18/VIIIa/0

4.002602

KEY

+2 ±4	**C** 6
	[12.0096; 12.0116] 12.011* 2-4

Atomic Weight Interval →

→ Conventional Atomic Weight

2 — Ne header region

| 13/IIIa | 14/IVa | 15/Va | 16/VIa | 17/VIIa | 18/VIIIa |

B 5, +3 — [10.806; 10.821] 10.81* 2-3

C 6, +2 ±4 — [12.0096; 12.0116] 12.011* 2-4

N 7, ±1 ±2 ±3 +4 +5 — [14.00643; 14.00728] 14.007* 2-5

O 8, −2 — [15.99903; 15.99977] 15.999* 2-6

F 9, −1 — 18.9984032 2-7

Ne 10, 0 — 20.1797 2-8

Al 13, +3 — 26.9815386 2-8-3

Si 14, +2 ±4 — [28.084; 28.086] 28.085* 2-8-4

P 15, ±3 +5 — 30.973762 2-8-5

S 16, +4 +6 −2 — [32.059; 32.076] 32.06* 2-8-6

Cl 17, ±1 +5 +7 — [35.446; 35.457] 35.45* 2-8-7

Ar 18, 0 — 39.948 2-8-8

Noble Gases

| 10/VIII | 11/Ib | 12/IIb |

Ni 28, +2 +3 — 58.6934 — -8-16-2
Cu 29, +1 +2 — 63.546 — -8-18-1
Zn 30, +2 — 65.38 — -8-18-2
Ga 31, +3 — 69.723 — -8-18-3
Ge 32, +2 +4 — 72.63 — -8-18-4
As 33, ±3 +5 — 74.92160 — -8-18-5
Se 34, +4 +6 −2 — 78.96 — -8-18-6
Br 35, ±1 +5 — 79.904 — -8-18-7
Kr 36, 0 — 83.798 — -8-18-8

Pd 46, +2 +4 — 106.42 — -18-18-0
Ag 47, +1 — 107.8682 — -18-18-1
Cd 48, +2 — 112.411 — -18-18-2
In 49, +3 — 114.818 — -18-18-3
Sn 50, +2 +4 — 118.710 — -18-18-4
Sb 51, ±3 +5 — 121.760 — -18-18-5
Te 52, +4 +6 −2 — 127.60 — -18-18-6
I 53, ±1 +5 +7 — 126.90447 — -18-18-7
Xe 54, 0 — 131.293 — -18-18-8

Pt 78, +2 +4 — 195.084 — -32-16-2
Au 79, +1 +3 — 196.966569 — -32-18-1
Hg 80, +1 +2 — 200.59 — -32-18-2
Tl 81, +1 +3 — [204.382; 204.385] 204.38* — -32-18-3
Pb 82, +2 +4 — 207.2 — -32-18-4
Bi 83, +3 +5 — 208.98040 — -32-18-5
Po 84, +2 +4 — (208.9824) — -32-18-6
At 85, ±1 +5 +7 — (209.9871) — -32-18-7
Rn 86, 0 — (222.0176) — -32-18-8

Ds 110 — (281.162)
Rg 111 — (280.164)
Cn 112 — (285.174)
Uut 113 (**) — (284.178)
Fl 114 — (289.187)
Uup 115 (**) — (288.192)
Lv 116 — (292.200)
Uus 117 (**)
Uuo 118 (**)

Eu 63, +2 +3 — 151.964 — -25-8-2
Gd 64, +3 — 157.25 — -25-9-2
Tb 65, +3 — 158.92535 — -27-8-2
Dy 66, +3 — 162.500 — -28-8-2
Ho 67, +3 — 164.93032 — -29-8-2
Er 68, +3 — 167.259 — -30-8-2
Tm 69, +3 — 168.93421 — -31-8-2
Yb 70, +2 +3 — 173.054 — -32-8-2
Lu 71, +3 — 174.9668 — -32-9-2

Am 95, +3 +4 +5 +6 — (243.0614) — -25-8-2
Cm 96, +3 — (247.0704) — -25-9-2
Bk 97, +3 +4 — (247.0703) — -27-8-2
Cf 98, +3 — (251.0796) — -28-8-2
Es 99, +3 — (252.0830) — -29-8-2
Fm 100, +3 — (257.0951) — -30-8-2
Md 101, +2 +3 — (258.0984) — -31-8-2
No 102, +2 +3 — (259.1010) — -32-8-2
Lr 103, +3 — (262.1096) — -32-9-2

Note: Elements with atomic numbers 113, 115, 117, and 118 have been reported but not fully authenticated.

* Conventional atomic weight; representative value for an element having an atomic weight interval.

** Symbols based on IUPAC systematic names.

View the Royal Society of Chemistry Visual Elements Periodic Table at www.rsc.org/periodic-table

THE MERCK INDEX

FIFTEENTH EDITION

1st Edition—1889
2nd Edition—1896
3rd Edition—1907
4th Edition—1930
5th Edition—1940
6th Edition—1952
7th Edition—1960
8th Edition—1968
9th Edition—1976
10th Edition—1983
11th Edition—1989
12th Edition—1996
13th Edition—2001
14th Edition—2006
15th Edition—2013

THE

MERCK INDEX

AN ENCYCLOPEDIA OF
CHEMICALS, DRUGS, AND BIOLOGICALS

FIFTEENTH EDITION

Maryadele J. O'Neil, *Editor-in-Chief*

Patricia E. Heckelman, *Senior Associate Editor*

Peter H. Dobbelaar, *Associate Editor*

Kristin J. Roman, *Assistant Editor*

Catherine M. Kenny, *Senior Editorial Assistant*

Linda S. Karaffa, *Technical Assistant*

Published by

The Royal Society of Chemistry

2013

The RSC is the largest organisation in Europe for advancing the chemical sciences. Supported by a worldwide network of members and an international publishing business, our activities span education, conferences, science policy and the promotion of chemistry to the public.

Our 47,500 members come from diverse areas of the chemical sciences and enjoy access to scientific information, careers advice and a wide range of exclusive benefits.

For further information see our web site at www.rsc.org

ISBN: 978-1-84973-670-1

A catalogue record for this book is available from the British Library

© The Royal Society of Chemistry, 2013. The name *The Merck Index* is owned by Merck Sharp & Dohme Corp., a subsidiary of Merck & Co., Inc., Whitehouse Station, N.J., U.S.A., and is licensed to The Royal Society of Chemistry for use in the U.S.A. and Canada.

Published by The Royal Society of Chemistry,
Thomas Graham House, Science Park, Milton Road,
Cambridge CB4 0WF, UK

Registered Charity Number 207890

RSCPublishing

FOREWORD

The Merck Index is an iconic reference work that has provided generations of professionals with comprehensive information on chemicals, drugs and biologicals for over 120 years. It is considered the standard chemistry reference work and has sold over one million copies worldwide—a major achievement for all who have been involved in its development.

The contents of this new edition have been extensively revised with particular attention paid to compounds used as standard laboratory reagents. More than 500 new monographs have been added, reflecting recent developments in chemistry, biomedicine, and environmental science. In response to requests from our readers, the number of graphical chemical structures was increased by more than 1000 depictions and physical property data were updated for 2800 compounds. Once again, all molecular weights have been recalculated to reflect the most recent IUPAC standards.

The Fifteenth Edition marks the beginning of a new chapter. The Royal Society of Chemistry is proud to become the new publisher of *The Merck Index*. We are very excited to be welcoming such a prestigious title to the RSC portfolio. *The Merck Index* is an excellent strategic fit with our other publishing activities and will help us achieve our stated aim to advance the chemical sciences. It is also emblematic of our organisation's rapidly expanding international presence and influence.

We eagerly look forward to developing *The Merck Index* for the digital future. We have already taken the first step in this development by offering online access together with the print version.

As we start to shape *The Merck Index* so that it continues to provide a valuable, trusted and relevant resource for future generations of scientists and researchers, we welcome your feedback and comments. If you would like to get in touch with the publishing team at the RSC, please email rscindex@rsc.org

Dr James Milne
Managing Director, RSC Publishing

TABLE OF CONTENTS

EXPLANATORY NOTES

The Fifteenth Edition of *The Merck Index* contains 10,400 monographs describing significant chemicals, drugs, and biological substances. The entries cover a wide range of compounds which have been selected on the basis of present or historic importance and interest. Since the publication of the Fourteenth Edition in 2006, over 5,500 monographs have been significantly revised and updated. Several hundred monographs have either been combined or deleted from the manuscript to make room for more than 500 new entries.

Entries are generally limited to single substances and related compounds such as isomers or salts. While multi-component drugs are, for the most part, excluded, there are a number of monographs devoted to families of natural products or biological substances. Monographs vary greatly in length. The length of a monograph, however, is not necessarily indicative of the importance of a compound, but rather may simply be an indication of the amount of relevant published information available for the compound.

For the purpose of illustrating the general monograph format, a typical monograph is depicted and the components are identified. While all possible categories of information in a monograph are described below, it must be emphasized that not all categories are present in every monograph.

Monograph Number. Sequential accession numbers are assigned to monographs which are alphabetized by title. Entries in the indices are referenced to these accession numbers, not to page numbers. (*Note:* Monograph numbers in the Fifteenth Edition do not correspond to monograph numbers of previous editions.)

Title. Titles are usually simple chemical names or in the case of drugs, the commonly used generic name such as the USAN (United States Adopted Name) or INN (International Nonproprietary Name). Registered trademarks, designated by ®, are used for a small number of entry titles, primarily when nonproprietary terms are not available. Plant monographs are titled using a common name rather than the full botanical name.

Chemical Abstracts Registry Number(s). The Chemical Abstracts Service (CAS) Registry Number appears following the title. These unique identifiers are provided for title substances and for selected derivatives. Where appropriate, numbers for isomeric and unspecified forms of the compound are listed. Descriptors are appended to the entry if more than one registry number has been associated with the compound.

Alternate Name(s). Other chemical names, trivial names, experimental drug codes, and trademarks that identify the entry are listed. Listing of trademarks is for information purposes only and it should not be assumed that the trademarks are in current use. The first letter of each trademark is capitalized; absence of capitalization, however, does not preclude that a name may either currently be a proprietary name, or may once have been the subject of proprietary rights. If known, the company associated with a particular trademark (as a manufacturer, distributor or trademark owner) is listed alongside the trademark in the Name Index.

Molecular Formula, Molecular Weight, % Composition. Elements in the molecular formula are listed according to the Hill convention (C, H, then other elements in alphabetical order). Formula and molecular weight are provided for

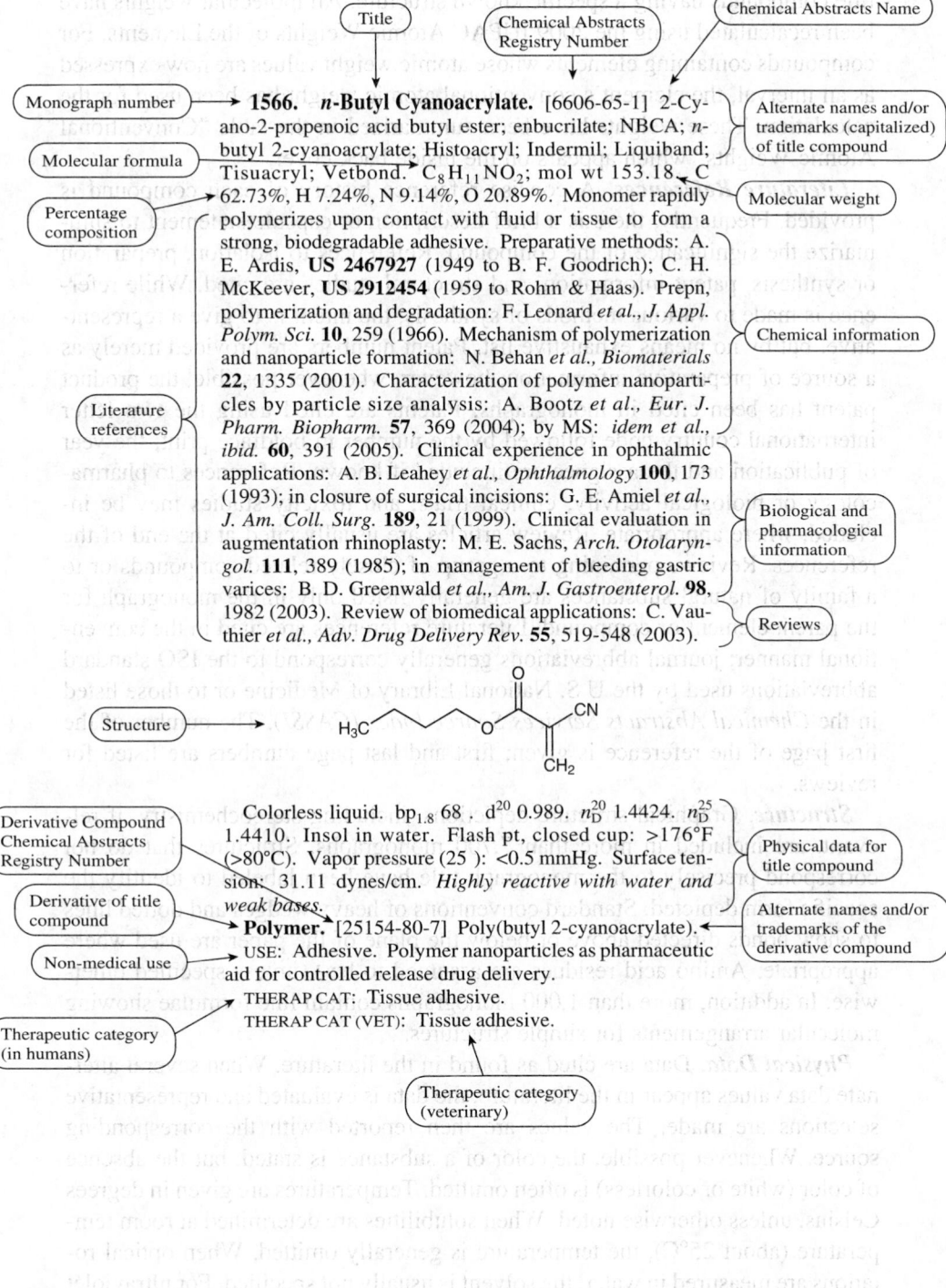

Title

Chemical Abstracts Registry Number

Chemical Abstracts Name

Monograph number → **1566. *n*-Butyl Cyanoacrylate.** [6606-65-1] 2-Cyano-2-propenoic acid butyl ester; enbucrilate; NBCA; *n*-butyl 2-cyanoacrylate; Histoacryl; Indermil; Liquiband; Tisuacryl; Vetbond. $C_8H_{11}NO_2$; mol wt 153.18. C 62.73%, H 7.24%, N 9.14%, O 20.89%. Monomer rapidly polymerizes upon contact with fluid or tissue to form a strong, biodegradable adhesive. Preparative methods: A. E. Ardis, **US 2467927** (1949 to B. F. Goodrich); C. H. McKeever, **US 2912454** (1959 to Rohm & Haas). Prepn, polymerization and degradation: F. Leonard *et al., J. Appl. Polym. Sci.* **10**, 259 (1966). Mechanism of polymerization and nanoparticle formation: N. Behan *et al., Biomaterials* **22**, 1335 (2001). Characterization of polymer nanoparticles by particle size analysis: A. Bootz *et al., Eur. J. Pharm. Biopharm.* **57**, 369 (2004); by MS: *idem et al., ibid.* **60**, 391 (2005). Clinical experience in ophthalmic applications: A. B. Leahey *et al., Ophthalmology* **100**, 173 (1993); in closure of surgical incisions: G. E. Amiel *et al., J. Am. Coll. Surg.* **189**, 21 (1999). Clinical evaluation in augmentation rhinoplasty: M. E. Sachs, *Arch. Otolaryngol.* **111**, 389 (1985); in management of bleeding gastric varices: B. D. Greenwald *et al., Am. J. Gastroenterol.* **98**, 1982 (2003). Review of biomedical applications: C. Vauthier *et al., Adv. Drug Delivery Rev.* **55**, 519-548 (2003).

Monograph number
Molecular formula
Percentage composition
Literature references

Alternate names and/or trademarks (capitalized) of title compound
Molecular weight
Chemical information
Biological and pharmacological information
Reviews

Structure →

Colorless liquid. $bp_{1.8}$ 68 . d^{20} 0.989. n_D^{20} 1.4424. n_D^{25} 1.4410. Insol in water. Flash pt, closed cup: >176°F (>80°C). Vapor pressure (25): <0.5 mmHg. Surface tension: 31.11 dynes/cm. *Highly reactive with water and weak bases.*

→ **Polymer.** [25154-80-7] Poly(butyl 2-cyanoacrylate).
USE: Adhesive. Polymer nanoparticles as pharmaceutic aid for controlled release drug delivery.
THERAP CAT: Tissue adhesive.
THERAP CAT (VET): Tissue adhesive.

Derivative Compound Chemical Abstracts Registry Number
Derivative of title compound
Non-medical use
Therapeutic category (in humans)

Physical data for title compound
Alternate names and/or trademarks of the derivative compound

Therapeutic category (veterinary)

title compounds having a specific known structure. All molecular weights have been recalculated using the 2009 IUPAC Atomic Weights of the Elements. For compounds containing elements whose atomic weight values are now expressed as an interval, the element's conventional atomic weight has been used for the calculation. These weights have been summarized in the table "Conventional Atomic Weights" which appears on the inside back cover.

Literature References. A concise reference history of each compound is provided. Frequently, there is a brief description or capsule statement to summarize the significance of the compound. References to isolation, preparation or synthesis, patent information, and structural studies are cited. While reference is made to various methods of synthesis, the intent is to give a representative, but by no means exhaustive list. Patent numbers are provided merely as a source of preparative information; however, whenever possible, the product patent has been cited in monographs. Patents are cited using the two letter international country code followed by the number in boldface print; the year of publication and the assignee are included if known. References to pharmacology or biological activity, clinical trials, and toxicity studies may be included, where appropriate. Review articles are usually cited at the end of the references. Reviews pertaining to a group of closely related compounds or to a family of natural substances are generally listed only in the monograph for the parent element or compound. Literature references are cited in the conventional manner; journal abbreviations generally correspond to the ISO standard abbreviations used by the U.S. National Library of Medicine or to those listed in the *Chemical Abstracts Services Source Index (CASSI)*. The number of the first page of the reference is given; first and last page numbers are listed for reviews.

Structure. Graphical structure depictions, including stereochemistry, if relevant, are included in more than 7,700 monographs. Structures that do not correspond precisely to the monograph title have been labeled to identify the specific form depicted. Standard conventions of heavy wedges and dotted lines to show bonds directed above or below the plane of the paper are used where appropriate. Amino acid residues are assumed to be L unless specified otherwise. In addition, more than 1,000 monographs contain line formulae showing molecular arrangements for simple structures.

Physical Data. Data are cited as found in the literature. When several alternate data values appear in the literature, the data is evaluated and representative selections are made. The values are then reported with the corresponding source. Whenever possible, the color of a substance is stated, but the absence of color (white or colorless) is often omitted. Temperatures are given in degrees Celsius, unless otherwise noted. When solubilities are determined at room temperature (about 25°C), the temperature is generally omitted. When optical rotations are measured in water, the solvent is usually not specified. For ultraviolet absorption measurements, the solvent is given within parentheses. Specific handling information, such as flammable or explosive, has been included where appropriate to draw attention to these characteristics.

An effort has been made to provide acute toxicity data (e.g. LD_{50}, LC_{50}) and to identify the source of this information. *Caution* and/or *Note* statements are also provided in a number of monographs. Specific statements are given for compounds on the U.S. Government's Schedules of Controlled Substances in Title 21 of the Code of Federal Regulations (CFR), for compounds listed as suspected or confirmed carcinogens in the *Twelfth Report on Carcinogens* issued in 2011 by the U.S. Department of Health and Human Services (USDHHS), and for chemicals considered potential occupational hazards as described in sources such as the *NIOSH Pocket Guide to Chemical Hazards (USDHHS)*. Compounds determined by the Stockholm Convention to be persistent organic pollutants (POPs) have also been identified. *Note:* Absence of toxicity data or specific cautions does not imply that toxic effects do not exist.

Derivatives. When derivatives (isomers, salts, etc.) of the title compound are described in a monograph, the information appears in the paragraph(s) directly following the physical data. These paragraphs may also be used to describe specific members of a large family of natural substances. Derivative data presentation mirrors that of the title compound and may include registry numbers, chemical and alternate names, molecular weights, percentage composition, literature references, and physical properties.

Use. Descriptions of specific uses, which are not medical or veterinary therapeutic applications, are summarized under this heading.

Therapeutic Category and Therapeutic Category (Veterinary). In most cases, therapeutic categories reflect the accepted terminology in the medical literature. When available, mode of action information is included in the literature references section of the monograph. Monographs for human drugs have been indexed by both therapeutic category and biological activity beginning on page THER-1.

The Merck Index is not intended as an official therapeutic guide. Inclusion of a drug or any other compound in this book is not an endorsement, but merely a statement of the fact that such a substance exists. THERAPEUTIC CATEGORY and THERAPEUTIC CATEGORY (VETERINARY) paragraphs are intended only as summary statements of major pharmacological properties or indications for the individual compounds. For additional information on uses, dosage, side effects, and adverse reactions, readers are directed to consult pertinent scientific and professional publications, product circulars, information sheets or material safety data sheets prepared or published by the respective manufacturers.

Indices. Four indices—Name, Formula, CAS Registry Number and Therapeutic Category—are included; each entry directs the reader to the number of the monograph in which the substance of interest is described. Approximately 60,000 synonyms, including titles, CAS names, alternate names, trademarks, and derivative names are contained in the **Name Index.** If known, trademarks have been matched with an associated company using an abbreviated form of the company name which appears in brackets following the trademark. Company names are provided as a source of additional information and do not necessarily imply trademark ownership. Due to reorganizations or mergers, some company names may have changed since the original matching process was completed.

The **Formula Index** contains molecular formulae for the title compounds and derivatives other than acid addition salts, hydrates or isomers.

The **CAS Registry Number Index** is arranged in ascending numerical order. A chemical name for the compound associated with each registry number is provided along with the number of the monograph in which it occurs.

In the **Therapeutic Category and Biological Activity Index,** monographs describing human drugs have been listed by one or more therapeutic indications and/or mechanisms of action. Cross references to closely related categories and mechanisms have been included. Whenever appropriate, subclassifications have been developed by grouping compounds according to chemical or pharmacological similarities.

Organic Name Reactions. This section is comprised of 485 named reactions and an index. A concise reference history and associated reaction schema are provided for each reaction or subreaction.

Tables. A compilation of over 80 pages of tables is provided to supplement the material presented in the monographs. Several tables have been extensively revised for this edition, including the addition of graphical structures to the Terms for Radicals and Groups used for Nonproprietary Names and the completely revised Periodic Chart of the Elements and Atomic Weights Tables.

Great care has been taken to assure the accuracy of the information contained in *The Merck Index*. However, the Editorial Staff and the Publisher cannot be responsible for errors incurred in publication or for any consequences arising from use of the information published in *The Merck Index*. Accordingly, reference to original sources is strongly encouraged.

Abbreviations

Included are abbreviations commonly used in *The Merck Index*. Please consult the Acronyms and Glossary (Suppl. Tables Section) for additional definitions.

α_D^{25}	specific optical rotation at 25° C for D (sodium) line; absence of brackets indicates optical rotation of a liquid in a 1 decimeter cell, neat	d	density; specific gravity (d_4^{19} specific gravity at 19° referred to water at 4°)
α_M	molar absorptivity	*d-*	*dextro* (rotatory); the opposite of *l*
Å	angstrom	D-	*Dextro* configuration; opposite of L
A	absorbance	Da	daltons
Ab	antibody	dec; decomp; decompn	decompose(s); decomposition
abs	absolute; absorption	deg	degree
abs config	absolute configuration	deliquesc	deliquescent
Ag	antigen	deriv	derivative
alc	alcohol; ethanol; ethyl alcohol	determn	determination
alk	alkali(ne)	dil	dilute; diluted; dilution
amps	ampules; amperes	distln	distillation
anhydr	anhydrous	*dl-*	racemic
Ann.	*Justus Liebig's Annalen der Chemie*	DL-	optically inactive by external compensation as contrasted with *meso-*
approx; ~	approximate(ly)	dyn	dynes
aq	aqueous	ε (epsilon)	molar extinction coefficient; dielectric constant
as-; asym-	asymmetrical; unsymmetrical	η (eta)	viscosity
at.	atomic	E.C. No.	Enzyme Commission Number
at. no.	atomic number	e.g.	(*exempli gratia*) for example
at. wt.	atomic weight	$E_{1cm}^{1\%}$	the absorbance of a solution containing one gram per 100 ml contained in a cell having an absorption path of one cm
atm	atmosphere(s)		
B.P.C.	British Pharmaceutical Codex		
Bé	Baumé (a specific gravity scale)	Ed(s).	editor(s)
Beilstein	*Beilstein's Handbuch der Organischen Chemie*	ed.	edition
		eidem	the same (authors); plural of *idem*
Ber.	*Chemische Berichte* (Berichte der Deutschen Chemischen Gesellschaft)	E_M	molar extinction coefficient (concn in g-moles/l)
bp	basepair; boiling point	equiv	equivalent
°C	Celsius degrees	*et al.*	(*et alii*) and others
c	concentration	etc.	(*et cetera*) and so forth
ca.	(*circa*) about	eV	electron volt
cal	calorie(s)	evac	evacuated
calc(d)	calculate; calculated	evapn	evaporation
cc	cubic centimeter(s) (milliliter)	exptl	experimental(ly)
cf.	(*confer*) compare	ext(d)	extract; extracted
Ci	curie	extern	externally
coll. vol.	collective volume	°F	Fahrenheit degrees
compd	compound	fp	freezing point
compn	composition	*Frdl.*	*P. Friedlander Fortschritte der Teer-farbenfabrikation*, a collection of patents
concd	concentrated		
concn	concentration	g	gram(s)
config	configuration	*Gmelin's*	*Gmelin's Handbuch der Anorganischen Chemie*
constit	constituent(s)		
contd	continued	habit.	habitat
contg	containing	*Houben*	a German collection of medicinal patents
corr	corrected	*Houben Weyl*	*Houben-Weyl Methoden der Organischen Chemie*
corresp	corresponding; corresponds		
C_p	heat capacity (constant pressure)	hr	hour
crit press	critical pressure	i.e.	(*id est*) that is
crit temp	critical temperature	i.g.	intragastric
cryst	crystalline; crystals	i.m.	intramuscular
crystn	crystallization	i.p.	intraperitoneal
		I.U.	international unit
Δ (delta)	indicates the locant of the double bond	i.v.	intravenous

ibid.	(*ibidem*) at the same place
idem	the same (author); plural: *eidem*, the same (authors)
incl	including
incompat	incompatibility
inorg	inorganic
insol	insoluble
Intl	International
isoln	isolation
K	dissociation constant; equilibrium constant; Kelvin temperature
λ (lambda)	wavelength; microliter
l	liter
l-	*levo*(rotatory); the opposite of *d*
L-	*Levo* configuration; opposite of D
LC_{50}	median lethal concentration; the concn of a chemical that is estimated to be fatal to 50% of the organisms tested
LD_{50}	median lethal dose; the quantity of a chemical that is estimated to be fatal to 50% of the organisms tested
loc. cit.	(*loco citato*) in the place cited
log	logarithm (common)
log *P*	logarithm of the partition coefficient
M	molar (concentration; moles/liter)
m-	*meta* chemical locant for ring substituents
Mab; mAb	monoclonal antibody
mass spec	mass spectrometry
MD	molecular rotation $\dfrac{[\alpha]_D \times \text{mol wt}}{100}$
Mellor's	*Mellor's Comprehensive Treatise on Inorganic and Theoretical Chemistry*
Mfg; manuf	manufacturing
mfr	manufacturer
misc	miscible
mixt	mixture
mol wt	molecular weight
Monatsh.	*Monatshefte für Chemie*
mp	melting point
M_r	relative molecular mass
N	normal concentration or nitrogen as a locant
n	index of refraction (n_D^{20} for 20° and sodium light); normal, as *n*-propyl
o-	*ortho* chemical locant for ring substituents
op. cit.	(*opere citato*) in the work cited
org	organic
OsM	osmolar; osmole
ψ (psi)	pseudo
P	poise
p-	*para* chemical locant for ring substituents
p; pp	page(s)
Pa	pascal
passim	here and there; scattered
pat.	patent
petr	petroleum
pH	acid-base scale; log of reciprocal of hydrogen ion concentration
pI	isoelectric point
pK	log of the reciprocal of the dissociation constant
ppm	parts per million
ppt; pptd	precipitate; precipitated
prepd; prepn	prepared; preparation
press.	pressure
pt	point
q.q.v.	(*quae vide*) which see, plural
q.v.	(*quod vide*) which see
r-	racemic
recryst(n)	recrystallize; recrystallization
ref	reference
rep [REP]	"roentgen equivalent physical" means a dose of ionizing radiation capable of producing energy absorption of 93 ergs per gram of tissue
resp	respectively
s.c.	subcutaneous
sapon(if)	saponification
satd	saturated
sec	second(s)
sepn	separation
sol	soluble
soly	solubility
solidif	solidifies; solidification
soln	solution
sp gr	specific gravity
sp.	species
spec	spectroscopy; spectrum; spectral
spp.	species (plural)
sqq	(*sequentia*) and following
subl	sublimes
suppl	supplement
sym-	symmetrical
$t_{1/2}$	half-life
tabl	tablet(s)
tech	technical
temp	temperature
uncor(r)	uncorrected
unsym-	unsymmetrical; asymmetrical
UV; uv	ultraviolet
v	volt(s)
v-	(*vicinal*) adjacent
v/v	percent "volume in volume" expresses the number of milliliters of an active constituent in 100 milliliters of solution
var	variety
viz.	(*videlicet*) that is to say; namely
vol	volume
vs	versus
w/v	percent "weight in volume" expresses the number of grams of an active constituent in 100 milliliters of solution, and is used regardless of whether water or another liquid is the solvent
w/w	percent "weight in weight" expresses the number of grams of an active constituent in 100 grams of solution or mixture
wt	weight

MONOGRAPHS

THE MERCK INDEX

OF CHEMICALS, DRUGS, AND BIOLOGICALS

A

1. Abacavir. [136470-78-5] (1*S*,4*R*)-4-[2-Amino-6-(cyclopropylamino)-9*H*-purin-9-yl]-2-cyclopentene-1-methanol; (−)-*cis*-4-[2-amino-6-(cyclopropylamino)-9*H*-purin-9-yl]-2-cyclopentene-1-methanol; 1592U89. $C_{14}H_{18}N_6O$; mol wt 286.34. C 58.73%, H 6.34%, N 29.35%, O 5.59%. Nucleoside reverse transcriptase inhibitor (NRTI). Prepn: S. M. Daluge, **EP 349242** (1990 to Wellcome Found.); *idem*, **US 5034394** (1991 to Burroughs Wellcome). Asymmetric synthesis: M. T. Crimmins, B. W. King, *J. Org. Chem.* **61**, 4192 (1996). Pharmacology and biological profile: S. M. Daluge *et al.*, *Antimicrob. Agents Chemother.* **41**, 1082 (1997). Review of antiviral activity and clinical evaluations: R. H. Foster, D. Faulds, *Drugs* **55**, 729-736 (1998). Clinical trial of triple nucleoside regimen in HIV patients: S. Staszewski *et al.*, *J. Am. Med. Assoc.* **285**, 1155 (2001).

White solid foam from acetonitrile, mp 165°. uv max (pH 1): 296, 255 nm (ε 14000, 10700); uv max (pH 7): 284, 259 nm (ε 15900, 9200); uv max (pH 13): 284, 259 nm (ε 15800, 9100). $[\alpha]_D^{20}$ −59.7°; $[\alpha]_{436}^{20}$ −127.8°; $[\alpha]_{365}^{20}$ −218.1° (c = 0.15 in methanol). Log P (1-octanol/0.1*M* sodium phosphate): 1.22 ±0.03 (pH 7.4). pKa 5.01. Soly in water (25°): >80 mM (pH 7).
Sulfate. [188062-50-2] Ziagen. $(C_{14}H_{18}N_6O)_2.H_2SO_4$; mol wt 670.75.

THERAP CAT: Antiretroviral.

2. Abamectin. [71751-41-2] Avermectin B₁; 5-*O*-demethylavermectin A₁ₐ and 5-*O*-demethyl-25-de(1-methylpropyl)-25-(1-methylethyl)avermectin A₁ₐ (4:1); avermectin B₁ₐ/ᵦ; MK-936; Agri-Mek; Avid; Zephyr. Mixture of avermectins, *q.v.*, that contains avermectin B₁ₐ ($C_{48}H_{72}O_{14}$) and not more than 20% of avermectin B₁ᵦ ($C_{47}H_{70}O_{14}$). Isoln from *Streptomyces avermitilis*: G. Albers-Schönberg *et al.*, **DE 2717040**; *eidem*, **US 4310519** (1977, 1982 both to Merck & Co.). Separation of components: T. W. Miller *et al.*, *Antimicrob. Agents Chemother.* **15**, 368 (1979); by semi-preparative HPLC: C. C. Ku *et al.*, *J. Liq. Chromatogr.* **7**, 2905 (1984). Structure determn: G. Albers-Schönberg *et al.*, *J. Am. Chem. Soc.* **103**, 4216 (1981). Absolute configuration: J. P. Springer *et al.*, *ibid.* 4221. Partial synthesis of B₁ₐ: K. C. Nicolaou *et al.*, *ibid.* **106**, 4189 (1984). Total synthesis: S. Hanessian *et al.*, *ibid.* **108**, 2776 (1986). Anthelmintic activity: L. S. Blair, W. C. Campbell, *J. Parasitol.* **64**, 1032 (1978); J. R. Egerton *et al.*, *Antimicrob. Agents Chemother.* **15**, 372 (1979); K. S. Todd *et al.*, *Am. J. Vet. Res.* **45**, 976 (1984). Pesticidal activity: I. Putter *et al.*, *Experientia* **37**, 963 (1981); R. A. Dybas, A. St. J. Green, *Proc. Br. Crop Prot. Conf. - Pests Dis.* **1984**,

947. Control of red imported fire ants: J. A. Greenblatt *et al.*, *J. Agric. Entomol.* **3**, 233 (1986). Interaction with GABA receptors: S. S. Pong *et al.*, *J. Neurochem.* **34**, 351 (1980); T. N. Mellin *et al.*, *Neuropharmacology* **22**, 89 (1983). Receptor binding studies: S. S. Pong, C. C. Wang, *ibid.* **19**, 311 (1980); G. Drexler, W. Sieghart, *Eur. J. Pharmacol.* **99**, 269 (1984). Fate in soil and plants: D. L. Bull *et al.*, *J. Agric. Food Chem.* **32**, 94 (1984); H. A. Moye *et al.*, *ibid.* **35**, 859 (1987); M. S. Maynard *et al.*, *ibid.* **37**, 178, 184 (1989). *Review*: J. R. Babu, *ACS Symp. Ser.* **380**, 91-108 (1988). Reviews of modes of action: C. C. Wang, S. S. Pong, *Prog. Clin. Biol. Res.* **97**, 373-395 (1982); D. J. Wright, *Biochem. Soc. Trans.* **15**, 65-67 (1987); of insecticidal activities and use: L. Strong, T. A. Brown, *Bull. Entomol. Res.* **77**, 357-389 (1987). *Book: Ivermectin and Abamectin*, W. C. Campbell, Ed. (Springer-Verlag, New York, 1989) 363 pp.

component B₁ₐ, R = C₂H₅   component B₁ᵦ, R = CH₃

Odorless, off-white to yellow crystals from methanol, mp 150-155° (dec). $[\alpha]_D$ +55.7 ±2° (c = 0.87 in CHCl₃). uv max (methanol): 237, 245, 253 nm (log ε 4.48, 4.53, 4.34). Vapor pressure: 1.5×10^{-9} torr. Soly at 21° (µg/l): water 10; (mg/ml): acetone 100; *n*-butanol 10; chloroform 25; cyclohexane 6; ethanol 20; isopropanol 70; kerosene 0.5; methanol 19.5; toluene 350. Hydrolysis does not occur in aq soln at pH 3, 5, 7. LD₅₀ (technical grade) orally in sesame oil in mouse, rat: 13.5, 10.0 mg/kg; dermally in rabbit: >2000 mg/kg. LD₅₀ in mallard duck, bobwhite quail: 84.6, >2000 mg/kg. LC₅₀ (96 hr) in rainbow trout, bluegill: 3.6, 9.6 µg/l; LC₅₀ (48 hr) in *Daphnia magna*: 0.34 µg/l (Merck Technical Data Sheet).

USE: Acaricide; insecticide.

THERAP CAT (VET): Anthelmintic.

3. Abarelix. [183552-38-7] *N*-Acetyl-3-(2-naphthalenyl)-D-alanyl-4-chloro-D-phenylalanyl-3-(3-pyridinyl)-D-alanyl-L-seryl-*N*-methyl-L-tyrosyl-D-asparaginyl-L-leucyl-*N*⁶-(1-methylethyl)-L-lysyl-L-prolyl-D-alaninamide; Ac-D-Nal¹-4-Cl-D-Phe²-D-Pal³-*N*-Me-Tyr⁵-D-Asn⁶-Lys(iPr)⁸-D-Ala¹⁰-LH-RH; PPI-149; Plenaxis. $C_{72}H_{95}ClN_{14}O_{14}$; mol wt 1416.09. C 61.07%, H 6.76%, Cl 2.50%, N 13.85%, O 15.82%. Decapeptide GnRH antagonist. Prepn: R. W. Roeske, **WO 9640757** (1996 to Indiana Univ. Found.); *idem*, **US 5843901** (1998 to Adv. Res. Technol. Inst.). Pharmacology and suppression of plasma gonadotropins: C. J. Molineaux *et al.*, *Mol. Urol.* **2**, 265 (1998). Clinical comparison with leuprolide in prostate cancer: J. Trachtenberg *et al.*, *J. Urol.* **167**, 1670 (2002). Clinical pharmacology: S. L. Wong *et al.*, *Clin. Pharmacol. Ther.* **73**, 304 (2003); of depot formulation: *eidem*, *J. Clin. Pharmacol.* **44**, 495 (2004). Review of clinical development: P. Mongiat-Artus, P. Teillac, *Expert Opin. Pharmacother.* **5**, 2171-2179 (2004).

White solid. Sol in water.

THERAP CAT: Antineoplastic (hormonal).

4. Abatacept. [332348-12-6] 1-25-Oncostatin M (human precursor) fusion protein with CTLA-4 (antigen)(human) fusion protein with immunoglobulin G1 (human heavy chain fragment), bimol. (146 → 146')-disulfide; CTLA-4Ig; BMS-188667; Orencia. Selective costimulation modulator that inhibits T cell activation via the CD28/B7 receptor pathway. Recombinant chimeric fusion protein composed of the extracellular domain of human cytotoxic lymphocyte-associated antigen 4 (CTLA-4) fused to the heavy constant region of human IgG. Prepn: P. S. Linsley *et al.*, *J. Exp. Med.* **174**, 561 (1991); *eidem*, **WO 9300431**; *eidem*, **US 5434131** (1993, 1995 both to Bristol-Myers Squibb). CE and HPLC analysis: K. F. Greve *et al.*, *J. Chromatogr. A* **723**, 273 (1996). Pharmacokinetics: N. R. Srinivas *et al.*, *J. Pharm. Sci.* **85**, 1 (1996). Clinical evaluation in psoriasis: J. R. Abrams *et al.*, *J. Clin. Invest.* **103**, 1243 (1999); in rheumatoid arthritis (RA): J. M. Kremer *et al.*, *Arthritis Rheum.* **52**, 2263 (2005); M. C. Genovese *et al.*, *N. Engl. J. Med.* **353**, 1114 (2005). Review of pharmacology: N. Najafian, M. H. Sayegh, *Expert Opin. Invest. Drugs* **9**, 2147-2157 (2000); of mechanism of action and clinical development in RA: G. G. Teng *et al.*, *Expert Opin. Biol. Ther.* **5**, 1245-1254 (2005).

THERAP CAT: Anti-inflammatory.

5. Abciximab. [143653-53-6] Anti-(human integrin α_{IIb}-β_3) immunoglobulin G1 Fab fragment (human-mouse monoclonal c7E3 clone p7E3V$_H$hC$_{\gamma1}$ γ_1-chain) disulfide with human-mouse monoclonal c7E3 clone p7E3V$_\kappa$hC$_\kappa$ κ-chain; c7E3 Fab; ReoPro. Platelet aggregation inhibitor. Fab fragment of chimeric human-murine monoclonal antibody c7E3 directed against the human platelet glycoprotein IIb/IIIa (GPIIb/IIIa) receptor. Chimeric fragment consists of the variable regions of murine *7E3* reconstructed with human constant regions of the Fab fragment. Mol wt ~47,600. Prepn: B. S. Coller, D. M. Knight, **WO 9512412** (1995 to Centocor and Res. Found. State Univ. New York); D. M. Knight *et al.*, *Mol. Immunol.* **32**, 1271 (1995). Pharmacology: C. Kohmura *et al.*, *Arterioscler. Thromb.* **13**, 1837 (1993). Clinical pharmacodynamics: J. E. Tcheng *et al.*, *Circulation* **90**, 1757 (1994). Clinical studies as adjunct in prevention of ischemic complications associated with coronary revascularization: EPIC Investigators, *N. Engl. J. Med.* **330**, 956 (1994); EPILOG Investigators, *ibid.* **336**, 1689 (1997). Review of development, binding studies and preclinical testing: R. E. Jordan *et al.* in *Adhesion Receptors as Therapeutic Targets*, M. A. Horton, Ed. (CRC Press, Boca Raton, 1996) p 281-305; of pharmacology and clinical efficacy: R. H. Foster, L. R. Wiseman, *Drugs* **56**, 629-665 (1998).

THERAP CAT: Antithrombotic.

6. Abetimus Sodium. [169147-32-4] d(C-A-C-A-C-A-C-A-C-A-C-A-C-A-C-A-C-A-C-A-C-A)DNA 5',5''',5''''',5'''''''-[[1,2-ethane-diylbis[oxy-2,1-ethanediyloxycarbonylnitrilobis[2,1-ethanediyl-imino(6-oxo-6,1-hexanediyl)imino(2-oxo-2,1-ethanediyl)thio-6,1-hexanediyl]]] tetrakis(hydrogen phosphate)] complex with d(T-G-T-G-T-G-T-G-T-G-T-G-T-G-T-G-T-G-T-G-T-G)DNA (1:4) hexapentacontahectasodium salt; LJP-394; Riquent. B-cell toleragen which prevents the expression of anti-ds-DNA. Consists of four identical ds-DNA oligonucleotides (20 basepairs) attached through a linker to a non-immunogenic modified triethylene glycol base. Relative mass of 54 kDa. Prepn: S. M. Coutts *et al.*, **EP 642798**; *eidem*, **US**

5552391 (1995, 1996 both to La Jolla Pharm.); D. S. Jones *et al.*, *J. Med. Chem.* **38**, 2138 (1995). Clinical evaluation in systemic lupus erythematosus (SLE): R. A. Furie *et al.*, *J. Rheumatol.* **28**, 257 (2001); in prevention of SLE renal flare: D. Alarcon-Segovia *et al.*, *Arthritis Rheum.* **48**, 442 (2003). Review of clinical pharmacology: D. J. Wallace, *Expert Opin. Invest. Drugs* **10**, 111-117 (2001); of clinical experience: M. H. Cardiel, *ibid.* **14**, 77-88 (2005).

Colorless aqueous solution with an osmolality of 255-310 mmol/kg at pH 6.8-7.8.

THERAP CAT: Immunosuppressant in treatment of SLE and associated nephritis.

7. Abietic Acid. [514-10-3] (1*R*,4a*R*,4b*R*,10a*R*)-1,2,3,4,-4a,4b,5,6,10,10a-Decahydro-1,4a-dimethyl-7-(1-methylethyl)-1-phenanthrenecarboxylic acid; 13-isopropylpodocarpa-7,13-dien-15-oic acid; sylvic acid. $C_{20}H_{30}O_2$; mol wt 302.46. C 79.42%, H 10.00%, O 10.58%. A widely available organic acid, prepared by isomerization of rosin: Harris, Sanderson, *Org. Synth.* **coll. vol. IV**, 1 (1963); Fieser, Fieser, *The Chemistry of Natural Products Related to Phenanthrene* (New York, 3rd ed., 1949). Synthesis from dehydroabietic acid: A. W. Burgstahler, L. W. Worden, *J. Am. Chem. Soc.* **83**, 2587 (1961); E. Wenkert *et al.*, *ibid.* **86**, 2038 (1964). Chromatographic study: A. G. Douglas, T. G. Powell, *J. Chromatogr.* **43**, 241 (1969). Metabolism in rabbits: Y. Asakawa *et al.*, *Xenobiotica* **16**, 753 (1986).

Monoclinic plates from alcohol + water, mp 172-175°. $[\alpha]_D^{24}$ −106° (c = 1 in abs alc). uv max: 235, 241.5, 250 nm (ε 19500, 22000, 14300). Insol in water. Sol in alc, benzene, chloroform, ether, acetone, carbon disulfide, dil NaOH soln. Commercial abietic acid made by heating rosin alone or with acids may be glassy or partly crystalline, usually of yellow color and melting as low as 85°.

Methyl ester *see* Methyl Abietate.

USE: Manufacture of esters (ester gums), e.g., methyl, vinyl and glyceryl esters for use in lacquers and varnishes. Manufacture of "metal resinates", soaps, plastics and paper sizes. Assists growth of lactic and butyric acid bacteria.

8. Abikoviromycin. [31774-33-1] (1a*R*,7*E*,7a*S*)-7-Ethylidene-1a,2,3,7-tetrahydrocyclopent[*b*]oxireno[*c*]pyridine; 4,4a-epoxy-5-ethylidene-2,3,4,4a-tetrahydro-5*H*-1-pyridine; abicoviromycin; latumcidin. $C_{10}H_{11}NO$; mol wt 161.20. C 74.51%, H 6.88%, N 8.69%, O 9.92%. Antiviral antibiotic produced by *Streptomyces abikoensis* and *Streptomyces rubescens*. Chromatographic isoln from broth cultures: Umezawa *et al.*, *Jpn. Med. J.* **4**, 331 (1951); *C.A.* **46**, 7167 (1952); Umezawa, **JP 54 6200** (1954 to Nippon). Identity with latumcidin: Sakagami *et al.*, *J. Antibiot.* **11A**, 231 (1958). Structure: Gurevich *et al.*, *Tetrahedron Lett.* **9**, 2209 (1968). Stereochemistry: Kono *et al.*, *J. Antibiot.* **23**, 572 (1970); Gurevich *et al.*, *Khim. Prir. Soedin.* **7**, 104 (1971); *C.A.* **75**, 5752e (1971). Crystal and molecular structure of the selenate: Y. Kono *et al.*, *Acta Crystallogr. B* **27**, 2341 (1971). *In vitro* antiviral activity: V. M. Roikhel, N. A. Zeitlenok, *Antibiotiki* **14**, 969 (1969); *C.A.* **72**, 19394q (1969).

Highly unstable and polymerizes promptly on isolation even at −50°; however, it can be handled in dilute solutions and in the form of its salts. uv max (neutral ethanol or 0.1*N* KOH): 218, 244, 289 nm (log ε 3.83, 3.99, 3.94); (0.1*N* HCl) 236, 341 nm (log ε 3.99, 4.05).

9. Abiraterone. [154229-19-3] (3β)-17-(3-Pyridinyl)androsta-5,16-dien-3-ol; 17-(3-pyridyl)androsta-5,16-dien-3β-ol; CB-7598. $C_{24}H_{31}NO$; mol wt 349.52. C 82.47%, H 8.94%, N 4.01%, O 4.58%. Androgen biosynthesis inhibitor; inhibits cytochrome P450c17 (17α-hydroxylase-$C_{17,20}$-lyase). Prepn: S. E. Barrie *et al.*, **WO 9320097**, *idem et al.*, **US 5604213** (1993, 1997 both to British Technol. Group); G. A. Potter *et al.*, *J. Med. Chem.* **38**, 2463 (1995). Large-scale synthesis of acetate ester prodrug: *idem*, *Org. Prep. Proced. Int.* **29**, 123 (1997). LC-MS/MS determn in plasma: V. Martins *et al.*, *J. Chromatogr. B* **843**, 262 (2006). Clinical pharmacokinetics and hormonal effects: A. O'Donnell *et al.*, *Br. J. Cancer* **90**, 2317 (2004). Review of mechanism of action and pharmacology: G. Attard *et al.*, *BJU Int.* **96**, 1241-1246 (2005); R. Aggarwal, D. J. Ryan, *Update Cancer Ther.* **2**, 171-175 (2007). Clinical trial in metastatic prostate cancer: J. S. deBono *et al.*, *N. Engl. J. Med.* **364**, 1995 (2011).

Crystals from toluene, mp 228-229°.

Acetate. [154229-18-2] (3β)-Acetoxy-17-(3-pyridyl)androsta-5,16-diene; CB-7630; Zytiga. $C_{26}H_{33}NO_2$; mol wt 391.56. Crystals from acetonitrile, mp 146-148°. Log P (octanol/water): 5.12. pKa 5.19 Practically insol in water.

THERAP CAT: Antineoplastic (hormonal).

10. Abrin. [1393-62-0] Agglutinin; toxalbumin. A toxic lectin and hemagglutinin obtained from seeds of *jequirity*, *Abrus precatorius* L., *Leguminosae*, a common vine of tropical countries, also found in central and southern Florida. Isoln and purification: J. Y. Lin *et al.*, *J. Formos. Med. Assoc.* **68**, 518 (1969), *C.A.* **72**, 98695 (1970); *eidem*, *Toxicon* **9**, 97 (1971). The high toxicity of abrin was originally believed to result from its hemagglutinating activity, but subsequent studies have shown that separate proteins are responsible for the toxicity and agglutination: S. Olsnes, A. Pihl, *Eur. J. Biochem.* **35**, 179 (1973). Five glycoproteins have been purified from the seeds of *A. precatorius*: *Abrus agglutinin* and the toxic principles, abrins a-d. *Abrus* agglutinin is a tetramer of 134,900 Da, is non-toxic to animal cells and a potent hemagglutinator. Abrins a through d (mol wt 63,000-67,000 Da) are composed of two disulfide-linked polypeptide chains. The smaller A-chain inhibits protein synthesis and causes cell death; the larger B-chain binds to the cell plasma membrane. Purification of major components: C. H. Wei *et al.*, *J. Biol. Chem.* **249**, 3061 (1974). Crystallographic study: C. H. Wei, J. R. Einstein, *ibid.* 2985. Improved purification, properties, crystallography of *Abrus* agglutinin: C. H. Wei *et al.*, *ibid.* **250**, 4790 (1975). Physical properties of the toxic principles: M. S. Herrmann, W. D. Behnke, *Biochim. Biophys. Acta* **621**, 43 (1980); *eidem*, *ibid.* **667**, 397 (1981). Isoln and purification of all five proteins: J. Y. Lin *et al.*, *Toxicon* **19**, 41 (1981). Amino acid sequence of the A-chain of abrin-a and a comparison with ricin: G. Funatsu *et al.*, *Agric. Biol. Chem.* **52**, 1095 (1988). Antitumor effects in animals: V. V. S. Reddy, M. Sirsi, *Cancer Res.* **29**, 1447 (1969); J. Y. Lin *et al.*, *Nature* **227**, 292 (1970); O. Fodstad *et al.*, *Cancer Res.* **37**, 4559 (1977). Immunoelectron microscopy studies of abrin toxic action on tumor cells: C. T. Lin *et al.*, *J. Ultrastruct. Res.* **73**, 310 (1980). Studies on toxicity and binding kinetics: M. Witten *et al.*, *Exp. Cell Biol.* **49**, 306 (1981); C. E. Bennett *et al.*, *ibid.* 319. Use of A-chain in cell-type-specific cytotoxic agents known as "immunotoxins": A.

J. Cumber *et al.*, *Methods Enzymol.* **112**, 207 (1985). Toxicity study: J. Y. Lin *et al.*, *J. Formos. Med. Assoc.* **68**, 322 (1969), *C.A.* **71**, 121926 (1969). *See also* Ricin, Lectins.

Yellowish-white powder. Sol in solns of sodium chloride, usually with turbidity. The toxic portion is heat-stable to incubation at 60° for 30 min; at 80°, most of the toxicity is lost in 30 min. LD_{50} i.p. in mice: 0.020 mg/kg (Lin p. 322).

Caution: Seeds of *A. precatorius* are extremely toxic; one seed, if thoroughly masticated, can cause fatal poisoning, *cf.* J. M. Kingsbury, *Poisonous Plants of the United States and Canada* (Prentice-Hall, New Jersey, 1964) p 303; K. Genest *et al.*, *Arzneim.-Forsch.* **21**, 888 (1971). *Note:* Do not confuse with abrine, *q.v.*

USE: Exptly in cancer research.

11. Abrine. [526-31-8] *N*-Methyl-L-tryptophan; (*S*)-3-(1*H*-indol-3-yl)-2-methylaminopropionic acid. $C_{12}H_{14}N_2O_2$; mol wt 218.26. C 66.04%, H 6.47%, N 12.84%, O 14.66%. Not to be confused with the albuminous substance abrin, *q.v.* Obtained from the seeds of *Abrus precatorius* L., *Leguminosae* (jequirity): Hoshino, *Ann.* **520**, 31 (1935). Synthesis: Miller, Robson, *J. Chem. Soc.* **1938**, 1910. Configuration: Cahill, Jackson, *J. Biol. Chem.* **126**, 29 (1938).

Prisms from water, dec 295°. $[\alpha]_D^{21}$ +44° (0.28 g in 10 ml 0.5*N* HCl). One gram dissolves in about 100 ml methanol, slightly sol in water, insol in ether. Sol in dil acids, alkalies.

Hydrochloride. $C_{12}H_{14}N_2O_2 \cdot HCl$. Needles, mp 222°, soluble in water.

Nitrate. $C_{12}H_{14}N_2O_2 \cdot HNO_3$. Needles, dec 143°.

Acetyl derivative. $C_{14}H_{16}N_2O_3$. mp 176°. $[\alpha]_D^{25}$ −148° (43 mg in 5 ml 0.1*N* NaOH).

12. Abscisic Acid. [21293-29-8] (2*Z*,4*E*)-5-[(1*S*)-1-Hydroxy-2,6,6-trimethyl-4-oxo-2-cyclohexen-1-yl]-3-methyl-2,4-pentadienoic acid; abscisin II; dormin; ABA. $C_{15}H_{20}O_4$; mol wt 264.32. C 68.16%, H 7.63%, O 24.21%. Abscission-accelerating plant hormone; naturally occurring as the (+)-*cis,trans*-form. Regulates seed maturation and germination and mediates the plant response to environmental stress. Isoln from young cotton fruit: K. Ohkuma *et al.*, *Science* **142**, 1592 (1963); from sycamore leaves: J. W. Cornforth *et al.*, *Nature* **205**, 1269 (1965). Identification in banana, birch, rose, cabbage, potato, lemon, avocado: *eidem*, *ibid.* **210**, 627 (1966). Structure and synthesis of (±)-*cis,trans*-form: *eidem*, *ibid.* **206**, 715 (1965); and enantiomers: F. Kienzle *et al.*, *Helv. Chim. Acta* **61**, 2616 (1978). Absolute configuration of naturally occurring form: G. Ryback, *Chem. Commun.* **1972**, 1190. Crystal and molecular structure: H. W. Schmalle *et al.*, *Acta Crystallogr. B* **33**, 2218 (1977). *Review:* F. T. Addicott, J. L. Lyon, *Annu. Rev. Plant Physiol.* **20**, 139 (1969). Review of role in root growth regulation: P. E. Pilet, P. W. Barlow, *Plant Growth Regul.* **6**, 217-265 (1987); of metabolism: J. A. D. Zeevaart, R. A. Creelman, *Annu. Rev. Plant Physiol. Plant Mol. Biol.* **39**, 439-473 (1988); of signal transduction: J. Leung, J. Giraudat, *ibid.* **49**, 199-222 (1998).

Crystals from ethyl acetate + hexane, mp 161-163°. Sublimes at 120°. $[\alpha]_D^{20}$ +411.1° (c = 1 in ethanol). $[\alpha]_D^{20}$ +426.5° (c = 1 in 0.005*N* methanolic H_2SO_4). Sol in aq $NaHCO_3$, chloroform, acetone, ethyl acetate, ether; slightly sol in benzene, water; sparingly sol in petr ether. uv max (methanol): 252 nm (ε 25200).

(−)-*cis,trans*-**Form.** [14398-53-9] (*R*)-(−)-Abscisic acid. mp 162-163°. $[\alpha]_D^{20}$ −426.2° (c = 1 in 0.005*N* H_2SO_4).

(±)-*cis,trans*-**Form.** [14375-45-2] Crystals, mp 188-190°.

13. Absinthin. [1362-42-1] (3*S*,3a*S*,6*S*,6a*R*,6b*S*,7*S*,7a*R*,-8*S*,10a*S*,11*S*,13a*S*,13b*R*,13c*R*,14b*S*)-3,3a,4,5,6,6a,6b,7,7a,8,9,10,-10a,13a,13c,14b-Hexadecahydro-6,8-dihydroxy-3,6,8,11,14,15-hexamethyl-2*H*-7,13b-ethenopentaleno[1″,2″:6,7;5″,4″:6′,7′]di-cyclohepta[1,2-b:1′,2′-b′]difuran-2,12(11*H*)-dione; absinthiin; ab-synthin. $C_{30}H_{40}O_6$; mol wt 496.64. C 72.55%, H 8.12%, O 19.33%. Chief bitter principle of wormwood, *Artemisia absinthium* L., *Compositae.* Isoln by chromatography: V. Herout *et al., Collect. Czech. Chem. Commun.* **21**, 1485 (1956); see also *Chem. Ind. (London)* **1955**, 569. Structural studies: L. Novotny *et al., ibid.* **1958**, 465; *Collect. Czech. Chem. Commun.* **25**, 1492 (1960); K. Vokác *et al., Tetrahedron Lett.* **9**, 3855 (1968). Structure: J. Beauhaire *et al., ibid.* **21**, 3191 (1980). Total synthesis: W. Zhang *et al., J. Am. Chem. Soc.* **127**, 18 (2005).

Colorless crystals from benzene, mp 165-166° (dec). $[\alpha]_D^{20}$ +107.0° (c = 1.9 in $CHCl_3$); $[\alpha]_D^{20}$ +103.5° (c = 1.0 in $CHCl_3$).

14. Acacetin. [480-44-4] 5,7-Dihydroxy-2-(4-methoxy-phenyl)-4*H*-1-benzopyran-4-one; 5,7-dihydroxy-4′-methoxyfla-vone; apigenin-4′-methyl ether. $C_{16}H_{12}O_5$; mol wt 284.27. C 67.60%, H 4.26%, O 28.14%. The aglycon of linarin, *q.v.*, and of acaciin. Isoln from linarin: Zemplén, Bognar, *Ber.* **74B**, 1818 (1941). From acaciin: Hattori, *Acta Phytochim.* **2**, 105 (1925). Isoln from *Robinia pseudoacacia* L., *Leguminosae:* Nakazawa, Matsuura, *J. Pharm. Soc. Jpn.* **73**, 481 (1953). Structure: Baker *et al., J. Chem. Soc.* **1951**, 691. Synthesis: Robinson, Venkataraman, *ibid.* **1926**, 2348; Zemplén, Bognar, *Ber.* **76B**, 452 (1943); Narasim-hachari, Seshadri, *Proc. Indian Acad. Sci.* **30A**, 151 (1949); Simp-son, *Sci. Proc. R. Dublin Soc.* **27**, 111 (1956),*C.A.* **51**, 8082a (1957).

Yellow needles from 95% alcohol, mp 263°. Sol in hot alc, prac-tically insol in ether. Sol in alkalies with yellow color.

7-Rhamnoglucoside. Acaciin. $C_{28}H_{32}O_{14}$. From *Robinia pseu-dacacia* L., *Leguminosae:* Freudenberg, Hartmann, *Ann.* **587**, 207 (1954). Structure: Zemplén, Mester, *Magy. Kem. Foly.* **56**, 2 (1950), *C.A.* **45**, 7977d (1951). Needles from pyridine + water, mp 263°. $[\alpha]$ −85.3° (pyridine); (−)99.5° (glacial acetic acid). Sparingly soluble in cold, more sol in boiling water; slightly sol in organic solvents.

15. Acacia. [9000-01-5] Gum arabic. Estimations of mol wt range from about 240,000: Oakley, *Trans. Faraday Soc.* **31**, 136 (1935); to 580,000: Anderson *et al., Carbohydr. Res.* **3**, 308 (1967). According to the U.S.P., acacia is the dried gummy exudation from the stems and branches of *Acacia senegal* (L.) Willd., *Leguminosae*, or other African species of Acacia. According to C. L. Mantell, *The Water-Soluble Gums* (New York, 1947), Kordofan gum (hashab ge-neina), the gum from *Acacia verek* Guill. & Perr. from plantations in the Kordofan province (Sudan) is considered the best commercial

variety. Grades of Kordofan gum which are clear, white (sun bleached) and tasteless are preferred for food prepns and pharmaceu-ticals. (There is a close relationship between color and flavor due to the presence of tannins.) Acacia was originally thought to be com-posed only of (−)-arabinose, (+)-galactose, (−)-rhamnose, (+)-gly-curonic acid. Revised composition and structural studies: Anderson *et al., J. Chem. Soc. C* **1966**, 1959. *See also:* Swenson *et al., J. Polym. Sci. Part A-2* **6**, 1593 (1968). General review: Anderson, Dea, *J. Soc. Cosmet. Chem.* **22**, 61-76 (1971). Review of use as food additive: D. M. W. Anderson, *Food Addit. Contam.* **3**, 225-230 (1986).

Occurs in spheroidal tears up to 32 mm in diameter. Also flakes and powder. Solns of gum from *Acacia verek* are levorotatory; other acacia species are dextrorotatory: Hamy, *Bull. Sci. Pharmacol.* **35**, 421 (1928). Specific gravity: 1.35-1.49 (samples dried at 100° are heavier). Moisture content usually varies from 13-15%. U.S.P. limit 15%. Material containing less than 12% chips easily and produces dust during transportation. Insol in alcohol. Almost completely sol in twice its weight of water. 100 grams of a satd soln contains 37 g at 25°; 38 g at 50°; 40 g at 90°: Taft, Malm, *Trans. Kans. Acad. Sci.* **32**, 49 (1929). Aq soln acid to litmus. Also sol in glycerol and in propylene glycol, but prolonged heating (several days) may be nec-essary for complete solution (about 5%).

USE: As mucilage, excipient for tablets, size, emulsifier, thickener, also in candy, other foods; as colloidal stabilizer. In the manufacture of spray-dried "fixed" flavors —stable, powdered flavors used in packaged dry-mix products (puddings, desserts, cake mixes) where flavor stability and long shelf life are important.

16. Acacic Acid. [1962-14-7] (3*β*,16*α*,21*β*)-3,16,21-Tri-hydroxyolean-12-en-28-oic acid. $C_{30}H_{48}O_5$; mol wt 488.71. C 73.73%, H 9.90%, O 16.37%. Isoln from pods of *Acacia concinna* D.C., *Leguminosae:* Varshney, Shamsuddin, *Tetrahedron Lett.* **1964**, 2055; from bark of *A. concinna* D.C.: R. Banerji, S. K. Nigam, *J. Indian Chem. Soc.* **57**, 1043 (1980). Structure and stereochemis-try: Varshney *et al., ibid.* **1965**, 1187. Revised structure: A. K. Barua *et al., Trans. Bose Res. Inst. (Calcutta)* **39**, 61 (1976); *C.A.* **87**, 53460c (1977).

Needles from methanol, mp 280-281°.

Methyl ester. $C_{31}H_{50}O_5$. Needles from methanol, mp 223-224°.

Diacetyl lactone. $C_{34}H_{50}O_6$. Crystals, mp 235-236°.

17. Acadesine. [2627-69-2] 5-Amino-1-*β*-D-ribofurano-syl-1*H*-imidazole-4-carboxamide; 5-amino-4-imidazolecarbox-amide ribonucleoside; AICA-riboside; Arasine; Protara. C_9H_{14}-N_4O_5; mol wt 258.23. C 41.86%, H 5.46%, N 21.70%, O 30.98%. Purine nucleoside analog. Prototype adenosine regulating agent; en-hances endogenous adenosine levels in ischemic tissue. Biosynthe-sis, isoln, and characterization: G. R. Greenberg, E. L. Spilman, *J. Biol. Chem.* **219**, 411 (1956). Chemical synthesis: J. Baddiley *et al., Proc. Chem. Soc. London* **1957**, 149; J. P. Ferris *et al., J. Org. Chem.* **50**, 747 (1985); T. Saito *et al., Chem. Pharm. Bull.* **42**, 2263 (1994). Pharmacology: Z.-Q. Zhao *et al., Cardiovasc. Res.* **29**, 495 (1995). Antiplatelet activity: D. A. Bullough *et al., J. Clin. Invest.* **94**, 1524 (1994). Disposition and metabolism: R. Dixon *et al., J. Clin. Pharmacol.* **33**, 955 (1993). HPLC determn in plasma: L.-S. Chen *et al., J. Liq. Chromatogr.* **18**, 1451 (1995). Review of phar-macology: K. Mullane, M. Young, *Drug Dev. Res.* **28**, 336-343 (1993); of development: K. Mullane *et al., Trends Cardiovasc. Med.* **3**, 227-234 (1993). Meta-analysis of clinical trials in coronary artery bypass graft surgery: D. T. Mangano *et al.*, *J. Am. Med. Assoc.* **277**, 325-332 (1997).

Crystals from methanol or water, mp 213-214° (dec). Also reported as slightly brownish prisms from 90% aq ethanol, mp 206-208° (dec) (Saito). uv max (pH 7 and 1*N* NaOH): 265 nm (ε 12400). uv max (1*N* HCl): 245, 265 nm (ε 8670, 10320).

THERAP CAT: Cardioprotective.

18. Acamprosate Calcium. [77337-73-6] 3-(Acetylamino)-1-propanesulfonic acid calcium salt (2:1); calcium acetyl homotaurinate; Ca-AOTA; calcium bisacetyl homotaurine; Aotal; Campral. $C_{10}H_{20}CaN_2O_8S_2$; mol wt 400.47. C 29.99%, H 5.03%, Ca 10.01%, N 7.00%, O 31.96%, S 16.01%. GABA (γ-aminobutyric acid, *q.v.*) agonist. Prepn: J. P. Durlach, **DE 3019350**; *idem*, **US 4355043** (1980, 1982 both to Lab. Meram). Physicochemical and pharmacological study: C. Chabenat *et al.*, *Methods Find. Exp. Clin. Pharmacol.* **10**, 311 (1988). Pharmacology: J. Durlach *et al.*, *ibid.* 437; A. Guiet-Bara *et al.*, *Alcohol* **5**, 63 (1988). Suppression of ethanol intake in rats: F. Boismare *et al.*, *Pharmacol. Biochem. Behav.* **21**, 787 (1984); J. Le Magnen *et al.*, *Alcohol* **4**, 97 (1987). Evaluation of abuse potential: K. A. Grant, W. L. Woolverton, *Pharmacol. Biochem. Behav.* **32**, 607 (1989). HPLC determn in plasma: C. Chabenat *et al.*, *J. Chromatogr.* **414**, 417 (1987). Clinical evaluation in relapse prevention in weaned alcoholics: J. P. L'Huintre *et al.*, *Lancet* **1**, 1014 (1985); J. P. L'Huintre *et al.*, *Alcohol Alcohol.* **25**, 613 (1990). Review of clinical efficacy in maintenance of abstinence in alcoholics: L. J. Scott *et al.*, *CNS Drugs* **19**, 445-464 (2005); of mechanism of action: P. De Witte *et al.*, *ibid.* 517-537.

Colorless crystalline powder, mp 270°. uv max (water): 192 nm (ε 7360). Freely sol in water. Practically insol in absolute ethanol, dichloromethane. LD_{50} i.p. in male mice: 1.87 g/kg (Durlach, 1982).

THERAP CAT: In treatment of alcoholism.

19. Acarbose. [56180-94-0] *O*-4,6-Dideoxy-4-[[(1*S*,4*R*,-5*S*,6*S*)-4,5,6-trihydroxy-3-(hydroxymethyl)-2-cyclohexen-1-yl]amino]-α-D-glucopyranosyl-(1 → 4)-*O*-α-D-glucopyranosyl-(1 → 4)-D-glucose; 4″,6″-dideoxy-4″-[(1*S*)-(1,4,6/5)-4,5,6-trihydroxy-3-hydroxymethyl-2-cyclohexenylamino]maltotriose; Bay g 5421; Glucobay; Prandase; Precose. $C_{25}H_{43}NO_{18}$; mol wt 645.61. C 46.51%, H 6.71%, N 2.17%, O 44.61%. Pseudotetrasaccharide containing an unsaturated cyclitol moiety. An α-glucosidase inhibitor that reduces sugar absorption in the gastrointestinal tract. Isoln from strains of *Actinoplanes:* W. Frommer *et al.*, **DE 2347782**; *eidem*, **US 4062950** (1975, 1977 both to Bayer). Total synthesis: S. Ogawa, Y. Shibata, *Chem. Commun.* **1988**, 605; *eidem*, *Carbohydr. Res.* **189**, 309 (1989). Biosynthetic studies: U. Degwert *et al.*, *J. Antibiot.* **40**, 855 (1987). Glucosidase inhibition studies: D. D. Schmidt *et al.*, *Naturwissenschaften* **64**, 535 (1977); W. Puls *et al.*, *ibid.* 536. Use in treatment of diabetic adults: D. Sailor, G. Roder, *Arzneim.-Forsch.* **30**, 2182 (1980); H. Laube *et al.*, *ibid.* 1154. Long-term study in sulfonylurea-treated diabetics: H. Vierhapper *et al.*, *Diabetologia* **20**, 586 (1981). Potential use in prophylaxis of dental caries: N. E. Fiehn, D. Moe, *Scand. J. Dent. Res.* **90**, 124 (1982). Review of

pharmacodynamics, pharmacokinetics and therapeutic potential: S. P. Clissold, C. Edwards, *Drugs* **35**, 214-243 (1988); of cardiovascular benefits in type 2 diabetes: U. Zeymer, *Int. J. Cardiol.***107**, 11-20 (2006).

White to off-white powder. $[\alpha]_D^{18}$ +165° (c = 0.4 in water). Sol in water. pKa 5.1.

THERAP CAT: Antidiabetic.

THERAP CAT (VET): Antidiabetic.

20. Acebutolol. [37517-30-9] *N*-[3-Acetyl-4-[2-hydroxy-3-[(1-methylethyl)amino]propoxy]phenyl]butanamide; 3′-acetyl-4′-[2-hydroxy-3-(isopropylamino)propoxy]butyranilide; 1-(2-acetyl-4-*n*-butyramidophenoxy)-2-hydroxy-3-isopropylaminopropane; 5′-butyramido-2′-(2-hydroxy-3-isopropylaminopropoxy)acetophenone. $C_{18}H_{28}N_2O_4$; mol wt 336.43. C 64.26%, H 8.39%, N 8.33%, O 19.02%. Cardioselective β-adrenergic blocker. Prepn: K. R. H. Wooldridge, B. Basil, **ZA 6808345**; *eidem*, **US 3857952** (1969, 1974 both to May & Baker). Pharmacology: Cuthbert, Owusu-Ankomah, *Br. J. Pharmacol.* **43**, 639 (1971); Basil *et al.*, *Br. J. Pharmacol.* **48**, 198 (1973); Lewis *et al.*, *Br. Heart J.* **35**, 743 (1973). HPLC determn in plasma and urine: M. Piquette-Miller *et al.*, *J. Chromatogr.* **526**, 129 (1990). Crystal structure: A. Carpy *et al.*, *Acta Crystallogr. B* **35**, 185 (1979). Review of pharmacology and therapeutic efficacy: B. N. Singh *et al.*, *Drugs* **29**, 531-569 (1985); G. DeBono *et al.*, *Am. Heart J.* **109**, 1211-1223 (1985). Comprehensive description: R. T. Foster, R. A. Carr, *Anal. Profiles Drug Subs.* **19**, 1-26 (1990).

Crystals, mp 119-123°.

Hydrochloride. [34381-68-5] M & B 17803A; IL-17803A; Acecor; Acetanol; Neptal; Prent; Sectral. $C_{18}H_{28}N_2O_4 \cdot HCl$; mol wt 372.89. Crystals from anhydr methanol-anhydr diethyl ether, mp 141-144°. Sol in alcohol, water. Very slightly sol in acetone, methylene chloride. Practically insol in ether. Soly at room temperature (mg/ml): water 200; ethanol 70.

THERAP CAT: Antihypertensive; antianginal; antiarrhythmic (class II).

21. Acecainide. [32795-44-1] 4-(Acetylamino)-*N*-[2-(diethylamino)ethyl]benzamide; 4′-[[2-(diethylamino)ethyl]carbamoyl]acetanilide; *N*-acetylprocainamide; NAPA. $C_{15}H_{23}N_3O_2$; mol wt 277.37. C 64.95%, H 8.36%, N 15.15%, O 11.54%. Metabolite of procainamide, *q.v.* Prepn: E. C. Schreiber, **DE 2062978** (1971 to Squibb), *C.A.* **75**, 76427 (1972). Pharmacokinetics: M. Wierzchowiecki *et al.*, *Int. J. Clin. Pharmacol. Ther. Toxicol.* **18**, 272 (1980). Clinical pharmacology and antiarrhythmic efficacy: J. Kluger *et al.*, *Am. J. Cardiol.* **45**, 1250 (1980); R. A. Winkle *et al.*, *ibid.* **47**, 123 (1981). HPLC determn in plasma: D. Raphanaud *et al.*, *Ther. Drug Monit.* **8**, 365 (1986). Review of pharmacology and therapeutic potential: D. W. G. Harron, R. N. Brogden, *Drugs* **39**, 720-740 (1990).

Hydrochloride. [34118-92-8] ASL-601. $C_{15}H_{23}N_3O_2 \cdot HCl$; mol wt 313.83. Crystals, mp 190-193°.

THERAP CAT: Antiarrhythmic.

22. Acecarbromal. [77-66-7] *N*-[(Acetylamino)carbonyl]-2-bromo-2-ethylbutanamide; *N*-acetyl-*N*-bromodiethylacetylurea; acetylbromodiethylacetylcarbamide; *N*-acetyl-*N'*-α-bromo-α-ethyl-butyrylcarbamide; acetylcarbromal; Abasin; Sedamyl. C_9H_{15}-BrN_2O_3; mol wt 279.13. C 38.73%, H 5.42%, Br 28.63%, N 10.04%, O 17.20%. Prepn: **DE 327129** (1920 to Bayer), *Frdl.* **13**, 809 (1923). HPLC determn in plasma, urine: M. Höbel, G. Bender, *Arch. Toxicol.* **37**, 307 (1977). GC-MS screening procedure in urine: H. H. Maurer, *J. Chromatogr.* **530**, 307 (1990).

Crystals, slightly bitter taste, mp 108-109°. Slightly sol in water; freely sol in alcohol, ethyl acetate.

THERAP CAT: Sedative, hypnotic.

23. Aceclofenac. [89796-99-6] 2-[(2,6-Dichlorophenyl)-amino]benzeneacetic acid carboxymethyl ester; 2-[(2,6-dichloro-phenyl)amino]phenylacetoxyacetic acid; glycolic acid [*o*-(2,6-di-chloroanilino)phenyl]acetate ester; PR-82/3; Airtal; Falcol; Gerbin; Preservex. $C_{16}H_{13}Cl_2NO_4$; mol wt 354.18. C 54.26%, H 3.70%, Cl 20.02%, N 3.95%, O 18.07%. Prepn: A. V. Casas, **ES 8404783**; *idem*, **US 4548952** (1984, 1985 both to Prodes). Gastrointestinal tolerance in rats in comparison with diclofenac, *q.v.*: V. Rimbau *et al.*, *Farmaco Ed. Prat.* **43**, 19 (1988). Clinical trial in comparison with acetaminophen, *q.v.*, in episiotomal pain: A. Yscla, *Drugs Exp. Clin. Res.* **14**, 491 (1988). Clinical evaluation in rheumatoid arthritis: R. Ballesteros *et al.*, *Clin. Trials J.* **27**, 12 (1990).

White crystals from cyclohexane, mp 149-150°. uv max (ethanol): 275 nm (log ε 4.14).

THERAP CAT: Anti-inflammatory; analgesic.

24. Acediasulfone. [80-03-5] *N*-[4-[(4-Aminophenyl)sul-fonyl]phenyl]glycine; *N*-*p*-sulfanilylphenylglycine; *p*-amino-*p'*-(carboxymethylamino)diphenyl sulfone; 4-carboxymethylamino-4'-aminodiphenylsulfone; diaminodiphenylsulfone-*N*-acetic acid. C_{14}-$H_{14}N_2O_4S$; mol wt 306.34. C 54.89%, H 4.61%, N 9.14%, O 20.89%, S 10.47%. Prepn: Jackson, *J. Am. Chem. Soc.* **70**, 680 (1948); **CH 254803** and **CH 278482** (1949, 1952, to Cilag Ltd.); Rawlins, **US 2589211** (1952 to Parke, Davis).

Crystals, mp 194°. Sol in methanol, dil sodium hydroxide, acetone.

Sodium salt. [127-60-6] Sulfon-Cilag. $C_{14}H_{13}N_2NaO_4S$; mol wt 328.32.

Morpholine salt. Bentrofene. $C_{18}H_{23}N_3O_5S$; mol wt 393.46. Prepn: Martin, Habicht, **US 2751382** (1956 to Cilag Ltd.). Glittering crystals, mp 133-135° (dec).

THERAP CAT: Antibacterial.

25. Acefylline. [652-37-9] 1,2,3,6-Tetrahydro-1,3-dimeth-yl-2,6-dioxo-7*H*-purine-7-acetic acid; carboxymethyltheophylline; 7-theophyllineacetic acid. $C_9H_{10}N_4O_4$; mol wt 238.20. C 45.38%, H 4.23%, N 23.52%, O 26.87%. Prepn: **DE 352980** (1922 to E. Merck); *Frdl.* **14**, 1320; S. M. Ride *et al.*, *Pharmazie* **32**, 672 (1977). Prepn of salts: J. Baisse, *Bull. Soc. Chim. Fr.* **1949**, 769; M. Milletti, F. Virgili, *Chimica* **6**, 394 (1951), *C.A.* **46**, 8615h (1952). GC determn in urine: J. Zuidema, H. Hilbers, *J. Chromatogr.* **182**, 445 (1980). HPLC determn in serum and pharmacokinetics: S. Sved *et al.*, *Biopharm. Drug Dispos.* **2**, 177 (1981).

Crystals from water, mp 271°.

Sodium salt. [837-27-4] $C_9H_9N_4NaO_4$. Silky needles, mp >300°.

Compd with piperazine. Acefylline piperazine; acepifylline; Dynaphylline; Etaphylline; Etafillina. Undefined mixture of the 1:1 and 2:1 salts; contains 75-78% theophylline acetic acid and 22-25% anhydrous piperazine.

THERAP CAT: Bronchodilator.

26. Aceglutamide. [2490-97-3] N^2-Acetyl-L-glutamine; α-*N*-acetyl-L-glutamine. $C_7H_{12}N_2O_4$; mol wt 188.18. C 44.68%, H 6.43%, N 14.89%, O 34.01%. Prepn: P. Karrer *et al.*, *Helv. Chim. Acta* **9**, 301 (1926); **GB 792576** (1958 to Merck & Co.); I. J. Maschler, N. Lichtenstein, *Biochim. Biophys. Acta* **57**, 252 (1962). Stability in soln: G. Sekules, G. Guadagnini, *Farmaco Ed. Prat.* **21**, 22 (1966). NMR study: W. Voelter *et al.*, *Z. Naturforsch.* **26B**, 213 (1971). HPLC-MS determn in urine: K. Sugahara *et al.*, *J. Chromatogr. B* **657**, 15 (1994). Clinical trial as component of total parenteral nutrition: I. Magnusson *et al.*, *Metabolism* **38**, Suppl. 1, 82 (1989).

Crystals from ethanol, mp 197°. $[\alpha]_D^{20}$ −12.5° (c = 2.9 in water). Stable in neutral aqueous solns.

Aluminum complex. [12607-92-0] Pentakis (N^2-acetyl-L-glu-taminato)tetrahydroxytrialuminum; aceglutamide aluminum; KW-110; Glumal. $C_{35}H_{59}Al_3N_{10}O_{24}$; mol wt 1084.85. Prepn: T. Kagawa *et al.*, **DE 2127176**; *eidem*, **US 3787466** (1971, 1974 both to Kyowa). Physico-chemical properties: E. Hayakawa *et al.*, *Yakugaku Zasshi* **97**, 731 (1977), *C.A.* **87**, 141198 (1977). Cytoprotective effect: H. Tanaka, *Arzneim.-Forsch.* **36**, 1485 (1986). White powder, mp 221° (dec). Sol in water. Practically insol in methanol, ethanol, acetone. LD_{50} in male mice, rats (g/kg): 14.3, >14.5 orally; 5.0, 4.2 i.p.; 0.46, 0.40 i.v. (Kagawa).

THERAP CAT: Nutritional supplement (glutamine source). Aluminum complex as antiulcerative.

27. Acemannan. Carrisyn. Long-chain, polydispersed, β-(1,4)-linked acetylated mannan obtained from the mucilage of *Aloe barbadensis*, Miller, *Liliaceae* (aloe vera). Isolation, purification, and characterization: B. H. McAnalley, **WO 8700052**; *idem*, **US 4735935** (1987, 1988 both to Carrington Labs.). Structural study:

S. Manna, B. H. McAnalley, *Carbohydr. Res.* **241**, 317 (1993). Immunostimulant effect *in vitro:* D. Womble, J. H. Helderman, *Int. J. Immunopharmacol.* **10**, 967 (1988). Inhibition of HIV-1 replication *in vitro:* J. B. Kahlon *et al.*, *Mol. Biother.* **3**, 127 (1991). Veterinary trial in feline-leukemia virus infection: M. A. Sheets *et al.*, *ibid.* 41; in canine and feline spontaneous tumors: C. Harris *et al.*, *ibid.* 207. Toxicological evaluation: R. W. Fogleman *et al.*, *Vet. Hum. Toxicol.* **34**, 201 (1992).

THERAP CAT: Antiviral; immunomodulator.

THERAP CAT (VET): Antiviral.

28. Acemetacin. [53164-05-9] 1-(4-Chlorobenzoyl)-5-methoxy-2-methyl-1*H*-indole-3-acetic acid carboxymethyl ester; [[1-(4-chlorobenzoyl)-5-methoxy-2-methylindol-3-yl]acetoxy]acetic acid; TV-1322; Acemix; Emflex; Rantudil; Solart. $C_{21}H_{18}ClNO_6$; mol wt 415.83. C 60.66%, H 4.36%, Cl 8.53%, N 3.37%, O 23.08%. Deriv of indomethacin, *q.v.* Prepn: K. H. Boltze *et al.*, **DE 2234651**; *eidem*, **US 3910952** (1972, 1975 both to Troponwerke). Series of articles on chemistry, analysis, pharmacodynamics, toxicology and clinical trials: *Arzneim.-Forsch.* **30**, 1313-1468 (1980). Toxicity data: H. Jacobi, H.-D. Dell, *ibid.* 1398.

Very fine pale yellow crystals from petr ether, mp 150-153°. LD_{50} in male, female mice, male, female rats (mg/kg): 55.5, 18.42, 24.2, 30.1 orally; 34.1, 51.1, 38.1, 28.3 i.v. (Jacobi, Dell).

THERAP CAT: Anti-inflammatory.

29. Acenaphthene. [83-32-9] 1,2-Dihydroacenaphthylene; *peri*-ethylenenaphthalene; 1,8-ethylenenaphthalene. $C_{12}H_{10}$; mol wt 154.21. C 93.46%, H 6.54%. Occurs in coal tar. Isoln: Ges. f. Teerverwertung, **DE 277110**; *Chem. Zentralbl.* **1914**, II, 597. In petroleum residues: Orloff *et al.*, *C.A.* **31**, 2800⁹ (1937). By passing ethylene and benzene or naphthalene though a red hot tube: Berthelot, *Bull. Soc. Chim.* [2] **7**, 274; **8**, 226, 245 (1867). By heating tetrahydroacenaphthene with sulfur at 180°: Braun *et al.*, *Ber.* **55**, 1694 (1922). From acenaphthenone or acenaphthenequinone by high pressure hydrogenation in decalin with nickel at 180-240°: Braun, Bayer, *Ber.* **59**, 921, 923 (1926). From acenaphthenone oxime: Morgan, Stanley, *J. Soc. Chem. Ind. London* **44**, 494T (1925). Review of toxicology and human exposure: *Toxicological Profile for Polycyclic Aromatic Hydrocarbons* (PB95-264370, 1995) 487 pp.

Orthorhombic bipyramidal needles from alcohol. d 1.189. mp 95°. bp 279°. uv spectrum: Seshan, *Proc. Indian Acad. Sci.* **A3**, 148 (1936). Factors influencing the uv spectrum: Jones, *J. Am. Chem. Soc.* **67**, 2127 (1945). Insol in water. One gram dissolves in 31 ml alcohol, 56 ml methanol, 25 ml propanol, 2.5 ml chloroform, 5 ml benzene or toluene. 3.2 g are sol in 100 ml glacial acetic acid. Forms water-sol, cryst complexes with desoxycholic acid, containing two molecules of the bile acid as a rule. The complexes crystallize when concentrated solns of the proper amount of the components in alcohol or dioxane are allowed to cool slowly.

USE: Dye intermediate; manuf plastics; insecticide; fungicide.

30. Acenocoumarol. [152-72-7] 4-Hydroxy-3-[1-(4-nitrophenyl)-3-oxobutyl]-2*H*-1-benzopyran-2-one; 3-(α-acetonyl-*p*-nitrobenzyl)-4-hydroxycoumarin; 3-(α-*p*-nitrophenyl-β-acetylethyl)-4-hydroxycoumarin; acenocoumarin; nicoumalone; G-23350; Mini-Sintrom; Sinthrome; Sintrom. $C_{19}H_{15}NO_6$; mol wt 353.33. C 64.59%, H 4.28%, N 3.96%, O 27.17%. Vitamin K antagonist;

structurally similar to warfarin, *q.v.* Prepn: W. Stoll, F. Litvan, **US 2648682** (1953 to Geigy); I. C. Ivanov *et al.*, *Arch. Pharm.* **323**, 521 (1990). Resolution and abs config of enantiomers: C. R. Wheeler, W. F. Trager, *J. Med. Chem.* **22**, 1122 (1979). Pharmacology and toxicity: M. Leroux, B. Jamain, *Therapie* **11**, 85 (1956). GC-MS determn in plasma: F. Pommier *et al.*, *J. Chromatogr. B* **654**, 35 (1994). Clinical trial: D. P. M. Brandjes *et al.*, *N. Engl. J. Med.* **327**, 1485 (1992).

Crystals, mp 196-199°. Sparingly sol in water and most organic solvents. Forms water-sol salt with alkalies. LD_{50} orally in mice, rats: 1470, 1000 mg/kg (Leroux, Jamain).

THERAP CAT: Anticoagulant.

31. Aceperone. [807-31-8] *N*-[[1-[4-(4-Fluorophenyl)-4-oxobutyl]-4-phenyl-4-piperidinyl]methyl]acetamide; *N*-[[1-[3-(*p*-fluorobenzoyl)propyl]-4-phenyl-4-piperidyl]methyl]acetamide; 1-(*p*-fluorophenyl)-4-(4-phenyl-4-acetamidomethylpiperidino)-1-butanone; 1-[γ-(4-fluorobenzoyl)propyl]-4-acetamidomethyl-4-phenylpiperidine; *N*-[1-[3-(*p*-fluorobenzoyl)propyl]-4-phenylpiperidin-4-ylmethyl]acetamide; 4'-fluoro-4-(4-acetamidomethyl-4-phenylpiperido)butyrophenone; acetabutone; acetobuton; R-3248. $C_{24}H_{29}FN_2O_2$; mol wt 396.51. C 72.70%, H 7.37%, F 4.79%, N 7.07%, O 8.07%. Prepd by reacting γ-chloro-4-fluorobutyrophenone with 4-acetamidomethyl-4-phenylpiperidine: P. A. J. Janssen, **BE 606849** (1961), corresp to **US 3083205** (1961, 1963 both to Janssen). Crystal structure: N. Van Opdenbosch *et al.*, *Acta Crystallogr. B* **33**, 171 (1977).

Crystals, mp 97-100°.

32. Acephate. [30560-19-1] *N*-Acetylphosphoramidothioic acid *O,S*-dimethyl ester; *O,S*-dimethyl acetylphosphoramidothioate; Ortho 12420; Orthene. $C_4H_{10}NO_3PS$; mol wt 183.16. C 26.23%, H 5.50%, N 7.65%, O 26.20%, P 16.91%, S 17.50%. Organophosphate insecticide; cholinesterase inhibitor. Prepn and activity: Magee, **US 3716600** and **US 3845172** (1973, 1974 both to Chevron). Physicochemical properties: A. K. Singh *et al.*, *Comp. Biochem. Physiol. C* **119**, 107 (1998). HPLC determn in aqueous soil extracts: E. Rodríguez-Gonzalo *et al.*, *J. Chromatogr.* **585**, 324 (1991). Dissipation in soil: J.-H. Yen *et al.*, *Ecotoxicol. Environ. Saf.* **45**, 79 (2000).

White solid, mp 88-90°. pKa 8.5. Soly in water (25°): 650 g/l. Moderately sol in acetone, alcohol, ethyl acetate. Low soly in aromatic solvents. Insol in *n*-hexane. Log P (octanol/water): −1.5 ± 0.2. LD_{50} orally in rats: 700 mg/kg (Magee, 1973).

USE: Contact and systemic insecticide.

33. Acepromazine. [61-00-7] 1-[10-[3-(Dimethylamino)propyl]-10*H*-phenothiazin-2-yl]ethanone; 10-[3-(dimethylamino)propyl]phenothiazin-2-yl methyl ketone; 2-acetyl-10-(3-dimethylaminopropyl)phenothiazine; 3-acetyl-10-(3-dimethylaminopropyl)-phenothiazine; 10-(3-dimethylaminopropyl)phenothiazine-3-ethyl-one; acetazine; acetopromazine; acetylpromazine; 1522-CB. $C_{19}H_{22}N_2OS$; mol wt 326.46. C 69.90%, H 6.79%, N 8.58%, O 4.90%, S 9.82%. **Prepn:** J. Schmitt *et al.*, *Bull. Soc. Chim. Fr.* **1957**, 938. Pharmacology and toxicology: W. Wirth *et al.*, *Arch. Int. Pharmacodyn.* **123**, 78 (1959). GLC determn in plasma: S. Ballard, T. Tobin, *J. Toxicol. Environ. Health* **7**, 745 (1981). Clinical trial in cats: D. H. Dyson *et al.*, *Can. Vet. J.* **33**, 462 (1992); in horses: A. Hashem, H. Keller, *J. Vet. Pharmacol. Ther.* **16**, 359 (1993). Brief review of clinical experience: N. Brock, *Can. Vet. J.* **35**, 458-459 (1994).

Yellow-colored oil, $bp_{0.5}$ 220-240°.

Maleate. [3598-37-6] Atravet; Calmivet; Plegicil; PromAce; Sedalin; Vetranquil. $C_{19}H_{22}N_2OS \cdot C_4H_4O_4$; mol wt 442.53. Yellow crystals from ethyl acetate, mp 135-136°. pKa 9.3. LD_{50} in rats (mg/kg): 400 orally, 59 i.v. (Wirth). Sol in water, alcohol, chloroform. pH of aq solns: ~4.0.

THERAP CAT (VET): Tranquilizer.

34. Acequinocyl. [57960-19-7] 2-(Acetyloxy)-3-dodecyl-1,4-naphthalenedione; AKD-2023; DPX-3792; DPX-T3792; AC-145; Kanemite. $C_{24}H_{32}O_4$; mol wt 384.52. C 74.97%, H 8.39%, O 16.64%. Naphthoquinone miticide; inhibits mitochondrial electron transfer. **Prepn:** NL 7505470; R. F. Bellina, D. L. Fost, US 4082848 (1975, 1978 both to Du Pont). Field study on mites: D. A. Marsden, W. W. Allen, *J. Econ. Entomol.* **73**, 168 (1980). Acaricidal mechanism: Y. Koura *et al.*, *J. Pestic. Sci.* **23**, 18 (1998). Review of physical properties, mode of action, and biological characteristics: F. Wakasa, S. Watanabe, *Agrochem. Jpn.* **75**, 17 (1999).

Fine yellowish powder, mp 59.6°. Soly in water at 25°: 6.7×10^{-6} g/l. Vapor pressure at 40°: 5.18×10^{-5} Pa. Log P (octanol/water): > 6.2. LD_{50} in rats (mg/kg): > 5000 orally, > 2000 dermally; in mice (mg/kg): > 5000 orally. LC_{50} (96hr) in carp, rainbow trout (mg/l): > 96.5, > 33 (Wakasa).

USE: Acaricide.

35. Acerin. [8001-08-9] An extract from the dried fruit of the Norway maple *Acer plantanoides* L., *Aceraceae*. Freshly prepd solutions of acerin destroy *E. coli* and vaccinia virus in 5 min but are less effective on the Staphylococcus Twort virus. Phagicidal action also exists even in the presence of the host cells. Extraction and properties: G. Fischer, *Acta Pathol. Microbiol. Scand.* **31**, 433 (1952); **34**, 482 (1954); Fischer *et al.*, *Experientia* **10**, 329 (1954).

Practically insol in methanol, ethanol, ether, acetone. Does not dialyze through a cellophane membrane at room temp but passes through a Seitz filter. Sodium chloride precipitates the active principle in acerin preparations as does also normal serum with 1:800 solns of acerin. Strong acerin preparations acquire reddish-brown color after treatment with alkali and oxidation in air. In general, the substance behaves similarly to vegetable tannins.

36. Acerola. The ripe fruit of *Malpighia punicifolia* L., *Malpighiaceae*, also called West Indian Cherry. *Habit.* Central America, Puerto Rico. Probably the richest natural source of ascorbic acid. Analysis of pitted fruit: Water 92.28%; ascorbic acid 1690

mg/100 g; vitamin A 11.0 mg/100 g; niacin 407 γ/100 g; vitamin B_6 8.7 γ/100 g; thiamine-HCl 30.0 γ/100 g; fluorine 10 γ/100 g; carbohydrates 6.4%. pH of juice 3.3. The pits comprise 19.25% of the fruit. *Ref:* Derse, Elvehjem, *J. Am. Med. Assoc.* **156**, 1501 (1954). Commercial aspects: *J. Agric. Food Chem.* **2**, 1155 (1954). *Review:* Moscoso, *Econ. Bot.* **10**, 280 (1956). Method of preparing acerola juice concentrates: Morse, US 3012942; US 3012943 (both 1961 to Nutrilite Products).

THERAP CAT: Nutrient.

37. ACES. [7365-82-4] 2-[(2-Amino-2-oxoethyl)amino]-ethanesulfonic acid; *N*-(carbamoylmethyl)taurine; *N*-(2-acetamido)-2-aminoethanesulfonic acid. $C_4H_{10}N_2O_4S$; mol wt 182.19. C 26.37%, H 5.53%, N 15.38%, O 35.13%, S 17.60%. One of several zwitterionic N-substituted amino acids known as "Good" buffers. Active in pH range 6-8. **Prepn:** N. E. Good *et al.*, *Biochemistry* **5**, 467 (1966). Thermodynamic parameters: D. C. McGlothlin, J. Jordan, *Anal. Lett.* **9**, 245 (1976). Metal complexation in solns: J. M. Pope *et al.*, *Anal. Biochem.* **103**, 214 (1980). Effect of temperature on pK values: R. N. Roy *et al.*, *J. Chem. Eng. Data* **42**, 41 (1997). Use as buffer in electrophoresis: D. J. Chappell, *Clin. Chem.* **31**, 1384 (1985); Q. Liu *et al.*, *Anal. Biochem.* **270**, 112 (1999); in chromatography: Z. Yu, D. Westerlund, *Chromatographia* **44**, 589 (1997).

Crystals from alcohol and water, dec 293°. pKa at 0.1M (0°): 7.32; (20°): 6.88; (37°): 6.56. $\Delta pKa/°C$: −0.020.

USE: Buffer for biological systems.

38. Acesulfame. [33665-90-6] 6-Methyl-1,2,3-oxathiazin-4(3*H*)-one 2,2-dioxide; 6-methyl-3,4-dihydro-1,2,3-oxathiazin-4-one 2,2-dioxide; acetosulfam. $C_4H_5NO_4S$; mol wt 163.15. C 29.45%, H 3.09%, N 8.59%, O 39.23%, S 19.65%. Non-nutritive artificial sweetener. **Prepn:** K. Clauss, H. Jensen, DE 2001017; *eidem*, US 3689486 (1971, 1972 to Hoechst); K. Clauss *et al.*, *Z. Lebensm.-Unters. Forsch.* **162**, 37 (1976). Crystal structure: E. F. Paulus, *Acta Crystallogr. B* **31**, 1191 (1975). HPLC analysis: H. Grosspietsch, H. Hachenberg, *Z. Lebensm.-Unters. Forsch.* **171**, 41 (1980). *Book: Acesulfame-K*, D. G. Mayer, F. H. Kemper, Eds. (Marcel Dekker, Inc., New York, 1991) 243 pp.

Needles from benzene or chloroform, mp 123-123.5°.

Potassium salt. [55589-62-3] Acesulfame-K; HOE-095K; Sunette. $C_4H_4KNO_4S$; mol wt 201.24. Colorless, odorless powder. Dec pt ~225°. d (solid) 1.81 g/cm³; d (bulk) 1.1-1.3 kg/dm³. uv max: 227 nm (ε 1.0762×10⁴). Very sol in DMF, DMSO. Sol in water, glycerin-water. Very slightly sol in acetone, alcohol. Soly in water (g/l): ~150 (0°); ~210 (10°); ~270 (20°); ~360 (30°); ~460 (40°); ~580 (50°); ~830 (70°); ~1300 (100°). Soly at 20° (g/l): methanol ~10; ethanol (anhydrous) ~1; glycerol (anhydrous) ~30; glycerol/water (80:20, v/v) ~82; glycerol/water (50:50, v/v) ~162; acetone ~0.8; glacial acetic acid ~130. Soly at 23° (g/l): ethanol/water (80:20, v/v) ~46; ethanol/water (60:40, v/v) ~100; ethanol/water (40:60, v/v) ~155; ethanol/water (20:80, v/v) ~221. LD_{50} in rats (mg/kg): 7431 orally, 2243 i.p. (Mayer, Kemper).

USE: Potassium salt as sweetener for foods, cosmetics.

39. Acetal. [105-57-7] 1,1-Diethoxyethane; diethylacetal; acetaldehyde diethyl acetal; ethylidene diethyl ether. $C_6H_{14}O_2$; mol wt 118.18. C 60.98%, H 11.94%, O 27.08%. Made from acetaldehyde and alcohol in the presence of anhydrous calcium chloride or of small quantities of mineral acid: H. Adkins, B. H. Nissen, *J. Am. Chem. Soc.* **44**, 2750 (1922); *eidem*, *Org. Synth.* coll. vol. I, 1 (2nd ed., 1941). Toxicity study: H. F. Smyth *et al.*, *J. Ind. Hyg. Toxicol.* **31**, 60 (1949).

Volatile liquid. *Flammable.* d_4^{20} 0.8254. bp_{760} 102.7°; bp_{200} 66.3°; bp_{60} 39.8°; bp_{40} 31.9°; bp_{20} 19.6°; bp_{10} 8.0°; bp_5 −2.3°; $bp_{1.0}$ −23°. n_D^{20} 1.38193. Flash pt, closed cup: 97°F (36°C). uv spectrum: Purvis, *J. Chem. Soc.* **127**, 9 (1925). 100 g water dissolve 5 g acetal. Misc with alcohol, 60% alcohol, ether. Sol in heptane, methylcyclohexane, propyl-, isopropyl-, butyl-, isobutyl alcohol, and ethyl acetate. Tends to polymerize on standing. Stable to alkalies. LD_{50} orally in rats: 4.57 g/kg (Smyth).

USE: Solvent; in synthetic perfumes such as jasmine; in organic syntheses.

40. Acetaldehyde. [75-07-0] Ethanal; "aldehyde"; acetic aldehyde; ethylaldehyde. C_2H_4O; mol wt 44.05. C 54.53%, H 9.15%, O 36.32%. CH_3CHO. Produced by oxidation of alcohol with $Na_2Cr_2O_7$ and H_2SO_4; usually from acetylene, dil H_2SO_4 and mercuric oxide as catalyst; also by passing alcohol vapor over a heated metallic catalyst. Lab procedure from ethanol: Wertheim, *J. Am. Chem. Soc.* **44**, 2658 (1922); Fricke, Havestadt, *Angew. Chem.* **36**, 546 (1923); Gattermann-Wieland, *Praxis des Organischen Chemikers* (de Gruyter, Berlin, 40th ed., 1961) p 180; from acetylene: Gattermann-Wieland, *op. cit.* 183; from paraldehyde: A. I. Vogel, *Practical Organic Chemistry* (Longmans, London, 3rd ed., 1959) p 324; by catalytic oxidation of ethylene in aq soln: J. Smidt *et al.*, *Angew. Chem.* **71**, 176 (1959); by oxidation of ethylene in fuel cells in the gas phase: K. Otsuka *et al.*, *Chem. Commun.* **1988**, 1272. Manuf: *Faith, Keyes & Clark's Industrial Chemicals*, F. A. Lowenheim, M. K. Moran, Eds. (Wiley-Interscience, New York, 4th ed., 1975) pp 1-7. Toxicity data: Smyth, *Arch. Ind. Hyg. Occup. Med.* **4**, 119 (1951). *Review:* H. J. Hagemeyer in *Kirk-Othmer Encyclopedia of Chemical Technology* **vol. 1** (John Wiley & Sons, New York, 4th ed., 1991) pp 94-109.

Liquid with characteristic, pungent odor. d_4^{16} 0.788. mp −123.5°. bp 21°. n_D^{20} 1.3316. Flash pt, closed cup: −36°F (−38°C). *Flammable.* Miscible with water, alcohol. *Keep cold.* LD_{50} orally in rats: 1930 mg/kg (Smyth).

Caution: Potential symptoms of overexposure are eye, nose and throat irritation; conjunctivitis; coughing; CNS depression; eye and skin burns; dermatitis; delayed pulmonary edema. *See NIOSH Pocket Guide to Chemical Hazards* (DHHS/NIOSH 97-140, 1997) p 2. *See also Clinical Toxicology of Commercial Products*, R. E. Gosselin *et al.*, Eds. (Williams & Wilkins, Baltimore, 5th ed., 1984) Section II, p 186. This substance is reasonably anticipated to be a human carcinogen: *Report on Carcinogens, Twelfth Edition* (PB2011-111646, 2011) p 21.

USE: Manuf paraldehyde, acetic acid, butanol, perfumes, flavors, aniline dyes, plastics, synthetic rubber; silvering mirrors, hardening gelatin fibers. Flavoring agent in foods and beverages. Fumigant for storage of apples and strawberries.

41. Acetaldehyde Ammonia. [75-39-8] 1-Aminoethanol; α-aminoethyl alcohol; aldehyde ammonia. C_2H_7NO; mol wt 61.08. C 39.33%, H 11.55%, N 22.93%, O 26.19%. $CH_3CH(OH)NH_2$. Prepd from acetaldehyde and ammonia: Aschan, *Ber.* **48**, 874 (1915).

Crystals; gradually turns yellow to brown in air. mp 97°. bp 110°, partly decomposing. Freely sol in water, slightly in ether. *Protect from light and air.*

Caution: Irritates eyes, mucous membranes.

USE: For preparing pure acetaldehyde; in organic syntheses.

42. Acetaldehyde Sodium Bisulfite. [918-04-7] 1-Hydroxyethanesulfonic acid sodium salt (1:1). $C_2H_5NaO_4S$; mol wt 148.11. C 16.22%, H 3.40%, Na 15.52%, O 43.21%, S 21.65%. $CH_3CH(OH)SO_3Na$.

Hemihydrate. Crystals, decomposed by acids. *Irritant.* Freely sol in water. Insol in alcohol.

USE: Making pure acetaldehyde; in organic synthesis.

43. Acetaldoxime. [107-29-9] Acetaldehyde oxime; aldoxime; ethylidenehydroxylamine. C_2H_5NO; mol wt 59.07. C 40.67%, H 8.53%, N 23.71%, O 27.08%. $CH_3CH=NOH$. Prepn: Dunstan,

Dymond, *J. Chem. Soc.* **61**, 470 (1892). Manuf: Donaruma, US 2763686 (1956 to du Pont).

Two crystalline modifications, mp 12° (β-form), mp 46.5° (α-form). d 0.966. bp 114.5°. n_D^{20} 1.415. *Flammable.* Dec by aq HCl into acetaldehyde and hydroxylamine. Very sol in water, alcohol, ether.

44. Acetamide. [60-35-5] Acetic acid amide. C_2H_5NO; mol wt 59.07. C 40.67%, H 8.53%, N 23.71%, O 27.08%. Prepd by fractional distillation of ammonium acetate: Coleman, Alvarado, *Org. Synth.* **coll. vol. I**, 3 (2nd ed., 1941); Gattermann-Wieland, *Praxis des Organischen Chemikers* (40th ed., 1961) p 118; Vogel, *Practical Organic Chemistry* (3rd ed., 1959) p 401. Prepn from methyl acetate, W. P. Munro *et al.*, US 2106697 (1936 to Calco Chem.); from ethyl acetate, Vogel, *op. cit.*, p 403. Studies of acetamide as an ionizing solvent: Jauder, Winkler, *J. Inorg. Nucl. Chem.* **9**, 24, 32, 39 (1959). Toxicological study: Weisburger *et al.*, *Toxicol. Appl. Pharmacol.* **14**, 163 (1969).

Deliquescent hexagonal crystals. Odorless when pure, but frequently has a mousy odor. d_4^3 1.159. mp 81°. bp_{760} 222°; bp_{100} 158°; bp_{40} 136°; bp_{20} 120°; bp_{10} 105°; bp_5 92°. n_D^{78} 1.4274. Neutral reaction. pKb (25°): 14.51. One gram dissolves in 0.5 ml water, 2 ml alcohol, 6 ml pyridine. Sol in chloroform, glycerol, hot benzene.

USE: Solvent; molten acetamide is an excellent solvent for many organic and inorganic compounds. Solubilizer; renders sparingly soluble substances more soluble in water by mere addition or by fusion. Plasticizer; stabilizer. Manuf methylamine, denaturing alcohol. In organic syntheses.

45. Acetamidine. [143-37-3] Ethanimidamide; ethanamidine; α-amino-α-iminoethane; ethenylamidine; acediamine; acetylamidine. $C_2H_6N_2$; mol wt 58.08. C 41.36%, H 10.41%, N 48.23%. Prepd by passing HCl into a soln of acetonitrile in abs alcohol, then passing NH_3 into the reaction mixture: Pinner, *Ber.* **16**, 1654 (1883); **17**, 178 (1884); A. W. Dox, *Org. Synth.* **8**, 1 (1928); by reduction of nitro-2-methylglyoxaline: R. G. Fargher, *J. Chem. Soc.* **117**, 668 (1920). Prepn from the hydrochloride: I. Crossland, F. S. Grevil, *Acta Chem. Scand. B* **35**, 605 (1981). Review of chemical behavior and use in syntheses: R. L. Shriner, F. W. Neumann, *Chem. Rev.* **35**, 351-425 (1944).

Colorless needles from methylene chloride, mp 66-67°. uv max: 224 nm (ε 4000). pK$_1$ (25°): 12.1. Has a strong alkaline reaction and on slight warming dissociates into ammonia and acetic acid.

Hydrochloride. [124-42-5] SN-4455. $C_2H_6N_2.HCl$. Long, colorless prisms from alcohol, mp 174° (Fargher). Also reported as mp 164-166° (Dox). *Irritant.* Very sol in water. Sol in alcohols. Practically insol in acetone, ether. Hygroscopic. Should be stored in a closed container and in a cool place. If alkali is added to an aq soln, the free base is liberated.

USE: In the synthesis of imidazoles, pyrimidines, triazines.

46. ε-Acetamidocaproic Acid. [57-08-9] 6-(Acetylamino)hexanoic acid; 6-acetamidohexanoic acid; acetaminocaproic acid; acexamic acid; N-acetyl-6-aminohexanoic acid. $C_8H_{15}NO_3$; mol wt 173.21. C 55.47%, H 8.73%, N 8.09%, O 27.71%. Prepn: Offe, *Z. Naturforsch.* **2b**, 182 (1947); FR M2332 (1964 to Rowa), *C.A.* **61**, 577b (1964). Anti-inflammatory activity of salts: O. Guillard *et al.*, *Pharmacology* **34**, 296 (1987). Clinical comparison with ranitidine in gastroduodenal ulcer: M. J. Varas Lorenzo, *Curr. Ther. Res.* **39**, 19 (1986).

Crystals from acetone, mp 104-105.5° (**FR M2332**), mp 112° (Offe).

Sodium salt. [7234-48-2] Plastenan. $C_8H_{14}NNaO_3$; mol wt 195.19.

Zinc salt. [70020-71-2] Zinc acexamate; Copinal. $C_{16}H_{28}N_2$-O_6Zn; mol wt 409.82.

THERAP CAT: Anti-inflammatory; zinc salt as antiulcerative.

47. Acetaminophen. [103-90-2] N-(4-Hydroxyphenyl)-acetamide; 4'-hydroxyacetanilide; p-hydroxyacetanilide; p-acetamidophenol; p-acetaminophenol; p-acetylaminophenol; N-acetyl-p-aminophenol; paracetamol; Acamol; Alpiny; Alvedon; Anhiba; Ben-u-ron; Calpol; Calradol; Captin; Dafalgan; Datril; Dirox; Disprol; Doliprane; Dolitabs; Dolviran; Efferalgan; Efferalganodis; Enelfa; Expandox; Fensum; Geluprane; Hedex; Malex; Mejoralito; Panadol; Panamax; Panodil; Pasolind N; Perfalgan; Sanipirina; Sedalito; Tempra; Tylenol. $C_8H_9NO_2$; mol wt 151.17. C 63.56%, H 6.00%, N 9.27%, O 21.17%. Synthetic non-opiate analgesic. Prepn from p-nitrophenol: Morse, Ber. **11**, 232 (1878); Tingle, Williams, Am. Chem. J. **37**, 63 (1907); from p-aminophenol: Lumière et al., Bull. Soc. Chim. Fr. [3] **33**, 785 (1905); Fierz-David, Kuster, Helv. Chim. Acta **22**, 94 (1939); Wilbert, De Angelis, US 2998450 (1961 to Warner-Lambert); Bergmann, DE 453577; Chem. Zentralbl. **1928**, I, 2663; Frdl. **16**, 238; from p-hydroxyacetophenone hydrazone: Pearson et al., J. Am. Chem. Soc. **75**, 5907 (1953). Evaluation of renal effects: D. P. Sandler et al., N. Engl. J. Med. **320**, 1238 (1989). Clinical trial in osteoarthritis: A. R. Temple et al., Clin. Ther. **28**, 222 (2006). Molecular toxicology: P. D. Josephy, Drug Metab. Rev. **37**, 581 (2005). Comprehensive description: J. E. Fairbrother, Anal. Profiles Drug Subs. **3**, 1-109 (1974). Review of pharmacology: B. Ameer, D. J. Greenblatt, Ann. Intern. Med. **87**, 202-209 (1977). Review of mechanism of hepatotoxicity: L. P. James et al., Drug Metab. Dispos. **31**, 1499-1506 (2003); of acetaminophen-induced acute liver failure: A. M. Larson et al., Hepatology **42**, 1364-1372 (2005).

Large monoclinic prisms from water, mp 169-170.5°. d_4^{21} 1.293. uv max (ethanol): 250 nm (ε 13800). Freely sol in alcohol. Sol in methanol, ethanol, dimethylformamide, ethylene dichloride, acetone, ethyl acetate, boiling water, 1 N sodium hydroxide. Slightly sol in ether. Very slightly sol in cold water. Practically insol in petr ether, pentane, benzene. LD_{50} in mice (mg/kg): 338 orally, 500 i.p. See: G. A. Stramer et al., Toxicol. Appl. Pharmacol. **19**, 20 (1971); D. C. Dahlin, S. D. Nelson, J. Med. Chem. **25**, 885 (1982).

USE: Manuf azo dyes, photographic chemicals.

THERAP CAT: Analgesic; antipyretic.

THERAP CAT (VET): Analgesic.

48. Acetamiprid. [160430-64-8] (1E)-N-[(6-Chloro-3-pyridinyl)methyl]-N'-cyano-N-methylethanimidamide; (E)-N-cyano-N'-(2-chloro-5-pyridylmethyl)-N'-methylacetamidine; NI-25; Assail; Intruder; Mospilan; Profil; Tri-Star. $C_{10}H_{11}ClN_4$; mol wt 222.68. C 53.94%, H 4.98%, Cl 15.92%, N 25.16%. Neonicotinoid insecticide for use in food crops and ornamental plants. Prepn: K. Ishimitsu et al., WO 9104965; eidem, US 5304566 (1991, 1994 both to Nippon Soda); of tritiated form: B. Latli et al., J. Labelled Compd. Radiopharm. **38**, 971 (1996). Physical properties, biological activity and field trials: H. Takahashi et al., Proc. Br. Crop Prot. Conf. - Pests Dis. **1992**, 89. HPLC determn in soil: M. Tokieda et al., J. Pestic. Sci. **23**, 296 (1998). LC-MS/MS determn in urine: A. Marín et al., J. Chromatogr. B **804**, 269 (2004). Stereochemistry and active conformation: A. Nakayama et al., Pestic. Sci. **51**, 157 (1997). Efficacy against diamondback moth: H. Takahashi et al., J. Pestic. Sci. **24**, 23 (1999); against green peach aphid: eidem, ibid. 270.

White crystals, mp 101.0-103.3°. Soly in water at 25°: 4200 mg/l. Sol in acetone, methanol, ethanol, dichloromethane, chloroform, acetonitrile, tetrahydrofuran. Vapor pressure at 25°: < 1 × 10⁻⁸ mmHg. LD_{50} in male, female rats, male, female mice (mg/kg): 217, 146, 198, 184 orally (Takahashi).

USE: Insecticide.

49. Acetanilide. [103-84-4] N-Phenylacetamide; antifebrin; acetylaniline; acetylaminobenzene. C_8H_9NO; mol wt 135.17. C 71.09%, H 6.71%, N 10.36%, O 11.84%. p-Aminophenol deriv related to acetaminophen, q.v., with analgesic and antipyretic properties. Usually prepd from aniline and acetic acids: A. I. Vogel, Practical Organic Chemistry (London, 3rd ed., 1959) p 577. From aniline and acetyl chloride: Gattermann-Wieland, Praxis des Organischen Chemikers (Berlin, 40th ed., 1961) p 114. From aniline, acetone and ketene: Hurd, Org. Synth. coll. vol. I (New York, 2nd ed., 1941) p 332. Chemistry: G. C. Derick, J. H. Bornman, J. Am. Chem. Soc. **35**, 1269 (1913). Toxicity data: P. K. Smith, W. E. Hambourger, J. Pharmacol. Exp. Ther. **54**, (1935). Monograph: Gross, Acetanilid (Hillhouse Press, New Haven, 1946).

Orthorhombic plates, scales from water. mp 113-115°; bp 304-305°. Slightly burning taste. Appreciably volatile at 95°. d_4^{15} 1.219. pK (28°): 13.0. One gram dissolves in 185 ml water, 20 ml boiling water, 3.4 ml alcohol, 3 ml methanol, 0.6 ml boiling alcohol, 3.7 ml chloroform, 4 ml acetone, 5 ml glycerol, 8 ml dioxane, 18 ml ether, 47 ml benzene. Very sparingly sol in petr ether. Chloral hydrate increases the soly of acetanilide in water. LD_{50} intragastric in rats: 800 mg/kg (Smith, Hambourger).

USE: Manuf of medicinals and dyes; stabilizer for H_2O_2 soln; as addition to cellulose ester varnishes.

50. p-Acetanisidine. [51-66-1] N-(4-Methoxyphenyl)acetamide; p-acetanisidide; methacetin; p-methoxyacetanilide. C_9H_{11}-NO_2; mol wt 165.19. C 65.44%, H 6.71%, N 8.48%, O 19.37%. Prepn by acetylation of p-anisidine: Reverdin, Bucky, Ber. **39**, 2679 (1906); from p-acetylanisole and hydroxylamine-O-sulfonic acid: Sanford et al., J. Am. Chem. Soc. **67**, 1941 (1945).

Cryst powder, feeble, bitter taste, mp 130-132°. Slightly sol in water; sol in alcohol, acetone, chloroform, dil acids, alkalies.

51. Acetarsone. [97-44-9] As-[3-(Acetylamino)-4-hydroxyphenyl]arsonic acid; N-acetyl-4-hydroxy-m-arsanilic acid; 3-acetamido-4-hydroxyphenylarsonic acid; 3-acetamido-4-hydroxybenzenearsonic acid; acetarsol; acetphenarsine; Ehrlich 594; Fourneau 190; F-190; Gynoplix; Orarsan; Spirocid; Stovarsol. C_8H_{10}-$AsNO_5$; mol wt 275.09. C 34.93%, H 3.66%, As 27.24%, N 5.09%, O 29.08%. Pentavalent arsenical formerly used in the treatment of syphilis. Also used in combination with arecoline, q.v. Prepn: Raiziss, Fisher, J. Am. Chem. Soc. **48**, 1323 (1926); Hewitt, King, J. Chem. Soc. **1926**, 823. Toxicity data: H. H. Anderson, C. D. Leake, Proc. Soc. Exp. Biol. Med. **27**, 267 (1930).

Stout prisms from water, dec 240-250°; slight acid taste. Slightly sol in water; freely sol in solns of alkalies and alkali carbonates. Stable at ordinary temps. MLD in rabbits, cats (mg/kg): 125-150, 150-175 orally (Anderson, Leake).

THERAP CAT: Antiprotozoal (Trichomonas).

THERAP CAT (VET): Antiprotozoal.

52. Acetazolamide. [59-66-5] N-[5-(Aminosulfonyl)-1,-3,4-thiadiazol-2-yl]acetamide; 5-acetamido-1,3,4-thiadiazole-2-sulfonamide; 2-acetylamino-1,3,4-thiadiazole-5-sulfonamide; 6063; Acetamox; Atenezol; Défiltran; Diamox; Didoc; Diuriwas; Donmox; Edemox; Fonurit; Glaupax. $C_4H_6N_4O_3S_2$; mol wt 222.24. C 21.62%, H 2.72%, N 25.21%, O 21.60%, S 28.85%. Carbonic anhydrase inhibitor. Prepn: R. O. Roblin, J. W. Clapp, *J. Am. Chem. Soc.* **72**, 4890 (1950); J. W. Clapp, R. O. Roblin, US 2554816 (1951 to Am. Cyanamid). HPLC determn in pharmaceuticals: Z. S. Gomaa, *Biomed. Chromatogr.* **7**, 134 (1993). Effect on retinal circulation: S. M. B. Rassam *et al.*, *Eye* **7**, 697 (1993). Clinical trial in postoperative elevation of intraocular pressure: I. D. Ladas *et al.*, *Br. J. Ophthalmol.* **77**, 136 (1993). Comprehensive description: J. Parasrampuria, *Anal. Profiles Drug Subs. Excip.* **22**, 1-32 (1993). Review of efficacy in acute mountain sickness: L. D. Ried *et al.*, *J. Wilderness Med.* **5**, 34-48 (1994).

Crystals from water, mp 258-259° (effervescence). Weak acid. pKa 7.2. Sparingly sol in practically boiling water. Slightly sol in alcohol, acetone. Very slightly sol in water. Practically insol in carbon tetrachloride, chloroform, ether. Soly (mg/ml): polyethylene glycol-400 87.81; propylene glycol 7.44; ethanol 3.93; glycerin 3.65; water 0.72.

Sodium salt. [1424-27-7] Vetamox.

THERAP CAT: Antiglaucoma; diuretic; in treatment of acute mountain sickness.

THERAP CAT (VET): Diuretic.

53. Acetiamine. [299-89-8] Ethanethioic acid S-[1-[2-(acetyloxy)ethyl]-2-[[(4-amino-2-methyl-5-pyrimidinyl)methyl]formylamino]-1-propen-1-yl] ester; thioacetic acid S-ester with N-[(4-amino-2-methyl-5-pyrimidinyl)methyl]-N-(4-hydroxy-2-mercapto-1-methyl-1-butenyl)formamide acetate; N-[(4-amino-2-methyl-5-pyrimidinyl)methyl]-N-(4-hydroxy-2-mercapto-1-methyl-1-butenyl)formamide O,S-diacetate; 3-acetylthio-4-[(4-amino-2-methyl-5-pyrimidinyl)methyl-N-formylamino]-3-pentenyl acetate; 5-acetoxy-3-acetylthio-2-[(4-amino-2-methyl-5-pyrimidinyl)methyl-N-formylamino]-2-pentene; diacethiamine; O,S-diacetylthiamine; vitamin B_1 O,S-diacetate; D.A.T.; Thianeuron. $C_{16}H_{22}N_4O_4S$; mol wt 366.44. C 52.44%, H 6.05%, N 15.29%, O 17.46%, S 8.75%. A fat-soluble deriv. of vitamin B_1. Prepn: Matsukawa, Kawasaki, *J. Pharm. Soc. Jpn.* **23**, 705, 709 (1953), *C.A.* **48**, 7017e (1954); Takamizawa *et al.*, *Bull. Chem. Soc. Jpn.* **36**, 1214 (1963). Synthesis and pharmacology: Gauthier *et al.*, *Ann. Pharm. Fr.* **21**, 655 (1963). Clinical studies: Wagner, Wagner, *Arzneim.-Forsch.* **16**, 1643 (1966); Blum, Thomas, *Pharmacol. Clin.* **2**, 177 (1970).

cis-form

Colorless prisms from benzene-petr ether, mp 122-123° (dec); from water, mp 123-124°. Soluble in water, methanol, ethanol.

Hydrochloride. [28008-04-0] Nevriton. $C_{16}H_{22}N_4O_4S \cdot HCl$; mol wt 402.89.

THERAP CAT: Vitamin (enzyme cofactor).

54. Acetic Acid. [64-19-7] Glacial acetic acid; ethanoic acid; Aci-Jel. $C_2H_4O_2$; mol wt 60.05. C 40.00%, H 6.71%, O 53.29%. Obtained in the destructive distillation of wood; from acetylene and water, via acetaldehyde by oxidation with air. Manuf pro-

cesses: Bhattacharyya, Sourirajan, *J. Appl. Chem.* **6**, 442 (1956); *eidem, ibid.* **9**, 126 (1959); Elce *et al.*, US 2800504 (1957 to Distillers Co.); Wirth, US 2818428 (1957 to British Petroleum); McKusick and Hoover, US 2940913 and US 2940914 (both 1960 to Du Pont); *Faith, Keyes & Clark's Industrial Chemicals*, F. A. Lowenheim, M. K. Moran, Eds. (Wiley-Interscience, New York, 4th ed., 1975) pp 8-15. Toxicity data: H. F. Smyth *et al.*, *Arch. Ind. Hyg. Occup. Med.* **4**, 119 (1951). *Review:* F. S. Wagner in *Kirk-Othmer Encyclopedia of Chemical Technology* vol. **1** (Wiley-Interscience, New York, 3rd ed., 1978) pp 124-147.

Liquid; pungent odor. $d^{16.67}$ (liq) 1.053; $d^{16.60}$ (solid) 1.266. d_{25}^{25} 1.049. bp 118°. mp 16.7°. n_D^{20} 1.3718. Flash pt, closed cup: 103°F (39°C). *Corrosive, flammable.* Contracts slightly on freezing. Excellent solvent for many organic compounds; also dissolves phosphorus, sulfur and halogen acids. Miscible with water, alcohol, glycerin, glycerol, ether, carbon tetrachloride. Practically insol in carbon disulfide. Weakly ionized in aq solns: pKa 4.74. pH of aq solns $1.0M$ = 2.4; $0.1M$ = 2.9; $0.01M$ = 3.4. LD$_{50}$ in rats (g/kg): 3.53 orally (Smyth).

Incompat. Carbonates, hydroxides, many oxides, phosphates.

Caution: Ingestion may cause severe corrosion of mouth and G.I. tract, with vomiting, hematemesis, diarrhea, circulatory collapse, uremia, death. Chronic exposure may cause erosion of dental enamel, bronchitis, eye irritation, *cf. Patty's Industrial Hygiene and Toxicology* vol. **2C**, G. D. Clayton, F. E. Clayton, Eds. (Wiley-Interscience, New York, 3rd ed., 1982) p 4909-4911.

USE: Manuf various acetates, acetyl compounds, cellulose acetate, acetate rayon, plastics and rubber in tanning; as laundry sour; printing calico and dyeing silk; as acidulant and preservative in foods; solvent for gums, resins, volatile oils and many other substances. Widely used in commercial organic syntheses. Aqueous and nonaqueous acid-base titrations. Trace metal analysis. Pharmaceutic aid (acidifier).

THERAP CAT (VET): Vesicant, caustic, destruction of warts.

55. Acetic Anhydride. [108-24-7] Acetic acid 1,1'-anhydride; acetic oxide; acetyl oxide; ethanoic anhydride. $C_4H_6O_3$; mol wt 102.09. C 47.06%, H 5.92%, O 47.01%. Equivalent to 117.64% acetic acid. Made formerly from sodium acetate and acetyl or sulfuryl chloride; can be obtained from acetaldehyde or acetic acid: *Faith, Keyes & Clark's Industrial Chemicals*, F. A. Lowenheim, M. K. Moran, Eds. (Wiley-Interscience, New York, 4th ed., 1975) pp 16-20. Of industrial importance is also the ketene process, starting with the thermal decompn of acetone: Schmidlin, Bergmann, *Ber.* **43**, 2821 (1910). Review of bulk manufacturing process: V. H. Agreda *et al.*, *Chemtech* **22**, 172-181 (1992). Toxicity study: H. F. Smyth *et al.*, *Arch. Ind. Hyg. Occup. Med.* **4**, 119 (1951). Review of synthetic uses: D. H. Kim, *J. Heterocycl. Chem.* **13**, 179-194 (1976); G. B. Baker *et al.*, *J. Pharmacol. Toxicol. Methods* **31**, 141-148 (1994).

Very refractive liquid; strong acetic odor. *Flammable, corrosive and combustible.* Reacts with alcohol, oxidizing materials, alkaline materials. Flash pt, closed cup: 120°F (49°C). d_4^{15} 1.080. mp −73°. bp 139°. n_D^{20} 1.3904. Slowly sol in water, forming acetic acid; with alcohol forms ethyl acetate; sol in chloroform, ether. LD$_{50}$ orally in rats: 1.78 g/kg (Smyth).

Caution: Potential symptoms of overexposure are conjunctivitis, lacrimation, corneal edema, opacity and photophobia; nasal, pharyngeal irritation; cough, dyspnea and bronchitis; skin burns, vesiculation and sensitization dermatitis. *See NIOSH Pocket Guide to Chemical Hazards* (DHHS/NIOSH 97-140, 1997) p 2.

USE: Manuf acetyl compounds, cellulose acetates. As acetulizer and solvent in examining wool fat, glycerol, fatty and volatile oils; resins; detection of rosin. Widely used in organic syntheses, e.g., as dehydrating agent in nitrations, sulfonations and other reactions where removal of water is necessary. In prepn of anhydrous acetic acid in nonaqueous titrimetry.

56. Acetimidoquinone. [50700-49-7] 4-(Acetylimino)-2,5-cyclohexadien-1-one; N-(4-oxo-2,5-cyclohexadien-1-ylidene)-acetamide; N-acetyl-p-benzoquinonimine; N-acetylimidoquinone; NAPQI. $C_8H_7NO_2$; mol wt 149.15. C 64.42%, H 4.73%, N 9.39%, O 21.45%. Proposed "ultimate" toxic metabolite of acetaminophen, q.v. Initially prepared by oxidation of acetaminophen with lead tetraacetate and characterized as a Diels-Alder adduct: I. C. Calder et al., J. Med. Chem. **16**, 499 (1973). Electrochemical generation in buffer: D. J. Miner, P. T. Kissinger, Biochem. Pharmacol. **28**, 3285 (1979). Synthesis in stable benzene soln and reactivity: I. A. Blair et al., Tetrahedron Lett. **21**, 4947 (1980). Prepn in cryst form, decomposition kinetics, preliminary toxicological studies: D. C. Dahlin, S. D. Nelson, J. Med. Chem. **25**, 885 (1982). Microsomal reactivity study: G. B. Corcoran, Adv. Exp. Med. Biol. **136B**, 1085 (1982). Review of acetaminophen-induced hepatotoxicity: J. A. Hinson et al., Life Sci. **29**, 107-116 (1981).

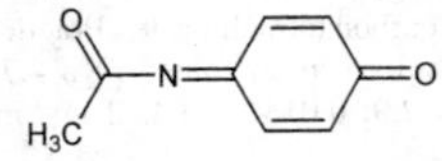

Yellow cubic cryst, mp 74-75°. Sublimes at 45-50° (0.07 mm Hg). uv max (n-hexane): 263, 376 nm (ε 3.3×10⁴, 1.6×10² M⁻¹ cm⁻¹). LD_{50} i.p. in male BALB/c mice: 20 mg/kg (Dahlin, Nelson).

57. Acetoacetanilide. [102-01-2] 3-Oxo-N-phenylbutanamide; α-acetylacetanilide; acetoacetic anilide; β-ketobutyranilide. $C_{10}H_{11}NO_2$; mol wt 177.20. C 67.78%, H 6.26%, N 7.90%, O 18.06%. Prepd by the reaction of ketene dimer with aniline: Chick, Wilsmore, J. Chem. Soc. **1908**, 946; Boese, Ind. Eng. Chem. **32**, 16 (1940); Williams, Krynitsky, Org. Synth. **coll. vol. III**, 10 (1955). By the reaction of aniline with ethyl acetoacetate: Knorr, Ann. **26**, 69 (1886); Roos, Ber. **21**, 624 (1883); Knorr, Reuter, Ber. **27**, 1169 (1894). Enol form (unstable) prepd by pouring an alkaline soln of the keto form in cold, dilute H_2SO_4.

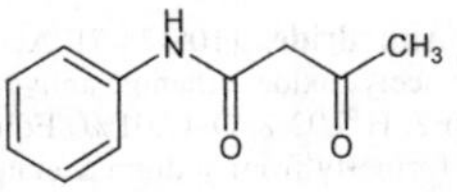

Ketone Form. Leaflets from dilute alcohol, mp 85°. Slightly soluble in water; sol in alcohol, chloroform, ether, hot benzene, hot petr ether, acids, or alkali hydroxide solns. Gives a violet color with ferric chloride.
USE: Manuf of yellow dyes, such as Hansa and benzidine yellows. In rubber compounding. In organic syntheses.

58. Acetoacetic Acid. [541-50-4] 3-Oxobutanoic acid; diacetic acid; acetylacetic acid; acetonecarboxylic acid; 3-ketobutyric acid; 2-ketobutyric acid. $C_4H_6O_3$; mol wt 102.09. C 47.06%, H 5.92%, O 47.01%. Prepd by hydrolysis of ethyl acetoacetate: Krueger, J. Am. Chem. Soc. **74**, 5536 (1952).

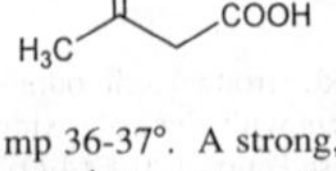

Crystals from ether, mp 36-37°. A strong, but unstable acid. At 100° violently decomposes into acetone and CO_2. Miscible with water, alc.
Caution: Severely irritating to skin, mucous membranes.
USE: In organic syntheses.

59. Acetobromglucose. [572-09-8] α-D-Glucopyranosyl bromide 2,3,4,6-tetraacetate; α-acetobromoglucose; 2,3,4,6-tetraacetyl-α-D-glucopyranosyl bromide. $C_{14}H_{19}BrO_9$; mol wt 411.20. C 40.89%, H 4.66%, Br 19.43%, O 35.02%. Prepd by the action of hydrogen bromide upon anhydrous glucose in acetic anhydride: Redemann, Niemann, Org. Synth. **coll. vol. III**, 11 (1955). See also:

Fischer, Ber. **49**, 584 (1916); Freudenberg, Ber. **60**, 241 (1927); Gattermann-Wieland, Praxis des Organischen Chemikers (de Gruyter, Berlin, 40th ed., 1961) p 340.

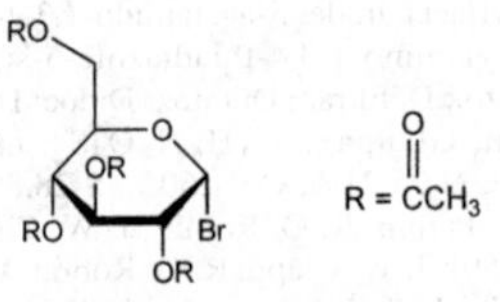

Crystals from isopropyl ether, mp 88-89°. $[\alpha]_D^{19}$ +199.3° (c = 3 in chloroform); $[\alpha]_D^{15}$ +230.3° (c = 9 in benzene). Best stored in a vacuum desiccator. Dec on contact with water. One gram dissolves in 20 ml absolute ethanol, more soluble in methanol. Freely sol in ether, chloroform, acetone, ethyl acetate, benzene. Slightly sol in petr ether.

60. Acetochlor. [34256-82-1] 2-Chloro-N-(ethoxymethyl)-N-(2-ethyl-6-methylphenyl)acetamide; 2-chloro-N-(ethoxymethyl)-6'-ethyl-o-acetotoluidide; 2'-ethyl-6'-methyl-N-(ethoxymethyl)-2-chloroacetanilide; MON-097; MG-02; Harness; Surpass; TopNotch; Trophy. $C_{14}H_{20}ClNO_2$; mol wt 269.77. C 62.33%, H 7.47%, Cl 13.14%, N 5.19%, O 11.86%. Pre-emergent herbicide. Prepn: J. F. Olin, US 3547620 (1970 to Monsanto). Description of physical and biological properties: R. P. Upchurch et al., Proc. North Cent. Weed Control Conf. **25**, 56 (1970). Synthesis: B. G. Zupancic, M. Sopcic, Synthesis **1982**, 942. Field trials on peanuts and soybeans: C. W. Derting et al., Proc. South. Weed Sci. Soc. **25**, 131 (1972); on peanuts: L. R. Hawf, K. R. Frost, Jr., ibid. **38**, 20 (1985). Soil adsorption and bioactivity: J. B. Weber, C. J. Peter, Weed Sci. **30**, 14 (1982). Soil transformation: P. C. C. Feng, Pestic. Biochem. Physiol. **40**, 136 (1991).

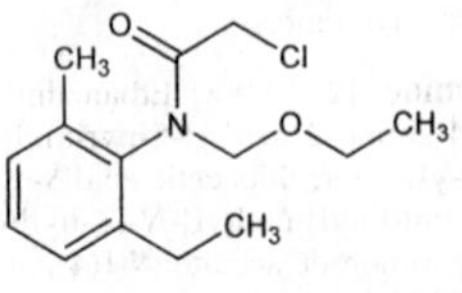

Pale straw-colored oil. $bp_{0.4\ torr}$ 134°. n_D^{20} 1.5272. Sol in water (25°): 400 ppm. LD_{50} orally in rats: 1160 mg/kg (Upchurch).
USE: Herbicide.

61. Acetohexamide. [968-81-0] 4-Acetyl-N-[(cyclohexylamino)carbonyl]benzenesulfonamide; 1-[(p-acetylphenyl)sulfonyl]-3-cyclohexylurea; 3-cyclohexyl-1-(p-acetylphenylsulfonyl)urea; N-(p-acetylbenzenesulfonyl)-N'-cyclohexylurea; cyclamide; tsiklamid; Dimelor; Dymelor; Dimeline; Ordimel. $C_{15}H_{20}N_2O_4S$; mol wt 324.40. C 55.54%, H 6.21%, N 8.64%, O 19.73%, S 9.88%. Prepn: GB 912789 (1962 to Lilly); Marshall et al., J. Med. Chem. **6**, 60 (1963). Comprehensive description: C. E. Shafer, Anal. Profiles Drug Subs. **1**, 1-14 (1972); A. A. Al-Badr, H. A. El-Obeid, Anal. Profiles Drug Subs. Excip. **21**, 1-41 (1992).

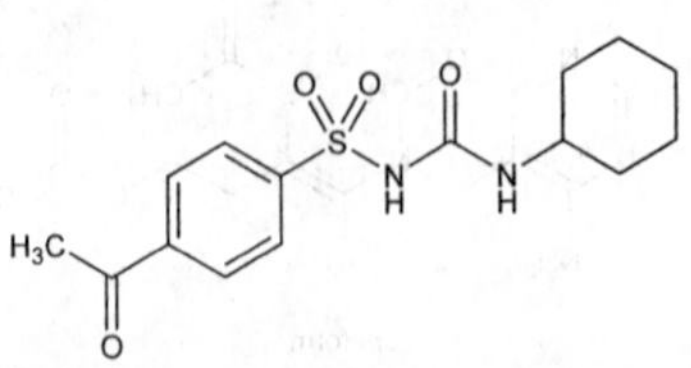

Crystals from 90% aq ethanol, mp 188-190° (**GB 912789**); from dil ethanol, mp 175-177° (Marshall). uv max (methanol): 247, 283 nm. Sol in pyridine, dilute solns of alkali hydroxides. Slightly sol in alcohol, chloroform. Practically insol in water, ether.
THERAP CAT: Antidiabetic.

62. Acetohydroxamic Acid. [546-88-3] N-Hydroxyacetamide; N-acetylhydroxylamine; acetic acid oxime; AHA; Lithostat. $C_2H_5NO_2$; mol wt 75.07. C 32.00%, H 6.71%, N 18.66%, O

42.62%. $CH_3CONHOH$. Urease inhibitor. Prepn: A. Miolati, *Ber.* **25**, 699 (1892); W. M. Wise, W. W. Brandt, *J. Am. Chem. Soc.* **77**, 1058 (1955); H. A. Staab *et al.*, *Ber.* **95**, 1275 (1962); G. Sosnovsky, J. A. Krogh, *Synthesis* **1980**, 654. Inhibition of urease activity: K. Kobashi *et al.*, *Biochim. Biophys. Acta* **65**, 380 (1962); W. N. Fishbein *et al.*, *Nature* **208**, 46 (1965); W. N. Fishbein, P. P. Carbone, *J. Biol. Chem.* **240**, 2407 (1965); D. P. Griffith *et al.*, *Invest. Urol.* **11**, 234 (1973). Metabolism: E. Wolpert *et al.*, *Proc. Soc. Exp. Biol. Med.* **136**, 592 (1971); W. N. Fishbein *et al.*, *J. Pharmacol. Exp. Ther.* **186**, 173 (1973). Pharmacokinetics: S. Feldman *et al.*, *Invest. Urol.* **15**, 498 (1978). Clinical studies in treatment of kidney stones: D. B. Griffith *et al.*, *J. Urol.* **119**, 9 (1978); A. Martelli *et al.*, *Urology* **17**, 320 (1981).

White crystalline powder. mp 89-92°. pKa 8.70. pH (aq soln): 9.39. Freely sol in water, alcohol. Very slightly sol in chloroform.

THERAP CAT: Antiurolithic. Antibacterial adjunct (urinary tract infection).

63. Acetoin. [513-86-0] 3-Hydroxy-2-butanone; 2,3-butanolone; acetyl methyl carbinol; dimethylketol; γ-hydroxy-β-oxobutane. $C_4H_8O_2$; mol wt 88.11. C 54.53%, H 9.15%, O 36.32%. A product of fermentation, also in cream ripened for churning. Obtained by the action of sorbose bacterium or *Mycoderma aceti* on 2,3-butanediol or by the action of fungi, such as *Aspergillus, Penicillium, Mycoderma* on sugar cane juice: Browne, *J. Am. Chem. Soc.* **28**, 467 (1906). By action of yeast on diacetyl: Nagelschmidt, *Biochem. Z.* **186**, 317 (1927). From diacetyl by partial reduction with Zn and acid: Diels, Stephan, *Ber.* **40**, 4338 (1907). Fermentation process: Vergnaud, **US 2529061** (1950 to Usines de Melle).

Liquid. Pleasant odor. mp 15°. d_4^{17} 0.9972. bp_{760} 148°. $n_D^{17.3}$ 1.4190. Miscible with water, alcohol. Sparingly sol in ether, petr ether. Reduces Fehling's soln forming acetic acid. Forms a solid dimer $C_8H_{16}O_4$ on standing or on treatment with granulated zinc. The dimer is easily converted back to the monomer by melting, distilling or dissolving.

64. Acetol. [116-09-6] 1-Hydroxy-2-propanone; hydroxyacetone; acetone alcohol; acetylcarbinol; acetylmethanol; 2-oxopropanol. $C_3H_6O_2$; mol wt 74.08. C 48.64%, H 8.16%, O 43.19%. Widely distributed in nature. Produced in animals as an intermediate in the intrahepatic metabolism of acetone. Also formed by the action of aldose reductase on methylglyoxal, *q.v.* and accumulates in uncontrolled diabetes. Prepn: W. H. Perkin, Jr., *J. Chem. Soc.* **59**, 786 (1891); T. Matsumoto *et al.*, *J. Org. Chem.* **50**, 603 (1985). Isoln from coffee extract: M. Stoll *et al.*, *Helv. Chim. Acta* **50**, 628 (1967). HPLC determn in serum: J. P. Casazza, J. L. Fu, *Anal. Biochem.* **148**, 344 (1985). Metabolism and potential role in diabetic complications: D. L. Vander Jagt *et al.*, *J. Biol. Chem.* **267**, 4364 (1992). Use in peptide synthesis: B. Kundu, *Tetrahedron Lett.* **33**, 3193 (1992).

Colorless oil with peculiar odor. bp_{20} 50°; bp_{200} 105-106°; bp_{760} 147° (dec). n_D^{20} 1.4235. d_4^{20} 1.0872. Misc with water.

USE: Reagent in organic synthesis; protecting group for the synthesis of peptides.

65. Acetone. [67-64-1] 2-Propanone; dimethylformaldehyde; dimethyl ketone; β-ketopropane; pyroacetic ether. C_3H_6O; mol wt 58.08. C 62.04%, H 10.41%, O 27.55%. Obtained by fermentation (by-product of butyl alcohol manufacture) or by chemical synthesis from isopropanol (as chief product); from cumene (by-product in phenol manufacture); from propane (by-product of oxidation-cracking); *Faith, Keyes & Clark's Industrial Chemicals*, F. A.

Lowenheim, M. K. Moran, Eds. (Wiley-Interscience, New York, 4th ed., 1975) pp 21-25. Toxicity: Smyth *et al.*, *Ind. Hyg. J.* **23**, 95 (1962). *Review:* W. L. Howard in *Kirk-Othmer Encyclopedia of Chemical Technology* vol. 1 (Wiley-Interscience, New York, 5th ed., 2004) pp 160-177. Review of toxicology and human exposure: *Toxicological Profile for Acetone* (PB95-100095, 1994) 276 pp; of use in chemical ionization MS: S. Prabhakar, M. Vairamani, *Mass Spectrom. Rev.* **16**, 259-281 (1997).

Volatile liquid; characteristic odor; pungent, sweetish taste. *Flammable.* d_{25}^{25} 0.788. bp 56.5°. mp −94°. n_D^{20} 1.3591. Flash pt, closed cup: 0°F (−18°C). Miscible with water, alcohol, DMF, chloroform, ether, most volatile oils. LD_{50} in rats: 10.7 ml/kg orally (Smyth).

Caution: Potential symptoms of overexposure are irritation of eyes, nose and throat; headache, dizziness, CNS depression; dermatitis. *See NIOSH Pocket Guide to Chemical Hazards* (DHHS/NIOSH 97-140, 1997) p 2.

USE: Solvent for fats, oils, waxes, resins, rubber, plastics, lacquers, varnishes, rubber cements. Versatile reagent in organic synthesis. Manuf of coatings, plastics, pharmaceuticals and cosmetics. In production of other solvents and intermediates including: methyl isobutyl ketone, mesityl oxide, acetic acid (ketene process), diacetone alcohol, bisphenol A, methyl methacrylate, explosives, rayon, photographic films, isoprene. In storing acetylene gas (takes up about 24 times its vol of the gas). In extraction of various principles from animal and plant substances; in paint and varnish removers; purifying paraffin; hardening and dehydrating tissues. Analytical uses in liquid chromatography; cleaning glassware. Pharmaceutic aid (solvent).

66. Acetone Cyanohydrin. [75-86-5] 2-Hydroxy-2-methylpropanenitrile; 2-methyllactonitrile; α-hydroxyisobutyronitrile. C_4H_7NO; mol wt 85.11. C 56.45%, H 8.29%, N 16.46%, O 18.80%. Prepd by adding acetone to sodium or potassium cyanide in water and treating with H_2SO_4 at below 20°: Welch, Clemo, *J. Chem. Soc.* **1928**, 2629; Cox, Stormont, *Org. Synth.* **coll. vol. II**, 7 (1943). Continuous production from HCN and aq acetone: G. Barsky, **US 2731490** (1956). Toxicity study: H. F. Smyth *et al.*, *Am. Ind. Hyg. Assoc. J.* **23**, 95 (1962).

Liquid. *Poisonous.* d_4^{25} 0.9267; d_4^{19} 0.932. mp −19°. bp_{15} 81°; bp_{20} 88-90°; bp_{23} 82°; bp_{760} 95°. n_D^{19} 1.40002; n_D^{15} 1.3980. Freely sol in water and in the usual organic solvents. Practically insol in petr ether and carbon disulfide. Decomposes readily to form hydrogen cyanide, *q.v.*, which is highly toxic. LD_{50} orally in rats: 0.17 g/kg (Smyth).

Caution: Potential symptoms of overexposure are irritation of eyes, skin, respiratory system; dizziness, weakness, headache, confusion, convulsions; liver, kidney injury; pulmonary edema, asphyxia. *See NIOSH Pocket Guide to Chemical Hazards* (DHHS/NIOSH 97-140, 1997) p 4.

USE: In preparative organic chemistry for transcyanohydrination, such as preparing the 17-monocyanohydrin of a 3,17-diketo steroid by hydrogen cyanide exchange with the reagent. *See:* Ercoli, de Ruggieri, *J. Am. Chem. Soc.* **75**, 650 (1953).

67. Acetonedicarboxylic Acid. [542-05-2] 3-Oxopentanedioic acid; 3-oxoglutaric acid; β-ketoglutaric acid. $C_5H_6O_5$; mol wt 146.10. C 41.11%, H 4.14%, O 54.75%. Prepd by the action of fuming sulfuric acid on citric acid: Willstätter, Pfannenstiel, *Ann.* **422**, 5 (1921); Ingold, Nickolls, *J. Chem. Soc.* **121**, 1642 (1922); R. Adams *et al.*, *Org. Synth.* **coll. vol. I**, 10 (2nd ed., 1941). By the action of *Aspergillus niger* on ammonium citrate mixed with some citric acid: Walker *et al.*, *J. Chem. Soc.* **1927**, 3050; also by B. *pyocyaneus:* Butterworth, Walker, *Biochem. J.* **23**, 931 (1929). Decomposition studies: Wiig, *J. Phys. Chem.* **32**, 961 (1928). Intermediate in synthesis of tropinone: R. Robinson, *J. Chem. Soc.* **111**, 762 (1917); C. Schöpf, G. Lehmann, *Ann.* **518**, 1 (1921).

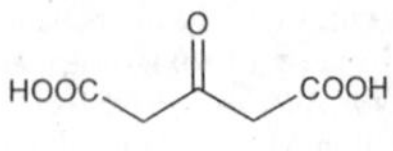

Needles from ethyl acetate, mp 138° (dec). The pure compound may be stored over P_2O_5 for several months. Crude material dec after a few hours. Decomposed by hot water, acids or alkalies to CO_2 and acetone. pK (25°) 3.10. Very sol in water and alcohol; slightly sol in ethyl acetate, ether. Insol in chloroform, ligroin, benzene.

USE: In organic synthesis.

68. Acetone Sodium Bisulfite. [540-92-1] 2-Hydroxy-2-propanesulfonic acid sodium salt (1:1); acetone sulfite. $C_3H_7NaO_4$-S; mol wt 162.13. C 22.22%, H 4.35%, Na 14.18%, O 39.47%, S 19.77%.

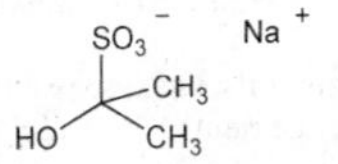

Crystals, fatty feel, slight SO_2 odor. Freely sol in water, sparingly in alcohol. Decomposed by acids.

USE: In photography; manuf of pure acetone; dyeing and printing texiles; in organic syntheses.

69. Acetonitrile. [75-05-8] Methyl cyanide; cyanomethane; ethanenitrile. C_2H_3N; mol wt 41.05. C 58.52%, H 7.37%, N 34.12%. CH_3CN. Small amounts occur in coal tar. Obtained commercially as a byproduct in manuf of acrylonitrile, *q.v.* Prepn by dehydration of acetamide: H. Adkins, B. H. Nissen, *J. Am. Chem. Soc.* **46**, 130 (1924); A. I. Vogel, *Practical Organic Chemistry* (Longmans, London, 3rd ed., 1959) p 407; Gattermann-Wieland, *Praxis des Organischen Chemikers* (de Gruyter, Berlin, 40th ed., 1961) p 125; or from acetylene and ammonia: **DE 365432**; *Chem. Zentralbl.* **1924**, I, 2398; **1925 II**, 1563. Review of purification methods: *Techniques of Chemistry* **vol. II**, entitled "Organic Solvents", J. A. Riddick, W. B. Bunger, Eds. (Wiley-Interscience, New York, 3rd ed., 1970) pp 798-805. Purification and use as a solvent for inorganic fluorides: J. M. Winfield, *J. Fluorine Chem.* **25**, 91 (1984).

Liquid. Ether-like odor. *Poisonous, flammable, combustible.* Burns with a luminous flame. mp −45.7°. bp$_{760}$ 81.6°. Flash pt, open cup: 6°C (42°F). d_4^{15} 0.78745; d_4^{30} 0.77125. n_D^{15} 1.34604, n_D^{30} 1.33934. Dielectric constant at 25° = 37.5. Surface tension at 20° = 29.04 dynes/cm. Misc with water, methanol, methyl acetate, ethyl acetate, acetone, ether, acetamide solutions, chloroform, carbon tetrachloride, ethylene chloride and many unsaturated hydrocarbons. Immiscible with many saturated hydrocarbons (petroleum fractions). Constant boiling mixture with water contains 16% H_2O and bp 76°. Log P (octanol/water): −0.34. Vapor pressure (20°): 74 mmHg. pKa: 25. LD_{50} orally in rats: 3800 mg/kg. *See:* H. F. Smyth, C. P. Carpenter, *J. Ind. Hyg. Toxicol.* **30**, 63 (1968).

Caution: Potential symptoms of overexposure are irritation of nose, throat; asphyxia; nausea, vomiting; chest pain; weakness; stupor, convulsions. *See NIOSH Pocket Guide to Chemical Hazards* (DHHS/NIOSH 97-140, 1997) p 4.

USE: In organic synthesis as starting material for acetophenone, α-naphthaleneacetic acid, thiamine, acetamidine. To remove tars, phenols, and coloring matter from petroleum hydrocarbons. To extract fatty acids from fish liver oils and other animal and vegetable oils. Polar solvent in non-aqueous titrations; non-aqueous solvent for inorganic salts. In HPLC, UV, and electrochemistry applications. Facilitates reactions between organic substrates and inorganic materials.

70. Acetonylacetone. [110-13-4] 2,5-Hexanedione; α,β-diacetylethane. $C_6H_{10}O_2$; mol wt 114.14. C 63.14%, H 8.83%, O 28.03%. Prepn by decarboxylation of diacetosuccinic ester (from sodium acetoacetic ester and iodine): Paal, *Ber.* **18**, 58 (1885); Knorr, *Ber.* **22**, 2100 (1889); by hydrolysis of 2,5-dimethylfuran:

Perkins, Toussaint, **US 2052652** (1936 to Union Carbide). Toxicity data: H. F. Smyth, C. P. Carpenter, *J. Ind. Hyg. Toxicol.* **26**, 269 (1944).

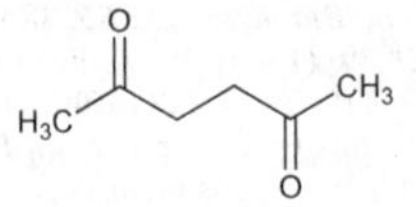

Liquid, gradually turns yellow. d_4^{20} 0.970. mp −9°. bp 188°, n_D^{20} 1.449. Misc with water, alcohol, ether. LD_{50} orally in rats: 2.7 g/kg (Smyth, Carpenter).

Caution: A mild local irritant. High concns cause narcosis.

71. Acetophenone. [98-86-2] 1-Phenylethanone; phenyl methyl ketone; acetylbenzene; hypnone. C_8H_8O; mol wt 120.15. C 79.97%, H 6.71%, O 13.32%. Made from benzene and acetylchloride in presence of aluminum chloride; catalytically from acetic and benzoic acids. Prepn from benzene and acetic anhydride: Adams, *J. Am. Chem. Soc.* **46**, 1889 (1924); A. I. Vogel, *Practical Organic Chemistry* (Longmans, London, 3rd ed., 1959) p 730; Gattermann-Wieland, *Die Praxis des Organischen Chemikers* (de Gruyter, Berlin, 40th ed., 1961) p 297. Toxicity study: H. F. Smyth, C. P. Carpenter, *J. Ind. Hyg. Toxicol.* **30**, 63 (1948).

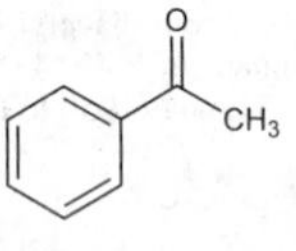

Liquid. Forms laminar crystals at low temp. mp 20.5°. d_{15}^{15} 1.033. bp 202°. n_D^{20} 1.5339. Flash pt, closed cup: 170.6°F (77°C); open cup: 180°F (82.2°C). Slightly sol in water; freely in alcohol, chloroform, ether, fatty oils, glycerol. Sol in concd H_2SO_4 with orange color. LD_{50} orally in rats: 0.90 g/kg (Smyth, Carpenter).

USE: In perfumery to impart an orange-blossom-like odor; catalyst for the polymerization of olefins; in organic syntheses, esp. as photosensitizer.

72. Acetotoluide. $C_9H_{11}NO$; mol wt 149.19. C 72.46%, H 7.43%, N 9.39%, O 10.72%. Prepn of *m*-, *o*-, and *p*-acetotoluidides, from *m*-, *o*-, or *p*-toluidine and acetic acid: Gasopoulos, *Ber.* **59B**, 2187 (1926).

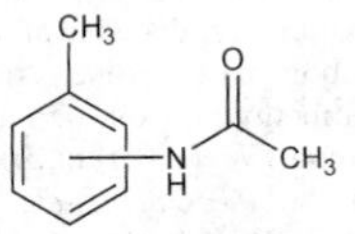

m-**Acetotoluide.** [537-92-8] *N*-(3-Methylphenyl)acetamide; *m*-acetotoluidide; *m*-tolylacetamide; aceto-*m*-aminotoluene; *m*-methylacetanilide; *N*-acetyl-*m*-toluidine. Needles, mp 65.5°, bp 303°. Slightly sol in water; freely sol in alc, ether.

o-**Acetotoluide.** [120-66-1] Crystals, mp 110°, bp 296°. Slightly sol in water; sol in alc, chloroform, ether. Determn of soly at various temps: Hall *et al.*, *J. Phys. Chem.* **37**, 1087 (1933).

p-**Acetotoluide.** [103-89-9] Crystals, mp 153°, bp 307°. Very slightly sol in water; sol in alc, ether, ethyl acetate, glacial acetic acid.

73. Acetoxime. [127-06-0] 2-Propanone oxime; acetone oxime; β-isonitrosopropane. C_3H_7NO; mol wt 73.10. C 49.29%, H 9.65%, N 19.16%, O 21.89%. Prepd by shaking an aqueous soln of hydroxylamine with acetone and extracting with ether: W. L. Semon, *Org. Synth.* **coll. vol. I**, 318 (2nd ed., 1941). Ionization constants: C. V. King, A. P. Marion, *J. Am. Chem. Soc.* **66**, 977 (1944).

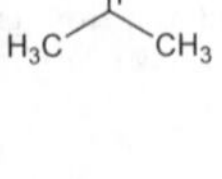

Columnar prisms. Volatilizes quickly in air. Odor resembling that of chloral hydrate. d_4^{62} 0.9113. mp 60°. bp_{728} 134.8°. Neutral reaction. pK (24.9°): 12.42. uv spectrum: Hartley, Dobbie, *J. Chem. Soc.* **77**, 326 (1900). Freely soluble in water, alcohol, ether, petr ether. Can be extracted with ether from neutral water solns, but not from alkaline or acidic solns.

74. Acetoxolone. [6277-14-1] ($3\beta,20\beta$)-3-(Acetyloxy)-11-oxoolean-12-en-29-oic acid; 3β-hydroxy-11-oxoolean-12-en-30-oic acid acetate; 3-*O*-acetyl-18β-glycyrrhetic acid; acetylglycyrrhetinic acid; glycyrrhetic acid acetate. $C_{32}H_{48}O_5$; mol wt 512.73. C 74.96%, H 9.44%, O 15.60%. Deriv of enoxolone, *q.v.;* homolog of carbenoxolone, *q.v.* Prepn: J. M. Beaton, F. S. Spring, *J. Chem. Soc.* **1955**, 3126. Prepn of the aluminum salt: A. Bonati, *ZA* **7006901**; *idem,* **US 3764618** (1971, 1973 both to Dott. Inverni & Della Beffa). NMR study of the acetate: M. Mousseron-Canet *et al., Bull. Soc. Chim. Fr.* **12**, 4668 (1967). Pharmacology of the aluminum salt: C. Capra, *Fitoterapia* **40**, 8 (1969), *C.A.* **73**, 64754 (1970). Absorption in rats: B. D. Cameron *et al., Arzneim.-Forsch.* **26**, 1680 (1976). Analysis of gastric protective effect: P. Altmayer *et al., ibid.* **31**, 853 (1981).

Crystals, mp 322-325°. $[\alpha]_D^{20}$ +141°.

Aluminum salt. [29728-34-5] Almacet. $C_{96}H_{141}AlO_{15}$; mol wt 1562.15. White powder, mp 286-290°. $[\alpha]_D^{20}$ +126±2° (c = 1 in chloroform). Insol in water. Sol in most organic solvents. LD_{50} in male rats (mg/kg): >3300 orally, and i.p. (Bonati).

THERAP CAT: Antiulcerative.

75. Aceturic Acid. [543-24-8] *N*-Acetylglycine; acetamidoacetic acid; acetylaminoacetic acid; acetylglycocoll; ethanoylaminoethanoic acid. $C_4H_7NO_3$; mol wt 117.10. C 41.03%, H 6.03%, N 11.96%, O 40.99%. Prepd by warming glycine with a slight excess of acetic anhydride in benzene: Radenhausen, *J. Prakt. Chem.* [2] **52**, 437 (1895); by warming glycine in glacial acetic acid with the stoichiometric amount of acetic anhydride: Dakin, *J. Biol. Chem.* **82**, 443 (1929).

Long needles from water, mp 206-208°. pK (25°) 3.64. Sol in water at 15° = 2.7%. Moderately soluble in alcohol. Slightly sol in acetone, chloroform, glacial acetic acid. Practically insol in ether, benzene. Forms stable salts with organic bases.

76. Acetylacetone. [123-54-6] 2,4-Pentanedione; diacetylmethane. $C_5H_8O_2$; mol wt 100.12. C 59.98%, H 8.05%, O 31.96%. Made by the action of sodium on ethyl acetate and acetone or by the action of anhydrous aluminum chloride on acetyl chloride in the presence of an inert solvent: Combes, *Compt. Rend.* **103**, 814 (1886); *Ann. Chim. (Paris)* **12**, 199 (1887); Claisen, *Ber.* **38**, 695 (1905); Hunt, **US 2737528** (1956 to Shawinigan Chem.); Georgieff, **US 2834811** (1958); Gattermann-Wieland, *Praxis des Organischen Chemikers* (de Gruyter, Berlin, 40th ed, 1961) p 219. Toxicity data: Carpenter, *J. Ind. Hyg. Toxicol.* **31**, 343 (1949).

Colorless or slightly yellow, flammable liquid; pleasant odor. d 0.976. bp 140.5°. mp −23°. n_D^{20} 1.4512. One part dissolves in about 8 parts of water. Miscible with alcohol, benzene, chloroform, ether, acetone, glacial acetic acid. LC_{50} (4 hrs) in rats: 1000 ppm (Carpenter).

Caution: Mild irritant to skin, mucous membranes.

USE: Forms organometallic complexes which are used as gasoline additives, lubricant additives, driers for varnishes and printer's inks, fungicides, insecticides, colors.

77. Acetyl Bromide. [506-96-7] C_2H_3BrO; mol wt 122.95. C 19.54%, H 2.46%, Br 64.99%, O 13.01%. Prepd from phosphorus tribromide and glacial acetic acid or acetic anhydride: Burton, Degering, *J. Am. Chem. Soc.* **62**, 227 (1940).

Fuming liquid. *Corrosive.* bp 76°. d^9 1.52. mp −96°. Violently decomposed by water, or by alcohol. Miscible with ether, chloroform, benzene. *Protect from water.*

78. α-Acetylbutyrolactone. [517-23-7] 3-Acetyldihydro-2(3*H*)-furanone; α-(2-hydroxyethyl)acetoacetic acid γ-lactone; α-acetyl-γ-hydroxybutyric acid γ-lactone; α-acetobutyrolactone. $C_6H_8O_3$; mol wt 128.13. C 56.24%, H 6.29%, O 37.46%. Prepn from sodium acetoacetate and ethylene oxide in abs alcohol: Knunyantz *et al., C. R. Acad. Sci. USSR* [N.S.] **1**, 312 (1934), *C.A.* **28**, 4383 (1934); Feofilaktov, Onishchenko, *J. Gen. Chem. USSR* **9**, 304 (1939), *C.A.* **34**, 378 (1940); Forman; Johnson, **US 2397134** and **US 2443827** (1946, 1948 both to U.S. Industrial Chemicals).

Liquid; ester-like odor. Soly in water: 20% v/v; soly of water in lactone: 12% v/v. d_4^{20} 1.1846; d_{20}^{20} 1.185-1.189; bp_{30} 142-143°; bp_{18} 130-132°; bp_5 107-108°. n_D^{20} 1.4562 (Knunyantz); n_D^{20} 1.4590 (Feofilaktov). Acquires a blue to bluish-purple color when in contact with iron.

Caution: Irritating to skin, mucous membranes.

USE: In synthesis of 3,4-disubstituted pyridines; of 5-(β-hydroxethyl)-4-methylthiazole.

79. Acetylcarnitine. [3040-38-8]; [14992-62-2] (DL-form). (2*R*)-2-(Acetyloxy)-3-carboxy-*N,N,N*-trimethyl-1-propanaminium inner salt; L-(3-carboxy-2-hydroxypropyl)trimethylammonium hydroxide inner salt acetate; acetyl-L-carnitine; ALCAR. $C_9H_{17}NO_4$; mol wt 203.24. C 53.19%, H 8.43%, N 6.89%, O 31.49%. Naturally occurring ester of carnitine, *q.v.*, important acetyl donor in metabolic processes. Prepn from carnitine extracted from meat: R. Engeland, *Ber.* **42**, 2457 (1909); R. Krimberg, W. Wittandt, *Biochem. Z.* **251**, 229 (1932); E. Strack *et al., Z. Physiol. Chem.* **238**, 183 (1936). Resolution of isomers: K. Brendel, R. Bressler, *Biochim. Biophys. Acta* **137**, 98 (1967). Extraction from calf brain tissue: E. A. Hosein, A. Orzeck, *Int. J. Neuropharmacol.* **3**, 71 (1964). Enzymatic determn in serum: K. Tomita *et al., J. Pharm. Biomed. Anal.* **24**, 1147 (2001). Clinical trial in cognitive impairment in the elderly: G. Salvioli, M. Neri, *Drugs Exp. Clin. Res.* **20**, 169 (1994). Review of pharmacology in the CNS: L. Janiri, E. Tempesta, *Int. J. Clin. Pharmacol. Res.* **3**, 295-306 (1983); of bioactivity and clinical studies in Alzheimer's disease and geriatric depression: J. W. Pettegrew *et al., Mol. Psychiatry* **5**, 616-632 (2000); of pharmacokinetics and metabolism: C. J. Rebouche, *Ann. N.Y. Acad. Sci.* **1033**, 30-41 (2004).

Hygroscopic crystals, mp 145° (dec). $[\alpha]_D^{20}$ −19.52° (c = 6). Very sol in water, alcohol. Practically insol in ether.

Hydrochloride. [5080-50-2] Acetyl-L-carnitine chloride; levacecarnine hydrochloride; ST-200; Branigen; Nicetile; Normobren; Zibren. $C_9H_{17}NO_4.HCl$; mol wt 239.70. Stable crystals, mp 187° (dec). Also reported as small rod-shaped crystals from butanol, mp 181° (Strack). $[\alpha]_D^{20}$ −26.9° (c = 9). $[\alpha]_D^{25}$ −28° (c = 2 in water). Very sol in water, sol in alcohol. Practically insol in ether.

THERAP CAT: Nootropic.

80. Acetyl Chloride. [75-36-5] Ethanoic acid chloride. C_2H_3ClO; mol wt 78.50. C 30.60%, H 3.85%, Cl 45.16%, O 20.38%. CH_3COCl. Prepd from acetic acid and chlorine in the presence of phosphorus; from acetic acid and salts of chlorosulfonic acid; from sodium acetate and sulfuryl chloride. Details of prepn: Gattermann-Wieland, *Praxis des Organischen Chemikers* (de Gruyter, Berlin, 40th ed., 1961) p 111; A. I. Vogel, *Practical Organic Chemistry* (Longmans, London, 3rd ed., 1959) p 367. Lab prepn from acetic anhydride and calcium chloride: Gmünder, *Helv. Chim. Acta* **36**, 2021 (1953).

Liquid; pungent odor. *Flammable; corrosive.* Imparts a green tinge to a colorless flame. d 1.104. mp −112°. bp 52°. n_D^{20} 1.3898. Decomposed violently by water or alc. Miscible with benzene, chloroform, ether, glacial acetic acid, petr ether. *Protect from water.*

Caution: May cause severe burns. Avoid contact with skin, eyes, mucous membranes.

USE: Acetylating agent; in testing for cholesterol, determn of H_2O in organic liquids.

81. Acetylcholine Bromide. [66-23-9] 2-(Acetyloxy)-*N*,*N*,*N*-trimethylethanaminium bromide (1:1); Pragmoline; Tonocholin B. $C_7H_{16}BrNO_2$; mol wt 226.11. C 37.18%, H 7.13%, Br 35.34%, N 6.19%, O 14.15%. From trimethylamine and β-bromoethyl acetate: Fourneau, Page, *Bull. Soc. Chim. Fr.* [4] **15**, 544 (1914).

Deliquesc crystals. Very soluble in cold water; decomposed by hot water or alkalies; soluble in alcohol; practically insol in ether. *Keep tightly closed.*

THERAP CAT: Cholinergic.

82. Acetylcholine Chloride. [60-31-1] 2-(Acetyloxy)-*N*,*N*,*N*-trimethylethanaminium chloride (1:1); Acecoline; Arterocoline; Miochol; Ovisot. $C_7H_{16}ClNO_2$; mol wt 181.66. C 46.28%, H 8.88%, Cl 19.51%, N 7.71%, O 17.61%. Prepd from trimethylamine and β-chloroethyl acetate: Fourneau, Page, *Bull. Soc. Chim. Fr.* [4] **15**, 544 (1914). Review of pharmacology: H. Molitor, *J. Pharmacol. Exp. Ther.* **58**, 337 (1936).

Crystalline powder, very deliquescent, mp 149-152°. Hygroscopic. Very sol in cold water. Freely sol in alcohol. Insol in ether. Decomposed by hot water, alkalies. *Keep tightly closed.* LD50 in mice, rats (mg/kg): 20, 22 i.v., 170, 250 s.c., 3000, 2500 i.p. (Molitor).

THERAP CAT: Cholinergic.

83. Acetylcysteine. [616-91-1] *N*-Acetyl-L-cysteine; L-α-acetamido-β-mercaptopropionic acid; *N*-acetyl-3-mercaptoalanine; Acetadote; Brunac; Fabrol; Fluimucil; Fluprowit; Mucocedyl; Mucolator; Mucomyst; Muco Sanigen; Mucosil; Mucret; Neo-Fluimucil; Parvolex; Tixair. $C_5H_9NO_3S$; mol wt 163.19. C 36.80%, H

5.56%, N 8.58%, O 29.41%, S 19.65%. Thiol-containing antioxidant. Prepn: Smith, Gorin, *J. Org. Chem.* **26**, 820 (1961). Prepn and use in treatment of respiratory diseases: Martin, Waller, **US 3184505** (1965 to Mead Johnson). Effect in corneal healing: F. Menna *et al.*, *Bull. Mem. Soc. Fr. Opthalmol.* **94**, 425 (1982); G. Petroutsos *et al.*, *Ophthalmic Res.* **14**, 241 (1982). Toxicity: E. I. Goldenthal, *Toxicol. Appl. Pharmacol.* **18**, 185 (1971). Clinical trial in acetaminophen overdose: M. J. Smilkstein *et al.*, *N. Engl. J. Med.* **319**, 1557 (1988); as renal protectant with radiocontrast agents: M. Tepel *et al.*, *N.Engl.J.Med.* **343**, 180 (2000). *Review:* G. R. McKinney, G. M. Sisson, in *Pharmacological and Biochemical Properties of Drug Substances* **vol. 2**, M. E. Goldberg, Ed. (Am. Pharm. Assoc., Washington, DC, 1979) pp 479-488. Review of use in clinical toxicology: R. J. Flanagan, T. J. Meredith, *Am. J. Med.* **91**, Suppl. 3C, 131S-139S (1991).

Crystals from water, mp 109-110°. Slight acetic odor. Freely sol in water, alcohol. Practically insol in chloroform, ether. LD50 orally in rats: 5050 mg/kg (Goldenthal).

THERAP CAT: Mucolytic; corneal vulnerary; antidote (acetaminophen poisoning).

THERAP CAT (VET): Expectorant.

84. Acetyldigitoxins. $C_{43}H_{66}O_{14}$; mol wt 806.99. C 64.00%, H 8.24%, O 27.76%. Obtained by enzymatic hydrolysis of lanatoside A, *q.v.* Composed of the aglycone digitoxigenin and 3 mols digitoxose, to one of which an acetyl group is attached. Acetyldigitoxin-α differs from acetyldigitoxin-β in the position of the acetyl group. The β-form is obtained either by splitting off the glucose residue from lanatoside A by means of enzymes, or by extraction from the leaves of *Digitalis ferruginea* L., *Scrophulariaceae*. The α-form can be obtained from acetyldigitoxin-β by heating in an anhydrous or aq organic solvent at pH 3.5 to 8: Stoll, Kreis, **US 2776963** (1957 to Sandoz).

α-Form. [1111-39-3] Acylanid. Platelets from methanol, mp 217-221°. $[\alpha]_D^{20}$ +5.0° (c = 0.7 in pyridine). Slightly sol in chloroform.

β-Form. Stout, solvated prisms from methanol losing their solvent of crystn in desiccator. When dry, dec 225°. $[\alpha]_D^{20}$ +16.7° (pyridine). One gram dissolves in 7 to 9 ml chloroform, in 150 ml methanol, in 220 ml amyl alc. Almost insol in ether, water (one gram dissolves in 200 liters H_2O at 20°).

THERAP CAT: Cardiotonic.

85. Acetylene. [74-86-2] Ethyne; ethine. C_2H_2; mol wt 26.04. C 92.25%, H 7.74%. HC≡CH. Manuf from calcium carbide and water: Eastman, **US 3017259** (1962 to Texaco); from methane: Anderson, **US 3051639** (1962 to Union Carbide). Toxicity data: Riggs, *Proc. Soc. Exp. Biol. Med.* **22**, 269 (1925). Review of manuf processes: *Faith, Keyes & Clark's Industrial Chemicals*, F. A. Lowenheim, M. K. Moran, Eds. (Wiley-Interscience, New York, 4th ed., 1975) pp 26-35. *Review:* Nieuwland, Vogt, *The Chemistry of Acetylene* (Reinhold, New York, 1945) pp 1-219. Comprehensive monograph in 2 vols: S. A. Miller, *Acetylene* (Academic Press, New York, 1965); several authors in *Kirk-Othmer Encyclopedia of Chemical Technology* **vol. 1** (Wiley-Interscience, New York, 3rd ed., 1978) pp 192-243.

Gas; odor not unpleasant when pure, but disagreeable when impure (due to phosphine). *Flammable.* mp −81° (subl). At 0° liquifies at 21.5 atm; below 37° (crit temp) liquifies at 68 atm. One liter at 0° and 760 mm weighs 1.165 g; d gas (air = 1) 0.90. Burns brilliantly in air with very sooty flame. Heat of combustion 313 cal. Not explosive at ordinary atmospheric pressure, but at 2 atms or more it is explosive by spark or decomposition. Mixture with air containing more than 3% or less than 65% gas is explosive, max being 1 vol gas and 12.5 vol air. Forms insoluble explosive compounds with copper and silver; hence copper or brass containers must be avoided. One vol dissolves in 1 vol water, in 6 vols glacial acetic acid or alcohol; soluble in ether, benzene. Acetone dissolves 25 vols acetylene at 15° and 760 mm; but 300 vols at 12 atm. LC in rats: 900000 ppm (Riggs).

Caution: Potential symptoms of overexposure are headache, dizziness; asphyxia; direct contact with liquid may cause frostbite. *See NIOSH Pocket Guide to Chemical Hazards* (DHHS/NIOSH 97-140, 1997) p 4.

USE: Illuminant, oxyacetylene welding, cutting, and soldering metals, signalling; pptg metals, particularly Cu; manuf acetaldehyde, acetic acid; fuel for motor boats.

86. Acetylene Dibromide. [540-49-8] 1,2-Dibromoethene; 1,2-dibromoethylene; *sym*-dibromoethylene. $C_2H_2Br_2$; mol wt 185.85. C 12.93%, H 1.08%, Br 85.99%. BrCH=CHBr. Prepd by reaction of tetrabromoethane with Zn and alc followed by sepn of *cis*- and *trans*-forms by fractional distillation: Noyes *et al., J. Am. Chem. Soc.* **72**, 33 (1950).

Liquid, gradually decomposed by air, moisture, or light. d_4^{17} 2.21. n_D^{20} 1.5428. Practically insol in water. Sol in many organic solvents.

cis-**Form.** n_D^{25} 1.5370.

trans-**Form.** n_D^{25} 1.5440.

Caution: Narcotic in high concns.

87. Acetylene Dichloride. [540-59-0] 1,2-Dichloroethene; 1,2-dichloroethylene; *sym*-dichloroethylene; dioform. $C_2H_2Cl_2$; mol wt 96.94. C 24.78%, H 2.08%, Cl 73.14%. ClCH=CHCl. Prepn: Bordner, US 2504919 (1950 to du Pont). Prepn of *trans*-form: Adler, US 2440997 (1948 to Stockholms Superfosfat Fabriks Aktiebolag). Sepn of *cis*- and *trans*-forms by fractional distillation: Wood, Dickinson, *J. Am. Chem. Soc.* **61**, 3259 (1939); Johnsen, Fitzpatrick, *Rec. Trav. Chim.* **70**, 823 (1951); Truce, Barney, *J. Org. Chem.* **27**, 128 (1962). Toxicity studies: D. Gradiski *et al., Eur. J. Toxicol.* **7**, 247 (1974); K. J. Freundt *et al., Toxicology* **7**, 141 (1977). *Review:* V. L. Stevens in *Kirk-Othmer Encyclopedia of Chemical Technology* Vol. 5 (Wiley-Interscience, New York, 3rd ed., 1979) pp 742-745. Review of toxicology and human exposure: *Toxicological Profile for 1,2-Dichloroethene* (PB97-121081, 1996) 198 pp.

Liquid; ethereal, slightly acrid odor; gradually decomposed by air, light and moisture, forming HCl. d ~1.28. bp ~55°. *Highly flammable; extremely corrosive.* Insol in water. Sol in alc, ether and most other organic solvents. LD$_{50}$ i.p. in mice: ~2150 mg/kg (Gradiski).

cis-**Form.** [156-59-2] Liquid. mp −81.5°. bp$_{745}$ 59.6°, bp$_{760}$ 60°. n_D^{25} 1.4435. d_4^{20} 1.2837. Log P (octanol/water): 1.86. Vapor pressure (mm Hg): 180 (20°), 250 (30°). Soly at 25°: water 3.5 g/l. Sol in ether, alc, benzene, acetone, chloroform.

trans-**Form.** [156-60-5] Liquid. mp −49.4°. bp$_{745}$ 47.2°. d_4^{20} 1.2565. Noticeably subject to air oxidation. Log P (octanol/water): 2.09. Vapor pressure (mm Hg): 265 (20°), 410 (30°). Soly at 25°: water 6.3 g/l. Sol in ether, alc, benzene, acetone, chloroform. LD$_{50}$ in rats (ml/kg): 1.0 orally; 60 i.p.; in mice (ml/kg): 3.2 i.p. (Freundt).

Caution: Potential symptoms of overexposure are irritation of eyes and respiratory system; CNS depression. *See NIOSH Pocket Guide to Chemical Hazards* (DHHS/NIOSH 97-140, 1997) p 98. *See also Patty's Industrial Hygiene and Toxicology* vol. **2B**, G. D. Clayton, F. E. Clayton, Eds. (Wiley-Interscience, New York, 3rd ed., 1981) pp 3550-3553.

USE: Solvent for fats, phenol, camphor, etc. Intermediate in synthesis of chlorinated solvents and compounds.

88. Acetyleneurea. [496-46-8] Tetrahydroimidazo[4,5-*d*]-imidazole-2,5-(1*H*,3*H*)-dione; glycoluril; acetylenediureine; acetylenediureine; glyoxaldiureine; acetylene carbamide. $C_4H_6N_4O_2$; mol wt 142.12. C 33.81%, H 4.26%, N 39.42%, O 22.51%. Prepd by the sodium amalgam reduction of allantoin: Biltz, Schiemann, *J. Prakt. Chem.* **113**, 77 (1926); from glyoxal and urea: Reibnitz, US 2731472 (1956 to BASF).

Crystals, dec at about 300°. Slightly sol in cold, more sol in hot water; sol in warm mineral acids, warm ammonia.

89. *N*-Acetyl-L-farnesylcysteine. [135304-07-3] *N*-Acetyl-*S*-[(2*E*,6*E*)-3,7,11-trimethyl-2,6,10-dodecatrien-1-yl]-L-cysteine; *N*-acetyl-*S*-*trans,trans*-farnesyl-L-cysteine; AFC; Arazine. $C_{20}H_{33}$-NO$_3$S; mol wt 367.55. C 65.36%, H 9.05%, N 3.81%, O 13.06%, S 8.72%. Anti-inflammatory G-protein modulator (GPM). Inhibitor of isoprenylcysteine methyltransferase, an enzyme involved in G-protein signaling. Prepn: J. B. Stock, US 5043268 (1991 to Princeton Univ.). Enzyme inhibiting activity: C. Volker *et al., J. Biol. Chem.* **266**, 21515 (1991). Structure-activity study: B. A. Gilbert *et al., J. Am. Chem. Soc.* **114**, 3966 (1992). Effect in mouse models of skin inflammation: J. S. Gordon *et al., J. Invest. Dermatol.* **128**, 643 (2008).

Colorless oil. [α]$_D$ −23.4° (c = 0.01 in methanol).

USE: In cosmetics as topical anti-inflammatory and antioxidant.

90. Acetyl Iodide. [507-02-8] C_2H_3IO; mol wt 169.95. C 14.13%, H 1.78%, I 74.67%, O 9.41%. Prepd from acetyl chloride and HI: Gustus, Stevens, *J. Am. Chem. Soc.* **55**, 374 (1933).

Liquid. *Corrosive.* Suffocating odor; fumes and turns brown in air. bp 108°, bp$_{735}$ 104-106°; d_4^{20} 2.0674; n_D^{20} 1.5491. Dec by water or alcohol; sol in benzene, ether. *Keep tightly closed and protected from light.*

Caution: A strong irritant. Avoid contact with skin. Vapors can cause pulmonary edema.

91. Acetylleucine. [99-15-0] *N*-Acetylleucine; acetyl-DL-leucine; α-acetamidoisocaproic acid; Tanganil. $C_8H_{15}NO_3$; mol wt 173.21. C 55.47%, H 8.73%, N 8.09%, O 27.71%. Prepn: E. Fisher, *Ber.* **34**, 433 (1901); W. C. Cahill, I. F. Burton, *J. Biol. Chem.* **132**, 161 (1940); H. R. Snyder *et al., J. Am. Chem. Soc.* **67**, 310 (1945). Effects on central vestibular neurons: N. Vibert, P.-P. Vidal, *Eur. J. Neurosci.* **13**, 735 (2001).

Large, flat needles, mp 160-161°. Sol in water, alcohol.

Monoethanolamine salt. [149-90-6] RP-7452. $C_8H_{15}NO_3 \cdot C_2$-H$_7$NO; mol wt 234.30. Prepn: P. Gailliot *et al.*, US 2941924 (1960 to Rhône-Poulenc). Crystals, mp ~150°. Soly in water: >20%; in alcohol: ~1%.

THERAP CAT: Antivertigo agent.

92. *N*-Acetylmethionine. [65-82-7] *N*-Acetyl-L-methionine. $C_7H_{13}NO_3S$; mol wt 191.25. C 43.96%, H 6.85%, N 7.32%, O 25.10%, S 16.76%. Naturally occurring derivative of methionine, *q.v.* Prepn of L-form and racemate: V. du Vigneaud, C. E. Meyer, *J. Biol. Chem.* **98**, 295 (1932); of D-form and racemate: G. P. Wheeler, A. W. Ingersoll, *J. Am. Chem. Soc.* **73**, 4604 (1951). Resolution of isomers: B. F. Tullar, US 3056799 (1962 to Sterling Drug). Evaluation of stability and organoleptic suitability for food supplementation: R. Damico, *J. Agric. Food Chem.* **23**, 30 (1975). Methionine replacement activity of isomers: D. H. Baker, *J. Nutr.* **109**, 970 (1979). Detection and stability in model food systems: K. L. Schleske, J. J. Warthesen, *J. Agric. Food Chem.* **30**, 1172 (1982).

Crystals, mp 104° (Tullar). $[\alpha]_D^{25}$ −20.3°(c = 4 in water). Also reported as crystals from acetone + ether, mp 102-103° (Damico).

DL-Form. [1115-47-5] Large prisms from water, mp 114-115°. Soly at 25° (g/100 ml): water, 9.12; acetone, 10.0; ethyl acetate, 2.29; chloroform, 1.33.

D-Form. [1509-92-8] Plates from water or ethyl acetate, mp 104-105°. $[\alpha]_D^{25}$ +20.3° (c = 4 in water). Soly at 25° (g/100 ml): water, 30.7; acetone, 29.6; ethyl acetate, 7.04; chloroform, 6.43.

USE: Food additive for methionine supplementation.

93. Acetyl Nitrate. [591-09-3] Acetic acid anhydride with nitric acid. $C_2H_3NO_4$; mol wt 105.05. C 22.87%, H 2.88%, N 13.33%, O 60.92%. CH_3COONO_2. Prepn from acetic anhydride and N_2O_5: Boh, *Ann. Chim.* [11] **20**, 437 (1945).

Fuming, mobile, hygroscopic liquid. Should be colorless. d_4^{15} 1.24. bp_{70} 22°. Although it may be stored in the dark over P_2O_5 for the weekend, it should be used in *statu nascendi* to avoid explosions. Always explodes when heated suddenly over 60° or when in contact with HgO. Explosions have occurred upon contact with ground glass surfaces: König, *Angew. Chem.* **67**, 157 (1955).

Caution: Irritant, corrosive.

USE: In nitrations, especially to introduce a single nitro group in an ortho position on an aromatic ring.

94. N-Acetylpenicillamine. [59-53-0] N-Acetyl-3-mercaptovaline; N-acetyl-D,L-penicillamine. $C_7H_{13}NO_3S$; mol wt 191.25. C 43.96%, H 6.85%, N 7.32%, O 25.10%, S 16.76%. Prepn of DL-form: J. C. Sheehan *et al.*, **US 2477148** and **US 2496416** (1949, 1950, both to Merck & Co.); and D-form: H. M. Crooks, Jr. in *The Chemistry of Penicillin* (Princeton Univ. Press, 1949) p 470. Use in treatment of mercury poisoning: T. W. Clarkson *et al.*, *J. Pharmacol. Exp. Ther.* **218**, 74 (1981).

Glistening colorless needles from hot water, mp 183°.

D-Form. [15537-71-0] N-Acetyl-3-mercapto-D-valine. Crystals from water, mp 189-190°; $[\alpha]_D^{25}$ +18° (50% ethanol).

THERAP CAT: Antidote (mercury poisoning).

95. N-Acetylsulfanilyl Chloride. [121-60-8] 4-(Acetylamino)benzenesulfonyl chloride; p-acetamidobenzenesulfonyl chloride; p-acetaminobenzenesulfonyl chloride; acetanilide-p-sulfonyl chloride; ASC. $C_8H_8ClNO_3S$; mol wt 233.67. C 41.12%, H 3.45%, Cl 15.17%, N 5.99%, O 20.54%, S 13.72%. Prepd by sulfonation of acetanilide: Stewart, *J. Chem. Soc.* **121**, 2558 (1922); Smiles, Stewart, *Org. Synth.* **coll. vol. I**, 8 (2nd ed., 1941).

Thick, light tan prisms from benzene, mp 149°. Has slight odor of acetic acid. Benzene and ether are poor solvents for purification by recrystn. Much better results are obtained using chloroform or ethylene dichloride. *Cf.* Northey, *Sulfonamides* (New York, 1948) p 12.

Caution: Irritant to skin, eyes, and mucous membranes of nose and throat.

USE: Intermediate in the prepn of sulfanilamide and its derivs.

96. Acetyltannic Acid. [1397-74-6] Diacetyltannic acid; tannyl acetate; Acetannin; Tannigen. Acetylation of tannin: Ciusa, Sollazzo, *Ann. Chim. Appl.* **33**, 72 (1942); *C.A.* **38**, 5794⁴ (1944).

Yellowish- or grayish-white odorless powder; darkens on exposure to light. Softens under water at about 70°. Slightly soluble in water or alcohol; sol in ethyl acetate, solns of borax or sodium phosphate, or with gradual decompn in solns of alkali hydroxides and carbonates. *Incompat.* Alkalies, iron salts.

THERAP CAT: Astringent (intestinal).

THERAP CAT (VET): Astringent (intestinal) for foals, calves, dogs.

97. Achillea. Milfoil; yarrow; thousand-leaf. Flowering herb, *Achillea millefolium* L., *Compositae. Habit.* Europe, Asia, naturalized in U.S. Extensively used as herbal remedy by many cultures. Constituents of plant and of the volatile oil may vary depending on origin of plant. Identification of constituents: Pailer, Kump, *Monatsh. Chem.* **90**, 395 (1959); *eidem*, *Arch. Pharm.* **293**, 646 (1960); A. J. Falk *et al.*, *J. Nat. Prod.* **37**, 598 (1974); R. F. Chandler *et al.*, *J. Pharm. Sci.* **71**, 690 (1982). Isoln of anti-inflammatory constituents: A. S. Goldberg, *J. Pharm. Sci.* **58**, 938 (1969). Phytopharmacology: J. P. Tewari *et al.*, *Indian J. Med. Sci.* **28**, 331 (1974). Review of medicinal uses and composition: R. F. Chandler *et al.*, *Econ. Bot.* **36**, 203-223 (1982); J. Gruenwald *et al.*, *PDR for Herbal Medicines* (Medical Economics, Montvale, 2nd Ed., 2000) pp 833-835.

Volatile oil. [8022-07-9] Oil of yarrow. Obtained by steam distillation of leaves and tops. *Constit.* Chamazulene (0-40%), camphor, β-pinene, cineol, caryophyllene. Blue liquid. d_{20}^{20} 0.905-0.925. Insol in water; very sol in alc, ether. *Keep well closed, cool and protected from light.*

USE: In hair preparations and skin lotions for cleansing and to promote healing.

THERAP CAT: Carminative; cholagogue; antiseptic.

98. Acibenzolar-S-methyl. [135158-54-2] 1,2,3-Benzothiadiazole-7-carbothioic acid S-methyl ester; 7-(methylthiocarbonyl)-benzo-1,2,3-thiadiazole; CGA-245704; Bion. $C_8H_6N_2OS_2$; mol wt 210.27. C 45.70%, H 2.88%, N 13.32%, O 7.61%, S 30.49%. Crop protection agent which activates the natural plant response to infection, known as systemic aquired resistance (SAR). Prepn: R. Schurter *et al.*, **EP 313512**; *eidem*, **US 5523311** (1989, 1996 both to Ciba-Geigy). Properties and biological activity: W. Ruess *et al.*, *Brighton Crop Prot. Conf. - Pests Dis.* **1996**, 53. Mode of action: H. Kessman *et al. ibid.* 961. Use as a seed treatment: B. D. Jensen *et al.*, *Pestic. Sci.* **52**, 63 (1998).

White to beige odorless powder, mp 132.9°. Vapor pressure at 25°: 4.6 × 10⁻⁴ Pa. Soly at 25°(g/l): n-hexane 1.3; toluene 36; n-octanol 5.4; acetone 28; dichloroethane 160. Soly in water (25°): 7.7 mg/l. Log P (n-octanol/water) at 25°: 3.1. LD_{50} in rats (mg/kg): >2000 orally; >2000 dermally (Ruess).

USE: Plant disease resistance activator.

99. Acid Fuchsin. [3244-88-0] 2-Amino-5-[(4-amino-3-sulfophenyl)(4-imino-3-sulfo-2,5-cyclohexadien-1-ylidene)methyl]-3-methylbenzenesulfonic acid sodium salt (1:2); C.I. Acid Violet 19; 2-amino-α⁵-(4-amino-3-sulfophenyl)-α⁵-(4-imino-3-sulfo-2,5-cyclohexadien-1-ylidene)-3,5-xylenesulfonic acid disodium salt; acid rubin; fuchsin(e) acid; acid roseine; C.I. 42685. $C_{20}H_{17}N_3Na_2O_9S_3$; mol wt 585.53. C 41.03%, H 2.93%, N 7.18%, Na 7.85%, O 24.59%, S 16.43%. The trisulfonated derivative of rosanilin, or pararosanilin. *See: Colour Index* **vol. 4** (3rd ed., 1971) p 4399. Use as stain: P. W. Benoit, *Stain Technol.* **48**, 177 (1973); E. Hals, *Scand. J. Dent. Res.* **85**, 542 (1977). Mechanism of action: L. F. Nielsen *et al.*, *Biotech. Histochem.* **73**, 71(1998). Brief review: *H. J. Conn's*

Biological Stains, R. D. Lillie, Ed. (Williams & Wilkins, Baltimore, 9th ed., 1977) pp 285-286.

Olive to dark olive-green coarse powder. Absorption max: 540-545 nm (10 mg/l of 0.01% HCl). Very sol in water, very slightly sol to insol in alcohol. Dil aq solns are purplish red. Very dil aq solns (1:10,000) are decolorized by drops of concd aq NaOH solns, but not returned by concd HCl. The decolorized soln neutralized with NaOH is called the *Andrade indicator*. Changes from red to colorless at pH 12-14.

USE: As pH indicator; biological stain.

100. Acid α-Glucosidase. Acid maltase; lysosomal α-glucosidase; GAA. Lysosomal enzyme that hydrolyzes α-1,4- and α-1,6-glucosidic linkages in oligosaccharides and α-glucans to liberate glucose at acid pH. Natural substrate is glycogen stored in lysosomes. Produced as a 110 kDa glycosylated precursor that is routed to lysosomes via the mannose-6-phosphate recognition marker and is proteolytically processed to the mature forms of 70 and 76 kDa. Genetic deficiency of the enzyme results in a glycogen storage disorder known as Pompe's disease. Identification and role in disease: H. G. Hers, *Biochem. J.* **86**, 11 (1963). Localization in liver lysosomes: N. Lejeune *et al., ibid.* 16. Purification: F. Auricchio, C. B. Bruni, *ibid.* **105**, 35 (1967); F. Auricchio *et al., ibid.* **108**, 161 (1968). *Review:* R. Hirschhorn in *The Metabolic and Molecular Bases of Inherited Disease*, C. R. Scriver *et al.*, Eds. (McGraw-Hill, New York, 7th Ed., 1995) pp 2443-2464.

Alglucosidase alfa. [420784-05-0] [199-arginine, 223-histidine]prepro-α-glucosidase (human); Lumizyme; Myozyme. Recombinant human precursor form. Production in Chinese hamster ovary cells: J. L. K. Van Hove *et al., Proc. Natl. Acad. Sci. USA* **93**, 65 (1996). Clinical evaluation in Pompe patients: A. Amalfitano *et al., Genet. Med.* **3**, 132 (2001).

THERAP CAT: Enzyme replacement therapy for Pompe's disease.

101. Acifluorfen. [50594-66-6] 5-[2-Chloro-4-(trifluoromethyl)phenoxy]-2-nitrobenzoic acid. $C_{14}H_7ClF_3NO_5$; mol wt 361.66. C 46.50%, H 1.95%, Cl 9.80%, F 15.76%, N 3.87%, O 22.12%. Selective pre- and post-emergence herbicide. Prepn: H. O. Bayer *et al.,* **DE 2311638**; *eidem,* **US 3928416, US 4063929** (1973, 1975, 1977 all to Rohm & Haas). Synthesis, activity and toxicity: W. O. Johnson *et al., J. Agric. Food Chem.* **26**, 285 (1978). Carcinogenicity studies: J. A. Quest *et al., Regul. Toxicol. Pharmacol.* **10**, 149 (1989).

Off-white solid, mp 151.5-157°.

Sodium salt. [62476-59-9] Scifluorfen; RH-6201; Blazer. $C_{14}H_6ClF_3NNaO_5$; mol wt 383.64. White powder, mp 124-125°. Soly in water >25%. LD_{50} orally in rats: 1300 mg/kg (Johnson).

Caution: Skin and eye irritant. Do not incorporate in soil. Do not mix with oils, surfactants, liquid fertilizers or other pesticides.

USE: Herbicide.

102. Acipimox. [51037-30-0] 5-Methyl-2-pyrazinecarboxylic acid 4-oxide; 2-carboxy-5-methylpyrazine 4-oxide; K-9321; Olbemox; Olbetam. $C_6H_6N_2O_3$; mol wt 154.13. C 46.76%, H 3.92%,

N 18.18%, O 31.14%. Prepn: V. Ambrogi *et al.,* **DE 2319834**; *eidem,* **US 4002750** (1973, 1977 both to Carlo Erba). Prepn and toxicology: *eidem, Eur. J. Med. Chem.* **15**, 157 (1980). Pharmacological profile: P. P. Lovisolo *et al., Pharmacol. Res. Commun.* **13**, 151, 163 (1981). Pharmacokinetics: L. M. Fuccella *et al., Clin. Pharmacol. Ther.* **28**, 790 (1980); L. Musatti *et al., J. Int. Med. Res.* **9**, 381 (1981). Mechanism of action: K. Aktories *et al., Arzneim.-Forsch.* **33**, 1525 (1983).

Crystals from water, mp 177-180°. LD_{50} orally in mice: 3500 mg/kg (Ambrogi).

THERAP CAT: Antilipemic.

103. Acitretin. [55079-83-9] (2E,4E,6E,8E)-9-(4-Methoxy-2,3,6-trimethylphenyl)-3,7-dimethyl-2,4,6,8-nonatetraenoic acid; etretin; Ro-10-1670; Neotigason; Soriatane. $C_{21}H_{26}O_3$; mol wt 326.44. C 77.27%, H 8.03%, O 14.70%. Synthetic retinoid; free acid form and major metabolite of etretinate, *q.v.* Prepn: W. Bollag *et al.,* **DE 2414619**; *eidem,* **US 4105681** (1974, 1978 both to Hoffmann-La Roche). Teratogenicity study: A. Kistler, H. Hummler, *Arch. Toxicol.* **58**, 50 (1985). HPLC determn in plasma: N. R. Al-Mallah *et al., Anal. Lett.* **21**, 1603 (1988). Pharmacokinetics in humans: F. G. Larsen *et al., Pharmacol. Toxicol.* **62**, 159 (1988). Clinical evaluation in cutaneous lupus erythematosus: T. Ruzicka *et al., Arch. Dermatol.* **124**, 897 (1988). Review of clinical pharmacology: A. Vahlquist, O. Rollman, *Dermatologica* **175**, Suppl. 1, 20-27 (1987). Review of clinical studies in psoriatic and nonpsoriatic dermatoses: J.-M. Geiger, B. M. Czarnetzki, *ibid.* **176**, 182-190 (1988); of clinical efficacy in psoriasis: C. S. Lee, J. Koo, *Expert Opin. Pharmacother.* **6**, 1725-1734 (2005).

Green-yellow crystalline powder. Crystals from hexane, mp 228-230°. Sparingly sol in tetrahydrofuran. Slightly sol in acetone, alcohol. Very slightly sol in cyclohexane. Practically insol in water (<0.1 mg/100 mg). pKa ~5. LD_{50} i.p. in mice (mg/kg): >4000 (1 day), 700 (10 days), 700 (20 days) (Bollag, 1978).

THERAP CAT: Antipsoriatic.

THERAP CAT (VET): In treatment of dermatologic conditions in small animals.

104. Aclacinomycins. Antitumor antibiotic complex of the anthracycline group, produced by *Streptomyces galilaeus.* Thirteen yellow and seven red-colored components have been identified. Isoln of the major components, aclacinomycins A and B: H. Umezawa *et al.,* **DE 3532568**; *eidem,* **US 3988315** (1974, 1976 both to Microbiochem. Res. Found.). *See also:* T. Oki *et al., J. Antibiot.* **28**, 830 (1975). Structure, taxonomy, production, properties: *eidem, ibid.* **32**, 791, 801 (1979). The aglycone portion of aclacinomycin A is known as *aklavinone.* Isoln of aklavinone: J. J. Gordon *et al., Tetrahedron Lett.* **8**, 28 (1960). Synthesis of racemic aklavinone: B. A. Pearlman *et al., J. Am. Chem. Soc.* **103**, 4248 (1981); P. A. Confalone, G. Pizzolato, *ibid.* 4251; R. K. Boeckman, F. W. Sum, *ibid.* **104**, 4604 (1982). Synthesis of optically active aklavinone: A. S. Kende, J. P. Rizzi, *ibid.* **103**, 4247 (1981); J. M. McNamara, Y. Kishi, *ibid.* **104**, 7371 (1982). *In vitro* metabolism of aclacinomycins: T. Komiyama *et al., Gann* **70**, 395 (1979), *C.A.* **92**, 15143x (1980). HPLC determn of aclacinomycin A and its metabolites: T. Ogasawara *et al., J. Antibiot.* **34**, 47, 52 (1981). Pharmacokinetics: M. J. Egorin *et al., Cancer Chemother. Pharmacol.* **8**, 41 (1982). Series of articles on absorption, excretion, distribution and general pharmacology: *Jpn. J. Antibiot.* **33**, 169-213 (1980), *C.A.* **93**, 19316,

125326, 125649, 142986 (1980). Immunological study: M. Ishizuka *et al.*, *J. Antibiot.* **34**, 331 (1981). Clinical studies: R. Maral, *Drugs Exp. Clin. Res.* **9**, 375 (1983); R. P. Warrell, Jr., S. J. Kempin, *Am. J. Clin. Oncol.* **6**, 81 (1983); A. Y. Bedikian *et al.*, *ibid.* 187 (1983). Review of pharmacology of aclacinomycin A: T. Oki, *Anthracyclines [Proc. Workshop]*, S. T. Crooke, S. D. Reich, Eds. (Academic Press, New York, 1980) pp 323-342, *C.A.* **93**, 160778h (1980).

Aclacinomycin A

Aclacinomycin A. [57576-44-0] [1*R*-(1α,2β,4β)]-2-Ethyl-1,2,-3,4,6,11-hexahydro-2,5,7-trihydroxy-6,11-dioxo-4-[[2,3,6-trideoxy-4-*O*-[2,6-dideoxy-4-*O*-[(2*R-trans*)-tetrahydro-6-methyl-5-oxo-2*H*-pyran-2-yl]-α-L-*lyxo*-hexopyranosyl]-3-(dimethylamino)-α-L-*lyxo*-hexopyranosyl]oxy]-1-naphthacenecarboxylic acid methyl ester; aclarubicin; antibiotic MA 144A1; NSC-208734; Jaclacin. $C_{42}H_{53}NO_{15}$; mol wt 811.88. Yellow microcryst powder from chloroform/hexane, mp 151-153° (dec). $[\alpha]_D^{24}$ −11.5° (c = 1 in methylene chloride). uv max (methanol): 229.5, 259, 289.5, 431 nm ($E_{1cm}^{1\%}$ 550, 326, 135, 161); (0.1*N* HCl) 229.5, 258.5, 290, 431 nm ($E_{1cm}^{1\%}$ 571, 338, 130, 161); (0.1*N* NaOH) 239, 287, 523 nm ($E_{1cm}^{1\%}$ 450, 113, 127). Sol in $CHCl_3$, ethyl acetate. Insol in ethyl ether, *n*-hexane, petr ether. Addn of alkali to aq solns gives an intense reddish-purple color; in conc HCl the soln is yellow. LD_{50} in mice (mg/kg): 22.6 i.p., 33.7 i.v. (Oki).

Aclacinomycin A hydrochloride. Aclarubicin hydrochloride; Aclacin; Aclacinon; Aclaplastin. $C_{42}H_{53}NO_{15}$.HCl; mol wt 848.34.

Aclacinomycin B. [57596-79-9] [1*R*-(1α,2β,4β)]-4-[[[2‴,3″-Anhydro]-*O*-3,6-dideoxy-α-L-*erythro*-hexopyranos-4-ulos-1-yl]-(1 → 4)-*O*-2,6-dideoxy-α-L-*lyxo*-hexopyranosyl]-(1 → 4)-2,3,6-trideoxy-3-(dimethylamino)-α-L-*lyxo*-hexopyranosyl]oxy]-2-ethyl-1,2,3,4,6,11-hexahydro-2,5,7-trihydroxy-6,11-dioxo-1-naphthacenecarboxylic acid methyl ester; antibiotic MA 144B1. $C_{42}H_{51}NO_{15}$; mol wt 809.86. Yellow microcryst powder from chloroform/hexane, mp 163-167° (dec). $[\alpha]_D^{24}$ +3° (c = 1 in methylene chloride). Other physical properties similar to aclacinomycin A. LD_{50} in mice (mg/kg): 13.7 i.p., 16.4 i.v. (Oki).

THERAP CAT: Antineoplastic.

105.   Aclatonium Napadisilate. [55077-30-0] 2-[2-(Acetyloxy)-1-oxopropoxy]-*N,N,N*-trimethylethanaminium 1,5-naphthalenedisulfonate (2:1); 1,5-naphthalenedisulfonic acid bis [2-[2-(acetyloxy)-1-oxopropoxy]-*N,N,N*-trimethylethanaminium]ion(2−); acetyllactoylcholine 1,5-naphthalenedisulfonate; choline 1,5-naphthalenedisulfonate (2:1), dilactate, diacetate; TM-723; Abovis. $C_{30}H_{46}N_2O_{14}S_2$; mol wt 722.82. C 49.85%, H 6.41%, N 3.88%, O 30.99%, S 8.87%. Spasmolytic agent related structurally to acetylcholine chloride, *q.v.* Prepn: K. Miura *et al.*, **DE 2425983**; *eidem*, **US 3903137** (1973, 1975 both to Toyama). Synthesis and smooth muscle activity: *eidem*, *Yakugaku Zasshi* **99**, 1245 (1979), *C.A.* **92**, 146197r (1980). Toxicity study: A. Takai *et al.*, *Oyo Yakuri* **19**, 93 (1980), *C.A.* **93**, 613367 (1980). Series of articles on metabolism,

pharmacology, toxicity studies: *Oyo Yakuri* **18**, 695-942 (1979), *C.A.* **92**, 191015r, 191281z, 191282a, 208995k-7n (1980).

Crystals, mp 189-191°. LD_{50} in mice: 15 g/kg orally, 826 mg/kg s.c. (Miura); in dogs: >10 g/kg orally (Takai).

THERAP CAT: Cholinergic.

106.   Aclidinium Bromide. [320345-99-1] (3*R*)-3-[(2-Hydroxy-2,2-di-2-thienylacetyl)oxy]-1-(3-phenoxypropyl)-1-azoniabicyclo[2.2.2]octane bromide (1:1); LAS-34273. $C_{26}H_{30}BrNO_4S_2$; mol wt 564.55. C 55.32%, H 5.36%, Br 14.15%, N 2.48%, O 11.34%, S 11.36%. Muscarinic M_3 antagonist. Prepn: D. Fernandez Forner *et al.*, **WO 0104118**; *eidem*, **US 6750226** (2001, 2004 both to Almirall Prodesfarma); and structure activity studies: M. Prat *et al.*, *J. Med. Chem.* **52**, 5076 (2009). Improved synthesis: N. Busquets Baque, F. Pajuelo Lorenzo, **WO 08009397** (2008 to Almirall). Pharmacology: A. Gavaldà *et al.*, *J. Pharmacol. Exp. Ther.* **331**, 740 (2009). Clinical pharmacokinetics: J. M. Jansat *et al.*, *Int. J. Clin. Pharmacol. Ther.* **47**, 460 (2009). Clinical evaluation in COPD: P. Chanez *et al.*, *Pulm. Pharmacol. Ther.* **23**, 15 (2010). Review of development and therapeutic potential: M. Cazzola, *Curr. Opin. Investig. Drugs* **10**, 482-490 (2009).

Crystals from acetonitrile, mp 230°.

THERAP CAT: Bronchodilator.

107.   Aconine. [509-20-6] (1α,3α,6α,14α,15α,16β)-20-Ethyl-1,6,16-trimethoxy-4-(methoxymethyl)aconitane-3,8,13,14,15-pentol. $C_{25}H_{41}NO_9$; mol wt 499.60. C 60.10%, H 8.27%, N 2.80%, O 28.82%. Obtained by hydrolysis of aconitine: Schulze, *Arch. Pharm.* **244**, 165 (1906); Schneider, *Ber.* **89**, 768 (1956). Structure: Schneider, Tausend, *Ann.* **628**, 114 (1959); McCaldin, Marion, *Can. J. Chem.* **37**, 1071 (1959).

Amorphous powder, mp 132°. Bitter taste. $[\alpha]_D$ +23°. pK 9.52. Absorption spectrum: Nath, *J. Indian Chem. Soc.* **32**, 75 (1955). Very sol in water, alcohol; moderately sol in chloroform; slightly sol in benzene. Practically insol in ether, petr ether.

Hydrochloride dihydrate. $C_{25}H_{42}ClNO_9.2H_2O$. Crystals, mp 175-176°. $[\alpha]_D$ −8°.

Hydrobromide sesquihydrate. $C_{25}H_{42}BrNO_9.1\frac{1}{2}H_2O$. Crystals from water, mp 225°.

108.   Aconite. Monkshood; wolf's bane; friar's cowl; mousebane. Dried tuberous root of *Aconitum napellus* L., *Ranunculaceae*. *Habit.* Mountainous regions of Europe, Asia, and North America. *Constit.* 0.4-0.8% aconitine; aconine; napelline (isoaconitine, pseudoaconitine); picraconitine; aconitic acid; itaconic acid; succinic acid; malonic acid; fat; levulose. *Quite toxic. Ref:* Freudenberg, *Ber.* **69**, 1962 (1936); Rogers, Freudenberg, *ibid.* **70**, 349 (1937); Lascombes, *Ann. Pharm. Fr.* **16**, 429 (1958). *See also* "Aconite," A. Ph. A. Monograph No. 1 (Am. Pharm. Assoc., Washington, D.C., 1938).

THERAP CAT: Antipyretic.

THERAP CAT (VET): Formerly used as an antihypertensive.

109. Aconitic Acid. [499-12-7] 1-Propene-1,2,3-tricarboxylic acid; equisetic acid; citridic acid; achilleic acid. $C_6H_6O_6$; mol wt 174.11. C 41.39%, H 3.47%, O 55.13%. Found in leaves and tubers of *Aconitum napellus* L., *Ranunculaceae*, in various species of *Achillea (Compositae)* and *Equisetum (Equisetaceae)*, in beet root, and in sugar cane. Can be prepd commercially from calcium magnesium aconitate recovered from sugar cane juice: McCalip, Seibert, *Ind. Eng. Chem.* **33**, 637 (1941); from molasses: Regna, Bruins, *ibid.* **48**, 1268 (1956). Most of the commercial aconitic acid is, however, manufactured by sulfuric acid dehydration of citric acid: Bruce, *Org. Synth.* **coll. vol. II**, p 12 (1943); by using methanesulfonic acid instead of sulfuric: Cranston, **US 2727066** (1955 to Daniel F. Kelly). Aconitic acid prepd by any of the above methods has the *trans*-configuration which is the form described here.

trans-form

Leaflets, plates from water. Decompn 198-199° (capillary inserted in oil bath at 190°); decompn 204-205° (capillary in oil bath at 195°); decompn 209° (electrically heated bar). K_1 at 25° = 1.58×10^{-3}; K_2 = 3.5×10^{-5}. One gram dissolves in 5.5 ml water at 13°, in 2 ml water at 25°. Soluble in 2 parts of 88% alcohol at 12°. Slightly sol in ether.

Triethyl ester. bp_5 155°.

Tributyl ester. bp_3 190°.

USE: Manuf itaconic acid. As plasticizer for buna rubber and plastics. Used in form of triethyl or tributyl ester.

110. Aconitine. [302-27-2] ($1\alpha,3\alpha,6\alpha,14\alpha,15\alpha,16\beta$)-20-Ethyl-1,6,16-trimethoxy-4-(methoxymethyl)aconitane-3,8,13,14,15-pentol 8-acetate 14-benzoate. $C_{34}H_{47}NO_{11}$; mol wt 645.75. C 63.24%, H 7.34%, N 2.17%, O 27.25%. Several isomers from *Aconitum napellus* L., *Ranunculaceae* and other aconites: Majima *et al.*, *Ber.* **57**, 1456 (1924); *Proc. Imp. Acad. Tokyo* **5**, 415 (1929); Freudenberg, *Ber.* **69**, 1964 (1936); Swanson *et al.*, *Aconite*, A. Ph. A. Monograph no. 1 (1938); *Methods of Analysis*, *A.O.A.C.*, 8th ed., 598, 651 (1955). Structure: Wiesner *et al.*, *Collect. Czech. Chem. Commun.* **28**, 2462 (1963); Wiesner *et al.*, *Can. J. Chem.* **47**, 2734 (1969). Stereochemistry: Bachelor *et al.*, *Tetrahedron Lett.* **1960**, no. 10, 1; Gilman, Marion, *ibid.* **1961**, 923; Tsuda, Marion, *Can. J. Chem.* **41**, 1634 (1963); Birnbaum *et al.*, *Tetrahedron Lett.* **1971**, 867. Pharmacology: H. Sato *et al.*, *Tohoku J. Exp. Med.* **128**, 175 (1979). Inotropic effects: P. Honerjäger, A. Meissner, *Arch. Pharmacol.* **322**, 49 (1983). Toxicity: Dybing *et al.*, *Acta Pharmacol. Toxicol.* **7**, 337 (1951).

Hexagonal plates, mp 204°. $[\alpha]_D$ +17.3° (chloroform). Aq soln is alkaline to litmus. pK 5.88. One gram dissolves in 2 ml chloroform, 7 ml benzene, 28 ml abs alcohol, 50 ml ether, 3300 ml water. Slightly sol in petr ether. *Poisonous.* LD_{50} in mice (mg/kg): 0.166

i.v.; 0.328 i.p.; approx 1 orally (Dybing); also reported as LD_{50} in mice (mg/kg): 1.8 orally, 0.270 s.c.; 0.380 i.p.; 0.12 i.v. (Sato).

Hydrobromide hemipentahydrate. $C_{34}H_{47}NO_{11}.HBr.2\frac{1}{2}H_2O$. Hexagonal tablets from water, mp 200-207° (sinters at 160°). Amorphous form mp 115-120°. Also crystallizes from ethanol + ether with $\frac{1}{2}H_2O$, mp 206-207°; $[\alpha]_D$ −30.8°. *Poisonous. Ref:* Paech, Tracey, *Modern Methods of Plant Analysis* vol. IV (Springer-Verlag, Berlin, 1955) p 375.

Hydrochloride hemipentahydrate. $C_{34}H_{47}NO_{11}.HCl.2\frac{1}{2}H_2O$. Crystals, mp 149-153°. mp 194-195° (dry). $[\alpha]_D$ −30.9°. *Poisonous. Ref:* Paech, Tracey.

Nitrate. $C_{34}H_{47}NO_{11}.HNO_3$. Crystals, mp about 200° (dec). $[\alpha]_D^{20}$ −35° (c = 2 in water). *Poisonous.* One gram dissolves in 10 ml boiling water. Less sol in cold water. Sol in alcohol.

Caution: Potential symptoms of overexposure are nausea, vomiting, diarrhea due to CNS stimulation; restlessness, ataxia, vertigo, slow and dyspneic breathing, hypothermia, convulsions. Direct contact may cause warm tingling sensation with subsequent numbness on mouth and mucous membranes. *See Clinical Toxicology of Commercial Products*, R. E. Gosselin *et al.*, Eds. (Williams & Wilkins, Baltimore, 5th ed., 1984) Section II, p 249.

USE: Used in producing heart arrhythmia in experimental animals: Boyadzhiev, *C.A.* **73**, 86256e (1970).

THERAP CAT: Has been used topically in neuralgia.

111. Aconitine, Amorphous. Mild aconitine. Highly toxic mixture of amorphous alkaloids from *Aconitum napellus* L., *Ranunculaceae*. Contains aconitine, mesaconitine, hypaconitine, neopelline, *l*-ephedrine, sparteine, neoline and napelline. *Ref:* Rogers, Freudenberg, *Ber.* **70**, 349 (1937); *eidem, J. Am. Chem. Soc.* **59**, 2572 (1937). Toxicity data: Munch, *J. Am. Pharm. Assoc.* **18**, 17 (1929).

Yellowish-white, amorphous powder. *Poisonous.* Insol in water. Sol in alcohol, chloroform, ether, dil acids. MLD in rats (mg/kg): 0.175 s.c., 0.1 i.p. (Munch).

112. Aconitum Ferox. Indian aconite; bish; visha; bishma; bikhroot. Tuber of *Aconitum ferox* Wall., *Ranunculaceae. Habit.* Nepal, Himalaya Mountains, India. *Constit.* Pseudoaconitine. Most powerful of all the aconites. Used by natives as arrow poison in big game hunting; in neuralgia and rheumatism. *Ref: Aconite*, A. Ph. A. Monograph no. 1, p 78 and *passim* (Am. Pharm. Assoc., Washington, D.C., 1938); N. B. Dutt, *Indian J. Pharm.* **1**, 81-84 (1939); *C.A.* **34**, 6768 (1940).

Caution: Extremely toxic. A very small dose can cause fatal cardiac depression.

113. Acotiamide. [185106-16-5] *N*-[2-[Bis(1-methylethyl)amino]ethyl]-2-[[2-hydroxy-4,5-dimethoxybenzoyl)amino]-4-thiazolecarboxamide; 2-[*N*-(4,5-dimethoxy-2-hydroxybenzoyl)amino]-4-[(2-diisopropylaminoethyl)aminocarbonyl]-1,3-thiazole; *N*-(*N'*,*N'*-diisopropylaminoethyl)-[2-(2-hydroxy-4,5-dimethoxybenzoylamino)-1,3-thiazole-4-yl]carboxyamide. $C_{21}H_{30}N_4O_5S$; mol wt 450.55. C 55.98%, H 6.71%, N 12.44%, O 17.75%, S 7.12%. Acetylcholinesterase inhibitor and muscarinic M_1- and M_2-receptor antagonist that enhances acetylcholine release in the enteric nervous system and stimulates gastric motility. Prepn: M. Nagasawa *et al.*, **WO 9636619**; *eidem*, **US 5981557** (1996, 1999 both to Zeria). HPLC determn by solid phase extraction in canine plasma: S. Furuta *et al.*, *J. Pharm. Biomed. Anal.* **25**, 599 (2001). Effect on gastric motility: T. Nakajima *et al.*, *J. Smooth Muscle Res.* **36**, 69 (2000). Receptor binding profile and effect on acetylcholine release: M. Ogishima *et al.*, *J. Pharmacol. Exp. Ther.* **294**, 33 (2000); Y. Doi *et al.*, *Eur. J. Pharmacol.* **505**, 31 (2004). Clinical evaluation in functional dyspepsia: K. Matsueda *et al.*, *Neurogastroenterol. Motil.* **22**, 618 (2010). Review of pharmacology and clinical experience: J. Tack, P. Janssen, *Expert Opin. Invest. Drugs* **20**, 701-712 (2011).

mp 179-182° (dec).

Hydrochloride. [185104-11-4] $C_{21}H_{30}N_4O_5S.HCl$; mol wt 487.01. Crystals from isopropyl alcohol + water, mp 160°.

Hydrochloride trihydrate. [773092-05-0] YM-443; Z-338. $C_{21}H_{30}N_4O_5S.HCl.3H_2O$; mol wt 541.06.

THERAP CAT: Gastroprokinetic; treatment of functional dyspepsia.

114. Acridine. [260-94-6] Dibenzo[b,e]pyridine; 10-azaanthracene. $C_{13}H_9N$; mol wt 179.22. C 87.12%, H 5.06%, N 7.82%. Occurs in coal tar. Isoln: Graebe, Caro, *Ann.* **158**, 265 (1871); from high boiling tar oils: Wirth, **DE 440771** (1926). Prepn from *N*-phenylanthranilic acid: Perkin, Clemo, **GB 214756** (1923); from Ca-anthranilate: Koller, Krakauer, *Monatsh. Chem.* **50**, 51 (1928); by passing benzylaniline vapor over red-hot platinum wire: Meyer, Hofmann, *ibid.* **37**, 698 (1916); *cf.* Ullmann, *Ber.* **40**, 2521 (1907). Absorption spectrum: Pinnow, *J. Prakt. Chem.* [2] **66**, 276 (1902). Toxicity: S. D. Rubbo, *Br. J. Exp. Pathol.* **28**, 1 (1947). *Reviews:* A. Albert, *The Acridines* (St. Martin's Press, New York, 2nd ed., 1966); R. M. Acheson, *Acridines* (Interscience, New York, 2nd ed., 1973).

Pale yellow orthorhombic plates, needles from diluted alcohol. *Poisonous.* There are five crystalline forms of acridine all of which melt below 111°. Purified acridine mp 110-111°; commercial samples mp 107° or mp 109°. Begins to sublime at 100°. bp_{760} 346°; also reported as bp_{760} >360°. Volatile with steam. Slightly sol in boiling water, liquid ammonia, liquid sulphur dioxide; sparingly sol in light petroleum; freely sol in alcohol, ether, hydrocarbons, carbon disulfide. 1 gm sol in <1 ml boiling benzene or alcohol; 5 ml benzene (20°); 6 ml alcohol (20°); 16 ml of ether (20°); 1.8 ml boiling cyclohexane. d 1.27-1.28. Dilute solns of acridine and its salts have a violet and green fluorescence, respectively. Weak base, colors litmus paper blue, forms yellow crystalline salts with mineral acids. Forms colored quaternary ammonium compounds (acridinium compounds), by the action of alkyl and aryl halides and sulfates. LD_{50} s.c. in mice: 0.40 g/kg (Rubbo).

Caution: Strongly irritating to skin, mucous membranes. *See: Clinical Toxicology of Commercial Products*, R. E. Gosselin *et al.*, Eds. (Williams & Wilkins, Baltimore, 5th ed., 1984) Section II, p 384.

USE: Manuf dyes and intermediates; some dyes derived from it are used as antiseptics, e.g. 9-aminoacridine, acriflavine and proflavine. The hydrochloride has been used as reagent for cobalt, iron and zinc.

115. Acriflavine. [86-40-8]; [65589-70-0] (unspecified composition). 3,6-Diamino-10-methylacridinium chloride (1:1); 2,8-diaminoacridine methochloride; neutral acriflavine; euflavine; trypaflavine; neutroflavine; gonacrine; C.I. 46000. $C_{14}H_{14}ClN_3$; mol wt 259.74. C 64.74%, H 5.43%, Cl 13.65%, N 16.18%. Fluorescent, acridine dye produced by the methylation of proflavine, *q.v.*, which is often present in varying amounts as a contaminant in commercial prepns. Prepn: L. Benda, *Ber.* **45**, 1787 (1912). Purification and properties: M. Gaillot, *Q. J. Pharm. Pharmacol.* **7**, 63 (1934); J. Marshall, *ibid.* 514; V. S. Gupta *et al.*, *J. Chromatogr.* **26**, 158 (1967). Bacteriostatic activity: C. H. Browning, W. Gilmour, *J. Pathol. Bacteriol.* **18**, 144 (1913); C. H. Browning *et al.*, *Br. Med. J.* **1**, 73 (1917); H. Berry, *Q. J. Pharm. Pharmacol.* **14**, 149 (1941). Toxicology: J. Ungar *et al.*, *J. Pharmacol. Exp. Ther.* **80**, 217 (1944). Use as DNA stain: D. Roth *et al.*, *Stain Technol.* **42**, 125 (1967); J. W. Levinson *et al.*, *J. Histochem. Cytochem.* **26**, 680 (1978). *Review: Conn's Biological Stains*, R. W. Horobin, J. A. Kiernan, Eds. (BIOS Scientific Publishers Ltd, Oxford, UK, 10th ed., 2002) pp 255-256.

Orange prisms from boiling methyl alcohol. Also reported as brick-red microcrystals from methanol-water, mp 295-298° (darkening at 285°). pK 10.1. Abs max: 452 nm (water); 465 nm (alcohol). Emission max: 510 nm (water). Soly in water (20°): 0.4%.

Mixture with 3,6-acridinediamine. [8048-52-0] Brick red powder. *Irritant.* Incompletely sol in alcohol; nearly insol in ether, chloroform, fixed oils. Aqueous solns are reddish orange and fluoresce on dilution. pH (1% soln) ~3.5.

Hydrochloride. [69235-50-3] Acid acriflavine; acriflavine dihydrochloride; Panflavin. Deep reddish-brown, crystalline powder. *Irritant.* pH (1% soln) ~1.5.

USE: Biological stain.

THERAP CAT: Antiseptic.

THERAP CAT (VET): Antiseptic, especially for aquarium fish.

116. Acrilan®. A brand of acrylic fiber, a copolymer of acrylonitrile and of a minor constituent with mildly basic character. Prepn of such copolymers: Mowry, Craig, **US 2744086** and Craig, **US 2749325** (both 1956 to Chemstrand). *Review:* R. W. Moncrieff, *Man-Made Fibres* (John Wiley & Sons, New York, 4th ed., 1963) pp 471-482.

Solid, d^{25} 1.17. Fiber decomposes before it melts. Sticks at 245°. Prolonged exposure in air at elevated temps causes some yellowing. Fabrics are not readily ignited and their rate of burning is less than that of cotton, viscose rayon, or acetate rayon. Fiber is practically insoluble in, and unaffected by common solvents. Has fairly good resistance to weak alkalies, very good resistance to mineral acids, sunlight, weathering. Has good dyeing properties. Its wear resistance is better than that of wool, but generally greatly inferior to that of nylon or Terylene. It is unaffected by mildew, molds, moth and carpet beetle larvae.

USE: Fabrics, sweaters, blankets, carpets.

117. Acrinathrin. [101007-06-1] (1R,3S)-2,2-Dimethyl-3-[(1Z)-3-oxo-3-[2,2,2-trifluoro-1-(trifluoromethyl)ethoxy]-1-propen-1-yl]cyclopropanecarboxylic acid (S)-cyano(3-phenoxyphenyl)-methyl ester; (S)-α-cyano-3-phenoxybenzyl (1R-cis)-2,2-dimethyl-3-[(Z)-3-oxo-3-[2-(1,1,1,3,3,3-hexafluoropropoxy)]-1-propenyl]cyclopropanecarboxylate; RU-38702; Rufast. $C_{26}H_{21}F_6NO_5$; mol wt 541.45. C 57.68%, H 3.91%, F 21.05%, N 2.59%, O 14.77%. Synthetic pyrethroid acaricide. Prepn: J. Martel *et al.*, **FR 2486073**; *eidem*, **US 4542142** (1982, 1985 both to Roussel-UCLAF); J. R. Tessier *et al.*, in *Pesticide Chemistry: Human Welfare and the Environment* vol. 1, J. Miyamoto, P. C. Kearney, Eds. (Pergamon Press, Oxford, 1983) pp 95-100. Field trial vs spider mite: B. H. Labanowska, C. Tkaczuk, *Fruit Sci. Rep.* **18**, 185 (1991). *Review:* J. J. Heller *et al.*, *Meded. Fac. Landbouwwet. Univ. Gent* **57**, 931-939 (1992).

mp 82°. $[\alpha]_D^{20}$ +23.5° (c = 0.5% in benzene). LD_{50} in rats (mg/kg): >5000 orally; >2000 dermally (Heller).

USE: Insecticide.

118. Acrivastine. [87848-99-5] (2E)-3-[6-[(1E)-1-(4-Methylphenyl)-3-(1-pyrrolidinyl)-1-propen-1-yl]-2-pyridinyl]-2-propenoic acid; (E)-6-[(E)-3-(1-pyrrolidinyl)-1-p-tolylpropenyl]-2-pyridineacrylic acid; BW-270C; BW-825C; BW-A825C; Semprex. $C_{22}H_{24}N_2O_2$; mol wt 348.45. C 75.83%, H 6.94%, N 8.04%, O 9.18%. Nonsedating type histamine H_1-receptor antagonist; analog of triprolidine, *q.v.* Prepn: G. G. Coker, J. W. A. Findlay, **EP 85959** (1983 to Wellcome); J. W. A. Findlay, G. G. Coker, **US 4501893** (1985). Pharmacodynamics and pharmacokinetics in humans: A. F. Cohen *et al.*, *Eur. J. Clin. Pharmacol.* **28**, 197 (1985). Evaluation of CNS effects: A. F. Cohen *et al.*, *Clin. Pharmacol. Ther.* **38**, 381 (1985). Clinical trials in idiopathic urticaria: J. G. Gibson *et al.*, *Dermatologica* **169**, 179 (1984); H. Neittaanmaki *et al.*, *ibid.* **177**, 98 (1988); in allergic rhinitis: T. G. Gibbs *et al.*, *J. Int. Med. Res.* **16**, 413 (1988). Review of pharmacology and therapeutic efficacy: R. N. Brogden, D. McTavish, *Drugs* **41**, 927-940 (1991).

Crystals from isopropanol, mp 222° (dec).
THERAP CAT: Antihistaminic.

119. Acrolein. [107-02-8] 2-Propenal; acrylic aldehyde; acrylaldehyde; acraldehyde; Aqualin; Magnacide. C_3H_4O; mol wt 56.06. C 64.28%, H 7.19%, O 28.54%. Prepd industrially by passing glycerol vapors over magnesium sulfate heated to 330-340°. Lab prepn by heating a mixture of anhydr glycerol, acid potassium sulfate and potassium sulfate in the presence of a small amount of hydroquinone and distilling in the dark: H. Adkins, W. H. Hartung, *Org. Synth.* **coll. vol. I**, 15 (1941). Formation from glycerol by the action of *B. amaracrylus:* Voisenet, *Compt. Rend.* **188**, 941, 1271 (1929); by *B. welchii:* Humphreys, *J. Infect. Dis.* **35**, 282; *Chem. Zentralbl.* **1925**, II, 309. Toxicity study: H. F. Smyth *et al.*, *Arch. Ind. Hyg. Occup. Med.* **4**, 119 (1951). *Review:* L. G. Hess *et al.*, in *Kirk-Othmer Encyclopedia of Chemical Technology* **vol. 1** (Wiley-Interscience, New York, 3rd ed., 1978) pp 277-297. Review of toxicology and human exposure: *Toxicological Profile for Acrolein* (PB2008-100001, 2007) 227 pp.

Liquid with pungent odor. *Flammable, poisonous.* mp −88°. d^0 0.8621; d^{20} 0.8389; d^{50} 0.8075. bp_{760} 52.5°; bp_{200} 17.5°; bp_{100} 2.5°; bp_{60} −7.5°; $bp_{1.0}$ −64.5°. n_D^{19} 1.4022. Sol in 2 to 3 parts water; in alcohol, ether. Flash pt, open cup: <0°F (−18°C). Vapor pressure at 20°: 210 mm Hg. Unstable, polymerizes (especially under light or in the presence of alkali or strong acid) forming disacryl, a plastic solid. Absorption spectrum: Lüthy, *Z. Phys. Chem.* **107**, 291, 298 (1923). Log P (octanol/water): −0.01. Soly at 25° (mg/l): water 2.12×10^5. Miscible with lower alcohols, ketones, benzene, diethyl ether, and other common organic solvents. LD_{50} orally in rats: 0.046 g/kg (Smyth).
Caution: Potential symptoms of overexposure are irritation of eyes, skin and mucous membranes; decreased pulmonary function; delayed pulmonary edema; chronic respiratory disease. *See NIOSH Pocket Guide to Chemical Hazards* (DHHS/NIOSH 97-140, 1997) p 6. *See also Clinical Toxicology of Commercial Products*, R. E. Gosselin *et al.*, Eds. (Williams & Wilkins, Baltimore, 5th ed., 1984) Section II, p 186.
USE: Manuf colloidal forms of metals; making plastics, perfumes; warning agent in methyl chloride refrigerant. Has been used in military poison gas mixtures. Used in organic syntheses. Aquatic herbicide.

120. Acrylamide. [79-06-1] 2-Propenamide. C_3H_5NO; mol wt 71.08. C 50.69%, H 7.09%, N 19.71%, O 22.51%. Prepd from acrylonitrile by treatment with H_2SO_4 or HCl: Bayer, *Angew. Chem.* **61**, 240 (1949); Weisgerber, US 2535245 (1950 to Hercules). *Reviews:* Carpenter, Davis, *J. Appl. Chem.* **7**, 671 (1957); C. E. Habermann in *Kirk-Othmer Encyclopedia of Chemical Technology* **vol. 1** (John Wiley & Sons, New York, 4th ed., 1991) pp 251-266. Toxicity: R. E. Peterson, N. K. Sheth, *Toxicol. Appl. Pharmacol.* **33**, 142 (1975). Review of carcinogenic risk: *IARC Monographs* **60**, 389-433 (1994). Identification in cooked carbohydrate-rich foods: E. Tareke *et al.*, *J. Agric. Food Chem.* **50**, 4998 (2002). Proposed formation from asparagine, *q.v.*, during cooking: D. S. Mottram *et al.*, *Nature* **419**, 448 (2002); R. H. Stadler *et al.*, *ibid.* 449.

Monomer, flake-like crystals from benzene. *Poisonous.* d_4^{30} 1.122. mp 84.5°. bp_2 87°; bp_5 103°; bp_{25} 125°. Solubilities in g/100 ml solvent at 30°: water 215.5; methanol 155; ethanol 86.2; acetone 63.1; ethyl acetate 12.6; chloroform 2.66; benzene 0.346; heptane 0.0068. The solid may be stored in a cool, dark place. Readily polymerizes at the mp or under uv light. Commercial solns of the monomer may be stabilized with hydroquinone, *tert*-butylpyrocatechol, *N*-phenyl-2-naphthylamine or other antioxidants. LD_{50} i.p. in mice: 170 mg/kg (Peterson, Sheth).
Polymer. Various forms, sol and insol in water, are obtained by heating with various polymerization catalysts: C. E. Schildknecht, *Vinyl and Related Polymers* (Wiley, New York, 1952) pp 314-322; D. Lipp, J. Kozakiewicz in *Kirk-Othmer Encyclopedia of Chemical Technology* **vol. 1** (John Wiley & Sons, New York, 4th ed., 1991) pp 266-287.
Caution: Potential symptoms of overexposure to the monomer are ataxia, numbness of limbs, paresthesia; muscle weakness; absence of deep tendon reflex; sweating of hands; fatigue, lethargy; irritation of eyes and skin; reproductive effects. *See NIOSH Pocket Guide to Chemical Hazards* (DHHS/NIOSH 97-140, 1997) p 6. Readily absorbed through intact skin from aqueous solutions. *See Clinical Toxicology of Commercial Products*, R. E. Gosselin *et al.*, Eds. (Williams & Wilkins, Baltimore, 5th ed., 1984) Section II, p 409. This substance is reasonably anticipated to be a human carcinogen: *Report on Carcinogens, Twelfth Edition* (PB2011-111646, 2011) p 25.
USE: Monomer as chemical intermediate in production of polyacrylamides; in synthesis of dyes; in copolymers for contact lenses; in construction of dam foundations, tunnels and sewers. Polymers as additives for water treatment, enhanced oil recovery, flocculants, papermaking aids, thickeners, soil conditioning agents, sewage and waste treatment, ore processing, permanent-press fabrics.

121. Acrylic Acid. [79-10-7] 2-Propenoic acid; vinylformic acid. $C_3H_4O_2$; mol wt 72.06. C 50.00%, H 5.60%, O 44.40%. Prepd by hydrolysis of acrylonitrile: Kaszuba, *J. Am. Chem. Soc.* **67**, 1227 (1945); by oxidation of acrolein: US 1911219 (1933 to Rohm & Haas); US 2288566 (1942 to Acrolein Corp.); US 2341339 (1944 to Distillers). Various other syntheses: *Org. Synth.* **coll. vol. III**, 30-34 (1955). Toxicity study: H. F. Smyth *et al.*, *Am. Ind. Hyg. Assoc. J.* **23**, 95 (1962). *Review:* J. W. Nemec, W. Bauer in *Kirk-Othmer Encyclopedia of Chemical Technology* **vol. 1** (Wiley-Interscience, New York, 3rd ed., 1978) pp 330-354.

Liquid; acrid odor and fumes. *Corrosive, flammable.* d_4^{16} 1.0621. mp 14°. bp 141.0°; bp_{400} 122.0°; bp_{200} 103.3°; bp_{100} 86.1°; bp_{40} 66.2°; bp_{10} 39.0°; bp_5 27.3°. n_D^{20} 1.4224. Flash pt, open cup: 155°F (68°C). pKa (25°): 4.25. Miscible with water, alc, ether. Polymerizes readily in the presence of oxygen. LD_{50} orally in rats: 2.59 g/kg (Smyth).
Caution: Potential symptoms of overexposure are irritation of eyes, skin, respiratory system; eye, skin burns; skin sensitization. *See NIOSH Pocket Guide to Chemical Hazards* (DHHS/NIOSH 97-140, 1997) p 8.
USE: In the manuf of plastics.

122. Acrylonitrile. [107-13-1] 2-Propenenitrile; vinyl cyanide; cyanoethylene; Ventox. C_3H_3N; mol wt 53.06. C 67.91%, H 5.70%, N 26.40%. Prepn by dehydration of ethylene cyanohydrin or acrylamide: Moureu, *Ann. Chim. Phys.* [7] **2**, 186 (1893). Manuf by ammoxidation of propylene: *Faith, Keyes & Clark's Industrial Chemicals*, F. A. Lowenheim, M. K. Moran, Eds. (Wiley-Interscience, New York, 4th ed., 1975) pp 46-49. Toxicity: H. F. Smyth, C. P. Carpenter, *J. Ind. Hyg. Toxicol.* **30**, 63 (1948). Causes acute and chronic adrenocortical insufficiency: S. Szabo *et al.*, *Lab. Invest.* **42**, 533 (1980); *eidem, J. Appl. Toxicol.* **4**, 131 (1984). Review of carcinogenic risk: *IARC Monographs* **19**, 73-133 (1979); of toxicology and human exposure: *Toxicological Profile for Acrylonitrile* (PB91-180489, 1990) 136 p. Comprehensive review: *The Chemistry of Acrylonitrile* (Am. Cyanamid, New York, 2nd ed., 1959) 272 pp; J. F. Brazdil in *Kirk-Othmer Encyclopedia of Chemical Technology* **vol. 1** (John Wiley & Sons, New York, 4th ed., 1991) pp 352-369.

Volatile liquid. *Flammable and combustible.* Should be stored and used in closed systems whenever possible. Work areas should be adequately ventilated, and should be free from open lights, flames, and equipment that is not explosion-proof. *Handle in hood.* May polymerize spontaneously, particularly in the absence of oxygen or on exposure to visible light. Polymerizes violently in the presence of concentrated alkali. On standing may slowly develop a yellow color particularly after excessive exposure to light. bp_{760} 77.3°; bp_{500} 64.7°; bp_{250} 45.5°; bp_{100} 23.6°; bp_{50} 8.7°. mp −83.55°. d_4^{20} 0.8060; d_4^{25} 0.8004. n_D^{25} 1.3888. Flash pt, open cup: 32°F (0°C). Explosive mixtures in air at 25°: 3.05% low limit; 17.0% upper limit. At 20° 7.35 parts dissolve in 100 parts water and 3.1 parts water dissolve in 100 parts acrylonitrile. Miscible with most organic solvents. LD_{50} orally in rats: 0.093 g/kg (Smyth, Carpenter).

Caution: Potential symptoms of overexposure are asphyxia; irritation of eyes and skin; headache; sneezing; nausea, vomiting; weakness, lightheadedness; skin vesiculation; scaling dermatitis. *See NIOSH Pocket Guide to Chemical Hazards* (DHHS/NIOSH 97-140, 1997) p 8. *See also Clinical Toxicology of Commercial Products,* R. E. Gosselin *et al.,* Eds. (Williams & Wilkins, Baltimore, 5th ed., 1984) Section II, p 215. This substance is reasonably anticipated to be a human carcinogen: *Report on Carcinogens, Twelfth Edition* (PB2011-111646, 2011) p 28.

USE: Manufacture of acrylic fibers. In the plastics, surface coatings, and adhesives industries. As a chemical intermediate in the synthesis of antioxidants, pharmaceuticals, dyes, surface-active agents, etc. In organic synthesis to introduce a cyanoethyl group. As a modifier for natural polymers. As a pesticide fumigant for stored grain. Experimentally to induce adrenal hemorrhagic necrosis in rats.

123. Actaplanins. A-4696; Kamoran. Complex of glycopeptide antibiotics produced by *Actinoplanes missouriensis.* Six actaplanins (A, B_1, B_2, B_3, C_1, G) have been isolated and characterized as having a central peptide core with the amino sugar *ristosamine* and up to four neutral sugars attached. Isoln: R. L. Hamill *et al.,* DE 2209018 (1972 to Lilly), *C.A.* 77, 138338n (1972); A. P. Raun, US 3816618; R. L. Hamill *et al.,* US 4115552 (1974, 1978 to Lilly). Chemical characterization: M. Debono *et al., J. Antibiot.* 37, 85 (1984). ^{1}H NMR studies and structures: A. H. Hunt *et al., J. Org. Chem.* 49, 635 (1984); *eidem, ibid.* 641. Growth promotant activity: C. Tsaltos *et al., Bull. Hellenic Vet. Med. Soc.* 33, 139 (1982). Use to increase milk production in ruminants: C. C. Scheifinger, EP 63491; *idem,* US 4430328 (1982, 1984 to Eli Lilly). Determn in milk: K. H. Hahne *et al., Milchwissenschaft* 39, 473 (1984).

	R	R'
Actaplanin A	mannosylglucose	mannose
Actaplanin B_1	rhamnosylglucose	mannose
Actaplanin B_2	glucose	mannose
Actaplanin B_3	mannosylglucose	H
Actaplanin C_1	rhamnosylglucose	H
Actaplanin G	glucose	H

Hydrochloride. White cryst solid, mp >220°. Approx mol wt 1158. $[\alpha]_D^{25}$ −42.3° (c = 1 in water). uv max (acidic and neutral

solns): 276 nm ($E_{1\,cm}^{1\%}$ 65). Sol in water. Insol in most organic solvents. Stable over pH 1.0 to 10.0 up to 27°.

THERAP CAT (VET): Growth stimulant.

124. Actarit. [18699-02-0] 4-(Acetylamino)benzeneacetic acid; (*p*-acetamidophenyl)acetic acid; MS-932; Mover; Orcl. C_{10}-$H_{11}NO_3$; mol wt 193.20. C 62.17%, H 5.74%, N 7.25%, O 24.84%. Immunomodulator. Prepn: S. Gabriel, *Ber.* 15, 834 (1882). *See also:* H. Munakata *et al.,* DE 3317107; *eidem,* US 4720506 (1983, 1988 both to Mitsubishi Chem. Ind. and Nippon Shinyaku). Pharmacology: H. Fujisawa *et al., Arzneim.-Forsch.* 40, 693 (1990). Pharmacokinetics: K. Sugihara *et al., ibid.* 800, 806. Antiarthritic effect in mice: H. Fujisawa *et al., ibid.* 44, 64 (1994). Acute toxicity: K. Toshida *et al., Oyo Yakuri* 40, 117 (1990), *C.A.* 114, 17276q (1991).

Odorless, white crystals or crystalline powder. mp 173-175° (Munakata); also reported as mp 168-170° (Gabriel). Freely sol in methanol; sol in ethanol; sparingly sol in acetone; slightly sol in water; very slightly sol in ether. LD_{50} in male, female mice, male, female rats (mg/kg): 1.06, 1.30, 1.95, 2.03 i.p.; 5.68, 5.48, 5.48, 6.12 s.c.; 15.3, 14.7, 14.8, 15.4 orally (Toshida).

THERAP CAT: Antiarthritic.

125. ACTH. [9002-60-2] Corticotropin; adrenocorticotrop-(h)in; corticotrophin; adrenocorticotrop(h)ic hormone of the pituitary gland; Acethropan; Acortan; Acthar; Acton; Cortiphyson; Cortrophin; Isactid. Pituitary hormone which stimulates the secretion of adrenal cortical steroids and induces growth of the adrenal cortex. Occurs also in female human urine and in serum of pregnant mares. Isoln procedure from swine pituitaries: Sayers *et al., J. Biol. Chem.* 149, 425 (1943); *Proc. Soc. Exp. Biol. Med.* 52, 199 (1943); from sheep pituitaries: Li *et al., J. Biol. Chem.* 149, 413 (1943); Li, *J. Am. Chem. Soc.* 74, 2124 (1952); from human pituitaries: Pickering *et al., Biochim. Biophys. Acta* 74, 763 (1963). Purification: Johnson, US 3124509 (1964 to Upjohn). Corticotropin is a single chain polypeptide containing 39 amino acids. The first 24 residues are identical in all species. *In vivo* studies show that this portion of the peptide chain is responsible for the biological activity and that the remaining residues, while not necessary for the hormonal action, are essential in the species immunological specificity of ACTH. The term α-ACTH is used to distinguish it from the pepsin- or acid-degraded product, β-ACTH. Prepn of active cleavage products by hydrolysis and simultaneous dialysis: Wettstein, Benz, US 2734015 (1956 to Ciba). Amino acid sequence of bovine ACTH: Li *et al., J. Am. Chem. Soc.* 80, 2587 (1958); proposed structure of human ACTH: Lee *et al., J. Biol. Chem.* 236, 2970 (1961). Revised amino acid sequences of porcine and human ACTH: Riniker *et al., Nature New Biol.* 235, 114 (1972). Revised sequences of ovine and bovine ACTH: Li, *Biochem. Biophys. Res. Commun.* 49, 835 (1972). Synthesis of the revised human corticotropin: Sieber *et al., Helv. Chim. Acta* 55, 1243 (1972); solid phase synthesis: Yamashiro, Li, *J. Am. Chem. Soc.* 95, 1310 (1973).

White powder. Freely sol in water. Partly precipitated at the isoelectric point (pH 4.65-4.80). Appreciably soluble in 60 to 70% alcohol or acetone. Almost completely precipitated in 2.5% trichloroacetic acid soln. Also precipitated from dilute soln by 20% sulfosalicylic acid and by 5% lead acetate soln. Solns are stable to heat. An aq soln buffered to pH 7.5 may be put in a boiling water bath for at least 120 minutes. In 0.10 molar HCl an 0.2% soln retains its biological potency when kept at 100° for 60 min, but in 0.10 molar NaOH the activity is lost within 30 min. At 60° at 0.2% soln at pH 10.8 maintains its potency for 60 min. In general the substance is more stable in acid soln. Pure ACTH has 150 to 200 potency units per mg. The biological activity is not destroyed by digestion with pepsin, and hydrolysis products have the biological activity of ACTH. One U.S.P. unit or one international unit or one Armour unit or one potency unit denotes the same activity.

Corticotrophin zinc hydroxide suspension. Cortrophine-Z. ACTH absorbed on zinc hydroxide. Used as an aqueous suspension.

THERAP CAT: Adrenocorticotropic hormone.

THERAP CAT (VET): Stimulates glucocorticoid production.

126. Actin. One of the major proteins of muscle and an important component of all eukaryotic cells. Formerly believed to be a single, highly conserved protein in all cell types, but multiple forms have been shown to exist. Muscle actin, or *α-actin*, is found in differentiated muscle cells; *β-actin* and *γ-actin* are present in all non-muscle cell types. All three forms contain equimolar amounts of *N*-methylhistidine, although α-actin differs from β- and γ-actins by several peptides; the two "non-muscle" actins are nearly identical. Several "minor actins" are also known. The various forms, although similar in activity, mol wt and amino acid composition, show immunological differences and are considered to be synthesized under the control of different genes. In addition to its role in muscle relaxation and contraction (together with myosin, *q.v.*), actin is involved in a variety of cellular events: cell movement, cytokinesis, phagocytosis, exocytosis and chromosome movement. Identification in muscle: F. B. Straub, *Stud. Szeged* **2**, 3 (1942). Isoln: Feuer *et al.*, *Hung. Acta Physiol.* **1**, 150 (1948). Separation of actin from muscle material: A. Szent-Gyorgyi, US 2742456 (1956 to Armour). In soln, actin and myosin combine to give **actomyosin**: Tonomura *et al.*, *J. Biol. Chem.* **237**, 1074 (1962). Actin, in the absence of salts, exists in a globular form, designated as *G-actin*; in the presence of ATP and potassium, sodium, or magnesium chlorides it polymerizes to a fibrous form, *F-actin*. Removal of bound ATP results in complete loss of polymerizability: Straub, Feuer, *Biochim. Biophys. Acta* **4**, 455 (1950). Structure of G-actin: Nagy, Jencks, *Biochemistry* **1**, 987 (1962); of F-actin: Hanson, Lowy, *J. Mol. Biol.* **6**, 46 (1963). Complete amino acid sequence of rabbit skeletal muscle actin: Elzinga *et al.*, *Proc. Natl. Acad. Sci. USA* **70**, 2687 (1973). Identification and characterization of multiple forms of actin: J. I. Garrels, W. Gibson, *Cell* **9**, 793 (1976). Partial amino acid sequence of calf brain actin: R. C. Lu, M. Elzinga, *Biochemistry* **16**, 5801 (1977). Comparison of actins from calf thymus, bovine brain, mouse cells and rabbit skeletal muscle: J. Vandeker, K. Weber, *Eur. J. Biochem.* **90**, 451 (1978). Isoln, characterization of porcine brain actin: J. P. Weir, D. W. Frederik, *Arch. Biochem.* **203**, 1 (1980). *In vivo* and *in vitro* synthesis of multiple forms of rat brain actin: E. Palmer, J. L. Saborio, *J. Biol. Chem.* **253**, 7482 (1978). Evidence for control of synthesis of human heart and platelet actins by different genes: M. Elzinga *et al.*, *Science* **191**, 94 (1976). Immunological differences between cardiac muscle, skeletal muscle and brain actins: J. L. Morgan *et al.*, *Proc. Natl. Acad. Sci. USA* **77**, 2069 (1980). History of discovery: Finck, *Science* **160**, 332 (1968). *Reviews:* Several authors, *Biochemistry of Muscle Contraction*, J. Gergely, Ed. (Little, Brown, Boston, 1964); Laki, "Actin" in *Contractile Proteins and Muscle*, K. Laki, Ed. (Marcel Dekker, New York, 1971) pp 97-133; E. D. Korn, *Proc. Natl. Acad. Sci. USA* **75**, 588-599 (1978); R. Kassab *et al.*, *Biochimie* **63**, 273-289 (1981); S. Highsmith, *Biochim. Biophys. Acta* **639**, 31-39 (1981); E. D. Korn, *Physiol. Rev.* **62**, 672-737 (1982).

127. Actinium. [7440-34-8] Ac; at. no. 89; valence 3. Group IIIB (3). No stable nuclides; known isotopes (mass numbers): 209-233; naturally occurring isotopes: 227, 228. Longest-lived isotope: ^{227}Ac (T½ 21.773 ±0.003 years, mode of decay α, β, γ-emission, rel. at. mass 227.0277). Daughter of ^{231}Pa; parent of ^{227}Th (radioactinium) and ^{223}Fr. Originally extracted from uranium ores where it occurs naturally as a member of the ^{235}U decay series; now prepd by neutron bombardment of ^{226}Ra in nuclear reactors. ^{228}Ac or *mesothorium II* (MsTh$_2$) occurs naturally as a member of the ^{232}Th decay series. T½ 6.13 ±0.03 hrs; decays by β and γ-emission. Daughter of ^{228}Ra (mesothorium I); parent of ^{228}Th (radiothorium). Discovery of ^{227}Ac: A. Debierne, *C. R. Hebd. Seances Acad. Sci.* **129**, 593 (1899); F. O. Geisel, *Ber.* **55**, 3608 (1902). Prepn of metal: J. G. Stites, Jr., *et al.*, *J. Am. Chem. Soc.* **77**, 237 (1955); J. D. Farr *et al.*, *J. Inorg. Nucl. Chem.* **18**, 42 (1961). Discovery of ^{228}Ac: O. Hahn, *Ber.* **40**, 1462 (1907). Review of actinium and the actinides: *Comprehensive Inorganic Chemistry* **vol. 5**, J. C. Bailar, Jr. *et al.*, Eds. (Pergamon Press, Oxford, 1973) *passim;* H. W. Kirby in *The Chemistry of the Actinide Elements* **vol. 1**, J. J. Katz *et al.*, Eds. (Chapman and Hall, New York, 1986) pp 14-40; G. T. Seaborg in *Kirk-Othmer Encyclopedia of Chemical Technology* **vol. 1** (Wiley-Interscience, New York, 4th ed., 1991) pp 412-444.

Cubic crystal, d^{25} 10.07. mp 1050 ±50°. bp ~3300°. Homologous with lanthanum; more basic than lanthanum. Strong electropositive element. Forms insol salts, e.g. carbonate, oxalate, phosphate etc. just as the lanthanides. Rapidly oxidizes to AcO$_3$ in moist air.

Caution: Radiation hazard; handling requires special equipment and shielding facilities (Katz *et al.*, *loc. cit.* **vol. 2**, p 1128).

128. Actinobolin. [24397-89-5] (2*S*)-2-Amino-*N*-[(3*R*,4*R*,-4a*R*,5*R*,6*R*)-3,4,4a,5,6,7-hexahydro-5,6,8-trihydroxy-3-methyl-1-oxo-1*H*-2-benzopyran-4-yl]propanamide; 4-(2-aminopropionamido)-3,4,4a,5,6,7-hexahydro-5,6,8-trihydroxy-3-methylisocoumarin. C$_{13}$H$_{20}$N$_2$O$_6$; mol wt 300.31. C 51.99%, H 6.71%, N 9.33%, O 31.96%. Antibiotic produced by *Streptomyces griseoviridus* var *atrofaciens;* (+)-form is naturally occurring. Isoln and characterization: T. H. Haskell, Q. R. Bartz, *Antibiot. Annu.* **1958-1959**, 505; T. H. Haskell *et al.*, US 3043830 (1962 to Parke, Davis). Structural studies: Struck *et al.*, *Tetrahedron Lett.* **8**, 1589 (1967); Munk *et al.*, *J. Am. Chem. Soc.* **89**, 4158 (1967). Structure: *eidem, ibid.* **90**, 1087 (1968). Absolute configuration and chemistry: Antosz *et al.*, *ibid.* **92**, 4933 (1970). Crystal structure: J. B. Wetherington, J. W. Moncrief, *Acta Crystallogr. B* **31**, 501 (1975). Total synthesis: M. Yoshioka *et al.*, *J. Am. Chem. Soc.* **106**, 1133 (1984); *eidem, Heterocycles* **21**, 151 (1984); R. S. Garigipati *et al.*, *J. Am. Chem. Soc.* **107**, 7790 (1985). Antibacterial activity and toxicity: R. F. Pittillo *et al.*, *Antibiot. Annu.* **1958-1959**, 497. Series of articles on antineoplastic activity: *ibid.* 515-532. Inhibition of protein synthesis: D. Smithers *et al.*, *Mol. Pharmacol.* **5**, 433 (1969). Experimental use as cariostat: D. E. Hunt *et al.*, *J. Dent. Res.* **50**, 371 (1971). Reduction of periodontal syndrome in mice, rats: J. H. Shaw, J. K. Ivimey, *Arch. Oral Biol.* **18**, 357 (1973).

Amorphous, fluffy, very hygroscopic powder. [α]$_D^{28}$ +59° (c = 0.5 in pH 7 phosphate buffer). pKa: 7.5, 8.8. Freely sol in water; mod sol in methanol, ethanol. Unstable in aqueous solns pH >7. uv max: 262 nm (0.1*N* HCl); 288 nm (0.1*N* NaOH). LD$_{50}$ in mice, rats (mg/kg): 800 ±27, 1550 ±26 i.v. (Pittillo).

Acetate. C$_{13}$H$_{20}$N$_2$O$_6$.C$_2$H$_4$O$_2$. Needles from ethanol; softens at 130°, resolidifies at 145°, mp 263-266°. [α]$_D^{26}$ +58°. pKa: 4.6, 7.5, 8.8. Very sol in water; sol in warm methanol, ethanol, acetone; sparingly sol in ethyl acetate. uv max: 264 nm (pH 7 phosphate buffer).

Hydrochloride. C$_{13}$H$_{20}$N$_2$O$_6$.HCl. Hygroscopic. [α]$_D^{22}$ +59° (c = 0.41) (natural); [α]$_D^{22}$ +55° (c = 0.47) (synthetic).

129. Actinodaphnine. [517-69-1] (7a*S*)-6,7,7a,8-Tetrahydro-11-methoxy-5*H*-benzo[*g*]-1,3-benzodioxolo[6,5,4-*de*]quinolin-10-ol; 10-methoxy-1,2-(methylenedioxy)-6aα-noraporphin-9-ol; 1,2-methylenedioxy-9-hydroxy-10-methoxynoraporphine. C$_{18}$H$_{17}$NO$_4$; mol wt 311.34. C 69.44%, H 5.50%, N 4.50%, O 20.56%. In bark of *Actinodaphne hookeri* Meissn., *Lauraceae*. Isoln: Krishna, Ghose, *J. Indian Chem. Soc.* **9**, 429 (1932). Structure: Ghose *et al.*, *Helv. Chim. Acta* **17**, 919 (1934). Photolytic synthesis: M. S. Premila *et al.*, *Indian J. Chem.* **13**, 945 (1975).

Needles from alcohol, mp 211°. $[\alpha]_D^{20}$ +33° (ethanol). Sol in acetone, alcohol, benzene, chloroform. Moderately sol in ether. Practically insol in water.

Hydrochloride. $C_{18}H_{17}NO_4.HCl$. Needles from alcohol-ether, mp 281° (dec). $[\alpha]_D^{20}$ +9°.

Methyl ether. $C_{19}H_{19}NO_4$. Needles from ether + alc, mp 114-115°. Synthesis: Hey, Lobo, *J. Chem. Soc.* **1954**, 2246.

130. Actinomycetin. [1402-37-5] Bacteriolytic cell-free fluid of culture filtrates of actinomycetes. Produced by most species of *Streptomyces;* studied mostly in *Streptomyces albus.* Protein-like in nature. Stated to consist of a bactericidal fatty acid fraction and an enzyme: Welsch, *J. Bacteriol.* **42**, 801 (1941); *ibid.* **44**, 571 (1942); *ibid.* **53**, 101 (1947). Dissolves dead gram-negative, and, with more difficulty, dead gram-positive organisms. Also dissolves living organisms in aq suspensions, such as *Staph. aureus.* Discussion of whether actinomycetin deserves the name "antibiotic": Hoogerheide, Welsch, *Bot. Rev.* **10**, 599 (1944); *J. Bacteriol.* **53**, 101 (1947). Purification and prepn from *Streptomyces albus:* Ghuysen, BE 517191 and BE 521114 (both 1953 to Soc. Belge de l'Azote), *C.A.* **53**, 5600g, 12593e (1959). *Review:* Caltrider in *Antibiotics* vol. 1, D. Gottlieb, P. D. Shaw, Eds. (Springer Verlag, New York, 1967) pp 681-683.

Sol in water. Pptd by alcohol, acetone, ammonium sulfate and other protein precipitants. Destroyed by strong acid. More stable at pH 10 than at pH 4. Rapidly inactivated by heat, 60-70° destroys activity.

131. Actinorhodine. [1397-77-9] $(1R,1'R,3S,3'S)$-3,3',4,-4',5,5',10,10'-Octahydro-6,6',9,9'-tetrahydroxy-1,1'-dimethyl-5,-5',10,10'-tetraoxo-[8,8'-bi-1*H*-naphtho[2,3-*c*]pyran]-3,3'-diacetic acid. $C_{32}H_{26}O_{14}$; mol wt 634.55. C 60.57%, H 4.13%, O 35.30%. Antibiotic pigment produced by *Streptomyces coelicolor*, found in woods near Göttingen, Germany: Brockmann, Pini, *Naturwissenschaften* **34**, 190 (1947); Brockmann *et al., Ber.* **83**, 161 (1950). Tentative structure: Brockmann, Hieronymus, *Ber.* **88**, 1379 (1955); Brockmann *et al., Naturwissenschaften* **49**, 131 (1962). Structure: Brockmann, *Angew. Chem.* **76**, 863 (1964); Brockmann *et al., Ann.* **698**, 209 (1966). Stereochemistry: Zeeck, Christiansen, *Ann.* **724**, 172 (1969). Biosynthesis and determn of point of dimerization: C. P. Gorst-Allman *et al., J. Org. Chem.* **46**, 455 (1981).

Fine red needles from dioxane, dec 270°. Absorption max (dioxane): 560, 523 nm. Sol in pyridine, piperidine, tetrahydrofuran, dioxane, phenol; slightly sol in alcohol, acetic acid, acetone. Red soln with acetone, blue soln with pyridine. Practically insol in aq acid; sol in aq alkali with bright blue color.

132. Activin. Polypeptide hormone, identified in ovarian follicular fluid, which selectively stimulates secretion of FSH, *q.v.* Dimer composed of β subunits of inhibin, *q.v.* (either β_A or β_B), mol wt 24,000 daltons. Isolation and characterization of the heterodimer, ***activin AB:*** N. Ling *et al., Nature* **321**, 779 (1986); of the homodimer, ***activin A:*** W. Vale *et al., ibid.* 776; N. Ling *et al., Biochem. Biophys. Res. Commun.* **138**, 1129 (1986). Comparison of FSH stimulation with gonadotropin releasing hormone (LH-RH, *q.v.*): R. H. Schwall *et al., ibid.* **151**, 1099 (1988). Approach to nonradiometric assays: R. H. Schwall *et al., Prog. Clin. Biol. Res.* **285**, 205 (1988). Brief review of chemical properties and biological activity: S.-Y. Ying, *Proc. Soc. Exp. Biol. Med.* **186**, 253-264 (1987). *Review:* S. A. Pangas, T. K. Woodruff, *Trends Endocrinol. Metab.* **11**, 309-314 (2000).

133. ACV. [32467-88-2] *N*-[(5*S*)-5-Amino-5-carboxy-1-oxopentyl]-L-cysteinyl-D-valine; *N*-[*N*-(L-5-amino-5-carboxyvaleryl)-L-cysteinyl]-D-valine; δ-(L-α-aminoadipyl)-L-cysteinyl-D-valine. $C_{14}H_{25}N_3O_6S$; mol wt 363.43. C 46.27%, H 6.93%, N 11.56%, O

26.41%, S 8.82%. Biosynthetic precursor of penicillins and cephalosporins. Isoln from mycelial extracts of *Penicillium chrysogenum:* H. R. V. Arnstein *et al., Biochem. J.* **76**, 353 (1960); structure determn of tripeptide: H. R. V. Arnstein, D. Morris, *ibid.* 357. Isoln from *Cephalosporium* sp., identification and optical configuration of constituent amino acids: P. B. Loder, E. P. Abraham, *ibid.* **123**, 471 (1971). Configuration of peptide from *P. chrysogenum*, identity with peptide from *Cephalosporium:* P. Adriaens *et al., Antimicrob. Agents Chemother.* **8**, 638 (1975). Synthesis of labelled tripeptide and role in penicillin biosynthesis: P. A. Fawcett *et al., Biochem. J.* **157**, 651 (1976). Synthesis of ACV as the disulfide: S. Wolfe, M. G. Jokinen, *Can. J. Chem.* **57**, 1388 (1979). Cellfree conversion of ACV directly into isopenicillin N: J. O'Sullivan *et al., Biochem. J.* **184**, 421 (1979); T. Konomi *et al., ibid.* 427. Confirmation by ^{13}C NMR: J. E. Baldwin *et al., Chem. Commun.* **1980**, 1271. *Reviews:* D. J. Aberhart, *Tetrahedron* **33**, 1545-1559 (1977); E. P. Abraham, *Jpn. J. Antibiot.* **30**, Suppl., S1-S26 (1977); S. Wolfe *et al., Science* **226**, 1386-1392 (1984).

134. Acyclovir. [59277-89-3] 2-Amino-1,9-dihydro-9-[(2-hydroxyethoxy)methyl]-6*H*-purin-6-one; acycloguanosine; 9-[(2-hydroxyethoxy)methyl]guanine; BW-248U; Wellcome 248U; Acicloftal; Avirase; Cycloviran; Maynar; Virmen; Viruseen; Zoliparin; Zovir; Zovirax. $C_8H_{11}N_5O_3$; mol wt 225.21. C 42.67%, H 4.92%, N 31.10%, O 21.31%. Orally active acyclic nucleoside with inhibitory activity towards several herpes viruses. Prepn: H. J. Schaeffer, DE 2539963; *idem,* US 4199574 (1976, 1980 to Wellcome). Convenient synthesis from guanine: H. Matsumoto *et al., Chem. Pharm. Bull.* **36**, 1153 (1988). Selectivity of action: G. B. Elion *et al., Proc. Natl. Acad. Sci. USA* **74**, 5716 (1977). Chemistry, antiviral activity, metabolism: H. J. Schaeffer *et al., Nature* **272**, 583 (1978). *In vitro* activity: P. Collins, D. J. Bauer, *J. Antimicrob. Chemother.* **5**, 431 (1979). Effect on herpes simplex infections in mice: H. J. Field *et al., Antimicrob. Agents Chemother.* **15**, 554 (1979); on herpes zoster in immunocompromised patients: H. H. Balfour *et al., N. Engl. J. Med.* **308**, 1448 (1983). Treatment of primary episodes of genital herpes simplex infection: Y. J. Bryson *et al., ibid.* 916; of recurrent genital herpes: S. E. Straus *et al., ibid.* **310**, 1545 (1984); J. M. Douglas *et al., ibid.* 1551. HPLC determn in serum and clinical pharmacokinetics: G. Bahrami *et al., J. Chromatogr. B* **816**, 327 (2005). Symposia on pharmacology and clinical studies: *Am. J. Med.* **73**, Suppl. 1A, 1-392 (1982); *J. Antimicrob. Chemother.* **12**, Suppl. B, 1-202 (1983); *Scand. J. Infect. Dis.* **Suppl. 47**, 1-176 (1985). *Review:* R. J. Whitley, J. W. Gnann, Jr., *N. Engl. J. Med.* **327**, 782-789 (1992).

Crystals from methanol, mp 256.5-257°. Maximum soly in water (37°): 2.5mg/ml. Sol in diluted hydrochloric acid; slightly sol in water. Insol in alcohol. pKa_1 2.27. pKa_2 9.25. LD_{50} in mice (mg/kg): >10,000 orally; 1000 i.p. (Schaeffer).

Sodium salt. [69657-51-8] $C_8H_{11}N_5NaO_3$; mol wt 248.20. White crystalline powder. Maximum soly in water (25°): >100 mg/ml.

THERAP CAT: Antiviral.

THERAP CAT (VET): Antiviral.

135. ADA. [26239-55-4] *N*-(2-Amino-2-oxoethyl)-*N*-(carboxymethyl)glycine; *N*-(2-acetamido)iminodiacetic acid. C_6H_{10}-N_2O_5; mol wt 190.16. C 37.90%, H 5.30%, N 14.73%, O 42.07%. One of several zwitterionic *N*-substituted amino acids known as "Good" buffers. Active in the pH range 6-8. Prepn: G. Schwarzenbach *et al.*, *Helv. Chim. Acta* **38**, 1147 (1955); and description: N. E. Good *et al.*, *Biochemistry* **5**, 467 (1966). Thermodynamic behavior: C. D. McGlothlin, J. Jordan, *Anal. Lett.* **9**, 245 (1976); F. Jumean, *J. Indian Chem. Soc.* **69**, 237 (1992). Use as buffer: Z. Pietrzkowski, W. Korohoda, *Folia Histochem. Cytobiol.* **26**, 143 (1988); as analytical reagent for chromium: C. Baluja-Santos *et al.*, *Rev. Roum. Chim.* **39**, 73 (1994); as chelator: M. C. Steele, J. Pichtel, *J. Environ. Eng.* **124**, 639 (1998).

Crystals from chloroacetamide, dec 220°. Slightly sol in water. pKa_1: <2; pKa_2: 2-3; pKa_3 when 0.1M (0°): 6.85; (20°): 6.60; (37°): 6.45. $\Delta pKa/°C = -0.011$.

USE: Buffer for biological systems and chelator for metals.

136. Adalimumab. [331731-18-1] Anti-(human tumor necrosis factor) immunoglobulin G1 (human monoclonal D2E7 heavy chain) disulfide with human monoclonal D2E7 light chain, dimer; D2E7; Humira. Fully human monoclonal antibody against tumor necrosis factor α (TNFα). Prepn: J. G. Salfeld *et al.*, **WO 9729131**; *eidem*, **US 6090382** (1997, 2000 both to BASF). Characterization: L. C. Santora *et al.*, *Anal. Biochem.* **275**, 98 (1999). Characterization of antigen binding: *idem et al.*, *ibid.* **299**, 119 (2001). Clinical trial in rheumatoid arthritis: A. A. den Broeder *et al.*, *Ann. Rheum. Dis.* **61**, 311 (2002); in combination with methotrexate: F. C. Breedveld *et al.*, *Arthritis Rheum.* **54**, 26 (2006); in psoriatic arthritis: P. J. Mease *et al.*, *ibid.* **52**, 3279 (2005); in Crohn's disease: K. A. Papadakis *et al.*, *Am. J. Gastroenterol.* **100**, 75 (2005). Review of pharmacology and clinical experience: K. P. Machold, J. S. Smolen, *Expert Opin. Biol. Ther.* **3**, 351-360 (2003). Review of safety and efficacy in psoriasis and rheumatoid arthritis: T. Patel, K. B. Gordon, *Dermatol. Ther.* **17**, 427-431 (2004); in psoriatic arthritis: P. J. Mease, *Expert Opin. Biol. Ther.* **5**, 1491-1504 (2005).

THERAP CAT: Anti-inflammatory.

137. Adamantane. [281-23-2] Tricyclo[3.3.1.1^{3,7}]decane; diamantane (obsolete). $C_{10}H_{16}$; mol wt 136.24. C 88.16%, H 11.84%. Isoln from petroleum from Moravia province in Czechoslovakia: Landa, Machacek, *Collect. Czech. Chem. Commun.* **3**, 1 (1933); from American petroleum: Mair *et al.*, *Anal. Chem.* **31**, 2082 (1959). Synthesis: Prelog, Seiwerth, *Ber.* **74**, 1769 (1941); Stetter *et al.*, *Ber.* **89**, 1922 (1956); by aluminum chloride-catalyzed isomerization of tetrahydrodicyclopentadiene: Yan, Shreh, *Bull. Inst. Chem. Acad. Sinica* **Dec. 1965**, no. 11, pp 79-81. Two step synthesis starting with dicyclopentadiene: Schleyer, *J. Am. Chem. Soc.* **79**, 3292 (1957); Schleyer, Donaldson, *ibid.* **82**, 4645 (1960); Ludwig **US 2937211** (1960 to du Pont). *Reviews:* Stetter, *Angew. Chem.* **66**, 217 (1954); Fort, Schleyer, *Chem. Rev.* **64**, 277 (1964).

Crystallizes at −30°; can be purified by recrystallization from acetone or by sublimation. mp 269.6-270.8°.

Note: The name diamantane has been abandoned as a synonym for adamantane and proposed as the name for the second member of the adamantane series, congressane: Vogl, Anderson, *Tetrahedron Lett.* **1966**, 415.

USE: The diamine as curing agent for epoxy resins [**US 3053907** (1962 to du Pont)].

138. Adapalene. [106685-40-9] 6-(4-Methoxy-3-tricyclo-[3.3.1.1^{3,7}]dec-1-ylphenyl)-2-naphthalenecarboxylic acid; 6-[3-(1-adamantyl)-4-methoxyphenyl]-2-naphthoic acid; CD-271; Differin. $C_{28}H_{28}O_3$; mol wt 412.53. C 81.52%, H 6.84%, O 11.63%. Retinoid selective for retinoic acid receptor (RAR) subtypes β and γ. Prepn: B. Shroot *et al.*, **EP 199636**; *eidem*, **US 4717720** (1986, 1988 both to Cent. Int. Recher. Dermatol.); and structure-activity study:

B. Charpentier *et al.*, *J. Med. Chem.* **38**, 4993 (1995). Pilot-scale synthesis: Z. Liu, J. Xiang, *Org. Process Res. Dev.* **10**, 285 (2006). HPLC determn in plasma and tissue: R. Ruhl, H. Nau, *Chromatographia* **45**, 269 (1997). Clinical pharmacology: C. E. M. Griffiths *et al.*, *J. Invest. Dermatol.* **101**, 325 (1993). Clinical trial in acne: A. Shalita *et al.*, *J. Am. Acad. Dermatol.* **34**, 482 (1996). Reviews of pharmacology and clinical potential: B. A. Bernard, *Skin Pharmacol.* **6**, Suppl. 1, 61-69 (1993); R. N. Brogden, K. L. Goa, *Drugs* **53**, 511-519 (1997); of clinical use in acne vulgaris: J. Waugh *et al.*, *Drugs* **64**, 1465-1478 (2004).

White crystals from THF and ethyl acetate, mp 319-322°. pK 4.2. Stable to light.

THERAP CAT: Antiacne.

139. Adefovir. [106941-25-7] *P*-[[2-(6-Amino-9*H*-purin-9-yl)ethoxy]methyl]phosphonic acid; 9-(2-phosphonylmethoxyethyl)-adenine; PMEA; GS-393. $C_8H_{12}N_5O_4P$; mol wt 273.19. C 35.17%, H 4.43%, N 25.64%, O 23.43%, P 11.34%. Acyclic nucleoside phosphonate (ANP) analog. Prepn: A. Holy, I. Rosenberg, **EP 206459**; *eidem*, *Collect. Czech. Chem. Commun.* **52**, 2801 (1987); of dipivoxil ester prodrug: J. E. Starrett, Jr. *et al.*, **EP 481214** (1992 to Bristol-Myers Squibb); *eidem*, **US 5663159** (1997 to Inst. Org. Chem. Biochem. Acad. Sci. Czech Rep.; Rega). Crystal structure: C. H. Schwalbe *et al.*, *J. Chem. Soc. Perkin Trans. 1* **1991**, 1348. HPLC determn: P. Augustijns *et al.*, *J. Liq. Chromatogr. Relat. Technol.* **19**, 2271 (1996). Clinical pharmacokinetics: K. C. Cundy *et al.*, *Antimicrob. Agents Chemother.* **39**, 2401 (1995). Review of pharmacology and therapeutic potential: L. Naesens *et al.*, *Antivir. Chem. Chemother.* **8**, 1-23 (1997). Clinical trial in HIV infected patients: J. Kahn *et al.*, *J. Am. Med. Assoc.* **282**, 2305 (1999); in hepatitis B virus (HBV): R. J. C. Gilson *et al.*, *J. Viral Hepat.* **6**, 387 (1999). Resistance surveillance in HBV infections: H. Yang *et al.*, *Hepatology* **36**, 464 (2002).

Off-white crystalline solid, mp >250°. pKa_1 2.0, pKa_2 6.8. uv max (H_2O) 208, 260 nm (ε 19600, 14100); (0.1*N* HCl) 210, 260 nm (ε 19000, 13700); (0.1*N* NaOH) 216, 262 nm (ε 9600, 14500).

Di(pivaloyloxymethyl) ester. [142340-99-6] Adefovir dipivoxil; bis(POM)-PMEA; GS-840; Hepsera; Preveon. $C_{20}H_{32}N_5O_8P$; mol wt 501.48. Off-white crystalline powder. Soly in water (mg/ml): 19 (pH 2.0); 0.4 (pH 7.2). Log P (octanol/aq phosphate buffer): 1.91.

THERAP CAT: Antiviral.

140. Adenine. [73-24-5] 9*H*-Purin-6-amine; 6-aminopurine; 6-amino-1*H*-purine; 6-amino-3*H*-purine; 6-amino-9*H*-purine; 1,6-dihydro-6-iminopurine; 3,6-dihydro-6-iminopurine; Leuco-4. $C_5H_5N_5$; mol wt 135.13. C 44.44%, H 3.73%, N 51.83%. Also referred to as *vitamin B₄*: Lecoq, *Int. Z. Vitaminforsch.* **27**, 291 (1957). Widespread throughout animal and plant tissues combined with niacinamide, D-ribose, and phosphoric acids; a constituent of nucleic acids and coenzymes, such as codehydrase I and II, adenylic acid, coalaninedehydrase. Isoln from bovine pancreas: Kossel, *Ber.* **18**, 79, 1928 (1885). Syntheses: Fischer, *ibid.* **30**, 2226 (1897); Traube, *Ann.* **331**, 64 (1904); Hoffer, *Jubilee Vol. Emil Barell* **1946**,

428-434; Taylor *et al.*, *Ciba Found. Symp. Chem. Biol. Purines* **1957**, 20, *C.A.* 53, 6238b (1959); Bredereck *et al.*, *Angew. Chem.* **71**, 524 (1959); Morita *et al.*, *Chem. Ind. (London)* **1968**, 1117; Sekiya, Suzuki, *Chem. Pharm. Bull.* **20**, 209 (1972); N. J. Kos *et al.*, *J. Org. Chem.* **44**, 3140 (1979). Toxicity study: Philips *et al.*, *J. Pharmacol. Exp. Ther.* **104**, 20 (1952). *Review:* Ts'o, "Bases, Nucleosides and Nucleotides" in *Basic Principles in Nucleic Acid Chemistry* **vol. 1**, P. O. P. Ts'o, Ed. (Academic Press, New York, 1974) pp 453-584.

Trihydrate, orthorhombic needles. Anhydr at 110°, dec 360-365°, subl 220°. uv max (pH 7.0): 207, 260.5 nm ($\varepsilon \times 10^{-3}$ 23.2, 13.4). One gram of anhydr compd dissolves in 2000 ml water, 40 ml boiling water. Aq solns are neutral. Slightly sol in alcohol. Practically insol in ether, chloroform. LD_{50} orally in rats: 745 mg/kg (Philips).

Hydrochloride hemihydrate. Monoclinic prisms. One gram dissolves in 42 ml water.

Sulfate dihydrate. Crystals. One gram dissolves in 150 ml water. Slightly sol in alc.

USE: In microbial determination of niacin; in research on heredity, virus diseases, and cancer.

141.   Adenosine. [58-61-7] 9-β-D-Ribofuranosyl-9*H*-purin-6-amine; 6-amino-9-β-D-ribofuranosyl-9*H*-purine; 9-β-D-ribofuranosidoadenine; adenine riboside; Adenocard; Adenocor; Adenoscan. $C_{10}H_{13}N_5O_4$; mol wt 267.25. C 44.94%, H 4.90%, N 26.21%, O 23.95%. Nucleoside; widely distributed in nature. From yeast nucleic acid: Levene and Bass, *Nucleic Acids* (New York, 1931) p 163. Structure: Levene, Tipson, *J. Biol. Chem.* **94**, 809 (1932); Bredereck, *Ber.* **66**, 198 (1933); *Z. Physiol. Chem.* **223**, 61 (1934); Gulland, Holiday, *J. Chem. Soc.* **1936**, 765. *See also:* Szent-Györgyi, *J. Physiol.* **68**, 213 (1930); Lythgoe *et al.*, *J. Chem. Soc.* **1947**, 355; **1948**, 965. Synthesis: Davoll *et al.*, *ibid.* **1948**, 967; H. Vorbrueggen, K. Krolikiewicz, *Angew. Chem. Int. Ed.* **14**, 421 (1975). Crystal structure: T. F. Lai, R. E. Marsh, *Acta Crystallogr. B* **28**, 1982 (1972). Conformational properties: D. B. Davies, A. Rabczenko, *J. Chem. Soc. Perkin Trans. 2* **1975**, 1703. Symposium on cardiac electrophysiology, pharmacology and clinical efficacy in supraventricular tachycardia: *Prog. Clin. Biol. Res.* **230**, 1-395 (1987). *Reviews: see* Adenine.

Crystals from water, mp 234-235°. $[\alpha]_D^{11}$ -61.7° (c = 0.706 in water); $[\alpha]_D^9$ -58.2° (c = 0.658 in water). uv max: 260 nm (ε 15100). Slightly sol in water. Practically insol in alcohol.

THERAP CAT: Antiarrhythmic.

142.   Adenosine Diphosphate. [58-64-0] Adenosine 5'-(trihydrogen diphosphate); ADP; adenosine 5'-pyrophosphoric acid; 5'-adenylphosphoric acid; adenosinediphosphoric acid. $C_{10}H_{15}N_5O_{10}P_2$; mol wt 427.20. C 28.12%, H 3.54%, N 16.39%, O 37.45%, P 14.50%. Formed from ATP in the muscle by the enzyme adenosinetriphosphatase upon stimulation of the muscle unless hydrolysis is prevented by injection of magnesium sulfate. Prepd by hydrolysis of ATP: LePage, *Biochem. Prep.* **I**, 1 (1949). Synthesis: Chambers *et al.*, *J. Am. Chem. Soc.* **82**, 970 (1960). *See also* Adenosine Triphosphate.

Barium salt. $Ba_3(C_{10}H_{12}N_5O_{10}P_2)_2$. The purity of the prepns can be checked by analyses for nitrogen, ribose (orcinol reaction), total phosphorus, easily hydrolyzable phosphorus, and inorganic phosphorus. ADP should give a ratio of 2:1 for total organic phosphorus to easily hydrolyzable phosphorus. uv max (pH 7.0): 259 nm (a_M 15.4 × 10³).

143.   Adenosine Triphosphate. [56-65-5] Adenosine 5'-(tetrahydrogen triphosphate); ATP; adenosine 5'-triphosphoric acid; Adetol; Atriphos; Striadyne; Triadenyl. $C_{10}H_{16}N_5O_{13}P_3$; mol wt 507.18. C 23.68%, H 3.18%, N 13.81%, O 41.01%, P 18.32%. Coenzyme valuable in the transfer of phosphate bond energy. Mammalian skeletal muscle at rest contains 350-400 mg ATP per 100 g. Upon stimulation of the muscle the ATP is hydrolyzed to ADP by the myosin-actin complex unless hydrolysis is prevented by injection of magnesium sulfate. Isoln from rabbit muscle: LePage, *Biochem. Prep.* **1**, 5 (1949); *cf.* Fiske, Subbarow, *Science* **70**, 381 (1929); Lohmann, *Biochem. Z.* **223**, 460 (1931); **254**, 381 (1932); Barrenscheen, Filz, *ibid.* **250**, 281 (1932); Kerr, *J. Biol. Chem.* **139**, 121 (1941); Needham, *Biochem. J.* **36**, 113 (1942). Synthetic routes: Tanaka, Honjo, US 3079379 (1963 to Takeda). Reviews of biosynthesis: Racker, *Adv. Enzymol.* **23**, 323-399 (1961); Deamer, *J. Chem. Educ.* **46**, 198-206 (1971). Reviews of nucleotide coenzymes: Bock in *The Enzymes* vol. 2A, P. D. Boyer *et al.*, Eds. (Academic Press, New York, 2nd ed., 1960) pp 3-38; A. M. Michelson, *The Chemistry of Nucleosides and Nucleotides* (Academic Press, New York, 1963) pp 153-250; D. W. Hutchinson, *Nucleotides and Coenzymes* (John Wiley, New York, 1964) pp 36-82.

Free ATP, isolated as a glass (by treatment of the Ba salt with H_2SO_4 and treating the concd aq soln with acetone) $[\alpha]_D^{22}$ -26.7° (c = 3.095). uv max (pH 7.0): 259 nm (a_M 15.4 × 10³). Freely sol in water. A 1% aq soln has a pH of about 2 and is stable at 0° for several hrs. ATP is a tetrabasic acid and after hydrolytic cleavage it is hexabasic. It is usually pptd as the dibarium salt with 4 or 6 mols of water of crystn which can be removed by prolonged drying at 100° over P_2O_5 and in a vacuum. The anhydr salt is stable, but the hydrated salt slowly decomp forming 5'-adenylic acid and barium pyrophosphate. For use the barium salt is converted to the sodium or potassium salt by treatment with sodium or potassium sulfate in HCl soln. At pH 6.8-7.4 the Na salt is stable in aq soln provided the product is pure. Ba^{2+} catalyzes breakdown.

Disodium salt. [987-65-5] Adetphos; Atenen; Circulen; Trinosin. $C_{10}H_{14}N_5Na_2O_{13}P_3$; mol wt 551.14.

USE: In biochemical research. To inhibit enzymatic browning of raw edible plant materials, such as sliced apples, potatoes, etc.

THERAP CAT: Coenzyme.

144.   S-Adenosylmethionine. [29908-03-0] 5'-[(3*S*)-(3-Amino-3-carboxypropyl)methylsulfonio]-5'-deoxyadenosine inner salt; *S*-(5'-desoxyadenosin-5'-yl)-L-methionine; active methionine; ademetionine; AdoMet; SAMe. $C_{15}H_{22}N_6O_5S$; mol wt 398.44. C 45.22%, H 5.57%, N 21.09%, O 20.08%, S 8.05%. Physiological methyl donor involved in enzymatic transmethylation reactions;

present in all living organisms. Enzymatic prepn: G. L. Cantoni, *J. Am. Chem. Soc.* **74**, 2942 (1952). Synthesis of racemate: J. Baddiley, G. A. Jamieson, *J. Chem. Soc.* **1954**, 4280; of optically active chloride: C. H. Shunk, J. W. Richter, US 2969353 (1961 to Merck & Co.). Review of enzymatic and microbial prepns: S. Shimizu, H. Yamada, *Trends Biotechnol.* **2**, 137-141 (1984). Pharmacology and toxicity: G. Stramentinoli *et al.*, *Minerva Med.* **66**, 1541 (1975). Radioenzymatic determn in biological fluids: P. Giulidori, G. Stramentinoli, *Anal. Biochem.* **137**, 217 (1984). Cytoprotective effects against bile-acid induced apoptosis: C. R. L. Webster *et al.*, *Vet. Ther.* **3**, 474 (2002). Clinical trial in arthritis: J. D. Bradley *et al.*, *J. Rheumatol.* **21**, 905 (1994); in depression: M. Fava *et al.*, *Psychiatry Res.* **56**, 295 (1995). Review of clinical efficacy in intrahepatic cholestasis: M. Coltorti *et al.*, *Methods Find. Exp. Clin. Pharmacol.* **2**, 69-78 (1990); in neurological disorders: T. Bottiglieri *et al.*, *Drugs* **48**, 137-152 (1994). Review of role in biological methylation: P. K. Chiang *et al.*, *FASEB J.* **10**, 471-480 (1996). Symposium on use in treatment of alcoholic liver disease: *Alcohol* **27**, 149-198 (2002). Review of clinical use in canine hepatitis: A. Honeckman, *Clin. Tech. Small Anim. Pract.* **18**, 239-244 (2003); in management of heptobiliary disorders: S. A. Center, *Vet. Clin. North Am. Small Anim. Pract.* **34**, 67-172 (2004).

1,4-Butanedisulfonate. [200393-05-1] Donamet; Samyr. $C_{15}H_{22}N_6O_5S.C_4H_{10}S_2O_6$.

Disulfate tosylate. [55722-12-8] 5′-[(3*S*)-(3-Amino-3-carboxypropyl)methylsulfonio]-5′-deoxyadenosine sulfate (salt) 4-methylbenzenesulfonate (salt) sulfate (salt) (1:1:2:1); Gumbaral; S-Amet; Zentonil. $C_{15}H_{23}N_6O_5S.2C_7H_8O_3S.H_2SO_4.HSO_4$; mol wt 938.98.

THERAP CAT: Antidepressant; antirheumatic; hepatoprotectant.

THERAP CAT (VET): Heptoprotectant.

145.   3′-Adenylic Acid. [84-21-9] Adenosine 3′-monophosphate; adenosine-3′-phosphoric acid; adenosine-3′-monophosphoric acid; adenylic acid b; yeast adenylic acid; synadenylic acid; h-adenylic acid. $C_{10}H_{14}N_5O_7P$; mol wt 347.22. C 34.59%, H 4.06%, N 20.17%, O 32.25%, P 8.92%. Early prepns from yeast nucleic acid: Levene, Bass, *Nucleic Acids* (Chemical Catalogue Co., New York, 1931). Early work probably done on mixtures of 2′- and 3′-adenylic acids; both compds isomerize readily to form an equilibrium mixture under acid conditions: Carter, Cohn, *Fed. Proc.* **8**, 190 (1949); Baddily, in *The Nucleic Acids* vol. **1**, E. Chargaff, J. N. Davidson, Eds. (Academic Press, New York, 1955) pp 165-168; A. M. Michelson, *The Chemistry of Nucleosides and Nucleotides* (Academic Press, New York, 1963) pp 100-106. Synthesis: Brown, Todd, *J. Chem. Soc.* **1952**, 44. Structure of dihydrate: Brown *et al.*, *Nature* **172**, 1184 (1953); Sundaralingam, *Acta Crystallogr.* **21**, 495 (1966). *Reviews: see* Adenosine.

Long, fine, colorless, needles. mp 197° (dec). Dihydrate, d (calc) 1.698. Absorption spectrum: Voet *et al.*, *Biopolymers* **1**, 193

(1963). Difficultly sol in boiling water. Gives quantitative yield of furfural when distilled with 20% HCl for 3 hrs (different from 5′-adenylic acid which yields only traces). *See:* Hoffman, *J. Biol. Chem.* **73**, 15 (1927); Embden, Schmidt, *Z. Physiol. Chem.* **181**, 130 (1929).

146.   5′-Adenylic Acid. [61-19-8] Muscle adenylic acid; ergadenylic acid; t-adenylic acid; adenosine 5′-monophosphate; adenosine phosphate; adenosine-5′-phosphoric acid; adenosine-5′-monophosphoric acid; A-5MP; AMP; NSC-20264; Adényl. $C_{10}H_{14}N_5O_7P$; mol wt 347.22. C 34.59%, H 4.06%, N 20.17%, O 32.25%, P 8.92%. Nucleotide; widely distributed in nature. Prepn from tissues: Embden, Zimmerman, *Z. Physiol. Chem.* **167**, 137 (1927); Embden, Schmidt, *ibid.* **181**, 130 (1929); *cf.* Kalckar, *J. Biol. Chem.* **167**, 445 (1947). Prepn by hydrolysis of ATP with barium hydroxide: Kerr, *J. Biol. Chem.* **139**, 131 (1941). Synthesis: Baddiley, Todd, *J. Chem. Soc.* **1947**, 648. Commercial prepn by enzymatic phosphorylation of adenosine. Monograph on synthesis of nucleotides: G. R. Pettit, *Synthetic Nucleotides* vol. **1** (Van Nostrand-Reinhold, New York, 1972) 252 pp. Crystal structure: Kraut, Jensen, *Acta Crystallogr.* **16**, 79 (1963). *Reviews: see* Adenosine.

Crystals from water + acetone, mp 196-200°. $[\alpha]_D^{20}$ −47.5° (c = 2, 2% NaOH); $[\alpha]_D^{20}$ −26.0° (c = 2, 10% HCl). $pK_1 = 3.8$; $pK_2 = 6.2$. uv max (pH 7.0): 259 nm (a_M 15.4 × 10³). Readily sol in boiling water. The compound is readily deaminated by nitrous acid to form inosinic acid; less rapidly hydrolyzed than 3′-adenylic acid by sulfuric acid. Furfural is formed only in traces on distillation with 20% HCl, *cf.* Levene, Bass, *Nucleic Acids* (New York, 1931) pp 230-232.

THERAP CAT: Nutrient.

147.   Adhatoda. Malabar nut; arusa; adulsa; vasaca; adhatodai; bakash. Leaves of *Adhatoda vasica* Nees, *Acanthacae. Habit.* East India. *Constit.* Adhatodic acid, vasicine. *Ref:* Krishnaswami, David, *Indian J. Pharm.* **2**, 141 (1940).

THERAP CAT: Expectorant.

148.   Adinazolam. [37115-32-5] 8-Chloro-*N*,*N*-dimethyl-6-phenyl-4*H*-[1,2,4]triazolo[4,3-*a*][1,4]benzodiazepine-1-methanamine; 8-chloro-1-[(dimethylamino)methyl]-6-phenyl-4*H*-*s*-triazolo[4,3-*a*][1,4]benzodiazepine; U-41123. $C_{19}H_{18}ClN_5$; mol wt 351.84. C 64.86%, H 5.16%, Cl 10.08%, N 19.91%. Triazolobenzodiazepine; dimethylamino derivative of alprazolam, *q.v.* Prepn: H. Allgeier, A. Gagneux, **DE 2201210**; **GB 1393256** (1972, 1975 both to Ciba-Geigy); J. B. Hester, Jr., US 4250094 (1984 to Upjohn); *idem, J. Heterocycl. Chem.* **17**, 575 (1980). Chromatographic determn in human plasma: G. W. Peng, *J. Pharm. Sci.* **73**, 1173 (1984). Pharmacological profile: R. A. Lahti *et al.*, *Neuropharmacology* **22**, 1277 (1984). Biological activity and metabolism: V. H. Sethy *et al.*, *J. Pharm. Pharmacol.* **36**, 546 (1984). Preliminary clinical study in severe depression: R. E. Pyke *et al.*, *Psychopharmacol. Bull.* **19**, 96 (1983).

Crystals from ethyl acetate, mp 171-172.5° (**US 4250094**). Also reported as mp 165-166° (**DE 2201210**).

Methanesulfonate. [57938-82-6] Adinazolam mesylate; Deracyn. $C_{19}H_{18}ClN_5\cdot CH_3SO_3H$; mol wt 447.94. Crystals from methanol-ether, mp 230-244°.

THERAP CAT: Antidepressant.

149. Adiphenine. [64-95-9] α-Phenylbenzeneacetic acid 2-(diethylamino)ethyl ester; 2-diethylaminoethyl diphenylacetate; diphenylacetyldiethylaminoethanol; diphacil; spasmolytin. $C_{20}H_{25}$ NO_2; mol wt 311.43. C 77.13%, H 8.09%, N 4.50%, O 10.27%. Anticholinergic. Prepn: **DE 626539** (1936 to Ciba); and pharmacology: T. Wagner-Jauregg *et al.*, *Ber.* **72**, 1551 (1939); R. R. Burtner, J. W. Cusic, *J. Am. Chem. Soc.* **65**, 262 (1943). Crystal and molecular structure: J. J. Guy, T. A. Hamor, *J. Chem. Soc. Perkin Trans. 2* **1973**, 942. Effect on TSH release: D. Jordan *et al.*, *Eur. J. Pharmacol.* **41**, 53 (1977). HPLC determn in tissues: J. Michelot *et al.*, *J. Chromatogr.* **257**, 395 (1983). Pharmacokinetics: *idem et al.*, *Xenobiotica* **11**, 123 (1981).

Hydrochloride. [50-42-0] Trasentine. $C_{20}H_{25}NO_2\cdot HCl$; mol wt 347.88. Crystals, mp 113-114°. Readily sol in water; very sparingly sol in alcohol, ether. LD_{50} i.p. in mice: 0.24 g/kg (Burtner, Cusic).

THERAP CAT: Antispasmodic.

150. Adipic Acid. [124-04-9] Hexanedioic acid; 1,4-butanedicarboxylic acid. $C_6H_{10}O_4$; mol wt 146.14. C 49.31%, H 6.90%, O 43.79%. Found in beet juice. Prepn from cyclohexanol: Bouveault, Locquin, *Bull. Soc. Chim.* [4] **3**, 438 (1908); Ellis, *Org. Synth.* **coll. vol. I**, 18 (2nd ed., 1941); Feagen, Copenhaver, *J. Am. Chem. Soc.* **62**, 869 (1940); **US 2191786** (1940); **US 2196357** (1940); Zilberman *et al.*, *J. Appl. Chem. USSR* **29**, 621 (1956). Convenient lab prepn from cyclohexanone: L. F. Fieser, *Organic Experiments* (Heath, Boston, 1964) pp 106-108. Prepn by one-step oxidation of cyclohexane: Onopchenko, Schulz, *J. Org. Chem.* **38**, 3729 (1973); Tanaka, 167th Am. Chem. Soc. Meet. (Los Angeles, March-April, 1974) *Abstracts of Papers*, p 29. Manuf: *Faith, Keyes & Clark's Industrial Chemicals*, F. A. Lowenheim, M. K. Moran, Eds. (Wiley-Interscience, New York, 4th ed., 1975) pp 50-54. *Review*: D. E. Danly, C. R. Campbell in *Kirk-Othmer Encyclopedia of Chemical Technology* **vol. 1** (Wiley-Interscience, New York, 3rd ed., 1978) pp 510-531.

Monoclinic prisms from ethyl acetate, from water or from acetone + petr ether. d_4^{25} 1.360. mp 152°. bp_{760} 337.5°; bp_{100} 265°; bp_{40} 240.5°; bp_{20} 222°; bp_{10} 205.5°; bp_5 191°; $bp_{1.0}$ 159.5°. K_1 (25°) = 3.90 × 10^{-5}; K_2 = 5.29 × 10^{-6}. Absorption spectrum: Ramart-Lucas, Salmon-Legagneur, *Compt. Rend.* **189**, 916 (1929). Freely sol in methanol, ethanol. Sol in boiling water, acetone. Slightly sol in water, cyclohexane. Practically insol in benzene, petr ether. 100 ml of a satd aq soln contains 1.44 g; 100 ml of boiling water dissolves 160 g; 100 parts of ether dissolve 0.633 parts (w/w) at 19°. pH of satd aq soln (25°) 2.7; pH of 0.1% soln 3.2.

Dimethyl ester. $C_8H_{14}O_4$. Liquid, solidifies at 0°, bp_{10} 112°.

Diethyl ester. Ethyl adipate. $C_{10}H_{18}O_4$. Liquid, bp 240-245°. d^{20} 1.009. Insol in water. Sol in alcohol and many other organic solvents.

USE: Manuf artificial resins, plastics (nylon), urethan foams. Buffering agent. Used as acidulant in baking powders instead of tartaric acid, cream of tartar, and phosphates because adipic acid is not hygroscopic. As an intermediate in lubricating oil additives.

151. Adiponectin. Acrp30; adipoQ; GBP28. Adipocyte hormone involved in glucose and lipid homeostasis. Consists of an N-terminal collagenous domain and a C-terminal globular domain with sequence homology to complement factor C1q. One of the most abundant proteins in human plasma, ranging from 0.5-30μg/ml; levels are reduced in diabetes and obesity. Discovery in murine adipocytes: P. E. Scherer *et al.*, *J. Biol. Chem.* **270**, 26746 (1995); in human adipose tissue: K. Maeda *et al.*, *Biochem. Biophys. Res. Commun.* **221**, 286 (1996). Isolation and characterization from human plasma: Y. Nakano *et al.*, *J. Biochem. (Tokyo)* **120**, 803 (1996). Review of structure and physiology: T.-S. Tsao *et al.*, *Eur. J. Pharmacol.* **440**, 213-221 (2002); of anti-inflammatory and atheroprotective effects in vascular tissue: B. J. Goldstein, R. Scalia, *J. Clin. Endocrinol. Metab.* **89**, 2563-2568 (2004); of role in insulin resistance: M. Haluzík *et al.*, *Physiol. Res.* **53**, 123-129 (2004).

152. Adiponitrile. [111-69-3] Hexanedinitrile; adipic acid dinitrile; 1,4-dicyanobutane; ADN. $C_6H_8N_2$; mol wt 108.14. C 66.64%, H 7.46%, N 25.91%. Chemical intermediate used to make hexamethylenediamine, *q.v.*, a raw material required for nylon production. Prepn: L. Henry, *Chem. Zentralbl.* **72**, II, 807 (1901); J. F. Thorpe, *J. Chem. Soc. Trans.* **95**, 1901 (1909). Manufacture from adipic acid: H. R. Arnold, W. A. Lazier, **US 2200734** (1940 to DuPont). Modified method: J. G. Mather, R. A. Williams, **US 3325531**; J. D. Rushton, R. A. Williams, **US 3325532** (both 1967 to Imperial Chem. Ind.). Manufacture from hydrocyanation of 1,3-butadiene: W. C. Drinkard, R. V. Lindsey, Jr., **US 3496215** (1970 to DuPont); from dimerization of acrylonitrile: D. E. Danly, *Hydrocarbon Process.* **60**, 161 (1981). Utility in prepn of hexamethylenediamine: O. R. Buehler *et al.*, **US 3461167** (1969 to DuPont); S. Alini *et al.*, *J. Mol. Catal. A* **206**, 363 (2003). Review of health and ecological effects: G. L. Kennedy, Jr., *Drug Chem. Toxicol.* **27**, 123-131 (2004).

Colorless oil. *Poisonous, irritant.* bp_{760} 295°; bp_{20} 180-182°; bp_{10} 147-148°; bp_1 124°. mp 0-1°. d^{19} 0.951; d_4^{20} 0.9623; d_4^{40} 0.9478; d_4^{60} 0.9335; d_4^{85} 0.9157. n_D^{20} 1.43854. Flash pt, closed cup: 325°F (163°C). Vapor pressure at 25°: 6.8 × 10^{-4} mm Hg. Sol in alcohol, chloroform; slightly sol in water. LD_{50} in fasted, non-fasted rats (mg/kg): 138, 301 orally; in rabbits (mg/kg): 2134 dermally. LC_{50} (4 hr) in rats (mg/l): 1.71 by inhalation. LC_{50} in minnows, sunfish, rainbow trout (mg/l): 1350, 1250, 750 (24 hr); 835-1300, 815, 750 (48 hr); 820-2140, 760 to >1000, 670 (96 hr) (Kennedy).

USE: Precursor in the manufacture of nylon.

153. Adipsin. [37213-56-2] Complement factor D. Serine protease homolog synthesized by mammalian adipocytes and secreted into the bloodstream; also produced by sciatic nerve. Developmentally regulated glycoprotein; mol wt ranges from 37-44 kilodaltons depending on the degree of glycosylation. Circulating levels of adipsin protein are reduced in certain genetic and experimental models of obesity. Originally identified as a 28 kDa protein encoded by a cDNA clone corresponding to messenger RNA specifically induced during adipocyte differentiation: B. M. Spiegelman *et al.*, *J. Biol. Chem.* **258**, 10083 (1983). Nucleotide analysis and predicted amino acid sequence: K. S. Cook *et al.*, *Proc. Natl. Acad. Sci. USA* **82**, 6480 (1985). Structure of adipsin gene: H. Y. Min, B. M. Spiegelman, *Nucleic Acids Res.* **14**, 8879 (1986). Isoln from adipose tissue, reportedly the primary site of synthesis, and from sciatic nerve: K. S. Cook *et al.*, *Science* **237**, 402 (1987). Studies in murine obesity models and potential biological role: J. S. Flier *et al.*, *ibid.* 405.

154. Adlumidine. [550-49-2] (6S)-6-[(5S)-5,6,7,8-Tetrahydro-6-methyl-1,3-dioxolo[4,5-g]isoquinolin-5-yl]furo[3,4-e]-1,3-benzodioxol-8(6H)-one. $C_{20}H_{17}NO_6$; mol wt 367.36. C 65.39%, H 4.66%, N 3.81%, O 26.13%. An alkaloid present in the *d*-form in the entire plant of *Adlumia fungosa* (Ait.) Greene (*A. cirrhosa* Raf.), *Corydalis thalictrifolia* Franch. and *C. incisa* (Thumb.) Pers., *Fumariaceae*. Schlotterbeck, Watkins, *J. Am. Chem. Soc.* **25**, 596 (1903); R. H. F. Manske, *Can. J. Res.* **21B**, 111 (1943). *l*-Adlumidine was isolated from *Corydalis sempervirens* (L.) Pers., *C. scouleri* Hook., and *C. crystallina* Engelm., *Fumariaceae*: R. H. F. Manske, *ibid.* **8**, 407 (1932); *ibid.* **14B**, 347 (1936); *ibid.* **17B**, 57 (1939). Structure: *idem*, *J. Am. Chem. Soc.* **72**, 3207 (1950). Stereochemistry: Blaha *et al.*, *Collect. Czech. Chem. Commun.* **29**, 2328 (1964); Snatzke *et al.*, *Tetrahedron* **25**, 5059 (1969). *Review*: Stanek,

Manske in *The Alkaloids* vol. **IV**, R. H. F. Manske, H. L. Holmes, Eds. (Academic Press, New York, 1954) pp 167-198. Stereoisomer of bicuculline, *q.v.*

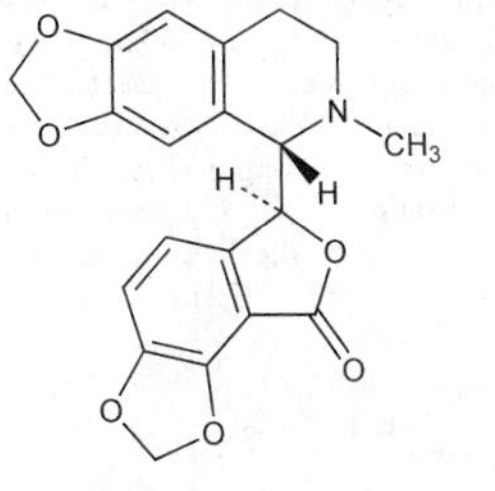

d-Adlumidine

Rhombic plates from chloroform + methanol, mp 236-237°. $[\alpha]_D^{25}$ +116.2° (c = 22 in chloroform). pKa 4.27. Practically insol in water. Very sparingly sol in alcohol, ether, hexane.

l-**Adlumidine.** [485-50-7] Capnoidine. Stout prisms from chloroform + methanol, mp 238°. $[\alpha]_D^{22}$ −113.2° (in chloroform). pKa 4.24. Practically insol in water. Sparingly sol in methanol; sol in chloroform.

155. Adlumine. [524-46-9] (6*S*)-6-[(1*S*)-1,2,3,4-Tetrahydro-6,7-dimethoxy-2-methyl-1-isoquinolinyl]-furo[3,4-*e*]-1,3-benzodioxol-8(6*H*)-one. $C_{21}H_{21}NO_6$; mol wt 383.40. C 65.79%, H 5.52%, N 3.65%, O 25.04%. An alkaloid present in the entire plant of some *Fumariaceae* species. *d*-Adlumine was isolated from *Adlumina fungosa* (Ait.) Greene (*A. cirrhosa* Raf.); *l*-adlumine was isolated from *Corydalis scouleri* Hook. and *C. sempervirens* (L.) Pers., *C. ophiocarpa* Hook. f. et Thoms *Fumariaceae:* Manske, *Can. J. Res.* **8**, 210 (1933); **14B**, 347 (1936); **16B**, 81 (1938); **17B**, 51 (1939). Preliminary stereochemical studies of *d*-form: Safe, Moir, *Can. J. Chem.* **42**, 160 (1964). Revised stereochemistry: Blaha *et al., Collect. Czech. Chem. Commun.* **29**, 2328 (1964); Snatzke *et al., Tetrahedron* **25**, 5059 (1969). *Review:* Stanek, Manske in *The Alkaloids* **Vol. IV**, R. H. F. Manske, H. L. Holmes, Eds. (Academic Press, New York, 1954) pp 167-198.

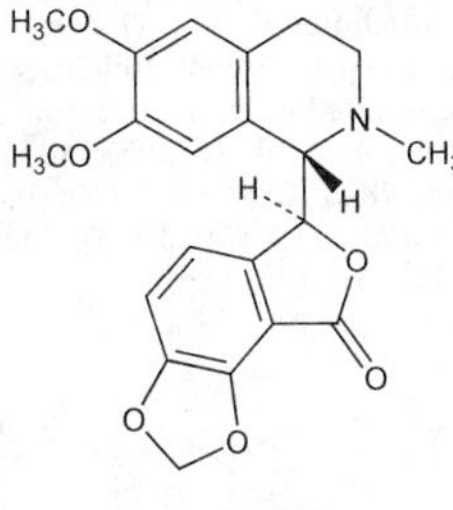

d-Adlumine

l-**Form.** Hexagonal plates from chloroform + methanol, mp 180°. $[\alpha]_D^{22}$ −42.5° (c = 0.8 in chloroform).

dl-**Form.** Stout short prisms, mp 190°.

156. Adonis vernalis. False hellebore; pheasant's eye; bird's eye. Herbaceous perennial, *Adonis vernalis* L., *Ranunculaceae*, used in traditional medicine as a cardiac stimulant. Medicinal formulations are derived from the aerial parts collected while flowering and dried. *Habit.* Central to eastern Europe and Siberia. *Constit.* Cardenolides, primarily adonitoxin, cymarin, k-strophanthoside, k-strophanthoside-β; flavonoids, including vitexin, luteolin. Isoln of constituents: F. W. Heyl *et al., J. Am. Chem. Soc.* **40**, 436 (1918); A. Pusz, S. Büchner, *Arzneim.-Forsch.* **13**, 409 (1963). Description and medicinal uses: J. Gruenwald *et al., PDR for Herbal Medicines* (Medical Economics, Montvale, 2nd Ed., 2000) pp 5-6.

THERAP CAT: Cardiotonic.

157. Adonitol. [488-81-3] Ribitol; adonite. $C_5H_{12}O_5$; mol wt 152.15. C 39.47%, H 7.95%, O 52.58%. A pentitol from *Adonis*

vernalis L., *Ranunculaceae; also from A. amurensis* L., *Ranunculaceae:* Santavy, Reichstein, *Pharm. Acta Helv.* **23**, 153 (1948).

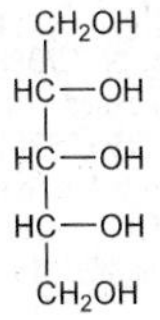

Large, optically inactive crystals. mp 102°. Does not reduce Fehling's soln. Freely soluble in water, hot alcohol. Insol in ether.

158. Adonitoxin. [17651-61-5] (3β,5β,16β)-3-[(6-Deoxy-α-L-mannopyranosyl)oxy]-14,16-dihydroxy-19-oxocard-20(22)-enolide. $C_{29}H_{42}O_{10}$; mol wt 550.65. C 63.26%, H 7.69%, O 29.05%. From *Adonis vernalis* L., *Ranunculaceae:* Katz, Reichstein, *Pharm. Acta Helv.* **22**, 437 (1947); *C.A.* **43**, 1790i (1949); Pitra, Cekan, *Collect. Czech. Chem. Commun.* **26**, 1551 (1961). Structure: Poláková, Cekan, *Chem. Ind. (London)* **1963**, 1766.

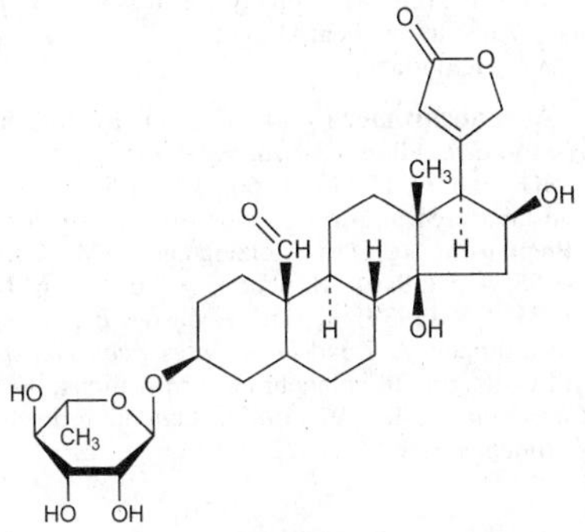

Crystals from chloroform + methanol + ether, decomp 251-258°.

3-*O*-Acetyladonitoxin. $C_{31}H_{44}O_{11}$. Needles from dil methanol, prisms from methanol+ ether, decomp 213-219°. $[\alpha]_D^{22}$ −19.9° (c = 0.33 in methanol).

159. Adrafinil. [63547-13-7] 2-[(Diphenylmethyl)sulfinyl]-*N*-hydroxyacetamide; 2-(benzhydrylsulfinyl)acetohydroxamic acid; CRL-40028; Olmifon. $C_{15}H_{15}NO_3S$; mol wt 289.35. C 62.27%, H 5.23%, N 4.84%, O 16.59%, S 11.08%. α-Adrenergic agonist with central stimulant and vigilence enhancing activity. Metabolized *in vivo* to modafinil, *q.v.* Prepn, pharmacology: L. Lafon, **BE 846880**; **US 4066686** (1977, 1978 both to Lab. Lafon). Synthesis and configuration of enantiomers: A. Osorio-Lozada *et al., Tetrahedron: Asymmetry* **15**, 3811 (2004). Mechanism of action study: J. Duteil *et al., Eur. J. Pharmacol.* **59**, 121 (1979). Psychopharmacology in mice: F. A. Rambert *et al., J. Pharmacol.* **17**, 37 (1986). Determn in urine for doping control analysis: J. Lu *et al., Rapid Commun. Mass Spectrom.* **23**, 1592 (2009). LC-MS determn in bulk drug and pharmaceutical formulations: R. N. Rao *et al., J. Sep. Sci.* **32**, 1312 (2009). Review of pharmacology and clinical experience: N. W. Milgram *et al., CNS Drug Rev.* **5**, 193-212 (1999).

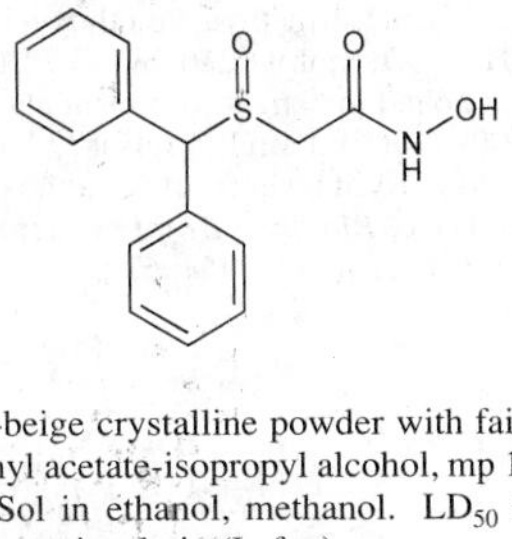

White to rosy-beige crystalline powder with faint odor of sulfur. Crystals from ethyl acetate-isopropyl alcohol, mp 159-160°. Soly in water: <1 g/l. Sol in ethanol, methanol. LD_{50} in mice (mg/kg): <2048 i.p.; 1950 gastric admin (Lafon).

THERAP CAT: CNS stimulant; nootropic.

160. Adrenalone. [99-45-6] 1-(3,4-Dihydroxyphenyl)-2-(methylamino)ethanone; 3′,4′-dihydroxy-2-(methylamino)aceto-phenone; 3,4-dihydroxy-α-methylaminoacetophenone; adrenone; 4-methylaminocetopyrocatechol; Stryphnon. $C_9H_{11}NO_3$; mol wt 181.19. C 59.66%, H 6.12%, N 7.73%, O 26.49%. Intermediate in some epinephrine manuf processes. Prepd from 4-chloroacetylpyro-catechol and methylamine: **DE 152814** (Hoechst); Stolz, *Ber.* **37**, 4152 (1904); by heating α-(*p*-toluene sulfonylmethylamino)-3,4-di-methoxyacetophenone with HCl under pressure: **DE 277540** (Bay-er); *Frdl.* **12**, 764; *Chem. Zentralbl.* **1914**, II, 740.

Needles, dec 235-236°. Sparingly sol in water, alc, ether.

Hydrochloride. [62-13-5] Stryphnasal. $C_9H_{11}NO_3 \cdot HCl$; mol wt 217.65. Crystals, mp 243°. Freely sol in water; sol in alcohol. Insol in ether. Aq solns are neutral to litmus.

THERAP CAT: Hemostatic.

161. Adrenochrome. [54-06-8] 2,3-Dihydro-3-hydroxy-1-methyl-1*H*-indole-5,6-dione; 3-hydroxy-1-methyl-5,6-indolinedi-one. $C_9H_9NO_3$; mol wt 179.18. C 60.33%, H 5.06%, N 7.82%, O 26.79%. Unstable, red pigment obtained by the oxidation of epi-nephrine. Prepn using Ag_2O as oxidizing agent: MacCarthy, *Chim. Ind. (Paris)* **55**, 435 (1946). Structure: D. E. Green, D. Richter, *Biochem. J.* **31**, 596 (1937); J. Harley-Mason, *Experientia* **4**, 307 (1948). Spectral data: A. Lund, *Acta Pharmacol. Toxicol.* **5**, 1218 (1949). HPLC determn in epinephrine formulations: E. C. Juenge *et al., J. Chromatogr.* **248**, 297 (1982). Oxidation mechanism: S. Adak *et al., Biochemistry* **37**, 16922 (1998).

Brilliant red crystals may exist as hemihydrate from methanol-formic acid, dec 115-120°. Absorption max (1 mg % aq sol): 220, 300, 485 nm (log ε 4.33, 4.01, 3.64). Well-formed and well-dried crystals can be kept in a vacuum desiccator for several weeks. Easily oxidized to melanin. Freely sol in water; fairly sol in alcohol. Al-most insol in benzene and ether. Solns are unstable. Optimum pH of water soln 4.0.

Oxime sesquihydrate. [6055-73-8] $C_9H_{10}N_2O_3 \cdot 1\frac{1}{2}H_2O$. Or-ange needles from water, mp 278°. Much more stable than adreno-chrome.

Monosemicarbazone see Carbazochrome.

Thiosemicarbazone. [113185-69-6] $C_{10}H_{12}N_4O_2S$. Descrip-tion: Fleischhacker, Barsel, **US 2712024** (1955 to International Hor-mones). mp 215-220°.

162. Adrenoglomerulotropin. [1210-56-6] 2,3,4,9-Tetra-hydro-6-methoxy-1-methyl-1*H*-pyrido[3,4-*b*]indole; aldosterone-stimulating hormone; ASH; 1-methyl-6-methoxy-1,2,3,4-tetrahydro-2-carboline. $C_{13}H_{16}N_2O$; mol wt 216.28. C 72.19%, H 7.46%, N 12.95%, O 7.40%. Found in extracts of pineal gland tissue: Farrell, *Circulation* **21**, 1009 (1960); Farrell, McIsaac, *Arch. Biochem. Bio-phys.* **94**, 543 (1961). Synthesis from 5-methoxytryptamine and acetaldehyde: McIsaac, *Biochim. Biophys. Acta* **52**, 607 (1961); Meek *et al., Chem. Ind. (London)* **1964**, 622.

Crystals, mp 150-151°. uv max (ethanol): 225, 280 nm (log ε 4.34, 3.86); (0.1*N* HCl): 220, 273 nm (log ε 4.40, 3.86).

163. Adrenolutin. [642-75-1] 1-Methyl-1*H*-indole-3,5,6-triol; *N*-methyl-5,6-dihydroxyindoxyl; 3,5,6-trihydroxy-*N*-methylin-dole. $C_9H_9NO_3$; mol wt 179.18. C 60.33%, H 5.06%, N 7.82%, O 26.79%. Alkaline rearrangement product of adrenochrome. Isoln and identification: Lund, *Acta Pharmacol. Toxicol.* **5**, 75 (1949). Structure and synthesis: Balsinger *et al., Helv. Chim. Acta* **36**, 708 (1953); Heacock, Mahon, *Can. J. Chem.* **36**, 1550 (1958). NMR studies and its existence in the ketonic form in solution: Powell, Heacock, *Chim. Ther.* **7**, 133 (1972).

Monohydrate, bright yellow prisms from water, mp 236° (dec); mp 195° (Balsinger). Anhydrous, bright yellow prisms, mp 245°.

164. Adrenomedullin. Hypotensive peptide originally iso-lated from human pheochromocytoma; subsequently found in all tis-sues with highest levels in heart, kidney, lung and adrenal gland. Highly conserved among species, human form is a 52-amino acid peptide; shares some sequence homology with calcitonin gene re-lated peptide and amylin, *q.v.* Plasma levels are elevated in patients with hypertension and heart failure. Biological effects include va-sodilation, diuresis and increased cardiac output; also exhibits anti-proliferative and anti-apoptotic effects in the myocardium and stim-ulates angiogenesis. Isoln and hypotensive activity: K. Kitamura *et al., Biochem. Biophys. Res. Commun.* **192**, 553 (1993). RIA determn in plasma: L. K. Lewis *et al., Clin. Chem.* **44**, 571 (1998). Review of biosynthesis, pharmacology and bioactivities: J. P. Hinson *et al., Endocr. Rev.* **21**, 138-167 (2000). Series of articles on physiological roles: *Regul. Pept.* **112** 1-196 (2003). Review of clinical relevance and therapeutic potential: D. C. Bunton *et al., Pharmacol. Ther.* **103**, 179-201 (2004); S. A. Hamid, G. F. Baxter, *ibid.* **105**, 95-112 (2005).

165. Adrenosterone. [382-45-6] Androst-4-ene-3,11,17-tri-one; Reichstein's substance G. $C_{19}H_{24}O_3$; mol wt 300.40. C 75.97%, H 8.05%, O 15.98%. Prepn: Reichstein, *Helv. Chim. Acta* **19**, 29, 1107 (1936); Kendall *et al., J. Biol. Chem.* **116**, 267 (1936); Reichstein, *Helv. Chim. Acta* **20**, 817, 953 (1937). Total synthesis: Velluz *et al., Compt. Rend.* **250**, 1293 (1960). Crystal structure: Ohrt *et al., Acta Crystallogr.* **19**, 479 (1965). Metabolism: Bradlow *et al., Steroids* **10**, 233 (1967).

Needles from alcohol, mp 220-224°. Sublimes in high vacuum. Soly in water: 9.85 (23°), 15.2 (37°) mg/100 ml. Sol in alcohol, acetone, ether. $[\alpha]_D^{20}$ +262° (abs alcohol); $[\alpha]_{5461}^{25}$ +364 ± 5° (c = 0.18 in abs alc). uv max: 235 nm.

166. Aequorin. Photoprotein from the luminescent jellyfish *Aequorea*, mol wt 20,000 Da. Contains a single chromophore, *coe-lentrazine*, which emits a blue light via an intramolecular reaction when traces of Ca^{2+} are present. The protein is converted to *apoae-quorin*, *coelenteramide*, and CO_2 with the emission of light. Isola-tion and purification: O. Shimomura *et al., J. Cell. Comp. Physiol.* **59**, 223 (1962). Mechanism of luminescence: O. Shimomura *et al., Biochemistry* **13**, 3278 (1974). Verification of coelentrazine as chro-mophore: S. Inoue *et al., Chem. Lett.* **1975**, 141. Characterization and properties of isoforms: O. Shimomura, *Biochem. J.* **234**, 271 (1986); of semisynthetic aequorin: O. Shimomura *et al., ibid.* **251**,

405 (1988); and recombinant forms: *eidem, ibid.* **270**, 309 (1990). Crystal structure: J. F. Head *et al., Nature* **405**, 372 (2000). Use in measurement of cellular calcium: A. Azzi, B. Chance, *Biochim. Biophys. Acta* **189**, 141 (1969); J. M. Kendall *et al., Anal. Biochem.* **221**, 173 (1994); M. N. Badminton *et al., Exp. Cell Res.* **216**, 236 (1995). Review: O. Shimomura in *Natural Products and Biological Activities,* H. Imura *et al.,* Eds (Elsevier Science Publishers, Amsterdam, 1986) p 33-44.

Sol in aq buffers >30 mg/ml. $E_{1cm}^{1\%}$ at 280 nm: 27.0; $E_{1cm}^{1\%}$ at 460 nm: 0.81. pI 4.2-4.9.

USE: For quantitation of free calcium in biological systems.

167. AET. [56-10-0] Carbamimidothioic acid 2-aminoethyl ester hydrobromide (1:2); 2-(2-aminoethyl)-2-thiopseudourea dihydrobromide; *S*-(2-aminoethyl)isothiuronium bromide hydrobromide; β-aminoethylisothiuronium bromide hydrobromide; Antiradon. $C_3H_{11}Br_2N_3S$; mol wt 281.01. C 12.82%, H 3.95%, Br 56.87%, N 14.95%, S 11.41%. Prepd by refluxing thiourea with 2-bromoethylamine HBr in isopropanol: Clinton *et al., J. Am. Chem. Soc.* **70**, 950 (1948); Funahashi, Miyano, *J. Agric. Chem. Soc. Jpn.* **27**, 775 (1953), *C.A.* **49**, 15737b (1955); Doherty *et al., J. Am. Chem. Soc.* **79**, 5667 (1957). Toxicity study: J. Pospisil *et al., Cas. Lek. Cesk.* **105**, 1165 (1966), *C.A.* **67**, 72261s (1967).

Crystals from abs ethanol + ethyl acetate, mp 194-195°. Hygroscopic, cakes together, and is converted in significant amounts of 2-aminothiazoline, a transformation that can be detected by a drop in melting point by as much as 30°. The animal organism appears to convert it to 2-mercaptoethylguanidine hydrobromide. LD_{50} in mice (mg/kg): 100 i.v., 280 s.c., 480 i.p., 1600 orally (Pospisil).

THERAP CAT: Radioprotective agent.

168. Afamelanotide. [75921-69-6] 4-L-Norleucine-7-D-phenylalanine-α-melanotropin (swine); [Nle⁴,D-Phe⁷]-α-MSH; CUV-1647; Melanotan; Scenesse. $C_{78}H_{111}N_{21}O_{19}$; mol wt 1646.87. C 56.89%, H 6.79%, N 17.86%, O 18.46%. Synthetic analog of human α-melanocyte-stimulating hormone, *see* MSH. Stimulates melanin production which protects skin from UV radiation. Prepn and bioactivity: T. K. Sawyer *et al., Proc. Natl. Acad. Sci. USA* **77**, 5754 (1980); *idem et al., J. Med. Chem.* **25**, 1022 (1982); V. J. Hruby *et al.,* **US 4457864** (1984 to University Patents). Review of pharmacology and development: M. E. Hadley *et al., Pharm. Biotechnol.* **11**, 575-595 (1998). Clinical induction of skin tanning: N. Levine *et al., J. Am. Med. Assoc.* **266**, 2730 (1991); R. T. Dorr *et al., Arch. Dermatol.* **140**, 827 (2004). Clinical evaluation in erythropoietic protoporphyria: J. Harms *et al., N. Engl. J. Med.* **360**, 306 (2009); in solar urticaria: A. K. Haylett *et al., Br. J. Dermatol.* **164**, 407 (2011). Review of clinical experience: E. I. Minder, *Expert Opin. Invest. Drugs* **19**, 1591-1602 (2010).

Ac–Ser–Tyr–Ser–Nle–Glu–His–D-Phe–Arg–Trp–Gly–Lys–Pro–Val–NH₂

THERAP CAT: Photoprotective agent; in treatment of solar urticaria.

169. Afatinib. [850140-72-6]; [439081-18-2] (unspecified double bond geometry). (2*E*)-*N*-[4-[(3-Chloro-4-fluorophenyl)amino]-7-[[(3*S*)-tetrahydro-3-furanyl]oxy]-6-quinazolinyl]-4-(dimethylamino)-2-butenamide; (*E*)-4-[(3-chloro-4-fluorophenyl)amino]-6-[[4-(*N*,*N*-dimethylamino)-1-oxo-2-buten-1-yl]amino]-7-((*S*)-tetrahydrofuran-3-yloxy)-quinazoline; BIBW-2992; Tomtovok. $C_{24}H_{25}ClFN_5O_3$; mol wt 485.94. C 59.32%, H 5.19%, Cl 7.30%, F 3.91%, N 14.41%, O 9.88%. EGFR and HER2 tyrosine kinase inhibitor. Prepn: F. Himmelsbach *et al.,* **WO 0250043**; *eidem,* **US 7019012** (2002, 2006 both to Boehringer Ingelheim); of (*E*)-isomer: R. Soyka *et al.,* **US 05085495** (2005 to Boehringer Ingelheim). Antitumor activity in lung cancer models: D. Li *et al., Oncogene* **27**, 4702 (2008); in colorectal tumor cell lines: S. A. Khelwatty *et al., Int. J. Oncol.* **39**, 483 (2011). Clinical pharmacokinetics and evaluation in solid tumors: T. A. Yap *et al., J. Clin. Oncol.* **28**, 3965 (2010).

Review of development: N. Minkovsky, A. Berezov, *Curr. Opin. Investig. Drugs* **9**, 1336-1346 (2008); and therapeutic potential in non-small-cell lung cancer: G. Metro, L. Crino, *Expert Rev. Anticancer Ther.* **11**, 673-682 (2011).

Crystals from butyl acetate + methylcyclohexane.
Dimaleate. [936631-70-8] $C_{24}H_{25}ClFN_5O_3 \cdot 2C_4H_4O_4$; mol wt 718.09. Crystals from ethanol, mp 178°.
THERAP CAT: Antineoplastic.

170. Affinin. [25394-57-4] (2*E*,6*Z*,8*E*)-*N*-(2-Methylpropyl)-2,6,8-decatrienamide; (*E*,*E*,*Z*)-*N*-isobutyl-2,6,8-decatrienamide; *N*-isobutyldeca-*trans*-2-*cis*-6-*trans*-8-trienamide; spilanthol. $C_{14}H_{23}NO$; mol wt 221.34. C 75.97%, H 10.47%, N 6.33%, O 7.23%. Insecticidal lipid amide isolated from *Heliopsis longipes* (A. Gray) Blake, *Compositae.* Isoln and structure: Acree *et al., J. Org. Chem.* **10**, 236, 449 (1945). Identity with spilanthol: Jacobson, *Chem. Ind. (London)* **1957**, 50. Stereochemistry and synthesis: Crombie, Krasinski, *ibid.* **1962**, 983; Crombie *et al., J. Chem. Soc.* **1963**, 4970.

Pale yellow viscous oil, bp₀.₂ 141°; bp₀.₃₋₀.₅ 160-165°. mp 23°. n_D^{25} 1.5134. Soluble in organic solvents. Practically insoluble in aq alkali and acid. uv max (95% ethanol): 228.5 nm (ε 33700).
USE: Insecticide synergist.

171. Aflatoxins B. Aflatoxins are a closely related group of secondary fungal metabolites shown to be mycotoxins. They are produced by *Aspergillus flavus* Link ex Fries, the causative principle of turkey "X" disease; and by *Aspergillus parasiticus*: Sargeant *et al., Nature* **192**, 1096 (1961); Hesseltine *et al., Proc. 1st U.S. Japan Conf. Toxic Microorganisms, Honolulu, Hawaii 1968*, p 202. Aflatoxins have been reported to naturally occur in peanuts, peanut meal, cottonseed meal, corn, dried chili peppers etc. However, the growth of the mold itself does not always indicate the presence of toxin since the yield of aflatoxin depends on growth conditions such as moisture, temperature, substrates, and aeration as well as genetic requirements. These heterocycles are now characterized as aflatoxins B_1, B_2, G_1, G_2, M_1 and M_2 (milk toxins) and B_{2a}, G_{2a}: Büchi, Rae in *Aflatoxin,* L. Goldblatt, Ed. (Academic Press, New York, 1969) pp 55-75. Toxic material is separated chromatographically into 4 distinct compounds based on fluorescent color (blue = B, green = G with subscripts relating to relative mobility): Nesbitt *et al., Nature* **195**, 1062 (1962); Sargeant *et al., Chem. Ind. (London)* **1963**, 53. B_1 is one of the most potent environmental mutagens and carcinogens known. B_2 and G_2 are the less toxic dihydro derivs of B_1 and G_1: Asao *et al., J. Am. Chem. Soc.* **85**, 1706 (1963); *ibid.* **87**, 882 (1965); and B_{2a} and G_{2a} are the 2-hydroxy derivs of B_2, G_2: Dutton, Heathcote, *Biochem. J.* **101**, 21P (1967); *Chem. Ind. (London)* **1968**, 418. Isoln of B_3 (*parasiticol*), a possible metabolite of G_1, from *A. flavus*: J. G. Heathcote, M. F. Dutton, *Tetrahedron* **25**, 1497 (1969). Aflatoxins R_0 (*aflatoxicol*), P_1, Q_1, RB_1, RB_2, and D_1 are also known: P. F. Schuda, *Top. Curr. Chem.* **91**, 77-106 (1980). Total syntheses of aflatoxins B_1 and G_1: Buchi *et al., J. Am. Chem. Soc.* **88**, 4534 (1966); Knight *et al., Chem. Commun.* **1966**, 706; Büchi, Weinreb, *J. Am. Chem. Soc.* **93**, 746 (1971). The extreme toxicity and carcinogenicity of aflatoxins may be due to their inhibition of nucleic acid synthesis by either direct interaction with enzymes involved or by a toxin-DNA template: Clifford, Rees, *Nature* **209**, 312 (1966); Sporn *et al., Science* **151**, 1539 (1966). *See also:* Wogan *et al., Food Cosmet. Toxicol.* **12**, 681 (1974). Inhibition of salt-induced conversion of B-DNA to Z-DNA by aflatoxin B_1: A. Nordheim *et al.*, *Science* **219**, 1434 (1983). Prepn of the *exo*-8,9-epoxide of B_1, the metabolic deriv thought to be responsible for B_1's carcinogenicity: S. W. Baertschi *et al.*, *J. Am. Chem. Soc.* **110**, 7929 (1988). uv spectrum of B_2: Hartley *et al.*, *Nature* **198**, 1056 (1963). Physicochemical data for B_1: A. E. Pohland *et al.*, *Pure Appl. Chem.* **54**, 2219 (1982). Toxicity data: Carnaghan *et al.*, *ibid.* **200**, 1101 (1963); G. Büchi *et al.*, *Life Sci.* **13**, 1143 (1973). Symposium on toxicology and synthesis: *J. Toxicol. Toxin Rev.* **8**, 1-416 (1991). Review and evaluation of studies of carcinogenic action in laboratory animals and humans: *IARC Monographs* **10**, 51-72 (1976). Comprehensive reviews: Goldblatt, *Econ. Bot.* **22**, 51-62 (1968); Detroy *et al.*, "Aflatoxin and Related Compounds" in *Microbial Toxins* **Vol. VI**, A. Ciegler *et al.*, Eds. (Academic Press, New York, 1971) pp 3-178; W. F. Busby, Jr., G. N. Wogan in *ACS Monograph Series* **no. 182**, entitled "Chemical Carcinogens," vol. 2, C. E. Searle, Ed. (American Chemical Society, Washington DC, 2nd ed., 1984) pp 945-1136. Review of chemistry and syntheses: P. F. Schuda, *loc. cit.*; of biosynthesis: M. W. Sinz, W. T. Shier, *J. Toxicol. Toxin Rev.* **10**, 87-121 (1991).

Aflatoxin B_1

Aflatoxin B_1. [1162-65-8] (6a*R*-*cis*)-2,3,6a,9a-Tetrahydro-4-methoxycyclopenta[*c*]furo[3′,2′:4,5]furo[2,3-*h*][1]benzopyran-1,11-dione. $C_{17}H_{12}O_6$; mol wt 312.28. Crystals, mp 268-269°. Exhibits blue fluorescence. $[\alpha]_D -558°$ (c = 0.1 in $CHCl_3$). $[\alpha]_D -480°$ (c = 0.1 in DMF). uv max (ethanol): 223, 265, 362 nm (ε 25600, 13400, 21800). LD_{50} orally in day old duckling: 18.2 μg/50 gm body wt (Carnaghan); i.p. in newborn mice: 9.50 mg/kg body wt (Büchi).

Aflatoxin B_2. [7220-81-7] (6a*R*-*cis*)-2,3,6a,8,9,9a-Hexahydro-4-methoxycyclopenta[*c*]furo[3′,2′:4,5]furo[2,3-*h*][1]benzopyran-1,11-dione. $C_{17}H_{14}O_6$; mol wt 314.29. The 8,9-dihydro deriv of aflatoxin B_1. Crystals, mp 286-289°. Exhibits blue fluorescence. $[\alpha]_D -492°$ (c = 0.1 in $CHCl_3$). uv max (ethanol): 265, 363 nm (ε 11700, 23400). LD_{50} orally in day old duckling: 84.8 μg/50 gm body wt (Carnaghan).

Caution: The aflatoxins are listed as known human carcinogens: *Report on Carcinogens, Twelfth Edition* (PB2011-111646, 2011) p 32.

172. Aflatoxins G. Toxic metabolites of fungi *Aspergillus flavus* Link ex Fries and *Aspergillus parasiticus*. Total synthesis of G_1: Büchi, Weinreb, *J. Am. Chem. Soc.* **93**, 746 (1971). Physical and chemical data: Hartley *et al.*, *Nature* **198**, 1056 (1963); Asao, *J. Am. Chem. Soc.* **87**, 882 (1965). Toxicity data: R. B. A. Carnaghan *et al.*, *Nature* **200**, 1101 (1963). For general refs *see* Aflatoxins B.

Aflatoxin G_1

Aflatoxin G_1. [1165-39-5] (7a*R*,10a*S*)-3,4,7a,10a-Tetrahydro-5-methoxy-1*H*,12*H*-furo[3′,2′:4,5]furo[2,3-*h*]pyrano[3,4-*c*][1]benzopyran-1,12-dione. $C_{17}H_{12}O_7$; mol wt 328.28. Crystals, mp 244-246°. Exhibits green fluorescence. $[\alpha]_D -556°$ (chloroform). uv max (ethanol): 243, 257, 264, 362 nm (ε 11500, 9900, 10000, 16100). LD_{50} orally in day old duckling: 39.2 μg/50 gm body wt (Carnaghan).

Aflatoxin G_2. [7241-98-7] (7a*R*,10a*S*)-3,4,7a,9,10,10a-Hexahydro-5-methoxy-1*H*,12*H*-furo[3′,2′:4,5]furo[2,3-*h*]pyrano[3,4-*c*]-[1]benzopyran-1,12-dione. $C_{17}H_{14}O_7$; mol wt 330.29. The 9,10-dihydro deriv of aflatoxin G_1. Crystals, mp 237-240°. Exhibits green-blue fluorescence. $[\alpha]_D -473°$ (c = 0.084 in chloroform). uv max (ethanol): 265, 363 nm (ε 9700, 21000). LD_{50} orally in day old duckling: 172.5 μg/50 gm body wt (Carnaghan).

Caution: The aflatoxins are listed as known human carcinogens: *Report on Carcinogens, Twelfth Edition* (PB2011-111646, 2011) p 32.

173. Aflatoxins M. Highly toxic 4-hydroxylated aflatoxin B derivs found in the milk of cows fed toxic meal. Initial reports: Allcroft, Carnaghan, *Vet. Rec.* **75**, 259 (1963). Isoln: De Iongh *et al.*, *Nature* **202**, 466 (1964); Allcroft *et al.*, *ibid.* **209**, 154 (1966). Isoln and structure of M_1 and M_2: C. W. Holzapfel, P. S. Steyn, *Tetrahedron Lett.* **1966**, 2799. Total synthesis of M_1: Büchi, Weinreb, *J. Am. Chem. Soc.* **91**, 5408 (1969); **93**, 746 (1971); G. Büchi *et al.*, *ibid.* **103**, 3497 (1981). Carcinogenicity studies: Wogan, Paglialunga, *Food Cosmet. Toxicol.* **12**, 381 (1974). For general refs *see* Aflatoxins B.

Aflatoxin M_1

Aflatoxin M_1. [6795-23-9] (6a*R*,9a*R*)-2,3,6a,9a-Tetrahydro-9a-hydroxy-4-methoxycyclopenta[*c*]furo[3′,2′:4,5]furo[2,3-*h*][1]benzopyran-1,11-dione; 4-hydroxyaflatoxin B_1. $C_{17}H_{12}O_7$; mol wt 328.28. Crystals from methanol, mp 299° (dec). Exhibits blue-violet fluorescence. $[\alpha]_D -280°$ (c = 0.1 in DMF). uv max (ethanol): 226, 265, 357 nm (ε 23100, 11600, 19000). LD_{50} orally in day old Pekin ducklings: 16.6 μg/duckling (Holzapfel, Steyn).

Aflatoxin M_2. [6885-57-0] 2,3,6a,8,9,9a-Hexahydro-9a-hydroxy-4-methoxycyclopenta[*c*]furo[3′,2′:4,5]furo[2,3-*h*][1]benzopyran-1,11-dione; 4-hydroxyaflatoxin B_2. $C_{17}H_{14}O_7$; mol wt 330.29. The 8,9-dihydro deriv of aflatoxin M_1. Crystals from methanol-chloroform, mp 293° (dec). uv max (ethanol): 221, 264, 357 nm (ε 20000, 10900, 21000). LD_{50} orally in day old Pekin ducklings: 62 μg/duckling (Holzapfel, Steyn).

Caution: The aflatoxins are listed as known human carcinogens: *Report on Carcinogens, Twelfth Edition* (PB2011-111646, 2011) p 32.

174. Aflibercept. [862111-32-8] Vascular endothelial growth factor receptor type VEGFR-1 (synthetic human immunoglobulin domain 2 fragment) fusion protein with vascular endothelial growth factor receptor type VEGFR-2 (synthetic human immunoglobulin domain 3 fragment) fusion protein with immunoglobulin G1 (synthetic Fc fragment), dimer; vascular endothelial growth factor trap; VEGF Trap; VEGF $Trap_{R1R2}$; VEGF Trap-Eye; Eylea. Antiangiogenic agent. Recombinant human decoy fusion protein produced in Chinese hamster ovary (CHO) cells; comprised of domain 2 of VEGF receptor-1 (Flt-1) and domain 3 of VEGF receptor-2 (Flk-1) fused to the Fc region of human IgG_1. Mol wt ~115 kDa including ~15% glycosylation. Prepn: N. J. Papadopoulos *et al.*, **WO 0075319**; T. J. Daly *et al.*, **US 7396664** (2000, 2008 both to Regeneron); J. Holash *et al.*, *Proc. Natl. Acad. Sci. USA* **99**, 11393 (2002). Effect on choroidal neovascularization: Y. Saishin *et al.*, *J. Cell. Physiol.* **195**, 241 (2003). Clinical evaluation in age-related macular degeneration: Q. D. Nguyen *et al.*, *Ophthalmology* **113**, 1522 (2006); in diabetic macular edema: D. V. Do *et al.*, *Br. J. Ophthalmol.* **93**, 144 (2009). Review of evaluation as tumor angiogenesis inhibitor: Q. S.-C. Chu, *Expert Opin. Biol. Ther.* **9**, 263-271 (2009). Clinical trial in metastatic melanoma: A. A. Tarhini *et al.*, *Clin. Cancer Res.* **17**, 6574 (2011).

THERAP CAT: In treatment of age-related macular degeneration.

175. Afloqualone. [56287-74-2] 6-Amino-2-(fluoromethyl)-3-(2-methylphenyl)-4(3*H*)-quinazolinone; 6-amino-2-fluoromethyl-3-(*o*-tolyl)-4(3*H*)-quinazolinone; HQ-495; Arofuto. $C_{16}H_{14}FN_3O$; mol wt 283.31. C 67.83%, H 4.98%, F 6.71%, N 14.83%, O 5.65%. Centrally acting muscle relaxant. Prepn: I. Inoue *et al.*, **DE 2449113**; *eidem*, **US 3966731** (1975, 1976 to Tanabe); J. Tani *et al.*, *J. Med. Chem.* **22**, 95 (1979). Pharmacology: T. Ochiai, R. Ishida, *Jpn. J. Pharmacol.* **31**, 491 (1981); **32**, 427 (1982). Metabolism: N. Otsuka *et al.*, *J. Pharmacobio-Dyn.* **6**, 708 (1983); S. Furuuchi *et al.*, *Drug Metab. Dispos.* **11**, 371 (1983). LC-MS/MS determn in plasma and clinical pharmacokinetics: H.-Y. Yun *et al.*, *Talanta* **73**, 635 (2007). Evaluation of phototoxic potential: H. Fujita, I. Matsuo, *Chem. Biol. Interact.* **64**, 139 (1987). Use in murine model of drug photoallergy: D. Nishio *et al.*, *J. Dermatol. Sci.* **55**, 34 (2009).

Pale yellow prisms from 2-propanol, mp 195-196°. uv max (ethanol): 293 nm (ε 14000). Sol in acetonitrile; sparingly sol in ethanol. Practically insol in water. LD_{50} in mice (mg/kg): 315.1 i.p. (Tani).

THERAP CAT: Muscle relaxant (skeletal).

176. Agar. [9002-18-0] Agar-agar; gelose; Japan agar; Bengal isinglass; Ceylon isinglass; Chinese isinglass; Japan isinglass; Layor Carang. A polysaccharide complex extracted from the agarocytes of algae of the *Rhodophyceae*. Predominant agar-producing genera are *Gelidium, Gracilaria, Acanthopeltis, Ceramium, Pterocladia* found in the Pacific and Indian Oceans and Japan Sea. Can be separated into a neutral gelling fraction, *agarose*, and a sulfated non-gelling fraction, *agaropectin*: Araki, *J. Chem. Soc. Jpn.* **58**, 1338 (1937). Structure believed to be a complex range of polysaccaride chains having alternating α-(1 → 3) and β-(1 → 4) linkages and varying in total charge content; three extremes of structure noted, namely neutral agarose, pyruvated agarose having little sulfation, and a sulfated galactan: Duckworth *et al.*, *Carbohydr. Res.* **16**, 189, 435, 446 (1971). *Reviews:* V. J. Chapman, *Seaweeds and Their Uses* (Pitman Publ., New York, 1952) pp 89-123; Humm, *Econ. Bot.* **1**, 17 (1947); Mori, *Adv. Carbohydr. Chem.* **8**, 317 (1953); Selby, Wynne, in *Industrial Gums*, R. L. Whistler, Ed. (Academic Press, N.Y., 2nd ed., 1973) pp 29-48.

Transparent, odorless, tasteless strips or coarse or fine powder. Sol in boiling water. Slowly sol in hot water to a viscid soln. A 1% soln forms a stiff jelly on cooling. Insol in cold water, alc.

USE: Substitute for gelatin, isinglass, etc. in making emulsions including photographic, gels in cosmetics, and as thickening agent in foods esp. confectionaries and dairy products; in meat canning; in production of medicinal encapsulations and ointments; as dental impression mold base; as corrosion inhibitor; sizing for silks and paper; in the dyeing and printing of fabrics and textiles; in adhesives. In nutrient media for bacterial cultures.

THERAP CAT: Cathartic.

THERAP CAT (VET): Laxative in dogs, cats. Demulcent.

177. Agaric. Larch agaric; touch wood; white agaric; purging agaric; amadou; German fungus. The dried fruit body of *Fomes laricis* (Jacq.) Murrill *(Polyporus officinalis* Fries), *Polyporaceae*, deprived of its outer rind. *Habit.* European and Asiatic Russia. Grows upon various species of *Pinus, Larix,* and *Picea. Constit.* 14-16% Agaricic acid; agaricoresin, agaricol, phytosterin, ricinoleic acid, cetyl alcohol, glucose, malic acid, carbohydrates. Isoln and separation of some constituents: Valentin, Knütter, *Pharm. Zentralhalle* **96**, 478 (1957). Impregnated with potassium nitrate and dried, it constitutes punk or tinder.

Light, fibrous, grayish-white to pale brown, spongy, friable pieces of irregular shape; feeble odor and bitter, acrid, yet somewhat sweetish taste.

USE: Anhidrotic.

178. Agaricic Acid. [666-99-9] 2-Hydroxy-1,2,3-nonadecanetricarboxylic acid; agaric acid; agaricin; agaricinic acid; laricic acid; *n*-hexadecylcitric acid; α-cetylcitric acid. $C_{22}H_{40}O_7$; mol wt 416.56. C 63.43%, H 9.68%, O 26.89%. Active principle of *Fomes laricis* (Jacq.) Murrill *(Polyporus officinalis* Fries), *Polyporaceae:* Thoms, Vogelsang, *Ann.* **357**, 145 (1907). Attempted syntheses: Evans, *J. Chem. Soc.* **1959**, 1313; Graf, Liu, *Arch. Pharm.* **306**, 366 (1973). Activity studies: Bacchi *et al.*, *J. Bacteriol.* **98**, 23 (1969).

Sesquihydrate, odorless, almost tasteless, crystalline powder. Anhydrous, mp 142° (dec). $[\alpha]_D^{19}$ −9° (NaOH). Slightly sol in cold water, chloroform or ether; freely sol in boiling water, alkalies, hot glacial acetic acid. One gram dissolves in 180 ml cold, 10 ml boiling alcohol.

Note: The name agaric acid has also been used for a lanostane-like mixture [*C.A.* **68**, 29894j (1968)].

USE: Has been used as antiperspirant.

179. Agaritine. [2757-90-6] L-Glutamic acid 5-[2-[4-(hydroxymethyl)phenyl]hydrazide]; β-*N*-[γ-L(+)-glutamyl]-4-hydroxymethylphenylhydrazine. $C_{12}H_{17}N_3O_4$; mol wt 267.29. C 53.92%, H 6.41%, N 15.72%, O 23.94%. Constituent of the commercial, edible mushroom *Agaricus bisporus* (Lange) Sing. *[Psalliota hortensis* Cooke var. *bispora], Agaricaceae (Agaricales).* Isoln and structure: Levenberg, *Fed. Proc.* **19**, 6 (1960); *J. Am. Chem. Soc.* **83**, 503 (1961); Daniels *et al.*, *ibid.* 3333; *see also* Levenberg in *Methods Enzymol.* **17** (Part A), 877 (1970). Synthesis: Kelly *et al.*, *J. Org. Chem.* **27**, 3229 (1962); Hinman, Kelly, **US 3274232**; **US 3288848** (both 1966 to Upjohn).

Glistening crystals from dil alcohol, dec 205-209°. $[\alpha]_D^{25}$ +7° (c = 0.8). uv max (water): 237.5, 280 nm (ε 12,000, 1,400). Very freely sol in water. Practically insol in the usual anhydr organic solvents. pKa in water: 3.4 and 8.86.

180. Agatolimod. [207623-20-9] d(*P*-Thio)(T-C-G-T-C-G-T-T-T-T-G-T-C-G-T-T-T-T-G-T-C-G-T-T)DNA. $C_{236}H_{326}N_{70}O_{133}P_{23}S_{23}$; mol wt 7721.34. C 36.71%, H 4.26%, N 12.70%, O 27.56%, P 9.23%, S 9.55%. Synthetic, 24 base phosphothioate oligodeoxynucleotide containing multiple cytosine-phosphate-guanosine (CpG) motifs. Toll-like receptor 9 (TLR9) agonist that stimulates B-cells and plasmacytoid dendritic cells. Prepn: A. M. Krieg, **WO 9818810**; *idem*, **US 7223741** (1998, 2007 both to Univ. Iowa Res. Foundn.). Immunostimulant effect on human dendritic cells: G. Hartmann *et al.*, *Proc. Natl. Acad. Sci. USA* **96**, 9305 (1999). CGE determn in plasma and metabolism study: B. O. Noll *et al.*, *Biochem. Pharmacol.* **69**, 981 (2005). Clinical evaluation as vaccine adjuvant: C. L. Cooper *et al.*, *J. Clin. Immunol.* **24**, 693 (2004); *eidem*, *Vaccine* **22**, 3136 (2004); as immunostimulant in metastatic melanoma: M. Pashenkov *et al.*, *J. Clin. Oncol.* **24**, 5716 (2006). Review of clinical development in cancer therapy: Y. M. Murad *et al.*, *Expert Opin. Biol. Ther.* **7**, 1257-1266 (2007).

Tricosasodium salt. [541547-35-7] Agatolimod sodium; CpG-2006; CPG-7909; PF-3512676; VaxImmune. $C_{236}H_{303}N_{70}Na_{23}O_{133}P_{23}S_{23}$; mol wt 8226.92.

USE: Vaccine adjuvant.

THERAP CAT: Antineoplastic; immunomodulator.

181. Agmatine. [306-60-5] *N*-(4-Aminobutyl)guanidine; 1-amino-4-guanidobutane. $C_5H_{14}N_4$; mol wt 130.20. C 46.13%, H 10.84%, N 43.03%. Decarboxylated arginine. Found in pollen of *Ambrosia artemisifolia* L., *Compositae*, in ergot, in sponges, in herring sperm, in octopus muscle: F. W. Heyl, *J. Am. Chem. Soc.* **41**,

670 (1919); A. Kossel, *Z. Physiol. Chem.* **66**, 257 (1910); J. L. Irvin, D. W. Wilson, *J. Biol. Chem.* **127**, 565 (1939). Prepn of salts: **DE 463576** (1928 to Schering-Kahlbaum AG). Synthesis: A. Kossel, *Z. Physiol. Chem.* **68**, 170 (1910); K. Odo, *J. Chem. Soc. Jpn.* **67**, 132 (1946); K. Dose, *Ber.* **90**, 1251 (1957). Use as marker for tumor cells: F. S. Steven *et al.*, *Anticancer Res.* **9**, 247 (1989); F. S. Steven *et al.*, *J. Enzyme Inhib.* **4**, 63 (1990).

Sulfate. [2482-00-0] $C_5H_{14}N_4.H_2SO_4$. Needles from dil methanol, mp 231°. Fairly sol in water. Nearly insol in alcohol.

Gold chloride. $C_5H_{14}N_4.2HCl.2AuCl_3$. Yellow needles from water, dec 223°.

USE: Biochemical probe.

182. Agnus Castus. Chaste tree; monk's pepper. Deciduous shrub, *Vitex agnus-castus* L., *Verbenaceae*. Used medicinally since ancient times for treatment of female conditions and for suppression of sexual desire. Medicinal parts include the dried leaves and the dried ripe fruit, known as **chasteberry**. The dried fruits are similar in appearance and odor to black pepper. *Habit.* Indigenous to the Mediterranean region; widely distributed in central Asia, the tropics and southern Europe. *Constit.* Iridoid glycosides, agnoside and aucubin; flavonoids including casticin and vitexin; labdan diterpenoids, rotundifuran, vitexilactone, and 6β,7β-diacetoxy-13-hydroxy-labda-8,14-diene; and volatile oil containing a complex mixture of monoterpenes and sesquiterpenes. Brief description: P. Houghton, *Pharm. J.* **253**, 720 (1994). Yield and composition of essential oil: J. M. Sorensen, S. T. Katsiotis, *Planta Med.* **66**, 245 (2000). HPLC analysis of diterpenoids: E. Hoberg *et al.*, *ibid.* 352. Inhibition of prolactin secretion: G. Sliutz *et al.*, *Horm. Metab. Res.* **25**, 253 (1993). Dopaminergic activity: H. Jarry *et al.*, *Exp. Clin. Endocrinol.* **102**, 448 (1994). Series of articles on phytochemistry, pharmacology and biological activity: *Z. Phytother.* **20**, 140-158 (1998). Historical overview: J. G. Mayer, F.-C. Czygan, *ibid.* 177. Clinical trial in treatment of hyperprolactinemia: A. Milewicz *et al.*, *Arzneim.-Forsch.* **43**, 752 (1993); in treatment of premenstrual syndrome: R. Schellenberg *et al.*, *Br. Med. J.* **322**, 134 (2001).

Dry extract. Agnolyt; Strotan; Valverde.

THERAP CAT: In treatment of premenstrual syndrome.

183. Agomelatine. [138112-76-2] *N*-[2-(7-Methoxy-1-naphthalenyl)ethyl]acetamide; S-20098; Valdoxan. $C_{15}H_{17}NO_2$; mol wt 243.31. C 74.05%, H 7.04%, N 5.76%, O 13.15%. Melatoninergic agonist and selective serotonin 5-HT_{2B} and 5-HT_{2C} receptor antagonist; metabolically stable analogue of melatonin, *q.v.* Prepn: J. Andrieux *et al.*, **EP 447285**; *eidem*, **US 5225442** (1991, 1993 both to Adir); and structure-activity studies: S. Yous *et al.*, *J. Med. Chem.* **35**, 1484 (1992); P. Depreux *et al.*, *ibid.* **37**, 3231 (1994). Mechanism of action: M. J. Millan *et al.*, *J. Pharmacol. Exp. Ther.* **306**, 954 (2003). Clinical evaluation in major depressive disorder: H. Lôo *et al.*, *Int. Clin. Psychopharmacol.* **17**, 239 (2002); S. H. Kennedy, R. Emsley, *Eur. Neuropsychopharmacol.* **16**, 93 (2006).

Crystals from toluene + hexane, 2:1, mp 109-110°.

THERAP CAT: Antidepressant.

184. Agroclavine. [548-42-5] 8,9-Didehydro-6,8-dimethyl-ergoline. $C_{16}H_{18}N_2$; mol wt 238.33. C 80.63%, H 7.61%, N 11.75%. A non-peptide ergot alkaloid obtained from cultures of fungi parasitic on *Elymus mollis* Trin.: Abe *et al.*, **JP 49 178336** (1949 to Takeda), *C.A.* **45**, 6352c (1951); *Annu. Rep. Takeda Res. Lab.* **10**, 145, 167, 171 (1951); **JP 54 7498** (1954), *C.A.* **50**, 6000ᵇ

(1956); **US 2835675** (1958). Found in fungi parasitic on *Pennisetum typhoideum* Rich.: Stoll *et al.*, *Helv. Chim. Acta* **37**, 1815 (1954). Structure and stereochemistry: Schreier, *ibid.* **41**, 1984 (1958). Biosynthesis: Floss *et al.*, *J. Am. Chem. Soc.* **90**, 6500 (1968). Synthesis: Plieninger *et al.*, *Ann.* **743**, 95 (1971). Metabolism: Ramstad, *Lloydia* **31**, 327 (1968).

Rods from ether, dec 198-203°; needles from acetone, dec 205-206°. $[\alpha]_D^{20}$ −155° (c = 0.9 in chloroform); $[\alpha]_D^{20}$ −182° (c = 0.5 in pyridine). uv max: 225, 284, 293 nm (ε 4.47, 3.88, 3.81). Freely sol in alc, chloroform, pyridine; sol in benzene, ether; very slightly sol in water.

185. Agrocybin. [544-44-5] 8-Hydroxy-2,4,6-octatriynamide. $C_8H_5NO_2$; mol wt 147.13. C 65.31%, H 3.43%, N 9.52%, O 21.75%. Phytotoxic acetylene responsible for killing of grass by fairy ring mushroom fungus. Isolation from basidiomycete *Agrocybe dura*: Kavanagh *et al.*, *Proc. Natl. Acad. Sci. USA* **36**, 102 (1950); from *Marasmius oreades* and activity: W. A. Ayer, P. A. Craw, *Can. J. Chem.* **67**, 1371 (1989). Structure and synthesis: Bu'Lock *et al.*, *Chem. Ind. (London)* **1954**, 990; and spectrum: Ashworth *et al.*, *J. Chem. Soc.* **1958**, 950. Origin of the carbon skeleton: E. R. H. Jones *et al.*, *J. Chem. Res. Miniprint* **1977**, 744.

Crystals from 20% alc or ether, mp 140° (conflagrates). Stable in air for about one day after which the crystals turn black and become insol in ether. Sol in alc, acetone, ether, chloroform, methyl isobutyl ketone; slightly sol in water. Practically insol in hexane.

186. Ajaconine. [545-61-9] (2*S*,4*R*,4a*S*,5*R*,6a*R*,9*R*,12a*S*,-12b*R*,13*R*)-Octahydro-4-hydroxy-9-methyl-3-methylene-5*H*,8*H*-2,4a-ethano-5,9,12a-ethanylylidene-2*H*-[2]benzopyrano[3,4-*b*]azocine-7(6a*H*)-ethanol. $C_{22}H_{33}NO_3$; mol wt 359.51. C 73.50%, H 9.25%, N 3.90%, O 13.35%. From seeds of larkspur, *Delphinium ajacis* L., *Ranunculaceae*: Keller, Volker, *Arch. Pharm.* **251**, 207 (1913); Goodson, *J. Chem. Soc.* **1945**, 245. Structure: Dvornik, Edwards, *Tetrahedron* **14**, 54 (1961); Nabors *et al.*, *Tetrahedron Lett.* **1969**, 2445. Stereochemistry: Solo, Pelletier, *Chem. Ind. (London)* **1960**, 1108; Whalley, *Tetrahedron* **18**, 43 (1962). Synthetic studies: Nabors *et al.*, *ibid.* **27**, 2385 (1971). Rearrangement: S. W. Pelletier, N. V. Mody, *J. Am. Chem. Soc.* **101**, 492 (1979).

Prisms from dil alcohol, mp 172°. $[\alpha]_D^{18}$ −119° (c = 2 in anhydr alcohol).

Sulfate heptahydrate. Hexagonal plates from dil acetone, mp 113°. $[\alpha]_D^{20}$ +5.5° (c = 2).

187. Ajmaline. [4360-12-7] (17*R*,21α)-Ajmalan-17,21-diol; rauwolfine; Aritmina; Gilurytmal; Ritmos; Tachmalin. $C_{20}H_{26}$-N_2O_2; mol wt 326.44. C 73.59%, H 8.03%, N 8.58%, O 9.80%. From roots of *Rauwolfia serpentina* (L.) Benth. (*Ophioxylon serpen-*

tinum L.), *Apocynaceae*. Isolation: S. Siddiqui, R. H. Siddiqui, *J. Indian Chem. Soc.* **8**, 667 (1931); **9**, 539 (1932); **12**, 37 (1935); L. van Itallie, A. J. Steenhauer, *Arch. Pharm.* **270**, 313 (1932). Structure: A. Chatterjee, S. Bose, *J. Indian Chem. Soc.* **31**, 17 (1954); F. A. L. Anet *et al.*, *J. Chem. Soc.* **1954**, 1242. Stereochemistry: M. F. Bartlett *et al.*, *J. Am. Chem. Soc.* **84**, 622 (1962). Synthesis: S. Masamune *et al.*, *ibid.* **89**, 2506 (1967); E. E. Van Tamelen, L. K. Oliver, *ibid.* **92**, 2136 (1970); K. Mashimo, Y. Sato, *Chem. Pharm. Bull.* **18**, 353 (1970). Physico-chemical properties: A. Petter, *Arzneim.-Forsch.* **24**, 874 (1974). Antiarrhythmic activity: A. Petter, K. Engelmann, *ibid.* 876. *Reviews:* R. Robinson in *Festschrift Arthur Stoll* (Birkhäuser-Verlag, Basel, 1957) pp 457-467; A. Koskinen, M. Lounasmaa in *Progress in the Chemistry of Natural Products* vol. **43**, W. Herz *et al.*, Eds. (Springer-Verlag, New York, 1983) pp 268-346.

Pale amber, solvated, tetragonal prisms from methanol, $C_{20}H_{26}N_2O_2 \cdot CH_3OH$, mp 158-160°. $[\alpha]_D^{18}$ +131° (c = 0.4 in chloroform). Anhydr mp 205-207°. $[\alpha]_D^{20}$ +144° (c = 0.8 in chloroform). uv max (ethanol): 247, 295 nm (log ε 3.94, 3.49). Sol in methanol, ethanol, ether, chloroform; slightly sol in water.

Hydrochloride dihydrate. $C_{20}H_{26}N_2O_2 \cdot 2HCl \cdot 2H_2O$. Hexagonal bipyramidal crystals from water, mp 140°. $[\alpha]_D^{18}$ +96.6°. One gram dissolves in 40 ml water.

THERAP CAT: Antihypertensive; antiarrhythmic.

188. Ajoene. [92285-01-3] 2-Propen-1-yl [3-(2-propen-1-ylsulfinyl)-1-propen-1-yl] disulfide; 4,5,9-trithiadodeca-1,6,11-triene 9-oxide. $C_9H_{14}OS_3$; mol wt 234.39. C 46.12%, H 6.02%, O 6.83%, S 41.03%. Antithrombotic principle found in garlic (*Allium sativum* L., *Liliaceae*). Formed from allicin, *q.v.*, in the approximate ratio of 4:1 (*E:Z*). The (*Z*)-isomer appears to be more bioactive. Isoln and effect on human platelet aggregation *in vitro*: R. Apitz-Castro *et al.*, *Thromb. Res.* **32**, 155 (1983). Structure initially assigned as *allyl 1,5-hexadienyltrisulfide*. Structural study, isoln of isomers and synthesis of (*E,Z*)-form: E. Block *et al.*, *J. Am. Chem. Soc.* **106**, 8295 (1984). Structure determn and structure-activity study: *eidem*, *ibid.* **108**, 7045 (1986). Prepd not claimed: E. Block, S. Ahmad, *US 4643994* (1987 to Res. Found. SUNY). Mechanism of antiplatelet activity: R. Apitz-Castro *et al.*, *Biochem. Biophys. Res. Commun.* **141**, 145 (1986). Synergism with other antiplatelet compounds: *eidem*, *Thromb. Res.* **42**, 303 (1986). *In vitro* antifungal activity: S. Yoshida *et al.*, *Appl. Environ. Microbiol.* **53**, 615 (1987).

(Z)–Ajoene

(E)–Ajoene

(*E*)-Form. [92284-99-6] Colorless, odorless oil. uv max: 240 nm.

189. Ajowan Oil. Ptychotis oil. Obtained by distillation of seeds of *Carum copticum* (DC.) Benth. & Hook. (*Ptychotis ajowan* DC.), *Umbelliferae*. Contains thymol, α-pinene, *p*-cymene, dipentene, γ-terpinene; fixed oil content includes petroselenic acid, oleic acid, linoleic acid, resin acids, and palmitic acid. *Ref:* Farook *et al. J. Sci. Food Agric.* **4**, 132 (1953); E. Guenther, *The Essential Oils* **IV**, 551-2 (Van Nostrand, New York, 1950).

Almost colorless or brownish liquid; thyme odor; sharp burning taste. d 0.910-0.930. Rotation 0° to +5°. n_D^{20} 1.498-1.504. Sol in 1-2.5 vol and more of 80% alcohol. The oil is toxic to earthworms, while alcoholic extracts are toxic to staphylococci and *Escherichia coli*. Dil extracts are useful as expectorants: Umanskii, Krutik, *Farmatsiya* **8**, no. 6, 19 (1945), *C.A.* **41**, 2209 (1947).

190. Ajugarins. Diterpenes isolated from the leaves of *Ajuga remota (Labiatae)*. Five different ajugarins have been isolated and identified; of these ajugarins I-III possess antifeedant activity against the African army worm, ajugarin-IV has insecticidal activity, ajugarin V is inactive. Isoln and structure of ajugarins I-III: I. Kubo *et al.*, *Chem. Commun.* **1976**, 949; of ajugarin IV: *eidem*, *ibid.* **1982**, 618; of ajugarin V: *eidem*, *Chem. Lett.* **1983**, 223. *See also:* **JP Kokai 82 48979** (1982 to Otsuka). X-ray crystal structure of 12-bromoajugarin-I and absolute configuration: I. Kubo *et al.*, *Chem. Commun.* **1980**, 897; ^{13}C-NMR shift data: J. M. Luteijn *et al.*, *Org. Magn. Reson.* **19**, 95 (1982). Synthetic approaches: D. J. Goldsmith *et al.*, *J. Org. Chem.* **43**, 3182 (1978); J. M. Luteijn, Ae. de Groot, *Tetrahedron Lett.* **1981**, 789. Total synthesis of ajugarin-I: S. V. Ley *et al.*, *Chem. Commun.* **1983**, 503.

Ajugarin-I   R = CH₂OCOCH₃

Ajugarin-I. [62640-05-5] [1*R*-(1α,4aβ,5β,6α,8α,8aα)]-4-[2-[8-(Acetyloxy)-8a-[(acetyloxy)methyl]octahydro-5,6-dimethylspiro-[naphthalene-1(2*H*),2'-oxiran]-5-yl]ethyl]-2(5*H*)furanone. $C_{24}H_{34}O_7$; mol wt 434.53. mp 155-157°. uv max (methanol): 212 nm (ε 10000).

Ajugarin-II. [62640-06-6] $C_{22}H_{32}O_6$; mol wt 392.49. 8-Hydroxy analog of ajugarin I. mp 188-189°.

Ajugarin-III. [62640-07-7] $C_{24}H_{36}O_8$; mol wt 452.54. 1,1-Dihydroxy analog of ajugarin I. mp 243-245°.

Ajugarin-IV. [82225-47-6] $C_{23}H_{34}O_6$; mol wt 406.52. mp 119-120.5°. $[\alpha]_D$ −57.5° (c = 0.06 in CHCl₃). uv max (ethanol): 215 nm (ε 17000).

Ajugarin-V. [82231-14-9] $C_{22}H_{32}O_5$; mol wt 376.49. mp 217-218°. $[\alpha]_D$ −13.5° (c = 0.18 in CHCl₃). uv max (ethanol): 210 nm (ε 11000).

191. Ajulemic Acid. [137945-48-3] (6a*R*,10a*R*)-3-(1,1-Dimethylheptyl)-6a,7,10,10a-tetrahydro-1-hydroxy-6,6-dimethyl-6*H*-dibenzo[*b*,*d*]pyran-9-carboxylic acid; 1',1'-dimethylheptyl-Δ⁸-tetrahydrocannabinol-11-oic acid; (3*R*,4*R*)-Δ⁶-THC-DMH-7-oic acid; CT-3; IP-751. $C_{25}H_{36}O_4$; mol wt 400.56. C 74.96%, H 9.06%, O 15.98%. Cannabinoid receptor agonist; synthetic, nonpsychoactive derivative of Δ⁹-tetrahydrocannabinol, *q.v.* Prepn: S. H. Burstein *et al.*, *J. Med. Chem.* **35**, 3135 (1992); S. H. Burstein, R. Mechoulam, **WO 9401429**; *eidem*, **US 5338753** (both 1994). GC-MS determn in plasma: C. Batista *et al.*, *J. Chromatogr. B* **820**, 77 (2005). Binding to CB-receptors: M.-H. Rhee *et al.*, *J. Med. Chem.* **40**, 3228 (1997); to PPAR$_\gamma$ receptor: J. Liu *et al.*, *Mol. Pharmacol.* **63**, 983 (2003). Clinical evaluation in chronic neuropathic pain: M. Karst *et al.*, *J. Am. Med. Assoc.* **290**, 1757 (2003). Reviews of development and pharmacology: S. Burstein *et al.*, *Life Sci.* **75**, 1513-1522 (2004); *idem*, *AAPS J.* **7**, E143-E148 (2005); J. L. Wiley, *IDrugs* **8**, 1002-1011 (2005).

Crystals from acetonitrile, mp 112-114° (sintering). $[\alpha]_D$ −275° (c = 3.8 in chloroform). Sol in most organic solvents. Practically insol in hexane.

Acetate. $C_{25}H_{36}O_4 \cdot C_2H_2O$; mol wt 442.60. Crystals from pentane, mp 120-122°. $[\alpha]_D$ −265° (c = 9.0 in chloroform).

THERAP CAT: Anti-inflammatory.

192. Akuammicine. [639-43-0] (19*E*)-2,16,19-20-Tetradehydrocuran-17-oic acid methyl ester. $C_{20}H_{22}N_2O_2$; mol wt 322.41. C 74.51%, H 6.88%, N 8.69%, O 9.92%. From the seeds of *Picralima klaineana*, Pierre, *Apocynaceae* found in the Gold Coast: Henry, Sharp, *J. Chem. Soc.* **1927**, 1950; Henry, *ibid.* **1932**, 2759. Structure: Aghoramurthy, Robinson, *Tetrahedron* **1**, 172 (1957); Bernauer *et al.*, *Helv. Chim. Acta* **43**, 717 (1960); Edwards, Smith, *J. Chem. Soc.* **1961** 152. Derivs: Robinson, Thomas, *ibid.* **1955**, 2049. Total synthesis: Kutney, Fuller, *Heterocycles* **3**, 197 (1975).

Plates from ethanol+ water, mp 182°. $[\alpha]_D^{16}$ −745° (c = 0.994 in ethanol). pKa 7.45. uv max (ethanol): 227, 300, 330 nm (log ε 4.09, 4.07, 4.24).

Hydrochloride dihydrate. $C_{20}H_{22}N_2O_2 \cdot HCl \cdot 2H_2O$. Leaflets from alcohol or water, mp 171°. $[\alpha]_D^{21}$ −610° (c = 1.430 in ethanol).

Perchlorate monohydrate. $C_{20}H_{22}N_2O_2 \cdot HClO_4 \cdot H_2O$. Needles from ethanol + water, mp 134-136°.

Hydriodide monohydrate. $C_{20}H_{22}N_2O_2 \cdot HI \cdot H_2O$. Square plates from water, mp 128°.

Methiodide. Crystals from water, mp 252°.

Nitrate. Needles from hot water, mp 182.5°.

193. Alacepril. [74258-86-9] 1-[(2*S*)-3-(Acetylthio)-2-methyl-1-oxopropyl]-L-prolyl-L-phenylalanine; 1-(D-3-acetylthio-2-methylpropanoyl)-L-prolyl-L-phenylalanine; *N*-[1-[(*S*)-3-mercapto-2-methylpropionyl]-L-prolyl]-3-phenyl-L-alanine acetate (ester); DU-1219; Cetapril. $C_{20}H_{26}N_2O_5S$; mol wt 406.50. C 59.09%, H 6.45%, N 6.89%, O 19.68%, S 7.89%. Angiotensin-converting enzyme (ACE) inhibitor. Prepn: T. Sawayama *et al.*, **JP Kokai 80 9058**; *eidem*, **US 4248883** (1980, 1981 both to Dainippon Pharm.). Pharmacology in animals: K. Takeyama *et al.*, *Arzneim.-Forsch.* **35**, 1502 (1985); *eidem, ibid.* 1507. Series of articles on pharmacology, metabolism, enzyme inhibiting activity: *ibid.* **36**, 47-83 (1986). Metabolism to captopril, *q.v.*: K. Matsumoto *et al.*, *ibid.* 40. HPLC determn of metabolites in plasma and urine: K. Hayashi *et al.*, *J. Chromatogr.* **338**, 161 (1985). Pharmacokinetics in humans: K. Onoyama *et al.*, *Clin. Pharmacol. Ther.* **38**, 462 (1985). Preclinical evaluation in essential hypertension: K. Mizuno *et al.*, *Res. Commun. Chem. Pathol. Pharmacol.* **49**, 175 (1985); H. Shionoiri *et al.*, *Curr. Ther. Res.* **38**, 537 (1985). Series of articles on toxicology: M. Iida *et al.*, *Yakuri to Chiryo* **13**, 7033-7121 (1985), *C.A.* **104**, 21888h-21890c; **105**, 396v (1986).

Crystals from ethanol/*n*-hexane, mp 155-156°. $[\alpha]_D^{25}$ −81.3° (c = 1.02 in ethanol). LD_{50} in rats, mice (mg/kg): >5000, >5000 orally; >3000, >3000 s.c.; ~2000, ~3000 i.p. (Iida, pp 7033-40).

THERAP CAT: Antihypertensive.

194. Alachlor. [15972-60-8] 2-Chloro-*N*-(2,6-diethylphenyl)-*N*-(methoxymethyl)acetamide; 2-chloro-2′,6′-diethyl-*N*-(methoxymethyl)acetanilide; metachlor; CP-50144; Lasso; Alanex. C_{14}-

$H_{20}ClNO_2$; mol wt 269.77. C 62.33%, H 7.47%, Cl 13.14%, N 5.19%, O 11.86%. Pre-emergence herbicide. Prepn: **NL 6602564** (1967 to Monsanto), *C.A.* **67**, 99832r (1967). Activity: D. M. Evans, *Chem. Ind. (London)* **1969**, 615. Soil degradn: R. S. Hargrove, M. G. Merkle, *Weed Sci.* **19**, 652 (1971). Mechanism of action: L. M. Deal, F. D. Hess, *ibid.* **28**, 168 (1980).

Cryst solid, mp 40-41°. $d_{15.6}^{25}$ 1.133. Soly in water at 23°: 140 mg/l. Sol in ether, acetone, benzene, ethanol, ethyl acetate. Hydrolyzed under strong acid or alkaline conditions. LD_{50} orally in rats: 1200 mg/kg (Evans).

USE: Herbicide.

195. Alafosfalin. [60668-24-8] *P*-[(1*R*)-1-[[(2*S*)-2-Amino-1-oxopropyl]amino]ethyl]phosphonic acid; 1*R*-1-(L-alanylamino)-ethylphosphonic acid; alaphosphin; Ro-3-7008. $C_5H_{13}N_2O_4P$; mol wt 196.14. C 30.62%, H 6.68%, N 14.28%, O 32.63%, P 15.79%. Synthetic phosphonodipeptide with antibacterial activity. Prepn: F. R. Atherton *et al.*, **DE 2602193**; *eidem*, **US 4016148** (1976, 1977 both to Hoffmann-La Roche); J. G. Allen *et al.*, *Nature* **272**, 56 (1978); F. R. Atherton *et al.*, *Antimicrob. Agents Chemother.* **15**, 677 (1979). Improved process: E. K. Baylis, **EP 10872**; *eidem*, **US 4331591** (1980, 1982 both to Ciba-Geigy). Separation of diastereoisomers: J. Szewczyk *et al.*, *Experientia* **38**, 983 (1982). Antibacterial spectrum: F. R. Atherton *et al.*, *Antimicrob. Agents Chemother.* **15**, 684 (1979); W. H. Traub, *Chemotherapy* **26**, 103 (1980). Synergism with β-lactams: H. B. Maruyama *et al.*, *Antimicrob. Agents Chemother.* **16**, 444 (1979); F. R. Atherton *et al.*, *ibid.* **20**, 470 (1981); M. Arisawa *et al.*, *ibid.* **21**, 706 (1982). Pharmacokinetics: J. D. Allen, L. J. Lees, *ibid.* **17**, 973 (1980). Comprehensive review: C. H. Hassall in *Antibiotics* **VI**, F. E. Hahn, Ed. (Springer-Verlag, New York, 1983) pp 1-11.

Crystals from ethanol-water, mp 295-296° (dec). $[\alpha]_D^{20}$ −44.0° (c = 1 in H_2O).

196. Alagebrium Chloride. [341028-37-3] 4,5-Dimethyl-3-(2-oxo-2-phenylethyl)thiazolium chloride (1:1); ALT-711. $C_{13}H_{14}$-ClNOS; mol wt 267.77. C 58.31%, H 5.27%, Cl 13.24%, N 5.23%, O 5.97%, S 11.97%. Catalytically cleaves crosslinks between advanced glycation end products (AGE) and extracellular matrix proteins. Prepn: A. Cerami *et al.*, **EP 808163**; *eidem*, **US 5853703** (1997, 1998 both to Alteon; Picower Inst. Med. Res.). In vivo AGE cross-link breaking activity: B. H. R. Wolffenbuttel *et al.*, *Proc. Natl. Acad. Sci. USA* **95**, 4630 (1998). Clinical pharmacology and effect on vascular stiffening: D. A. Kass *et al.*, *Circulation* **104**, 1464 (2001); in elderly patients with heart failure: W. C. Little *et al.*, *J. Card. Fail.* **11**, 191 (2005). Review of pharmacology and therapeutic potential: S. Vasan *et al.*, *Arch. Biochem. Biophys.* **419**, 89-96 (2003); of clinical evaluations in cardiovascular disease: G. L. Bakris *et al.*, *Am. J. Hypertens.* **17**, 23S-30S (2004).

THERAP CAT: In treatment of cardiovascular complications of diabetes and aging.

197. Alanine. [56-41-7] L-Alanine; Ala; A; L-α-alanine; L-α-aminopropionic acid; (S)-2-aminopropanoic acid. $C_3H_7NO_2$; mol wt 89.09. C 40.45%, H 7.92%, N 15.72%, O 35.92%. Non-essential amino acid for human development. First syllable of the name denotes its origin from aldehyde. First synthesized and named by: A. Strecker, *Ann.* **75**, 27 (1850); prior to identification in silk hydrolysate: P. Schützenberger, A. Bourgeous, *Compt. Rend.* **81**, 191 (1875); Th. Weyl, *Ber.* **21**, 1407 (1888). Early chemistry and biochemistry: *Amino Acids and Proteins*, D. M. Greenberg, Ed. (Charles C. Thomas, Springfield, IL, 1951) 950 pp., *passim*; J. P. Greenstein, M. Winitz, *Chemistry of the Amino Acids* **vols 1-3** (John Wiley and Sons, Inc., New York, 1961) pp. 1819-1840, *passim*. Enantiomeric selective synthesis: R. Amoroso *et al.*, *J. Org. Chem.* **57**, 1082 (1992). Review of metabolism in man: J. E. Liljenquist *et al.*, *Clin. Nutr. Update*, H. L. Greene *et al.*, Eds. (American Medical Association, Chicago, IL, 1977) pp 22-34; T. N. Palmer *et al.*, *Biosci. Rep.* **5**, 1015-1033 (1985). Review of industrial production by microorganisms: S. Takamatsu, T. Tosa, *Bioprocess Technol.* **16**, 25-35 (1993).

Orthorhombic crystals from water, dec 297°. d 1.401. $[\alpha]_D^{25}$ +2.42° (c = 10 in H_2O); +13.7° (c = 2.06 in 6N HCl). pK₁ 2.34; pK₂ 9.69. Soly (g/l) in water at 0°: 127.3; at 25°: 166.5; at 50°: 217.9; at 75°: 285.1; at 100°: 373.0. Soly in cold 80% ethanol: 0.2%. Freely sol in water. Insol in ether.

D-Form. [338-69-2] Occurs in bacterial cell walls and in some peptide antibiotics. *Review:* M. Bodanszky, D. Perlman, *Science* **163**, 352-358 (1969); C. T. Walsh, *J. Biol. Chem.* **264**, 2393-2396 (1989). Crystals, mp 289-291°. $[\alpha]_D$ −14.1° (c = 0.9 in 1N HCl).

DL-Form. [302-72-7] Orthorhombic bipyramidal needles or rods from water. d 1.424. Sublimes above 200°. Decomp 264-296° depending on rate of heating. pK₁ 2.35; pK₂ 9.87. Soly in water at 0°: 121 g/l, at 25°: 167 g/l, at 50°: 231 g/l, at 75°: 319 g/l, at 100°: 440 g/l; in ethanol 0.0087 g/100 g at 25°. Insol in ether.

198. β-Alanine. [107-95-9] β-Aminopropionic acid; 3-aminopropanoic acid; 3-aminopropanoic acid; Abufène. $C_3H_7NO_2$; mol wt 89.09. C 40.45%, H 7.92%, N 15.72%, O 35.92%. Prepd by the action of KOBr and KOH upon succinimide: Clarke, Behr, *Org. Synth.* **16**, 1 (1936). By the action of liq ammonia upon methyl acrylate: Morsch, *Monatsh. Chem.* **63**, 220 (1933), *C.A.* **41**, 4104 (1947); by the addition of NH_4OH to acrylonitrile: Ford *et al.*, *J. Am. Chem. Soc.* **69**, 844 (1947). By electrolytic oxidation of 3-amino-1-propanol in H_2SO_4 using Pb electrodes without diaphragm: *Jubilee Vol. Emil Barell* **1946**, 85-91. For industrial methods of prepn *see* several pats. by T. L. Gresham to B. F. Goodrich. Prepn from ethylene cyanohydrin (β-hydroxypropionitrile): Boatright, **US 2734081** (1956 to Am. Cyanamid); from β-aminopropionitrile: Ford, *Org. Synth.* **coll. vol. III**, 34 (1955). Improved process: Beutel, Klemchuk, **US 2956080** (1960 to Merck & Co.).

Orthorhombic bipyramidal crystals from water, decomp 207° (very rapid heating). Decomp 197-198° (Ford, *Org. Syn. loc. cit.*). Slightly sweet taste. pK₁ 3.60; pK₂ 10.19. pH of 5% aq soln: 6.0 to 7.3. Freely sol in water, slightly in alcohol. Practically insol in ether, acetone.

Hydrochloride. $C_3H_7NO_2$·HCl. Plates, leaflets, mp 122.5°. Freely sol in water, less sol in alcohol. Insol in ether.

Platinichloride. $2C_3H_7NO_2$·$2HCl$·$PtCl_4$. Yellow leaflets from alcohol + HCl, decomp 210°. Freely sol in water, sparingly in abs alcohol.

USE: In the synthesis of pantothenic acid and derivatives; as buffer in electroplating.

199. L-Alanosine. [5854-93-3] 3-(Hydroxynitrosoamino)-L-alanine; L-2-amino-3-(hydroxynitrosoamino)propanoic acid; L-2-amino-3-[(N-nitroso)hydroxylamino]propionic acid. $C_3H_7N_3O_4$; mol wt 149.11. C 24.17%, H 4.73%, N 28.18%, O 42.92%. Antibiotic substance from the fermentation of *Streptomyces alanosinicus* n. sp. Prepn: **NL 6509543**; J. Thiemann, Y. S. K. Murthy, **US 3676490** (1966, 1972 both to Lepetit). The first natural product found to have a N-nitrosohydroxylamino group on an aliphatic chain. Isoln and structure: Coronelli *et al.*, *Farmaco Ed. Sci.* **21**, 269 (1966). Characterization: Thiemann, Beretta, *J. Antibiot.* **19A**, 155 (1966). Structure and synthesis of L-, D- and DL-forms: Lancini *et al.*, *Tetrahedron Lett.* **1966**, 1769; *eidem, Farmaco Ed. Sci.* **24**, 169 (1969). Synthesis of L-form: Isowa *et al.*, *Bull. Chem. Soc. Jpn.* **46**, 1847 (1973). Improved synthesis of DL-form: Eaton *et al.*, *J. Med. Chem.* **16**, 289 (1973). Pharmacology: Murthy *et al.*, *Nature* **211**, 1198 (1966). Mode of action studies: Gale *et al.*, *Biochem. Pharmacol.* **17**, 363, 1823 (1968). Effect on insect growth: E. E. Kenaga, *J. Econ. Entomol.* **62**, 1006 (1969); S. Matsumoto *et al.*, *Agric. Biol. Chem.* **48**, 827 (1984).

Crystalline powder from slightly acidic water, dec 190°. $[\alpha]_D$ +8°, −46°, −37.8° (in 1N HCl, 0.1N NaOH, water). uv max (0.1N HCl): 228 nm ($E_{1cm}^{1\%}$ 505); in 0.1N NaOH: 250 nm ($E_{1cm}^{1\%}$ 630). pKa 4.8. Slightly sol in water. Practically insol in the common organic solvents. Sol in alkaline and acidic solns, from which it ppts by adjusting the pH between 4 and 6. LD_{50} in mice (mg/kg): 600 i.p.; 300 i.v. (Thiemann, Murthy).

D-Form. mp 183°. $[\alpha]_D$ +45° (c = 0.5N in 1N HCl).

USE: Experimental insect reproduction inhibitor.

200. Alantolactone. [546-43-0] (3aR,5S,8aR,9aR)-3a,5,6,-7,8,8a,9,9a-Octahydro-5,8a-dimethyl-3-methylenenaphtho[2,3-b]furan-2(3H)-one; 8β-hydroxy-4αH-eudesm-5-en-12-oic acid γ-lactone; helenin; alant camphor; elecampane camphor; inula camphor. $C_{15}H_{20}O_2$; mol wt 232.32. C 77.55%, H 8.68%, O 13.77%. A terpene from roots of *Inula helenium* L., *Compositae*: Kallen, *Ber.* **6**, 1506 (1873); **9**, 154 (1876); Ruzicka *et al.*, *Helv. Chim. Acta* **14**, 397, 1090 (1931); **16**, 268 (1933). Structure: Marshall, Cohen, *J. Org. Chem.* **29**, 3727 (1964). Stereoselective synthesis: Marshall *et al.*, *J. Am. Chem. Soc.* **88**, 3408 (1966).

Crystals from alcohol, mp 78-79°. bp 275°. $[\alpha]_D$ +175° (chloroform). uv max (ethanol): 212 nm (ε 9500). Volatile with steam. Freely sol in alcohol, chloroform, benzene, ether, oils. Practically insol in water.

201. Alazopeptin. [1397-84-8] 6-Diazo-5-oxo-N-2-propen-1-ylnorleucyl-6-diazo-5-oxonorleucine; L-alanyl-(6-diazo-5-oxo)-L-norleucyl-(6-diazo-5-oxo)-L-norleucine. $C_{15}H_{20}N_6O_5$; mol wt 364.36. C 49.45%, H 5.53%, N 23.07%, O 21.95%. Tumor-inhibiting antibiotic produced by *Streptomyces griseoplanus* from soil near Williamsburg, Iowa: De Voe *et al.*, *Antibiot. Annu.* **1956-7**, 730. Peptide consisting of one mole α-alanine and two moles of a C_6 diazo keto amino acid oxidizable to glutamic acid: E. L. Patterson *et al.*, *Antimicrob. Agents Chemother.* **1965**, 115-118. Antitumor activity: T. Hata *et al.*, *J. Antibiot.* **26**, 181 (1973). Toxicity: J. B. Thiersch, *Proc. Soc. Exp. Biol. Med.* **97**, 888 (1958).

Monohydrate. Crystals from dilute acetone. Poor stability. Has no definite melting point. $[\alpha]_D^{25}$ +9.5° (c = 4.7 in H_2O). uv max (pH 7.0 phosphate buffer): 242, 274 nm ($E_{1cm}^{1\%}$ 321, 549). Freely sol in water. Somewhat sol in acetic acid, formamide, DMSO, aq solns of methanol, ethanol, acetone. Practically insol in anhydr alcohols, acetone, ethyl acetate, ether. LD_{50} i.p. in rats: 150 mg/kg (Thiersch).

202.　Albaspidin. [58409-52-2] 2,2'-Methylenebis[6-butyryl-3,5-dihydroxy-4,4-dimethyl-2,5-cyclohexadien-1-one]; polystichalbin; methylenebis(butyrylfilicinic acid); albaspidin-BB. $C_{25}H_{32}O_8$; mol wt 460.52. C 65.20%, H 7.00%, O 27.79%. Found in the rhizomes of the male fern, *Aspidium filix mas* (L.) Schott., *Dryopteris filix mas* (L.) Schott., *A. spinulosum*, *Polypodiaceae* and other ferns. Isoln from *Aspidium* extract: Boehm, *Ann.* **318**, 305 (1901); McGookin *et al.*, *J. Chem. Soc.* **1953**, 1828; *see also:* Tryon *et al.*, *Phytochemistry* **12**, 683 (1973). Isoln as one component of a homologous mixture: Penttila, Sundman, *Acta Chem. Scand.* **18**, 344 (1964). Synthesis: Riedl, Mitteldorf, *Ber.* **89**, 2595 (1956); Inagaki *et al.*, *J. Pharm. Soc. Jpn.* **76**, 1258 (1956). Biosynthetic studies: Penttila, Fales, *J. Am. Chem. Soc.* **88**, 2327 (1966).

Crystals from ethanol or methanol, mp 149°. Freely sol in chloroform; moderately sol in ether, benzene; sparingly sol in alc, acetone, glacial acetic acid, sol in KOH solns; very sparingly sol in Na_2CO_3 solns. Practically insol in methanol.

203.　Albendazole. [54965-21-8] *N*-[6-(Propylthio)-1*H*-benzimidazol-2-yl]carbamic acid methyl ester; methyl 5-(propylthio)-2-benzimidazolecarbamate; 5-(propylthio)-2-carbomethoxy-aminobenzimidazole; SKF-62979; Albenza; Eskazole; Valbazen; Zentel. $C_{12}H_{15}N_3O_2S$; mol wt 265.33. C 54.32%, H 5.70%, N 15.84%, O 12.06%, S 12.08%. Prepn: R. J. Gyurik, V. J. Theodorides, **US 3915986** (1975 to SmithKline). Anthelmintic spectrum: V. J. Theodorides, *Experientia* **32**, 702 (1976). Metabolism: R. J. Gyurik *et al.*, *Drug Metab. Dispos.* **9**, 503 (1981). Efficacy in cattle vs gastrointestinal parasites: J. C. Williams *et al.*, *Am. J. Vet. Res.* **38**, 2037 (1977); vs liver flukes: T. M. Craig *et al.*, *ibid.* **53**, 1170 (1992). Clinical trial in mixed helminth infections: P. Pene *et al.*, *Am. J. Trop. Med. Hyg.* **31**, 263 (1982); in echinococcosis: R. J. Horton, *Trans. R. Soc. Trop. Med. Hyg.* **83**, 97 (1989).

Colorless crystals, mp 208-210°. Freely sol in anhydrous formic acid. Sol in DMSO, strong acids, strong bases. Slightly sol in methanol, chloroform, ethyl acetate, acetonitrile. Very slightly sol in ether, methylene chloride. Practically insol in water, alcohol.

Sulfoxide. [54029-12-8] Albendazole oxide; ricobendazole; rycobendazole; RS-8852; Rycoben. $C_{12}H_{15}N_3O_3S$; mol wt 281.33. Active metabolite of albendazole. Bioavailability: L. Dominguez *et al.*, *Farmaco* **50**, 697 (1995). mp 226-228° (dec).

　THERAP CAT: Anthelmintic (Cestodes).

　THERAP CAT (VET): Anthelmintic.

204.　Albiglutide. [782500-75-8] 7-36-Glucagon-like peptide I [8-glycine] (human) fusion protein with 7-36-glucagon-like peptide I [8-glycine] (human) fusion protein with serum albumin (human); albugon; naliglutide; GSK-716155; Syncria. Long acting glucagon-like peptide 1 (GLP-1) mimetic; created by genetic fusion of a DPP-IV resistant human GLP-1 dimer to recombinant human albumin. Prepn: C. A. Rosen, W. A. Haseltine, **WO 03059934**; C. A. Rosen *et al.*, **US 7141547** (2003, 2006 both to Human Genome Sciences). Pharmacology: L. L. Baggio *et al.*, *Diabetes* **53**, 2492 (2004). Clinical pharmacokinetics: J. E. Matthews *et al.*, *J. Clin. Endocrinol. Metab.* **93**, 4810 (2008). Clinical evaluation in type 2 diabetes: J. Rosenstock *et al.*, *Diabetes Care* **32**, 1880 (2009). Review of development and therapeutic potential: G. H. Tomkin, *Curr. Opin. Mol. Ther.* **11**, 579-588 (2009); E. L. St. Onge, S. A. Miller, *Expert Opin. Biol. Ther.* **10**, 801-806 (2010).

　THERAP CAT: Antidiabetic.

205.　Albizziin. [1483-07-4] 3-[(Aminocarbonyl)amino]-L-alanine; 2-amino-3-ureidopropionic acid. $C_4H_9N_3O_3$; mol wt 147.13. C 32.65%, H 6.17%, N 28.56%, O 32.62%. Constituent of several plants belonging to the family *Mimosaceae*. First obtained from the seeds of *Albizzia julibrissin* Durazz., *Mimosaceae*: Gmelin *et al.*, *Z. Naturforsch.* **13b**, 252 (1958); *Z. Physiol. Chem.* **314**, 28 (1959). Structure: Kjaer *et al.*, *Experientia* **15**, 253 (1959). Synthesis: Kjaer, Larsen, *Acta Chem. Scand.* **13**, 1565 (1959); Rudinger *et al.*, *Collect. Czech. Chem. Commun.* **25**, 2022 (1960).

Needles from dil alcohol, decomp 218-220°. $[\alpha]_D^{25}$ −66.2° (c = 4 in H_2O); $[\alpha]_D^{24}$ −22.2° (c = 4.24 in 1.0*N* HCl); $[\alpha]_D^{24}$ +3.2° (c = 4.7 in 1.0*N* NaOH). *See:* Kjaer *et al.*, *Acta Chem. Scand.* **18**, 2412 (1964).

206.　Albofungin. [37895-35-5] (1*S*,4*R*,8a*R*)-13-Amino-3,-4,8a,13-tetrahydro-1,15,16-trihydroxy-4-methoxy-12-methyl-1*H*-xantheno[4',3',2':4,5][1,3]benzodioxino[7,6-*g*]isoquinoline-14,17-(2*H*,9*H*)-dione. $C_{27}H_{24}N_2O_9$; mol wt 520.49. C 62.31%, H 4.65%, N 5.38%, O 27.66%. Antifungal antibiotic produced by *Streptomyces albus* var *fungistaticus (fungatus)* Solovyeva and Rudaya. Isoln procedure: Khokhlov, Liberman, *Proc. Symp. Antibiotics Prague* (May 1959) p 81. Structure: A. I. Gurevich *et al.*, *Tetrahedron Lett.* **1972**, 1751. Stereochemistry: A. I. Gurevich *et al.*, *ibid.* **1974**, 2801. Thought to be identical with *Ba-180265*: Liu *et al.*, in *Antimicrob. Agents Chemother.* **1962**, 767. Chemical and biological properties: A.I. Gurevich *et al.*, *Antibiotiki* **17**, 771 (1972).

Crystals from nitromethane, mp 304-307°. Also reported as mp 190° (dec). $[\alpha]_D^{20}$ −670° (DMF). uv max (ethanol): 228, 254, 303, 376 nm (log ε 4.58, 4.58, 4.19, 4.42). Practically insol in water, petr ether. Sparingly sol in alcohol; quite sol in chloroform, dichloroethane, acetone, chlorobenzene, formamide, dimethylformamide, glacial acetic acid.

207.　Albomycin. [1414-39-7] Iron-containing antibiotic produced by *Actinomyces subtropicus*: Gauze, Braznikova, *Nov. Med.* **23**, 3 (1951). Used as the sulfate. Consists of six components, α, β, γ, δ_1, δ_2, ε, of which δ_1 and δ_2 are the main components. δ_2 is the unstable, highly active albomycin A_1; the other components are its degradation products: Braznikova *et al.*, *Biokhimiya* **22**, 111 (1959); Turkovà *et al.*, *Antibiotiki* **7**, 878 (1962). Early studies proposed cyclic hexapeptide structures containing three serines and three ornithine derivatives for δ_1-, δ_2- and ε-albomycin (*cf.* Ferrichromes). Structural studies: *eidem.*, *Collect. Czech. Chem. Commun.* **30**, 118 (1965); Poddubnaya, el'Naggar, *Zh. Obshch. Khim.* **38**, 450 (1968). Conflicting structural analysis shows a 3:1 ornithine : serine ratio: Maehr, Pitcher, *J. Antibiot.* **24**, 830 (1971). Similarity to or identity with grisein, *q.v.:* Stapley, Ormond, *Science* **125**, 587 (1957); Turková *et al.*, *Collect. Czech. Chem. Commun.* **31**, 2444 (1966). Review and antibacterial spectrum of albomycin and other hydroxamic acids: Gauze, *Br. Med. J.* **2**, 1177 (1955); Bhuyan in *Antibiotics* **1**, D. Gottlieb, P. D. Shaw, Eds. (Springer-Verlag, New York, 1967) pp 153-155; Maehr, *Pure Appl. Chem.* **28**, 603-636 (1971); Emery, *Adv. Enzymol. Relat. Areas Mol. Biol.* **35**, 135-185 (1971).

Sulfate. Amorphous red powder. uv max: 283 nm ($E_{1cm}^{1\%}$ 880). Freely sol in water. Slightly sol in methanol. Practically insol in most other organic solvents. Aq solns have a bright orange color. Effective against penicillin-resistant pneumococci and staphylococci. The toxicity is comparable to that of penicillin.

Note: Not to be confused with albamycin.

208. Alborixin. [57760-36-8] ($\alpha R,2S,3S,5R,6S$)-6-[($2R,3S$)-3-[($2R,5S,6R$)-6-[[($2R,3S,5R,6R$)-6-[(S)-[($2S,2'R,3'R,5S,5'S$)-5'-[($2R,5R,6S$)-6-Ethyltetrahydro-5-hydroxy-5-methyl-2H-pyran-2-yl]-octahydro-2'-hydroxy-2,3',5'-trimethyl[2,2'-bifuran]-5-yl]hydroxy-methyl]tetrahydro-6-hydroxy-3,5-dimethyl-2H-pyran-2-yl]methyl]-tetrahydro-6-hydroxy-5-methyl-2H-pyran-2-yl]-2-hydroxybutyl]-tetrahydro-α,3,5-trimethyl-2H-pyran-2-acetic acid; Antibiotic S 14750A. $C_{48}H_{84}O_{14}$; mol wt 885.19. C 65.13%, H 9.57%, O 25.30%. Polycyclic polyether antibiotic ionophore, isolated from a strain of *Streptomyces albus*. Initial description and x-ray structure: M. Allèaume *et al.*, *Chem. Commun.* **1975**, 411. Production from *Streptomyces hygroscopicus* and use: M. Kuhn, H. D. King, **DE 2608337** (1976 to Sandoz), *C.A.* **84**, 28491k (1977). Isoln from *S. albus*, structure, properties: P. Gachon *et al.*, *J. Antibiot.* **29**, 603 (1976); C. Delhomme *et al.*, *ibid.* 692. NMR spectrum and solution conformation: N. A. Rodios, M. J. O. Anteunis, *Bull. Soc. Chim. Belg.* **88**, 279 (1979). Revised structure: H. Seto *et al.*, *J. Antibiot.* **32**, 970 (1979). *In vitro* study: M. Chapel *et al.*, *ibid.* 740. Effects on cardiovascular function and plasma cation concentration in the dog: N. Moins *et al.*, *J. Cardiovasc. Pharmacol.* **1**, 659 (1979).

White amorphous powder, mp 100-105°. $[\alpha]^{20}$ −7° (c = 4 in acetone). pKa 10.02 (25° in methanol). LD$_{50}$ in mice: 15 mg/kg s.c. (Delhomme). LD$_{50}$ originally reported as 150 mg/kg, corrected in *J. Antibiot.* **30**, Suppl. , facing p 77-12 (1977).

209. Albumen. Egg white; dried egg white. By convention the pure protein is spelled "albumin" and commercial egg white is called "albumen". The word "albumen" goes back to Roman times and is classical Latin for egg white: *Ref:* Plinius (Major), *Historia Naturalis* **28**, 6, 18 paragraph 66. Constitutes about six parts by weight of a hen's egg (wet basis). Average weight of raw egg white 32.9 g. Albumen contains 75% ovalbumin, *(q.v.)*, ovoconalbumin, ovomucoid, ovomucin, ovoglobulin, lysozyme, *(q.v.)* and avidin, *(q.v.)*. Monograph: A. L. Romanoff, A. J. Romanoff, *The Avian Egg* (Wiley, New York, 1949) 918 pp.

Raw egg albumen. Clear, colloidal, flowing, limpid mass. White rubbery solid when denatured. d_{25}^{25} 1.035. n_D^{25} 1.356. Coagulating temp 61°. mp −0.42°. pH 7.6.

Dried albumen. Yellow, transparent, amorphous lumps or scales, or yellow powder. Decomposes in moist air. In water it swells at first, then dissolves gradually. The soln coagulates at 61° (because of denaturation of the proteins). Albumen solns are also denatured on contact with the following chemicals: Salts of copper, iron, mercury and silver; H_2O_2, phenol, picric acid, alum, tannic acid, formaldehyde, ether, alcohol.

USE: For clarifying and refining wines and vinegars. As textile dye mordant, in lithography as vehicle for substances that sensitize plates, in gilding leather (Venetian decorations), stamping with gold

and bronze powder (especially in book binding). In adhesives and veneers. Ingredient of compositions used in sizing and in making papers. In fine color preparations for artist's paints. As activating agent for certain enzymes. Formerly in sugar refining. Ingredient of bakery products, confectionery, food preparations. In pharmaceutical compounding, i.e. to make various albuminates. As analytical reagent in testing for ionic mercury, foreign oil in olive oil, candy or wine colorings.

THERAP CAT: Antidote to mercury poisoning.

210. Albuterol. [18559-94-9] α^1-[[(1,1-Dimethylethyl)amino]methyl]-4-hydroxy-1,3-benzenedimethanol; α^1-[(*tert*-butylamino)methyl]-4-hydroxy-*m*-xylene-α,α'-diol; 2-(*tert*-butylamino)-1-(4-hydroxy-3-hydroxymethylphenyl)ethanol; 4-hydroxy-3-hydroxymethyl-α-[(*tert*-butylamino)methyl]benzyl alcohol; salbutamol; AH-3365. $C_{13}H_{21}NO_3$; mol wt 239.32. C 65.24%, H 8.85%, N 5.85%, O 20.06%. β-Adrenergic agonist. Prepn: L. H. C. Lunts *et al.*, **ZA 6705591**; *eidem*, **US 3644353** (1968, 1972 both to Allen & Hanburys). Synthesis and structure-activity studies: D. T. Collin *et al.*, *J. Med. Chem.* **13**, 674 (1970). Prepn of isomers: C. J. Hawkins, G. T. Klease, *ibid.* **16**, 856 (1973); R. P. Bakale, *Spec. Chem.* **15**, 249 (1995). Absolute configuration of isomers: D. Hartley, D. Middlemiss, *J. Med. Chem.* **14**, 895 (1971). Pharmacology: R. T. Brittain *et al.*, *Nature* **219**, 862 (1968); V. A. Cullum *et al.*, *Br. J. Pharmacol.* **35**, 141 (1969). Metabolism: L. E. Martin *et al.*, *Eur. J. Pharmacol.* **14**, 183 (1971). Clinical comparison of levalbuterol with racemate: H. S. Nelson *et al.*, *J. Allergy Clin. Immunol.* **102**, 943 (1998). Comprehensive description: H. Y. Aboul-Enein *et al.*, *Anal. Profiles Drug Subs.* **10**, 665-689 (1981). Review of clinical experience in asthma and in preterm labor: A. H. Price, S. P. Clissold, *Drugs* **38**, 77-122 (1989); of toxicology: S. E. Libretto, *Arch. Toxicol.* **68**, 213-216 (1994). Review of clinical trials for add-on therapy to inhaled steroids: S. Shrewsbury *et al.*, *Br. Med. J.* **320**, 1368-1373 (2000).

Crystalline powder from ethanol-ethyl acetate or ethyl acetate-cyclohexane, mp 151° (Lunts); 157-158° (Collin). Sol in alcohol, most organic solvents. Sparingly sol in water.

Sulfate. [51022-70-9] Aerolin; Airomir; Asmasal; Asmaven; Buventol; Ecovent; Loftan; Proventil; Salamol; Salbumol; Sultanol; Torpex; Venetlin; Ventilastin; Ventodisks; Ventolin; Volmax. (C_{13}H$_{21}$NO$_3$)$_2$.H$_2$SO$_4$; mol wt 576.70. White crystalline powder. Sol in water; slightly sol in ethanol.

(*R*)-Form. [34391-04-3] Levalbuterol; levosalbutamol.

(*R*)-Form hydrochloride. [50293-90-8] Xopenex. $C_{13}H_{21}$-NO$_3$.HCl; mol wt 275.77. Crystallized as the monohydrate; changes crystalline form at 175°; dec 188-195°. $[\alpha]_D^{20}$ −32.2° (c = 0.1 in water).

THERAP CAT: Bronchodilator; tocolytic.

THERAP CAT (VET): Bronchodilator.

211. Albutoin. [830-89-7] 5-(2-Methylpropyl)-3-(2-propen-1-yl)-2-thioxo-4-imidazolidinone; 3-allyl-5-isobutyl-2-thiohydantoin; 3-allyl-5-*sec*-butyl-2-thiohydantoin; CO-ORD; Euprax. C_{10}-H$_{16}N_2OS$; mol wt 212.31. C 56.57%, H 7.60%, N 13.19%, O 7.54%, S 15.10%. Prepn: Oba *et al.*, *C.A.* **46**, 3885f (1952).

Crystals, mp 210-211°.

USE: Fog inhibitor in photography.

THERAP CAT: Anticonvulsant.

212. Alcian Blue. [12040-44-7] Family of copper phthalocyanine dyes substituted with multiple cationic groups to enhance water solubility. Typical structure contains 2-4 alkyl-substituted guanidinium groups attached to to the phthalocyanine ring via thioether linkages. Prepn: N. H. Haddock, C. Wood, **GB 586340**; *eidem*, **GB 587636** (both 1947 to ICI); N. H. Haddock, *Research* **1**, 685 (1948). Composition and use as histochemical stain: J. E. Scott *et al.*, *Histochemie* **4**, 73 (1964). Structural studies: *idem, ibid.* **30**, 215 (1972). Review of staining methods: M. Karlsson, S. Björnsson, *Methods Mol. Biol.* **171**, 159-173 (2001). Brief review: R. W. Horobin in *Conn's Biological Stains* (Biological Stain Commission, 10th Ed., 2002) pp 385-387.

Alcian blue 8G

Alcian blue 8G. [33864-99-2]; [75881-23-1] (unspecified structure). [[*N,N',N'',N'''*-[(29*H*,31*H*-Phthalocyanine-*C,C,C,C*-tetrayl-κ*N*²⁹,κ*N*³⁰,κ*N*³¹,κ*N*³²)tetrakis[methylenethio[(dimethylamino)methylidyne]]]tetrakis[*N*-methylmethanaminiumato]](2−)] copper(4+) chloride (1:4); C.I. Ingrain Blue 1; C.I. 74240; Alcian blue 8GS; Alcian blue 8GX. $C_{56}H_{68}Cl_4CuN_{16}S_4$; mol wt 1298.86. Mixture of geometric isomers. Dark blue powder. Abs max (water): 623.4 nm ($E_{1cm}^{1\%}$ 425). Sol in water.

Alcian blue 7G. [37370-49-3] C.I. Ingrain Blue 3. Structurally similiar to Alcian blue 8G with fewer cationic substituents.

USE: Biological stain for peptidoglycans and mucopolysaccharides; bacterial stain; greenish-blue dye for textiles.

213. **Alclometasone.** [67452-97-5] (7α,11β,16α)-7-Chloro-11,17,21-trihydroxy-16-methylpregna-1,4-diene-3,20-dione; 7α-chloro-16α-methylprednisolone. $C_{22}H_{29}ClO_5$; mol wt 408.92. C 64.62%, H 7.15%, Cl 8.67%, O 19.56%. Non-fluorinated corticosteroid with low systemic effects. Prepn: M. J. Green *et al.*, **US 4076708**; M. J. Green, H. J. Shue, **US 4124707** (both 1978 to Schering); M. J. Green *et al., J. Steroid Biochem.* **11**, 61 (1979); H. J. Shue *et al., J. Med. Chem.* **23**, 430 (1980). Topical anti-inflammatory activity of the 17,21-dipropionate: B. Lutsky *et al., Arzneim.-Forsch.* **29**, 992 (1979); M. J. Green *et al., ibid.* **30**, 1618 (1980).

Crystals from acetone/hexane, mp 176-179°. $[\alpha]_D^{26}$ +47.5° (c = 0.3 in DMF). uv max (methanol): 242 nm (ε 15500).

17,21-Dipropionate. [66734-13-2] Sch-22219; Aclosone; Aclovate; Almeta; Delonal; Legederm; Modrasone; Perderm. $C_{28}H_{37}ClO_7$; mol wt 521.05. Crystals from acetone/methanol/isopropyl ether, mp 212-216°. $[\alpha]_D^{26}$ +42.6° (c = 0.3 in DMF). uv max (methanol): 242 nm (ε 15600).

THERAP CAT: Anti-inflammatory (topical).

214. **Alcuronium.** [23214-96-2] (1*R*,3a*S*,10*S*,11a*S*,12*R*,-14a*S*,19a*S*,20b*S*,21*S*,22a*S*,23*E*,26*E*)-2,3,11,11a,13,14,22,22a-Octahydro-23,26-bis(2-hydroxyethylidene)-1,12-di-2-propen-1-yl-10*H*,-21*H*-1,21:10,12-diethano-9a*H*,20b*H*-dipyrrolo[3,2-*f*:3',2'-*f'*][1,5]-diazocino[3,2,1-*jk*:7,6,5-*j'k'*]dicarbazolium; 4,4'-didemethyl-4,4'-di-2-propenyltoxiferine I; *N,N'*-diallylnortoxiferinium; diallylbis-

(nortoxiferine); diallylnortoxiferine; diallyltoxiferine. $[C_{44}H_{50}N_4-O_2]^{2+}$. Prepn of dichloride and diiodide: Boller *et al.*, **US 3080373** (1963 to Hoffmann-La Roche).

Dichloride. [15180-03-7] *N,N'*-Diallylnortoxiferinium dichloride; Ro-4-3816; Alloferin. $C_{44}H_{50}Cl_2N_4O_2$; mol wt 737.81. Crystals from methanol or ethanol. Compd contains 5 moles of water of crystn after equilibration in air; $[\alpha]_D^{22}$ −348° (methanol). uv max (methanol): 292 nm (ε 43000).

Diiodide. $C_{44}H_{50}I_2N_4O_2$. Solid. uv max (methanol): 291 nm (ε 39000).

THERAP CAT: Neuromuscular blocking agent.

THERAP CAT (VET): Neuromuscular blocking agent.

215. **Aldicarb.** [116-06-3] 2-Methyl-2-(methylthio)propanal *O*-[(methylamino)carbonyl]oxime; 2-methyl-2-(methylthio)propionaldehyde *O*-(methylcarbamoyl)oxime; UC-21149; Temik. $C_7H_{14}N_2O_2S$; mol wt 190.26. C 44.19%, H 7.42%, N 14.72%, O 16.82%, S 16.85%. Acetylcholinesterase inhibitor. Prepn: L. K. Payne, Jr., M. H. J. Weiden, **FR 1377474**; *eidem*, **US 3217037** (1963, 1965 both to Union Carbide); L. K. Payne, Jr. *et al., J. Agric. Food Chem.* **14**, 356 (1966). Metabolism: W. J. Bartley *et al., ibid.* **18**, 446 (1970). Crystal and molecular structure: F. Takusagawa, R. A. Jacobson, *ibid.* **25**, 333 (1977). HPLC-APCI-MS determn in fruits and vegetables: G. S. Nunes *et al., J. Chromatogr. A* **888**, 113 (2000). Efficacy for control of nematodes on potato: S. L. Hafez *et al., Nematol. Mediterranea* **30**, 227 (2002); of thrips on cotton: K. H. Lohmeyer *et al., J. Econ. Entomol.* **96**, 748 (2003). Efficacy on citrus and economic analysis: L. Blakeley *et al., HortTechnology* **13**, 694 (2003). *Review:* R. R. Romine, *Anal. Methods Pestic. Plant Growth Regul.* **7**, 147-162 (1973). Review of persistence in soil and root crops: A. C. Maheshwari, V. S. Kavadia, *Pesticides* **19**, 30-31 (1985); of groundwater contamination: H. A. Moye, C. J. Miles, *Rev. Environ. Contam. Toxicol.* **105**, 99-146 (1988). Review of toxicology: R. L. Baron, T. L. Merriam, *ibid.* **105**, 1-70 (1988); *idem, Environ. Health Perspect.* **102**, Suppl. 11, 23-27 (1994).

Crystals from isopropyl ether, mp 99-100° with dec >100°. d_4^{25} 1.195. Vapor pressure (25°): 1×10^{-4} mmHg. Sparingly sol in water and certain organic solvents, most sol in chloroform and acetone. Percent soly: acetone 28 (10°), 43 (30°), 50 (50°); benzene 9 (10°), 24 (30°), 49 (50°); carbon tetrachloride 2 (10°), 5 (30°), 25 (50°); chloroform 38 (10°), 44 (30°), 50 (50°); methyl isobutyl ketone 13 (10°), 24 (30°), 42 (50°); toluene 10 (10°), 12 (30°), 33 (50°); water 0.4 (10°), 0.9 (30°), 1.4 (50°). LD_{50} in male and female rats (mg/kg): 0.81-0.93, 0.67-1.20 orally. LC_{50} (96 hr) in rainbow trout, bluegill sunfish (ppb): 560-580, 50-100 (Baron, 1988).

USE: Insecticide, acaricide, nematocide.

216. **Aldioxa.** [5579-81-7] [*N*-(2,5-Dioxo-4-imidazolidinyl)ureato]dihydroxyaluminum; (allantoinato)dihydroxyaluminum; dihydroxy[(4,5-dihydro-2-hydroxy-5-oxo-1*H*-imidazol-4-yl)ureato]aluminum; aluminum dihydroxy allantoinate; RC-172; Alaneto-rin; Alusa; Arlanto; Ascomp; Chlokale; Isalon; Nische; Peptilate. $C_4H_7AlN_4O_5$; mol wt 218.10. C 22.03%, H 3.24%, Al 12.37%, N 25.69%, O 36.68%. Prepn: S. B. Mecca, US 2761867 (1956 to Schuylkill Chem.). Pharmacology: R. Cahen, J. F. Clement, *Ann. Pharm. Fr.* **20**, 693 (1962); R Cahen, A. Pessonnier, *ibid.* 704. Toxicology: R. Cahen, A. Pessonnier, *ibid.* 623. Metabolism studies: K. Fukawa *et al.*, *Oyo Yakuri* **11**, 421 (1976), *C.A.* **88**, 130701g (1978). HPLC determn in plasma: K. Hirota *et al.*, *J. Chromatogr.* **277**, 165 (1983). Clinical studies of combination therapy for peptic ulcer: M. Sakai *et al.*, *Jpn. Arch. Intern. Med.* **31**, 275 (1984); *idem et al.*, *ibid.* **33**, 61 (1986); *idem et al.*, *ibid.* **36**, 357 (1989). Review of cosmetic use and safety assessment: S. N. J. Pang, *J. Am. Coll. Toxicol.* **12**, 237-242 (1993).

White powder. mp 230°. Insol in polar and nonpolar solvents.
USE: Astringent and skin conditioning agent in cosmetics.
THERAP CAT: Antiulcerative.

217. Aldol. [107-89-1] 3-Hydroxybutanal; 3-hydroxybutyr-aldehyde; acetaldol. $C_4H_8O_2$; mol wt 88.11. C 54.53%, H 9.15%, O 36.32%. Manuf by condensation of acetaldehyde in aq NaOH: Alhéritière, Gobron, US 2713598 (1955 to Usines de Melle). Toxicity study: H. F. Smyth *et al.*, *J. Ind. Hyg. Toxicol.* **31**, 60 (1949).

Colorless, thick liquid. *Poisonous.* d^{16} 1.109. bp_{20} 83°; dec about 85°. Miscible with water, alcohol, ether. LD_{50} orally in rats: 2.18 g/kg (Smyth).
Caution: Direct contact may cause irritation of skin, eyes, mucous membranes and upper respiratory tract.
USE: Manuf rubber vulcanizers, accelerators and age resisters; in perfumes; ore flotation.

218. Aldosterone. [52-39-1] (11β)-11,21-Dihydroxy-3,20-dioxopregn-4-en-18-al; 3,20-diketo-11β,18-oxido-4-pregnene-18,21-diol; Aldocorten. $C_{21}H_{28}O_5$; mol wt 360.45. C 69.98%, H 7.83%, O 22.19%. Adrenocortical steroid which exerts regulatory influence on metabolism of electrolytes and water. Isoln: Simpson *et al.*, *Experientia* **9**, 333 (1953); *Helv. Chim. Acta* **37**, 1163 (1954); Mattox *et al.*, *J. Am. Chem. Soc.* **75**, 4869 (1953); Harman *et al.*, *ibid.* **76**, 5035 (1954). Solutions contain an equilibrium mixture of the aldehyde and the hemiacetal, the equilibrium favoring the latter. Structure: Tait *et al.*, *Experientia* **10**, 132 (1954); *Helv. Chim. Acta* **37**, 1200 (1954). Crystal structure and molecular conformation: Duax, Hauptmann, *J. Am. Chem. Soc.* **94**, 5467 (1972). ^{13}C-NMR spectrum: P. Gerard, *Org. Magn. Reson.* **11**, 478 (1978). Total synthesis: Schmidlin *et al.*, *Helv. Chim. Acta* **40**, 1438 (1957); Johnson *et al.*, *J. Am. Chem. Soc.* **80**, 2585 (1958); **85**, 1409 (1963). Three-step synthesis from corticosterone: Barton, Beaton, *ibid.* **82**, 2640 (1960); **83**, 4083 (1961). Alternate synthesis: D. H. R. Barton *et al.*, *J. Chem. Soc. Perkin Trans. 1* **1975**, 2243; M. Miyano, *J. Org. Chem.* **46**, 1846 (1981). Biosynthesized in the zona glomerulosa and transported chiefly by albumin. In man, 400 μg secreted normally in one day. Secretion influenced by ACTH, growth hormone, plasma sodium and potassium, and the renin-angiotensin system. Causes reabsorption of Na^+, Cl^-, and HCO_3^- and diuresis of K^+. *Review:* L. F. Fieser, M. Fieser, *Steroids* (Reinhold, New York, 1959) pp 701-720.

Hydrated crystals from dilute acetone, mp 108-112° (when anhydr mp 164°). $[\alpha]_D^{23}$ +152.2° (anhydr; c = 2 in acetone). $[\alpha]_D^{25}$ +161° (c = 0.1 in chloroform). uv max: 240 nm (log ε 4.20 for the monohydrate; ε_{mol} 15,000 for the anhydr).
21-Acetate. $C_{23}H_{30}O_6$. Synthesis: Wettstein *et al.*; Jeger, US **3002972** and US **3014029** (both 1958 to Ciba). Flat needles from acetone + ether, mp 198-199°. $[\alpha]_D^{24}$ +121.7° (c = 0.71 in chloroform).
THERAP CAT: Mineralocorticoid.
THERAP CAT (VET): Mineralocorticoid.

219. Aldrin. [309-00-2] *rel*-(1*R*,4*S*,4a*S*,5*S*,8*R*,8a*R*)-1,2,3,-4,10,10-Hexachloro-1,4,4a,5,8,8a-hexahydro-1,4:5,8-dimethano-naphthalene; HHDN; compd 118; Octalene. $C_{12}H_8Cl_6$; mol wt 364.90. C 39.50%, H 2.21%, Cl 58.29%. Chlorinated hydrocarbon insecticide; readily metabolized to dieldrin, *q.v.* Activity: C. W. Kearns *et al.*, *J. Econ. Entomol.* **42**, 127 (1949). Prepn of aldrin and *endo,endo*-isomer: Lidov, US **2635977** (1953 to Shell). Alternate syntheses: Schmerling, US **2911447** (1959 to Universal Oil Prod.); Korte, Rechmeier, *Ann.* **656**, 131 (1962). Toxicity study: T. B. Gaines, *Toxicol. Appl. Pharmacol.* **14**, 515 (1969). Review of toxicology and human exposure: *Toxicological Profile for Aldrin/Dieldrin* (PB2003-100134, 2002) 354 pp; of chemistry and environmental fate: V. Zitko in *The Handbook of Environmental Chemistry*, H. Fiedler, Ed. (Springer-Verlag, Berlin, 2003) pp 47-90.

Relative stereochemistry

Crystals, mp 104°. Vapor pressure at 20°: 7.5×10^{-5} mm Hg. Very sol in most organic solvents. Practically insol in water. Stable in presence of organic and inorganic alkalies; stable to the action of hydrated metal chlorides. LD_{50} in male, female rats (mg/kg): 39, 60 orally (Gaines).
endo,endo-**Isomer.** Isodrin; compd 711. Crystals, mp 240-242°. LD_{50} in male, female rats (mg/kg): 15, 7.0 orally (Gaines).
Caution: Potential symptoms of overexposure by ingestion, inhalation, or skin absorption are headache, dizziness; nausea, vomitting, malaise; myoclinic limb jerks; clonic and tonic convulsions; coma; hematuria, azotemia. Potential occupational carcinogen. See *NIOSH Pocket Guide to Chemical Hazards* (DHHS/NIOSH 97-140, 2003) p 8; *Patty's Industrial Hygiene and Toxicology* **vol. 2B**, G. D. Clayton, F. E. Clayton, Eds. (Wiley-Interscience, New York, 4th ed., 1994) pp 1517-1523.
Note: Aldrin is listed as a persistent organic pollutant (POP) in Annex A of the *Stockholm Convention on Persistent Organic Pollutants* (United Nations, Stockholm, 2001) 43 pp; amended (Geneva, 2009) 63 pp.
USE: Formerly as insecticide for soil insects and termites.

220. Alefacept. [222535-22-0] 1-92-LFA-3 (antigen) (human) fusion protein with immunoglobulin G1 (human hinge-C_H2-C_H3 γ1-chain), dimer; human LFA-3-IgG$_1$ fusion protein; LFA3TIP; BG-9273; BG-9712; Amevive. Recombinant fusion protein consisting of the first extracellular domain of human leukocyte function-associated antigen-3 (LFA-3) fused to the hinge and C_H2 and C_H3 sequences of human IgG$_1$. Expressed in Chinese hamster ovary cells; secreted as a glycosylated dimer, mol wt ~110 kDa. Antagonist at the CD2 receptor on T-lymphocytes and NK cells; inhibits T-cell proliferation. Prepn: B. P. Wallner *et al.*, **EP 503648**; *eidem*, US **5547853** (1992, 1996 both to Biogen); and binding study: G. T. Miller *et al.*, *J. Exp. Med.* **178**, 211 (1993). Pharmacology and ELISA determn in serum: P. L. Chisholm *et al.*, *Ther. Immunol.* **1**, 205 (1994). Effect of glycosylation on pharmacokinetics: W. Meier *et al.*, *ibid.* **2**, 159 (1995). Mechanism of action study: F. Chamian *et al.*, *Proc. Natl. Acad. Sci. USA* **102**, 2075 (2005). Clinical trial in psoriasis: C. N. Ellis *et al.*, *N. Engl. J. Med.* **345**, 248 (2001). Safety profile: B. Goffe *et al.*, *Clin. Ther.* **27**, 1912 (2005). Review of clinical development in psoriasis: R. G. Langley *et al.*, *Expert Opin. Pharmacother.* **6**, 2327-2333 (2005).
THERAP CAT: Antipsoriatic; immunosuppressant.

221. Aleglitazar. [475479-34-6] (αS)-α-Methoxy-4-[2-(5-methyl-2-phenyl-4-oxazolyl)ethoxy]benzo[*b*]thiophene-7-propanoic acid; (*S*)-2-methoxy-3-[4-[2-(5-methyl-2-phenyloxazol-4-yl)ethoxy]benzo[*b*]thiophen-7-yl]propionic acid; R-1439. $C_{24}H_{23}NO_5S$; mol wt 437.51. C 65.89%, H 5.30%, N 3.20%, O 18.28%, S 7.33%. Dual peroxisome proliferator-activated receptor (PPAR)-α/γ agonist. Prepn: A. Binggeli *et al.*, **WO 02092084**; *eidem*, **US 6642389** (2002, 2003 both to Hoffmann-La Roche). Prepn, structure-activity analysis, and receptor binding affinities: A. Bénardeau *et al.*, *Bioorg. Med. Chem. Lett.* **19**, 2468 (2009). Clinical pharmacokinetics and tolerability: P. Sanwald-Ducray *et al.*, *Clin. Pharmacol. Ther.* **88**, 197 (2010). Clinical evaluation in type 2 diabetes: R. R. Henry *et al.*, *Lancet* **374**, 126 (2009). Review of pharmacology and clinical experience in type 2 diabetes: L. M. Younk *et al.*, *Expert Opin. Drug Metab. Toxicol.* **7**, 753-763 (2011).

White crystals, mp 146-147°.
THERAP CAT: Antidiabetic.

222. Alemtuzumab. [216503-57-0] Anti-(human CD52 (antigen)) immunoglobulin G1 (human-rat monoclonal CAMPATH-1H γ_1-chain) disulfide with human-rat monoclonal CAMPATH-1H light chain, dimer; Campath-1H; LDP-03; Campath; MabCampath. Humanized monoclonal antibody directed against the pan-lymphocyte antigen, CD52. Induces complement-mediated cytolysis resulting in lymphocyte depletion. Prepn: L. Riechmann *et al.*, *Nature* **332**, 323 (1988); H. Waldmann *et al.*, **WO 8907452** (1989 to Medical Research Council); *eidem*, **US 5846534** (1998 to British Technology Group). Structure-activity study: J. Greenwood *et al.*, *Eur. J. Immunol.* **23**, 1098 (1993). Crystal structure of the antibody combining site: G. M. T. Cheetham *et al.*, *J. Mol. Biol.* **284**, 85 (1998). Clinical evaluation in chronic lymphocytic leukemia: A. Osterborg *et al.*, *Br. J. Haematol.* **93**, 151 (1996); in renal transplant patients: R. Calne *et al.*, *Transplantation* **68**, 1613 (1999). Clinical trial in early multiple sclerosis: A. J. Coles *et al.*, *N. Engl. J. Med.* **359**, 1786 (2008). *Review*: F. J. Dumont, *Curr. Opin. Investig. Drugs* **2**, 139-160 (2001).
THERAP CAT: Antineoplastic; immunosuppressant.

223. Alendronic Acid. [66376-36-1] *P,P'*-(4-Amino-1-hydroxybutylidene)bisphosphonic acid; α-hydroxy-δ-aminobutylidenediphosphonic acid; 4-amino-1-hydroxybutane-1,1-diphosphonic acid; ABDP. $C_4H_{13}NO_7P_2$; mol wt 249.10. C 19.29%, H 5.26%, N 5.62%, O 44.96%, P 24.87%. Bisphosphonate antiresorptive agent. Prepn: M. I. Kabachnik *et al.*, *Bull. Acad. Sci. USSR* **27**, 374 (1978). *See also*: **BE 903519**; G. Staibano, **US 4705651** (1986, 1987 both to Gentili). Improved prepn: G. R. Kieczykowski *et al.*, *J. Org. Chem.* **60**, 8310 (1995). Pharmacokinetics: J. H. Lin *et al.*, *Drug Metab. Dispos.* **19**, 926 (1991). Mechanism of action: M. Sato *et al.*, *J. Clin. Invest.* **88**, 2095 (1991). Determn by HPLC in urine: W. F. Kline *et al.*, *J. Chromatogr.* **534**, 139 (1990); by IC in pharmaceutical formulations: E. W. Tsai *et al.*, *ibid.* **596**, 217 (1992). Clinical evaluation in malignant hypercalcaemia: R. Rizzoli *et al.*, *Int. J. Cancer* **50**, 706 (1992). Clinical trial in postmenopausal osteoporosis: U. A. Liberman *et al.*, *N. Engl. J. Med.* **333**, 1437 (1995); in prevention of bone loss in women: D. Hosking *et al.*, *ibid.* **338**, 485 (1998). Review of clinical experience on bone mineral density in men: A. M. Sawka *et al.*, *J. Clin. Densitom.* **8**, 7-13 (2005); of long-term treatment: M. C. Hochberg, R. Rizzoli, *Expert Opin. Pharmacother.* **7**, 1201-1210 (2006).

Fine white powder, mp 233-235° (dec). pK_1 2.72 ±0.05; pK_2 8.73 ±0.05; pK_3 10.5 ±0.1; pK_4 11.6 ±0.1 at 25°, (0.1M KCl).
Monosodium salt trihydrate. [121268-17-5] Alendronate sodium; MK-217; G-704650; Adronat; Alendros; Bonalon; Dronal; Fosamac; Fosamax; Onclast. $C_4H_{12}NNaO_7P_2.3H_2O$; mol wt 325.12. White, crystalline, nonhygroscopic powder, mp 257-262.5°. Sol in water; very slightly sol in alcohol. Practically insol in chloroform.
THERAP CAT: Bone resorption inhibitor.

224. Aletris. Star grass; starwort; true unicorn root; blazing star; colic root. Rhizome of *Aletris farinosa* L., *Liliaceae*. *Habit.* Eastern U.S., Ontario. *Constit.* Starch, diosgenin. Isoln of sapogenin: Marker *et al.*, *J. Am. Chem. Soc.* **62**, 2620 (1940). Pharmacological studies: Butler, Costello, *J. Am. Pharm. Assoc. Sci. Ed.* **33**, 177 (1944).
THERAP CAT: Antiflatulent.

225. Aleuritic Acid. [17941-34-3]; [6949-98-0] (unspecified stereo). *rel*-(9*R*,10*R*)-9,10,16-Trihydroxyhexadecanoic acid; (±)-*threo*-9,10,16-trihydroxypalmitic acid; β-aleuritic acid. $C_{16}H_{32}O_5$; mol wt 304.43. C 63.13%, H 10.60%, O 26.28%. One of the constituent acids of shellac, *q.v.*; occurs naturally as a mixture of the *threo*-isomers. Isoln from shellac: W. Nagel, *Ber.* **60**, 605 (1927); R. G. Khurana *et al.*, *Tetrahedron* **26**, 4167 (1970). Synthesis and configuration: D. E. Ames *et al.*, *J. Chem. Soc.* **C 1968**, 268. Crystal structure: S. M. Prasad, M. P. Gupta, *Indian J. Phys.* **49**, 72 (1975). Resolution of *threo* isomers: J. F. McGhie *et al.*, *Chem. Ind. (London)* **1971**, 1074. Conversion to *erythro*-form: A. N. Singh *et al.*, *Tetrahedron* **34**, 595 (1978). Manuf from seed lac: L. J. M. Rao *et al.*, *J. Sci. Ind. Res.* **56**, 164 (1997). FTIR studies: P. C. Sarkar, A. K. Shrivastava, *J. Indian Chem. Soc.* **75**, 326 (1998). Review of synthetic applications in perfumery: G. B. V. Subramanian, K. H. Bhushan, *Perfum. Flavor.* **18**, 41-44 (1993).

Relative stereochemistry

Crystals from dilute ethanol or distilled water, mp 100-101°. d 1.15. Moderately sol in water; sol in methanol, ethanol, isopropanol.
Methyl ester. [57491-54-0]; [30009-42-8] (unspecified stereo). $C_{17}H_{34}O_5$. Fine feathery needles, mp 72-73°. $bp_{0.1}$ 235°. Sol in methanol, ethanol, chloroform, acetone. Less sol in benzene. Insol in petr ether.
DL-*erythro*-Aleuritic acid. [533-87-9] *rel*-(9*R*,10*S*)-9,10,16-Trihydroxyhexadecanoic acid; α-aleuritic acid. Crystals from ethyl acetate, mp 126-127°.
USE: Starting material in synthesis of fragrance compounds including musks and pheromones; ingredient in skin care formulations.

226. Alexidine. [22573-93-9] *N,N''*-Bis(2-ethylhexyl)-3,12-diimino-2,4,11,13-tetraazatetradecanediimidamide; 1,1'-hexamethylenebis[5-(2-ethylhexyl)biguanide]; Win-21904; Sterwin 904; Bisguadine. $C_{26}H_{56}N_{10}$; mol wt 508.80. C 61.38%, H 11.09%, N 27.53%. Prepn from 1,1'-hexamethylenebis(3-cyanoguanide) and 2-ethylhexylamine hydrochloride: **FR 1463818** (1965 to Sterling Drug). Evaluation as antimicrobial agent: McNamara *et al.*, *J. Soc. Cosmet. Chem.* **16**, 499 (1965).

Dihydrochloride. Crystals from methanol + ether, mp 220.6-223.4°.
THERAP CAT: Antibacterial.

227. Alfaprostol. [74176-31-1] (5*Z*)-7-[(1*R*,2*S*,3*R*,5*S*)-2-[(3*S*)-5-Cyclohexyl-3-hydroxy-1-pentyn-1-yl]-3,5-dihydroxycyclopentyl]-5-heptenoic acid methyl ester; 18,19,20-trinor-17-cyclohexyl-13,14-didehydro-PGF$_{2\alpha}$ methyl ester; Ro-22-9000; K-11941; Alfavet; Gabbrostim. $C_{24}H_{38}O_5$; mol wt 406.56. C 70.90%, H 9.42%, O 19.68%. Synthetic analog of prostaglandin $F_{2\alpha}$, *q.v.* Synthesis: C. Gandolfi *et al.*, **DE 2539116**; *eidem*, **US 4035415** (1976, 1977 both to Carlo Erba). Efficacy in bovine estrus synchronization: E. Schilling *et al.*, *Theriogenology* **18**, 413 (1982). Treatment of infertility in cows caused by persistent corpus luteum: G. Maffeo *et al.*, *Prostaglandins* **25**, 541 (1983). Efficacy in increasing calving rate in beef heifers and cows: E. R. doValle *et al.*, *J. Anim. Sci.* **75**, 897 (1997).

THERAP CAT (VET): Estrus control.

228. Alfaxalone. [23930-19-0] $(3\alpha,5\alpha)$-3-Hydroxypregnane-11,20-dione; alphaxalone; GR-2/234; Alfaxan. $C_{21}H_{32}O_3$; mol wt 332.48. C 75.86%, H 9.70%, O 14.44%. Steroid anesthetic. Prepn: Nagata *et al.*, *Helv. Chim. Acta* **42**, 1399 (1959); Browne, Kirk, *J. Chem. Soc. C* **1969**, 1653; Davis *et al.*, **DE 2030402**; **ZA 7003861** (both 1971 to Glaxo), *C.A.* **75**, 20793n, 64114w (1971). Mass spectral data: Ende, Spiteller, *Monatsh. Chem.* **102**, 929 (1971). Review of pharmacology and clinical efficacy of mixture with alfadolone acetate: *Postgrad. Med. J.* **48**, Suppl. 2, 1-139 (1972). *See also:* Child *et al.*, *Br. J. Anaesth.* **43**, 2-24 (1971); *Lancet* **1**, 888 (1972). Pharmacology and toxicity data: M. I. Al-Khawashki *et al.*, *J. Egypt. Med. Assoc.* **62**, 191 (1979). Mechanism of action study: N. L. Harrison *et al.*, *J. Neurosci.* **7**, 604 (1987). Veterinary trial in cats: S. Zaki *et al.*, *Aust. Vet. J.* **87**, 82 (2009).

Colorless prisms from ether, mp 172-174°. $[\alpha]_D^{26}$ +113.4° (c = 1.2 in chloroform). LD_{50} i.v. in mice: 43 mg/kg (Al-Khawashki).

Alfadolone acetate. [23930-37-2] $(3\alpha,5\alpha)$-21-(Acetyloxy)-3-hydroxypregnane-11,20-dione; GR-2/1574. $C_{23}H_{34}O_5$; mol wt 390.52. Crystals from acetone-hexane, mp 175-177°. $[\alpha]_D^{26}$ +97° (c = 1.02 in chloroform).

Mixture with alfadolone acetate. [8067-82-1] Alphadione; alfadione; CT-1341; Althesin; Saffan. LD_{50} in mice (mg/kg): 140 i.p., 47 i.v. (Al-Khawashki).

THERAP CAT (VET): Anesthetic (intravenous).

229. Alfentanil. [71195-58-9] N-[1-[2-(4-Ethyl-4,5-dihydro-5-oxo-1H-tetrazol-1-yl)ethyl]-4-(methoxymethyl)-4-piperidinyl]-N-phenylpropanamide; N-[1-[2-(4-ethyl-5-oxo-2-tetrazolin-1-yl)ethyl]-4-(methoxymethyl)-4-piperidyl]propionanilide. $C_{21}H_{32}$-N_6O_3; mol wt 416.53. C 60.56%, H 7.74%, N 20.18%, O 11.52%. Synthetic opioid analgesic; derivative of fentanyl, *q.v.* Prepn: F. Janssens, **DE 2819873**; *idem*, **US 4167574** (1978, 1979 both to Janssen); F. Janssens *et al.*, *J. Med. Chem.* **29**, 2290 (1986). Pharmacology and acute toxicity: C. J. E. Niemegeers, P. A. J. Janssen, *Drug Dev. Res.* **1**, 83 (1981). LC-MS/MS determn in urine: M. Thevis *et al.*, *Eur. J. Mass Spectrom.* **11**, 419 (2005). Review of pharmacology and clinical efficacy: G. E. Larijani, M. E. Goldberg, *Clin. Pharm.* **6**, 275-282 (1987). Comparative review of clinical pharmacokinetics: J. Scholz *et al.*, *Clin. Pharmacokinet.* **31**, 275-292 (1996); and clinical uses: J. S. Willens, N. R. Myslinski, *Heart Lung* **22**, 239-251 (1993).

Hydrochloride. [69049-06-5]; [70879-28-6] (monohydrate). R-39209; Alfenta; Brevafen; Fentalim; Limifen; Rapifen. $C_{21}H_{32}N_6$-$O_3 \cdot HCl$; mol wt 452.98. White to almost white powder. Monohydrate, crystals from acetone, mp 138.4°. pKa 6.5. Partition coefficient (*n*-octanol/water): 128 (pH 7.4). Freely sol in methanol, ethanol, chloroform; sol in water; sparingly sol in acetone. LD_{50} in rats, dogs (mg/kg): 47.5, 20 i.v. (Niemegeers, Janssen).

Note: This is a controlled substance (opiate): **21 CFR, 1308.12**.

THERAP CAT: Analgesic; anesthesia adjunct.

THERAP CAT (VET): Analgesic.

230. Alfimeprase. [259074-76-5] 3-203-Fibrolase [3-serine] (*Agkistrodon contortrix contortrix* recombinant). Truncated form of fibrolase, *q.v.*, containing 201 amino acid residues; mol wt ~22.6 kDa. Directly acting fibrinolytic metalloproteinase; preferentially degrades the Aα chain of fibrinogen, *q.v.* Inhibited by serum α_2-macroglobulin. Prepn by recombinant technology in *Pichia pastoris*: T. C. Boone *et al.*, **WO 0125445**; *eidem*, **US 6261820** (both 2001 to Amgen). Disulfide structure: G. Jones *et al.*, *Protein Sci.* **10**, 1264 (2001). Pharmacology: C. F. Toombs, *Haemostasis* **31**, 141 (2001). Review of biochemistry and pharmacology: *idem*, *Fundam. Clin. Cardiol.* **46**, 627-650 (2003). Clinical evaluation for peripheral arterial thrombolysis: K. Ouriel *et al.*, *J. Vasc. Interv. Radiol.* **16**, 1075 (2005).

THERAP CAT: Thrombolytic.

231. Alfuzosin. [81403-80-7] N-[3-[(4-Amino-6,7-dimethoxy-2-quinazolinyl)methylamino]propyl]tetrahydro-2-furancarboxamide; N^1-(4-amino-6,7-dimethoxyquinazol-2-yl)-N^1-methyl-N^2-(tetrahydrofuroyl-2)-propylenediamine; SL-77.499. $C_{19}H_{27}N_5O_4$; mol wt 389.46. C 58.60%, H 6.99%, N 17.98%, O 16.43%. α_1-Adrenoceptor antagonist structurally similar to prazosin, *q.v.* Prepn: P. M. J. Manoury, **DE 2904445**; *idem*, **US 4315007** (1979, 1982 both to Synthelabo); and antihypertensive activity in rats: P. M. Manoury *et al.*, *J. Med. Chem.* **29**, 19 (1986). Pharmacology: A. G. Ramage, *Eur. J. Pharmacol.* **129**, 307 (1986). HPLC determn in biological fluids: P. Guinebault *et al.*, *J. Chromatogr.* **353**, 361 (1986). Pharmacology in humans: A. H. Deering, *Br. J. Clin. Pharmacol.* **25**, 417 (1988). Clinical evaluation in essential hypertension: S. Leto Di Priolo *et al.*, *Eur. J. Clin. Pharmacol.* **35**, 25 (1988); A. K. Ghosh, S. Ghosh, *Ger. Cardiovasc. Med.* **1**, 81 (1988). Clinical trial in benign prostatic hyperplasia (BPH): C. G. Roehrborn *et al.*, *BJU Int.* **92**, 257 (2003). Review of clinical experience in BPH: D. M. Weiner, F. C. Lowe, *Expert Opin. Pharmacother.* **4**, 2057-2063 (2003).

Hydrochloride. [81403-68-1] SL-77.499-10; Mittoval; Urion; UroXatral; Xatral. $C_{19}H_{27}N_5O_4 \cdot HCl$; mol wt 425.91. Crystals from ethanol + ether, mp 225° (Manoury, 1986), also reported earlier as mp 235° (dec) (Manoury, 1982). pKa 8.13. Slightly hygroscopic. Freely sol in water; sparingly sol in alcohol. Practically insol in methylene chloride.

THERAP CAT: Antihypertensive. In treatment of benign prostatic hypertrophy.

232. Algestone. [595-77-7] (16α)-16,17-Dihydroxypregn-4-ene-3,20-dione; 16α,17-dihydroxyprogesterone; 4-pregnene-16α,-17α-diol-3,20-dione; alphasone. $C_{21}H_{30}O_4$; mol wt 346.47. C

72.80%, H 8.73%, O 18.47%. Prepn from 16-dehydroprogesterone: Inhoffen *et al.*, *Ber.* **87**, 593 (1954); Cooley *et al.*, *J. Chem. Soc.* **1955**, 4373; Allen, Bernstein, *J. Am. Chem. Soc.* **78**, 1909 (1956); Hydorn *et al.*, *Steroids* **3**, 493 (1964). Manuf from 16-dehydroprogesterone: Colton, US 2727909 (1955 to Searle); Hydorn *et al.*, US 3165541 (1965 to Olin Mathieson); from its 16α-acetate: Diassi, US 3027384 (1962 to Olin Mathieson).

Needles from ethanol + dichloromethane, mp 225°. $[\alpha]_D^{22}$ +95° (c = 0.81 in CHCl$_3$). uv max: 240 nm (ε 16600).

Cyclic acetal with acetone. [4968-09-6] 16α,17-[(1-Methylethylidene)bis(oxy)]pregn-4-ene-3,20-dione; algestone acetonide; alphasone acetonide; 16α,17α-isopropylidenedioxyprogesterone. C$_{24}$H$_{34}$O$_4$. Prepn: Cooley *et al.*, *loc. cit;* Fried *et al.*, *Chem. Ind. (London)* **1961**, 465. Needles from aq ethanol, mp 210°. $[\alpha]_D^{20}$ +137° (c = 0.7 in CHCl$_3$).

16α-Methyl ether. 17-Hydroxy-16α-methoxypregn-4-ene-3,20-dione. C$_{22}$H$_{32}$O$_4$. Prepd from the free diol via 16α,17-dihydroxypregn-4-ene-3,20-dione cyclic borate: Fried, US 3006930 (1961 to Olin Mathieson). Crystals from 95% ethanol, mp 142-143°. $[\alpha]_D^{23}$ +60° (c = 0.15 in CHCl$_3$). uv max: 234 nm (ε 15400).

THERAP CAT: Acetonide as anti-inflammatory (topical).

233. Algestone Acetophenide. [24356-94-3] (16α)-16,17-[[(1R)-1-Phenylethylidene]bis(oxy)]pregn-4-ene-3,20-dione; 16α,17-dihydroxypregn-4-ene-3,20-dione cyclic acetal with acetophenone; 16α,17α-dihydroxyprogesterone acetophenide; alphasone acetophenide; P-DHP; SQ-15101. C$_{29}$H$_{36}$O$_4$; mol wt 448.60. C 77.65%, H 8.09%, O 14.27%. Progestogen. Prepn: J. Fried, US 2941997 (1960 to Olin Mathieson); J. Fried *et al.*, *Chem. Ind. (London)* **1961**, 465. Improved prepn: S. J. Bramcato *et al.*, US 3488347 (1970 to Smith Kline & French). In estrus synchronization: J. N. Wiltbank *et al.*, *J. Anim. Sci.* **26**, 764 (1967). Clinical evaluations of combination with estradiol enanthate as injectable contraceptive: R. Plesner, *Acta Endocrinol.* **61**, 494 (1969); *idem, ibid.* **65**, 683 (1970); R. Recio *et al.*, *Contraception* **33**, 579 (1986).

Crystals from 95% ethanol, mp 150-151°. $[\alpha]_D^{23}$ +51° (CHCl$_3$). Stable to boiling mineral acids; readily cleaved by warming with formic acid, with subsequent deformylation.

Combination with estradiol enanthate. Perlutal; Perlutan; Unalmes; Yectames.

THERAP CAT: Progestogen; in combination with estradiol enanthate as injectable contraceptive.

234. Algin. [9005-38-3] Alginic acid sodium salt; sodium alginate; sodium polymannuronate; Alto; Alman; Alloid; Allose. A gelling polysaccharide extracted from giant brown seaweed (giant kelp, *Macrocystis pyrifera* (L.) Ag., *Lessoniaceae)* or from horsetail kelp (*Laminaria digitata* (L.) Lamour, *Laminariaceae*) or from sugar kelp (*Laminaria saccharina* (L.) Lamour). Process of manuf: Tseng, *Chem. Metall. Eng.* **52**, 97 (1945); Mantell, *The Water-Soluble Gums* (New York, 1947); Green, US 2036934 (1936 to Kelco); Gloahec, Herter, US 2128551 (1938 to Algin Corp. of America). Wound healing properties and use in hemostatic dressings: J. H. M. Miller, GB 1328088 (1973 to Wallace, Cameron & Co.), *C.A.* **80**,

6974u (1974). Series of articles on hemostatic effects: *Yakugaku Zasshi* **101**, 452-469 (1981), *C.A.* **95**, 35654e, 35655f, 35405z (1981). Immunoadjuvant effect: G. H. Scherr, A. S. Markowitz, US 3075883 (1963 to Consolidated Labs.). Clinical comparison with alum immunoadjuvant: G. Bruno *et al.*, *Ann. Allergy* **56**, 384 (1986). Review of structural studies: D. A. Rees, E. J. Welsh, *Angew. Chem. Int. Ed.* **16**, 214 (1977). Review of production, properties and use in the food industry: A. Askar, *Alimenta* **21**, 165-169 (1982). *Reviews:* McNeely, Pettitt, in *Industrial Gums*, R. L. Whistler, Ed. (Academic Press, New York, 2nd ed., 1973) pp 49-81; I. W. Cottrell, J. K. Baird, "Gums" in *Kirk-Othmer Encyclopedia of Chemical Technology* **vol. 12** (Wiley-Interscience, New York, 3rd ed., 1980) pp 48-51.

Cream-colored powder. Sol in water, forming a viscous, colloidal soln. Insol in alcohol and in hydro-alcoholic solns in which the alcohol content is >30% w/w. Insol in chloroform, ether, in aq acid solns when the pH is below 3.

USE: In the manufacture of ice cream where it serves as a stabilizing colloid, insuring creamy texture and preventing the growth of ice crystals. In drilling muds; in coatings; in the flocculation of solids in water treatment; as sizing agent; thickener; emulsion stabilizer; suspending agent in soft drinks; in dental impression preparations. Pharmaceutic aid (suspending agent).

THERAP CAT: Hemostatic.

235. Alginic Acid. [9005-32-7] Norgine; polymannuronic acid; Sazio. Mol wt about 240,000. A hydrophilic, colloidal polysaccharide obtained from seaweeds which, in the form of mixed salts of calcium, magnesium, and other bases, occurs as a structural component of the cell wall. Isoln from fronds of *Laminaria digitata* (L.) Edmonson, *Laminariaceae:* Stanford, *J. Chem. Soc.* **44**, 943 (1883); Bird, Haas, *Biochem. J.* **25**, 403 (1931); from *Macrocystis pyrifera* (L.) C. Ag., *Lessoniaceae:* Nelson, Cretcher, *J. Am. Chem. Soc.* **51**, 1914 (1929). Polysaccharides similar to alginic acid are also secreted by the bacteria *Pseudomonas aeruginosa* and *Azotobacter vinelandii:* G. H. Cohen, D. B. Johnstone, *J. Bacteriol.* **88**, 329 (1964); L. R. Evans, A. Linker, *ibid.* **116**, 915 (1973). Alginic acid is a linear polymer of β-(1 → 4)-D-mannosyluronic acid and α-(1 → 4)-L-gulosyluronic acid residues, the relative proportions of which vary with the botanical source and state of maturation of the plant. Clinical use of combination with antacid in gastric reflux esophagitis: D. Y. Graham *et al.*, *Curr. Ther. Res.* **22**, 653 (1977). Clinical evaluation of calcium salt as hemostatic dressing: T. Gilchrist, A. M. Martin, *Biomaterials* **4**, 317 (1983); A. R. Groves, J. C. Lawrence, *Ann. R. Coll. Surg. Engl.* **68**, 27 (1986). Review of structure studies and of secondary and tertiary structure in solutions and gels: D. A. Rees, E. J. Welsh, *Angew. Chem. Int. Ed.* **16**, 214 (1977). Review of production, properties and use in the food industry: A. Askar, *Alimenta* **21**, 165-169 (1982).

White to yellowish white, fibrous powder. Capable of absorbing 200-300 times its weight of water and salts to the extent of 60%. Resists hydrolysis. pH of a 3 in 100 suspension in water is between 2.0 and 3.4. Sol in alkaline solns. Insol in water, organic solvents.

Calcium salt. [9005-35-0] Sorbsan.

Sodium salt. *See* Algin.

USE: Sizing paper and textiles; as binder for briquettes; manuf artificial horn, ivory, celluloid; emulsionizing mineral oils; mucilage. Pharmaceutic aid (suspending and viscocity-increasing agent; tablet binder and disintegrant). Additional uses are described under Algin.

THERAP CAT: Calcium salt as hemostatic.

236. Alibendol. [26750-81-2] 2-Hydroxy-N-(2-hydroxyethyl)-3-methoxy-5-(2-propen-1-yl)benzamide; 5-allyl-2-hydroxy-N-(2-hydroxyethyl)-m-anisamide; 2-hydroxy-3-methoxy-5-allyl-N-(β-hydroxyethyl)benzamide; EB-1856; FC-54; H-3774; Cebera. C$_{13}$H$_{17}$NO$_4$; mol wt 251.28. C 62.14%, H 6.82%, N 5.57%, O 25.47%. Amide analog of eugenol, *q.v.* Prepn: FR 1584715; F. Clémence, O. Le Martret, US 3668238 (1967, 1972 both to Roussel). Synthesis and pharmacologic activity: F. Clémence *et al.*, *Chim. Ther.* **5**, 188 (1970).

Crystals from benzene, mp 95°. uv max (ethanol): 316, 218 nm. LD_{50} in Swiss male mice (mg/kg): >3000 orally; >2000 s.c.; 209 i.p.; 217 i.v. (Clémence).

THERAP CAT: Choleretic; antispasmodic.

237. Alicaforsen. [185229-68-9] d[(R)-P-Thio](G-C-C-C-A-A-G-C-T-G-G-C-A-T-C-C-G-T-C-A)DNA. Synthetic 20 base phosphorothioate antisense oligonucleotide designed to inhibit expression of intercellular adhesion molecule-1 (ICAM-1), a key mediator of inflammation. Prepn: C. F. Bennett, C. K. Mirabelli, **WO 9405333**; *eidem*, **US 5514788** (1994, 1996 both to Isis). Purifn and scale up: R. R. Deshmukh *et al.*, *Nucleosides Nucleotides Nucleic Acids* **20**, 567 (2001). ICAM-1 inhibition in cell culture: C. F. Bennett *et al.*, *J. Immunol.* **152**, 3530 (1994). Clinical pharmacokinetics: J. M. Glover *et al.*, *J. Pharmacol. Exp. Ther.* **282**, 1173 (1997). ELISA determn in plasma: R. Z. Yu *et al.*, *Anal. Biochem.* **304**, 19 (2002). Clinical evaluation in Crohn's disease: B. R. Yacyshyn *et al.*, *Gut* **51**, 30 (2002); in ulcerative colitis: S. J. H. van Deventer *et al.*, *ibid.* **53**, 1646 (2004). Review of toxicology: S. P. Henry *et al.*, *Toxicol. Pathol.* **27**, 95-100 (1999); A. A. Levin, *Biochim. Biophys. Acta* **1489**, 69-84 (1999); of pharmacology and clinical development: W. R. Shanahan, Jr., *Expert Opin. Invest. Drugs* **8**, 1417-1429 (1999); C. F. Barish, *Expert Opin. Biol. Ther.* **5**, 1387-1391 (2005).

Nonadecasodium salt. [331257-52-4] Alicaforsen sodium; ISIS-2302. $C_{192}H_{225}N_{75}Na_{19}O_{98}P_{19}S_{19}$; mol wt 6785.79.

THERAP CAT: Anti-inflammatory.

238. Alinidine. [33178-86-8] N-(2,6-Dichlorophenyl)-4,5-dihydro-N-2-propen-1-yl-1H-imidazol-2-amine; 2-(N-allyl-2,6-dichloroanilino)-2-imidazoline; 2-[N-allyl-N-(2,6-dichlorophenyl)-amino]-2-imidazoline; ST-567. $C_{12}H_{13}Cl_2N_3$; mol wt 270.16. C 53.35%, H 4.85%, Cl 26.24%, N 15.55%. Specific bradycardic agent; analog of clonidine, *q.v.* Prepn: H. Stähle *et al.*, **DE 1958201**; *eidem*, **US 3708485** (1971, 1973 both to Boehringer, Ing.); *eidem*, *J. Med. Chem.* **23**, 1217 (1980). Clinical pharmacology: D. W. G. Harron *et al.*, *J. Cardiovasc. Pharmacol.* **4**, 213 (1982). Mode of action study: J. S. Millar, E. M. Vaughan Williams, *Lancet* **1**, 1291 (1981). HPLC determn in human plasma: U.-W. Wiegand *et al.*, *J. Chromatogr.* **223**, 238 (1981). Clinical pharmacokinetics: *eidem*, *J. Cardiovasc. Pharmacol.* **4**, 59 (1982). Hemodynamic effects in angina or infarction: M. L. Simoons, P. G. Hugenholtz, *Eur. Heart J.* **5**, 227 (1984). Review of pharmacology and potential therapeutic uses: D. W. G. Harron, R. G. Shanks, *ibid.* **6**, 722-729 (1985).

Crystals, mp 130-131° (Stähle, 1973); also reported as mp 127-129° (Stähle, 1980). pKa 10.42.

Hydrobromide. $C_{12}H_{14}BrCl_2N_3$. Crystals from methanol + water, mp 193-194°.

239. Aliskiren. [173334-57-1] (αS,γS,δS,ζS)-δ-Amino-N-(3-amino-2,2-dimethyl-3-oxopropyl)-γ-hydroxy-4-methoxy-3-(3-methoxypropoxy)-α,ζ-bis(1-methylethyl)benzeneoctanamide; (2S,4S,5S,7S)-5-amino-N-(2-carbamoyl-2-methylpropyl)-4-hydroxy-2-isopropyl-7-[4-methoxy-3-(3-methoxypropoxy)benzyl]-8-methylnonanamide; 5S-amino-4S-hydroxy-2S,7S-diisopropyl-8-[4-methoxy-3-(3-methoxypropyloxy)phenyl]octanoic acid N-(2-carbamoyl-2,2-dimethylethyl)amide; CGP-60536. $C_{30}H_{53}N_3O_6$; mol wt 551.77. C 65.30%, H 9.68%, N 7.62%, O 17.40%. Orally active, synthetic nonpeptide renin inhibitor. Prepn: R. Göschke *et al.*, **EP 678503**; *eidem*, **US 5559111** (1995, 1996 both to Ciba-Geigy); H, Rüeger *et al.*, *Tetrahedron Lett.* **41**, 10085 (2000). Synthesis: A. Dondoni *et al.*, *ibid.* **42**, 4819 (2001). Determn by HPLC in biological fluids: G. Lefevre, S. Gauron, *J. Chromatogr. B* **738**, 129 (2000); by RIA in plasma: G. Lefevre *et al.*, *J. Immunoassay* **21**, 65 (2000). Structure activity study: J. M. Wood *et al.*, *Biochem. Biophys. Res. Commun.* **308**, 698 (2003). Clinical pharmacology: J. Nussberger *et al.*, *Hypertension* **39**, e1 (2002). Clinical study: A. H. Gradman *et al.*, *Circulation* **111**, 1012 (2005). Review of devel-

opment and clinical experience: E. O'Brien, *Expert Opin. Invest. Drugs* **15**, 1269-1277 (2006).

pKa 9.49. Log P (octanol/water): 2.45 (pH 7.4). Soly in water: >350 mg/ml (pH 7.4).

Hemifumarate. [173334-58-2] CGP-60536B; SPP-100; Rasilez; Tekturna. $2C_{30}H_{53}N_3O_6 \cdot C_4H_4O_4$; mol wt 1219.61. White to slightly yellowish crystalline powder. Sol in *n*-octanol; highly sol in water.

THERAP CAT: Antihypertensive.

240. Alitame. [80863-62-3] L-α-Aspartyl-N-(2,2,4,4-tetramethyl-3-thietanyl)-D-alaninamide; L-aspartyl-D-alanine-N-(2,2,4,4-tetramethylthietan-3-yl)amide; 3-(L-aspartyl-D-alaninamido)-2,2,-4,4-tetramethylthietane; CP-54802; Aclame. $C_{14}H_{25}N_3O_4S$; mol wt 331.43. C 50.74%, H 7.60%, N 12.68%, O 19.31%, S 9.67%. Dipeptide amide reported to be approx 2000 times sweeter than sucrose. Prepn: T. M. Brennan, M. E. Hendrick, **EP 34876**; *eidem*, **US 4411925** (1981, 1983 both to Pfizer). Conformational analysis: R. D. Feinstein *et al.*, *J. Am. Chem. Soc.* **113**, 3467 (1991). Review of development and uses: R. C. Glowaky *et al.*, *ACS Symp. Ser.* **450**, 57-67 (1991); of chemical and physical properties: M. E. Hendrick *et al.*, in *Advances in Sweeteners*, T. H. Grenby, Ed. (Blackie, Glasgow, 1996) pp 226-239; M. H. Auerbach *et al.*, in *Alternative Sweeteners*, L. O'Brien Nabors, Ed. (Marcel Dekker, Inc., New York, 2001) pp 31-40.

White, crystalline, non-hygroscopic powder, mp 136-147°. Faint, characteristic odor; intensely sweet taste. $[\alpha]_D^{25}$ +40 to +50° (c = 1 in water). pI 5.6. Soly at 25° (% w/v): water (pH 5.6) 13.1; methanol 41.9; ethanol 61.0; propylene glycol 53.7; chloroform 0.02; *n*-heptane 0.001.

USE: Non-nutritive sweetener.

241. Alitretinoin. [5300-03-8] 9-*cis*-Retinoic acid; 6-*cis*-retinoic acid; ALRT-1057; LGD-1057; Panretin; Toctino. $C_{20}H_{28}O_2$; mol wt 300.44. C 79.96%, H 9.39%, O 10.65%. Naturally occurring derivative of vitamin A, *q.v.* Morphogenic agent that modulates cell growth and differentiation; activates both retinoic acid (RAR) and retinoid X (RXR) receptors. Prepn: C. D. Robeson *et al.*, *J. Am. Chem. Soc.* **77**, 4111 (1955); M. Matsui *et al.*, *J. Vitaminol.* **4**, 178 (1958); M. F. Boehm *et al.*, *J. Med. Chem.* **37**, 408 (1994). Identification as physiological ligand of RXR: R. A. Heyman *et al.*, *Cell* **68**, 397 (1992); A. A. Levin *et al.*, *Nature* **355**, 359 (1992). Role in morphogenesis: C. Thaller *et al.*, *Development* **118**, 957 (1993). Biosynthesis from all-*trans*-retinoic acid: J. Urbach, R. R. Rando, *Biochem. J.* **299**, 459 (1994). HPLC determn in plasma: A. M. Dzerk *et al.*, *J. Pharm. Biomed. Anal.* **16**, 1013 (1998). Clinical pharmacokinetics: C. Weber, E. Dumont, *J. Clin. Pharmacol.* **37**, 566 (1997). Clinical evaluation in acute promyelocytic leukemia: S. L. Soignet *et al.*, *Leukemia* **12**, 1518 (1998); in advanced cancers: N. A. Rizvi *et al.*, *Clin. Cancer Res.* **4**, 1437 (1998); in chronic refractory eczema: T. Ruzicka *et al.*, *Br. J. Dermatol.* **158**, 808 (2008). Review of role as hormone: B. F. Tate *et al.*, *Trends Endocrinol. Metab.* **5**, 189-194 (1994); of clinical development: *J. Cell. Biochem. Suppl.* **26**, 158-167 (1996); of therapeutic use: C. Cheng *et al.*, *Expert Opin. Invest. Drugs* **17**, 437-443 (2008).

Yellow fine needles from ethanol, mp 190-191°. uv max (methanol): 343 nm (ε 39000).

THERAP CAT: Antineoplastic; in treatment of atopic eczema.

242. Alizapride. [59338-93-1] 6-Methoxy-N-[[1-(2-propen-1-yl)-2-pyrrolidinyl]methyl]-1H-benzotriazole-5-carboxamide; N-[(1-allyl-2-pyrrolidinyl)methyl]-6-methoxy-1H-benzotriazole-5-carboxamide. $C_{16}H_{21}N_5O_2$; mol wt 315.38. C 60.93%, H 6.71%, N 22.21%, O 10.15%. Prepn: **BE 825605**; G. Bulteau *et al.*, **US 4039672** (1975, 1977 both to Soc. d'Etudes Sci. Ind. de l'Ile-de-France). Series of articles on pharmacology and clinical studies in chemotherapy induced emesis: *Sem. Hop.* **58**, 323-374 (1982).

Cryst from acetone, mp 139°. LD_{50} i.v. in mice (5 days): 92.7 mg/kg (Bulteau).

Hydrochloride. [59338-87-3] Nausilen; Plitican; Vergentan. $C_{16}H_{21}N_5O_2$·HCl; mol wt 351.84. Cryst from methanol/methyl ethyl ketone, mp 206-208°.

THERAP CAT: Antiemetic.

243. Alizarin. [72-48-0] 1,2-Dihydroxy-9,10-anthracenedione; 1,2-dihydroxyanthraquinone; C.I. Mordant Red 11; C.I. Pigment Red 83; C.I. 58000. $C_{14}H_8O_4$; mol wt 240.21. C 70.00%, H 3.36%, O 26.64%. Occurs in the root of the madder plant (*Rubia tinctorum* L., *Rubiaceae;* Krappwurzel) in combination with 2 mols glucose, called ruberythric acid. Was known and used in ancient Egypt, Persia, and India. Synthesized from 2-anthraquinonesulfonic acid sodium salt: Caro *et al.*, *Ber.* **3**, 359 (1870); Perkin, *Ber.* **9**, 281 (1876). Historical review: Fieser, *J. Chem. Educ.* **7**, 2609 (1930). Laboratory prepn: Gattermann-Wieland, *Laboratory Methods of Organic Chemistry* (New York, 1937). Modern methods of manufacture: Pohl, *Ullmanns Encyklopädie der technischen Chemie* vol. **I**, p 200; Fierz-David and Blangey, *Grundlegende Operationen der Farbenchemie* (Vienna, 5th ed., 1943). *See also Colour Index* vol. **4**, (3rd ed., 1971) p 4513.

Orthorhombic, orange needles by sublimation or from abs alc. Solvated scales from dil alc or by evaporation from ether. Sublimes at 110° (2 mm Hg). mp 290°. bp 430°. Absorption spectrum: Moir, *J. Chem. Soc.* **1927**, 1810. Soly in water at 18°: 2.1×10^{-6} mols/l; at 25°: 2.5×10^{-6} mols/l. Sol in 300 parts boiling water; moderately sol in alcohol, freely in hot methanol and in ether at 25°. Also sol in benzene, toluene, xylene, pyridine, carbon disulfide, glacial acetic acid. Sol in water solns of alkalies with blue color, but without fluorescence. Fluorescent solns indicate unchanged 2-anthraquinone sodium sulfonate.

1-Methyl ether. $C_{15}H_{10}O_4$. Orange needles with $1H_2O$ from dil methanol. When dried at 100° mp 179°.

2-Methyl ether. $C_{15}H_{10}O_4$. Orange needles from alcohol, mp 231°.

Dimethyl ether. $C_{16}H_{12}O_4$. Golden-yellow needles from alcohol, mp 215°.

USE: In the manufacture of acid and chrome dyes for wool; acid-base indicator (in 0.5% alcoholic soln; pH: yellow 5.5, red 6.8); in spot tests as reagent for aluminum, indium, mercury, zinc, and zirconium; biological stain.

244. Alizarin Cyanine Green F. [4403-90-1] 2,2'-[(9,10-Dihydro-9,10-dioxo-1,4-anthracenediyl)diimino]bis(5-methylbenzenesulfonic acid) sodium salt (1:2); 6,6'-(1,4-anthraquinonylenediimino)di-*m*-toluenesulfonic acid disodium salt; D & C Green No. 5; C.I. Acid Green 25; C.I. 61570. $C_{28}H_{20}N_2Na_2O_8S_2$; mol wt 622.57. C 54.02%, H 3.24%, N 4.50%, Na 7.39%, O 20.56%, S 10.30%. Discovered by R. E. Schmidt in 1894: *Colour Index* vol. 4 (3rd ed., 1971) p 4541.

Green powder. Slightly sol in acetone, alc, pyridine. Insol in chloroform, toluene. Dull blue soln in conc H_2SO_4, turning turquoise on dilution. Absorption spectra: C. F. H. Allen *et al.*, *J. Org. Chem.* **7**, 63, 169 (1942).

USE: For nylon sutures: *Fed. Regist.* **42**, 52395 (1977). Permitted for use in drugs and cosmetics, excluding use in eye area: *ibid.* **47**, 49628 (1982).

245. Alizarine Blue. [568-02-5] 5,6-Dihydroxynaphtho-[2,3-*f*]quinoline-7,12-dione; 7,8-dihydroxy-5,6-phthalylquinoline; Alizarin Blue R; C.I. 67410. $C_{17}H_9NO_4$; mol wt 291.26. C 70.10%, H 3.11%, N 4.81%, O 21.97%. Prepn from 3-nitroalizarin, glycerol, and concd sulfuric acid: Auerbach, *J. Chem. Soc.* **35**, 799 (1879); *Colour Index* vol. 4 (3rd ed., 1971) p 4567.

Lustrous brownish violet needles from benzene, mp 268-270°. Practically insol in water; sparingly sol in alc, ether; slightly sol in cold benzene; sol in amyl alcohol, glacial acetic acid, hot benzene.

USE: As indicator in saturated alcoholic soln. pH: pink 0.0 to yellow 1.6; yellow 6.0 to green 7.6.

246. Alizarine Orange. [568-93-4] 1,2-Dihydroxy-3-nitro-9,10-anthracenedione; 1,2-dihydroxy-3-nitroanthraquinone; 3-nitroalizarin; C.I. Mordant Orange 14; C.I. 58015. $C_{14}H_7NO_6$; mol wt 285.21. C 58.96%, H 2.47%, N 4.91%, O 33.66%. From 3-bromoalizarin in acetic acid with nitric acid: Barnett, Cook, *J. Chem. Soc.* **121**, 1376 (1922). Additional prepns: *Colour Index* vol. 4 (3rd ed., 1971) p 4514.

Orange needles or plates from acetic acid, dec 244°. Sublimes with partial decompn. Yellow in organic solvents, purple-red in dil aq alkali, orange in H_2SO_4.

USE: Dyes cloth mordanted with Al orange, with Fe red to violet. As indicator in satd alc soln. pH 2.0-4.0, color change from golden orange to flat yellow (in water). pH 5.0-6.5 from yellow to purplish red.

247. Alizarine Yellow R. [2243-76-7] 2-Hydroxy-5-[2-(4-nitrophenyl)diazenyl]benzoic acid; 5-(p-nitrophenylazo)salicylic acid; p-nitrobenzeneazosalicylic acid; Alizarine Yellow RW; C.I. Mordant Orange 1; C.I. 14030. $C_{13}H_9N_3O_5$; mol wt 287.23. C 54.36%, H 3.16%, N 14.63%, O 27.85%. Prepd by coupling diazotized p-nitroaniline with salicylic acid: Armento, **US 2746955** (1956 to General Aniline and Film); *Colour Index* vol. **4** (3rd ed., 1971) p 4058.

Orange-brown needles from dil glacial acetic acid, dec 253-254°. Sol in water, alc; slightly sol in acetone, Cellosolve; practically insol in other org solvents.

Sodium salt. Mordant Yellow 3R. $C_{13}H_8N_3NaO_5$. Brownish yellow powder. Sol in water.

USE: Sodium salt used as indicator in 0.1% aq soln pH: yellow to red, range 10.2 to 12.0.

248. Alkanet. Alkanna; orcanette; dyer's alkanet; anchusa; orkanet. The root of *Alkanna tinctoria* Tausch, *Boraginaceae*. *Habit.* Asia Minor, Hungary, Greece, Mediterranean region. *Constit.* Alkannin (the coloring principle) and tannin. Extraction of roots: Betrabet, Chakravarti, *J. Indian Inst. Sci.* **16A**, 51 (1933); *C.A.* **28**, 1038 (1934); Majamdar, Chakravarti, *J. Indian Chem. Soc.* **17**, 272 (1940); *C.A.* **34**, 6262 (1940).

Alkannin paper. Anchusin paper; Boettger's paper. White paper impregnated with a 1% alc tincture of alkanet root and dried. A blue paper can be made from the red paper by treating the latter with a 1% Na_2CO_3 soln. This paper acts nearly like litmus paper. Alkalies = green or blue; acids = red. pH: 8.0 red; 10.0 blue.

USE: For coloring wines, cosmetics, confectionery. Alkannin paper as indicator.

THERAP CAT: Astringent.

249. Alkannin. [517-88-4] 5,8-Dihydroxy-2-[(1S)-1-hydroxy-4-methyl-3-penten-1-yl]-1,4-naphthalenedione; (−)-5,8-dihydroxy-2-(1-hydroxy-4-methyl-3-pentenyl)-1,4-naphthoquinone; anchusa acid; anchusin; alkanna red; alkanet extract; (1-hydroxy-3-isohexenyl)naphthazarine; 2-(1-hydroxy-4-methyl-3-pentenyl)-5,8-dihydroxy-1,4-naphthoquinone; C.I. Natural Red 20; C.I. 75530. $C_{16}H_{16}O_5$; mol wt 288.30. C 66.66%, H 5.59%, O 27.75%. Isoln from the root of *Alkanna tinctoria* Tausch, *Boraginaceae:* Brockmann, *Ann.* **521**, 1 (1936); Toribara, Underwood, *Anal. Chem.* **21**, 1352 (1949). Absolute configuration: Arakawa, Nakagaki, *Chem. Ind. (London)* **1961**, 947. Toxicity study: L. Majlathova, *Nahrung* **15**, 505 (1971); *C.A.* **76**, 122513j (1972).

Brownish-red prisms with a metallic sheen from benzene, mp 149°. Can be sublimed in high vac at 140-150°. $[\alpha]_{Cd}^{20}$ − 165° (benzene); −226° (chloroform). Also reported as −254 ± 7° (chloroform) (Toribara). Sol in organic solvents, sparingly sol in water. Buffered aq solns are red at pH 6.1; purple at pH 8.8; blue at pH 10.0. LD_{50} in male, female mice, rats (g/kg): 3.0 ± 1.0; 3.1 ± 0.1; >1.0 orally (Majlathova).

(+)-Form. Shikonin.

(±)-Form. Shikalkin.

USE: Red dye for cosmetics and food; spectrophotometric microdetermination of beryllium.

THERAP CAT: Astringent.

250. Allantoin. [97-59-6] N-(2,5-Dioxo-4-imidazolidinyl)-urea; 5-ureidohydantoin; glyoxyldiureide; cordianine. $C_4H_6N_4O_3$; mol wt 158.12. C 30.38%, H 3.82%, N 35.43%, O 30.35%. Major nitrogen storage and transport compound in a number of plant families. Product of purine metabolism; enzymatically converted from uric acid in most mammals with the exception of apes and humans. Occurs naturally as the S-(+)-form. Isoln from leaves of *Platanus orientalis*: R. Fosse *et al.*, *Compt. Rend.* **198**, 1953 (1934); from calf urine: P. E. Thomas, P. de Graeve, *ibid.*, 2205. Prepd synthetically by the oxidation of uric acid with alkaline potassium permanganate: *Org. Synth.* **coll. vol. II**, 23 (1943); by heating urea with dichloroacetic acid: C. N. Zellner, J. R. Stevens, **US 2158098** (1939 to Merck & Co.); from glyoxylic acid and urea using solid acid catalyst: C. Cativiela *et al.*, *Green Chem.* **5**, 275 (2003). Absolute configuration: E. J. 's-Gravenmade *et al.*, *Recl. Trav. Chim. Pays-Bas* **88**, 929 (1969). HPLC determn in urine: K. M. Kim *et al.*, *J. Chromatogr. B* **877**, 65 (2009); in cosmetics: T. Doi *et al.*, *ibid.*, 1005. Review of role in nitrogen metabolism in plants: R. G. Winkler *et al.*, *Trends Biochem. Sci.* **13**, 97-100 (1988). Phylogenetic comparison of degradation pathway: I. Ramazzina *et al.*, *Nat. Chem. Biol.* **2**, 144 (2006).

S-(+)-Form

Monoclinic plates or prisms from water, mp 238°. One gram dissolves in 190 ml water, 500 ml alcohol; more sol in hot water and hot alcohol. Almost insol in ether. pH of satd water soln: 5.5.

S-(+)-Form. [3844-67-5] Crystals. $[\alpha]_D^{22}$ +93° (water).

USE: Skin protectant in cosmetic and personal care formulations.

THERAP CAT: Vulnerary.

THERAP CAT (VET): Vulnerary; debriding agent.

251. Allenolic Acid. [553-39-9] 6-Hydroxy-2-naphthalenepropanoic acid; 2-hydroxy-6-naphthalenepropionic acid; amphihydroxynaphthyl-β-propionic acid. $C_{13}H_{12}O_3$; mol wt 216.24. C 72.21%, H 5.59%, O 22.20%. Prepn from the corresponding methoxy compd: Jacques, Horeau, *Bull. Soc. Chim. Fr.* **1948**, 714.

Crystals from aq methanol, mp 180-181°. Sol in alcohol, methanol, pyridine.

USE: In biochemical research, in the prepn of estrogenic compds.

252. Allethrins. Allyl cinerins. Synthetic analogues of the naturally occurring insecticides cinerins, jasmolins, and pyrethrins, *q.q.v.* *Review:* Barthel, *World Rev. Pest Control* **6**, 59 (1967); World Health Org., *Environ. Health Criter.* **87**, 1-75 (1989). Review of toxicology and human exposure: *Toxicological Profile for Pyrethrins and Pyrethroids* (PB2004-100004, 2003) 332 pp.

Allethrin I   R = CH₃
Allethrin II   R = COOCH₃

Allethrin I. [584-79-2] 2,2-Dimethyl-3-(2-methyl-1-propenyl)-cyclopropanecarboxylic acid 2-methyl-4-oxo-3-(2-propenyl)-2-cyclopenten-1-yl ester; allethrolone ester of chrysanthemummonocar-boxylic acid; Pynamin. $C_{19}H_{26}O_3$; mol wt 302.41. Synthesis: M. S. Schechter et al., J. Am. Chem. Soc. **71**, 1517, 3165 (1949); H. A. Stansbury, H. R. Guest, US 2768965 (1956 to UCC). Commercial process: H. J. Sanders, A. W. Taff, Ind. Eng. Chem. **46**, 414 (1954). The commercial product is a mixture of 8 optically active isomers. Clear, pale yellow oil. Poisonous. $bp_{0.1}$ 140°. Decomp by rapid pyrolysis at >400°. Vaporizes without dec when heated at 150°. Unstable in light, air, alkaline conditions. d_{20}^{20} 1.010; n_D^{20} 1.5040; n_D^{30} 1.5023. Practically insol in water. Sol in alcohol, petr ether, kerosene, carbon tetrachloride, ethylene dichloride, nitromethane. Incompatible with alkalies.

Allethrin II. [497-92-7] 3-(3-Methoxy-2-methyl-3-oxo-1-pro-penyl)-2,2-dimethylcyclopropanecarboxylic acid 2-methyl-4-oxo-3-(2-propenyl)-2-cyclopenten-1-yl ester; allethrolone ester of chrysan-themumdicarboxylic acid monomethyl ester; ENT-17510. C_{20}-$H_{26}O_5$; mol wt 346.42. Synthesis: M. Matsui, Y. Yamada, Agric. Biol. Chem. **27**, 373 (1963); M. Matsui, H. Meguro, ibid. **28**, 27 (1964); M. Elliott, N. F. Janes, Chem. Ind. (London) **1969**, 270; A. Kobayashi et al., Agric. Biol. Chem. **35**, 1961 (1971); T. Sugiyama et al., ibid. **36**, 565 (1972). Oily pale yellow liq. Poisonous. $n_D^{20.6}$ 1.5156. uv max (ethanol): 232 nm (ε 23000). Unstable in light, air, alkaline conditions. Sol in organic solvents. Practically insol in water.

Caution: Potential symptoms of toxicity are similar to those of the pyrethrins.

USE: Insecticide.

253. Allicin. [539-86-6] 2-Propene-1-sulfinothioic acid S-2-propen-1-yl ester; thio-2-propene-1-sulfinic acid S-allyl ester; diallyl disulfide-oxide. $C_6H_{10}OS_2$; mol wt 162.27. C 44.41%, H 6.21%, O 9.86%, S 39.51%. A biologically active constit of freshly crushed garlic (Allium sativum L., Liliaceae). Naturally formed by the action of the enzyme allicinase on alliin, q.v., when the tissue of the garlic bulb is disrupted. Isoln and antibacterial activity: C. J. Cavallito, J. H. Bailey, J. Am. Chem. Soc. **66**, 1950 (1944). Structure: Cavallito et al., ibid. 1952. Synthesis: Stoll, Seebeck, Experientia **6**, 330 (1950); Cavallito, Small, US 2508745 (1950 to Sterling Drug). Antifungal activity: Y. Yamada, K. Azuma, Antimicrob. Agents Chemother. **11**, 743 (1977). Pharmacological effects and prepn: P. R. Mayeux et al., Agents Actions **25**, 182 (1988). Stability in blood, solvents, and simulated physiological fluids: F. Freeman, Y. Kodera, J. Agric. Food Chem. **43**, 2332 (1995). Identification as the source of garlic's pungent burning and prickling sensations: L. J. Macpherson et al., Curr. Biol. **15**, 929 (2005).

Yellow liquid. True odor of garlic. Decomp on distilling. d_4^{20} 1.112. n_D^{20} 1.561. Soly in water at 10° about 2.5% w/w. pH about 6.5. Upon standing an oily precipitate forms from aq solns. Miscible with alcohol, ether, benzene; fairly insol in the Skellysolves; unstable to hot alkali; stable to acids. LD_{50} in mice (mg/kg): 60 i.v.; 120 s.c. (Cavallito, Bailey).

254. Alliin. [556-27-4] [S(S)]-S-2-Propen-1-yl-L-cysteine S-oxide; 3-[(S)-2-propenylsulfinyl]-L-alanine; 3-((S)-allylsulfinyl)-L-alanine; S-allyl-L-cysteine sulfoxide. $C_6H_{11}NO_3S$; mol wt 177.22. C 40.66%, H 6.26%, N 7.90%, O 27.08%, S 18.09%. Constituent of garlic, Allium sativum L., Liliaceae; also found in other Allium spp. Degraded by alliinase to produce the garlic flavor component, allicin, q.v. Isoln: Stoll, Seebeck, Helv. Chim. Acta **31**, 189 (1948). Synthesis: eidem, Experientia **6**, 330 (1950); Helv. Chim. Acta **34**, 481 (1951). Extraction from Allium plants and LC determn: E. Mochizuki et al., J. AOAC Int. **80**, 1052 (1997). In vitro immunomodulatory effects on peripheral blood cells: H. Salman et al., Int. J. Immunopharmacol. **21**, 589 (1999). Biosynthesis in garlic tissue cultures: J. Hughes et al., Phytochemistry **66**, 187 (2005).

Hemihydrate, odorless, bunched needles from dil acetone, mp 164-166° (effervescence). $[\alpha]_D^{20}$ +63.5° (c = 2). Freely sol in water. Practically insol in abs ethanol, chloroform, acetone, ether, benzene.

255. Allobarbital. [52-43-7] 5,5-Di-2-propen-1-yl-2,4,6-(1H,3H,5H)-pyrimidinetrione; 5,5-diallylbarbituric acid; allobarbitone. $C_{10}H_{12}N_2O_3$; mol wt 208.22. C 57.68%, H 5.81%, N 13.45%, O 23.05%. Prepn: US 1042265 (1912). Acute toxicity: F. Sandberg, Acta Physiol. Scand. **24**, 7 (1951).

Crystals, leaflets. Slightly bitter taste, mp 171-173°. One part dissolves in about 300 parts water, 50 parts boiling water, 20 parts cold alcohol, 20 parts ether; very sol in hot alcohol and in acetone; sol in ethyl acetate. Insol in aliphatic hydrocarbons. A satd aq soln is acid to litmus. LD_{50} i.p. in rats: 127.3 mg/kg (Sandberg).

Note: This is a controlled substance (depressant): **21 CFR,** 1308.13.

THERAP CAT: Sedative, hypnotic.

256. Allocholesterol. [517-10-2] (3β)-Cholest-4-en-3-ol; coprostenol; 4:5-coprosten-3-ol. $C_{27}H_{46}O$; mol wt 386.66. C 83.87%, H 11.99%, O 4.14%. Prepn: Windaus, Ann. **453**, 101 (1927); Schoenheimer, Evans, J. Biol. Chem. **114**, 567 (1936). Sepn from cholesterol: Stoll, Z. Physiol. Chem. **246**, 10 (1937).

Needles from ether-methanol, mp 132°. $[\alpha]_D^{23}$ +43.7° (c = 1 in benzene). Freely sol in benzene, acetone, ether, chloroform, dioxane, pyridine; less sol in methanol, alcohol. Pptd by digitonin. Deep-red color with 90% trichloroacetic acid.

Acetate. $C_{29}H_{48}O_2$. Long needles from dil methanol, mp 85°.

257. Allocryptopine. [485-91-6] 5,7,8,15-Tetrahydro-3,4-dimethoxy-6-methyl[1,3]benzodioxolo[5,6-e][2]benzazecin-14-(6H)-one; 5,7,8,15-tetrahydro-3,4-dimethoxy-6-methylbenzo-[e][1,3]dioxolo[4,5-k][3]benzazecin-14(6H)-one. $C_{21}H_{23}NO_5$; mol wt 369.42. C 68.28%, H 6.28%, N 3.79%, O 21.65%. Isomeric with cryptopine, q.v. Obtained from Chelidonium majus L., Bocconia cordata Willd., Sanguinaria canadensis L. and allied Papaveraceae: Manske in Manske, Holmes, The Alkaloids vol. IV (Academic Press, New York, 1954) p 159. Structure: Gadamer, Arch. Pharm. **257**, 298 (1919). Exists in two isomeric modifications designated as α- and β-forms. Identity of α-form with α-fagarine: Redemann et al., J. Am. Chem. Soc. **71**, 1030 (1949). Synthesis: Haworth, Perkin, J. Chem. Soc. **1926**, 445; Bentley, Murray, ibid. **1963**, 2497.

α-Allocryptopine. α-Fagarine. Crystals from ethanol, mp 160-161°. Soluble in alcohol, chloroform, ether, ethyl acetate and dil acids.

β-Allocryptopine. Crystallizes with ½ mol of alcohol or ethyl acetate. Melts after expelling the solvent, at 169-171°.

258. Allopregnane. [641-85-0] (5α)-Pregnane. $C_{21}H_{36}$; mol wt 288.52. C 87.42%, H 12.58%. Prepn from corticosterone: Steiger, Reichstein, *Helv. Chim. Acta* **21**, 161 (1938); from allopregnan-3-one: Ruzicka *et al.*, *ibid.* **22**, 1294 (1939); from 21-hydroxy-allopregnane: Plattner *et al.*, *ibid.* **27**, 1177 (1944); Casanova, Reichstein, *ibid.* **32**, 647 (1949); by degradation of conessine: Haworth *et al.*, *J. Chem. Soc.* **1949**, 831; by Hofmann decomposition of 3β- and 3α-dimethylaminopregnanes: Haworth *et al.*, *ibid.* **1953**, 1110.

Crystals from acetone + methanol, mp 84-85°. $[\alpha]_D^{19}$ +18.4° (c = 1.69 in chloroform).

259. Allopregnane-3α,20α-diol. [566-58-5] (3α,5α,20S)-Pregnane-3,20-diol; 5α-pregnane-3α,20α-diol; 3α,20α-dihydroxy-5α-pregnane. $C_{21}H_{36}O_2$; mol wt 320.52. C 78.69%, H 11.32%, O 9.98%. Progesterone metabolite. Isoln from human pregnancy urine: Beall, *Biochem. J.* **31**, 35 (1937); from bulls' urine: Marker *et al.*, *J. Am. Chem. Soc.* **60**, 2931 (1938). Prepn by reduction of 5α-pregn-1-ene-3,20-dione with lithium aluminum hydride: Schütt, Tamm, *Helv. Chim. Acta* **41**, 1751 (1958).
Crystals from methanol, mp 243-245°. $[\alpha]_D^{20}$ +17° (c = 0.148 in ethanol).
Diacetate. $C_{25}H_{40}O_4$. Crystals, mp 139.5-140.5°. $[\alpha]_D^{20}$ +18° (c = 0.408 in benzene).

260. Allopregnane-3α,20β-diol. [566-57-4] (3α,5α,20R)-Pregnane-3,20-diol; 3α,20β-dihydroxy-5α-pregnane. $C_{21}H_{36}O_2$; mol wt 320.52. C 78.69%, H 11.32%, O 9.98%. Progesterone metabolite. Prepn by hydrogenation of progesterone: Marker, Lawson, *J. Am. Chem. Soc.* **61**, 588 (1939); by catalytic hydrogenation of epiallopregnane-3-ol-20-one: Marker, **US 2231019** (1941 to Parke, Davis); by reduction of 5α-pregn-1-ene-3,20-dione with lithium aluminum hydride: Schütt, Tamm, *Helv. Chim. Acta* **41**, 1751 (1958).
Needles from acetone, mp 207-209°. $[\alpha]_D^{26}$ +12° (c = 1.132 in chloroform).

261. Allopregnane-3β,20α-diol. [566-56-3] (3β,5α,20S)-Pregnane-3,20-diol; 5α-pregnane-3,20-diol; 3β,20α-dihydroxy-5α-pregnane. $C_{21}H_{36}O_2$; mol wt 320.52. C 78.69%, H 11.32%, O 9.98%. Progesterone metabolite. Isoln from pregnant mares' urine: Brooks *et al.*, *Biochem. J.* **51**, 694 (1952). Prepn by heating 3α,20α-dihydroxy-5α-pregnane with sodium, or by hydrogenation of 5α-pregnane-20-ol-3-one in acid: Marker, **US 2196220** and **US 2250962** (1940, 1941, both to Parke, Davis); by reduction of allopregnan-3β-ol-20-one acetate with sodium: Klyne, Barton, *J. Am. Chem. Soc.* **71**, 1500 (1949); by reduction of 5,16-pregnadien-3β-ol-20-one with Zn + acetic acid: Ercoli, De Ruggieri, *Farm. Sci. Tec.* **7**, 11 (1952), *C.A.* **46**, 10186b (1952).
Crystals from acetone, mp 218-219°. $[\alpha]_D^{22}$ +23° (c = 0.93 in chloroform).
Diacetate. $C_{25}H_{40}O_4$. Crystals from petr ether, mp 164-165°. $[\alpha]_D^{19}$ +0.8° (c = 1.2 in chloroform).
Dibenzoate. $C_{35}H_{44}O_4$. Crystals, mp 170.5-172°. $[\alpha]_D^{20}$ +27.7° (c = 0.84 in chloroform).

262. Allopregnane-3β,20β-diol. [516-53-0] (3β,5α,20R)-Pregnane-3,20-diol; 3β,20β-dihydroxy-5α-pregnane. $C_{21}H_{36}O_2$; mol wt 320.52. C 78.69%, H 11.32%, O 9.98%. Progesterone metabolite. Isoln from pregnant mares' urine: Brooks *et al.*, *Biochem. J.* **51**, 694 (1952). Prepn by catalytic reduction of allopregnane-3,20-dione: Marker *et al.*, **GB 512940** (1939 to Parke, Davis); by hydrogenation of pregn-5-ene-3β-ol-20-one: Klyne, Barton, *J. Am. Chem.*

Soc. **71**, 1500 (1949); by hydrogenation of pregn-5-ene-3β,20β-diol: Klyne, Miller, *J. Chem. Soc.* **1950**, 1972.
Leaflets from ethyl acetate + petr ether, mp 194.5-195.5°. $[\alpha]_D$ +4.4° (c = 1.04 in chloroform).
Diacetate. $C_{25}H_{40}O_4$. Needles from ethyl acetate + petr ether, mp 141-142.5°. $[\alpha]_D^{20}$ +21° (c = 1 in chloroform).
Dibenzoate. $C_{35}H_{44}O_4$. Crystals, mp 237.5-239°. $[\alpha]_D^{20}$ −10.1° (c = 2.08 in chloroform).

263. Allopregnane-3β,17α-diol-20-one. [570-54-7] (3β,5α)-3,17-Dihydroxypregnan-20-one; 3β,17-dihydroxy-5α-pregnan-20-one; 3β,17-dihydroxy-20-oxo-5α-pregnane; 17-(1-ketoethyl)androstane-3,17-diol; Reichstein's substance L; Wintersteiner's compound G. $C_{21}H_{34}O_3$; mol wt 334.50. C 75.41%, H 10.25%, O 14.35%. Isoln from adrenal cortex: Wintersteiner, Pfiffner, *J. Biol. Chem.* **116**, 291 (1936). Structure: Reichstein, Gätzi, *Helv. Chim. Acta* **21**, 1497 (1938). Prepn from allopregnan-3β-ol-20-one 3-acetate: Rosenkranz *et al.*, *J. Am. Chem. Soc.* **72**, 4081 (1950); from pregn-5-ene-3β-ol-20-one: Ramirez, Stafie, *ibid.* **77**, 134 (1955).
Crystals from abs alcohol, mp 264-266°. $[\alpha]_D^{21}$ +30.6° (c = 0.54 in abs alcohol).
3-Acetate. $C_{23}H_{36}O_4$. Crystals from acetone, mp 187-189°. $[\alpha]_D^{20}$ +18° (acetone).

264. 3,20-Allopregnanedione. [566-65-4] (5α)-Pregnane-3,20-dione; 3,20-dioxo-5α-pregnane. $C_{21}H_{32}O_2$; mol wt 316.49. C 79.70%, H 10.19%, O 10.11%. From pregnancy urine: Hartmann, Locher, *Helv. Chim. Acta* **18**, 160 (1935); Lieberman *et al.*, *J. Biol. Chem.* **172**, 263 (1948). Prepn from pregna-4,11-diene-3,20-dione: Shoppee, Reichstein, *Helv. Chim. Acta* **24**, 356 (1941); by oxidation of allopregnan-3β-ol-20-one with chromium trioxide: Billeter, Miescher, *ibid.* **30**, 1409 (1947); by hydrogenation of pregnenolone: Pappas, Nace, *J. Am. Chem. Soc.* **81**, 4556 (1959); from funtumine: Janot *et al.*, *Bull. Soc. Chim. Fr.* **1960**, 1669.
Crystals from methylene chloride + hexane, mp 200°. $[\alpha]_D$ +125° (c = 1.2 in chloroform).
USE: In the synthesis of progesterone, **FR 845034** *C.A.* **34**, 8184² (1940).

265. Allopregnane-3α,11β,17α,21-tetrol-20-one. [302-91-0] (3α,5α,11β)-3,11,17,21-Tetrahydroxypregnan-20-one; 3α,11β,17,21-tetrahydroxy-5α-pregnan-20-one; 3α,11β,17,21-tetrahydroxy-20-oxo-5α-pregnane; 17-(1-keto-2-hydroxyethyl)androstane-3,11,17-triol; 3α-allotetrahydrocortisol; Kendall's compound C; Reichstein's substance C; Wintersteiner's compound D. $C_{21}H_{34}O_5$; mol wt 366.50. C 68.82%, H 9.35%, O 21.83%. Isoln from adrenal cortex: Mason *et al.*, *J. Biol. Chem.* **124**, 459 (1938); Kuizenga, Cartland, *Endocrinology* **24**, 526 (1939); v. Euw, Reichstein, *Helv. Chim. Acta* **25**, 988 (1942); v. Euw *et al.*, *ibid.* **41**, 1516 (1958). Prepn by hydrogenation of cortisol with rhodium: Caspi, *J. Org. Chem.* **24**, 669 (1959); from bismethylenedioxyhydrocortisone: Fukushima, Daum, *ibid.* **26**, 520 (1961).
Crystals from methanol, mp 244-245°. $[\alpha]_D^{21}$ +59.7° (c = 0.34 in methanol).
3,21-Diacetate. $C_{25}H_{38}O_7$. Crystals, dec 204-205°. $[\alpha]_D^{20}$ +73.8°; $[\alpha]_{542}^{20}$ +90.5° (dioxane).

266. Allopregnane-3β,11β,21-triol-20-one. [516-16-5] (3β,5α,11β)-3,11,21-Trihydroxypregnan-20-one; 3β,11β,21-trihydroxy-20-oxo-5α-pregnane; Reichstein's substance R. $C_{21}H_{34}O_4$; mol wt 350.50. C 71.96%, H 9.78%, O 18.26%. Isolation from adrenal glands: Reichstein, von Euw, *Helv. Chim. Acta* **21**, 1197 (1938); Reichstein, *ibid.* 1490. Partial synthesis by hydrogenation of corticosterone acetate: Pataki *et al.*, *J. Biol. Chem.* **195**, 751 (1952). Prepn from 3β,11β-dihydroxyalloetiocholanic acid: Lardon, Reichstein, *Helv. Chim. Acta* **37**, 443 (1954); from hydrocortisone: Mancera *et al*, *J. Am. Chem. Soc.* **77**, 5669 (1955); from 3β,21-diacetoxy-17(20)-allopregnene + osmium tetroxide and triethylamine oxide peroxide: Schneider, Hanze, **US 2769823** (1956 to Upjohn).
Needles from alc, mp 202-204°. $[\alpha]_D$ +110° (ethanol).
3,21-Diacetate. $C_{25}H_{38}O_6$. Crystals from acetone + ether, mp 170-172°. $[\alpha]_D^{22}$ +82.5° (c = 1.38 in dioxane); $[\alpha]_D^{20}$ +101° (acetone).

267. Allopregnan-3α-ol-20-one. [516-54-1] (3α,5α)-3-Hydroxypregnan-20-one; 3α-Hydroxy-5α-dihydroprogesterone; 5α-pregnan-3α-ol-20-one; epiallopregnanolone; allotetrahydroprogesterone. $C_{21}H_{34}O_2$; mol wt 318.50. C 79.19%, H 10.76%, O 10.05%. Naturally occurring metabolite of progesterone. Prototype of the neuroactive steroids known as *epalons* that allosterically modulate γ-aminobutyric acid type A ($GABA_A$) receptors in the CNS. Isoln from human pregnancy urine: R. E. Marker *et al.*, *J. Am. Chem. Soc.* **59**, 616 (1937). Prepn from pregnenolone: G. Fleischer *et al.*, *ibid.* **60**, 79 (1938). Chromatographic determn in urine: F. J. Bègue *et al.*, *J. Chromatogr. Sci.* **12**, 763 (1974). Synthesis, metabolism and pharmacology: R. H. Purdy *et al.*, *J. Med. Chem.* **33**, 1572 (1990). Review of pharmacology and therapeutic potential of epalons: K. W. Gee *et al.*, *Crit. Rev. Neurobiol.* **9**, 207-227 (1995).

Crystals from abs alc, mp 176-178°. $[α]_D$ +87.7° (abs alc).
Acetate. $C_{23}H_{36}O_3$. Crystals from aq ethanol, mp 141-142°. $[α]_D^{22}$ +94.5° (abs ethanol).

268. Allopregnan-3β-ol-20-one. [516-55-2] (3β,5α)-3-Hydroxypregnan-20-one; 3β-hydroxy-20-oxo-5α-pregnane. $C_{21}H_{34}O_2$; mol wt 318.50. C 79.19%, H 10.76%, O 10.05%. Isoln from adrenal cortex: von Euw, Reichstein, *Helv. Chim. Acta* **24**, 885 (1941); from corpus luteum: Butenandt, Mamoli, *Ber.* **68**, 1847 (1935); Prelog, Meister, *Helv. Chim. Acta* **32**, 2435 (1949); from human pregnancy urine: Lieberman *et al.*, *J. Biol. Chem.* **172**, 263 (1948); from human placenta: Pearlman, Cerceo, *ibid.* **194**, 807 (1952). Prepn from 20-methyl-Δ^{20}-allopregnen-3β-ol: Koechlin, Reichstein, *Helv. Chim. Acta* **27**, 549 (1944); from pregnenolone: Mancera *et al.*, *J. Org. Chem.* **16**, 192 (1951); Pappas, Nace, *J. Am. Chem. Soc.* **81**, 4556 (1959); by redn of allopregnane-3,20-dione with sodium borohydride: Mancera *et al.*, *ibid.* **75**, 1286 (1953); from progesterone 20-cycloethylene ketal: Sondheimer, Klibansky, *Tetrahedron* **5**, 15 (1959).

Plates from dil methanol, mp 194-195°. $[α]_D^{27}$ +91.2° (c = 0.4 in ethanol).
Acetate. $C_{23}H_{36}O_3$. Plates from methanol, mp 144-146°. $[α]_D^{25}$ +69° (chloroform).

269. Allopregnan-20α-ol-3-one. [516-59-6] (5α,20S)-20-Hydroxypregnan-3-one; 20α-hydroxy-5α-pregnan-3-one; 20α-hydroxy-3-oxo-5α-pregnane. $C_{21}H_{34}O_2$; mol wt 318.50. C 79.19%, H 10.76%, O 10.05%. Prepn from funtumidine: Janot *et al.*, *Bull. Soc. Chim. Fr.* **1960**, 1669.
Crystals, mp 179°. $[α]_D$ +36.6° (c = 0.8 in chloroform).
Acetate. $C_{23}H_{36}O_3$. Crystals from ethanol, mp 155°. $[α]_D$ +24° (c = 1.5 in chloroform).

270. Allopregnan-20β-ol-3-one. [516-58-5] (5α,20R)-20-Hydroxypregnan-3-one; 20β-hydroxy-5α-pregnan-3-one; 20β-hydroxy-3-oxo-5α-pregnane. $C_{21}H_{34}O_2$; mol wt 318.50. C 79.19%, H 10.76%, O 10.05%. Prepn from allopregnane-3β,20β-diol-3-acetate: Rubin *et al.*, *J. Am. Chem. Soc.* **73**, 2338 (1951); from isofuntumidine: Janot *et al.*, *Bull. Soc. Chim. Fr.* **1960**, 1669.
Crystals from heptane, mp 185°. $[α]_D$ +20° (c = 1.2 in chloroform).
Acetate. $C_{23}H_{36}O_3$. Crystals from ethanol, mp 148°. $[α]_D$ +57° (c = 1.9 in chloroform).

271. Allopurinol. [315-30-0] 1,5-Dihydro-4*H*-pyrazolo[3,4-*d*]pyrimidin-4-one; 1*H*-pyrazolo[3,4-*d*]pyrimidin-4-ol; 4-hydroxypyrazolo[3,4-*d*]pyrimidine; HPP; BW-56158; Adenock; Allurit; Alositol; Anoprolin; Anzief; Apurol; Apurin; Bleminol; Caplenal; Cellidrin; Epidropal; Foligan; Hexanurat; Remid; Takanarumin; Urosin; Zyloprim; Zyloric. $C_5H_4N_4O$; mol wt 136.11. C 44.12%, H 2.96%, N 41.16%, O 11.75%. Hypoxanthine analog that inhibits the enzyme xanthine oxidase, and decreases uric acid production. Prepn: R. K. Robins, *J. Am. Chem. Soc.* **78**, 784 (1956); **GB 798646** (1958 to Wellcome Found.); G. H. Hitchings, E. A. Falco, **US**

3474098 (1969 to Burroughs Wellcome); B. G. Hildick, G. Shaw, *J. Chem. Soc. C* **1971**, 1610; Y. Tominaga *et al.*, *J. Heterocycl. Chem.* **27**, 775 (1990). Crystal structure: P. Prusiner, M. Sundaralingam, *Acta Crystallogr.* **B28**, 2148 (1972). Comprehensive description: S. A. Benezra, T. R. Bennett, *Anal. Profiles Drug Subs.* **7**, 1-17 (1978). Review of clinical pharmacokinetics and pharmacodynamics: R. O. Day *et al.*, *Clin. Pharmacokinet.* **46**, 623-644 (2007); of pharmacology and clinical experience: P. Pacher *et al.*, *Pharmacol. Rev.* **58**, 87-114 (2006); of safety and efficacy in treatment of hyperuricemia in gout: J. Chao, R. Terkeltaub, *Curr. Rheumatol. Rep.* **11**, 135-140 (2009).

Colorless crystals from methanol, mp > 380°. Crystal density: 1.635. uv max (0.1*N* NaOH): 257 nm (ε 7200); (0.1*N* HCl): 250 nm (ε 7600); (methanol): 252 nm (ε 7600). pKa 9.4. Partition coefficient (octanol/water): 0.28. Soly in water (37°): 80 mg/dl. Sol in solns of potassium, sodium hydroxides; very slightly sol in alcohol. Practically insol in chloroform, ether.
Sodium salt. [17795-21-0] Aloprim. $C_5H_3N_4NaO$; mol wt 158.10. White amorphous mass. pKa 9.31.
THERAP CAT: Antiurolithic. In treatment of hyperuricemia and chronic gout.

272. D-Allose. [2595-97-3] D-Allose; β-D-allopyranose. $C_6H_{12}O_6$; mol wt 180.16. C 40.00%, H 6.71%, O 53.28%. An aldohexose sugar. Obtained from leaves of *Protea rubropilosa*: Beylis, Perold, *Chem. Commun.* **1971**, 597. Prepd by reduction of D-allonic lactone with sodium amalgam: Levene, Jacobs, *Ber.* **43**, 3147 (1910). Synthesis: Hughes, Speakman, *J. Chem. Soc.* **1965**, 2236; Baker *et al.*, *Carbohydr. Res.* **24**, 192 (1972).

Crystals from dilute methanol, mp 128-128.5°. Shows mutarotation: $[α]_D^{20}$ +0.58° (4 min) → +3.26° (10 min) → +14.41° (final value, 20 hr; c = 5). Soluble in water. Practically insol in alcohol.
Phenylosazone. $C_{18}H_{22}N_4O_4$. mp 178°.

273. Alloxan. [50-71-5] 2,4,5,6(1*H*,3*H*)-pyrimidinetetrone; 2,4,5,6-tetraoxohexahydropyrimidine; mesoxalylurea; mesoxalylcarbamide. $C_4H_2N_2O_4$; mol wt 142.07. C 33.82%, H 1.42%, N 19.72%, O 45.05%. Found by Liebig in mucus excreted during dysentery. Prepn by direct oxidation of uric acid: G. Brugnatelli, *Ann. Chim. Phys.* [2] **8**, 201 (1818); J. Liebig, F. Wöhler, *Ann.* **26**, 241 (1838); H. Biltz, M. Hehn, *ibid.* **413**, 60 (1917). Prepd from alloxantin: J. Liebig, *ibid.* **147**, 366 (1868); W. W. Hartman, O. E. Sheppard, *Org. Synth.* **coll. vol. III**, 37 (1955). *See also* A. V. Holmgren, W. W. Wenner, *ibid.* **coll. vol. IV**, 23 (1963). Produces diabetes in animals by selective necrosis of pancreatic islet β-cells: J. S. Dunn, N. G. B. McLetchie, *Lancet* **2**, 384 (1943); W. B. Kennedy, F. D. W. Lukens, *Proc. Soc. Exp. Biol. Med.* **57**, 143 (1944). Mechanism of action study: L. Boquist, *Acta Pathol. Microbiol. Scand. Sect. A* **88**, 201 (1980). *In vitro* antineoplastic activity: P. Grobon, *C. R. Seances Acad. Sci. Ser. D* **280**, 2413 (1975). Antibacterial and antifungal activity: J. D. Douros, A. F. Kerst, **JP Kokai 72 4900**; *eidem*, **US 3728454** (1972, 1973 both to Gates Rubber Co.). Toxicological study in mice: B. A. Waisbren, *Proc. Soc. Exp. Biol. Med.* **67**, 154 (1948).

Anhydrous, orthorhombic crystals from anhydr acetone or glacial acetic acid or by sublimation in vacuo. Turns pink at 230° and dec at 256°. Acid to litmus. pK (25°) 6.63. Absorption spectrum: Hartley, *J. Chem. Soc.* **87**, 1802, 1808 (1905). Freely sol in water. Hot aqueous solns are yellow and become colorless on cooling. Aqueous solns, after being in contact with the human skin for some time, give it a red color and a disagreeable odor. Sol in acetone, alcohol, methanol, glacial acetic acid; slightly sol in chloroform, petr ether, toluene, ethyl acetate and acetic anhydride. Insol in ether.

Tetrahydrate. Large triclinic prisms or oblique monoclinic rhombs from water.

Monohydrate. By heating the tetrahydrate at 100° or by exposing it to the air. Forms triclinic pinacoidal crystals.

Compound with urea. $CH_4N_2O.C_4H_2N_2O_4.H_2O$. Minute, yellow needles; red at 170°, decomp at 185-186°.

USE: In production of diabetes in experimental animals; in nutrition experiments; in organic syntheses.

274. Alloxantin. [76-24-4] 5,5'-Dihydroxy-[5,5'-bipyrimidine]-2,2',4,4',6,6'(1*H*,1'*H*,3*H*,5*H*,5'*H*)-hexone; 5,5'-dihydroxy-5,5'-bibarbituric acid; uroxin. $C_8H_6N_4O_8$; mol wt 286.16. C 33.58%, H 2.11%, N 19.58%, O 44.73%. Prepd from uric acid: Nightingale, *Org. Synth.* **coll. vol. III**, 42 (1955); from alloxan monohydrate: Tipson, *ibid.* **coll. vol. IV**, 25 (1963). Structure: Singh, *Acta Crystallogr.* **19**, 767 (1965).

Dihydrate. Cryst powder. On exposure to air becomes red. Becomes yellow at 225°; dec at 253-255°. Sparingly sol in cold water, alcohol or ether. Its aq soln is acid; reduces Ag salts and gives a blue ppt with $Ba(OH)_2$ soln. *Keep tightly closed.* Emits toxic fumes upon decompn.

Caution: Ingestion causes disturbed carbohydrate metabolism leading to diabetes, *cf. Dangerous Properties of Industrial Materials,* N. I. Sax, Ed. (Van Nostrand Reinhold, New York, 4th ed., 1975) p 366.

275. Allura Red AC. [25956-17-6] 6-Hydroxy-5-[(2-methoxy-5-methyl-4-sulfophenyl)diazenyl]-2-naphthalenesulfonic acid sodium salt (1:2); 6-hydroxy-5-[(2-methoxy-5-methyl-4-sulfophenyl)azo]-2-naphthalenesulfonic acid disodium salt; 6-hydroxy-5-[(6-methoxy-4-sulfo-*m*-tolyl)azo]-2-naphthalenesulfonic acid disodium salt; 1-[(6-methoxy-4-sulfo-*m*-tolyl)azo]-2-naphthol-6-sulfonic acid disodium salt; FD & C Red No. 40; C.I. 16035; C.I. Food Red 17. $C_{18}H_{14}N_2Na_2O_8S_2$; mol wt 496.42. C 43.55%, H 2.84%, N 5.64%, Na 9.26%, O 25.78%, S 12.92%. Prepn: **GB 1164249**; Rast, Steiner, **US 3519617** (1969, 1970 both to Allied Chem.).

Dark red powder. Soly at 25°: water, 22.5%; 50% alcohol, 1.3%.

USE: Color additive in foods, drugs and cosmetics.

276. Allyl Acetate. [591-87-7] Acetic acid 2-propen-1-yl ester; 3-acetoxy-1-propene. $C_5H_8O_2$; mol wt 100.12. C 59.98%, H 8.05%, O 31.96%. $CH_2=CHCH_2OCOCH_3$. Prepn: N. Zinin, *Ann.* **1855**, 361; K. N. Gurudutt *et al., Tetrahedron* **38**, 1843 (1982); and physical properties: G. H. Jeffery, A. I. Vogel, *J. Chem. Soc.* **1948**, 658. Vibrational spectra and thermodynamic functions: B. Singh *et al., Proc. Indian Acad. Sci.* **89**, 201 (1980). Acute toxicity: H. F.

Smyth, Jr. *et al., J. Ind. Hyg. Toxicol.* **31**, 60 (1949); P. M. Jenner *et al., Food Cosmet. Toxicol.* **2**, 327 (1964). Effect on respiratory function in mice: G. D. Nielsen *et al., Acta Pharmacol. Toxicol.* **54**, 292 (1984).

Liquid. *Flammable, poisonous.* bp_{760} 104° (Gurudutt); also reported as bp_{773} 104° (Jeffery, Vogel), bp 105° (Zinin). n_D 1.40396. d_4^{20} 0.9277. LD_{50}: 0.13 g/kg orally in rats; 1.1 ml/kg dermally in rabbits (Smyth). LD_{50} in rats, mice (mg/kg): 142, 170 orally (Jenner).

Caution: Direct contact may cause irritation of skin, eyes, mucous membranes and upper respiratory tract.

277. Allyl Alcohol. [107-18-6] 2-Propen-1-ol; 1-propenol-3; vinyl carbinol. C_3H_6O; mol wt 58.08. C 62.04%, H 10.41%, O 27.55%. Prepd by heating glycerol with formic acid: O. Kamm, C. S. Marvel, *Org. Synth.* **coll. vol. I**, 42 (2nd ed., 1941); *cf.* Delaby, Dubois, *Compt. Rend.* **188**, 710 (1929). Toxicity study: Smyth, Carpenter, *J. Ind. Hyg. Toxicol.* **30**, 63 (1948).

Colorless liquid; pungent, mustard-like odor. *Flammable, poisonous. Keep tightly closed.* d_4^{20} 0.8540. bp 96-97°. mp −50°. n_D^{20} 1.41345. Absorption spectrum: Lüthy, *Z. Phys. Chem.* **107**, 289 (1923). Flash pt 70°F (open cup); 75°F (closed cup); autoignition temp 713°F. 72.3% of the alcohol with 27.7% water forms a constant boiling mixture, bp at 87.5°. Miscible with water, alcohol, chloroform, ether, petr ether. Upon storage for several years allyl alcohol polymerizes and a thick syrup is formed (insol in water, sol in chloroform) which on treatment with ether yields a brittle resinoid mass: Blicke, *J. Am. Chem. Soc.* **45**, 1563 (1923). LD_{50} orally in rats: 64 mg/kg (Smyth, Carpenter).

Caution: Potential symptoms of overexposure are eye irritation, tissue damage; irritation of upper respiratory system and skin; pulmonary edema. *See NIOSH Pocket Guide to Chemical Hazards* (DHHS/NIOSH 97-140, 1997) p 10.

USE: Manuf allyl compds, war gas, resins, plasticizers.

278. Allylamine. [107-11-9] 2-Propen-1-amine; 3-aminopropylene. C_3H_7N; mol wt 57.10. C 63.11%, H 12.36%, N 24.53%. Manuf from allyl chloride and ammonia: Ploetz, **US 2915385** (1959 to Feldmühle Papier und Zellstoffwerke). Toxicity study: C. H. Hine *et al., Arch. Environ. Health* **1**, 34 (1960).

Liquid; burning taste; strong ammonia odor causing sneezing and tears. d_{20}^{20} 0.760. bp 55-58°. n_D^{20} 1.4186. Flash pt, closed cup: 10°F (−12°C). Misc with water, alc, chloroform, ether. *Poisonous; flammable. Keep tightly closed.* LD_{50} i.p. in mice: 49 mg/kg (Hine).

Caution: A strong eye and respiratory tract irritant; intolerable at 14 ppm. *See Patty's Industrial Hygiene and Toxicology* vol. **2B**, G. D. Clayton, F. E. Clayton, Eds. (Wiley-Interscience, New York, 3rd ed., 1981) pp 3157-3158.

USE: In the manuf of mercurial diuretics.

279. Allyl Bromide. [106-95-6] 3-Bromo-1-propene; 3-bromopropylene; bromallylene. C_3H_5Br; mol wt 120.98. C 29.78%, H 4.17%, Br 66.05%. Prepd from hydrobromic acid and allyl alcohol: O. Kamm, C. S. Marvel, *Org. Synth.* **coll. vol. I**, 27 (2nd ed., 1941); from triphenylphosphite, allyl alcohol and benzyl bromide: Landauer, Rydon, *J. Chem. Soc.* **1953**, 2224.

Colorless liquid; unpleasant, pungent odor. *Flammable, poisonous. Keep tightly closed.* d_4^{20} 1.398. bp_{760} 71.3°. mp −119°. n_D^{20} 1.46545. Slightly sol in water; miscible with alcohol, chloroform, ether, carbon disulfide, carbon tetrachloride.

Caution: Direct contact may cause burns and tissue destruction of eyes, skin, mucous membranes and upper respiratory tract.

USE: Manuf synthetic perfumes, other allyl compounds.

280. Allyl Chloride. [107-05-1] 3-Chloro-1-propene; 3-chloropropylene; chlorallylene. C_3H_5Cl; mol wt 76.52. C 47.09%, H 6.59%, Cl 46.33%. Manuf by chlorination of propylene: Samples, Hilbert, **US 3054831** (1962 to Union Carbide). Prepd from diphenylphosphite, allyl alcohol and benzyl chloride: Landauer, Rydon, *J. Chem. Soc.* **1953**, 2224. Toxicity data: H. F. Smyth, C. P. Carpenter, *J. Ind. Hyg. Toxicol.* **30**, 63 (1948).

Liquid; unpleasant, pungent odor. *Flammable, poisonous. Keep tightly closed.* d_4^{20} 0.938. bp 44-45°. mp $-134.5°$. n_D^{20} 1.4154. Flash pt, closed cup: $-25°F$ ($-31°C$). Slightly sol in water; misc with alcohol, chloroform, ether, petrol ether. LD_{50} orally in rats: 0.7 g/kg (Smyth, Carpenter).
Caution: Potential symptoms of overexposure are irritation of eyes, nose, skin and mucous membranes; pulmonary edema. *See NIOSH Pocket Guide to Chemical Hazards* (DHHS/NIOSH 97-140, 1997) p 10.
USE: In the synthesis of allyl compds.

281. Allyl Cyanide. [109-75-1] 3-Butenenitrile; vinylacetonitrile; β-butenonitrile. C_4H_5N; mol wt 67.09. C 71.61%, H 7.51%, N 20.88%. Found in some mustard oils. Prepd by treating dry CuCN with allyl bromide: Bruylants, *Bull. Soc. Chim. Belg.* **31**, 175 (1922); Supniewski, Salzberg, *Org. Synth.* **coll. vol. I**, 46 (2nd ed., 1941). Toxicity study: H. F. Smyth *et al.*, *Am. Ind. Hyg. Assoc. J.* **30**, 470 (1969).

Liquid. Agreeable onion-like odor. Stable to heat. d_4^{20} 0.8341. mp $-87°$. bp_{760} 119°; bp_{400} 98°; bp_{200} 78°; bp_{100} 60.2°; bp_{60} 48.8°; bp_{40} 40.0°; bp_{20} 26.6°; bp_{10} 14.1°; bp_5 2.9°; $bp_{1.0}$ $-19.6°$. n_D^{20} 1.4060. Absorption spectrum in hexane, water and dil alkali: Bruylants, Castille, *Bull. Soc. Chim. Belg.* **34**, 265; *Chem. Zentralbl.* **1926**, I, 1962. LD_{50} orally in rats: 0.115 g/kg (Smyth).

282. Allylestrenol. [432-60-0] (17β)-17-(2-Propen-1-yl)-estr-4-en-17-ol; 17-allylestr-4-en-17β-ol; 17α-allyl-17-hydroxy-19-nor-4-androstene; 17-hydroxy-17α-allyl-4-estrene; Gestanin; Gestanon; Gestanyn; Orageston; Turinal. $C_{21}H_{32}O$; mol wt 300.49. C 83.94%, H 10.73%, O 5.32%. Prepn: **GB 841411** (1960 to Organon). Determn in pharmaceutical formulations: S. Gorög *et al.*, *Analyst* **104**, 196 (1979). Clinical experience in pregnancy: J. Cortés-Prieto *et al.*, *Clin. Ther.* **3**, 200 (1980).

Crystals, mp 79.5-80°. Practically insol in water. Sol in alcohol, ether, acetone, chloroform. Sensitive to oxidizing agents.
THERAP CAT: Progestogen.

283. Allyl Ether. [557-40-4] 3,3'-Oxybis-1-propene; diallyl ether. $C_6H_{10}O$; mol wt 98.15. C 73.42%, H 10.27%, O 16.30%. $(CH_2=CHCH_2)_2O$. Prepn from allyl alcohol in the presence of $CuCl$-H_2SO_4: Stephenson, **GB 913919** (1962 to Monsanto); from allyl alcohol with K_2CO_3 and allyl bromide: Riemschneider, Kötzsch, *Monatsh. Chem.* **90**, 787 (1959). Toxicity: H. F. Smyth *et al.*, *J. Ind. Hyg. Toxicol.* **31**, 60 (1949).
Liquid; radish-like odor. d_0^{18} 0.805. bp 94°. n_D^{20} 1.4240. Practically insol in water. Miscible with alc, ether. LD_{50} orally in rats: 0.32 g/kg (Smyth).
Caution: An irritant. Can be absorbed through skin.

284. Allyl Ethyl Ether. [557-31-3] 3-Ethoxy-1-propene; ethyl 2-propenyl ether. $C_5H_{10}O$; mol wt 86.13. C 69.73%, H

11.70%, O 18.58%. Prepd from allyl bromide and sodium alcoholate: Brühl, *Ann.* **200**, 139 (1880).

Liquid. *Flammable, poisonous.* d 0.765. $bp_{742.9}$ 66-67°. n_D^{20} 1.3881. Practically insol in water. Miscible with alc, ether.
Caution: Direct contact may cause irritation to eyes, skin, mucous membranes.

285. Allyl Iodide. [556-56-9] 3-Iodo-1-propene; 3-iodopropylene. C_3H_5I; mol wt 167.98. C 21.45%, H 3.00%, I 75.55%. Prepd from allyl alcohol, methyl iodide, and triphenyl phosphite: Landauer, Rydon, *J. Chem. Soc.* **1953**, 2224; Rydon, Landauer, **GB 695468** (1953 to Nat. Res. Dev. Corp.).

Yellowish liquid; darkens on exposure to light and air, liberating iodine. Unpleasant, pungent odor. d^{12} 1.848. bp 103°. n_D^{20} 1.5540. Practically insol in water. Miscible with alc, chloroform, ether. *Flammable; corrosive. Keep tightly closed and protected from light.*
Caution: See Allyl Chloride.

286. Allyl Isothiocyanate. [57-06-7] 3-Isothiocyanato-1-propene; isothiocyanic acid allyl ester; allyl isosulfocyanate; "mustard oil"; allyl mustard oil. C_4H_5NS; mol wt 99.15. C 48.46%, H 5.08%, N 14.13%, S 32.33%. Principal constituent of *volatile oil of mustard*. Occurs in seeds as the glucoside, sinigrin, *q.v.* Isolated from *Brassica nigra* (L.) Koch, *Cruciferae* (black mustard seed), or prepd from allyl iodide and potassium thiocyanate: Dulière, *J. Pharm. Belg.* **2**, 981 (1920), *C.A.* **15**, 571[4] (1921). Description: E. Guenther, D. Althausen, *The Essential Oils* **vol. II** (Van Nostrand, New York, 1949) pp. 734-7. Review of isolation: E. Guenther, *ibid.* **vol VI**, 55-60 (1952); and synthesis: G. M. Dyson, *Perfum. Essent. Oil Rec.* **20**, 42 (1929). Evaluation in relieving muscle pain: J. G. Macarthur, S. Alstead, *Lancet* **2**, 1060 (1953). Toxicity study: P. M. Jenner *et al.*, *Food Cosmet. Toxicol.* **2**, 327 (1964). HPLC determn in brown mustard: K. Kanemaru *et al.*, *J. Jpn. Soc. Food Sci. Technol.* **37**, 565 (1990).

Colorless or pale yellow, very refractive liquid; very pungent, irritating odor; acrid taste. *Flammable, poisonous.* d_4^{15} 1.024. bp 152°. n_D^{25} 1.5248. Slightly sol in water. Misc with alcohol, carbon disulfide, ether. One ml dissolves in 8 ml 70% alcohol. LD_{50} orally in rats: 339 mg/kg (Jenner).
USE: Manuf flavors; war gas.
THERAP CAT: Counterirritant.
THERAP CAT (VET): Counterirritant.

287. Allylprodine. [25384-17-2] 1-Methyl-4-phenyl-3-(2-propen-1-yl)-4-piperidinol 4-propanoate; α-3-allyl-1-methyl-4-phenyl-4-propionoxypiperidine; α-3-allyl-1-methyl-4-phenyl-4-piperidinol propionate; NIH-7440; Ro-2-7113. $C_{18}H_{25}NO_2$; mol wt 287.40. C 75.23%, H 8.77%, N 4.87%, O 11.13%. Narcotic analgesic. Prepn: Ziering *et al.*, *J. Org. Chem.* **22**, 1521 (1957); Lee, Ziering, **US 2798073** (1957 to Hoffmann-La Roche).

Hydrochloride. $C_{18}H_{25}NO_2$·HCl. Crystals from acetone + methanol, mp 186-187°.

Note: This is a controlled substance (opiate): **21 CFR,** 1308.11.

288. Allyl Sulfide. [592-88-1] 3,3'-Thiobis-1-propene; diallyl sulfide; thioallyl ether; "oil garlic". $C_6H_{10}S$; mol wt 114.21. C 63.10%, H 8.83%, S 28.07%. Prepared from allyl bromide and sulfur in soln of Na in liquid NH_3: Brandsma, Wijers, *Rec. Trav. Chim.* **82**, 68 (1963).

$$H_2C\diagup\diagdown S\diagup\diagdown CH_2$$

Liquid, garlic odor. bp 139°. d_4^{27} 0.888. n_D^{27} 1.4877. Practically insoluble in water. Miscible with alc, chloroform, ether, carbon tetrachloride.

USE: Manuf flavors.

289. Allyltributylstannane. [24850-33-7] Tributyl-2-propen-1-ylstannane; allyltributyltin; tri-*n*-butylallyltin. $C_{15}H_{32}Sn$; mol wt 331.13. C 54.41%, H 9.74%, Sn 35.85%. Organotin reagent used to form carbon-carbon bonds. Prepn from allylmagnesium bromide and tri-*n*-butyltin chloride: W. J. Jones *et al., J. Chem. Soc.* **1947**, 1446; from allyllithium and tri-*n*-butyltin chloride: D. Seyferth, M. A. Weiner, *J. Org. Chem.* **26**, 4797 (1961); from allylmagnesium bromide and bis(tributyltin) oxide: N. G. Halligan, L. C. Blaszczak, *Org. Synth.* **coll. vol. VIII**, 23 (1993). NMR studies: R. G. Jones *et al., J. Organomet. Chem.* **35**, 291 (1972). Synthetic applications: M. Kosugi *et al., Chem. Lett.* **1977**, 301; G. E. Keck, E. J. Enholm, *J. Org. Chem.* **50**, 146 (1985); H. Nakamura *et al., J. Am. Chem. Soc.* **118**, 6641 (1996); Y. Niwa, M. Shimizu, *ibid.* **125**, 3720 (2003). Removal of organotin residues from reaction mixtures: C. J. Salomon *et al., J. Org. Chem.* **65**, 9220 (2000). Review: A. J. Borah, *Synlett* **2008**, 297-298; of allylstannanes: R. L. Marshall, *Sci. Synth.* **5**, 573-605 (2003).

$$H_3C\diagup\diagdown\diagup Sn \diagup\diagdown\diagup CH_3$$
$$H_3C\diagup\diagdown\diagup \quad\quad CH_2$$

Colorless oil. *Poisonous.* bp_{17} 155°; $bp_{1.6}$ 116°; $bp_{0.55}$ 94-95°; $bp_{0.28}$ 80°. n_D^{25} 1.4833. d_4^{25} 1.073. Stable towards air, heat, and hydrolysis.

USE: As reagent in synthetic organic chemistry.

290. Allyltrimethylsilane. [762-72-1] Trimethyl-2-propen-1-ylsilane; trimethylallylsilane; 3-(trimethylsilyl)propene. $C_6H_{14}Si$; mol wt 114.26. C 63.07%, H 12.35%, Si 24.58%. Nucleophilic reagent used to introduce allyl group functionality onto organic compds. Prepn from allylmagnesium bromide and trimethylchlorosilane: L. H. Sommer *et al., J. Am. Chem. Soc.* **70**, 2872 (1948); from methylmagnesium bromide and allyltrichlorosilane: C. A. Burkhard, *ibid.* **72**, 1078 (1950). NMR spectroscopy studies: L. Delmulle, G. P. Van Der Kelen, *J. Mol. Struct.* **66**, 315 (1980). Synthetic applications: A. Hosomi, H. Sakurai, *Tetrahedron Lett.* **17**, 1295 (1976); *idem et al., ibid.* **19**, 3043 (1978); J. R. Hwu *et al., Appl. Organomet. Chem.* **11**, 381 (1997); M. E. Jung, A. Maderna, *Tetrahedron Lett.* **45**, 5301 (2004). Reviews: I. Fleming in *Encyclopedia of Reagents for Organic Synthesis* **1**, L. A. Paquette, Ed. (John Wiley & Sons, New York, 1995) pp 133-136; P. K. Kalita, *Synlett* **2008**, 2080-2081.

$$H_3C\diagdown \underset{|}{\overset{CH_3}{Si}} \diagup\diagdown CH_2$$
$$H_3C\diagup$$

Colorless liquid. bp_{760} 85-86°; bp_{737} 84.9°. d^{20} 0.7193. n_D^{20} 1.4074. *Flammable. Irritant.* Flash pt, closed cup: 60.8°F (16°C). Freely sol in all organic solvents.

USE: Reagent in synthetic organic chemistry.

291. Allylurea. [557-11-9] *N*-2-Propen-1-ylurea; allylcarbamide. $C_4H_8N_2O$; mol wt 100.12. C 47.99%, H 8.05%, N 27.98%, O 15.98%. Prepn: Neville, McGee, *Can. J. Chem.* **41**, 2123 (1963).

$$\underset{H}{\overset{O}{\underset{||}{H_2N}}}\overset{O}{\underset{||}{N}}\diagup\diagdown CH_2$$

Crystals, mp 85°. Freely sol in water, alc. Practically insol in chloroform, ether, toluene, carbon disulfide.

USE: Manuf allylthiourea; corrosion inhibitors.

292. Almagate. [66827-12-1] Aluminum magnesium carbonate hydroxide $(AlMg_3(CO_3)(OH)_7)$ hydrate (1:2); carbonic acid aluminum-magnesium complex; [carbonato(2−)]heptahydroxy-(aluminum)trimagnesium dihydrate; Almax. $C_2H_{14}Al_2Mg_6O_{20}$·4H_2O; mol wt 629.97. C 3.81%, H 3.52%, Al 8.57%, Mg 23.15%, O 60.95%. $Al_2Mg_6(OH)_{14}(CO_3)_2$·4H_2O. A crystalline hydrated aluminum-magnesium hydroxycarbonate. Prepn: R. G. W. Spickett *et al.*, **BE 873843**; *eidem*, **US 4447417** (1979, 1984 both to Anphar). Crystal structure: J. Moragues *et al., Arzneim.-Forsch.* **34**, 1346 (1984). pH curves, acid consuming capacity: J. E. Beneyto *et al., ibid.* 1350. Effect of proteolytic enzymes, polypeptides: J. E. Beneyto, J. L. Fábregas, *ibid.* 1357. Serum levels of magnesium, aluminum not increased by high doses of almagate: J. Jauregui, J. Segura, *ibid.* 1364. Pharmacology: P. R. Beckett *et al., ibid.* 1367. Clinical trials: A. Suau *et al., ibid.* 1380.

Dehydrates at 510 K. USP acid neutralization capacity 28.3 mEq HCl/g.

THERAP CAT: Antacid.

293. Alminoprofen. [39718-89-3] α-Methyl-4-[(2-methyl-2-propen-1-yl)amino]benzeneacetic acid; 2-(*p*-methylallylamino-phenyl)propionic acid; EB-382; Minalfene. $C_{13}H_{17}NO_2$; mol wt 219.28. C 71.21%, H 7.81%, N 6.39%, O 14.59%. Prepn: **FR 2137211**; E. Bouchera, **US 3957850** (1971, 1976 both to Bouchera). Synthesis, pharmacology and toxicity data: B. Dumaitre *et al., Eur. J. Med. Chem.* **14**, 207 (1979). Pharmacokinetic study: A. Premel-Cabic *et al., Eur. J. Clin. Pharmacol.* **18**, 419 (1980).

$$H_2C\diagdown\diagup\underset{\underset{CH_3}{|}}{N}H\diagdown\diagdown\quad\overset{CH_3}{\underset{|}{CH}}\text{—COOH}$$

Crystals from cyclohexane, mp 107°. LD_{50} orally in mice: 2400 mg/kg (Dumaitre).

THERAP CAT: Anti-inflammatory.

294. Almitrine. [27469-53-0] 6-[4-[Bis(4-fluorophenyl)-methyl]-1-piperazinyl]-N^2,N^4-di-2-propen-1-yl-1,3,5-triazine-2,4-diamine; 2,4-bis[allylamino]-6-[4-[bis(*p*-fluorophenyl)methyl]-1-piperazinyl]-*s*-triazine; S-2620. $C_{26}H_{29}F_2N_7$; mol wt 477.56. C 65.39%, H 6.12%, F 7.96%, N 20.53%. Prepn: G. Regnier, R. Canevari, **DE 1947332**; *eidem*, **US 3647794** (1970, 1972 both to Sci. Union et Cie.-Soc. Franç. Recher. Méd.). Physical properties and toxicity: G. Regnier *et al., Experientia* **28**, 814 (1972). Pharmacology: M. Laubie, F. Diot, *J. Pharmacol.* **3**, 363 (1972); M. Laubie, H. Schmitt, *Eur. J. Pharmacol.* **61**, 125 (1980). Evaluation of combination with raubasine, *q.v.*, in cerebral ischemia in rats: M. G. Borzeix, J. Cahn, *Arzneim.-Forsch.* **37**, 491 (1987). Clinical effect on hypoxemia in chronic obstructive pulmonary disease: R. C. Bell *et al., Ann. Intern. Med.* **105**, 342 (1986); B. Gothe *et al., Am. J. Med.* **84**, 436 (1988).

Crystals, mp 181°.

Dimethanesulfonate. [29608-49-9] Almitrine dimesylate; Vectarion. $C_{26}H_{29}F_2N_7 \cdot 2CH_3SO_3H$; mol wt 669.76. mp 243° (dec). uv max (ethanol): 227, 246 nm (log ε 4.52, 4.53). LD_{50} in mice: 210 mg/kg i.v., 390 mg/kg i.p., >2 g/kg orally (Regnier).

THERAP CAT: Respiratory stimulant.

295. Almond, Bitter. Ripe seed of *Prunus amygdalus* Stokes var *amara* Focke *(P. communis* Arcang. var *amara* (Focke) Schneid.), *Rosaceae. Habit.* Italy, Spain, and Southern France. *Constit.* 35-50% fixed oil, about 3% amygdalin, proteins, emulsin (synaptase), sugar. *Ref:* E. W. Eckey, *Vegetable Fats and Oils* (Reinhold, New York, 1954) p 455.

Caution: Cyanide poisoning from ingestion of burnt bitter almonds has been reported: *C.A.* **50,** 6666a (1956).

USE: Preparing amygdalin, essential and expressed oils, almond and bitter-almond water; in perfumery and in manuf of liqueurs.

296. Almond, Sweet. Jordan almond. Ripe seed of *Prunus amygdalus* Stokes var *dulcis* (D.C.) Baill. *(P. communis* Arcang. var *dulcis* (Focke) Schneid.); *P. amygdalus* var *sativa* (Focke), *Rosaceae. Habit.* Italy, Spain and Southern France. *Constit.* About 50% fixed oil, proteins, emulsin, sugar. *Ref:* E. W. Eckey, *Vegetable Fats and Oils* (Reinhold, New York, 1954) p 455; Subrahmanyam, Achaya, *J. Sci. Food Agric.* **8,** 657 (1957).

USE: In perfumery and confectionery; preparing expressed oil of almond, almond milk, almond meal, etc.

297. Almorexant. [871224-64-5] $(\alpha R, 1S)$-3,4-Dihydro-6,7-dimethoxy-*N*-methyl-α-phenyl-1-[2-[4-(trifluoromethyl)phenyl]ethyl]-2(1*H*)-isoquinolineacetamide; (2*R*)-2-[(1*S*)-6,7-dimethoxy-1-[2-(4-trifluoromethylphenyl)ethyl]-3,4-dihydro-isoquinolin-2(1*H*)-yl]-*N*-methyl-2-phenylacetamide; ACT-078573. $C_{29}H_{31}F_3N_2O_3$; mol wt 512.57. C 67.96%, H 6.10%, F 11.12%, N 5.47%, O 9.36%. Dual orexin OX_1/OX_2-receptor antagonist. Prepn: T. Weller *et al.,* **WO 05118548;** H. Aissaoui *et al.,* **US 7763638** (2005, 2010 both to Actelion). Pharmacology, selectivity profile, and induction of somnolence: C. Brisbare-Roch *et al., Nat. Med.* **13,** 150 (2007). Receptor binding characteristics: P. Malherbe *et al., Mol. Pharmacol.* **76,** 618 (2009). Clinical pharmacokinetics and sleep-promoting potential: P. Hoever *et al., Clin. Pharmacol. Ther.* **87,** 593 (2010). Review of pharmacology and clinical evaluations for treatment of insomnia: D. N. Neubauer, *Curr. Opin. Investig. Drugs* **11,** 101-110 (2010).

White crystals from THF.

THERAP CAT: Sedative; hypnotic.

298. Almotriptan. [154323-57-6] *N,N*-Dimethyl-5-[(1-pyrrolidinylsulfonyl)methyl]-1*H*-indole-3-ethanamine; 1-[[[3-[2-(dimethylamino)ethyl]-1*H*-indol-5-yl]methyl]sulfonyl]pyrrolidine; 1-[[3-(2-dimethylaminoethyl)-5-indolyl]methanesulfonyl]pyrrolidine; LAS-31416. $C_{17}H_{25}N_3O_2S$; mol wt 335.47. C 60.87%, H 7.51%, N 12.53%, O 9.54%, S 9.56%. Serotonin $5HT_{1B/1D}$-receptor agonist. Prepn: D. F. Forner *et al.,* **WO 9402460;** *eidem,* **US 5565447** (1994, 1996 both to Almirall). Clinical pharmacokinetics: J. C. Fleishaker *et al., Clin. Pharmacol. Ther.* **67,** 498 (2000). Review of pharmacology: F. Kamali, *Curr. Opin. Cent. Peripher. Nerv. Syst. Invest. Drugs* **2,** 197-202 (2000); of clinical studies: D. W. Dodick, *Expert Opin. Pharmacother.* **4,** 1157-1163 (2003).

Maleate. [181183-52-8] PNU-180638E; Almogran; Axert. $C_{17}H_{25}N_3O_2S \cdot C_4H_6O_5$; mol wt 469.55. White to slightly yellow crystalline powder; sol in water.

THERAP CAT: Antimigraine.

299. Aloe. Genus of succulent plants (family *Liliaceae*) having triangular, spear-like leaves with thorny ridges. *Habit.* Africa, southern Arabia, Madagascar. Naturalized in the West Indies, southern U.S., Central America, Asia. Several species have been used medicinally, primarily *Aloe barbadensis* Miller (also known as *Aloe vera* Linné, Curaçao aloe, or Barbados aloe) and *Aloe ferox* Miller (also known as Cape aloe). "Drug aloe" or "aloes" is the dried, bitter, yellow exudate (latex) from the leaf bundle sheath cells and is used as a cathartic. The latex contains varying amounts of aloin, aloe-emodin, chrysophanic acid, aloesin, and other aloeresins. Aloe vera gel is the fresh mucilaginous gel obtained from the parenchymatous tissue in the leaf center, used for its emollient and wound healing activity. The gel consists primarily of water, sodium, potassium, and polysaccharides, such as acemannan. HPLC determn of phenolic constituents: N. Okamura *et al., J. Chromatogr. A* **746,** 225 (1996). Physical and chemical properties of the gel: Y.-T. Wang, K. J. Strong, *Phytother. Res.* **7,** S1 (1993). Effects in wound healing: J. P. Heggers *et al., ibid.* S48. Review of botany, constituents and biological activities: D. Grindlay, T. Reynolds, *J. Ethnopharmacol.* **16,** 117-151 (1986); M. D. Boudreau, F. A. Beland, *J. Environ. Sci. Health C* **24,** 103-154 (2006). Review of use in foods and cosmetics: K. Eshun, Q. He, *Crit. Rev. Food Sci. Nutr.* **44,** 91-96 (2004); of medicinal uses: C. Ulbricht *et al., J. Herb. Pharmacother.* **7,** 279-323 (2007); of safety assessment: *Int. J. Toxicol.* **26,** Suppl 2, 1-50 (2007). General review: I. A. Khan, E. A. Abourashed in *Leung's Encyclopedia of Common Natural Ingredients* (Wiley, Hoboken, 3rd Ed., 2010) pp 24-29.

USE: Gel as emollient and moisturizer in cosmetics and personal care products. Extracts in beverages as bitter flavoring agent

THERAP CAT: Latex as cathartic; gel as vulnerary.

300. Aloe-Emodin. [481-72-1] 1,8-Dihydroxy-3-(hydroxymethyl)-9,10-anthracenedione; 1,8-dihydroxy-3-(hydroxymethyl)-anthraquinone; 3-hydroxymethylchrysazin; rhabarberone. $C_{15}H_{10}O_5$; mol wt 270.24. C 66.67%, H 3.73%, O 29.60%. Occurs in the free state and as a glycoside in *Rheum* (rhubarb), in senna leaves and in various species of *Aloe* (*Liliaceae*). Isolation: Condo-Vissicchio, *Arch. Pharm.* **247,** 81 (1909); Mary *et al., J. Am. Pharm. Assoc.* **45,** 229 (1956). Prepn: Cahn, Simonsen, *J. Chem. Soc.* **1932,** 2573; Hay, Haynes, *ibid.* **1956,** 3141; Bapat *et al., Tetrahedron Letters* no. 5, 15 (1960). HPLC determn in plasma: M. Zaffaroni *et al., J. Chromatogr. B* **796,** 113 (2003).

Orange needles from toluene, mp 223-224°. Sublimes in CO_2 stream. Absorption spectrum: Stone, Furman, *J. Am. Chem. Soc.* **68,** 2742 (1946). Freely sol in hot alcohol, in ether, in benzene with yellow color, in ammonia water and in sulfuric acid with crimson color.

Trimethyl ether. $C_{18}H_{16}O_5$. Orange needles from acetic acid, mp 163°.

Triacetate. $C_{21}H_{16}O_8$. Yellow needles from benzene, mp 175-177°.

Note: See also Emodin.

THERAP CAT: Cathartic.

301. Aloesin. [30861-27-9] 8-β-D-glucopyranosyl-7-hydroxy-5-methyl-2-(2-oxopropyl)-4*H*-1-benzopyran-4-one; aloeresin B. $C_{19}H_{22}O_9$; mol wt 394.38. Major component of a family of naturally ocurring glycosides known as *aloeresins*; isolated from the latex of the aloe plant. Isoln: L. J. Haynes *et al.*, *J. Chem. Soc. C* **1970**, 2581. HPLC determn of aloeresins A and B: E. Graf, M. Alexa, *Arch. Pharm.* **313**, 285 (1980). Revised structure of A: P. Gramatica *et al.*, *Tetrahedron Lett.* **23**, 2423 (1982). HPLC determn in plasma: M. Baek *et al.*, *J. Chromatogr. B* **754**, 121 (2001); in commercial aloe preparations: M. Zahn *et al.*, *Phytochem. Anal.* **19**, 122 (2008).

Appl. **6**, 318 (1916); H. Böhme, J. Bertram, *Arch. Pharm.* **288**, 510 (1955). Structural studies: H. Mühlemann, *Pharm. Acta Helv.* **27**, 17 (1952); J. E. Hay, L. J. Haynes, *J. Chem. Soc.* **1956**, 3141. Separation of diastereomers: H. Auterhoff *et al.*, *Arch. Pharm.* **313**, 113 (1980); H.-W. Rauwald, *ibid.* **315**, 769 (1982). Crystal structure: *idem et al.*, *Angew. Chem. Int. Ed.* **28**, 1528 (1989). NMR spectra and absolute configuration: P. Manitto *et al.*, *J. Chem. Soc. Perkin Trans. 1* **1990**, 1297. Taxonomic distribution in aloe species: A. M. Viljoen *et al.*, *Biochem. Syst. Ecol.* **29**, 53 (2001). TLC determn in commercial formulations: R. Ramírez Durón *et al.*, *J. AOAC Int.* **91**, 1265 (2008).

Pale yellow with blue fluorescence; sinters at 143-144°. $[\alpha]_D^{27}$ +57.9° (ethanol). uv max (ethanol): 216, 248, 254, 297 nm (log ε 4.31, 4.21, 4.23, 3.96). Sol in water, acetonitrile.

Aloeresin A. [74545-79-2] 7-Hydroxy-8-[2-*O*-[(2*E*)-3-(4-hydroxyphenyl)-1-oxo-2-propen-1-yl]-β-D-glucopyranosyl]-5-methyl-2-(2-oxopropyl)-4*H*-1-benzopyran-4-one. $C_{28}H_{28}O_{11}$; mol wt 540.52. mp 148-150°. uv max (methanol): 228, 252, 300 nm (ε 34250, 25000, 37960). Sol in methanol.

302. Alogliptin. [850649-61-5] 2-[[6-[(3*R*)-3-Amino-1-piperidinyl]-3,4-dihydro-3-methyl-2,4-dioxo-1(2*H*)-pyrimidinyl]-methyl]benzonitrile; 6-((*R*)-3-aminopiperidin-1-yl)-1-(2-cyanobenzyl)-3-methyl-1*H*-pyrimidine-2,4-dione; (*R*)-2-((6-(3-aminopiperidin-1-yl)-3-methyl-2,4-dioxo-3,4-dihydropyrimidin-1(2*H*)-yl)methyl)benzonitrile. $C_{18}H_{21}N_5O_2$; mol wt 339.40. C 63.70%, H 6.24%, N 20.63%, O 9.43%. Dipeptidyl peptidase IV (DPP-IV) inhibitor. Prepn: J. Feng *et al.*, **JP 05263780**; *eidem*, **US 050261271** (both 2005 to Takeda). Discovery and optimization: J. Feng *et al.*, *J. Med. Chem.* **50**, 2297 (2007). Clinical pharmacokinetics: P. Covington *et al.*, *Clin. Ther.* **30**, 499 (2008). Clinical experience in combination with metformin in type 2 diabetes: M. A. Nauck *et al.*, *Int. J. Clin. Pract.* **63**, 46 (2009). Review of development and clinical experience: A. Glode, S. Abdelghany, *Formulary* **43**, 317-325 (2008); R. E. Pratley, *Expert Opin. Pharmacother.* **10**, 503-512 (2009).

Sol in THF, dioxane, acetonitrile, ethyl acetate, dichloromethane.
Benzoate. [850649-62-6] SYR-322; Nesina. $C_{18}H_{21}N_5O_2$.$C_7H_6O_2$; mol wt 461.52.
THERAP CAT: Antidiabetic.

303. Aloin. [8015-61-0] 1,8-Dihydroxy-3-hydroxymethyl-10-(6-hydroxymethyl-3,4,5-trihydroxy-2-pyranyl)anthrone; 10-(1′,5′-anhydroglucosyl)-aloe-emodin-9-anthrone. $C_{21}H_{22}O_9$; mol wt 418.40. C 60.28%, H 5.30%, O 34.41%. Bitter, purgative principle of aloe, *q.v.* Mixture of diastereoisomeric *C*-glucosides, aloins A and B, derived from aloe-emodin, *q.v.*; isomers readily interconvert in the presence of bases. Isoln from various species of aloe: E. Groenewold, *Arch. Pharm.* **228**, 115 (1890); E. Léger, *Ann. Chim.*

Lemon-yellow crystals, mp 148-149°. Soly at 18°: 57% in pyridine, 7.3% in glacial acetic acid, 5.4% in methanol, 3.2% in acetone, 2.8% in methyl acetate, 1.9% in ethanol, 1.8% in water, 1.6% in propanol, 0.78% in ethyl acetate, 0.27% in isopropanol.

Aloin A. [1415-73-2] (10*S*)-10-β-D-Glucopyranosyl-1,8-dihydroxy-3-(hydroxymethyl)-9(10*H*)-anthracenone; barbaloin. Crystals from methanol. mp 148°. $[\alpha]_D^{30}$ +10.2° (c = 0.5 in methanol).

Aloin B. [28371-16-6] (10*R*)-10-β-D-Glucopyranosyl-1,8-dihydroxy-3-(hydroxymethyl)-9(10*H*)-anthracenone; isobarbaloin. Yellow-brown, trigonal crystals from water/methanol. mp 138-140°. $[\alpha]_D^{30}$ −73.0° (c = 0.5 in methanol).

304. Alosetron. [122852-42-0] 2,3,4,5-Tetrahydro-5-methyl-2-[(4-methyl-1*H*-imidazol-5-yl)methyl]-1*H*-pyrido[4,3-*b*]indol-1-one; GR-68755. $C_{17}H_{18}N_4O$; mol wt 294.36. C 69.37%, H 6.16%, N 19.03%, O 5.44%. Serotonin 5HT₃-receptor antagonist. Prepn: I. H. Coates *et al.*, **EP 306323**; *eidem*, **US 5360800** (1989, 1994 both to Glaxo); of isotopically labelled compd: S. R. Prakash *et al.*, *J. Labelled Compd. Radiopharm.* **36**, 993 (1995). HPLC determn in plasma: T. L. Lloyd *et al.*, *J. Chromatogr. B* **678**, 261 (1996). Review of clinical pharmacology: M. D. Gunput, *Aliment. Pharmacol. Ther.* **13**, Suppl. 2, 70-76 (1999); of clinical studies: A. W. Mangel, A. R. Northcutt, *ibid.* 77-82. Clinical trial in irritable bowel syndrome: M. Camilleri *et al.*, *Lancet* **355**, 1035 (2000).

Hydrochloride. [122852-69-1] GR-68755C; Lotronex. $C_{17}H_{18}N_4O$.HCl; mol wt 330.82. mp 288-291°.
THERAP CAT: In treatment of irritable bowel syndrome.

305. Aloxiprin. [9014-67-9] Polyoxyaluminum acetylsalicylate; Lyman; Palaprin; Rumatral. A polymeric condensation product of aluminum oxide and aspirin. Prepd from aluminum isopropoxide and aspirin: Cummings *et al.*, *J. Pharm. Pharmacol.* **15**, 56 (1963).
White tasteless powder. Practically insol in water. Hyrolyzes rapidly in alkaline media.
THERAP CAT: Analgesic.

306. Alphaprodine. [77-20-3] *rel*-(3*R*,4*S*)-1,3-Dimethyl-4-phenyl-4-piperidinol 4-propanoate; (±)-α-1,3-dimethyl-4-phenyl-4-

piperidinyl propionate; dl-α-1,3-dimethyl-4-phenyl-4-propionoxy-piperidine; dl-α-prodine. $C_{16}H_{23}NO_2$; mol wt 261.37. C 73.53%, H 8.87%, N 5.36%, O 12.24%. Mixture of the two cis isomers of prodine; **betaprodine** is the mixture of $trans$ isomers. Prepn: J. Lee, A. Ziering, **US 2498433** (1950 to Hoffmann-La Roche); A. H. Beckett et al., J. Pharm. Pharmacol. **9**, 939 (1957); A. Ziering et al., J. Org. Chem. **22**, 1521 (1957). Configurational studies: F. R. Ahmed et al., Chem. Ind. (London) **1959**, 485; eidem, ibid. **1962**, 97. Stereostructure-activity studies: M. M. Abdel-Monem et al., J. Med. Chem. **15**, 494 (1972). Pharmacology and toxicology: G. M. Gruber et al., J. Pharmacol. Exp. Ther. **99**, 312 (1950). Review of clinical experience: R. C. Lunt, H. E. Howard, Pediatr. Dent. **10**, 121-126 (1988).

Relative stereochemistry

Hydrochloride. [561-78-4] Nu-1196; Nisentil. $C_{16}H_{23}NO_2$.HCl; mol wt 297.82. Crystals from acetone, mp 220-221°. Slightly saline taste. Freely sol in water, alc, chloroform. Practically insol in ether. pH of 1% aq soln 4.5-5.2. LD_{50} in mice: 54 mg/kg i.v., 73 mg/kg i.p.; in rats: 22 mg/kg i.p. (Gruber).
Betaprodine hydrochloride. [49638-23-5] (±)-β-Prodine hydrochloride; Nu-1779. Crystals from methyl ethyl ketone, mp 195-196° (Beckett); from acetone + methanol, mp 199-200° (Ziering).
Note: These are controlled substances (opiates): **21 CFR**, 1308.11 (betaprodine) and 1308.12 (alphaprodine).
THERAP CAT: Analgesic.

307. Alpiropride. [81982-32-3] 4-Amino-2-methoxy-5-[(methylamino)sulfonyl]-N-[[1-(2-propen-1-yl)-2-pyrrolidinyl]methyl]benzamide; (±)-N-[(1-allyl-2-pyrrolidinyl)methyl]-4-amino-5-(methylsulfamoyl)-o-anisamide; N-(1-allyl-2-pyrrolidinylmethyl)-2-methoxy-4-amino-5-methylsulfamoylbenzamide; RIV-2093; Rivistel. $C_{17}H_{26}N_4O_4S$; mol wt 382.48. C 53.39%, H 6.85%, N 14.65%, O 16.73%, S 8.38%. Dopamine D_2-receptor antagonist structurally similar to sulpiride, q.v. Prepn: J. Perrot, M. Thominet, **EP 47207**; eidem, **US 4550179** (1982, 1985 both to Soc. d'Etudes Sci. Ind. l'Ile-de-France). Dopamine receptor binding study: P. Sokoloff et al., Arch. Pharmacol. **327**, 221 (1984). HPLC determn in plasma and urine: F. Bressolle et al., J. Chromatogr. **343**, 443 (1985). Pharmacokinetics and metabolism in humans: F. Bressolle et al., J. Pharm. Clin. **4**, 261 (1985). Preliminary study in experimentally induced migraine: A. Bès et al., Int. J. Clin. Pharmacol. Res. **6**, 189 (1986).

Crystals from absolute ethanol, mp 168.5-169°. LD_{50} in male mice (mg/kg): 44 i.v.; 184 i.p.; 204 s.c.; 3600 orally (Perrot, Thominet, 1985).
THERAP CAT: Antimigraine.

308. Alprazolam. [28981-97-7] 8-Chloro-1-methyl-6-phenyl-4H-[1,2,4]triazolo[4,3-a][1,4]benzodiazepine; 8-chloro-1-methyl-6-phenyl-4H-s-triazolo[4,3-a][1,4]benzodiazepine; D-65MT; U-31889; Alplax; Cassadan; Esparon; Tafil; Tranquinal; Trankimazin; Xanax. $C_{17}H_{13}ClN_4$; mol wt 308.77. C 66.13%, H 4.24%, Cl 11.48%, N 18.15%. Benzodiazepine anxiolytic. Prepn: J. B. Hester, **DE 2012190**; idem, **US 3987052** (1970, 1976 both to Upjohn);

J. B. Hester et al., Tetrahedron Lett. **1971**, 1609; A. Walser, G. Zenchoff, J. Med. Chem. **20**, 1694 (1977). Central depressant activity: R. Nakajima et al., Jpn. J. Pharmacol. **21**, 497 (1971). Pharmacology: V. H. Sethy, Arch. Pharmacol. **301**, 157 (1978). Clinical studies: L. F. Fabre, Curr. Ther. Res. **19**, 661 (1976); J. B. Cohn, J. Clin. Psychiatry **42**, 347 (1981). Pharmacokinetics: D. R. Abernethy et al., ibid. **44**, 45 (1983). HPLC determn in tablets: P. Pérez-Lozano et al., J. Pharm. Biomed. Anal. **34**, 979 (2004); in plasma: A. Allqvist et al., J. Chromatogr. B **814**, 127 (2005). Review of pharmacokinetics, clinical efficacy, and mechanism of action: J. A. Fawcett, H. M. Kravitz, Pharmacotherapy **2**, 242-254 (1982); of pharmacology and efficacy in anxiety and depression: G. W. Dawson et al., Drugs **27**, 132-147 (1984); of extended release formulation in panic disorder: K. Rickels, Expert Opin. Pharmacother. **5**, 1599-1611 (2004).

Crystals from ethyl acetate, mp 228-228.5°. Freely sol in chloroform; sol in alc; sparingly sol in acetone; slightly sol in ethyl acetate. Insol in water. uv max (ethanol): 222 nm (ε 40250). LD_{50} in mice, rats (mg/kg): 1020, >2000 orally; 540, 610 i.p. (Nakajima).
Note: This is a controlled substance (depressant): **21 CFR**, 1308.14.
THERAP CAT: Anxiolytic.
THERAP CAT (VET): In treatment of behavioral disorders.

309. Alprenolol. [13655-52-2] 1-[(1-Methylethyl)amino]-3-[2-(2-propen-1-yl)phenoxy]-2-propanol; 1-(o-allylphenoxy)-3-(isopropylamino)-2-propanol; H-56/28. $C_{15}H_{23}NO_2$; mol wt 249.35. C 72.25%, H 9.30%, N 5.62%, O 12.83%. β-Adrenergic blocker. Prepn: Brandstrom et al., Acta Pharm. Suec. **3**, 303 (1966); **NL 6605692** (1966 to AB Hassle), C.A. **66**, 46214p (1967); **NL 6612958** (1967 to ICI), C.A. **67**, 99851w (1967). Pharmacology: Marmo, Clin. Ter. **56**, 121-176 (1971). Series of articles on clinical effect in myocardial infarction: Acta Med. Scand. **1984**, Suppl. 680, 1-64.

Hydrochloride. [13707-88-5] Apllobal; Aptine; Aptol Duriles; Gubernal; Regletin; Yobir. $C_{15}H_{23}NO_2$.HCl; mol wt 285.81. Crystals from ethyl acetate, mp 107-109°. LD_{50} in mice, rats, rabbits (mg/kg): 278.0, 597.0, 337.3 orally (Marmo).
THERAP CAT: Antihypertensive; antianginal; antiarrhythmic.

310. Alsactide. [34765-96-3] 1-β-Alanine-17-[N-(4-aminobutyl)-L-lysinamide]-α^{1-17}-corticotropin; 1-β-alanine-17-[L-2,6-diamino-N-(4-aminobutyl)hexanamide]-α^{1-17}-corticotropin; [β-ala^1,-lys^{17}]corticotropin-(1-17)-heptadecapeptide-4-amino-N-butylamide; [β-ala^1,lys^{17}]ACTH$^{1-17}$-4-amino-N-butylamide; alisactide; HOE-433; Synchrodyn. $C_{99}H_{155}N_{29}O_{21}S$; mol wt 2119.57. C 56.10%, H 7.37%, N 19.16%, O 15.85%, S 1.51%. Short-chain ACTH analog. Synthesis: R. Geiger, H. G. Schroeder, **DE 1954794**; eidem, **US 3749704** (1971, 1973, both to Hoechst); R. Geiger, Ann. **750**, 165 (1971). Pharmacology: J. Sandow et al., Arch. Pharmacol. **297**, Suppl. 2, R41 (1977). Dose dependent effects on cortisol and aldosterone plasma levels in man: A. Angeli et al., Horm. Metab. Res. **13**, 24 (1981). Review of chronobiological activity: idem et al., Chronobiologica **14**, 99-143 (1987).

β-Ala–Tyr–Ser–Met–Glu–His–Phe–Arg–Trp–Gly–Lys–Pro–Val–Gly–Lys–Lys–LysNH(CH₂)₄NH₂

Acetate. [33194-27-3] $C_{99}H_{155}N_{29}O_{21}S.xC_2H_4O_2$.
$[\alpha]_D^{20}$ −68.6° (c = 0.5 in 1% acetic acid).
THERAP CAT: Diagnostic aid (adrenal function).

311. Alstonidine. [25394-75-6] (2S,3S,4S)-3,4-Dihydro-3-(hydroxymethyl)-2-methyl-4-[(9-methyl-9H-pyrido[3,4-b]indol-1-yl)methyl]-2H-pyran-5-carboxylic acid methyl ester; 3β-(hydroxymethyl)-2α-methyl-4β-[(9-methyl-9H-pyrido[3,4-b]indol-1-yl)methyl]-2H-pyran-5-carboxylic acid methyl ester. $C_{22}H_{24}N_2O_4$; mol wt 380.44. C 69.46%, H 6.36%, N 7.36%, O 16.82%. From the bark or root of *Alstonia constricta* F. Muell., *Apocynaceae:* Hesse, *Ann.* **205**, 360 (1880); Svoboda, *J. Am. Pharm. Assoc.* **46**, 508 (1957). Structure: Boaz *et al., ibid.* 510. Stereochemistry: Crow *et al., Aust. J. Chem.* **23**, 2489 (1970).

Fine crystals from ether, mp 188-190°. Practically insol in water. Sol in alcohol, chloroform, ether, acetone, 5% HCl. uv max (methanol): 238, 291, 360 nm (log ε 4.66, 4.25, 3.74). The acid solutions are fluorescent.

Acetyl deriv trihydrate. $C_{24}H_{26}N_2O_5.3H_2O$. Fine needles from dil acetone, mp 92-96°.

312. Alstonine. [642-18-2] (19α,20α)-3,4,5,6,16,17-Hexadehydro-16-(methoxycarbonyl)-19-methyloxayohimbanium inner salt; (4S,4aS,14aS)-4,4a,5,13,14,14a-hexahydro-1-(methoxycarbonyl)-4-methylindolo[2,3-a]pyrano[3,4-g]quinolizin-6-ium inner salt. $C_{21}H_{20}N_2O_3$; mol wt 348.40. C 72.40%, H 5.79%, N 8.04%, O 13.78%. Indole alkaloid found in medicinal plants of the genus *Alstonia* and other *Apocynaceae*; stereoisomer of serpentine, *q.v.* Isoln from *A. constricta* F. Muell.: T. M. Sharp, *J. Chem. Soc.* **1934**, 287; from *Rauwolfia vomitoria* Afzel. and *R. obscura* K. Schum.: E. Schlittler *et al., Helv. Chim. Acta* **35**, 271 (1952); from *Vinca rosea* L.: P. P. Pillay, T. N. S. Kumari, *J. Sci. Ind. Res.* **20B**, 458 (1961). Structure: F. E. Bader, *Helv. Chim. Acta* **36**, 215 (1953). Stereochemistry: E. Wenkert *et al., J. Am. Chem. Soc.* **83**, 5037 (1961); M. Shamma, J. B. Moss, *ibid.* 5038; M. Shamma, J. M. Richey, *ibid.* **85**, 2507 (1963). Biosynthesis: R. B. Woodward, *Angew. Chem.* **68**, 13 (1956). Antipsychotic profile: L. Costa-Campos *et al., Pharmacol. Biochem. Behav.* **60**, 133 (1998). Review of pharmacology: E. Elisabetsky, L. Costa-Campos, *Evid. Based Complement. Alternat. Med.* **3**, 39-48 (2006).

Fine, yellow-orange needles from acetone, mp 205-210° (dec). uv max (methanol): 252, 289, 309, 369 nm (log ε 4.54, 4.08, 4.36, 3.39, 3.60). Quickly decomposes on standing in organic solvents, yielding red solns with blue fluorescence.

Hydrochloride. $C_{21}H_{20}N_2O_3.HCl$. Yellow plates from abs alcohol, mp 286° (dec). $[\alpha]_D$ +131.9° (c = 1.064 in water). Sol in water.

313. Althea. Marshmallow. Dried root of *Althaea officinalis* L., *Malvaceae* (marshmallow), deprived of brown, corky layer.

Habit. Europe, Western and Northern Asia, naturalized in Eastern U.S. *Constit.* Asparagine, 25-35% mucilage, sugar, pectin.
USE: Demulcent.

314. Althiazide. [5588-16-9] 6-Chloro-3,4-dihydro-3-[(2-propen-1-ylthio)methyl]-2H-1,2,4-benzothiadiazine-7-sulfonamide 1,1-dioxide; 3-[(allylthio)methyl]-6-chloro-3,4-dihydro-2H-1,2,4-benzothiadiazine-7-sulfonamide 1,1-dioxide; 3-allylthiomethyl-6-chloro-7-sulfamoyl-3,4-dihydrobenzothiadiazine 1,1-dioxide; 6-chloro-3,4-dihydro-7-sulfamoyl-3-(2-thiapent-4-enyl)-2H-1,2,4-benzothiadiazine 1,1-dioxide; altizide; P-1779. $C_{11}H_{14}ClN_3O_4S_3$; mol wt 383.88. C 34.42%, H 3.68%, Cl 9.23%, N 10.95%, O 16.67%, S 25.05%. Prepn: **GB 902658**; W. M. McLamore, G. D. Laubach, **US 3111517** (1962, 1963 both to Pfizer). HPLC determn in tablets: F. DeCroo *et al., J. Chromatogr.* **329**, 422 (1985). Fast GC-MS determn in urine: V. Morra *et al., J. Chromatogr. A* **1135**, 219 (2006). Series of articles on clinical experience of combination with spironolactone in hypertension: *Am. J. Cardiol.* **65**, K20-K35 (1990).

Solid, mp 206-207°.
Combination with spironolactone. Aldactazine.
THERAP CAT: Diuretic; antihypertensive.

315. Altrenogest. [850-52-2] (17β)-17-Hydroxy-17-(2-propen-1-yl)estra-4,9,11-trien-3-one; 17α-allyl-17-hydroxyestra-4,-9,11-trien-3-one; 13β-methyl-17α-allyl-Δ^{4,9,11}-gonatriene-17β-ol-3-one; allyltrenbolone; A-35957; RU-2267; Regu-Mate. $C_{21}H_{26}O_2$; mol wt 310.44. C 81.25%, H 8.44%, O 10.31%. Synthetic oral progestogen. Prepn: **NL 6401555**; G. Nomine *et al.,* **US 3257278** (1964, 1966 both to Roussel). Effect on estrous cycle and fertility of mares: E. L. Squires *et al., J. Anim. Sci.* **49**, 729 (1979). Synchronization of estrus in swine: R. R. Kraeling *et al., ibid.* **52**, 831 (1981); in gilts: V. G. Pursel *et al., ibid.* 130.

Cryst, mp 120°. $[\alpha]_D^{20}$ −72° (c = 0.5 in ethanol).
THERAP CAT (VET): Progestogen.

316. Altretamine. [645-05-6] N^2,N^2,N^4,N^4,N^6,N^6-Hexamethyl-1,3,5-triazine-2,4,6-triamine; 2,4,6-tris(dimethylamino)-s-triazine; hemel; hexamethylmelamine; HMM; ENT-50852; NSC-13875; Hexalen; Hexastat. $C_9H_{18}N_6$; mol wt 210.29. C 51.40%, H 8.63%, N 39.96%. An antitumor agent which also acts as a chemosterilant for male houseflies and other insects. Synthesis: A. W. Hofmann, *Ber.* **18**, 2755 (1885); D. W. Kaiser *et al., J. Am. Chem. Soc.* **73**, 2984 (1951); Y. Bessiere-Chretien, H. Serne, *Bull. Soc. Chim. Fr.* **6**, 2039 (1973). Production process: H. von Brachel, H. Kindler, **DE 1240870**; *eidem*, **US 3424752** (1967, 1969 both to Casella Farbwerke). Chemosterilant effect: S. C. Chang *et al., Science* **144**, 57 (1964); A. De Milo *et al., J. Econ. Entomol.* **65**, 1548 (1972). Chromatographic detn in plasma or serum: A. Hulshoff *et al., J. Chromatogr.* **181**, 363 (1980). Disposition and metabolism in rabbits and humans: M. M. Ames *et al., Cancer Res.* **39**, 5016 (1979). Evaluation in breast cancer: C. J. Fabian *et al., Cancer Treat. Rep.* **63**, 1359 (1979); in ovarian cancer: J. T. Wharton *et al., Am. J. Obstet. Gynecol.* **133**, 833 (1979). Mammalian toxicity study: R. L. Jasper *et al., Fed. Proc.* **24**, 641 (1965). Toxicological studies in dogs and mice: D. C. Thake *et al., Gov. Rep. Announce. Index (USA)*

79, 92 (1979). *In vitro* cytotoxicity study: M. D'Incalci *et al.*, *Br. J. Cancer* **41**, 630 (1980).

Needles from abs ethanol, mp 172-174°. uv max (ethanol): 226 nm (ε 49400), H. J. Anderson *et al.*, *Can. J. Chem.* **49**, 2315 (1971). Sol in chloroform. Insol in water. LD_{50} in rats, guinea pigs (mg/kg): 350, 255 orally (Jasper).

USE: As exptl insect chemosterilant.

THERAP CAT: Antineoplastic.

317. D-Altrose. [1990-29-0] D-Altropyranose. $C_6H_{12}O_6$; mol wt 180.16. C 40.00%, H 6.71%, O 53.28%. Prepd by the reduction of D-altronic lactone with sodium amalgam: Levene, Jacobs, *Ber.* **43**, 3143 (1910).

Prisms from dilute alcohol, mp 103-105°. $[\alpha]_D^{20}$ +32.6° (c = 7.6). Soluble in water. Practically insol in alcohol.

Phenylosazone. $C_{18}H_{22}N_4O_4$. mp 178°.

318. Aluminon. [569-58-4] 3,3'-[(3-Carboxy-4-oxo-2,5-cyclohexadien-1-ylidene)methylene]bis[6-hydroxybenzoic acid] ammonium salt (1:3); 5-[(3-carboxy-4-hydroxyphenyl)(3-carboxy-4-oxo-2,5-cylohexadien-1-ylidene)methyl]-2-hydroxybenzoic acid triammonium salt; 3-[bis(3-carboxy-4-hydroxyphenyl)methylene]-6-oxo-1,4-cyclohexadiene-1-carboxylic acid triammonium salt; ammonium salt of aurintricarboxylic acid; Lysofon. $C_{22}H_{23}N_3O_9$; mol wt 473.44. C 55.81%, H 4.90%, N 8.88%, O 30.41%. Prepd by reacting sodium nitrite with salicylic acid and adding formaldehyde, then treating with ammonia: G. B. Heisig, W. M. Lauer, *Org. Synth.* **coll. vol. I**, 54 (2nd ed., 1941).

Yellowish-brown, glassy powder. Freely sol in water.

USE: Forms brilliantly colored lakes with aluminum, chromium, iron, beryllium. Generally used for the detection and colorimetric estimation of aluminum in water, foods, tissues. *See:* Scherrer, Mogerman, *J. Res. Natl. Bur. Stand.* **21**, 105 (1938).

THERAP CAT: Pharyngeal aerosol spray.

319. Aluminum. [7429-90-5] Aluminium. Al; at. wt 26.9815386; at. no. 13; valence 3. Group IIIA (13). Naturally occurring isotope (mass number): 27 (100%); known artificial radioactive isotopes: 22-25, 26 ($T_{1/2}$ 7.2×10⁵ years, longest-lived known isotope, β^+ and γ emitter, found in meteors), 28-32. One of the most abundant metals in earth's crust: 8.3% by wt (83,000 ppm); occurs in nature primarily in combination with silica, also as oxide (*see* Aluminum Silicate; Aluminum Oxide). First obtained in impure form by Oersted in 1825; prepd as metal powder by Wöhler in 1827.

Commercially important source is bauxite. Reviews of aluminum, its alloys and compds: Brandt, "Aluminum and Aluminum Alloys" in *Proc. Met. Soc. Conf.* **vol. 40**, E. D. Verink, Ed. (Gordon & Breach, New York, 1966); *Aluminum* **3 vols.**, K. R. Van Horn, Ed. (American Society for Metals, Metal Park, Ohio, 1967); Wade, Bannister, "Aluminum, Gallium, Indium and Thallium" in *Comprehensive Inorganic Chemistry* **vol. 1**, J. C. Bailar, Jr. *et al.*, Eds. (Pergamon Press, Oxford, 1973) pp 993-1064; *Chemistry of the Elements*, N. N. Greenwood, A. Earnshaw, Eds. (Pergamon Press, New York, 1984) pp 243-295; J. T. Stanley, W. Haupin in *Kirk-Othmer Encyclopedia of Chemical Technology* **vol. 2** (John Wiley & Sons, 4th ed., 1992) pp 184-251; W. C. Sleppy *et al.*, *ibid.* 252-345. Review of clinical toxicology: C. D. Hewitt *et al.*, *Clin. Lab. Med.* **10**, 403-422 (1990); of toxicology and human exposure: *Toxicological Profile for Aluminum* (PB2009-100001, 2009) 357 pp. Book: *Chemistry of Aluminum, Gallium, Indium and Thallium*, A. J. Downs, Ed. (Blackie Academic & Professional, London, 1993) 515 pp.

Tin-white, malleable, ductile metal, with somewhat bluish tint; capable of taking brilliant polish which is retained in dry air. In moist air, oxide film forms which protects metal from corrosion. Available in bars, leaf, powder, sheets, or wire. d 2.70. mp 660°. bp 2327°. Does not vaporize even at high temps, but finely divided aluminum dust is easily ignited, and may cause explosions. Reacts with dil HCl, H_2SO_4, KOH and NaOH with evolution or hydrogen. Reduces the cations of many heavy metals to the metallic state E°(aq) Al^{3+}/Al −1.66 V. Solns of Al^{3+} in dil HCl or neutral or slightly acid solns of most aluminum salts, yield with Na_2S, a white ppt soluble in excess of Na_2S. Dil neutral soln of aluminum salts yields white gelatinous ppt on boiling with sodium acetate. Sol in HCl, H_2SO_4, hot water, alkalies. Insol in water.

Caution: Potential symptoms of overexposure are irritation of eyes, skin and respiratory system. *See NIOSH Pocket Guide to Chemical Hazards* (DHHS/NIOSH 97-140, 1997) p 12. *See also Patty's Industrial Hygiene and Toxicology* **vol. 2C**, G. D. Clayton, F. E. Clayton, Eds. (John Wiley & Sons, New York, 4th ed., 1994) pp 1881-1902.

USE: As pure metal or alloys (magnalium, aluminum bronze, etc.) for structural material in construction, automotive, electrical and aircraft industries. In cooking utensils, highway signs, fencing, containers and packaging, foil, machinery, corrosion resistant chemical equipment, dental alloys. The coarse powder in aluminothermics (thermite process); the fine powder as flashlight in photography; in explosives, fireworks, paints; for absorbing occluded gases in manuf of steel. In testing for Au, As, Hg; coagulating colloidal solns of As or Sb; pptg Cu; reducer for determining nitrates and nitrites; instead of Zn for generating hydrogen in testing for As. Forms complex hydrides with lithium and boron, such as $LiAlH_4$, which are used in preparative organic chemistry.

320. Aluminum Acetate. [139-12-8] Acetic acid aluminum salt (3:1); aluminum triacetate; neutral aluminum acetate. $C_6H_9AlO_6$; mol wt 204.11. C 35.31%, H 4.44%, Al 13.22%, O 47.03%. $Al(CH_3CO_2)_3$. Prepn from acetic acid + acetic anhydride with metallic aluminum: J. Lösch, US 2141477 (1938 to AG Stickstoffdunger, Knapsack); with solid anhydrous aluminum chloride: G. C. Hood, A. J. Ihde, *J. Am. Chem. Soc.* **72**, 2094 (1950). Antibacterial activity and clinical efficacy in treatment of ear infections: M. A. Thorp *et al.*, *J. Laryngol. Otol.* **112**, 925 (1998); M. Kashiwamura *et al.*, *Otol. Neurotol.* **25**, 9 (2004).

Minute white crystals. Must be prepd under strictly anhydrous conditions. Sol in water.

Aluminum acetate solution. Burow's solution; Domeboro. Prepd from aluminum subacetate solution and glacial acetic acid. May also be prepared by dissolving aluminum sulfate and calcium acetate in water. Colorless liquid; slight odor of acetic acid. Sweetish, astringent taste. d ~1.002. pH of 1:20 aq soln 4.2.

THERAP CAT: Astringent, antiseptic.

THERAP CAT (VET): Astringent, antiseptic.

321. Aluminum Acetotartrate. Alsol; Essitol. Consists of approx 70% basic aluminum acetate and 30% tartaric acid. Prepn: *Gmelins, Aluminum* (8th ed.) **35B**, 304 (1934).

Crystals or powder; slight acetic odor; astringent, acidulous taste. On long keeping loses acetic acid and becomes incompletely sol. Slowly sol in cold water. Practically insol in alcohol.

THERAP CAT: Astringent, antiseptic.

322. Aluminum Alkyls. Highly reactive compounds of the form R_3Al, *trialkylaluminum*; $[R_nAlX_{3-n}]_2$, n = 1, 2, *alkylaluminum halide*; or $R_3Al_2X_3$, *alkylaluminum sesquihalide*. (The sesquihalides are actually equilibrium mixtures of $R_3Al_2X_3$, $[R_2AlX]_2$ and $[RAlX_2]_2$.) First prepn of aluminum alkyls: ethylaluminum sesquiiodide $[(C_2H_5)_3Al_2I_3]$ prepd by Hallwachs, Schaferik, *Ann.* **109**, 207 (1859); trimethylaluminum, $[(CH_3)_3Al]_2$, and triethylaluminum, $[(C_2H_5)_3Al]_2$, prepd by Buckton, Odling, *Ann.* Suppl. **4**, 109 (1865). The alkylaluminum halides are prepd from alkyl halides and aluminum, or by halogenation of trialkylaluminums. The trialkylaluminums are prepd from aluminum, hydrogen and olefins; this improved "direct synthesis" was developed by Ziegler and co-workers. Comprehensive reviews of prepn, properties and chemistry: Schultz in *Adv. Chem. Ser.* **23**, entitled "Metal-Organic Compounds," M. Sittig, Ed. (ACS, Washington DC, 1959) pp 163-171; Ziegler in *ACS Monograph Series* no. **147**, entitled "Organometallic Chemistry," H. Zeiss, Ed. (Reinhold, New York, 1960) pp 194-269; Köster, Binger, *Adv. Inorg. Chem. Radiochem.* **7**, 263-348 (1965); T. Mole, E. A. Jeffery, *Organoaluminum Compounds* (Elsevier, New York, 1972) 465 pp; Wade, Banister in *Comprehensive Inorganic Chemistry* vol **1**, J. C. Bailar, Jr. *et al.*, Eds. (Pergamon Press, Oxford, 1973) pp 1058-1064.

Trialkylaluminum compounds. Colorless liquids at room temperature. Must be stored in an inert atm; sensitive to oxidation and hydrolysis in air; the lighter trialkylaluminums ignite spontaneously in air. The low mol wt, linear-chain alkyl compounds exist as dimers; the branched-chain alkyl compounds exist primarily as monomers. Among the industrially important trialkyl aluminums are *triethylaluminum*, dimeric liq, d^{25} 0.832, bp_{760} 194°, bp_{13} 100° and *triisobutylaluminum*, primarily monomeric liq, d^{25} 0.781; bp_{10} 86°, mp 6°.

Alkylaluminum halides. Low melting solids or colorless, volatile liquids. Strongly associated forming dimers; held together by bridging halogen bonds. Less sensitive than trialkylaluminums to oxidation upon exposure to air. Halogen aluminum bonds cleaved by water, alcohol. Industrially important halides include ethylaluminum dichloride, *q.v.*, and *chlorodiethylaluminum*, $[(C_2H_5)_2AlCl]_2$, liq, d^{25} 0.961, bp_{50} 127°, bp_{17} 100°, $bp_{1.9}$ 60°.

USE: Catalyst; with compds of early transition metals as Ziegler-Natta polymerization catalysts; intermediates in organic syntheses.

323. Aluminum Ammonium Sulfate. [7784-25-0] Sulfuric acid aluminum ammonium salt (2:1:1); ammonium aluminum sulfate; burnt ammonium alum; exsiccated ammonium alum. $AlH_4NO_8S_2$; mol wt 237.13. Al 11.38%, H 1.70%, N 5.91%, O 53.98%, S 27.04%. $AlNH_4(SO_4)_2$. Prepn: *Gmelins, Aluminum* (8th ed) **35B**, pp 508-515 (1934). Dodecahydrate crystal structure: Larson, Cromer, *Acta Crystallogr.* **22**, 793 (1967); enthalpy of soln and crystn: J. Fisher *et al.*, *J. Solution Chem.* **24**, 659 (1995); properties of mixed aqueous solns: J. W. Mullin, M. Sipek, *J. Chem. Eng. Data* **27**, 181 (1982); V. Hostomska, J. Hostomsky, *ibid.* **51**, 243 (2006).

About 97-98% pure, the balance is chiefly excess Al_2O_3. White powder. One gram dissolves in about 20 ml cold, 1.5 ml boiling water, usually incompletely. Practically insol in alcohol. *Keep well closed.*

Dodecahydrate. [7784-26-1] Ammonium alum. $AlH_4NO_8S_2\cdot12H_2O$; mol wt 453.31. Colorless crystals, white granules or powder; styptic taste. d 1.65. mp 94.5°. At about 250° becomes anhydr; decomposes above 280°. One gram dissolves in 7 ml water, 0.5 ml boiling water. Freely sol in glycerol. Practically insol in alcohol. The aq soln is acid to litmus; pH of 0.05 molar soln 4.6.

USE: Purifying drinking water; in baking powders; dyeing and printing fabrics; manuf pigments, lakes, artificial gems, paper, vegetable glue, marble and porcelain cements; fireproofing; tanning; electrolytic copperplating; in prepn of high-purity aluminum compounds; buffer in analytical chemistry. *See also* Aluminum Potassium Sulfate.

THERAP CAT: Astringent; hemostatic.

THERAP CAT (VET): Astringent; hemostatic.

324. Aluminum Antimonide. [25152-52-7] Aluminum compd. with antimony (1:1). AlSb; mol wt 148.74. Al 18.14%, Sb 81.86%. Prepn of high purity crystals, resistance measurements, activation energies: Justi, Lautz, *Abh. Braunschw. Wiss. Ges.* **5**, 36 (1953), *C.A.* **48**, 1756i (1954). Electrical properties: Willardson *et al.*, *J. Electrochem. Soc.* **101**, 354 (1954); and prepn: Kover, *Compt.*

Rend. **243**, 648 (1956). Zone melting to remove impurities and to improve electr properties: Scheel, *Z. Metallkd.* **46**, 58 (1955), *C.A.* **49**, 4486f (1955). Oxidation by water: Rudorff, Kohlmeyer, *Z. Metallkd.* **45**, 608 (1954), *C.A.* **49**, 7432c (1955).

Lattice constant: 6.1361×10^{-8} cm. mp 1050°. Absorption const, n, width of forbidden zone (opt & electr) = 0.78; 3.4; 16; 1.65 e.v.: Oswald, Schade, *Z. Naturforsch.* **9a**, 611 (1954). Width of forbidden band = 1.6 e.v. at 0 K: Blunt *et al.*, *Phys. Rev.* **96**, 578 (1954).

USE: In semiconductor research.

325. Aluminum Benzoate. [555-32-8] Benzoic acid, aluminum salt (3:1). $C_{21}H_{15}AlO_6$; mol wt 390.33. C 64.62%, H 3.87%, Al 6.91%, O 24.59%. $Al(C_6H_5COO)_3$. Prepn: Kränzlein *et al.*, **DE 569946** (1933 to I. G. Farbenind.); *Frdl.* **19**, 597.

Cryst powder. Very slightly sol in water.

326. Aluminum Borate. [11121-16-7] Boric acid aluminum salt. Occurs in nature as the mineral *eremeyevite* or *jeremejevite*, $Al_2O_3.B_2O_3$. Prepd by heating Al_2O_3 with B_2O_3; when heated to 1000° $2Al_2O_3.B_2O_3$ is formed; when heated to 1100° $9Al_2O_3.2B_2O_3$ is formed: Scholze, *Z. Anorg. Allg. Chem.* **284**, 272 (1956).

$2Al_2O_3.B_2O_3$. Needles, mp ~1050°. Practically insol in water.

$9Al_2O_3.2B_2O_3$. Needles, mp ~1440°. Practically insol in water.

USE: Catalyst for polymerizations; in glass manuf.

327. Aluminum Borohydride. [16962-07-5] Tetrahydroborate(1−) aluminum (3:1); aluminum tetrahydroborate. AlB_3H_{12}; mol wt 71.51. Al 37.73%, B 45.35%, H 16.92%. $Al(BH_4)_3$. Prepn from alkali metal hydride and aluminum halide: Schlesinger *et al.*, *J. Am. Chem. Soc.* **75**, 209 (1953); Hinkamp, Hnizda, *Ind. Eng. Chem.* **47**, 1560 (1955); Kollonitsch, Fuchs, *Nature* **176**, 1081 (1955); Hinkamp, **US 2854312** (1958 to Ethyl Corp.); Schechter, **US 2913306** (1959 to Callery Chem.); from trimethylaluminum and diborane: Schlesinger *et al.*, *J. Am. Chem. Soc.* **61**, 536 (1939).

Liquid. *Spontaneously combustible, dangerous when wet.* mp −64.5°; bp 44.5°; bp_{119} 0°. Reacts vigorously with water and hydrogen chloride to liberate hydrogen; ignites in air; decomposes slowly even at room temp evolving hydrogen. Forms addn products with dimethyl ether, trimethylamine and ammonia.

USE: Reducing agent; prepn of borohydrides of heavy metals; fuel for jet engines and rockets.

328. Aluminum Bromide. [7727-15-3] Aluminum tribromide. $AlBr_3$; mol wt 266.69. Al 10.12%, Br 89.88%. Prepd from aluminum and bromine: Nicholson *et al.*, *Inorg. Synth.* **3**, 30 (1950).

White to yellowish-red very hygroscopic lumps, mp 97°; bp reported within the range 250-270°; d_4^{18} 3.205. *Corrosive.* Fumes strongly in air; combines with water *with violence. Keep tightly closed and protect from moisture.* Sol in many organic solvents such as benzene, nitrobenzene, toluene, xylene, simple hydrocarbons.

Hexahydrate. Colorless to slightly yellow deliquesc crystals. mp 93°; d 2.5. Sol in water, alcohol, ether, carbon disulfide. *Keep well closed.*

USE: *Anhydrous* form as acid catalyst in organic syntheses. It is similar to anhydr $AlCl_3$ but is more reactive and more sol in organic media.

329. Aluminum *tert*-Butoxide. [556-91-2] 2-Methyl-2-propanol aluminum salt (3:1); *tert*-butyl alcohol aluminum salt; tris(2-methyl-2-propanolato)aluminum; aluminum tris(*tert*-butylate). $C_{12}H_{27}AlO_3$; mol wt 246.33. C 58.51%, H 11.05%, Al 10.95%, O 19.48%. $Al[OC(CH_3)_3]_3$. Prepd from aluminum *tert*-butyl alcohol and mercuric chloride: Wayne, Adkins, *Org. Synth.* **coll. vol. III**, 48 (1955).

Powder; can be recryst from benzene. Sublimes at 180°. Pure compd does not melt or decomp upon heating up to 300° in a sealed tube, but traces of moisture or *tert*-butyl alc cause it to melt at 160-200°. Very sol in organic solvents; approx 9 g dissolves in 5 g ethyl propionate at 120°.

USE: Reagent for oxidation of alcohols to ketones; in dealcoholation of orthoesters.

330. Aluminum Calcium Hydride. [16941-10-9] $(T$-4)-Tetrahydroaluminate(1−) calcium (2:1); calcium tetrahydroaluminate; calcium aluminum hydride. Al_2CaH_8; mol wt 102.11. Al 52.85%, Ca 39.25%, H 7.90%. $Ca(AlH_4)_2$. Prepd by the interaction of aluminum chloride and calcium hydride in tetrahydrofuran:

Schwab, Wintersberger, *Z. Naturforsch.* **8b**, 690 (1953); Conn, Taylor, **US 2999005** (1961 to Merck & Co.).

Slate-gray mass. The dry pulverized material can ignite spontaneously in moist air and is best handled under dry nitrogen. Reacts violently with water, the ensuing conflagration resembles a display of fireworks. Slightly less violent reaction with alcohols. Sol in dry tetrahydrofuran. Practically insol in dry ether, dioxane, benzene.

USE: Reducing agent for aldehydes, ketones, acid chlorides. Also in the reduction of esters to alcohols, nitriles to amines, aromatic nitro compounds to azo compounds.

331. Aluminum Carbide. [1299-86-1] Tetraaluminum tricarbide. C_3Al_4; mol wt 143.96. C 25.03%, Al 74.97%. Al_4C_3. Prepd by heating aluminum powder with carbon: Becher in *Handbook of Preparative Inorganic Chemistry* vol. **1**, G. Brauer, Ed. (Academic Press, New York, 2nd ed., 1963) p 832.

Yellow hexagonal crystals or powder. mp 2100°; decomposes above 2200°. d 2.36. *Dangerous when wet.* Decomposed by water with evolution of methane, *q.v.*

USE: In generating methane; reducing metal oxides; in manuf of aluminum nitride.

332. Aluminum Chlorate. [15477-33-5] Chloric acid, aluminum salt. $AlCl_3O_9$; mol wt 277.32. Al 9.73%, Cl 38.35%, O 51.92%. $Al(ClO_3)_3$. Occurs as hexahydrate and nonahydrate. Prepn: *Gmelins, Aluminum* (8th ed.) **35B**, 216-217 (1934).

Nonahydrate. [7784-15-8] Mallebrin. Deliquesc crystals. Freely sol in water; sol in alc. *Keep well closed.*

USE: Disinfectant; ClO_2 manuf; prevention of yellowing of acrylic fibers.

THERAP CAT: Antiseptic, astringent.

333. Aluminum Chloride. [7446-70-0] Aluminum trichloride. $AlCl_3$; mol wt 133.33. Al 20.24%, Cl 79.76%. Prepd from aluminum metal in a heated stream of HCl gas: Gattermann-Wieland, *Praxis des Organischen Chemikers* (de Gruyter, Berlin, 40th ed., 1961) p 295; H. J. Becher in *Handbook of Preparative Inorganic Chemistry* vol. **1**, G. Brauer, Ed. (Academic Press, New York, 2nd ed., 1963) p 812. Manufacture: *Faith, Keyes & Clark's Industrial Chemicals*, F. A. Lowenheim, M. K. Moran, Eds. (Wiley-Interscience, New York, 4th ed., 1975) pp 72-75. Monograph: *ACS Monograph Series* **no. 87**, entitled "Anhydrous Aluminum Chloride in Organic Chemistry," C. A. Thomas, Ed. (Reinhold, New York, 1941). Review of toxicology and human exposure: *Toxicological Profile for Aluminum* (PB2009-100001, 2009) 357 pp.

White when pure; ordinarily gray or yellow to greenish deliquescent, crystalline powder. mp 192.6°. Sublimes at 182.7° at 752 mm Hg. d 2.48. Fumes in air; strong odor of HCl; when heated in small quantities volatilizes without melting. *Corrosive. Keep tightly closed and protected from moisture.* Combines with water with explosive violence and liberation of much heat. Freely sol in many organic solvents, such as benzophenone, benzene nitrobenzene, carbon tetrachloride, chloroform.

Hexahydrate. [7784-13-6] Anhydrol; Driclor; Drysol; Xerac. Colorless crystals, or white or slightly yellow deliquesc, cryst powder; odorless or slight HCl odor. Sweet, very astrigent taste. One gram dissolves in 0.9 ml water, 4 ml alc; sol in ether, glycerol, propylene glycol. *Keep well closed.*

Caution: Direct contact with *anhydrous* form may cause skin, eye and respiratory system irritation; may cause severe eye and skin burns.

USE: The *anhydrous* form suitable as an acid catalyst, esp in Friedel-Crafts type reactions; in cracking of petroleum; in manuf rubbers, lubricants. The *hexahydrate* form used in preserving wood; disinfecting stables, slaughterhouses, etc.; in deodorants and antiperspirant preparations; refining crude oil; dyeing fabrics; manuf parchment paper.

THERAP CAT: *Hexahydrate* as anhidrotic.

334. Aluminum Ethoxide. [555-75-9] Ethanol aluminum salt (3:1); aluminum ethylate. $C_6H_{15}AlO_3$; mol wt 162.16. C 44.44%, H 9.32%, Al 16.64%, O 29.60%. $Al(OC_2H_5)_3$. Prepd by reacting aluminum powder with absolute ethanol in xylene using small amounts of mercuric chloride and iodine as catalysts: Meerwein, Schmidt, *Ann.* **444**, 232 (1925); *Newer Methods of Preparative Organic Chemistry* (Interscience, New York, 1948) p 132; *see also*

Farbwerke vorm. Meister, Lucius und Brüning, **DE 286596**; *J. Soc. Chem. Ind. London* **34**, 1168 (1915); Adkins, *J. Am. Chem. Soc.* **44**, 2178 (1922). Laboratory procedure: Gattermann-Wieland, *Praxis des Organischen Chemikers* (de Gruyter, Berlin, 40th ed., 1961) p 333.

Liquid. $bp_{6.8}$ 200°; bp_3 175-180°. Slowly solidifies to a whole white solid, mp 140°. May crystallize with alcohol of crystallization. Decomposed by water. Slightly sol in hot xylene, chlorobenzene, other high boiling solvents.

USE: In the reduction of aldehydes and ketones; as catalyst for polymerizations.

335. Aluminum Fluoride. [7784-18-1] Aluminum trifluoride. AlF_3; mol wt 83.98. Al 32.13%, F 67.87%. Prepd by heating $(NH_4)_3AlF_6$ to red heat in a stream of nitrogen: Witt, Barrow, *Trans. Faraday Soc.* **55**, 730 (1959); Kwasnik in *Handbook of Preparative Inorganic Chemistry* Vol. **1**, G. Brauer, Ed. (Academic Press, New York, 2nd ed., 1963) p 225. *Review:* Kemmitt, Sharp, *Adv. Fluorine Chem.* **4**, 154-155 (1965). Review of toxicology and human exposure: *Toxicological Profiles for Aluminum* (PB2009-100001, 2009) 357 pp.

Hexagonal crystals. d 3.10. Sublimes (760 mm) 1272°. Soly in water (25°) 0.559 g/100 ml. Sparingly sol in acids and alkalies, even hot concd H_2SO_4 has little effect. Hydrolyzed by superheated steam at 300-400°.

Monohydrate. Fluellite. Orthorhombic crystals, d 2.17.

Trihydrate. Usually $AlF_3.3.1H_2O$. Prepn: Ehret, Frere, *J. Am. Chem. Soc.* **67**, 64 (1945). Loses water at 100°, more at 200°. It does not seem possible to obtain the anhydrous compd free from oxides by dehydration of the hydrates.

Caution: Less toxic on ingestion than other fluorides because of slight solubility.

USE: In ceramics, as flux in metallurgy, in aluminum manufacture, as inhibitor of fermentation, as catalyst in organic reactions.

336. Aluminum Hexafluorosilicate. [17099-70-6] Hexafluorosilicate(2−) aluminum (3:2); aluminum fluosilicate; aluminum silicofluoride. $Al_2F_{18}Si_3$; mol wt 480.19. Al 11.24%, F 71.22%, Si 17.55%. $Al_2(SiF_6)_3$. Occurs in nature as *topaz*, $Al_2SiO_4(OH,F)_2$. Prepn: Sanfourche, Krapivine, *Compt. Rend.* **208**, 2080 (1939).

Nonahydrate. Hexagonal prisms. Easily sol in water; aq soln decomposes on heating or neutralization. Solid loses water on heating to temperatures below 500°, leaving a hexahydrate form; decomposes completely on heating to 1000°.

USE: Protection and preservation of construction materials; manuf of glass.

337. Aluminum Hydride. [7784-21-6] Aluminum trihydride. AlH_3; mol wt 30.01. Al 89.91%, H 10.08%. Mild, selective reducing agent. Prepd by treating lithium hydride with an ether solution of aluminum chloride: A. E. Finholt *et al.*, *J. Am. Chem. Soc.* **69**, 1199 (1947). Brief review: K. Lopinti, *Synlett* **2005**, 2265-2266.

Colorless solid, nonvolatile, probably highly polymerized and containing residual ether which cannot be completely removed. *Dangerous when wet.*

USE: As catalyst for polymerizations; reducing agent. Lithium aluminum hydride, *q.v.*, is a more powerful reagent because of its greater soly.

338. Aluminum Hydroxide. [21645-51-2] Aluminum hydrate; aluminum trihydrate; hydrated alumina. AlH_3O_3; mol wt 78.00. Al 34.59%, H 3.88%, O 61.53%. $Al(OH)_3$. Prepn and properties: Dominé-Berges, *Ann. Chim.* [12] **5**, 106 (1950); Hennig, *Chem. Tech.* **1**, 66 (1949); *Gmelins, Aluminum* (8th ed.) **35B**, pp 98-132 (1934); Becher in *Handbook of Preparative Inorganic Chemistry* vol. **1**, G. Brauer, Ed. (Academic Press, New York, 2nd ed., 1963) pp 820-821; Wagner, *ibid.* vol. **2** (1965) pp 1652-1654. Comparative pharmacology and clinical use of antacids: F. W. Green, Jr. *et al.*, *Am. J. Hosp. Pharm.* **32**, 425 (1975). Clinical comparison with calcium carbonate as phosphate binder in chronic renal failure: R. H. K. Mak *et al.*, *Br. Med. J.* **291**, 623 (1985). Review of use as vaccine adjuvant: W. Nicklas, *Res. Immunol.* **143**, 489-494 (1992). Review of toxicology and human exposure: *Toxicological Profile for Aluminum* (PB2009-100001, 2009) 357 pp.

Usually obtained as a white, bulky, amorphous powder. mp 300°. d 2.42. Sol in alkaline aq solns or in HCl, H_2SO_4 and other strong acids in the presence of some water. Practically insol in water. Forms gels on prolonged contact with water. Absorbs acids, CO_2.

Aluminum hydroxide gel. Aldrox; ALternaGEL; Aludyal; Amphojel; Cremorin; Pepsamar; Uracid. White, viscous suspension from which small amounts of clear liquid may separate on standing. May also be used as the dried gel. Sol in dilute mineral acids, solns of fixed alkali hydroxides. Insol in water, alc.

USE: Adsorbent; emulsifier; ion-exchanger; in chromatography; mordant in dyeing; filtering medium; manuf glass, fire clay, paper, pottery, printing inks, lubricating compositions, detergents; waterproofing fabrics; in antiperspirants, dentifrices. Vaccine adjuvant.

THERAP CAT: Antacid; antihyperphosphatemic.

THERAP CAT (VET): Antacid.

339. Aluminum Hydroxychloride. [1327-41-9] Basic aluminum chloride; aluminum chlorohydroxide; aluminum chlorohydrate; Chlorhydrol; Hyperdrol; Locron; Phosphonorm. The generally accepted empirical formula is $Al_2(OH)_5Cl.2H_2O$. Prepd by electrolyzing solns of suitable Al salts: **FR 837862** and **GB 509815** (both 1939 to I. G. Farben.); H. Huehn, W. Haufe, **US 2392531** (1946); Andersen, **US 2492085** (1949 to Elizabeth Arden). Structural studies and physicochemical properties: D. L. Teagarden et al., *J. Pharm. Sci.* **70**, 758, 762 (1981). Phosphate binding activity and use in hyperphosphatemia: **DE 3147869** (1983 to Nefro-Pharma), *C.A.* **99**, 110747a (1983).

Glassy solid. Dissolves in water, forming slightly turbid colloidal solns (up to 55% w/w). pH of 15% aq soln ~4.3.

USE: In antiperspirants.

THERAP CAT: Astringent; antihyperphosphatemic.

340. Aluminum Hypophosphite. [7784-22-7] $AlH_6O_6P_3$; mol wt 221.94. Al 12.16%, H 2.73%, O 43.25%, P 41.87%. Al-$(PO_2H_2)_3$. Prepd by heating $Al(OH)_3$ or a solution of an aluminum salt with hypophosphorous acid or sodium hypophosphite: Everest, *J. Chem. Soc.* **1952**, 2945.

Cryst powder; decomposes without melting at ~220° with evolution of phosphine. Practically insol in water. Sol in warm NaOH soln, dilute sulfuric and dilute or concd hydrochloric acid.

USE: In acrylonitrile polymer fiber finishes.

341. Aluminum Iodide. [7784-23-8] Aluminum triiodide. AlI_3; mol wt 407.69. Al 6.62%, I 93.38%. Prepn from aluminum and iodine: Watt, Hall, *Inorg. Synth.* **4**, 117 (1953); H. J. Becher in *Handbook of Preparative Inorganic Chemistry* vol. 1, G. Brauer, Ed. (Academic Press, New York, 2nd ed., 1963) p 814; Wilson, Worrall, *J. Chem. Soc. A* **1968**, 316.

White leaflets if pure; commercial grade yellowish- to blackish-brown lumps. mp 191°; bp 382°; d^{17} 3.948. Fumes in moist air; strong exothermic reaction with water. *Keep tightly closed and protected from light.* Sol in carbon disulfide, alcohol, ether, liquid ammonia.

Hexahydrate. Yellowish, deliquesc cryst powder. Sol in water, alcohol, ether. *Keep tightly closed.*

USE: Catalyst in organic reactions.

342. Aluminum Isopropoxide. [555-31-7] 2-Propanol aluminum salt (3:1); aluminum isopropylate. $C_9H_{21}AlO_3$; mol wt 204.25. C 52.90%, H 10.36%, Al 13.21%, O 23.50%. Al[OCH-$(CH_3)_2]_3$. Prepd from aluminum and isopropyl alcohol in the presence of mercuric chloride: Young et al., *J. Am. Chem. Soc.* **58**, 100 (1936); by adding excess isopropyl alcohol to a benzene soln of $AlCl_3$ at 6°: Teichner, *Compt. Rend.* **237**, 810 (1953). Forms trimers and tetramers: Shiner et al., *J. Am. Chem. Soc.* **85**, 2318 (1963); Oliver et al., *J. Inorg. Nucl. Chem.* **31**, 1609 (1969); Worrall, *J. Chem. Educ.* **46**, 510 (1969). Toxicity: Smyth et al., *Am. Ind. Hyg. Assoc. J.* **30**, 470 (1969). *Review:* Whitaker in *Adv. Chem. Ser.* **23**, entitled "Metal-Organic Compounds," M. Sittig, Ed. (ACS, Washington DC, 1959) pp 184-189.

Hygroscopic white solid, mp 119°. Solidifies rather slowly after distillation. bp_{10} 135°; $bp_{7.5}$ 131°; $bp_{5.5}$ 125.5°; $bp_{2.5}$ 113°; $bp_{1.5}$ 106°; $bp_{0.5}$ 94°. Sol in ethanol, isopropanol, benzene, toluene, chloroform, carbon tetrachloride, petroleum hydrocarbons. Decomposed by water. LD_{50} orally in rats: 11.3 g/kg (Smyth).

USE: Meerwein-Ponndorf reactions; alcoholysis and ester exchange; synthesis of higher alkoxides, chelates, and acylates; formation of aluminum soaps, formulation of paints; waterproofing finishes for textiles.

343. Aluminum Lactate. [18917-91-4] Tris[2-(hydroxy-κO)propanoato-κO]aluminum; Al(lact)$_3$. $C_9H_{15}AlO_9$; mol wt 294.19. C 36.74%, H 5.14%, Al 9.17%, O 48.94%. Prepn from lactic acid and aluminum isopropoxide or aluminum chloride: A. K. Rai et al., *J. Prakt. Chem.* **20**, 105 (1963); from lactic acid and aluminum foil: R. W. Jones, J. E. Cluskey, *Cereal Chem.* **40**, 589 (1963). Partition coefficients: A. Tapparo, M. Perazzolo, *Int. J. Environ. Anal. Chem.* **36**, 13 (1989). Molecular structure: G. G. Bombi et al., *Inorg. Chim. Acta* **171**, 79 (1990). Stability and NMR spectra: B. Corain et al., *J. Chem. Soc. Dalton Trans.* **1992**, 169. Optical rotation isotherms: I. G. Bratu et al., *COFrRoCA 2002, Actes Colloq. Fr.-Roum. Chim Appl.* **2**, 157 (2002). Reactant in sol-gel synthesis of aluminophosphate glass: L. Zhang et al., *Chem. Mater.* **15**, 2702 (2003). Inhibition of the biological effects of silica quartz: R. Bégin et al., *Exp. Lung Res.* **10**, 385 (1986); R. Duffin et al., *Toxicol. Appl. Pharmacol.* **176**, 10 (2001). Clinical evaluation as antiseptic throat spray: W. Klingbeil, *Fortschr. Med.* **100**, 146 (1982); in dental hypersensitivity: Y. Higuchi et al., *J. Clin. Dent.* **7**, 9 (1996). Review of toxicology and human exposure: *Toxicological Profile for Aluminum* (PB2009-100001, 2009) 357 pp.

White powder. $[\alpha]_{366}^{20}$ +48.31°; $[\alpha]_{436}^{20}$ +33.37°; $[\alpha]_{578}^{20}$ +10.20° (c= 0.1 M). $[\alpha]_{366}^{25}$ +48.40°; $[\alpha]_{436}^{25}$ +31.36°; $[\alpha]_{578}^{25}$ +10.22° (c = 0.1 M). Partition coefficient (n-octanol/water): 0.0037-0.0127. Soly in water (25°): 0.70 ±0.01 mol dm^{-3}. pH of saturated soln: 2.9.

USE: As a reagent in biological activity studies of aluminum; in sol-gel processing.

THERAP CAT: Antiseptic.

344. Aluminum Lithium Hydride. [16853-85-3] (T-4)-Tetrahydroaluminate(1−) lithium (1:1); lithium tetrahydroaluminate; lithium aluminum hydride; lithium aluminohydride; lithium alanate. AlH_4Li; mol wt 37.95. Al 71.10%, H 10.62%, Li 18.29%. $LiAlH_4$. Prepd by treating lithium hydride with an ether soln of $AlCl_3$: Finholt, et al., *J. Am. Chem. Soc.* **69**, 1199 (1947). Crystal structure: Sklar, Post, *Inorg. Chem.* **6**, 669 (1967). Review of chemistry: J. S. Pizey, *Synthetic Reagents* Vol. 1 (John Wiley, New York, 1974) pp 101-294.

Microcrystalline white powder when pure; gray when aluminum impurity present. Monoclinic crystals. d 0.92. Stable in dry air at room temperature, decomp above 125°, slowly loses hydrogen at 120°, decomp in moist air, may ignite on grinding in air. Soly (parts/100 parts solvent): 30 (ether); 13 (tetrahydrofuran); 10 (dimethylcellosolve); 2 (dibutyl ether); 0.1 (dioxane). Reacts rapidly with water and alcohols; reduces aldehydes, ketones, acid chlorides and esters to alcohols; nitriles to amines; aromatic nitro compounds to azo compounds. Does not attack olefinic double bonds unless they are conjugated with a phenyl group and a carbonyl or nitrile group.

USE: Reducing agent; in preparation of other hydrides.

345. Aluminum Magnesium Silicate. [12511-31-8] Silicic acid (H_4SiO_4) aluminum magnesium salt (2:2:1); magnesium aluminum silicate. $Al_2MgO_8Si_2$; mol wt 262.43. Al 20.56%, Mg 9.26%, O 48.77%, Si 21.40%. $MgAl_2(SiO_4)_2$. Occurs in nature in the minerals: *colerainite, leuchtenbergite, pyrope, saponite, sapphirine, sheridanite, zebedassite.* Prepn: **GB 834517** (1960 to Fuji Chem.)

Hydrate. Ervasil.

USE: As suspending agent, thickening agent.

THERAP CAT: Antacid.

346. Aluminum Nicotinate. Nicalex. Pharmaceutical composition consisting of aluminum hydroxydinicotinate and nicotinic acid. Manufacturing process: J. P. Miale, **US 2970082** (1961 to

Walker Labs.). Prepn, properties and clinical studies: *idem, Curr. Ther. Res.* **7**, 392 (1965). Clinical trial in hypercholesterolemia: E. S. McCabe, *Del. Med. J.* **38**, 49 (1966).

White, amorphous powder with very slight acidulous taste. Insol in water, alchol. Sol in diluted mineral acids.

THERAP CAT: Has been used as antilipemic.

347. Aluminum Nitrate. [13473-90-0] Nitric acid aluminum salt (3:1). AlN_3O_9; mol wt 212.99. Al 12.67%, N 19.73%, O 67.60%. $Al(NO_3)_3$. Occurs in several states of hydration of which the nonahydrate is the most stable. Prepn: *Gmelins, Aluminum* (8th ed.) **35B**, p 149-152 (1934). Toxicity data: Smyth *et al., Am. Ind. Hyg. Assoc. J.* **30**, 470 (1969). Review of toxicology and human exposure: *Toxicological Profile for Aluminum* (PB2009-100001, 2009) 357 pp.

Nonahydrate. [7784-27-2] $AlN_3O_9.9H_2O$; mol wt 375.13. Thermal decompn and γ-pyrolysis: E. El-Shereafy *et al., J. Radio-anal. Nucl. Chem.* **237**, 183 (1998). Deliquesc crystals; mp 73°; dec at 135°. *Oxidizer.* Very sol in water, alc; very slightly sol in acetone. Almost insol in ethyl acetate and pyridine. The aq soln is acid. *Keep well closed.* LD_{50} orally in rats: 4.28 g/kg (Smyth).

USE: Tanning leather; antiperspirant; corrosion inhibitor; extraction of uranium; nitrating agent. Nonahydrate as a metal standard solution in analytical chemistry.

348. Aluminum Nitride. [24304-00-5] AlN; mol wt 40.99. Al 65.82%, N 34.17%. Semiconductor material. Prepn: F. Briegleb, A. Geuther, *Ann.* **123**, 228 (1862). Lattice structure: H. Ott, *Z. Physik.* **22**, 201 (1924). Crystal structure: G. A. Jeffrey, G. S. Parry, *J. Chem. Phys.* **23**, 406 (1953). Prepn of AlN films by laser-induced chemical vapor deposition: X. Li, T. L. Tansley, *J. Appl. Phys.* **68**, 5369 (1990). Review of growth of high purity crystals: G. A. Slack, T. F. McNelly, *J. Cryst. Growth* **34**, 263-279 (1976); of prepn and properties: D. D. Marchant, T. E. Nemecek, *Adv. Ceram.* **26**, 19-54 (1989); of properties and applications: B. H. Mussler, *Am. Ceram. Soc. Bull.* **79**, 45-47 (2000).

Colorless, translucent material; crystallizes with a hexagonal wurtzite structure. Bandgap at room temp: 6.2 eV. d 3.26. Hardness no. 8 on Mohs' scale. mp 2400° (dec). Thermal conductivity at 300 K: 3.19 W/cm K. Specific heat: 738 ±20 J/kgK.

USE: In semiconductor electronics; in steel manuf. AlN components and substrates are used in electrical engines, microelectronics, naval radio and defense systems, railway transport systems, telecommunications and research satellites, and emission control systems.

349. Aluminum Oxalate. [814-87-9] [μ-[Ethanedioato-$(2-)-\kappa O^1, \kappa O'^2 : \kappa O^2, \kappa O'^1$]]bis[ethanedioato$(2-)-\kappa O^1, \kappa O^2$]di-aluminum; tris(oxalato)dialuminum. $C_6Al_2O_{12}$; mol wt 318.02. C 22.66%, Al 16.97%, O 60.37%. $Al_2(C_2O_4)_3$. Prepn: **GB 348789** and **GB 348790** (both 1930 to I.G. Farben).

Hydrate. Powder. Practically insol in water, alc. Sol in mineral acids.

USE: Mordant in printing textiles, dyeing cotton.

350. Aluminum Oxide. [1344-28-1] Alumina. Al_2O_3; mol wt 101.96. Al 52.93%, O 47.07%. Occurs in nature as the minerals: *bauxite, bayerite, boehmite, corundum, diaspore, gibbsite*. Prepn and properties: *Mellor's* vol. **V**, 263-273 (1929); *Gmelins, Aluminum* (8th ed.) **35B**, pp 7-98 (1934); Becher in *Handbook of Preparative Inorganic Chemistry* vol. **1**, G. Brauer, Ed. (Academic Press, New York, 2nd ed., 1963) pp 822-823; Wagner, *ibid.* vol. **2** (1965) pp 1660-1663. Use as column matrix in ion chromatography: W. Buchberger, K. Winsauer, *J. Chromatogr.* **482**, 401 (1989); in HPLC: M. T. Kelly, M. R. Smyth, *J. Pharm. Biomed. Anal.* **7**, 1757 (1989). Clinical evaluation in hip replacement: L. Sedel *et al., J. Bone Joint Surg. Br.* **72-B**, 658 (1990); of wear in hip replacement: L. P. Zichner, H.-G. Willert, *Clin. Orthop. Relat. Res.* **282**, 86 (1992). Review of properties, biocompatibility and clinical use: P. Boutin *et al., J. Biomed. Mater. Res.* **22**, 1203-1232 (1988); of biocompatibility: P. S. Christel, *Clin. Orthop. Relat. Res.* **282**, 10-18 (1992). Review of toxicology and human exposure: *Toxicological Profile for Aluminum* (PB2009-100001, 2009) 357 pp.

Approximate characteristics of native aluminum oxide: White cryst powder. mp ~2000°. bp 2980°. d_{20} 4.0. Very hard, about 8.8 on Moh's scale. An electrical insulator; electrical resistivity at 300° about 1.2×10^{13} ohms-cm. When heated above 800° it becomes

insol in acid and specific gravity increases from 2.8 to 4.0. Insol in water. Very hygroscopic.

Caution: Potential symptoms of overexposure by direct contact are irritation of eyes, skin and respiratory system. *See NIOSH Pocket Guide to Chemical Hazards* (DHHS/NIOSH 97-140, 1997) p 12.

USE: As adsorbent, desiccant, abrasive; as filler for paints and varnishes; in manuf of alloys, ceramic materials, electrical insulators and resistors, dental cements, glass, steel, artificial gems; in coatings for metals, etc.; as catalyst for organic reactions. As a chromatographic matrix; originally called *Brockmann aluminum oxide* when used for this purpose. The minerals *corundum* (hardness = 9) and *Alundum* (obtained by fusing bauxite in an electric furnace) are used as abrasives and polishes; in manuf of refractories.

351. Aluminum Palmitate. [555-35-1] Hexadecanoic acid aluminum salt (3:1); palmitic acid aluminum salt. $C_{48}H_{93}AlO_6$; mol wt 793.25. C 72.68%, H 11.82%, Al 3.40%, O 12.10%. $[CH_3-(CH_2)_{14}COO]_3Al$. Prepd from $AlCl_3$ and palmitic acid: Mehrotra, Rai, *J. Indian Chem. Soc.* **39**, 1 (1962).

White to yellow mass or powder. Practically insol in water or alcohol; when fresh, dissolves in petrol ether or oil turpentine.

USE: Thickening petroleum and lubricants; water-proofing fabrics; sizing and glazing paper and leather.

352. Aluminum Phosphate. [7784-30-7] Phosphoric acid aluminum salt (1:1); aluminum orthophosphate. AlO_4P; mol wt 121.95. Al 22.13%, O 52.48%, P 25.40%. $AlPO_4$. Occurs in nature as the minerals *angelite; coeruleolactite; evansite; lucinite; meta-variscite; sterretite; variscite; vashegyite; wavellite; zepharovichit*. Prepn from $NaAlO_2$ and H_3PO_4: Becher in *Handbook of Preparative Inorganic Chemistry* vol. **1**, G. Brauer, Ed. (Academic Press, New York, 2nd ed., 1963) p 831. Review of toxicology and human exposure: *Toxicological Profile for Aluminum* (PB2009-100001, 2009) 357 pp.

White, infusible powder. mp >1460°; d^{23} 2.56. Very slightly sol in concentrated hydrochloric acid and nitric acid. Practically insol in water or acetic acid. Isomorphous with quartz.

Aluminum phosphate gel. Fosfalugel; Phosphaljel; Phosphalugel. White, viscous suspension from which small amounts of water separate on standing.

USE: As cement in admixture with calcium sulfate and sodium silicate; as flux for ceramics; dental cements; for special glasses.

THERAP CAT: Antacid.

353. Aluminum Phosphide. [20859-73-8] Celphos; Phostoxin; Quickphos. AlP; mol wt 57.96. Al 46.55%, P 53.44%. Prepd from red phosphorus and aluminum powder: White, Bushey, *J. Am. Chem. Soc.* **66**, 1666 (1944); *Inorg. Synth.* **4**, 23 (1953); Montignie, *Bull. Soc. Chim. Fr.* **1946**, 276; from Al and Zn_3P_2: Wang *et al., J. Inorg. Nucl. Chem.* **25**, 326 (1963). Use as insecticidal fumigant: W. Freyberg, W. Haupt, **US 2117158** (1938 to Freyberg). Review of toxicology and human exposure: *Toxicological Profile for Aluminum* (PB2009-100001, 2009) 357 pp.

Dark gray or dark yellow crystals. Cubic zinc blende structure. *Poisonous. Dangerous when wet. Reacts readily with moist air to produce toxic phosphine gas.* d_4^{15} 2.85 (Montignie); d 2.40 (Wang *et al.*). Does not melt or decompose thermally at temps up to 1000°. Treatment with water and acid produces phosphine in quantitative yields.

USE: Source of phosphine, *q.v.*; in semiconductor research; as fumigant.

354. Aluminum Potassium Sulfate. [10043-67-1] Sulfuric acid aluminum potassium salt (2:1:1). $AlKO_8S_2$; mol wt 258.19. Al 10.45%, K 15.14%, O 49.57%, S 24.83%. $KAl(SO_4)_2$. Prepn: *Gmelins, Aluminum* (8th ed.) **35B**, 453-477 (1934). Crystal structure: Manoli *et al., Bull. Soc. Chim. Fr.* **1970**, 98; of dodecahydrate: Larson, Cromer, *Acta Crystallogr.* **22**, 793 (1967). Review of toxicology and human exposure: *Toxicological Profiles for Aluminum* (PB2009-100001, 2009) 357 pp.

Anhydrous. Burnt alum; exsiccated alum. Usually about 97-98% pure. White powder; styptic taste; attracts moisture from the air. One gram dissolves in about 20 ml cold or 1 ml boiling water, usually incompletely; practically insol in alc. *Keep well closed.*

Dodecahydrate. [7784-24-9] Alum; potassium alum; kalinite. $AlKO_8S_2.12H_2O$; mol wt 474.37. The technical product is also known as *alum flour*, *alum meal*, *cube alum*. It is about 99.5% pure. Colorless, odorless, hard, large, transparent crystals or cryst fragments or white, cryst powder; sweetish, astringent taste. Stable at ordinary temp; when kept for a long time at 60-65° (or over H_2SO_4) loses 9 H_2O which is reabsorbed on exposure to air. Becomes anhydr at about 200°; at higher temp loses SO_3 and becomes basic and incompletely sol in water. d 1.725. mp 92.5°. One gram dissolves in 7.2 ml water, in 0.3 ml boiling water; freely sol in glycerol. Insol in alc. The aq soln is acid; pH of 0.2 molar aq soln 3.3.

USE: In dyeing, printing fabrics; manuf dyes, lakes, paper, vegetable glue, marble cement, porcelain cement, explosives; in tanning, hardening gelatin, baking powders, purifying water, clarifying sugar, hardening plaster casts; electrolytic copperplating; as catalyst in synthesis of ammonia; in buffers; as flocculating reagent; as mordant in staining with carmine, eosine, hematoxylin; clarifier (as alumina cream); identifying coloring matters; hardening agent in microscopy.

THERAP CAT: Astringent.

THERAP CAT (VET): Astringent.

355. Aluminum Selenide. [1302-82-5] Dialuminum triselenide. Al_2Se_3; mol wt 290.84. Al 18.55%, Se 81.45%. Prepn from aluminum and selenium: Waitkins, Shutt, *Inorg. Synth.* **2**, 184 (1946); Becher in *Handbook of Preparative Inorganic Chemistry* **vol. 1**, G. Brauer, Ed. (Academic Press, New York, 2nd ed., 1963) p 825.

Yellowish to light-brown powder. d_4^{15} 3.437. Unstable in air. Decomp in water and acid.

USE: In prepn of hydrogen selenide; in semiconductor research.

356. Aluminum Silicate. [12141-46-7] Aluminum oxide silicate [$Al_2O(SiO_4)$]. Al_2O_5Si; mol wt 162.04. Al 33.30%, O 49.37%, Si 17.33%. Al_2SiO_5. Usually contains some water. Polymorphous; the three naturally occurring forms are *andalusite*, *cyanite*, *sillimanite*. Other aluminum silicate minerals are *anauxite*, *dickite*, *kaolinite*, *kochite*, *mullite*, *newtonite*, *pyrophyllite*, *takizolite*, *tmerite*, *ton*. For prepn and properties see *Gmelins, Aluminum* (8th ed.) **35B**, p 313-317 (1934).

USE: In dental cements, glass industry; manuf of semiprecious stones, enamels, ceramics, and colored lakes; paint filler; in washing compounds.

357. Aluminum Sodium Sulfate. [10102-71-3] Sulfuric acid aluminum sodium salt (2:1:1); sodium alum. $AlNaO_8S_2$; mol wt 242.08. Al 11.15%, Na 9.50%, O 52.87%, S 26.49%. NaAl(SO₄)₂. Prepn: *Gmelins, Aluminum* (8th ed) **35B**, p 378-384 (1934).

Dodecahydrate. Sodium alum; soda alum. Colorless crystals or white granules or powder. d 1.61. mp about 60°. Sol in 1 part water. Practically insol in alc.

USE: Industrially, like aluminum potassium sulfate.

THERAP CAT: Astringent.

THERAP CAT (VET): Astringent.

358. Aluminum Stearate. [637-12-7] Octadecanoic acid aluminum salt (3:1); stearic acid aluminum salt; aluminum tristearate. $C_{54}H_{105}AlO_6$; mol wt 877.41. C 73.92%, H 12.06%, Al 3.08%, O 10.94%. $[CH_3(CH_2)_{16}COO]_3Al$. Prepn: Gilmour *et al.*, *J. Chem. Soc.* **1956**, 1972.

Hard material, passes into plastic form on heating. mp 117-120°. Practically insol in water. When freshly made, sol in alcohol, benzene, oil turpentine, mineral oils. Forms cryst pyridine complex.

USE: Waterproofing fabrics, ropes; in paint and varnish driers; thickening lubricating oils; in cements; in light-sensitive photographic compositions.

359. Aluminum Subacetate. [142-03-0] Bis(acetato-κO)-hydroxyaluminum; bis(acetato)hydroxyaluminum; aluminum acetate, basic; aluminum diacetate; aluminum hydroxyacetate; aluminum hydroxydiacetate; basic aluminum acetate; hydroxyaluminum acetate; Lenicet. $C_4H_7AlO_5$; mol wt 162.08. C 29.64%, H 4.35%, Al 16.65%, O 49.36%. $Al(OH)(CH_3CO_2)_2$. Prepn from aluminum sulfate and lead acetate: W. Crum, *Q. J. Chem. Soc.* **6**, 216 (1854); from aluminum hydroxide and acetic acid or from sodium acetate and aluminum chloride hexahydrate: G. C. Hood, A. J. Ihde, *J. Am. Chem. Soc.* **72**, 2094 (1950). Large scale process: J. E. Jerome *et al.*, *US* 6498262 (2002 to Chattem Chem.).

White curdy precipitate or white amorphous powder. Material that has been oven-dried at 110° is practically insoluble in water.

Freshly prepared material forms numerous hydrates and is sol in water. Greatest soly is obtained by formation in solution. Incompatible with strong oxidizing agents.

Aluminum subacetate solution. Essigsäure Tonerde. Pharmacist's stock solution containing approx 8% aluminum subacetate in water. Prepd from aluminum sulfate, calcium carbonate, and acetic acid: *USP* **34**, 1822 (2011). Clear, colorless liquid. Odor of acetic acid. d 1.045. Acid to litmus. Gradually becomes turbid and colloidal. Heat promotes the decomposition. *Keep in a cool place and discard when cloudy.*

USE: Manuf color lakes; mordant in dyeing; in waterproofing and fireproofing fabrics; in antiperspirant formulations; in embalming fluid; in preparation of Burow's solution.

THERAP CAT: Astringent, antiseptic.

360. Aluminum Sulfate. [10043-01-3] Sulfuric acid aluminum salt (3:2); aluminum(III) sulfate. $Al_2O_{12}S_3$; mol wt 342.13. Al 15.77%, O 56.12%, S 28.11%. $Al_2(SO_4)_3$. Normally contains 14-18 molecules of water; anhydrous form is prepd by heating to 400-450°. Occurs in nature as the mineral *alunogen*. Prepn: *Gmelins, Aluminum* (8th ed.) **35B**, pp 267-269 (1934). Soly in aqueous ethanol: A. B. Gancy, C. A. Wamser, *J. Chem. Eng. Data* **24**, 192 (1979). Prepn and characterization of lower hydrated forms: A. B. Gancy, *Thermochim. Acta* **54**, 105 (1982). Thermal decompn studies: K.-S. Chou, C.-S. Soong, *ibid.* **78**, 285, 305 (1984); Y. Pelovski *et al.*, *ibid.* **205**, 219, 283 (1992). Evaluation as a first-aid treatment for stings and bites: D. Henderson, R. G. Easton, *Med. J. Aust.* **2**, 146 (1980). Review of toxicology and human exposure: *Toxicological Profiles for Aluminum* (PB2009-100001, 2009) 357 pp.

Hydrate. [17927-65-0] Stingose. $Al_2O_{12}S_3.xH_2O$. White, lustrous crystals, pieces, granules, or powder. Odorless with sweet taste, becoming mildly astringent. Hydrolyzes in water to form sulfuric acid. *Irritant*. d 1.61. Melts when gradually heated. Sol in 1 part water. Practically insol in alc. pH of 5% aq soln is 2.9.

Octadecahydrate. [7784-31-8] $Al_2O_{12}S_3.18H_2O$; mol wt 666.40. The commercial product is also known as *cake alum* or *patent alum*. White solid. *Irritant*.

USE: Tanning leather, sizing paper, mordant in dyeing; purifying water; manuf lakes, aluminum resinate; fireproofing and waterproofing cloth; clarifying oils and fats; treating sewage; waterproofing concrete; deodorizing and decolorizing petroleum; antiperspirants; agricultural pesticides; manuf aluminum salts; prepn of aluminum acetate solns; to lower pH of soil.

THERAP CAT: Astringent; in treatment of insect and marine organism stings and bites.

361. Aluminum Sulfide. [1302-81-4] Dialuminum trisulfide. Al_2S_3; mol wt 150.14. Al 35.94%, S 64.06%. Prepd by heating aluminum with sulfur: Flahaut, *Compt. Rend.* **232**, 334 (1951); Becher in *Handbook of Preparative Inorganic Chemistry* **vol. 1**, G. Brauer, Ed. (Academic Press, New York, 2nd ed., 1963) p 823.

Yellowish-gray, compact lumps; H_2S odor; decomposes in moist air to a gray powder. d 2.02. mp 1100°. Hydrolyzed by water to $Al(OH)_3$ and H_2S. *Keep tightly closed.*

362. Aluminum Tartrate. [815-78-1] (2*R*,3*R*)-2,3-Dihydroxybutanedioic acid aluminum salt (3:2). $C_{12}H_{12}Al_2O_{18}$; mol wt 498.17. C 28.93%, H 2.43%, Al 10.83%, O 57.81%. $Al_2(C_4H_4O_6)_3$. Usually contains some water. Prepn: Goldman, *Biochem. Z.* **133**, 459 (1922); **GB** 348790 (1930 to I. G. Farben.).

Odorless granules. Slowly sol in cold, readily in hot water; sol in ammonia.

USE: In textile dyeing.

363. Aluminum Thiocyanate. [538-17-0] Thiocyanic acid aluminum salt (3:1); aluminum sulfocyanate. $C_3AlN_3S_3$; mol wt 201.22. C 17.91%, Al 13.41%, N 20.88%, S 47.80%. $Al(CNS)_3$. Prepn: *Gmelins, Aluminum* (8th ed.) **35B**, p 306 (1934).

USE: Aq solution used as mordant in dye industry.

364. Aluminum Tris(8-hydroxyquinoline). [2085-33-8] Tris(8-quinolinolato-$\kappa N^1,\kappa O^8$)aluminum; tris(8-hydroxyquinoline)aluminium; Alq₃. $C_{27}H_{18}AlN_3O_3$; mol wt 459.44. C 70.59%, H 3.95%, Al 5.87%, N 9.15%, O 10.45%. Metal chelate; fluorescent solid-state material for organic light-emitting devices (OLED). Prepn: R. Berg, *Z. Anal. Chem.* **71**, 369 (1927); I. M. Kolthoff, E. B. Sandell, *J. Am. Chem. Soc.* **50**, 1900 (1928); of analogs: H. Jang et al., Synth. Met. **121**, 1667 (2001); of solid state: A. K. Saxena, Synth. React. Inorg. Met.-Org. Chem. **29**, 1747 (1999). Exists as two geometric isomers mer- and fac- with several crytalline forms. Crystal structure and optical properties of α-, β-, and γ-forms: M. Brinkmann et al., J. Am. Chem. Soc. **122**, 5147 (2000); of blue luminescent δ-form: M. Cölle et al., Adv. Funct. Mater. **13**, 108 (2003). Photoemission at thin-film faces: W. Zhao et al., Chem. Mater. **16**, 750 (2004). Reliability and degradation study: Z. D. Popovic, H. Aziz, IEEE J. Sel. Top. Quantum Electron. **8**, 362 (2002). Use in OLEDs: C. W. Tang, S. A. VanSlyke, Appl. Phys. Lett. **51**, 913 (1987); G. Baldacchini et al., Proc. 149th Int. School Phys. Enrico Fermi Rome 2002 561-567; as functionalized polymer: A. Meyers, M. Werk, Macromolecules **36**, 1766 (2003). Review as emitting material for electroluminescence: C. H. Chen, J. Shi, Coord. Chem. Rev. **171**, 161-174 (1998).

Light yellow complex, 330° (dec) (Jang); also reported as mp 340° (dec) (Saxena). Sublimes >500 K. uv max (0.1 mg/ml): 316, 372 nm; uv max (THF): 391 nm. Fluorescence max (0.09 mg/ml): 509 nm. Photoluminescence (THF): 516 nm; electroluminescence: 528 nm. Luminance max: 3400 cd/m² at 13 V.

USE: In OLEDs as an electron transport and/or light emitting layer.

365. Alverine. [150-59-4] N-Ethyl-N-(3-phenylpropyl)benzenepropanamine; N-ethyl-3,3′-diphenyldipropylamine; bis(γ-phenylpropyl)ethylamine; di(phenylpropyl)ethylamine. $C_{20}H_{27}N$; mol wt 281.44. C 85.35%, H 9.67%, N 4.98%. Prepn: Külz et al., Ber. **72**, 2165 (1939); Stühner, Elbrächter, Arch. Pharm. **287**, 139 (1954). Physical properties: E. F. Salim, W. R. Ebert, J. Pharm. Sci. **56**, 1162 (1967). Clinical trial in irritable bowel syndrome: G. J. Tudor, Br. J. Clin. Pract. **40**, 276 (1985).

Liquid, $bp_{0.3}$ 165-168°; bp_{13} 210-215°.

Citrate. [5560-59-8] Spasmaverine; Spasmonal. $C_{20}H_{27}N.C_6$-H_8O_7; mol wt 473.57. White to off-white powder having a sweet odor and slightly bitter taste, mp 100-102°. Slightly sol in water, chloroform; sparingly sol in alcohol; very slightly sol in ether.

THERAP CAT: Antispasmodic.

366. Alvimopan. [156053-89-3] N-[(2S)-2-[[(3R,4R)-4-(3-Hydroxyphenyl)-3,4-dimethyl-1-piperidinyl]methyl]-1-oxo-3-phenylpropyl]glycine; [[2(S)-[[4(R)-(3-hydroxyphenyl)-3(R),4-dimethyl-1-piperidinyl]methyl]-1-oxo-3-phenylpropyl]amino]acetic acid; ADL-8-2698; LY-246736; Entereg. $C_{25}H_{32}N_2O_4$; mol wt 424.54. C 70.73%, H 7.60%, N 6.60%, O 15.07%. Peripheral mu-opioid receptor antagonist. Prepn: B. E. Cantrell et al., **EP 506478**; eidem, **US 5250542** (1992, 1993 both to Lilly); D. M. Zimmerman et al., J. Med. Chem. **37**, 2262 (1994). Improved synthesis: J. A. Werner et al., J. Org. Chem. **61**, 587 (1996). Clinical pharmacology: S. S. Liu et al., Clin. Pharmacol. Ther. **68**, 66 (2000). Clinical trial in postoperative ileus: A. Taguchi et al., N. Engl. J. Med. **345**, 935 (2001); E. R. Viscusi et al., Surg. Endosc. **20**, 64 (2006). Review of clinical experience: W. K. Schmidt, Am. J. Surg. **182**, Suppl., 27S-38S (2001); and pharmacology: P. Neary, C. P. Delaney, Expert Opin. Invest. Drugs **14**, 479-488 (2005).

Dihydrate. [170098-38-1] Crystals, mp 210-213°. $[\alpha]_D^{25}$ +51.8° (c = 1.0 in DMSO).

THERAP CAT: Gastroprokinetic.

367. Amanitin. [11030-71-0] Group 1 mushroom toxin from the poisonous mushroom Amanita phalloides (Fr.) Secr., Agaricaceae. Inhibits protein synthesis of mammalian cells. Comprised of α-, β-, γ-amanitin and **amanin**. α-Amanitin is the major poisonous constituent of A. phalloides; it is 10-20 times more toxic than phalloidin, q.v. Isoln of α- and β-amanitin: Wieland, Ann. **564**, 152 (1949). Prepn of α-amanitin from β-amanitin: Wieland, Boehringer, ibid. **635**, 178 (1960). Structure: Wieland, Pure Appl. Chem. **9**, 145 (1964); Wieland, Gebert, Ann. **700**, 157 (1967). Review of chemistry and toxicology of the toxins of Amanita phalloides: Wieland, Wieland, Pharmacol. Rev. **11**, 87-107 (1959); T. Wieland, Fortschr. Chem. Org. Naturst. **25**, 214-250 (1967); T. Wieland, H. Faulstich, Crit. Rev. Biochem. **5**, 185-260 (1978). Book: H. Faulstich et al., Amanita Toxins and Poisoning: International Amanita Symposium (Lubrecht & Cramer, Heidelberg, 1980) 246 pp.

α-Amanitin   R = NH₂
β-Amanitin   R = OH

α-Amanitin. [23109-05-9] $C_{39}H_{54}N_{10}O_{14}S$; mol wt 918.98. Needles from methanol, mp 254-255°. $[\alpha]_D^{20}$ +191°. uv max: 302 nm. LD_{50} i.p. in albino mice: 0.1 mg/kg (Wieland, Wieland).

β-Amanitin. [21150-22-1] $C_{39}H_{53}N_9O_{15}S$; mol wt 919.96. Needles from methanol, mp 300°. uv max: 302 nm. Sol in water, methanol, ethanol, aqueous butanol. LD_{50} i.p. in albino mice: 0.4 mg/kg (Wieland, Wieland).

Caution: Highly toxic. Following a characteristic asymptomatic period of 6-15 hrs, potential symptoms of intoxication due to ingestion include violent gastroenteritis; fever, tachycardia, hyperglycemia, dehydration, electrolyte imbalance; liver dysfunction and necrosis; renal failure; may be fatal. See Clinical Toxicology of Commercial Products, R. E. Gosselin et al., Eds. (Williams & Wilkins, Baltimore, 5th ed., 1984) Section II, p 246; M. J. Ellenhorn, D. G. Barceloux, Medical Toxicology: Diagnosis and Treatment of Human Poisoning (Elsevier, New York, 1988) pp 1331-1338.

USE: As a tool in molecular biology.

368. Amantadine. [768-94-5] Tricyclo[3.3.1.1^{3,7}]decan-1-amine; 1-adamantanamine; 1-aminoadamantane; 1-aminodiamantane (obsolete); 1-aminotricyclo[3.3.1.1^{3,7}]decane. $C_{10}H_{17}N$; mol wt 151.25. C 79.41%, H 11.33%, N 9.26%. NMDA-receptor antagonist. Antiviral agent active vs influenza A and other RNA viruses; interacts with the M2 viral membrane matrix protein, blocking the release of viral RNA into the cytoplasm of infected cells. Prepn: H. Stetter et al., Ber. **93**, 226 (1960); W. Haaf, ibid. **97**, 3234 (1964); P. Kovacic, P. D. Roskos, Tetrahedron Lett. **1968**, 5833. LC-MS/MS determn in serum: T. Arndt et al., Clin. Chim. Acta **359**, 125 (2005). Antiviral activity: W. L. Davies et al., Science **144**, 862 (1964).

Pharmacology and toxicology: V. G. Vernier *et al.*, *Toxicol. Appl. Pharmacol.* **15**, 642 (1969). Comprehensive description: J. Kirschbaum, *Anal. Profiles Drug Subs.* **12**, 1-36 (1983). Review of use vs influenza A: R. L. Tominack, F. G. Hayden, *Infect. Dis. Clin. North Am.* **1**, 459-478 (1987); of pharmacokinetics: F. Y. Aoki, D. S. Sitar, *Clin. Pharmacokinet.* **14**, 35-51 (1988). Review of NMDA receptor binding and neuroprotective properties: J. Kornhuber *et al.*, *J. Neural Transm.* **43**, Suppl., 91-104 (1994). Series of articles on clinical experience in Parkinson's disease: *ibid.* **46**, Suppl., 399-421 (1995). Review of therapeutic potential in hepatitis C infection: J. K. Lim *et al.*, *J. Viral Hepat.* **12**, 445-455 (2005).

Crystals by sublimation, mp 160-190° (closed tube) (Stetter). Also reported as mp 180-192° (Haaf). pKa: 10.1. Sparingly sol in water.

Hydrochloride. [665-66-7] EXP-105-1; NSC-83653; Adekin; Lysovir; Mantadan; Mantadine; Mantadix; Symmetrel; Virofral. $C_{10}H_{17}N.HCl$; mol wt 187.71. Crystals from abs ethanol + anhydr ether, mp >360° (dec). Freely sol in water (at least 1:20); sol in alcohol, chloroform. Practically insol in ether. LD_{50} orally in mice, rats: 700, 1275 mg/kg (Vernier).

Sulfate. [31377-23-8] PK-Merz. $C_{10}H_{17}N.\frac{1}{2}H_2SO_4$; mol wt 200.29.

THERAP CAT: Antiviral; antiparkinsonian.

THERAP CAT (VET): Antiviral; adjunct for chronic pain.

369. Amaranth (Dye). [915-67-3] 3-Hydroxy-4-[2-(4-sulfo-1-naphthalenyl)diazenyl]-2,7-naphthalenedisulfonic acid sodium salt (1:3); C.I. Acid Red 27; C.I. 16185; FD & C Red No. 2; Red no. 2; 1-(4-sulfo-1-naphthylazo)-2-naphthol-3,6-disulfonic acid trisodium salt. $C_{20}H_{11}N_2Na_3O_{10}S_3$; mol wt 604.46. C 39.74%, H 1.83%, N 4.63%, Na 11.41%, O 26.47%, S 15.91%. Azo dye. Prepn: Knecht, *J. Soc. Dyers Colour.* **2**, 24 (1886). *See also Colour Index* vol. **4** (3rd ed., 1971) p 4093. Metabolism: J. L. Radomski, T. J. Mellinger, *J. Pharmacol. Exp. Ther.* **136**, 259 (1962). Use as indicator: N. K. Murty, K. R. Rao, *J. Indian Chem. Soc.* **53**, 532 (1976). Stability studies: A. Motawi, S. El-Gamal, *Sci. Pharm.* **48**, 377 (1980). Effect of pH and concentration on reduction: J. Maslowska, J. Janiak, *Zesz. Nauk. Politech. Lodz.* **1994**, 150. Spectrophotometric analysis: C. C. Blanco *et al.*, *Talanta* **43**, 1019 (1996). Review of carcinogenic risk: *IARC Monographs* **8**, 41-60 (1975).

Dark, reddish-brown powder. Absorption max at 25°: 520 nm (water); 528 nm (acetone); 520 nm (DMF). One gram dissolves in about 15 ml water. Also reported as 7.20 g/100 ml H_2O at 26°. Very slightly sol in ethanol, cellosolve. The aq soln is vivid red (1 cm layer). Discharge white by hydrosulfite on wool and silk. HCl does not change the color intensity of the soln, NaOH increases it. The aq soln is stable toward light.

USE: Dyeing wool and silk bright bluish-red from an acid bath. Food colorant. As indicator. In color photography.

370. Amaranth (Plant). Genus of the *Amaranthaceae* L. family which contains approx 60 species having worldwide distribution. Many species are considered weeds but *Amaranthus caudatus* L. (*love-lies-bleeding*), *A. hybridus* variety *hypochondriacus* L. (*prince's feather*), *A. tricolor* have been cultivated as ornamentals.

A. retroflexus L. and some of the other weedy species are known as *pigweed*, *redroot* and *water hemp*. *A. spinosus* L. has been used in the treatment of gonorrhea: W. H. Brown, *Useful Plants of the Philippines* **1** (Philippines Dept. Agr. and Natl. Resources, Manila, 1951) pp 510-515; as a poultice in the treatment of inflammation, bruises and eczema: T. H. P. de Tavera, *The Medicinal Plants of the Philippines* (P. Blakiston's Son, Philadelphia, 1901) pp 200-202. Most species are hardy, herbaceous and fast-growing cereal-like plants. Leaves and grain are used for food in parts of South America, Africa and Asia. Plants are high in protein; the amino acid composition is complementary to that of wheat. The grain was a basic food in pre-Columbian South and Central America and was important in Aztec ritual. Grain amaranths (*A. hypochondriacus*, *A. cruentas*, *A. caudatus*) produce large seedheads containing many edible seeds. The seed can be made into low gluten flour, cooked into gruel or popped like corn. Compositional study of grain: R. Becker *et al.*, *J. Food Sci.* **46**, 1175 (1981); K. C. Pant, *Nutr. Rep. Int.* **28**, 1445 (1983). Properties of seed starches: P. V. S. Rao, J. K. Goering, *Cereal Chem.* **47**, 655 (1970); Y. Tomita *et al.*, *J. Nutr. Sci. Vitaminol.* **27**, 471 (1981); saccharides: K. Lorenz, M. Gross, *Nutr. Rep. Int.* **29**, 721 (1984); lipids: F. I. Opute, *J. Exp. Bot.* **30**, 601 (1979). Baking potential of flour: K. Lorenz, *Starch/Staerke* **33**, 149 (1981). Most amaranth species have edible leaves with mild spinach-like flavor. *A. cruentas*, *A. dubius*, *A. hybridus*, *A. lividus* and *A. tricolor* are some of the species grown as vegetables. Magnesium and copper content of leaf: N. M. Guttiker *et al.*, *J. Nutr. Diet.* **3**, 4 (1966); vitamin A content: C. N. Rao, *ibid.* **4**, 10 (1967); amino acids: I. G. Vasi, V. P. Kalintha, *J. Inst. Chem. (India)* **52**, 13 (1980). Nutritive value of leaf protein concentrate: P. R. Cheeke *et al.*, *Can. J. Anim. Sci.* **61**, 199 (1981). Toxicology: R. M. Hill, P. D. Rawate: *J. Agric. Food Chem.* **30**, 465 (1982). Book: J. N. Cole, *Amaranth, From the Past for the Future* (Rodale Press, Emmaus, Pennsylvania, 1979) 311 pp; Nat'l. Academy of Sciences Report: *Amaranth, Modern Prospects for an Ancient Crop* (Nat'l. Academy Press, Washington, D.C., 1984) 80 pp.

371. Amarogentin. [21018-84-8] (4a*S*,5*R*,6*S*)-5-Ethenyl-4,-4a,5,6-tetrahydro-6-[[2-*O*-[(3,3',5-trihydroxy[1,1'-biphenyl]-2-yl)-carbonyl]-*β*-D-glucopyranosyl]oxy]-1*H*,3*H*-pyrano[3,4-*c*]pyran-1-one; sweroside 2'-(3'',5'',3'''-trihydroxydiphenyl)-2''-carboxylic acid ester; [4a*S*-(4aα,5β,6α)]-3,3',5-trihydroxy[1,1'-biphenyl]-2-carboxylic acid 2-ester with 5-ethenyl-6-(*β*-D-glucopyranosyloxy)-4,4a,-5,6-tetrahydro-1*H*,3*H*-pyrano[3,4-*c*]pyran-1-one. $C_{29}H_{30}O_{13}$; mol wt 586.55. C 59.38%, H 5.16%, O 35.46%. Strongly bitter glucoside. Isoln from *Swertia chirata* Buch-Ham., Gentianaceae: Korte, *Ber.* **88**, 704 (1955); **89**, 2404 (1956); from *Swertia japonica* Makino: Inouye *et al.*, *Chem. Pharm. Bull.* **18**, 1856 (1970). Occurrence in gentianaceous plants: Inouye, Nakamura, *J. Pharm. Soc. Jpn.* **91**, 755 (1971), *C.A.* **75**, 95431b (1971). Structure: *eidem*, *Tetrahedron Lett.* **9**, 4919 (1968); *eidem*, *Tetrahedron* **27**, 1951 (1971).

Colorless needles, mp 229-230° (monohydrate). $[\alpha]_D^{20}$ -116.6° (methanol). uv max: 230, 266, 306 nm (log ε 4.46, 4.07, 3.68). Slightly sol in benzene, water; freely sol in acetone, anhydr dioxane, tetrahydrofuran, methanol, ethanol. Practically insol in petr ether, ether, cyclohexane, $CHCl_3$.

372. Amarolide. [29913-86-8] $(2\alpha,11\alpha)$-2,11-Dihydroxy-picrasane-1,12,16-trione. $C_{20}H_{28}O_6$; mol wt 364.44. C 65.91%, H 7.74%, O 26.34%. Isoln from *Ailanthus altissima* (Mill.) Swingle (*A. glandulosa* Desf.), *Simaroubaceae*, and proposed structure: Casinovi *et al.*, *Tetrahedron Lett.* **1965**, 2273. Isoln from *Castela nicholsoni* Hook, *Simaroubaceae*, and revised structure: Stöcklin *et al.*, *ibid.* **1970**, 2399.

mp 253-255°. Practically insol in weak alkaline solns. Slightly sol in sodium carbonate, easily dissolved by sodium hydroxide.

Monoacetate. $C_{22}H_{30}O_7$. mp 264-265° for 11-acetate; mp 225° for 2-acetate.

Diacetate. $C_{24}H_{32}O_8$. mp 269-270° (from alcohol). Also reported as mp 260-262°.

373. Ambenonium Chloride. [115-79-7] N,N'-[(1,2-Dioxo-1,2-ethanediyl)bis(imino-2,1-ethanediyl)]bis[2-chloro-N,N-diethylbenzenemethanaminium] chloride (1:2); [oxalylbis(iminoethylene)]bis[(o-chlorobenzyl)diethylammonium chloride]; N,N'-bis[2-diethylaminoethyl]oxamide bis[2-chlorobenzyl chloride]; Win-8077; Mytelase. $C_{28}H_{42}Cl_4N_4O_2$; mol wt 608.47. C 55.27%, H 6.96%, Cl 23.30%, N 9.21%, O 5.26%. Reversible acetylcholinesterase inhibitor. Prepn: F. K. Kirchner, **DE 1024517** (1958 to Sterling Drug), *C.A.* **54**, 18366g (1960). Cholinesterase inhibition kinetics: A. S. Hodge *et al.*, *Mol. Pharmacol.* **41**, 937 (1992). HPLC determn in serum: K. Ohtsubo *et al.*, *J. Chromatogr.* **496**, 397 (1989). Clinical pharmacokinetics in patients with myasthenia gravis: C. Tharasse-Bloch *et al.*, *Eur. J. Drug Metab. Pharmacokinet.* **16**, 299 (1991).

Crystals, mp 196-199°. Freely soluble in water.

THERAP CAT: In treatment of myasthenia gravis.

374. Amber. [9000-02-6] Baltic amber; bernstein; succinite. A fossil resin from the extinct pine tree *Pinites succinifera* (Goepp.) Conway, *Pinaceae*. Found along the Baltic coast, also mined in Samland (East Prussia). Baltic amber contains: C 79%, H 10.5%, O 10.5%; succinic acid 3-8%; α-amyrin 20-30%. Refs: Plonait, *Angew. Chem.* **48**, 184 (1935); Schmid and co-workers: *Ann.* **503**, 269 (1933); *Monatsh. Chem.* **63**, 210 (1933); **65**, 348 (1935); **72**, 290, 311 (1939). Review: Berthelot, *Chim. Ind. (Paris)* **50**, 78-9 (1943); Frondel, *Econ. Bot.* **22**, 371 (1968). Infrared spectroscopy of different varieties of powdered amber: C. W. Beck *et al.*, *Nature* **201**, 256 (1964). Chemical constitution: J. B. Lambert, J. S. Frye, *Science* **217**, 55 (1982).

Pale-yellow to reddish-brown resin. Transparent or cloudy (due to enclosed air bubbles and free succinic acid). Brittle; conchoidal fracture. d 1.05 to 1.10. Harder than most other resins. n_D 1.539-1.545. Softens at 150°, mp 350-375° giving off a choking, aromatic odor. When rubbed it is a good generator of static electricity.

Oil of Amber, Rectified. Obtained by the destructive distillation of amber and purified by redistillation. Consists of a mixture of terpenes with resinous, oxygen-containing substances. Pale yellow to yellowish-brown, volatile oil; penetrating odor; burning acrid

taste. d 0.850-0.920. α_D^{20} +22 to +26°. Insol in water. Sol in about 10 vols alcohol; freely sol in chloroform, ether, carbon disulfide, oils.

USE: The best quality is machined into beads and other personal ornaments. For teething strings. Also used for making mouthpieces of tobacco pipes and cigarette holders. Small pieces are pressed into "ambroid" and then used for the same purpose. Impure material goes into the manufacture of "amber" varnishes.

375. Ambergris. [8038-65-1] Concretion from intestinal tract of the sperm whale, *Physeter catodon* L., *Physeteridae*. Found in tropical seas or seashores. Perfumers have used ambergris for centuries for its desirable odoriferous and fixative properties. Three major components isolated are the triterpene alcohol ambrein, epicoprostanol and coprostanone: Ruzicka, Lardon, *Helv. Chim. Acta* **29**, 912 (1946); Lederer *et al.*, *ibid.* 1354; Hardwick, Laws, *Analyst* **76**, 662 (1951). Ambergris falls under the Marine Mammal Protection Act of 1972 and is illegal to import in the U.S.A. Analytical method for the detection and identification of ambergris: T. F. Governo *et al.*, *J. Assoc. Off. Anal. Chem.* **60**, 160 (1977). Hypothesis on the biological origin of ambergris: P. A. Dubois, *Parfums Cosmet. Aromes* **19**, 35 (1978).

Gray to black, waxy mass; characteristic odor. d 0.8-0.92. mp ~60°. Flammable; almost completely volatile by heat. Insoluble in water or in alkali hydroxides. Sol in hot alcohol, chloroform, ether, fats, volatile oils.

USE: Chiefly in perfumery as tincture and essence for fixing delicate odors.

376. Amberlite®. [9079-25-8] Synthetic, high capacity cation and anion exchange resins and ion exchange-impregnated papers. Reviews: Nachod, Schubert, *Ion Exchange Technology* (Academic Press, New York, 1956); Kunin, *Ion Exchange Resins* (John Wiley, New York, 1958); Kunin, *Elements of Ion Exchange* (Reinhold, New York, 1960); Rieman, Walton, *Ion Exchange in Analytical Chemistry* (Pergamon Press, 1970); Dorfner, *Ion Exchangers Properties and Applications* (Ann Arbor Science, 1972).

377. Amberlyst 15®. [9037-24-5] Macroreticular sulfonic acid cation exchange resin based on a styrene-divinylbenzene copolymer. Prepn and characterization: R. Kunin *et al.*, *Ind. Eng. Chem. Prod. Res. Dev.* **1**, 140 (1962). Small-angle x-ray scattering structural study: B. Chu, D. M. T. Creti, *J. Phys. Chem.* **71**, 1943 (1967). Water sorption study: R. S. D. Toteja *et al.*, *Langmuir* **13**, 2980 (1997). Review of catalytic applications in synthetic organic chemistry: M. Kalesse, *Acros Org. Acta* **1**, 67-68 (1995).

Opaque beads. Moisture holding capacity: 50% H_2O. Apparent density: 1.012. True skeletal density: 1.513. Surface area: 42.5 m^2/g dry resin. Avg pore diameter: 288 Å.

USE: Ion exchange resin; reusable solid acid catalyst.

378. Ambrisentan. [177036-94-1] (αS)-α-[(4,6-Dimethyl-2-pyrimidinyl)oxy]-β-methoxy-β-phenylbenzenepropanoic acid; (+)-(2S)-2-[(4,6-dimethylpyrimidin-2-yl)oxy]-3-methoxy-3,3-diphenylpropanoic acid; BSF-208075; LU-208075; Letairis; Volibris. $C_{22}H_{22}N_2O_4$; mol wt 378.43. C 69.83%, H 5.86%, N 7.40%, O 16.91%. Nonpeptide endothelin ET_A receptor antagonist. Prepn: H. Riechers *et al.*, **WO 9611914**; *eidem*, **US 5932730** (1996, 1998 both to BASF); H. Riechers *et al.*, *J. Med. Chem.* **39**, 2123 (1996). Pharmacology: H. Vatter *et al.*, *Clin. Neuropharmacol.* **26**, 73 (2003). Review of pharmacology: H. Vatter, V. Seifert, *Cardiovasc. Drug Rev.* **24**, 63-76 (2006); of role in treatment of pulmonary arterial hypertension: R. J. Barst, *Vasc. Health Risk Manag.* **3**, 11-22 (2007).

White to off-white crystalline solid. pKa 4.0. Practically insol in water and in aq solns at low pH. Soly increases at higher pH.

(±)-Form. [713516-99-5] Crystals from diethylether, mp 190-191°.

THERAP CAT: In treatment of pulmonary arterial hypertension.

379. Ambrosin. [509-93-3] 3,3a,4,5,6,6a,9a,9b-Octahydro-6,9a-dimethyl-3-methyleneazuleno[4,5-*b*]furan-2,9-dione; 6β-hydroxy-4-oxo-10α*H*-ambrosa-2,11(13)-dien-12-oic acid γ-lactone. $C_{15}H_{18}O_3$; mol wt 246.31. C 73.15%, H 7.37%, O 19.49%. From herb of *Ambrosia maritima* L., *Compositae:* Abu-Shady, Soine, *J. Am. Pharm. Assoc.* **42**, 387 (1953); **43**, 365 (1954). Structure: Bernardi, Büchi, *Experientia* **13**, 466 (1957); Sorm *et al.*, *Collect. Czech. Chem. Commun.* **24**, 1548 (1959); Nerz *et al.*, *J. Am. Chem. Soc.* **84**, 2601 (1962). Absolute configuration: Emerson *et al.*, *Tetrahedron Lett.* **1966**, 6151. Stereospecific total synthesis of (±)-form: P. A. Grieco *et al.*, *J. Am. Chem. Soc.* **99**, 7393 (1977).

Crystals from alc, mp 146°. $[\alpha]_D^{22}$ −154.50° (c = 2 in alcohol). uv max (95% ethanol): 217, 324 nm (ε 13465; 36). Very sol in chloroform; sol in alc, benzene, sparingly sol in ether, petr ether. Practically insol in water, cold dil NaOH, dil acids.

(±)-Form. mp 188-190°.

380. Ambroxol. [18683-91-5] *trans*-4-[[(2-Amino-3,5-dibromophenyl)methyl]amino]cyclohexanol; *N*-(*trans-p*-hydroxycyclohexyl)-(2-amino-3,5-dibromobenzyl)amine; NA-872. $C_{13}H_{18}Br_2N_2O$; mol wt 378.11. C 41.30%, H 4.80%, Br 42.26%, N 7.41%, O 4.23%. Metabolite of bromhexine, *q.v.* Structure: E. Schraven *et al.*, *Eur. J. Pharmacol.* **7**, 445 (1967). Synthesis: J. Keck, *Ann.* **707**, 107 (1967); **FR 1522709**; J. Keck *et al.*, **US 3536713** (1968, 1970 both to Thomae). Toxicity: S. Püschmann, R. Engelhorn, *Arzneim.-Forsch.* **28**, 889 (1978). Series of articles on pharmacology, metabolism, and clinical studies: *ibid.* 889-935; on pharmacology and clinical efficacy of combination with amoxicillin, *q.v.*, in bronchopulmonary disease: *ibid.* **37**, 965-971 (1987). Symposium on pharmacology and efficacy in multicenter studies: *Respiration* **51**, Suppl. 1, 1-68 (1987).

Hydrochloride. [23828-92-4] Abramen; Ambril; Bronchopront; Duramucal; Fluibron; Fluixol; Frenopect; Lindoxyl; Motosol; Mucofar; Mucosan; Mucosolvan; Mucoclear; Mucovent; Pect; Solvolan; Stas-Hustenlöser; Surbronc; Surfactal. $C_{13}H_{18}Br_2N_2O\cdot HCl$; mol wt 414.57. Crystals from ethanol, mp 233-234.5° (dec). LD_{50} in mice, rats (mg/kg): 268, 380 i.p.; 2720, 13400 orally (Püschmann, Engelhorn).

THERAP CAT: Expectorant.

381. Ambutonium Bromide. [115-51-5] γ-(Aminocarbonyl)-*N*-ethyl-*N*,*N*-dimethyl-γ-phenylbenzenepropanaminium bromide; 3-(carbamoyl-3,3-diphenylpropyl)ethyldimethylammonium bromide; 4-dimethylamino-2,2-diphenylbutyramide ethyl bromide; R-100. $C_{20}H_{27}BrN_2O$; mol wt 391.35. C 61.38%, H 6.95%, Br 20.42%, N 7.16%, O 4.09%. Anticholinergic. Prepn: Janssen *et al.*, *Arch. Int. Pharmacodyn.* **103**, 82 (1955). Pharmacology: P. J. Southgate, *ibid.* **196**, 376 (1972).

Crystals, mp 228-229° (dec).

Combination with oxazepam. Praxiten SP.

THERAP CAT: Antispasmodic.

382. Amcinonide. [51022-69-6] (11β,16α)-21-(Acetyloxy)-16,17-[cyclopentylidenebis(oxy)]-9-fluoro-11-hydroxypregna-1,4-diene-3,20-dione; 9-fluoro-11β,16α,17,21-tetrahydroxypregna-1,4-diene-3,20-dione cyclic 16,17-acetal with cyclopentanone 21-acetate; CL-34699; Amciderm; Cyclocort; Penticort. $C_{28}H_{35}FO_7$; mol wt 502.58. C 66.92%, H 7.02%, F 3.78%, O 22.28%. Prepn: W. Shultz *et al.*, **DE 2437847** (1975 to Am. Cyanamid), *C.A.* **83**, 10608g (1975). Bioavailability study: R. Woodford, J. M. Haigh, *Curr. Ther. Res.* **26**, 301 (1979). Therapeutic use in eczematoid conditions: G. L. Rocha *et al.*, *ibid.* **19**, 538 (1976); E. W. Rosenberg, *Cutis* **24**, 642 (1979).

THERAP CAT: Glucocorticoid.

383. Amdinocillin. [32887-01-7] (2*S*,5*R*,6*R*)-6-[[(Hexahydro-1*H*-azepin-1-yl)methylene]amino]-3,3-dimethyl-7-oxo-4-thia-1-azabicyclo[3.2.0]heptane-2-carboxylic acid; 6-[(hexahydro-1*H*-azepin-1-yl)methyleneamino]penicillanic acid; mecillinam; FL-1060; Ro-10-9070; Selexid (inj). $C_{15}H_{23}N_3O_3S$; mol wt 325.43. C 55.36%, H 7.12%, N 12.91%, O 14.75%, S 9.85%. Semi-synthetic antibiotic related to penicillin. Prepn: F. J. Lund, **DE 2055531**; *idem*, **US 3957764** (1971, 1976 both to Lövens Kemiske Fabrik). Synthesis and chemical properties: H. B. König *et al.*, *Arzneim.-Forsch.* **33**, 88 (1983). X-ray structural study: J. W. Krajewski *et al.*, *J. Antibiot.* **34**, 282 (1981). *In vitro* study: B. Chattopadhyay, I. Hall, *J. Antimicrob. Chemother.* **5**, 549 (1979). Activity against gram-negative bacteria: D. S. Reeves, *J. Antimicrob. Chemother.* **3**, Suppl. B, 5 (1977). Studies on mechanism of action: B. G. Spratt, *ibid.* 13. Metabolism: A. P. Ball *et al.*, *ibid.* **4**, 241 (1978). Use in urinary tract infections and septicemia: N. Frimodt-Miller, T. J. Ravn, *Infection* **7**, 35 (1979). Determn in plasma and urine by HPLC: T. L. Lee, M. A. Brooks, *J. Chromatogr.* **227**, 137 (1982). Pharmacokinetics in man: B. R. Meyers *et al.*, *Antimicrob. Agents Chemother.* **23**, 827 (1983). Symposium on pharmacology, pharmacokinetics, clinical studies: *Am. J. Med.* **75**, no. 2A, 1-138 (1983). Review of pharmacology and clinical efficacy: H. C. Neu, *Pharmacotherapy* **5**, 1-10 (1985).

Crystals from methanol-acetone mp 156° (dec). $[\alpha]_D^{20}$ +285° (c = 1 in 0.1*N* HCl). Sol in water.

Pivaloyloxymethyl ester *see* Amdinocillin Pivoxil.

THERAP CAT: Antibacterial.

384.   Amdinocillin Pivoxil. [32886-97-8]   (2S,5R,6R)-6-[[[(Hexahydro-1H-azepin-1-yl)methylene]amino]-3,3-dimethyl-7-oxo-4-thia-1-azabicyclo[3.2.0]heptane-2-carboxylic acid (2,2-dimethyl-1-oxopropoxy)methyl ester; pivaloyloxymethyl 6-[(hexahydro-1H-azepin-1-yl)methyleneamino]penicillanate; pivamdinocillin; pivmecillinam; FL-1039; Selexid (susp.). $C_{21}H_{33}N_3O_5S$; mol wt 439.57. C 57.38%, H 7.57%, N 9.56%, O 18.20%, S 7.29%. Semisynthetic antibiotic related to penicillin. Pivaloyloxymethyl ester of amdinocillin, q.v. Prepn: F. J. Lund, **DE 2055531**; idem, **US 3957764** (1971, 1976 both to Lövens Kemiske Fabrik). HPLC stability analysis: R. B. Hagel, E. H. Waysek, J. Chromatogr. **178**, 97 (1979). Bacteriological and pharmacokinetic study: T. Damsgaard et al., J. Antimicrob. Chemother. **5**, 267 (1979). Metabolism: J. D. Anderson, M. A. Adams, Chemotherapy **25**, 1 (1979). Clinical study: B. T. Andersen et al., Infection **8**, 27 (1980). Toxicity study: S. Sato et al., Takeda Kenkyushoho **35**, 179 (1976), C.A. **86**, 115154w (1977).

Crystals from cyclohexane, mp 118.5-119.5°. $[\alpha]_D^{20}$ +231° (c = 1 in 96% ethanol). LD_{50} in mice, rats (mg/kg): 475-480, 465 i.v.; 1736-1930, 1935-2100 s.c.; 3020, 9500-10000 orally (Sato).

Hydrochloride. [32887-03-9] Selexid (tabl.); Melysin. $C_{21}H_{33}N_3O_5S$·HCl; mol wt 476.03. Crystals from methanol-diisopropyl ether, mp 172-173°. $[\alpha]_D^{20}$ +219° (c = 1 in 0.1N HCl).

THERAP CAT: Antibacterial.

385.   Amediplase. [151912-11-7] [173-Serine,174-tyrosine,175-glutamine]-173-275-plasminogen activator (human tissue-type reduced) fusion protein with urokinase (human urine β-chain reduced); K_2tu-PA; CGP-42935; MEN-9063. Chimeric, recombinant single chain glycoprotein of 365 amino acid residues composed of the kringle-2 domain of human tissue plaminogen activator, q.v., fused to the serine protease domain of single chain pro-urokinase, q.v. Prepn: B. Rajput et al., **EP 277313**; idem, **US 5242819** (1988, 1993 both to Ciba-Geigy); and primary structure: A. A. Bergwerff et al., Eur. J. Biochem. **212**, 639 (1993). Characterization of glycoforms: D. Müller et al., Biol. Mass Spectrom. **23**, 330 (1994). CHO cell-based production system: F. A. M. Asselbergs et al., J. Biotechnol. **42**, 221 (1995). Thrombolytic activity in vivo: G. Agnelli et al., Thromb. Haemostasis **68**, 331 (1992). Review of development and therapeutic potential: S. A. Doggrell, Curr. Opin. Investig. Drugs **5**, 344-347 (2004).

THERAP CAT: Thrombolytic.

386.   Americium. [7440-35-9] Am; at. no. 95; valences 3, 4, 5, 6. Man-made radioactive element. No stable nuclides; known isotopes (mass numbers): 234, 237-247. Longest-lived known isotope 243 ($T_{1/2}$ 7.38 × 10^3 years, rel. at. mass 243.0614, α-emitter). First isotope prepared: ^{241}Am ($T_{1/2}$ 432.7 years, α and γ-emitter, rel. at. mass 241.0568); prepd in 1944 by G. T. Seaborg et al. in The Transuranium Elements, G.T. Seaborg et al., Eds., (McGraw-Hill, New York, 1949) p 1525-1553; idem, Phys. Rev. **78**, 472 (1950). Isoln: Armstrong et al., AIChE J. **3**, 286 (1957); Coleman, J. Inorg. Nucl. Chem. **3**, 327 (1957). Prepn of metal: E. F. Westrum, Jr., L. Eyring, J. Am. Chem. Soc. **73**, 3396 (1951); Cunningham, Lohr, ibid. 2026. Superconducting properties: J. L. Smith, R. G. Haire, Science **200**, 535 (1978). Clinical application in bone mineral determn: E. G. De Puey et al., J. Nucl. Med. **16**, 891 (1975); in cancer radiotherapy: R. Nath et al., Int. J. Radiat. Oncol. Biol. Phys. **14**, 969 (1988). Reviews: C. Keller, The Chemistry of the Transuranium Elements (Verlag Chemie, Weinheim, English Ed., 1971) pp 485-527; Comprehensive Inorganic Chemistry vol. 5, J. C. Bailar, Jr. et al., Eds. (Pergamon Press, Oxford, 1973) passim; Handb. Exp. Pharmakol. **36**, 689-940 (1973); W. W. Schulz, R. A. Penneman in The Chemistry of the Actinide Elements vol. 2, J. J. Katz et al., Eds. (Chapman and Hall, New York, 1986) pp 887-961. See also metabolism study of internal contamination of ^{241}Am in man: N. Cohen et al., Science **206**, 64 (1979). Review of toxicology and human exposure: Toxicological Profile for Americium (PB2004-104396, 2004) 333 pp.

Silvery, ductile, very maleable, non-magnetic metal. Allotropic forms: double hexagonal close-packed α-form transforms to β-form at 658°; face-centered cubic β-form exists from 793-1004°; γ-form exists from ~1050° to mp. mp 1173°. bp (calc) 2067°. d 13.671. Dissolves readily in aq HCl; insol in liquid NH_3.

Trivalent americium. The most common in aq soln. Color light pink changing to yellow with increasing concn. Sharp absorption peak at 5027 Å. Americium dioxide, AmO_2, is obtained by ignition of most trivalent Am compounds.

Tetravalent americium. Known only in the solid state. When AmO_2 or AmF_3 is treated with fluoride, solid AmF_4 results.

Pentavalent and hexavalent americium compds. When obtained in soln are doubly oxygenated and have the general formula AmO_2^{+n}, where n = +1 for Am(V) and n = +2 for Am(VI). Hexavalent americium is yellow or light brown in dilute perchloric or nitric acid, green in fluoride solns and dark brown in sulfuric acid. A deep red ion complex is formed in bicarbonate-carbonate solns.

Caution: Radiation hazard; handling requires special equipment and shielding facilities (Katz et al., loc. cit. **vol. 2**, p 1128).

USE: ^{241}Am as radiation source for thickness gauging, density and radiographic measurements; parent for production of ^{242}Cf and ^{242}Pu; in ionization smoke detectors; for dissipation of static electrical charges. ^{241}Am-Be mixture as neutron source. ^{243}Am as target material for production of transcurium elements in high neutron-flux reactors.

THERAP CAT: ^{241}Am as diagnostic aid (bone mineral analyzer); as antineoplastic (radiation source).

387.   Ametryn. [834-12-8] N^2-Ethyl-N^4-(1-methylethyl)-6-(methylthio)-1,3,5-triazine-2,4-diamine; 2-(ethylamino)-4-(isopropylamino)-6-(methylthio)-s-triazine; 2-ethylamino-4-isopropylamino-6-methylmercapto-s-triazine; 2-methylthio-4-ethylamino-6-isopropylamino-s-triazine; ametryne; G-34162; Ametrex; Evik; Gesapax. $C_9H_{17}N_5S$; mol wt 227.33. C 47.55%, H 7.54%, N 30.81%, S 14.10%. Prepn: Gysin, Knuesli, **CH 337019**, C.A. **57**, 14226c (1962); Rufener et al., **US 3558622** (1959, 1971 both to Geigy). Acute toxicity: T. B. Gaines, R. E. Linder, Fundam. Appl. Toxicol. **7**, 299 (1986).

Crystals, mp 88-89°. Aqueous soly data: Ward, Weber, J. Agric. Food Chem. **16**, 959 (1968). LD_{50} in adult male, female rats (mg/kg): 508, 590 orally (Gaines, Linder).

USE: Herbicide.

388.   Amezinium Methyl Sulfate. [30578-37-1] 4-Amino-6-methoxy-1-phenylpyridazinium methyl sulfate (1:1); LU-1631; Regulton; Risumic; Supratonin. $C_{12}H_{15}N_3O_5S$; mol wt 313.33. C 46.00%, H 4.83%, N 13.41%, O 25.53%, S 10.23%. Sympathomimetic agent with vascular and cardiac activity. Prepn: F. Richenender, R. Kropp, **DE 1912941**; idem, **US 3631038** (1970, 1971 both to BASF). Series of articles on synthesis, pharmacology, mechanism of action, metabolism, pharmacokinetics, bioavailability, clinical trials: Arzneim.-Forsch. **31**, 1527-1671 (1981). Acute toxicity data: H. J. Teschendorf, ibid. 1568. HPLC determn in human plasma: D. Hotz, E. Brode, J. Chromatogr. **277**, 217 (1983). Disposition and identification of major metabolites in rats: K. Nambu et al., Arzneim.-Forsch. **38**, 909 (1988).

Crystals from water, mp 176° (dec). LD_{50} in mice, rats (mg/kg): 1630, 1410 orally; 40.4, 45.5 i.v. (Teschendorf).

THERAP CAT: Antihypotensive.

389. Amfenac. [51579-82-9] 2-Amino-3-benzoylbenzene-acetic acid; 2-amino-3-benzoylphenylacetic acid. $C_{15}H_{13}NO_3$; mol wt 255.27. C 70.58%, H 5.13%, N 5.49%, O 18.80%. Prepn: W. J. Welstead, H. W. Moran, **DE 2324768** *C.A.* **80**, 59708s (1974) and **US 4045576** (1973, 1977 both to A. H. Robbins). Synthesis and anti-inflammatory activity: W. J. Welstead *et al., J. Med. Chem.* **22**, 1074 (1979). Anti-inflammatory, analgesic, antipyretic activities: H. Fujimura *et al., Oyo Yakuri* **22**, 381 (1981), *C.A.* **97**, 49427m (1982). Platelet aggregation inhibition: *eidem, ibid.* 399, *C.A.* **97**, 49502g (1982). Effect on polymorphonuclear leukocytes: T. Matsumoto *et al., Pharmacol. Res. Commun.* **14**, 523 (1982). Toxicity data: L. F. Sancilio *et al.*, 170th Am. Chem. Soc. Meet. (Chicago, Aug. 1975), Abstracts of Papers, MEDI 17.

mp 121-123° (dec). LD_{50} in mice, rats (mg/kg): 615, 311 orally (Sancilio).

Sodium salt monohydrate. [61618-27-7] AHR-5850D; Fenazox. $C_{15}H_{12}NNaO_3 \cdot H_2O$; mol wt 295.27. Yellow solid from ethanol/isopropyl ether, mp 254-255.5°.

THERAP CAT: Anti-inflammatory.

390. Amicarbazone. [129909-90-6] 4-Amino-*N*-(1,1-dimethylethyl)-4,5-dihydro-3-(1-methylethyl)-5-oxo-1*H*-1,2,4-triazole-1-carboxamide; 4-amino-1-(*N*-*t*-butylcarbamoyl)-3-isopropyl-1,2,4-triazolin-5-one; 4-amino-*N*-*tert*-butyl-3-isopropyl-5-oxo-Δ²-1,2,4-triazoline-1-carboxamide; BAY-314666; BAY MKH 3586; Dinamic. $C_{10}H_{19}N_5O_2$; mol wt 241.30. C 49.78%, H 7.94%, N 29.02%, O 13.26%. Triazolinone herbicide for use in food crops. Prepn: K.-H. Müller *et al.*, **DE 3839206**; M. Lindig *et al.*, **US 5194085** (1990, 1993 both to Bayer). Comprehensive description: B. D. Philbrook *et al., Brighton Crop Prot. Conf. - Weeds* **1999**, 29-34.

Colorless crystals, mp 137.5°. d 1.12. Vapor pressure (Pa): 1.3 × 10⁻⁶ (20°); 3.0 × 10⁻⁶ (25°). Log P (octanol/water): 1.18 (pH 4); 1.23 (pH 7); 1.23 (pH 9). Soly in water (20°): 4.6 g/l (pH 4 - 9). LD_{50} in rats (mg/kg): 1015 orally; >2000 dermally. LC_{50} in rats (4 hr): 2.242 mg/l air. LC_{50} in bluegill sunfish, rainbow trout (96 hr): >129, >120 mg/l (Philbrook).

USE: Herbicide.

391. Amicetin. [17650-86-1] 4-[[(2*S*)-2-Amino-3-hydroxy-2-methyl-1-oxopropyl]amino]-*N*-[1-[(2*R*,5*S*,6*R*)-5-[[4,6-dideoxy-4-(dimethylamino)-α-D-glucopyranosyl]oxy]tetrahydro-6-methyl-2*H*-pyran-2-yl]-1,2-dihydro-2-oxo-4-pyrimidinyl]benzamide; $C_{29}H_{42}N_6O_9$; mol wt 618.69. C 56.30%, H 6.84%, N 13.58%, O 23.27%. Antibiotic substance produced by *Streptomyces vinaceus-drappus* isolated from soil near Kalamazoo, Mich.: C. DeBoer *et al., J. Am. Chem. Soc.* **75**, 499 (1953); J. W. Hinman *et al., ibid.* 5864; McKormich, Hoehn, *Antibiot. Chemother.* **3**, 718 (1953). Chemistry: E. H. Flynn *et al., J. Am. Chem. Soc.* **75**, 5867 (1953). Structural studies: C. L. Stevens *et al., J. Org. Chem.* **27**, 2991 (1962). Configuration: S. Hanessian, T. H. Haskell, *Tetrahedron Lett.* **5**, 2451 (1964). Inhibition of protein synthesis: A. Bloch, C. Coutsogeor-

gopoulos, *Biochemistry* **5**, 3345 (1966). Binding to 23S rRNA as mechanism of inhibition of peptide bond formation: I. G. Leviev *et al., EMBO J.* **13**, 1682 (1994).

Colorless granular crystals, mp 244-245°. pKa = 10.4, 7.0. $[\alpha]_D^{24}$ +116.5° (c = 0.5 in 0.1*N* HCl). uv max in water: 305 nm ($E_{1cm}^{1\%}$ 465); in 0.1*N* HCl: 316 nm ($E_{1cm}^{1\%}$ 433); in 0.1*N* NaOH: 322 nm ($E_{1cm}^{1\%}$ 470). Soly in water at 25° = 2 mg/ml. Slightly sol in common organic solvents. LD_{50} of citrate complex, pH 6 (mg/kg) in mice: ~90 i.v., 600-700 s.c.; in rats: ~200 i.v., ~600 s.c. (DeBoer).

392. Amicoumacin A. [78654-44-1] 3-Amino-2,3-dideoxy-*N*⁶-[1-(3,4-dihydro-8-hydroxy-1-oxo-1*H*-2-benzopyran-3-yl)-3-methylbutyl]hexaramide. $C_{20}H_{29}N_3O_7$; mol wt 423.47. C 56.73%, H 6.90%, N 9.92%, O 26.45%. Major component of antibiotics produced by *Bacillus pumilus* BN-103. Also exhibits anti-inflammatory activity *in vivo*. Isoln and biol activity: **JP Kokai 83 18379** (1983 to Meiji Seika), *C.A.* **99**, 20906x (1983); J. Itoh *et al., J. Antibiot.* **34**, 611 (1981); *eidem, Agric. Biol. Chem.* **46**, 1255 (1982). Structure: *eidem, ibid.* 2659. Use as acaricide: **JP Kokai 83 216107** (1983 to Meiji Seika), *C.A.* **100**, 116494b (1984).

Hydrochloride. Colorless powder, mp 132-135° (dec). uv max (methanol): 208, 247, 315 nm (ε 27300, 6400, 4380). $[\alpha]_D^{23}$ −97.2° (c = 1.0 in methanol). LD_{50} orally in mice: 132 mg/kg (Itoh).

393. Amidephrine. [37571-84-9] *N*-[3-[1-Hydroxy-2-(methylamino)ethyl]phenyl]methanesulfonamide; 3′-[1-hydroxy-2-(methylamino)ethyl]methanesulfonanilide; MJ-1996. $C_{10}H_{16}N_2O_3S$; mol wt 244.31. C 49.16%, H 6.60%, N 11.47%, O 19.65%, S 13.12%. α_1 Adrenoceptor agonist. Prepn: Larsen, Uloth, **FR M3027** (1965 to Mead Johnson), *C.A.* **62**, 13091a (1965); Uloth *et al., J. Med. Chem.* **9**, 88 (1966). Pharmacology of the racemate: Dungan *et al., Int. J. Neuropharmacol.* **4**, 219 (1965); Stanton *et al., ibid.* 235; of the isomers: Larsen, Lish, *Nature* **203**, 1283 (1964); Buchthal, Jenkinson, *Eur. J. Pharmacol.* **10**, 293 (1970). Toxicology: J. H. Weikel, Jr., K. H. Harper, *Toxicol. Appl. Pharmacol.* **23**, 589 (1972).

Crystals, mp 159-161°. pKa 9.1.

Methanesulfonate. [1421-68-7] Amidephrine mesylate; MJ-5190; Fentrinol. $C_{10}H_{16}N_2O_3S \cdot CH_3SO_3H$; mol wt 340.41. Crystals from ethanol, mp 207-209°. LD_{50} in mice (mg/kg): 2284 orally, 780 i.p., 190 i.v., 1990 s.c.; in male, female Mead Johnson rats (mg/kg): 229, 36 orally; 144, 25 i.p.; in male, female Sprague-Dawley rats (mg/kg): 24, 13 orally; in rabbits (mg/kg): 12 orally, 7.5 intradermally (Weikel, Harper).

THERAP CAT: Vasoconstrictor; decongestant (nasal).

394. Amidinomycin. [3572-60-9] (1*R*,3*S*)-3-Amino-*N*-(3-amino-3-iminopropyl)cyclopentanecarboxamide; (−)-*cis*-*N*-(2-amidinoethyl)-3-aminocyclopentanecarboxamide; myxoviromycin. C_9-$H_{18}N_4O$; mol wt 198.27. C 54.52%, H 9.15%, N 28.26%, O 8.07%. Antiviral antibiotic produced by *Streptomyces flavochromogenes* isolated from Japanese soil (Shiuoka Prefecture). Isoln and structure: S. Nakamura *et al.*, *J. Antibiot.* **14A**, 103 (1961); S. Nakamura, *Chem. Pharm. Bull.* **9**, 641 (1961). Identity with myxoviromycin: S. Nakamura *et al.*, *J. Antibiot.* **14A**, 163 (1961). Synthesis of (±)-*cis* and *trans* forms: H. Paul *et al.*, *Arch. Pharm.* **301**, 512 (1968). Crystal structure and absolute configuration: M. Kaneda *et al.*, *J. Antibiot.* **33**, 778 (1980). Enantioselective synthesis: H. Nagata *et al.*, *Tetrahedron: Asymmetry* **8**, 2679 (1997).

Sulfate. [74984-58-0] $C_9H_{18}N_4O.H_2SO_4$; mol wt 296.34. Plates or needles from water + methanol, dec 285-288°. $[\alpha]_D^{21}$ −3.9° (c = 3). Absorption spectra: S. Nakamura, *loc. cit.* Soluble in water. Prac insol in ether, benzene, ethyl acetate, methanol, ethanol, butan tone.

395. Amido-G-Acid. [86-65-7] 7-Amino-1,3-naphthalenedisulfonic acid; 2-naphthylamine-6,8-disulfonic acid; amino-G-acid. $C_{10}H_9NO_6S_2$; mol wt 303.30. C 39.60%, H 2.99%, N 4.62%, O 31.65%, S 21.14%. Prepd by sulfonation of β-naphthylamine: Fierz-David, Braunschweig, *Helv. Chim. Acta* **6**, 1146 (1923).

Tetrahydrate. Fine monoclinic needles. Sol in water, less sol in alc. Soly in water at 20°: 9.24 g in 100 g of satd soln.

USE: Manufacture of dyes.

396. Amido-R-Acid. [92-28-4] 3-Amino-2,7-naphthalenedisulfonic acid; 2-naphthylamine-3,6-disulfonic acid. $C_{10}H_9NO_6S_2$; mol wt 303.30. C 39.60%, H 2.99%, N 4.62%, O 31.65%, S 21.14%. Prepd by treating 2-hydroxy-3,6-naphthalenedisulfonic acid with ammonium sulfite and ammonium hydroxide: Petitcolas, Josué, *Bull. Soc. Chim. Fr.* **1952**, 89.

Crystals or powder. Soluble in water. Solutions show a violet-blue fluorescence.

USE: Manufacture of dyes.

397. Amidosulfuron. [120923-37-7] N-(4,6-Dimethoxy-2-pyrimidinyl)-4-methyl-3,5-dithia-2,4-diazahexanamide 3,3,5,5-tetraoxide; 3-(4,6-dimethoxypyrimidin-2-yl)-1-(N-methyl-N-methylsulfonyl)urea; HOE-75032; Gratil; Eagle. $C_9H_{15}N_5O_7S_2$; mol wt 369.37. C 29.27%, H 4.09%, N 18.96%, O 30.32%, S 17.36%. Postemergence herbicide for broad-leaf weed control in cereals. Prepn: L. Willms *et al.*, *EP 131258*; *eidem*, *US 4601747* (1985, 1986 both to Hoechst). Degradation in soil: G. Fent *et al.*, *Sci. Total Environ.* **132**, 201 (1993); K. L. Sagan *et al.*, *J. Agric. Food Chem.* **46**, 1205 (1998). Field trial in cereal crops: D. S. M. D'Souza *et al.*, *Brighton Crop Prot. Conf. - Weeds* **1993**, 567; for control of wild mustard in canola: K. J. Kirkland *Weed Technol.* **9**, 541 (1995).

mp 160-163°. Soly in water (mg/l): 3.3 at pH 3, 9.0 at pH 7, 13500 at pH 10. Vapor pressure at 20°: 1.3×10^{-5} Pa.

USE: Herbicide.

398. Amifampridine. [54-96-6] 3,4-Pyridinediamine; 3,4-DAP; 3,4-diaminopyridine. $C_5H_7N_3$; mol wt 109.13. C 55.03%, H 6.47%, N 38.51%. Selective blocker of potassium channels in nerve membranes. Prolongs activation of voltage-gated calcium channels and increases the release of acetylcholine in the neuromuscular synapse. Prepn: O. Bremer, *Ann.* **518**, 274 (1935); J. W. Clark-Lewis, R. P. Singh, *J. Chem. Soc.* **1962**, 2379; J. B. Campbell *et al.*, *J. Heterocycl. Chem.* **23**, 669 (1986). Prepn of phosphate: F. Guyon *et al.*, *WO 02062760*: *eidem*, *US 040106651* (2002, 2004). HPLC determn in serum: J. Leslie, C. T. Bever, *J. Chromatogr.* **496**, 214 (1989); in plasma: S. Goulay-Dufay *et al.*, *J. Chromatogr. B* **805**, 261 (2004). Acute toxicity: P. Lechat *et al.*, *Ann. Pharm. Fr.* **26**, 345 (1968). Effect on neurotransmitter release: J. Molgo *et al.*, *Eur. J. Pharmacol.* **61**, 25 (1980); R. H. Thomsen, D. F. Wilson, *J. Pharmacol. Exp. Ther.* **227**, 260 (1983). Review of clinical trials in Lambert-Eaton myasthenic syndrome (LEMS): P. W. Wirtz *et al.*, *Expert Rev. Clin. Immunol.* **6**, 867-874 (2010); A. Quarel *et al.*, *Curr. Med. Res. Opin.* **26**, 1363-1375 (2010).

Needles from water, mp 220° (Clark-Lewis, Singh); also reported as white to beige crystals from water, mp 218-219° (Campbell). pKa 0.49; 9.2. Unstable at high pH. Readily sol in water, alcohol; slightly sol in ether. LD_{50} i.v. in mice: 13 mg/kg (Lechat).

Phosphate. Firdapse; Zenas. $C_5H_7N_3.H_3O_4P$; mol wt 207.13. White crystalline powder, mp 229°. Sol in water; slightly sol in DMSO, glacial acetic acid, methanol; very slightly sol in ethanol, DMF. pH of 1% aq soln (25°): 4.6.

USE: Intermediate in synthesis of heterocyclic compds.

THERAP CAT: In treatment of Lambert-Eaton myasthenic syndrome.

399. Amifostine. [20537-88-6] 2-[(3-Aminopropyl)amino]-ethanethiol 1-(dihydrogen phosphate); phosphorothioic acid S-[2-[(3-aminopropyl)amino]ethyl] ester; aminopropylaminoethyl thiophosphate; ethiofos; gammaphos; SAPEP; NSC-296961; WR-2721; YM-08310; Ethyol. $C_5H_{15}N_2O_3PS$; mol wt 214.22. C 28.03%, H 7.06%, N 13.08%, O 22.41%, P 14.46%, S 14.97%. Thiophosphate derivative of cysteamine, *q.v.*; provides normal cells with selective protection against the toxic effects of cancer chemotherapy and radiation treatment. Prepn of monohydrate: J. R. Piper *et al.*, *J. Med. Chem.* **12**, 236 (1969); J. R. Piper, T. P. Johnston, *US 3892824* (1975 to Southern Res. Inst.). Differential radioprotective activity: J. M. Yuhas, J. B. Storer, *J. Natl. Cancer Inst.* **42**, 331 (1969). Mechanism of action study: G. D. Smoluk *et al.*, *Cancer Res.* **48**, 3641 (1988). Bioavailability: L. Fleckenstein *et al.*, *Pharmacol. Ther.* **39**, 203 (1988). Clinical pharmacokinetics: L. M. Shaw *et al.*, *ibid.* 195. HPLC determn in plasma: N. F. Swynnerton *et al.*, *Int. J. Radiat. Oncol. Biol. Phys.* **12**, 1495 (1986). Review of development as radioprotector: D. Q. Brown *et al.*, *Pharmacol. Ther.* **39**, 157-168 (1988); of role in chemotherapy: R. L. Capizzi *et al.*, *Cancer* **72**, 3495-3501 (1993); M. Treskes, W. J. M. van der Vijgh, *Cancer Chemother. Pharmacol.* **33**, 93-106 (1993).

Monohydrate. [63717-27-1] White solid from methanol/ether, mp 160-161° (dec). Freely sol in water. LD_{50} in mice (mg/kg): 700 i.p. (Piper, Johnston).

Trihydrate. [112901-68-5] Soly in water: >9 g/100 ml. pKa_1 <2.0; pKa_2 4.2; pKa_3 9.0; pKa_4 11.7.

THERAP CAT: Radioprotective agent.

400. Amikacin. [37517-28-5] O-3-Amino-3-deoxy-α-D-glucopyranosyl-(1 → 6)-O-[6-amino-6-deoxy-α-D-glucopyranosyl-(1 → 4)]-N^1-[(2S)-4-amino-2-hydroxy-1-oxobutyl]-2-deoxy-D-streptamine; 1-N-[L($-$)-4-amino-2-hydroxybutyryl]kanamycin A. $C_{22}H_{43}N_5O_{13}$; mol wt 585.61. C 45.12%, H 7.40%, N 11.96%, O 35.52%. Semisynthetic aminoglycoside antibiotic derived from kanamycin A. Prepn: Kawaguchi *et al.*, *J. Antibiot.* **25**, 695 (1972); H. Kawaguchi, T. Naito, **DE 2234315**; H. Kawaguchi *et al.*, **US 3781268** (both 1973 to Bristol-Myers). Biological formation from kanamycin A: L. M. Cappelletti, R. Spagnoli, *J. Antibiot.* **36**, 328 (1983). Microbiological evaluation: Price *et al.*, *ibid.* **25**, 709 (1972). Pharmacokinetics: Cabana, Taggart, *Antimicrob. Agents Chemother.* **3**, 478 (1973). *In vitro* studies: Yu, Washington, *ibid.* **4**, 133 (1973); Bodey, Stewart, *ibid.* 186. Pharmacology in humans: Bodey *et al.*, *ibid.* **5**, 508 (1974). Toxicity studies: Fujisawa *et al.*, *J. Antibiot.* **27**, 677 (1974). *Review*: K. A. Kerridge in *Pharmacological and Biochemical Properties of Drug Substances* **vol. 1**, M. E. Goldberg, Ed. (Am. Pharm. Assoc., Washington, DC, 1977) pp 125-153. Comprehensive description: P. M. Monteleone *et al.*, *Anal. Profiles Drug Subs.* **12**, 37-71 (1983).

White crystalline powder from methanol-isopropanol, mp 203-204° (sesquihydrate). $[\alpha]_D^{23}$ +99° (c = 1.0 in water). Sparingly sol in water. LD_{50} in mice of solns pH 6.6, pH 7.4 (mg/kg): 340, 560 i.v. (Kawaguchi).
Sulfate. [39831-55-5] Amiglyde-V; Amikin; Amiklin; BB-K8; Biklin; Lukadin; Mikavir; Novamin; Pierami. $C_{22}H_{43}N_5O_{13}.2H_2SO_4$; mol wt 781.75. Amorphous form, dec 220-230°. $[\alpha]_D^{22}$ +74.75° (water). Freely sol in water.
THERAP CAT: Antibacterial.
THERAP CAT (VET): Antibacterial.

401. Amiloride. [2609-46-3] 3,5-Diamino-N-(aminoiminomethyl)-6-chloro-2-pyrazinecarboxamide; N-amidino-3,5-diamino-6-chloropyrazinecarboxamide; N-amidino-3,5-diamino-6-chloropyrazinamide; 1-(3,5-diamino-6-chloropyrazinecarboxyl)guanidine; 1-(3,5-diamino-6-chloropyrazinoyl)guanidine; guanamprazine; amipramidin; amipramizide. $C_6H_8ClN_7O$; mol wt 229.63. C 31.38%, H 3.51%, Cl 15.44%, N 42.70%, O 6.97%. Sodium channel blocker. Prepn: E. J. Cragoe, Jr., **BE 639386**; *idem*, **US 3313813** (1964, 1967 both to Merck & Co.). NMR study on tautomerism and conformation: R. L. Smith *et al.*, *J. Am. Chem. Soc.* **101**, 191 (1979). Determn by isopotential fluorimetry: J. A. Murillo Pulgarín *et al.*, *Analyst* **122**, 247 (1997). Pharmacology: Baer *et al.*, *J. Pharmacol. Exp. Ther.* **157**, 472 (1967); Baba *et al.*, *Clin. Pharmacol. Ther.* **9**, 318 (1968); Lant *et al.*, *ibid.* **10**, 50 (1969). Metabolism: Weiss *et al.*, *ibid.* 401. Clinical evaluation in cystic fibrosis: M. R. Knowles *et al.*, *N. Engl. J. Med.* **322**, 1189 (1990); in combination with uridine 5′-triphosphate, *q.v.*: W. D. Bennett *et al.*, *Am. J. Respir. Crit. Care Med.* **153**, 1796 (1996). *Reviews*: H. L. Macfie *et al.*, *Drug Intell. Clin. Pharm.* **15**, 94-98 (1981); D. E. Hyams, *Int. Congr. Symp. Ser. - R. Soc. Med.* **44**, 65-73 (1981). Comprehensive description of hydrochloride: D. J. Mazzo, *Anal. Profiles Drug Subs.* **15**, 1-34 (1986). Review of mechanism of action: T. R. Kleyman, *et al.*, *Semin. Nephrol.* **19**, 524-532 (1999).

Solid, mp 240.5-241.5°. pKa 8.7.

Hydrochloride. [2016-88-8] $C_6H_8ClN_7O.HCl$; mol wt 266.09. mp 293.5°
Hydrochloride dihydrate. [17440-83-4] MK-870; Amikal; Midamor; Modamide. $C_6H_8ClN_7O.HCl.2H_2O$; mol wt 302.12. Crystalline solid, dec 285-288°. uv max (water): 212, 285, 362 nm ($E_{1cm}^{1\%}$ 642, 555, 617). Freely sol in DMSO, slightly sol in water, isopropanol, ethanol. Practically insol in acetone, chloroform, diethyl ether, ethyl acetate.
Hydrochloride dihydrate mixture with hydrochlorothiazide. Co-amilozide; Amilco; Aquaretic; Ecodurex; Hexarese; Moduretic; Moduretik; Normetic; Normorix.
THERAP CAT: Diuretic.

402. Aminacrine. [90-45-9] 9-Acridinamine; 5-aminoacridine; 9-aminoacridine. $C_{13}H_{10}N_2$; mol wt 194.24. C 80.39%, H 5.19%, N 14.42%. Prepn: A. Albert, B. Ritchie, *Org. Synth.* **coll. vol. III**, 53 (1955). Toxicity study: D. C. Brodie, E. Lowenhaupt, *J. Am. Pharm. Assoc.* **38**, 498 (1949). Spectrophotometric determn in pharmaceuticals: E. A. Bunch, *J. Assoc. Off. Anal. Chem.* **66**, 140 (1983). DNA intercalator used to induce exptl frameshift mutations: M. Conrad, M. D. Topal, *J. Biol. Chem.* **261**, 16226 (1986).

Sulfur-yellow needles from alcohol or acetone, mp 241°. Moderately strong base. pKa (25°): 4.53. Freely sol in alcohol; slightly sol in chloroform, toluene, pyridine; sol in acetone.
Hydrochloride. [134-50-9] Monacrin. $C_{13}H_{10}N_2.HCl$; mol wt 230.70. Pale yellow crystals. Neutral reaction. One of the most highly fluorescent substances. One gram dissolves in 300 ml water giving a faintly yellow soln showing bluish-violet fluorescence. LD_{50} orally in mice: 78 mg/kg (Brodie, Lowenhaupt).
THERAP CAT: Antiseptic.
THERAP CAT (VET): Antiseptic.

403. Amineptine. [57574-09-1] 7-[(10,11-Dihydro-5H-dibenzo[a,d]cyclohepten-5-yl)amino]heptanoic acid. $C_{22}H_{27}NO_2$; mol wt 337.46. C 78.30%, H 8.06%, N 4.15%, O 9.48%. Dopamine uptake blocker. Prepn: C. Malen *et al.*, **DE 2011806**; *eidem*, **US 3758528** (1970, 1973 both to Sci. Union et Cie Soc. Franc. Recher. Med.). Biochemical and pharmacological study: R. Samanin *et al.*, *J. Pharm. Pharmacol.* **29**, 555 (1977). Effect on dopamine uptake: N. B. Mercuri *et al.*, *Br. J. Pharmacol.* **104**, 700 (1991). Metabolism: L. Grislain *et al.*, *Eur. J. Drug Metab. Pharmacokinet.* **15**, 339 (1990). HPLC determn in plasma: P. P. Rop *et al.*, *J. Chromatogr.* **532**, 351 (1990). Clinical trial in depression: M. Paes de Sousa, J. Tropa, *Clin. Neuropharmacol.* **12**, Suppl. 2, S77 (1989).

Hydrochloride. [30272-08-3] S-1694; Maneon; Survector. $C_{22}H_{27}NO_2.HCl$; mol wt 373.92. Crystals from distilled water, mp 226-230°.
THERAP CAT: Antidepressant.

404. Aminitrozole. [140-40-9] N-(5-Nitro-2-thiazolyl)acetamide; 2-acetamido-5-nitrothiazole; 2-acetylamino-5-nitrothiazole; acinitrazole; nithiamide; Tritheon; Trichorad; Enheptin-A. $C_5H_5N_3O_3S$; mol wt 187.17. C 32.09%, H 2.69%, N 22.45%, O 25.64%, S 17.13%. Prepd by nitration of 2-acetamidothiazole: Ganapathi, Venkataraman, *Proc. Indian Acad. Sci.* **22A**, 343 (1945); Bellavita, *Ann. Chim. Appl.* **38**, 449 (1948); Hurd, Wehrmeister, *J. Am. Chem. Soc.* **71**, 4007 (1949); Yamamoto, *J. Pharm. Soc. Jpn.* **72**, 1017 (1952). As an antihistomonad: Waletzky, Marson, **US 2531756** (1950 to Am. Cyanamid).

Needles from alc, elongated plates from acetic acid. mp 264-265°. The commercial product may be yellow. Sol in aq solns of NaOH and NH_3 with deep orange color.

THERAP CAT: Antiprotozoal (Trichomonas).

THERAP CAT (VET): Antihistomonad for turkeys.

405.   *p*-Aminoacetanilide. [122-80-5] N-(4-Aminophenyl)-acetamide; 4′-aminoacetanilide; acetyl-*p*-phenylenediamine. C_8-$H_{10}N_2O$; mol wt 150.18. C 63.98%, H 6.71%, N 18.65%, O 10.65%. Prepd by catalytic hydrogenation of 4′-nitroacetanilide: Atkinson *et al.*, *J. Chem. Soc.* **1954**, 2023. Crystal and molecular structure: M. Haisa *et al.*, *Acta Crystallogr.* **33B**, 2449 (1977).

White or slightly reddish crystals, mp 163.5-166.0°; darkens in air. Slightly soluble in cold water; freely soluble in hot water, alc, ether.

USE: Intermediate in the manufacture of azo dyes, pharmaceuticals.

406.   Aminoacetonitrile. [540-61-4] Glycinonitrile; cyano-methylamine; glycine nitrile. $C_2H_4N_2$; mol wt 56.07. C 42.84%, H 7.19%, N 49.96%. H_2NCH_2CN. Prepd by the action of alcoholic HCl or H_2SO_4 on dimolecular or trimolecular methyleneaminoace-tonitrile: Anslow, King, *J. Chem. Soc.* **1929**, 2465.

Oily liquid. bp_{15} 58° (partial decompn).

Hydrochloride. [6011-14-9] $C_2H_4N_2$.HCl. Hygroscopic crystals from alcohol, dec 165°.

Sulfate (1:1). [151-63-3] $C_2H_4N_2$.H_2SO_4. Prepn: H. Stephen, *J. Chem. Soc.* **1931**, 871. Hygroscopic leaflets from ethanol + ether, mp 121°.

Sulfate (2:1). [5466-22-8] $(C_2H_4N_2)_2$.H_2SO_4. Long flat prisms from water/ethanol, mp 166° (dec).

407.   Aminoacetophenone. C_8H_9NO; mol wt 135.17. C 71.09%, H 6.71%, N 10.36%, O 11.84%. Prepn of *m*-, *o*-, and *p*-isomers: Grammaticakis, *Compt. Rend.* **235**, 546 (1952); Braude *et al.*, *J. Chem. Soc.* **1954**, 3586. Prepn of *m*-isomer: Tinsley, US 2797244 (1957 to Union Carbide); of *p*-isomer: Norman *et al.*, *Can. J. Chem.* **40**, 1547 (1962). Toxicity studies: J. M. Vandenbelt *et al.*, *J. Pharmacol. Exp. Ther.* **80**, 31 (1944); H. F. Smyth *et al.*, *Arch. Ind. Hyg. Occup. Med.* **10**, 61 (1954).

***m*-Aminoacetophenone.** 1-(3-Aminophenyl)ethanone; 3′-ami-noacetophenone; *m*-aminoacetylbenzene. Yellow leaflets, mp 98-99°. Partly volatile in steam. LD_{50} orally in rats: 1.87 g/kg (Smyth).

***o*-Aminoacetophenone.** Yellow oily liquid. bp_{760} 250-252° (some dec); bp_{17} 135°. Volatile with steam. Practically insol in water. Sol in alc.

***p*-Aminoacetophenone.** Yellow needles, pleasant, characteristic odor. mp 106°. bp 293-295°. Sparingly sol in cold, freely in hot water; sol in alc, ether, HCl; sparingly sol in benzene. LD_{50} i.p. in rats: 260 mg/kg (Vandenbelt).

408.   D-Amino Acid Oxidase. [9000-88-8] DAAO; EC 1.4.3.3. Flavoprotein that catalyzes the oxidative deamination of D-amino acids to the corresponding α-keto acids. Found in the kidney and liver of nearly all mammals studied. Isoln from pig kidney: Krebs, *Enzymologia* **7**, 53 (1939); Straub, *Nature* **141**, 603 (1938); Negelein, Brömel, *Biochem. Z.* **300**, 225 (1939); Massey *et al.*,

Biochim. Biophys. Acta **48**, 1 (1961); from sheep kidney: Burton, *Methods Enzymol.* **2**, 199 (1955). Physical properties: Yagi *et al.*, *J. Biochem.* **61**, 580 (1967). Partial structure: Kotaki *et al.*, *ibid.* 598. Studies on specificity and inhibition: Dixon, Kleppe, *Biochim. Biophys. Acta* **96**, 368 (1965). Reaction intermediates: K. Yagi, *Front. Physicochem. Biol., Proc. Int. Symp.* **1977**, B. Pullman, Ed. (Academic Press, New York, 1978) pp 299-308. *Review:* Meister, Wellner in *The Enzymes* **7**, P. Boyer *et al.*, Eds. (Academic Press, New York, 1963) pp 634-648.

409.   L-Amino Acid Oxidase. [9000-89-9] LAAO; EC 1.4.3.2. Flavoprotein that catalyzes the oxidative deamination of L-amino acids to the corresponding α-keto acids. Found in microor-ganisms and in animal tissue, esp in kidney and liver. Occurs also in many snake venoms. Isoln from rat kidney: Blanchard *et al.*, *J. Biol. Chem.* **161**, 583 (1945). Isoln of crystalline enzyme from the venom of the eastern diamondback rattlesnake (*Crotalus adamanteus*): Wellner, Meister, *J. Biol. Chem.* **235**, 2013 (1960). Structural stud-ies: DeKok, Rawitch, *Biochemistry* **8**, 1405 (1969). Studies on in-hibitors: DeKok, Veeger, *Biochim. Biophys. Acta* **167**, 35 (1968). Mechanism of reversible activation-deactivation: C. J. Coles *et al.*, *Flavins Flavoproteins, Proc. Int. Symp. 6th* **1978**, K. Yagi, T. Ya-mano, Eds. (Japan Sci. Soc. Press, Tokyo, 1980) pp 101-105. Mul-tiple conformational states: D. Wellner, L. A. Lichtenberg, *Dev. Biochem.* **21**, 78 (1982). *Review:* Meister, Wellner in *The Enzymes* **7**, P. Boyer *et al.*, Eds. (Academic Press, New York, 1963) pp 609-634.

410.   α-Aminoadipic Acid. [542-32-5] 2-Aminohexanedi-oic acid. $C_6H_{11}NO_4$; mol wt 161.16. C 44.72%, H 6.88%, N 8.69%, O 39.71%. An amino acid isolated from *Cholera vibrio*: Blass, Macheboeuf, *Helv. Chim. Acta* **29**, 1315 (1946). Occurrence in pro-teins or protein containing material: Windsor, *J. Biol. Chem.* **192**, 595 (1951). Synthesis: Dieckmann, *Ber.* **38**, 1656 (1905); T. P. Waalkes *et al.*, *J. Am. Chem. Soc.* **72**, 5760 (1952); A. I. Scott, T. J. Wilkinson, *Synth. Commun.* **10**, 127 (1980).

Forms a monohydrate if crystallized from water below 20°. An-hydrous platelets from water above 20°. mp 206° (effervescence). One gram dissolves in 450 ml water. Sparingly sol in alcohol, ether.

Diethyl ester. $C_{10}H_{19}NO_4$; mol wt 217.27. bp_{13} 155-156°.

411.   1-Aminoanthraquinone. [82-45-1] 1-Amino-9,10-an-thracenedione. $C_{14}H_9NO_2$; mol wt 223.23. C 75.33%, H 4.06%, N 6.27%, O 14.33%. Prepd by reduction of 1-nitroanthraquinone: Graham, Hort, US 2874168 (1959 to General Aniline & Film).

Ruby-red crystals. mp ~250°; also reported as mp 243°. Practi-cally insoluble in water. Freely soluble in alcohol, benzene, chloro-form, ether, glacial acetic acid, HCl.

USE: In the mfg of dyes and pharmaceuticals.

412.   1-Aminoanthraquinone-2-carboxylic Acid. [82-24-6] 1-Amino-9,10-dihydro-9,10-dioxo-2-anthracenecarboxylic acid. $C_{15}H_9NO_4$; mol wt 267.24. C 67.42%, H 3.39%, N 5.24%, O 23.95%. Prepd by reduction of 1-nitroanthraquinone-2-carboxylic acid with sodium sulfide: Terres, *Ber.* **46**, 1639 (1913); by boiling 1-nitro-2-methylanthraquinone with alcoholic potassium hydroxide: Scholl, *Monatsh. Chem.* **34**, 1011 (1913); by heating 1-nitro-2-bro-momethylanthraquinone with sodium acetate and *o*-dichloroben-zene: Locher, Fierz, *Helv. Chim. Acta* **10**, 667 (1927).

Red needles from nitrobenzene, mp 295-296°. Soluble in aniline, boiling nitrobenzene, in sodium hydroxide and aq pyridine with formation of a deep red soln, in concd sulfuric acid with formation of a yellowish-brown soln; slightly sol in ether, benzene, alcohol. Insol in water, ligroin.

Methyl ester. $C_{16}H_{11}NO_4$. Red needles from glacial acetic acid, mp 228°.

Ethyl ester. $C_{17}H_{13}NO_4$. mp 198°.

Phenyl ester. $C_{21}H_{13}NO_4$. Reddish-golden crystals from glacial acetic acid, mp 198°.

USE: For the detection of traces of aluminum, magnesium, or zinc.

413. 4-Aminoantipyrine. [83-07-8] 4-Amino-1,2-dihydro-1,5-dimethyl-2-phenyl-3*H*-pyrazol-3-one; 4-amino-2,3-dimethyl-1-phenyl-3-pyrazolin-5-one; ampyrone; 1,5-dimethyl-2-phenyl-4-aminopyrazolone; 1-phenyl-2,3-dimethyl-4-amino-4-pyrazolone. $C_{11}H_{13}N_3O$; mol wt 203.25. C 65.00%, H 6.45%, N 20.67%, O 7.87%. Reacts with phenols in the presence of alkaline oxidizing agents to form intensely colored compounds. Prepn: E. Waser, *Helv. Chim. Acta* **8**, 117 (1925); L. Freedman, A. E. Sherndal, **US 1877166** (1933 to H. A. Metz Labs.). Colorometric reaction with phenols: E. Emerson, *J. Org. Chem.* **8**, 417 (1943); D. Svobodová, J. Gasparic, *Mikrochim. Acta* **59**, 384 (1971); Y. Fiamegos *et al.*, *Anal. Chim. Acta* **467**, 105 (2002). ^{13}C-NMR spectrum: S. P. Singh *et al.*, *J. Pharm. Sci.* **68**, 470 (1979). Thermodynamic properties: I. Katime, F. A. Herraiz, *Thermochim. Acta* **49**, 139 (1981). Metabolite of aminopyrine, *q.v.*: E. S. Vesell *et al.*, *Clin. Pharmacol. Ther.* **20**, 661 (1976); G. F. Lockwood, J. B. Houston, *J. Pharm. Pharmacol.* **31**, 787 (1979).

Pale yellow crystals from benzene, mp 109°. Sol in water, ethanol, benzene; sparingly sol in ether.

Caution: Irritating to eyes, respiratory system and skin.

USE: Reagent for spectrophotometric and colorimetric determn of phenols; in analytical chemistry for trace phenol determn in water.

414. *p*-Aminoazobenzene. [60-09-3] 4-(2-Phenyldiazenyl)benzenamine; 4-(phenylazo)benzenamine; C.I. Solvent Yellow 1; *p*-(phenylazo)aniline; *p*-aminodiphenylimide; aniline yellow; C.I. 11000. $C_{12}H_{11}N_3$; mol wt 197.24. C 73.07%, H 5.62%, N 21.30%. Prepd from aniline, NaNO$_2$, and HCl: A. I. Vogel, *A Text-Book of Practical Organic Chemistry* (Longmans, Green & Co., New York, 3rd ed., 1956) p 627; W. R. Hydro, T. L. Willard, **US 2894942** (1959 to Goodrich); *Colour Index* vol. 4 (3rd ed., 1971) p 4014.

Brownish-yellow needles with a bluish cast. mp 128°. bp above 360°. Slightly sol in water; freely sol in alc, benzene, chloroform, ether. Reduces alcoholic ammoniacal AgNO$_3$.

USE: In form of its salts in dyeing; intermediate in manuf of Acid Yellow, diazo dyes and indulines.

415. *o*-Aminoazotoluene. [97-56-3] 2-Methyl-4-[2-(2-methylphenyl)diazenyl]benzenamine; 2-methyl-4-[(2-methylphenyl)azo]benzenamine; C.I. Solvent Yellow 3; 4-(*o*-tolylazo)-*o*-toluidine; 4′-amino-2,3′-dimethylazobenzene; 5-(*o*-tolylazo)-2-aminotoluene; toluazotoluidine; C.I. 11160. $C_{14}H_{15}N_3$; mol wt 225.30. C 74.64%, H 6.71%, N 18.65%. Prepd from *o*-toluidine, NaNO$_2$, and HCl: Shulman, **US 2538431** (1951 to Pfister Chem. Works); *Colour Index* vol. 4 (3rd ed., 1971) p 4017. Crystal and molecular structure: S. Kurosaki *et al.*, *Acta Crystallogr.* **32B**, 3160 (1976). Carcinogenic activity: Maini, Stich, *J. Natl. Cancer Inst.* **26**, 1413 (1961). *See also:* H. J. Conn's *Biological Stains*, R. D. Lillie, Ed. (Williams & Wilkins, Baltimore, 9th ed., 1977) p 87.

Golden crystals, mp 101-102°. Practically insoluble in water. Sol in alc, ether, chloroform, acetone, ethyl acetate, benzene, ligroin, linseed oil, oleic acid, stearic acid. Very sol in cellosolve. Absorption max (50% alcoholic 1*N* HCl): 326, 490 nm (ε 19000, 2500), Sawicki, *J. Org. Chem.* **21**, 605 (1956).

Caution: This substance is reasonably anticipated to be a human carcinogen: *Report on Carcinogens, Twelfth Edition* (PB2011-111646, 2011) p 37.

USE: Coloring oils, fats and waxes; manuf of pigments. Chemical intermediate for the production of dyes.

416. *m*-Aminobenzoic Acid. [99-05-8] 3-Aminobenzoic acid. $C_7H_7NO_2$; mol wt 137.14. C 61.31%, H 5.15%, N 10.21%, O 23.33%. Prepn: Toland, Heaton, **US 2878281** (1959 to California Res. Corp.); Neilson *et al.*, *J. Chem. Soc.* **1962**, 371.

Cryst, mp 174°. d 1.51. Aq soln turns brown on standing in air. Forms soluble salts with mineral acids; slightly sol in cold, freely in boiling water or alcohol; sol in ether.

417. *o*-Aminobenzoic Acid. [118-92-3] 2-Aminobenzoic acid; anthranilic acid. $C_7H_7NO_2$; mol wt 137.14. C 61.31%, H 5.15%, N 10.21%, O 23.33%. Found to be the same as *vitamin L$_1$*: Nakahara *et al.*, *Sci. Papers Inst. Phys. Chem. Res. Jpn.* (Tokyo) **42**, 39 (1945), *C.A.* **41**, 6317d (1947). Prepd by reduction of *o*-nitrobenzoic acid: Neilson *et al.*, *J. Chem. Soc.* **1962**, 371. Purification: Sugihara, Newman, *J. Org. Chem.* **21**, 1445 (1956).

White to pale yellow, cryst powder; sweetish taste. mp 144-146°. Sparingly sol in cold, freely in hot water, alcohol, ether. The solns in alcohol or ether and particularly in glycerol exhibit an amethyst fluorescence.

THERAP CAT (VET): The cadmium salt has been used as an ascaricide in swine.

418. *p*-Aminobenzoic Acid. [150-13-0] 4-Aminobenzoic acid; vitamin B$_x$; bacterial vitamin H^1; chromotrichia factor; antichromotrichia factor; trichochromogenic factor; anticanitic vitamin; PABA; Pabanol. $C_7H_7NO_2$; mol wt 137.14. C 61.31%, H 5.15%, N 10.21%, O 23.33%. Widely distributed in nature as a B complex factor. Baker's yeast contains 5 to 6 ppm, brewer's yeast from 10 to 100 ppm. Occurs free and in ester form. Prepn: Toland, Heaton, **US 2878281** (1959 to Calif. Research Corp.); Spiegler, **US 2947781**

(1960 to Du Pont); Nielson *et al.*, *J. Chem. Soc.* **1962**, 371. Purification: Lyding, **US 2735865** (1956 to Heyden Chem.). Antirickettsial activity: M. L. Robbins *et al.*, *J. Immunol.* **64**, 431 (1950). Toxicity studies: C. C. Scott, E. B. Robbins, *Proc. Soc. Exp. Biol. Med.* **49**, 184 (1942); R. K. Richards, *Fed. Proc.* **1**, 71 (1942); G. Cronheim, *ibid.* **10**, 289 (1952). Comprehensive description: H. A. El-Obeid, A. A. Al-Badr, *Anal. Profiles Drug Subs. Excip.* **22**, 33-106 (1993).

Monoclinic prisms from dil alcohol. May turn slightly yellow on prolonged exposure to light and air. mp 187.0-187.5°. pKa: 4.65, 4.80; pH (0.5% soln): 3.5. uv max (water): 266 nm ($E_{1cm}^{1\%}$ 1070); (isopropanol): 288 nm ($E_{1cm}^{1\%}$ 137). One gram dissolves in 170 ml water at 25°, in 90 ml boiling water; in 8 ml alcohol, in 50 ml ether. Sol in ethyl acetate, glacial acetic acid; slightly sol in benzene. Practically insol in petr ether. Incompatible with ferric salts and oxidizing agents. LD_{50} in mice, rats (g/kg): 2.85, >6.0 orally (Scott, Robbins). LD_{50} in rabbits (g/kg): 2.0 i.v. (Richards); 1.83 orally (Cronheim).

Diethylamine salt. [6018-84-4] $C_{11}H_{18}N_2O_2$. Crystals from acetone mp 170-173°. Very soluble in water.

Potassium salt *see* Potassium *p*-Aminobenzoate.

Note: p-Aminobenzoic acid is also known as ***anti-gray-hair factor*** (in rats only).

USE: Manuf various esters (local anesthetics), folic acid, and azo dyes; in sunburn preventives. Used in laboratories as sulfonamide antagonist.

THERAP CAT: Ultraviolet screen. Formerly as antirickettsial.

THERAP CAT (VET): Has been used in *eczema nasi* ("collie nose") in dogs.

419.   2-Aminobenzothiazole. [136-95-8] 2-Benzothiazolamine. $C_7H_6N_2S$; mol wt 150.20. C 55.98%, H 4.03%, N 18.65%, S 21.34%. Prepd from 2-chlorobenzothiazole by treatment with alcoholic ammonia at 150-160°: Hofmann, *Ber.* **12**, 1129 (1880); **13**, 11 (1881); from benzothiazole by boiling with hydroxylamine in water or in 2*N* NaOH: Skraup, *Ann.* **419**, 65 (1919). Toxicity study: E. F. Domino *et al.*, *J. Pharmacol. Exp. Ther.* **105**, 486 (1952).

Leaflets from water, mp 132°. Distills without decompn. Very sparingly sol in water; freely in alcohol, ether, chloroform; sol in concd acids. LD_{50} i.v. in mice: 126 mg/kg (Domino).

USE: In the prepn of azo dyes.

420.   6-Aminobenzothiazole. [533-30-2] 6-Benzothiazolamine. $C_7H_6N_2S$; mol wt 150.20. C 55.98%, H 4.03%, N 18.65%, S 21.34%. Prepd from 2,5-diaminothiophenol by boiling with concd formic acid or from 6-nitrobenzothiazole by reduction with tin and HCl: Mylius, *Thesis* (Berlin, 1883).

Prisms from water, mp 87°. Sol in alcohol. Practically insol in water and ether.

USE: In the prepn of azo dyes.

421.   *N*-(4-Aminobenzoyl)-L-glutamic acid. [4271-30-1] *N*-(*p*-Aminobenzoyl)glutamic acid; PABG. $C_{12}H_{14}N_2O_5$; mol wt 266.25. C 54.13%, H 5.30%, N 10.52%, O 30.05%. Catabolite of

folate. Prepn: J. Van der Scheer, K. Landsteiner, *J. Immunol.* **29**, 371 (1935). Improved synthesis: A. P. Reddy, C. P. R. Reddy, *Org. Prep. Proced. Int.* **22**, 117 (1990); P. Maunder *et al.*, *J. Chem. Soc. Perkin Trans. 1* 1311 (1999). HPLC determn: B. A. Allen, R. A. Newman, *J. Chromatogr.* **190**, 241 (1980); densitometric determn in folic acid tablets: J. Krzek, A. Kwiecien, *J. Pharm. Biomed. Anal.* **21**, 451 (1999). Use as marker of folate degradation: J. McPartlin, J. Scott, *Methods Enzymol.* **281**, 70 (1997); of turnover in pregnant women: J. F. Gregory, III *et al.*, *J. Nutr.* **131**, 1928 (2001).

Off white solid, mp 173-174° (dec). $[\alpha]_D^{23} -15°$ (c = 2 in 0.1 *M* HCl).

USE: Marker in folate metabolism studies.

422.   2-Amino-1-butanol. [96-20-8] 2-Amino-*n*-butyl alcohol. $C_4H_{11}NO$; mol wt 89.14. C 53.90%, H 12.44%, N 15.71%, O 17.95%. Prepd from 2-nitro-1-butanol by reduction or catalytic hydrogenation: Stiénon, *Chem. Zentralbl.* **1902**, I, 717; Vanderbilt, Haas, **US 2174242** (1940); Johnson, Degering, *J. Org. Chem.* **8**, 7 (1943).

Liquid. d_{20}^{20} 0.944. mp −2°. bp_{760} 178°; bp_{10} 79-80°. n_D^{20} 1.453. Miscible with water. Sol in alcohols. pH of 0.1 molar aq soln 11.1.

Hydrochloride. $C_4H_{11}NO.HCl$; mol wt 125.60. Deliquesc needles.

Platinichloride monohydrate. $(C_4H_{11}NO)_2.2HCl.PtCl_4.H_2O$; mol wt 606.09. Yellow leaflets, dec 189-190°. Moderately sol in water, freely sol in alcohol.

Oxalate. $(C_4H_{11}NO)_2.C_2H_2O_4$; mol wt 268.31. mp 176°.

USE: In the synthesis of surface-active agents, vulcanization accelerators, pharmaceuticals. As emulsifying agent for cosmetic creams and lotions, mineral oil and paraffin wax emulsions, leather dressings, textile specialties, polishes, cleaning compounds, so-called soluble oils.

423.   *α*-Aminobutyric Acid. [2835-81-6] 2-Aminobutanoic acid; *α*-amino-*n*-butyric acid. $C_4H_9NO_2$; mol wt 103.12. C 46.59%, H 8.80%, N 13.58%, O 31.03%. Prepn of the DL-form from *α*-bromobutyric acid and ammonia: Fisher, Mounegrat, *Ber.* **33**, 2388 (1900); from potassium cyanide, ammonium chloride and propionaldehyde: Zelinsky, Stadnikow, *Ber.* **41**, 2062 (1908); by reduction of *α*-oxobutyric acid: Knoop, Oesterlin, *Z. Physiol. Chem.* **148**, 305 (1925); from *α*-ethylacetoacetic ester and hydrazoic acid: Schmidt, *Ber.* **57**, 706 (1924). The L(+)-form has been isolated from proteins: Oikawa, *Chem. Zentralbl.* **1926**, 1, 148. Configuration: Clough, *J. Chem. Soc.* **113**, 544, 551; Levene, *Chem. Rev.* **2**, 203 (1926); Vogler, *Helv. Chim. Acta* **30**, 1766 (1947).

DL-Form. Crystals, mp 304° (begins to sublime when heated above 300°). Sol in water. One liter of water will dissolve 210.5 g at 25°. Sparingly sol in alcohol. One liter of boiling ethanol dissolves about 1.8 g. Insol in ether.

DL-Form ethyl ester. Viscous liquid. bp_{11} 61°. Sol in water and in organic solvents.

L-Form. Leaflets from dil alc. Sweet taste. mp 270-280° (depending on speed of heating, *see* Vogler, *loc. cit.*). $[M]_D$ +21.2° (5*N* HCl); $[M]_D$ +43.3° (glacial acetic acid); $[\alpha]_D^{16}$ +8.40° (c = 4); $[\alpha]_D^{16}$ +18.65° (c = 4.8 in 6*N* HCl).

L-Form hydrochloride. Needles; $[\alpha]_D^{19}$ +12.90° (c = 3.64). Readily sol in water.

424. β-Aminobutyric Acid. [541-48-0] 3-Aminobutanoic acid; β-amino-n-butyric acid. $C_4H_9NO_2$; mol wt 103.12. C 46.59%, H 8.80%, N 13.58%, O 31.03%. Prepd from pyrotartaric acid diamide: Weidel, Roithmer, *Monatsh. Chem.* **17**, 185 (1896); from β-chlorobutyric acid ethyl ester and alcoholic NH_3: Balbiano, *Ber.* **13**, 312 (1880); from crotonic acid and concd NH_3: Engel, *Bull. Soc. Chim.* [2] **50**, 102 (1888); Curtius, Gumlich, *J. Prakt. Chem.* [2] **70**, 204 (1904); Stadnikow, *Chem. Zentralbl.* **1909**, II, 1988; *see* Fischer, Roeder, *Ber.* **34**, 3755 (1901) footnote; Stoermer, Robert, *Ber.* **55**, 1038 (1922). Prepn of HCl salt from β-aminobutyronitrile: Bruylants, *Bull. Soc. Chim. Belg.* **32**, 259 (1923); *Chem. Zentralbl.* **1924**, I, 1668. By reducing acetoacetic ester phenylhydrazone and saponifying the resulting ester: Fischer, Groh, *Ann.* **383**, 338 (1911). The D(−)-form has been obtained by hydrolysis of its ester: Fischer, Scheibler, *ibid.* 346.

DL-Form. Crystals from alcohol, mp 193-194°. Practically tasteless. Sol in water. One liter of water dissolves 1250 g. Insol in cold absolute alcohol and ether.
Methyl ester. $C_5H_{11}NO_2$; mol wt 117.15. Odoriferous liquid; d^{20} 0.993; bp_{13} 54-55°. Sol in water, alcohol, ether, petr ether.
D-Form. Prisms from methanol. Dec near 220° without melting. $[\alpha]_D^{20}$ −35.20° (p = 10).

425. γ-Aminobutyric Acid. [56-12-2] 4-Aminobutanoic acid; γ-amino-n-butyric acid; piperidic acid; factor I; GABA. $C_4H_9NO_2$; mol wt 103.12. C 46.59%, H 8.80%, N 13.58%, O 31.03%. Ubiquitous nonprotein amino acid. Major inhibitory neurotransmitter in mammalian CNS; also has trophic role in neuronal development. Formed from glutamate by glutamic acid decarboxylase (GAD). Biological effects are mediated via 3 main types of GABA receptor complexes (A, B, C). The ionotropic GABA$_A$ receptor is the site of action of numerous pharmacological agents including benzodiazepines, barbiturates, anesthetics, and ethanol. Prepn from piperylurethan and fuming nitric acid: Schotten, *Ber.* **16**, 643 (1883); from succinimide: Tafel, Stern, *Ber.* **33**, 2224 (1900); from γ-chlorobutyronitrile and potassium phthalimide: Gabriel, *Ber.* **22**, 3335 (1889); **23**, 1771 (1890); C. C. DeWitt, *Org. Synth.* **coll. vol. II**, 25 (1943). Thermodynamic properties: E. J. King, *J. Am. Chem. Soc.* **76**, 1006 (1954). Identification in brain: J. Awapara *et al.*, *J. Biol. Chem.* **187**, 35 (1950); and enzymatic formation from glutamic acid: E. Roberts, S. Frankel, *ibid.* 55. Physiological properties: H. McLennan, *J. Physiol.* **139**, 79 (1957). Review of biochemical pharmacology: N. G. Bowery, T. G. Smart, *Br. J. Pharmacol.* **147**, Suppl. 1, S109-S119 (2006); of biosynthesis, metabolism, and homeostasis: H. S. Waagepetersen *et al.*, *J. Neurochem.* **73**, 1335-1342 (1999). Review of role in neurological disease: C. G. T. Wong *et al.*, *Ann. Neurol.* **54**, Suppl. 6, S3-S12 (2003); as developmental signal: D. F. Owens, A. R. Kriegstein, *Nat. Rev. Neurosci.* **3**, 715-727 (2002).

Leaflets from methanol + ether, needles from water + alcohol, mp 202° (dec on rapid heating). pK_1 4.031; pK_2 10.556 (25°). Log P (1-octanol/0.1M phosphate buffer): −3.33 (pH 5, 37°). Freely sol in water. Insol or poorly sol in other solvents.
Hydrochloride. [5959-35-3] $C_4H_9NO_2$·HCl. Crystals, mp 135-136°.
Ethyl ester. [5959-36-4] $C_6H_{13}NO_2$. Liquid, bp_{12} 76°.

426. 9-Aminocamptothecin. [91421-43-1] (4S)-10-Amino-4-ethyl-4-hydroxy-1H-pyrano[3′,4′:6,7]indolizino[1,2-b]quinoline-3,14(4H,12H)-dione; 9-amino-20(S)-camptothecin; 9-AC. $C_{20}H_{17}N_3O_4$; mol wt 363.37. C 66.11%, H 4.72%, N 11.56%, O 17.61%. Semisynthetic camptothecin which inhibits DNA topoisomerase I. Prepn: T. Miyasaka *et al.*, **JP Kokai 59 51289** (1984 to Yakult Honsha); M. C. Wani *et al.*, *J. Med. Chem.* **29**, 2358 (1986); and

activity: S. Sawada *et al.*, *Chem. Pharm. Bull.* **39**, 3183 (1991). Synthesis: W. Cabri *et al.*, *Tetrahedron Lett.* **36**, 9197 (1995). HPLC determn in plasma: W. J. Loos *et al.*, *J. Chromatogr. B* **694**, 435 (1997). Structure/activity study: M.-L. Li *et al.*, *Clin. Cancer Res.* **7**, 168 (2001). Clinical pharmacodynamics and pharmacokinetics: C. H. Takimoto *et al.*, *J. Clin. Oncol.* **15**, 1492 (1997). Clinical evaluation in colorectal cancer: H. C. Pitot *et al.*, *Cancer* **89**, 1699 (2000). Review of clinical developments: C. H. Takimoto, R. Thomas, *Ann. N.Y. Acad. Sci.* **922**, 224-236 (2000).

Orange-yellow solid from $CH_3OH/CHCl_3$ (13:87), mp 300° (dec). $[\alpha]_D^{23}$ +16° (c = 0.05 in $CH_3OH/CHCl_3$ 1:4). Insol in water.
THERAP CAT: Antineoplastic.

427. ε-Aminocaproic Acid. [60-32-2] 6-Aminohexanoic acid; epsilon-aminocaproic acid; epsilcapramin; EACA; CY-116; Amicar; Epsikapron; Hemocaprol. $C_6H_{13}NO_2$; mol wt 131.18. C 54.94%, H 9.99%, N 10.68%, O 24.39%. Antifibrinolytic; inhibits the activity of plasmin and plasminogen, q.q.v. Prepn: S. Gabriel, T. A. Maass, *Ber.* **32**, 1266 (1899); A. Galat, S. Mallin, *J. Am. Chem. Soc.* **68**, 2729 (1946); C. Y. Meyers, L. E. Miller, *Org. Synth.* **coll. vol. IV**, 39 (1963). Toxicity data: D. W. Hallesy *et al.*, *Pharmacologist* **3**, 62 (1961). Crystal structure: G. Bodor *et al.*, *Acta Crystallogr.* **23**, 482 (1967). Clinical trial and mode of action study: T. J. Vander Salm *et al.*, *J. Thorac. Cardiovasc. Surg.* **112**, 1098 (1996).

Crystals from alcohol, mp 202-203°. pK_1 4.43; pK_2 10.75. Freely sol in water; sparingly in methanol. Practically insol in ethanol, chloroform. LD_{50} in rats (g/kg): 7.0 i.p.; ~3.3 i.v. (Hallesy).
Hydrochloride. [4321-58-8] $C_6H_{13}NO_2$·HCl; mol wt 167.63. mp 128-129°.
THERAP CAT: Hemostatic.

428. Aminocarb. [2032-59-9] 4-(Dimethylamino)-3-methylphenol 1-(N-methylcarbamate); methylcarbamic acid 4-(dimethylamino)-m-tolyl ester; 4-dimethylamino-m-tolyl methylcarbamate; A-363; Bay 44646; ENT-25784; Matacil. $C_{11}H_{16}N_2O_2$; mol wt 208.26. C 63.44%, H 7.74%, N 13.45%, O 15.36%. Prepn: R. Heiss *et al.*, **GB 913439**; *eidem*, **US 3134806** (1962, 1964 both to Bayer). Metabolism: A. Strother, *Toxicol. Appl. Pharmacol.* **21**, 112 (1972). Photochemistry: J. B. Addison *et al.*, *Bull. Environ. Contam. Toxicol.* **11**, 250 (1974); J. B. Addison, *ibid.* **27**, 250 (1981). Toxicity study: T. B. Gaines, *Toxicol. Appl. Pharmacol.* **14**, 515 (1969).

Crystals, mp 94.5-95.5°. uv max (ethanol): 248.5 nm (ε 6.67 × 10^4). Slightly sol in water; moderately sol in aromatic solvents; sol in most polar organic solvents. LD_{50} orally in male, female rats: 40, 38 mg/kg (Gaines).
USE: Insecticide.

429. 7-Aminocephalosporanic Acid. [957-68-6] (6R,7R)-3-[(Acetyloxy)methyl]-7-amino-8-oxo-5-thia-1-azabicyclo[4.2.0]-

oct-2-ene-2-carboxylic acid; 7-amino-3-(hydroxymethyl)-8-oxo-5-thia-1-azabicyclo[4.2.0]oct-2-ene-2-carboxylic acid acetate ester; 3-acetoxymethyl-7-aminoceph-3-em-4-oic acid; 7-ACA. $C_{10}H_{12}N_2O_5S$; mol wt 272.28. C 44.11%, H 4.44%, N 10.29%, O 29.38%, S 11.77%. Starting material for semi-synthetic cephalosporins. Obtained by mild acid hydrolysis of cephalosporin C, *q.v.*: Loder *et al.*, *Biochem. J.* **79**, 408 (1961); Morin *et al.*, *J. Am. Chem. Soc.* **84**, 3400 (1962); Morin *et al.*, **BE 615955** (1962 to Lilly), *C.A.* **58**, 11373c (1963); by enzymatic hydrolysis of cephalosporin C: Walton, **US 3239394** (1966 to Merck & Co.). Improved prepn: Fechtig *et al.*, *Helv. Chim. Acta* **51**, 1108 (1968). Review of preparative methods: Huber *et al.*, in *Cephalosporins and Penicillins*, E. H. Flynn, Ed. (Academic Press, New York, 1972) pp 27-73.

Crystals. pI 3.5. R_f 0.14 in 1-butanol-ethanol-water (4:1:5 by vol).

430. **4-Amino-4′-chlorodiphenyl.** [135-68-2] 4′-Chloro-[1,1′-biphenyl]-4-amine; 4′-chloroxenylamine; 4′-chloro-4-amino-diphenyl; *p*′-chloro-*p*-phenylaniline; *p*-amino-*p*′-chlorobiphenyl. $C_{12}H_{10}ClN$; mol wt 203.67. C 70.77%, H 4.95%, Cl 17.41%, N 6.88%. Prepd from 4′-amino-4-biphenyldiazonium chloride by Sandmeyer's reaction: Gelmo, *Ber.* **39**, 4176 (1906); by chlorination of 4-nitrodiphenyl and reduction of 4-chloro-4′-nitrodiphenyl with iron and HCl in ethanol: Belcher *et al.*, *J. Chem. Soc.* **1953**, 1334.

Crystals from light petroleum, mp 134° (Gelmo); mp 128° (Belcher). uv max (0.1N HCl): 254 nm (ε 22090). Sol in warm alcohol, ether, benzene, acetone, glacial acetic acid. Practically insol in water and alkalies.

USE: As a reagent in the determination of sulfur, in coal, rubber, etc.

431. **Aminochromes.** Family of highly colored 2,3-dihydro-indole-5,6-quinones obtained by oxidative cyclization of catecholamines: Sobotka, Austin, *J. Am. Chem. Soc.* **73**, 3077 (1951). Best represented by a zwitterionic structure with the substituents determined by the catecholamine used, e.g. for epinephrine as starting material, R = CH_3, R′ = H, and R″ = OH. Numerous physiological activities, such as hallucinogenic, hemostatic, radioprotective, have been ascribed to aminochromes. *Reviews*: Sobotka *et al.*, *Fortschr. Chem. Org. Naturst.* **14**, 217 (1957); Heacock, *Adv. Heterocycl. Chem.* **5**, 205 (1965); Heacock, Powell, *Prog. Med. Chem.* **9**, 275 (1972). Review of toxicity: A. Bindoli *et al.*, *Toxicol. Lett.* **48**, 3-20 (1989).

See also adrenochrome, adrenolutin, carbazochrome salicylate.
THERAP CAT: Hemostatic.

432. **Aminocyclopyrachlor.** [858956-08-8] 6-Amino-5-chloro-2-cyclopropyl-4-pyrimidinecarboxylic acid; DPX-MAT28. $C_8H_8ClN_3O_2$; mol wt 213.62. C 44.98%, H 3.77%, Cl 16.59%, N 19.67%, O 14.98%. Synthetic auxin herbicide for broadleaf weeds in noncrop land. Prepn: D. A. Clark *et al.*, **WO 05063721**; *eidem*, **US 070197391** (2005, 2007 both to DuPont). HPLC/MS/MS determn in soil and water samples: S. C. Nanita *et al.*, *Anal. Chem.* **81**,

797 (2009). Absorption and translocation in Canada thistle: B. Bukun *et al.*, *Weed Sci.* **58**, 96 (2010). Volatility and vapor movement: S. D. Strachan *et al.*, *ibid.*, 103.

Solid, mp 144-146°. pKa 4.65. Log P (octanol/water): −2.48 (pH 7); −1.12 (pH 4). Vapor pressure: 4.89×10^{-6} Pa. Highly sol in water; sol in methanol. Degrades in solns of acetonitrile or acetone.

Methyl ester. [858954-83-3] Aminocyclopyrachlor-methyl; methyl 6-amino-5-chloro-2-cyclopropyl-4-pyrimidinecarboxylate; DPX-KJM44. $C_9H_{10}ClN_3O_2$; mol wt 227.65. Solid, mp 143-145°. Log P (octanol/water): 1.87. Vapor pressure: 4.46×10^{-4} Pa.

Potassium salt. [858956-35-1] Aminocyclopyrachlor-potassium; Imprelis. $C_8H_7ClKN_3O_2$; mol wt 251.71. Solid, mp 273° (dec).

USE: Herbicide.

433. **2-Amino-4,6-dichlorophenol.** [527-62-8] 2,4-Dichloro-6-aminophenol; 4,6-dichloro-*o*-aminophenol. $C_6H_5Cl_2NO$; mol wt 178.01. C 40.48%, H 2.83%, Cl 39.83%, N 7.87%, O 8.99%. Prepd by reduction of the corresponding nitrophenol: F. Fischer, *Z. Chem.* [N.F.] **4**, 386 (1868); *Ann. Suppl.* **7**, 189 (1870); Katz, Cohen, *J. Org. Chem.* **19**, 758 (1954). Purification: J. Meyer, *Helv. Chim. Acta* **41**, 1890 (1958).

Long needles from carbon disulfide, warts from benzene, mp 95-96°. Sublimes (0.06 torr) 70-80° (bath temp). Freely sol in benzene, somewhat less in carbon disulfide, much less in petr ether. The stability of the free base (snow-white when pure) seems to be impaired by impurities.

Hydrochloride. $C_6H_5Cl_2NO.HCl$. Crystals, dec 280-285°. Very stable when pure. The commercial product may be dark brown. Sol in water, alcohol. Precipitated from aq soln by the addition of concd HCl.

USE: Important azo-dye intermediate.

434. **Aminoethoxyvinylglycine.** [49669-74-1] (2S,3E)-2-Amino-4-(2-aminoethoxy)-3-butenoic acid; L-*trans*-2-amino-4-(2-aminoethoxy)butenoic acid. $C_6H_{12}N_2O_3$; mol wt 160.17. C 44.99%, H 7.55%, N 17.49%, O 29.97%. Inhibitor of 1-aminocyclopropane-1-carboxylate (ACC) synthase; applied to apples, pears, stone fruits, and walnuts to inhibit ethylene production and delay ripening. Amino acid antimetabolite produced by fermentation of *Streptomyces*. Prepn: J. Berger *et al.*, **US 3751459** (1973 to Hoffmann-La Roche); D. L. Pruess *et al.*, *J. Antibiot.* **27**, 229 (1974). Synthesis of DL-form: D. D. Keith *et al.*, *J. Org. Chem.* **43**, 3713 (1978). Crystal structure in complex with ACC synthase: G. Capitani *et al.*, *J. Biol. Chem.* **277**, 49735 (2002). Mechanism of action study: M. E. Saltveit, *Postharvest Biol. Technol.* **35**, 183 (2005).

Hydrochloride. [55720-26-8] ReTain. $C_6H_{12}N_2O_3.HCl$; mol wt 196.63. Crystals from methanol-H_2O, mp 193-195°. $[\alpha]_D^{25}$ +89.2° (c = 1 in 0.1M sodium phosphate buffer, pH 7). $[\alpha]_D^{25}$ +111.8° (c = 1 in 5N HCl).

DL-Form. [69257-01-8]; [67010-42-8] (hydrochloride). (3*E*)-2-Amino-4-(2-aminoethoxy)-3-butenoic acid. Hydrochloride as crystals from water/methanol, mp 187.5-189° (dec).

USE: Plant growth regulator.

435. 1-[(2-Aminoethyl)amino]-2-propanol. [123-84-2] *N*-(2-Hydroxypropyl)ethylenediamine; Monolene. $C_5H_{14}N_2O$; mol wt 118.18. C 50.82%, H 11.94%, N 23.70%, O 13.54%. Prepn: Kitchen, Pollard, *J. Org. Chem.* **8**, 342 (1943).

Viscous liquid; mild ammoniacal odor. $bp_{3.0}$ 94°; $bp_{10.0}$ 112°. d_4^{25} 0.9837. n_D^{25} 1.4738.

Dihydrochloride. $C_5H_{16}Cl_2N_2O$; mol wt 191.10. mp 184.7-185.0°.

USE: Rapid curing agent in the manuf of epoxy resins.

436. 2-Amino-2-ethyl-1,3-propanediol. [115-70-8] C_5H_{13}-NO_2; mol wt 119.16. C 50.40%, H 11.00%, N 11.75%, O 26.85%. An amino glycol prepd by reduction of catatyic hydrogenation of the corresp nitro compound: Vanderbilt, Hass, **US 2174242** (1940); Johnson, Degering, *J. Org. Chem.* **8**, 7 (1943). Manuf: McMillan, **US 2485982** (1949 to Comm. Solvents).

Crystalline mass, mp 37.5-38.5°. (The commercial product may be a viscous liquid.) d_{20}^{20} 1.099. n_D^{20} 1.490. bp_{10} 152-153°. Miscible with water. Sol in alcohols. pH of 0.1 molar aq soln 10.8.

USE: In the synthesis of surface-active agents, vulcanization accelerators, pharmaceuticals. As emulsifying agent for cosmetic creams and lotions, mineral oil and paraffin wax emulsions, leather dressings, textile specialties, polishes, cleaning compounds, so-called soluble oils. For absorbing CO_2 and H_2S from industrial gases.

437. Aminoglutethimide. [125-84-8] 3-(4-Aminophenyl)-3-ethyl-2,6-piperidinedione; 2-(*p*-aminophenyl)-2-ethylglutarimide; 3-ethyl-3-(*p*-aminophenyl)-2,6-dioxopiperidine; Cytadren; Orimeten. $C_{13}H_{16}N_2O_2$; mol wt 232.28. C 67.22%, H 6.94%, N 12.06%, O 13.78%. Aromatase inhibitor; also blocks adrenal steroidogenesis. Formerly used as anticonvulsant. Prepn: K. Hoffmann, E. Urech, **US 2848455** (1958 to Ciba). Mass spectrum: G. Ruecker, G. Bohn, *Arch. Pharm.* **302**, 204 (1969). Resolution and abs config of antipodes: N. Finch *et al.*, *Experientia* **31**, 1002 (1975). Comprehensive description: H. Y. Aboul-Enein, *Anal. Profiles Drug Subs.* **15**, 35-69 (1986). Review of chemistry and clinical use in Cushing's disease and breast cancer: R. J. Santen, R. I. Misbin, *Pharmacotherapy* **1**, 95-120 (1981); in prostate cancer: K. A. Havlin, D. L. Trump, *Cancer Treat. Res.* **39**, 83-96 (1988).

Crystals from methanol or ethyl acetate, mp 149-150°. Readily sol in most organic solvents. Poorly sol in ethyl acetate, 0.1*N* HCl, absolute ethanol. Practically insol in water.

THERAP CAT: Adrenocortical suppressant; antineoplastic.

438. Aminoguanidine. [79-17-4] Hydrazinecarboximidamide; guanylhydrazine; pimagedine. CH_6N_4; mol wt 74.09. C 16.21%, H 8.16%, N 75.62%. Nucleophilic hydrazine; nitric oxide synthase (NOS) inhibitor. Also inhibits the formation of advanced glycosylation end products (AGEs) that have been implicated in the etiology of diabetic complications. Prepn: J. Thiele, *Ann.* **270**, 1 (1892); G. B. L. Smith, E. Anzelmi, *J. Am. Chem. Soc.* **57**, 2730 (1935). Review of chemistry of aminoguanidine and related compounds: E. Lieber, G. B. L. Smith, *Chem. Rev.* **25**, 213-271 (1939); of preparative methods: F. Kurzer, L. E. A. Godfrey, *Chem. Ind. (London)* **1962**, 1584-1595. Prevention of glucose-derived aortic collagen cross-linking in diabetic rats: M. Brownlee *et al.*, *Science* **232**, 1629 (1986). Mechanistic study of AGE inhibition: D. Edelstein, M. Brownlee, *Diabetes* **41**, 26 (1992). Effect on NOS: T. P. Misko *et al.*, *Eur. J. Pharmacol.* **233**, 119 (1993). Review of pharmacology and clinical experience in prevention and treatment of chronic diabetic complications: E. Abdel-Rahman, W. K. Bolton, *Expert Opin. Invest. Drugs* **11**, 565-574 (2002).

Crystals. Sol in water, alc. Practically insol in ether. Aq soln is strongly alkaline and reddens on standing in air; ammonia is evolved on heating.

Hydrochloride. [1937-19-5] GER-11. $CH_6N_4 \cdot HCl$; mol wt 110.55. Large prisms from dil alc, mp 163°. Very sol in water; sol in alc. Practically insol in ether.

USE: NOS inhibitor in biochemical research.

439. *p*-Aminohippuric Acid. [61-78-9] *N*-(4-Aminobenzoyl)glycine; *N*-(*p*-aminobenzoyl)aminoacetic acid; PAH. C_9H_{10}-N_2O_3; mol wt 194.19. C 55.67%, H 5.19%, N 14.43%, O 24.72%. Prepn: J. B. Muenzen *et al.*, *J. Biol. Chem.* **67**, 469 (1926). Method to determine renal function: L. Schumann, P. W. Wüstenberg, *Clin. Nephrol.* **33**, 35 (1990); R. A. Gabel *et al.*, *J. Pharmacol. Toxicol. Methods* **36**, 189 (1996).

Short, irregular prisms from hot water, mp 199°. Soluble in alc, chloroform, benzene, acetone. Practically insol in cold water, ether, carbon tetrachloride.

Sodium salt. [94-16-6] Nephrotest. $C_9H_9N_2NaO_3$; mol wt 216.17. Soluble in water.

THERAP CAT: Diagnostic aid (renal function).

440. 3-Amino-4-hydroxybutyric Acid. [589-44-6] 3-Amino-4-hydroxybutanoic acid; γ-hydroxy-β-aminobutyric acid; GOBAB. $C_4H_9NO_3$; mol wt 119.12. C 40.33%, H 7.62%, N 11.76%, O 40.29%. Prepn: Jollès, Fromageot, *Bull. Soc. Chim. Fr.* **1951**, 862; Chibnall *et al.*, *Biochem. J.* **68**, 122 (1958); Piskov, *Zh. Obshch. Khim.* **32**, 3407 (1962); Nagai *et al.*, *Arzneim.-Forsch.* **17**, 1575 (1967); Kondo, Tanaka, **JP 68 12127** (1968 to Kaken Kagaku), *C.A.* **70**, 77328r (1969).

Prismatic crystals, mp 216°. Also reported as mp 228° (Nagai); mp 232-233.5° (Chibnall).

Hydrochloride. $C_4H_9NO_3 \cdot HCl$; mol wt 155.58. mp 156°.

THERAP CAT: Anti-inflammatory, antifungal.

441. 4-Amino-3-hydroxybutyric Acid. [352-21-6] 4-Amino-3-hydroxybutanoic acid; γ-amino-β-hydroxybutyric acid; buksamin; GABOB; Gabomade; Gamibetal. $C_4H_9NO_3$; mol wt 119.12. C 40.33%, H 7.62%, N 11.76%, O 40.29%. Prepn of DL-form: Tomita, *Z. Physiol. Chem.* **124**, 255 (1923); Balenovic *et al.*, *J. Org.*

Chem. **19**, 1589 (1964); Hayashi *et al.*, **JP 58 772** *C.A.* **53**, 1172d (1959); Sakai *et al.*, **JP 62 12264** (to Kaken), *C.A.* **59**, 9805e (1963); **ES 278780** (1963 to Antonio Gallardo, S.A.), *C.A.* **60**, 2779a (1964); Hayashi, **FR 1348105** (1964 to Kaken), *C.A.* **60**, 11956b (1964); D'Alo, Masserini, *Farmaco Ed. Sci.* **19**, 30 (1964); M. Pinza, G. Pifferi, *J. Pharm. Sci.* **67**, 120 (1978). Prepn of L-form: Tomita, Sendju, *Z. Physiol. Chem.* **169**, 270 (1927); Kaneko, Yoshida, *Bull. Chem. Soc. Jpn.* **35**, 1153 (1962); M. E. Jung, T. J. Shaw, *J. Am. Chem. Soc.* **102**, 6304 (1980); of D-form: Tomita, Sendju, *loc. cit.* Improved synthesis of L-form: S. Takano *et al.*, *Tetrahedron Lett.* **28**, 1783 (1987). Purification: Yamagiwa, Tanaka, **JP 63 24365** (to Kaken), *C.A.* **60**, 10515a (1964).

DL-Form. Crystals from dil alc, dec 218°. Fairly readily sol in water; very sparingly sol in methanol, alc, ether, chloroform, ethyl acetate.

D(+)-Form. Crystals from water, dec 214°. $[\alpha]_D^{20}$ +18.3° (c = 2 in H_2O).

L(−)-Form. Crystals from water or water + ethanol, dec 212°, also reported as dec 216-217°. $[\alpha]_D^{20}$ −21.06° (c = 2 in H_2O); −20.7° (c = 1.83 in H_2O).

THERAP CAT: Anticonvulsant.

442. α-Aminoisobutyric Acid. [62-57-7] 2-Methylalanine; 2-aminoisobutyric acid; 2-amino-2-methylpropanoic acid. C_4H_9-NO_2; mol wt 103.12. C 46.59%, H 8.80%, N 13.58%, O 31.03%. Prepd by the treatment of acetone with hydrocyanic acid and then with alcoholic ammonia (Strecker synthesis): Tiemann, Friedländer, *Ber.* **14**, 1970 (1881), *see also* p 1965; Marckwald *et al.*, *Ber.* **24**, 3283 (1891); Bailey, Randolph, *Ber.* **41**, 2507 (1908); Clarke, Bean, *Org. Synth.* **coll. vol. II**, 29 (1943); or directly with ammonium cyanide: Gulewitsch, *Ber.* **33**, 1900 (1900); or with a mixture of KCN and NH_4Cl: Zelinsky, Stadnikow, *Ber.* **39**, 1726 (1906); *cf.* Hellsing, *Ber.* **37**, 1921 (1904); and subsequent hydrolysis of the nitrile formed. By heating dimethylhydantoin (obtained from acetone, hydrocyanic and cyanic acids) with concd HCl: Urech, *Ann.* **164**, 268 (1872); Heilpern, *Monatsh. Chem.* **17**, 241 (1896).

Monoclinic prisms, tables, mp 335° (sealed capillary). Begins to sublime at 280°. Sweetish taste. Absorption spectrum: Ley, Arends, *Ber.* **61**, 219 (1928); Abderhalden, Rossner, *Z. Physiol. Chem.* **176**, 253 (1928). Freely sol in water. Difficultly sol in alcohol; insol in ether.

Hydrochloride. $C_4H_9NO_2$·HCl; mol wt 139.58. Platelets from water, dec 236-237°. Readily sol in water, methanol, alcohol.

443. δ-Aminolevulinic Acid. [106-60-5] 5-Amino-4-oxopentanoic acid; ALA. $C_5H_9NO_3$; mol wt 131.13. C 45.80%, H 6.92%, N 10.68%, O 36.60%. Naturally occurring amino acid; precursor of tetrapyrroles in the biosynthesis of chlorophyll and heme. Prepn: R. W. Wynn, A. H. Corwin, *J. Org. Chem.* **15**, 203 (1950); A. A. Marei, R. H. Raphael, *J. Chem. Soc.* **1958**, 2624; C. Herdeis, A. Dimmerlung, *Arch. Pharm.* **317**, 304 (1984). Role in tetrapyrrole biosynthesis: D. Shemin, C. S. Russell, *J. Am. Chem. Soc.* **75**, 4873 (1953). Enhancement of chlorophyll formation: E. C. Sisler, W. H. Klein, *Physiol. Plant.* **16**, 315 (1963). HPLC determn in plasma: K. Miyajima *et al.*, *J. Chromatogr. B* **654**, 165 (1994). Review of role in chlorophyll biosynthesis and potential as photodynamic herbicide: C. A. Rebeiz *et al.*, *Enzyme Microb. Technol.* **6**, 390-401 (1984). Review of role in heme biosynthesis and clinical experience in photodynamic therapy: Q. Peng *et al.*, *Cancer* **79**, 2282-2308 (1997).

Hydrochloride. [5451-09-2] Levulan. $C_5H_{10}ClNO_3$; mol wt 167.59. Needles from methanol-ether, mp 144-147° (dec).

THERAP CAT: Antineoplastic (photosensitizer).

444. β-Amino-α-methylphenethyl Alcohol. [52500-61-5] β-Amino-α-methylbenzeneethanol; 1-amino-1-phenyl-2-propanol; 1-phenyl-1-amino-2-propanol. $C_9H_{13}NO$; mol wt 151.21. C 71.49%, H 8.67%, N 9.26%, O 10.58%. Prepn from α-isonitroso-α-phenylacetone: Sichner, Pankova, *Collect. Czech. Chem. Commun.* **20**, 1419 (1955); from an oxazole: Viscontini, *Helv. Chim. Acta* **44**, 636 (1961).

DL-erythro-Form. So called **dl-norisoephedrine.** Crystals, mp 85°.

DL-erythro-Hydrochloride. $C_9H_{13}NO$·HCl. Crystals from ethanol + ether, mp 170-171°.

445. 2-Amino-2-methyl-1,3-propanediol. [115-69-5] 1,1-Di(hydroxymethyl)ethylamine; 1,3-dihydroxy-2-methyl-2-propylamine; AMPD. $C_4H_{11}NO_2$; mol wt 105.14. C 45.70%, H 10.55%, N 13.32%, O 30.43%. An amino glycol prepd by reduction of the corresponding nitro compd: O. Piloty, O. Ruff, *Ber.* **30**, 2057 (1897); B. M. Vanderbilt, H. B. Hass, *Ind. Eng. Chem.* **32**, 34 (1940); Johnson, Degering, *J. Org. Chem.* **8**, 7 (1943). Review of uses: G. A. Bergy, *J. Am. Pharm. Assoc. Pract. Pharm. Ed.* **3**, 358-364 (1942). Toxicological evaluation and review: Cosmetic, Toiletry and Fragrance Assoc., *J. Am. Coll. Toxicol.* **9**, 203-228 (1990).

Crystalline mass, mp 109-111°. bp$_{10}$ 151-152°. 250 grams dissolve in 100 ml water at 20°. Sol in alcohols. pH of 0.1M aq soln 10.8.

USE: In the synthesis of surface-active agents, vulcanization accelerators, pharmaceuticals. As emulsifying agent for cosmetic creams and lotions, mineral oil and paraffin wax emulsions, leather dressings, textile specialties, polishes, cleaning compounds, so-called sol oils. In hair sprays, wave sets and other noncoloring hair preparations. Absorbent for acidic gases. Biological buffer.

446. 2-Amino-2-methyl-1-propanol. [124-68-5] 2,2-Diethylethanolamine; 2-hydroxymethyl-2-propylamine; AMP. C_4H_{11}-NO; mol wt 89.14. C 53.90%, H 12.44%, N 15.71%, O 17.95%. An amino alcohol prepd by reduction of the corresp nitro compound: H. B. Hass, B. M. Vanderbilt, **US 2139122** (1938 to Purdue Res. Found.); Johnson, Degering, *J. Org. Chem.* **8**, 7 (1943). Toxicological evaluation and review: Cosmetic, Toiletry and Fragrance Assoc. (CTFA), *J. Am. Coll. Toxicol.* **9**, 203-228 (1990).

Crystalline mass, mp 30-31°. (The commercial product may be a viscous liquid.) d$_{20}^{20}$ 0.934. n$_D^{20}$ 1.449. bp$_{760}$ 165°; bp$_{10}$ 67.4°. Miscible with water. Sol in alcohols. pH of 0.1M aq soln 11.3. LD$_{50}$ in mice, rats (g/kg): 2.15 ± 0.2, 2.90 ± 0.14 orally (CTFA).

USE: In the synthesis of surface-active agents, vulcanization accelerators, pharmaceuticals. As emulsifying agent for cosmetic creams and lotions, mineral oil and paraffin wax emulsions, leather dressings, textile specialties, polishes, cleaning compounds, so-called soluble oils. In hair sprays, wave sets and hair dyes. Absorbent for acidic gases. For use as a drug *see* Pamabrom.

447. 2-Amino-4-methylthiazole. [1603-91-4] 4-Methyl-2-thiazolamine; 4-methyl-2-thiazolylamine; aminomethiazole. C_4H_6-

N_2S; mol wt 114.17. C 42.08%, H 5.30%, N 24.54%, S 28.08%. Prepn: Sprague *et al.*, *J. Am. Chem. Soc.* **68**, 2155 (1946); Rossi, *Gazz. Chim. Ital.* **85**, 898 (1955). Use in gravimetric determn and extraction of mercury: S. N. Tandon *et al.*, *Anal. Chim. Acta* **59**, 311 (1972).

Crystals, mp 45-46°. bp$_{20}$ 124-126°; bp$_{0.4}$ 70°. Very sol in water, alcohol, ether.
USE: Reagent.

448. 3-Amino-2-naphthoic Acid. [5959-52-4] 3-Amino-2-naphthalenecarboxylic acid; 3-aminoisonaphthoic acid. $C_{11}H_9NO_2$; mol wt 187.20. C 70.58%, H 4.85%, N 7.48%, O 17.09%. Prepd from 3-hydroxy-2-naphthoic acid: C. F. H. Allen, A. Bell, *Org. Synth.* coll. vol. III, 78 (1955).

Yellow scales from dil alcohol, mp 214°. Sol in alcohol, ether. Solns are yellow with greenish fluorescence.
Sodium salt. $C_{11}H_8NNaO_2$. Leaflets, very sparingly sol in water, alcohol.
Ethyl ester. $C_{13}H_{13}NO_2$. Yellow needles from dil alcohol. Sol in the usual organic solvents, mp 115-115.5°.
USE: In the determination of copper, nickel, cobalt.

449. 4-Amino-1-naphthol. [2834-90-4] 4-Amino-1-naphthalenol; 4-hydroxy-α-naphthylamine. $C_{10}H_9NO$; mol wt 159.19. C 75.45%, H 5.70%, N 8.80%, O 10.05%. Prepd by treating α-naphthol with benzenediazonium chloride and reducing the benzene-azo-α-naphthol with sodium hydrosulfite: Conant *et al.*, *Org. Synth.* **3**, 7 (1923); *cf.* Fieser, Fieser, *J. Am. Chem. Soc.* **57**, 493 (1935). From α-naphthylhydroxylamine by rearrangement in acetone: Neunhoffer, Liebich, *Ber.* **71B**, 2247 (1938).

Needles. Unless kept absolutely dry, it acquires a violet discoloration on storage and oxidizes to 1,4-naphthoquinone. Usually isolated as the hydrochloride, $C_{10}H_9NO.HCl$, needles, very sol in water.
N-Acetyl deriv. 4-Acetamido-1-naphthol; naphthacetol. $C_{12}H_{11}NO_2$. Needles from alcohol, mp 188°, sparingly sol in water, sol in alcohol.
USE: Polymerization inhibitor; prepn of 2-allyl-4-amino-1-naphthol hydrochloride (an antihemorrhagic cpd); prepn of 1,4-naphthoquinone.

450. 1-Amino-2-naphthol-4-sulfonic Acid. [116-63-2] 4-Amino-3-hydroxy-1-naphthalenesulfonic acid; 1,2,4-acid. $C_{10}H_9NO_4S$; mol wt 239.25. C 50.20%, H 3.79%, N 5.85%, O 26.75%, S 13.40%. Prepn: M. Schmidt, *J. Prakt. Chem. Chem.-Ztg.* **44**, 513 (1891); by treatment of nitroso-β-naphthol with sodium bisulfite and sulfuric acid: L. F. Fieser, *Org. Synth.* coll. vol. II, 42 (1943); *cf. idem, J. Am. Chem. Soc.* **57**, 494 (1935).

White or gray needles, usually contg 0.5 H_2O. May turn pink on exposure to light, especially when moist. Insol in water, alcohol, ether, benzene; sol in hot sodium bisulfite soln; sol in alkaline soln but such solns oxidize quickly on exposure to air yielding a brown substance which is sol in hot water giving a green soln. Controlled oxidation with HNO_3 yields ammonium 1,2-naphthoquinone-4-sulfonate, *see* Fieser, *loc. cit.*
Sodium salt. [5959-58-0] $C_{10}H_8NNaO_4S$; mol wt 261.23. Needles, sol in hot water with blue fluorescence
USE: Manuf of azo dyes, 1,2-naphthoquinone-4-sulfonic acid. In determn of phosphate.

451. 1-Amino-2-naphthol-6-sulfonic Acid. [5639-34-9] 5-Amino-6-hydroxy-2-naphthalenesulfonic acid. $C_{10}H_9NO_4S$; mol wt 239.25. C 50.20%, H 3.79%, N 5.85%, O 26.75%, S 13.40%. Prepd by reduction of sodium 1-nitroso-2-naphthol-6-sulfonate with zinc and acetic acid: Fierz-David *et al.*, *Helv. Chim. Acta* **29**, 1765 (1946).

Needles or prisms. Slightly sol in boiling water, less sol in alc. Practically insol in ether.
Sodium salt hemipentahydrate. Eikonogen. $C_{10}H_8NNaO_4S.2½H_2O$; mol wt 306.26. Powder. Sol in water. Practically insol in alc, ether. Strong reducing action on silver salts.
USE: Free acid formerly used in manuf of azo dyes. Sodium salt used as photographic developer and for detection of potassium.

452. 6-Aminonicotinic Acid. [3167-49-5] 6-Amino-3-pyridinecarboxylic acid; 6-amino-3-carboxypyridine. $C_6H_6N_2O_2$; mol wt 138.13. C 52.17%, H 4.38%, N 20.28%, O 23.17%. Prepd by heating 6-chloronicotinic acid with ammonia: Marckwald, *Ber.* **26**, 2188 (1893); **27**, 1319 (1894); Räth, Prauge, *Ann.* **467**, 4 (1928); Johnson *et al.*, *J. Biol. Chem.* **153**, 37 (1944).

Dihydrate. Crystals from dil acetic acid. Dec above 300° yielding 2-aminopyridine and CO_2. Sparingly sol in most solvents.
Potassium salt. $C_6H_5KN_2O_2$. Crystals, freely sol in water, insol in alcohol.
Hydrochloride. $C_6H_6N_2O_2.HCl$. Needles, freely sol in water, slightly sol in alcohol.

453. 2-Amino-5-nitrothiazole. [121-66-4] 5-Nitro-2-thiazolamine; Enheptin. $C_3H_3N_3O_2S$; mol wt 145.14. C 24.83%, H 2.08%, N 28.95%, O 22.05%, S 22.09%. Prepd by deacetylation of 2-acetamido-5-nitrothiazole: H. L. Hubbard, US 2573641; G. W. Steahly, US 2573656; *idem*, US 2573657 (all 1951 to Monsanto).

Greenish-yellow to orange-yellow fluffy powder, dec 202°. Slightly bitter taste. uv max (0.0005% in water): 386 nm (ε 0.540); min: 295 nm. pKa 0.61 (±0.03). Slightly soluble in water. At 20°, one gram dissolves in 150 g of 95% alcohol; in 250 g ether. Almost insol in chloroform. Sol in dilute mineral acids.
THERAP CAT (VET): Antihistomonad in turkeys. For trichomoniasis in pigeons.

454. 6-Aminopenicillanic Acid. [551-16-6] (2S,5R,6R)-6-Amino-3,3-dimethyl-7-oxo-4-thia-1-azabicyclo[3.2.0]heptane-2-carboxylic acid; 6-APA; penicin; penin. $C_8H_{12}N_2O_3S$; mol wt 216.26. C 44.43%, H 5.59%, N 12.95%, O 22.19%, S 14.82%. Obtained from cultures of *Penicillium chrysogenum* in the absence of side chain precursors: F. R. Batchelor *et al.*, *Nature* **183**, 257 (1959). Synthesis: J. C. Sheehan, K. R. Henery-Logan, *J. Am. Chem. Soc.* **81**, 5838 (1959); **84**, 2983 (1962). Intermediate in the manuf of synthetic penicillins: J. C. Sheehan, **US 3159617** (1964 to Arthur D. Little). HPLC determn: J. Haginaka, J. Wakai, *Anal. Biochem.* **158**, 146 (1986). Review of prepn: F. M. Huber *et al.* in *Cephalosporins Penicillins: Chem. Biol.*, E. H. Flynn, Ed. (Academic, New York, 1972) pp 27-73. Review of chemistry: E. J. Vandamme, J. P. Voets, *Adv. Appl. Microbiol.* **17**, 311-369 (1974). General review: G. N. Rolinson, *J. Antimicrob. Chemother.* **22**, 5-14 (1988).

Crystals from water + HCl, dec 208-209°, also reported as 207-208° (dried) (Sheehan). $[\alpha]_D^{31}$ +273° (c = 1.2 in 0.1*N* HCl).

USE: In manuf of synthetic penicillins.

455. Aminopentamide. [60-46-8] α-[2-(Dimethylamino)-propyl]-α-phenylbenzeneacetamide; 4-dimethylamino-2,2-diphenylvaleramide; α,α-diphenyl-γ-dimethylaminovaleramide; dimevamide; BL-139. $C_{19}H_{24}N_2O$; mol wt 296.41. C 76.99%, H 8.16%, N 9.45%, O 5.40%. Anticholinergic. Prepn similar to that of methadone, starts by condensing diphenylmethyl cyanide with chlorodimethylaminopropane: Walton, *et al.*, *J. Chem. Soc.* **1949**, 648; Cheney *et al.*, *J. Org. Chem.* **17**, 770 (1952); Specter, **US 2647926** (1953 to Bristol). Prepn of optical isomers: Wheatley *et al.*, *J. Org. Chem.* **19**, 794 (1954). *See also:* Moffett, Aspergren, *J. Am. Chem. Soc.* **79**, 4451 (1957). Toxicity study: Cazort, *J. Pharmacol. Exp. Ther.* **100**, 325 (1950).

***dl*-Form.** Long prisms from dilute alcohol, mp 183-184°. Practically insol in water.

***d*-Form.** Crystals from petr ether, mp 136.5-137.5° (sintering). $[\alpha]_D^{23}$ +98.9° (methanol).

***l*-Form.** Crystals from petr ether, mp 136.5-137.5° (sintering). $[\alpha]_D^{23}$ −101.9° (methanol).

***dl*-Form hydrochloride.** $C_{19}H_{24}N_2O.HCl$. Slightly deliquescent leaflets from alcohol + ether, dec 190-191°. Bitter taste. Soluble in water, alcohol. A 1% aq soln has a pH of 6.8. LD_{50} in mice (mg/kg): 34.7 i.v.; 396 orally (Cazort).

***dl*-Form acid sulfate.** Centrine. $C_{19}H_{24}N_2O.H_2SO_4$; mol wt 394.49. Deliquescent crystals from isopropanol + ethyl acetate, mp 185-187°. The commercial medicinal grade, mp 178-181°. Bitter taste. uv max (1% H_2SO_4): 258.5 nm ($A_{1cm}^{1\%}$ 10.3); min: 249 nm. Freely sol in water, alc. Very slightly sol in chloroform. Practically insol in ether. pH of 2.5% aq soln 1.3-2.2.

THERAP CAT: Antispasmodic.

THERAP CAT (VET): Antispasmodic; antiemetic.

456. *m*-Aminophenol. [591-27-5] 3-Aminophenol; 3-amino-1-hydroxybenzene; 3-hydroxyaniline. C_6H_7NO; mol wt 109.13. C 66.04%, H 6.47%, N 12.84%, O 14.66%. Manuf by reduction of *m*-nitrophenol: Freifelder, Robinson, **US 3079435** (1963 to Abbott). Toxicity study: Koelzer, Giesen, *Z. Naturforsch.* **6b**, 183 (1951).

Crystals, mp 122-123°. Sol in 40 parts cold water, freely in hot water, alcohol, ether, amyl alcohol; slightly in benzene, very slightly in petr ether. LD_{50} i.p. in mice: 4.5 mg/20g (Koelzer, Giesen).

USE: Dye intermediate, manuf *p*-aminosalicylic acid.

457. *o*-Aminophenol. [95-55-6] 2-Aminophenol; 2-amino-1-hydroxybenzene; 2-hydroxyaniline. C_6H_7NO; mol wt 109.13. C 66.04%, H 6.47%, N 12.84%, O 14.66%. Manuf by reduction of *o*-nitrophenol: Freifelder, Robinson, **US 3079435** (1963 to Abbott).

Crystals, rapidly becoming brown, mp 170-174°; sublimes. One gram dissolves in 50 ml cold water, 23 ml alcohol; freely soluble in ether, very slightly in benzene. *Keep tightly closed and protected from light.*

Hydrochloride. $C_6H_7NO.HCl$; mol wt 145.59. Crystals readily becoming gray on exposure to light. Freely sol in water or alcohol.

USE: Manuf azo and sulfur dyes; dyeing furs and hair. Hydrochloride used in dyeing fur, hair, leather, etc.

458. *p*-Aminophenol. [123-30-8] 4-Aminophenol; *p*-Hydroxyaniline; 4-amino-1-hydroxybenzene; Azol; Rodinal; Unal; Ursol P. C_6H_7NO; mol wt 109.13. C 66.04%, H 6.47%, N 12.84%, O 14.66%. Usually prepd by the reduction of *p*-nitrophenol: *BIOS Final Report* **986**; Freifelder, Robinson, **US 3079435** (1963 to Abbott).

Orthorhombic plates from water. *Deteriorates under the influence of air and light.* mp 189.6-190.2°. The commercial product is usually pink, mp 186°. Can be sublimed at 0.3 mm and 110° without decompn. bp_{760} 284° (dec); $bp_{8.0}$ 167°; $bp_{3.0}$ 150°; $bp_{0.3}$ 130.2°. Kb at 15° = 6.6 × 10^{-9}. Forms salts with acids and bases. Soly in water: 0.39% at 13°; 0.65% at 24°; 0.80% at 30°; 1.5% at 50°; 4.7% at 80°; 8.5% at 96°. Soly in ethyl methyl ketone: 9.3% at 58.5°; in abs ethanol: 4.5% at 0°. Practically insol in benzene, chloroform.

Hydrochloride. $C_6H_7NO.HCl$; mol wt 145.59. Cryst powder; gradually becomes darker. Dec about 306°. Very sol in water; sol in alc.

Caution: May cause skin sensitization, dermatitis. Inhalation can cause asthma, methemoglobin formation.

USE: Photographic developer; intermediate in the manufacture of sulfur and azo dyes; in dyeing furs and feathers.

459. *p*-Aminophenylacetic Acid. [1197-55-3] 4-Aminobenzeneacetic acid; *p*-amino-α-toluic acid. $C_8H_9NO_2$; mol wt 151.17. C 63.56%, H 6.00%, N 9.27%, O 21.17%. Prepd by the reduction of *p*-nitrophenylacetic acid with hydrogen sulfide in the presence of ammonia: G. R. Robertson, *Org. Synth.* **coll. vol. I**, 52 (2nd ed., 1941).

Plates, leaflets from water, mp 199-200° (dec). Moderately sol in hot water; sol in alcohol, in alkalies.

Hydrochloride. $C_8H_9NO_2.HCl$. Rods from HCl, freely sol in water, sol in alcohol (about 3% w/w).

Ethyl ester. $C_{10}H_{13}NO_2$. Platelets from water, mp 51°.

***N*-Benzoyl deriv.** Needles from alcohol, mp 205-206°.

460. 4-Amino-3-phenylbutyric Acid. [1078-21-3] β-(Aminomethyl)benzenepropanoic acid; β-(aminomethyl)hydrocinnamic acid; 4-amino-3-phenylbutanoic acid; β-phenyl-γ-aminobutyric acid; fenibut; phenibut; phenigamma; phenygam; PhGABA. $C_{10}H_{13}NO_2$; mol wt 179.22. C 67.02%, H 7.31%, N 7.82%, O 17.85%. Prepn starting with α,γ-diamino-β-phenylpropane: Jackson, Kenner, *J. Chem. Soc.* **1928**, 1657. Alternate prepn: Cologne, Pouchol, *Bull. Soc. Chim. Fr.* **1962**, 598. Pharmacology: R. A. Khaunina, *Bull. Exp. Biol. Med.* **57**, 52 (1964). Structure-activity studies: *idem, Farmakol. Toksikol.* **31**, 202 (1968). Activity studies of the isomers: *eidem, Byull. Eksp. Biol. Med.* **72**, 49 (1971), *C.A.* **76**, 81208t (1972). Review of pharmacology and clinical experience: I. Lapin, *CNS Drug Rev.* **7**, 471-481 (2001).

mp 250-253° (dec).

Hydrobromide. [103095-38-1] $C_{10}H_{13}NO_2$.HBr; mol wt 260.13. Irregular platelets from benzene, mp 114°.

Hydrochloride. [3060-41-1] $C_{10}H_{13}NO_2$.HCl; mol wt 215.68. LD_{50} in mice, rats (mg/kg): 900, 700 i.p. (Khaunina, 1964).

THERAP CAT: Anxiolytic.

461. Aminophylline. [317-34-0] 3,9-Dihydro-1,3-dimethyl-1*H*-purine-2,6-dione compd with 1,2-ethanediamine (2:1); theophylline compd with ethylenediamine; theophylline ethylenediamine; theophyllamine; Afonilum; Aminodur; Cardophylin; Euphyllina; Pecram; Phyllocontin; Phyllotemp; Planphylline; Pulmovet; Tefamin. $C_{16}H_{24}N_{10}O_4$; mol wt 420.43. C 45.71%, H 5.75%, N 33.32%, O 15.22%. Prepn: Grüter, **US 919161** (1909 to Byk). Toxicity data: C. R. Thompson, M. R. Warren, *J. Lab. Clin. Med.* **31**, 1337 (1946). Comprehensive description: K. D. Thakker, L. T. Grady, *Anal. Profiles Drug Subs.* **11**, 1-44 (1982). Reviews of clinical experience: J. A. Stirt, S. F. Sullivan, *Anesth. Analg.* **60**, 587-602 (1981); A. G. Perry, *AACN Clin. Issues* **6**, 297-306 (1995).

Occurs as the dihydrate. White or slightly yellowish granules or powder. Slight ammoniacal odor, bitter taste. Upon exposure to air, gradually loses ethylene diamine and absorbs carbon dioxide with liberation of theophylline. Insol in alcohol, ether. *Keep tightly closed.* LD_{50} orally in mice: 540 mg/kg (Thompson, Warren).

Caution: Potential symptoms of overexposure include acute restlessness, anorexia, nausea, fever, vomiting, dehydration; followed by tremors, delirium, convulsions and coma; may result in cardiovascular and respiratory collapse, shock, cyanosis and death. *See Clinical Toxicology of Commercial Products*, R. E. Gosselin et al., Eds. (Williams & Wilkins, Baltimore, 5th ed., 1984) Section III, pp 17-21.

THERAP CAT: Bronchodilator.

THERAP CAT (VET): Bronchodilator.

462. 2-Amino-4-picoline. [695-34-1] 4-Methyl-2-pyridinamine; 2-amino-4-methylpyridine; α-amino-γ-picoline; 4-methyl-2-aminopyridine; W-45; Askensil. $C_6H_8N_2$; mol wt 108.14. C 66.64%, H 7.46%, N 25.91%. Prepd by heating 4-picoline with sodamide in xylene: Seide, *Ber.* **57**, 791 (1924); **58**, 1733 (1925); Räth, *ibid.* 347. Formulations: von Haxthausen *et al.*, **US 2937118** (1960 to Raschig). Pharmacology: von Haxthausen, *Arch. Exp. Pathol. Pharmakol.* **226**, 163; **227**, 234 (1955); Marchetti *et al.*, *Arch. Int. Pharmacodyn. Ther.* **143**, 385 (1963).

Leaflets from petr ether, mp 100-100.5°. bp_{11} 115-117°. Sublimes on slow heating. Freely sol in water, lower alcohols, dimethylformamide, coal tar bases. Slightly sol in petr ether, aliphatic hydrocarbons.

THERAP CAT: Cardiotonic.

463. Aminopromazine. [58-37-7] N^1,N^1,N^2,N^2-Tetramethyl-3-(10*H*-phenothiazin-10-yl)-1,2-propanediamine; 10-[2,3-bis(dimethylamino)propyl]phenothiazine; proquamezine; tetrameprozine; RP-3828. $C_{19}H_{25}N_3S$; mol wt 327.49. C 69.68%, H 7.69%, N 12.83%, S 9.79%. Prepn: Jacob *et al.*, *Compt. Rend.* **243**, 1637 (1956); Horclois, **GB 800635** (1958 to Rhône-Poulenc).

Fumarate. [3688-62-8] Lispamol; Sedofarmolo. $(C_{19}H_{25}N_3S)_2.C_4H_4O_4$; mol wt 771.05. Crystals, sensitive to light, dec 166-170°. Soly at 20° (g/100 ml): water 9.0; methanol 5.0; ethanol 0.5. Very slightly sol in isopropanol, acetone. Practically insol in benzene, ether. pH of a 2% aq soln 5.0 to 7.0.

THERAP CAT: Antispasmodic.

THERAP CAT (VET): Antispasmodic.

464. 2-Aminopropanol. [78-91-1] 2-Amino-1-propanol; 2-aminopropyl alcohol; β-propanolamine; 2-hydroxyisopropylamine. C_3H_9NO; mol wt 75.11. C 47.97%, H 12.08%, N 18.65%, O 21.30%. Obtained together with isopropylamine on reduction of acetylcarbinoloxime with sodium amalgam in dilute acetic acid: Gabriel, *Ber.* **49**, 2121 (1916); from DL-alanylglycine by reduction with sodium and abs alcohol: Abderhalden, Schwab, *Z. Physiol. Chem.* **143**, 292 (1925); by boiling the ethyl ester of DL-acetylalanine with sodium in abs alcohol: Karrer, *Helv. Chim. Acta* **4**, 98 (1921); as hydrolytic cleavage product of ergonovine and ergometrinine: Stoll, Hofmann, *ibid.* **26**, 956 (1943); by catalytic hydrogenation of the ethyl ester of alanine: Adkins, Pavlic, *J. Am. Chem. Soc.* **69**, 3039 (1947).

***dl*-Form.** Liquid, fishy odor. bp 173-176°. Freely sol in water, alcohol, ether.

Hydrochloride. C_3H_9NO.HCl; mol wt 111.57. Leaflets from abs alcohol + acetone, dec 86-87.5°.

465. 3-Aminopropionitrile. [151-18-8] 3-Aminopropanenitrile; β-aminopropionitrile. $C_3H_6N_2$; mol wt 70.10. C 51.40%, H 8.63%, N 39.96%. Prepd by the reaction of acrylonitrile with ammonia: Buc *et al.*, *J. Am. Chem. Soc.* **67**, 93 (1945); *Org. Synth.* **coll. vol. III**, 93 (1955); Weijlard, Sullivan, **US 2742491** (1956 to Merck & Co.).

Liquid. Amine odor. bp_{760} 185°; bp_{20} 87-89°; bp_5 66-69°; bp_3 50-55°. n_D^{20} 1.4396. The pure, anhydrous material may be stored in tightly stoppered bottles for several months. Storing under refrigeration is recommended to avoid development of pressure in the bottles. Polymerization takes place slowly during storage in an open container, or very rapidly in the presence of acid or acidic compds.

USE: Intermediate in the manufacture of β-alanine and pantothenic acid.

466. *p*-Aminopropiophenone. [70-69-9] 1-(4-Aminophenyl)-1-propanone; ethyl *p*-aminophenyl ketone; PAPP. $C_9H_{11}NO$;

mol wt 149.19. C 72.46%, H 7.43%, N 9.39%, O 10.72%. Pretreatment antidote to cyanide; activity probably through the formation of methemoglobin, which has a higher affinity for cyanide than mitochondrial cytochrome oxidase. Prepd by the action of propionyl chloride on aniline in carbon bisulfide in the presence of aluminum chloride: Kunckell, *Ber.* **33**, 2641 (1900); Derrick, Bornemann, *J. Am. Chem. Soc.* **35**, 1283 (1913); Hartung, Foster, *J. Am. Pharm. Assoc.* **35**, 15 (1946). Antidotal activity: C. L. Rose *et al.*, *J. Pharmacol. Exp. Ther.* **89**, 109 (1947). Toxicity study: J. W. Scawin *et al.*, *Toxicol. Lett.* **23**, 359 (1984). Review of pharmacology of cyanide detoxification: S. I. Baskin, R. F. Fricke, *Cardiovasc. Drug Rev.* **10**, 358-375 (1992).

Yellow needles from water, mp 140°. Sol in water, alc, DMSO. LD_{50} in male, female mice (mg/kg): 145, 200 i.v.; in female mice, guinea pigs, rats, male rats (mg/kg): >5000, 1020, 223.7, 475 orally (Scawin).

Hydrochloride. [6170-25-8] $C_9H_{11}NO.HCl$. Needles, yellowish cast, mp 198-199°. Freely sol in water. LD_{50} i.v. in dogs: 7.15 ±0.89 mg/kg (Rose).

Sulfate. $(C_9H_{11}NO)_2.H_2SO_4$. Yellow plates from alcohol, dec 25°.

THERAP CAT: Antidote (cyanide).

467. α-(α-Aminopropyl)benzyl Alcohol. [5897-76-7] α-(1-Aminopropyl)benzenemethanol; 2-amino-1-phenyl-1-butanol; β-amino-α-phenylbutyl alcohol; 1-phenyl-1-hydroxy-2-amino-*n*-butane. $C_{10}H_{15}NO$; mol wt 165.24. C 72.69%, H 9.15%, N 8.48%, O 9.68%. Prepn of the *dl-threo*-form: Abrams, Kipping, *J. Chem. Soc.* **1936**, 1480; Rebstock *et al.*, *J. Am. Chem. Soc.* **73**, 3666 (1951). Prepn of the *dl-erythro*-form: Hartung *et al.*, *ibid.* **52**, 3317 (1930); Rebstock *et al.*, *loc. cit.* See also Beilstein **13**, suppl. 3, 1791-1792 (1973).

dl-threo-Form. Thick, shiny plates from benzene + petr ether, mp 79-80° (Abrams, Kipping); 78-79° (Rebstock *et al.*). Freely sol in alcohol; moderately sol in chloroform, benzene; sparingly sol in petr ether.

dl-threo-Form hydrochloride. $C_{10}H_{15}NO.HCl$. Lustrous prisms, mp 195-196° (Abrams, Kipping); 204-205° (Rebstock *et al.*). Freely sol in water; sol in alcohol; slightly sol in chloroform.

dl-erythro-Form. mp 80.5-81°.

dl-erythro-Form hydrochloride. mp 242°.

468. Aminopterin. [54-62-6] *N*-[4-[[(2,4-Diamino-6-pteridinyl)methyl]amino]benzoyl]-L-glutamic acid; 4-aminofolic acid; 4-aminopteroylglutamic acid; 4-amino-PGA; NSC-739. $C_{19}H_{20}N_8O_5$; mol wt 440.42. C 51.82%, H 4.58%, N 25.44%, O 18.16%. Folic acid antagonist. Prepn from 2,4,5,6-tetraminopyrimidine sulfate, 2,3-dibromopropionaldehyde and *p*-aminobenzoylglutamic acid: D. R. Seeger *et al.*, *J. Am. Chem. Soc.* **69**, 2567 (1947); from 6-(bromomethyl)-2,4-diaminopteridine HBr: J. R. Piper, J. A. Montgomery, *J. Heterocycl. Chem.* **11**, 279 (1974). Purification: T. L. Loo, *J. Med. Chem.* **8**, 139 (1965). Inhibition of dihydrofolate reductase: J. S. Erickson *et al.*, *J. Biol. Chem.* **247**, 5661 (1972). Clinical pharmacokinetics: A. F. Ratliff *et al.*, *J. Clin. Oncol.* **16**, 1458 (1998). Clinical evaluation in children with acute lymphoblastic leukemia (ALL): P. D. Cole *et al.*, *Cancer Chemother. Pharmacol.* **62**, 65 (2008).

Dihydrate. Clusters of yellow needles. uv max of the 0.75 hydrate (0.1*N* NaOH): 261, 282, 373 (log ε 4.41, 4.39, 3.91).

THERAP CAT: Antineoplastic.

469. α-Aminopyridine. [504-29-0] 2-Pyridinamine; 2-aminopyridine. $C_5H_6N_2$; mol wt 94.12. C 63.81%, H 6.43%, N 29.76%. Prepd from pyridine and sodamide: Tschitschibabin, *Chem. Zentralbl.* **1915**, I, 1065; Wibaut, *Rec. Trav. Chim.* **42**, 240 (1923); A. I. Vogel, *Practical Organic Chemistry* (Longmans, London, 3rd ed., 1959) p 1007; Gattermann-Wieland, *Praxis des Organischen Chemikers* (de Gruyter, Berlin, 40th ed., 1961) p 316.

Leaflets, or large crystals. mp 58.1°. bp 210.6°. Soluble in water, alcohol, benzene, ether, hot petrol ether.

β-Aminopyridine. [462-08-8] 3-Pyridinamine. Prepn: Allen, Wolf, *Org. Synth.* **coll. vol. IV**, 45 (1963). Crystals, mp 64°. bp 250-252°. Sol in water, alcohol, benzene, ether. Insol in petr ether.

Caution: Potential symptoms of overexposure to α-aminopyridine are irritation of eyes, nose, throat; headache, dizziness; excitement; nausea; high blood pressure; respiratory distress; weakness; convulsions; stupor. *See NIOSH Pocket Guide to Chemical Hazards* (DHHS/NIOSH 97-140, 1997) p 14.

USE: Manuf pharmaceuticals and dyes.

470. Aminopyrine. [58-15-1] 4-(Dimethylamino)-1,2-dihydro-1,5-dimethyl-2-phenyl-3*H*-pyrazol-3-one; 4-(dimethylamino)antipyrine; dimethylaminophenyldimethylpyrazolone; 4-dimethylamino-2,3-dimethyl-1-phenyl-3-pyrazolin-5-one; aminophenazone; amidazophen; dipyrine; dimethylaminophenazone; amidopyrine; Mamallet-A; Pyramidon; Netsusarin. $C_{13}H_{17}N_3O$; mol wt 231.30. C 67.51%, H 7.41%, N 18.17%, O 6.92%. Prepn: *Beilstein* **25**, 452 (1936); T. Takahashi *et al.*, *J. Pharm. Soc. Jpn.* **76**, 1180 (1956); T. Takahashi, K. Kenematsu, *Chem. Pharm. Bull.* **6**, 98 (1958). Metabolism: G. F. Lockwood, J. B. Houston, *J. Pharm. Pharmacol.* **31**, 787 (1979). Hepatotoxicity: E. Bien *et al.*, *Pharmazie* **36**, 492 (1981). Toxicity studies: Hart, *J. Pharmacol. Exp. Ther.* **89**, 205 (1947); Tubaro *et al.*, *Arzneim.-Forsch.* **20**, 1024 (1970). Use in breath test of hepatic function: G. W. Hepner, E. S. Vesell, *N. Engl. J. Med.* **291**, 1384 (1974).

Leaflets from ligroin, mp 107-109°. *Poisonous.* The aq soln is slightly alkaline to litmus. Soly in water is increased by the addition of sodium benzoate. One gram dissolves in 1.5 ml alcohol, 12 ml benzene, 1 ml chloroform, 13 ml ether, 18 ml water. Stable in air, but affected by light. Readily attacked by mild oxidizing agents in the presence of water. LD_{50} orally in rats: 1.7 g/kg (Hart).

Hydrochloride. [6170-29-2] $C_{13}H_{17}N_3O.HCl$. Deliquescent prisms, mp 143-144°. Freely sol in water with acid reaction.

Bicamphorate. [94442-12-3] $C_{13}H_{17}N_3O.2C_{10}H_{16}O_4$. Crystalline powder, mp 94°. Sol in water with gradual decompn; sol in alcohol.

Salicylate. [603-57-6] $C_{13}H_{17}N_3O.C_7H_6O_3$. Crystalline powder, mp 70°. One gram dissolves in 16 ml water, about 6 ml alcohol.

Caution: Agranulocytosis may occur: E. Urbach, H. L. Goldburgh, *J. Am. Med. Assoc.* **131**, 893 (1946); G. Discombe, *Br. Med. J.* **1**, 1270 (1952).

THERAP CAT: Antipyretic; analgesic. Diagnostic aid (hepatic function).

471. Aminoquinuride. [3811-56-1] *N*,*N'*-Bis(4-amino-2-methyl-6-quinolinyl)urea; bis(2-methyl-4-amino-6-quinolyl)urea; di(4-aminoquinald-6-yl)urea; aminoquincarbamide; aminochinuride; aminokinuride; Surfen. $C_{21}H_{20}N_6O$; mol wt 372.43. C 67.73%, H 5.41%, N 22.57%, O 4.30%. Prepn: Jensch, **DE 591480** (1934 to I. G. Farben). Oncogenic and heparin-neutralizing properties: D. T. Hunter, J. M. Hill, *Nature* **191**, 1378 (1961).

Crystals from butanol, dec 255° (effervescence).
THERAP CAT: Antiseptic.

472. Aminorex. [2207-50-3] 4,5-Dihydro-5-phenyl-2-oxazolamine; 2-amino-5-phenyl-2-oxazoline; aminoxafen; aminoxaphen; McN-742. $C_9H_{10}N_2O$; mol wt 162.19. C 66.65%, H 6.21%, N 17.27%, O 9.86%. Prepn: Poos *et al.*, *J. Med. Chem.* **6**, 266 (1963); Poos, **US 3161650** (1964 to McNeil).

Crystals from benzene, mp 136-138°.
Fumarate. [13425-22-4] Menocil; Apiquel. $C_9H_{10}N_2O.C_4H_4O_4$; mol wt 278.26.

Note: This is a controlled substance (stimulant): **21 CFR**, 1308.11.

THERAP CAT: Anorexic.

473. p-Aminosalicylic Acid. [65-49-6] 4-Amino-2-hydroxybenzoic acid; 4-aminosalicylic acid; PAS; PASER; Rezipas. $C_7H_7NO_3$; mol wt 153.14. C 54.90%, H 4.61%, N 9.15%, O 31.34%. Prepn: **DE 50835** (1889); **US 427564** (1890); Sheehan, *J. Am. Chem. Soc.* **70**, 1665 (1948); Erlenmeyer *et al.*, *Helv. Chim. Acta* **31**, 988 (1948); Wenis, Gardner, *J. Am. Pharm. Assoc.* **38**, 9 (1949); Centolella, **US 2844625** (1958 to Miles). Toxicity study: E. M. Bavin *et al.*, *J. Pharm. Pharmacol.* **2**, 764 (1950). Comprehensive description: M. M. A. Hassan *et al.*, *Anal. Profiles Drug Subs.* **10**, 1-27 (1981).

Minute crystals from alcohol, mp 150-151° with effervescence. uv max (0.1*N* HCl): 265, 300 nm. pKa 3.25. pH of 0.1% aq soln: 3.5. One gram dissolves in about 500 ml water, in 21 ml alcohol. Slightly sol in ether. Practically insol in benzene. Sol in dilute nitric acid, dil sodium hydroxide. At temps above 40°, aq solns of PAS and its hydrochloride are readily decarboxylated to give brown solns consisting mainly of *m*-aminophenol. LD_{50} orally in mice: 4 g/kg (Bavin).

Hydrochloride. $C_7H_7NO_3.HCl$. Crystals, dec 224°.
Sodium salt dihydrate. [6018-19-5] Aminacyl; Nemasol Sodium; Pamisyl Sodium; Paramisan Sodium; Pasalon. C_7H_6NNa-

$O_3.2H_2O$; mol wt 211.15. pH of 1% soln ~7. One gram dissolves in 2 ml water. Very sparingly sol in acetone. Practically insol in ether, chloroform, benzene. Solns of the sodium salt are more resistant to heat than PAS, but sterilization by bacterial filtration is recommended.

Calcium salt. [133-15-3] Nippas Calcium. $C_{14}H_{12}CaN_2O_6$; mol wt 344.34. Bittersweet crystals. One gram dissolves in about 7 ml water, slightly sol in alc. Aq solns dec slowly and darken in color.

Potassium salt. [133-09-5] Paskalium. $C_7H_6NO_3K$; mol wt 191.23. Crystals, freely sol in water. pH of 1% soln about 7. Reported to cause less gastric irritation than the free acid or the sodium salt.

Ethyl ester. $C_9H_{11}NO_3$. Needles from alc, mp 115°. Prepd by reduction of the ethyl ester of 4-nitrosalicylic acid.

Phenyl ester *see* Phenyl Aminosalicylate.

THERAP CAT: Antibacterial (tuberculostatic).

474. p-Aminosalicylic Acid Hydrazide. [6946-29-8] 4-Amino-2-hydroxybenzoic acid hydrazide; Apacizin; Apacizina. $C_7H_9N_3O_2$; mol wt 167.17. C 50.29%, H 5.43%, N 25.14%, O 19.14%. Prepn: Drain *et al.*, *J. Chem. Soc.* **1949**, 1498; Magrane, **ES 206645** (1952), *C.A.* **49**, 5529e (1955); **JP 54 7472** (1954), *C.A.* **50**, 9947c (1956).

Needles from alcohol, mp 190-200°. Slightly soluble in water, somewhat more in ethanol.
THERAP CAT: Antibacterial (tuberculostatic).

475. 2-Aminothiazole. [96-50-4] 2-Thiazolamine; Abadol; Basedol. $C_3H_4N_2S$; mol wt 100.14. C 35.98%, H 4.03%, N 27.97%, S 32.02%. Prepn from vinyl acetate: Christiansen, **US 2242237** (1941 to Squibb); Kyrides, **US 2330223** (1943 to Monsanto); *cf.* Skrimshire, **GB 540032** (1941 to B.D.H.). Prepd also by condensing tribromoparaldehyde with thiourea: Leitch, Brickman, **US 2230962** (1941); **US 2339083** (1944 to Mallinckrodt). Prepn from paraldehyde and thiourea: Erlenmeyer *et al.*, *Helv. Chim. Acta* **38**, 1293 (1955). Toxicology: W. B. Deichmann *et al.*, *J. Ind. Hyg. Toxicol.* **30**, 71 (1948).

Crystals from benzene + petr ether, mp 93°. Distills at 3 mm without decompn. Sol in hot water. Slightly sol in cold water, alc, ether. Freely sol in dil HCl and in 20% H_2SO_4. LD_{50} orally in rats: 0.48 g/kg (Deichmann).

Hydrochloride monohydrate. $C_3H_4N_2S.HCl.H_2O$. Needles, freely sol in water. Also used as the acid tartrate. Crystals, sol in water.

THERAP CAT: Thyroid inhibitor.

476. 2-Amino-1,1,3-tricyanopropene. [868-54-2] 2-Amino-1-propene-1,1,3-tricarbonitrile; 2-amino-1,3,3-tricyano-2-propene; malononitrile dimer. $C_6H_4N_4$; mol wt 132.13. C 54.54%, H 3.05%, N 42.40%. Prepn: Carboni *et al.*, *J. Am. Chem. Soc.* **80**, 2838 (1958); Carboni, **US 2719861** (1955 to du Pont); Coenen, **DE 922531** (1955 to Bayer).

Rod-like crystals from water, mp 170-173°.

477. Amioca. [9037-22-3] Amylopectin. A non-linear polymer of glucose, obtained from waxy corn: Caldwell, *Converter* **17**, No. 11, 12, 14 (1943); *Manuf. Confect.* **33**, No. 12, 15, 17 (1943); *C.A.* **38**, 1137[1] (1944); Schopmeyer, *Food Ind.* **17**, 1476 (1945), *C.A.* **40**, 2331[3] (1946). *Review:* Powell in *Industrial Gums*, R. L. Whistler, Ed. (Academic Press, New York, 2nd ed., 1973) pp 567-576.

478. Amiodarone. [1951-25-3] (2-Butyl-3-benzofuranyl)[4-[2-(diethylamino)ethoxy]-3,5-diiodophenyl]methanone; 2-butyl-3-benzofuranyl-4-[2-(diethylamino)ethoxy]-3,5-diiodophenyl ketone; 2-butyl-3-[3,5-diiodo-4-(β-diethylaminoethoxy)benzoyl]-benzofuran. $C_{25}H_{29}I_2NO_3$; mol wt 645.32. C 46.53%, H 4.53%, I 39.33%, N 2.17%, O 7.44%. Benzofuran derivative with multiple electrophysiological effects. Prepn: **FR 1339389**; R. Tondeur, F. Binon, **US 3248401** (1963, 1966 to Soc. Belge l'Azote Prod. Chim. Marly). Physicochemical properties: M. Bonati *et al.*, *J. Pharm. Sci.* **73**, 829 (1984). HPLC determn in plasma: M. De Smet, D. L. Massart, *J. Pharm. Biomed. Anal.* **6**, 277 (1988). Comprehensive description: T. A. Plomp, *Anal. Profiles Drug Subs.* **20**, 1-120 (1991). Review of pharmacology, clinical efficacy and safety: M. Chow, *Ann. Pharmacother.* **30**, 637-643 (1996); B. N. Singh, *Clin. Cardiol.* **20**, 608-618 (1997). Clinical trial in cardiac resuscitation: P. J. Kudenchuk *et al.*, *N. Engl. J. Med.* **341**, 871 (1999); to prevent atrial tachyarrythmias: D. Roy *et al.*, *ibid.* **342**, 913 (2000); L. B. Mitchell *et al.*, *J. Am. Med. Assoc.* **294**, 3093 (2005). Review of clinical experience: L. A. Siddoway, *Am. Fam. Physician* **68**, 2189-2196 (2003).

Hydrochloride. [19774-82-4] L-3428; Amiodar; Ancaron; Cordarex; Corbionax; Cordarone; Cornaron; Escodarone; Pacerone; Tachydaron; Trangorex. $C_{25}H_{29}I_2NO_3$·HCl; mol wt 681.78. Crystalline powder, mp 156°. Also reported as crystals from acetone, mp 159 ±2° (Bonati). Soly at 25° (g/100ml): chloroform 44.51; methylene chloride 19.20; methanol 9.98; ethanol 1.28; benzene 0.65; tetrahydrofuran 0.60; acetonitrile 0.32; 1-octanol 0.30; ether 0.17; 1-propanol 0.13; water 0.07; hexane 0.03 petroleum ether 0.001. Sparingly sol in isopropanol; slightly sol in acetone, dioxane, and carbon tetrachloride. pH (5% soln) 3.4-3.9. pK_a (25°C) 6.56 ±0.06. uv max (methanol): 208, 242 nm ($E_{1cm}^{1\%}$ 662 ±8, 623 ±10).

THERAP CAT: Antiarrhythmic (class III).
THERAP CAT (VET): Antiarrhythmic.

479. Amiphenazole. [490-55-1] 5-Phenyl-2,4-thiazolediamine; 2,4-diamino-5-phenylthiazole; DAPT; phenamizole; Dizol; Daptazole; Daptazile; Fenamizol. $C_9H_9N_3S$; mol wt 191.25. C 56.52%, H 4.74%, N 21.97%, S 16.76%. Prepd by the interaction of thiourea and α-bromobenzyl cyanide in alcohol: Davies *et al.*, *J. Chem. Soc.* **1950**, 3491; from thiourea and α-cyanobenzyl benzenesulfonate in acetone: Dodson, Turner, *J. Am. Chem. Soc.* **73**, 4517 (1951).

Flakes from water or dilute alcohol, dec 163-164°. Turns brown on exposure to light and air.
Hydrobromide. $C_9H_9N_3S$·HBr. Prisms from alc, dec >250°. Freely sol in hot water, moderately in cold water.
Benzenesulfonate. $C_{15}H_{15}N_3O_3S_2$. Crystals from alcohol + ether, dec 261-262°. Practically insol in water, most organic solvents. Slightly sol in alcohol.

THERAP CAT: Narcotic antagonist.
THERAP CAT (VET): Barbiturate and morphine antagonist.

480. Amiprilose. [56824-20-5] 3-*O*-[3-(Dimethylamino)-propyl]-1,2-*O*-(1-methylethylidene)-α-D-glucofuranose; 1,2-*O*-iso-propylidene-3-*O*-[3′-(*N*,*N*-dimethylamino)propyl]-α-D-glucofuranose. $C_{14}H_{27}NO_6$; mol wt 305.37. C 55.07%, H 8.91%, N 4.59%, O 31.44%. Synthetic, substituted monosaccharide with immunomodulatory activity. Prepn: **BE 823313**; P. Gordon, **US 3939146** (1975, 1976 both to Strategic Med. Res.); E. R. Garrett *et al.*, *J. Pharm. Sci.* **71**, 387 (1982). Absolute configuration: R. J. Linhardt *et al.*, *ibid.* **79**, 158 (1990). HPLC determn in plasma: S. T. Wu *et al.*, *J. Chromatogr. B* **692**, 149 (1997). Clinical pharmacokinetics: E. R. Garrett *et al.*, *J. Pharmacokinet. Biopharm.* **10**, 247 (1982). Clinical trial in rheumatoid arthritis: W. G. Riskin *et al.*, *Ann. Intern. Med.* **111**, 455 (1989). Review of immunoregulatory effects: D. M. Chang, *Immunopharmacol. Immunotoxicol.* **17**, 437-450 (1995).

Colorless, viscous oil. n^{25} 1.4687.
Hydrochloride. [60414-06-4] SM-1213; Therafectin. $C_{14}H_{27}$-NO_6·HCl; mol wt 341.83. Crystals from methanol, mp 181-183°. Sol in water, methanol, hot ethanol.
THERAP CAT: Immunomodulator; anti-inflammatory.

481. Amisulpride. [71675-85-9] 4-Amino-*N*-[(1-ethyl-2-pyrrolidinyl)methyl]-5-(ethylsulfonyl)-2-methoxybenzamide; 4-amino-*N*-[(1-ethyl-2-pyrrolidinyl)methyl]-5-(ethylsulfonyl)-*o*-anisamide; aminosultopride; DAN-2163; Deniban; Socian; Solian; Sulamid. $C_{17}H_{27}N_3O_4S$; mol wt 369.48. C 55.26%, H 7.37%, N 11.37%, O 17.32%, S 8.68%. Dopamine receptor antagonist. Prepn: M. Thominet *et al.*, **BE 872585**; *eidem*, **US 4401822** (1979, 1983 both to Soc. d'Etudes Sci. Ind. de l'Ile-de-France). Crystal structure: H. L. DeWinter *et al.*, *Acta Crystallogr.* **C46**, 313 (1990). Psychopharmacology: G. Perrault *et al.*, *J. Pharmacol. Exp. Ther.* **280**, 73 (1997). HPLC determn in plasma and urine: B. Malavasi *et al.*, *J. Chromatogr. B* **676**, 107 (1996). Series of articles on pharmacology and clinical efficacy in schizophrenia: *Int. Clin. Psychopharmacol.* **12**, Suppl. 2, S11-S36 (1997).

Crystals from acetone, mp 126-127°. LD_{50} in male mice (mg/kg): 56-60 i.v.; 175-180 i.p.; 224-250 s.c.; 1024-1054 orally (Thominet).
THERAP CAT: Antipsychotic.

482. Amitraz. [33089-61-1] *N*′-(2,4-Dimethylphenyl)-*N*-[[(2,4-dimethylphenyl)imino]methyl]-*N*-methylmethanimidamide; *N*-methyl-*N*′-2,4-xylyl-*N*-(*N*-2,4-xylylformimidoyl)formamidine; *N*,*N*-di-(2,4-xylyliminomethyl)methylamine; 2-methyl-1,3-di-(2,4-xylylimino)-2-azapropane; 1,5-di-(2,4-dimethylphenyl)-3-methyl-1,3,5-triazapenta-1,4-diene; *N*-methylbis(2,4-xylyliminomethyl)-amine; BTS-27419; U-36059; ENT-27967; Aludex; Ectodex; Mitaban; Mitac; Taktic; Topline. $C_{19}H_{23}N_3$; mol wt 293.41. C 77.78%, H 7.90%, N 14.32%. Formamide insecticide for use on deciduous fruit and citrus mites, and tick eradication in cattle. Prepn: I. R. Harrison *et al.*, **DE 2061132**; *eidem*, **US 3781355** (1971, 1973 both to Boots); and activity: *idem et al.*, *Pestic. Sci.* **4**, 901 (1973). Prepn and characterization of polymorphs: M. M. de Villiers *et al.*, *J.*

Agric. Food Chem. **52**, 7362 (2004). Bacterial degradation: E. R. Allcock, D. R. Woods, *J. Appl. Bacteriol.* **44**, 383 (1978). Properties and use as insecticide: V. Labonne *et al. Def. Veg.* **33**, 66 (1979). Pharmacokinetics and metabolism in animals: M. A. Pass, T. D. Mogg, *J. Vet. Pharmacol. Ther.* **18**, 210 (1995). Use in treatment of demodicosis in dogs: L. Medleau, T. Willemse, *J. Small Anim. Pract.* **36**, 3 (1995).

White monoclinic needles, mp 86-87°. Unstable to acidic pH. Soly at 20° (g/100 cc): water <10^{-4}; methanol 2.5; xylene 33; acetone 50. LD_{50} in male rats, female mice, rabbits, guinea pigs, bobwhite quail (mg/kg): 800, >1600, >100, >400, 788 orally; LD_{50} in rabbits, male rats (mg/kg): >200, >1600 dermally; LC_{50} (48 hr) in rainbow trout, Japanese carp: 3.3, 1.2 ppm (Labonne).

USE: Acaricide; insecticide.

THERAP CAT (VET): Ectoparasiticide.

483. Amitriptyline. [50-48-6] 3-(10,11-Dihydro-5H-dibenzo[a,d]cyclohepten-5-ylidene)-N,N-dimethyl-1-propanamine; 10,11-dihydro-N,N-dimethyl-5H-dibenzo[a,d]cycloheptene-$\Delta^{5,\gamma}$-propylamine; 5-(γ-dimethylaminopropylidene)-5H-dibenzo[a,d]-10,11-dihydrocycloheptene; 10,11-dihydro-5-(γ-dimethylaminopropylidene)-5H-dibenzo[a,d]cycloheptene; 5-(3-dimethylaminopropylidene)dibenzo[a,d][1,4]cycloheptadiene. $C_{20}H_{23}N$; mol wt 277.41. C 86.59%, H 8.36%, N 5.05%. Prepn: Hoffsommer *et al., J. Org. Chem.* **27**, 4134 (1962); *ibid.* **28**, 1751 (1963); *J. Med. Chem.* **8**, 555 (1965); Engelhardt *et al.,* **BE 584061** (1960 to Merck & Co.); **GB 858187**; **GB 858188** (both 1961 to Hoffmann-La Roche); Tristram, Tull, **US 3205264** (1965 to Merck & Co.). HPLC determn in plasma: A. Zarghi *et al., Boll. Chim. Farm.* **140**, 458 (2001). Toxicity data: A. Tobe *et al., Arzneim.-Forsch.* **31**, 1278 (1981). Comprehensive description: K. W. Blessel *et al., Anal. Profiles Drug Subs.* **3**, 127-148 (1974). Review of pharmacology and use in chronic pain: H. M. Bryson, M. I. Wilde, *Drugs Aging* **8**, 459-476 (1996).

Hydrochloride. [549-18-8] Ro-4-1575; Adepril; Amineurin; Domical; Elavil; Endep; Euplit; Laroxyl; Lentizol; Miketorin; Redomex; Saroten; Sarotex; Triptizol; Tryptanol; Tryptizol. $C_{20}H_{23}N \cdot HCl$; mol wt 313.87. Minute crystals, mp 196-197°. uv max (methanol): 240 nm (ε 13800). pKa 9.4. Freely sol in water, chloroform, alcohol. Insol in ether. LD_{50} in mice, rats (mg/kg): 350, 380 orally; 65, 75 i.p. (Tobe).

THERAP CAT: Antidepressant.

484. Amitriptylinoxide. [4317-14-0] 3-(10,11-Dihydro-5H-dibenzo[a,d]cyclohepten-5-ylidene)-N,N-dimethyl-1-propanamine N-oxide; 10,11-dihydro-N,N-dimethyl-5H-dibenzo[a,d]cycloheptene-$\Delta^{5,\gamma}$-propylamine N-oxide; amitriptyline N-oxide; Ambivalon; Equilibrin. $C_{20}H_{23}NO$; mol wt 293.41. C 81.87%, H 7.90%, N 4.77%, O 5.45%. Centrally acting metabolite of amitriptyline, *q.v.* Prepn: J. B. Pedersen, **GB 991651**; *idem*, **US 3299139** (1965, 1967 both to Dumex). Series of articles on pharmacology, pharmacokinetics, metabolism, clinical studies, toxicity studies, teratological studies: *Arzneim.-Forsch.* **28**, 1873-1926 (1978). HPLC determn: K. M. Jensen, *J. Chromatogr.* **183**, 321 (1980). Neuropharmacology: J. Hyttel *et al., Acta Pharmacol. Toxicol.* **47**, 53 (1980). Toxicity study: H. Friehe, R. Fontaine, *Arzneim.-Forsch.* **28**, 1898 (1978).

Crystals, mp 228-230°. (Dihydrate, mp 102-103°.) LD_{50} in mice, guinea pigs, rabbits, dogs (mg/kg): between 330-460 orally; in mice, rats: 87, 25 i.v.; 320, 110 i.p. (Friehe, Fontaine).

THERAP CAT: Antidepressant.

485. Amitrole. [61-82-5] 1H-1,2,4-Triazol-5-amine; 3-amino-1H-1,2,4-triazole; 3-amino-s-triazole; aminotriazole; ATA; ENT-25445; Amizol; Cytrol; Weedazol. $C_2H_4N_4$; mol wt 84.08. C 28.57%, H 4.80%, N 66.64%. Non-selective post-emergence, translocated herbicide. Prepn: G. Sjostedt, L. Gringas, *Org. Synth.* **coll. vol. III**, 95 (1955). Use as herbicide: Allen, **US 2670282** (1954 to Am. Chemical Paint Co.). Antithyroid activity: T. H. Jukes, C. B. Shaffer, *Science* **132**, 296 (1960). *Review:* E. Kröller, *Residue Rev.* **12**, 163-192 (1966). Review of carcinogenicity studies: *IARC Monographs* **7**, 31-43 (1974).

Crystals from abs ethanol, mp 159°. Soluble in water, methanol, ethanol, chloroform. Sparingly sol in ethyl acetate. Insol in ether, acetone. Aq solns are neutral. LD_{50} in mice, rats (g/kg): 14.7, 25.0 orally (Kröller).

Hydrochloride. $C_2H_4N_4 \cdot HCl$. Crystals from alcohol, mp 153°.

Caution: Potential symptoms of overexposure are irritation of eyes, skin; dyspnea, muscle spasms, ataxia, anorexia, salivation, increased body temperature; lassitude, skin dryness and depression associated with thyroid function suppression. *See NIOSH Pocket Guide to Chemical Hazards* (DHHS/NIOSH 97-140, 1997) p 14. Amitrole is reasonably anticipated to be a human carcinogen: *Report on Carcinogens, Twelfth Edition* (PB2011-111646, 2011) p 42.

USE: Herbicide.

486. Amlexanox. [68302-57-8] 2-Amino-7-(1-methylethyl)-5-oxo-5H-[1]benzopyrano[2,3-b]pyridine-3-carboxylic acid; 2-amino-7-isopropyl-5-oxo-5H-[1]benzopyrano[2,3-b]pyridine-3-carboxylic acid; 2-amino-7-isopropyl-1-azaxanthone-3-carboxylic acid; amoxanox; AA-673; CHX-3673; Aphthasol; Elics; Solfa. $C_{16}H_{14}N_2O_4$; mol wt 298.30. C 64.42%, H 4.73%, N 9.39%, O 21.45%. Inhibits release of allergic mediators from mast cells. Prepn: A. Nohara *et al.,* **BE 864647**; **US 4143042** (1978, 1979 both to Takeda); *eidem, J. Med. Chem.* **28**, 559 (1985). Mode of action studies: T. Saijo *et al., Int. Arch. Allergy Appl. Immunol.* **79**, 231 (1986); H. Makino *et al., ibid.* **82**, 66 (1987). Pharmacodynamics: G. Rankov *et al., Ophthalmic Res.* **22**, 359 (1990). Clinical trial in aphthous ulcers: R. O. Greer, Jr. *et al., J. Oral Maxillofac. Surg.* **51**, 243 (1993).

Crystals from DMF, mp >300°.

THERAP CAT: Antiallergic; antiasthmatic.

487. Amlodipine. [88150-42-9] 2-[(2-Aminoethoxy)methyl]-4-(2-chlorophenyl)-1,4-dihydro-6-methyl-3,5-pyridinedicarboxylic acid 3-ethyl 5-methyl ester; (±)-2-[(2-aminoethoxy)methyl]-4-(2-chlorophenyl)-3-ethoxycarbonyl-5-methoxycarbonyl-6-methyl-1,4-dihydropyridine; UK-48340. $C_{20}H_{25}ClN_2O_5$; mol wt 408.88. C 58.75%, H 6.16%, Cl 8.67%, N 6.85%, O 19.56%. Dihydropyridine calcium channel blocker; activity resides mainly in the $(-)$-isomer. Prepn: S. F. Campbell *et al.*, **EP 89167**; *eidem*, **US 4572909** (1983, 1986 both to Pfizer). Synthesis and pharmacology of racemate and enantiomers: J. E. Arrowsmith *et al.*, *J. Med. Chem.* **29**, 1696 (1986). Calcium antagonist activity: R. A. Burges *et al.*, *J. Cardiovasc. Pharmacol.* **9**, 110 (1987). GC determn in plasma: A. P. Beresford *et al.*, *J. Chromatogr.* **420**, 178 (1987). Metabolism: *idem et al.*, *Xenobiotica* **18**, 245 (1988). Review of pharmacology and therapeutic efficacy: M. Haria, A. J. Wagstaff, *Drugs* **50**, 560-586 (1995). Clinical potential in severe heart failure: M. Packer *et al.*, *N. Engl. J. Med.* **335**, 1107 (1996). Review of mechanism of plaque stabilization: R. P. Mason, *Atherosclerosis* **165**, 191-199 (2002); of clinical potential in combination with atorvastatin in atherosclerosis: J. W. Jukema, J. W. A. van der Hoorn, *Expert Opin. Pharmacother.* **5**, 459-468 (2004).

Maleate. [88150-47-4] UK-48340-11. $C_{20}H_{25}ClN_2O_5.C_4H_4O_4$; mol wt 524.95. White crystals from ethyl acetate, mp 178-179°.

Benzenesulfonate. [111470-99-6] Amlodipine besylate; UK-48340-26; Amlor; Antacal; Istin; Monopina; Norvasc. $C_{20}H_{25}ClN_2O_5.C_6H_5SO_3H$; mol wt 567.05. White crystalline powder. Slightly sol in water; sparingly sol in ethanol.

THERAP CAT: Antianginal; antihypertensive.

488. Ammonia. [7664-41-7] H_3N; mol wt 17.03. H 17.76%, N 82.25%. NH_3. Manufactured from water gas (obtained by blowing steam through incandescent coke) as source of hydrogen, and from producer gas (obtained from steam and air through incandescent coke), as source of nitrogen by the Haber-Bosch process. Manuf from natural gas: *Faith, Keyes & Clark's Industrial Chemicals*, F. A. Lowenheim, M. K. Moran, Eds. (Wiley-Interscience, New York, 4th ed., 1975) pp 83-92. Historical monograph: A. Mittasch, *Geschichte der Ammoniaksynthese* (Verlag Chemie, 1951). Reviews of prepn, properties and chemistry: Several authors in *Mellor's* **Vol. VIII**, supplement I, *Nitrogen* part 1 (1964) pp 240-369; Jones in *Comprehensive Inorganic Chemistry* **Vol. 2**, J. C. Bailar, Jr. *et al.*, Eds. (Pergamon Press, Oxford, 1973) pp 199-227; J. R. LeBlanc *et al.*, in *Kirk-Othmer Encyclopedia of Chemical Technology* **vol. 2** (Wiley-Interscience, New York, 3rd ed., 1978) pp 470-516. Review of toxicology and human exposure: *Toxicological Profile for Ammonia* (PB2004-107331, 2004) 269 pp.

Colorless, non-flammable gas; very pungent odor (characteristic of drying urine). Lower limit of human perception: 0.04 g/cubic meter or 53 ppm. *Poisonous, corrosive.* One liter of the gas weighs 0.7714 g. d 0.5967 (air = 1). mp $-77.7°$. bp_{760} $-33.35°$. Densities of liq NH_3 (temp; press.): 0.6818 ($-33.35°$; 1 atm); 0.6585 ($-15°$; 2.332 atm); 0.6386 (0°; 4.238 atm); 0.6175 (15°; 7.188 atm); 0.5875 (35°; 13.321 atm). Critical temp 132.4°; critical press. 111.5 atm. Heat capacity (25°) 8.38 cal/mole/deg. Mixtures of ammonia and air will explode when ignited under favorable conditions: *Angew. Chem.* **43**, 302 (1930), but ammonia is generally regarded as non-flammable. pH of 1.0N aq soln 11.6; of 0.1N aq soln 11.1; of 0.01N aq soln 10.6. Water at 0° holds 47%, at 15° 38%, at 20° 34%, at 25° 31%, at 30° 28%, at 50° 18%. d_4^{20} (aq solns): 0.9939 (1%); 0.9811 (4%); 0.9651 (8%); 0.9362 (16%); 0.9229 (20%); 0.9101 (24%); 0.8980 (28%). fp (aq solns): $-2.9°$ (4%); $-8.1°$ (8%); $-23.1°$ (16%); $-34.9°$ (20%); $-44.5°$ (24%); $-69.2°$ (28%). Solution of NH_3 in water is exothermic. 95% alcohol at 20° holds 15%, at 30° 11%. Abs ethanol at 0° 20%, at 25° 10%. Methanol at 25° 16%. It is also sol in chloroform and ether. Liquid ammonia produces low temps by its own evaporation. Heat of vaporization: 5.581 kcal/mole. It is a good solvent for many elements and compds. Usually marketed in liquefied form in steel cylinders or as ammonia water (aqua ammonia, ammonium hydroxide) in drums and bottles.

Caution: Potential symptoms of overexposure are eye, nose and throat irritation; dyspnea, bronchospasm and chest pain; pulmonary edema; pink frothy sputum; skin burns, vesiculation; direct contact with liquid may cause frostbite. See *NIOSH Pocket Guide to Chemical Hazards* (DHHS/NIOSH 97-140, 1997) p 14. *See also Patty's Industrial Hygiene and Toxicology* **vol. 2B**, G. D. Clayton, F. E. Clayton, Eds. (Wiley-Interscience, New York, 3rd ed., 1981) pp 3045-3052.

USE: Fertilizer, corrosion inhibitor, purification of water supplies, component of household cleaners, as refrigerant. Manuf nitric acid, explosives, synthetic fibers, fertilizers. In pulp and paper, metallurgy, rubber, food and beverage, textile and leather industries.

489. Ammonia Borane. [13774-81-7] $(T-4)$-Amminetrihydroboron; borane ammonia complex; borazane. BH_6N; mol wt 30.87. B 35.02%, H 19.59%, N 45.37%. H_3BNH_3. Storage medium for molecular hydrogen in clean fuel applications. Prepn: S. G. Shore, R. W. Parry, *J. Am. Chem. Soc.* **77**, 6084 (1955); *eidem, ibid.* **80**, 8 (1958); E. Mayer, *Inorg. Chem.* **11**, 866 (1972); M. G. Hu *et al.*, *J. Inorg. Nucl. Chem.* **39**, 2147 (1977). Thermal decompn studies: *idem et al.*, *Themochim. Acta* **23**, 249 (1978); F. Baitalow *et al.*, *ibid.* **445**, 121 (2006). Raman spectroscopy: S. Trudel, D. F. R. Gilson, *Inorg. Chem.* **42**, 2814 (2003). NMR structural studies: A. Lötz, J. Voitländer, *J. Magn. Reson.* **48**, 1 (1982); O. Gunaydin-Sen *et al.*, *J. Phys. Chem. B* **111**, 677 (2007). Dehydrogenation processes as chemical hydrogen storage methods: A. Gutowska *et al.*, *Angew. Chem. Int. Ed.* **44**, 3578 (2005); M. Chandra, Q. Xu, *J. Power Sources* **156**, 190 (2006); M. E. Bluhm *et al.*, *J. Am. Chem. Soc.* **128**, 7748 (2006); R. J. Keaton *et al.*, *ibid.* **129**, 1844 (2007).

White solid, mp 112-114° (dec). *Explosive.* d 0.74. Sol in ether, anhydrous liquid ammonia. Dissolves in water to form a colorless soln, pH 9.1.

USE: Hydrogen fuel storage compound.

490. Ammoniacum. [9000-03-7] Gum ammoniac. A gum-resin exuded from the flowering and fruiting stem of *Dorema ammoniacum*, D. Don, and probably other *Umbelliferae*. Habit: Persia, Northern India, Southern Siberia, Africa. *Constit.* 1.3-6.7% volatile oil; 50-70% resin; 18-26% gum; ash content about 2%, may be as high as 10%; salicylic acid; aminoresinol. *Ref:* W. Sandermann, *Naturharze, Terpentinöl, Tallöl* (Springer, 1960) pp 82-83.

Irregular rounded tears, yellowish or brownish outside and whitish within; brittle when cold, but soft when warm; also masses, darker in color and less homogeneous; peculiar odor; slightly sweetish, bitter, somewhat acrid taste. mp 45-55°. Acid no. 60-80. Sapon no. 97-114. d 1.207. Partly soluble in water, alcohol, ether, vinegar or alkali soln; forms emulsions with water.

USE: Ingredient of porcelain cements.

THERAP CAT: Diaphoretic; emmenagogue.

491. Ammonia Water. [1336-21-6] Ammonium hydroxide $((NH_4)(OH))$; ammonium hydroxide; aqua ammonia; spirit of hartshorn. H_5NO; mol wt 35.05. NH_4OH. A soln of 28-29% NH_3 in water.

Colorless liquid; intense, pungent, suffocating odor; acrid taste; strong alkaline reaction. d_{25}^{25} ~0.90, 26°Bé. Dissolves copper, zinc. Fumes are formed when brought near volatile acids. *Keep cool in strong glass, plastic or rubber-stoppered bottles not completely filled.* Reaction with H_2SO_4 or other strong mineral acids is exothermic; mixture becomes boiling hot.

Ammonia Water—10%. Colorless liquid. Very pungent odor. d_{25}^{25} 0.957. *Irritating to skin and mucous membranes.*

USE: Detergent, removing stains, bleaching, calico printing, extracting plant colors (cochineal, archil, etc.) and alkaloids; manuf ammonium salts and aniline dyes. Buffers; pH adjustment; trace metal analysis. Wide variety of other uses.

THERAP CAT: 10% Solution as reflex respiratory stimulant.

492. Ammonium Acetate. [631-61-8] Acetic acid ammonium salt (1:1). $C_2H_7NO_2$; mol wt 77.08. C 31.17%, H 9.15%, N 18.17%, O 41.51%. CH_3COONH_4. Commercial product contains 95-97% salt with acetic acid and some water. Prepd from acetic acid and NH_3: Zuffanti, *J. Am. Chem. Soc.* **63**, 3123 (1941). Toxicity data: Welch *et al.*, *J. Lab. Clin. Med.* **29**, 809 (1944).

Deliquesc crystals or crystalline masses. Slight acetous odor. d 1.07. mp 114° (Zuffanti, *loc. cit.*). Tends to lose NH_3. Sol in less than 1 part water; freely sol in alc; slightly sol in acetone. Very concd aq soln is slightly acid; a 0.5 molar aq soln has pH 7.0. *Keep cool and tightly closed.* LD i.v. in mice: 1.8 mg (NH_4^+)/20g (Welch).

Ammonium acetate solution. [8013-61-4] Mindererus spirit. Colorless, clear liquid; acid reaction to litmus. Prepd from 1 g ammonium carbonate, 20 ml dil acetic acid (6%). Contains 6.5-7.5% CH_3COONH_4.

USE: Preserving meats, dyeing, stripping; as a reagent in analytical chemistry, *e.g.*, for determining Pb, Fe; separating $PbSO_4$ from other sulfates. In buffer solutions.

THERAP CAT: Diuretic.

THERAP CAT (VET): Formerly as diuretic, antipyretic.

493. Ammonium Benzoate. [1863-63-4] Benzoic acid ammonium salt (1:1). $C_7H_9NO_2$; mol wt 139.15. C 60.42%, H 6.52%, N 10.07%, O 23.00%. It is about 99% pure. Manuf from benzoic acid and NH_3: J. A. Spina, **US 1704636** (1929 to Hooker Electrochem.). Analytical applications in metal cation separations: I. M. Kolthoff *et al.*, *J. Am. Chem. Soc.* **56**, 812 (1934); A. Jewsbury, S. H. Osborn, *Anal. Chim. Acta* **3**, 642 (1949). Structural studies: I. A. Oxton *et al.*, *Can. J. Chem.* **55**, 3831 (1977).

Lamellar crystals or crystalline powder; odorless or faint benzoic acid odor; gradually loses NH_3 on exposure to air. d 1.26. mp 198°. One gram dissolves in 4.7 ml water, 1.2 ml boiling water, 36 ml alcohol, 8 ml boiling alcohol, 8 ml glycerol. The aq soln is slightly acid. *Keep well closed. Incompat.* Ferric salts, acids, alkali hydroxides or carbonates.

USE: To preserve glue and latex. In determn of benzoate stabilizers in feeds.

494. Ammonium Bicarbonate. [1066-33-7] Carbonic acid ammonium salt (1:1); acid ammonium carbonate; ammonium hydrogen carbonate. CH_5NO_3; mol wt 79.06. C 15.19%, H 6.37%, N 17.72%, O 60.71%. NH_4HCO_3. Occurs in the urine of alligators: Coulson, Hernandez, *Proc. Soc. Exp. Biol. Med.* **88**, 682 (1955). Usually prepd by passing an excess of carbon dioxide through concd ammonia water. Manuf: **GB 304872** (1927 to I.G. Farben.); Brooks, **GB 742386** (1955 to I.C.I.). *See also* Ammonium Carbonate.

Shiny, hard, colorless or white prisms or crystalline mass. Faint odor of ammonia. Comparatively stable at room temp. Volatile with decompn at about 60°. The white fumes given off consist of NH_3 21.5%, CO_2 55.7%, H_2O vapor 22.8%. Rate of decompn increases as temp rises. mp 107.5° (very rapid heating). Soly in water: 14% (10°); 17.4% (20°); 21.3% (30°). Decomposed by hot water. One gram dissolves in 10 ml glycerol (pharmaceutical grade). pH of 0.1N soln in water at 25° = 7.8. Insol in alcohol and acetone. Negative heat of solution.

Pharmaceutical Incompat: Acids, caustic alkalies.

USE: In baking powder formulations; in cooling baths (one kg dissolved in 5 liters H_2O at 17° lowers it to 7°); in fire extinguishers; manuf porous plastics, ceramics; manuf dyes, pigments; in compost heaps to accelerate decompn; as fertilizer; for defatting textiles; in cold wave solns; in chrome leather tanning; to remove gypsum from heat exchanges and other processing equipment.

THERAP CAT: Expectorant.

THERAP CAT (VET): Expectorant. Used in bloat, colic.

495. Ammonium Bifluoride. [1341-49-7] Ammonium fluoride ((NH_4)(HF_2)); acid ammonium fluoride; ammonium hydrogen fluoride. F_2H_5N; mol wt 57.04. F 66.61%, H 8.84%, N 24.56%. NH_4HF_2. Prepn from hydrofluoric acid and NH_3: Hassel, Luzanski, *Z. Kristallogr.* **A83**, 449 (1932); *Gmelins, Ammonium* (8th ed.) **23**, 148 (1936).

Orthorhombic crystals which readily etch glass. d 1.5. mp 124.6°. Freely sol in water. *Keep in plastic, rubber, wood or paraffined containers and well closed. See also* Ammonium Fluoride.

USE: In manuf of Mg and Mg alloys; in brightening of Al; for purifying and cleansing various parts of beer-dispensing apparatus, tubes, etc., sterilizing dairy and other food equipment; in glass and porcelain industries; as mordant for aluminum; as a "sour" in laundering cloth. In lab production of HF.

496. Ammonium Binoxalate. [5972-72-5] Ethanedioic acid ammonium salt (1:1); ammonium acid oxalate; ammonium hydrogen oxalate. $C_2H_5NO_4$; mol wt 107.07. C 22.44%, H 4.71%, N 13.08%, O 59.77%. $NH_4OOCCOOH$. Prepn of monohydrate: Dehn, Heuse, *J. Am. Chem. Soc.* **29**, 1137 (1907). *Review: Gmelins, Ammonium* (8th ed.) **23**, pp 405-406 (1936).

Monohydrate. Rhombic crystals. *Poisonous.* d 1.56. Sol in 25 parts water; slightly sol in alcohol.

USE: To remove ink stains.

497. Ammonium Bisulfate. [7803-63-6] Sulfuric acid ammonium salt (1:1); acid ammonium sulfate; ammonium hydrogen sulfate. H_5NO_4S; mol wt 115.10. H 4.38%, N 12.17%, O 55.60%, S 27.85%. NH_4HSO_4. Prepn: *Gmelins, Ammonium* (8th ed.) **23**, 293-298 (1936).

Deliquesc crystals. mp about 147°. d 1.787. Freely sol in water. Practically insol in alcohol, acetone, pyridine. *Keep well closed.*

USE: In hair-waving preparations; as catalyst for organic reactions.

498. Ammonium Bisulfide. [12124-99-1] Ammonium sulfide ((NH_4)(SH)); ammonium hydrogen sulfide; ammonium hydrosulfide; ammonium sulfhydrate. H_5NS; mol wt 51.11. H 9.86%, N 27.41%, S 62.73%. NH_4HS. Prepd by mixing stoichiometric amts of NH_3 and H_2S gases at 0°: Thomas, Riding, *J. Chem. Soc.* **123**, 1181 (1923).

White, tetragonal or orthorhombic crystals. d 1.17. Sublimes *in vacuo.* Decomposes into H_2S and NH_3 rather easily at room temp when in crystal form. The commercial product is furnished in porcelain-like lumps which are more stable and can be stored in a closed bottle at a cool place. Dissociation press. at room temp about 350 mm Hg. Freely sol in water or alcohol giving colorless solns which turn yellow rapidly. Decomposed by boiling water. Soly (0°): 128.1 g/100 g H_2O. Slightly sol in acetone; almost insol in ether, benzene.

Caution: Very irritating to skin; penetrates more rapidly than hydrogen sulfide and may be fatal.

USE: In lubricants.

499. Ammonium Bisulfite. [10192-30-0] Sulfurous acid ammonium salt (1:1); ammonium acid sulfite; ammonium hydrogen sulfite. H_5NO_3S; mol wt 99.10. H 5.09%, N 14.13%, O 48.43%, S 32.35%. NH_4HSO_3. Prepn: *Gmelins, Ammonium* (8th ed.) **23**, 259-260 (1936).

Crystals. Soly in water (g/100 ml H_2O): 267 (10°); 620 (60°). *Keep well closed.*

USE: Preservative.

500. Ammonium Bitartrate. [3095-65-6] (2*R*,3*R*)-2,3-Dihydroxybutanedioic acid ammonium salt (1:1); L-tartaric acid monoammonium salt; ammonium acid tartrate; ammonium hydrogen tartrate. $C_4H_9NO_6$; mol wt 167.12. C 28.75%, H 5.43%, N 8.38%, O 57.44%. $NH_4OOCCH(OH)CH(OH)COOH$. Prepd from ammonium tartrate and tartaric acid: Dulk, *Ann.* **2**, 39 (1832).

Odorless crystals. d 1.68. Sol in 45.6 parts water at 15°; freely sol in hot water, alkalies, alkali carbonates. Practically insol in alc. $[\alpha]_D^{20}$ +26.0° (c = 1.5 in water): Long, *J. Am. Chem. Soc.* **23**, 813 (1901).

501. Ammonium Borate. [12007-58-8] Ammonium boron oxide ((NH_4)$_2B_4O_7$); ammonium tetraborate; ammonium biborate; ammonium metaborate. $B_4H_8N_2O_7$; mol wt 191.31. B 22.60%, H 4.22%, N 14.64%, O 58.54%. (NH_4)$_2B_4O_7$. Prepn: *Gmelins, Ammonium* (8th ed.) **23**, 323 (1936).

Tetrahydrate. Tetragonal crystals. Sol in water. Practically insol in alcohol.

USE: Fireproofing wood and textiles; in electrolytic condensers.

502. Ammonium Bromide. [12124-97-9] Ammonium bromide ((NH_4)Br). BrH_4N; mol wt 97.94. Br 81.58%, H 4.12%, N 14.30%. NH_4Br. Contains 99-99.5% NH_4Br. Prepn and properties: *Gmelins, Ammonium* (8th ed.) **23**, 203-218 (1936); Richards, "Ammonium Bromide" in *Mellor's* **Vol. VIII**, supplement I, *Nitrogen* (part 1) 433-447 (1964).

White, odorless, slightly hygroscopic crystals or granules; pungent, saline taste; slowly becomes yellowish in air; sublimes at high temp without melting. d^{25} 2.429. Freely sol in water, methanol, ethanol, acetone; slightly sol in ether. Practically insol in ethyl acetate. *Keep well closed.*

Incompat. Acids, acid salts, spirit nitrous ether, alkaloids; salts of lead, mercury, silver.

USE: Manuf of photographic films, plates, and papers; in process engraving and lithography; fireproofing of wood; in corrosion inhibitors; in photochemical reactions; precipitation of silver salts.

THERAP CAT: Sedative.

THERAP CAT (VET): Sedative.

503. Ammonium Caprylate. [5972-76-9] Octanoic acid ammonium salt (1:1); caprylic acid ammonium salt. $C_8H_{19}NO_2$; mol wt 161.25. C 59.59%, H 11.88%, N 8.69%, O 19.84%. C_7H_{15}-COONH$_4$. Prepn from the acid and ammonia: McMaster, Magill, *J. Am. Chem. Soc.* **38**, 1793 (1916); Stumpf, *Am. Paint J.* **38** (45), 60 (1954).

Monoclinic crystals from ether + alcohol. Somewhat hygroscopic. Decomposes on standing and develops odor of caprylic acid. mp 70-85°. Easily hydrolyzed by water. Freely sol in glacial acetic acid and ethanol; less sol in methanol; slightly sol in acetone and ethyl acetate. Practically insol in chloroform, benzene. *Keep well closed.*

USE: In photographic emulsions; as insecticide and nematocide; in manuf of zinc caprylate.

504. Ammonium Carbamate. [1111-78-0] Carbamic acid ammonium salt (1:1); "anhydride" of ammonium carbonate; ammonium aminoformate. $CH_6N_2O_2$; mol wt 78.07. C 15.38%, H 7.75%, N 35.88%, O 40.99%. NH$_2$COONH$_4$. Prepd from dry ice and liq ammonia: Brooks, Audrieth, *Inorg. Synth.* **2**, 85 (1946). Toxicology: R. P. Wilson *et al., Am. J. Vet. Res.* **29**, 897 (1968).

Cryst powder; ammonia odor; gradually loses ammonia in the air changing to ammonium bicarbonate. Volatilizes at about 60°. Freely sol in water; sol in alcohol. LD$_{50}$ i.v. in mice: 0.99 mmol/kg (Wilson).

USE: Ammoniating agent, less vigorous than NH$_3$.

505. Ammonium Carbonate. [506-87-6] Carbonic acid ammonium salt (1:2); bis(ammonium) carbonate; diammonium carbonate. $CH_8N_2O_3$; mol wt 96.09. C 12.50%, H 8.39%, N 29.15%, O 49.95%. (NH$_4$)$_2$CO$_3$. Not readily prepared in pure form. Commerical prepns are usually a mixture with ammonium carbamate and ammonium bicarbonate. Originally obtained by dry distillation of grated deer antlers. Prepn from gaseous ammonia, carbon dioxide and steam: J. Bueb, **US 1004361** (1911); V. M. Efimov *et al.*, **US 4335088** (1982). *Review:* Allen, "Ammonium Carbonate" in *Mellor's* **Vol. VIII**, supplement 1, *Nitrogen* (part 1) 459-468 (1964).

Flat, columnar, prismatic crystals or elongated flakes, mp 43°. *Irritant.*

Ammonium carbonate carbamate. [8000-73-5] Baker's ammonia; hartshorn salt; sal volatile. Contains 30-34% NH$_3$, about 45% CO$_2$. Colorless, hard, translucent, crystalline masses, white cubes or powder; strong odor of ammonia; sharp taste and alkaline reactions. Decomposes on exposure to air with loss of NH$_3$ and CO$_2$, becoming white and powdery and converting into ammonium bicarbonate. Volatilizes at about 60°. Freely sol in water. Decomposed by hot water. *Keep tightly closed in a cool place.*

Incompat. Acids and acid salts; salts of iron, zinc; alkaloids, alum, calomel, tartar emetic.

USE: In baking powders, smelling salts; as mordant in dyeing; foaming agent; as a reagent in analytical chemistry for pH adjustment. Pharmaceutic aid (source of ammonia; alkalinizing agent; buffering agent).

506. Ammonium Chloride. [12125-02-9] Ammonium chloride ((NH$_4$)Cl); ammonium muriate; sal ammoniac; salmiac. ClH$_4$N; mol wt 53.49. Cl 66.27%, H 7.54%, N 26.19%. NH$_4$Cl. Contains 99.5-99.8% NH$_4$Cl; principal impurity is NaCl; exists in two temperature dependent crystal modifications. Prepn and properties: *Gmelins, Ammonium* (8th ed.) **23**, pp 150-184 (1936); N. C. R. Kane, "Ammonium Chloride" in *Mellor's* **Vol. VIII**, supplement

1, *Nitrogen* (part 1), 378-432 (1964). Manuf: A. W. Bamforth, S. R. S. Sastry, *Chem. Process Eng.* **53**, 72 (1972). NMR study: K. H. Michel, *Proc. Int. Sch. Phys.* **1976**, 392. Toxicity data: E. M. Boyd, K. G. W. Seymour, *Exp. Med. Surg.* **4**, 223 (1946). Brief review: C. W. Weston in *Kirk-Othmer Encyclopedia of Chemical Technology* vol 2 (Wiley-Interscience, New York, 4th ed., 1992) pp 695-698. Review of properties and use in water treatment: P. Smeets, *Trib. Eau* **570**, 26-29 (1994).

Colorless, odorless crystals or cryst masses; or white, granular powder; cooling, saline taste; somewhat hygroscopic. Tendency to cake. Strongly endothermic. d_4^{20} 1.5274. Sublimes without melting. Soly in water (w/w): 22.9% (0°); 26.0% (15°); 28.3% (25°); 39.6% (80°). Freely sol in water, glycerin, more so in boiling water. HCl and NaCl decrease soly in water. Sparingly sol in alcohol. Almost insol in acetone, ether, ethyl acetate. pH of aq solns (25°): 1% 5.5; 3% 5.1; 10% 5.0. LD$_{50}$ in rats (mg/kg): 30 i.m. (Boyd, Seymour). LD$_{50}$ in rats (mg/kg): 1650 orally (Smeets).

Caution: Potential symptoms of overexposure to fumes are irritation of eyes, skin, respiratory system; cough, dyspnea, pulmonary sensitization. *See NIOSH Pocket Guide to Chemical Hazards* (DHHS/NIOSH 97-140, 1997) p 16.

USE: As a flux in zinc and tin plating; electroplating, electrolytic refining of zinc; etching solutions in manufacture of printed circuit boards; in dry and Leclanché batteries; as a nitrogen source for fertilization of rice and wheat, manufacturing of explosives; flame suppressant; hardener for formaldehyde-based adhesives; mordant for dyes and printing.

THERAP CAT: Acidifier.

THERAP CAT (VET): Expectorant; diaphoretic; acidifying diuretic.

507. Ammonium Chromate(VI). [7788-98-9] Chromic acid (H$_2$CrO$_4$) ammonium salt (1:2); neutral ammonium chromate; diammonium chromate. CrH$_8$N$_2$O$_4$; mol wt 152.07. Cr 34.19%, H 5.30%, N 18.42%, O 42.08%. (NH$_4$)$_2$CrO$_4$. Prepn: *Gmelins, Chromium* (8th ed.) **52**, part B, pp 707-712 (1962).

Yellow acicular crystals; loses some NH$_3$ in air; decomposes at 185°. d 1.8. Soly in water: 19.78% (0°); 41.20% (75°). Sparingly sol in liquid ammonia, acetone. Slightly sol in methanol. Practically insol in ethanol. The aq soln is alkaline. *Keep well closed.*

Caution: Chromium hexavalent (VI) compounds are listed as known human carcinogens: *Report on Carcinogens, Twelfth Edition* (PB2011-111646, 2011) p 106.

USE: Sensitizing gelatin in photography, in textile printing pastes, in fixing chromate dyes on wool, and as a reagent in analytical chemistry.

508. Ammonium Chromic Sulfate. [13548-43-1] Ammonium disulfatochromate(III); chromic ammonium sulfate. CrH$_4$NO$_8$S$_2$; mol wt 262.15. Cr 19.83%, H 1.54%, N 5.34%, O 48.82%, S 24.46%. NH$_4$Cr(SO$_4$)$_2$. Prepd by crystallization from a soln contg equimolar amounts of Cr$_2$(SO$_4$)$_3$ and (NH$_4$)$_2$SO$_4$: Howarth in *ACS Monograph Series* **no. 132**, entitled "Chromium," vol. 1, M. J. Udy, Ed. (Reinhold, New York, 1956) p 287; electrolytic prepn: Nishihara *et al.*, **JP 60 2164** (1960), *C.A.* **55**, 5200e (1961).

Dodecahydrate. Chrome alum ammonium. Small dark-violet or violet-blue, octahedral, cubic crystals; ruby-red by transmitted light. mp 94°; loses 9H$_2$O on melting and the remaining H$_2$O by 300°. d^{25} 1.72. Readily sol in water; slightly sol in alcohol. Aq soln is violet when cold, green when hot.

USE: Mordant in textile industry; in manuf of electrolytic Cr metal.

509. Ammonium Citrate, Dibasic. [3012-65-5] 2-Hydroxy-1,2,3-propanetricarboxylic acid ammonium salt (1:2); diammonium citrate; citric acid diammonium salt. $C_6H_{14}N_2O_7$; mol wt 226.19. C 31.86%, H 6.24%, N 12.39%, O 49.51%. Prepn: Heldt, *Ann.* **47**, 157 (1843).

$$H_4N^+\ ^-OOC\diagdown\diagup COO^-\ ^+NH_4$$
$$HO\quad COOH$$

Granules or crystals; acid reaction. d 1.48. Sol in about 1 part water, slightly in alc. pH of 0.1M soln in H$_2$O = 4.3.

USE: For the determn of phosphate, especially in fertilizers.

510. Ammonium Cobaltous Phosphate. [14590-13-7] Phosphoric acid ammonium cobalt(2+) salt (1:1:1); cobaltous ammonium phosphate. CoH_4NO_4P; mol wt 171.94. Co 34.28%, H 2.35%, N 8.15%, O 37.22%, P 18.01%. NH_4CoPO_4. Prepd by reaction of a cobaltous salt with $(NH_4)_3PO_4$, $(NH_4)_2HPO_4$, or H_3PO_4 and NH_3: Grat-Cabanac, *Bull. Microsc. Appl.* **8**, 97 (1958); Salutsky *et al.*; McCullogh, Salutsky, US 3126254; US 3141732 (both 1964 to W. R. Grace).

Hydrate, red to violet powder, or monoclinic rectangular lamellae. Practically insol in water. Sol in acids.

USE: As pigment in ceramic glazes and vitreous enamels; as temp indicator in textile industry; in fertilizers for plant nutrition; in Co analysis.

511. Ammonium Cupric Chloride. [15610-76-1] Tetrachlorocuprate(2−) ammonium (1:2); ammonium chlorocuprate; cupric ammonium chloride; diammonium tetrachlorocuprate(2−). $Cl_4CuH_8N_2$; mol wt 241.42. Cl 58.74%, Cu 26.32%, H 3.34%, N 11.60%. $(NH_4)_2CuCl_4$. Prepd by evaporation of a 2:1 soln of NH_4Cl and $CuCl_2$: Chrobak, *Bull. Int. Acad. Polonaise* **1929A**, 361; *C.A.* **24**, 3688[7] (1930); Willet, *J. Chem. Phys.* **41**, 2243 (1964).

Yellow, hygroscopic, orthorhombic crystals. Sol in water. *Keep well closed.*

Dihydrate. Ammonium tetrachlorodiaquocuprate(II). Blue to bluish-green tetragonal, rhombododecahedral crystals. d 2.0. Becomes anhyd at 110-120°; decomposes on stronger heating. Sol in water, alcohol, liquid NH_3; the aq soln is acid to litmus.

USE: Analytical reagent; formerly for determining carbon in iron and steel.

512. Ammonium Dichromate(VI). [7789-09-5] Chromic acid ($H_2Cr_2O_7$) ammonium salt (1:2); ammonium bichromate. $Cr_2H_8N_2O_7$; mol wt 252.06. Cr 41.26%, H 3.20%, N 11.11%, O 44.43%. $(NH_4)_2Cr_2O_7$. Prepd from ammonium sulfate and sodium dichromate or by the interaction of ammonia gas and chromic acid in soln: W. H. Hartford, *Ind. Eng. Chem.* **41**, 1993 (1949); *Chromium Chemicals* Vol. **52** (Mutual Chem. Div., Allied Chem.) p. 34-39. *Review: ACS Monograph Series* **no. 132**, entitled "Chromium," vol. 1, M. J. Udy, Ed. (Reinhold, New York, 1956) *passim.* Laboratory information profile: J. A. Young, *J. Chem. Educ.* **82**, 1617 (2005).

Combustible solid. Strong oxidizer. Swells dramatically; closed containers may rupture violently when heated. Bright orange-red monoclinic prisms. Odorless and non-hygroscopic. d_4^{25} 2.155. Bulk density: 82 lbs/cu ft. Dec at about 180°. Decomposition becomes self-sustaining at about 225° with spectacular swelling and evolution of heat and nitrogen, leaving Cr_2O_3. Heat of soln −23.0 cal/g. Very sol in water. Soly in water (w/w): 15.16% (0°); 26.67% (20°); 36.99% (40°); 46.14% (60°); 54.20% (80°); 60.89% (100°). Acid reaction. A 1% soln has a pH of 3.95 and a 10% soln has a pH of 3.45.

Caution: Irritating to eyes, skin and respiratory sytem. May cause ulcers of the nasal passages and pulmonary edema. Chromium hexavalent (VI) compounds are listed as known human carcinogens: *Report on Carcinogens, Twelfth Edition* (PB2011-111646, 2011) p 106.

USE: Source of pure nitrogen (especially in the laboratory); in pyrotechnics (Vesuvius fire); in lithography and photo engraving; in special mordant, catalysts, and porcelain finishes; intermediate in the manuf of pigments; of magnetic recording materials. Oxidimentric standard in analytical chemistry.

513. Ammonium Dithiocarbamate. [513-74-6] Carbamodithioic acid ammonium salt (1:1); ammonium sulfocarbamate; dithiocarbamic acid monoammonium salt. $CH_6N_2S_2$; mol wt 110.19. C 10.90%, H 5.49%, N 25.42%, S 58.19%. NH_2CSSNH_4. Prepd from CS_2 and NH_3: Mathes, *Inorg. Synth.* **3**, 48 (1950); Redemann *et al.*, *Org. Synth.* **coll. vol. III**, 763 (1955); Gatlow, Hahnkamm, *Z. Anorg. Allg. Chem.* **364**, 161 (1969). Crystal structure: Cappuchi *et al.*, *Chem. Commun.* **1966**, 441.

Yellow, lustrous almost odorless, orthorhombic crystals when fresh. Undergoes a reversible, exothermic transition at 63°; mp 99° (dec). d_4^{20} 1.451. Sol in water. Decomposes in air, and is then no longer clearly sol; acquires an odor of H_2S; the decomposition products contain ammonium thiocyanate, ammonium sulfide, etc. *Keep in tightly closed bottles.*

USE: Instead of H_2S or $(NH_4)_2S$ for pptg metals in chemical analysis; synthesis of heterocyclic compounds.

514. Ammonium Ferric Oxalate. [14221-47-7] (*OC-6-11*)-Tris[ethanedioato(2−)-$\kappa O^1, \kappa O^2$]ferrate(3−) ammonium (1:3); ammonium trioxalatoferrate(III); ferric ammonium oxalate. $C_6H_{12}FeN_3O_{12}$; mol wt 374.02. C 19.27%, H 3.23%, Fe 14.93%, N 11.23%, O 51.33%. $(NH_4)_3Fe(C_2O_4)_3$. Prepn: *Gmelins, Iron* (8th ed.) **59**, part B, pp 1020-1021 (1932).

Hydrate, bright-green, monoclinic, prismatic crystals. $d^{17.5}$ 1.78. Affected by light. Loses 3 H_2O by 100°, decomp at 160-170°. Very sol in water; practically insol in alcohol. *Protect from light.*

USE: In photography, blueprints; in coloring of Al and Al alloys.

515. Ammonium Ferric Sulfate. [10138-04-2] Sulfuric acid ammonium iron(3+) salt (2:1:1); ferric ammonium sulfate; iron ammonium disulfate. $FeH_4NO_8S_2$; mol wt 266.00. Fe 20.99%, H 1.52%, N 5.27%, O 48.12%, S 24.11%. $NH_4Fe(SO_4)_2$. Prepn: *Gmelins, Iron* (8th ed.) **59**, part B, pp 1010-1018 (1932).

Dodecahydrate. [7783-83-7] Ferric alum; iron alum. $FeH_4NO_8S_2.12H_2O$; mol wt 482.18. Colorless to pale-violet, transparent, efflorescent, octahedral crystals. Odorless; acid styptic taste. mp ~37°; d 1.71. Very sol in water; practically insol in alcohol; pH of $0.1M$ aq soln 2.5.

USE: As analytical reagent. Indicator for argentimetric titrations. Mordant in dyeing and printing textiles.

THERAP CAT: Astringent, styptic.

516. Ammonium Ferricyanide. [14221-48-8] (*OC-6-11*)-Hexakis(cyano-κC)ferrate(3−) ammonium (1:3); ammonium hexacyanoferrate(III). $C_6H_{12}FeN_9$; mol wt 266.07. C 27.09%, H 4.55%, Fe 20.99%, N 47.38%. $(NH_4)_3Fe(CN)_6$. Prepn: *Gmelins, Iron* (8th ed.) **59**, part B, p 1027 (1932).

Trihydrate, red crystals. Freely sol in water; practically insol in alcohol. *Protect from light.*

517. Ammonium Ferrocyanide. [14481-29-9] (*OC-6-11*)-Hexakis(cyano-κC)ferrate(4−) ammonium (1:4); ammonium hexacyanoferrate (II). $C_6H_{16}FeN_{10}$; mol wt 284.11. C 25.37%, H 5.68%, Fe 19.66%, N 49.30%. $(NH_4)_4Fe(CN)_6$. Prepn: *Gmelins, Iron* (8th ed.) **59**, part B, p 1024 (1932); Lux in *Handbook of Preparative Inorganic Chemistry* Vol. 2, G. Brauer, Ed. (Academic Press, New York, 2nd ed., 1965) pp 1509-1510.

Trihydrate. Yellow cryst powder. Loses NH_3 on exposure to air and light. Decomposes on heating. Freely sol in water. Practically insol in alcohol. *Protect from light.*

518. Ammonium Ferrous Sulfate. [10045-89-3] Sulfuric acid iron(2+) ammonium salt (2:1:2); ammonium iron(II) sulfate; ferrous ammonium sulfate; iron(II) ammonium sulfate; Mohr's salt. $FeH_8N_2O_8S_2$; mol wt 284.04. Fe 19.66%, H 2.84%, N 9.86%, O 45.06%, S 22.57%. $(NH_4)_2Fe(SO_4)_2$. Manuf from Fe, H_2SO_4 and NH_3: Demmerle *et al.*, *Ind. Eng. Chem.* **42**, 9 (1950); from pickling waste: Brundin, US 2694657 (1954 to Ekstrand and Tholand). Toxicity study: H. F. Smyth *et al.*, *Am. Ind. Hyg. Assoc. J.* **30**, 470 (1969).

Hexahydrate. [7783-85-9] $FeH_8N_2O_8S_2.6H_2O$; mol wt 392.13. Pale blue-green crystals or crystalline powder. Slowly oxidizes and effloresces in air. d_4^{20} 1.86. Soluble in water. Practically insol in alcohol. *Keep well closed and protected from light.* LD_{50} orally in rats: 3.25 g/kg (Smyth).

USE: In photography; as analytical standard; as polymerization catalyst; in dosimeters.

519. Ammonium Fluoride. [12125-01-8] Ammonium fluoride ((NH_4)F); neutral ammonium fluoride. FH_4N; mol wt 37.04. F 51.29%, H 10.89%, N 37.82%. NH_4F. Prepd by passing ammonia gas into ice-cooled 40% hydrofluoric acid or by heating 1 part NH_4Cl with 2.25 parts NaF and separating the ammonium fluoride by sublimation: Kwasnik in *Handbook of Preparative Inorganic Chemistry* vol. 1, G. Brauer, Ed. (Academic Press, New York, 2nd ed., 1963) p 183. *Review:* Steele, "Ammonium Fluoride" in *Mellor's* vol. VIII, supplement I, *Nitrogen* (part 1), 370-377 (1964).

Deliquescent, hygroscopic leaflets or needles. Hexagonal prisms by sublimation. Occurs commercially as a granular powder. d 1.015. On heating dec into NH_3 and HF. *Poisonous, corrosive.*

Soly in water (0°): 100 g/100 ml. Decomposed by hot water into NH_3 and ammonium bifluoride. ($NH_4F.HF$). Cannot be obtained by evapn of its aq soln. Slightly sol in alcohol. The aq soln is acid. May be stored in iron vessels. *Incompat.* Quinine salts; soluble calcium salts.

Caution: Potential symptoms of overexposure by ingestion are nausea, salivation, vomiting, abdominal pain, diarrhea, hemorrhagic gastroenteritis, muscular weakness, tremors, convulsions, vascular collapse. Increased respiration is followed by depression, death. *Chronic toxicity:* Fluoride poisoning characterized by mottling of enamel, generalized osteosclerosis, calcification in tendons and ligaments; synostoses.

USE: Etching and frosting glass; extracting agent; as antiseptic in brewing beer; preserving wood; in printing and dyeing textiles; as mothproofing agent.

520. Ammonium Formate. [540-69-2] Formic acid ammonium salt (1:1). CH_5NO_2; mol wt 63.06. C 19.05%, H 7.99%, N 22.21%, O 50.74%. $HCOONH_4$. Prepd from formic acid and NH_3: Zuffanti, *J. Am. Chem. Soc.* **63**, 3123 (1941); from methylformate and NH_3: Kelly, Cuthbert, **US 3122584** (1964 to Allied Chem.).

Deliquesc crystals or granules. d 1.27. mp 116°. Sol in less than its own wt of water; sol in alc. *Keep tightly closed.*

USE: In chemical analysis, especially to ppt base metals from salts of the "noble" metals.

521. Ammonium Hexafluoroaluminate. [7784-19-2] (*OC-6-11*)-Hexafluoroaluminate(3−) ammonium (1:3); ammonium cryolite; ammonium aluminum fluoride; ammonium fluoaluminate. $AlF_6H_{12}N_3$; mol wt 195.09. Al 13.83%, F 58.43%, H 6.20%, N 21.54%. $(NH_4)_3AlF_6$. Prepn from ammonium fluoride and aluminum hydroxide: v. Helmolt, *Z. Anorg. Allg. Chem.* **3**, 127 (1893); Petersen, *J. Prakt. Chem.* [2] **40**, 55 (1889); Kwasnik in *Handbook of Preparative Inorganic Chemistry* **vol. 1**, G. Brauer, Ed, (Academic Press, New York, 2nd ed.) 1963) p 236.

Cubic crystals, d 1.78. Thermally stable to above 100°. Freely sol in water. Does not attack glass.

USE: Prepn of pure aluminum fluoride.

522. Ammonium Hexafluorogallate. [14639-94-2] (*OC-6-11*)-Hexafluorogallate(3−) ammonium (1:3); ammonium hexafluorogallate. $F_6GaH_{12}N_3$; mol wt 237.83. F 47.93%, Ga 29.32%, H 5.09%, N 17.67%. $(NH_4)_3GaF_6$. Prepd from ammonium fluoride and NH_4F: Hannebohn, Klemm, *Z. Anorg. Allg. Chem.* **229**, 341 (1936); Kwasnik in *Handbook of Preparative Inorganic Chemistry* **vol. 1**, G. Brauer, Ed., (Academic Press, New York, 2nd ed.) 1963) p 228.

Octahedra. On heating in air changes to Ga_2O_3; on heating *in vacuo* at 200° for several hours forms GaN.

USE: In the prepn of GaF_3.

523. Ammonium Hexafluorophosphate. [16941-11-0] Hexafluorophosphate(1−) ammonium (1:1); ammonium hexafluorophosphate; ammonium phosphorus hexafluoride. F_6H_4NP; mol wt 163.00. F 69.93%, H 2.47%, N 8.59%, P 19.00%. NH_4PF_6. Prepn: Lange, Müller, *Ber.* **63**, 1063 (1930); v. Krueger, *Ber.* **65**, 1265 (1932); Woyski, *Inorg. Synth.* **3**, 111 (1950); Kwasnik in *Handbook of Preparative Inorganic Chemistry* **Vol. 1**, G. Brauer, Ed. (Academic Press, New York, 2nd ed.) 1963) p 195.

Square leaflets or tables, seldom rectangular plates; cubic system. d_4^{18} 2.180. Decomposes on heating to a relatively high temp without prior melting. Soly in water (20°): 74.8 g/100 ml. Sol in acetone, methanol, ethanol, methyl acetate. Does not etch glass at room temp. Slowly hydrolyzed by boiling with strong acids.

524. Ammonium Hexafluorosilicate. [16919-19-0] Hexafluorosilicate(2−) ammonium (1:2); ammonium fluosilicate; ammonium silicofluoride. $F_6H_8N_2Si$; mol wt 178.15. F 63.99%, H 4.53%, N 15.72%, Si 15.76%. $(NH_4)_2SiF_6$. Occurs in nature as the mineral **cryptohalite**. Prepn: *Gmelins, Ammonium* (8th ed.) **23**, pp 414-415 (1936). Crystal structure: Schlemper *et al., J. Chem. Phys.* **44**, 2499 (1966); **45**, 408 (1966). Toxicity study: Simonin, Pierron, *C. R. Seances Soc. Biol. Ses Fil.* **124**, 133 (1937).

Odorless crystalline powder. Two modifications at room temp: stable, cubic phase; metastable, trigonal phase. Freely sol in water. Practically insol in alcohol, acetone. LD orally in guinea pigs: 150 mg/kg (Simonin, Pierron).

Caution: Toxic symptoms similar to sodium fluoride.

USE: In pesticides; in soldering flux; etching glass.

525. Ammonium Hypophosphite. [7803-65-8] Phosphinic acid ammonium salt (1:1). H_6NO_2P; mol wt 83.03. H 7.28%, N 16.87%, O 38.54%, P 37.30%. $NH_4H_2PO_2$. Prepn: *Gmelins, Ammonium* (8th ed.) **23**, 416 (1936).

Hygroscopic and deliquesc crystals or white granules. Decomposes when heated with evolution of phosphine which ignites spontaneously. One gram dissolves in about 1 ml water, 0.2 ml boiling water, 20 ml alcohol; freely sol in boiling alcohol, practically insol in acetone. The aq soln is practically neutral. *Keep well closed.*

USE: As catalyst in polyamide manuf.

526. Ammonium Iodide. [12027-06-4] Ammonium iodide ($(NH_4)I$). H_4IN; mol wt 144.94. H 2.78%, I 87.56%, N 9.66%. NH_4I. Prepd from ammonia, iodine and hydrogen peroxide; from ammonia and hydrogen iodide; or from ammonium carbonate and hydrogen iodide: Wulff, Cameron, *Z. Phys. Chem.* **B10**, 350 (1930); Schmeisser in *Handbook of Preparative Inorganic Chemistry* **vol. 1**, G. Brauer, Ed. (Academic Press, New York, 2nd ed.) 1963) p 289. *Review:* Richards "Ammonium Iodide" in *Mellor's* **vol. VIII**, Supplement I, *Nitrogen* (part 1) 448-458 (1964). IR reflection spectra at high pressure: A. H. Abdullah, W. F. Sherman, *Vib. Spectrosc.* **13**, 155 (1997). Phase diagram: S. Salihoglu *et al., Mater. Chem. Phys.* **73**, 339 (2002). Use as iodine source in oxyiodination reactions: K. V. V. Krishna Mohan *et al., Tetrahedron Lett.* **45**, 8015 (2004).

White, odorless, very hygroscopic, tetragonal crystals or granular powder, sharp saline taste. Becomes yellow to brown on exposure to air and light because of liberation of iodine. When heated it partly decomposes and partly sublimes. d^{25} 2.5142. One gram dissolves in 0.6 ml water, 0.5 ml boiling water, 3.7 ml alc, 1.5 ml glycerol, 2.5 ml methanol. Except in the presence of a stabilizer, such as ammonium hypophosphite, the aq soln will quickly become yellow. The aq soln is nearly neutral to litmus. pH of 0.1M soln about 4.6. *Keep tightly closed and protected from light.*

USE: In photochemical reactions.

527. Ammonium Lactate. [515-98-0] 2-Hydroxy-propanoic acid ammonium salt (1:1); DL-lactic acid ammonium salt. $C_3H_9NO_3$; mol wt 107.11. C 33.64%, H 8.47%, N 13.08%, O 44.81%. $CH_3CH(OH)COONH_4$. Prepd by neutralizing DL-lactic acid with NH_4OH: Costello, Filachione, *J. Am. Chem. Soc.* **75**, 1242 (1953).

Crystals from propanol, mp 91-94°. Sol in water, glycerol, 95% alc; slightly sol in methanol; practically insol in ethyl, *n*-propyl, iso-propyl, and *n*-butyl alcohols, ether, acetone, ethyl acetate. For a 78.8% by wt soln: n_D^{20} 1.4543, n_D^{25} 1.4536, n_D^{40} 1.4503; d_4^{20} 1.2006, d_4^{25} 1.1984, d_4^{40} 1.1904.

THERAP CAT (VET): Has been used for bovine ketosis.

528. Ammonium Magnesium Chloride. [39733-35-2] Magnesium ammonium chloride. Cl_3H_4MgN; mol wt 148.69. Cl 71.52%, H 2.71%, Mg 16.35%, N 9.42%. $MgNH_4Cl_3$.

Hexahydrate, deliquesc crystals. Sol in 6 parts water. *Keep well closed.*

USE: Prepn of magnesia mixture and anhydrous magnesium chloride.

529. Ammonium Mandelate. [530-31-4] α-Hydroxybenzeneacetic acid ammonium salt (1:1); mandelic acid ammonium salt. $C_8H_{11}NO_3$; mol wt 169.18. C 56.80%, H 6.55%, N 8.28%, O 28.37%. $C_6H_5CH(OH)COONH_4$. Prepd from mandelic acid and NH_3 or excess of strong NH_3-water: Tabern *et al., US 2220692* (1941 to Abbott); Baker, **US 2209314** (1941 to Squibb).

Very deliquescent, cryst powder; odorless or with slight odor; discolors in light. Very sol in water, sparingly in alc. The aq soln is slightly acid to litmus. *Keep tightly closed and protected from light.*

THERAP CAT: Anti-infective (urinary).

THERAP CAT (VET): Has been used as a urinary antiseptic.

530. Ammonium Molybdate(VI). [12027-67-7] Ammonium molybdate ($Mo_7O_{24}^{6-}$) (6:1); hexaammonium molybdate ($Mo_7O_{24}^{6-}$); ammonium heptamolybdate; ammonium paramolybdate. $H_{24}Mo_7N_6O_{24}$; mol wt 1163.86. H 2.08%, Mo 57.71%, N 7.22%, O 32.99%. $(NH_4)_6Mo_7O_{24}$. Prepn and structure: Guiter, *Compt. Rend.* **220**, 146 (1945); Lindqvist, *Acta Chem. Scand.* **2**, 88 (1948). Toxicity studies: L. T. Fairhall *et al.,* "The Toxicity of

Molybdenum," *U.S. Public Health Service Bulletin No. 293*, Washington D.C. (1945) 39 pp. Synthetic use as a catalyst in oxidation reactions: B. Sur *et al.*, *Synthesis* **1985**, 652; B. M. Choudary *et al.*, *Synlett* **1994**, 450; N. Ismail, R. N. Rao, *Chem. Lett.* **7**, 844 (2000). Use in colorimetric determns: S. D. Carson, *Anal. Biochem.* **75**, 472 (1976); M. M. Abdel-Khalek, M. S. Mahrous, *Talanta* **31**, 635 (1984); T. Aman *et al.*, *Anal. Lett.* **35**, 1007 (2002); L. H. Marcolino-Junior *et al.*, *ibid.* **38**, 2315 (2005).

Tetrahydrate. [12054-85-2] $H_{24}Mo_7N_6O_{24}.4H_2O$; mol wt 1235.92. Colorless or slightly greenish or yellowish crystals. Loses one water at 90°; dec 190°. Sol in 2.3 parts water. Practically insol in alcohol. pH of 5% aq soln 5.0 to 5.5.

USE: In photography and for decorating ceramics; chromogenic reagent in detection and determn of phosphates, arsenates, lead; also as a colorimetic reagent for alkaloids and many other substances. Oxidant in organic synthesis.

531. Ammonium Nitrate. [6484-52-2] Nitric acid ammonium salt (1:1); Norway saltpeter. $H_4N_2O_3$; mol wt 80.04. H 5.04%, N 35.00%, O 59.97%. NH_4NO_3. First prepd in 1659 by reaction of ammonium carbonate with nitric acid. Review of uses in explosives and fertilizers, and physical properties: K. D. Shah, A. G. Roberts in *Nitric Acid and Fertilizer Nitrates*, C. Keleti, Ed. (Marcel Dekker, New York, 1985) pp 165-169, 171-196; of industrial production: G. D. Honti, *ibid.* pp 197-223; of use as a solid propellant oxidizer: C. Oommen, S. R. Jain, *J. Hazard. Mater.* **A67**, 253-281 (1999).

Odorless, transparent, hygroscopic, deliquesc crystals or white granules, mp 169.6°. Dec begins above 170°. Five solid phases exist at normal pressure. Orthorhombic at room temp. *Oxidizer in hot concentrated solns. Closed containers may rupture violently when heated.* d 1.725. One gram dissolves in 0.5 ml water, 0.1 ml boil. water, about 20 ml alc, about 8 ml methanol, pH of 0.1*M* soln in water: 5.43. Sol in anhydrous and aq ammonia, aq nitric acid, acetic acid.

USE: Major component in manuf of industrial and military explosives; in matches, pyrotechnics. In fertilizers; freezing mixtures. In manuf of nitrous oxide (laughing gas). When dissolved in liq NH_3 (Divers soln), used to strip ammonia from gases. Clean burning propellant oxidizer.

532. Ammonium Oleate. [544-60-5] (9Z)-9-Octadecenoic acid ammonium salt (1:1); oleic acid ammonium salt; ammonia soap. $C_{18}H_{37}NO_2$; mol wt 299.50. C 72.19%, H 12.45%, N 4.68%, O 10.68%. $CH_3(CH_2)_7CH{=}CH(CH_2)_7COONH_4$. Prepd from oleic acid and excess 28-30% NH_3 soln: Stumpf, *Am. Paint J.* **38**, no. 45, 60, 64, 68 (1954), *C.A.* **48**, 12428b (1954).

Yellowish-brown paste, softens at 50-55°F, mp 70-72°F. At 80°F: sol in water; slightly sol in acetone, ethanol, methanol, benzene, CCl_4, xylene, naphtha. Sol in water at 212°F, acetone at 134°F, ethanol at 172°F, methanol at 148°F, benzene at 176°F, CCl_4 at 170°F, xylene at 180°F, naphtha at 160°F.

USE: Detergent, solidifying alcohol, demonstrating liquid crystals according to Lehmann.

533. Ammonium Osmium Chloride. [12125-08-5] (*OC*-6-11)-Hexachloroosmate(2−) ammonium (1:2); ammonium hexachloroosmate(IV); ammonium chloroosmate; osmium ammonium chloride. $Cl_6H_8N_2Os$; mol wt 439.01. Cl 48.45%, H 1.84%, N 6.38%, Os 43.33%. $(NH_4)_2OsCl_6$. Prepn: Dwyer, Hogarth, *Inorg. Synth.* **5**, 206 (1957).

Red powder, or dark red, octahedral crystals. Soluble in water or alcohol.

534. Ammonium Oxalate. [1113-38-8] Ethanedioic acid ammonium salt (1:2). $C_2H_8N_2O_4$; mol wt 124.10. C 19.36%, H 6.50%, N 22.57%, O 51.57%. $(NH_4)_2C_2O_4$. Commercial product is 98-99% pure. Prepn of monohydrate from aq oxalic acid and NH_3 or ammonium carbonate: Bérard, *Ann. Chim.* **73**, 277 (1810). *Review: Gmelins, Ammonium* (8th ed.) **23**, pp 400-407 (1936). Crystal structure of monohydrate: Robertson, *Acta Crystallogr.* **18**, 410, 417 (1965). Photodecomposition study: M. N. R. Nair, V. R. P. Verneker, *J. Phys. Chem.* **80**, 2552 (1976). Heat capacity determn: S. V. Dalidovich *et al.*, *Thermochim. Acta* **89**, 387 (1985). Dehydration of monohydrate: J. E. House, Jr., R. P. Ralston, *ibid.* **214**, 255 (1993). Physical properties of aqueous solns: H. Frej *et al.*, *J. Chem. Eng. Data* **45**, 415 (2000).

Monohydrate. [6009-70-7] $C_2H_8N_2O_4.H_2O$; mol wt 142.11. Orthorhombic, odorless crystals or granules. *Poisonous, combustible.* d$^{18.5}$ 1.50. One gram dissolves in 20 ml water, 2.6 ml boil water; slightly sol in alcohol. The aq soln is practically neutral. pH of 0.1*M* soln 6.4.

USE: Manuf explosives; electrolytic detinning of iron; in dyeing, metal polishes; for detection and determn of Ca, Pb, and rare earth metals.

535. Ammonium Palmitate. [593-26-0] Hexadecanoic acid ammonium salt (1:1); palmitic acid ammonium salt. $C_{16}H_{35}NO_2$; mol wt 273.46. C 70.28%, H 12.90%, N 5.12%, O 11.70%. CH_3-$(CH_2)_{14}COONH_4$. Prepn from palmitic acid and excess 28-30% NH_3 soln: Stumpf, *Am. Paint J.* **38**, No. 45, 60, 64, 68 (1954), *C.A.* **48**, 12428b (1954); from palmitic acid and ammonium carbonate: Reiling, **US 3053867** (1962 to Boston Chem. Prods.).

Yellow-white powder, softens at 38-40°F, mp 70-73°F. At 80°F: sol in water; slightly sol in benzene, xylene; practically insol in acetone, ethanol, methanol, CCl_4, naphtha. Sol in water at 212°F, acetone at 134°F, ethanol at 172°F, methanol at 148°F, benzene at 176°F, CCl_4 at 170°F, xylene at 180°F, naphtha at 160°F.

USE: Waterproofing fabrics, thickening lubricants.

536. Ammonium Paratungstate. [11120-25-5] Tungstate $(W_{12}(OH)_2O_{40}{}^{10-})$ ammonium (1:10); ammonium tungstate; ammonium tungstate(VI); decaammonium tungstate; APT. $H_{42}N_{10}O_{42}$-W_{12}; mol wt 3060.44. H 1.38%, N 4.58%, O 21.96%, W 72.08%. $(NH_4)_{10}H_2W_{12}O_{42}$. Prepn from sodium tungstate solution: R. L. Pilloton, **US 3077379** (1963 to Union Carbide). Crystal structure of decahydrate: R. Allmann, *Acta Crystallogr. B* **27**, 1393 (1971). Industrial prepn from scheelite and wolframite ore: E. Lassner, *Int. J. Refract. Met. Hard Mater.* **13**, 35 (1995). Crystallization and processing in tungsten manufacturing: J. W. van Put, *ibid.*, 61. Thermal decomposition studies: M. S. Marashi *et al.*, *ibid.* **30**, 177 (2012).

White powder. Crystallizes in several hydrated forms with the tetrahydrate being the most stable. *Irritant.* Dec above 300° to form tungsten trioxide. Freely sol in water; practically insol in alcohol.

USE: Manuf of tungsten alloys, tungstic acid, or tungsten metal powder.

537. Ammonium Pentachlorozincate. [14639-98-6] Pentachlorozincate(3−) ammonium (1:3); zinc ammonium chloride. $Cl_5H_{12}N_3Zn$; mol wt 296.78. Cl 59.72%, H 4.08%, N 14.16%, Zn 22.04%. $ZnCl_2.3NH_4Cl$. Prepn: Klug, Alexander, *J. Am. Chem. Soc.* **66**, 1056 (1944).

Hygroscopic, orthorhombic, bipyramidal crystals. d 1.81. Sublimes at 340° without melting if absolutely dry. Very sol in water.

USE: In manuf of dry cells; as flux for welding, soldering, galvanizing.

538. Ammonium Perchlorate. [7790-98-9] Perchloric acid ammonium salt (1:1). ClH_4NO_4; mol wt 117.49. Cl 30.17%, H 3.43%, N 11.92%, O 54.47%. NH_4ClO_4. Prepn: *Gmelins, Ammonium* (8th ed.) **23**, pp 196-200 (1936). Review of decompn and combustion: Jacobs, Whitehead, *Chem. Rev.* **69**, 551-590 (1969).

Orthorhombic crystals. Dec on heating. d 1.95. Freely sol in water; sol in methanol; slightly sol in ethanol, acetone; almost insol in ethyl acetate, ether.

USE: In explosives, pyrotechnic compositions, jet and rocket propellants.

539. Ammonium Peroxydisulfate. [7727-54-0] Peroxydisulfuric acid ([(HO)S(O)$_2$]$_2$O$_2$) ammonium salt (1:2); ammonium persulfate. $H_8N_2O_8S_2$; mol wt 228.19. H 3.53%, N 12.28%, O 56.09%, S 28.10%. $(NH_4)_2S_2O_8$. Available oxygen 7.01%. Contains, when recently made, 95 to 98% $(NH_4)_2S_2O_8$. Prepd by anodic oxidation of a satd $(NH_4)_2SO_4$ soln: Feher in *Handbook of Preparative Inorganic Chemistry* **Vol. 1**, G. Brauer, Ed. (Academic Press, New York, 2nd ed., 1963) p 390. Toxicity study: Smyth *et al.*, *Am. Ind. Hyg. Assoc. J.* **30**, 470 (1969).

Odorless platelike or prismatic (monoclinic) crystals, or white granular powder. Stable for months when pure and dry; dec in presence of moisture, gradually evolving ozone-contg oxygen; dec on heating, evolving O_2 and forming $(NH_4)_2S_2O_7$. d 1.98. Strong oxidizing agent. Freely sol in water; aq soln is acid and dec slowly at room temp and rapidly at higher temp evolving O_2 and forming

NH₄HSO₄. Keep dry, in a cool place and protected from organic matter. LD_{50} orally in rats: 820 mg/kg (Smyth).

USE: As oxidizer and bleacher; to remove hypo; reducer and retarder in photography; in dyeing, manuf aniline dyes; oxidizer for copper; etching zinc; decolorizing and deodorizing oils; electroplating; washing infected yeast; removing pyrogallol stains; making soluble starch; depolarizer in electric batteries; in analytical chemistry chiefly for detection and determn of manganese and iron.

540. Ammonium Phosphate, Dibasic. [7783-28-0] Phosphoric acid ammonium salt (1:2); secondary ammonium phosphate; diammonium hydrogen phosphate; Fyrex. $H_9N_2O_4P$; mol wt 132.06. H 6.87%, N 21.21%, O 48.46%, P 23.45%. $(NH_4)_2HPO_4$. The grade used medicinally is 98-99% pure. Prepn: Gmelins, Ammonium (8th ed.) **23**, pp 422-426 (1936).

Odorless crystals or cryst powder; saline, cooling taste; gradually loses about 8% NH_3 on exposure to air. One gram dissolves in 1.7 ml water, 0.5 ml boil water; practically insol in alcohol, acetone. pH about 8. Keep well closed.

USE: Fireproofing textiles, paper, wood, and vegetable fibers; impregnating lamp wicks; preventing afterglow in matches; flux for soldering tin, copper, brass, and zinc; purifying sugar; in yeast cultures; in dentifrices; in corrosion inhibitors; in fertilizers; in buffers.

541. Ammonium Phosphate, Monobasic. [7722-76-1] Phosphoric acid ammonium salt (1:1); ammonium biphosphate; ammonium dihydrogen phosphate; primary ammonium phosphate. H_6NO_4P; mol wt 115.02. H 5.26%, N 12.18%, O 55.64%, P 26.93%. $(NH_4)H_2PO_4$. Prepn: Gmelins, Ammonium (8th ed.) **23**, pp 426-429 (1936).

Odorless crystals or white, cryst powder; stable in air. mp 193.3°. bp 376.1° (dec). d 1.80. One gram dissolves in about 2.5 ml water; slightly sol in alcohol; practically insol in acetone. pH of 0.2 molar aq soln 4.2.

USE: In buffer solutions. As baking powder with sodium bicarbonate; in fermentations (yeast cultures, etc.); fireproofing of paper, wood, fiberboard, etc.

542. Ammonium Phosphite. [51503-61-8] $H_9N_2O_3P$; mol wt 116.06. H 7.82%, N 24.14%, O 41.36%, P 26.69%. Prepn: Gmelins, Ammonium (8th ed.) **23**, pp 416-417 (1936).

Monohydrate, deliquesc crystals. Sol in water. Keep tightly closed.

USE: As reducing agent; corrosion inhibitor for lubricating grease.

543. Ammonium Phosphomolybdate. [12026-66-3] Tetracosa-μ-oxododecaoxo[μ₁₂-[phosphato(3−)-κO:κO:κO:κO':κO':-κO':κO'':κO'':κO''':κO''':κO''']]dodecamolybdate(3−) ammonium (1:3); ammonium molybdophosphate; $H_{12}Mo_{12}N_3O_{40}P$; mol wt 1876.45. H 0.64%, Mo 61.36%, N 2.24%, O 34.10%, P 1.65%. $(NH_4)_3PO_4.12MoO_3$. Prepn and constitution: Illingworth, Keggin, J. Chem. Soc. **1935**, 575; Thistlewaite, Analyst **72**, 531 (1947); Healy, Radiochim. Acta **3**, 100 (1964).

Yellow, heavy, cryst powder. Solubility in water (20°): 0.2 ±0.1 g/l. Practically insol in nitric acid; sol in fixed alkali hydroxides.

USE: In phosphorus analysis; as cation-exchanger.

544. Ammonium Picrate. [131-74-8] 2,4,6-Trinitrophenol ammonium salt (1:1); picric acid ammonium salt; ammonium picronitrate; ammonium carbazoate; $C_6H_6N_4O_7$; mol wt 246.14. C 29.28%, H 2.46%, N 22.76%, O 45.50%. Prepn: Berl, Berl, US 2350322 (1944). Crystal structure: Maartmann-Moe, Acta Crystallogr. B **25**, 1452 (1969).

Bright yellow, bitter scales or orthorhombic crystals. d 1.72. Explosive. Soly in water at 20°: about 1 g/100 ml. Slightly sol in alc. "Red modification" is not a distinct polymorph, but a slightly contaminated form of the yellow salt: Mitchell, Bryant, J. Am. Chem. Soc. **65**, 128 (1943). Physiologic effects and protective measures in cases of exposure to ammonium picrate: Foulger, U.S. Armed Forces Med. J. **4**, 1425 (1953).

USE: In explosives, fireworks, rocket propellants.

545. Ammonium Platinic Chloride. [16919-58-7] (OC-6-11)-Hexachloroplatinate(2−) ammonium (1:2); ammonium hexachloroplatinate(IV); ammonium chloroplatinate; platinic ammonium chloride. $Cl_6H_8N_2Pt$; mol wt 443.86. Cl 47.92%, H 1.82%, N 6.31%, Pt 43.95%. $(NH_4)_2PtCl_6$.

Orange-red crystals or yellow powder. d 3.06. Slightly sol in water. Practically insol in alcohol.

USE: Platinum plating; manuf spongy platinum.

546. Ammonium Platinous Chloride. [13820-41-2] (SP-4-1)-Tetrachloroplatinate(2−) ammonium (1:2); ammonium tetrachloroplatinate(II); ammonium chloroplatinite; ammonium platinochloride; platinous ammonium chloride. $Cl_4H_8N_2Pt$; mol wt 372.96. Cl 38.02%, H 2.16%, N 7.51%, Pt 52.31%. $(NH_4)_2PtCl_4$.

Dark ruby-red crystals. Sol in water.

USE: In photography.

547. Ammonium Salicylate. [528-94-9] 2-Hydroxybenzoic acid ammonium salt (1:1); salicylic acid monoammonium salt; Salicyl-Vasogen. $C_7H_9NO_3$; mol wt 155.15. C 54.19%, H 5.85%, N 9.03%, O 30.94%. Prepd from salicylic acid and aq NH_3: Cahours, Ann. **52**, 336 (1844).

Odorless, lustrous crystals or white, cryst powder. Discolors on exposure to light; loses some NH_3 on long exposure to air. Readily discolored by iron compounds. One gram dissolves in 1 ml water, 3 ml alcohol. The aq soln is slightly acid. Keep in the dark and protected from contamination by iron.

THERAP CAT: Analgesic; topically to loosen psoriatic scales.

548. Ammonium Selenate. [7783-21-3] Selenic acid ammonium salt (1:2). $H_8N_2O_4Se$; mol wt 179.03. H 4.50%, N 15.65%, O 35.75%, Se 44.10%. $(NH_4)_2SeO_4$. Prepd by treating a soln of selenic acid with ammonia: Retgers, Z. Phys. Chem. **8**, 36 (1891); King, J. Phys. Chem. **41**, 797 (1937). Crystal structure: Gatlow, Acta Crystallogr. **15**, 419 (1962).

Monoclinic crystals. d_4^{20} 2.194. Decomposed by heat. 117 parts dissolve in 100 parts water at 7°, 197 parts in 100 parts at 160°; sol in glacial acetic acid. Insol in alcohol, acetone, ammonia.

549. Ammonium Selenite. [7783-19-9] Selenious acid diammonium salt. $H_8N_2O_3Se$; mol wt 163.04. H 4.95%, N 17.18%, O 29.44%, Se 48.43%. $(NH_4)_2SeO_3$. Prepd by dissolving selenious acid in a slight excess of concd aqueous ammonia and evaporating: Berzelius, Acad. Handl. Stockholm **39**, 13 (1818); Nilson, Bull. Soc. Chim. [2] **21**, 253 (1874); ibid. [2] **23**, 262 (1875).

White or slightly reddish crystals. Sol in water. Dec by heat. Deliquescent.

USE: In manuf red glass; reagent for alkaloids.

550. Ammonium Stearate. [1002-89-7] Octadecanoic acid ammonium salt (1:1); stearic acid ammonium salt. $C_{18}H_{39}NO_2$; mol wt 301.52. C 71.70%, H 13.04%, N 4.65%, O 10.61%. Prepn from stearic acid and excess 28-30% NH_3 soln: Stumpf, Am. Paint J. **38**, No. 45, 60, 64, 68 (1954), C.A. **48**, 12428b (1954); from stearic acid and ammonium carbonate: Reiling, US 3053867 (1962 to Boston Chem. Prod.).

Yellow-white powder, softens at 35-40°F, mp 70-75°F. At 80°F: sol in methanol, ethanol; slightly sol in water, benzene, xylene, naphtha; practically insol in acetone, CCl_4. Sol in water at 212°F, acetone at 134°F, ethanol at 172°F, methanol at 148°F, benzene at 176°F, CCl_4 at 170°F, xylene at 180°F, naphtha at 160°F.

USE: In vanishing creams, in waterproofing cements.

551. Ammonium Sulfamate. [7773-06-0] Sulfamic acid ammonium salt (1:1); AMS; Ammate. $H_6N_2O_3S$; mol wt 114.12. H 5.30%, N 24.55%, O 42.06%, S 28.09%. $NH_4SO_3NH_2$. Weed-killing prepn: Cupery, Tanberg, US 2277744 (1941 to du Pont). Prepn from ammonia and sulfamic acid: Sisler, Audrieth, *Inorg. Synth.* **2**, 180 (1946). Toxicity study: Ambrose, *J. Ind. Hyg. Toxicol.* **25**, 26 (1943).

Hygroscopic crystals (large plates). mp 131°, dec 160°. Extremely soluble in water, liquid NH_3; slightly soluble in ethanol; moderately soluble in glycerol, glycol, formamide. pH of 0.27M soln in H_2O = 4.9. Aq solns are stable to boiling. LD_{50} orally in rats: 3.0 g/kg (Ambrose).

Caution: Potential symptoms of overexposure are irritation of eyes, nose and throat; coughing, dyspnea. *See NIOSH Pocket Guide to Chemical Hazards* (DHHS/NIOSH 97-140, 1997) p 16.

USE: In the manuf of fire-retardant compositions, for flameproofing textiles and paper products; in the manuf of weed killing compositions; in electroplating solns; for the generation of nitrous oxide gas.

552. Ammonium Sulfate. [7783-20-2] Sulfuric acid ammonium salt (1:2); diammonium sulfate; mascagnite. $H_8N_2O_4S$; mol wt 132.13. H 6.10%, N 21.20%, O 48.43%, S 24.26%. $(NH_4)_2$-SO_4. Prepn: *Gmelins, Ammonium* (8th ed.) **23**, 261-280 (1936). Manuf: *Faith, Keyes & Clark's Industrial Chemicals*, F. A. Lowenheim, M. K. Moran, Eds. (Wiley-Interscience, New York, 4th ed., 1975) pp 103-108. *Review:* Call, "Ammonium Sulfate" in *Mellor's* **Vol. VIII**, Supplement I, *Nitrogen* (part 1) 473-505 (1964). Solubility studies: A. C. D. Rivett, *J. Chem. Soc.* **121**, 379 (1922); R. M. Caven, T. C. Mitchell, *ibid.* **125**, 1428 (1924).

Odorless, orthorhombic crystals or white granules. d 1.77. Dec above 280°. One g sol in about 1.5 ml water. Soly in water (g $(NH_4)_2SO_4$ per 100 g of satd soln): 41.22 (0°); 43.47 (25°); 50.42 (100°) (Rivett). Soly in water (g/100 g H_2O): 70.6 (0°); 76.7 (25°); 103.8 (100°) (Caven, Mitchell). Insol in alcohol, acetone. The pH of 0.1 molar aq soln 5.5.

USE: Manuf ammonia alum; in the manuf of H_2SO_4 to free it from nitrogen oxides; analytical uses; freezing mixtures, flameproofing fabrics and paper; manuf viscose silk; tanning, galvanizing iron; in fractionation of proteins; as fertilizer.

553. Ammonium Sulfide. [12135-76-1] Ammonium sulfide $((NH_4)_2S)$; true ammonium sulfide. H_8N_2S; mol wt 68.14. H 11.83%, N 41.11%, S 47.05%. $(NH_4)_2S$. Prepn: *Gmelins, Ammonium* (8th ed.) **23**, p 246 (1936). Prepn of high purity soln: Johnson *et al., Chemist-Analyst* **53**, 46 (1964).

Forms crystals below −18°; at high temps it dec into NH_4HS, NH_3, polysulfides, etc. *Corrosive. Poisonous. Flammable.*

USE: To apply patina to bronze, in photographic developers, in textile manufacture, in trace metal analysis.

554. Ammonium Sulfite. [10196-04-0] Sulfurous acid ammonium salt (1:2). $H_8N_2O_3S$; mol wt 116.14. H 6.94%, N 24.12%, O 41.33%, S 27.60%. $(NH_4)_2SO_3$. Prepn: *Gmelins, Ammonium* (8th ed.) **23**, pp 256-258 (1936).

Monohydrate. [7783-11-1] $H_8N_2O_3S.H_2O$; mol wt 134.15. Efflorescent crystals. Under the influence of air and heat loses all its water of crystallization and is gradually oxidized to $(NH_4)_2SO_4$. Sol in water; almost insol in alcohol, acetone. Its aq soln is alkaline to litmus. *Keep tightly closed and in a cool place.*

USE: In photography; as reducing agent; in bricks for blast-furnace linings; in lubricants for metal cold-working.

555. Ammonium Tetrachloroaluminate. [7784-14-7] Aluminum ammonium chloride; ammonium chloroaluminate. $AlCl_4$-H_4N; mol wt 186.82. Al 14.44%, Cl 75.90%, H 2.16%, N 7.50%. NH_4AlCl_4. Prepn from $AlCl_3$ and NH_4Cl: Friedman, Taube, *J. Am. Chem. Soc.* **72**, 2236 (1950).

mp 304°. Sol in water and ether.

USE: In the processing of furs.

556. Ammonium Tetrachlorozincate. [14639-97-5] (T-4)-Tetrachlorozincate(2−) ammonium (1:2). $Cl_4H_8N_2Zn$; mol wt 243.29. Cl 58.28%, H 3.31%, N 11.51%, Zn 26.89%. $(NH_4)_2ZnCl_4$. Prepd by dissolving 70 g $ZnCl_2$ and 30 g NH_4Cl in 29 ml hot water and crystallizing: Meerburg, *Z. Anorg. Allg. Chem.* **37**, 199 (1903); Wagenknecht, Juza in *Handbook of Preparative Inorganic Chemistry* **Vol. 1**, G. Brauer, Ed. (Academic Press, New York, 2nd ed., 1963) p 1072.

White, thin, shiny platelets. Orthorhombic bipyramidal. Hygroscopic. d 1.879. mp ~150° (dec). Sublimes at 341° without melting, if absolutely dry. Very sol in water with absorption of heat.

557. Ammonium Tetrathiomolybdate. [15060-55-6] (T-4)-Tetrathioxomolybdate(2−) ammonium (1:2). $H_8MoN_2S_4$; mol wt 260.27. H 3.10%, Mo 36.87%, N 10.76%, S 49.27%. $(NH_4)_2$-MoS_4. Forms a complex with copper *in vivo*, blocking intestinal absorption. Prepn: G. Krüss, *Ann.* **225**, 1 (1884); and characterization: J. W. McDonald *et al., Inorg. Chim. Acta* **72**, 205 (1983). Thermal decomposition study: R. I. Walton *et al., Chem. Mater.* **10**, 3737 (1998). Mechanism of action study: G. N. George *et al., J. Am. Chem. Soc.* **125**, 1704 (2003). Use in treatment of copper poisoning in sheep: W. R. Humphries *et al., Vet. Rec.* **123**, 51 (1988). Clinical evaluation in Wilson's disease: G. J. Brewer *et al., Arch. Neurol.* **51**, 545 (1994); **53**, 1017 (1996); **60**, 379 (2003). Review of therapeutic potential: G. J. Brewer, *J. Cell. Mol. Med.* **7**, 11-20 (2003).

uv max (H_2O): 241, 316, 467 nm (ε 24700, 16750, 11850).

USE: In prepn of molybdenum disulfide, *q.v.*, and other transition metal sulfide clusters.

THERAP CAT: Chelating agent (copper); Wilson's Disease treatment.

THERAP CAT (VET): Antidote (copper poisoning).

558. Ammonium Thiocyanate. [1762-95-4] Thiocyanic acid ammonium salt (1:1); ammonium rhodanide; ammonium sulfocyanate; ammonium sulfocyanide. CH_4N_2S; mol wt 76.12. C 15.78%, H 5.30%, N 36.80%, S 42.12%. NH_4SCN. Commercial grade is 98-99% pure. Prepn: *Gmelins, Ammonium* (8th ed.) **23**, pp 372-383 (1936). ^{13}C and ^{15}N NMR study: R. M. Dickson *et al., Can. J. Chem.* **65**, 941 (1987). Reagent in determn of carboxyl termini of peptides: G. Stark, *Biochemistry* **7**, 1796 (1968); L. D. Cromwell, G. R. Stark, *ibid.* **8**, 4735 (1969). Review of toxicology and human exposure: *Toxicological Profile for Cyanide* (PB2007-100674, 2006) 341 pp. Synthetic applications in thiocyanation reactions: N. N. Karade *et al., Synth. Commun.* **35**, 1197 (2005); G. Wu *et al., Tetrahedron Lett.* **46**, 5831 (2005); P. Srihari *et al., Synthesis* **2006**, 2772. Laboratory information profile: J. A. Young, *J. Chem. Educ.* **82**, 1619 (2005).

Deliquesc crystals. mp ~149°. Dec at 170°. Freely sol in water, ethanol; sol in methanol, acetone. Practially insol in $CHCl_3$, ethyl acetate. *Reacts violently with oxidizing agents, esp potassium chlorate. Protect from light.*

Caution: Potential symptoms of overexposure by ingestion are hallucinations and distorted perceptions, nausea or vomiting, and other gastrointestinal effects.

USE: In matches; double-dyeing fabrics; photography; improving and increasing strength of silks weighted with tin salts; producing grayish-black coating on Zn; manuf transparent artificial resins, thiourea; in pesticides. Detection and determn of Fe, Ag, Hg.

559. Ammonium Thiosulfate. [7783-18-8] Thiosulfuric acid $(H_2S_2O_3)$ ammonium salt (1:2); Ammonium hyposulfite. H_8-$N_2O_3S_2$; mol wt 148.20. H 5.44%, N 18.90%, O 32.39%, S 43.27%. $(NH_4)_2S_2O_3$. Prepn: *Gmelins, Ammonium* (8th ed.) **23**, pp 304-306 (1936). Crystal structure: S. T. Teng *et al., Acta Crystallogr.* **B35**, 1682 (1979).

Crystals, dec at 150°. Sol in water; insol in alc, ether.

USE: To clean "white" metal; in photography; in lubricants for metal cold-working.

560. Ammonium Titanium Oxalate. [10580-02-6] (SP-5-21)-Bis[ethanedioata(2−)-$\kappa O^1,\kappa O^2$]oxotitanate(2−) ammonium (1:2); ammonium oxalatotitanate(IV); ammonium bis(oxalato)oxotitanate; bis(oxalato)oxotitanate(2−) diammonium; titanium ammonium oxalate; titanyl ammonium oxalate. $C_4H_8N_2O_9Ti$; mol wt

275.98. C 17.41%, H 2.92%, N 10.15%, O 52.17%, Ti 17.34%. $(NH_4)_2TiO(C_2O_4)_2$.

Monohydrate, crystals or crystalline powder. Very sol in water.

USE: As mordant in dyeing cotton and leather.

561. Ammonium 12-Tungstophosphate. [12026-93-6] Tetracosa-μ-oxododecaoxo[μ12-[phosphato(3−)-κO:κO;κO;$\kappa O'$:$\kappa O'$:$\kappa O'$:$\kappa O''$:$\kappa O''$:$\kappa O''$:$\kappa O'''$:$\kappa O'''$:$\kappa O'''$]]dodecatungstate(3−) ammonium (1:3); triammonium hexatriacontaoxo[phosphato(3−)]dodecatungstate(3−); tungstophosphoric acid ($H_3PW_{12}O_{40}$) triammonium salt; triammonium dodecatungstophophate(3−); ammonium tungstophosphate; ammonium phosphowolframate; AWP. $H_{12}N_3$-$O_{40}PW_{12}$; mol wt 2931.13. H 0.41%, N 1.43%, O 21.83%, P 1.06%, W 75.26%. $(NH_4)_3PW_{12}O_{40}$. Prepd from phosphotungstic acid with ammonium nitrate or other ammonium salt. Chemistry of phosphotungstic acid salts: H. Wu, *J. Biol. Chem.* **43**, 189 (1920). Prepn and properties: T. V. Healy. *Radiochim. Acta* **2**, 146 (1964). Use in chromatographic separation of aromatic amines: L. Lepri *et al.*, *J. Chromatogr.* **207**, 29 (1981); of lower hydrocarbons: V. S. Nayak, *ibid.* **498**, 349 (1990). Prepn from various ammonium salts and surface properties: V. S. Nayak, J. B. Moffat, *J. Mol. Catal.* **80**, 75 (1993).

Trihydrate. [1311-90-6] $H_{12}N_3O_{40}PW_{12}.3H_2O$; mol wt 2985.18. Microcryst powder. Soly in water (20°): 0.15 g/l. Freely sol in fixed alkali hydroxide solns.

USE: As ion-exchanger; chemical reagent.

562. Ammonium Uranate(VI). [7783-22-4] Ammonium uranium oxide $((NH_4)_2U_2O_7)$; ammonium diuranate; uranic acid diammonium salt. $H_8N_2O_7U_2$; mol wt 624.13. H 1.29%, N 4.49%, O 17.94%, U 76.28%. $(NH_4)_2U_2O_7$. Intermediate in the production of nuclear fuel. Prepn: Miller, Armstrong, US 2466118 (1949 to USAEC); Tridot, *Ann. Chim.* [12] **5**, 358 (1950). Reaction optimization: A. Rodríguez *et al.*, *J. Radioanal. Nucl. Chem.* **177**, 279 (1994); and conversion to uranium oxides: S. Rajogopal *et al.*, *J. Nucl. Mater.* **227**, 300 (1996). Thermal studies: *idem et al.*, *J. Therm. Anal. Calorim.* **61**, 99 (2000).

Reddish-yellow, amorphous powder. Practically insol in water, alkalies. Sol in acids or in ammonium carbonate soln.

USE: Prepn of uranium dioxide for nuclear fuels.

563. Ammonium Uranium Carbonate. [18077-77-5] (*HB*-8-22-111′1′1″)-Tris[carbonato(2−)-κO,$\kappa O'$]dioxouranate(4−) ammonium (1:4); ammonium dioxotricarbonatouranate(VI); uranium ammonium carbonate; uranyl ammonium carbonate. C_3H_{16}-$N_4O_{11}U$; mol wt 522.21. C 6.90%, H 3.09%, N 10.73%, O 33.70%, U 45.58%. May contain $2H_2O$.

Yellow crystals; decomposed on exposure to air. Sol in water.

USE: In uranium yellow glazes.

564. Ammonium Uranium Fluoride. [18433-40-4] (*PB*-7-22-11111)-Pentafluorodioxouranate(3−) ammonium (1:3); ammonium dioxopentafluorouranate(VI); triammonium pentafluorodioxouranate; uranium ammonium fluoride; uranyl ammonium fluoride. $F_5H_{12}N_3O_2U$; mol wt 419.14. F 22.66%, H 2.89%, N 10.03%, O 7.63%, U 56.79%. $UO_2(NH_4)_3F_5$. Usually contains some water of crystallization. Prepn and crystal structure: Nguyen-Quy-Dao, *Bull. Soc. Chim. Fr.* **1968**, 3543; *idem*, *Acta Crystallogr.* **25B**, 67 (1969).

Greenish-yellow, monoclinic, cryst powder. Freely sol in water; practically insol in alcohol.

USE: Has been used in x-ray work because of its fluorescence under these rays.

565. Ammonium Vanadate(V). [7803-55-6] Vanadate (VO_3^{1-}) ammonium (1:1); ammonium metavanadate; ammonium trioxovanadate; vanadic acid (HVO_3) ammonium salt. H_4NO_3V; mol wt 116.98. H 3.45%, N 11.97%, O 41.03%, V 43.55%. NH_4-VO_3. Prepn: Baker *et al.*, *Inorg. Synth.* **3**, 117 (1950). Hydrothermal prepn by reaction of vanadium pentoxide with NH_4OH or $(NH_4)_2CO_3$: L. Kótai *et al.*, *Chem. Lett.* **35**, 384 (2006). Toxicity study: H. F. Smyth *et al.*, *Am. Ind. Hyg. Assoc. J.* **30**, 470 (1969). Thermal decompn study: A. Arya *et al.*, *Thermochim. Acta* **244**, 257 (1994). Use as an oxidant in determn of penicillins: E. A. Ibrahim *et al.*, *Talanta* **24**, 328 (1977); in determn of tricyclic antidepressants: W. Misiuk, *J. Pharm. Biomed. Anal.* **22**, 189 (2000).

White or slightly yellow, crystalline powder. mp ~200°. *Poisonous.* d 2.33. Sol in 165 parts water; more sol in hot water, in dil ammonia. Loses water and ammonia on heating. LD_{50} orally in rats: 0.16 g/kg (Smyth).

USE: In dyeing and printing on woolens; staining wood black; manuf vanadium black and "indelible ink"; producing vanadium luster on pottery; as photographic developer; in hematoxylin staining in microscopy; in analytical chemistry as an oxidant for spectrophotometric determns and as a reagent in combustion analysis of carbon, hydrogen and nitrogen. Catalyst in organic and inorganic synthesis. In prepn and purification of vanadium compounds.

566. Ammonium Zirconyl Carbonate. [32535-84-5] (*T*-4)-Tris[carbonato(2−)-κO]hydroxyzirconate(3−) ammonium (1:3); ammonium tricarbonatozirconate. $C_3H_{13}N_3O_{10}Zr$; mol wt 342.37. C 10.52%, H 3.83%, N 12.27%, O 46.73%, Zr 26.64%. $(NH_4)_3$-$ZrOH(CO_3)_3$. Prepd from $ZrO_2CO_2.8H_2O$, NH_4HCO_3 and $(NH_4)_2$-CO_3: Blumenthal, *J. Chem. Educ.* **39**, 604 (1962).

Dihydrate. Large prisms from water. Unstable in air, gradually evolving carbon dioxide and ammonia. Sol in water. Marketed as aq soln: d_{24}^{24} 1.238. The aq soln is moderately stable at room temp, although, after standing a month or so, it is likely to deposit hydrous zirconia. At temps above 60° aq solns dec rapidly.

USE: Water-repellent. Leaves a zirconium residue after the evaporation of ammonia and carbon dioxide, without leaving other cations or anions.

567. Amobarbital. [57-43-2] 5-Ethyl-5-(3-methylbutyl)-2,-4,6(1*H*,3*H*,5*H*)-pyrimidinetrione; 5-ethyl-5-isopentylbarbituric acid; 5-ethyl-5-isoamylbarbituric acid; barbamil; amylobarbitone; 5-iso-amyl-5-ethylbarbituric acid; pentymal; Amal; Amasust; Amytal; Eunoctal; Isomytal; Mylodorm; Sednotic; Stadadorm. $C_{11}H_{18}N_2O_3$; mol wt 226.28. C 58.39%, H 8.02%, N 12.38%, O 21.21%. Prepn: US 1514573 (1924). Metabolism: Frey, Magnussen, *Arzneim.-Forsch.* **16**, 612 (1966). Toxicity data: K. Irrgang, *ibid.* **15**, 688 (1965). Comprehensive description: N. A. A. Mian *et al.*, *Anal. Profiles Drug Subs.* **19**, 27-58 (1990).

Slightly bitter crystals, mp 156-158°. One gram dissolves in 1300 ml water, in 5 ml alc, in 17 ml chloroform, in 6 ml ether. Freely sol in benzene; sol in alkaline solns. Insol in petr ether, aliphatic hydrocarbons. A satd aq soln is acid to litmus paper. Dissolves in aq solns of alkali hydroxides and carbonates. pH of a saturated soln in water about 5.6. pKa (25°C) 8.0. LD_{50} in mice (mg/kg): 212 s.c. (Irrgang).

Sodium salt. [64-43-7] Sodium Amytal. $C_{11}H_{17}N_2NaO_3$; mol wt 248.26. Hygroscopic, friable granules of powder. Slightly bitter taste. Very soluble in water; sol in alcohol (1:1). Practically insol in ether, chloroform.

Note: This is a controlled substance (depressant): **21 CFR,** 1308.12 and 1308.13.

THERAP CAT: Sedative, hypnotic.

THERAP CAT (VET): Sedative, hypnotic.

568. Amocarzine. [36590-19-9] 4-Methyl-*N*-[4-[(4-nitrophenyl)amino]phenyl]-1-piperazinecarbothioamide; 4-nitro-4′-[(*N*-methylpiperazinyl)thiocarbonylamino]diphenylamine; CGP-6140. $C_{18}H_{21}N_5O_2S$; mol wt 371.46. C 58.20%, H 5.70%, N 18.85%, O 8.61%, S 8.63%. Orally active macrofilaricide; derivative of amoscanate, *q.v.* Prepn: BE 772053 (1971 to Agripat); R. Spaun *et al.*, US 3781290 (1973 to Ciba-Geigy). Mechanism of action study: K. P. Davies *et al.*, *Exp. Parasitol.* **68**, 382 (1989). HPLC determn in biological fluids: S. C. Bhatia *et al.*, *J. Chromatogr.* **434**, 288 (1988). Pharmacokinetics: J. B. Lecaillon *et al.*, *Br. J. Clin. Pharmacol.* **30**, 625, 629 (1990). Clinical trial in onchocerciasis: A. A. Poltera *et al.*, *Lancet* **337**, 583 (1991).

mp 191-196°. Sol in acetonitrile.

THERAP CAT: Anthelmintic (Nematodes).

569. Amodiaquin. [86-42-0] 4-[(7-Chloro-4-quinolinyl)-amino]-2-[(diethylamino)methyl]phenol; 4-[(7-chloro-4-quinolyl)-amino]-α-(diethylamino)-o-cresol; 7-chloro-4-(3-diethylamino-methyl-4-hydroxyanilino)quinoline; 7-chloro-4-(3-diethylamino-methyl-4-hydroxyphenylamino)quinoline; 4-(3'-diethylaminometh-yl-4'-hydroxyanilino)-7-chloroquinoline; SN-10751. $C_{20}H_{22}ClN_3$-O; mol wt 355.87. C 67.50%, H 6.23%, Cl 9.96%, N 11.81%, O 4.50%. Prepd from 4,7-dichloroquinoline and 4-acetamido-α-dieth-ylamino-o-cresol: Burckhalter *et al.*, *J. Am. Chem. Soc.* **70**, 1363 (1948); **US 2474819**; **US 2474821** (1949 to Parke, Davis). Alternate synthesis from 2-aminomethyl-p-aminophenol and 4,7-dichloro-quinoline: Natarajan, Lan, *Arzneim.-Forsch.* **22**, 1230 (1972). Spectral study: J. Kracmar *et al*, *Pharmazie* **29**, 773 (1974). HPLC determn in plasma, blood and urine: E. Pussard *et al.*, *J. Chroma-togr.* **374**, 111 (1986). Comprehensive description: I. Ahmad *et al.*, *Anal. Profiles Drug Subs.* **21**, 43-73 (1992).

Crystals from abs. ethanol, mp 208° (dec).

Dihydrochloride dihydrate. [6398-98-7] CAM-AQ1; Camo-quin; Flavoquine. $C_{20}H_{22}ClN_3O.2HCl.2H_2O$; mol wt 464.81. Yel-low, bitter crystals, dec 150-160°. uv max (methanol): 342 nm ($E_{1cm}^{1\%}$ 349); (water): 341.5 nm ($E_{1cm}^{1\%}$ 389); (0.1N HCl): 342 nm ($E_{1cm}^{1\%}$ 396). Sol in water; sparingly sol in alc; very slightly sol in benzene, chloroform, ether. pH of 1% aq soln 4.0 to 4.8.

Dihydrochloride hemihydrate. Yellow crystals from methanol, mp 243°. Slightly sol in water, alc.

THERAP CAT: Antimalarial.

570. Amorolfine. [78613-35-1] *rel*-(2R,6S)-4-[3-[4-(1,1-Di-methylpropyl)phenyl]-2-methylpropyl]-2,6-dimethylmorpholine; *cis*-4-[3-(4-*tert*-amylphenyl)-2-methylpropyl]-2,6-dimethylmorpho-line; (±)-*cis*-2,6-dimethyl-4-[2-methylpropyl-3-(*p-tert*-pentylphenyl)pro-pyl]morpholine; Ro-14-4767/000. $C_{21}H_{35}NO$; mol wt 317.52. C 79.44%, H 11.11%, N 4.41%, O 5.04%. Antimycotic morpholine derivative; inhibits fungal ergosterol biosynthesis. Prepn (unspeci-fied stereochem): A. Pfiffner, K. Bohnen, **DE 2752096**; A. Pfiffner, **US 4202894** (1978, 1980 both to Hoffmann-La Roche); of *cis*-form: **NL 8004537** (1980 to Hoffmann-La Roche). *In vitro* comparative antifungal spectrum: S. Shadomy *et al.*, *Sabouraudia* **22**, 7 (1984). Mechanism of action: A. Polak-Wyss *et al.*, *ibid.* **23**, 433 (1985); A. Polak, *Ann. N.Y. Acad. Sci.* **544**, 221 (1988). LC determn in phar-maceutical formulations: M. A. Czech *et al.*, *J. Pharm. Biomed. Anal.* **9**, 1019 (1991). Series of articles on mode of action and clini-cal trials: *Clin. Exp. Dermatol.* **17**, Suppl. 1, 1-70 (1992). Review of pharmacology and clinical efficacy: M. Haria, H. M. Bryson, *Drugs* **49**, 103-120 (1995).

Relative stereochemistry

bp$_{0.1}$ 120°.

Hydrochloride. [78613-38-4] Ro-14-4767/002; Loceryl. C_{21}-$H_{35}NO.HCl$; mol wt 353.98.

THERAP CAT: Antifungal (topical).

571. Amoscanate. [26328-53-0] 4-Isothiocyanato-N-(4-ni-trophenyl)benzenamine; isothiocyanic acid p-(p-nitroanilino)phenyl ester; 4-isothiocyanato-4'-nitrodiphenylamine; nithiocyamine; C-9333-Go; CGP-4540. $C_{13}H_9N_3O_2S$; mol wt 271.29. C 57.56%, H 3.34%, N 15.49%, O 11.79%, S 11.82%. Analog of nitroscanate, *q.v.* Prepn: K. Antos *et al.*, **DE 1932690** (1970 to Cesk. Akad. Ved), *C.A.* **72**, 100265 (1970); S. Rajappa *et al.*, *J. Chem. Soc. Perkin Trans. 1* **1979**, 2001; N. Viswanathan, R. C. Desai, *Indian J. Chem.* **20B**, 308 (1981). Anthelmintic activity: H. P. Striebel, *Experientia* **32**, 457 (1976); K. R. Middleton *et al.*, *ibid.* **35**, 243 (1979); H. G. Sen, B. N. Deb, *Am. J. Trop. Med. Hyg.* **30**, 992 (1981). HPLC determn in human plasma: W. M. Kofi-Tsekpo, C. W. Karekezi, *Drugs Exp. Clin. Res.* **14**, 31 (1988). Clinical pharmacology: A. B. Vaidya *et al.*, *Br. J. Clin. Pharmacol.* **4**, 463 (1977). Mutagenicity study: B. S. Reddy *et al.*, *Antimicrob. Agents Chemother.* **22**, 707 (1982). Brief review: J. I. Bruce, *Int. J. Parasitol.* **17**, 131-140 (1987).

Crystals from acetone, mp 196-198°.

THERAP CAT: Anthelmintic (Schistosoma).

572. Amosulalol. [85320-68-9] 5-[1-Hydroxy-2-[[2-(2-methoxyphenoxy)ethyl]amino]ethyl]-2-methylbenzenesulfonamide. $C_{18}H_{24}N_2O_5S$; mol wt 380.46. C 56.83%, H 6.36%, N 7.36%, O 21.03%, S 8.43%. Sulfonamide-substituted phenylethanolamine with α- and β-adrenergic blocking activity. Prepn: K. Imai *et al.*, **DE 2843016**; *eidem*, **US 4217305** (1979, 1980 both to Yamanouchi). Pharmacology and receptor blocking activity: T. Takenaka *et al.*, *Eur. J. Pharmacol.* **85**, 35 (1982); of isomers: K. Honda *et al.*, *J. Pharmacol. Exp. Ther.* **236**, 776 (1986). GC determn in urine: H. Kamimura *et al.*, *J. Chromatogr.* **275**, 81 (1983); HPLC determn in plasma: H. S. Gwak *et al.*, *J. Chromatogr. B* **818**, 109 (2005). Clin-ical pharmacokinetics: M. Nakashima *et al.*, *Clin. Pharmacol. Ther.* **36**, 436 (1984); metabolism: H. Kamimura *et al.*, *Xenobiotica* **15**, 413 (1985). Clinical evaluation in hypertension: K. Ando *et al.*, *J. Cardiovasc. Pharmacol.* **20**, 7 (1992).

Monohydrochloride. [93633-92-2] YM-09538; Lowgan. C_{18}-$H_{24}N_2O_5S.HCl$; mol wt 416.92. Colorless crystals, mp 158-160°. pK$_1$ 7.4; pK$_2$ 10.2.

R-(−)-**Form hydrochloride.** mp 158°. $[\alpha]_D^{20}$ −30.4° (c = 1 in methanol).

S-(+)-**Form hydrochloride.** mp 158°. $[\alpha]_D^{20}$ +30.7° (c = 1 in methanol).

THERAP CAT: Antihypertensive.

573. Amoxapine. [14028-44-5] 2-Chloro-11-(1-piperazin-yl)dibenz[*b,f*][1,4]oxazepine; CL-67772; Asendin; Asendis; Defan-yl; Demolox; Moxadil. $C_{17}H_{16}ClN_3O$; mol wt 313.79. C 65.07%, H 5.14%, Cl 11.30%, N 13.39%, O 5.10%. Prepn: J. Schmutz *et al.*, *Helv. Chim. Acta* **50**, 245 (1967); *eidem*, *Chim. Ther.* **2**, 424 (1967); C. F. Howell *et al.*, **FR 1508536**; *eidem*, **US 3663696** (1968, 1972 both to Am. Cyanamid). Analysis of amoxapine and its metab-olites by GLC: T. B. Cooper, R. G. Kelly, *J. Pharm. Sci.* **68**, 216 (1979). X-ray crystallography: D. B. Cosulich, F. M. Lovell, *Acta*

Crystallogr. **B33**, 1147 (1977). Pharmacology: E. N. Greenblatt *et al., Arch. Int. Pharmacodyn. Ther.* **233**, 107 (1978). Clinical study: R. Takahashi *et al., J. Int. Med. Res.* **7**, 7 (1979). *Review:* T. A. Ban, *Psychopharmacol. Bull.* **15**, 22-25 (1979); E. N. Greenblatt *et al.,* in *Pharmacological and Biochemical Properties of Drug Substances* **vol. 2**, M. E. Goldberg, Ed. (Am. Pharm. Assoc., Washington, DC, 1979) pp 1-18. Review of pharmacology and therapeutic efficacy: S. G. Jue *et al., Drugs* **24**, 1-23 (1982).

Crystals from benzene/petr ether, mp 175-176°. Freely sol in chloroform; sol in tetrahydrofuran; sparingly sol in methanol, toluene; slightly sol in acetone. Practically insol in water. LD_{50} in mice (mg/kg): 122 i.p.; 112 orally (Howell, 1972).

THERAP CAT: Antidepressant.

574. Amoxicillin. [26787-78-0] (2*S*,5*R*,6*R*)-6-[[(2*R*)-2-Amino-2-(4-hydroxyphenyl)acetyl]amino]-3,3-dimethyl-7-oxo-4-thia-1-azabicyclo[3.2.0]heptane-2-carboxylic acid; (−)-6-[2-amino-2-(*p*-hydroxyphenyl)acetamido]-3,3-dimethyl-7-oxo-4-thia-1-azabicyclo[3.2.0]heptane-2-carboxylic acid; 6-[D(−)-α-amino-*p*-hydroxyphenylacetamido]penicillanic acid; α-amino-*p*-hydroxybenzylpenicillin; 6-(*p*-hydroxy-α-aminophenylacetamido)penicillanic acid; *p*-hydroxyampicillin; amoxycillin; AMPC; Helvamox; Pasetocin; Penimox; Supramox; Zamocilline. $C_{16}H_{19}N_3O_5S$; mol wt 365.40. C 52.59%, H 5.24%, N 11.50%, O 21.89%, S 8.77%. Semisynthetic antibiotic related to penicillin. Prepn: Nayler, Smith, **GB 978178** (1964 to Beecham); *eidem,* **US 3192198** (1965); Long, Nayler, **DE 1942693** and **GB 1241844** (1970 and 1971 to Beecham), *C.A.* **72**, 90447q (1970). Resolution of isomers: Long *et al., J. Chem. Soc. C* **1971**, 1920. Series of articles on activity, pharmacology, absorption and excretion: *Antimicrob. Agents Chemother.* **1970**, 407-430. Review of antibacterial activity, pharmacokinetics and therapeutic use: R. N. Brogden *et al., Drugs* **18**, 169-184 (1979). Comprehensive description: A. E. Bird, *Anal. Profiles Drug Subs. Excip.* **23**, 1-52 (1994).

Trihydrate. [61336-70-7] BRL-2333; Agram; Alfamox; Amocilline; Amodex; Amoram; Amoxidin; Amoxil; Amoxillat; Amoxypen; Ardine; Bactox; Betamox; Bristamox; Cuxacillin; Flemoxine; Gramidil; Hiconcil; Larotid; Ospamox; Polymox; Robamox; Sigamopen; Trimox; Velamox; Widecillin; Wymox; Zimox. Off-white crystalline powder. $[\alpha]_D^{20}$ +246° (c = 0.1). uv max (ethanol): 230, 274 nm (ε 10850, 1400); (0.1*N* HCl): 229, 272 nm (ε 9500, 1080); (0.1*N* KOH): 248, Insol in benzene, carbon tetrachloride, chloroform.291 (ε 2200, 3000). Solubility (mg/ml): water 4.0; methanol 7.5; abs ethanol 3.4. Insol in hexane, benzene, ethyl acetate, acetonitrile, carbon tetrachloride, chloroform.

Sodium salt. [34642-77-8] Clamoxyl; Ibiamox. $C_{16}H_{18}N_3$-NaO_5S; mol wt 387.39.

THERAP CAT: Antibacterial.

THERAP CAT (VET): Antibacterial.

575. AMPA. [77521-29-0] α-Amino-2,3-dihydro-5-methyl-3-oxo-4-isoxazolepropanoic acid; α-amino-3-hydroxy-5-methyl-4-isoxazolepropionic acid. $C_7H_{10}N_2O_4$; mol wt 186.17. C 45.16%, H 5.41%, N 15.05%, O 34.38%. Synthetic excitatory amino acid that characterizes a specific subset of ionotropic glutamate receptors in the CNS, consequently known as AMPA-receptors. The activity re-

sides primarily in the L-isomer. Prepn: J. J. Hansen, P. Krogsgaard-Larsen, *J. Chem. Soc. Perkin Trans. 1* **1980**, 1826; M. Begtrup, F. A. Slok, *Synthesis* **1993**, 861. Characterization of neuroexcitatory activity: P. Krogsgaard-Larsen *et al., Nature* **284**, 64 (1980). Resolution of enantiomers and stereospecific activity: J. J. Hansen *et al., J. Med. Chem.* **26**, 901 (1983). Review of AMPA receptors: K. Borges, R. Dingledine, *Prog. Brain Res.* **116**, 153-170 (1998); and potential for pharmacological intervention: G. J. Lees, *Drugs* **59**, 33-78 (2000).

L-AMPA

Crystals from water as the monohydrate, mp 252° (dec).

L-AMPA. [83643-88-3] *S*-AMPA. Crystals from water + ethanol as the hydrate, gradual decomp above ~200°. $[\alpha]_D^{28}$ −21 ±2° (c = 0.19 in water).

576. Ampelopsin. [27200-12-0] (2*R*,3*R*)-2,3-Dihydro-3,-5,7-trihydroxy-2-(3,4,5-trihydroxyphenyl)-4*H*-1-benzopyran-4-one; 3,3',4',5,5',7-hexahydroxyflavanone; ampeloptin; dihydromyricetin. $C_{15}H_{12}O_8$; mol wt 320.25. C 56.26%, H 3.78%, O 39.97%. From leaves of *Ampelopsis meliaefolia* Kudo, *Vitaceae:* Kotake, Kubota, *Ann.* **544**, 253 (1940); from bark of *Pinus contoria* Dougl., *Pinaceae:* Hergert, *J. Org. Chem.* **21**, 534 (1956); **US 2870165** (1959 to Rayonier); from *Erythrophleum africanum* (Welw.) Harms, *Caesalpiniaceae:* Hansel, Klaffenbach, *Arch. Pharm.* **294**, 158 (1961). Synthesis from myricetin, *q.v.:* Kotake, Kubota, *J. Inst. Polytech. Osaka City Univ.* **1**, no. 2, 47 (1950), *C.A.* **46**, 2052e (1952).

Hemipentahydrate. Needles from water, mp 245-246°.
Hexaacetate. $C_{27}H_{24}O_{14}$. Needles from alc, mp 174-175°.
Hexabenzoate. $C_{57}H_{42}O_{14}$. Needles from alc, mp 174°.

577. Amperozide. [75558-90-6] 4-[4,4-Bis(4-fluorophenyl)butyl]-*N*-ethyl-1-piperazinecarboxamide; FG-5606. $C_{23}H_{29}F_2$-N_3O; mol wt 401.50. C 68.81%, H 7.28%, F 9.46%, N 10.47%, O 3.98%. Serotonin 5-HT_{2A} receptor antagonist. Prepn: A. K. K. Björk *et al.,* **DE 2941880**; *eidem,* **US 4308387** (1980, 1981 both to Ferrosan). LC-MS determn in swine liver: G. Balizs *et al., Analyst* **119**, 2687 (1994). Mechanism of action study: J. T. Haskins *et al., Brain Res. Bull.* **19**, 465 (1987). Anti-aggressive effects in pigs: J. L. Barnett *et al., Appl. Anim. Behav. Sci.* **50**, 121 (1996).

Hydrochloride. [86725-37-3] Hogpax. $C_{23}H_{29}F_2N_3O\cdot HCl$; mol wt 437.96. mp 177-178°.

THERAP CAT (VET): Tranquilizer.

578. Amphenone B. [2686-47-7] 3,3-Bis(4-aminophenyl)-2-butanone; amphenone; 2-oxo-3,3-bis[*p*-aminophenyl]butane.

$C_{16}H_{18}N_2O$; mol wt 254.33. C 75.56%, H 7.13%, N 11.01%, O 6.29%. Prepn: Allen, Corwin, *J. Am. Chem. Soc.* **72**, 117 (1950); US 2539388 (1951). Structure: Bencze, Allen, *J. Org. Chem.* **22**, 352 (1957). Shows antiestrogenic activity in the chick oviduct test: Hertz *et al.*, *Recent Prog. Horm. Res.* **11**, 119-147 (1955). Decreases adrenal action. *Review: Subsidia Med.* **10**, 99-102 (1958).

Crystals, mp 137.5-138°.
Dihydrochloride. $C_{16}H_{20}Cl_2N_2O$. Crystals from ethanol, dec 272-275°. Soluble in water.
Note: Formerly a pinacolone structure was assigned to amphenone B: *1,2-Bis[p-aminophenyl]-2-methyl-1-propanone.*
USE: In biological research.

579. Amphetamine. [300-62-9] α-Methylbenzeneethanamine; (±)-α-methylphenethylamine; 1-phenyl-2-aminopropane; β-phenylisopropylamine; β-aminopropylbenzene; (±)-desoxynorephedrine. $C_9H_{13}N$; mol wt 135.21. C 79.95%, H 9.69%, N 10.36%. Prepn: L. Edeleano, *Ber.* **20**, 616 (1887); F. P. Nabenhauer, US 1921424 (1933 to SK & F); W. H. Hartung, J. C. Munch, *J. Am. Chem. Soc.* **53**, 1875 (1931); N. Kornblum, D. C. Iffland, *ibid.* **71**, 2137 (1949). Demonstration of binding sites in hypothalamic membranes: S. M. Paul *et al.*, *Science* **218**, 487 (1982). Toxicity data: M. R. Warren, H. W. Werner, *J. Pharmacol. Exp. Ther.* **85**, 119 (1945); W. A. Behrendt, R. Deininger, *Arzneim.-Forsch.* **13**, 711 (1963). Books: C. D. Leake, *The Amphetamines: Their Actions and Uses* (Thomas, Springfield, 1958) 167 pp; O. J. Kalant, *The Amphetamines: Toxicity and Addiction* (Thomas, Springfield, 1966) 151 pp. Review of use and abuse: J. P. Morgan, *Substance Abuse: Clinical Problems and Perspectives*, J. H. Lowinson, P. Ruiz, Eds. (Williams & Wilkins, Baltimore, 1981) pp 167-184. Review of pharmacology, behavioral effects and therapeutic uses: L. S. Seiden *et al.*, *Annu. Rev. Pharmacol. Toxicol.* **32**, 639-677 (1993). Review of HPLC determns in biological samples: P. Campins-Falco *et al.*, *J. Liq. Chromatogr.* **17**, 731-747 (1994).

Mobile liquid. Amine odor. Acrid, burning taste. Volatilizes slowly at room temp. d_4^{25} 0.913. bp_{760} 200-203°; bp_{13} 82-85°. Slightly soluble in water; sol in alc, ether; readily sol in acids. Aq solns are alkaline to litmus. LD_{50} in rats (mg/kg): 180 s.c. (Warren, Werner).
Sulfate. [60-13-9] Benzedrine. $(C_9H_{13}N)_2 \cdot H_2SO_4$; mol wt 368.49. White, odorless crystalline powder with slightly bitter taste. mp above 300° (dec). Freely sol in water (1:9); slightly sol in alcohol (about 1:500). Practically insol in ether. A soln of 1 g/10 ml water has a pH 5-6. LD_{50} in mice, rats (mg/kg): 24.2, 55 orally (Behrendt, Deininger).
Phosphate. [139-10-6] Actemin; Aktedron. $C_9H_{13}N \cdot H_3PO_4$; mol wt 233.20. Prepn: Goggin, US 2507468 (1950 to Clark & Clark). Crystals bitter taste. Sinters at about 150°. Dec around 300°. Freely sol in water. Slightly sol in alcohol. Practically insol in benzene, chloroform, ether. The pH of a 10% soln is about 4.6.
d-Form see Dextroamphetamine.
l-Form. [156-34-3] Levamphetamine; levamfetamine.
Note: This is a controlled substance (stimulant): **21 CFR**, 1308.12.
THERAP CAT: CNS stimulant; anorexic.

580. Amphetaminil. [17590-01-1] α-[(1-Methyl-2-phenylethyl)amino]benzeneacetonitrile; N-(α-methylphenethyl)-2-phenylglycinonitrile; α-phenyl-α-(β-phenylisopropylamino)acetonitrile; α-phenyl-α-(1-methyl-2-phenyl)ethylaminoacetonitrile; α-phenyl-α-N-(1-phenylisopropyl)aminoacetonitrile; AN 1; Aponeuron. C_{17}-

$H_{18}N_2$; mol wt 250.35. C 81.56%, H 7.25%, N 11.19%. Prepd by reaction of DL-β-phenylisopropylamine with sodium cyanide and benzaldehyde or with α-phenyl-α-bromoacetonitrile: Klosa, **DE 1112987** (1959), *C.A.* **56**, 3409d (1962); *idem*, *J. Prakt. Chem.* **20**, 275 (1963). Pharmacology: Dominok, Oelssner, *Acta Biol. Med. Ger.* **20**, 625 (1968); Beyer *et al.*, *Dtsch. Apoth. Ztg.* **111**, 677, 680 (1971). Metabolic studies: Remberg *et al.*, *Arch. Toxicol.* **29**, 153 (1972). Chemistry: Beyrich *et al.*, *Pharmazie* **27**, 28 (1972); Gloeckl, Beyrich, *ibid.* 95.

Crystals from ethanol-water, mp 85-87°.
Hydrochloride. $C_{17}H_{19}ClN_2$. Sinters at 100-104°, mp 134-136°.
THERAP CAT: Psychotropic.

581. Amphomycin. [1402-82-0] Amfomycin; glumamycin. $C_{58}H_{91}N_{13}O_{20}$; mol wt 1290.44. C 53.98%, H 7.11%, N 14.11%, O 24.80%. Polypeptide antibiotic modified at the amino terminus with a fatty acid moiety; active against gram positive bacteria. Produced by *Streptomyces canus* from soil collected near Syracuse, N.Y.: B. Heinemann *et al.*, *Antibiot. Chemother.* **3**, 1239 (1887). Production: *eidem*, US 3126317 (1964 to Bristol-Myers). Structure and identity with glumamycin: M. Bodanszky *et al.*, *J. Am. Chem. Soc.* **95**, 2352 (1973). Pharmacology and toxicity: D. E. Tisch *et al.*, *Antibiot. Annu.* **1954-1955**, 1011. Mechanism of action: H. Tanaka *et al.*, *Biochem. Biophys. Res. Commun.* **86**, 902 (1979). Use to improve feed efficiency in ruminant animals: M. Gordon, G. J. Christie, **DE 3027370**; *eidem*, US 4414206 (1981, 1983 both to Bristol-Myers).

R–Asp–MeAsp–Asp–Gly–Asp–Gly–Dab^e–Val–Pro
                                                    |
                              —Pip—Dab^t—

Dab^e = D-*erythro*-α,β–diaminobutyric acid
Dab^t = L-*threo*-α,β–diaminobutyric acid
Pip   = D-pipecolic acid
R   = (+)-3-anteisotridecenoic acid or (+)-3-isododecenoic acid

Acidic, surface-active polypeptide. $[\alpha]_D^{25}$ +7.5° (c = 1 at pH 6). Isoelectric point 3.5-3.6. Soluble in water and the lower alcohols. Insol in nonpolar solvents. Aq solns at neutral pH are stable at room temp for at least one month. Forms sodium and calcium salts. Induces hemolysis. Suggested as a topical agent for animal and plant infections.
Sodium salt. Amorphous solid, sol in water. LD_{50} in mice (mg/kg): 177.8 i.v. (Tisch).
Calcium salt. Crystalline solid, sol in water, methyl alcohol. Used in combination with other antibiotics as antibacterial ingredient in topical anti-inflammatory preparations. LD_{50} in mice (mg/kg): 120.2 i.v. (Tisch).
THERAP CAT: Antibacterial.

582. Amphotericin B. [1397-89-3] Fungizone; Fungilin; Ampho-Moronal. $C_{47}H_{73}NO_{17}$; mol wt 924.09. C 61.09%, H 7.96%, N 1.52%, O 29.43%. Polyene antibiotic produced by *Streptomycetes nodosus* M4575 obtained from soil of the Orinoco river region of Venezuela: Gold *et al.*, *Antibiot. Annu.* **1955-1956**, 579; Vandeputte *et al.*, *ibid.* 587; Dutcher *et al.*, *ibid.* **1956-1957**, 866; Walters *et al.*, *J. Am. Chem. Soc.* **79**, 5076 (1957); Dutcher *et al.*, **US 2908611** (1959 to Olin Mathieson). Structure studies: Borowski *et al.*, *Tetrahedron Lett.* **6**, 473 (1965). Carbon skeleton, ring size, and partial structure: Cope *et al.*, *J. Am. Chem. Soc.* **88**, 4228 (1966). Complete structure: Mechlinski *et al.*, *Tetrahedron Lett.* **11**, 3873 (1970); Borowski *et al.*, *ibid.* 3909; R. C. Pandey, K. L. Rinehart, *J. Antibiot.* **29**, 1035 (1976). Total synthesis: K. C. Nicolaou *et al.*, *J. Am. Chem. Soc.* **109**, 2821 (1987). Toxicity: G. R. Keim *et al.*, *Science* **179**, 584 (1973). Comprehensive description: I. M. Asher *et al.*, *Anal. Profiles Drug Subs.* **6**, 1-42 (1977). Review of clinical experience: H. A. Gallis *et al.*, *Rev. Infect. Dis.* **12**, 308-329 (1990); of lipid formulations: J. W. Hiemenz, T. J. Walsh, *Clin. Infect. Dis.* **22**, Suppl. 2, S133-S144 (1996). *Review:* K. M. Abu-Salah, *Br. J. Biomed. Sci.* **53**, 122-133 (1996).

Deep yellow prisms or needles from DMF. Dec gradually above 170°. uv max (methanol): 406, 382, 363, 345 nm. $[\alpha]_D^{24}$ +333° (acidic DMF); −33.6° (0.1N methanolic HCl). Sol in propylene glycol; slightly sol in methanol. Insol in anhydrous alcohol, ether, benzene, toluene. Insol in water at pH 6 to 7. Soly at pH 2 or pH 11 in water: about 0.1 mg/ml. Water soly increased by sodium desoxycholate. Soly in DMF 2 to 4 mg/ml; in DMF + HCl: 60 to 80 mg/ml; in DMSO: 30 to 40 mg/ml. Solids and solns appear stable for long periods between pH 4 and 10 when stored at moderate temps out of contact with light and air. LD$_{50}$ in mice (mg/kg): 88 i.p., 4 i.v. (Keim).

Compd with cholesteryl sulfate. [120895-52-5] Amphotericin B compd with (3β)-cholest-5-en-3-yl hydrogen sulfate (1:1); amphotericin B colloidal dispersion; ABCD; Amphocil. Reviews: L. S. S. Guo, P. K. Working, *J. Liposome Res.* **3**, 473-490 (1993); D. A. Stevens, *J. Infect.* **28**, Suppl.1, 45-49 (1994). Forms uniform discshaped particles in aq media.

Liposomal complex. AmBisome. Unilamellar liposomes containing phosphatidylcholine, cholesterol, distearoylphosphatidylglycerol and amphotericin B in a molar ratio of 2:1:0.8:0.4. Review: R. J. Hay, *J. Infect.* **28**, Suppl. 1, 35-43.

Amphotericin B lipid complex. ABLC; Abelcet. Complex with 2 phospholipids, L-α-dimyristoyl phosphatidylcholine and L-α-dimyristoyl phosphatidylglycerol, which are present in a 7:3 molar ratio. Review: J. Lister, *Eur. J. Haematol. Suppl.* **57**, 18-23 (1996).

THERAP CAT: Antifungal.

THERAP CAT (VET): Antifungal.

583. Ampicillin. [69-53-4] (2S,5R,6R)-6-[[(2R)-2-Amino-2-phenylacetyl]amino]-3,3-dimethyl-7-oxo-4-thia-1-azabicyclo-[3.2.0]heptane-2-carboxylic acid; 6-[D(−)-α-aminophenylacetamido]penicillanic acid; D(−)-α-aminobenzylpenicillin; AY-6108; BRL-1341; P-50; Albipen; Amfipen; Ampitab; Amplital; Omnipen; Pénicline. C$_{16}$H$_{19}$N$_3$O$_4$S; mol wt 349.41. C 55.00%, H 5.48%, N 12.03%, O 18.32%, S 9.18%. Orally active, semi-synthetic antibiotic; structurally related to penicillin. Prepn: Doyle *et al.*, US 2985648 (1961); *eidem*, GB 902703 (1962 to Beecham); Doyle *et al.*, *J. Chem. Soc.* **1962**, 1440. Prepn of the trihydrate: D. A. Johnson, G. A. Hardcastle, US 3157640 (1964 to Bristol-Myers); of the anhydrous crystalline form: N. H. Grant, H. E. Alburn, US 3144445 (1964 to Am. Home Prods.). Alternate syntheses: Dane, Dockner, *Ber.* **98**, 789 (1965); F. Kajfez *et al.*, *J. Heterocycl. Chem.* **13**, 561 (1976). LC determn: M. Margosis, *J. Assoc. Off. Anal. Chem.* **70**, 206 (1987). Series of articles on pharmacology and antibacterial activity: *Br. Med. J.* **2**, 193-206 (1961). Comprehensive description: E. Ivashkiv, *Anal. Profiles Drug Subs.* **2**, 1-61 (1973). Review of pharmacology and clinical efficacy in combination with sulbactam, *q.v.*: D. M. Campoli-Richards, R. N. Brogden, *Drugs* **33**, 577-609 (1987).

Crystals, dec 199-202°. $[\alpha]_D^{23}$ +287.9° (c = 1 in water). Slightly sol in water, methanol. Insol in benzene, carbon tetrachloride, chloroform.

Monohydrate. Crystals from water, dec 202°. $[\alpha]_D^{21}$ +281° (water). Sparingly sol in water at room temp.

Sesquihydrate. Dec 199-202°. $[\alpha]_D^{20}$ +283.1° (water).

Trihydrate. [7177-48-2] Amblosin; Amipenix; Ampilar; Cymbi; Doktacillin; Penbritin; Polycillin; Principen; Rosampline; Totapen; Vidopen.

Sodium salt. [69-52-3] Ampicin; Binotal; Cilleral; Penbristol; Pentrex; Pentrexyl; Polycillin-N; Synpenin; Totacillin N; Viccillin. C$_{16}$H$_{18}$NaN$_3$O$_4$S; mol wt 371.39. White crystalline powder. Hygroscopic. Very sol in water, isotonic sodium chloride, dextrose solutions.

L(+)-Form. Crystals, dec at about 205°. $[\alpha]_D^{20}$ +209° (c = 0.2 in water). Less active as an antibiotic than D(−)-form.

THERAP CAT: Antibacterial.

THERAP CAT (VET): Antibacterial.

584. Ampiroxicam. [99464-64-9] Carbonic acid ethyl 1-[[2-methyl-1,1-dioxido-3-[(2-pyridinylamino)carbonyl]-2H-1,2-benzothiazin-4-yl]oxy]ethyl ester; carbonic acid ethyl 1-[[2-methyl-3-[(2-pyridinylamino)carbonyl]-2H-1,2-benzothiazin-4-yl]oxy]ethyl ester S,S-dioxide; (±)-4-(1-hydroxyethoxy)-2-methyl-N-2-pyridyl-2H-1,2-benzothiazine-3-carboxamide ethyl carbonate (ester) 1,1-dioxide; 4-[1-(ethoxycarbonyloxy)ethoxy]-2-methyl-N-(2-pyridyl)-2H-1,2-benzothiazine-3-carboxamide 1,1-dioxide; CP-65703; Flucam. C$_{20}$H$_{21}$N$_3$O$_7$S; mol wt 447.46. C 53.69%, H 4.73%, N 9.39%, O 25.03%, S 7.16%. Prodrug of piroxicam, *q.v.* Prepn: A. Marfat, EP 147177; *idem*, US 4551452 (both 1985 to Pfizer). Clinical pharmacokinetics: F. C. Falkner *et al.*, *Xenobiotica* **20**, 645 (1990). Toxicity study: M. Iijima *et al.*, *Oyo Yakuri* **43**, 1 (1992); *C.A.* **116**, 187689k (1992). Pharmacology: T. J. Carty *et al.*, *Agents Actions* **39**, 157 (1993). Clinical trial in oral postoperative pain: K. Kurita *et al.*, *Oral. Ther. Pharmacol.* **10**, 138 (1991); in rheumatoid arthritis: S. Irimajiri *et al.*, *Jpn. J. Inflammation* **12**, 81 (1992).

White crystals from toluene, mp 159-161°. LD$_{50}$ orally in male, female rats: 1798, 747 mg/kg (Iijima).

THERAP CAT: Anti-inflammatory.

585. Amprenavir. [161814-49-9] N-[(1S,2R)-3-[[(4-Aminophenyl)sulfonyl](2-methylpropyl)amino]-2-hydroxy-1-(phenylmethyl)propyl]carbamic acid (3S)-tetrahydro-3-furanyl ester; 4-amino-N-((2syn,3S)-2-hydroxy-4-phenyl-3-((S)-tetrahydrofuran-3-yloxycarbonylamino)-butyl)-N-isobutylbenzene sulfonamide; 141W94; KVX-478; VX-478; Agenerase; Prozei. C$_{25}$H$_{35}$N$_3$O$_6$S; mol wt 505.63. C 59.39%, H 6.98%, N 8.31%, O 18.99%, S 6.34%. Peptidomimetic HIV protease inhibitor. Prepn: R. D. Tung *et al.*, WO 9405639; *eidem*, US 5585397 (1994, 1996 both to Vertex). Crystal structure of complex with HIV protease: E. E. Kim *et al.*, *J. Am. Chem. Soc.* **117**, 1181 (1995). Review of design strategy: M. A. Navia *et al.*, *Int. Antiviral News* **3**, 143-145 (1995); of pharmacology and clinical evaluation: J. C. Adkins, D. Faulds, *Drugs* **55**, 837-842 (1998).

White to cream-colored solid. Soly in water (25°): 0.04 mg/ml.

THERAP CAT: Antiretroviral.

586. Amprolium. [121-25-5] 1-[(4-Amino-2-propyl-5-pyrimidinyl)methyl]-2-methylpyridinium chloride (1:1); 1-[(4-amino-2-propyl-5-pyrimidinyl)methyl]-2-picolinium chloride; Corid. $C_{14}H_{19}ClN_4$; mol wt 278.78. C 60.32%, H 6.87%, Cl 12.72%, N 20.10%. Prepn: Rogers et al., J. Am. Chem. Soc. **82**, 2974 (1960); Rogers, Sarett, US **3020277** (1962 to Merck & Co.), see also US **3020200**. Formulations as poultry feed: Rogers, Sarett, US **3065132** (1962 to Merck & Co.).

Hydrochloride. Amprol. $C_{14}H_{19}ClN_4.HCl$; mol wt 315.24. Crystals from methanol + ethanol, dec 248-249°. Freely sol in water, methanol, 95% ethanol, dimethylformamide. Sparingly sol in abs ethanol. Practically insol in isopropanol, butanol, dioxane, acetone, ethyl acetate, acetonitrile, isooctane. pH of 10% aq soln 2.5-3.0.

THERAP CAT (VET): Coccidiostat.

587. Amrinone. [60719-84-8] 5-Amino-[3,4'-bipyridin]-6(1H)-one; 3-amino-5-(4-pyridinyl)-2(1H)-pyridinone; Win-40680; Cartonic; Inocor; Vesistol; Wincoram. $C_{10}H_9N_3O$; mol wt 187.20. C 64.16%, H 4.85%, N 22.45%, O 8.55%. Selective cAMP phosphodiesterase (PDE-3) inhibitor with positive inotropic and vasodilatory activity. Prepn: G. Y. Lesher, C. J. Opalka, US **4004012**; US **4072746** (1977, 1978 both to Sterling). Clinical hemodynamic assessment: J. R. Benotti et al., N. Engl. J. Med. **21**, 1373 (1978). Review of pharmacology and clinical efficacy: A. Ward et al., Drugs **26**, 468 (1983); M. B. Bottorff et al., Pharmacotherapy **5**, 227-237 (1985). Series of articles on pharmacology, mechanism of action and clinical experience: J. Cardiothorac. Anesth. **6**, Suppl. 2, 1-57 (1989).

Crystals from DMF, mp 294-297° (dec).

THERAP CAT: Cardiotonic.

588. Amrubicin. [110267-81-7] (7S,9S)-9-Acetyl-9-amino-7-[(2-deoxy-β-D-erythro-pentopyranosyl)oxy]-7,8,9,10-tetrahydro-6,11-dihydroxy-5,12-naphthacenedione; (+)-9-amino-4-demethoxy-9-deoxy-7-O-(2-deoxy-β-D-erythro-pentopyranosyl)daunomycinone. $C_{25}H_{25}NO_9$; mol wt 483.47. C 62.11%, H 5.21%, N 2.90%, O 29.78%. Synthetic anthracycline antibiotic; inhibits DNA topoisomerase II. Prepn: K. Ishizumi et al., EP **107486**; eidem, US **4673668** (1984, 1987 both to Sumitomo); eidem, J. Org. Chem. **52**, 4477 (1987). NMR study of complex with DNA: J. Igarashi, M. Sunagawa, Bioorg. Med. Chem. Lett. **5**, 2923 (1995). Mechanism of action: M. Hanada et al., Jpn. J. Cancer Res. **89**, 1229 (1998). Toxicology: S. Morisada et al., ibid. **80**, 77 (1989). Clinical pharmacokinetics: K. Inoue et al., Invest. New Drugs **7**, 213 (1989); Y. Matsunaga et al., Ther. Drug Monit. **28**, 76 (2006). Clinical evaluation in lung cancer: T. Sugiura et al., Invest. New Drugs **23**, 331 (2005); in combination with cisplatin: Y. Ohe et al., Ann. Oncol. **16**, 430 (2005).

mp 172-174°. $[\alpha]_D^{20}$ +119° (c = 0.02 in $CHCl_3$).

Hydrochloride. [110311-30-3] SM-5887; Calsed. $C_{25}H_{25}NO_9.HCl$; mol wt 519.93. mp 145-151°. LD_{50} i.v. in mice: 32-50 mg/kg (Morisada).

THERAP CAT: Antineoplastic.

589. Amsacrine. [51264-14-3] N-[4-(9-Acridinylamino)-3-methoxyphenyl]methanesulfonamide; 4'-(9-acridinylamino)methanesulfon-m-anisidide; m-AMSA; CI-880; NSC-249992; SN-11841; Amekrin; Amsidine; Amsidyl. $C_{21}H_{19}N_3O_3S$; mol wt 393.46. C 64.11%, H 4.87%, N 10.68%, O 12.20%, S 8.15%. Cytostatic agent with antiviral and immunosuppressive properties. Prepn: B. F. Cain et al., J. Med. Chem. **18**, 1110 (1975). Synthesis and biological properties of spin-labeled amsacrine: B. K. Sinha et al., ibid. **19**, 994 (1976). Mechanism of action: W. R. Wilson, Chem. N. Z. **37**, 148 (1973). Pharmacologic disposition: R. L. Cysyk et al., Drug Metab. Dispos. **5**, 579 (1977). Antiviral activity: D. M. Byrd, Ann. N.Y. Acad. Sci. **284**, 463 (1977). Immunosuppressive properties: B. C. Baguley et al., Eur. J. Cancer **10**, 169 (1974). Exptl antitumor properties: B. F. Cain, G. J. Atwell, ibid. 539. Efficacy in adult acute leukemia: S. S. Legha et al., Ann. Intern. Med. **93**, 17 (1980). Toxicologic studies: K. L. Pavkov et al., U.S. NTIS Report PB-298106 (1979) 284 pp. Review of pharmacology and clinical efficacy: J. Hornedo, D. A. Van Echo, Pharmacotherapy **5**, 78-90 (1985).

LD_{50} in male, female CDF_1 mice: 810 mg/m²; 729 mg/m² orally (Pavkov).

Hydrochloride. NSC-141549. $C_{21}H_{19}N_3O_3S.HCl$; mol wt 429.92. Crystals, mp 197-199°. LD_{50} i.p. in mice: ~60 mg/kg (Byrd).

Methanesulfonate. NSC-156303. $C_{21}H_{19}N_3O_3S.CH_3SO_3H$; mol wt 489.56. Crystals, mp 292-293°. LD_{50} i.p. in mice: ~24 mg/kg (Byrd).

THERAP CAT: Antineoplastic.

590. Amsonic Acid. [81-11-8] 2,2'-(1,2-Ethenediyl)bis[5-aminobenzenesulfonic acid]; 4,4'-diamino-2,2'-stilbenedisulfonic acid. $C_{14}H_{14}N_2O_6S_2$; mol wt 370.39. C 45.40%, H 3.81%, N 7.56%, O 25.92%, S 17.31%. Prepn: Bender, Schultz, Ber. **19**, 3234 (1886); H. E. Fierz-David, L. Blangey, Grundlegende Operationen der Farbenchemie (Springer-Verlag, Vienna, 7th ed., 1947) p 161; Spiegler, US **2784220** (1957 to du Pont).

Yellow needles; very slightly sol in water. Forms sparingly watersol, crystalline salts with many bisquaternary ammonium bases.

USE: In manuf of dyes, bleaching agents.

591. Amtolmetin Guacil. [87344-06-7] N-[2-[1-Methyl-5-(4-methylbenzoyl)-1H-pyrrol-2-yl]acetyl]glycine 2-methoxyphenyl ester; N-[(1-methyl-5-p-toluoylpyrrol-2-yl)acetyl]glycine o-methoxyphenyl ester; 1-methyl-5-p-toluoylpyrrole-2-acetamidoacetic acid guaicil ester; ST-679; MED-15; Eufans. $C_{24}H_{24}N_2O_5$; mol wt 420.47. C 68.56%, H 5.75%, N 6.66%, O 19.03%. Ester prodrug of tolmetin, q.v. Prepn: A. Baglioni, BE **896018**; idem, US **4578481** (1983, 1986 both to Sigma-Tau). Pharmacology: E. Arrigoni-Martelli, Drugs Exp. Clin. Res. **16**, 63 (1990); A. Caruso et al., ibid. **18**, 481 (1992). HPLC determn in plasma: A. Mancinelli et al., J. Chromatogr. **553**, 81 (1991). Series of articles on pharmacokinetics and clinical trials: Clin. Ter. **142** (1 pt 2) 3-59 (1993).

Crystals from cyclohexane-benzene, mp 117-120°. Sol in common organic solvents. LD_{50} in male mice, rats (mg/kg): 1370, 1100 i.p.; >1500, 1450 orally (Baglioni).

THERAP CAT: Anti-inflammatory.

592. Amygdalin. [29883-15-6] (αR)-α-[(6-O-β-D-Glucopyranosyl-β-D-glucopyranosyl)oxy]benzeneacetonitrile; amygdaloside; mandelonitrile-β-gentiobioside; D-mandelonitrile-β-D-glucosido-6-β-D-glucoside; NSC-15780. $C_{20}H_{27}NO_{11}$; mol wt 457.43. C 52.52%, H 5.95%, N 3.06%, O 38.47%. The name amygdalin is currently used interchangeably with **laetrile**. Cyanogenic glycoside which occurs in seeds of *Rosaceae;* principally in bitter almonds; also in peaches and apricots. Most common constituent of *Laetrile*® preparations. Structure and synthesis: W. N. Haworth, B. Wylam, *J. Chem. Soc.* **123**, 3120 (1923); Kuhn, *Ber.* **56**, 857 (1923); R. Campbell, W. N. Haworth, *J. Chem. Soc.* **125**, 1337 (1924); Hudson, *J. Am. Chem. Soc.* **46**, 483 (1924); Zemplén, Kunz, *Ber.* **57**, 1357 (1924); Kuhn, Sobotka, *ibid.* 1767; Baumann, Pigman, *The Carbohydrates*, W. Pigman, Ed. (Academic Press, New York, 1957) p 550. Enzymic hydrolysis studies: Haisman, Knight, *Biochem. J.* **103**, 528 (1967). The term Laetrile® has also been applied to *mandelonitrile β-glucuronide*. Purported prepn: E. T. Krebs, E. T. Krebs, Jr., **GB 788855** (1958) and **US 2985664** (1961). Synthesis, characterization and comparison of mandelonitrile β-glucuronide with amygdalin: C. Fenselau *et al., Science* **198**, 625 (1977). Pharmacology and cyanide toxicity studies of amygdalin (laetrile): C. G. Moertel *et al., J. Am. Med. Assoc.* **245**, 591 (1981); M. M. Ames *et al., Cancer Chemother. Pharmacol.* **6**, 51 (1981). Pharmacokinetics: A. G. Rauws *et al., Arch. Toxicol.* **49**, 311 (1982). Determn methods in tissues and fluids: J. Balkon, *J. Anal. Toxicol.* **6**, 244 (1982). Amygdalin (laetrile) is a toxic drug that is not effective as a cancer treatment: C. G. Moertel *et al., N. Engl. J. Med.* **306**, 201 (1982). Review of the controversial use of amygdalin (laetrile): V. Herbert, *Am. J. Clin. Nutr.* **32**, 1121-1158 (1979).

Trihydrate. Orthorhombic columns from water, mp 200°; mp about 220° when anhydr. The once melted and solidif substance remelts at 125-130°. $[\alpha]_D^{20}$ $-42°$ (anhydr basis). One gram dissolves in 12 ml water, in 900 ml alcohol, in 11 ml boiling alcohol. Very sol in boiling water; almost insol in ether. pH of satd aq soln ~7.

Note: The misleading term *vitamin B_{17}*, has sometimes been applied to amygdalin.

593. n-Amylamine. [110-58-7] 1-Pentanamine; pentylamine; 1-aminopentane. $C_5H_{13}N$; mol wt 87.17. C 68.89%, H 15.03%, N 16.07%. Prepd by reduction of valeronitrile with $LiAlH_4$ and with $LiAlH_4$-$AlCl_3$: Nystrom, *J. Am. Chem. Soc.* **77**, 2544 (1955).

Liquid. d^{19} 0.766. bp 104°. mp $-55°$. *Flammable, corrosive.* Very sol in water; sol in alcohol; miscible with ether.

Caution: A strong irritant.

594. Amylase. [9000-92-4] Enzymes catalyzing the hydrolysis of α-1 $\rightarrow$ 4 glucosidic linkages of polysaccharides such as glycogen, starch, or their degradation products. *Endoamylases* attack the α-1 $\rightarrow$ 4 linkage at random. A single type of endoamylase is known, *i.e.,* α-amylases (*α-1,4-glucan 4-glucanohydrolases*), so named, because the reducing hemiacetal group liberated by the hydrolysis has α optical configuration and mutarotates downward. The more common α-amylases include those isolated from human saliva, human, hog and rat pancreas, *Bacillus subtilis, B. coagulans, Aspergillus oryzae, A. candidus, Pseudomonas saccharophila*, and barley malt. *Exoamylases* attack the α-1 $\rightarrow$ 4 linkages only from the non-reducing outer polysaccharide chain ends. Those breaking every glucosidic bond to produce solely α-glucose are known as *glucoamylases (γ-amylases)*. Those breaking every alternate bond to produce maltose are known as β-amylases (*α-1,4-glucan maltohydrolases*). Exoamylases are exclusively of vegetable or microbial origin. *Reviews:* Fischer, Stein, "α-Amylases" and French, "β-Amylases" in *The Enzymes* Vol. **4**, P. D. Boyer *et al.,* Eds., (Academic Press, New York, 2nd ed., 1960) pp 313-343, 345-368; J. A. Thoma *et al., ibid.* Vol. **V** (3rd ed., 1971) pp 115-189; W. M. Fogarty, C. T. Kelly, *Microbial Enzymes and Bioconversions*, A. H. Rose, Ed. (Academic Press, New York, 1980) pp 115-170.

α-Amylase (porcine). [9000-90-2] Maxilase. Enzyme derived from swine pancrease. mol wt ~45,000. Prepn: Caldwell *et al., J. Am. Chem. Soc.* **74**, 4033 (1952).

α-Amylase (bacterial). [9000-85-5] Usually derived from *Bacillus subtilis.* Purification: Stein, Fischer, *Helv. Chim. Acta* **40**, 529 (1957).

β-Amylase (sweet potato). [9000-91-3] Mol wt ~152,000. Prepn: Balls *et al., J. Biol. Chem.* **173**, 9 (1948).

USE: In starch processing, brewing, distilling, baking, animal feed, sewage treatment.

THERAP CAT: Enzyme (digestive aid).

595. Amylbenzene. [538-68-1] Pentylbenzene; *n*-amylbenzene; 1-phenylpentane. $C_{11}H_{16}$; mol wt 148.25. C 89.12%, H 10.88%. Prepd by the action of benzylmagnesium chloride on *n*-butyl *p*-toluenesulfonate: Rossander, Marvel, *J. Am. Chem. Soc.* **50**, 1491 (1928); Gilman, Heck, *ibid.* **50**, 2223 (1928); Gilman, Robinson, *Org. Synth.* **coll. vol. II**, 47, (1943).

Liquid. mp $-78.5°$. bp_{760} 202.2°; bp_{10} 81°. d_4^{20} 0.8594. n_D^{20} 1.48849. *See:* Vogel, *J. Chem. Soc.* **1948**, 607. Insol in water. Sol in alcohol; miscible with ether, benzene.

596. n-Amyl Bromide. [110-53-2] 1-Bromopentane. $C_5H_{11}Br$; mol wt 151.05. C 39.76%, H 7.34%, Br 52.90%. Prepn: Fournier, *Bull. Soc. Chim. Fr.* **35**, 623 (1906); Lindstone, Morris, *Chem. Ind. (London)* **1958**, 560.

Liquid. d_4^{15} 1.2237. bp_{740} 129.7°. mp $-95°$. n_D^{20} 1.4444. Practically insol in water. Sol in alcohol; miscible with ether.

597. tert-Amyl Bromide. [507-36-8] 2-Bromo-2-methylbutane; 2-bromoisopentane. $C_5H_{11}Br$; mol wt 151.05. C 39.76%, H 7.34%, Br 52.90%. Prepd from trimethylethylene and dry HBr at $-78°$: Michael, Weiner, *J. Org. Chem.* **4**, 531 (1939).

Liquid, bp 107.4°. d_0^0 1.2439. n_D^{20} 1.4430.

598. n-Amyl Butyrate. [540-18-1] Butanoic acid pentyl ester. $C_9H_{18}O_2$; mol wt 158.24. C 68.31%, H 11.47%, O 20.22%.

Prepn: Gartenmeister, *Ann.* **233**, 269 (1886). Toxicity study: P. M. Jenner *et al.*, *Food Cosmet. Toxicol.* **2**, 327 (1964).

Liquid. Apricot-like odor. mp $-73.2°$. d_0^0 0.8832; d_4^{15} 0.8713. bp_{760} 185°. n_D^{20} 1.4110. *Flammable.* Soly in water (50°): 0.54 g/l; very sol in alc, ether. LD_{50} orally in rats: 12210 mg/kg (Jenner).

USE: Has been used in such flavors as apricot, pineapple, pear, plum, and sparingly in some perfume compositions.

599. *n*-Amyl Caproate. [540-07-8] Hexanoic acid pentyl ester; pentyl hexanoate; *n*-caproic acid *n*-amyl ester. $C_{11}H_{22}O_2$; mol wt 186.30. C 70.92%, H 11.90%, O 17.18%. Prepn: Simonini, *Monatsh. Chem.* **13**, 320 (1892).

Liquid, bp 222-227°.

600. Amyl Chloride. [543-59-9] 1-Chloropentane; *n*-amyl chloride; *n*-butylcarbonyl chloride. $C_5H_{11}Cl$; mol wt 106.59. C 56.34%, H 10.40%, Cl 33.26%. Prepd from 1-pentanol and concd HCl in sealed tube at 120°: Conant, Kirner, *J. Am. Chem. Soc.* **46**, 245 (1924); with HCl and zinc chloride: Clark, Streight, *Trans. R. Soc. Can. Sect. 3* **23 III**, 77 (1929); Vogel, *J. Chem. Soc.* **1943**, 638, 640; for prepn without admixed 2- and 3-chloropentanes *see* Whitmore's procedure from 1-pentanol with $SOCl_2$ and pyridine: Whitmore *et al.*, *J. Am. Chem. Soc.* **60**, 2540 (1938); Mixer, Young, *ibid.* **78**, 3382 (1956).

Liquid. d_4^{20} 0.8828. mp $-99°$. bp_{760} 107.8°. n_D^{20} 1.41280. Flash pt, closed cup: 55°F (13°C). *Flammable.* Miscible with alc, ether. Insol in water. Forms a constant boiling mixture with water, bp 82°, with ethanol bp 72.5°.

601. Amylene. [513-35-9] 2-Methyl-2-butene; β-isoamylene; trimethylethylene. C_5H_{10}; mol wt 70.14. C 85.62%, H 14.37%. Prepd by dehydration of *tert*-amyl alcohol in the presence of *p*-toluenesulfonic acid: Applequist, Babad, *J. Org. Chem.* **27**, 288 (1962); by disproportionation of isobutene with propylene or 2-butene: Banks, Regier, *Ind. Eng. Chem. Prod. Res. Dev.* **10**, 46 (1971).

Liquid, bp 37.5-38.5°. Highly flammable, flash pt 0°F. d_4^{15} 0.66. Disagreeable odor. Polymerizes on long standing. Practically insol in water. Miscible with alc, ether.

Caution: A simple asphyxiant.

602. *n*-Amyl Ether. [693-65-2] 1,1'-Oxybispentane; pentyl ether; amyl oxide; diamyl ether. $C_{10}H_{22}O$; mol wt 158.29. C 75.88%, H 14.01%, O 10.11%. Prepn from amyl alcohol with concd H_2SO_4: Hinton, Nieuwland, *Proc. Indiana Acad. Sci.* **42**, 109 (1933).

Liquid, bp 186.75°. mp $-69.43°$. Flash pt, closed cup: 57°C. d_4^{20} 0.78326, d_4^{25} 0.77924. n_D^{20} 1.41195, n_D^{25} 1.40985: Dreisbach, Martin, *Ind. Eng. Chem.* **41**, 2875 (1949). Practically insol in water. Miscible with alcohol, ether.

Caution: Vapors narcotic in high concns.

USE: Industrial solvent.

603. Amylin. [106602-62-4] Diabetes-associated peptide; insulinoma amyloid peptide; islet amyloid polypeptide; IAPP. Peptide hormone co-secreted with insulin by pancreatic beta cells in response to a meal or other nutrient stimuli. Affects carbohydrate absorption and disposition; modulates the effects of insulin. Major component of pancreatic amyloid deposits characteristic of noninsulin-dependent diabetes and thought to be involved in insulin resistance. Amylin deficiency occurs in insulin-dependent diabetes and may contribute to excessive insulin sensitivity and a heightened tendency to hypoglycemia. Structurally homologous with *calcitonin gene related peptide*, sharing certain bioactivities such as vasodilation, osteoclast inhibition, and appetite suppression. Isoln from human insulinoma amyloid: P. Westermark *et al.*, *Biochem. Biophys. Res. Commun.* **140**, 827 (1986); from diabetic islet amyloid: *idem et al.*, *Am. J. Pathol.* **127**, 414 (1987); G. J. S. Cooper *et al.*, *Proc. Natl. Acad. Sci. USA* **84**, 8628 (1987). Review of bioactivities: B. J. Edwards, J. E. Morley, *Life Sci.* **51**, 1899-1912 (1992); T. J. Rink *et al.*, *Trends Pharmacol. Sci.* **14**, 113-118 (1993). Review of role in insulin resistance: B. Leighton, G. J. S. Cooper, *Trends Biochem. Sci.* **15**, 295-299 (1990); in carbohydrate metabolism: A. Young *et al.*, *Biochem. Soc. Trans.* **23**, 325 (1995). Effects on appetite regulation and memory: J. E. Morley *et al.*, *Can. J. Physiol. Pharmacol.* **73**, 1042-1046 (1995). Review of role in age-related disease and therapeutic potential: G. J. S. Cooper, C. A. Tse, *Drugs Aging* **9**, 202-212 (1996).

Lys–Cys–Asn–Thr–Ala–Thr–Cys–Ala–Thr–Gln–Arg–Leu–Ala–Asn–Phe–Leu–Val–His

Ser

Ser

H_2N–Tyr–Thr–Asn–Ser–Gly–Val–Asn–Thr–Ser–Ser–Leu–Ile–Ala–Gly–Phe–Asn–Asn

Human Amylin

Human amylin. [122384-88-7] Amlintide; AC-001. $C_{165}H_{261}$ $N_{51}O_{55}S_2$; mol wt 3903.33. Composed of 37 amino acid residues; mol wt 3903.35.

THERAP CAT: In treatment of insulin-dependent diabetes.

604. *n*-Amyl Mercaptan. [110-66-7] 1-Pentanethiol; amyl thioalcohol. $C_5H_{12}S$; mol wt 104.21. C 57.63%, H 11.61%, S 30.76%. Prepn: Cossar *et al.*, *J. Org. Chem.* **27**, 93 (1962).

Liquid, bp 123-124°, n_D^{25} 1.4439. Penetrating, unpleasant odor. d_4^{20} 0.857. *Flammable.* Practically insol in water. Sol in alc.

Caution: Potential symptoms of overexposure are irritation of eyes, skin, nose, throat, respiratory system; headache, nausea, dizziness; vomiting, diarrhea; dermatitis, skin sensitization. *See NIOSH Pocket Guide to Chemical Hazards* (DHHS/NIOSH 97-140, 1997) p 244.

USE: In organic syntheses.

605. *tert*-Amyl Methyl Ether. [994-05-8] 2-Methoxy-2-methylbutane; methyl *tert*-pentyl ether; 1,1-dimethylpropyl methyl ether; methyl *tert*-amyl ether; TAME. $C_6H_{14}O$; mol wt 102.18. C 70.53%, H 13.81%, O 15.66%. Gasoline octane booster. Prepn: A. Reychler, *Bull. Soc. Chim. Belg.* **21**, 71 (1907); and properties: T. W. Evans, K. R. Edlund, *Ind. Eng. Chem.* **28**, 1186 (1936). Industrial synthesis and processing: H. Short, *Chem. Eng.* **93**, 34 (1986); S. Randriamahefa *et al.*, *J. Mol. Catal.* **49**, 85 (1988); W. J. Reagan, *Prepr. Pap. - Am. Chem. Soc. Div. Fuel Chem.* **39**, 337 (1994). Reaction equilibria in synthesis: L. K. Rihko *et al.*, *J. Chem. Eng. Data* **39**, 700 (1994). Chromatographic and spectroscopic analyses: J. S. Hardman *et al.*, *Fuel* **72**, 1563 (1993). Comparison with MTBE, *q.v.*, of effects on emissions: W. J. Koehl *et al.*, *Soc. Automot. Eng.* **SP-1000**, vol 2, 289 (1993). Reaction with nitrate radical (environmental implications): S. Langer, E. Ljungström, *J. Phys. Chem.* **98**, 5906 (1994).

Liquid, $bp_{760 \text{ mm}}$ 86.3°. d_4^{15} 0.7750; d_4^{20} 0.7703; d_4^{25} 0.7656; d_4^{30} 0.7607. n_D^{20} 1.3885. Vapor pressure (25°): 75 mm Hg. Soly in water (20°): 1.15g/100g soln.

USE: Fuel additive for gasoline.

606. Amyl Nitrite. Mixture of isomers containing not less than 97.0% and not more than 100.0% of $C_5H_{11}NO_2$. Consists chiefly of isoamyl nitrite [$(CH_3)_2CHCH_2CH_2ONO$], but other isomers are also present.

Clear, yellowish liquid having a peculiar, ethereal, fruity odor. Volatile even at low temperatures. bp 96°. The N.F. grade has d_{25}^{25} 0.870-0.876. Practically insol in water. Miscible with alcohol, ether. *Flammable. See* Isoamyl Nitrite.

607. Amylocaine. [644-26-8] 1-(Dimethylamino)-2-methyl-2-butanol 2-benzoate; 1-(dimethylaminomethyl)-1-methylpropyl benzoate; amyleine. $C_{14}H_{21}NO_2$; mol wt 235.33. C 71.45%, H 9.00%, N 5.95%, O 13.60%. Prepd by benzoylation of 1-(dimethylamino)-2-methyl-2-butanol: Fourneau, Ribas, *Bull. Sci. Pharmacol.* **35**, 273 (1928), *C.A.* **22**, 2919¹ (1928).

Hydrochloride. [532-59-2] Stovaine. $C_{14}H_{21}NO_2 \cdot HCl$; mol wt 271.79. Crystals, bitter taste, followed by temporary numbness of the tongue. Dec 177-179°. One gram dissolves in 2 ml water, in 3.3 ml abs ethanol. Practically insol in ether. A 5% aq soln is faintly acid to litmus and neutral to Congo red.

THERAP CAT: Anesthetic (local).

608. Amyloid β Peptide. Aβ; β/A4; β-amyloid protein. A 39-43 residue peptide deposited into the extracellular amyloid or senile plaques characterizing Alzheimer's disease and Down's syndrome. Mol wt approx 4,200. Multiple isoforms exist, the most predominant are 40 and 42 amino acids in length. Natural component of cerebrospinal fluid and blood plasma; sol under normal conditions. Forms insol, cytotoxic fibrillar plaques and aggregates in diseased brains. Derived from the proteolysis of *amyloid precursor protein* by proteases known as *secretases*. Purification and characterization: G. G. Glenner, C. W. Wong, *Biochem. Biophys. Res. Commun.* **120**, 885 (1984); C. L. Masters *et al., Proc. Natl. Acad. Sci. USA* **82**, 4245 (1985). Structure and neurotoxicity studies: L. K. Simmons *et al., Mol. Pharmacol.* **45**, 373 (1994); C. J. Pike *et al., J. Neurochem.* **64**, 253 (1995). Brain accumulation of Aβ during aging and in Alzheimer's disease: H. Funato *et al., Am. J. Pathol.* **152**, 1633 (1998). Role of soluble oligomers in Alzheimer's disease: R. Kayed *et al., Science* **300**, 486 (2003). Soly study of Aβ₄₀: P. Sengupta *et al., Biochemistry* **42**, 10506 (2003). Oligomerization study: G. Bitan *et al., J. Biol. Chem.* **278**, 34882 (2003). Evaluation of immunization with aggregated Aβ₄₂ in Alzheimer's disease: C. Hock *et al., Neuron* **38**, 547 (2003). Series of articles: *J. Struct. Biol.* **130**, 87-371 (2000). Review of solid state NMR studies: R. Tycko, *Biochemistry* **42**, 3151-3159 (2003). Reviews of role in Alzheimer's disease: C. L. Joachim, D. J. Selkoe, *Alzheimer Dis. Assoc. Disord.* **6**, 7-34 (1992), D. H. Small, C. A. McLean, *J. Neurochem.* **73**, 443-449 (1999); D. A. Butterfield *et al., Trends Mol. Med.* **7**, 548-554 (2001); W. I. Rosenblum, *Neurobiol. Aging* **23**, 225-230 (2002); T. E. Golde *et al., Biochim. Biophys. Acta* **1502**, 172-187 (2002).

Aβ₄₀. [131438-79-4] Human β-amyloid peptide (1-40).

Aβ₄₂. [107761-42-2] Human β-amyloid peptide (1-42).

609. Amylpenicillin. [4493-18-9] (2*S*,5*R*,6*R*)-3,3-Dimethyl-7-oxo-6-[(1-oxohexyl)amino]-4-thia-1-azabicyclo[3.2.0]heptane-2-carboxylic acid; dihydropenicillin F; hexanoylpenicillin; penicillin DF. $C_{14}H_{22}N_2O_4S$; mol wt 314.40. C 53.48%, H 7.05%, N 8.91%, O 20.35%, S 10.20%. Antibiotic produced by the mold *Aspergillus flavus* from a Czapek-Dox medium supplemented with corn steep liquor or by *Penicillium chrysogenum* Q176 or by *P. notatum* strains: McKee, MacPhillamy, *Proc. Soc. Exp. Biol. Med.* **53**, 247 (1943); Bush, Goth, *J. Pharmacol. Exp. Ther.* **78**, 164 (1943); McKee *et al., J. Bacteriol.* **47**, 187 (1944); Bush *et al., J. Pharmacol. Exp. Ther.* **84**, 264 (1945); Fried *et al., J. Biol. Chem.* **163**, 341 (1946); *see also* Wintersteiner under "Flavacidin" in *Chemistry of Penicillin* (Princeton, 1949). Prepn by hydrogenation of 2-pentenylpenicillin: Catch *et al.,* **GB 584852**; Cook, Heilbron in *Chemistry of Penicillin* (Princeton, 1949). Characterization and antibacterial activity: Leigh, *Nature* **163**, 95 (1949).

Sodium salt. [575-47-3] Sodium *n*-amylpenicillinate; flavacidin; flavicin. $C_{14}H_{21}N_2NaO_4S$. Flat, blunt-ended needles from moist acetone or moist ethyl acetate as the monohydrate. When anhydrous, mp 188° (dec). $[\alpha]_D^{23}$ +319°. Very soluble in water.

610. α-Amyrin. [638-95-9] (3β)-Urs-12-en-3-ol; α-amyrenol; viminalol. $C_{30}H_{50}O$; mol wt 426.73. C 84.44%, H 11.81%, O 3.75%. Occurs mostly as acetate in latex of rubber trees, in latex from *Ficus variegata* Blume, *Moraceae*, also in *Balanophora elongata* Blume, *Balanophoraceae*, and in *Erythroxylum coca* Lam. var. *novogranatense* Morris, and var. *spruceanum* Burck, *Erythroxylaceae*. Isoln from *Manila elemi:* Vesterberg, Westerlind, *Ann.* **428**, 247 (1922). Structural studies: Spring, Vickerstaff, *J. Chem. Soc.* **1937**, 249; Beynon *et al., ibid.* **1938**, 1233; Meisels *et al., Helv. Chim. Acta* **32**, 1075 (1949), **38**, 1298 (1955); Melera *et al., ibid.* **39**, 441 (1956). Identity with viminalol: Soldin, Marais, *J. Pharm. Soc.* **55**, 452 (1966). Formation from ursolic acid: Goodson, *J. Chem. Soc.* **1938**, 999; from boswellic acid: Ruzicka, Wirz, *Helv. Chim. Acta* **22**, 948 (1939). Partial synthesis from glycyrrhetic acid and stereochemistry: Corey, Cantrall, *J. Am. Chem. Soc.* **81**, 1745 (1959). *Review:* J. Simonsen, W. C. J. Ross, *The Terpenes* **vol. IV** (University Press, Cambridge, 1957) pp 116-148.

Needles from alcohol, mp 186°. $bp_{0.7}$ 243°. $[\alpha]_D^{17}$ +91.6° (c = 1.3 in benzene). Sol in 22 parts 98% alc. Sol in ether, benzene, chloroform, glacial acetic acid. Slightly sol in petr ether.

Acetate. $C_{32}H_{52}O_2$. Leaflets from petr ether, mp 227°. $[\alpha]_D^{20}$ +76.35° (c = 0.572 in CHCl₃).

Benzoate. $C_{37}H_{54}O_2$. Prisms from benzene + acetone, mp 195-196°. $[\alpha]_D^{10}$ +94.6° (c = 1.9 in CHCl₃).

611. β-Amyrin. [559-70-6] (3β)-Olean-12-en-3-ol; β-amyrenol. $C_{30}H_{50}O$; mol wt 426.73. C 84.44%, H 11.81%, O 3.75%. Occurs together with α-amyrin. Isoln and structural studies: *See:* α-Amyrin. *See also:* Vesterberg, *Bull. Soc. Chim.* **37**, 742 (1925);

Horrmann, Firzlaff, *Arch. Pharm.* **268**, 64 (1930); Ruzicka, Marxer, *Helv. Chim. Acta* **22**, 195 (1939); Jeger, Ruzicka, *ibid.* **28**, 209 (1945); Prelog *et al., ibid.* **29**, 360 (1946). Conversion of δ-amyrene to β-amyrin: Barton *et al., J. Chem. Soc. C* **1968**, 1031. Biogenetic-type total synthesis: van Tamelen *et al., J. Am. Chem. Soc.* **94**, 8229 (1972). Biosynthesis from squalene: Suga *et al., Chem. Lett.* **1972**, 129, 313.

Needles from petr ether or alc, mp 197-197.5°. bp$_{0.8}$ 260°. [α]$_D^{19}$ +99.8° (c = 1.3 in benzene). Somewhat less soluble than the α-form. Soluble in 37 parts of 98% alc.

Acetate. $C_{32}H_{52}O_2$. Prisms from petr ether, mp 241°. [α]$_D^{17}$ +79° (c = 0.9 in benzene).

Palmitate. Balanophorin. $C_{46}H_{80}O_2$. mp 77°. [α]$_D^{15}$ +54.5° (c = 1.1 in benzene). Occurs in *Balanophora elongata* Blume, *Balanophoraceae*, in *Erythroxylum coca* Lam. var *novogranatense* Morris, and var *spruceanum* Burck, *Erythroxylaceae*, in latex from *Ficus variegata* Blume, *Moraceae*.

Di-β-amyrin ether. $C_{60}H_{98}O$. mp 135-136°. *See:* Rollett, *Monatsh. Chem.* **47**, 437 (1926).

612. Anabasine. [494-52-0] 3-(2S)-2-Piperidinylpyridine; 2-(3-pyridyl)piperidine; neonicotine. $C_{10}H_{14}N_2$; mol wt 162.24. C 74.03%, H 8.70%, N 17.27%. In *Anabasis aphylla* L., *Chenopodiaceae:* Orechoff, Menschikoff, *Ber.* **64**, 266 (1931); in *Nicotiana glauca* Graham, *Solanaceae:* Smith, *J. Am. Chem. Soc.* **57**, 959 (1935); Pyriki, Oehler, *Pharmazie* **9**, 685 (1954). Synthesis: Späth, *Ber.* **70B**, 70 (1937). Industrial extraction processes: Sadykov, Timbekov, *J. Appl. Chem. USSR* **29**, 148 (1956). Abs config: Lukes *et al., Collect. Czech. Chem. Commun.* **27**, 751 (1962).

Liquid, bp 270-272°; bp$_{14}$ 145-147°; bp$_2$ 105°. Freezes at 9°. d$_4^{20}$ 1.0455. n$_D^{20}$ 1.5430. [α]$_D^{20}$ −83.1°. Sol in water and in most organic solvents.

Hydrochloride. [α]$_D$ +16.5° (c = 10 in water).

Caution: Acute and *subacute toxicity:* increased salivation, vertigo, confusion, disturbed vision and hearing, photophobia, cold extremities, nausea, vomiting, diarrhea, syncope, clonic spasms.

USE: Insecticide.

613. Anabsinthin. [6903-12-4] (3S,3aS,6S,6aR,6bS,7R,-7aR,8S,10aS,11S,13aS,13bS,13cR,14bS,15S)-3,3a,4,5,6,6a,6b,7,7a,-8,9,10,10a,13a,13c,14b-Hexadecahydro-6-hydroxy-3,6,8,11,14,15-hexamethyl-2H-8,15-epoxy-7,13b-ethanopentaleno[1″,2″:6,7; 5″,4″:6′,7′]dicyclohepta[1,2-b:1′,2′-b′]difuran-2,12(11H)-dione; anabsynthin. $C_{30}H_{40}O_6$; mol wt 496.64. C 72.55%, H 8.12%, O 19.33%. A bitter principle isolated from *Artemisia absinthium* L., *Compositae* (wormwood). Isoln: F. Sorm *et al., Chem. Ind. (London)* **1955**, 569; V. Herout *et al., Collect. Czech. Chem. Commun.* **21**, 1485 (1956). Structure: L. Novotny *et al., Chem. Ind. (London)* **1958**, 465; V. Herout *et al., Collect. Czech. Chem. Commun.* **25**, 1492 (1960); J. Beauhaire *et al., Tetrahedron Lett.* **21**, 3191 (1980).

Anhydrous Form. Crystals from benzene or isopropanol + diisopropyl ether, or by drying over phosphorus pentoxide in vacuo, mp 267°. [α]$_D^{20}$ +113° (c = 1.85 in chloroform).

Monohydrate. Crystals from methanol, mp 210°.

614. Anacardic Acid. Principal constituent of cashew nut-shell liquid, *Anacardium occidentalis* L., *Anacardiaceae*, member of the family of non-isoprenoid long-chain phenols. (*See also* Urushiol.) Anacardic acid is a mixture of *2-hydroxy-6-alkylbenzoic acids* in which the alkyl chain (C_{11} or higher) is fully saturated (I) or is a monoene (II), a diene (III) or a triene (IV). The name anacardic acid is also used in the literature to designate one component of the mixture: *6-pentadecyl-2-hydroxybenzoic acid* (I) (*6-pentadecylsalicylic acid*). Another component of the mixture *6-(8-pentadecenyl)-2-hydroxybenzoic acid* (II), is also known as **ginkgoic acid**. Isoln and structure: Städeler, *Ann.* **63**, 137 (1847); Ruhemann, Skinner, *Ber.* **20**, 1861 (1887); Haagen, Smit, *K. Akad. Wetensch.* **34**, 165 (1931); Backer, Haack, *Rec. Trav. Chim.* **660**, 61 (1941); Kremers, **US 2431127** (1947). Structure of side chain: Sletzinger, Dawson, *J. Org. Chem.* **14**, 670, 849 (1949). Structure of unsaturated components: J. H. P. Tyman, N. Jacobs, *J. Chromatogr.* **54**, 83 (1971). GLC analysis of components: J. H. P. Tyman, *ibid.* **111**, 285 (1975). Synthesis and prostaglandin synthetase inhibiting activity of I and II: Y. Yamagiwa *et al., Tetrahedron* **43**, 3387 (1987); of I: I. Kubo *et al., Chem. Lett.* **1987**, 1101. Antitumor activity of I: H. Itokawa *et al., Chem. Pharm. Bull.* **35**, 3016 (1987). *Review:* J. H. P. Tyman, *Chem. Soc. Rev.* **8**, 499-537 (1979).

I	R = $(CH_2)_7CH_3$
II	R = CH=CH$(CH_2)_5CH_3$
III	R = CH=CHCH$_2$CH=CH$(CH_2)_2CH_3$
IV	R = CH=CHCH$_2$CH=CHCH$_2$CH=CH$_2$

Mixture, crystals from acetone, mp 34-37°. Sparingly sol in water; freely sol in alc, ether, petr ether. Forms a water-soluble sodium salt.

6-Pentadecyl component (I). Needles from hexane, mp 90.2-91.5°.

6-(8-Pentadecenyl) component (II). Needles, mp 45.3-48° (subl).

615. Anacetrapib. [875446-37-0] (4S,5R)-5-[3,5-Bis(trifluoromethyl)phenyl]-3-[[4′-fluoro-2′-methoxy-5′-(1-methylethyl)-4-(trifluoromethyl)[1,1′-biphenyl]-2-yl]methyl]-4-methyl-2-oxazolidinone; (4S,5R)-5-[3,5-bis(trifluoromethyl)phenyl]-3-[[4′-fluoro-2′-methoxy-5′-(propan-2-yl)-4-(trifluoromethyl)[1,1′-biphenyl]-2-yl]methyl]-4-methyl-1,3-oxazolidin-2-one; MK-0859. $C_{30}H_{25}F_{10}NO_3$; mol wt 637.52. C 56.52%, H 3.95%, F 29.80%, N 2.20%, O 7.53%. Selective cholesteryl ester transfer protein (CETP) inhibitor. Prepn: A. Ali *et al.,* **WO 06014357**; *eidem,* **US 7652049** (2006, 2010 both to Merck & Co.). Clinical pharmacokinetics: R. Krishna *et al., Br. J. Clin. Pharmacol.* **68**, 535 (2009); and metabolism: S. Kumar *et al., Drug Metab. Dispos.* **38**, 474 (2010). Clinical evaluation in dyslipidemia: D. Bloomfield *et al., Am. Heart J.* **157**, 352 (2009). Mechanism of action study: M. Ranalletta *et al., J. Lipid Res.* **51**, 2739 (2010). Review of development and therapeutic potential: D. Masson, *Curr. Opin. Investig. Drugs* **10**, 980-987 (2009).

THERAP CAT: Antilipemic; antiatherosclerotic.

616. Anagrelide. [68475-42-3] 6,7-Dichloro-1,5-dihydro-imidazo[2,1-*b*]quinazolin-2(3*H*)-one; 6,7-dichloro-1,2,3,5-tetra-hydroimidazo[2,1-*b*]quinazolin-2-one. $C_{10}H_7Cl_2N_3O$; mol wt 256.09. C 46.90%, H 2.76%, Cl 27.69%, N 16.41%, O 6.25%. Phosphodiesterase inhibitor with antiplatelet activity. Prepn: W. N. Beverung, A. Partyka, **US 3932407**; **US RE 31617**; T. A. Jenks *et al.*, **US 4146718** (1976, 1984, 1979 all to Bristol-Myers); H. Yamaguchi, F. Ishikawa, *J. Heterocycl. Chem.* **18**, 67 (1981). Antithrombotic and platelet aggregation inhibiting properties: J. S. Fleming, J. P. Buyniski, *Thromb. Res.* **15**, 373 (1979). Mode of action studies: S. S. Tang, M. M. Frojmovic, *J. Lab. Clin. Med.* **95**, 241 (1980); S. Seiler *et al.*, *J. Pharmacol. Exp. Ther.* **243**, 767 (1987). GC-MS determn in human plasma: E. H. Kerns *et al.*, *J. Chromatogr.* **416**, 357 (1987). Clinical reduction of platelet counts: W. A. Andes *et al.*, *Thromb. Haemostasis* **52**, 325 (1984). Clinical trials to control thrombocytosis in chronic myeloproliferative diseases: M. N. Silverstein *et al.*, *N. Engl. J. Med.* **318**, 1292 (1988); Anagrelide Study Group, *Am. J. Med.* **92**, 69 (1992). Review of pharmacology and clinical experience: P. E. Petrides, *Expert Opin. Pharmacother.* **5**, 1781-1798 (2004).

Hydrochloride monohydrate. [58579-51-4] BL-4162A; BMY-26538-01; Agrylin; Thromboreductin; Xagrid. $C_{10}H_7Cl_2N_3O.$-$HCl.H_2O$; mol wt 310.56. Off-white powder. Very slightly sol in water; sparingly sol in DMSO, DMF. Also prepd as the hemihydrate; crystals from ethanolic HCl, mp >280°.

THERAP CAT: Antithrombocythemic.

617. Anagyrine. [486-89-5] (7*R*,14*R*,14a*R*)-1,3,4,6,7,13,-14,14a-Octahydro-7,14-methano-2*H*,11*H*-dipyrido[1,2-*a*:1',2'-*e*]-[1,5]diazocin-11-one; monolupine; rhombinin. $C_{15}H_{20}N_2O$; mol wt 244.34. C 73.74%, H 8.25%, N 11.47%, O 6.55%. Found in seeds of *Anagyris foetida* L., *Leguminosae* and in gorse *(Ulex europaeus* L., *Leguminosae)*. Isoln: Ing, *J. Chem. Soc.* **1933**, 504; Orekhov *et al.*, *Ber.* **67**, 1394 (1934); Couch, *J. Am. Chem. Soc.* **61**, 3327 (1939); Briggs, Russell, *J. Chem. Soc.* **1942**, 507; Galinsky, Stern, *Ber.* **77**, 132 (1944); Faugeras, Ann. *Pharm. Fr.* **29**, 241 (1971). Absolute configuration: Okuda *et al.*, *Chem. Ind. (London)* **1961**, 1116; Okuda *et al.*, *Chem. Pharm. Bull.* **13**, 491 (1965). Synthesis: van Tamelen, Baran, *J. Am. Chem. Soc.* **80**, 4659 (1958); Goldberg, Lipkin, *J. Org. Chem.* **37**, 1823 (1972).

Pale yellow glass. bp$_4$ 210-215°; bp$_{12}$ 260-270°. $[\alpha]_D^{25}$ −168° (c = 4.8 in ethanol). Sol in water, alcohol, chloroform; slightly sol in ether, benzene.

Hydrochloride trihydrate. $C_{15}H_{21}ClN_2O.3H_2O$. Crystals, mp 235-236° (mp 296° when dry). $[\alpha]_D^{25}$ −142.5° (c = 5). Freely sol in water.

618. Anandamide. [94421-68-8] *(5Z,8Z,11Z,14Z)-N-*(2-Hydroxyethyl)-5,8,11,14-eicosatetraenamide; arachidonylethanol-amide; *N*-arachidonoylethanolamide; *N*-(2-hydroxyethyl)arachi-donamide. $C_{22}H_{37}NO_2$; mol wt 347.54. C 76.03%, H 10.73%, N 4.03%, O 9.21%. Endogenous ligand for the mammalian cannabinoid receptor; derivative of arachidonic acid, *q.v.* Name coined from the Sanskrit word "ananda," meaning bliss. Identification in pig brain: W. A. Devane *et al.*, *Science* **258**, 1946 (1992). Identification in chocolate and cocoa: E. di Tomaso *et al.*, *Nature* **382**, 677 (1996). Pharmacology: P. B. Smith *et al.*, *J. Pharmacol. Exp. Ther.* **270**, 219 (1994). Biosynthetic studies: W. A. Devane, J. Axelrod, *Proc. Natl. Acad. Sci. USA* **91**, 6698 (1994); V. Di Marzo *et al.*, *Nature* **372**, 686 (1994). LC/MS/MS determn and distribution in human tissues: C. C. Felder *et al.*, *FEBS Lett.* **393**, 231 (1996). *Review:* V. Di Marzo, A. Fontana, *Prostaglandins Leukotrienes Essent. Fatty Acids* **53**, 1-11 (1995). Review of pharmacology: K. Smita *et al.*, *Fundam. Clin. Pharmacol.* **21**, 1-8 (2007); of regulatory role in food intake: R. Capasso, A. A. Izzo, *J. Neuroendocrinol.* **20**, Suppl. 1, 39-46 (2008).

619. Anastrozole. [120511-73-1] α1,α1,α3,α3-Tetramethyl-5-(1*H*-1,2,4-triazol-1-ylmethyl)-1,3-benzenediacetonitrile; 2,2'-[5-(1*H*-1,2,4-triazol-1-ylmethyl)-1,3-phenylene]di(2-methylpropio-nitrile); ZD-1033; ICI-D-1033; Arimidex. $C_{17}H_{19}N_5$; mol wt 293.37. C 69.60%, H 6.53%, N 23.87%. Aromatase inhibitor. Prepn: P. N. Edwards, M. S. Large, **EP 296749**; *eidem*, **US 4935437** (1989, 1990 both to ICI). Reviews of pharmacology, clinical pharmacokinetics: P. V. Plourde *et al.*, *Breast Cancer Res. Treat.* **30**, 103-111 (1994); *idem et al.*, *J. Steroid Biochem. Mol. Biol.* **53**, 175-179 (1995). Clinical trial in advanced breast cancer: A. U. Buzdar *et al.*, *Cancer* **79**, 730 (1997).

Crystals from ethyl acetate/cyclohexane, mp 81-82°.
THERAP CAT: Antineoplastic.

620. Anatabine. [581-49-7] (2*S*)-1,2,3,6-Tetrahydro-2,3'-bipyridine; 2-(3-pyridyl)-1,2,3,6-tetrahydropyridine; 1,2,3,6-tetra-hydro-2-(3-pyridyl)pyridine. $C_{10}H_{12}N_2$; mol wt 160.22. C 74.97%, H 7.55%, N 17.48%. The most abundant of the minor alkaloids of tobacco: Späth, Kesztler, *Ber.* **70**, 239, 704, 2450 (1937). Fresh *Nicotiana tabacum*, the species most commonly used for the production of cigarette tobacco, contains 3.9% anatabine. Configuration: Lukes *et al.*, *Collect. Czech. Chem. Commun.* **27**, 751 (1962). Total synthesis of *dl*-form: Quan *et al.*, *J. Org. Chem.* **30**, 2769 (1965). Biosynthesis: E. Leete, S. Slattery, *J. Am. Chem. Soc.* **98**, 6326 (1976). Biomimetic synthesis: E. Leete, M. E. Mueller, *ibid.* **104**, 6440 (1982).

Liquid. d_4^{19} 1.091. bp_{10} 145-146°. $[\alpha]_D^{17}$ −177.8°. n_D^{20} 1.5676. Misc with water. Sol in alc, ether, benzene.

dl-**Form.** Liquid. $bp_{6.5}$ 136°.

621. Anatoxins. ANTXS. First isolated, freshwater neurotoxins present in algal blooms associated with a number of blue-green cyanobacteria including *Anabaena*. spp., *Aphanizomenon* spp. and *Oscillatoria* spp. Responsible for fatal poisoning of cattle and wildlife as well as human illness worldwide. *Review:* W. W. Carmichael *et al., ACS Symp. Ser.* **418**, 87-106 (1990). Review of occurence, chemistry and analysis: K. Sivonen, *Food Sci. Technol.* **103**, 567-581 (2000).

Anatoxin A   Anatoxin A(s)

Anatoxin a. [64285-06-9] 1-(1*R*,6*R*)-9-Azabicyclo[4.2.1]non-2-en-2-ylethanone; 2-acetyl-9-azabicyclo[4.2.1]non-2ene; (+)-ANTX-a; Very Fast Death Factor. $C_{10}H_{15}NO$; mol wt 165.24. C 72.69%, H 9.15%, N 8.48%, O 9.68%. Alkaloid post-synaptic depolarizing neuromuscular blocker acting as a nicotinic cholinergic receptor agonist. Isoln and purification: J. P. Devlin *et al., Can. J. Chem.* **55**, 1367 (1977). Improved purification: K. Harada *et al., Toxicon* **27**, 1289 (1989). Synthesis: H. F. Campbell *et al., Can. J. Chem.* **55**, 1372 (1977); stereoselective synthesis: M. Skrinjar *et al., Tetrahedron: Asymmetry* **3**, 1263 (1992). TLC determn in algal material: I. Ojanpera *et al., Analyst* **116**, 265 (1991); by LC/MS in cyanobacteria and drinking water: A. Furey *et al., Rapid Commun. Mass Spectrom.* **17**, 583 (2003). Review of syntheses: H. L. Mansell, *Tetrahedron* **52**, 6025-6061 (1996). $[\alpha]_D^{25}$ +39.8° (c = 0.676 in abs. ethanol). pKa: 9.4 Degrades readily in sunlight and high pH to non-toxic products. LD_{50} i.p. in mice: 200 μg/kg (Carmichael).

Anatoxin a(s). [103170-78-1] (5*S*)-2-Amino-4,5-dihydro-1-[(hydroxymethoxyphosphinyl)oxy]-*N*,*N*-dimethyl-1*H*-imidazole-5-methanamine; ANTX-a(s). $C_7H_{17}N_4O_4P$; mol wt 252.21. C 33.34%, H 6.79%, N 22.21%, O 25.37%, P 12.28%. Irreversible acetylcholinesterase inhibitor; only known organophosphate toxin found in cyanobacteria. The (s) refers to the production of excess saliva upon ingestion. Identification and preliminary purification: N. A. Mahmood, W. W. Carmichael, *Toxicon* **24**, 425 (1986); anticholinesterase activity: *eidem, ibid.* **25**, 1221 (1987). Mechanism of action: E. G. Hyde, W. W. Carmichael, *J. Biochem. Toxicol.* **6**, 195 (1991). Pharmacology in rat: W. O. Cook *et al., J. Environ. Pathol. Toxicol. Oncol.* **9**, 393, (1989). Biosensor determn in water: F. Villatte *et al., Anal. Bioanal. Chem.* **372**, 322 (2002). Stable in neutral or acid conditions. LD_{50} i.p. in mice: 20 μg/kg (Carmichael).

Caution: Overexposure in animals has caused muscle fasciculation, gasping and convulsions and death due to respiratory arrest. Anatoxin a(s) also causes salivation. (Carmichael).

622. Anazolene Sodium. [3861-73-2] 4-Hydroxy-5-[2-[4-(phenylamino)-5-sulfo-1-naphthalenyl]diazenyl]-2,7-naphthalenedisulfonic acid sodium salt (1:3); 4′-anilino-8-hydroxy-1,1′-azonaphthalene-3,5′,6-trisulfonic acid trisodium salt; trisodium 4′-anilino-8-hydroxy-1,1′-azonaphthalene-3,6,5′-trisulfonate; 1-naphthol-3,6-disulfonic acid-8-azo-4′-[*N*-phenyl-1′-naphthylamine]-8′-sulfonic acid trisodium salt; C.I. Acid Blue 92; C.I. 13390; Sulfone Acid Blue R; Coomassie Blue RL. $C_{26}H_{16}N_3Na_3O_{10}S_3$; mol wt 695.57. C 44.90%, H 2.32%, N 6.04%, Na 9.92%, O 23.00%, S 13.83%. Prepn: Ulrich, **US 611664** (1897 to Bayer); **DE 108546** (1899 to Farbwerke Mülheim); *Frdl.* **5**, 497; *Beilstein* **16**, EII, 226; *Colour Index* **vol. 4** (3rd ed., 1971) p 4053. Properties and biological behavior: S. H. Taylor, J. M. Thorp, *Br. Heart J.* **21**, 492 (1959). Clinical applications: S. H. Taylor, J. P. Shillingford, *ibid.* 497; I. S. Menzies, *J. Clin. Pathol.* **19**, 179 (1966).

Reddish-black powder. Soluble in water, acetone, Cellosolve, giving a reddish-blue soln. Slightly sol in alc. Absorption max (water): 565-570 nm; in acetone 585 nm; in human plasma 580-590 nm ($E_{1cm}^{1\%}$ about 600). Aq solns are stable and are not affected by light. Solns up to 10% do not stain the skin appreciably. LD_{50} i.v. in mice: 450 mg/kg (Taylor, Thorp).

THERAP CAT: Diagnostic aid (cardiac output, blood volume determination).

623. Ancitabine. [31698-14-3] (2*R*,3*R*,3a*S*,9a*R*)-2,3,3a,9a-Tetrahydro-3-hydroxy-6-imino-6*H*-furo[2′,3′:4,5]oxazolo[3,2-*a*]pyrimidine-2-methanol; 2,2′-anhydro-(1β-D-arabinofuranosyl)cytosine; 2,2′-*O*-cyclocytidine; $O^{2,2′}$-cyclocytidine; ancytabine; anhydroara C. $C_9H_{11}N_3O_4$; mol wt 225.20. C 48.00%, H 4.92%, N 18.66%, O 28.42%. A cytostatic agent and intermediate in the synthesis of cytarabine, *q.v.* Prepn of the hydrochloride: E. R. Walwick *et al., Proc. Chem. Soc. London* **1959**, 84; T. Y. Shen, W. V. Ruyle, **US 3463850** (1969 to Merck & Co.); E. K. Hamamura *et al., J. Med. Chem.* **19**, 654 (1976). General pharmacological properties: H. Hirayama *et al., Oyo Yakuri* **6**, 1259 (1972), *C.A.* **79**, 49175f (1973). Metabolism: D. H. W. Ho, *Drug Metab. Dispos.* **1**, 752 (1973). Biochemical study: *idem, Biochem. Pharmacol.* **23**, 1235 (1974). Pharmacokinetic study: H. S. Chen, J. F. Gross, *Cancer Chemother. Pharmacol.* **2**, 85 (1979). Clinical studies: J. Z. Finklestein *et al., Cancer Treat. Rep.* **63**, 1331 (1979); T. Miale *et al., ibid.* 1913. Toxicity studies: K. Sugihara *et al., Oyo Yakuri* **8**, 1469 (1974); H. Hirayama *et al., ibid.* 1693, *C.A.* **83**, 71766, 37747 (1975). HPLC study: V. Reichelova *et al., J. Chromatogr.* **588**, 147 (1991).

Hydrochloride. [10212-25-6] NSC-145668; Cyclo-C. $C_9H_{11}N_3O_4$·HCl; mol wt 261.66. Cryst, mp 248-250° (dec). $[\alpha]_D^{}$ −21.8° (c = 2.0 in water). uv max (pH 1-7): 262, 231 nm (ε 10600, 9400).

THERAP CAT: Antineoplastic.

624. Ancrod. [9046-56-4] Agkistrodon serine proteinase; *Agkistrodon rhodostoma* venom proteinase; A-38414; Arvin; Viprinex. Defibrinating enzyme isolated from the venom of the Malayan pit-viper, *Agkistrodon rhodostoma* Boie (*Calloselasma rhodostoma* Boie). Glycosylated serine protease composed of 234 amino acid residues; mol wt ~35,400. Cleaves fibrinogen to form soluble, non-cross-linked fibrin and enhances the local release of tissue plasminogen activator, *q.v.* Isoln: **NL 6502120**; H. A. Reid *et al., US 3657416** (1965, 1972 to Nat. Res. Dev. Corp.); K. E. Chan *et al., Br. J. Haematol.* **11**, 646 (1965). Initial purification: M. P. Esnouf, G. W. Tunnah, *ibid.* **13**, 581 (1967). Improved purification, chemical composition: C. Nolan *et al., Methods Enzymol.* **45**, 205 (1976). Mechanism of action studies: W. R. Bell *et al., J. Lab. Clin. Med.* **91**, 592 (1978); C. R. M. Prentice *et al., Br. J. Haematol.* **83**, 276 (1993). Review of clinical pharmacology and efficacy: K. A. Illig, K. Ouriel, *Semin. Vasc. Surg.* **9**, 303-314 (1996). Review of use in anticoagulant therapy: R. L. Soutar, J. S. Ginsberg, *Crit. Rev.*

Oncol. Hematol. **15**, 23-33 (1993); in acute ischemic stroke: R. P. Atkinson, *Drugs* **54**, Suppl. 3, 100-108 (1997).

Colorless substance when pure, having a light powdery texture when in the freeze-dried state. Soluble in physiological saline. Absorbable on weakly basic anion exchange materials.

THERAP CAT: Anticoagulant.

625. Ancymidol. [12771-68-5] α-Cyclopropyl-α-(4-methoxyphenyl)-5-pyrimidinemethanol; α-cyclopropyl-4-methoxy-α-(pyrimidin-5-yl)benzyl alcohol; EL-531; A-Rest; Reducymol. C_{15}-$H_{16}N_2O_2$; mol wt 256.31. C 70.29%, H 6.29%, N 10.93%, O 12.48%. Prepn: J. D. Davenport *et al.*, **FR 1569940**; H. M. Taylor *et al.*, **US 3818009** (1969, 1974 to Lilly). Properties and efficacy: M. Snel, J. V. Gramlich, *Meded. Fac. Landbouwwet. Rijksuniv. Gent* **38**, 1033 (1973). GLC determn: R. Frank, E. W. Day, *Anal. Methods Pestic. Plant Growth Regul.* **8**, 475 (1976); S. D. West, E. W. Day, *J. Assoc. Off. Anal. Chem.* **60**, 904 (1977). Studies on the specificity and site of action: R. C. Coolbaugh *et al.*, *Plant Physiol.* **62**, 571 (1978).

Cryst. solid, mp 110-111°. Vapor pressure at 50°: $<1 \times 10^{-6}$ mm Hg. Thermally stable. Soly in water at 25°: ~650 mg/l. Readily sol in acetone, methanol, ethyl acetate, chloroform, acetonitrile. Moderately sol in aromatic hydrocarbons; slightly sol in saturated hydrocarbons. LD_{50} in rats, mice (mg/kg): 4500, 5000 orally (Snel, Gramlich).

USE: Plant growth regulator.

626. Andrographis. Chuan-xin-lian; Kalmegh. Herbaceous plant, *Andrographis paniculata* Nees, *Acanthaceae*, which has been used in Asian traditional medicine as an antipyretic, anti-inflammatory, and hepatoprotectant. *Habit.* Tropical regions of Asia; common ground flora in dry deciduous forests of southeastern India. *Constit.* Andrographolide, *q.v.*, deoxyandrographolide, neoandrographolide, andrographidines, paniculides. Brief description: R. D. Girach *et al.*, *Int. J. Pharmacognosy* **32**, 95 (1994). Review of constituents and pharmacology: W. Tang, G. Eisenbrand, *Chinese Drugs of Plant Origin* (Springer-Verlag, Berlin, 1992) pp 97-103. Chromatographic analysis of diterpenoids in leaves: S. Saxena *et al.*, *Phytochem. Anal.* **11**, 34 (2000). Clinical evaluation in respiratory tract infections: J. Melchior *et al.*, *Phytomedicine* **7**, 341 (2000); in familial Mediterranean fever: G. Amaryan *et al.*, *ibid.* **10**, 271 (2003). *In vitro* anticancer and immunostimulatory activities: R. A. Kumar *et al.*, *J. Ethnopharmacol.* **92**, 291 (2004).

THERAP CAT: In treatment of upper respiratory infections; hepatoprotectant.

627. Andrographolide. [5508-58-7] (3*E*,4*S*)-3-[2-[(1*R*,-4a*S*,5*R*,6*R*,8a*S*)-Decahydro-6-hydroxy-5-(hydroxymethyl)-5,8a-dimethyl-2-methylene-1-naphthalenyl]ethylidene]dihydro-4-hydroxy-2(3*H*)-furanone; 3α,14,15,18-tetrahydroxy-5β,9βH,10α-labda-8(20),12-dien-16-oic acid γ-lactone. $C_{20}H_{30}O_5$; mol wt 350.46. C 68.54%, H 8.63%, O 22.83%. Main bitter constituent and medicinally active principle isolated from the leaves of andrographis, *q.v.* Isoln: M. K. Gorter, *Rec. Trav. Chim.* **30**, 151 (1911). Rapid isoln method: M. Rajani *et al.*, *Pharm. Biol.* **38**, 204 (2000). Characterization: D. Chakravarti, R. N. Chakravarti, *J. Chem. Soc.* **1952**, 1697; R. Schwyzer *et al.*, *Helv. Chim. Acta* **34**, 652 (1951). Structure: M. P. Cava, B. Weinstein, *Chem. Ind. (London)* **1959**, 851. Revised structure: M. P. Cava *et al.*, *ibid.* **1963**, 167; M. P. Cava *et al.*, *Tetrahedron* **21**, 2617 (1965). X-ray crystallographic analysis: T. Fujita *et al.*, *Chem. Pharm. Bull.* **32**, 2117 (1984). HPLC determn in plant material: A. Sharma *et al.*, *Phytochem. Anal.* **3**, 129 (1992). Determn in Chinese medicinal prepns: Z. Yanfang *et al.*, *J. Pharm. Biomed. Anal.* **40**, 157 (2006). Mechanism of anti-inflammatory action: Y.-F. Xia *et al.*, *J. Immunol.* **173**, 4207 (2004). Clinical pharmacokinetics and bioavailability from herbal preparations: A. Panossian *et al.*, *Phytomedicine* **7**, 351 (2000).

Colorless plates from ethanol or methanol, mp 218° (dec) (Gorter). Also reported as mp 218-221° (Fujita). $[\alpha]_D^{25}$ −96.2° (c = 1.00 in C_5H_5N). d_4^{21} 1.2317. uv max (alcohol): 223 nm (log ε 4.09). Sparingly sol in water; sol in acetone, methanol, chloroform, ether.

Triacetyl derivative. $C_{26}H_{36}O_8$. Fine needles from alcohol + ether, mp 126-126.5°.

628. Androstane. [438-22-2] (5α)-Androstane; etioallocholane. $C_{19}H_{32}$; mol wt 260.47. C 87.61%, H 12.38%. From androstane-3,17-dione: Butenandt, Tscherning, *Z. Physiol. Chem.* **229**, 185 (1934). From androstane-3,17-diol: Steiger, Reichstein, *Helv. Chim. Acta* **20**, 817 (1937). From Δ^{16}-androstene: Prelog *et al.*, *ibid.* **27**, 66 (1944).

Leaflets from acetone-methanol, mp 50-50.5°. Sublimes at 60° and 0.003 mm Hg. $[\alpha]_D^{16}$ +2° (c = 1.2 in chloroform). Sol in acetone, alc, methanol, ether, petr ether, chloroform.

17-Amino-HCl. Dec 345°: Marker, *J. Am. Chem. Soc.* **58**, 480 (1936).

629. Androstane-3β,11β-diol-17-one. [514-17-0] (3β,5α,-11β)-3,11-Dihydroxyandrostan-17-one. $C_{19}H_{30}O_3$; mol wt 306.45. C 74.47%, H 9.87%, O 15.66%. First obtained by degradation of allopregnane-3β,11β,17α,20β,21-pentol (Reichstein's Substance A): Reichstein, *Helv. Chim. Acta* **19**, 402 (1936); later isolated in small quantities directly from extracts of the adrenal cortex: Reichstein, von Euw, *ibid.* **21**, 1197 (1938); *ibid.* **24**, 879 (1941). It is uncertain whether this substance occurs in the fresh adrenal gland; it may possibly occur by oxidation or decompn during the isolation procedure: Reichstein, Shoppee, *Vitam. Horm.* **I**, 368 (1943). Structure-activity study: S. Sassa *et al.*, *J. Biol. Chem.* **254**, 10011 (1979). Chromatographic studies: A. Kerebel *et al.*, *J. Chromatogr.* **140**, 229 (1977); J. T. Lin, E. Heftmann, *ibid.* **237**, 215 (1982).

Needles from acetone + ether, mp 235-238°. $[\alpha]_D^{20}$ +84.5° (ethanol); $[\alpha]_D^{19}$ +81.3° (dioxane); $[\alpha]_{545}^{19}$ +105° (dioxane). Precipitated by digitonin.

3-Acetate. $C_{21}H_{32}O_4$. Needles from actone + ether, mp 230-231°. $[\alpha]_D^{19}$ +70.5° (dioxane); $[\alpha]_{546}^{19}$ +87.1° (dioxane).

Diacetate. $C_{23}H_{34}O_5$. Crystals, mp 154-156°.

630. Androstenediol. [521-17-5] (3β,17β)-Androst-5-ene-3,17-diol; Δ^5-androstene-3β,17β-diol. $C_{19}H_{30}O_2$; mol wt 290.45. C 78.57%, H 10.41%, O 11.02%. Obtained from dehydroandrosterone: Butenandt, Hanisch, *Ber.* **68**, 1859 (1935); *Z. Physiol. Chem.* **237**, 89 (1935); Ruzicka, Wettstein, *Helv. Chim. Acta* **18**, 1264 (1935); *FIAT Final Report* 996, 45 (1947); Levy, Kapp, **US 2521586** (1950 to Nopco Chem.).

Leaflets from acetone + petr ether, or from methanol or ethyl acetate. Sublimes in high vacuum. mp 184°. $[\alpha]_D^{18}$ −55.5° (c = 0.4 in isopropanol). Insol in water.

3-Acetate. [1639-43-6] $C_{21}H_{32}O_3$. Crystals from hexane, mp 147-148°.

17-Acetate. [5937-72-4] $C_{21}H_{32}O_3$. Crystals from hexane, mp 146.5-148.5°, $[\alpha]_D^{18}$ −62.4° (alc).

Diacetate. [2099-26-5] $C_{23}H_{34}O_4$. Leaflets from hexane, mp 165-166°, $[\alpha]_D^{18}$ −56.5° (alc).

17-Benzoate. $C_{26}H_{34}O_3$. Crystals from methanol, mp 220-222°.

3-Acetate-17-benzoate. [5953-63-9] $C_{28}H_{36}O_4$. Crystals, mp 180-182°.

Dipropionate. [2297-30-5] $C_{25}H_{38}O_4$; mol wt 402.58.

Note: This is a controlled substance (anabolic steroid): **21 CFR**, 1308.13, as defined in 1300.01.

USE: Dietary supplement for body building.

631. Androstenedione. [63-05-8] Androst-4-ene-3,17-dione; Δ^4-androstenedione; Δ^4-etiocholendione-3,17. $C_{19}H_{26}O_2$; mol wt 286.42. C 79.68%, H 9.15%, O 11.17%. Weak androgen produced by the adrenal gland; converted in peripheral tissues to testosterone or estrone, *q.q.v.* Prepn: L. Ruzicka, A. Wettstein, *Helv. Chim. Acta* **18**, 986 (1935); E. S. Wallis, E. Fernholz, *J. Am. Chem. Soc.* **57**, 1511 (1935); and physiological activity: A. Butenandt, H. Kudszus, *Z. Physiol. Chem.* **237**, 75 (1935). Isoln from adrenal cortex: J. von Euw, T. Reichstein, *Helv. Chim. Acta* **24**, 879 (1941). Physiology: V. H. T. James, A. B. Goodall in *Proc. 9th Congr. Hung. Soc. Endocrinol. Metab.*, F. A. Laszlo, Ed. (Akad. Kiado, Budapest, 1979) pp 235-241. HPLC determn in plasma or serum: V. R. Walker *et al.*, *Anal. Biochem.* **234**, 194 (1996). Discussion of use as dietary supplement: L. Schnirring, *Physician Sportsmed.* **26**, 15-18 (1998). Clinical effect of oral supplement on blood chemistry: D. S. King *et al.*, *J. Am. Med. Assoc.* **281**, 2020 (1999).

Crystals from hexane, mp 173-174°. $[\alpha]_D^{30}$ +199° (in chloroform). *Note:* This is a controlled substance (anabolic steroid): **21 CFR**, 1308.13, as defined in 1300.01.

632. (3α,5α)-Androst-16-en-3-ol. [1153-51-1] 3α-Hydroxy-5α-androst-16-ene; Δ^{16}-androsten-3-ol. $C_{19}H_{30}O$; mol wt 274.45. C 83.15%, H 11.02%, O 5.83%. A major constituent of boar pheromone, having a pronounced musk-like odor. Isoln from swine testes: V. Prelog, L. Ruzicka, *Helv. Chim. Acta* **27**, 61 (1944). Prepn: V. Prelog *et al.*, *ibid.* 66; J. Fishman *et al.*, *J. Org. Chem.* **28**, 1443 (1963). Physiological role as a sex attractant for pigs: D. B. Gower, *J. Steroid Biochem.* **3**, 45 (1972). Use in pig artificial insemination: D. R. Melrose *et al.*, **DE 1937264**; *eidem*, **US 3681490** (1970, 1972 both to Nat. Res. Dev. Corp.). *In vivo* metabolism in boar testes: Y. A. Saat *et al.*, *Biochem. J.* **144**, 347 (1974). It has also been detected in human male axillary sweat, but has no androgenic activity: B. W. L. Brooksbank *et al.*, *Experientia* **30**, 864 (1974). Radioimmunoassay: D. C. Bickell, D. B. Gower, *J. Steroid Biochem.* **7**, 451 (1976). Receptor studies: J. N. Gennings *et al.*, *Biochim. Biophys. Acta* **496**, 547 (1977). Biosynthetic studies: E.

L. Hurden *et al.*, *J. Endocrinol.* **81**, 161P (1979); G. M. Cook, D. B. Gower, *ibid.* **88**, 409 (1981). Discovery of the presence of androst-16-en-3-ol in truffles *(Tuber melanosporum)* has been offered as an explanation for the ability of pigs to detect truffles growing as deep as 1 meter underground: R. Claus *et al.*, *Experientia* **37**, 1178 (1981).

Crystals, mp 142.5-143°. Purified by sublimation in high vacuum and recryst from acetone. $[\alpha]_D^{20}$ +13.1° (c = 0.957 in chloroform). Gives a blue color in the Kägi-Miescher test: *Helv. Chim. Acta* **22**, 683 (1939).

USE: As an aid to estrus determn in pig artificial insemination.

633. Androsterone. [53-41-8] (3α,5α)-3-Hydroxyandrostan-17-one; *cis*-androsterone; 3α-hydroxy-17-androstanone; androstan-3α-ol-17-one; 3α-hydroxyetioallocholan-17-one; 3-epihydroxyetioallocholan-17-one. $C_{19}H_{30}O_2$; mol wt 290.45. C 78.57%, H 10.41%, O 11.02%. Isolation from male urine after removal of the phenolic estrogen fraction: Butenandt, Tscherning, *Z. Physiol. Chem.* **229**, 167 (1934); v. Euw, Reichstein, *Helv. Chim. Acta* **25**, 988 (1942). Prepn from cholesterol: Ruzicka, *ibid.* **17**, 1389 (1934); Marker, *J. Am. Chem. Soc.* **57**, 1755 (1935); Schoeller *et al.*, **US 2232735** (1941 to Schering).

Crystals from acetone-ether, mp 185-185.5°. Sublimes in high vacuum. $[\alpha]_D^{20}$ +94.6° (c = 0.7 in abs alc). $[\alpha]_D^{15}$ +87.8° (c = 1.5 in dioxane). Not precipitated by digitonin. Barely soluble in water. Sol in most organic solvents.

Acetate. $C_{21}H_{32}O_3$. Crystals from ether, sublimes in high vac, mp 165°, $[\alpha]_D^{14}$ +76.7° (c = 2.04 in dioxane); $[\alpha]_D^{25}$ +86° (c = 2 in ethanol).

Propionate. $C_{22}H_{34}O_3$. mp 151-152°.

Benzoate. $C_{26}H_{34}O_3$. mp 178°.

634. Anecortave Acetate. [7753-60-8] 21-(Acetyloxy)-17-hydroxypregna-4,9(11)-diene-3,20-dione; $\Delta^{4,9(11)}$-pregnadien-17α,21-diol-3,20-dione-21-acetate; Al-3789; Retaane. $C_{23}H_{30}O_5$; mol wt 386.49. C 71.48%, H 7.82%, O 20.70%. Angiostatic steroid. Prepn: J. Fried, E. F. Sabo, *J. Am. Chem. Soc.* **75**, 2273 (1953); R. P. Graber *et al.*, *ibid.* **4722**; J. Fried, E. F. Sabo, *ibid.* **79**, 1130 (1957). Angiostatic activity *in vitro*: L. G. McNatt *et al.*, *J. Ocul. Pharmacol. Ther.* **15**, 413 (1999); *in vivo*: D. BenEzra *et al.*, *Invest. Ophthalmol. Visual Sci.* **38**, 1954 (1997). Clinical study in age-related macular degeneration: J. S. Slakter *et al.*, *Ophthalmology* **110**, 2372 (2003). Review of pharmacology: A. F. Clark, *Expert Opin. Invest. Drugs* **6**, 1867-1877 (1997); and clinical experience: S. A. Vinores, *IDrugs* **8**, 327-334 (2005).

Crystals from acetone, mp 236-237°. $[\alpha]_D$ +117° (c = 1.0 in $CHCl_3$). uv max (alcohol): 238 nm (ε 15500). Also reported as mp 231.5-234.5°. $[\alpha]_D^{22}$ +124° (c = 1.04 in $CHCl_3$).

THERAP CAT: In treatment of macular degeneration.

635. Anemonin. [508-44-1] rel-(5R,6R)-1,7-Dioxadispiro-[4.0.4.2]dodeca-3,9-diene-2,8-dione; β,β'-1,2-dihydroxy-1,2-cyclo-butanediacrylic acid di-γ-lactone; Anemone camphor; Pulsatilla camphor. $C_{10}H_8O_4$; mol wt 192.17. C 62.50%, H 4.20%, O 33.30%. Found in Anemone pulsatilla L. and other Ranunculaceae. Its precursor in plants is protoanemonin. Isoln from Ranunculus acer: Zecher, Wohlmuth, Sci. Pharm. 22, 95 (1954); C.A. 48, 13169b (1954). Structure: Moriarty et al., J. Am. Chem. Soc. 87, 3251 (1965); Romain, Diss. Abstr. B 27, 3867 (1967). Synthesis: Sugiyama et al., C.A. 67, 116604n (1967). Toxicity study: R. Brodersen, A. Kjaer, Acta Pharmacol. Toxicol. 2, 109 (1946).

S,S-Form

Crystals from petr ether, mp 157-158°. Volatile with steam. Slightly sol in cold, more in hot water; sol in hot alcohol, chloroform, alkalies with yellow color. Practically insol in ether. LD_{50} i.p. in mice: 150 mg/kg (Brodersen, Kjaer).

Note: Not to be confused with anemonine which is 5-(carboxymethyl)-1,1-dimethylimidazolium hydroxide inner salt.

636. Anethole. [4180-23-8]; [104-46-1] (unspecified stereo). 1-Methoxy-4-(1E)-1-propen-1-ylbenzene; trans-p-propenylanisole; trans-1-p-anisylpropene; anise camphor; isoestragole; Monasirup. $C_{10}H_{12}O$; mol wt 148.21. C 81.04%, H 8.16%, O 10.79%. Chief constituent of anise, star anise and fennel oils; responsible for the characteristic odor and flavor of aniseed. Isoln: A. Cahours, Ann. 41, 56 (1842). Description of prepns and properties: A. Wagner, Manuf. Chem. 23, 56 (1952). Prepn from estragole, q.v., and separation of isomers: Y.-R. Naves, Compt. Rend. 246, 1734 (1958). Improved synthesis: R. J. DePasquale, Synth. Commun. 10, 225 (1980). Characterization of isomers: Y.-R. Naves et al., Bull. Soc. Chim. Fr. 1958, 566; and comparative pharmacology: J.-R. Boissier et al., Therapie 22, 309 (1967). HPLC determn in alcoholic beverages: P. Curro et al., J. Chromatogr. 404, 273 (1987). Isoln by supercritical fluid extraction: L. K. Liu, Anal. Commun. 33, 175 (1996). Metabolism in humans: J. Caldwell, J. D. Sutton, Food Chem. Toxicol. 26, 87 (1988). Review of use as aroma chemical: G. S. Clark, Perfum. Flavor. 18, 11-18 (1993); of safety evaluation as flavoring substance: P. Newberne et al., Food Chem. Toxicol. 37, 789-811 (1999).

Colorless, slightly oily liquid; sweet, characteristic odor and taste. mp 21.4°. d_4^{20} 0.9883. bp 231-237°; $bp_{2.3}$ 81-81.5°. n_D^{20} 1.56145. uv max (ethanol): 259 nm (ε 22300). Freely sol in alcohol; 1 ml dissolves in 2 ml alc. Sol in benzene, ethyl acetate, acetone, carbon disulfide, petr ether. Very slightly sol in water. Misc with ether, chloroform. LD_{50} i.p. in rats: 900 mg/kg (Boissier).

cis-Anethole. [25679-28-1] Naturally occurs in trace amounts in oils of anise, star anise and fennel; by-product in prepn of synthetic anethole. Stereospecific prepn: Y.-R. Naves, Helv. Chim. Acta 43, 230 (1960). Camphoraceous, unpleasant, fennel-like odor. mp −22.5°. d_4^{20} 0.9878. $bp_{2.3}$ 79-79.5°. n_D^{20} 1.55455. uv max (ethanol): 253.5 nm (ε 18500). LD_{50} i.p. in rats: 93 mg/kg (Boissier).

USE: Flavoring agent in foods and beverages; in perfumery, particularly for soap and dentifrices; pharmaceutic aid (flavor).

THERAP CAT: Antitussive.

637. Anethole Trithione. [532-11-6] 5-(4-Methoxyphenyl)-3H-1,2-dithiole-3-thione; 3-(p-anisyl)trithione; 3-(p-methoxyphenyl)-4,5-dithiacyclopent-2-ene-1-thione; trithio-p-methoxyphenyl-

propene; 5-(p-methoxyphenyl)-1,2-dithiacyclopent-4-ene-3-thione; 3-(p-methoxyphenyl)trithione; trithioanethole; anethole dithiolthione; ADT; Felviten; Mucinol; Sialor; Sulfarlem. $C_{10}H_8OS_3$; mol wt 240.35. C 49.97%, H 3.36%, O 6.66%, S 40.02%. Organosulfur compound that enhances glutathione synthesis and stimulates secretion of bile and saliva. Prepn: B. Böttcher, A. Lüttringhaus, Ann. 557, 89 (1947); O. Gaudin, N. Lozac'h, Compt. Rend. 224, 557 (1947); A. Thuillier, J. Vialle, Bull. Soc. Chim. Fr. 1959, 1398. HPLC determn in urine: A. N. Masoud, E. Bueding, J. Chromatogr. 276, 111 (1983). Clinical trial in xerostomia: J. B. Epstein et al., Oral Surg. Oral Med. Oral Pathol. 56, 495 (1983). Clinical trial in chemoprevention of lung cancer: S. Lam et al., J. Natl. Cancer Inst. 94, 1001 (2002). Review of mechanism of chemoprotective and antioxidant effects: M.-O. Christen, Methods Enzymol. 252, 316-323 (1995).

Orange-colored prisms from butyl acetate. Very bitter taste. mp 111°. Practically insol in water. Sol in pyridine, chloroform, benzene, dioxane, carbon disulfide. Slightly sol in ether, acetone, ethyl acetate, acetic acid, alc, cyclohexane, petr ether.

THERAP CAT: Choleretic; sialogogue.

638. Angelica. Tall, perennial, herbaceous plant, Angelica archangelica L., Umbelliferae. Habit. Europe, Asia. Medicinal parts are the seed, leaves and root. Constit. Root: Volatile oil (0.35-1.9%); more than 20 furanocoumarins including bergapten, angelicin, archangelicin, isoimperatorin, xanthotoxin; osthol, umbelliferone, flavonoids, sugars. Fruit: Volatile oil, furanocoumarins, fatty oil, phytosterols. Review of constituents and uses: V. E. Tyler, The Honest Herbal (Pharmaceutical Products Press, New York, 3rd Ed., 1993) pp 29-30; N. G. Bisset, M. Wichtl, Herbal Drugs and Phytopharmaceuticals, English Ed. (CRC Press, Boca Raton, 1994) pp 70-72; J. Gruenwald et al., PDR for Herbal Medicines (Medical Economics, Montvale, 2nd Ed., 2000) pp 32-34.

Angelica root oil. [8015-64-3] Volatile oil obtained by steam distillation of the dried slender roots. Constit. Chiefly β-phellandrene (13-28%), α-phellandrene (2-14%), α-pinene (14-31%). Pale yellow to deep amber liquid with warm, pungent odor and bittersweet taste. d_{25}^{25} 0.850-0.880. n_D^{20} 1.473-1.487. Acid value not more than 7.0. Rotation: 0° to +46°. Sol in most fixed oils, slightly sol in mineral oil. Relatively insol in glycerin, propylene glycol. Keep well closed, cool, and protected from light.

Angelica seed oil. Volatile oil obtained by steam distillation of the fresh seeds. Light yellow liquid having a sweet, delicate aroma. d_{25}^{25} 0.853-0.876. n_D^{20} 1.480-1.488. Rotation: +4° to +16°. Acid value not more than 3.0. Keep well closed, cool, and protected from light.

USE: Flavoring for liqueurs and gin.

THERAP CAT: Carminative, diaphoretic, diuretic.

639. Angelic Acid. [565-63-9] (2Z)-2-Methyl-2-butenoic acid; cis-2-dimethylcrotonic acid; 2-methylisocrotonic acid; cis-2,3-dimethylacrylic acid. $C_5H_8O_2$; mol wt 100.12. C 59.98%, H 8.05%, O 31.96%. Stereoisomer of tiglic acid. Found in ester form in sumbul root, Angelica archangelica L., Umbelliferae and together with tiglic acid esters in the oil of the Roman camomile, Anthemis nobilis L., Compositae. Isoln from seeds of Schoenocaulon officinale (Lindl.) A. Gray, Liliaceae (cevadilla seeds) by alkaline hydrolysis of cevadine: Stoll, Seebeck, Helv. Chim. Acta 35, 1275 (1952). Synthesis by trans addition of bromine to tiglic acid: Buckles, Mock, J. Org. Chem. 15, 680 (1950). Review and bibliography: Buckles et al., Chem. Rev. 55, 659 (1955).

Monoclinic rods, needles, plates; mp 45°. Spicy odor. *Vesicant.* d_4^{47} 0.983. bp_{760} 185°; bp_{12} 86°. Sublimes. Volatile with steam. n_D^{47} 1.4434. pK (25°) 4.30. uv max (H_2O): 217 nm (ε 5.15 × 10³). Molar heat of combustion 626.6 kcal. Sparingly soluble in cold water, freely sol in hot water. Sol in alcohol, ether. Prolonged boiling of aq soln causes isomerization to tiglic acid; the process is speeded up by traces of bromine and sunlight, also by strong mineral acids or alks. Dry crystals of angelic acid have been stored in bottles for years without evidence of isomerization.

Calcium salt dihydrate. $Ca(C_5H_7O_2)_2 \cdot 2H_2O$. Leaflets. Much more soluble in water than calcium tiglate: 100 parts of aq soln satd at 17.5° contains 23 parts of anhydr calcium angelate.

Amide. C_5H_9NO. Crystals, mp 127-128°.

Methyl ester. $C_6H_{10}O_2$. Liquid; d_4^{20} 0.9413; bp_{764} 128°; n_D^{20} 1.4321.

Ethyl ester. $C_7H_{12}O_2$. Liquid; $d_4^{19.5}$ 0.9178; bp_{760} 141.5°, bp_{11} 49°. n_D^{20} 1.4304. Heat of formn at constant vol 963.1 kcal, at constant press. 964.2 kcal.

640. Angelica Lactone. $C_5H_6O_2$; mol wt 98.10. C 61.22%, H 6.17%, O 32.62%. Exists in three forms. Prepn of α and β-forms: Wolff, *Ann.* **229**, 250 (1885); Thiele, *Ann.* **319**, 184 (1901); v. Auwers, *Ber.* **56**, 1672 (1923); J. H. Helberger *et al.*, *Ann.* **561**, 215 (1949). Prepn of γ-form: J. P. Wineburg *et al.*, *J. Heterocycl. Chem.* **12**, 749 (1975); V. Jäger, H. J. Günther, *Tetrahedron Lett.* **1977**, 2543; R.A. Amos, J. A. Katzenellenbogen, *J. Org. Chem.* **43**, 560 (1978). Toxicity data for α-form: E. J. Moran *et al.*, *Drug Chem. Toxicol.* **3**, 249 (1980).

α-form   β-form   γ-form

α-Form. [591-12-8] 5-Methyl-2(3*H*)-furanone; Δ^2-angelica lactone; γ-methyl-β,γ-crotonolactone; 4-hydroxy-3-pentenoic acid γ-lactone. Volatile needles, mp 18°. Sweet, herbaceous odor; coconut-like taste. *Flammable.* d_4^{20} 1.084. bp_{12} 56°. n_{He}^{20} 1.4476. Flash point, closed cup: 154°F (68°C). One gram dissolves in 20 ml water at 15°. Readily isomerizes to the β-form. LD_{50} orally in mice: 2800 mg/kg (Moran).

β-Form. [591-11-7] 5-Methyl-2(5*H*)-furanone; Δ^1-angelica lactone; γ-methyl-α,β-crotonolactone; 4-hydroxy-2-pentenoic acid γ-lactone. Liquid. Not solidified at $-17°$. d_4^{20} 1.076. bp_{751} 208-209°. bp_{10} 87°. n_{He}^{20} 1.4603. Sol in water. Forms a dimer. More stable than α-form.

γ-Form. [10008-73-8] Dihydro-5-methylene-2(3*H*)-furanone; γ-methylene-γ-butyrolactone. bp_{17} 80°.

USE: As fragrance and flavoring agent.

641. Angeli's Salt. [13826-64-7] Hyponitric acid disodium salt; sodium *N*-nitrohydroxylaminate; sodium α-oxyhyponitrite; sodium trioxodinitrate. $N_2Na_2O_3$; mol wt 121.99. N 22.96%, Na 37.69%, O 39.35%. $Na_2(ONNO_2)$. Highly oxidizing compound; decomposes to produce nitroxyl (HNO). Prepn: A. Angeli, *Gazz. Chim. Ital.* **26**, 17 (1896); and uv spectrum: C. C. Addison *et al.*, *J. Chem. Soc.* **1952**, 338. Crystal structure of monohydrate: H. Hope, M. R. Sequeira, *Inorg. Chem.* **12**, 286 (1973). Reactivity studies: P. A. S. Smith, G. E. Hein, *J. Am. Chem. Soc.* **82**, 5731 (1960); L. Torun *et al.*, *Tetrahedron Lett.* **40**, 5279 (1999). Decomposition studies: F. T. Bonner, B. Ravid, *Inorg. Chem.* **14**, 558 (1975); C. M. Maragos *et al.*, *J. Med. Chem.* **34**, 3242 (1991). pH-dependent cytotoxicity: D. A. Stoyanovsky *et al.*, *ibid.* **47**, 210 (2004).

White, microcrystalline powder. Very sol in water. d 2.26 (monohydrate). uv max (0.1 *M* NaOH): 248 nm (ε 8300). uv max (pH 7.4 phosphate buffer): 237 nm (ε 6100).

USE: HNO/NO^- donor in biochemical model studies.

642. Angiogenin. Single-chain, basic protein of 123 amino acids that induces the *in vivo* formation of blood vessels. Mol wt ~14,000 Da. First isolated from human adenocarcinoma cells; subsequently found in normal human plasma and shown to be produced by the liver. Angiogenin exhibits a characteristic ribonucleolytic activity toward 28S and 18S ribosomal RNA. Its amino acid sequence is 35% identical with that of human pancreatic ribonuclease.

Isoln, characterization, and angiogenic activity: J. W. Fett *et al.*, *Biochemistry* **24**, 5480 (1985). Amino acid sequence: D. J. Strydom *et al.*, *ibid.* 5486. Cloning and DNA sequence of human angiogenin gene: K. Kurachi *et al.*, *ibid.* 5494. Structural study: K. A. Palmer *et al.*, *Proc. Natl. Acad. Sci. USA* **83**, 1965 (1986). Ribonucleolytic activity: R. Shapiro *et al.*, *Biochemistry* **25**, 3527 (1986). Isoln from normal human plasma: R. Shapiro *et al.*, *ibid.* **26**, 5141 (1987). Tissue distribution in neonatal and adult rats: H. L. Weiner *et al.*, *Science* **237**, 280 (1987); in human tumor and normal cells: S. M. Rybak *et al.*, *Biochem. Biophys. Res. Commun.* **146**, 1240 (1987). Inhibition of protein synthesis: D. K. St. Clair *et al.*, *Proc. Natl. Acad. Sci. USA* **84**, 8330 (1987). Inhibition of angiogenic and ribonucleolytic activities of angiogenin by placental ribonuclease inhibitor: R. Shapiro, B. L. Vallee, *ibid.* 2238; F. S. Lee, B. L. Vallee, *Biochemistry* **28**, 3556 (1989). *Reviews:* J. F. Riordan, B. L. Vallee, *Br. J. Cancer* **57**, 587-590 (1988); B. L. Vallee, J. F. Riordan, *Adv. Exp. Med. Biol.* **234**, 41-53 (1988).

643. Angiostatin. Naturally occurring inhibitor of angiogenesis that suppresses the growth of primary and metastatic tumors. 38 kDa protein generated by the cancer-mediated proteolysis of plasminogen, *q.v.*; amino acid sequence corresponds to the first four kringle domains. Isoln, amino acid sequence and bioactivity: M. S. O'Reilly *et al.*, *Cell* **79**, 315 (1994). Review of discovery: R. Vile, *Curr. Biol.* **5**, 10-13 (1995). Effect on human carcinomas in mice: M. S. O'Reilly *et al.*, *Nat. Med.* **2**, 689 (1996). Mechanism of angiostatin generation: S. Gately *et al.*, *Proc. Natl. Acad. Sci. USA* **94**, 10868 (1997).

644. Angiotensin. [1407-47-2] Formerly *hypertensin* and *angiotonin.* Pressor substance formed by the action of renin, *q.v.*, on a plasma substrate, *angiotensinogen (renin substrate, hypertensinogen).* The substance so formed is a decapeptide called *angiotensin I* which is converted to the active pressor agent, *angiotensin II*, by the splitting off of the *C*-terminal His-Leu residues by the *angiotensin converting enzyme* (ACE) or *angiotensinase.* The octapeptide angiotensin II differs among species only in the amino acid residue in position 5 being either Val or Ile. *See* reviews for refs to isoln and synthesis. Angiotensin II acts directly on the adrenal gland to stimulate the release of aldosterone, *q.v.* Rapid liquid phase synthesis of a protected angiotensin II: S. Nozaki, I. Muramatsu, *Bull. Chem. Soc. Jpn.* **55**, 2165 (1982). Extraction and characterization of angiotensins I and II from rat brain: D. Ganten *et al.*, *Science* **221**, 869 (1983). *Reviews:* M. Bodanszky, M. A. Ondetti, *Peptide Synthesis* (John Wiley, New York, 1966) pp 215-223; E. Schröder, K. Lübe, *The Peptides* vol. **II** (English ed., New York, 1966) pp 4-62; Bumpus, Smeby, "Angiotensin" in *Renal Hypertension*, I. Page, J. McCubbin, Eds. (Year Book Medical Publishers, Chicago, 1968) pp 62-98; Lee, "Angiotensin" in *Renin and Hypertension* (Williams & Wilkins, Baltimore, 1969) pp 32-94; G. M. Molinatti, P. Limone, *Minerva Med.* **72**, 715-732 (1981); J. A. Oliver *et al.*, *Ann. N.Y. Acad. Sci.* **394**, 275-277 (1982); M. Marin-Grez, *Biochem. Pharmacol.* **31**, 3941-3947 (1982). Symposium on hemodynamic effects of angiotensin II and its role in cardiovascular disease: *Am. J. Hypertens.* **15**, Suppl. 1, S1-S27 (2002). *See also* Tonin.

Asp–Arg–Val–Tyr–Ile–His–Pro–Phe

Angiotensin II (horse)

Angiotensins are very stable. Hydrolyzed by strong acids and bases and above pH 9.5. Sol in organic solvents, in aq solns pH 5-8. In high dilution, lost by absorption on walls of glass vessels.

Amide. [53-73-6] 5-Valine-angiotensin II amide; angiotensin II aspartic-β-amide 5-valine; 1-asparagine-5-valine-angiotensin II; val⁵-angiotensin II-asp¹-β-amide; Hypertensin. $C_{49}H_{70}N_{14}O_{11}$; mol wt 1031.19.

THERAP CAT: Amide as vasoconstrictor.

645. Angostura Bark. Cusparia bark; Carony bark; *Cortex Angosturae*; *Cortex Cuspariae.* The bark of the tree *Galipea officinalis* Hancock, habitat Venezuela, or the bark of the tree *Cusparia febrifuga* Humb., or *C. trifoliata* Engl., *Rutaceae*, habitat Brazil. *Ref:* Meyer, *Pharm. Ztg.* **80**, 120 (1935).

Unpleasant, musty odor. Bitter, slightly aromatic taste with pungent after-taste. The bitterness is due mainly to the bitter principle angosturin, $C_9H_{12}O_5$. The bark contains also the quinoline alkaloids cusparine, cuspareine; galipine, galipoidine, and galipoline.

THERAP CAT: Antipyretic.

646. Anhalamine. [643-60-7] 1,2,3,4-Tetrahydro-6,7-dimethoxy-8-isoquinolinol; 6,7-dimethoxy-8-hydroxy-1,2,3,4-tetrahydroisoquinoline. $C_{11}H_{15}NO_3$; mol wt 209.25. C 63.14%, H 7.23%, N 6.69%, O 22.94%. From *Lophophora williamsii* (Lemaire) Coutl. *(Anhalonium lewinii* Henn.), *Cactaceae:* Kauder, *Arch. Pharm.* **237**, 190 (1899); Späth, Becke, *Monatsh. Chem.* **66**, 327 (1935). Structure: *eidem, Ber.* **67**, 2100 (1934). Synthesis: Späth, Roder, *Monatsh. Chem.* **43**, 93 (1922); Brossi *et al., Helv. Chim. Acta* **47**, 2089 (1964); **49**, 403 (1966). Biosynthetic studies: Kapadia *et al., J. Am. Chem. Soc.* **92**, 6943 (1970). *Review:* Manske in R. H. F. Manske, H. L. Holmes, *The Alkaloids* **vol. IV** (Academic Press, New York, 1954) pp 8-14.

Crystals, mp 189-191°. uv max (ethanol): 274 nm (log ε 2.90). Almost insol in cold water, cold alcohol, ether. Sol in hot water, alcohol, acetone, dil acids.

Hydrochloride dihydrate. $C_{11}H_{15}NO_3 \cdot HCl \cdot 2H_2O$. Crystals from water, mp 258°.

647. Anhalonidine. [17627-77-9] (1*S*)-1,2,3,4-Tetrahydro-6,7-dimethoxy-1-methyl-8-isoquinolinol; 6,7-dimethoxy-8-hydroxy-1-methyl-1,2,3,4-tetrahydroisoquinoline. $C_{12}H_{17}NO_3$; mol wt 223.27. C 64.56%, H 7.68%, N 6.27%, O 21.50%. From mescal buttons, the buds of *Lophophora williamsii* (Lemaire) Coutl. *(Anhalonium lewinii* Henn.) *Cactaceae:* Heffter, *Ber.* **27**, 2975 (1894); **29**, 221 (1896); Kauder, *Arch. Pharm.* **237**, 190 (1899). Structure and synthesis from 3-acetoxy-4,5-dimethoxy-*N*-acetylphenethylamine: Späth, Passl, *Ber.* **65**, 1778 (1932); from mescaline: Brossi *et al., Helv. Chim. Acta* **47**, 2089 (1964). Biosynthesis: Kapadia *et al., J. Am. Chem. Soc.* **92**, 6943 (1970).

Small octahedra from benzene, mp 160-161°. uv max (ethanol): 270 nm (log ε 2.81). Strong base. Freely sol in water, alcohol, chlorofom, hot benzene. Sparingly sol in ether. Insol in petr ether. Solns of anhalonidine acquire a reddish color on standing.

648. Anhalonine. [519-04-0] (9*S*)-6,7,8,9-Tetrahydro-4-methoxy-9-methyl-1,3-dioxolo[4,5-*h*]isoquinoline. $C_{12}H_{15}NO_3$; mol wt 221.26. C 65.14%, H 6.83%, N 6.33%, O 21.69%. From mescal buttons *[Lophophora williamsii* (Lemaire) Coutl. *(Anhalonium lewinii* Henn.), *Cactaceae]* also in *Ariocarpus*, in *Gymnocalycium gibbosum*. Synthesis of *dl*-form and resolution: Späth, Kesztler, *Ber.* **68**, 1663 (1935); Brossi *et al., J. Am. Chem. Soc.* **93**, 6248 (1971). Configuration: Battersby, Edwards, *J. Chem. Soc.* **1960**, 1214.

Rhombic needles from petr ether, mp 86°, bp$_{0.02}$ 140°. $[\alpha]_D^{25}$ −63.8° (methanol); −56.3° (chloroform). Very sol in alcohol, ether, chloroform, benzene, petr ether.

Hydrochloride. $C_{12}H_{15}NO_3 \cdot HCl$. Orthorhombic prisms, dec 255°; freely sol in hot water. Aq soln is neutral.

649. Anidulafungin. [166663-25-8] 1-[(4*R*,5*R*)-4,5-Dihydroxy-N^2-[[4″-(pentyloxy)[1,1′:4′,1″-terphenyl]-4-yl]carbonyl]-L-ornithine]echinocandin B; V-echinocandin; LY-303366; VER-002; Ecalta; Eraxis. $C_{58}H_{73}N_7O_{17}$; mol wt 1140.25. C 61.10%, H 6.45%, N 8.60%, O 23.85%. Semisynthetic echinocandin antifungal; inhibits 1,3-β-D-glucan synthase. Prepn: F. J. Burkhardt *et al.,* **EP 561639** (1993 to Lilly); M. Debono *et al., J. Med. Chem.* **38**, 3271 (1995). HPLC determn in plasma: L. L. Zornes, R. E. Stratford, *J. Chromatogr. B* **695**, 381 (1997). Comparative *in vitro* susceptibility: M. A. Pfaller *et al., Diagn. Microbiol. Infect. Dis.* **30**, 251 (1998); M. Cuenca-Estrella *et al., J. Antimicrob. Chemother.* **46**, 475 (2000). Pharmacokinetics and antifungal activity: A. H. Groll *et al., Antimicrob. Agents Chemother.* **45**, 2845 (2001). Clinical evaluation in invasive candidiasis: M. A. Pfaller *et al., ibid.* **49**, 4795 (2005). Review of clinical experience: M. A. Pfaller, *Expert Opin. Invest. Drugs* **13**, 1183-1197 (2004); of pharmacology: J. A. Vazquez, *Clin. Ther.* **27**, 657-673 (2005).

White to off-white powder. Slightly sol in ethanol. Practically insol in water.

THERAP CAT: Antifungal.

650. Anilazine. [101-05-3] 4,6-Dichloro-*N*-(2-chlorophenyl)-1,3,5-triazin-2-amine; 2,4-dichloro-6-(*o*-chloroanilino)-*s*-triazine; (*o*-chloroanilino)dichlorotriazine; Dyrene. $C_9H_5Cl_3N_4$; mol wt 275.52. C 39.23%, H 1.83%, Cl 38.60%, N 20.34%. Prepn: C. N. Wolf, **US 2720480** (1955 to Ethyl Corp.); E. G. Hill, E. Clinton, **US 2820032** (1958); K. H. Rattenburg *et al.,* **US 3074946** (1963 to Chemagro). Toxicity study: S. D. Cohen, S. D. Murphy, *J. Agric. Food Chem.* **21**, 140 (1973).

White to tan crystals, mp 159-160°. Insol in water. Soly at 30° (g/100 ml): toluene, 5; xylene, 4; acetone, 10. Subject to hydrolysis; not compatible with oils and alkaline materials. LD$_{50}$ orally in rats: >5000 mg/kg, Mobay Technical Information Sheet, Jan. 1979.

USE: Fungicide.

651. Anileridine. [144-14-9] 1-[2-(4-Aminophenyl)ethyl]-4-phenyl-4-piperidinecarboxylic acid ethyl ester; 1-(*p*-aminophenethyl)-4-phenylisonipecotic acid ethyl ester; ethyl 1-(4-aminophenethyl)-4-phenylisonipecotate; *N*-[β-(*p*-aminophenyl)ethyl]-4-phenyl-4-carbethoxypiperidine; *N*-β-(*p*-aminophenyl)ethylnormeperidine; Leritine. $C_{22}H_{28}N_2O_2$; mol wt 352.48. C 74.97%, H 8.01%, N 7.95%, O 9.08%. Opioid analgesic. Synthesis: Weijlard *et al., J. Am. Chem. Soc.* **78**, 2342 (1956); **US 2966490** (1960 to Merck & Co.).

White crystalline powder. Oxidizes on exposure to air and light, becoming darker in color. mp 83°. Freely sol in alcohol, chloroform; sol in ether, although it may show turbidity. Very slightly sol in water.

Dihydrochloride. [126-12-5] $C_{22}H_{28}N_2O_2.2HCl$. Crystals from methanol + ether, mp 280-287° (dec). uv max (pH 7 in 90% methanol contg phosphate buffer): 235, 289 nm ($A_{1cm}^{1\%}$ 293, 34.5). Distribution coefficient (water, pH 3.6/n-butanol): 0.9. Freely soluble in water, methanol. Solubility in ethanol: 8 mg/g. pH of aq solns 2.0 to 2.5. Solns are stable at pH 3.5 and below. At pH 4 and higher the insol free base is precipitated.

Note: This is a controlled substance (opiate): **21 CFR,** 1308.12.

THERAP CAT: Analgesic.

652. Aniline. [62-53-3] Benzenamine; aniline oil; phenylamine; aminobenzene; aminophen; kyanol. C_6H_7N; mol wt 93.13. C 77.38%, H 7.58%, N 15.04%. First obtained in 1826 by Unverdorben from dry distillation of indigo. Runge found it in coal tar in 1834. Fritzsche, in 1841, prepared it from indigo and potash and gave it the name aniline. Manuf from nitrobenzene or chlorobenzene: *Faith, Keyes & Clark's Industrial Chemicals,* F. A. Lowenheim, M. K. Moran, Eds. (Wiley-Interscience, New York, 4th ed., 1975) pp 109-116. Procedures: A. I. Vogel, *Practical Organic Chemistry* (Longmans, London, 3rd ed., 1959) p 564; Gattermann-Wieland, *Praxis des Organischen Chemikers* (de Gruyter, Berlin, 40th ed., 1961) p 148. Brochure *"Aniline"* by Allied Chemical's National Aniline Division (New York, 1964) 109 pp, gives reactions and uses of aniline (877 references). Toxicity study: K. H. Jacobson, *Toxicol. Appl. Pharmacol.* **22**, 153 (1972).

Oily liquid; colorless when freshly distilled, darkens on exposure to air and light. *Poisonous.* Characteristic odor and burning taste; combustible; volatile with steam. d_{20}^{20} 1.022. bp 184-186°. Solidif −6°. Flash pt, closed cup: 169°F (76°C). n_D^{20} 1.5863. pKb 9.30. pH of 0.2 molar aq soln 8.1. One gram dissolves in 28.6 ml water, 15.7 ml boil. water; misc with alcohol, benzene, chloroform, and most other organic solvents. Combines with acids to form salts. It dissolves alkali or alkaline earth metals with evolution of hydrogen and formation of anilides, e.g., C_6H_5NHNa. *Keep well closed and protected from light. Incompat.* Oxidizers, albumin, solns of Fe, Zn, Al, acids, and alkalies. LD_{50} orally in rats: 0.44 g/kg (Jacobson).

Hydrobromide. [542-11-0] $C_6H_7N.HBr$; mol wt 174.04. White to slightly reddish, crystalline powder, mp 286°. Darkens in air and light. Sol in water, alc. *Protect from light.*

Hydrochloride. [142-04-1] $C_6H_7N.HCl$; mol wt 129.59. Crystals, mp 198°. d 1.222. Darkens in air and light. Sol in about 1 part water; freely sol in alc. *Protect from light.*

Hydrofluoride. [542-13-2] $C_6H_7N.HF$; mol wt 113.14. Crystalline powder. Turns gray on standing. Freely sol in water; slightly sol in cold alc, freely sol in hot alc.

Nitrate. [542-15-4] $C_6H_7N.HNO_3$; mol wt 156.14. Crystals, dec about 190°. d 1.36. Discolors in air and light. Sol in water, alc. *Protect from light.*

Hemisulfate. [542-16-5] $C_6H_7N.\frac{1}{2}H_2SO_4$; mol wt 142.17. Crystalline powder. d 1.38. Darkens on exposure to air and light. One gram dissolves in about 15 ml water; slightly sol in alc. Practically insol in ether. *Protect from light.*

Acetate. [542-14-3] $C_6H_7N.C_2H_4O_2$; mol wt 153.18. Prepd from aniline and acetic acid: Vignon, Evieux, *Bull. Soc. Chim. Fr.* [4] **3**, 1012 (1908). Colorless liquid. d 1.070-1.072. Darkens with age; gradually converted to acetanilide on standing. Misc with water, alc.

Oxalate. [591-43-5] $(C_6H_7N)_2.C_2H_2O_4$; mol wt 276.29. Prepd from aniline and oxalic acid in alc soln: Hofmann, *Ann.* **47**, 37 (1843). Triclinic rods from water, mp 174-175°. Readily sol in water; sparingly sol in abs alc. Practically insol in ether.

Caution: Poisoning may occur from inhalation, skin penetration or ingestion. Potential symptoms of overexposure are navy blue to black lips, tongue, mucous membranes; slate gray skin; headache, nausea, vomiting, confusion, ataxia, vertigo, tinnitus, weakness, disorientation, lethargy, drowsiness; dyspnea on effort; methemoglobinemia, cyanosis, coma; tachycardia, heart blocks, arryhthmia, shock; painful micturition, hematuria, hemoglobinuria, methemoglobinuria, oliguria, renal insufficiency; cirrhosis. Direct contact may cause eye irritation. Potential occupational carcinogen. *See NIOSH Pocket Guide to Chemical Hazards* (DHHS/NIOSH 97-140, 1997) p 18; *Clinical Toxicology of Commercial Products,* R. E. Gosselin *et al.,* Eds. (Williams & Wilkins, Baltimore, 5th ed., 1984) Section III, pp 31-36.

USE: Manuf dyes, medicinals, resins, varnishes, perfumes, shoe blacks; vulcanizing rubber; as solvent. Hydrochloride used in manuf of intermediates, aniline black and other dyes, in dyeing fabrics or wood black.

653. Aniline Mustard. [553-27-5] *N,N*-Bis(2-chloroethyl)-benzenamine; *N,N*-bis(2-chloroethyl)aniline; phenylbis[2-chloroethylamine]; β,β'-dichlorodiethylaniline. $C_{10}H_{13}Cl_2N$; mol wt 218.12. C 55.07%, H 6.01%, Cl 32.51%, N 6.42%. Prepd by the action of phosphorus pentachloride on *N,N*-bis-[2-hydroxyethyl]aniline (phenyldiethanolamine): Robinson, Watt, *J. Chem. Soc.* **1934**, 1538; Korshak, Strepikheev, *J. Gen. Chem. USSR* **14**, 312 (1944).

Stout prisms from methanol, mp 45°. bp_{14} 164°; $bp_{0.5}$ 110°. Sol in hot methanol, ethanol. Very slightly sol in ether.

Hydrochloride. $C_{10}H_{14}Cl_3N$. Crystals. *Vesicant.* Freely sol in water. Sol in alcohol.

USE: In cancer research.

654. 1-Anilino-8-naphthalenesulfonate. [82-76-8] 8-(Phenylamino)-1-naphthalenesulfonic acid; 8-anilino-1-naphthalenesulfonic acid; ANS; phenylperi acid. $C_{16}H_{13}NO_3S$; mol wt 299.34. C 64.20%, H 4.38%, N 4.68%, O 16.03%, S 10.71%. Hydrophobic fluorescent probe for protein studies; originally used in prepn of aniline dyes. Prepd from 8-aminonaphthalenesulfonic acid, aniline and anilinium hydrochloride. Fluorescence spectrum: L. Stryer, *Science* **162**, 526 (1968). Binding to bovine serum albumin: G. Weber, L. B. Young, *J. Biol. Chem.* **239**, 1415 (1964); to lipoproteins: R. A. Muesing, T. Nishida, *Biochemistry* **10**, 2952 (1971); to chymotrypsin: J. D. Johnson *et al., ibid.* **18**, 1292 (1979); L. D. Weber *et al., ibid.* 1297. Interaction with rod outer segment membrane: U. P. Andley, B. Chakrabarti, *ibid.* **20**, 1687 (1981).

Magnesium salt. $C_{32}H_{24}MgN_2O_6S_2$. Green crystals from water. uv max: 350 nm (ε 4.95 × 10^3).

Ammonium salt. $C_{16}H_{16}N_2O_3S$. mp 242-244°.

USE: Fluorescent probe. In protein conformation studies. Magnesium salt as visualization reagent for proteins.

655.   Aniracetam. [72432-10-1] 1-(4-Methoxybenzoyl)-2-pyrrolidinone; 1-p-anisoyl-2-pyrrolidinone; Ro-13-5057; Draganon; Sarpul. $C_{12}H_{13}NO_3$; mol wt 219.24. C 65.74%, H 5.98%, N 6.39%, O 21.89%. Cognition enhancer related to piracetam, q.v. Prepn: **JP Kokai 79 117468**; E. Kyburz, W. Aschwanden, **US 4369139** (1979, 1983 both to Hoffmann-La Roche). Effects on learning and memory in rats: R. Cumin et al., Psychopharmacology **78**, 104 (1982); K. Yamada et al., Pharmacol. Biochem. Behav. **22**, 645 (1985); in monkeys and pigeons: M. J. Pontecorvo, H. L. Evans, ibid. 745. Clinical evaluation in geriatric patients: P. Foltyn et al., Arzneim.-Forsch. **33**, 865 (1983); in Alzheimer's disease: L. B. Sourander et al., Psychopharmacology **91**, 90 (1987). Review of toxicology: B. Schläppi et al., Drug Invest. **5**, Suppl. 1, 50-67 (1993).

Crystals from ethanol, mp 121-122°. LD_{50} in rats, mice (mg/kg): ~4500, >5000 orally (Cumin).

THERAP CAT: Nootropic.

656.   p-Anisaldehyde. [123-11-5] 4-Methoxybenzaldehyde; anisic aldehyde. $C_8H_8O_2$; mol wt 136.15. C 70.58%, H 5.92%, O 23.50%. Metabolic product of the odoriferous fungus Lentinus lepidus Fr.: Birkinshaw et al., Biochem. J. **38**, 131 (1944); of wood-rotting fungus Polyporus benzoinus (Wahl.) Fr.: Birkinshaw et al., ibid. **50**, 509 (1952); of Daldalea juniperina Murr.: Birkinshaw, Chaplen, ibid. **60**, 255 (1955). Prepn: Niedzielski, Nord, J. Am. Chem. Soc. **63**, 1462 (1941); Sisti et al., J. Org. Chem. **27**, 279 (1962). Toxicity study: P. M. Jenner et al., Food Cosmet. Toxicol. **2**, 327 (1964).

Oily liquid, bp 248°, $bp_{1.5}$ 89-90°. mp 0°. d_4^{15} 1.119. n_D^{13} 1.5764. Volatile in steam. Very slightly sol in water; misc with alc, ether. LD_{50} orally in rats: 1510 mg/kg (Jenner).

USE: Perfumery and toilet soaps; odor resembles that of coumarin, but the aldehyde must be mixed with other odorous substances to yield an agreeable odor. Also used in organic syntheses.

657.   Anise. Aniseed. Dried ripe fruit of Pimpinella anisum L., Umbelliferae. Habit. Western Asia, Egypt; cultivated in Southern Europe, India and U.S. Constit. Coumarins including scopoletin, umbelliferone; flavonol and flavone glycosides; 2-6% volatile oil, 50% carbohydrate, 16% lipids, β-amyrin, stigmasterol. Description and uses: J. Barnes et al., Herbal Medicines (Pharmaceutical Press, London, 2nd Ed., 2002) pp 51-54.

Volatile oil. [8007-70-3] Aniseed oil; oil of anise. Constit. 80-95% trans-anethole, estragole (methylchavicol), anise ketone, β-caryophyllene, anisaldehyde. Colorless or pale yellow, refractive liquid. d_{25}^{25} 0.978-0.988. Solidif pt not below 15°. α_D +1 to −1°. n_D^{20} 1.553-1.560. Slightly sol in water; sol in about 3 vols alcohol, freely in chloroform, ether. Keep cool in well-closed and well-filled containers, protected from light.

USE: Pharmaceutic aid (flavor). As condiment and flavor in foods or beverages; in manuf of liqueurs.

THERAP CAT: Carminative; expectorant.

658.   Anise Alcohol. [105-13-5] 4-Methoxybenzenemethanol; p-methoxybenzyl alcohol; anisyl alcohol. $C_8H_{10}O_2$; mol wt 138.17. C 69.54%, H 7.30%, O 23.16%. Prepd by reduction of anisaldehyde with triisobutyl aluminum and with amine boranes: Ziegler et al., Ann. **623**, 9 (1959); Chamberlain, Schechter, **US 2898379** (1959 to Callery Chemical). Toxicity study: Woodart et al., J. Pharmacol. Exp. Ther. **93**, 26 (1948).

Liquid, bp 259°. mp 24-25°, solidif 17°. d_{15}^{15} 1.113. Practically insoluble in water. Freely soluble in alc, ether. LD_{50} orally in rats: 1.2 ml/kg (Woodart).

659.   p-Anisic Acid. [100-09-4] 4-Methoxybenzoic acid. $C_8H_8O_3$; mol wt 152.15. C 63.15%, H 5.30%, O 31.55%. Prepd from methoxybenzene: Gross et al., Ber. **96**, 1382 (1963). Manuf by catalytic oxidation of p-methoxytoluene: **GB 798619** (1958 to General Electric); Cotterill et al., **GB 842998** (1960 to I.C.I.).

Needles, mp 184° (subl). bp 275-280°. d 1.385. Soluble in 2500 parts water; more sol in boiling water; freely sol in alc, chloroform, ether, ethyl acetate.

660.   Anisidine. C_7H_9NO; mol wt 123.16. C 68.27%, H 7.37%, N 11.37%, O 12.99%. Prepn of p-isomer: Cahours, Ann. **74**, 298 (1850); of o-isomer: Mühlhäuser, Ann. **207**, 235 (1881); of m-isomer: Kadaba, Massie, J. Org. Chem. **22**, 333 (1957).

m-**Anisidine.** 3-Methoxybenzenamine; 3-methoxyaniline; 3-aminoanisole. Pale yellow, oily liquid. Poisonous. Remains fluid even at −10°. bp 251°, bp_2 81-86°. Sol in alc, acids; sparingly sol in water.

o-**Anisidine.** 2-Methoxybenzenamine. Yellowish liquid; becomes brownish on exposure to air. Poisonous. Volatile with steam. bp 225°. mp 5°. d_{15}^{15} 1.098. Practically insol in water. Miscible with alc, ether, acetone, benzene. Keep well closed and protected from light.

p-**Anisidine.** 4-Methoxybenzenamine. Crystals, mp 57°. bp 246°. Poisonous. Freely sol in methanol, ethanol; sparingly sol in water.

Caution: o-Anisidine hydrochloride is reasonably anticipated to be a human carcinogen: Report on Carcinogens, Twelfth Edition (PB2011-111646, 2011) p 43.

USE: In the manuf of azo dyes.

661.   Anisindione. [117-37-3] 2-(4-Methoxyphenyl)-1H-indene-1,3(2H)-dione; 2-p-anisyl-1,3-indandione; 2-(p-methoxyphenyl)-1,3-indandione; SPE-2792; Miradon; Unidone. $C_{16}H_{12}O_3$; mol wt 252.27. C 76.18%, H 4.79%, O 19.03%. Prepn: Koelsch, J. Am. Chem. Soc. **58**, 1331 (1936); Horeau, Jacques, Bull. Soc. Chim. Fr. **1948**, 53; Sperber, **US 2899358** (1959 to Schering).

Pale yellow crystals from acetic acid or ethanol, mp 156-157°.
THERAP CAT: Anticoagulant.

662. Anisole. [100-66-3] Methoxybenzene. C_7H_8O; mol wt 108.14. C 77.75%, H 7.46%, O 14.79%. Prepn from phenol and dimethyl sulfate: Ullmann, *Ann.* **327**, 114 (1903); Graebe, *Ann.* **340**, 204 (1905); G. S. Hiers, F. D. Hager, *Org. Synth.* **coll. vol. I**, 58 (2nd ed., 1941); from bromobenzene: Agfa, **DE 411052**; *Chem. Zentralbl.* **1925**, I, 2411; *Frdl.* **15**, 193; by passing methyl chloride into a suspension of sodium phenolate in liquid ammonia: White *et al.*, *J. Am. Chem. Soc.* **46**, 965 (1924); from phenol, methyl iodide and potassium carbonate in dimethylformamide: Brieger *et al.*, *J. Chem. Eng. Data* **13**, 581 (1968). Forms oils or resins by condensation with formaldehyde: **DE 403264**; **DE 406152**; *Chem. Zentralbl.* **1925**, I, 307, 1816; *Frdl.* **14**, 626, 627. Absorption spectrum: Scheibe, *Ber.* **59**, 2625 (1926). Soly in glycerol: McEwen, *J. Chem. Soc.* **123**, 2285 (1923). Toxicity studies: J. M. Taylor *et al.*, *Toxicol. Appl. Pharmacol.* **6**, 378 (1964).

Liquid. Agreeable aromatic odor. d_4^{18} 0.9956; d_4^{45} 0.9701. mp $-37.3°$. bp_{760} 155.5°; bp_{100} 93.0°; bp_{40} 70.7°; bp_{20} 55.8°; bp_{10} 42.2°; bp_5 30.0°; $bp_{1.0}$ 5.4°. n_D^{20} 1.51791. *Flammable*. Sol in alcohol and ether. Insol in water. LD_{50} orally in rats: 3700 mg/kg (Taylor).
USE: In perfumery, in organic syntheses.

663. Anisomycin. [22862-76-6] (2R,3S,4S)-2-[(4-Methoxyphenyl)methyl]-3,4-pyrrolidinediol 3-acetate; [2R-(2α,3α,4β)]-2-[(4-methoxyphenyl)methyl]-3,4-pyrrolidinediol 3-acetate; 2-p-methoxyphenylmethyl-3-acetoxy-4-hydroxypyrrolidine; 1,4,5-trideoxy-1,4-imino-5-(4-methoxyphenyl)-D-xylo-pentitol 3-acetate. $C_{14}H_{19}NO_4$; mol wt 265.31. C 63.38%, H 7.22%, N 5.28%, O 24.12%. Protein synthesis inhibiting antibiotic isolated from *Streptomyces*: B. A. Sobin, F. W. Tanner, Jr., *J. Am. Chem. Soc.* **76**, 4053 (1954); Tanner *et al.*, **US 2691618** (1954 to Pfizer). Activity: J. E. Lynch *et al.*, *Antibiot. Chemother.* **4**, 844, 899 (1954). Structure and stereochemistry: J. J. Beereboom *et al.*, *J. Org. Chem.* **30**, 2334 (1965); J. P. Schaefer, P. J. Wheatley, *ibid.* **33**, 166 (1968); K. Butler, *ibid.* 2136. Biosynthesis: *idem, ibid.* **31**, 317 (1966). Total synthesis: Oida, Ohki, *Chem. Pharm. Bull.* **16**, 2086 (1968); *ibid.* **17**, 1405 (1969); I. Felner, K. Schenker, *Helv. Chim. Acta* **53**, 754 (1970). Chiral synthesis: J. P. H. Verheyden *et al.*, *Pure Appl. Chem.* **50**, 1363 (1978). Stereospecific total synthesis: D. P. Schumacher, S. S. Hall, *J. Am. Chem. Soc.* **104**, 6076 (1982). Improved synthesis: A. N. Hulme, E. M. Rosser, *Org. Lett.* **4**, 265 (2002). Mechanism of action: A. Jiménez, D. Vázquez in *Antibiotics* **vol. 5**(pt. 2), F. E. Hahn, Ed. (Springer-Verlag, New York, 1979) pp 1-19. Soly and stability data: *Antibiot. Annu.* **1954-55**, pp 809-810. Memory impairment effects in mice: K. M. Lattal, T. Abel, *Proc. Natl. Acad. Sci. USA* **101**, 4667 (2004).

Long needles from ethyl acetate or water, mp 140-141°. $[\alpha]_D^{23}$ $-30°$ (methanol). uv max: 224, 277, 283 nm (ε 10800, 1800, 1600). pKa 7.9. Base is moderately sol in water; sol in lower alcohols,

esters, ketones, chloroform; slightly sol in benzene, toluene and hexane.
Hydrochloride. [1963-48-0] $C_{14}H_{19}NO_4 \cdot HCl$. Crystals from ethyl acetate + ethanol, mp 187-188°. Very sol in water.
Deacetylanisomycin. [27958-06-1] $C_{12}H_{17}NO_3$; mol wt 223.27. Prepn from anisomycin: Nickell *et al.*, **US 2935444** (1960 to Pfizer). mp 176-179°. $[\alpha]_D^{25}$ $-20.0°$ (methanol), pK 9.2.
USE: Agricultural fungicide.

664. o-(p-Anisoyl)benzoic Acid. [1151-15-1] 2-(4-Methoxybenzoyl)benzoic acid; S-23/46. $C_{15}O_4$; mol wt 256.26. C 70.31%, H 4.72%, O 24.97%. Prepn from phthalic anhydride and anisole: Meyer, Turnau, *Monatsh. Chem.* **30**, 486 (1909). Alternate route: Arcus, Marks, *J. Chem. Soc.* **1956**, 1627.

Leaflets from water. Stout crystals from alcohol or toluene, mp 146°. Very sparingly sol in water. Freely sol in alc, ether, toluene, chloroform, glacial acetic acid.
Sodium salt. $C_{15}H_{11}NaO_4$. Needles. Freely sol in water. Sol in alcohol. Rated approximately 150 times as sweet as cane sugar. Bitter taste if used in concns exceeding 0.2 g/liter. *Review:* Möhler, *Z. Lebensm.-Unters. Forsch.* **90**, 431 (1950), *C.A.* **44**, 8558d (1950).
USE: The sodium salt has been proposed as a sweetening agent.

665. p-Anisoyl Chloride. [100-07-2] 4-Methoxybenzoyl chloride. $C_8H_7ClO_2$; mol wt 170.59. C 56.33%, H 4.14%, Cl 20.78%, O 18.76%. Prepd from p-anisic acid and thionyl chloride: Vanderhaeghe *et al.*, *J. Pharm. Pharmacol.* **6**, 119 (1954).

Crystals or liquid. mp 22°. bp ~262-263° with slight decompn. Dec by water or alcohol. Sol in acetone, benzene.
Caution: Vapors can cause serious eye burns. Sealed containers may explode, because slow decompn at room temp may build up pressure. *Ref:* Carroll, *Chem. Eng. News* **38**, 40 (Aug. 22, 1960).

666. Anistreplase. [81669-57-0] Anisoylated plasminogen streptokinase activator complex; APSAC; BRL-26921; Eminase; Multilase. Binary complex of streptokinase and human plasminogen in which the catalytic site is reversibly blocked by a p-methoxybenzoyl group without affecting the fibrin-binding site. Prepn and fibrinolytic activity: R. A. G. Smith, J. G. Winchester, **EP 28489**; *eidem*, **US 4808405** (1981, 1989 both to Beecham). Deacylation of the fibrin-bound complex initiates thrombolysis: R. A. G. Smith *et al.*, *Nature* **290**, 505 (1981); D. Matsuo *et al.*, *Thromb. Res.* **24**, 347 (1981). Effect on hemostasis in humans: D. H. Staniforth *et al.*, *Eur. J. Clin. Pharmacol.* **24**, 751 (1983). Pharmacokinetics: M. Been *et al.*, *Int. J. Cardiol.* **11**, 53 (1986). Evaluation of thrombolytic efficacy in myocardial infarction: W. Kasper *et al.*, *Am. J. Cardiol.* **58**, 418 (1986); in major pulmonary embolism: J. H. Brett *et al.*, *Aust. N.Z. J. Med.* **17**, 77 (1987). Controlled clinical trial in acute myocardial infarction: S. Ikram *et al.*, *Br. Med. J.* **293**, 786 (1986). Review of mechanism of action, pharmacology and therapeutic use: J. P. Monk, R. C. Heel, *Drugs* **34**, 25-49 (1987).
THERAP CAT: Thrombolytic.

667. Annatto. [1393-63-1] C.I. Natural Orange 4; C.I. 75120; arnotta; annotta. Coloring matter from seeds of *Bixa orellana* L., *Bixaceae*. Extraction from seed: Barnett, Espoy, **US 2815287** (1957); Kocher, **US 2831775** (1958). Contains bixins, q.v., and several other yellow to orange-red pigments which give carotene reactions: Diemair *et al.*, *Naturwissenschaften* **39**, 211 (1933). Review of chemistry and applications: P. Collins, *Food Ingred. Proc. Int.*

13, 23-27 (1992). *See also*: *Colour Index* **vol. 3** (3rd ed., 1971) p. 3233.

Sol in alcohol, ether, oils.

USE: Food coloring in dairy products, esp butter and cheese, flour confectionary, fish, meat products, soft drinks, snack foods and dry mixes. In wood stains, polish, and varnishes. Spice or condiment. Insect repellent.

668. Annexins. Family of calcium and phopholipid binding proteins; virtually ubiquitous in both animal and plant cells. Characterized by a common structure composed of 4 or 8 homologous repeats of ~70 amino acids. Involved in membrane trafficking, transmembrane channel activity, inhibition of phospholipase A2, and inhibition of coagulation. Known variously by such names as *synexin*, *lipocortin*, *anchorin*, and *calpactin*, the term annexins was chosen for their ability to bind, or "annex", to cellular membranes. Description of nomenclature: M. J. Crumpton, J. R. Dedman, *Nature* **345**, 212 (1990). Isoln of synexin from bovine adrenal medullary tissue: C. E. Creutz *et al.*, *J. Biol. Chem.* **253**, 2858 (1978). Review of discovery, molecular biology and bioactivites: P. Raynal, H. B. Pollard, *Biochim. Biophys. Acta* **1197**, 63-93 (1994); of annexins in plant cells: G. B. Clark, S. J. Roux, *Plant Physiol.* **109**, 1133-1139 (1995). Review of three-dimensional structure: S. Liemann, R. Huber, *Cell. Mol. Life Sci.* **53**, 516-521 (1997). Review of role in plasminogen binding: H.-M. Kang et al., *Trends Cardiovasc. Med.* **9**, 92-102 (1999); K. A. Hajjar, S. Krishman, *ibid.* 128-138; of role in membrane fusion: H. Kubista *et al.*, *Subcell. Biochem.* **34**, 73-131 (2000); as nucleotide binding proteins: J. Bandorowicz-Pikula *et al.*, *BioEssays* **23**, 170-178 (2001). Comprehensive description: V. Gerke, S. E. Moss, *Physiol. Rev.* **82**, 331-371 (2002).

669. Annotinine. [559-49-9] $C_{16}H_{21}NO_3$; mol wt 275.35. C 69.79%, H 7.69%, N 5.09%, O 17.43%. Isoln from *Lycopodium annotinum* L., *Lycopodiaceae*: Manske, Marion, *Can. J. Res.* **21B**, 92 (1943). Structure: Przybylska, Marion, *Can. J. Chem.* **35**, 1075 (1957); Wiesner *et al.*, *Tetrahedron* **4**, 87 (1958); Przybylska, Ahmed, *Acta Crystallogr.* **11**, 718 (1958). Stereochemistry: Wiesner *et al.*, *Tetrahedron Lett.* **1961**, 187; Ho, *ibid.* **1969**, 1307. Synthesis: Wiesner *et al.*, *Can. J. Chem.* **47**, 433 (1969). Biogenesis: Leete, *Tetrahedron* **3**, 313 (1958).

Brilliant prisms from chloroform + methanol, mp 232°. Sol in chloroform, dil HCl. Sparingly sol in methanol.

670. p-Anol. [539-12-8] 4-(1-Propen-1-yl)phenol; *p*-propenylphenol; 4-hydroxy-1-propenylbenzene. $C_9H_{10}O$; mol wt 134.18. C 80.56%, H 7.51%, O 11.92%. Prepn from anethole: Stoermer, Kahlert, *Ber.* **34**, 1812 (1901); from 4-propenylphenylmagnesium bromide: Quelet, *Bull. Chim. Soc.* [4] **45**, 268 (1929).

Leaflets from boiling water. Weak odor of cloves, resembling that of eugenol. Spicy, pungent taste. mp 93-94°. bp_{760} 250° (dec); bp_{14} 138-140°. Slightly sol in hot water. Freely sol in dimethylformamide, other organic solvents. Sol in aq solns of KOH and NaOH.

USE: Intermediate in the synthesis of estrogens.

671. Anserine. [584-85-0] β-Alanyl-3-methyl-L-histidine; (*S*)-2-(3-aminopropanamido)-3-(1-methyl-1*H*-imidazol-5-yl)propanoic acid; 3-methyl-*N*,α-(β-alanyl)-L-histidine. $C_{10}H_{16}N_4O_3$; mol wt 240.26. C 49.99%, H 6.71%, N 23.32%, O 19.98%. Methylated analog of carnosine, *q.v.*, with similar bioactivity. First found in muscles of geese (*Anser* sp.), later in other animals, but is absent from human muscle. Acts as a physiological buffering agent and

antioxidant; present in higher concentrations in vigorously contracting muscle. Isoln: D. Ackermann *et al.*, *Z. Physiol. Chem.* **183**, 1 (1929); W. A. Wolff, D. W. Wilson, *J. Biol. Chem.* **109**, 565 (1935). Distribution in mammalian species: J. A. Zapp, Jr., D. W. Wilson, *ibid.* **126**, 19 (1938). Synthesis: O. K. Behrens, V. du Vigneaud, *ibid.* **120**, 517 (1937); H. Rinderknecht *et al.*, *J. Org. Chem.* **29**, 1968 (1964). FASI-CE determn in biological samples: A. Zinellu *et al.*, *Talanta* **84**, 931 (2011). Antioxidant activity: R. Kohen *et al.*, *Proc. Natl. Acad. Sci. USA* **85**, 3175 (1988); H. Fu *et al.*, *Radiat. Phys. Chem.* **78**, 1192 (2009). Review of distribution and bioactivity: A. A. Boldyrev, S. E. Severin, *Adv. Enzyme Regul.* **30**, 175-194 (1990).

Needles from dil alc, mp 240-242° (dec). $[\alpha]_D^{23}$ +11.4° (c = 5 in water). pK_1 2.64; pK_2 7.04; pK_3 9.49. The dried material is very hygroscopic. Freely soluble in water. Slightly sol in methanol, even less in ethanol.

Nitrate. $C_{10}H_{16}N_4O_3 \cdot HNO_3$. Needles from dil methanol, mp 226-228°.

672. Antazoline. [91-75-8] 4,5-Dihydro-*N*-phenyl-*N*-(phenylmethyl)-1*H*-imidazole-2-methanamine; 2-(*N*-benzylanilinomethyl)-2-imidazoline; phenazoline; 2-(*N*-phenyl-*N*-benzylaminomethyl)imidazoline; imidamine; 5512-M. $C_{17}H_{19}N_3$; mol wt 265.36. C 76.95%, H 7.22%, N 15.84%. Prepn: Miescher, Klarer, **US 2449241** (1948 to Ciba). HPLC determn in ophthalmic solutions: S. C. Ruckmick *et al.*, *J. Pharm. Sci.* **84**, 502 (1995). Clinical trial in combination with naphazoline in allergic conjunctivitis: M. B. Abelson *et al.*, *Arch. Ophthalmol.* **108**, 520 (1990).

Crystals, mp 120-122°.

Hydrochloride. [2508-72-7] Antistine; Antasten. $C_{17}H_{19}N_3 \cdot HCl$; mol wt 301.82. Bitter crystals producing temporary numbness of the tongue. mp 237-241°. uv max: 242 nm ($E_{1cm}^{1\%}$, 495 to 515); min 222 nm. One gram dissolves in 40 ml water, in 25 ml alc. Practically insol in ether, benzene, chloroform. pH (1% aq soln): 6.3.

Phosphate. [154-68-7] $C_{17}H_{19}N_3 \cdot H_3PO_4$. White crystalline powder. Bitter taste. mp 194-198°. Sol in water; sparingly sol in methanol. Practically insol in benzene, ether. pH (2% aq soln): 4.5.

THERAP CAT: Antihistaminic.

THERAP CAT (VET): Antihistaminic.

673. Antheridiol. [22263-79-2] (3β,22S,23R)-3,22,23-Trihydroxy-7-oxostigmasta-5,24(28)-dien-29-oic acid γ-lactone; 3β,-22,23-trihydroxy-24-(carboxymethylene)cholest-5-en-7-one 23,24-lactone. $C_{29}H_{42}O_5$; mol wt 470.65. C 74.01%, H 9.00%, O 17.00%. The first specific functioning sex hormone to be identified in the plant kingdom. A diffusible substance secreted by the female mycelium of the filamentous water molds *Achlya bisexualis* and *A. ambisexualis*, which induces the growth of antheridial hyphae in the male plant, thereby initiating sexual reproduction in the species. Isoln and preliminary chemical data: Raper, Haagen-Smit, *J. Biol. Chem.* **143**, 311 (1942). Isoln of crystalline material: McMorris, Barksdale, *Nature* **215**, 320 (1967). Structure: Arsenault *et al.*, *J. Am. Chem. Soc.* **90**, 5635 (1968). Synthesis: J. A. Edwards *et al.*, *ibid.* **91**, 1248 (1969); J. H. Fried, J. A. Edwards, **US 3547911** (1970 to Syntex). Stereochemical proposal: Green *et al.*, *Tetrahedron* **27**, 1199 (1971). Absolute stereochemistry and alternate syntheses: Edwards *et al.*, *Tetrahedron Lett.* **1972**, 791; T. C. McMorris *et al.*, *ibid.* 2673; *eidem*, *J. Org. Chem.* **39**, 669 (1974). *Reviews*: Barksdale, *Ann. N.Y. Acad. Sci.* **144**, 313-319 (1967); *idem, Science* **166**, 831-837 (1969); T. C. McMorris, *Lipids* **13**, 716-722 (1978).

Colorless crystals from methanol, mp 250-255°. uv max (ethanol): 220 nm (ε 17,000). Sol in hot methanol. Slightly sol in chloroform; very slightly sol in water.

USE: Control and regulation of plant fertility.

674. Anthracene. [120-12-7] $C_{14}H_{10}$; mol wt 178.23. C 94.35%, H 5.66%. Obtained from coal tar, *q.v.*: Dumas, Laurent, *Ann.* **5**, 10 (1833); Laurent, *Ann.* **34**, 287 (1840); Anderson, *Ann.* **122**, 294 (1862); *J. Chem. Soc.* **15**, 44 (1862); Auerbach, *Das Anthracen und seine Derivate* (Braunschweig, 1880); Perkin, *J. Soc. Arts* **27**, 572 (1879); Lunge, *Coal Tar and Ammonia* (1916); Barnett, *Anthracene and Anthraquinone* (London, 1921); Nanson, *Text. Color.* **48**, 605, 678, 751 (1926); **49**, 19, 246, 557, 593 (1927); Houben, Fischer, *Das Anthracen und die Anthrachinone* (Leipzig, 1929); Borrmann, *Der Teer* (Leipzig, 1940); Schumann, *Kokereiteer* (Stuttgart, 1940). Extensive patent literature on purification. Prepn of very pure anthracene from synthetic anthraquinone: Clar, *Ber.* **72**, 1645 (1939). *Review:* E. Clar, *Polycyclic Hydrocarbons* **Vol. 1 & 2** (Academic Press, New York, 1964). Review of toxicology and human exposure: *Toxicological Profile for Polycyclic Aromatic Hydrocarbons* (PB95-264370, 1995) 487 pp.

Monoclinic plates from alc. Sublimes. When pure, colorless with violet fluoresence; when impure (due to tetracene, naphthacene), yellow with green fluorescence. Strongly triboluminescent and triboelectric. d_4^{27} 1.25. mp 218°. bp$_{760}$ 342°. Absorption spectrum: Clar, *Ber.* **65**, 506 (1932). Less soluble than the isomeric phenanthrene. Insol in water; one gram dissolves in 67 ml abs alcohol, 70 ml methanol, 62 ml benzene, 85 ml chloroform, 200 ml ether, 31 ml carbon disulfide, 86 ml carbon tetrachloride, 125 ml toluene. Anthracene darkens in sunlight. According to Downs, **US 1303639** (1919), when solns of crude anthracene in coal tar naphtha are exposed to ultraviolet irradiation, the anthracene is precipitated as dianthracene (*para*-anthracene) which is reconverted to anthracene by sublimation. Forms molecular addn products with nitro compounds.

Picric acid complex. mp 139°.
sym-**Trinitrobenzene complex.** mp 164°.
Trinitrotoluene complex. mp 162°.

USE: Important source of dyestuffs (manuf anthraquinone, alizarin dyes).

675. Anthragallol. [602-64-2] 1,2,3-Trihydroxy-9,10-anthracenedione; 1,2,3-trihydroxyanthraquinone; anthragallic acid; anthracene brown. $C_{14}H_8O_5$; mol wt 256.21. C 65.63%, H 3.15%, O 31.22%. From gallic acid and benzoic acid with sulfuric acid at 125° or from phthalic anhydride and pyrogallol with sulfuric acid at 160°: Seuberlich, *Ber.* **10**, 39 (1877). Other methods: Kubota, Perkin, *J. Chem. Soc.* **127**, 1889 (1925); Perkin, Story, *ibid.* **1929**, 1399; Cross, Perkin, *ibid.* **1930**, 292. Absorption spectrum: Meyer, Fischer, *Ber.* **46**, 85 (1913).

Brown crystals, mp 312-313°. Sublimes 290°. Slightly sol in water, chloroform; sol in alcohol, ether, glacial acetic acid. Its soln in concd H_2SO_4 is reddish-brown. Greenish-brown soln in ammonia water changes to blue on heating. Forms salts with Na, K, Ba, Tl, Pb, La, Ce, Nd, Co.

Trimethyl ether. $C_{17}H_{14}O_5$. Yellow needles from benzene + petr ether, mp 168°. Insol in water solns of alkalies.

676. Anthralin. [1143-38-0] 1,8-Dihydroxy-9(10*H*)-anthracenone; 1,8-dihydroxyanthrone; dithranol; Anthraforte; Anthranol; Anthrascalp; Antraderm; Cignolin; Dithrocream; Drithocreme; Dritho-Scalp; Micanol; Psoradrate; Psoriderm. $C_{14}H_{10}O_3$; mol wt 226.23. C 74.33%, H 4.46%, O 21.22%. Prepn: Y. Hirosé, *Ber.* **45**, 2474 (1912); K. Zahn, H. Koch, *ibid.* **71B**, 172 (1938). Revised structure: H. M. Avdovich, G. A. Neville, *Can. J. Spectrosc.* **25**, 110 (1980); F. R. Ahmed, *Acta Crystallogr.* **B36**, 3184 (1980). Historical review: R. E. Ashton *et al., J. Am. Acad. Dermatol.* **9**, 173-192 (1983). Review of mechanism of action: L. Kemény *et al., Skin Pharmacol.* **3**, 1-20 (1990); of clinical pharmacology and efficacy in psoriasis: G. Mahrle, *Clin. Dermatol.* **15**, 723-737 (1997).

Lemon yellow leaflets or needles from ligroin, mp 176-181°. Sol in chloroform, acetone, benzene and in sols of alkali hydroxides. Slightly sol in alcohol, ether, glacial acetic acid. Sol in dil NaOH with yellow color and green fluorescence, becoming orange-red on exposure to air. Practically insol in water.

THERAP CAT: Antipsoriatic.

677. Anthramycin. [4803-27-4] (2*E*)-3-[(11*R*,11a*S*)-5,10,-11,11a-Tetrahydro-9,11-dihydroxy-8-methyl-5-oxo-1*H*-pyrrolo[2,1-*c*][1,4]benzodiazepin-2-yl]-2-propenamide; 5,10,11,11a-tetrahydro-9,11-dihydroxy-8-methyl-5-oxo-1*H*-pyrrolo[2,1-*c*][1,4]benzodiazepin-2-acrylamide. $C_{16}H_{17}N_3O_4$; mol wt 315.33. C 60.94%, H 5.43%, N 13.33%, O 20.29%. Antitumor antibiotic produced by *Streptomyces refuineus* var *thermotolerans*, NRRL 3143: Leimgruber *et al., J. Am. Chem. Soc.* **87**, 5791 (1965); Berger *et al.*, **US 3361742** (1968 to Hoffmann-La Roche). Also from *Streptomyces spadicogriseus*: N. Komatsu *et al., J. Antibiot.* **33**, 54 (1980). Structure: Leimgruber *et al., J. Am. Chem. Soc.* **87**, 5793 (1965). Synthesis: *eidem, ibid.* **90**, 5641 (1968); Batcho, Leimgruber, **US 3524849** (1970 to Hoffmann-La Roche). Activity studies: Horwitz, Grollman, *Antimicrob. Agents Chemother.* **1968**, 21. Biosynthesis: L. H. Hurley *et al., J. Am. Chem. Soc.* **97**, 4372 (1975). *Review:* Kohn in *Antibiotics* **vol. 3**, J. W. Corcoran, F. E. Hahn, Eds. (Springer-Verlag, New York, 1975) pp 3-11.

Yellow prisms from acetone + water, dec 188-194°. uv max (acetonitrile): 235, 333 nm (ε 18,200, 31,800). $[\alpha]_D^{25}$ +930° (DMF).

Methyl ether hydrate. $C_{17}H_{19}N_3O_4 \cdot H_2O$. Pale yellow needles from methanol + water, dec above 120°.

678. Anthranol. [529-86-2] 9-Anthracenol; 9-anthrol; 9-hydroxyanthracene. $C_{14}H_{10}O$; mol wt 194.23. C 86.57%, H 5.19%, O 8.24%. Prepd by dissolving anthrone in 5 to 10% boiling NaOH soln, cooling to −5° and pouring in cooled 5% H_2SO_4: Meyer, *Ann.* **379**, 56 (1911). Also prepared from anthraquinone.

Needles from glacial acetic acid, plates from dil alcohol. Orthorhombic, mp 120° (if bath is heated to 110°). Sinters at 120°, mp 152° (if bath is cold at start). Absorption spectrum (of acetate): Barnett *et al.*, *J. Chem. Soc.* **1928**, 885. Sol in most solvents with blue fluorescence. The solid is fairly stable when kept dry. In solution it changes quickly to anthrone. Recrystallizations of anthranol may lead to a less pure compound.

Acetate. $C_{16}H_{12}O_2$. Crystals from petr ether, mp 134°.

Benzoate. $C_{21}H_{14}O_2$. Crystals from pyridine + alcohol, mp 170-172°.

Methyl ether. $C_{15}H_{12}O$. Crystals from alcohol, mp 97-98°.

Ethyl ether. $C_{16}H_{14}O$. Crystals from methanol, mp 73°.

USE: In the manuf of dyes.

679. Anthraquinone. [84-65-1] 9,10-Anthracenedione; 9,10-anthraquinone; 9,10-dioxoanthracene; Flight Control. $C_{14}H_8O_2$; mol wt 208.22. C 80.76%, H 3.87%, O 15.37%. Prepn: A. Laurent, *Ann.* **34**, 287 (1840). Structural relationship with anthracene: C. Graebe, C. Liebermann, *Ber.* **2**, 332 (1869). Industrial prepn from phthalic anydride and benzene: Klipstein, *Ind. Eng. Chem.* **18**, 1327 (1926). From anthracene with vanadium pentoxide, sodium chlorate, glacial acetic and sulfuric acids: *Org. Synth.* **coll. vol. II**, 554 (1943). Convenient lab procedure: L. F. Fieser, *Organic Experiments* (Heath & Co., Boston, 1964) pp 195-200. Review of chemistry: M. Phillips, *Chem. Rev.* **6**, 157 (1929); of manufacture and uses: A. J. Cofrancesco in *Kirk-Othmer Encyclopedia of Chemical Technology* **Vol. 2** (Wiley-Interscience, New York, 4th ed., 1992) pp 801-814. Development as bird repellent: R. M. Poché, *Proc. 18th Vertebr. Pest Conf.* **1998**, 338. Field trial to control Canadian geese: P. Devers *et al. ibid.* 345.

Light yellow, slender monoclinic prisms by sublimation *in vacuo*. Almost colorless, orthorhombic, bipyramidal crystals from $H_2SO_4 + H_2O$. d_4^{20} 1.42-1.44. mp 286°. bp$_{760}$ 377°. Absorption spectrum: Flexser *et al.*, *J. Am. Chem. Soc.* **57**, 2103 (1935). Insol in water. Solubility (g/100 g) in alc at 18° 0.05; at 25° 0.44; in boiling alc 2.25; in ether at 25° 0.11; in chloroform at 20° 0.61; at 40° 1.00; at 60° 1.60; in benzene at 20° 0.26; at 40° 0.50; at 60° 1.00; at 80° 1.80; in toluene at 25° 0.30.

Caution: Low acute oral toxicity, but may cause skin irritation, sensitization. *See Clinical Toxicology of Commercial Products,* R. E. Gosselin *et al.*, Eds. (Williams & Wilkins, Baltimore, 5th ed., 1984) Section II, p 189.

USE: Starting material for the manufacture of dyes; bird repellent.

680. Anthrarufin. [117-12-4] 1,5-Dihydroxy-9,10-anthracenedione; 1,5-dihydroxyanthraquinone. $C_{14}H_8O_4$; mol wt 240.21. C 70.00%, H 3.36%, O 26.64%. Prepd from 1,5-anthraquinone potassium disulfonate: Fierz-David, Blangey, *Farbenchemie* (Vienna, 5th ed, 1943), pp 224-225.

Green to yellow crystals from acetic acid, mp 280°. Sublimes at 120°. Absorption spectrum in H_2SO_4: Meyer, Fischer, *Ber.* **46**, 85 (1913). Sol in concd H_2SO_4, in aq KOH soln (reddish color); insol in aq Na_2CO_3, NH_3, $Ba(OH)_2$. Moderately sol in alcohol, slightly in water.

USE: Important intermediate in the manuf of alizarin and indanthrene dyestuffs. Forms insol Ba and Ca lakes; has been proposed as analytical reagent for the detection of Ca.

681. Anthrimide. [82-22-4] 1,1′-Iminobis-9,10-anthracenedione; 1,1′-iminodianthraquinone; dianthraquinonylamine; dianthrimide. $C_{28}H_{15}NO_4$; mol wt 429.43. C 78.31%, H 3.52%, N 3.26%, O 14.90%. Prepd by heating 1-aminoanthraquinone with 1-chloroanthraquinone in nitrobenzene in the presence of copper: Bayer & Co., **DE 162823**; *Chem. Zentralbl.* **1905**, II, 1206; Eckert, Steiner, *Monatsh. Chem.* **35**, 1129 (1914); *FIAT Report* **no. 1313 vol. II**, (1948), p 137.

Deep red needles from chlorobenzene, rhombs from nitrobenzene. Also described as coppery red needles with metallic sheen. Practically insol in low boiling organic solvents. Sparingly sol in aniline, nitrobenzene, chlorobenzene, quinoline. Sol in concd sulfuric acid first giving a scarlet-red color which turns olive-green on standing.

USE: Determination of boron. Used as a 0.02% soln in concd H_2SO_4. The olive-green soln turns blue in the presence of boron. The blue soln has an absorption max at 620 nm. Also used in the determination of silicon.

682. Anthrone. [90-44-8] 9(10*H*)-Anthracenone; 9,10-dihydro-9-oxoanthracene; Carbothrone. $C_{14}H_{10}O$; mol wt 194.23. C 86.57%, H 5.19%, O 8.24%. Prepn from anthraquinone by reduction with tin, glacial acetic and hydrochloric acid: *Org. Synth.* **coll. vol. I**, 60 (1941). By cyclization of *o*-benzylbenzoic acid with liquid HF: Fieser, Hershberg, *J. Am. Chem. Soc.* **61**, 1278 (1939).

Orthorhombic needles from benzene + petr ether, mp 155°. Absorption spectrum: Martin, *Ann. Combustibles Liq.* **12**, 967 (1937). Sol in most organic solvents without fluorescence. Any fluorescence present is due to anthranol. Tendency to change to anthraquinone. Equilibrium in abs alc: 89% anthrone; 11% anthranol.

Addition compound with 4 mols desoxycholic acid. $C_{110}H_{170}O_{17}$. mp 179°.

USE: In organic syntheses; in the colorimetric determination of sugar and animal starch in body fluids.

683. Antimony. [7440-36-0] Stibium; regulus of antimony. Sb; at. wt 121.60; at. no. 51; valence 3, 5. Group VA (15). Two naturally occurring isotopes: 121 (57.25%); 123 (42.75%); artificial radioactive isotopes: 112-120; 122; 124-135; isotopes 122 and 124 are useful radioactive tracers. First accurate description of antimony by Thölde (Basil Valentine) in 1604. Antimony ore is mined in China, Mexico, and Bolivia. The antimony of commerce is about 99% pure. Prepn in the laboratory by reduction of Sb_2O_5 with KCN: Schenk in *Handbook of Preparative Inorganic Chemistry* **vol. 1**, G. Brauer, Ed. (Academic Press, New York, 2nd ed.), 1963) p 606. Toxicity study: W. R. Bradley, W. G. Fredrick, *Ind. Med.* **10**, 15 (1941); *Ind. Hyg. Sect.* **2**, 15 (1941). *Review:* Smith, "Arsenic, Antimony and Bismuth" in *Comprehensive Inorganic Chemistry* **vol. 2**, J. C. Bailar, Jr. *et al.*, Eds. (Pergamon Press, Oxford, 1973) pp 547-683; S. C. Carapella in *Kirk-Othmer Encyclopedia of Chemical Technology* **vol. 3** (Wiley-Interscience, New York, 3rd ed., 1978) pp 96-105. Review of toxicology and human exposure: *Toxicological Profile for Antimony and Compounds* (PB93-110641, 1992) 160 pp.

Silver-white, lustrous, hard, brittle metal; scale-like crystalline structure; or dark gray, lustrous powder. Is not tarnished in dry air and only slowly in moist air. d 6.68. mp 630°. bp 1635°; also reported to be 1440°: *Gmelins, Antimony* (8th ed.) **18B**, p 69 (1949). sp heat 0.049; electrical resistivity 39 μ-ohm-cm at 0°. Not affected by cold dil acids; attacked by hot concd H_2SO_4; readily by aqua regia. Nitric acid, depending on the concn, converts it to antimonous or antimonic oxide. When finely divided it reacts with hot concd HCl. Qualitative analysis for antimony: react with slight excess of HCl with aid of HNO_3; pour soln in large vol of water; white ppt forms which becomes orange-red on addn of H_2S and is sol in ammonium sulfide. LD_{50} in rats, guinea pigs (mg Sb/100 g): 10.0, 15.0 i.p. (Bradley, Fredrick).

Caution: Potential symptoms of overexposure are cough, dizziness, headache, nausea, vomiting, diarrhea, stomach cramps, insomnia, anorexia, inability to smell properly. Direct contact may cause irritation of eyes, skin, nose, throat and mouth. *See NIOSH Pocket Guide to Chemical Hazards* (DHHS/NIOSH 97-140, 2003) p 18.

USE: In manufacture of alloys, such as Britannia or Babbitt metal, hard lead, white metal, type, bullets and bearing metal; in fireworks; for thermoelectric piles, blackening iron, coating metals, etc.

684. Antimony Chloride Oxide. [7791-08-4] Chlorooxostibine; antimony oxychloride; basic antimony chloride; powder of Algaroth; Mercurius vitae. ClOSb; mol wt 173.21. Cl 20.47%, O 9.24%, Sb 70.30%. SbOCl. Prepd from $SbCl_3$ and water: Schenk in *Handbook of Preparative Inorganic Chemistry* vol 1, G. Brauer, Ed. (Academic Press, N.Y., 2nd ed., 1963) p 611.

Monoclinic crystals or crystalline powder. Hydrolyzed by water to Sb_2O_3. Sol in HCl, tartaric acid, CS_2. Practically insol in alcohol, ether. Heating to 250° results in formation of $Sb_2O_5Cl_2$; above 320° Sb_2O_3 is formed.

685. Antimony Dichlorotrifluoride. [7791-16-4] Antimony chloride fluoride; antimony dichlorofluoride; antimony trifluorodichloride. Cl_2F_3Sb; mol wt 249.66. Cl 28.40%, F 22.83%, Sb 48.77%. $SbCl_2F_3$. Prepd according to the exothermic reaction SbF_3 + Cl_2 → $SbCl_2F_3$: Henne, *Org. React.* **2**, 61 (1944); Kwasnik in *Handbook of Preparative Inorganic Chemistry* Vol. 1, G. Brauer, Ed. (Academic Press, New York, 2nd ed., 1963) pp 200-201. Prepd from $SbCl_5$ and ClF_3: Dehnicke, Weidlein, *Z. Anorg. Allg. Chem.* **323**, 267 (1963).

Viscous liquid. Can be stored in iron vessels.

Caution: Highly toxic.

USE: As catalyst in the manufacture of organic fluorine compds: Slesser, Schram, *Preparation, Properties, and Technology of Fluorine and Organic Fluoro Compounds* (New York, 1951) *passim.*

686. Antimony Pentachloride. [7647-18-9] Antimony chloride ($SbCl_5$); antimony perchloride; pentachloroantimony. Cl_5-Sb; mol wt 299.01. Cl 59.28%, Sb 40.72%. $SbCl_5$. Usually prepd by passing chlorine into molten antimony trichloride. Laboratory procedure: Schenk in *Handbook of Preparative Inorganic Chemistry* Vol. 1, G. Brauer, Ed. (Academic Press, New York, 2nd ed., 1963) p 610.

Colorless to yellow, oily liquid. Fumes in air. Cannot be distilled at atmospheric pressures without complete decompn. d_4^{16} 2.358; d_4^{36} 2.319; d_4^{52} 2.289; d_4^{78} 2.231. mp 3.5°. bp_{14} 68°; bp_{22} 79°; bp_{30} 92°; bp_{55} 85°; bp_{68} 102.5°. Dipole moment: 1.14 at 16.3° in CCl_4. Corrosive. Mono- and tetrahydrates are formed in the presence of small amounts of water. Large amounts of water cause hydrolysis to Sb_2-O_5. Sol in hydrochloric acid, chloroform, carbon tetrachloride.

USE: As catalyst when replacing a fluorine substituent with chlorine in organic compounds.

687. Antimony Pentafluoride. [7783-70-2] Antimony fluoride (SbF_5); pentafluoroantimony. F_5Sb; mol wt 216.75. F 43.83%, Sb 56.18%. SbF_5. Lewis acid. Prepd industrially (in aluminum apparatus) according to the equation $SbCl_5$ + $5HF$ → $5HCl$ + SbF_5: Ruff, Plato, *Ber.* **37**, 673 (1904); Perkins, Irwin, US 2410358 (1946). Laboratory procedure using SbF_3 and F_2: Woolf, Greenwood, *J. Chem. Soc.* **1950**, 2200; Kwasnik in *Handbook of Preparative Inorganic Chemistry* Vol. 1, G. Brauer, Ed. (Academic Press, New York, 2nd ed., 1963) p 200. *Reviews:* Burg in *Fluorine Chemistry* Vol. 1, J. Simons, Ed. (Academic Press, New York, 1950) pp 104-106; Kemmitt, Sharp, *Adv. Fluorine Chem.* **4**, 210-211 (1965).

Hygroscopic, moderately viscous liquid. *Poisonous, corrosive.* mp 8.3°; bp 141°. $d^{25.8}$ 3.097; density data: Hoffman, Jolly, *J. Phys. Chem.* **61**, 1574 (1957). Reacts violently with water. Also forms a solid dihydrate, which reacts violently with more water to form a clear soln. Slowly hydrolyzed in NaOH solns forming $Sb(OH)_6{}^-$. Forms solids with sulfur chloride, carbon disulfide, benzene, toluene, petr ether (resin formation), ether, alc, acetone, ethyl acetate. Glacial

acetic acid gives a clear soln. Slowly corrodes glass, copper, lead. May be stored in aluminum vessels.

USE: In the fluorination of organic compds, *see* the monograph *Preparation, Properties and Technology of Fluorine and Organic Fluoro Compounds,* C. Slesser, S. R. Schram, Eds. (McGraw-Hill, New York, 1951) 868 pp.

688. Antimony Pentasulfide. [1315-04-4] Antimony sulfide (Sb_2S_5); golden antimony sulfide; antimonic sulfide; antimonial saffron; antimony red. S_5Sb_2; mol wt 403.82. S 39.70%, Sb 60.30%. Sb_2S_5. Prepn: *Gmelins, Antimony* (8th ed.) **18B**, pp 534-539 (1949). Existence of Sb^{5+} sulfide doubted: Birchall, Della Valle, *Chem. Commun.* **1970**, 675. Toxicity study: W. R. Bradley, W. G. Fredrick, *Ind. Med.* **10**; *Ind. Hyg. Sect.* **2**, 15 (1941).

Orange-yellow, odorless powder. Insol in water, alc. Sol in concd HCl with evolution of H_2S; sol in solns of alkali hydroxides or sulfides, forming sulfantimoniates. *Incompat.* Acids, metal salts. LD_{50} i.p. in rats: 150.0 mg Sb/100 g (Bradley, Fredrick).

USE: As pigment; vulcanizing and coloring rubber; manuf matches and fireworks.

689. Antimony Pentoxide. [1314-60-9] Antimony oxide (Sb_2O_5); antimonic oxide; diantimony pentoxide; stibic anhydride. O_5Sb_2; mol wt 323.52. O 24.73%, Sb 75.27%. Sb_2O_5. Prepn: Schenk in *Handbook of Preparative Inorganic Chemistry* vol. 1, G. Brauer, Ed. (Academic Press, New York, 2nd ed., 1963) p 616. Toxicity study: W. R. Bradley, W. G. Fredrick, *Ind. Med.* **10**; *Ind. Hyg. Sect.* **2**, 15 (1941).

Yellowish powder. Cubic. d 3.78. Loses oxygen at 300° or higher. Studies have shown that the compound does not correspond fully to Sb_2O_5, but that it is always somewhat hydrated. Slightly soluble in water. Practically insol in HNO_3. Slowly dissolves in warm HCl or in warm KOH soln. LD_{50} i.p. in rats: 400.0 mg Sb/100 g (Bradley, Fredrick).

USE: Flame retardant.

690. Antimony Potassium Oxalate. [5965-33-3] Ethanedioic acid anhydride with antimonic acid (3:1) tripotassium salt; tripotassium tris(oxalato)antimonate(3−); potassium oxalatoantimonate(III); "antimony salt". $C_6K_3O_{12}Sb$; mol wt 503.11. C 14.32%, K 23.31%, O 38.16%, Sb 24.20%. $K_3[Sb(OOCCOO)_3]$. Prepn of trihydrate: Graddon, *J. Inorg. Nucl. Chem.* **3**, 308 (1956-1957).

Trihydrate, cryst powder. Sol in water. *Poisonous.*

USE: A mordant in dyeing and printing fabrics instead of tartar emetic.

691. Antimony Potassium Tartrate. [28300-74-5] Bis[μ-[2,3-di(hydroxy-κO)butanedioato(4−)-κO¹:κO⁴]]diantimonate(2−) potassium hydrate (1:2:3) stereoisomer; tartar emetic; tartrated antimony; tartarized antimony; potassium antimonyltartrate. $C_8H_4K_2O_{12}Sb_2.3H_2O$; mol wt 667.87. C 14.39%, H 1.51%, K 11.71%, O 35.93%, Sb 36.46%. Manuf from potassium bitartrate and metallic antimony in the presence of HNO_3 or solid antimony oxide: Davies, US 2335585 (1943 to Am. Cream Tartar); US 2391297 (1945 to Stauffer Chem.). Structural studies: P. Pfeiffer, E. Schmitz, *Pharmazie* **4**, 451 (1949); E. Chinoporos, N. Papathanasopoulos, *J. Phys. Chem.* **65**, 1643 (1961); D. Grdenic, B. Kamenar, *Acta Crystallogr.* **19**, 197 (1965); R. Iyer *et al.*, *J. Inorg. Nucl. Chem.* **34**, 3351 (1972). Toxicity: N. Ercoli, *Proc. Soc. Exp. Biol. Med.* **129**, 284 (1968). Anthelmintic activity: Z. Farid *et al.*, *Trans. R. Soc. Trop. Med. Hyg.* **66**, 119 (1972); R. R. C. New *et al.*, *Nature* **272**, 56 (1978). Oxidimetric determn: Y. A. Gawargious *et al.*, *Pharmazie* **41**, 59 (1980).

Transparent crystals (effloresce on exposure to air) or powder. Sweetish, metallic taste. d 2.6. $[\alpha]_D^{20}$ +140.69° (c = 2 in water),

+139.25° (c = 2 in glycerol). *Poisonous.* One gram dissolves in 12 ml water, 3 ml boiling water, 15 ml glycerol. Insol in alcohol. The aq soln is slightly acid. LD_{50} in mice (mg/kg): 55 s.c.; 65 i.v. (Ercoli). *Incompat.* Mineral acids, tannic acid, gallic acids, alkali hydroxides and carbonates, lead and silver salts, mercury bichloride, lime water, albumin, soap.

USE: As mordant in the textile and leather industry.

THERAP CAT: Anthelmintic (Schistosoma).

THERAP CAT (VET): Has been used as a parasiticide (gastrointestinal and blood parasites); as an expectorant, and as a ruminatoric.

692. Antimony Sodium Gluconate. General term referring to either a trivalent or pentavalent antimony complex with sodium gluconate. Pentavalent compound may require reduction *in vivo* to the trivalent form to be bioactive. Determn in biological materials and pharmacodynamics: R. J. Ward *et al., Clin. Chim. Acta* **99**, 143 (1979). Mechanism of action study: J. C. Mottram, G. H. Coombs, *Exp. Parasitol.* **59**, 151 (1985).

Sodium antimonylgluconate. [12550-17-3] $C_6H_8NaO_7Sb$; mol wt 336.87. Trivalent antimony compound. Prepn: Das Gupta, *Indian J. Pharm.* **15**, 84 (1953); A. Axon *et al.,* US 3306921 (1967 to Burroughs Wellcome). Antiparasitic activity: N. Ercoli *et al., Chemotherapy* **26**, 254 (1980). White, amorphous powder. Sol in water. Insol in organic solvents. pH of 2% aq soln 9-10; the soln at this pH is unstable.

Sodium stiboglucononate. [16037-91-5] 2,4:2′,4′-O-(Oxydistibylidyne)bis[D-gluconic acid] *Sb,Sb′*-dioxide trisodium salt nonahydrate; Pentostam; Solustibosan. $C_{12}H_{17}Na_3O_{17}Sb_2 \cdot 9H_2O$; mol wt 907.88. Pentavalent antimony compound. Prepn: Datta, Ghosh, *Sci. Cult.* **11**, 699 (1945-6); Bose, Ghosh, *Indian J. Pharm.* **11**, 155 (1949). Pharmacokinetics: J. D. Chulay *et al., Trans. R. Soc. Trop. Med. Hyg.* **82**, 69 (1988). Review of clinical safety and efficacy in leishmaniasis: B. L. Herwaldt, J. D. Berman, *Am. J. Trop. Med. Hyg.* **46**, 296 (1992). Crystals. Freely sol in water. pH of 10% aq soln 5.4-5.6.

THERAP CAT: Trivalent compd as an anthelmintic (Schistosoma); pentavalent compd as an antiprotozoal (Leishmania).

693. Antimony Sodium Tartrate. [34521-09-0] Bis[μ-[(2R,3R)-2,3-di(hydroxy-κO)butanedioato(4−)-κO¹:κO⁴]]diantimonate(2−) sodium (1:2) ; antimony sodium oxide L-(+)-tartrate; sodium antimonyl tartrate; Stibnal. $C_8H_4Na_2O_{12}Sb_2$; mol wt 581.61. C 16.52%, H 0.69%, Na 7.91%, O 33.01%, Sb 41.87%. Trivalent antimonial. Prepn: Fargher, Gray, *J. Pharmacol. Exp. Ther.* **18**, 341 (1921). Antiparasitic activity *in vitro:* N. Ercoli *et al., Chemotherapy (Basel)* **26**, 254 (1980).

Hygroscopic, transparent or whitish scales or powder. Sweetish taste. Sol in 1.5 parts water. Practically insol in 90% alc. Aq solns are slightly acid to litmus. LD i.v. in mice: 25 mg/kg (Fargher, Gray).

THERAP CAT: Anthelmintic (Schistosoma).

694. Antimony Sulfate. [7446-32-4] Antimonous sulfate; antimony trisulfate. $O_{12}S_3Sb_2$; mol wt 531.69. O 36.11%, S 18.09%, Sb 45.80%. Prepn: Schenk in *Handbook of Preparative Inorganic Chemistry* vol. 1, G. Brauer, Ed. (Academic Press, New York, 2nd ed., 1963) pp 618-619.

Crystalline powder or lumps; deliquesc in air. *Poisonous.* d 3.62. With a little water forms a solid mass which dissolves in more water, but excess of water changes it into an insol basic salt. Sol in dil acids. *Keep well closed.*

695. Antimony Tribromide. [7789-61-9] Br_3Sb; mol wt 361.47. Br 66.32%, Sb 33.68%. $SbBr_3$. Best prepd from the ele-

ments: Schenk in *Handbook of Preparative Inorganic Chemistry* vol. 1, G. Brauer, Ed. (Academic Press, New York, 2nd ed., 1963) p 613.

Orthorhombic bipyramidal needles. Less hygroscopic than $SbCl_3$. d_{23}^{23} 4.148. mp 96°. bp_{749} 288°. Cryoscopic constant: 26.7. Specific heat: 0.0709 at 33°. Critical temp: 904.5°; crit press. 56 atm. Dipole moment: 2.47. Dec by light, water, alc. Sol in dil HCl, HBr, carbon disulfide, acetone, benzene, chloroform.

696. Antimony Trichloride. [10025-91-9] Trichlorostibine; antimony(III) chloride; butter of antimony. Cl_3Sb; mol wt 228.11. Cl 46.62%, Sb 53.38%. $SbCl_3$. Prepd from the elements: Hensgen, *Rec. Trav. Chim.* **9**, 301 (1890); Kendall *et al., J. Am. Chem. Soc.* **45**, 967 (1923); Schenk in *Handbook of Preparative Inorganic Chemistry* vol. 1, G. Brauer, Ed. (Academic Press, New York, 2nd ed., 1963) p 608. Purification: Werner, *Z. Anorg. Allg. Chem.* **181**, 154 (1929); Schaafsma, US 2324240 (1943 to Shell). Toxicology profile: H. Huang *et al., Toxicology* **129**, 113 (1998).

Orthorhombic, deliquesc needles from carbon disulfide or by sublimation at 100°. Acrid pungent odor. Fumes in air. *Poisonous, corrosive.* d_4^{20} 3.14. mp 73°. bp 223.5°; bp_{70} 143.5°; bp_{11} 102°. Cryoscopic constant: 18.4. Heat of fusion at 73.2°: 13.29 cal/g. Specific heat at 33°: 0.110 cal/g/°C. Dipole moment: 3.12 at 25° in CS_2. One gram dissolves in 10.1 ml H_2O at 25°. Gradual hydrolysis to SbOCl. Considerably more sol in dil HCl. Sol in alcohol, benzene, carbon disulfide, dioxane, chloroform (about 22%), ether, acetone, carbon tetrachloride. A satd soln in carbon tetrachloride is about 1.1*M*. Insol in pyridine, quinoline, other organic bases. *Handle in an anhydrous, inert atomsphere.*

Antimony trichloride solution. Antimony chloride solution; liquid butter of antimony. Contains 36-38% $SbCl_3$ and 10-12% free HCl. Yellow to reddish-brown, clear, somewhat oily, strongly caustic liquid. d 1.5.

Caution: Potential symptoms of overexposure are skin and eye irritation or burns, irritation of the respiratory tract and breathing difficulties, pulmonary edema, liver, heart, and kidney damage.

USE: Reagent for chloral, aromatic hydrocarbons, and vitamin A. Lewis acid catalyst in organic synthesis; in making other antimony salts. As mordant; in bronzing iron.

697. Antimony Trifluoride. [7783-56-4] Trifluorostibine; antimony fluoride; antimonous fluoride. F_3Sb; mol wt 178.76. F 31.88%, Sb 68.11%. SbF_3. Prepd by dissolving Sb_2O_3 in aq HF and evaporating the water: Berzelius, *Pogg. Ann.* **1**, 34 (1824); Söll, *FIAT-Review* **23**, 276 (1947); Andersen *et al., Acta Chem. Scand.* **7**, 236 (1953); Kwasnik in *Handbook of Preparative Inorganic Chemistry* Vol. 1, G. Brauer, Ed. (Academic Press, New York, 2nd ed., 1963) p 199. Industrial prepn from $SbCl_3$ and HF: Midgeley *et al.,* US 2024654 (1935). *Reviews:* Burg in *Fluorine Chemistry* Vol. 1, J. H. Simons, Ed. (Academic Press, New York, 1950) p 106; Kemmitt, Sharp, *Adv. Fluorine Chem.* **4**, 211 (1965).

Orthorhombic, deliquesc crystals. *Poisonous. Irritates the skin.* d_{20}^{20} 4.379. mp 292°. bp 376°. Also reported as bp 319° (Andersen). Soly in water (g/100 ml): 443 (20°); 562 (30°). Dissolves in water with limited hydrolysis, readily forms complexes such as $[SbF_4]^-$ and many sol salts. For the purpose of halogen exchange fluorination, the compd must be dry: Henne, US 2082161 (1937). May be stored in glass vessels or steel drums. *Keep well closed.*

USE: To catalyze fluorinations by HF, manuf chlorofluorides, in dyeing, usually in form of double salts, e.g., antimony sodium fluoride or antimony fluoride and ammonium sulfate double salt, manuf pottery and porcelains.

698. Antimony Triiodide. [7790-44-5] Triiodostibine. I_3Sb; mol wt 502.47. I 75.77%, Sb 24.23%. SbI_3. Prepd by the interaction of antimony and iodine in boiling benzene or tetrachloroethane: Bailar, Cundy, *Inorg. Synth.* **1**, 104 (1939); Scattergood, *J. Chem. Educ.* **20**, 40 (1943).

Ruby-red, trigonal crystals (yellowish-green modifications have been observed). d_4^{17} 4.921. The tendency to sublime becomes noticeable at 100°. mp 168°. bp 420°. Critical temp 1101°; crit press. 55 atm. Dipole moment: 1.58. Dec by water and air to SbOI (antimony oxyiodide). Sol in alcohol, acetone, carbon disulfide, HCl, soln of KI. Insol in carbon tetrachloride.

Consult the Name Index before using this section.

699. Antimony Trioxide. [1309-64-4] Antimony oxide (Sb_2O_3); diantimony trioxide; flowers of antimony; exitelite; senarmontite; valentinite; weissspiessglanz. O_3Sb_2; mol wt 291.52. O 16.46%, Sb 83.53%. Sb_2O_3. Laboratory prepn from $SbCl_3$ and water: Schenk in *Handbook of Preparative Inorganic Chemistry* **vol. 1**, G. Brauer, Ed. (Academic Press, New York, 2nd ed., 1963) p 615. Obtained from antimony ore minerals by a volatilization (roasting) process: L. D. Freedman in *Kirk-Othmer Encyclopedia of Chemical Technology* **vol. 3** (Wiley-Interscience, New York, 3rd ed., 1978) pp 107-108. Toxicity study: H. F. Smyth *et al.*, *J. Ind. Hyg. Toxicol.* **30**, 63 (1948).

Crystals, polymorphic. mp 655°. bp 1425°; bp_{210} 870°. Sublimes in high vacuum at 400°. Exists in the vapor phase as Sb_4O_6. Heat capacity at 21° (294.4 K): 24.11 cal/g-atom/°C. *See:* Anderson, *J. Am. Chem. Soc.* **52**, 2712 (1930). Heat of vaporization: 17.82 kcal/mol. Slightly sol in water, dilute H_2SO_4, or dilute HNO_3. Soly in dil HCl (0.1 moles $HCl/kg H_2O$): ~1×10^{-4} g-atoms $Sb/kg H_2O$. *See:* Gayer, Garrett, *ibid.* **74**, 2353 (1952). Soly increases with increasing HCl concn: Lea, Wood, *J. Chem. Soc.* **125**, 137 (1924). Sol in solns of alkali hydroxides or sulfides, in warm soln of tartaric acid, or of bitartrates. LD_{50} orally in rats: >20 g/kg (Smyth).

USE: Manuf tartar emetic; as paint pigment; in enamels and glasses; as mordant; as fire retardant for textiles, canvas, rubber, adhesives, pigments, and paper.

700. Antimony Triselenide. [1315-05-5] Antimony selenide (Sb_2Se_3); diantimony triselenide. Sb_2Se_3; mol wt 480.40. Sb 50.69%, Se 49.31%. Prepd by the action of hydrogen selenide on a soln of potassium antimonyl tartrate: Moser, Atynsky, *Monatsh. Chem.* **45**, 235 (1924); Berzelius cited in *Mellor's* **Vol. 10**, 793 (1930); by direct union of the elements: Chretien, *Compt. Rend.* **142**, 1339, 1412 (1906); Parravano, *Gazz. Chim. Ital.* **43**, I, 210 (1913); Berzelius, *loc. cit. See also:* Chikashige, Fujita, *Mem. Coll. Sci. Kyoto Imp. Univ.* **2**, 233 (1917).

Gray powder. mp 605° (Berzelius); mp 611° (Chretien); mp 617° (Parravano); mp 572° (Chikashige, Fujite). Very slightly sol in water. Forms a brown soln with hot potash lye.

701. Antimony Trisulfide. [1345-04-6] Antimony sulfide (Sb_2S_3); antimonous sulfide; needle antimony; antimony glance. S_3Sb_2; mol wt 339.70. S 28.31%, Sb 71.69%. Sb_2S_3. Occurs in nature as the mineral **stibnite**. Prepn: Donges, Fricke, *Z. Anorg. Chem.* **253**, 2 (1945); Gagliardi, Pilz, *Z. Anal. Chem.* **136**, 344 (1952); *Gmelins, Antimony* (8th ed.) **18B**, pp 503-524 (1949). Toxicity study: W. R. Bradley, W. G. Fredrick, *Ind. Med.* **10**, Ind. Hyg. Sect. 2, 15 (1941).

Gray, lustrous, crystalline masses or grayish-black powder. Also exists in a red modification. mp 550°. Practically insol in water. Sol in concd HCl with evolution of H_2S; sol in solutions of the fixed alkali hydroxides. LD i.p. in rats: 100.0 mg Sb/100 g (Bradley, Fredrick).

USE: In pyrotechnics, Bengal fires; manuf ruby glass, matches, explosives; as a pigment in paints.

702. Antimycin A. [1397-94-0] Antifungal antibiotic produced by several *Streptomyces* spp. Complex of more than 24 components containing a characteristic nine-membered dilactone ring. Antimycins A_1 and A_3 are the most abundant. Inhibits mitochondrial electron transport between cytochromes *b* and c_1 and induces apoptosis. Isoln: B. R. Dunshee *et al.*, *J. Am. Chem. Soc.* **71**, 2436 (1949); J. L. Lockwood *et al.*, *Phytopathology* **44**, 438 (1954). Isoln of blastmycin from *Streptomyces blastmyceticus*: K. Watanabe *et al.*, *J. Antibiot.* **10**, 39 (1957). Separation of major components: Y. Harada *et al.*, *ibid.* **11**, 32 (1958); W.-C. Liu, F. M. Strong, *J. Am. Chem. Soc.* **81**, 4387 (1959). Structure: E. E. van Tamelen *et al.*, *ibid.* **83**, 1639 (1961); A. J. Birch *et al.*, *J. Chem. Soc.* **1961**, 889. Total synthesis of A_{3b}: M. Kinoshita *et al.*, *J. Antibiot.* **24**, 724 (1971). Absolute configuration: *eidem, ibid.* **25**, 373 (1972). Separation of subcomponents by HPLC: S. L. Abidi, *J. Chromatogr.* **447**, 65 (1988). Review of effect on mitochondrial respiratory chain: E. C. Slater, *Biochim. Biophys. Acta* **301**, 129-154 (1973). Mechanism of apoptotic activity: S.-P. Tzung *et al.*, *Nat. Cell Biol.* **3**, 183 (2001). Review of structures and syntheses: Y.-Q. Yang, Y. Wu, *Org. Prep. Proced. Int.* **39**, 135-152 (2007).

Antimycin A_{3b}

White to faint yellow powder. *Poisonous.* Soly in ethanol; 50 mg/ml. Sol in ether, acetone, chloroform. Practically insol in water.

Antimycin A_1. [642-15-9] 2(or 3)-Methylbutanoic acid (2*R*,3*S*,-6*S*,7*R*,8*R*)-3-[[3-(formylamino)-2-hydroxybenzoyl]amino]-8-hexyl-2,6-dimethyl-4,9-dioxo-1,5-dioxonan-7-yl ester. $C_{28}H_{40}N_2O_9$; mol wt 548.63. Mixture of isomers A_{1a} and A_{1b}. Crystals from ethyl acetate + Skellysolve B, mp 149-150°. $[\alpha]_D^{26}$ +76° (c = 1 in chloroform).

Antimycin A_3. [522-70-3] Blastmycin. Mixture of isomers A_{3a} and A_{3b}. Needles from benzene + petr ether, mp 170.5-171.5°. $[\alpha]_D^{26}$ +64.3° (c = 1 in chloroform).

Antimycin A_{3a}. [28068-14-6] (2*S*)-2-Methylbutanoic acid (2*R*,-3*S*,6*S*,7*R*,8*R*)-8-butyl-3-[[3-(formylamino)-2-hydroxybenzoyl]amino]-2,6-dimethyl-4,9-dioxo-1,5-dioxonan-7-yl ester. $C_{26}H_{36}N_2O_9$; mol wt 520.58. Total synthesis and configuration of natural isomer: T. Nishii *et al.*, *Tetrahedron Lett.* **44**, 7829 (2003). Colorless needles (rotamer mixture) from ether/petr ether, mp 173.1-174.0°. $[\alpha]_D^{21}$ +91.6° (c = 0.32 in chloroform).

Antimycin A_{3b}. [116095-17-1] 3-Methylbutanoic acid (2*R*,3*S*,-6*S*,7*R*,8*R*)-8-butyl-3-[[3-(formylamino)-2-hydroxybenzoyl]amino]-2,6-dimethyl-4,9-dioxo-1,5-dioxonan-7-yl ester. $C_{26}H_{36}N_2O_9$; mol wt 520.58. Enantioselective synthesis: Y. Wu, Y.-Q. Yang, *J. Org. Chem.* **71**, 4296 (2006). White needles, mp 173-174°. $[\alpha]_D^{23}$ +79.3° (c = 0.33 in chloroform).

USE: Biochemical reagent.

703. Antipyrine. [60-80-0] 1,2-Dihydro-1,5-dimethyl-2-phenyl-3*H*-pyrazol-3-one; 2,3-dimethyl-1-phenyl-3-pyrazolin-5-one; phenazone; 1,5-dimethyl-2-phenyl-3-pyrazolone; phenyldimethylpyrazolon(e); dimethyloxychinizin; dimethyloxyquinazine; oxydimethylquinizine; Antotalgin; Aurone; Migranin Phenazon; Tropex. $C_{11}H_{12}N_2O$; mol wt 188.23. C 70.19%, H 6.43%, N 14.88%, O 8.50%. Prepn: Müller *et al.*, *Monatsh. Chem.* **89**, 23 (1958); *Hagers Handb. Pharm. Praxis* **Vol. 6** (Springer Verlag, Berlin, 1977) p 571. Toxicity: Hart, *J. Pharmacol. Exp. Ther.* **89**, 205 (1947). Metabolism: H. Uchino *et al.*, *Xenobiotica* **13**, 155 (1983). Clinical comparison with paracetamol: H. Quiding *et al.*, *Int. J. Oral Surg.* **11**, 304 (1982). Use as indicator of hepatic drug metabolism: E. S. Vessel, *Clin. Pharmacol. Ther.* **26**, 275 (1979); G. C. Farrell, L. Zaluzny, *Br. J. Clin. Pharmacol.* **18**, 559 (1984). HPLC determn in urine: E. Brendel *et al.*, *J. Pharm. Biomed. Anal.* **7**, 1783 (1989). Clinical pharmacokinetics: H. A. Ali *et al.*, *Hum. Exp. Toxicol.* **13**, 658 (1994). Review of use in diagnosis of liver disease: J. V. St Peter, W. M. Awni, *Clin. Pharmacokinet.* **20**, 50-65 (1991).

Tabular crystals or white powder; slightly bitter taste. mp 111-113°. One gram dissolves in less than 1 ml water, 1.3 ml alcohol, 1 ml chloroform, 43 ml ether. The aq soln is neutral to litmus. LD_{50} orally in rats: 1.8 g/kg (Hart).

Acetylsalicylate. [569-84-6] $C_{20}H_{20}N_2O_5$; mol wt 368.39. Crystals, mp 63-65°. One part dissolves in 400 parts of water, in 20 parts of 2% aq sod. bicarbonate soln. Freely sol in hot water, cold alc, chloroform; sparingly sol in ether.

Mandelate. [603-64-5] Antipyrine amygdalate. $C_{11}H_{12}N_2O \cdot C_8H_8O_3$. Cryst powder, mp 52-55°. One gram dissolves in 15 ml water, 4 ml alc, 26 ml ether.

Methylethylglycolate. [5794-16-1] Antipyrine 2-hydroxy-2-methylbutyrate. $C_{16}H_{22}N_2O_4$; mol wt 306.36. Crystalline powder, mp 64-65.5°. Soluble in water, alc.

Salicylacetate. [603-59-8] α-Carboxy-o-anisic acid compd with antipyrine. $C_{20}H_{20}N_2O_6$; mol wt 384.39. Crystals, mp 149-150°. Bitter, acid taste. Sparingly soluble in water; soluble in alc.

Salicylate. [520-07-0] $C_{18}H_{18}N_2O_4$; mol wt 326.35. Prepd by fusing antipyrine and salicylic acid: *Hagers Handb. Pharm. Praxis* **Vol. 6**, (Springer Verlag, Berlin, 1977) p 574. Slightly sweet, cryst powder. mp 91-92°. One gram dissolves in 200 ml water, 40 ml boil. water; freely sol in alc, chloroform; sparingly sol in ether.

THERAP CAT: Analgesic. Diagnostic aid (liver function).

THERAP CAT (VET): Has been used as an antipyretic, analgesic and in laminitis of horses.

704. Antireticular Cytotoxic Serum. ACS; cytoxin-anticytotoxin serum; Bogomolets' serum; sarvinal. Biologic serum: Loiseleur, *Ann. Inst. Pasteur* **91**, 445 (1956). Purification: *C.A.* **52**, 6574e (1958). Stimulating effect on reticuloendothelial system: I. P. Miagkaia, *Zh. Mikrobiol. Epidemiol. Immunobiol.* **1978**, 82.

705. α₁-Antitrypsin. [9041-92-3] α₁-Trypsin inhibitor; α₁-proteinase inhibitor; α₁-protease inhibitor; alpha₁-antitrypsin; AAT; A1AT; A1PI; Aralast; Prolastin; Zemaira. Serum glycoprotein synthesized by the liver. Major serine protease inhibitor (serpin) in mammalian plasma; primarily inhibits neutrophil elastase. Consists of a single polypeptide chain of 394 amino acid residues and 3 carbohydrate side chains linked to asparagine residues. Mol wt ~52,-000 daltons. Highly pleomorphic protein with a number of genetic variants. Characterization of trypsin inhibitor in alpha₁-fraction of human serum: K. Jacobsson, *Scand. J. Clin. Invest.* **7**, Suppl. 14, 57-102 (1955). Isoln from human serum: F. C. Moll *et al., J. Biol. Chem.* **233**, 121 (1958); H. F. Bundy, J. W. Mehl, *ibid.* **234**, 1124 (1959). Purification: I. P. Crawford, *Arch. Biochem. Biophys.* **156**, 215 (1973); M. H. Coan, W. J. Brockway, **US 4379087** (1983 to Cutter). Structural study of active site: D. Johnson, J. Travis, *J. Biol. Chem.* **253**, 7142 (1978). Complete amino acid sequence and review of genetic variants: R. W. Carrell *et al., Nature* **298**, 329 (1982). Crystal structure: H. Loebermann *et al., J. Mol. Biol.* **177**, 531 (1984). Cloning and expression of human AAT gene: M. Courtney *et al., Proc. Natl. Acad. Sci. USA* **81**, 669 (1984). AAT deficiency has been associated with degenerative lung disease: C.-B. Laurell, S. Eriksson, *Scand. J. Clin. Lab. Invest.* **15**, 132 (1963); with hepatic cirrhosis: H. L. Sharp, *Gastroenterology* **70**, 611 (1976). Inhibition of leukocyte elastase: K. Ohlsson, *Scand. J. Clin. Lab. Invest.* **28**, 251 (1971); K. Beatty *et al., J. Biol. Chem.* **255**, 3931 (1980). Review of human phenotypes and relationship to disease: J. O. Morse, *N. Engl. J. Med.* **299**, 1045-1048 (1978); *ibid.* 1099-1105 (1978). Clinical evaluation in AAT deficiency: M. D. Werwers *et al., ibid.* **316**, 1055 (1987); N. Konietzko *et al., Dtsch. Med. Wochenschr.* **113**, 369 (1988); of the effect of cigarette smoking on the association rate constant with neutrophil elastase: F. Ogushi *et al., J. Clin. Invest.* **87**, 1060 (1991). Symposium on structure, function and clinical applications: *Am. J. Med.* **84**, Suppl. 6A, 1-90 (1988). Review of role in emphysema: P. J. Stone, *Clin. Chest Med.* **4**, 405-412 (1983); J. A. Pierce, *J. Am. Med. Assoc.* **259**, 2890-2895 (1988).

THERAP CAT: In treatment of emphysema associated with inherited α₁-antitrypsin deficiency.

706. Antivenin (Crotalidae) Polyvalent. North and South American antisnakebite serum. Immunoglobulin G (IgG) complex isolated from serum of horses that have been immunized with venoms of snakes from the *Crotalidae* family *(Crotalus atrox, C. adamanteus, C. durissus terrificus, Bothrops atrox). Ref:* Gingrich, Hohenadel, in *Venoms*, E. E. Buckley, N. Porges, Eds. (Publ. 44 of Am. Assoc. Adv. of Sci., Washington, 1956) p 381; Keegan, *ibid.* p 413. Purification: F. O. Cope, *Proc. Pa. Acad. Sci.* **53**, 43 (1979). Use in treatment of rattlesnake bites: T. M. Davidson, S. F. Schafer, *Postgrad. Med.* **96**, 107 (1994). ELISA determn in serum: C. L. Ownby *et al., South. Med. J.* **89**, 803 (1996).

THERAP CAT: Antidote to snake venom.

THERAP CAT (VET): Antidote to snake venom.

707. Antivenin (Latrodectus mactans). Antivenin. Black widow spider antivenin. Prepd from the serum of horses immunized against the venom of the New World spider *Latrodectus mactans* Fabr., *Araneae*. Review of clinical efficacy: R. F. Clark *et al., Ann. Emerg. Med.* **21**, 782-787 (1992).

THERAP CAT: In treatment of black widow spider bites.

708. ANTU. [86-88-4] *N*-1-Naphthalenylthiourea; 1-(1-naphthyl)-2-thiourea; α-naphthylthiourea; *N*-1-naphthylthiourea; α-naphthylthiocarbamide; krysid; chemical 109; Anturat; Bantu; Rattrack. $C_{11}H_{10}N_2S$; mol wt 202.28. C 65.32%, H 4.98%, N 13.85%, S 15.85%. Prepd from α-naphthylamine and ammonium or potassium or sodium thiocyanate: de Clermont, Wehrlin, *Bull. Soc. Chim.* [2] **26**, 125 (1876); Alvarez, *C.A.* **42**, 4560 (1948); from α-naphthylisothiocyanate by treatment with alcoholic ammonia: Dyson, Hunter, *Chem. News* **134**, 4 (1927). Rodenticide activity: Richter, *J. Am. Med. Assoc.* **129**, 927 (1945). Review of carcinogenic risk: *IARC Monographs* **30**, 347-357 (1983).

Prisms from alc. Bitter taste. mp 198°. Solubility in water at 25°: 0.06 g/100 ml; in acetone: 2.43 g/100 ml; in triethylene glycol: 8.6 g/100 ml. Fairly sol in hot alc.

Caution: Potential symptoms of toxicity following ingestion of large doses are vomiting, dyspnea, cyanosis and coarse pulmonary rales; liver damage. *See NIOSH Pocket Guide to Chemical Hazards* (DHHS/NIOSH 97-140, 1997) p 20. May cause glomerular injury in nonlethal poisonings. *See Clinical Toxicology of Commercial Products,* R. E. Gosselin *et al.,* Eds. (Williams & Wilkins, Baltimore, 5th ed., 1984) Section III, p 40-42.

USE: Rodenticide. Specific control for the adult Norway rat; less toxic to other rat species.

709. Apafant. [105219-56-5] 3-[4-(2-Chlorophenyl)-9-methyl-6*H*-thieno[3,2-*f*][1,2,4]triazolo[4,3-*a*][1,4]diazepin-2-yl]-1-(4-morpholinyl)-1-propanone; 4-[3-[4-(2-chlorophenyl)-9-methyl-6*H*-thieno[3,2-*f*][1,2,4]triazolo[4,3-*a*][1,4]diazepin-2-yl]-1-oxopropyl]morpholine; WEB-2086. $C_{22}H_{22}ClN_5O_2S$; mol wt 455.96. C 57.95%, H 4.86%, Cl 7.77%, N 15.36%, O 7.02%, S 7.03%. Platelet activating factor (PAF) antagonist. Prepn: K. H. Weber *et al,* **DE 3502392**; *eidem,* **US 5082839** (1986, 1992 both to Boehringer, Ing.). Pharmacology: J. Casals-Stenzel *et al., J. Pharmacol. Exp. Ther.* **241**, 974 (1987). Clinical pharmacology, pharmacokinetics: H. M. Brecht *et al., Arzneim.-Forsch.* **41**, 51 (1991). Clinical effect vs PAF-induced bronchoconstriction: W. S. Adamus *et al., Clin. Pharmacol. Ther.* **47**, 456 (1990).

Viscous, almost colorless oil. LD_{50} in mice (mg/kg): 540 i.v.; 4600 orally (Casals-Stenzel).

USE: Tool to evaluate the role of PAF in exptl models of human disease.

710. Apalcillin. [63469-19-2] (2*S*,5*R*,6*R*)-6-[[(2*R*)-2-[[(4-Hydroxy-1,5-naphthyridin-3-yl)carbonyl]amino]-2-phenylacetyl]-amino]-3,3-dimethyl-7-oxo-4-thia-1-azabicyclo[3.2.0]heptane-2-carboxylic acid. $C_{25}H_{23}N_5O_6S$; mol wt 521.55. C 57.57%, H

4.45%, N 13.43%, O 18.41%, S 6.15%. Semi-synthetic antibiotic related to penicillin. Prepn of the potassium salt: H. Tobiki *et al.*, **ZA 7205865**; *eidem*, **US 3864329** (1973, 1975 both to Sumitomo); of the sodium salt: Y. Hirotada *et al.*, **US 4005075** (1977 to Sumitomo); H. Tobiki *et al.*, *Yakugaku Zasshi* **100**, 49 (1980), *C.A.* **93**, 95179x (1980). Microbiological evaluation: H. Noguchi *et al.*, *Antimicrob. Agents Chemother.* **9**, 262 (1976). Analytical studies: I. Umeda *et al.*, *Yakugaku Zasshi* **99**, 717 (1979), *C.A.* **92**, 11287z (1980). Series of articles on pharmacology, exptl and clinical effects: *Chemotherapy (Tokyo)* **26**, Suppl. 2, 111-223 (1978). Pharmacokinetics in man: U. Busch *et al.*, *Arzneim.-Forsch.* **32**, 1131 (1982).

Sodium salt. [58795-03-2] PC-904; Lumota. $C_{25}H_{22}N_5NaO_6S$; mol wt 543.53. White cryst, sol in water.
THERAP CAT: Antibacterial.

711. Apamin. [24345-16-2] $C_{79}H_{131}N_{31}O_{24}S_4$; mol wt 2027.35. C 46.80%, H 6.51%, N 21.42%, O 18.94%, S 6.33%. Neurotoxic polypeptide consisting of 18 amino acid residues and 2 disulfide bridges. Comprises about 2% by wt of the dried venom of *Apis mellifica (mellifera)*, the honey bee. Isoln: E. Habermann, K. G. Reiz, *Naturwissenschaften* **51**, 61 (1964); *eidem*, *Biochem. Z.* **341**, 451 (1965). Structure: P. Haux *et al.*, *Z. Physiol. Chem.* **348**, 737 (1967); R. Shipolini *et al.*, *Chem. Commun.* **1967**, 679. Conformation: R. C. Hider, U. Ragnarsson, *FEBS Lett.* **111**, 189 (1980); B. Busetta, *ibid.* **112**, 138 (1980). Solution structure: J. H. B. Pease, D. E. Wemmer, *Biochemistry* **27**, 8491 (1988). Solid-phase synthesis: J. van Rietschoten *et al.*, *Eur. J. Biochem.* **56**, 35 (1975); B. E. B. Sandberg, U. Ragnarsson, *Int. J. Pept. Protein Res.* **11**, 238 (1978). Pharmacology: Wellheoner, *Arch. Pharmakol. Exp. Pathol.* **262**, 29 (1969). Biochemistry: E. Habermann, K. G. Reiz, *Biochem. Z.* **343**, 192 (1965). Action on CNS: E. Habermann, D. Cheng-Raude, *Toxicon* **13**, 465 (1975). Binding and toxicity studies: C. Labbé-Jullié *et al.*, *Eur. J. Biochem.* **196**, 639 (1991). *Review:* E. Habermann, *Science* **177**, 314-322 (1972); C. Granier, J. van Rietschoten, in *Natural Toxins*, D. Eaker, T. Wadström, Eds. (Pergamon, New York, 1980) pp 481-486; E. Habermann, *Pharmacol. Ther.* **25**, 255-270 (1984).

Cys–Asn–Cys–Lys–Ala–Pro–Glu–Thr–Ala–Leu–Cys–Ala–Arg–Arg–Cys–Gln–Gln–HisNH$_2$

Highly basic compd. Pharmacologic activity destroyed by oxidn with performic acid. LD_{50} i.v. in mice: 4 mg/kg (Habermann, Reiz); LD_{50} intracerebroventricularly in mice: 12 ng/animal (Labbé-Jullié).

712. Apaziquone. [114560-48-4] 5-(1-Aziridinyl)-3-(hydroxymethyl)-2-[(1*E*)-3-hydroxy-1-propen-1-yl]-1-methyl-1*H*-indole-4,7-dione; EO9; EOquin. $C_{15}N_{16}N_2O_4$; mol wt 496.29. C 36.30%, N 50.80%, O 12.89%. Bioreductive alkylating indoloquinone; structurally similar to mitomycin C, *q.v.* Prodrug activated by reductase enzymes to form the active cytotoxic agent. Prepn: W. N. Speckamp, E. A. Oostveen, **WO 8706227**; *eidem*, **US 5079257** (1987, 1992). Improved synthesis: A. S. Cotterill *et al.*, *Tetrahedron* **51**, 7223 (1995). Formulation and stability studies: S. C. van der Schoot *et al.*, *Int. J. Pharm.* **329**, 135 (2007). LC-MS/MS determn in plasma: L. D. Vainchtein *et al.*, *Rapid Commun. Mass Spectrom.* **22**, 462 (2008). Reduction by DT-diaphorase: S. M. Bailey *et al.*, *Biochem. Pharmacol.* **56**, 613 (1998); by cytochrome P450 reductase: *eidem*, *ibid.* **62**, 461 (2001). Clinical pharmacokinetics and safety following intravesical instillation in patients with bladder cancer: K. Hendricksen *et al.*, *J. Urol.* **180**, 116 (2008). Clinical evaluation in non-muscle invasive bladder cancer: *idem et al.*, *World J. Urol.* **27**, 337 (2009). Review of pharmacology and therapeutic potential: J. A. Witjes, P. S. Kolli, *Expert Opin. Invest. Drugs* **17**, 1085-1096 (2008).

Purple crystals, mp 160-169°. Low soly in water.
THERAP CAT: Antineoplastic.

713. Apazone. [13539-59-8] 5-(Dimethylamino)-9-methyl-2-propyl-1*H*-pyrazolo[1,2-*a*][1,2,4]benzotriazine-1,3(2*H*)-dione; 3-dimethylamino-7-methyl-1,2-(*n*-propylmalonyl)-1,2-dihydro-1,2,4-benzotriazine; azapropazone; AHR-3018; MI-85. $C_{16}H_{20}N_4O_2$; mol wt 300.36. C 63.98%, H 6.71%, N 18.65%, O 10.65%. Prepn: **FR 1440629**; I. Molnar *et al.*, **US 3349088** (1966, 1967, both to Siegfried); Mixich, *Helv. Chim. Acta* **51**, 532 (1968). Pharmacological and toxicological studies: Jahn, Adrian, *Arzneim.-Forsch.* **19**, 36 (1969). Metabolism: Mixich, *Helv. Chim. Acta* **55**, 1031 (1972). HPLC determn in plasma and urine: B. J. Kline *et al.*, *Arzneim.-Forsch.* **33**, 504 (1983).

mp 228°.
Dihydrate. [22304-30-9] MI-85Di; Prolixan; Rheumox. Almost colorless crystals, mp 247-248°.
THERAP CAT: Anti-inflammatory.

714. Aphidicolin. [38966-21-1] (3*R*,4*R*,4a*R*,6a*S*,8*R*,9*R*,-11a*S*,11b*S*)-Tetradecahydro-3,9-dihydroxy-4,11b-dimethyl-8,11a-methano-11a*H*-cyclohepta[*a*]naphthalene-4,9-dimethanol; (3α,4α,-5α,17α)-3,17-dihydroxy-4-methyl-9,15-cyclo-*C*,18-dinor-14,15-secoandrostane-4,17-dimethanol; ICI-69653; NSC-234714. C_{20}-$H_{34}O_4$; mol wt 338.49. C 70.97%, H 10.12%, O 18.91%. Novel tetracyclic diterpene antibiotic with antiviral and antimitotic properties, isolated as (+)-form from *Cephalosporium aphidicola* Petch. Specific inhibitor of DNA α-polymerase. Description and x-ray crystallographic determn of structure: K. M. Brundret *et al.*, *Chem. Commun.* **1972**, 1027. Prepn and antiviral activity: A. Borrow *et al.*, **US 3761512** (1973 to ICI). Structure and abs config: W. Dalziel *et al.*, *J. Chem. Soc. Perkin Trans. 1* **1973**, 2841. Total synthesis of (±)-aphidicolin: B. M. Trost *et al.*, *J. Am. Chem. Soc.* **101**, 1328 (1979); J. E. McMurry *et al.*, *ibid.* 1331; E. J. Corey *et al.*, *ibid.* **102**, 1743 (1980); J. E. McMurry *et al.*, *Tetrahedron* **37**, Suppl. 1, 319 (1981). Antiviral effects *in vitro* and *in vivo*: R. A. Bucknall *et al.*, *Antimicrob. Agents Chemother.* **4**, 294 (1973). Antimitotic activity: S. Ikegami *et al.*, *Nature* **275**, 458 (1978). Effects on DNA synthesis in mouse sarcoma: S. Seki *et al.*, *Biochim. Biophys. Acta* **610**, 413 (1980). Enzymatic determn method: G. Pedrali-Noy *et al.*, *J. Biochem. Biophys. Methods* **4**, 113 (1981). Induction of differentiation of human myeloid leukemia cells: J. Griffin *et al.*, *Exp. Hematol.* **10**, 774 (1982). Effects on cell proliferation: G. Iliakis *et al.*, *Int. J. Radiat. Biol.* **42**, 417 (1982); on γ-irradiated human fibroblasts: P. J. Smith, M. C. Paterson, *Biochim. Biophys. Acta* **739**, 17 (1983).

Needles from ethyl acetate, mp 227-232°. $[\alpha]_D^{27}$ +12° (c = 1 in methanol).

(±)-Form. White cryst, mp 218-220°.

USE: As a biological tool in studies of cell proliferation and differentiation.

715. Apholate. [52-46-0] 2,2,4,4,6,6-Hexakis(1-aziridinyl)-2λ5,4λ5,6λ5-1,3,5,2,4,6-triazatriphosphorine; 2,2,4,4,6,6-hexakis(1-aziridinyl)-2,2,4,4,6,6-hexahydro-1,3,5,2,4,6-triazatriphosphorine; hexakis(1-aziridinyl)phosphonitrile; 1-aziridinylphosphonitrile trimer. $C_{12}H_{24}N_9P_3$; mol wt 387.31. C 37.21%, H 6.25%, N 32.55%, P 23.99%. Prepn: Rätz, Grundmann, US 2858306 (1958 to Olin Mathieson). Toxicity study: T. B. Gaines, *Toxicol. Appl. Pharmacol.* **14**, 515 (1969).

Crystals from heptane, mp 147.5°. LD_{50} in male, female rats (mg/kg): 98, 113 orally (Gaines).

USE: Insect chemosterilant.

716. Aphylline. [577-37-7] (7R,7aS,14S,14aR)-Dodecahydro-7,14-methano-2H,6H-dipyrido[1,2-a:1',2'-e][1,5]diazocin-6-one. $C_{15}H_{24}N_2O$; mol wt 248.37. C 72.54%, H 9.74%, N 11.28%, O 6.44%. From *Anabasis aphylla* L., *Chenopodiaceae*. Extraction procedure and structure: Orechoff, Menschikow, *Ber.* **64**, 266 (1931); **65**, 234 (1932). Structure: Galinovsky, Jarish, *Monatsh. Chem.* **84**, 199 (1953). Stereochemistry: Edwards *et al.*, *Can. J. Chem.* **32**, 235 (1954); Galinovsky *et al.*, *Monatsh. Chem.* **86**, 1014 (1955). Synthesis: Bohlmann *et al.*, *Ber.* **90**, 653 (1957).

Crystals, mp 52-57°. bp_4 200°. $[\alpha]_D^{20}$ +10.3° (c = 20 in methanol). Soluble in the usual organic solvents.

Hydrochloride. $C_{15}H_{24}N_2O \cdot HCl$. Rhombohedra from alc, mp 209°. $[\alpha]_D^{20}$ +14° (c = 25). Freely soluble in water.

717. Apigenin. [520-36-5] 5,7-Dihydroxy-2-(4-hydroxyphenyl)-4H-1-benzopyran-4-one; 4',5,7-trihydroxyflavone; 2-(p-hydroxyphenyl)-5,7-dihydroxychromone; pelargidenon 1449. $C_{15}H_{10}O_5$; mol wt 270.24. C 66.67%, H 3.73%, O 29.60%. The aglucon of apiin and of apigenin-7-glucoside. From apiin by boiling with acids, from apigenin-7-glucoside by enzymatic hydrolysis with emulsin or by boiling with 15% H_2SO_4. Isoln and structure: Czajkowski *et al.*, *Ber.* **33**, 1992 (1900); Schmid, Waschkau, *Monatsh. Chem.* **49**, 83 (1928); Baker *et al.*, *J. Chem. Soc.* **1963**, 1477. Synthesis: Hutchins, Wheeler, *ibid.* **1939**, 91; Farooq *et al.*, *Arch. Pharm.* **292**, 792 (1959).

Yellow needles from aqueous pyridine, mp 345-350°. uv max (ethanol): 269, 340 nm (ε 18,800; 20,900). Practically insoluble in water; moderately sol in hot alcohol. Sol in dil KOH with intense yellow color.

USE: Has been used to dye Cr mordanted wool yellow. The color is fast to soap.

718. Apigetrin. [578-74-5] 7-(β-D-Glucopyranosyloxy)-5-hydroxy-2-(4-hydroxyphenyl)-4H-1-benzopyran-4-one; apigenin-7-D-glucoside; 7-D-glycosylapigenin; cossmetin. $C_{21}H_{20}O_{10}$; mol wt 432.38. C 58.34%, H 4.66%, O 37.00%. Isoln from flowers of *Anthemis nobilis* L., *Compositae*: Power, Browning, *J. Chem. Soc.* **105**, 1833 (1914); from parsley: Nordström, Swain, *Chem. Ind. (London)* **1953**, 85; *J. Chem. Soc.* **1953**, 2764.

Pale yellow, cryst powder, dec 178-180°. Astringent taste. uv max (ethanol): 335, 268 nm. Sol in water, dil alcohol.

719. Apiin. [26544-34-3] 7-[(2-O-D-Apio-β-D-furanosyl-β-D-glucopyranosyl)oxy]-5-hydroxy-2-(4-hydroxyphenyl)-4H-1-benzopyran-4-one; 4',5,7-trihydroxyflavone-7-apiosylglucoside; apigenin-7-apiosylglucoside; apioside. $C_{26}H_{28}O_{14}$; mol wt 564.50. C 55.32%, H 5.00%, O 39.68%. Isoln from parsley and from celery: Vongerichten, *Ber.* **33**, 2334, 2904 (1900); Gupta, Seshadri, *Proc. Indian Acad. Sci.* **35A**, 242 (1952), *C.A.* **47**, 3306c (1953); Rahman, *Z. Naturforsch.* **13b**, 201 (1958). Structure: Vongerichten, *Ann.* **318**, 121 (1901); Marchlewski, Skarzynski, *Biochem. Z.* **297**, 56 (1938); Hemming, Ollis, *Chem. Ind. (London)* **1953**, 85.

Crystals from ethanol, mp 230-232°. uv max (96% ethanol): 267.5, 342.5 nm (log ε 4.18, 4.30). Sol in hot water, hot alcohol; practically insol in ether. Sol in Na_2CO_3 or NH_3 solns with intense yellow color, in NaOH with pale yellow color.

720. Apiole (Dill). [484-31-1] 4,5-Dimethoxy-6-(2-propen-1-yl)-1,3-benzodioxole; 1-allyl-2,3-dimethoxy-4,5-(methylenedioxy)benzene; dill apiole. $C_{12}H_{14}O_4$; mol wt 222.24. C 64.85%, H 6.35%, O 28.80%. Occurs in dill oil, *Anethum graveolus* L., *Umbelliferae*. Isoln: G. Ciamician, P. Silber, *Ber.* **29**, 1799 (1896); D. B. Spoelstra, *Rec. Trav. Chim.* **48**, 372 (1929). Structure: H. Thoms, *Arch. Pharm.* **242**, 344 (1904). Synthesis: W. Baker *et al.*, *J. Chem. Soc.* **1934**, 1681; F. Dallacker, *Ber.* **102**, 2663 (1969); J. R. Cannon *et al.*, *J. Sci. Soc. Thailand* **6**, 59 (1980). Synergistic activity with insecticides: E. P. Lichtenstein *et al.*, *J. Agric. Food Chem.* **22**, 658 (1974); S.S. Tomar *et al.*, *Agric. Biol. Chem.* **43**, 1479 (1979).

Oil, mp 29.5°. bp 285°. n_D^{17} 1.5305; d_{15}^{15} 1.1598.

721. Apiole (Parsley). [523-80-8] 4,7-Dimethoxy-5-(2-propen-1-yl)-1,3-benzodioxole; 1-allyl-2,5-dimethoxy-3,4-methylenedioxybenzene; parsley apiole; apiol; apioline; parsley camphor. $C_{12}H_{14}O_4$; mol wt 222.24. C 64.85%, H 6.35%, O 28.80%. Occurs in parsley oil, *Petroselinum sativum*: Blanchet, Sell, *Ann.* **6**, 259 (1833); Vongerichten, *Ber.* **9**, 1477 (1876); Kolesnikov *et al.*, *Aptechnoe Delo* **7**(4), 27 (1958), *C.A.* **54**, 12491e (1960). Structure:

Ciamician, Silber, *Ber.* **23**, 2283 (1890); Thoms, *ibid.* **36**, 1714 (1903). Synthesis: Baker, Savage, *J. Chem. Soc.* **1938**, 1602; F. Dallacker, *Ber.* **102**, 2663 (1969). Synergistic activity with insecticides: E. P. Lichtenstein *et al., J. Agric. Food Chem.* **22**, 658 (1974).

Crystals; faint parsley odor. mp 29.5°. bp 294°. n_D^{20} 1.536-1.538. Insol in water. Sol in alcohol, benzene, chloroform, ether, acetone, oils. *Keep in a cool place.*

722. Apiose. [639-97-4] D-Apiose; tetrahydroxyisovaleraldehyde; 3-C-(hydroxymethyl)-D-glyceroaldotetrose. $C_5H_{10}O_5$; mol wt 150.13. C 40.00%, H 6.71%, O 53.28%. First found in parsley in which it occurs as the flavinoid glycoside apiin, *q.v.* Isoln from apiin: Vongerichten, *Ann.* **318**, 126 (1901); **321**, 74 (1902); Hemming, Ollis, *Chem. Ind. (London)* **1953**, 85. From the rubber plant, *Hevea brasiliensis, Euphorbiaceae:* Patrick, *Nature* **178**, 216 (1956). Discussion of structure and isoln from the Australian marine plant *Posidonia australis* Kon., *Potamogetonaceae:* Bell, *Methods in Carbohydrate Chemistry* vol. I (Academic Press, New York, 1962) pp 260-263. Synthesis: Gorin, Perlin, *Can. J. Chem.* **36**, 480 (1958); Khalique, *J. Chem. Soc.* **1962**, 2515; Ezekiel *et al., Tetrahedron Lett.* **1969**, 1635. Synthesis of L-form: Weygand, Schmiechen, *Ber.* **92**, 535 (1959); of DL-form: Kinoshita, Miwa, *Carbohydr. Res.* **28**, 175 (1973); Y. Araki *et al., ibid.* **58**, C4 (1977); of D- and L-forms: P. Ho, *Can. J. Chem.* **57**, 381 (1979). Chemistry, configuration and synthesis studies: Williams, Jones, *ibid.* **42**, 69 (1964); Hulyalker *et al., ibid.* **43**, 2085 (1965). *Review:* Watson, Orenstein, *Adv. Carbohydr. Chem. Biochem.* **31**, 135-184 (1975).

Syrup. $[\alpha]_D^{15}$ +5.6°; $[\alpha]_D^{19}$ +9.1°. Soluble in water.
D-Apiose di-*O*-isopropylidene. $C_{11}H_{18}O_5$. Plates from water containing a trace of NH_3, mp 81-83°. $[\alpha]_D^{20}$ +55.5° (c = 1.1 in ethanol).

723. Apixaban. [503612-47-3] 4,5,6,7-Tetrahydro-1-(4-methoxyphenyl)-7-oxo-6-[4-(2-oxo-1-piperidinyl)phenyl]-1*H*-pyrazolo[3,4-*c*]pyridine-3-carboxamide; 1-(4-methoxyphenyl)-7-oxo-6-(4-(2-oxopiperidin-1-yl)phenyl)-4,5,6,7-tetrahydro-1*H*-pyrazolo-[3,4-*c*]pyridine-3-carboxamide; BMS-562247; Eliquis. $C_{25}H_{25}$-N_5O_4; mol wt 459.51. C 65.35%, H 5.48%, N 15.24%, O 13.93%. Direct factor Xa inhibitor. Prepn: D. Pinto *et al.,* **WO 03026652**; *eidem,* **US 6967208** (2003, 2005 both to Bristol-Myers Squibb); and structure activity study: *eidem, J. Med. Chem.* **50**, 5339 (2007). Antithrombotic and antihemostatic properties: P. C. Wong *et al., J. Thromb. Haemost.* **6**, 820 (2008). Clinical evaluation for thromboprophylaxis after total knee replacement: M. R. Lassen *et al., ibid.* **5**, 2368 (2007); after hip replacement: M. R. Lassen *et al., N. Engl. J. Med.* **363**, 2487 (2010). Clinical trial for stroke prevention in patients with artrial fibrillation: S. J. Connolly *et al., ibid.* **364**, 806 (2011).

Colorless solid. Aq soly (0.9% saline soln): 40-50 $\mu g/ml$.
THERAP CAT: Antithrombotic.

724. Aplasmomycin. [61230-25-9] (*T*-4)-[(1*R*,2*R*,5*S*,6*R*,-8*S*,9*E*,12*R*,14*S*,17*R*,18*R*,19*R*,22*S*,23*R*,25*S*,26*E*,29*R*,31*S*,34*R*)-1,2,-18,19-Tetra(hydroxy-κO)-12,29-dihydroxy-6,13,13,17,23,30,30,34-octamethyl-4,7,21,24,35,37-hexaoxapentacyclo[29.3.1.15.8.114,-18.122,25]octatriaconta-9,26-diene-3,20-dionato(4−)]borate(1−) sodium (1:1); ICI-122378. $C_{40}H_{60}BNaO_{14}$; mol wt 798.71. C 60.15%, H 7.57%, B 1.35%, Na 2.88%, O 28.04%. Antibiotic produced by *Streptomyces griseus* strain SS-20, obtained from shallow sea mud: Y. Okami *et al., J. Antibiot.* **29**, 1019 (1976); *eidem,* **JP Kokai 77 108901** (1977 to Microbiochem. Res. Found.), *C.A.* **88**, 35843 (1978). It inhibits the growth of gram-positive bacteria *in vitro* and is active vs *Plasmodium berghei in vivo.* Aplasmomycin is a symmetrical dimer related to boromycin, *q.v.*, the only other known natural product that contains boron. Structure, x-ray crystallography: H. Nakamura *et al., J. Antibiot.* **30**, 714 (1977). Total synthesis of (+)-form: E. J. Corey *et al., J. Am. Chem. Soc.* **104**, 6816, 6818 (1982); T. Nakata *et al., Tetrahedron Lett.* **27**, 6341, 6345 (1986); J. D. White *et al., J. Am. Chem. Soc.* **108**, 8105 (1986). Use as growth promotant in ruminants: D. H. Davies *et al.,* **EP 2893**; *eidem,* **US 4225593** (1979, 1980 both to ICI). Two minor components, aplasmomycins B and C, have also been isolated from the fermentation; both aplasmomycin and aplasmomycin B show ionophoric properties, mediating net K^+ transport across a bulk phase: K. Sato *et al., J. Antibiot.* **31**, 632 (1978). NMR analysis of aplasmomycin and deboroaplasmomycin: T. S. S. Chen *et al., ibid.* **33**, 1316 (1980). Comparative anti-anaerobic activity of aplasmomycin: K. Watanabe *et al., Antimicrob. Agents Chemother.* **19**, 519 (1981). Structural studies: T. J. Stout *et al., Tetrahedron* **47**, 3511 (1991). Review of biosynthetic studies: H. G. Floss, C. Chang in *Antibiotics* IV, J. W. Corcoran, Ed. (Springer-Verlag, New York, 1981) pp 203-210.

(+)-Aplasmomycin

Colorless needles, mp 283-285° (dec). $[\alpha]_D^{22}$ +225° (c = 1.24 in chloroform). Very lipophilic. Practically insol in water. LD_{50} i.p. in mice: 125 mg/kg (Okami).
Silver salt. $C_{40}H_{60}AgBO_{14}$. Colorless needles, mp 218-220°. $[\alpha]_D^{23}$ +194° (c = 0.34 in chloroform).
USE: Growth promotant in ruminants.

725. Aplidine. [137219-37-5] N-[1-(1,2-Dioxopropyl)-L-prolyl]-didemnin A; dehydrodidemnin B; DDB; Aplidin. $C_{57}H_{87}$-N_7O_{15}; mol wt 1110.36. C 61.66%, H 7.90%, N 8.83%, O 21.61%. Cyclic depsipeptide analog of the didemnins, *q.v.*, isolated from the marine tunicate, *Aplidum albicans.* Induces rapid apoptotic death with or without previous cell cycle arrest. Isoln and synthesis: K. Rinehart, A. M. Lithgow-Bertelloni, **WO 9104985** (1991 to Pharma-Mar); G. Jou *et al., J. Org. Chem.* **62**, 354 (1997). Antiproliferative effects on Ehrlich carcinoma: J. L. Urdiales *et al., Cancer Lett.* **102**, 31 (1996); cytotoxic effect on acute lymphoblastic leukemia cells: E. Erba *et al., Br. J. Cancer* **89**, 763 (2003). Mechanism of action study: *idem et al., ibid.* **86**, 1510 (2002). Degradation kinetics in solution: J. C. M. Waterval *et al., J. Chromatogr. B* **754**, 161 (2001). 2D-NMR conformational analysis in solution: F. Cárdenas *et al., J.*

Org. Chem. **68**, 9554 (2003). LC/MS/MS determn in plasma: J. Yin *et al., Rapid Commun. Mass Spectrom.* **17**, 1909 (2003). *In vitro* toxicity on hematopoietic progenitors and stem cells: S. G. Gomez *et al., Exp. Hematol.* **31**, 1104 (2003). Clinical pharmacokinetics and evaluation in advanced malignancies: S. Faivre *et al., J. Clin. Oncol.* **23**, 7871 (2005).

White solid, mp 152-160°. $[\alpha]_D$ −95.9° (c = 1.8 in CHCl$_3$).
THERAP CAT: Antineoplastic.

726. Apoatropine. [500-55-0] α-Methylenebenzeneacetic acid (3-*endo*)-8-methyl-8-azabicyclo[3.2.1]oct-3-yl ester; 1α*H*,5α*H*-tropan-3α-ol atropate; atropamine; atropyltropeine. C$_{17}$H$_{21}$NO$_2$; mol wt 271.36. C 75.25%, H 7.80%, N 5.16%, O 11.79%. Anticholinergic. Occurs in root of *Atropa belladonna* L., *Solanaceae*. Also obtained from atropine by splitting off water or by total synthesis: Ladenburg, *Ann.* **217**, 102 (1883); Merck, *Arch. Pharm.* **230**, 134 (1892); **231**, 110 (1893); Hesse, *Ann.* **261**, 87 (1891); **271**, 124 (1892); **277**, 290 (1893). Isoln by chromatography: Steinegger, Phokas, *Pharm. Acta Helv.* **31**, 284 (1956).

Prisms from chloroform, mp 62°. Absorption spectrum: Gompel, Henri, *Compt. Rend.* **156**, 1543 (1913). Freely sol in alcohol, ether, chloroform, benzene, carbon disulfide; slightly in petr ether, isoamyl alcohol. Almost insol in water.
Hydrochloride. C$_{17}$H$_{21}$NO$_2$.HCl. Scales, mp 239°. Sol in hot water; sparingly sol in alc, acetone. Nearly insol in ether.
Sulfate pentahydrate. (C$_{17}$H$_{21}$NO$_2$)$_2$.H$_2$SO$_4$.5H$_2$O. Crystals, sparingly sol in water.

727. Apocodeine. [641-36-1] (6a*R*)-5,6,6a,7-Tetrahydro-10-methoxy-6-methyl-4*H*-dibenzo[*de,g*]quinolin-11-ol; 10-methoxy-6aβ-aporphin-11-ol. C$_{18}$H$_{19}$NO$_2$; mol wt 281.36. C 76.84%, H 6.81%, N 4.98%, O 11.37%. A monomethyl ether of apomorphine, *q.v.* Prepd by heating codeine with oxalic acid: Folkers, *J. Am. Chem. Soc.* **58**, 1814 (1936); by heating codeine with phosphoric acid: Small *et al., J. Org. Chem.* **5**, 344 (1940). Configuration: Corrodi, Hardegger, *Helv. Chim. Acta* **38**, 2038 (1955). Total synthesis and pharmacology: Neumeyer *et al., J. Med. Chem.* **16**, 1223 (1973).

Small prisms from methanol (lose solvent at 80°/2 mm), mp 124° (dry). $[\alpha]_D^{24}$ −97° (c = 0.45). Slightly soluble in water; sol in alcohol, ether, dilute acids.
Hydrochloride. C$_{18}$H$_{20}$ClNO$_2$. Crystals from alcohol-ether, softening 140°, dec 260-263°. $[\alpha]_D^{22}$ −43° (c = 0.51). Very sol in water; sol in alcohol.
THERAP CAT: Emetic.

728. Apocynin. [498-02-2] 1-(4-Hydroxy-3-methoxyphenyl)ethanone; acetovanillon; 4-hydroxy-3-methoxyacetophenone. C$_9$H$_{10}$O$_3$; mol wt 166.18. C 65.05%, H 6.07%, O 28.88%. From rhizome of Canadian hemp, *Apocynum cannabinum* L., *Apocynaceae* and from *A. androsaemifolium:* Finnemore, *J. Chem. Soc.* **93**, 1513, 1520 (1908). Also from the essential oil (butter) of the rhizomes of *Iris* spp., *Iridaceae:* Naves, *Helv. Chim. Acta* **32**, 1351 (1949).

Fine needles from water, mp 115°. bp 295-300°. Faint vanilla odor. Slightly sol in cold, freely in hot water, alc, benzene, chloroform, ether. Practically insol in petr ether.

729. Apocynum androsaemifolium. Dogbane; bitter root; spreading dogbane; milk ipecac; wild ipecac; rheumatism weed. Root of *Apocynum androsaemifolium* L., *Apocynaceae*. *Habit.* North America. *Constit.* Apocynin, apocynein, apocynamarin, volatile oil.
THERAP CAT: Cardiotonic.

730. Apocynum cannabinum. Canadian hemp; American Indian hemp; black Indian hemp; Indian physic; Indian dogbane. Dried rhizome and roots of *Apocynum cannabinum* L., *Apocynaceae*. *Habit.* U.S. *Constit.* Cynotoxin, apocyncein, apocynin, cymarin, resin, tannin, bitter extractive, starch.
Note: Not to be confused with *Apocynum androsaemifolium* which has few of its properties.
THERAP CAT: Cardiotonic.

731. Apo-β-erythroidine. [478-85-3] 4,5,7,8,9,12-Hexahydro-5*H*-pyrano[3,4-*d*]pyrrolo[3,2,1-*jk*][1]benzazepin-5-one. C$_{15}$H$_{15}$NO$_2$; mol wt 241.29. C 74.67%, H 6.27%, N 5.81%, O 13.26%. A degradation product of β-erythroidine. Prepd by heating β-erythroidine to 120° with phosphoric or sulfuric acid. Also prepd by reacting β-erythroidine with concd hydrobromic acid at 100°: Koniuszy, Folkers, *J. Am. Chem. Soc.* **73**, 333 (1951). Synthesis and structure: Blake *et al., J. Am. Chem. Soc.* **87**, 1397 (1965).

Crystals, mp 128-129°, also reported as mp 132-132.5°. uv max (ethanol): 345, 240 nm (ε 3500, 24500).

732. Apolipoprotein E. ApoE. Protein component of several classes of lipoproteins; plays a central role in plasma lipoprotein metabolism and cholesterol transport. Also involved in lipid-transport in the development and repair of nervous tissue. Polymorphic glycoprotein containing 299 amino acid residues; mol wt 34.2 kDa. Three major isoforms have been identified in humans (apo-E2, E3, E4) which are the product of 3 independent alleles (ε2, ε3, ε4) at a single gene locus. The isoforms differ from each other only in cysteine-arginine interchanges at 2 substitution sites. Allele ε3 is the most common; inheritance of ε4 has been associated with increased risk of Alzheimer's disease. Primarily synthesized in the liver by parenchymal cells and in the brain by astrocytes. Also produced in significant amounts by mature macrophages. Discovery as component of VLDL: V. G. Shore, B. Shore, *Biochemistry* **12**, 502 (1973). Purification and characterization: V. I. Zannis, J. L. Breslow, *Mol. Cell. Biochem.* **42**, 3 (1982). Amino acid sequence: S. C. Rall, Jr. *et al.*, *J. Biol. Chem.* **257**, 4171 (1982). Biosynthesis and activity: R. W. Mahley, *Science* **240**, 622 (1988). Association with Alzheimer's disease: A. M. Saunders *et al.*, *Neurology* **43**, 1467 (1993); L. A. Farrer *et al.*, *J. Am. Med. Assoc.* **278**, 1349 (1997). Review of role in lipoprotein metabolism: A. D. Dergunov, M. Rosseneu, *Biol. Chem. Hoppe-Seyler* **375**, 485-495 (1994); P. de Knijff *et al.*, *Hum. Mutat.* **4**, 178-194 (1994). Review of role in the nervous system: J. Poirier, *Trends Neurosci.* **17**, 525-530 (1994); K. H. Weisgraber *et al.*, *Curr. Opin. Lipidol.* **5**, 110-116 (1994); R. W. Mahley *et al.*, *ibid.* **6**, 86-91 (1995).

733. Apomorphine. [58-00-4] (6a*R*)-5,6,6a,7-Tetrahydro-6-methyl-4*H*-dibenzo[*de,g*]quinoline-10,11-diol; 6aβ-aporphine-10,11-diol. $C_{17}H_{17}NO_2$; mol wt 267.33. C 76.38%, H 6.41%, N 5.24%, O 11.97%. Dopamine (D$_1$ and D$_2$) receptor agonist. Synthetic opiate obtained by treating morphine with concd HCl: A. Matthiessen, C. R. A. Wright, *Proc. R. Soc. London Ser. B* **17**, 455 (1869). Structure: R. Pschorr *et al.*, *Ber.* **35**, 4377 (1902). Configuration: H. Corrodi, E. Hardegger, *Helv. Chim. Acta* **38**, 2038 (1955). Total synthesis of (±)-form: J. L. Neumeyer *et al.*, *J. Pharm. Sci.* **59**, 1850 (1970); of (+)- and (−)-forms: V. J. Ram, J. L. Neumeyer, *J. Org. Chem.* **46**, 2830 (1981). Toxicity data: J. Z. Ginos *et al.*, *J. Med. Chem.* **18**, 1194 (1975). Clinical evaluation in impotence: J. P. W. Heaton *et al.*, *Urology* **45**, 200 (1995). Historical review: J. L. Neumeyer *et al.*, in *Apomorphine and Other Dopaminomimetics* **vol. 1**, G. L. Gessa, G. U. Corsini, Eds. (Raven, New York, 1981) p 1-17. Comprehensive description: F. J. Muhtadi, M. S. Hifnawy, *Anal. Profiles Drug Subs.* **20**, 121-166 (1991). Review of pharmacology and clinical efficacy in Parkinson's disease: D. Muguet *et al.*, *Biomed. Pharmacother.* **49**, 197-209 (1995); in erectile dysfunction: F. Giuliano, J. Allard, *Int. J. Impot. Res.* **14**, Suppl. 1, S53-S56 (2002).

Hexagonal plates from chloroform and petr ether, dec 195°; subl in high vacuum. Oxidizes rapidly in air and becomes green. Sol in alcohol, acetone, chloroform. Slightly sol in water, benzene, ether, petr ether. Solns darken rapidly. pKb 7.0; pKa 8.92. uv max (98% alc): 336, 399 nm.

Hydrochloride. [314-19-2]; [41372-20-7] (hemihydrate). Apokinon; Apokyn; Apomine; Britaject; Ixense; Uprima. $C_{17}H_{17}NO_2$·HCl; mol wt 303.79. Small crystals (usually hemihydrate). Dec and turn green on exposure to light and air. $[\alpha]_D^{25}$ −48° (c = 1.2). uv spectrum: Csokan, *Z. Anal. Chem.* **124**, 344 (1942). pH of aq soln (1 in 300) = 4.8. One gram dissolves in 50 ml water, 17 ml water at 80°, 50 ml alcohol. Very slightly sol in chloroform and ether. LD$_{50}$ i.p. in mice: 145 μg/g (Ginos).

Diacetate (ester). [6191-56-6] Diacetylapomorphine. $C_{21}H_{21}$-NO$_4$. mp 127-128°, $[\alpha]_D^{24}$ −88° (c = 1.12 in 0.1*N* HCl).

Note: This is a controlled substance (opiate): **21 CFR**, 1300.01.

THERAP CAT: Antiparkinsonian; emetic. In treatment of male erectile dysfunction.

THERAP CAT (VET): Emetic.

734. Apoptolidin. [194874-06-1] (1*R*)-*O*-2,6-Dideoxy-3-*O*-methyl-β-D-*arabino*-hexopyranosyl-(1 → 4)-*O*-2,6-dideoxy-3-*C*-methyl-α-L-*arabino*-hexopyranosyl-(1 → 8)-3,5,7-trideoxy-1-*C*-[(2*S*,4*S*,5*S*,8*E*,10*E*,12*R*,13*R*,14*E*,16*E*,18*E*)-12-[(6-deoxy-4-*O*-methyl-α-L-glucopyranosyl)oxy]-5-hydroxy-4-methoxy-9,13,15,17,19-pentamethyl-20-oxooxacycloeicosa-8,10,14,16,18-pentaen-2-yl]-3,5-dimethyl-9-*O*-methyl-L-*glycero*-α-D-*galacto*-2-nonulopyranose; FU-40A. $C_{58}H_{96}O_{21}$; mol wt 1129.39. C 61.68%, H 8.57%, O 29.75%. Natural product isolated from *Nocardiopsis* sp.; specifically induces apoptosis in transformed tumor cells. Isoln, properties

and bioactivity: J. W. Kim *et al.*, *J. Antibiot.* **50**, 628 (1997). Structure: Y. Hayakawa *et al.*, *J. Am. Chem. Soc.* **120**, 3524 (1998). Partial synthesis: J. Schuppan *et al.*, *Tetrahedron Lett.* **41**, 621 (2000); K. C. Nicolaou *et al.*, *Chem. Commun.* **2000**, 307. Review of synthetic methods: P. T. Daniel *et al.*, *Angew. Chem. Int. Ed.* **45**, 872-893 (2006).

Colorless powder, mp 128-130°. $[\alpha]_D^{21}$ −5.2° (c = 1.0 in methanol). uv max (methanol): 234, 320 nm (ε 28300, 22000).

735. Aporeine. [2030-53-7] (7a*S*)-6,7,7a,8-Tetrahydro-7-methyl-5*H*-benzo[*g*]-1,3-benzodioxolo[6,5,4-*de*]quinoline; aporheine; 1,2-methylenedioxyaporphine; (+)-roemerine. $C_{18}H_{17}NO_2$; mol wt 279.34. C 77.40%, H 6.13%, N 5.01%, O 11.45%. From *Papaver dubium* L., *Papaveraceae*: Pavesi, *Gazz. Chim. Ital.* **37 I**, 629 (1907); **44 I**, 398 (1914). Structure and identity with (+)-roemerine: Slavik, *Collect. Czech. Chem. Commun.* **28**, 1738 (1963). Synthesis of *dl*-roemerine: Marion, Grassie, *J. Am. Chem. Soc.* **66**, 1290 (1944).

Needles from ether + petroleum ether, mp 102°. $[\alpha]_D^{22}$ +80° (c = 0.50 in ethanol). uv max: 262, 315 nm (log ε 4.3, 3.7). pK 6.1. Sol in ether, methanol, ethanol, chloroform; slightly sol in petr ether. Practically insol in water, alkali.

Hydrochloride. $C_{18}H_{17}NO_2$·HCl. Leaflets from ethanol or water, mp 266-267°. Slightly sol in ethanol, water.

Methiodide. $C_{18}H_{17}NO_2$·CH$_3$I. Crystals from boiling methanol, mp 232-233°.

736. Appel's Salt. [75318-43-3] 4,5-Dichloro-1,2,3-dithiazol-1-ium chloride (1:1). $C_2Cl_3NS_2$; mol wt 208.50. C 11.52%, Cl 51.01%, N 6.72%, S 30.75%. Readily prepd from chloroacetonitrile and disulfur dichloride; 5-chloro position is reactive to nucleophilic substitution. Prepn: R. Appel *et al.*, **DE 2848221** (1980 to Bayer); *eidem et al.*, *Ber.* **118**, 1632 (1985). Synthetic applications: J. J. Folmer, S. M. Weinreb, *Tetrahedron Lett.* **34**, 2737 (1993); A. M. Cuadro, J. Alvarez-Builla, *Tetrahedron* **50**, 10037 (1994); G. L'abbé *et al.*, *J. Chem. Soc. Perkin Trans. 1* **1995**, 2379; and crystal structure: C. W. Rees *et al.*, *ibid.* **2002**, 1535. Review of chemistry of polysulfur-nitrogen heterocycles: C. W. Rees, *J. Heterocycl. Chem.* **29**, 639-651 (1992).

Pale yellow-green crystals, mp 172° (dec). Insol in organic solvents. *Sensitive to water and other nucleophiles.*

USE: Synthon in heterocyclic chemistry; in conversion of alcohols and carboxylic acids to esters.

737. Appetize®. Blended cholesterol-reduced animal fat and vegetable oil free of *trans* fatty acids, wherein the final weight ratio of linoleic acid and myristic acid is between about 2 and 9: D. Perlman, K. C. Hayes, **WO 9322933;** *eidem,* **US 5382442** (1993, 1995 both to Brandeis University). Historical perspective: K. Kevin, *Food Process.* **55,** 81-85 (1994).

Contains approx 8 mg cholesterol per 100 g and as many calories as natural fats.

USE: Low cholesterol fat alternative.

738. Apraclonidine. [66711-21-5] 2,6-Dichloro-N^1-(4,5-di-hydro-1*H*-imidazol-2-yl)-1,4-benzenediamine; 2,6-dichloro-N'-2-imidazolidinylidene-1,4-benzenediamine; 2-[(4-amino-2,6-dichloro-phenyl)imino]imidazolidine; *p*-aminoclonidine; aplonidine; NC-14. $C_9H_{10}Cl_2N_4$; mol wt 245.11. C 44.10%, H 4.11%, Cl 28.93%, N 22.86%. α_2-Adrenergic agonist; structural analog of clonidine, *q.v.* Synthesis: B. Rouot, G. LeClerc, *Bull. Soc. Chim. Fr.* **Pt. 2,** 520 (1979). Prepn (not claimed) and use in treatment of intraocular pressure: B. M. York, Jr., **US 4517199** (1985 to Alcon). Pharmacology: B. Rouot *et al., C. R. Seances Acad. Sci. Ser. D* **286,** 909 (1978). Receptor binding studies: D. C. U'Prichard, *Prog. Clin. Biol. Res.* **71,** 53 (1981); D. C. Stump, D. E. MacFarlane, *J. Lab. Clin. Med.* **102,** 779 (1983). Clinical pharmacology: D. A. Abrams *et al., Arch. Ophthalmol.* **105,** 1205 (1987). Clinical evaluation in treatment of intraocular pressure: A. L. Robin *et al., ibid.* 1208; as additive to maximally tolerated medical therapy in glaucoma: A. L. Robin *et al., Am. J. Ophthalmol.* **120,** 423 (1995).

Solid, mp >230°.

Hydrochloride. [73218-79-8] ALO-2145; Iopidine. $C_9H_{10}Cl_2$-N_4·HCl; mol wt 281.57. White powder. Sol in methanol; sparingly sol in water, alcohol. Insol in chloroform, ethyl acetate, hexanes.

Dihydrochloride. [73217-88-6] $C_9H_{10}Cl_2N_4$·2HCl. uv max (ethanol): 254, 304 nm (ε 1800, 2500).

THERAP CAT: Treatment of post-surgical elevated intraocular pressure.

739. Apramycin. [37321-09-8] *O*-4-Amino-4-deoxy-α-D-glucopyranosyl-(1 → 8)-*O*-(8*R*)-2-amino-2,3,7-trideoxy-7-(methyl-amino)-D-glycero-α-D-*allo*-octodialdo-1,5:8,4-dipyranosyl-(1 → 4)-2-deoxy-D-streptamine; 4-*O*-[3α-amino-6α-[(4-amino-4-deoxy-α-D-glucopyranosyl)oxy]-2,3,4,4aβ,6,7,8aα-octahydro-8β-hydroxy-7β-(methylamino)pyranopyrano[3,2-*b*]pyran-2α-yl]-2-de-oxy-D-streptamine; nebramycin factor 2; EL-857; EL-857/820; 47657; Apralan. $C_{21}H_{41}N_5O_{11}$; mol wt 539.58. C 46.75%, H 7.66%, N 12.98%, O 32.62%. Broad spectrum aminocyclitol anti-biotic and component of the nebramycin complex, produced by a strain of *Streptomyces tenebrarius:* R. Q. Thompson, E. A. Presti, *Antimicrob. Agents Chemother.* **1967,** 332; W. M. Stark, **US 3691279** (1972 to Lilly). Structure, abs config and properties: S. O'Connor *et al., J. Org. Chem.* **41,** 2087 (1976). ^{13}C-NMR: E. Wenkert, E. W. Hagaman, *ibid.* 701. *In vitro* activity: R. Ryden *et al., J. Antimicrob. Chemother.* **3,** 609 (1977). Synthetic studies: H. C. Jarrell, W. A. Szarek, *Carbohydr. Res.* **67,** 43 (1978); *eidem, Can. J. Chem.* **56,** 144 (1978); **57,** 924 (1979). *See also* tobramycin.

Monohydrate. From aqueous ethanol, mp 245-247°. pKa (H_2O): 8.5, 7.8, 7.2, 6.2, 5.4. Very soluble in water; slightly sol in lower alcohols.

THERAP CAT (VET): Antibacterial.

740. Apremilast. [608141-41-9] *N*-[2-[(1*S*)-1-(3-Ethoxy-4-methoxyphenyl)-2-(methylsulfonyl)ethyl]-2,3-dihydro-1,3-dioxo-1*H*-isoindol-4-yl]acetamide; CC-10004. $C_{22}H_{24}N_2O_7S$; mol wt 460.50. C 57.38%, H 5.25%, N 6.08%, O 24.32%, S 6.96%. Phos-phodiesterase 4 (PDE4) inhibitor with tumor necrosis factor α (TNF-α) inhibitory activity. Prepn (stereochem unspec): G. W. Muller, H.-W. Man, **US 6020358** (2000 to Celgene). Prepn of *S*-isomer: P. H. Schafer *et al.,* **WO 03080049;** *eidem,* **US 7427638** (2003, 2008 both to Celgene). Synthesis and pharmacology of enantiomers: H.-W. Man *et al., J. Med. Chem.* **52,** 1522 (2009). Anti-inflammatory activity *in vitro:* P. H. Schafer *et al., Br. J. Pharmacol.* **159,** 842 (2010). Clinical evaluation in psoriasis: A. B. Gottlieb *et al., Curr. Med. Res. Opin.* **24,** 1529 (2008).

Crystals from ethanol + acetone. Soly in water: 0.012 mg/ml. **(±)-Form.** [253168-86-4] Yellow solid, mp 144.0°. Soly in wa-ter: 0.0034 mg/ml.

THERAP CAT: Antipsoriatic.

741. Aprepitant. [170729-80-3] 5-[[((2*R*,3*S*)-2-[(1*R*)-1-[3,5-Bis(trifluoromethyl)phenyl]ethoxy]-3-(4-fluorophenyl)-4-morpho-linyl]methyl]-1,2-dihydro-3*H*-1,2,4-triazol-3-one; (2*R*)-[(1*R*)-3,5-bis(trifluoromethyl)phenylethoxy]-(3*S*)-(4-fluoro)phenyl-4-(3-oxo-1,2,4-triazol-5-yl)methylmorpholine; MK-0869; Emend. $C_{23}H_{21}F_7$-N_4O_3; mol wt 534.43. C 51.69%, H 3.96%, F 24.88%, N 10.48%, O 8.98%. Selective substance P/neurokinin-1 (NK-1) receptor an-tagonist. Prepn: C. P. Dorn *et al.,* **WO 9516679;** *eidem,* **US 5719147** (1995, 1998 both to Merck & Co.); and pharmacology: J. J. Hale *et al., J. Med. Chem.* **41,** 4607 (1998). Improved synthesis: K. M. J. Brands *et al., J. Am. Chem. Soc.* **125,** 2129 (2003). LC/MS/ MS determn in plasma: C. M. Chavez-Eng *et al., J. Pharm. Biomed. Anal.* **35,** 1213 (2004). Clinical trial in prevention of cisplatin-in-duced emesis: S. Van Belle *et al., Cancer* **94,** 3032 (2002); with granisetron and/or dexamethasone: D. Campos *et al., J. Clin. Oncol.* **19,** 1759 (2001). Clinical pharmacokinetics: A. K. Majumdar *et al., J. Clin. Pharmacol.* **46,** 291 (2006). Review of clinical experience in prevention of chemotherapy-induced nausea and vomiting (CINV): T. M. Dando, C. M. Perry, *Drugs* **64,** 777-794 (2004); A. M. Massaro, K. L. Lenz, *Ann. Pharmacother.* **39,** 77-85 (2005).

White to off-white crystalline solid, mp 255°. $[\alpha]_D^{25}$ +69° (c = 1.00 in methanol). Sparingly sol in ethanol, isopropyl acetate; slightly sol in acetonitrile. Practically insol in water.

Fosaprepitant dimeglumine. [265121-04-8]; [172673-20-0] (free acid). 1-Deoxy-1-(methylamino)-D-glucitol [3-[[((2*R*,3*S*)-2-[(1*R*)-1-[3,5-bis(trifluoromethyl)phenyl]ethoxy]-3-(4-fluorophenyl)-4-morpholinyl]methyl]-2,5-dihydro-5-oxo-1*H*-1,2,4-triazol-1-yl]-phosphonate (2:1) (salt); MK-0517. $C_{23}H_{22}F_7N_4O_6P$·2$C_7H_{17}NO_5$;

mol wt 1004.84. Water-soluble prodrug. Prepn: C. P. Dorn *et al.*, **WO 9523798**; *eidem*, **US 5691336** (1995, 1997 both to Merck & Co.); J. J. Hale *et al.*, *J. Med. Chem.* **43**, 1234 (2000). Bioequivalence to aprepitant and tolerability: K. C. Lasseter *et al.*, *J. Clin. Pharmacol.* **47**, 834 (2007).

THERAP CAT: Antiemetic.

742. Apricitabine. [160707-69-7] 4-Amino-1-[(2*R*,4*R*)-2-(hydroxymethyl)-1,3-oxathiolan-4-yl]-2(1*H*)-pyrimidinone; 2*R*-hydroxymethyl-4*R*-(cytosin-1′-yl)-1,3-oxathiolane; (1′*R*,4′*R*)-2′-deoxy-3′-oxathiocytidine; (−)-2′-deoxy-3′-oxa-4′-thiocytidine; (−)-dOTC; AVX-754; BCH-10618; SPD-754. $C_8H_{11}N_3O_3S$; mol wt 229.25. C 41.91%, H 4.84%, N 18.33%, O 20.94%, S 13.98%. Nucleoside reverse transcriptase inhibitor (NRTI). Prepn: T. S. Mansour, H. Jin, **WO 9529176**; B. Belleau *et al.*,**US 5587480** (1995, 1996 both to BioChem Pharma); W. Wang *et al.*, *Tetrahedron Lett.* **35**, 4739 (1994). *In vitro* anti-HIV activity and intracellular metabolism: J.-M. de Muys *et al.*, *Antimicrob. Agents Chemother.* **43**, 1835 (1999). Clinical pharmacokinetics: P. Cahn *et al.*, *Clin. Drug Invest.* **28**, 129 (2008). Clinical evaluation in HIV infection: *eidem*, *AIDS* **20**, 1261 (2006). Review of development and therapeutic potential: S. Cox, J. Southby, *Expert Opin. Invest. Drugs* **18**, 199-209 (2009).

White solid from diethylether/methanol, mp 200° (dec). $[\alpha]_D^{25}$ −126.8° (c = 0.5 in methanol).

THERAP CAT: Antiretroviral.

743. Aprindine. [37640-71-4] N^1-(2,3-Dihydro-1*H*-inden-2-yl)-N^3,N^3-diethyl-N^1-phenyl-1,3-propanediamine; *N*,*N*-diethyl-*N*′-2-indanyl-*N*′-phenyl-1,3-propanediamine; *N*-[3-(diethylamino)propyl]-*N*-phenyl-2-indanamine; compd 99170; AC-1802; Lilly 99170. $C_{22}H_{30}N_2$; mol wt 322.50. C 81.94%, H 9.38%, N 8.69%. Membrane-stabilizing agent. Prepn: P. M. Vanhoof, P. M. Clarebout, **BE 760018**; *eidem*, **US 3923813** (1971, 1975 both to Christiaens). Toxicological study: A. Georges *et al.*, *Arzneim.-Forsch.* **23**, 519 (1973). Pharmacokinetics: T. Kobari *et al.*, *Eur. J. Clin. Pharmacol.* **26**, 129 (1984). HPLC determn in plasma: T. Kobari *et al.*, *J. Chromatogr.* **278**, 220 (1983). Clinical studies: D. P. Zipes *et al.*, *Am. J. Cardiol.* **40**, 586 (1977); P. J. Troup, D. P. Zipes, *Am. Heart J.* **97**, 322 (1979). *Reviews:* P. Danilo, *ibid.* 119; I. Stoel, S. F. Hagemeijer, *Eur. Heart J.* **1**, 147 (1980).

Hydrochloride. [33237-74-0] Compd 83846; Amidonal; Aspenon; Fiboran; Ritmusin. $C_{22}H_{30}N_2$.HCl; mol wt 358.95. Crystals from benzene, mp 120-121°.

THERAP CAT: Antiarrhythmic (class I).

744. Aprinocarsen. [151879-73-1] d(*P*-Thio)(G-T-T-C-T-C-G-C-T-G-G-T-G-A-G-T-T-T-C-A)DNA; CGP-64128A; ISI-641A; LY-900003. $C_{196}H_{249}N_{68}O_{105}P_{19}S_{19}$; mol wt 6435.16. C 36.58%, H 3.90%, N 14.80%, O 26.10%, P 9.15%, S 9.47%. Synthetic, 20 base phosphorothioate antisense oligonucleotide designed to inhibit protein kinase C-α (PKC-α) expression. Prepn: C. F. Bennett, N. Dean, **WO 9319203**; *eidem*, **US 5882927** (1993, 1999 both to Isis). Pharmacology: N. Dean *et al.*, *Cancer Res.* **56**, 3499 (1996). CGE determn in tissue: R. S. Geary *et al.*, *Anal. Biochem.* **274**, 241 (1999). Clinical pharmacology and evaluation in advanced cancer: J. Nemunaitis *et al.*, *J. Clin. Oncol.* **17**, 3586 (1999). Review of pharmacokinetics: P. L. Nicklin *et al.*, *Handb. Exp. Phar-*

macol. **131**, 141-168 (1998). Review of development and clinical evaluations: J. T. Holmlund *et al.*, *Curr. Opin. Mol. Ther.* **3**, 372-385 (1999); K. Li, J. Zhang, *Curr. Opin. Investig. Drugs* **2**, 1454-1461 (2001).

Nonadecasodium salt. [331257-53-5] Aprinocarsen sodium; ISIS-3521; Affinitac; Affinitak. $C_{196}H_{230}N_{68}Na_{19}O_{105}P_{19}S_{19}$; mol wt 6852.81.

THERAP CAT: Antineoplastic.

745. Aprobarbital. [77-02-1] 5-(1-Methylethyl)-5-(2-propen-1-yl)-2,4,6(1*H*,3*H*,5*H*)-pyrimidinetrione; 5-allyl-5-isopropylbarbituric acid; allypropymal; Alurate. $C_{10}H_{14}N_2O_3$; mol wt 210.23. C 57.13%, H 6.71%, N 13.33%, O 22.83%. Prepn: **US 1444802** (1923). Toxicity data: H. H. Frey, *Arzneim.-Forsch.* **12**, 389 (1962).

Slightly bitter crystals, mp 140-141.5°. Almost insoluble in water, petr ether, aliphatic hydrocarbons. Sol in alcohol, chloroform, ether, acetone, benzene, glacial acetic acid, also in solns of fixed alkali hydroxides. A satd aq soln is acid to litmus. LD_{50} i.p. in mice: 200 mg/kg (Frey).

Sodium salt. [125-88-2] Aprobarbital sodium; sodium 5-allyl-5-isopropylbarbiturate. $C_{10}H_{13}N_2NaO_3$. Hygroscopic powder; slightly bitter taste. Very soluble in water; slightly sol in alc. Practically insol in ether. Aq solns are alkaline to litmus.

Note: This is a controlled substance (depressant): **21 CFR**, 1308.13.

THERAP CAT: Sedative, hypnotic.

746. Aprotinin. [9087-70-1] Pancreatic basic trypsin inhibitor; pancreatic trypsin inhibitor (Kunitz); Bayer A 128; Riker 52G; RP-9921; Antagosan; Antikrein; Fosten; Iniprol; Kir Richter; Repulson; Trasylol; Trazinin; Zymofren. $C_{284}H_{432}N_{84}O_{79}S_7$; mol wt 6511.51. C 52.39%, H 6.69%, N 18.07%, O 19.41%, S 3.45%. Serine protease inhibitor which inhibits kallikrein, plasmin, trypsin, chymotrypsin and various intracellular proteases. Single chain polypeptide containing 58 amino acids; found in tissues and blood, highest concentration in bovine parotid gland, pancreas and lung. Initial description: H. Kraut *et al.*, *Z. Physiol. Chem.* **192**, 1 (1930). Isoln from bovine pancreas: M. Kunitz, J. H. Northrup, *J. Gen. Physiol.* **19**, 991 (1936); from bovine parotid glands: H. Kraut, R. Körbel, **US 2890986** (1959 to Bayer). Amino acid sequence: B. Kassell *et al.*, *Biochem. Biophys. Res. Commun.* **18**, 255 (1965); F. A. Anderer, S. Hörnle, *Z. Naturforsch.* **20b**, 457, 462 (1965); B. Kassell, M. Laskowski, *Biochem. Biophys. Res. Commun.* **20**, 463 (1966). Two-dimensional ^{1}H-NMR studies: K. Nagayama, K. Wüthrich, *Eur. J. Biochem.* **114**, 365 (1981); G. Wagner *et al.*, *ibid.* 375. Review of early literature: I. Trautschold *et al.*, *Biochem. Pharmacol.* **16**, 59-72 (1957). Review of mechanism of therapeutic action: G. Haberland, R. McConn, *Fed. Proc.* **38**, 2760-2767 (1979). Review of use as a proteolytic inhibitor in radioimmunoassays of polypeptide hormones: E. S. Zyzner, *Life Sci.* **28**, 1861-1866 (1981). Review of biochemistry and applications: H. Fritz, G. Wunderer, *Arzneim.-Forsch.* **33**, 479-494 (1983); of pharmacology and therapeutic efficacy in reducing blood loss associated with cardiac surgery: R. Davis, R. Whittington, *Drugs* **49**, 954-983 (1995).

```
                              ┌─Cys 55
1                                                14
Arg─Pro─Asp─Phe─Cys─Leu─Glu─Pro─Pro─Tyr─Thr─Gly─Pro─Cys─Lys─Ala─Arg─Ile─Ile─Arg─Tyr
                                                                                    │
                                                                                   Phe
                                                                                    │
     38                             30
Arg─Lys─Ala─Arg─Cys─Gly─Gly─Tyr─Val─Phe─Thr─Gln─Cys─Leu─Gly─Ala─Lys─Ala─Asn─Tyr
 │
Asn
 │      51                              58
Asn─Phe─Lys─Ser─Ala─Glu─Asp─Cys─Met─Arg─Thr─Cys─Gly─Gly─Ala
                              └─Cys 5
```

uv max (pH 5.9): 280 nm. Isoelec pt pH 10.5. Stable in neutral or acid media at high temp. Irreversible changes in molecular structure occur in strongly alkaline media (pH >12). Partially and reversibly denatured on treatment with 8M urea. May be kept at room temp in physiological saline soln for >1 yr without detrimental effects. LD$_{50}$ i.v. in mice: 2500000 kallikrein inhibitor units/kg (Trautschold).

USE: Proteolytic inhibitor in radioimmunoassays of polypeptide hormones.

THERAP CAT: Protease inhibitor.

747. Aptiganel. [137159-92-3] N-(3-Ethylphenyl)-N-methyl-N'-1-naphthalenylguanidine; N-(1-naphthyl)-N'-(3-ethylphenyl)-N'-methylguanidine; aptaguanal. C$_{20}$H$_{21}$N$_3$; mol wt 303.41. C 79.17%, H 6.98%, N 13.85%. Noncompetitive NMDA-receptor antagonist. Prepn: E. Weber, J. F. W. Keana, **WO 9112797**; *eidem*, **US 5262568** (1991, 1993 both to State of Oregon); N. L. Reddy *et al., J. Med. Chem.* **37**, 260 (1994). Clinical pharmacology and kinetics: K. W. Muir *et al., Ann. N.Y. Acad. Sci.* **765**, 279 (1995). Review of theoretical basis for use in stroke: K. W. Muir, K. R. Lees, *Stroke* **26**, 503-513 (1995). Clinical trial in acute ischemic stroke: G. W. Albers *et al., J. Am. Med. Assoc.* **286**, 2673 (2001).

Hydrochloride. [137160-11-3] CNS-1102; Cerestat. C$_{20}$H$_{21}$N$_3$·HCl; mol wt 339.87. Off-white needles from ethanol + diethyl ether, mp 223-225°.

THERAP CAT: Neuroprotective.

748. Apyrase. [9000-95-7] ATP diphosphohydrolase (phosphate-forming); ATP-diphosphatase; adenosine diphosphatase; adenylpyrophosphatase; nucleoside triphosphate diphosphohydrolase; NTPDase; EC 3.6.1.5. Family of ectonucleotidases responsible for the sequential hydrolysis of γ- and β-phosphates of tri- and diphosphonucleotides; yields nucleoside monophosphate (NMP) and inorganic phosphate. Requires Ca^{2+} as cofactor; although other divalent metal ions such as Mg^{2+} may be substituted. Widely distributed in plants and animals; multiple isoforms have been identified, including membrane bound and soluble extracellular forms. Isoln from potato: H. M. Kalckar, *J. Biol. Chem.* **153**, 355 (1944); from yeast: O. Meyerhoff, *ibid.* **157**, 105 (1945). Characterization of 2 isoenzymes in potatoes: J. Molnar, L. Lorand, *Arch. Biochem. Biophys.* **93**, 353 (1961); A. M. Kettlun *et al., Phytochemistry* **21**, 551 (1982). Identity with the human cell surface protein, CD39: T.-F. Wang, G. Guidotti, *J. Biol. Chem.* **271**, 9898 (1996); E. Kaczmarek *et al., ibid.* 33116. Comparison of plant and animal forms: M. A. Komoszynski, *Comp. Biochem. Physiol. B* **113**, 581 (1996). Review of biological role in plants: S. J. Roux, I. Steinebrunner, *Trends Plant Sci.* **12**, 522-527 (2007). Review of bioactivity and role in human disease: S. C. Robson *et al., Semin. Thromb. Hemostasis* **31**, 217-233 (2005); M. R. C. Schetinger *et al., BioFactors* **31**, 77-98 (2007).

Potato apyrase. Mol wt 49 kDa. Isoelectric point: 6.69 (isoform 1); 8.74 (isoform 2). Soly in water: 1 mg/ml.

USE: Biochemical reagent.

749. Aquaporins. Group of integral membrane proteins that function as molecular water channels to facilitate osmotic movement of water across the plasma membrane. Found in plant and animal cells characterized as having high osmotic water permeability. Part of the *MIP* family of channel proteins, typified by the major intrinsic protein of the eye, that are responsible for the permeability of ions and small molecules. Homologs exhibit ~40% sequence identity. Structure consists of six transmembrane domains with 5 connecting loops; the water pore is thought to be formed by the juxtaposition of 2 highly conserved Asn-Pro-Ala (NPA) sequences within the lipid bilayer. Isoln of aquaporin-1 from human erythrocytes and renal tubules: B. M. Denker *et al., J. Biol. Chem.* **263**, 15634 (1988). Cloning and sequence homology with MIP family: G. M. Preston, P. Agre, *Proc. Natl. Acad. Sci. USA* **88**, 11110 (1991). Identification

of function as membrane water channel: G. M. Preston *et al., Science* **256**, 385 (1992). Nomenclature: P. Agre *et al., Am. J. Physiol.* **265**, F461 (1993). Molecular structure: J. S. Jung *et al., J. Biol. Chem.* **269**, 14648 (1994). Review of structure and tissue distribution: C. H. van Os *et al., Biochim. Biophys. Acta* **1197**, 291-309 (1994); M. J. Chrispeels, P. Agre, *Trends Biochem. Sci.* **19**, 421-425 (1994). Review of aquaporins in plants: M. J. Chrispeels, C. Maurel, *Plant Physiol.* **105**, 9-13 (1994). Review of role in renal water transport: A. F. van Lieburg *et al., Pediatr. Nephrol.* **9**, 228-234 (1995); of model for water permeation: T. Zeuthen, *Trends Biochem. Sci.* **26**, 77-79 (2001); of function and physiology in the kidney: S. Nielsen *et al., Physiol. Rev.* **82**, 205-244 (2002).

Aquaporin-1. AQP1; CHIP28; AQP-CHIP; aquaporin-CHIP; channel forming integral protein. 269 amino acid residues; mol wt 28 kDa. Constitutes 2.4% of the plasma membrane protein in red blood cells; also found in renal proximal tubules and in other water-permeable epithelia.

Aquaporin-2. AQP2; AQP-CD; WCH-CD; water channel of collecting duct. 271 amino acids; mol wt 29 kDa. Vasopressin-regulated water channel in the kidney; has a primary role in concentrating the urine.

γ-TIP. γ-Tonoplast intrinsic protein; AQP.At1. Found in plant vacuolar membrane (tonoplast); facilitates rapid exchange of water between the vacuole and the cytoplasm.

750. Arabinose. [147-81-9] Aloe sugar; pectin sugar. C$_5$H$_{10}$O$_5$; mol wt 150.13. C 40.00%, H 6.71%, O 53.28%. Monosaccharide that is widespread in plants; also isolated from mycobacteria. Predominately found in nature in the L-form, either free or as a component of a complex polysaccharide; exists in the D-form to a lesser extent. Isoln of the L-form from mesquite gum: E. Anderson, L. Sands, *J. Am. Chem. Soc.* **48**, 3172 (1926); *eidem, Org. Synth.* **coll. vol. 1**, 67 (1941); E. V. White, *J. Am. Chem. Soc.* **69**, 715 (1947); from western red cedar *(Thuja plicata)*: A. B. Anderson, H. Erdtman, *ibid.* **71**, 2927 (1949); from sapote gum: E. V. White, *ibid.* **75**, 257 (1953); from heartwood of port orford cedar *(Chamaecyparis lawsoniana)*: G. Kritchevsky, A. B. Anderson, *ibid.* **77**, 3391 (1955). Prepn of the D-form: A. Wohl, *Ber.* **26**, 730 (1893); G. Braun, *Org. Synth.* **coll. vol. 3**, 101 (1955). Isolation of the (D)-form from aloins: J. E. Hay, L. J. Haynes, *J. Chem. Soc.* **1956**, 3141. Biosynthesis in mycobacteria: B. A. Wolucka, *FEBS J.* **275**, 2691 (2008).

β-L-Arabinose

L-form. [5328-37-0] Colorless orthorhombic bisphenoidal crystals, mp 157-160°. Shows mutarotation. $[\alpha]_D^{12}$ +173° (6 min); $[\alpha]_D^{20}$ +105.1° (equilibrium, 22½ hrs, c = 3 in water). One gram dissolves in about 1 ml water, about 250 ml 90% alc. K$_a$ (17°): -3.7×10^{-13}. Reduces Fehling's soln. Forms furfurol on heating to 200° in closed tube also contg water.

D-form. [10323-20-3] Colorless crystals from methanol, mp 159-160°. Shows mutarotation. $[\alpha]_D^{20}$ $-175°$ (initial, water); $[\alpha]_D^{20}$ $-105.0°$ (equilibrium, water). Molal soly at 25°C: 5.04 (water); 0.00853 (t-butanol).

USE: As culture medium for certain bacteria. Flavoring agent.

751. Arabitol. [2152-56-9] Arabinitol; 1,2,3,4,5-pentanepentol; arabite. C$_5$H$_{12}$O$_5$; mol wt 152.15. C 39.47%, H 7.95%, O 52.58%. D-Form obtained by reduction of D-arabinose or D-lyxose with sodium amalgam: Ruff, *Ber.* **32**, 555 (1899); Ruff, Ollendorff, *Ber.* **33**, 1802 (1900); Bertrand, *Bull. Soc. Chim.* [3] **15**, 593 (1896). Production by fermentation from molasses using *Saccharomyces rouxii* and *Saccharomyces mellis:* Lavin, Holloway, **US 2934474** (1960 to Comm. Solvents). L-Form obtained by reduction of L-arabinose with sodium amalgam: Kiliani, *Ber.* **20**, 1234, 1571 (1887). DL-Form obtained from equal parts of D- and L-arabitol: Ruff, *loc. cit.* Synthesis: Lespian, *Compt. Rend.* **206**, 1773 (1938).

D-Arabitol　　　　　L-Arabitol

Prisms from 90% alc. mp 105-106°. One part dissolves in 66 parts of 90% alc at 12°.

D-Form. Big prismatic crystals. Sweet taste. mp 103°. $[\alpha]_D^{20}$ +7.7° (c = 9.26 in satd borax soln). Freely sol in water. One part dissolves in 48 parts of 90% alc at 12°.

L-Form. Wart-like crystals, sweet taste. mp 102°. Weakly levorotatory in satd borax soln. Freely sol in water and in boiling 90% alc. One part dissolves in 46 parts of 90% alc at 12°. Does not reduce Fehling's soln.

Pentanitrate. $C_5H_7N_5O_{15}$; mol wt 377.13. Sirup, sol in alcohol, ether, acetone. Strongly reduces Fehling's soln.

752. D-Araboflavin. [5978-87-0] 1-Deoxy-1-(3,4-dihydro-7,8-dimethyl-2,4-dioxobenzo[g]pteridin-10(2H)-yl)-D-arabinitol; 7,8-dimethyl-10-(arabino-2,3,4,5-tetrahydroxypentyl)benzo[g]pteridine-2,4(3H,10H)-dione; 6,7-dimethyl-9-D-araboflavin. $C_{17}H_{20}$-N_4O_6; mol wt 376.37. C 54.25%, H 5.36%, N 14.89% O 25.51%. An antagonist of riboflavine. Prepd from D-(2-amino-4,5-dimethylphenyl)arabinamine and alloxan: Kuhn, Weygand, *Ber.* **68**, 1286 (1935); Sahashi *et al.*, *Bull. Inst. Phys. Chem. Res.* (Tokyo) **24**, 72 (1948), *C.A.* **42**, 5458 (1948).

Orange-yellow needles from dil acetic acid. Bitter taste, mp 302-303°. $[\alpha]_D^{20}$ +78.6° (c = 0.509 in 0.1N NaOH); $[\alpha]_D^{20}$ −441° (c = 0.253 in 0.2N NaOH satd with borax).

Tetraacetate. $C_{25}H_{28}N_4O_{10}$. Flat yellow prisms from ethyl acetate, mp 221-222°.

753. Arachidic Acid. [506-30-9] Eicosanoic acid; arachic acid. $C_{20}H_{40}O_2$; mol wt 312.54. C 76.86%, H 12.90%, O 10.24%. Fatty acid found in peanut oil, vegetable and fish oils, etc. Also obtained by hydrogenation of arachidonic acid: Ege *et al.*, *J. Am. Chem. Soc.* **83**, 3080 (1961). Synthesis: Linstead *et al.*, *J. Chem. Soc.* **1955**, 1097. Reviews: A. W. Ralston, *Fatty Acids and Their Derivatives* (Wiley & Sons, New York, 1948) pp 43-46; K. S. Markley, *Fatty Acids* Part I (Interscience, New York, 2nd ed., 1960) pp 164-167, 398-400.

Crystals from alc, mp 75.5°. bp_{760} ~328° with some decompn; $bp_{1.0}$ 205°. d_4^{100} 0.8240. n_D^{100} 1.4250. Practically insoluble in water; sparingly sol in cold water; freely sol in hot abs alcohol, benzene, chloroform, ether, petr ether.

Methyl ester. $C_{21}H_{42}O_2$. mp 47°; $bp_{1.0}$ 180°.
Ethyl ester. $C_{22}H_{44}O_2$. mp 41°; $bp_{0.3}$ 177°.

754. Arachidonic Acid. [506-32-1] (5Z,8Z,11Z,14Z)-5,8,-11,14-Eicosatetraenoic acid. $C_{20}H_{32}O_2$; mol wt 304.47. C 78.90%, H 10.59%, O 10.51%. An essential fatty acid and a precursor in the biosynthesis of prostaglandins, thromboxanes, and leukotrienes, *q.q.v.* Structure: Mowry *et al.*, *J. Biol. Chem.* **142**, 679 (1942); Arcus, Smedley-Maclean, *Biochem. J.* **37**, 1 (1943). Occurs in liver, brain, glandular organs, and depot fats of animals, in small amounts in human depot fats, and is a constituent of animal phosphatides. Isolation from liver lipids: Brown, *J. Biol. Chem.* **80**, 455 (1928); from beef suprarenal phosphatides: Ault, Brown, *ibid.* **107**, 615 (1934); Shinowara, Brown, *ibid.* **134**, 331 (1940). *See also:* Dolby *et al.*, *Biochem. J.* **34**, 1422 (1940). Synthesis from 2-propargyloxytetrahydropyran and 1-heptyne: Goldberg, Rachlin, US 2934570 (1960 to Hoffmann-La Roche); Rachlin *et al.*, *J. Org. Chem.* **26**, 2688 (1961). Alternate route: Osbond, Wickens, *Chem. Ind. (London)* **1959**, 1288; Ege *et al.*, *J. Am. Chem. Soc.* **83**, 3080 (1961). *Reviews:* K. S. Markley, *Fatty Acids* Part I (Interscience, New York, 2nd ed., 1960) pp 164-167, 398-400; T. K. Schaaf, *Annu. Rep. Med. Chem.* **12**, 182-190 (1977); B. B. Weksler, *N. Engl. Soc. Allergy Proc.* **2**, 56-61 (1981); N. A. Nelson *et al.*, *Chem. Eng. News* **60**, 30-44 (Aug. 16, 1982). Series of articles on metabolism, role in inflammation and therapeutic implications: *Drugs* **33**, Suppl. 1, 2-66 (1987).

Liquid. mp −49.5°. Neutralization value 184.20. Iodine value 333.50. n_D^{20} 1.4824.

Methyl ester. $C_{21}H_{34}O_2$. bp_2 200-205°. $bp_{0.001}$ 127-128°. n_D^{20} 1.4797 to 1.4810.

THERAP CAT: Nutrient (essential fatty acid).

THERAP CAT (VET): With linoleic and linolenic acids, in eczema and dermatitis in dogs and swine.

755. Aralia. Spikenard; American spikenard; spignet; pettymorrel; spice berry. Dried rhizome and roots of *Aralia racemosa* L., *Araliaceae*. *Habit.* Northeastern U.S. *Constit.* Starch, pectin, sugar, resin. The active constituent appears to be a volatile oil of unknown constitution.

Ingredient of Compound White Pine Syrup: *N. F. XI.*

756. Aranidipine. [86780-90-7] 1,4-Dihydro-2,6-dimethyl-4-(2-nitrophenyl)-3,5-pyridinedicarboxylic acid 3-methyl 5-(2-oxopropyl) ester; (±)-methyl 2-oxopropyl 1,4-dihydro-2,6-dimethyl-4-(2-nitrophenyl)-3,5-pyridinedicarboxylate; (±)-acetonyl methyl 1,4-dihydro-2,6-dimethyl-4-(o-nitrophenyl)-3,5-pyridinedicarboxylate; MPC-1304; Sapresta. $C_{19}H_{20}N_2O_7$; mol wt 388.38. C 58.76%, H 5.19%, N 7.21%, O 28.84%. Calcium channel blocker. Prepn: S. Ohno *et al.*, FR 2514761; *eidem*, US 4446325 (1983, 1984 both to Maruko Seiyaku); *idem et al.*, *Chem. Pharm. Bull.* **34**, 1589 (1986). Pharmacology and receptor binding studies: K. Miyoshi *et al.*, *Eur. J. Pharmacol.* **238**, 139 (1993). Determn by GC-MS in plasma and urine: Y. Umeno *et al.*, *J. Chromatogr.* **434**, 123 (1988); by HPLC in plasma: Y. Iida *et al.*, *ibid.* **571**, 277 (1991). Cardiovascular profile: A. Kanda *et al.*, *J. Cardiovasc. Pharmacol.* **22**, 167 (1993). Clinical pharmacology: S. Suzuki *et al.*, *Arzneim.-Forsch.* **43**, 1152 (1993). Toxicity studies: S. Nakano *et al.*, *Yakuri to Chiryo* **21**, S931 (1993), *C.A.* **119**, 131216 (1993).

Yellow prisms from ethyl acetate/hexane, mp 155°. LD_{50} in male, female mice, rats (mg/kg): 143, 193, 1982, 1459 orally; LD_{50} in male, female mice (mg/kg): 7.3, 9.1 i.p. (Nakano).

THERAP CAT: Antihypertensive.

757. Araroba. [1393-64-2] Goa powder; Bahia powder; Brazil powder. Dried latex found in cavities in the trunk and branches of *Andira araroba*, Aguiar *(Vouacapoua araroba* [Aguiar] Lyons), *Leguminosae. Habit.* In damp forests of Bahia, Brazil. *Constit.* Chrysarobin, chrysophanic acid, dichrysarobin, emodin anthrone methyl ether, gum, about 2% resin. The powder and extract have been used as a vermifuge, purgative and antipsoriatic. Description and use as emetic: J. A. Thompson, *Br. Med. J.* **1877**, 607. Prepn of the extract and constituents: F. Tutin, H. W. B. Clewer, *J. Chem. Soc. Trans.* **101**, 290 (1912). Description, constituents and uses: J. Gruenwald, *PDR for Herbal Medicines* (Thomson PDR, Montvale, 3rd Ed., 2004) p. 388. Historical use in psoriasis: C. Clark, S. Lewis-Jones, *Br. J. Dermatol.* **157**, Suppl. 1, 77 (2007).

Brownish-yellow powder; darkens with age. Mucilaginous, bitter taste; slight disagreeable odor. Insol in water.

Solvent extract. Purified Goa powder; purified araroba; crude chrysarobin. Complex mixture of anthrone derivatives prepd by recrystallization of dried latex in hot benzene or chloroform. Principle active ingredient is the anthrone chrysarobin, *q.v.*, which is also used as a name for the extract. Brownish to orange-yellow, microcrystalline, odorless, tasteless powder. *Irritant.* Very slightly sol in water; 1 g dissolves in 400 ml alcohol, 30 ml benzene, 15 ml chloroform, 160 ml ether; also sol in fats. Dissolves in alkali hydroxide solns or in sulfuric acid with a deep red color.

Caution: Ingestion may cause renal damage, severe gastroenteritis. Application to large areas of skin may cause renal irritation by percutaneous absorption. Severely irritating to skin and mucous membranes; stains the skin.

THERAP CAT: Antipsoriatic.

758. Arbaclofen Placarbil. [847353-30-4] (*βR*)-4-Chloro-*β*-[[[[(1*S*)-2-methyl-1-(2-methyl-1-oxopropoxy)propoxy]carbonyl]-amino]methyl]benzenepropanoic acid; (3*R*)-3-(4-chlorophenyl)-4-[[[(1*S*)-2-methyl-1-[(2-methylpropanoyl)oxy]propoxy]carbonyl]-amino]butanoic acid; 4-[[(1*S*)-isobutanoyloxyisobutoxy]carbonyl-amino]-(3*R*)-(4-chlorophenyl)butanoic acid; XP-19986. C$_{19}$H$_{26}$ClNO$_6$; mol wt 399.87. C 57.07%, H 6.55%, Cl 8.87%, N 3.50%, O 24.01%. Prodrug for the pharmacologically active *R*-isomer of baclofen, *q.v.* Prepn: M. A. Gallop *et al.*, **WO 05019163**; *eidem*, **US 7109239** (2005, 2006 both to Xenoport). Bioavailability, metabolism, and pharmacokinetics: R. Lai *et al.*, *J. Pharmacol. Exp. Ther.* **330**, 911 (2009). Clinical evaluation for spasticity due to spinal cord injury: P. W. Nance *et al.*, *Spinal Cord* **49**, 974 (2011).

White crystals from acetone/hexane or ethyl acetate/heptane.

THERAP CAT: Muscle relaxant (skeletal).

759. Arbaprostil. [55028-70-1] (5*Z*,11*α*,13*E*,15*R*)-11,15-Dihydroxy-15-methyl-9-oxoprosta-5,13-dien-1-oic acid; (15*R*)-15-methylprostaglandin E$_2$; (15*R*)-15-methyl-PGE$_2$; (*E,Z*)-7-[3-hydroxy-2-[(3*R*)-(3-hydroxy-3-methyl-1-octenyl]-5-oxocyclopentyl]-5-heptenoic acid; U-42842. C$_{21}$H$_{34}$O$_5$; mol wt 366.50. C 68.82%, H 9.35%, O 21.83%. Antisecretory, cytoprotective derivative of prostaglandin E$_2$, *q.v.* Prepn: G. L. Bundy *et al.*, **DE 2121980**; G. L. Bundy, **US 3804889** (1971, 1974 both to Upjohn). Synthetic studies: G. Bundy *et al.*, *Ann. N.Y. Acad. Sci.* **180**, 76 (1971); E. W. Yankee, G. Bundy, *J. Am. Chem. Soc.* **94**, 3651 (1972). Synthesis of *R*- and *S*-methyl esters: E. W. Yankee *et al.*, *ibid.* **96**, 5865 (1974). HPLC separation of epimers: R. K. Lustgarten, *J. Pharm. Sci.* **65**, 1533 (1976); G. E. Peng, V. K. Sood, *J. Liq. Chromatogr.* **6**, 1499 (1983). Pharmacology: J. R. Weeks *et al.*, *J. Pharmacol. Exp. Ther.* **186**, 67 (1973). Comparative activity of epimers: A. Robert, E. W. Yankee, *Proc. Soc. Exp. Biol. Med.* **148**, 1155 (1975). Kinetics of epimerization: M. V. Merritt, G. E. Bronson, *J. Am. Chem. Soc.* **100**, 1891 (1978). Clinical study in duodenal ulcers: G. Vantrappen *et al.*, *Gastroenterology* **83**, 357 (1982); in prevention of aspirin-induced gastric mucosal injury: D. A. Gilbert *et al.*, *ibid.* **86**, 339 (1984).

(15*S*)-Methyl ester. C$_{22}$H$_{36}$O$_5$. Colorless oil, [*α*]$_D$ −79° (c = 1.3 in chloroform). uv max (basic ethanol): 278 nm (*ε* = 25250).

(15*R*)-Methyl ester. C$_{22}$H$_{36}$O$_5$. Colorless oil, [*α*]$_D$ −74° (c = 1.0 in chloroform). uv max (basic ethanol): 278 nm (*ε* = 25200).

760. Arbekacin. [51025-85-5] *O*-3-Amino-3-deoxy-*α*-D-glucopyranosyl-(1 → 6)-*O*-[2,6-diamino-2,3,4,6-tetradeoxy-*α*-D-*erythro*-hexopyranosyl-(1 → 4)]-*N*^{1}-[(2*S*)-4-amino-2-hydroxy-1-oxobutyl]-2-deoxy-D-streptamine; 1-*N*-[(*S*)-4-amino-2-hydroxybutyryl]dibekacin; 1-*N*-[(*S*)-4-amino-2-hydroxybutyryl]-3′,4′-dideoxykanamycin B; HBK. C$_{22}$H$_{44}$N$_6$O$_{10}$; mol wt 552.63. C 47.82%, H 8.03%, N 15.21%, O 28.95%. Aminoglycoside antibiotic; derivative of kanamycin B, *q.v.* Prepn and antibacterial activity: S. Kondo *et al.*, *J. Antibiot.* **26**, 412 (1973); H. Umezawa *et al.*, **DE 2350169**; *eidem*, **US 4107424** (1974, 1978 both to Microbiochem. Res. Found.). Mechanism of action: N. Tanaka *et al.*, *Antimicrob. Agents Chemother.* **24**, 797 (1983); K. Matsunaga *et al.*, *J. Antibiot.* **37**, 596 (1984). Pharmacology: J. Stewens *et al.*, *Arzneim.-Forsch.* **35**, 1440 (1985); M. Kurebe *et al.*, *ibid.* **36**, 1511 (1986). Series of articles on antibacterial activity, pharmacology, pharmacokinetics, and clinical experience: *Chemotherapy* **34**, Suppl. 1, pp 1-670 (1986). *Review:* Y. Kobayashi *et al.*, *Int. J. Antimicrob. Agents* **5**, 227-230 (1995).

Sulfate. [104931-87-5] 1665-RB; Habekacin. C$_{22}$H$_{44}$N$_6$O$_{10}$·H$_2$SO$_4$; mol wt 650.70.

Dicarbonate. C$_{24}$H$_{48}$N$_6$O$_{16}$. Colorless, crystalline powder, mp 178° (dec). [*α*]$_D^{24}$ +86.8° (c = 0.77 in water). LD$_{50}$ i.v. in mice: >150 mg/kg (Umezawa, 1978).

THERAP CAT: Antibacterial.

761. Arborescin. [6831-14-7] (3a*R*,4a*S*,6a*S*,7*S*,9a*S*,9b*R*)-5,-6,6a,7,9a,9b-Hexahydro-1,4a,7-trimethyl-3*H*-oxireno[8,8a]azuleno[4,5-*b*]furan-8(4a*H*)-one; 1,10-epoxy-6*β*-hydroxy-1*β*,5*β*,7*α*-guiai-3-en-12-oic acid *γ*-lactone. C$_{15}$H$_{20}$O$_3$; mol wt 248.32. C 72.55%, H 8.12%, O 19.33%. Ancient Greek contraceptive. Isoln from *Artemisia arborescens* L. and *Matricaria globifera* (Thunb.) Druce, *Compositae:* Meisels, Weizmann, *J. Am. Chem. Soc.* **75**, 3865 (1953); Cekan *et al.*, *Collect. Czech. Chem. Commun.* **25**, 2553 (1960). Structure: Bates *et al.*, *Tetrahedron Lett.* **1963**, 1127; M. Suchy *et al.*, *Collect. Czech. Chem. Commun.* **29**, 1829 (1964). Synthesis and revised structure: M. Ando *et al.*, *Chem. Lett.* **1978**, 727. Stereochemistry: M. Ando *et al.*, *J. Org. Chem.* **47**, 3909 (1982).

Crystals from ethanol, mp 140-142°. [*α*]$_D^{20}$ +64° (c = 0.72 in chloroform).

762. Arbutamine. [128470-16-6] 4-[(1*R*)-1-Hydroxy-2-[[4-(4-hydroxyphenyl)butyl]amino]ethyl]-1,2-benzenediol; (*R*)-3,4-dihydroxy-α-[[[4-(p-hydroxyphenyl)butyl]amino]methyl]benzyl alcohol; 1-(3,4-dihydroxyphenyl)-2-(4-(4-hydroxyphenyl)butylamino)ethanol. $C_{18}H_{23}NO_4$; mol wt 317.39. C 68.12%, H 7.30%, N 4.41%, O 20.16%. Catecholamine cardiac stimulant. Prepn: R. R. Tuttle, C. E. Browne, III, **EP 329464**; *eidem*, **US 5395970** (1989, 1995 both to Gensia). Pharmacology and pharmacodynamics: M. Young *et al.*, *Drug Dev. Res.* **32**, 19 (1994). Efficacy as exercise stimulating agent (ESA) in comparison with dobutamine, *q.v.*: H. K. Hammond, M. D. McKirnan, *J. Am. Coll. Cardiol.* **23**, 475 (1994).

Hydrochloride. [125251-66-3] GP-2-121-3; GenESA. Off-white, amorphous solid. mp 55-58°. $[\alpha]_D^{23}$ −18.5° (in ethanol). Freely sol in water, ethanol. Practically insol in diethyl ether, hexane.

THERAP CAT: Stress agent for diagnosis of coronary artery disease.

763. Arbutin. [497-76-7] 4-Hydroxyphenyl-β-D-glucopyranoside; hydroquinone-β-D-glucopyranoside; hydroquinone glucose; arbutoside; ursin. $C_{12}H_{16}O_7$; mol wt 272.25. C 52.94%, H 5.92%, O 41.14%. Naturally occurring glycoside of hydroquinone, *q.v.*, found in the bark and leaves of various plants, usually occurring together with methylarbutin. Principal antibacterial constituent of the traditional medicine, uva ursi, *q.v.* Isoln from bearberry leaves *(Arctostaphylos uva-ursi* Spreng., *Ericaceae)*: A. Kawalier, *Ann.* **82**, 241 (1852); from leaves of blueberry, cranberry, and pear *(Pyrus communis* L., *Rosaceae)*: G. Urban, M. Rogowski, *Arch. Exp. Pathol. Pharmakol.* **211**, 194 (1950). Synthesis: C. Mannich, *Arch. Pharm.* **250**, 547 (1912); A. Robertson, R. B. Waters, *J. Chem. Soc.* **1930**, 2729. Production by plant cell culture: S. Inomata *et al.*, *Appl. Microbiol. Biotechnol.* **36**, 315 (1991). Physical properties: E. Lindpainter, *Arch. Pharm.* **277**, 398 (1940). Determn in drug prepns: L. Kraus, *Pharmazie* **19**, 41 (1964); E. Kenndler *et al.*, *J. Chromatogr.* **514**, 383 (1990). Inhibition of human melanin synthesis *in vitro*: K. Maeda, M. Fukuda, *J. Pharmacol. Exp. Ther.* **276**, 765 (1996). Review of human exposure: P. J. Deisinger *et al.*, *J. Toxicol. Environ. Health* **47**, 31-46 (1996).

Occurs as the monohydrate. Colorless elongated prisms from moist ethyl acetate, mp 199° after sintering at 163-164° (Robertson, Waters). Also reported as unstable form, mp 165°; stable form, mp 199.5-200° (Lindpaintner). $[\alpha]_D^{20}$ −60.3° (in water). Sol in water and in alc.

Methylarbutin. [6032-32-2] 4-Methoxyphenyl-β-D-glucopyranoside; methylarbutoside. $C_{13}H_{18}O_7$; mol wt 286.28. Crystallizes from water as the monohydrate, mp 158-160°; solidifies and melts again at 175° (Mannich). Also reported as unstable form, mp 160.5°; stable form, mp 176° (Lindpaintner). $[\alpha]_D^{20}$ −60.66° (in water). Sol in hot water or alcohol; slightly sol in ether.

USE: In cosmetics as skin lightening agent.

764. Arcitumomab. [154361-48-5] Anti-(human carcinoembryonic antigen) immunoglobulin G1 Fab′ fragment (mouse monoclonal IMMU-4 γ₁-chain) disulfide with mouse monoclonal IMMU-4 light chain. Murine monoclonal antibody fragment Fab′ directed against carcinoembryonic antigen (CEA), an antigen expressed by a majority of colorectal cancers and some adenocarcinomas. Radiolabeled form designed for imaging colorectal tumors. Series of articles on purifn and characterization of anti-CEA antibody: F. J. Primus *et al.*, *Cancer Res.* **43**, 679-701 (1983). Prepn of radiolabeled form: D. Shochat *et al.*, **EP 336678**; *eidem*, **US**

5061641 (1989, 1991 both to Immunomedics); and preclinical evaluation: H. J. Hansen *et al.*, *Cancer Res.* Suppl. **50**, 794s (1990). Clinical study of diagnostic use in colorectal immunoscintigraphy: K. Hughes *et al.*, *Ann. Surg.* **226**, 621 (1997); in breast cancer: D. M. Goldenberg *et al.*, *Cancer* **89**, 104 (2000). Review of pharmacology: S. A. Eccles, *Curr. Opin. Mol. Ther.* **1**, 737-744 (1999); of technical prepn and clinical use: D. A. Erb, H. A. Nabi, *J. Nucl. Med. Technol.* **28**, 12-18 (2000).

⁹⁹ᵐTc-Labeled form. [154361-49-6] Technetium Tc 99m arcitumomab; IMMU-4 ⁹⁹ᵐTc Fab′; CEA-Scan.

THERAP CAT: ⁹⁹ᵐTc-labeled form as diagnostic aid (radioactive tumor targeting agent).

765. Ardeparin. Low molecular weight heparin prepared by the depolymerization of porcine mucosal heparin in the presence of peroxides. Average mol wt 5500-6500 daltons. Manufacturing process: F. Fussi, **US 4281108** (1981 to Hepar Ind.). Description of properties and activity: *idem et al.*, *Dev. Biochem.* **12**, 535 (1981). Clinical pharmacokinetics: S. Troy *et al.*, *J. Clin. Pharmacol.* **35**, 1194 (1995). Clinical trials in prevention of postoperative venous thrombosis: M. N. Levine *et al.*, *Arch. Intern. Med.* **156**, 851 (1996); J. A. Heit *et al.*, *Thromb. Haemostasis* **77**, 32 (1997).

Sodium salt. WY-90493-RD; Normiflo. LD_{50} in mice, rats (mg/kg): >3000, >2000 i.p.; >1000, 354 ±30 i.v.; >6000, >6000 orally (Fussi *et al.*).

THERAP CAT: Antithrombotic.

766. Areca. Betel nuts; pinang. Nuts (seeds) of *Areca catechu* L., *Palmaceae*. *Habit.* East Indies. *Constit.* Arecoline, arecaidine, guvacine, guvacoline, arecolidine, choline, about 15% red tannin, about 14% fat. *Reviews:* L. Marion "The Alkaloids of Areca Nut" in Manske-Holmes, *The Alkaloids* **Vol. I** (Academic Press, New York, 1950) pp 171-175; Raghavan, Baruah, *Econ. Bot.* **12**, 315 (1958).

Hard and heavy; round-conical and depressed at base. Extern. brown, mottled with fawn color; intern. brownish-red with whitish veins; astringent taste; the fresh nuts have a faint, cheese-like odor.

THERAP CAT: Anthelmintic.

THERAP CAT (VET): Has been used for tapeworms, ascarids.

767. Arecaidine. [499-04-7] 1,2,5,6-Tetrahydro-1-methyl-3-pyridinecarboxylic acid; 1,2,5,6-tetrahydro-1-methylnicotinic acid; arecaine; methylguvacine. $C_7H_{11}NO_2$; mol wt 141.17. C 59.56%, H 7.85%, N 9.92%, O 22.67%. From seeds of *Areca catechu* L., *Palmaceae* (betel nuts). Synthesis: Wohl, Johnson, *Ber.* **40**, 4712 (1907); Hess, Leibbrandt, *ibid.* **51**, 806 (1918); Freudenberg, *ibid.* 976; Maurit, Preobrazhenskii, *Zh. Obshch. Khim.* **28**, 968 (1958), *C.A.* **52**, 17263 (1958); Merck, **DE 485139** *C.A.* **24**, 919 (1930).

Plates from dil alc, dec 232° (after drying at 102°). pH of 0.1 molar soln 5.6. Freely soluble in water, and dilute alc; almost insol in abs alc, chloroform, ether, benzene.

Hydrochloride. $C_7H_{11}NO_2$·HCl. mp 251°; needles, dec 263° (rapid heating). Freely sol in water.

Hydrobromide. $C_7H_{11}NO_2$·HBr. Crystals from methanol, dec 249°.

768. Arecoline. [63-75-2] 1,2,5,6-Tetrahydro-1-methyl-3-pyridinecarboxylic acid methyl ester; methyl 1,2,5,6-tetrahydro-1-methylnicotinate; methyl 1-methyl-Δ³,⁴-tetrahydro-3-pyridinecarboxylate; methyl N-methyltetrahydronicotinate; arecaline; arecholine; methylarecaidin. $C_8H_{13}NO_2$; mol wt 155.20. C 61.91%, H 8.44%, N 9.03%, O 20.62%. Cholinergic alkaloid found in betel nuts, the seeds of the areca palm, *Areca catechu* L., *Palmaceae*, *q.v.* Isoln: E. Johns, *Arch. Pharm.* **229**, 673 (1891). Synthesis: F. Chemnitius, *J. Prakt. Chem.* **117**, 147 (1926); Mannich, *Ber.* **75B**, 1480 (1942); Knox, **US 2506458** (1950 to Nopco). Improved synthesis from nicotinic acid: I. A. Kozello *et al.*, *Khim. Farm. Zh.* **10**, 90 (1976), *C.A.* **86**, 171205a (1977). HPLC determn in saliva: S.

Cox *et al.*, *J. Chromatogr. A* **1032**, 93 (2004). Effect on human serial learning: N. Sitaram *et al.*, *Science* **201**, 274 (1978). Clinical evaluation in Alzheimer's disease: T. T. Soncrant *et al.*, *Psychopharmacology* **112**, 421 (1993). *Review:* R. B. Burrows, *Progress in Drug Research* **vol. 17**, E. Jucker, Ed. (Birkhaüser Verlag, Basel, 1973) pp 108-210.

Oily liquid, bp 209°, bp$_7$ 92-93°, bp$_{12}$ 105°. Volatile with steam. Strong base. pK 6.84. n_D^{20} 1.4302. d^{20} 1.0495. Miscible with water, alc, ether. Soluble in chloroform. LD$_{50}$ in mice, dogs (mg/kg): 100, 5 s.c. (Burrows).

Hydrobromide. [300-08-3] $C_8H_{13}NO_2$.HBr. Bitter, optically inactive crystals, mp 169-171°. One gram dissolves in about 1 ml water, 10 ml alc, 2 ml boiling alc; slightly soluble in chloroform, ether. The aq soln is practically neutral.

THERAP CAT (VET): Anthelmintic (Cestodes); cathartic.

769.   Argatroban. [74863-84-6]   (2*R*,4*R*)-1-[(2*S*)-5-[(Aminoiminomethyl)amino]-1-oxo-2-[[(1,2,3,4-tetrahydro-3-methyl-8-quinolinyl)sulfonyl]amino]pentyl]-4-methyl-2-piperidinecarboxylic acid; (2*R*,4*R*)-4-methyl-1-[*N*^{2}-(3-methyl-1,2,3,4-tetrahydro-8-quinolinesulfonyl)-L-arginyl]-2-piperidinecarboxylic acid; (2*R*,4*R*)-4-methyl-1-[(*S*)-*N*^{2}-[[(*R*,*S*)-1,2,3,4-tetrahydro-3-methyl-8-quinolinyl]sulfonyl]arginyl]pipecolic acid; argipidine; MQPA. $C_{23}H_{36}$-N_6O_5S; mol wt 508.64. C 54.31%, H 7.13%, N 16.52%, O 15.73%, S 6.30%. Synthetic thrombin inhibitor. Prepn of argatroban and stereoisomers: R. Kikumoto *et al.*, **EP 8746**; S. Okamoto *et al.*, **US 4258192** (1980, 1981 both to Mitsubishi Chem. Ind.); S. Okamoto *et al.*, *Biochem. Biophys. Res. Commun.* **101**, 440 (1981). Comparison with heparin, *q.v.*, of antithrombotic effect in animals: T. Kumada, Y. Abiko, *Thromb. Res.* **24**, 285 (1981). Stereoselective inhibition of thrombin: R. Kikumoto *et al.*, *Biochemistry* **23**, 85 (1984). Clinical evaluation in hemodialysis: K. Ota *et al.*, *Proc. Eur. Dial. Transplant. Assoc.* **20**, 144 (1983); in disseminated intravascular coagulation: K. Kumon *et al.*, *Crit. Care Med.* **12**, 1039 (1984).

Crystals from ethanol, mp 188-191°.

Monohydrate. [141396-28-3]   DK-7419; MCI-9038; MD-805; OM-805; Novastan; Slonnon. Crystals from aq ethanol, mp 176-180°. $[\alpha]_D^{27}$ +76.1° (c = 1 in 0.2*N* HCl).

THERAP CAT: Antithrombotic.

770.   Arginine. [74-79-3]   L-Arginine; Arg; R; 2-amino-5-guanidinovaleric acid; (*S*)-2-amino-5-[(aminoiminomethyl)amino]-pentanoic acid. $C_6H_{14}N_4O_2$; mol wt 174.20. C 41.37%, H 8.10%, N 32.16%, O 18.37%. An essential amino acid for human development. Precursor for nitric oxide, *q.v.* Isoln from etiolated lupine seedlings: E. Schulze, E. Steiger, *Ber.* **19**, 1177 (1886). Early chemistry and biochemistry: *Amino Acids and Proteins,* D. M. Greenberg, Ed. (Charles C. Thomas, Springfield, IL, 1951) 950 pp., *passim*; J. P. Greenstein, M. Winitz, *Chemistry of the Amino Acids* vols **1-3** (John Wiley and Sons, Inc., New York, 1961) pp. 1841-1855, *passim*. Quick isoln from gelatin: R. M. Kothari, *J. Chromatogr.* **90**, 359 (1974). Fluorometric determn in serum: T. Miura *et al.*, *Anal. Biochem.* **139**, 432 (1984). Review of biosynthesis: M. E. Jones,

Adv. Enzyme Regul. **9**, 19-49 (1970). Review of biochemistry, physiology and therapeutic implication: A. Barbul, *J. Parenter. Enteral Nutr.* **10**, 227-238 (1986). Review of immune function: *idem, Nutrition* **6**, 53-62 (1990). Review of arginine:nitric oxide pathway: S. Moncada, A. Higgs, *N. Engl. J. Med.* **329**, 2002-2012 (1993); R. M. J. Palmer, *Curr. Opin. Nephrol. Hypertens.* **2**, 122-128 (1993). Review of structural role in proteins: C. L. Borders *et al.*, *Protein Sci.* **3**, 541-548 (1994). Review of role in health and renal disease: A. A. Reyes *et al.*, *Am. J. Physiol.* **267**, F331-F346 (1994).

Prisms containing 2 mol H_2O from water; anhydrous monoclinic plates from 66% alcohol. The dihydrate becomes anhydr at 105°, browns at 230°. Dec 244°. $[\alpha]_D^{20}$ +26.9° (c = 1.65 in 6.0*N* HCl); $[\alpha]_D^{20}$ +12.5° (c = 3.5 in water); $[\alpha]_D^{20}$ +11.8° (c = 0.87 in 0.5*N* NaOH). pK$_1$ 2.17; pK$_2$ 9.04; pK$_3$ 12.48. Absorption spectrum: Castille, Ruppol, *Bull. Soc. Chim. Biol.* **10**, 643 (1928); *Chem. Zentralbl.* **II**, 622 (1928). The satd aq soln contains 15% (w/w) at 21°. Freely sol in water; sparingly sol in alc. Insol in ether. Strongly alkaline, its aq solns absorb CO_2 from the air.

Hydrochloride. [1119-34-2]   R-Gene. $C_6H_{14}N_4O_2$.HCl. Plates, prisms from alc. Sinters at 218°, solidifies again at 225°. Dec 235°. $[\alpha]_D^{20}$ +12.0° (c = 4). $[\alpha]_D^{21}$ +21.9° (c = 12 in dil HCl). Freely sol in water, slightly sol in hot alc.

THERAP CAT: Ammonia detoxicant (hepatic failure); diagnostic aid (pituitary function).

771.   Arginine Glutamate. [4320-30-3]   L-Glutamic acid compd with L-arginine; Modumate. $C_{11}H_{23}N_5O_6$; mol wt 321.33. C 41.12%, H 7.22%, N 21.80%, O 29.87%. Prepd from L-arginine and L-glutamic acid: Barker, Chang, **US 2851482** (1958 to General Mills). Activity studies: Weisburger *et al.*, *Toxicol. Appl. Pharmacol.* **14**, 163 (1969); Yamamoto, Weisburger, *Life Sci.* **9(II)**, 285 (1970). Crystal and molecular structure: T. N. Bhat, M. Vijayan, *Acta Crystallogr. B* **33**, 1754 (1977). Reduction of blood ammonia: L. Zieve *et al.*, *Metab. Brain Dis.* **4**, 113 (1989).

Crystals, dec 193-194.5°. Supplied as a 25% (w/v) soln in water for injection; each 100 ml represents 13.5 g of arginine and 11.5 g of glutamic acid.

THERAP CAT: Ammonia detoxicant (hepatic failure).

772.   Argol. [8007-14-5]   Argilla vini; argil; arcilla; Weinstein; wine lees; crude cream of tartar; crude potassium bitartrate. Formed in the secondary fermentation of grapes for wine. Contains over 40% tartaric acid, potassium bitartrate, calcium. *Ref:* R. Pasternack in *Kirk-Othmer Encyclopedia of Chemical Technology* (Interscience, New York, 1954) **vol. 13**, p 649.

Gray or reddish-brown crystalline crusts.

USE: Mordant; manufacturing tartaric acid; vinegar from malt; in fertilizers.

773.   Argon. [7440-37-1]   Ar; at. wt 39.948; at. no. 18. Group VIIIA (18), also known as Group 0. A noble gas characterized by an electronic structure in which the outer *p* subshell is entirely filled: $1s^22s^22p^63s^23p^6$. Stable naturally occurring isotopes (mass numbers): 36 (0.337%); 38 (0.063%); 40 (99.600%); known artificial radioactive isotopes: 32-35; 37; 39; 41-46. Longest-lived known isotope: 39 (T$_½$ 269 years, β-emitter). Abundance in igneous rock of the earth's crust: 4×10^{-2} ppm by wt; concentration in the atmosphere: 93.40 ppm by vol. Discovered in 1894 by Rayleigh and Ramsay. Obtained commercially from the atmosphere by distillation-liquefaction process. Monograph: *Argon, Helium and the Rare Gases* **Vols. 1 2,**, G. A. Cook, Ed. (Interscience, New York,

1961) 818 pp. *Reviews:* A. H. Cockett, K. C. Smith, "The Monatomic Gases" in *Comprehensive Inorganic Chemistry* **vol. 1**, J. C. Bailar, Jr. *et al.*, Eds. (Pergamon Press, Oxford, 1973) pp 139-211; S.-C. Hwang, W. R. Weltmer, Jr. in *Kirk-Othmer Encyclopedia of Chemical Technology* **vol. 13** (John Wiley & Sons, 4th ed., 1995) pp 1-38; *Chemistry of the Elements*, N. N. Greenwood, A. Earnshaw, Eds. (Pergamon Press, New York, 1984) pp 1042-1059. Review of use in inductively coupled plasma-mass spectrometry: S. F. Durrant, *Fresenius J. Anal. Chem.* **347**, 389-392 (1993).

Colorless, odorless, tasteless, monatomic, inert gas; will form compds with highly electronegative elements such as O, F, Cl. Atomic radius 1.92×10^{-8} cm. Triple pt temp 83.80 K, press 68.90 kPa. Critical temp 150.86 K, critical press 4898 kPa, critical d 535.7 kg/m^3. *Non-flammable.* Gas: d^0 (101.3 kPa) 1.7838 kg/m^3, d (normal bp) 5.767 kg/m^3. Liquid: normal bp $-185.87°$, d (normal bp) 1393.9 kg/m^3, d (triple pt) 1415 kg/m^3, heat of vaporization (normal bp) 6469 J/mol. Solid: d (triple pt) 1623 kg/m^3, heat of vaporization (triple pt) 7.785 kJ/mol, heat of fusion (triple pt) 1.191 kJ/mol, exists as face-centered cubic crystals at normal pressure. Soly of gas in water (20°): 33.6 cc/kg water. Sol in organic liquids.

Caution: Can act as a simple asphyxiant by displacing air. *See: Matheson Gas Data Book* (Matheson Co., Inc., 4th ed., East Rutherford, NJ, 1966) pp 23-28.

USE: Gas as shield in gas metal-arc welding, in metal processing; carrier in gas-liquid and gas-solid chromatography; gas filler for incandescent light bulbs. Gas in fluorescent tubes analogous to neon lights, but produces a blue-purplish light; in rectifier tubes; in thermometers above mercury; in lasers; wherever an inert atmosphere is desired and the much cheaper nitrogen cannot be used; in ionization chambers and particle counters; in mixtures with He and Ne in Geiger counters; in argon-oxygen-decarburizing process for stainless steel; in manuf of semiconducting devices; in gas mixtures as the working fluid in plasma arc devices. Liquid as cryogen to produce low temps. The isotope ^{40}Ar is always found in minerals contg potassium, since it is a product of ^{40}K decay; measuring the amount of ^{40}Ar and ^{40}K can be used for determining the geologic age of minerals and meteors.

774. Aricine. [482-91-7] (19α,20α)-16,17-Didehydro-10-methoxy-19-methyloxayohimban-16-carboxylic acid methyl ester; quinovatine; cinchovatine; heterophylline. $C_{22}H_{26}N_2O_4$; mol wt 382.46. C 69.09%, H 6.85%, N 7.32%, O 16.73%. Originally found in Cusco bark, the bark of *Cinchona pelletierana* Wedd., *Rubiaceae:* Pelletier, Coriol, *J. Pharm. Chim.* [2] **15**, 565 (1829). Also from *Rauwolfia canescens* L., *R. heterophylla* Roem. & Schult., *R. sellowii* Muell. Argov. and *Aspidosperma marcgravianum* Woodson, *Apocynaceae:* Stoll *et al.*, *Helv. Chim. Acta* **38**, 270 (1955); Hochstein *et al.*, *J. Am. Chem. Soc.* **77**, 3551 (1955); Hochstein, *ibid.* **77**, 5744 (1955); Gilbert *et al.*, *J. Org. Chem.* **27**, 4702 (1962). Structure: Stoll *et al.*, *loc. cit.* Stereochemistry: Neuss, Boaz, *J. Org. Chem.* **22**, 1001 (1957); Shamma, Richey, *J. Am. Chem. Soc.* **85**, 2507 (1963).

Orthorhombic, elongated prisms from methanol, dec 188°. Sublimes at 0.01 mm and 180°. $[\alpha]_D^{20}$ $-91°$ (c = 1.4 in chloroform); $[\alpha]_D^{20}$ $-63°$ (c = 1.5 in pyridine); $[\alpha]_D^{20}$ $-57°$ (ethanol). uv max (ethanol): 229, 281 nm (log ε 4.54, 3.97). pKa in 80% methyl cellosolve 5.80; in 1:1 DMF-water 6.8. Practically insol in water. Sol in 100 parts of 90% alcohol, 33 parts ether. Very sol in chloroform.

Hydrochloride. $C_{22}H_{26}N_2O_4$·HCl. Square plates from methanol + acetone, dec 241-254°. $[\alpha]_D^{20}$ $-5°$ (c = 0.9 in 50% ethanol).

Hydrobromide. $C_{22}H_{26}N_2O_4$·HBr. Needles from methanol, dec 262-263°.

Note: Aricine does not seem to share the hypotensive and sedative properties of reserpine: Hochstein *et al.*, *J. Am. Chem. Soc.* **77**, 3551 (1955).

775. Arimoclomol. [289893-25-0] N-[(2R)-2-Hydroxy-3-(1-piperidinyl)propoxy]-3-pyridinecarboximidoyl chloride 1-oxide. $C_{14}H_{20}ClN_3O_3$; mol wt 313.78. C 53.59%, H 6.42%, Cl 11.30%, N 13.39%, O 15.30%. Co-inducer of heat shock protein expression involved in cellular stress response. Prepn: M. Kürthy *et al.*, **WO 0050403**; *eidem*, **US 6649628** (2000, 2003 both to Biorex). Manufacturing process: L. Urögdi *et al.*, **WO 0179174** (2001 to Biorex); *eidem*, **US 7126002** (2006 to CytRx). Neuroprotective effect in rat models of neuropathic pain: B. Kalmar *et al.*, *Exp. Neurol.* **184**, 636 (2003). Effect on disease progression in SOD1 mouse model of amyotrophic lateral sclerosis: D. Kieran *et al.*, *Nature Med.* **10**, 402 (2004). Mechanism of action study: B. Kalmar, L. Greensmith, *Cell. Molec. Biol. Lett.* **14**, 319 (2009). Clinical evaluation in patients with ALS: M. E. Cudkowicz *et al.*, *Muscle Nerve* **38**, 837 (2008). Review of development and clinical experience: V. Lanka *et al.*, *Expert Opin. Invest. Drugs* **18**, 1907-1918 (2009).

Crystals from hexane, mp 91-93°.

Citrate. [368860-21-3] $C_{14}H_{20}ClN_3O_3$·$C_6H_8O_7$; mol wt 505.91. Powder from methanol, mp 163-165°.

Maleate. [289893-26-1] BXR-220. $C_{14}H_{20}ClN_3O_3$·$C_4H_4O_4$; mol wt 429.85. Crystals from hot ethanol, mp 132.0-133.5°.

THERAP CAT: Neuroprotective.

776. Aripiprazole. [129722-12-9] 7-[4-[4-(2,3-Dichlorophenyl)-1-piperazinyl]butoxy]-3,4-dihydro-2(1H)-quinolinone; 7-[4-[4-(2,3-dichlorophenyl)-1-piperazinyl]butoxy]-3,4-dihydrocarbostyril; OPC-14597; OPC-31; Abilify. $C_{23}H_{27}Cl_2N_3O_2$; mol wt 448.39. C 61.61%, H 6.07%, Cl 15.81%, N 9.37%, O 7.14%. Dopamine-serotonin system stabilizer. Exhibits partial agonist activity at dopamine D$_2$-receptors and serotonin 5-HT$_{1A}$-receptors and antagonist activity at 5-HT$_{2A}$-receptors. Prepn: Y. Oshiro *et al.*, **EP 367141**; *eidem*, **US 5006528** (1990, 1991 both to Otsuka); and pharmacology: Y. Oshiro *et al.*, *J. Med. Chem.* **41**, 658 (1998). Description of action on dopamine receptors: S. M. Stahl, *J. Clin. Psychiatry* **62**, 841, 923 (2001); on serotonin receptors: S. Jordan *et al.*, *Eur. J. Pharmacol.* **441**, 137 (2002). LC-MS/MS determn in plasma: M. Song *et al.*, *Anal. Biochem.* **385**, 270 (2009). Clinical trial in schizophrenia: J. M. Kane *et al.*, *J. Clin. Psychiatry* **63**, 763 (2002). Review of pharmacology: J. K. McGavin, K. L. Goa, *CNS Drugs* **16**, 779-786 (2002); and clinical experience in schizophrenia: P. E. Keck Jr., S. L. McElroy, *Expert Opin. Invest. Drugs* **12**, 655-662 (2003); in bipolar disorder: A. Fagiolini *et al.*, *Expert Opin. Pharmacother.* **12**, 473-488 (2011).

Colorless, flake crystals from ethanol, mp 139.0-139.5°.
THERAP CAT: Antipsychotic.

777. Aristolochic Acid. [313-67-7] 8-Methoxy-6-nitrophenanthro[3,4-d]-1,3-dioxole-5-carboxylic acid; 3,4-methylenedioxy-8-methoxy-10-nitro-1-phenanthrenecarboxylic acid; aristo-lochic acid-I; aristolochine. $C_{17}H_{11}NO_7$; mol wt 341.28. C 59.83%, H 3.25%, N 4.10%, O 32.82%. One of a group of fourteen known, substituted 1-phenanthrenecarboxylic acids, aristolochic acids, that occur in *Aristolochiaceae* and in butterflies feeding on these plants. They are often accompanied by **aristololactams**, twelve of which have been characterized. Isoln: Gänshirt, *Pharmazie* **8**, 584 (1953); Pailer *et al.*, *Monatsh. Chem.* **86**, 676 (1955); Coutts *et al.*, *J. Pharm. Pharmacol.* **11**, 607 (1959); Kupchan, Doskotch, *J. Med. Pharm. Chem.* **5**, 657 (1962). Structure: Pailer *et al.*, *Monatsh. Chem.* **87**, 249 (1956). Synthesis: Kupchan, Wormser, *J. Org. Chem.* **30**, 3792 (1965). Biosynthesis: Spenser, Tiwari, *Chem. Commun.* **1966**, 55. Acute toxicity: U. Mengs, *Arch. Toxicol.* **59**, 328 (1987). *Review:* D. B. Mix *et al.*, *J. Nat. Prod.* **45**, 657-666 (1982).

Shiny brown leaflets from DMF + hot water, dec 281-286°. uv max (ethanol): 390, 318, 250 nm (ε 6500; 12000; 27000). Sol in alc, chloroform, ether, acetone, acetic acid, aniline, alkalies; slightly sol in water. Practically insol in benzene, carbon disulfide. LD_{50} in male, female mice, male, female rats (mg/kg): 38.4, 70.1, 82.5, 74.0 i.v.; 55.9, 106.1, 203.4, 183.9 orally (Mengs).

Methyl ester. $C_{18}H_{13}NO_7$. Orange-yellow rods from hot methanol, mp 285-286°.

Caution: Aristolochic acids are listed as known human carcinogens: *Report on Carcinogens, Twelfth Edition* (PB2011-111646, 2011) p 45.

778. Armepavine. [524-20-9] 4-[[(1*R*)-1,2,3,4-Tetrahydro-6,7-dimethoxy-2-methyl-1-isoquinolinyl]methyl]phenol; (−)-6,7-dimethoxy-1-*p*-hydroxybenzyl-2-methyl-1,2,3,4-tetrahydroisoquin-oline. $C_{19}H_{23}NO_3$; mol wt 313.40. C 72.82%, H 7.40%, N 4.47%, O 15.31%. Isoln from *Papaver armeniacum* (L.) DC., *Papaveraceae:* Konowalowa *et al.*, *Ber.* **68**, 2161 (1935); from *Euonymus europaea* L., *Celastraceae:* Bishay *et al.*, *J. Pharm. Pharmacol.* **23** (Suppl), 233S (1971); from *Rhamnus frangula* L., *Rhamnaceae:* Pailer, Haslinger, *Monatsh. Chem.* **103**, 1399 (1972). Structure: *eidem, J. Gen. Chem. USSR* **10**, 641 (1940). Synthesis: Marion *et al.*, *J. Org. Chem.* **15**, 216 (1950); Gibson *et al.*, *J. Heterocycl. Chem.* **3**, 99 (1966). Resolution: Knabe, Horn, *Arch. Pharm.* **300**, 547 (1967); Farber, Giacomazi, *Chem. Ind. (London)* **1968** (2), 57.

mp 149-150°. $[\alpha]_D^{20}$ −117.6° (CHCl$_3$) (Knabe); $[\alpha]_D^{27}$ −103.2° (Farber).

10-Camphorsulfonate hydrate. $C_{29}H_{39}NO_7S.H_2O$. mp 130°. $[\alpha]_D^{20}$ −65° (MeOH).

dl-**Form hydrate.** Needles from acetone + ether; loses its water around 100°, then melts at 140°. Freely sol in alc, acetone, chloroform; slightly sol in water. Almost insol in water.

dl-**Form hydrochloride.** $C_{19}H_{23}NO_3.HCl$. Needles from alcohol + ether, mp 152°. Freely soluble in water and alc.

779. Armstrong's Acid. [81-04-9] 1,5-Naphthalenedisulfonic acid. $C_{10}H_8O_6S_2$; mol wt 288.29. C 41.66%, H 2.80%, O 33.30%, S 22.24%. Prepd by sulfonation of naphthalene: Lynch, Scanlan, *Ind. Eng. Chem.* **19**, 1010 (1927).

Crystals. Soluble in water, alc; practically insol in ether.

780. Arnica. Arnica flowers; leopard's bane; wolf's bane; mountain tobacco. Dried flowerheads of *Arnica montana* L., *Compositae.* Habit. Northern Europe. Constit. 0.5-1% volatile oil; arnicin, arnisterol (arnidiol), anthoxanthine, tannin, resin. *Review:* Faber, *Pharmazie* **8**, 179, 286, 340 (1953).

THERAP CAT: Counterirritant (topical).

THERAP CAT (VET): Counterirritant.

781. Arogenic Acid. [53078-86-7] (αS,1α,4α)-α-Amino-1-carboxy-4-hydroxy-2,5-cyclohexadiene-1-propanoic acid; [1(*S*)-*cis*]-α-amino-1-carboxy-4-hydroxy-2,5-cyclohexadiene-1-propanoic acid; L-(8*S*)-β-(1-carboxy-4-hydroxy-2,5-cyclohexadien-1-yl)-alanine; arogenate; pretyrosine. $C_{10}H_{13}NO_5$; mol wt 227.22. C 52.86%, H 5.77%, N 6.16%, O 35.21%. Precursor in the biosynthesis of L-phenylalanine and L-tyrosine, distributed widely in nature. Pseudomonad bacteria and plants use both the prephenate and arogenate pathways for L-tyrosine biosynthesis. In cyanobacteria, coryneform bacteria and at least one yeast organism the arogenate pathway has been identified as the sole route of L-tyrosine biosynthesis. Identification of arogenate (pretyrosine) in blue-green algae biosynthesis of L-tyrosine: S. L. Stenmark *et al.*, *Nature* **247**, 290 (1974). Dual enzymic routes to L-tyrosine and L-phenylalanine via arogenate (pretyrosine) in *P. aeruginosa:* N. Patel *et al.*, *J. Biol. Chem.* **252**, 5839 (1977). Isoln and prepn: R. A. Jensen *et al.*, *J. Bacteriol.* **132**, 896 (1977). Confirmation that arogenate is an obligatory intermediate of L-tyrosine biosynthesis: A. M. Fazel *et al.*, *Proc. Natl. Acad. Sci. USA* **77**, 1270 (1980). Structure and config: L. O. Zamir *et al.*, *J. Am. Chem. Soc.* **102**, 4499 (1980). Synthesis via immobilized microbial proteins: *eidem, Bioorg. Chem.* **11**, 32 (1982). Arogenate pathway of tyrosine and phenylalanine biosynthesis in *P. aureofaciens:* B. Keller *et al.*, *J. Gen. Microbiol.* **128**, 1199 (1982).

Unstable. Quantitatively converted to phenylalanine at acidic pH. Forms a disodium salt that is stable in the lyophilized state at basic pH (7.5) and room temperature, but is thermally unstable.

782. Arotinolol. [68377-92-4] 5-[2-[[3-[(1,1-Dimethylethyl)amino]-2-hydroxypropyl]thio]-4-thiazolyl]-2-thiophenecarbox-amide; 2-(3′-*tert*-butylamino-2′-hydroxypropylthio)-4-(5′-carbamoyl-2′-thienyl)thiazole. $C_{15}H_{21}N_3O_2S_3$; mol wt 371.53. C 48.49%, H 5.70%, N 11.31%, O 8.61%, S 25.89%. Propanolamine deriv with α- and β-adrenergic blocking activity. Prepn: T. Hibino *et al.*, **DE 2341753**; *eidem*, **US 3932400** (1974, 1976 both to Sumitomo); Y. Hara *et al.*, *J. Pharm. Sci.* **67**, 1334 (1978). Characterization of adrenoceptor blocking effects *in vivo:* A. Miyagishi *et al.*, *Arch. Int. Pharmacodyn. Ther.* **271**, 249 (1984). Possible antianginal effects: M. Sakanashi *et al.*, *Oyo Yakuri* **28**, 709 (1984), *C.A.* **102**, 39669z (1985). Long-term antihypertensive effect in rats: K. Kishi *et al.*, *J. Pharmacobio-Dyn.* **8**, 50 (1985). Comparison with timolol, *q.v.*, of effect on intraocular pressure and hemodynamics: M. Nakashima *et al.*, *Eur. J. Clin. Pharmacol.* **28**, 391 (1985).

Crystals from chloroform/petroleum ether, mp 148-149°.

Consult the Name Index before using this section.

Hydrochloride. [68377-91-3] S-596; ARL; Almarl. $C_{15}H_{21}$-$N_3O_2S_3$.HCl; mol wt 407.99. Crystals from methanol/water, mp 234-235.5° (dec). LD_{50} in mice (mg/kg): 86 i.v., >360 i.p., >5000 orally (Hara).

THERAP CAT: Antihypertensive; antianginal; antiarrhythmic.

783. Arsanilic Acid. [98-50-0] As-(4-Aminophenyl)arsonic acid; p-aminobenzenearsonic acid; atoxylic acid; AS-101; Atoxyl; Pro-Gen. $C_6H_8AsNO_3$; mol wt 217.06. C 33.20%, H 3.72%, As 34.52%, N 6.45%, O 22.11%. Organoarsenical originally used in treatment of trypanosomiasis. Prepd by heating aniline and arsenic acid: A. Béchamp, *C. R. Hebd. Seances Acad. Sci.* **56**, 1172 (1863); W. L. Lewis, H. C. Cheetham, *Org. Synth.* **coll. vol. I**, 70 (1941); Hoffmann, Green, **US 3763201** (1973); from p-nitroaniline: H. Bart, *Ann.* **429**, 55 (1922); Gattermann-Wieland, *Praxis des Organischen Chemikers* (de Gruyter, Berlin, 40th ed., 1961) p 254. Metabolism in chickens: L. R. Overby, L. Straube, *Toxicol. Appl. Pharmacol.* **7**, 850 (1965). Toxicology in swine: Ledet *et al.*, *Clin. Toxicol.* **6**, 439 (1973). Toxicity data: E. I. Goldenthal, *Toxicol. Appl. Pharmacol.* **18**, 185 (1971). Use as a feed additive for Japanese quail: Q. Desheng, Z. Niya, *Poult. Sci.* **85**, 2097 (2006). Spectrophotometric determn in surface water: A. R. Roerdink, J. H. Aldstadt III, *Anal. Chim. Acta* **539**, 181 (2005). HPLC determn in animal feed: D. Chen *et al.*, *J. Chromatogr. B* **879**, 716 (2011). Review of discovery and role in chemotherapy: S. Riethmiller, *Chemotherapy* **51**, 234-242 (2005). Review of toxicology and human exposure: *Toxicological Profile for Arsenic* (PB2008-100002, 2007) 559 pp.

Needles from water or alc. Sol in hot water, pentanol, amyl alcohol, solns of alkali carbonates; sparingly sol in concd mineral acids; slightly sol in cold water, alc, acetic acid. Practically insol in acetone, benzene, chloroform, ether, and in moderately dil mineral acids. LD_{50} orally in male rats: >1000 mg/kg (Goldenthal).

Sodium salt. [127-85-5] Sodium arsanilate; sodium aminarsonate; sodium anilarsonate; Soamin. $C_6H_7AsNNaO_3$; mol wt 239.04.

Sodium salt tetrahydrate. [6696-54-4] $C_6H_7AsNNaO_3.4H_2O$; mol wt 311.10. White, odorless, cryst powder. *Poisonous.* Soluble in about 6 parts water, about 100 parts alcohol. The aq soln is moderately acid to litmus.

THERAP CAT (VET): Growth promotant; to improve feed efficiency. To control swine dysentery.

784. Arsenic. [7440-38-2] Grey arsenic; metallic arsenic; arsen (German). As; at. wt 74.92160; at. no. 33; valences 3, 5. Group VA (15) element, classified as a metalloid. Naturally occurring isotope (mass number): 75 (100%); known artificial, radioactive isotopes: 66-74; 76-87. Arsenic compds were described and used in antiquity, especially as poisons; their reduction to the element was known to medieval alchemists. Albertus Magnus credited with isolation of the element from the mineral orpiment in ~1250 A.D. First precise directions for the prepn of As found in Paracelsus' writings (*ca.* 1520). Arsenic probably occurs throughout the universe. Meteorites reported to contain from 0.0005 to 0.1% As. Occurrence in the earth's crust: 1.8 ppm. Found in nature to a small extent as the element; occurs mostly in minerals such as realgar (As_4S_4), orpiment (A_2S_3), arsenolite (As_2O_3). Commercial sources: as by-product in flue dusts from smelting copper, lead, cobalt and gold ores; by melting $FeAs_2$ or $FeAsS$ ores. Prepn of pure As by reduction with carbon (sugar charcoal) and sublimation in N_2: Krepelka, *Collect. Czech. Chem. Commun.* **2**, 255 (1930); E. H. Archibald, *The Preparation of Pure Inorganic Substances* (Wiley, New York, 1932) p 269. Other methods: Schenk in *Handbook of Preparative Inorganic Chemistry* **vol. 1**, G. Brauer, Ed. (Academic Press, New York, 2nd ed., 1963) pp 591-593. *Reviews: Gmelins, Arsenic* (8th ed.) **17**, 475 pp (1952); Smith, "Arsenic, Antimony and Bismuth" in *Comprehensive Inorganic Chemistry* **vol. 2**, J. C. Bailar, Jr. *et al.*, Eds. (Pergamon Press, Oxford, 1973) pp 547-683; *Chemistry of the Elements*, N. N. Greenwood, A. Earnshaw, Eds. (Pergamon Press, New York, 1984) pp 637-697; S. C. Carapella, Jr. in *Kirk-Othmer Encyclopedia of Chemical Technology* **vol. 3** (John Wiley & Sons, 4th ed., 1992) pp 624-633; G. O. Doak *et al.*, *ibid.* pp 633-659. Review of carcinogenicity

studies: *IARC Monographs* **23**, 39-141 (1980); of toxicology and human exposure: *Toxicological Profile for Arsenic* (PB2008-100002, 2007) 559 pp. Book: "The Chemistry of Organic Arsenic, Antimony and Bismuth Compounds," S. Patai, Ed. (John Wiley & Sons, New York, 1994) 962 pp.

Allotropic forms: α-form, metallic, steel-gray, shiny, brittle, rhombohedral crystal structure; β-form, dark gray, amorphous solid, d 4.700, transforms to metallic form at 280°. *Poisonous.* Can be heated to burn in air with bluish flame, giving off an odor of garlic and dense white fumes of As_2O_3. Stable in dry air; loses its luster on exposure to humid air as surface oxidizes, forming a black modification + As_2O_3. Brinell hardness: 147; Mohs' scale: 3.5. d_4^{25} 5.778. Sublimes$_{760}$ 615° without melting. mp 818° at 36 atm. Heat of vaporization 11.2 kcal/g-atom. Heat of sublimation 30.5 kcal/g-atom. Heat of fusion: 22.4 kcal/g-atom (*Gmelins, loc. cit.* pp 135-136). Also reported as: heat of fusion: 6.620 kcal/g-atom; heat of sublimation 7.63 kcal/g-atom: D. R. Stull, G. C. Sinke, "Thermodynamic Properties of the Elements" in *Advances in Chemistry Series* **18** (A.C.S., Washington, 1956) pp 11, 44. Latent heat of fusion: 27,940 J/mol K. Latent heat of sublimation: 31,974 J/mol K (Carapella). Specific heat (25°) 24.6 J/mol K. Dielectric constant = 10.23 at 20° and 60 cycles. Electrical and magnetic properties of crystalline As: Taylor *et al.*, *J. Phys. Chem. Solids* **26**, 69 (1965). Insol in water, caustic and non-oxidizing acids. Not attacked by cold H_2SO_4 or HCl; converted by HNO_3 or hot H_2SO_4 into arsenous or arsenic acid. A yellow modification which has no metallic properties has been reported from sudden cooling of As-vapor. This yellow arsenic is converted back to the gray modification upon very short exposure to ultraviolet light.

Note: In German and other languages *Arsenik* means arsenic trioxide.

Caution: Overexposure to arsenic and arsenic compounds has been associated with acute and chronic toxicity due to inhalation or ingestion. Organic forms are usually less harmful than inorganic forms. Direct contact can cause local irritation and dermatitis. Overexposure has been associated with an increased risk of skin, liver, bladder, kidney and lung cancer. *See Toxicological Profile, loc. cit.* Arsenic and inorganic arsenic compounds are listed as known human carcinogens: *Report on Carcinogens, Twelfth Edition* (PB2011-111646, 2011) p 794.

USE: In metallurgy for hardening copper, lead, nonferrous alloys; automotive body solder. In semiconductor materials. In the manufacture of low-melting glass. As wood preservative, herbicide, pesticide.

785. Arsenic Acid. [7778-39-4] Arsenic acid (AsH_3O_4); orthoarsenic acid. AsH_3O_4; mol wt 141.94. As 52.78%, H 2.13%, O 45.09%. H_3AsO_4. Exists as the hemihydrate. Excessive drying produces $As_2O_5.5/3H_2O$. Conveniently prepd from As_2O_3 and HNO_3: Simon, Thaler, *Z. Anorg. Allg. Chem.* **161**, 143 (1927); **246**, 19 (1941). Toxicity study: Joachimoglu, *Biochem. Z.* **70**, 144 (1915). Review of toxicology and human exposure: *Toxicological Profile for Arsenic* (PB2008-100002, 2007) 559 pp.

Hemihydrate. Hygroscopic crystals. $pKa_1 = 2.22$, $pKa_2 = 6.98$, $pKa_3 = 11.53$. *Poisonous.* Converted to As_2O_5 by heating above 300°. Freely sol in water, alcohol, glycerol. LD_{50} i.v. in rabbits: 6 mg/kg (Joachimoglu).

USE: In the manuf of arsenates.

786. Arsenicin A. [925705-41-5] 2,4,6-Trioxa-1,3,5,7-tetra-arsatricyclo[3.3.1.1^{3,7}]decane; 1,3,5,7-tetraarsa-2,4,6-trioxaadamantane. $C_3H_6As_4O_3$; mol wt 389.76. C 9.24%, H 1.55%, As 76.89%, O 12.31%. First known polyarsenical natural product. Isolated from the marine sponge, *Echinochalina bargibanti* Hooper & Lévi, *Peocilsclerida*. Isoln and characterization: I. Mancini *et al.*, *Chem. Eur. J.* **12**, 8989 (2006). Structural studies: P. Tähtinen *et al.*, *ibid.* **14**, 10445 (2008); G. Guella *et al.*, *Phys. Chem. Chem. Phys.* **11**, 2420 (2009). Synthesis and crystal structure: D. Lu *et al.*, *Organometallics* **29**, 32 (2010).

S-Enantiomer

Colorless prisms from benzene; crystallizes as a racemic compound. mp 182-184°. d (calc): 3.030. Stable in air.

787. Arsenic Pentafluoride. [7784-36-3] Pentafluoroarsorane. AsF$_5$; mol wt 169.91. As 44.09%, F 55.91%. Prepd from As + F$_2$: Ruff *et al.*, *Z. Anorg. Allg. Chem.* **206**, 59 (1932); Seel, Detmer, *ibid.* **301**, 113 (1959); Kwasnik in *Handbook of Preparative Inorganic Chemistry* vol. **1**, G. Brauer, Ed. (Academic Press, New York, 2nd ed., 1963) p 198. *Reviews:* Burg in *Fluorine Chemistry* vol. I, J. Simons, Ed. (Academic Press, New York, 1950) p 102; Kemmitt, Sharp, *Adv. Fluorine Chem.* **4**, 208 (1965).
Colorless gas. Forms white clouds in moist air. d$_{liq}^{25.8}$ 2.33. mp −79.8°. bp −53.2°. Instantly hydrolyzed by water. Sol in alcohol, ether, benzene. Dry AsF$_5$ does not attack glass, but a minute trace of moisture or HF catalyzes the etching reaction to the point of total destruction.

788. Arsenic Pentasulfide. [1303-34-0] Arsenic sulfide (As$_2$S$_5$); diarsenic pentasulfide. As$_2$S$_5$; mol wt 310.14. As 48.31%, S 51.69%. Prepd from H$_3$AsO$_4$ and H$_2$S: Schenk in *Handbook of Preparative Inorganic Chemistry* vol **1**, G. Brauer, Ed. (Academic Press, New York, 2nd ed., 1963) p 603.
Brownish-yellow, glassy, amorphous, highly refractive mass. Insol in water; sol in alkalies and alkali sulfides. Dec into As$_2$O$_3$, S, and As$_2$S$_3$ when boiled with water.
USE: In thin sheets as a light filter; in pigments.

789. Arsenic Pentoxide. [1303-28-2] Arsenic oxide (As$_2$O$_5$); arsenic(V) oxide; arsenic acid anhydride. As$_2$O$_5$; mol wt 229.84. As 65.19%, O 34.80%. Prepn: *Gmelins, Arsenic* (8th ed.) **17**, pp 273-277 (1952). Review of toxicology and human exposure: *Toxicological Profile for Arsenic* (PB2008-100002, 2007) 559 pp.
Amorphous lumps or powder. d 4.32. *Poisonous.* Gradually deliquesces on exposure to air. Freely sol in water or alc. Combines very slowly with H$_2$O to form H$_3$AsO$_4$. *Keep well closed.*
USE: Manuf of colored glass; in adhesives for metals; in wood preservatives; in weed control; as fungicide.

790. Arsenic Sulfide. [12279-90-2] 2,4,6,8-Tetrathia-1,3,-5,7-tetraarsatricyclo[3.3.0.0^{3,7}]octane; red arsenic sulfide; realgar; red orpiment; ruby arsenic; red arsenic glass; tetraarsenic tetrasulfide; C.I. Pigment Yellow 39; C.I. 77085. As$_4$S$_4$; mol wt 427.93. As 70.03%, S 29.97%. Prepn: *Gmelins, Arsenic* (8th ed.) **17**, pp 417-422 (1952); Schenk in *Handbook of Preparative Inorganic Chemistry* vol **1**, G. Brauer, Ed. (Academic Press, New York, 2nd ed., 1963) p 603.
Deep red, lustrous monoclinic crystals. mp 320°; bp 565°; d 3.5. Ignites at high temp. Practically insol in water; sol in alkali hydroxides; decomposed by HNO$_3$; very slightly sol in hot CS$_2$ and benzene.
USE: As pigment in painting; in fireworks as blue fire and to give an intense white flame; manuf shot; calico printing and dyeing; tanning and depilating hides.

791. Arsenic Tribromide. [7784-33-0] Arsenous tribromide. AsBr$_3$; mol wt 314.63. As 23.81%, Br 76.19%. Prepn: Oddo, Giachery, *Gazz. Chim. Ital.* **53**, 56 (1923); Schenk in *Handbook of Preparative Inorganic Chemistry* vol. **1**, G. Brauer, Ed. (Academic Press, New York, 2nd ed., 1963) p 597.
Deliquescent, orthorhombic prisms. *Poisonous.* d$_4^{25}$ 3.397; d$_4^{50}$ (liq) 3.3282; d$_4^{75}$ 3.2623; d$_4^{100}$ 3.1995. mp 31.1°. bp$_{760}$ 221°; bp$_{11}$ 89°. Dipole moment 1.66. Heat of fusion 8.93 cal/g. Fumes in moist air. Dec by water with the formation of As$_2$O$_3$ and HBr. Miscible with ether, benzene. Sol in hydrocarbons, chlorinated hydrocarbons, carbon disulfide, oils and fats.

792. Arsenic Trichloride. [7784-34-1] Arsenous trichloride; butter of arsenic; fuming liquid arsenic. AsCl$_3$; mol wt 181.27. As 41.33%, Cl 58.67%. Prepn: Smith, *Ind. Eng. Chem.* **11**, 109 (1919); Reisener in *Ullmanns Encyklopädie der technischen Chemie* vol. **3**, (Urban & Schwarzenberg, Munich, 1953) p 850; Schenk in *Handbook of Preparative Inorganic Chemistry* vol. **1**, G. Brauer, Ed. (Academic Press, New York, 2nd ed., 1963) p 596.
Oily liquid. *Poisonous.* Fumes in air. d$_4^{25}$ 2.1497. mp −16°. bp 130.21°; bp$_{11}$ 25°. Dipole moment: 2.17. Heat of vaporization 8.9 kcal/mol. Specific heat: 3.19 cal/mol/°C. n$_D^{20}$ 1.6006. Dec by water to form As(OH)$_3$ and HCl. One mol AsCl$_3$ can be dissolved in 9

mols H$_2$O, this soln (d 1.53) may be diluted again with another 9 mols H$_2$O giving a soln with d 1.346. Further dilution results in the precipitation of As$_2$O$_3$. Also dec by ultraviolet light. Miscible with, or solvent for, chloroform, carbon tetrachloride, ether, iodine, phosphorus, sulfur, alkali iodides, oils and fats.
USE: In the ceramic industry; in syntheses of chlorine-contg arsenicals, e.g., chloro derivs of arsine.

793. Arsenic Trifluoride. [7784-35-2] Arsenous trifluoride. AsF$_3$; mol wt 131.92. As 56.79%, F 43.20%. Prepn: Russell *et al.*, *J. Am. Chem. Soc.* **63**, 2825 (1941); Woolf, Greenwood, *J. Chem. Soc.* **1950**, 2200; Hoffman, *Inorg. Synth.* **4**, 151 (1953); Kwasnik in *Handbook of Preparative Inorganic Chemistry* vol. **1**, G. Brauer, Ed. (Academic Press, New York, 2nd ed., 1963) pp 197-198. *Review:* Kemmitt, Sharp, *Adv. Fluorine Chem.* **4**, 208-209 (1965).
Mobile liquid. *Poisonous.* Fumes in air. Etches glass. d$_{13}^{15}$ 2.73. mp −5.95°. bp 57.8°. Also reported as mp −8.5°; bp 63° (Kwasnik). Hydrolyzed by water. Sol in alcohol, ether, benzene. May be stored in iron vessels.
Caution: Extremely toxic.

794. Arsenic Triiodide. [7784-45-4] Arsenous triiodide. AsI$_3$; mol wt 455.64. As 16.44%, I 83.56%. Prepd from the elements or from AsCl$_3$ and KI: Bailar, *Inorg. Synth.* **I**, 103 (1939).
Orange-red, trigonal rhombohedra from acetone. Reacts slowly with O$_2$ from air, liberating iodine. d$_4^{25}$ 4.688. Some tendency to sublime below 100°. mp 140.9° forming a red liquid. bp$_{760}$ ~400°. One gram dissolves in 12 ml water forming a yellow soln. Does not hydrolyze rapidly and may be recovered from the water soln unchanged within 5 hrs. Aqueous solns are strongly acid (pH of 0.1N soln about 1.1) and ultimately form HI and As$_2$O$_3$, although an equilibrium AsI$_3$ + 3H$_2$O = H$_3$AsO$_3$ + 3HI has been observed. Freely sol in carbon disulfide, chloroform, benzene, toluene, xylene. Less sol in alc, ether.

795. Arsenic Trioxide. [1327-53-3] Arsenic oxide (As$_2$O$_3$); arsenic(III) oxide; arsenous acid; arsenous acid anhydride; arsenous oxide; arsenic sesquioxide; white arsenic; Trisenox. As$_2$O$_3$; mol wt 197.84. As 75.74%, O 24.26%. Commercially important form of arsenic, *q.v.* By-product in the smelting of copper. Has been used in traditional medicine; induces apoptosis in leukemic cells. Prepn: Schenk in *Handbook of Preparative Inorganic Chemistry* vol. **1**, G. Brauer, Ed. (Academic Press, New York, 2nd ed., 1963) p 600. Studies of antileukemic mechanisms: Z. Chen *et al.*, *Pharmacol. Ther.* **76**, 141 (1997); J. Dai *et al.*, *Blood* **93**, 268 (1999). Clinical evaluation in acute promyelocytic leukemia: S. L. Soignet *et al.*, *N. Engl. J. Med.* **339**, 1341 (1998). Acute toxicity: J. Harrison *et al.*, *Arch. Ind. Health* **17**, 118 (1958). Review of production, properties and uses: S. C. Carapella, Jr. in *Kirk-Othmer Encyclopedia of Chemical Technology* vol. **3** (John Wiley & Sons, 4th ed., 1992) pp 626-629; G. O. Doak, *ibid.* pp 633-638. Review of use in glass industry: J. T. Kohli, *Key Eng. Mater.* **94-95**, 363-378 (1994). Review of toxicology: A. K. Done, A. J. Peart, *Clin. Toxicol.* **4**, 343-355 (1971); *Cah. Notes Doc.* **136**, 543-548 (1989); and human exposure: *Toxicological Profile for Arsenic* (PB2008-100002, 2007) 559 pp. Review of clinical toxicology: W.-Y. Au, Y.-L. Kwong, *Acta Pharmacol. Sin.* **29**, 296-304 (2008).
Two crystalline modifications exist. *Claudetite*: thermodynamically stable, monoclinic form, mp 313°. *Arsenolite*: the commonly occurring, octahedral or cubic form; white solid, sublimes freely above 135°, mp 275°. bp 465°. *Poisonous.* Sol in 15 parts boiling water, in dil HCl, alkali hydroxide, carbonate solns, glycerol; sparingly and extremely slowly sol in cold water. Practically insol in alc, chloroform, ether. LD$_{50}$ in mice, rats (mg/kg): 39.4, 15.1 orally (Harrison).
Caution: Arsenic and inorganic arsenic compounds are listed as known human carcinogens: *Report on Carcinogens, Twelfth Edition* (PB2011-111646, 2011) p 50.
USE: Starting material for various arsenic compounds. Decolorizer and fining agent in manuf of glass. In wood preservatives, weed killers, rodenticides. As reductometric standard.
THERAP CAT: Antineoplastic.
THERAP CAT (VET): Has been used as ectoparasiticide.

796. Arsenic Triselenide. [1303-36-2] Arsenic selenide (As$_2$Se$_3$); arsenious selenide; arsenous selenide. As$_2$Se$_3$; mol wt

386.72. As 38.75%, Se 61.25%. Prepd by melting arsenic and selenium in the correct proportions: Uelsmann, *Ueber Selenverbindungen*, Göttingen (1860); *Ann.* **116**, 122 (1860); by the action of an arsenic salt on a soln of hydrogen selenide: Moser, Atynsky, *Monatsh. Chem.* **45**, 235 (1925).

Dark brown solid. d 4.75. mp 260°. Sol in nitric acid, alkali-lye, alkali sulfide solutions; insol in water.

797. Arsenic Trisulfide. [1303-33-9] Arsenic sulfide (As_2-S_3); yellow arsenic sulfide; orpiment; auripigment; arsenic yellow; King's yellow; King's gold. As_2S_3; mol wt 246.02. As 60.91%, S 39.09%. The article of commerce contains much less sulfur than theory. Prepn: *Gmelins, Arsenic* (8th ed.) **17**, pp 422-433 (1952).

Yellow or orange powder. mp reported from 300 to 325°. d 3.46. Practically insol in water. Sol in alkalies, alkali sulfides or carbonates; slowly sol in hot HCl; dec by HNO_3.

USE: Manuf of glass, particularly infrared-transmitting glass; manuf of oil cloth, linoleum; in electrical semiconductors, photoconductors; as pigment; for depilating hides; in pyrotechnics.

798. Arsenious Acid Solution. Arsenic chloride solution. Prepared with 1 g As_2O_3, 5 ml dil HCl and water to 100 ml. *Poisonous!*

THERAP CAT (VET): Has been used in skin and blood disorders.

799. Arsine. [7784-42-1] Arsenic trihydride; hydrogen arsenide. AsH_3; mol wt 77.95. As 96.11%, H 3.88%. Prepn: Wendt, Landauer, *J. Am. Chem. Soc.* **42**, 930 (1920); Jolly, Drake, *Inorg. Synth.* **7**, 41 (1963); Schenk in *Handbook of Preparative Inorganic Chemistry* **vol 1**, G. Brauer, Ed. (Academic Press, New York, 2nd ed., 1963) pp 593-595. Toxicity: Doig, *Lancet* **2**, 88 (1958); Vallee *et al., Arch. Ind. Health* **21**, 132 (1960).

Colorless, neutral gas. Disagreeable garlic odor. *Flammable, poisonous.* mp −117°. bp −62.5°. Dissociation pressure at 0° = 0.806 atm. Decomposes when heated at 300°, depositing arsenic which volatilizes at 400°. On exposure to light, moist arsine decomposes quickly depositing shiny black arsenic. Slightly sol in water. Aq solns are neutral. Traces are best removed by absorption in potassium permanganate soln or in bromine water.

Caution: Potential symptoms of overexposure are headache, malaise, weakness and dizziness; dyspnea; abdominal and back pain; nausea, vomiting; bronze skin; hematuria; jaundice; peripheral neuropathy; direct contact with liquid may cause frostbite. Potential occupational carcinogen. *See NIOSH Pocket Guide to Chemical Hazards* (DHHS/NIOSH 97-140, 1997) p 20.

800. Arsonoacetic Acid. [107-38-0] 2-Arsonoylacetic acid. $C_2H_5AsO_5$; mol wt 183.98. C 13.06%, H 2.74%, As 40.72%, O 43.48%. Prepn from sodium arsenite and sodium chloroacetate: Palmer, *J. Am. Chem. Soc.* **45**, 3023 (1923); *Org. Synth.* **coll. vol. I,** 73 (1941).

Shiny plates, mp 152°. Very sol in water, alc; sparingly sol in hot glacial acetic acid. Practically insol in petr ether, benzene, acetone, chloroform, ethyl acetate.

Barium salt. $C_4H_4As_2Ba_3O_{10}$. Feathery needles. Slightly sol in boiling water.

Sodium salt monohydrate. Disodium arsonoacetate; disodium acetarsenate; acetarsonic acid disodium salt; sodium acetoarsinate; Aricyl. $C_2H_3AsNa_2O_5.H_2O$; mol wt 245.96. Crystals. Freely sol in water.

THERAP CAT (VET): Disodium salt has been used to treat anaplasmosis, and as a stimulant in nervous diseases.

801. Arsphenamine. [139-93-5] Salvarsan; Ehrlich 606. Originally introduced as an antisyphilitic. Initially characterized as *4,4'-(1,2-diarsenediyl)bis[2-aminophenol] dihydrochloride;* composition was later determined to be of a mixture of cyclic 3-amino-4-hydroxyphenylarsenic species. Prepn: P. Ehrlich, A. Bertheim, **US 986148** (1911 to Hoechst); *eidem, Ber.* **45**, 756 (1912); W. A. Christiansen, *J. Am. Chem. Soc.* **42**, 2402 (1920). Toxicity data: J. F. Schamberg *et al., Am. J. Syph. Neurol.* **18**, 37 (1934). Review of

structure: A. S. Levinson, *J. Chem. Educ.* **54**, 98-99 (1977). MS determn of chemical composition: N. C. Lloyd *et al., Angew. Chem. Int. Ed.* **44**, 941 (2005).

Light yellow, somewhat hygroscopic powder. It is odorless, or has a slight odor of the precipitant used. Oxidizes on exposure to air, becoming darker and more toxic. *Poisonous.* Sol in water, alc or glycerol, very slightly sol in chloroform or ether. Its aq soln is acid; pH about 3. LD_{100} i.v. in rats: 140 mg/kg (Schamberg).

802. Arsthinol. [119-96-0] *N*-[2-Hydroxy-5-[4-(hydroxymethyl)-1,3,2-dithiarsolan-2-yl]phenyl]acetamide; 3-acetamido-4-hydroxydithiobenzenearsonous acid, cyclic (hydroxymethyl)ethylene ester; 2-(3'-acetamido-4'-hydroxyphenyl)-1,3-dithia-2-arsacyclopentane-4-methanol; 3-acetamido-4-hydroxydithiobenzenearsonous acid cyclic 3-hydroxypropylene ester; 2-acetylamino-4-(methylolcycloethylenedimercaptoarsine)phenol; Balarsen. $C_{11}H_{14}$-$AsNO_3S_2$; mol wt 347.28. C 38.04%, H 4.06%, As 21.57%, N 4.03%, O 13.82%, S 18.46%. Trivalent arsenical; derivative of acetarsone, *q.v.* Prepn: Friedheim, **US 2593434** (1952); **US 2772303** (1956).

Minute crystals, mp 163-166°. Soly in 95% ethanol about 2.7% w/v. Very slightly sol in water, ether.

THERAP CAT: Antiamebic.

803. Arteether. [75887-54-6] (3*R*,5a*S*,6*R*,8a*S*,9*R*,10*S*,12*R*,-12a*R*)-10-Ethoxydecahydro-3,6,9-trimethyl-3,12-epoxy-12*H*-pyrano[4,3-*j*]-1,2-benzodioxepin; dihydroartemisinin ethyl ether; dihydroqinghaosu ethyl ether; artemotil; SM-227. $C_{17}H_{28}O_5$; mol wt 312.41. C 65.36%, H 9.03%, O 25.61%. Derivative of artemisinin, *q.v.* Prepn: Y. Li *et al., Yaoxue Xuebao* **16**, 429 (1981). Synthesis and antimalarial activity: China Cooperative Research Group on Qinghaosu, *J. Tradit. Chin. Med.* **2**, 9 (1982); A. Brossi *et al., J. Med. Chem.* **31**, 645 (1988).

White crystalline solid, mp 80-82°. $[\alpha]_D^{21}$ +154.5° (c = 1.0 in $CHCl_3$).

THERAP CAT: Antimalarial.

804. Arteflene. [123407-36-3] (1*S*,4*R*,5*R*,8*S*)-4-[(1*Z*)-2-[2,4-Bis(trifluoromethyl)phenyl]ethenyl]-4,8-dimethyl-2,3-dioxabicyclo[3.3.1]nonan-7-one; (1*S*,4*R*,5*R*,8*S*)-4-[(*Z*)-2,4-bis(trifluoromethyl)styryl]-4,8-dimethyl-2,3-dioxabicyclo[3.3.1]nonan-7-one; Ro-42-1611. Synthetic sesquiterpene peroxide; structurally derived from the natural peroxides artemisinin, *q.v.* and yingzhaosu. Prepn: W. Hofheinz *et al.,* **EP 311955**; *eidem,* **US 4977184** (1989, 1990 both to Hoffmann-La Roche). Series of articles on prepn, biological activities, pharmacokinetics and clinical evaluations: *Trop. Med. Parasitol.* **45**, 261-291 (1994).

Crystalline stable material, mp 124°. Highly lipophilic, not sol in water. Stable in soln except in the presence of strong bases or strong reducing agents.

THERAP CAT: Antimalarial.

805.   Artemether. [71963-77-4] (3*R*,5a*S*,6*R*,8a*S*,9*R*,10*S*,-12*R*,12a*R*)-Decahydro-10-methoxy-3,6,9-trimethyl-3,12-epoxy-12*H*-pyrano[4,3-*j*]-1,2-benzodioxepin; dihydroartemisinin methyl ether; dihydroqinghaosu methyl ether; *o*-methyldihydroartemisinin; SM-224; Paluther. $C_{16}H_{26}O_5$; mol wt 298.38. C 64.41%, H 8.78%, O 26.81%. Derivative of artemisinin, *q.v.* Prepn: Y. Li *et al.*, *K'o Hsueh T'ung Pao* **24**, 667 (1979), *C.A.* **91**, 211376u (1979); *eidem*, *Yaoxue Xuebao* **16**, 429 (1981). Absolute configuration: X.-D. Luo *et al.*, *Helv. Chim. Acta* **67**, 1515 (1984). NMR spectral study: F. S. El-Feraly *et al.*, *Spectrosc. Lett.* **18**, 843 (1985). Inhibition of protein synthesis: H. M. Gu *et al.*, *Biochem. Pharmacol.* **32**, 2463 (1983). Antimalarial activity: S. Thaithong, G. H. Beale, *Bull. WHO* **63**, 617 (1985). Series of articles on chemistry, pharmacology and antimalarial efficacy: China Cooperative Research Group on Qinghaosu, *J. Tradit. Chin. Med.* **2**, 3-50 (1982). Toxicity data: *eidem*, *ibid.* 31. Clinical trial in cerebral malaria in children: M. B. van Hensbroek *et al.*, *N. Engl. J. Med.* **335**, 69 (1996). *Review:* R. N. Price, *Expert Opin. Invest. Drugs* **9**, 1815-1827 (2000).

Crystals, mp 86-88°. $[\alpha]_D^{19.5}$ +171° (c = 2.59 in $CHCl_3$). LD_{50} i.m. in mice: 263 mg/kg (China Cooperative Research Group on Qinghaosu).

THERAP CAT: Antimalarial.

806.   Artemisin. [481-05-0] (3*S*,3a*R*,4*S*,5a*S*,9b*S*)-3a,5,5a,-9b-Tetrahydro-4-hydroxy-3,5a,9-trimethylnaphtho[1,2-*b*]furan-2,8-(3*H*,4*H*)-dione; 6α,8α-dihydroxy-3-oxoeudesma-1,4-dien-12-oic acid 12,6-lactone; 8-hydroxysantonin. $C_{15}H_{18}O_4$; mol wt 262.31. C 68.68%, H 6.92%, O 24.40%. From the closed, unexpanded flower heads of several *Artemisia* spp, especially *Artemisia maritima* L., and *A. cina* Berg., *Compositae* ("wormseed"). Found in the mother liquors from the extraction of santonin. Isoln: E. Merck, *E. Merck's Jahresber.* **1894**, 3; *Chem. Zentralbl.* **1895**, I, 436. Structure: Sumi, *J. Am. Chem. Soc.* **80**, 4869 (1958); Cocker, McMurry, *Tetrahedron* **8**, 181 (1960). Stereochemistry: Bolt *et al.*, *J. Chem. Soc.* **1963**, 5235. Synthesis of (+)-artemisin: Nakazaki, Naemura, *Tetrahedron Lett.* **1966**, 2615.

Crystals from abs ethanol or ethyl acetate. Bitter taste. Turns yellow on exposure to light. mp 203°. $bp_{0.1}$ 260°. Sublimes$_{760}$ 170-175°. $[\alpha]_D^{23}$ −84.9° (c = 3 in 95% ethanol). One gram dissolves in 60 ml boiling water, in 3 ml boiling alc. Sol in ethyl acetate. Practically insol in petr ether. Somewhat sol in chloroform.

807.   Artemisinin. [63968-64-9] (3*R*,5a*S*,6*R*,8a*S*,9*R*,12*S*,-12a*R*)-Octahydro-3,6,9-trimethyl-3,12-epoxy-12*H*-pyrano[4,3-*j*]-1,2-benzodioxepin-10(3*H*)-one; artemisine; arteannuin; huanghuahaosu; QHS; qinghaosu; qing hau sau. $C_{15}H_{22}O_5$; mol wt 282.34. C 63.81%, H 7.85%, O 28.33%. Active antimalarial constituent of the traditional Chinese medicinal herb *Artemisia annua* L., *Compositae*, which has been known for almost 2000 years as **Qinghao**. Iso-

lated in 1972 and shown to be a sesquiterpene lactone with a peroxide moiety: *K'o Hsueh T'ung Pao* **22**, 142 (1977), *C.A.* **87**, 98788g (1977); L. Jing-Ming *et al.*, *Acta Chim. Sin.* **37**, 129 (1979), *C.A.* **92**, 94594w (1980). Total synthesis and absolute configuration: G. Schmid, W. Hofheinz, *J. Am. Chem. Soc.* **105**, 624 (1983). NMR spectral study: F. S. El-Feraly *et al.*, *Spectrosc. Lett.* **18**, 843 (1985). Antimalarial activity: Qinghaosu Antimalaria Coordinating Research Group, *Chin. Med. J.* **92**, 811 (1979); L. J. Bruce-Chwatt, *Br. Med. J.* **284**, 767 (1982). Clinical trials: J.-B. Jiang *et al.*, *Lancet* **2**, 285 (1982); W. Tongyin, X. Ruchang, *J. Tradit. Chin. Med.* **5**, 240 (1985). Toxicity data: China Cooperative Research Group on Qinghaosu, *ibid.* **2**, 31 (1982). Series of articles on chemistry, pharmacology, and antimalarial efficacy: *ibid.* 3-50. Brief reviews: H. Koch, *Pharm. Int.* **2**, 184-185 (1981); D. L. Klayman, *Science* **228**, 1049-1055 (1985). *Review:* R. N. Price, *Expert Opin. Invest. Drugs* **9**, 1815-1827 (2000). Review of chemistry, pharmacology and clinical applications: X. D. Luo, C. C. Shen, *Med. Res. Rev.* **7**, 29-52 (1987). Review of mechanism of action: S. R. Meshnick, *Int. J. Parasitol.* **32**, 1655-1660 (2002).

Needles, mp 156-157°. $[\alpha]_D^{17}$ +66.3°. (c = 1.64 in $CHCl_3$). Sol in most aprotic solvents. Slightly sol in oil. LD_{50} in mice (mg/kg): 5105 orally; 2800 i.m.; 1558 i.p. (Koch). LD_{50} in mice, rats (mg/kg): 4228, 5576 orally; 3840, 2571 i.m. (China Cooperative Research Group on Qinghaosu).

Dihydroartemisinin. [71939-50-9] Dihydroqinghaosu. $C_{15}H_{24}O_5$. Main metabolite of artemisinin, arteether, artemether, artesunate, *qqv.* HPLC determn: X.-D. Luo *et al.*, *Chromatographia* **23**, 112 (1987); in plasma: H. Naik *et al.*, *J. Chromatogr. B* **816**, 233 (2005).

THERAP CAT: Antimalarial.

808.   Artesunate. [88495-63-0] Butanedioic acid 1-[(3*R*,-5a*S*,6*R*,8a*S*,9*R*,10*S*,12*R*,12a*R*)-decahydro-3,6,9-trimethyl-3,12-epoxy-12*H*-pyrano[4,3-*j*]-1,2-benzodioxepin-10-yl] ester; artesunic acid; dihydroqinghaosu hemisuccinate. $C_{19}H_{28}O_8$; mol wt 384.43. C 59.36%, H 7.34%, O 33.29%. Derivative of artemisinin, *q.v.* Prepn: China Cooperative Research Group on Qinghaosu, *J. Tradit. Chin. Med.* **2**, 9 (1982). Absolute configuration: X.-D. Luo *et al.*, *Helv. Chim. Acta* **67**, 1515 (1984). GC/MS determn.: A. D. Theoharides *et al.*, *Anal. Chem.* **60**, 115 (1988); HPLC determn in plasma: H. Naik *et al.*, *J. Chromatogr. B* **816**, 233 (2005). Pharmacology: Y. Zhao, *J. Trop. Med. Hyg.* **88**, 391 (1985). Antimalarial activity: W. Peters *et al.*, *Ann. Trop. Med. Parasitol.* **80**, 483 (1986); A. J. Lin *et al.*, *J. Med. Chem.* **30**, 2147 (1987). Inhibition of cytochrome oxidase: Y. Zhao *et al.*, *J. Nat. Prod.* **49**, 139 (1986). Toxicology: China Cooperative Research Group on Qinghaosu, *J. Tradit. Chin. Med.* **2**, 31 (1982). Series of articles on chemistry, pharmacology, and antimalarial efficacy: *ibid.* 3-50. Clinical trial as add-on therapy in pediatric malaria: L. von Seidlein *et al.*, *Lancet* **355**, 352 (2000). *Review:* R. N. Price, *Expert Opin. Invest. Drugs* **9**, 1815-1827 (2000).

Fine white crystalline powder.

Sodium salt. SM-804. $C_{19}H_{27}NaO_8$; mol wt 406.41. Poor stability in aqueous solutions. LD_{50} in mice (mg/kg): 520 i.v.; 475 i.m. (China Cooperative Research Group); also reported as 699 ± 58.5 i.v. (Zhao, 1985).

THERAP CAT: Antimalarial.

809. Arylsulfatase B. [55354-43-3] Acetylgalactosamine 4-sulfatase; sulfatase B; EC 3.1.6.12. Lysosomal enzyme that hydrolyzes the 4-sulfate ester of N-acetylgalactosamine at the nonreducing terminus of dermatan sulfate and chondroitin-4-sulfate. Genetic deficiency of the enzyme results in the lysosomal storage disorder, mucopolysaccharidosis-VI (MPS-VI), also known as Maroteaux-Lamy syndrome. Purification from ox liver: A. B. Roy, *Biochem. J.* **57**, 30 (1954); from human liver: K. S. Dodgson, C. H. Wynn, *ibid.* **68**, 387 (1958). Identification of role in disease: D. A. Stumpf, J. H. Austin, *Trans. Am. Neurol. Assoc.* **97**, 29 (1972). Identification of natural substrate: A. A. Farooqui, *Experientia* **32**, 1242 (1976). Prepn of recombinant human form and uptake by human fibroblast cells: D. S. Anson *et al.*, *Biochem. J.* **284**, 789 (1992). Molecular structure and characterization of active site: C. S. Bond *et al.*, *Structure* **5**, 277 (1997).

Galsulfase. [552858-79-4] Acetylgalactosamine 4-sulfatase (human CSL4S-342 cell); aryplase; rhASB; BM-102; Naglazyme. Human arylsulfatase B produced by recombinant DNA technology in Chinese hamster ovary cells. Glycoprotein comprised of 495 amino acids; mol wt ∼56 kDa. Prepn: C. M. Starr, **WO 0183722** (2001 to BioMarin). Clinical evaluation in children with MPS-VI: P. Harmatz *et al.*, *J. Pediatr.* **144**, 574 (2004). *Review:* G. Yogalingam, *Curr. Opin. Investig. Drugs* **5**, 1111-1120 (2004).

THERAP CAT: Enzyme replacement therapy for mucopolysaccharidosis-VI.

810. Arzoxifene. [182133-25-1] 2-(4-Methoxyphenyl)-3-[4-[2-(1-piperidinyl)ethoxy]phenoxy]benzo[*b*]thiophene-6-ol; [6-hydroxy-3-[4-[2-(1-piperidinyl)ethoxy]phenoxy]2-(4-methoxyphenyl)]benzo[*b*]thiophene; LY-353381. $C_{28}H_{29}NO_4S$; mol wt 475.60. C 70.71%, H 6.15%, N 2.95%, O 13.46%, S 6.74%. Selective estrogen receptor modulator (SERM). Prepn: A. D. Palkowitz, K. J. Thrasher, **EP 729956**; *idem*, **US 5981765** (1996, 1999 both to Eli Lilly). Pharmacology: M. Sato *et al.*, *Endocrinology* **139**, 4642 (1998); N. Suh *et al.*, *Cancer Res.* **61**, 8412 (2001). Clinical pharmacokinetics: P. N. Münster *et al.*, *J. Clin. Oncol.* **19**, 2002 (2001). Review of development and clinical experience in hormone receptor-positive tumors: *idem*, *Expert Opin. Invest. Drugs* **15**, 317-326 (2006). Clinical effect on bone mineral density in postmenopausal women: M. Bolognese *et al.*, *J. Clin. Endocrinol. Metab.* **94**, 2284 (2009).

Solid from hexane, mp 174-176°.

Hydrochloride. [182133-27-3] $C_{28}H_{29}NO_4S \cdot HCl$; mol wt 512.06. Crystals from ethanol/ethyl acetate, mp 156-160°.

THERAP CAT: Antineoplastic (hormonal); antiosteoporotic.

811. Asafetida. [9000-04-8] Devil's dung; food of the gods; asafoetida; asant. Gum-resin obtained as an exudation of the decapitated rhizome and roots of *Ferula assafoetida* L., *F. foetida* Regel and some other species of *Ferula*, *Umbelliferae*. Habit. Iran, Turkestan, Afghanistan. *Constit.* 6-17% ethereal oil; 40-60% resin, consisting of ester of asaresinotannol and ferulic acid; pinene, vanillin, about 25% gum. Three sulfur-containing compounds isolated from asafetida resin are: *1-methylpropyl 1-propenyl disulfide*, *1-(methylthio)propyl 1-propenyl disulfide* and *1-methylpropyl 3-(methylthio)-2-propenyl disulfide*; the latter two have pesticidal properties. Studies of various constituents: Bézanger-Beauquesne,

Chosson, *Ann. Pharm. Fr.* **16**, 665 (1958); Caglioti *et al.*, *Helv. Chim. Acta* **41**, 2278 (1958); *ibid.* **42**, 2557 (1959). Isoln and structure elucidation of three sulfur-containing components: H. Naimie *et al.*, *Collect. Czech. Chem. Commun.* **37**, 1166 (1972). Synthesis and configuration of sulfur-containing constituents: J. Meijer, P. Vermeer, *Rec. Trav. Chim.* **93**, 242 (1974); A. Kjac *et al.*, *Acta Chem. Scand. B* **30**, 137 (1976). *Review:* Howes, *Econ. Bot.* **4**, 313 (1950); Subrahmanyam *et al.*, *J. Sci. Ind. Res.* **13A**, 382-386 (1954).

Soft mass or irregular lumps or "tears"; garlic-like odor; bitter acrid taste. When triturated with water, it makes a milky emulsion.

USE: In India, Iran, etc., as condiment and flavoring for foods; an ingredient in Worcestershire sauce. A 2% suspension as repellent against dogs, cats, rabbits, deer.

THERAP CAT: Carminative; expectorant.

THERAP CAT (VET): Has been used as a carminative, and externally to prevent bandage chewing by dogs.

812. Asarinin. [133-04-0] 5,5′-[(1R,3aS,4S,6aS)-Tetrahydro-1H,3H-furo[3,4-c]furan-1,4-diyl]bis-1,3-benzodioxole; *l*-asarinin; (−)-episesamin; xanthoxylin S. $C_{20}H_{18}O_6$; mol wt 354.36. C 67.79%, H 5.12%, O 27.09%. Naturally occurring *l*-form isolated from *Xanthoxylum clava-herculis* L. (*X. carolinianum* Lam.), *Rutaceae; Asarum sieboldi* Miguel var. *seulensis* Nakai, *A. blumei* Duch., *Aristolochiaceae:* Colton, *Am. J. Pharm.* **52**, 191 (1880); Eberhardt, *ibid.* **62**, 231 (1890); Gordin, *J. Am. Chem. Soc.* **28**, 1649 (1906); H. Dieterle *et al.*, *Arch. Pharm.* **269**, 384 (1931); Huang-Minlon, *Ber.* **70**, 951 (1937); T. Kaku *et al.*, *Keijo J. Med.* **9**, 1 (1934); *C.A.* **32**, 9090[1] (1938). Structure: H. Dieterle, K. Schwenger, *Arch. Pharm.* **277**, 33 (1939). Synthesis of *dl*-form: M. Beroza, M. S. Schechter, *J. Am. Chem. Soc.* **78**, 1242 (1956); K. Freudenberg, E. Fischer, *Ber.* **89**, 1230 (1956); D. R. Stevens, D. A. Whiting, *Tetrahedron Lett.* **27**, 4629 (1986). Diastereoisomeric with sesamin, *q.v.* Stereochemistry of the *d*-form: K. Freudenberg, G. S. Sidhu, *ibid.* **94**, 851 (1961). Antitubercular activity: Ramaswamy, *Naturwissenschaften* **44**, 380 (1957).

Crystals from alc, mp 121°. $[\alpha]_D^{20}$ −118.6°; $[\alpha]_D^{23}$ −122° (chloroform). Practically insol in water. Freely sol in boiling methanol, alcohol, chloroform, acetone, benzene.

***d*-Form.** Episesamin; (+)-episesamin. Crystals from ethanol, mp 121.5°. $[\alpha]_D^{25}$ +124° (chloroform).

***dl*-Form.** Crystals, mp 134-135°.

813. Asarones. 1,2,4-Trimethoxy-5-(1-propenyl)benzene; 2,4,5-trimethoxy-1-propenylbenzene; asarin; asarum camphor; asarabacca camphor. $C_{12}H_{16}O_3$; mol wt 208.26. C 69.21%, H 7.74%, O 23.05%. From root of *Asarum europaeum* L., *Aristolochiaceae* by distillation with water. Also found in the ethereal oils of *A. europaeum* and *A. arifolium* L., *Aristolochiaceae* and in *Acorus calamus* L., *Araceae.* Occurs in nature as a mixture of two isomeric forms, ***α-asarone*** being the (*E*)- or *trans*-isomer, the ***β-asarone***, the (*Z*)- or *cis*-isomer. The unqualified term asarone is often used synonymously with α-asarone. Isoln: Gattermann, Eggers, *Ber.* **32**, 289 (1899). Early syntheses: Seshadri, Thiruvengadam, *Proc. Indian Acad. Sci.* **32A**, 110 (1950); Sharma, Dandiya, *Indian J. Appl. Chem.* **32**, 236 (1969). Stereochemistry of isomers: Baxter *et al.*, *Can. J. Chem.* **40**, 154 (1962). Insect chemosterilant activity of β-asarone: B. P. Saxena *et al.*, *Nature* **270**, 512 (1977); *see also* G. Motolesy *et al.*, *Z. Naturforsch.* **35B**, 1449 (1980). Stereospecific synthesis of β-asarone: M. T. S. Hsia *et al.*, 177th Am. Chem. Soc. Meet. (Honolulu, April 1979), *Abstracts of Papers*, PEST 98. Synthetic and HPLC study of α- and β-asarone: L. Gracza, *Arch. Pharm.* **314**, 972 (1981).

α-Asarone   β-Asarone

α-Asarone. Needles from light petroleum, mp 62-63°. bp 296°. n_D^{11} 1.5719. Practically insol in water; sol in alcohol, ether, glacial acetic acid, carbon tetrachloride, chloroform, petr ether.

814. Asarum. Wild ginger; Canada snakeroot. Rhizomatous perennial herb, *Asarum canadense* L., *Aristolochiaceae*. Medicinal parts include the rhizomes and roots; used in Native American medicine for gastrointestinal discomfort, respiratory ailments and fever. *Habit.* North America, esp. Eastern U.S. and Canada. *Constit.* Flavonoids incl. kaempferol and quercetin glycosides; chalcononaringenin glycosides; volatile oil; aristolochic acid. Identification of flavonal and chalcone glycosides: T. Iwashina, J. Kitajima, *Phytochemistry* **55**, 971 (2000). Determn of aristolochic acids and safety of extracts: B. T. Schaneberg *et al.*, *Pharmazie* **57**, 10 (2002).
Volatile oil. [8016-69-1] *Constit.* Terpene (pinene), methyleugenol, borneol, linalool, geraniol. Yellowish-brown liq; aromatic odor and taste. d_{15}^{15} 0.93-0.96. α_D^{20} −1.4 to −3.5°. Sol in 2 vols 70% alc. Practically insol in water. *Keep well closed, cool and protected from light.*
USE: In perfumes; flavoring in foods.

815. Asbestos. Amianthus. A family of naturally occurring, flexible, fibrous mineral silicates. Divided into two groups: **serpentine** and **amphibole**. Most common form is **chrysotile** [$Mg_6(Si_4O_{10})(OH)_8$], the fibrous form of serpentine (*see also* magnesium silicates). Subdivisions of amphibole are **anthophyllite** [$(Mg,Fe)_7(Si_8O_{22})(OH)_2$] (low iron content); **amosite** [$Fe_5Mg_2(Si_8O_{22})(OH)_2$]; **actinolite** [$Ca_2(Mg,Fe)_5(Si_8O_{22})(OH)_2$]; **tremolite** [$Ca_2Mg_5(Si_8O_{22})(OH)_2$]; **crocidolite** or **blue asbestos** [$Na_2Fe_3^{2+}Fe_2^{3+}(Si_8O_{22})(OH)_2$]. Commercially important forms: chrysotile, anthophyllite, amosite, crocidolite. Review of occupational use and carcinogenic risk: *IARC Monographs* **14**, 1-106 (1977); of toxicology and human exposure: *Toxicological Profile for Asbestos* (PB2001-109101, 2001) 441 pp. Review of properties and industrial applications: C. R. Jolicoeur *et al.* in *Kirk-Othmer Encyclopedia of Chemical Technology* vol. 3 (John Wiley & Sons, New York, 4th ed., 1992) pp 659-688; of electron microscopic analysis of airborne fibers: P. N. Breysse, *Crit. Rev. Anal. Chem.* **22**, 201-277 (1991). Symposium on asbestos-related disease: P. J. Landrigan, H. Kazemi, Eds., *Ann. N.Y. Acad. Sci.* **643**, 1-628 (1991).
Fire resistant fibers. Chrysotile attacked by acid; amphiboles, acid resistant.
Caution: The effects of respiratory exposure to asbestos are subacute or chronic and exhibit a latent period. Nonmalignant respiratory diseases attributable to asbestos exposure include chronic pulmonary fibrosis (asbestosis), fibrotic pleural plaques, pleuritis, and diffuse pleural thickening. Neoplastic diseases associated with occupational exposure to airborne asbestos include lung cancer and mesothelioma. *See Patty's Industrial Hygiene and Toxicology* vol. **2A**, G. D. Clayton, F. E. Clayton, Eds. (Wiley-Interscience, New York, 4th ed., 1993) pp 849-864. Potential symptoms of overexposure are dyspnea, interstitial fibrosis, restricted pulmonary function and finger clubbing; eye irritation. *See NIOSH Pocket Guide to Chemical Hazards* (DHHS/NIOSH 97-140, 1997) p 22. Asbestos and all commercial forms of asbestos are listed as known human carcinogens: *Report on Carcinogens, Twelfth Edition* (PB2011-111646, 2011) p 53.
USE: Heat-resistant insulators, cements, furnace and hot pipe coverings, inert filler medium (laboratory & commercial), fireproof gloves, clothing, brake linings. NaOH treated asbestos, *Ascarite*, has been used to absorb CO_2 in combustion analysis.

816. Ascaridole. [512-85-6] 1-Methyl-4-(1-methylethyl)-2,3-dioxabicyclo[2.2.2]oct-5-ene; 1,4-peroxido-*p*-menthene-2. C_{10}-

$H_{16}O_2$; mol wt 168.24. C 71.39%, H 9.59%, O 19.02%. An organic peroxide which constitutes 60-80% of oil of chenopodium. Synthesis from α-terpinene by treatment with oxygen, chlorophyll, and light: Schenck, Ziegler, *Naturwissenschaften* **1944**, 157. Purification: Beckett *et al.*, *J. Pharm. Pharmacol.* **7**, 55 (1955). Isoln from *Chenopodium ambrosioides* L.: E. Okuyama *et al.*, *Chem. Pharm. Bull.* **41**, 1309 (1993). Review of use as anthelmintic: M. M. Kliks, *Soc. Sci. Med.* **21**, 879-886 (1985).

Liquid; unstable; prone to explode when heated or when treated with organic acids. d_4^{20} 1.0103; d_{20}^{20} 1.0113. mp 3.3°. $bp_{0.2}$ 39-40°. Sol in hexane, pentane, ethanol, toluene, benzene, castor oil.
THERAP CAT: Has been used as anthelmintic (Nematodes).
THERAP CAT (VET): Anthelmintic.

817. Asclepias. Pleurisy root; butterfly weed. Dried root of *Asclepias tuberosa* L., *Asclepiadaceae*. *Habit.* Ontario to Minnesota. *Constit.* Asclepiadin, resins, volatile oil.

818. Asclepias syriaca. Milkweed; silkweed; wild cotton. Root of *Asclepias syriaca* L. (*A. cornuti* Decaisne), *Asclepiadaceae*. *Habit.* Canada to North Carolina and Kansas. *Constit.* Asclepiadin, asclepion—a bitter principle; tannin, volatile oil.

819. Ascorbic Acid. [50-81-7] L-Ascorbic acid; vitamin C; 3-oxo-L-gulofuranolactone; *L-threo*-hex-2-enonic acid γ-lactone; L-3-keto-*threo*-hexuronic acid lactone; L-*xylo*-ascorbic acid; antiscorbutic vitamin; cevitamic acid; Ascorbicap; Ascorbin; Ascorvit; Cantan; Cebion; Cecon; Celaskon; Celin; Cetebe; Cevalin; Ce-Vi-Sol; Cevitan; Cewin; C-Vimin; Hicee; Hybrin; Laroscorbine; Redoxon; Ribena; Vitacimin; Vitacin; Vitascorbol. $C_6H_8O_6$; mol wt 176.12. C 40.92%, H 4.58%, O 54.50%. Physiological antioxidant. Coenzyme for a number of hydroxylation reactions; required for collagen synthesis. Widely distributed in plants and animals. Most primates (including humans), guinea pigs, and some birds and fish cannot synthesize ascorbic acid. Inadequate intake results in deficiency syndromes such as scurvy. Dietary sources include citrus fruits, potatoes, peppers, broccoli, cabbage, and rose hips. First isolated from the adrenal cortex of ox and later from lemons and paprika (originally called hexuronic acid): A. Szent-Györgyi, *Biochem. J.* **22**, 1387 (1928); W. N. Haworth, A. Szent-Györgyi, *Nature* **131**, 24 (1933). Structure studies: R. W. Herbert *et al.*, *J. Chem. Soc.* **1933**, 1270. Synthesis: R. G. Ault *et al.*, *ibid.* 1419; T. Reichstein *et al.*, *Helv. Chim. Acta* **16**, 561, 1019 (1933); **17**, 311, 510 (1934). Review of discovery, structure and synthesis: E. L. Hirst, *Fortschr. Chem. Org. Naturst.* **2**, 132-159 (1939). Crystal structure: J. Hvoself, *Acta Chem. Scand.* **18**, 841 (1964). Review of syntheses: T. C. Crawford, S. A. Crawford, *Adv. Carbohydr. Chem.* **37**, 79-155 (1980). HPLC determn in plasma: J. Lykkesfeldt *et al.*, *Anal. Biochem.* **229**, 329 (1995). Discussion of use in the treatment of the common cold: L. Pauling, *Proc. Natl. Acad. Sci. USA* **68**, 2678 (1971); H. Hemilä, *Scand. J. Infect. Dis.* **26**, 1 (1994). Clinical applications in immunology, lipid metabolism and cancer: *Int. J. Vitam. Nutr. Res.* **1982**, Suppl. 23, 294 pp. Comprehensive description: I. A. Al-Meshal, M. M. A. Hassan, *Anal. Profiles Drug Subs.* **11**, 45-78 (1982). Reviews: G. M. Jaffe in *Kirk-Othmer Encyclopedia of Chemical Technology* vol. **24** (John Wiley & Sons, New York, 3rd Ed., 1984) pp 8-40; "Vitamin C" in *Vitamins*, W. Friedrich, Ed. (Walter de Gruyter, Berlin, 1988) pp 929-1001. Review of pharmacology and clinical applications: M. Levine, *N. Engl. J. Med.* **314**, 892-902 (1986); H. E. Sauberlich, *Annu. Rev. Nutr.* **14**, 371-391 (1994).

White crystals or powder (usually plates, sometimes needles, monoclinic system). Pleasant, sharp acidic taste. mp 190-192° (some dec). d 1.65. $[\alpha]_D^{25}$ +20.5 to +21.5° (c = 1 in water); $[\alpha]_D^{23}$ +48° (c = 1 in methanol). pH = 3 (5 mg/ml); pH = 2 (50 mg/ml); pK_1 = 4.17; pK_2 = 11.57. uv max at pH 2: 245 nm ($E_{1cm}^{1\%}$ 695); at pH 6.4: 265 nm ($E_{1cm}^{1\%}$ 940). Redox potential (first stage): E_0^1 +0.166 V (pH 4). One gram dissolves in about 3 ml water, 30 ml alc, 50 ml abs alc, 100 ml glycerol, 20 ml propylene glycol. Soly in water: 80.0% at 100°; 40.0% at 45°. Insol in ether, chloroform, benzene, petr ether, oils, fats, fat solvents. Strong reducing agent. Stable to air when dry; aq solns are rapidly oxidized by air.

Calcium hypophosphite. Asphocalcium; Calscorbat. C_6H_9-CaO_8P; mol wt 280.18.

Calcium salt. [5743-27-1] Calcium ascorbate. $C_{12}H_{14}CaO_{12}$. Prepn: Ruskin, Merrill, *Science* **105**, 504 (1947). Dihydrate, triclinic crystals, $[\alpha]_D^{20}$ +95.6° (c = 2.4). Freely sol in water; slightly sol in alc. Insol in ether.

Sodium salt. [134-03-2] Sodium ascorbate; Cenolate; Xitix. C_6-H_7NaO_6; mol wt 198.11. Prepn: Holland, **US 2442005** (1948). Minute crystals, dec 218°. $[\alpha]_D^{20}$ +104.4°. Soly in water: 62 g/100 ml at 25°; 78 g/100 ml at 75°.

USE: As antimicrobial and antioxidant in foodstuffs.

THERAP CAT: Vitamin (antiscorbutic).

THERAP CAT (VET): Vitamin (antiscorbutic).

820. Ascorbigen. [8075-98-7] $C_{15}H_{15}NO_6$; mol wt 305.29. C 59.01%, H 4.95%, N 4.59%, O 31.44%. Isoln from savoy cabbage juice: Procházka, Sanda, *Collect. Czech. Chem. Commun.* **25**, 270 (1960). Prepn from L-ascorbic acid + glucobrassicin: Gmelin, Virtanan, *Suom. Kemistil. B* **34**, 15 (1961); *C.A.* **55**, 17774f (1961); from L-ascorbic acid + gramine methiodide: Procházka, *Collect. Czech. Chem. Commun.* **28**, 544 (1963). Prepn from L-ascorbic acid + 3-hydroxymethylindole; forms two diastereoisomers: *ascorbigen A* as the main product and *ascorbigen B* as a minor product: Kiss, Neukom, *Helv. Chim. Acta* **49**, 989 (1966); Kiss, *Angew. Chem.* **78**, 1066 (1966).

Ascorbigen A   Ascorbigen B

Ascorbigen A. [26676-89-1] 2-C-(1H-Indol-3-ylmethyl)-α-L-*xylo*-3-hexulofuranosonic acid γ-lactone. Amorphous powder, sinters about 65°. $[\alpha]_D^{25}$ +11.0° (c = 2 in ethanol). uv max (ethanol): 220, 273-274, 280, 290 nm.

Ascorbigen B. [26548-49-2] 2-C-(1H-Indol-3-ylmethyl)-α-L-*lyxo*-3-hexulofuranosonic acid γ-lactone. Yellow amorphous powder, sinters about 70°. $[\alpha]_D^{25}$ +12.5° (methanol).

821. Asenapine. [65576-45-6] *rel*-(3aR,12bR)-5-Chloro-2,-3,3a,12b-tetrahydro-2-methyl-1H-dibenz[2,3:6,7]oxepino[4,5-c]-pyrrole. $C_{17}H_{16}ClNO$; mol wt 285.77. C 71.45%, H 5.64%, Cl 12.41%, N 4.90%, O 5.60%. Combined serotonin ($5HT_2$) and dopamine (D_2) receptor antagonist. Mixture of *trans*-isomers; structurally related to mianserin, *q.v.* Prepn: **BE 854915** (1977 to Akzo); W. J. van der Burg, **US 4145434** (1979 to Akzona). Synthesis of labelled compd: J. Vader *et al., J. Labelled Compd. Radiopharm.* **34**, 845 (1994). Series of articles on properties, metabolism and pharmacology: *Arzneim.-Forsch.* **40**, 536-554 (1990). Receptor binding study: E. Richelson, T. Souder, *Life Sci.* **68**, 29 (2000). Clinical pharmacology: B. Andrée *et al., Psychopharmacology* **131**, 339 (1997). Clinical trial in bipolar disorder: R. S. McIntyre *et al., J. Affect. Disord.* **126**, 358 (2010). Review of clinical experience in schizophrenia and bipolar disorder: L. Citrome, *Int. J. Clin. Pract.* **63**, 1762-1784 (2009).

Relative stereochemistry

Log P at 37° (*n*-octanol/water): 6.33.

Maleate. [85650-56-2] Org-5222; Saphris; Sycrest. $C_{17}H_{16}$-$ClNO.C_4H_4O_4$; mol wt 401.84. Crystals, mp 141-145°. uv max (ethanol): 270 nm (ε 1900). Soly at 21° (g/l): water 3, ethanol 30, methanol 250, acetone 125, ethyl acetate 10, dichloromethane 250, hexane <1. pH at 20° (0.1% soln in water): 4.56 ±0.05. pH at 23° (saturated soln 5.8 g/l): 4.43. pK_1 <3, pK_2 7.52, pK_3 8.51.

THERAP CAT: Antipsychotic.

822. Asiaticoside. [16830-15-2] (2α,3β,4α)-2,3,23-Trihydroxyurs-12-en-28-oic acid O-6-deoxy-α-L-mannopyranosyl-(1 → 4)-O-β-D-glucopyranosyl-(1 → 6)-O-β-D-glucopyranosyl ester. $C_{48}H_{78}O_{19}$; mol wt 959.13. C 60.11%, H 8.20%, O 31.69%. Active principle of *Centella asiatica* (L.) Urban, *Umbelliferae*. Trisaccharide moiety linked to the aglycone: *asiatic acid*. Isoln: J.-E. Bontems, *Bull. Sci. Pharmacol.* **49**, 186 (1941), *C.A.* **38**, 4094 (1944); *idem, Gaz. Med. Madagascar* **5**, 29 (1942). Structure of asiatic acid: J. Polonsky, *Bull. Soc. Chim. Fr.* **1953**, 173. Structure of asiaticoside: P. Boiteau *et al., Nature* **163**, 258 (1949); J. Polonsky *et al., Bull. Soc. Chim. Fr.* **1959**, 880. Metabolism: L. F. Chasseaud *et al., Arzneim.-Forsch.* **21**, 1379 (1971). Wound healing properties: H. Rosen *et al., Proc. Soc. Exp. Biol. Med.* **125**, 279 (1967). Clinical study: J.-P. Bosse *et al., Ann. Plast. Surg.* **3**, 13 (1979). Review on asiatic acid: J. L. Simonsen, W. C. J. Ross, *The Terpenes* **vol. 5** (University Press, Cambridge, 1957) pp 58-67.

Minute needles from 60% methanol, mp 230-233°. Insol in water. Sol in alcohol, pyridine. $[\alpha]_D^{20} - 14°$ (alc).

823. Asimadoline. [153205-46-0] N-[(1S)-2-[(3S)-3-Hydroxy-1-pyrrolidinyl]-1-phenylethyl]-N-methyl-α-phenylbenzeneacetamide; N-methyl-N-[(1S)-1-phenyl-2-((3S)-3-hydroxy-1-pyrrolidino)ethyl]-2-diphenylacetamide. $C_{27}H_{30}N_2O_2$; mol wt 414.55. C 78.23%, H 7.29%, N 6.76%, O 7.72%. Selective κ-opioid receptor agonist (κ-ORA) with resticted ability to cross the blood-brain barrier. Prepn: R. Gottschlich *et al.,* **DE 4215213**; *eidem,* **US 5532266** (1993, 1996 both to Merck Patent GmbH); and structure activity study: *eidem, Bioorg. Med. Chem. Lett.* **4**, 677 (1994). Opioid activity of stereoisomers: R. Gottschlich *et al., Chirality* **6**, 685 (1994).

Receptor binding study and pharmacology: A. Barber *et al.*, *Br. J. Pharmacol.* **113**, 1317 (1994). Clinical trial in irritable bowel syndrome: A. W. Mangel *et al.*, *Aliment. Pharmacol. Ther.* **28**, 239 (2008). Review of pharmacology: M. Camilleri, *Neurogastroenterol. Motil.* **20**, 971-979 (2008); of development and therapeutic potential: A. W. Mangel, V. S. L. Williams, *Expert Opin. Invest. Drugs* **19**, 1257-1264 (2010).

Hydrochloride. [185951-07-9] EMD-61753. $C_{27}H_{30}N_2O_2$·HCl; mol wt 451.01. White powder, mp 197-198°. $[\alpha]_D^{20}$ +3.8° (c = 1 in methanol).

THERAP CAT: In treatment of irritable bowel syndrome.

824. Asoprisnil. [199396-76-4] [*C*(*E*)]-4-[(11β,17β)-17-Methoxy-17-(methoxymethyl)-3-oxoestra-4,9-dien-11-yl]benzaldehyde-1-oxime; 11β-[4-(hydroxyiminomethyl)phenyl]-17β-methoxy-17α-(methoxymethyl)estra-4,9-dien-3-one; J-867. $C_{28}H_{35}$-NO_4; mol wt 449.59. C 74.80%, H 7.85%, N 3.12%, O 14.23%. Selective progesterone receptor modulator. Prepn: G. Schubert *et al.*, **EP 648778**; *eidem*, **US 5693628** (1995, 1997 both to Jenpharm). Clinical effects on menstrual and ovarian cyclicity in premenopausal women: K. Chwalisz *et al.*, *Hum. Reprod.* **20**, 1090 (2005). Review of pharmacology and therapeutic potential: D. DeManno *et al.*, *Steroids* **68**, 1019-1032 (2003); of early clinical experience: K. Chwalisz *et al.*, *Semin. Reprod. Med.* **22**, 113-119 (2004).

Crystals from isopropanol + CH_2Cl_2, mp 118° (dec). $[\alpha]_D$ +197° (CHCl$_3$). uv max (methanol): 264, 299 nm (ε 20366, 20228).

THERAP CAT: In treatment of uterine fibroids.

825. Asoxime Chloride. [34433-31-3]; [124051-64-5] (free base). 1-[[[4-(Aminocarbonyl)pyridinio]methoxy]methyl]-2-[(hydroxyimino)methyl]pyridinium chloride (1:2); 4′-carbamoyl-2-formyl-1,1′-(oxydimethylene)dipyridinium chloride 2-oxime; 1-(2-hydroxyiminomethyl-1-pyridinio)-3-(4-carbamoyl-1-pyridinio)-2-oxapropane dichloride; HI-6. $C_{14}H_{16}Cl_2N_4O_3$; mol wt 359.21. C 46.81%, H 4.49%, Cl 19.74%, N 15.60%, O 13.36%. Cholinesterase reactivator. Prepn: I. Hagedorn, **US 3773775** (1973 to E. Merck); L. Y. Y. Hsiao, H. A. Musallam, **US 5130438** (1992 to U.S.A. Secy. of Army). Chemical stability study: P. Eyer *et al.*, *Arch. Toxicol.* **59**, 266 (1986). Dissolution properties: H. Thiermann *et al.*, *Int. J. Pharm.* **137**, 167 (1996). HPLC determn in plasma: M. P. McCluskey *et al.*, *J. Anal. Toxicol.* **14**, 239 (1990). Metabolism study: D. J. Ecobichon *et al.*, *Can. J. Physiol. Pharmacol.* **68**, 614 (1990). Clinical pharmacokinetics and safety: J. G. Clement *et al.*, *Biopharm. Drug Dispos.* **16**, 415 (1995). Clinical evaluation in insecticide poisoning: R. Kusic *et al.*, *Hum. Exp. Toxicol.* **10**, 113 (1991). Review of pharmacology: C. G. Rousseaux, A. K. Dua, *Can. J. Physiol. Pharmacol.* **67**, 1183-1189 (1989); of physical properties and acute toxicity: J. G. Clement *et al.*, *Arch. Toxicol.* **62**, 220-223 (1988).

Monohydrate. [82504-20-9] Crystals from water/ethanol (20:75), mp 145-147°. uv max (H_2O): 350, 300, 270, 218 nm (log ε 3.093, 4.018, 3.966, 4.181).

Dimethanesulfonate. [144252-71-1] $C_{14}H_{16}N_4O_3$·2CH_3O_3S; mol wt 478.49. Comparative studies of *in vitro* reactivation: S. Krummer *et al.*, *Arch. Toxicol.* **76**, 589 (2002); of *in vivo* pharmacokinetics and efficacy: P. M. Lundy *et al.*, *Toxicology* **208**, 399 (2005).

THERAP CAT: Antidote (organophosphate poisoning).

826. Asparaginase. [9015-68-3] L-Asparagine amidohydrolase; colaspase; L-asnase; L-asparaginase; EC 3.5.1.1; MK-965; NSC-109229; Crasnitin; Elspar; Kidrolase; Leunase. Enzyme which catalyzes the hydrolysis of L-asparagine to L-aspartic acid and ammonia. Inhibits the growth of tumor cells requiring exogenous asparagine. Widely distributed; occurs in bacteria, fungi, yeasts, plants, animal tissues and guinea pig serum. Commonly obtained from *E. coli* and *Erwinia* sp. L-Asparaginase from *E. coli* has a mol wt of ~136,000 and is composed of four identical subunits. Identification of antitumor activity: J. D. Broome, *Nature* **191**, 1114 (1961); *idem*, *J. Exp. Med.* **118**, 99 (1963). Purif, chemical properties of asparaginase from *E. coli*: D. H. Ho *et al.*, *J. Biol. Chem.* **245**, 3708 (1970). Structural studies: A. C. Greenquist, J. C. Wriston, *Arch. Biochem. Biophys.* **152**, 280 (1972). Amino acid sequence: T. Maita *et al.*, *J. Biochem.* **76**, 1351 (1974). Pharmacology and clinical effects: R. H. Adamson, S. Fabro, *Cancer Chemother. Rep. Part 1* **52**, 617 (1968); R. L. Capizzi *et al.*, *Annu. Rev. Med.* **21**, 433 (1970). *Reviews:* R. L. Capizzi, *Leuk. Lymphoma* **10**, Suppl., 147-150 (1993); M. J. Keating *et al.*, *ibid.* 153-157. Review of development, pharmacology, and clinical experience in leukemia: U. K. Narta *et al.*, *Crit. Rev. Oncol. Hematol.* **61**, 208-221 (2007); of sources, structural modification, and chemotherapeutic potential: N. Verma *et al.*, *Crit. Rev. Biotechnol.* **27**, 45-62 (2007).

White crystalline powder. Freely sol in water where it appears to be global in shape. $[\alpha]_D^{20}$ −30 to −32°. uv max (0.03M sodium phosphate, pH 7.3): 278 nm ($A_{1cm}^{1\%}$ 7.1 ±0.15). Active at pH 5-9. Practically insol in methanol, acetone, chloroform.

Compd with polyethylene glycol. [130167-69-0] Pegaspargase; PEG-asparaginase; (monomethoxypolyethylene glycol succinimidyl)$_{74}$-L-asparaginase; Oncaspar. L-Asparaginase from *E. coli* covalently coupled to polyethylene glycol moieties of mol wt ~5000 Da. Prepn: F. F. Davis *et al.*, **DE 2433883** (1976 to Research Corp.); *eidem*, **US 4179337** (1979). Prepn, pharmacology and antitumor efficacy: A. Abuchowski *et al.*, *Cancer Biochem. Biophys.* **7**, 175 (1984). Toxicology: A. T. Viau *et al.*, *Am. J. Vet. Res.* **47**, 1398 (1986). Review of clinical experience: F. Fuertges, A. Abuchowski, *J. Controlled Release* **11**, 139 (1990). Review of clinical experience and therapeutic potential of intravenous administration: C. H. Fu, K. M. Sakamoto, *Expert Opin. Pharmacother.* **8**, 1977-1984 (2007).

THERAP CAT: Antineoplastic (acute leukemia).

827. Asparagine. [70-47-3] L-Asparagine; Asn; N; L-β-asparagine; α-aminosuccinamic acid; aspartic acid β-amide; altheine; asparamide; agedoite; (*S*)-2,4-diamino-4-oxobutanoic acid. C_4H_8-N_2O_3; mol wt 132.12. C 36.36%, H 6.10%, N 21.20%, O 36.33%. Non-essential amino acid for human development. First amino acid noted in natural sources, was named for the asparagus juice from which it was isolated: L. N. Vauquelin, P. J. Robiquet, *Ann. Chim.* **57**, 88 (1806). Due to ease of conversion to aspartic acid, *q.v.*, it was not isolated from protein until 1932: M. Damodoran, *Biochem. J.* **26**, 235 (1932). Early chemistry and biochemistry: *Amino Acids and Proteins*, D. M. Greenberg, Ed. (Charles C. Thomas, Springfield, IL, 1951) 950 pp, *passim*; J. P. Greenstein, M. Winitz, *Chemistry of the Amino Acids* vols. 1-3 (John Wiley & Sons, Inc., New York, 1961) pp. 1856-1878, *passim*. As attachment point for saccharides: A. J. Mencke *et al.*, *Methods Enzymol.* **138**, 409 (1987). Colorimetric assay: S. Sheng *et al.*, *Anal. Biochem.* **211**, 242 (1993). Proposed formation of acrylamide, *q.v.*, in cooking: D. S. Mottram *et al.*, *Nature* **419**, 448 (2002); R. H. Stadler *et al.*, *ibid.* 449. Review of metabolism: D. A. Cooney, R. E. Handschumacher, *Annu. Rev. Pharmacol.* **10**, 421-440 (1970); in plants: K. A. Sieciechowicz *et al.*, *Phytochemistry* **27**, 663-671 (1988). Review of deamidation in proteins: R. Bischoff, H. V. J. Kolbe, *J. Chromatogr. B* **662**, 261-278 (1994).

White, orthorhombic bisphenoidal crystals, mp 234-235° (bath preheated to 226°). d_4^{15} 1.543. Acid to litmus. pK_1 2.02; pK_2 8.80. $[\alpha]_D^{20}$ −5.30° (c = 1.41); $[\alpha]_D^{20}$ +34.26° (c = 2.24 in 3.4N HCl); $[\alpha]_D^{20}$ −6.35° (c = 11.23 in 2.5N NaOH). Sol in water, acids, alkalies. Practically insol in methanol, ethanol, ether, benzene.

D-Form. [2058-58-4] In peptidoglycans of bacterial cell walls: D. A. Reaveley, R. E. Burge, *Adv. Microb. Physiol.* **7**, 1-81 (1972).

D-Form monohydrate. Crystals, mp 215°. $[\alpha]_D^{20}$ +5.41° (c = 1.3).

828. Asparagus. Shoot of *Asparagus officinalis* L., Liliaceae. *Habit.* Europe, cultivated everywhere. *Constit.* Asparagine, tyrosine, succinic acid, arginine, α-aminodimethyl-γ-butyrothetin (a methylsulfonium deriv of methionine), fat, sugar. In some humans, ingestion of asparagus is followed by excretion of a substance that produces a characteristic strong odor in the urine: M. Nencki, *Arch. Exp. Pathol. Pharmakol.* **28**, 206 (1891); C. Gautier, *C. R. Seances Soc. Biol. Ses Fil.* **89**, 239 (1923); A. C. Allison, K. G. McWhirter, *Nature* **178**, 748 (1956). The odor-causing substance was originally thought to be methyl mercaptan, but subsequent investigation using GC-mass spectrometry has suggested that *S*-methylthioacrylate and *S*-methyl 3-(methylthio)thiopropionate are the malodorous agents: R. H. White, *Science* **189**, 810 (1975).

829. Aspartame. [22839-47-0] L-α-Aspartyl-L-phenylalanine 2-methyl ester; 3-amino-*N*-(α-carboxyphenethyl)succinamic acid *N*-methyl ester; APM; SC-18862; Equal; NutraSweet; Sanecta. $C_{14}H_{18}N_2O_5$; mol wt 294.31. C 57.13%, H 6.16%, N 9.52%, O 27.18%. Dipeptide ester about 160 times sweeter than sucrose in aqueous solution. Prepn: Davey *et al.*, *J. Chem. Soc. C* **1966**, 555; J. M. Schlatter, **ZA 6702190**; *idem*, **US 3492131** (1968, 1970 both to Searle); H. Pietsch, *Tetrahedron Lett.* **17**, 4053 (1976); K. J. Vinick, S. Jung, *ibid.* **23**, 1315 (1982); utilizing immobilized enzyme technology: C. Fuganti, P. Grasselli, *ibid.* **27**, 3191 (1986). Structure-taste relationship: Mazur *et al.*, *J. Am. Chem. Soc.* **91**, 2684 (1969). Potential as a low-calorie sweetener: Cloninger, Baldwin, *Science* **170**, 81 (1970). Metabolism: Oppermann *et al.*, *J. Nutr.* **103**, 1454, 1460 (1973).

Colorless needles from water, mp 246-247°. Sweet taste. $[\alpha]_D^{22}$ −2.3° (1N HCl). Sparingly sol in water; slightly sol in alcohol.

USE: Non-nutritive sweetener.

830. Aspartic Acid. [56-84-8] L-Aspartic acid; Asp; D; aminosuccinic acid; asparagic acid; asparaginic acid; (*S*)-aminobutanedioic acid; 1-amino-1,2-carboxyethane; asparaginsäure (German). $C_4H_7NO_4$; mol wt 133.10. C 36.10%, H 5.30%, N 10.52%, O 48.08%. Non-essential amino acid for human development. First identified as the acid hydrolysis product of asparagine, *q.v.*: A. Plisson, *J. Pharm.* **13**, 477 (1827); later isolated from protein: H. Ritthausen, *J. Prakt. Chem.* **103**, 233, 239 (1868). Early chemistry and biochemistry: *Amino Acids and Proteins*, D. M. Greenberg, Ed. (Charles C. Thomas, Springfield, IL, 1951) 950 pp., *passim*; J. P. Greenstein, M. Winitz, *Chemistry of the Amino Acids* **vol 1-3** (John

Wiley and Sons, Inc., New York, 1961) pp. 1856-1878, *passim*. Synthesis of optically active forms: K. Harada, K. Matsumoto, *J. Org. Chem.* **31**, 2985 (1966); of labelled form: A. Ivanof *et al.*, *Anal. Biochem.* **110**, 267 (1981). Crystal structure: B. Dawson, *Acta Crystallogr.* **33B**, 882 (1977). Excitatory neurotransmitter: H. McLennan, H. V. Wheal, *Can. J. Physiol. Pharmacol.* **54**, 70 (1976). Role in protein activity: D. W. Urry *et al.*, *Biopolymers* **34**, 889 (1994); J. M. Denu *et al.*, *Biochemistry* **34**, 3396 (1995). Racemization with age in dentin: N. Saleh *et al.*, *Calcif. Tissue Int.* **53**, 103 (1993). Review of metabolism and pathology: L. D. Stegink, *J. Toxicol. Environ. Health* **2**, 215-242 (1976). Review of role as excitatory amino acid: P. J. Roberts, S. W. Davies, *Biochem. Soc. Trans.* **15**, 218-219 (1986); of use in forensics: E. R. Waite *et al.*, *Forensic Sci. Int.* **103**, 113-124 (1999); of pathobiology of age-related racemization in proteins: S. Ritz-Timme, M. J. Collins, *Ageing Res. Rev.* **1**, 43-59 (2002).

Orthorhombic bisphenoidal leaflets or rods, mp 270-271° (sealed capillary, preheated bath). $[\alpha]_D^{20}$ +25.0° (c = 1.97 in 6N HCl). $d^{12.5}$ 1.661. pK_1 1.88; pK_2 3.65; pK_3 9.60. Soly in water at 20°, 30°: 1 g/222.2 ml, 1 g/149.9 ml. Forms supersatd solns easily. More soluble in salt solns; sol in dilute solutions of mineral acids, alkalies. Practically insol in alcohol, ether.

D-Form. [1783-96-6] D-Asp. Occurs naturally although to a lesser extent than the L-form. Formed by non-enzymatic racemization in ageing protein. Purification and characterization from bovine lens: N. Fujii *et al.*, *Biochim. Biophys. Acta* **999**, 239 (1989); from tissue of the mollusc, *Aplysia kurodai*: M. Sato *et al.*, *Biochem. J.* **263**, 617 (1989). Review of industrial production by microorganisms: S. Takamatsu, T. Tosa, *Bioprocess Technol.* **16**, 25-35 (1993). $[\alpha]_D^{25}$ −2.0° (c = 3.93 in 5N HCl).

Compound with L-arginine. [3054-35-1] Arginine L-aspartate; Sargenor. $C_{10}H_{21}N_5O_6$; mol wt 307.31.

Compound with L-ornithine *see* Ornithine.

N-Acetyl-L-aspartic acid. [997-55-7] NAA. $C_6H_9NO_5$; mol wt 175.14. Found in nervous system; second most abundant free amino acid in mammals. Review: D. L. Birken, W. H. Oldendorf, *Neurosci. Biobehav. Rev.* **13**, 23-31 (1989).

USE: Nutritional supplement; in forensic determn of age.

831. Aspergillic Acid. [490-02-8] (+)-6-(1-methylpropyl)-3-(2-methylpropyl)pyrazinol 4-oxide; 1-hydroxy-6-(1-methylpropyl)-3-(2-methylpropyl)-2(1H)-pyrazinone; 6-*sec*-butyl-1-hydroxy-3-isobutyl-2(1H)-pyrazinone; 6-*sec*-butyl-3-isobutylpyrazinol 1-oxide; 2-hydroxy-3-isobutyl-6-(1-methylpropyl)pyrazine 1-oxide; 3-isobutyl-6-*sec*-butyl-2-hydroxypyrazine 1-oxide. $C_{12}H_{20}N_2O_2$; mol wt 224.30. C 64.26%, H 8.99%, N 12.49%, O 14.27%. Antibiotic substance produced by *Aspergillus flavus*: White, Hill, *J. Bacteriol.* **45**, 433 (1943); Dunn *et al.*, *J. Chem. Soc.* **1949**, Suppl. S126. Extraction and purification: Dutcher, *J. Biol. Chem.* **171**, 321 (1947). Structure: *idem*, *ibid.* **232**, 785 (1958). Biosynthesis: MacDonald, *ibid.* **236**, 512 (1961); **237**, 1977 (1962). Synthesis of racemate: Chigira *et al.*, *Bull. Chem. Soc. Jpn.* **39**, 632 (1966); Masaki *et al.*, *J. Org. Chem.* **31**, 4143 (1966); Ohta, Fujii, *Chem. Pharm. Bull.* **17**, 851 (1969). Industrial prepn: Omata, Ueno, **JP 65 13794** (1965 to Dainippon), *C.A.* **63**, 11589b (1965). Proposed as a hypotensive: Jones, Martin, **US 3720768** (1973 to Abbott). *Reviews*: MacDonald, "Aspergillic Acid and Related Compounds" in *Antibiotics* vol. **II**, D. Gottlieb, P. Shaw, Eds. (Springer-Verlag, New York, 1967) pp 43-51; Wilson, "Miscellaneous Aspergillus Toxins" in *Microbial Toxins* vol. VI, A. Ciegler *et al.*, Eds. (Academic Press, New York, 1971) pp 207-235.

Pale yellow needles having odor similar to black walnuts, mp 97-99° (methanol). pK′a 5.5. $[\alpha]_D^{18}$ +13.3° (c = 3.9 in ethanol). uv max (water pH 8): 328, 235 nm (ε 8500, 10500). Slightly sol in cold water; sol in dil acids and alkalies, alcohol, ether, acetone, benzene, chloroform, pyridine.

832. Aspergillin. The inappropriate designation of a number of different antibacterial agents produced by *Aspergilli* as "aspergillin" has resulted in confusion. The legitimate contender for this name by prior use is an allomelanin produced by mature spores of *A. niger:* Hugouneng, Florence, *Bull. Soc. Chim. Biol.* **2**, 133 (1920). This pigment is a mixture of chemically similar macromolecules of which the fundamental monomer is probably the perylene unit substituted with several oxygenated functions. *Review:* R. A. Nicolaus, *Melanins* (Hermann, Paris, 1968) pp 130-142.

Dried pigment, as shiny black blocks, does not melt but decomposes above 200°. Slightly sol in polar solvents. uv max (2% aqueous soln): 295, 450 nm.

833. Asperlicin. [93413-04-8] (7*S*)-6,7-Dihydro-7-[[(2*S*,-9*S*,9a*S*)-2,3,9,9a-tetrahydro-9-hydroxy-2-(2-methylpropyl)-3-oxo-1*H*-imidazo[1,2-*a*]indol-9-yl]methyl]quinazolino[3,2-*a*][1,4]-benzodiazepine-5,13-dione. $C_{31}H_{29}N_5O_4$; mol wt 535.60. C 69.52%, H 5.46%, N 13.08%, O 11.95%. Naturally occurring, nonpeptide cholecystokinin (CCK) antagonist. Produced by several strains of *Aspergillus alliaceus* along with minor related compounds known as asperlicin B, C, D and E. Isoln from fermentation cultures of *A. alliaceus* Thom and Church, ATCC 20655 and ATCC 20656: R. L. Monaghan *et al.*, **EP 116150**; *eidem,* **US 4530790** (1984, 1985 both to Merck & Co.). Fermentation, isoln and bioactivity: M. A. Goetz *et al., J. Antibiot.* **38**, 1633 (1985); of asperlicins B, C, D and E: *eidem, ibid.* **41**, 875 (1988). Structure elucidation: J. M. Liesch *et al., ibid.* **38**, 1638 (1985); of asperlicins B, C, D and E: *eidem, ibid.* **41**, 878 (1988). Biosynthetic study: D. R. Houck *et al., ibid.* **882**. Total synthesis of asperlicins C and E: M. G. Bock *et al., J. Org. Chem.* **52**, 1644 (1987). Pharmacology of asperlicin: R. S. L. Chang *et al., Science* **230**, 177 (1985). Effect on pancreatic enzyme secretion *in vitro:* K. A. Zucker *et al., Surgery* **102**, 163 (1987). Effect in exptl pancreatitis in rats: J. R. Wisner, Jr., I. G. Renner, *Pancreas* **3**, 174 (1988).

White crystals, mp 211-213°. $[\alpha]_D^{26.5}$ -185.3° (c = 1.10 in methanol). uv max (methanol): 310.5 nm (ε 4075). Sol in methylene chloride, acetone and lower alcohols. Insol in water.

Asperlicin B. [93413-08-2] $C_{31}H_{29}N_5O_5$; mol wt 551.60. Colorless powder. uv max (methanol): 310 nm (ε 4000).

Asperlicin C. [93413-06-0] $C_{25}H_{18}N_4O_2$; mol wt 406.45. Off-white powder. uv max (methanol): 222, 268, 278, 310 nm (ε 56300, 15100, 14000, 4650).

Asperlicin D. [93413-07-1] $C_{25}H_{18}N_4O_2$; mol wt 406.45. Colorless powder. uv max (methanol): 222, 290, 310 nm (ε 61060, 14600, 5920).

Asperlicin E. [93413-05-9] $C_{25}H_{18}N_4O_3$; mol wt 422.44. Off-white powder. uv max (methanol): 227, 268, 324 nm (ε 38450, 8680, 3200).

834. Asperuloside. [14259-45-1] (2a*S*,4a*S*,5*S*,7b*S*)-4-[(Acetyloxy)methyl]-5-(*β*-D-glucopyranosyloxy)-2a,4a,5,7b-tetrahydro-1*H*-2,6-dioxacyclopent[*cd*]inden-1-one; rubichloric acid. $C_{18}H_{22}O_{11}$; mol wt 414.36. C 52.18%, H 5.35%, O 42.47%. From

herb of *Asperula odorata* L., *Galium aparine* L., *Rubiaceae.* Occurs also in *Coprosma* spp. Isoln: Hérissey, *Compt. Rend.* **180**, 1695 (1925); Briggs, Nicholls, *J. Chem. Soc.* **1954**, 3940. Structure: Grimshaw, *Chem. Ind. (London)* **1961**, 403; Briggs *et al., J. Chem. Soc.* **1965**, 2595.

Needles from alcohol or acetone, mp 131-132°. $[\alpha]_D^{25}$ -198.6° (c = 1.44 in water). Absorption spectra: Briggs, Cain, *J. Chem. Soc.* **1954**, 4182. Sol in water, methanol, ethanol, acetone, ethylacetate, dioxane, pyridine, acetic acid. Practically insol in ether, benzene, chloroform, ligroin.

835. Asphalt. Asphaltum; mineral pitch; Judean pitch; bitumen. Bituminous substance resulting from petroleum by evaporation of lighter hydrocarbons and partial oxidation of the residue. Occurs in West Indies (chiefly Trinidad), Venezuela, Dead Sea, Switzerland, etc.

The "Syriac" asphalt (from the Dead Sea) forms deep black, shining, brittle masses of conchoidal fracture; faint, pitch-like odor and luster. d 1.00-1.18. *Flammable.* Insol in water, alc, acids, alkalies. Sol in oil turpentine, petroleum, CS_2, chloroform, ether, acetone.

Caution: Potential symptoms of overexposure to fumes are irritation of eyes and respiratory system. Potential occupational carcinogen. *See NIOSH Pocket Guide to Chemical Hazards* (DHHS/NIOSH 97-140, 1997) p 22.

USE: Making roads, roofs; making tanks watertight.

836. Aspidin. [584-28-1] 2-[[2,6-Dihydroxy-4-methoxy-3-methyl-5-(1-oxobutyl)phenyl]methyl]-3,5-dihydroxy-4,4-dimethyl-6-(1-oxobutyl)-2,5-cyclohexadien-1-one; 3′-[(5-butyryl-2,4-dihydroxy-3,3-dimethyl-6-oxo-1,4-cyclohexadien-1-yl)methyl]-2′,4′-dihydroxy-6′-methoxy-5′-methylbutyrophenone; polystichin. $C_{25}H_{32}O_8$; mol wt 460.52. C 65.20%, H 7.00%, O 27.79%. Active principle of fern root: Boehm, *Ann.* **302**, 171 (1898); **329**, 321 (1903); from *Dryopteris austriaca* (Jacq.) Woynar, *Polypodiaceae:* Aebi *et al., Helv. Chim. Acta* **40**, 266 (1957). Structure: *eidem, ibid.* 569. Synthesis: Riedl, Mitteldorf, *Ber.* **89**, 2595 (1956).

Crystals from ethanol, mp 124-125°. uv max (cyclohexane): 230, 290 nm (ε 25500, 21300). Soluble in ether, benzene, chloroform; less sol in petr ether; sparingly sol in methanol, ethanol, acetone.

THERAP CAT: Anthelmintic (Cestodes).

837. Aspidinol. [519-40-4] 1-(2,6-Dihydroxy-4-methoxy-3-methylphenyl)-1-butanone; 2′,6′-dihydroxy-4′-methoxy-3′-methyl-1-butyrophenone; 4-butyryl-2-methylphloroglucinol 1-methyl ether; 4-butyryl-3,5-dihydroxy-1-methoxy-2-methylbenzene. $C_{12}H_{16}O_4$; mol wt 224.26. C 64.27%, H 7.19%, O 28.54%. Occurs in extracts of male fern: Boehm, *Ann.* **318**, 247 (1901); Hausmann, *Arch. Pharm.* **237**, 559 (1899). Isoln from *Dryopteris austriaca* (Jacq.) Woynar, *Polypodiaceae:* Aebi *et al., Helv. Chim. Acta* **40**, 266 (1957). Synthesis from 2-methylphloroglucinol 1-methyl ether: Karrer, Widmer, *Helv. Chim. Acta* **3**, 392 (1920); Riedl, Mitteldorf, *Ber.* **89**, 2589 (1956).

Needles or prisms from benzene. mp 156-161°. Freely sol in alc, ether, chloroform, acetone; sparingly sol in water, benzene. Less soluble in ligroin than pseudoaspidinol, mp 116.5°. Sol in NaOH solns. Practically insol in Na_2CO_3 solns.

THERAP CAT: Anthelmintic (Cestodes).

838. **Aspidium.** Male fern; male shield-fern; filix mas (B.P.). Rhizome and stipes of *Dryopteris filix-mas* (L.) Schott., *Polypodiaceae. Habit.* North America, Northern Asia, Europe, Northern Africa. *Constit.* Filicic and flavaspidic acids, volatile oil, asbaspidin, filicin, filmaron, filix red, resin. It yields not less than 6.5% oleoresin (U.S.P.).

Oleoresin of Aspidium. Oleoresin of male fern. An ether extract of male fern containing not less than 24% crude filicin. Dark green, thick liq, usually depositing a granular, cryst substance. Bitter, unpleasant taste. d not less than 1.0. Insoluble in water. Partly sol in alcohol or chloroform. Almost completely sol in ether; not less than 85% sol in petr ether. *Shake well before dispensing.*

THERAP CAT: Anthelmintic (Cestodes).

THERAP CAT (VET): Anthelmintic (Cestodes).

839. **Aspidosperma.** Quebracho. Dried bark of *Aspidosperma quebracho-blanco* Schlecht., *Apocynaceae. Habit.* Argentina. *Constit.* 0.3-1.4% alkaloids—aspidospermine, aspidospermatine, aspidosamine, quebrachine, quebrachamine, hypoquebrachine, quebrachol, quebrachit; tannin.

USE: Chlorinated quebracho can be used to control nematodes and other parasitic worms in soils: Santmyer, **US 2799612** (1957 to Monsanto).

THERAP CAT: Respiratory stimulant.

840. **Aspidospermine.** [466-49-9] 1-Acetyl-17-methoxyaspidospermidine. $C_{22}H_{30}N_2O_2$; mol wt 354.49. C 74.54%, H 8.53%, N 7.90%, O 9.03%. In *Aspidosperma quebracho-blanco* Schlect., *Vallesia dichotoma* Ruiz & Pav., and *Vallesia glabra* (Cav.) Link, *Apocynaceae:* G. Fraude, *Ber.* **11**, 2189 (1878); Hesse, *Ann.* **211**, 249 (1882); Ewins, *J. Chem. Soc.* **105**, 2738 (1914); Deulofeu *et al., ibid.* **1940**, 1051; Holker *et al., J. Org. Chem.* **24**, 314 (1959). Structure: Conroy *et al., Tetrahedron Lett.* no. 11, 4 (1959). Crystal structure: Mills, Nyburg, *J. Chem. Soc.* **1960**, 1458. Stereochemistry: Smith, Wrobel, *ibid.* 1463; Craven, Zacharias, *Experientia* **24**, 770 (1968). Biogenesis: Robinson, *Tetrahedron Lett.* no. 18, 1F4 (1959). Synthesis of *dl*-form: Stork, Dolfini, *J. Am. Chem. Soc.* **85**, 2872 (1963); Ban *et al., Tetrahedron Lett.* **1965**, 2261; Stevens *et al., Chem. Commun.* **1971**, 857; S. F. Martin *et al., J. Am. Chem. Soc.* **102**, 3294 (1980). Aspidospermine has been reported as having a wide variety of pharmacological properties, including diuretic and respiratory stimulant activity. Biological and phytochemical evaluation: R. L. Lyon *et al., J. Pharm. Sci.* **62**, 218 (1973). Microbial transformation: S. K. Lin *et al., ibid.* **64**, 2021 (1975). NMR study: W. E. Campbell *et al., Spectrosc. Lett.* **26**, 707 (1993). Isoln: H. F. Deutsch *et al., J. Pharm. Biomed. Anal.* **12**, 1283 (1994). Acute toxicity data: V. Plzak, J. Doull, *Further Survey of Compounds for Radiation Protection* (AD691490, USAF Radiation Lab., 1969) 82 pp.

Needles or prisms from alc, needles from petr ether. mp 208°. Sublimes 180°. bp$_2$ 220°. $[\alpha]_D^{15}$ −100.2° (alc); $[\alpha]_D$ −93° (chloro-

form). uv max (methanol): 218, 255, 280-290 nm (log ε 4.52, 4.04, 3.53-3.40). One gram dissolves in 60 ml water, 50 ml alc, 100 ml ether. Also sol in benzene, chloroform, petr ether. LD$_{50}$ i.p. in mice: 40 mg/kg (Plzak, Doull).

N-Formyl-N-deacetylaspidospermine. Vallesine. $C_{21}H_{28}N_2O_2$. Structure: Taylor *et al., Helv. Chim. Acta* **42**, 2750 (1959). Long, fine needles from acetone, mp 154-156°. $[\alpha]_D^{24}$ −91 ± 2° (c = 1.814 in abs alc). uv max: 211, 250 nm (log ε 4.47, 3.94).

841. **Aspirin.** [50-78-2] 2-(Acetyloxy)benzoic acid; salicylic acid acetate; 2-acetoxybenzoic acid; acetylsalicylic acid; Acylpyrin; Angettes; Asatard; Aspro; Cardioaspirin; Cardiprin; Cemirit; Claragine; Ecotrin; Empirin; Encaprin; Rhonal; Solprin. $C_9H_8O_4$; mol wt 180.16. C 60.00%, H 4.48%, O 35.52%. Prepn: C. Gerhardt, *Ann.* **87**, 149 (1853). Manuf from salicylic acid and acetic anhydride: *Faith, Keyes & Clark's Industrial Chemicals*, F. A. Lowenheim, M. K. Moran, Eds. (Wiley-Interscience, New York, 4th ed., 1975) pp 117-120. Crystallization from acetone: Hamer, Phillips, **US 2890240** (1959 to Monsanto). Novel process involving distillation: Edmunds, **US 3235583** (1966 to Norwich Pharm.). Crystal structure: P. J. Wheatley, *J. Chem. Soc. (Suppl.)* **1964**, 6036. Characterization of a second polymorphic form: P. Vishweshwar *et al., J. Am. Chem. Soc.* **127**, 16802 (2005). Toxicity data: E. R. Hart, *J. Pharmacol. Exp. Ther.* **89**, 205 (1947). Evaluation as a risk factor in Reye's syndrome: P. J. Waldman *et al., J. Am. Med. Assoc.* **247**, 3089 (1982). Review of clinical trials in prevention of myocardial infarction and stroke: P. C. Elwood, *Drugs* **28**, 1-5 (1984). Symposium on aspirin therapy: *Am. J. Med.* **74**, no. 6A, 1-109 (1983). Clinical trials in prevention of colorectal cancer: R. S. Sandler *et al., N. Engl. J. Med.* **348**, 883 (2003); J. A. Baron *et al., ibid.* 891. Comprehensive description: K. Florey, *Anal. Profiles Drug Subs.* **8**, 1-46 (1979). Monograph: M. J. H. Smith, P. K. Smith, *The Salicylates* (Interscience, New York, 1966) 313 pp. Book: *Acetylsalicylic Acid*, H. J. M. Barnett *et al.*, Eds. (Raven, New York, 1982) 278 pp.

White monoclinic tablets or needle-like crystals. d 1.40. mp 135° (rapid heating); the melt solidifies at 118°. uv max (0.1N H_2SO_4): 229 nm (E$_{1cm}^{1\%}$ 484); (CHCl$_3$): 277 nm (E$_{1cm}^{1\%}$ 68). Is odorless, but in moist air it is gradually hydrolyzed into salicylic and acetic acids and acquires the odor of acetic acid. Stable in dry air. pK (25°) 3.49. One gram dissolves in 300 ml water at 25°, in 100 ml water at 37°, in 5 ml alcohol, 17 ml chloroform, 10-15 ml ether. Less soluble in anhydr ether. Decomp by boiling water or when dissolved in solns of alkali hydroxides and carbonates. Inorganic salts of acetylsalicylic acid are soluble in water (esp the Ca salt, *q.v.*), but are decomposed quickly. LD$_{50}$ orally in mice, rats: 1.1, 1.5 g/kg (Hart).

Guaiacol ester. [55482-89-8] Guacetisal; Guaiaspir. $C_{16}H_{14}O_5$; mol wt 286.28.

Methyl ester *see* Methyl Acetylsalicylate.

Caution: Potential symptoms of overexposure are increased blood clotting time; nausea, vomiting; liver, kidney injury. Direct contact may cause eye, skin and upper respiratory irritation. *See NIOSH Pocket Guide to Chemical Hazards* (DHHS/NIOSH 97-140, 1997) p 6.

THERAP CAT: Analgesic; antipyretic; anti-inflammatory; antithrombotic.

THERAP CAT (VET): Analgesic; antipyretic; anti-inflammatory; antithrombotic.

842. **Aspoxicillin.** [63358-49-6] (2R)-N-Methyl-D-asparaginyl-N-[(2S,5R,6R)-2-carboxy-3,3-dimethyl-7-oxo-4-thia-1-azabicyclo[3.2.0]hept-6-yl]-2-(4-hydroxyphenyl)glycinamide; 6-[D-2-(D-2-amino-3-N-methylcarbamoylpropionamido)-2-p-hydroxyphenylacetamido]penicillanic acid; (2S,5R,6R)-6-[(2R)-2-[(2R)-2-amino-3-(methylcarbamoyl)propionamido]-2-(p-hydroxyphenyl)acetamido]-3,3-dimethyl-7-oxo-4-thia-1-azabicyclo[3.2.0]heptane-2-carboxylic acid; N^4-methyl-D-asparaginylamoxicillin; ASPC; TA-058; Doyle. $C_{21}H_{27}N_5O_7S$; mol wt 493.54. C 51.11%, H 5.51%, N 14.19%, O

22.69%, S 6.50%. Semisynthetic penicillin. Prepn: M. Kawazu *et al.*, **DE 2638067**; *eidem*, **US 4053609** (both 1977 to Tanabe); prepn and antibacterial activity: M. Wagatsuma *et al.*, *J. Antibiot.* **36**, 147 (1985). Mechanism of action study: T. Nishino *et al.*, *Chemotherapy (Tokyo)* **33**, 132 (1985), *C.A.* **103**, 34792v (1985). Toxicological study: M. Takeshita *et al.*, *Oyo Yakuri* **30**, 687 (1985), *C.A.* **104**, 101990u (1986). HPLC determn in serum: J. Knöller *et al.*, *Zentralbl. Bakteriol. Mikrobiol. Hyg.* **265**, 176 (1987). Clinical evaluation in ocular infections: M. Ooishi *et al.*, *Acta Med. Biol.* **34**, 1 (1986). Series of articles on antibacterial activity, pharmacology and clinical efficacy: *Chemotherapy (Tokyo)* **32**, Suppl. 2, 1-791 (1984).

Colorless crystalline powder, mp 195-198° (dec).

THERAP CAT: Antibacterial.

843. Astacin. [514-76-1] β,β-Carotene-3,3′,4,4′-tetrone; 3,-4,3′,4′-tetraketo-β-carotene; 3,3′-dihydroxy-2,3,2′,3′-tetradehydro-β,β-carotene-4,4′-dione; astaceine. $C_{40}H_{48}O_4$; mol wt 592.82. C 81.04%, H 8.16%, O 10.80%. Red carotenoid pigment isolated from biological material originating from crustacea, algae, sponges, protozoa, fish and reptiles. Small amounts were isolated from the fat of mammals (whales, *Balaenoptera musculus).* Occurs together with astaxanthin from which it is formed by autoxidation. Appears to be an artifact rather than a natural product. Isoln from lobster shells: Kuhn, Lederer, *Ber.* **66**, 488 (1933). Structure: Karrer *et al.*, *Helv. Chim. Acta* **17**, 412, 745 (1934); **18**, 96 (1935); **19**, 479 (1936). Total synthesis: J. B. Davis, B. C. L. Weedon, *Proc. Chem. Soc. London* **1960**, 182; E. Widmer *et al.*, *Helv. Chim. Acta* **65**, 671 (1982). Prepn by autoxidation of canthaxanthin: R. D. G. Cooper *et al.*, *J. Chem. Soc. Perkin Trans. 1* **1975**, 2195.

Purple needles or leaflets with metallic luster, sometimes bent into sickle shape, esp when crystallized from pyridine + water, mp 240-243° (slow heating in evac tube, Karrer), mp 228° (Kuhn). Absorption max (pyridine): 500 nm. Practically insol in water; freely sol in chloroform, pyridine, dioxane, carbon disulfide and dil aq alkali; slightly sol in benzene, ethyl acetate, glacial acetic acid; nearly insol in ether, petr ether, methanol.

Diacetate. $C_{44}H_{52}O_6$. Black to violet needles from pyridine + water, dec 235°.

Dipalmitate (astacein). $C_{72}H_{108}O_6$. Almost square red leaflets from petr ether, mp 121°.

844. Astatine. [7440-68-8] At; at. no. 85. Radioactive halogen; Group VIIA (17). One of the rarest elements in nature. Known oxidation states: −1, 0, +5; existence of +1, +3, +7 oxidation states uncertain. No stable nuclides; radioisotopes range in mass numbers from 194 to 223. Naturally occurring isotopes produced in the natural decay series of uranium: 215, 218, 219 (longest-lived natural isotope, $T_{1/2}$ 56 sec). Most stable artificial isotopes: 209 ($T_{1/2}$ 5.41 hr); 210 ($T_{1/2}$ 8.1 hr, longest-lived isotope, rel. at. mass 209.9871); 211 ($T_{1/2}$ 7.21 hr, rel. at. mass 210.9875). Prepn and characterization of ^{211}At by cyclotron reaction ^{209}Bi (α,2n): D. R. Corson *et al.*, *Phys. Rev.* **58**, 672 (1940); of other artificial isotopes by α-particle bombardment of bismuth: E. L. Kelley, E. Segre, *Phys. Rev.* **75**, 999 (1949); Johnson *et al.*, *J. Chem. Phys.* **17**, 1, (1949); Neumann, *J. Inorg. Nucl. Chem.* **4**, 349 (1957). Prepn, purifn and

nuclear decay data for ^{211}At: R. M. Lambrecht, S. Mirzadeh, *Int. J. Appl. Radiat. Isot.* **36**, 443 (1985). Inorganic chemistry: G. W. M. Visser, E. L. Diemer, *Radiochim. Acta* **33**, 145 (1983); idem, ibid. **47**, 97 (1989). Toxicity studies of ^{211}At: L. M. Cobb *et al.*, *Hum. Toxicol.* **7**, 529 (1988). *Review:* A. J. Downs, C. J. Adams, "Chlorine, Bromine, Iodine and Astatine" in *Comprehensive Inorganic Chemistry* vol. 2, J. C. Bailar, Jr. *et al.*, Eds. (Pergamon Press, Oxford, 1973) pp 1573-1594. Review of organic chemistry and potential biomedical applications: I. Brown, *Adv. Inorg. Chem.* **31**, 43-88 (1987); K. Berei, L. Vasáros in *The Chemistry of Halides, Pseudohalides and Azides*, S. Patai, Z. Rappaport, Eds. (John Wiley & Sons, Chichester, 1995) pp 787-819.

More metallic than iodine. Soluble in organic solvents. E^0(aq) At/At⁻ 0.3 V.

845. Astaxanthin. [472-61-7] (3*S*,3′*S*)-3,3′-Dihydroxy-β,β-carotene-4,4′-dione; 3,3′-dihydroxy-4,4′-diketo-β-carotene; ovoester; AstaReal; BioAstin. $C_{40}H_{52}O_4$; mol wt 596.85. C 80.50%, H 8.78%, O 10.72%. Naturally occuring, red-orange carotenoid pigment with potent antioxidant activity. Primary carotenoid of salmon, rainbow trout, and crustaceans such as lobster and shrimp. Found as free or esterified forms or combined with lipids or proteins. In crustaceans, the pigment is bound to a protein carrier to form the slate-blue carotenoprotein, **β-crustacyanin**. The (*R*,*R*)-isomer is also naturally occuring and has been isolated from the red yeast, *Phaffia rhodozyma*. Like other carotenoids, the all-*trans*-forms readily convert into *cis*- isomers; particularly at the 9- and 13-positions. Isoln from lobster eggs and characterization of structure: R. Kuhn, N. A. Sörensen, *Angew. Chem.* **51**, 465 (1938); *eidem, Ber.* **71**, 1879 (1938). Distribution in plants and animals: R. Kuhn *et al.*, *ibid.* **72**, 1688 (1939). Isoln from flower petals of *Adonis annua* L., *Ranunculaceae:* A. Seybold, T. W. Goodwin, *Nature* **184**, 1714 (1959); from red feathers of birds of the *Lanarius* spp: O. Volker, *Z. Physiol. Chem.* **288**, 20 (1951). *Cis-trans*-isomerization and spectra: R. Grangaud, P. Chardenot, *Compt. Rend.* **242**, 1767 (1956). Abs config of astaxanthin from lobster and green algae: A. G. Andrewes *et al.*, *Acta Chem. Scand. B* **28**, 730 (1974); from red yeast: A. G. Andrewes, M. P. Starr, *Phytochemistry* **15**, 1009 (1976). Total synthesis: F. Kienzle, H. Mayer, *Helv. Chim. Acta* **61**, 2609 (1978); E. Widmer *et al.*, *ibid.* **64**, 2405 (1981). Commercial production using the green alga, *Haematococcus pluvialis:* R. T. Lorenz, G. R. Cysewski, *Trends Biotechnol.* **18**, 160 (2000). Review of chemistry, industrial sources, and uses: I. Higuera-Ciapara *et al.*, *Crit. Rev. Food Sci. Nutr.* **46**, 185-196 (2006); of antioxidant and anti-inflammatory activities and applications in human health and nutrition: M. Guerin *et al.*, *Trends Biotechnol.* **21**, 210-216 (2003); G. Hussein *et al.*, *J. Nat. Prod.* **69**, 443-449 (2006).

Dark violet crystals from methylene chloride/methanol, mp 223-225° (Widmer). Abs max: 492 nm (chloroform). Also reported as shiny purple platelets with gold luster from pyridine, mp 216° (some decompn). Readily sol in pyridine, from which it can be cryst by the addn of water (Kuhn, Sörensen).

Diacetate. [25494-88-6] $C_{44}H_{56}O_6$; mol wt 680.93. Stout, blue-black needles, mp 203-205° (vac) (Kuhn, Sörenson); also reported as mp 187-189° (Kienzle, Mayer).

(3*RS*,3′*RS*)-Form. [7542-45-2] *rac*-Astaxanthin; Carophyll Pink. Synthetic prepn consisting of the (*S*,*S*), (*R*,*S*), and (*R*,*R*) isomers in a 1:2:1 ratio. Synthesis: E. Widmer *et al.*, *Helv. Chim. Acta* **64**, 2436 (1981). Crystals from methanol, mp 216-218°. Abs max: 492 nm in chloroform.

USE: Nutritional supplement. Antioxidant in cosmetics and personal care products. In fish aquaculture to enhance pigmentation.

846. Astemizole. [68844-77-9] 1-[(4-Fluorophenyl)methyl]-*N*-[1-[2-(4-methoxyphenyl)ethyl]-4-piperidinyl]-1*H*-benzimidazol-2-amine; 1-(*p*-fluorobenzyl)-2-[[1-(*p*-methoxyphenethyl)-4-

piperidyl]amino]benzimidazole; R-43512; Astemisan; Hismanal; Histamine; Histaminos; Kelp; Laridal; Metodik; Novo-Nastizol A; Paralergin; Retolen; Waruzol. $C_{28}H_{31}FN_4O$; mol wt 458.58. C 73.34%, H 6.81%, F 4.14%, N 12.22%, O 3.49%. Nonsedating-type histamine H_1-receptor antagonist. Prepn: F. Janssens *et al.*, **EP 5318**; *eidem*, **US 4219559** (1979, 1980 both to Janssen). Pharmacology: J. Van Wauwe *et al.*, *Arch. Int. Pharmacodyn. Ther.* **251**, 39 (1981); A. Wauquier, C. J. E. Niemegeers, *Eur. J. Pharmacol.* **72**, 245 (1981). *In vitro* and *in vivo* binding characteristics: P. M. Laduron *et al.*, *Mol. Pharmacol.* **21**, 294 (1982). Effect on human psychomotor performance: T. Seppala, K. Savolainen, *Curr. Ther. Res.* **31**, 638 (1982). Clinical study in treatment of hay fever: J. Callier *et al.*, *ibid.* **29**, 24 (1981). Mutagenicity study: P. Vanparys *et al.*, *Arch. Toxicol.* **50**, 167 (1982). Review of pharmacology and clinical trials: D. M. Richards *et al.*, *Drugs* **28**, 38-61 (1984). Comprehensive description: A. M. Al-Obaid, M. S. Mian, *Anal. Profiles Drug Subs.* **20**, 173-208 (1991).

Crystals from 2,2'-oxybispropane, mp 172.9°. Freely sol in organic solvents. Practically insol in water. uv max (ethanol): 219, 249, 286 nm (ε 27250.229, 6480.293, 8634.280); (0.1N HCl): 209, 277 nm (ε 57889.908, 18073.394).

THERAP CAT: Antihistaminic.

847. Asulam. [3337-71-1] N-[(4-Aminophenyl)sulfonyl]carbamic acid methyl ester; N^1-methoxycarbonylsulfanilamide; sulfanilylcarbamic acid methyl ester; methyl p-aminobenzenesulfonylcarbamate; M & B 9057; Asulox. $C_8H_{10}N_2O_4S$; mol wt 230.24. C 41.73%, H 4.38%, N 12.17%, O 27.80%, S 13.92%. Carbamate herbicide; folic acid biosynthesis inhibitor. Prepn: **BE 622214** (1963 to May & Baker). Description of herbicidal properties: H. J. Cottrell, B. J. Heywood, *Nature* **207**, 655 (1965). Persistence in soil: A. E. Smith, A. Walker, *Pestic. Sci.* **8**, 449 (1977). Mode of action: P. Veerasekaran *et al.*, *Pestic. Sci.* **12**, 325, 330 (1981). Field study on sugarcane and torpedograss: M. A. Hossain *et al.*, *J. Weed Sci. Technol.* **43**, 10 (1998); in bracken control: R. J. Pakeman *et al.*, *J. Environ. Manage.* **53**, 255 (1998). Review of properties and analytical methods: C. H. Brockelsby, D. F. Muggleton, *Anal. Methods Pestic. Plant Growth Regul.* **7**, 497 (1973); A. Guardigli *et al.*, *ibid.* **13**, 197 (1984).

White crystals, mp 143-144° (dec). Soly: water 0.5%, hydrocarbons <2%, chlorinated hydrocarbons <2%, acetone 34%, methanol 28%. pKa 4.82. LD_{50} in rats, mice, dogs, rabbits (mg/kg): >5000, >5000, >5000, >2000 orally; in rats (mg/kg): >1200 dermally (Brockelsby).

USE: Herbicide.

848. Atacicept. [845264-92-8] Cytokine receptor TACI (synthetic human extracellular domain fragment) fusion protein with immunoglobulin G1 (synthetic human γ_1-chain Fc fragment), dimer; human transmembrane activator and CAML interactor (TACI) immunoglobulin G_1 Fc domain fusion protein (Fc5); TACI-Fc5; TACI-Ig. Soluble, fully human, recombinant fusion protein composed of the Fc portion of human immunoglobulin G and the extracellular portion of transmembrane activator and calcium-modulating ligand interactor (TACI) receptor. Acts as an inhibitor of B-lymphocyte stimulator (BLyS) and a proliferation-inducing ligand (APRIL) that are overexpressed in B-cell malignancies and autoimmune disorders. Prepn: M. W. Rixon, J. A. Gross, **WO 02094852**; *eidem*, **US**

7501497 (2002, 2009 both to ZymoGenetics); and receptor binding studies: J. A. Gross *et al.*, *Nature* **404**, 995 (2000). Pharmacology: S. Yaccoby *et al.*, *Leukemia* **22**, 406 (2008); and pharmacokinetics: M. Carbonatto *et al.*, *Toxicol. Sci.* **105**, 200 (2008). Clinical evaluation in multiple myeloma and Waldenström's macroglobulinemia: J.-F. Rossi *et al.*, *Br. J. Cancer* **101**, 1051 (2009). Clinical pharmacokinetics and activity in systemic lupus erythematosus: I. Nestorov *et al.*, *J. Pharm. Sci.* **99**, 524 (2010). Review of therapeutic potential in plasma cell diseases: B. Gatto *et al.*, *Curr. Opin. Investig. Drugs* **9**, 1216-1227 (2008); in rheumatoid arthritis: C. Bracewell *et al.*, *Expert Opin. Biol. Ther.* **9**, 909-919 (2009).

THERAP CAT: Immunosuppressant.

849. Atazanavir. [198904-31-3] (3S,8S,9S,12S)-3,12-Bis-(1,1-dimethylethyl)-8-hydroxy-4,11-dioxo-9-(phenylmethyl)-6-[[4-(2-pyridinyl)phenyl]methyl]-2,5,6,10,13-pentaazatetradecanedioic acid dimethyl ester; 1-[4-(pyridin-2-yl)phenyl]-5S-2,5-bis[[N-(methoxycarbonyl)-L-*tert*-leucinyl]amino]-4S-hydroxy-6-phenyl-2-azahexane; BMS-232632; CGP-73547. $C_{38}H_{52}N_6O_7$; mol wt 704.87. C 64.75%, H 7.44%, N 11.92%, O 15.89%. Azapeptide HIV protease inhibitor. Prepn: A. Fässler *et al.*, **WO 9740029**; *eidem*, **US 5849911** (1997, 1998 both to Novartis); G. Bold *et al.*, *J. Med. Chem.* **41**, 3387 (1998); of bisulfate salt: J. Singh *et al.*, **WO 9936404**; *eidem*, **US 6087383** (1999, 2000 both to Bristol-Myers Squibb); Z. Xu *et al.*, *Org. Process Res. Dev.* **6**, 323 (2002). Comparative anti-HIV activity: B. S. Robinson *et al.*, *Antimicrob. Agents Chemother.* **44**, 2093 (2000). HPLC determn in plasma: E. Cateau *et al.*, *J. Pharm. Biomed. Anal.* **39**, 791 (2005). Clinical evaluation in HIV: I. Sanne *et al.*, *J. Acquir. Immune Defic. Syndr.* **32**, 18 (2003). Review of pharmacology and clinical efficacy in HIV: C. Le Tiec *et al.*, *Clin. Pharmacokinet.* **44**, 1035-1050 (2005); T. S. Harrison, L. J. Scott, *Drugs* **65**, 2309-2336 (2005).

Crystals from ethanol/water, mp 207-209°. $[\alpha]_D$ −47° (c = 1 in ethanol). Insol in water (<1 μg/ml at 24°).

Sulfate. [229975-97-7] Reyataz. $C_{38}H_{52}N_6O_7.H_2SO_4$; mol wt 802.94. Crystals from ethanol + heptane, mp 195.0°, or acetone, mp 198-199° (dec). $[\alpha]_D^{22}$ −46.1° (c = 1 in 1:1 CH_3OH/H_2O, pH = 2.6). Soly in water: 4-5 mg/ml.

THERAP CAT: Antiretroviral.

850. Atenolol. [29122-68-7] 4-[2-Hydroxy-3-[(1-methylethyl)amino]propoxy]benzeneacetamide; 2-[p-[2-hydroxy-3-(isopropylamino)propoxy]phenyl]acetamide; 1-p-carbamoylmethylphenoxy-3-isopropylamino-2-propanol; ICI-66082; Atehexal; Ateno basan; Atenol; Cuxanorm; Myocord; Prenormine; Seles Beta; Tenoblock; Tenormin; Uniloc. $C_{14}H_{22}N_2O_3$; mol wt 266.34. C 63.14%, H 8.33%, N 10.52%, O 18.02%. Cardioselective β_1-adrenergic blocker. Prepn: Barrett *et al.*, **DE 2007751**; *eidem*, **US 3663607** and **US 3836671** (1970, 1972, 1974 all to I.C.I.). Pharmacology and clinical studies: Giudicelli *et al.*, *C. R. Seances Soc. Biol. Ses Fil.* **167**, 232 (1973); Hansson *et al.*, *Acta Med. Scand.* **194**, 549 (1973); Amery *et al.*, *N. Engl. J. Med.* **290**, 284 (1974). Clinical trial in treatment of alcohol withdrawal syndrome: M. L. Kraus *et al.*, *ibid.* **313**, 905 (1985). HPLC determn of enantiomers in plasma and urine: S. K. Chin *et al.*, *J. Chromatogr.* **489**, 438 (1989). Comprehensive description: V. Caplar *et al.*, *Anal. Profiles Drug Subs.* **13**, 1-25 (1984). *Review:* J. D. Fitzgerald in *Pharmacological and Biochemical Properties of Drug Substances* vol. 2, M. E. Goldberg, Ed. (Am. Pharm. Assoc., Washington, DC, 1979) pp 98-147; E. Marmo,

Drugs Exp. Clin. Res. **6**, 639-663 (1980). Symposium on clinical studies: *Drugs* **25**, Suppl. 2, 1-346 (1983).

Crystals from ethyl acetate, mp 146-148°. Also reported as mp 150-152° (Caplar). pKa 9.6. Partition coefficient (*n*-octanol/phosphate buffer): 0.008 (pH 7.0); 0.052 (pH 8.0). Dipole moment: 5.71 ±0.20 D at 20° in propionic acid. uv max (methanol): 225, 275, 283 nm. Freely sol in methanol; sol in acetic acid, DMSO; sparingly sol in 96% ethanol; slightly sol in water, isopropanol; very slightly sol in acetone, dioxane. Practically insol in acetonitrile, ethylacetate, chloroform. LD$_{50}$ in mice, rats (mg/kg): 2000, 3000 orally; 98.7, 59.24 i.v. (Fitzgerald).

THERAP CAT: Antihypertensive, antianginal, antiarrhythmic (class II).

THERAP CAT (VET): Antihypertensive; antiarrhythmic (class II).

851.   Athamantin. [1892-56-4] (8*S-cis*)-3-Methylbutanoic acid 8,9-dihydro-8-[1-methyl-1-(3-methyl-1-oxobutoxy)ethyl]-2-oxo-2*H*-furo[2,3-*h*]-1-benzopyran-9-yl ester; isovaleric acid diester with 8,9-dihydro-9-hydroxy-8-(1-hydroxy-1-methylethyl)-2*H*-furo-[2,3-*h*]-1-benzopyran-2-one; 2,3-dihydro-3,4-dihydroxy-2-(1-hydroxy-1-methylethyl)-5-benzofuranacrylic acid δ-lactone diisovalerate. C$_{24}$H$_{30}$O$_7$; mol wt 430.50. C 66.96%, H 7.02%, O 26.01%. From *Peucedanum oreoselinum* (L.) Moench (*Athamanta oreoselinum* L.), *Ammi visnaga* Lam., *Umbelliferae.* Isoln: Schnedermann, Winckler, *Ann.* **51**, 315 (1844). Structure: Halpern *et al.*, *Helv. Chim. Acta* **40**, 758 (1957). Absolute configuration: Nakazaki *et al.*, *Tetrahedron Lett.* **1966**, 4735.

Fine needles from petr ether, mp 58-60°; sublimes in high vacuum at 180-200°. [α]$_D^{20}$ +88° (c = 1.145 in glacial acetic acid); [α]$_{546}^{22}$ +129° (c = 0.5575 in methanol); [α]$_{546}^{21}$ +73.9° (c = 1.024 in chloroform). [M]$_D$ +440° (methanol); +258° (chloroform). uv max (96% ethanol): 217, 322 nm (log ε 4.18, 4.17). Practically insol in water; sol in alc, ether, chloroform.

852.   Atipamezole. [104054-27-5] 5-(2-Ethyl-2,3-dihydro-1*H*-inden-2-yl)-1*H*-imidazole; 4(5)-(2-ethyl-2-indanyl)imidazole. C$_{14}$H$_{16}$N$_2$; mol wt 212.30. C 79.21%, H 7.60%, N 13.20%. Specific α$_2$-adrenoceptor antagonist. Prepn: A. J. Karjalainen *et al.*, **EP 183492**; *eidem*, **US 4689339** (1986, 1987 both to Farmos). Behavioral and neurochemical effects: H. Scheinin *et al.*, *Eur. J. Pharmacol.* **151**, 35 (1988). Binding study in comparison with yohimbine, *q.v.*: R. Virtanen *et al.*, *Arch. Int. Pharmacodyn.* **297**, 190 (1989). LC-MS determn in pig plasma: H. Kanazawa *et al.*, *Biomed. Chromatogr.* **9**, 188 (1995). Pharmacodynamics and pharmacokinetics: X. Fargetton, T. Vähä-Vahe, *Tijdschr. Diergeneeskd.* **114**, Suppl. 1, 91S (1989). Preliminary report as detomidine antagonist in horses: L. Nilsfors, C. Kvart, *Acta Vet. Scand. Suppl.* **82**, 121 (1986). Use as medetomidine antagonist in zoo animals: H. Jalanka, *ibid.* **85**, 193 (1989); in dogs: M. Granholm *et al.*, *Vet. Rec.* **160**, 891 (2007).

Hydrochloride. [104075-48-1] MPV-1248; Antisedan; Atipam. C$_{14}$H$_{16}$N$_2$·HCl; mol wt 248.75. Crystals from dry HCl in ethyl acetate, mp 211-215°. LD$_{50}$ in male, female rats (mg/kg): >50, 44 s.c. (Scheinin).

THERAP CAT (VET): Antidote to α$_2$-adrenoceptor induced sedation.

853.   Atisine. [466-43-3] (6*R*,6a*R*,8a*S*,9*R*,11*S*,12a*R*,12b*S*,-12c*S*)-Decahydro-6-methyl-10-methylene-9*H*,12c*H*-8a,11-ethano-6,12b-propano-5*H*-benz[*h*]oxazolo[2,3-*a*]isoquinolin-9-ol; anthorine. C$_{22}$H$_{33}$NO$_2$; mol wt 343.51. C 76.92%, H 9.68%, N 4.08%, O 9.32%. From roots of the "atis" plant, *Aconitum heterophyllum* Wall., and from *A. anthora* L., *Ranunculaceae:* Broughton, *Blue Book of East India Cinchona Cultivation* **1877**, 133; Goris, Metin, *Compt. Rend.* **180**, 968 (1925); Lawson, Topps, *J. Chem. Soc.* **1937**, 1640. Structure: Wiesner *et al.*, *Chem. Ind. (London)* **1954**, 132; Pelletier, Jacobs, *J. Am. Chem. Soc.* **76**, 4496 (1954). Structure and stereochemistry: Dvornik, Edwards, *Can. J. Chem.* **42**, 137 (1964). Partial synthesis: Pelletier, Jacobs, *J. Am. Chem. Soc.* **78**, 4144 (1956); Pelletier, Parthasarathy, *Tetrahedron Lett.* **1963**, 205. Racemic synthesis and resolution: Nagata *et al.*, *J. Am. Chem. Soc.* **85**, 2342 (1963); *eidem, ibid.* **89**, 1499 (1967); Masamune, *ibid.* **86**, 291 (1964); Guthrie *et al.*, *Tetrahedron Lett.* **1966**, 4645. ^{13}C-NMR study of C-20 epimers: N. V. Mody, S. W. Pelletier, *Tetrahedron* **34**, 2421 (1978).

Solid, mp 57-60°. Distills in high vacuum at a bath temp of 140°. Strong base: pK 12.2.

Hydrochloride. C$_{22}$H$_{33}$NO$_2$·HCl. Flat needles from dil alc, dec 311-312°. [α]$_D^{25}$ +28°.

854.   Atomoxetine. [83015-26-3] (γ*R*)-*N*-Methyl-γ-(2-methylphenoxy)benzenepropanamine; (−)-*N*-methyl-3-(*o*-tolyloxy)-3-phenylpropylamine; tomoxetine. C$_{17}$H$_{21}$NO; mol wt 255.36. C 79.96%, H 8.29%, N 5.49%, O 6.27%. Norepinephrine reuptake inhibitor. Prepn (stereochem unspec): B. B. Molloy, K. K. Schmiegel, **DE 2500110**; *eidem*, **US 4314081** (1975, 1982 both to Lilly). Prepn of (*R*)-form: B. J. Foster, E. R. Lavagnino, **EP 52492** (1982 to Lilly); Y. Gao, K. B. Sharpless, *J. Org. Chem.* **53**, 4081 (1988). Clinical pharmacokinetics: N. A. Farid *et al.*, *J. Clin. Pharmacol.* **25**, 296 (1985). Binding study: D. R. Gehlert *et al.*, *J. Neurochem.* **64**, 2792 (1995). Biotransformation by human liver microsomes: B. J. Ring *et al.*, *Drug Metab. Dispos.* **30**, 319 (2002). Evaluation of abuse potential: S. H. Heil *et al.*, *Drug Alcohol Depend.* **67**, 149 (2002). Clinical trial in pediatric ADHD: C. J. Kratochvil *et al.*, *J. Am. Acad. Child Adolesc. Psychiatry* **41**, 776 (2002); and comorbid tic disorders: A. J. Allen *et al.*, *Neurology* **65**, 1941 (2005). Review of clinical pharmacokinetics: J.-M. Sauer *et al.*, *Clin. Pharmacokinet.* **44**, 571-590 (2005).

Hydrochloride. [82248-59-7] LY-139603; Strattera. C$_{17}$H$_{21}$-NO·HCl; mol wt 291.82. Crystals, mp 166-168° (Foster, Lavagnino). [α]$_D^{25}$ −38.01°; [α]$_{365}^{25}$ −177.26° (c = 1 in methanol). Also reported as white solid from acetonitrile, mp 162-164° (Gao, Sharpless). [α]$_D^{23}$ −41.37° (c = 1.02 in methanol); [α]$_D^{25}$ −40.3° (c = 0.94 in ethanol). Soly in water: 27.8 mg/ml. pKa 10.13.

THERAP CAT: In treatment of attention deficit hyperactivity disorder (ADHD).

855. Atorvastatin. [134523-00-5] ($\beta R,\delta R$)-2-(4-Fluoro-phenyl)-β,δ-dihydroxy-5-(1-methylethyl)-3-phenyl-4-[(phenylamino)carbonyl]-1H-pyrrole-1-heptanoic acid. $C_{33}H_{35}FN_2O_5$; mol wt 558.65. C 70.95%, H 6.32%, F 3.40%, N 5.01%, O 14.32%. HMG-CoA reductase inhibitor. Prepn: B. D. Roth, **EP 409281**; *idem,* **US 5273995** (1991, 1993 both to Warner-Lambert); K. L. Baumann *et al., Tetrahedron Lett.* **33**, 2283 (1992). Solubility and kinetic studies: A. S. Kearney *et al., Pharm. Res.* **10**, 1461 (1993). LC/MS/MS determn in serum: M. Jemal *et al., Rapid Commun. Mass Spectrom.* **13**, 1003 (1999). Clinical pharmacokinetics: D. D. Cilla, Jr. *et al., Clin. Pharmacol. Ther.* **60**, 687 (1996). Review of pharmacology and clinical experience: H. S. Yee, N. T. Fong, *Ann. Pharmacother.* **32**, 1030-1043 (1998); of efficacy in diabetic patients: K. F. Croom, G. L. Plosker, *Drugs* **65**, 137-152 (2005).

Calcium salt trihydrate. [344423-98-9]; [134523-03-8] (anhydrous). CI-981; Lipitor; Sortis; Torvast; Totalip. $C_{66}H_{68}CaF_2N_4$-$O_{10}.3H_2O$; mol wt 1209.41. White to off-white crystalline powder. $[\alpha]_D$ −7.4° (c = 1 in DMSO). Freely sol in methanol; slightly sol in ethanol; very slightly sol in acetonitrile, distilled water, phosphate buffer (pH 7.4). Insol in aq solns of pH 4 and below.
Sodium salt. [134523-01-6] $C_{33}H_{34}FN_2NaO_5$. pKa 4.46. Soly in water (30°): 20.4 μg/ml (pH 2.1); 1.23 mg/ml (pH 6.0).
Lactone. [125995-03-1] 5-(4-Fluorophenyl)-2-(1-methylethyl)-N,4-diphenyl-1-[2-[(2R,4R)-tetrahydro-4-hydroxy-6-oxo-2H-pyran-2-yl]ethyl]-1H-pyrrole-3-carboxamide. $C_{33}H_{33}FN_2O_4$; mol wt 540.64. mp 159.2-160.7°. $[\alpha]_D$ +26.05° (c = 1 in chloroform).

THERAP CAT: Antilipemic.

856. Atosiban. [90779-69-4] O-Ethyl-N-(3-mercapto-1-oxopropyl)-D-tyrosyl-L-isoleucyl-L-threonyl-L-asparaginyl-L-cysteinyl-L-prolyl-L-ornithylglycinamide cyclic (1 → 5)-disulfide; 1-(3-mercaptopropanoic acid)-2-(O-ethyl-D-tyrosine)-4-L-threonine-8-L-ornithineoxytocin; 1-(3-mercaptopropionic acid)-2-[3-(p-ethoxyphenyl)-D-alanine]-4-L-threonine-8-L-ornithineoxytocin; 1-deamino-2-D-Tyr-(OEt)-4-Thr-8-Orn-oxytocin; antocin; dTVT; ORF-22164; F-314; RWJ-22164; CAP-440; Tractocile. $C_{43}H_{67}N_{11}$-$O_{12}S_2$; mol wt 994.19. C 51.95%, H 6.79%, N 15.50%, O 19.31%, S 6.45%. Oxytocin antagonist. Prepn: P. O. R. Melin, J. A. Trojnar, **EP 112809**; *idem,* **US 4504469** (1984, 1985 both to Ferring). Mechanism of action: A. Lopez Bernal *et al., Br. J. Obstet. Gynaecol.* **96**, 1108 (1989). Effect on *in vitro* contractility of myometrium: V. A. Kinsler *et al., ibid.* **103**, 373 (1996). Clinical pharmacokinetics: T. M. Goodwin *et al., Am. J. Obstet. Gynecol.* **173**, 913 (1995). Clinical trial: T. M. Goodwin *et al., ibid.* **170**, 474 (1994).

THERAP CAT: Tocolytic.

857. Atovaquone. [95233-18-4] 2-[*trans*-4-(4-Chlorophenyl)cyclohexyl]-3-hydroxy-1,4-naphthalenedione; *trans*-2-[4-(4-chlorophenyl)cyclohexyl]-3-hydroxy-1,4-naphthoquinone; 566C80; BW-566C; BW-566C-80; Mepron; Wellvone. $C_{22}H_{19}ClO_3$; mol wt 366.84. C 72.03%, H 5.22%, Cl 9.66%, O 13.08%. Hydroxynaphthoquinone derivative that inhibits mitochondrial electron transport.

Prepn: A. T. Hudson, A. W. Randall, **EP 123238**; *eidem,* **US 5053432** (1984, 1991 both to Burroughs Wellcome). Antiprotozoal activity, metabolism, and pharmacokinetics: A. T. Hudson *et al., Drugs Exp. Clin. Res.* **17**, 427 (1991). Mechanism of action study: M. Fry, M. Pudney, *Biochem. Pharmacol.* **43**, 1545 (1992). Clinical trial for prevention of pneumocystis pneumonia in AIDS patients: W. M. El-Sadr *et al., N. Engl. J. Med.* **339**, 1889 (1998).

Relative stereochemistry

Crystals from acetonitrile, mp 216-219°. Freely sol in N-methyl-2-pyrrolidone, tetrahydrofuran; sol in chloroform; sparingly sol in acetone, di-n-butyl adipate, dimethyl sulfoxide, polyethylene glycol 400; slightly sol in alcohol, 1,3-butanediol, ethyl acetate, glycerin, octanol, polyethylene glycol 200; very slightly sol in 0.1 N sodium hydroxide. Insol in water.

THERAP CAT: Antipneumocystic.

858. Atractyloside. [17754-44-8] (2β,4α,15α)-15-Hydroxy-2-[[2-O-(3-methyl-1-oxobutyl)-3,4-di-O-sulfo-β-D-gluco-pyranosyl]oxy]-19-norkaur-16-en-18-oic acid potassium salt (1:2); potassium atractylate; atractylin (C_{30} glucoside). $C_{30}H_{44}K_2O_{16}S_2$; mol wt 802.98. C 44.87%, H 5.52%, K 9.74%, O 31.88%, S 7.99%. Toxic principle isolated from the thistle *Atractylis gummifera* L., *Compositae:* M. Lefranc, *Compt. Rend.* **67**, 1868 (1868). Structure and stereochemistry of the aglycone, *atractyligenin:* F. Piozzi *et al., Tetrahedron* Suppl. 8 **II**, 515 (1966); total synthesis of (±)-form: A. K. Singh *et al., J. Am. Chem. Soc.* **109**, 6187 (1987). Structure and stereochemistry of atractyloside: *eidem, Gazz. Chim. Ital.* **97**, 935 (1967). Toxicity studies: G. Cascio *et al., Boll. Soc. Ital. Biol. Sper.* **44**, 253 (1968). *Review: Atractyloside: Chemistry, Biochemistry, and Toxicology,* R. Santi, S. Luciani, Eds. (Piccin Medical Books, Padova, Italy, 1978) 136 pp.

Crystals, dec 174°. $[\alpha]_D^{20}$ −53° (water). *Highly toxic.* Has strychnine-like action; produces convulsion of a hypoglycemic nature: R. Santi, G. Cascio, *C.A.* **50**, 7320i (1956); R. Santi, *C.A.* **52**, 15733c (1958). LD$_{50}$ i.m. in rats: 431 mg/kg (Cascio).

859. Atracurium Besylate. [64228-81-5] 2,2'-[1,5-Pentanediylbis[oxy(3-oxo-3,1-propanediyl)]]bis[1-[(3,4-dimethoxyphenyl)methyl]-1,2,3,4-tetrahydro-6,7-dimethoxy-2-methylisoquinolinium] benzenesulfonate (1:2); 2-(2-carboxyethyl)-1,2,3,4-tetrahydro-6,7-dimethoxy-2-methyl-1-veratrylisoquinolinium benzenesulfonate pentamethylene ester; N,N'-dimethyl-N,N'-(4,10-dioxa-3,11-dioxotridecylene)-1,13-bis-tetrahydropapaverinium dibesylate; BW-33A; Wellcome 33-A-74; Tracrium. $C_{65}H_{82}N_2O_{18}S_2$; mol wt 1243.49. C 62.78%, H 6.65%, N 2.25%, O 23.16%, S 5.16%. Benzylisoquinolinium, competitive neuromuscular blocker. Prepn: J. B. Stenlake *et al.,* **DE 2655883**; *eidem,* **US 4179507** (1977, 1979 both to Burroughs Wellcome). Pharmacology: R. Hughes, D. J. Chapple, *Br. J. Anaesth.* **53**, 31 (1981). Clinical pharmacology: S. J. Basta *et* al., *Anesth. Analg.* **61**, 723 (1982). Neuromuscular effects in man: R. L. Katz *et al.*, *ibid.* 730. Metabolic studies: E. A. Neill, D. J. Chapple, *Xenobiotica* **12**, 203 (1982). Pharmacokinetics: S. Ward *et al.*, *Br. J. Anaesth.* **55**, 113 (1983). Use during halothane anesthesia in humans: J. A. Stirt *et al.*, *Anesth. Analg.* **62**, 207 (1983). Symposium on pharmacology, metabolism, and clinical studies: *Br. J. Anaesth.* **55**, Suppl. 1, 1S-139S (1983); on pharmacokinetics and clinical experience: *ibid.* **58**, Suppl. 1, 1S-113S (1986).

Off-white powder, mp 85-90°. Softens at 60°.

Cisatracurium besylate. [96946-42-8] $(1R,1'R,2R,2'R)$-2,2'-[1,5-Pentanediylbis[oxy(3-oxo-3,1-propanediyl)]]bis[1-[(3,4-dimethoxyphenyl)methyl]-1,2,3,4-tetrahydro-6,7-dimethoxy-2-methylisoquinolinium] benzenesulfonate (1:2); 51W89; Nimbex. Prepn: D. A. Hill, G. L. Turner, **WO 9200965** (1992 to Wellcome Found.). Clinical pharmacology: C. A. Lien *et al.*, *Anesthesiology* **82**, 1131 (1995); M. R. Belmont *et al.*, *ibid.* 1139. White solid.

THERAP CAT: Neuromuscular blocking agent.

THERAP CAT (VET): Neuromuscular blocking agent.

860. Atranorin. [479-20-9] 3-Formyl-2,4-dihydroxy-6-methylbenzoic acid 3-hydroxy-4-(methoxycarbonyl)-2,5-dimethylphenyl ester; atranoric acid. $C_{19}H_{18}O_8$; mol wt 374.35. C 60.96%, H 4.85%, O 34.19%. From a number of lichens. Belongs to the group of lichen acids. Isoln: Paterno, Ogliarori, *Gazz. Chim. Ital.* **1877**, 189; Hesse, *Ber.* **30**, 357, 1983 (1897); St. Pfau, *Helv. Chim. Acta* **9**, 650 (1926). Synthesis: Neelakantan *et al.*, *Tetrahedron Lett.* **1962**, 287. TLC analysis: R. Klee, L. Steubing, *J. Chromatogr.* **129**, 478 (1976); J. L. Ramaut *et al.*, *ibid.* **155**, 450 (1978).

Bitter crystals or cryst powder from chloroform, mp 195°. Practically insol in water; slightly sol in alc; sol in boiling benzene or in chloroform; also sol in alkalies giving a yellow soln.

861. Atrasentan. [173937-91-2] $(2R,3R,4S)$-4-(1,3-Benzodioxol-5-yl)-1-[2-(dibutylamino)-2-oxoethyl]-2-(4-methoxyphenyl)-3-pyrrolidinecarboxylic acid; *trans,trans*-2-(4-methoxyphenyl)-4-(1,3-benzodioxol-5-yl)-1-[[(dibutylamino)carbonyl]methyl]pyrrolidine-3-carboxylic acid; (+)-A-127722. $C_{29}H_{38}N_2O_6$; mol wt 510.63. C 68.21%, H 7.50%, N 5.49%, O 18.80%. Endothelin (ET) antagonist with selectivity for the ET_A subtype. Prepn: M. Winn *et al.*, **WO 9606905**; *eidem*, **US 5767144** (1996, 1998 both to Abbott); *eidem*, *J. Med. Chem.* **39**, 1039 (1996). Synthesis: S. J. Wittenberger, M. A. McLaughlin, *Tetrahedron Lett.* **40**, 7175 (1999). HPLC determn in pharmaceutical formulations: J. A. Morley *et al.*, *J. Pharm. Biomed. Anal.* **19**, 777 (1999); in plasma: P. D. Bryan *et al.*, *Biomed. Chromatogr.* **15**, 525 (2001). *In vitro* inhibition of ovarian carcinoma: D. Salani *et al.*, *Clin. Sci.* **103**, Suppl. 48, 318S (2002). Clinical pharmacokinetics and dynamics: M. C. Verhaar *et al.*, *Br. J. Clin. Pharmacol.* **49**, 562 (2000); and safety: M. A. Carducci *et al.*, *J. Clin. Oncol.* **20**, 2171 (2002). Clinical evaluation in prostate cancer: M. Fisher, *Clin. Prostate Cancer* **1**, 79 (2002). Review of clinical development in prostate cancer: A. Jimeno, M. Carducci, *Expert Opin. Invest. Drugs* **13**, 1631-1640 (2004).

mp 122-124°.

Hydrochloride. [195733-43-8] ABT-627; A-147627.1; Xinlay. $C_{29}H_{38}N_2O_6 \cdot HCl$; mol wt 547.09. Crystalline, non-hygroscopic solid. Readily sol in water.

THERAP CAT: Antineoplastic.

862. Atrazine. [1912-24-9] 6-Chloro-N^2-ethyl-N^4-(1-methylethyl)-1,3,5-triazine-2,4-diamine; 2-chloro-4-ethylamino-6-isopropylamino-*s*-triazine; G-30027; AAtrex; Atranex; Gesaprim. $C_8H_{14}ClN_5$; mol wt 215.69. C 44.55%, H 6.54%, Cl 16.44%, N 32.47%. Prepn: Gysin, Knüsli, **CH 342784**, **CH 342785** (both 1960 to Geigy), *C.A.* **55**, 5552d (1961); Mel'nikov *et al.*, *Khim. Prom.* **1961**, 703, *C.A.* **58**, 526c (1963); Andriska *et al.*, **HU 149189** (1962 to Nehézvegyipari Kutato Intézet), *C.A.* **58**, 13972c (1963); Mildner, **FR 1317812** (1963 to Radonja Kemijska Ind.), *C.A.* **59**, 8765h (1963). Toxicity study: S. Dalgaard-Mikkelsen, E. Poulsen, *Pharmacol. Rev.* **14**, 225 (1962). Review of toxicology and human exposure: *Toxicological Profile for Atrazine* (PB2004-100001, 2003) 262 pp.

Crystals, mp 171-174°. Soly at 25° in water 34.5 ppm; ether 12000 ppm; chloroform 52000 ppm; methanol 18000 ppm. Stable in slightly acidic or basic media; hydrolyzed to inactive hydroxy deriv by alkali or mineral acids. LD_{50} orally in mice: 1750 mg/kg (Dalgaard-Mikkelsen, Poulsen).

Caution: Potential symptoms of overexposure are irritation of eyes and skin; dermatitis, skin sensitization; dyspnea, weakness, incoordination, salivation; hypothermia; liver injury. *See NIOSH Pocket Guide to Chemical Hazards* (DHHS/NIOSH 97-140, 1997) p 22.

USE: Selective herbicide.

863. Atrial Natriuretic Peptide. [85637-73-6] Atriopeptin; ANF; ANP; atrial natriuretic factor; cardionatin. Peptide hormone involved in the regulation of salt and fluid balance and blood pressure homeostasis. Endogenous antagonist of the renin angiotensin aldosterone system; physiological effects include vasodilation, natriuresis and diuresis. Produced primarily by the cardiac atrium; also identified in the lung, brain and kidney. Synthesized as a species dependent, 149-153 residue precursor (preproANP) which is cleaved to produce a 126 residue prohormone that is stored in myocyte granules. The major circulating form, known as α-ANP, is a 28 amino acid peptide derived from the carboxy terminus of the prohormone. The amino acid sequence is highly conserved across species. Several bioactive forms of varying length have been identified. Removal of ANP from circulation is effected by clearance receptors and by *neutral endopeptidase*, also known as *atriopeptidase* or *enkephalinase*, that cleaves the disulfide bond required for activity. Demonstration of physiological effect: A. J. de Bold *et al.*, *Life Sci.* **28**, 89 (1981). Purification and amino acid sequence of rat ANP peptides: T. G. Flynn *et al.*, *Biochem. Biophys. Res. Commun.* **117**, 859 (1983); of human: K. Kangawa, H. Matsuo, *ibid.* **118**, 131 (1984). Synthesis of bioactive forms of ANP: N. G. Seidah *et al.*, *Proc. Natl. Acad.*

Sci. USA **81**, 2640 (1984); S. A. Atlas *et al.*, *Nature* **309**, 717 (1984). Series of articles on cloning and sequence analysis: *ibid.* 719-726. Identification of storage form: K. Kangawa *et al.*, *ibid.* **312**, 152 (1984); of major circulating form: A. Miyata *et al.*, *Biochem. Biophys. Res. Commun.* **129**, 248 (1985); D. Schwartz *et al.*, *Science* **229**, 397 (1985). Clinical pharmacology of synthetic peptide: A. M. Richards *et al.*, *Lancet* **1**, 545 (1985). Identification of clearance receptors: T. Maack *et al.*, *Science* **238**, 675 (1987). Role of endopeptidase to inactivate ANP: A. J. Kenny, S. L. Stephenson, *FEBS Lett.* **232**, 1 (1988). Determn in plasma by RIA: L. M. Burrell *et al.*, *J. Immunoassay* **11**, 159 (1990). Review of early literature: A. J. de Bold, *Science* **230**, 767-770 (1985); of physiological effects: P. Needleman, J. E. Greenwald, *N. Engl. J. Med.* **314**, 828-834 (1986); J. D. Baxter *et al.*, *Biotechnology* **6**, 529-546 (1988). Review of biosynthesis, release and metabolism: H. Ruskoaho, *Pharmacol. Rev.* **44**, 479-602 (1992). Review of clinical pharmacology and therapeutic potential: A. Deutsch *et al.*, *J. Clin. Pharmacol.* **34**, 1133-1147 (1994). Role in lung physiology and pathology: T. Perreault, J. Gutkowska, *Am. J. Respir. Crit. Care Med.* **151**, 226-242 (1995).

α-hANP

Carperitide. [89213-87-6] Human atriopeptin (1-28); human α-ANP; α-hANP; SUN-4936; Hanp. $C_{127}H_{203}N_{45}O_{39}S_3$; mol wt 3080.48. Peptide corresponding to the major circulating form of human ANP. Pharmacology: M. Yoshioka *et al.*, *Res. Commun. Chem. Pathol. Pharmacol.* **77**, 55 (1992).

Anaritide. [95896-08-5] Human atriopeptin (4-28); human atriopeptin (102-126); human α-ANP (4-28); Wy-47663; Auriculin. $C_{112}H_{175}N_{39}O_{35}S_3$; mol wt 2724.05. Peptide corresponding to residues 4-28 of human α-ANP. Clinical effects in congestive heart failure: M. A. Fifer *et al.*, *Am. J. Cardiol.* **65**, 211 (1990).

THERAP CAT: In treatment of acute heart failure; antihypertensive.

864. Atrolactic Acid. [515-30-0] α-Hydroxy-α-methylbenzeneacetic acid; α-methylmandelic acid; 2-phenyllactic acid; α-hydroxy-α-phenylpropionic acid; 2-hydroxy-2-phenylpropionic acid; 2-phenyl-2-hydroxypropionic acid. $C_9H_{10}O_3$; mol wt 166.18. C 65.05%, H 6.07%, O 28.88%. Prepd from acetophenone cyanohydrin: McKenzie, Clough, *J. Chem. Soc.* **101**, 393 (1912); Freudenberg *et al.*, *Ann.* **501**, 213 (1932); Eliel, Freeman, *Org. Synth.* **coll. vol. IV**, 58 (1963).

DL-Form, hemihydrate. Orthorhombic crystals from water, mp 88-90° (softening at 75°). Anhydr, mp 94.5° (obtained by drying the hemihydrate at 55° and 0.5 mm Hg). pKa (25°): 3.467. Soly of the anhydr acid in one liter of water at 18° = 17.04 g, at 25° = 21.17 g, at 30° = 25.65 g. Much more sol in boiling water. Slightly sol in petr ether.

865. Atropic Acid. [492-38-6] α-Methylenebenzeneacetic acid; α-phenylacrylic acid. $C_9H_8O_2$; mol wt 148.16. C 72.96%, H 5.44%, O 21.60%. Prepn: Normant, Maitte, *Bull. Soc. Chim. Fr.* **1956**, 1439.

Tabular or acicular crystals; volatilizes in steam. mp 106-107°. bp ~267° with partial decompn. Sol in 790 parts water; sol in alc, benzene, chloroform, ether, CS_2.

Caution: Irritating to skin, mucous membranes.

866. Atropine. [51-55-8] α-(Hydroxymethyl)benzeneacetic acid (3-*endo*)-8-methyl-8-azabicyclo[3.2.1]oct-3-yl ester; 1α*H*,5α*H*-tropan-3α-ol (±)-tropate (ester); *dl*-hyoscyamine; tropic acid ester with tropine; *dl*-tropyl tropate; tropine tropate. $C_{17}H_{23}NO_3$; mol wt 289.38. C 70.56%, H 8.01%, N 4.84%, O 16.59%. Parasympatholytic alkaloid isolated from *Atropa belladonna* L., *Datura stramonium* L., and other *Solanaceae*. Extraction procedure: Chemnitius, *J. Prakt. Chem.* **116**, 276 (1927). During extraction, partial racemization of the *l*-hyoscyamine takes place which is completed by treatment with dil alkali on heating in chloroform soln: Schneider, *Arch. Pharm.* **284**, 306 (1951). Structure and synthesis: Ladenburg, *Ann.* **217**, 75 (1883); Willstätter, *Ber.* **31**, 1537 (1898); *idem., Ann.* **326**, 23 (1903); Schwenker *et al.*, *Ber.* **99**, 2407 (1966). Prepn of the sulfate: DE 247455 (1912 to Hoffmann-La Roche); *Frdl.* **11**, 1022. Use as antidote to cholinesterase inhibitors: R. V. Brown, *Br. J. Pharmacol.* **15**, 170 (1960). Effect on cardiac arrhythmias: P. Schweitzer, H. Mark, *Am. Heart J.* **100**, 119, 255 (1980). Pharmacokinetics and pharmacodynamics: P. H. Hinderling *et al.*, *J. Pharm. Sci.* **74**, 703, 711 (1985). Toxicity: R. L. Cahen, K. Tvede, *J. Pharmacol. Exp. Ther.* **105**, 166 (1952); Goldenthal, *Toxicol. Appl. Pharmacol.* **18**, 185 (1971). Review of clinical use in anesthesia: L. E. Shutt, J. B. Bowes, *Anaesthesia* **34**, 476-490 (1979). Comprehensive description: A. A. Al-Badr, F. J. Muhtadi, *Anal. Profiles Drug Subs.* **14**, 325-389 (1985). Review of clinical toxicology: J. D. Truwit, *Crit. Care Clin.* **7**, 639-657 (1991).

Long, orthorhombic prisms from acetone, mp 114-116°. Sublimes in high vacuum at 93-110°. pK 4.35; pH of 0.0015 molar soln 10.0. Absorption spectra: Dobbie, Fox, *J. Chem. Soc.* **103**, 1194 (1913); Fischer, *Arch. Exp. Pathol. Pharmakol.* **170**, 623 (1933). One gram dissolves in 455 ml water, 90 ml water at 80°, 2 ml alc, 1.2 ml alc at 60°, 27 ml glycerol, 25 ml ether, 1 ml chloroform; also sol in benzene, dil acids. LD_{50} orally in rats: 750 mg/kg (Cahen, Tvede).

Methylbromide. [2870-71-5] Methylatropine bromide; Tropin. $C_{17}H_{23}NO_3 \cdot CH_3Br$; mol wt 384.31. Crystals, mp 222-223°. Sol in 1 part water; slightly sol in alc. Almost insol in chloroform, ether.

Methylnitrate. [52-88-0] Methylatropine nitrate; Eumydrin. $C_{17}H_{23}NO_3 \cdot CH_3NO_3$; mol wt 366.41. Crystals, mp 163°. Freely sol in water or alc; very slightly sol in chloroform, ether.

Sulfate monohydrate. [5908-99-6]; [55-48-1] (anhydrous). Atropisol; Atropt. $(C_{17}H_{23}NO_3)_2 \cdot H_2SO_4 \cdot H_2O$; mol wt 694.84. Granules or powder, mp 190-194°. Almost inactive optically. Very bitter. pH ~5.4. One gram dissolves in 0.4 ml water; 5 ml cold, 2.5 ml boil. alc; in 2.5 ml glycerol, 420 ml chloroform, 3000 ml ether. Bitterness threshold 1:10,000. Incompatible with alkalies, tannin, salts of mercury or gold, vegetable decoctions or infusions, borax, bromides, iodides, benzoates. LD_{50} orally in rats: 622 mg/kg (Goldenthal).

N-Oxide. [4438-22-6] Atropine aminoxide; aminoxytropine tropate; genatropine. $C_{17}H_{23}NO_4$; mol wt 305.37. Prepn: Polonovski, Polonovski, *Bull. Soc. Chim. Fr.* **39**, 1147 (1926). Crystalline powder, mp 127-128°, dec 135°. Very hygroscopic. Soluble in alc, chloroform. Practically insol in ether.

Caution: Potential symptoms of overexposure are blurred vision, suppressed salivation, vasodilation, hyperpyrexia, excitement, agitation, and delirium. *See: Clinical Toxicology of Commercial Products*, R. E. Gosselin *et al.*, Eds. (Williams & Wilkins, Baltimore, 5th Ed., 1984) Section III, pp 47-50.

THERAP CAT: Mydriatic; antispasmodic. In preanesthetic medication.

THERAP CAT (VET): Mydriatic; antispasmodic. Antidote to organophosphorus insecticides.

867. Attacins. Immune protein P5. A major class of nonspecific antibacterial proteins induced in the hemolymph of the pupae of silk moths, particularly *Hyalophora cecropia*, as part of the immune response. (*See also* Cecropins.) Originally designated as immune **protein P5**, attacins are a heterogeneous family consisting of four basic (A,B,C,D) and two acidic (E,F) forms with similar amino acid sequences and mol wts of 20,000-23,000 daltons. Initial identification and isolation of immune response proteins: I. Faye *et al.*, *Infect. Immun.* **12**, 1426 (1975). Purification and preliminary characterization: A. E. Pye, H. G. Boman, *ibid.* **17**, 408 (1977). Amino acid sequences and antibacterial activity: D. Hultmark *et al.*, *EMBO J.* **2**, 571 (1983). Structural study on attacin F: Å. Engström *et al.*, *ibid.* **3**, 2065 (1984). Mode of action: P. Engström *et al.*, *ibid.* 3347.

868.   Aucubin. [479-98-1] (1*S*,4a*R*,5*S*,7a*S*)-1,4a,5,7a-Tetra-hydro-5-hydroxy-7-(hydroxymethyl)cyclopenta[*c*]pyran-1-yl-*β*-D-glucopyranoside; rhinanthin; aucuboside. $C_{15}H_{22}O_9$; mol wt 346.33. C 52.02%, H 6.40%, O 41.58%. From leaves, roots, stalks and seeds of *Aucuba japonica* Thunb., *Cornaceae;* also occurs in 75 different plants: Paris, Chaslot, *Ann. Pharm. Fr.* **13**, 648 (1955). Isoln: Trim, Hill, *Biochem. J.* **50**, 310 (1952); Rombouts, Links, *Experientia* **12**, 78 (1956). Structure: Haegele *et al.*, *Tetrahedron Lett.* **2**, 110 (1961); Birch *et al.*, *J. Chem. Soc.* **1961**, 5194. Abs config: A. Bianco *et al.*, *Gazz. Chim. Ital.* **106**, 725 (1976). Brief review of structural studies: J. M. Bobbitt, K.-P. Segebarth, "The Iridoid Glycosides and Similar Substances" in *Cyclopentanoid Terpene Derivatives*, W. I. Taylor, A. R. Battersby, Eds. (Marcel Dekker, New York, 1969) pp 25-31.

Crystals from ethanol + ether, mp 181°. $[\alpha]_D^{21}$ −163.1° (c = 1.6). Soluble in water, alc, methanol. Practically insol in chloroform, ether, petr ether. Absorption spectrum: Trim, Hill, *loc. cit.*

869.   Auranofin. [34031-32-8] [1-(Thio-*κS*)-*β*-D-glucopy-ranose-2,3,4,6-tetraacetato](triethylphosphine)gold; (2,3,4,6-tetra-*O*-acetyl-1-thio-*β*-D-glucopyranosato-*S*)(triethylphosphine)gold; (1-thio-*β*-D-glucopyranosato)(triethylphosphine)gold 2,3,4,6-tetraace-tate; SKF-39162; Ridaura; Ridauran. $C_{20}H_{34}AuO_9PS$; mol wt 678.48. C 35.41%, H 5.05%, Au 29.03%, O 21.22%, P 4.57%, S 4.73%. Disease modifying antirheumatic drug (DMARD); orally active gold coordination complex. Prepn: E. R. McGusty, B. M. Sutton, **DE 2051495**; *eidem*, **US 3635945** (1971, 1972 both to SKF). Toxicity study: B. H. Payne, D. T. Walz, *Vet. Pathol.* **15**, Suppl. 5, 1 (1978). HPLC determn in urine: R. Kizu *et al.*, *Chem. Pharm. Bull.* **41**, 1261 (1993). Symposium on pharmacology and clinical experience: *J. Rheumatol.* **9**, Suppl. 8, 1-209 (1982). Clinical trial in rheumatoid arthritis: M. Itokazu *et al.*, *Clin. Ther.* **17**, 60 (1995); in steroid-dependent asthma: I. L. Bernstein *et al.*, *J. Allergy Clin. Immunol.* **98**, 317 (1996). *Review:* W. F. Kean *et al.*, *Br. J. Rheumatol.* **36**, 560-572 (1997).

R = COCH₃

White, odorless, crystalline powder, mp 112-115°. Very slightly sol in water; sol in alcohol. *Unstable to light and heat.* LD₅₀ in rats, mice (mg/kg): 265, 310 orally (Payne, Walz).

THERAP CAT: Antirheumatic.

THERAP CAT (VET): Antirheumatic.

870.   Aureothin. [2825-00-5] 2-Methoxy-3,5-dimethyl-6-[(2*R*,4*Z*)-tetrahydro-4-[(2*E*)-2-methyl-3-(4-nitrophenyl)-2-propen-1-ylidene]-2-furanyl]-4*H*-pyran-4-one; mycolutein. $C_{22}H_{23}NO_6$; mol wt 397.43. C 66.49%, H 5.83%, N 3.52%, O 24.15%. Antibi-otic by-product of aureothricin, *q.v.*, isolated from the culture of *Streptomyces thioluteus*. Isoln: K. Maeda, *J. Antibiot.* **6A**, 137

(1953); (as mycolutein): H. Schmitz, R. Woodside, *Antibiot. Chemother.* **5**, 652 (1955). Structure: Y. Hirata *et al.*, *Tetrahedron* **14**, 252 (1961). Identity with mycolutein: J. L. Schwartz *et al.*, *J. Antibiot.* **29**, 236 (1976). Absolute configuration: Y. Ishibashi *et al.*, *Tetrahedron Lett.* **33**, 521 (1992). Biosynthesis: R. Cardillo *et al.*, *Tetrahedron* **30**, 459 (1974); M. Yamazaki *et al.*, *Chem. Pharm. Bull.* **23**, 569 (1975). Synthesis of (±)-form: Y. Shizuri *et al.*, *Chem. Lett.* **1987**, 1381; M. F. Jacobsen *et al.*, *Org. Lett.* **7**, 641 (2005). Chemoenzymatic synthesis of naturally occurring (+)-form: M. Werneburg, C. Hertweck, *ChemBioChem* **9**, 2064 (2008).

Yellow prisms, mp 158°. $[\alpha]_D^{18}$ +51° (chloroform). Sol in meth-anol, ethanol, acetone, chloroform, tetrahydrofuran. Practically insol in water, nonpolar solvents. uv max (ethanol): 257, 346 nm (log ε 4.39, 4.27).

871.   Aureothricin. [574-95-8] *N*-(4,5-Dihydro-4-methyl-5-oxo-1,2-dithiolo[4,3-*b*]pyrrol-6-yl)propanamide; 4-methyl-6-pro-pionamido-1,2-dithiolo[4,3-*b*]pyrrol-5(4*H*)-one; 5-methyl-3-pro-pionamidopyrrolin-4-one[4,3-*d*]-1,2-dithiole; propionopyrrothine. $C_9H_{10}N_2O_2S_2$; mol wt 242.31. C 44.61%, H 4.16%, N 11.56%, O 13.21%, S 26.46%. Antibiotic substance produced by *Streptomyces sp* 26A (Mitaka, Tokyo, June 1947) which resembles *Nocardia far-cinicus* or *Streptomyces lipmanii:* Umezawa *et al.*, *Jpn. Med. J.* **1**, 512 (1948); Umezawa *et al.*, *J. Antibiot.* **2**, Suppl. A, 105 (1949); Maeda, *ibid.* **2**, 793 (1949); Maeda, *Jpn. Med. J.* **2**, 85 (1949); Celmer *et al.*, *J. Am. Chem. Soc.* **74**, 6304 (1952). Structure: Celmer, Solomons, *ibid.* **77**, 2861 (1955). Prepn: Celmer, **US 2752359** (1956 to Pfizer). Total synthesis: Schmidt, Geiger, *Angew. Chem.* **74**, 328 (1962).

Crystals, usually solvated. Stable in air. When anhydr dec 260-270° (sublimes near 200°). uv max: 248, 312, 388 nm (ε 6100, 3900, 11000). Practically insol in water. Slightly sol in ethyl ace-tate, butyl acetate, acetone, benzene, ether, alc.

872.   Aurin. [603-45-2] 4-[Bis(4-hydroxyphenyl)methyl-ene]-2,5-cyclohexadien-1-one; 4-(*p*,*p′*-dihydroxybenzhydrylidene)-2,5-cyclohexadien-1-one; *p*-rosolic acid; corallin; C.I. 43800. $C_{19}H_{14}O_3$; mol wt 290.32. C 78.61%, H 4.86%, O 16.53%. Prepd by heating a mixture of phenol, oxalic and sulfuric acids: Zulkowsky, *Ann.* **194**, 109, 122 (1878); **202**, 179 (1880); *Monatsh. Chem.* **16**, 358 (1895); *Colour Index* vol. 4 (3rd ed., 1971) pp 4407. By heating phenol with formic acid and stannous chloride: Nencki, Schmid, *J. Prakt. Chem.* [2] **23**, 549 (1881). By heating phenol and carbon tetrachloride to 140-160° in presence of $AlCl_3$ or $ZnCl_2$ and treating the reaction product with water: Heumann, **DE 68976**; *Frdl.* **3**, 103.

Deep red (garnet-like) crystals with metallic luster. Orthorhombic bipyramidal. Technical product may occur as yellowish-brown pieces with dark green metallic fracture. Dec 308-310°. Not volatile in vacuo at 180°. Absorption max (KOH): 534.6, 479.5 nm. Practically insol in water (0.12%) and in benzene. Freely sol in alc, giving a golden yellow solution, in aq or alc NaOH and KOH giving a carmine-red solution, in concd or 70% H_2SO_4, in HCl, in $HClO_4$ giving a yellow to orange soln. Moderately soluble in glacial acetic acid; slightly sol in ether, chloroform.

Sodium salt (yellow corallin). Yellowish pieces with greenish, metallic luster; sol in water, giving a carmine-red soln.

USE: Dye intermediate.

873. Aurodox. [12704-90-4] ($\alpha S,2R,3R,4R,6S$)-N-[(2E,4E,-6S,7R)-7-[(2S,3S,4R,5R)-5-[(1E,3E,5E)-7-(1,2-Dihydroxy-4-hydroxy-1-methyl-2-oxo-3-pyridinyl)-6-methyl-7-oxo-1,3,5-heptatrien-1-yl]tetrahydro-3,4-dihydroxy-2-furanyl]-6-methoxy-5-methyl-2,4-octadien-1-yl]-α-ethyltetrahydro-2,3,4-trihydroxy-5,5-dimethyl-6-(1E,3Z)-1,3-pentadien-1-yl-2H-pyran-2-acetamide; 1-methylmocimycin; antibiotic X-5108; goldinodox; goldinomycin; X-5108. $C_{44}H_{62}N_2O_{12}$; mol wt 810.98. C 65.17%, H 7.71%, N 3.45%, O 23.67%. Antibiotic produced by *Streptomyces goldiniensis* var. *goldiniensis:* J. Berger, **DE 2140322**; *idem*, **US 3708577** (1972, 1973 both to Hoffmann-La Roche). Isoln, production, properties and toxicity data: J. Berger *et al.*, *J. Antibiot.* **26**, 15 (1973). Improved production: J. Unowsky, D. C. Hoppe, *ibid.* **31**, 662 (1978). Structure: H. Maehr *et al.*, *J. Am. Chem. Soc.* **95**, 8449 (1973). Absolute stereochemistry: *eidem, ibid.* **96**, 4034 (1974). Biosynthesis: C.-M. Liu *et al.*, *J. Antibiot.* **30**, 416 (1977); **32**, 414 (1979). Stereospecific total synthesis: R. E. Dolle, K. C. Nicolaou, *J. Am. Chem. Soc.* **107**, 1691, 1695 (1985). Growth promoting activity: W. L. Marusich *et al.*, *Poult. Sci.* **53**, 636 (1974). Review of chemistry: H. Maehr *et al.*, *Can. J. Chem.* **58**, 501 (1980).

Yellow amorphous solid. Weakly acidic. Sol in methanol, ethyl acetate, $CHCl_3$, acetone, methylene chloride. Insol in water. pKa = 6.1. Unstable in acidic or basic soln, but can be kept in aq soln at pH 7-9 for 4 hrs at 25° without significant loss of activity. Solns kept at 25° or heated to 56° are 2-10 times more stable in methanol than in water or pH 9 borate buffer. LD_{50} in mice (mg/kg): >1000 s.c., >4000 orally (Berger *et al.*).

Sodium salt. [74219-54-8] $C_{44}H_{61}N_2NaO_{12}$. Yellow solid. $[\alpha]_D^{25}$ −82.8° (c = 0.52 in ethanol). uv max (0.1N HCl): 334, 233, 206 nm ($E_{1cm}^{1\%}$ 403, 610, 500); (0.1N KOH): 327, 231 nm ($E_{1cm}^{1\%}$ 416, 647). Sol in water, methanol, ethanol, isopropanol, butanol, DMF. Slightly sol in amyl alcohol, THF, dioxane. Insol in benzene, chloroform, ethyl ether, petr ether.

874. Aurothioglucose. [12192-57-3] [1-(Thio-κS)-D-glucopyranosato-κO^2]gold; (1-D-glucosylthio)gold; gold thioglucose; Aureotan; Solganal. $C_6H_{11}AuO_5S$; mol wt 392.18. C 18.38%, H 2.83%, Au 50.22%, O 20.40%, S 8.17%. Disease modifying antirheumatic drug (DMARD). Prepn: Lebeau-Janot, *Traité Pharm. Chim.* **II**, 661 (1956). Induction of obesity in exptl animals: A. F. Debons *et al.*, *Fed. Proc.* **36**, 143 (1977). Clinical comparison with auranofin, *q.v.*, in rheumatoid arthritis: P. L. C. M. Van Riel *et al.*, *Clin. Rheumatol.* **5**, 359 (1986); P. E. Prete *et al.*, *Clin. Rheumatol.* **13**, 60 (1994).

Yellow crystals. Slight mercaptan-like odor. Sol in water with decompn; slightly sol in propylene glycol. Practically insol in acetone, alcohol, chloroform, ether.

USE: To produce obesity in exptl animals.

THERAP CAT: Antirheumatic.

875. Avanafil. [330784-47-9] 4-[[(3-Chloro-4-methoxyphenyl)methyl]amino]-2-[(2S)-2-(hydroxymethyl)-1-pyrrolidinyl]-N-(2-pyrimidinylmethyl)-5-pyrimidinecarboxamide; (S)-2-(2-hydroxymethyl-1-pyrrolidinyl)-4-(3-chloro-4-methoxybenzylamino)-5-[N-(2-pyrimidinylmethyl)carbamoyl]pyrimidine; TA -1790. $C_{23}H_{26}ClN_7O_3$; mol wt 483.96. C 57.08%, H 5.42%, Cl 7.32%, N 20.26%, O 9.92%. Phosphodiesterase 5 (PDE5) inhibitor. Prepn: K. Yamada *et al.*, **WO 0119802**; *eidem*, **US 6656935** (2001, 2003 both to Tanabe Seiyaku). Clinical pharmacokinetics and tolerability: J. Jung *et al.*, *Clin. Ther.* **32**, 1178 (2010). Review of development and clinical trials in erectile dysfunction: M. Limin *et al.*, *Expert Opin. Invest. Drugs* **19**, 1427-1437 (2010).

Solid, mp 160-163°. Sol in methanol. Minimally sol in water.

Dibenzenesulfonate. [330784-48-0] $C_{23}H_{26}ClN_7O_3 \cdot 2C_6H_6O_3S$; mol wt 800.30. Colorless crystals from methanol + acetone, mp 158.5-161.5°.

THERAP CAT: In treatment of male erectile dysfunction.

876. Avasimibe. [166518-60-1] N-[2-[2,4,6-Tris(1-methylethyl)phenyl]acetyl]sulfamic acid 2,6-bis(1-methylethyl)phenyl ester; 2,6-diisopropylphenyl[(2,4,6-triisopropylphenyl)acetyl]sulfamate; CI-1011; PD-148515. $C_{29}H_{43}NO_4S$; mol wt 501.73. C 69.42%, H 8.64%, N 2.79%, O 12.76%, S 6.39%. Acyl-coenzyme A:cholesterol O-acyltransferase (ACAT) inhibitor. Prepn: H. T. Lee *et al.*, **WO 9426702**; *eidem*, **US 5491172** (1994, 1996 both to Warner Lambert). Scale up synthesis: G. J. Dozeman *et al*, *Org. Process Res. Dev.* **1**, 137 (1997). ACAT inhibition studies: H. T. Lee *et al.*, *J. Med. Chem.* **39**, 5031 (1996). HPLC/MS/MS determn in plasma: W. W. Bullen *et al.*, *J. Pharm. Biomed. Anal.* **17**, 1399 (1998). Clinical evaluation in hyperlipidemia: W. Insull Jr. *et al.*, *Atherosclerosis* **157**, 137 (2001). Toxicology: D. G. Robertson *et al.*, *Toxicol. Sci.* **59**, 324 (2001). Reviews of development and therapeutic potential: R. Jones, *Curr. Opin. Cardiovasc. Pulm. Renal Invest. Drugs* **1**, 263-269 (1999); *idem, ibid.* **2**, 274-278 (2000).

White solid, mp 178-180° (Lee). Also reported as white powder, mp 169.5-170.4° (Dozeman).

Sodium salt. [166518-61-2] White solid, mp 250-252°.

THERAP CAT: Antilipemic.

877. Avermectins. AVM. A group of broad-spectrum antiparasitic antibiotics which are derivs of pentacyclic 16-membered lactones related to the milbemycins, *q.v.* Isoln from a novel actinomycete, *Streptomyces avermitilis*, and separation of major components A_{1a}, A_{2a}, B_{1a}, B_{2a} and minor components A_{1b}, A_{2b}, B_{1b}, B_{2b}: G. Albers-Schönberg *et al.*, **DE 2717040**; *eidem*, **US 4310519** (1977, 1982 both to Merck & Co.); R. W. Burg *et al.*, *Antimicrob. Agents Chemother.* **15**, 361 (1979); T. W. Miller *et al.*, *ibid.* 368. Antiparasitic activity: S. R. Egerton *et al.*, *ibid.* 372. Mechanism of action: L. C. Fritz *et al.*, *Proc. Natl. Acad. Sci. USA* **76**, 2062 (1979). Pesticidal activity: I. Putter *et al.*, *Experientia* **37**, 963 (1981). Structure determn: G. Albers-Schönberg *et al.*, *J. Am. Chem. Soc.* **103**, 4216 (1981). Absolute configuration of avermectin B_{1a} and aglycon B_{2a}: J. P. Springer *et al.*, *ibid.* 4221. Prepn of aglycons: H. Mrozik *et al.*, *J. Org. Chem.* **47**, 489 (1982). Total synthesis of A_{1a} ($C_{49}H_{74}O_{14}$): S. J. Danishefsky *et al.*, *J. Am. Chem. Soc.* **109**, 8119 (1987). Biosynthetic studies: D. E. Cane *et al.*, *ibid.* **105**, 4110 (1983); M. D. Schulman *et al.*, *J. Antibiot.* **39**, 541 (1986); T. S. Chen *et al.*, *Arch. Biochem. Biophys.* **269**, 544 (1989). *Reviews:* W. C. Campbell *et al.*, *ACS Symp. Ser.* **255**, 1-20 (1984); M. H. Fisher, H. Mrozik, "The Avermectin Family of Macrolide Antibiotics" in *Macrolide Antibiotics*, S. Omura, Ed. (Academic Press, New York, 1984) pp 553-606; *Southwest. Entomol.* **1985**, Suppl. 7, 1-51; J. R. Babu, *ACS Symp. Ser.* **380**, 91-108 (1988). Review of syntheses: T. Blizzard *et al.*, in *Recent Prog. Synth. Antibiotics*, G. Lukacs, M. Ohno, Eds (Springer-Verlag, Berlin, 1990) pp 66-102. Review of structure-activity relationships: W. L. Shoop *et al.*, *Vet. Parasitol.* **59**, 139-156 (1995). For related structures *see* abamectin, ivermectin.

Avermectin A_{1a}

Avermectin $A_{1a/b}$. $[\alpha]_D^{27}$ +68.5 ± 2° (c = 0.77 in chloroform). uv max (methanol): 237, 243, 252 nm (ε 28700, 31275, 20290).

Avermectin $A_{2a/b}$. $[\alpha]_D^{27}$ +48.8 ± 2° (c = 1.64 in chloroform). uv max (methanol): 237, 243, 245 nm (ε 28800, 31740, 20425).

Avermectin $B_{1a/b}$ *see* Abamectin.

Avermectin $B_{2a/b}$. $[\alpha]_D^{27}$ +38.3 ± 2° (c = 0.87 in chloroform). uv max (methanol): 237, 243, 252 nm (ε 27580, 30590, 20060).

878.   Avibactam. [1192500-31-4] Sulfuric acid mono[(1R,-2S,5R)-2-(aminocarbonyl)-7-oxo-1,6-diazabicyclo[3.2.1]oct-6-yl] ester; (1R,2S,5R)-7-oxo-6-sulfoxy-1,6-diazabicyclo[3.2.1]octane-2-carboxamide. $C_7H_{11}N_3O_6S$; mol wt 265.24. C 31.70%, H 4.18%, N 15.84%, O 36.19%, S 12.09%. Non-β-lactam inhibitor of class A and class C beta-lactamase enzymes. Prepn of racemic *trans*-form: M. Lampilas *et al.*, **WO 0210172**; *eidem*, **US 7112592** (2002, 2006 both to Aventis). Prepn of active enantiomer as crystalline sodium salt: S. Bhattacharya *et al.*, **WO 11042560** (2011 to Novexel; Forest). β-lactamase inhibition and *in vitro* activity in combination with ceftazidime: A. Bonnefoy *et al.*, *J. Antimicrob. Chemother.* **54**, 410 (2004). Mechanistic studies: T. Stachyra *et al.*, *Antimicrob. Agents Chemother.* **54**, 5132 (2010). Activity *vs* carbapenemases: D. M. Livermore *et al.*, *ibid.* **55**, 390 (2011); *vs* clinical isolates of *Pseudomonas aeroginosa*: A. Walkty *et al.*, *ibid.* 2992.

Sodium salt. [1192491-61-4] AVE-1330A; NXL-104. C_7H_{10}-N_3NaO_6S; mol wt 287.22. Crystals from aq ethanol as the monohydrate.

THERAP CAT: Antibacterial adjunct (β-lactamase inhibitor).

879.   Avidin. Basic glycoprotein isolated from raw egg white. Exhibits high binding affinity for biotin and is capable of producing biotin deficiency in rats and chicks. Occurs in the white portion of

eggs and the oviducts of birds and amphibia. Destroyed by cooking or irradiation. Isoln: Eakin *et al.*, *J. Biol. Chem.* **136**, 801 (1940); Pennington *et al.*, *J. Am. Chem. Soc.* **64**, 469 (1942); Fraenkel-Conrat *et al.*, *Arch. Biochem. Biophys.* **39**, 80, 97 (1952). Improved purification and crystallization: Green *et al.*, *Biochem. J.* **118**, 67, 71 (1970). Structure is a glycoprotein containing four essentially identical subunits: Green, *ibid.* **92**, 16c (1964). The combined mol wt of the subunits is about 66,000. Each subunit is a single polypeptide chain containing 128 amino acid residues with alanine at the *N*-terminal, glutamic acid at the *C*-terminal, and a carbohydrate moiety attached at the asparaginyl residue, position 17. Complete amino acid sequence of the protein subunit: DeLange, Huang, *J. Biol. Chem.* **246**, 698 (1971). Studies on biotin inactivation by avidin: Becker, Wilchek, *Biochim. Biophys. Acta* **264**, 165 (1972). Review of chemistry and binding properties: N. M. Green, *Adv. Protein Chem.* **29**, 85-133 (1975). Review of use of avidin-biotin complex in molecular biology: E. A. Bayer, M. Wilchek, *Methods Biochem. Anal.* **26**, 1-45 (1980); in immunoassays: M. Wilchek, E. A. Bayer, *Immunol. Today* **5**, 39-43 (1984).

USE: Biochemical tool for affinity chromatography, affinity cytochemistry and immunoassay.

880.   Avilamycin. [11051-71-1] LY-048740; Maxus; Surmax. Polyether antibiotic complex produced by *Streptomyces viridochromogenes*, ETH 23575 (NRRL 2860); belongs to the orthosomycin family of antibiotics, members of which contain one or more ortho ester linkages associated with carbohydrate residues. Isoln: E. Gaeumann *et al.*, **DE 1116864**; *eidem*, **US 3131126** (1961, 1964 both to Ciba-Geigy); and characterization: F. Buzzetti *et al.*, *Experientia* **24**, 320 (1968). Mechanism of action studies: H. Wolf, *FEBS Lett.* **36**, 181 (1973). Avilamycin A is the main component; avilamycins B through N are also known and are derivatives of A differing at the C-45 linkage and/or the C-56 ketone adduct. Isoln of avilamycin C: W. Heilman *et al.*, *Helv. Chim. Acta* **62**, 1 (1979). Structural study of A and C: *eidem*, *ibid.*; W. Keller-Schierlein *et al.*, *ibid.* 7; E. Kupfer *et al.*, *ibid.* **65**, 3 (1982). Isoln and structure of avilamycins F through N: J. L. Mertz *et al.*, *J. Antibiot.* **39**, 877 (1986). GC determn of avilamycin residues in swine tissue and fluids: G. Formica, C. Giannone, *J. Assoc. Off. Anal. Chem.* **69**, 763 (1986). Synthesis of the A-B fragment of avilamycins A and C: P. Jütten *et al.*, *Tetrahedron* **43**, 4133 (1987); of the disaccharide C-D fragment: J.-M. Beau *et al.*, *Tetrahedron Lett.* **28**, 1105 (1987). Use as a feed additive: F. Knusel *et al.*, **US 4185091** (1980 to Ciba-Geigy); in prevention of swine dysentery: E. E. Ose, **US 4436734** (1984 to Lilly). Effect on feed conversion efficiency in swine: D. J. Jones *et al.*, *J. Anim. Sci.* **65**, 881 (1987). Metabolism in swine: J. D. Magnussen *et al.*, *J. Agric. Food Chem.* **39**, 306 (1991).

Avilamycin A

Colorless, needle-shaped crystals from acetone/ether, mp 188-189.5°. $[\alpha]_D^{20}$ +0.8° (c = 1.165 in abs ethanol) and −7.7° (c = 1.083 in chloroform).

Avilamycin A. [69787-79-7] $C_{61}H_{88}Cl_2O_{32}$; mol wt 1404.24. Colorless needles from chloroform/petr ether, mp 181-182° (1-2.-H_2O). uv max (methanol): 227, 286 nm (log ε 4.15, 3.33).

Avilamycin C. [69787-80-0] $C_{61}H_{90}Cl_2O_{32}$; mol wt 1406.26. Dihydrate, colorless fine plates from acetone/ether, mp 188-189°. $[\alpha]_D^{20}$ −4.8° (c = 1.44 in chloroform). uv max (methanol): 228, 284 nm (log ε 4.12, 3.33).

THERAP CAT (VET): Growth promotant.

881. Avobenzone. [70356-09-1] 1-[4-(1,1-Dimethylethyl)-phenyl]-3-(4-methoxyphenyl)-1,3-propanedione; butyl methoxydibenzoylmethane; 4-*tert*-butyl-4′-methoxydibenzoylmethane; Eusolex 9020; Neo Heliopan 357; Parsol 1789. $C_{20}H_{22}O_3$; mol wt 310.39. C 77.39%, H 7.14%, O 15.46%. UV-A blocker. Prepn: K.-F. De Polo, **DE 2945125**; *idem*, **US 4387089** (1980, 1983 both to Givaudan). Clinical efficacy as sunscreen: R. W. Gange *et al.*, *J. Am. Acad. Dermatol.* **15**, 494 (1986); K. Kaidbey, R. W. Gange, *ibid.* **16**, 346 (1987); N. J. Lowe *et al.*, *ibid.* **17**, 224 (1987). Assessment of photostability: A. Deflandre, G. Lang, *Int. J. Cosmet. Sci.* **10**, 53 (1988). Photodecomposition: N. M. Roscher *et al.*, *J. Photochem. Photobiol. A* **80**, 417 (1994). Photochemistry in solution: W. Schwack, T. Rudolph, *Photochem. Photobiol. B* **28**, 229 (1995). HPLC determn in cosmetic products: L. Gagliardi *et al.*, *J. Chromatogr.* **408**, 409 (1987); P. Wallner, *Dtsch. Lebensm. Rundsch.* **89**, 375 (1993). Effect of reflectance of light on protection of skin: G. J. Smith *et al.*, *Photochem. Photobiol.* **75**, 122 (2002).

Crystals from methanol, mp 83.5°.

THERAP CAT: Ultraviolet screen.

882. Avoparcin. [37332-99-3] AV-290; C-254; CL-81588; LL-AV290; Avotan. Veterinary, glycopeptide antibiotic complex produced by *Streptomyces candidus*. Prodn: M. P. Kunstmann, J. N. Porter, **US 3338786** (1967 to Am. Cyanamid). Fermentation, isoln, characterization: M. P. Kunstmann *et al.*, *Antimicrob. Agents Chemother.* **8**, 242 (1968). Isoln, purif of major components: W. J. McGahren *et al.*, *J. Antibiot.* **36**, 1671 (1983). HPLC separation of α-, β-avoparcin: F. Sztaricskai *et al.*, *ibid.* 1691. Chromatographic determn in tissue: M. S. S. Curren, J. W. King, *J. Chromatogr. A* **954** 41 (2002). Structural studies: J. J. Hlavka *et al.*, *Tetrahedron Lett.* **1974**, 175; W. J. McGahren *et al.*, *J. Am. Chem. Soc.* **101**, 2237 (1979). Structure: *eidem*, *ibid.* **102**, 1671 (1980). Stereochemistry and epimerization: G. A. Ellestad *et al.*, *J. Antibiot.* **36**, 1683 (1983). Relationship between structure, antibacterial activity: S. W. Fesik *et al.*, *Mol. Pharmacol.* **25**, 275 (1984). Antimicrobial spectrum: M. G. Cormican *et al.*, *Diagn. Microbiol. Infect. Dis.* **29**, 241 (1997). Effect on resistance patterns in chickens: J. R. Walton, *Zentralbl. Veterinaermed. (B)* **25**, 290 (1978), *C.A.* **89**, 173507 (1978). Prevention of exptly induced necrotic enteritis in chickens: J. F. Prescott, *Avian Dis.* **23**, 1072 (1979). Effect on feedlot performance: R. J. Johnson *et al.*, *J. Anim. Sci.* **48**, 1338 (1979).

α-Avoparcin R = H
β-Avoparcin R = Cl

White, hygroscopic, amorphous solid; no definitive mp. uv max: 280 nm in neutral or acidic solns; 300 nm in basic solns. Sol in water, DMF, DMSO. Max stability of aq solns is at pH 4-8. Moderately sol in methanol. LD_{50} in mice, rats, and chickens: >10000 mg/kg orally (Am. Cyanamid, company literature).

α-Avoparcin. [73957-86-5] $C_{89}H_{102}ClN_9O_{36}$. $[\alpha]_D^{25}$ −96 ±2° (c = 0.62 in 0.1N HCl). uv max (0.1N HCl): 280 nm ($E_{1cm}^{1\%}$ 42.0).

β-Avoparcin. [73957-87-6] $C_{89}H_{101}Cl_2N_9O_{36}$. $[\alpha]_D^{25}$ −102 ±2° (c = 0.65 in 0.1N HCl). uv max (0.1N HCl): 280 nm ($E_{1cm}^{1\%}$ 44.0).

THERAP CAT (VET): Antibacterial; growth promotant.

883. Axitinib. [319460-85-0] *N*-Methyl-2-[[3-[(1*E*)-2-(2-pyridinyl)ethenyl]-1*H*-indazol-6-yl]thio]benzamide; AG-013736. $C_{22}H_{18}N_4OS$; mol wt 386.47. C 68.37%, H 4.69%, N 14.50%, O 4.14%, S 8.30%. Receptor tyrosine kinase inhibitor of VEGFR, PDGFR, and c-KIT; induces anti-angiogenic response. Prepn: R. S. Kania *et al.*, **WO 0102369**; *eidem*, **US 6534524** (2001, 2003 both to Agouron). Anti-angiogenic activity in breast cancer xenografts: L. J. Wilmes *et al.*, *Magn. Reson. Imaging* **25**, 319 (2007). *In vivo* effect in combination with cyclophosphamide: J. Ma, D. J. Waxman, *Mol. Cancer Ther.* **7**, 79 (2008). Clinical pharmacokinetics in advanced solid tumors: H. S. Rugo *et al.*, *J. Clin. Oncol.* **23**, 5474 (2005). Clinical evaluation in renal cell carcinoma: O. Rixe *et al.*, *Lancet Oncol.* **8**, 975 (2007). Review of clinical experience in renal cell carcinoma: D. J. George, *Clin. Cancer Res.* **13**, Suppl. 2, 753s-757s (2007); G. Sonpavde *et al.*, *Expert Opin. Invest. Drugs* **17**, 741-748 (2008).

Sol in aq methanol.
THERAP CAT: Antineoplastic.

884. Azacitidine. [320-67-2] 4-Amino-1-β-D-ribofuranosyl-1,3,5-triazin-2(1*H*)-one; 5-azacytidine; 5-AzaC; ladakamycin; U-18496; NSC-102816; Mylosar (formerly); Vidaza. $C_8H_{12}N_4O_5$; mol wt 244.21. C 39.35%, H 4.95%, N 22.94%, O 32.76%. DNA methylation inhibitor; analog of the pyrimidine nucleoside, cytidine, *q.v.* Chemical synthesis: A. Piskala, F. Sorm, *Collect. Czech. Chem. Commun.* **29**, 2060 (1964); M. W. Winkley, R. K. Robins, *J. Org. Chem.* **35**, 491 (1970). Production by fermentation of *Streptoverticillium ladakanus* and activity: L. J. Hanka *et al.*, *Antimicrob. Agents Chemother.* **1966**, 619; M. E. Bergy, R. R. Herr, *ibid.* 625. HPLC determn in pharmaceutical prepns: L. D. Kissinger, N. L. Stemm, *J. Chromatogr.* **353**, 309 (1986). Toxicology study: P. E. Palm, C. J. Kensler, *U.S. Clearinghouse Fed. Sci. Tech. Inform.* (PB-194791, 1970) 191 pp. Review of clinical experience in acute nonlymphocytic leukemia: A. B. Glover *et al.*, *Cancer Treat. Rep.* **71**, 737-746 (1987); of mechanism of action: A. B. Glover, B. Leyland-Jones, *ibid.* 959-964. Review of carcinogenic risk: *IARC Monographs* **50**, 47-63 (1990). Clinical efficacy in β-thalassemia: C. H. Lowrey, A. W. Nienhuis, *N. Engl. J. Med.* **329**, 845 (1993); in myelodysplastic syndrome: L. R. Silverman *et al.*, *J. Clin. Oncol.* **20**, 2429 (2002); A. B. Kornblith *et al.*, *ibid.* 2441.

Crystals from aq ethanol, mp 235-237° (dec). $[\alpha]_D^{26}$ +22.4° (c = 1 in water). uv max (water): 241 nm (ε 8767); (0.01N HCl): 249 nm (ε 3077); (0.01N KOH): 223 nm (ε 24200). Soly (mg/ml): 40 warm water, 14 cold water, 28 0.1N HCl, 43 0.1N NaOH, 52.7 DMSO, 1 acetone, 1 chloroform, 1 hexane. LD_{50} in mice (mg/kg): 115.9 i.p.; 572.3 orally (Palm, Kensler).

Caution: This substance is reasonably anticipated to be a human carcinogen: *Report on Carcinogens, Twelfth Edition* (PB2011-111646, 2011) p 56.

THERAP CAT: Antineoplastic.

885. Azaconazole. [60207-31-0] 1-[[2-(2,4-Dichlorophenyl)-1,3-dioxolan-2-yl]methyl]-1H-1,2,4-triazole; R-28644; Rodewod; Safetray. $C_{12}H_{11}Cl_2N_3O_2$; mol wt 300.14. C 48.02%, H 3.69%, Cl 23.62%, N 14.00%, O 10.66%. Steroid demethylation inhibitor. Prepn: G. Van Reet *et al.*, **DE 2551660**; *eidem*, **US 4079062** (1975, 1978 both to Janssen). Wood fungicide for mushroom cultivation: L. Van Leemput *et al.*, *Meded. Fac. Landbouwwet. Rijksuniv. Gent* **52**, 703 (1987); A. Eicker, E. Strydom, *Bot. Bull. Acad. Sinica* **31**, 51 (1990). Use as a preservative for composite wood products: E. L. Schmidt, R. O. Gertjejansen, *Forest Prod. J.* **38**, 19 (1988); A. Jeihooni *et al.*, *Wood Fiber Sci.* **26**, 178 (1994).

Crystals from diisopropylether. mp 109.9°.

USE: Fungicide for cultivation on wood; preservative for composite wood products.

886. Azadirachtin. [11141-17-6] (2aR,3S,4S,4aR,5S,7aS,-8S,10R,10aS,10bR)-10-(Acetyloxy)octahydro-3,5-dihydroxy-4-methyl-8-[[(2E)-2-methyl-1-oxo-2-buten-1-yl]oxy]-4-[(1aR,2S,-3aS,6aS,7S,7aS)-3a,6a,7,7a-tetrahydro-6a-hydroxy-7a-methyl-2,7-methanofuro[2,3-b]oxireno[e]oxepin-1a(2H)-yl]-1H,7H-naphtho-[1,8-bc:4,4a-c']difuran-5,10a(8H)-dicarboxylic acid 5,10a-dimethyl ester. $C_{35}H_{44}O_{16}$; mol wt 720.72. C 58.33%, H 6.15%, O 35.52%. A tetranortriterpenoid isolated from the seeds of the neem tree, *Azadirachta indica* A. Juss. (*Melia azadirachta* L.), *Meliaceae*, and the chinaberry tree, *M. azedarach* L. Highly active insect feeding deterrent and growth regulator. Isoln from *A. indica* and identification as feeding inhibitor in locusts: J. H. Butterworth, E. D. Morgan, *Chem. Commun.* **1968**, 23; from *M. azedarach*: E. D. Morgan, M. D. Thornton, *Phytochemistry* **12**, 391 (1973). Total synthesis: G. E. Veitch *et al.*, *Angew. Chem. Int. Ed.* **46**, 7629 (2007). Structural studies: J. H. Butterworth *et al.*, *J. Chem. Soc. Perkin Trans. 2* **1972**, 2445. ^{1}H- and ^{13}C-NMR data and structure: P. R. Zanno *et al.*, *J. Am. Chem. Soc.* **97**, 1975 (1975); K. Nakanishi in *Recent Advances in Phytochemistry* **vol. 9**, V. C. Runeckles, Ed. (Plenum Press, New York, 1975) pp 283-298. Revised structure: W. Kraus *et al.*, *Tetrahedron Lett.* **26**, 6435 (1985); H. B. Broughton *et al.*, *Chem. Commun.* **1986**, 46. Isoln by HPLC: E. C. Uebel *et al.*, *J. Liq. Chromatogr.* **2**, 875 (1979); J. D. Warthen, Jr. *et al.*, *ibid.* **7**, 591 (1984). Antifeedant activity in locusts: J. S. Gill, C. T. Lewis, *Nature* **232**, 402 (1971); in fall army worms, cotton bollworms: J. A. Klocke, I. Kubo, *Entomol. Exp. Appl.* **32**, 299 (1982). Insect ecdysis and growth inhibition: H. Rembold, K. P. Sieber, *Z. Naturforsch.* **36C**, 466 (1981); I. Kubo, J. A. Klocke, *Agric. Biol. Chem.* **46**, 1951 (1982); K. P. Sieber, H. Rembold, *J. Insect Physiol.* **29**, 523 (1983). Series of articles on chemistry and activity: *Natural Pesticides from the Neem Tree*, Proc. 1st Int. Neem Conf., 1980, H. Schmutterer *et al.*, Eds. (German Agency for Technical Cooperation, Eschborn, 1981) 291 pp.

Microcrystalline powder from carbon tetrachloride, mp 154-158°. $[\alpha]_D$ −53° (c = 0.5 in CHCl$_3$). uv max (methanol): 217 nm (ε 9100).

USE: Experimentally as insect control agent.

887. Azafenidin. [68049-83-2] 2-[2,4-Dichloro-5-(2-propyn-1-yloxy)phenyl]-5,6,7,8-tetrahydro-1,2,4-triazolo[4,3-a]pyridin-3(2H)-one; DPX-R6447; Milestone; Evolus. $C_{15}H_{13}Cl_2N_3O_2$; mol wt 338.19. C 53.27%, H 3.87%, Cl 20.96%, N 12.43%, O 9.46%. Porphyrin biosynthesis inhibitor. Prepn: A. D. Wolf, **BE 862884**; *idem*, **US 4213773** (1978, 1980 both to Du Pont). Review of physical properties, mode of action, and activity: K. Amuti *et al.*, *Brighton Crop Prot. Conf. - Weeds* **1997**, 59-66.

White powdered solid, mp 168-168.5°. Vapor pressure at 20°: 1.0 × 10^{-11} torr. Partition coefficient (octanol/water): 229. LD_{50} in rats, mice, bobwhite quail, mallard duck (mg/kg): >5000, >5000, >2500, >2500 orally; in rabbits (mg/kg): >2000 dermally. LC_{50} in rats, rainbow trout, blue gill sunfish (mg/l): >5.3, 33, 48 (Amuti).

USE: Herbicide.

888. Azafrin. [507-61-9] (2E,4E,6E,8E,10E,12E,14E)-15-[(1R,2R)-1,2-Dihydroxy-2,6,6-trimethylcyclohexyl]-4,9,13-trimethyl-2,4,6,8,10,12,14-pentadecaheptaenoic acid; (5R,6R)-5,6-dihydro-5,6-dihydroxy-10′-*apo*-β,ψ-carotenoic acid; escobedin. $C_{27}H_{38}O_4$; mol wt 426.60. C 76.02%, H 8.98%, O 15.00%. Carotenoid-carboxylic acid from roots of the South American plant "Azafranillo," *Escobedia scabrifolia* Ruiz & Pav., and *Escobedia laevis* Cham. & Schlecht., *Scrophulariaceae*. Isoln: R. Kuhn *et al.*, *Ber.* **64**, 333 (1931); *ibid.* **65**, 1873 (1932). Structure: R. Kuhn, A. Deutsch, *ibid.* **66**, 883 (1933); R. Kuhn, H. Brockmann, *ibid.* **67**, 885 (1934); *Ann.* **516**, 104 (1935). Absolute configuration (5R:6R): W. Eschenmoser, C. H. Eugster, *Helv. Chim. Acta* **58**, 1722 (1975).

Orange-colored prisms from toluene, mp 213°. $[\alpha]_{6438}^{20}$ −75° (c = 0.28 in alcohol). Absorption max (chloroform): 458, 428 nm. Practically insol in water. Sol in dil NaOH or Na$_2$CO$_3$ solns, in chloroform, alc, acetic acid and benzene; sparingly sol in ether.

Methyl ester. $C_{28}H_{40}O_4$. Reddish-yellow leaflets from ether, mp 191°. $[\alpha]_{6438}^{22}$ −32° (chloroform).

889. 8-Azaguanine. [134-58-7] 5-Amino-3,6-dihydro-7H-1,2,3-triazolo[4,5-d]pyrimidin-7-one; 5-amino-1,6-dihydro-7H-v-triazolo[4,5-d]pyrimidin-7-one; 5-amino-1H-v-triazolo[d]pyrimidin-7-ol; 5-amino-7-hydroxy-1H-v-triazolo[d]pyrimidine; pathocidin; guanazolo. $C_4H_4N_6O$; mol wt 152.12. C 31.58%, H 2.65%, N 55.25%, O 10.52%. Triazolo analog of guanine. Prepd from 2,4,5-triamino-6-hydroxypyrimidine: Roblin, Jr. *et al.*, *J. Am. Chem. Soc.*

67, 290 (1945); from 2-amino-5-nitro-4-oxopyrimidine: H. U. Blank *et al.*, *J. Org. Chem.* **35**, 1131 (1970). Formation from guanine by *Streptomyces albus* and identity with pathocidin: K. Hirasawa, K. Isono, *J. Antibiot.* **31A**, 628 (1978). The first purine analog to show carcinostatic effects in murine malignancies; is readily incorporated into ribonucleic acids. *Review:* R. E. Parks, Jr., K. C. Agarwal, *Handb. Exp. Pharmacol.* **38**, pt. 2, 458 (1975); D. Grunberger, G. Grunberger in *Antibiotics* vol. 5(pt. 2), F. E. Hahn, Ed. (Springer-Verlag, New York, 1979) pp 110-123.

Crystals from dil aq NaOH. Dec above 300° without melting. Absorption spectrum: L. F. Cavalieri *et al.*, *J. Am. Chem. Soc.* **70**, 3875 (1948). Soluble in dil caustic, in dil acids. Practically insol in water, alcohol, ether.

USE: Purine antimetabolite.

890. Azanidazole. [62973-76-6] 4-[(1*E*)-2-(1-Methyl-5-nitro-1*H*-imidazol-2-yl)ethenyl]-2-pyrimidinamine; (*E*)-2-amino-4-[2-(1-methyl-5-nitroimidazol-2-yl)vinyl]pyrimidine; nitromidine; F-4; Triclose. $C_{10}H_{10}N_6O_2$; mol wt 246.23. C 48.78%, H 4.09%, N 34.13%, O 13.00%. Prepn and use: A. Garzia, **DE 2358483**; *eidem*, **US 3882105**; **US 3969520** (1974, 1975, 1976 all to Chemoterapico). Toxicological and teratological studies: R. Tammiso *et al.*, *Arzneim.-Forsch.* **28**, 2251 (1978).

Bright yellow odorless powder, mp 232-235°. Sol in DMF, DMSO, mineral oils and acids. Slightly sol in dioxane and acetone. LD_{50} in mice, rats (mg/kg): 5100, 7600 orally; 590, 860 i.p. (Tammiso).

THERAP CAT: Antiprotozoal (Trichomonas).

891. Azaperone. [1649-18-9] 1-(4-Fluorophenyl)-4-[4-(2-pyridinyl)-1-piperazinyl]-1-butanone; 4′-fluoro-4-[4-(2-pyridyl)-1-piperazinyl]butyrophenone; 1-[3-(4-fluorobenzoyl)propyl]-4-(2-pyridyl)piperazine; R-1929; Stresnil; Suicalm. $C_{19}H_{22}FN_3O$; mol wt 327.40. C 69.70%, H 6.77%, F 5.80%, N 12.83%, O 4.89%. Prepn: Janssen, **US 2979508** (1961). Synthesis of labeled compound: Soudijn, van Wijngaarden, *J. Lab. Comp.* **4**, 159 (1968). Distribution and metabolism studies in rat and pig: Heykants *et al.*, *Arzneim.-Forsch.* **21**, 982, 1263, 1357 (1971). Veterinary clinical studies: Symoens, van den Brande, *Vet. Rec.* **85**, 64 (1969). Crystal structure: M. H. J. Koch *et al.*, *Acta Crystallogr.* **33B**, 1975 (1977).

Crystals, mp 73-75°.

THERAP CAT (VET): Sedative; tranquilizer.

892. Azaserine. [115-02-6] *O*-(2-Diazoacetyl)-L-serine; L-serine diazoacetate (ester); CL-337; CN-15757; P-165. $C_5H_7N_3O_4$; mol wt 173.13. C 34.69%, H 4.08%, N 24.27%, O 36.96%. Antitumor antibiotic produced by *Streptomyces* sp. Isoln, characterization, and antitumor activity: C. C. Stock *et al.*, *Nature* **173**, 71 (1954); J. Ehrlich *et al.*, *ibid.* 72; Q. R. Bartz *et al.*, *ibid.* 72; S. A.

Fusari *et al.*, *J. Am. Chem. Soc.* **76**, 2878 (1954). Structural studies: *idem et al.*, *ibid.* 2881. Synthetic studies: J. A. Moore *et al.*, *ibid.* 2884; E. D. Nicolaides *et al.*, *ibid.* 2887; T. J. Curphey, D. S. Daniel, *J. Org. Chem.* **43**, 4666 (1978). Pathology and pharmacology studies: S. S. Sternberg, F. S. Philips, *Cancer* **10**, 889 (1957). Crystal and molecular structure: A. Fitzgerald, L. H. Jensen, *Acta Crystallogr. B* **34**, 828 (1978). Review of carcinogenicity studies: *IARC Monographs* **10**, 73-77 (1976). *Review:* Pettillo, Hunt, *Antibiotics* vol. 1, D. Gottlieb, P. Shaw, Eds. (Springer-Verlag, New York, 1967) pp 481-493. Use in study of the progression of pancreatic adenocarcinoma in rat models: K. Nagy *et al.*, *Histochem. Cell Biol.* **119**, 405 (2003).

Light yellow-green needles from aq ethanol, mp 146-162° (dec). $[\alpha]_D^{27.5}$ −0.5° (c = 8.46% in H_2O at pH 5.18). uv max (pH 7): 250.5 nm ($E_{1cm}^{1\%}$ 1140); in 0.1*N* NaOH: 252 nm ($E_{1cm}^{1\%}$ 1230). Very sol in water; slightly sol in abs methanol, abs ethanol, acetone, but sol in warm aq solns of these solvents. pKa 8.55. LD_{50} in mice, rats (mg/kg/day): 150, 170 orally (Sternberg, Philips).

USE: Reagent used to induce pancreatic cancer in experimental animal models.

893. Azasetron. [123040-69-7] *N*-1-Azabicyclo[2.2.2]oct-3-yl-6-chloro-3,4-dihydro-4-methyl-3-oxo-2*H*-1,4-benzoxazine-8-carboxamide; (±)-6-chloro-3,4-dihydro-4-methyl-3-oxo-*N*-(3-quinuclidinyl)-2*H*-1,4-benzoxazine-8-carboxamide; nazasetron. $C_{17}H_{20}ClN_3O_3$; mol wt 349.82. C 58.37%, H 5.76%, Cl 10.13%, N 12.01%, O 13.72%. Serotonin 5-HT$_3$ receptor antagonist. Prepn: T. Tahara *et al.*, **EP 313393**; *eidem*, **US 4892872** (1989, 1990 both to Yoshitomi); T. Kawakita *et al.*, *Chem. Pharm. Bull.* **40**, 624 (1992). Receptor binding study: N. Sato *et al.*, *Jpn. J. Pharmacol.* **59**, 443 (1992). Pharmacology: K. Haga *et al.*, *ibid.* **63**, 377 (1993). Toxicity study: S. Takagi *et al.*, *Oyo Yakuri* **42**, 255 (1991), *C.A.* **117**, 40208q (1991).

Hydrochloride. [123040-16-4] Y-25130; Serotone. $C_{17}H_{20}ClN_3O_3\cdot HCl$; mol wt 386.27. Crystals from ethanolic HCl, mp 281° (dec). Also reported as crystals from ethanol, mp 305° (dec) (Kawakita). LD_{50} in male, female rats (mg/kg): 135, 132 i.v. (Takagi).

THERAP CAT: Antiemetic.

894. Azatadine. [3964-81-6] 6,11-Dihydro-11-(1-methyl-4-piperidinylidene)-5*H*-benzo[5,6]cyclohepta[1,2-*b*]pyridine; 4-aza-5-(*N*-methyl-4-piperidinylidene)-10,11-dihydro-5*H*-dibenzo[*a,d*]cycloheptene. $C_{20}H_{22}N_2$; mol wt 290.41. C 82.72%, H 7.64%, N 9.65%. Prepn: F. J. Villani, **BE 647043** (1964 to Scherico); *idem*, **US 3326924** (1967 to Schering); *idem et al.*, *J. Med. Chem.* **15**, 750 (1972). Pharmacology: S. Tozzi *et al.*, *Agents Actions* **4**, 264 (1974). Toxicity studies: H. Tanaka *et al.*, *Jitchuken Zenrinsho Kenkyuho* **1**, 173 (1975), *C.A.* **84**, 25951h (1976). Clinical evaluation in allergic rhinitis: J. D. Wilson *et al.*, *N. Z. Med. J.* **94**, 79 (1981).

Crystals from isopropyl ether, mp 124-126°.
Dimaleate. [3978-86-7] Sch-10649; Bonamid; Idulian; Optimine; Zadine. $C_{20}H_{22}N_2.2C_4H_4O_4$; mol wt 522.55. Crystals from ethyl acetate-methanol, mp 152-154°.
THERAP CAT: Antihistaminic.

895. Azathioprine. [446-86-6] 6-[(1-Methyl-4-nitro-1H-imidazol-5-yl)thio]-1H-purine; 6-(1-methyl-4-nitro-5-imidazolyl)-mercaptopurine; azothioprine; BW-57-322; NSC-39084; Azanin; Azarek; Azasan; Imuran; Imurek; Imurel; Zytrim. $C_9H_7N_7O_2S$; mol wt 277.26. C 38.99%, H 2.54%, N 35.36%, O 11.54%, S 11.56%. Immunosuppressive antimetabolite; also active as disease modifying antirheumatic drug (DMARD). Metabolized *in vivo* to 6-mercaptopurine, *q.v.* Prepn: G. H. Hitchings, G. B. Elion, **US 3056785** (1962 to Burroughs Wellcome). Metabolism and pharmacokinetics: T. L. Ding, L. Z. Benet, *Drug Metab. Dispos.* **7**, 373 (1979). Carcinogenicity studies: P. S. Mitrou *et al., Arzneim.-Forsch.* **29**, 483, 662 (1979). HPLC determn in plasma: E. C. Van Os *et al., J. Chromatogr. B* **679**, 147 (1996). Clinical trial in rheumatoid arthritis: R. F. Wilkens, D. Stablein, *J. Rheumatol.* **23**, Suppl. 44, 64 (1996). Comprehensive description: W. P. Wilson, S. A. Benezra, *Anal. Profiles Drug Subs.* **10**, 29-53 (1981). Review of clinical efficacy in renal transplantation: G. L. C. Chan *et al., Pharmacotherapy* **7**, 165-177 (1987); in multiple sclerosis: C. Confavreux, T. Moreau, *Mult. Scler.* **1**, 379-384 (1996); in inflammatory bowel disease: W. J. Sandborn, *Scand. J. Gastroenterol.* **33**, Suppl. 225, 92-99 (1998).

Pale yellow crystals from 50% aq acetone, dec 243-244°. uv max (methanol): 276 nm (ε 1.82 × 10⁴); (0.1N HCl): 280 nm (ε 1.73 × 10⁴); (0.1N NaOH): 285 nm (ε 1.55 × 10⁴). pKa₂ 8.2. Sol in dilute solns of alkali hydroxides; sparingly sol in dilute mineral acids. Very slightly sol in alcohol, chloroform. Practically insol in water.
Sodium salt. [55774-33-9] $C_9H_6N_7NaO_2S$. Bright yellow amorphous mass or cake. Hygroscopic. Soly in water: 10 mg/ml.
Caution: Azathioprine is listed as a known human carcinogen: *Report on Carcinogens, Twelfth Edition* (PB2011-111646, 2011) p 57.
THERAP CAT: Immunosuppressant; antirheumatic.
THERAP CAT (VET): Immunosuppressant.

896. 6-Azathymine. [932-53-6] 6-Methyl-1,2,4-triazine-3,5(2H,4H)-dione; 5-methyl-6-azauracil; 3,5-dihydroxy-6-methyl-1,2,4-triazine. $C_4H_5N_3O_2$; mol wt 127.10. C 37.80%, H 3.97%, N 33.06%, O 25.18%. Prepn: Thiele, Bailey, *Ann.* **303**, 82 (1898); Bailey, *Am. Chem. J.* **28**, 386 (1902); Bougault, Daniel, *Compt. Rend.* **186**, 1216 (1928); Chang, *J. Org. Chem.* **23**, 1951 (1958).

Crystals from water. mp 210-212°. pKa 7.6. uv max (0.1N HCl): 261 nm (ε 5200); in 0.1N NaOH: 246 nm (ε 4770).

897. 6-Azauridine. [54-25-1] 2-β-D-Ribofuranosyl-1,2,4-triazine-3,5(2H,4H)-dione; 6-azauracil riboside; 3,5-dioxo-2,3,4,5-tetrahydro-1,2,4-triazine riboside; AzUR. $C_8H_{11}N_3O_6$; mol wt 245.19. C 39.19%, H 4.52%, N 17.14%, O 39.15%. Cytostatic nucleoside analog; inhibits orotidyl decarboxylase and blocks *de novo* pyrimidine biosynthesis. Production by *E. coli* in the presence of 6-azauracil: J. Skoda *et al., Experientia* **13**, 150 (1957). Chemical synthesis: M. Prystas *et al., Chem. Ind. (London)* **1961**, 947. Crystal structure and conformation: C. H. Schwalbe *et al., Biochem. Biophys. Res. Commun.* **44**, 57 (1971). Degradation kinetics: C. M. Riley *et al., Pharm. Res.* **12**, 1361 (1995). Clinical evaluation in recalcitrant psoriasis: J. C. Alper *et al., J. Am. Acad. Dermatol.* **13**, 567 (1985). Toxicity study: J. Plevova *et al., Toxicol. Appl. Pharmacol.* **17**, 511 (1970). Review of effect on vitamin B₆ and homocystinemia: W. Drell, A. D. Welch, *Pharmacol. Ther.* **41**, 195-206 (1989).

Crystals from ethanol, ether, mp 160-161°. $[\alpha]_D^{24}$ −132° (pyridine). uv max (water): 262 nm (ε 6100). pK 6.70.
2',3',5'-Triacetate. [2169-64-4] Azaribine; 2-(2,3,5-tri-O-acetyl-β-D-ribofuranosyl)-*as*-triazine-3,5(2H,4H)-dione; CB-304; Triazure. $C_{14}H_{17}N_3O_9$; mol wt 371.30. pKa (25°): 6.42. LD₅₀ in mice, rats (g/kg): 7.8, 12.0 orally (Plevova).
THERAP CAT: Antipsoriatic.

898. Azelaic Acid. [123-99-9] Nonanedioic acid; 1,7-heptanedicarboxylic acid; lepargylic acid; anchoic acid; Azelex; Finacea; Skinoren. $C_9H_{16}O_4$; mol wt 188.22. C 57.43%, H 8.57%, O 34.00%. Prepd by disruptive oxidation of ricinoleic acid: Hill, McEwen, *Org. Synth.* **coll. vol. II**, 53 (1943). Occurs in rancid oleic acid. Clinical comparison with tetracycline in acne: P. T. Bladon *et al., Br. J. Dermatol.* **114**, 493 (1986). Review of clinical studies in hyperpigmentary disorders: A. S. Breathnach, M. Nazzaro-Porro, *ibid.* **111**, 115-120 (1984); of pharmacology and clinical uses: M. Nazzaro-Porro, *J. Am. Acad. Dermatol.* **17**, 1033-1041 (1987). Clinical comparison with metronidazole, *q.v.*, in rosacea: S. Maddin, *J. Am. Acad. Dermatol.* **40**, 961 (1999).

HOOC———————COOH

Monoclinic prismatic needles, mp 106.5°. Distills above 360° with partial anhydride formation. bp₁₀₀ 286.5°; bp₅₀ 265°; bp₁₅ 237°; bp₁₀ 225°. d₄^{110.6} 1.0291. One liter of water dissolves 1.0 g at 1.0°; 2.4 g at 20°; 8.2 g at 50°; 22 g at 65°. Freely sol in boiling water, in alcohol. 1000 g of ether dissolves 18.8 g at 11° and 26.8 g at 15°. pK₁ (25°) 4.53; pK₂ 5.33.
Dimethyl ester. $C_{11}H_{20}O_4$. Liquid, d₄^{20} 1.0026. mp −3.9°. bp₈ 140°.
THERAP CAT: Antiacne.

899. Azelastine. [58581-89-8] 4-[(4-Chlorophenyl)methyl]-2-(hexahydro-1-methyl-1H-azepin-4-yl)-1(2H)-phthalazinone; 4-(p-chlorobenzyl)-2-(hexahydro-1-methyl-1H-azepin-4-yl)-1(2H)-phthalazinone; 4-(p-chlorobenzyl)-2-(N-methylperhydroazepin-4-yl)-1(2H)-phthalazinone. $C_{22}H_{24}ClN_3O$; mol wt 381.90. C 69.19%, H 6.33%, Cl 9.28%, N 11.00%, O 4.19%. Orally active H₁-histamine receptor antagonist. Prepn: **BE 778269**; D. Vogelsang *et al.,* **US 3813384** (1972, 1974 both to Asta-Werke AG). Synthesis and x-ray structure determn: G. Scheffler *et al., Arch. Pharm.* **321**, 205 (1988). Pharmacology: K. Tasaka, M. Akagi, *Arzneim.-Forsch.*

29, 488 (1979). Pharmacology and toxicology: H. J. Zechel *et al.*, *ibid.* **31**, 1184 (1981). Series of articles on pharmacokinetics, pharmacology and toxicology: *ibid.* 1184-1238. Mechanism of action studies: N. Chand *et al.*, *Int. J. Immunopharmacol.* **7**, 833 (1985); *eidem*, *Br. J. Pharmacol.* **87**, 443 (1986). HPLC determn in plasma: J. Pivonka *et al.*, *J. Chromatogr.* **420**, 89 (1987). Clinical evaluations in asthma: H. Magnussen, *Chest* **91**, 855 (1987); M. K. Albazzaz, K. R. Patel, *Thorax* **43**, 306 (1988); in allergic conjunctivitis: G. W. Canonica *et al.*, *Curr. Med. Res. Opin.* **19**, 321-329 (2003).

Oil. Sol in methylene chloride.

Monohydrate. $C_{22}H_{24}ClN_3O.H_2O$. Crystals from alcohol/water. Two distinct crystal forms have been identified (Scheffler).

Hydrochloride. [79307-93-0] A-5610; E-0659; W-2979M; Afluon; Allergodil; Astelin; Azeptin; Optilast; Rhinolast. $C_{22}H_{24}$- $ClN_3O.HCl$; mol wt 418.36. Crystals from alcohol, mp 225-229°. LD_{50} in male, female mice, male, female rats (mg/kg): 36.5, 35.5, 26.9, 30.3 i.v.; 56.4, 42.8, 43.2, 46.6 i.p.; 63.0, 54.2, 66.5, 59.6 s.c.; 124, 139, 310, 417 orally (Zechel).

THERAP CAT: Antihistaminic.

900. Azelnidipine. [123524-52-7] 2-Amino-1,4-dihydro-6-methyl-4-(3-nitrophenyl)-3,5-pyridinedicarboxylic acid 3-[1-(diphenylmethyl)-3-azetidinyl] 5-(1-methylethyl) ester; (±)-3-(1-diphenylmethylazetidin-3-yl) 5-isopropyl 2-amino-1,4-dihydro-6-methyl-4-(3-nitrophenyl)-3,5-pyridinedicarboxylate; CS-905; RS-9054; Calblock. $C_{33}H_{34}N_4O_6$; mol wt 582.66. C 68.03%, H 5.88%, N 9.62%, O 16.48%. Dihydropyridine L-type calcium channel blocker. Prepn: H. Koike *et al.*, **EP 266922**; *eidem*, **US 4772596** (both 1988 to Sankyo); T. Kobayashi *et al.*, *Chem. Pharm. Bull.* **43**, 797 (1995). Clinical evaluation in hypertension: M. Arita *et al.*, *J. Cardiovasc. Pharmacol.* **33**, 186 (1999). Review of pharmacology: Y. Yagil, A. Lustig, *Cardiovasc. Drug Rev.* **13**, 137-148 (1995); of development, pharmacokinetics, safety, and clinical experience: H. Koike *et al.*, *Annu. Rep. Sankyo Res. Lab.* **54**, 1-64 (2002); of pharmacology and clinical experience: K. Wellington, L. J. Scott, *Drugs* **63**, 2613-2621 (2003).

Pale yellow to yellow crystalline powder, mp 122-123°. Insol in water. pKa 7.89. LD_{50} in female, male mice, female, male rats (mg/kg): 785, 979, 1267, 1971 orally (Koike).

THERAP CAT: Antihypertensive.

901. 2-Azetidinecarboxylic Acid. [2517-04-6] $C_4H_7NO_2$; mol wt 101.11. C 47.52%, H 6.98%, N 13.85%, O 31.65%. From *Convallaria majalis* L. (lily-of-the-valley) and *Polygonatum officinale* Moensh., *Liliaceae*: Fowden, *Biochem. J.* **64**, 323 (1956); Fowden, Bryant, *ibid.* **70**, 626 (1958). Structure: Virtanen, *Angew. Chem.* **67**, 619 (1955); Fowden, *loc. cit.* Biosynthesis: Linko, *Acta Chem. Scand.* **12**, 101 (1958); Fowden, Bryant, *Biochem. J.* **71**, 210

(1959); Fowden, *ibid.* **71**, 643 (1959); Leete, *J. Am. Chem. Soc.* **86**, 3162 (1964). Shows growth-inhibitory activity on cultures of *Escherichia coli* and on germinating seeds of different species in which it does not normally occur; Fowden, Richmond, *Biochim. Biophys. Acta* **71**, 459 (1963). Acts as a proline analog where a stoichiometric replacement of proline occurred with production of abnormal proteins having impaired biological activity: Peterson, Fowden, *Nature* **200**, 148 (1963).

Crystals from 95% hot methanol, discolors at 200° and darkens until 310° when heating stopped. $[\alpha]_D^{20} -108°$ (c = 3.6). Unstable to mineral acids. Soluble in cold and hot water. Practically insol in abs ethanol.

902. Azidocillin. [17243-38-8] (2S,5R,6R)-6-[[(2R)-Azidophenylacetyl]amino]-3,3-dimethyl-7-oxo-4-thia-1-azabicyclo-[3.2.0]heptane-2-carboxylic acid; α-azidobenzylpenicillin; 6-[D-α-azidophenylacetamido]penicillanic acid; SPC-97D; BRL-2351. $C_{16}H_{17}N_5O_4S$; mol wt 375.40. C 51.19%, H 4.56%, N 18.66%, O 17.05%, S 8.54%. Semi-synthetic antibiotic related to penicillin. Prepn: B. O. H. Sjöberg, B. A. Ekström, **GB 940488**; *eidem*, **US 3293242** (1963, 1966 to Beecham); Ekström *et al.*, *Acta Chem. Scand.* **19**, 281 (1965). Pharmacology: Hanssen *et al.*, *Antimicrob. Agents Chemother.* **1967**, 568; Tunevall, Frisk, *ibid.* 573. Chemistry: Sjöberg *et al.*, *ibid.* 560. Metabolic studies: Ramsey *et al.*, *Arzneim.-Forsch.* **22**, 1962 (1972).

Sodium salt. [35334-12-4] Globacillin; Longatren; Syncillin. $C_{16}H_{16}N_5NaO_4S$; mol wt 397.38.

THERAP CAT: Antibacterial.

THERAP CAT (VET): Antimicrobial. (In mastitis, as feed supplement).

903. Azilsartan. [147403-03-0] 1-[[2'-(2,5-Dihydro-5-oxo-1,2,4-oxadiazol-3-yl) [1,1'-biphenyl]-4-yl]methyl]-2-ethoxy-1*H*-benzimidazole-7-carboxylic acid; 2-ethoxy-1-[[2'-(5-oxo-4,5-dihydro-1,2,4-oxadiazol-3-yl)biphenyl-4-yl]methyl]-1*H*-benzimidazole-7-carboxylic acid; TAK-536. $C_{25}H_{20}N_4O_5$; mol wt 456.46. C 65.78%, H 4.42%, N 12.27%, O 17.53%. Angiotensin II type-1 (AT_1) receptor antagonist; active moiety of the prodrug, azilsartan medoxomil. Prepn: T. Naka, Y. Inada, **EP 0520423**; *eidem*, **US 5243054** (1992, 1993 both to Takeda). Synthesis and pharmacology: Y. Kohara *et al.*, *J. Med. Chem.* **39**, 5228 (1996). Prepn of prodrug: T. Kuroita *et al.*, **US 7157584** (2007 to Takeda). *In vivo* effects on glucose intolerance: M. Iwai *et al.*, *Am. J. Hypertens.* **20**, 579 (2007). Receptor binding study and pharmacology: M. Ojima *et al.*, *J. Pharmacol. Exp. Ther.* **336**, 801 (2011). Clinical comparison with olmesartan and valsartan in hypertension: W. B. White *et al.*, *Hypertension* **57**, 413 (2011). Review of pharmacology and clinical experience: P. Singh *et al.*, *Formulary* **45**, 342-349 (2010).

Colorless prisms from EtOH, mp 212-214°. Log P (octanol/water) 0.90. pKa at 0.1 N NaOH (26°): 6.1.

Azilsartan medoxomil. [863031-21-4] (5-methyl-2-oxo-1,3-dioxol-4-yl)methyl 2-ethoxy-1-[[2'-(5-oxo-4,5-dihydro-1,2,4-oxadiazol-3-yl)biphenyl-4-yl]methyl]-1H-benzimidazole-7-carboxylate; TAK-491. $C_{30}H_{24}N_4O_8$; mol wt 568.54. Crystals from ethanol.

Azilsartan kamedoxomil. [863031-24-7] Azilsartan medoxomil potassium salt; Edarbi. $C_{30}H_{24}KN_4O_8$; mol wt 607.64. Crystals, mp 196° (dec). Freely sol in methanol. Practically insol in water.

THERAP CAT: Antihypertensive.

904. Azimilide. [149908-53-2] 1-[[[5-(4-Chlorophenyl)-2-furanyl]methylene]amino]-3-[4-(4-methyl-1-piperazinyl)butyl]-2,4-imidazolidinedione. $C_{23}H_{28}ClN_5O_3$; mol wt 457.96. C 60.32%, H 6.16%, Cl 7.74%, N 15.29%, O 10.48%. Potassium channel blocker. Prepn: S. S. Pelosi *et al.*, **WO 9304061** (1993 to Procter & Gamble); C.-N. Yu *et al.*, **US 5462940** (1995 to Norwich Eaton). Clinical pharmacokinetics: A. Corey *et al.*, *J. Clin. Pharmacol.* **37**, 946 (1997). Clinical trial to improve post-infarct survival: A. J. Camm *et al.*, *Am. J. Cardiol.* **81**, Suppl. 6A, 35D (1998). Review of electrophysiologic properties and pharmacology: R. Karam *et al.*, *ibid.* 40D-46D.

Dihydrochloride. [149888-94-8] NE-10064; Stedicor. $C_{23}H_{28}$-ClN_5O_3.2HCl; mol wt 530.88. Crystals from ethanol/water.

THERAP CAT: Antiarrhythmic (class III).

905. Azimsulfuron. [120162-55-2] N-[[(4,6-Dimethoxy-2-pyrimidinyl)amino]carbonyl]-1-methyl-4-(2-methyl-2H-tetrazol-5-yl)-1H-pyrazole-5-sulfonamide; 1-(4,6-dimethoxypyrimidin-2-yl)-3-[1-methyl-4-(2-methyl-2H-tetrazol-5-yl)-pyrazol-5-ylsulfonyl]-urea; DPX-A8947; IN-A8947. $C_{13}H_{16}N_{10}O_5S$; mol wt 424.40. C 36.79%, H 3.80%, N 33.00%, O 18.85%, S 7.55%. Broad spectrum, post-emergent sulfonylurea herbicide which inhibits acetolactate synthase. Prepn: G. Levitt, **US 4746353** (1988 to Du Pont). Description of chemical properties and biological activities: T. Marquez *et al.*, *Brighton Crop Prot. Conf. - Weeds* **1995**, 65. Environmental fate: A. C. Barefoot *et al.*, *ibid.* 713. Absorption, translocation and metabolism: S. Shirakura *et al.*, *Weed Res. Jpn.* **40**, 299 (1995). Efficacy in combination with bensulfuron methyl: *eidem, ibid.* 29.

Solid, mp 170°. Soly at 25° (ppm): methylene chloride 65900; acetone 26400; methanol 2100; acetonitrile 13900; ethyl acetate 13000. Aqueous soly at 20° (ppm): pH 5 72.3; pH 7 1050; pH 9 6540. pKa 3.6. Vapor pressure at 25°: 3.0×10^{-11} mm Hg. Partition coefficient (*n*-octanol/ water): 4.43 (pH 5); 0.043 (pH 7); 0.0083 (pH 9). LD_{50} in rats, bobwhite quail, mallard duck (mg/kg): >5000, >2250, >2250 orally; in rats (mg/kg): >2000 dermally (Marquez). LC_{50} (96 hr) in carp, bluegill sunfish, rainbow trout (ppm): >300, >1000, 154 (Marquez).

USE: Herbicide.

906. Azinphos-methyl. [86-50-0] Phosphorodithioic acid O,O-dimethyl S-[(4-oxo-1,2,3-benzotriazin-3(4H)-yl)methyl] ester; phosphorodithioic acid O,O-dimethyl ester, S-ester with 3-mercaptomethyl-1,2,3-benzotriazin-4(3H)-one; Bayer 17147; ENT-23233; R-1582; Cotnion; Gusathion M. $C_{10}H_{12}N_3O_3PS_2$; mol wt 317.32. C 37.85%, H 3.81%, N 13.24%, O 15.13%, P 9.76%, S 20.21%. Organophosphate insecticide; cholinesterase inhibitor. Prepn: W. Lorenz, **US 2758115** (1956 to Bayer). Activity: E. E. Ivy *et al., J.*

Econ. Entomol. **48**, 293 (1955). Multi-residue determn by GC-MS: K. S. Liapis *et al., J. Chromatogr. A* **996**, 181 (2003). Photodegradation study: M. Ménager *et al., J. Photochem. Photobiol. A* **192**, 41 (2007). Toxicity study: T. B. Gaines, *Toxicol. Appl. Pharmacol.* **14**, 515 (1969).

Crystals from methanol, mp 73-74°. d_4^{20} 1.44; n_D^{76} 1.6115. Soly in water at 25°: 33 mg/l. Sol in methanol, ethanol, propylene glycol, xylene, other organic solvents. Solns in ethanol and propylene glycol are stable for at least 3 weeks. Unstable at temperatures >200°. Hydrolyzes in acid or cold alkali. *Poisonous.* LD_{50} in female rats (mg/kg): 11 orally; 220 dermally (Gaines).

O,O-**Diethyl analog.** Azinphos-ethyl; Bayer 16259; ENT-22014; R-1513. $C_{12}H_{16}N_3O_3PS_2$; mol wt 345.37. Multi-residue determn by LC-MS/MS: B. Kmellár *et al., J. Chromatogr. A* **1215**, 37 (2008). Colorless needles, mp 53°, bp$_{0.001}$ 111°. *Poisonous.* d_4^{20} 1.284; n_D^{53} 1.5928. Vapor pressure at 20°: 2.2×10^{-7} mm Hg.

Caution: Potential symptoms of overexposure to azinphos-methyl are miosis; aching eyes; blurred vision, lacrimation and rhinorrhea; headache; chest tightness, wheezing, laryngeal spasms; salivation; cyanosis; anorexia; nausea, vomiting and diarrhea; sweating; twitching, paralysis and convulsions; low blood pressure, cardiac irregularities. *See NIOSH Pocket Guide to Chemical Hazards* (DHHS/ NIOSH 97-140, 1997) p 22.

USE: Insecticide; acaricide.

907. Azithromycin. [83905-01-5] (2R,3S,4R,5R,8R,10R,-11R,12S,13S,14R)-13-[(2,6-Dideoxy-3-C-methyl-3-O-methyl-α-L-*ribo*-hexopyranosyl)oxy]-2-ethyl-3,4,10-trihydroxy-3,5,6,8,10,-12,14-heptamethyl-11-[[3,4,6-trideoxy-3-(dimethylamino)-β-D-*xylo*-hexopyranosyl]oxy]-1-oxa-6-azacyclopentadecan-15-one; N-methyl-11-aza-10-deoxo-10-dihydroerythromycin A; 9-deoxo-9a-methyl-9a-aza-9a-homoerythromycin A. $C_{38}H_{72}N_2O_{12}$; mol wt 749.00. C 60.94%, H 9.69%, N 3.74%, O 25.63%. Semi-synthetic macrolide antibiotic; related to erythromycin A, *q.v.* Prepn: **BE 892357**; G. Kobrehel, S. Djokic, **US 4517359** (1982, 1985 both to Sour Pliva); of the crystalline dihydrate: D. J. M. Allen, K. M. Nepveux, **EP 298650**; *eidem*, **US 6268489** (1989, 2001 both to Pfizer). Antibacterial spectrum: S. C. Aronoff *et al., J. Antimicrob. Chemother.* **19**, 275 (1987); and mode of action: J. Retsema *et al., Antimicrob. Agents Chemother.* **31**, 1939 (1987). LC-MS determin in plasma: B.-M. Chen *et al., J. Pharm. Biomed. Anal.* **42**, 480 (2006). Series of articles on pharmacology, pharmacokinetics, and clinical experience: *J. Antimicrob. Chemother.* **31**, Suppl. E, 1-198 (1993). Clinical trial in prevention of *Pneumocystis carinii* pneumonia in AIDS patients: M. W. Dunne *et al., Lancet* **354**, 891 (1999); in cystic fibrosis: A. Clement *et al., Thorax* **61**, 895 (2006); in *P. vivax* malaria: M. W. Dunne *et al., Am. J. Trop. Med. Hyg.* **73**, 1108 (2005). Review of pharmacology and clinical efficacy in pediatric infections: H. D. Langtry, J. A. Balfour, *Drugs* **56**, 273-297 (1998). Review of clinical evaluation in lower respiratory tract infections: F. Blasi *et al., Expert Opin. Pharmacother.* **6**, 2335-2351 (2005).

Amorphous solid, mp 113-115°. $[\alpha]_D^{20}$ −37° (c = 1 in CHCl$_3$).

Dihydrate. [117772-70-0] CP-62993; XZ-450; Azitrocin; Ribotrex; Sumamed; Trozocina; Zithromax; Zitromax. White crystalline powder. mp 126°. $[\alpha]_D^{26}$ −41.4° (c = 1 in CHCl$_3$).

THERAP CAT: Antibacterial.

THERAP CAT (VET): Antibacterial.

908. Azlocillin. [37091-66-0] (2S,5R,6R)-3,3-Dimethyl-7-oxo-6-[[(2R)-2-[[(2-oxo-1-imidazolidinyl)carbonyl]amino]-2-phenylacetyl]amino]-4-thia-1-azabicyclo[3.2.0]heptane-2-carboxylic acid; D-α-[(imidazolidin-2-on-1-yl)carbonylamino]benzylpenicillin; Bay e 6905. C$_{20}$H$_{23}$N$_5$O$_6$S; mol wt 461.49. C 52.05%, H 5.02%, N 15.18%, O 20.80%, S 6.95%. Semi-synthetic, broad-spectrum acylureido penicillin. Prepn: H. Disselnkotter, K. G. Metzger, **FR 2100682**; *eidem*, **US 3933795** (1971, 1976 both to Bayer); H. B. Konig *et al.*, *Eur. J. Med. Chem. - Chim. Ther.* **17**, 59 (1982). *In vitro* studies: D. Stewart *et al.*, *Antimicrob. Agents Chemother.* **11**, 865 (1977). *In vitro* and *in vivo* activity: G. K. Daikos *et al.*, *Curr. Chemother., Proc. 10th Int. Congr. Chemother.*, 1977 (Amer. Soc. Microbiol., Washington, D.C., 1978) **1**, pp 626-8. Pharmacokinetics: P. Fiegel, K. Becker, *Antimicrob. Agents Chemother.* **14**, 288 (1978). Comparison with other penicillins: J. M. Andrews, K. A. Bedford, *ibid.* **559**. Clinical studies: H. Lode *et al.*, *Infection* **5**, 163 (1977); E. B. Helm *et al.*, *Dtsch. Med. Wochenschr.* **102**, 1211 (1977). Series of articles on antibacterial activity, pharmacology, and clinical trials: *Arzneim.-Forsch.* **29**, 1915-2032 (1979); *Infection* **10**, Suppl. 3, S121-S266 (1982); *J. Antimicrob. Chemother.* **11**, Suppl. B, 1-239 (1983).

Sodium salt. [37091-65-9] Azlin; Securopen. C$_{20}$H$_{22}$N$_5$NaO$_6$S; mol wt 483.47. Pale yellow crystals, sol in water, methanol, DMF. Slightly sol in ethanol, isopropanol.

THERAP CAT: Antibacterial.

909. Azobenzene. [103-33-3] 1,2-Diphenyldiazene; diphenyl diimide; azobenzide; azobenzol; benzeneazobenzene. C$_{12}$H$_{10}$N$_2$; mol wt 182.23. C 79.09%, H 5.53%, N 15.37%. Parent compound of azobenzene disperse dyes. Prepn: A. I. Vogel, *Practical Organic Chemistry* (Longmans, London, 3rd ed., 1959) p 631. Toxicity data: E. W. Schafer, Jr., W. A. Bowles, Jr., *Arch. Environ. Contam. Toxicol.* **14**, 111 (1985). Review of carcinogenicity studies: *IARC Monographs* **8**, 75 (1975). Review of photochemistry: G. S. Kumar, D. C. Neckers, *Chem. Rev.* **89**, 1915-1925 (1989). Review of use in artificial membranes: J.-I. Anzai, T. Osa, *Tetrahedron* **50**, 4039-4070 (1994); as modeling ligand: A. A. Pasynskii *et al.*, *Bull. Pol. Acad. Sci. Chem.* **42**, 23-33 (1994); in polymers: A. Natansohn, P. Rochon, *ACS Symp. Ser.* **672**, 236-250 (1997).

Orange-red leaflets. d 1.20. mp 68°. bp 293°. Insol in water. Sol in alcohol, ether, glacial acetic acid.

USE: Precursor for dyes, for polymers.

910. 1,1′-Azobis(cyclohexanecarbonitrile). [2094-98-6] 1,1′-Azobis(cyclohexane-1-carbonitrile); 1,1′-azobis[1-cyanocyclohexane]; 1,1′-azobis(cyclohexanenitrile); azobiscyclohexylnitrile; ACHN; ACN. C$_{14}$H$_{20}$N$_4$; mol wt 244.34. C 68.82%, H 8.25%, N 22.93%. Thermally stable initiator for free radical reactions. Prepn: H. Hartman, *Recl. Trav. Chim. Pays-Bas* **46**, 150 (1927); and decompn in soln: C. G. Overberger *et al.*, *J. Am. Chem. Soc.* **71**, 2661 (1949). Synthetic applications: G. E. Keck, D. A. Burnett, *J. Org. Chem.* **52**, 2958 (1987); W. A. Kinney, *Tetrahedron Lett.* **34**, 2715 (1993). *Review*: S. A. Kates, F. Albericio in *Encyclopedia of Reagents for Organic Synthesis* **1**, L. A. Paquette, Ed. (Wiley, New York, 1995) pp 228-229.

White solid from ethanol, mp 114-115°. *Flammable. Irritant.* Sol in benzene, toluene. Insol in water. Heating to temperatures above the mp results in exothermic decompn.

USE: Reagent in synthetic organic chemistry.

911. 2,2′-Azobisisobutyronitrile. [78-67-1] 2,2′-(1,2-Diazenediyl)bis[2-methylpropanenitrile]; AIBN; α,α′-azodiisobutyronitrile; 2,2′-dicyano-2,2′-azopropane; Porofor-57. C$_8$H$_{12}$N$_4$; mol wt 164.21. C 58.52%, H 7.37%, N 34.12%. Prepn by oxidation of α,α′-hydrazobutyric acid dinitrile: Overberger *et al.*, *J. Am. Chem. Soc.* **71**, 2661 (1949); Horner, Schwenk, *Ann.* **566**, 69 (1950); from 2-aminoisobutyronitrile + NaOCl: **GB 672106** (1952 to Rohm & Haas); Anderson, **US 2711405** (1955 to du Pont); from 2,2′-dichloro-2,2′-azopropane: Benzing, *Ann.* **631**, 1 (1960); **GB 929182** (1963 to Monsanto).

Crystals from ethanol + water, dec 107°. uv max (ethanol): 345 nm. Soly in methanol at 0°, 20°, 40°: 1.8, 4.96, 16.06 g/100 ml. Soly in ethanol at 0°, 20°, 40°: 0.58, 2.04, 7.15 g/100 ml. Can explode when dissolved in acetone: Carlisle, *Chem. Eng. News* **27**, 150 (1949).

Caution: In the organism, forms HCN which is found in blood, liver and brain: Rusin, *C.A.* **56**, 2682f (1962).

USE: Blowing agent for elastomers and plastics. Initiator for free radical reactions: Griesbaum *et al.*, *J. Org. Chem.* **30**, 261 (1965).

912. Azodicarbonamide. [123-77-3] 1,2-Diazenedicarboxamide; 1,1′-azobisformamide; azodicarboxamide; azobiscarbonamide; azobiscarboxamide; 1,1′-azobiscarbamide. C$_2$H$_4$N$_4$O$_2$; mol wt 116.08. C 20.69%, H 3.47%, N 48.27%, O 27.57%. Prepn: J. Thiele, *Ann.* **270**, 1 (1892); **271**, 127 (1892); T. Curtis, K. Heidenreich, *J. Prakt. Chem.* [2] **52**, 454 (1895); J. P. Picard, J. L. Boivin, *Can. J. Chem.* **29**, 223 (1951). Crystal structure: D. T. Cromer, A. C. Larson, *J. Chem. Phys.* **60**, 176 (1974). Safety assessment as a flour-maturing agent: B. L. Oser *et al.*, *Toxicol. Appl. Pharmacol.* **7**, 445 (1965).

Orange-red crystals, mp 225° (dec). Crystal d 1.68. *Flammable.* Sol in hot water. Insol in cold water, alc.

USE: As blowing and foaming agent for plastics; as maturing and bleaching agent in cereal flour.

913. Azolitmin. [1395-18-2] Purified coloring matter from litmus: I. M. Kolthoff, C. Rosenblum, *Acid-Base Indicators* (MacMillan Co., New York, 1937) pp 160-162, 174, 208, 355, 361, 365-366, 368-369, 373, 377, 387.

Dark violet scales or dark red powder. Sparingly soluble in water; freely sol in dil alkali hydroxides or carbonates. Insol in alc. The indicator solution is prepared by dissolving 0.5 g in 80 ml of warm H$_2$O, then adding 20 ml alcohol.

USE: Indicator instead of litmus. pH: 4.5 red, 8.3 blue. Usable with most mineral, some organic acids (not hydroxy acids) and some alkaloids; also used for preparing litmus media for bacteriologic purposes.

914. Azomycin. [527-73-1] 2-Nitro-1*H*-imidazole. C$_3$H$_3$N$_3$O$_2$; mol wt 113.08. C 31.87%, H 2.67%, N 37.16%, O 28.30%. Antibiotic substance produced by an unidentified *Streptomyces* sp. (resembling *Nocardia mesenterica* in some aspects) from soil collected at Shiba Shirokane Daimachi (Japan): Maeda *et al.*, *J. Antibiot.* **6A**, 182 (1953); Okami *et al.*, *ibid.* **7A**, 53 (1954); Nakamura,

Umezawa, *ibid.* **8A**, 66 (1955). Structure: Nakamura, *Pharm. Bull.* **3**, 379 (1955). Synthesis: Beaman *et al.*, *J. Am. Chem. Soc.* **87**, 389 (1965); Lancini, Lazzari, *Experientia* **21**, 83 (1965).

Crystals from methanol, dec 283°. uv max (ethanol): 313 nm ($E_{1cm}^{1\%}$ 915); in 0.1*N* NaOH: 374 nm (ε 12750). Soluble in methanol, ethanol, acetone, ethyl acetate, butyl acetate, alkaline water. Practically insol in ether, petr ether, chloroform, acidic water. LD_{50} i.v. in mice: 80 mg/kg (Maeda).

Caution: Irritating to skin.

915.   **Azosemide.** [27589-33-9] 2-Chloro-5-(2*H*-tetrazol-5-yl)-4-[(2-thienylmethyl)amino]benzenesulfonamide; 2-chloro-5-(2*H*-tetrazol-5-yl)-*N*⁴-2-thenylsulfanilamide; 5-(4′-chloro-2′-thenylamino-5′-sulfamoylphenyl)tetrazole; Ple-1053; Diart; Luret. $C_{12}H_{11}ClN_6O_2S_2$; mol wt 370.83. C 38.87%, H 2.99%, N 22.66%, O 8.63%, S 17.29%. Prepn: A. Popelak *et al.*, **DE 1815922**; *eidem*, **US 3665002** (1968, 1972 both to Boehringer, Mann.). Separation and analysis in blood and urine by HPLC: R. Seiwell, C. Brater, *J. Chromatogr.* **182**, 257 (1980). Sites of action: C. Brater, *Clin. Pharmacol. Ther.* **25**, 428 (1979). Clinical and pharmacological studies: F. Krueck *et al.*, *Eur. J. Clin. Pharmacol.* **14**, 153 (1978). Diuretic effect in animals: J. Greven, O. Heidenreich, *Arzneim.-Forsch.* **31**, 346, 350 (1981).

Crystals, mp 218-221°.
THERAP CAT: Diuretic.

916.   **Azoxybenzene.** [495-48-7] 1,2-Diphenyldiazene 1-oxide; azoxybenzide. $C_{12}H_{10}N_2O$; mol wt 198.23. C 72.71%, H 5.09%, N 14.13%, O 8.07%. Prepd by reduction of nitrobenzene with sodium arsenite: Bigelow, Palmer, *Org. Synth.* **coll. vol. II**, 57 (1943); also by treatment of nitrobenzene with glucose in alkaline medium: Opolonick, *Ind. Eng. Chem.* **27**, 1045 (1935). Industrial prepn by heating nitrobenzene with molasses and NaOH in high flash naphtha: **DE 228722**. Also prepd by catalytic reduction of nitrobenzene: Busch, Schulz, *Ber.* **62**, 1458 (1929). By peracetic acid oxidation of azobenzene: D'Ans, Kneip, *Ber.* **48**, 1145 (1915); *cf.* Greenspan, *Ind. Eng. Chem.* **39**, 847 (1947). Prepn of *cis*- and *trans*-forms by perbenzoic acid oxidation of azobenzene: G. M. Badger *et al.*, *J. Chem. Soc.* **1953**, 2143.

trans-Azoxybenzene

***trans*-Form.** [20972-43-4] Pale yellow orthorhombic needles. d_4^{26} 1.1590. d_4^{50} 1.1373. mp 36°. Slightly volatile with steam; easily volatile in superheated steam at 140-150°. Absorption spectrum: Müller, *Ann.* **493**, 166 (1932). Insol in water. Sol in alcohol, ether. 100 parts of abs alc satd at 16° contain 17.5 parts (w/w) azoxybenzene. At 15° 100 g ligroin dissolves 43.5 g azoxybenzene.

***cis*-Form.** [21650-65-7] mp 87°.
USE: In organic syntheses.

917.   **Azoxystrobin.** [131860-33-8] (α*E*)-2-[[6-(2-Cyanophenoxy)-4-pyrimidinyl]oxy]-α-(methoxymethylene)benzeneacetic acid methyl ester; methyl (*E*)-2-[2-[6-(2-cyanophenoxy)pyrimidin-4-yloxy]phenyl]-3-methoxyacrylate; ICI-A-5504; Amistar; Heritage; Quadris. $C_{22}H_{17}N_3O_5$; mol wt 403.39. C 65.51%, H 4.25%, N 10.42%, O 19.83%. Strobilurin fungicide; inhibits mitochondrial respiration by blocking electron transfer between cytochromes b and c_1. Prepn: J. M. Clough *et al.*, **EP 382375** (1990 to I.C.I.); *eidem*, **US 5395837** (1995 to Zeneca). Comprehensive description: J. R. Godwin *et al.*, *Brighton Crop Prot. Conf. - Pests Dis.* **1992**, 435-442. Field trials in turfgrass: J. A. Frank, P. L. Sanders, *ibid.* **1994**, 871. Environmental fate: E. D. Pilling *et al.*, *ibid.* **1996**, 315.

White crystalline solid, mp 118-119°. d 1.33. Partition coefficient (*n*-octanol/water): 440. Vapor pressure (20°): $<10^{-5}$ Pa. Soly in water (25°): 10 mg/l. LD_{50} in rats (mg/kg): >5000 orally; >2000 dermally (Godwin).
USE: Agricultural fungicide.

918.   **Aztreonam.** [78110-38-0] 2-[[(*Z*)-[1-(2-Amino-4-thiazolyl)-2-[[(2*S*,3*S*)-2-methyl-4-oxo-1-sulfo-3-azetidinyl]amino]-2-oxoethylidene]amino]oxy]-2-methylpropanoic acid; azthreonam; SQ-26776; Azactam; Primbactam. $C_{13}H_{17}N_5O_8S_2$; mol wt 435.43. C 35.86%, H 3.94%, N 16.08%, O 29.39%, S 14.73%. The first totally synthetic monocyclic β-lactam (monobactam) antibiotic. It has a high degree of resistance to β-lactamases and shows specific activity vs aerobic gram-negative rods. Prepn: R. B. Sykes *et al.*, **NL 8100571** (1981 to Squibb), *C.A.* **96**, 181062x (1982). Fast-atom-bombardment mass spectra: A. I. Cohen *et al.*, *J. Pharm. Sci.* **71**, 1065 (1982). Activity vs gram-negative bacteria: R. B. Sykes *et al.*, *Antimicrob. Agents Chemother.* **21**, 85 (1982). Series of articles on structure-activity, *in vitro* and *in vivo* properties, pharmacokinetics: *J. Antimicrob. Chemother.* **8**, Suppl. E, 1-148 (1981). Toxicology: G. R. Keim *et al.*, *ibid.* 141. Mechanism of action study: A. D. Russell, J. R. Furr, *ibid.* **9**, 329 (1982). Comparative stability to renal dipeptidase: H. Mikami *et al.*, *Antimicrob. Agents Chemother.* **22**, 693 (1982). Human pharmacokinetics: E. A. Swabb *et al.*, *ibid.* **21**, 944 (1982). Clinical evaluation in urinary tract infection: C. Donadio *et al.*, *Drugs Exp. Clin. Res.* **13**, 167 (1987). Clinical efficacy in neonatal sepsis: S. Sklavunu-Tsurutsoglu *et al.*, *Rev. Infect. Dis.* **13**, Suppl. 7, S591 (1991). Comprehensive description: K. Florey, *Anal. Profiles Drug Subs.* **17**, 1-39 (1988).

White crystalline, odorless powder, dec 227°. Very slightly sol in ethanol, slightly sol in methanol, sol in DMF, DMSO. Practically insol in toluene, chloroform, ethyl acetate.

Disodium salt. $C_{13}H_{15}N_5Na_2O_8S_2$. LD_{50} (mg/kg): 3300 i.v. in mice; 6600 i.p. in rats (Keim).

L-Lysine salt. [827611-49-4] Aztreonam lysinate; aztreonam lysine; Cayston. $C_{13}H_{17}N_5O_8S_2 \cdot C_6H_{14}N_2O_2$; mol wt 581.62. Prepn: V. Gyollai *et al.*, **WO 05005424** (2005 to Teva); *eidem*, **US 7262293** (2007 to Corus). Pharmacokinetics in patients with cystic fibrosis: R. L. Gibson *et al.*, *Pediatr. Pulmonol.* **41**, 656 (2006). Clinical trial for pseudomonas infection in cystic fibrosis: G. Z. Retsch-Bogart *et al.*, *Chest* **135**, 1223 (2009).

THERAP CAT: Antibacterial.

919. Azulene. [275-51-4] Cyclopentacycloheptene; bicyclo-[5.3.0]-deca-2,4,6,8,10-pentaene; bicyclo-[0.3.5]-deca-1,3,5,7,9-pentaene. $C_{10}H_8$; mol wt 128.17. C 93.71%, H 6.29%. Prepn from octahydronaphthalene: Plattner, Pfau, *Helv. Chim. Acta* **20**, 224 (1937); from indan: Plattner, Magyar, *ibid.* **25**, 581 (1942).

Intensely blue leaflets or monoclinic plates from alc. Odor of naphthalene. mp 98.5-99°. Absorption spectrum: Susz *et al.*, *Helv. Chim. Acta* **20**, 469 (1937); Plattner, *ibid.* **24**, 283 (1941). Sol in the usual organic solvents, in concd mineral acids with decompn. Insol in water.

Sodium sulfonate. [51873-93-9] Azunol; Azusalen.

THERAP CAT: Sodium sulfonate as antacid.

920. Azure A. [531-53-3] 3-Amino-7-(dimethylamino)phenothiazin-5-ium chloride (1:1); 7-(dimethylamino)-3-imino-3*H*-phenothiazine hydrochloride; 3-amino-7-dimethylaminophenazathionium chloride; *asym*-dimethyl-3,7-diaminophenazathionium chloride; *asym*-dimethylthionine chloride; methylene azure A; C.I. 52005. $C_{14}H_{14}ClN_3S$; mol wt 291.80. C 57.63%, H 4.84%, Cl 12.15%, N 14.40%, S 10.99%. One of the main constituents of methylene azure. Prepn: Bernthsen, *Ann.* **230**, 169 (1885); Kehrmann, *Ber.* **39**, 1804 (1906); MacNeal, Killian, *J. Am. Chem. Soc.* **48**, 740 (1926). *Review: H. J. Conn's Biological Stains*, R. D. Lillie, Ed. (Williams & Wilkins, Baltimore, 9th ed., 1977) pp 420-421, 603. *See also: Colour Index* **vol. 4** (3rd ed., 1971) p 4469.

Green glistening crystals or dark green powder. Soluble in water (blue soln); sparingly sol in alc. Absorption max: 620-634 nm (3 ml of a soln of 50 mg in 250 ml H_2O diluted to 200 ml; read in a Beckman spectrophotometer 1 cm cell).

USE: Biological stain.

921. Azure B. [531-55-5] 3-(Dimethylamino)-7-(methylamino)phenothiazin-5-ium chloride (1:1); 7-(dimethylamino)-3-(methylimino)-3*H*-phenothiazine hydrochloride; 3-methylamino-7-dimethylaminophenazathonium chloride; trimethyldiaminophenazathonium chloride; trimethylthionine chloride; methylene azure B; C.I. 52010. $C_{15}H_{16}ClN_3S$; mol wt 305.82. C 58.91%, H 5.27%, Cl 11.59%, N 13.74%, S 10.48%. One of the main constituents of methylene azure. Prepn: Bernthsen, *Ann.* **230**, 169 (1885); Kehrmann, *Ber.* **39**, 1804 (1906); Kehrmann, Duttenhöfer, *ibid.* 925, 1403; Kehrmann *et al.*, *ibid.* **46**, 2137 (1913); MacNeal, Killian, *J. Am. Chem. Soc.* **48**, 740 (1926). *Review: H. J. Conn's Biological Stains*, R. D. Lillie, Ed. (Williams & Wilkins, Baltimore, 9th ed., 1977) pp 421-423, 603-604. *See also: Colour Index* **vol. 4** (3rd ed., 1971) p 4470.

Green glistening crystals or dark green powder. Sol in water (blue soln); sparingly sol in alc. Absorption max: 648-655 nm (3 ml of a soln of 50 mg in 250 ml H_2O dil to 200 ml; read in a Beckman spectrophotometer 1 cm cell).

USE: Biological stain.

922. Azure C. [531-57-7] 3-Amino-7-(methylamino)phenothiazin-5-ium chloride (1:1); 3-imino-7-(methylamino)-3*H*-phenothiazine hydrochloride; 3-amino-7-methylaminophenazathonium chloride; monomethyldiaminodiphenazathionium chloride; monomethylthionine chloride; C.I. 52002. $C_{13}H_{12}ClN_3S$; mol wt 277.77. C 56.21%, H 4.35%, Cl 12.76%, N 15.13%, S 11.54%. Prepn: Holmes, French, *Stain Technol.* **1**, 17 (1926). *Review: H. J. Conn's Biological Stains*, R. D. Lillie, Ed. (Williams & Wilkins, Baltimore, 9th ed., 1977) pp 419-420, 604. *See also: Colour Index* **vol. 4** (3rd ed., 1971) p 4469.

Green glistening crystals or dark green powder. Sol in water (blue soln); sparingly sol in alc. Absorption max: 608-622 nm (3 ml of a soln of 50 mg in 250 ml H_2O dil to 200 ml; read in a Beckman spectrophotometer 1 cm cell).

USE: Biological stain.

B

923. Babbitt Metal. Originally an alloy of 69% Zn, 19% Sn, 4% Cu, 3% Sb, and 5% Pb. This name now covers a variety of formulations, such as lead base babbitt, silver base babbitt, tin base babbitt, cadmium base babbitt, arsenic base babbitt. Generally and better described as "white metal bearing alloys".
USE: For machinery bearings.

924. Bacampicillin. [50972-17-3] (2S,5R,6R)-6-[[(2R)-2-Amino-2-phenylacetyl]amino]-3,3-dimethyl-7-oxo-4-thia-1-azabi-cyclo[3.2.0]heptane-2-carboxylic acid 1-[(ethoxycarbonyl)oxy]ethyl ester; (2S,5R,6R)-6-[(R)-(2-amino-2-phenylacetamido)]-3,3-dimeth-yl-7-oxo-4-thia-1-azabicyclo[3.2.0]heptane-2-carboxylic acid ester with ethyl 1-hydroxyethyl carbonate; 6-[(R)-2-amino-2-phenylacet-amido]penicillanic acid [1-(ethoxycarbonyloxy)ethyl] ester; 1'-eth-oxycarbonyloxyethyl 6-(D-α-aminophenylacetamido)penicillanate. $C_{21}H_{27}N_3O_7S$; mol wt 465.52. C 54.18%, H 5.85%, N 9.03%, O 24.06%, S 6.89%. Semi-synthetic antibiotic related to penicillin. Prepn: B. A. Ekstrom, B. O. H. Sjoberg, **DE 2144457**; *eidem*, **US 3873521** and **US 3939270** (1972, 1975 and 1976 all to Astra). *In vitro* and *in vivo* study: N. O. Bodin *et al.*, *Antimicrob. Agents Chemother.* **8**, 518 (1975). Pharmacokinetics: M. Rozencweig *et al.*, *Clin. Pharmacol. Ther.* **19**, 592 (1976); T. Bergan, *Antimicrob. Agents Chemother.* **13**, 971 (1978). Animal studies: C. Carbon *et al.*, *Scand. J. Infect. Dis. Suppl.* **14**, 127 (1978); T. Bergan, I. Vers-land, *ibid.* 135; U. Forsgren *et al.*, *ibid.* 207. Clinical studies: P. V. Maesen *et al.*, *J. Antimicrob. Chemother.* **2**, 279 (1976); J. Sjövall, *J. Int. Med. Res.* **5**, 313 (1977); C. Ekedahl *et al.*, *Scand. J. Infect. Dis. Suppl.* **14**, 279 (1978). Toxicology: M. Edanaga *et al.*, *Chemotherapy (Tokyo)* **27**, Suppl. 4, 17 (1979).

Hydrochloride. [37661-08-8] Ambacamp; Ambaxin; Bacacil; Bacampicine; Penglobe. $C_{21}H_{27}N_3O_7S \cdot HCl$; mol wt 501.98. White crystals from acetone-petr ether, mp 171-176° (dec). $[\alpha]_D^{20}$ +161.5°, also reported as +173° (Bodin). Hygroscopic. Sol in water, methylene chloride. Freely sol in alc, chloroform. Very slightly sol in ether. LD_{50} in mice (mg/kg): 8529 orally; 176 i.p.; 9475 s.c.; 184 i.v. (Edanaga).
THERAP CAT: Antibacterial.

925. Bacillus thuringiensis. Bt; Sporeine. A gram-positive spore-forming bacterium which during sporulation forms a parasporal protein crystal with insecticidal properties. Various subspecies are used as bioinsecticides against Lepidoptera, Diptera, and Coleoptera pests. First isolated by Ishiwata in 1902 from dying silkworm larvae. Later isolated as *B. thuringiensis*, Berliner, from the larvae of the flour moth, *Ephestia kuehniella* Zell: Berliner, *Z. Angew. Entomol.* **2**, 29 (1915). Characterization of the protein crystal: K. W. Nickerson, *Biotechnol. Bioeng.* **22**, 1305 (1950); C. L. Hannay, *Nature* **172**, 1004 (1953). The protein crystals consist of one or more types of protoxin subunits known as δ-endotoxins. Bt also produces an exotoxin which is released in the culture medium during bacterial growth. Isoln and structure of the exotoxin: Sebesta *et al.*, *Collect. Czech. Chem. Commun.* **34**, 891 (1969); Farkas *et al.*, *ibid.* 1118. Review of ultrastructure, physiology and biochemistry: L. A. Bulla *et al.*, *Crit. Rev. Microbiol.* **8**, 147-204 (1980); of industrial production: X.-M. Yang, S. S. Wang, *Biotechnol. Appl. Biochem.* **28**, 95-98 (1998). Review of persistence in soil and effects on non-target organisms: J. A. Addison, *Can. J. For. Res.* **23**, 2329-2342 (1993); of mammalian safety and toxicity studies: J. T. McClintock *et al.*, *Pestic. Sci.* **45**, 95-105 (1995). Review of expression of toxin genes in plant cells: M. Mazier *et al.* in *Biotechnology Annual Review* vol. **3**, M. R. El-Gewely, Ed. (Elsevier Science, Amsterdam, 1997) pp

313-347; of use in transgenic plants for insect resistance: L. Jouanin *et al.*, *Plant Sci.* **131**, 1-11 (1998).
    Bacillus thuringiensis subsp. *israelensis.* Skeetal; Teknar.
    Bacillus thuringiensis subsp. *kurstaki.* Deliver; DiPel; Javelin; Thuricide.
USE: Bioinsecticide.

926. Bacilysin. [29393-20-2] L-Alanyl-3-[(1R,2S,6R)-5-oxo-7-oxabicyclo[4.1.0]hept-2-yl]-L-alanine; [1R-(1α,2β,6α)]-N-L-alanyl-3-(5-oxo-7-oxabicyclo[4.1.0]hept-2-yl)-L-alanine; α-[(2-ami-no-1-oxopropyl)amino]-5-oxo-7-oxabicyclo[4.1.0]heptane-2-propa-noic acid; α-(2-aminopropionamido)-5-oxo-7-oxabicyclo[4.1.0]hep-tane-2-propionic acid, stereoisomer; bacillin; tetaine. $C_{12}H_{18}N_2O_5$; mol wt 270.29. C 53.32%, H 6.71%, N 10.36%, O 29.60%. Antibiotic produced by the soil bacillus NCTC 7197: Gilliver *et al.*, in *Antibiotics* vol. I, Florey *et al.*, Eds. (Oxford, 1949) p 458. Production by *Bacillus subtilis* and purification: Rogers *et al.*, *Biochem. J.* **97**, 573 (1965). Identity with tetaine: K. Kaminski, T. Sololowska, *J. Antibiot.* **26**, 184 (1973). Identity with bacillin: K. Atsumi *et al.*, *ibid.* **28**, 77 (1975). Improved isoln: Walker, Abraham, *Biochem. J.* **118**, 557 (1970). Structural study: Rogers *et al.*, *ibid.* **97**, 579 (1965). Final structure: Walker, Abraham, *ibid.* **118**, 563 (1970).

White amorphous powder. Freely sol in water; sol in 80% alc; sparingly sol in abs alc. Stable in aq soln at 100° for 5 min at pH 7; becomes inactive at pH 2 or pH 9.

927. Bacitracin. [1405-87-4] Ak-Tracin; Baciim; Fortracin; Ocu-Tracin. Antibiotic polypeptide complex produced by *Bacillus subtilis* and *B. licheniformis*; mixture of at least nine bacitracins of which **bacitracin A** is the major component. Isoln: Johnson *et al.*, *Science* **102**, 376 (1945); Anker *et al.*, *J. Bacteriol.* **55**, 249 (1948). Purification of bacitracin by carrier displacement method: Porath, *Acta Chem. Scand.* **6**, 1237 (1952). Purification with ion exchange resin: Chaiet, Cochrane, **US 2915432** (1959 to Merck & Co.). Production: Johnson, Meleney, **US 2498165** (1950 to the U.S. Secy. of War); Freaney, Allen, **US 2828246** (1958 to Commercial Solvents). Solubilities: Weiss *et al.*, *Antibiot. Chemother.* **7**, 374 (1957). Preliminary structure studies: Hausmann *et al.*, *J. Am. Chem. Soc.* **77**, 723 (1955); Lockhart *et al.*, *Biochem. J.* **61**, 534 (1955); Stoffel, Craig, *J. Am. Chem. Soc.* **83**, 145 (1961). Structure of bacitracin A: Ressler, Kashelikar, *ibid.* **88**, 2025 (1966); Galardy *et al.*, *Biochemistry* **10**, 2429 (1971). Synthetic studies: Munekata *et al.*, *Bull. Chem. Soc. Jpn.* **46**, 3187, 3835 (1973). Mechanism of action: Storm, *Ann. N.Y. Acad. Sci.* **235**, 387 (1974). Comprehensive description: G. A. Brewen, *Anal. Profiles Drug Subs.* **9**, 1-69 (1980). *Reviews:* Craig *et al.*, "Bacitracin" in G. E. W. Wolstenholme, C. M. O'Connor, *Ciba Foundation Symposium on Amino Acids and Peptides with Antimetabolic Activity* (Little, Brown, Boston, 1958) pp 226-246; E. D. Weinberg in *Antibiotics*, D. Gottlieb, P. D. Shaw, Eds. (Springer-Verlag, New York, 1967) **I**, pp 90-99; **II**, pp 240-245; D. R. Storm, W. A. Toscano, Jr. in *Antibiotics* **vol. 5**, pt. 1, F. E. Hahn, Ed. (Springer-Verlag, New York, 1979) pp 1-17.

Bacitracin A

Grayish-white powder. Very bitter taste. Hygroscopic. Freely sol in water, alcohol, methanol, glacial acetic acid. Practically insol in ether, chloroform, acetone. Stable in acid soln; unstable in alkaline solns. Potency loss probably due to transformation of bacitracin A to bacitracin F, latter having little antimicrobial activity.

THERAP CAT: Antibacterial.

THERAP CAT (VET): Antibacterial.

928. Bacitracin Methylenedisalicylate. [55852-84-1] Bacitracin methylenebis[2-hydroxybenzoate] (salt); bacitracin methylenedisalicylic acid; BMD; Solu-Tracin. Prepn: Baron, US 2774712 (1956 to S. B. Penick & Co.). LC determn in animal feed: G. K. Webster, *J. AOAC Int.* **80**, 732 (1997). Efficacy as growth promotant: A. L. Izat *et al.*, *Poult. Sci.* **69**, 1787 (1990); M. D. Sims *et al.*, *ibid.* **83**, 1148 (2004); in combination with narasin vs necrotic enteritis: J. Brennan *et al.*, *ibid.* **82**, 360 (2003).

White to grayish-brown powder. Slightly disagreeable odor. Somewhat less bitter than bacitracin. uv max (dil acetic acid): 318 nm, min: 280 nm. Solubility in water about 50 mg/ml. pH of satd aq soln 3.5 to 5.0. Sol in pyridine, ethanol, less sol to insol in acetone, ether, chloroform, pentane, benzene. Sol in dil aq alkali at pH 6 and higher, increasingly insol at pH 6 to 3.

Sodium salt. Creamy-white powder. Sol in water. pH 2% aq soln 9.5.

THERAP CAT (VET): Animal growth promotant; antibacterial.

929. Bacitracin Zinc. [1405-89-6] Zinc bacitracin; Albac; Baciferm. Prepn: H. S. Anker *et al.*, *J. Bacteriol.* **55**, 249 (1948). Review of properties and formulatory characteristics: H. M. Gross, *Drug Cosmet. Ind.* **75**, 612 (1954); *idem et al.*, *J. Am. Pharm. Assoc. Sci. Ed.* **45**, 447 (1956). Solubility data: P. J. Weiss *et al.*, *Antibiot. Chemother.* **7**, 374 (1957). ELISA determn in animal feed: C. Williams *et al.*, *Analyst* **119**, 427 (1994). Field trials as performance enhancer in pigs: S. C. Kyriakis *et al.*, *Vet. Rec.* **138**, 489 (1996); in chickens: G. Huyghebaert, G. De Groote, *Poult. Sci.* **76**, 849 (1997). Clinical trial in giardiasis: B. J. Andrews *et al.*, *Am. J. Trop. Med. Hyg.* **52**, 318 (1995); in prevention of wound infection: D. J. Dire *et al.*, *Acad. Emerg. Med.* **2**, 4 (1995).

Creamy powder contg 1-4% H_2O. Less bitter and more stable than bacitracin. Hygroscopic. Soly at about 28° (mg/ml): water 5.1; methanol 6.55; ethanol 2.0; isopropanol 0.16; ethyl acetate 1.3; chloroform 0.01; petr ether 0.025.

THERAP CAT: Antibacterial.

THERAP CAT (VET): Antibacterial; growth promotant.

930. Baclofen. [1134-47-0] β-(Aminomethyl)-4-chlorobenzenepropanoic acid; β-(aminomethyl)-*p*-chlorohydrocinnamic acid; γ-amino-β-(*p*-chlorophenyl)butyric acid; β-(4-chlorophenyl)-GABA; Ba-34647; Lioresal. $C_{10}H_{12}ClNO_2$; mol wt 213.66. C 56.22%, H 5.66%, Cl 16.59%, N 6.56%, O 14.98%. Specific GABA-B receptor agonist; activity resides in the (*R*)-isomer. Prepn: NL 6407755; H. Keberle *et al.*, US 3471548 (1965, 1969 both to Ciba). Resolution and bioactivity of isomers: H.-R. Olpe *et al.*, *Eur. J. Pharmacol.* **52**, 133 (1978). Abs config of (*R*)-form: C. H. Chang *et al.*, *Acta Crystallogr. B* **38**, 2065 (1982). Stereoselective synthesis of (*R*)-form: C. Herdeis, H. P. Hubmann, *Tetrahedron: Asymmetry* **3**, 1213 (1992). Toxicity study: T. Tadokoro *et al.*, *Osaka Daigaku Igaku Zasshi* **28**, 265 (1976), *C.A.* **88**, 183016u (1978). Comprehensive description: S. Ahuja, *Anal. Profiles Drug Subs.* **14**, 527-548 (1985). Review of pharmacology and therapeutic efficacy in spasticity: R. N. Brogden *et al.*, *Drugs* **8**, 1-14 (1974); of intrathecal use in spinal cord injury: K. S. Lewis, W. M. Mueller, *Ann. Pharmacother.* **27**, 767-774 (1993). Clinical evaluation in reflex sympathetic dystrophy: B. J. van Hilten *et al.*, *N. Engl. J. Med.* **343**, 625 (2000). Benefit-risk assessment in severe spinal spasticity: A. Dario, G. Tomei, *Drug Saf.* **27**, 799-818 (2004).

Crystals from water, mp 206-208°. Slightly sol in water; very slightly sol in methanol. Insol in chloroform. LD_{50} in male mice, rats (mg/kg): 45, 78 i.v.; 103, 115 s.c.; 200, 145 orally (Tadokoro).

Hydrochloride. [28311-31-1] $C_{10}H_{12}ClNO_2.HCl$; mol wt 250.12. White tabular crystals from water. d (cryst) 1.386. mp 179-181°.

(*R*)-Form. [69308-37-8] Arbaclofen; *l*-baclofen; STX-209. White solid. mp 205-208°.

(*R*)-Form hydrochloride. [63701-55-3] Colorless crystals, mp 219°. $[\alpha]_D^{25} -2°$ (c = 0.2 in water).

THERAP CAT: Muscle relaxant (skeletal).

THERAP CAT (VET): Muscle relaxant (skeletal).

931. Bactericidal Permeability-Increasing Protein. BPI. Antibacterial protein found in azurophilic granules of mammalian neutrophils; plays a predominant role during phagocytosis. Selectively cytotoxic for gram negative bacteria; binds to both soluble and membrane lipopolysaccharide (LPS) and neutralizes endotoxin activity. Cationic protein containing 456 amino acid residues having a characteristic "boomerang" shape; mol wt ~55 kDa. Antibacterial activity resides in the amino terminal half while the C-terminal domain facilitates phagocytosis. Isoln from human polymorphonuclear leukocytes: J. Weiss *et al.*, *J. Biol. Chem.* **253**, 2664 (1978). Cloning and primary structure: P. W. Gray *et al.*, *ibid.* **264**, 9505 (1989). Crystal structure: L. J. Beamer *et al.*, *Science* **276**, 1861 (1997); *eidem*, *Biochem. Pharmacol.* **57**, 225 (1999). ELISA determn in body fluids: M. L. White *et al.*, *J. Immunol. Methods* **167**, 227 (1994). Review of role in host defense: P. Elsbach, *J. Leukocyte Biol.* **64**, 14-18 (1998); P. Elsbach, J. Weiss, *Curr. Opin. Immunol.* **10**, 45-49 (1998).

Opebacan. [158113-60-1] 1-193-Bactericidal/permeability-increasing protein [132-alanine] (human reduced); $rBPI_{21}$; Neuprex. Amino-terminal fragment of human BPI produced by recombinant DNA technology in Chinese hamster ovary cells. Biologically active mutein in which the cysteine at position 132 has been replaced with alanine; mol wt 25 kDa. Prepn: G. Theofan *et al.*, WO 9418323; *eidem*, US 5420019 (1994, 1995 both to Xoma); A. H. Horwitz *et al.*, *Protein Expression Purif.* **8**, 28 (1996). Clinical trial in meningococcal sepsis: M. Levin *et al.*, *Lancet* **356**, 961 (2000). Review of development and clinical potential in meningococcemia and hemorrhagic trauma: P. J. Scannon, *J. Endotoxin Res.* **5**, 209-212 (1999); O. Levy, *Expert Opin. Invest. Drugs* **11**, 159-167 (2002).

THERAP CAT: Antisepsis; in treatment of meningococcemia.

932. Bacteriorhodopsin. BR. The only protein of the purple membrane of *Halobacterium halobium* and other halophilic bacteria. It is a rhodopsin-like pigment containing retinal linked to lysine through a Schiff's base and has a mol wt of approx 26,000. Bacteriorhodopsin functions as an energy transducer or "proton pump"; unlike animal rhodopsins, it uses light energy to generate an electrochemical gradient and this stored energy is used by the cell for ATP synthesis and other important energy-requiring functions. Discovery in the purple membrane of *Halobacterium halobium:* D. Oesterhelt, W. Stoekenius, *Nature New Biol.* **233**, 149 (1971); A. E. Blaurock, W. Stoekenius, *ibid.* 152. Isoln and identification in *H. cutirubrum:* S. C. Kushwaha, M. Kates, *Biochim. Biophys. Acta* **316**, 235 (1973). Description of functions: D. Oesterhelt, W. Stoekenius, *Proc. Natl. Acad. Sci. USA* **70**, 2853 (1973). Proposed mechanism of the "proton pump": K. Schulten, P. Tavan, *Nature* **272**, 85 (1978). Structural elucidation: Y. A. Ovchinnikov *et al.*, *FEBS Lett.* **100**, 219 (1979); H. G. Khorana *et al.*, *Proc. Natl. Acad. Sci. USA* **76**, 5046 (1979). Three-dimensional crystallographic study: H. Michel, D. Oesterhelt, *ibid.* **77**, 1283 (1980). Series of articles on structure, biosynthesis, and energy transduction: *Photochem. Photobiol.* **33**, 417-608 (1981). Review of energy transduction: H. V. Westerhoff, Z. Dancshazy, *Trends Biochem. Sci.* **9**, 112 (1984). Comprehensive reviews: W. Stoekenius *et al.*, *Biochim. Biophys. Acta* **505**, 215-278 (1979); W. Stoekenius, R. A. Bogomolni, *Annu. Rev. Biochem.* **51**, 587-616 (1982). Review of crystallization, structure and function: J. Deisenhoffer, H. Michel, *Science* **245**, 1463-1473 (1989). Review of photophysical properties and optimization for use in bioelectronic devices: K. J. Wise *et al.*, *Trends Biotechnol.* **20**, 387-394 (2002).

Bacteriorhodopsin is not bleached by light in the same fashion as animal rhodopsins; only small spectral changes are caused by transfer from light to dark. When exposed to light, BR has a broad light absorption maximum at 568 nm that slowly shifts back to 558 nm in the dark. The 558 and 568 nm states are known as the dark-adapted and light-adapted states, respectively. Light-adapted BR, upon absorption of a light quantum, undergoes a series of reversible changes, having intermediate forms with absorption maxima of 590, 550, and 412 nm. uv max (2M HCl): 565 nm. Slow bleaching occurs when membrane suspensions are illuminated in the presence of hydroxylamine or sodium borohydride. When hydroxylamine is used, the uv max is shifted to 360 nm. The bleached membrane is called the "apomembrane"; the protein after loss of the chromophore is termed *bacterio-opsin*.

USE: As a tool in biological energy transduction research.

933. Badische Acid. [86-60-2] 7-Amino-1-naphthalenesulfonic acid; 2-naphthylamine-8-sulfonic acid. $C_{10}H_9NO_3S$; mol wt 223.25. C 53.80%, H 4.06%, N 6.27%, O 21.50%, S 14.36%. Prepd by sulfonation of 2-naphthylamine: Green, Vakil, *J. Chem. Soc.* **113**, 35 (1918); Hennion, Schmidle, *J. Am. Chem. Soc.* **65**, 2468 (1943).

Nec... or prisms. Sol in 1680 parts water, in alkalies; slightly sol in alc.

934. Bagasse. Fibrous residue from cane sugar mfg operations, which contains 50% cellulose, 25% pentosans, and 25% lignin: Wiggins, *Sugar J.* **16**, No. 8, pp 18, 22, 24 (1954), *C.A.* **48**, 9729h (1954). Review of properties and uses: Vo Tong Xuan, Samaniego, *Sugar News* **46**, 276, 284 (1970); M. A. Clarke in *Kirk-Othmer Encyclopedia of Chemical Technology* vol. 3 (Wiley-Interscience, New York, 3rd ed., 1978) pp 434-438.

Caution: Inhalation of bagasse dust may cause pneumonitis, asthma.

USE: Raw material for pulp, paper and board.

935. Baicalein. [491-67-8] 5,6,7-Trihydroxy-2-phenyl-4H-1-benzopyran-4-one; 5,6,7-trihydroxyflavone; noroxylin. $C_{15}H_{10}O_5$; mol wt 270.24. C 66.67%, H 3.73%, O 29.60%. From roots of *Scutellaria baicalensis*. Isoln and structure: Bargellini, *Gazz. Chim. Ital.* **49**, II, 47 (1919); Shibata *et al.*, *Acta Phytochim.* **1**, 109 (1923). Synthesis: Sastri, Seshadri, *Proc. Indian Acad. Sci.* **23A**, 262 (1946), *C.A.* **41**, 449 (1947); Schönberg *et al.*, *J. Am. Chem. Soc.* **77**, 5390 (1955); Jouanne, Mentzer, *Compt. Rend.* **254**, 727 (1962); Agasimundin, Siddappa, *J. Chem. Soc. Perkin Trans. 1* **1973**, 503. Pharmacology: Koda *et al.*, *C.A.* **75**, 47200d (1971).

Yellow prisms from alc, dec 264-265°. uv max (ethanol): 324, 276 nm (log ε 4.18, 4.42). Sol in alcohol, methanol, ether, acetone, ethyl acetate, hot glacial acetic acid. Sparingly sol in chloroform, nitrobenzene. Practically insol in water. Sol in dil NaOH with greenish-brown color. Concd H_2SO_4 gives yellow color, green fluorescence.

THERAP CAT: Astringent.

936. Bakankosin. [1398-17-0] (3S,4R,4aS)-4-Ethenyl-3-(β-D-glucopyranosyloxy)-3,4,4a,5,6,7-hexahydro-8H-pyrano[3,4-c]-pyridin-8-one; [3S-(3α,4β,4aα)]-4-ethenyl-3-(β-D-glucopyranosyloxy)-3,4,4a,5,6,7-hexahydro-8H-pyrano[3,4-c]pyridin-8-one; bacancosin; Bakankoside. $C_{16}H_{23}NO_8$; mol wt 357.36. C 53.78%, H 6.49%, N 3.92%, O 35.82%. Nitrogenous glucoside from seed of *Strychnos vacacoua* Baill., *Loganiaceae*. Ref: E. Bourquelot, H. Hérissey, *Compt. Rend.* **144**, 575 (1907); **147**, 750 (1908); *Arch. Pharm.* **247**, 56 (1909); K. Balenovic *et al.*, *Helv. Chim. Acta* **35**, 2519 (1952). Proposed structure: G. Büchi, R. E. Manning, *Tetrahedron Letters* no. 26, 5 (1960); H. C. Beyerman *et al.*, *Bull. Soc.*

Chim. Fr. **1961**, 1812; H. Inouye *et al.*, *Chem. Pharm. Bull.* **24**, 1406 (1976). Synthesis and structure: L. F. Tietze, *Tetrahedron Lett.* **29**, 2535 (1976).

Hydrate. Bitter crystals, mp 157°; when anhydr mp 200°, also reported as mp 211-212°. $[\alpha]_D -197°$. uv max (alcohol): 235 nm (log ε 4.2). Freely sol in water, alc; slightly sol in ethyl acetate.

937. Balofloxacin. [127294-70-6] 1-Cyclopropyl-6-fluoro-1,4-dihydro-8-methoxy-7-[3-(methylamino)-1-piperidinyl]-4-oxo-3-quinolinecarboxylic acid; Q-35; Baloxin. $C_{20}H_{24}FN_3O_4$; mol wt 389.43. C 61.69%, H 6.21%, F 4.88%, N 10.79%, O 16.43%. Fluorinated quinolone antibacterial. Prepn: H. Nagano *et al.*, **EP 342675**; *eidem*, **US 5051509** (1989, 1991 both to Chugai). Photostability: K. Marutani *et al.*, *Antimicrob. Agents Chemother.* **37**, 2217 (1993). Comparative antibacterial spectrum: H. Iwasaki *et al.*, *Chemotherapy (Basel)* **41**, 100 (1995). Pharmacokinetics: T. Nakagawa *et al.*, *Arzneim.-Forsch.* **45**, 719 (1995). HPLC determn in biological fluids: T. Nakagawa *et al.*, *ibid.* 716. Clinical evaluation in respiratory tract infection: O. Kunii, F. Matsumoto, *Proc. 18th Int. Congr. Chemother.*, *Stockholm 1993*, 22-23; in urinary tract infection: J. Kumazawa, F. Matsumoto, *ibid.* 69-70.

Colorless needles from acetonitrile-water, mp 134-135°.
THERAP CAT: Antibacterial.

938. Balsalazide. [80573-04-2] 5-[(1E)-2-[4-[[(2-Carboxyethyl)amino]carbonyl]phenyl]diazenyl]-2-hydroxybenzoic acid; (E)-5-[[p-[(2-carboxyethyl)carbamoyl]phenyl]azo]-2-salicylic acid; 5-[(1E)-[4-[[(2-carboxyethyl)amino]carbonyl]phenyl]azo]-2-hydroxybenzoic acid. $C_{17}H_{15}N_3O_6$; mol wt 357.32. C 57.14%, H 4.23%, N 11.76%, O 26.86%. Analog of sulfasalazine, *q.v.* Prodrug of 5-aminosalicylic acid where carrier molecule is 4-aminobenzoyl-β-alanine. Prepn: R. P. K. Chan, **GB 2080796**; *idem*, **US 4412992** (1982, 1983 both to Biorex). Toxicology study and clinical metabolism: *idem et al.*, *Dig. Dis. Sci.* **28**, 609 (1983). Review of pharmacology and clinical efficacy in ulcerative colitis: A. Prakash, C. M. Spencer, *Drugs* **56**, 83 (1998).

Crystals from hot ethanol, mp 254-255°.
Disodium salt dihydrate. [150399-21-6]; [82101-18-6] (anhydrous). BX-661A; Colazal; Colazide; Colazide. $C_{17}H_{13}N_3Na_2O_6 \cdot 2H_2O$; mol wt 437.32. Orange to yellow microcrystalline powder, mp >350°.

Nonhygroscopic. Freely sol in water, isotonic saline; sparingly sol in methanol, ethanol. Practically insol in organic solvents.

THERAP CAT: Anti-inflammatory (gastrointestinal).

939. Balsam Canada. Canada turpentine; balsam of fir. Improperly *"Balm of Gilead"*. Liquid oleoresin from *Abies balsamea* (L.), Mill., *Pinaceae*. *Habit*. Canada and Northern U.S. to Va., west to Minnesota. *Constit*. 27.5% Volatiles (pinene, nopinene, β-phellandrene), 44.5% resin acid (13% abietic, 8% neoabietic), 27% neutral resinous compounds. *Ref*: Lombard *et al.*, *Peint. Pigm. Vernis* **34**, 106 (1958), *C.A.* **52**, 12420 (1958).

Yellowish to greenish, viscid, transparent, slightly fluorescent liquid; agreeable, aromatic pine-like odor; bitter taste; on exposure to air gradually solidifies to a solid, noncryst mass. d 0.987-0.994. n_4^{20} 1.52-1.54. $[\alpha]_D^{20}$ +1 to +4°. Acid no. 84-87. Sapon no. 89.4-95.7 (2 g in xylene for 1 hr). Insoluble in water; miscible with benzene, chloroform, xylene, ethyl acetate, oil of cedar; completely sol or almost sol in ether, oil turpentine; about 90% dissolves in alcohol or petr ether.

USE: Cement for lenses; manuf fine lacquers; for mounting in microscopy.

940. Balsam Gurjun. Wood oil; "East Indian Copaiba". Oleoresin from various species of *Dipterocarpus*, *Dipterocarpaceae*. *Habit*. Eastern India, Burma. *Constit*. About 75% volatile oil, boiling at about 255°; gurjunic acid, resin, bitter substance sol in water. Isoln of an optically inactive sesquiterpene δ-elemene, from gurjun balsam: Gough, Powell, *Tetrahedron Lett.* **1961**, 763.

Clear liquid, light brown by transmitted light with greenish fluorescence; somewhat bitter, but not very acrid taste. Acid no. 5-15. Sapon no. 10-20 (2 g in 20 g xylene for 1 hr). d 0.95-0.97; $[\alpha]_D$ −23 to −70°; n_D^{28} 1.510-1.516. Insol in water. Completely sol in benzene, chloroform; incompletely in alcohol, ether, carbon disulfide, petr ether.

USE: As of copaiba; adapted for varnishes and lacquers, particularly for articles to be exposed to a temp of 80° or so.

941. Balsam Mecca. Balm of Gilead; balsam of Gilead; balsán-Katél; Duhnul-balsan. Balsam obtained from twigs of *Commiphora opobalsamum* (Kunth.) Engl. (*Balsamodendrum gilaedense* Kunth.), *Burseraceae*.

Light-colored, mobile to viscid, turbid, brownish-red liquid; aromatic odor. Insoluble in water; sol in alcohol, benzene, chloroform, acetone, glacial acetic acid, carbon disulfide, oil turpentine, ether.

USE: In perfumery.

942. Balsam Peru. Peruvian balsam; Indian balsam; China oil; Black balsam; Honduras balsam; Surinam balsam. From *Toluifera pereirae* (Klotzsch) Baill. (*Myroxylon pereirae* Klotzsch), *Leguminosae*. *Habit*. Central America (San Salvador) in forests near Pacific coast. *Constit*. 50-60% cinnamein—esters of cinnamic and benzoic acid; about 28% resin, styracine, vanillin. *Ref*: Bergemann, *Pharmazie* **5**, 494 (1950); Cortesi, *Bull. Soc. Pharm. Bordeaux* **89**, 141 (1951), *C.A.* **46**, 1718b (1952).

Dark brown, viscid liquid; pleasant aromatic odor; warm, bitter taste and persistent aftertaste. d 1.150-1.170. Insol in water, olive oil; sol in alcohol, chloroform and glacial acetic acid, usually with a slight opalescence. Partly sol in ether, petr ether.

USE: In perfumery and some chocolate flavorings; also in masking of odors.

THERAP CAT: Vulnerary; ectoparasiticide (scabies).

THERAP CAT (VET): Vulnerary; miticide.

943. Balsam Tolu. Thomas balsam; opobalsam; resin Tolu. From *Toluifera balsamum* L. (*Myroxylon toluiferum* H. B. K.), *Leguminosae*. *Habit*. South America (Venezuela, Colombia, Peru) on elevated plains and mountains. *Constit*. 12-15% free cinnamic and benzoic acids; about 40% benzyl, etc., esters of these acids (5.2-13.4% cinnamein); 1.5-3% volatile oil. *Ref*: Rosenthaler, *Pharm. Ztg.-Nachr.* **88**, 716 (1952), *C.A.* **47**, 5076 (1953).

Yellowish-brown or brown, semifluid or nearly solid, resinous mass; aromatic odor and taste; brittle when cold. Sol in alc, benzene, chloroform, ether, glacial acetic acid, partially in carbon disulfide or NaOH. Practically insol in water, petr ether, solvent hexane.

USE: In perfumery, confectionery and chewing gums. In pharmacy, as ingredient and vehicle for expectorants.

THERAP CAT: Expectorant.

THERAP CAT (VET): Has been used as an expectorant.

944. Balsam Traumatic. Friar's balsam; Turlington's balsam. Composed of 100 parts benzoin, 35 storax, 35 balsam Tolu, 16 balsam Peru, 8 aloe, 8 myrrh, 4 angelica, and alcohol to make 1000.

THERAP CAT: Topical protectant; expectorant.

THERAP CAT (VET): Has been used for chronic bronchitis. Wound antiseptic and styptic.

945. Bambermycins. [11015-37-5] Bambermycin; moenomycin; flavophospholipol; Flavomycin. Antibiotic complex comprised of at least 4 active components, moenomycins A, B₁, B₂ and C, with *moenomycin A* being the major component. Obtained from cultures of *Streptomyces bambergiensis*, *S. ghanaensis*, *S. ederensis*, *S. geysiriensis* and related strains: Wallhäuser *et al.*, *Antimicrob. Agents Chemother.* **1965**, 734; NL 6602132, *C.A.* **66**, 74946x (1967); Lindner, Wallhäuser, US 3674866 (1966, 1972 both to Hoechst). Structural studies: Tschesche *et al.*, *Ann.* **720**, 58 (1968); *Tetrahedron Lett.* **1968**, 2905; **1969**, 141; Huber, *J. Antibiot.* **25**, 1226 (1972); P. Welzel *et al.*, *Tetrahedron Lett.* **1973**, 227; F. J. Witteler *et al.*, *ibid.* **1979**, 3493; P. Welzel *et al.*, *Tetrahedron* **39**, 1583 (1983). Structure of moenomycin A: P. Welzel *et al.*, *Angew. Chem. Int. Ed.* **20**, 121 (1981). Synthesis of *moenocinol*, the lipid moiety: Tschesche, Reden, *Ann.* **1974**, 853; P. J. Kocienski, *J. Org. Chem.* **45**, 2037 (1980); R. M. Coates, M. W. Johnson, *ibid.* 2685. *Review*: G. Huber in *Antibiotics* vol. 5, pt. 1, F. E. Hahn, Ed. (Springer-Verlag, New York, 1979) pp 135-153.

Moenomycin A

Complex is a colorless, amorphous solid, no definite mp, decompn starting at 200°C. uv max (water pH 7): 258 nm ($E_{1cm}^{1\%}$ 60). Sol in water, methanol, DMF; less sol in ethanol, propanol; slightly sol in ether, ethyl acetate. Insol in benzene, chloroform. Stable in neutral aq and methanolic solns; slowly decomp in acid and alkaline solns. LD₅₀ in mice (mg/kg): >2000 orally, s.c. and i.p.; 1400 i.v. (Lindner, Wallhäuser).

THERAP CAT: Antibacterial.

THERAP CAT (VET): Antibacterial; feed additive (poultry, swine and calves).

946. Bambuterol. [81732-65-2] *N,N*-Dimethylcarbamic acid *C,C′*-[5-[2-[(1,1-dimethylethyl)amino]-1-hydroxyethyl]-1,3-phenylene] ester; 1-[bis(3′,5′-*N,N*-dimethylcarbamoyloxy)phenyl]-2-*N*-tert-butylaminoethanol; (±)-5-[2-(*tert*-butylamino)-1-hydroxyethyl]-*m*-phenylene bis(dimethylcarbamate); dimethylcarbamic acid 5-[2-[(1,1-dimethylethyl)amino]-1-hydroxyethyl]-1,3-phenylene ester; terbutaline bisdimethylcarbamate. C₁₈H₂₉N₃O₅; mol wt 367.45. C 58.84%, H 7.96%, N 11.44%, O 21.77%. Ester prodrug of the β₂-adrenergic agonist terbutaline, *q.v.* Prepn: O. A. T. Olsson *et al.*, **EP 43807**; *eidem*, **US 4419364** (1982, 1983 both to AB Draco).

Pharmacology: O. A. T. Olsson, L.-A. Svensson, *Pharm. Res.* **1984**, 19. Review of pharmacology and metabolism: L.-A. Svensson, *Acta Pharm. Suec.* **24**, 333 (1987). LC determn: O. Wannerberg, B. Persson, *J. Chromatogr.* **435**, 199 (1988). Clinical evaluation in asthma: T. Sandström *et al.*, *Respiration* **53**, 31 (1988). Clinical comparison with terbutaline: G. Persson *et al.*, *Eur. Respir. J.* **1**, 223 (1988).

Hydrochloride. [81732-46-9] KWD-2183; Bambec. $C_{18}H_{29}$-N_3O_5.HCl; mol wt 403.90.

THERAP CAT: Bronchodilator.

947.   Bamethan. [3703-79-5] α-[(Butylamino)methyl]-4-hydroxybenzenemethanol; α-[(butylamino)methyl]-*p*-hydroxybenzyl alcohol; 1-(*p*-hydroxyphenyl)-2-butylaminoethanol; 1-(4-hydroxyphenyl)-1-hydroxy-2-butylaminoethane; 2-butylamino-1-*p*-hydroxyphenylethanol. $C_{12}H_{19}NO_2$; mol wt 209.29. C 68.87%, H 9.15%, N 6.69%, O 15.29%. Prepn: Corrigan *et al.*, *J. Am. Chem. Soc.* **67**, 1894 (1945); Kovács, *Pharm. Zentralhalle* **92**, 193 (1953), *C.A.* **49**, 936 (1955). Metabolism of sulfate: Hajos, Szporny, *Arzneim.-Forsch.* **18**, 1212 (1968).

Crystals, mp 123.5-125°.

Hydrochloride. $C_{12}H_{19}NO_2$.HCl. mp 109-110°.

Sulfate. [5716-20-1] Vasculat; Vasculit; Bupatol; Garmian. $(C_{12}H_{19}NO_2)_2.H_2SO_4$; mol wt 516.65.

THERAP CAT: Vasodilator (peripheral).

948.   Bamifylline. [2016-63-9] 7-[2-[Ethyl(2-hydroxyethyl)-amino]ethyl]-3,7-dihydro-1,3-dimethyl-8-(phenylmethyl)-1*H*-purine-2,6-dione; 8-benzyl-7-[2-[ethyl(2-hydroxyethyl)amino]ethyl]-theophylline; 8-benzyl-7-[*N*-ethyl-*N*-(β-hydroxyethyl)aminoethyl]-theophylline; benzetamophylline; bamiphylline; 8102 CB. $C_{20}H_{27}$-N_5O_3; mol wt 385.47. C 62.32%, H 7.06%, N 18.17%, O 12.45%. Prepn: **BE 602888** (1961 to Christiaens), *C.A.* **56**, 5981c (1962). Pharmacology: Georges *et al.*, *Therapie* **17**, 211 (1962). Metabolism: Dodion *et al.*, *Arzneim.-Forsch.* **19**, 785 (1969). Toxicological studies: Georges *et al.*, *ibid.* **18**, 460 (1968).

Crystals, mp 80-80.5°.

Hydrochloride. [20684-06-4] AC-3810; BAX-2739Z; Briofil; Trentadil. $C_{20}H_{27}N_5O_3$.HCl; mol wt 421.93. Crystals, mp 185-186°. LD_{50} in mice, rats (mg/kg): 246, 1139 orally; 89, 131 i.p.; 67, 65 i.v. (Georges).

THERAP CAT: Bronchodilator.

949.   Bamipine. [4945-47-5] 1-Methyl-*N*-phenyl-*N*-(phenylmethyl)-4-piperidinamine; 4-*N*-benzylanilino-1-methylpiperidine; *N*-phenyl-*N*-benzyl-4-amino-1-methylpiperidine. $C_{19}H_{24}N_2$; mol

wt 280.42. C 81.38%, H 8.63%, N 9.99%. Prepn: Kallischnigg, **US 2683714** (1954 to Knoll). Toxicity studies: Hanna, *Toxicol. Appl. Pharmacol.* **3**, 393 (1961). Determn in pharmaceutical preparations: M. H. Barary, *Pharmazie* **39**, 706 (1984).

Crystals from butanone or methanol, mp 115°.

Hydrochloride. [1229-69-2] Soventol. $C_{19}H_{24}N_2$.HCl; mol wt 316.87.

THERAP CAT: Antihistaminic.

950.   Baobab. Upside-down tree; monkey's bread tree. Deciduous, fruit bearing tree, *Adansonia digitata* L., *Malvaceae*, characterized by its massive size and depressed, ovoid crown. The indehiscent fruit is globose to ovoid, covered by velvety yellowish or greenish hairs, and contains kidney-shaped seeds and powdery, off-white pulp. *Habit.* Sub-Saharan Africa; closely related species in Madagascar, India, Australia. Edible parts are the fruit pulp, seeds, and leaves which are a nutritional source of protein, amino acids, fiber, fatty acids, carbohydrate, minerals, particularly calcium, potassium, and iron, and vitamins, especially A, B and C. The bark is used in traditional medicine as a febrifuge and treatment for malaria. Chemical and nutrient analysis of fruit and seeds: M. A. Osman, *Plant Foods Hum. Nutr.* **59**, 29 (2004). Evaluation of antiviral and anti-inflammatory activity: S. Vimalanathan, J. B. Hudson, *J. Med. Plants Res.* **3**, 576 (2009). Comparison of phenolic content and antioxidant capacity: L. Nhukarume *et al.*, *J. Food Biochem.* **34**, 207 (2010). Review of constituents and nutritional value of food products: F. J. Chadare *et al.*, *Crit. Rev. Food Sci. Nutr.* **49**, 254-274 (2009). Review of phytochemistry and uses in food and traditional medicine: E. De Caluwé *et al.*, *ACS Symp. Ser.* **1021**, 51-84 (2009).

Baobab dried fruit pulp. BDFP. Free-flowing, coarsely milled, off-white to cream colored powder with tart, acidic taste. *Constit.* 79% carbohydrate, 14% moisture, 6% ash, 2% protein, 0.5% total fat. Appreciable concentrations of vitamins C, B_1, B_2, calcium, iron, magnesium, phosphorus, potassium, sodium; fatty acids, primarily oleic, linoleic; pectin, glucose, citric acid.

USE: Leaves, fruit, and seeds as food; dried fruit pulp as nutritional supplement and flavoring ingredient.

951.   Bapineuzumab. [648895-38-9] Anti-(human β-amyloid) immunoglobulin G1 (human-mouse monoclonal heavy chain) disulfide with human-mouse monoclonal light chain, dimer; AAB-001. Humanized monoclonal antibody directed against amyloid β peptide; designed to promote the clearance of cerebral β-amyloid. Prepn: G. Basi *et al.*, **WO 0246237**; G. Basi, J. W. Saldahna, **US 7179892** (2002, 2007 both to Neuralab; Wyeth). Clinical pharmacokinetics: R. S. Black *et al.*, *Alzheimer Dis. Assoc. Disord.* **24**, 198 (2010). Clinical evaluation of β-amyloid load changes in Alzheimer's disaese: J. O. Rinne *et al,. Lancet Neurol.* **9**, 363 (2010). Review of development and therapeutic potential: G. A. Kerchner, A. L. Boxer, *Expert Opin. Biol. Ther.* **10**, 1121-1130 (2010); F. Panza *et al. Immunotherapy* **2**, 767-782 (2010).

THERAP CAT: Antiamyloidogenic agent; in treatment of Alzheimer's disease.

952.   BAPTA. [85233-19-8] *N,N'*-[1,2-Ethanediylbis(oxy-2,1-phenylene)]bis[*N*-(carboxymethyl)glycine]; 1,2-bis(2-aminophenoxy)ethane-*N,N,N',N'*-tetraacetic acid. $C_{22}H_{24}N_2O_{10}$; mol wt 476.44. C 55.46%, H 5.08%, N 5.88%, O 33.58%. Calcium (Ca^{2+}) chelator derived from EGTA. Ca^{2+} binding causes a shift in the uv absorbance and fluorescence of the molecule. Parent compound of second generation fluorescent imaging agents such as Indo-1, Fura-2, *q.q.v.* Prepn: R. Y. Tsien, *Biochemistry* **19**, 2396 (1980). Dissociation constants: R. Pethig *et al.*, *Cell Calcium* **10**, 491 (1989). Use as chelator: P. W. Marks, F. R. Maxfield, *Anal. Biochem.* **193**, 61

(1991). Binding constants and fluorescence with heavy metals: A. Yuchi *et al.*, *Bull. Chem. Soc. Jpn.* **66**, 3377 (1993). Prepn of ^{19}F-forms and use for intracellular Ca^{2+} measurement: G. A. Smith *et al.*, *Proc. Natl. Acad. Sci. USA* **80**, 7178 (1983); L. A. Levy *et al.*, *Am. J. Physiol.* **252**, C441 (1987). NMR determn of Ca^{2+} in brain by *5F-BAPTA*: H. S. Bachelard *et al.*, *J. Neurochem.* **51**, 1311 (1988).

uv max (free dye): 209, 254 nm (ε 3.8 × 10^4, 1.6 × 10^4). uv max (complexed form): 203, 274 nm (ε 4.1 × 10^4, 4.2 ×10^3).

USE: As calcium chelator; ^{19}F labelled forms for measurement of intracellular calcium.

953. Baptigenin. [5908-63-4] 7-Hydroxy-3-(3,4,5-trihydroxyphenyl)-4*H*-1-benzopyran-4-one; 7,3',4',5'-tetrahydroxyisoflavone. $C_{15}H_{10}O_6$; mol wt 286.24. C 62.94%, H 3.52%, O 33.54%. The aglucon of baptisin. From baptisin by acid hydrolysis or vacuum sublimation. From *radix Baptisiae*: Fischer, Ehrlich, *C.A.* **31**, 4449^3 (1937). Structure: Böhm, *Arzneim.-Forsch.* **10**, 472 (1960). Synthesis: Farkas *et al.*, *Ber.* **96**, 1865 (1963). *See also* Pseudobaptigenin.

Needles from dil ethanol, mp 284-285°. Begins to sublime at 240°. Sublimes in oil pump vacuum at 180-200°. uv max (ethanol): 270.2, 247 nm. Practically insol in water, ammonia water; slightly sol in dil alc, hot glacial acetic acid; sol in acetone, in NaOH solns.

Tetraacetylbaptigenin. $C_{23}H_{18}O_{10}$. Needles from methanol, mp 214°.

Tetrabenzoylbaptigenin. $C_{43}H_{26}O_{10}$. Prisms from methanol, mp 191-192°.

Tetramethylbaptigenin. $C_{19}H_{18}O_6$. Crystals from methanol, mp 144-145°.

954. Baptisia. Wild indigo; indigo weed; false indigo; yellow indigo. Perennial herb. *Habit.* North America. Medicinal baptisia, from root of *Baptisia tinctoria* R. Br., *Leguminosae. Constit.* Baptin—a purgative glucoside, baptisin—a bitter glucoside, baptitoxine (identical with cytisine), an alkaloid. Also contains a blue dye which has been used as a substitute for indigo.

THERAP CAT: Anti-infective.

955. Barbasco. Name applied in the Spanish-speaking countries of the New World to many unrelated plants used to poison or stun fish. In Mexico it usually means roots of *Dioscorea composita* Hemsl., or of *Dioscorea tepinapensis* Uline, *Dioscoreaceae* which yield up to 5% of their dry weight in diosgenin: *Chem. Week* **79**, no. 2, p 20 (July 14, 1956). Isoln procedure: Julian, **US 3019220** (1962 to Julian Labs.). Book: D. G. Coursey, *Yams* (Longmans, London, 1967) 230 pp, an account of the nature, origins, cultivation and utilization of the useful members of the *Dioscoreaceae. Compare* Yam, Mexican.

956. Barberry Bark. Berberis bark; jaundice berry; woodsour; sowberry; pepperidge bush; sour-spine. Root bark of *Berberis vulgaris* L., *Berberidaceae. Habit.* Europe and Western Asia; also U.S. (New England States, Pennsylvania and Virginia). *Constit.*

Berberine, berbamine, oxyacanthine, tannin, wax, fat, resin. *Ref:* Neugebauer, Brunner, *Pharm. Zentralhalle* **80**, 113 (1939).

957. Barbital. [57-44-3] 5,5-Diethyl-2,4,6(1*H*,3*H*,5*H*)-pyrimidinetrione; 5,5-diethylbarbituric acid; barbitone; diethylmalonylurea; Veronal. $C_8H_{12}N_2O_3$; mol wt 184.20. C 52.17%, H 6.57%, N 15.21%, O 26.06%. Prototype of the barbiturate hypnotics. Prepn: M. Conrad, M. Guthzeit, *Ber.* **15**, 2841 (1882); E. Fischer, A. Dilthey, *Ann.* **335**, 334 (1904). Use as physiological buffer: L. Michaelis, *J. Biol. Chem.* **87**, 33 (1930); in enzyme determn: J. Bergerman, *Clin. Chem.* **12**, 797 (1966).

Faintly bitter needles (trigonal in the stable phase) from water, mp 188-192°. Can be sublimed *in vacuo*. pK (25°) 7.43. One gram dissolves in about 130 ml water, 13 ml boiling water, 14 ml alcohol, 75 ml chloroform, 35 ml ether. Sol in acetone, ethyl acetate, alkalies, petr ether, acetic acid, amyl alcohol, pyridine, aniline, nitrobenzene.

Sodium salt. [144-02-5] Barbital sodium; sodium 5,5-diethylbarbiturate; barbitone sodium; soluble barbital; sodium diethylmalonylurea; Veronal sodium; Medinal. $C_8H_{11}N_2NaO_3$; mol wt 206.18. Bitter crystals or powder. One gram dissolves in 5 ml water, 2.5 ml boiling water, 400 ml alc. Aq soln is alkaline to litmus and phenolphthalein. pH of 0.1 molar aq soln, 9.4.

Note: This is a controlled substance (depressant): **21 CFR**, 1308.14.

USE: Biological buffer.

THERAP CAT: Sedative, hypnotic.

THERAP CAT (VET): Sedative, hypnotic.

958. Barbituric Acid. [67-52-7] 2,4,6(1*H*,3*H*,5*H*)-Pyrimidinetrione; malonylurea; 2,4,6-trioxohexahydropyrimidine. $C_4H_4N_2O_3$; mol wt 128.09. C 37.51%, H 3.15%, N 21.87%, O 37.47%. Prepn from hydurilic acid + nitric acid: Baeyer, *Ann.* **127**, 199 (1863); *ibid.* **130**, 129 (1864). Structure: Mulder, *Ber.* **6**, 1233 (1873). Prepd from ethyl malonate and urea using sodium ethoxide as a condensing agent: Dickey, Gray, *Org. Synth.* **coll. vol. II**, 60 (1943). Crystal structure: Bolton, *Nature* **201**, 987 (1964). Toxicity study: E. I. Goldenthal, *Toxicol. Appl. Pharmacol.* **18**, 185 (1971). Unsubstituted barbituric acid has no hypnotic properties. *Review:* Carter, *J. Chem. Educ.* **28**, 524 (1951).

Dihydrate. Rhombs from water. mp ~248° when anhydrous, with some decompn. Strong acid. K at 25° = 9.9 × 10^{-5}. uv spectrum: Hartley, *J. Chem. Soc.* **87**, 1808 (1905). Difficultly sol in cold water; freely sol in hot water, in dil acids. Forms salts with metals. LD$_{50}$ orally in male rats: >5000 mg/kg (Goldenthal).

USE: Manuf plastics, pharmaceuticals.

959. Bardoxolone. [218600-44-3] 2-Cyano-3,12-dioxooleana-1,9(11)-dien-28-oic acid; CDDO; RTA-401; TP-151. $C_{31}H_{41}NO_4$; mol wt 491.67. C 75.73%, H 8.41%, N 2.85%, O 13.02%. Oleanane triterpenoid with cytoprotective and anti-inflammatory activity. Induces expression of Nrf2, a transcription factor with a key role in suppressing the inflammatory response and reducing oxidative stress. Prepn from oleanolic acid: G. W. Gribble *et al.*, **WO 9965478**; *eidem*, **US 6326507**; *eidem*, **US 7863327** (1999, 2001, 2011 all to Trustees Dartmouth Coll.); T. Honda *et al.*, *Bioorg. Med. Chem. Lett.* **8**, 2711 (1998). Structure-activity relationships: M. B. Sporn *et al.*, *J. Nat. Prod.* **74**, 537 (2011). Pharmacology in ischemic acute kidney injury: Q. Q. Wu *et al.*, *Am. J. Physiol. Renal Physiol.* **300**, F1180 (2011). Clinical effect on renal function in chronic kidney disease in diabetic patients: P. E. Pergola *et al., Am. J. Nephrol.* **33**, 469 (2011); *idem et al., N. Engl. J. Med.* **365**, 327 (2011).

Amorphous solid. $[\alpha]_D^{22}$ +33° (c = 0.28 in chloroform). uv max (ethanol): 240.4 (log ε 4.21).

Methyl ester. [218600-53-4] Methyl 2-cyano-3,12-dioxooleana-1,9(11)-dien-28-oate; CDDO-Me; RTA-402; TP-155; NSC-713200. $C_{32}H_{43}NO_4$; mol wt 505.70.

THERAP CAT: Antioxidant inflammation modulator in treatment of chronic kidney disease.

960. Barium. [7440-39-3] Ba; at. wt. 137.327; at. no. 56; valence 2. Group IIA (2). Alkaline earth metal. Abundance in earth's crust 0.05% by wt. Naturally occurring isotopes: 138 (71.70%); 137 (11.23%); 136 (7.85%); 135 (6.593%); 134 (2.42%); 132 (0.101%); 130 (0.106%). Known radioactive isotopes: 117, 119-129, 131, 133, 139-149. Does not occur free in nature as metal. Compds occur in minerals barite and witherite. Commercial production by thermal reduction of barium oxide with aluminum. First prepared as a mercury amalgam by Davy in 1808. Toxicity studies of barium compds: Syed, Hosain, *Toxicol. Appl. Pharmacol.* **22**, 150 (1972). *Reviews: Gmelins, Barium* (8th ed.) **30**, (1960); Goodenough, Stenger, "Magnesium, Calcium, Strontium, Barium and Radium" in *Comprehensive Inorganic Chemistry* **Vol. 1**, J. C. Bailar, Jr. *et al.*, Eds. (Pergamon Press, Oxford, 1973) pp 591-664; *Chemistry of the Elements*, N. N. Greenwood, A. Earnshaw, Eds. (Pergamon Press, New York, 1984) pp 117-154; C. Boffito in *Kirk-Othmer Encyclopedia of Chemical Technology* **vol. 3** (Wiley-Interscience, New York, 4th ed., 1992) pp 902-908. Review of toxicology and human exposure: *Toxicological Profile for Barium and Compounds* (PB2008-100003, 2007) 231 pp.

Silvery-white, slightly lustrous metal when pure; yellowish-white when contaminated with nitrogen. Body-centered cubic structure at atm pressure. Soft, ductile. Fairly volatile; extremely reactive, very easily oxidizable; must be kept under petroleum or other oxygen-free liquid to exclude air. Reacts vigorously with water, liberating hydrogen and creating an explosion hazard. Keep dry and tightly sealed. Pyrophoric when finely divided; store powder under dry argon or helium. d 3.6. mp ~710°. bp ~1600°. E^0 (aq) Ba^{2+}/Ba −2.91 V. Emits characteristic yellow-green color in flame. Dissolves in liq NH_3 to give deep blue-black solns. Solns of sol barium salts give a white ppt with H_2SO_4 or sol sulfates.

Caution: Potential symptoms of overexposure to soluble barium compounds are excessive salivation, vomiting, colic, diarrhea, convulsive tremors, slow, hard pulse, elevated blood pressure, confusion; hemorrhages in stomach, intestines, kidneys; respiratory failure, cardiac arrest. See: *Clinical Toxicology of Commercial Products*, R. E. Gosselin *et al.*, Eds. (Williams & Wilkins, Baltimore, 5th ed., 1984) Section III, pp 61-63; *Patty's Industrial Hygiene and Toxicology* **vol. 2C**, G. D. Clayton, F. E. Clayton, Eds. (Wiley-Interscience, New York, 4th ed., 1994) pp 1925-1930.

USE: Alloys with Al or Mg as "getters" to remove residual gases from vacuum systems and electronic tubes. Deoxidizer for steel and other metals. Carrier for radium. The β- and γ-radiation emitted by $^{140}Ba + ^{140}La$ makes a large contribution to the activity of the fission products of uranium rods during the first few weeks after their withdrawal from the reactor. The emissions from ^{133}Ba and ^{137m}Ba as standards in γ-spectrometry: Haissinsky, Adloff, *Radiochemical Survey of the Elements* (Elsevier, 1965) pp 12-14.

961. Barium Acetate. [543-80-6] Acetic acid barium salt (2:1). $C_4H_6BaO_4$; mol wt 255.42. C 18.81%, H 2.37%, Ba 53.77%, O 25.06%. $Ba(C_2H_3O_2)_2$. Prepn: *Gmelins, Barium* (8th ed.) **30**,

315 (1932) and supplement, 478 (1960). Toxicity study: Syed, Hosain, *Toxicol. Appl. Pharmacol.* **22**, 150 (1972). Crystal structure: I. Gautier-Luneau, A. Mosset, *J. Solid State Chem.* **73**, 473 (1988).

White crystals; slight acetic odor. d 2.51.

Monohydrate. [5908-64-5] $C_4H_6BaO_4.H_2O$; mol wt 273.43. *Poisonous.* d 2.19. Loses its H_2O of hydration at 110°. One gram dissolves in 1.5 ml cold or boiling water, in 700 ml alc. The aq soln is neutral or slightly acid to litmus. LD_{50} i.v. in ICR mice: 23.31 mg Ba^{2+}/kg (Syed, Hosain).

USE: Mordant for printing fabrics; in lubricating oil and grease; as catalyst for organic reactions.

962. Barium Benzenesulfonate. [515-72-0] Benzenesulfonic acid barium salt (2:1). $C_{12}H_{10}BaO_6S_2$; mol wt 451.65. C 31.91%, H 2.23%, Ba 30.41%, O 21.25%, S 14.20%. Prepn: Freund, *Ann.* **120**, 76 (1861).

Monohydrate. White, nacreous leaflets. *Poisonous.* Freely sol in water; slightly sol in alc.

USE: Lubricating oil additives.

963. Barium Bromate. [13967-90-3] Bromic acid barium salt (2:1). $BaBr_2O_6$; mol wt 393.13. Ba 34.93%, Br 40.65%, O 24.42%. $Ba(BrO_3)_2$. Prepd from potassium bromate and barium chloride: Pearce, Russell, *Inorg. Synth.* **2**, 20 (1946).

Monohydrate. Monoclinic crystals from hot water. *Oxidizer, poisonous.* May develop slight odor of bromine on long standing. d 3.99. Dec. at 260°. Sol in water (g/100 ml): 0.44 (10°); 0.96 (30°); 5.39 (100°). Sol in acetone. Practically insol in alc, most other organic solvents.

USE: In the prepn of rare earth bromates; as corrosion inhibitor for low-C steel.

964. Barium Bromide. [10553-31-8] $BaBr_2$; mol wt 297.14. Ba 46.22%, Br 53.78%. Prepn: *Gmelins, Barium* (8th ed.) **30**, 223 (1932) and supplement 380-381 (1960).

Dihydrate, crystals or granules; loses $1H_2O$ at 75° and all the H_2O at 120°. *Poisonous.* mp ~850° when anhyd. Very sol in water; sol in methanol. Almost insol in ethanol, ethyl acetate, acetone, dioxane.

USE: In the manuf of other bromides; in the prepn of phosphors.

965. Barium Carbonate. [513-77-9] Carbonic acid barium salt (1:1). $CBaO_3$; mol wt 197.34. C 6.09%, Ba 69.59%, O 24.32%. $BaCO_3$. Occurs in nature as the mineral **witherite**. Prepn: *Gmelins, Barium* (8th ed) **30**, 301-303 (1932) and supplement, 186-188, 461-466 (1960). Manuf: *Faith, Keyes & Clark's Industrial Chemicals*, F. A. Lowenheim, M. K. Moran, Eds. (Wiley-Interscience, New York, 4th ed., 1975) pp 121-125. Thermal transition from orthorhombic to hexagonal phase: C. M. Earnest, *Themochim. Acta* **137**, 365 (1989). Prepn of submicron particles by high gravity technique: C. Y. Tai *et al., Chem. Eng. Sci.* **61**, 7479 (2006). Polymer-controlled crystallization: T. Wang *et al., Angew. Chem. Int. Ed.* **45**, 4451 (2006).

White, heavy powder. *Poisonous.* d (witherite) 4.2865. Orthorhombic (γ), hexagonal (β), and cubic (α) polymorphs exist. Orthorhombic phase occurs under ambient conditions; transition to hexagonal phase occurs at ~810°; cubic at ~976°; dec into BaO and CO_2 at ~1300°. Almost insol in water, 0.024 g/l. Slightly sol (1:1000) in CO_2-water; sol in dil HCl, HNO_3 or acetic acid; also sol in soln NH_4Cl or NH_4NO_3.

Caution: Potential symptoms of overexposure are excessive salivation, vomiting, severe abdominal pain, violent diarrhea; increased blood pressure; tinnitus, giddiness, vertigo; muscle twitching, convulsions, paralysis; dilated pupils; confusion, somnolence; cardiac arrest; death due to respiratory failure. See *Clinical Toxicology of Commercial Products*, R. E. Gosselin *et al.*, Eds. (Williams & Wilkins, Baltimore, 5th ed., 1984) Section III, pp 61-63.

USE: Rat poison; in brick, glass, oil-drilling muds, ceramics, paints, enamels, marble substitutes, rubber; manuf of ceramics, photographic paper, barium salts, electrodes, optical glasses, electronic and magnetic materials; analytical reagent in the prepn of barium standard solutions.

966. Barium Chlorate. [13477-00-4] Chloric acid barium salt (2:1). $BaCl_2O_6$; mol wt 304.22. Ba 45.14%, Cl 23.31%, O 31.55%. $Ba(ClO_3)_2$. Prepn: Vanino, *Handb. Präp. Chem., Anorgan. Teil* (2.Aufl., Stuttgart, 1925) p 297; Schmeisser in *Handbook*

of Preparative Inorganic Chemistry **Vol. 1**, G. Brauer, Ed. (Academic Press, New York, 2nd ed., 1963) p 314. Large-scale process: Munroe, *Chem. Metall. Eng.* **23**, 188 (1920). Also prepd by electrolysis of barium chloride.

Monohydrate. Monoclinic prismatic crystals. *Oxidizer, poisonous.* d 3.179. Loses its water of hydration at 120°, begins to give off oxygen at 250°, mp 414°. Freely sol in water; sol in hydrochloric acid; moderately sol in ethylamine; very sparingly sol in alc, somewhat more in acetone. Practically insol in ethyl acetate, pyridine. Fire hazard when in contact with combustible material.

USE: In pyrotechnics (green fire); manuf of explosives and matches; mordant in dyeing.

967. Barium Chloride. [10361-37-2] Barium chloride ($BaCl_2$). $BaCl_2$; mol wt 208.23. Ba 65.95%, Cl 34.05%. Prepn: *Gmelins, Barium* (8th ed) **30**, 171-175 (1932) and supplement, 179-181, 324-325 (1960). Toxicity studies: I. B. Syed, F. Hosain, *Toxicol. Appl. Pharmacol.* **22**, 150 (1972). Induction of arrhythmia in exptl animals: F. W. Eichbaum, *Basic Res. Cardiol.* **68**, 73 (1973). Crystal structures of hydrated forms: A. Hasse, G. Brauer, *Z. Anorg. Allg. Chem.* **441**, 181 (1978); and spectroscopic and thermal properties: H. D. Lutz *et al.*, *ibid.* **457**, 84 (1979). Dehydration kinetics of dihydrate: R. K. Osterheld, P. R. Bloom, *J. Phys. Chem.* **82**, 1591 (1978); J. A. Lumpkin, D. D. Perlmutter, *Themochim. Acta* **249**, 335 (1995). Vapor pressure of hydrates: J. E. Tanner, *J. Chem. Eng. Data* **50**, 777 (2005). Phase diagrams of aqueous solns: J. Fenstad, D. J. Fray, *C. R. Chimie* **9**, 1235 (2006).

White, hygroscopic solid. mp 962°. d 3.913.

Dihydrate. [10326-27-9] $BaCl_2.2H_2O$; mol wt 244.26. Crystals or granules or powder; bitter salty taste. d 3.86. *Poisonous.* Very sol in water; sol in methanol. Practically insol in ethanol, acetone, ethyl acetate. LD_{50} i.v. in ICR mice: 19.2 mg Ba^{2+}/kg (Syed, Hosain).

Caution: Potential symptoms of overexposure are irritation of eyes, skin and upper respiratory system; gastroenteritis; muscle spasm; slow pulse, extrasystoles; hypokalemia; skin burns. *See NIOSH Pocket Guide to Chemical Hazards* (DHHS/NIOSH 97-140, 1997) p 24.

USE: Manuf pigments, color lakes, glass, mordant for acid dyes; weighting and dyeing textile fabrics; in Al refining; as pesticide; boiler compds for softening water; tanning and finishing leather. Dihydrate as an analytical reagent in sulfate determn.

968. Barium Chromate(VI). [10294-40-3] Chromic acid (H_2CrO_4) barium salt (1:1); C.I. 77103; C.I. Pigment Yellow 31; Baryta yellow; lemon yellow; permanent yellow; Steinbühl yellow; ultramarine yellow. $BaCrO_4$; mol wt 253.32. Ba 54.21%, Cr 20.53%, O 25.26%. Prepn: Beyer, Rieman, *J. Am. Chem. Soc.* **65**, 971 (1943); *Colour Index* **vol. 4** (3rd ed., 1971) p 4656.

Yellow, heavy, monoclinic, orthorhombic crystals. *Poisonous.* d 4.50. Practically insol in water, dil acetic or chromic acids. Dissolved or dec by mineral acids.

Caution: Chromium hexavalent (VI) compounds are listed as known human carcinogens: *Report on Carcinogens, Twelfth Edition* (PB2011-111646, 2011) p 106.

USE: As a pigment almost entirely in anticorrosion jointing pastes to prevent electro-chemical corrosion at junctions of dissimilar metals; some use in artists' colors and in coloring glass, ceramics, porcelain. Also used in metal primers, pyrotechnic compositions.

969. Barium Cyanide. [542-62-1] C_2BaN_2; mol wt 189.36. C 12.69%, Ba 72.52%, N 14.79%. $Ba(CN)_2$. Prepn: **GB 602393** (1948 to I.C.I.); *Gmelins, Barium* (8th ed.) **30**, 327 (1932) and supplement 483 (1960).

Crystals; slowly dec in air. *Poisonous.* Very sol in water; sol in alcohol.

USE: In electroplating processes; in metallurgy.

970. Barium Fluoride. [7787-32-8] BaF_2; mol wt 175.32. Ba 78.33%, F 21.67%. Prepd by dissolving $BaCO_3$ in excess HF, evaporating to dryness, and heating to red heat: W. Olbrich, *Thesis* (Technische Hochschule, Breslau, 1929), p 2; Kwasnik in *Handbook of Preparative Inorganic Chemistry* **vol. 1**, G. Brauer, Ed., (Academic Press, New York, 2nd ed., 1963) p 234.

Transparent cubic crystals (fluorite lattice). *Poisonous.* d 4.83. mp 1353°. bp 2260°. Soly in water (g/l): 1.586 (10°); 1.607 (20°);

1.620 (30°). Also sol in hydrochloric, nitric, hydrofluoric acids and in aq solns of ammonium chloride. May be stored in glass bottles.

USE: As a flux and opacifier in vitreous enamels; in the manuf of carbon brushes for DC motors and generators; in heat-treating metals; in embalming; in glass manuf.

971. Barium Formate. [541-43-5] $C_2H_2BaO_4$; mol wt 227.36. C 10.57%, H 0.89%, Ba 60.40%, O 28.15%. $Ba(HCOO)_2$. Prepn: *Gmelins, Barium* (8th ed) **30**, 311 (1932), and supplement, 477 (1960).

Crystals. *Poisonous.* d 3.21. Soluble in 4 parts cold, 3 parts boiling water. Practically insol in alcohol.

972. Barium Hexafluorosilicate. [17125-80-3] Hexafluorosilicate(2−) barium (1:1); barium fluosilicate; barium silicofluoride. BaF_6Si; mol wt 279.40. Ba 49.15%, F 40.80%, Si 10.05%. $BaSiF_6$. Prepd from $BaCl_2$ and H_2SiF_6: Truchot, *Compt. Rend.* **98**, 821 (1884); Hoffmann, Gutowsky, *Inorg. Synth.* **4**, 145 (1953).

Orthorhombic needles. d_4^{21} 4.29. Dec at 300°. Heat of formation (solid) −677.42 kcal/mol. Soly in water (g/100 ml H_2O): 0.015 (0°); 0.0235 (25°); 0.091 (100°). Prolonged contact with water produces hydrolysis which is much accelerated by the presence of alkali. Slightly sol in dil acids; sol in ammonium chloride soln; practically insol in alc.

Caution: Highly toxic, especially when brought into soln by alkali.

USE: Prepn of silicon tetrafluoride; as pesticide.

973. Barium Hydroxide. [17194-00-2] Barium hydroxide ($Ba(OH)_2$); barium dihydroxide; activated barium hydroxide; caustic baryta. BaH_2O_2; mol wt 171.34. Ba 80.15%, H 1.18%, O 18.68%. $Ba(OH)_2$. Reacts as base/nucleophile in organic syntheses. Prepn: *Gmelins, Barium* (8th ed) **30**, 106-111 and suppl. 175-177, 289 (1960). Thermal dehydration and decomposition study: G. M. Habashy, G. A. Kolta, *J. Inorg. Nucl. Chem.* **34**, 57 (1972). Use as a catalyst in organic synthesis: J. Barrios *et al.*, *J. Catal.* **112**, 528 (1988); I. Paterson *et al.*, *Synlett* **1993**, 774; Y.-D. Gong *et al.*, *J. Org. Chem.* **63**, 4854 (1998); R. S. Varma *et al.*, *Tetrahedron Lett.* **39**, 8437 (1998). Review of synthetic applications: S. Jeanmart, *Synlett* **2002**, 1739-1740. Review of toxicology and human exposure: *Toxicological Profile for Barium and Compounds* (PB2008-100003, 2007) 231 pp.

White powder, mp 408°. *n* 4.495.

Monohydrate. [22326-55-2] Dried barium hydroxide. $BaH_2O_2.H_2O$; mol wt 189.36. Obtained by heating commercial $Ba(OH)_2.8H_2O$ at 200°. White powder. *Poisonous.* d 3.743. Sol in dil acids; slightly sol in water.

Octahydrate. [12230-71-6] $BaH_2O_2.8H_2O$; mol wt 315.46. Most common form of the hydroxide, readily available commercially. Transparent crystals or white masses. Very alkaline; rapidly absorbs CO_2 from air, becoming incompletely sol in water. *Poisonous.* mp 78°. Freely sol in water, methanol; slightly sol in ethanol. Practically insol in acetone. *Keep tightly closed.*

USE: In manuf of alkali, glass; in synthetic rubber vulcanization, in corrosion inhibitors, pesticides, sugar industry; boiler scale remedy; refining animal and vegetable oils; softening water. Catalyst in organic synthesis. Carbonate-free base in alkalimetry.

974. Barium Hypophosphite. [14871-79-5] Phosphinic acid barium salt (2:1). $BaH_4O_4P_2$; mol wt 267.30. Ba 51.38%, H 1.51%, O 23.94%, P 23.18%. $Ba(H_2PO_2)_2$. Prepd by treating white phosphorus with barium hydroxide: $8P + 3Ba(OH)_2.8H_2O + H_2O \rightarrow 3Ba(H_2PO_2)_2.H_2O + 2PH_3$: Rose, *Pogg. Ann.* **9**, 370 (1827); Klement in *Handbook of Preparative Inorganic Chemistry* **Vol. 1**, G. Brauer, Ed. (Academic Press, New York, 2nd ed., 1963) p 557.

Monohydrate. Monoclinic platelets with nacreous sheen from hot water. *Poisonous.* d_4^{17} 2.90. Soly (g/100 ml H_2O): 28.6 (17°); 33.3 (100°). Practically insol in alcohol.

USE: In nickel plating.

975. Barium Iodate. [10567-69-8] Iodic acid (HIO_3) barium salt (2:1). BaI_2O_6; mol wt 487.13. Ba 28.19%, I 52.10%, O 19.71%. $Ba(IO_3)_2$. Prepn: Lambert, Yasada, *Inorg. Synth.* **7**, 13 (1963).

Monohydrate, crystals; becomes anhydr at 130°. d 5.00. *Poisonous.* Sol in 3350 parts water at 25°, 625 parts boiling water; sol in HCl or HNO_3. Practically insol in alcohol.

976. Barium Iodide. [13718-50-8] BaI_2; mol wt 391.14. Ba 35.11%, I 64.89%. Prepn: *Gmelins, Barium* (8th ed) **30**, 238-239 (1932) and supplement, 394 (1960).

Dihydrate, colorless, odorless, transparent crystals or white granules; rapidly becomes reddish in the air due to liberation of iodine. *Poisonous.* d 5.15. Freely sol in water; the aq soln is neutral or slightly alkaline. Sol in alcohol, acetone. *Keep well closed and protected from light.*

USE: In the manuf of other iodides.

977. Barium Manganate(VI). [7787-35-1] Manganic acid (H_2MnO_4) barium salt (1:1); manganese green; Cassel's green; Rosenstiehl's green. $BaMnO_4$; mol wt 256.26. Ba 53.59%, Mn 21.44%, O 24.97%. Prepn: Schlesinger, Stems, *J. Am. Chem. Soc.* **46**, 1965 (1924); Jellinek, *J. Inorg. Nucl. Chem.* **13**, 329 (1960); R. S. Nyholm, P. R. Woolliams, *Inorg. Synth.* **11**, 56 (1968); Bayan, Aymonino, *Ber.* **101**, 3337 (1968); H. Firouzabadi, Z. Mostafavi-poor, *Bull. Chem. Soc. Jpn.* **56**, 914 (1983).

Dark blue-green crystals. d 4.85. Disproportionates in water or dil acid to form $Ba(MnO_4)_2$ and MnO_2.

USE: As pigment in fresco painting instead of Scheele's green because not so poisonous as latter. As mild oxidizing agent in organic synthesis.

978. Barium Mercuric Iodide. [10048-99-4] (*T*-4)-Tetra-iodomercurate(2−) barium (1:1); barium tetraiodomercurate(II); mercuric barium iodide. $BaHgI_4$; mol wt 845.53. Ba 16.24%, Hg 23.72%, I 60.04%.

Yellow or reddish, deliquesc crystals; said to become red even when kept in sealed tubes. d 3.5 (conc aq soln). *Poisonous.* Very sol in water or alc. *Keep well closed!*

USE: As an aq soln known as *Rohrbach's Soln,* for separating minerals of different densities; also for microchemical detection of alkaloids.

979. Barium Nitrate. [10022-31-8] Nitric acid barium salt (2:1). BaN_2O_6; mol wt 261.34. Ba 52.55%, N 10.72%, O 36.73%. $Ba(NO_3)_2$. Prepn: *Gmelins, Barium* (8th ed) **30**, 149-151 (1932) and supplement, 178-179, 305 (1960). Toxicity study: Syed, Hosain, *Toxicol. Appl. Pharmacol.* **22**, 150 (1972). Crystal properties: I. Sunagawa *et al., Prog. Cryst. Growth Charact.* **30**, 153 (1995). Raman studies: P. G. Zverev *et al., Opt. Mater.* **11**, 315 (1999). AFM study: M. Plomp *et al., J. Cryst. Growth* **198/199**, 246 (1999). Molar volumes and heat capacities of aqueous solns: T. L. Niederhauser, E. M. Woolley, *J. Chem. Thermodyn.* **36**, 325 (2004).

Cubic crystals or crystalline powder. *Oxidizer, poisonous.* d 3.24. mp 592°; dec at higher temp. Freely sol in water; very slightly sol in alcohol, acetone. LD_{50} i.v. in ICR mice: 20.10 mg Ba^{2+}/kg (Syed, Hosain).

Caution: Potential symptoms of overexposure are upper respiratory irritation; gastroenteritis; muscle spasm; slow pulse, extrasystoles; hypokalemia; irritation of eyes and skin; skin burns. *See NIOSH Pocket Guide to Chemical Hazards* (DHHS/NIOSH 97-140, 1997) p 24.

USE: Manuf BaO_2; pyrotechnics for green fire; green signal lights; in the vacuum-tube industry. Precursor material in ceramics and superconductors; aids in sulfur fixation in coal combustion; promoter in catalytic synthesis of ammonia; in Raman spectroscopy optics; in sulfate determn.

980. Barium Nitrite. [13465-94-6] Nitrous acid barium salt (2:1). BaN_2O_4; mol wt 229.34. Ba 59.88%, N 12.22%, O 27.90%. $Ba(NO_2)_2$. Prepn: *Gmelins, Barium* (8th ed) **30**, 144-147 (1932) and suppl, 303 (1960).

Monohydrate, crystals. *Poisonous.* d 3.187. Sol in water. Practically insol in alc.

USE: In diazotization reactions; prevention of corrosion of steel bars; in explosives.

981. Barium Oxalate. [516-02-9] Ethanedioic acid barium salt (1:1). C_2BaO_4; mol wt 225.35. C 10.66%, Ba 60.94%, O 28.40%. BaC_2O_4. Prepn: *Gmelins, Barium* (8th ed) **30**, 320-323 (1932) and supplement, 480-482 (1960).

Monohydrate. Cryst powder. *Poisonous.* d 2.66. Sol in 10,000 parts cold, 5000 parts boiling water; sol in dil HNO_3 or HCl.

982. Barium Oxide. [1304-28-5] Barium monoxide; barium protoxide; calcined baryta. BaO; mol wt 153.33. Ba 89.56%, O

10.43%. Prepn: Ehrlich in *Handbook of Preparative Inorganic Chemistry* vol. **1**, G. Brauer, Ed. (Academic Press, New York, 2nd ed., 1963) p 933.

White to yellowish-white powder or lumps. Very alkaline; absorbs moisture and CO_2 on exposure to air. *Poisonous.* On contact with water it forms $Ba(OH)_2$ with evolution of much heat. At 450° combines with oxygen to form BaO_2, which is reduced to BaO above 600°. d 5.7. mp ~1920°. Sol in water, dil acids. Slowly, but considerably sol in methanol, ethanol forming barium alcoholate. *Keep tightly closed.*

USE: Porous grades are marketed especially for drying gases and solvents (particularly alcohols, aldehydes and petroleum solvents). Swells, but does not become sticky upon absorption of moisture. Used in manuf of lubricating oil detergents. Also used for making barium methoxide.

983. Barium Perchlorate. [13465-95-7] Perchloric acid barium salt (2:1). $BaCl_2O_8$; mol wt 336.22. Ba 40.84%, Cl 21.09%, O 38.07%. $Ba(ClO_4)_2$. Prepn: *Gmelins, Barium* (8th ed) **30**, 218 (1932) and supplement, 373 (1960).

Trihydrate, crystals. *Oxidizer, poisonous.* Sol in water, methanol; slightly sol in ethanol, ethyl acetate, acetone. Practically insol in ether.

USE: In the determination of ribonuclease; as absorbent for water in C and H analysis.

984. Barium Permanganate. [7787-36-2] Permanganic acid ($HMnO_4$) barium salt; barium manganate(VII). $BaMn_2O_8$; mol wt 375.20. Ba 36.60%, Mn 29.28%, O 34.11%. $Ba(MnO_4)_2$. Prepn: Lux in *Handbook of Preparative Inorganic Chemistry* vol. **1**, G. Brauer, Ed. (Academic Press, New York, 2nd ed., 1963) p 1462.

Brownish-violet to black crystals. *Poisonous, oxidizer.* d 3.77. Sparingly sol in water; dec by alcohol.

USE: As dry cell depolarizer.

985. Barium Peroxide. [1304-29-6] Barium dioxide; barium superoxide. BaO_2; mol wt 169.33. Ba 81.10%, O 18.90%. Prepn: *Gmelins, Barium* (8th ed) **30**, 92-98 (1932) and supplement, 177, 296 (1960).

White or grayish-white, heavy powder. *Poisonous, oxidizer.* Dec slowly in air. Insol in water, but slowly dec by contact with it; dec by dil acid or CO_2 in presence of H_2O forming H_2O_2. Combines with water to form octahydrate. *Keep well closed.* The article of commerce contains about 85% BaO_2; the remainder is chiefly BaO.

USE: Bleaching animal substances, vegetable fibers and straw; glass decolorizer; manuf H_2O_2 and oxygen; dyeing and printing textiles; with powdered aluminum in welding; in cathodes; in igniter compositions. Oxidizing agent in organic synthesis.

986. Barium Phosphate, Dibasic. [10048-98-3] Phosphoric acid barium salt (1:1); secondary barium phosphate. $BaHO_4P$; mol wt 233.30. Ba 58.86%, H 0.43%, O 27.43%, P 13.28%. $BaHPO_4$. Prepn: *Gmelins, Barium* (8th ed) **30**, 340 (1932) and supplement, 492 (1960).

Crystals. *Poisonous.* d 4.16. Practically insol in water. Sol in dil HCl or HNO_3.

USE: In fireproofing compositions; in prepn of phosphors.

987. Barium Platinous Cyanide. [562-81-2] (*SP*-4-1)-Tetrakis(cyano-*κ*C)platinate(2−) barium (1:1); barium tetracyanoplatinate(II); barium cyanoplatinate(II); barium platinocyanide; platinous barium cyanide. C_4BaN_4Pt; mol wt 436.48. C 11.01%, Ba 31.46%, N 12.84%, Pt 44.69%. $BaPt(CN)_4$.

Tetrahydrate. Large dichroic crystals; yellowish-green by transmitted, bluish-violet by reflected light. *Poisonous.* d 3.05. Sol in about 35 parts water; more sol in hot water.

USE: An aq soln mixed with some adhesive and painted on paper or wood exhibits luminescence when exposed to the invisible ultraviolet rays of the spectrum or to Roentgen, radium, or cathode rays; hence used in radiography for making x-ray screens.

988. Barium Selenide. [1304-39-8] BaSe; mol wt 216.29. Ba 63.49%, Se 36.51%. Prepn: Henglein, *Z. Anorg. Chem.* **120**, 77 (1921); *Z. Elektrochem.* **30**, 11 (1924); *Gmelins, Barium* (8th ed) **30**, 290 (1932) and supplement, 453 (1960).

Cubic microcrystalline powder. d 5.02. Turns red on exposure to air. Dec in water.

USE: In photocells, semiconductors.

989. Barium Silicide. [1304-40-1] BaSi$_2$; mol wt 193.50. Ba 70.97%, Si 29.03%. Prepn: *Gmelins, Barium* (8th ed) **30**, 330 (1932) and supplement, 486 (1960).

Metal-like, gray lumps; quite permanent in dry air, but dec by moisture with evolution of H$_2$. Melts at white heat.

USE: Deoxidizing and desulfurizing steel and for other metallurgical purposes.

990. Barium Sulfate. [7727-43-7] Sulfuric acid barium salt (1:1); blanc fixe; Baritop; Barosperse; Esophotrast; E-Z-Paque; Intestibar; Micropaque; Microtrast; Mixobar; Neobar; Oratrast; Prontobario; Radiopaque; Telebar; Tixobar. BaO$_4$S; mol wt 233.38. Ba 58.84%, O 27.42%, S 13.74%. BaSO$_4$. Occurs in nature as the mineral *barite*; also as *barytes, heavy spar*. Prepn: *Gmelins, Barium* (8th ed) **30**, 262-267 (1932) and supplement, 182-186, 412-414 (1960). Review of clinical use: R. F. Theoni, A. R. Margulis, *Radiology* **167**, 7 (1988).

Fine, heavy, odorless powder or polymorphous crystals, d 4.25-4.5. Dec above 1600°. Practically insol in water (one gram dissolves in 400,000 parts), organic solvents, dil acids, alc, alkalies. Sol in hot concd H$_2$SO$_4$.

Caution: Potential symptoms of overexposure by direct contact are irritation of eyes, nose and upper respiratory system; inhalation may cause benign pneumoconiosis. *See NIOSH Pocket Guide to Chemical Hazards* (DHHS/NIOSH 97-140, 1997) p 24.

USE: Manuf photographic papers, artificial ivory, cellophane; filler for rubber, linoleum, oil cloth, polymeric fibers and resins, paper, lithographic inks; as a water-color pigment for colored paper, in wallpaper; as a size for modifying the colors of other pigments; in heavy concrete for radiation shield.

THERAP CAT: Diagnostic aid (radiopaque medium).

THERAP CAT (VET): X-ray contrast medium.

991. Barium Sulfide. [21109-95-5] BaS; mol wt 169.39. Ba 81.07%, S 18.93%. The commercial article contains 80-90% BaS. Prepn: Ehrlich in *Handbook of Preparative Inorganic Chemistry* **vol. 1**, G. Brauer, Ed. (Academic Press, New York, 2nd ed., 1963) p 938.

Heavy, grayish-white or pale yellow powder. d 4.36. mp >2000°. *Poisonous.* Oxidizes in dry air; slowly dec in damp air into carbonate, etc., with evolution of H$_2$S. Slightly sol in cold, more in hot water; sol in NH$_4$Cl soln; decomposed by acids, even CO$_2$, forming H$_2$S. *Keep in well-closed containers.*

USE: As depilatory; in luminous paints; manuf lithopone; vulcanizing rubber, generating H$_2$S.

992. Barium Sulfide, Black. [8011-62-9] Barium sulfide (BaS), black; black ash. This is the crude barium sulfide obtained by strong heating of a mixture of barium sulfate mineral (barite) and charcoal. Contains 60-70% BaS; the remainder is chiefly barium sulfate and carbon.

USE: Prepn of barium salts.

993. Barium Sulfite. [7787-39-5] Sulfurous acid barium salt (1:1). BaO$_3$S; mol wt 217.38. Ba 63.17%, O 22.08%, S 14.75%. BaSO$_3$. Prepn: *Gmelins, Barium* (8th ed.) **30**, 261 (1932) and supplement, 411 (1960).

Odorless crystals or powder; gradually oxidizes in air to BaSO$_4$. *Poisonous.* Slightly sol in water. Practically insol in alcohol. Dec by acids with liberation of SO$_2$.

USE: In paper manufacture.

994. Barium Thiocyanate. [2092-17-3] Thiocyanic acid barium salt (2:1); barium sulfocyanate; barium sulfocyanide; barium rhodanide. C$_2$BaN$_2$S$_2$; mol wt 253.48. C 9.48%, Ba 54.18%, N 11.05%, S 25.30%. Ba(SCN)$_2$. Prepn: Herstein, *Inorg. Synth.* **3**, 24 (1950).

Deliquesc crystals. *Poisonous.* Very sol in water, sol in acetone, methanol, ethanol. *Keep well closed.*

Trihydrate. Needle-shaped crystals (from water).

USE: In dyeing; in photography; as dispersing agent for cellulose; in prepn of thiocyanates of other metals.

995. Barium Thiosulfate. [35112-53-9] Thiosulfuric acid (H$_2$S$_2$O$_3$) barium salt (1:1); barium hyposulfite. BaO$_3$S$_2$; mol wt 249.44. Ba 55.05%, O 19.24%, S 25.71%. BaS$_2$O$_3$. Prepn: *Gmelins, Barium* (8th ed.) **30**, 283 (1932) and supplement, 449 (1960).

Monohydrate. Cryst powder. *Poisonous.* Very slightly sol in water. Practically insol in alc, ether, acetone, CCl$_4$, CS$_2$.

USE: Manuf explosives, matches; as an iodometry standard; in photographic diffusion-transfer process.

996. Barium Titanate(IV). [12047-27-7] Barium titanium oxide (BaTiO$_3$); barium metatitanate. BaO$_3$Ti; mol wt 233.19. Ba 58.89%, O 20.58%, Ti 20.53%. BaTiO$_3$. Usually prepd by calcining (at about 1300°) an intimate mixture of titanium dioxide and barium carbonate. Prepn from Ti oxalate and a barium compd: Lynd, Merker, **US 2758911** (1956 to National Lead). Wet process by the addn of a titanium ester, such as tetrapropyl titanate, to an aq soln of barium hydroxide: Flaschen, *J. Am. Chem. Soc.* **77**, 6194 (1955). Prepn of high purity material by ignition of barium titanyloxalate: Clabaugh *et al.*, *J. Res. Natl. Bur. Stand.* **56**, 289 (1956). Prepn by ignition of Ba and Ti alcoholates in an organic solvent: DiVita, Fischer, **US 2985504** (1961 to U.S. Dept. of the Army).

Exists in five cryst modifications. The tetragonal form (obtained by the wet process) appears to have the most desirable electric properties and is described here: d 6.08. mp 1625°. Curie point 120°. Has ferroelectric and piezoelectric properties. Becomes permanently polarized when exposed to high voltage direct current, provided the temperature is never allowed to rise above Curie pt. Has high dielectric properties which can be influenced by temp, voltage, and frequency.

USE: In electronic devices, e.g., as voltage-sensitive dielectric in so-called dielectric amplifiers, in computer elements, magnetic amplifiers, memory devices.

997. Barium Uranium Oxide. [10380-31-1] Barium uranium oxide (BaU$_2$O$_7$); barium uranate(VI); barium diuranate; uranium barium oxide. BaO$_7$U$_2$; mol wt 725.38. Ba 18.93%, O 15.44%, U 65.63%. BaU$_2$O$_7$. Prepn: Allpress, *J. Inorg. Nucl. Chem.* **26**, 1847 (1964); Klima *et al.*, *ibid.* **28**, 1861 (1966).

Orange or yellow powder. Practically insol in water. Sol in acids.

USE: In painting on porcelain.

998. Barnidipine. [104713-75-9] (4S)-1,4-Dihydro-2,6-dimethyl-4-(3-nitrophenyl)-3,5-pyridinedicarboxylic acid 3-methyl 5-[(3S)-1-(phenylmethyl)-3-pyrrolidinyl] ester; (+)-(3′S,4S)-1-benzyl-3-pyrrolidinyl methyl 1,4-dihydro-2,6-dimethyl-4-(*m*-nitrophenyl)-3,5-pyridinedicarboxylate; (3S)-1-benzyl-3-pyrrolidinyl methyl (4S)-2,6-dimethyl-4-(*m*-nitrophenyl)-1,4-dihydropyridine-3,5-dicarboxylate; mepirodipine. C$_{27}$H$_{29}$N$_3$O$_6$; mol wt 491.54. C 65.98%, H 5.95%, N 8.55%, O 19.53%. Dihydropyridine calcium channel blocker. Prepn (stereochemistry unspecified): T. Kojima, T. Takenaka, **DE 2904552**; *eidem*, **US 4220649** (1979, 1980 both to Yamanouchi). Prepn and bioactivity of enantiomers: K. Tamazawa *et al.*, *J. Med. Chem.* **29**, 2504 (1986). GC-MS determn in plasma: T. Teramura *et al.*, *J. Chromatogr.* **528**, 191 (1990). Review of pharmacology and therapeutic efficacy: H. Satoh, *Cardiovasc. Drug Rev.* **9**, 340-356 (1991).

mp 137-139°. [α]$_D^{20}$ +64.8° (c = 1 in methanol).

Hydrochloride. [104757-53-1] YM-09730-5; Hypoca. C$_{27}$H$_{29}$N$_3$O$_6$·HCl; mol wt 528.00. mp 226-228°. [α]$_D^{20}$ +116.4° (c = 1 in methanol). Insol in water. LD$_{50}$ orally in male, female rats: 105, 113 mg/kg (Satoh).

THERAP CAT: Antihypertensive; antianginal.

999. Barrelene. [500-24-3] Bicyclo[2.2.2]octa-2,5,7-triene. C$_8$H$_8$; mol wt 104.15. C 92.26%, H 7.74%. Bicyclic hydrocarbon that is the formal Diels-Alder adduct of acetylene to benzene. Considered to be a Möbius-like molecule due to the unique arrangement of overlapping out-of-phase p-orbitals. The trivial name was suggested since the array of molecular orbitals can be envisioned as forming a barrel-shaped structure. Preliminary conception as a molecular target: J. Hine *et al.*, *J. Am. Chem. Soc.* **77**, 594 (1955). Prepn: H. E. Zimmerman, R. M. Paufler, *ibid.* **82**, 1514 (1960); and spectral properties: H. E. Zimmerman *et al.*, *ibid.* **91**, 2330 (1969). Improved prepn: W. G. Dauben *et al.*, *J. Org. Chem.* **41**, 887 (1976); S. Cossu *et al.*, *ibid.* **62**, 4162 (1997). Spectroscopic studies: F. A. Van-Catledge, C. E. McBride, Jr., *J. Am. Chem. Soc.* **98**, 304 (1976); *eidem*, *J. Phys. Chem.* **80**, 2987 (1976). Structural studies: S. Yamamoto *et al.*, *ibid.* **86**, 529 (1982). Synthesis of derivatives: M. W. Wagaman *et al.*, *J. Org. Chem.* **62**, 9076 (1997); S. I. Kozkushkov *et al.*, *Eur. J. Org. Chem.* **2006**, 2590.

Highly volatile liquid. mp 15-16°. uv max (ethanol): 208.0, 239 nm (log ε 3.05, 2.48). Sol in diethyl ether, *n*-pentane, THF. Heat of hydrogenation: −93.78±0.31 kcal/mole. Decomposes at 250° to benzene and acetylene.

1000. Barthrin. [70-43-9] 2,2-Dimethyl-3-(2-methyl-1-propen-1-yl)cyclopropanecarboxylic acid (6-chloro-1,3-benzodioxol-5-yl) methyl ester; 2,2-dimethyl-3-(2-methylpropenyl)cyclopropanecarboxylic acid 6-chloropiperonyl ester; 6-chloropiperonyl 2,2-dimethyl-3-(2-methylpropenyl)cyclopropanecarboxylate; 6-chloropiperonyl chrysanthemumate; chrysanthemummonocarboxylic acid 6-chloropiperonyl ester. $C_{18}H_{21}ClO_4$; mol wt 336.81. C 64.19%, H 6.28%, Cl 10.53%, O 19.00%. Prepn: Barthel, Alexander, *J. Org. Chem.* **23**, 1012 (1958); Barthel *et al.*, **US 2886485** (1959).

trans-form

Oily liquid, bp$_{0.2}$ 155-171°, bp$_{0.7}$ 184-206°. n_D^{25} 1.5383. Sol in kerosine.

USE: Insecticide.

1001. Basic Aluminum Carbonate Gel. [1339-92-0] Carbonic acid aluminum salt basic; Basaljel. An aluminum hydroxide/aluminum carbonate gel.

THERAP CAT: Phosphorus binding agent in prevention of recurrent renal phosphatic calculi; antacid.

1002. Basic Lead Carbonate. [1319-46-6] Lead carbonate hydroxide ($Pb_3(CO_3)_2(OH)_2$); C.I. 77597; C.I. Pigment White 1; lead subcarbonate; white lead; flake lead; ceruse; cerussa; bleiweiss (German). $C_2H_2O_8Pb_3$; mol wt 775.63. C 3.10%, H 0.26%, O 16.50%, Pb 80.14%. $(PbCO_3)_2 \cdot Pb(OH)_2$. Occurs in nature as the mineral *hydrocerussite*. *See: Mellor's* **vol. VII**, 837 (1927); *Colour Index* **vol. 4** (3rd ed., 1971) p 4676. Review of chemistry and use as pigment: E. J. Dunn, Jr., *Pigment Handbook* **1**, 65-84 (1973). Microdetermn by nuclear microprobe for use in analysis of artwork: B. Nens *et al.*, *Nucl. Instrum. Methods Phys. Res. B* **14**, 186 (1986); by laser breakdown: D. Anglos, *Appl. Spectrosc.* **55**, 186A (2001).

White, heavy powder. *Poisonous.* Dec at 400°, leaving a residue of PbO. Sol in acetic acid or dil HNO_3 with effervescence. Insol in water or alc.

Caution: Lead and all lead compounds are listed as reasonably anticipated to be human carcinogens: *Report on Carcinogens, Twelfth Edition* (PB2011-111646, 2011) p 251.

USE: Pigment in oil paints and water colors; in cements; for making putty and lead carbonate paper; in the processing of parchment.

1003. Basiliximab. [179045-86-4] Anti-(human interleukin 2 receptor) immunoglobulin G1 (human-mouse monoclonal CHI621 γ_1-chain) disulfide with human-mouse monoclonal CHI621 light chain, dimer; chRFT5; SDZ-CHI-621; Simulect. Human-murine chimeric monoclonal antibody directed against the interleukin-2 receptor α-chain (IL-2Rα), also known as CD25 antigen, on the sur-

face of activated T cells. Prepn: P. L. Amlot *et al.*, **EP 449769** (1991 to Sandoz; Roy. Free Hosp. Sch. Med.). Clinical pharmacology: *idem et al.*, *Transplantation* **60**, 748 (1995). Clinical pharmacokinetics: J. Kovarik *et al.*, *ibid.* **64**, 1701 (1997). Clinical trial in kidney transplantation: B. Nashan *et al.*, *Lancet* **350**, 1193 (1997).

THERAP CAT: Immunosuppressant.

1004. Basswood. Linden tree. Also called *Tilia americana* L., *Tiliaceae*. Grows in mountainous woods from Canada to Georgia and west to Texas. Various decoctions of flowers, trees, bark and wood are used in American folk medicine for disorders of the bile and liver. Preparation of active extracts from *Tilia alburnum*: Lafon, **US 3030271** (1962 to S. A. Orsymonde).

1005. Bathocuproine. [4733-39-5] 2,9-Dimethyl-4,7-diphenyl-1,10-phenanthroline; BCP. $C_{26}H_{20}N_2$; mol wt 360.46. C 86.64%, H 5.59%, N 7.77%. Copper(I) chelating agent. Prepn: F. H. Case, J. A. Brennan, *J. Org. Chem.* **19**, 919 (1954). Use in spectrophotometric determn of copper: G. F. Smith, D. H. Wilkins, *Anal. Chem.* **25**, 510 (1953); J. W. Moffett *et al.*, *Anal. Chim. Acta* **175**, 171 (1985); with the biuret reaction in protein determn: M. Matsushita *et al.*, *Clin. Chim. Acta* **216**, 103 (1993); in determn of copper in jet fuels: Q. Lu *et al.*, *Energy Fuels* **17**, 699 (2003). Use as an electron-transporting and hole-blocking material in electroluminescent devices: D. Troadec *et al.*, *Synth. Met.* **127**, 165 (2002); K. Ono *et al.*, *Chem. Lett.* **33**, 276 (2004); L. L. Chen *et al.*, *J. Lumin.* **122-123**, 667 (2007).

Off-white to buff colored solid. Crystals from benzene, mp 282-283°. Insol in water. Absorption max (CH_2Cl_2): 280, 312(sh) nm (log ε 4.63, 4.10). Absorption max (complex with Cu(I)): 479.0 nm (ε 14160).

USE: Reagent for colorimetric determn of copper. In organic semiconductor applications.

1006. Batimastat. [130370-60-4] $(2R,3S)$-N^4-Hydroxy-N^1-[(1S)-2-(methylamino)-2-oxo-1-(phenylmethyl)ethyl]-2-(2-methylpropyl)-3-[(2-thienylthio)methyl]butanediamide; $(2S,3R)$-5-methyl-3-[[(αS)-α-(methylcarbamoyl)phenethyl]carbamoyl]-2-[(2-thienylthio)methyl]hexanohydroxamic acid; [4-(*N*-hydroxyamino)-2*R*-isobutyl-3*S*-(2-thienylthiomethyl)succinyl]-L-phenylalanine-*N*-methylamide; BB-94. $C_{23}H_{31}N_3O_4S_2$; mol wt 477.64. C 57.84%, H 6.54%, N 8.80%, O 13.40%, S 13.42%. Synthetic matrix metalloproteinase (MMP) inhibitor. Prepn: C. Campion *et al.*, **WO 9005719**; *eidem*, **US 5240958** (1990, 1993 both to British Biotech.). Effect on transplanted human ovarian carcinoma: B. Davies *et al.*, *Cancer Res.* **53**, 2087 (1993). Inhibition of metastasis of transplanted human colorectal carcinoma: X. Wang *et al.*, *ibid.* **54**, 4726 (1994). Review of pharmacology and clinical studies as antimetastatic agent: H. S. Rasmussen, P. P. McCann, *Pharmacol. Ther.* **75**, 69-75 (1997).

Fine white powder. mp 236-238°.

USE: MMP inhibitor in pharmacological research.

1007. Batrachotoxin. [23509-16-2] 2,4-Dimethyl-1*H*-pyrrole-3-carboxylic acid (1*S*)-1-[(5a*R*,7a*R*,9*R*,11a*S*,11b*S*,12*R*,13a*R*)-1,2,3,4,7a,8,9,10,11,11a,12,13-dodecahydro-9,12-dihydroxy-2,11a-dimethyl-7*H*-9,11b-epoxy-13a,5a-propenophenanthro[2,1-*f*][1,4]-oxazepin-14-yl]ethyl ester; batrachotoxinin A 20-(2,4-dimethyl-1*H*-pyrrole-3-carboxylate); 3α,9α-epoxy-14β,18β-(epoxyethano-*N*-methylimino)-5β-pregna-7,16-diene-3β,11α,20α-triol 20α-ester with 2,4-dimethylpyrrole-3-carboxylic acid. $C_{31}H_{42}N_2O_6$; mol wt 538.69. C 69.12%, H 7.86%, N 5.20%, O 17.82%. One of a group of toxic steroidal alkaloids consisting of batrachotoxin, *pseudobatrachotoxin*, homobatrachotoxin, and batrachotoxinin A. Extracted orginally from the skin of South American poison-dart frogs, genus *Phyllobates*. Its name is derived from the Greek word "batrachos" meaning frog. Isoln: F. Märki, B. Witkop, *Experientia* **19**, 329 (1963); and preliminary characterization: J. W. Daly *et al.*, *J. Am. Chem. Soc.* **87**, 124 (1965). Structure: T. Tokuyama *et al.*, *ibid.* **91**, 3931 (1969). Synthetic studies: J. F. W. Keana, R. R. Schumaker, *J. Org. Chem.* **41**, 3840 (1976); P. Magnus *et al.*, *Chem. Commun.* **1985**, 1185; P. Hudson, *Tetrahedron Lett.* **34**, 7295 (1993). Mode of action: E. X. Albuquerque, *Fed. Proc.* **31**, 1133 (1972). Kinetics and gating action on Na^+ channels: J. A. Wasserstrom *et al.*, *Biophys. J.* **65**, 386 (1993). Review: J. Daly, B. Witkop, *Clin. Toxicol.* **4**, 331-342 (1971); of chemistry and pharmacology: E. X. Albuquerque *et al.*, *Science* **72**, 995-1002 (1971); J. W. Daly, *Fortschr. Chem. Org. Naturst.* **41**, 206-227 (1982). Review of sodium channel activation: G. B. Brown, *Int. Rev. Neurobiol.* **29**, 77-116 (1988).

[α]$_{584}^{24}$ −5 To −10°; [α]$_{300}^{24}$ −260° (c = 0.23 in methanol). uv max (0.1*N* HCl-MeOH): 234, 262 nm (log ε 3.99, 3.70). pKa 7.45. LD$_{50}$ s.c. in mice: 2 μg/kg (Tokuyama).

Batrachotoxinin A. [19457-37-5] $C_{24}H_{35}NO_5$; mol wt 417.55. Abs config: R. D. Gilardi, *Acta Crystallogr.* **B 26**, 440 (1970). LD$_{50}$ s.c. in mice: 1000 μg/kg (Tokuyama).

Homobatrachotoxin. [23509-17-3] Isobatrachotoxin. $C_{32}H_{44}$-N_2O_6; mol wt 552.71. Also isolated from the feathers and skin of hooded pitohui bird of New Guinea. First demonstration of chemical defense in birds: J. P. Dumbacher *et al.*, *Science* **258**, 799 (1992). uv max (0.1*N* HCl-MeOH): 233, 264 nm (log ε 3.95, 3.70). LD$_{50}$ s.c. in mice: 3 μg/kg (Tokuyama).

USE: Biochemical tool for study of Na^+ channels.

1008. Batroxobin. [9039-61-6] *Bothrops atrox* serine proteinase; Bothrops venom proteinase; Botropase; Defibrase. Thrombin-like enzyme from the venom of *Bothrops atrox* (Linn.), a pit viper found in several varieties in Southern and Central America. Prepn of conc extract from *B. jararaca* venom and coagulative properties: D. von Klobusitzky, *Arch. Exp. Pathol. Pharmakol.* **179**, 204 (1935). The enzyme from *B. atrox* venom is hemostatic at low doses and acts as a blood anti-coagulant at higher doses. Composition and manuf: E. E. Percs *et al.*, *DE 2201993*; *eidem*, *US 3849252* (1972, 1974 both to Pentapharm). Studies on the subunit structure of fibrin produced by batroxobin: P. Mattock, M. P. Esnouf, *Nature New Biol.* **233**, 277 (1971). *In vitro* blood-clotting activity: K. O. Wik *et al.*, *Br. J. Haematol.* **23**, 37 (1972). Electron microscopic study: S. Ishimaru *et al.*, *Thromb. Haemostasis* **45**, 276 (1981). Clinical pharmacology: K. Fukutake *et al.*, *Nippon Ketsueki Gakkai Zasshi* **44**, 1178 (1981), *C.A.* **96**, 62775x (1982). Review of characterization, properties: K. G. Stocker, G. H. Barlow, *Aktuel. Probl. Angiol.* **26**, 45-62 (1975).

Rectangular crystals. Sol in physiological saline. Practically insol in distilled water. Forms complexes with phenol and phenol derivs that are practically insol in water.

Mixture with factor-X activator. Hemocoagulase; Reptilase.
THERAP CAT: Hemostatic.

1009. Batyl Alcohol. [544-62-7] 3-(Octadecyloxy)-1,2-propanediol; monooctadecyl ether of glycerol. $C_{21}H_{44}O_3$; mol wt 344.58. C 73.20%, H 12.87%, O 13.93%. Isoln from shark liver oils: Heilbron, Owens, *J. Chem. Soc.* **1928**, 942; Davies *et al.*, *ibid.* **1933**, 165; Nakamiya, *C.A.* **33**, 8175 (1939); from yellow bone marrow: Holmes *et al.*, *J. Am. Chem. Soc.* **63**, 2607 (1941); from coral-reef-building animals: Kind, Bergmann, *J. Org. Chem.* **7**, 424 (1942). Synthesis from allyl octadecyl ether: Davies *et al.*, *J. Chem. Soc.* **1930**, 2542; Kornblum, Holmes, *J. Am. Chem. Soc.* **64**, 3045 (1942). Synthesis of optically active batyl alc: Baer, Fischer, *J. Biol. Chem.* **140**, 397 (1941).

Glistening plates from dilute acetone, mp 70.5-71°. [α]$_D^{20}$ +1.14° (c = 6.6 in CHCl$_3$). Sol in the usual fat solvents.

Bis(*p*-nitrobenzoate). [5908-83-8] $C_{35}H_{50}N_2O_9$; mol wt 642.79. Pale yellow needles from methanol, soft at 63°, mp 65-66°.

THERAP CAT (VET): Has been recommended for bracken fern poisoning in cattle.

1010. Bayberry Bark. Candleberry bark; myrica; myrtle wax; berry wax; tallow shrub. Dried root bark of *Myrica cerifera* L., *Myricaceae. Habit.* Maryland to Florida, west to Texas and Arkansas. *Constit.* Acrid and astringent resin, myricic acid, tannin, red coloring matter, gum, starch. Bayberry wax consists of palmitic, myristic and lauric acid esters.

THERAP CAT: Astringent; emetic.

1011. Bazedoxifene. [198481-32-2] 1-[[4-[2-(Hexahydro-1*H*-azepin-1-yl)ethoxy]phenyl]methyl]-2-(4-hydroxyphenyl)-3-methyl-1*H*-indol-5-ol; 1-[4-(2-azepan-1-ylethoxy)benzyl]-2-(4-hydroxyphenyl)-3-methyl-1*H*-indol-5-ol. $C_{30}H_{34}N_2O_3$; mol wt 470.61. C 76.57%, H 7.28%, N 5.95%, O 10.20%. Nonsteroidal selective estrogen receptor modulator (SERM). Prepn: C. P. Miller *et al.*, *EP 802183*; *eidem*, *US 5998402* (1997, 1999 both to American Home); C. P. Miller *et al.*, *J. Med. Chem.* **44**, 1654 (2001). Pharmacology and receptor binding study: B. S. Komm *et al.*, *Endocrinology* **146**, 3999 (2005). Clinical pharmacology: S. Ronkin *et al.*, *Obstet. Gynecol.* **105**, 1397 (2005). Review of pharmacology and clinical experience: A. L. Stump *et al.*, *Ann. Pharmacother.* **41**, 833-839 (2007).

mp 98-102°.
Acetate. [198481-33-3] TSE-424; WAY-140424; WAY-TSE-424; Conbriza; Viviant. $C_{30}H_{34}N_2O_3 \cdot C_2H_4O_2$; mol wt 530.67. Crystals from acetone + acetic acid, mp 174-178°.

THERAP CAT: Antiosteoporotic.

1012. BBOT. [7128-64-5] 2,2′-(2,5-Thiophenediyl)bis[5-(1,1-dimethylethyl)benzoxazole]; 2,5-bis(5-*tert*-butyl-2-benzoxazolyl)thiophene; C.I. Fluorescent Brightener 184; Uvitex OB. $C_{26}H_{26}N_2O_2S$; mol wt 430.57. C 72.53%, H 6.09%, N 6.51%, O 7.43%, S 7.45%. High mol wt fluorescent whitening agent. Prepn: P. Liechti *et al.*, *CH 427822* (1967 to Ciba); Z. Seha, C. D. Weis, *Helv. Chim. Acta* **63**, 413 (1980). Fluorescence depolarization studies: G. R. Fleming *et al.*, *Chem. Phys. Lett.* **51**, 399 (1977). Photostability study: G. Tury *et al.*, *J. Photochem. Photobiol. A* **114**, 51 (1998).

Consult the Name Index before using this section.

Yellow to green powder, mp 196-202°. Flash pt: >350°C. d^{20} 1.26. Vapor pressure (20°): $2.6×10^{-8}$ Pa. Soly at 20° (%w/w): water <0.01; acetone 0.5; chloroform 14; ethyl acetate 1; n-hexane 0.2; methanol <0.1. Absorption max (cyclohexane): 374 nm. Emission max (cyclohexane): 435 nm.

USE: Optical brightener additive in polymers, fibers, films, inks, laquers, and paints. In liquid scintillators.

1013. BCG. An abbreviation for *Bacillus Calmette-Guérin*, a strain of *Mycobacterium tuberculosis*. Prepn of BCG labeled with ^{14}C: Pasquier, Kurylowicz, *Rev. Immunol.* **20**, 245 (1956). Has the potential to act as a non-specific immunopotentiating agent that stimulates the whole range of immune responses. Role as an immunotherapeutic agent: Bartlett and Zbar *et al.*, *J. Natl. Cancer Inst.* **48**, 245, 1441, 1709 (1972); **49**, 119 (1972). Comparison of potentiation of specific tumor immunity in mice: M. T. Scott, R. Bomford, *ibid.* **57**, 555 (1976). Review of use in Hodgkin's disease: H. Zywicka-Lopaciuk *et al.*, *Arch. Immunol. Ther. Exp.* **29**, 739-755 (1981); in operable lung cancer: B. H. Stack *et al.*, *Thorax* **37**, 588-593 (1982). Review of use as anti-tumor agent and of the interaction of BCG with cells of the mammalian immune system: M. Davies, *Biochim. Biophys. Acta* **651**, 143-174 (1982). Review of pathogenesis of tuberculosis and the effectiveness of BCG vaccination: H. G. ten Dam, A. Pio, *Tubercle* **63**, 225-233 (1982).

THERAP CAT: BCG vaccine as active immunizing agent.

THERAP CAT (VET): Vaccination vs. tuberculosis (cattle). Use prohibited where bovine tuberculosis eradication programs are in progress.

1014. Bebeerine. [477-60-1] (13a*S*,25a*S*)-2,3,13a,14,15,16,-25,25a-Octahydro-18,29-dimethoxy-1,14-dimethyl-13*H*-4,6:21,24-dietheno-8,12-metheno-1*H*-pyrido[3′,2′:14,15][1,11]dioxacyclo-eicosino[2,3,4-*ij*]isoquinoline-9,19-diol; 1′α-6,6′-dimethoxy-2,2′-dimethyltubocuraran-7′,12′-diol; d-bebeerine; chondodendrine; pelosine. $C_{36}H_{38}N_2O_6$; mol wt 594.71. C 72.71%, H 6.44%, N 4.71%, O 16.14%. From bark of *Nectandra rodioei* Hook., *Lauraceae*: Maclagan, *Ann.* **48**, 106 (1843). From root of *Chondodendron microphyllum* (Eichl.) Moldenke and stem of *Ch. candicans* (Rick ex. D.C.) Sandwith, *Menispermaceae* and *radix pareirae bravae*: King, *J. Chem. Soc.* **1940**, 737. Structure: Faltis *et al.*, *Ber.* **69**, 1269 (1936); King, *J. Chem. Soc.* **1939**, 1157. Stereochemistry: King, *ibid.* **1948**, 265.

Needles from methanol, mp 215°. $[\alpha]_D^{20}$ +345.7° (c = 0.4 in 1*N* HCl). Sol in benzene, chloroform, pyridine.

Hydrochloride. $C_{36}H_{38}N_2O_6 \cdot HCl$. mp 260°. $[\alpha]_D^{20}$ +294° (c = 0.7). Sol in water and alcohol.

THERAP CAT: Antimalarial.

1015. Bebeeru Bark. Greenheart; bibiru bark; sipiri bark. From *Nectandra rodioei* Hook., *Lauraceae*. *Habit.* British Guiana. *Constit.* Bebeerine, sipirine, bebeeric acid, tannin, resin.

1016. Beclamide. [501-68-8] 3-Chloro-*N*-(phenylmethyl)-propanamide; *N*-benzyl-3-chloropropionamide; *N*-benzyl-β-chloro-propanamide; benzchlorpropamide; chloroethylphenamide; Neuracen; Nidrane; Posedrine; Seclar. $C_{10}H_{12}ClNO$; mol wt 197.66. C 60.77%, H 6.12%, Cl 17.93%, N 7.09%, O 8.09%. Prepd by the action of β-chloropropionyl chloride on benzylamine in cooled water

at pH 8: Cassell, Kushner, **US 2569288** (1951 to Am. Cyanamid); Kushner *et al.*, *J. Org. Chem.* **16**, 1283 (1951).

Large crystals from methanol, mp 94°. Slightly sol in water (0.005 to 0.01%). Moderately sol in the lower alcs. Dec in hot aq acid soln and in hot aq alkaline soln.

THERAP CAT: Anticonvulsant.

1017. Beclomethasone. [4419-39-0] (11β,16β)-9-Chloro-11,17,21-trihydroxy-16-methylpregna-1,4-diene-3,20-dione; 9α-chloro-16β-methyl-1,4-pregnadiene-11β,17α,21-triol-3,20-dione; 9α-chloro-16β-methylprednisolone. $C_{22}H_{29}ClO_5$; mol wt 408.92. C 64.62%, H 7.15%, Cl 8.67%, O 19.56%. Glucocorticoid. Prepn of free alcohol and 21-acetate: **GB 912378** (1962 to Merck & Co.); of 21-acetate: **GB 901093** (1962 to Schevico); of mono and diesters: J. Elks *et al.*, **BE 649170**; *eidem*, **US 3312590** (1964, 1967 both to Glaxo). Symposium on clinical studies: *Br. J. Clin. Pharmacol.* **4**, Suppl. 3, 249S-312S (1977). Use in chronic asthma in children: M. Rao *et al.*, *J. Asthma* **19**, 21 (1982); in treatment of asthma in steroid-independent adults: V. A. Malfitan, *Clin. Ther.* **4**, 472 (1982). Review of use in rhinitis: P. Small *et al.*, *Ann. Allergy* **49**, 127 (1982); of pharmacology, side effects and use in asthma and allergic rhinitis: R. N. Brogden *et al.*, *Drugs* **28**, 99-126 (1984).

Dipropionate. [5534-09-8] Sch-18020W; Aerobec; Aldecin; Anceron; Andion; Beclacin; Becloforte; Beclomet; Beclorhinol; Beclovent; Becodisks; Beconase; Beconasol; Becotide; Clenil-A; Entyderma; Inalone; Korbutone; Propaderm; Qvar; Rino-Clenil; Sanasthmax; Vancenase; Vanceril; Viarex; Viarox. $C_{28}H_{37}ClO_7$; mol wt 521.05. Crystals from acetone + ether, mp 117-120° (dec). $[\alpha]_D$ +98.0° (c = 1.0 in dioxane). Very sol in chloroform; freely sol in acetone, alc. Very slightly sol in water. uv max (ethanol): 238 nm (ε 15990).

THERAP CAT: Antiallergic, antiasthmatic (inhalant). Anti-inflammatory (topical).

1018. Bedaquiline. [843663-66-1] (α*S*,β*R*)-6-Bromo-α-[2-(dimethylamino)ethyl]-2-methoxy-α-1-naphthalenyl-β-phenyl-3-quinolineethanol; (1*R*,2*S*)-1-(6-bromo-2-methoxyquinolin-3-yl)-4-(dimethylamino)-2-(napthalen-1-yl)-1-phenylbutan-2-ol; R-207910; TCM-207. $C_{32}H_{31}BrN_2O_2$; mol wt 555.52. C 69.19%, H 5.63%, Br 14.38%, N 5.04%, O 5.76%. Prototype of the diarylquinoline antitubercular agents; specifically inhibits mycobacterial ATP synthase. Prepn: J. F. E. Van Gestel *et al.*, **WO 04011436**; *eidem*, **US 7498343** (2004, 2009 both to Janssen). Asymmetric synthesis: Y. Saga *et al.*, *J. Am. Chem. Soc.* **132**, 7905 (2010). Antimycobacterial activity, pharmacology, and pharmacokinetics: K. Andries *et al.*, *Science* **307**, 223 (2005). *In vitro* antimycobacterial spectrum: E. Huitric *et al.*, *Antimicrob. Agents Chemother.* **51**, 4202 (2007). Mechanism of action: A. Koul *et al.*, *Nat. Chem. Biol.* **3**, 323 (2007). Determn in biological samples: F. Cuyckens *et al.*, *Anal. Bioanal. Chem.* **390**, 1717 (2008). Clinical pharmacokinetics: R. Rustomjee *et al.*, *Antimicrob. Agents Chemother.* **52**, 2831 (2008). Clinical evaluation in multidrug-resistant tuberculosis: A. H. Diacon *et al.*, *N. Engl. J. Med.* **360**, 2397 (2009). Review of preclinical studies: A. Matteelli *et al.*, *Future Microbiol.* **5**, 849-858 (2010); of structure-activity studies: J. Guillemont *et al.*, *Future Med. Chem.* **3**, 1345-1360 (2011).

Consult the Name Index before using this section.

White solid, mp 118°. $[\alpha]_D^{20}$ −166.98° (c = 0.505 in DMF).

Fumarate. [845533-86-0] R-403323. $C_{32}H_{31}BrN_2O_2 \cdot C_4H_4O_4$; mol wt 671.59. Prepn: J. F. A. Hegyi et al., **WO 08068231**; eidem, **US 100028428** (2008, 2010 both to Janssen). White solid.

1019. Beeswax. [8012-89-3] Yellow beeswax. A substance obtained from bee honeycombs. Consists of esters of straight-chain monohydric alcohols with even-numbered carbon chains from C_{24} to C_{36} esterified with straight-chain acids also having even numbers of C atoms up to C_{36} (some C_{18} hydroxy acids). Examples of such esters are triacontanol hexadecanoate and hexacosanol hexacosanoate. These esters are mixed with about 20% (w/w) of hydrocarbons having odd-numbered straight carbon chains from C_{21} to C_{33}. Propolis, pigments and unidentified substances amount to about 6%. Composition: D. T. Downing et al., Aust. J. Chem. **14**, 253 (1961); Callow, Bee World **44**, 95 (1963). Brief review: C. S. Letcher in Kirk-Othmer Encyclopedia of Chemical Technology **Vol. 24** (Wiley-Interscience, New York, 3rd ed., 1984) pp 466-467.

Yellowish to brownish-yellow, soft to brittle; honey-like odor; slight balsamic taste. d 0.95-0.960. mp 62-65°. Saponification number 84. Acid number 20. Practically insol in water. Slightly sol in cold alc; sol in hot alc, chloroform, benzene, ether, carbon disulfide.

White beeswax. White wax; bleached yellow wax; bleached beeswax. Prepd by oxidizing yellow beeswax cakes with peroxide or in sunlight. Yellowish-white. Properties similar to those of yellow beeswax, except for a slightly different taste. Preferred to yellow beeswax in cosmetics.

USE: Manuf of wax paper, candles, cosmetics; modeling artificial fruits and flowers; in process engraving; shoe polish. Pharmaceutic aid (in ointments, plasters).

1020. Befloxatone. [134564-82-2] (5R)-5-(Methoxymethyl)-3-[4-[(3R)-4,4,4-trifluoro-3-hydroxybutoxy]phenyl]-2-oxazolidinone; MD-370503. $C_{15}H_{18}F_3NO_5$; mol wt 349.31. C 51.58%, H 5.19%, F 16.32%, N 4.01%, O 22.90%. Reversible monoamine oxidase type A (MAO-A) inhibitor. Prepn: F. X. Jarreau et al., **EP 424244**; eidem, **US 5036091** (1990, 1991 both to Delalande). Pharmacology: D. Caille et al., J. Pharmacol. Exp. Ther. **277**, 265 (1996). Biochemical profile: O. Curet et al., ibid. 253. Structure-activity study: J. Wouters et al., Bioorg. Med. Chem. **7**, 1683 (1999). Review of clinical pharmacology: P. Rosenzweig et al., J. Affect. Disord. **51**, 305-312 (1998).

Crystals from ethanol + isopropyl ether, mp 101°. $[\alpha]_D^{20}$ −11.5° (c = 1 in methylene chloride).

THERAP CAT: Antidepressant.

1021. Befunolol. [39552-01-7] 1-[7-[2-Hydroxy-3-[(1-methylethyl)amino]propoxy]-2-benzofuranyl]ethanone; 7-[2-hydroxy-3-(isopropylamino)propoxy]-2-benzofuranyl methyl ketone; 2-acetyl-7-(2-hydroxy-3-isopropylaminopropoxy)benzofuran. $C_{16}H_{21}NO_4$; mol wt 291.35. C 65.96%, H 7.27%, N 4.81%, O 21.97%. β-Adrenergic blocker. Prepn: K. Ito et al., **DE 2223184**; eidem, **US 3853923** (1972, 1974 both to Kakenyaku Kako). Prepn and activity

of optical isomers: J. Nakano et al., Chem. Pharm. Bull. **36**, 1399 (1988). Determn of befunolol and metabolites in human plasma: K. Kawahara, T. Ofuji, J. Chromatogr. **168**, 266 (1979). Absorption, distribution, excretion in rats: H. Kitagawa et al., Oyo Yakuri **17**, 383, 393 (1979), C.A. **91**, 13374w-5x (1979). General pharmacology: S. Masumoto et al., Iyakuhin Kenkyu **10**, 741 (1979), C.A. **92**, 88015s (1980). Pharmacodynamics: S. Harada et al., Arch. Int. Pharmacodyn. Ther. **252**, 262 (1981).

Crystals from cyclohexane/acetone, mp 115°. LD_{50} in mice: 100-105 mg/kg i.v. (Ito).

Hydrochloride. [39543-79-8] BFE-60; Benfuran; Bentos; Glauconex. $C_{16}H_{21}NO_4 \cdot HCl$; mol wt 327.81. Crystals from ethyl acetate, mp 163°.

(S)-(−)-Form hydrochloride. [66717-59-7] Pale yellow prisms from isopropyl alcohol, mp 151-152°. $[\alpha]_D$ −15.5° (c = 1 in methanol).

(R)-(+)-Form hydrochloride. [66685-79-8] Pale yellow prisms from isopropyl alcohol, mp 151°. $[\alpha]_D$ +15.3° (c = 1 in methanol).

THERAP CAT: Antiglaucoma.

1022. Behenic Acid. [112-85-6] Docosanoic acid. $C_{22}H_{44}O_2$; mol wt 340.59. C 77.58%, H 13.02%, O 9.39%. Minor constituent of most seed fats, animal milk fats and marine animal oils. Large amts (~50%) are found in (hydrogenated?) jamba oil, mustard seed oil and rape oil: Sudborough et al., J. Indian Inst. Sci. **9A**, 25 (1926). Prepn from erucic acid by catalytic reduction: Morgan, Holmes, J. Soc. Chem. Ind. London Trans. Commun. **44**, 491 (1925). Brief review: E. S. Lower, Manuf. Chem. **56**(7), 44-46 (1985).

Waxy solid, mp 79.95°. Neutralization value 164.73. bp_{60} 306°. d_4^{100} 0.8221; n_D^{100} 1.4270. One hundred grams of 90% ethanol dissolve 0.102 g of behenic acid at 17°; at 25° 0.218 g of the acid dissolves in 100 ml of 91.5% ethanol, 0.116 g in 100 ml of 86.2% ethanol, 0.011 g in 100 ml of 63.07% ethanol. One hundred grams of ether dissolves 0.1922 g of behenic acid at 16°.

Methyl ester. [929-77-1] $C_{23}H_{46}O_2$; mol wt 354.62. mp 54°.

Ethyl ester. [5908-87-2] $C_{24}H_{48}O_2$; mol wt 368.65. mp 50°. $bp_{0.2}$ 185°.

Amide. [3061-75-4] $C_{22}H_{45}NO$; mol wt 339.61. mp 111-112°.

USE: In lubricating oils, as solvent evaporation retarder in paint removers. Amide as anti-foam in the manuf of detergents, in floor polishes, in dripless candles.

1023. Belatacept. [706808-37-9] [29-Tyrosine,104-glutamic acid] CTLA-4 (antigen) (human extracellular domain-containing fragment) fusion protein with immunoglobulin G1 (human monoclonal Fc domain-containing fragment), bimol. (120 → 120')-disulfide; LEA29Y; BMS-224818; Nulojix. Selective costimulation blocker similar to abatacept, q.v. Soluble fusion protein combining a modified form of the extracellular portion of cytotoxic T-lymphocyte-associated antigen 4 (CTLA-4) with the constant region of human IgG1. Prepn: R. J. Peach et al., **WO 0192337**; eidem **US 05214313** (2001, 2005 both to Bristol-Myers Squibb). Pharmacology and development: C. P. Larsen et al., Am. J. Transplant. **5**, 443 (2005). Clinical trial in rheumatoid arthritis: L. W. Moreland et al., Arthritis Rheum. **46**, 1470 (2002); in renal transplantation: F. Vincenti et al., N. Engl. J. Med. **353**, 770 (2005).

THERAP CAT: Immunosuppressant.

1024. Belimumab. [356547-88-1] Anti-(human cytokine BAFF) immunoglobulin G1 (human monoclonal LymphoStat-B heavy chain) disulfide with human monoclonal LymphoStat-B λ-chain, dimer; Benlysta; LymphoStat-B. Fully human monoclonal antibody that specifically inhibits soluble B-lymphocyte stimulator (BLyS), also known as BAFF. Designed to treat autoimmune diseases associated with B-cell hyperactivity and increased levels of BLys. Prepn: S. Ruben *et al.*, **WO 0202641** (2002 to Human Genome Sciences, Cambridge Antibody Technology); *eidem*, **US 7138501** (2006 to Human Genome Sciences). Characterization: K. P. Baker *et al.*, *Arthritis Rheum.* **48**, 3253 (2003). Pharmacology and toxicology: W. G. Halpern *et al.*, *Toxicol. Sci.* **91**, 586 (2006). Clinical trial in systemic lupus erythematosus (SLE): S. V. Navarra *et al.*, *Lancet* **377**, 721 (2011). Review of pharmacology and clinical experience in SLE: C. Ding, *Expert Opin. Biol. Ther.* **8**, 1805 (2008); in rheumatoid arthritis and SLE: A. Goldberg, E. Katzap, *Int. J. Clin. Rheumatol.* **5**, 407-413 (2010).

THERAP CAT: In treatment of systemic lupus erythematosus.

1025. Belinostat. [414864-00-9] *N*-Hydroxy-3-[3-[(phenyl-amino)sulfonyl]phenyl]-2-propenamide; *N*-hydroxy-3-(3-phenylsul-famoylphenyl)acrylamide; PX-105684; PXD-101. $C_{15}H_{14}N_2O_4S$; mol wt 318.35. C 56.59%, H 4.43%, N 8.80%, O 20.10%, S 10.07%. Hydroxamate-type histone deacetylase (HDAC) inhibitor. Prepn: C. J. Watkins *et al.*, **WO 02030879** (2002 to Prolifix); *eidem*, **US 6888027** (2005 to Topotarget). HPLC determn in plasma: N. Zhang *et al.*, *Ther. Drug Monit.* **29**, 231 (2007). Growth inhibitory activity in prostate cell lines and xenografts: X. Qian *et al.*, *Int. J. Cancer* **122**, 1400 (2008). Clinical pharmacology in patients with advanced solid tumors: N. L. Steele *et al.*, *Clin. Cancer Res.* **14**, 804 (2008); with advanced hematological neoplasia: P. Gimsing *et al.*, *Eur. J. Haematol.* **81**, 170 (2008). Review of development and clinical experience: P. Gimsing, *Expert Opin. Invest. Drugs* **18**, 501-508 (2009).

White solid from acetonitrile and diethyl ether, mp 172°. Sol in ethanol.

THERAP CAT: Antineoplastic.

1026. Belladonna. Deadly nightshade; banewort; death's herb; dwale; poison black cherry. Dried leaves and root of *Atropa belladonna* L., Solanaceae. *Habit.* Southern and Central Europe, Asia Minor, Algeria; cultivated in North America. On extraction it yields atropine, hyoscyamine, scopolamine, asparagine, choline, chrysatropic acid, atroscine, leucatropic acid, phytosterol. Leaves usually contain 0.3-0.5% total alkaloids; dried root contains 0.4-0.7% total alkaloids. *See:* Romeike, *Pharmazie* **8**, 729 (1953); Phokas, Steinegger, *ibid.* **11**, 652 (1956); *Pharm. Acta Helv.* **31**, 284 (1956).

Caution: Toxicity based on content of atropine, *q.v.*, and related alkaloids.

THERAP CAT: Anticholinergic.

THERAP CAT (VET): Anticholinergic; antispasmodic.

1027. Belleau's Reagent. [88816-02-8] 2,4-Bis(4-phenoxy-phenyl)-1,3,2,4-dithiadiphosphetane 2,4-disulfide. $C_{24}H_{18}O_2P_2S_4$; mol wt 528.59. C 54.53%, H 3.43%, O 6.05%, P 11.72%, S 24.26%. *p*-Phenoxy derivative of Lawesson's reagent, *q.v.* Prepn: G. Lajoie *et al.*, *Tetrahedron Lett.* **24**, 3815 (1983). Synthetic applications: B. Yde *et al.*, *Tetrahedron* **40**, 2047 (1984); M. Caron, *J. Org. Chem.* **51**, 4075 (1986); M. M. Campbell *et al.*, *Tetrahedron Lett.* **30**, 1997 (1989); A. G. M. Barrett, A. C. Lee, *J. Org. Chem.* **57**, 2818 (1992); R. Shabana *et al.*, *Tetrahedron* **49**, 1271 (1993); *eidem, ibid.* **50**, 6975 (1994); S. V. Ley *et al.*, *Org. Lett.* **4**, 711 (2002).

Yellow crystals from toluene, mp 187-190°. Sol in $CHCl_3$, THF, toluene, acetonitrile.

USE: Mild thionation reagent in organic synthesis.

1028. Belotecan. [256411-32-2] (4*S*)-4-Ethyl-4-hydroxy-11-[2-[(1-methylethyl)amino]ethyl]-1*H*-pyrano[3′,4′:6,7]indolizino-[1,2-*b*]quinoline-3,14(4*H*,12*H*)-dione; (20*S*)-7-(2-isopropylamino)-ethylcamptothecin; 7-[2-(*N*-isopropylamino)ethyl]-(20*S*)-campto-thecin; (*S*)-7-[2-(*N*-propylamino)ethyl]camptothecin. $C_{25}H_{27}N_3O_4$; mol wt 433.51. C 69.27%, H 6.28%, N 9.69%, O 14.76%. DNA topoisomerase I inhibitor; semi-synthetic analog of camptothecin, *q.v.* Prepd (not claimed): C. I. Hong *et al.*, **WO 9902530**; *eidem*, **US 6310207** (1999, 2001 both to Chong Kun Dang). Improved synthesis: S. K. Ahn *et al.*, *J. Heterocycl. Chem.* **37**, 1141 (2000). Characterization of physicochemical properties: J.-H. Kim *et al.*, *Int. J. Pharm.* **239**, 207 (2002). HPLC determn in plasma: J.-Y. Cho *et al.*, *J. Chromatogr. B* **784**, 25 (2003). Inhibition of topoisomerase I in mammalian tumor cells and human tumor xenografts: L.-H. Lee *et al.*, *Arch. Pharmacal Res.* **21**, 581 (1998). Clinical pharmacology: D. H. Lee *et al.*, *Clin. Cancer Res.* **13**, 6182 (2007). Clinical evaluation in small cell lung cancer: D. H. Lee *et al.*, *Ann. Oncol.* **19**, 123 (2008).

Hydrochloride. [213819-48-8] CKD-602; Camtobell. $C_{25}H_{27}N_3O_4 \cdot HCl$; mol wt 469.97. Slightly hygroscopic, pale yellow, crystalline powder, mp 267-268° (dec). $[\alpha]_D$ +53.49° (c = 0.1 in water). pKa_1 2.32±0.05, pKa_2 9.15±0.02. Soly at 25° (mg/ml): acetic acid 10.70±0.08; methanol 4.13±0.32; ethanol 1.11±0.10; deionized water 8.22±0.18; acetonitrile 0.057±0.004.

THERAP CAT: Antineoplastic.

1029. Bemegride. [64-65-3] 4-Ethyl-4-methyl-2,6-piperi-dinedione; 3-ethyl-3-methylglutarimide; 4-ethyl-4-methyl-2,6-di-oxopiperidine; 2,6-dioxo-4-methyl-4-ethylpiperidine; methethari-mide; β,β-methylethylglutarimide; NP-13; Eukraton; Megimide. $C_8H_{13}NO_2$; mol wt 155.20. C 61.91%, H 8.44%, N 9.03%, O 20.62%. Prepn: Thole, Thorpe, *J. Chem. Soc.* **99**, 439 (1911); Sircar, *ibid.* **1937**, 602, 604; Benica, Wilson, *J. Am. Pharm. Assoc. Sci. Ed.* **39**, 451 (1950); Lukes, Ferles, *Chem. Listy* **49**, 510 (1955), *C.A.* **49**, 10290 (1955). Pharmacology: Delay *et al.*, *Presse Med.* **44**, 1525 (1956); Oberdorf, Meyer, *Arch. Exp. Pathol. Pharmakol.* **238**, 128 (1960); Kretzschmar *et al.*, *Arch. Int. Pharmacodyn. Ther.* **174**, 318 (1968). Metabolism: Nicholls, *Nature* **185**, 927 (1960). Toxicity study: F. Hahn, A. Oberdorf, *Arch. Int. Pharmacodyn. Ther.* **135**, 9 (1962). Reduction of barbiturate anesthesia in mice: V. K. Patel *et al.*, *Psychopharmacology* **71**, 21 (1980). Postsynaptic effects: P. W. Gage, P. Sah, *Br. J. Pharmacol.* **75**, 493 (1982).

Platelets from water or from acetone + ether, mp 127°. Sublimes at 100° and 2 mm press. Sol in water, acetone. LD_{50} in male mice (mg/kg): 20.1 ±1.41 i.v., 43.0 ±1.8 s.c.; LD_{50} in male rats (mg/kg): 16.3 ±1.24 i.v., 23.5 ±1.67 i.p (Hahn, Oberdorf).

THERAP CAT: CNS stimulant. Antidote (barbiturate poisoning).

THERAP CAT (VET): CNS stimulant. Antidote (barbiturate poisoning).

1030. Bemotrizinol. [187393-00-6] 2,2′-[6-(4-Methoxy-phenyl)-1,3,5-triazine-2,4-diyl]bis[5-[(2-ethylhexyl)oxy]phenol];

2,4-bis-[4-(2-ethylhexyloxy)-2-hydroxy]phenyl-6-(4-methoxyphen-yl)-1,3,5-triazine; bis-ethylhexyloxyphenol methoxyphenyl triazine; BEMT; Tinosorb S. $C_{38}H_{49}N_3O_5$; mol wt 627.83. C 72.70%, H 7.87%, N 6.69%, O 12.74%. Broad spectrum, photostable UV-A and UV-B filter in cosmetics and sunscreens. Prepn: D. Hüglin *et al.*, **EP 775698**; *idem*, **US 5955060** (1997, 1999 both to Ciba Specialty Chem.). Lack of estrogenic or androgenic activity: J. Ashby *et al.*, *Regul. Toxicol. Pharmacol.* **34**, 287 (2001). Photostabilizing effects on avobenzone and ethylhexyl methoxycinnamate, *q.q.v.*: E. Chatelain, B. Gabard, *Photochem. Photobiol.* **74**, 401 (2001). Comprehensive review: S. Mongiat *et al.*, *Cosmet. Toiletries* **118**, 47-54 (2003).

Slightly yellow crystals, mp 83-85° (Hüglin); also reported as mp 80° (Mongiat). Sol in oil. Absorption max (ethanol): 343 nm (ε 47000). Absorption max (isopropanol): 341 nm ($E_{1cm}^{1\%} \geq 790$).

THERAP CAT: Ultraviolet screen.

1031. Benactyzine. [302-40-9] α-Hydroxy-α-phenylbenzeneacetic acid 2-(diethylamino)ethyl ester; benzilic acid β-diethylaminoethyl ester; β-diethylaminoethyl benzilate; 2-diethylaminoethyl diphenylglycolate. $C_{20}H_{25}NO_3$; mol wt 327.42. C 73.37%, H 7.70%, N 4.28%, O 14.66%. An antagonist of acetylcholine in the central and peripheral nervous systems. Prepn: Horenstein, Pahlicke, *Ber.* **71**, 1654 (1938); Blicke, Maxwell, *J. Am. Chem. Soc.* **64**, 428 (1942); Hill, Holmes, **US 2394770** (1946 to Am. Cyanamid). Toxicity and pharmacodynamics: Fournier, Petit, *Therapie* **17**, 1245 (1962). Metabolism: Eldeson *et al.*, *Arch. Int. Pharmacodyn. Ther.* **187**, 139 (1970). Crystal and molecular structure determn by x-ray diffraction: T. J. Petcher, *J. Chem. Soc. Perkin Trans.* 2 **1974**, 1151.

Crystals, mp 51°.

Hydrochloride. [57-37-4] AY-5406-1; Cedad; Nutinal; Parasan; Suavitil. $C_{20}H_{25}NO_3 \cdot HCl$; mol wt 363.88. Crystals from acetone, mp 177-178°. Soly in water (25°): 14.9/100 ml. Practically insol in ether.

Methobromide. [3166-62-9] Spatomac. $C_{20}H_{25}NO_3 \cdot CH_3Br$; mol wt 422.36. Crystals from alcohol + ether, mp 169-170°.

THERAP CAT: Antispasmodic.

1032. Benalaxyl. [71626-11-4] N-(2,6-Dimethylphenyl)-N-(2-phenylacetyl)-alanine methyl ester; methyl N-phenylacetyl-N-2,6-xylyl-DL-alaninate; N-(2,6-dimethylphenyl)-N-(1-carbomethoxyethyl)phenylacetamide; N-(2,6-dimethylphenyl)-N-(phenylacetyl)-DL-alanine methyl ester; M-9834; Galben. $C_{20}H_{23}NO_3$; mol wt 325.41. C 73.82%, H 7.12%, N 4.30%, O 14.75%. Acylalanine fungicide systemically active against phytopathogens of the order *Peronosporales*. Prepn: E. Boson *et al.*, **DE 2903612**; *idem*, **US 4425357** (1978, 1984 both to Montedison). Activity and toxicology: P. Bergamaschi *et al.*, *Proc. Br. Crop Prot. Conf. - Pests Dis.* **1981**, 19. Residue accumulation in edible crops: P. Cabras *et al.*, *J. Agric. Food Chem.* **33**, 86 (1985). GC-MS study: B. D. Ripley, *ibid.* 560.

Crystals from ligroin, mp 78-80°. LD_{50} in rats (mg/kg): 4200 orally; 1100 i.p.; LC_{50} (96 hr) in rainbow trout: 3.75 mg/l (Bergamaschi).

USE: Agricultural fungicide.

1033. Benazepril. [86541-75-5] (3S)-3-[[(1S)-1-(Ethoxycarbonyl)-3-phenylpropyl]amino]-2,3,4,5-tetrahydro-2-oxo-1H-1-benzazepine-1-acetic acid; (3S)-1-(carboxymethyl)-[[(1S)-1-(ethoxycarbonyl)-3-phenylpropyl]amino]-2,3,4,5-tetrahydro-1H-[1]-benzazepin-2-one; (3S)-3-[[(1S)-1-carboxy-3-phenylpropyl]amino]-2,3,4,5-tetrahydro-2-oxo-1H-1-benzazepine-1-acetic acid 3-ethyl ester. $C_{24}H_{28}N_2O_5$; mol wt 424.50. C 67.91%, H 6.65%, N 6.60%, O 18.84%. Angiotensin converting enzyme (ACE) inhibitor. Hydrolyzed *in vivo* to the active diacid metabolite. Prepn: J. W. H. Watthey, **EP 72352**; *idem*, **US 4410520** (both 1983 to Ciba-Geigy); J. W. H. Watthey *et al.*, *J. Med. Chem.* **28**, 1511 (1985). Clinical pharmacology: M. D. Schaller *et al.*, *Eur. J. Clin. Pharmacol.* **28**, 267 (1985). GC-MS determn in plasma and urine: A. Sioufi *et al.*, *J. Chromatogr.* **434**, 239 (1988). Pharmacokinetics of benazepril and benazeprilat: G. Kaiser *et al.*, *Biopharm. Drug Dispos.* **10**, 365 (1989). Multicenter clinical trial in hypertension: M. Moser *et al.*, *ibid.* **49**, 322 (1991); in congestive heart failure: H. T. Colfer *et al.*, *Am. J. Cardiol.* **70**, 354 (1992).

mp 148-149°. $[\alpha]_D -159°$ (c = 1.2 in ethanol).

Hydrochloride. [86541-74-4] CGS-14824A; Briem; Cibacen; Fortekor; Lotensin. $C_{24}H_{28}N_2O_5 \cdot HCl$; mol wt 460.96. Crystals from 3-pentanone + methanol (10:1), mp 188-190°. $[\alpha]_D -141.0°$ (c = 0.9 in ethanol). Sol in water, ethanol, methanol.

Diacid. [86541-78-8] Benazeprilat; CGS-14831. $C_{22}H_{24}N_2O_5$; mol wt 396.44. mp 270-272°. $[\alpha]_D -200.5°$ (c = 1 in 3% aqueous NH_4OH).

THERAP CAT: Antihypertensive.

THERAP CAT (VET): In treatment of congestive heart failure in dogs.

1034. Bence-Jones Proteins. Low molecular weight, heat-sensitive proteins isolated from urine of multiple myeloma patients: Caputo, *Aminoacidosi* **7**, 163-77 (1957), *C.A.* **54**, 6841 (1960). Used as a model in structural studies of immunoglobulins, they are structurally homogeneous within each patient and antigenically related. Crystallographic structural study: E. E. Aeola *et al.*, *Biochemistry* **19**, 432 (1980). They constitute the light-chain component of myeloma globulin and have been implicated in the pathogenesis of amyloidosis: G. G. Glenner, *N. Engl. J. Med.* **302**, 1283 (1980); G. G. Glenner *et al.*, *Amyloid and Amyloidosis* (Excerpta Medica, Amsterdam, 1980) p 351.

Coagulates on heating at 45-55°. Redissolves partially or wholly on boiling. Amino acid composition in g-%: Arginine 5.6, histidine 3.3, lysine 9.5, tyrosine 6.4, tryptophan 3.4, phenylalanine 4.2, cystine 2.6, methionine 1.9, serine 8.4, threonine 8.1, leucine 7.3, isoleucine 2.2, valine 9.8, glycine 8.6, alanine 3.1, proline 6.9, glutamic acid 7.5, aspartic acid 9.2. The molecule also shows a carbohydrate content of 9.4% (mostly glucose, mannose, and glucosamine).

1035. Bencyclane. [2179-37-5] *N*,*N*-Dimethyl-3-[[1-(phenylmethyl)cycloheptyl]oxy]-1-propanamine; 3-[(1-benzylcycloheptyl)oxy]-*N*,*N*-dimethylpropylamine; 1-benzyl-1-(3-dimethylaminopropoxy)cycloheptane; *N*-[3-(1-benzylcycloheptyloxy)propyl]-*N*,*N*-dimethylamine; benzcyclan. $C_{19}H_{31}NO$; mol wt 289.46. C 78.84%, H 10.80%, N 4.84%, O 5.53%. Prepn: Pallos *et al.*, **HU 151865** (1965 to EGYT), *C.A.* **62**, 16125b (1965). Series of articles on pharmacology: *Arzneim.-Forsch.* **20**, 1337-1460 (1970). Toxicity: E. Komlos, L. E. Petöcz, *ibid.* 1338.

bp₃ 146-156°.

Fumarate. [14286-84-1] EGYT-201; Angiociclan; Dantrium; Dilangio; Fludilat; Fluxema; Halidor; Vasorelax. $C_{19}H_{31}NO.C_4H_4O_4$; mol wt 405.54. Crystals from 25% ethanol, mp 131-133°. Soly in water at 25°: 1 g/100 ml; in hot water: 2 g/100 ml. Poorly sol in acetone, easily in alc. uv spectrum: Simonyi *et al.*, *Acta Pharm. Hung.* **36**, 257 (1966). At pH 3.4-6.6, uv max (ethanol): 207 nm. LD_{50} in mice, rats (mg/kg): 445.6, 414 orally; 49.9, 41.3 i.v.; 132, 86.3 i.p.; 203, 257 s.c. (Komlos, Petöcz).

THERAP CAT: Vasodilator (peripheral, cerebral).

1036. Bendamustine. [16506-27-7] 5-[Bis(2-chloroethyl)amino]-1-methyl-1*H*-benzimidazole-2-butanoic acid; γ-[1-methyl-5-[bis(β-chloroethyl)amino]-2-benzimidazolyl]butyric acid. $C_{16}H_{21}Cl_2N_3O_2$; mol wt 358.26. C 53.64%, H 5.91%, Cl 19.79%, N 11.73%, O 8.93%. Bifunctional alkylating agent. Prepn: W. Ozegowski, D. Krebs, *J. Prakt. Chem.* **20**, 178 (1963); *eidem, Zentralbl. Pharm. Pharmakother. Laboratoriumsdiagn.* **110**, 1013 (1971). Antitumor activity: W. Jungstand *et al., ibid.* 1021. Capillary GC determn in plasma: H. Weber *et al., J. Chromatogr.* **525**, 459 (1990). Toxicity study: U. Horn *et al., Arch. Toxicol.* **Suppl. 8**, 504 (1985). Review of pharmacology and clinical development: K. Bremer, W. Roth, *Tumordiagn. Ther.* **17**, 1-6 (1996); J. A. Barman Balfour, K. L. Goa, *Drugs* **61**, 631-638 (2001). Review of clinical experience in chronic lymphocytic leukemia and non-Hodgkin's lymphoma: C. Ujjani, B. D. Cheson, *Expert Rev. Anticancer Ther.* **10**, 1353-1365 (2010).

Hydrochloride. [3543-75-7] IMET-3393; SDX-105; Cytostasan; Levact; Ribomustin; Treanda. $C_{16}H_{21}Cl_2N_3O_2.HCl$; mol wt 394.72. Monohydrate mp 152-156°. Sol in water. LD_{50} (monohydrate) in mice, rats (mg/kg): 400-500, 200-300 orally; 80, 40 i.v. (Horn).

THERAP CAT: Antineoplastic.

1037. Bendazac. [20187-55-7] 2-[[1-(Phenylmethyl)-1*H*-indazol-3-yl]oxy]acetic acid; [(1-benzyl-1*H*-indazol-3-yl)oxy]acetic acid; bendazolic acid; bindazac; AF-983; Zildasac. $C_{16}H_{14}N_2O_3$; mol wt 282.30. C 68.08%, H 5.00%, N 9.92%, O 17.00%. Principle action is inhibition of protein denaturation. Prepn: **BE 699226**; G. Palazzo, **US 3470194** (1967, 1969 both to Francesco Angelini). Pharmacology: B. Silvestrini *et al., Inflammation Biochem. Drug Interaction, Proc. Int. Symp.* **1968**, A. Bertelli, Ed. (Excerpta Med., Amsterdam, 1970) p 283. Mechanism of action: *eidem, Arzneim.-Forsch.* **20**, 250 (1970). Photoprotective capacity: Fuga *et al., Ann. Ital. Dermatol. Clin. Sper.* **24**, 205 (1970). NMR study of albumin, *q.v.*, binding: M. Delfini *et al., Bioorg. Chem.* **31**, 378 (2003). Toxicity: *Rx Bulletin* **3**, 147 (1972).

Odorless, crystalline powder, mp 158-159°. Insol in water. Sol in chloroform, ethanol, acetone. uv max: 306 nm ($E_{1cm}^{1\%}$ 191). LD_{50} in mice, rats (mg/kg): 380, 304 i.v.; 355, 388 i.p.; 440, 910 s.c.; 1105, ~1200 orally (*Rx Bulletin*).

Lysine salt. [81919-14-4] bendaline; bendazac lysine; Bendalina; Dogalina. $C_{16}H_{13}N_2O_3.C_6H_{14}N_2O_2$; mol wt 427.48. HPLC determn: S. Scalia, M. Massaccesi, *Int. J. Pharm.* **82**, 179 (1992). Clinical evaluation in senile cataract: P. Baraldi *et al., Graefe's Arch. Clin. Exp. Opthalmol.* **228**, 105 (1990); in *in vivo* contact lens cleaning: T. C. Evans *et al., Optom. Vis. Sci.* **70**, 210 (1993). Review of pharmacology and therapeutic potential: J. A. Balfour, S. P. Clissold, *Drugs* **39**, 575-596 (1990).

Sodium salt. [23255-99-4] Versus. $C_{16}H_{13}N_2NaO_3$; mol wt 304.28.

USE: Contact lens cleaning and wetting solution.

THERAP CAT: Anti-inflammatory; in treatment of cataracts.

THERAP CAT (VET): In treatment of cataracts.

1038. Bendiocarb. [22781-23-3] 2,2-Dimethyl-1,3-benzodioxol-4-ol 4-(*N*-methylcarbamate); 2,2-dimethyl-1,3-benzodioxol-4-ol methylcarbamate; methylcarbamic acid 2,3-(isopropylidenedioxy)phenyl ester; NC-6897; Ficam. $C_{11}H_{13}NO_4$; mol wt 223.23. C 59.19%, H 5.87%, N 6.27%, O 28.67%. Acetylcholinesterase inhibitor. Prepn: Gates, Gillon, **ZA 6800736** *C.A.* **71**, 38941m (1969), corresp to **US 3736338** (1968, 1973 both to Fisons). Insecticidal activity: Story, *Int. Pest Control* **14**, 6 (1972).

White solid, mp 129-130°. Soly in water: 40 ppm; in hexane: 350 ppm.

USE: Contact insecticide.

1039. Bendroflumethiazide. [73-48-3] 3,4-Dihydro-3-(phenylmethyl)-6-(trifluoromethyl)-2*H*-1,2,4-benzothiadiazine-7-sulfonamide 1,1-dioxide; 3-benzyl-6-trifluoromethyl-3,4-dihydro-7-sulfamoyl-2*H*-1,2,4-benzothiadiazine 1,1-dioxide; 3-benzyl-3,4-dihydro-7-sulfamyl-6-trifluoromethyl-1,2,4-benzothiadiazine 1,1-dioxide; bendrofluazide; benzydroflumethiazide; benzylhydroflumethiazide; Aprinox; Berkozide; Centyl; Naturetin; Neo-Naclex; Salures; Sinesalin. $C_{15}H_{14}F_3N_3O_4S_2$; mol wt 421.41. C 42.75%, H 3.35%, F 13.52%, N 9.97%, O 15.19%, S 15.22%. Prepn: C. T. Holdrege *et al., J. Am. Chem. Soc.* **81**, 4807 (1959); F. Lund, W. O. Godtfredsen, **GB 863474**; *eidem,* **US 3392168** (1961, 1968 both to Lövens Kemiske Fabrik). Comprehensive description: K. Florey, F. M. Russo-Alesi, *Anal. Profiles Drug Subs.* **5**, 1-19 (1976).

Crystals from dioxane, mp 224.5-225.5° (U.S. patent). Also reported as mp 221-223° (Holdrege). uv max (methanol): 208, 273, 326 nm ($E_{1cm}^{1\%}$ 745, 565, 96). Freely sol in acetone, alc. Insol in water, chloroform, benzene, ether.

THERAP CAT: Diuretic, antihypertensive.

1040. Benexate Hydrochloride. [78718-25-9] 2-[[[*trans*-4-[[(Aminoiminomethyl)amino]methyl]cyclohexyl]carbonyl]oxy]benzoic acid phenylmethyl ester monohydrochloride; benzyl salicylate *trans*-4-(guanidinomethyl)cyclohexanecarboxylate hydrochloride; (2'-benzyloxycarbonyl)phenyl *trans*-4-(guanidinomethyl)cyclohexanecarboxylate hydrochloride. $C_{23}H_{28}ClN_3O_4$; mol wt 445.94. C 61.95%, H 6.33%, Cl 7.95%, N 9.42%, O 14.35%. Synthetic protease inhibitor with antiulcer activity. Prepn: **BE 885263**; M. Muramatsu *et al.*, **US 4348410** (1981, 1982 to Nippon Chemiphar; Teikoku Chem. Ind.). Prepn and protease inhibition: T. Satoh *et al.*, *Chem. Pharm. Bull.* **33**, 647 (1985). Prepn of the clathrate compound with β-cyclodextrin: **JP Kokai 83 38250**; M. Shinoda *et al.*, **US 4478995** (1983, 1984 both to Teikoku Chem. Ind.). Effect on gastric secretion and exptl ulcers in rats: S. Okabe *et al.*, *Oyo Yakuri* **27**, 829 (1984), *C.A.* **101**, 143881c (1984); I. Tanaka, H. Tagami, *Nippon Yakurigaku Zasshi* **85**, 167 (1985), *C.A.* **103**, 64660t (1985); F. Hirose *et al.*, *Yakuri to Chiryo* **15**, 4749 (1987), *C.A.* **108**, 137769a (1988).

Relative stereochemistry

Crystals from methanol + ether, mp 83°.

Compd with β-cyclodextrin (1:1). [91574-91-3] Benexate-CD; TA-903; Lonmiel; Ulgut. $C_{65}H_{98}ClN_3O_{39}$; mol wt 1580.93.

THERAP CAT: Antiulcerative.

1041. Benfluorex. [23602-78-0] 2-[[1-Methyl-2-[3-(trifluoromethyl)phenyl]ethyl]amino]ethanol 1-benzoate; 2-[[α-methyl-*m*-(trifluoromethyl)phenethyl]amino]ethanol benzoate (ester); 1-(*m*-trifluoromethylphenyl)-2-(β-benzoyloxyethyl)aminopropane; *N*-(2-benzoyloxyethyl)norfenfluramine; benfluramate; S-780; SE-780. $C_{19}H_{20}F_3NO_2$; mol wt 351.37. C 64.95%, H 5.74%, F 16.22%, N 3.99%, O 9.11%. Analog of fenfluramine with anorectic activity. Prepn from 1-(*m*-trifluoromethyl)-2-(β-hydroxyethyl)aminopropane and benzoyl chloride: L. Beregi *et al.*, **FR 1517587**; *eidem*, **US 3607909** (1968, 1971 both to Sci. Union et Cie, Soc. Franc. Rech. Med.). Metabolism studies: A. H. Beckett *et al.*, *J. Pharm. Pharmacol.* **23**, 950 (1971); **24**, 281 (1972). Pharmacology: D. N. Brindley *et al.*, *ibid.* **28**, 676 (1976); P. Pritchard *et al.*, *ibid.* **29**, 343 (1977); *eidem*, *Biochem. J.* **166**, 639 (1977). Clinical trial in treatment of type 2 diabetes: P. Moulin *et al.*, *Diabetes Metab.* **35**, 64 (2009). Evaluation of risk for valvular heart disease: A. Weill *et al.*, *Pharmacoepidemiol. Drug Saf.* **19**, 1256 (2010); F. Le Ven *et al.*, *Eur. J. Echocardiogr.* **12**, 265 (2011).

Colorless oil.

Hydrochloride. [23642-66-2] S-992; JP-992; Minolip; Mediator; Mediaxal. $C_{19}H_{20}F_3NO_2 \cdot HCl$; mol wt 387.83. Crystals from ethyl acetate, mp 161-162°.

THERAP CAT: Antilipemic.

1042. Benfluralin. [1861-40-1] *N*-Butyl-*N*-ethyl-2,6-dinitro-4-(trifluoromethyl)benzenamine; *N*-butyl-*N*-ethyl-α,α,α-trifluoro-2,6-dinitro-*p*-toluidine; *N*-butyl-*N*-ethyl-2,6-dinitro-4-trifluoromethylaniline; benefin; bethrodine; EL-110; Balan; Balfin; Benefex; Quilan. $C_{13}H_{16}F_3N_3O_4$; mol wt 335.28. C 46.57%, H 4.81%, F 17.00%, N 12.53%, O 19.09%. Selective pre-emergence herbicide. Prepn: Q. F. Soper, **US 3257190** (1966 to Lilly). Activity: E. F. Alder *et al.*, *Proc. Northeast. Weed Control Conf.* **15**, 298 (1961). Environmental fate: T. Golab *et al.*, *J. Agric. Food Chem.* **18**, 838 (1970). Soil degradn: J. H. Miller *et al.*, *Weed Sci.* **23**, 211 (1975);

R. L. Zimdahl, S. M. Gwynn, *ibid.* **25**, 247 (1977). Toxicity study: E. I. Goldenthal, *Toxicol. Appl. Pharmacol.* **18**, 185 (1971).

Yellow-orange crystalline solid, mp 65-66.5°. Vapor pressure at 30°: 3.89×10^{-4} mm Hg. Soly in water at 25°: <1 mg/l. Sol in most organic solvents; less sol in ethanol. Decomp in uv light. LD_{50} orally in female rats: >10,000 mg/kg (Goldenthal).

USE: Herbicide.

1043. Benfotiamine. [22457-89-2] Benzenecarbothioic acid *S*-[2-[[(4-amino-2-methyl-5-pyrimidinyl)methyl]formylamino]-1-[2-(phosphonooxy)ethyl]-1-propen-1-yl] ester; thiobenzoic acid *S*-ester with *N*-[(4-amino-2-methyl-5-pyrimidinyl)methyl]-*N*-(4-hydroxy-2-mercapto-1-methyl-1-butenyl)formamide *O*-phosphate; *S*-benzoylthiamine monophosphate; BTMP; Biotamin; Vitanevril. $C_{19}H_{23}N_4O_6PS$; mol wt 466.45. C 48.92%, H 4.97%, N 12.01%, O 20.58%, P 6.64%, S 6.87%. Vitamin B_1 source. Prepn: A. Ito *et al.*, **DE 1130811** (1962), *C.A.* **57**, 13764h (1962). Exists as 3 temperature dependent crystalline forms, α, γ, δ: A. Ito *et al.*, *Takamine Kenkyusho Nempo* **14**, 64 (1962), *C.A.* **59**, 3920a (1963). Pharmacokinetics: H. Nogami *et al.*, *Chem. Pharm. Bull.* **18**, 1937 (1970). Clinical bioequivalence to thiamine: R. Bitsch *et al.*, *Ann. Nutr. Metab.* **35**, 292 (1991).

(Z)-form

Crystals, dec 165° (δ-form).

THERAP CAT: Vitamin (enzyme cofactor).

1044. Benfuracarb. [82560-54-1] 2-Methyl-4-(1-methylethyl)-7-oxo-8-oxa-3-thia-2,4-diazadecanoic acid 2,3-dihydro-2,2-dimethyl-7-benzofuranyl-ester; 2-methyl-4-(1-methylethyl)-7-oxo-8-oxa-3-thia-2,4-diazadecanoic acid 2,3-dihydro-2,2-dimethyl-7-benzofuranyl ester; 2,3-dihydro-2,2-dimethylbenzofuran-7-yl *N*-(*N*-isopropyl-*N*-ethoxycarbonylethylaminosulfenyl)-*N*-methylcarbamate; *N*-[[[[(2,3-dihydro-2,2-dimethyl-7-benzofuranyl)oxy]carbonyl]methylamino]thio]-*N*-(1-methylethyl)-β-alanine ethyl ester; 2,3-dihydro-2,2-dimethyl-7-benzofuranyl *N*-[*N*-[2-(ethoxycarbonyl)ethyl]-*N*-isopropylsulfenamoyl]-*N*-methylcarbamate; ethyl *N*-[2,3-dihydro-2,2-dimethylbenzofuran-7-yloxycarbonyl(methyl)aminothio]-*N*-isopropyl-β-alaninate; OK-174; Oncol. $C_{20}H_{30}N_2O_5S$; mol wt 410.53. C 58.51%, H 7.37%, N 6.82%, O 19.49%, S 7.81%. Broad spectrum soil and foliar insecticide; cholinesterase inhibitor. Prepn and insecticidal properties: A. Tanaka *et al.*, **BE 890162**; T. Goto *et al.*, **US 4413005** (1982, 1983 both to Otsuka). Biological activity and field trials: T. Goto *et al.*, *Proc. 10th Conf. Int. Congr. Plant Prot.* **1**, 360 (1983). Absorption and metabolism in plants: A. Tanaka *et al.*, *J. Agric. Food Chem.* **33**, 1049 (1985). Brief review, field trials: P. Cagnieul, *Def. Veg.* **38**, 3-10 (1984).

Viscous brown-red liquid. d^{20} 1.17. Almost insol in water (8 mg/l at 20°). Sol in organic solvents. LD_{50} in male rats, mice, dogs (mg/kg): 138, 175, 300 orally (Goto).

USE: Insecticide.

1045. Benidipine. [105979-17-7] *rel*-(4*R*)-1,4-Dihydro-2,6-dimethyl-4-(3-nitrophenyl)-3,5-pyridinedicarboxylic acid 3-methyl 5-[(3*R*)-1-(phenylmethyl)-3-piperidinyl] ester; (±)-(*R**)-3-[(*R**)-1-benzyl-3-piperidyl] methyl 1,4-dihydro-2,6-dimethyl-4-(*m*-nitrophenyl)-3,5-pyridinedicarboxylate; (±)-2,6-dimethyl-4-(3-nitrophenyl)-1,4-dihydropyridine-3,5-dicarboxylic acid-3-(1-benzyl-3-piperidyl) ester-5-methyl ester. $C_{28}H_{31}N_3O_6$; mol wt 505.57. C 66.52%, H 6.18%, N 8.31%, O 18.99%. Dihydropyridine calcium channel blocker. Prepn (stereochemistry unspecified): K. Muto *et al.*, **EP 63365**; *eidem*, **US 4448964** (1982, 1984 both to Kyowa). Prepn: *eidem*, **EP 106275** (1984 to Kyowa); and toxicity data: *eidem*, *Arzneim.-Forsch.* **38**, 1662 (1988). Structural studies: N. Hirayama, E. Shimizu, *Acta Crystallogr.* **C47**, 458 (1991). Series of articles on properties, pharmacology, determn and clinical evaluation: *Arzneim.-Forsch.* **38**, 1666-1763 (1988). *Review:* H. Suzuki, T. Saruta, *Cardiovasc. Drug Rev.* **7**, 25-38 (1989).

Relative stereochemistry

Hydrochloride. [91599-74-5] KW-3049; Coniel. $C_{28}H_{31}N_3O_6 \cdot HCl$; mol wt 542.03. Yellow crystalline powder, mp 199.4-200.4°. uv max (ethanol): 238, 359 nm (ε 2.80 × 10⁴, 6.68 × 10³). Soly at 25° (%): methanol 6.9; ethanol 2.2; water 0.19; chloroform 0.16; acetone 0.13; ethyl acetate 0.0056; toluene 0.0019; *n*-heptane 0.00009. pKa 7.34. Partition coefficient (*n*-octanol/water): 1230 (pH 6.4, 22°). LD_{50} orally in male mice: 218 mg/kg (Muto, 1988).

THERAP CAT: Antihypertensive.

1046. Benomyl. [17804-35-2] *N*-[1-[(Butylamino)carbonyl]-1*H*-benzimidazol-2-yl]carbamic acid methyl ester; 1-(butylcarbamoyl)-2-benzimidazolecarbamic acid methyl ester; methyl 1-(butylcarbamoyl)-2-benzimidazolecarbamate; F-1991; Benlate. $C_{14}H_{18}N_4O_3$; mol wt 290.32. C 57.92%, H 6.25%, N 19.30%, O 16.53%. Prepn: H. L. Klopping, **FR 1523597**; *eidem*, **US 3631176** (1968, 1971 both to du Pont). The degradation product, methyl 2-benzimidazolecarbamate, is thought to be the active component; *see* carbendazim. Antifungal activity and mode of action studies: Kilgore, White, *Bull. Environ. Contam. Toxicol.* **5**, 67 (1970); Maxwell, Brody, *Appl. Microbiol.* **21**, 944 (1971); Bartels-Schooley, MacNeill, *Phytopathology* **61**, 816 (1971). Activity as catalyst in biological oxidation of sewage and fertilizers: Kouba, **US 3649530** (1972 to du Pont). Toxicity: E. W. Schafer, *Toxicol. Appl. Pharmacol.* **21**, 315 (1972). Metabolism: J. A. Gardiner *et al.*, *J. Agric. Food Chem.* **22**, 419 (1974). Mutagenicity study: J. P. Seiler, *Mutat. Res.* **32**, 151 (1975); H. Sherman *et al.*, *Toxicol. Appl. Pharmacol.* **32**, 305 (1975). *Review:* W. E. Bleidner *et al.*, *Anal. Methods Pestic. Plant Growth Regul.* **10**, 157-171 (1978).

White crystalline solid. Sol in chloroform. Insol in water or oil. LD_{50} orally in rats: >9590 mg/kg (Schafer).

Caution: Potential symptoms of overexposure are irritation of eyes, skin and upper respiratory system; skin sensitization; possible reproductive and teratogenic effects. *See NIOSH Pocket Guide to Chemical Hazards* (DHHS/NIOSH 97-140, 1997) p 24.

USE: Fungicide; ascaricide.

THERAP CAT (VET): Anthelmintic.

1047. Benorylate. [5003-48-5] 2-(Acetyloxy)benzoic acid 4-(acetylamino)phenyl ester; salicylic acid acetate, ester with 4-hydroxyacetanilide; 2-acetoxy-4'-(acetamino)phenylbenzoate; *p*-acetamidophenyl acetylsalicylate; 4'-(acetamido)phenyl-2-acetoxybenzoate; fenasprate; Win-11450; Benoral; Benortan; Salipran. $C_{17}H_{15}NO_5$; mol wt 313.31. C 65.17%, H 4.83%, N 4.47%, O 25.53%. Prepn: **NL 6504517** (1965 to Sterwin), *C.A.* **64**, 8097c (1966), corresp to **GB 1101747** (1968); Robertson, **US 3431293** and Miller, **GB 1168289** (both 1969 to Sterling Drug). Pharmacological investigations: Rosner *et al.*, *Therapie* **23**, 525 (1968); Raab, *Arzneim.-Forsch.* **21**, 1662 (1971). Pharmacokinetic studies in animals: Liss, Palme, *ibid.* **19**, 1177 (1969). Clinical evaluation: Bain, Burt, *Clin. Trials J.* **7**, 307 (1970); Cardoe, *ibid.* 313.

Crystals from methanol or ethanol, mp 175-176°. LD_{50} in mice, rats (mg/ml): 2000, ~10000 orally; 1255, 1830 i.p. (Robertson).

THERAP CAT: Analgesic; anti-inflammatory; antipyretic.

1048. Benoxacor. [98730-04-2] 2,2-Dichloro-1-(2,3-dihydro-3-methyl-4*H*-1,4-benzoxazin-4-yl)ethanone; 4-(dichloroacetyl)-3,4-dihydro-3-methyl-2*H*-1,4-benzoxazine; CGA-154281. $C_{11}H_{11}Cl_2NO_2$; mol wt 260.11. C 50.79%, H 4.26%, Cl 27.26%, N 5.39%, O 12.30%. Protects maize from toxic effects of chloroacetanilide herbicides. Increases the expression of glutathione *S*-transferase (GST) isozymes in plants, accelerating herbicide metabolism. Prepn: H. Moser, **EP 149974**; *idem*, **US 4618361** (1985, 1986 both to Ciba-Geigy). Effects on metabolism of metolachlor, *q.v.*, in corn: C. K. Cottingham, K. K. Hatzios, *Z. Naturforsch.* **46c**, 846 (1991). Efficacy and mode of action studies: L. Rowe *et al.*, *Weed Sci.* **39**, 78 (1991); E. P. Fuerst *et al.*, *Pestic. Sci.* **43**, 242 (1995). Metabolism studies: K. D. Miller *et al.*, *J. Agric. Food Chem.* **44**, 3326, 3335 (1996). Environmental tolerances: *Fed. Regist.* **63**, 7299 (1998).

Crystals from diisopropyl ether, mp 105-107°. LD_{50} (mg/kg): >5000 orally in rats; >2010 dermally in rabbits. LC_{50} in rats (mg/l): >2000 by inhalation (Fed. Regist.).

USE: Herbicide safener.

1049. Benoxinate. [99-43-4] 4-Amino-3-butoxybenzoic acid 2-(diethylamino)ethyl ester; 3-butoxy-4-aminobenzoic acid 2-(diethylamino)ethyl ester; 2-(diethylamino)ethyl 4-amino-3-*n*-butoxybenzoate; oxibuprokain; oxybuprocaine. $C_{17}H_{28}N_2O_3$; mol wt 308.42. C 66.20%, H 9.15%, N 9.08%, O 15.56%. Prepn: **GB 654484** (1951 to Wander).

bp$_2$ 215-218°.

Hydrochloride. [5987-82-6] Conjuncain; Cebesine; Novesine; Benoxil; Lacrimin. C$_{17}$H$_{28}$N$_2$O$_3$.HCl; mol wt 344.88. Crystals, mp ~155° (*USP* **XVI**, 1960), also reported as mp 157-160° (*N.N.R.* 1955). Very sol in water, chloroform; sol in alc. Practically insol in ether. pH of aq solns, 4.5-5.2.

THERAP CAT: Anesthetic (local).

1050. Benperidol. [2062-84-2] 1-[1-[4-(4-Fluorophenyl)-4-oxobutyl]-4-piperidinyl]-1,3-dihydro-2*H*-benzimidazol-2-one; 1-[1-[3-(*p*-fluorobenzoyl)propyl]-4-piperidyl]-2-benzimidazolinone; 1-[1-[4-(*p*-fluorophenyl)-4-oxobutyl]piperidin-4-yl]-2-benzimidazolinone; benzperidol; McN-JR-4584; R-4584; Anquil; Frénactil; Glianimon. C$_{22}$H$_{24}$FN$_3$O$_2$; mol wt 381.45. C 69.27%, H 6.34%, F 4.98%, N 11.02%, O 8.39%. Prepn: **BE 626307** (1963 to Janssen), *C.A.* **60**, 10690c (1964), corresp. to **GB 989755**.

Solid, mp 170-171.8°.

Hydrochloride hydrate. Solid, mp 134-142°.

THERAP CAT: Antipsychotic.

1051. Benproperine. [2156-27-6] 1-[1-Methyl-2-[2-(phenylmethyl)phenoxy]ethyl]piperidine; 1-[1-methyl-2-[(α-phenyl-*o*-tolyl)oxy]ethyl]piperidine; 1-[2-(2-benzylphenoxy)-1-methylethyl]piperidine. C$_{21}$H$_{27}$NO; mol wt 309.45. C 81.51%, H 8.79%, N 4.53%, O 5.17%. Prepn: **GB 914008**; Rubinstein, **US 3117059** (1962, 1964 to Aktieselskabet Pharmacia). Pharmacology: Yamatsu *et al.*, *Jpn. J. Pharmacol.* **17**, 538 (1967); N. Tellini, V. De Fina, *Boll. Chim. Farm.* **109**, 476 (1970).

Liquid. bp$_{0.2}$ 159-161°. LD$_{50}$ orally in mice: 1087 mg/kg (Tellini, De Fina).

Trihydrogen phosphate. [19428-14-9] Tussafug. C$_{21}$H$_{27}$NO.H$_3$PO$_4$. Crystals from abs ethanol, mp 150-152°. LD$_{50}$ orally in mice: 1431 mg/kg (Tellini, De Fina).

Pamoate. [64238-92-2] Benzproperine embonate; Blascorid; Pirexyl. (C$_{21}$H$_{27}$NO)$_2$.C$_{23}$H$_{16}$O$_6$; mol wt 1007.28.

THERAP CAT: Antitussive.

1052. Benserazide. [322-35-0] Serine 2-[(2,3,4-trihydroxyphenyl)methyl] hydrazide; *N*-(DL-seryl)-*N*′-(2,3,4-trihydroxybenzyl) hydrazine. C$_{10}$H$_{15}$N$_3$O$_5$; mol wt 257.25. C 46.69%, H 5.88%, N 16.33%, O 31.10%. Peripheral decarboxylase inhibitor. Prepn: **BE 619015**; Hegedüs, Zeller, **US 3178476** (1962, 1965 both to Hoffmann-La Roche). Inhibition of dopa decarboxylase *in vitro* and *in vivo*: W. P. Burkard *et al.*, *Experientia* **18**, 411 (1962); *eidem*, *Arch. Biochem. Biophys.* **107**, 187 (1964). Synergy with dopa *in vivo*: G. Bartholini *et al.*, *Nature* **215**, 852 (1967); G. Bartholini, A. Pletscher, *J. Pharmacol. Exp. Ther.* **161**, 14 (1968). Clinical comparison with carbidopa, *q.v.*: I. Kuruma *et al.*, *J. Pharm. Pharmacol.* **24**, 289 (1972). Clinical and pharmacokinetic profile: M.-H. Marion *et al.*, *Adv. Neurol.* **45**, 493 (1987). Review of pharmacology and clinical efficacy of decarboxylase inhibitors: R. M. Pinder *et al.*, *Drugs* **11**, 329-377 (1976).

Hydrochloride. [14919-77-8] Ro-4-4602. C$_{10}$H$_{15}$N$_3$O$_5$.HCl; mol wt 293.70. White crystalline powder, mp 146-148°. Sol in water.

Combination with levodopa. [37270-69-2] Co-beneldopa; Madopar; Madopark; Prolopa.

THERAP CAT: In combination with levodopa as antiparkinsonian.

1053. Bensulfuron-methyl. [83055-99-6] 2-[[[[[(4,6-Dimethoxy-2-pyrimidinyl)amino]carbonyl]amino]sulfonyl]methyl]-benzoic acid methyl ester; methyl 2-[(4,6-dimethoxypyrimidin-2-yl)ureidosulfonylmethyl]benzoate; DPX-F5384; Londax; Mariner. C$_{16}$H$_{18}$N$_4$O$_7$S; mol wt 410.40. C 46.83%, H 4.42%, N 13.65%, O 27.29%, S 7.81%. Sulphonylurea herbicide primarily used on rice and aquatic weeds. Prepn: R. F. Sauers, **EP 51466**; *idem*, **US 4420325** (1982, 1983 both to DuPont). Conformation study: Y. K. Kang, D. W. Kim, *Bull. Korean Chem. Soc.* **11**, 144 (1990). Environmental persistence: A. Ferrero *et al.*, *Mobility Degrad. Xenobiot., Proc. 9th Simp. Pestic. Chem.* **1993**, 257. HPLC determn in soil: C. Salardi *et al.*, *Meded. Fac. Landbouwwet. Univ. Gent* **61**, 617 (1996). LC determn in rice and crayfish; M. Zhou *et al.*, *J. AOAC Int.* **79**, 791 (1996). Control of aquatic weeds: G. McCorkelle *et al.*, *Proc. 9th Aust. Weeds Conf.* **1990**, 549. Field trials on rice: J. L. Pacheco, C. L. Pope, *Proc. West. Soc. Weed Sci.* **39**, 147 (1986); M. P. Braverman, *Weed Technol.* **9**, 494 (1995).

Mp 179-183°.

USE: Herbicide.

1054. Bentazon. [25057-89-0] 3-(1-Methylethyl)-1*H*-2,1,3-benzothiadiazin-4(3*H*)-one 2,2-dioxide; 3-isopropyl-1*H*-2,1,3-benzothiadiazin-4(3*H*)-one 2,2-dioxide; bentazone; bendioxide; Basagran. C$_{10}$H$_{12}$N$_2$O$_3$S; mol wt 240.28. C 49.99%, H 5.03%, N 11.66%, O 19.98%, S 13.34%. Selective post-emergence herbicide. Prepn: A. Zeidler *et al.*, **ZA 6705164**; *eidem*, **US 3708277** (1968, 1973 both to BASF). Weed control: C. E. G. Mulder, G. B. Wortmann, *Proc. 3rd Natl. Weeds Conf. S. Afr.* **1980**, 159. HPLC determn in sugar maize and surface water: E. A. Hogendoorn, C. E. Goewie, *J. Chromatogr.* **475**, 432 (1989); in soil: F. S. Rasero *et al.*, *Sci. Total Environ.* **123/124**, 57 (1992). Stability in ground, surface, soil waters: C. Mouvet *et al.*, *Chemosphere* **35**, 1083 (1997). Toxicity data: G. Ugazio *et al.*, *Res. Commun. Chem. Pathol. Pharmacol.* **74**, 349 (1991). Review of synthesis: G. Hamprecht *et al.*, *Kem.-Kemi* **1**, 590-592 (1974); of environmental behavior: R. Huber, S. Otto, *Rev. Environ. Contam. Toxicol.* **137**, 111-134 (1994).

White crystalline powder, mp 137-139°. Soly (w/w) at 20°: water 0.05%; acetone 150.7%; benzene 3.3%; chloroform 18%; ethanol 86.1%. pKa (24°): 3.3. Vapor pressure at 20°: 1.7 × 10^{-6} hPa. Log P (octanol/water) at 22°: −0.456 (pH 7). LD$_{50}$ in male, female rats (mg/kg): 383.2, 433.6 orally (Ugazio).

USE: Herbicide.

1055. Benthiavalicarb-isopropyl. [177406-68-7] *N*-[(1*S*)-1-[[[(1*R*)-1-(6-Fluoro-2-benzothiazolyl)ethyl]amino]carbonyl]-2-

methylpropyl]carbamic acid 1-methylethyl ester; N^1[(R)-1-(6-fluoro-2-benzothiazolyl)ethyl]-N^2-isopropoxycarbonyl-L-valinamide; isopropyl [(S)-1-[(1R)-1-(6-fluorobenzothiazol-2-yl)-ethylcarbamoyl]-2-methylpropyl]carbamate; KIF-230. $C_{18}H_{24}FN_3O_3S$; mol wt 381.47. C 56.67%, H 6.34%, F 4.98%, N 11.02%, O 12.58%, S 8.40%. Amino acid amide carbamate fungicide active against oomycete pathogens. Prepn: M. Shibata *et al.*, **WO 96004232**; *eidem*, **US 5789428** (1996, 1998 both to Kumiai; Ihara). Comprehensive description: Y. Miyake *et al.*, *BCPC Int. Congr. - Crop Sci. Technol.* **2003**, 105. Field and greenhouse trials for control of downy mildew in grapevines: M. Reuveni, *Eur. J. Plant Pathol.* **109**, 243 (2003); of late blight diseases: Y. Miyake *et al.*, *J. Pestic. Sci.* **30**, 390 (2005). HPLC determn in soils: H. Mizutani *et al.*, *ibid.* **29**, 87 (2004). Soil degradation: M. Ikeda *et al.*, *ibid.* **30**, 22 (2005).

White, odorless powder, mp 169.2°. d 1.25. Log P (octanol/water): 2.52-2.59 (pH 5-9). Vapor pressure at 25°: <3.0 × 10⁻⁴ Pa. Soly in water at 20° (mg/l): 13.14, 10.96 (pH 5), 12.76 (pH 9). pH of 1% aq suspension (25°): 4.35. LD$_{50}$ in mice, rats (mg/kg): >5000 orally; in rats (mg/kg): >2000 dermally; in bobwhite and mallard (mg/kg): >2000 orally. LC$_{50}$ in rats (mg/kg): >4.6 by inhalation; LC$_{50}$ in rainbow trout, bluegill, *Daphnia* (mg/l): >10 (Miyake, 2003).

Mixture with folpet. [579508-24-0] Vincare.
Mixture with mancozeb. [192880-93-6] Valbon.
USE: Agricultural fungicide.

1056. Bentiromide. [37106-97-1] 4-[[(2S)-2-(Benzoylamino)-3-(4-hydroxyphenyl)-1-oxopropyl]amino]benzoic acid; (S)-*p*-(α-benzamido-*p*-hydroxyhydrocinnamamido)benzoic acid; *N*-benzoyl-L-tyrosyl-*p*-aminobenzoic acid; BTPABA; E-2663; Chymex; PFT Roche. $C_{23}H_{20}N_2O_5$; mol wt 404.42. C 68.31%, H 4.98%, N 6.93%, O 19.78%. Synthetic chymotrypsin-labile peptide, used in diagnosis of exocrine pancreatic disease. Prepn: P. L. De Benneville, N. J. Greenberger, **DE 2156835**; *eidem*, **US 3801562** (1972, 1974 both to Rohm & Haas). Synthesis, *in vitro* and *in vivo* data: P. L. De Benneville *et al.*, *J. Med. Chem.* **15**, 1098 (1972). Early clinical studies: K. Gyr *et al.*, *Schweiz. Med. Wochenschr.* **105**, 1717 (1975); W. Bornschein *et al.*, *Clin. Chim. Acta* **67**, 21 (1976). Pediatric study: G. Dockter *et al.*, *Eur. J. Pediatr.* **135**, 277 (1981). Human toxicity study, effects of renal insufficiency on pancreatic function test: C. Lang *et al.*, *J. Clin. Chem. Clin. Biochem.* **18**, 551 (1980).

Cryst from methanol/water, mp 240-242°. $[\alpha]_D^{25}$ +72.3° (c = 1 in DMF) (**US 3801562**); also reported as $[\alpha]_D^{25}$ +87° (c = 1 in DMF) (De Benneville).

THERAP CAT: Diagnostic aid (pancreatic function).

1057. Bentonite. [1302-78-9] Wilkinite. A colloidal, hydrated aluminum silicate (clay) found in the midwest of the U.S.A. and in Canada. Consists principally of montmorillonite, $Al_2O_3.4SiO_2.H_2O$. Usually contains some magnesium, iron, and cal-

cium carbonate. *Review:* J. Alexander, *Ind. Eng. Chem.* **16**, 1140 (1924).

The color in the massive condition varies from yellowish-white to almost black. The powder is cream colored to pale brown. It has the property of forming highly viscous suspensions or gels with not less than ten times its weight of water. The property of forming gels is very much increased by the addition of small amounts of alkaline substances such as magnesium oxide. Hygroscopic. Insol in water but swells to approximately twelve times its volume when added to water. Insol in and does not swell in organic solvents.

USE: As of Fuller's earth; as emulsifier for oils; as a base for plasters. Pharmaceutic aid (suspending and viscosity-increasing agent).

1058. Bentoquatam. [1340-69-8] Quaternium 18-bentonite; bis(hydrogenated tallow alkyl)dimethylammonium complex with sodium bentonite; Bentone 34; Ivy Block; Tixogel VP. Ion exchange addition product of the montmorillonite clay, bentonite, *q.v.*, and *quaternium-18*, a dimethyl, ditallow quaternary ammonium salt containing predominantly C_{18} alkyl chains. Acts as a barrier to block urushiol, *q.v.*, from the skin. Prepn of aliphatic ammonium bentonites: E. A. Hauser, **US 2531427** (1950); and organophilic properties: J. W. Jordan, *J. Phys. Colloid Chem.* **53**, 294 (1949); *idem et al.*, *ibid.* **54**, 1196 (1950). Use as stationary phase in GSC and HPLC: D. W. Grant *et al.*, *J. Chromatogr.* **99**, 721 (1974). Description of properties, toxicology and use in cosmetics: *J. Am. Coll. Toxicol.* **1**, 71-83 (1982). Clinical trial in poison ivy/oak dermatitis: J. G. Marks, Jr. *et al.*, *J. Am. Acad. Dermatol.* **33**, 212 (1995).

Hydrophobic, thixotropic organoclay with gel-like consistency. Expansible in water, methanol, ethanol, isopropanol, sorbitol, glycerine, acetone. Heat stable up to 500°. Resists base or acid attacks over pH range of 3-11.

USE: Emulsion stabilizer in cosmetics. Stationary phase in chromatography.

THERAP CAT: Barrier for the prevention of allergic contact dermatitis.

1059. Benzal Chloride. [98-87-3] (Dichloromethyl)benzene; benzylidene chloride; benzyl dichloride; α,α-dichlorotoluene; benzylene chloride. $C_7H_6Cl_2$; mol wt 161.03. C 52.21%, H 3.76%, Cl 44.03%. Obtained by chlorination of toluene. Lab procedure: A. I. Vogel, *Practical Organic Chemistry* (Longmans, London, 3rd ed., 1959) p 539; Gattermann-Wieland, *Praxis des Organischen Chemikers* (de Gruyter, Berlin, 40th ed., 1961) p 184.

Very refractive liquid; fumes in air; vapors irritate the eyes; pungent odor. d 1.26. bp 205°. mp −17°. Insol in water; freely sol in alcohol, ether.

USE: Manuf benzaldehyde, cinnamic acid.

1060. Benzaldehyde. [100-52-7] Benzoic aldehyde; artificial essential oil of almond. C_7H_6O; mol wt 106.12. C 79.23%, H 5.70%, O 15.08%. Occurs in kernels of bitter almonds; made synthetically from benzal chloride and lime or by oxidation of toluene. Laboratory prepn from benzal chloride: A. I. Vogel, *Practical Organic Chemistry* (Longmans, London, 3rd ed., 1959) p 693; Gattermann-Wieland, *Praxis des Organischen Chemikers* (de Gruyter, Berlin, 40th ed., 1961) p 184. Toxicity data: P. M. Jenner *et al.*, *Food Cosmet. Toxicol.* **2**, 327 (1964). *Review:* A. E. Williams in *Kirk-Othmer Encyclopedia of Chemical Technology* **vol. 3** (Wiley-Interscience, New York, 3rd ed., 1978) pp 736-743.

Strongly refractive liquid, becoming yellowish on keeping; characteristic odor of volatile oil of almond; burning aromatic taste. Oxidizes in air to benzoic acid; volatile with steam. d_4^{15} 1.050; 1.043 at 25°. bp 179°. mp −56.5°. Flash pt 62°C. n_D^{20} 1.5456. Sol in 350 parts water. Miscible with alcohol, ether, oils. It reduces ammoniacal $AgNO_3$, but not Fehling's soln. *Keep tightly closed and protected from light.* LD_{50} in rats, guinea pigs (mg/kg): 1300, 1000 orally (Jenner).

Note: **Benzaldehyde FFC** designates a grade of benzaldehyde free from chlorine.

Caution: Narcotic in high concns. May cause contact dermatitis.

USE: Manufacture of dyes, perfumery, cinnamic and mandelic acids, as solvent; in flavors.

1061. Benzalkonium Chloride. [8001-54-5] Alkylbenzyldimethyl quaternary ammonium compounds chlorides; alkyldimethyl-(phenylmethyl)ammonium chloride; Baktonium; Benirol; Bradosol; Callusolve; Capitol; Laudamonium; Osvan; Pharmatex; Roccal; Sagrotan; Zephiran; Zephirol. A mixture of alkyldimethylbenzylammonium chlorides; three major homologues consist of C_{12}, C_{14} and C_{16} straight chain alkyls. Prepn and biological activity: R. A. Cutler *et al.*, *Proc. Annu. Meet. Chem. Spec. Manuf. Assoc.* **1966**, 102. Acute toxicity data: L. M. Cummins, E. T. Kimura, *Toxicol. Appl. Pharmacol.* **20**, 89 (1971). Capillary electrophoretic determn in formulations: M. Jimidar *et al.*, *Biomed. Chromatogr.* **12**, 128 (1998); MS determn on skin: S. Kawakami *et al.*, *Analyst* **123**, 489 (1998). Efficacy as disinfecting wound irrigant: B. B. Tarbox *et al.*, *Clin. Orthop. Relat. Res.* **1998**, 255. Use as disinfectant against viruses: A. Wood, D. Payne, *J. Hosp. Infect.* **38**, 283 (1998). Brief review: A. C. S. Parr *et al.*, *Spec. Chem.* **16**, 157-160 (1996).

$$R = C_8H_{17} \text{ to } C_{18}H_{37}$$

White or yellowish-white, amorphous powder or gelatinous pieces. Very sol in water, alcohol, acetone; slightly sol in benzene. Almost insol in ether. The aq soln is slightly alkaline to litmus and foams strongly when shaken. LD_{50} orally in rats: 400 mg/kg (Cummins, Kimura).

USE: Cationic surface active agent and germicide. Pharmaceutic aid (preservative). Algicide.

THERAP CAT: Antiseptic, disinfectant.

THERAP CAT (VET): Antiseptic for skin preoperatively or for wounds, burns, etc. Udder wash.

1062. Benzamide. [55-21-0] Benzoylamide; C_7H_7NO; mol wt 121.14. C 69.40%, H 5.82%, N 11.56%, O 13.21%. Prepd from benzoyl chloride and ammonium carbonate. Lab prepn: Gattermann-Wieland, *Praxis des Organischen Chemikers* (de Gruyter, Berlin, 40th ed., 1961) p 119. Alternate procedure using concd ammonia soln: A. I. Vogel, *Practical Organic Chemistry* (Longmans, London, 3rd ed., 1959) p 797.

Crystals. d^4 1.341. mp 130°. bp 288°. One gram dissolves in 74 ml water, more sol in boiling water, in 6 ml alc, 3.3 ml pyridine. Sol in hot benzene, slightly in ether; sol in ammonia with formation of a small quantity of benzonitrile.

N-Chloro deriv. mp 116°. Sol in water; insol in alcohol or benzene.

1063. Benzanilide. [93-98-1] *N*-Phenylbenzamide; *N*-benzoylaniline; $C_{13}H_{11}NO$; mol wt 197.24. C 79.16%, H 5.62%, N 7.10%, O 8.11%. Prepd by the treatment of aniline with benzoic acid: Hübner, *Ann.* **208**, 291 (1881); Nägeli, *Bull. Soc. Chim.* [3] **11**,

892 (1894); Webb, *Org. Synth.* **7**, 6 (1927); **coll. vol. I** (2nd ed., 1941) p 82.

Leaflets from alc. d 1.315. mp 163°. Sublimes. Distills without decompn. bp_{10} 117-119°. Absorption spectrum: Crymble, *J. Chem. Soc.* **99**, 459 (1911). Insoluble in water. One gram dissolves in 60 ml alc, 7 ml boiling alc. Slightly sol in ether.

USE: Manuf dyes and perfumes.

1064. 1,2-Benzanthracene. [56-55-3] Benz[*a*]anthracene; 2,3-benzphenanthrene; tetraphene; benzanthrene; naphthanthracene. $C_{18}H_{12}$; mol wt 228.29. C 94.70%, H 5.30%. Occurs in coal tar, *q.v.*: Cook *et al.*, *J. Chem. Soc.* **1933**, 395. Synthesis from naphthalene and phthalic anhydride: Elbs, *Ber.* **19**, 2209 (1886). From *o*-toluylnaphthalene: Fieser, Dietz, *Ber.* **62**, 1827 (1929). From phenanthrene and succinic anhydride: Haworth, Mavin, *J. Chem. Soc.* **1933**, 1012. Absorption spectrum: Capper, Marsh, *J. Chem. Soc.* **1926**, 726; Clar, *Ber.* **65**, 507 (1932); Mayneord, Roe, *Proc. R. Soc. London* **A152**, 299 (1935). *Review:* E. Clar, *Polycyclic Hydrocarbons* **Vol. 1 & 2** (Academic Press, New York, 1964). Review of toxicology and human exposure: *Toxicological Profile for Polycyclic Aromatic Hydrocarbons* (PB95-264370, 1995) 487 pp.

Plates from glacial acetic acid or alc. Greenish-yellow fluorescence. Sublimes. mp 155-157° (Fieser, Dietz); mp 160° [Fieser, Hershberg, *J. Am. Chem. Soc.* **59**, 2502 (1937)]; mp 167° (I. G. Farbenind., **DE 481819**; **DE 486766**). Difficultly sol in boiling alc; sol in most other organic solvents. Insol in water.

Caution: This substance is reasonably anticipated to be a human carcinogen: *Report on Carcinogens, Twelfth Edition* (PB2011-111666, 2011) p 353.

1065. Benzanthrone. [82-05-3] 7*H*-Benz[*de*]anthracen-7-one. $C_{17}H_{10}O$; mol wt 230.27. C 88.67%, H 4.38%, O 6.95%. Prepd by heating a reduction product of anthraquinone with sulfuric acid and glycerol: Macleod, Allen, *Org. Synth.* **14**, 4 (1934); *cf.* **US 1626392** (1927). Review of toxicology: G. B. Singh *et al.*, *J. Sci. Ind. Res.* **49**, 288-296 (1990).

Pale yellow needles from alcohol or xylene. mp 170°. Absorption spectrum: Clar, *Ber.* **65**, 846 (1932). Solubility at 20° = 0.52 g in 100 g glacial acetic acid; 1.61 g in 100 g benzene; 2.05 g in 100 g chlorobenzene. Solution in H_2SO_4 is orange with green fluorescence. LD_{50} in albino rabbits: >3 g/kg dermally; LD_{50} in rats, mice (g/kg): 1.5, 0.29 i.p. (Singh).

USE: Manufacture of dyes.

1066. Benzathine. [140-28-3] N^1,N^2-Bis(phenylmethyl)-1,2-ethanediamine; *N,N'*-dibenzylethylenediamine; 1,2-bis(benzylamino)ethane; DBED. $C_{16}H_{20}N_2$; mol wt 240.35. C 79.96%, H 8.39%, N 11.66%. Prepd from *N,N'*-dibenzenesulfonyl-*N,N'*-dibenzylethylenediamine by heating with concd HCl in bomb tube at 170-180°: Bleier, *Ber.* **32**, 1829 (1899); from ethylene chloride and benzylamine: Frost *et al.*, *J. Am. Chem. Soc.* **71**, 3842 (1949); by reduction of *N,N'*-dibenzylideneethylenediamine: **DE 98031** (1898 to Schering); *Chem. Zentralbl.* **1898**, II, 743; Van Alphen, *Rec. Trav. Chim.* **54**, 93 (1935); Lob, *ibid.* **55**, 859 (1936); Szabo *et al.*, *Antibiot. Chemother.* **1**, 499 (1951); Rebenstorf, **US 2773098** (1956 to Upjohn).

Oily liquid, mp 26°. d_4^{20} 1.024. bp_{12} 212-213°; bp_4 195°. n_D^{20} 1.5624. Insol in water. Freely sol in the usual organic solvents except CS_2, with which it gives a solid addition product.

Diacetate. $C_{20}H_{28}N_2O_4$. Needles, mp 111°. Soly in water: 253 mg/ml.

Dihydrobromide. $C_{16}H_{22}Br_2N_2$. Shiny leaflets, dec 300°. Soly in water: 30 mg/ml.

Dihydrochloride. $C_{16}H_{22}Cl_2N_2$. Shiny leaflets, dec 298°. Soly in water: 23.9 mg/ml.

Sulfate. $C_{16}H_{22}N_2O_4S$. Crystals, dec 247-250°. Soly in water: 15.8 mg/ml.

Disalicylate. $C_{30}H_{34}N_2O_6$. Crystals, mp 85°. Soly in water: 2.43 mg/ml.

USE: Manuf of a repository form of penicillins G and V, *q.q.v.*

1067. Benzbromarone. [3562-84-3] (3,5-Dibromo-4-hydroxyphenyl)(2-ethyl-3-benzofuranyl)methanone; 3,5-dibromo-4-hydroxyphenyl 2-ethyl-3-benzofuranyl ketone; 2-(3,5-dibromo-4-hydroxybenzoyl)-2-ethylbenzofuran; 2-ethyl-3-benzofuranyl 4-hydroxy-3,5-dibromophenyl ketone; 2-ethyl-3-(3,5-dibromo-4-hydroxybenzoyl)benzofuran; 2-ethyl-3-(3,5-dibromo-4-hydroxybenzoyl)oxaindene; 2-ethyl-3-benzofuryl 3,5-dibromo-4-hydroxyphenyl ketone; L-2214; MJ-10061; Azubromaron; Desuric; MaxUric; Minuric; Narcaricin; Normurat; Uricovac; Urinorm. $C_{17}H_{12}$-Br_2O_3; mol wt 424.09. C 48.15%, H 2.85%, Br 37.68%, O 11.32%. Prepn: **BE 553621** (1957 to Labaz); Buu-Hoi *et al.*, *J. Chem. Soc.* **1957**, 625; Buu-Hoi, Beaudet, **US 3012042** (1961 to Soc. Belge l'Azote Prod. Chim. Marly). Structure-activity study: Delbarre *et al.*, *Chim. Ther.* **3**, 470 (1968). Pharmacology of hypouricemic effect: D. S. Sinclair, I. H. Fox, *J. Rheumatol.* **2**, 437 (1975). Pharmacokinetic and clinical studies: T. F. Yu, *ibid.* **3**, 305 (1976). Pharmacokinetics and biotransformation in man: H. Ferber *et al.*, *Eur. J. Clin. Pharmacol.* **19**, 431 (1981). HPLC determn in serum: H. Vergin *et al.*, *J. Chromatogr.* **183**, 383 (1980).

Yellowish prisms, mp 151°.
THERAP CAT: Uricosuric.

1068. Benzene. [71-43-2] Benzol; cyclohexatriene. C_6H_6; mol wt 78.11. C 92.26%, H 7.74%. Natural component of petroleum, usually <1.0% by weight. Discovered by Faraday in compressed oil gas in 1825. Obtained in the coking of coal and in the production of illuminating gas from coal. Manuf by catalytic reforming and separation of aromatic compounds, thermal or catalytic dealkylation of toluene, toluene disproportionation; from pyrolysis gasoline. Purification by washing with water: **GB 863711** (1961 to Schloven-Chemie and H. Koppers GmbH), *C.A.* **55**, 16971f (1961). Lab prepn from aniline: Gattermann-Wieland, *Praxis des Organischen Chemikers* (de Gruyter, Berlin, 40th ed., 1961) p 247. Production of pure benzene: French, *Ind. Chem.* **39**, 9-12 (1963). Manuf: *Faith, Keyes & Clark's Industrial Chemicals*, F. A. Lowenheim, M. K. Moran, Eds. (Wiley-Interscience, New York, 4th ed., 1975) pp

126-137. Physical properties: Thorne *et al.*, *Ind. Eng. Chem. Anal. Ed.* **17**, 481 (1945). Solubility studies: F. P. Schwarz, *Anal. Chem.* **52**, 10 (1980). Toxicity data: Kimura *et al.*, *Toxicol. Appl. Pharmacol.* **19**, 699 (1971). *Review:* W. Fruscella in *Kirk-Othmer Encyclopedia of Chemical Technology* vol. 4 (Wiley-Interscience, New York, 4th ed., 1992) pp 73-103. Reviews of toxicology: R. Snyder *et al.*, *Rev. Biochem. Toxicol.* **3**, 123-154 (1981); D. J. Paustenbach *et al.*, *Environ. Health Perspect.* **101**, Suppl. 6, 177-200 (1993); and human exposure: *Toxicological Profile for Benzene* (PB2008-10004, 2007) 438 pp. Symposia on metabolism, toxicity and carcinogenesis: *Environ. Health Perspect.* **82**, 3-310 (1989); *ibid.* **104**, Suppl. 6, 1121-1441 (1996).

Clear, colorless, volatile liquid; characteristic odor. d_4^{15} 0.8787. bp 80.1°. mp 5.5°. n_D^{20} 1.50108. Flash pt, closed cup: 12°F (−11°C). Soly in water at 23.5°C (w/w): 0.188%. *Flammable.* Miscible with alc, chloroform, ether, carbon disulfide, carbon tetrachloride, glacial acetic acid, acetone, oils. *Keep in well-closed containers in a cool place and away from fire.* LD$_{50}$ orally in young adult rats: 3.8 ml/kg (Kimura).

Sodium salt. [1623-99-0] Phenyl sodium. C_6H_5Na; mol wt 100.10. Prepn: Schlosser, *Angew. Chem.* **76**, 267 (1964). Solid mass, dec by water, acids, alkalies. Sol in liquid ammonia, tetrahydrofuran.

Caution: Potential symptoms of overexposure by inhalation or ingestion are dizziness, headache, vomiting, visual disturbances, staggering gait, hilarity, fatigue, anorexia, lassitude, CNS depression, loss of consciousness, respiratory arrest. Chronic exposure has been associated with bone marrow depression and leukemia. Direct contact may cause irritation of eyes, nose, respiratory system and skin; dermititis may develop due to defatting action. Aspiration into the lung may lead to chemical pneumonitis. *See Patty's Industrial Hygiene and Toxicology* vol. 2B, G. D. Clayton, F. E. Clayton, Eds. (Wiley-Interscience, New York, 4th ed., 1994) pp 1306-1326; *NIOSH Pocket Guide to Chemical Hazards* (DHHS/NIOSH 97-140, 1997) p 26. Benzene is listed as a known human carcinogen: *Report on Carcinogens, Twelfth Edition* (PB2011-111646, 2011) p 60.

USE: Manuf of industrial chemicals such as polymers, detergents, pesticides pharmaceuticals, dyes, plastics, resins. Organic solvent for waxes, resins, oils, natural rubber, etc. Reference for quantitating compds. Gasoline additive.

THERAP CAT (VET): Has been used as a disinfectant.

1069. Benzenearsonic Acid. [98-05-5] *As*-Phenylarsonic acid; phenylarsonic acid. $C_6H_7AsO_3$; mol wt 202.04. C 35.67%, H 3.49%, As 37.08%, O 23.76%.

Cryst powder. mp 158-162° with dec. Sol in 40 parts water, 50 parts alcohol. Insol in chloroform.
USE: Reagent for tin.

1070. Benzeneboronic Acid. [98-80-6] Phenylboronic acid; phenylboric acid; phenylboron dihydroxide. $C_6H_7BO_2$; mol wt 121.93. C 59.10%, H 5.79%, B 8.87%, O 26.24%. Prepd by the reaction of phenylmagnesium bromide with methyl borate: Washburn *et al.* in *Adv. Chem. Ser.* **23**, entitled "Metal-Organic Compounds," M. Sittig, Ed. (ACS, Washington DC, 1959) pp 102-128. Crystal and molecular structure: S. J. Rettig, J. Trotter, *Can. J. Chem.* **55**, 3071 (1977).

Crystals from water. Spontaneous conversion to C_6H_5BO, *benzene boronic anhydride* or *phenylboroxide* on standing in dry air, although the best method is by azeotropic dehydration with toluene. mp 215-216° (anhydride). pKa 13.7. Dipole moment 1.72 (dioxane). Soly in water at 25°: 2.5%, benzene: 1.75%, xylene: 1.2%, ether: 30.2%, methanol: 178%.

1071. Benzenesulfonic Acid. [98-11-3] $C_6H_6O_3S$; mol wt 158.17. C 45.56%, H 3.82%, O 30.35%, S 20.27%. Made by treating benzene with fuming H_2SO_4: Gattermann-Wieland, *Praxis des Organischen Chemikers* (de Gruyter, Berlin, 40th ed., 1961) p 168.

Sesquihydrate. [5928-72-3] $C_6H_6O_3S.1\frac{1}{2}H_2O$; mol wt 185.19. Deliquescent plates, tablets, mp 43-44°. When anhydrous mp 50-51°, also reported as mp 65-66°. pKa (25°): 0.699. Freely sol in water, alc; slightly sol in benzene; insol in ether, carbon disulfide. *Keep well closed.*

Ethyl ester. [515-46-8] Ethyl benzenesulfonate. $C_8H_{10}O_3S$; mol wt 186.23. Colorless to slightly yellow, almost odorless liq. bp_{15} 156°. d_4^{17} 1.219. Slightly sol in water; misc with alc, benzene, chloroform, ether.

Sodium salt. [515-42-4] Sodium benzenesulfonate; sodium benzosulfonate. $C_6H_5NaO_3S$; mol wt 180.15. Crystals. Sol in water.

Caution: Highly irritant to skin, eyes, mucous membranes.

USE: Manuf phenol by fusion with NaOH.

1072. Benzenesulfonic Anhydride. [512-35-6] Benzenesulfonic acid 1,1'-anhydride. $C_{12}H_{10}O_5S_2$; mol wt 298.33. C 48.31%, H 3.38%, O 26.81%, S 21.49%. Prepd by heating benzenesulfonic acid with excess phosphorus pentoxide mixed with inert support: Field, *J. Am. Chem. Soc.* **74**, 394 (1952).

Light tan solid, mp 60-85° (softens about 55°); may be recrystallized from ether, mp 88-91° (softens about 75°). Liquefies upon exposure to air for 2.5 hrs. Explosion occurs when mixed with 90-95% H_2O_2.

USE: In Friedel-Crafts sulfone synthesis; in other sulfonylation reactions.

1073. Benzenesulfonyl Chloride. [98-09-9] Benzene sulfonechloride; benzenesulfonic (acid) chloride. $C_6H_5ClO_2S$; mol wt 176.61. C 40.81%, H 2.85%, Cl 20.07%, O 18.12%, S 18.15%. Prepd from benzene and chlorosulfonic acid or from the sodium salt of benzenesulfonic acid and PCl_5 or $POCl_3$: R. Adams, C. S. Marvel, *Org. Synth.* **coll. vol. I**, (2nd ed.), 1964) p 84; H. T. Clarke *et al., ibid.* p 85.

Colorless, oily liquid. d_{15}^{15} 1.3842. Solidifies 0°, mp 14.5°. bp_{760} 251-252° (decompn); bp_{100} 177°; bp_{10} 120°. *Corrosive.* Insol in cold water. Sol in ether, alc.

1074. Benzenesulfonyl Hydrazide. [80-17-1] Benzenesulfonic acid hydrazide; benzenesulfohydrazide; phenylsulfohydrazide; Porofor BSH. $C_6H_8N_2O_2S$; mol wt 172.20. C 41.85%, H 4.68%, N 16.27%, O 18.58%, S 18.62%. *Ref:* Lober, *Angew. Chem.* **64**, 65 (1952).

Crystals, dec 103-104° with the evolution of nitrogen. May be stored indefinitely at temps up to 80°. Sensitive to moist oxidizing agents.

USE: Gas generating agent for use in making foam rubber and foam plastics.

1075. 1,2,4-Benzenetriol. [533-73-3] Hydroxyhydroquinone; hydroxyquinol. $C_6H_6O_3$; mol wt 126.11. C 57.15%, H 4.80%, O 38.06%. Prepd by hydrolysis of the triacetate with H_2SO_4 in methanol, the triacetate being formed by the action of acetic anhydride on quinone: Vliet, *Org. Synth.* **coll. vol. I** (2nd ed, 1941) p 317.

Monoclinic prismatic leaflets from ether, mp 141°. Freely sol in water, alcohol, ether, ethyl acetate; almost insol in chloroform, carbon disulfide, ligroin, benzene.

Trimethyl ether. $C_9H_{12}O_3$. bp 247°.

Triacetate. $C_{12}H_{12}O_6$. Needles from abs alc, mp 96.5-97.0°. Readily hydrolyzed by acids or alkalies.

USE: In gas analysis; 1,2,4-benzenetriol in alkaline soln is just as good an absorbent for oxygen as is pyrogallol.

1076. Benzethonium Chloride. [121-54-0] *N,N*-Dimethyl-*N*-[2-[2-[4-(1,1,3,3-tetramethylbutyl)phenoxy]ethoxy]ethyl]benzenemethanaminium chloride (1:1); benzyldimethyl[2-[2-(*p*-1,1,-3,3-tetramethylbutylphenoxy)ethoxy]ethyl]ammonium chloride; diisobutylphenoxyethoxyethyl dimethyl benzyl ammonium chloride; Hyamine 1622; Phemeride; Phemerol Chloride. $C_{27}H_{42}ClNO_2$; mol wt 448.09. C 72.37%, H 9.45%, Cl 7.91%, N 3.13%, O 7.14%. Quaternary ammonium salt with antimicrobial activity. Prepn: H. A. Bruson, US 2115250 (1938 to Rohm & Haas). Germicidal activity and toxicity data: P. Weiss *et al., J. Am. Pharm. Assoc. Sci. Ed.* **40**, 267 (1951). HPLC determn in fish: T. B. A. Reuvers *et al., J. Chromatogr.* **467**, 321 (1989). Use for determn of total protein in cerebrospinal fluid (CSF): R. W. Luxton *et al., Clin. Chem.* **35**, 1731 (1989). Review of properties, commercial uses and toxicology studies: *J. Am. Coll. Toxicol.* **4**, 65-106 (1985).

Odorless, colorless crystals with very bitter taste. mp 160-165°. Sol in water giving a foamy, soapy soln. Sol in alcohol, chloroform; slightly sol in ether. Incompatible with soap, anionic detergents. LD_{50} i.v. in mice: 29.5 mg/kg (Weiss).

USE: In cosmetics as preservative; cationic surfactant. As disinfectant in dairies and food industries. Clinical reagent for determn of protein in CSF; pharmaceutic aid (preservative).

THERAP CAT: Antiseptic, disinfectant.

THERAP CAT (VET): Antiseptic, disinfectant.

1077. Benzetimide. [14051-33-3] 3-Phenyl-1'-(phenylmethyl)-[3,4'-bipiperidine]-2,6-dione; 2-(1-benzyl-4-piperidyl)-2-phenylglutarimide; 1-benzyl-4-(2,6-dioxo-3-phenyl-3-piperidyl)piperidine. $C_{23}H_{26}N_2O_2$; mol wt 362.47. C 76.21%, H 7.23%, N 7.73%, O 8.83%. Anticholinergic. Prepn: Janssen, US 3125578 (1964 to Janssen); Hermans *et al., J. Med. Chem.* **11**, 797 (1968). Synthesis of labelled compound: van Wijngaarden, Soudijn, *J. Labelled Compd.* **1**, 207 (1965). Resolution of isomers and pharmacology:

van Wijngaarden, *Life Sci.* **8**, 517 (1969); *ibid.* **9**, part 1, 489 (1970); Janssen *et al.*, *Arzneim.-Forsch.* **21**, 1365 (1971). Metabolism: van Wijngaarden, Soudijn, *Life Sci.* **7**, 225 (1968).

Crystals, mp 156-159°.

Hydrochloride. [5633-14-7] R-4929; Dioxatrine; Spasmentral. $C_{23}H_{26}N_2O_2 \cdot HCl$; mol wt 398.93. Crystals, mp 299-301.5°. LD_{50} in rats, mice (mg/kg): 37.6, 46.0 i.v. (Janssen).

l-**Form.** mp 180.5-182° (from toluene). $[\alpha]_D^{20}$ −124° (chloroform). LD_{50} i.v. in mice: 38.5 mg/kg (Janssen).

d-**Form** *see* Dexetimide.

THERAP CAT: Antiparkinsonian.

THERAP CAT (VET): Antidiarrheal.

1078. Benzhydrylamine. [91-00-9] α-Phenylbenzenemethanamine; 1,1-diphenylmethylamine; α-aminodiphenylmethane. $C_{13}H_{13}N$; mol wt 183.25. C 85.21%, H 7.15%, N 7.64%. Prepd by boiling benzophenone oxime with zinc dust and ammonia: Scholl, *Ber.* **60**, 1247 (1927).

Hexagonal plates from water, mp 34°. d_4^{22} 1.0635 (supercooled liq). bp_{763} 304.1°; bp_{12} 166°. n_D^{99} 1.59631 (supercooled liq). Strong base. Absorbs CO_2 from the air forming a substance that melts at 91°. Slightly sol in water.

Hydrochloride. $C_{13}H_{13}N \cdot HCl$. Needles, mp 293°. Sparingly sol in cold water.

1079. Benzidine. [92-87-5] [1,1′-Biphenyl]-4,4′-diamine; *p*-diaminodiphenyl. $C_{12}H_{12}N_2$; mol wt 184.24. C 78.23%, H 6.57%, N 15.21%. Discovered by Zinin in 1845. Produced by reduction of nitrobenzene with Zn and NaOH; the resultant hydrazobenzene is heated with acid. Lab proc: A. I. Vogel, *Practical Organic Chemistry* (Longmans, London, 3rd ed., 1959) p 633; Gattermann-Wieland, *Praxis des Organischen Chemikers* (de Gruyter, Berlin, 40th ed., 1961) p 165. Review of carcinogenic risk: *IARC Monographs* **1**, 80-86 (1972); *ibid.* **29**, 149-183 (1982); M. Boeniger, *Carcinogenicity and Metabolism of Azo Dyes, Especially Those Derived from Benzidine* (DHHS/NIOSH 80-119, 1980) 141 pp. Review of toxicology and human exposure: *Toxicological Profile for Benzidine* (PB2001-109102, 2001) 242 pp.

White or slightly-reddish, cryst powder; darkens on exposure to air and light. *Poisonous.* mp 115-120° when slowly heated; 128° when anhydr and rapidly heated. bp ~400°. Soly: one gram dissolves in 2500 ml cold, 107 ml boiling water, 5 ml boiling alcohol, 50 ml ether. *Keep well closed and protected from light.*

Dihydrochloride. $C_{12}H_{12}N_2 \cdot 2HCl$. Crystals. Sol in water, alc.

Caution: Potential symptoms of overexposure are hematuria; secondary anemia from hemolysis; acute cystitis; acute liver disorders; dermatitis; painful and irregular urination. *See NIOSH Pocket Guide to Chemical Hazards* (DHHS/NIOSH 97-140, 1997) p 26. Benzidine and dyes metabolized to benzidine are listed as known human carcinogens: *Report on Carcinogens, Twelfth Edition* (PB2011-111646, 2011) p 62, 64.

USE: Manuf direct azo dyes; as a reagent for H_2O_2 in milk; for detection of blood. Dihydrochloride used for quantitative determn of sulfates, and as reagent for metals.

1080. Benzil. [134-81-6] 1,2-Diphenyl-1,2-ethanedione; bibenzoyl; dibenzoyl; diphenylglyoxal; diphenyl-α,β-diketone. $C_{14}H_{10}O_2$; mol wt 210.23. C 79.99%, H 4.79%, O 15.22%. Prepd by the oxidation of benzoin, $C_6H_5CH(OH)COC_6H_5$, with HNO_3 or with a copper sulfate-pyridine mixture: Adams, Marvel, *Org. Synth.* **vol. I**, p 25 (1921); Clarke, Dreger, *ibid.* **coll. vol. I**, 80 (87, 2nd ed); Hatt *et al.*, *J. Chem. Soc.* **1936**, 93; L. F. Fieser, *Experiments in Organic Chemistry* (Boston, 3rd ed, 1955) p 173; *Organic Experiments* (Boston, 1964) p 214.

Yellow prisms (trigonal trapezohedral) from alcohol. d_4^{15} 1.23. mp 95°. bp_{760} 346-348°; bp_{12} 188°. Absorption spectrum: Hantzsch, Schwiete, *Ber.* **49**, 216 (1916). uv max (ethanol): 260 nm (ε 22000). Infrared in chloroform: 5.93; 6.22; 6.85 μ. Insol in water; sol in alcohol, ether, chloroform, ethyl acetate, benzene, toluene, nitrobenzene.

α-Benzilmonoxime. $C_{14}H_{11}NO_2$. Leaflets from 30% alc, mp 137-138°.

β-Benzilmonoxime. $C_{14}H_{11}NO_2$. Solvated needles from benzene, when dry mp 113-114°.

Disemicarbazone. $C_{16}H_{16}N_6O_2$. Leaflets from alc, dec 243-244°.

USE: In organic syntheses.

1081. Benzil Dioxime. [23873-81-6] 1,2-Diphenyl-1,2-ethanedione 1,2-dioxime; diphenylglyoxime. $C_{14}H_{12}N_2O_2$; mol wt 240.26. C 69.99%, H 5.03%, N 11.66%, O 13.32%. Exists in three isomeric forms: α or *anti*, β or *syn*, γ or *amphi*. Prepn: F. W. Atack, *J. Chem. Soc. Trans.* **103**, 1317 (1913); J. Meisenheimer, W. Lamparter, *Ber.* **57**, 276 (1924); J. H. Boyer *et al.*, *J. Am. Chem. Soc.* **77**, 5688 (1955). Use in determn of nickel: F. W. Atack, *Analyst* **38**, 316 (1913); and palladium: F. J. Welcher, *Organic Analytical Reagents* **vol. III** (Van Nostrand, New York, 1947) pp 224-228.

α-Form

α-Form. [522-34-9] (*E,E*)-Benzil dioxime; α-benzildioxime. White microcrystalline leaflets from acetone, mp 235-237° (dec) (Welcher). Also reported as bright white scales from methanol, mp 243-244° (Meisenheimer, Lamparter). Practically insol in water, ether, glacial acetic acid. Slightly sol in alc; readily sol in NaOH solns.

β-Form. [572-45-2] (*Z,Z*)-Benzil dioxime; β-benzildioxime. Crystals, mp 212-214°.

γ-Form. [572-43-0] (*E,Z*)-Benzil dioxime; γ-benzildioxime. Crystals, mp 164-165°.

USE: Reagent for determn of nickel and other metals.

1082. Benzilic Acid. [76-93-7] α-Hydroxy-α-phenylbenzeneacetic acid; diphenylglycolic acid. $C_{14}H_{12}O_3$; mol wt 228.25. C 73.67%, H 5.30%, O 21.03%. Prepd from benzil by the action of concd aq or alc KOH: Adams, Marvel, *Org. Synth.* **vol. 1**, p 29 (1921); Ballard, Dehn, *ibid.* **coll. vol. I**, 82; Kao, Ma, *J. Chem. Soc.* **1931**, 443.

Monoclinic needles from water. Bitter taste. mp 150°. Melt is deep red at higher temp. pKa (25°): 3.036. Slightly sol in cold water; freely sol in hot water, alc, ether.

Potassium salt. $C_{14}H_{11}O_3K$. Crystals, very sol in water, alc.

Lead salt. $(C_{14}H_{11}O_3)_2Pb$. Amorphous precipitate. Upon heating it becomes a red liquid.

Methyl ester. $C_{15}H_{14}O_3$. mp 74-75°. bp$_{13}$ 187°.

1083. Benzimidazole. [51-17-2] 1*H*-Benzimidazole; benzimidazole; 1,3-benzodiazole; azindole; benzoglyoxaline; *N,N'*-methenyl-*o*-phenylenediamine. $C_7H_6N_2$; mol wt 118.14. C 71.17%, H 5.12%, N 23.71%. Prepd by the reaction of *o*-phenylenediamine with formic acid: Wundt, *Ber.* **11**, 826 (1878); Wagner, Millett, *Org. Synth. coll. vol.* **II**, 65 (1943). Absorption spectrum: Steck *et al., J. Am. Chem. Soc.* **70**, 3406 (1948). *Reviews:* J. B. Wright, *Chem. Rev.* **48**, 397-541 (1951); K. Hofmann, *Imidazole and Its Derivatives* (Interscience, New York, 1953) p 247 sqq.

Tabular crystals. Orthorhombic and monoclinic modifications. mp 170.5°. bp$_{760}$ >360°. Weak base. pKa = 5.48 at 25°. Dipole moment: 3.93 (dioxane). Sparingly sol in cold water, more sol in hot water. Freely sol in alcohol, sparingly sol in ether. Practically insol in benzene, petr ether. One gram dissolves in 2 g of boiling xylene. Sol in aq solns of acids and strong alkalies. High degree of chemical stability.

1084. 2-Benzimidazolethiol. [583-39-1] 1,3-Dihydro-2*H*-benzimidazole-2-thione; 2-mercaptobenzimidazole. $C_7H_6N_2S$; mol wt 150.20. C 55.98%, H 4.03%, N 18.65%, S 21.34%. Prepn from *o*-phenylenediamine and potassium ethyl xanthate: Van Allan, Deacon, *Org. Synth. coll. vol.* **IV**, 569 (1963).

Glistening platelets from 95% ethanol, mp 303-304°. Slightly sol in water; sol in methanol, ethanol.

1085. Benziodarone. [68-90-6] (2-Ethyl-3-benzofuranyl) (4-hydroxy-3,5-diiodophenyl) methanone; 2-ethyl-3-(3',5'-diiodo-4'-hydroxybenzoyl)benzofuran; 2-ethyl-3-(3',5'-diiodo-4'-hydroxybenzoyl)oxaindene; 2-ethyl-3-benzofuryl 3',5'-diiodo-4'-hydroxyphenyl ketone; 2-ethyl-3-(3,5-diiodo-4-hydroxybenzoyl)coumarone; 2-ethyl-3-(4-hydroxy-3,5-diiodobenzoyl)benzofuran; 2329 Labaz; L-2329; Amplivix; Dilafurane. $C_{17}H_{12}I_2O_3$; mol wt 518.09. C 39.41%, H 2.33%, I 48.99%, O 9.26%. Description: Charlier, *Acta Cardiol. Suppl.* **7**, 1-60 (1959). Synthesis: Beaudet, Henaux, **BE 553621**; **GB 836272** (1957, 1960, both to Labaz); Buu-Hoi, Beaudet, **US 3012042** (1961 to Soc. Belge Azote Prod. Chim. Marly). Clinical trial in hyperuricemia in renal transplant patients: F. Perez-Ruiz *et al., Nephrol. Dial. Transplant.* **18**, 603 (2003).

Yellowish powder, mp 167°. Soly in water at 25° about 0.2%, at 45° about 1.0%. Sol in chloroform, acetone.

THERAP CAT: Uricosuric.

1086. Benznidazole. [22994-85-0] 2-Nitro-*N*-(phenylmethyl)-1*H*-imidazole-1-acetamide; *N*-benzyl-2-nitroimidazole-1-acetamide; Ro-7-1051; Radanil. $C_{12}H_{12}N_4O_3$; mol wt 260.25. C 55.38%, H 4.65%, N 21.53%, O 18.44%. Prepn: **GB 1138529**; A. G. Beaman *et al.,* **US 3679698** (1966, 1972 both to Hoffmann-La Roche). *In vitro* effect on *T. cruzi:* S. Yoneda *et al., Experientia* **33**, 1201 (1977). Pharmacokinetics: J. Raaflaub, W. H. Ziegler, *Arzneim.-Forsch.* **29**, 1611 (1979). Mutagenicity study: C. E. Voogd *et al., Mutat. Res.* **66**, 207 (1979). Clinical trial in Chagas disease: A. L. S. Sgambatti de Andrade *et al., Lancet* **348**, 1407 (1996); R. Viotti *et al., Ann. Intern. Med.* **144**, 724 (2006).

Crystals from ethanol, mp 188.5-190°. uv max (ethanol): 313 nm (ε 7,600). Soly in water at 37°: 40 mg/100 ml.

THERAP CAT: Antiprotozoal (Trypanosoma).

1087. Benzo Azurine G. [2429-71-2] 3,3'-[(3,3'-Dimethoxy[1,1'-biphenyl]-4,4'-diyl)bis(2,1-diazenediyl)]bis[4-hydroxy-1-naphthalenesulfonic acid] sodium salt (1:2); disodium *o*-dianisidine-diazobis(1-naphthol-4-sulfonate); C.I. Direct Blue 8; C.I. 24140. $C_{34}H_{24}N_4Na_2O_{10}S_2$; mol wt 758.68. C 53.83%, H 3.19%, N 7.38%, Na 6.06%, O 21.09%, S 8.45%. Prepared from diazotized *o*-dianisidine and sodium 4-hydroxy-1-naphthalenesulfonate: F. J. Welcher, *Organic Analytical Reagents* vol. 4 (D. Van Nostrand Co., New York, 1948) pp 337-338. *See also Colour Index* vol. 4 (3rd ed., 1971) p 4202.

Bluish-black powder. Sol in water, aq sodium hydroxide (red soln), sulfuric acid (blue soln), Cellosolve; very slightly sol in ethanol. Practically insol in other organic solvents.

USE: As a dye: *Colour Index* vol. 2 (3rd ed., 1971) p 2223. As analytical reagent in detection of Mg.

1088. Benzocaine. [94-09-7] 4-Aminobenzoic acid ethyl ester; ethyl aminobenzoate; Americaine; Anaesthesin; Flavamed; Subcutin. $C_9H_{11}NO_2$; mol wt 165.19. C 65.44%, H 6.71%, N 8.48%, O 19.37%. Prepd by the esterification of *p*-aminobenzoic acid: Salkowski, *Ber.* **28**, 1921 (1895); Vorländer, Meyer, *Ann.* **320**, 135 (1902); by the reduction of ethyl *p*-nitrobenzoate: Limpricht, *ibid.* **303**, 278 (1898); R. Adams, F. L. Cohen, *Org. Synth. coll. vol.* **I**, 240 (2nd ed.), 1941). In industrial practice the reducing agent is usually iron and water in the presence of a little acid. Comprehensive description: S. L. Ali, *Anal. Profiles Drug Subs.* **12**, 73-104 (1983).

Rhombohedra from ether, mp 88-90°. Stable in air. One gram dissolves in about 2500 ml water, 5 ml alcohol, 2 ml chloroform, in about 4 ml ether, and in 30 to 50 ml of expressed almond oil or olive oil. Also sol in dil acids. pKa 2.5.

THERAP CAT: Anesthetic (local).

THERAP CAT (VET): Local (usually surface) anesthetic.

1089. Benzoctamine. [17243-39-9] *N*-Methyl-9,10-ethano-anthracene-9(10*H*)-methanamine; 1-(methylaminomethyl)di-benzo[*b,e*]bicyclo[2.2.2]octadiene; 9-(methylaminomethyl)-9,10-di-hydro-9,10-ethanoanthracene. $C_{18}H_{19}N$; mol wt 249.36. C 86.70%, H 7.68%, N 5.62%. Prepn: **BE 610863**; P. Schmidt *et al.*, **US 3399201** (1962, 1968 to Ciba); Wilhelm, Schmidt, *Helv. Chim. Acta* **52**, 1385 (1969). Pharmacology: Keberle *et al.*, *Present Status Psychotropic Drugs* (Excerpta Medica, New York, 1969) p 123; Baltzer, Bein, *Arch. Int. Pharmacodyn. Ther.* **201**, 25 (1973). Toxicity: E. I. Goldenthal, *Toxicol. Appl. Pharmacol.* **18**, 185 (1971).

Hydrochloride. [10085-81-1] Ba-30803; Tacitin. $C_{18}H_{19}N.$-HCl; mol wt 285.82. Crystals, mp 320-322°. pKa: 7.6. LD_{50} orally in rats: 700 ±170 mg/kg (Goldenthal).

THERAP CAT: Anxiolytic.

1090. Benzofuran. [271-89-6] Coumarone; cumarone. C_8-H_6O; mol wt 118.14. C 81.33%, H 5.12%, O 13.54%. Constituent of coal tar, *q.v.* Isoln from coal tar oils: Kraemer, Spilker, *Ber.* **23**, 78 (1890); *eidem, ibid.* **33**, 2261 (1900); Breston, Gauger, *Am. Gas Assoc. Proc.* **28**, 492 (1946). Synthesis by heating phenoxyacetaldehyde with zinc chloride and acetic acid: Stoermer, *Ber.* **30**, 1703 (1897); *Ann.* **312**, 261 (1900). Review of toxicology and human exposure: *Toxicological Profile for 2,3-Benzofuran* (PB93-110666, 1992) 95 pp.

Oil. Aromatic odor. Not solid at −18°. bp_{760} 173-175°; bp_{15} 62-63°. Volatile with steam. $d_4^{22.7}$ 1.0913. $n_D^{16.3}$ 1.56897; $n_D^{22.7}$ 1.565. Insol in water, aq alkaline solns; miscible with benzene, petr ether, abs alcohol, ether. Slowly polymerizes on standing.

USE: Manuf of coumarone-indene resins.

1091. Benzoguanamine. [91-76-9] 6-Phenyl-1,3,5-triazine-2,4-diamine; 2,4-diamino-6-phenyl-*s*-triazine; 4,6-diamino-2-phenyl-*s*-triazine. $C_9H_9N_5$; mol wt 187.21. C 57.74%, H 4.85%, N 37.41%. Prepd from benzonitrile and dicyandiamide in the presence of Na and liquid ammonia: Jones, **US 2735850** (1956 to British Oxygen Co.); from benzonitrile and dicyandiamide in the presence of KOH and methyl Cellosolve: Simons, Saxton, *Org. Synth.* **33**, 13 (1953).

Crystals. d_4^{25} 1.40. mp 227-228°. uv max (ethanol): 249 nm (ε 25,000). Sol in alcohol, ether, dil HCl; partially sol in dimethylformamide. Practically insol in acetone, chloroform, ethyl acetate. Soly in water at 22° 0.06%; at 100° 0.6%.

USE: In the manuf of thermosetting resins, pesticides, pharmaceuticals and dyestuffs.

1092. Benzohydrol. [91-01-0] α-Phenylbenzenemethanol; benzhydrol; diphenylcarbinol. $C_{13}H_{12}O$; mol wt 184.24. C 84.75%, H 6.57%, O 8.68%. Prepd by reducing benzophenone with zinc dust

in strongly alkaline soln: Wiselogle, Sonneborn, *Org. Synth.* **coll. vol. I** (2nd ed, 1941) p 90. By reducing benzophenone with magnesium in methanol: Zechmeister, Rom, *Ann.* **468**, 123 (1929).

Needles from ligroin, mp 69°. bp_{748} 298°; bp_{20} 180°; bp_{13} 176°. Absorption spectrum: Orndorff *et al.*, *J. Am. Chem. Soc.* **49**, 1542 (1927). One gram dissolves in 2 liters of water at 20°. Freely sol in alcohol, ether, chloroform and carbon bisulfide; almost insol in cold ligroin.

Diphenylmethyl ether. [574-42-5] $C_{26}H_{22}O$; mol wt 350.46. Monoclinic crystals from benzene, mp 110°; bp_{15} 267°.

USE: In organic syntheses.

1093. Benzoic Acid. [65-85-0] Benzenecarboxylic acid; phenylformic acid; dracylic acid. $C_7H_6O_2$; mol wt 122.12. C 68.85%, H 4.95%, O 26.20%. Occurs in nature in free and combined forms. Gum benzoin may contain as much as 20%. Most berries contain appreciable amounts (around 0.05%). Excreted mainly as hippuric acid by almost all vertebrates, except fowl. Mfg processes include the air oxidation of toluene, the hydrolysis of benzotrichloride, and the decarboxylation of phthalic anhydride: *Faith, Keyes & Clark's Industrial Chemicals*, F. A. Lowenheim, M. K. Moran, Eds. (Wiley-Interscience, New York, 4th ed., 1975) pp 138-144. Lab prepn from benzyl chloride: A. I. Vogel, *Practical Organic Chemistry* (Longmans, London, 3rd ed, 1959) p 755; from benzaldehyde: Gattermann-Wieland, *Praxis des Organischen Chemikers* (de Gruyter, Berlin, 40th ed, 1961) p 193. Prepn of ultra-pure benzoic acid for use as titrimetric and calorimetric standard: Schwab, Wicher, *J. Res. Natl. Bur. Stand.* **25**, 747 (1940). *Review:* A. E. Williams in *Kirk-Othmer Encyclopedia of Chemical Technology* **vol. 3** (Wiley-Interscience, New York, 3rd ed., 1978) pp 778-792.

Monoclinic tablets, plates, leaflets. d 1.321 (also reported as 1.266). mp 122.4°. Begins to sublime at ~100°. bp_{760} 249.2°; bp_{400} 227°; bp_{200} 205.8°; bp_{100} 186.2°; bp_{60} 172.8°; bp_{40} 162.6°; bp_{20} 146.7°; bp_{10} 132.1°. Volatile with steam. Flash pt 121°C. pK (25°) 4.19. pH of satd soln at 25°: 2.8. Soly in water (g/l) at 0° = 1.7; at 10° = 2.1; at 20° = 2.9; at 25° = 3.4; at 30° = 4.2; at 40° = 6.0; at 50° = 9.5; at 60° = 12.0; at 70° = 17.7; at 80° = 27.5; at 90° = 45.5; at 95° = 68.0. Mixtures of excess benzoic acid and water form two liquid phases beginning at 89.7°. The two liquid phases unite at the critical soln temp of 117.2°. Composition of critical mixture (32.34% benzoic acid, 67.66% water): Ward, Cooper, *J. Phys. Chem.* **34**, 1484 (1930). One gram dissolves in 2.3 ml cold alc, 1.5 ml boiling alc, 4.5 ml chloroform, 3 ml ether, 3 ml acetone, 30 ml carbon tetrachloride, 10 ml benzene, 30 ml carbon disulfide, 23 ml oil of turpentine; also sol in volatile and fixed oils, slightly in petr ether. The soly in water is increased by alkaline substances, such as borax or trisodium phosphate, *see also* Sodium Benzoate.

Barium salt dihydrate. [5908-68-9] Barium benzoate. $C_{14}H_{10}$-$BaO_4.2H_2O$; mol wt 415.59. Nacreous leaflets. *Poisonous.* Soluble in about 20 parts water; slightly sol in alc.

Calcium salt trihydrate. [5743-30-6] Calcium benzoate. C_{14}-$H_{10}CaO_4.3H_2O$; mol wt 336.35. Orthorhombic crystals or powder. d 1.44. Soluble in 25 parts water; very sol in boiling water.

Cerium salt trihydrate. [105546-51-8] Cerous benzoate. C_{21}-$H_{15}CeO_6.3H_2O$; mol wt 557.51. White to reddish-white powder. Sol in hot water or hot alc.

Copper salt dihydrate. [6046-97-5] Cupric benzoate. $C_{14}H_{10}$-$CuO_4.2H_2O$; mol wt 341.81. Light blue, cryst powder. Slightly soluble in cold water, more in hot water; sol in alc or in dil acids with separation of benzoic acid.

Lead salt monohydrate. [6080-57-5] Lead benzoate. $C_{14}H_{10}$-$O_2Pb.H_2O$; mol wt 435.45. Cryst powder. *Poisonous.* Slightly sol in water.

Manganese salt tetrahydrate. [6146-97-0] Manganese benzoate. $C_{14}H_{10}MnO_4.4H_2O$; mol wt 369.23. Pale-red powder. Sol in water, alc. Also occurs with $3H_2O$.

Nickel salt trihydrate. [6018-91-3] Nickel benzoate. $C_{14}H_{10}$-$NiO_4.3H_2O$; mol wt 354.97. Light-green odorless powder. Slightly sol in water; sol in ammonia; dec by acids.

Potassium salt trihydrate. [6100-02-3] Potassium benzoate. $C_7H_5KO_2.3H_2O$; mol wt 214.26. Crystalline powder. Sol in water, alc.

Silver salt. [532-31-0] Silver benzoate. $C_7H_5AgO_2$; mol wt 228.98. Light-sensitive powder. Sol in 385 parts cold water, more sol in hot water; very slightly sol in alc.

Uranium salt. [532-60-5] Uranyl benzoate. $C_{14}H_{10}O_6U$; mol wt 512.26. Yellow powder. Slightly sol in water, alc.

Caution: Potential symptoms of overexposure may include irritation to the skin, eyes, nose, throat, and mucous membranes.

USE: Preserving foods, fats, fruit juices, alkaloidal solns, etc; manuf benzoates and benzoyl compds, dyes; as a mordant in calico printing; for curing tobacco. As standard in volumetric and calorimetric analysis. Pharmaceutic aid (antifungal).

THERAP CAT (VET): Has been used with salicylic acid as a topical antifungal.

1094. Benzoic Anhydride. [93-97-0] Benzoic acid 1,1'-anhydride. $C_{14}H_{10}O_3$; mol wt 226.23. C 74.33%, H 4.46%, O 21.22%. Prepd by heating benzoic acid with acetic anhydride in the presence of a trace of phosphoric acid: Clarke, Rahrs, *Org. Synth.* **3**, 2 (1923); **coll. vol. I** (Wiley, New York, 2nd ed., 1941) p 91.

Orthorhombic bipyramidal prisms from benzene + petr ether. d_4^{15} 1.1989. mp 42°. bp$_{760}$ 360°; bp$_{200}$ 299.1°; bp$_{60}$ 252.7°; bp$_{40}$ 239.8°; bp$_{20}$ 218°; bp$_{10}$ 198°; bp$_5$ 180°; bp$_{1.0}$ 143.8°. n_D^{15} 1.57665. Almost insol in water (0.01 g/l). Sol in alcohol, chloroform, acetone, ethyl acetate, benzene, toluene, xylene, ether, glacial acetic acid, acetic anhydride; moderately sol in petr ether; stable in water and in cold alkaline solns.

USE: In organic syntheses, as benzoylating agent in the manufacture of pharmaceuticals, dyes, and intermediates.

1095. Benzoin. [119-53-9] 2-Hydroxy-1,2-diphenylethanone; benzoylphenylcarbinol; α-hydroxy-α-phenylacetophenone; bitter-almond-oil camphor. $C_{14}H_{12}O_2$; mol wt 212.25. C 79.22%, H 5.70%, O 15.08%. Prepd by treating an alcoholic soln of benzaldehyde with an alkali cyanide: Adams, Marvel, *Org. Synth.* **vol. 1**, p 33 (1921); **coll. vol. I**, 88; Arnold, Fuson, *J. Am. Chem. Soc.* **58**, 1295 (1936); L. F. Fieser, *Organic Experiments* (D. C. Heath & Co., Boston, 1964) pp 211-214.

dl-**Form.** Six-sided monoclinic prisms from alcohol, mp 137°. bp$_{768}$ 344°. bp$_{12}$ 194°. Reduces Fehling's soln. uv max (ethanol): 247 nm (ε 14500); infrared in chloroform: 2.88; 5.93; 6.21; 6.28, 6.85 μ. Soluble in 3335 parts water, more in hot water, in 5 parts pyridine; sol in acetone, in boiling alc; slightly in ether.

Methyl ether. $C_{15}H_{14}O_2$. Needles, mp 49°.
Ethyl ether. $C_{16}H_{16}O_2$. Needles, mp 62°.
l-**Form.** Needles, mp 132°. $[\alpha]_D^{12}$ −118° (c = 1.2 in acetone).
d-**Form.** Needles, mp 132°. $[\alpha]_D^{12}$ +120.5° (c = 1.2 in acetone).
Note: Not to be confused with Gum Benzoin.
USE: In organic syntheses.

1096. Benzoin Oxime. [441-38-3] 2-Hydroxy-1,2-diphenylethanone oxime; 2-hydroxy-2-phenylacetophenone oxime. $C_{14}H_{13}$-NO_2; mol wt 227.26. C 73.99%, H 5.77%, N 6.16%, O 14.08%. Two isomers occur: α or *anti* and β or *syn*. Both have been prepd from benzoin and hydroxylamine hydrochloride. Prepn of α-form: Werner, Detscheff, *Ber.* **38**, 69 (1905); F. J. Welcher, *Organic Analytical Reagents* **vol. III** (Van Nostrand, New York, 1947) pp 239-251. Prepn of β-form: Werner, Detscheff, *loc. cit.* Configuration of α- and β-forms: Meisenheimer, Meis, *Ber.* **57**, 289 (1924).

α-form   β-form

α-**Form.** Cupron. Prisms from benzene, mp 151-152°. Darkens on exposure to light. Slightly sol in water; sol in alc, aq ammonium hydroxide soln. *Protect from light.*

β-**Form.** Prismatic crystals from ether. Ether of crystn is lost on standing in air. Ether-free compd, mp 99°.

USE: The α-form is used in the detection and determination of Cu, Mo, and W: Welcher, *loc. cit.*

1097. 6,7-Benzomorphan. [575-19-9] 1,2,3,4,5,6-Hexahydro-2,6-methano-3-benzazocine; 6,7-benzmorphan. $C_{12}H_{15}N$; mol wt 173.26. C 83.19%, H 8.73%, N 8.08%. Parent compound of a series of potent analgetics. Synthesis: Kanematsu *et al.*, *J. Am. Chem. Soc.* **90**, 1064 (1968); Kanematsu *et al.*, *J. Med. Chem.* **12**, 405 (1969). Review of derivative syntheses and pharmacology: May, Sargent, "Morphine and its Modifications", in *Analgetics*, G. de Stevens, Ed. (Academic Press, New York, 1965) pp 145-171.

Hydrochloride. $C_{12}H_{15}N.HCl$. Needles from methanol-acetone, mp 261-262°.

1098. Benzonatate. [104-31-4] 4-(Butylamino)benzoic acid 3,6,9,12,15,18,21,24,27-nonaoxaoctacos-1-yl ester; nonaethyleneglycol monomethyl ether *p-n*-butylaminobenzoate; *p*-butylaminobenzoic acid ω-O-methylnonaethyleneglycol ester; benzononatine; Exangit; Tessalon. $C_{30}H_{53}NO_{11}$; mol wt 603.75. C 59.68%, H 8.85%, N 2.32%, O 29.15%. Prepn: Matter, **US 2714608** (1955 to Ciba).

Colorless to faintly yellow oil. Miscible with water in all proportions. Freely sol in chloroform, alcohol, benzene. Sol in most organic solvents except aliphatic hydrocarbons.

THERAP CAT: Antitussive.

1099. Benzonitrile. [100-47-0] Phenyl cyanide; cyanobenzene. C_7H_5N; mol wt 103.12. C 81.53%, H 4.89%, N 13.58%. Prepd by heating Na benzenesulfonate with NaCN or by adding benzenediazonium chloride soln to a hot aq NaCN soln contg $CuSO_4$ and distilling. Lab prepn: A. I. Vogel, *Practical Organic Chemistry* (Longmans, London, 3rd ed, 1959) p 608.

Consult the Name Index before using this section.

Liquid, odor of volatile oil of almond. bp_{760} 190.7°, bp_{100} 123.5°, bp_{10} 69.2°, bp_1 28.2°. mp −12.75°. d_{15}^{15} 1.010. n_D^{20} 1.5289. Fire pt 167°F. Slightly sol in cold water; sol to the extent of 1% in water at 100°; miscible with common organic solvents.

USE: Solvent.

1100. Benzophenone. [119-61-9] Diphenylmethanone; diphenyl ketone; benzoylbenzene. $C_{13}H_{10}O$; mol wt 182.22. C 85.69%, H 5.53%, O 8.78%. Prepd by the Friedel-Crafts ketone synthesis from benzene and benzoyl chloride in the presence of $AlCl_3$: Marvel, Sperry, *Org. Synth.* **coll. vol. I** (Wiley, New York, 2nd ed., 1941) p 95. By decarboxylation of *o*-benzoylbenzoic acid in the presence of copper catalyst: L. F. Fieser, *Organic Experiments* (D. C. Heath & Co., Boston, 1964) pp 201-203.

Stable form (there are two other labile forms), orthorhombic bisphenoidal prisms from alcohol or ether. Geranium-like odor. d_4^{18} 1.1108. d_4^{50} 1.0869. $n_D^{45.2}$ 1.5975. Absorption spectrum: Purvis, McCleland, *J. Chem. Soc.* **101**, 1516 (1912). mp 48.5°. bp_{760} 305.4°; bp_{400} 276.8°; bp_{200} 249.8°; bp_{100} 224.4°; bp_{60} 208.2°; bp_{40} 195.7°; bp_{20} 175.8°; bp_{10} 157.6°; bp_5 141.7°; $bp_{1.0}$ 108.2°. Insoluble in water. One gram dissolves in 7.5 ml alcohol, 6 ml ether; sol in chloroform.

Oxime. [574-66-3] $C_{13}H_{11}NO$; mol wt 197.24. Prepn: *Org. Synth.* **10**, 10. Crystals from ligroin, mp 143-144°. Freely sol in ether, acetone.

USE: Fixative for heavy perfumes, such as geranium, new-mown hay, especially when used in soaps. In the manuf of antihistamines, hypnotics, insecticides.

1101. Benzophenone-6. [131-54-4] Bis(2-hydroxy-4-methoxyphenyl)methanone; 2,2′-dihydroxy-4,4′-dimethoxybenzophenone; Uvinul D49. $C_{15}H_{14}O_5$; mol wt 274.27. C 65.69%, H 5.15%, O 29.17%. Prepn: Grover *et al.*, *J. Chem. Soc.* **1955**, 3982; Hardy, Forster, **US 2773903** (1956 to Am. Cyanamid); Dayan, Roberts; Hahn, Stanley, **US 2853522** and **US 2853523** (both 1958 to Gen. Aniline); Hosler, Storfer, **US 2928878** (1960 to Am. Cyanamid).

Crystals, mp 139-140°. uv max: 284, 340 nm (log ε 4.12, 4.12).

USE: As ultraviolet light absorber, esp in paints, plastics.

1102. Benzophenone Imine. [1013-88-3] α-Phenylbenzenemethanimine; diphenyl ketimine; 1,1-diphenylmethylenimine. $C_{13}H_{11}N$; mol wt 181.24. C 86.15%, H 6.12%, N 7.73%. Reagent for protection of primary amines. Prepn of salt forms: A. Hantzsch, F. Kraft, *Ber.* **24**, 3511 (1891). Prepn by addn of phenylmagnesium bromide to benzonitrile, followed by hydrolysis with CH_3OH: P. L. Pickard, T. L. Tolbert, *J. Org. Chem.* **26**, 4886 (1961); *eidem*, *Org. Synth.* **coll. vol. V**, 520 (1973); by reaction of benzophenone with NH_3: G. Verardo *et al.*, *Synth. Commun.* **18**, 1501 (1988). Electrochemical properties: S. Zhan, M. D. Hawley, *J. Electroanal. Chem.* **319**, 275 (1991). Synthetic applications: M. J. O'Donnell, *Aldrichim. Acta* **34**, 3 (2001); Y. Liu *et al.*, *J. Heterocycl. Chem.* **40**, 713 (2003). *Review*: A. Crespo, *Synlett* **2006**, 2345-2346.

Liquid, $bp_{3.5}$ 127-128°. bp_8 151-152°. n_D^{20} 1.6180-1.6191. d^{20} 1.0849. *Light sensitive. Store under Nitrogen.*

Hydrochloride. [5319-67-5] $C_{13}H_{11}N.HCl$; mol wt 217.70. mp 139°. Sol in cold water; slightly sol in chloroform. Insol in ether, benzene.

USE: Protecting group for primary amines in peptide chemistry; ammonia surrogate in catalyzed coupling reactions.

1103. Benzopinacol. [464-72-2] 1,1,2,2-Tetraphenyl-1,2-ethanediol; benzpinacone; tetraphenylethylene glycol. $C_{26}H_{22}O_2$; mol wt 366.46. C 85.22%, H 6.05%, O 8.73%. Prepd by photochemical reduction of benzophenone: Bachmann, *Org. Synth.* **coll. vol. II**, 71 (1943); L. F. Fieser, *Organic Experiments* (D. C. Heath & Co., Boston, 1964) p 203. By the action of phenylmagnesium bromide on benzil: Acree, *Ber.* **37**, 2761 (1904).

Monoclinic prisms, may contain 1 mol C_6H_6 when crystallized from benzene. mp 197° (open capillary, rapid heating); mp 222° (copper block). On heating to mp it dec to benzophenone and benzohydrol. Soluble in 11.5 parts boiling glacial acetic acid; in 26 parts boiling benzene; in 39 parts boiling 95% alcohol. Freely sol in ether, carbon bisulfide, chloroform.

1104. Benzopurpurine 4B. [992-59-6] 3,3′-[(3,3′-Dimethyl[1,1′-biphenyl]-4,4′-diyl)bis(2,1-diazenediyl)]bis[4-amino-1-naphthalenesulfonic acid] sodium salt (1:2); disodium *o*-tolidinediazobis(1-naphthylamine-4-sulfonate); C.I. Direct Red 2; C.I. 23500; azamin 4B; eclipse red; fast scarlet; Paper Red 4B; Sultan Red 4B; Cotton Red 4B. $C_{34}H_{26}N_6Na_2O_6S_2$; mol wt 724.72. C 56.35%, H 3.62%, N 11.60%, Na 6.34%, O 13.25%, S 8.85%. Prepn from diazotized *o*-tolidine and sodium 4-amino-1-naphthalenesulfonate: **DE 35615** (1885 to A. G. für Anilinfabrikation Berlin), *Frdl.* **1**, 473; F. J. Welcher, *Organic Analytical Reagents* **vol. 4** (Van Nostrand Co., New York, 1948) pp 338-339; *Colour Index* **vol. 4** (3rd ed., 1971) p 4189. *See also*: H. J. Conn's *Biological Stains*, R. D. Lillie, Ed. (Williams & Wilkins, Baltimore, 9th ed., 1977) p 154.

Brown powder. Soluble in water, sodium hydroxide, sulfuric acid, ethanol, acetone, Cellosolve. Practically insol in other organic solvents.

USE: For dyeing primarily cotton and viscose rayon: *Colour Index* **vol. 2** (3rd ed., 1971) p 2101. As an analytical reagent in detection of Al, Mg, Hg, Ag, U; as a biological stain; as pH indicator, violet 1.2 to red 4.0.

1105. 1,2-Benzopyran. [254-04-6] 2*H*-1-Benzopyran; α-5:6-benzopyran; 1,2-chromene; 3-chromene. C_9H_8O; mol wt 132.16. C 81.79%, H 6.10%, O 12.11%. Prepn from *cis-o*-hydroxycinnamyl alcohol: Chatterjea, *J. Indian Chem. Soc.* **36**, 76 (1959), *C.A.* **54**,

519b (1960); from phenyl propargyl ether in diethylphenylamine: Iwai, Ide, *Chem. Pharm. Bull.* **11**, 1042 (1963); Iwai, Iwade, *JP 63 22587* (1963 to Sankyo), *C.A.* **60**, 2901e (1964). Earlier prepns: Maitte, *Ann. Chim. (Paris)* **9**, 431 (1954); Normant, Maitte, *Compt. Rend.* **234**, 1787 (1952).

Liquid; bp_{102} 132°, bp_{15} 92-92.5°, bp_{13} 91°, bp_9 77°. d^{16} 1.0993. n_D^{24} 1.5869; n_D^{16} 1.5923 (Maitte); also reported as n_D 1.5837 (**JP 63 22587**).

1106. Benzo[a]pyrene. [50-32-8] 3,4-Benzpyrene. $C_{20}H_{12}$; mol wt 252.32. C 95.20%, H 4.79%. Formerly called *1,2-benzpyrene*. Occurs in coal tar, *q.v.* Isoln by fractionation: Cook *et al., J. Chem. Soc.* **1933**, 395; by adsorption and fluorimetric determn: Hieger, *Am. J. Cancer* **29**, 705 (1937); Winterstein *et al., Z. Physiol. Chem.* **230**, 158, 169 (1934); *Naturwissenschaften* **22**, 237 (1934). Synthesis from pyrene and succinic anhydride: Cook, Hewett, *J. Chem. Soc.* **1933**, 398; Fieser, Fieser, *J. Am. Chem. Soc.* **57**, 782 (1935); Winterstein *et al., Ber.* **68**, 1079 (1935); Fieser *et al., J. Am. Chem. Soc.* **57**, 1509 (1935). Absorption spectrum: Mayneord, Roe, *Proc. R. Soc. London* **A152**, 299 (1935). *Review:* Clar, *Polycyclic Hydrocarbons* **Vol. 1 & 2** (Academic Press, New York, 1964). Review of carcinogenicity studies: *IARC Monographs* **3**, 91-136 (1973). Inhibition of the mutagenicity of the ultimate carcinogenic metabolite of benzo[a]pyrene by ellagic acid: A. W. Wood *et al., Proc. Natl. Acad. Sci. USA* **79**, 5513 (1982). Study of the reaction between this metabolite (***benzo[a]pyrene-7,8-diol 9,10-epoxide***) and ellagic acid: J. M. Sayer *et al., J. Am. Chem. Soc.* **104**, 5562 (1982). *Review:* D. H. Phillips, *Nature* **303**, 468-472 (1983). Review of toxicology and human exposure: *Toxicological Profile for Polycyclic Aromatic Hydrocarbons* (PB95-264370, 1995) 487 pp.

Yellowish plates, needles from benzene + methanol, mp 179-179.3°. Crystals may be monoclinic or orthorhombic. bp_{10} 310-312°. Dil benzene solns exhibit violet fluorescence. Soluble in benzene, toluene, xylene; sparingly sol in alc, methanol. Practically insol in water.

Caution: This substance is reasonably anticipated to be a human carcinogen: *Report on Carcinogens, Twelfth Edition* (PB2011-111646, 2011) p 353.

1107. Benzo[e]pyrene. [192-97-2] 1,2-Benzpyrene; 4,5-benzpyrene. $C_{20}H_{12}$; mol wt 252.32. C 95.20%, H 4.79%. Constituent of coal tar, *q.v.* Isoln: Cook *et al., J. Chem. Soc.* **1933**, 396. Synthesis: Cook, Hewett, *ibid.* 398; Clar, *Ber.* **76**, 609 (1943); Buchta, Kröger, *Ann.* **705**, 190 (1967). NMR spectra: Cobb, Memory, *J. Chem. Phys.* **47**, 2020 (1967). *Review:* Clar, *Polycyclic Hydrocarbons* **Vol. 2**, (Academic Press, New York, 1964) p 127. Review of toxicology and human exposure: *Toxicological Profile for Polycyclic Aromatic Hydrocarbons* (PB95-264370, 1995) 487 pp.

Prisms or plates from benzene, mp 178-179°.

1108. Benzo[f]quinoline. [85-02-9] β-Naphthoquinoline; 5,6-benzoquinoline; naphthopyridine. $C_{13}H_9N$; mol wt 179.22. C

87.12%, H 5.06%, N 7.82%. Prepn from from β-naphthylamine by the Skraup reaction: Knueppel, *Ber.* **29**, 703 (1896); Clem, Hamilton, *J. Am. Chem. Soc.* **62**, 2349 (1940); Uhle, Jacobs, *J. Org. Chem.* **10**, 76 (1945).

Crystals from alcohol + water, mp 93°. bp_{721} 349-350°. uv max (ethanol): 347, 331, 316, 266, nm (log ε 3.54, 3.41, 3.18, 4.06). Practically insol in water. Sol in dil acids; very sol in alcohol, ether, benzene.

USE: As a reagent for the determination of cadmium which is pptd as $(C_{13}H_9N)_2H_2(CdI_4)$ from dil nitric or sulfuric acid soln in the presence of potassium iodide.

1109. Benzoresorcinol. [131-56-6] (2,4-Dihydroxyphenyl)-phenylmethanone; 2,4-dihydroxybenzophenone; 4-benzoylresorcinol; resbenzophenone; benzisotriazole. Uvinul 400. $C_{13}H_{10}O_3$; mol wt 214.22. C 72.89%, H 4.71%, O 22.41%. Prepd from benzoyl chloride and resorcinol: Dobner, *Ann.* **210**, 246 (1881); by Fries rearrangement: Amin, Shah, *J. Indian Chem. Soc.* **29**, 351 (1952).

Needles from hot water, mp 144-145°. Practically insol in cold water. Easily sol in alcohol, ether, glacial acetic acid; scarcely sol in cold benzene.

USE: Ultraviolet light absorber, esp in paints and plastics.

1110. Benzothiazole. [95-16-9] Benzosulfonazole; 1-thia-3-azaindene. C_7H_5NS; mol wt 135.18. C 62.20%, H 3.73%, N 10.36%, S 23.72%. Prepd from *N,N*-dimethylaniline and sulfur: Knowles, Watt, *J. Org. Chem.* **7**, 56 (1942). Chemistry of benzothiazoles, their use as carbonyl equivalents and in carbon-carbon bond formation: E. J. Corey, D. L. Boger, *Tetrahedron Lett.* **19**, 5 (1978); *eidem, ibid.* 9; *eidem, ibid.* 13. Toxicity study: E. F. Domino *et al., J. Pharmacol. Exp. Ther.* **105**, 486 (1952).

Liquid. Odor similar to that of quinoline. Volatile with steam. d_4^{20} 1.246. bp_{765} 227-228°; bp_{34} 131°. n_D^{20} 1.6379. Slightly sol in water. Freely sol in alcohol, carbon disulfide. LD_{50} i.v. in mice: 95±3 mg/kg (Domino).

USE: In organic synthesis.

1111. 1H-Benzotriazole. [95-14-7] 1,2,3-Benzotriazole; benztriazole; azimidobenzene; benzisotriazole. $C_6H_5N_3$; mol wt 119.13. C 60.49%, H 4.23%, N 35.27%. Prepd by the action of nitrous acid on *o*-phenylenediamine: Ladenburg, *Ber.* **9**, 219 (1876); Damschroder, Peterson, *Org. Synth.* **coll. vol. III**, 106 (1955).

Needles from benzene, mp 98.5°. bp_{15} 204°; $bp_{2.0}$ 159°. *Caution:* May explode during vacuum distillation, *see Chem. Eng. News* **34**, 2450 (1956). Sparingly sol in water. Sol in alc, benzene, toluene, chloroform, DMF.

1112. Benzotrichloride. [98-07-7] (Trichloromethyl)benzene; α,α,α-trichlorotoluene; phenylchloroform; ω,ω,ω-trichlorotoluene; benzenyl trichloride; toluene trichloride. $C_7H_5Cl_3$; mol wt 195.47. C 43.01%, H 2.58%, Cl 54.41%. Produced by chlorination of boiling toluene in the presence of light and of 2% phosphorus trichloride: Swarts, *Bull. Soc. Chim. Belg.* **31**, 375 (1922); Conklin, US 1828858; US 1828859 (both 1931 to Solvay Process Co.). Commercial grades may contain hydrochloric acid, benzylidene chloride, and benzyl chloride. Purification procedures: Holleman, deMooy, *Rec. Trav. Chim.* **33**, 25, 33 (1914); Britton, US 1804458 (1931 to Dow); *Chem. Zentralbl.* **1931**, II, 497. Toxicity data: H. F. Smyth *et al.*, *Arch. Ind. Hyg. Occup. Med.* **4**, 119 (1951). *Review:* H. C. Lin in *Kirk-Othmer Encyclopedia of Chemical Technology* **vol. 6** (John Wiley & Sons, New York, 4th ed., 1993) pp 113-126.

Colorless, oily liquid with pungent odor. Fumes in air. mp −5.0°. d_4^{20} 1.3756. bp_{760} 220.8°. bp_{60} 129°; bp_{25} 105°; bp_{10} 89°. n_D^{20} 1.55789. Autoignition temp: 211°. Unstable. *Corrosive.* Hydrolyzes in the presence of moisture, forming benzoic and hydrochloric acids. Sol in alc, benzene, ether, many other organic solvents. Insol in water. LD_{50} orally in rats: 6.0 g/kg (Smyth).

Caution: Vapors may be stongly irritating and lacrimatory (Lin). This substance is reasonably anticipated to be a human carcinogen: *Report on Carcinogens, Twelfth Edition* (PB2011-111646, 2011) p 66.

USE: In dye chemistry. In organic syntheses (source of benzenyl group).

1113. Benzotrifluoride. [98-08-8] (Trifluoromethyl)benzene; α,α,α-trifluorotoluene; phenylfluoroform. $C_7H_5F_3$; mol wt 146.11. C 57.54%, H 3.45%, F 39.01%. Prepd by the action of hydrogen fluoride on benzotrichloride: Simons, Lewis, *J. Am. Chem. Soc.* **60**, 492 (1938); Simons, *Ind. Eng. Chem.* **32**, 178 (1940); by the action of antimony trifluoride on benzotrichloride: Swarts, *Bull. Sci. Acad. R. Belg.*, [4] **35**, 375 (1898); [5] **6**, 389 (1920); [5] **13**, 175 (1927); Holt *et al.*, US 2058453 (1937 to Kinetic Chemicals).

Liquid. mp −29.05°. d^{20} 1.1886. bp 103.46°. $n_D^{13.3}$ 1.41486. *Flammable.* Soluble in many organic solvents.

USE: In dye chemistry; in the manuf of substituted benzotrifluorides contg an ethylenic group, used in high polymer chemistry; in dielectric fluids, such as transformer oils.

1114. Benzoxonium Chloride. [19379-90-9] *N*-Dodecyl-*N,N*-bis(2-hydroxyethyl)benzenemethanaminium chloride (1:1); benzyldodecylbis(2-hydroxyethyl)ammonium chloride; dodecyldi-(β-hydroxyethyl)benzylammonium chloride; D-301; ZY-15021; Absonal; Bactofen; Bialcol; $C_{23}H_{42}ClNO_2$; mol wt 400.04. C 69.06%, H 10.58%, Cl 8.86%, N 3.50%, O 8.00%. Quaternary ammonium salt with antimicrobial activity. Prepn: CH 306648 (1955 to Ciba), *C.A.* **51**, 2023g (1957). Antimicrobial activity and toxicology: H. Goeth *et al.*, *Arzneim.-Forsch.* **9**, 622 (1959). Potentiometric determn: S. Pinzauti, E. La Porta, *Analyst* **102**, 938 (1977). Clinical evaluation in dental plaque control: U. P. Saxer *et al.*, *J. Clin. Periodontol.* **9**, 162 (1982). *In vitro* antibacterial spectrum: M. Cortat, P. Fels, *Arzneim.-Forsch.* **37**, 463 (1987). Clinical evaluation in pharyngeal infections: M. A. Weibel *et al.*, *ibid.* 467.

Colorless powder from ethyl ether, mp 107-109°. Sol in water, alcohol, benzene, toluene, chlorobenzene. LD_{50} orally in rats: 750 mg/kg (Goeth).

THERAP CAT: Antiseptic.

THERAP CAT (VET): Antiseptic.

1115. Benzoyl Chloride. [98-88-4] Benzenecarbonyl chloride. C_7H_5ClO; mol wt 140.57. C 59.81%, H 3.59%, Cl 25.22%, O 11.38%. Prepd by partial hydrolysis of benzotrichloride: Davies, Dick, *J. Chem. Soc.* **1932**, 2808; by chlorination of benzaldehyde: Wöhler, *Ann.* **3**, 262 (1832). Lab prepn from benzoic acid and thionyl chloride: A. I. Vogel, *Practical Organic Chemistry* (Longmans, London, 3rd ed., 1959) p 792; Gattermann-Wieland, *Praxis des Organischen Chemikers* (de Gruyter, Berlin, 40th ed., 1961) p 112. Microwave structural study: M. Onda *et al.*, *J. Mol. Struct.* **162**, 183 (1987). Derivatization of polyamines for HPLC determn: Y.-H. Mei, *J. Liq. Chromatogr.* **17**, 2413 (1994).

Liquid. Penetrating odor. d_4^{25} 1.2070. mp −1.0°. bp_{760} 197.2°; bp_{35} 100°; bp_{15} 82.3°; bp_9 71°; bp_3 49°. n_D^{20} 1.55369. Dipole moment 3.28. Flash pt 88°C (190.4°F). Parachor 289.8. *Corrosive.* Dec by water and alc. Miscible with ether, benzene, carbon disulfide, oils.

Caution: Lacrimator. Potential symptoms of overexposure may include irritation and burns of skin, eyes, mucous membranes, and respiratory tract.

USE: For acylation, i.e., introduction of the benzoyl group into alcohols, phenols, and amines (Schotten-Baumann reaction); in the manuf of benzoyl peroxide and of dye intermediates. In organic analysis for making benzoyl derivatives for identification purposes.

1116. Benzoylecgonine. [519-09-5] (1*R*,2*R*,3*S*,5*S*)-3-(Benzoyloxy)-8-methyl-8-azabicyclo[3.2.1]octane-2-carboxylic acid; 3β-hydroxy-1α*H*,5α*H*-tropane-2β-carboxylic acid benzoate; ecgonine benzoate. $C_{16}H_{19}NO_4$; mol wt 289.33. C 66.42%, H 6.62%, N 4.84%, O 22.12%. Major metabolite of cocaine, *q.v.* Isoln from coca leaves: De Jong, *Rec. Trav. Chim.* **42**, 980 (1923). Prepn from *l*-ecgonine and benzoic anhydride: *idem, ibid.* **66**, 544 (1947).

mp 199-201°.

Tetrahydrate. Orthorhombic prisms or needles from water, mp 86-92° (dec 195°, dry). $[\alpha]_D^{15}$ −45° (c = 3 in abs alc). Soluble in alc, hot H_2O.

1117. Benzoyl Isothiocyanate. [532-55-8] Benzoylthiocarbimide. C_8H_5NOS; mol wt 163.19. C 58.88%, H 3.09%, N 8.58%, O 9.80%, S 19.65%. Prepd by refluxing a mixture of KSCN and benzoyl chloride in benzene at 110-120°: Ambelang, Johnson, *J. Am. Chem. Soc.* **61**, 632 (1939). From NH_3SCN, acetone and benzoyl chloride: Frank, Smith, *Org. Synth.* **28**, 89 (1948).

Liquid; bp_{18} 133-137°; bp_{10} 119°. $d_4^{18.3}$ 1.2142; d_4^{16} 1.197. $n_D^{18.3}$ 1.6382. Reacts with a primary or secondary amine to form a benzoylthiourea, which is readily hydrolyzed by alkalies to the free thiourea.

1118. Benzoylpas. [13898-58-3] 4-(Benzoylamino)-2-hydroxybenzoic acid; *N*-benzoyl-*p*-aminosalicylic acid; 4-benzamidosalicylic acid. $C_{14}H_{11}NO_4$; mol wt 257.25. C 65.37%, H 4.31%, N

Consult the Name Index before using this section.

5.44%, O 24.88%. Prepn: Drain *et al., J. Chem. Soc.* **1949**, 1498; **GB 676363** (1952 to Wander). Prepn of sodium and calcium salts: Suddaby, Sumpter, **GB 711163** (1954 to Herst).

Crystals, mp 260-261°.
Calcium salt pentahydrate. [5631-00-5] Benzoylpas calcium; Therapas. $C_{28}H_{20}CaN_2O_8.5H_2O$; mol wt 642.63. White powder or crystals.
Mixture of calcium salt with isoniazid. Iso-Benzacyl.
THERAP CAT: Antibacterial (tuberculostatic).

1119.   Benzoyl Peroxide. [94-36-0] Dibenzoyl peroxide; benzoyl superoxide; Acetoxyl; Acnegel; Akneroxid L; Benoxyl; Benzac; Benzagel; Benzaknen; Brevoxyl; Debroxide; Desanden; Lucidol; Nericur; Oxy-5; PanOxyl; Peroxyderm; Persa-gel; Sanoxit; Theraderm; Xerac BP. $C_{14}H_{10}O_4$; mol wt 242.23. C 69.42%, H 4.16%, O 26.42%. Prepd by interaction of benzoyl chloride and a cooled soln of sodium peroxide. Laboratory procedure: A. I. Vogel, *Practical Organic Chemistry* (Longmans, London, 3rd ed., 1954) p 807; Gattermann-Wieland, *Praxis des Organischen Chemikers* (de Gruyter, Berlin, 40th ed., 1961) p 115. Comparative clinical study with clindamycin in acne vulgaris: L. J. Swinyer *et al., Br. J. Dermatol.* **119**, 615 (1988). Review of carcinogenic and allergenic potential: D. J. Hogan, *Int. J. Dermatol.* **30**, 467-470 (1991).

Crystals. mp 103-106°. *May explode when heated.* Sparingly sol in water or alcohol; sol in benzene, chloroform, ether. One gram dissolves in 40 ml carbon disulfide, in about 50 ml olive oil.
Caution: Potential symptoms of overexposure are irritation of skin, eyes and mucous membranes; sensitization dermatitis. *See NIOSH Pocket Guide to Chemical Hazards* (DHHS/NIOSH 97-140, 1997) p 26.
USE: Source of free radicals for industrial processes. Oxidizing agent in bleaching oils, flour, etc.; catalyst in the plastics industry; initiator in polymerization.
THERAP CAT: Keratolytic.
THERAP CAT (VET): Keratolytic.

1120.   3,4-Benzphenanthrene. [195-19-7] Benzo[c]phenanthrene. $C_{18}H_{12}$; mol wt 228.29. C 94.70%, H 5.30%. Synthesis by a Pschorr reaction from diazotized α-(2-naphthyl)-2-aminocinnamic acid: Cook, *J. Chem. Soc.* **1931**, 2524. From diphenylmethylsuccinic anhydride: Hewett, *J. Chem. Soc.* **1936**, 599. By double ring closure of β-benzohydrylglutaric acid: Newman, Joshel, *J. Am. Chem. Soc.* **60**, 487 (1938). From 4-keto-1,2,3,4-tetrahydrophenanthrene: Bachmann, Edgerton, *ibid.* **62**, 2970 (1940).

Needles from alcohol, mp 68° (needles or leaflets from petr ether, fine needles from alcohol + acetone). Absorption maxima: Mayneord, Roe, *Proc. R. Soc. London* **A158**, 63 (1937).

1121.   Benzphetamine. [156-08-1] (αS)-N,α-Dimethyl-N-(phenylmethyl)benzeneethanamine; N-benzyl-N,α-dimethylphenethylamine; d-N-methyl-N-benzyl-β-phenylisopropylamine. $C_{17}H_{21}N$; mol wt 239.36. C 85.31%, H 8.84%, N 5.85%. Prepd from d-desoxyephedrine: Heinzelman, Aspergren, **US 2789138** (1957 to Upjohn).

Liquid. $bp_{0.02}$ 127°. n_D^{19} 1.5515. Practically insol in water. Sol in methanol, ethanol, ether, chloroform, acetone, benzene.
Hydrochloride. [5411-22-3] Didrex; Inapetyl. $C_{17}H_{21}N.HCl$; mol wt 275.82. Crystals from ethyl acetate, mp 129-130°. Dextrorotatory. Sol in water, 95% ethanol.
Note: This is a controlled substance (stimulant): **21 CFR,** 1308.13.
THERAP CAT: Anorexic.

1122.   Benzquinamide. [63-12-7] 2-(Acetyloxy)-N,N-diethyl-1,3,4,6,7,11b-hexahydro-9,10-dimethoxy-2H-benzo[a]quinolizine-3-carboxamide; N,N-diethyl-1,3,4,6,7,11b-hexahydro-2-hydroxy-9,10-dimethoxy-2H-benzo[a]quinolizine-3-carboxamide acetate; 2-acetoxy-3-(N,N-diethylcarboxamido)-9,10-dimethoxy-1,2,-3,4,6,7-hexahydro-11bH-benzopyridocoline; 2-hydroxy-3-diethylcarbamyl-9,10-dimethoxy-1,2,3,4,6,7-hexahydro-11bH-benzoquinolizine acetate; BZQ; NSC-64375; P-2647; Emete-Con; Promecon; Quantril. $C_{22}H_{32}N_2O_5$; mol wt 404.51. C 65.32%, H 7.97%, N 6.93%, O 19.78%. Prepn: Trettner, **US 3053845** (1962 to Pfizer). Pharmacology: Scriabine *et al., J. Am. Med. Assoc.* **184**, 276 (1963); Kadzielawa, Gumulka, *Arch. Int. Pharmacodyn. Ther.* **163**, 139 (1966). Metabolic studies: Koe, Pinson, *J. Med. Chem.* **7**, 635 (1964); Wiseman *et al., Biochem. Pharmacol.* **13**, 1421 (1964). Toxicity: E. I. Goldenthal, *Toxicol. Appl. Pharmacol.* **18**, 185 (1971).

Crystals from diisopropyl ether, mp 130-131.5°. LD_{50} orally in rats: 990 mg/kg; i.p. in mice: 376 mg/kg (Goldenthal).
THERAP CAT: Antipsychotic, antiemetic.

1123.   Benzthiazide. [91-33-8] 6-Chloro-3-[[(phenylmethyl)-thio]methyl]-2H-1,2,4-benzothiadiazine-7-sulfonamide 1,1-dioxide; 3-[(benzylthio)methyl]-6-chloro-2H-1,2,4-benzothiadiazine-7-sulfonamide 1,1-dioxide; 3-[(benzylthio)methyl]-6-chloro-7-sulfamoyl-2H-benzo-1,2,4-thiadiazine 1,1-dioxide; 6-chloro-7-sulfamoyl-3-benzylthiomethyl-2H-1,2,4-benzothiadiazine 1,1-dioxide; benzothiazide; Exna; Fovane. $C_{15}H_{14}ClN_3O_4S_3$; mol wt 431.92. C 41.71%, H 3.27%, Cl 8.21%, N 9.73%, O 14.82%, S 22.27%. Prepn: J. M. McManus *et al., 136th Am. Chem. Soc. Meet.* (Atlantic City, Sept. 1959) *Abstr. of Papers,* pp 13-O; W. M. McLamore, G. D. Laubach, **DE 1137740**; *eidem,* **US 3440244** (1962, 1969 both to Pfizer). Pharmacology, toxicity: S. Y. P'an *et al., J. Pharmacol. Exp. Ther.* **128**, 122 (1960).

Crystals from acetone, mp 231-232° (U.S. patent); also reported as mp 238-239° (P'an). Bitter taste. Practically insol in water. Sol in alkaline solns. LD_{50} in mice, rats (mg/kg): >5000, >10000 orally; 410, 422 i.v. (P'an).

THERAP CAT: Diuretic, antihypertensive.

1124. Benztropine. [86-13-5] (3-*endo*)-3-(Diphenylmethoxy)-8-methyl-8-azabicyclo[3.2.1]octane; 3α-(diphenylmethoxy)-1α*H*,5α*H*-tropane; tropine benzohydryl ether. $C_{21}H_{25}NO$; mol wt 307.44. C 82.04%, H 8.20%, N 4.56%, O 5.20%. Anticholinergic. Prepd by the action of diphenyldiazomethane on tropine: Phillips, **US 2595405** (1952 to Merck & Co.). Crystal structure: P. G. Jones *et al.*, *Acta Crystallogr. B* **34**, 3125 (1978). Comparison with clozapine for Parkinson's related tremor: J. H. Friedman *et al.*, *Neurology* **48**, 1077 (1997). Evaluation of transdermal formulation: N. T. Hai *et al.*, *Int. J. Pharm.* **357**, 55 (2008).

Viscous colorless oil. Log P (octanol/water): 2.21.

Methanesulfonate. [132-17-2] Benztropine mesylate; Cogentin. $C_{21}H_{25}NO.CH_4O_3S$; mol wt 403.54. Crystals from acetone + ether, mp 143°. uv max: 259 nm (E_M = 437). Slightly hygroscopic. Freely sol in alc; very sol in water. Very slightly sol in ether. pH about 6.

THERAP CAT: Antiparkinsonian.

1125. Benzydamine. [642-72-8] *N,N*-Dimethyl-3-[[1-(phenylmethyl)-1*H*-indazol-3-yl]oxy]-1-propanamine; 1-benzyl-3-[3-(dimethylamino)propoxy]-1*H*-indazole; 1-benzyl-1*H*-indazol-3-yl 3-(dimethylamino)propyl ether; benzindamine. $C_{19}H_{23}N_3O$; mol wt 309.41. C 73.76%, H 7.49%, N 13.58%, O 5.17%. Prepn: **FR 1382855**; Palazzo, **US 3318905** (1964, 1967 both to Angelini Francesco); Palazzo *et al.*, *J. Med. Chem.* **9**, 38 (1966). Pharmacology: Lisciani *et al.*, *Eur. J. Pharmacol.* **3**, 157 (1968). Metabolism: Catanese *et al.*, *Arzneim.-Forsch.* **16**, 1354 (1966); Kataoka *et al.*, *Chem. Pharm. Bull.* **19**, 1511 (1971). Toxicology: B. Silvestrini *et al.*, *Toxicol. Appl. Pharmacol.* **10**, 148 (1967). Series of articles on pharmacology: *Arzneim.-Forsch.* **37**, 587-646 (1987).

$bp_{0.05}$ 160°.

Hydrochloride. [132-69-4] Afloben; Andolex; Benzyrin; Difflam; Enzamin; Imotryl; Opalgyne; Riripen; Salyzoron; Saniflor; Tamas; Tantum; Verax. $C_{19}H_{23}N_3O.HCl$; mol wt 345.87. Crystals, mp 160°. uv max: 306 nm ($E_{1cm}^{1\%}$ 160). Very sol in water; rather sol in ethanol, chloroform, *n*-butanol. LD_{50} in mice, rats (mg/kg): 110, 100 i.p.; 515, 1050 orally (Silvestrini).

THERAP CAT: Analgesic; anti-inflammatory.

THERAP CAT (VET): Anti-inflammatory.

1126. Benzyl Acetate. [140-11-4] Acetic acid phenylmethyl ester; acetic acid benzyl ester. $C_9H_{10}O_2$; mol wt 150.18. C 71.98%,

H 6.71%, O 21.31%. Occurs in a number of plants, particularly jasmine: S. Arctander, *Perfume and Flavor Materials of Natural Origin* (Elizabeth, N.J., 1960) pp 313-314. Prepd from benzyl chloride, acetic acid and sodium acetate and triethylamine: Merker, Scott, *J. Org. Chem.* **26**, 5180 (1961); Hennis *et al.*, *Ind. Eng. Chem. Prod. Res. Dev.* **6**, 193 (1967). Toxicity study: P. M. Jenner *et al.*, *Food Cosmet. Toxicol.* **2**, 327 (1964).

Liquid; pear-like odor. bp 213°, bp_{102} 134°. mp −51°. d_4^{25} 1.050. n_D^{20} 1.5232, n_D^{25} 1.4998. Flash pt, closed cup: 216°F (102°C). Practically insol in water. Misc with alcohol, ether. LD_{50} orally in rats: 2490 mg/kg (Jenner).

Caution: Ingestion can cause G.I. irritation with vomiting and diarrhea. Also irritating to skin, eyes, respiratory tract.

USE: In perfumery, solvent for cellulose acetate and nitrate.

1127. Benzyl Alcohol. [100-51-6] Benzenemethanol; phenylcarbinol; phenylmethanol; α-hydroxytoluene. C_7H_8O; mol wt 108.14. C 77.75%, H 7.46%, O 14.79%. Constituent of jasmine, hyacinth, ylang-ylang oils, Peru and Tolu balsams, storax, where it occurs in ester form also. Originally prepd by the Cannizzaro reaction from benzaldehyde + KOH: Cannizzaro, *Ann.* **88**, 129 (1853); Hickinbottom, *Reactions of Organic Compds.* (Longmans, London, 3rd ed., 1957) p 251; A. I. Vogel, *Practical Organic Chemistry* (Longmans, London, 3rd ed., 1959) p 711; Gattermann-Wieland, *Praxis des Organischen Chemikers* (de Gruyter, Berlin, 40th ed., 1961) p 193. Produced on a large scale by the action of sodium or potassium carbonate on benzyl chloride: **DE 484662**; *Chem. Zentralbl.* **1930**, I, 1052; *Frdl.* **16**, 426; *Kirk-Othmer Encyclopedia of Chemical Technology* **vol. 3** (Interscience, New York, 1964) pp 442-449. Toxicity: Smyth *et al.*, *Arch. Ind. Hyg. Occup. Med.* **4**, 119 (1951).

Liquid. Faint aromatic odor. Sharp burning taste. d_4^{20} 1.04535; d_4^{25} 1.04156. mp −15.19°. bp_{760} 204.7°; bp_{400} 183.0°; bp_{200} 160.0°; bp_{100} 141.7°; bp_{60} 129.3°; bp_{40} 119.8°; bp_{20} 105.8°; bp_{10} 92.6°; bp_5 80.8°; $bp_{1.0}$ 58.0°. n_D^{20} 1.54035; n_D^{25} 1.53837. *See:* Dreisbach, Martin, *Ind. Eng. Chem.* **41**, 2875 (1941). Absorption spectrum: Brode, *J. Phys. Chem.* **30**, 61 (1926). Vapor density 3.72 (air = 1.00). Flash pt, closed cup 213°F, open cup 220°F. Autoignition temp 817°F. One gram dissolves in about 25 ml water. One volume dissolves in 1.5 vols of 50% ethyl alcohol. Freely sol in 50% alc; sparingly sol in water. Misc with abs and 94% alcohol, ether, chloroform. Practically insol in glycerin. LD_{50} orally in rats: 3.1 g/kg (Smyth).

USE: Manuf other benzyl compds. Pharmaceutic aid (antimicrobial). Organic solvent for gelatin, casein (when hot), solvent for cellulose acetate, shellac. Used in perfumery and in flavoring (mostly in form of its aliphatic esters). In microscopy as embedding material.

THERAP CAT (VET): Has been used for relief from pruritis.

1128. Benzylamine. [100-46-9] Benzenemethanamine; aminotoluene; phenylmethylamine; moringine. C_7H_9N; mol wt 107.16. C 78.46%, H 8.47%, N 13.07%. Prepn from benzylchloride and ammonia: Mason, *J. Chem. Soc.* **63**, 1311 (1893); by redn of benzonitrile: Carothers, Jones, *J. Am. Chem. Soc.* **47**, 3051 (1925); from benzyl bromide + acetamide: Erikson, *Ber.* **59**, 2665 (1926); from *N*-benzylphthalimide + hydrazine hydrate: Ing, Manske, *J. Chem. Soc.* **129**, 2348 (1926). Identity with moringine: Chakravarti, *Bull. Calcutta Sch. Trop. Med.* **3**, 162 (1955); *C.A.* **50**, 16891e (1956).

335; by the action of bromine on dibenzyl ether: Lachman, *J. Am. Chem. Soc.* **45**, 2359 (1923).

Liquid; strongly alkaline reaction. bp 185°; bp$_{12}$ 90°. d$_4^{19}$ 0.983. n$_D^{20}$ 1.5401. Miscible with water, alcohol, ether.

Hydrochloride. C$_7$H$_9$N.HCl. Crystals, mp 253°.

Hydroiodide. C$_7$H$_9$N.HI. Leaflets, mp 162°.

Caution: Highly irritating to skin, mucous membranes.

USE: In organic synthesis.

1129. Benzylaniline. [103-32-2] *N*-Phenylbenzenemethanamine; *N*-phenylbenzylamine; benzylphenylamine. C$_{13}$H$_{13}$N; mol wt 183.25. C 85.21%, H 7.15%, N 7.64%. Prepn from benzyl alc and aniline in the presence of KOH: Sprinzak, *J. Am. Chem. Soc.* **78**, 3207 (1956); from benzaldehyde and aniline in the presence of NaBH$_4$: Schellenberg, *J. Org. Chem.* **28**, 3259 (1963).

Prisms, mp 37-38°. bp 306-307°. Practically insol in water; sol in alcohol, chloroform, ether.

1130. Benzyl Benzoate. [120-51-4] Benzoic acid phenylmethyl ester; benzoic acid benzyl ester; benzylbenzenecarboxylate; Acarosan; Antiscabiosum; Ascabiol. C$_{14}$H$_{12}$O$_2$; mol wt 212.25. C 79.22%, H 5.70%, O 15.08%. Contained in Peru and Tolu balsams. Prepd by the action of sodium benzylate on benzaldehyde: Kamm, Kamm, *Org. Synth.* **coll. vol. I**, 104 (2nd ed., 1941); by the dry esterification of sodium benzoate and benzyl chloride in the presence of triethylamine: Thorp, Nottorf, *Ind. Eng. Chem.* **39**, 1300 (1947). Toxicity studies: Graham, Kuizenga, *J. Pharmacol. Exp. Ther.* **84**, 358 (1945); Draize *et al.*, *J. Pharmacol. Exp. Ther.* **93**, 26 (1948). Comprehensive description: M. M. A. Hassan, J. S. Mossa, *Anal. Profiles Drug Subs.* **10**, 55-74 (1981).

Leaflets or oily liq; faint, pleasant, aromatic odor; sharp burning taste. mp 21°. d$_4^{25}$ 1.118. bp 323-324°. bp$_{16}$ 189-191°. bp$_{4.5}$ 156°. Sparingly volatile with steam. n$_D^{21}$ 1.5681. Misc with alc, chloroform, ether, oils. Practically insol in water, glycerin. LD$_{50}$ in rats, mice, rabbits, guinea pigs (g/kg): 1.7, 1.4, 1.8, 1.0 orally (Draize).

Caution: Direct contact may cause skin irritation. Potential symptoms of overexposure by ingestion in exptl animals are progressive incoordination, CNS excitation, convulsions. *See Clinical Toxicology of Commercial Products*, R. E. Gosselin *et al.*, Eds. (Williams & Wilkins, Baltimore, 5th ed., 1984) Section II, p 203.

USE: As solvent of cellulose acetate, nitrocellulose and artificial musk; substitute for camphor in celluloid and plastic pyroxylin compds; perfume fixative; in confectionery and chewing gum flavors.

THERAP CAT: Scabicide, pediculicide.

THERAP CAT (VET): Acaricide, pediculicide. *Contraindicated* in cats.

1131. Benzyl Bromide. [100-39-0] (Bromomethyl)benzene; α-bromotoluene; ω-bromotoluene. C$_7$H$_7$Br; mol wt 171.04. C 49.16%, H 4.13%, Br 46.72%. Prepd by the action of bromine on toluene in ultraviolet light: v. Konek, Loczka, *Ber.* **57**, 679 (1924); Zelinsky, **DE 478084**; *Chem. Zentralbl.* **1929**, II, 1216; *Frdl.* **16**,

Lacrimatory liquid. mp −3.9°. bp 198-199°. bp$_{80}$ 127°. d$_0^{22}$ 1.4380; d^{17} 1.443; d$_4^{64}$ 1.3886. *Poisonous, corrosive.* Slowly decomp by water.

Caution: Intensely irritating to skin, eyes, mucous membranes. Large doses cause CNS depression.

1132. Benzyl Chloride. [100-44-7] (Chloromethyl)benzene; α-chlorotoluene. C$_7$H$_7$Cl; mol wt 126.58. C 66.42%, H 5.57%, Cl 28.01%. Made by cautious chlorination of toluene: A. I. Vogel, *Practical Organic Chemistry* (Longmans, London, 3rd ed., 1959) p 538; Gattermann-Wieland, *Praxis des Organischen Chemikers* (de Gruyter, Berlin, 40th ed., 1961) p 92. Manuf: *Faith, Keyes & Clark's Industrial Chemicals*, F. A. Lowenheim, M. K. Moran, Eds. (Wiley-Interscience, New York, 4th ed., 1975) pp 145-148.

Very refractive liquid; rather unpleasant, irritating odor. d$_{20}^{20}$ 1.100. bp 179°. mp −48 to −43°. n$_D^{15}$ 1.5415. *Poisonous, corrosive.* Insol in water. Miscible with alcohol, chloroform, ether. Rapidly dec when heated in the presence of iron.

Caution: Potential symptoms of overexposure are irritation of eyes, skin and nose; weakness; irritability; headache; skin eruption; pulmonary edema. *See NIOSH Pocket Guide to Chemical Hazards* (DHHS/NIOSH 97-140, 1997) p 28.

USE: Manuf benzyl compds, perfumes, pharmaceutical products, dyes, synthetic tannins, artificial resins.

1133. Benzyl Cinnamate. [103-41-3]; [78277-23-3] (*trans*-form). 3-Phenyl-2-propenoic acid phenylmethyl ester; cinnamein. C$_{16}$H$_{14}$O$_2$; mol wt 238.29. C 80.65%, H 5.92%, O 13.43%. Constituent of storax, Peru and Tolu balsams: Tschirch, Trog, *Arch. Pharm.* **232**, 70 (1894); Tschirch, Oberländer, *ibid.* 559. Prepn: E. H. Volwiler, E. B. Vliet, *J. Am. Chem. Soc.* **43**, 1672 (1921); E. L. Eliel, R. P. Anderson, *ibid.* **74**, 547 (1952). Stereoselective synthesis of *trans*-form: K. Zeitler, *Org. Lett.* **8**, 637 (2006). Toxicity study: P. M. Jenner *et al.*, *Food Cosmet. Toxicol.* **2**, 327 (1964). Review of properties and use as fragrance ingredient: S. P. Bhatia *et al.*, *Food Chem. Toxicol.* **45**. S40-S48 (2007).

trans-form

Crystals from 95% ethanol; sweet odor of balsam. mp 39°; also reported as mp 33-34° (Volwiler, Vliet). Dec on distillation at ordinary pressure; bp$_{0.5}$ 154-157°, bp$_5$ 195-200°, bp$_{22}$ 228-230°. Practically insol in water, propylene glycol and glycerin. Sol in alc, ether, oils. LD$_{50}$ orally in rats: 5530 mg/kg (Jenner).

USE: Fragrance ingredient; in foods as flavoring agent.

1134. Benzyl Cyanide. [140-29-4] Benzeneacetonitrile; phenylacetonitrile; α-tolunitrile; ω-cyanotoluene. C$_8$H$_7$N; mol wt 117.15. C 82.02%, H 6.02%, N 11.96%. Occurs in garden cress and other plants; made from benzyl chloride, and NaCN: Adams, Thal, *Org. Synth.* **vol. 2**, 9 (1922); **coll. vol. I**, 101 (107 in 2nd ed.).

Oily liquid, aromatic odor. d_{15}^{15} 1.0214. mp $-23.8°$. bp_{760} 233.5°; bp_{100} 161.8°; bp_{20} 119.4°; $bp_{1.0}$ 60°. n_D^{25} 1.52105. Insoluble in water, miscible with alc, ether.

1135. Benzyl Ether. [103-50-4] 1,1'-[Oxybis(methylene)]-bis[benzene]; dibenzyl ether. $C_{14}H_{14}O$; mol wt 198.27. C 84.81%, H 7.12%, O 8.07%. Prepd: Lachman, *J. Am. Chem. Soc.* **45**, 2356 (1923); Staab, Wendel, *Ber.* **93**, 2902 (1960); Lichtenberger, Tritsch, *Bull. Soc. Chim. Fr.* **1961**, 363. Manuf by reduction of benzaldehyde in the presence of $[Co(CO)_4]_2$: Wender, Orchin, **US 2614107** (1952 to U.S.A. as represented by the Secy. of Agr.). Physical properties: Svirbely *et al.*, *J. Am. Chem. Soc.* **71**, 507 (1949); Dreisbach, Martin, *Ind. Eng. Chem.* **41**, 2875 (1949). Miscibility: Jackson, Drury, *ibid.* **51**, 1491 (1959).

Unstable liquid, bp 295-298° (with dec), bp_{21} 173-174°; bp_2 125.5-126.5°. Appears to dec slowly at ordinary temps. d^{35} 1.0341; d_4^4 0.99735; d_4^{20} 1.00142; d^{15} 1.0482. n_D^{25} 1.5601 (Svirbely *et al.*), 1.53851 (Dreisbach, Martin); n_D^{20} 1.54057 (Dreisbach, Martin), 1.566 (Lichtenberger, Tritsch). Practically insol in water; miscible with ethanol, ether, chloroform, acetone.

USE: Plasticizer for nitrocellulose; solvent in perfumery.

1136. Benzyl Ethyl Ether. [539-30-0] (Ethoxymethyl)benzene. $C_9H_{12}O$; mol wt 136.19. C 79.37%, H 8.88%, O 11.75%. Preparation from sodium ethoxide and benzyl bromide: Letsinger, Pollart, *J. Am. Chem. Soc.* **78**, 6079 (1956); by reduction of benzaldehyde diethyl acetal with LiAlH$_4$-AlCl$_3$: Eliel, Rerick, *J. Org. Chem.* **23**, 1088 (1958).

Oily liquid, aromatic odor. bp 186°; bp_{10} 65°. d 0.949, n_D^{20} 1.4955. Volatile with steam. Practically insol in water; miscible with alcohol, ether.

1137. Benzyl Formate. [104-57-4] Formic acid phenylmethyl ester; formic acid benzyl ester. $C_8H_8O_2$; mol wt 136.15. C 70.58%, H 5.92%, O 23.50%. Prepn from formic acid and benzyl alcohol: Mailhe, *Chem. Ztg.* **35**, 508 (1911).

Liquid; pleasant fruity odor. d 1.081. bp 203°. Practically insol in water. Sol in alcohol.

USE: Solvent for cellulose esters; in perfumery.

1138. Benzyl Fumarate. [538-64-7] (2E)-2-Butenedioic acid 1,4-bis(phenylmethyl) ester; fumaric acid dibenzyl ester; dibenzyl fumarate. $C_{18}H_{16}O_4$; mol wt 296.32. C 72.96%, H 5.44%, O 21.60%. Prepd from fumaric acid and benzyl alcohol: Volwiler, Vliet, *J. Am. Chem. Soc.* **43**, 1672 (1921).

Cryst powder, mp 58.5-59.5°. bp_5 210-211°. Practically insol in water. Sol in alcohol, chloroform, ether, oils.

USE: In room spray deodorant: Kulka, **US 3077457** (1963 to Fritzsche Bros.).

1139. Benzylhydrochlorothiazide. [1824-50-6] 6-Chloro-3,4-dihydro-3-(phenylmethyl)-2*H*-1,2,4-benzothiadiazine-7-sulfonamide 1,1-dioxide; 3-benzyl-6-chloro-3,4-dihydro-2*H*-1,2,4-benzothiadiazine-7-sulfonamide 1,1-dioxide; 6-chloro-7-sulfamoyl-3-benzyl-3,4-dihydro-1,2,4-benzothiadiazine 1,1-dioxide; 3-benzyl-6-chloro-3,4-dihydro-7-sulfamoyl-1,2,4-benzothiadiazine 1,1-dioxide; Behyd. $C_{14}H_{14}ClN_3O_4S_2$; mol wt 387.85. C 43.36%, H 3.64%, Cl 9.14%, N 10.83%, O 16.50%, S 16.53%. Prepn: Werner *et al.*, *J. Am. Chem. Soc.* **82**, 1161 (1960); Novello *et al.*, *J. Org. Chem.* **25**, 970 (1960); Ugi, **US 3108097** (1963).

Crystals from acetic acid + water, mp 260-262°. Also reported as crystals from water, mp 269°.

THERAP CAT: Antihypertensive; diuretic.

1140. Benzylideneacetone. [122-57-6]; [1896-62-4] ((*E*)-form); [937-53-1] ((*Z*)-form). 4-Phenyl-3-buten-2-one; acetocinnamone; benzalacetone; cinnamyl methyl ketone; methyl styryl ketone. $C_{10}H_{10}O$; mol wt 146.19. C 82.16%, H 6.90%, O 10.94%. Prepn from acetone and benzaldehyde: L. Claisen, A. Claparede, *Ber.* **14**, 2460 (1881); K. V. Auwers, *ibid.* **45**, 2764 (1912); N. L. Drake, P. Allen, Jr., *Org. Synth.* **3**, 17 (1923). Purification by steam distillation: Fromm, Haas, *Ann.* **394**, 291 (1912). Absorption spectrum: F. Baker, *J. Chem. Soc., Trans.* **91**, 1490 (1907); E. C. C. Baly, K. Schafer, *ibid.* **93**, 1808 (1908). Prepn of isomers: G. Gamboni *et al.*, *Helv. Chim. Acta* **38**, 255 (1955).

Lustrous plates on vacuum distillation. Coumarin type odor; sweet, pungent taste. *Irritant*. mp 41.5°. d_{15}^{15} 1.0377; $d_4^{45.2}$ 1.0097. bp_{760} 261°; bp_{200} 211°; bp_{100} 187.8°; bp_{40} 161.3°; bp_{20} 143.8°; bp_{10} 127.4°; bp_5 112.2°; $bp_{1.0}$ 81.7°. $n_D^{45.9}$ 1.5836. Freely sol in alcohol, benzene, chloroform, ether; sparingly sol in water, petr ether.

USE: In perfumery; as flavoring agent. Reagent in organic syntheses.

1141. Benzylideneaniline. [538-51-2]; [1750-36-3] ((*E*)-form); [33993-35-0] ((*Z*)-form). N-(Phenylmethylene)benzenamine; benzalaniline; benzaldehyde N-phenylimine; N-phenylbenzaldimine. $C_{13}H_{11}N$; mol wt 181.24. C 86.15%, H 6.12%, N 7.73%. Prepn from aniline and benzaldehyde: O. Hinsberg, *Ber.* **20**, 1585 (1887); A. Hantzsch, O. Schwab, *ibid.* **34**, (1901); L. A. Bigelow, H. Eatough, *Org. Synth.* **coll. vol. I** (2nd ed., 1941) p 80. Molecular structure: M. Traetteberg *et al.*, *J. Mol. Struct.* **48**, 395 (1978). Absorption spectrum: Baly *et al.*, *J. Chem. Soc.* **97**, 590 (1910).

Crystals from 85% ethanol, mp 52°. *Irritant*. d_4^{50} 1.045. bp_{760} 300°. Sol in alcohol, chloroform, acetic anhydride, carbon disulfide.

USE: Reagent in organic synthesis.

1142. Benzylimidobis(*p*-methoxyphenyl)methane. [524-96-9] N-[Bis(4-methoxyphenyl)methylene]benzenemethanamine. $C_{22}H_{21}NO_2$; mol wt 331.42. C 79.73%, H 6.39%, N 4.23%, O 9.65%. Prepd by heating *p,p'*-dimethoxybenzophenone, thionyl chloride and benzylamine: Schönberg, Urban, *Ber.* **67**, 1999 (1934).

Pale-yellow crystals, mp 89-91°. Soluble in ether, chloroform. Slightly sol in petr ether.

USE: Detection of elementary sulfur.

1143. Benzyl Methyl Ether. [538-86-3] (Methoxymethyl)-benzene; methyl benzyl ether. $C_8H_{10}O$; mol wt 122.17. C 78.65%, H 8.25%, O 13.10%. Prepd from benzyl chloride, methanol, and NaOH: Olson *et al.*, *J. Am. Chem. Soc.* **69**, 2451 (1947).

Liquid. d 0.987. bp 174°. Practically insol in water. Sol in alcohol, ether.

1144. *o*-Benzylphenol. [28994-41-4] 2-(Phenylmethyl)phenol; α-phenyl-*o*-cresol; (2-hydroxydiphenyl)methane. $C_{13}H_{12}O$; mol wt 184.24. C 84.75%, H 6.57%, O 8.68%. Prepn from benzyl bromide and sodium phenoxide: Kornblum, Lurie, *J. Am. Chem. Soc.* **81**, 2711 (1959).

Crystals or liquid. mp 20.2-20.9°. bp$_{10}$ 154-156°; bp$_{1.0}$ 121-123°. n_D^{20} 1.59945. Practically insol in water. Sol in organic solvents, in fixed alkali hydroxide solns.

1145. *p*-Benzylphenol. [101-53-1] 4-(Phenylmethyl)phenol; α-phenyl-*p*-cresol; (4-hydroxydiphenyl)methane. $C_{13}H_{12}O$; mol wt 184.24. C 84.75%, H 6.57%, O 8.68%. Prepd from phenol and benzyl chloride in the presence of zinc chloride: Ziegenbein *et al.*, *Ber.* **88**, 1906 (1955).

Crystals, mp 84°. bp 322°; bp$_4$ 154-157°. Slightly sol in cold water; moderately sol in hot water; sol in organic solvents, glacial acetic acid, alkali hydroxide solns.

USE: Germicide, antiseptic, preservative; also in organic syntheses.

1146. Benzyl Salicylate. [118-58-1] 2-Hydroxybenzoic acid phenylmethyl ester; salicylic acid benzyl ester. $C_{14}H_{12}O_3$; mol wt 228.25. C 73.67%, H 5.30%, O 21.03%. Prepd from sodium salicylate and benzyl chloride: Volwiler, Vliet, *J. Am. Chem. Soc.* **43**, 1672 (1921).

Thick liquid, slight, pleasant odor. d^{20} 1.175. bp$_{25}$ 208°. Slightly sol in water; miscible with alcohol or ether.

USE: As fixer in perfumery; in sunscreen preparations.

1147. Benzyl Sulfide. [538-74-9] 1,1'-[Thiobis(methylene)]-bisbenzene; dibenzylsulfide. $C_{14}H_{14}S$; mol wt 214.33. C 78.46%, H 6.58%, S 14.96%. Prepn: Runge *et al.*, *J. Prakt. Chem.* **11**, 284 (1960). Manuf from benzyl chloride and Na$_2$S: Stucker, Brennan, **US 2755305** (1956 to Pure Oil Co.).

Plates, mp 49°. Practically insol in water. Sol in alc, ether.

1148. *S*-Benzylthiourea. [621-85-2] Carbamimidothioic acid phenylmethyl ester; 2-benzyl-2-thiopseudourea; *S*-benzylisothiourea. $C_8H_{10}N_2S$; mol wt 166.24. C 57.80%, H 6.06%, N 16.85%, S 19.29%. Prepn from benzyl chloride and thiocarbamide: E. A. Werner, *J. Chem. Soc., Trans.* **57**, 283 (1890); and use in separation of organic acids: J. J. Donleavy, *J. Am. Chem. Soc.* **58**, 1004 (1936).

White, microcrystalline powder, mp 88° (dec). Freely sol in chloroform, benzene, ether, alcohol.

Hydrochloride. [538-28-3] *S*-Benzylthiuronium chloride. $C_8H_{10}N_2S\cdot HCl$; mol wt 202.70. Crystals from alc or dil HCl, mp 172-174°. Metastable form, mp 146-148°.

USE: Reagent for cobalt and nickel; identification and separation of carboxylic, sulfinic, and sulfonic acid.

1149. Benzylurea. [538-32-9] *N*-(Phenylmethyl)urea; benzylcarbamide. $C_8H_{10}N_2O$; mol wt 150.18. C 63.98%, H 6.71%, N 18.65%, O 10.65%. Prepn: Neville, McGee, *Can. J. Chem.* **41**, 2123 (1963).

Crystals, mp 147-148°; dec at 200°. One gram dissolves in 60 ml warm water, 33 ml acetone; slightly sol in benzene, ether.

1150. Bephenium Hydroxynaphthoate. [3818-50-6] *N*,*N*-Dimethyl-*N*-(2-phenoxyethyl)benzenemethanaminium salt with 3-hydroxy-2-naphthalene carboxylic acid (1:1); bephenium embonate; Lecibis. $C_{28}H_{29}NO_4$; mol wt 443.54. C 75.82%, H 6.59%, N 3.16%, O 14.43%. Prepn: F. C. Copp, **US 2918401** (1959 to Burroughs Wellcome). Mode of action: G. C. Coles *et al.*, *Experientia* **30**, 1265 (1974). Spectrophotometric determn: C. S. P. Sastry, M. Aruna, *Pharmazie* **43**, 361 (1988). Efficacy against strongylosis in elephants: K. Chandrasekharan *et al.*, *Kerala J. Vet. Sci.* **13**, 15 (1982). Clinical trial in hookworm infections: J. Nahmias *et al.*, *Ann. Trop. Med. Parasitol.* **83**, 625 (1989).

Crystals, mp 170-171°. Absorption max (water): 600 nm.

THERAP CAT: Anthelmintic (Nematodes).

THERAP CAT (VET): Anthelmintic (Nematodes).

1151. Bepotastine. [190786-43-7] 4-[(S)-(4-Chlorophenyl)-2-pyridinylmethoxy]-1-piperidinebutanoic acid; betotastine. C_{21}-$H_{25}ClN_2O_3$; mol wt 388.89. C 64.86%, H 6.48%, Cl 9.12%, N 7.20%, O 12.34%. Histamine H_1-receptor antagonist. Prepn (stereochem. unspec.): A. Koda et al., **EP 335586**; eidem, **US 4929618** (1989, 1990 both to Ube). Prepn of optically active salts: J. Kita et al., **EP 949260** (1999 to Ube; Tanabe Seiyaku). Pharmacology: M. Kato et al., Arzneim.-Forsch. **47**, 1116 (1997). Suppression of IL-5 production: O. Kaminuma et al., Biol. Pharm. Bull. **21**, 411 (1998). Antiallergic activity in animal models: M. Ueno et al., Pharmacology **57**, 206 (1998). Clinical evaluation in pollen-induced allergic rhinitis: K. Hashiguchi et al., Expert Opin. Pharmacother. **10**, 523 (2009).

Benzenesulfonate salt. [190786-44-8] Bepotastine besilate; TAU-284; Bepreve; Talion. $C_{21}H_{25}ClN_2O_3 \cdot C_6H_6O_3S$; mol wt 547.06. Pale grey prisms from acetonitrile, mp 161-163°. $[\alpha]_D^{20}$ +6.0° (c = 5 in methanol).

THERAP CAT: Antihistaminic.

1152. Bepridil. [64706-54-3] β-[(2-Methylpropoxy)methyl]-N-phenyl-N-(phenylmethyl)-1-pyrrolidineethanamine; 1-[2-(N-benzylanilino)-1-(isobutoxymethyl)ethyl]pyrrolidine; 1-isobutoxy-2-pyrrolidino-3-N-benzylanilinopropane; 3-isobutoxy-2-pyrrolidino-N-phenyl-N-benzylpropylamine. $C_{24}H_{34}N_2O$; mol wt 366.55. C 78.64%, H 9.35%, N 7.64%, O 4.36%. Calcium channel blocker with antianginal and antiarrhythmic properties. Prepn: R. Y. Mauvernay et al., **DE 2310918**; eidem, **US 3962238**; reissued with structure correction: N. Busch et al., **US RE 30577** (1973, 1976, 1981 all to CERM). See also: R. Y. Mauvernay et al., **DE 2802864** (1978 to CERM), C.A. **89**, 179852s (1978). Cardiovascular pharmacology in dogs: M.-T. Michelin et al., Thérapie **32**, 485 (1977). Mechanism of action studies: S. Vogel et al., J. Pharmacol. Exp. Ther. **210**, 378 (1979); C. Labrid et al., ibid. **211**, 546 (1979). Clinical comparison of hemodynamic and coronary changes: J. P. Merillon et al., Thérapie **36**, 123 (1981). Clinical evaluation in angina: M. K. Sharma et al., Am. J. Cardiol. **61**, 1210 (1988).

Viscous liquid, $bp_{0.1}$ 184°, $bp_{0.5}$ 192°. n_D^{20} 1.5538.

Hydrochloride monohydrate. [74764-40-2] CERM-1978; Cordium; Vascor; Unicordium. $C_{24}H_{34}N_2O \cdot HCl \cdot H_2O$; mol wt 421.02. Crystals, mp 91 ±2°. LD_{50} in mice (mg/kg): 1955 orally, 23.5 i.v. (Mauvernay, 1978).

THERAP CAT: Antianginal.

1153. Beractant. [108778-82-1] Surfactant TA; A-60386X; Surfacten; Survanta. Organic solvent extract of bovine lung, consisting mostly of phospholipids, modified by the addition of colfosceril palmitate, palmitic acid and tripalmitin, q.q.v. Prepn: **JP 83 45299**; Y. Tanaka et al., **US 4397839** (both 1983 to Tokyo Tanabe). Effect of protein component on surface activity: S. B. Hall et al., Am. Rev. Respir. Dis. **145**, 24 (1992). Pharmacology: J. J. Cummings et al., ibid. 999. Metabolism in rabbits: J. F. Lewis et al., ibid. 19. Evaluation of establishment of normal surfactant function in comparison with Exosurf: E. M. Scarpelli et al., Am. J. Perinatol. **9**, 414 (1992). Clinical trial in neonatal respiratory distress syn-

drome: M. D. Reller et al., Am. J. Dis. Child. **145**, 1017 (1991); E. M. Zola et al., Pediatrics **91**, 546 (1993).

THERAP CAT: Pulmonary surfactant.

1154. Beraprost. [88430-50-6] 2,3,3a,8b-Tetrahydro-2-hydroxy-1-(3-hydroxy-4-methyl-1-octen-6-ynyl)-1H-cyclopenta(b)-benzofuran-5-butanoic acid; dl-16-methyl-18,19-tetradehydro-5,6,7-trinor-4,8-inter-m-phenyleneprostaglandin I_2; dl-4-[(1R,2R,3aS,8bS)-1,2,3a,8b-tetrahydro-2-hydroxy-[(3S,4RS)-3-hydroxy-4-methyl-oct-6-yne-(E)-1-enyl]-5-cyclopenta(b)benzofuranyl]butanoic acid; befaprost; ML-1229. $C_{24}H_{30}O_5$; mol wt 398.50. C 72.34%, H 7.59%, O 20.07%. Platelet aggregation inhibitor; stable analog of prostacyclin, q.v. Prepn: K. Ohno et al., **EP 84856**; eidem, **US 4474802** (1983, 1984 both to Toray); eidem, Adv. Prostaglandin Thromboxane Leukotriene Res. **15**, 279 (1985). PGI$_2$ receptor binding: N. Kajikawa et al., Arzneim.-Forsch. **39**, 495 (1989). Series of articles on pharmacology: ibid. 856-876. Clinical pharmacokinetics: J.-L. Demolis et al., J. Cardiovasc. Pharmacol. **22**, 711 (1993). Clinical trial in intermittent claudication: M. Lievre et al., ibid. **27**, 788 (1996); in diabetic angiopathy: Y. Okuda et al., Prostaglandins **52**, 375 (1996).

Sodium salt. [88475-69-8] ML-1129; TRK-100; Dorner; Procylin. $C_{24}H_{29}NaO_5$; mol wt 420.48.

THERAP CAT: Antithrombotic; vasodilator (peripheral).

1155. Berbamine. [478-61-5] (4aS,16aR)-3,4,4a,5,16a,17,-18,19-Octahydro-21,22,26-trimethoxy-4,17-dimethyl-16H-1,24:6,9-dietheno-11,15-metheno-2H-pyrido[2',3':17,18][1,11]dioxacycloeicosino[2,3,4-ij]isoquinolin-12-ol; 6,6',7-trimethoxy-2,2'-dimethylberbaman-12-ol; berbenine. $C_{37}H_{40}N_2O_6$; mol wt 608.74. C 73.00%, H 6.62%, N 4.60%, O 15.77%. Alkaloid found in various members of Berberis family. Used as a traditional medicine in China as antipyretic, antidiarrheoa and as a tonic. Identification in Berberis vulgaris: C. Rüdel, Arch. Pharm. **229**, 631 (1895); isoln from Atherosperma moschatum Labill., Monimiaceae: I. R. C. Bick et al., Aust. J. Chem. **9**, 111 (1956). Structure: F. von Bruchhausen et al., Ann. **507**, 144 (1933); configuration: Y. Inubishi, J. Pharm. Soc. Jpn. **72**, 220 (1952), C.A. **47**, 6429e (1953). Immunopharmacology and toxicology: C. W. Wong et al., Int. J. Immunopharmacol. **13**, 579 (1991). Brief review: C.-X. Liu et al., Phytother. Res. **5**, 228-230 (1991).

Colorless plates from aq ethanol as the monohydrate, sinters 147°. pKa (20°): 7.33 in aq methanol. $[\alpha]_D^{20}$ +114.6° (c = 1.0 in chloroform). uv max (ethanol): 282 nm (log ε 3.89; E $_{1cm}^{1\%}$ 126).

1156. Berberine. [2086-83-1] 5,6-Dihydro-9,10-dimethoxy-benzo[g]-1,3-benzodioxolo[5,6-a]quinolizinium; 7,8,13,13a-tetradehydro-9,10-dimethoxy-2,3-(methylenedioxy)berbinium; umbellatine. $[C_{20}H_{18}NO_4]^+$. Tautomeric alkaloid widely distributed in nine or more botanical famiies but occurs most frequently in *Berberidaceae*. A component of many traditional medicines, first isolated in 1826. Structure: Perkin, Robinson, *J. Chem. Soc.* **97**, 305 (1910). Biosynthesis: J. R. Gear, I. D. Spenser, *Can. J. Chem.* **41**, 783 (1963). Pharmacology: Fukuda *et al., Chem. Pharm. Bull.* **18**, 1299 (1970). Total synthesis: Kametani *et al., J. Chem. Soc. C* **1969**, 2036. NMR: G. Blasko *et al., Heterocycles* **27**, 911 (1988). HPLC determn in biological fluids: C.-M. Chen, H.-C. Chang, *J. Chromatogr. B* **665**, 117 (1995). Clinical evaluation in malaria: W. D. Sheng *et al., East Afr. Med. J.* **74**, 283 (1997). Review of biological activities: F. E. Hahn, J. Ciak, in *Antibiotics* vol. 3, D. Gottlieb *et al.*, Eds. (Springer-Verlag, New York, 1975) pp 577-584; of chemistry, distribution and use: R. S. Thakur, S. K. Srivastava, *Curr. Res. Med. Aromat. Plants* **4**, 249-272 (1982). Review of antidiarrheal action: A. W. Baird *et al., Adv. Drug Delivery Rev.* **23**, 111-120 (1997).

Yellow needles from ether, mp 145°: Gadamer, *Arch. Pharm.* **243**, 33 (1905). uv max: 265, 343 nm. pK 2.47.

Acid sulfate. Berberine bisulfate. $C_{20}H_{19}NO_8S$. Yellow needles. Sol in about 100 parts water; slightly sol in alc.

Chloride dihydrate. $C_{20}H_{18}NO_4.Cl.2H_2O$. Yellow crystals. Slightly sol in cold, freely in boiling water. Practically insol in cold alcohol, chloroform, ether.

Sulfate trihydrate. Yellow needles. Sol in about 30 parts water; sol in alcohol. Activity studies: Amin *et al., Can. J. Microbiol.* **15**, 1067 (1969). Pharmacology: Sabir, Bhide, *Indian J. Physiol. Pharmacol.* **15**, 111 (1971).

THERAP CAT: Antiprotozoal (Leishmania); antimalarial; antibacterial; antidiarrheal.

1157. Berberis. Holly-leaved barberry; Oregon grape root; mountain grape. Dried rhizome and roots of species of section of *Mahonia* DC. of the genus *Berberis* L., *Berberidaceae. Habit.* U.S. and British Columbia. *Constit.* Berberine, berbamine, oxyacanthine, phytosterol, sugar.

THERAP CAT: Bitter. Antipyretic.

1158. Berberis Aristata. Indian barberry; ruswut; rusat. Dried stem of *Berberis aristata* DC., *Berberidaceae. Habit.* India, Ceylon. *Constit.* Berberine and other alkaloidal substances, tannin, resin, starch.

THERAP CAT: Bitter. Antipyretic.

1159. Bergamottin. [7380-40-7] 4-[[(2E)-3,7-Dimethyl-2,6-octadien-1-yl]oxy]-7H-furo[3,2-g][1]benzopyran-7-one; bergamotine; bergaptin; 5-geranoxypsoralen. $C_{21}H_{22}O_4$; mol wt 338.40. C 74.54%, H 6.55%, O 18.91%. Furanocoumarin found in bergamot, grapefruit, and other citrus fruits. Inactivates cytochrome P450 3A4 (CYP3A4) and affects the oral bioavailability of drugs metabolized by this enzyme. Isoln from bergamot oil: E. Späth, P. Kainrath, *Ber.* **70**, 2272 (1937); from lemon oil: W. L. Stanley, S. H. Vannier, *J. Am. Chem. Soc.* **79**, 3488 (1957); from grapefruit peel oil: J. F. Fisher, H. E. Nordby, *J. Food Sci.* **30**, 869 (1965). Synthesis: A. Chatterjee, B. Chaudhury, *J. Chem. Soc.* **1961**, 2246. Configuration of side chain: R. B. Bates *et al., Tetrahedron Lett.* **4**, 1683 (1963). HPLC determn in grapefruit tissues and juice: W. V. De Castro *et al., J. Agric. Food Chem.* **54**, 249 (2006). LC-MS/MS determn in plasma: Y. Uesawa *et al., Pharmazie* **63**, 110 (2008). Photodegradation and photosensitizing properties: P. Morlière *et al., J. Photochem. Photobiol. B* **7**, 199 (1990); idem *et al., Photochem. Photobiol.* **53**, 13 (1991). Identification of photodegradation products: M.-T. Martin *et al., ibid.* **57**, 222 (1993). Inactivation of CYP3A4: K. He

et al., Chem. Res. Toxicol. **11**, 252 (1998). Effect on felodipine bioavailability: D. G. Bailey *et al., Clin. Pharmacol. Ther.* **68**, 468 (2000); T. C. Goosen *et al., ibid.* **76**, 607 (2004).

Stout rods from ethanol, mp 60°. uv max: 220, 250, 309 nm (log ε 4.41, 4.28, 4.17). Fluorescence excitation max: 320 nm; emission max: 485 nm. Readily sol in alcohol; poorly sol in water. Degraded by uv light.

Dihydroxybergamottin. [145414-76-2] 6′,7′-Dihydroxybergamottin; 6,7-DHB. $C_{21}H_{24}O_6$; mol wt 372.42. Synthesis and activity as CYP3A4 inhibitor: F. H. Bellevue III *et al., Bioorg. Med. Chem. Lett.* **7**, 2593 (1997); E. C. Row *et al., Org. Biomol. Chem.* **4**, 1604 (2006). Small colorless needles, mp 112-113°.

1160. Bergapten. [484-20-8] 4-Methoxy-7H-furo[3,2-g][1]-benzopyran-7-one; 5-methoxypsoralen; bergaptan; bergaptene; heraclin; majudin; 5-MOP; Psoraderm-5. $C_{12}H_8O_4$; mol wt 216.19. C 66.67%, H 3.73%, O 29.60%. Naturally occurring analog of psoralen and isomer of methoxsalen, *q.q.v.*, found in a wide variety of plants. It was first isolated from oil of bergamot from *Citrus bergamia* Risso, *Aurantiodiae:* Pomeranz, *Monatsh. Chem.* **12**, 379 (1891), **14**, 28 (1893). Isoln from *Fagara xanthoxyloides* Lam., *Rutaceae:* H. Thoms, E. Baetcke, *Ber.* **44**, 3326 (1911); *ibid.* **45**, 3705 (1912). Synthesis: E. Späth *et al., Ber.* **70**, 478 (1937); W. N. Howell, R. Robertson, *J. Chem. Soc.* **1937**, 293; G. Caporale, *Farmaco Ed. Sci.* **13**, 784 (1958); V. K. Ahluwalia *et al., Indian J. Chem.* **7**, 831 (1969). Use in photochemotherapy of psoriasis: H. Hönigsmann *et al., Br. J. Dermatol.* **101**, 369 (1979). Mutagenicity studies: B. R. Scott *et al., Mutat. Res.* **39**, 29 (1976); M. J. Ashwood-Smith *et al., Nature* **285**, 407 (1980). Phototoxicity study: A. Kornhauser *et al., Science* **217**, 733 (1982).

Needles from alcohol, mp 188° (sublimes). Practically insol in boiling water. Slightly sol in glacial acetic acid, chloroform, benzene, warm phenol. Sol in abs alcohol: 1 part in 60. Its soln in sulfuric acid is yellow-gold.

USE: Has been used to promote tanning in suntan preparations.

THERAP CAT: Antipsoriatic.

1161. Berkelium. [7440-40-6] Bk; at. no. 97; valence 3, 4. Man-made radioactive element; second element in the curide series. No stable nuclides, known isotopes (mass numbers): 240, 242-251. Longest-lived known isotope: 247 ($T_{1/2}$ 1.38×10^3 years, α-decay, rel. at. mass 247.0703). Prepn of first isotope, ^{243}Bk ($T_{1/2}$ 4.5 hrs, electron capture and α-decay) by helium ion bombardment of ^{241}Am: S. G. Thompson *et al., Phys. Rev.* **77**, 838 (1950); *eidem, ibid.* **80**, 781. Prepn of macroscopic quantities as ^{249}Bk ($T_{1/2}$ 320 days, β-decay, rel. at. mass 249.0750): B. B. Cunningham, *J. Chem. Educ.* **36**, 32-37 (1959). Prepn of metal and determn of crystal structure: J. A. Fahey, *U.S. Atomic Energy Commission* TID-25741 (1971) 119 pp, *C.A.* **76**, 9882r (1972); J. R. Peterson *et al., J. Inorg. Nucl. Chem.* **33**, 3345 (1971). *Reviews:* M. Haissinsky, J.-P. Adloff, *Radiochemical Survey of the Elements* (Elsevier, New York, 1965) pp 14-15; C. Keller, *The Chemistry of the Transuranium Elements* (Verlag Chemie, Weinheim, English Ed., 1971) pp 553-566; Silva, "Trans-Curium Elements" in *MTP Int. Rev. Sci.: Inorg. Chem., Ser. One* vol. 8, A. G. Maddock, Ed. (University Park Press, Baltimore, 1972) pp 71-105; *Comprehensive Inorganic Chemistry* vol. 5, J. C. Bailar,

Jr. *et al.*, Eds. (Pergamon Press, Oxford, 1973) *passim; Handb. Exp. Pharmakol.* **36**, 689-928 (1973); D. E. Hobart, J. R. Peterson in *The Chemistry of the Actinide Elements* vol. **1**, J. J. Katz *et al.*, Eds. (Chapman and Hall, New York, 1986) pp 989-1024.

Metal. Two allotropic forms: double hexagonal close-packed α-form, d (calc) 14.78, transforms to β-form at 930°; face-centered cubic β-form, d (calc) 13.25, exists from 930-986°. mp 986 ± 25° (Fahey). Also reported as mp 1050° (Katz *et al., loc. cit.* vol. 2, p 1150). Changes from the trivalent to the tetravalent state under the influence of oxidizing agents. In the trivalent state, its chemical properties are very close to those of curium. Can be separated from other transuranium elements by ion-exchange or by extraction of Bk(IV) with dioctylphosphoric acid in heptane (Haissinsky, Adloff). Metal dissolves rapidly in aqueous mineral acids.

Caution: Radiation hazard; handling requires special equipment and shielding facilities (Katz *et al., loc. cit.* vol. 2, p 1128).

1162. Berninamycin. [58798-97-3] Threonyl-2-(1-amino-1-propen-1-yl)-5-methyl-4-oxazolecarbonyl-2,3-didehydroalanyl-3-methylthreonyl-2-(1-aminoethenyl)-5-methyl-4-oxazolecarbonyl-2,3-didehydroalanyl-6-[2-(1-aminoethenyl)-4-oxazolyl]-5-(4-carboxy-2-thiazolyl)-2-pyridinecarbonyl-2,3-didehydroalanyl-2,3-didehydroalaninamide (7 → 1)-lactam; berninamycin A. $C_{51}H_{51}N_{15}O_{15}S$; mol wt 1146.12. C 53.45%, H 4.49%, N 18.33%, O 20.94%, S 2.80%. Major component of cyclic peptide antibiotic complex also containing minor component *Berninamycin B* ($C_{51}H_{51}N_{15}O_{14}S$). Isoln from *Streptomyces bernesis* Dietz: M. E. Bergy *et al.*, US 3689639 (1972). Inhibition of protein synthesis: F. Reusser, *Biochemistry* **8**, 3303 (1969); J. Thompson *et al., J. Gen. Microbiol.* **128**, 875 (1982). Structural studies and characterization of degradation product berninamycinic acid, a novel sulfur containing moiety: J. M. Liesch *et al., J. Am. Chem. Soc.* **98**, 299 (1976); J. M. Liesch *et al., ibid.* 8237; J. M. Liesch, K. L. Rinehart, Jr., *ibid.* **99**, 1645 (1977). Revised structure: H. Abe *et al., Tetrahedron Lett.* **29**, 1401 (1988). Confirmation of structures: R. C. M. Lau, K. L. Rinehart, *J. Antibiot.* **47**, 1466 (1994). Biosynthetic study: C. J. Pearce, K. L. Rinehart, Jr., *J. Am. Chem. Soc.* **101**, 5069 (1979). Use as growth permittant: R. M. Pellegrino, EP 112233 (1984 to Merck & Co.). Synthesis of berninamycinic acid: T. R. Kelly *et al., Tetrahedron Lett.* **25**, 2127 (1984).

Berninamycin A

White crystals, mp >290°. Sol in DMF, methanol, ethanol, propanol, butanol. Relatively insol in water, ether, cyclohexane, benzene, acetone, ethyl acetate. uv max (methanol): 208, 236 nm (A = 62.4, 64.4).

Berninamycinic acid. $C_{12}H_6N_2O_5S$. Golden needles, mp 210° (dec). pKa 5.8. Insol in water. uv max (0.01N HCl): 228, 272 nm (ε 13500, 13500); uv max (0.01N NaOH): 232, 294 nm (ε 9500, 13000).

Note: The characteristics of berninamycin are similar to those previously reported in the literature for *theiomycetin*: M. Shibata, Takeda Kenkyusho Nempo **18**, 44 (1959), *C.A.* **54**, 19840c (1960).

1163. Beryllium. [7440-41-7] Glucinium. Be; at. wt 9.012182; at. no. 4; valence 2. Group IIA (2). Alkaline earth metal. Estimated abundance in earth's crust 2 to 6 ppm. Natural isotopes (mass number): 9 (100%); known artificial radioactive isotopes: 6-8; 10-12. Oxide discovered by Vauquelin in 1797; free metal isolated by Wöhler and Bussy in 1828. Produced industrially from *beryl* (3BeO.Al$_2$O$_3$.6SiO$_2$) and *bertrandite* (4BeO.2SiO$_2$.H$_2$O); also found in *phenacite* (Be$_2$SiO$_4$), *chrysoberyl* (BeO.Al$_2$O$_3$). Precious forms of beryl: *emerald, aquamarine.* Reviews of beryllium and its compounds: Kjellgren, "Beryllium" in *Rare Metals Handbook*, C. A. Hampel, Ed. (Reinhold, New York, 1954) pp 31-55; D. A. Everest, *Chemistry of Beryllium* (Elsevier, New York, 1964) 151 pp. Review of properties and use: N. P. Pinto, J. Greenspan, "Beryllium" in *Modern Materials* vol. **6**, B. W. Gonser, Ed. (Academic Press, New York, 1968) pp 319-372; D. A. Everest, "Beryllium" in *Comprehensive Inorganic Chemistry*, J. C. Bailar, Jr. *et al.*, Eds. (Pergamon Press, Oxford, 1973) pp 531-590; A. J. Stonehouse *et al.*, in *Kirk-Othmer Encyclopedia of Chemical Technology* vol. **4** (Wiley-Interscience, New York, 4th ed., 1992) pp 126-146; A. J. Stonehouse, M. N. Emly, *ibid.* 147-153. Review of carcinogenic risk of beryllium and its compds: *IARC Monographs* **1**, 17-28 (1972); *ibid.* **23**, 143-204 (1980); of toxicology and human exposure: *Toxicological Profile for Beryllium* (PB2003-100135, 2002) 290 pp.; of immunotoxicology: A. L. Reeves, O. P. Preuss in *Immunotoxicology and Immunopharmacology*, J. H. Dean *et al.*, Eds. (Raven Press, New York, 1985) pp 441-455; of human and environmental toxicology: WHO, *Environ. Health Criter.* **106**, 1-210 (1990).

Gray, brittle metal; close-packed hexagonal structure; anisotropic; high permeability to X-rays; can serve as neutron source through either (α,n) or (n,2n) reactions. Lightest of all solid and chemically-stable substances. Displays excellent electrical and thermal conductivities. *Poisonous, flammable.* mp 1287°. bp 2500° (extrapolated). d 1.8477. Heat capacity at constant pressure (30°) 0.437 cal/g/°C: Walker *et al., J. Chem. Eng. Data* **7**, 595 (1962). Latent heat of fusion: 3.5 kcal/ mole. Brinell hardness: 60-125. Chemical properties similar to aluminum; metal resistant to attack by acid due to the formation of a thin oxide film. E° (aq) Be/Be^{2+} 1.85 V (calc.). Finely divided or amalgamated metal reacts with HCl, dil H$_2$SO$_4$ and dil HNO$_3$; attacked by strong bases with evolution of H$_2$.

Caution: Overexposure to beryllium and beryllium compounds has been associated with acute and chronic toxicity. Toxicity due to chronic inhalation is called "berylliosis" and appears after a latent period as an immunologically mediated, progressive, systemic disease leading to the formation of a characteristic granulomatous lesion. Potential symptoms of overexposure are anorexia, weight loss, weakness, chest pains, cough, clubbing of fingers, cyanosis, pulmonary insufficiency; direct contact may cause eye irritation. Dermal exposure may cause acute contact dermatitis; chronic dermal exposure may cause granulomatous skin ulceration. *See Patty's Industrial Hygiene and Toxicology* vol. **2C**, G. D. Clayton, F. E. Clayton, Eds. (John Wiley & Sons, Inc., New York, 4th ed., 1994) pp 1930-1948; *NIOSH Pocket Guide to Chemical Hazards* (DHHS/NIOSH 97-140, 1997) p 28. Beryllium and beryllium compounds are listed as known human carcinogens: *Report on Carcinogens, Twelfth Edition* (PB2011-111646, 2011) p 67.

USE: Source of neutrons when bombarded with alpha particles according to the equation $^9_4Be + ^4_2He \rightarrow ^{12}_6C + ^1_0n$. This yields about 30 neutrons per million alpha particles. Also as neutron reflector and neutron moderator in nuclear reactors. In beryllium copper and beryllium aluminum alloys (by direct reduction of beryllium oxide with carbon in the presence of Cu or Al). In aerospace, aircraft and satellite structures; x-ray transmission windows; missile parts; nuclear weapons; fuel containers; precision instruments; rocket propellants; navigational systems; heat shields; and mirrors. For fiber optics and cellular network communications systems.

1164. Beryllium Acetate. [543-81-7] Acetic acid beryllium salt. $C_4H_6BeO_4$; mol wt 127.10. C 37.80%, H 4.76%, Be 7.09%, O 50.35%. (CH$_3$COO$^-$)$_2$Be^{2+}. Prepn: Besson, Hardt, *Compt. Rend.* **237**, 1525 (1953).

Crystals, dec at 60-100° when heated slowly, 150-180° when heated rapidly. Dissolves slowly with hydrolysis in boiling water. Practically insol in abs alc and other common organic solvents.

Caution: Beryllium and beryllium compounds are listed as known human carcinogens: *Report on Carcinogens, Twelfth Edition* (PB2011-111646, 2011) p 67.

1165. Beryllium Acetylacetonate. [10210-64-7] $(T$-4)-Bis-(2,4-pentanedionato-$\kappa O^2,\kappa O^4$)beryllium. $C_{10}H_{14}BeO_4$; mol wt 207.23. C 57.96%, H 6.81%, Be 4.35%, O 30.88%. Prepd by the action of 2,4-pentanedione on an soln of beryllium hydroxide in dil acetic acid or on a soln of beryllium chloride in presence of ammonia: Biltz, *Ann.* **331**, 336 (1904); Parsons, *J. Am. Chem. Soc.* **26**, 732 (1904); Arch, Young, *Inorg. Synth.* **2**, 17 (1946); from 2,4-pentanedione and beryllium sulfate in NaOH soln: Jones, *J. Am. Chem. Soc.* **81**, 3188 (1959).

Monoclinic crystals. mp 108°. bp 270°. d_4^{20} 1.168. Practically insol in water; hydrolyzed by boiling water. Freely sol in alc, acetone, ether, benzene, CS_2, other organic solvents.

Caution: Beryllium and beryllium compounds are listed as known human carcinogens: *Report on Carcinogens, Twelfth Edition* (PB2011-111646, 2011) p 67.

1166. Beryllium Bromide. [7787-46-4] Beryllium dibromide. $BeBr_2$; mol wt 168.82. Be 5.34%, Br 94.66%. Prepn: Ehrlich in *Handbook of Preparative Inorganic Chemistry* **vol. 1**, G. Brauer, Ed. (Academic Press, New York, 2nd ed., 1963) p 891. Review of beryllium halides: Bell, *Adv. Inorg. Chem. Radiochem.* **14**, 255-332 (1972).

Orthorhombic crystals, d 3.465. mp 506-509°; also reported as 488°: Bell, *loc. cit.* Sublimes at 473°. bp 520°. Very hygroscopic. Freely sol in water. By saturating the concd viscous soln with HBr, the tetrahydrate is formed. Sol in ethanol, in pyridine (185.6 g/l), in ethyl bromide (1.0 g/l). Forms addition compounds with amines, alcohols. *Keep tightly closed.*

Caution: Beryllium and beryllium compounds are listed as known human carcinogens: *Report on Carcinogens, Twelfth Edition* (PB2011-111646, 2011) p 67.

1167. Beryllium Carbide. [506-66-1] CBe_2; mol wt 30.04. C 39.98%, Be 60.00%. Be_2C. Prepn: Coobs, Koshuba, *J. Electrochem. Soc.* **99**, 115 (1952); Mallett *et al., ibid.* **101**, 298 (1954); Ehrlich in *Handbook of Preparative Inorganic Chemistry* **vol. 1**, G. Brauer, Ed. (Academic Press, New York, 2nd ed., 1963) p 899.

Brick-red or yellow-red octahedra, d 1.90, dec above 2100°. Very slowly dec by water, somewhat faster by mineral acids and quickly by alkalies with the evolution of methane.

Caution: Beryllium and beryllium compounds are listed as known human carcinogens: *Report on Carcinogens, Twelfth Edition* (PB2011-111646, 2011) p 67.

USE: Nuclear reactor core material: Schwartz, **US 3170812** (1965 to USAEC).

1168. Beryllium Chloride. [7787-47-5] Beryllium dichloride. $BeCl_2$; mol wt 79.91. Be 11.28%, Cl 88.72%. Prepn from the elements: Tannenbaum, *Inorg. Synth.* **5**, 22 (1957); from BeO, Cl_2 and C: Ehrlich in *Handbook of Preparative Inorganic Chemistry* **vol. 1**, G. Brauer, Ed. (Academic Press, New York, 2nd ed., 1963) p 889. Toxicity data: K. W. Cochran *et al., Fed. Proc.* **9**, 264 (1950). Review of beryllium halides: Bell, *Adv. Inorg. Chem. Radiochem.* **14**, 255-332 (1972).

White to faintly yellow, very deliquesc, orthorhombic crystals or cryst mass. Reported mp ranges from 399.2° to 440°. 399.2° is considered to be the most reliable (Bell). bp 482.3°. Sublimes *in vacuo* at 300°. d 1.90. Very sol in water with evolution of heat; the aq soln is strongly acid. Sol in alc, ether, pyridine, CS_2. Insol in benzene, toluene. *Keep tightly closed.*

Tetrahydrate. Monoclinic deliquesc platelets. Has been reported to have 4½H₂O: Semenenko, Turova, *Russ. J. Inorg. Chem.* **10**, 42 (1965). LD_{50} in guinea pigs, rats (mg Be/kg): 63, 0.6 i.p. (Cochran).

Caution: Beryllium and beryllium compounds are listed as known human carcinogens: *Report on Carcinogens, Twelfth Edition* (PB2011-111646, 2011) p 67.

USE: Manuf of beryllium. *Anhydrous* form used as acid catalyst in organic reactions, similar to $AlCl_3$.

1169. Beryllium Fluoride. [7787-49-7] Beryllium difluoride. BeF_2; mol wt 47.01. Be 19.17%, F 80.83%. Prepd by heating ammonium fluoroberyllate $(NH_4)_2BeF_4$: Lebeau, *Compt. Rend.* **126**, 1418 (1898); Kwasnik in *Handbook of Preparative Inorganic Chemistry* **vol. 1**, G. Brauer, Ed. (Academic Press, New York, 2nd ed., 1963) p 231. Review of prepn and properties of beryllium halides: Bell, *Adv. Inorg. Chem. Radiochem.* **14**, 255-332 (1972).

Glassy hygroscopic mass (tetragonal system). True mp 555°; becomes free-flowing about 800°. Sublimes at 1036° under 1 mm press. in the presence of beryllium. d_4^{25} 1.986. Very freely sol in water; sparingly sol in alc; more sol in a mixture of alc and ether. Insol in anhydr HF.

Caution: Beryllium and beryllium compounds are listed as known human carcinogens: *Report on Carcinogens, Twelfth Edition* (PB2011-111646, 2011) p 67.

USE: Manuf of Be and Be alloys; manuf of glass; in nuclear reactors.

1170. Beryllium Hydride. [7787-52-2] Beryllium dihydride. BeH_2; mol wt 11.03. Be 81.71%, H 18.28%. Lower purity material prepd by treating dimethylberyllium with $LiAlH_4$ in ether: Barbaras *et al., J. Am. Chem. Soc.* **73**, 4585 (1951); higher purity by pyrolysis of di-*tert*-butylberyllium: Coates, Glocking, *J. Chem. Soc.* **1954**, 2526; Head *et al., J. Am. Chem. Soc.* **79**, 3687 (1957); from triphenyl phosphine and beryllium borohydride: Banford, Coates, *J. Chem. Soc.* **1964**, 5591.

White solid. Higher purity material is inert to laboratory air. Loss of hydrogen at 190-200° negligible, rapid at 220°. Reacts slowly with water, rapidly with dil acids. Insol in ether, toluene, isopentane. Reacts with diborane to form beryllium borohydride.

Caution: Beryllium and beryllium compounds are listed as known human carcinogens: *Report on Carcinogens, Twelfth Edition* (PB2011-111646, 2011) p 67.

1171. Beryllium Hydroxide. [13327-32-7] Beryllium dihydroxide. BeH_2O_2; mol wt 43.03. Be 20.94%, H 4.69%, O 74.36%. $Be(OH)_2$. Prepn: Ehrlich in *Handbook of Preparative Inorganic Chemistry* **vol. 1**, G. Brauer, Ed. (Academic Press, New York, 2nd ed., 1963) p 894.

Amorphous powder or crystals. d 1.92. Amphoteric. Sol in hot concd NaOH soln and acids; very slightly sol in water and dil alkali.

Caution: Beryllium and beryllium compounds are listed as known human carcinogens: *Report on Carcinogens, Twelfth Edition* (PB2011-111646, 2011) p 67.

USE: Manuf of beryllium and beryllium oxide.

1172. Beryllium Nitrate. [13597-99-4] Nitric acid beryllium salt (2:1). BeN_2O_6; mol wt 133.02. Be 6.78%, N 21.06%, O 72.17%. $Be(NO_3)_2$. Prepn: *Gmelins, Beryllium* (8th ed.) **26**, 102-104 (1930). Review of toxicology: B. Venugopal, T. D. Luckey, *Environ. Qual. Saf.* **1**, Suppl. 1, 4-73 (1975); of toxicity: *Dangerous Prop. Ind. Mater. Rep.* **9**, 2-7 (1989).

Trihydrate. [7787-55-5] White to slightly yellow, deliquesc cryst mass. mp ~60°. *Poisonous, oxidizer.* Very sol in water, ethanol. *Keep well closed in a cool place.* LD_{50} i.p. in mice: 0.5 mg/kg (Venugopal, Luckey).

Caution: Beryllium and beryllium compounds are listed as known human carcinogens: *Report on Carcinogens, Twelfth Edition* (PB2011-111646, 2011) p 67.

USE: Stiffening mantles in gas and acetylene lamps.

1173. Beryllium Nitride. [1304-54-7] Be_3N_2; mol wt 55.05. Be 49.11%, N 50.89%. Prepn: Ehrlich in *Handbook of Preparative Inorganic Chemistry* **vol. 1**, G. Brauer, Ed. (Academic Press, New York, 2nd ed., 1963) p 898; Langsdorf, Jr., **US 2567518** (1951 to USAEC).

White crystals to grayish white powder; mp 2200 ±40°. Volatile at bp, on further heating it dissociates into Be and N_2. Oxidized in air at 600°. Dec slowly by water, quickly by acids and alkalies with the evolution of ammonia.

Caution: Beryllium and beryllium compounds are listed as known human carcinogens: *Report on Carcinogens, Twelfth Edition* (PB2011-111646, 2011) p 67.

1174. Beryllium Oxide. [1304-56-9] Beryllia. BeO; mol wt 25.01. Be 36.03%, O 63.97%. Prepn: *Gmelins, Beryllium* (8th ed.) **26,** 82-91 (1930); Ehrlich in *Handbook of Preparative Inorganic Chemistry* vol. **1,** G. Brauer, Ed. (Academic Press, New York, 2nd ed., 1963) p 893. *Review:* Lillie, *USAEC* **UCRL 6457,** 23 pp (1961).

Light, amorphous powder. mp 2530°. Very sparingly sol in water; slowly sol in concd acids or solns of fixed alkali hydroxides. After ignition it is almost insol in these solvents. Pure (100%) BeO insulates electrically like a ceramic, but conducts heat like a metal. Electrical resistivity in ohm-cm: $>10^{16}$. Dielectric const at 8.5 giga-cycles: 6.57.

Caution: Beryllium and beryllium compounds are listed as known human carcinogens: *Report on Carcinogens, Twelfth Edition* (PB2011-111646, 2011) p 67.

USE: Manuf of beryllium oxide ceramics, glass, electrical insulators, gyroscopes, thermocouple tubing; in nuclear reactor fuels and moderators; catalyst for organic reactions.

1175. Beryllium Potassium Sulfate. [53684-48-3] BeK_2-O_8S_2; mol wt 279.32. Be 3.23%, K 28.00%, O 45.82%, S 22.96%. $BeSO_4.K_2SO_4$. Prepn: *Gmelins, Beryllium* (8th ed) **26,** 174 (1930).

Dihydrate. Brilliant crystals. Sol in water, concd K_2SO_4 solns; practically insol in alc.

Caution: Beryllium and beryllium compounds are listed as known human carcinogens: *Report on Carcinogens, Twelfth Edition* (PB2011-111646, 2011) p 67.

USE: In chromium- and silver-plating.

1176. Beryllium Sodium Fluoride. [13871-27-7] Sodium tetrafluoroberyllate. BeF_4Na_2; mol wt 130.99. Be 6.88%, F 58.01%, Na 35.10%. Na_2BeF_4. Prepn: *Gmelins, Beryllium* (8th ed.) **26,** 169 (1930). Review of prepn and properties of beryllium halides: Bell, *Adv. Inorg. Chem. Radiochem.* **14,** 255-332 (1972).

Orthorhombic or monoclinic crystals. mp ~350°. Sol in water.

Caution: Beryllium and beryllium compounds are listed as known human carcinogens: *Report on Carcinogens, Twelfth Edition* (PB2011-111646, 2011) p 67.

1177. Beryllium Sulfate. [13510-49-1] Sulfuric acid beryllium salt (1:1). BeO_4S; mol wt 105.07. Be 8.58%, O 60.91%, S 30.51%. $BeSO_4$. Prepn: *Gmelins, Beryllium* (8th ed.) **26,** 130-141 (1930). Toxicity study: White *et al., J. Pharmacol. Exp. Ther.* **102,** 88 (1951).

Tetrahydrate. Crystals. d 1.71. At about 100° loses $2H_2O$. Very sol in water. Practically insol in alc. LD_{50} i.v. in mice: 0.5 mg Be/kg (White).

Caution: Beryllium and beryllium compounds are listed as known human carcinogens: *Report on Carcinogens, Twelfth Edition* (PB2011-111646, 2011) p 67.

1178. Besifloxacin. [141388-76-3] 7-[(3*R*)-3-Aminohexahydro-1*H*-azepin-1-yl]-8-chloro-1-cyclopropyl-6-fluoro-1,4-dihydro-4-oxo-3-quinolinecarboxylic acid. $C_{19}H_{21}ClFN_3O_3$; mol wt 393.84. C 57.94%, H 5.37%, Cl 9.00%, F 4.82%, N 10.67%, O 12.19%. Fluorinated quinolone antibacterial; inhibits both DNA gyrase and topoisomerase-IV, which are involved in coiling and uncoiling of bacterial DNA. Prepn: F. Konno *et al.,* **WO 9201676;** *eidem,* **US 5447926** (1992, 1995 both to SS Pharma). HPLC determn in tears: D. R. Arnold *et al., J. Chromatogr. B* **867,** 105 (2008). Anti-inflammatory activity in corneal epithelial cells: J.-Z. Zhang *et al., Curr. Eye Res.* **33,** 923 (2008); in THP-1 monocytes: J.-Z. Zhang, K. W. Ward, *J. Antimicrob. Chemother.* **61,** 111 (2008). Pharmacokinetics and pharmacodynamics: K. W. Ward *et al., J. Ocul. Pharmacol. Ther.* **23,** 243 (2007); K. W. Proksch *et al., ibid.* **25,** 335 (2009). Comparative clinical trial with moxifloxacin in bacterial conjunctivitis: M. B. McDonald *et al., Ophthalmology* **116,** 1615 (2009). Review of pharmacology and clinical experience: J.

S. Bertino, J.-Z. Zhang, *Expert Opin. Pharmacother.* **10,** 2545-2554 (2009).

Colorless powder, mp 255-257° (dec).

Hydrochloride. [405165-61-9] BOL-303224-A; SS-734; Besivance. $C_{19}H_{21}ClFN_3O_3.HCl$; mol wt 430.30. White to pale yellowish-white powder.

THERAP CAT: Antibacterial.

1179. Bestmann-Ohira Reagent. [90965-06-3] *P*-(1-Diazo-2-oxopropyl)phosphonic acid dimethyl ester; dimethyl (1-diazo-2-oxopropyl)phosphonate. $C_5H_9N_2O_4P$; mol wt 192.11. C 31.26%, H 4.72%, N 14.58%, O 33.31%, P 16.12%. Diazoalkylphosphonate reagent; converted *in situ* to the Seyferth-Gilbert reagent, *q.v.* Prepn: P. Callant *et al., Synth. Commun.* **14,** 155 (1984). Synthetic applications: S. Ohira, *ibid.* **19,** 561 (1989); S. Müller *et al., Synlett* **1996,** 521; G. J. Roth *et al., Synthesis* **2004,** 59; E. Quesada, R. J. K. Taylor, *Tetrahedron Lett.* **46,** 6473 (2005).

USE: Reagent for transformation of aldehydes to terminal alkynes and ketones to methyl enol ethers.

1180. Betahistine. [5638-76-6] *N*-Methyl-2-pyridineethanamine; 2-[2-(methylamino)ethyl]pyridine; [2-(2-pyridyl)ethyl]methylamine. $C_8H_{12}N_2$; mol wt 136.20. C 70.55%, H 8.88%, N 20.57%. Prepn: Löffler, *Ber.* **37,** 161 (1904); Walter *et al., J. Am. Chem. Soc.* **63,** 2771 (1941).

Liquid. bp_{30} 113-114°. Soluble in water, alcohol, ether, chloroform.

Dihydrochloride. [5579-84-0] Betaserc; Serc; Vasomotal. C_8-$H_{12}N_2.2HCl$; mol wt 209.11. Crystals from alc, mp 148-149°.

Maleate. Suzutolon. $C_8H_{12}N_2.C_4H_4O_4$; mol wt 252.27.

Mesylate. Medan; Menitazine; Merislon; Remark; Tenyl. C_8-$H_{12}N_2.CH_3SO_3H$; mol wt 232.30.

Dimesylate. Aequamen; Melopat; Ribrain. $C_8H_{12}N_2.(CH_3SO_3$-$H)_2$; mol wt 328.40.

THERAP CAT: Vasodilator.

1181. Betaine. [107-43-7] 1-Carboxy-*N,N,N*-trimethylmethanaminium inner salt; (carboxymethyl)trimethylammonium hydroxide inner salt; glycine betaine; glycocoll betaine; lycine; oxyneurine; trimethylglycine hydroxide inner salt; trimethylglycocoll anhydride; Cystadane. $C_5H_{11}NO_2$; mol wt 117.15. C 51.26%, H 9.46%, N 11.96%, O 27.31%. Widely distributed in plants and animals: M. Guggenheim, *Die biogenen Amine* (S. Karger, Basel, 4th ed., 1951) pp 240-242. Prepn via or from hydrochloride: Stoltzenberg, *Z. Physiol. Chem.* **92,** 445 (1914); Edsall, *J. Am. Chem. Soc.* **65,** 1767 (1943). Prepn of the hydrochloride: Kuhn, Ruelius, *Ber.* **83,** 420 (1950); of the hydrate: Vassel, **US 2800502** (1957 to IMC). Structure of hydrate: Leifer, Lippincott, *J. Am. Chem. Soc.* **79,** 5098 (1957). HPLC determn in urine: M. D. Laryea *et al., Clin. Chim. Acta* **230,** 169 (1994). Clinical evaluations in homocystinuria: D.

E. Wilcken *et al.*, *N. Engl. J. Med.* **309**, 448 (1983); N. P. B. Dudman *et al.*, *J. Nutr.* **126**, Suppl. 4, 1295S (1996).

Deliquescent scales or prisms dec around 310°. (Isomerizes at the mp to methyl ester of dimethylaminoacetic acid.) Sweet taste. Solubility (g/100 g solvent): water, 160; methanol, 55; ethanol, 8.7. Sparingly sol in ether. Betaine yields trimethylamine with concd KOH.

Monohydrate. [590-47-6] (Carboxymethyl)trimethylammonium hydroxide; trimethylglycine hydroxide. Formed by crystn of betaine from aq solvents. pH of satd soln about 8.0. Loses water at 100° forming inner salt again.

Hydrochloride. [590-46-5] 1-Carboxy-*N*,*N*,*N*-trimethylmethanaminium chloride; (carboxymethyl)trimethylammonium chloride; pluchine; Acidol. $C_5H_{11}NO_2 \cdot HCl$; mol wt 153.61. Monoclinic crystals from alc, dec 227-228° (Stoltzenberg); also reported as dec 232° (Kuhn, Ruelius). Soly in water at 25° = 64.7 g/100 ml, in ethanol 5.0 g/100 ml. pH of 5% aq soln 1.0. Practically insol in chloroform, ether.

Sodium aspartate. [93227-64-6] Somatyl. $C_9H_{17}N_2NaO_6$; mol wt 272.23. Prepn: R. Cote, **FR 1356945** (1964), *C.A.* **61**, 7098 (1964). Prepd as trihydrate, dec 160-170°. Hygroscopic. Sol in water.

USE: In soldering, resin curing fluxes, organic synthesis.

THERAP CAT: In treatment of homocystinuria; hepatoprotectant. Hydrochloride as gastric acidifier.

1182. Betamethasone. [378-44-9] (11β,16β)-9-Fluoro-11,-17,21-trihydroxy-16-methylpregna-1,4-diene-3,20-dione; 9α-fluoro-16β-methylprednisolone; 16β-methyl-9α-fluoro-Δ¹-hydrocortisone; 16β-methyl-9α-fluoroprednisolone; betadexamethasone; flubenisolone; β-methasone; Sch-4831; NSC-39470; beta-Corlan; Becort; Betasolon; Betnelan; Celestene; Celestone; Dermabet; Visubeta. $C_{22}H_{29}FO_5$; mol wt 392.47. C 67.33%, H 7.45%, F 4.84%, O 20.38%. Prepn: Taub *et al.*, *J. Am. Chem. Soc.* **80**, 4435 (1958); Oliveto *et al.*, *ibid.* 6688; Taub *et al.*, *ibid.* **82**, 4012 (1960); **US 3053865** (1962 to Merck & Co.); Amiard *et al.*, **US 3104246** (1963 to Roussel-UCLAF). Also prepared from hecogenin. Comprehensive description of the dipropionate ester: M. G. Ferrante, B. C. Rudy, *Anal. Profiles Drug Subs.* **6**, 43-60 (1977).

Crystals from ethyl acetate, mp 231-234° (dec). Sparingly sol in acetone, alcohol, dioxane, methanol; very slightly sol in chloroform, ether. Insol in water. $[\alpha]_D$ +108° (acetone). uv max (methanol): 238 nm (ε 15200).

21-Acetate. [987-24-6] Betafluorene; Celestovet. $C_{24}H_{31}FO_6$; mol wt 434.50. Hexagonal prisms from acetone + ether, mp 205-208° (Taub); also reported as mp 196-201° (Oliveto). $[\alpha]_D$ +140° (chloroform). uv max (methanol): 238 nm (ε 14800). Freely sol in acetone; sol in alc, chloroform. Practically insol in water.

17-Benzoate. [22298-29-9] W-5975; Bebate; Beben; Benisone; Euvaderm; Flurobate; Uticort. $C_{29}H_{33}FO_6$; mol wt 496.58. Crystals from acetone-ether, mp 225-228°. $[\alpha]_D^{24}$ +63.5° (dioxane). Sol in alc, methanol, chloroform. Insol in water. Synthesis and activity: Ercoli *et al.*, *J. Med. Chem.* **15**, 783 (1972). *See also:* Cullen, *Curr. Ther. Res.* **15**, 243 (1973).

17,21-Dipropionate. [5593-20-4] Sch-11460; Diproderm; Diprolene; Diprophos; Diprosis; Diprosone; Maxivate; Rinderon-DP. $C_{28}H_{37}FO_7$; mol wt 504.60. Powder, mp 170-179° (dec). $[\alpha]_D^{26}$ +65.7° (dioxane). uv max (methanol): 238 nm (ε 15700). Freely sol in acetone, chloroform; sparingly sol in alcohol. Insol in water.

17-Valerate. [2152-44-5] Bedermin; Betnesol-V; Betneval; Betnovate; Bextasol; Celestan-V; Celestoderm-V; Dermosol; Dermovaleas; Ecoval 70; Hormezon; Tokuderm; Valisone. $C_{27}H_{37}FO_6$; mol wt 476.59. Needles from acetone + petr ether, mp 183-184°. $[\alpha]_D$ +77° (dioxane). uv max: 239 nm (ε 15920). **NL 6406615** (1964 to Glaxo). Freely sol in acetone, chloroform; sol in alc; slightly sol in benzene, ether. Practically insol in water.

21-Phosphate disodium salt. [151-73-5] Betamethasone 21-(dihydrogen phosphate) disodium salt; Bentelan; Betnesol; Celestan; Durabetason; Vista-Methasone. $C_{22}H_{28}FNa_2O_8P$; mol wt 516.41. White odorless powder. Hygroscopic. Freely sol in water, methanol. Practically insol in acetone, chloroform.

THERAP CAT: Glucocorticoid.

THERAP CAT (VET): Glucocorticoid.

1183. Betamipron. [3440-28-6] *N*-Benzoyl-β-alanine; 3-(benzoylamino)propionic acid; β-benzamidopropionic acid; CS-443. $C_{10}H_{11}NO_3$; mol wt 193.20. C 62.17%, H 5.74%, N 7.25%, O 24.84%. Acyl amino acid that inhibits active transport of carbapenem antibiotics in the renal cortex. *See also* panipenem. Prepn: F. H. Holm, *Arch. Pharm.* **242**, 590 (1904); C. C. Barker, *J. Chem. Soc.* **1954**, 317; J. Barluenga *et al.*, *Tetrahedron* **45**, 2183 (1989). Use as renal protectant: T. Shiokari *et al.*, **EP 226304**; *eidem*, **US 4757066** (1987, 1988 both to Sankyo). Crystal structure: N. Feeder, W. Jones, *Acta Crystallogr.* **C50**, 813 (1994). Nephroprotective effect: Y. Hirouchi *et al.*, *Jpn. J. Pharmacol.* **63**, 487 (1993). Mode of action: *eidem*, *ibid.* **66**, 1 (1994). Pharmacokinetics: A. Kurihara *et al.*, *Antimicrob. Agents Chemother.* **36**, 1810 (1992).

Colorless prisms from hot water, mp 120° (Holm); also reported as crystals from water, mp 133° (Barker). Readily sol in warm water, chloroform. Very easily sol in alcohol, ether, acetone. LD_{50} in rats (mg/kg): >3000 i.v. (Hirouchi, 1994).

THERAP CAT: Antibacterial adjunct (renal protectant).

1184. Betaxolol. [63659-18-7] 1-[4-[2-(Cyclopropylmethoxy)ethyl]-phenoxy]-3-[(1-methylethyl)amino]-2-propanol; (±)-1-(isopropylamino)-3-[*p*-(cyclopropylmethoxyethyl)phenoxy]-2-propanol. $C_{18}H_{29}NO_3$; mol wt 307.43. C 70.32%, H 9.51%, N 4.56%, O 15.61%. Cardioselective β₁-adrenergic blocker. Prepn: P. M. J. Manoury *et al.*, **DE 2649605**; *eidem*, **US 4252984** (1977, 1981 both to Synthelabo). Blood concn and pharmacodynamic effects: S. J. Warrington *et al.*, *Br. J. Clin. Pharmacol.* **10**, 449 (1980). Pharmacokinetics: G. Bianchetti *et al.*, *Arzneim.-Forsch.* **30**, 1912 (1980). Cardiovascular effects in normal volunteers: P. J. Cadigan *et al.*, *Br. J. Clin. Pharmacol.* **9**, 569 (1980). Efficacy and pharmacokinetics: K. Balnave *et al.*, *ibid.* **11**, 171 (1981). Use in treatment of glaucoma: A. R. Berrospi, H. M. Leibowitz, *Arch. Ophthalmol.* **100**, 943 (1982). Antihypertensive effect: M. Pathe *et al.*, *Therapie* **37**, 75 (1982). *Book: Betaxolol and Other β₁-Adrenoceptor Antagonists*, P. L. Morselli *et al.*, Eds. (Raven Press, New York, 1983) 385 pp.

Crystals from petr ether, mp 70-72°.

Hydrochloride. [63659-19-8] SLD-212; SL-75.212; Betoptic; Betoptima; Kerlone. $C_{18}H_{29}NO_3 \cdot HCl$; mol wt 343.89. Crystals from acetone, mp 116°. Freely sol in water, alcohol, chloroform, methanol. LD_{50} in mice (mg/kg): 944 orally; 37 i.v. (Manoury).

THERAP CAT: Antihypertensive; antiglaucoma.

1185. Betazole. [105-20-4] 1*H*-Pyrazole-3-ethanamine; 3-(2-aminoethyl)pyrazole; 3-(β-aminoethyl)pyrazole; ametazole; gastramine. $C_5H_9N_3$; mol wt 111.15. C 54.03%, H 8.16%, N 37.81%. Prepd by the catalytic reduction of 3-pyrazoleacetaldehyde hydrazone: Jones, Mann, *J. Am. Chem. Soc.* **75**, 4048 (1953); Jones, **US 2785177** (1957 to Lilly).

Viscous liquid. $bp_{0.5}$ 118-123°.

Dihydrochloride. [138-92-1] Histimin; Histalog. $C_5H_9N_3$.·2HCl; mol wt 184.06. Crystals from ethanol, mp 224-226°. Sol in water. Practically insol in chloroform. Aq solns are acid to litmus.

THERAP CAT: Diagnostic aid (gastric secretion stimulant).

1186. Betel. Dried leaves of *Piper betle* L., *Piperaceae*. *Habit.* India, Ceylon, Malay Archipelago. *Constit.* 0.2-1% volatile oil, chavibetol, chavicol, cadinene, allylpyrocatechol. *Ref:* Ueda, Sasaki, *J. Pharm. Soc. Jpn.* **71**, 559 (1951), *C.A.* **45**, 9137g (1951).

THERAP CAT: Counterirritant.

1187. Bethanechol Chloride. [590-63-6] 2-[(Aminocarbonyl)oxy]-N,N,N-trimethyl-1-propanaminium chloride (1:1); (2-hydroxypropyl)trimethylammonium chloride carbamate; 2-carbamoyloxypropyltrimethylammonium chloride; carbamylmethylcholine chloride; β-methylcholine chloride urethan; Duvoid; Myocholine; Myotonine; Urecholine. $C_7H_{17}ClN_2O_2$; mol wt 196.68. C 42.75%, H 8.71%, Cl 18.02%, N 14.24%, O 16.27%. Prepn: Dalmer, Diehl, **US 1894162** (1933); Major, Bonnett, **US 2322375** (1943 to Merck & Co.). ¹H-NMR determn in tablets: G. M. Hanna, C. A. Lau-Cam, *J. Assoc. Off. Anal. Chem.* **70**, 557 (1987). Clinical trial in postoperative urinary retention: L. Gottesman *et al.*, *Dis. Colon Rectum* **32**, 867 (1989).

Hygroscopic crystals, slight amine odor, dec 218-219°. One gram dissolves in 0.6 ml of water, in 12.5 ml of 95% alc. pH of an 0.5% aq soln 5.5-6.0. Aq solns may be sterilized by autoclaving at 120° for 20 minutes.

THERAP CAT: Cholinergic; in treatment of urinary retention.

THERAP CAT (VET): Cholinergic.

1188. Bethanidine. [55-73-2] N,N'-Dimethyl-N''-(phenylmethyl)guanidine; 1-benzyl-2,3-dimethylguanidine; N-benzyl-N',N''-dimethylguanidine. $C_{10}H_{15}N_3$; mol wt 177.25. C 67.76%, H 8.53%, N 23.71%. Adrenergic neuron blocking agent. Prepn: Walton, Ruffell, **US 3168562** (1965 to Wellcome Found.); **GB 1084461** and **GB 1111564** (1967 and 1968 to Wellcome Found.). Pharmacology: A. L. A. Boura, A. F. Green, *Br. J. Pharmacol.* **20**, 36 (1963); J. A. Oates *et al.*, *Ann. N.Y. Acad. Sci.* **179**, 302 (1971). Pharmacokinetics: C. N. Corder, *J. Clin. Pharmacol.* **19**, 428 (1979). HPLC determn in plasma: J. R. Shipe *et al.*, *Clin. Chem.* **29**, 1793 (1983). Anti-arrhythmic activity: M. B. Bacaner, D. G. Benditt, *Am. J. Cardiol.* **50**, 728 (1982); J. C. Somberg *et al.*, *ibid.* **54**, 343 (1984). *Review:* A. F. Green, *Br. J. Clin. Pharmacol.* **13**, 25-34 (1982).

Crystals from methanol + ether, mp 195-197°.

Hydriodide. $C_{10}H_{15}N_3$·HI. Crystals from ethanol + ether, mp 141-146°.

Sulfate. [114-85-2] Benzaidin; BW-467C60; Bendogen; Benzoxine; Betaling; Betanidol; Esbatal; Eusmanid; Hypersin; Tena-

than. $(C_{10}H_{15}N_3)_2$·H_2SO_4; mol wt 452.57. LD_{50} in mice (mg/kg): 12 i.v.; 150 i.p.; 260 s.c.; 520 by stomach intubation (Boura, Green).

THERAP CAT: Antihypertensive.

1189. Betoxycaine. [3818-62-0] 3-Amino-4-butoxybenzoic acid 2-[2-(diethylamino)ethoxy]ethyl ester; 2-diethylaminoethoxyethyl 3-amino-4-butoxybenzoate. $C_{19}H_{32}N_2O_4$; mol wt 352.48. C 64.74%, H 9.15%, N 7.95%, O 18.16%. Prepn and properties: E. Cuingnet, **US 3209022** (1965 to Corbiere). Ionization potential: A. Cier *et al.*, *Therapie* **26**, 941 (1971). Investigations of possible antifibrillatory properties: P. Amaud *et al.*, *C. R. Seances Soc. Biol. Ses Fil.* **159**, 2427 (1965); G. Faucon *et al.*, *Therapie* **21**, 1253 (1966).

Monohydrochloride. [5003-47-4] Millicaine. $C_{19}H_{32}N_2O_4$.·HCl; mol wt 388.93. White crystals from acetone, mp 117°. Soluble in methanol.

THERAP CAT: Anesthetic (local).

1190. Betti Base. [481-82-3] 1-(Aminophenylmethyl)-2-naphthalenol; 1-(α-aminobenzyl)-2-naphthol; β-naphthol-phenylaminomethane. $C_{17}H_{15}NO$; mol wt 249.31. C 81.90%, H 6.06%, N 5.62%, O 6.42%. The single enantiomers and their derivatives serve as chiral auxiliaries and ligands in asymmetric synthesis. Prepn: M. Betti, *Gazz. Chim. Ital.* **31**, I, 377 (1901). Large scale prepn: *idem, Org. Synth.* **coll. vol. I**, 381 (1941). Resolution procedures: *idem, Gazz. Chim. Ital.* **36**, II, 392 (1906); Y. Dong *et al.*, *J. Org. Chem.* **70**, 8617 (2005); and absolute configuration determn: C. Cardellicchio *et al.*, *Tetrahedron: Asymmetry* **9**, 3667 (1998). Synthetic applications: *eidem, Tetrahedron* **55**, 14685 (1999); Y. Dong *et al.*, *Tetrahedron: Asymmetry* **15**, 1667 (2004); X. Wang *et al.*, *J. Org. Chem.* **70**, 1897 (2005). Review of chemistry of derivatives: I. Szatmári, F. Fülöp, *Curr. Org. Synth.* **1**, 155-165 (2004).

mp 124-125°.

Hydrochloride. [219897-32-2] $C_{17}H_{15}NO$·HCl; mol wt 285.77. Light pink or white needles, mp 190-220° (dec).

(R)-Form. [219897-35-5] mp 137°. $[\alpha]_D^{30}$ -60 ± 0.02°.

(S)-Form. [219897-38-8] mp 137°. $[\alpha]_D^{30}$ $+60\pm0.02$°.

USE: Reagent in synthetic organic chemistry.

1191. Betula. European white birch. Bark and leaves of *Betula alba* L., *Betulaceae*. *Habit.* Europe and Northern Asia, also America, north of Pennsylvania. *Constit.* 10-15% Betulin (betula camphor), betuloresinic acid, essential oil, saponins, betulol (sesquiterpene alcohol), apigenin dimethyl ether, betuloside, gaultherin, methyl salicylate, ascorbic acid. *Ref:* Kreitmair, *Pharmazie* **8**, 534 (1953).

USE: Pharmaceutic aid (flavor).

1192. Betulin. [473-98-3] (3β)-Lup-20(29)-ene-3,28-diol; lup-20(30)-ene-3β,28-diol; trochol; betulinol; betulol. $C_{30}H_{50}O_2$; mol wt 442.73. C 81.39%, H 11.38%, O 7.23%. In the outer portion of the bark of white birch (up to 24%), in other barks, and in lignite. Isoln: Lowitz, *Crell's Chem. Ann.* **1**, 312 (1788); L. Ruzicka, O. Isler, *Helv. Chim. Acta* **19**, 506 (1936); and botanical distribution: Steiner, *Molisch Festschrift* (1936). Isoln from *Lemaireocereus griseus* Britton et Rose, *Cactaceae*: C. Djerassi *et al.*, *J. Am. Chem. Soc.* **78**, 2312 (1956). Structure: Ames *et al.*, *J. Chem. Soc.* **1951**, 450; Davy *et al.*, *ibid.* **1951**, 2696, 2702. Stereochemistry: Guider *et al., ibid.* **1953**, 3024; Das, *Chem. Ind. (London)* **1971**, 1331. *Reviews:* J. Simonsen, W. C. J. Ross, *The Terpenes* vol. **IV** (Cambridge Univ. Press, 1957) pp 187-328; E. W. H. Hayek *et al., Phytochemistry* **28**, 2229-2242 (1989).

Crystals from methanol-chloroform, mp 248-251°; sublimes at 240° at 0.01 mm. Solvated needles from alc contg one mol EtOH. After drying sublimes at 170-180° (bath temp) at 0.08 mm. uv max (H$_2$SO$_4$): 316 nm. [α]$_D^{15}$ +20° (c = 2 in pyridine). Sparingly sol in cold water, petr ether, carbon disulfide. One part is sol in 149 parts alc, 251 ether, 113 chloroform, 417 benzene. Sol in acetic acid.

Diacetate. [1721-69-3] C$_{34}$H$_{54}$O$_4$. mp 223-224°. [α]$_D^{20}$ +22° (c = 1.2 in CHCl$_3$). d$_4^{228.5}$ 0.9635; n$_D^{228.5}$ 1.4661.

USE: Light stabilizer for cellulose and wood pulp. In mfr of resins, laquers, emulsifiers and polyurethanes. Cosmetics additive.

1193. Betulinic Acid. [472-15-1] (3β)-3-Hydroxylup-20-(29)-en-28-oic acid; betulic acid; ALS-357. C$_{30}$H$_{48}$O$_3$; mol wt 456.71. C 78.90%, H 10.59%, O 10.51%. Naturally occurring triterpene found in many plant sources, notably the bark of white birch trees, *Betula alba* L., *Betulaceae*. Melanoma-specific cytotoxic agent. Isoln: L. Ruzicka *et al., Helv. Chim. Acta* **21**, 1706 (1938); and characterization: A. Robertson *et al., J. Chem. Soc.* **1939**, 1267. Synthesis from betulin, *q.v.*: D. S. H. Kim *et al., Synth. Commun.* **27**, 1607 (1997). Isoln and crystal structure of benzyl ester: G. Bringmann *et al., Planta Med.* **63**, 255 (1997). Inhibition of HIV replication: T. Fujioka *et al., J. Nat. Prod.* **57**, 243 (1994). Induction of apoptosis in human melanoma cells: E. Pisha *et al., Nat. Med.* **1**, 1046 (1995). Mechanism of action study: S. Fulda *et al., Cancer Res.* **57**, 4956 (1997). Pharmacokinetics in mice: G. O. Udeani *et al., Biopharm. Drug Dispos.* **20**, 379 (1999). Tumor cell-specific cytotoxicity: V. Zuco *et al., Cancer Lett.* **175**, 17 (2002). LC/MS determn in plasma: X. Cheng *et al., Rapid Commun. Mass Spectrom.* **17**, 2089 (2003). Review of natural distribution: G. Pavanasasivam, M. U. S. Sultanbawa, *Phytochemistry* **13**, 2002-2006 (1974); of anticancer activity: D. A. Eiznhamer, Z.-Q. Xu, *IDrugs* **7**, 359-373 (2004); of chemistry, biology and therapeutic potential: R. H. Cichewicz, S. A. Kouzi, *Med. Res. Rev.* **24**, 90-114 (2004); of biological properties: P. Yogeeswari, D. Sriram, *Curr. Med. Chem.* **12**, 657-666 (2005).

Colorless needles from methanol, mp 290-293°. [α]$_D^{25}$ +7.5° (c = 0.37 in pyridine). Highly sol in pyridine, acetic acid. Limited soly in methanol, ethanol, CHCl$_3$, ether. Low soly in H$_2$O, petroleum ether, DMF, DMSO, benzene.

1194. Bevacizumab. [216974-75-3] Anti-(human vascular endothelial growth factor) immunoglobulin G1 (human-mouse monoclonal rhuMAb-VEGF γ$_1$-chain) disulfide with human-mouse monoclonal rhuMAb-VEGF light chain, dimer; rhuMAb-VEGF;

Avastin. Recombinant humanized monoclonal IgG1 antibody directed against vascular endothelial growth factor (VEGF). Prepn: L. G. Presta *et al., Cancer Res.* **57**, 4593 (1997); M. Baca *et al.,* **WO 9845331** (1998 to Genentech). Biodistribution study: Y. S. Lin *et al., J. Pharmacol. Exp. Ther.* **288**, 371 (1999). Clinical pharmacokinetics: M. S. Gordon *et al., J. Clin. Oncol.* **19**, 843 (2001). Clinical trial in renal cancer: J. C. Yang *et al., N. Engl. J. Med.* **349**, 427 (2003); in combination therapy in colorectal cancer: F. Kabbinavar *et al., J. Clin. Oncol.* **21**, 60 (2003). Symposium on antiangiogenic and antitumor activity: *Semin. Oncol.* **29**, Suppl. 16, 1-18 (2002). Review of development and clinical experience: N. Ferrara *et al., Nat. Rev. Drug Discovery* **3**, 391-400 (2004); of clinical efficacy and safety: R. E. Sanborn, A. B. Sandler, *Expert Opin. Drug Saf.* **5**, 289-301 (2006).

THERAP CAT: Antineoplastic.

1195. Bevantolol. [59170-23-9] 1-[[2-(3,4-Dimethoxyphenyl)ethyl]amino]-3-(3-methylphenoxy)-2-propanol; 1-[(3,4-dimethoxyphenethyl)amino]-3-(m-tolyloxy)-2-propanol. C$_{20}$H$_{27}$NO$_4$; mol wt 345.44. C 69.54%, H 7.88%, N 4.05%, O 18.53%. Cardioselective β$_1$-adrenergic blocker. Prepn: **BE 790165**; A. Holmes, R. F. Meyer, **US 3857891** (1973, 1974 both to Parke, Davis & Co.); M. L. Hoefle *et al., J. Med. Chem.* **18**, 148 (1975). Pharmacology in animals: S. G. Hastings *et al., Arch. Int. Pharmacodyn. Ther.* **226**, 81 (1977). Cardiovascular effects in animals: I. D. Dukes, E. M. Vaughan Williams, *Br. J. Pharmacol.* **84**, 365 (1985). Cardioselectivity in asthmatic patients: C.-G. Löfdahl *et al., Pharmacotherapy* **4**, 205 (1984). GC determn in plasma: E. J. Randinitis *et al., J. Chromatogr.* **308**, 345 (1984). Pharmacokinetics in humans: P. Vermeij, P. van Brummelen, *Eur. J. Clin. Pharmacol.* **30**, 375 (1986). Comparative clinical trial with hydrochlorothiazide in hypertension: C. P. Lucas *et al., Clin. Ther.* **8**, 49 (1985). Symposium on pharmacology and clinical efficacy: *Am. J. Cardiol.* **58**, 1E-44E (1986).

Hydrochloride. [42864-78-8] CI-775; Ranestol; Sentiloc; Vantol. C$_{20}$H$_{27}$NO$_4$.HCl; mol wt 381.90. Crystals from acetonitrile, mp 137-138°.

THERAP CAT: Antianginal; antihypertensive; antiarrhythmic.

1196. Bevirimat. [174022-42-5] (3β)-3-(3-Carboxy-3-methyl-1-oxobutoxy)-lup-20(29)-en-28-oic acid; 3-O-(3′,3′-dimethylsuccinyl)betulinic acid; PA-457; YK-FH312. C$_{36}$H$_{56}$O$_6$; mol wt 584.84. C 73.93%, H 9.65%, O 16.41%. HIV maturation inhibitor derived from the natural product, betulinic acid, *q.v.* Disrupts processing of HIV Gag protein and blocks the conversion to mature capsid. Prepn: K.-H. Lee *et al.,* **WO 9639033** (1996 to UNC Chapel Hill); *eidem,* **US 5679828** (1997 to Biotech; UNC Chapel Hill); Y. Kashiwada *et al., J. Med. Chem.* **39**, 1016 (1996). Anti-HIV activity: T. Kanamoto *et al., Antimicrob. Agents Chemother.* **45**, 1225 (2001). Mechanism of action: F. Li *et al., Proc. Natl. Acad. Sci. USA* **100**, 13555 (2003); *eidem, Virology* **356**, 217 (2006). Clinical pharmacokinetics: D. E. Martin *et al., Antimicrob. Agents Chemother.* **51**, 3063 (2007). *Review:* Z. Temesgen, J. E. Feinberg, *Curr. Opin. Investig. Drugs* **7**, 759-765 (2006).

Colorless needles from methanol, mp 274-276°. $[\alpha]_D^{19}$ +23.5° (c = 0.71 in chloroform/methanol).

Dimeglumine salt. [823821-85-8] PA-457N; PA-103001. $C_{36}H_{56}O_6 \cdot 2C_7H_{17}NO_5$; mol wt 975.27. Prepn: M. D. Power, D. E. Martin, **WO 05090380**; *eidem*, **US 050239748** (both 2005 to Panacos). White solid from diethyl ether. Soly (mg/ml): propylene glycol >255, methanol 37.03, ethanol 24.99, deionized water >7.

THERAP CAT: Antiretroviral.

1197. Bevonium Methyl Sulfate. [5205-82-3] 2-[[(2-Hydroxy-2,2-diphenylacetyl)oxy]methyl]-1,1-dimethylpiperidinium methyl sulfate (1:1); 2-(hydroxymethyl)-1,1-dimethylpiperidinium methyl sulfate benzilate; piribenzil methyl sulfate; benzilic acid ester with 2-(hydroxymethyl)-1,1-dimethylpiperidinium methyl sulfate; CG-201; Acabel. $C_{23}H_{31}NO_7S$; mol wt 465.56. C 59.34%, H 6.71%, N 3.01%, O 24.06%, S 6.89%. Anticholinergic. Prepn: **BE 616951** (1962 to Grünenthal), *C.A.* **58**, 7914d (1963); Beckmann, *Arzneim.-Forsch.* **16**, 910 (1966). Series of publications on pharmacology, toxicology, clinical trials: *ibid.* 901-988.

Crystals from petr ether, mp 134-135°.

THERAP CAT: Antispasmodic.

1198. Bexarotene. [153559-49-0] 4-[1-(5,6,7,8-Tetrahydro-3,5,5,8,8-pentamethyl-2-naphthalenyl)ethenyl]benzoic acid; SR-11247; LGD-1069; Targretin. $C_{24}H_{28}O_2$; mol wt 348.49. C 82.72%, H 8.10%, O 9.18%. Selective retinoid X receptor (RXR) agonist. Prepn: M. F. Boehm *et al.*, **WO 9321146** (1993 to Ligand Pharm.); M. L. Dawson *et al.*, **US 5466861** (1995 to SRI Int.; La Jolla Cancer Res. Found.); and structure-activity study: M. F. Boehm *et al.*, *J. Med. Chem.* **37**, 2930 (1994); M. I. Dawson *et al.*, *ibid.* **38**, 3368 (1995). Chemopreventive effect in rat mammary carcinoma: M. M. Gottardis *et al.*, *Cancer Res.* **56**, 5566 (1996). Effect on glucose homeostasis in exptl diabetes: R. Mukherjee *et al.*, *Nature* **386**, 407 (1997). Clinical pharmacokinetics and evaluation in advanced cancers: V. A. Miller *et al.*, *J. Clin. Oncol.* **15**, 790 (1997). Clinical evaluation in cutaneous T-cell lymphoma: D. Breneman *et al.*, *Arch. Dermatol.* **138**, 325 (2002).

White solid from methylene chloride, mp 230-231° (Dawson, 1995); also reported as white crystals, mp 234° (Boehm, 1994). uv max (methanol): 264 nm (ε 16400).

THERAP CAT: Antineoplastic.

1199. Bezafibrate. [41859-67-0] 2-[4-[2-[(4-Chlorobenzoyl)amino]ethyl]phenoxy]-2-methylpropanoic acid; 2-[p-[2-(p-chlorobenzamido)ethyl]phenoxy]-2-methylpropionic acid; α-[4-(4-chlorobenzoylaminoethyl)phenoxy]isobutyric acid; BM-15075; Befizal; Bezalip; Bezatol; Cedur; Difaterol. $C_{19}H_{20}ClNO_4$; mol wt 361.82. C 63.07%, H 5.57%, Cl 9.80%, N 3.87%, O 17.69%. Prepn: E. Witte *et al.*, **DE 2149070**; *eidem*, **US 3781328** (both 1973 to Boehringer, Mann.). Pharmacology: R. Zimmerman *et al.*, *Atherosclerosis* **29**, 477 (1978). Clinical studies: A. G. Olsson, P. D. Lang, *ibid.* **31**, 421, 429 (1978); P. Wahl *et al.*, *Dtsch. Med. Wochenschr.* **103**, 1233 (1978). Review of pharmacodynamics and therapeutic use: J. P. Monk, P. A. Todd, *Drugs* **33**, 539-576 (1987).

Crystals from acetone, mp 186°.

THERAP CAT: Antilipemic.

1200. Bezitramide. [15301-48-1] 4-[2,3-Dihydro-2-oxo-3-(1-oxopropyl)-1H-benzimidazol-1-yl]-α,α-diphenyl-1-piperidinebutanenitrile; 1-[1-(3-cyano-3,3-diphenylpropyl)-4-piperidinyl]-1,3-dihydro-3-(1-oxopropyl)-2H-benzimidazol-2-one; 1-[1-[(3-cyano-3,3-diphenylpropyl)-4-piperidyl]-3-propionyl-2-benzimidazolinone; 1-(3-cyano-3,3-diphenylpropyl)-4-(2-oxo-3-propionyl-1-benzimidazolinyl)piperidine; benzitramide; R-4845; Burgodin. $C_{31}H_{32}N_4O_2$; mol wt 492.62. C 75.58%, H 6.55%, N 11.37%, O 6.50%. Opioid analgesic. Prepn: **BE 633495**; P. A. J. Janssen, **US 3196157** (1963, 1965 both to Janssen). Pharmacology and clinical data: P. A. J. Janssen *et al.*, *Arzneim.-Forsch.* **21**, 862 (1971); W. K. P. Amery *et al.*, *ibid.* 868; H. Knape, *Anaesthesist* **21**, 251 (1972).

White crystalline powder, mp 145-149°. Also reported as pale yellow amorphous powder, mp 124.5-126°. Solubility of >1 g/100 ml in ethyl acetate, acetone, benzene, chloroform. Almost insol in water, dilute acids. LD_{50} orally in mice, rats: 2101, 141 mg/kg (Janssen).

Note: This is a controlled substance (opiate): **21 CFR**, 1308.12.

THERAP CAT: Analgesic.

1201. Biapenem. [120410-24-4] 6-[[(4R,5S,6S)-2-Carboxy-6-((1R)-1-hydroxyethyl)-4-methyl-7-oxo-1-azabicyclo[3.2.0]hept-2-en-3-yl]thio]-6,7-dihydro-5H-pyrazolo[1,2-a][1,2,4]triazol-4-ium inner salt; (1R,5S,6S)-2-[(6,7-dihydro-5H-pyrazolo[1,2-a][1,2,4]-triazolium-6-yl)thio]-6-[(R)-1-hydroxyethyl]-1-methylcarbapen-2-em-3-carboxylate; LJC-10627; L-627; CL-186815. $C_{15}H_{18}N_4O_4S$; mol wt 350.39. C 51.42%, H 5.18%, N 15.99%, O 18.26%, S 9.15%. 1-β-methyl-carbapenem antibiotic. Prepn: T. Kumagai *et al.*, **EP 289801**; *eidem*, **US 4990613** (1988, 1991 both to Lederle). Total synthesis: Y. Nagao *et al.*, *J. Org. Chem.* **57**, 4243 (1992). Renal dehydropeptidase stability: M. Hikida *et al.*, *Antimicrob. Agents Chemother.* **36**, 481 (1992). *In-vitro* antimicrobial spectrum: H. M. Wexler *et al.*, *J. Antimicrob. Chemother.* **33**, 629 (1994); H. Y. Chen, D. M. Livermore, *ibid.* 949. Clinical pharmacokinetics: M. Nakashima *et al.*, *Int. J. Clin. Pharmacol. Ther. Toxicol.* **31**, 70 (1993). Clinical evaluation: H. Meguro *et al.*, *Jpn. J. Antibiot.* **47**, 903 (1994).

Amorphous, pale yellow powder from acetone/water, occurs as hemihydrate. $[\alpha]_D^{20}$ −32.9° (c = 0.5).

THERAP CAT: Antibacterial.

1202. Bibenzyl. [103-29-7] 1,1'-(1,2-Ethanediyl)bisbenzene; dibenzyl; *sym*-diphenylethane; 1,2-diphenylethane. $C_{14}H_{14}$; mol wt 182.27. C 92.26%, H 7.74%. Prepn by the reduction of benzil or benzoin: Clemmensen, *Ber.* **47**, 688 (1914); Gattermann-Wieland, *Praxis des Organischen Chemikers* (de Gruyter, Berlin, 40th ed., 1961) p 333; from stilbene: Kleiderer, Kornfeld, *J. Org. Chem.* **13**, 455 (1948).

Monoclinic prisms from methanol, mp 52.0-52.5°. bp_{760} 284°. d_4^0 1.104; d_4^{25} 0.9782; d_4^{58} 0.958. Heat of combustion (25°): 9909.9 cal (15°)/g. Moderately sol in alcohol; freely sol in carbon disulfide, ether, chloroform, amyl acetate; sol in liquid sulfur dioxide. Practically insol in water, liquid ammonia.

1203. Bibrocathol. [6915-57-7] 4,5,6,7-Tetrabromo-2-hydroxy-1,3,2-benzodioxabismole; tetrabromopyrocatechol bismuth derivative; bismuth tetrabromopyrocatechol; bibrocathin; Noviform; Posiformin. $C_6HBiBr_4O_3$; mol wt 649.67. C 11.09%, H 0.16%, Bi 32.17%, Br 49.20%, O 7.39%. Prepn: Hundrup, *Arch. Pharm. Chemi* **54**, 537 (1947), *C.A.* **42**, 2727a (1948). Antibacterial activity: Frank, Stark, *Pharm. Acta Helv.* **29**, 283 (1954). Identity tests: Hakkesteegt, *Pharm. Weekbl.* **99**, 922 (1964).

Yellow, odorless, tasteless powder. Practically insol in water. Slightly sol in alcohol, ether. Dec in acid, alkalies.

THERAP CAT: Antiseptic.

1204. Bicalutamide. [90357-06-5] *N*-[4-Cyano-3-(trifluoromethyl)phenyl]-3-[(4-fluorophenyl)sulfonyl]-2-hydroxy-2-methylpropanamide; 4-cyano-3-trifluoromethyl-*N*-(3-*p*-fluorophenylsulfonyl-2-hydroxy-2-methylpropionyl)aniline; (±)-4'-cyano-α,α,α-trifluoro-3-[(*p*-fluorophenyl)sulfonyl]-2-methyl-*m*-lactotoluidide; ICI-176334; Casodex. $C_{18}H_{14}F_4N_2O_4S$; mol wt 430.37. C 50.24%, H 3.28%, F 17.66%, N 6.51%, O 14.87%, S 7.45%. Non-steroidal, peripherally active antiandrogen. Prepn: H. Tucker, **EP 100172**; *idem*, **US 4636505** (1984, 1987 both to ICI); and structure-activity: H. Tucker *et al.*, *J. Med. Chem.* **31**, 954 (1988). Resolution and activity of enantiomers: H. Tucker, G. J. Chesterson, *ibid.* 885. Pharmacokinetics: I. D. Cockshott *et al.*, *Eur. Urol.* **18**, Suppl. 3, 10 (1990). Clinical trials in prostate cancer: G. Lunglmayr, *Horm. Res.* **32**, Suppl. 1, 77 (1989); D. W. W. Newling, *Eur. Urol.* **18** Suppl. 3, 18 (1990). Review of pharmacology: B. J. A. Furr in *Hormonal Therapy of Prostatic Diseases: Basic and Clinical Aspects*, M. Motta, M. Serio, Eds. (Mediocom Europe, Milan, 1987) 148-161.

Crystals from a 1:1 (v/v) mix of ethyl acetate and petroleum ether, mp 191-193°. Freely sol in tetrahydrofuran, acetone; sol in acetonitrile; sparingly sol in methanol; slightly sol in alcohol.

THERAP CAT: Antiandrogen; antineoplastic (hormonal).

1205. Bicine. [150-25-4] *N*,*N*-Bis(2-hydroxyethyl)glycine; di(hydroxyethyl)glycine; *N*,*N*-bis(hydroxyethyl)aminoacetic acid;

N,*N*-di(hydroxyethyl)aminoacetic acid; diethylolglycine; 2-HxG. $C_6H_{13}NO_4$; mol wt 163.17. C 44.17%, H 8.03%, N 8.58%, O 39.22%. One of the zwitterionic amino acids known as "Good" buffers, active in the pH range 6-8.5. Prepn by hydrolysis of its lactone (obtained from glycine and ethylene oxide): Pascal, *Compt. Rend.* **245**, 1318 (1957); from diethanolamine and bromoacetic acid: N. E. Good *et al.*, *Biochemistry* **5**, 467 (1966). Crystal structure: V. Cody *et al.*, *Acta Crystallogr.* **33B**, 905 (1977). Use as sequestering agent: A. Grawitz, *Rev. Tech. Ind. Cuir* **65**, 187, 190 (1973), *C.A.* **80**, 49281h (1974). Temperature effects on pKa: M. L. Soni, R. C. Kapoor, *Int. J. Quantum Chem.* **20**, 385 (1981). Use as a buffer: A. L. Remisov, *Biokhimiya* **25**, 323 (1960); S. Ito *et al.*, *Histochem. J.* **16**, 489 (1984).

Crystals from dil ethanol, mp 193-195° (slight decompn). pKa (20°): 8.35. pKa_2 (0.1*M*): 0°, 8.7; 20°, 8.35; 37°, 8.2. $\Delta pKa/°C$ −0.018. Saturated aq soln is 1.1*M* at 0°. Slightly sol in water. Forms a water-soluble sodium salt.

USE: Biological buffer and chelating agent.

1206. Bicuculline. [485-49-4] (6*R*)-6-[(5*S*)-5,6,7,8-Tetrahydro-6-methyl-1,3-dioxolo[4,5-*g*]isoquinolin-5-yl]furo[3,4-*e*]-1,3-benzodioxol-8(6*H*)-one. $C_{20}H_{17}NO_6$; mol wt 367.36. C 65.39%, H 4.66%, N 3.81%, O 26.13%. Alkaloid naturally occurring in the *d*-form; found in *Dicentra cucullaria* (L.) Bernh., *Adlumia fungosa* (Ait.) Greene, *Fumariaceae*, and several *Corydalis* species: Manske, *Can. J. Res.* **7**, 265 (1932); **8**, 210, 407 (1933); **9**, 436 (1933); Edwards, Handa, *Can. J. Chem.* **39**, 1801 (1961). Synthesis of *dl*-form: Groenewoud, Robinson, *J. Chem. Soc.* **1936**, 199. Resolution of isomers: Haworth *et al.*, *Nature* **165**, 529 (1950). Stereoisomer of adlumidine, *q.v.*, and of capnoidine: Manske, *J. Am. Chem. Soc.* **72**, 3207 (1950). Preliminary stereochemical studies: Safe, Moir, *Can. J. Chem.* **42**, 160 (1964). Revised stereochemistry: Blaha *et al.*, *Collect. Czech. Chem. Commun.* **29**, 2328 (1964); Snatzke *et al.*, *Tetrahedron* **25**, 5059 (1969). Crystal and molecular structure: Gorinsky, Moss, *J. Cryst. Mol. Struct.* **3**, 299 (1973). Shows GABA (*q.v.*) antagonist activity: Curtis *et al.*, *Nature* **226**, 1222 (1970).

Elongated plates from chloroform-methanol, mp 215°; also reported as mp 177°, solidifies and remelts 193-195°: Manske, *Can. J. Res.* **21B**, 13 (1943). uv max (acidified ethanol): 225, 296, 324 nm (ε 36700, 6390, 5870). $[α]_D^{25}$ +130.5° (CHCl₃). pKa 4.84. Sol in benzene, chloroform, ethyl acetate. Sparingly sol in alc and ether.

1207. Bicyclomycin. [38129-37-2] (1*S*,6*R*)-6-Hydroxy-5-methylene-1-[(1*S*,2*S*)-1,2,3-trihydroxy-2-methylpropyl]-2-oxa-7,9-diazabicyclo[4.2.2]decane-8,10-dione; bicozamycin; aizumycin; 8,10-diaza-6-hydroxy-5-methylene-1-(2-methyl-1,2,3-trihydroxypropyl)-2-oxabicyclo[4.2.2]decan-7,9-dione; antibiotic 5879; WS-4545 antibiotic. $C_{12}H_{18}N_2O_7$; mol wt 302.28. C 47.68%, H 6.00%, N 9.27%, O 37.05%. Structurally unique antibiotic; exhibits broad spectrum activity vs Gram-negative bacteria and the Gram-positive bacterium, *Micrococcus luteus*. Prepn by fermentation of *Streptomyces sapporensis*: T. Miyoshi *et al.*, **DE 2150593**; H. Imanaka *et al.*, **US 3923790** (1972, 1975 both to Fujisawa). Isoln and characterization: T. Miyoshi *et al.*, *J. Antibiot.* **25**, 569 (1972). Isoln from *S.*

aizunensis and identity with antibiotic 5879: S. Miyamura *et al.*, *ibid.* **26**, 479 (1973). Structural elucidation: T. Kamiya *et al.*, *ibid.* **25**, 576 (1972). Crystal and molecular structure: Y. Tokuma *et al.*, *Bull. Chem. Soc. Jpn.* **47**, 18 (1974). *In vitro* and *in vivo* activity studies: M. Nishida *et al.*, *J. Antibiot.* **25**, 582 (1972). Metabolism: *eidem, ibid.* 594. Total synthesis of (±)-bicyclomycin: S. Nakatsuka *et al.*, *Tetrahedron Lett.* **24**, 5627 (1983); of (+)-bicyclomycin: R. M. Williams *et al.*, *J. Am. Chem. Soc.* **106**, 5749 (1984); *eidem, ibid.* **107**, 3253 (1985). Efficacy vs Shiga toxin-producing *E. coli* in calves: N. Misawa *et al.*, *Microbiol. Immunol.* **44**, 891 (2000). Review of synthetic, mechanistic and biological studies: R. M. Williams, C. A. Durham, *Chem. Rev.* **88**, 511-540 (1988); of mechanism of action: H. Kohn, W. Widger, *Curr. Drug Targets Infect. Disord.* **5**, 273-295 (2005).

Monoclinic crystals from ethanol, mp 188-191° (dec); rhombic crystals from methanol + acetone, mp 187-189° (dec) (Imanaka), also reported as mp 166-170° (Nakatsuka *et al.*). $[\alpha]_D^{23}$ +63.5° (methanol). Weakly basic substance. Soly in water: 192 mg/ml. Sol in methanol; sparingly sol in ethanol. Slightly sol in acetone. Practically insol in chloroform, ethyl acetate, benzene, *n*-hexane. Unstable in alkaline soln. LD$_{50}$ orally, s.c., i.p. in mice: >4 g/kg (Williams, Durham).

Benzoate. [37134-40-0] Bicomarin. $C_{19}H_{22}N_2O_8$; mol wt 406.39.

THERAP CAT (VET): Antibacterial; feed additive (livestock).

1208. Bidisomide. [116078-65-0] α-[2-[Acetyl-(1-methylethyl)amino]ethyl]-α-(2-chlorophenyl)-1-piperidinebutanamide; (±)-α-(*o*-chlorophenyl)-α-[2-(*N*-isopropylacetamido)ethyl]-1-piperidinebutyramide; SC-40230. $C_{22}H_{34}ClN_3O_2$; mol wt 407.98. C 64.77%, H 8.40%, Cl 8.69%, N 10.30%, O 7.84%. Sodium channel blocker. Prepn: B. N. Desai *et al.*, **EP 170901**; *eidem*, **US 4639524** (1986, 1987 both to Searle); *eidem, J. Med. Chem.* **31**, 2158 (1988). Cardiovascular pharmacology: L. G. Frederick *et al.*, *Cardiovasc. Res.* **23**, 897 (1989). Electrophysiologic study: S. M. Garthwaite *et al.*, *Drug Dev. Res.* **27**, 329 (1992). Bioavailability and tolerability in humans: R. L. Page *et al.*, *Clin. Pharmacol. Ther.* **51**, 371 (1992). Clinical pharmacokinetics: C. S. Cook *et al.*, *Pharm. Res.* **10**, 1675 (1993).

Crystals from ethyl acetate, mp 140-141°.

THERAP CAT: Antiarrhythmic (class I).

1209. Bietamiverine. [479-81-2] α-Phenyl-1-piperidineacetic acid 2-(diethylamino)ethyl ester; β-diethylaminoethyl phenyl-piperidinoacetate; β-diethylaminoethyl α-(1-piperidyl)phenylacetate. $C_{19}H_{30}N_2O_2$; mol wt 318.46. C 71.66%, H 9.50%, N 8.80%, O 10.05%. Prepd by reacting piperidine with phenylchloroacetic acid ethyl ester in chloroform: Reetz, **DE 859892** (1952 to Nordmark-Werke); from β-diethylaminoethyl phenylbromoacetate and piperidine in chloroform: Blicke *et al.*, *J. Am. Chem. Soc.* **76**, 3161 (1954); from ethyl α-(1-piperidyl)phenylacetate and β-diethylaminoethyl chloride: Moffett, Hart, *J. Am. Pharm. Assoc. Sci. Ed.* **42**, 717 (1953). Clinical tests: Ciravegna *et al.*, *Clin. Ter.* **57**, 227 (1971).

Liquid, d_4^{25} 1.0184. bp$_1$ 65°. n_D^{25} 1.5070.

Hydrochloride. [1477-10-7] Novosparol. $C_{19}H_{30}N_2O_2 \cdot HCl$; mol wt 354.92. Crystals from methyl isobutyl ketone, mp 187-189°.

Dihydrochloride. [2691-46-5] $C_{19}H_{30}N_2O_2 \cdot 2HCl$. Crystals from ethanol + ether, mp 194-195°.

THERAP CAT: Antispasmodic.

1210. Bifemelane. [90293-01-9] *N*-Methyl-4-[2-(phenylmethyl)phenoxy]-1-butanamine; 4-(*o*-benzylphenoxy)-*N*-methylbutylamine; 2-(4-methylaminobutoxy)diphenylmethane; 2-benzyl-1-[4-(methylamino)butoxy]benzene. $C_{18}H_{23}NO$; mol wt 269.39. C 80.25%, H 8.61%, N 5.20%, O 5.94%. Monoamine oxidase inhibitor. Prepn: R. Kikumoto *et al.*, **DE 2627227**; *eidem*, **US 4091114** (1976, 1978 both to Mitsubishi); and antidepressant activity: R. Kikumoto *et al.*, *J. Med. Chem.* **24**, 145 (1981). Pharmacology: A. Tobe *et al.*, *Arzneim.-Forsch.* **31**, 1278 (1981). Effects on experimental amnesia in rats: A. Tobe *et al.*, *Jpn. J. Pharmacol.* **39**, 153 (1985). Effects on neuronal activity in cats: M. Egawa *et al.*, *Neuropharmacology* **26**, 379 (1987). Inhibition of MAO: M. Naoi *et al.*, *J. Neurochem.* **50**, 243 (1988).

Hydrochloride. [62232-46-6] E-0687; MCI-2016; Alnert; Celeport. $C_{18}H_{23}NO \cdot HCl$; mol wt 305.85. Crystals from acetone, mp 117-121°. LD$_{50}$ in mice, rats (mg/kg): 1000, 1080 orally; 173, 130 i.p. (Tobe, 1981).

THERAP CAT: Nootropic.

1211. Bifenazate. [149877-41-8] 2-(4-Methoxy[1,1'-biphenyl]-3-yl)hydrazinecarboxylic acid 1-methylethyl ester; isopropyl 3-(4-methoxy-3-biphenylyl)carbazate; D-2341; Floramite. $C_{17}H_{20}N_2O_3$; mol wt 300.36. C 67.98%, H 6.71%, N 9.33%, O 15.98%. Hydrazinecarboxylate miticide for use on fruits and ornamentals. Prepn: M. A. Dekeyser, P. T. McDonald, **WO 9310083**; *eidem*, **US 5367093** (1993, 1994 both to Uniroyal). Review of physical properties, biological activity, and field trials: M. A. Dekeyser *et al.*, *Brighton Crop Prot. Conf. - Pests Dis.* **1996**, 487-492.

White crystals, mp 120-121°. Soly in water at 20°: 3.76 mg/l. Log P (*n*-octanol/water) at 25°: 3.4 (pH 7). LD$_{50}$ in rats (mg/kg): 5000 orally, >2000 dermally (Dekeyser).

USE: Acaricide.

1212. Bifenox. [42576-02-3] 5-(2,4-Dichlorophenoxy)-2-nitrobenzoic acid methyl ester; methyl 5-(2,4-dichlorophenoxy)-2-nitrobenzoate; 2,4-dichlorophenyl 3-(methoxycarbonyl)-4-nitrophenyl ether; MC-4379; Modown. $C_{14}H_9Cl_2NO_5$; mol wt 342.13. C 49.15%, H 2.65%, Cl 20.72%, N 4.09%, O 23.38%. Pre-emergence herbicide. Prepn and herbicidal activity: **BE 749444**; R. J. Thiessen, **US 3652645** (1970, 1972 both to Mobil). Comparison with other herbicides in corn: W. M. Dest *et al.*, *Proc. Annu. Meet. Northeast. Weed Sci. Soc.* **27**, 31 (1973). Metabolism in soil, plants: G. R.

Leather, C. L. Foy, *Pestic. Biochem. Physiol.* **7**, 437 (1977). Brief description: R. H. Dreger, *Weeds Today* **8**(2), 18 (1977). *Review:* P. J. Kruger *et al., Proc. 12th Br. Weed Control Conf.* **2**, 839-845 (1974).

Yellow tan crystals, mp 84-86°. Practically insol in water (0.35 ppm at 25°). Soly in xylene at 25°: 30%. Vapor pressure at 30°: 2.4 × 10⁻⁶ mm Hg. LD₅₀ orally in rats, mice: >6400, 4556 mg/kg; LC₅₀ in pheasants, wild ducks: >5000 ppm (Kruger).

USE: Herbicide.

1213. Bifenthrin. [82657-04-3] *rel*-(1*R*,3*R*)-3-[(1*Z*)-2-Chloro-3,3,3-trifluoro-1-propen-1-yl]-2,2-dimethylcyclopropanecarboxylic acid (2-methyl[1,1'-biphenyl]-3-yl)methyl ester; 2-methylbiphenyl-3-ylmethyl-(*Z*)-(1*RS*)-*cis*-3-(2-chloro-3,3,3-trifluoroprop-1-enyl)-2,2-dimethylcyclopropanecarboxylate; biphenate; biphenthrin; biphentrin; FMC-54800; Brigade; Talstar; Capture; Wisdom. C₂₃H₂₂ClF₃O₂; mol wt 422.87. C 65.33%, H 5.24%, Cl 8.38%, F 13.48%, O 7.57%. Third generation synthetic pyrethroid. Prepn: J. F. Engel, **EP 3336**; *idem,* **US 4238505** (1979, 1980 both to FMC); E. L. Plummer *et al., Pestic. Sci.* **14**, 560 (1983). Physical properties, toxicology and review of field studies: H. J. H. Doel *et al., Meded. Fac. Landbouwwet. Rijksuniv. Gent* **49**, 929 (1984). Insecticidal activity: M. S. Mulla, H. A. Darwazeh, *Bull. Soc. Vector Ecol.* **10**, 1 (1985); M. S. Hamed, C. O. Knowles, *J. Econ. Entomol.* **81**, 1295 (1988). Field studies: Y. Antignus *et al., Ann. Appl. Biol.* **110**, 557 (1987); J. T. Trumble *et al., J. Econ. Entomol.* **81**, 608 (1988). Review of toxicology and human exposure: *Toxicological Profile for Pyrethrins and Pyrethroids* (PB2004-100004, 2003) 332 pp.

Relative stereochemistry

Light brown viscous oil, mp 51-66°. d²⁵ 1.212 g/ml. Vapor pressure at 25°: 1.81 × 10⁻⁷ Torr. Sol in methylene chloride, chloroform, acetone, ether, toluene. Slightly sol in heptane, methanol. Soly in water: <0.1 ppb. LD₅₀ orally in rats: 54.5 mg/kg; LD₅₀ dermally in rabbits: >2000 mg/kg (Doel).

USE: Insecticide, acaricide.

1214. Bifeprunox. [350992-10-8] 7-[4-([1,1'-Biphenyl]-3-ylmethyl)-1-piperazinyl]-2(3*H*)-benzoxazolone. C₂₄H₂₃N₃O₂; mol wt 385.47. C 74.78%, H 6.01%, N 10.90%, O 8.30%. Exhibits partial agonist activity at dopamine D₂ and serotonin 5-HT₁ₐ receptors. Prepn: R. W. Feenstra *et al.,* **WO 9736893**; *idem,* **US 6225312** (1997, 2001 both to Duphar); and receptor binding studies: *idem et al., Bioorg. Med. Chem. Lett.* **11**, 2345 (2001). Analysis of binding affinity vs *in vitro* efficacy: A. Newman-Tancredi *et al., Int. J. Neuropsychopharmacol.* **8**, 341 (2005). Review of development: W. Wolf, *Curr. Opin. Investig. Drugs* **4**, 72-76 (2003).

Methanesulfonate. [350992-13-1] Bifeprunox mesylate; DU-127090. C₂₄H₂₃N₃O₂.CH₄O₃S; mol wt 481.57.

THERAP CAT: Antipsychotic.

1215. Bifidus Factor. [9007-03-8] *Lactobacillus bifidus* factor; *Lactobacillus bifidus* growth factor. A factor found in human milk and causing a predominant occurrence of *L. bifidus* in the intestinal tract of breast-fed infants: Petuely, Kristen, *Oesterr. Z. Kinderheilkd. Kinderfuersorge* **6**, 173 (1951); Petuely, *Naturwissenschaften* **40**, 349 (1953). Essential growth factor for *L. bifidus* var *Penn:* György in Ciba Found. Symp., *Chemistry and Biology of Mucopolysaccharides* (Little, Brown, Boston, 1958) pp 140-156. Isoln from human milk: György *et al., Arch. Biochem. Biophys.* **48**, 193, 202, 209, 214 (1954); György *et al.,* **US 2786051** (1957 to Am. Home Prod.); from *L. bifidus* cultured together with *Escherichia coli:* Kludas, **US 2962424** (1960 to J. Carl Pflüger). Isoln from carrots and identification of the major and minor bifidus factor: Samejima *et al., Chem. Pharm. Bull.* **19**, 166, 178, 186 (1971); Z. Tamura *et al., Proc. Jpn. Acad.* **48**, 138, 144 (1972). Prepn from porcine gastric mucosa and use as dietetic adjuvant in infant food: **FR 2101032**, *C.A.* **78**, 28221g (1970); P. C. Wirth, **DE 2040268**, *C.A.* **78**, 41859r (1970) (both 1972 to Sogeras).

USE: As adjuvant in powdered milk formulas for infants.

1216. Bifonazole. [60628-96-8] 1-([1,1'-Biphenyl]-4-yl-phenylmethyl)-1*H*-imidazole; (±)-1-(*p*,α-diphenylbenzyl)imidazole; Bay h 4502; Amycor; Azolmen; Bedriol; Mycospor; Mycosporan. C₂₂H₁₈N₂; mol wt 310.40. C 85.13%, H 5.85%, N 9.03%. Antimycotic deriv of imidazole. Prepn: E. Regel *et al.,* **DE 2461406**; *eidem,* **US 4118487** (1976, 1978 both to Bayer). Series of articles on *in vitro* and *in vivo* antimycotic efficacy, microscopic studies, pharmacokinetics, efficacy in dermatomycoses and comparison with clotrimazole and miconazole, *q.q.v.: Arzneim.-Forsch.* **33**, 517-551, 745-754 (1983). Toxicology: G. Schlüter, *ibid.* 739.

Crystals from acetonitrile, mp 142°. Very lipophilic. Sol in alcohols, DMF, DMSO. Soly in water at pH 6: <0.1 mg/100 ml. Stable in aq soln at pH 1-12. LD₅₀ in male mice, rats (mg/kg): 2629, 2854 orally (Schlüter).

THERAP CAT: Antifungal.

1217. BIGCHAP. [86303-22-2] (3α,5β,7α,12α)-*N*,*N*-Bis[3-(D-gluconoylamino)propyl]-3,7,12-trihydroxycholan-24-amide. C₄₂H₇₅N₃O₁₆; mol wt 878.07. C 57.45%, H 8.61%, N 4.79%, O 29.15%. Homogeneous nonionic detergent; derivative of cholic acid, *q.v.* Prepn: L. M. Hjelmeland *et al., Anal. Biochem.* **130**, 485 (1983). QSAR study as surfactant: P. Campbell *et al., Biotechnol. Bioeng.* **64**, 527 (1999). Use in protein purification via hydrophobic interaction displacement chromatography: K. M. Sunasara *et al., ibid.* **82**, 330 (2003). Clinical evaluation as a transduction enhancing agent in gene therapy: J. Kuball *et al., J. Clin. Oncol.* **20**, 957 (2002).

BIGCHAP R = OH
deoxy-BIGCHAP R = H

Crystals at 0° from 2-propanol. Critical micelle conc: 2.9 mM; also reported as 3.4 mM. Log P (octanol/water): 2.450. Dipole moment: 7.44 D.

Deoxy-BIGCHAP. [86303-23-3] N,N'-[[[(3α,5β,12α)-3,12-Di-hydroxy-24-oxocholan-24-yl]imino]di-3,1-propandiyl]bis-D-glu-conamide. $C_{42}H_{75}N_3O_{15}$; mol wt 862.07. Use as mobile phase additive in microcolumn LC: W. Hu *et al.*, *J. High Resolut. Chromatogr.* **15**, 275 (1992). Critical micelle conc: 1.4 mM.

USE: Surfactant and solubilizer for biological systems.

1218. Biguanide. [56-03-1] Imidodicarbonimidic diamide; guanylguanidine; amidinoguanidine; diguanide. $C_2H_7N_5$; mol wt 101.11. C 23.76%, H 6.98%, N 69.27%. Preparation from dicyanodiamide: Rackmann, *Ann.* **376**, 169 (1910); Karipides, Fernelius, *Inorg. Synth.* **7**, 56, 58 (1963). Crystal structure: S. R. Ernst, F. W. Cagle, *Acta Crystallogr.* **33B**, 235 (1977); S. R. Ernst, *ibid.* 237.

Crystals from alc, mp 130°; dec rapidly at about 142°. Sol in water, alcohol. Insol in ether, benzene, chloroform. The aq soln dec on standing or heating. Aq solns are alkaline. Biguanide is a stronger base (pK_1 = 11.52; pK_2 = 2.93) than ammonia (pK = 9.61).

Hydrochloride. $C_2H_7N_5$.HCl. Needles, mp 235°, very sol in water, slightly sol in alc.

Neutral sulfate. $(C_2H_7N_5)_2$.H_2SO_4.$2H_2O$. Large crystals, sol in water.

Acid sulfate. $C_2H_7N_5$.H_2SO_4.H_2O. Rhombic prisms, sol in water, the soln giving an acid reaction.

USE: Sulfates in the determination of copper and nickel.

1219. Bikhaconitine. [6078-26-8] (1α,6α,14α,16β)-20-Ethyl-1,6,16-trimethoxy-4-(methoxymethyl)aconitane-8,13,14-triol 8-acetate 14-(3,4-dimethoxybenzoate); acetylveratroylbikhaconine. $C_{36}H_{51}NO_{11}$; mol wt 673.80. C 64.17%, H 7.63%, N 2.08%, O 26.12%. From roots of *Aconitum spicatum* Stapf., *Ranunculaceae*: Dunstan, Andrews, *J. Chem. Soc.* **87**, 1636 (1905). Structure: Tsuda, Marion, *Can. J. Chem.* **41**, 3055 (1963).

Prisms from *n*-hexane, mp 163.5-164°.

Monohydrate. Needles from ether or dil methanol, mp 105-110°. $[\alpha]_D$ +16° (c = 1.6 in ethanol). Sol in alcohol, chloroform, ether, dil acids. Practically insol in water, petr ether.

1220. Bilastine. [202189-78-4] 4-[2-[4-[1-(2-Ethoxyethyl)-1*H*-benzimidazol-2-yl]-1-piperidinyl]ethyl]-α,α-dimethylbenzene-acetic acid; 2-(4-(2-(4-(1-(2-ethoxyethyl)-1*H*-benzo[*d*]imidazol-2-yl)piperidin-1-yl)ethyl)phenyl)-2-methylpropanoic acid. $C_{28}H_{37}$-N_3O_3; mol wt 463.62. C 72.54%, H 8.04%, N 9.06%, O 10.35%. Non-sedating, nonbrain penetrant H_1 receptor antagonist. Prepn: A. Orjales *et al.*, **EP 0818454**; *eidem*, **US 5877187** (1998, 1999 both to FAES). Alternative synthesis: S. J. Collier *et al.*, *Synth. Commun.* **41**, 1394 (2011). HPLC determn in feces: L. A. Berrueta *et al.*, *J. Chromatogr. B* **760**, 185 (2001). Clinical pharmacokinetics: N. Jauregizar *et al.*, *Clin. Pharmacokinet.* **48**, 543 (2009). Pharmacology: C. Bachert *et al.*, *Allergy* **64**, 158 (2009). Clinical trial in seasonal allergic rhinitis: P. Kuna *et al.*, *Clin. Exp. Allergy.* **39**, 1338 (2009); in chronic idiopathic urticaria: T. Zuberbier *et al.*, *Allergy* **65**, 516 (2010). Review of pharmacology and clinical experience in allergic rhinoconjunctivitis and urticaria: C. Bachert *et al.*, *ibid.* **65**, Suppl. 93, 1-13 (2010); M. K. Church *Expert Opin. Drug Saf.* **10**, 779-793 (2011).

Yellow solid, mp 199-201°.

THERAP CAT: Antihistaminic.

1221. Bilberry. Heidelberry; huckleberry; European blueberry; whortleberry. Deciduous, dwarf shrub, *Vaccinium myrtillus* L., *Ericaceae*, bearing edible blue-black berries. Medicinal parts include the leaves and fruit. *Habit.* Northern and central Europe. *Constit.* Berries: Anthocyanins, particularly glycosides of delphinidin, cyanidin, petunidin, peonidin, malvidin; quercetin, catechin, epicatechin, tannins, pectins, vitamin C. Leaves: Catechol tannins (0.8-6.7%); leucoanthocyans; flavonoids esp. quercetin glycosides; phenolic acids incl. caffeic, *p*-coumaric, *p*-hydroxybenzoic, protocatechuic, melilotic; iridoids; manganese; chromium. Berries have been used in traditional medicine to treat diarrhea and mouth and throat inflammations; leaves used in antidiabetic teas. Comprehensive description and medicinal uses: J. Barnes *et al.*, *Herbal Medicines* (Pharmaceutical Press, London, 2nd Ed., 2002) pp 73-77. Anthocyanin content and radical scavenging activity: J. Nakajima *et al.*, *J. Biomed. Biotechnol.* **5**, 241-247 (2004). Phenolic profile and antioxidant activity: S. Ehala *et al.*, *J. Agric. Food Chem.* **53**, 6484 (2005). Review of clinical trials in impaired night vision: P. H. Carter, E. Ernst, *Surv. Ophthalmol.* **49**, 38-50 (2004).

Bilberry extract. [84082-34-8] Alcodin; Difrarel; Tegens. Anthocyanin-enriched extract from the fruit.

USE: In jams and jellies; food flavoring. Extracts as astringent and capillary protectant in eyecare and skin products.

THERAP CAT: In treatment of night blindness.

1222. Bilirubin. [635-65-4] 2,17-Diethenyl-1,10,19,22,-23,24-hexahydro-3,7,13,18-tetramethyl-1,19-dioxo-21*H*-biline-8,12-dipropanoic acid; 1,10,19,22,23,24-hexahydro-2,7,13,17-tetra-methyl-1,19-dioxo-3,18-divinylbiline-8,12-dipropionic acid; 1,3,-6,7-tetramethyl-4,5-dicarboxyethyl-2,8-divinyl-(*b*-13)-dihydrobil-enone; bilirubin IXα. $C_{33}H_{36}N_4O_6$; mol wt 584.67. C 67.79%, H 6.21%, N 9.58%, O 16.42%. Principal pigment of bile and constituent of many biliary calculi. Major end-product of the biological breakdown of heme, *q.v.* Bilirubin is the chromophore responsible for coloration in various forms of jaundice. Also found in blood serum, where it exists in four major forms: unconjugated bilirubin, the monoglucuronide, the diglucuronide, and albumin-bound bilirubin. Most easily obtained from ox gallstones which are largely calcium bilirubinate: Städeler, *Ann.* **132**, 323 (1864); Küster, *Z. Physiol. Chem.* **94**, 136 (1915); **99**, 86 (1917); **121**, 80 (1922); Küster, Haas, *ibid.* **141**, 279 (1924); Fischer, *ibid.* **3**, 204 (1911); Fischer, Hess, *ibid.* **194**, 193 (1931). Isoln from pig bile: Gibson, Lowe, *J. Biol. Chem.* **123**, XLI (1938); Gray *et al.*, *J. Chem. Soc.* **1961**, 2264, 2268; from ox bile: Libowitzky, *Z. Physiol. Chem.* **263**, 267 (1940). Industrial isoln from ox bile using chlorobenzene as extractant: **US 2166073**; **US 2331574**; **US 2363471**; **US 2386716**. Structure and synthesis: Fischer, Plieninger, *Naturwissenschaften* **30**, 382 (1942); *Z. Physiol. Chem.* **274**, 231 (1942); *cf.* Fischer-Orth, *Die Chemie des Pyrrols* **II**, 1, 621 (Leipzig, 1937); Gray *et al.*, *Nature* **181**, 183 (1958); *eidem*, *J. Chem. Soc.* **1961**, 2276. Structure: Fog, Jellum, *Nature* **198**, 88 (1963). Configuration: Kuenzle *et al.*, *Biochem. J.* **133**, 364 (1973). X-ray analysis and structure: Bonnett *et al.*, *J. Chem. Soc. Perkin Trans. 2* **1972**, 902, 1335; *Nature* **262**, 326 (1976). NMR conformation studies: D. Kaplan, G. Navon, *J. Chem. Soc. Perkin Trans. 2* **1981**, 1374. Separation of bilirubin species in serum and bile by reversed-phase HPLC: J. J. Lauff *et al.*, *J. Chromatogr.* **226**, 391 (1981); *eidem*, *ibid.* **3**, 800 (1983). Clinical importance of albumin-bound bilirubin: J. S. Weiss *et al.*, *N. Engl. J. Med.* **309**, 148 (1983). Comprehensive reviews: Lemberg, Legge, *Hematin Compounds and Bile Pigments* (New York, 1949);

With, *Bile Pigments* (Academic Press, New York, 1968). Reviews of toxicity: D. Bratlid, *N. Y. State J. Med.* **91**, 489-493 (1991); R. P. Wennberg, *ibid.* 493-496.

Light orange to deep reddish-brown monoclinic rhomboid, prisms, plates from chloroform. Gradually blackens on heating and does not melt. The greenish solns show a red fluorescence in ultraviolet light. A 0.001% soln in chloroform shows selective abs from 490 to 400 nm with a max at 453 nm; ε mM 60.7 ± 0.8. Practically insol in water. Sol in benzene, chloroform, chlorobenzene, carbon disulfide, acids, alkalies; slightly sol in alcohol, ether. Spreads on water. Penetrates into cholesterol, octadecylamine, and protein monolayers: Stenhagen, Rideal, *Biochem. J.* **33**, 1591 (1939).

1223. Biliverdine. [114-25-0] 3,18-Diethenyl-1,19,22,24-tetrahydro-2,7,13,17-tetramethyl-1,19-dioxo-21*H*-biline-8,12-dipropanoic acid; 1,19,22,24-tetrahydro-2,7,13,17-tetramethyl-1,19-dioxo-3,18-divinylbiline-8,12-dipropionic acid; 1,3,6,7-tetramethyl-4,5-dicarboxyethyl-2,8-divinylbilenone; 4,5-di(2-carboxyethyl)-1,-3,6,7-tetramethyl-2,8-divinylbilatriene; dehydrobilirubin; uteroverdine; oöcyan. $C_{33}H_{34}N_4O_6$; mol wt 582.66. C 68.03%, H 5.88%, N 9.62%, O 16.48%. Precursor of bilirubin. Formed in the body from hemoglobin. The bile of amphibia and of birds contains biliverdine only. Does not occur in normal human bile or normal human serum, but regularly accompanies bilirubin in the serum of patients with carcinomatous obstruction of the bile duct, and frequently in that of patients with liver cirrhosis, catarrhal jaundice, and bile duct occlusion by gallstones. Can be obtained by autooxidation of bilirubin in alkaline soln but the yield is poor: Lemberg, *Biochem. J.* **28**, 978 (1934); better yields by oxidation of bilirubin with ferric chloride in glacial acetic acid: Lemberg, *Ann.* **499**, 25 (1932); by coupled oxidation of hemoglobin and ascorbic acid: Lemberg *et al.*, *Biochem. J.* **35**, 363 (1941); by oxidation of bilirubin with ferric chloride in methanol: Gray *et al.*, *J. Chem. Soc.* **1961**, 2264. Crystal and molecular structure: W. S. Sheldrick, *J. Chem. Soc. Perkin Trans. 2* **1976**, 1457. Comprehensive reviews: Lemberg, Legge, *Hematin Compounds and Bile Pigments* (New York, 1949); With, *Bile Pigments* (Academic Press, New York, 1968).

Dark green plates or prisms with violet surface color from methanol. Does not melt, blackens and dec above 300°. Absorption spectrum: Gray *et al.*, *loc. cit.* Soluble in methanol, ether, chloroform, carbon disulfide, benzene, soln of alkali hydroxides. Gives the later color changes of the Gmelin test starting with green.

Dimethyl ester. $C_{35}H_{38}N_4O_6$. Green crystals from chloroform + petr ether. When produced from bilirubin, mp 215-223°; synthetic 206-209°; from hemin 208°; from hemoglobin 216°.

Dimethyl ester ferrichloride. $C_{35}H_{39}Cl_4FeN_4O_6$. (Green hemin ester), pleochroitic elongated platelets, no definite mp. Ferric chloride and HCl can be removed by washing the chloroform soln with water.

1224. Bilobalide. [33570-04-6] (3a*S*,5a*R*,8*R*,8a*S*,9*R*,10a*S*)-9-(1,1-Dimethylethyl)-10,10a-dihydro-8,9-dihydroxy-4*H*,5a*H*,9*H*-furo[2,3-*b*]furo[3′,2′:2,3]cyclopenta[1,2-*c*]furan-2,4,7(3*H*,8*H*)trione. $C_{15}H_{18}O_8$; mol wt 326.30. C 55.21%, H 5.56%, O 39.23%. Sesquiterpene isolated from the leaves of *Ginko biloba* L., *Ginkgoaceae*; closely related to the ginkgolides, *q.v.* Component of the *Ginkgo biloba* extract, EGb 761, *q.v.* Isoln: K. Weinges, W. Bähr, *Ann.* **724**, 214 (1969). NMR and MS analysis: *eidem, ibid.* **759**, 158 (1972). Structure determn: K. Nakanishi *et al.*, *J. Am. Chem. Soc.* **93**, 3544 (1971). Total synthesis of (±)-form: E. J. Corey, W. Su, *ibid.* **109**, 7534 (1987). Enantioselective synthesis of (−)-form: *eidem, Tetrahedron Lett.* **29**, 3423 (1988). Separation from *Ginkgo biloba* extracts: S. Jaracz *et al.*, *Phytochemistry* **65**, 2897 (2004). Neuroprotective effects in animal models of brain ischemia: K. Chandrasekaran *et al.*, *Brain Res.* **922**, 282 (2001). GABA$_A$ antagonist activity: S. H. Huang *et al.*, *Eur. J. Pharmacol.* **464**, 1 (2003). Review of neuroprotective properties: F. V. Defeudis, *Pharmacol. Res.* **46**, 565-568 (2002); of physical properties: T. A. van Beek, *Bioorg. Med. Chem.* **13**, 5001-5012 (2005).

Colorless crystals from water, mp >300°. $[\alpha]_{578}^{20}$ −64.3° (c = 2 in acetone).

1225. Bimatoprost. [155206-00-1] (5*Z*)-7-[(1*R*,2*R*,3*R*,5*S*)-3,5-Dihydroxy-2-[(1*E*,3*S*)-3-hydroxy-5-phenyl-1-penten-1-yl]-cyclopentyl]-*N*-ethyl-5-heptenamide; (5*Z*,9*α*,11*α*,13*E*,15*S*)-9,11,15-trihydroxy-17-phenyl-18,19,20-trinorprosta-5,13-dienoic acid ethylamide; 17-phenyl-18,19,20-trinorprostaglandin F$_{2\alpha}$ ethylamide; AGN-192024; Latisse; Lumigan. $C_{25}H_{37}NO_4$; mol wt 415.57. C 72.26%, H 8.97%, N 3.37%, O 15.40%. Synthetic prostamide; structurally related to prostaglandin F$_{2\alpha}$, *q.v.* Prepn: D. F. Woodward *et al.*, **WO 9406433**; *eidem*, **US 6403649** (1994, 2002 both to Allergan). Pharmacology: D. F. Woodward *et al.*, *J. Pharmacol. Exp. Ther.* **305**, 772 (2003). Clinical potential for treatment of eyelash hypotrichosis: A. Tosti *et al.*, *J. Am. Acad. Dermatol.* **51**, S149 (2004). Clinical trial in glaucoma and ocular hypertension: R. Quinones *et al.*, *J. Ocul. Pharmacol. Ther.* **20**, 115 (2004). Review of pharmacology and clinical efficacy: L. B. Cantor, *Expert Opin. Invest. Drugs* **10**, 721-731 (2001); of mechanism of action: A. H.-P. Krauss, D. F. Woodward, *Surv. Ophthalmol.* **49**, Suppl. 1, S5-S11 (2004). Long-term safety and efficacy in glaucoma and ocular hypertension: R. D. Williams *et al.*, *Br. J. Ophthalmol.* **92**, 1387 (2008). Review of pharmacology and clinical use for enhancement of eyelash growth: J. L. Cohen, *Dermatol. Surg.* **36**, 1361-1371 (2010).

Powder. Very sol in ethanol, methanol; slightly sol in water.
THERAP CAT: Antiglaucoma; in treatment of hypotrichosis of the eyelashes.
THERAP CAT (VET): Antiglaucoma in dogs.

1226. Bimosiamose. [187269-40-5] 3′,3‴-(1,6-Hexanediyl)-bis[6′-(α-D-mannopyranosyloxy)[1,1′-biphenyl]]-3-acetic acid; 1,6-bis[3-(3-carboxymethylphenyl)-4-(2-α-D-mannopyranosyloxy)-phenyl]hexane; TBC-1269. $C_{46}H_{54}O_{16}$; mol wt 862.92. C 64.03%, H 6.31%, O 29.66%. Pan-selectin receptor antagonist. Prepn: T. P. Kogan *et al.*, **WO 9701335**; *eidem*, **US 5919768** (1997, 1999 both to Texas Biotechnol.); *eidem*, *J. Med. Chem.* **41**, 1099 (1998). Stereospecific synthesis: I. L. Scott *et al.*, *Carbohyd. Res.* **317**, 210 (1999). Review of pharmacology and clinical development: L. Pradella, *Curr. Opin. Anti-Inflam. Immunomod. Invest. Drugs* **1**, 56-60 (1999); E. Aydt, G. Wolff, *Pathobiology* **70**, 297-301 (2003). Clinical pharmacokinetics: M. Meyer *et al.*, *Int. J. Clin. Pharmacol. Ther.* **43**, 463 (2005). Clinical evaluation in allergic asthma: K. M. Beeh *et al.*, *Pulm. Pharmacol. Ther.* **19**, 233 (2006).

White solid, mp 115-117°.

Disodium salt. [187269-60-9] TBC-1269z. $C_{46}H_{52}Na_2O_{16}$; mol wt 906.89. Slightly hygroscopic white powder, mp 243-245°.

THERAP CAT: Anti-inflammatory

1227. BINAP. [98327-87-8] 1,1′-[1,1′-Binaphthalene]-2,2′-diylbis[1,1-diphenylphosphine]; 2,2′-bis(diphenylphosphino)-1,1′-binaphthyl. $C_{44}H_{32}P_2$; mol wt 622.69. C 84.87%, H 5.18%, P 9.95%. Axially disymmetric bis(triaryl)phosphine ligand. Synthesis: A. Miyashita *et al.*, *J. Am. Chem. Soc.* **102**, 7932 (1980); preparative scale synthesis of enantiomers: *eidem*, *Tetrahedron* **40**, 1245 (1984). Review of asymmetric synthesis by metal-BINAP catalysts: S. Akutagawa, *Appl. Catal. A* **128**, 171-207 (1995). Review of industrial applications: H. Kumobayashi, *Rec. Trav. Chim.* **115**, 201-210 (1996); *idem et al.*, *Synlett* **2001**, S1 1055-1064.

(R)-form

(R)-(+)-Form. [76189-55-4] Crystals from benzene/ethanol, mp 240-241°. $[\alpha]_D^{25}$ +229° (c = 0.32 in benzene).

(S)-(−)-Form. [76189-56-5] Crystals from benzene/ethanol, mp 241-242°. $[\alpha]_D^{25}$ −229° (c = 0.31 in benzene).

USE: Chiral ligand for metal mediated asymmetric catalysis.

1228. Binifibrate. [69047-39-8] 3-Pyridinecarboxylic acid 3,3′-[2-[2-(4-chlorophenoxy)-2-methyl-1-oxopropoxy]-1,3-propanediyl] ester; trihydroxypropane 2-p-chlorophenoxyisobutyrate-1,3-dinicotinate; 2-(p-chlorophenoxy)-2-methylpropionic acid ester with 1,3-dinicotinoyloxy-2-propanol; glyceryl 2-p-chlorophenoxy-isobutyrate-1,3-dinicotinate; WAC-104; Biniwas. $C_{25}H_{23}ClN_2O_7$; mol wt 498.92. C 60.18%, H 4.65%, Cl 7.11%, N 5.61%, O 22.45%. Antilipemic agent structurally related to clofibrate, *q.v.* Prepn: R. Andreoli, X. Cicera, **ES 463218** (1978 to Soc. Espan. Espec. Farm.-Ter.), *C.A.* **90**, 72067h (1979); and pharmacology: *eidem*, **BE 884722** (1980 to Soc. Espan. Espec. Farm.-Ter.). Microhemorheological properties in animals: L. Bruseghini *et al.*, *Arzneim.-Forsch.* **33**, 854 (1983); in patients: M. Dalmau *et al.*, *ibid.* 858. Clinical evaluation: M. Labios *et al.*, *Clin. Hemorheol. Microcirc.* **21**, 79 (1999).

Yellowish-white crystals from alcohol-isopropyl ether, mp 100°. Saponification number 166. LD_{50} orally in mice and rats: >4000 mg/kg (Andreoli).

THERAP CAT: Antilipemic.

1229. Binodenoson. [144348-08-3] 2-[2-(Cyclohexylmethylene)hydrazinyl]adenosine; MRE-0470; WRC-0470; CorVue. $C_{17}H_{25}N_7O_4$; mol wt 391.43. C 52.16%, H 6.44%, N 25.05%, O 16.35%. Selective adenosine A_{2A}-receptor agonist. Prepn: R. A. Olsson, R. D. Thompson, **EP 567094**; *eidem*, **US 5278150** (1993, 1994 both to Whitby Research); K. Niiya *et al.*, *J. Med. Chem.* **35**, 4557 (1992). Dose ranging and clinical evaluation as coronary vasodilator in myocardial imaging: J. E. Udelson *et al.*, *Circulation* **109**, 457 (2004). Clinical pharmacokinetics: R. J. Barrett *et al.*, *J. Nucl. Cardiol.* **12**, 166 (2005). *Review:* A. Zaza, *Curr. Opin. Cardiovasc. Pulm. Renal Invest. Drugs* **1**, 301-306 (1999).

Crystals from methanol/water, mp 154-157°.

THERAP CAT: Diagnostic aid (cardiac stress testing).

1230. BINOL. [602-09-5] [1,1′-Binaphthalene]-2,2′-diol; β-binaphthol; 2,2′-dihydroxy-1,1′-binaphthyl. $C_{20}H_{14}O_2$; mol wt 286.33. C 83.90%, H 4.93%, O 11.18%. C_2-symmetric ligand for metal-mediated stereoselective catalysis. Prepn: H. Walder, *Ber.* **15**, 2166 (1882). Resolution of enantiomers: E. P. Kyba *et al.*, *J. Org. Chem.* **42**, 4173 (1977). Elucidation of catalyst structure: H. Sasai *et al.*, *J. Am. Chem. Soc.* **115**, 10372 (1993). Effect of substituents on enantioselectivity: C. Qian *et al.*, *J. Chem. Soc. Perkin Trans. 1* **1998**, 2097. Photochromism: M. Cavazza *et al.*, *J. Am. Chem. Soc.* **118**, 9990 (1996). Review of role in catalysis: R. Zimmer, J. Suhrbier, *J. Prakt. Chem.* **339**, 758-762 (1997); with titanium: K. Mikami, *Pure Appl. Chem.* **68**, 639-644 (1996); with rare-earth metals: H. Sasai *et al.*, *Chemtracts Org. Chem.* **11**, 264-267 (1998); of synthesis and use as a chiral reagent and ligand: J. M. Brunel, *Chem. Rev.* **105**, 857-897 (2005).

Long flat lustrous white needles from alcohol, mp 216-218°.

(R)-(+)-Form. [18531-94-7] Crystals from benzene, mp 207.5-208.5°. $[\alpha]_{546}^{25}$ +50.9°; $[\alpha]_{589}^{25}$ +34.3° (c = 1.0 in THF).

(S)-(−)-Form. [18531-99-2] Crystals from benzene, mp 207-208°. $[\alpha]_{546}^{25}$ −51.3°; $[\alpha]_{589}^{25}$ −33.3° (c = 1.1 in THF).

USE: Chiral ligand.

1231. Biochanin A. [491-80-5] 5,7-Dihydroxy-3-(4-methoxyphenyl)-4*H*-1-benzopyran-4-one; 5,7-dihydroxy-4′-methoxyisoflavone; genestein 4′-methyl ether; olmelin. $C_{16}H_{12}O_5$; mol wt 284.27. C 67.60%, H 4.26%, O 28.14%. Phytoestrogen found in trifolium (red clover), *q.v.*, and other legumes; metabolic precursor of genistein, *q.v.* Prepn from genistein: E. Walz, *Ann.* **489**, 118 (1931). Synthesis: R. L. Shriner, C. J. Hull, *J. Org. Chem.* **10**, 288 (1945); W. Baker et al., *J. Chem. Soc.* **1953**, 1852. Isoln from red clover: G. S. Pope et al., *Chem. Ind. (London)* **1953**, 1092. Metabolism in rats: A. Nilsson, *Nature* **192**, 358 (1961). HPLC determn in red clover samples: E. Gikas et al., *J. Liq. Chromatogr. Relat. Technol.* **31**, 1181 (2008); in nutritional supplements: H. Schwartz, G. Sontag, *Monatsh. Chem.* **139**, 865 (2008). Clinical cholesterol-lowering effect: P. Nestel et al., *Eur. J. Clin. Nutr.* **58**, 403 (2004). Review of estrogenic activity, bioavailability, and clinical pharmacology: V. Beck et al., *J. Steroid Biochem. Mol. Biol.* **94**, 499-518 (2005).

Long, white needles from ethanol, mp 214.5-215°. uv max: 262.5 nm (log ε 4.56).

Sissotrin. [5928-26-7] 7-(β-D-Glucopyranosyloxy)-5-hydroxy-3-(4-methoxyphenyl)-4*H*-1-benzopyran-4-one; biochanin A 7-*O*-glucoside. $C_{22}H_{22}O_{10}$; mol wt 446.41. Isoln from garbanzo beans: E. Wong et al., *Phytochemistry* **4**, 89 (1965). HPLC determn in red clover: X. He et al., *J. Chromatogr. A* **755**, 127 (1996). Synthesis: P. Lewis et al., *J. Chem. Soc. Perkin Trans. 1* **1998**, 2481. White solid from methanol-water, mp 212-214°. $[\alpha]_D^{24}$ −32.7° (c = 0.001 g/ml in DMSO).

USE: Dietary supplement.

1232. Biocytin. [576-19-2] N^6-[5-[(3a*S*,4*S*,6a*R*)-Hexahydro-2-oxo-1*H*-thieno[3,4-*d*]imidazol-4-yl]-1-oxopentyl]-L-lysine; ε-*N*-biotinyl-L-lysine; biotin complex of yeast. $C_{16}H_{28}N_4O_4S$; mol wt 372.48. C 51.59%, H 7.58%, N 15.04%, O 17.18%, S 8.61%. A naturally occurring complex of biotin. Contains 65.6% biotin. Characterized microbiologically by its availability as a source of biotin to *Lactobacillus casei*, *L. delbrückii* LD 5, *L. acidophilus*, *Streptococcus fecalis* R, *Neurospora crassa*, and *Saccharomyces carlsbergensis* and by its unavailability as a source of biotin to *Lactobacillus arabinosus*, *L. pentosus*, and *Leuconostoc mesenteroides* P-60. Isoln: Wright et al., *J. Am. Chem. Soc.* **72**, 1048 (1950); **74**, 1996 (1952). Structure: Peck et al., *ibid.* **72**, 1048 (1950); **74**, 1999 (1952). Synthesis: Wolf et al., *ibid.* **74**, 2002 (1952); Weijlard et al., *ibid.* **76**, 2505 (1954); Wolf, Folkers, US 2710298 (1955 to Merck & Co.). Purification: McCormick, Föry, *Methods Enzymol.* **18** (pt A), 413 (1970). Infrared absorption spectrum: *J. Am. Chem. Soc.* **74**, 2001 (1952). *Review:* A. F. Wagner, K. Folkers, *Vitamins and Coenzymes* (Wiley, New York, 1964) pp 138-159.

Crystals, mp 241-243°. Upon rapid crystn from dil methanol or dil acetone, mp 228-230° (dec); upon slow crystn sinters at 227°. mp 245-252° (dec, microblock). Crystals from water, mp 228.5°. $[\alpha]_D^{25}$ +53° (c = 1.05 in 0.1*N* NaOH). Freely sol in water, glacial acetic acid. Less sol in alc. Practically insol in acetone and most other organic solvents. When subjected to strong acid hydrolysis (at least 3*N* at 120° for one hour) biocytin yields biotin and L-lysine. Forms a crystalline hydrochloride.

1233. Bioflavonoids. Vitamin P complex; citrus flavonoid compounds; Arliflav; Pecitrol Veinogène; C.V.P.. A group of compounds which contribute to the maintenance of normal blood vessel conditions by decreasing capillary permeability and fragility. Widely distributed among plants: J. B. Harborne, *Comparative Biochemistry of the Flavonoids* (Academic Press, New York, 1967). Biosynthesis: Grisebach, Barz, *Naturwissenschaften* **56**, 538 (1969). High concentrates can be obtained from all citrus fruits, rose hips, and black currants. Commercial methods extract the rinds of oranges, tangerines, lemons, limes, kumquats, and grapefruits. Solvents used in the extraction processes are aqueous alkalies, hot water, or water-miscible organic solvents, such as isopropanol: Freedman et al., US 2888381 (1959 to U.S. Vitamin). Early interest in the compounds developed because of their synergistic effect with ascorbic acid. Other pharmacologic effects such as inhibition of adrenaline autooxidation and of enzyme action are also under study: several authors in *The Pharmacology of Plant Phenolics*, J. W. Fairbairn, Ed. (Academic Press, New York, 1959); several authors in *Angiologica* 9, 133-446 (1972). Metabolism: DeEds, "Flavonoid Metabolism" in *Comprehensive Biology* Vol. 20, M. Florkin, E. H. Stotz, Eds. (Elsevier, New York, 1968). *Reviews:* Scarborough, Bacharach, *Vitam. Horm.* **7**, 1-55 (1949); Baier et al., *Ann. N.Y. Acad. Sci.* **61**, (Art. 3), 637-736 (1955); T. Robinson, *The Organic Constituents of Higher Plants* (Burgess, Minneapolis, 1967) pp 178-209; H. Geiger, C. Quinn in *Flavonoids*, J. B. Harborne et al., Eds. (Academic Press, New York, 1975) pp 692-742.

THERAP CAT: Capillary protectant.

1234. Biopterin. [22150-76-1] 2-Amino-6-[(1*R*,2*S*)-1,2-dihydroxypropyl]-4(1*H*)-pteridinone; 1-(2-amino-4-hydroxy-6-pteridinyl)-1,2-propanediol; 2-amino-4-hydroxy-6-(1,2-dihydroxypropyl)pteridine; pterin HB_2. $C_9H_{11}N_5O_3$; mol wt 237.22. C 45.57%, H 4.67%, N 29.52%, O 20.23%. A pteridine widely distributed in nature; naturally occurring as the L-*erythro*-form. Considered as a growth factor for some insects; *see also* Neopterin. Isoln from human urine: Patterson et al., *J. Am. Chem. Soc.* **77**, 3167 (1955); **78**, 5871 (1956); GB 814462 (1959 to Am. Cyanamid); from drosophila: H. S. Forrest, H. K. Mitchell, *J. Am. Chem. Soc.* **77**, 4865 (1955); from queen bee jelly: Butenandt, Rembold, *Z. Physiol. Chem.* **311**, 79 (1958). Absolute configuration: E. L. Patterson et al., *J. Am. Chem. Soc.* **78**, 5871 (1956). Synthesis: E. L. Patterson et al., *ibid.* 5868; Tschesche et al., *Ann.* **658**, 193 (1962); Rembold, Metzger, *Ber.* **96**, 1395 (1963); Viscontini et al., *Helv. Chim. Acta* **55**, 570, 574 (1972); B. Schircks et al., *ibid.* **60**, 211 (1977); T. Sugimoto et al., *Bull. Chem. Soc. Jpn.* **53**, 2344 (1980). Synthesis from neopterin: A. Kaiser, H. P. Wessel, *Helv. Chim. Acta* **70**, 766 (1987). Biosynthesis: G. Kapatos et al., *Science* **213**, 1129 (1981).

Minute, yellow, spherical crystals from water. Chars without melting at 250-280°. $[\alpha]_D^{24}$ −50° (c = 0.4 in 0.1*N* HCl); $[\alpha]_D^{24}$ −26° (c = 0.92 in 0.1*N* NaOH). uv max (0.08*N* HCl): 247 nm (ε 11000). Soly in water: 0.7 mg/ml (20°); 4 mg/ml (90°). Soly in alc, ether, acetone, benzene: <0.1 mg/ml; in 1*N* NaOH, 1*N* HCl: >25 mg/ml. Fluoresces with a blue color in alkaline soln.

1235. Bioresmethrin. [28434-01-7] (1*R*,3*R*)-2,2-Dimethyl-3-(2-methyl-1-propen-1-yl)cyclopropanecarboxylic acid [5-(phenylmethyl)-3-furanyl]methyl ester; *trans*-(+)-2,2-dimethyl-3-(2-methylpropenyl)cyclopropanecarboxylic acid (5-benzyl-3-furyl)methyl ester; 5-benzyl-3-furylmethyl-(+)-*trans*-chrysanthemate; *d-trans*-[(5-benzyl-3-furyl)methyl]chrysanthemumate; (+)-*trans*-resmethrin; NRDC-107; NIA-18739; SBP-1390; Resbuthrin; Biobenzyfuroline. $C_{22}H_{26}O_3$; mol wt 338.45. C 78.07%, H 7.74%, O 14.18%. Synthetic pyrethroid insecticide. Prepn: M. Elliott, N. F. Janes, FR 1503260; eidem, US 3465007 and US 3542928 (1967, 1969, 1970, all to Nat. Res. Dev. Corp.). Insecticidal activity: M. Elliott et al., *Nature* **213**, 493 (1967); idem, idem, *Pestic. Sci.* **2**, 243 (1971). Mammalian metabolism: C. O. Abernathy, J. E. Casida, *Science* **179**, 1235 (1973). Control of cockroaches: P. R. Chadwick et al., *J. Med. Entomol.* **13**, 625 (1977). Acute toxicity: T. B. Gaines, R. E. Linder, *Fundam. Appl. Toxicol.* **7**, 299 (1986). Brief review: D. S.

Gunew, *Anal. Methods Pestic. Plant Growth Regul.* **10**, 19-29 (1978). Review of toxicology and human exposure: *Toxicological Profile for Pyrethrins and Pyrethroids* (PB2004-100004, 2003) 332 pp.

bp$_{0.0008}$ 174°. n_D^{20} 1.5346. $[\alpha]_D^{20}$ −7.8° (c = 5 in acetone). Sol in most organic solvents. Insol in water. LD$_{50}$ in adult male, female rats (mg/kg): 1244, 1721 orally (Gaines, Linder).

USE: Insecticide.

1236. Biotin. [58-85-5] (3a*S*,4*S*,6a*R*)-Hexahydro-2-oxo-1*H*-thieno[3,4-*d*]imidazole-4-pentanoic acid; *cis*-tetrahydro-2-oxo-thieno[3,4-*d*]imidazoline-4-valeric acid; *cis*-hexahydro-2-oxo-1*H*-thieno[3,4]imidazole-4-valeric acid; vitamin H; coenzyme R; bios II; Biodermatin. C$_{10}$H$_{16}$N$_2$O$_3$S; mol wt 244.31. C 49.16%, H 6.60%, N 11.47%, O 19.65%, S 13.12%. Growth factor present in minute amounts in every living cell. Plays an indispensable role in numerous naturally occurring carboxylation reactions. Occurs mainly bound to proteins or polypeptides. The richest sources are liver, kidney, pancreas, yeast, and milk. The biotin content of cancerous tumors is higher than that of normal tissue. Biotin combines with the proteinaceous substance, avidin, in raw egg-white and becomes inactive. When on diets contg large amounts of raw egg-white the rat or chick develops characteristic skin lesions and growth is retarded; these symptoms can be prevented by feeding additional biotin. Isoln from egg yolk: Kögl, Tönnis, *Z. Physiol. Chem.* **242**, 43 (1936); from liver: György *et al., J. Biol. Chem.* **131**, 745 (1939); du Vigneaud *et al., ibid.* **140**, 643 (1941). Identity of biotin with vitamin H: du Vigneaud *et al., Science* **92**, 62 (1940). Structure: du Vigneaud *et al., J. Am. Chem. Soc.* **146**, 475 (1942). First synthesis: Harris *et al., J. Am. Chem. Soc.* **67**, 2096 (1945). Configuration: Traub, *Nature* **178**, 649 (1956). Stereospecific total synthesis of the naturally occurring *d*-form: P. N. Confalone *et al., J. Am. Chem. Soc.* **97**, 5936 (1975); *eidem, J. Org. Chem.* **42**, 1630 (1977); T. Ogawa *et al., Carbohydr. Res.* **57**, C31 (1977); F. G. M. Vogel *et al., Ann.* **1980**, 1972; R. R. Schmidt, M. Maier, *Synthesis* **1982**, 747. Synthesis of the *dl*-form: P. N. Confalone *et al., Helv. Chim. Acta* **59**, 1005 (1976); M. Marx *et al., J. Am. Chem. Soc.* **99**, 6794 (1977); A. Fliri, K. Hohenlohe-Oehringen, *Ber.* **113**, 607 (1980); Ph. Rossy *et al., Tetrahedron Lett.* **22**, 3493 (1981). Biosynthetic studies: R. J. Parry, M. G. Kunitani, *J. Am. Chem. Soc.* **98**, 4024 (1976); R. J. Parry, M. V. Naidu, *Tetrahedron Lett.* **1980**, 4783. Coordination properties: H. Siegel, *Experientia* **37**, 789 (1981). Mechanism of action: M. J. Cravey, H. Kohn, *J. Am. Chem. Soc.* **102**, 3928 (1980); D. L. Vesely, *Science* **216**, 1329 (1982). Reviews: A. F. Wagner, K. Folkers, *Vitamins and Coenzymes* (Wiley, New York, 1964) pp 138-159; Harris in *The Vitamins* vol. II, W. H. Sebrell, R. S. Harris, Eds. (Academic Press, New York, 2nd ed., 1968) pp 261-359; D. B. McCormick, *Nutr. Rev.* **33**, 97-102 (1975). Review of assay methods: P. György, *The Vitamins* vol. VII, P. Gyorgy, W. N. Pearson, Eds. (1967) pp 303-313.

Characteristics of biotin isolated from liver or milk (Kögl's β-biotin): fine long needles, mp 232-233°. $[\alpha]_D^{21}$ +91° (c = 1 in 0.1*N* NaOH). Isoelec pt pH 3.5. pH of 0.01% aq soln 4.5. Soly at 25° (mg/100 ml): water ~22; 95% alc ~80. Very slightly sol in water, alcohol. More sol in hot water and in dil alkali. Insol in other common organic solvents. The pure compd is stable to air and temp. Moderately acid and neutral solns are stable several months; alkaline solns are less stable, but appear reasonably stable up to a pH of about

9. Aq solns are very susceptible to mold growth. Acidic solns can be heat sterilized.

THERAP CAT: Vitamin (enzyme cofactor).

THERAP CAT (VET): Vitamin (enzyme cofactor).

1237. Biotin *l*-Sulfoxide. [3376-83-8] (3a*S*,4*S*,5*R*,6a*R*)-Hexahydro-2-oxo-1*H*-thieno[3,4-*d*]imidazole-4-pentanoic acid 5-oxide; [3a*S*-(3aα,4β,5β,6aα)]-hexahydro-2-oxo-1*H*-thieno[3,4-*d*]-imidazole-4-pentanoic acid 5-oxide; AN factor. C$_{10}$H$_{16}$N$_2$O$_4$S; mol wt 260.31. C 46.14%, H 6.20%, N 10.76%, O 24.58%, S 12.32%. Isolated from *Aspergillus niger* culture filtrate where growth had taken place in the presence of pimelic acid; also from milk residue concentrates or by synthesis: Melville, *J. Biol. Chem.* **208**, 495 (1954); Wright *et al., J. Am. Chem. Soc.* **76**, 4163 (1954).

Polymorphic plates from water, mp 238° (some dec). $[\alpha]_D^{20}$ −39.5° (c = 1.01 in 0.1*N* NaOH). Ineffective in curing the biotin deficiency syndrome when administered to rats at a level 100 times the effective dose of biotin.

1238. Biperiden. [514-65-8] α-Bicyclo[2.2.1]hept-5-en-2-yl-α-phenyl-1-piperidinepropanol; α-5-norbornen-2-yl-α-phenyl-1-piperidinepropanol; 3-piperidino-1-phenyl-1-bicycloheptenyl-1-propanol; 1-bicycloheptenyl-1-phenyl-3-piperidinopropanol; 3-piperidino-1-phenyl-(Δ^5-bicyclo[2.2.1]hepten-2-yl)-1-propanol; KL-373. C$_{21}$H$_{29}$NO; mol wt 311.47. C 80.98%, H 9.39%, N 4.50%, O 5.14%. Prepn: W. Klavehn, US 2789110 (1957 to Knoll). Toxicology study: C. Hanna, *Toxicol. Appl. Pharmacol.* **2**, 379 (1960).

White, practically odorless, crystalline powder; mp 101°. Freely sol in chloroform; sparingly sol in alcohol. Practically insol in water. LD$_{50}$ in mice (mg/kg): 195 s.c., 56 i.v. (Klavehn).

Hydrochloride. [1235-82-1] Akineton; Akinophyl. C$_{21}$H$_{29}$-NO.HCl; mol wt 347.93. Fluffy white powder, mp 238°. Soly in water (mg/100 ml) at 30°: 220 (pH 3.8), 280 (pH 5.5), 290 (pH 6.7), 264 (pH 7.9), 244 (9.2-10.1). Sparingly sol in methanol; slightly sol in water, ether, alcohol, chloroform. LD$_{50}$ in rats, dogs (mg/kg): 750, 340 orally (Hanna).

THERAP CAT: Antiparkinsonian.

1239. *p*-Biphenylamine. [92-67-1] [1,1'-Biphenyl]-4-amine; 4-aminobiphenyl; *p*-aminodiphenyl; anilinobenzene; xenylamine. C$_{12}$H$_{11}$N; mol wt 169.23. C 85.17%, H 6.55%, N 8.28%. Prepn from diazoaminobenzene: Heusler, *Ann.* **260**, 232 (1890). Review of carcinogenicity studies: *IARC Monographs* **1**, 74-79 (1972).

Leaflets from alc or water. mp 53°. Volatile with steam. Slightly sol in cold water, readily sol in hot water, alcohol, chloroform.

Caution: Potential symptoms of overexposure are headache, dizziness; lethargy, dyspnea; ataxia, weakness; methemoglobinemia; urinary burning; acute hemorrhagic cystitis. *See NIOSH Pocket Guide to Chemical Hazards* (DHHS/NIOSH 97-140, 1997) p 14. This substance is listed as a known human carcinogen: *Report on Carcinogens, Twelfth Edition* (PB2011-111646, 2011) p 38.

USE: In the detection of sulfates. Formerly as rubber antioxidant. As carcinogen in cancer research.

1240. 2,4′-Biphenyldiamine. [492-17-1] [1,1′-Biphenyl]-2,4′-diamine; diphenyline; 2,4′-diphenyldiamine; *o,p*′-dianiline; 2,4′-diaminodiphenyl. $C_{12}H_{12}N_2$; mol wt 184.24. C 78.23%, H 6.57%, N 15.21%. Preparation by reducing azobenzene with tin and hydrochloric acid: Fischer, *Monatsh. Chem.* **6**, 547 (1885).

Needles from dil alc, mp 45°. bp 363°. Very slightly sol in alc or ether.

USE: Detection of tungsten; manuf azo dyes.

1241. 2,2′-Bipyridine. [366-18-7] 2,2′-Bipyridyl; α,α′-dipyridyl; CI-588. $C_{10}H_8N_2$; mol wt 156.19. C 76.90%, H 5.16%, N 17.94%. Bidentate metal chelating ligand. Prepn from pyridine: H. Meyer, A. Hofmann-Meyer, *J. Prakt. Chem. Chem.-Ztg.* **102**, 287 (1921); C. R. Smith, *J. Am. Chem. Soc.* **46**, 414 (1924); using Raney Ni: W. H. F. Sasse, *Org. Synth.* **coll. vol. V**, 102 (1973). Manuf: P. F. H. Freeman, R. Ghosh, US 2962502 (1960 to I.C.I.). Absorption spectrum: P. Krumholz, *J. Am. Chem. Soc.* **73**, 3487 (1951). Crystal structure: L. L. Merritt, Jr., E. D. Schroeder, *Acta Crystallogr.* **9**, 801 (1956). Pharmacologic study: P. Bass *et al.*, *J. Pharmacol. Exp. Ther.* **152**, 104 (1966). Review of chemistry of complex formation: E. König, *Coord. Chem. Rev.* **3**, 471-495 (1968); of properties of metal complexes: G. Nord, *Comments Inorg. Chem.* **4**, 193-212 (1985); of ligands containing multiple 2,2′-bipyridine units: C. Kaes *et al.*, *Chem. Rev.* **100**, 3553-3590 (2000); of synthesis methods: G. R. Newkome *et al.*, *Eur. J. Org. Chem.* **2004**, 235-254.

Crystals from petroleum ether, mp 70-71°. bp 272-273°. d 1.28. *Poisonous.* uv max (water): 233, 280 nm (ε 10200, 13300). Soly in water (mg/ml): 6.4; in 0.1 *N* HCl: 25.7. Insol in sodium hydroxide. Very sol in alcohol, ether, benzene, chloroform, petr ether. LD$_{50}$ i.p. in mice: 200 mg/kg. *See:* R. W. Grady *et al.*, *J. Pharmacol. Exp. Ther.* **196**, 478 (1976).

USE: Reagent for the determn of iron. Ligand in coordination chemistry; building block in supramolecular chemistry.

1242. Birch Tar Oil, Empyreumatic. Oil white birch; oleum rusci. Obtained by destructive distillation of the bark and wood of *Betula alba* L., *Betulaceae* (white birch). *Constit.* Oil turpentine, other isomeric hydrocarbons, various empyreumatic resins, guaiacol, cresol, pyrocatechol, betulin.

Black, viscid liquid; characteristic odor. d 0.926-0.955. Partly sol in alcohol; sol in chloroform, fats, oils.

USE: For preserving leather and wood.

1243. Birch Tar Oil, Rectified. "Essential oil of birch wood". Obtained from the empyreumatic oil by steam distillation. *Constit.* Phenol, cresol, xylenol, guaiacol, creosol, pyrocatechol.

Dark brown liquid. d 0.886-0.950. One ml dissolves in about 3 ml abs alcohol; sol in benzene, chloroform, ether, glacial acetic acid, carbon disulfide, oil turpentine.

THERAP CAT: Dermatologic.

THERAP CAT (VET): Has been used extern. in skin diseases.

1244. Biricodar. [159997-94-1] (2*S*)-1-[2-Oxo-2-(3,4,5-trimethoxyphenyl)acetyl]-2-piperidinecarboxylic acid 4-(3-pyridinyl)-1-[3-(3-pyridinyl)propyl]butyl ester. $C_{34}H_{41}N_3O_7$; mol wt 603.72. C 67.64%, H 6.85%, N 6.96%, O 18.55%. Chemosensitizer that restores sensitivity to cytotoxic agents in multidrug resistant (MDR) cells. Prepn: D. M. Armistead *et al.*, WO 9407858; *eidem*, US 5620971 (1994, 1997 both to Vertex). Effect on MDR mediated by P-glycoprotein: U. A. Germann *et al.*, *Anti-Cancer Drugs* **8**, 125 (1997); by multidrug resistance-associated protein (MRP): *eidem*,

ibid. 141. Clinical pharmacology, pharmacokinetics: E. K. Rowinsky *et al.*, *J. Clin. Oncol.* **16**, 2964 (1998).

Dicitrate. [174254-13-8] VX-710; Incel. $C_{34}H_{41}N_3O_7.2C_6H_8O_7$; mol wt 987.96. Sol in water.

THERAP CAT: Antineoplastic adjunct (chemosensitizer).

1245. α-Bisabolol. [515-69-5] *rel*-(α*R*,1*R*)-α,4-Dimethyl-α-(4-methyl-3-penten-1-yl)-3-cyclohexene-1-methanol; 6-methyl-2-(4-methyl-3-cyclohexen-1-yl)-5-hepten-2-ol; 1-methyl-4-(1,5-dimethyl-1-hydroxyhex-4(5)-enyl)cyclohexen-1; Camilol; Dragosantol; Hydagen B. $C_{15}H_{26}O$; mol wt 222.37. C 81.02%, H 11.79%, O 7.19%. Fragrant sesquiterpene isolated from the essential oils of a variety of plants, shrubs and trees. Both isomers occur in nature; the (−)-form is the most widespread and is an active anti-inflammatory component of chamomile. The (+) and (−)-epi isomers are also naturally occurring. Isoln of (−)-form from *Matricaria chamomilla*: F. Sorm *et al.*, *Collect. Czech. Chem. Commun.* **16**, 626 (1951); of (+)-form from *Populus balsamifera*: *idem et al.*, *Chem. Listy* **46**, 364 (1952). Synthesis of racemic mixture: L. Ruzicka, M. Liguori, *Helv. Chim. Acta* **15**, 3 (1932); C. D. Gutsche *et al.*, *Tetrahedron* **24**, 859 (1968). Enantioselective synthesis of (−)-form: H. Nemoto *et al.*, *Tetrahedron Lett.* **34**, 4939 (1993). Separation of four stereoisomers: K. Günther *et al.*, *Fresenius J. Anal. Chem.* **345**, 787 (1993). Pharmacology and toxicology of (−)-form: V. Jakovlev, A. von Schlichtegroll, *Arzneim.-Forsch.* **19**, 615 (1969); S. Habersang *et al.*, *Planta Med.* **37**, 115 (1979). Comparative pharmacology of racemate and enantiomers: V. Jakovlev *et al.*, *ibid.* **35**, 125 (1979). Determn by HPLC in chamomile extracts: R. Herrmann, *Dtsch. Apoth. Ztg.* **36**, 1797 (1982); by GC in cosmetic products: D. Andre *et al.*, *Int. J. Cosmet. Sci.* **13**, 137 (1991). Evaluation of use as dermal penetration enhancer: R. Kadir, B. W. Barry, *Int. J. Pharm.* **70**, 87 (1991). Review of anti-inflammatory activity and use in cosmetics: N. Jellinek, *Parfums Cosmet. Aromes* **57**, 55-57 (1984); of absolute configuration studies: G. W. O'Donnell, M. D. Sutherland, *Aust. J. Chem.* **42**, 2021-2034 (1989).

(−)-form

bp$_{12}$ 155-157°. d$_4^{23}$ 0.9223. n_D^{23} 1.4917. Also reported as colorless oil, bp$_{1.5}$ 128-130° (Gutsche). Miscible with alcohols, oils and lipophilic substances.

(−)-**Form.** [23089-26-1] (α*S*,1*S*)-α-Bisabolol; levomenol. bp$_{12}$ 153°. d^{20} 0.9211. n_D^{20} 1.4936. $[\alpha]_D$ −55.7°. Also reported as $[\alpha]_D^{20}$ −51.02° (Günther). LD$_{50}$ in mice, rats (mg/kg): 11350, 14850 orally (Jakovlev, von Schlichtegroll). LD$_{50}$ in male, female mice, rats (ml/kg): 15.2, 15.0, 14.9, 15.6 orally (Habersang).

(+)-**Form.** [23178-88-3] (α*R*,1*R*)-α-Bisabolol. bp$_{1.0}$ 120-122°. d^{20} 0.9213. n_D^{20} 1.4919. $[\alpha]_D^{20}$ +51.7°. Also reported as $[\alpha]_D^{20}$ +57.04° (Günther).

(−)-**Epi-α-bisabolol.** [78148-59-1] (α*R*,1*S*)-α-Bisabolol; anymol. Isoln from *Myoporum crassifolium*: K. G. O'Brien *et al.*, *Aust. J. Chem.* **6**, 166 (1953). $[\alpha]_D^{20}$ −67.6°. Also reported as $[\alpha]_D^{20}$ −67.48° (Günther).

(+)-Epi-α-bisabolol. [76738-75-5] (αS,1R)-α-Bisabolol. Isoln from *Salvia stenophylla:* E. J. Brunke, F. J. Hammerschmidt, *Proc. 15th Int. Symp. Essent. Oils Aromat. Plants* 1985, 145. $[\alpha]_D^{20}$ +67.4°. Also reported as $[\alpha]_D^{20}$ +70.11° (Günther).

USE: In cosmetics.

THERAP CAT: Anti-inflammatory.

1246. Bisacodyl. [603-50-9] 4,4'-(2-Pyridylmethylene)bis-phenol 1,1'-diacetate; bis(p-acetoxyphenyl)-2-pyridylmethane; (4,4'-diacetoxydiphenyl)(2-pyridyl)methane; 2-(4,4'-diacetoxydi-phenylmethyl)pyridine; Contalax; Correctol; Dulcolax; Durolax; Ex-Lax; Gentlax; Perilax; Pyrilax; Stadalax. $C_{22}H_{19}NO_4$; mol wt 361.40. C 73.12%, H 5.30%, N 3.88%, O 17.71%. Diphenylmeth-ane laxative. Prepn: **GB 730243**; A. Kottler, E. Seeger, **US 2764590** (1955, 1956 both to Thomae). Pharmacology and toxicology: Schmidt, *Arzneim.-Forsch.* **3**, 19 (1953). Comparison of efficacy and safety with other stimulant laxatives: A. Wald, *J. Clin. Gas-troenterol.* **36**, 386 (2003). Clinical evaluation in constipation: S. Kienzle-Horn *et al., Aliment. Pharmacol. Ther.* **23**, 1479 (2006).

White tasteless crystals; mp 138°. Sol in acids, acetone, chloro-form, benzene, propylene glycol, and other organic solvents. Spar-ingly sol in alcohol, methanol. Slightly sol in ether. Practically insol in water. LD_{50} orally in rats: >3 g/kg (Schmidt).

THERAP CAT: Cathartic.

THERAP CAT (VET): Cathartic.

1247. Bis(4-amino-1-anthraquinonyl)amine. [128-87-0] 1,1'-Iminobis[4-amino-9,10-anthracenedione]; 4,4'-diamino-1,1'-dianthraquinonylamine; 4,4'-diamino-1,1'-anthrimide; 4,4'-diami-no-1,1'-iminobisanthraquinone. $C_{28}H_{17}N_3O_4$; mol wt 459.46. C 73.20%, H 3.73%, N 9.15%, O 13.93%. Prepn: **DE 255822** (1913); *Frdl.* **11**, 615; Tinker, Stallmann, **US 2420022** (1947), *C.A.* **41**, 5730 (1947).

On heating with fuming sulfuric acid (30% SO_3) at 100° yields a dark powder, sol in water or alc with bluish-black coloration, used as dyestuff for wool. On heating with molten potassium hydroxide at 190° and oxidizing the reaction product with sodium hypochlorite, an olive-green vat dye is obtained.

USE: In the textile industry; in the determination of boron: Beck-ett, Webster, *Analyst* **68**, 306 (1943), *C.A.* **38**, 38 (1944).

1248. Bisantrene. [78186-34-2] 9,10-Anthracenedicarbox-aldehyde bis[(4,5-dihydro-1H-imidazol-2-yl)hydrazone]; 9,10-an-thracenedicarboxaldehyde bis(2-imidazolin-2-ylhydrazone). C_{22}-$H_{22}N_8$; mol wt 398.47. C 66.31%, H 5.57%, N 28.12%. Prepn: K. C. Murdock *et al.,* **DE 2850822**; K. C. Murdock, F. E. Durr, **US 4258181** (1979, 1981 both to Am. Cyanamid); and antitumor activ-ity: K. C. Murdock *et al., J. Med. Chem.* **25**, 505 (1982). Crystal-lographic characterization of hydrochloride: R. B. Bates *et al., Acta Crystallogr.* **C42**, 186 (1986); C. G. Pierpont, S. A. Lang, Jr., *ibid.* 1085. HPLC determn in plasma: Y.-M. Peng *et al., J. Chromatogr.* **233**, 235 (1982). Antitumor activity *in vitro:* J. D. Cowan *et al., Invest. New Drugs* **1**, 139 (1983); *in vivo:* R. V. Citarella *et al., Cancer Res.* **42**, 440 (1982). Clinical pharmacokinetics: K. Lu *et al., Cancer Chemother. Pharmacol.* **16**, 156 (1986). Clinical trials

in breast cancer: H.-Y. Yap *et al., Cancer Res.* **43**, 1402 (1983); C. K. Osborne *et al., Cancer Treat. Rep.* **68**, 357 (1984).

Dihydrochloride. [71439-68-4] ADAH; ADCA; NSC-337766; CL-216942; Orange Crush; Zantrène. $C_{22}H_{22}N_8.2HCl$; mol wt 471.39. Crystalline orange solid from ethanol. Hemihydrate, mp 288-289° (dec). uv max (H_2O) 260, 415 nm (ε 72700, 16300).

THERAP CAT: Antineoplastic.

1249. 2,5-Bis(1-aziridinyl)-3,6-bis(2-methoxyethoxy)-1,4-benzoquinone. [800-24-8] 2,5-Bis(1-aziridinyl)-3,6-bis(2-meth-oxyethoxy)-2,5-cyclohexadiene-1,4-dione; Bayer E 39 soluble. C_{16}-$H_{22}N_2O_6$; mol wt 338.36. C 56.80%, H 6.55%, N 8.28%, O 28.37%. Prepn: Gauss, Petersen, *Angew. Chem.* **69**, 252 (1957); **GB 793796** (1958 to Bayer). Antineoplastic activity *in vitro:* M. Akhtar *et al., Can. J. Chem.* **53**, 2891 (1975); J. S. Driscoll *et al., J. Pharm. Sci.* **68**, 185 (1979).

Gray needles from petr ether, mp 79-80.5°: Gauss, Petersen, *Med. Chem.* (Bayer) **7**, 649 (1963), *C.A.* **60**, 2875i (1964).

1250. Bisbenzimide. [23491-44-3] 4-[5-(4-Methyl-1-pipera-zinyl)[2,5'-bi-1H-benzimidazol]-2'-yl]phenol; bisbenzimidazole; 2-[2-(4-hydroxyphenyl)-6-benzimidazoyl]-6-(1-methyl-4-piperazyl)-benzimidazole; 2'-(4-hydroxyphenyl)-5-(4-methyl-1-piperazinyl)-2-5'-bi(1H-benzimidazole); pibenzimol. $C_{25}H_{24}N_6O$; mol wt 424.51. C 70.73%, H 5.70%, N 19.80%, O 3.77%. Membrane-permeable dye and nuclear stain utilized in numerous cytological applications. Binds to A-T rich DNA sequences with blue fluorescence. Prepn: **FR 1519964**; H. Loewe *et al.,* **US 3538097** (1967, 1970 both to Hoechst). DNA staining: I. Hilwig, A. Gropp, *Exp. Cell Res.* **75**, 122 (1972). Counterstaining applications: S. A. Schnell, M. W. Wessendorf, *Histochemistry* **103**, 111 (1995). Fluorescence spectro-scopic DNA binding studies: Y. Kubota, *Bull. Chem. Soc. Jpn.* **63**, 758 (1990); A. Adhikary *et al., Nucleic Acids Res.* **31**, 2178 (2003); and photophysical properties: G. Cosa *et al., Photochem. Photobiol.* **73**, 585 (2001). DNA quantification assays: C. F. Cesarone *et al., Anal. Biochem.* **100**, 188 (1979); D. L. Stout, F. F. Becker, *ibid.* **127**, 302 (1982).

mp 235° (dec). pKa_1 ~ −2; pKa_2 3.49±0.03; pKa_3 5.68±0.03; pKa_4 7.94±0.05; pKa_5 8.90±0.03; pKa_6 11.69±0.06; pKa_7 12.91±0.04. Sol in methanol.

Trihydrochloride. [23491-45-4] Bisbenzimide H 33258; Hoechst 33258; HOE 33258. $C_{25}H_{24}N_6O.3HCl$; mol wt 533.88. mp 280° (dec). *Irritant.* Light sensitive. Sol in DMF, methanol.

Soly in water: 2%. Will precipitate in phosphate buffer solns. As pentahydrate, abs max: 337 nm; 349 nm (complexed to ssDNA); 349 nm (complexed to dsDNA). Emission max: 508 nm; 466 nm (complexed to ssDNA); 458 nm (complexed to dsDNA).

Ethoxide. [23491-52-3] 2'-(4-Ethoxyphenyl)-5-(4-methyl-1-piperazinyl)-2,5'-bi-1H-benzimidazole. $C_{27}H_{28}N_6O$; mol wt 452.56. mp 268° (sesquihydrate).

Ethoxide Trihydrochloride. [875756-97-1] Bisbenzimide H 33342; Hoechst 33342; HOE 33342. $C_{27}H_{28}N_6O.3HCl$; mol wt 561.94. Fluorescent probe for membrane permeability studies: M. E. Lalande et al., Proc. Natl. Acad. Sci. USA **78**, 363 (1981); for stem cell differentiation studies: I. Bertoncello, B. Williams, Methods Mol. Biol. **263**, 181 (2004). Dark yellow powder. Irritant. Light sensitive. Hygroscopic. Sol in water, DMF. As pentahydrate, abs max: 340 nm; 350 nm (complexed to ssDNA); 350 nm (complexed to dsDNA). Emission max: 471 nm; 466 nm (complexed to ssDNA); 456 nm (complexed to dsDNA).

USE: Fluorescent cytological stain for DNA, flow cytometry, microscopy, immunostaining, autoradiography, nuclear counterstaining, and in situ hybridization. In assays to quantify DNA and to detect apoptosis. Probe for cellular differentiation and for membrane permeability.

1251. 1,2-Bis(chlorodimethylsilyl)ethane. [13528-93-3] 1,1'-(1,2-Ethanediyl)bis[1-chloro-1,1-dimethylsilane]; 2,5-dichloro-2,5-dimethyl-2,5-disilahexane; 1,1,4,4-tetramethyl-1,4-dichlorodisilylethylene. $C_6H_{16}Cl_2Si_2$; mol wt 215.26. C 33.48%, H 7.49%, Cl 32.94%, Si 26.09%. Silylating reagent used to protect primary amines; the resulting derivative is the tetramethyldisilylazacyclopentane (Stabase) adduct. Protection of primary amines: W. A. Piccoli et al., J. Am. Chem. Soc. **82**, 1883 (1960); H. Sakurai et al., Tetrahedron Lett. **7**, 5493 (1966). Use in protection of primary amines: S. Djuric et al., ibid. **22**, 1787 (1981); T. L. Guggenheim, ibid. **25**, 1253 (1984). Additional synthetic applications: W. B. Motherwell, L. R. Roberts, ibid. **36**, 1121 (1995); T. Uchida et al., Tetrahedron **62**, 3103 (2006). Review: F. Z. Basha, S. W. Elmore in Encyclopedia of Reagents for Organic Synthesis **1**, L. A. Paquette, Ed. (Wiley, New York, 1995) pp 444-446.

White solid, fp 37°. bp$_{734}$ 198°. bp$_{13}$ 80°. Flammable. Corrosive. Flash point, closed cup: 149°F (65°C). Moisture sensitive.

USE: Reagent in synthetic organic chemistry.

1252. Bis(p-chlorophenoxy)methane. [555-89-5] 1,1'-[Methylenebis(oxy)]bis[4-chloro]benzene; di-(p-chlorophenoxy)-methane; DCPM; K-1875; Neotran. $C_{13}H_{10}Cl_2O_2$; mol wt 269.12. C 58.02%, H 3.75%, Cl 26.35%, O 11.89%. Prepd from p-chlorophenol and dichloromethane: Moyl, US 2503207 (1950 to Dow); from sodium phenolate and dichloromethane: Miron, Lowry, J. Am. Chem. Soc. **73**, 1872 (1951). Toxicology: H. C. Spencer et al., Arch. Ind. Hyg. Occup. Med. **1**, 341 (1950).

Crystals from petr ether. mp 69.7-70.2°. bp$_6$ 189-194°. Solubilities at 25° in g/100 ml: acetone 189; benzene 40; carbon tetrachloride 28; methanol 0.5; ether 87. Very sol in ethyl ether. Practically insol in water, petr oils, naphtha. LD$_{50}$ orally in rats: 5.8 g/kg (Spencer).

USE: Miticide.

1253. Bis(1,2-dimethylpropyl)borane. [1069-54-1] Disiamylborane; di-sec-isoamylborine; bis(3-methyl-2-butyl)borane. $C_{10}H_{23}B$; mol wt 154.10. C 77.94%, H 15.04%, B 7.01%. Prepn: Brown, Zweifel, J. Am. Chem. Soc. **83**, 1241 (1961). Use in selective reductions: G. W. Kabalka, C. F. Lane, Chem. Tech. **6**, 324 (1976).

Crystals, mp 35-40°. Unstable to air.

USE: Selective reagent for steric control of hydroboration of olefins.

1254. Bis(diphenylphosphine) Ligands. Class of phosphorus-based bidentate ligands with an alkyl chain backbone; form catalytic complexes upon chelation with transition metals. Prepn of DPPM and DPPE: K. Issleib, D.-W. Müller, Ber. **92**, 3175 (1959); W. Hewertson, H. R. Watson, J. Chem. Soc. **1962**, 1490; of DPPP: G. R. Van Hecke, W. D. Horrocks, Jr., Inorg. Chem. **5**, 1960 (1966); of DPPB: P. W. Clark, Org. Prep. Proceed. Int. **11**, 103 (1979). Improved prepn of DPPM, DPPE, and DPPB: E. N. Tsvetkov et al., Synthesis **1986**, 198. Utility of DPPE in allylic alkylations: J.-C. Fiaud, J.-L. Malleron, J. Chem. Soc. Chem. Commun. **1981**, 1159; of DPPP in Heck reactions: W. Cabri et al., J. Org. Chem. **57**, 3558 (1992); of DPPB in alkyne hydroesterifications: B. E. Ali et al., Tetrahedron Lett. **42**, 2385 (2001). Use of ligands in carbonylative and direct Suzuki-Miyaura cross-couplings: A. Petz et al., J. Mol. Catal. A **255**, 97 (2006). Review of DPPM chemistry: R. J. Puddephatt, Chem. Soc. Rev. **12**, 99-127 (1983).

DPPM: n = 1
DPPE: n = 2
DPPP: n = 3
DPPB: n = 4

DPPM. [2071-20-7] 1,1'-Methylenebis[1,1-diphenylphosphine]; 1,1-bis(diphenylphosphino)methane. $C_{25}H_{22}P_2$; mol wt 384.40. C 78.12%, H 5.77%, P 16.12%. Colorless needles from propanol, mp 120.5-121.5°

DPPE. [1663-45-2] 1,1'-(1,2-Ethanediyl)bis[1,1-diphenylphosphine]; 1,2-bis(diphenylphosphino)ethane; diphos. $C_{26}H_{24}P_2$; mol wt 398.43. C 78.38%, H 6.07%, P 15.55%. White solid, mp 143-144° (Hewertson); also reported as mp 159-161° (Issleib). Sol in tetrahydrofuran, chloroform, dichloromethane, diethyl ether. Air stable. Solns oxidize readily and should be handled under inert gas.

DPPP. [6737-42-4] 1,1'-(1,3-Propanediyl)bis[1,1-diphenylphosphine]; 1,3-bis(diphenylphosphino)propane. $C_{27}H_{26}P_2$; mol wt 412.45. C 78.63%, H 6.35%, P 15.02%. White powder from methanol. mp 63.3-65.3°. Air stable. Solns oxidize readily and should be handled under inert gas.

DPPB. [7688-25-7] 1,1'-(1,4-Butanediyl)bis[1,1-diphenylphosphine]; 1,4-bis(diphenylphosphino)butane. $C_{28}H_{28}P_2$; mol wt 426.48. C 78.86%, H 6.62%, P 14.53%. White crystalline solid, mp 133.8-133.9°. Irritant. Air stable. Solns oxidize readily and should be handled under inert gas.

USE: Ligands in organic synthesis.

1255. Bis(2-ethylhexyl) Sebacate. [122-62-3] Decanedioic acid 1,10-bis(2-ethylhexyl) ester; di(2-ethylhexyl) sebacate; Octoil S; Plexol 201. $C_{26}H_{50}O_4$; mol wt 426.68. C 73.19%, H 11.81%, O 15.00%. Prepn: Bruno, US 2628249 (1953 to Pittsburgh Coke & Chemical); GB 747260 (1956 to Chemische Werke Hüls).

Liquid, d$_{25}^{25}$ 0.9119, n_D^{25} 1.4496.

USE: In vacuum pumps.

1256. Bismark Brown R. [5421-66-9]; [8005-78-5] (unspecified structure). 4,4′-[(4-Methyl-1,3-phenylene)bis(azo)]bis[6-methyl-1,3-benzenediamine] dihydrochloride; C.I. Basic Brown 4; 5,5′-[(4-methyl-*m*-phenylene)bis(azo)]bis[toluene-2,4-diamine] dihydrochloride; C.I. 21010; Bismarck Brown R; Bismark Brown 53; Vesuvine. $C_{21}H_{26}Cl_2N_8$; mol wt 461.40. C 54.67%, H 5.68%, Cl 15.37%, N 24.29%. Prepn: P. Griess, *Ber.* **11**, 624 (1878). Prepd by reaction of toluene-2,4-diamine HCl with nitrous acid: *Colour Index* **vol. 4** (3rd ed., 1971) p 4154. Use in protein staining: J.-L. Choi *et al., Anal. Biochem.* **236**, 82 (1996).

2HCl

Dark brown solid. mp 222°. Very sol in water (yellowish-brown soln); sol in ethanol, Cellosolve; slightly sol in acetone; practically insol in benzene. In concd H_2SO_4 gives brown soln which on dilution turns reddish-brown; in concd HNO_3 gives violet soln which turns brown.

Free base. [4482-25-1] C.I. Solvent Brown 12; C.I. 21010:1. $C_{21}H_{24}N_8$. mp 130-135°. Very slightly sol in water; sol in ethanol, acetone.

USE: As textile dye, leather dye, biological stain.

1257. Bismark Brown Y. [10114-58-6] 4,4′-[1,3-Phenylenebis(2,1-diazenediyl)]bis[1,3-benzenediamine] hydrochloride (1:2); C.I. Basic Brown 1; 4,4′-[*m*-phenylenebis(azo)]bis[*m*-phenylenediamine] dihydrochloride; C.I. 21000; Bismarck Brown Y; phenylene brown. $C_{18}H_{20}Cl_2N_8$; mol wt 419.31. C 51.56%, H 4.81%, Cl 16.91%, N 26.72%. Prepd by reaction of *m*-phenylenediamine HCl and nitrous acid: F. J. Welcher, *Organic Analytical Reagents* (D. Van Nostrand Co., New York, 1948) p 339; *Colour Index* **vol. 4** (3rd ed., 1971) p 4154.

2 HCl

Blackish-brown powder. Very sol in water; slightly sol in ethanol, Cellosolve. Practically insol in acetone, benzene carbon tetrachloride. In concd H_2SO_4 gives brown soln; in concd nitric acid gives orange soln which turns yellow.

USE: As a textile dye, biological stain.

1258. Bis(2-mercaptoethyl)sulfone. [145626-87-5] 2,2′-Sulfonylbis(ethanethiol); BMS. $C_4H_{10}O_2S_3$; mol wt 186.30. C 25.79%, H 5.41%, O 17.18%, S 51.63%. Reducing agent for proteins. Prepn: G. V. Lamoureux, G. M. Whitesides, *J. Org. Chem.* **58**, 633 (1993). Enhanced reduction rate in comparison with dithiothreitol, *q.v.*: R. Singh, G. M. Whitesides, *Bioorg. Chem.* **22**, 109 (1994). Use in reduction of disulfide bonds: D. Winitz *et al.,J. Biol. Chem.* **271**, 27645 (1996); A. Brunella, O. Ghisalba, *J. Mol. Catal. B* **10**, 215 (2000); S. Lee, J. P. N. Rosazza, *Org. Lett.* **6**, 365 (2004).

White fluffy crystals from hexane, mp 57-58°. pKa (aqueous soln at 25°): 7.9, 9.0. Keq (for the reduction of DTT): 0.35.

USE: Reagent for reduction of disulfide bonds in aqueous solutions.

1259. Bis(1-methylamyl) Sodium Sulfosuccinate. [6001-97-4] 2-Sulfobutanedioic acid 1,4-bis(1-methylpentyl) ester sodium salt (1:1); dihexyl sodium sulfosuccinate; Aerosol MA; Alphasol MA. $C_{16}H_{29}NaO_7S$; mol wt 388.45. C 49.47%, H 7.53%, Na 5.92%, O 28.83%, S 8.25%. The bis(1-methylamyl) ester of sulfosuccinic acid monosodium salt, perhaps in admixture with the dihexyl ester. Prepd by the action of the appropriate alcohols on maleic anhydride followed by addition of sodium bisulfite: A. O. Jaeger, **US 2028091**; **US 2176423** (1936, 1939 both to Am. Cyanamid); **FR 776495** (1935 to Selden Co.).

Available as white, slightly hygroscopic, wax-like pellets. Must be soaked to dissolve in cold water. Dissolves rapidly in hot water. Solubility in water at 25° = 343 g/l; at 70° = 447 g/l. Maximum concn of electrolyte soln in which 1% of the wetting agent is sol: 2% NaCl; 2% NH_4Cl; 14% $(NH_4)_2HPO_4$; 3% $NaNO_3$ (slightly turbid); 3% Na_2SO_4. Also sol in pine oil, oleic acid, acetone, kerosene, carbon tetrachloride, 2B ethanol, benzene, hot olive oil, glycerol. Insol in liquid petrolatum. Stable in acid and neutral solns, hydrolyzes in alkaline soln.

USE: Wetting agent.

1260. Bis(methylthio)methane. [1618-26-4] Bis[methylmercapto]methane; methylenebis[methyl sulfide]. $C_3H_8S_2$; mol wt 108.22. C 33.30%, H 7.45%, S 59.25%. Odorous principle of the white truffle *Tuber magnatum* (Pico) Vitt., *Tuberaceae*. Isoln and mass spectra: A. Fiecchi *et al., Tetrahedron Lett.* **8**, 1681 (1967). Synthesis from methyl mercaptan: Böhme, Marx, *Ber.* **74**, 1672 (1941).

Oily liquid. Odor in high dilutions reminiscent of white truffles. The odor of the neat liquid resembles that of freshly prepd mustard without its acrid and irritating qualities. bp 148-149°. Insol in aq alkali solns.

1261. Bismuth. [7440-69-9] Bi; at. wt 208.98040; at. no. 83; valence 3, 5. Group VA (15). One naturally occurring isotope: 209; artificial radioactive isotopes: 199-208; 210-215. Confused with tin until 1450. First isolated by Hillot in 1737. It was, however, Geoffrey the Younger who clearly proved its individuality in 1753. Pott and Bergmann are named as the scientific discoverers. Occurrence in the earth's crust: approx 0.2 ppm. Obtained as a byproduct from the processing of lead, copper, and tin ores. *Reviews: Nouveau Traité de Chimie Minérale* **11**, P. Pascal, Ed. (Masson, Paris, 1958); *Gmelins, Bismuth* (8th ed.) **19**, pp 1-104 (1927); supplement, pp 1-621 (1964); Smith, "Arsenic, Antimony and Bismuth" in *Comprehensive Inorganic Chemistry* **vol. 2**, J. C. Bailar, Jr. *et al.*, Eds. (Pergamon Press, Oxford, 1973) pp 547-683; S. C. Carapella, H. E. Howe in *Kirk-Othmer Encyclopedia of Chemical Technology* **vol. 3** (Wiley-Interscience, New York, 3rd ed., 1978) pp 912-921.

Grayish-white with reddish tinge and bright metallic luster; soft and brittle; superficially oxidized by air, frequently becoming iridescent. mp 271°; contracts when melted. bp 1420°; bp 1490°: *Gmelins*, p. 43. d_4^{20} 9.78; d_4^{271} 10.07. Poor conductor of electricity. Has the greatest Hall effect of any metal, *i.e.*, its resistance increases when placed in a magnetic field. Attacked by dil HNO_3, hot H_2SO_4, concd HCl. Cold solns of Bi give a white ppt with NaOH, turning yellow on boiling; with HCl a white ppt sol in excess of acid. The solns in HCl or HNO_3 yield with much water a white ppt blackened by H_2S (different from Sb).

Precipitated bismuth. Prepd by treating a soln of bismuth chloride in HCl with hypophosphorous acid, washing and drying. Contains not less than 98.5% metallic bismuth. Dull gray powder. The particles are of no greater diameter than 15 microns (0.015 mm). Easily dispersed in water.

D'Arcet metal-fusible. An alloy of 49.2% Bi, 32.2% Pb and 18.4% Sn. Whitish-gray metal, mp 96-97°.

USE: Manuf Bi salts, fusible alloys, stereotype metal, fusible boiler plugs, electric fuses, low-melting solders; tempering baths for steel; "silvering" mirrors; in dental technique.

THERAP CAT (VET): Has been used externally in dusting powders for indolent, moist or suppurating lesions; internally as a protectant of the gastrointestinal lining, and as an x-ray contrast medium. Has also been recommended to treat buccal warts in dogs.

1262. Bismuth Bromide. [7787-58-8] Tribromobismuthine; bismuth tribromide. $BiBr_3$; mol wt 448.69. Bi 46.58%, Br 53.42%. Prepn from the elements: Schenk in *Handbook of Preparative Inorganic Chemistry* vol. 1, G. Brauer, Ed. (Academic Press, New York, 2nd ed., 1963) p 623.

Yellowish crystals; odor of HBr. d 5.7. mp 218°. bp 441°. Dec by water, forming BiOBr; sol in solns of KI, KBr, KCl, and in dil HCl. Practically insol in alc. *Keep tightly closed.*

1263. Bismuth Bromide Oxide. [7787-57-7] Bromooxobismuthine; basic bismuth bromide; bismuth oxybromide; bismuthyl bromide; bismuth "subbromide". BiBrO; mol wt 304.88. Bi 68.55%, Br 26.21%, O 5.25%. BiOBr. Prepn: Schenk in *Handbook of Preparative Inorganic Chemistry* vol. 1, G. Brauer, Ed. (Academic Press, New York, 2nd ed., 1963) p 624.

Crystals or amorphous powder. Very stable, melts at red heat. Practically insol in water, alc; sol in HCl, HBr, HNO_3.

USE: Manuf of dry cell cathodes.

1264. Bismuth Chloride. [7787-60-2] Trichlorobismuthine; bismuth trichloride. $BiCl_3$; mol wt 315.33. Bi 66.27%, Cl 33.73%. Prepn from the elements: Schenk in *Handbook of Preparative Inorganic Chemistry* vol. 1, G. Brauer, Ed. (Academic Press, New York, 2nd ed., 1963) p 621.

White to yellowish, deliquesc crystals; HCl odor. d 4.75. mp ~230°. Sublimes at about 430°. bp 447°. Dec by water or aq alc into BiOCl; sol in HCl, HNO_3, abs alcohol, acetone, ethyl acetate. *Keep tightly closed.*

USE: In the manuf of other bismuth salts; as catalyst for organic reactions.

1265. Bismuth Fluoride. [7787-61-3] Trifluorobismuthine; bismuth trifluoride. BiF_3; mol wt 265.98. Bi 78.57%, F 21.43%. Prepd by dissolving bismuth oxide or hydroxide in aqueous HF. May also be prepd by reduction of BiF_5 with CO_2 in very dil H_2 at 80-150° in a Pt tube: Muir *et al.*, *J. Chem. Soc.* **39**, 33 (1881); v. Wartenberg, *Z. Anorg. Allg. Chem.* **244**, 344 (1940); Aurivillius, *Acta Chem. Scand.* **9**, 1206 (1955). *Review:* Kemmitt, Sharp, *Adv. Fluorine Chem.* **4**, 213 (1965).

White to gray dimorphic crystals. d 8.3. mp 725-730°. Volatilizes at higher temps slowly and without decompn. Practically insol in water. Sol in concd HF with the formation of complexes.

USE: In the prepn of bismuth pentafluoride, *q.v.*

1266. Bismuth Germanate. [12233-56-6] Bismuth germanium oxide ($Bi_4Ge_3O_{12}$); BGO. $Bi_4Ge_3O_{12}$; mol wt 1245.80. Bi 67.10%, Ge 17.49%, O 15.41%. $Bi_4(GeO_4)_3$. High density scintillation material; γ-ray absorber. Single crystal growth and characterization: R. Nitsche, *J. Appl. Phys.* **36**, 2358 (1965); R. V. A. Murthy *et al.*, *J. Cryst. Growth* **197**, 865 (1999); J. B. Shim *et al.*, *ibid.* **243**, 157 (2002). Luminescence properties: M. J. Weber, R. R. Monchamp, *J. Appl. Phys.* **44**, 5495 (1973). Temperature-dependent scintillation properties: H. V. Piltingsrud, *J. Nucl. Med.* **20**, 1279 (1979). Laser-excited optical properties: F. Rogemond *et al.*, *J. Lumin.* **33**, 455 (1984); in radiation detectors: Z. S. Macedo *et al.*, *Nucl. Instrum. Methods Phys. Res. B* **218**, 153 (2004). Optical and photoelastic properties: P. A. Williams *et al.*, *Appl. Opt.* **35**, 3562 (1996). Thermophysical properties: V. D. Golyshev *et al.*, *J. Cryst. Growth* **262**, 202, 212 (2004). Performance in PET applications: G. Gervino, E. Monticone, *Sens. Actuators A* **41-42**, 487 (1994).

Colorless, non-hygroscopc, cubic crystals, mp ~1050°. n_{550} 2.1204. n_{600} 2.1058. n_{750} 2.0786. d 7.13. Emission max: 480 nm.

USE: Scintillator crystal in nuclear physics, high-energy physics, geology, and nuclear medical imaging applications including PET, X-ray CT and positron CT; in electromagnetic calorimeters.

1267. Bismuth Hydroxide. [10361-43-0] Bismuth trihydroxide; bismuth hydrate. BiH_3O_3; mol wt 260.00. Bi 80.38%, H

1.16%, O 18.46%. $Bi(OH)_3$. Prepn: *Gmelins, Bismuth* (8th ed.) **19**, pp 119-122 (1927).

White to yellowish-white, amorphous powder. d^{15} 4.962. When freshly pptd it is soluble in glycerol in presence of NaOH; sol in acids. Readily loses one mol H_2O forming the metahydroxide which is yellow. Practically insol in water.

USE: As absorbent for rutin and quercetin; in hydrolysis of ribonucleic acid; in separation of plutonium from irradiated uranium.

1268. Bismuthine. [18288-22-7] Bismuth trihydride; Bismutan. BiH_3; mol wt 212.00. Bi 98.58%, H 1.43%. Prepn: Wiberg, Mödritzer, *Z. Naturforsch.* **12b**, 123 (1957); Amberger, *Ber.* **94**, 1447 (1961).

Thermally unstable liquid, bp 16.8° (est).

USE: In manuf of Ge or Si semiconductors, Scott *et al.*, **US 2910394** (1959 to Int'l Standard Electric).

1269. Bismuth Iodide. [7787-64-6] Triiodobismuthine; bismuth triiodide. BiI_3; mol wt 589.69. Bi 35.44%, I 64.56%. Best prepd from the elements: Watt *et al.*, *Inorg. Synth.* **4**, 114 (1953).

Black, minute, hexagonal crystals with metallic sheen. d_4^{17} 5.778, also reported as d 5.64. Sublimes at 439° (760 mm), dec at 500°. Practically insol in cold water, dec slowly in hot water. Sol in liquid ammonia, abs ethanol (about 3.5% at 20°), in aq solns of KI, HI, and HCl. The solid is slowly converted to $Bi(IO_3)_3$ upon prolonged exposure to air.

1270. Bismuth Iodide Oxide. [7787-63-5] Basic bismuth iodide; bismuth oxyiodide; bismuthyl iodide; bismuth "subiodide". BiIO; mol wt 351.88. Bi 59.39%, I 36.06%, O 4.55%. BiOI. Prepn: *Gmelins, Bismuth* (8th ed) **19**, p 159 (1927); Schenk in *Handbook of Preparative Inorganic Chemistry* vol. 1, G. Brauer, Ed. (Academic Press, New York, 2nd ed., 1963) p 625.

Brick-red, heavy, odorless powder or copper-colored crystals. d 7.92. Fuses at red heat with partial decompn. Practically insol in water, alcohol, $CHCl_3$; sol in HCl; dec by HNO_3 or alkali.

USE: Manuf of dry cell cathodes.

THERAP CAT: Anti-infective.

1271. Bismuth Nitrate. [10361-44-1] Nitric acid bismuth-(3+) salt (3:1); bismuth trinitrate. BiN_3O_9; mol wt 394.99. Bi 52.91%, N 10.64%, O 36.45%. $Bi(NO_3)_3$. Prepn: *Gmelins, Bismuth* (8th ed.) **19**, pp 126-129 (1927).

Pentahydrate. [10035-06-0] $BiN_3O_9.5H_2O$; mol wt 485.07. Lustrous, hygroscopic crystals; acid reaction; odor of nitric acid. d 2.83. Sol in water containing nitric acid; dec by water alone into subnitrate; sol in glycerol, dil acids including acetic acid, acetone. Practically insol in alcohol, ethyl acetate.

USE: Prepn of other bismuth salts, luminous paints; pptn of alkaloids. Trace metal analysis.

1272. Bismuth Oxalate. [6591-55-5] Ethanedioic acid bismuth(3+) salt (3:2); oxalic acid bismuth salt. $C_6Bi_2O_{12}$; mol wt 682.01. C 10.57%, Bi 61.28%, O 28.15%. $Bi_2(C_2O_4)_3$. Prepn: Vanino, Zumbusch, *Ber.* **41**, 3994 (1908); Soerbye, Kruse, *Acta Chem. Scand.* **16**, 1662 (1962).

Powder. *Poisonous.* Practically insol in water or alcohol. Sol in moderately dil HCl or HNO_3.

1273. Bismuth Oxide. [1304-76-3] Bismuth trioxide; bismuthous oxide; bismuth yellow; bismuth(3+) oxide. Bi_2O_3; mol wt 465.96. Bi 89.70%, O 10.30%. Occurs in nature as the mineral *bismite.* Prepn: *Gmelins, Bismuth* (8th ed.) **19**, p 109-113 (1927).

Yellow, heavy, odorless powder or monoclinic crystals; stable in air. Sol in HCl, HNO_3. Practically insol in water.

USE: In disinfectants, magnets, glass, rubber vulcanization; in fireproofing of papers and polymers; in catalysts.

THERAP CAT: Astringent.

1274. Bismuth Oxychloride. [7787-59-9] Chlorooxobismuthine; bismuth chloride oxide; basic bismuth chloride; bismuthyl chloride; bismuth subchloride; pearl white; blanc d'Espagne; blanc de perle; chlorbismol; C.I. Pigment White 14; C.I. 77163. BiClO; mol wt 260.43. Bi 80.24%, Cl 13.61%, O 6.14%. BiOCl. Pearlescent white pigment commonly prepared by hydrolysis of acidic bismuth solns in the presence of chlorine. Wide band-gap semiconductor with photocatalytic properties. Prepn: Schenk in *Handbook of Preparative Inorganic Chemistry* vol. 1, G. Brauer, Ed.

Consult the Name Index before using this section.

(Academic Press, New York, 2nd ed., 1963) p 622. Crystal structure: K. G. Keramidas *et al.*, *Z. Kristallogr.* **205**, 35 (1993). Electronic structure and photocatalytic activity: K.-L. Zhang *et al.*, *Appl. Catal. B* **68**, 125 (2006). Synthesis of crystalline nanostructures: H. Peng *et al.*, *Chem. Mater.* **21**, 247 (2009). Use in treatment of syphilis: R. R. Willcox, *Practitioner* **161**, 203 (1948). Review of use in cosmetics: Q. Peng, M. Tellefsen, *Cosmet. Toiletries* **118**, 53-62 (2003).

Fine powder or tetragonal crystals. Dec slowly above 400°. d 7.72. Practically insol in water, alcohol. Sol in HCl, HNO_3.

USE: Pigment in face powders, nail enamels, and other cosmetics; manuf artificial pearls, dry-cell cathodes, photochemical cells; photocatalyst.

1275. Bismuth Pentafluoride.

[7787-62-4] BiF_5; mol wt 303.97. Bi 68.75%, F 31.25%. Prepd according to the equation $BiF_3 + F_2 \rightarrow BiF_5$: v. Wartenberg, *Z. Anorg. Allg. Chem.* **244**, 344 (1940); Kwasnik in *Handbook of Preparative Inorganic Chemistry* **vol. 1**, G. Brauer Ed. (Academic Press, New York, 2nd ed., 1963) pp 202-203. Prepn from Bi and F_2: Fischer, Rodzitis, *J. Am. Chem. Soc.* **81**, 6375 (1959); Kemmitt, Sharp, *Adv. Fluorine Chem.* **4**, 212 (1965); Hebecker, *Z. Anorg. Allg. Chem.* **384**, 111 (1971).

Body-centered tetragonal crystals, sublimes at 120°. d 5.55. Very sensitive to moisture, discolors quickly in moist air. Violent reaction with water forming BiF_3 and ozone. Reacts with liquid petrolatum above 50°.

Caution: Highly toxic and irritating to mucous membranes, skin, eyes, respiratory tract.

USE: Fluorinating agent.

1276. Bismuth Phosphate.

[10049-01-1] Phosphoric acid bismuth(3+) salt (1:1); Bismugel. BiO_4P; mol wt 303.95. Bi 68.75%, O 21.05%, P 10.19%. $BiPO_4$. Prepn: Schenk in *Handbook of Preparative Inorganic Chemistry* **vol. 1**, G. Brauer, Ed. (Academic Press, New York, 2nd ed., 1963) p 626.

Odorless powder or monoclinic crystals. Does not melt on heating. d^{15} 6.323. Slightly sol in water and dil acids; not hydrolyzed by boiling water. Practically insol in alcohol, acetic acid; sol in concd HNO_3 and HCl.

USE: In the separation of plutonium from fission products; manuf of optical flint glass.

THERAP CAT: Antacid; protectant.

1277. Bismuth Potassium Iodide.

[41944-01-8] Potassium heptaiodobismuthate(4−) (4:1). BiI_7K_4; mol wt 1253.70. Bi 16.67%, I 70.86%, K 12.47%. K_4BiI_7. Prepn: *Gmelins, Potassium* (8th ed.) **22**, p 1068 (1936).

Red crystals. Partly dec by water; completely sol in alkali iodide solns.

USE: Pptn of vitamins, particularly thiamine HCl, and antibiotics from soln.

1278. Bismuth Potassium Tartrate.

[5798-41-4] (2*R*,3*R*)-2,3-Dihydroxybutanedioic acid bismuth(3+) potassium salt (1:1:1); potassium bismuthotartrate; potassium bismuth tartrate; potassium bismuthyl tartrate; tartaric acid bismuth complex potassium salt. The bismuth content of the complex salts of bismuth and alkali tartrate may be made to range within wide limits (35-75% Bi). The medicinal grade contains 60-64% bismuth. Prepn: Kober, US 1663201 (1928 to Searle); *Hagers Handb. Pharm. Praxis* **1st Suppl.**, 316 (Berlin, 1949).

Odorless powder; sweetish taste; darkens on exposure to light. Sol in 2 parts water. Practically insol in alc or other organic solvents. The aq soln is slightly alkaline. Dec by mineral acids.

THERAP CAT: Formerly as antisyphilitic.

1279. Bismuth Selenide.

[12068-69-8] Bismuth triselenide; dibismuth triselenide. Bi_2Se_3; mol wt 654.84. Bi 63.83%, Se 36.17%. Prepn by heating a stoichiometric mixture at 475° in evacuated tube: Dönges, *Z. Anorg. Allg. Chem.* **265**, 56 (1951); Konorov, *Zh. Tekh. Fiz.* **26**, 1394 (1956); from BiO, Se and sodium oxalate at high press.: Cambi, Elli, *Chim. Ind. (Milan)* **50**, 94 (1968).

Black crystals. Rhombohedral and hexagonal crystal structure. d_4^{15} 7.70 (also reported as 6.82). mp 710°. Heat of formation: −13.9 kcal/mol. Heat of combustion: 296.4 kcal/mol. Dec when heated in air. Dec by concd nitric acid and aqua regia. Insol in water.

USE: In semiconductor research.

1280. Bismuth Sodium Tartrate.

[31586-77-3] (2*R*,3*R*)-2,3-Dihydroxybutanedioic acid bismuth sodium salt (1:?:?); sodium bismuth tartrate; sodium bismuthyl tartrate; tartaric acid bismuth complex sodium salt; Natrol; Tartrol. The bismuth content of the complex salts of bismuth and alkali tartrate may be made to range within wide limits (35-75% Bi). The article generally used contains 70-74% bismuth. Prepn: Kober, US 1663201 (1928 to Searle); *Hagers Handb. Pharm. Praxis* **1st Suppl.**, 316 (Berlin, 1949).

Odorless, tasteless powder; discolors in light. Sol in about 3 parts water. Insol in alc or other organic solvents. The aq soln is slightly alkaline.

THERAP CAT: Formerly as antisyphilitic.

1281. Bismuth Subacetate.

[5142-76-7] Acetic acid oxido-bismuthinyl ester; acetic acid basic bismuth salt; (acetyloxy)oxobismuthine; bismuth acetate basic; bismuth acetate oxide; bismuth oxide acetate; bismuthyl acetate. $C_2H_3BiO_3$; mol wt 284.02. C 8.46%, H 1.06%, Bi 73.58%, O 16.90%. $CH_3COOBiO$. Prepn: Hofmann, *Ann.* **223**, 110 (1884); and crystal structure: Aurivillius, *Acta Chem. Scand.* **9**, 1213 (1955).

Thin crystal plates. Slight acetic odor. Practically insol in water. Sol in glacial acetic acid, dil HCl or HNO_3.

1282. Bismuth Subcarbonate.

[5892-10-4] Bismuth carbonate oxide ($Bi_2(CO_3)O_2$); bismuth oxycarbonate; bismutite. CBi_2O_5; mol wt 509.97. C 2.36%, Bi 81.96%, O 15.69%. $(BiO)_2CO_3$. Prepn: *Gmelins, Bismuth* (8th ed.) **19**, p 178 (1927). Crystal structure, stability and spectral studies: P. Taylor *et al.*, *Can. J. Chem.* **62**, 2863 (1984). Thermal decompn study: H. Henmi *et al.*, *Thermochim. Acta* **114**, 393 (1987). Powder neutron diffraction structural study: C. Greaves, S. K. Blower, *Mater. Res. Bull.* **23**, 1001 (1988). Vibrational spectrum of hemihydrate: G. E. Tobon-Zapata *et al.*, *J. Mater. Sci. Lett.* **16**, 656 (1997). *In vitro* efficacy vs *Campylobacter pylori*: K. Vogt *et al.*, *Zentralbl. Bakteriol.* **273**, 33 (1990). Review of pharmaceutical prepns: M. Wesolowski, *Arzneim.-Forsch.* **28**, 372-378 (1978).

White or almost white powder. Practically insol in water, alc, ether. Dissolves in dilute acids with effervescence.

Hemihydrate. [5798-45-8] Odorless, tasteless powder. Practically insol in water or alc. Sol in mineral acids, in concd acetic acid.

USE: In admixture with other substances in glazes on ceramics; for pearly surfaces for plastics.

THERAP CAT: Antacid; antidiarrheal.

THERAP CAT (VET): Antidiarrheal.

1283. Bismuth Subgallate.

[99-26-3] 2,7-Dihydroxy-1,3,2-benzodioxabismole-5-carboxylic acid; gallic acid bismuth basic salt; bismuth gallate, basic; B.S.G.; Dermatol. $C_7H_5BiO_6$; mol wt 394.09. C 21.33%, H 1.28%, Bi 53.03%, O 24.36%. Prepd from bismuth nitrate and gallic acid in an acetic acid medium. The acetic acid may be replaced with a glycol or mannitol: Pfeiffer, Schmitz, *Pharmazie* **5**, 517 (1950). Effect on hemostasis: H. Thorisdottir *et al.*, *J. Lab. Clin. Med.* **112**, 481 (1988); on wound healing: L.-M. Mai *et al.*, *Biomaterials* **24**, 3005 (2003). Clinical experience in tonsillectomy: J. E. Fenton *et al.*, *J. Laryngol. Otol.* **109**, 203 (1995); R. C. Hatton, *Ann. Pharmacother.* **34**, 522 (2000).

Bright yellow, odorless, tasteless powder. Stable in air but is affected by light. Sol in dil alkali hydroxide solns, in hot mineral acids with decompn. Forms a water-sol sodium salt which has an alkaline reaction. Practically insol in water, alc, chloroform, ether. Insol in very dilute mineral acids.

THERAP CAT: Hemostatic; vulnerary.

1284. Bismuth Subnitrate.

[1304-85-4] Bismuth hydroxide nitrate oxide ($Bi_5(OH)_9(NO_3)_4O$); bismuth nitrate basic; bismuth oxynitrate; bismuth subnitricum; bismuthyl nitrate; bismuth white; magistery of bismuth; novismuth; paint white; Spanish white. $Bi_5H_9N_4O_{22}$; mol wt 1461.98. Bi 71.47%, H 0.62%, N 3.83%, O 24.08%. A basic salt, the compn of which varies with the conditions of preparation. Contains 70 to 74% Bi or 79 to 82% Bi_2O_3. Prepd by partial hydrolysis of $Bi(NO_3)_3$: *Gmelins, Bismuth* (8th ed.) **19**, pp 132-135 (1927); *Traité Pharm. Chim.* **vol. 1**, P. Lebeau, M. M. Janot, Eds. (Masson, Paris, 1956) p 371; *Handbuch der Pharmazie* **vol. 4**(1), H. Thoms, Ed. (Urban & Schwarzenberg, Berlin, 1927) p 325. Mechanism of action study: S. Pugh, M. R. Lewin, *J. Gastroenterol. Hepatol.* **5**, 382 (1990). Clinical studies in ulcer caused by *Helicobacter pylori* infection: A. F. Carvalho *et al.*, *Aliment. Pharmacol. Ther.* **12**, 557 (1998); M. W. Whitehead *et al.*, *Helicobacter* **5**, 169 (2000); R. H. Phillips *et al.*, *ibid.* 176.

Odorless, tasteless, heavy, slightly hygroscopic, microcrystalline powder. Dec to Bi_2O_3 and nitrogen oxides when heated to red heat. Sol in dil HCl and HNO_3. Practically insol in water, alc. *Keep well closed and protect from light. Incompat.* Alkaline bicarbonates, soluble iodides, gallic acid, calomel, salicylic acid, tannin, sulfur.

USE: Manuf bismuth fluxes for enamels; in cosmetics.

THERAP CAT: Antacid; antiulcerative.

THERAP CAT (VET): See Bismuth.

1285. Bismuth Subsalicylate. [14882-18-9] 2-Hydroxy-4*H*-1,3,2-benzodioxabismin-4-one; (2-hydroxybenzoato-O^1,O^2)-oxobismuth; oxo(salicylato)bismuth; basic bismuth salicylate; bismuth oxysalicylate; Bismed; Jatrox; Pepto-Bismol. $C_7H_5BiO_4$; mol wt 362.09. C 23.22%, H 1.39%, Bi 57.72%, O 17.67%. Prepn: Fischer, Grützner, *Arch. Pharm.* **231**, 680 (1893). Clinical trial in infant diarrhea: D. Figueroa-Quintanilla *et al.*, *N. Engl. J. Med.* **328**, 1653 (1993). Symposium on pharmacology and therapeutic use in gastrointestinal infections: *Rev. Infect. Dis.* **12**, Suppl. 1, S1-S119 (1990). Mechanism of gastroprotection: D. Bagchi *et al.*, *Dig. Dis. Sci.* **44**, 2419 (1999).

White or nearly white amorphous or microcrystalline, odorless powder. Practically insol in water, alc, ether. Reacts with alkalies and mineral acids.

THERAP CAT: Antidiarrheal; antiemetic.

THERAP CAT (VET): Antidiarrheal.

1286. Bismuth Sulfate. [7787-68-0] $Bi_2O_{12}S_3$; mol wt 706.13. Bi 59.19%, O 27.19%, S 13.62%. $Bi_2(SO_4)_3$. Prepn: *Gmelins, Bismuth* (8th ed.) **19**, p 170 (1927).

White crystals. Dec by water or alc into a basic salt; sol in acids.

USE: In analysis of other metallic sulfates.

1287. Bismuth Sulfide. [1345-07-9] Bismuth(3+) sulfide. Bi_2S_3; mol wt 514.14. Bi 81.29%, S 18.71%. Occurs in nature as the mineral *bismuthinite*. Prepn: *Gmelins, Bismuth* (8th ed.) **19**, pp 162-167 (1927).

Blackish-brown, orthorhombic, bipyramidal crystals. Sol in HNO_3, HCl. Practically insol in water, ethyl acetate.

1288. Bismuth Tannate. Tanbismuth; tannic acid bismuth derivative. Contains about 36% bismuth equivalent to 40% Bi_2O_3. Prepd from freshly pptd bismuth hydroxide and tannin: *Hagers Handb. Pharm. Praxis* **Band 1**, 685 (Berlin, 1930).

Yellow or brownish-yellow powder. Practically insol in water, alcohol, ether.

THERAP CAT: Astringent; protective.

1289. Bismuth Telluride. [1304-82-1] Tellurobismuthite; bismuth(3+) telluride; dibismuth tritelluride. Bi_2Te_3; mol wt 800.76. Bi 52.20%, Te 47.80%. Prepd by heating stoichiometric amounts of the elements to 475° for several days in an evacuated glass or quartz tube: Dönges, *Z. Anorg. Allg. Chem.* **265**, 56 (1951). Prepn of single crystals: Ainsworth, *Proc. Phys. Soc. London* **B69**, 606 (1956); in zone-melting apparatus: Harmon *et al.*, *J. Phys. Chem. Solids* **2**, 181 (1957). Review of different methods: Minden, *Sylvania Technol.* **11** (no. 1), 13-25 (1958).

Gray hexagonal platelets. d 7.642. mp 585°. Single crystals have been grown by the Czochralski technique in which a hydrogen atmosphere was used to minimize the evaporation of tellurium. Since the crystals cleave readily along the (0001) basal hexagonal plane, it is mechanically easier to orient the seed so that the growth direction is in this plane rather than normal to it. The resulting crystals grow more readily along the basal plane, so that they have an oval cross section, often with a characteristic notch. All crystals so pulled are the *P* type. Heat of formation: −8 kcal/mol. Resistivity: 0.00033 ohm-cm. Thermal conductivities at room temp: $\lambda_0 = 0.015$ watt/

cm-deg; $\lambda_e = 1.4 \times 10^{-3}$ watt/cm-deg. Energy gap: 0.15 ev. Electron mobility: 800 cm^2/volt-sec. Hole mobility: 400 cm^2/volt-sec.

Caution: Potential symptoms of overexposure are irritation of eyes, skin, upper respiratory system; garlic breath. *See NIOSH Pocket Guide to Chemical Hazards* (DHHS/NIOSH 97-140, 1997) p 28.

USE: In electronics as semiconductor.

1290. Bismuth Tetroxide. [12048-50-9] "Bismuth peroxide". Bi_2O_4; mol wt 481.96. Bi 86.72%, O 13.28%. It may contain 1 or 2 mols of H_2O. Prepn: *Gmelins, Bismuth* (8th ed.) **19**, 115 (1927); Schenk in *Handbook of Preparative Inorganic Chemistry* **vol. 1**, G. Brauer, Ed. (Academic Press, New York, 2nd ed., 1963) p 629.

Orange-red to yellowish-brown heavy powder. Slowly dec by water without dissolving. Dec and dissolved by hot mineral acids.

USE: In lubricants for metal extrusion.

1291. Bismuth Tribromophenate. [5175-83-7] 2,4,6-Tribromophenol bismuth(3+) salt (3:1); bismuth tribromophenol; tribromophenolbismuth; tris(2,4,6-tribromophenoxy)bismuthine; Sigmaform; Xeroform. $C_{18}H_6BiBr_9O_3$; mol wt 1198.36. C 18.04%, H 0.50%, Bi 17.44%, Br 60.01%, O 4.01%. $(Br_3C_6H_2O)_3Bi$. Prepd from sodium tribromophenolate and bismuth nitrate: Kollo, *Pharm. Post.* **45**, 1013-1014, *C.A.* **7**, 2831[1] (1913); *Hagers Handb. Pharm. Praxis* **Band 1**, 686 (Berlin, 1930); *N.N.R.* **1951**, pp 66, 505. [In early publications bismuth tribromophenate was described as a basic bismuth salt $(Br_3C_6H_2O)_2BiOH.Bi_2O_3$.]

Amorphous yellow powder. Stable below 120°. Neutral to moistened litmus paper. Slightly sol in water, alcohol, chloroform, vegetable oils. Dec by alkalies and strong acid.

THERAP CAT: Anti-infective.

THERAP CAT (VET): See Bismuth.

1292. Bismuth Triflate. [88189-03-1] 1,1,1-Trifluoromethanesulfonic acid bismuth(3+) salt (3:1); bismuth(III) tris(trifluoromethanesulfonate); $Bi(OTf)_3$. $C_3BiF_9O_9S_3$; mol wt 656.17. C 5.49%, Bi 31.85%, F 26.06%, O 21.94%, S 14.66%. Lewis acid catalyst. Prepn: S. Singh *et al.*, *Indian J. Chem.* **22A**, 814 (1983); M. Labrouillère *et al.*, *Tetrahedron Lett.* **40**, 285 (1999). Catalysis of large scale acetylations of alcohols and diols: M. D. Carrigan *et al.*, *Synthesis* **2001**, 2091; of Biginelli Reaction: R. Varala *et al.*, *Synlett* **2003**, 67; of epoxide opening with aromatic amines: T. Ollevier, G. Lavie-Compin, *Tetrahedron Lett.* **45**, 49 (2004). Review as catalyst for acylations and sulfonylations: C. Le Roux, J. Dubac, *Synlett* **2002**, 181-200; brief review of catalytic use in a variety of organic reactions: S. Antoniotti, *ibid.* **2003**, 1566-1567.

White hygroscopic solid, stable in air to 100 °C. Decomposes in two steps: between 100-400 °C to $Bi_2O_2SO_4$; between 400-600 °C to Bi_2O_3. Sol in DMSO and DMF.

USE: Catalyst for organic syntheses.

1293. Bis(1-naphthylmethyl)amine. [5798-49-2] *N*-(1-Naphthalenylmethyl)-1-naphthalenemethanamine; di(α-naphthylmethyl)amine; α-dinaphthomethylamine. $C_{22}H_{19}N$; mol wt 297.40. C 88.85%, H 6.44%, N 4.71%. Prepd by catalytic hydrogenation of α-naphthonitrile: Rupe, Becherer, *Helv. Chim. Acta* **6**, 880 (1923).

Pale yellow crystals from petr ether, mp 62°.

USE: In determination of nitrates in fertilizers: Konek, *Z. Anal. Chem.* **97**, 416 (1934), *C.A.* **28**, 5779 (1934). The acetate has been proposed as a sensitive reagent for nitric acid.

1294. Bisoctrizole. [103597-45-1] 2,2′-Methylenebis[6-(2*H*-benzotriazol-2-yl)-4-(1,1,3,3-tetramethylbutyl)phenol]; MBBT; Tinosorb M; Tinuvin 360. $C_{41}H_{50}N_6O_2$; mol wt 658.89. C 74.74%, H 7.65%, N 12.76%, O 4.86%. UV-A filter with dual mechanism of action. Photostable organic molecule with broad UV absorption; microfine structure causes light scattering and reflection. Prepn and use in light stabilizer formulations: N. Kubota, A. Nishimura, **EP 180992**; *eidem*, **US 4681905** (1986, 1987 both to Adeka Argus). Light stabilization of polymers: G. Rytz *et al.*, *Angew. Makromol. Chem.* **247**, 213 (1997). Skin photoprotection study: C. Gélis *et al.*, *Photodermatol. Photoimmunol. Photomed.* **19**, 242 (2003). HPLC determn in suncare formulations: C. G. Smyrniotakis, H. A. Archontaki, *J. Chromatogr. A* **1031**, 319 (2004). UV attenuating properties of microparticles: B. Herzog *et al.*, *J. Colloid Interface Sci.* **276**, 354 (2004).

Slightly yellow powder, mp 195°. Flash pt: >200°C. d^{20} 1.2. Vapor pressure (25°): 6×10^{-13} Pa. Soly at 20° (%w/w): water <0.001; acetone 0.05; chloroform 100; *n*-hexane 0.03; methylene chloride 75; toluene 34. Absorption max (chloroform): 308, 349 nm (ε 31895). Absorption max (*n*-heptane): 348 nm (ε 31600).

USE: Light stabilizer in polymers and resins.

THERAP CAT: Ultraviolet screen.

1295. Bisoprolol. [66722-44-9] 1-[4-[[2-(1-Methylethoxy)-ethoxy]methyl]phenoxy]-3-[(1-methylethyl)amino]-2-propanol; (±)-1-[[α-(2-isopropoxyethoxy)-*p*-tolyl]oxy]-3-(isopropylamino)-2-propanol; (±)-1-[*p*-(2-isopropoxyethoxymethyl)phenoxy]-3-(isopropylamino)-2-propanol; EMD-33512. $C_{18}H_{31}NO_4$; mol wt 325.45. C 66.43%, H 9.60%, N 4.30%, O 19.66%. Cardioselective β_1-adrenergic blocker. Prepn: **BE 859425**; R. Jonas *et al.*, **US 4258062** (1978, 1981 both to E. Merck). Clinical pharmacology: A. E. Tattersfield *et al.*, *Br. J. Clin. Pharmacol.* **18**, 343 (1984); P. Dorow, U. Tönnesmann, *Eur. J. Clin. Pharmacol.* **27**, 135 (1984). Binding study of (−)-form: A. J. Kaumann, H. Lemoine, *Arch. Pharmacol.* **331**, 27 (1985). HPLC determn of enantiomers in plasma and urine: T. Suzuki *et al.*, *J. Chromatogr.* **619**, 267 (1993). Clinical evaluation in angina: R. S. Kohli *et al.*, *Eur. Heart J.* **6**, 845 (1985); in hypertension: F. R. Bühler *et al.*, *J. Cardiovasc. Pharmacol.* **8**, Suppl. 11, S122 (1986). Review of pharmacokinetics: J. Grevel *et al.*, *Clin. Pharmacokinet.* **17**, 53-63 (1989); of clinical efficacy in combination with hydrochlorothiazide: P. K. Zachariah *et al.*, *Clin. Ther.* **15**, 779-787 (1993). Clinical trial in chronic heart failure: CIBIS-II Investigators, *Lancet* **353**, 9 (1999).

Hemifumarate. [104344-23-2] Concor; Detensiel; Emconcor; Emcor; Euradal; Isoten; Monocor; Soprol; Zebeta. $C_{18}H_{31}NO_4 \cdot \frac{1}{2}C_4H_4O_4$; mol wt 383.49. Crystals, mp 100°. Sol in ethanol.

Mixture with hydrochlorothiazide. Ziac.

THERAP CAT: Antihypertensive.

1296. Bisoxatin Acetate. [14008-48-1] 2,2-Bis[4-(acetyloxy)phenyl]-2*H*-1,4-benzoxazin-3(4*H*)-one; 2,2-bis(4-acetoxyphenyl)-3-oxo-2,3-dihydrobenz-1,4-oxazine; Wy-8138; Laxonalin; Maratan; Talsis; Tasis. $C_{24}H_{19}NO_6$; mol wt 417.42. C 69.06%, H 4.59%, N 3.36%, O 23.00%. Prepd from 2,2-dichloro-3-oxo-2,3-dihydrobenz-1,4-oxazine and phenylacetate or by acetylation of the

diphenol: Seeger, **US 3006917** (1961 to Thomae). Toxicity: E. I. Goldenthal, *Toxicol. Appl. Pharmacol.* **18**, 185 (1971).

Crystals from ethanol, mp 190°. LD_{50} orally in rats, mice: 8000, >10,000 mg/kg (Goldenthal).

THERAP CAT: Cathartic.

1297. Bis(oxazoline) Ligands. Box ligands. C_2-symmetric ligands; when coordinated with metal cations, the bidentate complexes serve as enantioselective catalysts for a wide range of reactions. Pyridine bis(oxazoline) ligands, *q.v.*, are a similar class of compds that also serve as chiral catalysts when bound to metal ions. Prepn of *t*-Bu-box and evaluation of the copper complex as a catalyst in cyclopropanations: D. A. Evans *et al.*, *J. Am. Chem. Soc.* **113**, 726 (1991). Design of Ph-box and utility of the iron-ligand complex in Diels-Alder reactions: E. J. Corey *et al.*, *ibid.* 728. Improved prepn of *t*-Bu-box: D. A. Evans *et al.*, *J. Org. Chem.* **63**, 4541 (1998); of box ligands: A. Cornejo *et al.*, *Synlett* **2005**, 2321. X-ray analysis of copper-ligand complexes: J. Thorhauge *et al.*, *Chem. Eur. J.* **8**, 1888 (2002). Synthetic utility of metal-box ligand complexes as chiral catalysts for olefin aziridinations: D. A. Evans *et al.*, *J. Am. Chem. Soc.* **115**, 5328 (1993); for Diels-Alder reactions: *idem et al.*, *ibid.* **121**, 7559 (1999); for Mukaiyama-Michael reactions: *idem et al.*, *ibid.* **123**, 4480 (2001); for olefin cyclopropanations: N. Ostergaard *et al.*, *Tetrahedron* **57**, 6083 (2001); for C-H carbene insertions: J. M. Fraile *et al.*, *Org. Lett.* **9**, 731 (2007). Brief review: R. Basak, *Synlett* **2003**, 1223-1224.

(*S,S*)-*t*-Bu-box: R = C(CH₃)₃

(*S,S*)-Ph-box: R = phenyl

(*S,S*)-*t*-**Bu-box.** [131833-93-7] (4*S*,4′*S*)-2,2′-(1-Methylethylidene)bis[4-(1,1-dimethylethyl)-4,5-dihydrooxazole]; 2,2-bis[2-[4-(*S*)-*tert*-butyl-1,3-oxazolinyl]]propane; (*S,S*)-2,2′-isopropylidenebis[4-*tert*-butyl-2-oxazoline]. $C_{17}H_{30}N_2O_2$; mol wt 294.44. White crystalline solid from pentane, mp 88.9-89.8°. $[\alpha]_{365}$ −394° (c = 0.97 in dichloromethane). Sol in common organic solvents. Insol in water.

(*R,R*)-*t*-**Bu-box.** [131833-97-1] White powdery solid, mp 87.8-88.6°. $[\alpha]_{365}$ +384° (c = 1.22 in dichloromethane).

(*S,S*)-**Ph-box.** [131457-46-0] (4*S*,4′*S*)-2,2′-(1-Methylethylidene)bis[4,5-dihydro-4-phenyloxazole]; 2,2-bis[2-[4-(*S*)-phenyl-1,3-oxazolinyl]]propane; (*S,S*)-2,2′-isopropylidenebis[4-phenyl-2-oxazoline]. $C_{21}H_{22}N_2O_2$; mol wt 334.42. Viscous oil, bp$_{0.03}$ 193°. mp 36-40°. $[\alpha]_D^{23}$ −171.3° (c = 1.0 in ethanol) (Corey). Flash pt, closed cup: 230°F (110°C). Store at −20°C.

(*R,R*)-**Ph-box.** [150529-93-4] Solid. *Poisonous*. Store at −20°C.

USE: Ligands in asymmetric synthesis.

1298. Bisphenol A. [80-05-7] 4,4′-(1-Methylethylidene)bisphenol; 4,4′-isopropylidenediphenol; 2,2-bis(4-hydroxyphenyl)propane. $C_{15}H_{16}O_2$; mol wt 228.29. C 78.92%, H 7.06%, O 14.02%. Monomer used for polycarbonate and epoxy resins; exhibits estrogenic activity. Manuf from phenol and acetone: Jansen, **US 2468982** (1949); *Faith, Keyes & Clark's Industrial Chemicals*, F. A.

Lowenheim, M. K. Moran, Eds. (Wiley-Interscience, New York, 4th ed., 1975) pp 149-152. Identification as xenoestrogen released from polycarbonate: A. V. Krishnan *et al.*, *Endocrinology* **132**, 2279 (1993). HPLC determn in serum: K. Inoue *et al.*, *J. Chromatogr. B* **749**, 17 (2000); in food: H. Petersen *et al.*, *Eur. Food Res. Technol.* **216**, 355 (2003). Review of bioactivity and uses: N. Ben-Jonathan, R. Steinmetz, *Trends Endocrinol. Metab.* **9**, 124-128 (1998); of properties and environmental fate: C. A. Staples *et al.*, *Chemosphere* **36**, 2149-2173 (1998).

Crystals or flakes. Mild phenolic odor. mp 150-155° (solidification range). bp₄ 220°. Dec above 8 mm pressure when heated above 220°. Log P (*n*-octanol/water): 3.40. Soly in water: 120-300 mg/l. Sol in alkaline solns, alcohol, acetone. Slightly sol in carbon tetrachloride. LC₅₀ (96 hr) in fathead minnow, rainbow trout: 4600, 3000-3500 μg/l (Staples).

Diglycidyl ether. [1675-54-3] BADGE. $C_{21}H_{24}O_4$; mol wt 340.42. Monohydrate as white solid, mp 40°. uv max (acetonitrile/ water): 226.5, 277.1 nm.

USE: In the manuf of epoxy resins and polycarbonates for food packaging.

1299. Bisphenol B. [77-40-7] 4,4′-(1-Methylpropylidene)-bisphenol; *p,p′-sec*-butylidenediphenol; 2,2-bis(4-hydroxyphenyl)-butane. $C_{16}H_{18}O_2$; mol wt 242.32. C 79.31%, H 7.49%, O 13.20%. May be prepd from phenol and ethyl methyl ketone: Jansen, **US 2468982** (1949 to Goodrich).

Crystals or tan granules, mp 118.9-121.7°. Approximate soly per 100 g: acetone 266 g; benzene 2.3 g; carbon tetrachloride <0.1 g; ether 133 g; methanol 166 g; V.M.P. naphtha <0.1 g; water <0.1 g.

USE: In the manufacture of phenolic resins.

1300. 9,10-Bis(2-phenylethynyl)anthracene. [10075-85-1] BPEA. $C_{30}H_{18}$; mol wt 378.47. C 95.21%, H 4.79%. Fluorescent dye. Prepn: W. Ried *et al.*, *Ber.* **94**, 1051 (1961). Scintillation and fluorescence properties: A. Heller, G. Rio, *Bull. Soc. Chim. Fr.* **1963**, 1707. Absorption and emission properties: D. R. Maulding, B. G. Roberts, *J. Org. Chem.* **34**, 1734 (1969). ¹H NMR and ¹³C NMR of dianion: M. Hirayama *et al.*, *Bull. Chem. Soc. Jpn.* **61**, 1501 (1988). Photophysics and electronic spectroscopy: M. Levitus, M. A. Garcia-Garibay, *J. Phys. Chem. A* **104**, 8632 (2000).

Orange needles from benzene + petroleum ether, mp 249-250°. Absorption max (benzene): 439, 455 nm (log ε 4.50, 4.52).

USE: Bright yellow-green fluorescer in chemiluminescent formulations including *Cyalume* light sticks.

1301. Bis(pinacolato)diborane. [73183-34-3] 4,4,4′,4′,5,5,-5′,5′-Octamethyl-2,2′-bi-1,3,2-dioxaborolane; Miyaura's reagent; pinacol diborane; 4,4,5,5-tetramethyl-1,3,2-dioxaborolan-1-yl-4′,-4′,5′,5′-tetramethyl-1′,3′,2′-dioxaborolan; B₂pin₂. $C_{12}H_{24}B_2O_4$; mol wt 253.94. C 56.76%, H 9.53%, B 8.51%, O 25.20%. Boron

source for organic syntheses. Prepn and crystal structure: H. Nöth, *Z. Naturforsch.* **39b**, 1463 (1984). Synthesis: T. Ishiyama *et al.*, *Org. Synth.* **77**, 176 (2000). NMR spectroscopic data: W. Biffar *et al.*, *Ber.* **113**, 333 (1980). First use as boron source in Pt-catalyzed diboration of alkynes: T. Ishiyama *et al.*, *J. Am. Chem. Soc.* **115**, 11018 (1993). Stereospecific synthesis via cross coupling with aromatic amine: C. Malan, C. Morin, *J. Org. Chem.* **63**, 8019 (1998). Rh-catalyzed activation of the C-H bond: Y. Kondo *et al.*, *J. Am. Chem. Soc.* **124**, 1164 (2002). Brief review: X. Liu, *Synlett* **2003**, 2442-2443.

Colorless plates, mp 138°. Crystalline form can be handled in air and stored in capped bottle.

USE: As a boron source in Pt-mediated diborations, coupling reactions; in Rh-or Ir-mediated borylations of alkanes and arenes, and in carbenoid chemistry.

1302. Bispyribac. [125401-75-4] 2,6-Bis[(4,6-dimethoxy-2-pyrimidinyl)oxy]benzoic acid. $C_{19}H_{18}N_4O_8$; mol wt 430.37. C 53.03%, H 4.22%, N 13.02%, O 29.74%. Post-emergence herbicide; turfgrass growth regulator. Prepn: N. Wada *et al.*, **EP 321846**; *eidem*, **US 4906285** (1989, 1990 both to Kumiai Chem. Ind. and Ihara Chem. Ind.). Metabolism in rats: Y. Fukai *et al.*, *J. Pestic. Sci.* **20**, 479 (1995). Field studies in rice: M. P. Braverman, D. L. Jordan, *Weed Technol.* **10**, 876 (1996); in turfgrass: M. J. Fagerness, D. Penner, *Weed Technol.* **12**, 436 (1998). Review of physical properties, mode of action and efficacy: M. Yokoyama *et al.*, *Brighton Crop Prot. Conf. - Weeds* **1993**, 61-66.

White powder, mp 148-150°.

Sodium salt. [125401-92-5] KIH-2023; KUH-911; V-10029; Nominee. $C_{19}H_{17}N_4NaO_8$; mol wt 452.35. White powder, mp 228.0° (dec). Soly in water at 25°: 733 g/l. Vapor pressure at 25°: 5.05 × 10⁻⁹ Pa. LD₅₀ in male, female rats (mg/kg): 4111, 2635 orally; in rats (mg/kg): >2000 dermally. LC₅₀ in bluegill sunfish, rainbow trout (ppm): >100, >100 (Yokoyama).

USE: Herbicide

1303. Bis(pyridine)iodonium Tetrafluoroborate. [15656-28-7] Bis(pyridine)iodine(1+) tetrafluoroborate(1−) (1:1); Barluenga's reagent; IPy₂BF₄. $C_{10}H_{10}IN_2 \cdot BF_4$; mol wt 371.91. C 32.30%, H 2.71%, I 34.12%, N 7.53%, B 2.91%, F 20.43%. Mild source of iodonium ions in organic synthesis. Prepn: J. Barluenga *et al.*, *Angew. Chem. Int. Ed.* **24**, 319 (1985). Oxidation of alcohols: *idem et al.*, *Chem. Eur. J.* **10**, 4206 (2004). Brief review: K. Muñiz, *Synlett* **1999**, 1679. Review of iodination chemistry: J. Barluenga, *Pure Appl. Chem.* **71**, 431-436 (1999).

Crystals from dichloromethane, mp 149-151° (dec). *Store under inert atmosphere at 4° in the dark.*

USE: Iodination and oxidation reagent.

1304. 1,4-Bis(trichloromethyl)benzene. [68-36-0] α,α,α,-α′,α′,α′-Hexachloro-*p*-xylene; *p*-bis(perchloromethyl)benzene; Bitriben; Hetol. $C_8H_4Cl_6$; mol wt 312.82. C 30.72%, H 1.29%, Cl

67.99%. Usually obtained by photochlorination of *p*-xylene: Mc-Bee *et al.*, *Ind. Eng. Chem.* **39**, 298 (1947); *eidem* in Slesser, Schram, *Preparation, Properties, and Technology of Fluorine and Organic Fluoro Compounds* (McGraw-Hill, New York, 1951) pp 207-221; Ross *et al.*, *J. Am. Chem. Soc.* **75**, 4697 (1953); Rabjohn, *ibid.* **76**, 5479 (1954); Harvey *et al.*, *J. Appl. Chem.* **4**, 319 (1954); Ligett, **US 2654789** (1953 to Ethyl Corp.).

Cl₃C—⬡—CCl₃

Crystals from hexane or ether, mp 108-110°.
Note: The name Hetol was previously used to designate sodium cinnamate.
USE: Insecticide.
THERAP CAT (VET): Anthelmintic (Trematodes).

1305. Bis(triphenylphosphine)dicarbonylnickel. [13007-90-4] (*T*-4)-Dicarbonylbis(triphenylphosphine)nickel; dicarbonyldi-(triphenylphosphino)nickel. C₃₈H₃₀NiO₂P₂; mol wt 639.30. C 71.39%, H 4.73%, Ni 9.18%, O 5.01%, P 9.69%. [(C₆H₅)₃P]₂Ni-(CO)₂. Obtained by reacting one mole of nickel carbonyl with two moles of triphenylphosphine: Reppe, Schweckendiek, *Ann.* **560**, 104 (1948); Rose, Statham, *J. Chem. Soc.* **1950**, 69. Also by reacting CO with a methanol soln of triphenylphosphine and reduced Ni at 170° and 60 atm: Yamamoto, Oku, *Bull. Chem. Soc. Jpn.* **27**, 382 (1954), *C.A.* **49**, 6856 (1955).
Crystals from benzene, mp 210-215° (Rose); mp 206-209° (Reppe). Catalytic activity decreases with storage.
USE: Catalyst in polymerization of acetylene to benzene and styrene, trimerization of ethynyl compds, cyclization of butadiene.

1306. Bitertanol. [55179-31-2] β-[(1,1′-Biphenyl)-4-yloxy]-α-(1,1-dimethylethyl)-1*H*-1,2,4-triazol-1-ethanol; 1-(biphenyl-4-yl-oxy)-3,3-dimethyl-1-(1*H*-1,2,4-triazol-1-yl)butan-2-ol; biloxazol; BAY KWG 0599; Baycor; Sibutol. C₂₀H₂₃N₃O₂; mol wt 337.42. C 71.19%, H 6.87%, N 12.45%, O 9.48%. Prepn: **BE 814831**; W. Kramer *et al.*, **US 3952002** (1974, 1976 both to Bayer AG). Physical properties and fungicidal activity: W. Brandes *et al.*, *Pflanzen-schutz-Nachr.* **32**, 1 (1979). Mechanism of action: P. Kraus, *ibid.* 17; S. V. Overton *et al.*, *J. Hortic. Sci.* **63**, 183 (1988). GC determn in plants, soil and water: R. Brennecke, *Pflanzenschutz-Nachr.* **38**, 33 (1985). Resolution of diastereomers: R. S. Burden *et al.*, *J. Chromatogr.* **391**, 273 (1987). *In vitro* activity: T. B. Sutton *et al.*, *Plant Dis.* **69**, 700 (1985). Field use in control of apple diseases: K. S. Yoder, *ibid.* **66**, 580 (1982); W. F. S. Schwabe, *Pflanzenschutz-Nachr.* **35**, 125 (1982); in wheat seed pretreatment: J. A. Hoffmann, D. V. Sisson, *Plant Dis.* **71**, 839 (1987).

[structure]

The marketed product is a mixture of diastereomers. Colorless crystals, mp 125-129°. Vapor pressure at 20°: 10⁻⁵mbar. Soly at 20° (g/100 g solvent): water 0.0005, ligroin (80-110°) 0-1, propan-2-ol 1-5, toluene 1-5, cyclohexane 5-10, methylene chloride 10-20. Stable in aqueous acid and alkaline solns. LD₅₀ in rats, male and female mice (mg/kg): >5000, 4488, 4202 orally (Brandes).
USE: Agricultural fungicide.

1307. Bithionol. [97-18-7] 2,2′-Thiobis[4,6-dichlorophenol]; TBP; bis(2-hydroxy-3,5-dichlorophenyl)sulfide; XL-7; Bithin; Lo-rothidol. C₁₂H₆Cl₄O₂S; mol wt 356.04. C 40.48%, H 1.70%, Cl 39.83%, O 8.99%, S 9.00%. Prepn from 2,4-halo substituted phenol with S halides: Muth, **DE 583055** (1933 to I. G. Farbenind.), *C.A.* **28**, 179 (1934); Copper, Godfrey, **US 2849494** (1958 to Monsanto). Comprehensive review and bibliography: Shumard *et al.*, *Soap Sanit. Chem.* **29**, no. 1, 34-37, 90 (1953). Anthelmintic activity of

the sulfoxide: J. Guilhon, M. Graber, *Bull. Acad. Vet. Fr.* **52**, 225 (1979). Metabolism of the sulfoxide in humans: M. Sakamoto *et al.*, *J. Toxicol. Sci.* **6**, 307 (1981).

[structure]

Crystals, mp 188°. d₄²⁵ 1.73. pK₁ 4.82; pK₂ 10.50. Vapor press. at 37°: 1.1 × 10⁻⁹ mm Hg. Practically insol in water (0.0004% at 25°). Sol in dil caustic solns. A 4% NaOH soln will dissolve 16.2% bithionol. Soly (g/100 ml): acetone 15.0; polysorbate 80 19.0; di-methylacetamide 72.5; lanolin at 42° 5.0; pine oil 4.0; corn oil 1.0; propylene glycol 0.5; 70% ethanol 0.3.
Sodium salt. [6385-58-6] Bithionolate sodium; Vancide BN. C₁₂H₄Cl₄Na₂O₂S; mol wt 400.00. Harvey *et al.*, **US 3024163** (1962 to Vanderbilt Co.).
Sulfoxide. BTS; Bitin-S; Disto-5. C₁₂H₆Cl₄O₃S; mol wt 372.04.
USE: Surfactant-formulated antimicrobial against bacteria, molds and yeast. In cosmetics. Proposed as agricultural fungicide.
THERAP CAT: Anti-infective (topical).
THERAP CAT (VET): Anthelmintic; antiseptic.

1308. Bitolterol. [30392-40-6] 4-Methylbenzoic acid 1,1′-[4-[2-[(1,1-dimethylethyl)amino]-1-hydroxyethyl]-1,2-phenylene] es-ter; *p*-toluic acid 4-[2-(*t*-butylamino)-1-hydroxyethyl]-*o*-phenylene ester; α-[(*tert*-butylamino)methyl]-3,4-dihydroxybenzyl alcohol 3,4-di-*p*-toluate. C₂₈H₃₁NO₅; mol wt 461.56. C 72.86%, H 6.77%, N 3.03%, O 17.33%. β₂-Adrenergic agonist; a diester of *N-tert*-butyl-norepinephrine. Prepn: H. Minatoya *et al.*, **DE 2015573**; *eidem*, **US 4138581** (1970, 1979 both to Sterling); B. F. Tullar *et al.*, *J. Med. Chem.* **19**, 834 (1976). Pharmacology: H. Minatoya, *J. Pharmacol. Exp. Ther.* **206**, 515 (1978). Metabolism and excretion: T. Aimoto *et al.*, *Xenobiotica* **9**, 173 (1979). Comparative clinical studies with isoproterenol, *q.v.*, in asthma: J. L. Pinnas *et al.*, *J. Allergy Clin. Immunol.* **79**, 768 (1987); R. A. Nathan *et al.*, *ibid.* 822. *Review:* S. B. Walker *et al.*, *Pharmacotherapy* **5**, 127-137 (1985).

[structure]

Methanesulfonate. [30392-41-7] Bitolterol mesylate; Win-32784; Biterol; Effectin; Tornalate. C₂₈H₃₁NO₅.CH₃SO₃H; mol wt 557.66. White crystalline powder, mp 170-172°. Sol in DMSO.
THERAP CAT: Bronchodilator.

1309. Bitoscanate. [4044-65-9] 1,4-Diisothiocyanatoben-zene; phenylene-1,4-diisothiocyanate; isothiocyanic acid *p*-phenyl-ene ester; C₈H₄N₂S₂; mol wt 192.25. C 49.98%, H 2.10%, N 14.57%, S 33.35%. Prepn: Lieber, Slutkin, *J. Org. Chem.* **27**, 2214 (1962). Prepn and use as anthelmintic: **FR M1652** and **GB 1001314** (1963, 1965 to Hoechst). Purification: Soeder, Laemmer, **DE 1172664** (1964 to Hoechst). Series of articles on clinical studies: *Progress in Drug Research* vol. 19, E. Jucker, Ed. (Birkhäuser Verlag, Basel and Stuttgart, 1975) pp 2-107.

S=C=N—⬡—N=C=S

Tasteless, odorless, colorless needles, from acetic acid or acetone, mp 132°.

THERAP CAT: Anthelmintic (Nematodes).

1310. Biuret. [108-19-0] Imidodicarbonic diamide; carbamylurea. $C_2H_5N_3O_2$; mol wt 103.08. C 23.30%, H 4.89%, N 40.77%, O 31.04%. Prepd on a large scale by the action of heat on urea. Other syntheses are based on the treatment of urea with inorganic halides, such as thionyl chloride, on the interaction of urea and cyanic acid, and on the ammonolysis of allophanic esters and related compds. Has weak bacteriostatic and diuretic properties in rats, also causes a fall in blood pressure, but produces strong irritation of the urinary tract. Increases the activity of pepsin. Comprehensive review and bibliography: Kurzer, *Chem. Rev.* **56**, 95-197 (1956).

Hygroscopic, elongated plates from ethanol. d_4^{-5} 1.467. When crystallized from water, $5C_2H_5N_3O_2.4H_2O$ is formed, which becomes anhydr at 110°, then dec at about 193°. Pyrolysis at higher temps yields melamine. Soly (g/100 g soln) in water at 25° = 2.01; at 50° = 7; at 75° = 20; at 105.5° = 53.5. Freely sol in alc, very slightly sol in ether. Aq solns treated with cupric sulfate and NaOH give a reddish-violet color (biuret reaction, details: Kurzer, *loc. cit.*, p 181).

1311. Bivalirudin. [128270-60-0] D-Phenylalanyl-L-prolyl-L-arginyl-L-prolylglycylglycylglycylglycylglycyl-L-aspariginylglycyl-L-α-aspartyl-L-phenylalanyl-L-α-glutamyl-L-α-glutamyl-L-isoleucyl-L-prolyl-L-α-glutamyl-L-glutamyl-L-tyrosyl-L-leucine; Hirulog-1; hirulog-8; BG-8967; Angiomax; Angiox; Hirulog. $C_{98}H_{138}N_{24}O_{33}$; mol wt 2180.32. C 53.99%, H 6.38%, N 15.42%, O 24.22%. Synthetic peptide of 20 amino acids; a specific and reversible bivalent direct thrombin inhibitor. Characterized by an anion binding exosite (ABE) derived from hirudin, *q.v.*, linked to a thrombin catalytic site inhibitor moiety by tetraglycine. Prepn and structure activity study: J. M. Maraganore *et al.*, **WO 9102750**; *eidem*, **US 5196404** (1991, 1993 both to Biogen); *eidem*, *Biochemistry* **29**, 7095 (1990). Determn in plasma and whole blood: T. J. Reid, B. M. Alving, *Thromb. Haemostasis* **70**, 608 (1993). Clinical pharmacology: I. Fox *et al.*, *ibid.* **69**, 157 (1993). Review of pharmacology and clinical development: N. W. Shammas, *Cardiovasc. Drug Rev.* **23**, 345-360 (2005); of clinical experience in coronary syndromes: T. E. Warkentin, A. Koster, *Expert Opin. Pharmacother.* **6**, 1349-1371 (2005).

D-Phe–Pro–Arg–Pro–Gly–Gly–Gly–Gly–Asn–Gly–Asp
|
Phe
|
Leu–Tyr–Glu–Glu–Pro–Ile–Glu–Glu

THERAP CAT: Antithrombotic.

1312. Bixins. Carotenoid carboxylic acids isolated from seeds of *Bixa orellana* L., *Bixaceae*; principal pigments of the crude coloring extract, annatto, *q.v.* The natural product is the *cis*-form. Isomerization to the more stable all-*trans*-form occurs upon heating or commercial extraction. Isoln: J. B. Boussingault, *Ann. Chim. Phys.* **28**, 440 (1825). Isomerization studies: J. Herzig, F. Faltis, *Ann.* **431**, 40 (1923); P. Karrer *et al.*, *Helv. Chim. Acta* **12**, 741 (1929). Structural studies: R. Kuhn *et al.*, *ibid.* **11**, 427 (1928); **12**, 64, 904 (1929); *Ber.* **64**, 1732 (1931); **65**, 646, 1873 (1932). NMR determn of stereochemistry: M. S. Barber *et al.*, *J. Chem. Soc.* **1961**, 1625. Properties of bixin and norbixin: J. F. Reith, J. W. Gielen, *J. Food Sci.* **36**, 861 (1971). ^{1}H and ^{13}C NMR, MS, and x-ray crystal structure determn: D. R. Kelly *et al.*, *J. Chem. Res. Miniprint* **10**, 2640 (1996). HPLC analysis of food coloring formulations: M. J. Scotter *et al.*, *J. Agric. Food Chem.* **46**, 1031 (1998). Biosynthesis and prodn in genetically engineered *E. coli* (stereo unspecified): F. Bouvier *et al.*, *Science* **300**, 2089 (2003). Review of *cis*-*trans* isomerization and stereochemistry: L. Zechmeister, *Chem. Rev.* **34**, 267-344 (1944); of chemistry and extraction from annatto: H. D. Preston, M. D. Rickard, *Food Chem.* **5**, 47-56 (1980).

all-*trans* Bixin

cis Bixin

***cis*-Form.** [6983-79-5] (2E,4Z,6E,8E,10E,12E,14E,16E,18E)-4,-8,13,17-Tetramethyl-2,4,6,8,10,12,14,16,18-eicosanonaenedioic acid monomethyl ester; 9-*cis*-6,6′-diapo-ψ,ψ-carotenedioic acid monomethyl ester; 6′-methyl hydrogen (9′Z)-6,6′-diapocarotene-6,6′-dioate; α-bixin; labile bixin. $C_{25}H_{30}O_4$; mol wt 394.51. C 76.11%, H 7.67%, O 16.22%. Dark purple, lustrous crystals from ethanol + chloroform or dichloromethane + acetone, mp 190 ± 1°; dec 200.3 ± 0.6°. Absorption max (chloroform): 502, 471 nm (log ε = 5.02, 5.07).

***trans*-Form.** [39937-23-0] (2E,4E,6E,8E,10E,12E,14E,16E,-18E)-4,8,13,17-Tetramethyl-2,4,6,8,10,12,14,16,18-eicosanonaenedioic acid monomethyl ester; all-*trans* bixin; β-bixin; isobixin; stable bixin. $C_{25}H_{30}O_4$; mol wt 394.51. Orange to purple plates from acetone, mp 204-206° (dec). Absorption max (chloroform): 507, 476 nm ($E_{1cm}^{1\%}$ 2970 ±40, 3240 ±50). Sol in oil.

α-Norbixin. [626-76-6] (2E,4E,6E,8E,10E,12E,14E,16Z,18E)-4,8,13,17-Tetramethyl-2,4,6,8,10,12,14,16,18-eicosanonaenedioic acid; 9-*cis*-6,6′-diapo-ψ,ψ-carotenedioic acid. $C_{24}H_{28}O_4$; mol wt 380.48. C 75.76%, H 7.42%, O 16.82%. No melting observed ≤ 280°; yellow coloring observed at 240°; black coloring (carbonization) at ≥280°. Absorption max (chloroform): 499, 468 nm ($E_{1cm}^{1\%}$ 2200 ±30, 2470 ±30).

β-Norbixin. [542-40-5] (all-E)-4,8,13,17-Tetramethyl-2,4,6,8,-10,12,14,16,18-eicosanonaenedioic acid; 6,6′-diapo-ψ,ψ-carotenedioic acid; all-*trans* norbixin. $C_{24}H_{28}O_4$; mol wt 380.48. No melting observed ≤300°; red color darkens ≥250°. Absorption max (chloroform): 506, 475 nm; (0.1 N NaOH): 486, 457 nm. Sol in H_2O.

USE: Food coloring; imparts a golden yellow color to butter and cheese.

1313. Black Cohosh. Black snake root; bugbane; bugwort; cimicifuga; rattle weed. Flowering perennial plant, *Actaea racemosa* L., also known as *Cimicifuga racemosa* (L.) Nutt., *Ranunculaceae*. Traditionally used in Native American medicine for gynecological disorders, rheumatism and snake bite. Medicinal formulations are prepared from the dried rhizome and roots. *Habit.* Eastern North America; cultivated in Europe. *Constit.* Triterpene glycosides such as actein, 27-deoxyactein, cimicifugosides; isoferulic and salicylic acids; 15-20% cimicifugin; tannin, volatile oils, resin. Description of botany, constituents and medical uses: D. J. McKenna *et al.*, *Altern. Ther. Health Med.* **7**, 93 (2001). LC/MS determn of active triterpene glycosides in commercial formulations: K. He *et al.*, *Planta Med.* **66**, 635 (2000). *Review:* J. Barnes *et al.*, *Herbal Medicines* (Pharmaceutical Press, London, 2nd Ed., 2002) pp 141-146.

Alcoholic aqueous extract. Remifemin. Review of clinical experience: E. Liske, *Adv. Ther.*, **15**, 45-53 (1998).

Note: Should not be confused with blue cohosh, *q.v.*

THERAP CAT: In treatment of menopausal symptons.

1314. Black Pepper. Dried, unripe fruit of *Piper nigrum* L., *Piperaceae*. White pepper consists of the seed only, with the ripe fruit removed. Widely used as a cooking spice and in traditional medicine for dyspepsia. *Habit.* Southern India; cultivated in tropical Asia and the Caribbean. *Constit.* Pungent principles, predominantly piperine, *q.v.*, piperylin, piperolein A and B, cumaperine; 3,4-dihydroxyphenylethanol glycosides; volatile oil. HPLC determn of piperine content: M. Rathnawathie, K. A. Buckle, *J. Chromatogr.* **264**, 316 (1983). GC-MS analysis of constituents and antifungal and antioxidant properties: G. Singh *et al.*, *J. Sci. Food Agric.* **84**, 1878 (2004). Insecticidal activity: H. C. F. Su, *J. Econ. Entomol.* **70**, 18

(1977); W. P. Scott, G. H. McKibben, *ibid.* **71**, 343 (1978). Review of medicinal uses: J. Gruenwald *et al.*, *PDR for Herbal Medicines* (Thomson PDR, Montvale, 3rd Ed., 2004) pp 105-106.

Volatile oil. [8006-82-4] Black pepper oil; oil of pepper. Obtained from dried, unripened fruit. *Constit.* Complex mixture containing predominantly β-caryophyllene, limonene, sabinene, β-bisabolene, α-copaene, β-pinene. Colorless or slightly green liquid. d_{25}^{25} 0.864-0.884. Angular rotation: −1 to −23°. n_D^{20} 1.479-1.488. Sol in most fixed oils, mineral oil, propylene glycol; sparingly sol in glycerin. Sol in 3 vols 95% alcohol.

USE: Pungently aromatic condiment in foods.

1315. Blackstrap Molasses. The final, unpurified mother liquor in the cane sugar industry from which no more sugar can be crystallized by factory methods. Rich in inorganic constituents, may analyze as much as 10% ash upon ignition. Sucrose content about 30%.

1316. Blankophor® R. [2606-93-1] 2,2′-(1,2-Ethenediyl)-bis[5-[[(phenylamino)carbonyl]amino]benzenesulfonic acid] sodium salt (1:2); 4,4′-bis(3-phenylureido)-2,2′-stilbenedisulfonic acid disodium salt; [stilbene-(4,4′)]bis[ω-phenylurea]-2,2′-disulfonic acid disodium salt; C.I. Fluorescent Brightener 30; C.I. 40600; Blancol C; Leucophor R; Lumisol RV; Phorwite RN; Photine R; Pontamine White BR; Tintophen X. $C_{28}H_{22}N_4Na_2O_8S_2$; mol wt 652.60. C 51.53%, H 3.40%, N 8.59%, Na 7.05%, O 19.61%, S 9.83%. Prepd by treating 4,4′-diamino-2,2′-stilbenedisulfonic acid with phenyl isocyanate in aq soln at 40°: **DE 746569** and **FR 878155** (to I. G. Farbenind.); **GB 683895** (1952 to Bayer); *FIAT Final Rept.* No. 1302 (Sept. 15, 1947); *Colour Index* Vol. 4 (3rd ed., 1971) p 4371.

trans-form

USE: Fluorescent dye for cellulose, protein fibers, nylon, wool or paper.

1317. Blasticidin S. [2079-00-7] (S)-4-[[3-Amino-5-[(amino-iminomethyl)methylamino]-1-oxopentyl]amino]-1-(4-amino-2-oxo-1(2H)-pyrimidinyl)-1,2,3,4-tetradeoxy-β-D-*erythro*-hex-2-enopyranuronic acid; 4-[3-amino-5-(1-methylguanidino)valeramido]-1-(4-amino-2-oxo-1(2H)-pyrimidinyl)-1,2,3,4-tetradeoxy-β-D-*erythro*-hex-2-enopyranuronic acid; 1-(1′-cytosinyl)-4-[L-3′-amino-5′-(1″-N-methylguanidino)-valerylamino]-1,2,3,4-tetradeoxy-β-D-*erythro*-hex-2-enuronic acid. $C_{17}H_{26}N_8O_5$; mol wt 422.45. C 48.33%, H 6.20%, N 26.53%, O 18.94%. Nucleoside antibiotic produced by *Streptomyces griseochromogenes.* Isoln and antimicrobial activity: Takeuchi *et al.*, *J. Antibiot.* **11A**, 1 (1958); Sumiki, Umezawa, **JP 60 16449** (1960 to Japan Antibiot. Res. Assoc.), *C.A.* **55**, 21474e (1961). Structure: Otake *et al.*, *Agric. Biol. Chem.* **30**, 132 (1966); Fox, Watanabe, *Tetrahedron Lett.* **1966**, 897. Abs config: Yonehara *et al.*, *ibid.* **1966**, 3785. Synthesis of the unsaturated carbohydrate: Goody *et al.*, *ibid.* **1970**, 293; of the cytosinine moiety: Fox, Watanabe, *Pure Appl. Chem.* **28**, 475 (1971); Kondo *et al.*, *Tetrahedron Lett.* **1972**, 1881; *eidem*, *Tetrahedron* **29**, 1801 (1973). Crystal and molecular structure: V. Swaminathan *et al.*, *Biochim. Biophys. Acta* **655**, 335 (1981).

Blastidic Acid　　　　　　Cytosinine

Needles from water, dec 235-236°. $[\alpha]_D^{11}$ +108.4° (water). uv max (0.1N HCl): 275 nm ($E_{1cm}^{1\%}$ 349); (0.1N NaOH): 266-270 nm ($E_{1cm}^{1\%}$ 266). Sol in water, acetic acid. Practically insol in methanol, ethanol, acetone, benzene, ether, ethyl acetate, chloroform, carbon tetrachloride, cyclohexane, xylene, pyridine, dioxane. LD_{50} i.v. in mice: 2.82 mg/kg (Takeuchi).

Hydrochloride. Crystals, dec 224-225°.

USE: Antifungal against rice blast disease in Japan.

1318. Blebbistatin. [674289-55-5] 1,2,3,3a-Tetrahydro-3a-hydroxy-6-methyl-1-phenyl-4H-pyrrolo[2,3-b]quinolin-4-one; 1-phenyl-1,2,3,4-tetrahydro-4-hydroxypyrrolo[2,3-b]-7-methylquinolin-4-one. $C_{18}H_{16}N_2O_2$; mol wt 292.34. C 73.95%, H 5.52%, N 9.58%, O 10.95%. Small molecule inhibitor of the ATPase activity of myosin II; affects cell blebbing during cell division. Prepn and inhibition of non-muscle myosin II: A. F. Straight *et al.*, *Science* **299**, 1743 (2003). Prepn and crystal structure of (−)-form: C. Lucas-Lopez *et al.*, *Eur. J. Org. Chem.* **2005**, 1736. Crystal structure in complex with myosin II: J. S. Allingham *et al.*, *Nat. Struct. Mol. Biol.* **12**, 378 (2005). Kinetics of non-muscle myosin IIB inhibition: B. Ramamurthy *et al.*, *Biochemistry* **43**, 14832 (2004). Mechanism of myosin II inhibition: M. Kovács *et al.*, *J. Biol. Chem.* **279**, 35557 (2004). Photochemistry studies: J. Kolega, *Biochem. Biophys. Res. Commun.* **320**, 1020 (2004); T. Sakamoto *et al.*, *Biochemistry* **44**, 584 (2005).

(S)-Form

Bright yellow solid, mp 192-193°. Absorption max (phosphate-buffered saline): 265, 430 nm.

(S)-Form. (−)-Blebbistatin. Crystals from acetonitrile, $[\alpha]_D^{26}$ −464° (c = 0.2 in dichloromethane).

USE: In studies of cytoskeletal dynamics and the biological roles of class-II myosins.

1319. Bleomycins. NSC-125066. A group of related glycopeptide antibiotics. Variations in the terminal amine account for differing activities. Isolated from *Streptomyces verticillus:* Umezawa, *Antimicrob. Agents Chemother.* **1965**, 1079. Purification and separation into bleomycins A and B and their components: Umezawa *et al.*, *J. Antibiot.* **19**, 200, 210 (1966); T. Takita *et al.*, *ibid.* **21**, 79 (1968); **22**, 237 (1969). Bleomycin A_2 is the main component of the bleomycin employed clinically. Total structure elucidation: T. Takita *et al.*, *ibid.* **25**, 755 (1972). Revised structure: *eidem*, *ibid.* **31**, 801 (1978). Terminal amines: Fujii *et al.*, *ibid.* **26**, 398 (1973). Synthesis of new bleomycins: T. Takita *et al.*, *ibid.* 254. Total synthesis of bleomycin A_2: *eidem*, *Tetrahedron Lett.* **23**, 521 (1982). Improved total synthesis: S. Saito *et al.*, *J. Antibiot.* **36**, 92 (1983). Biosynthesis: Fujii *et al.*, *ibid.* **27**, 73 (1974). Bleomycins are believed to react with DNA and cause strand scission; they have also been shown to have a type of oxygen transferase activity. Mechanism of action studies: R. M. Burger *et al.*, *Life Sci.* **28**, 715 (1981); N. Marugesan *et al.*, *J. Biol. Chem.* **257**, 8600 (1982). Coordination chemistry: J. C. Dabrowiak, *J. Inorg. Biochem.* **13**, 317 (1980). Clinical pharmacology: S. T. Crooke, *Cancer Chemother.* **3**, 343 (1981). Characterization of analogs: N. J. Oppenheimer *et al.*, *J. Biol. Chem.* **257**, 1606 (1982). *Reviews:* H. Umezawa, *Pure Appl. Chem.* **28**, 665-680 (1971); C. W. Haidle, R. S. Lloyd, *Antibiotics* **vol. 5**(pt. 2), F. E. Hahn, Ed. (Springer-Verlag, New York, 1979) pp 124-154; H. Umezawa, *Anticancer Agents Based on Natural Product Models*, J. M. Cassady, J. D. Douros, Eds. (Academic Press, New York, 1980) pp 147-166.

Colorless or yellowish powder which becomes bluish depending on Cu content. Very sol in water, methanol; slightly sol in ethanol. Practically insol in acetone, ethyl acetate, butyl acetate, ether. uv max: 244-248, 289-294 nm ($E_{1cm}^{1\%}$ 121-148, 102-121.5).

Sulfate. [9041-93-4] Blenoxane; Bleo. Cream-colored, amorphous powder. Very sol in water.

R = terminal amine

Bleomycin A₂. [11116-31-7] N^1-[3-(Dimethylsulfonio)propyl]-bleomycinamide. $C_{55}H_{84}N_{17}O_{21}S_3$; mol wt 1415.56.
THERAP CAT: Antineoplastic.

1320. Blonanserin. [132810-10-7] 2-(4-Ethyl-1-piperazinyl)-4-(4-fluorophenyl)-5,6,7,8,9,10-hexahydrocycloocta[*b*]pyridine; AD-5423; Lonasen. $C_{23}H_{30}FN_3$; mol wt 367.51. C 75.17%, H 8.23%, F 5.17%, N 11.43%. Dopamine D_2 and serotonin 5-HT$_2$ receptor antagonist. Prepn: K. Hino *et al.*, **EP 385237**; *eidem*, **US 5021421** (1990, 1991 both to Dainippon). HPLC determn in plasma: M. Matsuda *et al.*, *J. Pharm. Biomed. Anal.* **15**, 1449 (1997). X-ray crystal structure: K. Suzuki *et al.*, *Anal. Sci.* **18**, 1289 (2002). *In vitro* receptor binding study and *in vivo* pharmacology in animals: M. Oka *et al.*, *J. Pharmacol. Exp. Ther.* **264**, 158 (1993). Evaluation in mouse models of schizophrenia: T. Nagai *et al.*, *NeuroReport* **14**, 269 (2003). Review of pharmacology and clinical experience in schizophrenia: E. D. Deeks, G. M. Keating, *CNS Drug Rev.* **24**, 65-84 (2010).

Crystals from acetonitrile, mp 123-124°.
THERAP CAT: Antipsychotic.

1321. Blue Cohosh. Caulophyllum; papoose root; squaw root. Perennial herb, *Caulophyllum thalictroides* Michx., *Berberidaceae*. Mature plant is a peculiar bluish green color and bears dark blue fruit. Traditionally used in Native American medicine as a uterine stimulant, emmenagogue, antispasmodic. Medicinal formulations are prepared from the dried rhizome and roots. *Habit.* Damp woods of eastern North America. *Constit.* Alkaloids, primarily baptifoline, methylcytisine, anagyrine, magnoflorine, *q.q.v.*; 2 glycosides, caulosaponin, cauloside D; citrullol, gum, resins, phosphoric acid, phytosterol. GC determn of alkaloids: J. M. Betz *et al.*, *Phytochem. Anal.* **9**, 232 (1998). Tetratogenicity study of constituents: E. J. Kennelly *et al.*, *J. Nat. Prod.* **62**, 1385 (1999). Brief review of clinical use in labor stimulation: B. L. McFarlin *et al.*, *J. Nurse-Midwifery* **44**, 205-216 (1999). Reviews of medicinal uses: V. E. Tyler, *Herbs of Choice* (Pharmaceutical Products Press, New York, 1994) pp 47-48; J. Barnes *et al.*, *Herbal Medicines* (Pharmaceutical Press, London, 2nd Ed., 2002) pp 147-148.
Note: Do not confuse with black cohosh, *q.v.*
THERAP CAT: Emmenogogue; oxytocic.

1322. Boceprevir. [394730-60-0] (1*R*,2*S*,5*S*)-*N*-[3-Amino-1-(cyclobutylmethyl)-2,3-dioxopropyl]-3-[(2*S*)-2-[[[(1,1-dimethylethyl)amino]carbonyl]amino]-3,3-dimethyl-1-oxobutyl]-6,6-dimethyl-3-azabicyclo[3.1.0]hexane-2-carboxamide; SCH-503034; Victrelis. $C_{27}H_{45}N_5O_5$; mol wt 519.69. C 62.40%, H 8.73%, N 13.48%, O 15.39%. Inhibitor of hepatitis C virus serine protease NS3. Prepn: A. K. Saksena *et al.*, **WO 0208244** (2002 to Schering; Corvas); *eidem*, **US 7012066** (2006 to Schering; Dendreon); S. Venkatraman *et al.*, *J. Med. Chem.* **49**, 6074 (2006). Structure-based optimization: A. J. Prongay *et al.*, *ibid.* **50**, 2310 (2007). *In vitro* antiviral activity: B. A. Malcolm *et al.*, *Antimicrob. Agents Chemother.* **50**, 1013 (2006). LC/MS/MS determn in plasma: H. Farnik *et al.*, *J. Chromatogr. B* **877**, 4001 (2009). Clinical evaluation in combination with PEG interferon α-2b vs hepatitis C virus: C. Sarrazin *et al.*, *Gastroenterology* **132**, 1270 (2007). Review of development: F. G. Njoroge *et al.*, *Acc. Chem. Res.* **41**, 50-59 (2008); and antiviral efficacy: K. Berman, P. Y. Kwo, *Clin. Liver Dis.* **13**, 429-439 (2009).

White to off-white powder. Freely sol in methanol, ethanol, isopropanol; slightly sol in water.
THERAP CAT: Antiviral.

1323. 2-(Boc-oxyimino)-2-phenylacetonitrile. [58632-95-4] Carbonic acid (cyanophenylmethylene)azanyl 1,1-dimethylethyl ester; 2-(*tert*-butoxycarbonyloxyimino)-2-phenylacetonitrile; α-[[[(1,1-dimethylethoxy)carbonyl]oxy]imino]benzeneacetonitrile; Boc-ON. $C_{13}H_{14}N_2O_3$; mol wt 246.27. C 63.40%, H 5.73%, N 11.38%, O 19.49%. Reagent used to introduce the *tert*-butoxycarbonyl (Boc) group to protect amines; also used to protect alcohols. Prepn and use in protection of amines: M. Itoh *et al.*, *Tetrahedron Lett.* **16**, 4393 (1975); *eidem*, *Bull. Chem. Soc. Jpn.* **50**, 718 (1977). Additional protection application: X. Ariza *et al.*, *Tetrahedron Lett.* **39**, 9101 (1998). *Review:* M. S. Wolfe, J. Aubé in *Encyclopedia of Reagents for Organic Synthesis* **2**, L. A. Paquette, Ed. (Wiley, New York, 1995) 838-839.

White needles or plates from methanol, mp 84-86°. *Irritant. Protect from light.* Very sol in ethyl acetate, diethyl ether, benzene, chloroform, dioxane, acetone; sol in methanol, 2-propanol, *tert*-butanol. Insol in water, petr ether. Store at −20°C. Slowly dec with evolution of CO_2
USE: Reagent in synthetic organic chemistry.

1324. Bohrium. [54037-14-8] Element 107; nielsbohrium; unnilseptium. Bh, Ns, Uns; at. no. 107. Group VIIB(7). No stable nuclides. Known isotopes: 261, 262, 262m, 266, 267. Prepn of α-emitting isotope [261]107 by [209]Bi (^{54}Cr,2n); decay by spontaneous fission, T$_{1/2}$ 1-2 msec: Y. T. Organessian *et al.*, *Nucl. Phys. A* **273**, 505 (1976); G. N. Flerov, *C.A.* **87**, 173830v (1977); A. S. Iljinov *et al.*, *Report* **JINR-E-7-9686** (1976), *C.A.* **90**, 62354k (1979). Prepn of isotopes including [262]107 (T$_{1/2}$ 102 ±26 msec, α-emitter) by [209]Bi (^{54}Cr,1n): G. Münzenberg, *et al.*, *Z. Phys.* **A300**, 107 (1981); and revised T$_{1/2}$ 11.8 msec for [261]107: *eidem, ibid.* **A333**, 163 (1989). Chemical characterization of 6 isotopes as bohrium oxychloride: R. Eichler *et al.*, *Nature* **407**, 63 (2000). Review of history, prepn and properties: R. J. Silva in *The Chemistry of the Actinide Elements* **vol. 2**, J. J. Katz *et al.*, Eds. (Chapman and Hall, New York, 1986) pp 1110-1112; N. Flerov, G. M. Ter-Akopian *Prog. Particle Nucl. Phys.* **19**, 197-239 (1987); G. T. Seaborg, W. D. Loveland, *The Elements Beyond Uranium* (John Wiley & Sons, Inc., New York, 1990) p 56-57. Brief review of properties and prepn of ^{267}Bh (T$_{1/2}$ 17 sec): R. F. Service, *Science* **289**, 1270 (2000).

1325. Bolandiol. [19793-20-5] (3β,17β)-Estr-4-ene-3,17-diol; 3β,17β-dihydroxyestr-4-ene. C$_{18}$H$_{28}$O$_2$; mol wt 276.42. C 78.21%, H 10.21%, O 11.58%. Prepn: Colton, US 2843608 (1958 to Searle).

Crystals from dil acetone and from ethyl acetate + petr ether, mp 169-172°.
Dipropionate. [1986-53-4] Norpropandrolate; 3β,17β-dipropionyloxy-4-estrene; SC-7525; Anabiol; Storinal. C$_{24}$H$_{36}$O$_4$; mol wt 388.55.
THERAP CAT: Anabolic.

1326. Bolasterone. [1605-89-6] (7α,17β)-17-Hydroxy-7,17-dimethylandrost-4-en-3-one; 7α,17α-dimethyltestosterone. C$_{21}$-H$_{32}$O$_2$; mol wt 316.49. C 79.70%, H 10.19%, O 10.11%. Synthetic anabolic; epimeric with calusterone, *q.v.* Prepn: BE 610385; Babcock, Campbell, US 3341557 (1962, 1967 both to Upjohn); Campbell, Babcock, *Hormonal Steroids, Proc. 1st Int. Congr.*, **2**, L. Martini, A. Pecile, Eds. (Academic Press, New York, 1965) pp 59-67. Activity studies: Stucki *et al., ibid.* 119-132. HPLC determn in urine: R. Gonzalo-Lumbreras, R. Izquierdo-Hornillos, *J. Chromatogr. B* **742**, 47 (2000).

Crystals, mp 163-165°.
Note: This is a controlled substance (anabolic steroid): **21 CFR,** 1308.13, as defined in 1300.01.

1327. Boldenone. [846-48-0] (17β)-17-Hydroxyandrosta-1,4-dien-3-one; 1,4-androstadien-17β-ol-3-one; 3-oxo-17β-hydroxy-1,4-androstadiene; dehydrotestosterone. C$_{19}$H$_{26}$O$_2$; mol wt 286.42. C 79.68%, H 9.15%, O 11.17%. Anabolic steroid. Prepn: Meystre *et al., Helv. Chim. Acta* **39**, 734 (1956); Florey, US 2875196 (1949 to Olin Mathieson); Nobile, US 2837464 (1958 to Schering). Metabolism in humans and GC/MS determn in urine: W. Schänzer, M. Donike, *Biol. Mass Spectrom.* **21**, 3 (1992). LC/MS determn in horse urine: F. Pu *et al., J. Chromatogr. B* **813**, 241 (2004). Review of metabolism in animal species: H. F. De Brabander *et al., Food Addit. Contam.* **21**, 515-525 (2004).

Crystals, mp 164-166°. [α]$_D^{25}$ +25° (in chloroform).
Acetate. C$_{21}$H$_{28}$O$_3$. Crystals, mp 151-153°.
10-Undecenoate. [13103-34-9] Boldenone undecylenate; Ba-29038; Equipoise; Parenabol. C$_{30}$H$_{44}$O$_3$; mol wt 452.68.

Note: This is a controlled substance (anabolic steroid): **21 CFR,** 1308.13, as defined in 1300.01.
THERAP CAT (VET): Anabolic.

1328. Boldine. [476-70-0] (6aS)-5,6,6a,7-Tetrahydro-1,10-dimethoxy-6-methyl-4H-dibenzo[de,g]quinoline-2,9-diol; 1,10-dimethoxy-6aα-aporphine-2,9-diol; 1,10-dimethoxy-2,9-dihydroxyaporphine; 2,6-dihydroxy-3,5-dimethoxyaporphine. C$_{19}$H$_{21}$NO$_4$; mol wt 327.38. C 69.71%, H 6.47%, N 4.28%, O 19.55%. Major alkaloid from boldo *(Peumus boldus* Molina, *Monimiaceae).* Isoln: Bourgoin, Verne, *J. Pharm. Chim.* **16**, 191 (1872); from *Laurelia novaezelandiae* A. Cunn.: Bernauer, *Helv. Chim. Acta* **50**, 1583 (1967). Structure: Warnat, *Ber.* **58**, 2768; **59**, 85 (1926); Späth, Tharrer, *ibid.* **66**, 904 (1933); Schlittler, *ibid.* 988. Synthesis of *dl*-form: S. M. Kupchan *et al., Chem. Commun.* **1976**, 91. Biosynthetic study: D. S. Bhakuni *et al., J. Chem. Soc. Perkin Trans. 1* **1977**, 706.

Crystals from ether. mp 212-220°. [α]$_D^{25}$ +127° (c = 0.1 in alcohol). Very slightly sol in water or ether; sol in alcohol, chloroform, dil acids.
Hydrochloride. Crystals from methanol-ether and methanol, mp 212-220°.
dl-Form. mp 159-162°.
Boldine dimethyl ether *see* Glaucine.
USE: Ingredient in choleretics and laxatives.

1329. Boldo. Boldu; boldea; boldus; boldoa. Aromatic, evergreen shrub, *Peumus boldus* Molina, *Monimiaceae.* Leaves are used in traditional medicine to treat dyspepsia, gastrointestinal spasms, and cholelithiasis. *Habit.* Peru, Chile. *Constit.* Isoquinoline alkaloids, especially boldine, isoboldine; flavonoids, such as isorhamnetin, and their glycosides; 2-3% volatile oil containing *p*-cymene, cineole, ascaridole. Description and constituents: H. Schindler, *Arzneim.-Forsch.* **7**, 747 (1957). Review of constituents and medicinal uses: J. Barnes *et al., Herbal Medicines* (Pharmaceutical Press, London, 3rd Ed., 2007) pp 91-93.
Alkaloid extract. Grayish-white to grayish-yellow, bitter powder. Almost insol in water. Sol in alc, chloroform, slightly in ether.

1330. Bole, Armenian. Bolus armena; bolus rubra; red bole. A red variety of clay (aluminum silicate) contg naturally occurring ferric oxide (hematite). Originally found in Armenia.
Reddish, soft, unctuous pieces; adheres to the tongue; easily reduced to powder. Insol in usual solvents. d 1.9-2.0.
USE: For coloring powders, cements, and as a pigment.
THERAP CAT: Adsorbant; protectant.

1331. Boleko Oil. Isano oil. Oil extracted from the nuts of the tree *Ongokea gore* (Hua) Pierre, *Olacaceae,* growing in equatorial Africa. Contains fatty acids (as glycerides): isanic acid 46%, isanolic acid 44%. *Refs:* Scher, *Arch. Pharm.* **287**, 548 (1954); Dupont *et al., Bull. Soc. Chim. Fr.* **1957**, 1495; De Vries, *Oleagineux* **12**, 427 (1957); Kneeland *et al., J. Am. Oil Chem. Soc.* **35**, 361 (1958). Extraction procedure: Lambert, US 2800492 (1957 to UCB). Structure of acids: Gunstone, Sealy, *J. Chem. Soc.* **1963**, 5772; Morris, *ibid.* 5779.
Typical characteristics: Acid no. 1-10; ester no. 183.5; sapon no. 185-200; iodine no. 254-259; d$_4^{20}$ 0.973-0.983; n_D^{20} 1.505-1.509; viscosity (25°): 7-10 P. Soluble in acetone, benzene, ethyl ether, carbon tetrachloride, chloroform. Slightly sol in petr ether, ethanol, hexane. The viscosity increases upon heating and the oil is transformed into a rubber-like mass. Does not polymerize when heated to a moderate temp; when heated rapidly to temp above 200° polymerization becomes so rapid as to become explosive. Polymerization is explained by the acetylenic structure of isanic acid.
USE: In fire retardant paints.

1332. Bombesin. [31362-50-2] 2-L-Glutamine-6-L-aspara-ginealytesin. $C_{71}H_{110}N_{24}O_{18}S$; mol wt 1619.87. C 52.65%, H 6.84%, N 20.75%, O 17.78%, S 1.98%. Pharmacologically active tetradecapeptide found in skins of European amphibians of the family Discoglossidae, principally *Bombina bombina* and *Bombina variegata variegata*. Isoln and structure: A. Anastasi *et al.*, *Experientia* **27**, 166 (1971). Pharmacological activity: V. Erspamer *et al.*, *J. Pharm. Pharmacol.* **22**, 875 (1970). Synthesis: L. Bernardi *et al.*, *Experientia* **27**, 873 (1971). It is a potent stimulant of gastric and pancreatic secretions in mammals; a bombesin-like immunoreactive peptide is found in both brain and gut, *cf.* J. M. Polak *et al.*, *Lancet* **1**, 1109 (1976). Other actions include hypertensive reactions, antidiuresis, and hyperglycemic activity. Bombesin has been shown to have a strong effect on core temperature lowering in rats: M. Brown *et al.*, *Science* **196**, 996 (1977). High levels of intracellular bombesin have also been found in human small-cell lung carcinoma: T. W. Moody *et al.*, *ibid.* **214**, 1246 (1981). Brief review of chemistry, isoln, purification: V. Mutt, *Biochem. Soc. Trans.* **8**, 11 (1980). Effect on suppression of food intake: J. Gibbs *et al.*, *Nature* **282**, 208 (1979). Pharmacological reviews: G. Bertaccini, *Pharmacol. Rev.* **28**, 127 (1976); S. R. Bloom, J. M. Polak, *Adv. Clin. Chem.* **21**, 177 (1980).

5-oxoPro–Gln–Arg–Leu–Gly–Asn–Gln–Trp–Ala–Val–Gly–His–Leu–MetNH₂

Hydrochloride. $C_{71}H_{112}Cl_2N_{24}O_{18}S$. Cryst from 99% ethanol, mp 185° (dec). $[\alpha]_D^{24}$ −20.6° (c = 0.65 in DMF-HMPT 8:2).

1333. Bone Morphogenetic Proteins. Bone morphogenic factors; bone morphogenic proteins; BMPs. Multifunctional cytokines that induce formation of cartilage and bone; originally isolated from demineralized bone matrix. Structurally related to the transforming growth factor-β (TGF-β) superfamily; at least 15 closely related forms exist. All are dimers linked by 7 disulfide bonds. Prepn from rabbit bone: M. R. Urist *et al.*, *Proc. Natl. Acad. Sci. USA* **76**, 1828 (1979); from human bone: *idem et al.*, *Proc. Soc. Exp. Biol. Med.* **173**, 194 (1983). Clinical evaluations in bone repair: E. E. Johnson *et al.*, *Clin. Orthop. Relat. Res.* **230**, 249, 257 (1988). Review of purification and structures of BMP-1 through BMP-7: J. M. Wozney, *Prog. Growth Factor Res.* **1**, 267-280 (1989); and osteogenic activity: *idem*, *Mol. Reprod. Dev.* **32**, 160-167 (1992). Review of potential clinical applications and delivery strategies: C. A. Kirker-Head, *Adv. Drug Delivery Rev.* **43**, 65-92 (2000); of clinical studies in human bone regeneration: E. H. J. Groeneveld, E. H. Burger, *Eur. J. Endocrinol.* **142**, 9-21 (2000); of clinical efficacy and safety: P. J. Harwood, P. V. Giannoudis, *Expert Opin. Drug Saf.* **4**, 75-89 (2005).
BMP-2. BMP-2A. Mature human form is a homodimer, mol wt 32 kDa. Each monomer consists of 114 amino acid residues. Review of bioactivity: E. H Riley *et al.*, *Clin. Orthop. Relat. Res.* **324**, 39-46 (1994).
BMP-7. Osteogenic protein-1; OP-1. Mature human form is a homodimer, mol wt 32-36 kDa. Each monomer consists of 139 amino acid residues. Prepn of recombinant human BMP-7 (*rhBMP-7*): T. K. Sampath *et al.*, *J. Biol. Chem.* **267**, 20352 (1992). Review of bioactivity and potential clinical applications: S. D. Cook, D. C. Rueger, *Clin. Orthop. Relat. Res.* **324**, 29-38 (1996).
Dibotermin alfa. [246539-15-1] Bone morphogenetic protein 2 (human recombinant rhBMP-2); rhBMP-2; InductOs. Recombinant human form of BMP-2 prepared in Chinese hamster ovary (CHO) cells. Glycosylated, disulfide-linked dimer, mol wt ~28-33 kDa. Prepn: E. A. Wang *et al.*, *WO 8800205*; *eidem*, *US 5013649* (1988, 1991 both to Genetics Institute); E. A. Wang *et al.*, *Proc. Natl. Acad. Sci. USA* **87**, 2220 (1990). Clinical trial of collagen sponge implant in tibial fractures: S. Govender *et al.*, *J. Bone Joint Surg.* **84-A**, 2123 (2002). Review of clinical trials in spinal fusion: S. N. Khan, J. M. Lane, *Expert Opin. Biol. Ther.* **4**, 741-748 (2004).
THERAP CAT: Osteoinductive agent.

1334. Bongkrekic Acid. [11076-19-0] (2E,4Z,6R,8Z,10E,-14E,17S,18E,20Z)-20-(Carboxymethyl)-6-methoxy-2,5,17-trimethyl-2,4,8,10,14,18,20-docosaheptaenedioic acid; 3-carboxymethyl-17-methoxy-6,18,21-trimethyldocosa-2,4,8,12,14,18,20-heptaenedioic acid; BA. $C_{28}H_{38}O_7$; mol wt 486.61. C 69.11%, H 7.87%, O 23.01%. One of the two toxic antibiotic principles produced by *Pseudomonas cocovenenans* on partially defatted coconut; the other

being toxoflavin, *q.v.* Name derived from "bongkrek", a molded coconut product from Indonesia which becomes highly poisonous when *P. cocovenenans* outgrows the mold. Isoln: van Veen, Mertens, *Rec. Trav. Chim.* **53**, 257 (1934); **54**, 373 (1935); Nugteren, Berends, *ibid.* **76**, 13 (1957). Purification and properties: Lijmbach *et al.*, *Tetrahedron* **26**, 5993 (1970). Structural studies: *eidem*, *ibid.* **27**, 1839 (1971). Revised structure: De Bruijn *et al.*, *ibid.* **29**, 1541 (1973). Absolute configuration: Zylber *et al.*, *Experientia* **29**, 387, 648 (1973). Influence on carbohydrate metabolism: van Veen, Mertens, *Arch. Neerl. Physiol.* **21**, 73 (1936), *C.A.* **30**, 3880⁹ (1936); inhibition of adenine nucleotide translocation: Henderson, Lardy, *J. Biol. Chem.* **245**, 1319 (1970); Klingenberg *et al.*, *Biochem. Biophys. Res. Commun.* **39**, 363 (1970).

White, amorphous solid, mp 50-60°. uv max (methanol): 237, 267 nm (ε 32000, 36700). $[\alpha]_D^{25}$ +162.5°. LD₅₀ i.v. in mice: 1.41 mg/kg (Lijmbach).

1335. Bopindolol. [62658-63-3] 1-[(1,1-Dimethylethyl)amino]-3-[(2-methyl-1*H*-indol-4-yl)oxy]-2-propanol 2-benzoate; (±)-1-(*tert*-butylamino)-3-[(2-methylindol-4-yl)oxy]-2-propanol benzoate (ester). $C_{23}H_{28}N_2O_3$; mol wt 380.49. C 72.60%, H 7.42%, N 7.36%, O 12.61%. Nonselective β-adrenergic blocker. Prepn: F. Troxler, F. Seemann, *DE 2635209*; *eidem*, *US 4340541* (1977, 1982 both to Sandoz). Clinical pharmacology: P. van Brummelen *et al.*, *Eur. J. Clin. Pharmacol.* **22**, 491 (1982). HPLC determn in plasma: C. J. Oddie *et al.*, *J. Chromatogr.* **273**, 469 (1983). Effect on plasma lipid fractions: P. van Brummelen *et al.*, *Br. J. Clin. Pharmacol.* **17**, 86 (1984). Pharmacokinetics: R. Platzer *et al.*, *Clin. Pharmacol. Ther.* **36**, 5 (1984). Clinical trials in hypertension: U. L. Hulthen *et al.*, *J. Cardiovasc. Pharmacol.* **5**, 426 (1983); W. J. Schiess *et al.*, *Eur. J. Clin. Pharmacol.* **27**, 529 (1984).

Sol in ether, methylene chloride. LD₅₀ i.v. in mice: 17 mg/kg (Troxler, Seemann, 1977).
Malonate. [82857-38-3] LT-31-200; Sandonorm; Wandonorm. $C_{23}H_{28}N_2O_3 \cdot C_3H_4O_4$; mol wt 484.55.
THERAP CAT: Antihypertensive.

1336. Borane Complexes. Complexes formed from diborane, *q.v.*, with the Lewis bases, tetrahydrofuran and dimethyl sulfide. Used synthetically in hydroborations and reductions of numerous functional groups. Determn of the tetrahydrofuran complex: J. R. Elliott *et al.*, *J. Am. Chem. Soc.* **74**, 5211 (1952); H. E. Wirth *et al.*, *J. Phys. Chem.* **62**, 870 (1958). Prepn and use in hydroborations: H. C. Brown *et al.*, *J. Am. Chem. Soc.* **82**, 4233 (1960). Utility in hydroborations: H. C. Brown, G. Zweifel, *ibid.* **83**, 2544 (1961); in functional group reductions: H. C. Brown *et al.*, *ibid.* **92**, 1637 (1970). Soln stability studies: M. Potyen *et al.*, *Org. Process Res. Dev.* **11**, 210 (2007). Prepn of the dimethyl sulfide complex: A. B. Burg, R. I. Wagner, *J. Am. Chem. Soc.* **76**, 3307 (1954). Utility in hydroborations: L. M. Braun *et al.*, *J. Org. Chem.* **36**, 2388 (1971); C. F. Lane, *ibid.* **39**, 1437 (1974); in functional group reductions: H. C. Brown *et al.*, *ibid.* **47**, 3153 (1982). Review of the synthetic applications of the dimethylsulfide complex: R. O Hutchins, F. Cistone, *Org. Prep. Proced. Int.* **13**, 225-240 (1981).

Borane-tetrahydrofuran complex

Borane-dimethyl sulfide complex

Borane-tetrahydrofuran complex. [14044-65-6] (T-4)-Trihydro(tetrahydrofuran)boron; borane-THF; BTHF. $C_4H_{11}BO$; mol wt 85.94. C 55.90%, H 12.90%, B 12.58%, O 18.62%. Liquid. Sol in THF. *Flammable.* Air sensitive. Reacts violently with water. Commercial solns in THF are stable over prolonged periods of time when stored at 0°C under inert gas.

Borane-dimethyl sulfide complex. [13292-87-0] (T-4)-Trihydro[thiobis[methane]]boron; borane-methyl sulfide; borane-DMS; BMS. C_2H_9BS; mol wt 75.96. C 31.62%, H 11.94%, B 14.23%, S 42.21%. Liquid with a stench. mp −40 to −38°. d^{23} 0.80. Sol in benzene, dichloromethane, dimethyl ether, diethyl ether, diglyme, ethyl acetate, hexane, toluene, xylene. Reacts with alcohols, acetone. *Flammable.* Air and moisture sensitive. Store under inert gas. Stable indefinitely when refrigerated and for prolonged periods at room temp.

USE: Reagents in synthetic organic chemistry.

1337. Borazine. [6569-51-3] Inorganic benzene; hexahydro-s-triazaborine; borazane; borazole; triborine triamine; triboron nitride; s-triazaborane. $B_3H_6N_3$; mol wt 80.50. B 40.29%, H 7.51%, N 52.20%. Isoelectronic with benzene. Prepn: Stock, Pohland, *Ber.* **59**, 2215 (1926); Wiberg, Bolz, *Ber.* **73**, 209 (1940). Structural study: H. Watanabe, W. Kubo, *J. Am. Chem. Soc.* **82**, 2428 (1960); W. Harshbarger *et al.*, *Inorg. Chem.* **8**, 1683 (1969). Electronic structure: A. Lötz *et al.*, *Chem. Phys.* **103**, 317 (1986). Raman spectroscopy: J. V. Kainnady, A. Weber, *J. Raman Spectrosc.* **5**, 35 (1976). Electric and magnetic properties: G. R. Dennis, G. L. D. Ritchie, *J. Phys. Chem.* **97**, 8403 (1993). Use in synthesis of ceramic composites: K. Su *et al.*, *Chem. Mater.* **5**, 547 (1993). *Review:* M. Smolin, L. Rapoport, *s-Triazines and Derivatives* (Interscience, New York, 1959) pp 597-626.

Mobile liquid; mp −58°; bp 53°. Stable up to 500° when pure and totally anhydrous. d_4^0 0.824; d_4^{57} 0.898. n_D^{20} 1.3821. Critical temp 252°. Heat of vaporization 7.0 kcal. Mol vol at bp = 99.7° ml. Surface tension at mp = 31.09° dynes/cm. Parachor 207.9. Trouton's constant 21.4. Eötvös' constant 2.0. Dipole moment in benzene (25°) = 0.50. Dissolves in water, giving a soln with strong reducing properties. Hydrolyzes slowly in aq soln yielding hydrogen, boric acid, and ammonia.

1338. Bordeaux Mixture. [8011-63-0] Mixture of calcium hydroxide (hydrated lime) and copper(II) sulfate, q.q.v. Prepn and use vs downy mildew on grapes: Millardet, *J. Agric. Prat. (Paris)* **49**, 513 (1885); *see also*: D. E. H. Frear, *Pesticide Index* (College Science, 4th ed., 1969) p 72. Field trial vs pink disease in mandarin oranges: P. M. Pradhanang, *Crop Prot.* **13**, 550 (1994); vs Persian walnut blight: A. Belisario, A. Zoina, *Eur. J. Forest Pathol.* **25**, 224 (1995). Effects on Sauvignon grape composition and wine aroma: E. Hatzidimitriou *et al.*, *J. Int. Sci. Vigne Vin* **30**, 133 (1996).

Fine, blue pptd powder. Insol in water.

Burgundy Mixture. [11125-96-5] Carbonic acid disodium salt, mixt. with copper(2+) sulfate (1:1). Analog to Bordeaux mixture produced from copper(II) sulfate and sodium carbonate, q.q.v. Prepn and use vs downy mildew on grapes: E. Masson, *J. Agric. Prat. (Paris)* **51**, 814 (1887); *see also*: Frear, *Chemistry of the Pesticides* (Van Nostrand, 3rd ed., 1955) p 322. Review of chemistry: R. L. Mond, C. Heberlein, *J. Chem. Soc.* **115**, 908-922 (1919). Blue to green colloidal ppt. Practically insol in water.

USE: Fungicide.

1339. Boric Acid. [10043-35-3] Boric acid (H_3BO_3); boracic acid; orthoboric acid; Borofax. BH_3O_3; mol wt 61.83. B 17.48%,

H 4.89%, O 77.63%. H_3BO_3. Occurs in nature as the mineral *sassolite.* Manuf: *Faith, Keyes & Clark's Industrial Chemicals*, F. A. Lowenheim, M. K. Moran, Eds. (Wiley-Interscience, New York, 4th ed., 1975) pp 153-158. Toxicity study: Smyth *et al.*, *Am. Ind. Hyg. Assoc. J.* **30**, 470 (1969). Review of toxicology and human exposure: *Toxicological Profile for Boron* (PB2010-100001, 2010) 249 pp.

Colorless, odorless, transparent crystals, or white granules or powder; slightly unctuous to the touch. mp ∼171°. d 1.48. pKa = 9.42. Log P (octanol/water): 0.175. Phase diagram for the $B_2O_3.H_2O$ system: Kracek *et al.*, *Am. J. Sci.* **35A**, 143 (1938). Volatile with steam. pH: 5.1 (0.1 molar). Stable in air. One gram dissolves in 18 ml cold, 4 ml boiling water, in 18 ml cold, 6 ml boiling alc, in 4 ml glycerol; soly in water is increased by HCl, citric or tartaric acids. Soly of boric acid in glycerol solns of various concns: Sciarra, Elliott, *J. Am. Pharm. Assoc. Sci. Ed.* **49**, 116 (1960). LD_{50} orally in rats: 5.14 g/kg (Smyth).

Caution: Potential symptoms of overexposure by ingestion or absorption are vomiting, diarrhea, abdominal cramps, erythematous lesions on skin and mucous membranes, circulatory collapse, tachycardia, cyanosis, delirium, convulsions, coma. Chronic use may cause borism (dry skin, eruptions, gastric disturbances). *See* E. Browning, *Toxicity of Industrial Metals* (Appleton-Century-Crofts, New York, 2nd ed., 1969) pp 90-97.

USE: For weatherproofing wood and fireproofing fabrics; as a preservative; manuf cements, crockery, porcelain, enamels, glass, borates, leather, carpets, hats, soaps, artificial gems; in nickeling baths; in buffers; cosmetics; printing and dyeing, painting; photography; for impregnating wicks; electric condensers; hardening steel. Also used as insecticide for cockroaches and black carpet beetles. Pharmaceutic aid (buffering agent).

THERAP CAT: Astringent, antiseptic.

THERAP CAT (VET): Antibacterial and antifungal. Used chiefly in aqueous solution or powders for external use.

1340. Borneol. [507-70-0] rel-(1R,2S,4R)-1,7,7-Trimethylbicyclo[2.2.1]heptan-2-ol; *endo*-1,7,7-trimethylbicyclo[2.2.1]heptan-2-ol; *endo*-2-bornanol; *endo*-2-camphanol; *endo*-2-hydroxycamphane; (±)-borneol; *dl*-borneol; bornyl alcohol; Baros camphor; Sumatra camphor; Borneo camphor; Dryobalanops camphor; Bhimsaim camphor; Malayan camphor; camphol. $C_{10}H_{18}O$; mol wt 154.25. C 77.87%, H 11.76%, O 10.37%. Constituent of essential oils of many plants. Both *d*- and *l*- forms are naturally occurring. The *exo*-isomer is known as isoborneol, q.v. Occurrence of *d*-form in oil from *Dryobalanops aromatica* Gaertn., *Dipterocarpaceae*, and *l*-form in *Blumea balsamifera* (L.) DC., *Compositae*: E. Gildemeister, F. Hoffman, *Die ätherischen Ole* (Schimmel, Leipzig, 3rd ed., 1928) pp 475-481. Racemic borneol is prepd synthetically by reduction of camphor: Truett, Moulton, *J. Am. Chem. Soc.* **73**, 5913 (1951); Ziegler *et al.*, *Ann.* **623**, 9 (1959); Ziegler, **GB 803178** (1958); from pinene: Schwyzer, *Pharm. Ztg.* **75**, 1275 (1930). Configuration: Toivonen *et al.*, *Acta Chem. Scand.* **3**, 991 (1949). Toxicity study: R. C. Beier, *Rev. Environ. Contam. Toxicol.* **113**, 47-127 (1990). *Review:* J. L. Simonsen, *The Terpenes* **vol. II** (University Press, Cambridge, 2nd ed., 1949) pp 349-365.

d-form

mp 206-207°.

d-Form. [464-43-7] (1R,2S,4R)-borneol. Hexagonal plates from petr ether, mp 208°. Peculiar peppery odor and burning taste somewhat resembling that of mint. Sublimes, but is less volatile than camphor. d_4^{20} 1.011. bp 212°. $[\alpha]_D^{20}$ +37.7° (c = 5 in alc); $[\alpha]_{546}^{22}$ +44.4° (c = 0.5 in toluene). *Flammable.* Almost insol in water. Sol in alc (176 parts dissolve in 100 parts w/w of abs alc), ether, petr ether (about 1:6), benzene (about 1:5), toluene, acetone, decalin, tetralin. LD orally in rabbits: 2000 mg/kg (Beier).

l-Form. [464-45-9] (1*S*,2*R*,4*S*)-borneol. Hexagonal plates, mp 204°. bp$_{779}$ 210°. $[\alpha]_D^{20}$ −37.7° (c = 5 in alc); $[\alpha]_{546}^{22}$ −44.4° (c = 0.5 in toluene).

Caution: May cause nausea, vomiting, mental confusion, dizziness, convulsions.

USE: In flavors and fragrances; in the manuf of its esters.

1341. Bornyl Acetate. [76-49-3] *rel*-(1*R*,2*S*,4*R*)-1,7,7-Trimethylbicyclo[2.2.1]heptan-2-ol 2-acetate; borneol acetate; *dl*-bornyl acetate; (±)-bornyl acetate. $C_{12}H_{20}O_2$; mol wt 196.29. C 73.43%, H 10.27%, O 16.30%. Constituent of essential oils of various plants; both *d*- and *l*- forms are naturally occurring. Prepn by acylation of borneol: M. A. Haller, *Compt. Rend.* **109**, 29 (1889); K. Sisido *et al.*, *J. Am. Chem. Soc.* **82**, 125 (1960). Properties: W. J. Considine, *J. Org. Chem.* **25**, 671 (1960).

d-form

Crystals, mp 7.0°.

d-Form. [20347-65-3] (1*R*,2*S*,4*R*)-Bornyl acetate. Crystals, mp 24° (Haller); also reported as mp 26.5°(Considine). bp 225-226°; bp$_8$ 92-93°. d 0.99. n_D^{22} 1.4623. $[\alpha]_D$ +44.38° (Haller); $[\alpha]_D$ +41.2° (Considine); $[\alpha]_D^{14}$ +44.72°, neat (Sisido). Very slightly sol in water; sol in alcohol, ether.

l-Form. [5655-61-8] (1*S*,2*R*,4*S*)-Bornyl acetate. Crystals, mp 24° (Haller); also reported as mp 27° (Considine). bp$_{14}$ 103°. $[\alpha]_D$ −44.45° (Haller); $[\alpha]_D$ −42.0° (Considine).

USE: In flavors and fragrances.

1342. Bornyl Chloride. [464-41-5] *rel*-(1*R*,2*S*,4*R*)-2-Chloro-1,7,7-trimethylbicyclo[2.2.1]heptane; pinene hydrochloride; 2-chlorobornane; 2-chlorocamphane; "terpene" hydrochloride; "turpentine camphor". $C_{10}H_{17}Cl$; mol wt 172.70. C 69.55%, H 9.92%, Cl 20.53%. Prepd from α-pinene: Zeiss, Zwanzig, *J. Am. Chem. Soc.* **79**, 1733 (1957); Hückel, Gelchsheimer, *Ann.* **625**, 12 (1959); by chlorination of camphane: Gandini, *Gazz. Chim. Ital.* **66**, 357 (1936). Configuration: Kwart, *J. Am. Chem. Soc.* **75**, 5942 (1953); Kwart, Null, *ibid.* **78**, 5943 (1956). *Review:* J. L. Simonsen, *The Terpenes* vol. II (University Press, Cambridge, 2nd ed., 1949) pp 340-349.

Relative stereochemistry

Crystals from alc, mp 132°. Odor resembling camphor. bp 207-208°. When prepd from optically active sources it has the same sign as the hydrocarbon from which it is prepd: Thurber, Thielke, *J. Am. Chem. Soc.* **53**, 1032 (1931). Practically insol in water. Sol in alc, ether.

THERAP CAT: Antiseptic.

1343. Boromycin. [34524-20-4] Hydrogen (*T*-4)-[(1*R*)-1-[(1*S*,2*R*,5*S*,6*R*,8*R*,12*R*,14*S*,17*R*,18*S*,19*R*,22*S*,24*Z*,28*S*,30*S*,33*R*)-1,-2,18,19-tetra(hydroxy-κ*O*)-12,28-dihydroxy-6,13,13,17,29,29,33-heptamethyl-3,20-dioxo-4,7,21,34,35-pentaoxatetracyclo-[28.3.1.15,8.114,18]hexatriacont-24-en-22-yl]ethyl D-valinato-(4−)]borate(1−). $C_{45}H_{74}BNO_{15}$; mol wt 879.89. C 61.43%, H 8.48%, B 1.23%, N 1.59%, O 27.27%. Antibiotic produced by *Streptomyces antibioticus* ETH 28829: Hütter *et al.*, *Helv. Chim. Acta* **50**, 1533 (1967). The first known natural product in which boron has been found. It is a complex of boric acid with a tetraden-

tate organic complexing agent and yields D-valine, boric acid and a polyhydroxy macrolide-type compound upon hydrolysis. Structure: Dunitz *et al.*, *ibid.* **54**, 1709 (1971). Mechanism of action studies: Pache, Zähner, *Arch. Mikrobiol.* **67**, 156 (1969). Activity as a coccidiostat: Miller, Burg, **US 3864479** (1975 to Merck & Co.). Biosynthesis: T. S. S. Chen *et al.*, *J. Org. Chem.* **46**, 2661 (1981). Partial synthesis: M. A. Avery *et al.*, *Tetrahedron Lett.* **22**, 3123 (1981). Synthesis of C-3′ to C-17′: S. Hanessian *et al.*, *J. Am. Chem. Soc.* **103**, 6243 (1981); of C-3 to C-17: *eidem*, *Can. J. Chem.* **61**, 634 (1983); of C-1 to C-17: J. D. White *et al.*, *J. Am. Chem. Soc.* **105**, 6517 (1983). Total synthesis: *eidem*, *ibid.* **111**, 790 (1989). *Review:* Pache in *Antibiotics* vol. **3**, J. W. Corcoran, F. E. Hahn, Eds. (Springer-Verlag, New York, 1975) pp 585-587.

Colorless crystals from methanol, mp 223-228° (dec). $[\alpha]_D$ +63.5° (c = 0.55 in CHCl$_3$). No uv absorption between 210 and 400 nm. LD$_{50}$ orally in mice: 180 mg/kg (Hütter).

1344. Boron. [7440-42-8] B; at. wt [10.806; 10.821]; conventional at. wt 10.81; at. no. 5; valence 3. Group IIIA (13). Two naturally-occurring isotopes: 10; 11; three short-lived, artificial isotopes: 8, 12, 13. Occurrence in the earth's crust about 0.001% in the form of its compounds, never as the element. First obtained by Moissan in 1895 by reduction of boric anhydride (B$_2$O$_3$) with magnesium in a thermite-type reaction: Moissan, *Ann. Chim. Phys.* [7] **6**, 296 (1895). Prepn of high purity crystalline boron by vapor phase reduction of boron trichloride with hydrogen on electrically heated filaments in a flow system: D. R. Stern, L. Lynds, *J. Electrochem. Soc.* **105**, 676 (1958). Sublimation pressure: A. W. Searcy, C. E. Myers, *J. Phys. Chem.* **61**, 957 (1957). Reviews of prepn and properties of boron and its compds: *Boron, Metallo-Boron Compounds and Boranes*, R. M. Adams, Ed. (Interscience, New York, 1964) 765 pp; *The Chemistry of Boron and Its Compounds*, E. L. Muetterties, Ed. (John Wiley, New York, 1967) 699 pp; Greenwood, "Boron" in *Comprehensive Inorganic Chemistry* vol. 1, J. C. Bailar, Jr. *et al.*, Eds. (Pergamon Press, Oxford, 1973) pp 665-991; L. H. Jansen in *Kirk-Othmer Encyclopedia of Chemical Technology* vol. 4 (Wiley-Interscience, New York, 5th ed., 2004) pp 132-138. Review of synthesis and applications of vinylic organoboranes: H. C. Brown, J. B. Campbell, *Aldrichim. Acta* **14**, 3-11 (1981). Review of synthesis and properties of crystalline modifications and of boron-rich borides: B. Albert, H. Hillebrecht, *Angew. Chem. Int. Ed.* **48**, 8640-8668 (2009). Review of toxicology and human exposure: *Toxicological Profile for Boron* (PB2010-100001, 2010) 248 pp.

Polymorphic; exists as amorphous powder or in multiple crystalline forms. Crystals are almost as hard as diamond. α-Rhombohedral form (α-B$_{12}$): red or maroon crystals, 12 atoms/unit cell, d 2.46. mp 2180°. Mohs' hardness 9.3. β-Rhombohedral form (β-B): grayish-black crystals, 105 atoms/unit cell, d 2.33. Heat capacity at 25°: 2.650 cal/g-atom/°C. Tetragonal form (T-192): black crystals, 192 atoms/unit cell, d 2.36. Amorphous form: black or dark brown powder, mp 2300°. d 2.35. Vapor pressure at 2413 K (2140°): 1.56 × 10^{-5} atm. Heat capacity at 25°: 2.858 cal/g-atom/°C. Feeble conductor of electricity at room temp, good conductor at high temps. Admixture of traces of carbon improves conductivity. Self-limiting reaction with oxygen due to formation of B$_2$O$_3$ film; oxide coating evaporates above 1000°. Reacts with fluorine at room temp. Insol in water. Unaffected by aq hydrochloric and hydrofluoric acids. When finely divided, it is sol in boiling nitric and sulfuric acids and in most molten metals, such as copper, iron, magnesium, aluminum, calcium. Reacts vigorously with fused sodium peroxide, or with a fusion mixture of sodium carbonate and potassium nitrate.

USE: In nuclear chemistry as neutron absorber; as shielding in nuclear reactors. In alloys, usually to harden other metals; as dopant for semiconductors.

1345. Boron Carbide. [12069-32-8] Norbide; tetraboron carbide. CB_4; mol wt 55.25. C 21.74%, B 78.26%. B_4C. Prepd in an electric furnace at 2500° according to the equation $2B_2O_3 + 7C \rightarrow B_4C + 6CO$: Ridgway, *Trans. Electrochem. Soc.* **66**, 117-133 (1934); also formed by reducing boric anhydride with magnesium in the presence of carbon: Dawihl, **DE 752324** (1942 to Krupp); *BIOS* rept. no. 925, p 22 (1947). Lab prepn by the reduction of boron trichloride with hydrogen in the presence of carbon or hydrocarbons: *Bell Labs. Record* **28**, 477 (1950). Comprehensive monograph: P. W. Gilles in *Adv. Chem. Ser.* **32**, entitled "Borax to Boranes," D. L. Martin, Ed. (ACS, Washington, DC, 1961).

Black shiny rhombohedra or octahedra. d_4^{25} 2.508-2.512. mp 2350° (no decompn); bp >3500°. Its hardness is less than that of industrial diamonds, but higher than the hardness of silicon carbide: ca 5,000 kg/mm², on Mohs' hardness scale = 9.3. Less brittle than most ceramics. Remarkably resistant to chemical action. Not attacked by hot HF, HNO_3 or $HCrO_4$. Decomposed by molten alkalis at red heat. Does not burn in oxygen flame.

USE: Abrasive. In the manuf of hard and chemicals-resistant ceramics or wear-resistant tools. Finely pulverized B_4C can be molded under (considerable) pressure and heat.

1346. Boron Monoxide. [12505-77-0] Boron oxide (BO). BO; mol wt 26.81. B 40.32%, O 59.68%. Exists in multiple polymeric forms, $(BO)_x$. Prepn: E. Zintl *et al.*, *Z. Anorg. Allg. Chem.* **245**, 8 (1940); T. Wartik, E. F. Apple, *J. Am. Chem. Soc.* **77**, 6400 (1955); and proposed structure: A. L. McCloskey *et al.*, *ibid.* **83**, 4750 (1961). VUV absorption spectrum: H. Bredohl *et al.*, *Mol. Phys.* **101**, 2145 (2003).

Fluffy white solid. Hygroscopic, converts back to tetrahydroxyboron on reaction with water. Sol in methanol, warm absolute ethanol, warm isopropyl alcohol. Sparingly sol in water. Insol in dimethylamine and methyl borate.

1347. Boron Nitride. [10043-11-5] BN; mol wt 24.82. B 43.55%, N 56.43%. Prepd by igniting compds of boron with compds of nitrogen: Taylor, **US 2855316** (1958 to Carborundum Co.). Reviews: Giardini, *Bur. Mines Inf. Circ.* **7664**, 13 pp (1953); K. Niedenzu, J. W. Dawson, *Boron-Nitrogen Compounds* (Academic Press, New York, 1965) pp 147-153.

Crystals with hexagonal, graphite lattice is most common form. *Borazon*, cubic crystalline modification, is probably the hardest substance known. There exists also an amorphous modification. mp 3000°. Begins to sublime at a temp slightly below 3000°. Begins to dissociate *in vacuo* at about 2700°. The chemical behavior of BN is dependent on the method of prepn. Not attacked by mineral acids, water; in general resistant to chemical attack. Hot concd alkali cleaves boron-nitrogen bond. Oxidation in air begins above 1200°. *See* Niedenzu, Dawson, *loc. cit.*

USE: Manuf of alloys; in semiconductors, nuclear reactors, lubricants.

1348. Boron Oxide. [1303-86-2] Boron oxide (B_2O_3); boron sesquioxide; boron trioxide; boric anhydride; boric oxide. B_2O_3; mol wt 69.62. B 31.05%, O 68.94%. Prepn of crystalline form: McCulloch, *J. Am. Chem. Soc.* **59**, 2650 (1937). Review of toxicology and human exposure: *Toxicological Profile for Boron* (PB2010-100001, 2010) 249 pp.

Colorless, brittle, vitreous, semitransparent, hygroscopic lumps or hard, white crystals. d (amorph) 1.8; d (cryst) 2.46. mp (cryst) 450°. Sol in alc, glycerol; slowly sol in 30 parts cold, or 5 parts boiling water. *Keep dry.*

Note: Has been improperly called **anhydrous boric acid** or **fused boric acid.**

Caution: Potential symptoms of overexposure are irritation of eyes, skin, respiratory system; cough; conjunctivitis; skin erythema.

See NIOSH Pocket Guide to Chemical Hazards (DHHS/NIOSH 97-140, 1997) p 30.

USE: In metallurgy; in analysis of silicates to determine SiO_2 and alkalies; in blowpipe analysis.

1349. Boron Tribromide. [10294-33-4] Tribromoborane. BBr_3; mol wt 250.52. B 4.32%, Br 95.69%. Prepn: Gamble, *Inorg. Synth.* **3**, 27 (1950); Becher in *Handbook of Preparative Inorganic Chemistry* Vol. 1, G. Brauer, Ed. (Academic Press, New York, 2nd ed., 1963) p 781. Review of boron halides: Massey, *Adv. Inorg. Chem. Radiochem.* **10**, 1-152 (1967). Review of toxicology and human exposure: *Toxicological Profile for Boron* (PB2010-100001, 2010) 249 pp. Brief review of synthetic applications: E. G. Doyagüez, *Synlett* **2005**, 1636-1637.

Colorless, fuming liquid; dec by water or alc. Sharp, irritating odor. mp −46.0°; bp 90°. d^0 2.698. *Poisonous, corrosive. Store under dry inert atmosphere. Reacts violently with water, alcohols.* Vapor pressure data: Barber *et al.*, *J. Chem. Eng. Data* **9**, 137 (1964).

Caution: Potential symptoms of overexposure are irritation of eyes, skin, respiratory system; skin and eye burns; dyspnea, pulmonary edema. *See NIOSH Pocket Guide to Chemical Hazards* (DHHS/NIOSH 97-140, 1997) p 32.

USE: Manuf of diborane; ultra high purity boron. Reagent for cleavage of ethers, amines, thiols; addition of allenes and alkynes.

1350. Boron Trichloride. [10294-34-5] Trichloroborane. BCl_3; mol wt 117.16. B 9.23%, Cl 90.77%. Prepn: Gamble, *Inorg. Synth.* **3**, 27 (1950). *Reviews:* Gerrard, Lappert, *Chem. Rev.* **58**, 1081-1111 (1958); Massey, *Adv. Inorg. Chem. Radiochem.* **10**, 1-152 (1967). Review of toxicology and human exposure: *Toxicological Profile for Boron* (PB2010-100001, 2010) 249 pp.

Colorless, fuming liquid at low temp; dec by water or alcohol. *Poisonous, corrosive.* bp 12.5°. mp −107°. d_4^{12} 1.35. d^0 1.3728. *See:* Ward, *J. Chem. Eng. Data* **14**, 167 (1969).

USE: Manuf and purification of boron; as catalyst for organic reactions; in semiconductors; in bonding of iron, steels; in purification of metal alloys to remove oxides, nitrides and carbides.

1351. Boron Trifluoride. [7637-07-2] Trifluoroborane. BF_3; mol wt 67.81. B 15.94%, F 84.05%. A strong Lewis acid. Prepn: Swinehart, **US 2148514, US 2196907** (1939, 1940 both to Harshaw Chemical); Booth, Wilson, *Inorg. Synth.* **1**, 21 (1939); Kwasnik in *Handbook of Preparative Inorganic Chemistry* Vol. 1, G. Brauer, Ed. (Academic Press, New York, 2nd ed., 1963) pp 219-222; Wiesboeck, **US 3690821** (1972 to U.S. Steel). Dihydrate: McGrath *et al.*, *J. Am. Chem. Soc.* **66**, 1263 (1944). *Reviews:* Booth, Martin, *Boron Trifluoride and Its Derivatives* (John Wiley & Sons, 1949), 296 pp; Booth in *Fluorine Chemistry* Vol. 1, J. Simons, Ed. (Academic Press, New York, 1950) pp 201-224; Topchiev *et al.*, *Boron Fluoride and Its Compounds as Catalysts in Organic Chemistry* (Pergamon Press, 1959) 326 pp; Martin in *Kirk-Othmer Encyclopedia of Chemical Technology* Vol. 9 (Interscience, New York, 2nd ed., 1966) pp 554-562; Massey, *Adv. Inorg. Chem. Radiochem.* **10**, 1-152 (1967). Review of toxicology and human exposure: *Toxicological Profile for Boron* (PB2010-100001, 2010) 249 pp.

Colorless gas. Pungent, suffocating odor. Forms dense white fumes in moist air. mp −127.1°. bp −100.4°. d_4 (−100.4°; liq) 1.57. d (gas at STP) 3.07666 g/l. *Poisonous.* Soly in water (0°): 332 g/100 g; some hydrolysis occurs to form fluoboric and boric acids. Soly in anhydrous H_2SO_4: 1.94 g/100 g acid. Forms solid complex with nitric acid ($HNO_3.2BF_3$). Sol in most saturated and halogenated hydrocarbons and in aromatic compds. Polymerizes unsaturated molecules. Easily forms coordination complexes with molecules having at least one unshared pair of electrons. Reacts with incandescence when heated with alkali metals or alkaline earth metals except magnesium.

Caution: Potential symptoms of overexposure are irritation of eyes, skin, nose, respiratory system; epistaxis; burns to eyes and skin. *See NIOSH Pocket Guide to Chemical Hazards* (DHHS/NIOSH 97-140, 1997) p 32. *See also Patty's Industrial Hygiene and Toxicology* **vol. 2B**, G. D. Clayton, F. E. Clayton, Eds. (Wiley-Interscience, New York, 3rd ed., 1981) pp 2996-2999.

USE: To protect molten magnesium and its alloys from oxidation; as a flux for soldering magnesium; as a fumigant; in ionization chambers for the detection of weak neutrons. By far the largest application of boron trifluoride is in catalysis with and without promoting agents.

1352. Boron Trifluoride Etherate. [109-63-7] (*T*-4)-Trifluoro[1,1'-oxybis[ethane]]boron; boron fluoride etherate; boron fluoride ethyl ether; ethyl ether-boron trifluoride complex. C_4H_{10}-BF_3O; mol wt 141.93. C 33.85%, H 7.10%, B 7.62%, F 40.16%, O 11.27%. $(CH_3CH_2)_2O.BF_3$. Prepd by vapor-phase reaction of anhydr ether with BF_3: Laubengayer, Finlay, *J. Am. Chem. Soc.* **65**, 884 (1943).

Fuming liquid, immediately hydrolyzed by moisture in air. d_4^{25} 1.125. bp 125.7°. mp −60.4°. n_D^{20} 1.348. Heat of formation: 12.5 kcal. Heat of soln at 0° in ether: 2.7 kcal.

Caution: On decomposition forms highly toxic fumes of fluorides.
USE: Catalyst in acetylation, alkylation, polymerization, dehydration, and condensation reactions.

1353. Bortezomib. [179324-69-7] *B*-[(1*R*)-3-Methyl-1-[[(2*S*)-1-oxo-3-phenyl-2-[(2-pyrazinylcarbonyl)amino]propyl]amino]butyl]boronic acid; *N*-(2-pyrazine)carbonyl-L-phenylalanine-L-leucine boronic acid; MG-341; PS-341; Velcade. $C_{19}H_{25}BN_4O_4$; mol wt 384.24. C 59.39%, H 6.56%, B 2.81%, N 14.58%, O 16.66%. Dipeptide boronic acid proteasome inhibitor. Prepn: J. Adams *et al.*, **WO 9613266**; *eidem*, **US 5780454** (1996, 1998 both to ProScript); *idem et al.*, *Bioorg. Med. Chem. Lett.* **8**, 333 (1998). Pharmacology and cytotoxic activity: *idem et al.*, *Cancer Res.* **59**, 2615 (1999). Crystal structure in complex with the yeast 20S proteasome: M. Groll *et al.*, *Structure* **14**, 451 (2006). Mechanism of action study: J. Codony-Servat *et al.*, *Mol. Cancer Ther.* **5**, 665 (2006). Clinical trial in refractory myeloma: P. G. Richardson *et al.*, *N. Engl. J. Med.* **348**, 2609 (2003). Review of clinical development: *idem et al.*, *Annu. Rev. Med.* **57**, 33-47 (2006).

Soly in water: 3.3-3.8 mg/ml (pH 2-6.5).
THERAP CAT: Antineoplastic.

1354. Boscalid. [188425-85-6] 2-Chloro-*N*-(4'-chloro[1,1'-biphenyl]-2-yl)-3-pyridinecarboxamide; 2-chloro-*N*-(4'-chlorobiphenyl-2-yl)nicotinamide; nicobifen; BAS-510F; Cantus; Endura. $C_{18}H_{12}Cl_2N_2O$; mol wt 343.21. C 62.99%, H 3.52%, Cl 20.66%, N 8.16%, O 4.66%. Broad spectrum carboxamide fungicide used to control diseases in turf, field, row, orchard, and vineyard crops. Prepn: K. Eicken *et al.*, **EP 545099**; *eidem*, **US 5589493** (1993, 1996 both to BASF). Terrestrial field dissipation study: S. H. Jackson, *Pest Manage. Sci.* **60**, 8 (2003). Comparative field trial vs sclerotinia drop in lettuce: M. E. Matheron, M. Porchas, *Plant Dis.* **88**, 665 (2004). Review of chemistry and toxicology: *Boscalid Pesticide Fact Sheet* (U.S. EPA, July 2003) 18 pp.

White crystalline solid, mp 142.8-143.8°. d 1.381. Partition coefficient (octanol/water): 915 at 21°. Soly at 20° (g/100 ml): acetone 16-20; acetonitrile 4-5; methanol 4-5; ethyl acetate 6.7-8; dichloromethane 20-25; toluene 2-2.5; 1-octanol <1. Soly in water at

20°: 6 mg/l. Vapor pressure (hPa): 7×10^{-9} at 20°; 2×10^{-8} at 25°. pH 5.5 at 23° (1% soln). LD_{50} in rats (mg/kg): >5000 orally; >2000 dermally (U.S. EPA).
USE: Fungicide.

1355. Bosentan. [147536-97-8] 4-(1,1-Dimethylethyl)-*N*-[6-(2-hydroxyethoxy)-5-(2-methoxyphenoxy)[2,2'-bipyrimidin]-4-yl]-benzenesulfonamide; *p-tert*-butyl-*N*-[6-(2-hydroxyethoxy)-5-(*o*-methoxyphenoxy)-2-(2-pyrimidinyl)-4-pyrimidinyl]benzenesulfonamide; Ro-47-0203. $C_{27}H_{29}N_5O_6S$; mol wt 551.62. C 58.79%, H 5.30%, N 12.70%, O 17.40%, S 5.81%. Mixed endothelin receptor antagonist. Prepn: K. Burri *et al.*, **CA 2071193**; *eidem*, **US 5292740** (1992, 1994 both to Hoffmann-La Roche). Receptor binding study: M. Clozel *et al.*, *J. Pharmacol. Exp. Ther.* **270**, 228 (1994). LC-MS determn in plasma: B. Lausecker, G. Hopfgartner, *J. Chromatogr. A* **712**, 75 (1995). Clinical pharmacology: G. Sütsch *et al.*, *Cardiovasc. Drugs Ther.* **10**, 717 (1996). Clinical pharmacokinetics: C. Weber *et al.*, *Clin. Pharmacol. Ther.* **60**, 124 (1996). Clinical trial in hypertension: H. Krum *et al.*, *N. Engl. J. Med.* **338**, 784 (1998); in pulmonary arterial hypertension: L. J. Rubin *et al.*, *ibid.* **346**, 896 (2002). Review of pharmacology and clinical experience in pulmonary hypertension: S. G. Raja, G. D. Dreyfus, *Ann. Card. Anaesth.* **11**, 6-14 (2008).

Monohydrate. [157212-55-0] Tracleer. $C_{27}H_{29}N_5O_6S.H_2O$; mol wt 569.63. White to yellowish powder. Soly (mg/100 ml): water 1.0; aq soln 0.1 (pH 1.1); 0.1 (pH 4.0); 0.2 (pH 5.0); 43 (pH 7.5).
Sodium salt. [150726-52-6] $C_{27}H_{28}N_5NaO_6S$. mp 195-198°.
THERAP CAT: Antihypertensive; in treatment of pulmonary hypertension.

1356. Bostrycoidin. [4589-33-7] 6,9-Dihydroxy-7-methoxy-3-methylbenz[*g*]isoquinoline-5,10-dione; 5,8-dihydroxy-6-methoxy-3-methyl-2-aza-9,10-anthraquinone. $C_{15}H_{11}NO_5$; mol wt 285.26. C 63.16%, H 3.89%, N 4.91%, O 28.04%. Antibiotic substance produced by *Fusarium bostrycoides*: Hamilton *et al.*, *Antibiot. Chemother.* **3**, 853 (1953); Cajori *et al.*, *J. Biol. Chem.* **208**, 107 (1954). Structure: Arsenault, *Tetrahedron Lett.* **1965**, 4033. Synthesis: D. W. Cameron *et al.*, *ibid.* **21**, 5089 (1980).

Dark red crystals; mp 243-244°, changes to purple in alkaline medium, yellow in acid medium. Stable at room temp and withstands autoclaving. Insol in water. Sol in abs ethanol, 60% ethanol; moderately sol in dioxane, benzene, acetone, chloroform, carbon tetrachloride. Soluble in aq sodium carbonate solns. Partly sol in corn oil. Active *in vitro* against *Mycobacterium tuberculosis*. Serum seems to interfere.

1357. Bosutinib. [380843-75-4] 4-[(2,4-Dichloro-5-methoxyphenyl)amino]-6-methoxy-7-[3-(4-methyl-1-piperazinyl)propoxy]-3-quinolinecarbonitrile; 4-[(2,4-dichloro-5-methoxyphenyl)-amino]-6-methoxy-7-[3-(4-methylpiperazin-1-yl)propoxy]quinoline-3-carbonitrile; SKI-606; WAY-173606. $C_{26}H_{29}Cl_2N_5O_3$; mol wt 530.45. C 58.87%, H 5.51%, Cl 13.37%, N 13.20%, O 9.05%. Tyrosine kinase inhibitor with dual activity against Src and Abl kinases. Prepn: A. Wissner *et al.*, **WO 9843960**; *eidem*, **US 6002008**

(1998, 1999 both to American Cyanamid). Structure-activity relationship: D. H. Boschelli *et al.*, *J. Med. Chem.* **44**, 3965 (2001). Improved synthesis: *eidem, ibid.* **47**, 1599 (2004). Protonation constants and characterization of micro-speciation: K. J. Box *et al., J. Pharm. Biomed. Anal.* **47**, 303 (2008). *In vitro* and *in vivo* evaluation in imatinib-resistant chronic myelogenous leukemia models: M. Puttini *et al., Cancer Res.* **66**, 11314 (2006). *In vitro* and *in vivo* inhibition of Src-mediated signaling pathways in breast cancer models: H. Jallal *et al., ibid.* **67**, 1580 (2007). Clinical trial in chronic myeloid leukemia: J. E. Cortes *et al., Blood* **118**, 4567 (2011).

Light pink solid from diethyl ether, mp 116-120°. Protonation constants (log K): 11.2, 7.92, 4.75, 3.79.

THERAP CAT: Antineoplastic.

1358. β-Boswellic Acid. [631-69-6] (3α,4β)-3-Hydroxyurs-12-en-23-oic acid. $C_{30}H_{48}O_3$; mol wt 456.71. C 78.90%, H 10.59%, O 10.51%. Occurs as the acetate in frankincense *(olibanum)* from *Boswellia carterii, Burseraceae.* The β-form is predominant and is accompanied by small amounts of α- and γ-boswellic acid. Isoln from olibanum tears: Winterstein, Stein, *Z. Physiol. Chem.* **208**, 9 (1932). Early structural studies: Simpson, Williams, *ibid.* **1938**, 686, 1712; Ruzicka, Wirz, *Helv. Chim. Acta* **22**, 948 (1939); **23**, 132 (1940); Ruzicka *et al., ibid.* **27**, 1859 (1944). Revised structure and stereochemistry: Beton *et al., J. Chem. Soc.* **1956**, 2904; Allan, *Chimia* **17**, 382 (1963); *idem, Phytochemistry* **7**, 963 (1968). *Review:* J. Simonsen, W. C. J. Ross, *The Terpenes* vol. **5** (University Press, Cambridge, 1957) pp 68-74.

Long prisms from methanol, mp 228-232° with preliminary sintering. $[\alpha]_D$ +107° (c = 0.75 in CHCl$_3$). 100 ml of boiling methanol will dissolve 8 grams of β-boswellic acid. Sol in chloroform, ether, acetone, alc.

Acetate. $C_{32}H_{50}O_4$. Prisms, mp 275-278°, $[\alpha]_D$ +63° (c = 1.88 in CHCl$_3$).

Methyl ester. $C_{31}H_{50}O_3$. mp 195-196°, $[\alpha]_D$ +111° (c = 1.6 in CHCl$_3$).

1359. Botrydial. [54986-75-3] (1S,3aR,4S,6R,7S,7aS)-4-(Acetyloxy)octahydro-7a-hydroxy-1,3,3,6-tetramethyl-*1H*-indene-1,7-dicarboxaldehyde. $C_{17}H_{26}O_5$; mol wt 310.39. C 65.78%, H 8.44%, O 25.77%. Sesquiterpenoid phytotoxin produced by *Botrytis cinerea* Pers., *Sclerotiniaceae*, the pathogenic gray mold fungus of grapevines, tobacco, vegetables, and fruits. Isoln from *B. cinerea* cultures: H.-W. Fehlhaber *et al., Ber.* **107**, 1720 (1974). Biosynthetic study and ^{13}C-NMR: J. R. Hanson, R. Nyfeler, *Chem. Commun.* **1976**, 72. Phytotoxic activity: L. Rebordinos *et al., Phytochemistry* **42**, 383 (1996). Identification in infected plant tissue: N. Deighton *et al., ibid.* **57**, 689 (2001). Characterization of biosynthetic gene cluster: C. Pinedo *et al., ACS Chem. Biol.* **3**, 791 (2008).

Crystals from petroleum ether, mp 108-110°. $[\alpha]_D^{20}$ +34° (c = 0.14 in chloroform).

1360. Botryococcene. [42719-34-6] (3S,7S,10S,11E,13R,-16S,20S)-10-Ethenyl-2,3,7,10,13,16,20,21-octamethyl-6,17-bis-(methylene)-1,11,21-docosatriene; (−)-C_{34}-botryococcene. $C_{34}H_{58}$; mol wt 466.84. C 87.48%, H 12.52%. Most abundant member of a family of terpenoid hydrocarbons (botryococcenes) having the general structure C_nH_{2n-10}, n=30-37. Produced by the B race of *Botryococcus braunii*, a common green microalga that has been found in oil-rich geological deposits. Isoln from *B. braunii* (Kützing): J. R. Maxwell *et al., Phytochemistry* **7**, 2157 (1968). Structure: R. E. Cox *et al., J. Chem. Soc. Chem. Commun.* **1973**, 284; of various isomers: P. Metzger *et al., Phytochemistry* **24**, 2995 (1985). Abs config: J. D. White *et al., J. Am. Chem. Soc.* **108**, 5352 (1986). Total synthesis: *idem et al., ibid.* **110**, 1624 (1988). Recovery of hydrocarbons by solvent extraction from algal cultures: J. Frenz *et al., Biotechnol. Bioeng.* **34**, 755 (1989). Review of biosynthesis, production methods, and conversion to biofuels: A. Banerjee *et al., Crit. Rev. Biotechnol.* **22**, 245-279 (2002).

$[\alpha]_D^{22}$ −3.5° (c = 6.1 in chloroform).

Isobotryococcene. [11054-04-9] $C_{34}H_{58}$; mol wt 466.84.

USE: Renewable source of hydrocarbon fuels.

1361. Bottromycin. [1393-68-6] B-mycin. A complex of five antibiotics of which the main active component is bottromycin A$_2$. Produced by *Streptomyces bottropensis* and *S. canadensis:* Waisvisz *et al., J. Am. Chem. Soc.* **79**, 4520, 4522, 4524 (1957); Miller *et al., Antimicrob. Agents Chemother.* **1967**, 407. Production: **GB 762736** (1956 to Koninklijke Nederlandsche Gist en Spiritus-Fabriek); Umesawa *et al.,* **JP 68 10998** (1968 to Microbiochem. Res. Found.), *C.A.* **69**, 85439x (1968); Hata *et al.,* **US 3650904** (1972). Partial structure: Waisvisz, Hoeven, *J. Am. Chem. Soc.* **80**, 38 (1958). Sepn and structures of bottromycins A$_1$, A$_2$, and B: Nakamura *et al., J. Antibiot.* **18A**, 47, 60 (1965); **19A**, 10 (1966). Isoln and characterization of bottromycins A$_2$, B$_2$, C$_2$: Nakamura *et al., ibid.* **20A**, 1 (1967). Revised structure of bottromycin A$_2$: Takahashi *et al., ibid.* **29A**, 1120 (1976); D. Schipper, *ibid.* **36A**, 1076 (1983). Mode of action of bottromycin A$_2$: T. Otaka, A. Kaji, *J. Biol. Chem.* **251**, 2299 (1976); *eidem, FEBS Lett.* **123**, 173 (1981).

Bottromycin A$_2$

Bottromycin A₂. Bottromycic A_2 acid methyl ester. $C_{42}H_{62}N_8O_7S$; mol wt 823.07.

1362. Botulin Toxins. Botulinum toxins; botulinus toxins. Potent neurotoxins produced by *Clostridium botulinum*. Seven serologically distinct types are known, designated A-G; mol wt ~150 kDa. Zinc-proteases synthesized as non-toxic single-chain precursors which become toxic upon proteolytic cleavage. Active form consists of one heavy and one light chain linked by a disulfide bond. Prevents release of acetylcholine from presynaptic nerve terminals at the neuromuscular junction. Identification of the causative agent of botulism: E. P. Van Ermengem, *Z. Hyg. Infektionskrankh.* **26**, 1 (1897). Isoln: P. T. Snipe, H. Sommer, *J. Infect. Dis.* **43**, 152 (1928). Review of isoln and purification: L. L. Simpson *et al.*, *Methods Enzymol.* **165**, 76-85 (1988); of structure and function: K. Oguma *et al.*, *Microbiol. Immunol.* **39**, 161-168 (1995). Review of pharmacology: J. L. Middlebrook, *J. Toxicol. Toxin Rev.* **5**, 177-190 (1986); of mode of action and detection: M. Wictome, C. C. Shone, *J. Appl. Microbiol. Symp. Suppl.* **84**, 87S-97S (1998). Review of clinical uses in focal dystonias: K. R. Kessler, R. Benecke, *Neurotoxicology* **18**, 761-770 (1997); in cosmetic dermatology: A. Carruthers, J. Carruthers, *Basic Clin. Dermatol.* **15**, 207-236 (1998).

MLD in mouse: 0.0003 μg/kg (Middlebrook).

Botulin A. [93384-43-1] Botulinum toxin type A; oculinum; AGN-191622; BoTox; Dysport. Clinical trial in blepharospasm: B. Arthurs *et al.*, *Can. J. Ophthalmol.* **22**, 24 (1987); in achalasia: P. J. Pasricha *et al.*, *N. Engl. J. Med.* **332**, 774 (1995); in hyperhidrosis: M. Heckmann *et al.*, *ibid.* **344**, 488 (2001); in spasticity following stroke: A. Brashear *et al.*, *ibid.* **347**, 395 (2002).

Botulin B. [93384-44-2] Botulinum toxin type B; NeuroBloc. Clinical trial in cervical dystonia: P. A. Cullis *et al.*, *Adv. Neurol.* **78**, 227 (1998).

THERAP CAT: Muscle relaxant (skeletal).

1363. BPE Ligands. Bidentate, C_2-symmetric bis(phospholane) ligands with an alkyl chain backbone. Catalytic complexes formed from transition metals are used in asymmetric transformations. Prepn of Me-BPE and use in enantioselective hydrogenations: M. J. Burk *et al.*, *Organometallics* **9**, 2653 (1990); *eidem, Tetrahedron: Asymmetry* **2**, 569 (1991); M. J. Burk, *US 5008457* (1991 to Du Pont). Improved prepn of Me-BPE and Et-BPE and use in enantioselective hydrogenations: *idem, J. Am. Chem. Soc.* **113**, 8518 (1991); *idem et al., ibid.* **115**, 10125 (1993). Utility of ligands in enantioselective hydrogenations: *idem et al., ibid.* **118**, 5142 (1996); *idem et al., Angew. Chem. Int. Ed.* **37**, 1931 (1998); *idem et al., Tetrahedron Lett.* **40**, 3093 (1999). Review of phospholane ligand chemistry: M. J. Burk, *Acc. Chem. Res.* **33**, 363-372 (2000).

(R,R)-Me-BPE: R = CH₃
(R,R)-Et-BPE: R = CH₂CH₃

(R,R)-Me-BPE. [129648-07-3] (2R,2'R,5R,5'R)-1,1'-(1,2-Ethanediyl)bis[2,5-dimethylphospholane]; (+)-1,2-bis[(2R,5R)-2,5-dimethylphospholano]ethane. $C_{14}H_{28}P_2$; mol wt 258.33. C 65.09%, H 10.93%, P 23.98%. Colorless oil, $bp_{0.06}$ 64-67°. $[\alpha]_D^{25}$ +263 ± 3° (c = 1 in hexane). *Spontaneously combustible.* Reacts violently with water. Handle and store under inert gas.

(S,S)-Me-BPE. [136779-26-5] Colorless liquid. *Spontaneously combustible.* Reacts violently with water. Handle and store under inert gas.

(R,R)-Et-BPE. [136705-62-9] (2R,2'R,5R,5'R)-1,1'-(1,2-Ethanediyl)bis[2,5-diethylphospholane]; (+)-1,2-bis[(2R,5R)-2,5-diethylphospholano]ethane. $C_{18}H_{36}P_2$; mol wt 314.43. C 68.76%, H 11.54%, P 19.70%. Colorless oil, $bp_{0.05}$ 104-106°. $[\alpha]_D^{25}$ +320 ± 4° (c = 1 in hexane). Air and moisture sensitive. Handle and store under inert gas.

(S,S)-Et-BPE. [136779-27-6] Liquid. *Spontaneously combustible.* Reacts violently with water. Handle and store under inert gas.

USE: Ligands in organic synthesis.

1364. Bradykinin. [58-82-2] Kallidin I; kallidin-9; callidin I; BRS-640. $C_{50}H_{73}N_{15}O_{11}$; mol wt 1060.23. C 56.64%, H 6.94%, N 19.82%, O 16.60%. A tissue hormone belonging to a group of hypotensive peptides known as plasma kinins. First obtained by incubation with the venom of *Bothrops jararaca* or with crystalline trypsin: Rocha e Silva *et al.*, *Cien Cult. (Sao Paulo)* **1**, 32 (1949); *eidem, Am. J. Physiol.* **156**, 261 (1949); Prado *et al.*, *Arch. Biochem.* **27**, 410 (1950); Werle *et al.*, *Biochem. Z.* **320**, 372 (1950). Large scale prepn from whole bovine plasma: Hamberg, Deutsch, *Arch. Biochem. Biophys.* **76**, 262 (1958). Formed by proteolysis of a precursor in the globulin fraction of plasma referred to as kininogen by the action of enzymes such as trypsin, plasmin, and plasma kallikrein, *q.q.v.* Acts on smooth muscle, dilates peripheral vessels, increases capillary permeability. Also is a potent pain-producing agent. Structure: Elliott *et al.*, *Biochem. Biophys. Res. Commun.* **3**, 87 (1960); Werle *et al.*, *Z. Physiol. Chem.* **326**, 174 (1961). Synthesis: Boissonnas *et al.*, *Helv. Chim. Acta* **43**, 1349 (1960); **45**, 170 (1962); Merrifield, *Biochemistry* **3**, 1385 (1964); Young *et al.*, *J. Chem. Soc. C* **1971**, 46; Bajusz *et al.*, *HU 3840* (1972 to Gyogyszerkutato Intézet), *C.A.* **77**, 20027g (1972); F. Sipos, *DE 2212787*; *idem, US 3714140* (1972, 1973 both to Squibb); Corley *et al.*, *Biochem. Biophys. Res. Commun.* **47**, 1353 (1972); N. S. S. Kumari *et al.*, *Indian J. Chem.* **17B**, 152 (1979). *Review:* Schröder, Hempel, *Experientia* **20**, 529 (1964).

Arg–Pro–Pro–Gly–Phe–Ser–Pro–Phe–Arg

Amorphous precipitate. $[\alpha]_D^{25}$ −76.5° (c = 1.37 in $1N$ acetic acid). Sol in glacial acetic acid, in a 10% soln of trichloroacetic acid, 70% ethanol, in hot methanol; less sol in 90% ethanol or cold methanol. Almost insol in acetone, chloroform, ethyl ether, ethyl methyl ketone, petr ether, butanol, amyl alcohol, ethyl acetate.

THERAP CAT: Vasodilator.

1365. Brain Natriuretic Peptide. [114471-18-0] Brain natriuretic factor; B-type natriuretic peptide; BNP. Peptide hormone structurally and physiologically similar to atrial natriuretic peptide, *q.v.* Originally discovered in mammalian brain, subsequently found to be produced primarily by the cardiac ventricle in response to hemodynamic stress. Amino acid sequence shows a marked interspecies diversity; the normal circulating form in humans contains 32 amino acid residues. Isoln from porcine brain: T. Sudoh *et al.*, *Nature* **332**, 78 (1988). Cloning of human BNP precursor: *idem et al.*, *Biochem. Biophys. Res. Commun.* **159**, 1427 (1989). Direct isoln from human heart: Y. Kambayashi *et al.*, *FEBS Lett.* **259**, 341 (1990). Use in prognosis of risk in acute coronary syndromes: J. A. de Lemos *et al.*, *N. Engl. J. Med.* **345**, 1014 (2001); in diagnosis of congestive heart failure: A. S. Maisel *et al.*, *ibid.* **347**, 161 (2002). *Review:* N. C. Davidson, A. D. Struthers, *J. Hypertens.* **12**, 329-336 (1994); T. Takeda, M. Kohno, *Hypertens. Res.* **18**, 259-266 (1995); of pathophysiological significance: M. Yoshibayashi *et al.*, *Eur. J. Endocrinol.* **135**, 265-268 (1996); of clinical experience: N. Valli *et al.*, *J. Lab. Clin. Med.* **134**, 437-444 (1999).

```
1
Ser–Pro–Lys–Met–Val–Gln–Gly–Ser–Gly–Cys–Phe–Gly–Arg–Lys–Met–Asp–Arg
                                        |                            |
                                        |                           Ile
  32                                    |
His–Arg–Arg–Leu–Val–Lys–Cys–Gly–Leu–Gly–Ser–Ser–Ser–Ser
```

human BNP

Nesiritide. [124584-08-3] Human brain natriuretic peptide-32; natriuretic factor-32 (human brain clone λhBNP57); Natrecor. Clinical hemodynamic effects: L. S. Marcus *et al.*, *Circulation* **94**, 3184 (1996). Clinical trial in congestive heart failure: W. S. Colucci *et al.*, *N. Engl. J. Med.* **343**, 246 (2000).

THERAP CAT: In treatment of congestive heart failure.

1366. Brallobarbital. [561-86-4] 5-(2-Bromo-2-propen-1-yl)-5-(2-propen-1-yl)-2,4,6(1H,3H,5H)-pyrimidinetrione; 5-allyl-5-

(2-bromoallyl)barbituric acid; Vesperone. $C_{10}H_{11}BrN_2O_3$; mol wt 287.11. C 41.83%, H 3.86%, Br 27.83%, N 9.76%, O 16.72%. Prepn: Morren, **BE 497501** (1950), *C.A.* **49**, 1100e (1955). Metabolism: Keding, Schmidt, *Arzneim.-Forsch.* **19**, 342 (1969).

Solid, mp 168-169°.

Note: This is a controlled substance (depressant): **21 CFR,** 1308.13.

THERAP CAT: Sedative, hypnotic.

1367. Brandy. A potable alcoholic liquid distilled from wine or from the fermented juices of peaches, cherries, apples, or other fruit.

To meet the specifications of the National Formulary, brandy must have been obtained by distillation of fermented juice from sound ripe grapes and contain between 48 and 54% ethanol (v/v), d_4^{15} 0.921-0.933. It must have been stored in wooden containers for a period of not less than 2 years. Unlike whisky, brandy is never made from cereal mash or from potatoes (as some brands of vodka).

THERAP CAT: Sedative, vasodilator (peripheral).

1368. Brassard's Diene. [90857-62-8]; [74272-66-5] (unspecified stereo). [[(1E)-1,3-Dimethoxy-1,3-butadien-1-yl]oxy]trimethylsilane; 1,3-dimethoxy-1-trimethylsiloxy-1,3-butadiene. $C_9H_{18}O_3Si$; mol wt 202.33. C 53.43%, H 8.97%, O 23.72%, Si 13.88%. Highly reactive nucleophilic silyloxydiene; *cf.* Danishefsky's diene. Prepn: J. Savard, P. Brassard, *Tetrahedron Lett.* **20**, 4911 (1979); *eidem, Tetrahedron* **40**, 3455 (1984). Synthetic applications: M. M. Midland, R. W. Koops, *J. Org. Chem.* **55**, 5058 (1990); T. Ito *et al., Tetrahedron Lett.* **34**, 6583 (1993); Q. Fan *et al., Eur. J. Org. Chem.* **2005**, 3542; of polymer-supported derivatives: C. Pierres *et al.,Tetrahedron Lett.* **44**, 3645 (2003).

Oil. $bp_{0.5}$ 54°. $bp_{0.8}$ 57-62°.

USE: Electron-rich diene in hetero Diels-Alder reactions.

1369. Brassidic Acid. [506-33-2] (13E)-13-Docosenoic acid. $C_{22}H_{42}O_2$; mol wt 338.58. C 78.04%, H 12.50%, O 9.45%. Prepn by isomerization of its *cis*-isomer (erucic acid): Reimer, Will, *Ber.* **19**, 3320 (1886); Skellon, *Manuf. Chem.* **33**, 405 (1962). Synthesis and separation from erucic acid by crystn: Bowman, *Nature* **163**, 95 (1949).

Thin, cryst plates from alc, mp 61-62°. bp_{30} 282°. n_D^{57} 1.448. Practically insol in water; sparingly sol in cold alcohol; sol in ether. Solubility at 10°, 0°, −10°, −20°, and −30° in methanol, ethyl acetate, ether, acetone, toluene, *n*-heptane: Kobl, *Diss. Abstr.* **20**, 82 (1959).

1370. Brassinolide. [72962-43-7] (1R,3aS,3bS,6aS,8S,9R,-10aR,10bS,12aS)-1-[(1S,2R,3R,4S)-2,3,Dihydroxy-1,4,5-trimethylhexyl]hexadecahydro-8,9-dihydroxy-10a,12a-dimethyl-6*H*-benz[c]-indeno[5,4-e]oxepin-6-one; (2α,3α,5α,22R,23R,24S)-2,3,22,23-tetrahydroxy-*B*-homo-7-oxaergostan-6-one; 2α,3α,22,23-tetrahydroxy-24-methyl-*B*-homo-7-oxa-5α-cholestan-6-one. $C_{28}H_{48}O_6$; mol wt 480.69. C 69.96%, H 10.07%, O 19.97%. Plant hormone; natural steroid containing a seven-membered B-ring lactone, that promotes both cell elongation and cell division. Over ten brassinosteroids have been isolated and characterized from sources such as pollen, seedling, leaf. Isoln, structure and activity of brassinolide

from rape pollen, *Brassica napus* L.: M. D. Grove *et al., Nature* **281**, 216 (1979). Stereoselective synthesis: S. Fung, J. B. Siddall, *J. Am. Chem. Soc.* **102**, 6580 (1980). Synthesis of two stereoisomers: M. J. Thompson *et al., J. Org. Chem.* **44**, 5002 (1979). Improved synthesis: T. Kametani *et al.,J. Org. Chem.* **53**, 1982 (1988). Structure-activity relationship of brassinosteroids: S. Takatsuto *et al., Phytochemistry* **22**, 2437 (1983); interaction with cytokinin: C. Schlagnhaufer *et al., Physiol. Plant.* **60**, 347 (1984); bioassay: K. Wada *et al., Agric. Biol. Chem.* **48**, 719 (1984).

Crystals from methanol, mp 274-275°. $[\alpha]_D^{27}$ +16°.

USE: Plant growth regulator.

1371. Brayera. Kousso; kosso; cusso; koso; cousso; kouso; kusso. Dried panicles of pistillate flowers of *Hagenia abyssinica* J. F. Gmel. (*Brayera anthelmintica* Kunth), *Rosaceae. Habit.* Abyssinia. *Constit.* Kosin, kossein, kosidin, protokosin, kosotoxin, volatile oil, tannin. *Ref:* Leisenring, *Arch. Pharm.* **232**, 50 (1894); Kondakow, *ibid.* **237**, 481 (1899); Hess, Todd, *J. Chem. Soc.* **1937**, 562; Birch, Todd, *ibid.* **1952**, 3102.

THERAP CAT: Anthelmintic (Cestodes).

THERAP CAT (VET): Has been used as a teniacide.

1372. Brazilin. [474-07-7] (6aS,11bR)-7,11b-Dihydrobenz-[b]indeno[1,2-d]pyran-3,6a,9,10(6H)-tetrol; brasilin; C.I. Natural Red 24; C.I. 75280. $C_{16}H_{14}O_5$; mol wt 286.28. C 67.13%, H 4.93%, O 27.94%. From *Caesalpinia echinata* Lam. (Brazil-wood), or *C. sappan* L. (sappan-wood), *Leguminosae.* Isoln and structure: Perkin *et al., J. Chem. Soc.* **1928**, 1504; Pfeiffer *et al., Ber.* **63**, 1301 (1930). Synthesis of (±)-form: Dann, Hofmann, *Ann.* **667**, 116 (1963); Kirkiacharian, Billet, *Bull. Soc. Chim. Fr.* **1972**, 3292. Synthesis, resolution: Morsingh, Robinson, *Tetrahedron* **26**, 281 (1970). Stereochemistry: Craig *et al., J. Chem. Soc.* **30**, 1573 (1965); *Colour Index* **vol. 4** (3rd ed., 1971) p 4628. *Review:* Robinson, *Bull. Soc. Chim. Fr.* **1958**, 125-134.

Amber-yellow crystals; turn orange in air and light. Dec above 130°. May crystallize as the mono- or hemihydrate. Sol in water, freely in alcohol, ether, also in alkali hydroxide solns with carmine-red color. *Protect from air and light.*

USE: Chiefly as a dye. Has also been recommended as indicator in acid-base titrations; acids = yellow, alkalies = carmine-red.

1373. Bredereck's Reagent. [5815-08-7] 1-(1,1-Dimethyl-ethoxy)-*N,N,N',N'*-tetramethylmethanediamine; bis(dimethylamino)-*tert*-butyloxymethane; *tert*-butoxybis(dimethylamino)methane; *tert*-butoxy-*N,N,N',N'*-tetramethylmethanediamine. $C_9H_{22}N_2O$; mol wt 174.29. C 62.02%, H 12.72%, N 16.07%, O 9.18%. Aminal ester used in condensation reactions. Prepn: H. Bredereck *et al., Angew. Chem. Int. Ed.* **6**, 356 (1967); *idem et al., Ber.* **101**, 41 (1968). Synthetic applications: B. M. Trost, M. Preckel, *J. Am. Chem. Soc.* **95**, 7862 (1973); W. Haefliger, H. Knecht, *Tetrahedron Lett.* **25**, 285 (1983); J.-P. Vors, *J. Heterocycl. Chem.* **28**, 1043 (1991); A. C. Spivey *et al., J. Org. Chem.* **65**, 5253 (2000); E. Morera *et al., Org. Lett.* **4**, 1139 (2002); S. Csihony *et al., Adv. Synth.*

Catal. **346**, 1081 (2004). *Review:* W. Kantlehner, *J. Prakt. Chem. Chem.-Ztg.* **337**, 418-421 (1995).

bp$_{10-12}$ 48-52°. bp$_{15}$ 50-55°. n_D^{20} 1.4260.

USE: Mild enamination reagent; catalyst in ring-opening polymerization reactions.

1374. Brefeldin A. [20350-15-6] (1*R*,2*E*,6*S*,10*E*,11a*S*,13*S*,-14a*R*)-1,6,7,8,9,11a,12,13,14,14a-Decahydro-1,13-dihydroxy-6-methyl-4*H*-cyclopent[*f*]oxacyclotridecin-4-one; γ,4-dihydroxy-2-(6-hydroxy-1-heptenyl)-4-cyclopentanecrotonic acid λ-lactone; ascotoxin; cyanein; decumbin. $C_{16}H_{24}O_4$; mol wt 280.36. C 68.55%, H 8.63%, O 22.83%. A fungal metabolite which is a macrocyclic lactone exhibiting a wide range of antibiotic activity. Produced by *Penicillium brefeldianum* Dodge: E. Haerri *et al.*, *Helv. Chim. Acta* **46**, 1235 (1963). Also produced by *P. decumbens:* V. L. Singleton *et al.*, *Nature* **181**, 1072 (1958); *P. cyaneum:* V. Betina *et al.*, *Folia Microbiol.* **7**, 353 (1962). Structure: H. P. Sigg, *Helv. Chim. Acta* **47**, 1401 (1964). Abs configuration: H. P. Weber *et al.*, *ibid.* **54**, 2763 (1971). Synthesis of (±)-form: E. J. Corey, R. H. Wollenberg, *Tetrahedron Lett.* **1976**, 4705; E. J. Corey *et al.*, *ibid.* **1977**, 2243; R. Baudouy *et al.*, *ibid.* 2973; P. A. Bartlett, F. R. Green, *J. Am. Chem. Soc.* **100**, 4548 (1978); A. E. Greene *et al.*, *ibid.* **102**, 7583 (1980); M. Honda *et al.*, *Tetrahedron Lett.* **1981**, 2679. Total synthesis of (+)-form: T. Kitahara *et al.*, *ibid.* **1979**, 3021. Biosynthesis: B. E. Cross, P. Hendley, *Chem. Commun.* **1975**, 124; C. R. Hutchinson *et al.*, *J. Am. Chem. Soc.* **103**, 2474, 2477 (1981); M. Sunagawa *ibid. J. Antibiot.* **36**, 25 (1983). Antifungal activity: V. Betina *et al.*, *ibid.* **17A**, 93 (1964); anti-HeLa cell effect: *eidem*, *Naturwissenschaften* **49**, 241 (1964). *See also:* W. Keller-Schierlein, "Chemistry of Macrolide Antibiotics" in *Fortschr. Chem. Org. Naturst.* **30**, 313-445 (1973).

Colorless prisms from methanol/ether, mp 204-205°. uv max (ethanol): 215 nm (log ε 4.05). [α]$_D^{22}$ +96 ±2° (c = 1.08 in methanol). LD$_{50}$ i.p. in mice: >200 mg/kg (Haerri).

1375. Brentuximab Vedotin. [914088-09-8] Anti-(human CD30 (antigen)) immunoglobulin G1 (human-mouse monoclonal CAC10 γ$_1$-chain) disulfide with human-mouse monoclonal CAC10 κ-chain, dimer complex with *N*-[[[4-[[*N*-[6-(2,5-dihydro-2,5-dioxo-1*H*-pyrrol-1-yl)-1-oxohexyl]-L-valyl-*N* ^{5}-(aminocarbonyl)-L-ornithyl]amino]phenyl]methoxy]carbonyl]-*N*-methyl-L-valyl-*N*-[(1*S*,-2*R*)-4-[(2*S*)-2-[(1*R*,2*R*)-3-[[(1*R*,2*S*)-2-hydroxy-1-methyl-2-phenylethyl]amino]-1-methoxy-2-methyl-3-oxopropyl]-1-pyrrolidinyl]-2-methoxy-1-[(1*S*)-1-methylpropyl]-4-oxobutyl]-*N*-methyl-L-valinamide; cAC10-vcMMAE; SGN-35; Adcetris. Antibody drug conjugate (ADC) designed to deliver the tubulin polymerization inhibitor, monomethylauristatin E (MMAE), to CD-30 expressing tumor cells. Composed of chimeric anti-CD30 monoclonal antibody (cAC10) conjugated to MMAE via the enzyme-cleavable dipeptide linker, maleimidocaproyl-valine-citrulline. Prepn: P. D. Senter *et al.*, **WO 04010957**; *eidem*, **US 7851437** (2004, 2010 both to Seattle Genetics); and antitumor activity: J. A. Francisco *et al.*, *Blood* **102**, 1458 (2003); S. O. Doronina *et al.*, *Nat. Biotechnol.* **21**, 778 (2003). Stability in serum: R. J. Sanderson *et al.*, *Clin. Cancer Res.* **11**, 843 (2005). Intracellular activation: N. M. Okeley *et al.*, *Clin. Cancer Res.* **16**, 888 (2010). Clinical evaluation in CD30-positive lymphomas: A. Younes *et al.*, *N. Engl. J. Med.* **363** 1812 (2010). Review of development and therapeutic potential: K. V. Foyil, N. L. Bartlett, *Immunotherapy* **3**, 475-485 (2011).

Brentuximab. [775303-41-8] Anti-(human CD30 (antigen)) immunoglobulin G1 (human-mouse monoclonal SGN-30 γ$_1$-chain) disulfide with human-mouse monoclonal SGN-30 κ-chain, dimer; SGN-30. Prepn: J. A. Francisco *et al.*, **WO 0243661**; *eidem*, **US 7090843** (2002, 2006 both to Seattle Genetics); and antitumor activity: A. F. Wahl *et al.*, *Cancer Res.* **62**, 3736 (2002).

THERAP CAT: Antineoplastic.

1376. Brequinar. [96187-53-0] 6-Fluoro-2-(2′-fluoro[1,1′-biphenyl]-4-yl)-3-methyl-4-quinolinecarboxylic acid. $C_{23}H_{15}F_2$-NO_2; mol wt 375.37. C 73.59%, H 4.03%, F 10.12%, N 3.73%, O 8.52%. Inhibitor of dihydroorotate dehydrogenase, the fourth enzyme in the *de novo* pyrimidine biosynthetic pathway; depresses lymphocyte proliferation. Prepn: D. P. Hesson, **EP 133244**; *idem*, **US 4680299** (1985, 1987 both to DuPont). Mechanisms of action: T. L. Forrest *et al.*, *Transplantation* **58**, 920 (1994). Clinical pharmacokinetics: M. de Forni *et al.*, *Eur. J. Cancer* **29A**, 983 (1993). Symposium on pharmacology, toxicology: *Transplant. Proc.* **25**, Suppl. 2, 1-83 (1993). Review of development and clinical experience: L. Makowka *et al.*, *Immunol. Rev.* **136**, 51-70 (1993).

White crystals from DMF + water, mp 315-317°.
Sodium salt. [96201-88-6] DUP-785; NSC-368390. $C_{23}H_{14}F_2$-$NNaO_2$; mol wt 397.36. White solid, mp >360°. Sol in water.

THERAP CAT: Immunosuppressant.

1377. Bretylium Tosylate. [61-75-6] 2-Bromo-*N*-ethyl-*N*,*N*-dimethylbenzenemethanaminium 4-methylbenzenesulfonate (1:1); (*o*-bromobenzyl)ethyldimethylammonium *p*-toluenesulfonate; *N*-ethyl-*N*-*o*-bromobenzyl-*N*,*N*-dimethylammonium tosylate; Bretylate; Bretylol. $C_{18}H_{24}BrNO_3S$; mol wt 414.36. C 52.18%, H 5.84%, Br 19.28%, N 3.38%, O 11.58%, S 7.74%. Prepn: Copp, Stephenson, **US 3038004** (1962 to Burroughs Wellcome). Metabolism: R. Kuntzman *et al.*, *Clin. Pharmacol. Ther.* **11**, 829 (1970). Toxicity study: E. I. Goldenthal, *Toxicol. Appl. Pharmacol.* **18**, 185 (1971). Review of pharmacology: R. H. Heissenbuttel, J. T. Bigger, *Ann. Intern. Med.* **91**, 229-238 (1979); R. J. Lee *et al.*, in *Pharmacological and Biochemical Properties of Drug Substances* vol. **2**, M. E. Goldberg, Ed. (Am. Pharm. Assoc., Washington, DC, 1979) pp 148-164. Comprehensive description: J. E. Carter *et al.*, *Anal. Profiles Drug Subs.* **9**, 71-86 (1980).

Crystalline powder, mp 97-99°. Extremely bitter taste. uv max: 278, 271, 264 nm. Freely sol in water, methanol, ethanol. Practically insol in ether, ethyl acetate, hexane. LD$_{50}$ in mice (mg/kg): 400 orally; 250 i.m. (Goldenthal).

THERAP CAT: Antiarrhythmic (class III).

1378. Brevetoxins. BTX. Structurally unique marine neurotoxins produced by the "red-tide" dinoflagellate *Ptychodiscus brevis* Davis (*Karenia brevis*, formerly *Gymnodinium breve* Davis). Dense growths of these algae have been responsible for massive fish kills, mollusk poisoning and human food poisoning in the Gulf of Mexico and along the Florida coast, *cf. Marine Natural Products*, P. J. Scheuer, Ed. (Academic Press, New York, 1978). Unlike previously isolated dinoflagellate toxins, such as saxitoxin, *q.v.*, which are water-soluble sodium channel blockers, the brevetoxins are lipid-soluble sodium channel activators. Isoln of brevetoxins A, B, and C and structure of B, the major component: Y. Y. Lin *et al.*, *J. Am. Chem. Soc.* **103**, 6773 (1981). Structure of C: J. Golik *et al.*, *Tetrahedron Lett.* **23**, 2535 (1982). Structure of A, the most potent toxin: Y. Shimizu *et al.*, *J. Am. Chem. Soc.* **108**, 514 (1986). Synthetic approaches to brevetoxin B: K. C. Nicolau *et al.*, *Chem. Commun.* **1985**, 1359. Absolute configuration: Y. Shimizu *et al.*, *Chem. Commun.* **22**, 1656 (1987). Biosynthetic study: H. N. Chou, Y. Shimizu, *J. Am. Chem. Soc.* **109**, 2184 (1987). ELISA determn in seawater, shellfish and biological fluids: J. Naar *et al.*, *Environ. Health Perspect.* **110**, 179 (2002). Series of articles on pharmacology of brevetoxins: *Toxicon* **23**, 469-524 (1985); on exposures and health effects of aerosolized toxins: *Environ. Health Perspect.* **113**, 618-657 (2005). Review of chemistry: K. Nakanishi, *ibid.* 473; of pharmacology, toxicokinetics and determn: M. A. Poli, *Recent Adv. Marine Biotechnol.* **7**, 1-3 (2002). For additional information on red tide algae, *see Toxic Dinoflagellate Blooms*, D. L. Taylor, H. H. Seliger, Eds. (Elsevier, New York, 1979) pp 327-354.

Brevetoxin B : R =

Brevetoxin C : R =

Brevetoxin A. GB-1. $C_{49}H_{70}O_{13}$; mol wt 867.09. Fine prisms from acetonitrile, mp 197-199°; 218-220° (double melting point). LC_{100} in guppies: 4 ng/ml (Shimizu).

Brevetoxin B. BTX-B; GB-2; T-34; T-47. $C_{50}H_{70}O_{14}$; mol wt 895.10. Needles from acetonitrile, mp 270° (dec). uv max (methanol): 208 nm (ε 16000, enal). LC_{50} (1 hr) in fresh water "zebra" fish, *Brachydanio rerio:* 16 ng/ml (Lin).

Brevetoxin C. BTX-C. $C_{49}H_{69}ClO_{14}$; mol wt 917.53. uv max (methanol): 208 nm (ε 11300, ene-lactone). LC_{50} (1 hr) in fresh water "zebra" fish: 30 ng/ml (Lin).

USE: As tools in neurochemical research.

1379. Brilliant Blue FCF. [3844-45-9] *N*-Ethyl-*N*-[4-[[4-[ethyl[(3-sulfophenyl)methyl]amino]phenyl](2-sulfophenyl)methylene]-2,5-cyclohexadien-1-ylidene]-3-sulfobenzenemethanaminium inner salt, sodium salt (1:2); FD & C Blue No. 1; C.I. Acid Blue 9; C.I. Food Blue 2; C.I. 42090. $C_{37}H_{34}N_2Na_2O_9S_3$; mol wt 792.84. C 56.05%, H 4.32%, N 3.53%, Na 5.80%, O 18.16%, S 12.13%. Discovered by Sandmeyer in 1896: *Colour Index* vol. 4 (3rd ed., 1971) p 4385. Also prepared as the diammonium salt. Metabolism: S. M. Hess, O. G. Fitzhugh, *J. Pharmacol. Exp. Ther.* **114**, 38 (1955); J. P. Brown *et al.*, *Food Cosmet. Toxicol.* **18**, 1 (1980). Toxicology: E. Gross, *Z. Krebsforsch.* **64**, 287 (1961); W. A. Mannell, H. C. Grice, *J. Pharm. Pharmacol.* **16**, 56 (1964); W. H. Hansen *et al.*, *Toxicol. Appl. Pharmacol.* **8**, 29 (1966). Chronic toxicity and carcinogenicity: J. F. Borzelleca *et al.*, *Food Chem. Toxicol.* **28**, 221 (1990). Review of carcinogenicity studies: *IARC Monographs* **16**, 171-186 (1978).

Reddish-violet powder or granules with a metallic lustre. Absorption max: 630 nm. Sol in water, ethanol. Practically insol in vegetable oils. Pale amber soln in conc H_2SO_4, changing to yellow then greenish blue on dilution. LD_{50} s.c. in mice: 4.6 g/kg (Gross).

USE: Colorant in food, drugs and cosmetics. Biological stain; textile dye; wood stain; indicator.

1380. Brilliant Green. [633-03-4] *N*-[4-[[4-(Diethylamino)-phenyl]phenylmethylene]-2,5-cyclohexadien-1-ylidene]-*N*-ethylethanaminium sulfate (1:1); C.I. Basic Green 1; C.I. 42040; malachite green G; emerald green; diamond green G; fast green J; solid green. $C_{27}H_{34}N_2O_4S$; mol wt 482.64. C 67.19%, H 7.10%, N 5.80%, O 13.26%, S 6.64%. Prepn: Doebner, *Ber.* **13**, 2222 (1880); Fischer, *ibid.* **14**, 2521 (1881); *Colour Index* vol. 4 (3rd ed., 1971) p 4382. Toxicity: Anderson, *Proc. Soc. Exp. Biol. Med.* **31**, 825 (1934). Antibacterial activity: O. H. Paetzold, *Arch. Klin. Exp. Dermatol.* **224**, 90 (1966); *idem*, *Arch. Dermatol. Forsch.* **243**, 1 (1972). Review: *H. J. Conn's Biological Stains*, R. D. Lillie, Ed. (Williams & Wilkins, Baltimore, 9th ed., 1977) pp 251-252, 580.

Minute, glistening, golden crystals. Soluble in water or alcohol with green color. Absorption max: 623 nm. Soln changes color from yellow to green at pH 0.0 to 2.6. LD_{100} i.v. in mice: 3 mg/kg (Anderson).

USE: Dyeing silk, wool, leather, jute and cotton yellowish-green; manuf green ink; biological stain; indicator.

THERAP CAT: Antiseptic.

THERAP CAT (VET): Antiseptic for external and internal (oral) use. In wounds and scours.

1381. Brimonidine. [59803-98-4] 5-Bromo-*N*-(4,5-dihydro-1*H*-imidazol-2-yl)-6-quinoxalinamine; 5-bromo-6-(2-imidazolin-2-ylamino)quinoxaline; UK-14304; AGN-190342. $C_{11}H_{10}BrN_5$; mol wt 292.14. C 45.23%, H 3.45%, Br 27.35%, N 23.97%. α_2-Adrenoceptor agonist. Prepn: J. C. Danielewicz *et al.*, *DE 2309160*; *eidem*, *US 3890319* (1973, 1975 both to Pfizer). HPLC determn in serum and aqueous humour of the eye: N. K. Karamanos *et al.*, *Biomed. Chromatogr.* **13**, 86 (1999). Clinical study in control of intraocular pressure: R. David *et al.*, *Arch. Ophthalmol.* **111**, 1387 (1993); L. J. Katz *et al.*, *Am. J. Ophthalmol.* **127**, 20 (1999). Review of pharmacology and mechanisms of action: J. C. Adkins, J. A. Balfour, *Drugs Aging* **12**, 225-241 (1998); of pharmacotherapeutic profile: L. B. Cantor, *Expert Opin. Pharmacother.* **1**, 815-834 (2000).

Yellow crystals from ethanol, mp 252°.

D-Tartrate. [70359-46-5] UK-14304-18; AGN-190342-LF; Alphagan. $C_{11}H_{10}BrN_5 \cdot C_4H_6O_6$; mol wt 442.23. Off-white to pale yellow powder. Crystals from acetone, mp 207.5°. Soly in water: 0.6 mg/ml.

THERAP CAT: Antiglaucoma.

1382. Brinzolamide. [138890-62-7] (4*R*)-4-(Ethylamino)-3,4-dihydro-2-(3-methoxypropyl)-2*H*-thieno[3,2-*e*]-1,2-thiazine-6-sulfonamide 1,1-dioxide; AL-4682; Azopt. $C_{12}H_{21}N_3O_5S_3$; mol wt 383.50. C 37.58%, H 5.52%, N 10.96%, O 20.86%, S 25.08%. Carbonic anhydrase inhibitor. Prepn: T. R. Dean *et al.*, *WO 9115486*; *eidem*, *US 5378703* (1991, 1995 both to Alcon). Clinical trial: L. H. Silver, *Am. J. Ophthalmol.* **126**, 400 (1998).

White or almost white powder. Slightly sol in alc, methanol. Insol in water.

Hydrochloride. [150937-43-2] $C_{12}H_{21}N_3O_5S_3 \cdot HCl$. Crystals from methanol + methylene chloride, mp 175-177°. $[\alpha]_D$ +10.35° (c = 1 in water).

THERAP CAT: Antiglaucoma agent.

1383. Brivanib. [649735-46-6] $(2R)$-1-[[4-[(4-Fluoro-2-methyl-1H-indol-5-yl)oxy]-5-methylpyrrolo[2,1-f][1,2,4]triazin-6-yl]oxy]-2-propanol; BMS-540215 . $C_{19}H_{19}FN_4O_3$; mol wt 370.38. C 61.61%, H 5.17%, F 5.13%, N 15.13%, O 12.96%. Dual vascular endothelial growth factor receptor-2 (VEGFR-2) and fibroblast growth factor receptor-1 (FGFR-1) tyrosine kinase inhibitor. L-alanine ester prodrug hydrolyzed to the active moiety by cytochrome p450 (CYP3A4) in the liver. Prepn: R. Bhide *et al.*, **WO 04009784**; *eidem*, **US 6869952** (2004, 2005 both to BMS); and structure-activity study: R. S. Bhide *et al.*, *J. Med. Chem.* **49**, 2143 (2006). Prepn and bioavailability of ester prodrug: Z. Cai *et al.*, *ibid.* **51**, 1976 (2008). Metabolism and distribution: J. Gong *et al.*, *Drug Metab. Dispos.* **39**, 891 (2011). Pharmacokinetics in patients with advanced tumors: T. Mekhail *et al.*, *ibid.* **38**, 1962 (2010). Antiangiogenic activity in human tumor xenograft models: R. S. Bhide *et al.*, *Mol. Cancer Ther.* **9**, 369 (2010). Clinical evaluation in hepatocellular carcinoma: J.-W. Park *et al.*, *Clin. Cancer Res.* **17**, 1973 (2011). Review of pharmacology: W. C. M. Dempke, R. Zippel, *Anticancer Res.* **30**, 4477-4484 (2010); of development and clinical experience I. Diaz-Padilla, L. L. Siu, *Expert Opin. Invest. Drugs* **20**, 577-586 (2011).

Off-white solid, mp 208-210°. pKa 2.8. Soly in water: <1 μg/ml (pH 6.5).

L-Alanine ester. [649735-63-7] L-Alanine $(1R)$-2-[[4-[(4-fluoro-2-methyl-1H-indol-5-yl)oxy]-5-methylpyrrolo[2,1-f][1,2,4]triazin-6-yl]oxy]-1-methylethyl ester; $(1R,2S)$-2-aminopropionic acid 2-[4-(4-fluoro-2-methyl-1H-indol-5-yloxy)-5-methylpyrrolo[2,1-f][1,2,-4]triazin-6-yloxy]-1-methylethyl ester; brivanib alaninate; BMS-582664. $C_{22}H_{24}FN_5O_4$; mol wt 441.46. White solid, mp 136-142°.

THERAP CAT: Antineoplastic.

1384. Brivaracetam. [357336-20-0] $(\alpha S,4R)$-α-Ethyl-2-oxo-4-propyl-1-pyrrolidineacetamide; $(2S)$-2-[$(4R)$-2-oxo-4-propylpyr-rolidinyl]butanamide; UCB-34714; Rikelta. $C_{11}H_{20}N_2O_2$; mol wt 212.29. C 62.24%, H 9.50%, N 13.20%, O 15.07%. Orally active synaptic vesicle protein 2 (SV2A) ligand; analog of levetiracetam, *q.v.* Prepn: E. Differding *et al.*, **WO 0162726**; *eidem*, **US 6911461** (2001, 2005 both to UCB); B. M. Kenda *et al.*, *J. Med. Chem.* **47**, 530 (2004). Clinical trial in epilepsy: J. A. French *et al.*, *Neurology* **75**, 519 (2010). Review of pharmacology and clinical experience: B. Malawska, K. Kulig, *Curr. Opin. Investig. Drugs* **6**, 740-746 (2005).

Colorless, monoclinic crystals from isopropyl ether, mp 76.38°. $[\alpha]_D^{25}$ −60.57° (c = 1 in methanol). Crystal density: 1.093. Log P (octanol/water, pH 7.4 PBS): 1.00. Soly in water: 681 mg/ml.

THERAP CAT: Anticonvulsant; in treatment of progressive myoclonic epilepsies.

1385. Brivudine. [69304-47-8] 5-[(1E)-2-Bromoethenyl]-2'-deoxyuridine; (E)-5-(2-bromovinyl)-2'-deoxyuridine; brivudin; BVDU; Brivex; Brivirac; Nervinex; Zecovir; Zostex. $C_{11}H_{13}$-BrN_2O_5; mol wt 333.14. C 39.66%, H 3.93%, Br 23.99%, N 8.41%, O 24.01%. Analog of thymidine, *q.v.*, with selective activity against herpes simplex virus type 1 and varicella-zoster virus. Prepn: A. S. Jones *et al.*, **DE 2915254**; *eidem*, **US 4424211** (1979, 1984 both to University of Birmingham and Rega Institut); and antiviral activity: E. De Clercq *et al*, *Proc. Natl. Acad. Sci. USA* **76**, 2947 (1979). Mechanism of action studies: H. S. Allaudeen *et al.*, *ibid.* **78**, 2698 (1981); J. Balzarini, E. De Clercq, *Methods Find. Exp. Clin. Pharmacol.* **11**, 379 (1989). Cytotoxic properties vs viral tumor cells: C. Grignet-Debrus *et al.*, *Cancer Gene Ther.* **7**, 215 (2000). CE determn in plasma and urine: J. Olgemöller *et al.*, *J. Chromatogr. B* **726**, 261 (1999). Clinical evaluation in herpetic keratitis: P. C. Maudgal, E. De Clercq, *Curr. Eye Res.* **10**, Suppl., 193 (1991). Clinical comparison with acyclovir, *q.v.*, in herpes zoster: S. W. Wassilew *et al.*, *Antiviral Res.* **59**, 49, 57 (2003). Review of pharmacology and clinical efficacy in herpes zoster: S. J. Keam *et al.*, *Drugs* **64**, 2091-2097 (2004); of antiviral activity, mechanism of action, and clinical efficacy: E. De Clercq, *Med. Res. Rev.* **25**, 1-20 (2005).

White needles from methanol-water, mp 123-125° (dec). uv max: 253, 295 nm (ε 13100, 10300).

THERAP CAT: Antiviral.

1386. Brodifacoum. [56073-10-0] 3-[3-(4'-Bromo[1,1'-bi-phenyl]-4-yl)-1,2,3,4-tetrahydro-1-naphthalenyl]-4-hydroxy-2H-1-benzopyran-2-one; 3-[3-(4'-bromobiphenyl-4-yl)-1,2,3,4-tetrahy-dro-1-naphthyl]-4-hydroxycoumarin; PP-581; WBA-8119; Talon. $C_{31}H_{23}BrO_3$; mol wt 523.43. C 71.13%, H 4.43%, Br 15.27%, O 9.17%. Coumarin analog. Prepn: M. R. Hadler, R. S. Shadbolt, **DE 2424806**; *eidem*, **US 3957824** (1975, 1976 to Ward, Blenkinsop & Co.); *eidem; J. Chem. Soc. Perkin Trans. 1* **1976**, 1190. Anticoagulant activity: M. R. Hadler, R. S. Shadbolt, *Nature* **253**, 275 (1975). Field trials: B. D. Rennison, A. C. Dubock, *J. Hyg.* **80**, 77 (1978); F. P. Rowe *et al.*, *ibid.* **81**, 197 (1978). Efficacy and toxicity: L. Yuet-Ming, *Malays. Agric. J.* **52**, 1 (1980).

Off-white powder, mp 228-230°. Insol in water. Slightly sol in alc, benzene; sol in acetone, chloroform. LD_{50} orally in male, female field rats: 0.16, 0.18 mg/kg (Yuet-Ming).

USE: Rodenticide.

1387. Brodimoprim. [56518-41-3] 5-[(4-Bromo-3,5-dimeth-oxyphenyl)methyl]-2,4-pyrimidinediamine; 2,4-diamino-5-(4-bro-mo-3,5-dimethoxybenzyl)pyrimidine; Ro-10-5970. $C_{13}H_{15}$-

BrN_4O_2; mol wt 339.19. C 46.03%, H 4.46%, Br 23.56%, N 16.52%, O 9.43%. Dihydrofolate reductase inhibitor; structural analog of trimethoprim, *q.v.* Prepn: M. Hoffer, I. Kompis, **DE 2452889** (1975 to Hoffmann-La Roche), *C.A.* **83**, 97361 (1975); I. Kompis, **US 4024145** (1977 to Hoffmann-La Roche); I. Kompis, A. Wick, *Helv. Chim. Acta* **60**, 3025 (1977). Antibacterial activity: G. Giammanco *et al.*, *Drugs Exp. Clin. Res.* **9**, 721 (1983). Comparison with trimethoprim as inhibitor of dihydrofolate reductase: R. L. Then *et al.*, *Rev. Infect. Dis.* **4**, 372 (1982); R. L. Then, F. Hermann, *Chemotherapy (Basel)* **30**, 18 (1984). Antimycobacterial activity *in vitro*: J. K. Seydel *et al.*, *ibid.* **29**, 249 (1983). Pharmacokinetics in human serum, skin blister fluid: T. Kalager *et al.*, *Chemotherapy (Basel)* **31**, 405 (1985). Clinical evaluation in respiratory infections: H. A. Salmi *et al.*, *Drugs Exp. Clin. Res.* **12**, 349 (1986); in urinary tract infections: F. Scaglione *et al.*, *Int. J. Clin. Pharmacol. Res.* **15**, 121 (1995); in bacterial cystitis: E. V. Cosmi *et al.*, *Eur. J. Obstet. Gynecol. Reprod. Biol.* **64**, 207 (1996).

Crystals from methanol, mp 225-228°. pKa 7.15.

THERAP CAT: Antibacterial.

1388.　Bromacil. [314-40-9] 5-Bromo-6-methyl-3-(1-methylpropyl)-2,4(1*H*,3*H*)-pyrimidinedione; 5-bromo-3-*sec*-butyl-6-methyluracil; 5-bromo-6-methyl-3-(1-methylpropyl)uracil; Du Pont Herbicide 976; Hyvar; Uragon; Urox B. $C_9H_{13}BrN_2O_2$; mol wt 261.12. C 41.40%, H 5.02%, Br 30.60%, N 10.73%, O 12.25%. Prepn: H. M. Loux, **US 3235357** (1966 to du Pont). Mode of action: C. E. Hoffman, *Pestic. Chem., Proc. 2nd Int. Congr. Pestic.* **5**, A. S. Tahori, Ed, (Gordon and Breach Science Publishers, New York, 1972) pp 65-85. Toxicology: H. Sherman, A. M. Kaplan, *Toxicol. Appl. Pharmacol.* **34**, 189 (1975). *Review:* Pease, Deye, *Anal. Methods Pestic. Plant Growth Regul. Food Addit.* **5**, 335 (1967).

White crystalline solid, mp 157.5-160°. Vapor pressure at 100°: 8×10^{-4} mm Hg. Soly in water at 20°: 815 mg/l. Moderately sol in strong aq bases, acetone, acetonitrile, ethanol. LD_{50} orally in rats: 5200 mg/kg (Sherman, Kaplan).

Caution: Potential symptoms of overexposure are irritation of eyes, skin and upper respiratory system. *See NIOSH Pocket Guide to Chemical Hazards* (DHHS/NIOSH 97-140, 1997) p 32.

USE: Herbicide.

1389.　Bromadiolone. [28772-56-7] 3-[3-(4'-Bromo[1,1'-biphenyl]-4-yl)-3-hydroxy-1-phenylpropyl]-4-hydroxy-2*H*-1-benzopyran-2-one; 3-[α-[*p*-(*p*-bromophenyl)-β-hydroxyphenethyl]benzyl]-4-hydroxycoumarin; LM-637; Bromone; Maki; Rozol; Super-Caid. $C_{30}H_{23}BrO_4$; mol wt 527.41. C 68.32%, H 4.40%, Br 15.15%, O 12.13%. Anticoagulant rodenticide. Prepn: E. Boschetti *et al.*, **DE 1959317**; *eidem*, **US 3764693** (1970, 1973 both to Lipha); *eidem*, *Chim. Ther.* **7**, 20 (1972). Toxicological studies: M. Grand, *Phytiatr.-Phytopharm.* **25**, 69 (1976). Evaluation of efficacy: R. Redfern, J. E. Gill, *J. Hyg.* **84**, 263 (1980); of risk to non-target species: P. Giraudoux *et al.*, *Environ. Res.* **102**, 291 (2006). HPLC determn in animal tissues: K. Hunter *et al.*, *J. Chromatogr.* **435**, 83 (1988); in blood: M.-C. Jin *et al.*, *Forensic Sci. Int.* **171**, 52 (2007).

White to off-white powder, mp 200-210°. uv max (ethanol): 260 nm ($E_{1cm}^{1\%}$ 538-582). pKa (21°) 4.04. Soly at 20-25° (g/l): dimethylformamide 730.0; ethyl acetate 25.0; acetone 22.3; chloroform 10.1; ethanol 8.2; methanol 5.6; ethyl ether 3.7; hexane 0.2; water 0.019. LD_{50} in rats, mice (mg/kg): 1.125, 1.75 orally (Grand).

USE: Rodenticide.

1390.　Bromal. [115-17-3] 2,2,2-Tribromoacetaldehyde. C_2-HBr_3O; mol wt 280.74. C 8.56%, H 0.36%, Br 85.39%, O 5.70%. Br_3CCHO. Preparation from ethanol and bromine: Löwig, *Ann.* **3**, 288 (1832); from chloral and a bromide: Müller, **US 2053964** (1936 to Winthrop); from paraldehyde and bromine: Long, Howard, *Org. Synth.* **17**, 18 (1937).

Yellowish, oily liquid; forms with water bromal hydrate which is solid at temps below 50°. d 2.66. bp ~174° with decompn. Sol in water, alcohol or ether.

1391.　Bromal Hydrate. [507-42-6] 2,2,2-Tribromo-1,1-ethanediol; tribromoacetaldehyde hydrate. $C_2H_3Br_3O_2$; mol wt 298.76. C 8.04%, H 1.01%, Br 80.24%, O 10.71%. $Br_3CCH(OH)_2$. Prepd from bromal and water: Löwig, *Ann.* **3**, 288 (1832). Structure: Jain, Soundararajan, *Tetrahedron* **20**, 1589 (1964).

Deliquescent crystals, mp 53.5°. Odor of chloral and pungent taste. Dipole moment in benzene, 2.56D. Sol in water, alcohol, chloroform, ether, glycerol. *Keep tightly closed in a cool place.*

1392.　Bromazepam. [1812-30-2] 7-Bromo-1,3-dihydro-5-(2-pyridinyl)-2*H*-1,4-benzodiazepin-2-one; 7-bromo-5-(2-pyridyl)-3*H*-1,4-benzodiazepin-2(1*H*)-one; Ro-5-3350; Compendium; Creosedin; Durazanil; Lectopam; Lexomil; Lexotan; Lexotanil; Normoc. $C_{14}H_{10}BrN_3O$; mol wt 316.16. C 53.19%, H 3.19%, Br 25.27%, N 13.29%, O 5.06%. Prepn: Berger *et al.*, **BE 619101**; *eidem*, Fryer *et al.*, **US 3100770** (1962, 1963 to Hoffmann-La Roche); *eidem*, *J. Pharm. Sci.* **53**, 264 (1964); **US 3182065**; **US 3182067** (both 1965 to Hoffmann-La Roche). Pharmacology: Korol, Brown, *Pharmacology* **1**, 115 (1968). Metabolism: M. A. Schwartz *et al.*, *J. Pharm. Sci.* **62**, 1776 (1973); *Drug Metab. Dispos.* **2**, 31 (1974). Evaluation as pre-anesthesia medication: P. Chalmers, J. N. Horton, *Anaesthesia* **39**, 370 (1984). Multicenter clinical comparison with lorazepam, *q.v.*: G. J. Cordingley *et al.*, *Curr. Med. Res. Opin.* **9**, 505 (1985). Evaluation of adverse effects and withdrawal reactions: R. Fontaine *et al.*, *Psychopharmacol. Bull.* **21**, 91 (1985). Toxicity: E. I. Goldenthal, *Toxicol. Appl. Pharmacol.* **18**, 185 (1971). Comprehensive description: M. M. Hassan, M. A. Abounassif, *Anal. Profiles Drug Subs.* **16**, 1-51 (1987).

Colorless prisms from acetone, mp 237-238.5° (dec). LD_{50} orally in rats: 3050 ±405 mg/kg (Goldenthal).

Note: This is a controlled substance (depressant): **21 CFR,** 1308.14.

THERAP CAT: Anxiolytic.

1393. Bromcresol Green. [76-60-8] 4,4'-(1,1-Dioxido-3*H*-2,1-benzoxathiol-3-ylidene)bis[2,6-dibromo-3-methylphenol]; 4,4'-(3*H*-2,1-benzoxathiol-3-ylidene)bis[2,6-dibromo-3-methylphenol] *S,S*-dioxide; α,α-bis(3,5-dibromo-4-hydroxy-*o*-tolyl)-α-hydroxytoluenesulfonic acid, γ-sultone; 3,3',5,5'-tetrabromo-*m*-cresolsulfonphthalein; bromocresol green. $C_{21}H_{14}Br_4O_5S$; mol wt 698.01. C 36.14%, H 2.02%, Br 45.79%, O 11.46%, S 4.59%. Prepd by adding bromine to a suspension of *m*-cresolsulfonphthalein in glacial acetic acid: Clark, Lubs, *J. Wash. Acad. Sci.* **5**, 610 (1915); **6**, 481 (1916); *J. Bacteriol.* **2**, 110 (1917); Cohen, *Biochem. J.* **16**, 31 (1922); **17**, 535 (1923); Cohen, *Public Health Rep.* **38**, 814 (1923); **41**, 3051 (1926); *Proc. Soc. Exp. Biol. Med.* **20**, 124 (1922); Orndorff, Purdy, *J. Am. Chem. Soc.* **48**, 2216 (1926).

Minute, slightly yellow crystals from acetic acid, mp 218-219°. Sparingly sol in water. Readily sol in alcohol, ether, ethyl acetate. Fairly sol in benzene. Very sensitive to alkalies, tap water being sufficiently alkaline to give the characteristic blue-green color. pKa = 4.7. pH 3.8 yellow; pH 5.4 blue-green. To prepare a soln for use as pH indicator, dissolve 0.10 g in 7.15 ml *N*/50 NaOH and dil with water to 250 ml. To prepare a soln for use as indicator in volumetric work, dissolve 0.1 g in 250 ml alc.

Sodium salt. [62625-32-5] $C_{21}H_{13}Br_4NaO_5S$; mol wt 720.00.

USE: As pH indicator. As selective inhibitor of anion transport in kidney physiology. In lithography.

1394. Bromcresol Purple. [115-40-2] 4,4'-(1,1-Dioxido-3*H*-2,1-benzoxathiol-3-ylidene)bis[2-bromo-6-methylphenol]; 4,4'-(3*H*-2,1-benzoxathiol-3-ylidene)bis[2-bromo-6-methylphenol] *S,S*-dioxide; α-(5-bromo-4-hydroxy-*m*-tolyl)-α-(3-bromo-5-methyl-4-oxo-2,5-cyclohexadien-1-ylidene)-*o*-toluenesulfonic acid; 5,5'-dibromo-*o*-cresolsulfonphthalein; bromocresol purple. $C_{21}H_{16}Br_2O_5S$; mol wt 540.22. C 46.69%, H 2.99%, Br 29.58%, O 14.81%, S 5.93%. Prepd by treating *o*-cresol red with bromine in glacial acetic acid: *See refs under* Bromcresol Green.

Minute, slightly yellow crystals, mp 241-242°. Practically insol in water. Sol in alcohol, dil alkalies. pKa 6.3. pH 5.2 yellow; pH 6.8 purple. To prepare a soln for use as pH indicator, dissolve 0.10 g in 9.25 ml *N*/50 NaOH and dil with water to 250 ml. To prepare a soln for use as indicator in volumetric work, dissolve 0.05 g in 250 ml alc.

Sodium salt. [62625-30-3] $C_{21}H_{15}Br_2NaO_5S$; mol wt 562.20.

USE: As pH indicator, especially in microbiological culture media. In instrumentation systems for ammonia detection. Vital stain in differentiation between live and dead yeast. Analytical stain for carbonic anhydrase isoenzymes.

1395. Bromelain. Ananase; Extranase; Traumanase. Mixture of proteolytic enzymes found in pineapple, *Ananas comosus* L., and other plants of the *Bromeliaceae*. First isolated from pineapple fruit, closely related but structurally different proteases were subsequently discovered in stem tissue. Commercial extracts prepared from pineapple stem contain primarily stem bromelain wih minor components of ananain and fruit bromelain. Isoln of fruit bromelain: Marcano, *Bull. Pharm.* **5**, 77 (1891); Chittenden, *Trans. Conn. Acad. Sci.* **8**, 281 (1892). Discovery in stem tissue: R. M. Heinicke, W. A. Gortner, *Econ. Bot.* **11**, 225 (1957). Commercial prodn of stem extract: R. M. Heinicke, US 3002891 (1961 to Pineapple Res. Inst.). Purification and characterization of fruit and stem bromelains: T. Murachi *et al.*, *Biochemistry* **3**, 48 (1964); S. Ota *et al.*, *ibid.* 180. Review of purification and properties: T. Murachi, *Methods Enzymol.* **19**, 273-284 (1970); W. M. Cooreman *et al.*, *Pharm. Acta Helv.* **51**, 73-97 (1976). Isolation of ananain from pineapple stem extract: A. D. Rowan *et al.*, *Arch. Biochem. Biophys.* **267**, 262 (1988). Amino acid sequence of stem bromelain: A. Ritonja *et al.*, *FEBS Lett.* **247**, 419 (1989). Constituents, activity, and stability of bromelain extracts: L. P. Hale *et al.*, *Int. Immunopharmacol.* **5**, 783 (2005). Review of biochemistry and medicinal use: H. R. Maurer, *Cell. Mol. Life Sci.* **58**, 1234-1245 (2001). Review of clinical experience with medicinal extract: *Altern. Med. Rev.* **15**, 361-368 (2010).

Fruit bromelain. [9001-00-7] Bromelin; juice bromelain; E.C. 3.4.22.33. Nonglycosylated, acidic peptide; mol wt 23 kDa. Isoelectric point: 4.6. uv max: 280 nm ($A_{1cm}^{1\%}$ 19.2). Unlike papain, fruit bromelain does not disappear as the fruit ripens.

Stem bromelain. [37189-34-7] E.C. 3.4.22.32. Major proteinase in pineapple stem. Basic glycoprotein consisting of a single chain with 212 residues; mol wt 23.8 kDa. Isoelectric point: 9.55. uv max: 280 nm ($A_{1cm}^{1\%}$ 20.1).

Ananain. [119129-70-3] E.C. 3.4.22.31. Cysteine proteinase occuring in pineapple stem. Non-glycosylated peptide with 216 residues, mol wt 23.4 kDa. Isoelectric point: >10.

USE: As meat tenderizer.

THERAP CAT: Digestive aid; debriding agent; to reduce pain and swelling following injury; in treatment of osteoarthritis.

1396. Bromethalin. [63333-35-7] *N*-Methyl-2,4-dinitro-*N*-(2,4,6-tribromophenyl)-6-(trifluoromethyl)benzenamine; EL-614; Cykill; Fastrac; Gladiator; Rampage; Talpirid. $C_{14}H_7Br_3F_3N_3O_4$; mol wt 577.93. C 29.10%, H 1.22%, Br 41.48%, F 9.86%, N 7.27%, O 11.07%. Rodenticide which acts by uncoupling oxidative phosphorylation. Single-feed toxicant, effective against warfarin-resistant strains of mice, rats. Prepn and identification as rodenticide: B. A. Dreikorn, US 4187318 (1980 to Eli Lilly). HPLC determn in commercial formulations: M. Z. Mesmer, R. A. Flurer, *J. Chromatogr. Sci.* **39**, 49 (2001). Toxicology and activity: B. A. Dreikorn *et al.*, *Proc. Br. Crop Prot. Conf. - Pests Dis.* **1979**, 491. Review of structure-activity and mode of action: B. A. Dreikorn, G. O. P. O'Doherty, *ACS Symp. Ser.* **255**, 45-63 (1984); of toxic symptoms in dogs and cats: E. Dunayer *Vet. Med.* **98**, 732-736 (2003).

Pale yellow odorless crystals from ethanol, mp 150-151°. *Poisonous.* Sol in chloroform, acetone; moderately sol in aromatic hydrocarbons. Insol in water. LD_{50} in mice, rats, cats, dogs (mg/kg): 2, 5, 2, 5 orally (Dreikorn).

USE: Rodenticide.

1397. Bromfenac. [91714-94-2] 2-Amino-3-(4-bromobenzoyl)benzeneacetic acid; AHR-10282. $C_{15}H_{12}BrNO_3$; mol wt 334.17. C 53.91%, H 3.62%, Br 23.91%, N 4.19%, O 14.36%. Prostaglandin synthetase inhibitor; analog of amfenac, *q.v.* Prepn: D. A. Walsh *et al.*, *J. Med. Chem.* **27**, 1379 (1984). *See also*: R. G. Poser, US 4683242 (1987 to A. H. Robins). HPLC determn in plasma: M. A. Osman *et al.*, *J. Chromatogr.* **489**, 452 (1989). Pharmacology: L. F. Sancilio *et al.*, *Arzneim.-Forsch.* **37**, 513 (1987). Clinical pharmacokinetics: P. Högger, P. Rohdewald, *ibid.* **43**, 1114 (1993). Evaluation of hepatotoxicity: R. J. Fontana *et al.*, *Liver Transpl.*

Surg. **5**, 480 (1999). Clinical trial in allergic conjunctivitis: M. Mi-yake-Kashima *et al.*, *Jpn. J. Ophthalmol.* **48**, 587 (2004); for post-operative ocular pain and inflammation: E. D. Donnenfeld *et al.*, *Ophthalmology* **114**, 1653 (2007).

Monosodium salt sesquihydrate. [120638-55-3]; [91714-93-1] (monosodium salt). AHR-10282B; Bromday; Xibrom; Yellox. C_{15}-$H_{11}BrNNaO_3.1\frac{1}{2}H_2O$; mol wt 383.17. Bright orange-yellow crys-talline powder, mp 284-286° (dec). pKa 4.29. Sol in water, meth-anol, dilute base. Insol in chloroform, dilute acid.

THERAP CAT: Anti-inflammatory (ophthalmic).

1398. Bromhexine. [3572-43-8] 2-Amino-3,5-dibromo-*N*-cyclohexyl-*N*-methylbenzenemethanamine; 3,5-dibromo-*N*$^\alpha$-cyclo-hexyl-*N*$^\alpha$-methyltoluene-α,2-diamine; *N*-cyclohexyl-*N*-methyl-2-(2-amino-3,5-dibromo)benzylammonium; *N*-(2-amino-3,5-dibromo-benzyl)-*N*-methylcyclohexylammonium. $C_{14}H_{20}Br_2N_2$; mol wt 376.14. C 44.71%, H 5.36%, Br 42.49%, N 7.45%. Prepn: Keck, *Ann.* **662**, 171 (1963); K. Thomae, **BE 625022** (1963), *C.A.* **61**, 5564 (1964). Pharmacology: R. Engelhorn, S. Püschmann, *Arzneim.-Forsch.* **13**, 474 (1963). Metabolism: R. Jauch *et al.*, *ibid.* **25**, 1954 (1975). Toxicity: Boyd, Sheppard, *Arch. Int. Pharmacodyn. Ther.* **163**, 284 (1966).

Crystals from abs ethanol, dec 237.5-238°. One gram dissolves in 250 ml of water or 10% ethanol. Approx LD_{50} orally in rabbits: >10 g/kg (Boyd, Sheppard).

Hydrochloride. [611-75-6] NA-274; Auxit; Bisolvon; Ophtosol; Quentan. $C_{14}H_{20}Br_2N_2.HCl$; mol wt 412.59.

THERAP CAT: Expectorant; mucolytic.

THERAP CAT (VET): Expectorant.

1399. Bromic Acid. [7789-31-3] $BrHO_3$; mol wt 128.91. Br 61.98%, H 0.78%, O 37.23%. $HBrO_3$. Prepd from $Ba(BrO_3)_2$ and H_2SO_4: Burchard, *Z. Phys. Chem.* **2**, 814 (1888); Schmeisser in *Handbook of Preparative Inorganic Chemistry* **vol. 1**, G. Brauer, Ed. (Academic Press, 2nd ed, 1963) p 315.

Known in aq soln only. *Keep refrigerated.* Colorless soln, turns yellow on standing at room temp, even in the dark. Careful evapn *in vacuo* at $-12°$ yields about a 50% soln. Dec on heating to 100°. Can be dil with cold water.

Caution: Highly irritating to skin, eyes, mucous membranes.

USE: Oxidizing agent.

1400. Bromindione. [1146-98-1] 2-(4-Bromophenyl)-1*H*-in-dene-1,3(2*H*)-dione; 2-(4-bromophenyl)-1,3-dioxohydrindene; HL-255; MG-2555; Fluidane; Halinone. $C_{15}H_9BrO_2$; mol wt 301.14. C 59.83%, H 3.01%, Br 26.53%, O 10.63%. Prepn: Cavallini *et al.*, *Farmaco Ed. Sci.* **10**, 710 (1955); Freedman *et al.*, **US 2847474** (1958 to U.S. Vitamin).

Crystals from ligroin, mp 137-139°.

THERAP CAT: Anticoagulant.

1401. Bromine. [7726-95-6] Br; at. wt 79.904; at. no. 35; valences 1, 3, 5, 7. A halogen; Group VIIA (17). Does not exist as elemental state, Br, in nature. Occurs as diatomic molecule, Br_2. Abundance in igneous rock: 1.6×10^{-4}% by wt; in seawater: 0.0065% by wt. Extracted commercially from natural brines (salt lakes) and seawater. Naturally occurring stable isotopes (mass num-bers): 79 (50.69%), 81 (49.31%); known artificial radioactive iso-topes: 69-76, 77 (longest-lived known isotope, $T_{\frac{1}{2}}$ 57.036 hr; EC decay), 78, 79m, 80, 80m, 82-94. Discovery: A. J. Balard, *Ann. Chim. Phys.* **32**, 337 (1826). Books: *Bromine and its Compounds*, Z. E. Jolles, Ed. (E. Benn, London, 1966) 940 pp; *Bromine Com-pounds: Chemistry and Applications*, D. Price *et al.*, Eds. (Elsevier, Amsterdam, 1988) 422 pp. *Reviews: MTP Int. Rev. Sci.: Inorg. Chem., Ser. One* vol. 3, V. Gutmann, Ed. (Butterworths, London, 1972); A. J. Downs, C. J. Adams, "Chlorine, Bromine, Iodine and Astatine" in *Comprehensive Inorganic Chemistry* **vol. 2**, J. C. Bailar, Jr. *et al.*, Eds. (Pergamon Press, Oxford, 1973) pp 1107-1573; *Chem-istry of the Elements*, N. N. Greenwood, A. Earnshaw, Eds. (Perga-mon Press, New York, 1984) pp 920-1041; P. F. Jackish in *Kirk-Othmer Encyclopedia of Chemical Technology* **vol. 4** (Wiley-Inter-science, New York, 4th ed., 1992) pp 536-560.

Dark reddish-brown, volatile, mobile, diatomic liquid; suffocating odor; vaporizes rapidly at room temp. Only nonmetallic element liquid at standard conditions. Nonflammable, but may ignite com-bustibles on contact. *Poisonous, corrosive.* mp $-7.25°$ (265.90 K); bp 59.47° *(JANAF Thermochemical Tables)*; 58.78° (Mellor's Suppl. II, Part I, "The Halogens"); d_4^{25} 3.1023; crit temp: 315°; crit pressure: 102 atm. Heat capacity at constant pressure (liq, 25°) 18.089 cal/mole deg: Hildenbrand *et al.*, *J. Am. Chem. Soc.* **80**, 4129 (1958). Vapor pressure data: A. N. Nesmeyanov, *Vapor Pressure of the Chemical Elements*, R. Gary, Ed. (Elsevier, New York, 1963) pp 354-58. Total soly in water (25°): 0.2141 moles/l with formation of 0.00115 moles/l of HOBr; freely sol in alc, ether, $CHCl_3$, CCl_4, CS_2, concd HCl, aq solns of bromides. Oxidizing agent; less reactive than chlorine; E^0 (aq) $\frac{1}{2}Br_2/Br^-$ 1.065 V; dissociation energy (25°): 46.072 kcal. *Keep sealed or glass-stoppered.*

Caution: Potential symptoms of overexposure are dizziness, head-ache; lacrimation, epistaxis; coughing, feeling of oppression, pul-monary edema and pneumonia; abdominal pain, diarrhea; measle-like eruptions; direct contact may cause severe burns of eyes and skin. *See NIOSH Pocket Guide to Chemical Hazards* (DHHS/NIOSH 97-140, 1997) p 32. *See also Patty's Industrial Hygiene and Toxicology* vol. **2B**, G. D. Clayton, F. E. Clayton, Eds. (Wiley-Inter-science, New York, 4th ed., 1994) pp 4505-4513.

USE: Manuf of organic and inorganic chemicals, such as fuel additives, fire retardants, pesticides, oil well drilling fluids, pharma-ceuticals, dyestuffs. Brominating agent. In water disinfection; as bleaching agent, surface disinfectant.

1402. Bromine Pentafluoride. [7789-30-2] Bromine fluoride (BrF_5). BrF_5; mol wt 174.90. Br 45.69%, F 54.31%. Prepd by fluorination of bromine at 200° in iron or Monel metal apparatus: Ruff, Menzel, *Z. Anorg. Allg. Chem.* **202**, 49 (1931); Kwasnik in *Handbook of Preparative Inorganic Chemistry* **Vol. 1**, G. Brauer, Ed. (Academic Press, New York, 2nd ed., 1963) pp 158-159; from F_2 and KBr: Hyde, Boudakian, *Inorg. Chem.* **7**, 2648 (1968). *Re-views:* Kemmitt, Sharp, *Adv. Fluorine Chem.* **4**, 243-244 (1965); Stein, "Physical and Chemical Properties of Halogen Fluorides" in *Halogen Chemistry* **Vol. 1**, V. Gutmann, Ed. (Academic Press, New York, 1967) pp 133-224; Meinert, *Z. Chem.* **7**, 41-57 (1967).

Liquid. Fumes in air. mp $-60.5°$. bp 40.76°. d^{25} 2.4604. Trou-ton constant 23.7. Thermostable up to 460°. Does not attack quartz when dry. *Produces an explosion on contact with water. Oxidizer, poisonous, corrosive.* Very reactive, usually with conflagration.

Caution: Potential symptoms of overexposure are irritation of eyes, skin, respiratory system; corneal necrosis; skin burns; cough, dyspnea, pulmonary edema; liver, kidney injury. *See NIOSH Pocket Guide to Chemical Hazards* (DHHS/NIOSH 97-140, 1997) p 34.

1403. Bromine Trifluoride. [7787-71-5] Bromine fluoride (BrF_3). BrF_3; mol wt 136.90. Br 58.37%, F 41.63%. Prepd by fluorination of bromine at +80°: Lebeau, *Compt. Rend.* **141**, 1018 (1905); Prideaux, *J. Chem. Soc.* **89**, 316 (1906); Ruff, Braida, *Z. Anorg. Allg. Chem.* **206**, 59 (1932); **214**, 91 (1933); Simons, *Inorg. Synth.* **3**, 184 (1950); Kwasnik in *Handbook of Preparative Inor-ganic Chemistry* **Vol. 1**, G. Brauer, Ed. (Academic Press, 2nd ed.,

1963) pp 156-157. *Reviews:* Kemmitt, Sharp, *Adv. Fluorine Chem.* **4**, 244-245 (1965); Stein, "Physical and Chemical Properties of Halogen Fluorides" in *Halogen Chemistry* **Vol. 1**, V. Gutmann, Ed. (Academic Press, New York, 1967) pp 133-224; Meinert, *Z. Chem.* **7**, 41-57 (1967).

Colorless liquid; also reported to be pale yellow. Long prisms when solid. mp 8.77°. bp 125.75°. d^{25} 2.8030. *Smokes in air. Attacks skin. Very reactive.*

Caution: Corrosive and irritating to skin, eyes, mucous membranes, respiratory tract.

USE: Solvent for fluorides.

1404. Bromisovalum. [496-67-3] *N*-(Aminocarbonyl)-2-bromo-3-methylbutanamide; (α-bromoisovaleryl)urea; 2-monobromo-isovalerylurea; α-bromo-β-dimethylpropanoylurea; bromvaletone; B.V.U.; Bromural; Bromisoval; Uvaleral; Bromuvan; Somnurol; Brovalurea; Dormigene; Isobromyl; Alluval; Pivadorm. C_6H_{11}-BrN_2O_2; mol wt 223.07. C 32.31%, H 4.97%, Br 35.82%, N 12.56%, O 14.34%. Prepn from urea and α-bromoisovaleryl bromide: US 914518 (1909). Toxicity: R. I. Mrongovius *et al., Clin. Exp. Pharmacol. Physiol.* **3**, 443 (1976).

Practically tasteless needles, mp 147-149°. Sublimes. Slightly sol in cold water, but freely sol in hot water. Readily sol in alcohol, ether and in alkaline solns. LD_{50} in male mice (mmoles/kg): 3.25 i.p. (Mrongovius).

THERAP CAT: Sedative, hypnotic.

1405. *N*-Bromoacetamide. [79-15-2] Acetobromamide. C_2-H_4BrNO; mol wt 137.96. C 17.41%, H 2.92%, Br 57.92%, N 10.15%, O 11.60%. $CH_3CONHBr$. Prepd by treating a cooled (0-5°) soln of acetamide dissolved in bromine with ice cold aq 50% KOH, allowing to stand for 2-3 hr at 0-5°, treating with NaCl, and extracting with chloroform: Oliveto, Gerold, *Org. Synth.* **31**, 17 (1951).

Needles from chloroform + hexane, mp 102-105°. Sol in warm water, freely sol in cold ether. Unstable to light and heat.

Monohydrate. Rectangular plates from hydr ether, mp 70-80°. Freely sol in cold water, alcohol, ether; less freely in chloroform.

USE: Brominating agent; in oxidation of primary and secondary alcohols.

1406. *p*-Bromoacetanilide. [103-88-8] *N*-(4-Bromophenyl)-acetamide; 4'-bromoacetanilide; monobromoacetanilide; bromoanilide. C_8H_8BrNO; mol wt 214.06. C 44.89%, H 3.77%, Br 37.33%, N 6.54%, O 7.47%. Prepd by bromination of acetanilide: Remmers, *Ber.* **7**, 346 (1874); Merker, Vona, *J. Chem. Educ.* **26**, 613 (1949); Knowles, Alt, US 3012035 (1959 to Monsanto).

Crystals, from 95% alc, mp 168° (with previous softening). d 1.72. Practically insol in cold water. Sparingly sol in hot water; sol in benzene, chloroform, ethyl acetate; moderately sol in alcohol.

1407. Bromoacetic Acid. [79-08-3] 2-Bromoacetic Acid. $C_2H_3BrO_2$; mol wt 138.95. C 17.29%, H 2.18%, Br 57.51%, O 23.03%. $BrCH_2COOH$. Prepn by bromination of acetic acid: Perkin, Duppa, *Ann.* **108**, 106 (1858); from chloroacetic acid and HBr: Lake, Asadorian; Asadorian, Burk, US 2553518; US 3130222 (1951, 1964, both to Dow); from glycolic acid and HBr: Johnston, US 2876255 (1959 to Ethyl Corp.).

Hygroscopic crystals, mp 50°. bp 208°. d 1.93. Very sol in water, alcohol. *Corrosive. Protect from air and moisture.*

1408. Bromoacetone. [598-31-2] 1-Bromo-2-propanone. C_3H_5BrO; mol wt 136.98. C 26.31%, H 3.68%, Br 58.33%, O

11.68%. Prepn by bromination of acetone: Emmerling, Wagner, *Ann.* **204**, 29 (1880); Catch *et al., J. Chem. Soc.* **1948**, 272; Ross US 2452154 (1948 to Colgate-Palmolive-Peet); by bromination acetone enol acetate: Magerlein, US 2752341 (1956 to Upjohn).

Liquid, mp −36.5°. bp 137°. bp_{50} 63.5-64°. d^{23} 1.634. n_D^{15} 1.4697. Turns violet rapidly even in absence of air. Sparingly sol in water; sol in alcohol, acetone. *Poisonous, flammable. Keep tightly closed and protected from light.*

Caution: Violent lacrimator.

USE: Chemical war gas.

1409. ω-Bromoacetophenone. [70-11-1] 2-Bromo-1-phen-ylethanone; phenacyl bromide. C_8H_7BrO; mol wt 199.05. C 48.27%, H 3.54%, Br 40.14%, O 8.04%. Prepn from acetophenone and bromine: Rother, Reid, *J. Am. Chem. Soc.* **41**, 77 (1919); Cowper, Davidson, *Org. Synth.* **19**, 24 (1939); Shevchuk, Dombrovskii, *Zh. Obshch. Khim.* **33**(4), 1135 (1963).

Crystals, mp 50°. d 1.65. bp_{20} 133-135°. Irritating vapors. *Lacrimator.* Practically insol in water. Freely sol in alcohol, benzene, chloroform, ether.

Caution: Highly irritating to skin, eyes, mucous membranes.

1410. *p*-Bromoacetophenone. [99-90-1] 1-(4-Bromophen-yl)ethanone; 1-acetyl-4-bromobenzene; 4'-bromoacetophenone; methyl *p*-bromophenyl ketone. C_8H_7BrO; mol wt 199.05. C 48.27%, H 3.54%, Br 40.14%, O 8.04%. Prepd from bromobenzene and acetic anhydride in CS_2 in the presence of anhyd $AlCl_3$: Adams, Noller, *Org. Synth.* **5**, 17 (1925).

Leaflets from alc. mp 54°. bp_{736} 255.5°; bp_{15} 130°; bp_7 117°. Easily volatile with steam. Sol in alcohol, ether, glacial acetic acid, benzene, petr ether, carbon disulfide.

Oxime. [5798-71-0]; [59862-55-4] ((*E*)-form); [73744-33-9] ((*Z*)-form). C_8H_8BrNO; mol wt 214.06. Needles from dil alc, mp 128.5°.

1411. *p*-Bromoaniline. [106-40-1] 4-Bromobenzeneamine; 4-bromoaniline. C_6H_6BrN; mol wt 172.03. C 41.89%, H 3.52%, Br 46.45%, N 8.14%. Prepd by steam distilling sodium hydroxide and *p*-bromoacetanilide: Scott, *J. Chem. Soc.* **123**, 3199 (1923); by direct bromination of aniline: Kosolapoff, *J. Am. Chem. Soc.* **75**, 3596 (1953).

Rhombic crystals from dil alc; mp 66-66.5°. $d_4^{99.6}$ 1.4970 (liq). Very sol in alcohol and ether. Insol in cold water. pKb (25°): 10.28 also reported as 9.98.

USE: In prepn of azo dyes; condensed with formaldehyde in prepn of dihydroquinazolines.

1412. 5-Bromoanthranilic Acid. [5794-88-7] 2-Amino-5-bromobenzoic acid; 5-bromo-2-aminobenzoic acid. $C_7H_6BrNO_2$; mol wt 216.03. C 38.92%, H 2.80%, Br 36.99%, N 6.48%, O

14.81%. Prepd by bromination of anthranilic acid: Wheeler, *J. Am. Chem. Soc.* **31**, 565 (1909); Wheeler, Oates, *ibid.* **32**, 771 (1910).

Crystals, mp 218-219°. Very slightly sol in water, moderately sol in alcohol, ether, chloroform, benzene, acetic acid, freely sol in acetone.

USE: Determination of cobalt, copper, nickel, and zinc.

1413. Bromobenzene. [108-86-1] Monobromobenzene; phenyl bromide. C_6H_5Br; mol wt 157.01. C 45.90%, H 3.21%, Br 50.89%. Prepd industrially by the action of bromide on benzene in the presence of iron powder: Gattermann-Wieland, *Praxis des Organischen Chemikers* (de Gruyter, Berlin, 40th ed., 1961) p 95; alternate procedure using pyridine as halogen carrier: A. I. Vogel, *Practical Organic Chemistry* (Longmans, London, 3rd ed., 1959) p 535.

Mobile liquid. Aromatic odor. d_4^0 1.5220; d_4^{10} 1.5083; d_4^{15} 1.5017; d_4^{20} 1.4952; d_4^{30} 1.4815; d_4^{71} 1.426. mp $-30.6°$. bp_{760} 156.2°; bp_{400} 132.3°; bp_{200} 110.1°; bp_{40} 68.6°; bp_{20} 53.8°; bp_{10} 40.0°; bp_5 27.8°; $bp_{1.0}$ 2.9°. n_D^{15} 1.5625; n_D^{20} 1.5602. Flash pt 51°C. Fire pt 155°. Critical temp 397°; crit press. 33,912 mm (44.6 atm). Viscosity at 20° = 1.124 cP. Vapor density (air = 1): 5.41. Specific heat at 26.84° = 0.2368. Heat of melting 16.186 cal/g at 15°. *Flammable.* Practically insol in water (0.045 g/100 g at 30°). Miscible with chloroform, benzene, petr hydrocarbons. Sol in alc (10.4 g/100 g at 25°), in ether (71.3 g/100 g at 25°).

Caution: Irritating to skin.

USE: In organic synthesis, especially to make phenyl magnesium bromide; as solvent, especially for crystns on a large scale and where a heavy liquid is desirable; as additive to motor oils.

1414. 4-Bromobenzenesulfonyl Chloride. [98-58-8] C_6H_4-$BrClO_2S$; mol wt 255.51. C 28.20%, H 1.58%, Br 31.27%, Cl 13.87%, O 12.52%, S 12.55%. Prepd from sodium *p*-bromobenzenesulfonate and phosphorous pentachloride: Marvel, Smith, *J. Am. Chem. Soc.* **45**, 2696 (1923).

Needles from ligroin, mp 74.5°.
USE: Identification of amines.

1415. 4-Bromobenzoic Acid. [586-76-5] $C_7H_5BrO_2$; mol wt 201.02. C 41.83%, H 2.51%, Br 39.75%, O 15.92%. Prepd by oxidation of *p*-bromotoluene with potassium permanganate: Hale, Thorp, *J. Am. Chem. Soc.* **35**, 269 (1913).

Needles from ether, leaflets from water or 90% alcohol. mp 251-253°. Slightly sol in hot water. Sol in alc, ether.

Methyl ester. $C_6H_4BrCO_2CH_3$. Needles from ether or leaflets from dil alcohol, mp 81°. d 1.689. Sol in alc, ether.

Ethyl ester. $C_6H_4BrCO_2C_2H_5$. Liquid; $bp_{737.4}$ 262°.

Phenyl ester. $C_{13}H_9BrO_2$. Scales, mp 117°. Sol in alc, ether, chloroform, CS_2, benzene; less sol in petr ether.

USE: For the detection of strontium; in org syntheses.

1416. *p*-Bromobenzyl Bromide. [589-15-1] 1-Bromo-4-(bromomethyl)benzene; *p*,α-dibromotoluene. $C_7H_6Br_2$; mol wt 249.93. C 33.64%, H 2.42%, Br 63.94%. Prepd by photo-bromination of *p*-bromotoluene: Weizmann, Patai, *J. Am. Chem. Soc.* **68**, 150 (1946). Alternate procedure: Goerner, Nametz, *ibid.* **73**, 2940 (1951).

Crystals from alc. Agreeable aromatic odor. mp 61°. bp_{12} 115-124°. Sol in water, cold alc, more sol in hot alc, ether, carbon disulfide, benzene, glacial acetic acid.

Caution: Irritating to eyes, mucous membranes of nose and throat. Action similar to, but less intense than, benzylbromide, *q.v.*

USE: Identification of aromatic carboxylic acids.

1417. *p*-Bromobenzyl Chloride. [589-17-3] 1-Bromo-4-(chloromethyl)benzene; *p*-bromo-α-chlorotoluene; α-chloro-4-bromotoluene. C_7H_6BrCl; mol wt 205.48. C 40.92%, H 2.94%, Br 38.89%, Cl 17.25%. Prepd from *p*-bromobenzene, paraformaldehyde and $SnCl_4$: Quelet, *Bull. Soc. Chim.* **41**, 329 (1927); by benzoyl peroxide catalyzed chlorination of *p*-bromotoluene with sulfhydryl chloride: Goerner, Nametz, *J. Am. Chem. Soc.* **73**, 2940 (1951).

Needles from alc, mp 40-41°. bp_{12} 105-115°; bp_{27} 136-139°. Freely sol in hot alc.

1418. α-Bromobenzyl Cyanide. [5798-79-8] α-Bromobenzeneacetonitrile; α-bromophenylacetonitrile; α-bromo-α-tolunitrile; B.B.C.; C.A.; Camite. C_8H_6BrN; mol wt 196.05. C 49.01%, H 3.08%, Br 40.76%, N 7.14%. The practical industrial prepn consists of three steps: (1) Chlorination of toluene to form benzyl chloride, (2) conversion of benzyl chloride to benzyl cyanide by the action of sodium cyanide in alcoholic soln, (3) bromination of the benzyl cyanide with bromine vapor in the presence of sunlight: Steinkopf *et al.*, *Ber.* **53**, 1146 (1920); Nekrassov, *J. Prakt. Chem.* **119**, 108 (1928).

Crystalline mass, mp 29°. Odor of soured fruit. bp_{760} 242° (dec); bp_{12} 132-134°. d_4^{29} 1.539. Vapor density 6.8 (air = 1). Vapor pressure at 20° = 0.012 mm; at 30° = 0.028 mm. Slightly sol in water. Freely sol in alcohol, ether, chloroform, acetone, and other common organic solvents; also sol in phosgene, chloropicrin, benzyl cyanide. LC (30 min.): 0.90 mg/l (A. M. Prentiss, *Chemicals in War* (McGraw-Hill, New York, 1937) p 141).

Caution: Highly potent lacrimator. *See Patty's Industrial Hygiene and Toxicology* vol. **2D**, G. D. Clayton, F. E. Clayton, Eds. (Wiley-Interscience, New York, 4th ed., 1994) p 3163.

USE: War gas.

1419. Bromobutide. [74712-19-9] 2-Bromo-3,3-dimethyl-*N*-(1-methyl-1-phenylethyl)butanamide; 2-bromo-*N*-(α,α-dimethylbenzyl)-3,3-dimethylbutyramide; *N*-(1-methyl-1-phenylethyl)-2-bromo-3,3-dimethylbutanamide; S-4347; S-47; Sumiherb. $C_{15}H_{22}$-$BrNO$; mol wt 312.25. C 57.70%, H 7.10%, Br 25.59%, N 4.49%, O 5.12%. 4-Hydroxyphenylpyruvate dioxygenase (4-HPPD) inhibitor; developed as a selective herbicide for weed control. Prepn: **NL 7906809**; O. Kirino *et al.*, **US 4288244** (1980, 1981 both to Sumitomo); *eidem, Agric. Biol. Chem.* **45**, 2669 (1981). Synthesis: H. Nakatsuji *et al.*, *Tetrahedron* **63**, 12071 (2007). Herbicidal activity: S. Hashimoto *et al.*, *J. Pestic. Sci.* **8**, 493 (1983). Photodegradation study: N. Takahashi *et al. ibid.* **10**, 247 (1985). Multiresidue GC/MS determn in river water: A. Tanabe *et al.*, *J. Chromatogr. A* **754**, 159 (1996); in sediment: K. Kawata *et al.*, *J. AOAC Int.* **88**, 1440 (2005); in agricultural products: Y. Hirahara *et al.*, *J. Health Sci.* **51**, 617 (2005). Bioconcentration studies in fish and shellfish: T. Tsuda *et al.*, *Bull. Environ. Contam. Toxicol.* **82**, 716 (2009).

Crystals from ethanol, mp 182-183°.
USE: Herbicide.

1420. α-Bromobutyric Acid. [80-58-0] 2-Bromobutanoic acid; *dl*-2-bromobutyric acid. $C_4H_7BrO_2$; mol wt 167.00. C 28.77%, H 4.23%, Br 47.85%, O 19.16%. Prepd by bromination of butyric acid: Naumann, *Ann.* **119**, 115 (1861); Stevens, Holland, *J. Org. Chem.* **18**, 1112 (1953); Smissman, *J. Am. Chem. Soc.* **76**, 5805 (1954). Toxicity study: J. L. Morrison, *J. Pharmacol. Exp. Ther.* **86**, 336 (1946).

Oily liquid, mp −4°. bp_{250} 181-182°, bp_{25} 127-128°. d_4^4 1.5855, d_{15}^{15} 1.5735, d_{20}^{20} 1.5669, d_{25}^{25} 1.5620: Perkin, *J. Chem. Soc.* **65**, 402 (1894). Boils with decompn at ordinary pressure. Sol in 15 parts water; sol in alcohol, ether. *Corrosive.* LD_{50} orally in mice: 310 mg/kg (Morrison).

Ethyl ester. [533-68-6] Ethyl α-bromobutyrate. $C_6H_{11}BrO_2$; mol wt 195.06. Liquid, bp 177-178° with slight dec. d_{20}^{20} 1.329. Insol in water. Misc with alc, ether. *Protect from light. Lacrimator.*

1421. 3-Bromo-*d*-camphor. [76-29-9] 3-Bromo-1,7,7-trimethylbicyclo[2.2.1]heptan-2-one; 3-bromo-*d*-2-bornanone. $C_{10}H_{15}BrO$; mol wt 231.13. C 51.97%, H 6.54%, Br 34.57%, O 6.92%. Of the two configurations found, the *endo*-form is more stable than the *exo*-form: Lowry *et al.*, *J. Chem. Soc.* **121**, 633 (1922); Cookson, *ibid.* **1954**, 282. Prepn of the *endo*-form by bromination of *d*-camphor: Kipping, Pope, *ibid.* **63**, 548 (1893); *cf.* Woods, Roberts, *J. Org. Chem.* **22**, 1124 (1957). Prepn of *exo*-form by isomerization of *endo*-form: Lowry *et al.*, *loc. cit.* Configuration: Wiebenga, Krom, *Rec. Trav. Chim.* **65**, 663 (1946); Cookson, *loc. cit. Review:* J. L. Simonsen, Ed., *The Terpenes* vol. II (University Press, Cambridge, 2nd ed., 1949), pp 401-404.

endo-form   *exo*-form

endo-**Form.** α-Bromo-*d*-camphor; 3α-bromo-*d*-camphor; bromated camphor; camphor monobromated. Crystals from benzene, mp 76°. Camphor-like odor and taste. Discolors on prolonged exposure to light. d 1.449. $[\alpha]_D^{20}$ +122.7° (14.5 g/100 g benzene soln), Cutter *et al.*, *J. Chem. Soc.* **127**, 1260 (1925). uv max (cyclohexane): 307.5 nm (log ε 1.98), Cookson, *loc. cit.* Sublimes, bp 274°. Almost insol in water: 1 g dissolves in 6.5 ml alcohol, 0.5 ml chloroform, 1.6 ml ether; sol in olive oil, slightly in glycerol. When phenol, chloral hydrate, salol, menthol, or thymol is triturated with bromocamphor the mixture melts; these compds, however, are not incompatible.

exo-**Form.** α'-Bromo-*d*-camphor; 3β-bromo-*d*-camphor. Needles from methanol or ethanol, mp 78.5°. d 1.484. $[\alpha]_D^{20}$ −42.1° (14.5 g/100 g benzene soln), Cutter *et al.*, *loc. cit.* uv max (cyclohexane): 312 nm (log ε 1.95), Cookson, *loc. cit.*

THERAP CAT: Topical counterirritant.

1422. α-Bromo-*n*-caproic Acid. [616-05-7] 2-Bromohexanoic acid; $C_6H_{11}BrO_2$; mol wt 195.06. C 36.95%, H 5.68%, Br 40.96%, O 16.40%. Prepd by heating bromine with *n*-caproic acid in the presence of phosphorus trichloride as catalyst: Clarke, Taylor, *Org. Synth.* **4**, 9 (1924). Resoln into optically active forms by means of the strychnine salt: Levene *et al.*, *J. Biol. Chem.* **75**, 352 (1927); Levene, Mardashew, *ibid.* **117**, 707 (1937).

Liquid. bp_{760} 240°; bp_{30} 148-153°; bp_{17} 128-136°; bp_8 116-125°. Soluble in alcohol, ether.

Ethyl ester. [615-96-3] $C_8H_{15}BrO_2$; mol wt 223.11. Liquid, odor of anise oil. bp_{760} 205-210°; bp_{32} 99-102°; bp_{11} 103°; bp_9 95-96°.

1423. *B*-Bromocatecholborane. [51901-85-0] 2-Bromo-1,3-2-benzodioxaborole; catechol boron bromide; *o*-phenylene bromoboronate; BrBcat. $C_6H_4BBrO_2$; mol wt 198.81. C 36.25%, H 2.03%, B 5.44%, Br 40.19%, O 16.09%. Reagent used primarily in deprotection chemistry to cleave ester, ether, and carbamate protecting groups. Prepn: W. Gerrard *et al.*, *J. Chem. Soc.* **1959**, 1529. Crystal structure: R. B. Coapes *et al.*, *J. Chem. Soc. Dalton Trans.* **2001**, 1201. Protecting group cleavage applications: R. K. Boeckman, Jr., J. C. Potenza, *Tetrahedron Lett.* **26**, 1411 (1985); P. F. King, S. G. Stroud, *ibid.* 1415; C. Yu *et al.*, *ibid.* **41**, 819 (2000). Use in catalysis of Diels-Alder reactions: T. R. Kelly *et al.*, *ibid.* **30**, 1357 (1989). *Review:* P. C. Anderson, Y. Guindon in *Encyclopedia of Reagents for Organic Synthesis* **1**, L. A. Paquette, Ed. (Wiley, New York, 1995) pp 709-710.

Colorless solid. *Flammable. Corrosive.* mp 47°. bp_9 76°. Flash pt, closed cup: 107°F (42°C). Sol in dichloromethane. Light and moisture sensitive. Fumes in air.

USE: Reagent in synthetic organic chemistry.

1424. Bromocriptine. [25614-03-3] (5′α)-2-Bromo-12′-hydroxy-2′-(1-methylethyl)-5′-(2-methylpropyl)ergotaman-3′,6′,18-trione; 2-bromoergocriptine; 2-bromo-α-ergokryptin; CB-154. $C_{32}H_{40}BrN_5O_5$; mol wt 654.61. C 58.71%, H 6.16%, Br 12.21%, N 10.70%, O 12.22%. Dopamine D_2 receptor agonist; derivative of the ergotoxin group of ergot alkaloids. Prepn: E. Flückiger *et al.*, **DE 1926045**; *eidem*, **US 3752814** (1969, 1973 both to Sandoz). Pharmacology: E. Flückiger, H. R. Wagner, *Experientia* **24**, 1130 (1968); E. Del Pozo *et al.*, *Schweiz. Med. Wochenschr.* **103**, 847 (1973). Relationship of stereochemistry and biological activity: H. P. Weber, *Adv. Biochem. Psychopharmacol.* **23**, 25 (1980); N. Camerman, A. Camerman, *Mol. Pharmacol.* **19**, 517 (1981). LC/MS/MS determn in plasma: A. Salvador *et al.*, *J. Chromatogr. B* **820**, 237 (2005). Long term clinical trial in Parkinson's disease: T. Nakanishi *et al.*, *Eur. Neurol.* **32**, Suppl. 1, 9 (1992). Clinical effect on body weight and glucose tolerance in obesity: A. H. Cincotta, A. H. Meier, *Diabetes Care* **19**, 667 (1996). Clinical study in type 2 diabetes: H. Pijl *et al.*, *ibid.* **23**, 1154 (2000). Effects of prolactin concentration and estrus in beagles: N. J. Beijerink *et al.*, *Theriogenology* **60**, 1379 (2003). Comprehensive description: D. A. Giron-Forest, W. D. Schönlein, *Anal. Profiles Drug Subs.* **8**, 47-81 (1979). Review of pharmacology, toxicology and therapeutic uses: D. Parkes, *Adv. Drug Res.* **12**, 247-344 (1977); of therapeutic applications in endocrine and neurological diseases: K. Y. Ho, M. O. Thorner, *Drugs* **36**, 67-82 (1988); of clinical experience in rheumatic and autoimmune diseases: R. W. McMurray, *Semin. Arthritis Rheum.* **31**, 21-32 (2001).

Crystals from methyl ethyl ketone-isopropyl ether, mp 215-218° (dec). $[\alpha]_D^{20}$ −195° (c = 1 in methylene chloride).

Methanesulfonate. [22260-51-1] CB-154 mesylate; Bromo-Kin; Cycloset; Parlodel; Pravidel. $C_{32}H_{40}BrN_5O_5.CH_3SO_3H$; mol wt 750.71. Crystals from methyl ethyl ketone, mp 192-196° (dec). $[\alpha]_D^{20}$ +95° (c = 1 in methanol-methylene chloride). Soly at 25° (mg/ml): methanol 910; ethanol 23.0; water 0.8; chloroform 0.45; benzene <0.1; hexane <0.1. pKa 4.90. LD_{50} in mice, rats, rabbits (mg/kg): 190, 72, 12.5 i.v. (Parkes).

THERAP CAT: Prolactin inhibitor; antiparkinsonian; antidiabetic.

THERAP CAT (VET): Prolactin inhibitor.

1425. Bromodichloromethane. [75-27-4] Dichlorobromomethane; BDCM; NCI-C55243. $CHBrCl_2$; mol wt 163.82. C 7.33%, H 0.62%, Br 48.78%, Cl 43.28%. Trihalomethane contaminant found in chlorinated water. Prepn: O. Jacobsen, R. Neumeister, *Ber.* **15**, 599 (1882); M. Fedorynski *et al., Synth. Commun.* **7**, 287 (1977). Disposition and metabolism: J. M. Mathews *et al., J. Toxicol. Environ. Health* **30**, 15 (1990). Toxicity studies: F. J. Bowman *et al., Toxicol. Appl. Pharmacol.* **44**, 213 (1978); Y. Aida *et al., J. Toxicol. Sci.* **17**, 51 (1992). Review of quantitative analyses, toxicity and carcinogenicity: *IARC Monographs* **52**, 179-212 (1991); of toxicology and human exposure: *Toxicological Profile for Bromodichloromethane* (PB90-167461, 1989) 89 pp.

Colorless liquid. bp 88.4-88.6° (Fedorynski); also reported as bp 91-92° (Jacobsen). d^{15} 1.9254. n_D^{20} 1.4967. LD_{50} in male, female mice (mg/kg): 450, 900 orally (Bowman).

Caution: Potential symptoms of acute oral overexposure in animals are liver and kidney damage; CNS depression, incoordination, sleepiness, lethargy, sedation. Chronic exposure has been associated with cancer of liver, kidney and intestines. *See: Toxicological Profile for Bromodichloromethane* (PB90-167461, 1989). This substance is reasonably anticipated to be a human carcinogen: *Report on Carcinogens, Twelfth Edition* (PB2011-111646, 2011) p 73.

USE: Chemical reagent, intermediate in organic synthesis.

1426. Bromodimethylborane. [5158-50-9] Dimethylboron bromide; dimethylbromoborane; Me_2BBr. C_2H_6BBr; mol wt 120.78. C 19.89%, H 5.01%, B 8.95%, Br 66.16%. Organoboron bromide reagent utilized for the cleavage of ethers, acetals, and ketals. Prepn: P. I. Paetzold, H.-J. Hansen, *Z. Anorg. Allg. Chem.* **345**, 79 (1966); and synthetic applications: Y. Guindon *et al., J. Org. Chem.* **49**, 3912 (1984). Additional synthetic applications: *idem et al., ibid.* 4538; *idem et al., ibid.* **52**, 1680 (1987); R. L. Dorow, D. E. Gingrich, *ibid.* **60**, 4986 (1995). *Reviews:* Y. Guindon *et al., Pure Appl. Chem.* **60**, 1705-1714 (1988); Y. Guindon, P. C. Anderson in *Encyclopedia of Reagents for Organic Synthesis* vol. 1, L. A. Paquette, Ed. (Wiley, New York, 1995) pp 715-718; B. Y. Michel, *Synlett* **2008**, 2893-2894.

Colorless liquid. bp 30°. *Flammable. Corrosive.* Flash pt, closed cup: −35°F (−37°C). Sol in dichloromethane, 1,2-dichloroethane, hexane. Moisture sensitive. Stable for months when stored as soln in dichloromethane or 1,2-dichloroethane at −15°C.

USE: Reagent in synthetic organic chemistry.

1427. Bromodimethylsulfonium Bromide. [50450-21-0] Dimethylsulfur dibromide; BDMS. $C_2H_6Br_2S$; mol wt 221.94. C 10.82%, H 2.73%, Br 72.01%, S 14.45%. Brominating reagent in synthetic organic chemistry. Several forms of the dimethyl sulfide-bromine adduct exist; structure is dependent upon reaction conditions. Although the charge transfer-complex is more stable, the ionic species is often presumed when considering its synthetic applications. Prepn from dimethyl sulfide and bromine: A. Hantzsch, H. Hibbert, *Ber.* **40**, 1508 (1907); P. Haas, *Biochem. J.* **29**, 1297 (1935); A. H. Fenselau, J. G. Moffatt, *J. Am. Chem. Soc.* **88**, 1762 (1966); G. A. Olah *et al., Synthesis* **1979**, 720; from DMSO and bromotrimethylsilane: G. Megyeri, T. Keve, *Synth. Commun.* **19**, 3415 (1989); *in situ* from DMSO and aq hydrobromic acid: G. Majetich *et al., J. Org. Chem.* **62**, 4321 (1997). Crystal and molecular structure studies: G. B. M. Vaughan *et al., J. Chem. Soc. Dalton Trans.* **1999**, 79. Raman spectroscopy studies of structural forms: H. F. Askew *et al., J. Raman Spectrosc.* **22**, 265 (1991). Synthetic applications in bromination reactions: G. Olah *et al., Tetrahedron Lett.*

20, 3653 (1979); Y. L. Chow, B. H. Bakker, *Can. J. Chem.* **60**, 2268 (1982); in solvent free catalysis: A. T. Khan *et al., J. Mol. Catal. A* **239**, 158 (2005); *idem et al., Tetrahedron Lett.* **48**, 3805 (2007). Review: M. L. H. Choudhury, *Synlett* **2006**, 1619-1620.

Yellow needles from carbon tetrachloride, mp 90-91° (Fenselau). Also reported as light orange crystals, mp 80° (dec) (Olah). Sol in water, alcohol, ether, and chloroform.

USE: In synthetic organic chemistry as a brominating reagent and catalyst.

1428. Bromodiphenhydramine. [118-23-0] 2-[(4-Bromophenyl)phenylmethoxy]-*N,N*-dimethylethanamine; 2-(*p*-bromo-α-phenylbenzyloxy)-*N,N*-dimethylethylamine; β-(*p*-bromobenzhydryloxy)ethyldimethylamine; β-dimethylaminoethyl *p*-bromobenzhydryl ether; bromdiphenhydramine; bromazine; bromanautine; Bromo-Benadryl; Deserol; Histabromamine. $C_{17}H_{20}BrNO$; mol wt 334.26. C 61.09%, H 6.03%, Br 23.90%, N 4.19%, O 4.79%. Prepn: Rieveschl, US 2527963 (1950 to Parke, Davis).

Hydrochloride. [1808-12-4] Ambodryl. $C_{17}H_{20}BrNO.HCl$; mol wt 370.72. Crystals from isopropanol, mp 144-145°. Freely sol in water, alc; sol in isopropyl alc. Insol in ether, solvent hexane.

THERAP CAT: Antihistaminic.

1429. Bromofenofos. [21466-07-9] 3,3',5,5'-Tetrabromo-[1,1'-biphenyl]-2,2'-diol 2-(dihydrogen phosphate); 3,3',5,5'-tetrabromo-2,2'-biphenyldiol mono(dihydrogen phosphate); 4,4',6,6'-tetrabromobiphenyl-2,2'-diol mono(dihydrogen phosphate); bromophenophos; bromphenphos; PH-1882. $C_{12}H_7Br_4O_5P$; mol wt 581.77. C 24.77%, H 1.21%, Br 54.94%, O 13.75%, P 5.32%. Anthelmintic used to control fascioliasis in ruminants. Prepn: **NL 6505635**; S. Van der Meer *et al.,* **US 3662035** (1966, 1972 both to Chemiefarma); S. Van der Meer, H. Pouwels, *J. Med. Chem.* **12**, 534 (1969). Activity: J. Guilhon *et al., Bull. Acad. Vet. Fr.* **43**, 67 (1970). Toxicology: J. M. Poul, M. Dagorn, *Recl. Med. Vet.* **158**, 363 (1982). HPLC determn in plasma: Y. S. Endoh *et al., J. Chromatogr.* **426**, 202 (1988).

Cryst, no mp, dec >350°.

THERAP CAT (VET): Anthelmintic (Trematodes).

1430. Bromoform. [75-25-2] Tribromomethane. $CHBr_3$; mol wt 252.73. C 4.75%, H 0.40%, Br 94.85%. Prepd from acetone and sodium hypobromite: Günther, *Jahresber. Fortschr. Chem.* **1887**, 741; *Beilstein* vol. 1, 68; Kergomard, *Bull. Soc. Chim. Fr.* **1961**, 2360. Toxicity data: Kutob, Plaa, *Toxicol. Appl. Pharmacol.* **4**, 354 (1962). Review of toxicology and human exposure: *Toxicological Profile for Bromoform and Dibromochloromethane* (PB2006-100001, 2005) 273 pp.

Heavy liquid; chloroform odor; sweetish taste. bp 149-150°. mp 7.5°. d_4^{15} 2.9035. n_D^{15} 1.6005. Sol in about 800 parts water; miscible with alc, benzene, chloroform, ether, petr ether, acetone, oils. Gradually dec, acquiring a yellow color; air and light accelerate the decompn. May be preserved by the addition of 3-4% alc. d 2.6-2.7. *Poisonous. Keep in well-closed containers, protected from light. Incompat.* Caustic alkalies. LD_{50} s.c. in mice: 7.2 mmol/kg (Kutob, Plaa).

Caution: Potential symptoms of overexposure are irritation of skin, eyes and respiratory system; CNS depression; liver, kidney damage. *See NIOSH Pocket Guide to Chemical Hazards* (DHHS/NIOSH 97-140, 1997) p 34.

USE: In separating mixtures of minerals.

THERAP CAT: Has been used as sedative, hypnotic; antitussive.

1431. α-Bromoisobutyric Acid. [2052-01-9] 2-Bromo-2-methylpropanoic acid. $C_4H_7BrO_2$; mol wt 167.00. C 28.77%, H 4.23%, Br 47.85%, O 19.16%. Prepd by bromination of isobutyric acid: Markownikow, *Ann.* **153**, 228 (1870); Smissman, *J. Am. Chem. Soc.* **76**, 5805 (1954).

Crystals, mp 48-49°. bp 198-200°, bp_{20} 110-116°. d 1.52. Sparingly sol in cold water; sol in alcohol, ether. Dec by hot water into the hydroxy acid.

1432. α-Bromoisovaleric Acid. [565-74-2] 2-Bromo-3-methylbutanoic acid. $C_5H_9BrO_2$; mol wt 181.03. C 33.17%, H 5.01%, Br 44.14%, O 17.68%. Prepn of *dl*-form: Ley, Popow, *Ann.* **174**, 61 (1874); Smissman, *J. Am. Chem. Soc.* **76**, 5805 (1954); Marvel, *Org. Synth.* **coll. vol. III**, 848 (1955). Prepn of *d*- and *l*-forms: Berlingozzi, Furia, *Gazz. Chim. Ital.* **56**, 82 (1926); Berlingozzi, Lenoci, *ibid.* **68**, 721 (1938). Configuration of *d*-form: P. Brewster *et al., Nature* **166**, 179 (1950).

Prisms from ether or chloroform, mp 44°. bp ~230° with some decompn. Sparingly sol in water; sol in alcohol, ether.

(*R*)-Form. [76792-22-8] Crystals from petroleum ether, mp 43-44°. $[\alpha]_D^{20}$ +22.6° (c = 4 in benzene).

(*S*)-Form. [26782-75-2] Crystals from petroleum ether, mp 43-44°. $[\alpha]_D^{20}$ −22.4° (c = 4 in benzene).

1433. Bromolysergide. [478-84-2] (8β)-2-Bromo-9,10-didehydro-*N,N*-diethyl-6-methylergoline-8-carboxamide; D-2-bromolysergic acid diethylamide; 2-bromo-*N,N*-diethyl-D-lysergamide; bromo-LSD; BOL-148. $C_{20}H_{24}BrN_3O$; mol wt 402.34. C 59.71%, H 6.01%, Br 19.86%, N 10.44%, O 3.98%. Prepd by bromination of lysergide, *q.v.*: Troxler, Hofmann, *Helv. Chim. Acta* **40**, 2160 (1957). Serotonin antagonist without the hallucinogenic activity of LSD: J. P. Bennett, S. H. Snyder, *Brain Res.* **94**, 523 (1975).

Needles from ether, mp 120-127°. $[\alpha]_D^{20}$ +15° (c = 0.5 in pyridine); $[\alpha]_D^{20}$ +53° (c = 0.5 in chloroform). uv max: 240, 301 nm (log ε 4.28, 3.95).

1434. p-Bromomandelic Acid. [6940-50-7] 4-Bromo-α-hydroxybenzeneacetic acid; *p*-bromophenylglycolic acid. $C_8H_7BrO_3$; mol wt 231.05. C 41.59%, H 3.05%, Br 34.58%, O 20.77%. Prepd by condensing bromobenzene with ethyl oxomalonate in presence of boron trifluoride: Riebsomer *et al., J. Am. Chem. Soc.* **60**, 2974 (1938).

Crystals, mp 117-118°. Slightly sol in water.
Zirconium salt. Soly in water at 25°: 0.0446 g/l.
USE: Analytical reagent for zirconium.

1435. (2S)-1-Bromo-2-methylbutane. [534-00-9] (*S*)-(+)-1-bromo-2-methylbutane; (*S*)-2-methyl-1-bromobutane; *d*-amyl bromide. $C_5H_{11}Br$; mol wt 151.05. C 39.76%, H 7.34%, Br 52.90%. Prepn: W. Marckwald, *Ber.* **37**, 1038 (1904); H. O. Jones, *J. Chem. Soc.* **87**, 135 (1905). Prepn and use in asymmetric synthesis: R. L. Letsinger, *J. Am. Chem. Soc.* **70**, 406 (1948); G. Y. Brokaw, W. R. Brode, *J. Org. Chem.* **13**, 194 (1948).

Colorless oil. *Flammable; irritant.* d_4^{20} 1.221. bp 120-121°. n_D^{25} 1.4425. $[\alpha]_D^{25}$ +3.84°. Flash pt, closed cup: 71.6°F (22°C). Practically insol in water. Sol in alcohol, ether.
USE: Reagent in organic synthesis.

1436. 1-Bromonaphthalene. [90-11-9] α-Bromonaphthalene. $C_{10}H_7Br$; mol wt 207.07. C 58.00%, H 3.41%, Br 38.59%. Prepd by dropping bromine into a mixture of naphthalene and carbon tetrachloride, distilling the CCl_4, heating the residue with NaOH, and fractionating the liquid under reduced pressure: Clarke, Schram, *Org. Synth.* **1**, 35 (1921); Clarke, Brethen, *ibid.* **coll. vol. I**, 121 (2nd ed., 1941). Also prepd by heating naphthalene (liquid or gaseous) with bromine; 2-bromonaphthalene also being produced, esp with increasing temps: Wibaut, *Chem. Weekbl.* **39**, 326, 328 (1942); Suyver, Wibaut, *Rec. Trav. Chim.* **64**, 65 (1945).

Oily liquid at room temp. More pungent odor than naphthalene. d_4^{20} 1.4834; d_4^{25} 1.4785; d_4^{30} 1.4732. When solid it exists in two forms: mp 0.2-0.7° and mp 6.2°. Darkens on standing when distilled at 760 mm, but remains colorless when distilled at 16 mm. Volatile with steam. bp_{760} 281.1°; bp_{400} 252.0°; bp_{200} 224.2°; bp_{100} 198.8°; bp_{60} 183.5°; bp_{40} 170.2°; bp_{20} 150.2°; bp_{10} 133.6°; bp_5 117.5°; $bp_{1.0}$ 84.2°. $n_D^{16.5}$ 1.66011. Slightly sol in water; miscible with alcohol, ether, benzene, chloroform. Absorption spectrum: DeLaszlo, *Proc. R. Soc. London Ser. A* **111**, 356 (1926).
Compd with 1,3,5-trinitrobenzene. $C_{10}H_7Br.C_6H_3N_3O_6$. Lemon-yellow needles, mp 137°.
USE: Immersion fluid in the determination of the refractive index of crystals. For the determination of water in alc by the cloud point method. For refractometric fat determinations. Mixed with polymerized castor oil as a general immersion oil in microscopy.

1437. 2-Bromonaphthalene. [580-13-2] β-Bromonaphthalene. $C_{10}H_7Br$; mol wt 207.07. C 58.00%, H 3.41%, Br 38.59%. Prepn: Liebermann, Palm, *Ann.* **183**, 267 (1876); Vingiello *et al., J. Chem. Educ.* **40**, 544 (1963).

Crystals from alcohol, mp 59°; also reported as mp 54-56° (Vingiello *et al.*, *loc. cit.*). bp 281-282°, bp$_{4.5}$ 122-127°. d 1.60. Slightly sol in water; sol in 8 parts alcohol; very sol in ether, chloroform, benzene.

1438. *p*-Bromophenacyl Bromide. [99-73-0] 2-Bromo-1-(4-bromophenyl)ethanone; 2,4'-dibromoacetophenone; *p*,α-dibromoacetophenone. C$_8$H$_6$Br$_2$O; mol wt 277.94. C 34.57%, H 2.18%, Br 57.50%, O 5.76%. Prepd by adding bromine to *p*-bromoacetophenone in glacial acetic acid at below 20°: Langley, *Org. Synth.* **coll. vol. I,** 127 (2nd ed., 1941).

Crystals from ethanol, mp 109-110°. Very sol in warm alcohol.
USE: Identification of carboxylic acids.

1439. Bromophenol. C$_6$H$_5$BrO; mol wt 173.01. C 41.65%, H 2.91%, Br 46.18%, O 9.25%. Prepn of *m*-isomer: Wurster, Nölting, *Ber.* **7,** 904 (1874); Carpenter *et al.*, *J. Org. Chem.* **16,** 586 (1951); of *o*- and *p*-isomers: Medola, Streatfield, *J. Chem. Soc.* **73,** 681 (1898); Kaeding, Lindstrom, US **2805263** (1957 to Dow).

m-**Bromophenol.** [591-20-8] Crystals, mp 33°, also reported as mp 31°. bp 235-236°, bp$_3$ 88-89°. Sol in alcohol, ether, alkalies.

o-**Bromophenol.** [95-56-7] Yellow to red oily liquid; unpleasant odor. d ~1.5. bp 194°. mp 6°. Sol in water; miscible in chloroform, ether. Fusion with NaOH gives resorcinol.

p-**Bromophenol.** [106-41-2] Tetragonal bipyramidal crystals from chloroform or ether. mp 64°. bp 238°. d^{15} 1.840; d^{80} 1.5875. Small amounts of water depress the mp considerably and may prevent crystallization. Absorption spectrum: Ley, *Z. Phys. Chem.* **94,** 412 (1920). Soluble in about 7 parts water; freely sol in alc, chloroform, ether, glacial acetic acid.
p-**Form methyl ether.** *p*-Bromoanisole. C$_7$H$_7$BrO. Crystals, mp 9-10°. bp 223°.
p-**Form ethyl ether.** *p*-Bromophenetole. C$_8$H$_9$BrO. Crystals, mp 4°. bp 233°.
USE: *p*-Bromophenol as disinfectant.

1440. *p*-Bromophenylhydrazine. [589-21-9] (4-Bromophenyl)hydrazine. C$_6$H$_7$BrN$_2$; mol wt 187.04. C 38.53%, H 3.77%, Br 42.72%, N 14.98%. Prepn from 4-bromophenyldiazonium chloride by modified Fischer phenylhydrazine synthesis: Chattaway, Humphrey, *J. Chem. Soc.* **1927,** 1323.

Needles from water, mp 108-109°. Sol in alcohol, ether, chloroform, benzene; moderately sol in ligroin; slightly sol in water.
USE: In prepn of indoleacetic acid derivatives; in the study of transosazonation of sugar phenylosazones.

1441. *p*-Bromophenyl Isocyanate. [2493-02-9] 1-Bromo-4-isocyanatobenzene; 4-bromophenylcarbimide; *p*-bromophenylcarbonimide; *p*-bromocarbanil. C$_7$H$_4$BrNO; mol wt 198.02. C 42.46%, H 2.04%, Br 40.35%, N 7.07%, O 8.08%. Prepd by heating phenyl isocyanate dibromide: Curtius, *J. Prakt. Chem.* **87,** 517

(1913); by distilling *p*-bromophenylurethan and P$_2$O$_5$: Dennstedt, *Ber.* **13,** 228 (1880).

Needles, pungent odor, mp 42°. bp$_{14}$ 158°. Very sol in ether.
USE: Prepn of bromophenylurea and urethan derivatives.

1442. Bromophos. [2104-96-3] Phosphorothioic acid *O*-(4-bromo-2,5-dichlorophenyl) *O,O*-dimethyl ester; *O*-(4-bromo-2,5-dichlorophenyl)-*O,O*-dimethylphosphorothioate; *O,O*-dimethyl-*O*-(2,5-dichloro-4-bromophenyl)thiophosphate; bromofos; bromophosmethyl; ENT-27162; OMS-658; S-1942; Nexion. C$_8$H$_8$BrCl$_2$O$_3$PS; mol wt 365.99. C 26.25%, H 2.20%, Br 21.83%, Cl 19.37%, O 13.11%, P 8.46%, S 8.76%. Organophosphate insecticide; cholinesterase inhibitor. Prepn: **BE 625198;** R. Sehring, K. Zeile, **US 3275718** (1963, 1966 both to Boehringer, Ing.). Crystal structure: R. G. Baughman, R. A. Jacobson, *J. Agric. Food Chem.* **24,** 1036 (1976). Metabolism in rats: M. Stiasni *et al.*, *ibid.* **15,** 474 (1967). Toxicity study: T. B. Gaines, *Toxicol. Appl. Pharmacol.* **14,** 515 (1969). TLC determn in food crops: S. Traore, J. J. Aaron, *Talanta* **28,** 765 (1981). *Review:* D. Eichler, *Residue Rev.* **41,** 65-112 (1972).

Yellow crystals, mp 53-54°. bp$_{0.01}$ 140-142°. Vapor pressure at 20°C: 1.3 × 10^{-4} mm Hg. Soly in water at 25°: 40 mg/l. Sol in CCl$_4$, ether, toluene. Stable in soln up to pH 9. Non-corrosive. LD$_{50}$ in male, female rats (mg/kg): 1600, 1730 orally (Gaines).
O,O-**Diethyl analog.** Bromophos-ethyl; ENT-27258; OMS-659; S-2225; Nexagan. C$_{10}$H$_{12}$BrCl$_2$O$_3$PS; mol wt 394.04. Colorless to pale yellow oil, bp$_{0.001}$ 122-133°. Vapor pressure at 30°: 4.6 × 10^{-5} mm Hg. Soly in water at 25°: 2 mg/l.
USE: Insecticide; acaricide.

1443. Bromopride. [4093-35-0] 4-Amino-5-bromo-*N*-[2-(diethylamino)ethyl]-2-methoxybenzamide; 4-amino-5-bromo-*N*-[2-(diethylamino)ethyl]-*o*-anisamide; Emepride; Emoril; Viadil. C$_{14}$H$_{22}$BrN$_3$O$_2$; mol wt 344.25. C 48.85%, H 6.44%, Br 23.21%, N 12.21%, O 9.29%. Prepn: M. L. Thominet, **BE 620543;** *idem,* **US 3177252** (1962, 1965 both to Soc. d'Etudes Sci. Ind. de l'Ile-de-France); H. Mori, K. Shibata, **DE 2119724** (1971 to Teikoku Hormone Manuf. Co.), *C.A.* **76,** 99375e (1972). Action on guinea-pig ileum: J. Fontaine, J. J. Reuse, *Arch. Int. Pharmacodyn. Ther.* **213,** 322 (1975). Effects on intestinal peristalsis in dogs: P. Sava *et al.*, *Arzneim.-Forsch.* **29,** 799 (1979). Clinical study: M. Fischer *et al.*, *Fortschr. Med.* **97,** 883 (1979). Pharmacokinetics: P. W. Lücker *et al.*, *Arzneim.-Forsch.* **33,** 453 (1983).

Hydrochloride. [52423-56-0] Cascapride; Plesium; Praiden; Valopride; Viaben. C$_{14}$H$_{22}$BrN$_3$O$_2$.HCl; mol wt 380.71.
Dihydrochloride monohydrate. Opridan. C$_{14}$H$_{22}$BrN$_3$O$_2$.2HCl.H$_2$O; mol wt 435.18.
THERAP CAT: Antiemetic.

1444. β-Bromopropionic Acid. [590-92-1] 3-Bromopropanoic acid. C$_3$H$_5$BrO$_2$; mol wt 152.98. C 23.55%, H 3.29%, Br 52.23%, O 20.92%. Prepd by the action of hydrobromic acid on acrylic acid, on hydracrylic acid, and on ethylene cyanohydrin: Kendall, McKenzie, *Org. Synth.* **3**, 25 (1923). Prepn of the ethyl ester: *eidem, ibid.* 51; of the methyl ester: Mozingo, Patterson, *ibid.* **20**, 64 (1940).

Br⁀⁀COOH

Plates from CCl₄. mp 62.5°. pK (25°): 4.00. Sol in water, alcohol, ether, chloroform, benzene. Aq alkalies hydrolyze β-bromopropionic acid to hydracrylic acid.

Ethyl ester. [539-74-2] $C_5H_9BrO_2$. Liquid. Pungent odor. Becomes yellow on exposure to light. d_4^{18} 1.4123. bp₄₄ 112°; bp₁₂ 70°. n_D^{18} 1.4569. Insol in water; miscible with alchol, ether. *Protect from light.*

Methyl ester. [3395-91-3] $C_4H_7BrO_2$. Liquid. d¹⁵ 1.4897. bp₂₇ 80°. bp₁₈ 65°.

1445. Bromopropylate. [18181-80-1] 4-Bromo-α-(4-bromophenyl)-α-hydroxybenzeneacetic acid 1-methyl ethyl ester; 4,4′-dibromobenzilic acid isopropyl ester; isopropyl 4,4′-dibromobenzilate; phenisobromolate; ENT-27552; GS-19851; Acarol; Folbex VA; Neoron. $C_{17}H_{16}Br_2O_3$; mol wt 428.12. C 47.69%, H 3.77%, Br 37.33%, O 11.21%. Contact acaricide with residual activity. Prepn: **FR 1504969**; K. Gubler, **US 3639446** (1967, 1972 to Geigy). Toxicity data: E. E. Kenaga, W. E. Allison, *Bull. Entomol. Soc. Am.* **15**, 85 (1969). Persistence in soil: W. B. Wheeler *et al., J. Environ. Qual.* **2**, 115 (1973). Control of bee mites: A. Rojahn, *Dtsch. Tieraerztl. Wochenschr.* **92**, 103 (1985). GC determn in honey: L. Torreti, A. Simonella, *J. High Resolut. Chromatogr.* **13**, 142 (1990).

Cryst solid, mp 77°. Vapor pressure at 20°: 5.1 × 10⁻⁸ mm Hg. Soly in water at 20°: <5 mg/l. Readily sol in most organic solvents. LD₅₀ orally in rats: 5000 mg/kg (Kenaga, Allison).
USE: Acaricide.

1446. 5-Bromosalicylhydroxamic Acid. [5798-94-7] 5-Bromo-*N*,2-dihydroxybenzamide; Brosalamid; Bromocyl. C_7H_6-BrNO₃; mol wt 232.03. C 36.24%, H 2.61%, Br 34.44%, N 6.04%, O 20.69%. Prepd by direct bromination of salicylhydroxamic acid in acetic acid: Urbanski *et al., Nature* **170**, 753 (1952); *Bull. Acad. Pol. Sci. Cl. 3* **1**, 319 (1953); Hornung, Krakowska, *Gruzlica* **20**, 469 (1952).

Crystals from alcohol, dec 232°. Very sparingly sol in water. Forms a water-soluble sodium salt.
THERAP CAT: Antibacterial (tuberculostatic).

1447. Bromosuccinic Acid. [923-06-8] 2-Bromobutanedioic acid; monobromosuccinic acid. $C_4H_5BrO_4$; mol wt 196.98. C 24.39%, H 2.56%, Br 40.56%, O 32.49%. Prepn of *dl*-form by bromination of succinic acid: Kekulé, *Ann.* **117**, 120 (1861); from fumaric acid and HBr: Fittig, *Ann.* **188**, 42 (1877); by bromination of succinyl bromide or succinic anhydride: Hughes, Watson, *J. Chem. Soc.* **1930**, 1733. Prepn of *l*-form from *l*-aspartic acid: Walden, *Ber.* **29**, 133 (1896); Karrer *et al., Helv. Chim. Acta* **30**, 271 (1947).

dl-**Form.** Crystals, mp 161°. d 2.07. Soluble in 5.5 parts water; sol in alc.

d-**Form.** [3972-41-6] (2*R*)-2-Bromobutanedioic acid. Crystals, mp 172°. $[\alpha]_D^{15}$ +41.9° (c = 5); $[\alpha]_D^{20}$ +67.9° (ether). *Ref:* Levene, Mikeska, *J. Biol. Chem.* **55**, 795 (1923); Clough, *J. Chem. Soc.* **129**, 1674 (1926).

l-**Form.** [20859-23-8] (2*S*)-2-Bromobutanedioic acid. Crystals, dec 177-178°. $[\alpha]_D$ −43.8° (c = 6), −65.0° (c = 6 in abs alc), −73.5° (c = 6 in acetone).

1448. *N*-Bromosuccinimide. [128-08-5] 1-Bromo-2,5-pyrrolidinedione; succinbromimide; NBS. $C_4H_4BrNO_2$; mol wt 177.99. C 26.99%, H 2.27%, Br 44.89%, N 7.87%, O 17.98%. Prepn: K. Ziegler *et al., Ann.* **551**, 109 (1942). In bromination of olefins: *eidem, ibid.* 80. The method has been extended to other classes of compds by the use of catalysts: H. Schmid, P. Karrer, *Helv. Chim. Acta* **29**, 573 (1946); H. Schmid, *ibid.* 1144. In oxidation of aldehydes: Y. F. Cheung, *Tetrahedron Lett.* **1979**, 3809. Review of uses: C. Djerassi, *Chem. Rev.* **43**, 271-317 (1948); R. Filler, *ibid.* **63**, 21-43 (1963); J. S. Pizey, *Synthetic Reagents* **vol. 2** (John Wiley, New York, 1974) pp 1-63.

Orthorhombic bisphenoidal crystals, mp 173-175° (slight decompn). Faint odor of bromine. d 2.098. Soly (g/100 g of solvent at 25°): water 1.47; *tert*-butanol 0.73; acetone 14.40; carbon tetrachloride 0.02; hexane 0.006; glacial acetic acid 3.10.
Caution: Highly irritating to eyes, skin, mucous membranes.
USE: In bromination of olefins; in oxidation of alcohols to aldehydes and ketones and of aldehydes to acid bromides.

1449. Bromotoluene. C_7H_7Br; mol wt 171.04. C 49.16%, H 4.13%, Br 46.72%. Prepn of *o*-isomer: Bourgeois, *Ber.* **28**, 2312 (1895). Prepn of *m*-isomer: Acree, *Ber.* **37**, 994 (1904); Kohn, Bum, *Monatsh. Chem.* **33**, 924 (1912); Feitler, *Z. Phys. Chem.* **4**, 77 (1889). Prepn of *m*-, *o*- and *p*-isomers: Bigelow *et al., Org. Synth.* **coll. vol. I** (2nd ed., 1941) pp 133, 135, 136 resp. Isomerization of *p*-isomer yields mixture of monobromotoluenes containing 48.2% of *o*-isomer: Crump, **US 3077503** (1963 to Dow). Separation of *o*- and *p*-isomers by clathration: Coscia, **US 3114784** (1963 to Am. Cyanamid).

m-**Bromotoluene.** 1-Bromo-3-methylbenzene; 3-bromotoluene; *m*-tolylbromide. Liquid. d_4^{20} 1.4099; d_4^{58} 1.309; d_4^{184} 1.201. mp −39.8°. bp₇₆₀ 183.7°; bp₄₀₀ 160°; bp₂₀₀ 138°; bp₁₀₀ 117.8°; bp₆₀ 104.1°; bp₄₀ 93.9°; bp₂₀ 78.1°; bp₁₀ 64.0°; bp₅ 50.8°; bp₁.₀ 14.8°. n_D^{20} 1.551. Absorption spectrum: Purvis, *J. Chem. Soc.* **99**, 1706, 1710 (1911). Sol in alc, ether, benzene.

o-**Bromotoluene.** Colorless liq. d_{15}^{15} 1.431. bp 181°. mp −26°. n_D^{20} 1.555. Practically insol in water; misc with alc, benzene.

p-**Bromotoluene.** Crystals from abs alc. d_{55}^{35} 1.3959; d_{50}^{50} 1.3856; d_{100}^{100} 1.3637; d¹⁸⁴ 1.1931. mp 28.5°. bp₇₆₀ 184.5°; bp₁₀₀ 116.4°; bp₆₀ 102.3°; bp₄₀ 91.8°; bp₂₀ 75.2°; bp₁₀ 61.1°; bp₅ 47.5°; bp₁.₀ 10.3°. n_D^{20} 1.5490. Sol in alc, ether, benzene.
Caution: Irritants!

1450. 5-Bromouracil. [51-20-7] 5-Bromo-2,4(1*H*,3*H*)-pyrimidinedione. $C_4H_3BrN_2O_2$; mol wt 190.98. C 25.16%, H 1.58%, Br 41.84%, N 14.67%, O 16.75%. Major chemical mutagen; incorporates into DNA, altering base-pair sequencing by replacing thymine. Prepn: H. L. Wheeler, H. F. Merriam, *Am. Chem. J.* **29**, 478 (1903); P. A. Levene, F. B. LaForge, *Ber.* **45**, 608 (1912). UV effects on DNA containing 5-bromouracil: A. G. Skavronskaya *et al., Mutat. Res.* **6**, 319 (1968); F. Hutchinson, *Q. Rev. Biophys.* **6**, 201 (1973). Mechanisms of mutagenesis: I. Pietrzykowska, *Mutat. Res.* **19**, 1 (1973); E. M. Witkin, E. C. Parisi, *ibid.* **25**, 407 (1974); B.

Rydberg, *Mol. Gen. Genet.* **152**, 19 (1977). Crystal structure and base stacking properties: H. Sternglanz, C. E. Bugg, *Biochim. Biophys. Acta* **378**, 1 (1975).

Prisms from water, mp 293°.
USE: Exptly as mutagen.

1451. Bromoxynil. [1689-84-5] 3,5-Dibromo-4-hydroxybenzonitrile; 3,5-dibromo-4-hydroxyphenyl cyanide; 2,6-dibromo-4-cyanophenol; broxynil; ENT-20852; MB-10064. $C_7H_3Br_2NO$; mol wt 276.92. C 30.36%, H 1.09%, Br 57.71%, N 5.06%, O 5.78%. Selective contact herbicide. Prepn: K. Auwers, J. Reis, *Ber.* **29**, 2359 (1896); E. Müller *et al.*, *Ber.* **92**, 2278 (1959); K. Carpenter *et al.*, *Weed Res.* **4**, 175 (1964); **FR 1375311** (1964 to May & Baker), *C.A.* **62**, 3982h (1965). Herbicidal activity: K. Carpenter, B. J. Heywood, *Nature* **200**, 28 (1963). HPLC analysis: J. C. Van Damme, M. Galoux, *J. Chromatogr.* **190**, 401 (1980). Persistence in soil: A. E. Smith, *Pestic. Sci.* **11**, 341 (1980).

Colorless solid, mp 194-195°. pKa 4.06. Subl at 135° (0.15 mm Hg). Slightly volatile in steam. Soly in (g/l) at 25°: water 0.13; methanol 90; acetone 170; tetrahydrofuran 410. LD_{50} orally in mice: 111 mg/kg (Carpenter, 1964).
Octanoate. [1689-99-2] MB-10731; NPH-1320; RP-16272; Brominal; Buctril. $C_{15}H_{17}Br_2NO_2$; mol wt 403.11. Waxy solid, mp 45-46°. Low volatility; subl at 90° (0.1 mm Hg). Insol in water.
USE: Herbicide.

1452. Bromperidol. [10457-90-6] 4-[4-(4-Bromophenyl)-4-hydroxy-1-piperidinyl]-1-(4-fluorophenyl)-1-butanone; 4-[4-(p-bromophenyl)-4-hydroxypiperidino]-4'-fluorobutyrophenone; R-11333; Azurene; Impromen; Tesoprel. $C_{21}H_{23}BrFNO_2$; mol wt 420.32. C 60.01%, H 5.52%, Br 19.01%, F 4.52%, N 3.33%, O 7.61%. Bromine analog of haloperidol, *q.v.* Prepn (no data): P. A. J. Janssen, **GB 895309**; *idem*, **US 3438991** (1962, 1969 both to Janssen); C. J. E. Niemegeers, P. A. J. Janssen, *Arzneim.-Forsch.* **24**, 45 (1974). Prepn of ^{82}Br bromperidol and preliminary tissue distribution studies: S. H. Vincent *et al.*, *J. Med. Chem.* **23**, 75 (1980). Pharmacological study: C. J. E. Niemegeers, P. A. J. Janssen, *Life Sci.* **24**, 2201 (1979). Preliminary clinical study: B. Woggon *et al.*, *Int. Pharmacopsychiatry* **14**, 213 (1979). Radioimmunoassay: E. Van Den Eeckhout *et al.*, *Eur. J. Drug Metab. Pharmacokinet.* **5**, 45 (1980).

Off-white amorphous or microcrystalline powder, mp 155-158°. uv max: 245 nm. pKa 8.6-8.7. Soly in water: 0.09 mg/ml; in 0.1M tartaric, lactic, citric and acetic acids: ≥10 mg/ml.
THERAP CAT: Antipsychotic.

1453. Brompheniramine. [86-22-6] γ-(4-Bromophenyl)-N,N-dimethyl-2-pyridinepropanamine; 2-[p-bromo-α-(2-dimethylaminoethyl)benzyl]pyridine; 1-(p-bromophenyl)-1-(2-pyridyl)-3-dimethylaminopropane; 3-(p-bromophenyl)-3-(2-pyridyl)-N,N-dimethylpropylamine; parabromdylamine. $C_{16}H_{19}BrN_2$; mol wt 319.25. C 60.20%, H 6.00%, Br 25.03%, N 8.77%. Prepn from α-(p-bromophenyl)-α-(β-dimethylaminoethyl)-2-pyridylacetonitrile: Sperber *et al.*, **US 2567245** (1951); **US 2676964** (1954 to Schering). Prepn of d-form: L. A. Walter, **US 3061517** (1962 to Schering).

Oily liquid. Slightly yellow color. Characteristic amine-like odor. $bp_{0.5}$ 147-152°. Soluble in dilute acids.
Maleate. [980-71-2] Dimegan; Dimetane; Dimotane; Ilvin; Nagemid; Symptom 3; Veltane. $C_{16}H_{19}BrN_2 \cdot C_4H_4O_4$; mol wt 435.32. Crystals, mp 132-134°. Freely sol in water; sol in alcohol, chloroform. pH of a 2% aq soln about 5. Slightly sol in ether, benzene.
d-Form. [132-21-8] Dexbrompheniramine; (+)-parabromdylamine. Oily liquid. $[\alpha]_D^{25}$ +42.7° (c = 1 in DMF).
d-Form maleate. [2391-03-9] Disomer; Ebalin. $C_{16}H_{19}BrN_2 \cdot C_4H_4O_4$; mol wt 435.32. Crystals, mp 103-113°.
THERAP CAT: Antihistaminic.

1454. Bromphenol Blue. [115-39-9] 4,4'-(1,1-Dioxido-3H-2,1-benzoxathiol-3-ylidene)bis[2,6-dibromophenol]; 4,4'-(3H-2,1-benzoxathiol-3-ylidene)bis[2,6-dibromophenol] S,S-dioxide; α,α-bis(3,5-dibromo-4-hydroxyphenyl)-α-hydroxy-o-toluenesulfonic acid γ-sultone; 3,3',5,5'-tetrabromophenolsulfonphthalein; bromophenol blue; Albutest. $C_{19}H_{10}Br_4O_5S$; mol wt 669.96. C 34.06%, H 1.50%, Br 47.71%, O 11.94%, S 4.79%. Prepd by slow addition of excess bromine to a hot soln of phenolsulfonphthalein in glacial acetic acid: White, Acree, *J. Am. Chem. Soc.* **41**, 1205 (1919); Orndorff, Sherwood, *ibid.* **45**, 495 (1923). *See also* Bromcresol Green.

Elongated hexagonal prisms from acetic acid + acetone. Dec 279°. (An orange discoloration with formation of a green sublimate sets in at 210°). Soluble in water (about 0.4 g/100 ml); more sol in methyl and ethyl alcohol, and in benzene. Freely sol in NaOH solns with the formation of a water-sol sodium salt. pKa = 4.0. pH 3.0 yellow; pH 4.6 purple. To prepare a soln for use as pH indicator, dissolve 0.10 g in 7.45 ml N/50 NaOH and dil with water to 250 ml.
Sodium salt. [34725-61-6] $C_{19}H_9Br_4NaO_5S$; mol wt 691.94.
USE: As pH indicator. With mercuric chloride as a semiquantitative protein stain.

1455. Bromthymol Blue. [76-59-5] 4,4'-(1,1-Dioxido-3H-2,1-benzoxathiol-3-ylidene)bis[2-bromo-3-methyl-6-(1-methylethyl)phenol]; 4,4'-(3H-2,1-benzoxathiol-3-ylidene)bis[2-bromo-3-methyl-6-(1-methylethyl)phenol] S,S-dioxide; α,α-bis(6-bromo-5-hydroxycarvacryl)-α-hydroxy-o-toluenesulfonic acid γ-sultone;

3,3′-dibromothymolsulfonphthalein; bromothymol blue. $C_{27}H_{28}$-Br_2O_5S; mol wt 624.38. C 51.94%, H 4.52%, Br 25.59%, O 12.81%, S 5.13%. Prepd by the action of bromine on thymol blue in glacial acetic acid: Clark, Lubs, *J. Wash. Acad. Sci.* **5**, 609 (1915); *ibid.* **6**, 482 (1916); *Chem. Zentralbl.* **1916**, I, 175; *ibid.* II, 1068. *See also* Bromcresol Green.

Cream-colored crystals. Sparingly sol in water; sol in alcohol and in aq solns of alkalies. Also sol in ether. Less sol in benzene, toluene, xylene. Practically insol in petr ether. pKa = 7.0. pH 6.0 yellow; pH 7.6 blue. To prepare a soln for use as pH indicator, dissolve 0.10 g in 8.0 ml *N*/50 NaOH and dil with water to 250 ml. To prepare a soln for use as indicator in volumetric work, dissolve 0.1 g in 100 ml of 50% alc.

Sodium salt. [34722-90-2] $C_{27}H_{27}Br_2NaO_5S$; mol wt 646.37.

USE: As pH indicator, especially in microbiological culture media. In fiber-optic instrumentation systems for chloride and potassium determn.

1456. Bromuconazole. [116255-48-2] 1-[[4-Bromo-2-(2,4-dichlorophenyl)tetrahydro-2-furanyl]methyl]-1*H*-1,2,4-triazole; 1-[(2*RS*,4*RS*;2*RS*,4*SR*)-4-bromo-2-(2,4-dichlorophenyl)tetrahydrofurfuryl]-1*H*-1,2,4-triazole; LS-860263; Granit. $C_{13}H_{12}BrCl_2N_3O$; mol wt 377.06. C 41.41%, H 3.21%, Br 21.19%, Cl 18.80%, N 11.14%, O 4.24%. Ergosterol biosynthesis inhibiting triazole. Prepn: A. Greiner, R. Pepin, **EP 258161** (1988 to Rhone Poulenc), *C.A.* **109**, 110440v (1988). Properties and antifungal activity: R. Pepin *et al.*, *Brighton Crop Prot. Conf. - Pests Dis.* **1990**, 439. Effect on fungus ultrastructure: M. Mangin-Peyrard, R. Pepin, *Z. Pflanzenkrankh. Pflanzenschutz* **103**, 142 (1996). Determn by TLC in water: S. Butz, H.-J. Stan, *Anal. Chem.* **67**, 620 (1995); by GC with atomic emission detection in foodstuffs: H.-J. Stan, M. Linkerhägner, *J. Chromatogr. A* **750**, 369 (1996). Field trials in combination with iprodione, *q.v.*: P. Duvert *et al.*, *Agro-Food-Ind. Hi-Tech* **7**, 34 (1996); in combination with prochloraz, *q.v.*: *eidem*, *Phytoma* **490**, 32 (1997).

White to off-white odorless powder, mp 84°. Moderate to high soly in organic solvents; soly in water 50 mg/l. Vapor pressure (25°): 0.3×10^{-7} mm Hg. LD$_{50}$ orally in rats, mice: 365, 1151 mg/kg; LD$_{50}$ dermally in rats: >2000 mg/kg; LD$_{50}$ by inhalation in rabbits: >5 mg/l; LC$_{50}$(96 hr) in rainbow trout, bluegill sunfish (mg/l): 1.7, 3.1 (Pepin).

USE: Agricultural fungicide.

1457. Bronopol. [52-51-7] 2-Bromo-2-nitro-1,3-propanediol; β-bromo-β-nitrotrimethyleneglycol; Bronosol. $C_3H_6BrNO_4$; mol wt 199.99. C 18.02%, H 3.02%, Br 39.95%, N 7.00%, O 32.00%. Prepn: E. Schmidt, R. Wilkendorf, *Ber.* **52**, 389 (1919); R. Wessendorf, **DE 1804068**; *idem*, **DE 1954173**; *idem*, **US 3658921**;

idem, **US 3711561** (1970, 1971, 1972, 1973 all to Henkel & Cie). Antibacterial, antifungal and toxic properties: Z. Eckstein *et al.*, *Bull. Acad. Pol. Ser. Sci. Chim.* **11**, 687 (1963); B. Croshaw *et al.*, *J. Pharm. Pharmacol.* **16** (Suppl), 127T (1964); N. G. Clark *et al.*, *eidem*, **GB 1057131**; *eidem*, **US 3558788** (1967, 1971 both to Boots); R. J. Stretton, T. W. Mason, *J. Appl. Bacteriol.* **36**, 61 (1973). Biotransformation and distribution in rats: H. S. Buttar, R. H. Downie, *Toxicol. Lett.* **6**, 101 (1980). As a nitrosating agent for diethanolamine: I. Schmeltz, A. Wenger, *Food Cosmet. Toxicol.* **17**, 105 (1979); R. L. Elder, *J. Environ. Pathol. Toxicol.* **4**, 47 (1980). Review of bronopol and other preservatives: B. Croshaw, *J. Soc. Cosmet. Chem.* **28**, 3-16 (1977).

Odorless crystals from ethyl acetate-chloroform, mp 120-122°. Sol in water, alcohol, ethyl acetate. Slightly sol in chloroform, acetone, ether, benzene. Insol in ligroin. LD$_{50}$ in mice, rats (mg/kg): 350, 400 orally (Croshaw).

USE: Preservative in cosmetics and toiletries. Antiseptic.

1458. Brostallicin. [203258-60-0] *N*-[5-[[[5-[[[2-[(Aminoiminomethyl)amino]ethyl]amino]carbonyl]-1-methyl-1*H*-pyrrol-3-yl]amino]carbonyl]-1-methyl-1*H*-pyrrol-3-yl]-4-[[[4-[(2-bromo-1-oxo-2-propen-1-yl)amino]-1-methyl-1*H*-pyrrol-2-yl]carbonyl]amino]-1-methyl-1*H*-pyrrole-2-carboxamide. $C_{30}H_{35}BrN_{12}O_5$; mol wt 723.59. C 49.80%, H 4.88%, Br 11.04%, N 23.23%, O 11.06%. DNA minor groove binder; bromoacryloyl derivative of distamycin A, *q.v.* Prepn: P. Cozzi *et al.*, **WO 9804524**; *eidem*, **US 6482920** (1998, 2002 both to Pharmacia & Upjohn). LC/MS/MS determn in plasma: S. Calderoli *et al.*, *J. Pharm. Biomed. Anal.* **32**, 601 (2003). Mechanism of action studies: C. Geroni *et al.*, *Cancer Res.* **62**, 2332 (2002); A. Fedier *et al.*, *Br. J. Cancer* **89**, 1559 (2003). Clinical pharmacokinetics and evaluation in advanced solid tumors: A. J. ten Tije *et al.*, *Clin. Cancer Res.* **9**, 2957 (2003); A. C. Lockhart *et al.*, *ibid.* **10**, 468 (2004). Review of clinical development: M. Broggini *et al.*, *Anti-Cancer Drugs* **15**, 1-6 (2004).

Hydrochloride. [203258-38-2] PNU-166196. $C_{30}H_{35}BrN_{12}$-O_5.HCl; mol wt 760.05. Yellow solid. uv max (95% ethanol): 312.0 nm (ε 48792).

THERAP CAT: Antineoplastic.

1459. Brotizolam. [57801-81-7] 2-Bromo-4-(2-chlorophenyl)-9-methyl-6*H*-thieno[3,2-*f*][1,2,4]triazolo[4,3-*a*][1,4]diazepine; 2-bromo-4-(*o*-chlorophenyl)-9-methyl-6*H*-thieno[3,2-*f*]-*s*-triazolo-[4,3-*a*][1,4]diazepine; 8-bromo-6-(*o*-chlorophenyl)-1-methyl-4*H*-s-triazolo[3,4-*c*]thieno[2,3-*e*]-1,4-diazepine; WE-941-BS; Lendorm; Lendormin; Mederantil; Nimbisan; Sintonal. $C_{15}H_{10}BrClN_4S$; mol wt 393.69. C 45.76%, H 2.56%, Br 20.30%, Cl 9.00%, N 14.23%, S 8.14%. One of a class of triazolo-1,4-thienodiazepines having psychotropic activity. Prepn: K. H. Weber *et al.*, **DE 2410030**; *eidem*, **US 4094984** (1975, 1978 to Boehringer, Ing.); *eidem*, *Ann.* **8**, 1257 (1978). Pharmacodynamics: J. Gruenberger *et al.*, *Curr. Ther. Res.* **24**, 427 (1978). Bioavailability: B. Saletu *et al.*, *Arzneim.-Forsch.* **29**, 700 (1979). Pharmaco-EEG study: M. Fink, P. Irwin, *Clin. Pharmacol. Ther.* **30**, 336 (1981). Toxicity studies: C. Hewett *et* al., Arzneim.-Forsch. **36**, 592 (1986). Series of articles on pharmacology, pharmacokinetics, toxicology: *ibid.* 517-620. Veterinary trial as appetite stimulant: A. van Miert *et al.*, *J. Vet. Pharmacol. Ther.* **12**, 147 (1989). Review of pharmacology and clinical efficacy: M. S. Langley, S. P. Clissold, *Drugs* **35**, 104-122 (1988).

Colorless crystals from ethanol, mp 212-214°. LD_{50} in mice, rats (mg/kg): >10000, >10000 orally; 920, 1000 i.p. (Hewett).

THERAP CAT: Sedative, hypnotic.

THERAP CAT (VET): Appetite stimulant.

1460. Brovincamine. [57475-17-9] $(3\alpha,14\beta,16\alpha)$-11-Bromo-14,15-dihydro-14-hydroxyeburnamenine-14-carboxylic acid methyl ester; *cis*-11-bromovincamine. $C_{21}H_{25}BrN_2O_3$; mol wt 433.35. C 58.20%, H 5.82%, Br 18.44%, N 6.46%, O 11.08%. Cerebrovascular agent, vincamine deriv. Prepn: P. Pfaeffli, **DE 2458164**; *idem*, **US 4146643** (1975, 1979 both to Sandoz). HPLC determn in plasma: R. R. Brodie, L. F. Chasseaud, *J. Chromatogr.* **228**, 413 (1982). Effect on autonomic nervous system in laboratory animals: K. Kushiku *et al.*, *J. Pharmacobio-Dyn.* **7**, 177 (1984). Cardiovascular effects, mechanistic studies: *eidem*, *Clin. Exp. Pharmacol. Physiol.* **12**, 121 (1985). Mode of action studies: T. Katsuragi *et al.*, *Gen. Pharmacol.* **15**, 43 (1984). Clinical trials in patients with multi-infarct dementia: S. Hagstadius *et al.*, *Psychopharmacology* **83**, 321 (1984).

Crystals from isopropanol, mp 214° (dec). $[\alpha]_D^{20}$ +8.7° (1% in $CHCl_3$).

Hydrogen fumarate. [84964-12-5] BV-26-723; Sabromin; Zabromin. $C_{21}H_{25}BrN_2O_3 \cdot C_4H_4O_4$; mol wt 549.42. mp 144°. $[\alpha]_D^{20}$ +4.7° (0.388% in H_2O).

THERAP CAT: Vasodilator (peripheral).

1461. Broxuridine. [59-14-3] 5-Bromo-2′-deoxyuridine; 5-bromouracil deoxyriboside; BrdUrd; BUdR; NSC-38297; Broxine; Neomark; Radibud. $C_9H_{11}BrN_2O_5$; mol wt 307.10. C 35.20%, H 3.61%, Br 26.02%, N 9.12%, O 26.05%. Thymidine, *q.v.*, analog. Preferentially incorporated into cellular DNA in place of thymidine during replication causing increased radiosensitivity of the cell. Prepn and inhibition of DNA biosynthesis: R. E. Beltz, D. W. Visser, *J. Am. Chem. Soc.* **77**, 736 (1955); T. J. Bardos *et al.*, *ibid.* 4279. X-ray crystallography: J. Iball *et al.*, *Nature* **209**, 1230 (1966). HPLC determn in plasma: D. A. Ganes, J. G. Wagner, *J. Chromatogr.* **432**, 233 (1988). Clinical studies in brain tumors: H. S. Greenberg *et al.*, *Neurology* **44**, 1715 (1994); T. L. Phillips *et al.*, *Int. J. Radiat. Oncol. Biol. Phys.* **21**, 709 (1991). Review of radiobiology and therapeutic potential: T. J. Kinsella *et al.*, *ibid.* **10**, 1399-1406 (1984); of genetic toxicology: S. M. Morris, *Mutat. Res.* **258**, 161-188 (1991); of role in management of CNS tumors: A. Freese *et al.*, *J. Neuro-Oncol.* **20**, 81-95 (1994). Series of reviews on diagnostic and research use: F. Dolbeare, *Histochem. J.* **27**, 339-369, 923-964 (1995); *ibid.* **28**, 531-575 (1996).

Crystals from abs ethanol, mp 187-189° (Beltz, Visser); also reported as mp 181-183° (Bardos). uv max (in HCl): 280 nm (ε 9.9 × 10^{-3}); uv max (in NaOH): 277 nm (ε 7.2 × 10^{-3}).

USE: Research tool for measuring DNA synthesis.

THERAP CAT: Antineoplastic adjunct (radiosensitizer); diagnostic aid (tumor cell label for cytokinetic analysis).

1462. Broxyquinoline. [521-74-4] 5,7-Dibromo-8-quinolinol; 5,7-dibromo-8-hydroxyquinoline; Brodiar; Broxykinolin; Colepur; Fenilor; Intensopan. $C_9H_5Br_2NO$; mol wt 302.95. C 35.68%, H 1.66%, Br 52.75%, N 4.62%, O 5.28%. Prepn by bromination of 5-formyl-8-quinolinol: Matsumura, Ito, *J. Am. Chem. Soc.* **77**, 6671 (1955); by bromination of 8-hydroxyquinaldine: Irving, Pinnington, *J. Chem. Soc.* **1957**, 285; by bromination of 8-quinolinol: Luis, Palomo, *Afinidad* **28**, 163 (1951), *C.A.* **47**, 10533d (1953); Zinnei, Fiedler, *Arch. Pharm.* **291**, 493 (1958); Aristov, Kostina, *Zh. Obshch. Khim.* **34**, 3421 (1964). Crystal structure: Kashino, Haisa, *Bull. Chem. Soc. Jpn.* **46**, 1094 (1973). Metabolism: Rodriguez, Close, *Biochem. Pharmacol.* **17**, 1647 (1968).

Monoclinic needles from alc. d 2.189. mp 196°. Freely sol in chloroform, alcohol, benzene, acetic acid; slightly sol in ether. Practically insol in water.

USE: In testing for Cu, Fe, Ti.

THERAP CAT: Antiseptic; disinfectant.

1463. Bruceantin. [41451-75-6] $[11\beta,12\alpha,15\beta]$-15-[[(2E)-3,4-Dimethyl-1-oxo-2-penten-1-yl]oxy]-13,20-epoxy-3,11,12-trihydroxy-2,16-dioxopicras-3-en-21-oic acid methyl ester; NSC-165563. $C_{28}H_{36}O_{11}$; mol wt 548.59. C 61.30%, H 6.61%, O 32.08%. Antileukemic quassinoid from the simaroubaceous tree *Brucea antidysenterica* J. F. Mill. Isoln and structure: S. M. Kupchan, **DE 2347576**; S. M. Kupchan, R. W. Britton, **US 3969369** (1975, 1976 both to Research Corp.); S. M. Kupchan *et al.*, *J. Org. Chem.* **40**, 648 (1975). Mode of action: L. L. Liao *et al.*, *Mol. Pharmacol.* **12**, 167 (1976). *In vivo* study: R. K. Johnson *et al.*, *Cancer Treat. Rep.* **62**, 1535 (1978). Pharmacology: S. M. Sieber *et al.*, *ibid.* **60**, 1127 (1976); M. Fresno *et al.*, *Biochim. Biophys. Acta* **518**, 104 (1978). Clinical study: A. Y. Bedikian *et al.*, *Proc. Am. Assoc. Cancer Res.* **20**, 193 (1979). Toxicologic evaluation: T. R. Castles *et al.*, *U.S. NTIS Report* **PB-257175** (1976) 348 pp. Synthetic studies: O. D. Dailey, P. L. Fuchs, *J. Org. Chem.* **45**, 216 (1980); R. J. Pariza, P. L. Fuchs, *ibid.* **48**, 2306 (1983). Total synthesis: M. Sasaki *et al.*, *ibid.* **55**, 528 (1990).

Crystals from ether, mp 225-226°. $[\alpha]_D^{25}$ −43° (c = 0.31 in pyridine). uv max (ethanol): 280, 221 nm (ε 8680, 18000); (ethanol, NaOH): 328, 221 nm (ε 7290, 28600). LD_{50} in male, female mice (mg/kg): 1.95, 2.58 i.v. (Castles).

1464. Brucine. [357-57-3] 2,3-Dimethoxystrychnidin-10-one; 10,11-dimethoxystrychnine. $C_{23}H_{26}N_2O_4$; mol wt 394.47. C 70.03%, H 6.64%, N 7.10%, O 16.22%. Toxic alkaloid resembling strychnine. Isoln from Strychnos seeds *(Strychnos nux-vomica* L. and *S. ignatii* Berg., *Loganiaceae):* C. Hartwick, P. Geiger, *Arch. Pharm.* **239**, 491 (1901). Prepn from strychnine, *q.v.*: E. Tedeschi *et al., Tetrahedron* **24**, 4573 (1968). ^{13}C-NMR study: E. Wenkert *et al., J. Org. Chem.* **43**, 1099 (1978). Electrophoretic determn in seeds: Y.-Y. Zong, C. Che, *Planta Med.* **61**, 456 (1995). Toxicity data: M. H. Malone *et al., J. Ethnopharmacol.* **35**, 295 (1992). Use as reagent for sulfur compounds: M. K. Tummuru *et al., Analyst* **109**, 1105 (1984); for nitrates: J. Masini *et al., Anal. Chem.* **69**, 1077 (1997). Review of structural elucidation: H. L. Holmes, Elucidation of the Structure of Strychnine and Brucine in *The Alkaloids* **vol. I**, R. H. F. Manske, Ed. (Academic Press, New York, 1950) pp 377-420.

Needles from acetone + water, mp 178°. $[\alpha]_D$ −127° (chloroform), −85° (in abs alcohol). uv max (ethanol): 263, 301 nm (log ε 4.09, 3.93). *Poisonous.*

Tetrahydrate. [5892-11-5] $C_{23}H_{26}N_2O_4.4H_2O$; mol wt 466.53. Monoclinic prisms. Also forms a dihydrate. Very bitter taste. Bitterness threshold 1:220,000. *Handle dry powder in hood only.* Becomes anhydr at 100°. One gram dissolves in 0.8 ml methanol, 1.3 ml alcohol, 5 ml chloroform, 25 ml ethyl acetate, 36 ml glycerol, about 100 ml benzene, 187 ml ether, 1320 ml water, 750 ml boiling water. pH of satd water soln 9.5. pK_1 6.04, pK_2 11.7. LD_{50} in mice (mg/kg): 12.0 i.v.; 62.0 i.p.; 150.0 orally (Malone).

Sulfate heptahydrate. [60583-39-3] $(C_{23}H_{26}N_2O_4)_2.H_2SO_4.7H_2O$; mol wt 1013.12.

USE: Denaturing alcohol and oils; for separating racemic mixtures. Addition agent to lubricants. In analytical chemistry for determn of nitrate.

THERAP CAT: CNS stimulant.

1465. Bryonia. Bryony. Dried root of *Bryonia alba* L., or of *B. dioica* Jacq., *Cucurbitaceae. Habit.* Europe. *Constit.* Bryonin, bryonidin, bryonicine, bryoamarid glycoside, Δ^7-stigmastenol, volatile oil, resin. Isoln of constituents: Tunmann, Wienecke, *Arch. Pharm.* **293**, 195 (1960); Biglino, *Farmaco Ed. Sci.* **14**, 673 (1959). Bryogenine structure studies: Biglino *et al., Tetrahedron Lett.* **1963**, 1651. *Poisonous!*

THERAP CAT: Cathartic.

THERAP CAT (VET): Formerly as a purgative.

1466. Bryostatins. Family of biologically active macrolides isolated from the marine bryozoans, *Bugula neritina*, L. and related organisms. Protein kinase C activators. Isoln and structure of bryostatin 1, one of the most abundant of the series: G. R. Pettit *et al., J. Am. Chem. Soc.* **104**, 6846 (1982). Large-scale isoln of 1: D. E. Schaufelberger *et al., J. Nat. Prod.* **54**, 1265 (1991). Abs config of 1, 2 and 3: G. R. Pettit *et al., J. Org. Chem.* **56**, 1337 (1991). ^{1}H- and ^{13}C- NMR analyses of 1: D. E. Schaufelberger *et al., Magn. Reson. Chem.* **29**, 366 (1991). Approaches to synthesis: R. Roy *et al., Chem. Commun.* **1989**, 1308; K. Ohmori *et al., Tetrahedron Lett.* **34**, 4981 (1993). Total synthesis of 7: M. Kageyama *et al., J. Am. Chem. Soc.* **112**, 7407 (1990). Conversion of 2 to 1: G. R. Pettit *et al., Can. J. Chem.* **69**, 856 (1991). Pharmacodynamics: R. L. Berkow *et al., Cancer Res.* **53**, 2810 (1993); C. Stanwell *et al., Int. J. Cancer* **56**, 585 (1994). Clinical evaluation of 1 in advanced cancer: J. Prendiville *et al., Br. J. Cancer* **68**, 418 (1993). Review of pharmacology: P. M. Blumberg, G. R. Pettit, "The Bryostatins, a Family

of Protein Kinase C Activators with Therapeutic Potential", in *New Leads and Targets in Drug Research*, P. Krogsgaard-Larsen *et al.*, Eds. (Munksgaard, Copenhagen, 1992) pp 273-285. Review of isolation, structure elucidation and biological activity: G. R. Pettit, *Fortschr. Chem. Org. Naturst.* **57**, 153-195 (1991); of syntheses and clinical biology: R. Mutter, M. Wills, *Bioorg. Med. Chem.* **8**, 1841-1860 (2000).

Bryostatin 1

Bryostatin 1. [83314-01-6] $C_{47}H_{68}O_{17}$. Colorless crystals from methylene chloride/methanol, mp 230-235°. $[\alpha]_D^{25}$ +34.1° (c = 0.044 in methanol). uv max (methanol): 233, 263 nm (ε 25700, 28700). Relatively insol in water. Soly in ethanol >4000 μg/ml. LD_{50} in mice, rats (mg/kg): 0.038, 0.068 i.v. (Prendiville).

1467. Bucetin. [1083-57-4] *N*-(4-Ethoxyphenyl)-3-hydroxybutanamide; 3-hydroxy-*p*-butyrophenetidide; β-hydroxybutyric acid *p*-phenetidide; *p*-ethoxy-*N*-(β-hydroxybutyryl)aniline. $C_{12}H_{17}NO_3$; mol wt 223.27. C 64.56%, H 7.68%, N 6.27%, O 21.50%. Phenetidine derivative with analgesic and antipyretic activity. Prepn: G. Ehrhart *et al., US 2830087* (1958 to Hoechst); *eidem, Arzneim.-Forsch.* **15**, 727 (1965). Toxicity data: H. Fujimura, *Folia Pharmacol. Jpn.* **62**, 123 (1966). Metabolism in rabbits: J. Shibasaki *et al., Chem. Pharm. Bull.* **16**, 1726, 2269 (1968). Synergistic effect with ethenzamide, *q.v.*: H. Kojima *et al., Oyo Yakuri* **16**, 549 (1978), *C.A.* **90**, 81035y (1979). TLC determn in saliva: P. Rohdewald, G. Drehsen, *J. Chromatogr.* **225**, 427 (1981).

Crystals from isopropanol, mp 160°. Sparingly sol in water. LD_{50} in mice (mg/kg): 790 i.p., 2800 orally (Fujimura).

Note: Component in combination analgesics also containing ethenzamide, caffeine and vitamin B_1, e.g. *Butylon, Bucetalon.*

THERAP CAT: Analgesic.

1468. Buchu. Bucco; bucku; buku. Dried leaves of *Barosma betulina* Bartl. & Wendl. (short buchu) or of *B. crenulata* (Linné) Hooker (oval buchu), or of *B. serratifolia* Willd. (long buchu), *Rutaceae. Habit.* Southern Africa (Cape of Good Hope). *Constit.* Diosphenol (barosma camphor), diosmin, bitter extractive, resin, *l*-menthone, mucilage, hesperidin, 1-2% volatile oil. *Review:* Feldman, Youngken, *J. Am. Pharm. Assoc.* **33**, 277 (1944).

THERAP CAT: Antiseptic (urinary).

THERAP CAT (VET): Has been used as a urinary antiseptic, diuretic.

1469. Buchwald Phosphine Ligands. Dialkylbiaryl phosphine ligands; biaryl phosphane ligands. Class of structurally related phosphine ligands. Utilized in a variety of palladium-catalyzed carbon-carbon and carbon-nitrogen bond forming reactions, including cross-couplings and aminations. Prepn of DavePhos and use in cross-coupling reactions: D. W. Old *et al., J. Am. Chem. Soc.* **120**, 9722 (1998). Improved prepn and synthesis of related ligands: H. Tomori *et al., J. Org. Chem.* **65**, 5334 (2000). Prepn of JohnPhos and use in diaryl ether synthesis: A. Aranyos *et al.*, *J. Am. Chem. Soc.* **121**, 4369 (1999); of XPhos and use in aminations and amidations: X. Huang *et al.*, *ibid.* **125**, 6653 (2003). Design of SPhos and use in Suzuki-Miyaura couplings: S. D. Walker *et al.*, *Angew. Chem. Int. Ed.* **43**, 1871 (2004). Prepn and use of SPhos: T. E. Barder *et al.*, *J. Am. Chem. Soc.* **127**, 4685 (2005). Prepn of RuPhos and use in Negishi cross-couplings: J. E. Milne, S. L. Buchwald, *J. Am. Chem. Soc.* **126**, 13028 (2004); of BrettPhos and use in amination reactions: B. P. Fors *et al.*, *ibid.* **130**, 13552 (2008). Review of phosphine ligands in Suzuki-Miyaura cross-coupling reactions: R. Martin, S. L. Buchwald, *Acc. Chem. Res.* **41**, 1461-1473 (2008); in amination reactions: D. S. Surry, S. L Buchwald, *Angew. Chem. Int. Ed.* **47**, 6338-6361 (2008); *eidem*, *Chem. Sci.* **2**, 27-50 (2011).

DavePhos: Y = cyclohexyl; R^1, R^2, R^4, R^5 = H; R^3 = N(CH$_3$)$_2$
JohnPhos: Y = C(CH$_3$)$_3$; R^1, R^2, R^3, R^4, R^5 = H
XPhos: Y = cyclohexyl; R^1, R^2 = H; R^3, R^4, R^5 = CH(CH$_3$)$_2$
SPhos: Y = cyclohexyl; R^1, R^2, R^4, R^5 = H; R^3 = OCH$_3$
RuPhos: Y = cyclohexyl; R^1, R^2, R^4 = H; R^3, R^5 = OCH(CH$_3$)$_2$
BrettPhos: Y = cyclohexyl; R^1, R^2 = OCH$_3$; R^3, R^4, R^5 = CH(CH$_3$)$_2$

DavePhos phosphine ligand. [213697-53-1] 2'-(Dicyclohexylphosphino)-*N*,*N*-dimethyl-[1,1'-biphenyl]-2-amine; 2-dicyclohexylphosphino-2'-(*N*,*N*-dimethylamino)biphenyl; 1-(*N*,*N*-dimethylamino)-1'-(dicyclohexylphosphino)biphenyl. C$_{26}$H$_{36}$NP; mol wt 393.55. White solid from hot ethanol, mp 110°. *Irritant.* Sol in most organic solvents. Store under inert gas.

JohnPhos phosphine ligand. [224311-51-7] [1,1'-Biphenyl]-2-ylbis(1,1-dimethylethyl)phosphine; (2-biphenyl)di-*tert*-butylphosphine; 2-(di-*tert*-butylphosphino)biphenyl. C$_{20}$H$_{27}$P; mol wt 298.41. White solid from methanol, mp 86-86.5°. Sol in most organic solvents. Store under inert gas.

XPhos phosphine ligand. [564483-18-7] Dicyclohexyl[2',4',6'-tris(1-methylethyl)[1,1'-biphenyl]-2-yl]phosphine; 2-dicyclohexylphosphino-2',4',6'-triisopropylbiphenyl. C$_{33}$H$_{49}$P; mol wt 476.73. White crystalline solid from diethyl ether + dichloromethane + methanol, mp 182-184°.

SPhos phosphine ligand. [657408-07-6] Dicyclohexyl(2',6'-dimethoxy[1,1'-biphenyl]-2-yl)phosphine; 2-dicyclohexylphosphino-2',6'-dimethoxybiphenyl. C$_{26}$H$_{35}$O$_2$P; mol wt 410.54. White solid from acetone, mp 162-162.5°. Sol in most organic solvents. Store under inert gas.

RuPhos phosphine ligand. [787618-22-8] [2',6'-Bis(1-methylethoxy)[1,1'-biphenyl]-2-yl]dicyclohexylphosphine; 2-dicyclohexylphosphino-2',6'-diisopropoxybiphenyl; 2-(2',6'-diisopropoxyphenyl)phenyldicyclohexylphosphine. C$_{30}$H$_{43}$O$_2$P; mol wt 466.65. White solid from ethanol, mp 123-124°. Sol in most organic solvents. Insol in water.

BrettPhos phosphine ligand. [1070663-78-3] Dicyclohexyl[3,6-dimethoxy-2',4',6'-tris(1-methylethyl)[1,1'-biphenyl]-2-yl]phosphine; 2-(dicyclohexylphosphino)-3,6-dimethoxy-2',4',6'-triisopropyl-1,1'-biphenyl. C$_{35}$H$_{53}$O$_2$P; mol wt 536.78. White crystals from acetone, mp 191-193°.

USE: Ligands in organic synthesis.

1470. Bucillamine. [65002-17-7] *N*-(2-Mercapto-2-methyl-1-oxopropyl)-L-cysteine; *N*-(2-mercapto-2-methylpropanoyl)-L-cysteine; *N*-(2-mercaptoisobutyryl)-L-cysteine; tiobutarit; thiobutarit; DE-019; SA-96; Rimatil. C$_7$H$_{13}$NO$_3$S$_2$; mol wt 223.31. C 37.65%, H 5.87%, N 6.27%, O 21.49%, S 28.71%. Disease modifying antirheumatic drug (DMARD) which modulates the immune response; structurally related to tiopronin, *q.v.* Prepn: T. Fujita *et al.*, *DE*

2709820; *eidem*, *US* **4305958** (1977, 1981 both to Santen); M. Oya *et al.*, *Chem. Pharm. Bull.* **29**, 940 (1981). GC-MS determn in blood: K. Matsuura *et al.*, *J. Mass Spectrom. Soc. Jpn.* **46**, 25 (1998). Clinical immunopharmacology: H. Matsuno *et al.*, *Drugs Exp. Clin. Res.* **19**, 205 (1993). Clinical trials in rheumatoid arthritis: M. Yasuda *et al.*, *Clin. Rheumatol.* **13**, 446 (1994); H. A. Kim, Y. W. Song, *Rheumatol. Int.* **17**, 5 (1997).

Crystals from ethyl acetate, mp 139-140°. $[\alpha]_D^{25}$ +32.3° (c = 1.0 in ethanol). LD$_{50}$ in mice (mg/kg): 2285 i.p.; 989.6 i.v. (Fujita, 1981).

THERAP CAT: Antirheumatic.

1471. Bucindolol. [71119-11-4] 2-[2-Hydroxy-3-[[2-(1*H*-indol-3-yl)-1,1-dimethylethyl]amino]propoxy]benzonitrile. C$_{22}$H$_{25}$N$_3$O$_2$; mol wt 363.46. C 72.70%, H 6.93%, N 11.56%, O 8.80%. Non-selective β-adrenergic blocker with vasodilating activity. Prepn: **BE 868943** (1979 to Bristol-Myers); W. E. Kreighbaum, W. T. Comer, *US* **4234595** (1980 to Mead Johnson); W. E. Kreighbaum *et al.*, *J. Med. Chem.* **23**, 285 (1980). Crystal structure: J.-M. Léger *et al.*, *Acta Crystallogr.* **C40**, 706 (1984). Conformation: E. Coutinho, *Indian J. Chem.* **34B**, 553 (1995). Review of chemistry and pharmacology: D. Deitchman *et al.*, in *New Drugs Annual: Cardiovascular Drugs* vol. 1, A. Scriabine, Ed. (Raven Press, New York, 1983) pp 1-18. Clinical pharmacokinetics: P. A. Meredith *et al*, *Xenobiotica* **15**, 979 (1985). Clinical pharmacology: N. Bett *et al.*, *Am. J. Cardiol.* **57**, 678 (1986). GC-MS determn in plasma: M. J. Bartek *et al.*, *J. Chromatogr.* **377**, 183 (1986). Clinical trial in chronic heart failure: E. J. Eichhorn *et al.*, *N. Engl. J. Med.* **344**, 1659 (2001); C. M. O'Connor *et al.*, *Am. J. Cardiol.* **95**, 558 (2005).

White crystals from abs ethanol, mp 125-127°.

Hydrochloride. [70369-47-0] MJ-13105-1; Gencaro. C$_{22}$H$_{25}$N$_3$O$_2$.HCl; mol wt 399.92. White crystals from abs ethanol, mp 185-187°. Sol in water. Nonhygroscopic. pKa 8.86. Partition coefficient (chloroform/0.01*M* phosphate buffer): 83.4 (pH 7.19); 361 (pH 7.78). LD$_{50}$ in mice, rats (mg/kg): ~100, ~100 orally (Deitchmann).

THERAP CAT: Antihypertensive; in treatment of congestive heart failure.

1472. Buckminsterfullerene. [99685-96-8] [5,6]Fullerene-C$_{60}$-Ih; buckyball; soccerballene. C$_{60}$; mol wt 720.66. C 100.00%. Named after R. Buckminster Fuller, the American engineer, whose geodesic domes possess the same symmetry as the C$_{60}$ truncated icosohedron typified by a soccerball. The most abundant of a class of carbon molecules characterized by a hollow cage structure and known as the *fullerenes*. One of the allotropic forms of carbon, *q.v.*, and the only molecular form, these molecules, described by C$_{20+2m}$, form stable closed nets composed of 12 pentagons and 'm' hexagons. Prepn: H. W. Kroto *et al.*, *Nature* **318**, 162 (1985); in macroscopic quantities: W. Krätschmer *et al.*, *ibid.* **347**, 354 (1990). Rational synthesis: L. T. Scott *et al.*, *Science* **295**, 1500 (2002). Identification in shungite, a carbonaceous rock: P. R. Buseck *et al.*, *Science* **257**, 215 (1992). Review of early work: R. F. Curl, R. E. Smalley, *Sci. Am.* **265**, 54-63 (1991). Review of synthesis, properties, chemistry including metal-doping: *Acc. Chem. Res.* **25**, 98-175 (1992); *Fullerenes*, G. S. Hammond, V. J. Kuck, Eds. *ACS Symp. Ser.* **481**, 1-195 (1992). Review of solids chemistry: A. Penicaud, *Fullerene Sci. Tech.* **6**, 731-741 (1998). Book: *The Chemical Physics of Fullerenes 10 (and 5) Years Later*, W. Andreoni, Ed. (Kluwer Academic Publishers, Dordrecht, 1996) 498 pp.

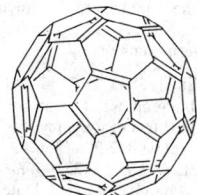

1473. Bucladesine. [362-74-3] *N*-(1-Oxobutyl)adenosine cyclic 3′,5′-(hydrogen phosphate) 2′-butanoate; *N*-(9-β-D-ribofuranosyl-9*H*-purin-6-yl)butyramide cyclic 3′,5′-(hydrogen phosphate) 2′-butyrate; N^6,2′-*O*-dibutyryladenosine 3′,5′-cyclic monophosphate; N^6,2′-*O*-dibutyryl cAMP; DBcAMP. $C_{18}H_{24}N_5O_8P$; mol wt 469.39. C 46.06%, H 5.15%, N 14.92%, O 27.27%, P 6.60%. Vasodilating cyclic nucleotide derivative which can permeate the cell membrane. Mimics the action of endogenous cyclic AMP, *q.v.* Prepn: Th. Posternak *et al.*, *Biochim. Biophys. Acta* **65**, 558 (1962); O. Nagase *et al.*, **JP Kokai 76 113896**; T. Hirayama *et al.*, **JP Kokai 77 39698**; **JP 77 39699** (1976, 1977, 1977 all to Daiichi Seiyaku); *see:* *C.A.* **86**, 140409r (1977); *C.A.* **87**, 136330m-136331n (1977). Pharmacology: H. S. White *et al.*, *Eur. J. Pharmacol.* **57**, 107 (1979); J. D. Johnson *et al.*, *ibid.* **91**, 343 (1983); N. Bondar *et al.*, *J. Physiol.* **355**, 33 (1984). Clinical trial in congestive heart failure: S. Matsui *et al.*, *Am. J. Cardiol.* **51**, 1364 (1983); after cardiopulmonary bypass surgery: T. Yoshitake *et al.*, *Prog. Clin. Biol. Res.* **111**, 211 (1983).

Sodium salt. [16980-89-5] DC-2797; Actosin. $C_{18}H_{23}N_5$-NaO_8P; mol wt 491.37.

Barium salt. [18837-96-2] $C_{36}H_{46}BaN_{10}O_{16}P_2$. uv max (ethanol-0.5*M* ammonium acetate, 5:2) 270 nm.

THERAP CAT: Cardiotonic.

1474. Buclizine. [82-95-1] 1-[(4-Chlorophenyl)phenylmethyl]-4-[[4-(1,1-dimethylethyl)phenyl]methyl]piperazine; 1-(*p-tert*-butylbenzyl)-4-(*p*-chloro-α-phenylbenzyl)piperazine; 1-(*p*-chlorobenzhydryl)-4-(*p-tert*-butylbenzyl)diethylenediamine; 1-(*p-tert*-butylbenzyl)-4-(*p*-chlorodiphenylmethyl)piperazine; histabutizine; histabutyzine. $C_{28}H_{33}ClN_2$; mol wt 433.04. C 77.66%, H 7.68%, Cl 8.19%, N 6.47%. Prepn: Morren, **US 2709169** (1955 to U.C.B.), *see also* **DE 964040**, Lui *et al.*, *C.A.* **62**, 2776a (1965). Clinical evaluation as antiemetic in migraine: P. B. Jorgensen *et al.*, *Curr. Ther. Res.* **16**, 1276 (1974); E. I. Adam, *J. Int. Med. Res.* **15**, 71 (1987).

bp$_{0.001}$ 217-220°.

Dihydrochloride. [129-74-8] UCB-4445; Aphilan; Buclina; Longifene; Postafen. $C_{28}H_{33}ClN_2$.2HCl; mol wt 505.95. mp 230-240°.

THERAP CAT: Antiemetic.

1475. Bucolome. [841-73-6] 5-Butyl-1-cyclohexyl-2,4,6-(1*H*,3*H*,5*H*)pyrimidinetrione; 5-butyl-1-cyclohexylbarbituric acid; 5-*n*-butyl-1-cyclohexyl-2,4,6-trioxoperhydropyrimidine; BCP; Paramidin. $C_{14}H_{22}N_2O_3$; mol wt 266.34. C 63.14%, H 8.33%, N 10.52%, O 18.02%. Prepn: Senda *et al.*, **US 3243344** and **US 3274195** (both 1966 to Takeda). Metabolic studies in man: Yashiki *et al.*, *Chem. Pharm. Bull.* **19**, 468, 869 (1971).

Needles from methanol, mp 84°. bp$_{0.8}$ 185-187°.

THERAP CAT: Anti-inflammatory.

1476. Budesonide. [51333-22-3] (11β,16α)-16,17-[Butylidenebis(oxy)]-11,21-dihydroxypregna-1,4-diene-3,20-dione; (*R,S*)-11β,16α,17,21-tetrahydroxypregna-1,4-diene-3,20-dione cyclic 16,17-acetal with butyraldehyde; S-1320; Bidien; Budeson; Cortivent; Desonax; Entocort; Miflonil; Novopulmon; Pulmaxan; Preferid; Pulmicort; Rhinocort; Spirocort. $C_{25}H_{34}O_6$; mol wt 430.54. C 69.74%, H 7.96%, O 22.30%. Non-halogenated glucocorticoid related to triamcinolone hexacetonide, *q.v.* with a high ratio of topical to systemic activity. Prepn: R. L. Brattsand *et al.*, **DE 2323215**; *eidem*, **US 3929768** (1973, 1975 both to Bofors). Synthesis and anti-inflammatory properties: A. Thalén, R. L. Brattsand, *Arzneim.-Forsch.* **29**, 1787 (1979). Pharmacokinetic study: A. Ryrfeldt *et al.*, *J. Steroid Biochem.* **10**, 317 (1979). HPLC determn of epimers, impurities, content: G. Roth *et al.*, *J. Pharm. Sci.* **69**, 766 (1980). Clinical experience in collagenous colitis: S. Miehlke *et al.*, *Aliment. Pharmacol. Ther.* **22**, 1115 (2005). Safety profile in asthma patients: A. L. Sheffer *et al.*, *Ann. Allergy Asthma Immunol.* **94**, 48 (2005). Review of physical and pharmocokinetic properties: E. J. O'Connell, *Clin. Ther.* **25**, C42-C60 (2003); of pharmacokinetics and clinical experience in Crohn's disease: S. Edsbäcker, T. Andersson, *Clin. Pharmacokinet.* **43**, 803-821 (2004); in allergic rhinitis: B. E. Stanaland, *Clin. Ther.* **26**, 473-492 (2004); in pediatric asthma: W. E. Berger, *Drugs* **65**, 1973-1989 (2005).

Crystals; mp 221-232° (dec). Mixture of two isomers; the content of the *S*-isomer in the mixture varies between 40-51%. $[\alpha]_D^{25}$ +98.9° (c = 0.28 in methylene chloride). Partition coefficient (octanol/water): 1.6×10^3. Freely sol in chloroform; sparingly sol in ethanol, alc. Practically insol in water, heptane.

THERAP CAT: Anti-inflammatory.

THERAP CAT (VET): Anti-inflammatory in treatment of inflammatory intestinal diseases.

1477. Budipine. [57982-78-2] 1-(1,1-Dimethylethyl)-4,4-diphenylpiperidine; 1-*tert*-butyl-4,4-diphenylpiperidine. $C_{21}H_{27}N$; mol wt 293.45. C 85.95%, H 9.27%, N 4.77%. Centrally acting deriv of diphenylpiperidine. Prepn: H. G. Menge, J. Klosa, **DE 1936452**; *eidem*, **US 4016280** (1969, 1977 both to Byk Gulden). Pharmacological studies: U. Brand, H. G. Menge, *Arzneim.-Forsch.* **30**, 1242 (1980); M. Eltze, *ibid.* 1129; H. G. Menge, U. Brand, *ibid.*

32, 85 (1982). Comparison to other drugs in treatment of dyskinesias: J. Siegfried, *Can. J. Neurol. Sci.* **6**, 89 (1979). Clinical evaluation with levodopa in Parkinson's disease: K. Jellinger, H. Bliesath, *J. Neurol.* **234**, 280 (1987).

Hydrochloride. [63661-61-0] Parkinsan. $C_{21}H_{27}N.HCl$; mol wt 329.91. LD_{50} in male mice, rats (mg/kg): 120, 165 orally; 33, 28 i.v. (Menge, Brand).
THERAP CAT: Antiparkinsonian.

1478. Bufalin. [465-21-4] $(3\beta,5\beta)$-3,14-Dihydroxybufa-20,22-dienolide. $C_{24}H_{34}O_4$; mol wt 386.53. C 74.58%, H 8.87%, O 16.56%. Cardiotonic steroid constituent of Ch'an Su (also called Senso), a galenical preparation of the dried venom of the Chinese toad, *Bufo bufo gargarizans.* Isoln and characterization: M. Kotake, K. Kuwada, *Sci. Papers Inst. Phys. Chem. Res. Jpn.* **36**, 106 (1939), *C.A.* **33**, 7304⁹ (1939); K. Meyer, *Helv. Chim. Acta* **32**, 1238 (1949). Structure: K. Kuwada, *J. Chem. Soc. Jpn.* **60**, 335 (1939), *C.A.* **35**, 51123⁹ (1941). Prepn from resibufogenin, *q.v.:* H. Kondo, S. Ohno, **US 3134772** (1964); from digitoxigenin, *q.v.:* G. R. Pettit *et al.*, **US 3687944** (1972). Total synthesis: F. Sondheimer *et al.*, *J. Am. Chem. Soc.* **91**, 1228 (1969). Additional syntheses: G. R. Pettit *et al.*, *J. Org. Chem.* **35**, 2895 (1970); Y. Kamano, G. R. Pettit, *ibid.* **38**, 222 (1973); E. Yoshii *et al.*, *Chem. Pharm. Bull.* **25**, 2249 (1977); K. Wiesner *et al.*, *Helv. Chim. Acta* **66**, 2632 (1983). Pharmacology: S. Yoshida, T. Sakai, *Jpn. J. Pharmacol.* **23**, 859 (1973); **24**, 97 (1974).

Needles from methanol/chloroform, mp 242-243°. $[\alpha]_D$ −20°. uv max (ethanol): 298 nm (log ε 3.77).
Sodium sulfate. Sodium bufalin-3-sulfate. $C_{24}H_{33}NaO_7S$. Isoln from Japanese toad *Bufo vulgaris formosus:* K. Shimada *et al.*, *Tetrahedron Lett.* **1974**, 2767. Colorless amorphous solid, mp 165.5-166.5°. $[\alpha]_D^{23}$ −33.1° (c = 0.05 in methanol).

1479. Bufexamac. [2438-72-4] 4-Butoxy-N-hydroxybenzeneacetamide; 2-(*p*-butoxyphenyl)acetohydroxamic acid; 2-[*p*-(butyloxy)phenyl]acetohydroxamic acid; CP-1044-J3; Droxarol; Droxaryl; Feximac; Malipuran; Mofenar; Norfemac; Parfenal. $C_{12}H_{17}NO_3$; mol wt 223.27. C 64.56%, H 7.68%, N 6.27%, O 21.50%. Synthesis: N. P. Buu-Hoï *et al.*, *Compt. Rend.* **261**, 2259 (1965); *eidem*, **BE 661226**; *eidem*, **US 3479396** (1965, 1969 to Madan). Pharmacology and toxicology: Lambelin *et al.*, *Med. Pharmacol. Exp.* **15**, 545 (1966).

Needles from acetone, mp 153-155°. Practically insol in water. LD_{50} orally in mice, rats: >8, >4 g/kg (Lambelin).
THERAP CAT: Anti-inflammatory.

1480. Buflomedil. [55837-25-7] 4-(1-Pyrrolidinyl)-1-(2,4,6-trimethoxyphenyl)-1-butanone; 2′,4′,6′-trimethoxy-4-(1-pyrrolidinyl)butyrophenone; (2,4,6-trimethoxyphenyl)-(3-pyrrolidinopropyl) ketone. $C_{17}H_{25}NO_4$; mol wt 307.39. C 66.43%, H 8.20%, N 4.56%, O 20.82%. Prepn: L. Lafon, **DE 2122144**; *idem*, **US 3895030** (both 1971 to Orsymonde). Pharmacology: J. Duteil *et al.*, *Therapie* **30**, 207 (1975); C. Debray *et al.*, *ibid.* 259. HPLC determn: J. A. Badmin *et al.*, *J. Chromatogr.* **172**, 319 (1979). Clinical pharmacokinetics: U. Gundert-Remy *et al.*, *Eur. J. Clin. Pharmacol.* **20**, 459 (1981). Comparative safety and efficacy study: G. Rosas *et al.*, *Angiology* **32**, 291 (1981). Review of pharmacology and therapeutic effects: S. P. Clissold *et al.*, *Drugs* **33**, 430-460 (1987).

Hydrochloride. [35543-24-9] LL-1656; Bufedil; Buflan; Buflocit; Buflonat; Fonzylane; Irrodan; Lofton; Loftyl; Provas. $C_{17}H_{25}NO_4.HCl$; mol wt 343.85. White crystals from isopropanol, mp 192-193°. LD_{50} in mice (mg/kg): 80 ±4.6 i.v. (Lafon).
THERAP CAT: Vasodilator (peripheral).

1481. Buformin. [692-13-7] N-Butylimidodicarbonimidic diamide; 1-butylbiguanide; *n*-butylbiguanide; butyldiguanide; butformin; W-37. $C_6H_{15}N_5$; mol wt 157.22. C 45.84%, H 9.62%, N 44.55%. Prepn: Shapiro *et al.*, *J. Am. Chem. Soc.* **81**, 3728 (1959); Shapiro, Freedman, **US 2961377** (1960 to USV). Toxicity study: Rikimaru *et al.*, *J. Antibiot.* **18A**, 196 (1965).

Strong base. Very sol in water.
Hydrochloride. [1190-53-0] Andere; Biforon; Bigunal; Bufonamin; Bulbonin; Diabrin; Dibetos; Gliporal; Insulamin; Krebon; Panformin; Silubin; Sindiatil; Tidemol; Ziavetine. $C_6H_{15}N_5.HCl$; mol wt 193.68. Crystals, mp 174-177°. Freely sol in water, alcohol. LD_{50} i.p. in mice: 380 mg/kg (Rikimaru).
Nitrate. $C_6H_{15}N_5.HNO_3$. Crystals from ethanol, dec 125-126°.
THERAP CAT: Antidiabetic.

1482. Bufotalin. [471-95-4] $(3\beta,5\beta,16\beta)$-16-(Acetyloxy)-3,14-dihydroxybufa-20,22-dienolide; $3\beta,14,16\beta$-trihydroxy-5β-bufa-20,22-dienolide 16-acetate. $C_{26}H_{36}O_6$; mol wt 444.57. C 70.24%, H 8.16%, O 21.59%. One of the genins found in the venom of the common European toad *(Bufo vulgaris).* Isoln and structure: Faust, *Arch. Exp. Pathol. Pharmakol.* **47**, 279 (1902); **49**, 1 (1902); Wieland, Weil, *Ber.* **46**, 3315 (1913); Wieland, Weyland, *Sitzungsber. Bayer. Akad. Wiss.* (math.-physikal. Klasse) **1920**, 329; Wieland *et al.*, *Ann.* **524**, 203 (1936); Meyer, *Helv. Chim. Acta* **32**, 1993 (1949); Pettit *et al.*, *Chem. Commun.* **1970**, 1566. Synthesis: G. R. Pettit *et al.*, *J. Org. Chem.* **52**, 3573 (1987).

Solvated crystals from alc. Sinters at 154°, dec 223°. Sublimes in high vacuum at 225-230°. uv max: 300 nm. $[\alpha]_D^{20}$ +5.4° (c = 0.5 in chloroform). Sol in alcohol, chloroform.

3-Acetate. mp 269-272°.

1483. Bufotenine. [487-93-4] 3-[2-(Dimethylamino)ethyl]-1H-indol-5-ol; 3-(2-dimethylaminoethyl)-5-indolol; 5-hydroxy-N,N-dimethyltryptamine; N,N-dimethylserotonin; 3-(β-dimethyl-aminoethyl)-5-hydroxyindole; mappine. $C_{12}H_{16}N_2O$; mol wt 204.27. C 70.56%, H 7.90%, N 13.71%, O 7.83%. Isoln from toads: Wieland et al., Ann. **513**, 1 (1934); Wieland, Wieland, ibid. **528**, 239 (1937); from toadstools: Wieland, Motzel, ibid. **581**, 10 (1953). Isoln from Piptadenia peregrina Benth., Leguminosae: Stromberg, J. Am. Chem. Soc. **76**, 1707 (1954). Synthesis: Hoshino, Shimo-daira, Ann. **520**, 19 (1935); Harley-Mason, Jackson, Chem. Ind. (London) **1952**, 954; see also serotonin; and Speetor, US **2708197** (1955 to Upjohn); Stoll et al., Helv. Chim. Acta **38**, 1452 (1955). Activity: Bhattacharya, Sanyal, Indian J. Physiol. Pharmacol. **15**, 133 (1971). Crystal and molecular structure: G. Falkenberg, Acta Crystallogr. B **28**, 3219 (1972).

Stout prisms from ethyl acetate, mp 146-147°, $bp_{0.1}$ 320°. uv max: 220, 265 nm (log ε 4.0, 3.7). Almost insol in water. Freely sol in alcohol, less sol in ether. Sol in dil acids and alkalies.

Methyliodide. $C_{13}H_{19}IN_2O$. Stout prisms from methanol, dec 214-215°.

Note: This is a controlled substance (hallucinogen): **21 CFR, 1308.11.**

1484. Bufotoxin. [464-81-3] (3β,5β,16β)-16-(Acetyloxy)-3-[[8-[[(1S)-4-[(aminoiminomethyl)amino]-1-carboxybutyl]amino]-1,8-dioxooctyl]oxy]-14-hydroxybufa-20,22-dienolide; vulgarobufo-toxin; bufotalin 3-suberoylarginine ester. $C_{40}H_{60}N_4O_{10}$; mol wt 756.94. C 63.47%, H 7.99%, N 7.40%, O 21.14%. Principal toxin of the venom of the common European toad, Bufo vulgaris. Isoln and structure: Wieland, Alles, Ber. **55**, 1789 (1922); Wieland et al., Ann. **524**, 203 (1936); Meyer, Helv. Chim. Acta **32**, 1993 (1949); Urscheler et al., ibid. **38**, 883 (1955). Alternate structure: Kamano et al., Tetrahedron Lett. **1968**, 5673; Linde-Tempel, Helv. Chim. Acta **53**, 2188 (1970). Partial synthesis: Pettit, Kamano, Chem. Commun. **1972**, 45.

Monohydrate, bitter needles from alc, dec 205°. uv max: 295 nm (log ε 3.74). Freely sol in methanol, pyridine, sparingly sol in abs alc. Practically insol in water, ether, acetone, chloroform, petr ether.

1485. Bufuralol. [54340-62-4] α-[[(1,1-Dimethylethyl)ami-no]methyl]-7-ethyl-2-benzofuranmethanol; α-[(tert-butylamino)-

methyl]-7-ethyl-2-benzofuranmethanol; 2-(2-tert-butylamino-1-hy-droxyethyl)-7-ethylbenzofuran; 1-(7-ethylbenzofuran-2-yl)-2-tert-butylamino-1-hydroxyethane. $C_{16}H_{23}NO_2$; mol wt 261.37. C 73.53%, H 8.87%, N 5.36%, O 12.24%. β-Adrenergic blocker with peripheral vasodilating activity. Prepn: **NL 6606441**; G. A. Foth-ergill et al., **US 3929836** (1966, 1975 both to Hoffmann-La Roche). Prepn and resolution of isomers: eidem, Experientia **31**, 1322 (1975); eidem, Arzneim.-Forsch. **27**, 981 (1977). Pharmacology: T. C. Hamilton, M. W. Parkes, ibid. 1410. HPLC determn in plasma: P. Haefelfinger, J. Chromatogr. **221**, 327 (1980). Metabolism of isomers and racemate: R. J. Francis et al., Eur. J. Clin. Pharmacol. **23**, 529 (1982). Pharmacokinetics in hypertensive patients: M. Eck-ert et al., ibid. **24**, 479 (1983). Hemodynamic effects in patients with angina: M. Pfisterer et al., J. Cardiovasc. Pharmacol. **6**, 417 (1984).

Hydrochloride. [59652-29-8] Ro-3-4787; Angium. $C_{16}H_{23}$-NO_2.HCl; mol wt 297.82. Fine, white powder from acetone, mp 146°. LD_{50} in mice (mg/kg): 29.7 i.v.; 88.0 i.p.; 177 orally; in rats (mg/kg): 1400 s.c.; 750 orally (Hamilton, Parkes).

(+)-Hydrochloride. [57704-11-7] Crystals from ethyl acetate-ether, mp 122-123°. $[\alpha]_{365}^{20}$ +135.0° (c = 1.0 in ethanol).

(−)-Hydrochloride. [57704-10-6] Crystals from ethyl acetate-ether, mp 122-123°. $[\alpha]_{365}^{20}$ −136.0° (c = 1.0 in ethanol).

THERAP CAT: Antianginal; antihypertensive.

1486. Bulbocapnine. [298-45-3] (7aS)-6,7,7a,8-Tetrahydro-11-methoxy-7-methyl-5H-benzo[g]-1,3-benzodioxolo[6,5,4-de]-quinolin-12-ol; 10-methoxy-1,2-(methylenedioxy)-6aα-aporphin-11-ol. $C_{19}H_{19}NO_4$; mol wt 325.36. C 70.14%, H 5.89%, N 4.31%, O 19.67%. From root of Corydalis cava (L.) Schweigg. & Körte (C. tuberosa DC), Fumariaceae and Dicentra canadensis Walp., Papa-veraceae. Isoln: Freund, Josephi, Ber. **25**, 2411 (1892); Manske, Can. J. Res. **7**, 258 (1932). Pharmacology and toxicity study: H. Molitar, J. Pharmacol. Exp. Ther. **56**, 85 (1936). Structure: Gada-mer, Chem. Ztg. **34**, 1004 (1910). Configuration: Ayer, Taylor, J. Chem. Soc. **1956**, 472; Corrodi, Hardegger, Helv. Chim. Acta **39**, 889 (1956). Synthesis: Kikkawa, C.A. **54**, 4649b (1960). Periph-eral-dopamine-receptor blocking activity: R. G. Pendleton et al., Arch. Pharmacol. **289**, 171 (1975).

Columns, mp 201-203°. $[\alpha]_D^{22}$ +231°. Practically insol in water, sol in alcohol, chloroform. LD_{50} s.c. in mice: 195 mg/kg (Molitar).

dl-Form. Columns, mp 213-214°. Absorption max: Brustier, C.A. **49**, 12127b (1955).

l-Form. Columns, mp 202-203°. $[\alpha]_D^{22}$ −239°.

Methyl ether. $C_{20}H_{21}NO_4$. Crystals from petr ether, mp 129°. $[\alpha]_D^{20}$+260°. Absorption max: Girardet, J. Chem. Soc. **1931**, 2630.

1487. Bullvalene. [1005-51-2] Tricyclo[3.3.2.0²,⁸]deca-3,-6,9-triene; tricyclo[3.3.2.0⁴,⁶]deca-2,7,9-triene. $C_{10}H_{10}$; mol wt 130.19. C 92.26%, H 7.74%. Molecular structure is highly fluxion-al; spontaneously undergoes a 3-fold structurally degenerate Cope rearrangement at room temp, resulting in 10!/3 structurally equiva-lent geometric isomers. ¹H NMR spectrum is a sharp singlet peak at high temp. Predicted as a hypothetical molecule: W. von E. Doer-ing, W. R. Roth, Tetrahedron **19**, 715 (1963). Prepn: G. Schröder, Angew. Chem. **75**, 722 (1963). ¹H NMR spectrum: R. Merényi et al., Ber. **97**, 3150 (1964). Vapor phase structure determn: B. Andersen, A. Marstrander, *Acta Chem. Scand.* **25**, 1271 (1971). Microwave and vibrational spectral study: W. M. Stigliani, V. W. Laurie, *J. Mol. Spectrosc.* **60**, 188 (1976). Calorimetry study: M. Mansson, S. Sunner, *J. Chem. Thermodyn.* **13**, 671 (1981). Review of chemistry: G. Schröder, J. F. M. Oth, *Angew. Chem. Int. Ed.* **6**, 414-423 (1967); of discovery and characterization: A. Ault, *J. Chem. Educ.* **78**, 924-927 (2001).

mp 96°.

1488. Bumadizone. [3583-64-0] 2-Butylpropanedioic acid-1-(1,2-diphenylhydrazide); butylmalonic acid mono(1,2-diphenylhydrazide); *N*-(2-carboxycaproyl)hydrazobenzene; α-carboxycaproyl-*N,N*′-diphenylhydrazine. $C_{19}H_{22}N_2O_3$; mol wt 326.40. C 69.92%, H 6.79%, N 8.58%, O 14.70%. Non-steriodal anti-inflammatory drug (NSAID). Prepn: **NL 6406412**; Pfister *et al.*, **US 3455999** (1964, 1969 to Geigy); **NL 6600685** (1966 to Byk-Gulden). Main product of the hydrolysis of phenylbutazone, *q.v.:* Schmid, *Helv. Chim. Acta* **53**, 2239 (1970). Chemical characterization: Pawelczyk, Wachowiak, *Acta Pol. Pharm.* **26**, 433 (1969), *C.A.* **72**, 136355p (1970). Series of articles on pharmacology: *Arzneim.-Forsch.* **23**, 1226-1251, 1813-1822 (1973). Toxicity: R. Riedel, W. Schoetensack, *ibid.* 1215.

Crystals from ether-petr ether, mp 116-117°. Also reported as mp 77-79° (dependent on speed of crystallization). uv max (0.1*N* NaOH): 234, 264 nm (ε 16200, 3700).
Calcium salt hemihydrate. [69365-73-7] Desflam; Eumotol. $C_{38}H_{42}CaN_4O_6 \cdot \frac{1}{2}H_2O$; mol wt 699.86. Dec 154°. Sol in chloroform, alcohol, ether. Slightly sol in water. LD_{50} in mice, rats (mg/kg): 2500, 1250 orally; 258, 263 i.v. (Riedel, Schoetensack).
THERAP CAT: Anti-inflammatory.

1489. Bumetanide. [28395-03-1] 3-(Aminosulfonyl)-5-(butylamino)-4-phenoxybenzoic acid; 3-(butylamino)-4-phenoxy-5-sulfamoylbenzoic acid; PF-1593; Ro-10-6338; Bumex; Burinex; Butinat; Fontego; Fordiuran; Lunetoron. $C_{17}H_{20}N_2O_5S$; mol wt 364.42. C 56.03%, H 5.53%, N 7.69%, O 21.95%, S 8.80%. Prepn: P. W. Feit, **DE 1964503** and **DE 1964504**; *idem*, **US 3806534** (1970, 1970, 1974 all to Lövens Kemiske Fabrik); *idem, J. Med. Chem.* **14**, 432 (1971). Pharmacology: E. H. Oestergaard *et al.*, *Arzneim.-Forsch.* **22**, 66 (1972). HPLC and GC/MS determn in urine: C. Y. Gradeen *et al., J. Anal. Toxicol.* **14**, 123 (1990). Review of pharmacology and therapeutic use: A. Ward, R. C. Heel, *Drugs* **28**, 426-464 (1984). Series of articles on pharmacokinetics and comparative clinical studies in combination therapy: *Curr. Ther. Res.* **50**, Suppl. A, 1-82 (1991). Comprehensive description: P. N. V. Tata *et al., Anal. Profiles Drug Subs. Excip.* **22**, 107-144 (1993).

Crystals from aq ethanol, mp 230-231°. pK_1 3.6, pK_2 7.7. Sol in alkaline solns; slightly sol in water. Soly (mg/ml): water 0.1; ethanol 30.6; propylene glycol 18.7; dimethylacetamide >500; methanol

76.5; benzene 0.4; benzyl alcohol 21.6; acetone 50.2. uv max (water): 260, 220 nm ($E^{1\%}$ 18.9, 17.1); (0.1 *N* NaOH): 326 nm; (methanol): 270, 345 nm. LD_{50} i.v. in mice: 330 mg/kg (Oestergaard).
THERAP CAT: Diuretic.

1490. Bunamiodyl. [1233-53-0] 2-[[2,4,6-Triiodo-3-[(1-oxobutyl)amino]phenyl]methylene]butanoic acid; 3-butyramido-α-ethyl-2,4,6-triiodocinnamic acid; 3-(3-butyrylamino-2,4,6-triiodophenyl)-2-ethylacrylic acid; α-ethyl-β-(2,4,6-triiodo-3-butyramidophenyl)acrylic acid; α-(2,4,6-triiodo-3-butyrylaminobenzylidene)butyric acid; buniodyl. $C_{15}H_{16}I_3NO_3$; mol wt 639.01. C 28.19%, H 2.52%, I 59.58%, N 2.19%, O 7.51%. Prepn: H. Cassebaum, K. Dierbach, *Pharmazie* **16**, 389 (1961). Description and use in cholecystography: J. P. Arcomano *et al., Am. J. Dig. Dis.* **4**, 466 (1959). Toxicity studies: I. Levenstein *et al., J. Pharm. Sci.* **50**, 957 (1961).

mp 105-120°.
Sodium salt. [1923-76-8] Orabilex; Orabilix. $C_{15}H_{15}I_3NNaO_3$; mol wt 660.99. Microcrystalline white powder. Slightly sol in water. LD_{50} in rats: 2.5 g/kg orally (Levenstein).
THERAP CAT: Diagnostic aid (radiopaque medium).

1491. Bunazosin. [80755-51-7] 1-[4-(4-Amino-6,7-dimethoxy-2-quinazolinyl)hexahydro-1*H*-1,4-diazepin-1-yl]-1-butanone; 1-(4-amino-6,7-dimethoxy-2-quinazolinyl)hexahydro-4-(1-oxobutyl)-1*H*-1,4-diazepine; 1-(4-amino-6,7-dimethoxy-2-quinazolinyl)-4-butyrylhexahydro-1*H*-1,4-diazepine; 4-amino-6,7-dimethoxy-2-[4-(*n*-butyryl)homopiperazin-1-yl]quinazoline; DDQ. $C_{19}H_{27}N_5O_3$; mol wt 373.46. C 61.11%, H 7.29%, N 18.75%, O 12.85%. Aminoquinazoline derivative with potent hypotensive properties. Prepn: T. Takahashi, H. Sugimoto, **BE 806626**; *eidem*, **US 3920636** (1974, 1975 to Eisai); alternative prepn: *eidem*, **JP Kokai 75 140474** (1975 to Eisai), *C.A.* **85**, 46769r (1976). *In vitro* and *in vivo* studies of adrenergic blocking action and antihypertensive properties: T. Shoji *et al., Jpn. J. Pharmacol.* **30**, 763 (1980); K. Hoshi, S. Fujino, *ibid.* 427. Antihypertensive effect in volunteers: T. Kawasaki *et al., Eur. J. Clin. Pharmacol.* **20**, 399 (1981). Pharmacokinetics: C. Yamato *et al., Xenobiotica* **12**, 549 (1982).

Hydrochloride. [52712-76-2] E-643; Detantol. $C_{19}H_{27}N_5O_3 \cdot$HCl; mol wt 409.92. Crystals from methanol/ethanol, mp 280-282°.
THERAP CAT: Antihypertensive.

1492. Bungarotoxins. Constituent proteins of the venom of the *Elapidae* snake, *Bungarus multicinctus* (Southeast Asian banded krait). Separation of the crude venom yields several fractions, the most important being **α-bungarotoxin (α-Bgt)** and **β-bungarotoxin (β-BuTX)**: C. C. Chang, C. Y. Lee, *Arch. Int. Pharmacodyn. Ther.* **144**, 241 (1963). α-Bgt is a single polypeptide chain of mol wt about 8,000 containing 74 amino acid residues with 5 disulfide bridges: D. Mebs *et al., Biochem. Biophys. Res. Commun.* **44**, 711 (1971); *eidem, Z. Physiol. Chem.* **353**, 243 (1972). Purification, characterization, immunochemical studies: D. G. Clark *et al., Biochemistry* **11**, 1663 (1972). It is a post-synaptic neurotoxin with curare-like action that binds irreversibly to acetylcholine receptor sites, producing neuromuscular blockade and skeletal muscle paralysis. Activity and use as a probe for acetylcholine receptors: C. C. Chang, *Nature* **215**,

1177 (1967); J. P. Changeux *et al.*, *Proc. Natl. Acad. Sci. USA* **67**, 1241 (1970); R. E. Oswald, J. A. Freeman, *Neuroscience* **6**, 1 (1981). β-Bungarotoxin has been shown to contain several components, the major protein being designated as β_1- or β-bungarotoxin. It is composed of 2 subunits of mol wts of about 13,000 and 7,000, linked by disulfide bonds. The larger chain contains 120 amino acid residues including 13 half-cystine; the smaller chain contains 60 residues including 7 half cystine. Purification: T. Abe *et al.*, *Eur. J. Biochem.* **80**, 1 (1977); *eidem*, *Proc. R. Soc. London Ser. B* **207**, 487 (1980). Chemical properties, and amino acid sequence of the two polypeptide chains: K. Kondo *et al.*, *J. Biochem.* **83**, 91, 101 (1978). Complete purification and characterization of its action on synaptosomal accumulation and release of acetylcholine: J. W. Spokes, J. O. Dolly, *Biochim. Biophys. Acta* **596**, 81 (1980). β-BuTX is a presynaptic neurotoxin that prevents acetylcholine release at skeletal neuromuscular junctions without affecting the sensitivity of the postsynaptic membrane. Its proposed phospholipase A_2 activity is believed to be responsible for some of its effects at motor nerve terminals. Activity studies: P. N. Strong *et al.*, *Proc. Natl. Acad. Sci. USA* **75**, 1029 (1976); T. Abe, R. Miledi, *Proc. R. Soc. London Ser. B* **200**, 225 (1978); M. T. Alderdice, R. L. Volle, *Arch. Pharmacol.* **316**, 126 (1981). *Review:* A. T. Tu, *Venoms: Chemistry and Molecular Biology* (Wiley, New York, 1977) pp 185-187, 240-251.

The crude venom of *Bungarus multicinctus* is quite toxic, having an LD$_{50}$ in mice (μg/g): 0.16 s.c. (Chang, Lee); also reported as 0.33 s.c. (Mebs, 1972). LD$_{50}$ of α-Bgt in mice (μg/g): 0.21 s.c.; 0.15 i.p. (Mebs, 1972). LD$_{50}$ of β-BuTX in mice (μg/g): 0.019 i.p. (Kondo).

USE: As exptl tools in investigating neuromuscular processes.

1493. Bunitrolol. [34915-68-9] 2-[3-[(1,1-Dimethylethyl)-amino]-2-hydroxypropoxy]benzonitrile; *o*-[3-(*tert*-butylamino)-2-hydroxypropoxy]benzonitrile; 1-(2-cyanophenoxy)-2-hydroxy-3-*tert*-butylaminopropane; Ko-1366. $C_{14}H_{20}N_2O_2$; mol wt 248.33. C 67.71%, H 8.12%, N 11.28%, O 12.89%. β-Adrenergic blocker. Prepn: H. Koeppe *et al.*, **ZA 6803783**; *eidem*, **US 3541130** (1968, 1970 both to Boehringer, Ing.). Pharmacology: T. Kimura, *Experientia* **28**, 813 (1972). Studies on absorption, distribution, metabolism, excretion in rats and mice: T. Suzuki, T. Rikihisa, *Arzneim.-Forsch.* **29**, 1707 (1979). Clinical studies: A. Reale *et al.*, *J. Int. Med. Res.* **4**, 338 (1976); J. J. C. Jonker *et al.*, *Arzneim.-Forsch.* **31**, 1140 (1981). Toxicology study: M. Kanda *et al.*, *Oyo Yakuri* **9**, 457, 465, 499 (1975), *C.A.* **83**, 172616 (1975).

Hydrochloride. [23093-74-5] Betriol; Stresson. $C_{14}H_{20}N_2O_2 \cdot$HCl; mol wt 284.78. Crystals from ethanol, mp 163-165°. LD$_{50}$ in mice, rats (mg/kg): 1344-1440, 639-649 orally; 264-265, 222-225 i.p. (Kanda).

THERAP CAT: Antihypertensive, antiarrhythmic, antianginal.

1494. Bunte Salts. Water-sol salts of certain alkyl or aralkyl thiosulfuric acids of the general formula: $RSSO_2ONa$. Prepn: Lecher, Hardy, **US 2712547** (1955 to Am. Cyanamid); El-Heweki, Taeger, *J. Prakt. Chem.* **7**, 191 (1958); Tabushi *et al.*, *Kogyo Kagaku Zasshi* **67**, 478 (1964).

THERAP CAT (VET): Have been used as coccidiostats.

1495. Buparvaquone. [88426-33-9] 2-[[4-(1,1-Dimethylethyl)cyclohexyl]methyl]-3-hydroxy-1,4-naphthalenedione; 2-[(4-*tert*-butylcyclohexyl)methyl]-3-hydroxy-1,4-naphthoquinone; 3-(4-*t*-butylcyclohexyl)methyl-2-hydroxy-1,4-naphthoquinone; BW-720C; Butalex. $C_{21}H_{26}O_3$; mol wt 326.44. C 77.27%, H 8.03%, O 14.70%. Analog of antimalarial hydroxynaphthoquinones; *t*-butyl deriv of parvaquone, *q.v.* Prepn: A. T. Hudson, A. W. Randall, **EP 77550**; *eidem*, **US 4485117** (1983, 1984 both to Wellcome). *In vivo* and *in vitro* antiprotozoal activity: A. T. Hudson *et al.*, *Parasitology* **90**, 45 (1985); N. McHardy *et al.*, *Res. Vet. Sci.* **39**, 29 (1985). Treatment of bovine theileriosis: S. Dhar *et al.*, *Vet. Rec.* **119**, 635 (1986). Efficacy as a chemoimmunoprophylactic of theileriosis: *eidem, ibid.* **120**, 375 (1987).

Crystals, mp 124-125°. LD$_{50}$ orally in rats: >2 g/kg (Hudson).
THERAP CAT (VET): Antiprotozoal (Theileria).

1496. Buphanamine. [6793-24-4] (1*R*,4a*R*,5*S*,11b*S*)-4,4a-Dihydro-7-methoxy-1*H*,6*H*-5,11b-ethano[1,3]dioxolo[4,5-*j*]phenanthridin-1-ol; (1α)-2,3-didehydro-7-methoxycrinan-1-ol. $C_{17}H_{19}$NO$_4$; mol wt 301.34. C 67.76%, H 6.36%, N 4.65%, O 21.24%. Isoln from bulb of *Buphane (Boöphane) disticha* Herb., Appleg. *(Haemanthus toxicarius* Herb.) *Amaryllidaceae:* Humber, Taylor, *Can. J. Chem.* **33**, 1268 (1955); Hauth, Stauffacher, *Helv. Chim. Acta* **44**, 491 (1961). From *B. fischeri* Baker: Renz *et al., ibid.* **38**, 1209 (1955). Identity with "oily" haemanthine: Goosen, Warren, *J. Chem. Soc.* **1960**, 1094. Structure and stereochemistry: Fales, Wildman, *J. Org. Chem.* **26**, 881 (1961). Revised structure: Wildman in *The Alkaloids* vol. **XI**, R. H. F. Manske, Ed. (Academic Press, New York, 1968) p 361. NMR of revised structure: Crain *et al., J. Am. Chem. Soc.* **93**, 990 (1971). High-resolution mass spectrum: P. Longevialle *et al., Org. Mass Spectrom.* **7**, 401 (1973).

Prisms from ethyl acetate, mp 183-185° (Fales); from acetone, mp 192-194° (Goosen). $[\alpha]_{589}^{24}$ −195°; $[\alpha]_{436}^{24}$ −408° (c = 0.97); $[\alpha]_D^{20}$ −205° (c = 0.69, 95% ethanol); $[\alpha]_D^{21}$ −194° (c = 0.247, chloroform). uv max: 287 nm (ε 1495).

Hydrochloride. $C_{17}H_{19}NO_4 \cdot$HCl. Prisms from ethanol + ether, mp 180°.

1497. Buphanitine. [4673-18-1] (1*R*,3*R*,4a*R*,5*S*,11b*S*)-2,3,-4,4a-Tetrahydro-7-methoxy-1*H*,6*H*-5,11b-ethano[1,3]dioxolo[4,5-*j*]phenanthridine-1,3-diol; (1α,3α)-7-methoxycrinan-1,3-diol; hemanthine; nerbowdine. $C_{17}H_{21}NO_5$; mol wt 319.36. C 63.94%, H 6.63%, N 4.39%, O 25.05%. Major alkaloid from *Buphane (Boöphane) disticha* Herb., Appleg. *(Haemanthus toxicarius* Herb.), *Amaryllidaceae.* Isoln: Lewin, *Arch. Exp. Pathol. Pharmakol.* **68**, 333 (1912); Tutin, *ibid.* **69**, 314 (1912). Identity with hemanthine: Goosen, Warren, *Chem. Ind.* **1957**, 267; with nerbowdine: Lyle *et al., J. Am. Chem. Soc.* **82**, 2620 (1960). Structure studies: Goosen, Warren, *J. Chem. Soc.* **1960**, 1097; Goosen *et al., ibid.* **1961**, 4038. High-resolution mass spectrum: P. Longevialle *et al., Org. Mass Spectrom.* **7**, 401 (1973).

Needles from chloroform + ether (changes crystalline form at 210°), mp 232°. Prisms from acetone. $[\alpha]_D^{20}$ −102° (c = 1 in chloroform).

Hydrochloride. $C_{17}H_{22}ClNO_5$. mp 265°.
Nitrate. $C_{17}H_{22}N_2O_8$. mp 222-224°.

1498. Bupirimate. [41483-43-6] *N,N*-Dimethylsulfamic acid 5-butyl-2-(ethylamino)-6-methyl-4-pyrimidinyl ester; 5-butyl-2-

(ethylamino)-6-methyl-4-pyrimidinyl dimethylsulfamate; PP-588; Nimrod. $C_{13}H_{24}N_4O_3S$; mol wt 316.42. C 49.35%, H 7.65%, N 17.71%, O 15.17%, S 10.13%. Prepn: A. M. Cole et al., **DE 2246645**; eidem, **US 3880852** (1973, 1975 both to ICI). Activity: J. R. Finney et al., Proc. 8th Br. Insectic. Fungic. Conf. **2**, 667 (1975).

Pale tan waxy solid, mp 50-51°. Vapor pressure at 20°: 5×10^{-7} mm Hg. Soly in water at 25°: 22 mg/l. Sol in most organic solvents except paraffins. Easily hydrolyzed by dil acids. LD_{50} orally in rats: 4000 mg/kg (Finney).

USE: Fungicide.

14 Bupivacaine. [38396-39-3] 1-Butyl-N-(2,6-dimethylpheny 2-piperidinecarboxamide; dl-1-butyl-2',6'-pipecoloxylidide; 1-n-butyl-2',6'-dimethyl-2-piperidinecarboxanilide; dl-N-n-butylpipecolic acid 2,6-xylidide; 1-butyl-2-(2,6-xylylcarbamoyl)piperidine; dl-1-n-butylpiperidine-2-carboxylic acid 2,6-dimethylanilide. $C_{18}H_{28}N_2O$; mol wt 288.44. C 74.95%, H 9.79%, N 9.71%, O 5.55%. Prepn: B. Ekenstam et al., Acta Chem. Scand. **11**, 1183 (1957); B. T. Ekenstam, B. G. Pettersson, **US 2955111** (1960 to AB Bofors). Resolution of isomers: B. F. Tullar, J. Med. Chem. **14**, 891 (1971). Stereospecific synthesis: B. Adger et al., Tetrahedron Lett. **37**, 6399 (1996). Pharmacology of racemate: F. Henn, R. Brattsand, Acta Anaesthesiol. Scand. Suppl. **21**, 9 (1966), C.A. **66**, 17863u (1967); of isomers: F. P. Luduena et al., Arch. Int. Pharmacodyn. **200**, 359 (1972). Clinical pharmacokinetics: D. W. Blake et al., Anaesth. Intensive Care **22**, 522 (1994). Comprehensive description: T. D. Wilson, Anal. Profiles Drug Subs. **19**, 59-94 (1990). Review of use in spinal anesthesia: Acta Anaesthesiol. Scand. **35**, 1-10 (1991). Review of pharmacology and clinical efficacy of levobupivacaine: K. J. McClellan, C. M. Spencer, Drugs **56**, 355-362 (1998).

mp 107.5-108°. pKa 8.09; also reported as 8.17. Partition coefficient: (oleyl alcohol/water) 1565; (n-heptane/pH 7.4 buffer) 27.5.

Hydrochloride. [18010-40-7]; [73360-54-0] (monohydrate). AH-2250; LAC-43; Carbostesin; Marcaine; Sensorcaine. $C_{18}H_{28}N_2O.HCl$; mol wt 324.89. White, odorless crystalline powder. mp 258.5°. Freely sol in water, alc. Soly (mg/ml): water 40; alcohol 125. Slightly sol in acetone, chloroform, ether. LD_{50} in mice (mg/kg): 7.8 i.v., 82 s.c. (Henn, Brattsand).

(−)-**Form.** [27262-47-1] Levobupivacaine; (S)-bupivacaine. Crystals from isopropanol, mp 135-137°. $[\alpha]_D^{25}$ −80.9° (c = 5 in methanol).

(−)-**Form hydrochloride.** [27262-48-2] Chirocaine. $C_{18}H_{28}N_2O.HCl$; mol wt 324.89. mp 255-257°. $[\alpha]_D^{25}$ −12.3° (c = 2 in water).

THERAP CAT: Anesthetic (local).

1500. Bupranolol. [14556-46-8] 1-(2-Chloro-5-methylphenoxy)-3-[(1,1-dimethylethyl)amino]-2-propanol; 1-(tert-butylamino)-3-[(6-chloro-m-tolyl)oxy]-2-propanol; 1-(6-chloro-3-methylphenoxy)-3-tert-butylaminopropan-2-ol; 1-tert-butylamino-3-(2-chloro-5-methylphenoxy)-2-propanol; bupranol; Ophtorenin. $C_{14}H_{22}ClNO_2$; mol wt 271.79. C 61.87%, H 8.16%, Cl 13.04%, N 5.15%, O 11.77%. β-Adrenergic blocker. Prepn: Kunz et al., **DE 1236523** (1967 to Sanol-Arzneimittel Dr. Schwarz), C.A. **67**, 64046k (1967); **US 3309406** (1967). Pharmacology: Waterloh et al., Arzneim.-

Forsch. **19**, 153, 330, 1710 (1969); Pendleton et al., Arch. Int. Pharmacodyn. Ther. **187**, 75 (1970); P. Montastruc et al., Arch. Farmacol. Toxicol. **3**, 93 (1977).

Hydrochloride. [15148-80-8] KL-255; Betadran; Betadrenol; Looser; Panimit. $C_{14}H_{22}ClNO_2.HCl$; mol wt 308.24. Crystals, mp 220-222°.

THERAP CAT: Antihypertensive; antianginal; antiarrhythmic; antiglaucoma.

1501. Buprenorphine. [52485-79-7] (αS,5α,7α)-17-(Cyclopropylmethyl)-α-(1,1-dimethylethyl)-4,5-epoxy-18,19-dihydro-3-hydroxy-6-methoxy-α-methyl-6,14-ethenomorphinan-7-methanol; 21-cyclopropyl-7α-[(S)-1-hydroxy-1,2,2-trimethylpropyl]-6,14-endo-ethano-6,7,8,14-tetrahydrooripavine; 21-cyclopropyl-7α-(2-hydroxy-3,3-dimethyl-2-butyl)-6,14-endo-ethano-6,7,8,14-tetrahydrooripavine; RX-6029-M. $C_{29}H_{41}NO_4$; mol wt 467.65. C 74.48%, H 8.84%, N 3.00%, O 13.68%. Semisynthetic derivative of thebaine, q.v. with mixed opioid agonist-antagonist properties. Prepn: K. W. Bentley, **GB 1136214**; idem, **US 3433791** (1968, 1969 to Reckitt & Sons). See also: K. W. Bentley, "The Morphine Alkaloids" in The Alkaloids vol. **13**, R. F. Manske, Ed. (Academic Press, New York, 1971) pp 75-120. Review of synthesis and analgesic activity: J. W. Lewis, "Ring C-Bridged Derivatives of Thebaine and Oripavine" in Advan. Biochem. Psychopharmacol. vol. **8**, M. C. Braude et al., Eds. (Raven Press, New York, 1974) pp 123-137. NMR and stereochemistry: B. C. Uff et al., Magn. Reson. Chem. **23**, 454 (1985). LC/ESI-MS/MS determn in urine, blood and hair: D. Favretto et al., Rapid Commun. Mass Spectrom. **20**, 1257 (2006). Review of pharmacology and therapeutic efficacy in cancer pain: M. P. Davis, Support. Care Cancer **13**, 878-887 (2005); in opioid addiction: S. E. Robinson, CNS Drugs **20**, 697-712 (2006).

Crystals, mp 209°.

Hydrochloride. [53152-21-9] CL-112302; NIH-8805; UM-952; Buprenex; Lepetan; Subutex; Temgesic. $C_{29}H_{41}NO_4.HCl$; mol wt 504.11. White powder. Highly lipophilic. Slightly acidic with limited solubility in water.

Note: This is a controlled substance (narcotic): **21 CFR**, 1308.13.

THERAP CAT: Analgesic; in treatment of opioid dependence.

1502. Buprofezin. [69327-76-0] 2-[(1,1-Dimethylethyl)imino]tetrahydro-3-(1-methylethyl)-5-phenyl-4H-1,3,5-thiadiazin-4-one; 2-tert-butylimino-3-isopropyl-5-phenylperhydro-1,3,5-thiadiazin-4-one; NNI-750; NNK-758; NN-29285; PP-618; Applaud. $C_{16}H_{23}N_3OS$; mol wt 305.44. C 62.92%, H 7.59%, N 13.76%, O 5.24%, S 10.50%. Insect growth regulator which inhibits chitin synthesis. Prepn: Z. Grünecker et al., **DE 2824126**; K. Ikeda et al., **US 4159328** (1978, 1979 both to Nihon Nohyaku); H. Kanno, Pure Appl. Chem. **59**, 1027 (1987). Mode of action study: T. Asai et al., Appl. Entomol. Zool. **20**, 111 (1985). Control of whiteflies and scale insects: I. Ishaaya et al., Meded. Fac. Landbouwwet. Univ. Gent **54**, 1003 (1989). GC-MS determn in clementine citrus: P. Cabras et al., J. Agric. Food Chem. **46**, 4255 (1998). Persistence in olives and olive oil: *idem et al., Food Addit. Contam.* **17**, 855 (2000). Review of physical properties, activity and field trials: H. Kanno *et al., Proc. Br. Crop Prot. Conf. - Pests Dis.* **1981**, 59-66.

Crystals from isopropyl alcohol, mp 106.1°. Soly at 25° (g/l): acetone 240, chloroform 520, ethanol 80, toluene 320; in water 0.9 mg/l. Vapor pressure at 25°: 9.4×10^{-6} mmHg. LD_{50} in mice, rats (mg/kg): 10000, 8740 orally. LC_{50} (48 hr) in carp: 2-10 mg/l (Kanno, 1981).

USE: Insecticide.

1503. Bupropion. [34911-55-2] 1-(3-Chlorophenyl)-2-[(1,1-dimethylethyl)amino]-1-propanone; (±)-2-(*tert*-butylamino)-3'-chloropropiophenone; *m*-chloro-α-(*tert*-butylamino)propiophenone; amfebutamone. $C_{13}H_{18}ClNO$; mol wt 239.74. C 65.13%, H 7.57%, Cl 14.79%, N 5.84%, O 6.67%. Unicyclic aminoketone with noradrenergic and dopaminergic activity. Prepn: N. B. Mehta, D. A. Yeowell, **DE 2059618** (1971 to Wellcome Found.); N. B. Mehta, **US 3819706** (1974 to Burroughs Wellcome). Pharmacology: F. Soroko *et al., J. Pharm. Pharmacol.* **29**, 767 (1977). Preclinical toxicology: W. E. Tucker, *J. Clin. Psychiatry* **44**, 60-62 (1983). Pharmacokinetics in smokers: P.-H. Hsyu *et al., J. Clin. Pharmacol.* **37**, 737 (1997). HPLC determn in plasma: K. K. Loboz *et al., J. Chromatogr. B* **823**, 115 (2005). Clinical trial in attention deficit disorder with hyperactivity: C. K. Conners *et al., J. Am. Acad. Child Adolesc. Psychiatry* **34**, 1314 (1996); in smoking cessation: D. E. Jorenby *et al., N. Engl. J. Med.* **340**, 685 (1999). Review of pharmacology and clinical efficacy in depression: S. G. Bryant *et al., Clin. Pharm.* **2**, 525-537 (1983); of mechanism of action studies: J. A. Ascher *et al., J. Clin. Psychiatry* **56**, 395-401 (1995); of clinical use in smoking cessation: R. West, *Expert Opin. Pharmacother.* **4**, 533-540 (2003).

Pale yellow oil, bp.005 52°. Sol in methanol, ethanol, acetone, ether, benzene. Very hygroscopic and susceptible to decompn.

Hydrochloride. [31677-93-7] Budeprion; Wellbutrin; Zyban. $C_{13}H_{18}ClNO.HCl$; mol wt 276.20. Crystals from isopropanol and abs ethanol, mp 233-234°. Bitter, anesthetizing taste. Soly (mg/ml): water 312; alcohol 193; 0.1*N* HCl 333. pKa: 7.9. LD_{50} in mice, rats (mg/kg): 230, 210 i.p.; 575, 600 orally (Soroko).

Hydrobromide. [905818-69-1] Aplenzin. $C_{13}H_{18}ClNO.HBr$; mol wt 320.66. White or almost white crystalline powder with bitter taste. Sol in water.

THERAP CAT: Antidepressant; aid in smoking cessation.

1504. Buquinolate. [5486-03-3] 4-Hydroxy-6,7-bis(2-methylpropoxy)-3-quinolinecarboxylic acid ethyl ester; 4-hydroxy-6,7-diisobutoxy-3-quinolinecarboxylic acid ethyl ester; ethyl 6,7-diisobutoxy-4-hydroxyquinoline-3-carboxylate; Bonaid. $C_{20}H_{27}NO_5$; mol wt 361.44. C 66.46%, H 7.53%, N 3.88%, O 22.13%. Prepn: **BE 659237**; E. J. Watson, Jr. **US 3267106** (1965, 1966 both to Norwich). Efficacy in chickens: D. K. McLoughlin, *Avian Dis.* **14**, 126 (1970).

Crystals, mp 288-291°.

THERAP CAT (VET): Coccidiostat.

1505. Burgess Reagent. [29684-56-8] 3,3-Diethyl-6-oxo-7-oxa-4-thia-3,5-diazaoctane-2,4-diaminium inner salt 4,4-dioxide; *N,N*-diethyl-*N*-[[(methoxycarbonyl)amino]sulfonyl]ethanaminium inner salt; (methoxycarbonylsulfamoyl)triethylammonium hydroxide inner salt; methyl(carboxysulfamoyl)triethylammonium hydroxide inner salt; methyl-*N*-(triethylammoniumsulfonyl)carbamate. C_8-$H_{18}N_2O_4S$; mol wt 238.30. C 40.32%, H 7.61%, N 11.76%, O 26.86%, S 13.45%. Dehydrating agent. Synthesis method: G. M. Atkins, Jr., E. M. Burgess, *J. Am. Chem. Soc.* **90**, 4744 (1968). Prepn and applications in dehydration reactions: E. M. Burgess *et al., ibid.* **92**, 5224 (1970); *eidem, J. Org. Chem.* **38**, 26 (1973). Prepn and reaction with primary alcohols to form urethanes: E. M. Burgess *et al., Org. Synth.* **56**, 40 (1977). Reaction with epoxides: U. Rinner *et al., Synlett* **2003**, 1247. Review of chemistry: C. Lamberth, *J. Prakt. Chem. Chem.-Ztg.* **342**, 518-522 (2000); K. C. Nicolaou *et al., Chem. Eur. J.* **10**, 5581-5606 (2004).

Colorless needles from toluene, mp 71-72°.

USE: Reagent for dehydration of secondary and tertiary alcohols; source of heteroatoms (N, O, S) in organic synthesis.

1506. Buserelin. [57982-77-1] 6-[*O*-(1,1-Dimethylethyl)-D-serine]-9-(*N*-ethyl-L-prolinamide)-1-9-luteinizing hormone-releasing factor (swine); 6-[*O*-(1,1-dimethylethyl)-D-serine]-9-(*N*-ethyl-L-prolinamide)-10-deglycinamideluteinizing hormone-releasing factor (pig); [D-Ser(Bu^t)^6-des-Gly^{10}-NH$_2$]-LH-RH ethylamide. $C_{60}H_{86}$-$N_{16}O_{13}$; mol wt 1239.45. C 58.14%, H 6.99%, N 18.08%, O 16.78%. Synthetic nonapeptide agonist analog of gonadotropin-releasing hormone, *q.v.* Synthesis: W. Konig *et al.,* **DE 2438350**; *eidem,* **US 4024248** (1976, 1977 both to Hoechst); A. S. Dutta *et al., J. Med. Chem.* **21**, 1018 (1978). Clinical pharmacology: A. Lemay *et al., Fertil. Steril.* **37**, 193 (1982). Radioimmunoassay in plasma and urine: S. Saito *et al., J. Immunol. Methods* **79**, 173 (1985). Veterinary use to increase conception rate: K. Moller, E. D. Fielden, *N. Z. Vet. J.* **29**, 214 (1981). Clinical evaluation in prostatic carcinoma: J. H. Waxman, *Br. J. Urol.* **55**, 737 (1983); as ovulatory stimulant for *in vitro* fertilization: V. MacLachlan *et al., N. Engl. J. Med.* **320**, 1233 (1989). Review of pharmacokinetics and clinical profile: R. N. Brogden *et al., Drugs* **39**, 399-437 (1990); of efficacy in prostatic carcinoma: H. J. de Voogt *et al., Scand. J. Urol. Nephrol. Suppl.* **138**, 131-136 (1991).

5-oxoPro–His–Trp–Ser–Tyr–D-Ser(*t*-Bu)–Leu–Arg–ProNHCH$_2$CH$_3$

$[\alpha]_D^{20}$ −40.4° (c = 1 in dimethylacetamide).

Monoacetate. [68630-75-1] HOE-766; Receptal; Suprecur; Suprefact. $C_{60}H_{86}N_{16}O_{13}.C_2H_4O_2$; mol wt 1299.50.

THERAP CAT: Antineoplastic (hormonal). Gonad-stimulating principle.

THERAP CAT (VET): Gonad-stimulating principle.

1507. Buspirone. [36505-84-7] 8-[4-[4-(2-Pyrimidinyl)-1-piperazinyl]butyl]-8-azaspiro[4.5]decane-7,9-dione. $C_{21}H_{31}N_5O_2$; mol wt 385.51. C 65.43%, H 8.11%, N 18.17%, O 8.30%. Nonbenzodiazepine anxiolytic; 5-hydroxytryptamine (5-HT$_1$) receptor agonist. Prepn: Y. H. Wu *et al., J. Med. Chem.* **15**, 477 (1972); Y. H. Wu, J. W. Rayburn, **DE 2057845** (1971 to Bristol-Myers); *eidem,* **US 3717634** (1973 to Mead-Johnson). Pharmacology: L. E. Allen *et al., Arzneim.-Forsch.* **24**, 917 (1974). Comparison with diazepam in treatment of anxiety: H. L. Goldberg, R. J. Finnerty, *Am. J. Psychiatry* **136**, 1184 (1979). Disposition and metabolism: S. Caccia *et al., Xenobiotica* **13**, 147 (1983). Series of articles on chemistry, pharmacology, addictive potential, and clinical trials: *J. Clin. Psychiatry* **43**, pp 1-116 (1982); on pharmacology, safety and clinical comparison with clorazepate: *Am. J. Med.* **80**, Suppl. 3B, 1-51

(1986). Review of pharmacology and therapeutic efficacy: K. L. Goa, A. Ward, *Drugs* **32**, 114-129 (1986); M. W. Jann, *Pharmacotherapy* **8**, 100-116 (1988); of discovery and clinical development: J. S. New, *Med. Res. Rev.* **10**, 283-326 (1990); of pharmacokinetics and pharmacodynamics: I. Mahmood, C. Sahajwalla, *Clin. Pharmacokinet.* **36**, 277-287 (1999).

Hydrochloride. [33386-08-2] Ansial; Anxut; Bespar; Buspar. $C_{21}H_{31}N_5O_2 \cdot HCl$; mol wt 421.97. Crystals from abs ethanol, mp 201.5-202.5°. Very sol in water; freely sol in methanol, methylene chloride; sparingly sol in ethanol, acetonitrile; very slightly sol in ethyl acetate. Practically insol in hexanes. pKa_1: 4.12. pKa_2: 7.32. LD_{50} i.p. in rats: 136 mg/kg (Allen).

THERAP CAT: Anxiolytic.

THERAP CAT (VET): In treatment of behavioral disorders in dogs and cats.

1508. Busulfan. [55-98-1] 1,4-Butanediol 1,4-dimethanesulfonate; 1,4-bis(methanesulfonoxy)butane; 1,4-di(methanesulfonyloxy)butane; 1,4-di(methylsulfonoxy)butane; methanesulfonic acid tetramethylene ester; tetramethylene bis(methanesulfonate); busulphan; CB-2041; GT-41; Busulfex; Myleran. $C_6H_{14}O_6S_2$; mol wt 246.29. C 29.26%, H 5.73%, O 38.98%, S 26.03%. Cytotoxic alkylating agent with antileukemic activity; conditioning agent in bone marrow transplants. Discovery: A. Haddow, G. M. Timmis, *Lancet* **1**, 207 (1953). Prepn: G. M. Timmis, **US 2917432** (1959 to Burroughs Wellcome). Comprehensive description: M. Tariq, A. A. Al Badr, *Anal. Profiles Drug Subs.* **16**, 53-83 (1987). Pharmacokinetics: H. Ehrsson *et al.*, *Clin. Pharmacol. Ther.* **34**, 86 (1983). Review of pharmacology: C. D. R. Dunn, *Exp. Hematol.* **2**, 101-117 (1974); of toxicology: J. B. Bishop, J. S. Wassom, *Mutat. Res.* **168**, 15-45 (1986). Chemosterilant effect in boll weevils: J. W. Haynes *et al.*, *J. Econ. Entomol.* **66**, 619 (1973). LC/MS determn in plasma and saliva: M. Rauh *et al.*, *Clin. Pharmacokinet.* **45**, 305 (2006). Clinical pretreatment with cyclophosphamide for bone marrow transplants: G. W. Santos *et al.*, *N. Engl. J. Med.* **309**, 1347 (1983). Clinical pharmacokinetics in stem cell transplant patients: L. Nguyen *et al.*, *Cancer Chemother. Pharmacol.* **57**, 191 (2006); in bone marrow transplantation in thalassemia: M. Chandy *et al.*, *Bone Marrow Transplant.* **36**, 839 (2005).

Crystals, mp 114-118°. Soly in acetone at 25°: 2.4 g/100 ml; in alcohol: 0.1 g/100 ml. Very slightly sol in water, but will dissolve slowly as hydrolysis takes place. LD_{50} i.v. in rats: 1.8 mg/kg. See: H. R. Scherf *et al.*, *Arzneim.-Forsch.* **20**, 1467 (1970).

Caution: This substance is listed as a known human carcinogen: *Report on Carcinogens, Twelfth Edition* (PB2011-111646, 2011) p 77.

USE: Insect sterilant.

THERAP CAT: Antineoplastic.

THERAP CAT (VET): Antineoplastic.

1509. Butabarbital. [125-40-6] 5-Ethyl-5-(1-methylpropyl)-2,4,6(1*H*,3*H*,5*H*)-pyrimidinetrione; 5-*sec*-butyl-5-ethylbarbituric acid; 5-ethyl-5-(1-methylpropyl)barbituric acid; secbutabarbital; sec-butobarbitone. $C_{10}H_{16}N_2O_3$; mol wt 212.25. C 56.59%, H 7.60%, N 13.20%, O 22.61%. Prepn: Shonle, **US 1856792** (1932 to Lilly).

LC determn in urine: Y. Martín-Biosca *et al.*, *J. Pharm. Biomed. Anal.* **21**, 331 (1999).

White, odorless, crystalline powder. mp 165-168°. Sol in alc, chloroform, ether and in solutions of alkali hydroxides and carbonates. Very slightly sol in water.

Sodium salt. [143-81-7] Sodium 5-*sec*-butyl-5-ethylbarbiturate; Asturidon; Butisol Sodium. $C_{10}H_{15}N_2NaO_3$; mol wt 234.23. White powder, having a bitter taste. One gram dissolves in 2 ml water, in about 7 ml alcohol. Practically insol in ether, benzene. The pH of 10% aq soln 10.0-11.2.

Note: This is a controlled substance (depressant): **21 CFR, 1308.13.**

THERAP CAT: Sedative, hypnotic.

1510. Butacaine. [149-16-6] 3-(Dibutylamino)-1-propanol 1-(4-aminobenzoate); 3-(*p*-aminobenzoxy)-1-di-*n*-butylaminopropane; dibutylaminopropyl-*p*-aminobenzoate; *p*-aminobenzoyldibutylaminopropanol; Butelline. $C_{18}H_{30}N_2O_2$; mol wt 306.45. C 70.55%, H 9.87%, N 9.14%, O 10.44%. Prepd from *p*-nitrobenzoyl chloride and γ-di-*n*-butylaminopropanol followed by reduction of the NO_2 group to NH_2: O. Kamm *et al.*, **US 1358751**; Adams, Volwiler, **US 1676470**; A. Weston, **US 2437984** (1920, 1928, 1948 all to Abbott); Burnett *et al.*, *J. Am. Chem. Soc.* **59**, 2248 (1937); Kaye, Roberts, *ibid.* **73**, 4762 (1951). Toxicity study: Schmidt *et al.*, *Toxicol. Appl. Pharmacol.* **1**, 454 (1956).

Liquid, $bp_{0.11}$ 178-182°.

Sulfate. [149-15-5] Butyn Sulfate. $(C_{18}H_{30}N_2O_2)_2 \cdot H_2SO_4$; mol wt 710.97. Crystals from *n*-propanol, mp 138.5-139.5°. Also reported as mp 100-103°. Produces numbness of tongue on tasting. Affected by light. One gram dissolves slowly in somewhat less than 1 ml water, more rapidly on heating. Quite sol in warm alc, in acetone; slightly sol in chloroform. Practically insol in ether. Aq soln is practically neutral to litmus and may be boiled for sterilization without dec. LD_{50} i.v. in mice: 12.4 mg/kg (Schmidt). *Pharmaceutical Incompat.* Alkalies and alkaline-reacting substances liberate the free base as an oily liquid from solns. Bicarbonates produce a precipitate of butacaine carbonate. Iodine gives a brown precipitate. Chlorides form the almost insol butacaine chloride which may precipitate.

Hydrochloride. $C_{18}H_{30}N_2O_2 \cdot HCl$. Crystals from ethanol, mp 157-158.5°.

THERAP CAT: Anesthetic (local).

THERAP CAT (VET): Anesthetic (local).

1511. Butachlor. [23184-66-9] *N*-(Butoxymethyl)-2-chloro-*N*-(2,6-diethylphenyl)acetamide; *N*-(butoxymethyl)-2-chloro-2',6'-diethylacetanilide; 2-chloro-2,6-diethyl-*N*-(butoxymethyl)acetanilide; CP-53619; Machete; Butanex. $C_{17}H_{26}ClNO_2$; mol wt 311.85. C 65.48%, H 8.40%, Cl 11.37%, N 4.49%, O 10.26%. Selective preemergent and pre-plant herbicide. Prepn: J. F. Olin, **US 3442945** (1969 to Monsanto). Toxicity data: A. Strateva, *Exp. Med. Morphol.* **13**, 123 (1974). Degradation: Y.-L. Chen, T.-C. Wu, *J. Pestic. Sci.* **3**, 411 (1978). Environmental persistence: T. Ohyama *et al.*, *Bull. Environ. Contam. Toxicol.* **39**, 555 (1987); and soil absorption: H.-C. Chiang *et al.*, *ibid.* **58**, 758 (1997). HPLC determn in soil: X. Weibing, *J. Environ. Sci. (China)* **8**, 504 (1996). Control of barnyardgrass: T. C. Miller *et al.*, *MAFES Res. Highlights* **46**, 4 (1983); in rice: B. S. Azad *et al.*, *Seed Farms* **15**, 29 (1989).

Light yellow oil, bp$_{0.5}$ 196°. d$_4^{30}$ 1.0695. Soly in water at 20°: 20 mg/l. Sol in most organic solvents. LD$_{50}$ orally in rats: 1740 mg/kg (Strateva).

USE: Herbicide.

1512. 1,3-Butadiene. [106-99-0] α,γ-Butadiene; bivinyl; divinyl; erythrene; vinylethylene; biethylene; pyrrolylene. C$_4$H$_6$; mol wt 54.09. C 88.82%, H 11.18%. Manuf as a coproduct of hydrocarbon steam cracking to produce ethylene. Also produced from ethanol; from petroleum gases, i.e., by catalytic dehydrogenation of butene or butene-butane mixtures; by cracking naphtha and light oil. Manuf: *Faith, Keyes & Clark's Industrial Chemicals*, F. A. Lowenheim, M. K. Moran, Eds. (Wiley-Interscience, New York, 4th ed., 1975) pp 164-172. Toxicity study: Carpenter *et al.*, *J. Ind. Hyg. Toxicol.* **26**, 69 (1944). *Reviews:* Norton, *Chem. Rev.* **31**, 319 (1942); Alder, "The Diene Synthesis" in *Newer Methods of Preparative Organic Chemistry* (Interscience, New York, 1948); Konrad, *Angew. Chem.* **62**, 491 (1950); Hillyer, Stallings, *Pet. Refin.* **35**(12), 157 (1956); A. S. Onishchenko, *Diene Synthesis* (New York, 1964); Bailey, "Butadiene" in *Vinyl and Diene Monomers* (part 2), E. C. Leonard, Ed. (Interscience, New York, 1971) pp 757-995; H. N. Sun, J. P. Wristers in *Kirk-Othmer Encyclopedia of Chemical Technology* **vol. 4** (John Wiley & Sons, New York, 4th ed., 1992) pp 663-690. Review of toxicology: L. S. Birnbaum, *Environ. Health Perspect.* **101**, Suppl. 6, 161-167 (1993); and human exposure: *Toxicological Profile for 1,3-Butadiene* (PB93-110690, 1992) 148 pp; of carcinogenic risk: *IARC Sci. Publ.* **127**, 1-412 (1993).

Colorless gas. Mild aromatic odor. *Flammable and combustible.* mp −108.966°: Glasgow *et al.*, *Anal. Chem.* **22**, 1521 (1950). bp$_{760}$ −4.5°. d$_4^{-6}$ 0.650. Freezing pt −108.902°. Densities from −17.8° to 60°: *Ind. Eng. Chem. Anal. Ed.* **16**, 7 (1944). n$_D^{-6}$ 1.4223. bp (at pressures greater than 1 atm): 2 atm: 15.3°; 5 atm: 47.0°; 10 atm: 76.0°; 20 atm: 114.0°; 30 atm: 139.8°; 40 atm: 160.0°. Critical temperature 161.8°; critical pressure 42.6 atm. Infrared absorption spectrum: *ibid.* 422. Stability: *Ind. Eng. Chem.* **36**, 3 (1944). Stabilization with *o*-dihydroxybenzene: **GB 569412**; with aliphatic mercaptans: **US 2373754**. Explosion limits in air, vol%: lower 2.0, upper 11.5. Polymerizes and copolymerizes easily, e.g. under the influence of sodium, thereby forming synthetic rubbers. Sparingly soluble in water; slightly soluble in methanol, ethanol; soluble in organic solvents such as diethyl ether, benzene, carbon tetrachloride. Alcohol dissolves about 40 vols at room temp. LC$_{50}$ in rats (4 hr inhalation): 129,000 ppm; in mice (2 hr inhalation): 122,000 ppm (Birnbaum).

Caution: Potential symptoms of overexposure are irritation of eyes, nose and throat; drowsiness, lightheadedness; teratogenic and reproductive effects; direct contact with liquid may cause frostbite. *See NIOSH Pocket Guide to Chemical Hazards* (DHHS/NIOSH 97-140, 1997) p 34; *Patty's Industrial Hygiene and Toxicology* **vol. 2B**, G. D. Clayton, F. E. Clayton, Eds. (John Wiley & Sons, Inc., New York, 4th ed., 1994) p 1242, 1250-1252. This substance is listed as a known human carcinogen: *Report on Carcinogens, Twelfth Edition* (PB2011-111646, 2011) p 75.

USE: As component in the manuf of polymers such as synthetic rubbers, plastics, resins. As a chemical intermediate for the production of many industral materials; in the manuf of adiponitrile.

1513. Butalamine. [22131-35-7] N^1,N^1-Dibutyl-N^2-(3-phenyl-1,2,4-oxadiazol-5-yl)-1,2-ethanediamine; 5-[[2-(dibutylamino)ethyl]amino]-3-phenyl-1,2,4-oxadiazole; 3-phenyl-5-(dibutyl-

aminoethylamino)-1,2,4-oxadiazole. C$_{18}$H$_{28}$N$_4$O; mol wt 316.45. C 68.32%, H 8.92%, N 17.71%, O 5.06%. Prepn: **FR M3334**; Aron-Samuel, Sterne, **US 3338899** (1965, 1967 to Aron-Samuel). Pharmacology and toxicity studies: J. Sterne *et al.*, *Therapie* **24**, 735 (1969); J. Sterne, *ibid.* 745; J. L. Fontaine, R. Fontaine, *ibid.* **25**, 961 (1970).

Hydrochloride. [56974-46-0] LA-1221; Adrevil; Hemotrope; Surem; Surheme. C$_{18}$H$_{28}$N$_4$O.HCl; mol wt 352.91. mp 145°. LD$_{50}$ in mice (mg/kg): 43 i.v., 2500 s.c., 625 orally; LD$_{50}$ in rats (mg/kg): >4000 s.c., 1600 orally (Sterne).

THERAP CAT: Vasodilator (peripheral).

1514. Butalbital. [77-26-9] 5-(2-Methylpropyl)-5-(2-propen-1-yl)-2,4,6(1*H*,3*H*,5*H*)-pyrimidinetrione; 5-allyl-5-isobutylbarbituric acid; 5-allyl-5-(2-methylpropyl)barbituric acid; 5-isobutyl-5-allylbarbituric acid; alisobumal; allylbarbital; itobarbital; tetrallobarbital; Sandoptal. C$_{11}$H$_{16}$N$_2$O$_3$; mol wt 224.26. C 58.91%, H 7.19%, N 12.49%, O 21.40%. Prepn: Volwiler, *J. Am. Chem. Soc.* **47**, 2236 (1925).

Prisms, having a slightly bitter taste. mp 138-139°. Stable in air. A satd aq soln is acid to litmus. Freely sol in alcohol, chloroform, ether; sol in boiling water, acetone, glacial acetic acid, and in solns of fixed alkali hydroxides and alkali carbonates; slightly sol in cold water. Practically insol in water, petr ether.

Note: This is a controlled substance (depressant): **21 CFR**, 1308.13.

THERAP CAT: Sedative, hypnotic.

1515. Butamben. [94-25-7] 4-Aminobenzoic acid butyl ester; butyl aminobenzoate; *n*-butyl *p*-aminobenzoate; Butesin; Butoform; Planoform; Scuroforme. C$_{11}$H$_{15}$NO$_2$; mol wt 193.25. C 68.37%, H 7.82%, N 7.25%, O 16.56%. Prepn: Brill, *J. Am. Chem. Soc.* **43**, 1322 (1921); Adams, Volwiler, **US 1440652** (1923 to Abbott); **GB 252870**; *C.A.* **17**, 1243 (1923); *C.A.* **21**, 2478 (1927).

Crystals from alc, mp 57-59°. bp$_8$ 174°. One gram dissolves in about 7 liters of water. Sol in dil acids, alcohol, chloroform, ether, and fatty oils. It is hydrolyzed slowly when boiled with water.

THERAP CAT: Anesthetic (local).

THERAP CAT (VET): Anesthetic (local).

1516. Butamirate. [18109-80-3] α-Ethylbenzeneacetic acid 2-[2-(diethylamino)ethoxy]ethyl ester; 2-phenylbutyric acid 2-[2-(diethylamino)ethoxy]ethyl ester; 2-[2-(diethylamino)ethoxy]ethyl 2-phenylbutyrate; butamyrate. C$_{18}$H$_{29}$NO$_3$; mol wt 307.43. C 70.32%, H 9.51%, N 4.56%, O 15.61%. Prepd by the esterification of phenylethylacetyl chloride with diethylaminoethoxyethanol: **DE 1151515**; Heusser, **US 3349114** (1963, 1967, both to Hommel A.G.).

Practically colorless liquid with peculiar odor. bp_1 140-155°. Practically insol in water. Very sol in alcohol, acetone, ether.

Citrate. [18109-81-4] Abbott 36581; HH-197; Acodeen; Panatus; Sincodix; Sinecod. $C_{18}H_{29}NO_3 \cdot C_6H_8O_7$; mol wt 499.56. White, hygroscopic crystals from acetone, mp 75°.

THERAP CAT: Antitussive.

1517. Butane. [106-97-8] *n*-Butane. C_4H_{10}; mol wt 58.12. C 82.66%, H 17.34%. Occurrence in petroleum: Mabery, *J. Am. Chem. Soc.* **30**, 143 (1908); in natural gas and in refinery cracking products. Prepd from C_2H_5I and sodium amalgam: Löwig, *Jahresber. Fortschr. Chem.* **1860**, 397; *Beilstein* vol. **1**, 118. Recovery of butanes from natural and refinery gases: Kirkbride, Bertelli, *Ind. Eng. Chem.* **35**, 1242 (1943); Walters, *ibid.* **47**, 2544 (1955); Gilmore, Bauer, *Oil Gas J.* **50**, 84, 90, 94, 119 (1951), *C.A.* **46**, 1743d (1952). Separation of butane and isobutane: Stone, *Pet. Refin.* **25** (4), 164 (1946), *C.A.* **43**, 2414 (1949). Handbook: *Butane-Propane Gases*, L. C. Denny et al., Eds. (Chilton Co., Los Angeles, 4th ed., 1962) 383 pp.

Flammable gas. bp −0.50°. d(gas) 2.046 (air = 1). One vol of water dissolves 0.15 vol and 1 vol of alcohol 18 vols of the gas at 17° and 770 mm; 1 vol of ether or chloroform at 17° dissolves 25 or 30 vols of the gas, resp.

Caution: Potential symptoms of overexposure are drowsiness, narcosis, asphyxia; direct contact with liquid may cause frostbite. See *NIOSH Pocket Guide to Chemical Hazards* (DHHS/NIOSH 97-140, 1997) p 34.

USE: As producer gas; raw material for motor fuels, in the manuf of synthetic rubbers.

1518. 1,4-Butanediol. [110-63-4] 1,4-Butylene glycol; 1,4-dihydroxybutane; tetramethylene glycol; BDO. $C_4H_{10}O_2$; mol wt 90.12. C 53.31%, H 11.19%, O 35.51%. Commercial chemical used in the manufacture of polymer fibers and plastics. Metabolically converted to the CNS depressant, γ-hydroxybutyrate, (GHB) *q.v.* Prepn from *N,N'*-dinitrobutanediyldiamine: M. P. J. Dekkers, *Recl. Trav. Chim. Pays-Bas* **9**, 92 (1890); from 4-benzyloxybutan-1-ol: W. R. Kirner, G. H. Richter, *J. Am. Chem. Soc.* **51**, 2503 (1929). LC-MS/MS determn in urine: M. Wood et al., *J. Chromatogr. A* **1056**, 83 (2004). Toxicological profile: R. D. Irwin, *J. Appl. Toxicol.* **26**, 72 (2006). Clinical pharmacokinetics, metabolism, and toxicology: D. Thai et al., *Clin. Pharm. Ther.* **81**, 178 (2007). Review of abuse as substitute for GHB: R. B. Palmer, *Toxicol. Rev.* **23**, 21-31 (2004). Review of industrial prepns and uses: M. L. Morgan, *Chem. Ind. (London)* **3**, 166-168 (1997); T. Haas et al., *Appl. Catal. A* **280**, 83-88 (2005).

Colorless, viscous liquid. mp 19-19.5°. bp 230°. d_4^{20} 1.0171. n_D^{20} 1.4467. Sol in water, DMSO, acetone, 95% ethanol. LD_{50} in albino rats (mg/kg): 1000 i.p.; 1550 orally (Irwin).

Caution: When ingested, may cause dizziness, alcohol-like inebriation, sedation. Symptoms of overexposure include vomiting, myoclonus, respiratory depression, coma, or death (Thai).

USE: Industrial solvent; intermediate in organic synthesis; polymer feedstock.

1519. Butaperazine. [653-03-2] 1-[10-[3-(4-Methyl-1-piperazinyl)propyl]-10*H*-phenothiazin-2-yl]-1-butanone; 2-butyryl-10-[3-(4-methyl-1-piperazinyl)propyl]phenothiazine; *N*-[γ-(4'-methyl-1'-piperazinyl)propyl]-3-butyrylphenothiazine; butyrylperazine; Bayer 1362; Repoise; Tyrylen. $C_{24}H_{31}N_3OS$; mol wt 409.59. C

70.38%, H 7.63%, N 10.26%, O 3.91%, S 7.83%. Prepn: Hoerlein et al., **DE 1120451** (1961 to Bayer), *C.A.* **57**, 4677c (1962).

$bp_{0.05}$ 270-280°.

Dimaleate. [1063-55-4] Randolectil. $C_{24}H_{31}N_3OS \cdot 2C_4H_4O_4$; mol wt 641.74.

Maleate. $C_{24}H_{31}N_3OS \cdot C_4H_4O_4$. Crystals from carbon tetrachloride, mp 180-182°.

THERAP CAT: Antipsychotic.

1520. Butedronic Acid. [51395-42-7] 2-(Diphosphonomethyl)butanedioic acid; (diphosphonomethyl)succinic acid; 2,3-dicarboxypropane-1,1-diphosphonic acid; DPD. $C_5H_{10}O_{10}P_2$; mol wt 292.07. C 20.56%, H 3.45%, O 54.78%, P 21.21%. Prepn: A. Heins et al., **DE 2217692** (1973 to Henkel); eidem, **US 3923876** (1975 to Bayer A.G.); K. H. Worms, H. Blum, *Z. Anorg. Allg. Chem.* **457**, 219 (1979); and physical-chemical properties: N. Vanlic-Razumenic et al., *J. Serb. Chem. Soc.* **51**, 63 (1986). Clinical pharmacokinetics of ^{99m}Tc complex: C. Schümichen, H. Schmidt, *Nuklearmedizin Suppl.* **19**, 930 (1982). Clinical evaluation of ^{99m}Tc complex as skeletal imaging agent: G. Godart et al., *Clin. Nucl. Med.* **11**, 92 (1986).

Monohydrate. White crystalline powder from glacial acetic acid and water (15:1), mp 150°. uv max (water): 208 nm (ε 274).

Tetrasodium salt. [97772-98-0] Tc-924 (DPD); Teceos. C_5H_6-$Na_4O_{10}P_2$; mol wt 380.00.

THERAP CAT: ^{99m}Tc complex as diagnostic aid (radioactive imaging agent).

1521. Butenafine. [101828-21-1] *N*-[[4-(1,1-Dimethylethyl)-phenyl]methyl]-*N*-methyl-1-naphthalenemethanamine; *N*-(*p-tert*-butylbenzyl)-*N*-methyl-1-naphthalenemethylamine. $C_{23}H_{27}N$; mol wt 317.48. C 87.01%, H 8.57%, N 4.41%. Benzylamine antifungal; squalene epoxidase inhibitor. Prepn: T. Maeda et al., **EP 164697**; eidem, **US 5021458** (1985, 1991 both to Kaken). Antifungal activity *in vivo*: T. Arika et al., *Antimicrob. Agents Chemother.* **34**, 2250 (1990). Review of antifungal activity, mechanism of action, and clinical trials: R. Fukushiro et al., in *Recent Progress in Antifungal Chemotherapy*, H. Yamaguchi et al., Eds. (Dekker, New York, 1992) pp. 147-157. Clinical trial in treatment of athlete's foot: T. A. Syed et al., *Clin. Drug Invest.* **9**, 393 (2000).

Hydrochloride. [101827-46-7] KP-363; Mentax. $C_{23}H_{27}N$·HCl; mol wt 353.93. Crystals from acetone + ethanol, mp 211-213°. Easily sol in methanol, ethanol, dichloromethane, chloroform. Slightly sol in water.

THERAP CAT: Antifungal.

1522. 1-Butene. [106-98-9] α-Butylene; ethylethylene. C_4H_8; mol wt 56.11. C 85.62%, H 14.37%. Occurs in oil and coal gas. Obtained by cracking of petr oils and by thermal decompn of butane or pentane or isopentane: Egloff *et al.*, *Ind. Eng. Chem.* **28**, 1283 (1936); Calingaert, *J. Am. Chem. Soc.* **45**, 130 (1923); from butyl alcohol by treatment with conc H_2SO_4: *Compt. Rend.* **176**, 813 (1923).

Gas. Not solid at −190°. bp_{760} −6.47°. $d_4^{-6.47}$ 0.6255. Explodes in mixtures with oxygen. Description of properties: *Oil Gas J.* **44**, 119 (1945), *C.A.* **39**, 3783; *J. Am. Chem. Soc.* **50**, 1427 (1928). *Caution:* May be narcotic in high concns. A simple asphyxiant.

1523. 2-Butene. [107-01-7] Pseudo-butylene; *sym*-dimethylethylene; β-butylene. C_4H_8; mol wt 56.11. C 85.62%, H 14.37%. Occurs in coal gas. Obtained by cracking of petroleum oils. From isobutanol by the action of hot $ZnCl_2$: LeBel, Greene, *Am. Chem. J.* **2**, 24 (1880). Configuration of the stereoisomeric forms: Kistiakowsky, *J. Am. Chem. Soc.* **57**, 879 (1935).

(E)-form

Flammable gas.
(E)-**Form.** [624-64-6] mp −105.8°; bp_{744} 0.3-0.4°.
(Z)-**Form.** [590-18-1] mp −139.3°; bp_{760} 3.73°.

1524. Butethal. [77-28-1] 5-Butyl-5-ethyl-2,4,6(1H,3H,5H)-pyrimidinetrione; 5-butyl-5-ethylbarbituric acid; butobarbitone; Soneryl; Neonal; Butobarbital; Etoval. $C_{10}H_{16}N_2O_3$; mol wt 212.25. C 56.59%, H 7.60%, N 13.20%, O 22.61%. Prepn: **US 1609520** (1926). Pharmacology and toxicity: G. A. Alles *et al.*, *J. Pharmacol. Exp. Ther.* **89**, 356 (1947).

Crystals. Slightly bitter taste. mp 124-127°. One gram dissolves in about 5 ml alcohol, 10 ml ether. Practically insol in water; insol in petr ether, aliphatic hydrocarbons. LD_{50} i.p. in mice: 1.506 mM/kg (Alles).
Note: This is a controlled substance (depressant): **21 CFR**, 1308.13.
THERAP CAT: Sedative, hypnotic.

1525. Butethamate. [14007-64-8] α-Ethylbenzeneacetic acid 2-(diethylamino)ethyl ester; 2-phenylbutyric acid 2-(diethylamino)-ethyl ester; β-diethylaminoethyl ethylphenylacetate; 2-diethylaminoethyl 2-phenylbutyrate. $C_{16}H_{25}NO_2$; mol wt 263.38. C 72.97%, H 9.57%, N 5.32%, O 12.15%. Prepn: Di Paco, Tauro, *Farmaco Ed. Sci.* **11**, 540 (1956); **CH 291375** (1953 to AG Hommel's Haematogen) and **CH 292596** (1953 to Chem. Fabrik "PARA"), *C.A.* **49**, 2505a,c (1955). Pharmacology: Jordan, *Arzneim.-Forsch.* **8**, 716 (1958); Fleisch *et al.*, *ibid.* **11**, 1119 (1961).

bp_{11} 167-169°. n_D^{20} 1.4909.

Citrate. [3639-12-1] Abuphenine; Convenil; Hicoseen; Phenesin; Phenetin. $C_{16}H_{25}NO_2.C_6H_8O_7$; mol wt 455.50. Crystals from abs alcohol, mp 109-110°. Freely sol in water, alcohol.
THERAP CAT: Antitussive.

1526. Butethamine. [2090-89-3] 2-[(2-Methylpropyl)amino]ethanol 1-(4-aminobenzoate); 2-(isobutylamino)ethanol *p*-aminobenzoate (ester); 2-(isobutylamino)ethyl *p*-aminobenzoate. C_{13}-$H_{20}N_2O_2$; mol wt 236.32. C 66.07%, H 8.53%, N 11.85%, O 13.54%. Prepn of the hydrochloride: Goldberg, **US 2139818** (1938 to Novocol Chem.); J. Büchi *et al.*, *Arzneim.-Forsch.* **14**, 161 (1964); **16**, 1657 (1966).

Formate. $C_{13}H_{20}N_2O_2.CH_2O_2$. mp 136-139°. Freely sol in water and alcohol. Slightly sol in chloroform, ether; very slightly sol in benzene. pH (1% aq soln): about 6.1.
Hydrochloride. [553-68-4] Ibylcaine; Monocaine. $C_{13}H_{20}N_2$-O_2.HCl; mol wt 272.77. mp 192-196°. Sol in water, slightly sol in alcohol, chloroform, benzene. Practically insol in ether. pH (1% aq soln): about 4.7.
***meta*-Isomer hydrochloride.** [553-58-2] Metabutethamine hydrochloride. $C_{13}H_{20}N_2O_2$.HCl. Bitter crystals, mp 181-184°. Sol in water. pH of 1:50 aq soln about 6.2. Slightly sol in alcohol, acetone, chloroform.
THERAP CAT: Anesthetic (local).
THERAP CAT (VET): Local anesthetic for nerve block.

1527. Buthiazide. [2043-38-1] 6-Chloro-3,4-dihydro-3-(2-methylpropyl)-2H-1,2,4-benzothiadiazine-7-sulfonamide 1,1-dioxide; 6-chloro-3,4-dihydro-3-isobutyl-7-sulfamoyl-1,2,4-benzothiadiazine 1,1-dioxide; thiabutazide; butizide; isobutylhydrochlorothiazide; Su-6187; S-3500; Eunephran; Saltucin. $C_{11}H_{16}ClN_3O_4S_2$; mol wt 353.84. C 37.34%, H 4.56%, Cl 10.02%, N 11.88%, O 18.09%, S 18.12%. Prepd from 5-chloro-2,4-disulfamoylaniline and isovaleraldehyde: Werner *et al.*, *J. Am. Chem. Soc.* **82**, 1161 (1960); **GB 861367**; **GB 885078**; G. de Stevens, L. H. Werner, **US 3163645** (1961, 1961, 1964 all to Ciba); Topliss *et al.*, *J. Org. Chem.* **26**, 3842 (1961).

Crystals, mp 241-245° (Werner, 1960); from methanol + chloroform, mp 228° (Topliss). An ingredient of *Modenol*.
THERAP CAT: Diuretic, antihypertensive.

1528. Buthionine Sulfoximine. [5072-26-4] 2-Amino-4-(*S*-butylsulfonimidoyl)butanoic acid; *S*-(3-amino-3-carboxypropyl)-*S*-butylsulfoximine; *S*-(*n*-butyl)homocysteine sulfoximine; BSO. C_8-$H_{18}N_2O_3S$; mol wt 222.30. C 43.22%, H 8.16%, N 12.60%, O 21.59%, S 14.42%. Selective inhibitor of γ-glutamyl cysteine synthetase, an enzyme in the glutathione, *q.v.*, biosynthetic pathway. BSO-mediated depletion of intracellular glutathione has been associated with increased sensitivity of tumor cells to antineoplastic agents. Prepn: K. Hayashi, *Chem. Pharm. Bull.* **8**, 177 (1960). Synthesis and *in vitro* and *in vivo* enzyme inhibition: O. W. Griffith, A. Meister, *J. Biol. Chem.* **254**, 7758 (1979). Mechanism of action and metabolism in animals: O. W. Griffith, *ibid.* **257**, 13704 (1982). Chemosensitization by BSO of melphalan-resistant murine leukemia cells *in vivo*: S. Somfai-Relle *et al.*, *Biochem. Pharmacol.* **33**, 485 (1984); R. A. Kramer *et al.*, *Cancer Res.* **47**, 1593 (1987). Enhanced melphalan, *q.v.*, cytotoxicity in human cancer cells *in vitro*: J. A. Green *et al.*, *ibid.* **44**, 5427 (1984); and in mice bearing human tumor xenografts: H. S. Friedman *et al.*, *J. Natl. Cancer Inst.* **81**, 524 (1989).

Crystals from aq ethanol, mp 214-215.5° (dec). Sol in water. Partition coefficient (octanol/water): 4.34 ±0.0004.

USE: Biological tool for depletion of glutathione.

1529. Butibufen. [55837-18-8] α-Ethyl-4-(2-methylpropyl)-benzeneacetic acid; 2-(4-isobutylphenyl)butyric acid; Butilopan. $C_{14}H_{20}O_2$; mol wt 220.31. C 76.33%, H 9.15%, O 14.52%. Prepn: L. Aparicio *et al.*, **DE 2505813**; *eidem*, **US 4031243** (1976, 1977 both to Juste); J. M. Carretero *et al.*, *Eur. J. Med. Chem.* **13**, 77 (1978). Pharmacology: L. Aparicio, *Arch. Int. Pharmacodyn. Ther.* **227**, 130 (1977). Pharmacokinetics: R. Revilla De Granda *et al.*, *An. R. Acad. Farm.* **43**, 419 (1977), *C.A.* **88**, 83349 (1978). HPLC determination: L. González Tavares *et al.*, *Arzneim.-Forsch.* **42**, 818 (1992).

Solid, mp 51-53°. LD_{50} orally in mice: 810 mg/kg (Carretero).

THERAP CAT: Anti-inflammatory.

1530. Butirosin. [12772-35-9] Ambutyrosin. $C_{21}H_{41}N_5O_{12}$; mol wt 555.58. C 45.40%, H 7.44%, N 12.61%, O 34.56%. Aminoglycosidic antibiotic complex obtained from fermentation filtrates of mucoid strains of *Bacillus circulans* (NRRL B-3312 and B-3313). Consists of two components, butirosin A (80-85%) and butirosin B (15-20%), isomers which differ only in the configuration at one carbon atom in the pentose moiety. Prepn: P. W. K. Woo *et al.*, **DE 1914527**; *eidem*, **US 3541078** (1969, 1970 both to Parke, Davis). Isoln and characterization: Dion *et al.*, *Antimicrob. Agents Chemother.* **2**, 84 (1972). Structures: Woo *et al.*, *Tetrahedron Lett.* **1971**, 2617, 2621, 2625. Activity: Howells *et al.*, *Antimicrob. Agents Chemother.* **2**, 79 (1972); Heifetz *et al.*, *ibid.* 89. Synthesis of butirosin B: Ikeda *et al.*, *J. Antibiot.* **25**, 741 (1972); Akita *et al.*, *ibid.* **26**, 365 (1973). Proposed biosynthetic pathway to the butirosins: K. Takeda *et al.*, *ibid.* **31**, 250 (1978).

Butirosin A: R = OH   R' = H
Butirosin B: R = H   R' = OH

Butirosin sulfate dihydrate. [51022-98-1] $C_{21}H_{41}N_5O_{12}.2H_2SO_4.2H_2O$. No sharp mp, dec ~225°. $[\alpha]_D^{25}$ +29° (c = 2 in water). pKa' (water): 5.5, 7.2, 8.5, 9.4. Very sol in water; moderately sol in methanol; slightly sol in ethanol. LD_{50} i.v. in mice: 450-500 mg/kg (Howells).

Butirosin A. [34291-02-6] O-2,6-Diamino-2,6-dideoxy-α-D-glucopyranosyl-(1 → 4)-O-[β-D-xylofuranosyl-(1 → 5)]-N¹-[(2S)-4-amino-2-hydroxy-1-oxobutyl]-2-deoxy-D-streptamine; White, amorphous solid, melts with dec over wide range, beginning at ~149°. $[\alpha]_D^{25}$ +26° (c = 1.46 in water). pKa' (water): 5.6, 7.3, 8.7, 9.8.

Butirosin B. [34291-03-7] O-2,6-Diamino-2,6-dideoxy-α-D-glucopyranosyl-(1 → 4)-O-[β-D-ribofuranosyl-(1 → 5)]-N¹-[(2S)-

4-amino-2-hydroxy-1-oxobutyl]-2-deoxy-D-streptamine; 1-N-[(S)-4-amino-2-(hydroxybutyryl)]ribostamycin. Occurs as a dihydrate, melts over wide range beginning at 146°. $[\alpha]_D^{25}$ +33° (c = 1.5 in water). pKa' (water): 5.3, 7.1, 8.6, 9.8.

THERAP CAT: Antibacterial.

1531. Butoconazole. [64872-76-0] 1-[4-(4-Chlorophenyl)-2-[(2,6-dichlorophenyl)thio]butyl]-1H-imidazole. $C_{19}H_{17}Cl_3N_2S$; mol wt 411.77. C 55.42%, H 4.16%, Cl 25.83%, N 6.80%, S 7.79%. Imidazole derivative with antifungal properties. Prepn: K. A. M. Walker, **US 4078071** (1978 to Syntex). Prepn, toxicity, activity vs *Candida albicans* in mice: K. A. M. Walker *et al.*, *J. Med. Chem.* **21**, 840 (1978). *In vitro* comparison with other antifungal agents: F. C. Odds *et al.*, *J. Antimicrob. Chemother.* **14**, 105 (1984). Clinical trials in treatment of vulvovaginal candidiasis: W. Droegemueller *et al.*, *Obstet. Gynecol.* **64**, 530 (1984); J. B. Jacobson *et al.*, *Acta Obstet. Gynecol. Scand.* **64**, 241 (1985). Comparison with miconazole, *q.v.*: C. S. Bradbeer *et al.*, *Genitourin. Med.* **61**, 270 (1985).

Crystals from cyclohexane, mp 68-70.5°.

Nitrate. [64872-77-1] RS-35887; Femstat; Gynomyk. $C_{19}H_{17}Cl_3N_2S.HNO_3$; mol wt 474.78. Colorless blades from acetone + ethyl acetate, mp 162-163°. Sparingly sol in methanol; slightly sol in acetonitrile, acetone, dichloromethane, tetrahydrofuran; very slightly sol in ethyl acetate. Practically insol in water. LD_{50} in mice, male, female rats (mg/kg): >3200, >3200, 1720 orally; >1600, 940, 940 i.p. (Walker).

THERAP CAT: Antifungal (topical).

1532. Butorphanol. [42408-82-2] 17-(Cyclobutylmethyl)-morphinan-3,14-diol; (−)-N-cyclobutylmethyl-3,14-dihydroxymorphinan; *levo*-BC-2627. $C_{21}H_{29}NO_2$; mol wt 327.47. C 77.02%, H 8.93%, N 4.28%, O 9.77%. Mixed opioid agonist-antagonist. Prepn: I. Pachter *et al.*, **DE 2243961**; *eidem*, **US 3819635** (1973, 1974 to Bristol-Myers); I. Monkovic, T. Thomas, **US 3775414** (1973 to Bristol-Myers). Total synthesis and pharmacology: I. Monkovic *et al.*, *J. Am. Chem. Soc.* **95**, 7910 (1973); *eidem*, *Can. J. Chem.* **53**, 3094 (1975). Clinical study: F. Vargas-Arreola *et al.*, *Curr. Ther. Res.* **22**, 186 (1977). Antitussive effect: R. L. Cavanaugh *et al.*, *Arch. Int. Pharmacodyn. Ther.* **220**, 258 (1976). *Review:* R. C. Heel *et al.*, *Drugs* **16**, 473-505 (1978); F. S. Caruso *et al.*, in *Pharmacological and Biochemical Properties of Drug Substances* vol. 2, M. E. Goldberg, Ed. (Am. Pharm. Assoc., Washington, DC, 1979) pp 19-57.

Solid, mp 215-217°. $[\alpha]_D$ −70.0° (c = 0.1 in methanol).

Tartrate. [58786-99-5] Stadol; Torbugesic; Torbutrol. $C_{21}H_{29}NO_2.C_4H_6O_6$; mol wt 477.55. White powder. mp 217-219°. $[\alpha]_D^{22}$ −64.0° (c = 0.4 in methanol). Sol in dilute acids; sparingly sol in water; slightly sol in methanol. Insol in alc, ethanol, chloroform, ethyl acetate, ethyl ether, hexane. LD_{50} in mice, rats (mg/kg): 40-57, 17-20 i.v.; 395-527, 570-756 orally (Heel).

Note: This is a controlled substance: **21 CFR**, 1308.14.

THERAP CAT: Analgesic.

THERAP CAT (VET): Analgesic; antitussive.

1533. Butoxycaine. [3772-43-8] 4-Butoxybenzoic acid 2-(diethylamino)ethyl ester; 2-diethylaminoethyl p-butoxybenzoate. C_{17}-

$H_{27}NO_3$; mol wt 293.41. C 69.59%, H 9.28%, N 4.77%, O 16.36%. Prepd from *p*-butoxybenzoyl chloride and β-diethylaminoethanol: Christiansen, Harris, **US 2412966** (1946 to Squibb). Alternate prepn and activity: Reynaud *et al.*, *Chim. Ther.* **2**, 25 (1967). *See also:* Büchi *et al.*, *Arzneim.-Forsch.* **18**, 610 (1968).

Hydrochloride. [2350-32-5] Stadacain. $C_{17}H_{27}NO_3 \cdot HCl$; mol wt 329.87. Heavy crystals, mp 146°.

THERAP CAT: Anesthetic (local).

1534. Butralin. [33629-47-9] 4-(1,1-Dimethylethyl)-*N*-(1-methylpropyl)-2,6-dinitrobenzenamine; *N*-*sec*-butyl-4-*tert*-butyl-2,6-dinitroaniline; dibutalin; Amchem 70-25; A-820; Amexine; Tamex. $C_{14}H_{21}N_3O_4$; mol wt 295.34. C 56.94%, H 7.17%, N 14.23%, O 21.67%. Pre-emergence herbicide. Prepn: J. J. Damiano, **DE 2058201**; *idem*, **US 3672866** (1971, 1972 both to Amchem). Activity: S. R. McLane *et al.*, *Proc. South. Weed Sci. Soc.* **24**, 58 (1971). Soil persistence and metabolism: P. C. Kearney *et al.*, *J. Agric. Food Chem.* **22**, 856 (1974). Photochemistry: J. R. Plimmer, U. I. Klingebiel, *ibid.* 689.

Yellow-orange crystals, mp 60-61°, bp$_{0.5}$ 134-136°. Vapor pressure at 25°: 1.3×10^{-5} mm Hg. Flash pt, open cup: 97°F (36°C). Soly in water at 25°: 1 mg/l. Soly at 25° (kg/kg): methanol 0.125; acetone 4.48; benzene 2.7; xylene 3.88; butanone 9.55; carbon tetrachloride 1.46. LD$_{50}$ orally in rats: 2500 mg/kg (McLane).

USE: Herbicide.

1535. Butriptyline. [35941-65-2] 10,11-Dihydro-*N*,*N*,β-trimethyl-5*H*-dibenzo[*a*,*d*]cycloheptene-5-propanamine; 5-(3-dimethylamino-2-methylpropyl)-10,11-dihydro-5*H*-dibenzo[*a*,*d*]cycloheptene; 5-(2-methyl-3-dimethylaminopropyl)dibenzo[*a*,*d*][1,4]cycloheptadiene; butriptylene. $C_{21}H_{27}N$; mol wt 293.45. C 85.95%, H 9.27%, N 4.77%. Prepn: Winthrop, Davis, **BE 613750** (1962 to Ayerst, McKenna & Harrison); *eidem*, *J. Org. Chem.* **27**, 230 (1962); Villani, **US 3409640** (1968 to Schering). Review of pharmacology and clinical data: *J. Med.* **2**, 249-343 (1971). Metabolic studies: Cameron *et al.*, *Arzneim.-Forsch.* **24**, 93 (1974). Toxicity: Voith, Herr, *Arch. Int. Pharmacodyn. Ther.* **182**, 318 (1969).

Oil, bp$_1$ 180-185°.
Hydrochloride. [5585-73-9] AY-62014; Evadene; Evadyne. $C_{21}H_{27}N \cdot HCl$; mol wt 329.91. Crystals from isopropyl alcohol-ether, mp 188-190° (dec). uv max (methanol): 273, 270, 266 nm (ε 460, 441, 552). Freely sol in water; moderately sol in aliphatic alc, chloroform. Insol in ether, paraffinic hydrocarbons. LD$_{50}$ in mice (mg/kg): 120 i.p.; 345 orally (Voith, Herr).

THERAP CAT: Antidepressant.

1536. Butropium Bromide. [29025-14-7] [3(*S*)-*endo*]-8-[(4-Butoxyphenyl)methyl]-3-(3-hydroxy-1-oxo-2-phenylpropoxy)-8-methyl-8-azoniabicyclo[3.2.1]octane bromide; 8-(*p*-butoxybenzyl)-3α-hydroxy-1α*H*,5α*H*-tropanium bromide (−)-tropate; *l*-[1-(*p*-*n*-butoxybenzyl)hyoscyaminium] bromide; BHB; Coliopan. $C_{28}H_{38}$-BrNO$_4$; mol wt 532.52. C 63.15%, H 7.19%, Br 15.00%, N 2.63%, O 12.02%. Anticholinergic. Prepn: S. Tanaka, K. Hashimoto, **DE 1950378**; *eidem*, **US 3696110** (1970, 1972 both to Eisai); Tanaka *et al.*, *J. Pharm. Soc. Jpn.* **92**, 510 (1972). Prepn of the labelled compound: Fujita *et al.*, *J. Labelled Compd.* **9**, 149, 555 (1972). Activity: Akutsu, Ichikawa, *Showa Igakkai Zasshi* **32**, 494 (1972), *C.A.* **78**, 119243g (1973).

Crystals from ethanol-acetone, mp 166-168°; also reported as white needles from isopropanol, mp 158-160° (Tanaka, Hashimoto). $[\alpha]_D^{20}$ −21.7° (c = 0.5 in water). Freely sol in glacial acetic acid; sol in chloroform, DMF. Sparingly sol in ethanol; slightly sol in water, 0.1*N* HCl, 0.1*N* NaOH. Practically insol in acetone, ether, benzene. LD$_{50}$ in male mice (mg/kg): 1500 orally; 660 s.c.; 12.0 i.v. (Tanaka, Hashimoto, 1972).

THERAP CAT: Antispasmodic.

1537. *n*-Butyl Acetate. [123-86-4] Acetic acid butyl ester. $C_6H_{12}O_2$; mol wt 116.16. C 62.04%, H 10.41%, O 27.55%. Prepd from acetic acid and butyl alcohol: Leyes, Othmer, *Ind. Eng. Chem.* **37**, 968 (1945); Vogel, *J. Chem. Soc.* **1948**, 624; Zettlemoyer *et al.*, **US 2644839** (1953 to FMC); *Faith, Keyes & Clark's Industrial Chemicals*, F. A. Lowenheim, M. K. Moran, Eds. (Wiley-Interscience, New York, 4th ed., 1975) pp 171-177. Toxicity data: H. F. Smyth *et al.*, *Arch. Ind. Hyg. Occup. Med.* **10**, 61 (1954).

Liquid. d$_{20}^{20}$ 0.8826. bp 125-126°. mp −77°. n$_D^{20}$ 1.3951. Flash pt, closed cup: 72°F (22°C). Sol in about 120 parts water at 25°; misc with alcohol, ether; sol in most hydrocarbons. LD$_{50}$ orally in rats: 14.13 g/kg (Smyth).

Caution: Potential symptoms of overexposure are headache, drowsiness, narcosis; irritation of eyes, upper respiratory system and skin. *See NIOSH Pocket Guide to Chemical Hazards* (DHHS/NIOSH 97-140, 1997) p 36.

USE: Manuf lacquer, artificial leather, photographic films, plastics, safety glass. Organic solvent.

1538. *sec*-Butyl Acetate. [105-46-4] Acetic acid 1-methylpropyl ester; acetic acid *sec*-butyl ester. $C_6H_{12}O_2$; mol wt 116.16. C 62.04%, H 10.41%, O 27.55%. Prepd from *sec*-butanol and acetic anhydride: R. Altschul, *J. Am. Chem. Soc.* **68**, 2605 (1946). Prepn of *d*- and *l*-form: J. Kenyon *et al.*, *J. Chem. Soc.* **1935**, 1072. Manuf: *Faith, Keyes & Clark's Industrial Chemicals*, F. A. Lowenheim, M. K. Moran, Eds. (Wiley-Interscience, New York, 4th ed., 1975) pp 171-177.

dl-**Form.** Liquid. d$_{27}$ 0.865. bp$_{761}$ 111-111.5°. n$_D^{27}$ 1.3848. Flash pt, open cup: 88°F (31°C). Slightly sol in water; sol in alcohol, ether.

d-**Form.** [66610-38-6] Acetic acid (1*S*)-1-methylpropyl ester. $[\alpha]_{546.1}^{18}$ +10.52° (neat).

l-**Form.** [54657-08-8] Acetic acid (1*R*)-1-methylpropyl ester. Liquid, bp 116-117°. d$_4^{19}$ 0.873. $[\alpha]_{546.1}^{19}$ −20.19° (neat). $[\alpha]_{546.1}^{19}$ −18.86° (c = 5.046 in ethanol). n$_D^{18}$ 1.3899.

Caution: Potential symptoms of overexposure are irritation of eyes; headache; drowsiness; narcosis; dryness of upper respiratory system and skin. *See NIOSH Pocket Guide to Chemical Hazards* (DHHS/NIOSH 97-140, 1997) p 38.

1539. *tert*-Butyl Acetate. [540-88-5] Acetic acid 1,1-dimethylethyl ester; acetic acid *tert*-butyl ester. $C_6H_{12}O_2$; mol wt 116.16. C 62.04%, H 10.41%, O 27.55%. Prepn: Baker, Bordwell; Hauser *et al.*, *Org. Synth.* **coll. vol. III**, 141, 142 (1955). Manuf from acetic acid and isobutylene: Young, Pare, **US 3031495** (1962 to Sinclair); Wheeler *et al.*, **US 3102905** (1963 to Celanese); Heisler *et al.*, **US 3096365** (1963 to Texaco); *Faith, Keyes & Clark's Industrial Chemicals*, F. A. Lowenheim, M. K. Moran, Eds. (Wiley-Interscience, New York, 4th ed., 1975) pp 171-177.

Liquid, bp 97.8°. d_4^{20} 0.8665, d_4^{25} 0.8593. n_D^{20} 1.3870. Practically insol in water. Miscible with alcohol, ether.

Caution: Potential symptoms of overexposure are itching and inflammation of eyes; irritation of upper respiratory tract; headache; narcosis; dermatitis. *See NIOSH Pocket Guide to Chemical Hazards* (DHHS/NIOSH 97-140, 1997) p 38.

USE: As gasoline additive (Wheeler *et al.*).

1540. *tert*-Butylacetic Acid. [1070-83-3] 3,3-Dimethylbutanoic acid; 3,3-dimethylbutyric acid. $C_6H_{12}O_2$; mol wt 116.16. C 62.04%, H 10.41%, O 27.55%. Prepn: Homeyer *et al.*, *J. Am. Chem. Soc.* **55**, 4209 (1933); Botterson, Shulman, *J. Org. Chem.* **27**, 1059 (1962); A. Nilsson, R. Carlson, *Acta Chem. Scand.* **B34**, 621 (1980).

Liquid, bp_{26} 96°, bp_{739} 183.0-183.3°, bp 190°. mp 6-7°. n_D^{20} 1.4115 (Botterson, Shulman), also reported as 1.4096 (Homeyer *et al.*). d_4^{20} 0.9124.

1541. *n*-Butyl Acrylate. [141-32-2] 2-Propenoic acid butyl ester; acrylic acid *n*-butyl ester. $C_7H_{12}O_2$; mol wt 128.17. C 65.60%, H 9.44%, O 24.97%. Prepn from *n*-butanol and methyl acrylate: Rehberg, *Org. Synth.* **coll. vol. III**, 146 (1955). Toxicity study: H. F. Smyth *et al.*, *Arch. Ind. Hyg. Occup. Med.* **4**, 119 (1951).

Liquid. d_4^{20} 0.8986; d_4^{15} 0.9110; d_4^{12} 0.9117; d_4^{0} 0.9202. bp_{760} 145°; also reported bp_{760} 138°; bp_{101} 84-86°; bp_{25} 59°; bp_{10} 39°; bp_8 35°. n_D^{20} 1.4190; n_D^{12} 1.4254. Sp heat $(-60°)$: 0.467 cal/g/°C; heat of vaporization 8.11 kcal/mol; heat of combustion 974.46 kcal/mol. Soly in water at 20°: 0.14 g/100 ml; at 40°: 0.12 g/100 ml. Soly of water in *n*-butyl acrylate at 20°: 0.8 ml/100 g. LD_{50} orally in rats: 3.73 g/kg (Smyth).

Polymer. Elastic, tacky substance. Brittle temp $-45°$.

Caution: Potential symptoms of overexposure to the monomer are irritation of eyes, skin, upper respiratory system; sensitization dermatitis; dyspnea. *See NIOSH Pocket Guide to Chemical Hazards* (DHHS/NIOSH 97-140, 1997) p 38.

USE: The monomer in the manuf of polymers and resins for textile and leather finishes, paint formulations, etc.

1542. *n*-Butyl Alcohol. [71-36-3] 1-Butanol; butyl alcohol; propyl carbinol. $C_4H_{10}O$; mol wt 74.12. C 64.82%, H 13.60%, O 21.59%. Prepn by reduction of butyraldehyde with sodium borohydride: Chaikin, Brown, *J. Am. Chem. Soc.* **71**, 122 (1949). Manuf from ethylene oxide and triethylaluminum: Rudner, **US 3091627** (1963 to Koppers); by oxidation of tributylborane: Mirviss, **US**

3067235 (1962 to Esso). Manuf by carbohydrate fermentation, by hydrogenation of butyraldehyde, from crotonaldehyde: *Faith, Keyes & Clark's Industrial Chemicals*, F. A. Lowenheim, M. K. Moran, Eds. (Wiley-Interscience, New York, 4th ed., 1975) pp 178-185. Purification and vapor pressure: Biddiscombe *et al.*, *J. Chem. Soc.* **1963**, 1954. Toxicity study: H. F. Smyth *et al. Arch. Ind. Hyg. Occup. Med.* **4**, 119 (1951).

Highly refractive liquid; burns with a strongly luminous flame; leaves a transitory greasy spot on paper. Odor similar to that of fusel oil, but weaker. d_4^{20} 0.810. bp 117-118°. mp $-90°$. Flash pt 36-38°C. n_D^{20} 1.3993. A mixture of 63% of the alcohol and 37% water forms a constant boiling mixture, boiling at 92°. Soly at 25°, 9.1 ml/100 ml H_2O: Booth, Everson, *Ind. Eng. Chem.* **40**, 1491 (1948). Miscible with alc, ether and many other organic solvents. LD_{50} orally in rats: 4.36 g/kg (Smyth).

Caution: Potential symptoms of overexposure are irritation of eyes, nose and throat; headache, vertigo and drowsiness; corneal inflammation, blurred vision, lacrimation and photophobia; dermatitis; possible auditory nerve damage, hearing loss; CNS depression. *See NIOSH Pocket Guide to Chemical Hazards* (DHHS/NIOSH 97-140, 1997) p 38.

USE: As organic solvent for fats, waxes, resins, shellac, varnish, gums etc.; manuf lacquers, rayon, detergents, other butyl compds; in microscopy for preparing paraffin imbedding materials.

1543. *sec*-Butyl Alcohol. [78-92-2] 2-Butanol; butylene hydrate; 2-hydroxybutane; methyl ethyl carbinol. $C_4H_{10}O$; mol wt 74.12. C 64.82%, H 13.60%, O 21.59%. Prepn by reduction of 2-butanone: S. W. Chaikin, W. G. Brown, *J. Am. Chem. Soc.* **71**, 122 (1949); R. F. Nystrom *et al.*, *ibid.* 3245. Manuf by hydration of 2-butene or hydrocarbons contg butene: C. B. Dale *et al.*, *Ind. Eng. Chem.* **48**, 913 (1956); F. M. Archibald, H. O. Mottern, **US 2543820** (1951 to Standard Oil); D. A. Limerick, O. L. Wylie, Jr., **US 2776324** (1957 to Shell). Purification and vapor pressure: Biddiscombe *et al.*, *J. Chem. Soc.* **1963**, 1954. Prepn of *l*-form from D$(-)$-2,3-butanediol: P. J. Leroux, H. J. Lucas, *J. Am. Chem. Soc.* **73**, 41 (1951). Prepn of *l*- or *d*-form by hydroboration of *cis*-2-butene: H. C. Brown *et al.*, *J. Am. Chem. Soc.* **86**, 397 (1964). Resolution of *dl*-form: S. W. Kantor, C. R. Hauser, *ibid.* **75**, 1744 (1953). Absolute configuration of *l*-form: H. C. Brown *et al.*, *ibid.* **86**, 1071 (1964). Toxicity: H. F. Smyth *et al.*, *Arch. Ind. Hyg. Occup. Med.* **10**, 61 (1954).

dl-**Form.** Liquid, bp 99.5°. mp $-114.7°$. d_4^{20} 0.808. n_D^{25} 1.3949. Flash pt, open cup: 88°F (31°C). Sol in 12 parts water; misc with alcohol, ether. LD_{50} orally in rats: 6.48 g/kg (Smyth).

d-**Form.** [4221-99-2] (2*S*)-2-Butanol. Liquid, bp 98-99.5°. d_4^{27} 0.8025. n_D^{20} 1.3954. $[\alpha]_D^{22}$ +13.1. $[\alpha]_D^{27}$ +13.52°.

l-**Form.** [14898-79-4] (2*R*)-2-Butanol. Liquid, bp_{744} 98°. d_4^{25} 0.8042. n_D^{20} 1.3970, n_D^{25} 1.3949. $[\alpha]_D^{25}$ $-13.51°$.

Caution: Potential symptoms of overexposure are eye, skin, nose, throat irritation; narcosis. *See NIOSH Pocket Guide to Chemical Hazards* (DHHS/NIOSH 97-140, 1997) p 40.

USE: In the synthesis of flotation agents, flavors, perfumes, dyestuffs, wetting agents. In industrial cleaners, paint removers. Solvent for many natural resins, linseed and castor oils.

1544. *tert*-Butyl Alcohol. [75-65-0] 2-Methyl-2-propanol; trimethyl carbinol; *tert*-butanol. $C_4H_{10}O$; mol wt 74.12. C 64.82%, H 13.60%, O 21.59%. Prepd from acetyl chloride and dimethylzinc: Butlerow, *Ann.* **144**, 1 (1867). Manuf by catalytic hydration of isobutylene: Kreps, Nachod, **US 2477380** (1949 to Atlantic Refining); Serniuk, Vanderbilt, **US 2534304** (1950 to Standard Oil); by reduction of *tert*-butyl hydroperoxide: Lorand, **US 2484841** (1949 to Hercules Powder); De Jong, **US 2853532** (1958 to Shell). Purification: Biddiscombe *et al.*, *J. Chem. Soc.* **1963**, 1954. Physical properties: *eidem, ibid.*; Dreisbach, Martin, *Ind. Eng. Chem.* **41**, 2875 (1949). Toxicity: Schaffarzick, Brown, *Science* **116**, 663 (1952).

Crystals, camphor-like odor, mp 25.6°. At 99.69 mol-% purity, bp 82.41°. mp 25.7°. d_4^{20} 0.78581, d_4^{25} 0.78086, d_4^{25} (calcd) 0.78080; n_D^{20} 1.38468, n_D^{25} 1.38231. Flash pt, closed cup: 52°F (11.1°C). Sol in water. Miscible with alcohol, ether. LD_{50} orally in rats: 3.5 g/kg (Schaffarzick, Brown).

Caution: Potential symptoms of overexposure are drowsiness, narcosis; irritation of skin, eyes, nose, throat. *See NIOSH Pocket Guide to Chemical Hazards* (DHHS/NIOSH 97-140, 1997) p 40.

USE: Denaturant for ethanol, mfg flotation agents, flavors, perfumes; as organic solvent; in paint removers. Octane booster in gasoline.

1545. n-Butylamine. [109-73-9] 1-Butanamine; 1-aminobutane. $C_4H_{11}N$; mol wt 73.14. C 65.69%, H 15.16%, N 19.15%. Prepn by reduction of butyraldoxime: Lycan *et al.*, *Org. Synth.* **coll. vol. II**, 319 (1943). Usually manuf by catalytic alkylation of ammonia with butyl alcohol: Davies *et al.*, US **2609394** (1952 to I.C.I.); Hindley, Fisher, US **2782237** (1957 to British Celanese); Lemon, Myerly, US **3022349** (1962 to Union Carbide); Shirley, Speranza, US **3128311** (1964 to Jefferson Chem.). Manuf from butyraldehyde and ammonia in the presence of Raney nickel: Brimer *et al.*, US **2518659** (1950 to Eastman Kodak). Toxicity study: C. H. Hine *et al.*, *Arch. Environ. Health* **1**, 343 (1960).

Liquid, ammoniacal odor, bp 78°. mp −50°. d_4^{25} 0.7327. n_D^{20} 1.4010. Flash pt, open cup: 30°F (−1°C). *Flammable, corrosive.* Misc with water, alcohol, ether. LD_{50} orally in rats: 500 mg/kg (Hine).

Caution: Potential symptoms of overexposure are irritation of eyes, skin, nose and throat; headache; skin flushing and burns. *See NIOSH Pocket Guide to Chemical Hazards* (DHHS/NIOSH 97-140, 1997) p 40.

USE: Intermediate for pharmaceuticals, dyestuffs, rubber chemicals, emulsifying agents, insecticides, synthetic tanning agents.

1546. sec-Butylamine. [13952-84-6] 2-Butanamine; 2-aminobutane; Frucote; Deccotane; Tutane. $C_4H_{11}N$; mol wt 73.14. C 65.69%, H 15.16%, N 19.15%. Prepn by reduction of ethyl methyl ketoxime: Lycan *et al.*, *Org. Synth.* **coll. vol. II**, 319 (1943). Manuf: Taylor *et al.*, US **2636902** (1953 to I.C.I.); Thurston, US **2689868** (1954 to Am. Cyanamid). Resolution of *dl*-form: Leithe, *Ber.* **63**, 800 (1930); Bruck *et al.*, *J. Chem. Soc.* **1956**, 921. Toxicity: Goldenthal, *Toxicol. Appl. Pharmacol.* **18**, 204 (1971).

dl-Form. Liquid, bp 63°. mp −104°. d_4^{20} 0.724. n_D^{20} 1.394. Miscible with water, alc. LD_{50} orally in rats: 380 mg/kg (Goldenthal).
d-Form. [513-49-5] (2*S*)-2-Butanamine. Liquid, bp 63°. d_4^{15} 0.7308. n_4^{15} 1.3963. $[\alpha]_D^{15}$ +7.80° (neat).
l-Form. [13250-12-9] (2*R*)-2-Butanamine. Liquid, bp 63°. d_4^{19} 0.728. $[\alpha]_D^{19}$ −7.64°.
Caution: Irritating to skin and mucous membranes.
USE: Agricultural fungistat.

1547. tert-Butylamine. [75-64-9] 2-Methyl-2-propanamine; 2-aminoisobutane; 2-amino-2-methylpropane. $C_4H_{11}N$; mol wt 73.14. C 65.69%, H 15.16%, N 19.15%. Prepn: Campbell *et al.*, *Org. Synth.* **coll. vol. III**, 148 (1955). Manuf: Gresham *et al.*, US **2501509** (1950 to du Pont); Albert, Kibler, US **2773097** (1956 to Firestone Tire & Rubber).

Liquid, bp 44-46°. mp −72.65°. d_4^{20} 0.6951, d_4^{25} 0.6867. n_D^{20} 1.37. Miscible with alcohol.
Hydrochloride. [10017-37-5] $C_4H_{11}N.HCl$; mol wt 109.60. mp 310°. Sol in cold methanol, in boiling isopropyl alcohol.

1548. Butylate. [2008-41-5] *N*,*N*-Bis(2-methylpropyl)carbamothioic acid *S*-ethyl ester; diisobutylthiocarbamic acid *S*-ethyl ester; *S*-ethyl *N*,*N*-diisobutylthiocarbamate; butilate; diisocarb; R-1910; Sutan Plus; Anelda. $C_{11}H_{23}NOS$; mol wt 217.37. C 60.78%, H 10.67%, N 6.44%, O 7.36%, S 14.75%. Selective pre-planting herbicide. Prepn: H. Tilles, *J. Am. Chem. Soc.* **81**, 714 (1959); H. Tilles, J. Antognini, US **2913327** (1959 to Stauffer). Metabolism: J. P. Hubbell, J. E. Casida, *J. Agric. Food Chem.* **25**, 404 (1977). Soil degradation: F. W. Roeth *et al.*, *Weed Technol.* **3**, 24 (1989). HPLC determn in water: C. H. Marvin *et al.*, *Anal. Chem.* **62**, 1495 (1990). Weed control in alfalfa: N. Sarpe *et al.*, *Meded. Fac. Landbouwwet. Univ. Gent* **57**, 1117 (1992); in combination with atrazine, *q.v.*, in corn: G. D. Hoyt, *Crop Prot.* **14**, 75 (1995).

Clear liquid, bp_{21} 138°. n_D^{30} 1.4701; d 0.9417. Vapor press at 25°: 1.3×10^{-3} mm Hg. Soly in water at 25°: 45 mg/l.
USE: Herbicide.

1549. Butylated Hydroxyanisole. [25013-16-5] (1,1-Dimethylethyl)-4-methoxyphenol; 2(3)-*tert*-butyl-4-hydroxyanisole; BHA; Antrancine 12; Embanox; Nipantiox 1-F; Sustane 1-F; Tenox BHA. $C_{11}H_{16}O_2$; mol wt 180.25. C 73.30%, H 8.95%, O 17.75%. A mixture of **2-tert-butyl-4-methoxyphenol** (also called 3-*tert*-butyl-4-hydroxyanisole) and **3-tert-butyl-4-methoxyphenol** (also called 2-*tert*-butyl-4-hydroxyanisole). Prepn from *p*-methoxyphenol and isobutene: R. H. Rosenwald, US **2459540**; US **2470902** (both 1949 to Universal Oil Products). Toxicity study: A. J. Lehman *et al.*, *Adv. Food Res.* **3**, 197 (1951). Review of carcinogenic risk: *IARC Monographs* **40**, 123-159 (1986); antimicrobial activity: M. Riccach, *J. Food Saf.* **6**, 141-170 (1984); of physicochemical properties, antioxidant activity and toxicology: H. Verhagen *et al.*, *Chem. Biol. Interact.* **80**, 109-134 (1991); of safety assessment as food additive: G. M. Williams *et al.*, *Food Chem. Toxicol.* **37**, 1027-1038 (1999).

White or slightly yellow waxy solid, with a faint characteristic odor. mp 48-55°. bp_{733} 264-270°. Freely sol in alc, propylene glycol, chloroform, ether; sol in petr ether (Skellysolve H), fats, oils. Insol in water. Exhibits antioxidant properties and synergism with acids, BHT, propyl gallate, hydroquinone, methionine, lecithin, thiodipropionic acid, etc. LD_{50} in mice, rats (mg/kg): 2000, 2200 orally (Lehman).

Caution: This substance is reasonably anticipated to be a human carcinogen: *Report on Carcinogens, Twelfth Edition* (PB2011-111646, 2011) p 78.

USE: Antioxidant and preservative, esp in foods, cosmetics, pharmaceuticals; also rubber and petroleum products.

1550. Butylated Hydroxytoluene. [128-37-0] 2,6-Bis(1,1-dimethylethyl)-4-methylphenol; 2,6-di-*tert*-butyl-*p*-cresol; 2,6-di-*tert*-butyl-4-methylphenol; BHT; Antrancine 8; Ionol CP; Sustane; Dalpac; Impruvol; Vianol. $C_{15}H_{24}O$; mol wt 220.36. C 81.76%, H 10.98%, O 7.26%. Prepn from *p*-cresol and isobutylene: Stillson, US **2428745** (1947 to Gulf); McConnell, Davis, **US**

3082258 (1963 to Eastman Kodak). Toxicity study: W. A. McOmie *et al.*, *J. Am. Pharm. Assoc.* **38**, 366 (1949). Review of physicochemical properties, antioxidant activity and toxicity: H. Babich, *Environ. Res.* **29**, 1-29 (1982); of antimicrobial activity: M. Raccach, *J. Food Saf.* **6**, 141-170 (1984); of safety assessment as food additive: G. M. Williams *et al.*, *Food Chem. Toxicol.* **37**, 1027-1038 (1999).

Crystals, mp 70°. d_4^{20} 1.048. bp 265°. Flash pt (open cup): 260°F (127°C). Freely sol in alc, chloroform, ether, toluene. Sol in methanol, isopropanol, methyl ethyl ketone, acetone, Cellosolve, petr ether, benzene, most other hydrocarbon solvents. Soly in liquid petrolatum (white oil): 0.5% w/w. More sol in food oils and fats than butylated hydroxyanisole. Good soly in linseed oil. Insol in water, propylene glycol. LD_{50} orally in mice: 1040 mg/kg (McOmie).
Caution: Potential symptoms of overexposure are irritation of eyes and skin. *See NIOSH Pocket Guide to Chemical Hazards* (DHHS/NIOSH 97-140, 1997) p 124.
USE: Antioxidant for food, animal feed, petroleum products, synthetic rubbers, plastics, animal and vegetable oils, soaps. Antiskinning agent in paints and inks.

1551. *n*-**Butylbenzene.** [104-51-8] Butylbenzene; 1-phenylbutane. $C_{10}H_{14}$; mol wt 134.22. C 89.49%, H 10.51%. Prepn: Radziszewski, *Ber.* **9**, 261 (1876); Balbiano, *Ber.* **10**, 296 (1877); Read, Foster, *J. Am. Chem. Soc.* **48**, 1606 (1926).

Liquid, mp −88.5°. d_4^{20} 0.8604. bp_{760} 183.1°; bp_{400} 159.2°; bp_{200} 136.9°; bp_{100} 116.2°; bp_{60} 102.6°; bp_{40} 92.4°; bp_{20} 76.3°; bp_{10} 62.0°; bp_5 48.8°; $bp_{1.0}$ 22.7°. n_D^{20} 1.49040. Flash pt, open cup: 160°F (71°C). Insol in water. Miscible with alcohol, ether, benzene.

1552. *sec*-**Butylbenzene.** [135-98-8] (1-Methylpropyl)benzene; 2-phenylbutane. $C_{10}H_{14}$; mol wt 134.22. C 89.49%, H 10.51%. Prepd from benzene and *n*-butyl chloride in presence of AlCl$_3$: Schramm, *Monatsh. Chem.* **9**, 621 (1888); by the action of sodium on γ-chloro-*sec*-butylbenzene: Braun *et al.*, *Ber.* **46**, 1277 (1913); with other products by heating *n*- or *sec*-butyl alcohol with 80% H_2SO_4: Meyer, Bernhauer, *Monatsh. Chem.* **53**, 727 (1929).

Liquid, mp −82.7°. d_4^{20} 0.8608. bp_{760} 173.5°; bp_{400} 150.3°; bp_{200} 128.8°; bp_{100} 109.5°; bp_{60} 96.0°; bp_{40} 86.2°; bp_{20} 70.6°; bp_{10} 57.0°; bp_5 44.2°; $bp_{1.0}$ 18.6°. n_D^{20} 1.48980. Flash pt, closed cup: 126°F (52°C). Insol in water. Misc with alcohol, ether, benzene.
d-**Form.** [5787-28-0] [(1*S*)-1-Methylpropyl]benzene. $[\alpha]_D^{25}$ +26.6°: Bonner, Greenlee, *J. Am. Chem. Soc.* **81**, 3336 (1959).
l-**Form.** [5787-29-1] [(1*R*)-1-Methylpropyl]benzene. $[\alpha]_D^{25}$ −27.3°.
USE: Solvent; in organic syntheses.

1553. *tert*-**Butylbenzene.** [98-06-6] (1,1-Dimethylethyl)benzene; 2-methyl-2-phenylpropane; trimethylphenylmethane; pseudobutylbenzene. $C_{10}H_{14}$; mol wt 134.22. C 89.49%, H 10.51%. Prepn: Konowalow, *Bull. Soc. Chim.* [3] **16**, 865 (1896); Shoesmith, Mackie, *J. Chem. Soc.* **1928**, 2336; Meyer, Bernhauer, *Monatsh. Chem.* **53**, 727 (1929); Wilt, Abegg, *J. Org. Chem.* **33**, 923 (1968).

See also: Groose, Ipatieff, *J. Am. Chem. Soc.* **57**, 2415 (1935); Ipatieff, Pines, *ibid.* **58**, 1056 (1936).

Liquid, mp −58.1°. d_4^{20} 0.8669. bp_{760} 168.5°; bp_{400} 145.8°; bp_{200} 123.7°; bp_{100} 103.8°; bp_{60} 90.6°; bp_{40} 80.8°; bp_{20} 65.6°; bp_{10} 51.7°; bp_5 39.0°; $bp_{1.0}$ 13.0°. n_D^{20} 1.49235. Flash pt, open cup: 140°F (60°C). Insol in water. Misc with alcohol, ether, benzene.

1554. *n*-**Butyl Benzoate.** [136-60-7] Benzoic acid butyl ester. $C_{11}H_{14}O_2$; mol wt 178.23. C 74.13%, H 7.92%, O 17.95%. Prepn: Newman, Fones, *J. Am. Chem. Soc.* **69**, 1046 (1947); Justoni, **GB 719891** (1954 to Vismara). Toxicity study: H. F. Smyth *et al.*, *Arch. Ind. Hyg. Occup. Med.* **10**, 61 (1954).

Thick, oily liquid. d 1.00. mp −22°. bp 250°. Practically insoluble in water. Sol in alcohol or ether. LD_{50} orally in rats: 5.14 g/kg (Smyth).

1555. *n*-**Butyl Bromide.** [109-65-9] 1-Bromobutane. C_4H_9Br; mol wt 137.02. C 35.06%, H 6.62%, Br 58.32%. Prepd from *n*-butyl alc and a hydrobromic-sulfuric acid mixture: Kamm, Marvel, *Org. Synth.* **vol. 1**, 5 (1921); Skau, McCullough, *J. Am. Chem. Soc.* **57**, 2440 (1935).

Colorless liquid. d_4^{25} 1.2686. bp_{760} 101.3°. mp −112°. n_D^{20} 1.4398. *Flammable.* Insol in water. Sol in alcohol, ether.

1556. *sec*-**Butyl Bromide.** [78-76-2] 2-Bromobutane; methylethylbromomethane. C_4H_9Br; mol wt 137.02. C 35.06%, H 6.62%, Br 58.32%. Prepn: Levene, Marker, *J. Biol. Chem.* **91**, 405 (1931); Kenyon *et al.*, *J. Chem. Soc.* **1935**, 1080; Skau, McCullough, *J. Am. Chem. Soc.* **57**, 2439 (1935); Colson *et al.*, *J. Chem. Soc.* **1965**, 2364. Prepn of optically pure isomers: Goodwin, Hudson, *J. Chem. Soc. B* **1968**, 1333.

dl-**Form.** Colorless liquid, pleasant odor. d_4^{25} 1.2530. bp 91.2°. mp −112°. n_D^{25} 1.4344. *Flammable.* Insol in water. Freely sol in alcohol, ether.
d-**Form.** [5787-32-6] (2*S*)-2-Bromobutane. n_D^{20} 1.4359-1.4362. α_D^{20} +42.64°.
l-**Form.** [5787-33-7] (2*R*)-2-Bromobutane. n_D^{20} 1.4368. α_D^{20} −43.7°.
Caution: Narcotic in high concns.

1557. *tert*-**Butyl Bromide.** [507-19-7] 2-Bromo-2-methylpropane; 2-bromoisobutane; trimethylbromomethane. C_4H_9Br; mol wt 137.02. C 35.06%, H 6.62%, Br 58.32%. Prepn: Brunel, *J. Am. Chem. Soc.* **39**, 1978 (1917); Bryce-Smith, Howlett, *J. Chem. Soc.* **1951**, 1141; Coe *et al.*, *ibid.* **1954**, 2281.

Colorless liquid. d_4^{25} 1.2125. bp 73.3°. mp −16.3°. At 210° changes to isobutyl bromide. n_D^{25} 1.4249. Insol in water. Miscible with organic solvents.

1558. *n*-Butyl *n*-Butyrate. [109-21-7] Butanoic acid butyl ester; butyric acid butyl ester. $C_8H_{16}O_2$; mol wt 144.21. C 66.63%, H 11.18%, O 22.19%. Prepn from butyl alcohol: Robertson, *Org. Synth.* **coll. vol. I**, 138 (1941); Horton, **US 2522676** (1950 to Socony-Vacuum Oil).

Liquid, bp 165°. d_4^{20} 0.8692. n_D^{20} 1.4064. Practically insol in water. Miscible with alcohol, ether.

1559. *n*-Butyl Carbonate. [542-52-9] Carbonic acid dibutyl ester; dibutyl carbonate. $C_9H_{18}O_3$; mol wt 174.24. C 62.04%, H 10.41%, O 27.55%. Prepn from ethyl carbonate, butyl alcohol and ethylmagnesium bromide: Frank *et al., J. Am. Chem. Soc.* **66**, 1509 (1944); from butyl alcohol and CO in the presence of Pd and $CuCl_2$: Mador, Blackham, **US 3114762** (1963 to National Distillers).

Liquid, bp 206.6°. d_4^{20} 0.9251, d_4^{25} 0.9388. n_D^{20} 1.4117. Practically insol in water. Miscible with ethanol, benzene, chloroform, acetone, ether and other organic solvents, *see:* Jackson, Drury, *Ind. Eng. Chem.* **51**, 1491 (1959).

1560. Butyl Cellosolve®. [111-76-2] 2-Butoxyethanol; ethylene glycol monobutyl ether. $C_6H_{14}O_2$; mol wt 118.18. C 60.98%, H 11.94%, O 27.08%. Prepn: L. H. Cretcher, W. H. Pittenger, *J. Am. Chem. Soc.* **46**, 1503 (1924); W. W. Carlson, **US 2448767** (1948 to Mellon Inst. Ind. Res.); R. Riemschneider, P. Gross, *Monatsh. Chem.* **90**, 783 (1959). Toxicity: H. F. Smyth *et al., J. Ind. Hyg. Toxicol.* **23**, 259 (1941); C. P. Carpenter *et al., Arch. Ind. Health* **14**, 114 (1956). Series of articles on toxicology: *Environ. Health Perspect.* **57**, 1-275 (1984). Review of toxicology and human exposure: *Toxicological Profile for 2-Butoxyethanol and 2-Butoxyethanol Acetate* (PB99-102527, 1998) 404 pp.

Liquid, bp 171-172°. d_4^{20} 0.9012, d_{20}^{20} 0.9019. n_D^{20} 1.4196. Flash pt, closed cup: 141°F (60°C). Sol in water, mineral oil, most organic solvents. LD_{50} orally in rats: 1.48 g/kg (Smyth).
Caution: Potential symptoms of overexposure are hemolysis, hemoglobinuria; CNS depression, headache; vomiting. Direct contact may cause irritation of skin, eyes, nose and throat. *See NIOSH Pocket Guide to Chemical Hazards* (DHHS/NIOSH 97-140, 1997) p 36. *See also Patty's Industrial Hygiene and Toxicology* **vol. 2D**, G. D. Clayton, F. E. Clayton, Eds. (Wiley-Interscience, New York, 4th ed., 1994) pp 2765, 2795-2804.
USE: Solvent for nitrocellulose, resins, grease, oil, albumin; dry cleaning.

1561. *n*-Butyl Chloride. [109-69-3] 1-Chlorobutane; *n*-propylcarbinyl chloride; butyl chloride. C_4H_9Cl; mol wt 92.57. C 51.90%, H 9.80%, Cl 38.30%. Prepd from *n*-butyl alcohol by heating with HCl and anhyd $ZnCl_2$: Whaley, Copenhaver, *J. Am. Chem. Soc.* **60**, 2497 (1938); *Org. Synth.* **coll. vol. I**, 142 (2nd ed., 1941). Toxicity study: H. F. Smyth *et al., Arch. Ind. Hyg. Occup. Med.* **10**, 61 (1954).

Liquid. *Flammable.* d_4^{15} 0.89197; d_4^{20} 0.88648; d_4^{25} 0.88098. One gallon weighs 7.35 pounds. mp −123.1°. bp_{760} 78.5°. n_D^{20} 1.40223. Flash pt −6.7°C (20°F). Dipole moment: 1.95. Practically insol in water (0.066% at 12°). Misc with alcohol, ether. LD_{50} orally in rats: 2.67 g/kg (Smyth).
USE: As butylating agent in organic synthesis, e.g., in the manuf of butyl cellulose.
THERAP CAT (VET): Anthelmintic.

1562. *sec*-Butyl Chloride. [78-86-4] 2-Chlorobutane; 2-chloro-3-methylpropane. C_4H_9Cl; mol wt 92.57. C 51.90%, H 9.80%, Cl 38.30%. Prepn from *sec*-butyl alcohol, hydrochloric acid and $ZnCl_2$: Norris, Taylor, *J. Am. Chem. Soc.* **46**, 753 (1924); Copenhaver, Whaley, *Org. Synth.* **coll. vol. I**, 143 (1941); from *sec*-butyl alcohol and PCl_3: Coulson *et al., J. Chem. Soc.* **1965**, 2364. Prepn of optically pure isomers: Goodwin, Hudson, *J. Chem. Soc. B* **1968**, 1333. Toxicity data: H. F. Smyth *et al., Am. Ind. Hyg. Assoc. J.* **30**, 470 (1969).

dl-Form. Liquid; pleasant, ethereal odor. d_4^{20} 0.871. bp 68°. n_D^{20} 1.3960; n_D^{25} 1.3953. *Flammable.* One gram dissolves in 1000 ml water at 25°; misc with alcohol, ether. LD_{50} orally in rats: 20.0 ml/kg (Smyth).
d-Form. [22156-91-8] (2*S*)-2-Chlorobutane. n_D^{20} 1.3963. α_D^{20} +30.8°. *Flammable.*
l-Form. [22157-31-9] (2*R*)-2-Chlorobutane. n_D^{20} 1.3968. α_D^{20} −31.0°. *Flammable.*

1563. *tert*-Butyl Chloride. [507-20-0] 2-Chloro-2-methylpropane; 2-chloroisobutane; trimethylchloromethane. C_4H_9Cl; mol wt 92.57. C 51.90%, H 9.80%, Cl 38.30%. Prepd by shaking *tert*-butyl alcohol with concd HCl and distilling: Norris, Olmsted, *Org. Synth.* **8**, 50 (1928). Prepn from *tert*-butyl alcohol and PCl_3: Gerrard *et al., J. Chem. Soc.* **1953**, 1920.

Liquid. d_4^{15} 0.847. n_D^{18} 1.38686. mp −26.5°. bp_{760} 51.0°; bp_{400} 32.6°; bp_{200} 14.6°; bp_{100} −1.0°; bp_{60} −11.4°; bp_{40} −19.0°. Sparingly sol in water, miscible with alcohol and ether. Boiling with water yields *tert*-butyl alcohol.

1564. *tert*-Butyl Chloroacetate. [107-59-5] 2-Chloroacetic acid 1,1-dimethylethyl ester; *t*-butyl chloroacetate; chloroacetic acid *tert*-butyl ester. $C_6H_{11}ClO_2$; mol wt 150.60. C 47.85%, H 7.36%, Cl 23.54%, O 21.25%. Prepd by reacting monochloroacetic acid and isobutylene in dioxane in the presence of sulfuric acid: Johnson *et al., J. Am. Chem. Soc.* **75**, 4995 (1953); from *tert*-butyl alcohol, chloroacetyl chloride and dimethylaniline: Baker, *Org. Synth.* **24**, 21 (1944).

Liquid, bp 155° (dec); bp_{11} 48-49°; bp_{16-17} 56-57°. n_D^{25} 1.4204-1.4210; n_D^{20} 1.4259-1.4260. Hydrolyzes to *tert*-butyl alcohol and chloroacetic acid.
USE: In the glycidic ester condensation.

1565. Butyl Citrate. [77-94-1] 2-Hydroxy-1,2,3-propanetricarboxylic acid 1,2,3-tributyl ester; citric acid tributyl ester; *n*-butyl citrate; tributyl citrate. $C_{18}H_{32}O_7$; mol wt 360.45. C 59.98%, H 8.95%, O 31.07%. Prepn: Fodor, *Arch. Biochem.* **28**, 274 (1950). Synthesis from *n*-butyl alcohol and citric acid: Benedict, *Chemistry* **47**, 27 (1974).

Colorless or pale yellow, odorless liquid. d_4^{20} 1.045. bp_{22} about 233°. mp −20°. n_D^{20} 1.4460 (Benedict). Insol in water. Miscible with most organic liquids.
USE: Plasticizer and solvent for nitrocellulose lacquers; in polishes, inks and similar prepns; also as anti-foam agent.

1566. *n*-Butyl Cyanoacrylate. [6606-65-1] 2-Cyano-2-propenoic acid butyl ester; enbucrilate; NBCA; *n*-butyl 2-cyanoacrylate; Histoacryl; Indermil; Liquiband; Tisuacryl; Vetbond. $C_8H_{11}NO_2$; mol wt 153.18. C 62.73%, H 7.24%, N 9.14%, O 20.89%. Monomer rapidly polymerizes upon contact with fluid or tissue to form a strong, biodegradable adhesive. Preparative methods: A. E. Ardis, US 2467927 (1949 to B. F. Goodrich); C. H. McKeever, US 2912454 (1959 to Rohm & Haas). Prepn, polymerization and degradation: F. Leonard *et al.*, *J. Appl. Polym. Sci.* **10**, 259 (1966). Mechanism of polymerization and nanoparticle formation: N. Behan *et al.*, *Biomaterials* **22**, 1335 (2001). Characterization of polymer nanoparticles by particle size analysis: A. Bootz *et al.*, *Eur. J. Pharm. Biopharm.* **57**, 369 (2004); by MS: *idem et al.*, *ibid.* **60**, 391 (2005). Clinical experience in ophthalmic applications: A. B. Leahey *et al.*, *Ophthalmology* **100**, 173 (1993); in closure of surgical incisions: G. E. Amiel *et al.*, *J. Am. Coll. Surg.* **189**, 21 (1999). Clinical evaluation in augmentation rhinoplasty: M. E. Sachs, *Arch. Otolaryngol.* **111**, 389 (1985); in management of bleeding gastric varices: B. D. Greenwald *et al.*, *Am. J. Gastroenterol.* **98**, 1982 (2003). Review of biomedical applications: C. Vauthier *et al.*, *Adv. Drug Delivery Rev.* **55**, 519-548 (2003).

Colorless liquid. $bp_{1.8}$ 68°. d^{20} 0.989. n_D^{20} 1.4424. n_D^{25} 1.4410. Insol in water. Flash pt, closed cup: >176°F (>80°C). Vapor pressure (25°): <0.5 mmHg. Surface tension: 31.11 dynes/cm. *Highly reactive with water and weak bases.*

Polymer. [25154-80-7] Poly(butyl 2-cyanoacrylate).

USE: Adhesive. Polymer nanoparticles as pharmaceutic aid for controlled release drug delivery.

THERAP CAT: Tissue adhesive.

THERAP CAT (VET): Tissue adhesive.

1567. *tert*-Butyldimethylchlorosilane. [18162-48-6] Chloro(1,1-dimethylethyl)dimethylsilane; *tert*-butylchlorodimethylsilane; *tert*-butyldimethylsilyl chloride; TBDMS chloride; TBDMS-Cl; TBS-Cl; *t*-BuMe₂SiCl. $C_6H_{15}ClSi$; mol wt 150.72. C 47.81%, H 10.03%, Cl 23.52%, Si 18.63%. Reagent used to introduce the *tert*-butyldimethylsilyl (TBDMS) group, especially in the protection of alcohols. Prepn: L. H. Sommer, L. J. Tyler, *J. Am. Chem. Soc.* **76**, 1030 (1954). Use in protection of hydroxyl groups: E. J. Corey, A. Venkateswarlu, *ibid.* **94**, 6190 (1972); S. K. Chaudhary, O. Hernandez, *Tetrahedron Lett.* **20**, 99 (1979); in preparation of TBDMS enol ethers: J. Orban *et al.*, *ibid.* **25**, 5099 (1984). *Review*: B. E. Huff in *Encyclopedia of Reagents for Organic Synthesis* **2**, L. A. Paquette, Ed. (Wiley, New York, 1995) pp 863-868.

White crystalline solid, mp 92.5°. bp 125°. *Corrosive.* Very sol in most organic solvents. Hygroscopic. Many decompose on exposure to moisture.

USE: Reagent in synthetic organic chemistry.

1568. *tert*-Butyldiphenylchlorosilane. [58479-61-1] 1,1'-[Chloro(1,1-dimethylethyl)silylene]bisbenzene; *tert*-butylchlorodiphenylsilane; *tert*-butyldiphenylsilyl chloride; TBDPS-Cl; *t*-BuPh₂SiCl. $C_{16}H_{19}ClSi$; mol wt 274.86. C 69.92%, H 6.97%, Cl 12.90%, Si 10.22%. Reagent used to introduce the *tert*-butyldiphenylsilyl (TBDPS) group, especially in the protection of alcohols. Prepn and use in protection of hydroxyl groups: S. Hanessian, P. Lavallee, *Can. J. Chem.* **53**, 2975 (1975). Synthetic method for the protection of hydroxyl groups: S. A. Hardinger, N. Wijaya, *Tetrahedron Lett.* **34**, 3821 (1993); for the protection of primary amines: L. E. Overman *et al.*, *ibid.* **27**, 4391 (1986). *Review*: S. Hanessian

in *Encyclopedia of Reagents for Organic Synthesis* **2**, L. A. Paquette, Ed. (Wiley, New York, 1995) pp 878-879.

Colorless liquid. $bp_{0.05}$ 120-122°; $bp_{0.015}$ 90°. *Corrosive.* Flash point, closed cup: 233.6°F (112°C). Miscible in most organic solvents. May decompose on exposure to moisture.

USE: Reagent in synthetic organic chemistry.

1569. α-Butylene Dibromide. [533-98-2] 1,2-Dibromobutane. $C_4H_8Br_2$; mol wt 215.92. C 22.25%, H 3.73%, Br 74.01%. Prepn by bromination of butene: Wurtz, *Ann.* **152**, 21 (1869); of bromobutane: Kharasch *et al.*, *J. Org. Chem.* **20**, 1430 (1955).

Yellowish liquid, bp 166°. mp −65°. d_4^{20} 1.7946. n_D^{20} 1.5144. Practically insol in water; miscible with alcohol.

1570. 1,3-Butylene Glycol. [107-88-0] 1,3-Butanediol; 1,3-dihydroxybutane; β-butyleneglycol; methyltrimethylene glycol; butane-1,3-diol. $C_4H_{10}O_2$; mol wt 90.12. C 53.31%, H 11.19%, O 35.51%. Usually prepd by catalytic hydrogenation of aldol using Raney nickel: Hancock, Henson, *Ind. Eng. Chem.* **45**, 629 (1953); GB 853266 (1960 to Celanese); F. S. Wagner in *Kirk-Othmer Encyclopedia of Chemical Technology* vol. 11 (Wiley-Interscience, New York, 3rd ed., 1980) pp 956-962. Toxicity: Smyth *et al.*, *Arch. Ind. Hyg. Occup. Med.* **4**, 119 (1951).

Viscous liquid. The pure compd is colorless. d_{20}^{20} 1.004-1.006. One gallon weighs 8.398 lbs at room temp. Viscosity (cSt): 24.6 (50°); 96 (25°); 590 (0°); 3253 (−17.7°); 6059 (−23°). mp < 50°. Because of the high viscosity at low temps heating is necessary for pumping. Very hygroscopic: Will absorb 38.5 wt % of water within 144 hrs. at 81% relative humidity. bp 207.5°. n_D^{20} 1.4401. Flash pt (tag open cup): 250°F (121°C). Dielectric constant 28.8 at 25°. Surface tension: 37.8 dynes/cm at 25°. Sol in water, acetone, methyl ethyl ketone, ethanol, dibutyl phthalate, castor oil. Practically insol in aliphatic hydrocarbons, benzene, toluene, carbon tetrachloride, ethanolamines, mineral oil, linseed oil. LD_{50} orally in rats: 22.8 g/kg (Smyth).

USE: Intermediate in the manufacture of polyester plasticizers, humectant for cellophane, tobacco. Has some mold inhibiting action.

1571. 2,3-Butylene Glycol. [513-85-9] 2,3-Butanediol; 2,3-dihydroxybutane; dimethylethylene glycol. $C_4H_{10}O_2$; mol wt 90.12. C 53.31%, H 11.19%, O 35.51%. Occurs in 3 isomeric forms: *meso*- or *erythro*-, D(−)-*threo*-, and L(+)-*threo*-forms. The commercial product is usually either the *meso*- or the D(−)-form. Prepn of *meso*-form from *trans*-2,3-epoxybutane and of DL-form from *cis*-2,3-epoxybutane: Wilson, Lucas, *J. Am. Chem. Soc.* **58**, 2396 (1936). Prepn of D(−)- and L(+)-forms from corresponding D- and L-mannitols: Rubin *et al.*, *ibid.* **74**, 425 (1952). Manuf of D(−)-form by fermentation of carbohydrate solns with organisms of the *Bacillus subtilis* group: Vergnaud, US 2529061 (1950 to Usines de Melle). Manuf from 2-butene: Cosby *et al.*, US 2808429 (1957 to Allied Chem.); Keith *et al.*, US 2974161 (1961 to Sinclair); from 2-butyne: Saegebarth, US 3157704 (1964 to du Pont). Configuration:

Morell, Auernheimer, *J. Am. Chem. Soc.* **66**, 792 (1944); Leroux, Lucas, *ibid.* **73**, 41 (1951); Rubin *et al., loc. cit.*

D(-)-*threo*-form

meso-Form (erythro-Form). [5341-95-7] *rel*-(2*R*,3*S*)-2,3-Butanediol. Hygroscopic crystals from dry diisopropyl ether, mp 34.4°. bp$_{742}$ 181.7°, bp$_{16}$ 89°. d$_4^{25}$ 0.9939. n$_D^{35}$ 1.4324. Moderately sol in diisopropyl ether.

DL-threo-Form. [6982-25-8] *rel*-(2*R*,3*R*)-2,3-Butanediol. Hygroscopic crystals from diisopropyl ether, mp 7.6°. bp$_{742}$ 172.7°, bp$_{16}$ 86°. n$_D^{25}$ 1.4310. Very sol in diisopropyl ether.

D(−)-threo-Form. [24347-58-8] (2*R*,3*R*)-2,3-Butanediol. mp 19.7°. bp$_{745}$ 179-180°, bp$_{10}$ 77.5-78°. d$_4^{25}$ 0.9869. n$_D^{25}$ 1.4315. [α]$_D^{25}$ −13.0° (neat).

L(+)-threo-Form. [19132-06-0] (2*S*,3*S*)-2,3-Butanediol. bp 179-182°. d^{25} 0.9872. n$_D^{25}$ 1.4306.

1572. n-Butyl Ether. [142-96-1] 1,1′-Oxybisbutane; butyl ether; *n*-dibutyl ether. C$_8$H$_{18}$O; mol wt 130.23. C 73.78%, H 13.93%, O 12.29%. Prepn: Smyth, *J. Chem. Educ.* **39**, 212 (1962); d'Engenieres *et al., Bull. Soc. Chim. Fr.* **1964**(10), 2471. Toxicity study: H. F. Smyth *et al., Arch. Ind. Hyg. Occup. Med.* **10**, 61 (1954).

Liquid. d$_{20}^{20}$ 0.769. mp −98°. bp 142-143°. Flash pt, closed cup: 100°F (37°C). Almost insol in water. Misc with alcohol or ether. Tends to form explosive peroxides, especially when anhydr. LD$_{50}$ orally in rats: 7.4 g/kg (Smyth).

1573. tert-Butyl Hydroperoxide. [75-91-2] 1,1-Dimethylethylhydroperoxide. C$_4$H$_{10}$O$_2$; mol wt 90.12. C 53.31%, H 11.19%, O 35.51%. Prepn from *tert*-butyl alcohol and 30% H$_2$O$_2$: N. A. Milas, S. A. Harris, *J. Am. Chem. Soc.* **60**, 2434 (1938); N. A. Milas, Surgenor, *ibid.* **68**, 205 (1946); **US 2573947** (1951 to Shell). By oxidation of *tert*-butylmagnesium chloride: Walling, Buckler, *J. Am. Chem. Soc.* **75**, 4372 (1953); **77**, 6032 (1955).

Liquid. Stable to 75°. d$_4^{20}$ 0.896. mp −8°. bp$_{20}$ 35°. n$_D^{20}$ 1.4007. Sol in organic solvents. *Flammable.* Slow first-order decompn can be accelerated by the presence of 1 mol-% of Cu, Co and Mn salts.

USE: Catalyst in polymerization reactions. To introduce peroxy group into org molecules, in radical substitution reactions: Kharasch, Fono, *J. Org. Chem.* **23**, 325 (1948); *see also* Kharasch, Sosnovsky, *Tetrahedron* **3**, 97, 105 (1958).

1574. tert-Butyl Hypochlorite. [507-40-4] Hypochlorous acid 1,1-dimethylethyl ester; *t*-butyl hypochlorite. C$_4$H$_9$ClO; mol wt 108.57. C 44.25%, H 8.36%, Cl 32.65%, O 14.74%. Prepd by adding aq *tert*-butyl alcohol to a cooled aq soln of NaOH and passing chlorine into the mixture: Teeter, Bell, *Org. Synth.* **32**, 20 (1952). By adding CO$_2$ to neutral aq *tert*-butyl alcohol and NaOCl satd with NaCl: Katz, **US 2694722** (1954 to Bjorksten Res. Labs.).

Pale yellow liquid, irritating odor. d$_{20}^{20}$ 0.910; d$_4^{18}$ 0.9583. bp 77-78°. n$_D^{20}$ 1.403. Hydrolyzes in presence of aq alkali but not aq acid. Dec upon exposure to bright light to methyl chloride and acetone

with considerable evolution of heat. *Spontaneously combustible; corrosive.* Should be stored in an inert atmosphere in a dark, refrigerated place. Violent reaction occurs if exposed to rubber, strong light or overheating. Has far greater stability than the corresponding *n*-butyl and *sec*-butyl hypochlorites.

USE: Dehydration of alcohols; *N*- and *C*-chlorinations.

1575. Butylidene Chloride. [541-33-5] 1,1-Dichlorobutane. C$_4$H$_8$Cl$_2$; mol wt 127.01. C 37.83%, H 6.35%, Cl 55.82%. Prepn from butyraldehyde and PCl$_5$: Henne *et al., J. Am. Chem. Soc.* **61**, 938 (1939).

Oily liquid. bp$_{752}$ 114.8-115.1°. d^{25} 1.0797; n$_D^{25}$ 1.4305: Brown, Ash, *J. Am. Chem. Soc.* **77**, 4019 (1955). Practically insol in water. Sol in alcohol, chloroform, ether.

1576. n-Butyl Iodide. [542-69-8] 1-Iodobutane. C$_4$H$_9$I; mol wt 184.02. C 26.11%, H 4.93%, I 68.96%. Prepn: Franzen, *Ber.* **87**, 1148 (1954); Landauer, Rydon, *J. Chem. Soc.* **1954**, 2281; Stone, Shechter, *Org. Synth.* **coll. vol. IV**, 321 (1963). Manuf: Huber, Schenck, **US 2899471** and **US 3053910** (1959 and 1962, both to GAF).

Liquid, bp 130.4°. mp −103.0°. d$_4^{20}$ 1.616. n$_D^{20}$ 1.4998: Vogel, *J. Chem. Soc.* **1943**, 636. Practically insol in water. Sol in alcohol, ether. *Protect from light.*

1577. sec-Butyl Iodide. [513-48-4] 2-Iodobutane. C$_4$H$_9$I; mol wt 184.02. C 26.11%, H 4.93%, I 68.96%. Prepn of *dl*-form from 2-butanol: Kornblum *et al., J. Am. Chem. Soc.* **77**, 5528 (1955); from dibutyl ether: Long, Freeguard, *Chem. Ind. (London)* **1965**, 223. Prepn of *d*-form from *l*-butanol: Kenyon *et al., J. Chem. Soc.* **1935**, 1072. Prepn of *l*-form from *d*-2-butanol: Levene, Rothen, *J. Biol. Chem.* **115**, 415 (1936); Kornblum *et al., J. Am. Chem. Soc.* **70**, 746 (1948).

dl-Form. Liquid, rapidly turns brown on exposure to light. bp 120°. mp −104°. d$_4^{20}$ 1.592, n$_D^{20}$ 1.4991: Vogel, *J. Chem. Soc.* **1943**, 636. Practically insol in water. Sol in alc, ether. *Protect from light.*

l-Form. [22156-92-9] (2*R*)-2-Iodobutane. Liquid, bp 118°. d$_4^{20}$ 1.596, n$_D^{20}$ 1.495, [α]$_D^{24}$ −15.9° (neat); also reported as [α]$_D^{17}$ −31.98° (Kornblum *et al.*).

1578. Butyl Isocyanate. [111-36-4] 1-Isocyanatobutane; isocyanic acid butyl ester; *n*-butyl isocyanate; BIC. C$_5$H$_9$NO; mol wt 99.13. C 60.58%, H 9.15%, N 14.13%, O 16.14%. Prepn: R. J. Slocombe *et al., J. Am. Chem. Soc.* **72**, 1888 (1950); T. I. Bieber, *ibid.* **74**, 4700 (1952); E. T. Shawl *et al.*, **EP 408277** (1991 to Arco). Toxicity study: J. Pauluhn, A. Eben, *Arch. Toxicol.* **66**, 118 (1992). Analytical methods development: R. J. Rando *et al., J. Liq. Chromatogr.* **16**, 3977 (1993).

Flammable liquid. bp 113-116°. n$_D^{20}$ 1.4060. d$_4^{20}$ 0.880. LC$_{50}$ (4hr) in male, female rats (mg/m^3): 60, 55 by inhalation (Pauluhn, Eben).

USE: Reagent in organic synthesis.

1579. n-Butylmalonic Acid. [534-59-8] 2-Butylpropanedioic acid; pentane-1,1-dicarboxylic acid. C$_7$H$_{12}$O$_4$; mol wt 160.17. C 52.49%, H 7.55%, O 39.96%. Prepd by the interaction of sodium *n*-

amyl chloride and carbon dioxide under pressure with ligroin as solvent: Morton *et al.*, *J. Am. Chem. Soc.* **58**, 754 (1936).

H₃C —— COOH / COOH

Prisms from water, mp 102°. pK (5°): 2.96. Sol in water: 100 g of a satd aq soln contain 11.6 g at 0°; 79.3 g at 50°. Sol in alc, ether. Heating to 150° yields caproic acid.

Diethyl ester. [133-08-4] $C_{11}H_{20}O_4$; mol wt 216.28. Prepared by the interaction of *n*-butyl bromide, diethyl malonate, and sodium alcoholate: Adams, Kamm, *Org. Synth.* **4**, 11 (1925). Liquid, bp$_{760}$ 235-240°; bp$_{40}$ 140-145°; bp$_{20}$ 130-135°. n_D^{20} 1.425. Very sol in alcohol, ether.

1580. *n*-Butyl Mercaptan. [109-79-5] 1-Butanethiol; normal butyl thioalcohol; thiobutyl alcohol. $C_4H_{10}S$; mol wt 90.18. C 53.28%, H 11.18%, S 35.55%. Formed by the action of yeast on a product obtained when hydrogen sulfide is made to react with butyraldehyde in alcoholic ammonia: Nord, *Ber.* **52**, 1209 (1919); by passing vapors of butanol and hydrogen sulfide over thorium oxide catalyst: Kramer, Reid, *J. Am. Chem. Soc.* **43**, 880 (1921); together with dibutyl sulfide in distilling an aq soln of sodium butyl sulfate and sodium sulfide: Gray, Gutekunst, *J. Am. Chem. Soc.* **42**, 858 (1920); by slightly warming dithiocarbamic butyl ester with aq KOH: v. Braun, Engelbertz, *Ber.* **56**, 1574 (1923); the sodium salt is formed when dibutyl disulfide is treated with sodium in ether or alcohol: Moses, Reid, *J. Am. Chem. Soc.* **48**, 777 (1926). Physical properties: Mathias, *ibid.* **72**, 1897 (1950); W. E. Haines *et al.*, *J. Phys. Chem.* **60**, 549 (1956). Once cited to occur in "skunk" fluid: Beckmann, *Pharm. Zentralhalle* **37**, 557 (1896); T. B. Aldrich, *J. Exp. Med.* **1**, 323 (1896); however, recent literature rebuts this: K. K. Anderson, D. T. Bernstein, *J. Chem. Ecol.* **1**, 493 (1975); *eidem*, *J. Chem. Educ.* **55**, 159 (1978).

H₃C —— SH

Mobile liquid. Heavy skunk odor. mp −115.9°. bp$_{766}$ 98.2°; bp$_{760}$ 98.4°. *Flammable.* d$_4^{25}$ 0.83679. n_D^{25} 1.44014. Slightly sol in water. Very sol in alcohol, ether, liquid hydrogen sulfide. Forms azeotropic mixtures with butyl alcohol (bp 97.8°; 85.16% butanethiol) and with butyl alcohol and water.

Caution: Potential symptoms of overexposure are irritation of eyes, skin; muscle weakness, malaise, sweating, nausea, vomiting, headache, confusion. *See NIOSH Pocket Guide to Chemical Hazards* (DHHS/NIOSH 97-140, 1997) p 42.

1581. *sec*-Butyl Mercaptan. [513-53-1] 2-Butanethiol; *sec*-butyl thioalcohol. $C_4H_{10}S$; mol wt 90.18. C 53.28%, H 11.18%, S 35.55%. Prepd from 2-iodobutane by means of alcoholic KSH soln: Reymann, *Ber.* **7**, 1287 (1874); from 2-bromobutane in the same manner: Ellis, Reid, *J. Am. Chem. Soc.* **54**, 1674 (1932).

H₃C —— SH / CH₃

Mobile liq. Heavy skunk odor. mp −165°. bp 84-85°. *Flammable.* d$_4^{17}$ 0.8299; d$_4^{25}$ 0.8246. n_D^{25} 1.4338. Slightly sol in water. Very sol in alc, ether, liquid H₂S.

1582. *tert*-Butyl Mercaptan. [75-66-1] 2-Methyl-2-propanethiol. $C_4H_{10}S$; mol wt 90.18. C 53.28%, H 11.18%, S 35.55%. Prepd from *tert*-butyl iodide, ZnS and alcohol: Dobbin, *J. Chem. Soc.* **57**, 641 (1890). *See also* US 2020421, *C.A.* **30**, 489 (1936); US 2051806, *C.A.* **30**, 6760 (1936).

SH / H₃C—C—CH₃ / CH₃

Mobile liquid. Heavy skunk odor. mp −0.5°. bp$_{760}$ 63.7-64.2°. d$_4^{25}$ 0.79426; n_D^{25} 1.41984. *See:* Mathias, *J. Am. Chem. Soc.* **72**, 1897

(1950). *Flammable.* Remarkably stable to oxidizing agents. Slightly sol in water. Very sol in alcohol, ether, liquid H₂S.

1583. *n*-Butylmercuric Chloride. [543-63-5] Butylchloromercury. C_4H_9ClHg; mol wt 293.16. C 16.39%, H 3.09%, Cl 12.09%, Hg 68.42%. Prepn: Slotta, Jacobi, *J. Prakt. Chem.* [2] **120**, 249 (1929); R. C. Larock, H. C. Brown, *J. Am. Chem. Soc.* **92**, 2467 (1970). Distribution in mice: M. Yonaha *et al.*, *Chem. Pharm. Bull.* **23**, 1718 (1975). Prepn of bromide: Slotta, Jacobi, *J. Prakt. Chem.* [2] **120**, 249 (1929); Marvel *et al.*, *J. Am. Chem. Soc.* **47**, 3009 (1925). Review and bibliography: Krause, von Grosse, *Die Chemie der Metallorganischen Verbindungen* (Berlin, 1937).

H₃C —— Cl / Hg

White needles from ethanol, sometimes leaflets, mp 130°. Soly (g/100 ml) in water at 18°: 1.4×10^{-4}; at 100°: 3.3×10^{-4}; soly (g/100 g) in ethanol at 18°: 1.5; at 78°: 9.0; in chloroform at 18°: 5.6.

1584. Butylmethylimidazolium. [80432-08-2] 3-Butyl-1-methyl-1*H*-imidazolium; 1-butyl-3-methyl-1*H*-imidazolium; BMIM; C₄mim. $C_8H_{15}N_2^+$. Cation used in prepn of room temperature ionic liquids, *q.v.* Prepn of hexafluorophosphate and tetrafluoroborate: Y. Chauvin *et al.*, *Angew. Chem. Int. Ed.* **34**, 2698 (1995); and physical properties: P. A. Z. Suarez *et al.*, *J. Chim. Phys.* **95**, 1626 (1998). Prepn and properties of ionic liquids containing the imidazolium cation: J. G. Huddleston *et al.*, *Green Chem.* **3**, 156 (2001).

H₃C—N⁺⧸⧹N—CH₃

Chloride. [79917-90-1] $C_8H_{15}ClN_2$; mol wt 174.67. Prepn: J. S. Wilkes *et al.*, *Inorg. Chem.* **21**, 1263 (1982). Slightly yellow, may be crystalline at room temp depending on water content. mp 41° (dried) (Huddleston). Also reported as mp 65-69° (Wilkes). d^{25} 1.08 (dried). Miscible with water.

Hexafluorophosphate. [174501-64-5] $C_8H_{15}F_6N_2P$; mol wt 284.19. Use as extraction medium: J. G. Huddleston *et al.*, *Chem. Commun.* **16**, 1765 (1998); as a solvent for alkylation reactions: M. J. Earle *et al.*, *ibid.* **20**, 2245 (1998). Use in environmentally benign manufacturing processes: L. A. Blanchard *et al.*, *Nature* **339**, 28 (1999). Microscopic solvent properties: S. N. Baker *et al.*, *Green Chem.* **4**, 165 (2002). Viscous liquid, mp 4° (water equilibrated); 10° (dried). Glass transition temperature: −83° (water equilibrated), −80° (dried). Viscosity (25°): 397 cP (water equilibrated), 450 cP (dried). d^{25} 1.35 (water equilibrated), 1.36 (dried). n^{25} 1.409 (dried). Miscible with water. Imiscible with toluene, hexane, other non-polar solvents.

Tetrafluoroborate. [174501-65-6] $C_8H_{15}BF_4N_2$; mol wt 226.03. Use as a solvent for nucleophilic substitution reactions: D. W. Kim *et al.*, *J. Org. Chem.* **68**, 4281 (2003). Viscous liquid. Glass transition temperature: −97° (dried). Viscosity (25°): 219 cP (dried). d^{25} 1.12(dried). Miscible with water. Sol in methanol, ethanol. Insol in hydrocarbons.

USE: Solvent for organic reactions.

1585. *n*-Butyl Nitrite. [544-16-1] Nitrous acid butyl ester. $C_4H_9NO_2$; mol wt 103.12. C 46.59%, H 8.80%, N 13.58%, O 31.03%. Prepd by the action of nitrous acid on butyl alcohol: Noyes, *Org. Synth.* **coll. vol. II**, 108 (1943); Miller, Audrieth, *Inorg. Synth.* **2**, 139 (1946).

H₃C —— O—N=O

Oily liquid. Characteristic odor. Breathing of vapor causes headache and vasodilation. d$_4^0$ 0.9114. bp$_{760}$ 78.2° (some decompn). Miscible with alcohol, ether. Dec on storage. Polymerization products of butyraldehyde have been found in five-month-old samples.

Caution: Lowers blood pressure through vasodilation, causing headache, throbbing, weakness. Effects resemble those produced by amyl nitrate.

USE: In the manuf of rare earth azides.

1586. *tert*-Butyl Nitrite. [540-80-7] Nitrous acid 1,1-dimethyl ethyl ester; α,α-dimethylethyl nitrite; nitrous acid *tert*-butyl ester. $C_4H_9NO_2$; mol wt 103.12. C 46.59%, H 8.80%, N 13.58%, O 31.03%. Prepd by adding *t*-butyl chloride to a cooled slurry of silver nitrite and anhydr ether: Kornblum *et al., J. Am. Chem. Soc.* **77**, 5528 (1955); by passing NO_2 or NO_2 + NO into *t*-butyl alcohol at 25-30°: Treacy, **US 2739166** (1956); from *t*-butanol, sodium nitrite and sulfuric acid: Coe, Doumani, *J. Am. Chem. Soc.* **70**, 1516 (1948).

Yellow liquid, agreeable odor. d_4^0 0.8941; d_4^{20} 0.8671. bp_{760} 63°; bp_{250} 34°; n_D^{20} 1.3687. Very sol in alcohol, ether, chloroform, carbon disulfide; slightly sol in water; practically insol in glycerol.

USE: Jet propellant.

1587. Butylparaben. [94-26-8] 4-Hydroxybenzoic acid butyl ester; *n*-butyl *p*-hydroxybenzoate; Butyl Chemosept; Butyl Parasept; Tegosept B. $C_{11}H_{14}O_3$; mol wt 194.23. C 68.02%, H 7.27%, O 24.71%. Prepn of calcium and magnesium salts: Engels, Weijlard, **US 2046324** and **US 2056176** (both 1936 to Merck & Co.). Prepn of analogous derivatives: *See* ethyl-, methyl- and propylparaben.

Crystalline powder, mp 68-69°. Freely sol in acetone, alc, ether, chloroform, propylene glycol. Very slightly soluble in water (1:6500), glycerin. Preserve in well-closed containers.

Calcium salt. $C_{22}H_{26}CaO_6$. Cryst powder. Soly in water: approx 1:125.

Magnesium salt. $C_{22}H_{26}MgO_6$. Cryst powder. Soly in water: approx 1:110.

USE: Pharmaceutic aid (antifungal). Preservative in foods.

1588. 4-*tert*-Butylphenol. [98-54-4] 4-(1,1-Dimethylethyl)-phenol; butylphen. $C_{10}H_{14}O$; mol wt 150.22. C 79.96%, H 9.39%, O 10.65%. Prepd by heating phenol with isobutanol in the presence of zinc chloride: **DE 17311**; *Frdl.* **1**, 22; also from phenol, *tert*-butyl chloride and excess alkali in alcohol: Lewis, *J. Chem. Soc.* **83**, 329 (1903). *See also* Smith, *J. Am. Chem. Soc.* **55**, 3718 (1933); Natelson, *ibid.* **56**, 1583 (1934); Huston, Hsieh, *ibid.* **58**, 439 (1936); **US 2039344** (1936 to Dow); Isagulyants, Bagryantseva, *C.A.* **33**, 8183 (1939). Toxicity study: H. F. Smyth *et al., Am. Ind. Hyg. Assoc. J.* **30**, 470 (1969).

Needles from water, mp 98°. d_4^{114} 0.9081. bp 237°. Volatile with steam. Practically insol in cold water. Sol in alcohol, ether. LD_{50} orally in rats: 3.25 ml/kg (Smyth).

Sodium salt. $C_{10}H_{13}NaO$. Deliquescent leaflets. Sol in water.

USE: Intermediate in the manuf of varnish and lacquer resins; as a soap antioxidant; ingredient in de-emulsifiers for oil field use; in motor oil additives.

1589. 4-*tert*-Butylphenyl Salicylate. [87-18-3] 2-Hydroxybenzoic acid 4-(1,1-dimethylethyl)phenyl ester; salicylic acid *p-tert*-butylphenyl ester; TBS. $C_{17}H_{18}O_3$; mol wt 270.33. C 75.53%, H 6.71%, O 17.75%. Prepd from salicylic acid and 4-*tert*-butylphenol in the presence of $POCl_3$: Stoesser, Sommerfield, **US 2606920** (1952 to Dow).

Crystals, slight odor resembling that of salol, mp 62-64°. Maximum light absorption at 290-330 nm. Soly (w/w): in water <0.1%; abs ethanol 79%; ethyl acetate 153%; methyl ethyl ketone 197%; toluene 158%; Stoddard solvent 39%.

USE: Light absorber in plastic food wrappings.

1590. *n*-Butyl Propionate. [590-01-2] Propanoic acid butyl ester; propionic acid butyl ester. $C_7H_{14}O_2$; mol wt 130.19. C 64.58%, H 10.84%, O 24.58%. Prepn from 1-butanol and propionic acid: Vogel, *J. Chem. Soc.* **1948**, 616; Spindt, Stevens, **US 2470876** (1949 to Gulf).

Liquid, bp 146.8°. mp −89°. d_4^{20} 0.8754. n_D^{20} 1.401. Very slightly sol in water; very sol in alcohol, ether.

1591. *N*-Butylscopolammonium Bromide. [149-64-4] (1α,-2β,4β,5α,7β)-9-Butyl-7-[(2*S*)-3-hydroxy-1-oxo-2-phenylpropoxy]-9-methyl-3-oxa-9-azoniatricyclo[3.3.1.0^{2,4}]nonane bromide (1:1); α-(hydroxymethyl)benzeneacetic acid 9-butyl-9-methyl-3-oxa-9-azoniatricyclo[3.3.1.0^{2,4}]non-7-yl ester bromide; 8-butyl-6β,7β-epoxy-3α-hydroxy-1αH,5αH-tropanium bromide (−)-tropate; butylscopolamine bromide; hyoscine-*N*-butyl bromide; scopolamine-*N*-butyl bromide; scopolamine bromobutylate; Buscapina; Buscolysin; Buscopan; Scoburen. $C_{21}H_{30}BrNO_4$; mol wt 440.38. C 57.28%, H 6.87%, Br 18.14%, N 3.18%, O 14.53%. Anticholinergic. Prepn: F. Adickes *et al.*, **DE 856890**; *eidem*, **US 2872452** (1952, 1959 both to Boehringer Ing.). Biological activity studies: Bauer *et al., Arzneim.-Forsch.* **18**, 1132 (1968). Toxicological studies: K. Stockhaus, H. Wick, *Arch. Int. Pharmacodyn. Ther.* **180**, 155 (1969). Use in endoscopic examinations: J. R. Lee, *Clin. Radiol.* **33**, 273 (1982); D. N. Hupscher, O. Dommerholt, *Diagn. Imag. Clin. Med.* **53**, 77 (1984); A. C. Steger *et al., Am. J. Gastroenterol.* **81**, 615 (1986).

Crystals from methanol, mp 142-144°. $[\alpha]_D^{20}$ −20.8° (c = 3 in water). LD_{50} in mice (mg/kg): 15.6 i.v.; 74 i.p.; 570 s.c.; 3000 orally (Stockhaus, Wick).

THERAP CAT: Antispasmodic.

THERAP CAT (VET): Antispasmodic.

1592. Butyl Stearate. [123-95-5] Octadecanoic acid butyl ester. $C_{22}H_{44}O_2$; mol wt 340.59. C 77.58%, H 13.02%, O 9.39%. Prepn from silver stearate and *n*-butyl iodide: Whitby, *J. Chem. Soc.* **1926**, 1458; from stearic acid and *n*-butanol: Smith, *ibid.* **1931**, 802. Physical properties: *Beilstein* **vol. 2**, Suppl. 3, 1016; Smith, *loc. cit.*; *Ind. Eng. Chem.* **32**, 880 (1940).

Crystals from alcohol, propanol, or ether, mp 27°. Also reported as mp 16°. bp 343°. Closed-cup flash pt 160°C (320°F). d_{25}^{25} 0.855-0.875. Slightly sol in water; sol in alcohol, ether.

USE: Solvent, spreading and softening agent in plastics, textiles, cosmetics, rubber industries.

1593. *n*-Butyl Sulfide. [544-40-1] 1,1′-Thiobisbutane; butylthiobutane; dibutyl sulfide. $C_8H_{18}S$; mol wt 146.29. C 65.68%, H 12.40%, S 21.92%. Prepd by refluxing a soln contg sodium sulfide and sodium *n*-butyl sulfate for several hrs: Gray, Gutekunst, *J. Am. Chem. Soc.* **42**, 856 (1920).

$$H_3C\diagup\diagdown\diagup S\diagdown\diagup\diagdown CH_3$$

Liquid. mp −79.7°. d_0^{16} 0.839. d_4^0 0.852. bp 182°. Insol in water. Very sol in alcohol, ether.

1594. Butyltin Trichloride. [1118-46-3] Butyltrichlorostannane; butyltrichlorotin; monobutyltin trichloride; trichlorobutyltin; BuSnCl_3. $C_4H_9Cl_3Sn$; mol wt 282.18. C 17.03%, H 3.21%, Cl 37.69%, Sn 42.07%. Organotin halide that serves as a Lewis acid catalyst and reagent for polymerization, dehydration, esterification, and acetylation reactions. Prepd (not claimed): E. W. Johnson, J. M. Church, **US 2599557** (1952 to Metal & Thermit). Improved process: **GB 739883** (1955 to Metal & Thermit); W. P. Neumann, G. Burkhardt, *Ann.* **663**, 11 (1963). Applications as catalyst: G. Tagliavini *et al.*, *Tetrahedron* **45**, 1187 (1989); D. Marton *et al.*, *ibid.* 7099; L. A. Hobbs, P. J. Smith, *Appl. Organomet. Chem.* **6**, 95 (1992). *Review*: G. Tagliavini, *J. Organomet. Chem.* **437**, 15-22 (1992).

$$\underset{Cl}{\overset{Cl}{\underset{|}{\overset{|}{Sn}}}}\diagup\diagdown\diagup CH_3$$

Colorless liquid. *Flammable. Corrosive.* bp_{11} 97.5-98.5°; $bp_{0.1}$ 44-45°. Flash point, closed cup: 178°F (81°C). Sol in most organic solvents. Dissolves in water to give strongly acidic soln.
USE: Reagent and catalyst in synthetic organic chemistry.

1595. Butyraldehyde. [123-72-8] Butanal. C_4H_8O; mol wt 72.11. C 66.63%, H 11.18%, O 22.19%. Prepn from butyryl chloride: Brown, Tsukamoto, *J. Am. Chem. Soc.* **83**, 2016 (1961); by reduction of corresponding nitrile: Gaiffe, Pallaud, *Compt. Rend.* **254**, 496 (1962); by alkali aluminum hydride reduction of methyl butyrate: Zakharkin *et al.*, *Tetrahedron Lett.* **1963**, 208; by oxidation of butanol: Harrison, *Proc. Chem. Soc. London* **1964**, 110. Usually manuf by catalytic dehydrogenation of butanol, catalytic hydrogenation of crotonaldehyde, or by the oxo process from propene: Dunbar, Arnold, *J. Org. Chem.* **10**, 501 (1945); Horn, *Ind. Eng. Chem.* **51**, 655 (1959); W. L. Faith *et al.*, *Industrial Chemicals* (John Wiley, New York, 3rd ed., 1965) pp 183, 304-305. Toxicity study: Smyth *et al.*, *Arch. Ind. Hyg. Occup. Med.* **4**, 119 (1951). *Review:* P. D. Sherman in *Kirk-Othmer Encyclopedia of Chemical Technology* vol. 4 (Wiley-Interscience, New York, 3rd ed., 1978) pp 376-386.

$$H_3C\diagup\diagdown\diagup\overset{\overset{\displaystyle H}{|}}{C}\diagdown_O$$

Flammable. bp 74.8°. mp −99°. Closed-cup flash pt −6.67°C (20°F). d_4^{20} 0.8016. n_D^{20} 1.379. Soly in water at 25°, 7.1 (wt-%): Smith, Bonner, *Ind. Eng. Chem.* **43**, 1169 (1951). Miscible with ethanol, ether, ethyl acetate, acetone, toluene, many other organic solvents and oils. Single-dose LD_{50} orally in rats: 5.89 g/kg (Smyth).
Caution: May act as irritant, narcotic.
USE: Chiefly in the manuf of rubber accelerators, synthetic resins, solvents, plasticizers.

1596. *n*-Butyramide. [541-35-5] Butanamide. C_4H_9NO; mol wt 87.12. C 55.15%, H 10.41%, N 16.08%, O 18.36%. Prepn by a modified Willgerodt reaction: Jelinek, **US 2572809**; **US 2572810** (both 1951 to GAF); from butyryl chloride: Philbrook, *J. Org. Chem.* **19**, 623 (1954); from butyraldoxime: Huber, **US**

2721199 (1955 to du Pont); from butyric acid: Rahman, *Rec. Trav. Chim.* **79**, 188 (1960); from butyronitrile: Gilbert, Rumanowski, **US 3062883** (1962 to Allied Chem.).

$$H_2N\overset{\overset{\displaystyle O}{||}}{C}\diagup\diagdown\diagup CH_3$$

Crystals, mp 115-116°. bp 216°. Sol in water, alcohol; slightly sol in ether.

1597. Butyric Acid. [107-92-6] Butanoic acid; *n*-butyric acid; ethylacetic acid. $C_4H_8O_2$; mol wt 88.11. C 54.53%, H 9.15%, O 36.32%. Present as an ester to the extent of 4-5%. Obtained by suitable fermentation of carbohydrates; prepn from *n*-propanol + CO at 200 atm in the presence of $Ni(CO)_4$ and NiI_2: Reppe *et al.*, *Ann.* **582**, 83 (1953); lab prepn from ethylmalonic acid: Gattermann-Wieland, *Praxis des Organischen Chemikers* (de Gruyter, Berlin, 40th ed., 1961) p 221. Toxicity study: Smyth *et al.*, *Arch. Ind. Hyg. Occup. Med.* **10**, 61 (1954).

$$H_3C\diagup\diagdown\diagup\overset{\overset{\displaystyle O}{||}}{C}\diagdown_{OH}$$

Oily liq; unpleasant, rancid odor. d_4^{20} 0.959. bp 163.5°. mp −7.9°. n_D^{20} 1.3991. Flash pt, closed cup: 170°F (77°C). *Corrosive.* Neutralization value 636.79. Miscible with water, alcohol, ether. The calcium salt of this acid is less soluble in hot than in cold water (difference from isobutyric acid). LD_{50} orally in rats: 8.79 g/kg (Smyth).
Magnesium salt. [556-45-6] Magnesium butyrate. C_8H_{14}-MgO_4; mol wt 198.50. Deliquesc leaflets. Sol in water. *Keep well closed.*
USE: Manuf of esters, some of which serve as bases of artificial flavoring ingredients of certain liqueurs, soda-water syrups, candies; also for varnishes; as decalcifier of hides.

1598. Butyric Anhydride. [106-31-0] Butanoic acid 1,1′-anhydride; butyryl oxide. $C_8H_{14}O_3$; mol wt 158.20. C 60.74%, H 8.92%, O 30.34%. Prepn from butyric acid: Williams, Krynitsky, *Org. Synth.* **coll. vol. III**, 165 (1955); Kuwajima, Mukaiyama, *J. Org. Chem.* **29**, 1385 (1964). Manuf by catalytic hydrogenation of crotonic acid: Smith, Hunter, **US 2492403** (1949 to Celanese); by catalytic carbonylation of the corresponding acid ester: Reppe, Friederich, **US 2730546** (1956 to BASF).

$$H_3C\diagup\diagdown\diagup\overset{\overset{\displaystyle O}{||}}{C}\diagdown_O\diagup\overset{\overset{\displaystyle O}{||}}{C}\diagup\diagdown CH_3$$

Liq, bp 199.4-201.4°. mp −75°. d_4^{20} 0.9668. n_D^{20} 1.4070. Flash pt, open cup: 190°F (88°C). *Corrosive.* Sol in water and in alc with decompn; sol in ether.

1599. Butyroin. [496-77-5] 5-Hydroxy-4-octanone; 5-octanol-4-one. $C_8H_{16}O_2$; mol wt 144.21. C 66.63%, H 11.18%, O 22.19%. Prepd by the reaction between an ether soln of ethyl butyrate and sodium or potassium: Bouveault, Locquin, *Bull. Soc. Chim.* [3] **35**, 629 (1906); Corson *et al.*, *J. Am. Chem. Soc.* **52**, 3988 (1930); Scheibler, Emden, *Ann.* **434**, 265 (1923); Snell, McElvain, *Org. Synth.* **coll. vol. II**, 114 (1943).

$$H_3C\diagup\diagdown\diagup\overset{\overset{\displaystyle O}{||}}{C}\diagup\underset{\underset{\displaystyle OH}{|}}{C}H\diagup\diagdown CH_3$$

Liquid. d_4^0 0.9367; $d_4^{16.7}$ 0.91075. bp_{760} 180-190°; bp_{155} 150-154°; bp_{20} 95°; bp_{10} 85°. $n_D^{16.7}$ 1.43455. Reduces Fehling's soln.

Oxime. [5787-53-1] $C_8H_{17}NO_2$; mol wt 159.23. Viscous liquid, bp_{10} 143°.

1600. Butyrolactone. [96-48-0] Dihydro-2(3H)-furanone; γ-butyrolactone; 1,2-butanolide; 1,4-butanolide; γ-hydroxybutyric acid lactone; 3-hydroxybutyric acid lactone; 4-hydroxybutanoic acid lactone. $C_4H_6O_2$; mol wt 86.09. C 55.81%, H 7.03%, O 37.17%. Prepd from acetylene and formaldehyde: Reppe, *Chem. Ing. Tech.* **1950**, 365; *Chem. Eng.* **58**, no. 6, 176 (1951); also prepd from ethylene chlorohydrin, glutaric acid, ν-hydroxybutyric acid solns, tetrahydrofuran, or vinylacetic acid: F. C. Whitmore, *Organic Chemistry* (Van Nostrand, New York, 2nd ed., 1951). Alternate synthesis: Y. Ogata *et al.*, *J. Org. Chem.* **45**, 1320 (1980). Physical properties: McKinley, Copes *J. Am. Chem. Soc.* **72**, 5331 (1950). Toxicity: H. F. Smyth *et al.*, *Am. Ind. Hyg. Assoc. J.* **30**, 470 (1969).

Oily liquid. d_0^0 1.1441; d_0^{15} 1.1286. mp -43.53°. bp_{760} 204°. bp_{12} 89°. n_D^{25} 1.4348. Flash pt, open cup: 209°F (98°C). Volatile with steam. Misc with water. Sol in methanol, ethanol, acetone, ether, benzene. Hydrolyzed by hot alkaline solns. LD_{50} orally in rats: 17.2 ml/kg (Smyth).

USE: Intermediate in the synthesis of polyvinylpyrrolidone, DL-methionine, piperidine, phenylbutyric acid, thiobutyric acids. Solvent for polyacrylonitrile, cellulose acetate, methyl methacrylate polymers, polystyrene. Constituent of paint removers, textile aids, drilling oils.

1601. Butyronitrile. [109-74-0] Butanenitrile; propyl cyanide; butyric acid nitrile. C_4H_7N; mol wt 69.11. C 69.52%, H 10.21%, N 20.27%. Prepd from 1-butanol by controlled cyanation with NH_3 at 300° in the presence of Ni—Al_2O_3 catalysts: Popov, Shuikin, *Izv. Akad. Nauk SSSR Otd. Khim. Nauk* **1958**, 713-718, *C.A.* **52**, 19924 (1958). Toxicity: H. F. Smyth *et al.*, *Am. Ind. Hyg. Assoc. J.* **23**, 95 (1962).

Liquid. d_4^0 0.8091; d_4^{15} 0.7954; d_4^{30} 0.7817. mp -112°. bp_{760} 117.5°; bp_{400} 96.8°; bp_{200} 76.7°; bp_{100} 59.0°; bp_{60} 47.3°; bp_{40} 38.4°; bp_{20} 25.7°; bp_{10} 13.4°; bp_5 2.1°; $bp_{1.0}$ -20.0°. n_D^{20} 1.38385. Flash pt, open cup: 85°F (29°C). Viscosity ($\eta \times 10^5$) at 15° = 624; at 30° = 515. Dipole moment 3.5. *Flammable, poisonous.* Sparingly sol in water. Misc with alcohol, ether, dimethylformamide. LD_{50} orally in rats: 0.14 g/kg (Smyth).

Caution: Potential symptoms of overexposure are irritation of eyes, skin, respiratory system; headache, dizziness, weakness, giddiness, confusion, convulsions; dyspnea; abdominal pain, nausea, vomiting. *See NIOSH Pocket Guide to Chemical Hazards* (DHHS/NIOSH 97-140, 1997) p 44.

1602. *n*-Butyryl Chloride. [141-75-3] Butanoyl chloride; butyryl chloride. C_4H_7ClO; mol wt 106.55. C 45.09%, H 6.62%, Cl 33.27%, O 15.02%. Prepn from butyric acid: Helferich, Schaefer, *Org. Synth.* **coll. vol. I**, 147 (1941).

Liquid, bp 101-102°. mp -89°. $d_4^{20.6}$ 1.0263. n_D^{20} 1.412. *Flammable; corrosive.* Dissolves slowly with decompn in water, alcohol. Miscible with ether.

1603. Buzepide. [3691-21-2] Hexahydro-α,α-diphenyl-1H-azepine-1-butanamide; 2,2-diphenyl-4-hexamethyleneiminobutyramide; R-658. $C_{22}H_{28}N_2O$; mol wt 336.48. C 78.53%, H 8.39%, N 8.33%, O 4.75%. Anticholinergic. Prepn: Janssen *et al.*, *J. Med. Pharm. Chem.* **1**, 187 (1959); of methiodide as well as free base: Janssen, de Jongh, **US 2881165** (1959 to N.V. Nederlandsche Combinatie Chem. Ind.). Clinical trial in irritable bowel syndrome: J. Ph. Barbier *et al.*, *Ann. Gastroenterol. Hepatol.* **25**, 123 (1989).

Crystals from isopropanol, mp 141.5-143.5°.

Methiodide. [15351-05-0] Metazepium iodide; 1-(3-carbamoyl-3,3-diphenylpropyl)hexahydro-1-methylazepinium iodide; N-(3,3-diphenyl-3-carbamoylpropyl)-N-methylperhydroazepinium iodide; diphexamide methiodide. $C_{22}H_{28}N_2O.CH_3I$. Crystals, dec 212-213°.

Mixture with haloperidol. Vésadol.

THERAP CAT: Antispasmodic.

C

1604. Cabazitaxel. [183133-96-2] $(\alpha R,\beta S)$-β-[[(1,1-Dimethylethoxy)carbonyl]amino]-α-hydroxybenzenepropanoic acid (2aR,-4S,4aS,6R,9S,11S,12S,12aR,12bS)-12b-(acetyloxy)-12-(benzoyloxy)-2a,3,4,4a,5,6,9,10,11,12,12a,12b-dodecahydro-11-hydroxy-4,6-dimethoxy-4a,8,13,13-tetramethyl-5-oxo-7,11-methano-1H-cyclodeca[3,4]benz[1,2-b]oxet-9-yl ester; 1-hydroxy-7β,10β-dimethoxy-9-oxo-5β,20-epoxytax-11-ene-2α,4,13α-triyl 4-acetate 2-benzoate 13-[(2R,3S)-3-[[(tert-butoxy)carbonyl]amino]-2-hydroxy-3-phenylpropanoate]; dimethoxydocetaxel; RPR-116258; TXD-258; XRP-6258; Jevtana. $C_{45}H_{57}NO_{14}$; mol wt 835.94. C 64.66%, H 6.87%, N 1.68%, O 26.79%. Semisynthetic taxane; structurally similar to docetaxel, q.v. Antimitotic agent that stabilizes microtubules against cold-induced depolymerization; exhibits low affinity for the drug efflux pump, P-glycoprotein. Prepn: H. Bouchard et al., **WO 9630355**; eidem, **US 5847170** (1996, 1998 both to Rhone Poulenc Rorer). Pharmacokinetics and brain penetration: S. Cisternino et al., Br. J. Pharmacol. **138**, 1367 (2003). Clinical evaluation in patients with taxane-resistant breast cancer: X. Pivot et al., Ann. Oncol. **19**, 1547 (2008); with advanced solid tumors: A. C. Mita et al., Clin. Cancer Res. **15**, 723 (2009). Clinical trial in prostate cancer: J. S. de Bono et al., Lancet **376**, 1147 (2010).

$[\alpha]_D^{20}$ $-32.9°$ (c = 0.5 in methanol). Log P (octanol/water): 3.88.
THERAP CAT: Antineoplastic.

1605. Cabenegrins. Orally active antidotes against snake venoms; isolated from the root of a South American plant called "Cabeca de Negra" and structurally related to pterocarpin, q.v. Isoln: L. L. Darko et al., **EP 89229**; eidem, **US 4429141** (1983, 1984 both to Richter, Budapest). Structure determn: M. Nakagawa et al., Tetrahedron Lett. **1982**, 3855. Synthesis of (±)-cabenegrins A-I and A-II: M. Ishiguro et al., ibid. 3859.

Cabenegrin A-I   Cabenegrin A-II

Cabenegrin A-I. [84297-59-6] [6aR-[4(E),6aα,12aα]]-6a,12a-Dihydro-4-(4-hydroxy-3-methyl-2-butenyl)-6H-[1,3]dioxolo[5,6]-benzofuro[3,2-c][1]benzopyran-3-ol. $C_{21}H_{20}O_6$; mol wt 368.39. White crystalline solid, mp 167-168°. uv max (ethanol): 309 nm (ε 13000); uv max (methanol): 209, 233, 309 nm (ε 75000, 24000, 13000).
Cabenegrin A-II. [84297-60-9] 6a,12a-Dihydro-3-hydroxy-β-methyl-6H-[1,3]dioxolo[5,6]benzofuro[3,2-c][1]benzopyran-2-butanol. $C_{21}H_{22}O_6$; mol wt 370.40. uv max (methanol): 204, 230, 292, 308 nm (ε 116000, 8000, 9400, 11800).

1606. Cabergoline. [81409-90-7] (8β)-N-[3-(Dimethylamino)propyl]-N-[(ethylamino)carbonyl]-6-(2-propen-1-yl)ergoline-8-carboxamide; 1-ethyl-3-(3′-dimethylaminopropyl)-3-(6′-allylergoline-8′β-carbonyl)urea; 1-[(6-allylergoline-8β-yl)carbonyl]-1-[3-

(dimethylamino)propyl]-3-ethylurea; FCE-21336; Dostinex. $C_{26}H_{37}N_5O_2$; mol wt 451.62. C 69.15%, H 8.26%, N 15.51%, O 7.09%. Dopamine D_2-receptor agonist. Prepn: P. Salvati et al., **BE 888243**; eidem, **US 4526892** (1981, 1985 both to Farmitalia Carlo Erba). Prepn and bioactivity: E. Brambilla et al., Eur. J. Med. Chem. **24**, 421 (1989). Clinical pharmacology: C. Ferrari et al., J. Clin. Endocrinol. Metab. **63**, 941 (1986). Veterinary trial as abortifacient in dogs: K. Post et al., Theriogenology **29**, 1233 (1988). Clinical evaluation to prevent puerperal lactation: G. B. Melis et al., Obstet. Gynecol. **71**, 311 (1988); in hyperprolactinemic disorders: C. Ferrari et al., J. Clin. Endocrinol. Metab. **68**, 1201 (1989). Clinical trial in Parkinson's disease: J. T. Hutton et al., Neurology **46**, 1062 (1996).

White crystals from diethyl ether, mp 102-104°. Sol in ethyl alcohol, chloroform, DMF; slightly sol in 0.1 N HCl; very slightly sol in n-hexane. Insol in water. LD_{50} orally in male mice: >400 mg/kg (Brambilla).
Diphosphate. [85329-89-1] $C_{26}H_{37}N_5O_2 \cdot 2H_3PO_4$. mp 153-155°.
THERAP CAT: Prolactin inhibitor; antiparkinsonian.
THERAP CAT (VET): Prolactin inhibitor.

1607. Cacao Shell. Cocoa shells. Shells of the seed of Theobroma cacao L., Sterculiaceae. Habit. Brazil, Central America, Mexico, West Indies, and most tropical countries. Constit. Theobromine, caffeine, cacao red, protein, pentosans, pectic acid, starch. Ref: Dittmar, Engenharia e quim **5**, no. 1, 1 (1953), C.A. **48**, 2949e (1954). Monographs: E. M. Chatt, Cocoa (Interscience, New York, 1953); D. H. Urquhart, Cocoa (Longmans, London, 1961); Powell, Harris, Chocolate and Cocoa in Kirk-Othmer Encyclopedia of Chemical Technology vol. **5** (Interscience, New York, 2nd ed., 1964) pp 363-402.
Thin, papery, reddish-brown, concavo-convex shells; weak chocolate-like odor and taste.
Caution: Occasionally causes allergic dermatitis from handling.
USE: In the manuf of caffeine, theobromine.

1608. Cacodyl. [471-35-2] 1,1,2,2-Tetramethyldiarsine; tetramethyldiarsenic; dicacodyl. $C_4H_{12}As_2$; mol wt 209.98. C 22.88%, H 5.76%, As 71.36%. Prepn from cacodyl chloride by heating with zinc in an atm of carbon dioxide: Bunsen, Ann. **42**, 14 (1842); from dimethylarsine by the action of oxides of nitrogen, aqueous chromic acid, lead peroxide, cacodyl chloride or potassium ferricyanide: Dehn, Wilcox, Am. Chem. J. **35**, 1 (1906); by the action of free methyl on arsenic: Paneth, Loleit, J. Chem. Soc. **1935**, 366; by reduction of cacodyl oxide: Witten, **US 2531487** (1950 to U.S.A.); Fuson, Shive, **US 2756245** (1956 to U.S.A.).

Oily liquid. Solidifies to large quadratic plates, mp $-6°$. Almost intolerable garlicky odor. Inflames spontaneously in dry air. bp_{760} 165°. Slightly sol in water. Controlled oxidation with moist air yields cacodyl oxide and cacodylic acid. Reduction with tin and HCl yields Erytrarsin $(CH_3As)_4As_2O_3$.

1609. Cacodylic Acid. [75-60-5] As,As-Dimethylarsinic acid; dimethylarsinic acid; hydroxydimethylarsine oxide; Phytar. C_2H_7-AsO_2; mol wt 138.00. C 17.41%, H 5.11%, As 54.29%, O 23.19%.

Consult the Name Index before using this section.

$(CH_3)_2As(O)OH$. Prepn: Guinot, *J. Pharm. Chim.* **27**, 55 (1923), *C.A.* **17**, 2103[8] (1923); Inverni, *Boll. Chim. Farm.* **62**, 129 (1923), *C.A.* **17**, 2413[7] (1923); Challenger, Ellis, *J. Chem. Soc.* **1935**, 396; Fioretti, Portelli, *Ann. Chim. (Rome)* **53**, 1869 (1963), *C.A.* **60**, 10712c (1964). Use as herbicide: Sprague, US 3056668 (1962 to Ansul). Toxicity study: G. W. Bailey, J. L. White, *Residue Rev.* **10**, 97 (1965). Review of toxicology and human exposure: *Toxicological Profile for Arsenic* (PB2008-100002, 2007) 559 pp.

Crystals from alcohol + ether. mp 195-196°. Hygroscopic. *Poisonous*. Sol in 0.5 part water; very sol in alcohol; sol in acetic acid. Practically insol in ether. *Keep well closed*. LD_{50} orally in rats: 1350 mg/kg (Bailey, White).

Mercury salt. Mercuric cacodylate. $C_4H_{12}As_2HgO_4$. Hygroscopic, somewhat unstable cryst powder. Soluble in water, alcohol. Practically insol in ether. *Keep well closed*.

USE: Herbicide.

THERAP CAT: Dermatologic.

THERAP CAT (VET): Has been used in chronic eczema, anemia and as a tonic.

1610. Cacotheline. [561-20-6] $(\beta S, 16\beta)$-19,20-Didehydro-β,18-epoxy-10,11-dihydro-12-nitro-10,11-dioxo-17-norcuran-16-propanoic acid; 2,3-dihydro-4-nitro-2,3-dioxo-9,10-secostrychnidin-10-oic acid. $C_{21}H_{21}N_3O_7$; mol wt 427.41. C 59.01%, H 4.95%, N 9.83%, O 26.20%. A nitro derivative of brucine made by treating brucine with 10% nitric acid at 60-70°: Leuchs *et al.*, *Ber.* **43**, 1042 (1910). Structure: Teuber, *ibid.* **86**, 232 (1953). Used as reversible redox indicator for Sn^{+2} titrations: P. Szarvas, J. Lantos, *Talanta* **10**, 477 (1963).

Yellow crystals; sparingly sol in water.

USE: Indicator.

1611. Cactinomycin. [8052-16-2] Actinomycin C; HBF-386; Sanamycin. Antibiotic complex produced by *Streptomyces chrysomallus*: Brockmann, Grubhofer, *Naturwissenschaften* **36**, 376 (1949); **37**, 494 (1950); *Ber.* **84**, 260 (1951); US 2953495 (1960 to Shenley Inds.); Lindenmann, *Arch. Mikrobiol.* **17**, 361 (1952). Mixture of actinomycins C_1 (dactinomycin, *q.v.*), C_2 and C_3, 10%, 45% and 45%, resp: Brockmann, Pfennig, *Naturwissenschaften* **39**, 429 (1952); Brockmann, Gröne, *ibid.* **40**, 222 (1953). Description of other actinomycins: Waksman *et al.*, *Proc. Natl. Acad. Sci. USA* **44**, 602 (1958). Structures: Brockmann *et al.*, *Angew. Chem.* **68**, 70 (1956); Brockmann, Boldt, *Naturwissenschaften* **50**, 19 (1963). Synthesis of actinomycin C_3: Brockmann, Lackner, *ibid.* **47**, 230 (1960); **48**, 555 (1961); **51**, 407 (1964); Brockmann *et al.*, DE 1172680 (1964 to Bayer); Brockmann, Lackner, *Ber.* **100**, 353 (1967); **101**, 1312 (1968). Synthesis of actinomycin C_2: Brockmann, Lackner, *Tetrahedron Lett.* **1964**, 3517. Comprehensive review: H. Brockmann, in *Fortschr. Chem. Org. Naturst.* **18**, 1-54 (1960).

Actinomycin C_2	R = D-Valine	R' = D-Alloisoleucine
Actinomycin C_3	R = D-Alloisoleucine	R' = D-Alloisoleucine

Alizarin-red hexagonal bipyramids from ethyl acetate, mp 252°. $[\alpha]_D^{25}$ -325 to -349° (c = 0.25 in ethanol). Sparingly sol in water; moderately sol in ethanol; sol in chloroform, ethyl acetate, benzene, acetone. *Protect from light*.

Actinomycin C_2. $C_{63}H_{88}N_{12}O_{16}$; mol wt 1269.47. Red bipyramids, prisms or needles from ethyl acetate, mp 237-239°. $[\alpha]_D^{21}$ -325 ±10° (c = 0.23 in methanol). Abs max (methanol): 443 nm (ϵ 25400).

Actinomycin C_3. $C_{64}H_{90}N_{12}O_{16}$; mol wt 1283.49. Red hexagonal bipyramids from ethyl acetate or methanol, dec 235°. $[\alpha]_D^{17}$ -328° (c = 0.5 in ethanol). Absorption max (methanol): 443 nm (ϵ 24100). Weak base.

THERAP CAT: Antineoplastic.

1612. Cactus Grandiflorus. Night-blooming cereus; large-flowered cereus. Fresh, succulent stems of *Selenicereus grandiflorus* (L.) Britt. and Rose, Cactaceae. *Habit*. Tropical America. *Constit.* Cactine(?), acrid resinous glucoside, fat, wax. Extract containing combined principles from freshly tinctured stems and petals is known as *cactoid*.

THERAP CAT: Cardiotonic.

THERAP CAT (VET): Has been used as circulatory stimulant.

1613. Cadalene. [483-78-3] 1,6-Dimethyl-4-(1-methylethyl)-naphthalene; 4-isopropyl-1,6-dimethylnaphthalene; cadalin. $C_{15}H_{18}$; mol wt 198.31. C 90.85%, H 9.15%. Obtained from cadinene and other sesquiterpenes or sesquiterpene alcohols by dehydrogenation: Ruzicka, Meyer, *Helv. Chim. Acta* **4**, 505 (1921); Ruzicka, Stoll, *ibid.* **7**, 84 (1924). Synthesis: Ruzicka, Seidel, *ibid.* **5**, 369 (1922); Barnett, Cook, *J. Chem. Soc.* **1933**, 22; Johnson, Jones, *J. Am. Chem. Soc.* **69**, 792 (1947); Kohli *et al.*, *Experientia* **28**, 131 (1972).

Liquid. d_4^{25} 0.9667. bp_{720} 291-292°; bp_{10} 149°. n_D^{25} 1.5785. uv max: 228, 232, 280, 284, 295, 310, 317, 325 nm. Insol in water. Sol in fat solvents, oils.

1614. Cadaverine. [462-94-2] 1,5-Pentanediamine; pentamethylenediamine; animal coniine. $C_5H_{14}N_2$; mol wt 102.18. C 58.77%, H 13.81%, N 27.42%. Biogenic polyamine and homolog of putrescine, *q.v.*, produced by decarboxylation of lysine. Found in cholera discharge. Isoln: Bocklisch, *Ber.* **18**, 1922 (1885); Ackermann, *Z. Physiol. Chem.* **54**, 16 (1907). Prepn: Ladenburg, *Ber.* **18**, 2956 (1885). GC determn in foods: W. F. Staruszkiewicz, J. F. Bond, *J. Assoc. Off. Anal. Chem.* **64**, 584 (1981). Metabolism study of ^{14}C-cadaverine in rat brain: S. K. Salzman, M. Stepita-Klauco, *J. Neurochem.* **37**, 1308 (1981). Biosynthetic study in *S. ruminantium*: Y. Kamio *et al.*, *J. Biol. Chem.* **257**, 3326 (1982).

Colorless, syrupy liquid; characteristic odor. Strong base, fumes and attracts CO_2 on exposure to air. pKa$_1$ 10.25; pKa$_2$ 9.13. *Poisonous*. d_4^{20} 0.873. mp 9°. bp 178-180°. n_D^{20} 1.463. Sol in water, alcohol; slightly sol in ether. *Keep well closed*.

Dihydrochloride. [1476-39-7] $C_5H_{14}N_2$·2HCl; mol wt 175.10. Needles from water, mp 225-230°. Sol in water. Practically insol in abs alc.

Caution: Skin irritant and possible sensitizer.

1615. Cadexomer Iodine. Iodosorb. A hydrophilic modified starch polymer containing 0.9% (w/w) iodine within a helical matrix. Produced by the reaction of dextrin with epichlorohydrin coupled with ion exchange groups and iodine. Clinical use in venous ulcers: E. Skog *et al.*, *Br. J. Dermatol.* **109**, 77 (1983); M. C. Ormiston *et al.*, *Br. Med. J.* **291**, 308 (1985); L. Hillström, *Acta Chir. Scand. Suppl.* **544**, 53 (1988).

THERAP CAT: Vulnerary.

1616. Cadherins. Family of calcium-dependent cell adhesion molecules thought to be key regulators of morphogenesis in vertebrates. Transmembrane glycoproteins; mol wt ~120-140 kDa. Mediate tissue development by connecting cells bearing identical cadherin types via homophilic interactions. Loss of cadherin function has been associated with tumor metastasis. Two subfamilies have been identified: classic cadherins and the desmosomal cadherins, known as ***desmocollins*** and ***desmogleins***, which differ from the classic cadherins primarily in the structure of the cytoplasmic domains. A number of classic cadherins have been identified; nomenclature is based on the cell from which they were first identified, *e.g.* E-, N- and P-cadherins from epithelial, neural, and placental cells, resp. Among subclasses, ~50% of amino acids are conserved within a particular species. Classic cadherins are complexed with cytoplasmic proteins known as ***catenins*** which anchor them to the actin-based cytoskeleton and are required for adhesive function. Review of discovery and characterization of cadherins: M. Takeichi, *Development* **102**, 639-655 (1988); *idem, Annu. Rev. Biochem.* **59**, 237-252 (1990); of properties and role in morphogenesis: *idem, Science* **251**, 1451-1455 (1991); J. Behrens, *Acta Anat.* **149**, 165-169 (1994). Review of interaction with catenins: R. Kemler, *Trends Genet.* **9**, 317-321 (1993); L. Hinck *et al., Trends Biochem. Sci.* **19**, 538-542 (1994). Review of role in tumor metastasis: M. Takeichi, *Curr. Opin. Cell Biol.* **5**, 806-811 (1993); W. Birchmeier, J. Behrens, *Biochim. Biophys. Acta* **1198**, 11-26 (1994); O. W. Blaschuk *et al., Endocrine* **3**, 83-89 (1995). Review of desmosomal cadherins: P. J. Koch, W. W. Franke, *Curr. Opin. Cell Biol.* **6**, 682-687 (1994).

1617. Cadinenes. $C_{15}H_{24}$; mol wt 204.36. C 88.16%, H 11.84%. Sesquiterpenes occurring in essential oils from Juniper species and cedars (oil of cade). Nine possible isomers differing in stereochemistry and position of the double bonds, the principal isomer being β-cadinene: Sykora *et al., Chem. Listy* **52**, 1314 (1958). Prepn and structure: Campbell, Soffer, *J. Am. Chem. Soc.* **64**, 417 (1942); Campbell *et al., ibid.* **425**; Rao *et al., Tetrahedron Lett.* **1960**, 27; Herout *et al., Collect. Czech. Chem. Commun.* **31**, 3012 (1966). Synthesis: Soffer, Günay, *Tetrahedron Lett.* **1965**, 1355. Of the nine possible isomers, all able to yield (−)-cadinene dihydrochloride, six are known: Herout, Sykora, *Tetrahedron* **4**, 246 (1958); Kartha *et al., ibid.* **19**, 241 (1963). Structure and configuration of isomers: Sykora *et al., Collect. Czech. Chem. Commun.* **23**, 2181 (1958). Isoln of **α-cadinene** from Japanese hop: Y. Naya, M. Kotake, *Bull. Chem. Soc. Jpn.* **42**, 1468 (1969). Synthesis: O. P. Vig, *Indian J. Chem.* **21B**, 145 (1982). Brief review in *Rodd's Chemistry of Carbon Compounds* vol. 2, part C, S. Coffey, Ed. (Elsevier, New York, 1969) pp 268-270.

β-form

β-Cadinene. Oil. Slight pleasant odor. bp_9 124°. d_4^{20} 0.9239. n_D^{20} 1.5059. $[α]_D^{20}$ −251°.

1618. Cadmium. [7440-43-9] Cd; at. wt 112.411; at. no. 48; valence 2. Group IIB (12) element. Abundance in earth's crust: 0.1 to 0.2 ppm. Naturally occurring isotopes (mass numbers): 114 (28.73%); 112 (24.13%); 111 (12.80%); 110 (12.49%); 113 (12.22%), $T_{1/2}$ 9.3 × 10^15 yrs, β-emitter; 116 (7.49%); 106 (1.25%); 108 (0.89%); known artificial radioactive isotopes: 97-105, 107, 109, 115, 117-122, 124, 126. Found in zinc ores; also as CdS, greenockite; $CdCO_3$, otavite. Obtained in vapor form when roasting zinc ores, as sludge from zinc sulfate purification. Lab prepns from $CdSO_4$: Treadwell, *Helv. Chim. Acta* **4**, 551 (1921). Implicated as causative agent in Itai-Itai ("ouch-ouch") disease in Japan: Flick *et al., Environ. Res.* **4**, 71-85 (1971); Fassett, *Annu. Rev. Pharmacol.* **15**, 425-435 (1975). *Review:* Aylett "Group IIB" in *Comprehensive Inorganic Chemistry* vol. 3, J. C. Bailar Jr. *et al.*, Eds. (Pergamon Press, Oxford, 1973) pp 187-328; D. S. Carr in *Kirk-Othmer Ency-*

clopedia of Chemical Technology vol. 4 (John Wiley & Sons, New York, 4th ed., 1992) pp 748-760; N. Herron, *ibid.* 760-776. Review of carcinogenic risk: *IARC Monographs* **11**, 39-74 (1976); and toxicity: *Cadmium in the Human Environment: Toxicity and Carcinogenicity, IARC Scientific Publ.* **118**, G. F. Nordberg *et al.*, Eds. (1992) 469 pp; of toxicology and human exposure: *Toxicological Profile for Cadmium* (PB99-166621, 1999) 439 pp.

Silver-white, blue-tinged, lustrous metal; distorted hexagonal close-packed structure; easily cut with a knife; available in the form of bars, sheets or wire or a gray, granular powder. mp 321°. bp 765°. d^{25} 8.65. Specific heat at constant pressure (25°) 6.22 cal/mole deg. Slowly oxidized by moist air to form CdO. E° (aq) Cd/Cd^{2+} 0.4025 V. Insol in water. Reacts readily with dil HNO_3; reacts slowly with hot HCl; does not react with alkalies. Other reactions similar to those of zinc. Solns of cadmium salts and H_2S or Na_2S yield a yellow ppt insol in excess Na_2S.

Caution: Overexposure to cadmium and cadmium compounds has been associated with acute and chronic toxicity. Potential symptoms of acute poisoning from the inhalation of cadmium dusts or fumes include headache, chest pains, cough, metal fume fever, weakness. Potential symptoms of acute poisoning from ingestion of cadmium salts are GI disturbances, headache, muscular cramps, vertigo, convulsions. Chronic inhalation may cause pulmonary emphysema and chronic bronchitis. *See Clinical Toxicology of Commercial Products,* R. E. Gosselin *et al.*, Eds. (Williams & Wilkins, Baltimore, 5th ed., 1984) Section II, p 99; Section III, p 77-84. Potential toxic effects due to chronic overexposure by inhalation or ingestion are anemia; kidney damage; osteomalacia and osteoporosis. Itai-itai disease is a skeletal disease associated with a cadmium-induced renal disorder and is attributed to high oral intake of cadmium in food and water; characterized by progressive bone demineralization with painful joints and bones. *See Patty's Industrial Hygiene and Toxicology* vol. 2C, G. D. Clayton, F. E. Clayton, Eds. (John Wiley & Sons, New York, 4th ed., 1994) p 1954-1967. *See also NIOSH Pocket Guide to Chemical Hazards* (DHHS/NIOSH 97-140, 1997) p 44. Cadmium and cadmium compounds are listed as known human carcinogens: *Report on Carcinogens, Twelfth Edition* (PB2011-111646, 2011) p 80.

USE: Batteries, including Ni-Cd storage batteries; coating and electroplating steel and cast iron; pigments; plastic stabilizers; constituent of low melting or easily fusible alloys, e.g., Lichtenberg's, Abel's, Lipowitz', Newton's, and Wood's metal; electronics and optics; soft solder and solder for aluminum; reactor control rods; hardener for copper; catalytsts.

THERAP CAT (VET): Many cadmium salts, especially the oxide and anthranilate, are used or have been suggested as anthelmintics in swine and poultry.

1619. Cadmium Acetate. [543-90-8] Acetic acid cadmium salt (2:1). $C_4H_6CdO_4$; mol wt 230.50. C 20.84%, H 2.62%, Cd 48.77%, O 27.76%. $Cd(CH_3COO)_2$. Prepd by treating cadmium nitrate with acetic anhydride: Späth, *Monatsh. Chem.* **33**, 241 (1912); Wagenknecht, Juza, in *Handbook of Preparative Inorganic Chemistry* vol. 2, G. Brauer, Ed. (Academic Press, New York, 2nd ed., 1965) p 1105.

Dihydrate. Crystals. Slight acetic acid odor; becomes anhydr at about 130°. d 2.01. d (anhydr) 2.341. mp (anhydr) 255°. Freely sol in water; sol in alcohol. pH of 0.2 molar aq soln 7.1.

USE: Producing iridescent effects on porcelains and pottery; as a reagent for determination of S, Se, and Te; in cadmium electroplating.

1620. Cadmium Bromide. [7789-42-6] Br_2Cd; mol wt 272.22. Br 58.71%, Cd 41.29%. $CdBr_2$. Usually prepd from the elements: Honigschmid, Schlee, *Z. Anorg. Allg. Chem.* **227**, 184 (1936); Wagenknecht, Juza, in *Handbook of Inorganic Chemistry* vol. 2, G. Brauer, Ed. (Academic Press, New York, 2nd ed., 1965) p 1096.

Hexagonal, pearly flakes; highly hygroscopic. mp 566°; bp 963°; d 5.192. Freely sol in water, alc; moderately sol in acetone; slightly sol in ether. Crystallizes as the monohydrate below 36°, as the tetrahydrate above 36°.

USE: In photography, process engraving, and lithography.

1621. Cadmium Carbonate. [513-78-0] Carbonic acid cadmium salt (1:1). $CCdO_3$; mol wt 172.42. C 6.97%, Cd 65.20%, O

27.84%. CdCO₃. Occurs in nature as the mineral, *otavite*. Prepn: de Schulten, *Bull. Soc. Chim. Fr.* [3] **19**, 34 (1898); Biltz, *Z. Anorg. Allg. Chem.* **220**, 312 (1934).

Powder or rhombohedral leaflets. d 4.26. Practically insol in water. Sol in concd solns of ammonium salts, dil acids.

1622. Cadmium Chloride. [10108-64-2] Cadmium chloride (CdCl₂); Caddy; Vi-Cad. CdCl₂; mol wt 183.31. Cd 61.32%, Cl 38.68%. Prepn: *Gmelins, Cadmium* (8th ed.) **33**, pp 82-83 (1925); supplement, pp 464-465 (1959); Pray, *Inorg. Synth.* **5**, 153 (1957). Book: *Cadmium in the Environment, Part I: Ecological Cycling*, J. O. Nriagu, Ed. (Wiley-Interscience, New York, 1980) 696 pp. Crystal structure of hemipentahydrate: H. Leligny, J. C. Monier, *Acta Crystallogr.* **B31**, 728 (1975). Evaluation of carcinogenic risk: *IARC Monographs* **58**, 119-237 (1993).

Hygroscopic, rhombohedral crystals. mp 568°; bp 960°; d 4.05. Freely sol in water; sol in acetone; slightly sol in methanol, ethanol. Practically insol in ether. LD₅₀ orally in rats: 88 mg/kg (Nriagu).

Hemipentahydrate. [7790-78-5] CdCl₂.2½H₂O; mol wt 228.35. Efflorescent granules or rhombohedral leaflets. d²⁰ 2.84. Freely sol in water.

Caution: Cadmium and cadmium compounds are listed as known human carcinogens: *Report on Carcinogens, Twelfth Edition* (PB2011-111646, 2011) p 80.

USE: Photography; dyeing and calico printing; in the vacuum tube industry; manuf of cadmium yellow; galvanoplasty; manuf of special mirrors; as ice-nucleating agent; as lubricant; in analysis of sulfides to absorb the H₂S; in testing for pyridine bases; as fungicide; in pyrotechnics; in electroplating.

1623. Cadmium Cyanide. [542-83-6] C₂CdN₂; mol wt 164.45. C 14.61%, Cd 68.36%, N 17.03%. Cd(CN)₂. Prepn from Cd(OH)₂ and HCN: Biltz, *Z. Anorg. Allg. Chem.* **170**, 161 (1928); Wagenknecht, Juza in *Handbook of Preparative Inorganic Chemistry* **vol. 2**, G. Brauer, Ed. (Academic Press, New York, 2nd ed., 1965) p 1105.

Crystals or white powder. Turns brown on heating in air. *Poisonous.* d 2.226. Soly in water (15°) 1.71 g/100 ml. Slightly sol in alc; sol in solns of alkali cyanides or hydroxides. Not appreciably attacked by organic acids, but readily dec by dil mineral acids with evolution of hydrogen cyanide. Readily forms complex cyanides.

USE: In copper bright electroplating.

1624. Cadmium Fluoride. [7790-79-6] CdF₂; mol wt 150.41. Cd 74.74%, F 25.26%. Prepn from CdCl₂ and NH₄F: Kurtenacker *et al.*, *Z. Anorg. Chem.* **211**, 89 (1933); from CdCO₃ and HF: Kwasnik in *Handbook of Preparative Inorganic Chemistry* **vol. 1**, G. Brauer, Ed. (Academic Press, New York, 2nd ed., 1963) p 243. Cubic crystals. mp 1049°; bp 1748°; d 6.33. *Poisonous.* Soly in water (25°) 4.3 g/100 ml. Sol in HF and other mineral acids; practically insol in alcohol and liquid ammonia.

USE: Manufacture of phosphors, glass; in nuclear reactor controls.

1625. Cadmium Hydroxide. [21041-95-2] Cadmium hydrate. CdH₂O₂; mol wt 146.43. Cd 76.77%, H 1.38%, O 21.85%. Cd(OH)₂. Usually prepd from a Cd salt by treatment with KOH: de Schulten, *Compt. Rend.* **101**, 72 (1885); Fricke, Blaschke, *Z. Elektrochem.* **46**, 46 (1940); Scholder, Staufenbiel, *Z. Anorg. Allg. Chem.* **247**, 271 (1941).

Powder or trigonal and hexagonal crystals. d 4.79. Dehydration starts at 130° and is complete at 200°. Absorbs CO₂ from the air. Practically insol in water; slightly sol in NaOH soln; sol in dil acids, in NH₄OH or NH₄Cl solns.

USE: In storage battery electrodes.

1626. Cadmium Iodide. [7790-80-9] CdI₂; mol wt 366.22. Cd 30.69%, I 69.31%. Prepd from the elements or from the sulfate and KI: Wagenknecht, Juza in *Handbook of Preparative Inorganic Chemistry* **vol. 2**, G. Brauer, Ed. (Academic Press, New York, 2nd ed., 1965) p 1096.

Hexagonal, lustrous, flake-like crystals; becomes yellow on long exposure to air and light. mp 388°; bp 787°; d 5.67. Sol in water, alc, ether, acetone.

USE: In electrodeposition of Cd; as nematocide; in manuf of phosphors; as lubricant; in photoconductors; in photography, process engraving; lithography; in analytical chemistry.

1627. Cadmium Nitrate. [10325-94-7] Nitric acid cadmium salt (2:1). CdN₂O₆; mol wt 236.42. Cd 47.55%, N 11.85%, O 40.60%. Cd(NO₃)₂. Exists most commonly as tetrahydrate. Prepn: *Gmelins, Cadmium* (8th ed.) **33**, pp 76-78 (1925); suppl. p 446 (1959). Raman spectra of hydrated and anhydrous forms: D. W. James *et al.*, *Aust. J. Chem.* **31**, 1189 (1978); M. T. Carrick *et al.*, *ibid.* **36**, 223 (1983); and X-ray scattering in aqueous soln: R. Caminiti *et al.*, *J. Phys. Chem.* **88**, 2382 (1984). Thermal decompn of tetrahydrate: K. T. Wojciechowski, A. Malecki, *Thermochim. Acta* **331**, 73 (1999).

Tetrahydrate. [10022-68-1] CdN₂O₆.4H₂O; mol wt 308.48. Hygroscopic, orthorhombic crystals. mp 59.5°. Sol in 0.6 part water; sol in alc, acetone, ethyl acetate; practically insol in concd HNO₃. *Poisonous. Emits toxic Cd and NOx fumes when heated to dec.*

Caution: Cadmium and cadmium compounds are listed as known human carcinogens: *Report on Carcinogens, Twelfth Edition* (PB2011-111646, 2011) p 80. Direct contact may cause severe skin and moderate eye irritation.

USE: In making other Cd salts; in photographic emulsions.

1628. Cadmium Oxide. [1306-19-0] Aska-Rid. CdO; mol wt 128.41. Cd 87.54%, O 12.46%. Prepn: *Gmelins, Cadmium* (8th ed.) **33**, pp 69-70 (1925); supplement, pp 419-420 (1959). Toxicity data: Barrett *et al.*, *J. Ind. Hyg. Toxicol.* **29**, 279 (1947).

Dark-brown, infusible powder or cubic crystals. d 8.15. Practically insol in water. Sol in dil acids; slowly sol in ammonium salts. LC₅₀ in rats, monkeys (mg/m³): 500, ~15000 (Barrett).

Caution: Potential symptoms of overexposure to fumes are pulmonary edema, dyspnea, cough, chest tightness, substernal pain; headache; chills, muscle aches; nausea, vomiting, diarrhea; emphysema, proteinuria, anosmia; mild anemia. See *NIOSH Pocket Guide to Chemical Hazards* (DHHS/NIOSH 97-140, 1997) p 44. Cadmium and cadmium compounds are listed as known human carcinogens: *Report on Carcinogens, Twelfth Edition* (PB2011-111646, 2011) p 80.

USE: In phosphors, semiconductors; manuf of silver alloys, glass; in storage battery electrodes; as nematocide; as catalyst for organic reactions, in cadmium electroplating; in ceramic glazes.

THERAP CAT (VET): Has been used as an ascaricide in swine.

1629. Cadmium Potassium Cyanide. [14402-75-6] (*T*-4)-Tetrakis(cyano-*κC*)cadmate(2−) potassium (1:2); potassium tetracyanocadmate. C₄CdK₂N₄; mol wt 294.68. C 16.30%, Cd 38.15%, K 26.54%, N 19.01%. K₂Cd(CN)₄. Prepn: Dickinson, *J. Am. Chem. Soc.* **44**, 774 (1922); Biltz, *Z. Anorg. Allg. Chem.* **170**, 161 (1928).

Highly refractive, cubic crystals. d 1.846. mp ~450°. When heated, melts to a colorless liquid solidifying to a gray, cryst mass on cooling. Sol in 3 parts cold, 1 part boiling water; slightly sol in alc.

1630. Cadmium Salicylate. [19010-79-8] (*T*-4)-Bis[2-(hydroxy-*κO*)benzoato-*κO*] cadmium. C₁₄H₁₀CdO₆; mol wt 386.64. C 43.49%, H 2.61%, Cd 29.07%, O 24.83%. Cd(C₇H₅O₃)₂. Prepn: Prasad *et al.*, *J. Indian Chem. Soc.* **35**, 267 (1958).

Monohydrate. Small needles or plates. mp 242° (dec). Slightly sol in cold water, freely sol in boiling water; very slightly sol in methanol, ethanol.

THERAP CAT: Antiseptic.

1631. Cadmium Selenide. [1306-24-7] CdSe; mol wt 191.37. Cd 58.74%, Se 41.26%. Prepd by heating cadmium in a current of hydrogen selenide and subliming the product in hydrogen at a dull red heat: Margottet, *Recherches sur les Sulfures, les Séléniures, et les Tellures Métalliques*, Paris (1879); by passing hydrogen selenide over heated cadmium chloride and washing with warm water: Grzenkowsky, *Ueber Selenide und Erdalkaliferrocyanide* (Danzig, 1925); from cadmium sulfate and hydrogen selenide: Wagenknecht, Juza in *Handbook of Preparative Inorganic Chemistry* **vol. 2**, G. Brauer, Ed. (Academic Press, New York, 2nd ed., 1965) p 1099. Prepn of high purity CdSe: Taylor, Conn, US 3540859 (1970 to Merck & Co.).

White to brown cubic or hexagonal crystals; turns red in sunlight. d 5.8; mp 1350°. Dec in air or acids. Practically insol in water.

USE: In photoconductors, semiconductors, photoelectric cells, and rectifiers; in phosphors.

1632. Cadmium Sulfate. [10124-36-4] Sulfuric acid cadmium salt (1:1). CdO₄S; mol wt 208.47. Cd 53.92%, O 30.70%, S

15.38%. CdSO₄. Prepn: *Gmelins, Cadmium* (8th ed.) **33**, p 121 (1925); suppl. pp 609-610. Book: *Cadmium in the Environment, Part I: Ecological Cycling,* J. O. Nriagu, Ed. (Wiley-Interscience, New York, 1980) 696 pp. Crystal structure of hydrate: R. Caminiti, G. Johansson, *Acta Chem. Scand. A* **35**, 373, 451 (1981). ¹¹³Cd NMR study: P. D. Murphy, B. C. Gerstein, *J. Am. Chem. Soc.* **103**, 3282 (1981). Thermal decompn: H. Tagawa, K. Kawabe, *Thermochim. Acta* **158**, 293 (1990). Evaluation of carcinogenic risk: *IARC Monographs* **58**, 119-237 (1993). Use as a catalyst in organic synthesis: S. Tu *et al., J. Chem. Res. Synop.* **2003**, 544.

Hydrate. [7790-84-3] 8/3 Moles water per mole cadmium sulfate, odorless, monoclinic crystals. On heating loses water above 40°, forming monohydrate by 80°. Does not become anhydrous on further heating. d 3.09. Freely sol in water. Almost insol in ethanol, acetone, ammonia, ethyl acetate. *Poisonous. Emits toxic Cd and SOx fumes when heated to dec.*

Caution: Cadmium and cadmium compounds are listed as known human carcinogens: *Report on Carcinogens, Twelfth Edition* (PB2011-111646, 2011) p 80.

USE: In electrodeposition of Cd, Cu, and Ni; in phosphors; manuf of standard cadmium elements; catalyst in the Marsh test for As; determining H₂S and detecting fumaric acid; as nematocide, fungicide, bactericide; lubricant; electrolyte in Weston cell.

1633. Cadmium Sulfide. [1306-23-6] C.I. 77199; Capsebon. CdS; mol wt 144.47. Cd 77.81%, S 22.19%. Occurs in nature as the mineral *greenockite.* Prepd from CdSO₄ + H₂S: Milligan, *J. Phys. Chem.* **38**, 797 (1934); Frerichs, *Naturwissenschaften* **33**, 281 (1946); prepn of single crystals: Grillot, *Compt. Rend.* **230**, 1280 (1950); Czyzak *et al., J. Appl. Phys.* **23**, 932 (1952); Wagenknecht, Juza in *Handbook of Preparative Inorganic Chemistry* vol. **II**, G. Brauer, Ed. (Academic Press, New York, 2nd ed., 1965) pp 1098-1099. *See also Colour Index* vol. **4** (3rd ed., 1971) p 4659.

Light-yellow or orange-colored cubic or hexagonal crystals. Cubic structure: d 4.50; hexagonal structure: d 4.82. The light-yellow variety is also known as **Cadmium Yellow** or **Jaune Brilliant.** d 4.82. Sublimes at 980°. Soly in water (18°): 0.13 mg/100 g. Sol in concd or warm dil mineral acids with evolution of H₂S; readily dec and dissolved by moderately dil HNO₃.

Caution: Cadmium and cadmium compounds are listed as known human carcinogens: *Report on Carcinogens, Twelfth Edition* (PB2011-111646, 2011) p 80.

USE: As a pigment being fast to light and not affected by H₂S; color for soaps; coloring glass yellow; coloring textiles, paper, rubber; in printing inks, ceramic glazes, fireworks; in phosphors and fluorescent screens; in scintillation counters, semiconductors, photoconductors.

THERAP CAT: Dermatologic.

1634. Cadmium Telluride. [1306-25-8] CdTe; mol wt 240.01. Cd 46.84%, Te 53.16%. Prepd by fusion of the elements or by the action of H₂Te on CdCl₂ solns: Dennis, Anderson, *J. Am. Chem. Soc.* **36**, 887 (1914); Kroger, de Nobel, *J. Electron.* **1**, 190 (1955); Kretschmar, Schilberg, *J. Appl. Phys.* **28**, 865 (1957); from aluminum telluride and a cadmium salt: Nitsche, US **2767049** (1956 to du Pont). Prepn of high purity CdTe: Taylor, Conn, US **3540859** (1970 to Merck & Co.); prepn of single crystals: Kyle, US **3519399** (1970 to Hughes Aircraft).

Brownish-black, cubic crystals by sublimation in hydrogen. d₄¹⁵ 6.2. mp 1041°. Oxidizes upon prolonged exposure to moist air. Practically insol in water and acids, except nitric, in which it is sol with decompn.

USE: In semiconductor research, in phosphors.

1635. Cadmium Tungstate(VI). [7790-85-4] Cadmium tungsten oxide (CdWO₄). CdO₄W; mol wt 360.25. Cd 31.20%, O 17.76%, W 51.03%. CdWO₄. Prepn: Karl, *Compt. Rend.* **196**, 1403 (1933); prepn of single crystals: Uitert, Soden, *J. Appl. Phys.* **31**, 328 (1960).

White or yellowish monoclinic crystals or powder. Practically insol in water or dil acids. Sol in solns of alkali cyanides.

USE: In x-ray screens; in scintillation counters; in phosphors; as catalyst for organic reactions.

1636. Cadusafos. [95465-99-9] Phosphorodithioic acid O-ethyl S,S-bis(1-methylpropyl) ester; S,S-di-sec-butyl O-ethyl phos-

phorodithioate; ebufos; FMC-67825; Apache; Rugby; Taredan. C₁₀H₂₃O₂PS₂; mol wt 270.39. C 44.42%, H 8.57%, O 11.83%, P 11.46%, S 23.71%. Organophosphate insecticide structurally similar to ethoprop, *q.v.* Manufacturing process: J. M. Brochard *et al.,* EP **235056** (1987 to Rhone-Poulenc Agrochimie). Field trial as nematocide for bananas: P. Quénéhervé *et al., Rev. Nematol.* **14**, 251 (1991). Behavior in soils: S. Q. Zheng *et al., Sci. Total Environ.* **156**, 1 (1994).

Colorless to yellow liquid. Vapor press (25°): 120 mPa. Soly in water: 248 mg/l.

USE: Insecticide; nematocide.

1637. Cafestol. [469-83-0] (3bS,5aS,7R,8R,10aR,10bS)-3b,-4,5,6,7,8,9,10,10a,10b,11,12-Dodecahydro-7-hydroxy-10b-methyl-5a,8-methano-5aH-cyclohepta[5,6]naphtho[2,1-b]furan-7-methanol; cafesterol. C₂₀H₂₈O₃; mol wt 316.44. C 75.91%, H 8.92%, O 15.17%. Diterpenoid constituent of coffee. Isoln from green coffee oil: Slotta, Neisser, *Ber.* **71**, 1991, 2342 (1938); C. Djerassi *et al., J. Org. Chem.* **18**, 1449 (1953). Prepn and purification: R. Bertholet, US **4692534** (1987 to Nestec). Structure: C. Djerassi *et al., J. Am. Chem. Soc.* **81**, 2386 (1959); R. A. Finnegan, C. Djerassi, *ibid.* **82**, 4342 (1960). Stereochemical studies: R. A. Finnegan, *J. Org. Chem.* **26**, 3057 (1961); A. I. Scott *et al., J. Am. Chem. Soc.* **84**, 3197 (1962); A. I. Scott *et al., Tetrahedron* **20**, 1339 (1964). Stereospecific total synthesis of (±)-form: E. J. Corey *et al., J. Am. Chem. Soc.* **109**, 4717 (1987).

Crystals from hexane, mp 158°-160°. [α]D −101°. uv max: 222 nm (log ε 3.78).

Acetate. C₂₂H₃₀O₄. Needles from petr ether, mp 167-168°. [α]D −89°. uv max: 222 nm (log ε 3.80).

Tetrahydrocafestol. C₂₀H₃₂O₃. Crystals from dil methanol, mp 154.5-157°.

1638. Caffeic Acid. [331-39-5] 3-(3,4-Dihydroxyphenyl)-2-propenoic acid; 3,4-dihydroxycinnamic acid. C₉H₈O₄; mol wt 180.16. C 60.00%, H 4.48%, O 35.52%. Constituent of plants, probably occurs in plants only in conjugated forms, e.g., chlorogenic acid. Isoln from green coffee: Wolfrom *et al., J. Agric. Food Chem.* **8**, 58 (1960); from roasted coffee: Krasemann, *Arch. Pharm.* **293**, 721 (1960). Formation by acid hydrolysis of chlorogenic acid: Fiedler, *Arzneim.-Forsch.* **4**, 41 (1954); Whiting, Carr, *Nature* **180**, 1479 (1957); Guern, *C.A.* **61**, 9965h (1964). Synthesis: Hayduck, *Ber.* **36**, 2935 (1903); Posner, *J. Prakt. Chem.* **82**, 432 (1910); Mauthner, *ibid.* **142**, 33 (1935); Pandya *et al., Proc. Indian Acad. Sci.* **9A**, 511 (1939); Neish, *Can. J. Biochem. Physiol.* **37**, 1431 (1959). *Review:* Herrmann, *Pharmazie* **11**, 433 (1956).

Yellow crystals from concd aq solns. Monohydrate from dil solns. Dec 223-225° (softens at 194°). Rf values: Fiedler, *loc. cit.* Sparingly sol in cold water. Freely sol in hot water, cold alc. Alkaline solns turn from yellow to orange.

Methyl ester. C₁₀H₁₀O₄. Colorless needles from water, mp 152-153°.

1639. Caffeine. [58-08-2] 3,7-Dihydro-1,3,7-trimethyl-1*H*-purine-2,6-dione; 1,3,7-trimethylxanthine; 1,3,7-trimethyl-2,6-dioxopurine; coffeine; thein; guaranine; methyltheobromine; No-Doz. $C_8H_{10}N_4O_2$; mol wt 194.19. C 49.48%, H 5.19%, N 28.85%, O 16.48%. Naturally occurring alkaloid commonly found in tea leaves, coffee beans, cocoa beans, maté leaves, guarana paste and kola nuts. Originally isolated from coffee in 1821. Synthesis: E. Fischer, *Ber.* **15**, 453 (1882); E. Fischer, L. Ach, *ibid.* **28**, 3135 (1895); W. Traube, *ibid.* **33**, 1371, 3035 (1900); H. Bredereck *et al.*, *ibid.* **83**, 201 (1950). Crystal structure: D. J. Sutor, *Acta Crystallogr.* **11**, 453 (1958). Arrhythmogenic effects in humans: D. J. Dobmeyer *et al.*, *N. Engl. J. Med.* **308**, 814 (1983). Teratogenicity study: P. E. Palm *et al.*, *Toxicol. Appl. Pharmacol.* **44**, 1 (1978). HPLC determn in serum: D. T. Holland *et al.*, *J. Chromatogr. B* **707**, 105 (1998). Comprehensive description: M. U. Zubair *et al.*, *Anal. Profiles Drug Subs.* **15**, 71-150 (1986). Review of dietary sources: D. M. Graham, *Nutr. Rev.* **36**, 97-102 (1978); of clinical pharmacology: N. L. Benowitz, *Annu. Rev. Med.* **41**, 277-288 (1990); of CNS effects: A. Nehlig *et al.*, *Brain Res. Rev.* **17**, 139-170 (1992); of therapeutic uses: J. Sawynok, *Drugs* **49**, 37-50 (1996).

Hexagonal prisms by sublimation, mp 238°. Odorless with bitter taste. Sublimes 178°. Fast sublimation is obtained at 160-165° under 1 mm press. at 5 mm distance. d_4^{18} 1.23. pH of 1% soln 6.9. Absorption spectrum: Hartley, *J. Chem. Soc.* **87**, 1802 (1905). One gram dissolves in 46 ml water, 5.5 ml water at 80°, 1.5 ml boiling water, 66 ml alcohol, 22 ml alcohol at 60°, 50 ml acetone, 5.5 ml chloroform, 530 ml ether, 100 ml benzene, 22 ml boiling benzene. Freely sol in chloroform, pyrrole, tetrahydrofuran contg about 4% water; sol in ethyl acetate; sparingly sol in water, alc; slightly sol in petr ether. Soly in water is increased by alkali benzoates, cinnamates, citrates or salicylates. LD_{50} orally in mice, hamsters, rats, rabbits (mg/kg): 127, 230, 355, 246 (males); 137, 249, 247, 224 (females) (Palm).

Monohydrate. Felted needles, contg 8.5% H_2O. Efflorescent in air; complete dehydration takes place at 80°.

Mixture with citric acid. [69-22-7] Caffeine citrate; citrated caffeine; Cafcit. Clinical trial in treatment of apnea in premature neonates: P. B. Larsen *et al.*, *Acta Paediatr.* **84**, 360 (1995). White, crystalline powder; acid reaction. Sol in about 4 parts warm water.

THERAP CAT: CNS stimulant; respiratory stimulant.

THERAP CAT (VET): Cardiac and respiratory stimulant; diuretic.

1640. Calamus. Sweet flag; calmus; sweet cane; sweet grass. Perennial, iris-like plant, *Acorus calamus* L., *Araceae*. *Habit.* Europe, North America, Western Asia; cultivated in Burma and Ceylon. Several varieties exist and are characterized based on β-asarone content. Medicinal portion is the rhizome. *Constit.* Volatile oil (1.5-3.5%), resins (2.5%), tannins (1.5%); bitter principles such as acorin; mucilage, starch, sugars. Review of toxicology: M. J. Cupp, *Toxicology and Clinical Pharmacology of Herbal Products* (Humana Press, Totowa, 2000) pp 171-175; of medicinal uses and constituents: J. Barnes *et al.*, *Herbal Medicines* (Pharmaceutical Press, London, 2nd Ed., 2002) pp 100-102.

Volatile oil. [8015-79-0] Oil of sweet flag. Obtained by steam distillation of the fresh or unpeeled dried rhizome. *Constit.* β-Asarone, highly variable, 90-96% in Indian variety; calamenol (5%), calamene (4%), calamone, methyl eugenol, eugenol, sesquiterpenes. Toxicity study: J. M. Taylor *et al.*, *Toxicol. Appl. Pharmacol.* **10**, 405 (1967). Fragrance monograph: *Food Cosmet. Toxicol.* **15**, 623-626 (1977). Review of composition and uses: C. Singh *et al.*, *J. Med. Aromat. Plant Sci.* **23**, 687-708 (2001). Yellow to yellowish-brown viscid liquid; bitter taste. Soft, mellow fragrance reminiscent of patchouli. d_{20}^{20} 0.960-0.970. α_D^{20} +9 to +31°. n_D^{20} 1.507-1.515. Sapon no. 16-20. Very slightly sol in water; miscible with alcohol. LD_{50} orally in rats (Jammu variety): 777 mg/kg (Taylor). *Keep well closed, cool and protected from light.*

USE: In perfumery; as moisturizer in skin and hair preparations. Has been used as a flavoring ingredient and cooking spice.

THERAP CAT: Carminative, spasmolytic, diaphoretic.

1641. Calcifediol. [19356-17-3] (ε*R*,1*R*,3a*S*,4*E*,7a*R*)-4-[(2*Z*)-2-[(5*S*)-5-Hydroxy-2-methylenecyclohexylidene]ethylidene]-octahydro-α,α,ε,7a-tetramethyl-1*H*-indene-1-pentanol; 25-hydroxy-vitamin D₃; (3β,5*Z*,7*E*)-9,10-secocholesta-5,7,10(19)-triene-3,25-diol; 25-hydroxycholecalciferol; 25-HCC; U-32070E; Dedrogyl; Didrogyl; Hidroferol. $C_{27}H_{44}O_2$; mol wt 400.65. C 80.94%, H 11.07%, O 7.99%. The principal circulating form of vitamin D₃, formed in the liver by hydroxylation at C-25: Ponchon, DeLuca, *J. Clin. Invest.* **48**, 1273 (1969). It is the intermediate in the formation of 1α,25-dihydroxycholeciferol, *q.v.*, the biologically active form of vitamin D₃ in the intestine. Identification in rat as an active metabolite of vitamin D₃: Lund, DeLuca, *J. Lipid Res.* **7**, 739 (1966); Morii *et al.*, *Arch. Biochem. Biophys.* **120**, 513 (1967). Evaluation of biological activity in comparison with vitamin D₃: Blunt *et al.*, *Proc. Natl. Acad. Sci. USA* **61**, 717 (1968); *ibid.* 1503. Isoln from porcine plasma and establishment of structure: Blunt *et al.*, *Biochemistry* **7**, 3317 (1968). Synthesis: Blunt, DeLuca, *ibid.* **8**, 671 (1969). Review of isoln, identification and synthesis: DeLuca, *Am. J. Clin. Nutr.* **22**, 412 (1969). Review of bioassays: J. G. Haddad Jr., *Basic Clin. Nutr.* **2**, 579-597 (1980).

uv max (ethanol): 265 nm (ε 18000) (Blunt, DeLuca).

THERAP CAT: Calcium regulator.

1642. Calcimycin. [52665-69-7] 5-(Methylamino)-2-[[(2*R*,-3*R*,6*S*,8*S*,9*R*,11*R*)-3,9,11-trimethyl-8-[(1*S*)-1-methyl-2-oxo-2-(1*H*-pyrrol-2-yl)ethyl]-1,7-dioxaspiro[5.5]undec-2-yl]methyl]-4-benzox-azolecarboxylic acid; antibiotic A-23187; A-23187. $C_{29}H_{37}N_3O_6$; mol wt 523.63. C 66.52%, H 7.12%, N 8.02%, O 18.33%. Polyether antibiotic produced by a strain of *Streptomyces chartreusensis* Calhoun and Johnson NRRL 3882. Activity as a divalent cation ionophore in isolated mitochondria: P. W. Reed, H. A. Lardy, *J. Biol. Chem.* **247**, 6970 (1972). Prepn and antimicrobial activity: R. M. Gale *et al.*, *US* 3923823 (1975 to Lilly). Elucidation of structure: M. O. Chaney *et al.*, *J. Am. Chem. Soc.* **96**, 1932 (1974). Spectral studies of ionophore and metal ion complexes: D. R. Pfeiffer *et al.*, *Biochemistry* **13**, 4007 (1974). Total synthesis and absolute configuration: D. A. Evans *et al.*, *J. Am. Chem. Soc.* **101**, 6789 (1979); P. A. Grieco *et al.*, *J. Org. Chem.* **45**, 3537 (1980). Stereospecific synthesis: G. R. Martinez *et al.*, *J. Am. Chem. Soc.* **104**, 1436 (1982); D. P. Negri, Y. Kishi, *Tetrahedron Lett.* **28**, 1063 (1987). Review of cation binding and transport properties: D. R. Pfeiffer *et al.*, *Ann. N.Y. Acad. Sci.* **307**, 402-423 (1978). Use in model systems of calcium transport: M. Takamori *et al.*, *J. Neurol. Sci.* **50**, 89 (1981); M. Takamori *et al.*, *ibid.* **51**, 207 (1981); M. H. Freedman *et al.*, *Cell. Immunol.* **58**, 134 (1981); G. Thomas, *Eur. J. Pharmacol.* **81**, 35 (1982); V. L. Lew, J. Garcia-Sancho, *Cell Calcium* **6**, 15 (1985).

Crystalline solid, mp 181-182°. $[\alpha]_D^{25}$ −56° (c = 1 in chloroform). uv max (ethanol): 204, 225, 278, 378 nm (E 28200, 26200, 18200, 8200). pKa_1 6.9 in 90% DMSO. Slightly sol in water, readily sol in ethyl acetate, chloroform, methanol, DMSO. Also reported as mp 184.5-186° (Evans, 1979). LD_{50} i.p. in mice: 10 mg/kg (Gale).

Mixed calcium-magnesium salt. Colorless crystalline solid, mp 230-250° (dec). uv max (ethanol, neutral): 202, 228, 303, 370 nm ($E_{1cm}^{1\%}$ 425, 490, 278, 109). Insol in water, pentane, hexane, heptane. Very slightly sol in methanol, DMSO. Very sol in methylene chloride, chloroform, acetone, methyl ethyl ketone, diethyl ketone, ethyl acetate.

USE: Biochemical tool used to study the role of divalent cations in various biological systems.

1643. Calcineurin. [9025-75-6] Protein phosphoserine/phosphothreonine phosphatase; EC 3.1.3.16; phosphoprotein phosphatase. Ca^{2+}/calmodulin dependent ser-thr phosphatase that participates in many signalling pathways involved in gene regulation or biological responses to external stimuli in various organisms and cell types. Highly conserved heterodimer from yeast to humans comprised of a 58-69 kDa catalytic and calmodulin binding subunit A, *calcineurin A*, and a 16-19 kDa regulatory subunit B, *calcineurin B*. Isoln: J. H. Wang, R. Desai, *Biochem. Biophys. Res. Commun.* **72**, 926 (1976). Purification: C. B. Klee, M. H. Krinks, *Biochemistry* **17**, 120 (1978); and phosphatase activity: C. B. Klee *et al.*, *Proc. Natl. Acad. Sci. USA* **76**, 6270 (1979). Review of early work: C. B. Klee, J. Haiech, *Ann. N.Y. Acad. Sci.* **1980**, 43-54; of role in neuronal cells and brain injury: M. Morioka *et al.*, *Prog. Neurobiol.* **58**, 1-30 (1999). Review of structure and function: C. B. Klee *et al.*, *J. Biol. Chem.* **273**, 13367-13370 (1998); J. Aramburu *et al.*, *Curr. Top. Cell. Regul.* **36**, 237-295 (2000).

1644. Calcipotriene. [112965-21-6] (1R,3S,5Z)-5-[(2E)-2-[(1R,3aS,7aR)-1-[(1R,2E,4S)-4-Cyclopropyl-4-hydroxy-1-methyl-2-buten-1-yl]octahydro-7a-methyl-4H-inden-4-ylidene]ethylidene]-4-methylene-1,3-cyclohexanediol; (1α,3β,5Z,7E,22E,24S)-24-cyclopropyl-9,10-secochola-5,7,10(19),22-tetraene-1,3,24-triol; (1S,1'E,-3R,5Z,7E,20R)-9,10-seco-20-(3'-cyclopropyl-3'-hydroxyprop-1'-enyl)-1,3-dihydroxypregna-5,7,10(19)-triene; calcipotriol; MC-903; Daivonex; Dovonex; Psorcutan. $C_{27}H_{40}O_3$; mol wt 412.61. C 78.60%, H 9.77%, O 11.63%. Vitamin D_3 analog with low calcemic activity. Prepn: M. J. Calverley, E. T. Binderup, **WO 8700834**; *eidem*, **US 4866048** (1987, 1989 both to Leo Pharm.); M. J. Calverley, *Tetrahedron* **43**, 4609 (1987). Pharmacology: L. Binderup, E. Bramm, *Biochem. Pharmacol.* **37**, 889 (1988); P. J. Marie *et al.*, *Bone* **11**, 171 (1990). Receptor binding: T. Valaja *et al.*, *Biochem. Pharmacol.* **40**, 1827 (1990). Review of clinical trials in psoriasis: D. M. Ashcroft *et al.*, *Br. Med. J.* **320**, 963-967 (2000).

Crystals from methyl formate, mp 166-168°. uv max (96% ethanol): 264 nm (ε 17200).

THERAP CAT: Antipsoriatic.

1645. Calciseptine. [134710-25-1] Calciseptin (*Dendroaspis polylepis polylepis*). Peptide toxin, 60 amino acids long with 4 disulfide bridges, isolated from the venom of the black mamba snake (*Dendroaspis polylepis polylepis*); specific blocker of L-type Ca^{2+} channels. Isoln and characterization: J. R. De Weille *et al.*, *Proc. Natl. Acad. Sci. USA* **88**, 2437 (1991). Solution synthesis: H. Kuroda *et al.*, *Pept. Res.* **5**, 265 (1992). NMR 3d structure determn: H.

Haruyama *et al.*, *Bull. Magn. Reson.* **18**, 125 (1996). Analysis of residues involved in binding: R. M. Kini *et al.*, *Biochemistry* **37**, 9058 (1998). Differential sensitivities to L-channel subtypes: C. M. Rogers, E. R. Brown, *Exp. Physiol.* **86**, 689 (2001). Comparison with 1,4-dihydropyridines of effects on smooth muscle: T. X. Watanbe *et al.*, *Jpn. J. Pharmacol.* **68**, 305 (1995); on skeletal muscle: M. C. Garcia *et al.*, *J. Membr. Biol.* **184**, 121 (2001).

USE: Biochemical probe for L-type Ca^{2+} channels.

1646. Calcitonin. [9007-12-9] Thyrocalcitonin; TCA; TCT. Calcium regulating hormone secreted from the mammalian thyroid gland and in non-mammalian species from the ultimobranchial gland. Postulation of a plasma-calcium lowering substance: Copp *et al.*, *Endocrinology* **70**, 638 (1962). Recognition as a hormone: Hirsch *et al.*, *ibid.* **73**, 244 (1963); of thyroid origin: G. V. Foster *et al.*, *Nature* **202**, 1303 (1964). Over-all action is to oppose the bone and renal effects of parathyroid hormone, *q.v.*; inhibits bone resorption of Ca^{2+}, with accompanying hypocalcemia and hypophosphatemia and decreased urinary Ca^{2+} concentrations. Also abolishes the osteolytic effect of toxic doses of vitamins A and D. Calcitonin structures are single polypeptide chains containing 32 amino acid residues. Structure of porcine: J. T. Potts *et al.*, *Proc. Natl. Acad. Sci. USA* **59**, 1321 (1968); P. H. Bell *et al.*, *J. Am. Chem. Soc.* **90**, 2704 (1968); *eidem*, *Biochemistry* **9**, 1665 (1970). Synthesis of porcine: St. Guttmann *et al.*, *Helv. Chim. Acta* **51**, 1155 (1968). Isoln of human calcitonin from non-pathological thyroid glands: A. Haymovits, J. F. Rosen, *Endocrinology* **81**, 993 (1967); from medullary carcinoma of the thyroid: R. Neher *et al.*, *Nature* **220**, 984 (1968); B. Riniker *et al.*, *Helv. Chim. Acta* **51**, 1738 (1968). Structure of human: R. Neher *et al.*, *ibid.* 1900. Synthesis of human: P. Sieber *et al.*, *ibid.* 2057; J. Hirt *et al.*, *Rec. Trav. Chim.* **98**, 143 (1979). Amino acid sequence differs among mammalian species, salmon calcitonin showing a marked difference from that of the higher vertebrae as well as a more potent biological activity. Mechanism of action: E. M. Brown, G. D. Aurbach, *Vitam. Horm.* **38**, 236 (1980). Review of early literature: Munson, Hirsch, *Clin. Orthop.* **49**, 209 (1966). Review of isoln, structure, synthesis: Behrens, Grinnan, *Annu. Rev. Biochem.* **38**, 83 (1969); Potts *et al.*, *Vitam. Horm.* **29**, 41 (1971). Comprehensive review: *Calcitonin, Proc. Symp. on Thyrocalcitonin and the C Cells*, S. Taylor, Ed. (Springer-Verlag, New York, 1968); Foster *et al.*, "Calcitonin" in *Clinics in Endocrinology and Metabolism*, I. MacIntyre, Ed. (W. B. Saunders, Philadelphia, 1972) pp 93-124. Review of pharmacology and therapeutic use: J. C. Stevenson, I. M. A. Evans, *Drugs* **21**, 257-272 (1981).

Calcitonin, human synthetic. [21215-62-3] Cibacalcin.

Calcitonin, salmon synthetic. [47931-85-1] Salcatonin; Calcimar; Calsyn; Calsynar; Karil; Miacalcic; Miacalcin; Miadenil; Salmotonin; Tonocalcin. Clinical trial in postmenopausal osteoporosis: C. H. Chesnut *et al.*, *Am. J. Med.* **109**, 267 (2000). LC determn in biological fluids: M. Aguiar *et al.*, *J. Chromatogr. B* **818**, 301 (2005). Review of therapeutic potential in acute pain: E. J. Visser, *Acute Pain* **7**, 185-189 (2005). *See also* Elcatonin.

THERAP CAT: Calcium regulator. Treatment of Paget's disease.

THERAP CAT (VET): Calcium regulator.

1647. Calcitriol. [32222-06-3] (1R,3S,5Z)-4-Methylene-5-[(2E)-2-[(1R,3aS,7aS)-octahydro-1-[(1R)-5-hydroxy-1,5-dimethylhexyl]-7a-methyl-4H-inden-4-ylidene]ethylidene]-1,3-cyclohexanediol; (1α,3β,5Z,7E)-9,10-secocholesta-5,7,10(19)-triene-1,3,25-triol; 1α,25-dihydroxycholecalciferol; 1α,25-dihydroxyvitamin D_3; 1,25-DHCC; Ro-21-5535; Calcijex; Rocaltrol; Silkis. $C_{27}H_{44}O_3$; mol wt 416.65. C 77.83%, H 10.64%, O 11.52%. The biologically active form of vitamin D_3, *q.v.*, in intestinal calcium transport and bone calcium resorption. First obtained from chick intestine and designated as metabolite 4B: J. F. Myrtle *et al.*, *J. Biol. Chem.* **245**, 1190 (1970); M. R. Haussler *et al.*, *Proc. Natl. Acad. Sci. USA* **68**, 177 (1971). Isoln and identification: M. F. Holick *et al.*, *ibid.* 803; D. E. M. Lawson *et al.*, *Nature* **230**, 228 (1971); A. W. Norman *et al.*, *Science* **173**, 51 (1971). Synthesis: D. H. R. Barton *et al.*, *J. Chem. Soc. Chem. Commun.* **1974**, 203; T. Sato *et al.*, *Chem. Pharm. Bull.* **26**, 2933 (1978). Effect of calcium on *in vivo* synthesis: I. T. Boyle *et al.*, *Proc. Natl. Acad. Sci. USA* **68**, 2131 (1971). Stimulation of bone resorption in tissue culture: L. G. Raisz *et al.*, *Science* **175**, 768 (1972). Classification as a steroid hormone: J. S. Emtage *et al.*, *Nature* **246**, 100 (1973). Pharmacokinetics: J. R. Muindi *et al.*, *Clin. Pharmacol. Ther.* **72**, 648 (2002). Clinical evaluation in prostate cancer: C. Gross *et al., J. Urol.* **159**, 2035 (1998); in secondary hyperparathyroidism: S. Koshikawa *et al., Nephron* **90**, 413 (2002). Comprehensive description: T. Suda, *Vitamins* **45**, 175-188 (1972); E. Debesis, *Anal. Profiles Drug Subs.* **8**, 83-100 (1979). Review of clinical efficacy in osteoporosis: K. L. Dechant, K. L. Goa, *Drugs Aging* **5**, 300-317 (1994); of mechanism of action in renal failure: C. H. Hsu *et al., Kidney Int.* **46**, 605-612 (1994); of clinical experience in treatment of psoriasis: L. Kowalzick, *Br. J. Dermatol.* **144**, Suppl. 58, 21-25 (2001).

White crystalline powder, mp 111-115°. uv max (abs ethanol): 264 nm (ε 19000). $[\alpha]_D^{25}$ +48° (methanol). Freely sol in alc; sol in ether, fatty oils; slightly sol in methanol, ethanol, ethyl acetate, THF. Practically insol in water. Air, heat and light sensitive.

THERAP CAT: Calcium regulator; vitamin (antirachitic); antihyperparathyroid; antineoplastic; antipsoriatic.

1648. Calcium. [7440-70-2] Ca; at. wt 40.078; at. no. 20; valence 2. Group IIA (2). Alkaline earth metal. Occurrence in the earth's crust 3.64% (fifth element in order of abundance). Sea water contains about 400 g/ton. Naturally occurring isotopes: 40 (96.941%), 44 (2.086%), 42 (0.647%), 48 (0.187%), 43 (0.135%), 46 (0.004%). Known radioactive isotopes: 35-39, 41, 45, 47-51. Found naturally only in the form of its compds, never uncombined, in minerals such as limestone, dolomite, marble, chalk, iceland spar, gypsum, anhydrite, fluorite, apatite. Principal commercial source is limestone, *q.v.* Major commercial production by high temperature vacuum reduction of calcium oxide in aluminothermal process; less commonly by electrolysis followed by redistillation. Essential constituent of bones, shells, teeth, coral, pearls. Essential nutrient for animal life. First isolated by Davy in 1808. Produced by electrolysis of calcium chloride: Rathenau, Suter, **DE 155433** (1903); *Z. Elektrochem.* **10**, 502 (1904); Goodwin, *J. Am. Chem. Soc.* **27**, 1403 (1905); also by thermal reduction of lime with silicon, or with aluminum. Prepn of the pure metal for laboratory use: Whaley, *Inorg. Synth.* **6**, 18 (1960). Purifn of commercial material: Marshall, Whaley, *ibid.* 24. *Reviews:* Schaufler in *Ullmanns Encyklopädie der technischen Chemie* **vol. 4** (Munich, 3rd ed., 1953) pp 830-836; Mantell in C. A. Hampel, *Rare Metals Handbook* (Reinhold, New York, 1954) p 17-29. Review of calcium and its compounds: Goodenough, Stenger, "Magnesium, Calcium, Strontium, Barium and Radium" in *Comprehensive Inorganic Chemistry* **Vol. 1**, J. C. Bailar Jr. *et al.*, Eds. (Pergamon Press, Oxford, 1973) pp 591-664; *Chemistry of the Elements*, N. N. Greenwood, A. Earnshaw, Eds. (Pergamon Press, New York, 1984) pp 117-154; R. L. Petersen, M. B. Freilich in *Kirk-Othmer Encyclopedia of Chemical Technology* **vol. 4** (Wiley-Interscience, New York, 4th ed., 1992) pp 777-796. Book: R. P. Rubin, *Calcium and Cellular Secretion* (Plenum, New York, 1982) 276 pp.

Lustrous, silver-white surface (when freshly cut); face-centered cubic structure, transforms to body-centered cubic form at 428 ±2°. Soft, ductile. Brinell hardness: 17. Much harder than sodium, but softer than aluminum or magnesium. Oxidizes and acquires bluish-gray tarnish on exposure to moist air. d_4^{20} 1.54. mp 839°. bp 1484°. *Dangerous when wet.* Electrical resistivity at 20°: 3.5 μohm cm. Heat of combustion 151.9 cal/g. Specific heat (0-100°) 0.149 cal/g. Considerably less reactive than sodium. E° (aq) Ca^{2+}/Ca −2.87 V. Reacts with water, alcohols, and dil acids with evolution of hydrogen, creating explosion hazard. Reacts with halogens. Dissolves in liquid ammonia to form a blue black soln. Ignites in air when finely

divided. Emits characteristic orange-red color in flame. Insol in and inert towards benzene, kerosene.

USE: Reducing agent for production of less common metals; alloying agent to increase strength and corrosion resistance in lead, to improve mechanical and electrical properties in aluminum; refining agent to remove bismuth from lead. In metallurgy as scavenger to deoxidize, desulfurize and degas steel and cast iron; to control nonmetallic inclusions in steel; to promote uniform microstructure in gray iron. As anode material in thermal batteries; as "getter" for oxygen and nitrogen.

1649. Calcium Acetate. [62-54-4] Acetic acid calcium salt (2:1); Phos-ex; PhosLo. $C_4H_6CaO_4$; mol wt 158.17. C 30.37%, H 3.82%, Ca 25.34%, O 40.46%. $Ca(CH_3COO)_2$. The technical product was originally known as *brown acetate of lime* or *gray acetate of lime*. Prepn and analysis of acetate of lime: Stillwell, Gladding, *J. Am. Chem. Soc.* **4**, 94 (1882); T. S. Gladding, *Ind. Eng. Chem.* **1**, 250 (1909). Distillation to produce acetone: E. G. R. Ardagh *et al., ibid.* **16**, 1133 (1924). Electrolytic prepn: H. Schmidt, *Z. Anorg. Allg. Chem.* **270**, 188 (1952). Analysis of hydrated forms: J. Panzer, *J. Chem. Eng. Data* **7**, 140 (1962). Toxicity study: Smyth *et al., Am. Ind. Hyg. Assoc. J.* **30**, 470 (1969). Clinical evaluation in hyperphosphatemia: W. Y. Qunibi *et al., Kidney Int.* **65**, 1914 (2004). Very hygroscopic, rod-shaped crystals. On heating above 160° dec to acetone and $CaCO_3$. d 1.50. Freely sol in water; slightly sol in methanol. Practically insol in ethanol, dehydrated alc, acetone, benzene. *Keep well closed.*

Monohydrate. [5743-26-0] $C_4H_6CaO_4.H_2O$; mol wt 176.18. Needles, granules or powder. Does not lose all its water below 150°. Sol in water; slightly sol in alcohol. pH of 0.2 molar aq soln 7.6. LD_{50} orally in rats: 4.28 g/kg (Smyth).

USE: Manuf of acetic acid, acetone; in dyeing, tanning, and curing skins; in lubricants; as food stabilizer; as corrosion inhibitor. Monohydrate used in precipitation of oxalates.

THERAP CAT: Antihyperphosphatemic.

1650. Calcium Acetylsalicylate. [69-46-5] 2-(Acetyloxy)benzoic acid calcium salt (2:1); salicylic acid acetate calcium salt; acetylsalicylic acid calcium salt; calcium aspirin; soluble aspirin; Cal-Aspirin; Dispril; Disprin; Kalmopyrin; Kalsetal; Solaspin; Tylcalsin. $C_{18}H_{14}CaO_8$; mol wt 398.38. C 54.27%, H 3.54%, Ca 10.06%, O 32.13%. $Ca(OOCC_6H_4OCOCH_3)_2$. Prepd from calcium carbonate and acetylsalicylic acid: Lawrence, **US 2003374** (1935 to Lee Labs.).

Dihydrate. Amorphous, nonhygroscopic powder. One gram dissolves in 6 ml water, 80 ml alcohol.

Complex with urea. [5749-67-7] Carbasalate calcium; acetylsalicylic acid calcium salt complex with urea; calcium acetylsalicylate carbamide; urea calcium acetylsalicylate; carbaspirin calcium; Alcacyl; Ascal; Calurin; Cardiosolupsan; Solupsan. $C_{19}H_{18}$-CaN_2O_9; mol wt 458.44. Description and hydrolytic stability: Parrott, *J. Pharm. Sci.* **51**, 897 (1962). Clinical trial in migraine: C. Le Jeunne *et al., Eur. Neurol.* **41**, 37 (1999). Amorphous powder, dec 243-245°. Soly in water at 37°: 231 mg/ml, pH 4.8.

THERAP CAT: Analgesic; antipyretic; anti-inflammatory.

1651. Calcium Aluminosilicate. [1327-39-5] Silicic acid aluminum calcium salt ; aluminum calcium silicate. Many different forms of calcium aluminosilicate are known, the most common of which are $CaAl_2Si_2O_8$ and $Ca_2Al_2SiO_7$. It occurs in nature as the minerals: *anorthite, bavenite, clinozoisite, didymolite, epistilbite, gehlenite, gismondite, grossularite, heulandite, hibschite, laubanite, laumontite, lawsonite, levynite, margarite, meionite, plazolite, pumpellyite, scolecite, stellerite, vesuvianite, zoisite*. Prepn and properties: *Gmelins, Aluminum* (8th ed.) **35B**, 576-586 (1934). Summary and references for minerals: Hey, *An Index of Mineral Species and Varieties* (British Museum, London, 2nd ed., 1962) pp 159-162.

USE: Constituent of cement; in refractories.

1652. Calcium Arsenate. [7778-44-1] Arsenic acid (H_3-AsO_4) calcium salt (2:3); tricalcium arsenate. $As_2Ca_3O_8$; mol wt 398.07. As 37.64%, Ca 30.20%, O 32.15%. $Ca_3(AsO_4)_2$. Prepn: Les Veaux, **US 2715562** (1955 to FMC). Toxicity study: T. B. Gaines, *Toxicol. Appl. Pharmacol.* **14**, 515 (1969). *Review:* Guerin, *Chim. Ind. (Paris)* **77**, 1288 (1957). Review of toxicology and human exposure: *Toxicological Profile for Arsenic* (PB2008-100002, 2007) 559 pp.

Powder. *Poisonous.* d 3.620. Sol in dil acids; slightly sol in water. Insol in organic acids. LD_{50} orally in female rats: 298 mg/kg (Gaines).

Caution: Potential symptoms of overexposure are weakness; GI disturbance; peripheral neuropathy; skin hyperpigmentation, palmar planter hyperkeratoses; dermatitis. Potential occupational carcinogen. *See NIOSH Pocket Guide to Chemical Hazards* (DHHS/NIOSH 97-140, 1997) p 46. Arsenic and inorganic arsenic compounds are listed as known human carcinogens: *Report on Carcinogens, Twelfth Edition* (PB2011-111646, 2011) p 50.

USE: Insecticide; molluscicide.

1653. Calcium Arsenite. [52740-16-6] Arsonic acid calcium salt (1:1). Variable composition. Prepd by passing steam over a dry mixture of CaO and As_2O_3: Altwegg, Dutel, **US 1700756** (1929).

White, granular powder. *Poisonous.* Slightly sol in water; sol in acids.

USE: As insecticide, germicide, molluscicide.

1654. Calcium Bisulfite, Solution. [13780-03-5] Sulfurous acid calcium salt (2:1). $Ca(HSO_3)_2$ is known only in soln. Prepn from sulfite liquor: Arend, *Chem. Prod.* **10**, 53 (1947); Lougheed, *Pulp Paper Mag. Can.* **49**(3), 215 (1948); Schoeffel, **US 2696424** (1954 to Sterling Drug). The product here described is substantially a soln of calcium sulfite in an aq sulfur dioxide soln.

Colorless or slightly yellow liquid; strong SO_2 odor. On standing in the air, crystals of $CaSO_3.2H_2O$ form. d about 1.06. This soln corrodes metals.

USE: As germicide, preservative, and disinfectant; for washing (1:1000) casks in brewing to prevent souring and cloudiness of beer and to prevent secondary fermentation; as antichlor in bleaching fabrics; largely in manuf sulfite cellulose from wood for paper-making.

1655. Calcium Borate. [12007-56-6] Boron calcium oxide (B_4CaO_7); calcium pyroborate; calcium tetraborate. B_4CaO_7; mol wt 195.31. B 22.14%, Ca 20.52%, O 57.34%. CaB_4O_7. Prepn by direct fusion of B_2O_3 with $CaCO_3$: Griveau, *Compt. Rend.* **166**, 993 (1918). Various calcium borate minerals occur in nature. These include: *colemanite, ginorite, inyoite, meyerhofferite, pandermite, priceite.*

Hexahydrate. Powder. Almost insol in cold, moderately sol in hot water; sol in dil acids.

USE: As flux in heavy-metal metallurgy; mfr of forsterite porcelain insulators; in glycol antifreeze; in fire-retardant paint.

1656. Calcium Borogluconate. [5743-34-0] D-Gluconic acid cyclic 4,5-ester with boric acid (H_3BO_3) calcium salt (2:1); calcium diborogluconate. $C_{12}H_{20}B_2CaO_{16}$; mol wt 481.97. C 29.90%, H 4.18%, B 4.49%, Ca 8.32%, O 53.11%. Prepn and use in milk fever: H. Dryerre, J. R. Greig, *Vet. Rec.* **15**, 456 (1935); H. T. Macpherson, J. Stewart, *Biochem. J.* **32**, 76 (1938). Physical properties: L. Seekles, E. Havinga, *Nor. Veterinaertidsskr.* **58**, No. 11, 433-445 (1946), *C.A.* **43**, 1907 (1949). Effect on plasma calcium concentration in sheep: D. A. H. Farningham, *Res. Vet. Sci.* **39**, 70 (1985). Spectrophotometric determn in plasma: D. J. Lyons, K. P. Spann, *J. Assoc. Off. Anal. Chem.* **68**, 160 (1985). Evaluation in bovine milk fever: P. A. Mullen, *Vet. Rec.* **97**, 87 (1975).

Crystals, freely sol in water. Soly in water 1:1 at 15°: 2.8:1 at 100°. A 20% aq soln has a pH of 3.5. If desired, the pH may be adjusted to 7.0 by the addition of CaO which is very sol in Ca borogluconate solns.

THERAP CAT (VET): In hypocalcemic states including bovine milk fever.

1657. Calcium Bromide. [7789-41-5] Br_2Ca; mol wt 199.89. Br 79.95%, Ca 20.05%. $CaBr_2$. Prepn: *Gmelins, Calcium* (8th ed.) **28B**, 100-102, 584-599 (1958). The N.F. article is a hydrated salt, contg not less than 84% and not more than 94% $CaBr_2$.

Odorless, deliquesc granules or rhombic crystals; sharp, saline taste. Becomes yellow on long exposure to air. When anhyd mp 730°; d_4^{25} 3.353. When strongly heated in air, becomes alkaline due to loss of bromine and formation of lime. Very sol in water, methanol, ethanol; sol in acetone; practically insol in dioxane, chloroform, ether. The aq soln is neutral or only slightly alkaline to litmus. *Keep well closed.*

USE: In photography for making dry plates and light-sensitive papers; manuf mineral waters, NH_4Br, fire-extinguishing compositions.

THERAP CAT: Sedative, anticonvulsant.

THERAP CAT (VET): Has been used in hypocalcemic states such as canine eclampsia.

1658. Calcium Carbide. [75-20-7] Acetylenogen. C_2Ca; mol wt 64.10. C 37.48%, Ca 62.52%. CaC_2. Prepn: Ehrlich in *Handbook of Preparative Inorganic Chemistry* **Vol. 1**, G. Brauer, Ed. (Academic Press, New York, 2nd ed., 1963) p 943. *Review:* Brennan, *J. Electrochem. Soc.* **99**, 61c (1952).

Grayish-black, irregular lumps or orthorhombic crystals; dec by water with evolution of acetylene leaving a residue of lime. d 2.22; mp 2300°. *Dangerous when wet.*

USE: Generating acetylene gas for lighting purposes (1 kg of carbide yields ~300 liters acetylene); as reducing agent, e.g., for direct reduction of copper sulfide to metallic copper; signal fires for marine service; manuf of calcium, iron, alloys, lampblack, cyanamide; welding and cutting metals.

1659. Calcium Carbonate. [471-34-1] Carbonic acid calcium salt (1:1); Cacit; Calcichew; Calcidia; Caltrate; Citrical; Fixical; Maalox Quick Dissolve; Maalox Regular Strength; Os-Cal; Tums Smooth Dissolve. $CCaO_3$; mol wt 100.09. C 12.00%, Ca 40.04%, O 47.95%. $CaCO_3$. Exists in nature as the minerals *aragonite, calcite* and *vaterite.* Solubility and scale formation study: J.-Y. Gal *et al., Talanta* **43**, 1497 (1996). Clinical evaluation in protease inhibitor-induced diarrhea in HIV: M. J. Turner *et al., HIV Clin. Trials* **5**, 19 (2004). Review of commercial applications in paints, coatings and adhesives: R. D. Athey, *Eur. Coatings J.* **5**, 256-263 (1994).

Odorless, tasteless powder or crystals. Two crystal forms are of commercial importance: Aragonite, orthorhombic, mp 825° (dec), d 2.83, formed at temps above 30°; Calcite, hexagonal-rhombohedral, mp 1339° (102.5 atm), $d^{25.2}$ 2.711, formed at temps below 30°. At about 825° is dec into CaO and CO_2. Sol in 1N acetic acid, 3N hydrochloric acid, 2N nitric acid. Soly in water increased by the presence of ammonium salt or carbon dioxide. Soly in water reduced by presence of alkali hydroxide. Practically insol in water. Insol in alc.

Precipitated calcium carbonate. Precipitated chalk; Aeromatt; Albacar; Purecal. Commercial $CaCO_3$ produced by chemical means. It is 98-99% pure. The byproduct process, the carbonation process, and the calcium chloride process of manuf from limestone are outlined in *Kirk-Othmer Encyclopedia of Chemical Technology* **vol. 4** (Interscience, New York, 2nd ed., 1964) pp 7-11. *Review:* Woerner in *Pigment Handbook* **vol. 1**, T. C. Patton, Ed. (John Wiley, New York, 1973) pp 119-128.

Prepared calcium carbonate. Drop chalk; prepared chalk; whiting; English white; Paris white. Native $CaCO_3$ purified by elutriation.

Caution: Potential symptoms of overexposure are irritation of eyes, skin, respiratory system; cough. *See NIOSH Pocket Guide to Chemical Hazards* (DHHS/NIOSH 97-140, 1997) p 46.

USE: Manuf of paint, rubber, plastics, paper, dentifrices, ceramics, putty, polishes, insecticides, inks, shoe dressings; as a filler in production of adhesives, matches, pencils; crayons, linoleum, insulating compds, welding rods. In foods, cosmetics, antibiotics; in pharmaceuticals as tablet and capsule diluent; removing acidity of wines. In analytical chemistry for detecting and determining halogens in organic combinations; with NH_4Cl for decomposing silicates; preparing $CaCl_2$ soln for standardizing soap solns; for water analyses; as a chelometric standard.

THERAP CAT: Antacid. Calcium supplement.

THERAP CAT (VET): Antacid, calcium supplement, antidiarrheal agent.

1660. Calcium Chlorate. [10137-74-3] Chloric acid calcium salt (2:1). $CaCl_2O_6$; mol wt 206.97. Ca 19.36%, Cl 34.26%, O 46.38%. $Ca(ClO_3)_2$. Prepn: Wilderman, **GB 183671** (1921); Duveau, *Bull. Soc. Chim.* **10**, 374 (1943).

Dihydrate. [10035-05-9] Monoclinic, hygroscopic crystals. d 2.711. mp 100° when rapidly heated. Sol in 0.6 part water; sol in alcohol. *Oxidizer. Keep well closed.*

USE: Herbicide, insecticide, seed disinfectant.

1661. Calcium Chloride. [10043-52-4] Intergravin-orales. $CaCl_2$; mol wt 110.98. Ca 36.11%, Cl 63.89%. Forms mono-, di-, tetra- and hexahydrates. Obtained as a byproduct from the ammonia-soda (Solvay) process and as a joint product from natural salt brines: *Faith, Keyes & Clark's Industrial Chemicals*, F. A. Lowenheim, M. K. Moran, Eds. (Wiley-Interscience, New York, 4th ed., 1975) pp 186-190. Crystal structure of dihydrate: A. Leclaire, M. M. Borel, *Acta Crystallogr.* **B33**, 1608 (1977). Acute toxicity: I. B. Syed, F. Hosain, *Toxicol. Appl. Pharmacol.* **22**, 150 (1972).

Cubic crystals, granules or fused masses. Very hygroscopic. mp 772°. bp >1600°; d_4^{15} 2.152. Freely sol in water (with liberation of much heat), alcohol. The commercial product is about 94-97% $CaCl_2$, the chief impurity being $Ca(OH)_2$. *Keep well closed.* LD_{50} i.v. in mice: 42.2 mg/kg (Syed, Hosain).

Dihydrate. [10035-04-8] $CaCl_2.2H_2O$; mol wt 147.01. Hygroscopic granules, flakes or powder. Deliquescent. d 1.86. Freely sol in water, alcohol. Commercial grades contain 73-80% $CaCl_2$. *Keep well closed.*

Hexahydrate. [7774-34-7] $CaCl_2.6H_2O$; mol wt 219.07. Deliquesc trigonal crystals. mp 30°. d^{17} 1.68. Loses all H_2O at 200°. Extremely sol in water, alcohol. *Keep well closed.*

USE: The *anhydrous* form used as a drying and dehydrating agent for organic liquids and gases, and in desiccators. The *dihydrate* and *hexahydrate* forms are used for antifreeze and refrigerating solns, in fire extinguishers, etc. (a 40% soln freezes at −41°); to preserve wood, stone; manuf ice, glues, cements; fireproofing fabrics; automobile antifreeze mixtures; to melt ice and snow; as coagulant in rubber manuf, as size in admixture with starch paste; in concrete mixes to give quicker initial set and greater strength; freezeproofing of coal and ores; dust control on unpaved roads; sizing and finishing cotton fabrics; as brine for filling inflatable tires on tractors to increase traction.

THERAP CAT: Electrolyte replacement. Has been used as diuretic, urinary acidifier, antiallergic.

THERAP CAT (VET): May be used intravenously in hypocalcemic states such as milk fever.

1662. Calcium Chromate(VI). [13765-19-0] Chromic acid (H_2CrO_4) calcium salt (1:1); calcium chrome yellow; gelbin; yellow ultramarine; C.I. 77223; C.I. Pigment Yellow 33. $CaCrO_4$; mol wt 156.07. Ca 25.68%, Cr 33.32%, O 41.00%. Prepn: Mylius, Wrochem, *Ber.* **33**, 3689 (1900); Udy, **US 2493789**; **US 2494215** (both 1950); Dunn, O'Brien, **US 2745764/5** (both 1956 to Vanadium Corp. of America).

Yellow monoclinic or rhombic crystals. Sparingly sol in water; sol in dil acids; practically insol in alcohol. Also occurs as hemihydrate, monohydrate, and dihydrate.

Caution: Chromium hexavalent compounds are listed as known human carcinogens: *Report on Carcinogens, Twelfth Edition* (PB2011-111646, 2011) p 106.

USE: As a pigment, corrosion inhibitor; manuf of chromium; in oxidizing reactions; in battery depolarization.

1663. Calcium Citrate. [813-94-5] 2-Hydroxy-1,2,3-propanetricarboxylic acid calcium salt (2:3). $C_{12}H_{10}Ca_3O_{14}$; mol wt 498.43. C 28.92%, H 2.02%, Ca 24.12%, O 44.94%. $Ca_3(C_6H_5O_7)_2$. Prepn from citrus fruit: Cole, **US 2389766** (1945 to California Fruit Growers Exchange). Use in foods: R. Labin-Goldscher, S. Edelstein, *Food Technol.* **50**, 96 (1996). Bioavailability compared with calcium carbonate: K. Sakhaee *et al.*, *Am. J. Ther.* **6**, 313 (1999). Clinical effect on bone density: L. A. Ruml *et al.*, *ibid.* 303;

on serum lipid concentrations: I. R. Reid *et al.*, *Am. J. Med.* **112**, 343 (2002).

Tetrahydrate. [5785-44-4] Calcimax; Citracal. Odorless powder. Loses most of its water at 100° and all at 120°. Sol in 1050 parts cold water; somewhat more sol in hot water. Freely sol in diluted 3N hydrochloric acid, diluted 2N nitric acid. Insol in alcohol.

USE: In foods as calcium fortifier; as anticaking agent in dry mixes.

THERAP CAT: Calcium supplement.

1664. Calcium Cyanamide. [156-62-7] Cyanamide calcium salt (1:1); calcium carbimide; "cyanamide"; nitrolime. $CCaN_2$; mol wt 80.10. C 15.00%, Ca 50.03%, N 34.97%. Prepn: Kastens, McBurney, *Ind. Eng. Chem.* **43**, 1020 (1951); Franck, Heimann, *Angew. Chem.* **44**, 372 (1931); Owen, *Trans. Faraday Soc.* **57**, 670 (1961); Dedman, Owen, *ibid.* 678.

Commercial grades may occur as grayish-black lumps of powder. While pure calcium cyanamide is nonvolatile and noncombustible, commercial grades may contain small amounts of calcium carbide which will produce acetylene in containers and processing vessels. Other contaminants are carbon, $Ca(OH)_2$, CaO, and $CaCO_3$. Pure calcium cyanamide occurs as glistening, hexagonal crystals belonging to the rhombohedral system. mp ~1340°; d^{20} 2.29. Sublimes at 1150-1200°. Heat of formn from $CaC_2 + N_2$: −69.0 kcal/mole (25°). Heat of fusion 1.29 cal/g. Essentially insol in water, but undergoes partial hydrolysis to the sol calcium hydrogen cyanamide, a source of cyanamide ions. No known solvent will bring about soln without decompn.

Caution: Potential symptoms of overexposure are irritation of eyes, skin, respiratory system; headache, vertigo, rapid breathing, low blood pressure, nausea, vomiting; skin burns and sensitization; cough; Antabuse-like effects. *See NIOSH Pocket Guide to Chemical Hazards* (DHHS/NIOSH 97-140, 1997) p 46.

USE: As fertilizer, defoliant, herbicide, pesticide; manuf and refining of iron; manuf of calcium cyanide, melamine, dicyandiamide.

THERAP CAT (VET): Has been used as an anthelmintic.

1665. Calcium Cyanamide Citrated. [8013-88-5] 2-Hydroxy-1,2,3-propanetricarboxylic acid mixt. with cyanamide calcium salt (1:1); Citrated calcium carbimide; CCC; carbimide; Colme; Dipsan; Abstem; Temposil. Contains citric acid in two parts by weight to one part of calcium cyanamide suitably purified for drug use. Used in treatment of alcoholism. Formulations: de Grunigen, Ferguson, **US 2998350** (1961 to Cyanamid and Alcoholism Res. Found., Toronto). Acts by inhibiting aldehyde dehydrogenase: J. A. Smith *et al.*, *J. Am. Med. Assoc.* **165**, 2181 (1957). Effect on cardiovascular system: J. E. Peachey *et al.*, *Clin. Pharmacol. Ther.* **29**, 40 (1981); on liver cells: J. J. Vázquez, S. Cervera, *Lancet* **1**, 361 (1980). Comparison with disulfiram, *q.v.:* M. S. Levy *et al.*, *Am. J. Psychiatry* **123**, 1018 (1967). Review of drug therapy for alcoholism: E. M. Sellers *et al.*, *N. Engl. J. Med.* **305**, 1255 (1981).

THERAP CAT: Alcohol deterrent.

1666. Calcium Cyanide. [592-01-8] Calcium cyanide (Ca(CN)₂); Cyanogas. C_2CaN_2; mol wt 92.11. C 26.08%, Ca 43.51%, N 30.41%. $Ca(CN)_2$. Prepn: *Gmelins, Calcium* (8th ed.) **28B**, 173-178, 958-960 (1958). Commercial prepns contain 40-50% $Ca(CN)_2$. Toxicity study: H. F. Smyth *et al.*, *Am. Ind. Hyg. Assoc. J.* **30**, 470 (1969). Review of toxicology and human exposure: *Toxicological Profile for Cyanide* (PB2007-100674, 2006) 341 pp.

Rhombohedric crystals or powder. dec in moist air liberating hydrogen cyanide. *Poisonous.* Sol in water with gradual liberation of HCN; even very weak acid (CO_2) liberates HCN; sol in alc. *Keep dry.* LD_{50} orally in rats: 39 mg/kg (Smyth).

USE: Fumigant; rodenticide; in stainless-steel manuf; in leaching ores of precious metals; stabilizer for cement.

1667. Calcium Cyclamate. [139-06-0] N-Cyclohexylsulfamic acid calcium salt (2:1); cyclohexanesulfamic acid calcium salt; calcium cyclohexanesulfamate; calcium cyclohexylsulfamate; cyclamate calcium; Cyclan; Sucaryl Calcium. $C_{12}H_{24}CaN_2O_6S_2$; mol wt 396.53. C 36.35%, H 6.10%, Ca 10.11%, N 7.06%, O 24.21%, S 16.17%. $[C_6H_{11}NHSO_3]_2Ca^{2+}$. Prepn: Cummins, Johnson; McQuaid, **US 2799700**; **US 2804477** (both 1957 to du Pont); Freifelder, **US 3082247** (1963 to Abbott); Birsten, Rosin, **US 3361798**; **US 3366670** (both 1968 to Baldwin-Montrose). Metabolism: Wallace *et al.*, *J. Pharmacol. Exp. Ther.* **175**, 325 (1970); Prosky, O'Dell, *J. Pharm. Sci.* **60**, 1341 (1971); Renwick, Williams, *Biochem. J.* **129**, 869 (1972). *See also* Cyclamic Acid.

Dihydrate. Crystals with pleasant, very sweet taste. Freely sol in water. Practically insol in alc, benzene, chloroform, ether. pH of 10% aq soln 5.5-7.5. Said to be more resistant to cooking temps than saccharin.

Note: Consult latest Government regulations on use in foods.

USE: Non-nutritive sweetener.

1668. Calcium Dichromate(VI). [14307-33-6] Chromic acid ($H_2Cr_2O_7$) calcium salt (1:1); calcium bichromide. $CaCr_2O_7$; mol wt 256.06. Ca 15.65%, Cr 40.61%, O 43.74%. Prepn: Hartford *et al.*, *J. Am. Chem. Soc.* **72**, 3353 (1950).

Trihydrate. Bipyramidal orange-red crystals. Nonhygroscopic if pure. Dec on heating above 100° to $CaCrO_4$ and CrO_3. d_4^{30} 2.370. Very sol in water; insol in ether, CCl_4, hydrocarbons; dissolves in alc with immediate reduction of the dichromate and pptn of brown hydrous chromic chromate; dissolves in acetone with subsequent pptn of $CaCrO_4$.

Caution: Chromium hexavalent (VI) compounds are listed as known human carcinogens: *Report on Carcinogens, Twelfth Edition* (PB2011-111646, 2011) p 106.

USE: As catalyst; in manuf of $CrCl_3$ and CrO_3; corrosion inhibitor.

1669. Calcium Fluoride. [7789-75-5] CaF_2; mol wt 78.07. Ca 51.34%, F 48.67%. Occurs in nature as the mineral *fluorite* or *fluorspar*. Prepd from $CaCO_3$ + HF: O. Ruff, *Die Chemie des Fluors* (Berlin, 1920) p 89; Emeleus in *Fluorine Chemistry* vol. **I**, J. H. Simons, Ed. (Academic Press, 1950) p 36; Kwasnik in *Handbook of Preparative Inorganic Chemistry* **Vol. 1**, G. Brauer, Ed. (Academic Press, New York, 2nd ed., 1963) p 233. Toxicity study: K.-R. Stratmann, *Dtsch. Zahnaerztl. Z.* **34**, 484 (1979).

White powder or cubic crystals. When F ions are pptd with Ca^{2+} in the absence of CO_3 ions, a gel is obtained. For the crystals: d 3.18; mp 1403°; bp 2500°; Mohs' hardness: 4. Becomes luminous when heated. Practically insol in water (soly at 18°: 0.0015 g/100 ml). Slightly sol in dil mineral acids; dissolved by concd mineral acids with liberation of HF. LD_{50} i.p. in mice: 2638.27 mg/kg (Stratmann).

USE: Fluorspar is the main primary source of fluorine and its compds. In ferrous metallurgy it is used as a flux to increase the fluidity of the slag. The steel industry is the largest consumer; the chemical industry, second and glass and ceramics, third. Synthetic fluorspar is used in the optical industry (transmits u.v. rays), and pure calcium fluoride is used as catalyst in dehydration and dehydrogenations. Used to fluoridate drinking water.

1670. Calcium Formate. [544-17-2] Formic acid calcium salt (2:1). $C_2H_2CaO_4$; mol wt 130.11. C 18.46%, H 1.55%, Ca 30.80%, O 49.19%. $Ca(HCOO)_2$. Prepn from $Ca(OH)_2$ and CO at high temp and pressure: Enderli, **US 1920851**; **US 1995607** (1933, 1935 to Rudulf Koepp); Erasmus, Hamby, **US 2913318** (1959 to Union Carbide); from $CaCl_2$ and formic acid: Funk, Romer, *Z. Anorg. Allg. Chem.* **239**, 288 (1938).

Orthorhombic crystals or cryst powder. Slight acetic acid-like odor. d 2.02. Sol in water, practically insol in alc.

USE: Preservative for food, silage; as binder for fine-ore briquets; in drilling fluids and lubricants.

1671. Calcium Gluconate. [299-28-5] D-Gluconic acid calcium salt (2:1); Calciofon; Calglucon; Ebucin; Glucal; Glucobiogen. $C_{12}H_{22}CaO_{14}$; mol wt 430.37. C 33.49%, H 5.15%, Ca 9.31%, O 52.04%. $Ca[HOCH_2(CHOH)_4COO]_2$. Description of injectable solns with sodium ascorbate: **GB 495675**; **DE 702185**. General directions and stability data: Siegrist, *Pharm. Acta Helv.* **24**, 430 (1949).

Odorless, tasteless crystals, granules, or powder. $[\alpha]_D^{20}$ about +6°. Does not lose its water on drying without some decomposition. Freely sol in about 5 parts boiling water; sparingly and slowly sol in 30 parts cold water. Insol in alc or other organic solvents. pH aq soln: 6-7. More concd (20 to 30%) aq solns are easily obtained by the addition of boric acid or similar complex-forming acids. Supersatd injectable solns prepd with calcium D-saccharate.

USE: In sewage purification; in coffee powders to prevent caking.

THERAP CAT: Calcium replenisher.

THERAP CAT (VET): In hypocalcemic states, including bovine milk fever.

1672. Calcium Glycerophosphate. [27214-00-2] 1,2,3-Propanetriol mono(dihydrogen phosphate) calcium salt (1:1); calcium glycerinophosphate; calcium phosphoglycerate; Neurosin. $C_3H_7CaO_6P$; mol wt 210.13. C 17.15%, H 3.36%, Ca 19.07%, O 45.68%, P 14.74%. Three isomers exist: The *β-glycerophosphoric acid calcium salt*, D(+)- and L(−)-*α-glycerophosphoric acid calcium salt*. Commercial product is a mixture of calcium β- and DL-α-glycerophosphates: Toal, Phillips, *J. Pharm. Pharmacol.* **1**, 869 (1949). Prepn of the calcium α- and β-acids: King, Pyman, *J. Chem. Soc.* **105**, 1238 (1914); Toal, Phillips, *loc. cit.* Sepn of the α-acid from β- and polyglycerophosphoric acids via the α-acid calcium salt: Carrara, **IT 460219** (1950), *C.A.* **46**, 5077a (1952). Protective action against demineralization of dental enamel: T. H. Grenby, J. M. Bull, *Caries Res.* **14**, 210 (1980).

Commercial product, fine, odorless, almost tasteless, slightly hygroscopic powder; dec >170°. Sol in about 50 parts water; almost insol in alc, boiling water. Soly in water is increased by citric or lactic acid. The aq soln is alkaline.

Mixture with calcium lactate. Calphosan.

USE: In dentifrices, baking powder, as food stabilizer.

THERAP CAT: Calcium and phosphorus source. Tonic.

THERAP CAT (VET): Has been used as dietary supplement.

1673. Calcium Hexafluorosilicate. [16925-39-6] Calciumhexafluorosilicate(2−) (1:1); calcium fluosilicate; calcium silicofluoride. CaF_6Si; mol wt 182.15. Ca 22.00%, F 62.58%, Si 15.42%. $CaSiF_6$. Prepn from Ca salt and H_2SiF_6: Moller, Kreth, **GB 263780** (1925). Review of toxicology of fluoride compounds: G. L. Waldbott, *Acta Med. Scand. Suppl.* **400**, 1-44 (1963).

Dihydrate, powder. d 2.25. Almost insol in cold water. Partially dec by hot water. Practically insol in acetone. LD in guinea pigs (mg/kg): 250 orally, 450 s.c. (Waldbott).

USE: In wood, rubber, textile industries; flotation agent; insecticide.

1674. Calcium Hydride. [7789-78-8] CaH_2; mol wt 42.09. Ca 95.22%, H 4.79%. Prepd by direct combination of calcium and hydrogen at 300-400°: **GB 597055** (1948); by reduction of lime with magnesium in the presence of hydrogen: Gibb, *Trans. Electrochem. Soc.* **93**, 198-211 (1948); from $CaCl_2$ and hydrogen in the presence of sodium: Wade, Alexander, **US 2702740** (1955 to Metal Hydrides). *Reviews:* Halls, *Ind. Chem.* **22**, 680 (1946); Kilb, *USAEC* APEX-485, 57 pp (1959).

Orthorhombic crystals or powder; the commercial product is gray. d 1.7. mp 1.86°. *Dangerous when wet.* Decomposes with water, lower alcohols and carboxylic acids to form hydrogen; moderately powerful condensing agent with ketones and acid esters; more powerful reducing agent toward metal oxides than lithium or sodium hydrides.

USE: To prepare rare metals by reduction of their oxides; as a drying agent for liquids and gases; to generate hydrogen: 1 g of calcium hydride in water liberates 1 liter of hydrogen at STP; in organic syntheses.

1675. Calcium Hydroxide. [1305-62-0] Calcium hydroxide ($Ca(OH)_2$); calcium hydrate; slaked lime. CaH_2O_2; mol wt 74.09. Ca 54.09%, H 2.72%, O 43.19%. $Ca(OH)_2$. Contains at least 95% $Ca(OH)_2$. Commercial prepn by hydration of lime: W. L. Faith *et al.*, *Industrial Chemicals* (John Wiley, New York, 3rd ed., 1965) pp 483-484. Laboratory prepn by treating an aq soln of a calcium salt with alkali: Ehrlich in *Handbook of Preparative Inorganic Chemistry* vol. **1**, G. Brauer (Academic Press, New York, 2nd ed., 1963) p 934. Toxicity study: Smyth *et al.*, *Am. Ind. Hyg. Assoc. J.* **30**, 470 (1969).

Crystals or soft, odorless, granules or powder. Slightly bitter, alkaline taste. Readily absorbs CO_2 from air forming $CaCO_3$. Loses water when ignited; forms CaO. d 2.08-2.34. Sol in glycerol, sugar or NH$_4$Cl solns, acids with evolution of much heat. Slightly sol in water; very slightly sol in boiling water. Insol in alc. pH of aq soln satd at 25°: 12.4. *Keep well closed.* LD$_{50}$ orally in rats: 7.34 g/kg (Smyth).

Caution: Potential symptoms of overexposure are irritation of eyes, skin, upper respiratory system; eye and skin burns; skin vesiculation; cough, bronchitis, pneumonia. *See NIOSH Pocket Guide to Chemical Hazards* (DHHS/NIOSH 97-140, 1997) p 46.

USE: In mortar, plaster, cement and other building and paving materials; in lubricants, drilling fluids, pesticides, fireproofing coatings, water paints; as egg preservative; manuf of paper pulp; in SBR rubber vulcanization; in water treatment; as absorbant for carbon dioxide; dehairing hides.

THERAP CAT: Astringent.

1676. Calcium Hypochlorite. [7778-54-3] Hypochlorous acid calcium salt (2:1); chlorinated lime; Losantin; Induchlor; Zappit. CaCl$_2$O$_2$; mol wt 142.98. Ca 28.03%, Cl 49.59%, O 22.38%. Ca-(OCl)$_2$. Solid, stable form of hypochlorite. Prepd by the chlorination of lime. Commercial products normally contain various impurities such as CaCl$_2$, CaCO$_3$, Ca(OH)$_2$ and water. Preparation of solid product contg 90-94% Ca(OCl)$_2$: Cady, *Inorg. Synth.* **5**, 161 (1957).

White powder or flat plates, chlorine odor. Dec 100°. d 2.35. *Corrosive. Oxidizer.* Soly in water (25°): 21.4%.

USE: Disinfectant for drinking water and swimming pools; algicide; oxidizing agent; as household and industrial bleaching agent and sanitizer.

1677. Calcium Hypophosphite. [7789-79-9] Phosphinic acid calcium salt (2:1). CaH$_4$O$_4$P$_2$; mol wt 170.05. Ca 23.57%, H 2.37%, O 37.63%, P 36.43%. Ca(H$_2$PO$_2$)$_2$. Prepn: *Gmelins, Calcium* (8th ed.) **28B**, 1119-1121 (1958).

Monoclinic, prismatic crystals or granular powder. When heated above 300° it evolves spontaneously-inflammable phosphine. Sol in water; slightly sol in glycerol. Practically insol in alcohol. The aq soln is slightly acid.

USE: As corrosion inhibitor; in nickel plating. Pharmaceutic aid (retards oxidation of ferrous salts).

THERAP CAT: Calcium source.

THERAP CAT (VET): Has been used as a dietary supplement and also as a "nerve tonic".

1678. Calcium Iodate. [7789-80-2] Iodic acid (HIO$_3$) calcium salt (2:1); lautarite. CaI$_2$O$_6$; mol wt 389.88. Ca 10.28%, I 65.10%, O 24.62%. Ca(IO$_3$)$_2$. Prepd by passing chlorine into a hot soln of lime in which iodine has been dissolved: Bahl, Singh, *J. Indian Chem. Soc.* **17**, 397 (1940).

Nonhygroscopic, monoclinic-prismatic crystals. d$_4^{15}$ 4.519. Stable up to 540°. Sensitive to reducing agents. Soly in water (g/100 ml): 0.10 (0°); 0.95 (100°). More sol in aq solns of iodides and in amino acid solns. Sol in nitric acid. Insol in alcohol.

Monohydrate. Cubic crystals. Slightly sol in water.

Hexahydrate. Orthorhombic crystals. Slightly sol in water.

USE: Nutritional source of iodine in foods and feedstuffs. More stable in table salts than iodides: *Food Field Reporter*, Aug. 8, 1956; Daum, *C.A.* **51**, 5324 (1957); to improve properties of yeast-leavened bakery products.

THERAP CAT: Antiseptic.

1679. Calcium Iodide. [10102-68-8] CaI$_2$; mol wt 293.89. Ca 13.64%, I 86.36%. Prepn: Farr, US 2415346 (1947 to Mallinckrodt); Chaigneau, *Bull. Soc. Chim. Fr.* **1957**, 886; *Gmelins, Calcium* (8th ed.) **28B**, 102, 610-622 (1958). The commercial product usually contains 16-20% water.

Very hygroscopic hexagonal lamella. Becomes yellow and completely insol on exposure to air due to liberation of I$_2$ and absorption of CO$_2$. mp 740°; bp 1100°. Very sol in water, methanol, ethanol, acetone; practically insol in ether, dioxane. The aq soln is neutral or slightly alkaline. *Keep tightly closed and protected from light.*

Hexahydrate. Hexagonal, thick needles, or plates, or lumps, or powder. Very hygroscopic; becomes yellow in air, mp ~42°. Freely sol in water, alcohol. *Keep tightly closed and protected from light.*

THERAP CAT: Expectorant.

1680. Calcium Lactate. [814-80-2] 2-Hydroxypropanoic acid calcium salt (2:1). C$_6$H$_{10}$CaO$_6$; mol wt 218.22. C 33.02%, H

4.62%, Ca 18.37%, O 43.99%. Ca[CH$_3$CH(OH)COO]$_2$. Commercial prepn usually contains about 25% water, and on the anhydr basis it is at least 98% pure. Prepd commercially by neutralization of lactic acid, from fermentation of dextrose, molasses, starch, sugar or whey, with CaCO$_3$: Inskeep *et al.*, *Ind. Eng. Chem.* **44**, 1955 (1952). Clinical use as a supplement: G. M. Day *et al.*, *Pediatr. Res.* **9**, 568 (1975). Evaluation as anticaries agent: B. M. Shrestha *et al.*, *Caries Res.* **16**, 12 (1982); M. J. M. Schaeken, J. S. van der Hoeven, *ibid.* **24**, 376 (1990). Clinical pharmacology: M. Goddard *et al.*, *Am. J. Clin. Nutr.* **44**, 653 (1986).

Pentahydrate, odorless, white, slightly efflorescent granules or powder. Becomes anhydr at 120°. pH: 6-8. Sol water. Practically insol in alcohol.

USE: As a preservative in foods and beverages; in dentifrices.

THERAP CAT: Calcium replenisher.

THERAP CAT (VET): Calcium replenisher.

1681. Calcium Levulinate. [591-64-0] 4-Oxopentanoic acid calcium salt (2:1); levulinic acid calcium salt. C$_{10}$H$_{14}$CaO$_6$; mol wt 270.29. C 44.44%, H 5.22%, Ca 14.83%, O 35.52%. (CH$_3$COCH$_2$-CH$_2$COO)$_2$Ca. Prepn: Cox, Dodds, US 2033909 (1936 to Niacet Chemicals).

Dihydrate, crystals or granular powder. Bitter, salty taste. mp 125°. Loses 1 H$_2$O on drying *in vacuo* at room temp and all H$_2$O at 50°. Freely sol in water; the aq soln is practically neutral. Slightly sol in alc. Insol in ether, chloroform.

THERAP CAT: Calcium replenisher.

THERAP CAT (VET): May be used in hypocalcemic states.

1682. Calcium Magnesium Acetate. [76123-46-1] Acetic acid, calcium magnesium salt; CMA; Ice-B-Gon. C$_2$H$_4$-O$_2$.xCa.xMg. Prepn and use as deicer: S. A. Dunn, R. U. Schenk, *Transport. Res. Rec.* **776**, 12 (1980); *eidem, U.S. Federal Highway Administration Report* **FHWA-RD-79-109** (1980) pp 17. Prepn by fermentation: C. W. Marynowski *et al.*, *Ind. Eng. Chem. Prod. Res. Dev.* **24**, 457 (1985); via extraction and sorption: H. Reisinger, C. J. King, *Ind. Eng. Chem. Res.* **34**, 845 (1995). Acute toxicity: G. A. Rausina *et al.*, *Acute Toxic. Data* **1**, 87 (1990). Environmental and toxicological evaluation: B. L. McFarland, K. T. O'Reilly, *Chemical Deicers and the Environment*, F. M. D'Itri, Ed. (Lewis Publishers, Chelsea, MI, 1992) pp 193-227. Anti-icing effect: E. E. Adams *et al.*, *ibid.* pp. 481-493. Degradation in soil: D. W. Ostendorf *et al.*, *J. Environ. Qual.* **22**, 299 (1993). Effects on cement mortar: O. Peterson, *Cem. Concr. Res.* **25**, 617 (1995). Book: *Calcium Magnesium Acetate*, D. L. Wise *et al.*, Eds. (Elsevier Science Publishers, Amsterdam, 1991) pp 511.

Non-flammable. Shows no significant corrosion of steel, aluminum or zinc. LD$_{50}$ orally in male, female rats: 3240, 3071 mg/kg (Rausina).

USE: Deicing agent for roadways; in fossil fuel combustion for removal of sulfur.

1683. Calcium Molybdate(VI). [7789-82-4] (*T*-4)-Molybdate (MoO$_4^{2-}$) calcium (1:1); calcium molybdenum oxide. Ca-MoO$_4$; mol wt 200.02. Ca 20.04%, Mo 47.97%, O 31.99%. Prepn from sodium molybdate and CaSO$_4$: Carosella, US 2460974 (1949 to U.S. Vanadium); by heating a stoichiometric mixture of CaO or CaCO$_3$ and molybdic acid: Kroger, *Nature* **159**, 674 (1947). Toxicity studies: L. T. Fairhall *et al.*, "The Toxicity of Molybdenum," *U.S. Public Health Service Bulletin No. 293*, Washington D.C. (1945) 36 pp.

Tetragonal crystals. d. 4.35. Practically insol in water, alcohol. Sol in concd mineral acids.

USE: In phosphors and luminescent materials.

1684. Calcium Nitrate. [10124-37-5] Nitric acid calcium salt (2:1). CaN$_2$O$_6$; mol wt 164.09. Ca 24.42%, N 17.07%, O 58.50%. Ca(NO$_3$)$_2$. Prepn: *Gmelins, Calcium* (8th ed.) **28B**, 59-69, 341-382 (1956).

Deliquesc granules, mp ~560°. Very sol in water, heat being evolved; freely sol in methanol, ethanol, acetone. Almost insol in concd HNO$_3$. pH of 5% aq soln 6.0. *Oxidizer. Keep well closed.* *Note:* Ca(NO$_3$)$_2$ crystallizes also with 4H$_2$O (30.5%), mp 45°. Technical flake usually contains 28.6% H$_2$O.

Tetrahydrate. [13477-34-4] CaN$_2$O$_6$.4H$_2$O; mol wt 236.15.

USE: In explosives, fertilizers, matches, pyrotechnics; manuf of incandescent mantles, radio tubes, HNO_3; corrosion inhibitor in diesel fuels.

1685. Calcium Nitrite. [13780-06-8] Nitrous acid calcium salt (2:1). CaN_2O_4; mol wt 132.09. Ca 30.34%, N 21.21%, O 48.45%. $Ca(NO_2)_2$. Prepd by reaction of nitric oxide with a mixture of calcium ferrate(III) and calcium nitrate: Ray, Ogg, Jr., *J. Am. Chem. Soc.* **79**, 265 (1957).

White or yellowish, deliquesc, hexagonal crystals. d 2.23. Freely sol in water; slightly sol in alc. *Keep well closed.*

USE: Corrosion inhibitor in lubricants, concrete.

1686. Calcium Oleate. [142-17-6] (9Z)-9-Octadecenoic acid calcium salt (2:1); oleic acid calcium salt. $C_{36}H_{66}CaO_4$; mol wt 603.00. C 71.71%, H 11.03%, Ca 6.65%, O 10.61%. $Ca(C_{18}H_{33}-O_2)_2$. Prepn: Harrison, *Biochem. J.* **18**, 1222 (1924); Pink, *J. Chem. Soc.* **1939**, 619.

Pale-yellow transparent solid. Dec above 140°. Slowly absorbs moisture from the air to form the monohydrate. Practically insol in water, alcohol, ether, acetone, petr ether; sol in chloroform, benzene.

USE: Thickening lubricating grease; waterproofing concrete; emulsifier for benzene, kerosene, etc.; in modeling waxes to vary hardness.

1687. Calcium Oxalate. [563-72-4] Ethanedioic acid calcium salt (1:1). C_2CaO_4; mol wt 128.10. C 18.75%, Ca 31.29%, O 49.96%. CaC_2O_4. Prepn from calcium formate: Bredt, **US 1622991** (1927); from calcium cyanamide: Barsky, Buchanan, *J. Am. Chem. Soc.* **53**, 1270 (1931).

Monohydrate. Cubic crystals. Loses all of its water at 200°. When ignited is converted into $CaCO_3$ or CaO without appreciable charring. d 2.2. Practically insol in water and acetic acid; sol in dil HCl or HNO_3.

USE: In ceramic glazes; as carrier for separation of rare earth metals; analysis for calcium: Ingols, Murray, *Anal. Chem.* **21**, 525 (1949).

1688. Calcium Oxide. [1305-78-8] Lime; burnt lime; calx; quicklime. CaO; mol wt 56.08. Ca 71.47%, O 28.53%. Properly stored lime of commerce contains 90-95% free CaO. Commercial production from limestone: W. L. Faith *et al.*, *Industrial Chemicals* (John Wiley, New York, 3rd ed., 1965) pp 482-487. Lab prepn by ignition of $CaCO_3$: Ehrlich in *Handbook of Preparative Inorganic Chemistry* vol. **1**, G. Brauer, Ed. (Academic Press, New York, 2nd ed., 1963) p 931. *Review:* R. S. Boynton in *Kirk-Othmer Encyclopedia of Chemical Technology* vol. **14** (Wiley-Interscience, New York, 3rd ed., 1981) pp 343-382.

Crystals, white or grayish-white lumps, or granular powder; commercial material sometimes has a yellowish or brownish tint, due to iron. mp 2572°; bp 2850°; d 3.32-3.35. Readily absorbs CO_2 and H_2O from air, becoming air-slaked. Sol in water forming $Ca(OH)_2$ and generating a large quantity of heat; sol in acids, glycerol, sugar soln; practically insol in alc. *Corrosive. Keep tightly closed and dry.*

Caution: Potential symptoms of overexposure are irritation of eyes, skin, upper respiratory tract; ulceration or perforation of nasal septum; pneumonia; dermatitis. *See NIOSH Pocket Guide to Chemical Hazards* (DHHS/NIOSH 97-140, 1997) p 48.

USE: In bricks, plaster, mortar, stucco and other building and construction materials; manuf of steel, aluminum, magnesium, and flotation of non-ferrous ores; manuf of glass, paper, Na_2CO_3 (Solvay process), Ca salts and many other industrial chemicals; dehairing hides; clarification of cane and beet sugar juices; in fungicides, insecticides, drilling fluids, lubricants; water and sewage treatment; in laboratory to absorb CO_2 (the combination with NaOH is known as soda-lime, *q.v.*).

1689. Calcium Palmitate. [542-42-7] Hexadecanoic acid calcium salt (2:1); palmitic acid calcium salt. $C_{32}H_{62}CaO_4$; mol wt 550.92. C 69.77%, H 11.34%, Ca 7.27%, O 11.62%. $Ca(C_{16}H_{31}-O_2)_2$. Prepn: Harrison, *Biochem. J.* **18**, 1222 (1924).

Powder or rhombic crystals. Dec above 155°. Practically insol in water, alcohol, ether, acetone, petr ether; slightly sol in chloroform, benzene, acetic acid.

USE: Thickening lubricating oils; waterproofing fabrics and lubricating greases; as corrosion inhibitor in halohydrocarbons.

1690. Calcium Permanganate. [10118-76-0] Permanganic acid ($HMnO_4$) calcium salt. $CaMn_2O_8$; mol wt 277.95. Ca 14.42%, Mn 39.53%, O 46.05%. $Ca(MnO_4)_2$. Prepn from $KMnO_4$ and $CaCl_2$: **GB 624885** (1949 to Boots Pure Drug Co. and T. Hagyard); from $Al(MnO_4)_3$ and $Ca(OH)_2$: Jaskowiak, **US 2504130** (1950 to Carus Chemical).

Violet or dark-purple, deliquesc crystals. Freely sol in water; dec in alcohol. *Oxidizer. Keep tightly closed.*

USE: Antiseptic, disinfectant, deodorizer; with CaF_2 as binder for welding electrode coatings and fluxes.

1691. Calcium Peroxide. [1305-79-9] Calcium dioxide. CaO_2; mol wt 72.08. Ca 55.60%, O 44.39%. The commercial product usually contains about 60% CaO_2, water, and some $Ca(OH)_2$ and $CaCO_3$. Prepn: Young, **US 2533660** (1950 to du Pont); Ehrlich in *Handbook of Preparative Inorganic Chemistry* vol. **1**, G. Brauer, Ed. (Academic Press, New York, 2nd ed., 1963) p 936.

White or yellowish, odorless, almost tasteless powder. Dec in moist air. Slightly sol in water; sol in acids with formation of H_2O_2. *Oxidizer. Keep well closed.*

USE: Stabilizer for rubber.

THERAP CAT: Antiseptic.

1692. Calcium Phenolsulfonate. [127-83-3] 4-Hydroxybenzenesulfonic acid calcium salt (2:1); calcium sulfocarbolate; calcium sulphophenolate. $C_{12}H_{10}CaO_8S_2$; mol wt 386.40. C 37.30%, H 2.61%, Ca 10.37%, O 33.12%, S 16.59%. $Ca[C_6H_4(OH)SO_3]_2$. Prepn: *Hagers Handb. Pharm. Praxis* vol. **2**, 420 (Berlin, 1930).

Hydrate, odorless cryst powder. Sol in water or alcohol. The aq soln is neutral, and has a bitter, astringent taste.

THERAP CAT (VET): Has been used as an intestinal antiseptic, in dusting powders for ulcers and in ophthalmic solns.

1693. Calcium Phenoxide. [5793-84-0] Phenol calcium salt (2:1); calcium carbolate; calcium phenate; calcium phenolate; calcium phenylate. $C_{12}H_{10}CaO_2$; mol wt 226.29. C 63.69%, H 4.45%, Ca 17.71%, O 14.14%. $Ca(OC_6H_5)_2$. Prepn: Kluge, Drake, **US 2870134** (1959 to Texas Co.).

Reddish powder. Dec in air. Slightly sol in water or alcohol. *Keep well closed.*

USE: Detergent; additive for motor oils.

1694. Calcium Phosphate, Dibasic. [7757-93-9] Phosphoric acid calcium salt (1:1); calcium monohydrogen phosphate; dicalcium orthophosphate; secondary calcium phosphate. $CaHO_4P$; mol wt 136.06. Ca 29.46%, H 0.74%, O 47.04%, P 22.76%. $CaHPO_4$. Occurs in nature as the mineral *monetite*. Prepn from $CaCl_2$ and Na_2HPO_4: Jensen, Rathley, *Inorg. Synth.* **4**, 19, 20 (1953); from $Ca_3(PO_4)_2$ and H_3PO_4: Perloff, Posner, *ibid.* **6**, 16 (1960), where it is an intermediate in the preparation of hydroxyapatite.

Triclinic crystals. At red heat dehydrated to calcium pyrophosphate. Sol in $3N$ HCl or $2N$ HNO_3. Practically insol in water, alcohol.

Dihydrate. Brushite. Monoclinic crystals. Loses water of crystn slowly below 100°. Dehydr at red heat to calcium pyrophosphate. d 2.31. Practically insol in water, alcohol. Sol in dil HCl or HNO_3; slightly sol in dil acetic acid.

USE: Chiefly in animal feeds; mineral supplement in cereals and other foods; manuf of glass; in dental products, fertilizers (*see also* Calcium Phosphate, Monobasic). Pharmaceutic aid (tablet and capsule diluent).

THERAP CAT: Calcium replenisher.

THERAP CAT (VET): Has been used as a dietary supplement, and as an antacid.

1695. Calcium Phosphate, Monobasic. [7758-23-8] Phosphoric acid calcium salt (2:1); acid calcium phosphate; calcium biphosphate; monocalcium orthophosphate; monocalcium phosphate; primary calcium phosphate; "calcium superphosphate". $CaH_4O_8P_2$; mol wt 234.05. Ca 17.12%, H 1.72%, O 54.69%, P 26.47%. $Ca(H_2PO_4)_2$. Commercial prepn for fertilizers by treating pulverized phosphate rock with H_2SO_4 or H_3PO_4: *Faith, Keyes & Clark's Industrial Chemicals*, F. A. Lowenheim, M. K. Moran, Eds. (Wiley-Interscience, New York, 4th ed., 1975) pp 191-200. Laboratory prepn from $CaCO_3$ and H_3PO_4: Jensen, Kathley, *Inorg. Synth.* **4**, 18 (1953).

Monohydrate, large, shining, triclinic plates, cryst powder or granules. Non-hygroscopic when pure, but traces of impurities such as

H_3PO_4 cause material to be deliquesc. Strong acid taste. Loses H_2O at 100°, dec at 200°. d_4^{18} 2.220. Moderately sol in water; sol in dil HCl or HNO_3 or acetic acid.

Note: The products obtained from commercial processes are not pure monobasic calcium phosphate. The superphosphate obtained from the H_2SO_4 treatment is about 30% $CaH_4(PO_4)_2.H_2O$, 10% $CaHPO_4$, 45% $CaSO_4$, 10% iron oxide, silica, alumina, etc. and 5% water; it contains 18-21% available P_2O_5. The triple superphosphate obtained from the H_3PO_4 treatment contains from 43 to 50% available P_2O_5.

USE: Chiefly in fertilizers; as acidulant in baking powders and in wheat flours; mineral supplement for foods and feeds; in enameling.

1696. Calcium Phosphate, Tribasic. [7758-87-4] Phosphoric acid calcium salt (2:3); tricalcium orthophosphate; tricalcium phosphate; tertiary calcium phosphate; Calcigenol Simple. $Ca_3O_8P_2$; mol wt 310.17. Ca 38.76%, O 41.27%, P 19.97%. $Ca_3(PO_4)_2$. It is about 96% pure, usually contg an excess of CaO. Occurs in nature as the minerals: *oxydapatit, voelicherite, whitlockite.* The technical product is also known as *"bone ash"*. Commercial prepn from phosphate rock: Hignett, Hubbard, *Ind. Eng. Chem.* **38**, 1208 (1946); Elmore, US 2474831 (1949 to T.V.A.); Hollingsworth, US 2556541 and US 2562718 (both 1951 to Coronet Phosphate); Brosheer, Hignett, *Chem. Eng. Rept.* no. 7, 143 pp (1953).

Amorphous, odorless, tasteless powder. mp 1670°. d 3.14. readily sol in 3*N* HCl, 2*N* HNO_3. Practically insol in water, alcohol, acetic acid.

USE: Manuf of fertilizers, H_3PO_4 and P compds; manuf milkglass, polishing and dental powders, porcelains, pottery; enameling; clarifying sugar syrups; in animal feeds; as noncaking agent; in the textile industry. Pharmaceutic aid (tablet and capsule diluent).

THERAP CAT: Calcium replenisher.

THERAP CAT (VET): Has been used as a dietary supplement, and as an antacid.

1697. Calcium Phosphide. [1305-99-3] Photophor. Ca_3P_2; mol wt 182.18. Ca 66.00%, P 34.00%. Prepn: Ehrlich in *Handbook of Preparative Inorganic Chemistry* **vol. 1**, G. Brauer, Ed. (Academic Press, New York, 2nd ed., 1963) p 943.

Red-brown cryst powder or gray lumps. Dec by moist air or water, evolving spontaneously-flammable phosphine. d 2.51; mp ~1600°. *Dangerous when wet. Keep tightly closed. Poisonous.*

USE: For signal fires; in purification of Cu and Cu alloys; as rodenticide.

1698. Calcium Phosphite. [21056-98-4] Phosphonic acid calcium salt (1:1). $CaHO_3P$; mol wt 120.06. Ca 33.38%, H 0.84%, O 39.98%, P 25.80%. $CaHPO_3$. Prepn: *Gmelins, Calcium* (8th ed.) **28B**, 1121 (1958).

Monohydrate. Crystals. Loses water at 200°; dec above 300°. Slightly sol in water; practically insol in alcohol.

USE: Fertilizers; polymerization catalyst.

1699. Calcium Polycarbophil. [126040-58-2] Polycarbophil calcium salt; calcium polycarbophil; WL-140; Equalactin; FiberCon; Fiber-Lax. Calcium salt of a synthetic loosely crosslinked hydrophilic resin of the polycarboxylic type. Review of pharmacology, toxicology, clinical efficacy, and adverse effects: I. E. Danhof, *Pharmacotherapy* **2**, 18-28 (1982).

White or creamy white powder. Insol in water, dilute acids, dilute alkalies, common organic solvents.

THERAP CAT: Laxative; bulking agent.

1700. Calcium Propionate. [4075-81-4] Propionic acid calcium salt (2:1). $C_6H_{10}CaO_4$; mol wt 186.22. C 38.70%, H 5.41%, Ca 21.52%, O 34.37%. $Ca(CH_3CH_2COO)_2$. Occurs as mono- or trihydrate. Prepn: *Beilstein* **vol. 2**, 238, 2nd suppl., 218, 3rd suppl., 516.

Powder or monoclinic crystals. Sol in water; slightly sol in methanol, ethanol; practically insol in acetone, benzene.

USE: As an inhibitor of molds and other microorganisms in foods, tobacco, pharmaceuticals; in butyl rubber to improve processability and scorching resistance.

THERAP CAT: Antifungal.

1701. Calcium Pyrophosphate. [7790-76-3] Diphosphoric acid calcium salt (1:2); calcium diphosphate. $Ca_2O_7P_2$; mol wt

254.10. Ca 31.55%, O 44.07%, P 24.38%. Prepn by ignition of $CaHPO_4$: St. Pierre, *J. Am. Chem. Soc.* **77**, 2197 (1955).

Polymorphous crystals or powder. d 3.09. mp 1353°. Practically insol in water. Sol in dil HCl or HNO_3.

USE: Abrasive; fertilizer; feed supplement; in dentifrices, ceramic ware, china, glass, phosphors.

1702. Calcium D-Saccharate. [5793-88-4] D-Glucaric acid calcium salt (1:1). $C_6H_8CaO_8$; mol wt 248.20. C 29.04%, H 3.25%, Ca 16.15%, O 51.57%. $CaC_6H_8O_8$. The normal calcium salt of D-saccharic acid, a dicarboxylic sugar acid derived from the oxidation of D-gluconic acid. Calcium D-saccharate is a true chemical compd and should not be confused with saccharated lime, formerly called "calcium saccharate" and produced by the action of lime upon sugar. Prepn: *Beilstein* **vol. 3**, 2nd suppl., 378; *Hagers Handb. Pharm. Praxis* **vol. 1**, 755 (Berlin, 1930).

Tetrahydrate. Odorless, tasteless crystals or fine white powder. Stable to air. Becomes anhydr upon heating at 100° *in vacuo*. Very slightly sol in water, alcohol. Sol in dil mineral acids and in calcium gluconate solns. Practically insol in ether, chloroform.

USE: Pharmaceutic aid (stabilizer for calcium gluconate solns). As plasticizer in cement, concrete, mortar.

1703. Calcium Selenide. [1305-84-6] CaSe; mol wt 119.04. Ca 33.67%, Se 66.33%. Prepd by reducing $CaSeO_4$ in a stream of H_2 at 400-500°: Ehrlich in *Handbook of Preparative Inorganic Chemistry* **vol. 1**, G. Brauer, Ed. (Academic Press, New York, 2nd ed., 1963) p 939.

White powder. In air may turn red within a few minutes and light brown in a few hours. d 3.82. Decomposed by water. Treatment with HCl produces H_2Se gas, and red Se separates.

USE: In electron emitters, photoconductors, photoelectric cells, rectifiers.

1704. Calcium Silicate. [1344-95-2] Silicic acid calcium salt. Many different forms of calcium silicate are known. Among the most common forms are $CaSiO_3$, Ca_2SiO_4 and Ca_3SiO_5. Usually occur in hydrated form contg various percentages of water of crystallization. Names of calcium silicate minerals are: *afwillite; akermanite; calcium pectolith; centrallasite; crestmoreite; eaklite; foshagite; foshallasite; gjellebaekite; grammite; gyrolite; hillebrandite; larnite; okenite; parawollastonite; pseudo-wollastonite; riverside-ite; table spate; tobermorite; wollastonite; xonaltite; xonotlite.* Commercial calcium silicate sold for industrial use, such as *Micro-Cell* and *Silene*, is prepared synthetically to control its absorbing power. The usual method of prepn is from lime and diatomaceous earth under carefully controlled conditions: Boss, *Chem. Eng. News* **27**, 677 (1949); Steinour, *Chem. Rev.* **40**, 391 (1947). The commercial product is described here.

White or slightly cream-colored, free-flowing powder. Approximate analysis: CaO 19%, SiO_2 67%, H_2O 6 to 8%. d^{25} 2.10. Bulk density: 15-16 lb/cu ft. Absorbs 1 to 2.5 times its weight of liquids and still remains a free-flowing powder. Total absorption power for water about 600%, for mineral oil about 500%. Available surface area: 95 to 175 m^2/g. Ultimate particle size: 0.02 to 0.07 μ. pH of aq slurry 8.0 to 10.0. Practically insol in water. Forms a siliceous gel with mineral acids.

Caution: Potential symptoms of overexposure are irritation of eyes, skin, upper respiratory system. *See NIOSH Pocket Guide to Chemical Hazards* (DHHS/NIOSH 97-140, 1997) p 48.

USE: Constituent (produced *in situ*) of lime glass, portland cement; reinforcing filler in elastomers and plastics; absorbent for liquids, gases, vapors; as anti-caking agent, suspension agent, pigment and pigment extender; binder for refractory material; in chromatography; in road construction.

1705. Calcium Stearate. [1592-23-0] Octadecanoic acid calcium salt (2:1); stearic acid calcium salt. $C_{36}H_{70}CaO_4$; mol wt 607.03. C 71.23%, H 11.62%, Ca 6.60%, O 10.54%. The commercial prepn also contains palmitate. Prepn: Harrison, *Biochem. J.* **18**, 1222 (1924); Kebrich, Petrot, **US 2650932** (1953 to National Lead).

Granular, fatty powder. Bulk density ~20 lb/cu ft, mp 147-149° (determined by gradient bar). Practically insol in water, ether, chloroform, acetone, cold alcohol; slightly sol in hot alcohol, in hot vegetable and mineral oils; quite sol in hot pyridine.

USE: For waterproofing fabrics, cement, stucco, explosives; as a releasing agent for plastic molding powders; as a stabilizer for polyvinyl chloride resins; lubricant; in pencils and wax crayons. Food grade calcium stearate, derived from edible tallow, is used as a conditioning agent in certain food and pharmaceutical products.

1706. Calcium Stearyl-2 Lactylate. [5793-94-2] Octadecanoic acid 2-(1-carboxyethoxy)-1-methyl-2-oxoethyl ester calcium salt (2:1); stearic acid ester with lactate of lactic acid calcium salt; calcium stelate; Verv-Ca. $C_{48}H_{86}CaO_{12}$; mol wt 895.28. C 64.40%, H 9.68%, Ca 4.48%, O 21.44%. Use in improving the mixing characteristics of flour: Thompson, Buddemeyer, *Cereal Chem.* **31**, 296 (1954); in improving whipping and baking properties of dried egg whites: Gorman, Keith, **US 2919992** (1960 to Seymour Foods). Metabolism: J. C. Phillips *et al.*, *Food Cosmet. Toxicol.* **19**, 7 (1981).

Free flowing, nonhygroscopic powder. Sparingly sol in water. pH of a 2% aq suspension 4.7.

USE: Dough conditioner in yeast-leavened bakery products; emulsifier in cosmetic and pharmaceutical industry.

1707. Calcium Succinate. [140-99-8] Butanedioic acid calcium salt (1:1); succinic acid calcium salt; Artume. $C_4H_4CaO_4$; mol wt 156.15. C 30.77%, H 2.58%, Ca 25.67%, O 40.98%. $CaC_4H_4O_4$. Prepn: *Beilstein* vol. **2**, 607, 2nd suppl., 548, 3rd suppl., 1657.

Trihydrate. Needles or granules. Slightly sol in water; practically insol in alcohol; sol in dil acids.

THERAP CAT: Combined with salicylates for rheumatic fever and rheumatoid arthritis.

1708. Calcium Sulfate. [7778-18-9] Sulfuric acid calcium salt (1:1). CaO_4S; mol wt 136.13. Ca 29.44%, O 47.01%, S 23.55%. $CaSO_4$. *Review:* R. J. Wenk, P. L. Henkels in *Kirk-Othmer Encyclopedia of Chemical Technology* vol. **4** (Wiley-Interscience, New York, 3rd ed., 1978) pp 437-448.

The natural form of anhydrous calcium sulfate is known as the mineral **anhydrite**; also as **karstenite, muriacite, anhydrous sulfate of lime, anhydrous gypsum**. Crystals are orthorhombic, color varies, e.g., white with blue, gray or reddish tinge, or brick red. d 2.96. Hardness 3-3.5 (Mohs'). Sol in water (18.75°) 0.2 pts/100 pts. **Insoluble anhydrite** or **dead-burned gypsum** which has the same crystal structure as the mineral is obtained upon complete dehydration of gypsum at above 650°. **Soluble anhydrite** is obtained in granular or powder form by complete dehydration of gypsum at below 300° in an electric oven. Estimated pore space is 38% by volume. Possesses high affinity for water and will absorb 6.6% of its weight of water forming the stable hemihydrate.

Hemihydrate. [26499-65-0] Dried calcium sulfate; dried gypsum; plaster of Paris; Annalin. $CaO_4S.\frac{1}{2}H_2O$; mol wt 145.14. Fine, odorless, tasteless powder. When mixed with water, sets to a hard mass. *Keep well closed.*

Dihydrate. [13397-24-5] Native calcium sulfate; precipitated calcium sulfate; gypsum; alabaster; selenite; terra alba; satinite; mineral white; satin spar; light spar. $CaO_4S.2H_2O$; mol wt 172.16. Fine, white to slightly yellow-white, odorless powder. d 2.32. It loses only part of its water at 100-150°. Sol in 3*N* HCl; slightly sol in water; very slowly sol in glycerol. Practically insol in most organic solvents.

Caution: Potential symptoms of overexposure to anhydrous compd, dihydrate or hemihydrate are irritation of eyes, skin, mucous membranes, respiratory system; cough. Potential symptoms of overexposure to anhydrous compd also include conjunctivitis; rhinitis, epistaxis. Potential symptoms of overexposure to dihydrate also include sneezing, rhinorrhea. *See NIOSH Pocket Guide to Chemical Hazards* (DHHS/NIOSH 97-140, 1997) pp 48, 154, 260.

USE: *Anhydrous:* Insol anhydrite is used in cement formulations and as a paper filler. Soluble anhydride, because of its strong tendency to absorb moisture, is useful as a drying agent for solids, organic liquids and gases; the desiccant used in laboratory and industry is known under the name **Drierite**. This material can be regenerated repeatedly and reused without noticeable decrease in its desiccating efficiency. The *hemihydrate* is used for wall plasters; wallboard; tiles and blocks for the building industry; moldings; statuary; in the paper industry. The *dihydrate* is used in the manuf of portland cement; in soil treatment to neutralize alkali carbonates and to prevent loss of volatile and dissolved nitrogenous compounds by volatilization and leaching; for the manuf of plaster of Paris, artificial marble; as a white pigment, filler or glaze in paints, enamels, pharmaceuticals, paper, insecticide dusts, yeast manuf, water treatment, polishing powders; in the manuf of sulfuric acid, CaC_2, $(NH_4)_2SO_4$, porous polymers; in the determn of oxalates. Pharmaceutic aid (in plaster casts).

1709. Calcium Sulfide. [20548-54-3] CaS; mol wt 72.14. Ca 55.56%, S 44.44%. Pure CaS prepd in the laboratory by heating pure $CaCO_3$ in a stream of $H_2S + H_2$ at 1000°: Ehrlich in *Handbook of Preparative Inorganic Chemistry* vol. 1, G. Brauer, Ed. (Academic Press, New York, 2nd ed., 1963) p 938. Crude calcium sulfide, erroneously called **sulfurated lime, calcic liver of sulfur, liver of lime, hepar calcis**, made by igniting calcium sulfate with carbonaceous matter. Contains not less than 55% CaS; the balance is calcium sulfate, sulfite and carbonate, and the "ash" from the carbonaceous material. *See Mellor's* vol III, p 740 (1928). **Luminous calcium sulfide** or **Canton's phosphorus** made by igniting a mixture of $CaCO_3$ and S with very small quantities of Bi or Mn salts, etc.: Verneuil, *Compt. Rend.* **103**, 600 (1886); *Mellor's*, *loc. cit.*

White powder if pure; crude and luminous calcium sulfide may be yellowish to pale-gray. Odor of H_2S in moist air; unpleasant alkaline taste. Oxidizes in dry air and dec in moist air. mp >2000°. d 2.59. Slightly sol in cold, more sol in hot water with partial decompn; freely sol in solns of ammonium salts; practically insol in alcohol; dec even by weak acids, evolving H_2S. *Keep well closed.*

USE: In phosphors; as lubricant additive. Pure CaS used in electron emitters. Luminous CaS used for making luminous paints or varnishes.

THERAP CAT (VET): Has been used in chronic suppurative lesions.

1710. Calcium Sulfite. [10257-55-3] Sulfurous acid calcium salt (1:1). CaO_3S; mol wt 120.14. Ca 33.36%, O 39.95%, S 26.69%. $CaSO_3$. Prepn: *Gmelins, Calcium* (8th ed.) **28B**, 107-108, 660-674 (1958).

Dihydrate. Crystals or powder. Slowly oxidizes in air to $CaSO_4$. Slightly sol in water, alcohol; sol in SO_2 solns, acids with liberation of SO_2.

USE: Preserving cider and other fruit juices; as disinfectant of brewing vats; antichlor in bleaching textiles; in sugar manuf; in paper pulp cooking; in cement.

1711. Calcium Tartrate. [3164-34-9] (2R,3R)-2,3-Dihydroxybutanedioic acid calcium salt (1:1). $C_4H_4CaO_6$; mol wt 188.15. C 25.53%, H 2.14%, Ca 21.30%, O 51.02%. $CaC_4H_4O_6$. A byproduct of the wine industry. Prepn from wine dregs: Dabul, **US 3114770** (1963 to Orandi & Massera). *See also* the processes mentioned under L-tartaric acid.

Tetrahydrate. Powder. Slightly sol in water (from about 0.04% at 10° to about 0.2% at 85°) or in alcohol; sol in dil HCl or HNO_3.

USE: As preservative for fruits, vegetables, seafoods; in deodorization of fish; as antacid.

1712. Calcium Thiocyanate. [2092-16-2] Thiocyanic acid calcium salt (2:1); calcium rhodanate; calcium sulfocyanate. C_2-CaN_2S_2; mol wt 156.23. C 15.38%, Ca 25.65%, N 17.93%, S 41.04%. $Ca(SCN)_2$. Prepn: *Gmelins, Calcium* (8th ed.) **28B**, 972-976 (1958).

Tetrahydrate. Hygroscopic crystals or cryst powder. Dec on heating above 160°. Very sol in water; sol in methanol, ethanol, acetone. *Keep well closed.*

USE: In manuf of acrylonitrile polymers; for parchmentizing; for stiffening of textiles; soln as a solvent for textiles.

1713. Calcium Thioglycollate. [814-71-1] 2-Mercaptoacetic acid calcium salt (2:1); Depil. $C_4H_6CaO_4S_2$; mol wt 222.29. C 21.61%, H 2.72%, Ca 18.03%, O 28.79%, S 28.85%. Prepn: Hoshall, *J. Assoc. Off. Agric. Chem.* **23**, 727 (1940).

Trihydrate. [5793-98-6] Prismatic rod crystals. Odorless or faint mercaptan odor; somewhat astringent and fetid taste. Slowly loses H_2O above 95°, darkens at 220° and partially fuses with decompn at 280-290°. Sol in water; very slightly sol in alcohol, chloroform; practically insol in ether, petr ether, benzene. Solns readily absorb CO_2 from air, pptg $CaCO_3$. *Keep well closed.*

USE: Depilatory; tanning leather; in hair-waving prepns.

1714. Calcium Tungstate(VI). [7790-75-2] Calcium tungsten oxide ($CaWO_4$). CaO_4W; mol wt 287.91. Ca 13.92%, O 22.23%, W 63.85%. $CaWO_4$. Occurs in nature as the mineral *scheelite*. The article of commerce is usually made by pptn: Boericke, Boericke, US 2390687 (1945); prepn by heating a stoichiometric mixture of CaO or $CaCO_3$ and tungstic acid: Kroger, *Nature* **159**, 674 (1947); prepn of single crystals: Uitert, Soden, *J. Appl. Phys.* **31**, 328 (1960).

Tetragonal crystals. d 6.06. Practically insol in water. Dec by hot HCl or HNO_3.

USE: Very small crystals have been used for injection into malignant tumors, etc., thus affording by transillumination a means of x-ray treatment; for preparing screens for x-ray observations and photographs; in luminous paints; in scintillation counters.

1715. Calendula. Marigold; Mary-bud; gold-bloom; holligold. Dried, ligulate florets of *Calendula officinalis* L., *Compositae*. *Habit.* Southern Europe and Levant; cultivated everywhere in gardens. *Constit.* Volatile oil; calendulin, carotenoid pigments, a saponin which on hydrolysis yields oleanolic acid, bitter principle (caledin). *Refs:* Zimmerman, *Helv. Chim. Acta* **29**, 445 (1946); Gedeon, *Pharmazie* **6**, 547 (1951); **9**, 922 (1954), Kasprzyk, *C.A.* **47**, 6918c (1953); Suchy, Herout, *Collect. Czech. Chem. Commun.* **26**, 890 (1961).

THERAP CAT: Anti-inflammatory (topical).

1716. Calfactant. [183325-78-2] CLSE; CLL; Infasurf. Chloroform/methanol extract of calf lung lavage fluid. Consists of 35 mg/ml phospholipid, 55-70% of which is dipalmitoylphosphatidylcholine, and of low mol wt surfactant proteins. Also contains 0.9 mg/ml fatty acids and 1.8 mg/ml neutral lipids. *See also:* Beractant, Poractant Alfa. Prepn: R. H. Notter *et al.*, *Pediatr. Res.* **16**, 515 (1982). Physiologic effects in preterm lambs: J. J. Cummings *et al.*, *Am. Rev. Respir. Dis.* **145**, 999 (1992). Surface tension: E. M. Scarpelli *et al.*, *Am. J. Perinatol.* **9**, 414 (1992). Separation of hydrophobic constituents: S. B. Hall *et al.*, *J. Lipid Res.* **35**, 1386 (1994). Clinical trial in neonates: M. S. Kwong *et al.*, *Pediatrics* **76**, 585 (1985). Efficacy of prophylactic administration to neonates: J. Katwinkel *et al.*, *ibid.* **92**, 90 (1993). Clinical trial in pediatric acute lung injury: D. F. Wilson *et al.*, *J. Am. Med. Assoc.* **293**, 470 (2005).

THERAP CAT: Pulmonary surfactant. In treatment of respiratory distress syndrome.

1717. Calicheamicins. LL-E33288; CLM. Family of 15-20 enediyne antitumor antibiotics produced by *Micromonospora echinospora* ssp. *calichensis*, a bacterium isolated from chalky soil or a caliche. Naming is based on TLC mobility using Greek letters with subscripts and on the halogen substitution which is indicted by the superscript. Binds to the minor groove of DNA and initiates double stranded DNA cleavage via a radical abstraction process. This type of damage is usually lethal as it is not repairable by the cell. Isoln: W. M. Maiese *et al.*, *J. Antibiot.* **42**, 558 (1989). Structure elucidation: M. D. Lee *et al.*, *J. Am. Chem. Soc.* **109**, 3464, 3466 (1987);

and NMR studies: *eidem, ibid.* **114**, 985 (1992). DNA cleavage behavior: N. Zein *et al.*, *Science* **240**, 1198 (1988); S. Walker *et al.*, *Proc. Natl. Acad. Sci. USA* **89**, 4608 (1992); M. Chatterjee *et al.*, *J. Am. Chem. Soc.* **118**, 1938 (1996). Series of articles on total synthesis of γ_1^I: *ibid.* **115**, 7593-7635 (1993). Book on calicheamicins and related compounds: *Enediyne Antibiotics As Antitumor Agents,* D. B. Borders, T. W. Doyle, Eds. (Marcel Dekker, Inc., New York, 1995) 478 pp.

Calicheamicin γ_1

Calicheamicin γ_1^I. [108212-75-5] CLM γ_1^I. $C_{55}H_{74}IN_3O_{21}S_4$; mol wt 1368.34. White amorphous powder, $[\alpha]_D^{26} -124°$ (c = 0.98%, EtOH).

1718. Californium. [7440-71-3] Cf; at. no. 98; valence 4, 3, 2. Man-made, radioactive element. No stable nuclides; known isotopes (mass numbers): 239-256. Longest-lived known isotope: 251 ($T_{1/2}$ 898 years, α-emitter, rel. at. mass 251.0796). Prepn of isotope ^{245}Cf ($T_{1/2}$ 43.6 min) by bombarding ^{242}Cm with helium ions: S. G. Thompson *et al.*, *Phys. Rev.* **78**, 298 (1950); *eidem, ibid.* **80**, 790. Prepn of ^{249}Cf metal ($T_{1/2}$ 351 years, α-emitter, rel. at. mass 249.0748) by reduction of Cf_2O_3 with lanthanum metal: Haire, Baybarz, *J. Inorg. Nucl. Chem.* **36**, 1295 (1974). Medical uses of ^{252}Cf (α-decay, $T_{1/2}$ half-life 2.645 years, rel. at. mass 252.0816; spontaneous fission $T_{1/2}$ half-life 85.5 years): Seaborg, *Handb. Exp. Pharmakol.* **36**, 929 (1973). *Reviews:* Cunningham, *J. Chem. Educ.* **36**, 32-37 (1959); M. Haissinsky, J. P. Adloff, *Radiochemical Survey of the Elements* (Elsevier, New York, 1965) pp 28-29; C. Keller, *The Chemistry of the Transuranium Elements* (Verlag Chemie, Weinheim, English Ed., 1971) pp 567-581; Silva, "Trans-Curium Elements" in *MTP Int. Rev. Sci.: Inorg. Chem., Ser. One* **vol. 8**, A. G. Maddock, Ed. (University Park Press, Baltimore, 1972) pp 71-105; *Comprehensive Inorganic Chemistry* **vol.5**, J. C. Bailar, Jr. *et al.*, Eds. (Pergamon Press, Oxford, 1973) *passim*; *Handb. Exp. Pharmakol.* **36**, 689-928 (1973); R. G. Haire in *The Chemistry of the Actinide Elements* vol. 2, J. J. Katz *et al.*, Eds. (Chapman and Hall, New York, 1986) pp 1025-1070. Review of radiobiology and therapeutic applications: Y. Maruyama *et al.*, *Oncology* **35**, 172 (1978); of use in brachytherapy for cervical cancer: *idem et al.*, *Cancer* **68**, 1189-1197 (1991). Review of production, distribution and uses of ^{252}Cf: R. C. Martin *et al.*, *Appl. Radiat. Isot.* **53**, 785-792 (2000).

Metal; two crystalline forms: double hexagonal close-packed α-form, d 15.10, exists below 900°; face-centered cubic β-form, d 8.74, (Katz *et al., loc. cit.* **vol. 2**, p. 1150). mp 900±30° (Haire, Baybarz). Slowly oxidized in air at room temp; rate increases with increased moisture in air. Oxidized when warmed in air; reacts when heated with nitrogen, hydrogen or a chalogen. Reacts rapidly with dry hydrogen halides and with aq mineral acids.

Caution: Radiation hazard; handling requires special equipment and shielding facilities (Katz *et al., loc. cit.* **vol. 2**, p. 1128).

USE: ^{252}Cf as neutron source; startup source for nuclear reactors; in nuclear reactor fuel rod scanners; for neutron radiography of weapons components.

THERAP CAT: ^{252}Cf as antineoplastic (radiation source).

1719. Calixarenes. Molecular baskets into which smaller molecules can fit. Name is derived from the Greek "calix" for "chal-

---

ice" or "cup" and arene for the aryl residues; the name originally referred to the phenolic tetramer linked by methylene groups. The nomenclature uses calix[*n*]arene where *n* is the number of the aryl groups. Modifications to the phenol are indicated by prefixes; whereas total changes of the aryl group are indicated in front of the "arene". Prepn: A. Zinke, E. Ziegler, *Ber.* **77**, 264 (1944); A. Zinke *et al.*, *Monatsh. Chem.* **79**, 438 (1948). Modification of basic structures: J. L. Atwood *et al.*, *Angew. Chem. Int. Ed.* **32**, 1093 (1993). Book: *Calixarenes*, C. D. Gutsche, Ed. (Royal Society of Chemistry, Cambridge, U.K., 1989) 222 pp.

1720.   Calmagite. [3147-14-6] 3-Hydroxy-4-[2-(2-hydroxy-5-methylphenyl)diazenyl]-1-naphthalenesulfonic acid; 3-hydroxy-4-[(2-hydroxy-5-methylphenyl)azo]-1-naphthalenesulfonic acid; 1-(1-hydroxy-4-methyl-2-phenylazo)-2-naphthol-4-sulfonic acid. $C_{17}H_{14}N_2O_5S$; mol wt 358.37. C 56.98%, H 3.94%, N 7.82%, O 22.32%, S 8.95%. Prepn from 1-amino-2-naphthol-4-sulfonic acid and *p*-cresol: Lindstrom, Diehl, *Anal. Chem.* **32**, 1123 (1960).

Red crystals from acetone. Sol in water. Absorption max (pH 10.10): 610 nm (ε 20300). Functions as acid-base indicator: Aq solns are bright red at low pH, red at pH 7.1 to 9.1, blue at pH 9.1 to 11.4. The blue color at pH 10 is changed to red by the addition of calcium or magnesium.

USE: As indicator in titration of Ca or Mg with EDTA.

1721.   Calmodulin. [77107-46-1] CaM; calcium-dependent regulator protein; CDR. A calcium-binding multifunctional regulatory protein, ubiquitously distributed in eukaryotic cells. It functions as an intracellular intermediary for calcium ions and activates a number of enzymes involved in fundamental cell processes, such as protein phosphorylation, contractile processes, and metabolism of cyclic nucleotides, of glycogen and of calcium, as well as in other metabolic reactions. Originally discovered as a protein activator of cyclic 3′,5′-nucleotide phosphodiesterase: W. Y. Cheung, *Biochem. Biophys. Res. Commun.* **29**, 478 (1967); idem, *Biochim. Biophys. Acta* **191**, 303 (1969); idem, *Biochem. Biophys. Res. Commun.* **38**, 533 (1970); S. Kakiuchi *et al.*, *ibid.* **41**, 1104 (1970). Calcium-binding activity: T. S. Teo, J. H. Wang, *J. Biol. Chem.* **248**, 5950 (1973). Calmodulin is a relatively small, acidic, stable monomer of mol wt 15,000-19,000, lacking cysteine, hydroxyproline, and tryptophan. It has a high content of acidic amino acids and low tyrosine content; almost all calmodulins isolated also contain a single, fully trimethylated lysyl residue. Purification from bovine brain: Y. M. Lin *et al.*, *ibid.* **249**, 4943 (1974); eidem, *Methods Enzymol.* **39**, Pt. C, 262 (1974). Amino acid sequence of CaM from bovine brain: T. C. Vanaman in *Calcium Binding Proteins and Calcium Function*, R. H. Wasserman *et al.*, Eds. (Elsevier, New York, 1977) pp 107-116; D. M. Watterson *et al.*, *J. Biol. Chem.* **255**, 962 (1980); from rat testis: J. R. Dedman *et al.*, *ibid.* **253**, 343 (1978); from sea invertebrate, *Renilla reniformis:* T. C. Vanaman, F. Sharief, *Fed. Proc.* **38**, 788 (1979). Radioimmunoassay: R. W. Wallace, W. Y. Cheung, *J. Biol. Chem.* **254**, 6564 (1979); J. C. Chafouleas *et al.*, *ibid.* 10262. CaM has four Ca^{2+}-binding sites; binding to any one of the sites results in a conformational change, which is required for calmodulin to regulate enzyme systems. Conformational transition study: C. B. Klee, *Biochemistry* **16**, 1017 (1977). An important feature of calmodulin's structure is the presence of four homologous internal amino acid sequences, called domains. Each of these domains (one for each bound calcium) are reportedly "E-F Hands", a structural concept described by R. H. Kretsinger in *Calcium Transport in Contraction and Secretions*, E. Carafoli *et al.*, Eds. (Elsevier, Amsterdam, 1975) pp 469-478. Calmodulins obtained from a wide range of phylogenetically different sources are similar in amino acid sequence and in physicochemical and biological properties; hence, the protein lacks both species and tissue specificity and appears to be structurally and functionally conserved throughout evolution. *Reviews:* W. Y. Cheung, *Science* **207**, 19-27 (1980); A. R. Means, J. R. Dedman, *Nature* **285**, 73-77 (1980); C. B. Klee *et al.*, *Annu. Rev. Biochem.* **49**, 489-515 (1980); A. R. Means, *Recent Prog. Horm. Res.* **37**, 333-367 (1981); Y. M. Lin, *Mol. Cell. Biochem.* **45**, 101-112 (1982). Books: *Ann. N.Y. Acad. Sci.* **356**, entitled "Calmodulin and Cell Functions", D. M. Watterson, F. F. Vincenzi, Eds. (1980) 455 pp; *Calcium and Cell Function* vol. 1, W. Y. Cheung, Ed. (Academic Press, New York, 1980) 395 pp.

Isoelectric pt 3.9-4.3. $\varepsilon_{276nm}^{1\%}$ 1.8; $\varepsilon_{280nm}^{1\%}$ 2.1. Stable when subjected to heat at neutral or acidic pH.

1722.   Calotropin. [1986-70-5] [2α,3β,5α]-14-Hydroxy-19-oxo-2,3-[[(2*S*,3*S*,4*S*,6*R*)-tetrahydro-3,4-dihydroxy-6-methyl-2*H*-pyran-3,2-diyl]bis(oxy)]card-20(22)-enolide. $C_{29}H_{40}O_9$; mol wt 532.63. C 65.40%, H 7.57%, O 27.03%. African arrow poison isolated from milk sap of *Calotropis procera* Dryand., *Asclepiadaceae*. Isolation: Lewin, *Arch. Exp. Pathol. Pharmakol.* **71**, 142 (1913); Hesse *et al.*, *Ann.* **526**, 252 (1936); **566**, 130 (1950); Rajagopalan *et al.*, *Helv. Chim. Acta* **38**, 1809 (1955). Isoln from *Asclepias curassavica* L., *Asclepiadaceae:* S. M. Kupchan *et al.*, *Science* **146**, 1685 (1964). Structure: G. Hesse, G. Lettenbauer, *Ann.* **623**, 142 (1959); Hesse *et al.*, *ibid.* **625**, 157, 161 (1959); D. G. H. Crout *et al.*, *Tetrahedron Lett.* **1963**, 63; *J. Chem. Soc.* **1964**, 2187. Extraction from *C. procera* R.Br. and toxicity: F. Brüschweiler *et al.*, *Helv. Chim. Acta* **52**, 2086 (1969). Revised structure: *eidem, ibid.* 2276. Sequestration by larvae of Monarch butterfly *Danaus plexippus* L.: J. N. Sieber *et al.*, *J. Chem. Ecol.* **6**, 321 (1980). Quantitative analysis of cardenolides in latex and leaves of *C. procera: eidem, Phytochemistry* **21**, 2343 (1982). Biosynthesis of labelled compd: M. S. Lee, J. N. Sieber, *ibid.* **22**, 923 (1983).

Rectangular platelets from alcohol or ethyl acetate, mp 223° (dec). $[\alpha]_D^{18}$ +66.8° (in methanol). Sol in water, alc. Practically insol in ether. uv max: 217, 310 nm (log ε 4.21, 1.49). MLD i.v. in cats: 0.12 mg/kg (Brüschweiler).

1723.   Calumba. Colombo. Root of *Jatrorrhiza palmata* (DC.) Miers (*J. columba* Miers), *Menispermaceae. Habit.* Eastern Africa. *Constit.* Columbin, chasmanthin, palmarin (isomer of chasmanthin), jatrorrhizine, columbic acid, columbamine; contains no tannin: Barton, Elad, *J. Chem. Soc.* **1956**, 2085, 2090. *Review:* Feist, *Arzneim.-Forsch.* **1**, 418 (1951).

THERAP CAT (VET): Has been used as a stomachic.

1724.   Calusterone. [17021-26-0] (7β,17β)-17-Hydroxy-7,17-dimethylandrost-4-en-3-one; 7β,17α-dimethyltestosterone; methosarb; U-22550. $C_{21}H_{32}O_2$; mol wt 316.49. C 79.70%, H 10.19%, O 10.11%. Synthetic androgen epimeric with bolasterone, *q.v.* Has been used in treatment of breast cancer. Prepn: Campbell, Babcock, US 3029263; US 3341557 (1962, 1967 both to Upjohn). Clinical studies: Gordan *et al.*, *J. Am. Med. Assoc.* **219**, 483 (1972).

Crystals from acetone, mp 127-129°. $[\alpha]_D$ +57° (CHCl$_3$). uv max (alcohol): 243 nm.

Note: This is a controlled substance (anabolic steroid): **21 CFR, 1308.13**, as defined in 1300.01.

1725. Calycanthine. [595-05-1] (4bS,5R,10bS,11R)-5,6,-11,12-Tetrahydro-13,18-dimethyl-5,10b:11,4b-bis(iminoethano)-dibenzo[c,h][2,6]naphthyridine. C$_{22}$H$_{26}$N$_4$; mol wt 346.48. C 76.26%, H 7.56%, N 16.17%. In *Calycanthus floridus* L., *C. glaucus* Willd.; *Chimonanthus praecox* (L.) Link, Calycanthaceae. First isoln: Eccles, *Proc. Am. Pharm. Assoc.* **84**, 382 (1888). Extraction procedure: Manske-Marion, *Can. J. Res.* **17B**, 293 (1939). Structure: Woodward *et al.*, *Proc. Chem. Soc. London* **1960**, 76; Hamor *et al.*, *ibid.* 78; Hamor, Robertson, *J. Chem. Soc.* **1962**, 194. Configuration: Clayton *et al.*, *Tetrahedron* **18**, 1495 (1962). Synthesis of DL-form: Hendrickson *et al.*, *ibid.* **20**, 565 (1964); Hall *et al.*, *ibid.* **23**, 4131 (1967).

Crystals, mp 245° (evac tube). $[\alpha]_D^{18}$ +684° (c = 1.2 in abs alc). uv max (95% ethanol): 250, 309 nm (log ε 4.28, 3.80). Alkaline reaction to litmus. Freely sol in alc, chloroform; sol in ether, acetone, pyridine; slightly sol in water.

Monohydrate. Orthorhombic bipyramidal crystals from water, mp 220°.

Caution: Highly toxic. Can cause violent convulsions, paralysis, cardiac depression.

1726. Calyculin A. [101932-71-2] N-[(3S)-3-[4-[[(1E)-3-[(2R,3R,5R,7S,8S,9R)-2-[(1S,3S,4S,5R,6R,7E,9E,11E,13Z)-14-Cyano-3,5-dihydroxy-1-methoxy-4,6,8,9,13-pentamethyl-7,9,11,13-tetradecatetraen-1-yl]-9-hydroxy-4,4,8-trimethyl-3-(phosphonooxy)-1,6-dioxaspiro[4.5]dec-7-yl]-1-propen-1-yl]-2-oxazolyl]butyl]-4-deoxy-4-(dimethylamino)-5-O-methyl-L-ribonamide; ($-$)-calyculin A. C$_{50}$H$_{81}$N$_4$O$_{15}$P; mol wt 1009.18. C 59.51%, H 8.09%, N 5.55%, O 23.78%, P 3.07%. One of a family of antitumor marine toxins which inhibits protein phosphatases 1 and 2a. Isoln from the marine sponge, *Discodermia calyx*, and structure determn: Y. Kato *et al.*, *J. Am. Chem. Soc.* **108**, 2780 (1986); of calyculins B-D: *eidem, J. Org. Chem.* **53**, 3930 (1988); of E-H: S. Matsunaga *et al.*, *Tetrahedron* **47**, 2999 (1991). Revised stereochemistry: S. Matsunaga, N. Fusetani, *Tetrahedron Lett.* **32**, 5605 (1991). Synthesis: N. Tanimoto *et al.*, *Angew. Chem. Int. Ed.* **33**, 673 (1994). Inhibition of protein phosphatases: H. Ishihara *et al.*, *Biochem. Biophys. Res. Commun.* **159**, 871 (1989); selectivity *in vivo*: B. Favre *et al.*, *J. Biol. Chem.* **272**, 13856 (1997). Binding model: M. K. Lindvall *et al.*, *ibid.* 23312. Use as inhibitor in biological studies: M. C. Arufe *et al.*, *Endocrine* **11**, 235 (1999); I. Bize *et al.*, *Am. J. Physiol.* **277**, C926 (1999).

Colorless needles from a mixture of hexane, ether and acetone, mp 247-249°. $[\alpha]_D$ $-60°$ (c = 0.1 in ethanol). uv max (ethanol): 230, 342 nm (ε 12000, 19000).

USE: Inhibitor of phosphatases 1 and 2a.

1727. Camazepam. [36104-80-0] N,N-Dimethylcarbamic acid 7-chloro-2,3-dihydro-1-methyl-2-oxo-5-phenyl-1H-1,4-benzodiazepin-3-yl ester; 7-chloro-1,3-dihydro-3-hydroxy-1-methyl-5-phenyl-2H-1,4-benzodiazepin-2-one dimethylcarbamate (ester); 7-chloro-1,3-dihydro-3-(N,N-dimethylcarbamoyl)-1-methyl-5-phenyl-2H-1,4-benzodiazepin-2-one; SB-5833; Albego. C$_{19}$H$_{18}$-ClN$_3$O$_3$; mol wt 371.82. C 61.38%, H 4.88%, Cl 9.53%, N 11.30%, O 12.91%. Prepn: G. Ferrari, C. Casagrande, **DE 2142181**; *eidem*, **US 3799920** (1972, 1974 both to Siphar). Metabolism: F. Marcucci *et al.*, *J. Pharm. Sci.* **67**, 1470 (1978). Pharmacology: L. Merlo *et al.*, *Arzneim.-Forsch.* **24**, 1759 (1974); R. Ferrini *et al.*, *ibid.* 2029. Clinical studies: A. Tammaro *et al.*, *ibid.* **27**, 2177 (1978); S. Carrara *et al.*, *Eur. J. Clin. Pharmacol.* **13**, 335 (1978).

White crystalline powder from ethyl acetate, mp 173-174°. Sol in alcohol, moderately sol in water. LD$_{50}$ in mice, rats (mg/kg): 970, >4000 orally (Ferrini).

Note: This is a controlled substance (depressant): **21 CFR, 1308.14.**

THERAP CAT: Anxiolytic.

1728. Cambendazole. [26097-80-3] N-[2-(4-Thiazolyl)-1H-benzimidazol-6-yl]carbamic acid 1-methylethyl ester; 2-(4-thiazolyl)-5-benzimidazolecarbamate isopropyl ester; 5-isopropoxycarbonylamino-2-(4-thiazolyl)benzimidazole; 5-isopropoxycarbonylaminothiabendazole; isopropyl 2-(4-thiazolyl)-5-benzimidazolecarbamate; MK-905; Camvet; Noviben. C$_{14}$H$_{14}$N$_4$O$_2$S; mol wt 302.35. C 55.62%, H 4.67%, N 18.53%, O 10.58%, S 10.60%. Prepn and activity studies: Hoff, Fisher, **ZA 6800351** (1969 to Merck & Co.), *C.A.* **72**, 90461q (1970); Hoff *et al.*, *Experientia* **26**, 550 (1970). Clinical studies with sheep, cattle and swine: Egerton *et al.*, *Res. Vet. Sci.* **11**, 193, 495, 590 (1970).

Odorless, white crystalline solid, mp 238-240° (dec). Sol in alcohol, dimethylformamide; sparingly sol in acetone; slightly sol in benzene; very slightly sol in 0.1M HCl. Practically insol in isooctane and water (0.02 mg/ml). Stable in acid and base in range of pH 1 to 12. uv max (0.1N HCl): 319, 232 nm (A$_{1cm}^{1\%}$ 740, 670).

THERAP CAT (VET): Anthelmintic.

1729. Cameleons. [205599-45-7] Yellow cameleons; YCs. Engineered proteins designed to measure intracellular calcium concentrations via FRET. Characterized by four domains: a modified green fluorescent protein, *q.v.*; a calmodulin; a cyan fluorescent protein; the calmodulin binding peptide, M13. Prepn: A. Miyawaki *et al.*, *Nature* **388**, 882 (1997); modification: *eidem, Proc. Natl. Acad. Sci. USA* **96**, 2135 (1999). Use in monitoring calcium changes at vesicle surface: E. Emmanouilidou *et al.*, *Curr. Biol.* **9**, 915 (1999); in the endoplasmic reticulum: R. Yu, P. M. Hinkle, *J. Biol. Chem.* **275**, 23648 (2000). Use as a tool for nervous system function in model organisms: R. Kerr *et al.*, *Neuron* **26**, 583 (2000).

USE: Indicator for intracellular calcium.

1730. Camostat. [59721-28-7] 4-[[4-[(Aminoiminomethyl)-amino]benzoyl]oxy]benzeneacetic acid 2-(dimethylamino)-2-oxo-ethyl ester; *N,N*-dimethylcarbamoylmethyl-*p*-(*p*-guanidinobenzoyloxy)phenylacetate. $C_{20}H_{22}N_4O_5$; mol wt 398.42. C 60.29%, H 5.57%, N 14.06%, O 20.08%. Orally active, non-peptide proteolytic enzyme inhibitor with anti-trypsin and anti-plasmin activities, related structurally to gabexate, *q.v.* Prepn: S. Fujii *et al.*, **DE 2548886**; *eidem*, **US 4021472** (1976, 1977 both to Ono). Enzyme inhibition study: Y. Tamura *et al.*, *Biochim. Biophys. Acta* **484**, 417 (1977). Metabolism: I. Midgley *et al.*, *Xenobiotica* **24**, 79 (1994). Clinical trial in acute pancreatitis: N. Tanaka *et al.*, *Adv. Exp. Med. Biol.* **120B**, 367 (1979); in treatment of dyspepsia in mild pancreatic disease: J. K. Sai *et al.*, *J. Gastroenterol.* **45**, 335 (2010).

Methanesulfonate. [59721-29-8] Camostat mesylate; FOY-305; Foipan. $C_{20}H_{22}N_4O_5.CH_3SO_3H$; mol wt 494.52. Solid from methanol/ether, mp 150-155°. Sol in water.

THERAP CAT: Protease inhibitor.

1731. Campesterol. [474-62-4] (3β,24R)-Ergost-5-en-3-ol. $C_{28}H_{48}O$; mol wt 400.69. C 83.93%, H 12.08%, O 3.99%. Small amounts are found in rape-seed oil derived from *Brassica campestris* L., *Cruciferae*, in soybean oil, and in wheat germ oil. Isoln: Fernholz, MacPhillamy, *J. Am. Chem. Soc.* **63**, 1155 (1941). Structure: Fernholz, Ruigh, *ibid.* 1157. Synthesis: Tarzia *et al.*, *Gazz. Chim. Ital.* **97**, 102 (1967), *C.A.* **67**, 32883q (1967).

Crystals from acetone, mp 157-158°. $[\alpha]_D^{23}$ −33° (22.5 mg in 5 ml chloroform).

Acetate. $C_{30}H_{50}O_2$. Crystals from alc, mp 137-138°. $[\alpha]_D^{23}$ −35° (28.8 mg in 1 ml chloroform).

1732. Camphene. [79-92-5] 2,2-Dimethyl-3-methylenebicyclo[2.2.1]heptane; 2,2-dimethyl-3-methylenenorbornane; 3,3-dimethyl-2-methylenenorcamphane. $C_{10}H_{16}$; mol wt 136.24. C 88.16%, H 11.84%. Occurs in many essential oils, such as turpentine (*levo* and *dextro* forms), in cypress oil (*dextro* form), in camphor oil from species of *Lauraceae* (*dextro*), in bergamot oil, in oil of citronella, neroli, ginger, valerian. Reviews on isolation, preparation and properties: J. L. Simonsen, *The Terpenes* **vol. II** (Cambridge Univ. Press, 1949) pp 280-322; E. Guenther, *The Essential Oils* **vol. II** (Van Nostrand, 1949) pp 66-70. Synthesis: G. W. Hana, H. Koch, *Ber.* **111**, 2527 (1968).

***dl*-Form.** Cubic crystals from alcohol. Large dodecahedra by slow sublimation. Volatilizes on exposure to air. Insipid odor. mp 51-52°. bp_{760} 158.5-159.5°; bp_{100} 92.4°; bp_{16} 55-56°. d_4^{54} 0.8422.

n_D^{54} 1.45514. Practically insol in water. Moderately sol in alcohol; sol in ether, cyclohexane, cyclohexene, dioxane, chloroform.

***d*-Form.** mp 52°. $[\alpha]_D^{17}$ +103.5° (c = 9.67 in ether). d_4^{50} 0.8486. n_D^{50} 1.4605.

***l*-Form.** mp 52°. $[\alpha]_D^{21}$ −119.11° (c = 2.33 in benzene). d_4^{54} 0.8422. n_D^{40} 1.4620.

1733. *d*-Camphocarboxylic Acid. [18530-30-8] (1R,2S,4R)-4,7,7-Trimethyl-3-oxobicyclo[2.2.1]heptane-2-carboxylic acid; *d*-2-oxo-3-bornanecarboxylic acid; *d*-3-camphorcarboxylic acid; *d*-2-oxo-3-camphanecarboxylic acid; *d*-3-carboxy-2-bornanone; *d*-3-carboxy-2-camphanone. $C_{11}H_{16}O_3$; mol wt 196.25. C 67.32%, H 8.22%, O 24.46%. Prepd by carboxylation of *d*-camphor: Brühl, *Ber.* **24**, 3373 (1891).

Crystals from benzene, water, ether, or 50% alcohol. mp 127-128°. Sparingly sol in cold water, more sol in warm water; sol in alcohol, ether, chloroform, in about 2 parts boiling benzene. Sparingly sol in cold benzene. Practically insol in cold petr ether; very slightly sol in boiling petr ether. For soln contg 0.38 g in 25 ml solvent: $[\alpha]_D$ +18° (benzene), +60° (alcohol), +73.3° (water).

Ammonium salt. [6799-26-4] Camphydryl; camphor solubilized; Canfoxil. $C_{11}H_{19}NO_3$; mol wt 213.28.

Basic bismuth salt. [4154-53-4] Angimuth. $C_{33}H_{46}Bi_2O_{11}$; mol wt 1036.68. Prepn: Raiziss, Clemence, **US 1921638** (1933 to Abbott). Powder. Odor of camphor. Practically insoluble in water. Soluble in methanol, ether, benzene, oils.

THERAP CAT: Basic bismuth salt formerly as antisyphilitic.

1734. Camphor. [76-22-2] 1,7,7-Trimethylbicyclo[2.2.1]-heptan-2-one; 2-bornanone; 2-camphanone; 2-keto-1,7,7-trimethyl-norcamphane; gum camphor; Japan camphor; Formosa camphor; laurel camphor. $C_{10}H_{16}O$; mol wt 152.24. C 78.90%, H 10.59%, O 10.51%. Naturally occuring in both the *d*- and *l*-forms; originally obtained commercially as the *d*-form from the camphor tree, *Cinnamomum camphora* T. Nees & Ebermeier, *Lauraceae*. Primarily manufactured from pinene as the racemate. History of isolation and production of natural and synthetic forms: I. Gubelmann, H. W. Elley, *Ind. Eng. Chem.* **26**, 589 (1934); J. M. Derfer, M. M. Derfer in *Kirk-Othmer Encyclopedia of Chemical Technology* **vol. 23** (John Wiley & Sons, New York, 4th ed., 1997) pp 865-866. Enantiomeric composition in oils of coriander, sage, and basil: F. Tateo *et al.*, *Anal. Commun.* **36**, 149 (1999). GC determn in human plasma: J. S. Valdez *et al.*, *J. Chromatogr. B* **729**, 163 (1999); in pharmaceutical formulation: E. Gonzálea-Penas *et al.*, *Chromatographia* **52**, 245 (2000). *Review: Camphor and Camphor Containing Products* (PB293503, 1979) 65 pp; J. S. Mossa, M. M. A. Hassan, *Anal. Profiles Drug Subs.* **13**, 28-93 (1984). Review of use as starting material for syntheses: T. Money, *Org. Synth.* **3**, 1-83 (1996).

d-form   *l*-form

White or colorless crystals or crystalline masses; also colorless to white translucent masses. Characteristic fragrant and penetrating odor. Pungent, aromatic taste. d_4^{25} 0.992. mp 179°. $bp_{101.3 kPa}$ 209°. Volatilizes slowly. uv max (CHCl₃): 292 nm. *Flammable.* Freely sol in carbon disulfide, solvent hexane, petr. benzin, fixed and volatile oils. At 25° one gram dissolves in about 800 ml water, in 1 ml alcohol, 1 ml ether, 0.5 ml chloroform. Sol in concd mineral acids, phenol, liquid NH₃, liquid SO₂. LD₅₀ orally in mice: 1.3 g/kg (PB293505).

d-Form. [464-49-3] (1R,4R)-(+)-Camphor. Colorless, transparent crystals, mp 179.8°, sublimes 204°. $[\alpha]_D^{25}$ +41 to +43° (c = 10 in U.S.P. alcohol) according to U.S.P. specif.

l-Form. [464-48-2] (1S,4S)-(−)-Camphor.

Spirit of Camphor. A soln of camphor in alcohol contg 10 g camphor per 100 ml soln. Colorless liquid; camphor odor.

Caution: Potential symptoms of overexposure to synthetic camphor are irritation of eyes, skin, mucous membranes; nausea, vomiting, diarrhea; headache, dizziness, confusion, vertigo, excitement, restlessness, delerium, hallucinations; epileptic convulsions; CNS depression, coma. *See NIOSH Pocket Guide to Chemical Hazards* (DHHS/NIOSH 97-140, 1997) p 48; *Clinical Toxicology of Commercial Products*, R. E. Gosselin *et al.*, Eds. (Williams & Wilkins, Baltimore, 5th ed., 1984) Section III, pp 84-86.

USE: Starting reagent for organic syntheses. Used as an odorant and flavorant and as a moth repellant. Plasticizer in cosmetics and as a preservative.

THERAP CAT: Topical analgesic; topical antipruritic.

THERAP CAT (VET): Has been used internally as a stimulant and carminative; externally as an antipruritic, counterirritant and antiseptic.

1735. Camphoric Acid. [5394-83-2] *rel*-(1R,3S)-1,2,2-Trimethyl-1,3-cyclopentanedicarboxylic acid. $C_{10}H_{16}O_4$; mol wt 200.23. C 59.99%, H 8.05%, O 31.96%. Prepn by oxidation of camphor: Bredt, *Ber.* **26**, 3047 (1893). Prepn of isomers: O. Aschan, *Ann.* **316**, 196 (1901); J. D. M. Ross, I. C. Somerville, *J. Chem. Soc.* **1926**, 2770; A. N. Campbell, *J. Am. Chem. Soc.* **53**, 1661 (1931). Synthesis: N. J. Toivonen *et al.*, *Acta Chem. Scand.* **2**, 597 (1948).

d-form

d-Form. [124-83-4] (1R,3S)-Camphoric acid; dextrocamphoric acid. Leaflets from water, monoclinic prisms from alc, mp 186-188°. d 1.186. $[\alpha]_D^{20}$ +47 to +48° (alc). One gram dissolves in 125 ml water, 10 ml boiling water, 1 ml alc, 20 ml glycerol; sol in chloroform, ether, fats, oils.

l-Form. [560-09-8] (1S,3R)-Camphoric acid. mp 187.5°. $[\alpha]_D^{16}$ −48.12° (c = 8.24 in alc).

USE: Reagent in organic synthesis.

1736. d-Camphorsulfonic Acid. [3144-16-9] (1S,4R)-7,7-Dimethyl-2-oxobicyclo[2.2.1]heptane-1-methanesulfonic acid; (+)-2-oxo-10-bornanesulfonic acid; d-10-camphorsulfonic acid; (+)-β-camphorsulfonic acid; Reychler's acid. $C_{10}H_{16}O_4S$; mol wt 232.29. C 51.71%, H 6.94%, O 27.55%, S 13.80%. Prepn from powdered camphor, concd H_2SO_4 and acetic anhydride: Reychler, *Bull. Soc. Chim. Fr.* [3] **19**, 120 (1898); Armstrong, Lowry, *J. Chem. Soc.* **81**, 1447 (1902); Lipp, Knapp, *Ber.* **73B**, 915 (1940). Structure: Loudon, *J. Chem. Soc.* **1933**, 823; Komppa, *J. Prakt. Chem.* **162**, 19 (1943).

Prisms from glacial acetic acid or ethyl acetate, dec 193-195°. $[\alpha]_D^{20}$ +43.5° (c = 4.3 in alcohol); $[\alpha]_D^{20}$ +21.5° (c = 4.3 in water). Deliquesc in moist air. Practically insol in ether; slightly sol in glacial acetic acid, ethyl acetate.

Ammonium salt. $C_{10}H_{19}NO_4S$. Needles from water. $[\alpha]_D^{16}$ +20.5° (c = 5 in water). Very sol in water.

Potassium salt. $C_{10}H_{15}KO_4S$. Needles from alcohol. $[\alpha]_D^{16}$ +18.4° (c = 4.4 in water).

USE: Resolution of optically active isomers.

1737. Camptothecin. [7689-03-4] (4S)-4-Ethyl-4-hydroxy-1H-pyrano[3′,4′:6,7]indolizino[1,2-b]quinoline-3,14(4H,12H)-dione. $C_{20}H_{16}N_2O_4$; mol wt 348.36. C 68.96%, H 4.63%, N 8.04%, O 18.37%. Antitumor alkaloid; prototype DNA topoisomerase I inhibitor. Isoln from the stem wood of the Chinese tree, *Camptotheca acuminata* Decsne., *Nyssaceae*, and structure: M. E. Wall *et al.*, *J. Am. Chem. Soc.* **88**, 3888 (1966). Approach to synthesis: Kepler *et al.*, *J. Org. Chem.* **34**, 3853 (1969). Total synthesis: E. J. Corey *et al.*, *ibid.* **40**, 2140 (1975). Total synthesis of racemate: G. Stork, A. G. Schultz, *J. Am. Chem. Soc.* **93**, 4074 (1971); Volkmann *et al.*, *ibid.* 5576; C. S. F. Tang *et al.*, *ibid.* **97**, 159 (1975); J. C. Bradley, G. Buchi, *J. Org. Chem.* **41**, 699 (1976); T. Kametani *et al.*, *J. Chem. Soc. Perkin Trans. 1* **1981**, 1563. Pharmacologic and clinical evaluation: Gottlieb *et al.*, *Cancer Chemother. Rep. Part 1* **54**, 461 (1970); Gallo *et al.*, *J. Natl. Cancer Inst.* **46**, 789 (1971); S. M. Sieber *et al.*, *Cancer Treat. Rep.* **60**, 1127 (1976). Mechanism of action: Y. H. Hsiang *et al.*, *J. Biol. Chem.* **260**, 14873 (1985). HPLC determn in plasma: J. H. Beijnen *et al.*, *J. Chromatogr.* **617**, 111 (1993). *Reviews:* M. Potmesil, *Cancer Res.* **54**, 1431-1439 (1994); M. E. Wall, M. C. Wani, *ibid.* **55**, 753-760 (1995); C. J. Thomas *et al.*, *Bioorg. Med. Chem.* **12**, 1585-1604 (2004).

Pale yellow needles from methanol + acetonitrile, dec 264-267°. Also reported as mp 275-277° (Volkmann); 287-288° (Stork, Schultz). $[\alpha]_D^{25}$ +31.3° (in chloroform-methanol, 8:2). Exhibits intense blue fluorescence under uv light. uv max: 220, 254, 290, 370 nm (ε 37320, 29230, 4980, 19900). Does not form stable salts with acids. Poorly sol in water.

Acetate. [7688-64-4] $C_{22}H_{18}N_2O_5$. Crystals, dec 271-274°. uv max: 220, 254, 290, 360-370 nm (ε 39010, 28740, 6160, 22000).

Chloroacetate. [7688-65-5] $C_{22}H_{17}ClN_2O_5$. Crystals, dec 245-248°.

1738. Canadine. [522-97-4] 5,8,13,13a-Tetrahydro-9,10-dimethoxy-6H-benzo[g]-1,3-benzodioxolo[5,6-a]quinolizine; 9,10-dimethoxy-2,3-(methylenedioxy)berbine; 5,6,13,13a-tetrahydro-9,10-dimethoxy-2,3-(methylenedioxy)-8H-dibenzo[a,g]quinolizine; tetrahydroberberine; xanthopuccine. $C_{20}H_{21}NO_4$; mol wt 339.39. C 70.78%, H 6.24%, N 4.13%, O 18.86%. A protoberberine alkaloid from *Corydalis cava* (L.), Schweigg & Korte (*C. tuberosa* DC.), *Fumariaceae*: Späth, Julian, *Ber.* **64**, 1131 (1931). Prepn by reduction of berberine: Bersch, Seufert, *ibid.* **70**, 1121 (1937); Awe, Hertel, *Arch. Pharm.* **288**, 516 (1955); Russell, *J. Am. Chem. Soc.* **78**, 3115 (1956). Configuration: Corrodi, Hardegger, *Helv. Chim. Acta* **39**, 889 (1956). Other syntheses: Kametani *et al.*, *J. Chem. Soc. C* **1969**, 2036; **1971**, 2709; M. Cushman, F. W. Dekon, *J. Org. Chem.* **44**, 407 (1979); K. Iwasa *et al.*, *ibid.* **46**, 4744 (1981); N. S. Narasimhan *et al.*, *Tetrahedron Lett.* **22**, 2797 (1981). Pharmacology: F. Sadritdinov, M. B. Sultanov, *C.A.* **66**, 74714v (1967). *Review:* R. H. F. Manske, H. L. Holmes, *The Alkaloids* vol. 4 (Academic Press, New York, 1954) pp 91-92.

LD$_{50}$ in mice (mg/kg): 940 orally; 790 s.c.; 100 i.v. (Sadritdinov, Sultanov).

d-Form. [2086-96-6] mp 132°. $[\alpha]_D^{15}$ +299° (chloroform). Sol in aq methanol.

l-Form. [5096-57-1] Crystals from methanol, mp 135°. $[\alpha]_D^{22}$ −308°; −317° (c = 0.28; 0.4, both in methanol). Sol in aq methanol.

dl-Form. [29074-38-2] mp 172° (Späth, Julian), also reported as 163-165° (Cushman, Dekon). uv max (95% ethanol): 209, 284 nm (ε 28,300; 5200). Sol in aq methanol.

1739. Canakinumab. [914613-48-2] Anti-(human interleukin 1β) immunoglobulin G1 (human clone ACZ885); anti-(human interleukin-1β (IL-1β)) immunoglobulin G1 human monoclonal ACZ885; (1Glu>Glp)-γ$_1$ heavy chain (221-214')-disulfide with kappa light chain, dimer (227-227":230-230")-bisdisulfide; ACZ-885; Ilaris. Fully human monoclonal antibody directed against human interleukin 1β; developed for the treatment of autoinflammatory disorders. Prepn: H. Gram, F. E. DiPadova, **WO 0216436**; *eidem*, **US 7446175** (2002, 2008 both to Novartis); and pharmacology: R. Alten *et al.*, *Arthritis Res. Ther.* **10**, R67 (2008). Clinical study in cryopyrin-associated periodic syndrome (CAPS): H. J. Lachmann *et al.*, *N. Engl. J. Med.* **360**, 2416 (2009); in gouty arthritis: A. So *et. al.*, *Arthritis Rheum.* **62**, 3064 (2010). Review of development and clinical experience in CAPS: L. D. Church, M. F. McDermott, *Expert Rev. Clin. Immunol.* **6**, 831-841 (2010); in CAPS and rheumatoid arthritis: E. Dhimolea, *mAbs* **2**, 3-13 (2010).

THERAP CAT: Immunomodulator.

1740. Canavanine. [543-38-4] *O*-[(Aminoiminomethyl)amino]-L-homoserine; L-2-amino-4-(guanidinooxy)butyric acid. C$_5$H$_{12}$N$_4$O$_3$; mol wt 176.18. C 34.09%, H 6.87%, N 31.80%, O 27.24%. Naturally occurring, non-protein amino acid; guanidinooxy analog of arginine, *q.v.* Predominantly found in leguminous plants; serves as nitrogen storage compound and defensive mechanism. Cytotoxic antimetabolite; substrate for arginyl-tRNA synthetase. Isoln from jack bean, *Canavalia ensiformis* (L.) DC., *Leguminosae:* M. Kitagawa, T. Tomiyama, *J. Biochem. (Tokyo)* **11**, 265 (1929); from alfalfa and clover: E. A. Bell *Biochem. J.* **70**, 617 (1958). Distribution in the plant kingdom: *idem, ibid.* **75**, 618 (1960); B. L. Turner, J. B. Harborne, *Phytochemistry* **6**, 863 (1967). Structure: J. M. Gulland, C. J. O. R. Morris, *J. Chem. Soc.* **1935**, 763; M. Kitagawa, A. Takani, *J. Biochem. (Tokyo)* **23**, 181 (1936). Synthesis of racemate: D. D. Nyberg, B. E. Christensen, *J. Am. Chem. Soc.* **79**, 1222 (1957). HPLC determn in plant tissues: C. Oropeza *et al.*, *J. Chromatogr.* **456**, 405 (1988). Toxicity and pharmacokinetics in rats: D. A. Thomas, G. A. Rosenthal, *Toxicol. Appl. Pharmacol.* **91**, 395 (1987). Review of biological properties: G. A. Rosenthal, *Q. Rev. Biol.* **52**, 155-178 (1977); of anticancer activity: A. K. Bence, P. A. Crooks, *J. Enzyme Inhib. Med. Chem.* **18**, 383-394 (2003); of association with autoimmunity: J. Akaogi *et al.*, *Autoimmun. Rev.* **5**, 429-435 (2006).

Crystals from abs alcohol, mp 184°. $[\alpha]_D^{20}$ +7.9° (c = 3.2). pK$_1$ 2.35; pK$_2$ 7.01; pK$_3$ 9.22. LD$_{50}$ in rats (g/kg): 5.9 ±1.8 s.c. (Thomas, Rosenthal).

Sulfate. [2219-31-0] C$_5$H$_{12}$N$_4$O$_3$.H$_2$SO$_4$; mol wt 274.25. Crystals from dil alcohol, dec 172°. $[\alpha]_D^{17}$ +19.4° (c = 2). Freely sol in water.

DL-Form. [13269-28-8] White to slightly gray crystals from ethanol, mp 180-184° (dec on rapid heating).

1741. Candelilla Wax. [8006-44-8] Obtained from the candelilla plants; *Euphorbia antisyphilitica* Zucc., *Euphorbiaceae* and *E. cerifera* Alcocer (which is only doubtfully distinct from *E. antisyphilitica*) are now the principal sources of candelilla wax. *Pedilanthus pavonis* Boiss. and *P. aphyllus* Boiss. are secondary sources, yielding a wax of lower melting range and lower saponification num-

ber. Most of the wax is produced in Mexico by immersing the plants in boiling water containing sulfuric acid and skimming off the wax which rises to the surface as described by Dickinson, *Am. J. Pharm.* **91**, 808 (1919); Hodge, Sineath, *Econ. Bot.* **10**, 134 (1956). The main constituent is the hydrocarbon hentriacontane. Brief review: C. S. Letcher in *Kirk-Othmer Encyclopedia of Chemical Technology* vol. 24 (Wiley-Interscience, New York, 3rd ed., 1984) pp 468-469.

Brownish to yellowish-brown, hard, brittle, easily pulverizable lumps. d 0.950-0.990. mp 68-70°. Saponification number 50-65. Acid number 10-20. Iodine number 30-35. Sol in chloroform, toluene, acetone, benzene, carbon disulfide, decalin, hot petr ether, gasoline, oils, turpentine, hot chloroform, carbon tetrachloride. Sparingly sol in alcohol. Practically insoluble in water.

USE: Manuf cosmetics, rubber substitutes, furniture and leather polishes, candles, sealing wax, phonograph records; for waterproofing boxes and fabrics; electric insulations; lithographic, printing, stamping and writing inks; molding compositions; sizing paper; hardening other waxes; protective coating for citrus fruits; formerly in chewing gum.

1742. Candesartan. [139481-59-7] 2-Ethoxy-1-[[2'-(2H-tetrazol-5-yl)[1,1'-biphenyl]-4-yl]methyl]-1H-benzimidazole-7-carboxylic acid; 2-ethoxy-1-[4-[2-(1H-tetrazol-5-yl)phenyl]benzyl]-7-benzimidazolecarboxylic acid; CV-11974. C$_{24}$H$_{20}$N$_6$O$_3$; mol wt 440.46. C 65.45%, H 4.58%, N 19.08%, O 10.90%. Angiotensin II type-1 receptor antagonist. Prepn: T. Naka *et al.*, **EP 459136**; *eidem*, **US 5196444** (1991, 1993 both to Takeda); K. Kubo *et al.*, *J. Med. Chem.* **36**, 2182, 2343 (1993). Pharmacology: Y. Shibouta *et al.*, *J. Pharmacol. Exp. Ther.* **266**, 114 (1993). Clinical pharmacology, pharmacokinetics: T. Ogihara *et al.*, *Clin. Ther.* **16**, 74 (1994). Series of clinical trials in chronic heart failure: *Lancet* **362**, 759-781 (2003). Review of clinical experience in hypertension: S. E. Easthope, B. Jarvis, *Drugs* **62**, 1253-1287 (2002); in heart failure: R. S. McKelvie, *Expert Rev. Cardiovasc. Ther.* **7**, 9-16 (2009).

Colorless crystals from ethyl acetate + methanol, mp 183-185°.

1-[[(Cyclohexyloxy)carbonyl]oxy]ethyl ester. [145040-37-5] Candesartan cilexetil; TCV-116; Atacand; Amias; Blopress; Kenzen. C$_{33}$H$_{34}$N$_6$O$_6$; mol wt 610.67. Ester prodrug; hydrolyzed *in vivo* to the active carboxylic acid. Colorless crystals from ethanol + water, mp 163° (dec). Also reported as white to off-white powder, sparingly sol in methanol. Practically insol in water.

THERAP CAT: Antihypertensive. In treatment of congestive heart failure.

1743. Candicidin. [1403-17-4] Levorin; Vanobid. Heptaene macrolide antifungal antibiotic complex composed of candicidins A, B, C and D (major component). Produced by a strain of *Streptomyces griseus* (Rutgers no. 3570): Lechevalier *et al.*, *Mycologia* **45**, 155 (1953). Methods of isoln: Siminoff, **US 2872373** (1959 to Penick); Waksman, Lechevalier, **US 2992162** (1961 to Rutgers Res. & Ed. Found.). Activity: Kligman, Lewis, *Proc. Soc. Exp. Biol. Med.* **82**, 399 (1953); *Lancet* **1**, 507 (1954); Lechevalier, *Presse Med.* **61**, 1327 (1953). Structure of candicidin D: J. Zielinski *et al.*, *Tetrahedron Lett.* **1979**, 1791. Biosynthesis studies: Liu *et al.*, *J. Antibiot.* **25**, 116 (1972). Anticholesteremic property and mode of action studies: I. H. Kwon, H. Fisher, *Nutr. Rep. Int.* **9**, 245 (1974); A. K. Singhal *et al.*, *Lipids* **16**, 423 (1981). HPLC comparison of candicidin, hamycin, trichomycin, *q.q.v.:* P. Helboe *et al.*, *J. Chromatogr.* **189**, 249 (1980). Toxicity study: L. C. Vining *et al.*, *Antibiot. Annu.* **1954-55**, 980.

Candicidin D

Small yellow needles or rosettes from aq tetrahydrofuran or pyridine/acetic acid/water soln (when pure). Absorption max: 403, 380 ($E_{1cm}^{1\%}$ 1150), 360 nm. Practically insol in water, alcohols, ketones, esters, ethers, hydrocarbons, and other lipophilic solvents. Sol in DMSO, DMF, and lower aliphatic acids. Very sol in 80% aq tetrahydrofuran soln. The addn of 5%-25% water to alcohols greatly increases soly. Forms sol salts in alkaline solns. Soly: Marsh, Weiss, *J. Assoc. Off. Anal. Chem.* **50**, 457 (1967). LD_{50} i.p. in mice: 14 mg/kg (Vining).

Candicidin D. Levorin A_2. $C_{59}H_{84}N_2O_{18}$; mol wt 1109.32. Identity with levorin A_2: Bosshardt, Bickel, *Experientia* **24**, 422 (1968); J. Zielinski *et al.*, *loc. cit.*

THERAP CAT: Antifungal (topical).

1744. Candidin. [1405-90-9] (10R)-8,9-Dideoxy-10-hydroxy-7-oxoamphotericin B; 7-oxo-7-deoxy-28,29-didehydronystatin. $C_{47}H_{71}NO_{17}$; mol wt 922.08. C 61.22%, H 7.76%, N 1.52%, O 29.50%. Main component of an antifungal antibiotic complex produced by the soil actinomycete *Streptomyces viridoflavus:* W. A. Taber *et al.*, *Antibiot. Chemother.* **4**, 455 (1954); L. C. Vining *et al.*, *Antibiot. Annu.* **2**, 980 (1954-1955); L. C. Vining, W. A. Taber, *Can. J. Chem.* **34**, 1163 (1956). The two other active principles are designated as *candidinin* and *candidoin*. Prepn of pure crystalline candidin: Preud'homme, Vuillemin, FR 1298345 (1962 to Rhone-Poulenc), *C.A.* **58**, 420c (1963). Synthesis: M. Subramanian *et al.*, *J. Nat. Prod.* **55**, 1213 (1992). Structure: Borowski *et al.*, *Tetrahedron Lett.* **12**, 1987 (1971); and NMR characterization: J. Pawlak *et al.*, *J. Antibiot.* **46**, 1598 (1993). Antifungal activity studies: Solotorovsky *et al.*, *Antibiot. Chemother.* **8**, 364 (1958).

Golden-yellow needles from methanol + chloroform + water. Does not melt, but darkens slowly above 180°. $[\alpha]_D^{27}$ +363° (c = 0.3 in DMF); $[\alpha]_D^{27}$ +205° (c = 0.3 in glacial acetic acid). uv max (methanol): 406, 383, 362, 347 nm. Practically insol in water and most organic solvents. Moderately sol in glacial acetic acid, pyridine, DMF. LD_{50} in mice (mg/kg): 1.5 i.v., >100 orally, 30 s.c. (Vining *et al.*).

1745. Candoxatril. [123122-55-4] *cis*-4-[[[1-[(2S)-3-[(2,3-Dihydro-1H-inden-5-yl)oxy]-2-[(2-methoxyethoxy)methyl]-3-oxopropyl]cyclopentyl]carbonyl]amino]cyclohexanecarboxylic acid; (αS)-1-[(cis-4-carboxycyclohexyl)carbamoyl]-α-[(2-methoxyethoxy)methyl]cyclopentanepropionic acid α-5-indanyl ester; (S)-cis-4-[1-[2-(5-indanyloxycarbonyl)-3-(2-methoxyethoxy)propyl]-1-cyclopentanecarboxamido]-1-cyclohexanecarboxylic acid; UK-79300. $C_{29}H_{41}NO_7$; mol wt 515.65. C 67.55%, H 8.01%, N 2.72%, O 21.72%. Neutral endopeptidase inhibitor. Prepn: I. T. Barnish *et al.*, EP 274234; *eidem*, US 5030654 (1988, 1991 both to Pfizer). Enzyme inhibition and pharmacology: J. C. Danilewicz *et al.*, *Biochem. Biophys. Res. Commun.* **164**, 58 (1989). Clinical pharmacokinetics and effect on plasma ANP: J. G. Motwani *et al.*, *Clin. Phar-*

macol. Ther. **54**, 661 (1993). Clinical evaluation in congestive heart failure: C. D. Kimmelstiel *et al.*, *Cardiology* **87**, 46 (1996).

White crystals, mp 107-109°. $[\alpha]_D$ −5.8° (c = 1 in methanol).

Diacid. [123122-54-3] Candoxatrilat; UK-73967. $C_{20}H_{33}NO_7$; mol wt 399.48. White crystals from ethyl acetate, mp 108.5-109.1°. $[\alpha]_D$ +1.4° (c = 1 in methanol).

THERAP CAT: In treatment of congestive heart failure.

1746. Canella. White cinnamon; wild cinnamon; false Winter's bark; Bahama white wood; wild canilla. Bark of *Canella alba* Murr. (*C. winterana* Gaertn.), *Canellaceae*. *Habit*. W. Indies and Florida. *Constit*. Eugenol, cineol, terpenes, caryophyllene, mannitol, resin, cancellin.

USE: As spice and as an addition to smoking tobacco.

1747. Canertinib. [267243-28-7] N-[4-[(3-Chloro-4-fluorophenyl)amino]-7-[3-(4-morpholinyl)propoxy]-6-quinazolinyl]-2-propenamide; N-[4-(3-chloro-4-fluorophenylamino)-7-(3-morpholin-4-ylpropoxy)quinazolin-6-yl]acrylamide. $C_{24}H_{25}ClFN_5O_3$; mol wt 485.94. C 59.32%, H 5.19%, Cl 7.30%, F 3.91%, N 14.41%, O 9.88%. Irreversible pan-erbB tyrosine kinase inhibitor. Prepn: A. J. Bridges *et al.*, WO 0031048; *eidem*, US 6344455 (2000, 2002 both to Warner-Lambert); J. B. Smaill *et al.*, *J. Med. Chem.* **43**, 1380 (2000). Clinical pharmacokinetics in patients with solid malignancies: E. Calvo *et al.*, *Clin. Cancer Res.* **10**, 7112 (2004); and tolerability in refractory cancer: J. Nemunaitis *et al.*, *ibid.* **11**, 3846 (2005). Review of pharmacology and mechanism of action: L. F. Allen *et al.*, *Semin. Oncol.* **30**, Suppl. 16, 65-78 (2003); of development and clinical experience: C. M. Galmarini, *IDrugs* **7**, 58-63 (2004).

Crystals from methanol, mp 188-190°.

Dihydrochloride. [289499-45-2] CI-1033. $C_{24}H_{25}ClFN_5O_3$·2HCl; mol wt 558.86. Sol in water.

THERAP CAT: Antineoplastic.

1748. Canfosfamide. [158382-37-7] (2R)-L-γ-Glutamyl-3-[[2-[[bis[bis(2-chloroethyl)amino]phosphinyl]oxy]ethyl]sulfonyl]-L-alanyl-2-phenylglycine; γ-glutamyl-α-amino-β-((2-ethyl-N,N,-N',N'-tetrakis(2-chloro)ethylphosphoramidate)sulfonyl)propionyl-(R)-(−)phenylglycine; TER-286. $C_{26}H_{40}Cl_4N_5O_{10}PS$; mol wt 787.46. C 39.66%, H 5.12%, Cl 18.01%, N 8.89%, O 20.32%, P 3.93%, S 4.07%. Nitrogen mustard prodrug preferentially activated by glutathione S-transferase P1-1. Prepn: M. H. Lyttle *et al.*, *J. Med. Chem.* **37**, 1501 (1994); L. M. Kauvar *et al.*, WO 9509866; *eidem*, US 5556942 (1995, 1996 both to Terrapin). Mechanism of activation: A. Satyam *et al.*, *J. Med. Chem.* **39**, 1736 (1996). *In vitro* and *in vivo* antitumor and myelosuppressive effects: A. S. Morgan *et al.*, *Cancer Res.* **58**, 2568 (1998). Clinical pharmacokinetics and toxicology in advanced solid malignancies: L. S. Rosen *et al.*, *Clin. Cancer Res.* **9**, 1628 (2003); *idem et al.*, *ibid.* **10**, 3689 (2004).

White solid, mp 82-90°.

Hydrochloride. [439943-59-6] TLK-286; Telcyta. $C_{26}H_{40}Cl_4$-$N_5O_{10}PS.HCl$; mol wt 823.92.

THERAP CAT: Antineoplastic.

1749.   Cangrelor.   [163706-06-7]   N-[2-(Methylthio)ethyl]-2-[(3,3,3-trifluoropropyl)thio]-5'-adenylic acid monoanhydride with (dichloromethylene)bis[phosphonic acid]; N^6-(2-methylthioethyl)-2-(3,3,3-trifluoropropylthio)-β,γ-dichloromethylene ATP; AR-C69931XX. $C_{17}H_{25}Cl_2F_3N_5O_{12}P_3S_2$; mol wt 776.35. C 26.30%, H 3.25%, Cl 9.13%, F 7.34%, N 9.02%, O 24.73%, P 11.97%, S 8.26%. Specific P2Y$_{12}$ purinoceptor antagonist; inhibits ADP-induced platelet aggregation. Prepn: A. H. Ingall et al., **WO 9418216** (1994 to Fisons); eidem, **US 5721219** (1998 to Astra); and in vivo antithrombotic activity: idem et al., J. Med. Chem. **42**, 213 (1999). In vivo antithrombotic effects in canine arterial thrombosis: J. Huang et al., J. Pharmacol. Exp. Ther. **295**, 492 (2000). Mechanism of action study: A. Ishii-Watabe et al., Biochem. Pharmacol. **59**, 1345 (2000). Clinical safety assessment and evaluation in acute coronary syndromes: R. F. Storey et al., Thromb. Haemostasis **85**, 401 (2001); in angina pectoris and non-Q-wave myocardial infarction: F. Jacobsson et al., Clin. Ther. **24**, 752 (2002). Clinical pharmacodynamics compared with clopidogrel: R. F. Storey et al., Platelets **13**, 407 (2002). Review of pharmacology and clinical development: M. Ueno et al., Expert Rev. Cardiovasc. Ther. **8**, 1069-1077 (2010).

Tetrasodium salt. [163706-36-3] AR-C69931MX. $C_{17}H_{21}Cl_2$-$F_3N_5Na_4O_{12}P_3S_2$; mol wt 864.27. Freely sol in water.

THERAP CAT: Antithrombotic.

1750.   Cannabidiol.   [13956-29-1]   2-[(1R,6R)-3-Methyl-6-(1-methylethenyl)-2-cyclohexen-1-yl]-5-pentyl-1,3-benzenediol; trans-(−)-2-p-mentha-1,8-dien-3-yl-5-pentylresorcinol. $C_{21}H_{30}O_2$; mol wt 314.47. C 80.21%, H 9.62%, O 10.18%. Major nonpsychoactive constituent of cannabis, q.v. (Cannabis sativa L., Cannabinaceae). Exhibits multiple bioactivities including anticonvulsant, anxiolytic and anti-inflammatory effects. Isoln from wild hemp: R. Adams et al., J. Am. Chem. Soc. **62**, 196, 2194 (1940); from hashish: A. Jacob, A. R. Todd, J. Chem. Soc. **1940**, 649. Structure: R. Mechoulam, Y. Shvo, Tetrahedron **19**, 2073 (1963). Crystal and molecular structure: T. Ottersen et al., Acta Chem. Scand. B **31**, 807 (1977). Abs config: Y. Gaoni, R. Mechoulam, J. Am. Chem. Soc. **93**, 217 (1971). Synthesis of (±)-form: eidem, ibid. **87**, 3273 (1965); of (−)-form: T. Petrzilka et al., Helv. Chim. Acta **52**, 1102 (1969); H. J. Kurth et al., Z. Naturforsch. **36B**, 275 (1981). LC-IT-MS determn in cannabis products: A. A. M. Stolker et al., J. Chromatogr. A **1058**, 143 (2004). Review of isoln, chemistry and metabolism: R. Mechoulam, L. Hanus, Chem. Phys. Lipids **121**, 35-43 (2002). Review of pharmacology and bioactivity: R. Mechoulam et al., J. Clin. Phar-

macol. **42**, 11S-19S (2002); and therapeutic potential in neurodegenerative disorders: T. Iuvone et al., CNS Neurosci. Ther. **15**, 65-75 (2009).

Pale yellow resin or crystals, mp 66-67°. bp$_2$ 187-190° (bath temp 220°). bp$_{0.001}$ 130°. d$_4^{40}$ 1.040. n_D^{20} 1.5404. [α]$_D^{27}$ −125° (0.066 g in 5 ml 95% ethanol). [α]$_D^{18}$ −129° (c = 0.45 in ethanol). uv max (ethanol): 282, 274 nm (log ε 3.10, 3.12). Practically insol in water or 10% NaOH. Sol in ethanol, methanol, ether, benzene, chloroform, petr ether.

1751.   Cannabinol.   [521-35-7]   6,6,9-Trimethyl-3-pentyl-6H-dibenzo[b,d]pyran-1-ol; 3-amyl-1-hydroxy-6,6,9-trimethyl-6H-dibenzo[b,d]pyran; CBN. $C_{21}H_{26}O_2$; mol wt 310.44. C 81.25%, H 8.44%, O 10.31%. Nonpsychoactive constituent of cannabis, q.v. (Cannabis sativa L. Cannabinaceae); weak cannabinoid receptor ligand. Isoln from cannabis resin: T. B. Wood et al., J. Chem. Soc. **69**, 539 (1896); R. S. Cahn, J. Chem. Soc. **1931**, 630; T. S. Work et al., Biochem. J. **33**, 123 (1939). Structural studies: R. S. Cahn, J. Chem. Soc. **1932**, 1342; **1933**, 1400; F. Bergel, K. Vögele, Ann. **493**, 250 (1932). Structure and synthesis: R. Adams et al., J. Am. Chem. Soc. **62**, 2204 (1940). Crystal structure: T. Ottersen et al., Acta Chem. Scand. B **31**, 781 (1977). Improved syntheses: P. C. Meltzer et al., Synthesis **1981**, 985; J. Novák, C. A. Salemink, Tetrahedron Lett. **23**, 253 (1982). Pharmacology: I. Yamamoto et al., Chem. Pharm. Bull. **35**, 2144 (1987); F. Petitet et al., Life Sci. **63**, 1 (1998). Review of chromatographic determn methods in biological samples: C. Staub, J. Chromatogr. B **733**, 119-126 (1999). Comparison of pharmacology with other cannabinoids: I. Yamamoto et al., J. Toxicol. Toxin Rev. **22**, 577-589 (2003).

Leaflets from petr ether, mp 76-77°. Sublimes at 4 mm with a bath temp of 180-190°. bp$_{0.05}$ 185°. Insol in water. Sol in methanol, ethanol, aq alkaline solns.

1752.   Cannabis.   Hemp; Indian hemp. Annual, dioecious plant, Cannabis sativa L. Cannabinaceae. Used since antiquity for its edible seed, fiber to produce rope and cloth, and medicinally as an analgesic, anti-emetic, hypnotic and intoxicant. Habit. Temperate to tropical regions, originally in central Asia, China and India. Constit. More than 60 known cannabinoids, primarily isomeric tetrahydrocannabinols, cannabidiol, cannabinol, q.q.v.; other constituents include alkaloids, proteins, sugars, steroids, flavonoids and vitamins. Seeds and seed oil contain fatty acids, including linoleic, oleic, stearic, and palmetic acids, vitamin E, phytosterols, carotenes. Pistillate plants secrete a cannabinoid containing resin from which **hashish** or **charas** is prepared. Preparations of dried flowering tops from these plants are known as **bhang, ganja**, or **marijuana**. Comprehensive description of constituents: C. E. Turner et al., J. Nat. Prod. **43**, 169-234 (1980). Review of analytical methods: T. J. Raharjo, R. Verpoorte, Phytochem. Anal. **15**, 79-94 (2004); of pharmacology and toxicology: I. B. Adams, B. R. Martin, Addiction **91**, 1585-1614 (1996). Series of articles on psychiatric effects, pharmacology and therapeutic uses: Br. J. Psychiatry **178**, 101-128 (2001). Book:

Cannabis and Cannabinoids: Pharmacology, Toxicology, and Therapeutic Potential, F. Grotenhermen, E. Russo, Eds. (Haworth Press, New York, 2002) 439 pp.

Extract. GW-1000; Sativex. Medicinal preparation containing approximately equal amounts of Δ^9-tetrahydrocannabinol and cannabidiol. Prepn of extracts from dried leaf and flowerhead: B. Whittle, G. Guy, **WO 02064109** (2002 to GW Pharma); *eidem,* **US 04192760** (2004). Clinical evaluation for relief of neuropathic pain: J. S. Berman *et al., Pain* **112**, 299 (2004); in multiple sclerosis: C. M. Brady *et al., Mult. Scler.* **10**, 425 (2004). Review of development and clinical experience: P. F. Smith, *Curr. Opin. Investig. Drugs* **5**, 748-754 (2004).

Caution: This is a controlled substance (hallucinogen): **21 CFR,** 1308.11. Acute intoxication is frequently due to recreational use by ingestion or by inhalation of smoke. Psychological responses include euphoria, feelings of detachment and relaxation, visual and auditory hallucinations, anxiety, panic, paranoia, depression, drowsiness, psychotic symptoms. Other effects include impairment of cognitive and psychomotor performance, tachycardia, vasodilation, reddening of the conjuctivae, dry mouth, increased appetite. Chronic inhalation of smoke causes respiratory tract irritation and bronchoconstriction, and may be a significant risk factor for lung cancer. *See* Grotenhermen, Russo, *loc. cit.*

THERAP CAT: Analgesic.

1753. Canrenone. [976-71-6] (17α)-17-Hydroxy-3-oxopregna-4,6-diene-21-carboxylic acid γ-lactone; 17α-(2-carboxyethyl)-17β-hydroxyandrosta-4,6-dien-3-one lactone; 17α-(2-carboxyethyl)-17β-hydroxy-3-oxoandrosta-4,6-diene lactone; 6-dehydrotestosterone-17α-propionic acid γ-lactone; 3-(3-oxo-17β-hydroxy-4,6-androstadien-17α-yl)propionic acid γ-lactone; Phanurane. $C_{22}H_{28}O_3$; mol wt 340.46. C 77.61%, H 8.29%, O 14.10%. Aldosterone antagonist. Prepd by dehydrogenation of 17-hydroxy-3-oxo-17α-pregn-4-ene-21-carboxylic acid γ-lactone: Cella, Tweit, *J. Org. Chem.* **24**, 1109 (1959); Cella, **US 2900383** (1959 to Searle).

Crystals from ethyl acetate, mp 149-151°, solidifies and remelts at 165°. $[\alpha]_D$ +24.5° (chloroform). uv max: 283 nm (ε 26700).

Free acid potassium salt. [2181-04-6] Potassium canrenoate; Kanrenol; Soldactone; Venactone. $C_{22}H_{29}KO_4$; mol wt 396.57.

THERAP CAT: Diuretic.

1754. Cantharides. Spanish fly; blistering fly; blistering beetle. Preparation of dried blister beetles, *Lytta vesicatoria,* also known as *Cantharis vesicatoria, Meloidae;* traditionally used as a vesicant, rubifacient, and aphrodisiac. *Habit.* Southern and Central Europe, mainly on plants of the family *Oleaceae* and *Caprifoliaceae. Constit.* 0.6-1% cantharidin, *q.v.,* 10-15% fat, resinous substances, acetic and uric acids. *Review:* Ude, Heeger, *Pharm. Zentralhalle* **82**, 193 (1941); A. Kar, *Pharmacognosy and Pharmacobiotechnology* (New Age Intl., New Delhi, 2003) pp 199-200.

Caution: Extreme irritant and vesicant. Direct contact with skin may cause intense blister formation. Potential symptoms of overexposure are burning sensation of oral cavity and throat; diarrhea, hemorrhagic necrosis in upper GI tract, abdominal pain; urinary urgency, strangury, hematuria, priapism, oliguria, tubular necrosis, renal failure. *See Clinical Toxicology of Commercial Products,* R. E. Gosselin *et al.,* Eds. (Williams & Wilkins, Baltimore, 5th ed., 1984) Section II, p 270.

1755. Cantharidin. [56-25-7] *rel-*(3aR,4S,7R,7aS)-Hexahydro-3a,7a-dimethyl-4,7-epoxyisobenzofuran-1,3-dione; (3aα,4β,-7β,7aα)-hexahydro-3a,7a-dimethyl-4,7-epoxyisobenzofuran-1,3-dione; *exo-*1,2-*cis-*dimethyl-3,6-epoxyhexahydrophthalic anhydride;

cantharides camphor. $C_{10}H_{12}O_4$; mol wt 196.20. C 61.22%, H 6.17%, O 32.62%. Defensive toxin produced by blister beetles of the family *Meloidae.* Active principle of cantharides, *q.v.* and notorious *"Spanish Fly"* aphrodisiac. Synthesis and stereochemistry: R. B. Woodward, R. B. Loftfield, *J. Am. Chem. Soc.* **63**, 3167 (1941); K. Ziegler *et al., Ann.* **551**, 1 (1942). Stereospecific synthesis: G. Stork *et al., J. Am. Chem. Soc.* **75**, 384 (1953). Simple efficient synthesis: W. G. Dauben *et al., ibid.* **102**, 6893 (1980). Crystal structure: M. Zehnder, U. Thewalt, *Helv. Chim. Acta* **60**, 740 (1977). GC-MS determn in blister beetles: D. Mebs *et al., Toxicon* **53**, 466 (2009). Review of biological source and medicinal uses: L. Moed *et al., Arch. Dermatol.* **137**, 1357-1360 (2001); of clinical experience in pediatric molluscum contagiosum: J. Coloe, D. S. Morrell, *Pediatr. Dermatol.* **26**, 405-408 (2009).

Relative stereochemistry

Orthorhombic plates, scales, mp 218°. Sublimes at about 110° (12 mm Hg, 3-5 mm distance). Insol in cold water, somewhat sol in hot water. One gram dissolves in 40 ml acetone, 65 ml chloroform, 560 ml ether, 150 ml ethyl acetate. Sol in oils.

Caution: Extreme irritant and vesicant. Direct contact may cause intense blister formation. Potential symptoms of overexposure are burning sensation of throat and oral cavity; diarrhea, hemorrhagic necrosis of upper GI tract, abdominal pain; urinary urgency, strangury, hematuria, priapism, oliguria, tubular necrosis, renal failure. *See Clinical Toxicology of Commercial Products,* R. E. Gosselin *et al.,* Eds. (Williams & Wilkins, Baltimore, 5th ed., 1984) Section II, 270.

THERAP CAT: Vesicant; in treatment of warts and molluscum contagiosum.

1756. Canthaxanthin. [514-78-3] β,β-Carotene-4,4′-dione; 4,4′-dioxo-β-carotene; Food Orange 8; C.I. 40850; Carophyll Red; Orobronze; Roxanthin Red 10; Carotaben plus. $C_{40}H_{52}O_2$; mol wt 564.85. C 85.06%, H 9.28%, O 5.66%. All *trans-*carotenoid pigment widely distributed in nature. Isoln from the edible mushroom *Cantharellus cinnabarinus* Adans. ex Fr., *Agaricaceae:* Haxo, *Bot. Gaz.* **112**, 228 (1950); also isolated from flamingo feathers. Structure and synthesis: Isler *et al., Verh. Naturforsch. Ges. Basel* **67**, 379 (1956); Zeller *et al., Helv. Chim. Acta* **42**, 841 (1959). Alternate syntheses: M. Akhtar, B. C. L. Weedon, *J. Chem. Soc.* **1959**, 4058; R. Rüegg, G. Saucy, **US 2983752** and J. D. Surmatis, **US 3311656** (1961, 1967 both to Hoffmann-La Roche); J. D. Surmatis *et al., Helv. Chim. Acta* **53**, 974 (1970); M. Rosenberger *et al., Pure Appl. Chem.* **51**, 871 (1979); *eidem, J. Org. Chem.* **47**, 2130 (1982).

Violet crystals from methylene chloride, dec 217°. Absorption max (cyclohexane): 470 nm ($E_{1cm}^{1\%}$ 2250). Sol in chloroform, oils. Oil solns are more red than those of β-carotene.

USE: Permissible color additive for food and drugs (exempt from certification): *Fed. Regist.* **34**, no. 5 (Jan. 8, 1969). Oral suntanning agent.

1757. Capecitabine. [154361-50-9] 5′-Deoxy-5-fluoro-*N*-[(pentyloxy)carbonyl]cytidine; [1-(5-deoxy-β-D-ribofuranosyl)-5-fluoro-1,2-dihydro-2-oxo-4-pyrimidinyl]carbamic acid pentyl ester; pentyl 1-(5-deoxy-β-D-ribofuranosyl)-5-fluoro-1,2-dihyro-2-oxo-4-pyrimidinecarbamate; Ro-9-1978; Xeloda. $C_{15}H_{22}FN_3O_6$; mol wt

359.35. C 50.14%, H 6.17%, F 5.29%, N 11.69%, O 26.71%. Orally active fluoropyrimidine. Prodrug of doxifluridine, *q.v.*, designed to be metabolized to 5-fluorouracil, *q.v.*, at the tumor site. Prepn: M. Arasaki *et al.*, **EP 602454**; *eidem*, **US 5472949** (1994, 1995 both to Hoffmann-La Roche). Metabolism by tumor enzymes: T. Ishikawa *et al.*, *Cancer Res.* **58**, 685 (1998). Clinical comparison with i.v. fluorouracil in colorectal cancer: P. M. Hoff *et al.*, *J. Clin. Oncol.* **19**, 2282 (2001); E. Van Cutsem *et al.*, *ibid..* 4097. Review of pharmacology and clinical experience: J. McKendrick, J. Coutsouvelis, *Expert Opin. Pharmacother.* **6**, 1231-1239 (2005).

Crystals from ethyl acetate, mp 110-121°. Freely sol in methanol; sol in acetonitrile, alc. Sparingly sol in water; soly in water (20°): 26 mg/ml.

THERAP CAT: Antineoplastic.

1758. Caprenin. [138184-95-9] Reduced calorie fat with functional properties similar to cocoa butter. Synthetic triglyceride composed primarily of caprylic and capric acids (derived from coconut and palm-kernel oils) and behenic acid (derived from rapeseed oil). Prepn: A. M. Erhman *et al.*, **US 4888196** and **US 5066510** (1989, 1991, both to Procter & Gamble). Digestion and absorption in rats: D. R. Webb, R. A. Sanders, *J. Am. Coll. Toxicol.* **10**, 325 (1991). Toxicology and metabolism in animals: D. R. Webb *et al.*, *ibid.* 341. Absorption and caloric value in humans: J. C. Peters *et al.*, *ibid.* 357. Brief description: C. O'Donnell, D. Best, *Prepared Foods* **162**, 41-42 (February, 1993). Clinical effect on serum lipoproteins: J. T. Snook *et al.*, *Nutr. Res.* **16**, 925 (1996).

Caloric value in humans: 3-5 kcal/g.

USE: Reduced calorie fat substitute in soft candies and confectionary coatings.

1759. Capreomycin. [11003-38-6] Capastat. Cyclic peptide antibiotic similar to viomycin; produced by *Streptomyces capreolus* NRRL 2773; Herr *et al.*, 140th Am. Chem. Soc. Meet. (Chicago, Sept. 1961), *Abstracts of Papers*, p 49C; *Chem. Eng. News* **39**, 57 (Sept. 18, 1961); **GB 920563**; **US 3143468** (1963, 1964 both to Lilly). Mixture of capreomycins IA, IB, IIA, and IIB in the approx percentages, 25%, 67%, 3%, 6%, resp.: Herr, Redstone, *Ann. N.Y. Acad. Sci.* **135**, 940 (1966). Activity studies: Sutton *et al.*, *ibid.* 947; Lucchesi, *Antibiot. Chemother.* **16**, 27 (1970). Proposed structure: Bycroft *et al.*, *Nature* **231**, 301 (1971). Revised structure and total synthesis: T. Shiba *et al.*, *Tetrahedron Lett.* **1976**, 3907; S. Nomoto *et al.*, *Tetrahedron* **34**, 921 (1978). Structure of IA and IB: *eidem*, *J. Antibiot.* **30**, 955 (1977). Toxicity study: Welles *et al.*, *Ann. N.Y. Acad. Sci.* **135**, 960 (1966).

Capreomycin IA   R = OH
Capreomycin IB   R = H

The mixture is a white solid. Sol in water. Practically insol in most organic solvents. pKa in 66% aq DMF: 6.2, 8.2, 10.1, 13.3. Stable in aq soln at pH 4-8; unstable in strongly acidic or strongly basic solns.

Disulfate. [1405-36-3] Caprocin; Ogostal. Solubility data: March, Weiss, *J. Assoc. Off. Anal. Chem.* **50**, 457 (1967). LD_{50} in mice, rats (mg/kg): 250, 325 i.v.; 514, 1191 s.c. (Welles).

Capreomycin IA. [37280-35-6] $C_{25}H_{44}N_{14}O_8$; mol wt 668.72. mp 246-248° (dec). $[\alpha]_D^{22}$ −21.9° (c = 0.5 in water). uv max (0.1N HCl): 269 nm (ε 24000); (H$_2$O): 268 nm (ε 23900); (0.1N NaOH): 287 nm (ε 15900) (all for hydrochloride ethanolate).

Capreomycin IB. [33490-33-4] $C_{25}H_{44}N_{14}O_7$; mol wt 652.72. mp 253-255° (dec). $[\alpha]_D^{22}$ −44.6° (c = 0.5 in water). uv max (0.1N HCl): 268 nm (ε 22700); (H$_2$O): 268 nm (ε 22300); (0.1N NaOH): 290 nm (ε 14400) (all for hydrochloride ethanolate).

Note: **Capreomycin IIA, capreomycin IIB.** Structures corresp to capreomycins IA and IB but lack β-lysine residues.

THERAP CAT: Antibacterial (tuberculostatic).

1760. *n*-Capric Acid. [334-48-5] Decanoic acid. $C_{10}H_{20}O_2$; mol wt 172.27. C 69.72%, H 11.70%, O 18.57%. Prepn from octyl bromide: Shishido *et al.*, *J. Am. Chem. Soc.* **81**, 5817 (1959); Closson, De Pree, **US 2918494** (1959 to Ethyl Corp.). Recovery from *Cuphea llavea* Llave et Lex, *Lythaceae* seed oil: Miwa *et al.*, **US 2964546** (1960 to U.S.D.A.). Toxicity study: L. Orö, A. Wretlind, *Acta Pharmacol. Toxicol.* **18**, 141 (1961). *Review: Fatty Acids* Part I, K. S. Markley, Ed. (Interscience, New York, 2nd ed., 1960) pp 34, 39.

Crystalline solid, mp 31.4°. Rancid odor. bp 270°. d_4^{50} 0.8782. n_D^{40} 1.4288. Practically insol in water (0.015 g/100 g at 20°). Sol in ethanol, ether, chloroform, benzene, carbon disulfide. Also sol in dil HNO_3 from which it precipitates unchanged by addition of water. LD_{50} i.v. in mice: 129 ±5.4 mg/kg (Orö, Wretlind).

USE: Manuf of esters for artificial fruit flavors and perfumes; as an intermediate in other chemical syntheses.

1761. *n*-Caproic Acid. [142-62-1] Hexanoic acid. $C_6H_{12}O_2$; mol wt 116.16. C 62.04%, H 10.41%, O 27.55%. Occurs in milk fats (about 2%), in coconut oil (<1%), various palm and other oils. Prepn: Vliet *et al.*, *Org. Synth.* **coll. vol. II**, 417 (1943); Reid, Ruhoff, *ibid.* 475. Manuf by catalytic reduction of corresponding β-lactone: Caldwell, **US 2484486** (1949 to Kodak); from oleic acid: Follett, Murray, **US 2580417** (1952 to Arthur D. Little); from castor oil or a ricinoleate: Steadman, Peterson, **US 2847432** (1958 to National Res. Corp.); by ozonolysis of tall oil unsaturated fatty acids: Maggiolo, **US 2865937** (1958 to Welsbach); from 1,3-butadiene and potassium acetate in presence of $NaNH_2$: Schmerling, Toekelt, **US 3075010** (1963 to Universal Oil Prod.); from cyclohexanol: Bartlett, Lippincott, **US 3121728** (1964 to Esso); by catalytic oxidation of *n*-hexanol: Hay, **US 3173933** (1965 to General Electric). Toxicity study: H. F. Smyth, C. P. Carpenter, *J. Ind. Hyg. Toxicol.* **26**, 269 (1944). *Review: Fatty Acids* Part 1, K. S. Markley, Ed. (Interscience, New York, 2nd ed., 1960) pp 34, 37.

Oily liquid, bp 205°. Characteristic goat-like odor. mp −3.4°. d_4^{20} 0.9265. n_D^{20} 1.4163. Slightly soluble in water (1.082 g/100 g); readily soluble in ethanol, ether. LD_{50} orally in rats: 3.0 g/kg (Smyth, Carpenter).

USE: Manuf of esters for artificial flavors, and of hexyl derivatives, especially hexylphenols, hexylresorcinol, etc.

1762. Caproic Aldehyde. [66-25-1] Hexanal; caproaldehyde; hexaldehyde. $C_6H_{12}O$; mol wt 100.16. C 71.95%, H 12.08%, O 15.97%. Prepn: Bagard, *Bull. Soc. Chim.* **1**, 307 (1907). Toxicity study: Smyth *et al.*, *Arch. Ind. Hyg. Occup. Med.* **10**, 61 (1954).

Liquid. d_4^{20} 0.8335. bp_{760} 131°; bp_{12} 28°. Autooxidizes and polymerizes, especially in the presence of traces of acid. LD_{50} orally in rats: 4.89 g/kg (Smyth).

1763. Caprolactam. [105-60-2] Hexahydro-2*H*-azepin-2-one; ε-caprolactam; 2-oxohexamethylenimine; 2-ketohexamethylenimine; aminocaproic lactam. $C_6H_{11}NO$; mol wt 113.16. C 63.69%, H 9.80%, N 12.38%, O 14.14%. Prepn: Wallach, *Ann.* **312**, 187 (1900); **343**, 43 (1905); Ruzicka *et al., Helv. Chim. Acta* **4**, 477 (1921); Eck, Marvel, *J. Biol. Chem.* **106**, 387 (1934); Marvel, Eck, *Org. Synth.* **coll. vol. II**, 371 (1943); Lazier, Rigby, **US 2234566** (1941 to du Pont); Schlack, **US 2249177** (1941 to I. G. Farben); **DE 739953** (1943); **DE 745224** (1943); P. Smith, *J. Am. Chem. Soc.* **70**, 320 (1948); E. Schmitz *et al., J. Prakt. Chem.* **319**, 274 (1977). Environmentally benign synthesis from cyclohexanone: J. M. Thomas, R. Raja, *Proc. Natl. Acad. Sci. USA* **102**, 13732 (2005). Purification: Kampschmidt, **US 2786052** (1957 to Stamicarbon N. V.). Stabilization with alkalies: Indest *et al.,* **US 2884414** (1959 to Vereinigte Glanzstoff-Fabriken). Toxicity study: H. F. Smyth *et al., Am. Ind. Hyg. Assoc. J.* **30**, 470 (1969). *Reviews: CIOS Repts.* **no. 22** and **31**, File XXXIII Synthetic Fiber Developments in Germany, parts I & II; K. Kahr *et al.* in *Ullmanns Encyklopädie der technischen Chemie* **vol. 9**, E. Bartholome *et al.,* Eds. (Verlag Chemie, Weinheim, 4th ed., 1975) pp 96-114.

Hygroscopic leaflets from petr ether, mp 70°. d_4^{75} (liq) 1.02. bp_{50} 180°; bp_3 100°. Viscosity (78°): 9 cP. Flash pt, open cup: 257°F (125°C). Freely sol in water, methanol, ethanol, ether, tetrahydrofurfuryl alcohol, dimethylformamide. Also sol in chlorinated hydrocarbons, cyclohexene, petroleum fractions. A 70% aq soln has d_4^{25} 1.05; n_D^{31} 1.4965; n_D^{40} 1.4935. LD_{50} orally in rats: 2.14 g/kg (Smyth).

Caution: Potential symptoms of overexposure are irritation of eyes, skin, respiratory system; epistaxis; dermatitis, skin sensitization; asthma; irritability, confusion, dizziness, headache; abdominal cramps, diarrhea, nausea, vomiting; liver and kidney injury. *See NIOSH Pocket Guide to Chemical Hazards* (DHHS/NIOSH 97-140, 1997) p 50.

USE: Manuf of synthetic fibers of the polyamide type (Perlon); solvent for high mol wt polymers; precursor of nylon-6, *q.v.*

1764. Capromab. [151763-64-3] Anti-(human prostatic carcinoma cell) immunoglobulin G1 (mouse monoclonal 7E11-C5.3 γ₁-chain) disulfide with mouse monoclonal 7E11-C5.3 light chain, dimer. Murine monoclonal antibody directed against prostate specific membrane antigen (PSMA), a transmembrane glycoprotein expressed by prostate epithelium. Designed for the detection of primary and metastatic prostate carcinoma. Prepn: J. S. Horoszewicz *et al., Anticancer Res.* **7**, 927 (1987); *idem,* **US 5162504** (1992 to Cytogen). Characterization and specificity of immunoconjugate: A. D. Lopes *et al., Cancer Res.* **50**, 6423 (1990). Diagnostic use of labeled form in metastatic prostate cancer: M. J. Manyak *et al., Urology* **54**, 1058 (1999). Review of clinical studies: R. T. Maguire *et al., Cancer* **72**, 3453-3462 (1993); and pharmacology: H. M. Lamb, D. Faulds, *Drugs Aging* **12**, 293-304 (1998). Review of diagnostic use: M. K. Haseman *et al., Cancer Biother. Radiopharm.* **15**, 131-140 (2000).

Capromab pendetide. [145464-28-4] 7E11-C5.3-GYK-DTPA; 7E11-C5.3-glycyl-tyrosyl-(*N*-ε-diethylenetriaminepentaacetic acid)-lysine; CYT-356; ProstaScint. Monoclonal antibody conjugated to a linker-chelator to enable radiolabeling. Prepd as hydrochloride and labeled with Indium ^{111}In immediately before administration.

THERAP CAT: ^{111}In-labeled form as diagnostic aid (radioactive tumor-targeting agent).

1765. Caproyl Chloride. [142-61-0] Hexanoyl chloride. $C_6H_{11}ClO$; mol wt 134.60. C 53.54%, H 8.24%, Cl 26.34%, O 11.89%. Prepn: Brown, *J. Am. Chem. Soc.* **60**, 1325 (1938). Manuf: Wygant, **US 2806061** (1957 to Monsanto).

Liquid, bp 151-153°. mp −87.3°. d_4^{15} 0.9805. n_D^{15} 1.4286. Dec by water or alcohol. Sol in ether, chloroform.

1766. Caprylene. [111-66-0] 1-Octene; octylene. C_8H_{16}; mol wt 112.22. C 85.62%, H 14.37%. Prepn from appropriate alkylmagnesium bromide and allyl bromide or chloride: Geisler, Pilz, *Ber.* **95**, 96 (1962); from formaldehyde or paraformaldehyde and triphenyl(phenylmethylene)phosphorane: Hauser *et al., J. Org. Chem.* **28**, 372 (1963); by catalytic dehydration of 2-octanol: Lundeen, Hoozer, *J. Am. Chem. Soc.* **85**, 2180 (1963).

Liquid, bp 121°, bp_{100} 61.5-61.7°. mp −102°. d_4^{20} 0.7149, d_4^{25} 0.7109. n_D^{20} 1.4087, n_D^{25} 1.4062. Flash pt, open cup: 70°F (21°C). Practically insol in water; misc with alcohol, ether.

1767. Caprylic Acid. [124-07-2] Octanoic acid. $C_8H_{16}O_2$; mol wt 144.21. C 66.63%, H 11.18%, O 22.19%. Prepn from 1-heptene: Dupont *et al., Compt. Rend.* **240**, 628 (1955); by oxidation of octanol: Langenbeck, Richter, *Ber.* **89**, 202 (1956). Manuf: Alexander, **US 2821534** (1958 to GAF); McAlister *et al.,* **US 3053869** (1962 to Standard Oil Co., Indiana). Antifungal properties: O. Wyss *et al., Arch. Biochem.* **7**, 418 (1945). Toxicity: P. M. Jenner *et al., Food Cosmet. Toxicol.* **2**, 327 (1964). *Review: Fatty Acids* Part 1, K. S. Markley, Ed. (Interscience, New York, 2nd ed., 1960) pp 34, 38.

Oily liquid, bp 239.7°. Slightly unpleasant rancid taste. mp 16.7°. d_4^{20} 0.910. n_D^{20} 1.4280. Very slightly sol in water (0.068 g/100 g at 20°); freely sol in alcohol, chloroform, ether, carbon disulfide, petr ether, glacial acetic acid. LD_{50} orally in rats: 10,080 mg/kg (Jenner).

USE: An intermediate in manuf of esters used in perfumery; in manuf of dyes, etc.

1768. Caprylic Aldehyde. [124-13-0] Octanal; caprylaldehyde; octaldehyde. $C_8H_{16}O$; mol wt 128.22. C 74.94%, H 12.58%, O 12.48%. Prepn: Stephen, *J. Chem. Soc.* **127**, 1874 (1925).

Liquid. d_4^{20} 0.821. bp_{760} 163.4°; bp_{20} 72°; bp_9 60°. n_D^{26} 1.41667. Slightly sol in water; misc with alc, ether.

1769. CAPS. [1135-40-6] 3-(Cyclohexylamino)-1-propanesulfonic acid. $C_9H_{19}NO_3S$; mol wt 221.32. C 48.84%, H 8.65%, N 6.33%, O 21.69%, S 14.49%. One of the zwitterionic amino acids known as "Good" buffers; active in the pH range of 9.7-11.1. Prepn: H. Dorn, K. Walter, *Z. Chem.* **7**, 151 (1967). Thermodynamic parameters: C. D. McGlothlin, J. Jordan, *Anal. Lett.* **9**, 245 (1976). Effects on protein measurement: V. Kaushal, L. D. Barnes, *Anal. Biochem.* **157**, 291 (1986); H. M. Himmel, W. Heller, *J. Clin. Chem. Clin. Biochem.* **25**, 909 (1987). Use as buffer: I. I. Koukli *et al., Analyst* **113**, 603 (1988); R. M. Smith *et al., J. Chromatogr.* **514**, 97 (1990).

Colorless needles from methanol, mp 302-303°. pK (25°): 10.35.
USE: Biological buffer.

1770. Capsaicin. [404-86-4] (6*E*)-*N*-[(4-Hydroxy-3-meth-oxyphenyl)methyl]-8-methyl-6-nonenamide; *trans*-8-methyl-*N*-van-illyl-6-nonenamide; *N*-(4-hydroxy-3-methoxybenzyl)-8-methylnon-*trans*-6-enamide; Axsain; Qutenza; Zacin; Zostrix. $C_{18}H_{27}NO_3$; mol wt 305.42. C 70.79%, H 8.91%, N 4.59%, O 15.72%. Pungent principle in fruit of hot, red peppers, such as *Capsicum annuum* or *C. frutescens*, *Solanaceae*. Powerful irritant; initial administration causes burning pain. Prolonged treatment induces desensitization to painful stimuli. Agonist of the transient receptor potential vanilloid 1 receptor (TRPV1); depletes substance P from sensory neurons. Isoln from paprika and cayenne: Thresh, *Pharm. J. Trans.* **7**, 21 (1876); Micko, *Z. Nahr. Genussm.* **1**, 818 (1898). Early structure study: E. K. Nelson, *J. Am. Chem. Soc.* **42**, 597 (1920). Synthesis: Späth, Darling, *Ber.* **63**, 737 (1930); L. Crombie *et al.*, *J. Chem. Soc.* **1955**, 1025; O. P. Vig *et al.*, *Indian J. Chem.* **17B**, 558 (1979). Desensitization effect: N. Jancso *et al.*, *Br. J. Pharmacol. Chemother.* **31**, 138 (1967). Neuronal depletion of substance P: T. M. Jessell *et al.*, *Brain Res.* **152**, 183 (1978). Identification of capsaicin receptor (TRPV1): M. J. Caterina *et al.*, *Nature* **389**, 816 (1997). LC/MS/MS determn in biological samples: F. Beaudry, P. Vachon, *Biomed. Chromatogr.* **23**, 204 (2009). Clinical trial for treatment of postherpetic neuralgia: M. Backonja *et al.*, *Lancet Neurol.* **7**, 1106 (2008). Review of chemistry and pharmacology: J. Molnar, *Arzneim.-Forsch.* **15**, 718-727 (1965); R. M. Virus, G. F. Gebhart, *Life Sci.* **25**, 1273-1284 (1979); S. J. Conway, *Chem. Soc. Rev.* **37**, 1530-1545 (2008). Review of safety assessment for use in cosmetics: *Int. J. Toxicol.* **26**, Suppl.1, 3-106 (2007); of clinical experience in treatment of chronic pain: L. Mason *et al.*, *Br. Med. J.* **328**, 991-995 (2004).

Monoclinic, rectangular plates, scales from petr ether, mp 65°. bp$_{0.01}$ 210-220° (air-bath temp). uv max: 227, 281 nm (ε 7000, 2500). Burning taste, one part in 100,000 can be detected by tasting. Practically insol in cold water. Freely sol in alc, ether, benzene, chloroform; slightly sol in CS_2. *Irritant*.

USE: Food flavoring; in pepper spray self-defense devices; as a tool in neurobiological research.

THERAP CAT: Topical analgesic.

THERAP CAT (VET): Topical analgesic.

1771. Capsanthin. [465-42-9] (3*R*,3'*S*,5'*R*)-3,3'-Dihydroxy-β,κ-caroten-6'-one. $C_{40}H_{56}O_3$; mol wt 584.89. C 82.14%, H 9.65%, O 8.21%. Carotenoid pigment isolated from paprika *(Capsicum annuum* L., *Solanaceae)*: L. Zechmeister, L. Cholnoky, *Ann.* **454**, 54 (1927); L. Cholnoky *et al.*, *Ann.* **606**, 194 (1957); Warren, Weedon, *J. Chem. Soc.* **1958**, 3972. Structure: Entschel, Karrer, *Helv. Chim. Acta* **43**, 89 (1960); Faigle, Karrer, *ibid.* **44**, 1257, 1904 (1961). Absolute configuration: Bartlett *et al.*, *J. Chem. Soc. C* **1969**, 2527. Crystal structure: I. Ueda, W. Nowacki, *Verh. Schweiz. Naturforsch. Ges.* **1971**, 152, *C.A.* **77**, 106449h (1972).

Deep carmine-red needles from petr ether, mp 181-182°. Absorption max 483 nm (ε × 10⁻³ 121). [α]$_{Cd}$ +36° (chloroform). Freely sol in acetone, chloroform. Sol in methanol, ethanol, ether, benzene. Slightly sol in petr ether, CS_2.

Diacetate. $C_{44}H_{60}O_5$. Red plates from methanol, mp 150°.

Dipalmitate. $C_{72}H_{116}O_5$. Bordeaux-red plates from benzene + methanol, mp 95°.

1772. Capsicum. Cayenne pepper; chilli pepper; red pepper. Genus of herbaceous shrubs of the family, *Solanaceae*. Several species are cultivated, the most common being *Capsicum annuum* and *Capsicum fructescens*; varieties produce a pungent red fruit that is widely used as a spice. Mild fruits, commonly known as **paprika** or **bell pepper**, are produced by varieties of *C. annuum*. *Habit*. Native to tropical Americas, but cultivated worldwide in tropical to temperate climates. *Constit*. Pungent principles, primarily capsaicin (0.1-1.5%), dihydrocapsaicin; carotenoid pigments, such as capsanthin, capsorubin, cryptoxanthin, zeaxanthin; oleoresin; volatile oil. Overview of botany and cultivation: V. S. Govindarajan, *Crit. Rev. Food Sci. Nutr.* **22**, 109-176 (1985); of products and uses: *idem, ibid.* **23**, 207-288 (1986); of color, aroma, and pungency constituents: *idem, ibid.* **24**, 245-355 (1986). Determn of capsaicin in peppers and pepper sauces by GC-MS: A. Pena-Alvarex *et al.*, *J. Chromatogr. A* **1216**, 2843 (2009); of non-pungent capsinoids by HPLC: S. Singh *et al.*, *J. Agric. Food Chem.* **57**, 3452 (2009). Identification of aroma constituents by GC-MS: A. Rodríguez-Burreuzo *et al.*, *ibid.* **58**, 4388 (2010). Review of safety assessment for use in cosmetics: *Int. J. Toxicol.* **26**, Suppl. 1, 3-106 (2007). General reviews: T. G. Berke, S. C. Shieh in *Handbook of Herbs and Spices*, **Vol. 1**, K. V. Peter, Ed., (Woodhead Publishing, Cambridge, 2001) pp 111-122; I. A. Khan, E. A. Abourashed in *Leung's Encyclopedia of Common Natural Ingredients* (Wiley, Hoboken, 3rd ed., 2010) pp 132-138.

Oleoresin capsicum. [8023-77-6] Obtained by solvent extraction of dried, ripe fruits of pungent varieties of *C. annuum* or *C. frutescens*. Clear red to dark red, viscous liquid. Characteristic odor, flavor, and bite. d²⁵ 1.0073-1.073. mp < −60°. bp >187°. Partly sol in alcohol, with oily separation and sediment. Sol in most fixed oils.

Oleoresin paprika. [68917-78-2] C.I. Natural Red 34; C.I. 75133. Obtained by solvent extraction of pods of *C. annuum*. Deep red to purple-red, viscous liquid. Characteristic odor and flavor. Parly sol in alcohol with oily separation. Sol in most fixed oils.

USE: Colorant, flavor, source of pungency in foods. Nutrient; dietary source of vitamins A, C, E. Resins and extracts in cosmetics as skin conditioning agent, external analgesic, flavoring agent, and fragrance component.

THERAP CAT: Counterirritant.

1773. Captafol. [2425-06-1] 3a,4,7,7a-Tetrahydro-2-[(1,1,-2,2-tetrachloroethyl)thio]-1*H*-isoindole-1,3(2*H*)-dione; *N*-(1,1,2,2-tetrachloroethylthio)-4-cyclohexene-1,2-dicarboximide; *N*-(1,1,2,2-tetrachloroethylmercapto)-4-cyclohexene-1,2-dicarboximide; *N*-(1,-1,2,2-tetrachloroethylthio)-Δ⁴-tetrahydrophthalimide; *N*-(1,1,2,2-tetrachloroethylsulfenyl)-*cis*-4-cyclohexene-1,2-dicarboximide; Difolatan. $C_{10}H_9Cl_4NO_2S$; mol wt 349.05. C 34.41%, H 2.60%, Cl 40.62%, N 4.01%, O 9.17%, S 9.18%. *N*-tetrachloroethyl analog of captan, *q.v.* Prepn: Kohn, **BE 633205** (1963 to California Res. Corp.), *C.A.* **60**, 15789cd (1964). Toxicology: R. Ben-Dyke *et al.*, *World Rev. Pest Control* **9**, 119 (1970); G. L. Kennedy, Jr. *et al.*, *Food Cosmet. Toxicol.* **13**, 55 (1975). Carcinogenicity study: S. Tamano *et al.*, *Jpn. J. Cancer Res.* **81**, 1222 (1990).

Crystals, mp 160-161°. LD$_{50}$ in rats, rabbits (mg/kg): 2500-6200 orally; 15400 dermally (Ben-Dyke).

Caution: Potential symptoms of overexposure are irritation of eyes, skin, respiratory system; dermatitis, skin sensitization; conjunctivitis; bronchitis, wheezing; diarrhea, vomiting; liver and kidney injury; high blood pressure. See *NIOSH Pocket Guide to Chemical Hazards* (DHHS/NIOSH 97-140, 1997) p 50. This substance is listed as reasonably anticipated to be a human carcinogen: *Report on Carcinogens, Twelfth Edition* (PB2011-111646, 2011) p 83.

USE: Agricultural fungicide, especially for potatoes.

1774. Captan. [133-06-2] 3a,4,7,7a-Tetrahydro-2-[(trichloromethyl)thio]-1*H*-isoindole-1,3(2*H*)-dione; *N*-(trichloromethyl-thio)-4-cyclohexene-1,2-dicarboximide; *N*-trichloromethylthio-3a,-4,7,7a-tetrahydrophthalimide; *N*-trichloromethylmercapto-4-cyclohexene-1,2-dicarboximide; *N*-(trichloromethylmercapto)-Δ^4-tetrahydrophthalimide; ENT-26538; SR-406; Merpan; Orthocide-406; Vancide 89. $C_9H_8Cl_3NO_2S$; mol wt 300.58. C 35.96%, H 2.68%, Cl 35.38%, N 4.66%, O 10.65%, S 10.67%. Prepn: A. R. Kittleson, **US 2553771**; **US 2653155**; **US 2713058** (1951, 1953, 1955 all to Standard Oil); *idem, Science* **115**, 84 (1952); *idem, J. Agric. Food Chem.* **1**, 677 (1953). Colorimetric determination: A. R. Kittleson, *Anal. Chem.* **24**, 1173 (1952). Review of mutagenicity studies: B. A. Bridges, *Mutat. Res.* **32**, 3 (1975); of carcinogenic risk: *IARC Monographs* **30**, 295-318 (1983).

Odorless crystals from CCl_4, mp 178°. d 1.74. Practically insol in water. Soly at 26° in g/100 ml: chloroform 7.78; tetrachloroethane 8.15; cyclohexanone 4.96; dioxane 4.70; benzene 2.13; toluene 0.69; heptane 0.04; ethanol 0.29; ether 0.25. LD_{50} orally in rats: 9000 mg/kg (Bridges).

Caution: Potential symptoms of overexposure are irritation of eyes, skin, upper respiratory system; blurred vision; dermatitis, skin sensitization; dyspnea; diarrhea, vomiting. Potential occupational carcinogen. *See NIOSH Pocket Guide to Chemical Hazards* (DHHS/NIOSH 97-140, 1997) p 50.

USE: Fungicide; bacteriostat in soap.

1775. Captodiamine. [486-17-9] 2-[[[4-(Butylthio)phenyl]-phenylmethyl]thio]-*N*,*N*-dimethylethanamine; 2-[*p*-(butylthio)-α-phenylbenzylthio]-*N*,*N*-dimethylethylamine; *p*-butylmercaptobenzhydryl β-dimethylaminoethyl sulfide; *p*-butylthiodiphenylmethyl 2-dimethylaminoethyl sulfide; captodiam; captodramin. $C_{21}H_{29}NS_2$; mol wt 359.59. C 70.14%, H 8.13%, N 3.90%, S 17.83%. Prepn: Hübner, Petersen, **US 2830088** (1958). Pharmacological studies: R. Kopf, I. Moller-Nielsen, *Arzneim.-Forsch.* **8**, 154 (1958).

Hydrochloride. [904-04-1] Covatine; Covatix; Suvren. $C_{21}H_{29}$-NS_2.HCl; mol wt 396.05. Crystals, mp 131-132°. LD_{50} (96 hr) in mice, rats (mg/kg): 180, 343 i.p.; 1630, 3800 orally (Kopf, Moller-Nielsen).

THERAP CAT: Anxiolytic.

1776. Captopril. [62571-86-2] 1-[(2*S*)-3-Mercapto-2-methyl-1-oxopropyl]-L-proline; (2*S*)-1-(3-mercapto-2-methylpropionyl)-L-proline; D-2-methyl-3-mercaptopropanoyl-L-proline; SQ-14225; Acepril; Acepress; Capoten; Capotena; Capotolane; Captoril; Cesplon; Lopirin; Lopril; Tensobon. $C_9H_{15}NO_3S$; mol wt 217.28. C 49.75%, H 6.96%, N 6.45%, O 22.09%, S 14.76%. First orally active angiotensin-converting enzyme (ACE) inhibitor. Prepn: M. A. Ondetti, D. W. Cushman, **DE 2703828**; *eidem*, **US 4046889** and **US 4105776** (1977, 1977, 1978 all to Squibb). Design and synthesis: M. A. Ondetti *et al., Science* **196**, 441 (1977); D. W. Cushman *et al., Biochemistry* **16**, 5484 (1977). Improved synthesis: D. H. Nam *et al., J. Pharm. Sci.* **73**, 1843 (1984). Pharmacology: B. Rubin *et al., Eur. J. Pharmacol.* **51**, 377 (1978); *eidem, Prog. Cardiovasc. Dis.* **21**, 183 (1978). Clinical studies: D. B. Case *et al., ibid.* 195; H. R. Brunner *et al., Ann. Intern. Med.* **90**, 19 (1979). Toxicology and metabolism: G. R. Keim in *Captopril and Hypertension*, D. B. Case, Ed. (Plenum, New York, 1980) p 137. GC/MS determn in biological fluids: T. Ito, Y. Matsuki, *J. Chromatogr.* **417**, 79 (1987). HPLC determn in pharmaceutical formulations: T. Mirza, H. S. I. Tan, *J. Pharm. Biomed. Anal.* **25**, 39 (2001). Thermal analysis: Y. Huang

et al., Thermochim. Acta **367-368**, 43 (2001). Historical review and comprehensive bibliography: Z. P. Horovitz in *Pharmacological and Biochemical Properties of Drug Substances* vol. 3, M. E. Goldberg, Ed. (Am. Pharm. Assoc., Washington, DC, 1981) pp 148-175. Comprehensive description: H. Kadin, *Anal. Profiles Drug Subs.* **11**, 79-137 (1982). *Reviews: Am. Heart J.* **104**, 1125-1228 (1982); *Br. J. Clin. Pharmacol.* **14**, Suppl. 2, 69S-252S (1982). Series of articles on pharmacology and therapeutic efficacy: *Postgrad. Med. J.* **62**, Suppl. 1, 1-191 (1986).

Crystals from ethyl acetate/hexane, mp 103-104° (Ondetti, Cushman). Generally regarded as polymorphic: stable form, mp 106°; unstable form, mp 86° (Florey); also reported as mp 87-88°, resolidifies, second mp 104-105° (Cushman). Slight sulfurous odor. $[\alpha]_D^{22}$ −131.0° (c = 1.7 in ethanol). pK_1 3.7, pK_2 9.8. Freely sol in water (~160mg/ml), methylene chloride, methanol, chloroform, alc; sparingly sol in ethyl acetate. LD_{50} in mice (mg/kg): 1040 i.v.; 6000 orally (Keim).

THERAP CAT: Antihypertensive.

THERAP CAT (VET): Antihypertensive.

1777. Caramel. Burnt sugar coloring; burnt sugar. Made by heating sugar or glucose, adding small quantities of alkali, alkaline carbonate or a trace of mineral acid during the heating.

Dark-brown, thick liq; pleasant, bitter taste; odor of burnt sugar. d about 1.35. Misc with water. Sol in dil alc. Insol in benzene, chloroform, ether, acetone, petr ether, oil turpentine, solvent hexane.

USE: Coloring foods, confectionery, galenicals.

1778. Caramiphen. [77-22-5] 1-Phenylcyclopentanecarboxylic acid 2-(diethylamino)ethyl ester; diethylaminoethyl 1-phenylcyclopentane-1-carboxylate. $C_{18}H_{27}NO_2$; mol wt 289.42. C 74.70%, H 9.40%, N 4.84%, O 11.06%. Anticholinergic. Prepn: **CH 234452**; H. Martin, F. Häfliger, **US 2404588** (1945, 1946 both to Geigy). Prepn of the ethanedisulfonate: **CH 272708** (1951 to Geigy), *C.A.* **46**, 4563i (1952); *Chem. Zentralbl.* **1952**, 1571. Pharmacology and toxicology of the hydrochloride: C. P. Kraatz *et al., J. Pharmacol. Exp. Ther.* **96**, 42 (1949); of the ethanedisulfonate: J. J. Toner, E. Macko, *ibid.* **106**, 246 (1952). GLC determn in blood: P. Levandoski, T. Flanagan, *J. Pharm. Sci.* **69**, 1353 (1980). Mechanism of antitussive action: E. F. Domino *et al., J. Pharmacol. Exp. Ther.* **233**, 249 (1985).

$bp_{0.07}$ 112-115°.

Ethanedisulfonate. [125-86-0] $(C_{18}H_{27}NO_2)_2.C_2H_6O_6S_2$; mol wt 769.02. Crystals from acetone, mp 115-116°. More sol in water than the hydrochloride. Sol in alc, pharmaceutical syrups. Mixture with phenylpropanolamine hydrochloride, *Tuss-Ornade*.

Hydrochloride. [125-85-9] $C_{18}H_{27}NO_2$.HCl; mol wt 325.88. Crystals, mp 145-146°. Sol in alc; slightly sol in water. LD_{50} i.p. in rats: 209 mg/kg (Kraatz).

THERAP CAT: Antitussive.

1779. Caraway. Dried ripe fruit of *Carum carvi* L., *Umbelliferae. Habit.* Europe, Central and Western Asia; cultivated in England, Russia, U.S. *Constit.* 3-7% volatile oil; terpenes, chiefly *d*-limonene; 10-18% fixed oil; protein; carbohydrate; flavonoids. Composition studies: Von Schantz, Ek, *Sci. Pharm.* **39**, 82 (1971). *Review:* Arctander, *Perfume and Flavor Materials of Natural Origin* (S. Arctander, Elizabeth, N.J., 1960) pp 124-125; of medicinal uses: N. G. Bisset, M. Wichtl, *Herbal Drugs and Phytopharmaceuticals*, English Ed. (CRC Press, Boca Raton, 1994) pp 128-129.

Volatile oil. [8000-42-8] Oil of caraway. *Constit.* 53-63% carvone (by vol), *d*-limonene. Colorless or pale yellow liq; darkens and thickens with age. d_{23}^{25} 0.900-0.910. a_D^{25} +70 to +80°. n_D^{20} 1.484-1.488. Almost insol in water. Sol in 8 vols 80% or in 1 vol 90% alcohol. *Keep well closed, cool, and protected from light.*

USE: Pharmaceutic aid (flavor). In manuf liqueurs and perfuming soaps; as a spice in baking.

THERAP CAT: Carminative.

1780. Carazolol. [57775-29-8] 1-(9*H*-Carbazol-4-yloxy)-3-[(1-methylethyl)amino]-2-propanol; BM-51052; Conducton; Suacron. $C_{18}H_{22}N_2O_2$; mol wt 298.39. C 72.45%, H 7.43%, N 9.39%, O 10.72%. β-Adrenergic blocker. Prepn: H. Leinert *et al.*, **DE 2240599** (1974 to Boehringer, Mann.), *C.A.* **80**, 133455a (1974). Comparative study of cardiac action: W. Bartsch *et al.*, *Arzneim.-Forsch.* **27**, 1022 (1977). Initial clinico-pharmacological study: E. Chorianopoulos *et al.*, *Herz/Kreislauf* **9**, 965 (1977), *C.A.* **88**, 164460t (1978). Receptor binding studies: G. Kaiser *et al.*, *Arch. Pharmacol.* **305**, 41 (1978); R. B. Innis *et al.*, *Life Sci.* **24**, 2255 (1979). Use in treatment of stress syndrome in pigs: G. Ballarini, F. Guizzardi, *Tieraerztl. Umsch.* **36**, 171 (1981).

Hydrochloride. $C_{18}H_{23}ClN_2O_2$. Crystals, mp 234-235°.

THERAP CAT: Antihypertensive, antianginal, antiarrhythmic.

THERAP CAT (VET): Treatment of stress in pigs.

1781. Carbachol. [51-83-2] 2-[(Aminocarbonyl)oxy]-*N,N,N*-trimethylethanaminium chloride (1:1); (2-hydroxyethyl)trimethyl ammonium chloride carbamate; choline chloride carbamate; carbamylcholine chloride; carbocholine; Doryl; Isopto Carbachol; Jestryl; Miostat. $C_6H_{15}ClN_2O_2$; mol wt 182.65. C 39.46%, H 8.28%, Cl 19.41%, N 15.34%, O 17.52%. Parasympathomimetic. Prepn: O. Dalmer, C. Diehl, **DE 539329** (1930 to E. Merck); *eidem*, **US 1894162** (1933); Kreitmair, *Arch. Exp. Pathol. Pharmakol.* **164**, 346 (1932); R. D. Haworth *et al.*, *J. Chem. Soc.* **1947**, 176. Toxicity: Molitor, *J. Pharmacol. Exp. Ther.* **58**, 337 (1936). Evaluation in bronchoprovocation test for asthma: G. Rosati *et al.*, *Eur. J. Respir. Dis.* **64**, Suppl. 128, 417 (1983). Clinical effect on post-surgical intraocular pressure: R. S. Ruiz *et al.*, *Am. J. Ophthalmol.* **107**, 7 (1989).

Prismatic crystals from alcohol-ether, mp 207° (Haworth); also reported as mp 204-205° (Kreitmair). Hygroscopic. Odorless or slight amine-like odor. Freely sol in water; sparingly sol in alc. One gram dissolves in 1 ml water, 50 ml alcohol. Practically insol in chloroform, ether. LD_{50} in mice (mg/kg): 15 orally, 0.3 i.v. (Molitor).

THERAP CAT: Cholinergic; miotic.

THERAP CAT (VET): Cholinergic; miotic.

1782. Carbadox. [6804-07-5] 2-[(1,4-Dioxido-2-quioxalinyl)methylene]hydrazinecarboxylic acid methyl ester; (2-quinoxalinylmethylene)hydrazinecarboxylic acid methyl ester *N,N'*-dioxide; 3-(2-quinoxalinylmethylene)carbazic acid methyl ester *N,N'*-dioxide; methyl 3-(2-quinoxalinylmethylene)carbazate N^1,N^4-dioxide; 2-formylquinoxaline-1,4-dioxide carbomethoxyhydrazone; GS-6244; Fortigro; Mecadox. $C_{11}H_{10}N_4O_4$; mol wt 262.23. C 50.38%, H 3.84%, N 21.37%, O 24.40%. Prepn: Johnston, **BE 669353**; *idem*, **US 3371090**; **US 3433871** (1964, 1968, 1969, all to Pfizer). Animal studies: Thrasher *et al.*, *J. Anim. Sci.* **26**, 911 (1967); Kornegay *et al.*, *ibid.* **27**, 1134 (1968).

Minute yellow crystals, mp 239.5-240°. uv max (water): 236, 251, 303, 366, 373 nm (ε 11000, 10900, 36400, 16100, 16200). Practically insol in water.

THERAP CAT (VET): Antimicrobial.

1783. Carbamazepine. [298-46-4] 5*H*-Dibenz[*b,f*]azepine-5-carboxamide; 5-carbamoyl-5*H*-dibenz[*b,f*]azepine; G-32883; Biston; Carbatrol; Epitol; Finlepsin; Equetro; Sirtal; Stazepine; Tegretol; Telesmin; Timonil. $C_{15}H_{12}N_2O$; mol wt 236.27. C 76.25%, H 5.12%, N 11.86%, O 6.77%. Tricyclic iminostilbene derivative. Prepn: W. Schindler, **US 2948718** (1960 to Geigy). Comprehensive description: H. Y. Aboul-Enein, A. A. Al-Badr, *Anal. Profiles Drug Subs.* **9**, 87-106 (1980). HPLC determn in plasma: E. Oh *et al.*, *Anal. Bioanal. Chem.* **386**, 1931 (2006). Clinical pharmacokinetics: R. H. Levy, B. M. Kerr, *J. Clin. Psychiatry* **49**, 58 (1988). Review of GLC methods, pharmacokinetics and metabolism: S. Pynnönen, *Ther. Drug Monit.* **1**, 409-431 (1979). Review of clinical use in pediatric seizures: J. T. Gilman, *DICP Ann. Pharmacother.* **25**, 1109-1112 (1991); in trigeminal neuralgia: A. Sidebottom, S. Maxwell, *J. Clin. Pharm. Ther.* **20**, 31-35 (1995); in bipolar disorders: P. W. Wang, T. A. Ketter, *Expert Opin. Pharmacother.* **6**, 2887-2902 (2005).

Crystals from abs ethanol + benzene, mp 190-193°. Sol in alcohol, acetone, propylene glycol. Practically insol in water. LD_{50} orally in mice, rats: 3750, 4025 mg/kg. See: E. G. Stenger, F. C. Roulet, *Med. Exp. Int. J. Exp. Med.* **11**, 191 (1964).

THERAP CAT: Anticonvulsant; antimanic. In treatment of pain associated with trigeminal neuralgia.

1784. Carbamide Peroxide. [124-43-6] Urea compd with hydrogen peroxide (H_2O_2) (1:1); urea hydrogen peroxide; Debrox; Exterol; Gly-Oxide; Hyperol. $CH_6N_2O_3$; mol wt 94.07. C 12.77%, H 6.43%, N 29.78%, O 51.02%. $CO(NH_2)_2.H_2O_2$. Prepn: **GB 1555** (1911 to Bayer); M. Strauss, *Chem. Zentralbl.* **11**, 820 (1913). Use for prepn of H_2O_2: J. Milbauer, *Chem. Ztg.* **35**, 871 (1911), *C.A.* **5**, 3391 (1911). Clinical trial in aphthous stomatitis: M. F. Miller, N. W. Chilton, *Pharmacol. Ther. Dent.* **5**, 55 (1980); as cerumenolytic: S. Fahmy, M. Whitefield, *Br. J. Clin. Pract.* **36**, 197 (1982); as dental bleach: V. B. Haywood *et al.*, *J. Am. Dent. Assoc.* **125**, 1219 (1994).

White crystals or cryst powder. Dec in air into urea, oxygen, and water. Sol in 2.5 parts water; partly dec by alcohol and ether into H_2O_2 and urea.

USE: For extemporaneous prepn of H_2O_2.

THERAP CAT: Antiseptic; cerumenolytic; dental bleach.

1785. Carbamyl Chloride. [463-72-9] Carbamic chloride; chloroformamide. CH_2ClNO; mol wt 79.48. C 15.11%, H 2.54%, Cl 44.60%, N 17.62%, O 20.13%. Prepd by passing HCl gas over heated cyanuric acid: Gattermann, Rossolymo, *Ber.* **23**, 1190 (1890); Gattermann, *Ber.* **32**, 1117 (1899). From ammonia and phosgene at 400°: Rupe, Labhard, *Ber.* **33**, 236 (1900); Gattermann, *Ann.* **244**, 30 (1888).

Liquid. Acrid, offensive odor. Has been obtained cryst, mp about 50°, bp 61-62° (decompn). Reacts violently on contact with water, forming ammonium chloride and carbon dioxide. During storage it gives off HCl and slowly changes to cyanuric acid.

1786. Carbanilide. [102-07-8] N,N'-Diphenylurea; diphenylcarbamide; 1,3-diphenylurea; sym-diphenylurea. $C_{13}H_{12}N_2O$; mol wt 212.25. C 73.57%, H 5.70%, N 13.20%, O 7.54%. Obtained during the preparation of phenylurea from aniline hydrochloride and urea: Davis, Blanchard, *Org. Synth.* coll. vol. I, 453 (2nd ed., 1941). Crystal structure: W. Dannecker *et al.*, *Cryst. Struct. Commun.* **8**, 429 (1979).

Orthorhombic prisms from alc. d 1.239. mp 238°. bp 260° (decompn). Sublimes in current of hydrogen at 220°. Sol in ether, glacial acetic acid. Sparingly sol in water (0.15 g/l), acetone, alcohol, chloroform. Moderately sol in pyridine (69.0 g/l).

1787. Carbarsone. [121-59-5] [4-[(Aminocarbonyl)amino]phenyl]arsonic acid; N-carbamoylarsanilic acid; p-ureidobenzenearsonic acid; N-carbamylarsanilic acid; p-carbamidobenzenearsonic acid; 4-ureido-1-phenylarsonic acid; 4-carbamylaminophenylarsonic acid; p-arsonophenylurea; Carb-O-Sep. $C_7H_9AsN_2O_4$; mol wt 260.08. C 32.33%, H 3.49%, As 28.81%, N 10.77%, O 24.61%. Pentavalent organic arsenical with antiprotozoal activity. Prepd from the sodium salt of arsanilic acid by treatment with potassium cyanate or cyanogen bromide: Meister *et al.*, **DE 213155**; Stickings, *J. Chem. Soc.* **1928**, 3131; from arsanilic acid by treatment with phosgene: Nakatsu, Kawase, *Annu. Rep. Takamine Lab.* **8**, 44-47 (1956); **JP 58 4418** (1958 to Sankyo). Toxicity study: N. A. David *et al.*, *Proc. Soc. Exp. Biol. Med.* **29**, 125 (1931). Effect on bodyweight gain and food conversion in turkeys: A. N. Worden, E. C. Wood, *J. Sci. Food Agric.* **24**, 35 (1973). HPLC determn in animal feed: D. Chen *et al.*, *J. Chromatogr. B* **879**, 716 (2011).

White powder. mp 174°. Slightly sol in water, alcohol; sol in solns of alkali hydroxides and carbonates. The satd aq soln is acid to litmus. Nearly insol in ether or chloroform.

THERAP CAT (VET): Antihistomonad in turkeys.

1788. Carbaryl. [63-25-2] 1-Naphthalenol 1-(N-methylcarbamate); methyl carbamic acid 1-naphthyl ester; 1-naphthyl N-methylcarbamate; ENT-23969; OMS-29; UC-7744; Carylderm; Derbac; Ravyon; Sevin. $C_{12}H_{11}NO_2$; mol wt 201.23. C 71.63%, H 5.51%, N 6.96%, O 15.90%. Cholinesterase inhibitor. Prepn and description: Haynes *et al.*, *Contrib. Boyce Thompson Inst.* **18**, 507 (1957); Lambrech, **US 2903478** (1959 to Union Carbide). Metabolism: W. E. Whitehurst *et al.*, *J. Agric. Food Chem.* **11**, 167 (1963); J. B. Houston *et al.*, *Xenobiotica* **5**, 637 (1975). Degradation: D. G. Crosby *et al.*, *J. Agric. Food Chem.* **13**, 204 (1965); D. L. Heywood, *Environ. Qual. Saf.* **4**, 128 (1975). Toxicology: I. Nisbet, D. Miner, *Environment* **13**, 10 (1971). Toxicity: M. Vandekar *et al.*, *Bull. WHO* **44**, 241 (1971). Clinical trial in pediculosis: J. W. Maunder, *Clin. Exp. Dermatol.* **6**, 605 (1981). *Review: Carbamate Insecticides: Chemistry, Biochemistry and Toxicology*, R. J. Kuhr, H. W. Dorough, Eds. (CRC Press, Cleveland, 1976) 301 pp.

Crystals, mp 142°. d_{20}^{20} 1.232. Moderately sol in DMF, acetone, isophorone, cyclohexanone. Soly in water at 30°: 120 mg/l. Vapor pressure at 25°: $<4 \times 10^{-5}$ mm Hg. Stable to heat, light, acids; hydrolyzed in alkalies; noncorrosive. LD_{50} orally in rats: 250 mg/kg (Vandekar).

Caution: Potential symptoms of overexposure are skin irritation; miosis, blurred vision, tearing; rhinorrhea, excessive salivation, sweating; abdominal cramps, nausea, vomiting, diarrhea; tremor, muscle twitching, slurred speech; cyanosis, convulsions; coma, respiratory failure; possible reproductive effects. *See NIOSH Pocket Guide to Chemical Hazards* (DHHS/NIOSH 97-140, 1997) p 50; *Clinical Toxicology of Commercial Products*, R. E. Gosselin *et al.*, Eds. (Williams & Wilkins, Baltimore, 5th ed., 1984) Section II p 305, Section III, pp 86-90.

USE: Contact insecticide.

THERAP CAT: Ectoparasiticide.

THERAP CAT (VET): Ectoparasiticide.

1789. Carbazochrome. [69-81-8] 2-(1,2,3,6-Tetrahydro-3-hydroxy-1-methyl-6-oxo-5H-indol-5-ylidene)hydrazinecarboxamide; 3-hydroxy-1-methyl-5,6-indolinedione 5-semicarbazone; adrenochrome monosemicarbazone; Adrenoxyl; Styptocid; Styptochrome. $C_{10}H_{12}N_4O_3$; mol wt 236.23. C 50.84%, H 5.12%, N 23.72%, O 20.32%. Stable derivative of adrenochrome, *q.v.* Prepn: G. Dechamps *et al.*, **US 2506294** (1950 to Société Belge de l'Azote); C. Beaudet *et al.*, *Experientia* **7**, 293 (1951). Structural studies: F. Ramirez, P. von Ostwalden, *J. Org. Chem.* **20**, 1676 (1955). Clinical trial of combination with troxerutin in treatment of hemorrhoids: M. Basile *et al.*, *Curr. Med. Res. Opin.* **17**, 256 (2001).

Orange-red crystals from dil alcohol, mp 203° (dec).

Compd with sodium salicylate. [13051-01-9] Carbazochrome salicylate; Sigmachrome. $C_{17}H_{17}N_4NaO_6$; mol wt 396.33. Prepd by refluxing carbazochrome and Na salicylate in 30% methanol: Iwao *et al.*, **JP 57 546** (1957 to Tanabe), *C.A.* **52**, 4693 (1958). Orange-red powder, mp 196-197.5° (dec). Soly in water (25°): 0.61 mg/ml. Practically insol in ether, chloroform. pH of 10% aq soln 6.7 to 7.3.

THERAP CAT: Hemostatic.

THERAP CAT (VET): Hemostatic.

1790. Carbazochrome Sodium Sulfonate. [51460-26-5] 5-[2-(Aminocarbonyl)hydrazinylidene]-2,3,5,6-tetrahydro-1-methyl-6-oxo-1H-indole-2-sulfonic acid sodium salt (1:1); 5-[(aminocarbonyl)hydrazono]-2,3,5,6-tetrahydro-1-methyl-6-oxo-1H-indole-2-sulfonic acid monosodium salt; 5,6-dihydro-1-methyl-5,6-dioxo-2-indolinesulfonic acid 5-semicarbazone sodium salt; AC-17; Adona; Odanon. $C_{10}H_{11}N_4NaO_5S$; mol wt 322.27. C 37.27%, H 3.44%, N 17.39%, Na 7.13%, O 24.82%, S 9.95%. Prepn: Iwao, *Pharm. Bull.* **4**, 251 (1956), *C.A.* **51**, 6860f (1957); **GB 795184** (1958 to Gohei Tanabe). Revision of structure from a 3- to a 2-sulfonic acid: Kawazu *et al.*, *J. Heterocycl. Chem.* **10**, 1059 (1973). Mode of action study: T. Seno *et al.*, *Arch. Pharmacol.* **368**, 175 (2003).

Yellow-orange needles from aq methanol, dec 227-228° (free acid dec 195°). Soluble in water.

THERAP CAT: Hemostatic.

1791. Carbazole. [86-74-8] 9H-Carbazole; 9-azafluorene; dibenzopyrrole; diphenylenimine. $C_{12}H_9N$; mol wt 167.21. C

86.20%, H 5.43%, N 8.38%. Prepn: Bunyan, Cadogan, *J. Chem. Soc.* **1963**, 42. Manuf from 2-biphenylamine: Conover, **US 2481292** (1949 to Monsanto); Voltz *et al.*, **US 2891965** (1959 to Houdry Process Corp.); Nevitt, Seelig, **US 3085095** (1963 to Standard Oil, Indiana); from 2-nitrobiphenyl: Larrison, **US 2508791** (1950 to Allied Chem.); by hydrogenolysis of coal-hydrogenation products: Murray *et al.*, **US 2913397** (1959 to Union Carbide); from diphenylamine: Grotta, **US 2921942** (1960 to American-Marietta); Bearse *et al.*, **US 3041349** (1962 to Martin-Marietta); Nevitt, Seelig, *loc. cit.* Purification: Rottschaefer, **US 2459135** (1949 to GAF); Insinger, **US 2464811** (1949 to Koppers Co.). Toxicity study: Eagle, Carlson, *J. Pharmacol. Exp. Ther.* **99**, 450 (1950). Review of prepn and properties: W. C. Sumpter, F. M. Miller, *Heterocyclic Compounds* (Interscience, New York, 1954) pp 70-109.

Crystals from alcohol, benzene, toluene, glacial acetic acid, mp 245°. Sublimes. bp 355°, bp$_{147}$ 200°. d$_4^{18}$ 1.10. Exhibits strong fluorescence and long phosphorescence on exposure to ultraviolet light. Extremely weak base. Insol in water. One gram dissolves in 3 ml quinoline, 6 ml pyridine, 9 ml acetone, 2 ml acetone at 50°, 35 ml ether, 120 ml benzene, 135 ml abs alcohol. Slightly sol in petr ether, chlorinated hydrocarbons, acetic acid. Dissolves in concd H$_2$SO$_4$ without dec. KOH fusion yields *N*-potassium salt. LD$_{50}$ orally in rats: >5 g/kg (Eagle, Carlson).

USE: Important dye intermediate. Used in making photographic plates sensitive to ultraviolet light. Reagent for lignin, carbohydrates, and formaldehyde.

1792. Carbendazim. [10605-21-7] *N*-1*H*-Benzimidazol-2-ylcarbamic acid methyl ester; 2-benzimidazolecarbamic acid methyl ester; 2-(methoxycarbonylamino)benzimidazole; methyl 2-benzimidazolecarbamate; carbendazole; BMC; MBC; BCM; BAS-3460; BAS-67054; CTR-6669; HOE-17411; Bavistin; Derosal; Delsene. C$_9$H$_9$N$_3$O$_2$; mol wt 191.19. C 56.54%, H 4.75%, N 21.98%, O 16.74%. Degradn product of benomyl, *q.v.* Prepn: H. M. Loux, **US 3010968** (1961 to du Pont); H. A. Selling *et al.*, *Chem. Ind. (London)* **1970**, 1625. Activity: G. P. Clemons, H. D. Sisler, *Pestic. Biochem. Physiol.* **1**, 32 (1971). Degradn in soil: A. Helweg, *Pestic. Sci.* **8**, 71 (1977). Efficacy in controlling powdery mildew: G. R. Singh, T. B. Anilkumar, *Indian J. Mycol. Plant Pathol.* **16**, 30 (1986). Review of environmental and toxicological effects: *Carbendazim: Environ. Health Criteria* 149, (World Health Organization, Geneva, 1993) 132 pp.

White powder, mp 302-307° (dec). pKa 4.48. Soly 20° (mg/ml): water, pH 7 8; pH 4 29. Soly 20° (mg/l): hexane 0.5; benzene 36; dichloromethane 68; ethanol 300; dimethylformamide 5000, chloroform 100, acetone 300. Slowly decomp in alkaline soln.

USE: Fungicide.

1793. Carbenicillin. [4697-36-3] (2*S*,5*R*,6*R*)-6-[(2-Carboxy-2-phenylacetyl)amino]-3,3-dimethyl-7-oxo-4-thia-1-azabicyclo[3.2.0]heptane-2-carboxylic acid; *N*-(2-carboxy-3,3-dimethyl-7-oxo-4-thia-1-azabicyclo[3.2.0]hept-6-yl)-2-phenylmalonamic acid; α-carboxybenzylpenicillin; 6-(α-carboxyphenylacetamido)penicillanic acid; α-phenyl(carboxymethylpenicillin). C$_{17}$H$_{18}$N$_2$O$_6$S; mol wt 378.40. C 53.96%, H 4.79%, N 7.40%, O 25.37%, S 8.47%. Semisynthetic antibiotic related to penicillin. Prepn of monopotassium salt: Hobbs, **US 3142673** (1964 to Pfizer); of disodium salt: **BE 646991**; Brain, Nayler, **US 3282926** (1964, 1966 both to Beecham). Activity studies and pharmacology: Naumann, Kempf, *Arzneim.-Forsch.* **19**, 1222 (1969). Chemistry and mode of action: Butler *et al.*, *J. Infect. Dis.* **122**, *Suppl.*, 81 (1970). Clinical data: Hoffler *et*

al., ibid. 1233; Gritz, Naumann, *ibid.* 1237. HPLC determn of epimers in plasma and urine: M. Ishida *et al.*, *J. Chromatogr. B* **652**, 43 (1994). Review of antibacterial activity, pharmacology, and clinical use: H. C. Neu, *Med. Clin. North Am.* **66**, 61-76 (1982).

Disodium salt. [4800-94-6] Carbenicillin disodium; BRL-2064; CP-15639-2; Geopen; Pyopen. C$_{17}$H$_{16}$N$_2$Na$_2$O$_6$S; mol wt 422.36. White powder. Freely sol in water; sol in alc. Practically insol in chloroform, ether. LD$_{50}$ i.p. in rats: >2000 mg/kg *See:* E. I. Goldenthal, *Toxicol. Appl. Pharmacol.* **18**, 185 (1971).

Phenyl sodium. [21649-57-0] Carfecillin sodium; BRL-3475. C$_{23}$H$_{21}$N$_2$NaO$_6$S; mol wt 476.48. Prepn and activity studies: Clayton *et al.*, *J. Med. Chem.* **18**, 172 (1975). Crystals from ethanol, [α]$_D^{20}$ +216.2° (H$_2$O).

THERAP CAT: Antibacterial.

THERAP CAT (VET): Antimicrobial in reptiles.

1794. Carbenoxolone. [5697-56-3] (3β,20β)-3-(3-Carboxy-1-oxopropoxy)-11-oxoolean-12-en-29-oic acid; 3β-hydroxy-11-oxoolean-12-en-30-oic acid hydrogen succinate; 3-*O*-(β-carboxypropionyl)-11-oxo-18β-olean-12-en-30-oic acid; 18β-glycyrrhetic acid hydrogen succinate; carbenoxalone. C$_{34}$H$_{50}$O$_7$; mol wt 570.77. C 71.55%, H 8.83%, O 19.62%. Anti-inflammatory glucocorticoid related to enoxolone, *q.v.* Prepn: Gottfried, Baxendale, **GB 843133**; **US 3070623** (1960, 1961 to Biorex). Monograph: *Carbenoxolone Sodium*, J. M. Robson, F. M. Sullivan, Eds. (Butterworths, London, 1969) 263 pp. Symposium on clinical efficacy: *Scand. J. Gastroenterol.* **15**, Suppl. 65, 1-121 (1980). Effect on gastric prostaglandin levels in humans: J. Rask-Madsen *et al.*, *Eur. J. Clin. Invest.* **13**, 351 (1983); P. Minuz *et al.*, *Pharmacol. Res. Commun.* **16**, 875 (1984). Comprehensive description: S. Pindado *et al.*, *Anal. Profiles Drug Subs. Excip.* **24**, 1-43 (1996).

Cream-colored crystals, mp 291-294°. [α]$_D^{20}$ +128° (chloroform).

Disodium salt. [7421-40-1] Bioplex; Bioral; Neogel; Sanodin; Ulcus-Tablinen. C$_{34}$H$_{48}$Na$_2$O$_7$; mol wt 614.73. Creamy-white solid. Freely sol in water. Insol in chloroform, ether. pKa$_1$ 4.18; pKa$_2$ 5.52. uv max (water): 260 nm (E$_{cm}^{1\%}$ 172). LD$_{50}$ in male mice (mg/kg): 198 i.v.; 120 i.p.; in male rats (mg/kg): 3200 orally (Robson, Sullivan).

THERAP CAT: Antiulcerative.

1795. Carbetapentane. [77-23-6] 1-Phenylcyclopentanecarboxylic acid 2-[2-(diethylamino)ethoxy]ethyl ester; 2-(diethylaminoethoxy)ethyl 1-phenyl-1-cyclopentanecarboxylate; 2-(diethylaminoethoxy)ethyl 1-phenylcyclopentyl-1-carboxylate; 1-phenylcyclopentane-1-carboxylic acid diethylaminoethoxyethyl ester; pentoxyverine; pentoxiverin. C$_{20}$H$_{31}$NO$_3$; mol wt 333.47. C 72.04%, H 9.37%, N 4.20%, O 14.39%. Prepn: H. G. Morren, **GB 753779**; idem, **US 2842585** (1956, 1958 both to Union Chimique Belge). Antispasmodic activity: D. Wellens, *Arzneim.-Forsch.* **17**, 495 (1967). Clinical effect on lung function: E. Krieger, *ibid.* **22**, 389 (1972).

$bp_{0.01}$ 165-170°.

Citrate. [23142-01-0] UCB-2543; Antees; Calnathal; Carbetane; Cossym; Fustpentane; Germapect; Pencal; Sedotussin; Toclase; Tosnone; Tuclase. $C_{20}H_{31}NO_3.C_6H_8O_7$; mol wt 525.60. Crystals, mp 93°. Freely sol in water, chloroform; sol in alcohol, acetone, ethyl acetate. Practically insol in ether, petr ether, benzene.

THERAP CAT: Antitussive.

1796. Carbetocin. [37025-55-1] 1-Butanoic acid-2-(O-methyl-L-tyrosine)-1-carbaoxytocin; 1-butyric acid-2-[3-(p-methoxyphenyl)-L-alanine]oxytocin; deamino-2-O-methyltyrosine-1-carbaoxytocin; 1-thia-4,7,10,13,16-pentaazacycloeicosane cyclic peptide deriv; 1-desamino-1-monocarba-[2-tyr(OMe)]-OT; (2-O-methyltyrosine)deamino-1-carbaoxytocin; d(COMOT); Depotocin; Duratocin; Lonactene; Pabal. $C_{45}H_{69}N_{11}O_{12}S$; mol wt 988.17. C 54.70%, H 7.04%, N 15.59%, O 19.43%, S 3.24%. Synthetic longacting analog of oxytocin, q.v. Prepn: I. Fric et al., Collect. Czech. Chem. Commun. **39**, 1290 (1974); J. H. Cort et al., DE 2732175 (1976 to Czech. Akad. Ved.). Chromatographic properties: M. Lebl, Collect. Czech. Chem. Commun. **45**, 2927 (1980). Uterotonic and galactogogic activity: T. Barth et al., ibid. 3045; T. Barth et al., ibid. **46**, 2441 (1981). Pharmacokinetics in lactating sows: N. Cort et al., Am. J. Vet. Res. **42**, 1804 (1981). Use in regulation of bovine labor: Z. Veznik et al., ibid. **40**, 425 (1979). Effect on milk letdown in sows: N. Cort et al., ibid. **43**, 1283 (1982). Clinical pharmacology: D. J. S. Hunter et al., Clin. Pharmacol. Ther. **52**, 60 (1992). Clinical trials in prevention of postpartum hemorrhage: J. Dansereau et al., Am. J. Obstet. Gynecol. **180**, 670 (1999); S. W. Leung et al., BJOG **113**, 1459 (2006).

Solid from methanol with ether $[\alpha]_D$ −69.0° (c = 0.25 in 1M acetic acid).

THERAP CAT: Oxytocic.

THERAP CAT (VET): Oxytocic, stimulates milk let-down.

1797. Carbic Anhydride. [129-64-6] rel-(3aR,4S,7R,7aS)-3a,4,7,7a-Tetrahydro-4,7-methanoisobenzofuran-1,3-dione; cis-endo-5-norbornene-2,3-dicarboxylic anhydride; endo-cis-bicyclo-[2.2.1]hept-5-ene-2,3-dicarboxylic anhydride; 3,6-endo-methylene-1,2,3,6-tetrahydro-cis-phthalic anhydride; 3,6-endo-methylene-Δ⁴-tetrahydrophthalic anhydride; nadic anhydride. $C_9H_8O_3$; mol wt 164.16. C 65.85%, H 4.91%, O 29.24%. Prepd by the reaction of maleic anhydride with cyclopentadiene in benzene: Diels, Alder, Ann. **460**, 98 (1928). Crystal structure: Destro et al., Acta Crystallogr. B **25**, 2465 (1969).

Relative stereochemistry

Shiny, orthorhombic crystals from petr ether, mp 164-165°. d 1.417. Converted to equilibrium mixtures with exo-cis isomers

when heated above mp. Sol in benzene, toluene, acetone, carbon tetrachloride, chloroform, ethanol, ethyl acetate. Slightly sol in petr ether. Reacts with water to form the corresponding acid. Forms the γ-lactone of 5-hydroxy-2,3-norcamphanedicarboxylic acid in 50% H_2SO_4.

1798. Carbidopa. [38821-49-7]; [28860-95-9] (anhydrous). (αS)-α-Hydrazinyl-3,4-dihydroxy-α-methylbenzenepropanoic acid hydrate (1:1); (−)-L-α-hydrazino-3,4-dihydroxy-α-methylhydrocinnamic acid monohydrate; α-hydrazino-α-methyl-β-(3,4-dihydroxyphenyl)propionic acid monohydrate; L-α-(3,4-dihydroxybenzyl)-α-hydrazinopropionic acid monohydrate; α-methyldopahydrazine; HMD; MK-486; Lodosyn. $C_{10}H_{14}N_2O_4.H_2O$; mol wt 244.25. C 49.18%, H 6.60%, N 11.47%, O 32.75%. Peripheral decarboxylase inhibitor. Prepn of DL-form: Pfister, FR M1553 (1962 to Merck & Co.), C.A. **59**, 12921e (1963); Sletzinger et al., J. Med. Chem. **6**, 101 (1963); GB **940596**; Chemerda et al., US **3462536** (1963, 1969 both to Merck & Co.). Synthesis of the L-form: Karady et al., DE 2062285; DE 2062332 (both 1971 to Merck & Co.), C.A. **75**, 118122t, 118120r (1971); eidem, J. Org. Chem. **36**, 1946, 1949 (1971). Inhibition of dopa decarboxylase: Porter et al., Biochem. Pharmacol. **11**, 1067 (1962); Moran, Sourkes, J. Pharmacol. Exp. Ther. **148**, 252 (1962); Watanabe et al., Clin. Pharmacol. Ther. **11**, 740 (1970). Only the L-form is pharmacologically active: Lotti, Porter, J. Pharmacol. Exp. Ther. **172**, 406 (1970).

Crystals from hot water, mp 203-205° (dec). $[\alpha]_D$ −17.3° (methanol). Also reported as mp 208°. Freely sol in 3N hydrochloric acid; slightly sol in water, methanol. Practically insol in alc, acetone, chloroform, ether.

Combination with levodopa. [57308-51-7] Co-careldopa; Isicom; Nacom; Sinemet.

DL-Form. [302-53-4] Tan fluffy crystals, mp 206-208° (dec). uv max (methanol): 282.5 nm (ε 2940).

THERAP CAT: In combination with levodopa as antiparkinsonian.

1799. Carbimazole. [22232-54-8] 2,3-Dihydro-3-methyl-2-thioxo-1H-imidazole-1-carboxylic acid ethyl ester; 1-ethoxycarbonyl-3-methyl-2-thio-4-imidazoline; ethyl 3-methyl-2-thioimidazoline-1-carboxylate; 1-methyl-3-carbethoxy-2-thioglyoxalone; athyromazole; Neo-mercazole; Neo-Thyreostat. $C_7H_{10}N_2O_2S$; mol wt 186.23. C 45.15%, H 5.41%, N 15.04%, O 17.18%, S 17.22%. Thyroid inhibitor. Prepn: Rimington et al., US **2671088**, reissued as US RE **24505**; US **2815349** (1954, 1958, 1957, all to Natl. Res. Dev. Corp.); Baker, J. Chem. Soc. **1958**, 2387. Clinical efficacy in Graves' disease: J. Duprey et al., Presse Med. **17**, 1124 (1988).

Crystalline powder with characteristic odor; tasteless at first, followed by bitter taste; mp 122-125°. Sol (at 20°) in 500 parts of water; in 50 parts of ethanol; in 330 parts of ether; in 3 parts of chloroform; in 17 parts of acetone. uv max in 0.1N HCl : water (1:8): 291 nm; in 0.1N H_2SO_4: 227 nm and 291 nm ($E_{1cm}^{1\%}$ 557).

THERAP CAT: Antihyperthyroid.

1800. Carbinoxamine. [486-16-8] 2-[(4-Chlorophenyl)-2-pyridinylmethoxy]-N,N-dimethylethanamine; 2-[p-chloro-α-(2-dimethylaminoethoxy)benzyl]pyridine; paracarbinoxamine. $C_{16}H_{19}$-ClN_2O; mol wt 290.79. C 66.09%, H 6.59%, Cl 12.19%, N 9.63%, O 5.50%. Prepn: Tilford, Shelton, US **2606195** (1952 to Wm. S. Merrell); Swain, US **2800485** (1957 to McNeil Labs.). Prepn of l-form: GB **905993** (1962 to McNeil Labs.), C.A. **58**, 5644a (1962). Abs config of l-form: V. Barouh et al., J. Med. Chem. **14**, 834 (1971). Pharmacology and toxicology: R. Cahen, Ann. Pharm. Fr.

20, 463 (1962). GLC determn in serum: D. J. Hoffman *et al., J. Pharm. Sci.* **72**, 1342 (1983).

Liquid, $bp_{0.1}$ 158-162°.

Hydrochloride. $C_{16}H_{19}ClN_2O.HCl$. Crystals from isopropanol + ethyl acetate, dec 162-164°. Sol in water.

Maleate. [3505-38-2] Allergefon; Clistin; Ciberon; Lergefin; Polistin T-Caps. $C_{16}H_{19}ClN_2O.C_4H_4O_4$; mol wt 406.86. Bitter crystals from ethyl acetate, mp 117-119°. Very sol in water; freely sol in alcohol, chloroform. Very slightly sol in ether. pH of 1% aq soln 4.6-5.1. LD_{50} in mice (mg/kg): 166 i.p. (Cahen).

l-**Form.** Levocarbinoxamine; rotoxamine; McN-R-73-Z. $bp_{0.5}$ 143-144°. n_D^{20} 1.5522. $[\alpha]_D^{25}$ −6.8° (c = 2 in methanol).

l-**Form *d*-tartrate.** [49746-00-1] Twiston. $C_{16}H_{19}ClN_2O.C_4H_6O_6$; mol wt 440.88. Crystals from isopropanol, mp 143-144.5°. $[\alpha]_D^{25}$ +37.2° (c = 20 in methanol).

THERAP CAT: Antihistaminic.

1801. Carbobenzoxy Chloride. [501-53-1] Carbonochloridic acid phenylmethyl ester; benzyl chloroformate; chloroformic acid benzyl ester; benzylcarbonyl chloride. $C_8H_7ClO_2$; mol wt 170.59. C 56.33%, H 4.14%, Cl 20.78%, O 18.76%. Prepn by action of phosgene absorbed in toluene on benzyl alcohol: Carter *et al., Org. Synth.* **23**, 13 (1943); by reacting carbonyl chloride and benzyl alcohol at −20 to −30°: Farthing, *J. Chem. Soc.* **1950**, 3213.

Oily liquid; acrid odor; lacrimator. bp_{20} 103°; bp_7 85-87°. Dec to CO_2 and benzyl chloride upon heating at 100-155°.

USE: In peptide synthesis to block the amino group.

1802. Carbocysteine. [638-23-3] *S*-(Carboxymethyl)-L-cysteine; (2*R*)-2-amino-3-[(carboxymethyl)thio]propionic acid; 3-[(carboxymethyl)thio]alanine; AHR-3053; LJ-206; Carbocit; Fluifort; Lisomucil; Loviscol; Mucilar; Mucocis; Mucodyne; Mucolase; Mucolex; Mucopront; Pectox; Pulmoclase; Reomucil; Rhinathiol; Siroxyl; Transbronchin. $C_5H_9NO_4S$; mol wt 179.19. C 33.51%, H 5.06%, N 7.82%, O 35.71%, S 17.89%. Prepn: M. D. Armstrong, J. D. Lewis, *J. Org. Chem.* **16**, 749 (1951); Schöberl, Wagner, *Z. Physiol. Chem.* **304**, 97 (1956); Foye, Verderame, *J. Am. Pharm. Assoc.* **46**, 273 (1957); Goodman *et al., J. Org. Chem.* **23**, 1251 (1958). Pharmacology: Huyen-Vu-Ngoc *et al., C. R. Seances Soc. Biol. Ses Fil.* **160**, 1849 (1966); Quevauviller *et al., Therapie* **22**, 485 (1967). Clinical trial in chronic bronchitis: M. Grillage, K. Bernard-Jones, *Br. J. Clin. Pract.* **39**, 395 (1985). Review of pharmacology and clinical uses: D. T. Brown, *Drug Intell. Clin. Pharm.* **22**, 603-608 (1988).

mp 204-207°. $[\alpha]_D^{24}$ +0.5° (1*N* HCl).

DL-Form. [25390-17-4] Spherical aggregates of needles.

THERAP CAT: Mucolytic; expectorant.

1803. Carbofuran. [1563-66-2] 2,3-Dihydro-2,2-dimethyl-7-benzofuranol 7-(*N*-methylcarbamate); methyl carbamic acid 2,3-dihydro-2,2-dimethyl-7-benzofuranyl ester; 2,2-dimethyl-2,3-dihydro-7-benzofuranyl-*N*-methylcarbamate; 2,2-dimethyl-7-coumaranyl *N*-methylcarbamate; Bay 70143; NIA-10242; Furadan. C_{12}-

$H_{15}NO_3$; mol wt 221.26. C 65.14%, H 6.83%, N 6.33%, O 21.69%. Cholinesterase inhibitor. Prepn and use as insecticide: **NL 6407316** (1964 to Bayer), *C.A.* **63**, 583a (1965); **NL 6500340**; W. G. Scharpf, **US 3474170**; **US 3474171** (1965, 1969, 1969 all to FMC); E. F. Orwoll, **US 3356690** (1967 to FMC). Metabolism: H. W. Dorough, *J. Agric. Food Chem.* **16**, 319 (1968); J. B. Knaak, *ibid.* **18**, 832 (1970). Toxicity studies: J. S. Tobin, *J. Occup. Med.* **12**, 16 (1970); M. A. Fahmy *et al., J. Agric. Food Chem.* **18**, 793 (1970). Teratogenicity study: K. D. Courtney *et al., J. Environ. Sci. Health* **B20**, 373 (1985).

White crystalline solid, mp 150-153°. Soly in water at 25°: 700 ppm. Unstable in alk. LD_{50} orally in mice: 2 mg/kg (Fahmy, 1970).

Caution: Potential symptoms of overexposure are miosis, blurred vision; sweating, salivation, abdominal cramps, diarrhea, headache, nausea, vomiting, weakness; muscle twitching, incoordination, convulsions. See *NIOSH Pocket Guide to Chemical Hazards* (DHHS/NIOSH 97-140, 1997) p 52; *Clinical Toxicology of Commercial Products*, R. E. Gosselin *et al.*, Eds. (Williams & Wilkins, Baltimore, 5th ed., 1984) Section II, p 305.

USE: Systemic insecticide, acaricide, nematocide.

1804. Carbohydrazide. [497-18-7] Carbonic dihydrazide; 1,3-diaminourea. CH_6N_4O; mol wt 90.09. C 13.33%, H 6.71%, N 62.19%, O 17.76%. Prepd by refluxing diethyl carbonate with hydrazine hydrate: Mohr *et al., Inorg. Synth.* **4**, 32 (1953).

Crystals from water + ethanol, dec 153-154°. Freely sol in water. pH of 1% aq soln ~7.4. Practically insol in alcohol, ether, chloroform, benzene. Forms salts with acids. *With nitrous acid it forms the highly explosive carbonyl azide* $CO(N_3)_2$.

1805. γ-Carboline. [244-69-9] 5*H*-Pyrido[4,3-*b*]indole; 2*H*-pyrid[4,3-*b*]indole; 5-carboline. $C_{11}H_8N_2$; mol wt 168.20. C 78.55%, H 4.79%, N 16.66%. Prepn: Robinson, Thornley, *J. Chem. Soc.* **125**, 2169 (1924). Prepn of derivs: Hörlein, *Ber.* **87**, 463 (1954); C. Ducrocq *et al., J. Heterocycl. Chem.* **12**, 963 (1975). NMR studies: F. Balkau, M. L. Heffernan, *Aust. J. Chem.* **26**, 1501, 1523 (1973).

Monoclinic needles from water, mp 225°. d 1.352. Can be distilled at atmospheric pressure without dec. Strong base. Freely sol in methanol; somewhat less sol in ethanol. Slightly sol in benzene, water.

1806. Carbomycin. Sixteen-membered-ring macrolide antibiotic complex similar to leucomycin, *q.v.* and erythromycin, *q.v.*, produced by *Streptomyces halstedii*. Isoln and antibacterial activity: F. W. Tanner *et al., Antibiot. Chemother.* **2**, 441 (1952). Two components have been isolated: Carbomycin A (major) and carbomycin B. Isoln of A: Friedman *et al.*, **US 2960438** (1960 to Pfizer); of B: F. A. Hochstein, K. Murai, *J. Am. Chem. Soc.* **76**, 5080 (1954). Structure of A and B: R. B. Woodward, *Angew. Chem.* **69**, 50 (1957); revised structure: M. Kuehne, B. W. Benson, *J. Am. Chem. Soc.* **87**, 4660 (1965); R. B. Woodward *et al., ibid.* 4662. Abs config of A and B: W. D. Celmer, *ibid.* **88**, 5028 (1966). Identity of A with deltamycin A$_4$: Y. Shimauchi *et al.*, *J. Antibiot.* **31**, 270 (1978). Synthesis of B: K. Tatsuta *et al.*, *J. Am. Chem. Soc.* **99**, 5826 (1977). Stereospecific total synthesis of B: *eidem*, *Tetrahedron Lett.* **1980**, 2837. Retrosynthetic studies: K. C. Nicolaou *et al.*, *J. Am. Chem. Soc.* **103**, 1222 (1981). *Reviews:* D. Vazquez, in *Antibiotics* **Vol. 1**, D. Gottlieb, P. D. Shaw, Eds. (Springer-Verlag, New York, 1967) pp 366-377; W. Keller-Schierlein in *Fortschr. Chem. Org. Naturst.* **30**, 314-460 (1973).

Carbomycin A

Carbomycin A. [4564-87-8] (12S,13S)-9-Deoxy-12,13-epoxy-12,13-dihydro-9-oxoleucomycin V 3-acetate 4^B-(3-methylbutanoate); deltamycin A$_4$; M-4209. C$_{42}$H$_{67}$NO$_{16}$; mol wt 841.99. Blunt needles from ethanol, mp 214°. [α]$_D^{25}$ −58.6° (chloroform). uv max (abs ethanol): 238, 327 nm (E$_{1cm}^{1\%}$ 585, 0.9). Carbomycin standard is the free base having a potency of 1080 units/mg. Stability of soln data: H. L. Martin, *Antibiot. Chemother.* **3**, 865 (1953). Weak base, pKb 7.2. Solubility determn: Weiss *et al.*, *ibid.* **7**, 374 (1957). Soly (mg/ml) at about 28°: water 0.295; methanol >20; ethanol >20. LD$_{50}$ i.v. in mice: 550 mg/kg (Tanner).
Carbomycin B. [21238-30-2] 9-Deoxy-9-oxoleucomycin V 3-acetate 4^B-(3-methylbutanoate). C$_{42}$H$_{67}$NO$_{15}$; mol wt 825.99. Colorless anisotropic plates from acetone/water, mp 141-144° (dec), softens at 138°. [α]$_D^{25}$ −35° (c = 1 in chloroform). uv max (abs ethanol): 278 nm (E$_{1cm}^{1\%}$ 276). pKb 7.56. Solubilities in mg/ml at 25°: ethanol 450; water 0.1-0.2.

1807. Carbon. [7440-44-0] C; at. wt [12.0096; 12.0116]; conventional at. wt 12.011; at. no. 6; valence 4. Group IVA(14). Stable isotopes: 12 (98.892%); 13 (1.108%); radioactive isotopes: 9-11; 14-16. Abundance in earth's crust: approx 0.027%. Cosmic abundance: 6 atoms/atom Si. Occurs in 4 allotropic forms: (1) diamond, *q.v.;* (2) graphite, *q.v.* or black lead; (3) amorphous carbon such as coal, lampblack; (4) fullerenes, *see* Buckminsterfullerene, the only molecular form. Comprehensive reviews: P. L. Walker, *Am. Sci.* **50**, 259-293 (June 1962); Holliday *et al.* in *Comprehensive Inorganic Chemistry* **vol. 1**, J. C. Bailar, Jr. *et al.*, Eds. (Pergamon Press, Oxford, 1973) pp 1173-1294; several authors in *Kirk-Othmer Encyclopedia of Chemical Technology* **vol. 4** (Wiley-Interscience, New York, 3rd ed., 1978) pp 556-709.
^{14}**C isotope.** Continuously formed in the earth's atm by the bombardment of nitrogen with cosmic neutrons according to the reaction $^{14}_{7}$N + $^{1}_{0}$n → $^{14}_{6}$C + $^{1}_{1}$H. The ^{14}C is rapidly oxidized to CO$_2$, in this form it penetrates into animals and plants by photosynthesis and metabolism. The ^{14}C content of living matter is estimated at 15.3 disintegrations per minute and per gram of carbon, corresponding to the equilibrium reached between formation of ^{14}C and its exchange with ^{12}C. This equilibrium stops when the plant or animal dies, and the ^{14}C content begins to decrease, because the ^{14}C decays with a half-life of 5760 years. This fact can be used to date organic matter (not more than 40,000 years old) by comparison with the standard 15.3 disintegrations per min per gram: M. Haissinsky, J. P. Adloff, *Radiochemical Survey of the Elements* (Elsevier, New York, 1965) pp 30-32.

1808. Carbon, Amorphous. Carbon black; activated carbon; decolorizing carbon. A quasi-graphitic form of carbon of small particle size. By the term "carbon black" several forms of artificially prepared carbon or charcoal are designated, *e.g.*: (1) *Animal charcoal*, obtained by charring bones, meat, blood, etc.; (2) *Gas black*; *furnace black; channel black; C.I. 77266*, obtained by incomplete combustion of natural gas; (3) *Lamp black*, obtained by burning various fats, oils, resins, etc., under suitable conditions; (4) *Activated*

charcoal, *e.g.* **Carbomix, Carboraffin, Medicoal, Norit, Opocarbyl, Ultracarbon**, prepd from wood and vegetables. Monograph: H. W. Davidson *et al.*, *Manufactured Carbon* (Pergamon Press, New York, 1968). *Reviews:* Cohan in *Science of Petroleum* **vol. V**, Pt 2, B. T. Brooks, A. E. Dunstan, Eds. (Oxford Univ. Press, 1953), pp 79-89; Smisek, Cerny, *Active Carbon* (Elsevier Publishing Co., Amsterdam, 1970).
Caution: Potential symptoms of overexposure are cough; eye irritation. Potential occupational carcinogen in the presence of polycyclic aromatic hydrocarbons. *See NIOSH Pocket Guide to Chemical Hazards* (DHHS/NIOSH 97-140, 1997) p 52.
USE: Pigment for rubber tires; for printing, stenciling and drawing inks; for leather; stove polish, phonograph records, electrical insulating apparatus. Activated charcoal for clarifying, deodorizing, decolorizing and filtering. As a color additive in foods and cosmetics.
THERAP CAT: Activated charcoal as antidote; adsorptive.
THERAP CAT (VET): Internally as an adsorptive in diarrhea; externally in foul wounds.

1809. Carbon Dioxide. [124-38-9] Carbonic acid gas; carbonic anhydride. CO$_2$; mol wt 44.01. C 27.29%, O 72.71%. Occurs in the atms of many planets. In our solar system, e.g., on Venus, the optical layer thickness due to CO$_2$ is 100,000 cm/atm, but only 220 cm/atm on Earth. Analyses of air in the temperate zones of the Earth show 0.027 to 0.036% (v/v) of CO$_2$: G. P. Kuiper, *The Atmospheres of the Earth and the Planets* (Univ. of Chicago Press, 1949); Landolt-Bornstein, *Zahlenwerte* **vol. III** (Springer-Verlag, 6th ed., 1952) pp 59 and 585. Constituent of carbonate type of minerals and products of animal metabolism. Necessary for the respiration cycle of plants and animals. Obtained industrially as a by-product in the manuf of lime during the "burning" of limestone (CaCO$_3$). Also produced by burning coke or other carbonaceous material. In the U.S.A. large amounts are produced by fermentation (Backus process and Reich process). When glucose is fermented by yeast, the chief products are ethyl alcohol and CO$_2$. Prepd in the laboratory by dropping acid on a carbonate: E. H. Archibald, *The Preparation of Pure Inorganic Substances* (Wiley, New York, 1932) p 196; Loomis, Walters, *J. Am. Chem. Soc.* **48**, 3103 (1926). Purification: Glemser in *Handbook of Preparative Inorganic Chemistry* G. Brauer, Ed. (Academic Press, New York, 2nd ed., 1963) p 647. Discovery of a second polymorph of dry ice: L.-G. Liu, *Nature* **303**, 508 (1983). *Reviews:* E. L. Quinn, *J. Chem. Educ.* **7**, 151-162 and 403-419 (1930); J. Kuprianoff, *Die feste Kohlensäure (Trockeneis)* (Enke, Stuttgart, 1939); E. L. Quinn, C. L. Jones, *Carbon Dioxide* (Reinhold, New York, 1947); W. R. Ballou, in *Kirk-Othmer Encyclopedia of Chemical Technology* **vol. 4** (Interscience, New York, 3rd ed., 1978) pp 725-742. Reviews of supercritical CO$_2$ (*sc*CO$_2$) solvent uses and environmental benefits in polymer chemistry: S. L. Wells, J. De-Simone, *Angew. Chem. Int. Ed.* **40**, 518-527 (2001); in catalysis: W. Leitner, *Acc. Chem. Res.* **35**, 746-756 (2002).
Colorless, odorless gas. *Non-flammable.* Faint acid taste. Usually a nonsupporter of combustion, although burning magnesium continues to burn when transferred into a CO$_2$ atm. Usually marketed in steel cylinders (under sufficient pressure to keep it liquid) or in solid form as *Dry Ice* (compressed carbon dioxide snow, d 1.35). When shipped in steel cylinders, CO$_2$ is in the form of gas over liquid and at 20° exerts a pressure of 830 psi. Use gloves when handling dry ice, as its temp is at least −78.5°; momentary skin contact with dry ice has caused serious frostbites and blisters. At atmospheric pressures the solid form changes into the gaseous phase without liquefaction. d (gas) 1.527 (air = 1); d (gas) 1.557 (N$_2$ = 1); abs d 0.1146 lb/cu ft at 25°; vol at 25°: 8.76 cu ft/lb. d (gas, 0°) 1.976 g/l at 760 mm; d (liq, 0°) 0.914 at 34.3 atm; d (solid, −56.6°) 1.512. Sublimes at −78.48° (760 mm). mp$_{5.2\ atm}$ −56.6°. The gas is not affected by heat until temp reaches about 2000°. Crit temp 31.3°; crit press 72.9 atm; crit density 0.464. Triple point −56.6° at 5.11 atm. Vapor press at −120°: 10.5 mm; at −100°: 104.2 mm; at −82°: 569.1 mm. Heat of formation 94.05 kcal/mol. Latent heat of vaporization 83.12 g cal/g. Specific heat 0.19 to 0.21 Btu/lb. Soly in water (ml CO$_2$/100 ml H$_2$O at 760 mm): 0° = 171; 20° = 88; 60° = 36. One vol dissolves in about 1 vol of water. More sol at higher pressures. Less sol in alcohol, other neutral organic solvents. Absorbed by alkaline solns with the formation of carbonates.
Caution: Potential symptoms of overexposure are headache, dizziness, restlessness and paresthesia; dyspnea; sweating, malaise; in-

creased heart rate and cardiac output; elevated blood pressure; coma; asphyxia; convulsions; direct contact with liquid or dry ice may cause frostbite. *See NIOSH Pocket Guide to Chemical Hazards* (DHHS/NIOSH 97-140, 1997) p 52.

USE: In the carbonation of beverages; manuf of carbonates; in fire prevention and extinction; for inerting flammable materials during manuf, handling and transfer; as propellant in aerosols; as dry ice for refrigeration; to produce harmless smoke or fumes on stage; as rice fumigant; as antiseptic in bacteriology and in the frozen food industry. Supercritical or liquid CO_2 used in extraction of caffeine and hops aroma; dry cleaning; metal degreasing; cleaning semiconductor chips; paint spraying; polymer modification. Environmentally benign alternative to potentially hazardous solvents in organic and polymer chemistry.

THERAP CAT: Respiratory stimulant.

THERAP CAT (VET): Respiratory stimulant (inhalant).

1810. Carbon Diselenide. [506-80-9] Carbon selenide (CSe_2). CSe_2; mol wt 169.93. C 7.07%, Se 92.93%. Prepd by the action of methylene chloride vapor on heated selenium: Ives *et al.*, *J. Chem. Soc.* **1947**, 1080; or from a mixture of CCl_4 and H_2Se in a stream of N_2 at 500°: Grimm, Metzger, *Ber.* **69**, 1356 (1936); from the elements by electrical discharge on Se vapor in the presence of sugar charcoal: Steudel, *Z. Anorg. Allg. Chem.* **361**, 195 (1968).

Light-sensitive, golden yellow, strongly refractive, liquid. Odor of rotten radishes. Turns brown to black on storage. d_4^{20} 2.6824; d_4^{25} 2.6626. mp −45.5°. bp 125-126°; $bp_{8.0}$ 10.0°. n_D^{20} 1.845. Heat of formation: 34 kcal/mol. Miscible with carbon tetrachloride, carbon disulfide, toluene, other organic solvents. Practically insol in water. Dec by alc, pyridine.

1811. Carbon Disulfide. [75-15-0] Carbon bisulfide; dithiocarbonic anhydride. CS_2; mol wt 76.13. C 15.78%, S 84.22%. Minute amounts occur in coal tar and in crude petroleum. Prepd on an industrial scale by heating charcoal with vaporized sulfur; from sulfur and natural gas: *Faith, Keyes & Clark's Industrial Chemicals*, F. A. Lowenheim, M. K. Moran, Eds. (Wiley-Interscience, New York, 4th ed., 1975) pp 224-229. Laboratory purification: Glemser in *Handbook of Preparative Inorganic Chemistry* vol. 1, G. Brauer, Ed. (Academic Press, New York, 2nd ed., 1963) p 652. Review of production and uses: Bushell, *Chem. Ind. (London)* **1961**, 1465; R. W. Timmerman in *Kirk-Othmer Encyclopedia of Chemical Technology* vol. 4 (Wiley-Interscience, New York, 3rd ed., 1978) pp 742-757; of toxicology and human exposure: *Toxicological Profile for Carbon Disulfide* (PB97-121073, 1996) 252 pp.

Highly refractive, mobile liquid. *Flammable, poisonous. May be ignited by hot steam pipes.* The purest distillates ever obtained are reported to have a sweet, pleasing, and ethereal odor, while the usual commercial and reagent grades are foul smelling. Dec on standing for a long time. Burns with a blue flame to CO_2 and SO_2. Flash pt, closed cup: −30°C. Ignition pt: 100°. Explosive range: 1 to 50% (v/v) in air. d_4^0 1.29272; d_4^{15} 1.27055; d_4^{20} 1.2632; d_4^{30} 1.24817. Vapors sink to the ground. Vapor density 2.67 (air = 1). mp −111.6°. $bp_{1.0}$ −73.8°; bp_{10} −44.7°; bp_{100} −5.1°; bp_{400} 28.0°; bp_{760} 46.5°; $bp_{(2\ atm)}$ 69.1°; $bp_{(5\ atm)}$ 104.8°. Crit temp 280.0°; crit press. 72.9 atm. n_D^{15} 1.63189; $n_D^{20.1}$ 1.62803; $n^{23.5}$ 1.62543. Surface tension at 20°: 32.25. Coefficient of viscosity at 20°: 0.363. Heat of vaporization at bp: 84.1 cal/g. Heat of fusion: 1.049 kcal/mole. Heat capacity at 24.3°: 18.17 cal/mole/deg: Brown, Manov, *J. Am. Chem. Soc.* **59**, 500 (1937). Ebullioscopic constant: 2.35°. Dielectric constant at low frequencies: 2.641. Dipole moment: 0.0. Soly in water at 20°: 0.294%. Soly of water in CS_2: <0.005%. Azeotrope with water bp 42.6°, contains 97.2% CS_2. Misc with anhydr methanol, ethanol, ether, benzene, chloroform, carbon tetrachloride, oils. Can be stored in iron, aluminum, glass, porcelain, Teflon.

Caution: Poisoning usually occurs from inhalation but also may be caused by ingestion and skin absorption. Direct contact with liquid or concentrated vapors may cause irritation of skin, eyes, mucous membranes; eye and skin burns; dermatitis. Potential symptoms of acute overexposure are headache, garlicky breath, nausea, vomiting, diarrhea, fatigue, weakness, vertigo, dizziness, poor sleep; CNS depression with respiratory paralysis. Potential symptoms of chronic overexposure are nervousness, anorexia, weight loss; psychosis; polyneuropathy; Parkinson-like syndrome; ocular changes;

coronary heart disease; gastritis; kidney and liver injury; reproductive effects; marked psychic disturbances ranging from extreme irritability to mania with hallucinations, tremors, auditory and visual disturbances. *See NIOSH Pocket Guide to Chemical Hazards* (DHHS/NIOSH 97-140, 1997) p 52; *Clinical Toxicology of Commercial Products*, R. E. Gosselin *et al.*, Eds. (Williams & Wilkins, Baltimore, 5th ed., 1984) Section III, pp 90-93.

USE: In the manuf of rayon, carbon tetrachloride, xanthogenates, soil disinfectants, electronic vacuum tubes. Solvent for phosphorus, sulfur, selenium, bromine, iodine, fats, resins, rubbers.

1812. Carbonic Anhydrase. [9001-03-0] Carbonate dehydratase; carbonate hydro-lyase; EC 4.2.1.1. Mol wt ~30,000. A small zinc-contg enzyme which catalyzes the hydration of CO_2. Found in higher concns in erythrocytes, renal cortex, and gastric mucosa of mammals; also found in other animal tissues, in plants and in some bacteria. Isoln from bovine erythrocytes: Lindskog, *Biochim. Biophys. Acta* **39**, 218 (1960); from human erythrocytes: Nyman, *ibid.* **52**, 1 (1961); from renal cortex: Höber, *Proc. Soc. Exp. Biol. Med.* **49**, 87 (1942); from gastric mucosa: Davenport, *Physiol. Rev.* **26**, 560 (1946). Human carbonic anhydrase consists of two isoenzymes with distinctly different amino acid sequences and specific activities. The high activity form is called carbonic anhydrase C; the low activity form is called B; modified forms of these two isoenzymes exist: Funakoshi, Deutsch, *J. Biol. Chem.* **243**, 6474 (1968); **244**, 3438 (1969). Amino acid sequence of carbonic anhydrase B: Anderson *et al.*, *Biochem. Biophys. Res. Commun.* **48**, 670 (1972); Lin, Deutsch, *J. Biol. Chem.* **248**, 1885 (1973); sequence of carbonic anhydrase C: Henderson, Henriksson, *Biochem. Biophys. Res. Commun.* **52**, 1388 (1973); Lin, Deutsch, *J. Biol. Chem.* **249**, 2329 (1974). Crystal structure of carbonic anhydrase C: Liljas *et al.*, *Nature New Biol.* **235**, 131 (1972). Catalyzes the reversible reaction of CO_2 and H_2O to HCO_3^- and H^+. Permits CO interchange between blood and tissues. In gastric mucosa, reaction rate is sufficient to neutralize the excess alkalinity produced by the ionization of water and secretion of hydrogen ions: Roughton, Clark in *The Enzymes* vol. 1, part 2, J. B. Sumner, K. Myrbäck, Eds. (Academic Press, New York, 1951) pp 1250-1265. In the kidney, participates in Na^+ transport. Review of physiology: Maren in *Oxygen Affinity of Hemoglobin and Red Cell Acid Base Status, Alfred Benzon Symposium* IV, P. Astrup, M. Roerth, Eds. (Academic Press, New York, 1972) pp 418-433. Review of metal ion function: Prince, Woolley, *Angew. Chem. Int. Ed.* **11**, 408-417 (1972). *Review:* Lindskog *et al.*, "Carbonic Anhydrase" in *The Enzymes* vol. **5**, P. D. Boyer, Ed. (Academic Press, New York, 1971) pp 587-665.

1813. Carbon Monoxide. [630-08-0] CO; mol wt 28.01. C 42.88%, O 57.12%. Produced on an industrial scale by partial oxidation of hydrocarbon gases from natural gas or by the gasification of coal and coke. Conveniently prepd in the laboratory by heating calcium carbonate with Zn dust: Weinhouse, *J. Am. Chem. Soc.* **70**, 442 (1948); by dehydration of formic acid with H_2SO_4: Gilliland, Blanchard, *Inorg. Synth.* **2**, 81 (1946). Purification of carbon monoxide bought in steel cylinders: A. Klemenc, *Die Behandlung und Reindarstellung von Gasen* (Vienna, 1948) p 160; Glemser in *Handbook of Preparative Inorganic Chemistry* vol. 1, G. Brauer, Ed. (Academic Press, New York, 2nd ed., 1963) p 646. *Review:* R. Pierantozzi in *Kirk-Othmer Encyclopedia of Chemical Technology* vol. **5** (Wiley-Interscience, New York, 4th ed., 1993) pp 97-122. Review of clinical toxicology: Stewart, *Annu. Rev. Biochem.* **15**, 409-423 (1975); D. Gorman *et al.*, *Toxicology* **187**, 25-38 (2003); of industrial toxicology: *Patty's Industrial Hygiene and Toxicology* vol. **2F**, G. D. Clayton, F. E. Clayton, Eds. (Wiley-Interscience, New York, 4th ed., 1994) pp 4523-4552.

Odorless, colorless, tasteless gas. Ignition pt in air: 700°. mp −205.0°. bp −191.5°. d_4^{-195} (liq) 0.814. d (gas) 0.968 (air = 1.000). d_4^0 at 760 mm: 1.250 g/liter. The top pressure is 500 psi. *Poisonous, flammable.* Flammable limits in air: 12 to 75 vol %. Crit press 35 atm, crit temp −139°. Heat capacity at 20°: 6.95 cal/mole/°C. Heat value per m^3: 3033 kcal. Heat of formation: −26.39 kcal/mol. Dec into carbon and carbon dioxide between 400 and 700°, at lower temp when in contact with catalytic surfaces. Above 800° the equilibrium reaction favors CO formation. Hopcalite, a mixture of the oxides of manganese and copper, catalyzes the decompn at room temp, as does Pd on silica gel. Sparingly sol in water: 3.3 ml/100 ml H_2O at 0°; 2.3 ml/100 ml H_2O at 20°; freely absorbed by a concd soln of cuprous chloride in HCl or in NH_4OH. Appreciably sol in some organic solvents, such as ethyl acetate, $CHCl_3$, acetic acid. The soly in methanol and ethanol is about 7 times as great as the soly in water.

Caution: Combines with the hemoglobin in the blood to form carboxyhemoglobin which disrupts oxygen transport and delivery throughout the body. Potential symptoms of overexposure by inhalation are headache, tachypnea, nausea, vomiting, lassitude, fatigue, dizziness, confusion, hallucinations; dimness of vision; irritability, impaired judgement; cyanosis; depression of S-T segment of electrocardiogram, angina, syncope; coma. *See NIOSH Pocket Guide to Chemical Hazards* (DHHS/NIOSH 97-140, 2003) p 54; *Clinical Toxicology of Commercial Products*, R. E. Gosselin *et al.*, Eds. (Williams & Wilkins, Baltimore, 5th ed., 1984) Section III, pp 94-101.

USE: As reducing agent in metallurgical operations especially in the Mond process for the recovery of nickel; in organic synthesis especially in the Fischer-Tropsch processes for petroleum-type products and in the oxo reaction; in the manuf of metal carbonyls.

1814. Carbon Nitride. [143334-20-7] C_3N_4; mol wt 92.06. C 39.14%, N 60.86%. Covalent crystalline solid which exists in two forms α and β. β-form may be harder than diamond. Model of structure and hardness: A. Y. Liu, M. L. Cohen, *Science* **245**, 841 (1989). Synthesis: E. E. Haller *et al.*, **WO 9116196**; **US 5110679** (1991, 1992 both to Regents U. California, Berkeley); C. Niu *et al.*, *Science* **261**, 334 (1993). Hardness evaluation: O. Fukunaga, *Kagaku to Kyoiku* **41**, 325 (1993), *C.A.* **119**, 54572g (1993).

1815. Carbon Suboxide. [504-64-3] 1,2-Propadiene-1,3-dione; tricarbon dioxide. C_3O_2; mol wt 68.03. C 52.97%, O 47.04%. O=C=C=C=O. Prepn by thermal decompn of malonic acid: Glemser in *Handbook of Preparative Inorganic Chemistry* vol. 1, G. Brauer, Ed. (Academic Press, New York, 2nd ed., 1963) p 648. Reactions in organic synthesis: Dashkevich, Beilin, *Russ. Chem. Rev.* **36**, 391 (1967). Comprehensive reviews: Reyerson, Kobe, *Chem. Rev.* **7**, 479 (1930); Vol'kenshtein, *Usp. Khim.* **4**, 610 (1935); Grauer, *Chimia* **14**, 11 (1960); T. Kappe, E. Ziegler, *Angew. Chem. Int. Ed.* **13**, 491-504 (1974); reprinted in *New Synthetic Methods* vol. 1 (Verlag Chemie, Weinheim, 1975) pp 29-69.

Colorless, highly refractive liquid or colorless gas which burns with a blue, sooty flame. Odor like acrolein and mustard oil. mp $-111.3°$. bp_{760} 6.8°. d_4^0 1.114. n_D^0 1.45384; n_D^{-12} 1.46757. Vapor pressure at 0°: 587-589 mm. Explosive limits, 6 to 30 vol % in air. Dipole moment: 0.7D. Thermodynamic constants: Thompson, *Trans. Faraday Soc.* **37**, 249 (1941). The gas can be stored at pressures of up to 100 mm, but even at these pressures polymerization may occur, giving a red, water-sol product. This invariably occurs at higher pressure or in the liquid state. Polymerization facilitated by presence of P_2O_5. Dec when passed through heated glass tubes, forming a mirror surface. Difficultly sol in carbon disulfide, xylene. With water forms malonic acid quantitatively. Forms malonamide with ammonia.

Caution: In small amounts acts as a lacrimator; in high concns attacks eyes, nose, respiratory organs, producing a feeling of suffocation.

USE: Prepn of malonates; improving dye affinity of fibers.

1816. Carbon Tetrachloride. [56-23-5] Tetrachloromethane; perchloromethane; Necatorina; Benzinoform. CCl_4; mol wt 153.81. C 7.81%, Cl 92.19%. Obtained from carbon disulfide and chlorine in presence of a catalyst, e.g., $SbCl_5$, Fe filings, or by the chlorination of hydrocarbons: *Faith, Keyes & Clark's Industrial Chemicals*, F. A. Lowenheim, M. K. Moran, Eds. (Wiley-Interscience, New York, 4th ed., 1975) pp 230-234; H. D. DeShon in *Kirk-Othmer Encyclopedia of Chemical Technology* vol. 5 (Wiley-Interscience, New York, 3rd ed., 1979) pp 704-714. Toxicity data: Svirbely, *J. Ind. Hyg. Toxicol.* **29**, 382 (1947); E. Browning, *Toxicity and Metabolism of Industrial Solvents* (Elsevier, New York, 1965) pp 173-188. Use in induction of experimental liver disease: P. Trivedi, A. P. Mowat, *Br. J. Exp. Pathol.* **64**, 25 (1983). Review of carcinogenic risk: *IARC Monographs* **20**, 371-399 (1979); of toxicity studies: R. O. Recknagel *et al.*, *Pharmacol. Ther.* **43**, 139-154 (1989); of toxicology and environmental exposure: *Environ. Health Criter.*

208, 1-177 (1999); of toxicology and human exposure: *Toxicological Profile for Carbon Tetrachloride* (PB2006-100002, 2005) 361 pp.

Colorless, clear, nonflammable, heavy liquid; characteristic odor. May form phosgene when used to put out electrical fires. *Poisonous. Use only when adequate ventilation is possible.* d_{25}^{25} 1.589. bp 76.7°. mp $-23°$. n_D^{20} 1.4607. One ml dissolves in 2000 ml water; misc with alcohol, benzene, chloroform, ether, carbon disulfide, petr ether, oils. LC_{50} for mice: 9528 ppm (Svirbely). LD_{50} in rats, mice, dogs (g/kg): 2.92, 12.1-14.4, 2.3 orally; LD_{50} in mice (g/kg): 4.1 i.p., 30.4 s.c. (IARC, 1979).

Caution: Potential symptoms of overexposure are CNS depression; drowsiness, dizziness, incoordination; nausea, vomiting; liver and kidney injury. Direct contact may cause skin and eye irritation; dermatitis through defatting action. *See NIOSH Pocket Guide to Chemical Hazards* (PB2003-100121, 2003) p 54; *Patty's Industrial Hygiene and Toxicology* vol. 2E, G. D. Clayton, F. E. Clayton, Eds. (Wiley-Interscience, New York, 4th ed., 1994) p 4071-4080. This substance is reasonably anticipated to be a human carcinogen: *Report on Carcinogens, Twelfth Edition* (PB2011-111646, 2011) p 86.

USE: As solvent for oils, fats, lacquers, varnishes, rubber waxes, resins; starting material in manuf of organic compds. Pharmaceutic aid (solvent). Formerly used as dry cleaning agent, fire extinguisher and grain fumigant.

THERAP CAT: Formerly as anthelmintic (Nematodes).

THERAP CAT (VET): Anthelmintic.

1817. Carbon Tetrafluoride. [75-73-0] Tetrafluoromethane; Freon 14. CF_4; mol wt 88.00. C 13.65%, F 86.36%. Prepd from carbon or carbon monoxide and fluorine: Yost, *Inorg. Synth.* **1**, 34 (1939); Simons, Block, *J. Am. Chem. Soc.* **61**, 2962 (1939); Kwasnik in *Handbook of Preparative Inorganic Chemistry* vol 1, G. Brauer, Ed. (Academic Press, New York, 2nd ed., 1963) p 203. May also be prepd from SiC + F_2: Priest, *Inorg. Synth.* **3**, 178 (1950).

Colorless, odorless gas. Thermally stable. Chemically very inert. d (solid, $-195°$) 1.98. d (liq, $-183°$) 1.89. mp $-183.6°$. bp $-127.8°$. May be stored in steel cylinders.

Caution: Narcotic in high concns.

USE: Low temp refrigerant; gaseous insulator.

1818. Carbon Tetraiodide. [507-25-5] Tetraiodomethane. CI_4; mol wt 519.63. C 2.31%, I 97.69%. Prepd by the interaction of carbon tetrachloride and aluminum or calcium iodide: Gustavson, *Ann.* **172**, 173 (1874); boron iodide: Moissan, *C. R. Hebd. Seances Acad. Sci.* **113**, 19 (1891); lithium or calcium iodide: Lantenois, *ibid.* **156**, 1385 (1913); ethyl iodide in the presence of aluminum chloride: Walker, *J. Chem. Soc.* **85**, 1090 (1904); McArthur, Simons, *Inorg. Synth.* **3**, 37 (1950).

Red cubic crystals. Odor of iodine. Dec to iodine and tetraiodoethylene under the influence of light or heat. d_4^{20} 4.32. mp 171°. Sol in benzene, chloroform. Dec by hot alcohol. Practically insol in water, but hydrolyzes slowly in contact with water, forming iodoform and iodine.

1819. *trans*-Carbonylchlorobis(triphenylphosphine)rhodium(I). [15318-33-9]; [13938-94-8] (unspecified stereo). (*SP*-4-3)-Carbonylchlorobis(triphenylphosphine)rhodium; *trans*-bis(triphenylphosphine)rhodiumcarbonyl chloride; *trans*-carbonylbis(triphenylphosphine)chlororhodium(I); *trans*-chlorocarbonylbis(triphenylphosphine)rhodium(I); *trans*-rhodium carbonyl bis(triphenylphosphine) chloride; *trans*-RhCl(CO)(PPh$_3$)$_2$. $C_{37}H_{30}ClOP_2Rh$; mol wt 690.95. C 64.32%, H 4.38%, Cl 5.13%, O 2.32%, P 8.97%, Rh 14.89%. Square-planar rhodium complex; catalyst for a variety of synthetic transformations. Rhodium analogue of Vaska's Compound, *q.v.* Prepn: L. Vallarino, *J. Chem. Soc.* **1957**, 2287. Improved prepn: D. Evans *et al.*, *Inorg. Synth.* **28**, 79 (1990). Crystal structure: A. Del Pra *et al.*, *Cryst. Struct. Commun.* **8**, 959 (1979); Y.-J. Chen *et al.*, *Acta Crystallogr. C* **47**, 2441 (1991). Catalytic decarbonylation studies: J. Blum *et al.*, *J. Am. Chem. Soc.* **89**, 2338 (1967); K. Ohno, J. Tsuji, *ibid.* **90**, 99 (1968). Synthetic applications: L. S. Hegedus *et al.*, *ibid.* **97**, 5448 (1975); A. J. Kunin, R. Eisenberg, *Organometallics* **7**, 2124 (1988). *Review:* K. Kikukawa, S. A. Westcott in *Encyclopedia of Reagents for Organic Synthesis* **2**, L. A. Paquette, Ed. (Wiley, New York, 1995) pp 1001-1004.

Yellow crystalline solid, mp 209-210° (dec). Partially isomerizes when heated with chlorobenzene in sealed tube at 220°. Crystal density 1.41. uv max: 365 nm. Very sol in dichloromethane, chloroform; sol in carbon tetrachloride, benzene, nitrobenzene; slightly sol in acetone, diethyl ether. Insol in methanol, ethanol, hexane. Air stable. Readily reacts with oxygen in soln.

cis-**Form.** [16353-77-8] mp 204-205° (dec).

USE: Catalyst.

1820. N,N′-Carbonyldiimidazole. [530-62-1] Di-1*H*-imidazol-1-ylmethanone; 1,1′-carbonylbis-1*H*-imidazole; diimidazol-1-yl ketone. $C_7H_6N_4O$; mol wt 162.15. C 51.85%, H 3.73%, N 34.55%, O 9.87%. Prepd from phosgene and imidazole in dry tetrahydrofuran or dry benzene: Staab, *Ann.* **609**, 75 (1957); Anderson, Paul, *J. Am. Chem. Soc.* **80**, 4423 (1958).

Crystals from tetrahydrofuran or benzene, mp 115.5-116°. Should be handled under exclusion of atmospheric moisture. Hydrolyzed by water in a few sec with evolution of CO_2.

USE: In the synthesis of peptides. Reacts readily with carboxylic acids to form acyl imidazoles; subsequent reaction with amines to form amides goes smoothly.

1821. Carbonyl Fluoride. [353-50-4] Carbonic difluoride; fluophosgene. CF_2O; mol wt 66.01. C 18.20%, F 57.56%, O 24.24%. COF_2. Prepd from CO and F_2 or BrF_3 and CO: Ruff, Miltschitzky, *Z. Anorg. Allg. Chem.* **221**, 154 (1935); Kwasnik in *Handbook of Preparative Inorganic Chemistry* vol. 1, G. Brauer, Ed. (Academic Press, New York, 2nd ed., 1963) p 206. Alternate route from CO + AgF_2: Farlow *et al.*, *Inorg. Synth.* **6**, 155 (1960).

Pungent, very hygroscopic gas. d (solid, −190°): 1.388. d (liq, −114°): 1.139. mp −114.0°. bp −83.1°. Heat of formation: 166.6 kcal. Instantly hydrolyzed by water. *Poisonous, corrosive.*

Caution: Potential symptoms of overexposure are irritation of eyes, skin, mucous membranes, respiratory system; eye, skin burns; lacrimation; cough, pulmonary edema, dyspnea; direct contact with liquid may cause frostbite. Chronic exposure may cause GI pain, muscular fibrosis, skeletal fluorosis. *See NIOSH Pocket Guide to Chemical Hazards* (DHHS/NIOSH 97-140, 1997) p 54.

1822. Carbonyl Sulfide. [463-58-1] Carbon oxide sulfide (COS); carbon oxysulfide. COS; mol wt 60.07. C 20.00%, O 26.63%, S 53.37%. Fluid found in petroleum and coal refinery gases and in coal gassification streams; impurity in commercial sources of propane. Prepn: C. Than, *Ann.* **5**, 236 (1867). Thermodynamic properties: J. D. Kemp, W. F. Giauque, *J. Am. Chem. Soc.* **59**, 79 (1937). Photolysis and extinction coefficient measurements: G. S. Forbes, J. E. Cline, *ibid.* **61**, 151 (1939). Potentiometric determn in petroleum gases: D. B. Bruss *et al.*, *Anal. Chem.* **29**, 807 (1957). Hydrolysis in water-propane systems: W. C. Andersen, T. J. Bruno, *Ind. Eng. Chem. Res.* **42**, 963 (2003). Use as a grain fumigant: H. J. Banks *et al.*, **WO 9313659**; *eidem*, **US 6203824** (1993, 2001 both to CSIRO). Uptake in grain products: P. C. Annis *et al.*, *J. Agric. Food Chem.* **48**, 3646 (2000). Review of chemistry: R. J. Ferm, *Chem. Rev.* **57**, 621-640 (1957); P. D. N. Svoronos, T. J. Bruno, *Ind. Eng. Chem. Res.* **41**, 5321-5336 (2002); of development as a fumigant: E. J. Wright, *Proc. Austral. Postharvest Tech. Conf., Canberra 2003* 224-229 (2003); of toxicology: A. R. Bartholomaeus, V. S. Haritos, *Food Chem. Toxicol.* **43**, 1687-1701 (2005).

$$O=C=S$$

Colorless gas, bp −50.2°. mp −138.8°. Odorless when very pure. *n* 1.3785. Liquid density (174 K): 1.274 g/cm³. Vapor density (25°, 1 atm): 2.4849 ±0.0005 g/l. Critical temp 105°. Critical press 61 atm. Soly at 1 atm (ml/ml): water 0.80 (13.5°); toluene 15.0 (22.0°). Log P (octanol/water): ~0.8. uv max: 208, 225 nm. *Poisonous, flammable.* LD_{50} i.p. in rats: 22.5 mg/kg. LC_{50} by inhalation (mg/m³): 2940 in mice (35 min); 2650 in rats (4 h) (Bartholomaeus, Haritos).

Caution: Potential symptoms of overexposure are related to carbonyl sulfide hydrolysis to form hydrogen sulfide, *q.v.*

USE: Grain fumigant.

1823. Carboplatin. [41575-94-4] (*SP*-4-2)-Diammine[1,1-cyclobutanedicarboxylato(2−)-*κO,κO″*]platinum; 1,1-cyclobutanedicarboxylic acid platinum complex; *cis*-diammine(1,1-cyclobutanedicarboxylato)platinum(II); CBDCA; JM-8; NSC-241240; Carboplat; Paraplatin. $C_6H_{12}N_2O_4Pt$; mol wt 371.26. C 19.41%, H 3.26%, N 7.55%, O 17.24%, Pt 52.55%. Analog of cisplatin, *q.v.*, with reduced nephrotoxicity. Prepn and antitumor activity: **NL 7307863**; M. J. Cleare *et al.*, **US 4140707** (1973, 1979 both to Research Corp.). Improved prepn: R. C. Harrison *et al.*, *Inorg. Chim. Acta* **46**, L15 (1980). Crystal structure: S. Neidle *et al.*, *J. Inorg. Biochem.* **13**, 205 (1980). Comparison with other antitumor platinum complexes: M. J. Cleare *et al.*, *Biochimie* **60**, 835 (1978). Early clinical studies: A. H. Calvert *et al.*, *Cancer Chemother. Pharmacol.* **9**, 140 (1982). Clinical pharmacokinetics: S. J. Harland *et al.*, *Cancer Res.* **44**, 1693 (1984); M. J. Egorin *et al.*, *ibid.* 5432. Toxicity, activity in mice, rats, dogs: P. Lelieveld *et al.*, *Eur. J. Cancer Clin. Oncol.* **20**, 1087 (1984). Comparison with cisplatin chemotherapy in advanced seminoma: M. J. Peckham *et al.*, *Br. J. Cancer* **52**, 7 (1985). In treatment of small cell lung cancer: I. E. Smith *et al.*, *Cancer Treat. Rep.* **69**, 43 (1985); in metastatic breast cancer: E. A. Perez, *Oncologist* **9**, 518 (2004). Clinical pharmacokinetics: S. B. Duffull, B. A. Robinson, *Clin. Pharmacokinet.* **33**, 161 (1997). Review of veterinary studies: L. E. Fox, *J. Am. Anim. Hosp. Assoc.* **36**, 13-14 (2000).

White crystals, sol in water. LD_{50} in mice (mg/kg): 150 i.p., 140 i.v.; in rats (mg/kg): 85 i.v. (Lelieveld).

THERAP CAT: Antineoplastic.

THERAP CAT (VET): Antineoplastic.

1824. Carboprost. [35700-23-3] (5*Z*,9*α*,11*α*,13*E*,15*S*)-9,-11,15-Trihydroxy-15-methylprosta-5,13-dien-1-oic acid; 7-[3,5-dihydroxy-2-(3-hydroxy-3-methyl-1-octenyl)cyclopentyl]-5-heptenoic acid; (15*S*)-15-methyl PGF$_{2α}$; U-32921. $C_{21}H_{36}O_5$; mol wt 368.51. C 68.45%, H 9.85%, O 21.71%. Analog of prostaglandin F$_{2α}$, *q.v.* Prepn: G. L. Bundy *et al.*, **DE 2121980**; G. L. Bundy, **US 3728382** (1971, 1973 both to Upjohn); *eidem, Ann. N.Y. Acad. Sci.* **180**, 76 (1971); E. W. Yankee *et al.*, *J. Am. Chem. Soc.* **96**, 5865 (1974). Biological activity: J. R. Weeks *et al.*, *J. Pharmacol. Exp. Ther.* **186**, 67 (1973). Mechanism of action: A. I. Csapo, M. O. Pulkkinen, *Prostaglandins* **18**, 479 (1979). Clinical studies: P. C. Schwallie, K. R. Lamborn, *J. Reprod. Med.* **23**, 289 (1979); M. P. Mapa *et al.*, *Int. J. Gynaecol. Obstet.* **20**, 125 (1982). Teratological study: G. M. Szczech *et al.*, *Adv. Prostaglandin Thromboxane Res.* **4**, 157 (1978).

Tromethamine salt. [58551-69-2] Carboprost trometamol; U-32921E; Hemabate; Prostin/15M. $C_{25}H_{47}NO_8$; mol wt 489.65. Sol in water.

Methyl ester. [35700-21-1] Carboprost methyl; U-36384. $C_{22}H_{38}O_5$; mol wt 382.54. Crystals from ether/hexane, mp 55-56°. $[α]_D$ +24° (c = 0.81 in ethanol).

THERAP CAT: Oxytocic.

1825. Carboquone. [24279-91-2] 2-[2-[(Aminocarbonyl)-oxy]-1-methoxyethyl]-3,6-bis(1-aziridinyl)-5-methyl-2,5-cyclo-hexadiene-1,4-dione; 2,5-bis(1-aziridinyl)-3-(2-hydroxy-1-meth-oxyethyl)-6-methyl-*p*-benzoquinone carbamate (ester); 2,5-bis(1-az-iridinyl)-3-(2-carbamoyloxy-1-methoxyethyl)-6-methyl-1,4-benzo-quinone; carbazilquinone; Esquinon. $C_{15}H_{19}N_3O_5$; mol wt 321.33. C 56.07%, H 5.96%, N 13.08%, O 24.89%. Prepn and antitumor activity: Nakao *et al.*, *Chem. Pharm. Bull.* **20**, 1968 (1972); T. Nak-amura *et al.*, **DE 1905224**; H. Nakao *et al.*, **US 3631026** (1969, 1971 both to Sankyo). Effect on tumors in mice: Arakawa *et al.*, *Gann* **61**, 535 (1970). Acute toxicity: H. Masuda, Y. Suzuki, *Oyo Yakuri* **8**, 501 (1974), *C.A.* **81**, 163349g (1974).

Red to reddish-brown crystals, mp 202° (dec). Slightly sol in chlo-roform, acetone, abs alcohol. Practically insol in water. LD_{50} in male mice, male rats (mg/kg): 6.09, 3.88 i.v.; 3.84, 3.16 i.p.; 30.8, 28.0 orally (Masuda, Suzuki).

THERAP CAT: Antineoplastic.

1826. Carbostyril. [59-31-4] 2(1*H*)-Quinolinone; 2-hy-droxyquinoline; 2-quinolinol; 2(1*H*)-quinolone; *o*-aminocinnamic acid lactam. C_9H_7NO; mol wt 145.16. C 74.47%, H 4.86%, N 9.65%, O 11.02%. Prepn from quinoline by heating with potassium hydroxide to 225° under anhydr conditions: Tschitschibabin, *Ber.* **56**, 1883 (1923); **DE 406208**; *Chem. Zentralbl.* **1925**, I, 1536; *Frdl.* **14**, 515.

Prisms from methanol, mp 199-200°. Sublimes at atmospheric pressure without dec. A monohydrate has been obtained from a satd aq soln. pKb (18°) 8.71. Very sparingly sol in water: One gram dissolves in 950 ml H_2O at 22°. Sol in alcohol, ether, dil HCl. Forms easily hydrolyzed Na and K salts.

Compd with 1,3,5-trinitrobenzene. $(C_9H_7NO)_2 \cdot C_6H_3N_3O_6$. Yellow needles, mp 178°.

1827. Carbosulfan. [55285-14-8] *N*-[(Dibutylamino)thio]-*N*-methylcarbamic acid 2,3-dihydro-2,2-dimethyl-7-benzofuranyl es-ter; FMC-35001; Marshal; Posse. $C_{20}H_{32}N_2O_3S$; mol wt 380.55. C 63.12%, H 8.48%, N 7.36%, O 12.61%, S 8.42%. Carbamate cholin-esterase inhibitor which is hydrolyzed to carbofuran in soil. Prepn: A. L. Black, T. R. Fukuto, **BE 817517**; *eidem*, **US 4006231** (1975, 1977 both to Regents Univ. Calif.). Metabolism in orange tree leaves and fruit: V. E. Clay, T. R. Fukuto, *Arch. Environ. Contam. Toxicol.* **13**, 53 (1984). Degradation and persistence in soil: A. Sa-hoo *et al.*, *Bull. Environ. Contam. Toxicol.* **50**, 29 (1993). HPLC determn in oranges: M. W. Brooks, A. Barros, *Analyst* **120**, 2479 (1995). Spectrophotometric determn in formulations and water sam-ples: K. R. Mohan *et al.*, *Asian J. Chem.* **10**, 457 (1998). Review of physical properties and field trials: E. G. Maitlen, N. A. Sladen, *Proc. Br. Crop Prot. Conf. - Pests Dis.* **1979**, 557-564.

Viscous, brown liquid. Soly in water at 25°: 0.3 ppm. Soly in organic solvents: > 50%. Vapor pressure: 0.31×10^{-6} mmHg.

LD_{50} in male, female rats (mg/kg): 250, 185 orally; in male, female rabbits (mg/kg): >2000, >2000 dermally; in pheasant, mallard, quail (ppm): 26.2, 8.1, 81.6 orally. LC_{50} (96 hr) in bluegill, trout (ppb): 14.9, 42.4 (Maitlen).

USE: Insecticide.

1828. Carboxin. [5234-68-4] 5,6-Dihydro-2-methyl-*N*-phen-yl-1,4-oxathiin-3-carboxamide; 2,3-dihydro-5-carboxanilido-6-methyl-1,4-oxathiin; carbathiin; DCMO; Vitavax. $C_{12}H_{13}NO_2S$; mol wt 235.30. C 61.25%, H 5.57%, N 5.95%, O 13.60%, S 13.63%. Systemic fungicide, effective against loose smut in cereals. Prepn: B. von Schmeling *et al.*, **US 3249499** (1966 to U.S. Rubber); M. Kulka *et al.*, **US 3393202** (1968 to Uniroyal). Fungicidal activity: B. von Schmeling, M. Kulka, *Science* **152**, 659 (1966). Inhibitor of succinate oxidation: D. E. Mathre, *Phytopathology* **60**, 671 (1970); G. A. White, *Biochem. Biophys. Res. Commun.* **44**, 1212 (1971); P. C. Mowery *et al.*, *ibid.* **71**, 354 (1976). Residue determn in crops: J. R. Lane, *J. Agric. Food Chem.* **18**, 409 (1970); H. R. Sisken, J. E. Newell, *ibid.* **19**, 738 (1971). Toxicity studies: T. V. Dyadicheva, *Gig. Tr. Prof. Zabol.* **1979**(2), 55, *C.A.* **90**, 146673b (1979).

Crystals, from ethanol or methanol, mp 93-95°. LD_{50} in rats, mice (mg/kg): 430, 3200 orally (Dyadicheva).

USE: Systemic plant fungicide.

1829. γ-Carboxyglutamic Acid. [53861-57-7] (3*S*)-3-Ami-no-1,1,3-propanetricarboxylic acid; γ-carboxy-L-glutamic acid; L-γ-carboxyglutamic acid; Gla. $C_6H_9NO_6$; mol wt 191.14. C 37.70%, H 4.75%, N 7.33%, O 50.22%. Amino acid found in blood coagu-lation proteins (prothrombin, Factor VII, Factor IX, Factor X, *q.q.v.*), plasma proteins and proteins from calcified tissue. Presence of Gla confers metal binding properties to the protein. Discovery in pro-thrombin: J. Stenflo, *Proc. Natl. Acad. Sci. USA* **71**, 2730 (1974); G. L. Nelsestuen *et al.*, *J. Biol. Chem.* **249**, 6347 (1974). Synthesis: S. Danishefsky *et al.*, *J. Am. Chem. Soc.* **101**, 4385 (1979); of DL-form: H. R. Morris *et al.*, *Biochem. Biophys. Res. Commun.* **62**, 856 (1975); W. Märki, R. Schwyzer, *Helv. Chim. Acta* **58**, 1471 (1975); B. Weinstein *et al.*, *J. Org. Chem.* **41**, 3634 (1976). Resolution of D- and L-forms: W. Märki *et al.*, *Helv. Chim. Acta* **60**, 798 (1977). Biosynthesis by vitamin K-dependent carboxylation of glutamic acid: C. T. Esmon *et al.*, *J. Biol. Chem.* **250**, 4744 (1975). Metal binding properties: G. L. Nelsestuen *et al.*, *Biochem. Biophys. Res. Commun.* **65**, 233 (1975); G. L. Nelsestuen, *J. Biol. Chem.* **251**, 5648 (1976); R. Robertson *et al.*, *ibid.* **253**, 5880 (1978); S. P. Bajaj *et al.*, *ibid.* **257**, 3726 (1982). HPLC determn in proteins, bone, urine: M. Kuwada, K. Katayama, *Anal. Biochem.* **117**, 259 (1981). *Reviews:* J. Stenflo, J. W. Suttie, *Annu. Rev. Biochem.* **46**, 157-172 (1977); R. E. Olson, J. W. Suttie, *Vitam. Horm.* **35**, 59-108 (1977); J. W. Suttie, *Crit. Rev. Biochem.* **8**, 191-223 (1980); J. P. Burnier *et al.*, *Mol. Cell. Biochem.* **39**, 191-207 (1981).

Crystals, mp 167-167.5° (dec). $[\alpha]_D^{20}$ +35.3° (c = 1 in 6*N* HCl). **DL-Form.** White powder, mp 90-92°.

1830. Carboxymethylcellulose Sodium. [9004-32-4] Cellu-lose carboxymethyl ether sodium salt; sodium carboxymethylcellu-lose; sodium cellulose glycolate; Cethylose; CMC; Cel-O-Brandt; Glykocellon; Carbose D; Xylo-Mucine; Tylose MGA; Cellolax; Po-lycell. Prepd by treating alkali cellulose with sodium chloroacetate: *Faith, Keyes & Clark's Industrial Chemicals*, F. A. Lowenheim, M. K. Moran, Eds. (Wiley-Interscience, New York, 4th ed., 1975) pp 235-238. Review and bibliography: Ott, *Cellulose and Cellulose Derivatives*, New York, 1946 (2nd ed., 1955).

Consult the Name Index before using this section. **Page 319**

White granules or powder. Powder is hygroscopic. Soly in water depends on degree of substitution. Water-soluble CMC is available in various viscosities (5-2000 cP in 1% soln), and the soly is equally good in hot and cold water (difference from methyl cellulose). Easily dispersed in water to form colloidal solns. Insol in alc, ether, most organic solvents. Also the presence of metal salts has little effect on the viscosity. Solns are stable between pH 2 and 10. Below pH 2 precipitation of a solid occurs, above pH 10 the viscosity decreases rapidly. pKa 4.30. The free acid is obtained from aq soln at pH 2.5 and may be precipitated with alcohol.

USE: In drilling muds, in detergents as a soil-suspending agent, in resin emulsion paints, adhesives, printing inks, textile sizes, as protective colloid in general. As stabilizer in foods. Pharmaceutic aid (suspending agent; tablet excipient; viscosity-increasing agent).

1831. Carbutamide. [339-43-5] 4-Amino-*N*-[(butylamino)-carbonyl]benzenesulfonamide; 1-butyl-3-sulfanilylurea; *N*^{1}-(butyl-carbamoyl)sulfanilamide; *N*^{1}-sulfanilyl-*N*^{2}-butylurea; *N*^{1}-sulfanilyl-*N*^{2}-butylcarbamide; *N*-(4-aminobenzenesulfonyl)-*N'*-butylurea; aminophenurobutane; BZ-55; U-6987; Glucidoral; Invenol; Nadisan. C$_{11}$H$_{17}$N$_3$O$_3$S; mol wt 271.34. C 48.69%, H 6.32%, N 15.49%, O 17.69%, S 11.82%. First generation sulfonylurea antidiabetic; originally investigated as an antibacterial agent. Prepn: E. Haack, A. Hagedorn, US 2907692 (1959 to Boehringer, Mann.). Description of clinical hypoglycemic activity: H. Franke, J. Fuchs, *Dtsch. Med. Wochenschr.* **80**, 1449 (1955). Pharmacology: J. D. Achelis, K. Hardebeck, *ibid.*, 1452; J. D. Achelis *et al.*, *Arch. Exp. Pathol. Pharmakol.* **228**, 63 (1956). Clinical trial in diabetes: A. Markkanen *et al.*, *Br. Med. J.* **1**, 1089 (1960). Determn in plasma by LC-MS/MS: G. Hoizey *et al.*, *Clin. Chem.* **51**, 1666 (2005). Historical review of discovery: H. Kleinsorge, *Exp. Clin. Endocrinol. Diabetes* **106**, 149-151 (1998).

Crystals, mp 140-142°. Sol in water at pH 5 to 8. LD$_{50}$ s.c. in mice: 3 g/kg (Haack, Hagedorn).

THERAP CAT: Antidiabetic.

1832. Carbuterol. [34866-47-2] *N*-[5-[2-[(1,1-Dimethyleth-yl)amino]-1-hydroxyethyl]-2-hydroxyphenyl]urea; α-(*t*-butylamino-methyl)-4-hydroxy-3-ureidobenzyl alcohol. C$_{13}$H$_{21}$N$_3$O$_3$; mol wt 267.33. C 58.41%, H 7.92%, N 15.72%, O 17.95%. A β-adrenergic agonist related to isoproterenol, *q.v.*, with selectivity for airway smooth muscle receptors. Prepn: C. Kaiser, S. T. Ross, DE 2106620; *eidem*, US 3763232 (1971, 1973 both to SKF); C. Kaiser *et al.*, *J. Med. Chem.* **17**, 49 (1974). Pharmacology, mechanism of action, toxicity study: J. R. Wardell *et al.*, *J. Pharmacol. Exp. Ther.* **189**, 167 (1974). Analysis in aq soln: L. J. Ravin *et al.*, *J. Pharm. Sci.* **67**, 1523 (1978). Clinical study: T. D. James, H. A. Lyons, *J. Am. Med. Assoc.* **241**, 704 (1979).

Crystals, mp 174-176°. LD$_{50}$ in mice (mg/kg): 32.8 i.v.; 3134.6 orally; in rats: 77.2 i.v. (Wardell).

Hydrochloride. [34866-46-1] SKF-40383; Bronsecur; Pirem. C$_{13}$H$_{21}$N$_3$O$_3$.HCl; mol wt 303.79. Crystals from ethanol/ether, mp 205-207° (dec).

THERAP CAT: Bronchodilator.

1833. Cardamom. Perennial, ginger-like plant, *Elettaria cardamomum* L. Maton, *Zingiberaceae*; seeds and volatile oil have been used since ancient times in cooking and traditional medicine. Numerous cultivars are grown, paticularly the Malabar and Mysore varieties. Plants of *Amomum sp.* and *Afromomum sp.* yield seeds of

similar flavor, but are considered inferior to true cardamom. *Habit:* India; cultivated in Sri Lanka, Guatemala, Tanzania. *Constit:* Mature seeds contain volatile oil (6.5-10.5%); crude fiber (6.7-12.8%); protein (8.8-10.5%); starch (38-47%); ash (3.8-6.7%); fixed oil containing glycerides of oleic, stearic, linoleic, palmitic, caprylic and caproic acids. Comprehensive review: V. S. Govindarajan *et al.*, *Crit. Rev. Food Sci. Nutr.* **16**, 229-326 (1982). Review of medicinal uses: A. Y. Leung, S. Foster, *Encyclopedia of Common Natural Ingredients*, (Wiley-Interscience, Hoboken, 2nd Ed., 2003) pp 121-122; J. Gruenwald *et al.*, *PDR for Herbal Medicines* (Thomson PDR, Montvale, 3rd Ed., 2004) pp 157-158.

Volatile oil. [8000-66-6] Oil of cardamom. Usually obtained by steam distillation from the ripe seeds. *Constit.* Complex mixture, mainly α-terpinyl acetate (28-42%); 1,8-cineole (21-41%); linalyl acetate, limonene, linalool. Description: E. Guenther, *The Essential Oils* vol. V (New York, 1952) pp 85-106. Composition studies: Y. S. Lewis *et al.*, *Perfum. Essent. Oil Rec.* **57**, 623 (1966). Supercritical fluid extraction and analysis: B. Marongiu *et al.*, *J. Agric. Food Chem.* **52**, 6278 (2004). Colorless or pale yellow liquid. Warm and spicy flavor with camphoraceous and lemony undertones. d$_4^{25}$ 0.917-0.947. Angular rotation: +22 to +44°. n$_D^{20}$ 1.462-1.466. Sapon. value: 91.6 to 144.7. Acid value: 1.7 to 7.1. Sol in benzyl alcohol, diethyl phthalate, fixed oil, mineral oil, in 5 vols 70% alcohol. Insol in glycerin, propylene glycol. *Keep well closed, cool and protected from light.*

USE: Flavoring in baked goods, confectionery, coffee, tea; component of curry powder and garam masala. Pharmaceutic aid (flavor). Fragrance component of perfumes, soaps, lotions.

THERAP CAT: Carminative.

1834. Cardiotoxins. [12663-44-4] Family of cytotoxins found in the venom of elapid snakes, particularly in cobras of the genus, *Naja*. Basic proteins which cause irreversible depolarization of cell membranes, contraction of skeletal and smooth muscle, and cardiac arrest. Structures consist of single polypeptide chains of approx 60 amino acids cross-linked by four disulfide bridges. Mol wts < 7000. Identification in *Naja tripudians* venom: N. K. Sarkar, *J. Indian Chem. Soc.* **21**, 227 (1947). Purification from venom of Taiwan cobra, *Naja naja atra* and pharmacology: C. Y. Lee *et al.*, *Arch. Exp. Pathol. Pharmakol.* **259**, 360 (1968). Amino acid sequence: K. Narita, C. Y. Lee, *Biochem. Biophys. Res. Commun.* **41**, 339 (1970). Sequence of 2 cardiotoxins from Indian cobra, *Naja naja*: K. Hayashi *et al.*, *ibid.* **45**, 1357 (1971). HPLC separation from cobra venoms: N. Kaneda, K. Hayashi, *J. Chromatogr.* **281**, 389 (1983). Sequence comparison of 6 cardiotoxins from Taiwan cobra: S.-H. Chiou *et al.*, *Biochem. Biophys. Res. Commun.* **206**, 22 (1995). Review of isolation and activity: E. Condrea, *Experientia* **30**, 121-129 (1974); of mechanisms of action: J. E. Fletcher, M.-S. Jiang, *Toxicon* **31**, 669-695 (1993).

Cardiotoxin III. [11061-96-4] Cytotoxin D1 (Naja naja atra); cardiotoxin A3 (Naja atra); CTX-III. First isolated and most abundant cardiotoxin in venom of the Taiwan cobra, *Naja naja atra*. 60 amino acid residues; mol wt 6747. Isoelectric point: 10.181.

1835. 3-Carene. [13466-78-9] 3,7,7-Trimethylbicyclo-[4.1.0]hept-3-ene; Δ^3-carene; 4,7,7-trimethyl-3-norcarene; isodiprene. C$_{10}$H$_{16}$; mol wt 136.24. C 88.16%, H 11.84%. Constituent of turpentine. The turpentine from *Pinus sylvestris* L. may contain as much as 42%, turpentine from *Pinus longifolia* Roxb., *Pinaceae* about 30%. Isoln and structure: Simonsen, *The Terpenes* vol. II (Cambridge, 1949) pp 64-72; Guenther, *The Essential Oils* vol. II (Van Nostrand, 1949) pp 49-51. Conformation: Acharya, *Tetrahedron Lett.* **1966**, 4117. Absorption spectrum: Cole, *J. Chem. Soc.* **1954**, 3807.

d-**Form.** Mobile liquid. Readily oxidized on exposure to air. Sweet and pungent odor, more agreeable than the odor of turpentine.

d_{15}^{15} 0.8668; d_{30}^{30} 0.8586. bp$_{705}$ 168-169°. bp$_{200}$ 123-124°. $[\alpha]_D^{20}$ +7.69°. n_D^{30} 1.468. Practically insol in water. Miscible with fat solvents and oils.

d-Form nitrosate. $C_{10}H_{16}N_2O_4$. Prepd from *d*-Δ^3-carene with amyl nitrite, acetic and nitric acid, prisms, dec 147.5°.

1836. Carfentrazone-ethyl. [128639-02-1] α,2-Dichloro-5-[4-(difluoromethyl)-4,5-dihydro-3-methyl-5-oxo-1*H*-1,2,4-triazol-1-yl]-4-fluorobenzenepropanoic acid ethyl ester; ethyl 2-chloro-3-[2-chloro-4-fluoro-5-(4-difluoromethyl-4,5-dihydro-3-methyl-5-oxo-1*H*-1,2,4-triazol-1-yl)phenyl]propionate; F8426; Aurora. $C_{15}H_{14}$-$Cl_2F_3N_3O_3$; mol wt 412.19. C 43.71%, H 3.42%, Cl 17.20%, F 13.83%, N 10.19%, O 11.64%. Cereal herbicide that disrupts membranes by inhibition of protoporphyrinogen oxidase. Prepn: K. M. Poss, **WO 9002120** (1990 to FMC). Properties: W. A. Van Saun *et al.*, *Brighton Crop Prot. Conf. - Weeds* **1993**, 19. Synthesis and structure-activity: G. Theodoridis *et al.*, *ACS Symp. Ser.* **584**, 90 (1995). Mode of action: F. E. Dayan *et al.*, *Pestic. Sci.* **51**, 65 (1997). Control of broadleaf weeds in maize: S. F. Tutt *et al.*, *Brighton Crop Prot. Conf. - Weeds* **1995**, 731; in wheat and barley: S. W. Shires *et al.*, *ibid.* **1997**, 117.

Viscous yellow liquid, mp $-22.1°$. bp$_{760}$ 350-355°. d^{20} 1.457 g/cm^3. Flash point: >110°C. Log P (octanol/water): 3.36. Soly in water (μg/ml): 12 (20°); 22 (25°); 23 (30°). Vapor pressure at 25°: 1.2×10^{-7}mm Hg; at 20°: 5.4×10^{-8}mm Hg. LD$_{50}$ in rats (mg/kg): 5143 orally (female); >4000 dermally (Van Saun).

USE: Post-emergent herbicide.

1837. Carfilzomib. [868540-17-4] (αS)-α-[[2-(4-Morpholinyl)acetyl]amino]benzenebutanoyl-L-leucyl-N-[(1S)-3-methyl-1-[[(2R)-2-methyl-2-oxiranyl]carbonyl]butyl]-L-phenylalaninamide; PR-171. $C_{40}H_{57}N_5O_7$; mol wt 719.92. C 66.74%, H 7.98%, N 9.73%, O 15.56%. Synthetic tetrapeptide irreversible proteasome inhibitor; epoxyketone structure similar to epoxomicin, *q.v.* Prepn: M. S. Smyth *et al.*, **WO 05105827**; M. S. Smyth, G. J. Laidig, **US 7417042** (2005, 2008 both to Proteolix). Proteasome inhibition and antitumor activity: S. D. Demo *et al.*, *Cancer Res.* **67**, 6383 (2007). Selectively inhibits chymotrypsin-like activity of proteasome: F. Parlati *et al.*, *Blood* **114**, 3439 (2009). Clinical pharmacokinetics and tolerability in patients with hematologic malignancies: O. A. O'Connor *et al.*, *Clin. Cancer Res.* **15**, 7085 (2009). Review of pharmacology and clinical experience: S. Jain *et al.*, *Core Evid.* **6**, 43-57 (2011).

THERAP CAT: Antineoplastic.

1838. Carglumic Acid. [1188-38-1] *N*-(Aminocarbonyl)-L-glutamic acid; carbamylglutamic acid; *N*-carbamoyl-L-glutamic acid; *l*-uramidoglutaric acid; ureidoglutaric acid; Carbaglu. C_6H_{10}-N_2O_5; mol wt 190.16. C 37.90%, H 5.30%, N 14.73%, O 42.07%. Metabolically stable analog of *N*-acetylglutamate, a physiological activator of the first enzyme of the urea cycle, carbamylphosphate

synthetase (CAPS). Prepn: H. McIlwain, *Biochem. J.* **33**, 1942 (1939). Effect on blood urea and ammonia levels and potential clinical application: J.-E. O'Connor *et al.*, *Eur. J. Pediatr.* **143**, 196 (1985). Evaluation in treatment of CAPS deficiency: G. Kuchler *et al.*, *J. Inherit. Metab. Dis.* **19**, 220 (1996); of *N*-acetylglutamate synthetase (NAGS) deficiency: B. Plecko *et al.*, *Eur. J. Pediatr.* **157**, 996 (1998).

mp 174°.
THERAP CAT: In treatment of inherited urea cycle disorders.

1839. Carindacillin. [35531-88-5] (2S,5R,6R)-6-[[3-[(2,3-Dihydro-1*H*-inden-5-yl)oxy]-1,3-dioxo-2-phenylpropyl]amino]-3,3-dimethyl-7-oxo-4-thia-1-azabicyclo[3.2.0]heptane-2-carboxylic acid; *N*-(2-carboxy-3,3-dimethyl-7-oxo-4-thia-1-azabicyclo[3.2.0]-hept-6-yl)-2-phenylmalonamic acid 1-(5-indanyl) ester; 6-[2-(5-indanyloxycarbonyl)phenylacetamido]-3,3-dimethyl-7-oxo-4-thia-1-azabicyclo[3.2.0]heptane-2-carboxylic acid; 1-(5-indanyl) *N*-(2-carboxy-3,3-dimethyl-7-oxo-4-thia-1-azabicyclo[3.2.0]hept-6-yl)-2-phenylmalonamate; 6-[2-phenyl-2-(5-indanyloxycarbonyl)acetamido]penicillanic acid; 6-(2-carboxy-2-phenylacetamido)-3,3-dimethyl-7-oxo-4-thia-1-azabicyclo[3.2.0]heptane-2-carboxylic acid 6-(5-indanyl ester); α-(5-indanyloxycarbonyl)benzylpenicillin; carbenicillin indanyl ester; CP-15464. $C_{26}H_{26}N_2O_6S$; mol wt 494.56. C 63.14%, H 5.30%, N 5.66%, O 19.41%, S 6.48%. Semi-synthetic antibiotic related to penicillin. Prepn: K. Butler, **ZA 6900060**; *idem*, **US 3679801** (1969, 1972 both to Pfizer). New process: S. Nakanishi, **US 3759898** (1973 to Pfizer). Activity studies: Butler, *Del. Med. J.* **43**, 366 (1971); English *et al.*, *Antimicrob. Agents Chemother.* **1**, 185 (1972). Clinical evaluation: Wallace *et al.*, *Antimicrob. Agents Chemother.* **1970**, 223; Bran *et al.*, *Clin. Pharmacol. Ther.* **12**, 525 (1971); *Indanyl Carbenicillin*, H. Swarz, F. E. Storari, Eds. (Am. Elsevier, New York, 1974) 100 pp.

Sodium salt. [26605-69-6] CP-15464-2; Carindapen; Geocillin; G.U.-Pen. $C_{26}H_{25}N_2NaO_6S$; mol wt 516.54. mp 207-213°.
THERAP CAT: Antibacterial.

1840. Cariporide. [159138-80-4] *N*-(Aminoiminomethyl)-4-(1-methylethyl)-3-(methylsulfonyl)benzamide; 4-isopropyl-3-(methanesulfonylbenzoyl)guanidine; $C_{12}H_{17}N_3O_3S$; mol wt 283.35. C 50.87%, H 6.05%, N 14.83%, O 16.94%, S 11.31%. Selective inhibitor of the sodium-hydrogen exchanger subtype 1 (NHE-1); protects against myocardial injury following ischemia and reperfusion. Prepn: H. J. Lang *et al.*, **EP 589336**; *eidem*, **US 5591754** (1994, 1997 both to Hoechst); A. Weichert *et al.*, *Arzneim.-Forsch.* **47**, 1204 (1997). Pharmacology: W. Scholz *et al.*, *Cardiovasc. Res.* **29**, 260 (1995). Symposium on pharmacology and clinical experience: *Am. J. Cardiol.* **83**, Suppl. 10A, 1G-25G (1999). Review: W. Scholz *et al.*, *J. Thromb. Thrombolysis* **8**, 61-70 (1999).

Colorless solid, mp 90-94° (dec).

Methanesulfonate. [159138-81-5] Cariporide mesylate; HOE-642. $C_{12}H_{17}N_3O_3S.CH_3SO_3H$; mol wt 379.45. Colorless needles from isopropanol + water, mp 227-230°.

THERAP CAT: Cardioprotective.

1841. Carisbamate. [194085-75-1] (1S)-1-(2-Chlorophenyl)-1,2-ethanediol 2-carbamate; (S)-(2-(2-chlorophenyl)-2-hydroxyethyl)oxocarboxamide; (S)-2-O-carbamoyl-1-o-chlorophenylethanol; JNJ-10234094; RWJ-333369; YKP-509. $C_9H_{10}ClNO_3$; mol wt 215.63. C 50.13%, H 4.67%, Cl 16.44%, N 6.50%, O 22.26%. Monocarbamate antiepileptic drug with neuroprotective activity. Prepn: Y. M. Choi, M. W. Kim, **WO 97026241** (1997 to Yukong); *eidem*, **US 6103759** (2000 to SK Corp.). Effects in hippocampal neuronal culture: L. S. Deshpande *et al.*, *Epilepsy Res.* **79**, 158 (2008); in models of genetic epilepsy: J. François *et al.*, *Epilepsia* **49**, 393 (2008). Disposition in humans: G. S. J. Mannens *et al.*, *Drug Metab. Dispos.* **35**, 554 (2007). Clinical pharmacokinetics: C. Yao *et al.*, *Epilepsia* **47**, 1822 (2006); in elderly: R. Levy *et al.*, *Epilepsy Res.* **79**, 22 (2008). Clinical evaluation in idiopathic photosensitive epilepsy: D. G. A. Kasteleijn-Nolst Trenité *et al.*, *ibid.* **74**, 193 (2007). Review of pharmacology in epilepsy: G. P. Novak *et al.*, *Neurotherapeutics* **4**, 106-109 (2007); of development and clinical experience in epilepsy: K. Kulig, B. Malawska, *IDrugs* **10**, 720-727 (2007).

From methanol, mp 133°. Sol in methanol. $[\alpha]_D$ +64.9° (c=2.69 in methanol).

THERAP CAT: Anticonvulsant.

1842. Carisoprodol. [78-44-4] N-(1-Methylethyl)carbamic acid 2-[[(aminocarbonyl)oxy]methyl]-2-methylpentyl ester; N-isopropyl-2-methyl-2-propyl-1,3-propanediol dicarbamate; isopropyl meprobamate; isobamate; carisoprodate; Carisoma; Flexartal; Rela; Sanoma; Soma; Somadril; Somalgit. $C_{12}H_{24}N_2O_4$; mol wt 260.33. C 55.37%, H 9.29%, N 10.76%, O 24.58%. Prepn: Berger, Ludwig, **US 2937119** (1960 to Carter Prod.). Pharmacology: Berger *et al.*, *J. Pharmacol. Exp. Ther.* **127**, 66 (1959).

Crystals, mp 92-93°. Slightly bitter taste. Freely sol in alc, chloroform, acetone. Sol in many common organic solvents. Very slightly sol in water. Practically insol in vegetable oils. Stable in dil acids and alkalies. LD_{50} in mice, rats (mg/kg): 2340, 1320 orally; 980, 450 i.p. (Berger).

THERAP CAT: Muscle relaxant (skeletal).

1843. Carminic Acid. [1260-17-9] 7-β-D-Glucopyranosyl-9,10-dihydro-3,5,6,8-tetrahydroxy-1-methyl-9,10-dioxo-2-anthracenecarboxylic acid; C.I. Natural Red 4; C.I. 75470. $C_{22}H_{20}O_{13}$; mol wt 492.39. C 53.67%, H 4.09% O 42.24%. Red anthraquinone pigment produced by the scale insect, *Dactylopius coccus* Costa, (*Coccus cacti* L.), *Homoptera*; commonly known as cochineal. The essential constituent of the natural dye, carmine. Isoln: E. Schunck, L. Marchlewski, *Ber.* **27**, 2979 (1894); O. Dimroth, S. Scheuer, *Ann.* **399**, 43 (1913). Structure: O. Dimroth, H. Kammerer, *Ber.* **53**, 471 (1920); M. A. Ali, L. J. Haynes, *J. Chem. Soc.* **1959**, 1033; revised structure: S. B. Bhatia, K. Venkataraman, *Indian J. Chem.* **3** (2), 92 (1965). Configuration of the C-glycosidic bond: A. Fiecchi *et al.*, *J. Org. Chem.* **46**, 1511 (1981); and crystal structure: T. Ishida *et al.*, *Acta Crystallogr. C* **43**, 1541 (1987). Total synthesis: P. Allevi *et*

al., *J. Chem. Soc. Perkin Trans. 1* **1998**, 575. HPLC determn in foods: F. E. Lancaster, J. F. Lawrence, *J. Chromatogr. A* **732**, 394 (1996); in historical documents: R. Blanc *et al.*, *ibid.* **1122**, 105 (2006).

Red prisms from alc, no distinct melting point, darkens at 120°. Has a deep red color in water and is yellow to violet in acid solns. uv max (water): 500 nm (ε 6800); (0.02N HCl): 490-500 nm (ε 5800); (0.0001N NaOH): 540 nm (ε 3450). $[\alpha]_{654}^{15}$ +51.6° (water). Sol in water, alc, concd H_2SO_4; slightly sol in ether. Practically insol in petr ether, benzene, chloroform.

Methyl tetra-O-methylcarminate. [71013-43-9] $C_{27}H_{30}O_{13}$. Yellow needles from benzene + petr ether, mp 185-188°.

USE: In color photography; pigment for artists' paints; as bacteriological stain; food colorant.

1844. Carmofur. [61422-45-5] 5-Fluoro-N-hexyl-3,4-dihydro-2,4-dioxo-1(2H)-pyrimidinecarboxamide; 1-(n-hexylcarbamoyl)-5-fluorouracil; HCFU; Mifurol; Yamaful. $C_{11}H_{16}FN_3O_3$; mol wt 257.27. C 51.35%, H 6.27%, F 7.38%, N 16.33%, O 18.66%. Orally active cytostatic deriv of fluorouracil, *q.v.* Prepn: S. Ozaki *et al.*, **JP Kokai 77 78886**, *C.A.* **87**, 152265 (1977); *eidem*, **US 4071519** (1977, 1978 both to Mitsui); *eidem*, *Bull. Chem. Soc. Jpn.* **50**, 2406 (1977). Anti-tumor activity: A. Hoshi *et al.*, *Chem. Pharm. Bull.* **26**, 161 (1978). Determn in body fluids: S. Watanabe *et al.*, *Chemotherapy (Tokyo)* **27**, 778 (1979); O. Nakajima *et al.*, *J. Chromatogr.* **225**, 91 (1981). General pharmacology: Z. Henmi *et al.*, *Oyo Yakuri* **19**, 369 (1980), *C.A.* **93**, 125639s (1980). Pharmacokinetics: M. Iigo *et al.*, *Cancer Chemother. Pharmacol.* **4**, 189 (1980). Metabolism: *eidem*, *Xenobiotica* **10**, 847 (1980); T. Kobari *et al.*, *ibid.* **11**, 57 (1981). Phase I clinical study: Y. Koyama, Y. Koyama, *Cancer Treat. Rep.* **64**, 861 (1980). Mutagenicity study: Y. Seino *et al.*, *Cancer Res.* **38**, 2148 (1978). Review: T. Taguchi, *Recent Results Cancer Res.* **70**, 125-132 (1980).

White cryst from ethanol, mp 110-111°. uv max (chloroform): 258 nm (ε 1.16×10^4) (S. Ozaki *et al.*, *Bull. Chem. Soc. Japan*). Also reported as mp 283° (dec) (**US 4071519**).

THERAP CAT: Antineoplastic.

1845. Carmustine. [154-93-8] N,N'-Bis(2-chloroethyl)-N-nitrosourea; BCNU; NSC-409962; Becenun; Bicnu; Carmubris. $C_5H_9Cl_2N_3O_2$; mol wt 214.05. C 28.06%, H 4.24%, Cl 33.12%, N 19.63%, O 14.95%. Chloroethylnitrosourea derivative with antitumor activity. Similar to chlorozotocin, lomustine, nimustine, ranimustine, *q.q.v.* Synthesis: Johnston *et al.*, *J. Med. Chem.* **6**, 669 (1963). Properties: Loo *et al.*, *J. Pharm. Sci.* **55**, 492 (1966). Decompn studies as related to antileukemic activity: Montgomery *et al.*, *J. Med. Chem.* **10**, 668 (1967). Antifungal action: Hunt, Pittilo, *Antimicrob. Agents Chemother.* **1965**, 710. Toxicology studies: Thompson, Larson, *Toxicol. Appl. Pharmacol.* **21**, 405 (1972). Review of pulmonary toxicity: A. C. Smith, *Pharmacol. Ther.* **41**, 443-460 (1989).

Light yellow powder that melts to an oily liquid; mp 30-32°. Both powder and liquid are stable. Dec rapidly in acid and in soln above pH 7. Most stable in petroleum ether or aqueous soln at pH 4. Nonionized at pH 7 with consequent high lipid solubility. Sol in water up to 4 mg/ml and in 50% ethanol up to 150 mg/ml: DeVita *et al.*, *Cancer Res.* **25**, 1876 (1965). LD_{50} in mice (mg/kg): 19-25 orally, 26 i.p., 24 s.c.; in rats (mg/kg): 30-34 orally (Thompson, Larson).

Caution: This substance is reasonably anticipated to be a human carcinogen: *Report on Carcinogens, Twelfth Edition* (PB2011-111646, 2011) p 325.

THERAP CAT: Antineoplastic.

1846. Carnauba Wax. [8015-86-9] Brazil wax. An exudate from the pores of the leaves of the Brazilian wax palm tree *Copernicia prunifera* (Muell.) H. E. Moore [*Copernicia cerifera* (Arruda da Camara) Mart.], *Palmae.* The botany of the tree and the native wax-collecting procedures are described adequately by A. H. Warth, *The Chemistry and Technology of Waxes* (Reinhold, New York, 1947). The hardness and high-polish capability of this important wax can be ascribed to the presence of esters of hydroxylated unsaturated fatty acids having about 12 carbon atoms in the acid chain. The usual names for the constituents, *i.e.* cerotic acid, melissyl cerotate, carnaubic acid etc., are meaningless. Brief review: C. S. Letcher in *Kirk-Othmer Encyclopedia of Chemical Technology* vol. 24 (Wiley-Interscience, New York, 3rd ed., 1984) pp 469-470.

Hard greenish solid, cryst fracture. Sharp, characteristic, not unpleasant odor upon melting. mp 82-85.5°. d 0.990 to 0.999. Saponification number 78 to 89. Iodine number about 13. n_D^{90} 1.4500. Sparingly soluble in fat solvents at 25°, quite sol at 45°.

USE: Wherever a hard, high-polish wax is desired, *e.g.* in automobile waxes, floor wax emulsions, high quality shoe polishes, in the paper industry (especially for making carbon papers). As a plasticizer in dental impression compounds. To raise the melting point of other waxes; often used together with candelilla wax. The presence of the lower-melting ouricury wax is considered as an adulteration. Purified and bleached carnauba wax is used for cosmetic materials, such as depilatories and deodorant sticks. In pharmacy as the last stage in tablet coating. Skin sensitization or irritation by carnauba wax seems infrequent.

1847. Carnegine. [490-53-9] 1,2,3,4-Tetrahydro-6,7-dimethoxy-1,2-dimethylisoquinoline; 6,7-dimethoxy-1,2-dimethyl-1,2,3,4-tetrahydroisoquinoline; pectenine. $C_{13}H_{19}NO_2$; mol wt 221.30. C 70.56%, H 8.65%, N 6.33%, O 14.46%. In *Carnegiea gigantea* (Engelm.) Britt. & Rose *(Cereus giganteus* Engelm.), *Cactaceae,* a cactus of Arizona and Mexico. Extraction procedure: Heyl, *Arch. Pharm.* **266**, 668 (1928). Synthesis: Späth, *Ber.* **62**, 1021 (1929); Nakada, Nisgihara, *J. Pharm. Soc. Jpn.* **64**, 74 (1944). Pharmacology: E. Santi-Soncin, M. Furlanut, *Fitoterapia* **43**, 21 (1972), *C.A.* **80**, 66667f (1974). Review of synthetic methods: A. B. J. Bracca, T. S. Kaufman, *Tetrahedron* **60**, 10575-10610 (2004).

Viscous liquid, dec on standing. Distills at 1 mm pressure and 170° air bath temp. Sol in ether, alc, chloroform. LD_{50} i.p. in mice: 15.23 mg/kg (Santi-Soncin, Furlanut).

Hydrochloride monohydrate. [5864-18-6] $C_{13}H_{19}NO_2$.HCl.-H_2O. Clusters from dil alc, mp 207° (mp 211° when anhydr). Sol in water, slightly sol in alc.

Hydrobromide monohydrate. [5853-25-8] $C_{13}H_{19}NO_2$.HBr.-H_2O. Needles from alc, mp 228°.

Methyliodide. [5911-58-0] $C_{13}H_{19}NO_2$.CH$_3$I. Needles from methanol, mp 211° (evac tube) after drying at 100° and 10 mm.

1848. Carnidazole. [42116-76-7] *N*-[2-(2-Methyl-5-nitro-1*H*-imidazol-1-yl)ethyl]carbamothioic acid *O*-methyl ester; *O*-methyl [2-(2-methyl-5-nitroimidazol-1-yl)ethyl]thiocarbamate; R-25831; Spartrix. $C_8H_{12}N_4O_3S$; mol wt 244.27. C 39.34%, H 4.95%, N 22.94%, O 19.65%, S 13.12%. Prepn and antiprotozoal

activity: J. Heeres *et al.*, **DE 2429755**; *eidem,* **US 3928374** (both 1975 to Janssen); *eidem, Eur. J. Med. Chem. - Chim. Ther.* **11**, 237 (1976). Crystal structure of monohydrate: N. M. Blaton *et al., Acta Crystallogr.* **B35**, 753 (1979). Clinical efficacy in vaginal trichomoniasis: A. Notowicz *et al., Br. J. Vener. Dis.* **53**, 129 (1977); P. Chaudhuri, A. C. Drogendijk, *Eur. J. Obstet. Gynecol. Reprod. Biol.* **10**, 325 (1980).

Crystals from ethanol, mp 142.4° (dec).

THERAP CAT (VET): Antiprotozoal (Trichomonas).

1849. Carnitine. [541-15-1] (2*R*)-3-Carboxy-2-hydroxy-*N*,-*N*,*N*-trimethyl-1-propanaminium inner salt; L-(3-carboxy-2-hydroxypropyl)trimethylammonium hydroxide inner salt; (−)-β-hydroxy-γ-trimethylaminobutyric acid; L-carnitine; levocarnitine; vitamin B$_T$; Carnicor; Carnitor; Levocarnil; Nefrocarnit. $C_7H_{15}NO_3$; mol wt 161.20. C 52.16%, H 9.38%, N 8.69%, O 29.77%. Essential cofactor of fatty acid metabolism; required for the transport of fatty acids through the inner mitochondrial membrane. Synthesized primarily in the liver and kidney; highest concentrations found in heart and skeletal muscle. Name derived from the Latin word, *carnis,* meaning meat or flesh. Dietary sources include red meat, dairy products, beans, avocado. Isoln from meat extract: W. Gulewitsch, R. Krimberg, *Z. Physiol. Chem.* **45**, 326 (1905); H. E. Carter, P. K. Bhattacharyya, *Methods Enzymol.* **III**, 660 (1957). Identification as vitamin B$_T$: H. E. Carter *et al., Arch. Biochem. Biophys.* **38**, 405 (1952). Synthesis: M. Tomita, Y. Sendju, *Z. Physiol. Chem.* **169**, 263 (1927); R. Voeffray *et al., Helv. Chim. Acta* **70**, 2058 (1987). Enantioselective synthesis from glycerol: M. Marzi *et al., J. Org. Chem.* **65**, 6766 (2000). Biosynthesis: G. Wolf, C. R. A. Berger, *Arch. Biochem. Biophys.* **92**, 360 (1961). Absolute configuration: T. Kaneko, R. Yoshida, *Bull. Chem. Soc. Jpn.* **35**, 1153 (1962). Metabolism in humans: M. E. Mitchell, *Am. J. Clin. Nutr.* **31**, 293 (1978). Chromatographic determn in food supplement formulations: A. Kakou *et al., J. Chromatogr. A* **1069**, 209 (2005). Use in canine dilated cardiomyopathy: S. L. Sanderson, *Vet. Clin. North Am. Small Anim. Pract.* **36**, 1325 (2006). Historical review: G. Fraenkel, S. Friedman, *Vitam. Horm.* **XV**, 73-118 (1957). Review of physiological significance and deficiency syndromes: C. J. Rebouche, D. J. Paulson, *Annu. Rev. Nutr.* **6**, 41-66 (1986); of clinical pharmacology: J. J. Bahl, R. Bressler, *Annu. Rev. Pharmacol. Toxicol.* **27**, 257-277 (1987); of biosynthesis in mammals: F. M. Vaz, R. J. A. Wanders, *Biochem. J.* **361**, 417-429 (2002). Review of effect on myocardial metabolism and clinical experience in ischemic heart disease: R. Lango *et al., Cardiovasc. Res.* **51**, 21-29 (2001); of use as dietary supplement in athletes: H. Karlic, A. Lohninger, *Nutrition* **20**, 709-715 (2004). Symposium on physiology, pharmacology and therapeutic potential: *Ann. N.Y. Acad. Sci.* **1033**, 1-197 (2004).

Crystals from anhydrous ethanol + acetone, dec 196-198° (Carter, Bhattacharyya); also reported as crystals from isopropanol, mp 200° (dec) (Marzi). Very hygroscopic solid. $[\alpha]_D^{25}$ −31.3° (c = 10 in water); $[\alpha]_{546}^{25}$ −37.0° (c = 10 in water). Sol in water, hot alcohol. Practically insol in acetone, ether, benzene.

Hydrochloride. [6645-46-1] Levocarnitine chloride. C_7H_{15}-NO_3.HCl; mol wt 197.66. Crystals, dec 142°.

DL-Form. [406-76-8] γ-Amino-β-hydroxybutyric acid trimethylbetaine; γ-trimethyl-β-hydroxybutyrobetaine. Synthesis: E. Strack *et al., Ber.* **86**, 525 (1953); H. E. Carter, P. K. Bhattacharyya, *J. Am. Chem. Soc.* **75**, 2503 (1953). Hygroscopic crystalline solid, dec 195-197°. Sol in water, ethanol.

DL-Form hydrochloride. [461-05-2] Needles from ethanol, mp 196° (dec). Very sol in water; sol in hot ethanol; slightly sol in cold ethanol. Practically insol in acetone, ether.

D-Form. [541-14-0] Prepn: S. Friedman *et al.*, *Arch. Biochem. Biophys.* **66**, 10 (1957). Crystals, dec 210-212°. $[\alpha]_D$ +30.9°. Very sol in water and alcohol. Practically insol in acetone and ether.

D-Form hydrochloride. [10017-44-4] Crystals, dec 142°.

THERAP CAT: Vitamin (enzyme cofactor).

THERAP CAT (VET): In treatment of feline hepatic lipidosis and canine dilated cardiomyopathy.

1850. Carnosic Acid. [3650-09-7] (4a*R*,10a*S*)-1,3,4,9,10,-10a-Hexahydro-5,6-dihydroxy-1,1-dimethyl-7-(1-methylethyl)-4a-(2*H*)-phenanthrenecarboxylic acid; 11,12-dihydroxy-13-isopropyl-podocarpa-8,11,13-trien-17-oic acid; salvin. $C_{20}H_{28}O_4$; mol wt 332.44. C 72.26%, H 8.49%, O 19.25%. Phenolic diterpene antioxidant principally found in rosemary and sage. Readily cyclizes to form the lactone, carnosol, *q.v.*; precursor for structurally similar antioxidants found in these plants. Isoln from *Salvia officinalis* L., Lamiaceae: H. Linde, *Helv. Chim. Acta* **47**, 1234 (1964); from *Rosmarinus officinalis* L., *Lamiaceae*: E. Wenkert *et al.*, *J. Org. Chem.* **30**, 2931 (1965). Configuration: C. R. Narayanan, H. Linde, *Tetrahedron Lett.* **6**, 3647 (1965). Antioxidant activity in lipid systems: C. H. Brieskorn, H.-J. Dömling, *Z. Lebensm.-Unters. Forsch.* **141**, 10 (1969); A. I. Hopia *et al.*, *J. Agric. Food Chem.* **44**, 2030 (1996). Mechanism of antioxidant action: T. Masuda *et al.*, *ibid.* **49**, 5560 (2001). HPLC determn in rosemary extracts and comparison with derivative antioxidant compds: S. L. Richheimer *et al.*, *J. Am. Oil Chem. Soc.* **73**, 507 (1996). HPLC determn in foods: W. Ternes, K. Schwarz, *Z. Lebensm.-Unters. Forsch.* **201**, 548 (1995); in antioxidant food packaging material: K. Bentayeb *et al.*, *Anal. Bioanal. Chem.* **389**, 1989 (2007).

Prisms from ether-pentane, mp 196-215° (dec); $[\alpha]_D^{22}$ +139.8 ±2° (c = 1.15 in chloroform) (Linde). Also reported as colorless crystals from hexane, mp 185-190° (dec); $[\alpha]_D^{23}$ +191° (c = 1.07 in MeOH) (Wenkert).

USE: Antioxidant in edible oils, meats, and other fat containing foods.

1851. Carnosine. [305-84-0] β-Alanyl-L-histidine; ignotine. $C_9H_{14}N_4O_3$; mol wt 226.24. C 47.78%, H 6.24%, N 24.76%, O 21.22%. Naturally occurring dipeptide found in large amounts in skeletal muscle. Also present in other tissues such as brain, cardiac muscle, kidney. Water soluble antioxidant; functions as a free-radical scavenger. Isoln: Gulewitsch, Amiradzibi, *Ber.* **33**, 1902 (1900); Wolff, Wilson, *J. Biol. Chem.* **95**, 495 (1932); **109**, 565 (1935). Synthesis from histidine and β-iodo- or β-nitropropionyl chloride: Baumann, Ingvaldsen, *ibid.* **35**, 271 (1918); Barger, Tutin, *Biochem. J.* **12**, 406 (1918). Later syntheses: Sifford, du Vigneaud, *J. Biol. Chem.* **108**, 753 (1935); R. A. Turner, *J. Am. Chem. Soc.* **75**, 2388 (1953); F. J. Vinick, S. Jung, *J. Org. Chem.* **48**, 392 (1983). Crystal structure: H. Itoh *et al.*, *Acta Crystallogr.* **33B**, 2959 (1977). Possible role in wound healing: D. E. Fischer *et al.*, *Proc. Soc. Exp. Biol. Med.* **158**, 402 (1978). Review of physiological properties and therapeutic potential: S. E. Gariballa, A. J. Sinclair, *Age Ageing* **29**, 207-210 (2000).

Crystals from aqueous ethanol, mp 262° (dec) (Vinick, Jung); also reported as mp 260° (capillary tube) and as mp 308-309° (Dennis

bar) (Sifford, du Vigneaud). $[\alpha]_D^{25}$ +21.0° (c = 1.5 in water). pK$_1$ 2.64; pK$_2$ 6.83; pK$_3$ 9.51. Alkaline reaction. One gram dissolves in 3.1 ml water at 25°.

Nitrate. [5852-98-2] $C_9H_{15}N_5O_6$. Crystals, dec 222°. $[\alpha]_D^{20}$ +24.1° (c = 1.5 in water). Very sol in water.

Hydrochloride. [5852-99-3] $C_9H_{15}ClN_4O_3$. Crystals, dec 245°. Very sol in water.

D-Form. [5853-00-9] Crystals, mp 260°. $[\alpha]_D^{28}$ −20.4° (c = 1.5).

1852. Carnosol. [5957-80-2] (4a*R*,9*S*,10a*S*)-1,3,4,9,10,10a-Hexahydro-5,6-dihydroxy-1,1-dimethyl-7-(1-methylethyl)-2*H*-9,4a-(epoxymethano)phenanthren-12-one; 7β,11,12-trihydroxy-13-isopropylpodocarpa-8,11,13-trien-17-oic acid 17,7-lactone; picrosalvin. $C_{20}H_{26}O_4$; mol wt 330.42. C 72.70%, H 7.93%, O 19.37%. Phenolic diterpene antioxidant principally occurring in rosemary and sage; lactone of carnosic acid, *q.v.* Isoln from *Salvia carnosa*, Dougl., *Lamiaceae*: A. I. White, G. L. Jenkins, *J. Am. Pharm. Assoc.* **31**, 33, 37 (1942). Isoln as picrosalvin from *S. officinalis*, L.: C. H. Brieskorn, A. Fuchs, *Ber.* **95**, 3034 (1962). Isoln from rosemary and identity with picrosalvin: *idem et al.*, *J. Org. Chem.* **29**, 2293 (1964). Prepn by oxidation of carnosic acid: J. G. Marrero *et al.*, *J. Nat. Prod.* **65**, 986 (2002). Crystal structure and spectroscopic properties: M. Gajhede *et al.*, *J. Crystallogr. Spectrosc. Res.* **20**, 165 (1990). Antioxidant activity in lipids: C. H. Brieskorn, H.-J. Dömling, *Z. Lebensm.-Unters. Forsch.* **141**, 1431 (1969); in biological systems: O. I. Aruoma *et al.*, *Xenobiotica* **22**, 257 (1992); in oil-in-water emulsion: E. N. Frankel *et al.*, *J. Agric. Food Chem.* **44**, 131 (1996). Anticancer effects in murine model of colonic tumorigenesis: A. E. Moran *et al.*, *Cancer Res.* **65**, 1097 (2005).

Long, white, odorless needles, mp 219.5° (dec) (White, Jenkins). Also reported as mp 221-226° (Brieskorn, Fuchs). Intensely bitter taste. $[\alpha]_D$ −66°. uv max (ethanol): 210, 284 nm (log ε 4.42, 3.40). Soly (mg/ml): 8 in ethanol, 35 in DMF, 250 in DMSO. Sol in methanol, ether, chloroform; slightly sol in petr ether. Sol in dil alkali with red-brown color.

USE: Antioxidant in edible oils, meats, and other fat containing foods.

1853. Caro's Acid. [7722-86-3] Peroxymonosulfuric acid; sulfomonoperacid; persulfuric acid. H_2O_5S; mol wt 114.07. H 1.77%, O 70.13%, S 28.11%. Dry reagent is prepd by stirring 10 g potassium persulfate into 11 g concd H_2SO_4 for 10 min and adding 30 g finely powdered potassium sulfate; liquid reagent is obtained by triturating potassium persulfate with three times as much (by weight) of H_2SO_4; dil reagent is prepd by stirring 10 g potassium persulfate into 11 g concd H_2SO_4 and adding 50 cc ice: Baeyer, Villiger, *Ber.* **32**, 3625 (1899); another prepn by reacting 90% H_2O_2 with chlorosulfonic acid at −40 to −50°: Ball, Edwards, *J. Am. Chem. Soc.* **78**, 1125 (1956).

The product is a sirupy liquid consisting of about equal amounts of Caro's acid and H_2SO_4. pK$_2$ of Caro's acid 9.4 ± 0.1. Oxygen is evolved at room temp; should be stored at dry ice temp.

Caution: Can be dangerously unstable, like most peroxides. Description of explosion at Brown University: J. O. Edwards, *Chem. Eng. News* **33**, 3336 (1955); of explosion at Sun Oil: *ibid.* **38**, 59 (Nov. 21, 1960). May be highly irritating to skin, eyes, mucous membranes.

USE: In prepn of dyes; oxidation of olefins to α-glycols; oxidation of ketones to lactones or esters; treating woolens to prevent felting and shrinking; in bleaching compositions.

1854. **α-Carotene.** [7488-99-5] (6′R)-β,ε-Carotene. $C_{40}H_{56}$; mol wt 536.89. C 89.49%, H 10.51%. About as widely distributed as its β-isomer, but in smaller amounts. Best sources for both the α- and β-isomers are carrots, palm oil, and green leaves of various species. As a provitamin A it is half as active as β-carotene. Found in the mother liquors after crystallizing β-carotene. Isoln by chromatography: Karrer, Walker, *Helv. Chim. Acta* **16**, 641 (1933). Structure: Kuhn, Lederer, *Ber.* **64**, 1349 (1931); Karrer *et al.*, *Helv. Chim. Acta* **14**, 614 (1931); **16**, 975 (1933). Natural (+)-α-carotene has 6′R configuration: Eugster *et al.*, *ibid.* **52**, 1729 (1969). Synthesis of dl-α-carotene: Eugster, Karrer, *ibid.* **38**, 610 (1955); Tscharner *et al.*, *ibid.* **40**, 1676 (1957); Ruegg *et al.*, *ibid.* **44**, 985 (1961).

Deep purple prisms, polyhedra from petr ether or from benzene + methanol, mp 187.50° (evacuated tube). $[\alpha]_{643}^{18}$ +385° (c = 0.08 in benzene). Absorption max ($CHCl_3$): 485, 454 nm. More sol than β-carotene. Freely sol in carbon disulfide, chloroform; sol in ether, benzene. Slightly sol in petr ether, alcohol. 100 ml hexane dissolves 294 mg at 0°. Practically insol in water, acids, alkalies. Absorbs oxygen from the air, giving rise to inactive colorless oxidation products. The oxidation in light is autocatalytic. *Store in darkness in sealed ampoules and at low temp (−20°C).*

THERAP CAT: Vitamin A precursor.

THERAP CAT (VET): Vitamin A precursor for all species except cats.

1855. **β-Carotene.** [7235-40-7] β,β-Carotene; Carotaben; Provatene; Solatene. $C_{40}H_{56}$; mol wt 536.89. C 89.49%, H 10.51%. Most important of the provitamins A. Widely distributed in the plant and animal kingdoms. In plants it occurs almost always together with chlorophyll. Isoln from carrots: Willstätter, Escher, *Z. Physiol. Chem.* **64**, 47 (1910); Kuhn, Lederer, *Ber.* **64**, 1349 (1931); Barnett *et al.*, US 2848508 (1958). Chromatography: Karrer, Walker, *Helv. Chim. Acta* **16**, 641 (1933). Structure: Willstätter, Mieg, *Ann.* **355**, 1 (1907); Zechmeister *et al.*, *Ber.* **61**, 566 (1928); **66**, 123 (1933); Karrer *et al.*, *Helv. Chim. Acta* **12**, 1142 (1929); **13**, 1084 (1930); **14**, 1033 (1931); Kuhn, Brockmann, *Ber.* **65**, 894 (1932); **66**, 1319 (1933); **67**, 1408 (1934); *Ann.* **516**, 95 (1935). Crystal structure: Sterling, *Acta Crystallogr.* **17**, 1224 (1964). Synthesis: Milas *et al.*, *J. Am. Chem. Soc.* **72**, 4844 (1950); Karrer, Eugster, *Compt. Rend.* **250**, 1920 (1950); Inhoffen *et al.*, *Chem. Ztg.* **74**, 285, 309 (1950); Surmatis, Ofner, *J. Org. Chem.* **26**, 1171 (1961); Rüegg *et al.*, *Helv. Chim. Acta* **44**, 985 (1961); Bestmann *et al.*, *Ann.* **1973**, 760; Fischli, Mayer, *Helv. Chim. Acta* **58**, 1584 (1975). Industrial mfg procedure: Isler *et al.*, *Helv. Chim. Acta* **39**, 249 (1956); Isler *et al.*, US 2917539 (1959 to Hoffmann-La Roche). Microbial production by *Choanephora trispora*: Zajic, US 2959521; US 2959522 and US 3123026 (1960, 1960 and 1964, all to Grain Processing); Miescher, US 3001912 (1961 to C.S.C.). *Review:* Fleming, *Selected Organic Syntheses* (John Wiley, London, 1973) pp 70-74.

Deep-purple, hexagonal prisms from benzene + methanol. Red, rhombic, almost square leaflets from petr ether. mp 183° (evacuated tube). Absorption max (chloroform): 497, 466 nm. Less sol than α-carotene. Sol in CS_2, benzene, chloroform. Moderately sol in ether, petr ether, oils. 100 ml hexane dissolve 109 mg at 0°. Very

sparingly sol in methanol and ethanol. Practically insol in water, acids, alkalies. Dil solns are yellow. Absorbs oxygen from the air giving rise to inactive, colorless oxidation products. *Keep tightly closed and protected from light. Store at low temp (−20°C).* Commercial crystalline β-carotene has a vitamin A activity of 1.67 million U.S.P. units per gram. The I.U. of 0.6 μg β-carotene is almost exactly equivalent to 0.3 μg vitamin A.

USE: Yellow coloring agent for foods.

THERAP CAT: Vitamin A precursor. Ultraviolet screen.

THERAP CAT (VET): Vitamin A precursor for all species except cats.

1856. **γ-Carotene.** [472-93-5] β,ψ-Carotene; (all-E)-2-(3,7,-12,16,20,24-hexamethyl-1,3,5,7,9,11,13,15,17,19,23-pentacosa-undecaenyl)-1,3,3-trimethylcyclohexene. $C_{40}H_{56}$; mol wt 536.89. C 89.49%, H 10.51%. A rare carotenoid. Has provitamin A activity. Best source is *Penicillium sclerotiorum*: Mase *et al.*, *Arch. Biochem. Biophys.* **68**, 150 (1957). Also present in small proportions in many plant materials, esp fruits containing β-carotene. Chromatographic isoln from crude carotenes: Kuhn, Brockmann, *Ber.* **66**, 407 (1933); Winterstein, *Z. Physiol. Chem.* **219**, 249 (1933). Structure: Kuhn, Brockmann, *loc. cit.; Naturwissenschaften* **21**, 44 (1933). Synthesis: Garbers *et al.*, *Helv. Chim. Acta* **36**, 1783 (1953); Ruegg *et al.*, *ibid.* **44**, 985 (1961).

Probably crystallizes in polymorphous forms. Synthetic form: red plates, mp 152-153.5°. Absorption max (petr ether): 437, 462, 494 nm ($E_{1cm}^{1\%}$ 2055, 3100, 2720). Natural form: minute, deep-red prisms with bluish luster from benzene + methanol. mp 177.5°. Absorption max (chloroform): 508.5, 475, 446 nm. Somewhat less sol than β-carotene. *Store in darkness in sealed ampoules at low temps (0°C).*

THERAP CAT: Vitamin A precursor.

THERAP CAT (VET): Vitamin A precursor for all species except cats.

1857. **δ-Carotene.** [472-92-4] ε,ψ-Carotene; (all-E)-6-(3,7,-12,16,20,24-hexamethyl-1,3,5,7,9,11,13,15,17,19,23-pentacosa-undecaenyl)-1,5,5-trimethylcyclohexene. $C_{40}H_{56}$; mol wt 536.89. C 89.49%, H 10.51%. Extracted from the fruit of *Gonocaryum pyriforme* Mig., *Icacinaceae*: Winterstein, *Z. Physiol. Chem.* **219**, 249 (1933). Occurs also in carrots and certain varieties of tomatoes. Isoln from tomato mutants: Porter, Murphey, *Arch. Biochem. Biophys.* **32**, 21 (1951). Structure: Kargl, Quackenbush, *ibid.* **88**, 59 (1960). Synthesis: Manchand *et al.*, *J. Chem. Soc.* **1965**, 2019. Absolute configuration: Buchecker, Eugster, *Helv. Chim. Acta* **54**, 327 (1971).

Long orange-red needles from CS_2 + hexane + ethanol, mp 140.5°. $[\alpha]_{Cd}$ +317°; $[\alpha]_D^{25}$ +352° ±16% (hexane). Absorption spectrum: Kargl, Quackenbush, *loc. cit.*

1858. **Carotol.** [465-28-1] (3R,3aS,8aR)-2,3,4,5,8,8a-Hexahydro-6,8a-dimethyl-3-(1-methylethyl)-3a(1H)-azulenol; 2,3,4,5,8,-8a-hexahydro-3-isopropyl-6,8a-dimethyl-3a(1H)-azulenol. $C_{15}H_{26}$O; mol wt 222.37. C 81.02%, H 11.79%, O 7.19%. From oil of carrot seeds, *Daucus carota* L., *Umbelliferae*: Asahina, Tsukamoto, *J. Pharm. Soc. Jpn.* **525**, 961 (1925), *C.A.* **20**, 2845 (1926); Sorm *et al.*, *Collect. Czech. Chem. Commun.* **16**, 47 (1951); Pigulevskii, Kovaleva, *Zh. Prikl. Khim.* **32**, 2703 (1959). Structure: Sykora *et al.*, *Collect. Czech. Chem. Commun.* **26**, 788 (1961); Zalkow *et al.*, *J. Org. Chem.* **26**, 981 (1961). Stereochemistry: Levisalles, Rudler, *Bull. Soc. Chim. Fr.* **1964**, 2020; **1967**, 2059. Synthesis of (+)-form: DeBroissia *et al.*, *ibid.* **1972**, 4314; *eidem*, *Chem. Commun.* **1972**, 855.

Liquid. bp$_{2.5}$ 126°. [α]$_D^{20}$ +30.4°. n_D^{20} 1.4964. d^{20} 0.9624.

1859. Caroverine. [23465-76-1] 1-[2-(Diethylamino)ethyl]-3-[(4-methoxyphenyl)methyl]-2(1H)-quinoxalinone; 1-(diethylaminoethyl)-3-(p-methoxybenzyl)dihydro-2-quinoxalone. C$_{22}$H$_{27}$N$_3$O$_2$; mol wt 365.48. C 72.30%, H 7.45%, N 11.50%, O 8.76%. Smooth muscle relaxant with calcium channel blocking and glutamate antagonist activity. Prepn: Zellner et al., US 3028384 (1962 to Donau-Pharm.). Polarographic study: P. Pflegel, Pharmazie 24, 667 (1969). Pharmacological study: F. Hahn et al., Arch. Int. Pharmacodyn. Ther. 199, 108 (1972). Antioxidant activity: N. Udilova et al., Biochem. Pharmacol. 65, 59 (2003). Review of clinical experience in treatment of tinnitus and other inner ear diseases: K. Ehrenberger, Adv. Otorhinolaryngol. 59, 156-162 (2002).

Crystals from isopropyl alcohol, mp 69°. bp$_{0.01}$ 202°.
Hydrochloride. Spasmium; Tinnitin. Dec 188°.
THERAP CAT: Antispasmodic.

1860. Carpaine. [3463-92-1] (1S,11R,13S,14S,24R,26S)-13,26-Dimethyl-2,15-dioxa-12,25-diazatricyclo[22.2.2.2^{11,14}]triacontane-3,16-dione. C$_{28}$H$_{50}$N$_2$O$_4$; mol wt 478.72. C 70.25%, H 10.53%, N 5.85%, O 13.37%. Found in Carica papaya L. and in Vasconcellosia hastata Caruel, Caricaceae. Isoln from papaya leaves: Greshoff, Mededeel. uit's Lands. Plant., Buitenzorg. No. 7, 5 (1890); Rapoport, Baldridge, J. Am. Chem. Soc. 73, 343 (1951). Reportedly causes bradycardia, CNS depression: Kakowski, Arch. Int. Pharm. 15, 84 (1905). Structure and stereochemistry: Govindachari et al., Tetrahedron Lett. 1965, 1907. Absolute configuration: Coke, Rice, J. Org. Chem. 30, 3420 (1965). Review of structural studies: Govindachari, J. Indian Chem. Soc. 45, 945 (1968).

Monoclinic prisms from acetone, mp 119-120°. Sublimes at 120° under 0.05 mm pressure. [α]$_D^{12}$ +24.7° (c = 1.07 in ethanol). Slightly sol in water. Sol in most organic solvents except petr ether.

1861. Carpetimycins. Carbapenem antibiotics related to thienamycin and olivanic acids, q.q.v. Carpetimycins A, B, C and D are known. Prodn of A and B by Streptomyces strain No. C-19393, now named Streptomyces griseus subsp. cryophilus, and antibacterial properties: NL 8000628; A. Imada et al., US 4518529 (1980, 1985 to Takeda); A. Imada et al., J. Antibiot. 33, 1417 (1980); by Streptomyces strain KC-6643: M. Nakayama et al., ibid. 1388. Structure and abs config: S. Harada et al., J. Antibiot. 33, 1425 (1980); M. Nakayama et al., ibid. 34, 818 (1981). Mode of action

study: Y. Nozaki et al., ibid. 206. Total synthesis of (±)-carpetimycin A: H. Natsugari et al., J. Chem. Soc. Perkin Trans. 1 1983, 403; M. Ihara et al., Heterocycles 20, 2182 (1983); J. D. Buynak, M. N. Rao, J. Org. Chem. 51, 1571 (1986). Synthesis of (−)-carpetimycin A: T. Iimori et al., J. Am. Chem. Soc. 105, 1659 (1983); M. Aratani et al., Tetrahedron Lett. 26, 223 (1985); M. Shibasaki et al., ibid. 2217.

Carpetimycin A: R = H
Carpetimycin B: R = SO$_3$H

Carpetimycin A. [76025-73-5] (5R,6R)-3-[(R)-[(1E)-2-(Acetylamino)ethyl]sulfinyl]-6-(1-hydroxy-1-methylethyl)-7-oxo-1-azabicyclo[3.2.0]hept-2-ene-2-carboxylic acid; antibiotic Ab 651; antibiotic C-19393-H$_2$; antibiotic KA-6643-A; C-19393-H$_2$. C$_{14}$H$_{18}$N$_2$O$_6$S; mol wt 342.37. Colorless solid, mp >145° (dec). [α]$_D^{24}$ −27° (c = 1.7 in water). uv max (water): 240, 288 nm (E$_{1cm}^{1\%}$ 369, 300) (Nakayama, 1980).
Carpetimycin B. [76094-36-5] Antibiotic C-19393-S$_2$; antibiotic KA-6643-B; C-19393-S$_2$. C$_{14}$H$_{18}$N$_2$O$_9$S$_2$; mol wt 422.42. Colorless solid melting above 130° (dec). [α]$_D^{24}$ −145° (c = 1 in water). uv max (water): 240, 285 nm (E$_{1cm}^{1\%}$ 357, 305) (Nakayama, 1980).

1862. Carprofen. [53716-49-7] 6-Chloro-α-methyl-9H-carbazole-2-acetic acid; C-5720; Ro-20-5720/000; Imadyl; Rimadyl. C$_{15}$H$_{12}$ClNO$_2$; mol wt 273.72. C 65.82%, H 4.42%, Cl 12.95%, N 5.12%, O 11.69%. Prepn: L. Berger, A. J. Corraz, DE 2337340; eidem, US 3896145 (1974, 1975 both to Hoffmann-La Roche). Pharmacology: L. O. Randall, H. Baruth, Arch. Int. Pharmacodyn. 220, 94 (1976). Stereospecific assay and disposition: J. M. Kemmerer et al., J. Pharm. Sci. 68, 1274 (1979). Metabolism: F. Rubio et al., ibid. 69, 1245 (1980). Pharmacokinetic and clinical studies in gout: T. F. Yu, J. Perel, J. Clin. Pharmacol. 20, 347 (1980). Use in treatment of rheumatoid arthritis: S. Jalava, Scand. J. Rheumatol. 1983, Suppl. 48, 5-12. Review of pharmacology and clinical efficacy: W. M. O'Brien, G. F. Bagby, Pharmacotherapy 7, 16 (1987).

Cryst from chloroform, mp 197-198°. LD$_{50}$ orally in mice: 400 mg/kg (Berger, Corraz). Freely sol in ether, acetone, ethyl acetate, sodium hydroxide TS, sodium carbonate TS. Practically insol in water.
THERAP CAT: Anti-inflammatory.
THERAP CAT (VET): Anti-inflammatory.

1863. Carpropamid. [104030-54-8] 2,2-Dichloro-N-[1-(4-chlorophenyl)ethyl]-1-ethyl-3-methylcyclopropanecarboxamide; KTU 3616; Win. C$_{15}$H$_{18}$Cl$_3$NO; mol wt 334.67. C 53.83%, H 5.42%, Cl 31.78%, N 4.19%, O 4.78%. Melanin inhibiting fungicide for control of rice blast. Prepn: Y. Kurahashi et al., EP 170842; eidem, US 4710518 (1985, 1987 both to Nihon Tokushu Noyaku Seizo). Synthesis of stereoisomers: U. Kraatz, M. Littmann, Pflanzenschutz-Nachr. 51, 201 (1998). Inhibition of melanin biosynthesis: G. Tsuji et. al., Pestic. Biochem. Physiol. 57, 211 (1997). Suppression of secondary infection in rice: Y. Kurahashi et. al., Pestic. Sci. 55, 31 (1999). Review of physical properties, mode of action, and field trials: T. Hattori et. al., Brighton Crop Prot. Conf. - Pests Dis. 1994, 517-524.

Technical grade is a mixture of (1RS, 3SR, 1'RR) isomers, mp 152°. Sol in dichloromethane, slightly sol in toluene, very low sol in water. Vapor pressure at 20°: 2.70 × 10⁻⁷ Pa.

USE: Fungicide.

1864. Carrageenan. [9000-07-1] Carrageen; carrageenin. Mixture of sulfated polysaccharides extracted from red seaweed *(Rhodophyceae)*. Chief sources are *Chondrus crispus* (L.) Stackhouse, and *Gigartina stellata* (G. mammillosa) (Goodenough and Woodward) J. Aghardh, *Gigartinaceae*, found most abundantly in North Atlantic coastal regions from Norway to North Africa. The name carrageenan is derived from the Irish coastal town of Carragheen. The κ and λ structural families are identified based on position of sulfate and the presence/absence of anhydrogalactose. Kappa (κ) family consists of κ,ι,μ,ν carrageenans of which κ and ι are the most prevalent. Characterized by a repeating unit of 4-sulfate-β-D-galactopyranosyl(1 → 4)-α-D-galactose linked (1 → 3). The galactose unit varies: 3,6-anhydro-α-D-galactose for κ-form; 3,6-anhydro-α-D-galactose-2-sulfate for ι-form. Due to helical tertiary structure which allows for gelling, κ-family is of major commercial importance. Lambda (λ) family consists of λ and ξ; characterized by a repeating (1 → 3) linked disaccharide of 2-sulfate-β-D-galactopyranosyl(1 → 4)-α-D-galactose. The galactose residue is 2,6-sulfated for λ form and 2-sulfated for ξ form. These carrageenans are non-gelling. Conformational study: N. S. Anderson *et al.*, *J. Mol. Biol.* **45**, 85 (1969); and structural analysis: D. A. Rees, E. J. Welsh, *Angew. Chem. Int. Ed.* **16**, 214 (1977); C. Bodeau-Bellion, *Physiol. Veg.* **21**, 785 (1983). Use as phlogistic agent: C. A. Winter *et al.*, *Proc. Soc. Exp. Biol. Med.* **111**, 544 (1962); L. Levy, *Life Sci.* **8**, 601 (1969). IR determn of κ, ι, and λ forms: E. Tojo, J. Prado, *Carbohydr. Res.* **338**, 1309 (2003). Review of effect on inflammation and immunity: H. J. Schwartz in *Inadvert. Modif. Immune Resp.* **FDA-80-1074**, 109-114 (1980); S. Nicklin, K. Miller, *Food Addit. Contam.* **6**, 425-436 (1989). Review of toxicology: M. L. Weiner, *Agents Actions* **32**, 46-51 (1991). Review of industrial and non-food uses: R. J. Tye, *Carbohydr. Polym.* **10**, 259-280 (1989). *Reviews:* G. H. Therkelsen in *Industrial Gums*, R. L. Whistler, J. N. BeMiller, Eds. (Academic Press, New York, 3rd ed., 1993) pp 145-180; J. K. Baird, "Gums" in *Kirk-Othmer Encyclopedia of Chemical Technology* **vol. 12** (Wiley-Interscience, New York, 4th ed., 1994) pp 842-862.

Yellowish or tan to white, coarse to fine powder. Practically odorless; mucilaginous taste. Sol in water at 80°, forming a viscous, clear or slightly opalescent solution that flows readily. Disperses in water more readily if first moistened with alc, glycerin or saturated soln of sucrose in water.

κ-Form. Ability to form thermoreversible gels. Gelation depends on temperature, concentration of counterions and other polysaccharides. Sol in very polar solvents. Sol in hot milk, hot conc. sugar soln; above 60° in water, as Na⁺ salt in cold water. K⁺ and Ca²⁺ salts are insol in cold water. Insol in in conc. salt soln; 35% alcohol soln; insol with swelling in cold milk. Optimum stability occurs in pH 9. Degree of sulphation varys from 25-30%.

ι-Form. Ability to form thermoreversible gels. Gelation depends on temperature, concentration of counterions and other polysaccharides. Sol in very polar solvents. Sol in hot milk, hot conc. salt soln; above 60° in water, as Na⁺ salt in cold water; slightly sol hot conc. sugar soln. K⁺ and Ca²⁺ salts are insol in cold water. Insol in 35% alcohol soln; insol in cold milk. Optimum stability occurs in pH 9. Degree of sulfation varies from 28-35%. LD₅₀ orally in rat: >5000 mg/kg; dermally in rabbit: >2000 mg/kg (Weiner). LC50 (4 hr) in rat: >930.8 ±74.4 mg/m³ (Weiner).

λ-Form. Sol in very polar solvents. Sol in water, milk, hot conc. sugar soln, conc. salt soln., as Na⁺salt in 35% alcohol soln. Optimum stability occurs in pH 9. The most highly sulfated from 32-39%.

USE: Gelling, emulsifying, and stabilizing agent and viscosity builder in foods and non-foods, but esp in milk or water systems. Demulcent. To induce experimental edema in laboratory animals.

1865. Cartap. [15263-53-3] Carbamothioic acid $S^C,S^{C'}$-[2-(dimethylamino)-1,3-propanediyl] ester; 1,3-bis(carbamoylthio)-2-N,N-(dimethylamino)propane. $C_7H_{15}N_3O_2S_2$; mol wt 237.34. C 35.42%, H 6.37%, N 17.70%, O 13.48%, S 27.02%. Insecticide modeled on a toxin isolated from the marine annelid *Lumbrineris heteropoda*. Prepn and insecticidal activity: K. Konishi *et al.*, **FR 1452338**; *eidem*, **US 3332943** (1966, 1968 both to Takeda); K. Konishi, *Agric. Biol. Chem.* **34**, 935 (1970). Crystal structure of hydrochloride: C. J. Cheer, F. J. Pickles, *J. Chem. Soc. Perkin Trans. 2* **1980**, 1805. Comparative study of insecticidal activity: B. Gumey, N. W. Hussey, *Plant Pathol.* **23**, 127 (1974). Effect on soil enzyme activity: T. Endo *et al.*, *J. Pestic. Sci.* **7**, 101 (1982). Aquatic toxicity study: S. Lakota *et al.*, *Acta Hydrobiol.* **23**, 183 (1981). Chronic toxicity study in mice: Y. Tsubura *et al.*, *Nara Igaku Zasshi* **26**, 368 (1975), *C.A.* **84**, 174837c (1976). *Review:* Y. Kono, *Jpn. Pestic. Inf.* **34**, 22 (1978).

Colorless prisms from ethyl acetate, mp 130.5-131° (dec).

Monohydrochloride. NTD-2; Padan. $C_7H_{16}ClN_3O_2S_2$; mol wt 273.79. Colorless rods from methanol, mp 176° (dec). Sol in water. LD₅₀ in mice, rats (mg/kg): 165, 250 orally (Konishi, 1970).

USE: Insecticide.

1866. Carteolol. [51781-06-7] 5-[3-[(1,1-Dimethylethyl)-amino]-2-hydroxypropyl]-3,4-dihydro-2(1H)-quinolinone; 5-[3-(tert-butylamino)-2-hydroxypropoxy]-3,4-dihydrocarbostyril. $C_{16}H_{24}N_2O_3$; mol wt 292.38. C 65.73%, H 8.27%, N 9.58%, O 16.42%. β-Adrenergic blocker. Prepn: K. Nakagawa *et al.*, **DE 2302027**; Y. Tamura *et al.*, **US 3910924** (1973, 1975 both to Otsuka); K. Nakagawa, *J. Med. Chem.* **17**, 529 (1974). Pharmacological evaluation in rats: S. Morita *et al.*, *Biochem. Pharmacol.* **25**, 1836 (1976). HPLC determn in plasma and urine: S. Y. Chu, *J. Pharm. Sci.* **67**, 1623 (1978). *In vitro* study: S. Chiba, *Arzneim.-Forsch.* **29**, 895 (1979). Carcinogenicity study: M. Kurosomi *et al.*, *Farmaco Ed. Prat.* **34**, 202 (1979). Series of articles on pharmacology, toxicology and clinical studies: *Arzneim.-Forsch.* **33**, 277-345 (1983). Acute toxicity data: W. Lang, *ibid.* 290. Review of clinical experience in glaucoma and ocular hypertension: P. Chrisp, E. M. Sorkin, *Drugs Aging* **2**, 58-77 (1992).

Hydrochloride. [51781-21-6] Abbott 43326; OPC-1085; Arteoptic; Carteol; Endak; Mikelan; Ocupress; Teoptic. $C_{16}H_{24}N_2O_3$·HCl; mol wt 328.84. Crystals from ethanol, mp 278°. LD₅₀ in male mice, rats (mg/kg): 810, 1380 orally; 54.5, 158 i.v.; 380, 400 i.p. (Lang).

THERAP CAT: Antihypertensive; antianginal; antiarrhythmic; antiglaucoma.

1867. Carthamin. [36338-96-2] (2Z,6S)-6-β-D-Glucopyranosyl-2-[[(3S)-3-β-D-glucopyranosyl-2,3,4-trihydroxy-5-[(2E)-3-(4-hydroxyphenyl)-1-oxo-2-propen-1-yl]-6-oxo-1,4-cyclohexadien-1-yl]methylene]-5,6-dihydroxy-4-[(2E)-3-(4-hydroxyphenyl)-1-oxo-2-propen-1-yl]-4-cyclohexene-1,3-dione; carthamic acid; safflor carmine; safflor red; C.I. Natural Red 26; C.I. 75140. $C_{43}H_{42}O_{22}$; mol wt 910.79. C 56.71%, H 4.65%, O 38.65%. Coloring principle from safflower, *Carthamus tinctorius* L., *Compositae*. Purification and review of early efforts: T. Kametaka, A. G. Perkin, *J. Chem. Soc.* **97**, 1415 (1910). Structure: C. Kuroda, *J. Chem. Soc.* **1930**, 752,

765. Revised structure: H. Obara, J. Onodera, *Chem. Lett.* **1979**, 201. Absolute configuration: S. Sato *et al.*, *ibid.* **1996**, 833. Total synthesis as acetate: S. Sato *et al.*, *ibid.* **2001**, 1318. Biogenesis: M. Shimokoriyama, S. Hattori, *Arch. Biochem. Biophys.* **54**, 93 (1955). Determn of levels in *Carthamus Red*: T. Morimoto *et al.*, *Nippon Shokuhin Kagaku Gakkaishi* **5**, 236 (1998). Color stability in aqueous soln: J.-B. Kim, Y.-S. Paik, *Arch. Pharmacal Res.* **20**, 643 (1997); with polysaccharides: K. Saito, *Acta Bot. Croat.* **57**, 123 (1998). Review of prepn protocols: K. Saito, Y. Fukaya, *Acta Alimentaria* **26**, 141-152 (1997). *See also: Colour Index* vol. **4** (3rd ed., 1971) p 4625.

Bright scarlet prismatic needles, mp 228-230° (dec). Hygroscopic. uv max (ethanol): 244, 373, 515 nm (log ε 4.33, 4.48, 4.99). Absorption max (DMF): 530 nm; $E_{1cm}^{1\%}$ 992 (ε 90370). Sparingly sol in water, ethanol and methanol; sol in dil alkali hydroxides, sodium carbonate and ammonia with an orange color. Insol in acetone.

USE: Food color and dye in cosmetics.

1868. Carthamus. Safflower; African saffron; thistle saffron; American saffron; false saffron; bastard saffron; Dyer's saffron. Florets of *Carthamus tinctorius* L. *Compositae. Habit.* Levant; Orient; cultivated extensively in Europe and America. *Constit.* Carthamin, safflor yellow.

USE: In dyeing; surrogate for Spanish saffron; coloring butter, liqueurs, confectionery, cosmetics.

THERAP CAT: Dietary supplement.

1869. Carticaine. [23964-58-1] 4-Methyl-3-[[1-oxo-2-(propylamino)propyl]amino]-2-thiophenecarboxylic acid methyl ester; 3-propylamino-α-propionylamino-2-carbomethoxy-4-methylthiophene; methyl 4-methyl-3-(2-propylaminopropionamido)thiophene-2-carboxylate; articaine. $C_{13}H_{20}N_2O_3S$; mol wt 284.37. C 54.91%, H 7.09%, N 9.85%, O 16.88%, S 11.27%. Prepn: H. Ruschig *et al.*, ZA **6804265**; *eidem*, US **3855243** (1969, 1974 both to Farbwerke Hoechst). Pharmacological studies: R. Muschaweck, R. Rippel, *Prakt. Anaesth.* **9**, 135 (1974); H. Hofer *et al.*, *ibid.* 157; A. Den Hertog, *Eur. J. Pharmacol.* **26**, 175 (1974). Toxicity studies: C. Baeder *et al.*, *Prakt. Anaesth.* **9**, 147 (1974). Mechanism of action: U. Borchard, H. Drouin, *Eur. J. Pharmacol.* **62**, 73 (1980). *Review:* H. Schneider, *Schweiz. Monatsschr. Zahnheilkd.* **86**, 1188-1194 (1976).

$bp_{0.3}$ 162-167°.

Hydrochloride. [23964-57-0] HOE-045; HOE-40045; Ultracain. $C_{13}H_{20}N_2O_3S$·HCl; mol wt 320.83. White, fine crystals, mp 177-178°. LD_{50} i.v. in mice: 37 mg/kg (Muschaweck, Rippel).

THERAP CAT: Anesthetic (local).

1870. Carubicin. [50935-04-1] (8*S*,10*S*)-8-Acetyl-10-[(3-amino-2,3,6-trideoxy-α-L-*lyxo*-hexopyranosyl)oxy]-7,8,9,10-tetrahydro-1,6,8,11-tetrahydroxy-5,12-naphthacenedione; (1*S*,3*S*)-3-acetyl-1,2,3,4,6,11-hexahydro-3,5,10,12-tetrahydroxy-6,11-dioxo-1-naphthacenyl 3-amino-2,3,6-trideoxy-α-L-*lyxo*-hexopyranoside; 4-*O*-demethyldaunorubicin; carminomycin; carminomycin I; karminomycin. $C_{26}H_{27}NO_{10}$; mol wt 513.50. C 60.82%, H 5.30%, N 2.73%, O 31.16%. Anthracycline antitumor antibiotic, related to daunorubicin and doxorubicin, *q.q.v.* Isoln from *Actinomadura car-*

minata: G. F. Gauze *et al.*, *Antibiotiki* **18**, 675 (1973); M. G. Brazhnikova *et al.*, *ibid.* 678. Antitumor activity: V. A. Shorin *et al.*, *ibid.* 681. Physico-chemical characteristics, structure: M. G. Brazhnikova *et al.*, *J. Antibiot.* **27**, 254 (1974). Pharmacokinetics, pharmacodynamics, toxicity study: L. E. Goldberg *et al.*, *Antibiotiki* **19**, 57 (1974). Production: M. G. Brazhnikova *et al.*, SU **508076** (1976 to Inst. Antibiot. Res., USSR), *C.A.* **86**, 15215 (1977). Stereochemistry: *eidem*, *J. Antibiot.* **29**, 469 (1976). Synthesis from daunomycinone: G. Cassinelli *et al.*, *ibid.* **31**, 178 (1978). Molecular pharmacology: V. H. DuVernay *et al.*, *Cancer Res.* **40**, 387 (1980). Analysis in human serum: S. E. Fandrich, K. A. Pittman, *J. Chromatogr.* **223**, 155 (1981). Early clinical studies: L.H. Baker *et al.*, *Cancer Treat. Rep.* **63**, 899 (1979). Embryotoxicity and teratogenicity study: I. Damjanov, A. Celluzzi, *Res. Commun. Chem. Pathol. Pharmacol.* **28**, 497 (1980).

Hydrochloride. [52794-97-5] NSC-180024. $C_{26}H_{27}NO_{10}$·HCl; mol wt 549.96. Crystals from ethanol/benzene. $[\alpha]_D^{20}$ +289°. uv max (ethanol): 236, 255, 462, 478, 492 ($E_{1cm}^{1\%}$ 300), 510, 525 nm. pKa_1 8.00; pKa_2 10.16. Sol in water, methanol. Practically insol in other organic solvents. LD_{50} in mice (mg/kg): 7.3 orally; 1.3 i.v.; 3.7 s.c. (Goldberg).

THERAP CAT: Antineoplastic.

1871. Carumonam. [87638-04-8] 2-[[(Z)-[2-[(2*S*,3*S*)-2-[[(Aminocarbonyl)oxy]methyl]-4-oxo-1-sulfo-3-azetidinyl]amino]-1-(2-amino-4-thiazolyl)-2-oxoethylidene]amino]oxy]acetic acid; (Z)-[[[(2-amino-4-thiazolyl)[[(2*S*,3*S*)-2-(hydroxymethyl)-4-oxo-1-sulfo-3-azetidinyl]carbamoyl]methylene]amino]oxy]acetic acid carbamate (ester); (3*S*,4*S*)-*cis*-3-[2-(2-amino-4-thiazolyl)-2-(Z)-carboxymethoxyiminoacetamido]-4-carbamoyloxymethyl-2-azetidinone-1-sulfonic acid. $C_{12}H_{14}N_6O_{10}S_2$; mol wt 466.40. C 30.90%, H 3.03%, N 18.02%, O 34.30%, S 13.75%. Synthetic monocyclic β-lactam (monobactam) antibiotic. Prepn: S. Kishimoto *et al.*, EP **93376**; T. Matsuo *et al.*, US **4572801** (1983, 1986 both to Takeda); M. Sendai *et al.*, *J. Antibiot.* **38**, 346 (1985). Alternate synthesis: P. S. Manchand *et al.*, *J. Org. Chem.* **53**, 5507 (1988). Comparative *in vitro* antimicrobial activity: R. J. Fass, V. L. Helsel, *Antimicrob. Agents Chemother.* **28**, 834 (1985); B. R. Smith *et al.*, *ibid.* **29**, 346 (1986); I. M. Hoepelman *et al.*, *Chemotherapy (Basel)* **33**, 103 (1987). β-Lactamase stability: R. L. Then, *ibid.* **30**, 398 (1984). Pharmacokinetics in humans: E. Weidekamm *et al.*, *Antimicrob. Agents Chemother.* **26**, 898 (1984); C. A. M. McNulty *et al.*, *ibid.* **28**, 425 (1985).

Colorless powder. $[\alpha]_D^{26}$ −45° (c = 1 in DMSO).

Disodium salt. [86832-68-0] AMA-1080; Ro-17-2301; Amasulin; Mobactam. $C_{12}H_{12}N_6Na_2O_{10}S_2$; mol wt 510.36.

THERAP CAT: Antibacterial.

1872. Carvacrol. [499-75-2] 2-Methyl-5-(1-methylethyl)-phenol; 2-*p*-cymenol; 2-hydroxy-*p*-cymene; isopropyl-*o*-cresol; isothymol. $C_{10}H_{14}O$; mol wt 150.22. C 79.96%, H 9.39%, O 10.65%. Found in oil of origanum, thyme, marjoram, summer savory: E.

Guenther, *The Essential Oils* **vol. 2** (Van Nostrand, New York, 1949) p 503; Carpenter, Easter, *J. Org. Chem.* **20**, 401 (1955). Prepn by chlorination of α-pinene with *tert*-butyl hypochlorite: Ritter, Ginsburg, *J. Am. Chem. Soc.* **72**, 2381 (1950); from 2-bromo-*p*-cymol: Strubell, Baumgartel, *Arch. Pharm.* **291**, 66 (1958). Toxicity data: Kochmann, *Arch. Exp. Pathol. Pharmakol.* **161**, 196 (1931).

Liquid; thymol odor. d_4^{20} 0.976; d_{25}^{25} 0.9751. bp_{760} 237-238°; bp_{18} 118-122°; bp_3 93°. mp ~0°. n_D^{20} 1.52295. uv max (95% ethanol): 277.5 nm (log ε 3.262). Volatile with steam. Practically insol in water. Freely soluble in alc or ether. LD orally in rabbits: 100 mg/kg (Kochmann).

USE: As disinfectant; in organic syntheses.

THERAP CAT: Has been used as anti-infective; anthelmintic (Nematodes).

1873. Carvedilol. [72956-09-3] 1-(9*H*-Carbazol-4-yloxy)-3-[[2-(2-methoxyphenoxy)ethyl]amino]-2-propanol; BM-14190; DQ-2466; Coreg; Dilatrend; Dimitone; Eucardic; Kredex; Querto. $C_{24}H_{26}N_2O_4$; mol wt 406.48. C 70.92%, H 6.45%, N 6.89%, O 15.74%. Nonselective β-adrenergic blocker with $α_1$-blocking activity. Prepn: F. Wiedemann *et al.*, **DE 2815926**; *eidem*, **US 4503067** (1979, 1985 both to Boehringer Mannheim). General pharmacological profile: M. Hirohashi *et al.*, *Arzneim.-Forsch.* **40**, 735 (1990). HPLC determn in biological fluids: K. Reiff, *J. Chromatogr.* **413**, 355 (1987). Series of articles on clinical pharmacology and efficacy in hypertension: *J. Cardiovasc. Pharmacol.* **19**, Suppl. 1, S1-S146 (1992). Clinical trial in chronic heart failure: M. Packer *et al.*, *N. Engl. J. Med.* **334**, 1349 (1996). *Review:* W. H. Frishman, *ibid.* **339**, 1759-1765 (1998).

Colorless crystals from ethyl acetate, mp 114-115°. Freely sol in DMSO; sol in methylene chloride, methanol; sparingly sol in ethanol, isopropanol; slightly sol in ethyl ether. Practically insol in water.

THERAP CAT: Antihypertensive; in treatment of congestive heart failure.

1874. Carvone. [99-49-0] 2-Methyl-5-(1-methylethenyl)-2-cyclohexen-1-one; *p*-mentha-6,8-dien-2-one; 1-methyl-4-isopropenyl-Δ⁶-cyclohexen-2-one; carvol. $C_{10}H_{14}O$; mol wt 150.22. C 79.96%, H 9.39%, O 10.65%. *d*-Carvone is found in caraway seed and dill seed oils: Schweizer, *J. Prakt. Chem.* **24**, 257 (1841); Gladstone, *J. Chem. Soc.* **25**, 1 (1872). Isoln of *d*-carvone from mandarin peel oil *(Citrus reticulata* Blanco, *Rutaceae)*: Kugler, Kováts, *Helv. Chim. Acta* **46**, 1480 (1963). *l*-Carvone is found in spearmint and kuromoji oils: Kwaenick, *Ber.* **24**, 82 (1891). *dl*-Carvone is found in gingergrass oil: Walbaum, Hüthig, *J. Prakt. Chem.* **71**, 459 (1905). Prepn from α-pinene: Booth, Klein, **US 2796428** (1957 to Glidden). Structure: G. Wagner, *Ber.* **27**, 2270 (1894); F. Tiemann, F. W. Semmler, *ibid.* **28**, 2141 (1895). Synthesis of *l*-carvone: Royals, Horne, *J. Am. Chem. Soc.* **73**, 5856 (1951); Shono *et al.*, *Chem. Lett.* **1975**, 4330. Synthesis of *dl*-carvone: Suga, *Bull. Chem. Soc. Jpn.* **31**, 569 (1958); I. Fleming, I. Patterson, *Synthesis* **1979**, 736. Toxicity study: P. M. Jenner *et al.*, *Food Cosmet. Toxicol.* **2**, 327 (1964). *Review:* J. L. Simonsen, *The Terpenes* **vol. I** (University Press, Cambridge, 2nd ed., 1947) pp 394-408; B. Singaram, J. Verghese, *Perfum. Flavor.* **2**, 47 (1977).

d-**Form.** Liquid, bp_{755} 230°; bp_{5-6} 91°. d_4^{20} 0.965. n_D^{20} 1.4989. $[α]_D^{20}$ +61.2°.

l-**Form.** Liquid, bp_{763} 230-231°. d_{15}^{15} 0.9652. n_D^{20} 1.4988. $[α]_D^{20}$ −62.46°.

dl-**Form.** Liquid, bp 230-231°. d_{15}^{15} 0.9645. n_D^{20} 1.5003. Practically insol in water. Misc with alc. LD_{50} orally in rats: 1640 mg/kg (Jenner).

USE: As oil of caraway; also for flavoring liqueurs; in perfumery and soaps.

THERAP CAT: Carminative.

1875. Caryophyllene. [87-44-5] (1*R*,4*E*,9*S*)-4,11,11-Trimethyl-8-methylenebicyclo[7.2.0]undec-4-ene; β-caryophyllene; *trans*-caryophyllene. $C_{15}H_{24}$; mol wt 204.36. C 88.16%, H 11.84%. Sesquiterpenoid occurring in many essential oils and especially in clove oil, the oils from stems and flowers of *Syzygium aromaticum* (L.) Merrill & Perry *(Jambrosa caryophyllus* Niedenzu; *Eugenia caryophyllata* Thunb.), *Myrtaceae.* Occurs in nature as a mixture with isocaryophyllene and α-caryophyllene (humulene, *q.v.*). Isolation of mixture: Schreiner, Kremers, *Pharm. Arch.* **2**, 273, 293 (1899). Existence of isomers: Deussen, *Ann.* **356**, 1 (1907). Structural studies: A. Aebi *et al.*, *J. Chem. Soc.* **1953**, 3124; G. R. Ramage, R. Whitehead, *ibid.* **1954**, 4336. Abs config: D. H. R. Barton, A. Nickon, *ibid.* 4665. Total synthesis of racemic *trans/cis*-forms: E. J. Corey *et al.*, *J. Am. Chem. Soc.* **86**, 485 (1964). Rearrangement to isocaryophyllene: Rachlin, **DE 2044018** (1971 to I.F.F.), *C.A.* **75**, 49364j (1971). *Reviews:* Simonsen, *The Terpenes* **vol. III** (University Press, Cambridge, 1952) pp 39-71; Barton, de Mayo, *Q. Rev. Chem. Soc.* **11**, 197 (1957); Halsall, *ibid.* **16**, 101 (1962).

Caryophyllene   Isocaryophyllene

Liquid. Has a terpene odor about midway between odor of cloves and turpentine. bp_{14} 129-130°; $bp_{9.7}$ 118-119°. $[α]_D$ −8 to −9° (chloroform). n_D^{17} 1.5009; n_D^{15} 1.5030. d_4^{17} 0.9052.

Isocaryophyllene. [118-65-0] γ-Caryophyllene; *cis*-caryophyllene. Liquid. bp_{19} 130°; $bp_{14.5}$ 125-125.5°. $[α]_D$ −24° (chloroform). n_D^{19} 1.4966. d^{19} 0.8995.

USE: In perfumery.

1876. Carzenide. [138-41-0] 4-(Aminosulfonyl)benzoic acid; *p*-sulfamoylbenzoic acid; 4-carboxybenzenesulfonamide; *p*-sulfonamidobenzoic acid; benzoic acid 4-sulfamide. $C_7H_7NO_4S$; mol wt 201.20. C 41.79%, H 3.51%, N 6.96%, O 31.81%, S 15.93%. Byproduct in the production of saccharin. Prepn from toluene *p*-sulfonamide: W. A. Noyes, *Am. Chem. J.* **7**, 145 (1885); A. A. Goldberg, A. H. Wragg, *J. Chem. Soc.* **1960**, 1408. Metabolism: L. M. Ball *et al.*, *Xenobiotica* **8**, 183 (1978). HPLC determn in saccharin samples: A. E. Mooser, *J. Chromatogr.* **287**, 113 (1984). Toxicity data: D. Appenroth, H. Bräunlich, *Pharmazie* **38**, 102 (1983).

Flat, shiny prisms from water, dec 280°. pKa 3.50. Practically insol in cold water, ether, benzene. Freely sol in alcohol. LD$_{50}$ i.p. in rats: 350 ± 22 mg/kg (Appenroth, Bräunlich).

Sodium salt. C$_7$H$_6$NNaO$_4$S. Crystals. Neutral. Soly in water: ~20 g/100 ml.

1877. Carzinophilin A. [106486-76-4] 3-Methoxy-5-methyl-1-naphthalenecarboxylic acid (1S)-2-[[(1E)-1-[(3R,4R,5S)-3-(acetyloxy)-4-hydroxy-1-azabicyclo[3.1.0]hex-2-ylidene]-2-[[(1Z)-1-(hydroxymethylene)-2-oxopropyl]amino]-2-oxoethyl]amino]-1-[(2S)-2-methyloxiranyl]-2-oxoethyl ester; cardinophyllin; CZ; azinomycin B. C$_{31}$H$_{33}$N$_3$O$_{11}$; mol wt 623.62. C 59.71%, H 5.33%, N 6.74%, O 28.22%. Antitumor antibiotic; alkylates DNA by interstrand cross-linking of the major groove. Isoln from *Streptomyces sahachiroi*: T. Hata *et al., J. Antibiot.* **7A**, 107 (1954). Purification: H. Kamada *et al., ibid.* **8A**, 187 (1955). Mechanism of action: J. W. Lown, K. C. Majumdar, *Can. J. Biochem.* **55**, 630 (1977); T. Fujiwara, I. Saito, *Tetrahedron Lett.* **40**, 315 (1999). Proposed structure: J. W. Lown, C. C. Hanstock, *J. Am. Chem. Soc.* **104**, 3213 (1982). Identity with azinomycin B: E. J. Moran, R. W. Armstrong, *Tetrahedron Lett.* **32**, 3807 (1991). Purification: K. Nagaoka *et al., J. Antibiot.* **39**, 1527 (1986). Total structure as azinomycin B: K. Yokoi *et al., Chem. Pharm. Bull.* **34**, 4554 (1986).

Amorphous white solid, mp 190° (dec). [α]$_D^{20}$ +48° (c = 0.48 in CHCl$_3$). uv max (methanol): 217, 250, 340 nm (ε 65500, 30400, 8500). Sol in acetone, chloroform, ethyl acetate, methanol. Insol in diethyl ether, *n*-hexane, water.

1878. Casanthranol. Cantralax; Peristim. Complex mixture of the purified anthranol glycosides derived from *Cascara sagrada, q.v.*, practically devoid of free anthraquinones. Two active fractions have been identified as *casanthranol A* and *casanthranol B*. Upon oxidative hydrolysis, casanthranol A yields emodin, while casanthranol B under similar treatment yields the aglycone, aloe-emodin, and traces of chrysophanic acid. Prepn: Lee, Berger, US 2552896 (1951 to Hoffmann-La Roche).

Light tan to brown, amorphous, hygroscopic powder. Freely sol in water, with some residue; partially sol in methanol, hot isopropyl alc. Practically insol in acetone.

Component of *Peri-Colace, Casakol.*

THERAP CAT: Cathartic.

1879. Cascara Amara. Honduras bark. Bark of *Picramnia antidesma* Sw., *Simaroubaceae* and related species. *Habit.* West Indies, Mexico.

1880. Cascara Sagrada. Sacred bark; Chittem bark; Chittim bark; Purshiana bark; Persian bark; bearberry bark; bearwood. Dried bark of *Rhamnus purshiana* DC., *Rhamnaceae*, from which a naturally occurring cathartic is extracted. *Habit.* Northern Idaho, west to Northern California. The cathartic properties are primarily due to the presence of *cascarosides*, anthraglycosides which are related to glycosides found in aloe, *q.v.* and buckthorn. *See:* A. Y. Leung, *Drug Cosmet. Ind.* **121**, 42 (December, 1977). Other constituents are aloins (C-glycosides), O-glycosides, and free anthraquinones: *Analyst* **93**, 749 (1968); *ibid.* **98**, 830 (1973) (Joint Committee Reports). Isoln of anthraquinone aglycones and glycosides: S. C. Yung Su, N. M. Ferguson, *J. Pharm. Sci.* **63**, 899 (1973). Structure of cascarosides A and B: J. W. Fairbairn *et al., J. Pharm. Pharmacol.* **15** (Suppl.) 292T (1963); F. J. Evans *et al., ibid.* **27** (Suppl.), 91P (1975). Biological evaluation of cascara bark prepns: J. W. Fairbairn, G. E. D. H. Mahran, *ibid.* **5**, 827 (1953).

Commercial prepn, *Cas-Evac.*

THERAP CAT: Cathartic.

THERAP CAT (VET): Laxative.

1881. Cascarilla. Eleuthera bark; sweet-wood bark. Dried bark of *Croton eluteria* (L.) Sw., *Euphorbiaceae. Habit.* Bahama Islands, Cuba, Haiti. *Constit.* About 25% resins, 1.5-3% volatile oil, diterpene bitter principles including cascarillin A (15%). Description and medicinal use: J. Gruenwald *et al., PDR for Herbal Medicines* (Medical Economics, Montvale, 2nd Ed., 2000) p 156.

Volatile oil. [8007-06-5] Oil of cascarilla; oil of sweetwood bark. *Constit. l*-Limonene, *p*-cymene, dipentene, eugenol, cascarillic acid. Chromatographic separation of constituents: A. Claude-Lafontaine *et al., Bull. Soc. Chim. Fr.* **1973** part 2, 2866. Fragrance monograph: D. L. J. Opdyke, *Food Cosmet. Toxicol.* **14**, Suppl. 1, 707 (1976). Light yellow to brown amber liquid with a pleasant, spicy odor. d$_{25}^{25}$ 0.892-0.914. n$_D^{20}$ 1.488-1.494. Rotation +2° to +5° in a 100-mm tube at 20°. Very sol in alcohol, ether. Sol in most fixed oils, mineral oil. Practically insol in glycerin, propylene glycol. *Keep well closed, cool, and protected from light.*

USE: Flavoring agent; as an addition to smoking tobacco for flavoring.

THERAP CAT: Aromatic bitter.

1882. Cascarillin. [10118-56-6] (1S,2S,3R,4aR,5R,6R,8aS)-3-(Acetyloxy)-1-[(2R)-2-(3-furanyl)-2-hydroxyethyl]decahydro-5,6-dihydroxy-2,4a,5-trimethyl-1-naphthalenecarboxaldehyde. C$_{22}$H$_{32}$O$_7$; mol wt 408.49. C 64.69%, H 7.90%, O 27.42%. Major constituent of a family of bitter diterpenoids isolated from the bark of the cascarilla plant, *Croton eluteria* (L.) Sw., *Euphorbiaceae.* Isoln: Duval, *J. Pharm.* **8**, 91 (1845); Mylius, *Ber.* **6**, 1051 (1873); Naylor, Littlefield, *Pharm. J.* **57**, 95 (1896). Structure and stereochemistry: McEachan *et al., J. Chem. Soc. B* **1966**, 633; of cascarillin A: T. G. Halsall *et al., Chem. Commun.* **1965**, 218. Isoln of cascarillins B-D: C. Vigor *et al., Phytochemistry* **57**, 1209 (2001); of E-I: *eidem, J. Nat. Prod.* **65**, 1180 (2002).

Needles from alc, mp 205°. Very bitter taste. Freely sol in ether, hot alc; slightly sol in water, cold alc, chloroform.

THERAP CAT: Aromatic bitter.

1883. Casein. [9000-71-9] A mixture of related phosphoproteins occurring in milk and cheese. Present to the extent of 3% in bovine milk. One of the most nutritive milk proteins in that it contains all of the common amino acids and is rich in the essential ones. Produced in mammary tissue from amino acids supplied by the blood. Obtained from milk by removing the cream and acidifying the skimmed milk which causes casein to precipitate. In cheese manufacture, casein is precipitated by the lactic acid formed from the same milk by fermentation. Precipitation by rennet is favored for casein intended for plastics manuf. Curdling by electricity has been described. The prepn of pure casein is described by Hammarsten, *Textbook of Physiological Chemistry* (New York, 7th ed., 1911) p 619; *Abderhalden's Handbuch* **vol. II** (Berlin, 1910) p 384. Alternate prepn: van Slyke, Baker, *J. Biol. Chem.* **35**, 127 (1918); cf. Cohn, Hendry, *Org. Synth.* **coll. vol. II**, 120 (1943). Prepn of casein free from vitamin B$_{12}$ for nutrition experiments: Kissel, US 2853479 (1958 to Natl. Dairy Prods.). The major casein components may be distinguished by electrophoresis and are designated as α-, β-, γ- and K-caseins, in order of decreasing mobility at pH 7. The complete amino acid sequence of bovine β-casein is known and contains 209 residues with an approx. mol wt of 23,600: Ribadeaudumas *et al., Eur. J. Biochem.* **25**, 505 (1972). *Review:* McKenzie, *Adv. Protein Chem.* **22**, pp 75-135 (1967).

White, amorphous powder or granules without odor or taste. Very sparingly sol in water and in nonpolar organic solvents; sol in aqueous solns of alkalies, levorotatory. The isoelectric zone is around

Consult the Name Index before using this section.

pH 4.7; sol in concd HCl with light violet color. Amphoteric; forms salts with both acids and bases. Present in bovine milk as neutral calcium caseinate and in human milk as potassium caseinate. Precipitated from solns satd with metallic salts. Forms a hard, insol plastic with formaldehyde.

Casamino acids. Commercial acid-hydrolyzed casein. Hydrolysis is carried out until all the nitrogen in the casein is converted to amino acids or other compounds of relative chemical simplicity. Prepn: Mueller, Miller, *J. Immunol.* **40**, 21 (1941); Mueller, Johnson, *ibid.* 33. Typical analysis: N 10%, NaCl 14%, ash 20%, phosphorus as PO_4 2%, Fe 15 mg/3 g.

Note: Legumin, also known as *avenin* or *vegetable casein*, occurring in beans and nuts, is a globulin resembling casein. Isoln from *Avena sativa* L., *Graminea*: Sanson, *Compt. Rend.* **96**, 75 (1883). Structure: C. J. Bailey, D. Boulter, *Eur. J. Biochem.* **17**, 460 (1970).

USE: In the manuf of molded plastics, adhesives, paints, textile finishes, paper coatings, man-made fibers. Vitamin-free casein is used in diets of animals employed for the biological assay of vitamins. Medicinal grades are used in dietetic prepns and for determining the effectiveness of digestive enzyme prepns contg. pepsin, trypsin, papain. Casamino acids are used in microbial assays and in the prepn of microbiological media.

THERAP CAT: Nutrient.

1884. Caspase-1. [122191-40-6] Interleukin 1β precursor proteinase; interleukin-1β converting enzyme; IL-1β converting enzyme; ICE; EC 3.4.22.36. Primary protease responsible for cleavage of the proinflammatory cytokines *interleukin-1β (IL-1β)* and *interleukin-18 (IL-18)* from their biologically inactive precursors. IL-1β is the predominant form of interleukin-1, *q.v.*, produced by human monocytes. Caspase-1 is synthesized as an inactive 45 kDa precursor protein, *procaspase-1*, that is cleaved to form an active heterodimer composed of 10 and 20 kDa subunits. Identification in monocytes and IL-1β activation: M. J. Kostura *et al.*, *Proc. Natl. Acad. Sci. USA* **86**, 5227 (1989); R. A. Black *et al.*, *FEBS Lett.* **247**, 386 (1989). Purification, primary structure and catalytic mechanism: N. A. Thornberry *et al.*, *Nature* **356**, 768 (1992). Molecular cloning of human cDNA: D. P. Cerretti *et al.*, *Science* **256**, 97 (1992). Induction of apoptosis in fibroblasts: M. Miura *et al.*, *Cell* **75**, 653 (1993). Crystal structure of recombinant human form: N. P. C. Walker *et al.*, *ibid.* **78**, 343 (1994); and mechanism of action: K. P. Wilson *et al.*, *Nature* **370**, 270 (1994). Activation of IL-18 (*interferon-γ-inducing factor*): T. Ghayur *et al.*, *ibid.* **386**, 619 (1997). Overexpression in chronic pancreatitis: M. Ramadani *et al.*, *Pancreas* **22**, 383 (2001); in multiple sclerosis plaques: X. Ming *et al.*, *J. Neurol. Sci.* **197**, 9 (2002). Review of assay methods, purification and inhibition: N. A. Thornberry, *Methods Enzymol.* **244**, 615-631 (1994); of structure and function: M. J. Tocci, *Vitam. Horm.* **53**, 27-63 (1997); of role in intestinal inflammation: B. Siegmund, *Biochem. Pharmacol.* **64**, 1-8 (2002).

1885. Caspase-3. [169592-56-7] Apopain; cysteine protease P32; CPP32; EC 3.4.22.56. Primary effector caspase required for certain distinctive biochemical and morphological events in cells undergoing apoptosis. Necessary for the dATP/cytochrome *c*, *q.v.*, inducible cleavage of many cellular targets, including activation of an endonuclease, *caspase-activated DNase (CAD)*, that is responsible for DNA fragmentation. CAD activation is achieved by caspase-3 cleavage and inactivation of the *inhibitor of CAD (ICAD)*. Synthesized as a 32 kDa inactive zymogen, *procaspase-3*, that is activated by a proteolytic event generating a heterodimer consisting of a 20 kDa (*N*-terminal) and a 12 kDa (*C*-terminal) subunit. Identification: T. Fernandes-Alnemri *et al.*, *J. Biol. Chem.* **269**, 30761 (1994). Identification of poly(ADP-ribose) polymerase (PARP)-cleaving activity: M. Tewari *et al.*, *Cell* **81**, 801 (1995); and characterization of caspase-3: D. W. Nicholson *et al.*, *Nature* **376**, 37 (1995). Role in Fas-mediated apoptosis: J. Schlegel *et al.*, *J. Biol. Chem.* **271**, 1841 (1996). Crystal structure in complex with a tetrapeptide inhibitor: P. R. E. Mittl *et al.*, *ibid.* **272**, 6539 (1997). Cleavage of ICAD: H. Sakahira *et al.*, *Nature* **391**, 96 (1998); and activation of DNA fragmentation: B. B. Wolf *et al.*, *J. Biol. Chem.* **274**, 30651 (1999). Role in the execution phase of apoptosis: R. U. Jänicke *et al.*, *ibid.* **273**, 9357 (1998); E. A. Slee *et al.*, *ibid.* **276**, 7320 (2001). Effect of pH, ionic charge, and osmolality on activity: M. S. Segal, E. Beem, *Am. J. Physiol. Cell Physiol.* **281**, C1196 (2001). Fluorescence studies characterizing the active site: M. Kyoung *et al.*, *Biochim. Bio-*

phys. Acta **1598**, 74 (2002). Review of role in apoptosis: A. G. Porter, R. U. Jänicke, *Cell Death Differ.* **6**, 99-104 (1999).

1886. Caspase-8. [179241-78-2] FLICE proteinase; FADD-like ICE; FLICE; MACH; MORT1-associated CED-3 homolog; Mch5; EC 3.4.22.61. Initiator caspase required for Fas/CD95/APO-1-mediated apoptosis. Synthesized as a 55 kDa inactive zymogen, *procaspase-8*. Activated by association with the *death-inducing signaling complex (DISC)*, leading to the release of active ~12 and ~20 kDa subunits into the cytosol. Isoln, characterization, and apoptotic function: M. P. Boldin *et al.*, *Cell* **85**, 803 (1996); M. Muzio *et al.*, *ibid.* 817. DISC binding and activation: J. P. Medema *et al.*, *EMBO J.* **16**, 2794 (1997). Expression and characterization of isoforms: C. Scaffidi *et al.*, *J. Biol. Chem.* **272**, 26953 (1997). Activation of caspase-3, *q.v.*: H. R. Stennicke *et al.*, *ibid.* **273**, 27084 (1998). Cleavage of the proapoptotic Bcl2 protein, *BID*: H. Li *et al.*, *Cell* **94**, 491 (1998). Role in Fas-mediated apoptosis: P. Juo *et al.*, *Curr. Biol.* **8**, 1001 (1998). X-ray crystal structure of complexes with peptide inhibitors: H. Blanchard *et al.*, *Structure Fold. Des.* **7**, 1125 (1999); W. Watt *et al.*, *ibid.* 1135. Role in lymphocyte activation: H. J. Chun *et al.*, *Nature* **419**, 395 (2002). *Review*: M. Kruidering, G. I. Evan, *IUBMB Life* **50**, 85-90 (2000).

1887. Caspase-9. [180189-96-2] ICE-LAP6 proteinase; Mch6; EC 3.4.22.62. Central initiator caspase. Synthesized as an inactive 46 kDa precursor protein, *procaspase-9*, which consists of three domains: an *N*-terminal pro-domain, a 20 kDa large subunit, and a 10 kDa small subunit. Activated in the cytoplasm by cleavage of the prodomain, in response to mitochondrial damage. Activation occurs by binding to an Apaf-1/cytochrome *c* complex known as the apoptosome, in the presence of dATP. Isoln and characterization: H. Duan *et al.*, *J. Biol. Chem.* **271**, 16720 (1996); S. M. Srinivasula *et al.*, *ibid.* 27099. Purification, binding to Apaf-1, and activation of caspase-3, *q.v.*: P. Li *et al.*, *Cell* **91**, 479 (1997). Role in initiation of the caspase cascade: E. A. Slee *et al.*, *J. Cell Biol.* **144**, 281 (1999). Crystal structure of the prodomain in complex with the pro-domain of Apaf-1: H. Qin *et al.*, *Nature* **399**, 549 (1999). Human gene structure and chromosomal location: S. Hadano *et al.*, *Mammal. Genome* **10**, 757 (1999). Activation mechanism: J. Rodriguez, Y. Lazebnik, *Genes Dev.* **13**, 3179 (1999); E. N. Shiozaki *et al.*, *Proc. Natl. Acad. Sci. USA* **99**, 4197 (2002). Review of structure and function: K. Kuida, *Int. J. Biochem. Cell Biol.* **32**, 121-124 (2000); of activation mechanism: Y. Shi, *Structure* **10**, 285-288 (2002).

1888. Caspases. Class of cysteine proteases present as inactive zymogens (procaspases) in cytosol of healthy cells. Activated during apoptosis to cleave specific protein targets, usually after aspartic acid residues, leading to organized cellular destruction. At least 14 distinct mammalian caspases have been identified. Apoptotic caspases are divided into two categories, initiator caspases: caspase-2, -8, -9, and -10, and effector caspases: caspase-3, -6, and -7. Caspase-1, and likely caspases-4, -5, -11 and -12, are cytokine activators, responsible for generation of inflammatory response. Identification of interleukin-1β-converting enzyme (caspase-1) as a cysteine protease: N. A. Thornberry *et al.*, *Nature* **356**, 768 (1992). Isolation and apoptotic role of *CED-3*, a caspase homolog from *Caenorhabditis elegans*: J. Yuan *et al.*, *Cell* **75**, 641 (1993). Nomenclature: E. S. Alnemri *et al.*, *Cell* **87**, 171 (1996). Substrate specificites: N. A. Thornberry *et al.*, *J. Biol. Chem.* **272**, 17907 (1997). Induced proximity activation mechanism: G. S. Salvesen, V. M. Dixit, *Proc. Natl. Acad. Sci. USA* **96**, 10964 (1999). Analysis of active sites: D. Chéreau *et al.*, *Biochemistry* **42**, 4151 (2003). Review of role in apoptosis: N. A. Thornberry, Y. Lazebnik, *Science* **281**, 1312-1316 (1998); V. Cryns, J. Yuan, *Genes Dev.* **13**, 1551-1570 (1998); of mammalian caspases: W. C. Earnshaw *et al.*, *Annu. Rev. Biochem.* **68**, 383-424 (1999); of activation pathways: I. Budihardjo *et al.*, *Annu. Rev. Cell Dev. Biol.* **15**, 269-290 (1999); of structure, function and inhibition: J.-B. Denault, G. S. Salvesen, *Chem. Rev.* **102**, 4489-4499 (2002); of mechanisms of activation and inhibition: Y. Shi, *Mol. Cell* **9**, 459-470 (2002); of apoptotic role in neurodegenerative diseases: R. M. Friedlander, *N. Engl. J. Med.* **348**, 1365-1375 (2003).

1889. Caspofungin. [162808-62-0] 1-[(4*R*,5*S*)-5-[(2-Amino-ethyl)amino]-*N*²-[(10*R*,12*S*)-10,12-dimethyl-1-oxotetradecyl]-4-hydroxy-L-ornithine]-5-[(3R)-3-hydroxy-L-ornithine]pneumocandin B₀. $C_{52}H_{88}N_{10}O_{15}$; mol wt 1093.33. C 57.13%, H 8.11%, N 12.81%, O 21.95%. Semisynthetic echinocandin antifungal that inhibits 1,3-β-D-glucan synthase which is required for cell wall synthesis. Prepn: J. M. Balkovec *et al.*, **WO 9421677**; *eidem*, **US 5378804** (1994, 1995 both to Merck & Co.). Abs config of dimethylmyristoyl side chain: W. R. Leonard, Jr. *et al.*, *Org. Lett.* **4**, 4201 (2002). Synthesis: *idem et al.*, *J. Org. Chem.* **72**, 2335 (2007). HPLC/MS determn in plasma: C. M. Chavez-Eng *et al.*, *J. Chromatogr. B* **721**, 229 (1999). Clinical study vs amphotericin B in *Candida* esophagitis: A. Villanueva *et al.*, *Clin. Infect. Dis.* **33**, 1529 (2001). Review of pharmacology and clinical experience: A. Hoang, *Am. J. Health Syst. Pharm.* **58**, 1206-1214 (2001); W. W. Hope *et al.*, *Expert Opin. Drug Metab. Toxicol.* **3**, 263-274 (2007).

Acetate. [179463-17-3] MK-0991; Cancidas. $C_{52}H_{88}N_{10}O_{15}$·- $2C_2H_4O_2$; mol wt 1213.44. Hygroscopic white to off-white powder. $[\alpha]_{405}^{25}$ −105° (c = 1.0 in water). Freely sol in water, methanol; slightly sol in ethanol.

THERAP CAT: Antifungal.

1890. Cassaidine. [26296-41-3] (2E)-2-[(1R,4aS,4bR,7S,- 8aR,10S,10aS)-Dodecahydro-7,10-dihydroxy-1,4b,8,8-tetramethyl- 2(1H)-phenanthrenylidene]acetic acid 2-(dimethylamino)ethyl ester; (E)-3β,7β-dihydroxy-14α-methyl-8β-podocarpane-Δ¹³,ᵅ-acetic acid 2-(dimethylamino)ethyl ester. $C_{24}H_{41}NO_4$; mol wt 407.60. C 70.72%, H 10.14%, N 3.44%, O 15.70%. Cardiotonic principle from the bark of *Erythrophleum guineense* G. Don., *Leguminosae.* Isoln and structure: Ruzicka, Dalma, *Helv. Chim. Acta* **23**, 753 (1940); Engel, *ibid.* **42**, 1127 (1959). Synthesis: Turner *et al.*, *J. Am. Chem. Soc.* **88**, 1766 (1966). Stereochemistry: Clarke *et al.*, *ibid.* **88**, 5865 (1966).

Prisms from acetone + ether, mp 139.5°. $[\alpha]_D^{20}$ −98° (ethanol); $[\alpha]_D^{20}$ −104° (0.1N HCl). Slightly sol in methanol, ethanol, acetone, acetic, chloroform. Practically insol in ether, benzene.

Oxalate. $C_{24}H_{41}NO_4$·(COOH)₂. Crystals from methanol+ acetone, mp 198-201°. $[\alpha]_D^{17}$ −84° (c = 1.39 in water).

1891. Cassaine. [468-76-8] (2E)-[(1R,4aS,4bR,7S,8aR,- 10aS)-Dodecahydro-7-hydroxy-1,4b,8,8-tetramethyl-10-oxo-2(1H)- phenanthrenylidene]acetic acid 2-(dimethylamino)ethyl ester; (E)-

3β-hydroxy-14α-methyl-7-oxopodocarpane-Δ¹³,ᵅ-acetic acid 2-(di-methylamino)ethyl ester. $C_{24}H_{39}NO_4$; mol wt 405.58. C 71.07%, H 9.69%, N 3.45%, O 15.78%. Cardiotonic principle from the bark of *Erythrophleum guineense* G. Don., *Leguminosae*: Dalma, *Helv. Chim. Acta* **22**, 1497 (1939). Structure: R. B. Turner *et al.*, *Tetrahedron Lett.* **1**(2), 7 (1959); Gensler *et al.*, *J. Am. Chem. Soc.* **81**, 5217 (1959). Absolute configuration: Hauth *et al.*, *Helv. Chim. Acta* **48**, 1087 (1965); Clarke *et al.*, *J. Am. Chem. Soc.* **88**, 5865 (1966). Synthesis: Turner *et al.*, *ibid.* 1766.

Glossy flakes from ether, mp 142.5°. $[\alpha]_D^{23}$ −110.5° (alc); $[\alpha]_D^{23}$ −101° (0.1N HCl). Absorption spectrum: Ruzicka, Dalma, *Helv. Chim. Acta* **22**, 1516 (1939). Sol in methanol, ethanol, acetone, acetic acid, chloroform, ether, benzene.

Hydrochloride hydrate. $C_{24}H_{40}ClNO_4.H_2O$. Minute crystals from alcohol + methyl ethyl ketone + ether (1:1:4), mp 220°.

1892. Cassella's Acid. 7-Hydroxy-2-naphthalenesulfonic acid; 2-naphthol-7-sulfonic acid; β-naphthol-δ-monosulfonic acid; β-naphtholsulfonic acid F; F Acid. $C_{10}H_8O_4S$; mol wt 224.23. C 53.57%, H 3.60%, O 28.54%, S 14.30%. Prepn: Bayer, Duisberger, *Ber.* **20**, 1426 (1887); Green, *J. Chem. Soc.* **55**, 33 (1889).

Needles from HCl, mp 89°. Readily sol in water, alcohol; practically insol in ether, benzene.

USE: Dyestuff intermediate.

1893. Cassella's Acid F. [494-44-0] 7-Amino-2-naphtha-lenesulfonic acid; 2-naphthylamine-7-sulfonic acid; β-naphthyl-amine-δ-sulfonic acid. $C_{10}H_9NO_3S$; mol wt 223.25. C 53.80%, H 4.06%, N 6.27%, O 21.50%, S 14.36%. Prepn by sulfonation of β-naphthylamine and separation from the 6-amino isomer: Green, *J. Chem. Soc.* **55**, 33 (1889); from 7-hydroxy-2-naphthalenesulfonic acid and ammonia: Green, *loc. cit.*; Wait, **US 1492497** (1924).

Monohydrate. Crystals. Sol in 5040 parts cold water, 350 parts boiling water; sol in glacial acetic acid.

Copper salt. Orange-yellow crystals. Sparingly sol in water. *Ref*: Green, Vakil, *J. Chem. Soc.* **113**, 35 (1918).

Note: **Bronner's acid** was first described as *6-amino-2-naphtha-lenesulfonic acid* or *2-naphthylamine-6-sulfonic acid*. However, this product obtained by sulfonation of β-naphthylamine, was subsequently shown to be a mixture of about equal parts of 6- and 7-amino-2-naphthalenesulfonic acids: Green, *loc. cit.*

USE: Both the title compd and its 6-amino isomer are used in the manuf of azo dyes, *e.g.*, **GB 810246** (1959 to Bayer).

1894. Cassia Fistula. Cassia pods; drumstick; Indian laburnum; pudding-stick; pudding pipe; purging cassia. Dried fruit of *Cassia fistula* L. (*Cathartocarpus fistula* [L.] Pers.), *Leguminosae*. *Habit*. Upper Egypt, E. India; cultivated in tropical America and Africa. The pulp of the ripe fruit, *cassia pulp*, is an almost black, viscid mass with a sweetish taste. *Constit.* Hydroxymethylanthra-quinones, gum, tannin, albuminoids, about 60% sugars.

THERAP CAT: Cathartic.

1895. Castanea. Chestnut. Leaves of *Castanea dentata* (Marsh.) Borkh., *Fagaceae*, collected in September and October. *Habit.* Southern Europe. There are hardly any chestnut trees left in the U.S. *Constit.* Tannin, gum, albumin, resin.

1896. Castanospermine. [79831-76-8] (1*S*,6*S*,7*R*,8*R*,8a*R*)-Octahydro-1,6,7,8-indolizinetetrol; 1,6,7,8-tetrahydroxyoctahydroindolizine; (1*S*,6*S*,7*R*,8*R*,8a*R*)-1,6,7,8-tetrahydroxyindolizidine. $C_8H_{15}NO_4$; mol wt 189.21. C 50.78%, H 7.99%, N 7.40%, O 33.82%. Polyhydroxy alkaloid isolated from the seeds of the Australian leguminous tree, *Castanospermum australe*. Inhibits enzymatic glycoside hydrolysis. Isoln of the naturally occurring (+)-form: L. D. Hohenschutz *et al.*, *Phytochemistry* **20**, 811 (1981). Total synthesis and absolute configuration: R. C. Bernotas, B. Ganem, *Tetrahedron Lett.* **25**, 165 (1984). Alternate synthesis: H. Hamana *et al.*, *J. Org. Chem.* **52**, 5492 (1987). Inhibition of α- and β-glucosidases: R. Saul *et al.*, *Arch. Biochem. Biophys.* **221**, 593 (1983); *eidem, ibid.* **230**, 668 (1984). Insect antifeedant activity: D. L. Dreyer *et al.*, *J. Chem. Ecol.* **11**, 1045 (1985). Inhibition of HIV infectivity *in vitro*: B. D. Walker *et al.*, *Proc. Natl. Acad. Sci. USA* **84**, 8120 (1987); R. A. Gruters *et al.*, *Nature* **330**, 74 (1987).

Crystals from aq ethanol, mp 212-215° (dec). $[\alpha]_D^{25}$ +79.7° (c = 0.93 in water). pK 6.09.

1897. Castle's Intrinsic Factor. Intrinsic factor; IF. A thermolabile mucoprotein with mol wt about 60,000. Promotes vitamin B_{12} absorption by transporting it through the intestinal wall. Occurs in normal gastric juice, but is deficient in patients with pernicious anemia. Prepd from hog mucosa: Castle *et al.*, *Am. J. Med. Sci.* **178**, 748, 764 (1929); **180**, 305 (1930); Glass *et al.*, *Science* **115**, 101 (1952); Heinle *et al.*, *Trans. Assoc. Am. Physicians* **65**, 214 (1952); Latner *et al.*, *Biochem. J.* **55**, XXIII (1953); Callender *et al.*, *Br. Med. J.* **I**, 10 (1954); Latner *et al.*, *Lancet* **I**, 497 (1954); Baum, Federman, **US 2912360** (1959 to Lilly). Purification: Robbins, **US 3008877** (1961 to Armour); Highley, Ellenbogen, **US 3434927** and **US 3591678** (1969, 1971, both to Am. Cyanamid). In approx 30% of pernicious anemia patients, antibodies are produced in the serum which combine with IF, thus inhibiting its biological activity. In clinical tests diminished excretion of vitamin B_{12} in the feces is taken as evidence of intrinsic factor activity. Function in the metabolism of vitamin B_{12}: Glass, *Physiol. Rev.* **43**, 529 (1963). *Review:* Gräsbeck, *Prog. Hematol.* **6**, 233 (1969).

Combination with vitamin B_{12}. Gastrhéma.
THERAP CAT: Adjuvant in vitamin B_{12} utilization.

1898. Castor Oil. Ricinus oil; oil of Palma Christi; tangantangan oil; Neoloid. Fixed oil obtained by cold-pressing the seeds of *Ricinus communis* L., *Euphorbiaceae*. Triglyceride of fatty acids. Fatty acid composition is approx ricinoleic 87%, oleic 7%, linoleic 3%, palmitic 2%, stearic 1% and dihydroxystearic trace amounts: Binder *et al.*, *J. Am. Oil Chem. Soc.* **39**, 513 (1962). Review and bibliography: Anderson, *J. Philipp. Pharm. Assoc.* **42**, 5-16 (1955); Dominguez *et al.*, *J. Chem. Educ.* **20**, 446 (1952); F. C. Naughton *et al.*, in *Kirk-Othmer Encyclopedia of Chemical Technology* **vol. 5** (Wiley-Interscience, New York, 3rd ed., 1979) pp 1-15.

Pale yellow, viscous oil. Slight somewhat characteristic odor. The crude oil tastes slightly acrid with a decidedly nauseating aftertaste. Has excellent keeping qualities, does not turn rancid unless subjected to excessive heat. Dextrorotatory (undil. in sodium light). $d_{15.5}^{15.5}$ 0.961-0.963. Wt of tech grades: 8.1 to 8.9 lbs/gallon. n_D^{25} 1.473-1.477. n_D^{40} 1.466-1.473. Solidif −10° to −18°. Viscosity (25°): 6-8 P, also expressed as U ± ½ (Gardner-Holdt Scale). Flash pt 445°F (230°C); ignition temp 840°F (449°C). Surface tension (dynes/cm): at 20°, 39.0; at 80°, 35.2. Acid value <4. Sapon no. 176-187. Iodine no. (Wijs') 81-91. Reichert-Meissl value <0.5. Polenske value <0.5. Acetyl value 144-150. Hydroxyl value 161-169. Miscible with abs ethanol, methanol, ether, chloroform, glacial acetic acid. Dissolves in its own vol of petr ether or 95% alcohol. Does not dissolve to any extent in mineral oil, unless mixed with another vegetable oil. When heated to 300° for several hours it polymerizes and becomes miscible with mineral oil.

USE: As an industrial raw material for the prepn of chemical derivs used in coatings, urethane derivs, surfactants and dispersants, cosmetics, lubricants; chief raw material for the production of sebacic acid, a basic ingredient in the production of synthetic resins and fibers; as lubricant in metal drawing, machine lubrication and 2-cycle engine fuels, in hydraulic fluids, rubber preservative and mold lubricants; constituent of embalming fluids; in soap manuf; to impart emollient and lubricant properties to cosmetic prepns; as Turkey-red oil (sulfated castor oil) for dyeing and finishing textiles; as dehydrated castor oil in alkyds, resinous copolymers, varnishes, oil-based paints, enamels, calks and putties; as blown oil (oxidized oil) for plasticizing oilcloth, artificial leather, coated fabrics, and lacquers; to plasticize rosin in the manuf of sticky fly-paper, for nitrocellulose and similar coating systems, hot melts, duplicating and stencil inks, adhesives and laminants; as release and anti-sticking agent in hard candy manuf.

THERAP CAT: Cathartic.
THERAP CAT (VET): Mild purgative, but considered unreliable in adult horses. Emollient.

1899. Castor Oil, Hydrogenated. Opalwax; Castorwax.
Mol wt about 932. A hard, white wax, mp 86-88°. Iodine number (Wijs') <5.0. Extremely insol in water and in the more common organic solvents.
USE: In water-repellent coatings, candles, shoe polish, carbon paper, ointments and cosmetics; for impregnating paper, wood, cloth; for electrical condenser impregnation; as solid lubricant; as a pressure mold release agent in the manuf of formed plastics and rubber goods.

1900. Catalase. [9001-05-2] Caperase; Equilase; Optidase. Enzymes which promote reactions involving the decompn of hydrogen peroxide to water and oxygen. Although widely distributed among animals, plants, bacteria, and fungi, catalase in mammalian liver and blood has been most intensively studied. Catalase for commercial use obtained from animal liver, bacterial (*Micrococcus lysodeikticus*), and fungal (*Aspergillus niger*) sources. Isoln from mammalian livers and kidneys: Lolli, Cavanaugh, **US 2703779** (1955 to Armour); Schroeder *et al.*, *Biochim. Biophys. Acta* **58**, 611 (1962); Dan, **US 2992167** (1961 to Chr. Hansen's Lab.); from *Aspergillus niger*: Faucett *et al.*, **US 3102081** (1963 to Miles Labs.). All catalases so far isolated contain four tetrahedrally arranged subunits of equal size giving an approximate mol wt of 240,000. Each subunit consists of a single polypeptide chain associated with a single prosthetic group, ferric protoporphyrin IX. Amino acid sequence of bovine liver catalase subunit: Schroeder *et al.*, *Arch. Biochem. Biophys.* **131**, 653 (1969). Bovine **hepatocatalase**, a term for catalase obtained from liver, was reported to lower serum cholesterol; the active form was found to contain a catalytic amount of zinc (0.32%): Azarnoff, Curran, *J. Lab. Clin. Med.* **60**, 856 (1962); Laporte *et al.*, *Biochem. Pharmacol.* **11**, 670 (1962). *Reviews:* Nicholls, Schonbaum, in *The Enzymes* **vol. 8**, P. D. Boyer *et al.*, Eds. (Academic Press, New York, 1963) pp 147-225; Deisseroth, Daunce, *Physiol. Rev.* **50**, 319 (1970); Schonbaum, Chance, in *The Enzymes* **vol. 13** (part C), P. D. Boyer, Ed. (Academic Press, New York, 3rd ed., 1976) pp 363-408.

USE: In combination with glucose oxidase, for treatment of food wrappers to prevent oxidative deterioration of food: Sarett, Scott, **US 2765233** (1956 to Ben L. Sarett). In the removal of traces of peroxide in the process of cold sterilization (preservation of milk and cheese by treatment with hydrogen peroxide). With glucose oxidase, *q.v.*, in food preservation.

1901. Catalposide. [6736-85-2] (1a*S*,1b*S*,2*S*,5a*R*,6*S*,6a*S*)-1a,1b,2,5a,6,6a-Hexahydro-6-[(4-hydroxybenzoyl)oxy]-1a-(hydroxymethyl)oxireno[4,5]cyclopenta[1,2-*c*]pyran-2-yl-β-D-glucopyranoside; Catalpin. $C_{22}H_{26}O_{12}$; mol wt 482.44. C 54.77%, H 5.43%, O 39.80%. From unripe fruit of *Catalpa bignonioides* Walt. or *C. ovata* G. Don, *Bignoniaceae*: Claassen, *Am. Chem. J.* **10**, 228 (1888). Historical summary: Bobbitt *et al.*, *J. Org. Chem.* **26**, 3090 (1961). Structure: Lunn *et al.*, *Can. J. Chem.* **40**, 104 (1962). Stereochemistry: Bobbitt *et al.*, *J. Org. Chem.* **31**, 500 (1966); **32**, 1459 (1967).

Crystals from methanol + ethyl acetate, mp 215-216.5°. $[\alpha]_D^{23.2}$ −184° (c = 0.87 in methanol). uv max (ethanol): 260 nm (log ε 4.27); in sodium hydroxide: 303 nm (log ε 4.35). Soluble in water, alcohol; slightly sol in benzene, chloroform, ether.

Hexaacetate. $C_{34}H_{38}O_{18}$. Crystals from ethanol, mp 141.5°. $[\alpha]_D^{21.7}$ −106° (c = 0.75 in chloroform).

1902. Catechin. [154-23-4] (2R,3S)-2-(3,4-Dihydroxyphenyl)-3,4-dihydro-2H-1-benzopyran-3,5,7-triol; catechol; trans-(+)-3,-3',4',5,7-flavanpentol; catechinic acid; catechuic acid; (+)-cyanidanol-3; dexcyanidanol; cyanidol; Catergen. $C_{15}H_{14}O_6$; mol wt 290.27. C 62.07%, H 4.86%, O 33.07%. Flavonoid found primarily in higher woody plants as (+)-catechol along with (−)-*epicatechin* (cis form). From catechu (gambir and acacia), mahogany wood, etc.: Perkin, Yoshitake, J. Chem. Soc. **81**, 1160 (1902); Freudenberg et al., Ber. **55**, 1734 (1922). Structure: Freudenberg et al., Ann. **444**, 135 (1925). Stereochemistry: Clark-Lewis, Chem. Ind. (London) **1955**, 1218; Hardegger et al., Helv. Chim. Acta **40**, 1819 (1957); Clark-Lewis, J. Chem. Soc. **1960**, 2433. Pharmacology: Van Cauwenberge et al., C. R. Seances Soc. Biol. Ses Fil. **165**, 1195 (1971). Metabolism: Das, Sothy, Biochem. J. **125**, 417 (1971); and absorption in human: N. P. Das, Biochem. Pharmacol. **20**, 3435 (1971). Biosynthesis of epicatechins: Zaprometov, Grisebach, Z. Naturforsch. **28c**, 113 (1973). Other catechins such as *afzelechin* and *gallocatechin* and catechin oligomers (*procyanidins*) also exist in nature: Thompson et al., J. Chem. Soc. Perkin Trans. 1 **1972**, 1287. See also Bioflavonoids.

Needles from water + acetic acid, occurs as hydrate. mp 93-96°; 175-177° when anhydr. $[\alpha]_D^{18}$ +16 to +18.4°.

dl-Form. [7295-85-4] Needles from water + acetic acid, mp 212-216°. Slightly sol in cold water, ether; sol in hot water, alcohol, glacial acetic acid, acetone. Practically insol in benzene, chloroform, petr ether.

l-Form. [18829-70-4] Needles from water + acetic acid, mp 93-96°; 175-177° when anhydr. $[\alpha]_D$ −16.8°.

Note: Catechin is also called catechol (flavan) to distinguish it from catechol (pyrocatechol, q.v.).

USE: In dyeing and tanning.

THERAP CAT: Antidiarrheal.

1903. Catecholborane. [274-07-7] 1,3,2-Benzodioxaborole; 1,3,2-benzodioxaborolane. $C_6H_5BO_2$; mol wt 119.91. C 60.10%, H 4.20%, B 9.02%, O 26.69%. Versatile boron hydride reagent used in synthesis to perform hydroborations and reductions. Prepn from 2-chloro-1,3,2-benzodioxaborole: H. C. Newsom, W. G. Woods, Inorg. Chem. **7**, 177 (1968); from catechol: H. C. Brown, S. K. Gupta, J. Am. Chem. Soc. **93**, 1816 (1971); S. W. May et al., ibid. **99**, 2017 (1977). Improved prepn: J. V. B. Kanth et al., Org. Pro-

cess Res. Dev. **4**, 550 (2000). Synthetic applications: H. C. Brown, S. K. Gupta, J. Am. Chem. Soc. **97**, 5249 (1975); D. A. Evans, A. H. Hoveyda, J. Org. Chem. **55**, 5190 (1990); D. A. Evans, G. C. Fu, ibid. 5678; D. J. Harrison et al., Tetrahedron Lett. **45**, 8493 (2004); R. R. Huddleston et al., J. Org. Chem. **68**, 11 (2003). Reviews: C. F. Lane, G. W. Kabalka, Tetrahedron **32**, 981-990 (1976); G. W. Kabalka, Org. Prep. Proced. Int. **9**, 133-147 (1977); B. Zhang, Synlett **2007**, 666-667.

Colorless liquid. Flammable. Corrosive. Moisture sensitive. Stable in dry air. bp_{156} 88°; bp_{100} 76-77°; bp_{80} 66°; bp_{26} 50°. mp ~10-12°. n_D^{20} 1.5070. d 1.27. Flash point, closed cup: 35.6°F (2°C). Sol in most aprotic organic solvents.

USE: Reagent in organic synthesis.

1904. Catechu Black. Cutch; cachou; pegu catechu; cashoo. Extract prepd from heartwood of Acacia catechu Willd., Leguminosae. Habit. India, Hindustan, Ceylon; naturalized in Jamaica. Constit. 25-35% catechutannic acid, 2-10% catechin; catechu red, quercetin, gum.

Incompat. Iron compds, gelatin, lime water, zinc sulfate.

USE: Tanning, dyeing fabrics brown and black; staining wood; in toilet preparations.

THERAP CAT: Astringent (diarrheal).

THERAP CAT (VET): Has been used as intestinal astringent.

1905. Catharanthine. [2468-21-5] (2α,5β,6α,18β)-3,4-Didehydroibogamine-18-carboxylic acid methyl ester; 7-ethyl-9,10,-12,13-tetrahydro-6,9-methano-5H-pyrido[1',2':1,2]azepino[4,5-b]-indole-6(6aH)-carboxylic acid methyl ester. $C_{21}H_{24}N_2O_2$; mol wt 336.44. C 74.97%, H 7.19%, N 8.33%, O 9.51%. Precursor of vinblastine-type alkaloids. Isoln from Vinca rosea Linn. (Catharanthus roseus G. Don.) Apocynaceae: M. Gorman et al., J. Am. Pharm. Assoc. Sci. Ed. **48**, 256 (1959); G. H. Svoboda et al., ibid. 659. Structure: N. Neuss, M. Gorman, Tetrahedron Lett. **1961**, 206. Abs config: K. Bláha et al., ibid. **1972**, 2763. Synthesis of (±)-form: A. A. Qureshi, A. I. Scott, Chem. Commun. **1968**, 947, 948; A. R. Battersby et al., ibid. 951; G. Büchi et al., J. Am. Chem. Soc. **92**, 999 (1970); J. P. Kutney, F. Bylsma, Helv. Chim. Acta **58**, 1672 (1975); B. M. Trost et al., J. Org. Chem. **44**, 2052 (1979); T. Imanishi et al., Tetrahedron Lett. **21**, 3285 (1980).

Crystals from methanol, mp 126-128°. uv max (ethanol): 226, 284, 292 nm (log ε 4.56, 3.92, 3.88). $[\alpha]_D^{27}$ +29.8° (CHCl₃). pKa' 6.8.

1906. Cathepsins. Intracellular proteinases obtained from animal tissue extracts, the richest sources being liver, kidney and spleen. Located primarily in the lysosomal fraction within the cell. Part of the general enzymic apparatus of animal cells; in most cases they do not specialize in functions characteristic of individual tissues. Review of cathepsins A-C: J. S. Fruton in The Enzymes vol. 4, P. D. Boyer et al., Eds. (Academic Press, New York, 1960) pp 233-241; of cathepsins A-E: M. J. Mycek, Methods Enzymol. **19**, 285-315 (1970); of cathepsins B, D and G: several authors, Res. Monogr. Cell Tissue Physiol. **2**, 57-89, 181-248 (1977); of cathepsins B, D, G, H, L, N and S: several authors, Ciba Found. Symp. **75**, 1-68 (1980).

Cathepsin C. Dipeptidyl transferase; dipeptidyl amino peptidase I. Isoln from beef spleen: Tallan et al., J. Biol. Chem. **194**, 793 (1952); de la Haba et al., ibid. **234**, 316 (1959). Hydrolyzes dipeptidyl amides or esters bearing a free α-amino (or α-imino) group in the N-terminal position, esp. those containing an aromatic amino acid adjacent to the free α-amino group: Planta, Gruber, *Biochim. Bio-phys. Acta* **53**, 443 (1961); Wurz *et al.*, *Biochemistry* **1**, 19 (1962). Enhances the proteolysis of prothrombin to thrombin and thus plays an important role in blood clotting: Purcell, Barnhart, *Biochim. Bio-phys. Acta* **78**, 800 (1963).

Cathepsin D. Isoln from bovine spleen: Press *et al.*, *Biochem. J.* **74**, 501 (1960); Webb, *ibid.* **76**, 538 (1960); **84**, 455 (1962). In-volved in the catabolism of cartilage and connective tissue: Weston *et al.*, *Nature* **222**, 285 (1969).

Cathepsin G. Isoln from human spleen: G. Starkey *et al.*, *Bio-chem. J.* **155**, 255 (1976). Chymotrypsin-like enzyme. Catalytic and immunological properties: P. M. Starkey, A. J. Barrett, *ibid.* 273.

1907. Cathinone. [71031-15-7] (2*S*)-2-Amino-1-phenyl-1-propanone; (−)-α-aminopropiophenone; norephedrone. $C_9H_{11}NO$; mol wt 149.19. C 72.46%, H 7.43%, N 9.39%, O 10.72%. Psycho-active alkaloid found in the leaves of the khat plant, *Catha edulis* Forsk., *Celastraceae*. Prepn of racemic compd: L. Behr-Bergowski, *Ber.* **30**, 1515 (1897); S. Gabriel, *ibid.* **41**, 1127 (1908). Isoln from *C. edulis*: X. Schorno, E. Streinegger, *Experientia* **35**, 572 (1979); and identification as active constituent of khat: D. W. Peterson *et al.*, *Life Sci.* **27**, 2143 (1980). Stereospecific synthesis from norephe-drine: B. D. Berrang *et al.*, *J. Org. Chem.* **47**, 2643 (1982). Metab-olism in humans: R. Brenneisen *et al.*, *J. Pharm. Pharmacol.* **38**, 298 (1986). HPLC determn in urine: K. Mathys, R. Brenneisen, *J. Chromatogr.* **593**, 79 (1992). Quantitative determn in khat: A. M. Al-Obaid *et al.*, *J. Pharm. Biomed. Anal.* **17**, 321 (1998). Review of pharmacology: P. Kalix, *Pharmacol. Toxicol.* **70**, 77-86 (1992).

Sol in methanol (with racemization), methylene chloride. $[\alpha]_D$ −46.8° (c = 0.24 in methanol).

Hydrochloride. [76333-53-4] $C_9H_{11}NO.HCl$. Crystals from isopropanol-THF, mp 189-190°.

Note: This is a controlled substance (stimulant): **21 CFR,** 1308.11.

1908. Catnep. Cataria; catnip; catmint. Herb of *Nepeta ca-taria* L., *Labiatae*. *Habit.* Europe, Asia; naturalized in U.S. *Con-stit.* Volatile oil, nepetalactone, *q.v.*, nepetalic acid and related compds, tannin: McElvain, Eisenbraun, *J. Am. Chem. Soc.* **77**, 1599 (1955). Its odor is very attractive to all members of the cat family.

THERAP CAT: Aromatic.

1909. Catumaxomab. [509077-98-9] Anti-(human antigen 17-1A) immunoglobulin G2a (mouse monoclonal Ho-3/TP-A-01/TPBs01 heavy chain) disulfide with mouse monoclonal Ho-3/TP-A-01/TPBs01 light chain disulfide with anti-(human CD3 (antigen)) immunoglobulin G2b (rat monoclonal 26/II/6-1.2/TPBs01 heavy-chain) disulfide with rat monoclonal 26/II/6-1.2/TPBs01 light chain; Removab. Heterologous (mouse/rat), bispecific monoclonal anti-body directed against human epithelial cell adhesion molecule (Ep-CAM) and human CD3 T-cell surface antigen; elicits a trifunctional immune response that targets tumor cells which cause malignant ascites. Consists of a mouse anti-human EpCAM heavy/light chain Ig pair combined with a rat anti-human CD3 heavy/light chain Ig pair. Also binds via the intact Fc-region to Fcγ-receptors on immune accesory cells. Description of preparative method using quadroma technology: H. Lindhofer, S. Thierfelder, **DE 4419399**; *eidem*, **US 5945311** (1995, 1999 both to GSF). Use in treatment of malignant ascites: H. Lindhofer, **US 030223999** (2003). Structural and func-tional characterization: D. Chelius *et al.*, *mAbs* **2**, 309 (2010). Tri-functional mode of action: F. Hirschhaeuser *et al.*, *Cancer Immunol. Immunother.* **59**, 1675 (2010). Clinical pharmacology: M. Sebas-tian *et al.*, *ibid.* **56**, 1637 (2007). Clinical trial in treatment of malig-nant acites from epithelial cancers: M. M. Heiss *et al.*, *Int. J. Cancer* **127**, 2209 (2010). Review of development and clinical experience: D. Seimetz *et al.*, *Cancer Treat. Rev.* **36**, 458-467 (2010); C. Boke-meyer *Expert Opin. Biol. Ther.* **10**, 1259-1269 (2010).

THERAP CAT: Antineoplastic.

1910. 2C-B. [66142-81-2] 4-Bromo-2,5-dimethoxyben-zeneethanamine; 4-bromo-2,5-dimethoxyphenethylamine; α-des-methyl DOB; BDMPEA; MFT; nexus. $C_{10}H_{14}BrNO_2$; mol wt 260.13. C 46.17%, H 5.42%, Br 30.72%, N 5.38%, O 12.30%. Psy-choactive analog of mescaline, *q.v.* Synthesis and activity: A. T. Shulgin, M. F. Carter, *Psychopharmacol. Commun.* **1**, 93 (1975). Qualitative analysis: F. A. Ragan, Jr. *et al.*, *J. Anal. Toxicol.* **9**, 91 (1985). Pharmacology: R. A. Glennon *et al.*, *Pharmacol. Biochem. Behav.* **30**, 597 (1988); M. Lobos *et al.*, *Gen. Pharmacol.* **23**, 1139 (1992).

Hydrochloride. [56281-37-9] $C_{10}H_{14}BrNO_2.HCl$. Pale pink crystals from ethanol, mp 237-239° (dec). uv max (dil H_2SO_4): 294 nm.

Note: This is a controlled substance (hallucinogen): **21 CFR,** 1308.11.

1911. Ceanothic Acid. [21302-79-4] (3β,3′β,4α,4′α,5β,8α,-9β,10α,13α,14β,15β)-Tetrahydro-4′-hydroxy-4,5,9-tetrameth-yl-15-(1-methylethyl)-3′*H*-cyclopenta[3,4]-18-norandrost-3-ene-3′,13-dicarboxylic acid; (2α,3β)-3-hydroxy-A(1),28-dinorlup-20-(29)-ene-2,17-dicarboxylic acid; emmolic acid. $C_{30}H_{46}O_5$; mol wt 486.69. C 74.04%, H 9.53%, O 16.44%. From *Ceanothus ameri-canus* L., *Rhamnaceae* (New Jersey tea). Isoln: Julian *et al.*, *J. Am. Chem. Soc.* **60**, 77 (1938). Identity of ceanothic acid and emmolic acid: Mechoulam, *Chem. Ind. (London)* **1961**, 1835. Structure: de Mayo, Starratt, *Tetrahedron Lett.* **1961**, 259; Mechoulam, *J. Org. Chem.* **27**, 4070 (1962). Stereochemistry: de Mayo, Starratt, *Can. J. Chem.* **40**, 788 (1962). Revised stereochemistry: Eade *et al.*, *Aust. J. Chem.* **24**, 621 (1971).

Crystals from methanol + ether, mp 356-357°. $[\alpha]_D$ +38° (c = 1.20 in ethanol).

Dimethyl ester. $C_{32}H_{50}O_5$. Crystals from ether + petr ether, mp 221-223°. $[\alpha]_D$ +41° (chloroform).

1912. Cecropins. A major class of non-specific antibacterial proteins induced in hemolymph of some insects as part of the im-mune response. First characterized in *Lepidoptera*, particularly *Hy-alophora cecropia*. (*See also* attacins.) Small, basic polypeptides which possess similar amino acid sequences with relatively long hy-drophobic regions, mol wt approx 4000 Da. Originally designated *protein P9*, three major peptides have been characterized, cecropins A, B, D; three minor peptides C, E, F are also known. Identification of cell free immune response in *Lepidoptera*: H. Boman *et al.*, *In-fect. Immun.* **10**, 136 (1974). Initial identification and isolation of immune response proteins in *H. cecropia*: I. Faye *et al.*, *ibid.* **12**, 1426 (1975). Purification and preliminary characterization of cecro-pins A and B: D. Hultmark *et al.*, *Eur. J. Biochem.* **106**, 7 (1980); antibacterial spectrum and amino acid sequence: H. Steiner *et al.*, *Nature* **292**, 246 (1981). Partial synthesis of A: R. B. Merrifield *et al.*, *Biochemistry* **21**, 5020 (1982). Total synthesis of A: Andreu *et al.*, *Proc. Natl. Acad. Sci. USA* **80**, 6475 (1983); of B: P. van Hofsten *et al.*, *ibid.* **82**, 2240 (1985). Isolation, characterization of cecropins D, E, F, and comparison with A and B: D. Hultmark *et al.*, *Eur. J. Biochem.* **127**, 207 (1982); X. Qu *et al.*, *ibid.* 219. Isoln of cecropin-like proteins from flesh fly larvae: M. Okada, S. Natori,

Biochem. J. **211**, 727 (1983); from tsetse fly: G. P. Kaaya *et al.*, *Insect Biochem.* **17**, 309 (1987). Cecropin-like proteins designated as *lepidopterans* have been isolated from *Bombyx mori* silkworms: T. Teshima *et al.*, *Tetrahedron Lett.* **28**, 4705 (1987). Mechanism of action: H. Steiner *et al.*, *Biochim. Biophys. Acta* **939**, 260 (1988); B. Christensen *et al.*, *Proc. Natl. Acad. Sci. USA* **85**, 5072 (1988). Review of cecropins A and B: H. G. Boman, H. Steiner in *Current Topics in Microbiology and Immunology* **94/95**, W. Henle *et al.*, Eds. (Springer-Verlag, N.Y., 1981) pp 75-91.

1913. Cedrol. [77-53-2] (3*R*,3a*S*,6*R*,7*R*,8a*S*)-Octahydro-3,- 6,8,8-tetramethyl-1*H*-3a,7-methanoazulen-6-ol; 8*βH*-cedran-8-ol; cedar camphor; cypress camphor. $C_{15}H_{26}O$; mol wt 222.37. C 81.02%, H 11.79%, O 7.19%. Occurs in cedar wood oil from *Juniperus virginiana* L. (*J. sabina* Kook), *Cupressaceae*, in cypress oil from *Cupressus sempervirens*: *Schimmel's Report* **1904 II**, 20; in the oil from *Juniperus chinensis*: Kondo, *J. Pharm. Soc. Jpn.* **1907**, 236; and from *Origanum smyrnaeum* L., *Labiatae*: *Schimmel's Report* **1906**, Oct., 72. Structure: Stork, Breslow, *J. Am. Chem. Soc.* **75**, 3291, 3292 (1953). Stereochemistry and total synthesis: Stork, Clarke, *ibid.* **83**, 3114 (1961). Synthesis of *dl*-form: Corey *et al.*, *ibid.* **91**, 1557 (1969); *eidem*, *Tetrahedron Lett.* **1973**, 3153; E. G. Breitholle, A. G. Fallis, *J. Org. Chem.* **43**, 1964 (1978).

Needles from dil methanol, mp 86-87°. $[α]_D^{28}$ +9.9° (c = 5 in chloroform).

USE: In fragrances.

1914. Cefaclor. [53994-73-3] (6*R*,7*R*)-7-[[(2*R*)-2-Amino-2-phenylacetyl]amino]-3-chloro-8-oxo-5-thia-1-azabicyclo[4.2.0]oct-2-ene-2-carboxylic acid; 7-(D-2-amino-2-phenylacetamido)-3-chloro-3-cephem-4-carboxylic acid; 3-chloro-7-D-(2-phenylglycinamido)-3-cephem-4-carboxylic acid. $C_{15}H_{14}ClN_3O_4S$; mol wt 367.80. C 48.98%, H 3.84%, Cl 9.64%, N 11.42%, O 17.40%, S 8.72%. Semi-synthetic cephalosporin antibiotic, related to cephalexin, *q.v.* Prepn: R. R. Chauvette, **DE 2408698**; *idem*, **US 3925372** (1974, 1975 both to Lilly); R. R. Chauvette, P. A. Pennington, *J. Med. Chem.* **18**, 403 (1975). Comparative antibacterial spectrum: M. S. Silver *et al.*, *Antimicrob. Agents Chemother.* **12**, 591 (1977). Clinical pharmacokinetics: G. R. Hodges *et al.*, *ibid.* **14**, 454 (1978); A. Glynne *et al.*, *J. Antimicrob. Chemother.* **4**, 343 (1978). Clinical trial in otitis media: J. D. Nelson *et al.*, *Am. J. Dis. Child.* **132**, 992 (1978). Comprehensive description: L. J. Lorenz, *Anal. Profiles Drug Subs.* **9**, 107-123 (1980). Review of pharmacology and clinical experience: B. R. Meyers, *Clin. Ther.* **22**, 154-166 (2000).

Monohydrate. [70356-03-5] Compd 99638; Alfacet; Alfatil; Ceclor; Distaclor; Haxifal; Keflor; Panacef; Panoral; Raniclor. $C_{15}H_{14}ClN_3O_4S.H_2O$; mol wt 385.82. Crystalline solid. uv max (pH 7 buffer): 265 nm (ε 6800). pKa: 2.43, 7.16. Soly in water (pH5): 8.59 g/L. Practically insol in methanol, chloroform, benzene. Solns are stable at pH 2.5-4.5.

THERAP CAT: Antibacterial.

THERAP CAT (VET): Antibacterial.

1915. Cefadroxil. [50370-12-2] (6*R*,7*R*)-7-[[(2*R*)-2-Amino-2-(4-hydroxyphenyl)acetyl]amino]-3-methyl-8-oxo-5-thia-1-azabicyclo[4.2.0]oct-2-ene-2-carboxylic acid; 7-[D-(−)-α-amino-α-(4-hydroxyphenyl)acetamido]-3-methyl-3-cephem-4-carboxylic acid;

p-hydroxycephalexine. $C_{16}H_{17}N_3O_5S$; mol wt 363.39. C 52.88%, H 4.72%, N 11.56%, O 22.01%, S 8.82%. Semi-synthetic cephalosporin antibiotic. Prepn: **NL 6812382**; L. B. Crast, Jr., **US 3489752** (1969, 1970 both to Bristol-Myers); T. Takahashi *et al.*, **DE 2216113**; *eidem*, **US 3816253** (1972, 1974 both to Takeda). Prepn of crystalline monohydrate: D. Bouzard *et al.*, **US 4504657** (1985 to Bristol-Myers). Antimicrobial activity: R. E. Buck, K. E. Price, *Antimicrob. Agents Chemother.* **11**, 324 (1977). Pharmacology: M. Pfeffer *et al.*, *ibid.* 331; A. I. Hartstein *et al.*, *ibid.* **12**, 93 (1977). HPLC determn in biological fluids and pharmaceutical prepns: V. F. Samanidou *et al.*, *J. Chromatogr. B* **788**, 147 (2003). *Review: J. Antimicrob. Chemother.* **10**, Suppl. B, 1-162 (1982). Series of articles on clinical trials in respiratory tract infections: *Drugs* **32**, Suppl. 3, 1-56 (1986).

Monohydrate. [66592-87-8] BL-S578; MJF-11567-3; Baxan; Cefadril; Cefa-Drops; Cefamox; Ceforal; Duracef; Duricef; Oracéfal; Ultracef. $C_{16}H_{17}N_3O_5S.H_2O$; mol wt 381.40. White crystals, mp 197° (dec). Slightly sol in water. Practically insol in alc, chloroform, ether.

THERAP CAT: Antibacterial.

THERAP CAT (VET): Antibacterial.

1916. Cefamandole. [34444-01-4] (6*R*,7*R*)-7-[[(2*R*)-2-Hydroxy-2-phenylacetyl]amino]-3-[[(1-methyl-1*H*-tetrazol-5-yl)thio]-methyl]-8-oxo-5-thia-1-azabicyclo[4.2.0]oct-2-ene-2-carboxylic acid; 7-D-mandelamido-3-[[(1-methyl-1*H*-tetrazol-5-yl)thio]methyl]-3-cephem-4-carboxylic acid; 7-D-mandelamido-3-(1-methyl-1,-2,3,4-tetrazole-5-thiomethyl)-Δ³-cephem-4-carboxylic acid; CMT; compd 83405. $C_{18}H_{18}N_6O_5S_2$; mol wt 462.50. C 46.75%, H 3.92%, N 18.17%, O 17.30%, S 13.86%. Broad-spectrum semisynthetic cephalosporin antibiotic. Prepn: C. W. Ryan, **DE 2018600**; *idem*, **US 3641021** (1970, 1972 both to Lilly). Prepn of nafate: J. M. Greene, J. M. Indelicato, **US 3928592** (1975 to Lilly). Comparative antibacterial spectrum: W. E. Wick, D. A. Preston, *Antimicrob. Agents Chemother.* **1**, 221 (1972); H. C. Neu, *ibid.* **6**, 177 (1974). Pharmacokinetics: R. S. Griffith *et al.*, *ibid.* **10**, 814 (1976). HPLC determn in plasma and urine: H. S. Lee, O. P. Zee, *J. Chromatogr.* **528**, 425 (1990). Comprehensive description: R. H. Bishara, E. C. Rickard, *Anal. Profiles Drug Subs.* **9**, 125-154 (1980). Clinical trial for prophylaxis of post-surgical infection: T. R. Townsend *et al.*, *J. Thorac. Cardiovasc. Surg.* **106**, 664 (1993).

Nafate. [42540-40-9] *O*-Formylcefamandole sodium; Cefam; Cemado; Kefadol; Mancef; Mandokef; Mandol. $C_{19}H_{17}N_6NaO_6S_2$; mol wt 512.49. White, odorless needles, mp 190° (dec). uv max (H_2O): 269 nm (ε 10800). pKa 2.6-3.0. Sol in water, methanol. Practically insol in ether, chloroform, benzene, cyclohexane.

THERAP CAT: Antibacterial.

1917. Cefatrizine. [51627-14-6] (6*R*,7*R*)-7-[[(2*R*)-2-Amino-2-(4-hydroxyphenyl)acetyl]amino]-8-oxo-3-[(1*H*-1,2,3-triazol-5-yl-thio)methyl]-5-thia-1-azabicyclo[4.2.0]oct-2-ene-2-carboxylic acid; 7-[D-α-amino-α-(4-hydroxyphenyl)acetamido]-3-(1,2,3-triazol-4(5)-ylthiomethyl)-3-cephem-4-carboxylic acid; BL-S640; SKF-60771; S-640P. $C_{18}H_{18}N_6O_5S_2$; mol wt 462.50. C 46.75%, H 3.92%, N 18.17%, O 17.30%, S 13.86%. Orally active semi-synthetic cephalosporin antibiotic. Prepn: G. L. Dunn, J. R. E. Hoover, **DE 2316866**; *eidem*, **US 3867380** (1974, 1975 both to SK&F); G. L. Dunn *et al.*, *J. Antibiot.* **29**, 65 (1976). Prepn of the propylene glycolate: M. A. Kaplan, A. P. Granatek, **DE 2500385**; *eidem*, **US 3970651** (1975, 1976 both to Bristol-Myers). *In vitro* and *in vivo* antibacterial activity and pharmacokinetic behavior: P. Actor *et al.*, *J. Antibiot.* **28**, 594 (1975); F. Leitner *et al.*, *Antimicrob. Agents Chemother.* **7**, 298 (1975). Determn in serum and urine by HPLC: E. Crombez *et al.*, *J. Chromatogr.* **173**, 165 (1979). LC determn in pharmaceutical formulations: L. Manna, L. Valvo, *Chromatographia* **60**, 645 (2004). Acute toxicity: M. Matsuzaki *et al.*, *Jpn. J. Antibiot.* **29**, 612 (1976), *C.A.* **85**, 171724y (1976). Series of articles on pharmacokinetics and clinical efficacy: *Drugs Exp. Clin. Res.* **11**, 441-462 (1985).

Zwitterionic. LD_{50} in male, female mice, male, female rats (mg/kg): 6880, 6410, 4325, 4325 i.p. (Matsuzaki).

Compd with propylene glycol. [64217-62-5] Bricef; Cefaperos; Faretrizin; Trizina. $C_{18}H_{18}N_6O_5S_2 \cdot C_3H_8O_2$; mol wt 538.59. Rod-like crystals. $[\alpha]_D^{23}$ +55.9° (c = 1% in 1N HCl). uv max: 227, 272 nm.

THERAP CAT: Antibacterial.

1918. Cefazolin. [25953-19-9] (6R,7R)-3-[[(5-Methyl-1,3,4-thiadiazol-2-yl)thio]methyl]-8-oxo-7-[[2-(1H-tetrazol-1-yl)acetyl]amino]-5-thia-1-azabicyclo[4.2.0]oct-2-ene-2-carboxylic acid; CEZ. $C_{14}H_{14}N_8O_4S_3$; mol wt 454.50. C 37.00%, H 3.10%, N 24.65%, O 14.08%, S 21.16%. Semi-synthetic antibiotic derived from 7-aminocephalosporanic acid, *q.v.* Prepn: T. Takano *et al.*, **ZA 6804513**; *eidem*, **US 3516997** (1969, 1970 to Fujisawa). Synthesis and properties: Kariyone *et al.*, *J. Antibiot.* **23**, 131 (1970). Activity and clinical studies: Nishida *et al.*, *ibid.* **137**, 184; Shibata, Fujii, *Antimicrob. Agents Chemother.* **1970**, 467. HPLC determn in plasma and tissue: S. M. Bayoumi *et al.*, *Int. J. Pharm.* **30**, 57 (1986); in ophthalmic ointment: C. F. Martin *et al.*, *J. Chromatogr.* **402**, 376 (1987). Antibacterial activity of crystalline modifications: G. Opalchenova, G. N. Kalinkova, *Int. J. Pharm.* **189**, 235 (1999). Metabolic studies: Kozatani *et al.*, *Chem. Pharm. Bull.* **20**, 1105 (1972). Toxicology: H. A. Birkhead *et al.*, *J. Infect. Dis.* **128**, Suppl., 379 (1973). Comprehensive description: A. E. Zappala *et al.*, *Anal. Profiles Drug Subs.* **4**, 1-20 (1975). Clinical trial for prophylaxis of postoperative wound infection: R. Saginur *et al.*, *J. Thorac. Cardiovasc. Surg.* **120**, 1120 (2000).

Needles from aq acetone, mp 198-200° (dec). uv max (buffer pH 6.4): 272 nm (ε 13150). Sol in DMF, pyridine; sparingly sol in acetone; slightly sol in alc, methanol, water; very slightly sol in ethyl acetate, isopropyl alc, methyl isobutyl ketone. Practically insol in chloroform, benzene, ether, methylene chloride.

Sodium salt. [27164-46-1] Sodium CEZ; SKF-41558; Acef; Ancef; Cefacidal; Cefamedin; Cefamezin; Cefazil; Gramaxin; Kefzol; Totacef; Zolicef. $C_{14}H_{13}N_8NaO_4S_3$; mol wt 476.48. White to yellowish-white, odorless crystalline powder with a bitter, salty taste. Crystallizes in α, β, and γ-forms (Kariyone). Freely sol in water, saline TS, dextrose solutions; slightly sol in methanol; very slightly sol in alc. Practically insol in benzene, acetone, chloroform, ether. LD_{50} in mice, rats (mg/kg): 3.9, 3.18 i.v.; 6.2, 7.4 i.p. (Birkhead).

THERAP CAT: Antibacterial.

THERAP CAT (VET): Antibacterial.

1919. Cefbuperazone. [76610-84-9] (6R,7S)-7-[[(2R,3S)-2-[[(4-Ethyl-2,3-dioxo-1-piperazinyl)carbonyl]amino]-3-hydroxy-1-oxobutyl]amino]-7-methoxy-3-[[(1-methyl-1H-tetrazol-5-yl)thio]methyl]-8-oxo-5-thia-1-azabicyclo[4.2.0]oct-2-ene-2-carboxylic acid; (6R,7S)-7-[(2R,3S)-2-(4-ethyl-2,3-dioxo-1-piperazinecarboxamido)-3-hydroxybutyramido]-7-methoxy-3-[[(1-methyl-1H-tetrazol-5-yl)thio]methyl]-8-oxo-5-thia-1-azabicyclo[4.2.0]oct-2-ene-2-carboxylic acid; 7β-[D-α-(4-ethyl-2,3-dioxo-1-piperazinecarboxamido)-β-(S)-hydroxybutanamido]-7α-methoxy-3-[5-(1-methyl-1,-2,3,4-tetrazolyl)thiomethyl]-Δ³-cephem-4-carboxylic acid. $C_{22}H_{29}N_9O_9S_2$; mol wt 627.65. C 42.10%, H 4.66%, N 20.08%, O 22.94%, S 10.22%. Broad spectrum injectable cephamycin. Prepn: I. Saikawa *et al.*, **BE 879217**; *eidem*, **US 4263292** (1980, 1981 both to Toyama). *In vitro* and *in vivo* antibacterial activity: M. Tai *et al.*, *Antimicrob. Agents Chemother.* **22**, 728 (1982). β-lactamase stability and activity vs anaerobic bacteria: V. E. Del Bene *et al.*, *ibid.* **27**, 817 (1985). Pharmacokinetics and efficacy in exptl biliary infection: J. Kameyama *et al.*, *Curr. Med. Res. Opin.* **11**, 576 (1989). Clinical evaluation for prophylaxis of post-surgical infection: L. Kager *et al.*, *Drugs Exp. Clin. Res.* **12**, 983 (1986).

Solid, mp 118-120° (dec).

Sodium salt. [76648-01-6] T-1982; Keiperazon; Tomiporan. $C_{22}H_{28}N_9NaO_9S_2$; mol wt 649.63.

THERAP CAT: Antibacterial.

1920. Cefcapene. [135889-00-8] (6R,7R)-3-[[(Aminocarbonyl)oxy]methyl]-7-[[(2Z)-2-(2-amino-4-thiazolyl)-1-oxo-2-penten-1-yl]amino]-8-oxo-5-thia-1-azabicyclo[4.2.0]oct-2-ene-2-carboxylic acid; (6R,7R)-7-[(Z)-2-(2-aminothiazol-4-yl)-2-pentenoylamino]-3-(carbamoyloxymethyl)-3-cephem-4-carboxylic acid; S-1006. $C_{17}H_{19}N_5O_6S_2$; mol wt 453.49. C 45.03%, H 4.22%, N 15.44%, O 21.17%, S 14.14%. Cephalosporin antibiotic. Prepn: M. Boberg, K. G. Metzger, **EP 49448**; *eidem*, **US 4416880** (1982, 1983 both to Bayer); of orally active esters: Y. Hamashima *et al.*, **BE 904517**; *idem et al.*, **US 4731361** (1986, 1988 both to Shionogi); K. Ishikura *et al.*, *J. Antibiot.* **47**, 466 (1994). Comparative antibacterial spectrum and β-lactamase stability: H. C. Neu *et al.*, *Antimicrob. Agents Chemother.* **36**, 1336 (1992). Metabolism: K. Totsuka *et al.*, *ibid.* 757. Safety and pharmacokinetics: M. Nakashima *et al.*, *ibid.* 762. Clinical trial in chronic respiratory infection: A. Saito *et al.*, *J. Int. Med. Res.* **32**, 590 (2004).

Pivaloyloxymethyl ester. [105889-45-0] Cefcapene pivoxil. $C_{23}H_{29}N_5O_8S_2$; mol wt 567.63.

Pivoxil hydrochloride hydrate. [147816-24-8]; [147816-23-7] (anhydrous). S-1108; Flomox. $C_{23}H_{29}N_5O_8S_2 \cdot HCl \cdot H_2O$; mol wt 622.11. White to pale yellowish white, crystalline powder; slight characteristic odor. Freely sol in DMF, methanol; sol in ethanol, slightly sol in water. Practically insol in diethyl ether.

THERAP CAT: Antibacterial.

1921. Cefdinir. [91832-40-5] (6R,7R)-7-[[(2Z)-2-(2-Amino-4-thiazolyl)-2-(hydroxyimino)acetyl]amino]-3-ethenyl-8-oxo-5-thia-1-azabicyclo[4.2.0]oct-2-ene-2-carboxylic acid; syn-7-[2-(2-amino-4-thiazolyl)-2-hydroxyiminoacetamido]-3-vinyl-3-cephem-4-carboxylic acid; FK-482; BMY-28488; Omnicef. $C_{14}H_{13}N_5O_5S_2$; mol wt 395.41. C 42.53%, H 3.31%, N 17.71%, O 20.23%, S 16.22%. Cephalosporin antibiotic structurally similar to cefixime, *q.v.* Prepn: T. Takaya *et al.*, **BE 897864**; *eidem*, **US 4559334** (1985 both to Fujisawa); Y. Inamoto *et al.*, *J. Antibiot.* **41**, 828 (1988); H. Kamachi *et al.*, *ibid.* 1602. Improved prepn: M. González *et al.*, *Farmaco* **58**, 409 (2003). LC determn in pharmaceutic formulations: T. N. Mehta *et al.*, *J. AOAC Int.* **88**, 1661 (2005); and in urine: G. M. Hadad *et al.*, *Chromatographia* **70**, 1593 (2009). Review of pharmacokinetics and *in vitro* antimicrobial activity: D. R. P. Guay, *Pediatr. Infect. Dis. J.* **19**, S141-S146 (2000); of antibacterial activity, pharmacology and clinical efficacy: C. M. Perry, L. J. Scott, *Drugs* **64**, 1433-1464 (2004); H. S. Sader, R. N. Jones, *Expert Rev. Anti Infect. Ther.* **5**, 29-33 (2007).

White to slightly brownish-yellow solid, mp 170° (dec). uv max (pH 7 phosphate buffer): 223, 286 nm (ε 17400, 19700). pKa 9.70. Soluble in 0.1 M pH 7.0 phosphate buffer solution; slightly sol in dilute HCl. Practically insol in water, alc, diethyl ether.

THERAP CAT: Antibacterial.

1922. Cefditoren. [104145-95-1] (6*R*,7*R*)-7-[[(2*Z*)-2-(2-Amino-4-thiazolyl)-2-(methoxyimino)acetyl]amino]-3-[(1*Z*)-2-(4-methyl-5-thiazolyl)ethenyl]-8-oxo-5-thia-1-azabicyclo[4.2.0]oct-2-ene-2-carboxylic acid; (+)-(6*R*,7*R*)-7-[2-(2-amino-4-thiazolyl)gly-oxylamido]-3-[(*Z*)-2-(4-methyl-5-thiazolyl)vinyl]-8-oxo-5-thia-1-azabicyclo[4.2.0]oct-2-ene-2-carboxylic acid 7²-(*Z*)-(*O*-methyloxime); ME-1206. $C_{19}H_{18}N_6O_5S_3$; mol wt 506.57. C 45.05%, H 3.58%, N 16.59%, O 15.79%, S 18.99%. Broad spectrum, third generation cephalosporin; active metabolite of orally absorbed pivaloyloxymethyl ester prodrug. Prepn: K. Atsumi *et al.*, **EP 175610**; *eidem*, **US 4839350** (1986, 1989 both to Meiji); K. Sakagami *et al.*, *J. Antibiot.* **43**, 1047 (1990); *idem et al.*, *Chem. Pharm. Bull.* **39**, 2433 (1991). Comparative antibacterial spectrum: D. Felmingham *et al.*, *Drugs Exp. Clin. Res.* **20**, 127 (1994). HPLC determn in plasma: R. V. S. Nirogi *et al.*, *Arzneim.-Forsch.* **56**, 309 (2006). Review of pharmacokinetics and therapeutic efficacy: E. A. Balbisi, *Pharmacotherapy* **22**, 1278-1293 (2002).

Sodium salt sesquihydrate. $C_{19}H_{17}N_6NaO_5S_3 \cdot 1\frac{1}{2}H_2O$; mol wt 555.57. Pale yellow crystals from water, mp 195-200° (dec). $[\alpha]_D^{20}$ +121.6° (c = 0.5 in methanol).

Pivaloyloxymethyl ester. [117467-28-4] Cefditoren pivoxil; ME-1207; Giasion; Meiact; Spectracef. $C_{25}H_{28}N_6O_7S_3$; mol wt 620.71. Pale yellow powder, mp 127-129°. $[\alpha]_D^{20}$ −48.5° (c = 0.5 in methanol).

THERAP CAT: Antibacterial.

1923. Cefepime. [88040-23-7] 1-[[(6*R*,7*R*)-7-[[(2*Z*)-2-(2-Amino-4-thiazolyl)-2-(methoxyimino)acetyl]amino]-2-carboxy-8-oxo-5-thia-1-azabicyclo[4.2.0]oct-2-en-3-yl]methyl]-1-methylpyrrolidinium inner salt; 1-[[(6*R*,7*R*)-7-[2-(2-amino-4-thiazolyl)glyoxylamido]-2-carboxy-8-oxo-5-thia-1-azabicyclo[4.2.0]oct-2-en-3-yl]methyl]-1-methylpyrrolidinium hydroxide inner salt 7²-(*Z*)-2-(*O*-methyloxime); 7-[(*Z*)-2-(2-aminothiazol-4-yl)-2-methoxyiminoacetamido]-3-(1-methylpyrrolidinio)methyl-3-cephem-4-carboxylate; BMY-28142. $C_{19}H_{24}N_6O_5S_2$; mol wt 480.56. C 47.49%, H 5.03%, N 17.49%, O 16.65%, S 13.34%. Semisynthetic, fourth generation cephalosporin antibiotic. Prepn: S. Aburaki *et al.*, **DE 3307550**; *eidem*, **US 4406899** (both 1983 to Bristol-Myers); and antibacterial activity: T. Naito *et al.*, *J. Antibiot.* **39**, 1092 (1986). *In vitro* comparative antimicrobial spectrum: N. J. Khan *et al.*, *Antimicrob.*

Agents Chemother. **26**, 585 (1984); and β-lactamase stability: H. C. Neu *et al.*, *J. Antimicrob. Chemother.* **17**, 441 (1986). HPLC determn in plasma and urine: R. H. Barbhaiya *et al.*, *Antimicrob. Agents Chemother.* **31**, 55 (1987). Review of clinical pharmacokinetics: M. P. Okamoto *et al.*, *Clin. Pharmacokinet.* **25**, 88-102 (1993); of clinical experience and safety assessment: D. Yahav *et al.*, *Lancet Infect. Dis.* **7**, 338-348 (2007).

Colorless powder, mp 150° (dec). uv max (pH 7 phosphate buffer): 235, 257 nm (ε 16700, 16100).

Sulfate. $C_{19}H_{24}N_6O_5S_2 \cdot H_2SO_4$. mp 210° (dec). uv max (pH 7 phosphate buffer): 236, 258 nm (ε 17200, 16900).

Hydrochloride monohydrate. [123171-59-5] 1-[[(6*R*,7*R*)-7-[[(2*Z*)-2-(2-Amino-4-thiazolyl)-2-(methoxyimino)acetyl]amino]-2-carboxy-8-oxo-5-thia-1-azabicyclo[4.2.0]oct-2-en-3-yl]methyl]-1-methylpyrrolidinium chloride hydrochloride hydrate (1:1:1:1); cefepime hydrochloride; Axepim; Cepimax; Cepimex; Maxipime. $C_{19}H_{25}ClN_6O_5S_2 \cdot HCl \cdot H_2O$; mol wt 571.49. Nonhygroscopic solid. Freely sol in water.

THERAP CAT: Antibacterial.

THERAP CAT (VET): Antibacterial.

1924. Cefetamet. [65052-63-3] (6*R*,7*R*)-7-[[(2*Z*)-2-(2-Amino-4-thiazolyl)-2-(methoxyimino)acetyl]amino]-3-methyl-8-oxo-5-thia-1-azabicyclo[4.2.0]oct-2-ene-2-carboxylic acid; (6*R*,7*R*)-7-[2-(2-aminothiazol-4-yl)-(*Z*)-2-(methoxyimino)acetamido]-3-methyl-3-cephem-4-carboxylic acid; (6*R*,7*R*)-7-[2-(2-amino-4-thiazolyl)-glyoxylamido]-3-methyl-8-oxo-5-thia-1-azabicyclo[4.2.0]oct-2-ene-2-carboxylic acid 7²-(*Z*)-(*O*-methyloxime); deacetoxycefotaxime; LY-097964; Ro-15-8074. $C_{14}H_{15}N_5O_5S_2$; mol wt 397.42. C 42.31%, H 3.80%, N 17.62%, O 20.13%, S 16.13%. Third generation cephalosporin antibiotic. Prepn: R. Heymes, A. Lutz, **DE 2713272**; *eidem*, **US 4396618** (1977, 1983 both to Roussel-UCLAF); R. Bucourt *et al.*, *Tetrahedron* **34**, 2233 (1978). Prepn of the pivaloyloxymethyl ester: M. Ochiai *et al.*, **DE 2715385**; *eidem*, **US 4680390** (1977, 1987 both to Takeda). Antibacterial spectrum: M. G. Thomas, S. D. R. Lang, *Antimicrob. Agents Chemother.* **29**, 945 (1986). Determn in biological fluids by HPLC: R. Wyss, F. Bucheli, *J. Chromatogr.* **430**, 81 (1988); by chemiluminescence: X. Shao *et al.*, *J. Anal. Chem.* **64**, 71 (2009). Pharmacokinetics: K. Stoeckel *et al.*, *Curr. Med. Res. Opin.* **11**, 432 (1989). Multicenter clinical trial: L. Bernstein-Hahn *et al.*, *ibid.* 442. Review of clinical experience: H. M. Bryson, R. N. Brogden, *Drugs* **45**, 589-621 (1993).

Cefetamet pivoxil. [65243-33-6] Cefetamet pivaloyloxymethyl ester; Ro-15-8075. $C_{20}H_{25}N_5O_7S_2$; mol wt 511.57. White powder. Lipophilic.

Cefetamet pivoxil hydrochloride. [111696-23-2] Cefyl; Globocef. $C_{20}H_{25}N_5O_7S_2 \cdot HCl$; mol wt 548.03. Prepn: A. Furlenmeier, **US 4716227** (1987 to Roche). Crystals from ethanol/hexane, mp 169-170° (dec). Log *P* (octanol/water): 2.8.

THERAP CAT: Antibacterial.

THERAP CAT (VET): Antibacterial.

1925. Cefixime. [79350-37-1] (6*R*,7*R*)-7-[[(2*Z*)-2-(2-Amino-4-thiazolyl)-2-[(carboxymethoxy)imino]acetyl]amino]-3-ethenyl-8-oxo-5-thia-1-azabicyclo[4.2.0]oct-2-ene-2-carboxylic acid; 7β-[(*Z*)-

2-(2-amino-4-thiazolyl)-2-(carboxymethoxyimino)acetamido]-3-vinyl-3-cephem-4-carboxylic acid; FK-027; FR-17027; CL-284635. $C_{16}H_{15}N_5O_7S_2$; mol wt 453.44. C 42.38%, H 3.33%, N 15.45%, O 24.70%, S 14.14%. Orally active, third generation cephalosporin antibiotic. Prepn: T. Takaya *et al.*, **EP 30630**; *eidem*, **US 4409214** (1981, 1983 both to Fujisawa); H. Yamanaka *et al.*, *J. Antibiot.* **38**, 1738 (1985). Synthesis and activity of (*E*)-isomer: K. Kawabata *et al.*, *Chem. Pharm. Bull.* **34**, 3458 (1986). Mechanism of action: Y. Shigi *et al.*, *J. Antibiot.* **37**, 790 (1984). Comparative antibacterial spectrum *in vitro* and *in vivo*: T. Kamimura *et al.*, *Antimicrob. Agents Chemother.* **25**, 98 (1984). *In vitro* activity and β-lactamase stability: H. C. Neu *et al.*, *ibid.* **26**, 174 (1984). HPLC determn in human plasma and urine: Y. Tokuma *et al.*, *J. Chromatogr.* **311**, 339 (1984). Clinical pharmacokinetics: D. R. P. Guay *et al.*, *Antimicrob. Agents Chemother.* **30**, 485 (1986). Review of clinical experience in respiratory and urinary tract infections: D. H. Wu, *Clin. Ther.* **15**, 1108-1119 (1993).

Trihydrate. [125110-14-7] Cefixoral; Cefspan; Cephoral; Denvar; Oroken; Suprax; Unixime. $C_{16}H_{15}N_5O_7S_2 \cdot 3H_2O$; mol wt 507.49. White to light yellow crystalline powder. Sol in methanol, propylene glycol; slightly sol in ethanol, acetone, glycerol; very slightly sol in 70% sorbitol and octanol. Practically insol in water, ether, ethyl acetate, hexane.
Disodium salt. [79350-82-6] $C_{16}H_{13}N_5Na_2O_7S_2$. mp >250°.
(*E*)-**Form trihydrate.** Pale yellow solid, mp 218-225° (dec).
THERAP CAT: Antibacterial.
THERAP CAT (VET): Antibacterial.

1926. Cefmenoxime. [65085-01-0] (6*R*,7*R*)-7-[[(2*Z*)-2-(2-Amino-4-thiazolyl)-2-(methoxyimino)acetyl]amino]-3-[[(1-methyl-1*H*-tetrazol-5-yl)thio]methyl]-8-oxo-5-thia-1-azabicyclo[4.2.0]oct-2-ene-2-carboxylic acid; (6*R*,7*R*)-7-[2-(2-amino-4-thiazolyl)glyoxylamido]-3-[[(1-methyl-1*H*-tetrazol-5-yl)thio]methyl]-8-oxo-5-thia-1-azabicyclo[4.2.0]oct-2-ene-2-carboxylic acid 7²-(*Z*)-(*O*-methyloxime); SCE-1365. $C_{16}H_{17}N_9O_5S_3$; mol wt 511.55. C 37.57%, H 3.35%, N 24.64%, O 15.64%, S 18.80%. Third generation cephalosporin antibiotic; related structurally to cefotaxime and ceftizoxime, *q.q.v.* Prepn (unspecified stereochemistry): M. Ochiai *et al.*, **DE 2556736**; *eidem*, **US 4098888** (1976, 1978 both to Takeda); of the *syn*-isomer: R. Heymes, A. Lutz, **DE 2713272**; *eidem*, **US 4476122** (1983, 1984 both to Roussel-Uclaf). Comparative antibacterial spectrum: K. Tsuchiya *et al.*, *Antimicrob. Agents Chemother.* **19**, 56 (1981). β-Lactamase stability: K. Okonogi *et al.*, *ibid.* **20**, 171 (1981). Clinical pharmacokinetics study: D. Höffler, P. Koeppe, *Arzneim.-Forsch.* **33**, 269 (1983). HPLC determn in plasma and urine: D. P. Reitberg, J. J. Schentag, *Clin. Chem.* **29**, 1415 (1983). Review of antibacterial activity and therapeutic use: D. M. Campoli-Richards, P. A. Todd, *Drugs* **34**, 188-221 (1987).

Hydrochloride. [75738-58-8] Abbott 50192; Bestcall; Tacef. $(C_{16}H_{17}N_9O_5S_3)_2 \cdot HCl$; mol wt 1059.56. White to light orange-yellow crystals or crystalline powder. Freely sol in formamide; slightly sol in methanol; very slightly sol in water. Practically insol in dehydrated alc, ether.
Sodium salt. [65085-02-1] $C_{16}H_{16}N_9NaO_5S_3$. $[\alpha]_D^{20}$ −13.5 ±1° (c = 1 in water).
THERAP CAT: Antibacterial.

1927. Cefmetazole. [56796-20-4] (6*R*,7*S*)-7-[[2-[(Cyanomethyl)thio]acetyl]amino]-7-methoxy-3-[[(1-methyl-1*H*-tetrazol-5-yl)thio]methyl]-8-oxo-5-thia-1-azabicyclo[4.2.0]oct-2-ene-2-carboxylic acid; CS-1170; SKF-83088. $C_{15}H_{17}N_7O_5S_3$; mol wt 471.53. C 38.21%, H 3.63%, N 20.79%, O 16.96%, S 20.40%. Semi-synthetic antibiotic derived from cephamycin C, *q.v.* Prepn: H. Nakao *et al.*, **DE 2455884** (1975 to Sankyo); *eidem*, *J. Antibiot.* **29**, 554 (1976). *See also:* J. E. Dolfini, **US 3920639** (1975 to Squibb). Toxicological studies: H. Masuda *et al.*, *Sankyo Kenkyusho Nempo* **30**, 112 (1978), *C.A.* **90**, 180268h (1979). Comparative antibacterial spectrum and β-lactamase stability: R. N. Jones *et al.*, *J. Clin. Microbiol.* **24**, 1055 (1986). HPLC determn in serum and urine: J. C. Garcia-Glez *et al.*, *J. Chromatogr. A* **812**, 197 (1998). LC/MS/MS determn in manufacturing environments: N. Fukutsu *et al.*, *J. Pharm. Biomed. Anal.* **41**, 1243 (2006). Review of antibacterial activity, pharmacokinetics and clinical experience: J. J. Schentag, *Pharmacotherapy* **11**, 2-19 (1991).

Sodium salt. [56796-39-5] Cefmetazon; Metafar. $C_{15}H_{16}N_7NaO_5S_3$; mol wt 493.51. White powder. uv max (water): 272 nm (ε 10900). $[\alpha]_D^{20}$ +73 to +85° (c = 1 in water). Very sol in water, methanol; sol in acetone; slightly sol in ethanol; very slightly sol in THF. Practically insol in chloroform. LD_{50} i.v. in rats: >5000 mg/kg (Masuda).
THERAP CAT: Antibacterial.

1928. Cefminox. [84305-41-9] (6*R*,7*S*)-7-[[2-[[(2*S*)-2-Amino-2-carboxyethyl]thio]acetyl]amino]-7-methoxy-3-[[(1-methyl-1*H*-tetrazol-5-yl)thio]methyl]-8-oxo-5-thia-1-azabicyclo[4.2.0]oct-2-ene-2-carboxylic acid; 7β-(2-D-amino-2-carboxyethylthioacetamido)-7α-methoxy-3-[[(1-methyl-1*H*-tetrazol-5-yl)thio]methyl]-3-cephem-4-carboxylic acid. $C_{16}H_{21}N_7O_7S_3$; mol wt 519.57. C 36.99%, H 4.07%, N 18.87%, O 21.55%, S 18.51%. Semisynthetic broad spectrum cephamycin antibiotic. Prepn: **BE 880656**, K. Iwamatsu *et al.*, **US 4357331** (1980, 1982 both to Meiji Seika). Synthesis, biological activity: *eidem*, *J. Antibiot.* **36**, 229 (1983). Structure-activity studies: S. Inouye *et al.*, *ibid.* **37**, 1403 (1984). Comparative antibacterial spectrum: *idem et al.*, *Antimicrob. Agents Chemother.* **26**, 722 (1984). Toxicity studies: M. Kurebe *et al.*, *Jpn. J. Antibiot.* **37**, 847 (1984). Series of articles on pharmacokinetics and clinical trials: *Chemotherapy (Tokyo)* **32**, Suppl. 5 (1984) 561 pp. Pharmacology: S. Watanabe, S. Omoto, *Drugs Exptl. Clin. Res.* **16**, 461 (1990). Clinical trial in intra-abdominal infections: A. J. Torres *et al.*, *Infection* **28**, 318 (2000).

Sodium salt heptahydrate. [88641-36-5]; [75498-96-3] (anhydrous). MT-141; Meicelin. $C_{16}H_{20}N_7NaO_7S_3 \cdot 7H_2O$; mol wt 667.65. White to light-yellow crystalline powder. Crystals from ethanol-water, mp 90-91°. Freely sol in water; sparingly sol in methanol. Practically insol in ethanol. LD_{50} in male, female mice (mg/kg): 6100, 5200 i.v.; in male, female rats (mg/kg): 6600, 5700 i.v.; 8600, 8550 i.p.; >15000, >15000 s.c. or orally (Kurebe).
THERAP CAT: Antibacterial.

1929. Cefodizime. [69739-16-8] (6*R*,7*R*)-7-[[(2*Z*)-2-(2-Amino-4-thiazolyl)-2-(methoxyimino)acetyl]amino]-3-[[[5-(carboxymethyl)-4-methyl-2-thiazolyl]thio]methyl]-8-oxo-5-thia-1-azabicyclo[4.2.0]oct-2-ene-2-carboxylic acid; (6*R*,7*R*)-7-[2-(2-amino-4-thiazolyl)glyoxylamido]-3-[[[5-(carboxymethyl)-4-methyl-2-thiazol-

yl]thio]methyl]-8-oxo-5-thia-1-azabicyclo[4.2.0]oct-2-ene-2-car-
boxylic acid $7^2(Z)$-$(O$-methyloxime). $C_{20}H_{20}N_6O_7S_4$; mol wt
584.66. C 41.09%, H 3.45%, N 14.37%, O 19.16%, S 21.93%.
Third generation cephalosporin antibiotic with immunomodulating
activity; derivative of cefotaxime, q.v. Prepn: **BE 865632**; W.
Dürckheimer et al., **US 4278793** (1978, 1981 both to Hoechst AG);
and antibacterial activity: J. Blumbach et al., J. Antibiot. **40**, 29
(1987). In vitro comparative antibacterial spectrum: R. N. Jones et
al., Antimicrob. Agents Chemother. **20**, 760 (1981); and β-lactamase
stability: M. Limbert et al., J. Antibiot. **37**, 892 (1984). Determn in
biological fluids by HPLC: T. Marunaka et al., J. Chromatogr. **420**,
329 (1987); by capillary zone electrophoresis: Y. Mrestani et al.,
Anal. Chim. Acta **349**, 207 (1997). Clinical study in urogenital gonorrhea: A. H. van der Willigen et al., Antimicrob. Agents Chemother. **32**, 426 (1988). Physical properties and preclinical studies: A.
Bryskier et al., J. Antimicrob. Chemother. **26**, Suppl. C, 1 (1990).
Series of articles on pharmacokinetics and clinical studies: ibid. 9-
134. Review of clinical experience: L. B. Barradell, R. N. Brogden,
Drugs **44**, 800-834 (1992).

$[\alpha]_D^{25}$ −55.9°. pK$_1$ 2.85; pK$_2$ 3.37; pK$_3$ 4.18. uv max (water):
228, 260, 288 nm (log ε 4.25, 4.25, 4.20).
Disodium salt. [86329-79-5] HR-221; THR-221; Diezime; Kenicef; Modivid; Timecef. $C_{20}H_{18}N_6Na_2O_7S_4$; mol wt 628.62. White
to yellow crystalline powder. Soly in water: ~270 g/L. LD$_{50}$ in
mice, rabbits (mg/kg): 4000-8000 i.v. both species; in rats (mg/kg):
4000-8000 i.v., 15000-17500 s.c., 8000-11000 i.p. (Bryskier).
THERAP CAT: Antibacterial.

1930. Cefonicid. [61270-58-4] $(6R,7R)$-7-[[(2R)-2-Hydroxy-
2-phenylacetyl]amino]-8-oxo-3-[[[1-(sulfomethyl)-1H-tetrazol-5-
yl]thio]methyl]-5-thia-1-azabicyclo[4.2.0]oct-2-ene-2-carboxylic
acid; $(6R,7R)$-7-[(R)-mandelamido]-8-oxo-3-[[[1-(sulfomethyl)-1H-
tetrazol-5-yl]thio]methyl]-5-thia-1-azabicyclo[4.2.0]oct-2-ene-2-
carboxylic acid; 7-[(R)-mandelamido]-3-(1-sulfomethyl-1H-tetra-
zol-5-ylthiomethyl)-3-cephem-4-carboxylic acid. $C_{18}H_{18}N_6O_8S_3$;
mol wt 542.56. C 39.85%, H 3.34%, N 15.49%, O 23.59%, S
17.73%. Injectable semi-synthetic cephalosporin antibiotic related
to cefamandole, q.v. Prepn: D. A. Berges, **DE 2611270**; idem, **US
4048311** (1976, 1977 both to Smith Kline). Comparative antibacterial spectrum, pharmacokinetics: P. Actor et al., Antimicrob. Agents
Chemother. **13**, 784 (1978). Stability towards β-lactamases: R.
Mehta et al., J. Antibiot. **34**, 202 (1981). Kinetics and renal handling: D. Pitkin et al., Clin. Pharmacol. Ther. **30**, 587 (1981). Review of antibacterial activity, pharmacology, therapeutic use: E. Saltiel, R. N. Brogden, Drugs **32**, 222-259 (1986).

Disodium salt. [61270-78-8] SKF-75073; Cefodie; Chefir; Day-
cef; Emidoxin; Fonicid; Monocid; Praticef; Sofarcid; Valecid. C_{18}-
$H_{16}N_6Na_2O_8S_3$; mol wt 586.52. White to off-white solid. Freely
sol in water, 0.9% sodium chloride solution, 5% dextrose solution;
sol in methanol; very slightly sol in dehydrated alcohol.
THERAP CAT: Antibacterial.

1931. Cefoperazone. [62893-19-0] $(6R,7R)$-7-[[(2R)-2-[[(4-
Ethyl-2,3-dioxo-1-piperazinyl)carbonyl]amino]-2-(4-hydroxyphen-
yl)acetyl]amino]-3-[[(1-methyl-1H-tetrazol-5-yl)thio]methyl]-8-

oxo-5-thia-1-azabicyclo[4.2.0]oct-2-ene-2-carboxylic acid; 7-[D-
(−)-α-(4-ethyl-2,3-dioxo-1-piperazinecarboxamido)-α-(4-hydroxy-
phenyl)acetamido]-3-[[(1-methyl-1H-tetrazol-5-yl)thio]methyl]-3-
cephem-4-carboxylic acid. $C_{25}H_{27}N_9O_8S_2$; mol wt 645.67. C
46.51%, H 4.22%, N 19.52%, O 19.82%, S 9.93%. Broad spectrum
third generation cephalosporin antibiotic. Prepn: I. Saikawa et al.,
BE 837682; eidem, **US 4410522** (1976, 1983 both to Toyama); eidem, Yakugaku Zasshi **99**, 929 (1979). Stability in aq soln: eidem,
ibid. 1207. In vitro activity: M. V. Borobio et al., Antimicrob.
Agents Chemother. **17**, 129 (1980). Clinical pharmacokinetics: A.
F. Allaz, Schweiz. Med. Wochenschr. **109**, 1999 (1979). Antibacterial activity in combination with sulbactam: J. D. Williams, Clin.
Infect. Dis. **24**, 494 (1997). Determn in pharmaceutical formulations
by HPLC: T.-L. Tsou et al., J. Sep. Sci. **30**, 2407 (2007); in plasma
by capillary electrophoresis: P. Puig et al., J. Chromatogr. B **856**,
365 (2007). Clinical trial with sulbactam for intra-abdominal infection: A. Chandra et al., Surg. Infect. **9**, 367 (2008). Review of
pharmacology and therapeutic efficacy: R. N. Brogden et al., Drugs
22, 423-460 (1981).

Crystals from acetonitrile/water, mp 169-171° (hydrated). Stable
at pH 4.0-7.0; slightly unstable in acid; highly unstable in alkaline
soln.
Sodium salt. [62893-20-3] CP-52640-2; T-1551; Bioperazone;
Cefazone; Cefobid; Cefobis; Cefoneg; Dardum; Farecef. $C_{25}H_{26}$-
$N_9NaO_8S_2$; mol wt 667.65. Freely sol in water, methanol; slightly
sol in dehydrated alc. Insol in acetone, ethyl acetate, ether.
THERAP CAT: Antibacterial.
THERAP CAT (VET): Antibacterial.

1932. Ceforanide. [60925-61-3] $(6R,7R)$-7-[[2-[2-(Aminomethyl)phenyl]acetyl]amino]-3-[[[1-(carboxymethyl)-1H-tetrazol-
5-yl]thio]methyl]-8-oxo-5-thia-1-azabicyclo[4.2.0]oct-2-ene-2-car-
boxylic acid; $(6R,7R)$-7-[2-(α-amino-o-tolyl)acetamido]-3-[[[1-(carboxymethyl)-1H-tetrazol-5-yl]thio]methyl]-8-oxo-5-thia-1-azabicy-
clo[4.2.0]oct-2-ene-2-carboxylic acid; BL-S786. $C_{20}H_{21}N_7O_6S_2$;
mol wt 519.55. C 46.24%, H 4.07%, N 18.87%, O 18.48%, S
12.34%. Injectable, semi-synthetic cephalosporin antibiotic. Prepn:
M. A. Kaplan et al., **DE 2538804**; W. J. Gottstein et al., **US 4100346**,
(1976, 1978, both to Bristol-Myers); W. J. Gottstein et al., J. Antibiot. **29**, 1226 (1976). Comparative antibacterial spectrum and pharmacology: F. Leitner et al., Antimicrob. Agents Chemother. **10**, 426
(1976). Comparative tissue distribution: F. H. Lee et al., ibid. **19**,
625 (1981). Clinical trial in pneumonia: R. J. Wallace et al., ibid.
20, 648 (1981). Spectrophotometric determn in pharmaceutical formulations: P. B. Issopoulos, Analyst **114**, 237 (1989). Review of
antibacterial spectrum, pharmacokinetics and clinical experience: D.
M. Campoli-Richards et al., Drugs **34**, 411-437 (1987).

White solid, mp >150° (dec). Very sol in 1N sodium hydroxide.
Practically insol in water, methanol, chloroform, ether.
L-Lysine salt. [63767-79-3] Precef; Radacef. $C_{20}H_{21}N_7O_6S_2$.-
$C_6H_{14}N_2O_2$; mol wt 665.74.
THERAP CAT: Antibacterial.

1933. Cefoselis. [122841-10-5] $(6R,7R)$-7-[[(2Z)-2-(2-Ami-
no-4-thiazolyl)-2-(methoxyimino)acetyl]amino]-3-[[2,3-dihydro-2-

Cefotiam

1936

(2-hydroxyethyl)-3-imino-1H-pyrazol-1-yl]methyl]-8-oxo-5-thia-1-azabicyclo[4.2.0]oct-2-ene-2-carboxylic acid; (syn)-7β-[2-(2-aminothiazol-4-yl)-2-methoxyiminoacetamido]-3-[3-amino-2-(2-hydroxyethyl)-1-pyrazolio]methyl-3-cephem-4-carboxylate; (−)-5-amino-2-[[(6R,7R)-7-[2-(2-amino-4-thiazolyl)glyoxylamido]-2-carboxy-8-oxo-5-thia-1-azabicyclo[4.2.0]oct-2-en-3-yl]methyl]-1-(2-hydroxyethyl)pyrazolium hydroxide inner salt 7^2-(Z)-(O-methyloxime). $C_{19}H_{22}N_8O_6S_2$; mol wt 522.56. C 43.67%, H 4.24%, N 21.44%, O 18.37%, S 12.27%. Fourth generation cephalosporin antibiotic. Prepn: K. Sakane et al., **EP 307804**; eidem, **US 4952578** (1989, 1990 both to Fujisawa); H. Ohki et al., J. Antibiot. **46**, 359 (1993). In vitro activity and β-lactamase stability: H. C. Neu et al., Antimicrob. Agents Chemother. **37**, 566 (1993). Clinical pharmacokinetics: R. Wise et al., ibid. **38**, 2369 (1994). Series of articles on activity, clinical pharmacology and therapeutic efficacy: Chemotherapy (Tokyo) **42**, Suppl. 3, 1-464 (1994).

Sulfate. [122841-12-7] FK-037; Wincef. $C_{19}H_{22}N_8O_6S_2 \cdot H_2SO_4$; mol wt 620.63.

THERAP CAT: Antibacterial.

1934. Cefotaxime. [63527-52-6] (6R,7R)-3-[(Acetyloxy)methyl]-7-[[(2Z)-2-(2-amino-4-thiazolyl)-2-(methoxyimino)acetyl]amino]-8-oxo-5-thia-1-azabicyclo[4.2.0]oct-2-ene-2-carboxylic acid; (6R,7R)-7-[2-(2-amino-4-thiazolyl)glyoxylamido]-3-(hydroxymethyl)-8-oxo-5-thia-1-azabicyclo[4.2.0]oct-2-ene-2-carboxylate 7^2-(Z)-(O-methyloxime) acetate. $C_{16}H_{17}N_5O_7S_2$; mol wt 455.46. C 42.19%, H 3.76%, N 15.38%, O 24.59%, S 14.08%. Broad spectrum third generation cephalosporin antibiotic. Prepn (unspecified stereochemistry): M. Ochiai et al., **DE 2556736**; eidem, **US 4098888** (1976, 1978 both to Takeda); of the syn-isomer: R. Heymes, A. Lutz, **DE 2702501**; eidem, **US 4152432** (1977, 1979 both to Roussel-UCLAF). Antibacterial activity: R. Wise, Antimicrob. Agents Chemother. **14**, 807 (1978); H. W. Van handuyt, M. Pyckavet, ibid. **16**, 109 (1979). Clinical pharmacology: R. Lüthy et al., ibid. 127. Determn in urine by HPLC and voltammetric methods: M. M. Aleksic et al., Talanta **77**, 131 (2008). Comprehensive description: F. J. Muhtadi, M. M. A. Hassan, Anal. Profiles Drug Subs. **11**, 139-168 (1982). Series of articles on clinical efficacy: J. Antimicrob. Chemother. **26**, Suppl. A, 1-83 (1990). Review of antibacterial activity, pharmacokinetics, and clinical efficacy: R. N. Brogden, C. M. Spencer, Drugs **53**, 483-510 (1997). Clinical trial in combination with sulbactam for respiratory infections: A. Pareek et al., Expert Opin. Pharmacother. **9**, 2751 (2008).

Sodium salt. [64485-93-4] HR-756; RU-24756; Cefotax; Claforan. $C_{16}H_{16}N_5NaO_7S_2$; mol wt 477.44. White to slightly yellow powder. Hygroscopic. $[\alpha]_D^{20}$ +55±2° (c = 0.8 in water). Freely sol in water; slightly sol in methanol. Practically insol in organic solvents.

THERAP CAT: Antibacterial.

THERAP CAT (VET): Antibacterial.

1935. Cefotetan. [69712-56-7] (6R,7S)-7-[[[4-(2-Amino-1-carboxy-2-oxoethylidene)-1,3-dithietan-2-yl]carbonyl]amino]-7-methoxy-3-[[(1-methyl-1H-tetrazol-5-yl)thio]methyl]-8-oxo-5-thia-1-azabicyclo[4.2.0]oct-2-ene-2-carboxylic acid; (6R,7S)-7-[4-(car-

bamoylcarboxymethylene)-1,3-dithietane-2-carboxamido]-7-methoxy-3-[[(1-methyl-1H-tetrazol-5-yl)thio]methyl]-8-oxo-5-thia-1-azabicyclo[4.2.0]oct-2-ene-2-carboxylic acid. $C_{17}H_{17}N_7O_8S_4$; mol wt 575.60. C 35.47%, H 2.98%, N 17.03%, O 22.24%, S 22.28%. Broad-spectrum injectable semi-synthetic antibiotic derived from cephamycin C, q.v. Prepn: M. Iwanami et al., **DE 2824559**; eidem, **US 4263432** (1979, 1981 both to Yamanouchi); eidem, Chem. Pharm. Bull. **28**, 2629 (1980). Commercial-scale synthesis: M. Fujimoto et al., Org. Process Res. Dev. **8**, 915 (2004). In vitro comparison of antibacterial activity: B. Chattopadhyay, J. C. Teli, J. Antimicrob. Chemother. **10**, 151 (1982). Toxicity study: K. Imamura et al., Chemotherapy (Tokyo) **30**, Suppl. 1, 212 (1982). Series of articles on pharmacology, clinical studies, toxicology: ibid. 1-947; J. Antimicrob. Chemother. **11**, Suppl. A, 1-236 (1983). Review of antibacterial activity, pharmacokinetics, therapeutic use: A. Ward, D. M. Richards, Drugs **30**, 382-426 (1985).

White or light yellowish white powder. Sparingly sol in methanol; slightly sol in ethanol, water.

Disodium salt. [74356-00-6] ICI-156834; YM-09330; Apatef; Cefotan. $C_{17}H_{15}N_7Na_2O_8S_4$; mol wt 619.57. LD_{50} in male mice, rats (g/kg): 6.35, 8.48 i.v.; 8.12, 8.37 i.p.; >10 orally and s.c. both species (Imamura).

THERAP CAT: Antibacterial.

THERAP CAT (VET): Antibacterial.

1936. Cefotiam. [61622-34-2] (6R,7R)-7-[[2-(2-Amino-4-thiazolyl)acetyl]amino]-3-[[[1-[2-(dimethylamino)ethyl]-1H-tetrazol-5-yl]-thio]methyl]-8-oxo-5-thia-1-azabicyclo[4.2.0]oct-2-ene-2-carboxylic acid; 7β-[2-(aminothiazol-4-yl)acetamido]-3-[[[1-(2-dimethylaminoethyl)-1H-tetrazol-5-yl]thio]methyl]ceph-3-em-4-carboxylic acid; SCE-963. $C_{18}H_{23}N_9O_4S_3$; mol wt 525.62. C 41.13%, H 4.41%, N 23.98%, O 12.18%, S 18.30%. Semi-synthetic cephalosporin antibiotic. Prepn: S. Terao et al., **DE 2458695**; M. Numata et al., **US 4080498** (1975, 1978 both to Takeda); eidem, J. Antibiot. **31**, 1262 (1978). Prepn of the crystalline dihydrochloride: K. Naito et al., **DE 2738711**; eidem, **US 4146710** (1978, 1979 both to Takeda). Antibacterial spectrum: K. Tsuchiya et al., Antimicrob. Agents Chemother. **14**, 557 (1978). Clinical pharmacokinetics: F. D. Daschner et al., ibid. **22**, 958 (1982). Clinical trial in respiratory infections: H. M. Beumer et al., Int. J. Clin. Pharmacol. Ther. Toxicol. **23**, 105 (1985); for prophylaxis of post-surgical infection: K. Ishizaka et al., J. Infect. Chemother. **13**, 324 (2007).

Dihydrochloride. [66309-69-1] Abbott 48999; CGP-14221/E; Pansporin; Spizef. $C_{18}H_{23}N_9O_4S_3 \cdot 2HCl$; mol wt 598.54. White to light yellow crystals. Sol in methanol. Slightly sol in ethanol.

1-(Cyclohexyloxycarbonyloxy)ethyl ester dihydrochloride. [95789-30-3] Cefotiam hexetil hydrochloride; SCE-2174; Pansporin T; Taketiam; Texodil. $C_{27}H_{37}N_9O_7S_3 \cdot 2HCl$; mol wt 768.75. Prepn: T. Nishimura et al., **EP 182029**; eidem, **US 5120841** (1984, 1992 both to Takeda); eidem et al., J. Antibiot. **40**, 81 (1987). Acute toxicity: S. Chiba et al., Oyo Yakuri **35**, 179 (1988), C.A. **109**, 31615c (1988). Clinical trial: H. Maier et al. Arzneim.-Forsch. **42**, 980 (1992). Colorless powder. Soly in water at pH 4.5: >1 mg/ml. LD_{50} in male, female mice, male, female rats (mg/kg): 1290, 1150, 1470, 1620 i.p.; >2000 s.c.; >5000 orally (Chiba).

THERAP CAT: Antibacterial.

Consult the Name Index before using this section.

Page 341

1937. Cefovecin. [234096-34-5] (6R,7R)-7-[[(2Z)-2-(2-Amino-4-thiazolyl)-2-(methoxyimino)acetyl]amino]-8-oxo-3-[(2S)-tetrahydro-2-furanyl]-5-thia-1-azabicyclo[4.2.0]oct-2-ene-2-carboxylic acid. $C_{17}H_{19}N_5O_6S_2$; mol wt 453.49. C 45.03%, H 4.22%, N 15.44%, O 21.17%, S 14.14%. Semi-synthetic cephalosporin antibiotic. Prepn: J. H. Bateson et al., **WO 9201696** (1992 to Beecham); eidem, **US 6020329** (2000 to Pfizer). Antimicrobial spectrum: M. R. Stegemann et al., Antimicrob. Agents Chemother. **50**, 2286 (2006). Pharmacokinetics in dogs: M. R. Stegemann et al., J. Vet. Pharmacol. Ther. **29**, 501 (2006); in cats: eidem, ibid., 513. Efficacy in skin infections in dogs: R. Six et al., J. Am. Vet. Med. Assoc. **233**, 433 (2008); in cats: eidem, ibid. **234**, 81 (2009). Review of manufacturing process: T. Norris, Process Chemistry in the Pharmaceutical Industry, **Vol. 2**, K. Gadamasetti, T. Braish, Ed. (Taylor & Francis, Boca Raton, 2008) pp 191-204.

Sodium salt. [141195-77-9] Convenia. $C_{17}H_{18}N_5NaO_6S_2$; mol wt 475.47. Off white to yellow powder. Sol in water.

THERAP CAT (VET): Antibacterial.

1938. Cefoxitin. [35607-66-0] (6R,7S)-3-[[(Aminocarbonyl)oxy]methyl]-7-methoxy-8-oxo-7-[[2-(2-thienyl)acetyl]amino]-5-thia-1-azabicyclo[4.2.0]oct-2-ene-2-carboxylic acid; (6R,7S)-3-(hydroxymethyl)-7-methoxy-8-oxo-7-[2-(2-thienyl)acetamido]-5-thia-1-azabicyclo[4.2.0]oct-2-ene-2-carboxylic acid carbamate (ester); 3-carbamoyloxymethyl-7α-methoxy-7-[2-(2-thienyl)acetamido]-3-cephem-4-carboxylic acid. $C_{16}H_{17}N_3O_7S_2$; mol wt 427.45. C 44.96%, H 4.01%, N 9.83%, O 26.20%, S 15.00%. Semi-synthetic, broad spectrum antibiotic derived from cephamycin C, q.v., possessing high resistance to β-lactamase inactivation. Synthesis: B. G. Christensen et al., **DE 2129675**; **DE 2203653**; eidem, **US 4297488** (1971, 1972, 1981 all to Merck & Co.); Karady et al., J. Am. Chem. Soc. **94**, 1410 (1972); Ratcliffe, Christensen, Tetrahedron Lett. **1973**, 4653. Biological evaluation: Wallick, Hendlin, Antimicrob. Agents Chemother. **5**, 25 (1974); Miller et al., ibid. 33; Onishi et al., ibid. 38; Hamilton, Miller et al., J. Antibiot. **27**, 42 (1974). HPLC determn in serum and tissue: S. K. Cox et al., J. Chromatogr. B **705**, 145 (1998). Mode of action: Onishi et al., Ann. N.Y. Acad. Sci. **235**, 406 (1974). Toxicity: S. Takayama et al., Chemotherapy (Tokyo) **26**, Suppl. 1, 150 (1978). Comprehensive reviews: J. Antimicrob. Chemother. **4**, Suppl. B, 1-256 (1978); R. N. Brogden et al., Drugs **17**, 1-37 (1979); E. O. Stapley, K. R. Brown, in Pharmacological and Biochemical Properties of Drug Substances **vol. 3**, M. E. Goldberg, Ed. (Am. Pharm. Assoc., Washington, DC, 1981) pp 262-290. Comprehensive description: G. S. Brenner, Anal. Profiles Drug Subs. **11**, 169-195 (1982). Symposium on therapeutic use in anaerobic and aerobic infectons: Hosp. Pract. **25**, Suppl. 4, 1-56 (1990).

Crystals, mp 149-150° (dec). pKa 2.2. Very sol in acetone; sol in aq NaHCO₃; very slightly sol in water. Practically insol in ether and chloroform.

Sodium salt. [33564-30-6] MK-306; Mefoxin; Mefoxitin. $C_{16}H_{16}N_3NaO_7S_2$; mol wt 449.43. White crystals with characteristic odor. $[\alpha]_{589nm}^{25}$ +210° (c = 1 in methanol). Hygroscopic. Very sol in water; sol in methanol; sparingly sol in ethanol, DMF; slightly sol in acetone. Insol in ether, chloroform, aromatic and aliphatic hydrocarbons. LD₅₀ in mice, rats, dogs (g/kg): 5.10, 8.98, >10.0 i.v. (Takayama).

THERAP CAT: Antibacterial.
THERAP CAT (VET): Antibacterial.

1939. Cefozopran. [113359-04-9] 1-[[(6R,7R)-7-[[(2Z)-2-(5-Amino-1,2,4-thiadiazol-3-yl)-2-(methoxyimino)acetyl]amino]-2-carboxy-8-oxo-5-thia-1-azabicyclo[4.2.0]oct-2-en-3-yl]methyl]-imidazo[1,2-b]pyridazinium inner salt; 7β-[2-(5-amino-1,2,4-thiadiazol-3-yl)-2(Z)-methoxyiminoacetamido]-3-[(imidazo[1,2-b]pyridazinium-1-yl)methyl]-3-cephem-4-carboxylate. $C_{19}H_{17}N_9O_5S_2$; mol wt 515.52. C 44.27%, H 3.32%, N 24.45%, O 15.52%, S 12.44%. Prepn: A. Miyake et al., **WO 8605183**; eidem, **US 4864022** (1986, 1989 both to Takeda); A. Miyake et al., J. Antibiot. **45**, 709 (1992). Antibacterial spectrum: T. Iwahi et al., Antimicrob. Agents Chemother. **36**, 1358 (1992). In vivo activity: Y. Iizawa et al., ibid. **37**, 100 (1993). HPLC determn in plasma: K. Ikeda et al., J. Pharm. Biomed. Anal. **45**, 811 (2007). Clinical pharmacokinetics: K. Nomura et al., J. Antimicrob. Chemother. **61**, 892 (2008). Clinical trial in febrile neutropenia: T. Sato et al., Pediatr. Blood Cancer **51**, 774 (2008).

Hydrochloride. SCE-2787; Firstcin. $C_{19}H_{17}N_9O_5S_2 \cdot HCl$; mol wt 551.98. White to pale yellow crystals or crystalline powder. Freely sol in DMSO, formamide; slightly sol in water, methanol, ethanol. Practically insol in acetonitrile, diethyl ether.

THERAP CAT: Antibacterial.

1940. Cefpiramide. [70797-11-4] (6R,7R)-7-[[(2R)-2-[[(4-Hydroxy-6-methyl-3-pyridinyl)carbonyl]amino]-2-(4-hydroxyphenyl)acetyl]amino]-3-[[(1-methyl-1H-tetrazol-5-yl)thio]methyl]-8-oxo-5-thia-1-azabicyclo[4.2.0]oct-2-ene-2-carboxylic acid; 7-[(R)-2-(4-hydroxy-6-methylnicotinamido)-2-(p-hydroxyphenyl)acetamido]-3-[[(1-methyl-1H-tetrazol-5-yl)thio]methyl]-8-oxo-5-thia-1-azabicyclo[4.2.0]oct-2-ene-2-carboxylic acid; D-7-[(4-hydroxy-6-methylnicotinamido)-4-hydroxyphenylacetamido]-3-(1-methyltetrazol-5-yl)thiomethylcephem-4-carboxylic acid. $C_{25}H_{24}N_8O_7S_2$; mol wt 612.64. C 49.01%, H 3.95%, N 18.29%, O 18.28%, S 10.47%. Broad spectrum semi-synthetic cephalosporin antibiotic. Prepn: H. Yamada et al., **BE 833063**; eidem, **US 4156724** (1975, 1979 both to Sumitomo); I. Isaka et al., **JP Kokai 79 30197** (1979 to Yamanouchi), C.A. **91**, 57037a (1979). Prepn, NMR data, antibacterial activity: H. Yamada et al., J. Antibiot. **36**, 522, 543 (1983). HPLC determn in plasma: T. Ohshima et al., J. Liq. Chromatogr. **11**, 3457 (1988). Comparative antibacterial spectrum: C. Quentin et al., Eur. J. Clin. Microbiol. Infect. Dis. **7**, 544 (1988). Clinical pharmacokinetics: J. E. Conte, Antimicrob. Agents Chemother. **31**, 1585 (1987). Series of articles on in vitro, in vivo activity, determn in body fluids, pharmacokinetics, clinical studies: Chemotherapy (Tokyo) **31**, Suppl. 1, 1-842 (1983).

Yellow crystals, mp 213-215° (dec).

Sodium salt. [74849-93-7] SM-1652; Wy-44635; Cefpiran; Suncefal; Sepatren. $C_{25}H_{23}N_8NaO_7S_2$; mol wt 634.62.

THERAP CAT: Antibacterial.

1941. Cefpirome. [84957-29-9] 1-[[(6R,7R)-7-[[(2Z)-2-(2-Amino-4-thiazolyl)-2-(methoxyimino)acetyl]amino]-2-carboxy-8-

oxo-5-thia-1-azabicyclo[4.2.0]oct-2-en-3-yl]methyl]-6,7-dihydro-5H-cyclopenta[b]pyrindinium inner salt; 1-[[(6R,7R)-7-[2-(2-amino-4-thiazolyl)glyoxylamido]-2-carboxy-8-oxo-5-thia-1-azabicyclo-[4.2.0]oct-2-en-3-yl]methyl]-6,7-dihydro-5H-1-pyrindinium hydroxide, inner salt, 7²-(Z)-(O-methyloxime); HR-810. $C_{22}H_{22}N_6$-O_5S_2; mol wt 514.58. C 51.35%, H 4.31%, N 16.33%, O 15.55%, S 12.46%. Fourth generation cephalosporin antibiotic. Prepn: R. Lattrell et al., GB 2098216 (1982 to Hoechst); Prepn of crystalline salts: W. Dürckheimer, R. Lattrell, US 4609653 (1986 to Hoechst); and in vitro antibacterial activity: G. Seibert et al., Arzneim.-Forsch. 33, 1084 (1983). Physical properties: R. Lattrell et al., Proc. 13th Int. Congr. Chemother. 4, part 99, 1 (1983). Comparative in vitro antibacterial spectrum and β-lactamase stability: H. C. Neu et al., Infection 13, 146 (1985). HPLC determn in serum and urine: H. Uihlein et al., ibid. 16, 135 (1988). Clinical pharmacokinetics: J. Kavi et al., J. Antimicrob. Chemother. 22, 911 (1988). Toxicity studies: H. H. Donaubauer, D. Mayer, ibid. 29, Suppl. A, 71 (1992). Clinical trial in bacteremia or sepsis: S. R. Norrby et al., ibid. 42, 503 (1998).

Sulfate. [98753-19-6] Cefrom. $C_{22}H_{22}N_6O_5S_2 \cdot H_2SO_4$; mol wt 612.65. White to pale yellowish-white, crystalline powder. Hygroscopic. Crystals as the monohydrate, dec 198-202°. $[\alpha]_D^{25}$ −4.7° (c = 5 in H_2O). uv max: 265 nm (ε 21100). Soly (aq buffer, pH 6.5) >50%. Sol in water. Practically insol in ethanol, diethyl ether.
THERAP CAT: Antibacterial.

1942. Cefpodoxime. [80210-62-4] (6R,7R)-7-[[(2Z)-2-(2-Amino-4-thiazolyl)-2-(methoxyimino)acetyl]amino]-3-(methoxymethyl)-8-oxo-5-thia-1-azabicyclo[4.2.0]oct-2-ene-2-carboxylic acid; 7β-[2-(2-aminothiazol-4-yl)-(Z)-2-methoxyiminoacetamido]-3-methoxymethyl-3-cephem-4-carboxylic acid; R-3763; U-76253. $C_{15}H_{17}N_5O_6S_2$; mol wt 427.45. Third generation cephalosporin antibiotic. Prepn: T. Takaya et al., EP 29557; eidem, US 4409215 (1981, 1983 both to Fujisawa); of orally active esters: H. Nakao et al., EP 49118; eidem, US 4486425 (1982, 1984 both to Sankyo). Prepn, pharmacokinetics and NMR analysis: K. Fujimoto et al., J. Antibiot. 40, 370 (1987). Antimicrobial spectrum and β-lactamase stability: R. N. Jones, A. L. Barry, Diagn. Microbiol. Infect. Dis. 8, 245 (1987); of the proxetil ester: Y. Utsui et al., Antimicrob. Agents Chemother. 31, 1085 (1987). HPLC determn in plasma and sinus mucosa: F. Camus et al., J. Chromatogr. B 656, 383 (1994); in pharmaceutical formulations and stability study: N. Fukutsu et al., J. Chromatogr. A 1129, 153 (2006). Review of chemistry, antibacterial activity, toxicity and clinical studies: H. Nakao et al., Sankyo Kenkyusho Nempo 39, 1-44 (1987), C.A. 109, 27503x (1988). Comprehensive review: W. M. Todd, Int. J. Antimicrob. Agents 4, 37-62 (1994).

Sodium salt. [82619-04-3] R-3746; U-76253A. $C_{15}H_{16}N_5$-NaO_6S_2; mol wt 449.43.
1-(Isopropoxycarbonyloxy)ethyl ester. [87239-81-4] (6R,7R)-7-[[(2Z)-2-(2-Amino-4-thiazolyl)-2-(methoxyimino)acetyl]amino]-3-(methoxymethyl)-8-oxo-5-thia-1-azabicyclo[4.2.0]oct-2-ene-2-carboxylic acid 1-[[(1-methylethoxy)carbonyl]oxy]ethyl ester; cefpodoxime proxetil; CS-807; U-76252; Cefodox; Orelox; Otreon; Podomexef; Simplicef; Vantin. $C_{21}H_{27}N_5O_9S_2$; mol wt 557.59. White to slightly brownish-white powder with bitter taste. Odorless

or slight characteristic smell. No defined mp. Freely sol in dehydrated ethanol; sol in acetonitrile, methanol; slightly sol in ether; very slightly sol in water. pKa 3.20±0.13. Partition coefficient (octanol/water): 0.08 (pH 1.2); 1.53 (pH 5); 1.50 (pH 9). Partition coefficient (chloroform/water): 1.60 (pH 1.2); 3.08 (pH 5); 3.18 (pH 9). LD_{50} in male, female mice, male, female rats (mg/kg): >10000, >10000, >2000, >2000 s.c., 3502, 2535, >4000, >4000 i.p.; >8000, >8000, >4000, >4000 orally (Nakao, 1988).
THERAP CAT: Antibacterial.
THERAP CAT (VET): Antibacterial.

1943. Cefprozil. [92665-29-7] (6R,7R)-7-[[(2R)-2-Amino-2-(4-hydroxyphenyl)acetyl]amino]-8-oxo-3-(1-propen-1-yl)-5-thia-1-azabicyclo[4.2.0]oct-2-ene-2-carboxylic acid; BMY-28100. C_{18}-$H_{19}N_3O_5S$; mol wt 389.43. C 55.52%, H 4.92%, N 10.79%, O 20.54%, S 8.23%. Semisynthetic oral cephalosporin consisting of ~90:10 Z/E isomeric mixture. Prepn and antibacterial activity: H. Hoshi et al., DE 3402642; eidem, US 4520022 (1984, 1985 both to Bristol-Myers); T. Naito et al., J. Antibiot. 40, 991 (1987). Separation of isomers: M. A. Kaplan et al., US 4727070 (1988 to Bristol-Myers). Comparative in vitro antibacterial spectrum and β-lactamase stability: R. N. Jones et al., Diagn. Microbiol. Infect. Dis. 9, 11 (1988). Clinical pharmacokinetics and tissue penetration: R. H. Barbhaiya et al., Antimicrob. Agents Chemother. 34, 1204 (1990). Comparative clinical trials: A. Iravani, ibid. 35, 1940 (1991); A. G. Arguedas et al., Pediatr. Infect. Dis. J. 10, 375 (1991). HPLC determn in plasma and urine: W. C. Shyu et al., Pharm. Res. 8, 992 (1991). Review of clinical experience: S. Bhargava et al., Indian J. Pediatr. 70, 395-400 (2003).

Z - isomer

Monohydrate. [121123-17-9] BMY-28100-03-800; Cefzil; Cronocef; Procef; Radacefe; Rozicel; Serozil; Zamalin. $C_{18}H_{19}N_3$-$O_5S \cdot H_2O$; mol wt 407.44.
Z-Form. [121412-77-9] BMY-28100. Isolated as the hemihydrate, crystals from acetone, mp 218-220° (dec). uv max (pH 7 phosphate buffer): 228, 279 nm (ε 12300, 9800).
E-Form. [92676-86-3] BMY-28167; BBS-1067. Colorless prisms from methanol, mp 230° (dec). uv max (pH 7 phosphate buffer): 228, 292 nm (ε 13000, 16900).
THERAP CAT: Antibacterial.

1944. Cefquinome. [84957-30-2] 1-[[(6R,7R)-7-[[(2Z)-2-(2-Amino-4-thiazolyl)-2-(methoxyimino)acetyl]amino]-2-carboxy-8-oxo-5-thia-1-azabicyclo[4.2.0]oct-2-en-3-yl]methyl]-5,6,7,8-tetrahydroquinolinium inner salt; HOE-111. $C_{23}H_{24}N_6O_5S_2$; mol wt 528.60. C 52.26%, H 4.58%, N 15.90%, O 15.13%, S 12.13%. Broad-spectrum fourth generation injectable aminothiazolyl cephalosporin. Prepn: BE 893163; R. Lattrell et al., US 5071979 (1982, 1991 both to Hoechst); R. F. Brown et al., J. Med. Chem. 33, 2114 (1990). Antibacterial spectrum and pharmacokinetics in animals: M. Limbert et al., Antimicrob. Agents Chemother. 35, 14 (1991). In vitro activity vs bovine bacterial isolates: A. Böttner et al., J. Vet. Med. B 42, 377 (1995); vs equine bacterial isolates: E. Thomas et al., Vet. Microbiol. 115, 140 (2006). LC/MS/MS determn in biological fluids: A. Maes et al., J. Mass Spectrom. 42, 657 (2007). Efficacy in the treatment of mastitis in dairy cows: N. Y. Shpigel et al., J. Dairy Sci. 80, 318 (1997); of bovine endometritis: G. S. Amiridis et al., J. Vet. Pharmacol. Ther. 26, 387 (2003).

Sulfate. [118443-89-3] HR-111V; Cobactan. $C_{23}H_{24}N_6O_5S_2$.H_2SO_4; mol wt 626.67.

THERAP CAT (VET): Antibacterial.

1945. Cefroxadine. [51762-05-1] (6*R*,7*R*)-7-[[(2*R*)-2-Amino-2-(1,4-cyclohexadien-1-yl)acetyl]amino]-3-methoxy-8-oxo-5-thia-1-azabicyclo[4.2.0]oct-2-ene-2-carboxylic acid; 7-β-[D-2-amino-2-(1,4-cyclohexadienyl)acetamido]-3-methoxy-3-cephem-4-carboxylic acid; CGP-9000; Oraspor. $C_{16}H_{19}N_3O_5S$; mol wt 365.40. C 52.59%, H 5.24%, N 11.50%, O 21.89%, S 8.77%. Orally active cephalosporin derivative. Prepn: R. Scartazzini, H. Bickel, **DE 2331133**; *eidem*, **US 4073902** (1974, 1978 both to Ciba-Geigy). Antibiotic activity: O. Zak *et al.*, *J. Antibiot.* **29**, 653 (1976). *In vivo* and *in vitro* microbiological evaluation: K. Yasuda *et al.*, *Antimicrob. Agents Chemother.* **18**, 105 (1980). Pharmacokinetics: T. Bergan, *Chemotherapy* **26**, 225 (1980). HPLC determn in plasma: Y.-S. Kang *et al.*, *J. Pharm. Biomed. Anal.* **40**, 369 (2006). Clinical trial: L. Bertoli *et al.*, *Curr. Ther. Res.* **44**, 975 (1988); in children: F. Scaglione *et al.*, *Drugs Exp. Clin. Res.* **15**, 71 (1989).

Dihydrate. [95615-72-8] $C_{16}H_{19}N_3O_5S$.$2H_2O$; mol wt 401.43. Pale yellowish white to yellow crystalline powder. $[\alpha]_D^{20}$ +87° (c = 1.093 in 0.1*N* HCl). uv max (0.1*N* HCl): 268 nm (ε 6700). Very sol in formic acid; slightly sol in water, methanol; very slightly sol in acetonitrile, ethanol. LD_{50} in mice (mg/kg): >6000 orally; 7090 i.p. (Scartazzini, Bickel, 1978).

THERAP CAT: Antibacterial.

1946. Cefsulodin. [62587-73-9] 4-(Aminocarbonyl)-1-[[(6*R*,7*R*)-2-carboxy-8-oxo-7-[[(2*R*)-2-phenyl-2-sulfoacetyl]amino]-5-thia-1-azabicyclo[4.2.0]oct-2-en-3-yl]methyl]pyridinium inner salt; 7-(D-α-sulphophenylacetamido)-3-(4'-carbamoylpyridinium)-methyl-3-cephem-4-carboxylic acid. $C_{22}H_{20}N_4O_8S_2$; mol wt 532.54. C 49.62%, H 3.79%, N 10.52%, O 24.03%, S 12.04%. Third generation cephalosporin antibiotic. Prepn: S. Morimoto *et al.*, **DE 2234280**; *eidem*, **US 4065619** (1973, 1977 both to Takeda). Prepn and separation of isomers: H. Nomura *et al.*, *J. Med. Chem.* **17**, 1312 (1974). Comparative antipseudomonal activity: K. Tsuchiya *et al.*, *Antimicrob. Agents Chemother.* **13**, 137 (1978). Antibacterial spectrum and β-lactamase stability: A. King *et al.*, *Antimicrob. Agents Chemother.* **17**, 165 (1980). Review of activity and pharmacology: H. C. Neu, B. E. Scully, *Rev. Infect. Dis.* **6**, Suppl. 3, S667-S677 (1984); and therapeutic use: D. B. Wright, *Drug Intell. Clin. Pharm.* **20**, 845-849 (1986).

Sodium salt. [52152-93-9] Sulcephalosporin; Abbott 46811; CGP-7174/E; SCE-129; Monaspor; Pseudocef; Takesulin; Tilmapor. $C_{22}H_{19}N_4NaO_8S_2$; mol wt 554.52. Colorless needles from ethanol/water, mp 175° (dec). Hygroscopic. $[\alpha]_D^{23}$ + 16.5° (c = 1.08 in water). uv max (water): 263 nm (ε 14600). Freely sol in water, formamide; slightly sol in methanol; very slightly sol in ethanol. LD_{50} in mice (mg/kg): >4000 i.p.; >15000 orally (Bryskier).

THERAP CAT: Antibacterial.

1947. Ceftaroline. [189345-04-8] 4-[2-[[(6*R*,7*R*)-7-[[(2*Z*)-2-(5-Amino-1,2,4-thiadiazol-3-yl)-2-(ethoxyimino)acetyl]amino]-2-carboxy-8-oxo-5-thia-1-azabicyclo[4.2.0]oct-2-en-3-yl]thio]-4-thiazolyl]-1-methylpyridinium inner salt; 7β-[2-(5-amino-1,2,4-thiadi-

azol-3-yl)-2(Z)-ethoxyiminoacetamido]-3-[4-(1-methyl-4-pyridinio)-1,3-thiazol-2-yl]thio-3-cephem-4-carboxylate; T-91825. $C_{22}H_{20}N_8O_5S_4$; mol wt 604.69. C 43.70%, H 3.33%, N 18.53%, O 13.23%, S 21.21%. Fourth generation cephalosporin; active form of the *N*-phosphono prodrug, ceftaroline fosamil. Prepn: H. Tawada, K. Okonogi, **JP 9100283** (1997 to Takeda); of *N*-phosphono prodrug: T. Ishikawa *et al.*, **WO 9932497**; *eidem*, **US 6417175** (1999, 2002 both to Takeda). Synthesis and physiochemical properties: T. Ishikawa *et al.*, *Bioorg. Med. Chem.* **11**, 2427 (2003). Stability study: Y. Ikeda *et al.*, *Chem. Pharm. Bull.* **56**, 1406 (2008). Comparative antibacterial spectrum *in vitro*: Y. Iizawa *et al.*, *J. Infect. Chemother.* **10**, 146 (2004). Comparative *in vivo* activity in experimental methicillin resistant *Staphylococcus aureus* (MRSA) infection: C. Jacqueline *et al.*, *ibid.* **51**, 3397 (2007). Review of antibacterial activity, pharmacokinetics and therapeutic use: G. G. Zhanel *et al.*, *Drugs* **69**, 809-831 (2009). Review of clinical trials in complicated skin and skin-structure infections: E. C. Nannini *et al.*, *Expert Opin. Pharmacother.* **11**, 1197-1206 (2010).

Soly in water: 2.3 mg/ml.

Ceftaroline fosamil. [229016-73-3] 4-[2-[[(6*R*,7*R*)-2-carboxy-7-[[(2*Z*)-2-(ethoxyimino)-2-[5-(phosphonoamino)-1,2,4-thiadiazol-3-yl]acetyl]amino]-8-oxo-5-thia-1-azabicyclo[4.2.0]oct-2-en-3-yl]thio]-4-thiazolyl]-1-methylpyridinium inner salt. $C_{22}H_{21}N_8O_8PS_4$; mol wt 684.67.

Ceftaroline fosamil monoacetate. [400827-46-5]; [400827-55-6] (monohydrate). PPI-0903; TAK-599; Teflaro. $C_{22}H_{21}N_8O_8PS_4$.$C_2H_4O_2$; mol wt 744.72. Crystals from aq acetic acid as the monohydrate, mp 221-223° (dec). Soly in water at 25°: >100 mg/ml.

THERAP CAT: Antibacterial.

1948. Ceftazidime. [72558-82-8] 1-[[(6*R*,7*R*)-7-[[(2*Z*)-2-(2-Amino-4-thiazolyl)-2-[(1-carboxy-1-methylethoxy)imino]acetyl]-amino]-2-carboxy-8-oxo-5-thia-1-azabicyclo[4.2.0]oct-2-en-3-yl]methyl]pyridinium inner salt; 1-[[(6*R*,7*R*)-7-[2-(2-amino-4-thiazol-yl)glyoxylamido]-2-carboxy-8-oxo-5-thia-1-azabicyclo[4.2.0]oct-2-en-3-yl]methyl]pyridinium hydroxide inner salt 7^2-(*Z*)-[*O*-(1-carboxy-1-methylethyl)oxime]; GR-20263. $C_{22}H_{22}N_6O_7S_2$; mol wt 546.57. C 48.35%, H 4.06%, N 15.38%, O 20.49%, S 11.73%. Third generation cephalosporin antibiotic. Prepn: C. H. O'Callaghan *et al.*, **DE 2921316**; *eidem*, **US 4258041** (1979, 1981 both to Glaxo). Prepn of crystalline pentahydrate: A. Brodie, L. A. Wetherill, **DE 3037102**; *eidem*, **US 4329453** (1981, 1982 both to Glaxo). Antibacterial properties: C. H. O'Callaghan, *Antimicrob. Agents Chemother.* **17**, 876 (1980). Comparative antibacterial spectrum: R. Wise *et al.*, *ibid.* 884. Comprehensive description: M. A. Abounassif *et al.*, *Anal. Profiles Drug Subs.* **19**, 95-121 (1990). Review of antibacterial activity, pharmacokinetics and therapeutic use: C. P. Rains *et al.*, *Drugs* **49**, 577-617 (1995).

Pentahydrate. [78439-06-2] Fortam; Fortaz; Fortum; Fortumset; Glazidim; Modacin; Panzid; Spectrum; Starcef; Tazidime. $C_{22}H_{22}N_6O_7S_2$.$5H_2O$; mol wt 636.65. Crystalline powder. uv max (pH 6): 257 nm ($E_{1cm}^{1\%}$ 348). Sol in alkali, DMSO; slightly sol in

DMF, methanol, water. Insol in acetone, alcohol, chloroform, dioxane, ether, ethyl acetate, toluene.

THERAP CAT: Antibacterial.

THERAP CAT (VET): Antibacterial, esp in reptiles.

1949. Cefteram. [82547-58-8] (6R,7R)-7-[[(2Z)-2-(2-Amino-4-thiazolyl)-2-(methoxyimino)acetyl]amino]-3-[(5-methyl-2H-tetrazol-2-yl)methyl]-8-oxo-5-thia-1-azabicyclo[4.2.0]oct-2-ene-2-carboxylic acid; (6R,7R)-7-[2-(2-aminothiazol-4-yl)-2-syn-methoxyiminoacetamido]-3-[(5-methyl-2H-tetrazol-2-yl)methyl]-3-cephem-4-carboxylic acid; (+)-(6R,7R)-7-[2-(2-amino-4-thiazolyl)glyoxylamido]-3-[(5-methyl-2H-tetrazol-2-yl)methyl]-8-oxo-5-thia-1-azabicyclo[4.2.0]oct-2-ene-2-carboxylic acid 7^2-(Z)-(O-methyloxime); ceftetrame; Ro-19-5247; T-2525. $C_{16}H_{17}N_9O_5S_2$; mol wt 479.49. C 40.08%, H 3.57%, N 26.29%, O 16.68%, S 13.37%. Third generation, orally active cephalosporin antibiotic. Prepn: **BE 890499**; H. Sadaki et al., **US 4489072** (1982, 1984 both to Toyama). Comparative antibacterial spectrum: R. Wise et al., Antimicrob. Agents Chemother. **29**, 1067 (1986); P. Y. Chau et al., ibid. **31**, 473 (1987). Series of articles on activity, toxicology, pharmacology and clinical efficacy: Chemotherapy (Tokyo) **34**, Suppl. 2, 1-984 (1986). Acute toxicity: S. Sato et al., ibid. 166. LC determn in plasma: J. J. Zou et al., Chromatographia **68**, 817 (2008). Clinical pharmacokinetics: idem et al., Clin. Ther. **30**, 654 (2008). Clinical trial in respiratory infections: A. Saito et al., J. Int. Med. Res. **32**, 590 (2004).

Crystals from acetone, mp >200°.

Pivaloyloxymethyl ester. [82547-81-7] Cefteram pivoxil; T-2588; Tomiron. $C_{22}H_{27}N_9O_7S_2$; mol wt 593.63. Oral prodrug of cefteram. White to pale yellowish white powder. Bitter taste. mp 127-128° (dec). Very sol in acetonitrile; freely sol in methanol, ethanol, chloroform. Practically insol in water. LD_{50} in male, female mice, rats (g/kg): >6.00, 5.86, 5.63, 5.09 i.p.; in both species >6.00 s.c.; in male mice, rats, dogs (g/kg): >6.00, >6.00, >2.00 orally (Sato).

THERAP CAT: Antibacterial.

1950. Ceftibuten. [97519-39-6] (6R,7R)-7-[[(2Z)-2-(2-Amino-4-thiazolyl)-4-carboxy-1-oxo-2-buten-1-yl]amino]-8-oxo-5-thia-1-azabicyclo[4.2.0]oct-2-ene-2-carboxylic acid; 7β-[(Z)-2-(2-aminothiazol-4-yl)-4-carboxy-2-butenoylamino]-3-cephem-4-carboxylic acid; 7432-S; Sch-39720. $C_{15}H_{14}N_4O_6S_2$; mol wt 410.42. C 43.90%, H 3.44%, N 13.65% O 23.39%, S 15.62%. Orally active third generation cephalosporin antibiotic. Prepn: Y. Hamashima, **EP 136721**; idem, **US 4634697** (1985, 1987 both to Shionogi); and in vitro antibacterial activity: Y. Hamashima et al., J. Antibiot. **40**, 1468 (1987). Industrial synthesis utilizing penicillin G: M. Yoshioka, Pure Appl. Chem. **59**, 1041 (1987); utilizing cephalosporin C: E. Bernasconi et al., Org. Process Res. Dev. **6**, 152; 158; 169 (2002). LC determn in pharmaceutical formulations: L. Manna, L. Valvo, Chromatographia **60**, 645 (2004). Comparative antibacterial spectrum: R. N. Jones, A. L. Barry, Antimicrob. Agents Chemother. **32**, 1576 (1988). Pharmacokinetics in humans: M. Nakashima et al., J. Clin. Pharmacol. **28**, 246 (1988). Review of pharmacology and clinical efficacy: R. C. Owens et al., Pharmacotherapy **17**, 707-720 (1997); D. R. P. Guay, Ann. Pharmacother. **31**, 1022-1033 (1997). Clinical trial in pediatric urinary tract infection: S. Marild et al., Pediatr. Nephrol. **24**, 521 (2009).

Dihydrate. [118081-34-8] Cedax; Isocef; Keimax; Seftem. C_{15}-$H_{14}N_4O_6S_2 \cdot 2H_2O$; mol wt 446.45. White to pale yellowish white crystalline powder with slight characteristic odor. mp ~ 235° (dec). Partition coefficient (n-octanol/pH 7 buffer): 0.004. Freely sol in DMF, DMSO. Practically insol in water, ethanol, diethyl ether.

THERAP CAT: Antibacterial.

1951. Ceftiofur. [80370-57-6] (6R,7R)-7-[[(2Z)-2-(2-Amino-4-thiazolyl)-2-(methoxyimino)acetyl]amino]-3-[[(2-furanylcarbonyl)thio]methyl]-8-oxo-5-thia-1-azabicyclo[4.2.0]oct-2-ene-2-carboxylic acid; (6R,7R)-7-[2-(2-amino-4-thiazolyl)glyoxylamido]-3-(mercaptomethyl)-8-oxo-5-thia-1-azabicyclo[4.2.0]oct-2-ene-2-carboxylic acid 7^2-(Z)-(O-methyloxime) 2-furoate (ester); Excede. C_{19}-$H_{17}N_5O_7S_3$; mol wt 523.55. C 43.59%, H 3.27%, N 13.38%, O 21.39%, S 18.37%. Broad spectrum, third generation cephalosporin antibiotic. Prepn: B. Labeeuw, A. Salhi, **EP 36812**; eidem, **US 4464367** (1981, 1984 both to Sanofi). In vivo and in vitro antibacterial activity and β-lactamase stability: R. J. Yancey et al., Am. J. Vet. Res. **48**, 1050 (1987). HPLC determn in swine tissue: M. G. Beconi-Barker et al., J. Chromatogr. B **673**, 231 (1995); in bovine plasma: G. A. Jacobson et al., J. Pharm. Biomed. Anal. **40**, 1249 (2006). Safety assessment in horses: C. R. Mahrt, Am. J. Vet. Res. **53**, 2201 (1992). Review of use in food animals: R. E. Hornish, S. F. Kotarski, Curr. Top. Med. Chem. **2**, 717-731 (2002).

Monosodium salt. [104010-37-9] CM-31916; U-64279E; Naxcel; Spectramast. $C_{19}H_{16}N_5NaO_7S_3$; mol wt 545.53.

Monohydrochloride. [103980-44-5] U-67279A; Excenel. C_{19}-$H_{17}N_5O_7S_3 \cdot HCl$; mol wt 560.01.

THERAP CAT (VET): Antibacterial.

1952. Ceftizoxime. [68401-81-0] (6R,7R)-7-[[(2Z)-2-(2-Amino-4-thiazolyl)-2-(methoxyimino)acetyl]amino]-8-oxo-5-thia-1-azabicyclo[4.2.0]oct-2-ene-2-carboxylic acid; (6R,7R)-7-[2-(2-imino-4-thiazolyl)glyoxylamido]-8-oxo-5-thia-1-azabicyclo[4.2.0]oct-2-ene-2-carboxylic acid 7^2-(Z)-(O-methyloxime). $C_{13}H_{13}N_5$-O_5S_2; mol wt 383.40. C 40.73%, H 3.42%, N 18.27%, O 20.86%, S 16.72%. Injectable third generation cephalosporin antibiotic related to cefotaxime, q.v. Exhibits broad spectrum activity and resistance to β-lactamase hydrolysis. Prepn: T. Takaya et al., **DE 2810922**; eidem, **US 4427674** (1978, 1984 both to Fujisawa). Comparative antibacterial spectrum and β-lactamase stability: T. Kamimura et al., Antimicrob. Agents Chemother. **16**, 540 (1979); M. Nishida et al., J. Antibiot. **32**, 1319 (1979). Series of articles on metabolism, pharmacology, laboratory evaluation: Arzneim.-Forsch. **30**, 1662-1687 (1980). Toxicity and reproduction studies: K. Fukuhara et al., ibid. 1669. HPLC determn in biological fluids: F. Péhourcq, C. Jarry, J. Chromatogr. A **812**, 159 (1998); analysis of impurities in bulk drug: C. Bharathi et al., J. Pharm. Biomed. Anal. **43**, 733 (2007). Clinical pharmacokinetics: B. Facca et al., Antimicrob. Agents Chemother. **42**, 1783 (1998). Clinical trial for prophylaxis of post-surgical infection: J. A. McGregor et al., J. Am. Coll. Surg. **178**, 123 (1994). Review: D. M. Richards, R. C. Heel, Drugs **29**, 281-329 (1985).

Crystals, mp 227° (dec).

Sodium salt. [68401-82-1] FK-749; FR-13479; SKF-88373; Cefizox; Ceftix; Epocelin; Eposerin. $C_{13}H_{12}N_5NaO_5S_2$; mol wt

405.38. White to pale yellow crystalline powder. Freely sol in water; sparingly sol in methanol. Practically insol in ethanol. LD_{50} in mice, rats (mg/kg): ~6000 i.v. (Fukuhara).

THERAP CAT: Antibacterial.

1953. Ceftobiprole Medocaril. [376653-43-9] $(6R,7R)$-7-[[(2Z)-2-(5-Amino-1,2,4-thiadiazol-3-yl)-2-(hydroxyimino)acetyl]-amino]-3-[[(E,3'R)-1'-[[(5-methyl-2-oxo-1,3-dioxol-4-yl)methoxy]-carbonyl]-2-oxo[1,3'-bipyrrolidin]-3-ylidene]methyl]-8-oxo-5-thia-1-azabicyclo[4.2.0]oct-2-ene-2-carboxylic acid. $C_{26}H_{26}N_8O_{11}S_2$; mol wt 690.66. C 45.22%, H 3.79%, N 16.22%, O 25.48%, S 9.28%. Fourth generation cephalosporin antibiotic prodrug; rapidly cleaved in plasma to the active metabolite, ceftobiprole. Prepn: P. Hebeisen *et al.*, **WO 9965920**; *eidem*, **US 6232306** (1999, 2001 both to Hoffmann-La Roche). Activity against methicillin-resistant *Staphylococcus aureus* (MRSA): *idem et al.*, *Antimicrob. Agents Chemother.* **45**, 825 (2001). Comparative antibacterial spectrum *in vitro*: R. N. Jones *et al.*, *J. Antimicrob. Chemother.* **50**, 915 (2002). Clinical pharmacokinetics and safety: A. Schmitt-Hoffmann *et al.*, *ibid.* **48**, 2576 (2004). Clinical trial in skin infections: G. J. Noel *et al.*, *Clin. Infect. Dis.* **46**, 647 (2008). Review of use in treatment of skin and skin structure infections: P. L. Schirmer, S. C. Deresinski, *Expert Rev. Anti Infect. Ther.* **7**, 777-791 (2009).

Sodium salt. [252188-71-9] BAL-5788; Ro-65-5788; Zeftera. $C_{26}H_{25}N_8NaO_{11}S_2$; mol wt 712.64. White to yellowish or slightly brownish powder. Freely sol in water. pKa (25°) 2.8. pH (1% aq soln): 4.52.

Ceftobiprole. [209467-52-7] $(6R,7R)$-7-[[(2Z)-2-(5-Amino-1,2,4-thiadiazol-3-yl)-2-(hydroxyimino)acetyl]amino]-8-oxo-3-[(E)-[(3'R)-2-oxo[1,3'-bipyrrolidin]-3-ylidene]methyl]-5-thia-1-azabicyclo[4.2.0]oct-2-ene-2-carboxylic acid; BAL-9141; Ro-63-9141. $C_{20}H_{22}N_8O_6S_2$; mol wt 534.57. Prepn: P. Angehrn *et al.*, **EP 849269**; *eidem*, **US 5981519** (1997, 1999 both to Hoffmann-La Roche).

THERAP CAT: Antibacterial.

1954. Ceftolozane. [689293-68-3] $(6R,7R)$-3-[[4-[[[(2-Aminoethyl)amino]carbonyl]amino]-2,3-dihydro-3-imino-2-methyl-1H-pyrazol-1-yl]methyl]-7-[[(2Z)-2-(5-amino-1,2,4-thiadiazol-3-yl)-2-[(1-carboxy-1-methylethoxy)imino]acetyl]amino]-8-oxo-5-thia-1-azabicyclo[4.2.0]oct-2-ene-2-carboxylic acid; 7β-[(Z)-2-(5-amino-1,2,4-thiadiazol-3-yl)-2-(1-carboxy-1-methylethoxyimino)acetamido]-3-[3-amino-4-[3-(2-aminoethyl)ureido]-2-methyl-1-pyrazolio]-methyl-3-cephem-4-carboxylic acid. $C_{23}H_{30}N_{12}O_8S_2$; mol wt 666.69. C 41.44%, H 4.54%, N 25.21%, O 19.20%, S 9.62%. Cephalosporin antibiotic with anti-pseudomonal activity. Prepn: H. Ohki *et al.*, **WO 04039814** (2004 to Fujisawa; Wakunaga); *eidem*, **US 7129232** (2006 to Astellas; Wakunaga); and structure-activity study: A. Toda *et al.*, *Bioorg. Med. Chem. Lett.* **18**, 4849 (2008). β-Lactamase stability: S. Takeda *et al.*, *Int. J. Antimicrob. Agents* **30**, 443 (2007). Comparative antibacterial spectrum: H. S. Sader *et al.*, *Antimicrob. Agents Chemother.* **55**, 2390 (2011). Clinical pharmacokinetics and safety: Y. Ge *et al.*, *ibid.* **54**, 3427 (2010).

Amorphous solid.

Sulfate. [936111-69-2] CXA-101; FR-264205. $C_{23}H_{30}N_{12}O_8S_2 \cdot H_2SO_4$; mol wt 764.76. White crystals from ethanol. pKa 7.95.

THERAP CAT: Antibacterial.

1955. Ceftriaxone. [73384-59-5] $(6R,7R)$-7-[[(2Z)-2-(2-Amino-4-thiazolyl)-2-(methoxyimino)acetyl]amino]-8-oxo-3-[[(1,2,5,6-tetrahydro-2-methyl-5,6-dioxo-1,2,4-triazin-3-yl)thio]methyl]-5-thia-1-azabicyclo[4.2.0]oct-2-ene-2-carboxylic acid; $(6R,7R)$-7-[2-(2-amino-4-thiazolyl)glyoxylamido]-3-[[(2,5-dihydro-6-hydroxy-2-methyl-5-oxo-*as*-triazin-3-yl)thio]methyl]-8-oxo-5-thia-1-azabicyclo[4.2.0]oct-2-ene-2-carboxylic acid 7^2-(Z)-(O-methyloxime); cefatriaxone. $C_{18}H_{18}N_8O_7S_3$; mol wt 554.57. C 38.98%, H 3.27%, N 20.21%, O 20.19%, S 17.34%. Parenteral third generation cephalosporin antibiotic. Prepn: M. Montavon, R. Reiner, **GB 2022090**; *eidem*, **US 4327210** (1979, 1982 both to Hoffmann-La Roche); R. Reiner *et al.*, *J. Antibiot.* **33**, 783 (1980). Comparative antibacterial spectrum: P. Angehrn *et al.*, *Antimicrob. Agents Chemother.* **18**, 913 (1980). Mechanism of action study: R. B. Wright *et al.*, *J. Antibiot.* **34**, 590 (1981). Determn in plasma by HPLC: D. B. Bowman *et al.*, *J. Chromatogr.* **309**, 209 (1984); in pharmaceutical formulations by HPTLC: S. Eric-Jovanovic *et al.*, *J. Pharm. Biomed. Anal.* **18**, 893 (1998). Toxicology: K. Teelman *et al.*, *Proc. Hahnenklee Symp.*, *1981*, R. Grieshalber, Ed. (Editiones Roche, Basel, 1982) pp 91-111. Review of pharmacokinetics and pharmacodynamics: T. R. Perry, J. J. Schentag, *Clin. Pharmacokinet.* **40**, 685-694 (2001). Review of clinical experience: T. R. Beam, *Pharmacotherapy* **5**, 237-253 (1985); in community acquired and nosocomial infections: H. M. Lamb *et al.*, *Drugs* **62**, 1041-1089 (2002).

Disodium salt hemiheptahydrate. [104376-79-6]; [74578-69-1] (anhydrous). Ceftriaxone sodium; Ro-13-9904/001; Cefraden; Megion; Rocefin; Rocephin. $C_{18}H_{16}N_8Na_2O_7S_3 \cdot 3\frac{1}{2}H_2O$; mol wt 661.59. White crystalline powder, mp >155° (dec). $[\alpha]_D^{25}$ −165° (c = 1 in water) (calc as anhydrous). uv max (water): 242, 272 nm (ε 32300, 29530). pKa: ~3 (COOH), 3.2 (NH_3^+), 4.1 (enolic OH). Sparingly sol in methanol; very slightly sol in alc. Soly in water at 25°: ~40 g/100 ml. LD_{50} in male, female, mice, rats (mg/kg): 3000, 2800, 2175, 2175 i.v.; >10000 all species orally; >5000 all species s.c. (Teelman).

THERAP CAT: Antibacterial.

THERAP CAT (VET): Antibacterial.

1956. Cefuroxime. [55268-75-2] $(6R,7R)$-3-[[(Aminocarbonyl)oxy]methyl]-7-[[(2Z)-2-(2-furanyl)-2-(methoxyimino)acetyl]amino]-8-oxo-5-thia-1-azabicyclo[4.2.0]oct-2-ene-2-carboxylic acid; $(6R,7R)$-3-carbamoyloxymethyl-7-[2-(2-furyl)-2-(methoxyimino)acetamido]-8-oxo-5-thia-1-azabicyclo[4.2.0]oct-2-ene-2-carboxylic acid; $(6R,7R)$-3-carbamoyloxymethyl-7-[2-(2-furyl)-2-(methoxyimino)acetamido]ceph-3-em-4-carboxylic acid. $C_{16}H_{16}N_4O_8S$; mol wt 424.38. C 45.28%, H 3.80%, N 13.20%, O 30.16%, S 7.55%. Prepn: M. C. Cook *et al.*, **DE 2439880**; *eidem*, **US 3974153** (1973, 1976 both to Glaxo). Prepn of the 1-acetoxyethyl ester: M. Gregson, B. Sykes, **DE 2706413**; *eidem*, **US 4267320** (1977, 1981 both to Glaxo). Antibacterial spectrum and β-lactamase stability: C. H. O'Callaghan *et al.*, *Antimicrob. Agents Chemother.* **9**, 511 (1976); and pharmacology: *idem et al.*, *J. Antibiot.* **29**, 29 (1976). Comprehensive description of the sodium salt: T. J. Wozniak, J. R. Hicks, *Anal. Profiles Drug Subs.* **20**, 209-236 (1991). LC-MS/MS determn in plasma: P. Partani *et al.*, *J. Chromatogr. B* **878**, 428 (2010). Review of antibacterial activity, pharmacology and therapeutic efficacy: R. N. Brogden *et al.*, *Drugs* **17**, 233-266 (1979); of the axetil ester: L. J. Scott *et al.*, *ibid.* **61**, 1455-1500 (2001).

White crystalline solid. $[\alpha]_D^{20}$ +63.7° (c = 1.0 in 0.2M pH 7 phosphate buffer). uv max (pH 6 phosphate buffer): 274 nm (ε 17600).

Sodium salt. [56238-63-2] Anaptivan; Biociclin; Cefamar; Cefoprim; Cefurin; Curoxim; Zinacef. $C_{16}H_{15}N_4NaO_8S$; mol wt 446.37. White or faintly yellow powder. $[\alpha]_D^{20}$ +60° (c = 0.91 in water). uv max (water): 274 nm (ε 17400). Freely sol in water and buffered solutions; sol in methanol; very slightly sol in ethanol, ethyl acetate, diethyl ether, octanol, benzene and chloroform. Soly in water: 500 mg/2.5 ml. pKa (water): 2.5; (DMF): 5.1. Solns are stable at room temp for 13 hrs; <10% decompn in 48 hrs at 25° (O'Callaghan, *J. Antibiot.*).

1-Acetoxyethyl ester. [64544-07-6] Cefuroxime axetil; CCI-15641; Ceftin; Cefurax; Cepazine; Cetoxil; Curocef; Elobact; Novador; Novocef; Oraxim; Zinat; Zinnat. $C_{20}H_{22}N_4O_{10}S$; mol wt 510.47. White powder. Freely sol in acetone; sparingly sol to sol in chloroform, ethyl acetate, methanol; slightly sol in dehydrated alcohol. Insol in ether, water.

THERAP CAT: Antibacterial.

THERAP CAT (VET): Antibacterial.

1957. Celastrol. [34157-83-0] (9β,13α,14β,20α)-3-Hydroxy-9,13-dimethyl-2-oxo-24,25,26-trinoroleana-1(10),3,5,7-tetraen-29-oic acid; tripterine. $C_{29}H_{38}O_4$; mol wt 450.62. C 77.30%, H 8.50%, O 14.20%. Anti-inflammatory triterpenoid isolated from the Chinese medicinal herb, thunder god vine, *Tripterygium wilfordii* Hook. f., and other plants of the family, *Celastraceae*. Identification as red pigment in *T. wilfordii*: T. Q. Chou, P. F. Mei, *Chin. J. Physiol.* **10**, 529 (1936); *eidem, Chin. Med. J.* **54**, 37 (1938). Isoln from *Celastrus scandens*: O. Gisvold, *J. Am. Pharm. Assoc.* **28**, 440 (1939). Identity of tripterine and celastrol: M. S. Schechter, H. L. Haller, *J. Am. Chem. Soc.* **64**, 182 (1942). Structure: R. Harada *et al., Tetrahedron Lett.* **3**, 603 (1962). Isoln from *Reissantia buchananii*: F.-R. Chang *et al., J. Nat. Prod.* **66**, 1416 (2003). HPLC/MS/MS determn in human blood: Q. Xu *et al., Chromatographia* **66**, 735 (2007). Antiangiogenic and antitumor activity: Y. Huang *et al., Cancer Lett.* **264**, 101 (2008). Efficacy in rat models of arthritis: H. Li *et al., J. Ethnopharmacol.* **118**, 479 (2008).

Ruby red, cubic crystals, mp 205° (dec) (Gisvold). Also reported as orange powder, mp 196-199° (Chang). $[\alpha]_D$ −61.22° (c = 0.049 in methanol). abs max (methanol): 209, 424 nm. Sol in methanol, ethanol, DMSO, DMF. Sparingly sol in aq buffers.

Methyl ester. [1258-84-0] Pristimerin. $C_{30}H_{40}O_4$; mol wt 464.65. Orange needles from ether + petr ether, mp 217.5-218°. $[\alpha]_D^{24.5}$ −157.43° (c = 0.101 in methanol). abs max (methanol): 211, 252, 428 nm.

1958. Celecoxib. [169590-42-5] 4-[5-(4-Methylphenyl)-3-(trifluoromethyl)-1H-pyrazol-1-yl]benzenesulfonamide; SC-58635; YM-177; Celebrex; Onsenal. $C_{17}H_{14}F_3N_3O_2S$; mol wt 381.37. C 53.54%, H 3.70%, F 14.94%, N 11.02%, O 8.39%, S 8.41%. Selective cyclooxygenase-2 (COX-2) inhibitor. Prepn: J. J. Talley *et al.*, **WO 9515316**; *eidem*, **US 5466823** (both 1995 to Searle); T. D. Pen-

ning *et al., J. Med. Chem.* **40**, 1347 (1997). Clinical pharmacology: P. E. Lipsky, P. C. Isakson, *J. Rheumatol.* **24**, Suppl. 49, 9 (1997). Clinical trials in arthritis: L. S. Simon *et al., Arthritis Rheum.* **41**, 1591 (1998). Evaluation of risk of gastrointestinal toxicity in patients with arthritis: F. E. Silverstein *et al., J. Am. Med. Assoc.* **284**, 1247 (2000). Clinical trial in familial adenomatous polyposis: G. Steinbach *et al., N. Engl. J. Med.* **342**, 1946 (2000).

Pale yellow solid, mp 157-159°.

THERAP CAT: Anti-inflammatory. In treatment of familial adenomatous polyposis.

1959. Celery Seed. Dried, ripe fruit of *Apium graveolens* L., *Umbelliferae*. *Habit.* Southern Europe; cultivated everywhere. *Constit.* Volatile and fixed oils; bitter extractive; resin.

1960. Celesticetin. [2520-21-0] 2-[(2-Hydroxybenzoyl)oxy]ethyl 6,8-dideoxy-7-O-methyl-6-[[[(2S)-1-methyl-2-pyrrolidinyl]carbonyl]amino]-1-thio-D-*erythro*-α-D-*galacto*-octopyranoside; α-2-hydroxyethyl 6,8-dideoxy-7-O-methyl-6-(1-methyl-L-2-pyrrolidinecarboxamido)-1-thio-D-*erythro*-D-*galacto*-octopyranoside monosalicylate; S-demethyl-3′-depropyl-S-[2-[(2-hydroxybenzoyl)oxy]ethyl]-7-O-methyllincomycin. $C_{24}H_{36}N_2O_9S$; mol wt 528.62. C 54.53%, H 6.86%, N 5.30%, O 27.24%, S 6.06%. Antibiotic related to lincomycin, *q.v.*; produced by *Streptomyces caelestis* from a soil obtained from Emigration Canyon near Salt Lake City: De-Boer *et al., Antibiot. Annu.* **1954-1955**, 831; Hoeksema *et al., ibid.* 837. Structure: Hoeksema, *J. Am. Chem. Soc.* **86**, 4224 (1964); **90**, 755 (1968).

Amphoteric, amorphous base. $[\alpha]_D^{24}$ +126.6° (c = 0.5 in chloroform). uv max (0.01N alcoholic H_2SO_4): 240, 310 nm ($E_{1cm}^{1\%}$ 183.7, 80.6). Soluble in acidic or strongly basic aq solns, practically insol in the pH range of 7 to 10. Soluble in the more polar organic solvents, but not in ether or light hydrocarbons. Active against gram-positive organisms and plant pathogens such as *B. stewartii, P. fascians, X. pruni* at 0.5 γ/ml. The antibacterial activity is rapidly destroyed in aq solns above pH 9 at 24°, but remains at pH 5 to 7 for at least 60 days. Celesticetin is more stable in acid than in basic solns, but is destroyed rapidly in 1N HCl at 100°. Cross resistance with erythromycin.

Salicylate. Monoclinic tablets, mp 139°. $[\alpha]_D^{24}$ +90.2° (c = 0.5 in H_2O). Sol in water.

Oxalate. Crystals, mp 149-154°. $[\alpha]_D^{24}$ +106.6° (c = 0.5 in H_2O). Sol in water.

1961. Celestin Blue. [1562-90-9] 1-(Aminocarbonyl)-7-(diethylamino)-3,4-dihydroxyphenoxazin-5-ium chloride (1:1); 1-carbamoyl-7-(diethylamino)-3,4-dihydroxyphenazoxonium chloride; C.I. Mordant Blue 14; C.I. 51050; michrome no. 66; Corein 2R. $C_{17}H_{18}ClN_3O_4$; mol wt 363.80. C 56.13%, H 4.99%, Cl 9.74%, N 11.55%, O 17.59%. Prepd by the action of gallamide on N,N-diethyl-p-nitrosoaniline HCl: Gnehm, Bauer, *J. Prakt. Chem.* **72**, 249 (1905); Grandmougin, Bodmer, *ibid.* **77**, 502 (1908). *See also:* **US**

534809 (1895); **US 1227407** (1917); *Colour Index* **vol. 4** (3rd ed., 1971). *Review:* H. J. Conn's *Biological Stains*, R. D. Lillie, Ed. (Williams & Wilkins, Baltimore, 9th ed., 1977) pp 407-408.

Greenish-black powder. With water it yields a reddish-purple soln, which appears blue on extreme diln. The alcoholic soln is blue in all concns. Approximate soly in water 2.0%, abs ethanol 1.5%, Cellosolve 2.25%, ethylene glycol 6.5%, xylene 0.005%.

USE: To dye fabrics a navy-blue color with faint reddish tinge. As nuclear and connective tissue stain. One of the ingredients of Picro-Mallory stain: Lendrum, McFarlane, *J. Pathol. Bacteriol.* **50**, 381 (1940).

1962. Celiprolol. [56980-93-9] N'-[3-Acetyl-4-[3-[(1,1-dimethylethyl)amino]-2-hydroxypropoxy]phenyl]-N,N-diethylurea; N-[3-acetyl-4-(3'-*tert*-butylamino-2'-hydroxy)propoxy]phenyl-N'-diethylurea; ST-1396. $C_{20}H_{33}N_3O_4$; mol wt 379.50. C 63.30%, H 8.77%, N 11.07%, O 16.86%. Cardioselective β_1-adrenergic blocker. Prepn: **BE 823411**; G. Zölss *et al.*, **US 4034009** (1975, 1977 both to Chemie Linz). Hemodynamic effects: J. Bonelli *et al.*, *Wien. Klin. Wochenschr.* **90**, 350 (1978); H. Pittner, *Arch. Pharmacol.* **311**, Suppl., 180 (1980). Series of articles on determn in biological material, pharmacology, toxicology, clinical studies: *Arzneim.-Forsch.* **33**, 1-79 (1983). Toxicity: W. Wendtlandt, H. Pittner, *ibid.* 41. Symposium on pharmacology, clinical efficacy and comparison with other β-blockers: *J. Cardiovasc. Pharmacol.* **8**, Suppl. 4, S1-S152 (1986). Comprehensive description: D. J. Mazzo *et al.*, *Anal. Profiles Drug Subs.* **20**, 237-301 (1991).

Crystals, mp 110-112°.
Hydrochloride. [57470-78-7] Celectol; Corliprol; Selectol. $C_{20}H_{33}N_3O_4 \cdot HCl$; mol wt 415.96. White, odorless crystals, mp 197-200° (dec). Soly at ~25°C (g/100 ml): water 15.1, methanol 18.2, ethanol 1.61, chloroform 0.42. uv max (water): 231, 324 nm ($E^{1\%}$ 652, 57); (0.01N HCl): 231, 324 nm ($E^{1\%}$ 660, 60); (0.01N NaOH): 231, 324 nm ($E^{1\%}$ 640, 60); (methanol): 232, 329 nm ($E^{1\%}$ 775, 58). LD_{50} in male mice, rats (mg/kg): 56.2, 68.3 i.v.; 1834, 3826 orally (Wendtlandt, Pittner).

THERAP CAT: Antihypertensive, antianginal.

1963. Cellobiose. [528-50-7] 4-*O*-β-D-Glucopyranosyl-D-glucose; β-cellobiose; cellose; 4-(β-D-glucosido)-D-glucose. $C_{12}H_{22}O_{11}$; mol wt 342.30. C 42.11%, H 6.48%, O 51.41%. Unit of cellulose and lichenin. Does not occur free in nature, or as glucoside. Prepn from cotton: Braun, *Org. Synth.* **coll. vol. II**, 122, 124 (1943). Prepn from cell-free enzymatic hydrolyzate of cellulose: Whistler, Smart, *J. Am. Chem. Soc.* **75**, 1916 (1953). Structure: Haworth, Hirst, *J. Chem. Soc.* **119**, 193 (1923); Charlton *et al.*, *ibid.* **1926**, 89; Zemplén, *Ber.* **59**, 1254 (1926); Haworth *et al.*, *J. Chem. Soc.* **1927**, 2809; Peterson, Spencer, *J. Am. Chem. Soc.* **49**, 2822 (1927); Helferich *et al.*, *Ber.* **63**, 992 (1930); Hess, Dziengel, *ibid.* **68**, 1594 (1935); Hassid, Ballou in *The Carbohydrates*, W. Pigman, Ed. (Academic Press, New York, 1957) p 490. Synthesis: Haskins *et al.*, *J. Am. Chem. Soc.* **64**, 1289 (1942). *Review:* Pazur in *The Carbohydrates* **vol. 2A**, W. Pigman *et al.*, Eds. (Academic Press, New York, 2nd ed., 1970) pp 109-110; R. G. Edwards, *Dev. Food Carbohydr.* **2**, 229-273 (1980).

Minute crystals from dil alcohol which retain 0.25 to 0.50 mol water after drying in vacuo. Indifferent taste. Dec 225°. Shows mutarotation. $[\alpha]_D^{20}$ +14.2° → +34.6° (15 hrs, c = 8). One gram dissolves in 8 ml water, in 1.5 ml boiling water. Almost insol in abs alc and ether. Reduces Fehling's soln. Hydrolysis with acid or emulsin yields 2 mols β-D-glucose. Not fermented by brewers' yeast, maltase, or invertase.

Octaacetyl-aldehydro-cellobiose. $C_{28}H_{38}O_{19}$; mol wt 678.59. mp 105-110°. $[\alpha]_D^{20}$ +17.7° (c = 3 in chloroform).
Octaacetyl-α-cellobiose. $C_{28}H_{38}O_{19}$; mol wt 678.59. mp 229°. $[\alpha]_D^{20}$ +41° (c = 6 in chloroform).
Octaacetyl-β-cellobiose. mp 202°. $[\alpha]_D^{20}$ −14.7° (c = 5 in chloroform).

1964. Cellocidin. [543-21-5] 2-Butynediamide; acetylenedicarboxamide; acetylenedicarboxylic acid diamide; aquamycin; lenamycin. $C_4H_4N_2O_2$; mol wt 112.09. C 42.86%, H 3.60%, N 24.99%, O 28.55%. Antibiotic substance with antibacterial activity. Produced by *Streptomyces chibaensis* from soil collected at Chiba City, Japan: Suzuki *et al.*, *J. Antibiot.* **11**, 81 (1958). Synthesis from dimethyl acetylenedicarboxylate and concd ammonium hydroxide at −10°: Saggiomo, *J. Org. Chem.* **22**, 1171 (1957); Suzuki, Okuma, *J. Antibiot.* **11**, 84 (1958). Identity with lenamycin: Y. Sekizawa, *Meiji Seika Kenkyu Nempo* **1960**, 42, *C.A.* **56**, 14609a (1962). Biosynthesis: E. R. H. Jones, *J. Chem. Soc. Perkin Trans. 1* **1973**, 148.

Crystals from dil methanol, dec. 216-218°. uv max (0.1N NaOH): 299 nm ($E^{1\%}_{1cm}$ 290). Sparingly sol in water, methanol, ethanol, acetone, chloroform, glacial acetic acid. Relatively stable in neutral or acid solns, showing no loss of activity at pH 2 to 7 when heated for 10 min at 100°. Unstable in alkaline soln evolving ammonia. LD_{50} i.v. in mice: 11 mg/kg (Suzuki).

1965. Celluloid. [8050-88-2] Pyralin; Zylonite. Prepd from nitrocellulose and camphor.
Colorless, amorphous mass. *Flammable.* Prone to spontaneous decompn which is retarded or prevented by the addition of urea, ZnO, $MgCO_3$, diphenylamine, etc. It is rendered less flammable by addition of ammonium phosphate. Softens in boiling water; sol in acetone.

USE: Plastic material for manuf of toilet articles, toys, photographic films; substitute for amber, ivory, ebonite, tortoise shell; also in surgery for bandages and in dentistry as substitute for rubber.

1966. Cellulose. [9004-34-6] $(C_6H_{10}O_5)_n$. Polysaccharide with the glucose units linked as in cellobiose. Chief constituent of the fiber of plants; cotton is the purest natural form, contg about 90%. Rayon is regenerated cellulose. Books: C. Dorée, *The Methods of Cellulose Chemistry* (Chapman & Hall, London, 1947); T. Lieser, *Kurzes Lehrbuch der Cellulosechemie* (Gebrüder Borntraeger, Berlin, 1953); S. D. Antonovskii, *Chemistry of Wood and Cellulose* (Vsesoyuz. Zaochnyi Lesotekh Instit., Leningrad, 1954); E. Ott *et al.*, *Cellulose and Cellulose Derivatives* **vols. 1-3** (Interscience, New York, 1954, 1955). *Reviews:* Several authors in *Encyclopedia of Polymer Science and Technology* **vol. 3**, N. M. Bikales, Ed. (Interscience, New York, 1965) pp 131-539; Shafizadeh, *Pure Appl. Chem.* **35**, 195-208 (1973); A. F. Turbak *et al.* in *Kirk-Othmer Encyclopedia of Chemical Technology* **vol. 5** (Wiley-Interscience, New York, 3rd ed., 1979) pp 70-89. Comprehensive review on constitution, conformation, size of molecule, fine structure and superstructure: H. Krässig, *Papier (Darmstadt)* **33**, 9-20 (1979). Review of toxicity studies: R. L. Anderson *et al.*, *Cancer Lett.* **63**, 83-92 (1992).

White substance. Practically insol in water or other usual solvents. Dissolved by concd soln of zinc chloride, by ammoniacal copper hydroxide soln; also by caustic alkali with carbon disulfide.

Microcrystalline Form. Avicel. Prepn and manuf of crystallite cellulosic aggregates: Battista, *Ind. Eng. Chem.* **42**, 502 (1950); Battista, Smith, **US 2978446** (1961 to Am. Viscose); **US 3141875** (1964 to FMC). Non-fibrous powder. Particle shape: rigid rods. Refractive index: 1.55. Bulk density: 18-19 lb/cubic foot. Practically insol, but dispersible in water; practically insol in and resistant to dil acid; practically insol and inert in organic acids. Partially sol with swelling in dil alkali.

Caution: Potential symptoms of overexposure are irritation of eyes, skin, mucous membranes. *See NIOSH Pocket Guide to Chemical Hazards* (DHHS/NIOSH 97-140, 1997) p 56.

USE: Fibrous form is the basic material for the textile and paper industries. Nitrated it yields nitrocellulose used for manuf of explosives, collodion, lacquers. Basic material also for cellulose acetate, cellulose xanthate. Also used in chromatography and as ion exchange material especially in the form of derivatives such as **DEAE-cellulose** (diethylaminoethyl cellulose) and ECTEOLA-cellulose, *q.v.* Microcrystalline forms of cellulose are used as combination binder-disintegrants in tableting, as separatory medium in thin-layer and column chromatography. Colloidal cellulose particles aid in stabilization and emulsification of liquid and foam systems. May be used as pure cellulose raw-material. Incorporation of cellulose crystallite aggregates in foods to reduce caloric content: Battista, **US 3023104** (1962 to American Viscose); also used in food industry as stabilizer, thickener, texturizer.

1967. Cellulose Acetates. Partially acetylated cellulose, *q.v.* Several acetates of cellulose are known, which differ from one another only in the degree of acetylation. In triacetates, no less than 92% of the hydroxyl groups are acetylated. In characterizing the degree of acetylation, percent acetyl value and percent combined acetic acid are used. All cellulose acetates are obtained by treating cellulose with acetic anhydride at various temps for different lengths of time to produce amorphous white solid material in granular, flake, or powder form from which fibers may be formed by extrusion. In the plastics industry, it is usual to acetylate fully and then to lower the acetyl value to 52-56% by partial hydrolysis. Such material when compounded with suitable plasticizers gives a tough thermoplastic product. Manuf: *Faith, Keyes & Clark's Industrial Chemicals,* F. A. Lowenheim, M. K. Moran, Eds. (Wiley-Interscience, New York, 4th ed., 1975) pp 239-243. Toxicity studies: W. C. Thomas *et al., Food Chem. Toxicol.* **29**, 453 (1991). *Review:* G. A. Serad, J. R. Sanders, "Cellulose Acetate and Triacete Fibers" in *Kirk-Othmer Encyclopedia of Chemical Technology* **Vol. 5** (Wiley-Interscience, New York, 3rd ed., 1979) pp 89-117.

R = –OOCCH₃

Cellulose Triacetate

Commercial products do not have sharp melting points. Solubility is affected by the acetyl value; the triacetate is insol in water, alcohol, ether, but sol in glacial acetic acid; the tetraacetate is insol in water, alcohol, ether, glacial acetic acid, methanol; the pentaacetate is insol in water, but sol in alcohol.

USE: Manuf rubber and celluloid substitutes, nonflammable photographic and cinema films, airplane dopes, varnishes and lacquers, filaments, phonograph records; waterproofing fabrics and rendering balloons gas-tight; sizing and finishing fabrics; coating skins; insulating electric wires; tow for cigarette smoke filters.

1968. Cellulose Ethyl Hydroxyethyl Ether. [9004-58-4] Ethyl 2-hydroxyethyl ether cellulose; ethyl hydroxyethyl cellulose; Etulos. Prepn: Jullander, *C.A.* **48**, 6114g (1954); *idem,* **DE 1000367** (1957 to Mo och Domsjö Ab.), *C.A.* **54**, 5088f (1960). Use as laxative: Alm, *C.A.* **50**, 2122i (1956); *idem, Am. J. Dig. Dis.* **2**, 493 (1957).

R is H, –CH₂CH₃, or –CH₂CH₂OH

THERAP CAT: Cathartic.

1969. Centaurein. [35595-03-0] 7-(β-D-Glucopyranosyl-oxy)-5-hydroxy-2-(3-hydroxy-4-methoxyphenyl)-3,6-dimethoxy-4H-1-benzopyran-4-one; 3′,5,7-trihydroxy-3,4′,6-trimethoxyflavone 7-β-D-glucoside. C₂₄H₂₆O₁₃; mol wt 522.46. C 55.17%, H 5.02%, O 39.81%. Isoln from root of *Centaurea jacea* L., *Compositae:* Bridel, Charaux, *Compt. Rend.* **175**, 833, 1168 (1922). Structure: Farkas *et al., Ber.* **97**, 1666 (1964).

Monohydrate. Yellow crystals, mp 208-209°. $[\alpha]_D^{20}$ −76.6° (c = 1.4 in methanol). uv max (methanol): 349, 258 nm (log ε 4.31, 4.30). Sol in hot water, hot alcohol, and hot acetone; practically insol in cold water, chloroform, ether.

1970. Centaury. Minor centaury; lesser centaury; bitter herb. Dried flowering plant of *Centaurium erythraea* Rafn. (syn. *C. umbellatum* Gilib., *Erythraea centaurium* Pers.), Gentianaceae. *Habit.* Europe, North America, Middle East, North Africa. *Constit.* Gentiopicrin, amarogentin, gentisin: Korte, *Ber.* **87**, 1357 (1954); **88**, 704 (1955). *Review:* A. Y. Leung, *Encyclopedia of Common Natural Ingredients* (John Wiley & Sons, New York, 1980) pp 109-110.

Chilean centaury. Chanchalagua; canchalagua. Dried flowering plant of *C. cachanlahuen* (Mol) Robins (syn *Gentiana peruviana* Lam.). *Habit.* Chile, Peru.

THERAP CAT: Bitter tonic.

1971. Centaury, American. Rose-pink; bitter-bloom; square-stem rose-gentian. Dried flowering plant of *Sabatia angularis* (L.) Pursh, Gentianaceae. *Habit.* Florida to Louisiana and Oklahoma, northern to southeastern New York; southern Ontario, Michigan, Wisconsin, Missouri. *Constit.* Gentiopicrin, amarogentin, gentisin: Korte, *Ber.* **87**, 1357 (1954); **88**, 704 (1955). Prepn of sedative extracts: Huneker, *Am. J. Pharm.* **43**, 207 (1871).

THERAP CAT: Bitter tonic.

1972. Centchroman. [31477-60-8] *rel*-1-[2-[4-[(3R,4R)-3,4-Dihydro-7-methoxy-2,2-dimethyl-3-phenyl-2H-1-benzopyran-4-yl]phenoxy]ethyl]pyrrolidine; *(trans)*-1-[2-[p-(7-methoxy-2,2-dimethyl-3-phenyl-4-chromanyl)phenoxy]ethyl]pyrrolidine; 3,4-*trans*-2,2-dimethyl-3-phenyl-4-[p-(β-pyrrolidinoethoxy)phenyl]-7-methoxychroman; *trans*-centchroman; ormeloxifene; Centron; Saheli. C₃₀H₃₅NO₃; mol wt 457.61. C 78.74%, H 7.71%, N 3.06%, O 10.49%. Estrogen antagonist; synthetic, nonsteroidal, postcoital antifertility agent. Pharmacology: V. P. Kamboj *et al., Indian J. Exp. Biol.* **9**, 103 (1971); *idem et al., ibid.* **15**, 1144 (1977). Prepd (not claimed): J. W. Bolger, **DE 2329201** (1974 to Riker); *idem,* **US 3822287** (1974 to Rexall); S. Ray *et al., J. Med. Chem.* **19**, 276 (1976). Mechanism of action: M. S. Sankaran, M. R. N. Prasad, *Contraception* **9**, 279 (1974). Clinical pharmacology: R. Vaidya *et al., ibid.* **1173**. Physicochemical properties: R. K. Seth *et al., Indian J. Pharm. Sci.* **45**, 14 (1983). Resolution and stereoselective binding of enantiomers: M. Salman *et al., J. Med. Chem.* **29**, 1801 (1986). Absolute configuration: N. Srivastava *et al., Bioorg. Med. Chem. Lett.* **6**, 1747 (1996). HPLC determn in serum and milk: J. Lal *et al., J. Chroma-*

togr. B **658**, 193 (1994). Clinical pharmacokinetics: *idem et al.,* *Contraception* **52**, 297 (1995). Toxicity: I. M. Chak *et al., Indian J. Exp. Biol.* **15**, 1159 (1977). Review of clinical experience: M. M. Singh, *Med. Res. Rev.* **21**, 302-347 (2001).

Relative stereochemistry

Crystals from ether/petr ether, mp 99-101°, sinters ~50°.
Hydrochloride. [51023-56-4] 6720 CDRI. $C_{30}H_{35}NO_3 \cdot HCl$; mol wt 494.07. White crystals, mp 165-166°. uv max (methanol): 232, 278 nm (ε 3701 at 278 nm). pKa 2.1. Sol in 10 parts chloroform, 20 parts acetone, 60 parts 95% ethanol, 20 parts methanol. Practically insol in water, isobutanol, 0.1*N* HCl, 0.1*N* NaOH. LD_{50} i.p. in mice: 400 mg/kg (Chak).
(−)-Form. [78994-23-7] Levormeloxifene. Prepd as hydrochloride, mp 197°. $[\alpha]_D^{20}$ −192.9° (c = 1.0 in $CHCl_3$).
THERAP CAT: Oral contraceptive.

1973. Cephaeline. [483-17-0] (1*R*)-1-[[(2*S*,3*R*,11b*S*)-3-Ethyl-1,3,4,6,7,11b-hexahydro-9,10-dimethoxy-2*H*-benzo[*a*]quinolizin-2-yl]methyl]-1,2,3,4-tetrahydro-7-methoxy-6-isoquinolinol; 7′,-10,11-trimethoxyemetan-6′-ol; desmethylemetine; dihydropsychotrine. $C_{28}H_{38}N_2O_4$; mol wt 466.62. C 72.07%, H 8.21%, N 6.00%, O 13.71%. Next to emetine, the most important alkaloid of ipecac, the ground roots of *Uragoga ipecacuanha* (Brot.) Baill. [*Cephaelis ipecacuanha* (Brot.) A. Rich.], *Rubiaceae*: Carr, Pyman, *J. Chem. Soc.* **105**, 1591 (1914); Hesse, *Ann.* **405**, 1 (1914). Structure: Pailer, Porschinski, *Monatsh. Chem.* **80**, 101 (1949). Stereochemistry: Van Tamelen *et al., J. Am. Chem. Soc.* **79**, 4817 (1957). Partial synthesis: A. K. Garg, J. R. Gear, *Tetrahedron Lett.* **50**, 4377 (1969). Biosynthetic studies: *eidem, ibid.* **1968**, 141. HPLC determn in biological fluids: D. J. Crouch *et al., J. Anal. Toxicol.* **8**, 63 (1984). *Review:* M. Janot in Manske, Holmes, *The Alkaloids* **vol. 3** (Academic Press, New York, 1953) pp 363-394.

Needles from ether. Faintly bitter taste. mp 115-116°. $[\alpha]_D^{20}$ −43.4° (c = 2 in chloroform). Practically insol in water. Freely sol in dil hydrochloric acid, dil sulfuric acid, acetic acid, methanol, ethanol, acetone, chloroform. Less sol in ether, petr ether.
Dihydrochloride heptahydrate. $C_{28}H_{40}Cl_2N_2O_4 \cdot 7H_2O$. Prisms, sinters at 245°, slowly melts up to 270°. $[\alpha]_D^{20}$ +25.0° (c = 2). Sol in water. Solns turn yellow. Less sol in alc, acetone, chloroform. Practically insol in benzene, ligroin.
Dihydrobromide heptahydrate. $C_{28}H_{40}Br_2N_2O_4 \cdot 7H_2O$. Prisms from dil hydrobromic acid, sinters at 266°, slowly melts up to 293°. Sol in water; moderately sol in alcohol, acetone. Practically insol in benzene.

Methyl ether *see* Emetine.
THERAP CAT: Emetic; antiamebic.
THERAP CAT (VET): Has been used as an emetic and expectorant.

1974. Cephalexin. [15686-71-2] (6*R*,7*R*)-7-[[(2*R*)-2-Amino-2-phenylacetyl]amino]-3-methyl-8-oxo-5-thia-1-azabicyclo[4.2.0]-oct-2-ene-2-carboxylic acid; cefalexin; 7-(D-α-aminophenylacetamido)desacetoxycephalosporanic acid; 7-(D-2-amino-2-phenylacetamido)-3-methyl-Δ³-cephem-4-carboxylic acid. $C_{16}H_{17}N_3O_4S$; mol wt 347.39. C 55.32%, H 4.93%, N 12.10%, O 18.42%, S 9.23%. Semi-synthetic cephalosporin antibiotic. Prepn: R. B. Morin, B. G. Jackson, **US 3275626**; **US 3507861** (1966, 1970 both to Lilly); Ryan *et al., J. Med. Chem.* **12**, 310 (1969). Pharmacology and toxicology: Muggleton *et al., Antimicrob. Agents Chemother.* **1968**, 353; Kind *et al.*, Welles *et al., ibid.* 361, 489. Comprehensive description: L. P. Marrelli, *Anal. Profiles Drug Subs.* **4**, 21-46 (1975). Clinical pharmacology, efficacy, adverse reactions: *Postgrad. Med. J.* **59**, Suppl. 5, 1-56 (1983). Veterinary trial in canine superficial pyoderma: S. Toma *et al., J. Small Anim. Pract.* **49**, 384 (2008).

Crystals. uv max: 260 nm (ε 7750). pKa 5.2, 7.3.
Monohydrate. [23325-78-2] Ceporex; Ceporexin; Ibilex; Keforal; Keflex; Larixin; Oracef; Servispor. White to off-white crystalline powder. Slightly sol in water. Practically insol in alc, chloroform, ether. LD_{50} in mice, rats (g/kg): 1.6-4.5, >5.0 orally; 0.4-1.3, >3.7 i.p. (Welles).
Monohydrochloride monohydrate. [105879-42-3] Cephalexin hydrochloride; LY-061188. $C_{16}H_{17}N_3O_4S \cdot HCl \cdot H_2O$; mol wt 401.86. White to off-white crystalline powder. Sol in water, acetone, acetonitrile, alcohol, DMF, methanol. Practically insol in chloroform, ether, ethyl acetate, isopropyl alcohol.
Sodium salt. [38932-40-0] Alfaspoven. $C_{16}H_{16}N_3NaO_4S$; mol wt 369.37.
Pivalate ester *see* Pivcefalexin.
THERAP CAT: Antibacterial.
THERAP CAT (VET): Antibacterial.

1975. Cephalins. Kephalins; phosphatidylethanolamine. Group of phospholipids found in all living organisms. Significant constituent of nervous tissue and brain substance. Cephalins consist of glycerophosphoric acid in which the two free hydroxyls are esterified with long-chain fatty acid residues, and ethanolamine forms an ester linkage with the phosphate group. α-Isomers are derivatives of α-glycerophosphoric acid, *q.v.;* β-isomers are derivatives of β-glycerophosphoric acid, *q.v.* Natural products occur in the α-form, while the β-form is now recognized to be an artifact. Prepn from commercial "soybean lecithin": Scholfield, Dutton, *Biochem. Prep.* **5**, 5 (1957); **US 2801255** (1957 to USDA). Diagnostic use: C. R. Ratliff *et al., Am. J. Gastroenterol.* **55**, 589 (1971). Reviews on natural and synthetic cephalins: E. Baer in *Progress in the Chemistry of Fats and Other Lipids* **vol. 6**, (MacMillan, New York, 1963) pp 39-44; Van Deenen, de Haas in *Advances in Lipid Research* **vol. 2** (Academic Press, New York, 1964) pp 183-189; Verkade, *Bull. Soc. Chim. Fr.* **1963**, 1993.

R can be, but is not necessarily, the same fatty acid.

α - Cephalin

Yellowish amorphous substances; characteristic odor and taste. Practically insol in water, acetone. Freely sol in chloroform, ether; slightly sol in ethanol.

USE: Clinical reagent (liver function test).

THERAP CAT: Hemostatic (local).

1976. Cephalonic Acid. [18456-04-7] 8-Hydroxy-5-oxo-ophiobola-3,6,19-trien-25-oic acid; ophiobolin D. $C_{25}H_{36}O_4$; mol wt 400.56. C 74.96%, H 9.06%, O 15.98%. Minor antibiotic metabolite produced by *Cephalosporium caerulens*. Exhibits weak activity against *Staphylococcus aureus*. Isolation, structure elucidation and physical data: Itai *et al.*, *Tetrahedron Lett.* **1967**, 4111; Nozoe *et al.*, *ibid.* 4113. Crystal structure: Itai *et al.*, *Acta Crystallogr.* **25B**, 872 (1969). Synthetic studies on related compounds: P. C. Dutta, T. K. Das, *Synth. Commun.* **6**, 253 (1976); W. G. Dauben, D. S. Hart, *J. Org. Chem.* **42**, 922 (1977); R. K. Boeckman *et al.*, *ibid.* 3630.

Solid, mp 139°. $[\alpha]_D$ +76.2° (c = 0.54 in chloroform). uv max: 259 nm (ε 11700).

1977. Cephalonium. [5575-21-3] 4-(Aminocarbonyl)-1-[[(6R,7R)-2-carboxy-8-oxo-7-[[2-(2-thienyl)acetyl]amino]-5-thia-1-azabicyclo[4.2.0]oct-2-en-3-yl]methyl]pyridinium inner salt; cefalonium; 2-thienylmethyl isonicotinamide cephalosporin C_A; Cepravin Dry Cow. $C_{20}H_{18}N_4O_5S_2$; mol wt 458.51. C 52.39%, H 3.96%, N 12.22%, O 17.45%, S 13.98%. Cephalosporin antibiotic. Prepn: E. H. Flynn, **GB 1067644** (1967 to Lilly); and antibacterial activity: J. L. Spencer *et al.*, *Antimicrob. Agents Chemother.* **1966**, 573. Determn in milk samples: J. E. Hillerton *et al.*, *J. Dairy Sci.* **82**, 704 (1999). Prophylactic use in dry udder infections: R. Curtis *et al.*, *Vet. Rec.* **100**, 557 (1977); in mastitis: J. H. Williamson *et al.*, *N. Z. Vet. J.* **43**, 228 (1995); in combination with a teat sealer: M. W. Woolford *et al.*, *ibid.* **46**, 12 (1998).

Crystals from hot water, mp 147-150° (dec). uv max: 233, 262 nm (ε 16650, 13550). pKa: 3.3.

THERAP CAT (VET): Prophylactic use in mastitis and dry udder therapy.

1978. Cephalosporin C. [61-24-5] (6R,7R)-3-[(Acetyloxy)-methyl]-7-[[(5R)-5-amino-5-carboxy-1-oxopentyl]amino]-8-oxo-5-thia-1-azabicyclo[4.2.0]oct-2-ene-2-carboxylic acid; 7-(D-5-amino-5-carboxyvaleramido)-3-(hydroxymethyl)-8-oxo-5-thia-1-azabicyclo[4.2.0]oct-2-ene-2-carboxylic acid acetate. $C_{16}H_{21}N_3O_8S$; mol wt 415.42. C 46.26%, H 5.10%, N 10.12%, O 30.81%, S 7.72%. First naturally occurring antibiotic substance known to contain the 7-aminocephalosporanic acid nucleus. Produced along with penicillin N by a strain of *Cephalosporium acremonium* (now known as *Acremonium chrysogenum*) that was cultivated from sea water near a sewage outlet on the coast of Sardinia: G. Brotzu, *Lav. Ist. Igiene Cagliari* (1948). Isoln: G. G. F. Newton, E. P. Abraham, *Nature* **175**, 548 (1955); *eidem*, *Biochem. J.* **62**, 651 (1956); *eidem*, **GB 810196** (1959 to Natl. Res. Dev. Corp). Structure: E. P. Abraham, G. G. F. Newton, *Biochem. J.* **79**, 377 (1961); D. C. Hodgkin, E. N. Maslen, *ibid.* 393. Conversion to 7-aminocephalosporanic acid (7-

ACA): R. B. Morin *et al.*, *J. Am. Chem. Soc.* **84**, 3400 (1962). Total synthesis: R. B. Woodward *et al.*, *ibid.* **88**, 852 (1966). Overview of discovery and prepn of derivatives: E. P. Abraham, *Q. Rev. Chem. Soc.* **21**, 231 (1967). Review of biosynthesis: A. L. Demain, J. Zhang, *Crit. Rev. Biotechnol.* **18**, 283-294 (1998). Review of culture techniques and industrial production methods: C. Tollnick *et al.*, *Adv. Biochem. Eng. Biotechnol.* **86**, 1-45 (2004). Review of enzymatic conversion methods to 7-ACA and industrial significance: A. Parmar *et al.*, *Crit. Rev. Biotechnol.* **18**, 1-12 (1998); V. C. Sonawane, *ibid.* **26**, 95-120 (2006).

Sodium salt dihydrate. $C_{16}H_{20}N_3NaO_8S.2H_2O$. Monoclinic crystals from dil alc. $[\alpha]_D^{20}$ +103° (H_2O). uv max: 260 nm ($E_{1cm}^{1\%}$ 200). pKa values: <2.6, 3.1, 9.8. Sol in water. Practically insol in ethanol, ether.

USE: Starting material for the production of semi-synthetic cephalosporin antibiotics.

1979. Cephalosporin P$_1$. [13258-72-5] (3α,4α,6α,7β,8α,-9β,13α,14β,16β,17Z)-6,16-Bis(acetyloxy)-3,7-dihydroxy-29-nor-dammara-17(20),24-dien-21-oic acid. $C_{33}H_{50}O_8$; mol wt 574.76. C 68.96%, H 8.77%, O 22.27%. Antibiotic substance produced by a *Cephalosporium* sp. cultivated from the sea near Sardinia. Crude cephalosporin P contains at least 5 components: P$_1$, P$_2$, P$_3$, P$_4$, P$_5$, the major active substance being cephalosporin P$_1$. Isoln: Burton, Abraham, *Biochem. J.* **50**, 168 (1951). Structure: Baird *et al.*, *Proc. Chem. Soc. London* **1961**, 257; Halsall *et al.*, *ibid.* **1963**, 16; Melera, *Experientia* **19**, 565 (1963); Halsall *et al.*, *Chem. Commun.* **1966**, 685.

Orthorhombic crystals from 50% aq ethanol, $C_{33}H_{50}O_8.H_2O.\frac{1}{2}C_2$-$H_5OH$. mp 147°. $[\alpha]_D^{20}$ +28° (c = 2.7 in chloroform). uv max: 211 nm (ε 9140). Readily sol in chloroform and ethanol; sparingly sol in hexane and water.

1980. Cephalotaxine. [24316-19-6] (1S,3aR,14bS)-1,5,6,8,-9,14b-Hexahydro-2-methoxy-4H-cyclopenta[a][1,3]dioxolo[4,5-h]-pyrrolo[2,1-b][3]benzazepine-1-ol. $C_{18}H_{21}NO_4$; mol wt 315.37. C 68.55%, H 6.71%, N 4.44%, O 20.29%. Predominant alkaloid found in several varieties of Japanese plum yew, *Cephalotaxus harringtonia*, *Cephalotaxacea*, a coniferous shrub used in Chinese traditional medicine as a treatment for cancer. Biologically inactive, biosynthetic precursor for the antitumor cephalotaxine esters such as homoharringtonine, q.v. Isoln: W. W. Paudler *et al.*, *J. Org. Chem.* **28**, 2194 (1963). Structure: R. G. Powell *et al.*, *Tetrahedron Lett.* **10**, 4081 (1969). Crystal structure: S. K. Arora *et al.*, *J. Org. Chem.* **41**, 551 (1976). Biosynthetic studies: R. J. Parry *et al.*, *J. Am. Chem. Soc.* **102**, 1099 (1980). Total synthesis of (±)-form: J. Auerbach, S. M. Weinreb, *ibid.* **94**, 7172 (1972); of naturally occurring (−)-form: N. Isono, M. Mori, *J. Org. Chem.* **60**, 115 (1995). Enantioselective synthesis: L. F. Tietze, H. Schirok, *J. Am. Chem. Soc.* **121**, 10264 (1999); W. R. Esmieu *et al.*, *Org. Lett.* **10**, 3045 (2008).

Crystals from benzene, mp 131-132°. pKa (ethanol): 8.95. $[\alpha]_D^{25}$ −204° (c = 1.8 in chloroform). uv max (ethanol): 290, 238 nm (log ε 3.55, 3.56).

USE: Synthetic precursor.

1981. Cephalothin. [153-61-7] (6R,7R)-3-[(Acetyloxy)methyl]-8-oxo-7-[[2-(2-thienyl)acetyl]amino]-5-thia-1-azabicyclo[4.2.0]-oct-2-ene-2-carboxylic acid; (7R)-3-(acetoxymethyl)-7-[2-(2-thienyl)acetamido]-3-cephem-4-carboxylic acid; cefalotin. $C_{16}H_{16}N_2$-O_6S_2; mol wt 396.43. C 48.48%, H 4.07%, N 7.07%, O 24.21%, S 16.17%. Semi-synthetic cephalosporin antibiotic. Prepn: Chauvette *et al.*, *J. Am. Chem. Soc.* **84**, 3401 (1962); E. H. Flynn, **US 3218318** (1965 to Lilly); **FR 1384197** (1965 to Glaxo), *C.A.* **63**, 11591c (1965). Bacteriology and pharmacology: Lee, Anderson, *Antimicrob. Agents Chemother.* **1962**, 695; Walters *et al.*, *ibid.* 706; Naumann, *Arzneim.-Forsch.* **16**, 1099 (1966). Acute toxicity: M. Kuramoto *et al.*, *Jpn. J. Antibiot.* **27**, 746 (1974), *C.A.* **83**, 71972t (1975). Comprehensive description: R. J. Simmons, *Anal. Profiles Drug Subs.* **1**, 319-341 (1972). HPLC determn in serum: J. S. Wold, S. A. Turnipseed, *Clin. Chim. Acta* **78**, 203 (1977).

mp 160-160.5°. $[\alpha]_D^{20}$ +50° (c = 1.03 in acetonitrile), *see:* R. B. Woodward *et al.*, *J. Am. Chem. Soc.* **88**, 852 (1966).

Sodium salt. [58-71-9] Ceporacin; Keflin. $C_{16}H_{15}N_2NaO_6S_2$; mol wt 418.41. White crystalline powder. Freely sol in water. Insol in most organic solvents. mp 204-205°. $[\alpha]_D$ +135° (c = 1.0 in water). uv max: 236, 260 nm (ε 12950, 9350). LD_{50} in mice, rats (mg/kg): >20000, >10000 orally; 5670, 7716 i.p. (Kuramoto).

THERAP CAT: Antibacterial.

THERAP CAT (VET): Antibacterial.

1982. Cephamycins. A family of β-lactam antibiotics produced by various *Streptomyces* species. Detection and production: Stapley *et al.*, *Antimicrob. Agents Chemother.* **2**, 122 (1972). Chemical characterization: Miller *et al.*, *ibid.* 132. Cephamycins A, B, and C have been isolated, the latter being identical to a *Streptomyces clavuligerus* metabolite: Nagarajan *et al.*, *J. Am. Chem. Soc.* **93**, 2308 (1971). Structures: Albers-Schoenberg *et al.*, *Tetrahedron Lett.* **1972**, 2911. Antibacterial activity studies: Miller *et al.*, *Antimicrob. Agents Chemother.* **2**, 281, 287 (1972); Daoust *et al.*, *ibid.* **3**, 254 (1973). Review of syntheses: T. Hiraoka *et al.*, *Heterocycles* **8**, 719 (1977).

Cephamycin C

1983. Cephapirin. [21593-23-7] (6R,7R)-3-[(Acetyloxy)-methyl]-8-oxo-7-[[(4-pyridinylthio)acetyl]amino]-5-thia-1-azabicyclo[4.2.0]oct-2-ene-2-carboxylic acid; 7-[α-(4-pyridylthio)acetamido]cephalosporanic acid; cefapirin. $C_{17}H_{17}N_3O_6S_2$; mol wt 423.46.

C 48.22%, H 4.05%, N 9.92%, O 22.67%, S 15.14%. Broad spectrum, cephalosporin antibiotic. Prepn: L. B. Crast, **ZA 6707783**; *eidem*, **US 3422100** (1968, 1970 both to Bristol-Myers). Alternate processes: H. H. Silvestri, D. A. Johnson, **US 3503967**; R. E. Havranek, L. B. Crast, **US 3578661** (1970, 1971 both to Bristol-Myers); L. B. Crast *et al.*, *J. Med. Chem.* **16**, 1413 (1973). Activity studies: Chisholm *et al.*, *Antimicrob. Agents Chemother.* **1969**, 244; Axelrod *et al.*, *Appl. Microbiol.* **22**, 904 (1971); Bran *et al.*, *Antimicrob. Agents Chemother.* **1**, 35 (1972); Wiesner *et al.*, *ibid.* 303. LC-MS/MS determn in milk: D. M. Holstege *et al.*, *J. Agric. Food Chem.* **50**, 406 (2002). Veterinary trial in lactating dairy cows: J.-P. Roy *et al.*, *Can. Vet. J.* **50**, 1257 (2009).

Sodium salt. [24356-60-3] Cephapirin sodium; BL-P-1322; Brisfirina; Cefadyl; Cefa-Lak; Cefaloject; Cefatrex; Cefatrexyl; Lopitrex; Piricef. $C_{17}H_{16}N_3NaO_6S_2$; mol wt 445.44. White crystalline powder. Very sol in water. Insol in most organic solvents.

Compd with N,N'-dibenzylethylenediamine. [97468-37-6] Cephapirin benzathine; Cefa-Dri; Metricure. $(C_{17}H_{17}N_3O_6S_2)_2$·$C_{16}H_{20}N_2$; mol wt 1087.27. White to off-white crystalline powder. Practically insol in water, ether, toluene, alcohol. Sol in 0.1N HCl.

THERAP CAT: Antibacterial.

THERAP CAT (VET): Antibacterial.

1984. Cepharanthine. [481-49-2] (14aS,26aR)-2,3,13,14,-14a,15,26,26a-Octahydro-22,30-dimethoxy-1,14-dimethyl-1H-4,-6:16,19-dietheno-21,25-metheno-12H-[1,3]dioxolo[4,5-g]pyrido-[2',3':17,18][1,10]dioxacycloeicosino[2,3,4-ij]isoquinoline; 6',12'-dimethoxy-2,2'-dimethyl-6,7-[methylenebis(oxy)]oxyacanthan. $C_{37}H_{38}N_2O_6$; mol wt 606.72. C 73.25%, H 6.31%, N 4.62%, O 15.82%. From tubers of *Stephania cephalantha* Hayata, and *Stephania sasahii* Hayata, *Menispermaceae:* Kondo *et al.*, **US 2206407** (1940); Kondo, Hasegawa, **US 2248241** (1941). Structure: Kondo, Keimatsu, *Ber.* **71**, 2553 (1938); Tomita, Sasaki, *Pharm. Bull.* **2**, 89, 375 (1954). Total synthesis (*dl*-form): Tomita *et al.*, *Tetrahedron Lett.* **8**, 1201 (1967).

Yellow powder, mp 145-155°. Obtained by drying solvated needles from acetone + benzene. $[\alpha]_D^{20}$ +277° (c = 2 in chloroform). Soluble in the usual organic solvents except petr ether.

1985. Cephradine. [38821-53-3] (6R,7R)-7-[[(2R)-2-Amino-2-(1,4-cyclohexadien-1-yl)acetyl]amino]-3-methyl-8-oxo-5-thia-1-azabicyclo[4.2.0]oct-2-ene-2-carboxylic acid; 7-[D-2-amino-2-(1,4-cyclohexadienyl)acetamido]desacetoxycephalosporanic acid; cefradine; SQ-11436; Eskacef; Velocef; Zeefra. $C_{16}H_{19}N_3O_4S$; mol wt 349.41. C 55.00%, H 5.48%, N 12.03%, O 18.32%, S 9.18%. First generation, semisynthetic cephalosporin antibiotic. Prepn and activity data: Weisenborn *et al.*, **US 3485819** (1969 to Squibb); Dolfini *et al.*, *J. Med. Chem.* **14**, 117 (1971). Comprehensive description:

K. Florey, *Anal. Profiles Drug Subs.* **5**, 21-59 (1976). LC-MS/MS determn in plasma: S.-J. Choi *et al.*, *J. Chromatogr. B* **877** 4059 (2009). Determn of degradation products in pharmaceutical formulations: C. Lu *et al.*, *Talanta* **81**, 698 (2010). Review of pharmacology: W. E. Brown, J. R. Knill, in *Pharmacological and Biochemical Properties of Drug Substances* **vol. 2**, M. E. Goldberg, Ed. (Am. Pharm. Assoc., Washington, DC, 1979) pp 279-304. Clinical trial in urinary tract infections: W. Brumfitt, J. M. T. Hamilton-Miller, *Antimicrob. Agents Chemother.* **34**, 1803 (1990); for prophylaxis of postsurgical infection: D. Skipper *et al.*, *J. Hospital Infect.* **17**, 303 (1991).

White or slightly yellow, hygroscopic powder. Sparingly sol in water. Practically insol in alcohol, *n*-hexane. *Store in airtight containers. Protect from light.*

Monohydrate. [75975-70-1] Small, colorless crystals. mp 140-142° (dec). pK_1 2.63; pK_2 7.57. Sol in propylene glycol; slightly sol in acetone, ethanol. Insol in ether, chloroform, benzene, hexane.

Dihydrate. [31828-50-9] SQ-22022. mp 183-185°.

THERAP CAT: Antibacterial.

1986. Ceresin. [8001-75-0] Purified ozokerite; earth wax; mineral wax; cerosin; cerin. A mixture of hydrocarbons of complex composition purified by treatment with concd H_2SO_4 and filtration through bone-black. Found in Ukraine, Lake Baikal, also in Utah, Texas.

White or yellow, tasteless, waxy cakes. The white is odorless, the yellow has a slight odor. Fracture is very much like that of white wax. d 0.91-0.92. mp 61-78°. It is very stable toward oxidizing agents. Insol in water; sol in 30 parts abs alcohol; sol in benzene, chloroform, petr ether, hot oils.

USE: Substitute for beeswax; for making candles, wax figures; for waxed paper and cloth; in polishes, electrical insulators; waterproofing fabrics; for bottles for hydrofluoric acid; in dentistry for impression and inlay waxes and modeling compounds.

1987. Ceric Fluoride. [10060-10-3] Cerium fluoride (CeF₄); cerium tetrafluoride. CeF₄; mol wt 216.11. Ce 64.84%, F 35.16%. Prepd from CeF_3 and F_2: Klemm, Henkel, *Z. Anorg. Allg. Chem.* **220**, 181 (1934); von Wartenberg, *ibid.* **244**, 343 (1940); Kwasnik in *Handbook of Preparative Inorganic Chemistry* **vol. 1**, G. Brauer, Ed. (Academic Press, New York, 2nd ed., 1963) pp 247-248; from CeO_2 and F_2: Asker, Wylie, *Aust. J. Chem.* **18**, 959 (1965).

Minute crystals. d 4.77. mp >650°. Practically insol in water; very slowly hydrolyzed by cold water. Thermally stable below 550°. May be reduced to CeF_3 by H_2 at 300°.

Monohydrate. White powder. Sol in acids.

USE: Fluorinating agent.

1988. Ceric Oxide. [1306-38-3] Cerium oxide (CeO₂); cerium dioxide; ceria. CeO₂; mol wt 172.11. Ce 81.41%, O 18.59%. Prepn from Ce salts: Duval, *Anal. Chim. Acta* **1**, 341 (1947); other prepns: Warf, *US 2564241* (1951 to USAEC); Wilke, *Z. Anorg. Allg. Chem.* **330**, 164 (1964). Purification: Wetzel in *Handbook of Preparative Inorganic Chemistry* **vol. 2**, G. Brauer, Ed. (Academic Press, New York, 2nd ed., 1965) pp 1132-1135. *Review:* Davis, Wayman, *Can. Chem. Process Ind.* **29**, 230 (1945). Toxicity data: C. E. Lambert *et al.*, *J. Am. Coll. Toxicol.* **12**, 617 (1993).

Heavy powder or cubic crystals; white if pure, but usually pale yellow; traces of lanthanum, praseodymium, etc., give a pink to reddish-brown color. Practically insol in water, acids. LD_{50} orally in rats: >5.0 g/kg (Lambert).

USE: Polishing and decolorizing glass; as opacifier for vitreous enamels; in heat-resistant alloy coatings; in coatings for infrared filters to prevent reflection; in analysis for Ce and oximetry; as catalyst for organic reactions.

1989. Ceric Sulfate. [13590-82-4] Sulfuric acid cerium(4+) salt (2:1); cerium disulfate. CeO₈S₂; mol wt 332.23. Ce 42.17%, O 38.53%, S 19.30%. Ce(SO₄)₂. Prepd by heating CeO_2 with concd H_2SO_4: Vanino, *Handbuch der Präparativen Chemie* **vol. 1** (Stuttgart, 3rd ed., 1925) p 755. *Review:* Smith, *Ceric Sulfate* (G. Frederick Smith Chem. Co., Columbus, Ohio, 2nd ed., 1935) 51 pp. The commercial salt usually contains small amounts of associated rare earths.

Tetrahydrate. Yellow to orange powder or orthorhombic crystals. On heating loses water, becoming anhydr at 180-200°; dec above 350°, forming CeOSO₄. Sol in a small quantity of water, but dec with much water with separation of a basic salt. Slowly sol in cold, more readily sol in hot mineral acids; sol in dil H_2SO_4.

USE: Oxidizing agent; analytical reagent; in radiation dosimeters.

1990. Cerium. [7440-45-1] Ce; at. wt 140.116; at. no. 58; valences 3, 4. A rare earth metal, most abundant member of the lanthanide series. Naturally occurring isotopes (mass numbers): 140 (88.48%); 142 (11.08%), radioactive, $T_½ >5 \times 10^{16}$ years; 138 (0.25%); 136 (0.19%). Known artificial radioactive isotopes: 124-135; 137; 139; 141; 143-152. Estimated abundance in the earth's crust: 46-66 ppm. Commercially important sources are the rare earth minerals monazite and bastnaesite; also found in cerite. Discovered by Klaproth, Hisinger and Berzelius in 1803. Can be separated from other rare earths by selective precipitation of ceric (4+) salts from buffered solns (pH 3-4); also by ion exchange techniques. Prepn of metal: Spedding *et al.*, *Ind. Eng. Chem.* **44**, 553 (1952). Review of prepn, properties and compds: *The Rare Earths*, F. H. Spedding, A. H. Daane, Eds. (Krieger, Huntington, N.Y., 1971, reprint of 1961 ed.) 641 pp; Hulet, Bode, "Separation Chemistry of the Lanthanides and Transplutonium Actinides" in *MTP Int. Rev. Sci.: Inorg. Chem., Ser. One* **vol. 7**, K. W. Bagnall, Ed. (University Park Press, Baltimore, 1972) pp 1-45; Moeller, "The Lanthanides" in *Comprehensive Inorganic Chemistry* **vol. 4**, J. C. Bailar, Jr. *et al.*, Eds. (Pergamon Press, Oxford, 1973) pp 1-101; B. T. Kilbourn in *Kirk-Othmer Encyclopedia of Chemical Technology* **vol. 5** (John Wiley & Sons, New York, 4th ed., 1993) pp 728-749; *Chemistry of the Elements*, N. N. Greenwood, A. Earnshaw, Eds. (Pergamon Press, New York, 1984) pp 1423-1449. Brief review of properties: G. T. Seaborg, *Radiochim. Acta* **61**, 115-122 (1993).

Iron-gray, ductile, malleable metal. Only material known to have a solid-solid critical point. Crystalline forms: face-centered cubic α-form, d 6.770, transforms to β-form at −150°; hexagonal β-form transforms to γ-form at −10°; face-centered cubic γ-form transforms to δ-form at 730°; body-centered cubic δ-form exists at >730°. mp 798°. bp 3433°. Heat of fusion: 5.179 kJ/mol. Heat of sublimation (25°): 422.6 kJ/mol. E^0(aq) Ce³⁺/Ce −2.48 V (calc). *Flammable as slabs, ingots or rods. Turnings or gritty powder dangerous when wet.* Stable in dry air, but superficially oxidized in moist air; when finely divided may ignite spontaneously. Slowly dec by cold, rapidly by hot water; sol in dil mineral acids. Ceric salts usually are yellow to orange-red in color and liberate iodine from KI. Cerous salts are usually white and give a white ppt with alkali hydroxides or sulfides, insol in excess of reagent; they also are pptd by ammonium oxalate from cold dil acid sols.

USE: In metallurgy as stabilizers in alloys and as an alternative to thorium oxide in welding electrodes. In glass as polishing agent, decolorizer to stabilize impurities, to render glass opaque to near uv radiation, to resist discoloration from strong light or high energy electron bombardment (as in television screens). In ceramics as an opacifying and strengthening agent. Catalysts to impart high cracking activity for crude oil processing, in automotive exhaust control devices, as combustion additive, polymerization initiator, paint drier, polymer stabilizer. As phosphor in fluorescent lamps, cathode ray tubes and thorium dioxide gas mantles.

1991. Cerium(IV) Ammonium Nitrate. [16774-21-3]; [10139-51-2] (nitric acid ammonium cerium salt). Ammonium (*OC*-6-11)-hexakis(nitrato-*κO*)cerate(2−) (2:1); ammonium ceric nitrate; ammonium hexanitratocerate; ammonium nitratocerate. CeH₈N₈O₁₈; mol wt 548.22. Ce 25.56%, H 1.47%, N 20.44%, O 52.53%. (NH₄)₂Ce(NO₃)₆. Versatile one-electron oxidizing agent. Prepn: G. F. Smith *et al.*, *Ind. Eng. Chem. Anal. Ed.* **8**, 449 (1936). Use as a primary standard in oximetry: G. F. Smith, W. H. Fly, *Anal. Chem.* **21**, 1233 (1949). Improved prepn: G. F. Smith, *Talanta* **10**,

709 (1963). Review of synthetic applications: J. R. Hwu, K.-Y. King, *Curr. Sci.* **81**, 1043-1053 (2001); V. Nair *et al., Acc. Chem. Res.* **37**, 21-30 (2004).

Orange, non-hygroscopic solid. Soly in water (g/ml): 1.41 at 25°; 2.27 at 80°. Sol in sulfuric, nitric, perchloric, and hydrochloric acids. Extremely limited soly in organic solvents.

USE: As a standard in oxidimetry. Oxidant in redox reactions. In reactions involving carbon-carbon bond formation, oxidative carbon-carbon bond cleavage, nitration, and removal of protecting groups. Soln in dilute perchloric acid as etchant for prepn of printed circuits, metal samples, and surface cleaning prior to fabrication by soldering.

1992. Cerivastatin. [145599-86-6] (3*R*,5*S*,6*E*)-7-[4-(4-Fluorophenyl)-5-(methoxymethyl)-2,6-bis(1-methylethyl)-3-pyridinyl]-3,5-dihydroxy-6-heptenoic acid; (3*R*,5*S*,6*E*)-7-[4-(*p*-fluorophenyl)-2,6-diisopropyl-5-(methoxymethyl)-3-pyridyl]-3,5-dihydroxy-6-heptenoic acid. $C_{26}H_{34}FNO_5$; mol wt 459.56. C 67.95%, H 7.46%, F 4.13%, N 3.05%, O 17.41%. HMG-CoA reductase inhibitor. Prepd and claimed as methyl ester: R. Angerbauer *et al.,* **EP 325130**; *eidem,* **US 5006530**; as (+)-form sodium salt: *eidem,* **US 5177080** (1989, 1991, 1993 all to Bayer). Determn in plasma by fluorescence linked LC: G. J. Krol *et al., J. Pharm. Biomed. Anal.* **11**, 1269 (1993); by HPLC: *eidem, Methodol. Surv. Bioanal. Drugs* **23**, 147 (1994). Effects on LDL receptor levels: G. C. Ness *et al., Arch. Biochem. Biophys.* **325**, 242 (1996). Pharmacokinetics in dogs and rats: W. Steinke *et al., Jpn. Pharmacol. Ther.* **24** Suppl. 9, S1217 (1996). Clinical trial in hyperlipidemia: N. Nakaya *et al., ibid.* S1381.

Sodium salt. [143201-11-0] Rivastatin; Bay w 6628; Baycol; Lipobay. $C_{26}H_{33}FNNaO_5$; mol wt 481.54. Solid, $[\alpha]_D^{20}$ +24.1° (c = 1 in ethanol).

THERAP CAT: Antilipemic.

1993. Ceronapril. [111223-26-8] 1-[(2*S*)-6-Amino-2-[[hydroxy(4-phenylbutyl)phosphinyl]oxy]-1-oxohexyl]-L-proline; 1-[(2*S*)-6-amino-2-hydroxyhexanoyl]-L-proline, hydrogen (4-phenylbutyl)phosphonate (ester); ceranapril; SQ-29852. $C_{21}H_{33}N_2O_6P$; mol wt 440.48. C 57.26%, H 7.55%, N 6.36%, O 21.79%, P 7.03%. Hydroxylphosphonate angiotensin converting enzyme (ACE) inhibitor. Prepn: D. S. Karanewsky, E. W. Petrillo, Jr., **EP 97534**; *eidem,* **US 4452790** (both 1984 to Squibb); *eidem et al., J. Med. Chem.* **31**, 204 (1988). HPLC determn in body fluids: H. Kadin *et al., J. Chromatogr.* **487**, 135 (1989). Pharmacology: J. M. DeForrest *et al., J. Cardiovasc. Pharmacol.* **16**, 121 (1990). Radioimmunoassay and pharmacokinetics: J. Tu *et al., Ther. Drug Monit.* **14**, 209 (1992). Clinical evaluation in hypertension: J. Sasaki *et al., Int. J. Clin. Pharmacol. Ther. Toxicol.* **31**, 83 (1993).

Prepd as the monohydrate. Crystals from water, mp 190-195° (dec). $[\alpha]_D$ −47.5° (c = 1 in methanol).

THERAP CAT: Antihypertensive.

1994. Cerous Bromide. [14457-87-5] Cerium bromide (CeBr$_3$); cerium tribromide. Br$_3$Ce; mol wt 379.83. Br 63.11%, Ce 36.89%. CeBr$_3$. Prepn: Jantsch, Wein, *Monatsh. Chem.* **69**, 161 (1936).

Heptahydrate. Colorless, deliquesc needles. mp 732°. Sol in water, alcohol. *Keep well closed.*

1995. Cerous Carbonate. [537-01-9] Carbonic acid cerium-(3+) salt (3:2); cerium tricarbonate. $C_3Ce_2O_9$; mol wt 460.26. C 7.83%, Ce 60.89%, O 31.28%. $Ce_2(CO_3)_3$. Prepn: Head, Holley, Jr., *USAEC* **LADC-5579**, 8 pp (1962).

Pentahydrate. Powder, or microcryst prisms. Almost insol in water; sol in dil mineral acids and in soln of ammonium salts.

1996. Cerous Chloride. [7790-86-5] Cerium chloride (CeCl$_3$); cerium trichloride. CeCl$_3$; mol wt 246.47. Ce 56.85%, Cl 43.15%. Prepn: Kleinheksel, Kremers, *J. Am. Chem. Soc.* **50**, 959 (1928); Didtschenko, **US 2932553** (1960 to Union Carbide); Harmon, Wickers, **US 2982603** (1961 to USAEC); Taylor, Carter, *J. Inorg. Nucl. Chem.* **24**, 387 (1962). Toxicity study: Graca *et al., Arch. Environ. Health* **5**, 437 (1962).

Very fine powder, mp 822°. d^{25} 3.97. Sol in water, alcohol (exothermic). LD$_{50}$ in mice, guinea pigs (mg/kg): 352, 110 i.p. (Graca).

Heptahydrate. Colorless to yellow deliquesc orthorhombic crystals. Begins to lose water above 90°, becomes anhydr by 230°. Very sol in H$_2$O, alcohol. *Keep well closed.*

USE: In manuf of Ce and Ce salts; with Al or Mg salts as catalyst in the polymerization of olefins.

1997. Cerous Fluoride. [7758-88-5] Cerium fluoride (CeF$_3$); cerium trifluoride. CeF$_3$; mol wt 197.11. Ce 71.09%, F 28.92%. Prepd by the action of HF on CeO$_2$: von Wartenberg, *Z. Anorg. Allg. Chem.* **244**, 343 (1940); by the action of F$_2$ or concd HF on CeSi$_2$: Sterba, *Ann. Chim. Phys.* [8] **2**, 193 (1904); from CCl$_3$F$_2$ and CeO$_2$: Pausewang, Rüdorff, *Z. Anorg. Allg. Chem.* **369**, 89 (1969).

Hexagonal crystals or powder. d 6.157. mp 1460°. Practically insol in, but slowly hydrolyzed by water.

1998. Cerous Iodide. [7790-87-6] Cerium iodide (CeI$_3$); cerium triiodide. CeI$_3$; mol wt 520.83. Ce 26.90%, I 73.10%. Prepn from Ce and HgI$_2$: Asprey *et al., Inorg. Chem.* **3**, 1137 (1964); from the oxide and HI in the presence of NH$_4$I: Taylor, Curtis, *J. Inorg. Nucl. Chem.* **24**, 387 (1962).

Bright yellow, orthorhombic crystals. mp 752°. Dec in moist air. Sol in water.

Nonahydrate. White to reddish-white crystals. Very sol in water, the soln readily decomposing with separation of iodine; sol in alc. *Keep well closed and protected from light.*

1999. Cerous Nitrate. [10108-73-3] Nitric acid cerium(3+) salt (3:1); cerium nitrate. CeN$_3$O$_9$; mol wt 326.13. Ce 42.96%, N 12.88%, O 44.15%. Ce(NO$_3$)$_3$. Prepn from Ce sulfate and Ca-(NO$_3$)$_2$: Blumenfeld, **US 2166702** (1939 to Soc. de Produits Chimiques des Terres Rares); in separation of cerium from other rare earths: Pierce, Butler, *Inorg. Synth.* **2**, 51 (1946); Scargill *et al., J. Inorg. Nucl. Chem.* **4**, 304 (1957). For physical and chemical properties *see* USAEC **IDO-14504**, 126 pp (1961). Toxicity study: Bruce *et al., Toxicol. Appl. Pharmacol.* **5**, 750 (1963).

Hexahydrate. LD$_{50}$ in rats: 290 mg/kg i.p.; 4.2 g/kg orally (Bruce).

USE: Separation of Ce from other rare earths; catalyst in hydrolysis of phosphoric acid esters; mixture with Th(NO$_3$)$_4$ (1:99) formerly used in manuf incandescent mantles.

2000. Cerous Oxalate. [139-42-4] [μ-[Ethanedioato(2−)-$\kappa O^1,\kappa O^{2'}:\kappa O^{1'},\kappa O^2$]]bis[ethanedioato(2−)-$\kappa O^1,\kappa O^2$]dicerium; oxalic acid cerium(3+) salt (3:2); tris(oxalato)dicerium; cerium oxalate; sedemesis. $C_6Ce_2O_{12}$; mol wt 544.29. C 13.24%, Ce 51.49%, O 35.27%. $Ce_2(C_2O_4)_3$. Prepn from Ce(NO$_3$)$_3$.6H$_2$O and oxalic acid: Wylie, *J. Chem. Soc.* **1947**, 1687. The article of commerce contains variable amounts of the oxalates of cerium, lanthanum, and other associated elements.

Nonahydrate. White or slightly pink, tasteless powder. Practically insol in water; sol in hot, moderately dil H$_2$SO$_4$ or HCl.

THERAP CAT: Antiemetic.

2001. Cerous Sulfate. [13454-94-9] Sulfuric acid cerium(3+) salt (3:2); cerium sulfate; dicerium trisulfate. $Ce_2O_{12}S_3$; mol wt

568.40. Ce 49.30%, O 33.78%, S 16.92%. $Ce_2(SO_4)_3$. Prepn: Blandin, Rerat, *Compt. Rend.* **239**, 1055 (1954); Wendlandt, *J. Inorg. Nucl. Chem.* **7**, 51 (1958).

Octahydrate. Orthorhombic octahedral crystals. d 2.87. Loses water on heating, becoming anhydr by 250°; loses SO_3 to form a basic salt above 650°. Soluble in water.

USE: Developing aniline black; said to be superior to vanadium for this purpose.

2002. Certolizumab Pegol. [428863-50-7] Anti-(human tumor necrosis factor α) immunoglobulin Fab′ fragment (human-mouse monoclonal CDP870 heavy chain) disulfide with human-mouse monoclonal CDP870 light chain, pegylated; CDP-870; Cimzia. Recombinant, humanized monoclonal antibody Fab′ fragment directed against tumor necrosis factor α (TNF-α) and linked to poly-ethylene glycol. Mol wt ~91 kDa. Prepn: D. S. Athwal *et al.*, **WO 0194585**; *eidem*, **US 7012135** (2001, 2006 both to Celltech). Clinical trial in Crohn's disease: S. Schreiber *et al.*, *Gastroenterology* **129**, 807 (2005). Review of clinical safety and efficacy in rheumatoid arthritis: R. Fleischmann, *Expert Opin. Biol. Ther.* **10**, 773-786 (2010).

THERAP CAT: Anti-inflammatory.

2003. Cerulenin. [17397-89-6] (2*R*,3*S*)-3-[(4*E*,7*E*)-1-Oxo-4,7-nonadien-1-yl]-2-oxiranecarboxamide; 2,3-epoxy-4-oxo-7,10-dodecadienamide; 2,3-epoxy-4-oxo-7,10-dodecadienoylamide; helicocerin. $C_{12}H_{17}NO_3$; mol wt 223.27. C 64.56%, H 7.68%, N 6.27%, O 21.50%. Antifungal antibiotic isolated from *Cephalosporium caerulens; Acryocylindrium oryzae; Helicoceras oryzae*: T. Hata *et al.*, *Jpn. J. Bacteriol.* **15**, 1075 (1960); Matsumae *et al.*, *J. Antibiot.* **16A**, 236, 239 (1963); Sano *et al.*, *ibid.* **20A**, 344 (1967); Furuya, Shirasaka, **JP 70 21638** (1970 to Sankyo), *C.A.* **73**, 108271k (1970). Biological characteristics: Matsumae *et al.*, *J. Antibiot.* **17A**, 1 (1964). Structure: Omura *et al.*, *ibid.* **20A**, 349 (1967). Abs config: *eidem*, *Chem. Pharm. Bull.* **17**, 2361 (1969); Arison, Omura, *J. Antibiot.* **27**, 28 (1974). Stereoselective synthesis of (+)- and (−)-*tetrahydrocerulenin* and corrected abs config of (+)-cerulenin: H. Ohrui, S. Emoto, *Tetrahedron Lett.* **1978**, 2095; J.-R. Pougny, P. Sinay, *ibid.* 3301. Stereoselective synthesis of the (+)-form from D-glucose: N. Sueda *et al.*, *ibid.* **1979**, 2039; M. Pietraszkiewicz, P. Sinay, *ibid.* 4741. Interrupts yeast-type fungi growth by inhibiting the biosynthesis of sterols and fatty acids. Mechanism of action studies: Nomura *et al.*, *J. Biochem.* **71**, 783 (1972); *eidem*, *J. Antibiot.* **25**, 365 (1972); *see also* D. Vance, *Biochem. Biophys. Res. Commun.* **48**, 649 (1972). Total synthesis of (±)-cerulenin: R. K. Boeckman, Jr., E.W. Thomas, *J. Am. Chem. Soc.* **99**, 2805 (1977); A. A. Jakubowski *et al.*, *Tetrahedron Lett.* **1977**, 2399; E. J. Corey, D. R. Williams, *ibid.* 3847; K. Mikami *et al.*, *Chem. Lett.* **1981**, 1721; A. A. Jakubowski *et al.*, *J. Org. Chem.* **47**, 1221 (1982).

White needles from benzene, mp 93-94°. bp 120° $(10^{-8}$ mm). $[\alpha]_D^{16}$ +63° (c = 2 in methanol). Stable in neutral and acidic solns. Sol in ethanol, acetone, benzene and most common solvents. Slightly sol in water. Practically insol in petr ether. LD_{50} in mice (mg/kg): 154 i.v.; 211 i.p.; 547 orally (Matsumae 1964).

DL-Form. mp 40-43°.

USE: Biochemical tool.

2004. Ceruletide. [17650-98-5] Caerulein; cerulein; FI-6934. $C_{58}H_{73}N_{13}O_{21}S_2$; mol wt 1352.41. C 51.51%, H 5.44%, N 13.46%, O 24.84%, S 4.74%. Cholecystokinin analog; decapeptide discovered in the skins of Australian amphibians: V. Erspamer *et al.*, *Nature* **212**, 204 (1966). Isoln from *Hyla caerulea* and structure: A. Anastasi *et al.*, *Experientia* **23**, 699 (1967); *eidem*, *Arch. Biochem. Biophys.* **125**, 57 (1968). Synthesis: L. Bernardi *et al.*, *Experientia* **23**, 700 (1967); *eidem*, **ZA 6704716**; *eidem*, **US 3472832** (1968, 1969 to Soc. Farm. Italia). Shows hypotensive activity, stimulates smooth muscle and increases gastric, pancreatic, and biliary secretions: V. Erspamer *et al.*, *Experientia* **23**, 702 (1967). Series of articles on metabolism: *Eur. J. Biochem.* **91**, 21-48 (1978). Pharmacodynamics: N. Iwatsuki, O. H. Petersen, *J. Cell Biol.* **79**, 533

(1978); T. Fujita *et al.*, *Adv. Exp. Med. Biol.* **106**, 147 (1978). Diagnostic use: R. Fujita *et al.*, *Acta Gastroenterol. Belg.* **40**, 167 (1977); C. Monti *et al.*, *Radiology* **129**, 611 (1978). Toxicology: T. Chieli *et al.*, *Toxicol. Appl. Pharmacol.* **23**, 480 (1972). Reviews of pharmacology and clinical applications: G. Bertaccini, *Pharmacol. Rev.* **28**, 127 (1976); and toxicology: M. E. Vincent *et al.*, *Pharmacotherapy* **2**, 223 (1982).

5-oxoPro–Gln–Asp–Tyr–Thr–Gly–Trp–Met–Asp–PheNH₂
|
SO₃H

mp 224-226° (dec). $[\alpha]_D^{20}$ −26° (c = 1 in DMF).

Diethylamine salt. [71247-25-1] Ceosunin; Takus; Tymtran. Off-white hygroscopic powder. Sol in DMF and DMSO. Insol in acetone, diethyl ether. LD_{50} i.v. in mice: 1012 mg/kg (Chieli).

THERAP CAT: Stimulant (gastric secretory). Diagnostic aid (pancreatic function; cholecystokinetic agent in cholecystography).

2005. Ceruloplasmin. [9031-37-2] Caeruloplasmin; ferroxidase. Intensely blue colored copper-containing glycoprotein of the α₂-globulin fraction of mammalian blood; it is the principal copper transport protein and is believed to play an important role in iron mobilization via its ferroxidase activity. First reported by C. G. Holmberg, *Acta Physiol. Scand.* **8**, 227 (1944). Isoln, purification and description of properties of porcine and human ceruloplasmin: C. G. Holmberg, C. B. Laurell, *Acta Chem. Scand.* **2**, 550 (1948). Ceruloplasmin accounts for 90-95% of the circulating copper in normal mammals. Its concentration increases by a factor of 2 to 3 during pregnancy and varies significantly in several diseases and hormonal conditions. Prepd by Cohn cold ethanol fractionation: Cohn *et al.*, *J. Am. Chem. Soc.* **68**, 459 (1946); Steinbuch, Quentin, *Nature* **183**, 323 (1959); and further purified from fraction IV: Sanders *et al.*, *Arch. Biochem. Biophys.* **84**, 60 (1959); **US 3003918** (1961 to Merck & Co.). Different mol wts have been reported for human ceruloplasmin, ranging from 126,000 to 160,000; the most generally accepted is 134,000 ±3000, *cf.* L. Ryder, I. Björk, *Biochemistry* **15**, 3411 (1976). Chemical and structural studies of porcine ceruloplasmin: Osaki *et al.*, *J. Biochem. (Tokyo)* **48**, 190 (1960); **50**, 24, 29 (1961); Mukasa *et al.*, *Biochim. Biophys. Acta* **168**, 132 (1968); Matsunaga, Nosoh, *ibid.* **215**, 280 (1970); of human: Kasper, Deutsch, *J. Biol. Chem.* **238**, 2325 (1963); Jamieson, *ibid.* **240**, 2019 (1965); Poillon, Bearn, *Biochim. Biophys. Acta* **127**, 407 (1966); Simons, Bearn, *ibid.* **175**, 260 (1969); Ryden, *Eur. J. Biochem.* **26**, 380 (1972); T. G. Samsonidze *et al.*, *Int. J. Pept. Protein Res.* **14**, 161 (1979); V. N. Zaitsev *et al.*, *Kristallografiya* **25**, 174 (1980), *C.A.* **92**, 210415q (1980). Human metabolism: Kekki *et al.*, *Nature* **209**, 1252 (1966). Reviews of biological role: E. Frieden, H. S. Hsieh, *Adv. Exp. Med. Biol.* **74**, 505-529 (1976); J. M. C. Gutteridge, *Ann. Clin. Biochem.* **15**, 293-296 (1978). Comprehensive review: S. H. Laurie, E. S. Mohammed, *Coord. Chem. Rev.* **33**, 279-312 (1980).

Absorption max: 280, 610 nm ($E_{1cm}^{1\%}$ 14.9, 0.68). Electrophoretic mobility (cm/volt/sec): −5.05 × 10^{-5} at pH 7.0 (0.1*M* phosphate buffer); −5.32 × 10^{-5} at pH 8.6 (0.1*M* barbital sodium buffer).

2006. Cervicarcin. [18700-78-2] 1,2,3,4-Tetrahydro-1,3,4,-5,10-pentahydroxy-2-methyl-3-[(3-methyloxiranyl)carbonyl]-4a,9a-epoxyanthracen-9(10*H*)-one; 4a,9a-epoxy-3-(2,3-epoxybutyryl)-1,-2,3,4,4a,9a-hexahydro-1,3,4,5,10-pentahydroxy-2-methylanthrone. $C_{19}H_{20}O_9$; mol wt 392.36. C 58.16%, H 5.14%, O 36.70%. Antineoplastic antibiotic produced by *Streptomyces ogaensis*: Okuma *et al.*, *J. Antibiot.* **15A**, 152, 247 (1962). Prepn: Sumiki *et al.*, **JP 64 7400** (1964 to Inst. Phys. & Chem. Res.). Structure: Marumo *et al.*, *J. Am. Chem. Soc.* **86**, 4507 (1964); *eidem*, *Agric. Biol. Chem.* **32**, 209 (1968). Stereochemistry: *eidem*, *ibid.* **35**, 1931 (1971). Activity studies: C. Itakura *et al.*, *J. Antibiot.* **16A**, 231 (1963).

Needles, mp 203-205° (dec). $[\alpha]_D^{20}$ −144° (in ethanol). uv max: 227, 264, 323 nm (ε 14200, 7800, 2600). pKa 9.0. Sol in acetone, lower alcohols, acetic acid, pyridine. Moderately sol in ethyl acetate, ethyl ether, chloroform, carbon tetrachloride. Slightly sol in water, benzene. Insol in petr ether, ligroin. Stable in neutral or acidic soln; unstable in alkaline soln. LD_{50} i.p. in mice: 48.5 mg/kg (Itakura).

2007. Cesium. [7440-46-2] Caesium. Cs; at. wt 132.9054519; at. no. 55; valence 1. Group IA (1). Alkali metal. Occurrence in earth's crust: 2.6 ppm. Naturally occurring isotope: ^{133}Cs (100%); artificial isotopes (mass nos.): 112-132, 134-148. Occurs in nature in the aluminosilicates, ***pollucite*** and ***lepidolite*** and in the borate, ***rhodizite***. Acid digestion of pollucite ore is primary commercial source. Discovered by Bunsen and Kirchhoff in 1860. Prepn from pollucite: *Inorg. Synth.* **4**, 5 (1953). ^{137}Cs (T½ 30.07 years; β^- 0.514, 1.176 MeV) is a product of atomic fission of uranium and an important constituent of radioactive fallout; decays to and reaches radioactive equilibrium with ^{137m}Ba (T½ 2.552 min; γ 0.662 MeV). Review of cesium and its compounds: Whaley, "Sodium, Potassium, Rubidium, Cesium, and Francium" in *Comprehensive Inorganic Chemistry* **vol. 1**, J. C. Bailar, Jr. *et al.*, Eds. (Pergamon Press, Oxford, 1973) pp 369-529; *Chemistry of the Elements*, N. N. Greenwood, A. Earnshaw, Eds. (Pergamon Press, New York, 1984) pp 75-116; R. O. Burt in *Kirk-Othmer Encyclopedia of Chemical Technology* **vol. 5** (Wiley-Interscience, New York, 4th ed., 1993) pp 749-764. Review of toxicology and human exposure: *Toxicological Profile for Cesium* (PB2004-104397, 2004) 310 pp. Clinical use of ^{131}Cs in prostate brachytherapy: W. S. Bice *et al.*, *Brachytherapy* **7**, 290 (2008); R. Yang *et al.*, *Cancer Biother. Pharmaceut.* **24**, 701 (2009).

Silver-white, ductile metal; body-centered cubic structure. mp 28.5°. bp 705°. d^{20} 1.90. Specific heat (25°) 0.057 cal/g/°C. E°/v (aq) Cs/Cs$^+$ −2.92. Electrical resistivity 36.6 micro-ohm cm at 30°. Mohs' hardness 0.2. Oxidizes rapidly in air; in moist air may ignite spontaneously to produce nonluminous reddish-violet flame. Emits characteristic blue color (455.5 nm) in flame. Reacts with water to form hydroxide with evolution of hydrogen which ignites spontaneously. Sol in liquid ammonia. *Dangerous when wet. Keep immersed in mineral oil.*

USE: In photoelectric cells, as a "getter" in vacuum tubes; in photoemitter devices, scintillation counters; in industrial radiography. Adsorbent in CO_2 purifn; scavenger of gases and impurities in metallurgy. For doping catalysts. As plasma seeding agent in magnetohydrodynamic power generators. For construction and operation of one type of atomic clock based on the vibrational frequency (9,192.76 megacycles/sec) of ^{133}Cs. ^{137}Cs as gamma ray source for sterilization of food, medical instruments, sewage and sludge

THERAP CAT: ^{131}Cs as antineoplastic (radiation source).

2008. Cesium Bromide. [7787-69-1] BrCs; mol wt 212.81. Br 37.55%, Cs 62.45%. CsBr. Prepn from pollucite: Thomas, Steton, *Compt. Rend.* **241**, 56 (1955). Toxicity study: Cochran *et al.*, *Arch. Ind. Hyg.* **1**, 637 (1950).

Crystals. d 4.44. mp 636°. bp ~1300°. Very sol in water; sol in alc. Practically insol in acetone. LD_{50} i.p. in rats: 1.4 g/kg (Cochran).

USE: In x-ray fluorescent screens, spectrometer prisms, absorption-cell windows.

2009. Cesium Carbonate. [534-17-8] Carbonic acid cesium salt (1:2); dicesium carbonate. CCs$_2$O$_3$; mol wt 325.82. C 3.69%, Cs 81.58%, O 14.73%. Cs$_2$CO$_3$. Prepn: Suhrmann, Clusius, *Z. Anorg. Allg. Chem.* **152**, 52 (1926); Thomas, *Ann. Chim.* **6**, 367 (1957); Bernard, *Compt. Rend.* **243**, 1528 (1956).

Very deliquesc crystals; melts at red heat. Extremely sol in water, alcohol; sol in ether.

USE: Catalyst in ethylene oxide polymerization; in coating for spatter-free welding of steel in CO_2; in oxide cathode.

2010. Cesium Chloride. [7647-17-8] ClCs; mol wt 168.36. Cl 21.06%, Cs 78.94%. CsCl. Prepn from carnallite and from pollucite: Donges in *Handbook of Preparative Inorganic Chemistry* **vol. 1**, G. Brauer, Ed. (Academic Press, New York, 2nd ed., 1963) pp 951, 955. Toxicity study: Cochran *et al.*, *Arch. Ind. Hyg.* **1**, 637 (1950).

Deliquesc cubic crystals. d 3.99. mp 646°. bp 1303°. Very sol in water; sol in alc. *Keep well closed.* LD_{50} i.p. in rats: 1.5 g/kg (Cochran).

^{131}CsCl. Cescan-131.

USE: In the final evacuation of radio and television vacuum tubes; in x-ray fluorescent screens; in manuf of cesium. ^{131}CsCl as radiographic contrast medium.

2011. Cesium Fluoride. [13400-13-0] CsF; mol wt 151.90. Cs 87.50%, F 12.51%. Basic catalyst in fluoride chemistry and oxidation reactions. Prepn from pulverized pollucite: R. V. Winsor, G. H. Cady, *J. Am. Chem. Soc.* **70**, 1500 (1948). Properties as a scintillator: M. Moszynski *et al.*, *Nucl. Instrum. Methods* **179**, 271 (1981); and use as a detector in positron cameras: R. Allemand *et al.*, *J. Nucl. Med.* **21**, 153 (1980). Vapor pressure determn: H. Kawano *et al.*, *Bull. Chem. Soc. Jpn.* **57**, 581 (1984). Infrared spectroscopy: A. G. Maki, W. B. Olson, *J. Mol. Spectrosc.* **140**, 185 (1990). Use in malonic ester synthesis: T. Sato, J. Otera, *J. Org. Chem.* **60**, 2627 (1995); in trifluoromethylation of esters, aldehydes and ketones: R. P. Singh *et al.*, *ibid.* **64**, 2873 (1999); in synthesis of esters and ethers: S. T. A. Shah *et al.*, *Tetrahedron* **61**, 6652 (2005); in Stille coupling reactions: S. P. H. Mee *et al.*, *Chem. Eur. J.* **11**, 3294 (2005).

Hygroscopic crystalline solid, mp 682°. d 4.61.

USE: Catalyst in oxidation reactions; as a solid base when adsorbed onto Celite, *q.v.* Fluoride source in organic synthesis. Inorganic scintillant.

2012. Cesium Hydroxide. [21351-79-1] Cesium hydrate. CsHO; mol wt 149.91. Cs 88.66%, H 0.67%, O 10.67%. CsOH. Prepd by electrolysis of Cs salts: Winslow *et al.*, *J. Phys. Colloid Chem.* **51**, 967 (1947); Jolibois, Berger, *Compt. Rend.* **224**, 78 (1947). Alternate prepn: H. Jacobs, B. Harbrecht, *Z. Naturforsch.* **B36**, 270 (1981). Toxicity study: Cochran *et al.*, *Arch. Ind. Hyg.* **1**, 637 (1950).

White or yellowish, fused, very deliquescent, crystalline mass; strongly alkaline reaction. mp 272°. d 3.68. *Corrosive.* Sol in about 0.25 part water, much heat being evolved; sol in alcohol. *Keep tightly closed.* LD_{50} i.p. in rats: 100 mg/kg (Cochran).

Caution: Potential symptoms of overexposure are irritation of eyes, skin, upper respiratory tract; eye and skin burns. *See NIOSH Pocket Guide to Chemical Hazards* (DHHS/NIOSH 97-140, 1997) p 56.

USE: In storage-battery electrolytes; as catalyst in the polymerization of cyclic siloxanes.

2013. Cesium Iodide. [7789-17-5] CsI; mol wt 259.81. Cs 51.15%, I 48.85%. Prepn: *Gmelins, Cesium* (8th ed.) **25**, pp 188-204 (1938). Toxicity study: Cochran *et al.*, *Arch. Ind. Hyg.* **1**, 637 (1950).

Deliquesc crystals or crystalline powder. d 4.5. mp 621°. bp about 1280°. Very sol in water; sol in ethanol; slightly sol in methanol. Practically insol in acetone. *Keep well closed.* LD_{50} i.p. in rats: 1.4 g/kg (Cochran).

USE: Prisms for infrared spectroscopy; in x-ray fluorescent screens, scintillation counters.

2014. Cesium Nitrate. [7789-18-6] Nitric acid cesium salt (1:1). CsNO$_3$; mol wt 194.91. Cs 68.19%, N 7.19%, O 24.63%. Prepn from pollucite: Watt, *Inorg. Synth.* **4**, 6 (1953). Toxicity study: Cochran *et al.*, *Arch. Ind. Hyg.* **1**, 637 (1950).

White, lustrous hexagonal or cubic prisms; saltpeter taste. d$_4^{20}$ 3.64-3.68. mp 414°; dec at higher temp. *Oxidizer.* Sol in 5 parts cold, 0.5 part boiling water; sol in acetone; very slightly sol in alc. LD_{50} i.p. in rats: 1.2 g/kg (Cochran).

USE: Prepn of other Cs salts.

2015. Cesium Sulfate. [10294-54-9] Sulfuric acid cesium salt (1:2); dicesium sulfate. Cs$_2$O$_4$S; mol wt 361.87. Cs 73.45%, O 17.68%, S 8.86%. Cs$_2$SO$_4$. Prepn: *Gmelins, Cesium* (8th ed.) **25**, pp 218-225 (1938).

Orthorhombic or hexagonal prisms. d 4.24. mp 1019°. Very sol in water. Practically insol in alc, acetone, pyridine.

USE: With V or V$_2$O$_5$ as catalyst for SO$_2$ oxidation.

2016. Cetalkonium Chloride. [122-18-9] *N*-Hexadecyl-*N*,*N*-dimethylbenzenemethanaminium chloride (1:1); benzylhexadecyldimethylammonium chloride; cetyldimethylbenzylammonium chloride; hexadecyldimethylbenzylammonium chloride; Banicol; Acetoquat CDAC; Acquat CDAC; Ammonyx G; Zettyn; Ammonyx T; Cetol. C$_{25}$H$_{46}$ClN; mol wt 396.10. C 75.81%, H 11.71%, Cl 8.95%, N 3.54%. Prepn: **FR 771746** (1934 to I. G. Farben); Piggott, **US 2075958** (1937 to I.C.I.); Westphal, Jerchel, *Ber.* **73**, 1011

(1940). *Review* of this type of quaternary ammonium compounds: J. P. Sisley, P. J. Wood, *Encyclopedia of Surface-Active Agents* (Chem. Publishing Co., New York, 1952) p 96 sqq.

Leaflets from ethyl acetate + petr ether, mp 59°. Sol in water, alc, acetone, ethyl acetate, propylene glycol, sorbitol solns, glycerol, ether, CCl_4. pH of aq solns 7.2.

USE: Cationic quaternary ammonium surfactant germicide and fungicide. Used in leather processing, textile dyeing. A mildew preventive in silicone-based water repellents. Compatible with many nonionic detergents. Active in moderately alkaline solns.

THERAP CAT: Anti-infective (topical).

2017. Cethexonium Bromide. [1794-74-7] *N*-Hexadecyl-2-hydroxy-*N*,*N*-dimethylcyclohexanaminium bromide (1:1); hexadecyl(2-hydroxycyclohexyl)dimethylammonium bromide; cetyldimethyl(2-hydroxycyclohexyl)ammonium bromide; dimethylcetyl-2-cyclohexanolammonium bromide; *N*-cetyl-*N*,*N*-dimethyl-2-cyclohexanolammonium bromide; Biocidan. $C_{24}H_{50}BrNO$; mol wt 448.57. C 64.26%, H 11.24%, Br 17.81%, N 3.12%, O 3.57%. Prepd by the action of cetyl bromide upon 2-dimethylaminocyclohexanol: Mousseron *et al.*, *Bull. Soc. Chim. Biol.* **33**, 369 (1951).

White powder, mp 75°. Sol in water, alcohol, chloroform. Practically insol in petr ether.

THERAP CAT: Antiseptic.

2018. Cethromycin. [205110-48-1] (3a*S*,4*R*,7*R*,9*R*,10*R*,-11*R*,13*R*,15*R*,15a*R*)-4-Ethyloctahydro-3a,7,9,11,13,15-hexamethyl-11-[[3-(3-quinolinyl)-2-propenyl]oxy]-10-[[3,4,6-trideoxy-3-(dimethylamino)-β-D-*xylo*-hexopyranosyl]oxy]-2*H*-oxacyclotetradecino[4,3-*d*]oxazole-2,6,8,14(1*H*,7*H*,9*H*)-tetrone; A-195773; ABT-773; Restanza. $C_{42}H_{59}N_3O_{10}$; mol wt 765.95. C 65.86%, H 7.76%, N 5.49%, O 20.89%. Semisynthetic macrolide antibiotic of the ketolide class; structurally similar to telithromycin, *q.v.* Prepn: Y. S. Or *et al.*, **WO 9809978**; *eidem*, **US 5866549** (1998, 1999 both to Abbott); *eidem*, *J. Med. Chem.* **43**, 1045 (2000). Improved prepn: D. J. Plata *et al.*, *Tetrahedron* **60**, 10171 (2004). Comparative *in vitro* antibacterial spectrum: A. M. Nilius *et al.*, *Antimicrob. Agents Chemother.* **45**, 2163 (2001). Mechanism of action study: W. S. Champney, J. Pelt, *Curr. Microbiol.* **45**, 155 (2002). HPLC-MS determn in biological samples: Q. Ren *et al.*, *J. Chromatogr. Sci.* **41**, 494 (2003). Clinical pharmacokinetics: M. W. Pletz *et al.*, *Antimicrob. Agents Chemother.* **47**, 1129 (2003). Review of pharmacology, antibacterial spectrum, and clinical experience: M. R. Hammerschlag, R. Sharma, *Expert Opin. Invest. Drugs* **17**, 387-400 (2008).

White crystalline solid, mp 211-213°.

THERAP CAT: Antibacterial.

2019. Cetiedil. [14176-10-4] α-Cyclohexyl-3-thiopheneacetic acid 2-(hexahydro-1*H*-azepin-1-yl)ethyl ester; α-cyclohexyl-α-(3-thienyl)acetic acid 2-hexamethyleneiminoethyl ester. $C_{20}H_{31}NO_2S$; mol wt 349.53. C 68.73%, H 8.94%, N 4.01%, O 9.15%, S 9.17%. Prepn: Pons, Robba, **FR 1460571** and Pons *et al.*, **FR M5504** (1966, 1967, both to Innothera), *C.A.* **68**, 59429d (1968); **71**, 91286c (1969). Prepn and activity: Robba, LeGuen, *Chim. Ther.* **2**, 120 (1967). Antisickling effect: T. Asakura *et al.*, *Proc. Natl. Acad. Sci. USA* **77**, 2955 (1980); L. R. Berkowitz, E. P. Orringer, *J. Clin. Invest.* **68**, 1215 (1981).

Citrate. [16286-69-4] Stratene; Vasocet. $C_{20}H_{31}NO_2S.C_6H_8O_7$; mol wt 541.66. Crystals from ethanol-ether, mp 115°.

Hydrochloride. $C_{20}H_{31}NO_2S.HCl$. Crystals from acetonitrile, mp 152° (Robba, LeGuen); also mp 143° (Pons, Robba).

THERAP CAT: Vasodilator (peripheral).

2020. Cetilistat. [282526-98-1] 2-(Hexadecyloxy)-6-methyl-4*H*-3,1-benzoxazin-4-one; ATL-962. $C_{25}H_{39}NO_3$; mol wt 401.59. C 74.77%, H 9.79%, N 3.49%, O 11.95%. Oral pancreatic lipase inhibitor. Prepn: H. F. Hodson *et al.*, **WO 0040569**; *eidem*, **US 6624161** (2000, 2003 both to Alizyme). Pharmacology: Y. Yamada *et al.*, *Horm. Metab. Res.* **40**, 539 (2008). Clinical study in obese diabetics: P. Kopelman *et al.*, *Obesity* **18**, 108 (2010). Review of development and therapeutic potential: R. Padwal, *Curr. Opin. Investig. Drugs* **9**, 414-421 (2008).

White solid, mp 72-73°.

THERAP CAT: Antiobesity agent.

2021. Cetirizine. [83881-51-0] 2-[2-[4-[(4-Chlorophenyl)phenylmethyl]-1-piperazinyl]ethoxy]acetic acid; [2-[4-(*p*-chloro-α-phenylbenzyl)-1-piperazinyl]ethoxy]acetic acid. $C_{21}H_{25}ClN_2O_3$; mol wt 388.89. C 64.86%, H 6.48%, Cl 9.12%, N 7.20%, O 12.34%. Nonsedating type histamine H_1-receptor antagonist; major metabolite of hydroxyzine, *q.v.* Pharmacological activity resides primarily in the (*R*)-isomer. Prepn: E. Baltes *et al.*, **EP 58146**; *eidem*, **US 4525358** (1982, 1985 both to UCB); of enantiomers: E. Cossement *et al.*, **GB 2225321** (1989 to UCB). *See also*: *eidem*, **US 5478941** (1995 to UCB). Enantioselective HPLC determn in urine: S. O. Choi *et al.*, *Arch. Pharmacal Res.* **23**, 178 (2000). Stereoselective binding study: M. Gillard *et al.*, *Mol. Pharmacol.* **61**, 391 (2002). Clinical pharmacokinetics: E. Baltes *et al.*, *Fundam. Clin. Pharmacol.* **15**, 269 (2001). Clinical comparison with enantiomers on cutaneous response: J. L. Devalia *et al.*, *Allergy* **56**, 50 (2001); on nasal response: D. Y. Wang *et al.*, *ibid.* 339. Review of pharmacology and clinical experience in allergic disorders: J. M. Portnoy, C. Dinakar, *Expert Opin. Pharmacother.* **5**, 125-135 (2004).

Crystals from ethanol, mp 110-115°.

Dihydrochloride. [83881-52-1] P-071; Alerlisin; Formistin; Reactine; Virlix; Zirtek; Zyrtec; Zyrlex. $C_{21}H_{25}ClN_2O_3.2HCl$; mol wt 461.81. Crystals from isopropanol, mp 225°. Sol in water.

(*R*)-Form. [130018-77-8] Levocetirizine; (−)-cetirizine.

(*R*)-Form dihydrochloride. [130018-87-0] Levocetirizine hydrochloride; UCB-28556; Xusal; Xyzal. mp 229.3°. $[\alpha]_D^{25}$ −12.79° (c = 1 in water).

THERAP CAT: Antihistaminic.

2022. Cetotiamine. [137-76-8] Carbonic acid 4-[[(4-amino-2-methyl-5-pyrimidinyl)methyl]formylamino]-3-[(ethoxycarbonyl)-thio]-3-penten-1-yl ethyl ester; *O,S*-bis(ethoxycarbonyl)thiamine; *O,S*-dicarbethoxythiamine; thiocarbonic acid *O*-ethyl ester, *S*-ester with *N*-[(4-amino-2-methyl-5-pyrimidinyl)methyl]-*N*-(4-hydroxy-2-mercapto-1-methyl-1-butenyl)formamide ethyl carbonate (ester); DCET. $C_{18}H_{26}N_4O_6S$; mol wt 426.49. C 50.69%, H 6.15%, N 13.14%, O 22.51%, S 7.52%. Prepn: Takamizawa, Hirai, *Chem. Pharm. Bull.* **10**, 1102 (1962); Takamizawa *et al.*, *ibid.* 1107; Yamamoto *et al.*, *Bitamin* **25**, 472 (1962), *C.A.* **60**, 9773e (1964); **GB 944641** (1963 to Shionogi).

cis-form

Prisms from ethyl acetate + petr ether, mp 113.5-114.5°.

Hydrochloride monohydrate. Dicethiamin; Dicetamin. $C_{18}H_{26}N_4O_6S.HCl.H_2O$; mol wt 480.96. Crystals from ethyl acetate, dec 122-124°. Sol in water, methanol. Practically insol in ether, benzene.

THERAP CAT: Vitamin B_1 source.

2023. Cetraxate. [34675-84-8] 4-[[[*trans*-4-(Aminomethyl)-cyclohexyl]carbonyl]oxy]benzenepropanoic acid; *p*-hydroxyhydrocinnamic acid *trans*-(4-aminomethyl)cyclohexanecarboxylate; tranexamic acid *p*-(2-carboxyethyl)phenyl ester. $C_{17}H_{23}NO_4$; mol wt 305.37. C 66.87%, H 7.59%, N 4.59%, O 20.96%. Deriv of tranexamic acid, *q.v.* Prepn: Y. Yamamura *et al.*, **DE 1951061**; *eidem*, **US 3699149** (1970, 1972 both to Daiichi); O. Atsuji *et al.*, *J. Med. Chem.* **15**, 247 (1972); S. Kitahara, **JP Kokai 73 75547** (1973 to Daiichi), *C.A.* **80**, 59727x (1974). Mechanism of action: Y. Suzuki *et al.*, *Jpn. J. Pharmacol.* **29**, 829 (1979), *C.A.* **92**, 88029 (1980). Anti-ulcer effects in rats: T. Hashizume *et al.*, *Arch. Int. Pharmacodyn. Ther.* **240**, 314 (1979). Clinical study: A. Ishimori *et al.*, *Arzneim.-Forsch.* **29**, 1625 (1979); S. Yamagata, K. Miura, *ibid.* **33**, 1191 (1983).

Relative stereochemistry

Crystals from methanol, melts over a range of 200-280°.

Hydrochloride. [27724-96-5] DV-1006; Neuer. $C_{17}H_{23}NO_4$.HCl; mol wt 341.83. Crystals from methanol/ether, mp 238-240°.

THERAP CAT: Antiulcerative.

2024. Cetrimonium Bromide. [57-09-0] *N,N,N*-Trimethyl-1-hexadecanaminium bromide (1:1); hexadecyltrimethylammonium bromide; cetyltrimethylammonium bromide; Bromat; Cetab; Cetavlon; Cetylamine; C.T.A.B.; Lissolamine V; Micol; Quamonium. $C_{19}H_{42}BrN$; mol wt 364.46. C 62.62%, H 11.62%, Br 21.92%, N 3.84%. Prepd from cetyl bromide and trimethylamine: Shelton *et al.*, *J. Am. Chem. Soc.* **68**, 753 (1946). Toxicity and pharmacology: B. Isomaa, K. Bjondahl, *Acta Pharmacol. Toxicol.* **47**, 17 (1980). RP-HPLC determn in commercial formulations: A. Malenovic *et al.*, *Farmaco* **60**, 157 (2005).

Crystals, mp 237-243°. Soluble in about 10 parts water. Freely sol in alc; sparingly sol in acetone. Practically insol in ether, benzene. Stable in acid soln. LD_{50} in mice, rats (mg/kg): 32.0, 44.0 i.v. (Isomaa, Bjondahl).

***p*-Toluenesulfonate analog.** [138-32-9] Cetrimonium tosylate; Cetats. $C_{26}H_{49}NO_3S$; mol wt 455.74.

Note: **Cetrimide** is a mixture consisting chiefly of tetradecyltrimethylammonium bromide together with smaller amounts of dodecyltrimethylammonium bromide and cetrimonium bromide.

USE: As cationic detergent and antiseptic; as laboratory reagent.

THERAP CAT: Antiseptic, disinfectant.

THERAP CAT (VET): Antiseptic, cleansing agent.

2025. Cetrorelix. [120287-85-6] *N*-Acetyl-3-(2-naphthalenyl)-D-alanyl-4-chloro-D-phenylalanyl-3-(3-pyridinyl)-D-alanyl-L-seryl-L-tyrosyl-N^5-(aminocarbonyl)-D-ornithyl-L-leucyl-L-arginyl-L-prolyl-D-alaninamide; Ac-D-Nal(2)^1-D-Phe(4Cl)^2-D-Pal(3)^3-D-Cit^6-D-Ala^{10}-LH-RH; SB-75. $C_{70}H_{92}ClN_{17}O_{14}$; mol wt 1431.06. C 58.75%, H 6.48%, Cl 2.48%, N 16.64%, O 15.65%. Decapeptide GnRH antagonist. Prepn: A. V. Schally, S. Bajusz, **EP 299402** (1989 to Asta); *eidem*, **US 4800191** (1989); S. Bajusz *et al.*, *Int. J. Pept. Protein Res.* **32**, 425 (1988). Structural study: A. Müller *et al.*, *ibid.* **43**, 264 (1994). HPLC determn in plasma: H. H. Raffel *et al.*, *J. Chromatogr. B* **653**, 102 (1994). Pharmacology: T. Reissmann *et al.*, *Eur. J. Cancer* **32A**, 1574 (1996). Clinical evaluation in prostatic cancer: D. Gonzalez-Barcena *et al.*, *Urology* **45**, 275 (1995); in assisted fertilization: C. Albano *et al.*, *Fertil. Steril.* **67**, 917 (1997). Review of development and clinical experience in female infertility and endometriosis: D. Finas *et al.*, *Expert Opin. Pharmacother.* **7**, 2155-2168 (2006).

Ser-Tyr-D-Cit-Leu-Arg-Pro-D-AlaNH₂

Acetate. [145672-81-7] D-20761; Cetrotide. Sol in water.

THERAP CAT: Antineoplastic (hormonal). In treatment of female infertility.

2026. Cetuximab. [205923-56-4] Anti-(human epidermal growth factor receptor) immunoglobulin G1 (human-mouse monoclonal C225 γ_1-chain), disulfide with human-mouse monoclonal C225 κ-chain, dimer; C225; IMC-C225; Erbitux. Chimeric monoclonal antibody directed against human epithelial growth factor receptor (EGFR). Prepn: N. I. Goldstein *et al.*, **WO 9640210** (1995 to Imclone and MRC); *eidem*, *Clin. Cancer Res.* **1**, 1311 (1995). ELISA determn in serum: N. Cézé *et al.*, *Ther. Drug Monit.* **31**, 597 (2009). Clinical pharmacokinetics: C. Delbaldo *et al.*, *Eur. J. Cancer* **41**, 1739 (2005). Review of development, mechanisms of action, and biological properties: B. Vincenzi *et al.*, *Curr. Cancer Drug Targets* **10**, 80-95 (2010). Review of clinical experience in squamous cell carcinoma of head and neck: P. Specenier, J. B. Vermorken, *Expert Rev. Anticancer Ther.* **11**, 511-524 (2011); in colorectal cancer: C. R. Garrett, C. Eng, *Expert Opin. Biol. Ther.* **11**, 937-949 (2011).

THERAP CAT: Antineoplastic.

2027. Cetyl Alcohol. [36653-82-4] 1-Hexadecanol; ethal; ethol; palmityl alcohol. $C_{16}H_{34}O$; mol wt 242.45. C 79.26%, H 14.14%, O 6.60%. Discovered by Chevreul in 1813. Obtained from spermaceti by saponification: Spada, Gavioli, *Farm. Sci. Tec.* (Pavia) **7**, 435 (1952), *C.A.* **47**, 891c (1953). Prepn from palmitoyl chloride + NaBH₄: Caikin, Brown, *J. Am. Chem. Soc.* **71**, 122 (1949); from methylthiopalmitate + Raney Ni: Ruzicka, Prelog, **US 2509171** (1950 to Ciba); from hexadecyl bromide: Levine, Clippinger, **US 3018308** (1962 to California Res. Corp.).

White crystals. d 0.811. mp 49°. bp 344°; bp$_{15}$ 190°. n_D^{79} 1.4283. Practically insol in water. Sol in alcohol, chloroform, ether.

Hexadecyl alcohol. Primary, branched chain, C$_{16}$ alcohol, made up of an array of isomeric compds maintained in constant proportion by a complex manufacturing process (not from spermaceti): Edman, Lowden, *Drug Cosmet. Ind.* **93**, 631 (Nov. 1963). Liquid, d$_{20}^{20}$ 0.842. bp$_{50}$ 195-205°. Freezes at <−60°. Miscible with most alcohols, glycols, esters, ketones, cosmetic oils and aromatics. Immiscible with water.

USE: In cosmetics as emollient, emulsion modifier, coupling agent. Pharmaceutic aid (emulsifying and stiffening agent).

2028. Cetyldimethylethylammonium Bromide. [124-03-8] *N*-Ethyl-*N*,*N*-dimethyl-1-hexadecanaminium bromide (1:1); ethylhexadecyldimethylammonium bromide; ethyl cetab; CDA; Ammonyx DME. C$_{20}$H$_{44}$BrN; mol wt 378.48. C 63.47%, H 11.72%, Br 21.11%, N 3.70%. Cationic germicidal detergent. Prepn and antibacterial activity: R. S. Shelton *et al., J. Am. Chem. Soc.* **68**, 753 (1946).

White powder, mp 178-186°. Soluble in water, alcohol; slightly sol in chloroform, benzene, ether.

USE: Disinfectant; laboratory reagent.

2029. Cetyl Lactate. [35274-05-6] 2-Hydroxypropanoic acid hexadecyl ester; 1-hexadecanol lactate; lactic acid cetyl ester; lactic acid hexadecyl ester; Ceraphyl 28. C$_{19}$H$_{38}$O$_3$; mol wt 314.51. C 72.56%, H 12.18%, O 15.26%. Preparation: Rehberg, Marion, *J. Am. Chem. Soc.* **72**, 1918 (1950).

Waxy solid. mp 41°. bp$_{0.1}$ 132°; bp$_1$ 170°; bp$_{10}$ 219°. n_D^{40} 1.4410; n_D^{50} 1.4370.

USE: Nonionic emollient. To improve feel and texture of cosmetic and pharmaceutical prepns.

2030. Cetyl Palmitate. [540-10-3] Hexadecanoic acid hexadecyl ester; palmitic acid hexadecyl ester; palmitic acid palmityl ester; hexadecyl hexadecanoate; hexadecyl palmitate; palmityl palmitate. C$_{32}$H$_{64}$O$_2$; mol wt 480.86. C 79.93%, H 13.42%, O 6.65%. Naturally occuring wax; constituent of spermaceti, *q.v.*, also produced by reef corals and other marine organisms. Isoln from spermaceti: Heintz, *Ann.* **80**, 293 (1851); F. Krafft, *Ber.* **16**, 3018 (1883). Isoln from the staghorn coral, *Madrepora cervicornis*: D. Lester, W. Bergmann, *J. Org. Chem.* **6**, 120 (1941). Prepn from palmitoyl chloride and cetyl alcohol in the presence of Mg: Paquot, Bouquet, *Bull. Soc. Chim. Fr.* **1947**, 321; by CrO$_3$H$_2$SO$_4$ oxidation of cetyl alcohol: Cymerman-Craig, Horning, *J. Org. Chem.* **25**, 2098 (1960). Biosynthesis using inoculum of *Nocardia salmonicolor:* Davis, **US 3169099** (1965 to Socony Mobil Oil). Synthesis by esterification of cetyl alcohol using H$_2$O$_2$ and HBr: M. G. Kulkarni, S. B. Sawant, *Eur. J. Lipid Sci. Technol.* **104**, 387 (2002). Review of safety assessment for use in cosmetics: *J. Am. Coll. Toxicol.* **1**, 13-35 (1982).

Monoclinic leaflets, mp 54°. d^{20} 0.989. n_D^{70} 1.4398. Soly in chloroform: 100 mg/ml. Freely sol in alc, ether. Practically insol in water.

USE: In cosmetics as emollient; base for lotions, creams, and emulsions. Pharmaceutic aid (stiffening agent).

2031. Cetylpyridinium Chloride. [123-03-5] 1-Hexadecylpyridinium chloride (1:1); Ceepryn; Medilave; Merocet; Pristacin; Pyrisept. C$_{21}$H$_{38}$ClN; mol wt 339.99. C 74.19%, H 11.27%, Cl 10.43%, N 4.12%. Pharmacology and toxicology: *J. Pharmacol. Exp. Ther.* **74**, 401 (1942). Review of early literature: C. L. Huyck, *Am. J. Pharm.* **116**, 50 (1944). Toxicity data: J. W. Nelson, S. C. Lyster, *J. Am. Pharm. Assoc.* **35**, 89 (1946).

Monohydrate. [6004-24-6] Dobendan; Halset. White powder, mp 77-83°. Freely sol in water, alcohol, chloroform; slightly sol in benzene, ether. pH (1% aq soln): 6.0 to 7.0. Surface tension (25°): 43 dyn/cm (0.1% aq soln); 41 dyn/cm (1.0%); 38 dyn/cm (10%). LD$_{50}$ in rats (mg/kg): 250 s.c.; 6 i.p.; 30 i.v.; 200 orally (Nelson, Lyster).

USE: Pharmaceutic aid (preservative).

THERAP CAT: Antiseptic; disinfectant.

THERAP CAT (VET): Topical antiseptic; disinfectant.

2032. Cevadine. [62-59-9] [3β,4α,16β]-4,9-Epoxycevane-3,4,12,14,16,17,20-heptol 3-[(2Z)-2-methyl-2-butenoate]; veratrine. C$_{32}$H$_{49}$NO$_9$; mol wt 591.74. C 64.95%, H 8.35%, N 2.37%, O 24.33%. From seeds of *Schoenocaulon officinale* (Schlecht & Cham.) A. Gray *(Sabadilla officinarum* Brandt), *Liliaceae:* Poetsch *et al., J. Am. Pharm. Assoc.* **38**, 525 (1949); Ringel, *ibid.* **45**, 433 (1956). Evaluation as insecticide: Ikawa, Link *et al., J. Biol. Chem.* **159**, 517 (1945). Structure: Kupchan, Alfonso, *ibid.* **49**, 242 (1960). Toxicity study: Swiss, Bauer, *Proc. Soc. Exp. Biol. Med.* **76**, 847 (1951). *Review:* Wintersteiner in Graff, *Essays in Biochemistry* (Wiley, New York, 1956) pp 308-321.

Flat needles from ether, decomp 213-214.5°. [α]$_D^{20}$ +12.8° (c = 3.2 in alc). One gram dissolves in about 15 ml alc or ether; slightly sol in water. LD$_{50}$ i.p. in mice: 3.5 mg/kg (Swiss, Bauer).

Aurichloride. Fine yellow needles from alc, dec 190°.

Mercurichloride. C$_{32}$H$_{49}$NO$_9$.HCl.HgCl$_2$. Silvery scales, dec 172°.

Caution: Extremely irritating locally, particularly to mucous membranes. Serious poisoning has resulted from local application. Caution must be used in handling. *See also* Veratrine (Mixture).

2033. Cevimeline. [107233-08-9] *rel*-(2'*R*,3*R*)-2'-Methylspiro[1-azabicyclo[2.2.2]octane-3,5'-[1,3]oxathiolane]; (±)-*cis*-2-methylspiro[1,3-oxathiolane-5,3'-quinuclidine]. C$_{10}$H$_{17}$NOS; mol wt 199.31. C 60.26%, H 8.60%, N 7.03%, O 8.03%, S 16.09%. Muscarinic M$_1$ and M$_3$ receptor agonist. Prepn: A. Fisher *et al.,* **JP Kokai 61 280497**; *eidem,* **US 4855290**; (1986, 1989 both to State of Israel). Improved process: K. Hayashi *et al.,* **US 5571918** (1996 to Ishihara Sangyo Kaisha). Sialogogic effect in animals: H. Masunaga *et al., Eur. J. Pharmacol.* **339**, 1 (1997). General pharmacology: H. Arisawa *et al., Arzneim.-Forsch.* **52**, 14, 81 (2002). Clinical experience in Sjögren's syndrome dry eye: M. Ono *et al., Am. J. Ophthalmol.* **138**, 6 (2004); in dry mouth: K. Suzuki *et al., Pharmacology* **74**, 100 (2005). Review of clinical pharmacokinetics and efficacy in Sjögren's syndrome: H. Yasuda, H. Niki, *Clin. Drug Invest.* **22**, 67-73 (2002).

Relative stereochemistry

Hydrochloride hemihydrate. [153504-70-2]; [107220-28-0] (anhydrous). AF-102B; SNI-2011; Evoxac. $C_{10}H_{17}NOS.HCl.$½H_2O; mol wt 244.78. White to off white crystalline powder, mp 201-203°. Freely sol in alcohol, chloroform; very sol in water. Practically insol in ether.

THERAP CAT: Sialagogue.

2034. Cevine. [124-98-1] (3α,4α,16β)-4,9-Epoxycevane-3,-4,12,14,16,17,20-heptol. $C_{27}H_{43}NO_8$; mol wt 509.64. C 63.63%, H 8.50%, N 2.75%, O 25.11%. By hydrolysis of cevadine. Structure and stereochemistry: Barton *et al.*, *Experientia* **10**, 81 (1954); Kupchan *et al.*, *J. Am. Chem. Soc.* **80**, 1769 (1958); Kupchan *et al.*, *Tetrahedron* **7**, 47 (1959); Eeles, *Tetrahedron Letters* no. 7, 24 (1960). Comparative toxicity: O. Krayer *et al.*, *J. Pharmacol. Exp. Ther.* **82**, 167 (1944); K. Tanaka, *ibid.* **113**, 89 (1955).

Occurs as the hemiheptahydrate, triclinic prisms from dil alc. After drying at 110° it sinters at 165° and becomes a transparent resin at 165-170°: M. Ikawa *et al.*, *J. Biol. Chem.* **159**, 517 (1945). $[\alpha]_D^{17}$ −17.5° (aq alc). Sol in water, alc; slightly sol in ether. LD_{50} i.p. in rats: 87.0 mg/kg (Krayer). LD_{50} s.c. in male mice: 160 mg/kg (Tanaka).

Hydrochloride. mp 247°.

2035. Chalcomycin. [20283-48-1] (1S,2R,3R,6E,8S,9S,10S,-12S,14E,16S)-2[[(6-Deoxy-2,3-di-O-methyl-β-D-allopyrano-syl)oxy]methyl]-9-[(4,6-dideoxy-3-O-methyl-β-D-xylo-hexopyrano-syl)oxy]-12-hydroxy-3,8,10,12-tetramethyl-4,17-dioxabicyclo-[14.1.0]heptadeca-6,14-diene-5,13-dione. $C_{35}H_{56}O_{14}$; mol wt 700.82. C 59.98%, H 8.05%, O 31.96%. Antibiotic substance produced from *Streptomyces bikiniensis* NRRL 2737: Frohardt *et al.*, **DE 1109835** (1960 to Parke, Davis). Structure: Woo *et al.*, *J. Am. Chem. Soc.* **86**, 2726 (1964). *Review:* Jordan, "Chalcomycin" in *Antibiotics* **I**, D. Gottlieb, P. Shaw, Eds. (Springer-Verlag, New York, 1967) pp 446-450.

Crystals from butanol, mp 121-123°. $[\alpha]_D^{27}$ −43.5° (ethanol). uv max: 218 nm (ε 22770). Sol in ethyl acetate, methanol, ethanol, benzene, toluene, chloroform; sparingly sol in carbon tetrachloride, ether, water. Practically insol in petr ether.

2036. Chalcone. [94-41-7] 1,3-Diphenyl-2-propen-1-one; chalkone; benzylideneacetophenone; benzalacetophenone; phenyl styryl ketone. $C_{15}H_{12}O$; mol wt 208.26. C 86.51%, H 5.81%, O 7.68%. Naturally occuring aromatic ketone. Intermediate in the biosynthesis of flavones and template for numerous bioactive compounds known collectively as chalcones. Usually exists as the thermodynamically more stable (E)-isomer. Prepd by the condensation of benzaldehyde and acetophenone: E. P. Kohler, H. M. Chadwell, *Org. Synth.* **coll. vol. I**, 78 (1941); D. S. Breslow, D. R. Hauser, *J. Am. Chem. Soc.* **62**, 2385 (1940). Prepn of *cis*-chalcone by irradiation of *trans*-isomer: R. E. Lutz, R. H. Jordan, *ibid.* **72**, 4090 (1950); and reactivity of isomers: R. E. Lutz, J. O. Weiss, *ibid.* **77**, 1814 (1955). Conformational studies: P. Baas, H. Cerfontain, *Tetrahedron* **33**, 1509 (1977).

(E)-form

(E)-Form. [614-47-1] *trans*-Chalcone. Pale yellow crystals from alcohol, mp 56-57°. bp$_{25}$ 208°. d_4^{62} 1.0712. uv max (isooctane): 298 nm (ε 23600).

(Z)-Form. [614-46-0] *cis*-Chalcone. Deep yellow rosettes from *n*-pentane, mp 45-46°. uv max (isooctane): 248, 290 nm (ε 14000, 8950).

Caution: Skin irritant.

2037. Chalcopyrite. [1308-56-1] Cupric ferrous sulfide; copper pyrites; yellow copper; Chalkopyrite; Kupferkies. $CuFeS_2$; mol wt 183.51. Cu 34.63%, Fe 30.43%, S 34.94%. Occurs as a mineral. It is an important copper ore found in Canada, Montana, Utah, Arizona, Tennessee, Chile, Peru, Bolivia, Europe, Japan. Prepd synthetically by treating $KFeS_2$ with an ammoniacal soln of Cu^+ ions: Schneider, *J. Prakt. Chem.* [2] **38**, 569 (1888); **56**, 415 (1897); Boon, *Rec. Trav. Chim.* **63**, 69 (1954).

Yellow brass-colored or bronze-colored crystals. Metallic, greenish sheen. Tetragonal crystal system. d 4.1-4.3. mp 950°. Specific heat 0.1291. Hardness 3.5-4.0. Soluble in nitric acid, but dissolves faster in aqua regia. Practically insol in hydrochloric acid.

Note: Copper iron sulfide of the formula $FeS.2Cu_2S.CuS$ occurs as the mineral **bornite**.

USE: The ore as source of copper. The synthetic material in semiconductor research.

2038. D-Chalcose. [3150-28-5] 4,6-Dideoxy-3-O-methyl-D-xylo-hexose; 4,6-dideoxy-3-O-methyl-D-glucose; lankavose. $C_7H_{14}O_4$; mol wt 162.19. C 51.84%, H 8.70%, O 39.46%. Component of chalcomycin and lankamycin, *q.q.v.* Prepn from chalcomycin and structure: Woo *et al.*, *J. Am. Chem. Soc.* **83**, 3352 (1961). Stereochemistry: Woo *et al.*, *ibid.* **84**, 1066 (1962); Foster *et al.*, *J. Chem. Soc.* **1965**, 2318. Synthesis: Kochetkov, Usov, *Tetrahedron Lett.* **1963**, 519; McNally, Overend, *Chem. Ind. (London)* **1964**, 2021; Lawton *et al.*, *Can. J. Chem.* **47**, 2899 (1969). Synthesis of DL-form: R. M. Srivastava, R. K. Brown, *Can. J. Chem.* **48**, 830 (1970); K.Torssell, M. P. Tyagi, *Acta Chem. Scand. B* **31**, 1 (1977); S. Danishefsky, J. F. Kerwin, *J. Org. Chem.* **47**, 1597 (1982).

Crystals, mp 96-99°. $[\alpha]_D^{24}$ +120° (2 min) → +97° (10 min) → +76° (3 hr and 26 hr) (c = 1.5 in water).

Methyl chalcoside. $C_8H_{16}O_4$. Crystals, mp 101.5-102°. $[\alpha]_D^{27}$ −21° (c = 2.04 in chloroform).

2039. Chamazulene. [529-05-5] 7-Ethyl-1,4-dimethylazulene; 1,4-dimethyl-7-ethylazulene; dimethulene. $C_{14}H_{16}$; mol wt 184.28. C 91.25%, H 8.75%. Anti-inflammatory principle obtained from chamazulenogenic compds found in chamomile (*Matricaria chamomilla* L.), wormwood (*Artemisia absinthium* L.), and yarrow (*Achillea millefolium* L., Compositae). Isoln: Ruzicka, Rudolph, *Helv. Chim. Acta* **9**, 118 (1926); Ruzicka, Haagen-Smit, *ibid.* **14**, 1104 (1931). Structure: Sorm *et al.*, *Collect. Czech. Chem. Commun.* **18**, 527 (1953); Meisels, Weizmann, *J. Am. Chem. Soc.* **75**, 3865 (1953); Suchy *et al.*, *Collect. Czech. Chem. Commun.* **21**, 477 (1956). Synthesis: Mangoni, Bandiera, *Gazz. Chim. Ital.* **90**, 947 (1960); White, Winter, *Tetrahedron Lett.* **1963**, 137; D. Mukherjee *et al.*, *J. Am. Chem. Soc.* **101**, 251 (1979). Pharmacology: I. G. Boldina, *Farmakol. Toksikol.* **29**, 672 (1966); I. G. Boldina, M. V. Nazarenko, *C.A.* **67**, 42441h (1967).

Blue oil. bp_{12} 161°; bp_{11} 145°. d_4^{20} 0.9883. uv max: 370 nm (log ε 3.7). LD_{50} i.m. in white mice: 3 g/kg (Boldina, Nazarenko).

Trinitrobenzene derivative. $C_{20}H_{19}N_3O_6$. Dark violet needles from abs ethanol, mp 131.5-132.5°.

2040. Chamomile. Camomile. The term chamomile refers to the dried flowerheads of either of two distinct plants of the family *Asteraceae (Compositae)*: German chamomile, *Chamomilla recutita* (L.) Rauschert, also known as *Matricaria recutita* L. or *M. chamomilla*, and Roman chamomile, *Chamaemelum nobile* (L.) All. (formerly *Anthemis nobilis* L.). Both chamomiles have been used in traditional medicines as antispasmodic and anti-inflammatory agents. Evaluation of hydroalcoholic extracts for antispasmodic activity: H. B. Forster *et al.*, *Planta Med.* **40**, 309 (1980); for anti-inflammatory activity: A. Tubaro *et al.*, *ibid.* **50**, 359 (1984). Stability of extracts: R. Carle *et al.*, *ibid.* **55**, 540 (1989). HPLC determn of flavones and coumarins in extracts: P. Pietta *et al.*, *J. Chromatogr.* **404**, 279 (1987). Reviews: A. Y. Leung, *Encyclopedia of Common Natural Ingredients* (Wiley-Interscience, New York, 1980) pp 110-112; C. Mann, E. J. Staba in *Herbs, Spices, and Medicinal Plants: Recent Advances in Botany, Horticulture, and Pharmacology* vol. 1 (Oryx Press, 1986) pp 235-280.

German chamomile. Hungarian chamomile; wild chamomile; Matricaria. *Habit*. Europe, northern and western Asia; naturalized in North America. Cultivated in Hungary, Romania, Germany. *Constit*. 0.24-1.9% volatile oil; flavonoids including apigenin, luteolin, quercetin; coumarins such as herniarin, umbelliferone; proazulenes (matricin, matricarin, etc.); phenolic carboxylic acids (anisic, vanillic, syringic, caffeic); polysaccharide mucilage.

German chamomile oil. [8002-66-2] Matricaria oil. Extracted from flower heads by steam distillation. *Constit*. $(-)$-α-bisabolol (up to 50%), bisabolol oxides, chamazulene, farnesene. Review of constituents: B. M. Lawrence, *Perfum. Flavor.* **12**, 35-52 (1987). Deep ink-blue oil. d 0.91-0.95. Ester value: <40. Acid value: 5 to 50. Sol in propylene glycol. Insol in glycerin, mineral oil.

Roman chamomile. English chamomile; garden chamomile. *Habit*. Southern and western Europe; naturalized in North America. Cultivated in Great Britain, Belgium, France, U.S. *Constit*. up to 1.75% volatile oil; up to 0.6% sesquiterpene lactones of the germacranolide type (mainly nobilin, 3-epinobilin); flavonoids (apigenin, luteolin, apiin, etc.); phenolic carboxylic acids (caffeic, ferulic); coumarins; and thiophene derivatives.

Roman chamomile oil. [8015-92-7] Oil of anthemis. Extracted from flowerheads by steam distillation. *Constit*. esters of angelic and tiglic acids (~85%), chamazulene, α-pinene, farnesol. Blue oil. n^{20} 1.44-1.45. $[\alpha]^{20}$ -1 to $+4°$. Ester value: 250 to 310. Acid value <15. Sol in mineral oil and most fixed oils; sol with haziness in propylene glycol. Practically insol in glycerin.

USE: Flowers in herbal teas. Extracts and oils in perfumery; as flavor in foods and alcoholic beverages; hair dye.

THERAP CAT: Carminative; vulnerary. Extracts and oils as topical anti-inflammatory.

2041. Chanoclavine. [2390-99-0] (2*E*)-2-Methyl-3-[(4*R*,5*R*)-1,3,4,5-tetrahydro-4-(methylamino)benz[*cd*]indol-5-yl]-2-propen-1-ol; (*E*)-8,9-didehydro-6-methyl-6,7-secoergoline-8-methanol; chanoclavin-I; secaclavine. $C_{16}H_{20}N_2O$; mol wt 256.35. C 74.97%, H 7.86%, N 10.93%, O 6.24%. Precursor of the tetracyclic ergolines, agroclavine, elymoclavine, and lysergic acid amide, q.q.v.: Gröger *et al.*, *Z. Naturforsch.* **21b**, 827 (1966); Floss *et al.*, *Chem. Commun.* **1967**, 105. Occurs in sclerotia of *Claviceps purpurea* (Fries) Tul., *Hypocreaceae*, in *Ipomea tricolor* Cav., *Convolvulaceae* and is one of the active principles of the ancient Aztec drug "Ololiuqui," *Rivea corryubosa* (L.) Hall. f., *Convolvulaceae*: Hofmann, Tscherter, *Experientia* **16**, 414 (1960). Isoln and structure: Hofmann *et al.*, *Helv. Chim. Acta* **40**, 1358 (1957). Isoln of stereoisomers: Stauffacher, Tscherter, *ibid.* **47**, 2186 (1964). Stereochemistry: Acklin *et al.*, *Chem. Commun.* **1966**, 799. Biosynthetic studies: Floss *et al.*, *J. Am. Chem. Soc.* **90**, 6500 (1968). Total synthesis of (±)-chanoclavine: H. Plieninger *et al.*, *Tetrahedron Lett.* **1975**, 1827; H. Plieninger, D. Schmalz, *Ber.* **109**, 2140 (1976); M. Natsume, H. Muratake, *Heterocycles* **16**, 375 (1981).

Thick prisms and polyhedra from acetone or methanol, mp 220-222°. $[\alpha]_D^{20}$ $-240°$ (pyridine), $[\alpha]_D^{20}$ $-205°$ (c = 0.75 in alcohol). uv max: 225, 284, 293 nm (ε 4.44, 3.82, 3.72). One gram dissolves in 25 ml boiling methanol, 140 ml boiling acetone, 170 ml boiling ethyl acetate or 350 ml boiling chloroform. Practically insol in water.

Epimer at position 10. Chanoclavine II.

2042. Chaperonins. Cpns. Subset of molecular chaperone proteins, including the nucleoplasmins and heat shock proteins, that guide the in-vivo transport, folding and assembly of other protein structures but are not themselves components of these structures as part of their biological function. "Chaperonin" coined to describe the function of these proteins: S. M. Hemmingsen *et al.*, *Nature* **333**, 330 (1988). Characterized by large oligomeric cage-like structures with multiple rings of subunits possessing 7,8 or 9-fold symmetry. Ubiquitous conserved proteins which have been identified in most microorganisms and cellular compartments, require the interaction of two chaperonins to ensure correct folding: J. Martin *et al.*, *Nature* **366**, 228, 279 (1993). Molecular study of mechanism: S. G. Burston *et al.*, *J. Mol. Biol.* **249**, 138 (1995). Review: R. J. Ellis, S. M. van der Vies, *Annu. Rev. Biochem.* **60**, 321-347 (1991). Review of structures and conformations: H. Saibil, S. Wood, *Curr. Opin. Struct. Biol.* **3**, 207-213 (1993). Review of *E. coli* chaperonins, cpn60-cpn10: F. Baneyx, *Ann. N.Y. Acad. Sci.* **745**, 383-394 (1994).

2043. CHAPS. [75621-03-3] *N,N*-Dimethyl-*N*-(3-sulfopropyl)-3-[[(3α,5β,7α,12α)-3,7,12-trihydroxy-24-oxocholan-24-yl]amino]-1-propanaminium inner salt; 3-[(3-cholamidopropyl)dimethylammonio]-1-propanesulfonate. $C_{32}H_{58}N_2O_7S$; mol wt 614.88. C 62.51%, H 9.51%, N 4.56%, O 18.21%, S 5.21%. Sulfobetaine zwitterionic derivative of cholic acid, q.v. Prepn: L. M. Hjelmeland, *Proc. Natl. Acad. Sci. USA* **77**, 6368 (1980); idem, US **4372888** (1983 to U.S. Secy. Health and Human Services); and physical properties: L. M. Hjelmeland *et al.*, *Anal. Biochem.* **130**, 72 (1983). Physical characterization: R. E. Stark *et al.*, *J. Phys. Chem.* **88**, 6063 (1984). Micellar properties: M. A. Partearroyo *et al.*, *Biochem. Int.* **16**, 259 (1988). Solubilization of membrane proteins: A. J. Bitonti *et al.*, *Biochemistry* **21**, 3650 (1982); W. Wouters *et al.*, *Eur. J. Pharmacol.* **115**, 1 (1985); L. Bancells *et al.*, *Biochem. Pharmacol.* **36**, 2539 (1987); of opioid receptors: J. Simon *et al.*, *J. Neurochem.* **46**, 695 (1986).

Crystals from absolute methanol at 0°.

Hydroxy analog. [82473-24-3] CHAPSO. $C_{32}H_{58}N_2O_8S$. White granular detergent.

USE: Nondenaturing biological detergent.

2044. Chartreusin. [6377-18-0] 10-[[6-Deoxy-2-O-(6-deoxy-3-O-methyl-α-D-galactopyranosyl)-β-D-galactopyranosyl]-oxy]-6-hydroxy-1-methylbenzo[h][1]benzopyrano[5,4,3-cde][1]-benzopyran-5,12-dione; antibiotic X-465A; lambdamycin; NSC-5159. $C_{32}H_{32}O_{14}$; mol wt 640.59. C 60.00%, H 5.04%, O 34.97%. Antibiotic substance produced by *Streptomyces chartreusis* from African soil, also by another *Streptomyces* sp. from North American soil: B. E. Leach *et al.*, *J. Am. Chem. Soc.* **75**, 4011 (1953). Identity with Antibiotic X-465A: J. Berger *et al.*, *ibid.* **80**, 1636 (1958). Structure: L. H. Sternbach *et al.*, *ibid.* 1639; E. Simonitsch *et al.*, *Helv. Chim. Acta* **47**, 1459 (1964); W. Eisenhuth *et al.*, *ibid.* 1475. Anticancer activity: J. P. McGovern *et al.*, *Cancer Res.* **37**, 1666 (1977). Synthesis of *chartarin*, the chartreusin aglycone: T. R. Kelly *et al.*, *J. Am. Chem. Soc.* **102**, 798 (1980); F. M. Hauser, D.W. Combs, *J. Org. Chem.* **45**, 4071 (1980).

Yellow plates from acetone or methylene chloride + ethanol, mp 184-186°. Weak acid. $[\alpha]_D^{25}$ +132.5° (c = 0.2 in pyridine); $[\alpha]_D^{25}$ −33° (c = 0.3 in glacial acetic acid). uv max: 237, 262, 332, 382, 405, 422 nm. Practically insol in water. Sol in acetone.

Sodium salt. Gold-colored needles or plates from water. Soly in water: at least 20 mg/ml (pH 9.5). LD_{50} i.v. in mice: 250 mg/kg (Leach).

2045. Charybdotoxin. [95751-30-7] ChTX; CTX. $C_{176}H_{277}N_{57}O_{55}S_7$; mol wt 4295.92. C 49.21%, H 6.50%, N 18.59%, O 20.48%, S 5.22%. Peptide inhibitor of K^+ channels isolated from the venom of the scorpion *Leiurus quinquestriatus* var. *hebraeus*. Single chain of 4.3 kDa comprised of 37 amino acids with 3 disulfide linkages. Identification: C. Miller *et al.*, *Nature* **313**, 316 (1985). Purification and amino acid sequence: G. Gimenez-Gallego *et al.*, *Proc. Natl. Acad. Sci. USA* **85**, 3329 (1988). Synthesis: E. E. Sugg *et al.*, *J. Biol. Chem.* **265**, 18745 (1990); P. Lambert *et al.*, *Biochem. Biophys. Res. Commun.* **170**, 684 (1990). NMR structural study: F. Bontems *et al.*, *Eur. J. Biochem.* **196**, 19 (1991). Mechanism of channel block: R. MacKinnon, C. Miller, *J. Gen. Physiol.* **91**, 335 (1988). Characterization of binding sites: J. Vázquez *et al.*, *J. Biol. Chem.* **264**, 20902 (1989); J. Vázquez *et al.*, *ibid.* **265**, 15564 (1990). Use as probe for K^+ channels: M. Garcia, G. Gimenez-Gallego, US **4960867** (1990 to Merck & Co.).

pyroE-F-T-N-V-S-C-T-T-S-K-E-C-W-S-V-C-Q-R-L-H-N-T-S-R-G-K-C-M-N-K-K-C-R-C-Y-S

Extinction coefficient: 1.52 $(mg/ml)^{-1}cm^{-1}$ at 280 nm; 15.05 $(mg/ml)^{-1}cm^{-1}$ at 215 nm.

USE: Biochemical probe.

2046. Chaulmoogra Oil. Hydnocarpus oil; gynocardia oil. Fixed oil expressed from chaulmoogra, the seeds of *Taraktogenos kurzii* King, *Bixaceae; Hydnocarpus wightiana* Blume or *H. anthelmintica* Pierre, *Flacourtiaceae*. The seeds contain about 50% chaulmoogra oil. *Constit.* Glycerides of chaulmoogric and hydnocarpic acids with small amounts of glycerides of palmitic acid, etc.

Yellow or brownish-yellow oil. Below 25° it is a soft solid; characteristic odor. About d_{25}^{25} 0.95; d_{15}^{45} 0.940. $[\alpha]_D^{20}$ +48 to +60°. Sol in benzene, chloroform, ether, petr ether, slightly in cold alcohol; almost entirely sol in hot alcohol, carbon disulfide.

THERAP CAT: Antibacterial (leprostatic).

2047. Chaulmoogric Acid. [29106-32-9] (1S)-2-Cyclopentene-1-tridecanoic acid; D-13-(2-cyclopenten-1-yl)tridecanoic acid; hydnocarpylacetic acid; $C_{18}H_{32}O_2$; mol wt 280.45. C 77.09%, H 11.50%, O 11.41%. Isoln from chaulmoogra oil and characterization: Power, Gornall, *J. Chem. Soc.* **85**, 838, 851 (1904); Barrowcliff, Power, *ibid.* **91**, 557 (1907); Hashimoto, *J. Am. Chem. Soc.* **47**, 2325 (1925); Cole, Cardoso, *ibid.* **61**, 2349, 2351, 3442 (1939). Structure: Shriner, Adams, *ibid.* **47**, 2727 (1925). Synthesis: Noller, Adams, *ibid.* **48**, 1080 (1926); Perkins, Cruz, *ibid.* **49**, 1070 (1927). Synthesis and stereochemistry: Mislow, Steinberg, *ibid.* **77**, 3807 (1955). Activity against *Mycobacterium leprae:* L. Levy, *Am. Rev. Respir. Dis.* **111**, 703 (1975). Biosynthesis: T. Kaneda, *Biochem. Biophys. Res. Commun.* **99**, 1226 (1981).

Shiny leaflets from petr ether or alcohol. mp 68.5°. bp_{20} 247-248°. $[\alpha]_D^{20}$ +60.3° (c = 4 in chloroform). Iodine value 90.5. Freely sol in ether, chloroform, ethyl acetate; sol in other organic fat solvents.

Methyl ester. [24828-59-9] $C_{19}H_{34}O_2$. Prepd by passing HCl gas into a methanol soln of the acid. Needles, mp 22°. bp_{20} 227°. d_{25}^{25} 0.9119. $[\alpha]_D^{15}$ +50° (c = 5 in chloroform).

Ethyl ester. [623-32-5] Chaulmestrol; Moogrol. $C_{20}H_{36}O_2$; mol wt 308.51. Essentially a mixt of the ethyl esters of the unsaturated fatty acids (chaulmoogric and hydnocarpic) of chaulmoogra oil. Pale yellow, clear liquid; slight fruity odor; not unpleasant taste. Iodine no. 90-100. d_{25}^{25} ~0.904. $[\alpha]_D^{25}$ +44.5° (chloroform). Insol in water; misc with alc, chloroform.

THERAP CAT: Ethyl ester in treatment of leprosy, sarcoidosis.

2048. Chavicine. [495-91-0] (2Z,4Z)-5-(1,3-Benzodioxol-5-yl)-1-(1-piperidinyl)-2,4-pentadien-1-one; (Z,Z)-1-[5-(1,3-benzodioxol-5-yl)-1-oxo-2,4-pentadienyl]piperidine; (Z,Z)-1-piperoylpiperidine. $C_{17}H_{19}NO_3$; mol wt 285.34. C 71.56%, H 6.71%, N 4.91%, O 16.82%. One of the most active constituents of black pepper: Buchheim, *Arch. Exp. Pathol. Pharmakol.* **5**, 455 (1876); Ott, Eichler, *Ber.* **55**, 2653 (1922); Ott, Lüdemann, *Ber.* **57**, 214 (1924). Stereoisomer of piperine: Lohaus, Gall, *Ann.* **517**, 282 (1935). Loss of pungency of ground pepper during storage attributed to gradual isomerization of chavicine into piperine, *q.v.*: Newman, *Chem. Prod.* **16**, 379 (1953). Synthesis of chavicine and questioning of its existence as constituent of black pepper: R. Grewe *et al.*, *Ber.* **103**, 3752 (1970). Spectroscopic structural elucidation and prepn: R. DeCleyn, M. Verzele, *Bull. Soc. Chim. Belg.* **84**, 435 (1975).

Yellowish, oily mass; sharp, peppery taste. Sol in alc, ether, petr ether. uv max (methanol): 318 nm (ε 16200).

2049. Chavicol. [501-92-8] 4-(2-Propen-1-yl)phenol; *p*-allylphenol; γ-(*p*-hydroxyphenyl)-α-propylene. $C_9H_{10}O$; mol wt 134.18. C 80.56%, H 7.51%, O 11.92%. Found together with terpenes in volatile betel oils (from leaves of *Piper betle* L., *Pipera-*

ceae): Eijkman, *Ber.* **22**, 2739 (1889). Prepd from a mixture of estragole, ethyl bromide and magnesium in benzene: Grignard, *Compt. Rend.* **151**, 323 (1910); from 4-allylphenylmagnesium bromide in ether by the action of oxygen: Quelet, *Bull. Soc. Chim. Fr.* [4] **45**, 265 (1929).

Liquid. mp 15.8°. d_4^{15} 1.0203; d_4^{20} 1.0175. bp_{760} 238°; bp_{16} 123°. n_D^{18} 1.5441. Miscible with alcohol, ether, chloroform, petr ether.

Methyl ether *see* Estragole.

2050. Chelerythrine. [34316-15-9] 1,2-Dimethoxy-12-methyl[1,3]dioxolo[4′,5′:4,5]benzo[1,2-*c*]phenanthridinium; 1,2-dimethoxy-12-methyl[1,3]benzodioxolo[5,6-*c*]phenanthridinium; toddaline. $[C_{21}H_{18}NO_4]^+$. From root of *Chelidonium majus* L., *Papaveraceae*: Kratzmann, *Pharm. Monatsh.* **5**, 161 (1924), *C.A.* **18**, 3406[2] (1924); Platonova *et al.*, *J. Gen. Chem. USSR* **26**, 181 (1956). Structure: Späth, Kuffner, *Ber.* **64**, 1123 (1931); N. Decaudain *et al.*, *Ann. Pharm. Fr.* **35**, 521 (1977). Identity with toddaline: Govindachari, Thyagarajan, *J. Chem. Soc.* **1956**, 769. Synthesis of chelerythrine chloride: Bailey, Worthing, *ibid.* 4535; S. V. Kessar *et al.*, *J. Org. Chem.* **53**, 1708 (1988). Pharmacology: Chelombit'o, Murav'eva, *Aktual. Probl. Farmakol. Farm. Vses. Nauchn. Konf.* **1971**, 183, *C.A.* **76**, 112g (1972).

Crystals from chloroform + methanol, mp 207°. *See:* Manske, *Can. J. Res.* **21B**, 140 (1943). The free base is colorless, but its quaternary salts are yellow. Aq solns of base and salts show violet fluorescence. uv spectrum of chelerythrine chloride: Hruban *et al.*, *Collect. Czech. Chem. Commun.* **35**, 3420 (1970).

2051. Chelidonic Acid. [99-32-1] 4-Oxo-4*H*-pyran-2,6-dicarboxylic acid; 4-oxo-1,4-pyran-2,6-dicarboxylic acid; jerva acid; jervasic acid. $C_7H_4O_6$; mol wt 184.10. C 45.67%, H 2.19%, O 52.14%. Presence in various plants: Stransky, *Arch. Pharm.* **258**, 56 (1920); Ramstad, *Pharm. Acta Helv.* **28**, 45 (1953). Structure: Verkade, *Recl. Trav. Chim. Pays-Bas* **43**, 879 (1924). Synthesis: Riegel, Reinhard, *J. Am. Chem. Soc.* **48**, 1334 (1926); Riegel, Zwilgmeyer, *Org. Synth.* **coll. vol. II**, 126 (1943); Toomey, Riegel, *J. Org. Chem.* **17**, 1492 (1952).

Crystals, dec 257°. uv max (water): 270 nm (ε 11500). One gram dissolves in 65 ml water, 26 ml boiling water. Sparingly sol in alc.

Monohydrate. Monoclinic crystals from hot alcohol + water. The water of crystn is given up when heated at 102° and then at 160° to constant weight.

Sodium salt. Sparingly sol in water.

2052. Chelidonine. Stylophorin. $C_{20}H_{19}NO_5$; mol wt 353.37. C 67.98%, H 5.42%, N 3.96%, O 22.64%. Hexahydrobenzophenanthridine alkaloid. Occurs in nature in (+)-form, (−)-form and racemic form. Isoln of (+)-form from root of *Chelidonium majus* L., *Stylophorum diphyllum* (Michx.) Nutt., and *Dicranostigma franchetianum* (Prain) Fedde, *Papaveraceae*: J. M. Probst, *Ann.* **29**, 113 (1839); E. Schmidt, F. Selle, *Arch. Pharm.* **228**, 441 (1890); F. Selle, *ibid.* 96; Manske, *Can. J. Res.* **20B**, 53 (1942); J. Slavik, *Collect.*

Czech. Chem. Commun. **20**, 198 (1955); from *Symphoricarpos albus* L., Blake, *Caprifoliaceae:* M. Szaufer *et al.*, *Phytochemistry* **17**, 1446 (1978). Isoln of (−)-form from *Glaucium corniculatum* Curt., *Papaveraceae:* J. Slavik, L. Slavikova, *Collect. Czech. Chem. Commun.* **22**, 279 (1957). Structure: F. von Bruchhausen, H. W. Bersch, *Ber.* **63**, 2520 (1930); E. Späth, F. Kuffner, *ibid.* **64**, 370 (1931); H. W. Bersch, *Arch. Pharm.* **291**, 491 (1958). Identity of (±)-form with diphylline: J. Slavik, *Collect. Czech. Chem. Commun.* **26**, 2933 (1961). Absolute configuration of (+)-*p*-bromobenzoate: N. Takao *et al.*, *Tetrahedron Lett.* **1979**, 495. Conformation: M. Cushman, T.-C. Choong, *Heterocycles* **14**, 1935 (1980); M. Sugiura *et al.*, *J. Chem. Soc. Perkin Trans.* 2 **1986**, 175. Chiroptic properties and absolute configuration of (+)-chelidonine: N. Takao *et al.*, *Arch. Pharm.* **317**, 223 (1984). Biosynthesis: E. Leete, *J. Am. Chem. Soc.* **85**, 473 (1963). Total synthesis of (±)-form: W. Oppolzer, K. Keller, *ibid.* **93**, 3836 (1971); M. Cushman *et al.*, *J. Org. Chem.* **45**, 5067 (1980); W. Oppolzer, C. Rabbiani, *Helv. Chim. Acta* **66**, 1119 (1983); M. Hanaoka *et al.*, *Chem. Lett.* **1986**, 736. Effect on smooth muscle: P. J. Hanzlik: *J. Pharmacol. Exp. Ther.* **7**, 99 (1915). Inhibition of reverse transcriptase activity: M. L. Sethi, *Can. J. Pharm. Sci.* **16**, 29 (1981). Cytotoxic effects: M. Cushman *et al.*, *loc. cit.* Review of pharmacological effects: V. Preininger, "The Biology of Papaveraceae Alkaloids" in *The Alkaloids* **vol. 15**, R. H. F. Manske, Ed. (Academic Press, Orlando, 1975) pp 241-242. General review: V. Simanek, "Benzophenanthridine Alkaloids" *ibid.* **vol. 26**, A. Brossi, Ed. (1985) pp 185-240. Toxicity data: R. C. Anderson, K. K. Chen, *Fed. Proc.* **5**, 163 (1946).

(+)-form

(+)-Form. [476-32-4] [5b*R*-(5bα,6β,12bα)]-5b,6,7,12b,13,14-Hexahydro-13-methyl[1,3]benzodioxolo[5,6-*c*]-1,3-dioxolo[4,5-*i*]-phenanthridin-6-ol. Monoclinic prisms from methanol, ethanol or ethanol + chloroform, mp 135-136°. $bp_{0.002}$ 220° (air-bath temp). $[\alpha]_D^{22}$ +115 ±3° (ethanol); $[\alpha]_D^{20}$ +117° (c = 3 in CHCl₃). uv max (methanol): 289, 239, 208 nm (log ε 3.89, 3.88, 4.69). Sol in alc, chloroform, ether, amyl alc. Practically insol in water. LD₅₀ in mice (mg/kg): 34.6 ±2.44 i.v. (Anderson, Chen).

(+)-*O*-Acetylchelidonine. $C_{22}H_{21}NO_6$. Crystals from chloroform, mp 184-186°. $[\alpha]_D$ +110°.

(+)-Benzoylchelidonine. $C_{27}H_{23}NO_6$. Crystals from chloroform, mp 210-211°.

(−)-Form. [88200-01-5] Crystals from aqueous ethanol, mp 136°. $[\alpha]_D^{22}$ −112 ±3° (c = 0.47 in ethanol).

(±)-Form. [20267-87-2] Diphylline. Crystals from ethanol, mp 215-216°.

2053. Chemerin. [635677-02-0] Chemoattractant protein that induces migration of macrophages and dendritic cells; also promotes adipogenesis and angiogenesis. Elevated serum levels have been associated with inflammatory disease, obesity, and metabolic syndrome. Originally identified as the protein product of the tazarotene-induced gene 2 (TIG2) and as the natural ligand of the ChemR23 receptor, also known as chemokine-like receptor 1 (CMKLR1), which is present on antigen-presenting cells, adipocytes, and various other cells. Transcribed as a 163 amino acid preproprotein and secreted as the inactive 143 amino acid protein, prochemerin. The fully bioactive form contains 137 amino acids that correspond to residues 21-157 of preprochemerin. Identification of the TIG2 gene and protein transcript from psoriatic skin lesions: S. Nagpal *et al.*, *J. Invest. Dermatol.* **109**, 91 (1997). Identification as ligand for ChemR23 receptor: V. Wittamer *et al.*, *J. Exp. Med.* **198**, 977 (2003). Role as chemoattractant for antigen-presenting cells: W. Vermi *et al.*, *ibid.* **201**, 509 (2005). Cleavage to active form by proteases released upon neutrophil degranulation: V. Wittamer *et al.*, *J. Immunol.* **175**, 487 (2005). Identification in adipose tissue and role in adipogenesis: S. Roh *et al.*, *Biochem. Biophys. Res. Commun.*

362, 1013 (2007). Identification in human endothelial cells and role in angiogenesis: J. Kaur *et al.*, *ibid.* **391**, 1762 (2010). Serum levels in patients with metabolic syndrome: B. Dong *et al.*, *Intern. Med.* **50**, 1093 (2011). Review of role in inflammation and obesity: M. C. Ernst, C. J. Sinal, *Trends Endocrinol. Metab.* **21**, 660-667 (2010).

2054. **Chenodiol.** [474-25-9] (3α,5β,7α)-3,7-Dihydroxycholan-24-oic acid; 3α,7α-dihydroxy-5β-cholanic acid; anthropodesoxycholic acid; gallodesoxycholic acid; 17β-(1-methyl-3-carboxypropyl)etiocholane-3α,7α-diol; chenic acid; chenodeoxycholic acid; CDC; Chendol; Chenocol; Chenofalk; Chenossil; Cholanorm; Fluibil. $C_{24}H_{40}O_4$; mol wt 392.58. C 73.43%, H 10.27%, O 16.30%. A major bile acid in many vertebrates, occurring as the *N*-glycine and/or *N*-taurine conjugate. With other bile acids, forms mixed micelles with lecithin in bile which solubilize cholesterol and thus facilitates its excretion. Facilitates fat absorption in the small intestine by micellar solubilization of fatty acids and monoglycerides. Has cathartic properties since it induces fluid secretion from large intestine. Main constituent of the bile of hens, geese and other fowl; occurs in appreciable amounts in the bile of hamster, hog, guinea pig, bear and man. Epimeric with ursodiol, *q.v.* Isoln: Windhaus *et al.*, *Z. Physiol. Chem.* **140**, 177 (1924); Wieland, Reveney, *ibid.* 186. Configuration: Lettré, *Ber.* **68**, 766 (1935). Prepn from cholic acid: Fieser, Rajagopalan, *J. Am. Chem. Soc.* **72**, 5530 (1950); Hauser *et al.*, *Helv. Chim. Acta* **43**, 1595 (1960); Hofmann, *Acta Chem. Scand.* **17**, 173 (1963). Alternate prepns: Sato, Ikekawa, *J. Org. Chem.* **24**, 1367 (1959); T. Iida, F. C. Chang, *ibid.* **46**, 2786 (1981). Stereoselective total synthesis: T. Kametani *et al.*, *J. Am. Chem. Soc.* **103**, 2890 (1981). Asymmetric total synthesis of (+)-form: *eidem*, *J. Org. Chem.* **47**, 2331 (1982). Dissolution of cholesterol gallstones: Danzinger *et al.*, *N. Engl. J. Med.* **286**, 1 (1972); Bell *et al.*, *Lancet* **II**, 1213 (1972). Use in long-term treatment of cerebrotendinous xanthomatosis: V. M. Berginer *et al.*, *N. Engl. J. Med.* **311**, 1649 (1984). Monograph on bile acids: *The Bile Acids*, 2 vols., P. P. Nair, D. Kritchevsky, Eds. (Plenum Press, New York, 1971, 1973). Review of pharmacology and therapeutic use of chenodeoxycholic acid: J. H. Iser, A. Sali, *Drugs* **21**, 90-119 (1981). Effect on cholesterol and bile acid metabolism: G. S. Tint *et al.*, *Gastroenterology* **91**, 1007 (1986).

Needles from ethyl acetate + heptane, mp 119°. $[\alpha]_D^{20}$ +11.5° (dioxane). Freely sol in methanol, alc, acetone, acetic acid; more sol in ether and ethyl acetate than deoxycholic acid. Practically insol in water, petr ether, benzene. High solvent power for alkali soaps, but does not form "choleic" acid addition compds as does deoxycholic acid. Forms beautiful cryst salts of Na, K and Ba. While the acid is tasteless, the Na salt tastes slightly sweet at first, then bitter.
Diformate. $C_{25}H_{40}O_6$. Clusters of needles from alc; mp with slight effervescence at 137°, upon further heating solidifies again, and finally melts around 172°.
Methyl ester. $C_{25}H_{42}O_4$. Fine needles from benzene + heptane, mp 90-91°. $[\alpha]_D^{25}$ +20°.
THERAP CAT: Anticholelithogenic.

2055. **CHES.** [103-47-9] 2-(Cyclohexylamino)ethanesulfonic acid; *N*-cyclohexyltaurine. $C_8H_{17}NO_3S$; mol wt 207.29. C 46.35%, H 8.27%, N 6.76%, O 23.15%, S 15.47%. Zwitterionic *N*-substituted aminosulfonic acid in the style of the "Good" buffers; active in the pH range 8.6-10.0. Synthesis: A. Champseix *et al.*, *Bull. Soc. Chim. Fr.* **1985**, 463. Effects of freezing on buffering: D. L. Williams-Smith *et al.*, *Biochem. J.* **167**, 593 (1977). Effects on Lowry protein assay: C. Cookson, *Anal. Biochem.* **88**, 340 (1978). Dissociation constants and buffering capacity: M. J. Taylor, Y. Pignat, *Cryobiology* **19**, 99 (1982). Use as buffer: M. G. N. Hartmanis,

T. C. Stadtman, *Proc. Natl. Acad. Sci. USA* **84**, 76 (1987); C. Engstrand *et al.*, *Biochim. Biophys. Acta* **1122**, 321 (1992).

Crystals, mp 320°. pKa in water (25°) 9.27 ±0.01; in 20% (w/w) DMSO (25°) 9.10, (0°) 9.76, (−5.5°) 10.01; in 30% (w/w) DMSO (25°) 9.11, (0°) 9.89, (−12°) 10.27.
USE: Biological buffer.

2056. **Chicle.** The partially evaporated, milky juice from *Manilkara zapotilla* (Jacq.) Gilly *(Achras sapota* L.), *Sapotaceae. Habit.* West Indies, Mexico and Central America.
USE: In the chewing gum industry.

2057. **Chimaphila.** Pipsissewa; Prince's pine; bitter wintergreen; rheumatism weed; ground holly; pyrola; pine tulip. Dried leaves of *Chimaphila umbellata* (L.) Nutt., *Ericaceae. Habit.* Europe, Asia, North America. *Constit.* Chimaphilin, arbutin, ericolin, urson, tannin, resin.
THERAP CAT: Urinary antiseptic.

2058. **Chimaphilin.** [482-70-2] 2,7-Dimethyl-1,4-naphthalenedione; 2,7-dimethyl-1,4-naphthoquinone. $C_{12}H_{10}O_2$; mol wt 186.21. C 77.40%, H 5.41%, O 17.18%. From *Chimaphila carymbosa* Pursh. and *Pyrola incarnata* (Fisch.) Fernald, *Ericaceae:* DiModica *et al.*, *Gazz. Chim. Ital.* **83**, 393 (1953); Inouye, *J. Pharm. Soc. Jpn.* **76**, 976 (1956). Structure: DiModica, Tira, *Gazz. Chim. Ital.* **86**, 234 (1956).

Yellow needles, mp 113.5-114.5°. uv max (alc): 233, 256, 339 nm. Polymerizes in sunlight to needles, mp 232-233°.
Diacetate. Crystals from ethanol + water, mp 90.5-91.5°.

2059. **Chimonanthine.** [5545-89-1] (3a*S*,3'a*S*,8a*S*,8'a*S*)-2,-2',3,3',8,8',8a,8'a-Octahydro-1,1'-dimethyl-3a,3'a(1*H*,1'*H*)-bipyrrolo[2,3-*b*]indole; 1-demethylcalycanthidine. $C_{22}H_{26}N_4$; mol wt 346.48. C 76.26%, H 7.56%, N 16.17%. From leaves of *Chimonanthus fragrans* Lindle, *Calicanthaceae:* Hodson *et al.*, *Proc. Chem. Soc. London* **1961**, 465. Structure: Clayton *et al.*, *Tetrahedron* **18**, 1495 (1962). Synthesis: Hendrickson *et al.*, *Proc. Chem. Soc. London* **1962**, 383; Scott *et al.*, *J. Am. Chem. Soc.* **86**, 302 (1964); Hall *et al.*, *Tetrahedron* **23**, 4131 (1967).

Crystals, mp 188-189°. Weak diacidic base of equivalent weight 173. uv max: 246, 304 nm.
Dimethiodide. Crystals, dec 235-236°.

2060. **Chimyl Alcohol.** [506-03-6] (2*S*)-3-(Hexadecyloxy)-1,2-propanediol; *d*-α-(*n*-hexadecyl)glycerol; (*S*)-(+)-chimyl alcohol; testriol. $C_{19}H_{40}O_3$; mol wt 316.53. C 72.10%, H 12.74%, O 15.16%. Isoln from bull and boar testes: Hirano, *J. Pharm. Soc.*

Jpn. **56**, 122 (1936); Prelog *et al., Helv. Chim. Acta* **27**, 674 (1944). Synthesis: E. Baer, H. O. L. Fischer, *J. Biol. Chem.* **140**, 397 (1941).

Leaflets from hexane, mp 64°. bp$_{0.005}$ 120°. $[\alpha]_D^{20}$ +3.0° (c = 1.16 in chloroform). Sol in acetone, hexane, chloroform.

Bis(phenylurethan). C$_{33}$H$_{50}$N$_2$O$_5$; mol wt 554.77. Leaflets from petr ether, mp 97.5-98.5°.

2061. Chinese Wax. The excretion of an insect, *Coccus ceriferus* Fabr., or *C. pela* Westwood, deposited on the twigs and branches of a species of ash tree in Western China. Chief constituent is ceryl cerotate.

White to yellowish-white, practically odorless, tasteless solid. d ~0.93. mp ~92°. Solidif 80-81°. Sapon no. 80-92. Iodine no. 1.4. Acid no. 63. Insol in water; freely sol in benzene, slightly in alcohol or ether.

USE: Manuf candles, leather and furniture polish; treating silk and cotton fabrics; sizing and glazing papers.

2062. Chirald. [38345-66-3] (αS)-α-[(1*R*)-2-(Dimethylamino)-1-methylethyl]-α-phenylbenzeneethanol; (+)-oxyphene; (2*S*,3*R*)-(+)-4-dimethylamino-3-methyl-1,2-diphenyl-2-butanol; Darvon alcohol. C$_{19}$H$_{25}$NO; mol wt 283.42. C 80.52%, H 8.89%, N 4.94%, O 5.64%. Asymmetric reducing agent. Prepn of racemate: W. G. Stoll *et al., Helv. Chim. Acta* **33**, 1194 (1950). Resolution of enantionmers: A. Pohland, H. R. Sullivan, *J. Am. Chem. Soc.* **77**, 3400 (1955). Asymmetric synthesis: T. J. Erickson, *J. Org. Chem.* **51**, 934 (1986). HPLC determn as an impurity in dextropropoxyphene, *q.v.*: S. Kryger, P. Helboe, *J. Chromatogr.* **539**, 186 (1991). Use as an enantioselective reducing agent: M. N. Nefedova *et al., J. Organomet. Chem.* **425**, 125 (1992); D. R. Williams *et al., Tetrahedron Lett.* **41**, 9397 (2000).

mp 55-57°. bp$_{0.1 mm}$ 138-139°. Flash point: >230°F . $[\alpha]^{21}$ +8.2° (c = 10 in ethanol).

Note: Also referred to in the literature as Mosher's reagent, *q.v.*

USE: Reducing reagent in organic synthesis.

2063. (S,S)-Chiraphos. [64896-28-2] 1-1'-[(1*S*,2*S*)-1,2-Dimethyl-1,2-ethanediyl]bis[1,1-diphenylphosphine]; (2*S*,3*S*)-(−)-bis-(diphenylphosphino)butane. C$_{28}$H$_{28}$P$_2$; mol wt 426.48. C 78.86%, H 6.62%, P 14.53%. Chiral, bidentate phosphine ligand in asymmetric catalysis. Prepn: M. D. Fryzuk, B. Bosnich, *J. Am. Chem. Soc.* **99**, 6262 (1977); J. F. G. Jansen, B. L. Feringa, *Tetrahedron: Asymmetry* **1**, 719 (1990); S. H. Bergens *et al., Inorg. Synth.* **31**, 131 (1997). ^{1}H and ^{31}P NMR of platinum complexes: N. C. Payne, D. W. Stephan, *J. Organomet. Chem.* **221**, 203 (1981). Synthetic applications: C. Moinet, J.-C. Fiaud, *Tetrahedron Lett.* **36**, 2051 (1995); H. Frauenrath *et al., Tetrahedron Asymmetry* **9**, 1103 (1998); K. Akiyama *et al., Adv. Synth. Catal.* **347**, 1569 (2005).

Large, colorless plates from absolute ethanol, mp 108-109°. $[\alpha]_D^{27}$ −211° (c = 1.5 in CHCl$_3$). Sol in acetone, THF, dichloromethane, chloroform, diethyl ether, toluene, hot ethanol. Insol in hexanes, methanol, cold ethanol.

USE: Ligand for palladium-catalyzed carbon-carbon bond formation.

2064. Chirata. Chiretta; chirayita; bitter stick; East Indian balmony. Dried plant, *Swertia (Ophelia) chirata* (Roxb.) Buch.-Ham., *Gentianaceae. Habit.* India (Himalaya). *Constit.* Chiratin, ophelic acid.

THERAP CAT: Bitter tonic, antimalarial.

2065. Chitin. [1398-61-4] (C$_8$H$_{13}$NO$_5$)$_n$. C 47.29%, H 6.45%, N 6.89%, O 39.37%. Cellulose-like biopolymer consisting predominantly of unbranched chains of β-(1 → 4)-2-acetamido-2-deoxy-D-glucose (also named *N*-acetyl-D-glucosamine) residues. Found in fungi, yeasts, marine invertebrates and arthropods, where it is a principal component in the exoskeletons. May be regarded as a derivative of cellulose, in which the C-2 hydroxyl groups have been replaced by acetamido residues. Occurrence in lower animals: A. G. Richards, *The Integument of Arthropods* (University of Minnesota Press, Minneapolis, 1951). Occurrence in the plant kingdom: F. von Wettstein, *Handbuch der systematischen Botanik* (F. Deuticke, Leipzig and Vienna, 4th ed., 1933). Typical isoln: Hackman, *Aust. J. Biol. Sci.* **7**, 168 (1954); Horowitz *et al., J. Am. Chem. Soc.* **79**, 5046 (1957). Occurrence in the anthozoan *Stylobates aeneus* Dall, the first sea anemone proved capable of synthesizing chitin: D. F. Dunn, M. H. Liberman, *Science* **221**, 157 (1983). Structure: Dweltz, *Biochim. Biophys. Acta* **44**, 416 (1960); **51**, 283 (1961); Carlstrom, *ibid.* **59**, 361 (1962); Ramachadran, Ramakrishnan, *ibid.* **63**, 307 (1962). *Review:* Foster, Webber, *Adv. Carbohydr. Chem.* **15**, 371-393 (1960); C. Jeuniaux, "Chitinous Structures" in *Comprehensive Biochemistry* **vol. 26c**, M. Florkin, E. H. Stotz, Eds. (Elsevier, New York, 1971) pp 595-632. Book: R. A. A. Muzzarelli, *Chitin* (Pergamon Press, New York, 1977). Review of properties and possible novel applications: P. R. Austin *et al., Science* **212**, 749-753 (1981).

Amorphous solid. Practically insol in water, dil acids, dil and concd alkalies, alcohol and other organic solvents; sol in concd HCl, H$_2$SO$_4$, 78-97% H$_3$PO$_4$, anhydr HCOOH. There are substantial variations in solubility, molecular weight, acetyl values, specific rotation among chitins of different origins and prepared by different methods.

Acetate. Prepn: Shorigin, Hait, *Ber.* **68**, 971 (1935). Sol in HCOOH, 50% resorcinol, concd HCl, H$_2$SO$_4$, HNO$_3$; practically insol in organic solvents.

Sulfate. Process for sulfating chitin: Cushing, Kratovil, US **2755275** (1956 to Abbott). Sulfated chitin has been found to possess anticoagulant properties in laboratory animals: Roth *et al., Am. J. Physiol.* **171**, 761 (1952); *Proc. Soc. Exp. Biol. Med.* **86**, 315 (1954).

USE: Deacylated chitin, *chitosan*, used in water treatment; in photographic emulsions; in improving the dyeability of synthetic fibers and fabrics.

THERAP CAT: Vulnerary.

2066. Chlophedianol. [791-35-5] 2-Chloro-α-[2-(dimethylamino)ethyl]-α-phenylbenzenemethanol; 2-chloro-α-(2-dimethylaminoethyl)benzhydrol; 1-*o*-chlorophenyl-1-phenyl-3-dimethylamino-1-propanol; 1-phenyl-1-(*o*-chlorophenyl)-3-dimethylaminopropanol; α-(dimethylaminoethyl)-*o*-chlorobenzhydrol; clofedanol; Tussistop. C$_{17}$H$_{20}$ClNO; mol wt 289.80. C 70.46%, H 6.96%, Cl 12.23%, N 4.83%, O 5.52%. Prepn: Henecka, Lorenz, **DE 1083277** (1960); Lorenz *et al., US* **3031377** (1962, both to Bayer).

mp 120°. LD_{50} i.v. in mice: 70 mg/kg (Lorenz).

Hydrochloride. [511-13-7] Coldrin; Pectolitan; Refugal; Ulone; ULO. $C_{17}H_{20}ClNO \cdot HCl$; mol wt 326.26. Crystals, mp 190-191°. Freely sol in water, methanol, ethanol. Sparingly sol in ether, benzene, ethyl acetate. LD_{50} orally in rats: 350 mg/kg; s.c. in mice: 95 mg/kg (Lorenz).

THERAP CAT: Antitussive.

2067. Chloracetyl Chloride. [79-04-9] 2-Chloroacetyl chloride. $C_2H_2Cl_2O$; mol wt 112.94. C 21.27%, H 1.79%, Cl 62.78%, O 14.17%. Prepn from chloroacetic acid and benzoyl chloride: Brown, *J. Am. Chem. Soc.* **60**, 1325 (1938); by chlorination of ketene: Erickson, Prill, *J. Org. Chem.* **23**, 141 (1958); Prill, US **2889365** (1959 to Monsanto); from chloroacetic acid and pyrocatechylphosphorus trichloride: Gross, Gloede, *Ber.* **96**, 1387 (1963). Manuf from chloroacetic acid and $POCl_2$: MacKenzie, Morris, US **2848491** (1958 to Dow); from ketene: Prill, US **2889365** (1959 to Monsanto). Physical data: McDonald *et al.*, *J. Chem. Eng. Data* **4**, 311 (1959).

Liquid, very pungent odor. bp 106°. mp −21.77°. d_4^{20} 1.4202. n_D^{20} 1.4541. Dec by water. *Protect from moisture.*

Caution: Potential symptoms of overexposure are irritation of eyes, skin, respiratory system; eye and skin burns; cough, wheezing, dyspnea, lacrimation. *See NIOSH Pocket Guide to Chemical Hazards* (DHHS/NIOSH 97-140, 1997) p 60.

USE: In the synthesis of organic compounds.

2068. Chloracizine. [800-22-6] 1-(2-Chloro-10*H*-phenothiazin-10-yl)-3-(diethylamino)-1-propanone; 2-chloro-10-[3-(diethylamino)-1-oxopropyl]-10*H*-phenothiazine; 2-chloro-10-(*N,N*-diethyl-β-alanyl)phenothiazine; 10-(β-diethylaminopropionyl)-2-chlorophenothiazine; 2-chloro-10-(β-diethylaminopropionyl)phenothiazine; chloracysin; chlorocyzine; chlorocizin; khloratsizin; G-020. $C_{19}H_{21}ClN_2OS$; mol wt 360.90. C 63.23%, H 5.87%, Cl 9.82%, N 7.76%, O 4.43%, S 8.88%. Prepn: Horclois, Metivier, **FR 1060715** (1954 to Rhône Poulenc); **GB 740932** (1955); Zhuravlev, Gritsenko, *Zh. Obshch. Khim.* **26**, 3385 (1956), *C.A.* **51**, 9623h (1957). For other prepns in Russian journals *see C.A.* **55**, 6484b, 9425a (1961).

Hydrochloride. $C_{19}H_{21}ClN_2OS \cdot HCl$. Crystals from isopropanol, mp 168-169°. Sol in water.

THERAP CAT: Vasodilator (coronary).

2069. Chloral Alcoholate. [515-83-3] 2,2,2-Trichloro-1-ethoxyethanol; chloral ethylalcoholate; trichloroacetaldehyde monoethylacetal. $C_4H_7Cl_3O_2$; mol wt 193.45. C 24.84%, H 3.65%, Cl 54.98%, O 16.54%. Prepd by refluxing trichloroacetaldehyde or chloral hydrate with alcohol: Personne, *Compt. Rend.* **69**, 1363 (1869); Magnani, McElvain, *J. Am. Chem. Soc.* **60**, 2212 (1938); Post, *J. Org. Chem.* **6**, 830 (1941).

Crystals, d 1.143. mp 47.5°. bp_{760} 116°. Less sol in water than chloral hydrate. Sol in organic solvents. With concd H_2SO_4 it forms trichloroacetaldehyde and ethyl sulfate.

2070. Chloral Formamide. [515-82-2] *N*-(2,2,2-Trichloro-1-hydroxyethyl)formamide; chloralamide; chloramide. $C_3H_4Cl_3NO_2$; mol wt 192.42. C 18.73%, H 2.10%, Cl 55.27%, N 7.28%, O 16.63%. Prepn or manuf from chloral and formamide: Bennett, Campbell, *Q. J. Pharm. Pharmacol.* **8**, 398 (1955); and herbicidal activity: Reuter, Sehring, **DE 1168161** (1964 to Boehringer, Ing.), *C.A.* **62**, 1024g (1965).

Crystals from water of 30% alc, mp 124-126°, also reported as 115-116° (Bennett, Campbell). Soluble in water; readily sol in alcohol, ether, glycerol, acetone, ethyl acetate. Dec by hot solvents.

THERAP CAT: Sedative, hypnotic.

2071. Chloral Hydrate. [302-17-0] 2,2,2-Trichloro-1,1-ethanediol; trichloroacetaldehyde monohydrate; Escre; Noctec; Nycton; Chloraldurat. $C_2H_3Cl_3O_2$; mol wt 165.39. C 14.52%, H 1.83%, Cl 64.30%, O 19.35%. Made by adding the required amount of water to trichloroacetaldehyde, *q.v.* First synthesized by Liebig in 1832. Introduced as hypnotic by Liebreich in 1869. Comprehensive description: J. E. Fairbrother, *Anal. Profiles Drug Subs.* **2**, 85-143 (1973). LC determn in drinking water: M. C. Bruzzoniti *et al.*, *J. Chromatogr. A* **920**, 283 (2001). Dissociation in water: Piguet *et al.*, *Helv. Chim. Acta* **46**, 406 (1963). Toxicity data: E. Goldenthal, *Toxicol. Appl. Pharmacol.* **18**, 185 (1971). Evaluation of carcinogenic risk: *IARC Monographs* **63**, 245-269 (1995). Review of pharmacology: T. L. Sourkes, *Mol. Chem. Neuropathol.* **17**, 21-30 (1992); and use in pediatric dentistry: P. A. Moore, *Anesth. Prog.* **31**, 191-196 (1984); V. Avalos-Arenas *et al.*, *Curr. Med. Res. Opin.* **14**, 219-226 (1998).

Colorless, transparent, or white crystals with aromatic, penetrating and sightly acrid odor, slightly bitter, caustic taste. *Irritant.* d 1.91. Slowly volatilizes on exposure to air. mp 57° when heated in an open vessel. bp 98° with dissociation into chloral and water. One gram dissolves in 1.3 ml alcohol, in 2 ml chloroform, in 1.5 ml ether, in 1.4 ml olive oil, in 0.5 glycerol, in 68 g carbon disulfide. Very sol in water, olive oil. Freely sol in acetone, methyl ethyl ketone. One ml of water dissolves the following amounts of chloral hydrate: 2.4 g at 0°, 8.3 g at 25°, 14.3 g at 40°. pH of 10% aq soln is 3.5 to 5.5. LD_{50} orally in rats: 479 mg/kg (Goldenthal). *Store in airtight containers.*

Note: This is a controlled substance (depressant): **21 CFR,** 1308.14.

USE: Manuf DDT and other insecticides.

THERAP CAT: Sedative, hypnotic.

THERAP CAT (VET): Anesthetic, sedative, hypnotic.

2072. α-Chloralose. [15879-93-3] 1,2-*O*-[(1*R*)-2,2,2-Trichloroethylidene]-α-D-glucofuranose; α-D-glucochloralose; glucochloral; anhydroglucochloral; chloralosane; Alphakil; Dorcalm; Somio. $C_8H_{11}Cl_3O_6$; mol wt 309.52. C 31.04%, H 3.58%, Cl 34.36%, O 31.01%. Prepd from anhydr glucose and trichloroacetaldehyde (anhydr chloral); β-chloralose is also formed. Prepn: M. Hanroit, C. Richet, *C. R. Hebd. Seances Acad. Sci.* **116**, 63 (1893); and structure: Pictet, Reichel, *Helv. Chim. Acta* **6**, 621 (1923); Freudenberg, Vajda, *J. Am. Chem. Soc.* **59**, 1955 (1937); Coles *et al.*, *ibid.* **51**, 519 (1929); White, Hixon, *ibid.* **55**, 2438 (1933). Crystal structure: T. Taga *et al.*, *Acta Crystallogr.* **B38**, 1874 (1982). Immobilizing agent in control of depredating birds: Redpath *et al.*, *Ann. Appl. Biol.* **49**, 77 (1961). Anesthetic for dogs in experimental work: B. G. Bass, N. M. Buckley, *Am. J. Physiol.* **210**, 854 (1966); R. H. Cox, *ibid.* **223**, 660 (1972). CNS depressant activity: J. L. Barker, *Nature* **252**, 52 (1974). GLC determn in rodenticides: J. Theobald, *J. Chromatogr.* **129**, 444 (1976); in tissue: E. Odam *et al.*, *Analyst* **109**, 1335 (1984). Acute toxicity: E. W. Schafer, *Toxicol. Appl. Pharmacol.* **21**, 315 (1972). Review of pharmacology: G. U. Balis, R. R. Monroe, *Psychopharmacologia* **6**, 1-30 (1964); and toxicology: P. Lees, *Vet. Rec.* **91**, 330-333 (1972). Review as canine anesthetic: H. Holzgrefe *et al.*, *Lab. Anim. Sci.* **37**, 587-595 (1987).

Needles from alcohol or ether, mp 187°. $[\alpha]_D^{22}$ +19° (c = 5 in 98% alc). One gram dissolves in 225 ml water at 15°, in 120 ml at 37°, in 15 ml alc at 25°. Sol in ether, glacial acetic acid; slightly in chloroform. Almost insol in petr ether. LD_{50} orally in mice: 400 mg/kg (Schafer).

β-Chloralose. mp 227-230°, is much less sol in water, alcohol, and ether.

USE: Rodenticide. Control of avian pests.

THERAP CAT: Sedative, hypnotic.

THERAP CAT (VET): Surgical anesthetic for laboratory animals.

2073. Chlorambucil. [305-03-3] 4-[Bis(2-chloroethyl)amino]benzenebutanoic acid; 4-[*p*-[bis(2-chloroethyl)amino]phenyl]-butyric acid; γ-[*p*-di(2-chloroethyl)aminophenyl]butyric acid; *N,N*-di-2-chloroethyl-γ-*p*-aminophenylbutyric acid; chloraminophene; chloroambucil; CB-1348; Leukeran. $C_{14}H_{19}Cl_2NO_2$; mol wt 304.21. C 55.28%, H 6.30%, Cl 23.31%, N 4.60%, O 10.52%. Nitrogen mustard deriv. Prepn: W. C. J. Ross *et al.*, **GB 727336**; *eidem*, **US 2944079** (1955, 1960 both to Nat. Res. Dev. Corp.); Everett *et al.*, *J. Chem. Soc.* **1953**, 2386. Improved process: Phillips, Mentha, **US 3046301** (1962 to Burroughs Wellcome). Chemotherapeutic activity: Van Putten *et al.*, *Eur. J. Cancer* **7**, 11 (1971). Toxicity: W. C. J. Ross, *Biochem. Pharmacol.* **13**, 969 (1964). Review of carcinogenicity studies: *IARC Monographs* **9**, 125-134 (1975). Comprehensive description: M. Tariq, A. A. Al-Badr, *Anal. Profiles Drug Subs.* **16**, 85-118 (1987).

Flattened needles from petr ether, mp 64-66°. Sol in ether, dilute alkali. Sol (20°) in 1.5 parts alc, in 2.5 parts chloroform, in 2 parts acetone. Very slightly sol in water. LD_{50} i.p. in rats: 58.2 μmole/kg (Ross).

Caution: This substance is listed as a known human carcinogen: *Report on Carcinogens, Twelfth Edition* (PB2011-111646, 2011) p 90.

THERAP CAT: Antineoplastic.

THERAP CAT (VET): Antineoplastic agent, esp for leukemia.

2074. Chloramine-B. [127-52-6] *N*-Chlorobenzenesulfonamide sodium salt (1:1); (*N*-chlorobenzenesulfonamido)sodium; sodium benzenesulfochloramine; Neomagnol. $C_6H_5ClNNaO_2S$; mol wt 213.61. C 33.74%, H 2.36%, Cl 16.60%, N 6.56%, Na 10.76%, O 14.98%, S 15.01%. Prepd via benzenesulfonamide: Chattaway, *J. Chem. Soc.* **87**, 145 (1905); Cuiban *et al.*, *Pharmazie* **13**, 407 (1958).

Trihydrate. [17440-73-2] $C_6H_5ClNNaO_2S.3H_2O$; mol wt 267.66. Prisms. *Corrosive.* Soluble in 20 parts of water, more sol in hot water. Sol in 25 parts of ethanol, yielding a turbid soln. Very sparingly sol in ether, chloroform. An aq soln first turns red litmus paper blue, then gradually bleaches it. Gives a red color wth phenolphthalein.

THERAP CAT: Antibacterial.

THERAP CAT (VET): Antiseptic (topical).

2075. Chloramine-T. [127-65-1] *N*-Chloro-4-methylbenzenesulfonamide sodium salt (1:1); (*N*-chloro-*p*-toluenesulfonamido)sodium; sodium *p*-toluenesulfonchloramide; chloramine; tosylchloramide sodium; Chlorazene; Euclorina; Halamid. $C_7H_7Cl-NNaO_2S$; mol wt 227.64. C 36.93%, H 3.10%, Cl 15.57%, N 6.15%, Na 10.10%, O 14.06%, S 14.08%. Prepd via *p*-toluenesulfonamide: Chattaway, *J. Chem. Soc.* **87**, 145 (1905); Inglis, *J. Soc. Chem. Ind. London* **37**, 288 (1918); F. J. Welcher, *Organic Analytical Reagents* vol. 4 (Van Nostrand, New York, 1948) pp 316-320. Reviews of synthetic applications: M. M. Campbell, G Johnson, *Chem. Rev.* **78**, 65-79 (1978); G. Agnihotri, *Synlett* **2005**, 2857-2858.

Trihydrate. [7080-50-4] $C_7H_7ClNNaO_2S.3H_2O$; mol wt 281.68. Prisms. Loses water on drying. Dec slowly on exposure to air. *Keep well closed. Corrosive.* Fairly sol in water. Practically insol in benzene, chloroform, ether. Dec by alc.

USE: Detection of bromate and halogens. Source of halonium ions and nitrogen anions in organic synthesis.

THERAP CAT: Antibacterial.

THERAP CAT (VET): Antiseptic (topical).

2076. Chloraminophenamide. [121-30-2] 4-Amino-6-chloro-1,3-benzenedisulfonamide; 4-amino-6-chloro-*m*-benzenedisulfonamide; 5-chloro-2,4-bis(sulfonamido)aniline; 6-amino-4-chlorobenzene-1,3-disulfonamide; 5-chloro-2,4-disulfamylaniline; Idorese. $C_6H_8ClN_3O_4S_2$; mol wt 285.72. C 25.22%, H 2.82%, Cl 12.41%, N 14.71%, O 22.40%, S 22.44%. Prepn: Novello, Sprague, *J. Am. Chem. Soc.* **79**, 2028 (1957); Novello, **US 2809194**; **US 2965655**; **US 2965656** (1957, 1960, 1960 all to Merck & Co.); Novello *et al.*, *J. Org. Chem.* **25**, 965 (1960).

Crystals from aq ethanol, mp 251-252°. uv max (ethanol): 223.5-224.5, 265-266, 312-314 nm (ε 41776, 18633, 3874). Slightly sol in water; more freely sol in alkalies.

THERAP CAT: Diuretic.

2077. Chloramphenicol. [56-75-7] 2,2-Dichloro-*N*-[(1*R*,-2*R*)-2-hydroxy-1-(hydroxymethyl)-2-(4-nitrophenyl)ethyl]acetamide; D-*threo-N*-dichloroacetyl-1-*p*-nitrophenyl-2-amino-1,3-propanediol; D(−)-*threo*-2-dichloroacetamido-1-*p*-nitrophenyl-1,3-propanediol; D-*threo-N*-(1,1′-dihydroxy-1-*p*-nitrophenylisopropyl)dichloroacetamide; Aquamycetin; Chlorocid; Chloromycetin; Chloroptic; Fenicol; Pantovernil; Paraxin; Quemicetina; Sintomicetina; Tifomycine; Viceton. $C_{11}H_{12}Cl_2N_2O_5$; mol wt 323.13. C 40.89%, H 3.74%, Cl 21.94%, N 8.67%, O 24.76%. Broad spectrum antibiotic obtained from cultures of the soil bacterium *Streptomyces vene-*

zuelae: Bartz, *J. Biol. Chem.* **172**, 445 (1948); Gottlieb *et al., J. Bacteriol.* **55**, 409 (1948); Ehrlich *et al., ibid.* **56**, 467 (1948). Isoln from the moon snail, *Lunatia heros:* C. A. Price *et al., J. Antibiot.* **34**, 118 (1981). Structure: Rebstock *et al., J. Am. Chem. Soc.* **71**, 2458 (1949). Synthesis: Controulis *et al., ibid.* 2463; Long, Troutman, *ibid.* 2469, 2473. *See also* **US 2483871**; **US 2483884**; **US 2483892**. Alternate synthesis: Ehrhart *et al., Ber.* **90**, 2088 (1957); **GB 795131**; **GB 796901**, *C.A.* **53**, 2161 (1959) (both to Chinoin); **US 2839577** (1958 to Chinoin). Review and evaluation of toxicity studies: *IARC Monographs* **10**, 85-98 (1976). Review of pharmacology and clinical efficacy: I. Shalit, M. I. Marks, *Drugs* **28**, 281-291 (1984). Comprehensive description: D. Szulczewski, F. Eng, *Anal. Profiles Drug Subs.* **4**, 47-90 (1975); A. A. Al-Badr, H. A. El-Obeid, *ibid.* **15**, 701-760 (1986). *Reviews:* Hahn in *Antibiotics* **vol. 1**, D. Gottlieb, P. D. Shaw, Eds. (Springer-Verlag, New York, 1967) pp 308-330; Pestka, *ibid.* **vol. 3**, J. W. Corcoran, F. E. Hahn, Eds. (1975) pp 370-395; O. Pongs, *ibid.* **vol. 5**, pt. 1, F. E. Hahn, Ed. (1979) pp 26-42.

Needles or elongated plates from water or ethylene dichloride. mp 150.5-151.5°. Sublimes in high vacuum. $[\alpha]_D^{27}$ +18.6° (c = 4.86 in ethanol). $[\alpha]_D^{25}$ −25.5° (ethyl acetate). uv max: 278 nm ($E_{1cm}^{1\%}$ 298). Soly (25°) in water: 2.5 mg/ml; in propylene glycol: 150.8 mg/ml. Very sol in methanol, ethanol, butanol, ethyl acetate, acetone. Fairly sol in ether. Insol in benzene, petr ether, vegetable oils. Soly in 50% acetamide soln about 5%. Additional soly data: Weiss *et al., Antibiot. Chemother.* **7**, 374 (1957). Aq solns are neutral. Neutral and acid solns are stable on heating.

Monosuccinate sodium salt. [982-57-0] Chloramphenicol sodium succinate; Globenicol; Kemicetine. $C_{15}H_{15}Cl_2N_2NaO_8$; mol wt 445.18. Freely sol in alc, water (about 50% w/w).

Palmitate. [530-43-8] Chloropal. Prepn: Edgerton, **US 2662906** (1953 to Parke, Davis). Structure: Edgerton *et al., J. Am. Chem. Soc.* **77**, 27 (1955). Description: Glazko *et al., Antibiot. Chemother.* **2**, 234 (1952). Soly data: Weiss *et al., ibid.* **7**, 374 (1957). Crystals from benzene, mp 90°. Practically tasteless. $[\alpha]_D^{26}$ +24.6° (c = 5 in ethanol). uv max (ethanol): 271 nm ($E_{1cm}^{1\%}$ 179). Freely sol in acetone, methanol, chloroform, benzene; sol in ether; sparingly sol in alc; very slightly sol in solvent hexane. Insol in water.

Monosuccinate arginine salt. [34327-18-9] Chloramphenicol arginine succinate. $C_{21}H_{30}Cl_2N_6O_{10}$; mol wt 597.40. mp 135-145° (dec). *See: Jpn. Med. Gaz.* **7**(10), 15 (1970).

Pantothenate calcium complex (4:1). [31342-36-6] Chloramphenicol pantothenate. $C_{62}H_{80}CaCl_8N_{10}O_{30}$; mol wt 1769.04. Prepn: I. Villax, **GB 866787**; **GB 866788**; **GB 866789** (all 1961); I. Villax, **US 3078300** (1963).

Caution: Chloramphenicol is reasonably anticipated to be a human carcinogen: *Report on Carcinogens, Twelfth Edition* (PB2011-111646, 2011) p 92.

THERAP CAT: Antibacterial; antirickettsial.

THERAP CAT (VET): Antibacterial.

2078. Chloranil. [118-75-2] 2,3,5,6-Tetrachloro-2,5-cyclohexadiene-1,4-dione; 2,3,5,6-tetrachloro-*p*-benzoquinone; tetrachloroquinone; Spergon; Vulklor. $C_6Cl_4O_2$; mol wt 245.86. C 29.31%, Cl 57.68%, O 13.01%. Prepd from *p*-phenylenediamine or phenol by treating with $KClO_3$ and HCl. Because of its great resistance to further oxidation, chloranil is formed as the final product of the chlorate-HCl oxidation of many aromatic compds. Comprehensive list: Huntress, *Organic Chlorine Compounds* (New York, 1948). Laboratory procedure starting with phenol or *p*-chlorophenol: Fierz-David, Blangey, *Grundlegende Operationen der Farbenchemie* (Vienna, 5th ed., 1943) p 140. Use as reagent for pamaquine, *q.v.,* in urine: Schulemann *et al., Chem. Zentralbl.* **1928**, I, 2193. Aqueous decomposition study: D. H. Sarr *et al., Environ. Sci. Technol.* **29**, 2735 (1995). *Review: Health and Environmental Effects Profile for Chloranil* (EPA PB88-219696, 1986) 59 pp.

Golden-yellow platelets from acetic acid or acetone. Monoclinic prisms from benzene or toluene or by sublimation *in vacuo*, mp 290°. Absorption spectrum in chloroform: Lifschitz *et al., Rec. Trav. Chim.* **43**, 276, 658 (1924). Insol in water; almost insol in cold petr ether, cold alcohol. Sol in ether; sparingly sol in chloroform, carbon tetrachloride, carbon disulfide. Solubility data: Dimroth, Bamberger, *Ann.* **438**, 103, 106 (1924).

Caution: Potential symptoms of overexposure are watery diarrhea, CNS depression, coma. Not absorbed percutaneously; direct contact may cause skin irritation. *See Clinical Toxicology of Commercial Products*, R. E. Gosselin *et al.,* Eds. (Williams & Wilkins, Baltimore, 5th ed., 1984) p II-317.

USE: Agricultural fungicide, dye intermediate, reagent. Manuf of electrodes for pH measurement.

2079. Chloranilic Acid. [87-88-7] 2,5-Dichloro-3,6-dihydroxy-2,5-cyclohexadiene-1,4-dione; 2,5-dichloro-3,6-dihydroxy-*p*-benzoquinone. $C_6H_2Cl_2O_4$; mol wt 208.98. C 34.48%, H 0.96%, Cl 33.93%, O 30.62%. Prepd from chloranil by alkaline hydrolysis: Conant, Fieser, *J. Am. Chem. Soc.* **46**, 1866 (1924).

Red crystals, mp 283-284°. Tendency to sublime. Relatively strong dibasic acid. Absorption max (pH 4 in water): 290-340 nm, 520-555 nm.

USE: Reacts with metal cations to form stable salts. Used in spectrophotometry: Hart, *Org. Chem. Bull.* **33**, no. 3 (1961).

2080. Chlorantraniliprole. [500008-45-7] 3-Bromo-*N*-[4-chloro-2-methyl-6-[(methylamino)carbonyl]phenyl]-1-(3-chloro-2-pyridinyl)-1*H*-pyrazole-5-carboxamide; DPX-E2Y45; Altacor; Coragen; Dermacor; Rynaxypyr. $C_{18}H_{14}BrCl_2N_5O_2$; mol wt 483.15. C 44.75%, H 2.92%, Br 16.54%, Cl 14.67%, N 14.50%, O 6.62%. Anthranilic diamide pesticide; insect-specific ryanodine receptor activator. Prepn: G. P. Lahm *et al., WO 03015519*; *eidem*, **US 7232836** (2003, 2007 both to DuPont); and ryanodine receptor activity: *idem et al., Bioorg. Med. Chem. Lett.* **17**, 6274 (2007). Physical properties and activity: A. Bassi *et al., Proc. 16th Int. Plant Prot. Congr.* **1**, 52 (2007). HPLC-MS/MS determn in fruits and vegetables: P. Caboni *et al., J. Agric. Food Chem.* **56**, 7696 (2008). Field trials vs white grubs: A. M. Koppenhöfer, E. M. Fuzy, *Biol. Control* **45**, 93 (2008); vs fruit flies: L. A. F. Teixeira *et al., Pest Manag. Sci.* **65**, 137 (2009).

Fine crystalline off-white powder, mp 208-210°. Log P (octanol/water): 2.76 (pH 7). Vapor pressure (20°): 6.3×10^{-12} Pa. pKa (20°): 10.88. Soly (mg/l) at 20°: acetone 3446, methanol 1714, ethyl acetate 1144, water 1.0. LD_{50} in rats (mg/kg): >5000 orally; >5000 dermally. LC_{50} (4 hr) in rats (mg/l): >5.1 by inhalation. LC_{50} in rainbow trout, bluegill sunfish (mg/l): >13.8, >15.7 (Bassi).

USE: Insecticide.

2081. Chlorbenzoxamine. [522-18-9] 1-[2-[(2-Chlorophenyl)phenylmethoxy]ethyl]-4-[(2-methylphenyl)methyl]piperazine; 1-[2-(o-chloro-α-phenylbenzyloxy)ethyl]-4-o-methylbenzylpiperazine; 1-[2'-(o-chlorobenzhydryloxy)ethyl]-4-(o-methylbenzyl)diethylenediamine; 1-(o-chlorobenzhydryloxyethyl)-4-(o-methylbenzyl)-piperazine; chlorbenzoxyethamine; chlorobenzoxamine. $C_{27}H_{31}$-ClN_2O; mol wt 435.01. C 74.55%, H 7.18%, Cl 8.15%, N 6.44%, O 3.68%. Anticholinergic. Prepn: Morren et al., Ind. Chim. Belge **22**, 409 (1957); Morren, **GB 837986** (1960). Toxicity study: Levis et al., Arch. Int. Pharmacodyn. Ther. **118**, 167 (1959).

bp$_{0.01}$ 234-236°; bp$_{0.005}$ 235°.
Dihydrochloride. [5576-62-5] UCB-1474; Libratar; Gastomax. $C_{27}H_{31}ClN_2O.2HCl$; mol wt 507.92. Bitter crystals, mp 197-200°. Freely sol in methanol; sol in ethanol, chloroform, acetic acid; slightly sol in water, acetone. Practically insol in acetonitrile, ether, benzene. LD$_{50}$ in rats (mg/kg): 4000 s.c., 66 i.v.; in mice (mg/kg): 1400 orally (Levis).
THERAP CAT: Antispasmodic.

2082. Chlorcyclizine. [82-93-9] 1-[(4-Chlorophenyl)phenylmethyl]-4-methylpiperazine; 1-(p-chloro-α-phenylbenzyl)-4-methylpiperazine; 1-(4-chlorobenzhydryl)-4-methylpiperazine; N-methyl-N'-(4-chlorobenzhydryl)piperazine; chlorocyclizine; compd 47-282; Alergicide; Perazyl; Trihistan. $C_{18}H_{21}ClN_2$; mol wt 300.83. C 71.87%, H 7.04%, Cl 11.78%, N 9.31%. Prepn: Baltzly et al., J. Org. Chem. **14**, 775 (1949); Murfitt, Dewing, **GB 656043** (1951 to Wellcome Found.); **US 2630435** (1953 to Burroughs Wellcome). Toxicity study: Castillo et al., J. Pharmacol. Exp. Ther. **96**, 388 (1949).

Oil, bp$_{0.1-0.15}$ 137-145°.
Hydrochloride. [1620-21-9] AH-289; Di-Paralene; Perazil; Histantin. $C_{18}H_{21}ClN_2.HCl$; mol wt 337.29. Crystalline powder, mp 226-227°. One gram dissolves in about 2 ml water, in 11 ml alc, in about 4 ml chloroform. Practically insol in ether, benzene.
Dihydrochloride. $C_{18}H_{21}ClN_2.2HCl$. Prisms from alc + ether, mp 216-216.5°. Freely sol in water; sol in alc. Aq soln is acid to litmus. LD$_{50}$ i.p. in mice: 137 mg/kg (Castillo).
Di[(tert-butyl)naphthalene sodium sulfonate]. Bexedan. C_{18}-$H_{21}ClN_2.2C_{14}H_{15}NaOS$; mol wt 809.48.
THERAP CAT: Antihistaminic.
THERAP CAT (VET): The hydrochloride as antihistaminic.

2083. Chlordane. [57-74-9] 1,2,4,5,6,7,8,8-Octachloro-2,3,-3a,4,7,7a-hexahydro-4,7-methano-1H-indene; 1,2,4,5,6,7,8,8-octachloro-3a,4,7,7a-tetrahydro-4,7-methanoindan; 1,2,4,5,6,7,8,8-octachloro-4,7-methane-3a,4,7,7a-tetrahydroindane; chlordan; CD-68; Velsicol 1068; Toxichlor; Niran; Octachlor; Ortho-Klor; Synklor; Corodane; Belt. $C_{10}H_6Cl_8$; mol wt 409.76. C 29.31%, H 1.48%, Cl 69.21%. Organochlorine pesticide. The commercial product is a mixture containing 60 to 75% of the pure compound and 25 to 40% of related compds. Chlorine content: 64-67%. Prepn: Hyman, **GB 618432** (1949); Kleiman, **US 2598561** (1952 to Velsicol). Insecti-

cidal activity: Bussart, Schor, Soap Sanit. Chem. **24**, 126 (1948). Biodegradation in river sediment: T. Hirano et al., Chemosphere **67**, 428 (2007). GC-MS determn in sediment: M.-S. Kim et al., J. Chromatogr. A **1208**, 25 (2008). Toxicity study: R. B. Harbison, Toxicol. Appl. Pharmacol. **32**, 443 (1975). Review of toxicology and human exposure: Toxicological Profile for Chlordane (PB95-100111, 1994) 262 pp.

Viscous, amber-colored liquid. Viscosity (25°): 69 P (about that of 95% glycerol). Viscosity reduced considerably by heating to 120-140°F when it may be sprayed directly. d^{25} 1.59-1.63. n_D^{25} 1.56-1.57. Insol in water. Miscible with aliphatic and aromatic hydrocarbon solvents, including deodorized kerosene. Loses its chlorine in presence of alkaline reagents, and should not be formulated with any solvent, carrier, diluent or emulsifier, which has an alkaline reaction. LD$_{50}$ i.p. in male rats: 343 mg/kg (Harbison).
Caution: Potential symptoms of overexposure are blurred vision; confusion; ataxia, delirium; coughing; abdominal pain, nausea, vomiting and diarrhea; irritability, tremor, convulsions; anuria. Potential occupational carcinogen. See NIOSH Pocket Guide to Chemical Hazards (DHHS/NIOSH 97-140, 1997) p 56. Direct contact may cause skin, eye and mucous membrane irritation. See Patty's Industrial Hygiene and Toxicology vol. **2B**, G. D. Clayton, F. E. Clayton, Eds. (Wiley-Interscience, New York, 3rd ed., 1981) pp 3718-3725. See also Clinical Toxicology of Commercial Products, R. E. Gosselin et al., Eds. (Williams & Wilkins, Baltimore, 5th ed., 1984) Section III, p 108-109.
Note: Chlordane is listed as a persistent organic pollutant (POP) in Annex A of the Stockholm Convention on Persistent Organic Pollutants (United Nations, Stockholm, 2001) 43 pp; amended (Geneva, 2009) 63 pp.
USE: Insecticide, fumigant.
THERAP CAT (VET): Insecticide, acaricide.

2084. Chlordecone. [143-50-0] 1,1a,3,3a,4,5,5,5a,5b,6-Decachlorooctahydro-1,3,4-metheno-2H-cyclobuta[cd]pentalen-2-one; GC-1189; Kepone. $C_{10}Cl_{10}O$; mol wt 490.61. C 24.48%, Cl 72.26%, O 3.26%. Organochlorine pesticide. Prepn: Gilbert, Giolito, **US 2616825** and **US 2616928** (1952 to Allied Chem.); reissued as **US RE 24435** (1958). Structure: McBee et al., J. Am. Chem. Soc. **78**, 1511 (1956). Review: Ungnade, McBee, Chem. Rev. **58**, 249-320 (1958). GC-MS determn in soil: L. Amalric et al., Int. J. Environ. Anal. Chem. **86**, 15 (2006). Toxicity data: T. B. Gaines, Toxicol. Appl. Pharmacol. **14**, 515 (1969). Occupational exposure: Chem. Eng. News **54**, 19 (Feb. 2, 1976). Comparative toxicity in man and animals: P. S. Guzelian, Annu. Rev. Pharmacol. Toxicol. **22**, 89-113 (1982). Review of toxicology and human exposure: Toxicological Profile for Mirex and Chlordecone (PB95-264354, 1995) 362 pp.

Crystals, dec 350°. Slightly sol in water and hydrocarbon solvents. Sol in alcohols, ketones, acetic acid. LD$_{50}$ orally in rats: 125 mg/kg (Gaines).
Caution: Potential symptoms of overexposure are headache, anxiety, tremor; liver and kidney damage; visual disturbances; ataxia, chest pain, skin erythema; testicular atrophy, low sperm count. See NIOSH Pocket Guide to Chemical Hazards (DHHS/NIOSH 97-140, 1997) p 184. See also Patty's Industrial Hygiene and Toxicology

vol. 2B, G. D. Clayton, F. E. Clayton, Eds. (John Wiley & Sons, New York, 4th ed., 1994) 1504-1506, 1545-1550. This substance is reasonably anticipated to be a human carcinogen: *Report on Carcinogens, Twelfth Edition* (PB2011-111646, 2011) p 250.

Note: Chlordecone is listed as a persistent organic pollutant (POP) in Annex A of the *Stockholm Convention on Persistent Organic Pollutants* (United Nations, Stockholm, 2001) 43 pp; amended (Geneva, 2009) 63 pp.

USE: Formerly as insecticide, fungicide, miticide.

2085. **Chlordiazepoxide.** [58-25-3] 7-Chloro-*N*-methyl-5-phenyl-3*H*-1,4-benzodiazepin-2-amine 4-oxide; 7-chloro-2-methyl-amino-5-phenyl-3*H*-1,4-benzodiazepine 4-oxide; methaminodiazepoxide; clopoxide; Helogaphen; Libritabs; Multum; Risolid; Silibrin; Tropium. $C_{16}H_{14}ClN_3O$; mol wt 299.76. C 64.11%, H 4.71%, Cl 11.83%, N 14.02%, O 5.34%. Prototype of the benzodiazepine anxiolytics. Prepn: Sternbach, **US 2893992** (1959 to Hoffmann-La Roche); Sternbach, Reeder, *J. Org. Chem.* **26**, 1111 (1961). Comprehensive description: A. MacDonald *et al.*, *Anal. Profiles Drug Subs.* **1**, 15-51 (1972). Review of pharmacokinetics: D. J. Greenblatt *et al.*, *Clin. Pharmacokinet.* **3**, 381-394 (1978); of pharmacology and effect on cognitive function: J. B. Murray, *Genet. Psychol. Monogr.* **109** (2), 167-197 (1984). Clinical trial in alcohol withdrawal: A. K. Burroughs *et al.*, *Alcohol Alcohol.* **20**, 263 (1985).

Light yellow plates from ethanol, mp 236-236.5°. pK 4.8. Sparingly sol in chloroform, alc. Insol in water.

Hydrochloride. [438-41-5] Ansiacal; A-Poxide; Balance; Benzodiapin; Cebrum; Corax; Disarim; Elenium; Equibral; Labican; Lentotran; Librium; O.C.M.; Psichial; Psicoterina; Reliberan; Seren Vita; SK-Lygen; Viansin. $C_{16}H_{14}ClN_3O\cdot HCl$; mol wt 336.22. Crystals from methanol, mp 213°. Sol in water; sparingly sol in alc. Insol in solvent hexane.

Note: This is a controlled substance (depressant): **21 CFR,** 1308.14.

THERAP CAT: Anxiolytic.

THERAP CAT (VET): Tranquilizer.

2086. **Chlordimeform.** [6164-98-3] *N'*-(4-Chloro-2-methyl-phenyl)-*N*,*N*-dimethylmethanimidamide; *N'*-(4-chloro-*o*-tolyl)-*N*,*N*-dimethylformamidine; chlorophenamidine; chlorphenamidine; spanon; CDM; Ciba 8514; Schering 36268; Fundal; Galecron. $C_{10}H_{13}ClN_2$; mol wt 196.68. C 61.07%, H 6.66%, Cl 18.02%, N 14.24%. Member of a class of insecticides possessing the formamidine moiety, effective vs many organophosphate and carbamate resistant pests. Prepn: H. Arndt, W. Steinhausen, **DE 1172081**; *eidem*, **US 3502720** (1964, 1970, both to Schering AG). Acts by interfering with amine regulatory mechanisms: Beeman, Matsumura, *Nature* **242**, 273 (1973); Aziz, Knowles, *ibid.* 417. Review of metabolism: Knowles, *J. Agric. Food Chem.* **18**, 1038 (1970). Degradation products: Witkonton, Ercegovich, *ibid.* **20**, 569 (1972). Toxicity data: C. P. Robinson, P. W. Smith, *J. Toxicol. Environ. Health* **3**, 565 (1977). Toxicological studies: D. S. Folland *et al.*, *J. Am. Med. Assoc.* **13**, 1052 (1978); K. T. Maddy *et al.*, *Toxicol. Lett.* **33**, 37 (1986). Review of carcinogenicity studies: *IARC Monographs* **30**, 61-72 (1983).

Colorless crystals, mp 35°. $bp_{0.4}$ 156-157°. n_D^{25} 1.5885. d_4^{25} 1.105. Vapor pressure at 20°: 3.5×10^{-4} mm. Slightly sol in water; easily sol in organic solvents. LD_{50} i.p. in rats: 238 mg/kg (Robinson, Smith).

USE: Acaricide, insecticide.

2087. **Chlorendic Anhydride.** [115-27-5] 4,5,6,7,8,8-Hexa-chloro-3a,4,7,7a-tetrahydro-4,7-methanoisobenzofuran-1,3-dione; 1,4,5,6,7,7-hexachloro-*endo*-5-norbornene-2,3-dicarboxylic anhydride; hexachloroendomethylenetetrahydrophthalic anhydride; 1,4,-5,6,7,7-hexachloro-*endo*-bicyclo[2.2.1]hept-5-ene-2,3-dicarboxylic anhydride. $C_9H_2Cl_6O_3$; mol wt 370.81. C 29.15%, H 0.54%, Cl 57.36%, O 12.94%. Prepd from hexachlorocyclopentadiene and maleic anhydride: Herzfeld *et al.*, **US 2606910** (1952 to Velsicol); Baranauckas *et al.*, **US 2903463** (1959 to Hooker Chem.). Purification: Zimmer *et al.*, **DE 1113451** (1961 to Hooker Chem.), *C.A.* **57**, 697a (1962).

Crystals, mp 231-235°.

USE: In prepn of polyester resins.

2088. **Chlorethoxyfos.** [54593-83-8] Phosphorothioic acid *O,O*-diethyl *O*-(1,2,2,2-tetrachloroethyl) ester; *O,O*-diethyl *O*-(1,2,-2,2-tetrachloroethyl)phosphorothioate; DPX-43898; SD-208304; Fortress. $C_6H_{11}Cl_4O_3PS$; mol wt 335.98. C 21.45%, H 3.30%, Cl 42.20%, O 14.29%, P 9.22%, S 9.54%. Organophosphate insecticide; cholinesterase inhibitor. Prepn: M. Schnell *et al.*, **DD 107581** (1974); *eidem*, *J. Prakt. Chem.* **319**, 723 (1977). Physical properties and field trials: I. A. Watkinson, D. W. Sherrod, *Brit. Crop Prot. Conf. - Pests Dis.* **1986**, 107. Persistence in soil: R. A. Chapman *et al.*, *J. Environ Sci. Health* **B28**, 151 (1993). Assessment of volatility losses from soil: B. W. Fuller *et al.*, *J. Agric. Entomol.* **14**, 399 (1997). Field trials for control of corn rootworm: *eidem, J. Econ. Entomol.* **90**, 1332 (1997).

Colorless liquid, $bp_{0.05\,torr}$ 80°. n_D^{25} 1.4980. d_4^{20} 1.464. *Poisonous.* Also reported as white, crystalline, powder (Watkinson). Soly in water < 1 mg/l. Sol in hexane, ethanol, xylene, acetonitrile, chloroform. Vapor pressure at 20°: 8×10^{-4} mm Hg. LD_{50} in rats, mice (mg/kg): 1-10, 20-50 orally; in rabbits (mg/kg): 20-200 dermally (Watkinson).

Caution: Potential symptoms of overexposure by ingestion, inhalation or absorption through eye or skin are headache, nausea, vomiting, diarrhea, abdominal cramps; excessive sweating, salivation, tearing; constricted pupils, blurred vision; chest tightening, weakness, muscle twitching, confusion; unconsciousness; convulsions; severe respiratory depression.

USE: Insecticide.

2089. **Chlorfenapyr.** [122453-73-0] 4-Bromo-2-(4-chloro-phenyl)-1-(ethoxymethyl)-5-(trifluoromethyl)-1*H*-pyrrole-3-carbonitrile; AC-303630; CL-303630; Kotetsu; Pylon. $C_{15}H_{11}BrClF_3N_2O$; mol wt 407.62. C 44.20%, H 2.72%, Br 19.60%, Cl 8.70%, F 13.98%, N 6.87%, O 3.92%. Halogenated pyrrole insecticide that uncouples oxidative phosphorylation in mitochondria. Prepn: D. G. Brown *et al.*, **BR 8803788**; *eidem*, **US 5010098** (1988, 1991 both to Am. Cyanamid). Chemical and biological properties: J. B. Lovell *et al.*, *Brighton Crop Prot. Conf. - Pests Dis.* **1990**, 37. Summary of field trials: T. P. Miller *et al.*, *ibid.* 43. Field trials on apples: Y.-J. Ahn *et al.*, *Appl. Entomol. Zool.* **31**, 67 (1996). Mechanism of action: B. C. Black *et al.*, *Pestic. Biochem. Physiol.* **50**, 115 (1994). ELISA determn in fruit: E. Watanabe *et al.*, *J. Chromatogr. A* **1074**, 145 (2005). GC determn in water samples: Y. Wu, Z. Zhou, *Microchim Acta* **162**, 161 (2008).

White solid, mp 91-92°. Sol in acetone, diethyl ether, DMSO, THF, acetonitrile, alcohols. Insol in water. LD_{50} (single dose) in male, female rats (mg/kg): 223, 459 orally; in rabbits: >2000 dermally (Lovell). LC_{50} in Japanese carp: 0.5 ppm (Lovell).

USE: Insecticide; acaricide.

2090. Chlorfenvinphos. [470-90-6] Phosphoric acid 2-chloro-1-(2,4-dichlorophenyl)ethenyl diethyl ester; 2-chloro-1-(2,4-dichlorophenyl)vinyl diethyl phosphate; *O,O*-diethyl *O*-[2-chloro-1-(2,4-dichlorophenyl)vinyl] phosphate; 2,4-dichloro-α-(chloromethylene)benzyl alcohol diethyl phosphate; CVP; SD-7859; GC-4072; Birlane; Steladone; Supona. $C_{12}H_{14}Cl_3O_4P$; mol wt 359.56. C 40.09%, H 3.92%, Cl 29.58%, O 17.80%, P 8.61%. Organophosphate insecticide; cholinesterase inhibitor. Commercial prepn contains at least 90% of the active (Z)-isomer. Prepn: E. E. Gilbert *et al.*, *US 3003916* (1961 to Allied Chemical); R. R. Whetstone *et al.*, *J. Agric. Food Chem.* **14**, 352 (1966). Dermal absorption and metabolism in cattle: W. F. Chamberlain, D. E. Hopkins, *J. Econ. Entomol.* **55**, 86 (1962). GLC determn in animal tissue: M. C. Ivey *et al.*, *J. Agric. Food Chem.* **21**, 822 (1973). Toxicity studies: A. M. Ambrose *et al.*, *Toxicol. Appl. Pharmacol.* **17**, 323 (1970). Persistence in soil: K. I. Beynon, A. N. Wright, *J. Sci. Food Agric.* **18**, 143 (1967). Review of toxicology and human exposure: *Toxicological Profile for Chlorfenvinphos* (PB98-101116, 1997) 220 pp.

(Z)-isomer

Amber liquid, mild odor. *Poisonous.* $bp_{0.001}$ 120°, $bp_{0.5}$ 167-170°. Vapor press at 25°C: 7.5×10^{-6} mm Hg. n_D^{25} 1.5272. Soly in water at 23°: 145 ppm. Miscible with acetone, ethanol, propylene glycol. Slowly hydrolyzed by water. Corrosive to metal. LD_{50} in rats (mg/kg): 6.6 i.v.; 9.66 orally (Ambrose).

Caution: Potential symptoms of overexposure are headache, dizziness, weakness, feeling of anxiety, confusion, runny nose, constriction of pupils, blurred vision, nausea, vomiting, abdominal cramps, slow pulse, diarrhea, difficulty breathing, fainting (PB98-101116).

USE: Insecticide; acaricide.

THERAP CAT (VET): Ectoparasiticide.

2091. Chlorguanide. [500-92-5] *N*-(4-Chlorophenyl)-*N'*-(1-methylethyl)imidodicarbonimidic diamide; 1-(*p*-chlorophenyl)-5-isopropylbiguanide; N^1-*p*-chlorophenyl-N^5-isopropyldiguanide; chloroguanide; proguanil. $C_{11}H_{16}ClN_5$; mol wt 253.73. C 52.07%, H 6.36%, Cl 13.97%, N 27.60%. Prepn: Curd, Rose, *J. Chem. Soc.* **1946**, 729; Curd *et al.*, *ibid.* **1948**, 1630; Ainley *et al.*, *ibid.* **1949**, 98. Manuf of acetate: Gailliot, *FR 1001548* (1952 to Rhône-Poulenc), *C.A.* **51**, 7411e (1957). Toxicity study: Schmidt *et al.*, *J. Pharmacol. Exp. Ther.* **90**, 233 (1947).

Rectangular plates from toluene, mp 129°.

Acetate. $C_{11}H_{16}ClN_5 \cdot C_2H_4O_2$. Crystals from acetone, mp 189-190°.

Hydrochloride. [637-32-1] M-4888; RP-3359; SN-12837; Paludrine. $C_{11}H_{16}ClN_5 \cdot HCl$; mol wt 290.19. Crystals from water or aq ethanol, mp 243-244°. uv max (alc): 259 nm. Sol in alc; slightly sol in water. Practically insol in chloroform and ether. pH of satd aq soln 5.8-6.3. LD_{50} orally in rats: 200 mg/kg (Schmidt).

THERAP CAT: Antimalarial.

2092. Chlorhexidine. [55-56-1] *N,N''*-Bis(4-chlorophenyl)-3,12-diimino-2,4,11,13-tetraazatetradecanediimidamide; 1,1'-hexamethylenebis[5-(*p*-chlorophenyl)biguanide]; 1,6-bis[*N'*-(*p*-chlorophenyl)-N^5-biguanido]hexane; 1,6-bis(N^5-*p*-chlorophenyl-*N'*-diguanido)hexane; 1,6-di(4'-chlorophenyldiguanido)hexane; 10040. $C_{22}H_{30}Cl_2N_{10}$; mol wt 505.45. C 52.28%, H 5.98%, Cl 14.03%, N 27.71%. Bisbiguanide with bacteriostatic activity. Prepn: Rose, Swain, *J. Chem. Soc.* **1956**, 4422; *eidem*, *US 2684924* (1954 to I.C.I.). Antibacterial activity and acute toxicity: G. E. Davies *et al.*, *Br. J. Pharmacol.* **9**, 192 (1954). LC-MS determn in hemolyzed blood: K. Usui *et al.*, *J. Chromatogr. B* **831**, 105 (2006). Review of toxicology and clinical uses: D. M. Foulkes, *J. Periodontal Res.* **8**, Suppl. 12, 55-60 (1973). Series of articles on clinical efficacy in gingivitis and plaque control: *ibid.* **21**, Suppl. 16, 1-89 (1986). Clinical trial in prevention of nosocomial infection: P. Segers *et al.*, *J. Am. Med. Assoc.* **296**, 2460 (2006).

Crystals from methanol, mp 134°. Strong alkaline reaction. Soly in water at 20°: 0.08% (w/v).

Dihydrochloride. [3697-42-5] C.E.T. HEXtra; Nolvasan. $C_{22}H_{30}Cl_2N_{10} \cdot 2HCl$; mol wt 578.37. Crystals, dec 260-262°. Soly in water at 20°: 0.06 g/100 ml.

Diacetate. [56-95-1] Chlorasept 2000. $C_{22}H_{30}Cl_2N_{10} \cdot 2C_2H_4O_2$; mol wt 625.56. Crystals, mp 154-155°. Neutral reaction. Soly in water at 20°: 1.9 g/100 ml. Aq solns dec when heated above 70°. Soluble in alcohol, glycerol, propylene glycol, polyethylene glycols. LD_{50} orally in mice: 2 g/kg (Davies).

D-Digluconate. [18472-51-0] Biorgasept; Chlorhexamed; Collunovar; Corsodyl; Diaseptyl; Dosiseptive; Exoseptoplix; Hibident; Hibidil; Hibiscrub; Hibisprint; Hibitane; Merfene; Paroex; Peridex; Plurexid; Prexidine; Secalan; Septeal; Unisept. Soly in water at 20°: >50% (w/v). LD_{50} in mice (mg/kg): 22 i.v.; 1800 orally (Foulkes).

USE: Ingredient in mouth rinses, contact lens cleaners, burn creams and cosmetics.

THERAP CAT: Antiseptic; disinfectant.

THERAP CAT (VET): Antiseptic; disinfectant.

2093. Chloric Acid. [7790-93-4] $ClHO_3$; mol wt 84.46. Cl 41.97%, H 1.19%, O 56.83%. $HClO_3$. Prepd from barium chlorate and sulfuric acid: Lamb *et al.*, *J. Am. Chem. Soc.* **42**, 1643 (1920); from sodium chlorate using ion-exchange resins: Klement, *Z. Anorg. Allg. Chem.* **260**, 271 (1949).

Known in aq soln only. Aq solns are stable if pure and protected from light. *Oxidizer.* 1% aq soln d_4^{18} 1.0044; 6% soln d_4^{18} 1.0344; 10% soln d_4^{18} 1.0594; 16% soln d_4^{18} 1.0991; 20% soln d_4^{18} 1.1273; 24% soln d_4^{18} 1.1568. A 40% soln corresponds to $HClO_3 \cdot 7H_2O$, d_4^{20} 1.282. If higher concns are attempted by evaporation the soln begins to dec with evolution of chlorine and oxygen and formation of perchloric acid.

Caution: Strongly irritating to skin, mucous membranes.

USE: Oxidizing agent; with H_2SO_3 as catalyst in acrylonitrile polymerization.

2094. Chlorimuron-ethyl. [90982-32-4] 2-[[[[(4-Chloro-6-methoxy-2-pyrimidinyl)amino]carbonyl]amino]sulfonyl]benzoic acid ethyl ester; ethyl 2-[[[[(4-chloro-6-methoxypyrimidin-2-yl)amino]carbonyl]amino]sulfonyl]benzoate; DPX-F6025; Classic. $C_{15}H_{15}ClN_4O_6S$; mol wt 414.82. C 43.43%, H 3.64%, Cl 8.55%, N 13.51%, O 23.14%, S 7.73%. Selective sulfonylurea herbicide.

Prepn: A. D. Wolf, **AT 8316181**; *idem*, **US 4547215** (1984, 1985 both to Du Pont). Effect on plant growth and pigment synthesis: R. M. Devlin, Z. K. Koszanski, *Proc. Annu. Meet. Northeast. Weed Sci. Soc.* **40**, 115 (1986). Metabolism by plants: H. M. Brown, S. M. Neighbors, *Pestic. Biochem. Physiol.* **29**, 112 (1987). Field trial in soybeans: G. N. Rhodes *et al.*, *Tenn. Farm Home Sci.* **142**, 21 (1987). HPLC determn in crops: J. L. Prince, R. A. Guinivan, *J. Agric. Food Chem.* **36**, 63 (1988). Brief review: J. S. Claus, *Weed Technol.* **1**, 114-115 (1987).

Crystals from butyl chloride, mp 198-201°. Soly (ppm): acetone 71000, acetonitrile 31000, benzene 8000, methylene chloride 153000, water (pH 7) 1200, (pH 6.5) 450, (pH 5) 11. LD$_{50}$ in male, female rats (mg/kg): 4102, 4236 orally (Claus).

USE: Herbicide.

2095. Chlorine. [7782-50-5] Cl; at. wt [35.446; 35.457]; conventional at. wt 35.45; at. no. 17; valences 1, 3, 5, 7. A halogen; Group VIIA (17). Does not exist as elemental state, Cl, in nature. Occurs as diatomic molecule, Cl$_2$. Abundance in igneous rock (95% of earth's crust): 0.031% by wt; in seawater: 1.9% by wt (primarily as NaCl). Naturally occurring stable isotopes (mass numbers): 35 (75.77%), 37 (24.23%); known artificial radioactive isotopes: 31-34, 36 (longest-lived known isotope, T$_{½}$ 3.0 × 10^5 yrs; β^-, EC decay), 38-46, 48. Discovered in 1774 by C. W. Scheele; recognized as an element in 1810 by H. Davy. Commercial sources: seawater, ocean derived mineral deposits, brines from lakes, wells and springs. Industrial prepn from brine in electrolytic cells. Lab prepn from MnO$_2$ and HCl: Schmeisser in *Handbook of Preparative Inorganic Chemistry* **vol. 1**, G. Brauer, Ed. (Academic Press, New York, 2nd ed., 1963) p 272. Cosmogenic production and determn of ^{36}Cl for geological dating: H. E. Gove, *Philos. Trans. R. Soc. London Ser. A* **323**, 103 (1987); M. G. Zreda *et al.*, *Earth Planet. Sci. Lett.* **105**, 94 (1991). *Reviews:* Ciba Review **vol. 12**, no. 139 (Aug. 1960); *ACS Monograph Series* **no. 154**, entitled "Chlorine," J. S. Sconce, Ed. (Reinhold, New York, 1962) 901 pp; *MTP Int. Rev. Sci.: Inorg. Chem., Ser. One* **vol. 3**, V. Gutmann, Ed. (Butterworths, London, 1972); A. J. Downs, C. J. Adams, "Chlorine, Bromine, Iodine and Astatine" in *Comprehensive Inorganic Chemistry* **vol. 2**, J. C. Bailar, Jr. *et al.*, Eds. (Pergamon Press, Oxford, 1973) pp 1107-1594; *Chemistry of the Elements*, N. N. Greenwood, A. Earnshaw, Eds. (Pergamon Press, New York, 1984) pp 920-1041; L. C. Curlin, T. V. Bommaraju in *Kirk-Othmer Encyclopedia of Chemical Technology* **vol. 1** (Wiley-Interscience, New York, 4th ed., 1994) pp 938-1025. Review of potential human health and environmental adverse effects of chlorine and its compounds: E. Delzell *et al.*, *Regul. Toxicol. Pharmacol.* **2**, S1-S1056 (1994).

Greenish-yellow, diatomic gas; suffocating odor. mp −101.00°. bp −34.05°. d^{20} at 6.864 atm 1.4085 (liq); d^{-35} at 0.9949 atm 1.5649 (liq). d (relative to air) 2.48. Heat capacity at constant pressure (gas, 25°) 8.11 cal/mole/°C. Vapor pressure data: Giauque, Powell, *J. Am. Chem. Soc.* **61**, 1970 (1939). Critical temp 144°; critical pressure 76.1 atm; critical density 0.573. *Poisonous, corrosive.* Sol in water (25°) with formation of aqueous Cl$_2$ (0.062 moles/l), HOCl (0.030 moles/l) and Cl$^-$ (0.030 moles/l); total soly: 0.092 moles/l. More sol in alkalies. Oxidizing agent. Very reactive; E^0 (aq) ½Cl$_2$/ Cl$^-$ 1.356 V; dissociation energy (25°): 57.978 kcal. Forms halides with all elements except the rare gases helium, neon and argon. Noncombustible in air; most combustible materials will burn in chlorine. Forms explosive mixtures with flammable gases and vapors. Reacts explosively or forms explosive compounds with many common chemicals, especially acetylene, turpentine, ether, ammonia gas, fuel gas, hydrocarbons, hydrogen and finely divided metals. LC$_{50}$ (1 hr) in rats, mice (ppm): 293, 137 (K. C. Back *et al.*, *Reclassification*

of Materials Listed as Transportation Health Hazards (TSA-20-72-3; PB 214-270, 1972) pp A-182-183).

Caution: Potential symptoms of overexposure are burning of eyes, nose and mouth; lacrimation, rhinorrhea; coughing, choking and substernal pain; nausea, vomiting; headache, dizziness; syncope; pulmonary edema; pneumonia; hypoxemia; dermatitis; direct contact with liquid may cause frostbite. *See NIOSH Pocket Guide to Chemical Hazards* (DHHS/NIOSH 97-140, 1997) p 58. *See also Patty's Industrial Hygiene and Toxicology* **vol. 2F**, G. D. Clayton, F. E. Clayton, Eds. (Wiley-Interscience, New York, 4th ed., 1994) pp 4483-4505.

USE: Manuf of organic and inorganic chemicals. As oxidizing and bleaching agent in pulp and paper industry, and for textiles. As disinfectant for water purification, industrial waste, sewage, swimming pools. In the extraction and refining of metals. ^{36}Cl for determining geological age of samples such as meteorites, surface rocks, polar ice and ground water. Has been used as a military poison gas under the name *bertholite*.

2096. Chlorine Dioxide. [10049-04-4] Chlorine oxide (ClO$_2$); chlorine peroxide. ClO$_2$; mol wt 67.45. Cl 52.56%, O 47.44%. Prepn from chlorine and sodium chlorite: Derby, Hutchinson, *Inorg. Synth.* **4**, 152 (1953); from potassium chlorate and sulfuric acid: Bodenstein *et al.*, *Z. Anorg. Allg. Chem.* **147**, 233 (1925); by passing NO$_2$ through a column of sodium chlorate: Hutchinson, Derby, *Cereal Chem.* **24**, 372 (1947); alternate methods of prepn: Schmeisser in *Handbook of Preparative Inorganic Chemistry* **vol. 1**, G. Brauer, Ed. (Academic Press, New York, 2nd ed., 1963) p 301. *Review:* Bedumeau, *Rev. Prod. Chim.* **57**, 173-177, 257-261 (1954). Book: W. J. Masschelein, *Chlorine Dioxide*, R. G. Rice, Ed (Ann Arbor Science Publishers Inc., Ann Arbor, MI, 1979) 189 pp. Review of toxicology and human exposure: *Toxicological Profile for Chlorine Dioxide and Chlorite* (PB2004-107332, 2004) 191 pp.

Strongly oxidizing, yellow to reddish-yellow gas at room temp. Unpleasant odor similar to that of chlorine and reminiscent of that of nitric acid. Unstable in light; stable in dark if pure, but chlorides catalyze its decompn even in the dark. *Oxidizer, poisonous. Reacts violently with organic materials.* In concns in excess of 10% at atm pressure easily detonated by sunlight, heat, contact with mercury or carbon monoxide. mp −59°; bp 11°; d^0(liq) 1.642: Cheesman, *J. Chem. Soc.* **1930**, 35. Sol in water (3.01 g/l at 25° and 34.5 mm Hg), with slight hydrolysis to chlorous and chloric acid; sol in alkaline and sulfuric acid solns. Solid is yellowish-red crystalline mass; liquid is reddish-brown.

Caution: Potential symptoms of overexposure are irritation of eyes, nose and throat; coughing, wheezing, bronchitis, pulmonary edema; chronic bronchitis. *See NIOSH Pocket Guide to Chemical Hazards* (DHHS/NIOSH 97-140, 1997) p 58.

USE: Bleaching cellulose, paper-pulp, flour, leather, fats and oils, textiles, beeswax; purification of water; taste and odor control of water; cleaning and detanning leather; manuf of chlorite salts; oxidizing agent; bactericide, antiseptic and deodorizer.

2097. Chlorine Heptoxide. [12015-53-1] Perchloric acid trioxidochloryl ester; dichlorine heptoxide; perchloric anhydride. Cl$_2$O$_7$; mol wt 182.89. Cl 38.77%, O 61.24%. Prepd by dehydration of perchloric acid with P$_2$O$_5$: Michael, Conn, *Am. Chem. J.* **23**, 445 (1900); *eidem, ibid.* **25**, 92 (1901); Meyer, Kessler, *Ber.* **54**, 566 (1921); Goodeve, Powney, *J. Chem. Soc.* **1932**, 2078.

Colorless, very volatile oily liquid. d$^0_4$ 1.86. mp −91.5°; bp 82°; bp$_{23.7}$ 0°. Trouton constant 23.4. The most stable oxide of chlorine. *Explodes violently upon concussion or on contact with a flame or iodine.* Dipole moment in CCl$_4$ at 20° is 0.72. Does not attack wood or paper. Slowly hydrolyzed by water, forming perchloric acid.

Caution: May be irritating to skin, mucous membranes.

USE: Catalyst in cellulose esterification.

2098. Chlorine Monofluoride. [7790-89-8] Chlorine fluoride. ClF; mol wt 54.45. Cl 65.11%, F 34.89%. Prepd from Cl$_2$ and F$_2$ at 400°: Ruff *et al.*, *Z. Anorg. Allg. Chem.* **176**, 256 (1928); Kwasnik in *Handbook of Preparative Inorganic Chemistry* **vol. 1**, G. Brauer, Ed. (Academic Press, New York, 2nd ed., 1963) p 153.

Colorless gas. Slightly yellow when liquid. mp −155.6°. bp −100.1°. Crit temp −14°. d (liq; −108°) 1.67. *Extremely reactive.* Destroys glass instantly, attacks quartz readily in presence of moisture; organic matter bursts into flame on contact; violent reaction with water.

Caution: Extremely corrosive and irritating to skin, eyes, mucous membranes, respiratory tract.

2099. Chlorine Monoxide. [7791-21-1] Chlorine oxide (Cl$_2$O); dichlorine monoxide; dichloromonoxide; dichloroxide; hypochlorous anhydride; oxygen dichloride. Cl$_2$O; mol wt 86.90. Cl 81.59%, O 18.41%. First preparation by J. L. Gay-Lussac in 1842 from yellow mercuric oxide and chlorine. See also: Cady, *Inorg. Synth.* **5**, 156 (1957); Schmeisser in *Handbook of Preparative Inorganic Chemistry* **vol. 1**, G. Brauer, Ed. (Academic Press, New York, 2nd ed., 1963) p 299. Use as a powerful and selective chlorinating agent: F. D. Marsh *et al.*, *J. Am. Chem. Soc.* **104**, 4680 (1982).

Yellowish-brown gas. Disagreeable, penetrating odor. *Explodes on contact with organic matter.* Can also be caused to explode by a spark or by heating. Dec at moderate rate at room temp. mp −120.6°. bp 2.2°. Trouton constant: 22.5. Suffers photochemical and thermal decompn: at 100-140° there is an induction period, followed by a second order reaction, *see* N. V. Sidgwick, *The Chemical Elements and Their Compounds* **vol. II** (Oxford, 1950) p 1201. One vol of water at 0° dissolves more than 100 vols Cl$_2$O with formation of HClO; sol in CCl$_4$. Stored as a liquid or solid at temps below −80°.

Caution: Intensely irritating to eyes, skin, mucous membranes, respiratory tract.

USE: Chlorinating agent.

2100. Chlorine Trifluoride. [7790-91-2] Chlorine fluoride (ClF$_3$); trifluorochlorine. ClF$_3$; mol wt 92.45. Cl 38.35%, F 61.65%. Prepd from F$_2$ and Cl$_2$ or ClF. May be purified by distillation in suitable steel apparatus: Ruff, Krug, *Z. Anorg. Allg. Chem.* **190**, 270 (1930); Grisard *et al.*, *J. Am. Chem. Soc.* **73**, 5724 (1951); Kwasnik in *Handbook of Preparative Inorganic Chemistry* **vol. 1**, G. Brauer, Ed. (Academic Press, New York, 2nd ed., 1963) p 156. *Reviews:* Kemmitt, Sharp, *Adv. Fluorine Chem.* **4**, 240-241 (1965); Stein, "Physical and Chemical Properties of Halogen Fluorides" in *Halogen Chemistry* **vol. 1**, V. Gutmann, Ed. (Academic Press, New York, 1967) pp 133-224; Meinert, *Z. Chem.* **7**, 41-57 (1967).

Corrosive, colorless gas. Somewhat sweet, suffocating odor. mp −76.34°. bp 11.75°. *Poisonous, oxidizer, corrosive. Extremely reactive.* Glass wool and organic matter burst into flames on contact even with dil vapors. Violently hydrolyzed by water. Attacks quartz if traces of moisture are present. Liquid is yellow-green in color and solid is white.

Caution: Potential symptoms of overexposure to liquid or high vapor concentrations are eye and skin burns; respiratory irritation. *See NIOSH Pocket Guide to Chemical Hazards* (DHHS/NIOSH 97-140, 1997) p 60.

USE: Fluorinating agent; in nuclear reactor fuel processing; incendiary; igniter and propellant for rockets; pyrolysis inhibitor for fluorocarbon polymers.

2101. Chlorisondamine Chloride. [69-27-2] 4,5,6,7-Tetrachloro-2,3-dihydro-2-methyl-2-[2-(trimethylammonio)ethyl]-1*H*-isoindolium chloride (1:2); 4,5,6,7-tetrachloro-2-(2-dimethylamino-ethyl)-2-methylisoindolinium chloride methochloride; 4,5,6,7-tetrachloro-2-(2-dimethylaminoethyl)isoindoline dimethylchloride; *N*-[(2-dimethylammonium)ethyl]-4,5,6,7-tetrachloroisoindolinium dimethochloride; chlorisondamine dimethochloride; Ecolid. C$_{14}$H$_{20}$Cl$_6$N$_2$; mol wt 429.03. C 39.19%, H 4.70%, Cl 49.58%, N 6.53%. Nicotinic receptor antagonist of the ganglion blocking class. Prepn: Allen, Ocampo, *J. Electrochem. Soc.* **103**, 452, 682 (1956); Huebner, US 3025294 (1962 to Ciba). Pharmacology: P. B. S. Clarke *et al.*, *Br. J. Pharmacol.* **111**, 397 (1994). Use in studies of neuronal nicotinic acetylcholine receptors: A. S. Woods *et al.*, *J. Proteome Res.* **2**, 207 (2003).

Non-hygroscopic crystals contg ethanol of crystallization, mp 258-265° (dec). Sol in water, methanol, ethanol. pH of aq solns 4.7 to

6.2. Forms yellow, chloroform-soluble complex with bromcresol green, absorption max 420 nm.

USE: Biochemical probe in nicotinic receptor studies.

2102. Chlormadinone Acetate. [302-22-7] 17-(Acetyloxy)-6-chloropregna-4,6-diene-3,20-dione; 6-chloro-17-hydroxypregna-4,6-diene-3,20-dione acetate; 6-chloro-6-dehydro-17α-hydroxyprogesterone acetate; 6-chloro-6-dehydro-17α-acetoxyprogesterone; 17α-acetoxy-6-chloro-6,7-dehydroprogesterone; Chronosyn; Cyclonorm; Fertiletten; Gestafortin; Lormin; Luteran; Matrol; Normenon; Menstridyl; Prostal; Traslan. C$_{23}$H$_{29}$ClO$_4$; mol wt 404.93. C 68.22%, H 7.22%, Cl 8.75%, O 15.80%. Orally active progestogen with antiandrogenic activity; has been used in combinations as an oral contraceptive. Prepn: Brückner, **DE 1075114** (1960 to E. Merck, AG); Brückner *et al.*, *Ber.* **94**, 1225 (1961); Sciaky, *Gazz. Chim. Ital.* **91**, 545 (1961); **GB 932153**; H. J. Ringold, A. Bowers, **US 3485852** (1963, 1969 both to Syntex). Endocrinological activities: D. M. Brennan, R. J. Kraay, *Acta Endocrinol.* **44**, 367 (1963). Metabolism: S. Honma *et al.*, *Chem. Pharm. Bull.* **25**, 2019 (1977). Clinical evaluation in prostatic carcinoma: R. Nishimura, K. Shida, *Prostate*, Suppl. 1, 27 (1981); in benign prostatic hypertrophy: T. Usui *et al.*, *Acta Urol. Jpn.* **27**, 327 (1981), *C.A.* **73**, 27225 (1982). Review of carcinogenicity studies: *IARC Monographs* **21**, 365-375 (1979).

Crystals from methanol or ether, mp 212-214°. [α]$_D$ +6° (c = 1 in CHCl$_3$). uv max: 283.5, 286 nm (ε 23400, 22100).

Mixture with ethinyl estradiol. [37301-55-6] Amenyl; Lutestral; Menova.

Mixture with mestranol. [8065-91-6] C-Quens; Gestamestrol; Sequens.

THERAP CAT: Progestogen; antineoplastic (hormonal).

THERAP CAT (VET): Progestogen; estrus regulator.

2103. Chlormephos. [24934-91-6] *S*-(Chloromethyl)phosphorodithioic acid *O,O*-diethyl ester; chlormethylfos; *O,O*-diethyl *S*-(chloromethyl)dithiophosphate; MC-2188; Dotan. C$_5$H$_{12}$ClO$_2$PS$_2$; mol wt 234.69. C 25.59%, H 5.15%, Cl 15.11%, O 13.63%, P 13.20%, S 27.32%. Non-systemic organophosphate insecticide; cholinesterase inhibitor. Prepn: O. Scherer *et al.*, **DE 1015794**; *eidem*, **GB 817360** (1957, 1959 both to Hoechst); W. W. Brand *et al.*, *Phosphorus Sulfur* **10**, 183 (1981). Photodegradation: S. J. Buckland, R. S. Davidson, *Pestic. Sci.* **19**, 61 (1987). TLC/MS determn in biological samples: H. Brzezinka, N. Bertram, *J. Chromatogr. Sci.* **40**, 609 (2002). Review of physical properties and field trials: F. Colliot *et al.*, *Proc. 7th Br. Insectic. Fungic. Conf.* **2**, 557-565 (1973); physical properties and residue analysis: V. P. Lynch, *Anal. Methods Pestic. Plant Growth Regul.* **10**, 49-55 (1978).

Water-white oil. bp$_{1mm}$ 93-95°; bp$_{0.1 Torr}$ 81-85°. Vapor pressure (30°): 5.7 x 10^{-2}. Miscible with most organic solvents. Soly in water (20°): 60 ppm. d^{20} 1.260. n$_D^{20}$ 1.5244. uv max (ethanol): 208 nm. *Poisonous.* LD$_{50}$ in rats (mg/kg): 7 orally; 27 dermally. LC$_{50}$ in fish: 1.5 mg/l (Lynch).

USE: Insecticide primarily for soil applications.

2104. Chlormequat Chloride. [999-81-5] 2-Chloro-*N,N,N*-trimethylethanaminium chloride (1:1); (2-chloroethyl)trimethyl-ammonium chloride; chlorocholine chloride; choline dichloride; CCC; AC-38555; Cycocel; Cycogan. $C_5H_{13}Cl_2N$; mol wt 158.07. C 37.99%, H 8.29%, Cl 44.85%, N 8.86%. Prepn: Kauffmann, Vorlander, *Ber.* **43**, 2740 (1910); Freiss, Carville, *J. Am. Chem. Soc.* **76**, 2260 (1954); H. Linser *et al.*, **AT 222145** (1962 to OSSW), *C.A.* **57**, 7660f (1962). Activity: N. E. Tolbert, *J. Biol. Chem.* **235**, 475 (1960); J. Namokar, *Pesticides* **11**, 53 (1977). Metabolism: R. C. Blinn, *J. Agric. Food Chem.* **15**, 984 (1967). Toxicity study: G. Hennighausen, B. Wiegershausen, *Acta Biol. Med. Ger.* **27**, 663 (1971).

White crystalline solid, mp 245° (dec). Fish-like odor; very hygroscopic. Sol in water, lower alcs. Insol in ether, hydrocarbons. Aq solns corrosive to metal. LD_{50} in mice (mg/kg): 7 i.v., 54 orally (Hennighausen, Wiegershausen).

USE: Plant growth regulator.

2105. Chlormerodrin. [62-37-3] [3-[(Aminocarbonyl-*κO*)-amino]-2-methoxypropyl-*κC*]chloromercury; 1-[3-(chloromercuri)-2-methoxypropyl]urea; chlormeroprin; Katonil; Mercoral; Diurone; Mercloran; Percapyl; Merilid; Oricur. $C_5H_{11}ClHgN_2O_2$; mol wt 367.20. C 16.35%, H 3.02%, Cl 9.65%, Hg 54.63%, N 7.63%, O 8.71%. Prepn: R. L. Rowland, **US 2635982** (1953 to Lakeside); R. L. Rowland *et al.*, *J. Am. Chem. Soc.* **72**, 3595 (1950); E. L. Foreman, **US 2635983** (1953 to Lakeside). Toxicity data: E. I. Goldenthal, *Toxicol. Appl. Pharmacol.* **18**, 185 (1971).

Crystals from ethanol, mp 152-153°. Bitter, metallic taste. Stable to light and air. Soly in water, methanol: both 1.1 g/100 ml; in ethanol: 0.56 g/100 ml. Sparingly sol in chloroform. Freely sol in alkaline solns. pH of a 0.5% aq soln 4.3-5.0. LD_{50} orally in rats: ~82 mg/kg (Goldenthal).

THERAP CAT: Diuretic.

THERAP CAT (VET): Diuretic.

2106. Chlormezanone. [80-77-3] 2-(4-Chlorophenyl)tetrahydro-3-methyl-4*H*-1,3-thiazin-4-one 1,1-dioxide; 2-(*p*-chlorophenyl)perhydro-3-methyl-1,3-thiazin-4-one 1,1-dioxide; 2-(4-chlorophenyl)-3-methyl-4-metathiazanone 1,1-dioxide; dichloromethazanone; chlormethazanone; Alinam; Banabin-Sintyal; Fenarol; Lobak; Mio-Sed; Rexan; Rilansyl; Rilaquil; Rilassol; Supotran; Suprotan; Tanafol; Trancopal; Trancote; Transanate. $C_{11}H_{12}ClNO_3S$; mol wt 273.73. C 48.27%, H 4.42%, Cl 12.95%, N 5.12%, O 17.53%, S 11.71%. Prepn starting with 4-chlorobenzylidenemethylamine: Surrey *et al.*, *J. Am. Chem. Soc.* **80**, 3469 (1958); **GB 815203** (1959 to Sterling Drug).

Crystals, mp 116.2-118.2°. Soly in water at 25° less than 0.25% (w/v); in 95% ethanol at 25° less than 1.0% (w/v).

THERAP CAT: Anxiolytic; muscle relaxant (skeletal).

2107. Chlormidazole. [3689-76-7] 1-[(4-Chlorophenyl)methyl]-2-methyl-1*H*-benzimidazole; 1-(*p*-chlorobenzyl)-2-methyl-benzimidazole; 2-methyl-1-(*p*-chlorobenzyl)benzimidazole; clomi-dazole. $C_{15}H_{13}ClN_2$; mol wt 256.73. C 70.18%, H 5.10%, Cl 13.81%, N 10.91%. Prepn: Herrling *et al.*, **US 2876233** (1959 to Grünenthal).

bp$_{12}$ 240-242°.

Monohydrate. mp 67-68°.

Hydrochloride. [74298-63-8] H-115; Diamyceline; Futrican. $C_{15}H_{13}ClN_2 \cdot HCl$; mol wt 293.19. mp 227-228°.

THERAP CAT: Antifungal.

2108. Chlornaphazine. [494-03-1] *N,N*-Bis(2-chloroethyl)-2-naphthylamine; dichloroethyl-*β*-naphthylamine; *β*-naphthyldi(2-chloroethyl)amine; *β*-naphthylbis(*β*-chloroethyl)amine; di(2-chloroethyl)-*β*-naphthylamine; CB-1048; R-48; Cloronaftina; Erysan. $C_{14}H_{15}Cl_2N$; mol wt 268.18. C 62.70%, H 5.64%, Cl 26.44%, N 5.22%. Prepd from 2-$C_{10}H_7N(C_2H_4OH)_2$ by treatment with $POCl_3$: Ross, *J. Chem. Soc.* **1949**, 183; Ghielmetti, *Farmaco Ed. Sci.* **5**, 275 (1950); **11**, 603 (1956). Mechanism of action studies: Jeney *et al.*, *Kiserl. Orvostud.* **20**, 369 (1968), *C.A.* **70**, 18766j (1969). Review of carcinogenicity studies: *IARC Monographs* **4**, 119-124 (1974).

Platelets from petr ether, mp 54-56°. bp$_5$ 210°. Very sparingly sol in water, glycerol. More sol (in ascending degree) in petr ether, ethanol, olive oil, ether, acetone, benzene.

Caution: This substance has been listed as a known carcinogen: *Fourth Annual Report on Carcinogens* (NTP 85-002, 1985) p 44; delisted because no U.S. residents exposed: *Fifth Annual Report on Carcinogens* (NTP 89-239, 1989) p 340.

THERAP CAT: Antineoplastic.

2109. Chloroacetaldehyde. [107-20-0] 2-Chloroacetaldehyde; 2-chloro-1-ethanal; monochloroacetaldehyde. C_2H_3ClO; mol wt 78.50. C 30.60%, H 3.85%, Cl 45.16%, O 20.38%. Prepd industrially by carefully controlled chlorination of acetaldehyde: Söll, **DE 844595** (1943 to Knapsack-Griesheim); Shawinigan Chem. Ltd., **GB 644914** (1947); Guinot, Tabuteau, *Compt. Rend.* **231**, 234 (1950); **FR 1012991** (1950 to Comp. des Prod. Chim. et Electrometallurg.).

Liquid. Acrid, penetrating odor. bp$_{760}$ 85-86°. The anhydr substance polymerizes on standing but reverts to the monomer on distillation.

Hemihydrate. Probably $ClCH_2CH(OH)OCH(OH)CH_2Cl$, platelets from water, mp 43-50°, bp 85.5° with decompn into water and chloraldehyde. Sol in water, alcohol, ether.

Caution: Potential symptoms of overexposure are irritation of eyes, skin and mucous membranes; skin burns; eye damage; pulmonary edema; skin, respiratory system sensitization. *See NIOSH Pocket Guide to Chemical Hazards* (DHHS/NIOSH 97-140, 1997) p 60.

USE: In the manufacture of 2-aminothiazole; to facilitate bark removal from tree trunks.

2110. Chloroacetamide. [79-07-2] 2-Chloroacetamide. C_2H_4ClNO; mol wt 93.51. C 25.69%, H 4.31%, Cl 37.91%, N 14.98%, O 17.11%. Prepn from ethyl chloroacetate and ammonia: Jacobs,

Heidelberger, *Org. Synth.* **coll. vol. I**, 153 (2nd ed., 1941); from chloroacetyl chloride and ammonium acetate: Finan, Fothergill, *J. Chem. Soc.* **1962**, 2824.

Crystals from water, mp 119-120°. bp ~225° (dec). Sol in 10 parts water, 10 parts abs alc; very slightly sol in ether. Two isomorphous crystalline modifications: α-form, "stable form", obtained by sublimation and by crystallization from nonpolar solvents; β-form, "unstable form", obtained by quenching the melt or by crystallization from polar solvents, *see* Katayama, *Acta Crystallogr.* **9**, 986 (1956); **10**, 468 (1957); B. Kalyanaraman *et al.*, *J. Cryst. Mol. Struct.* **8**, 175 (1978).

2111. Chloroacetanilide. C_8H_8ClNO; mol wt 169.61. C 56.65%, H 4.75%, Cl 20.90%, N 8.26%, O 9.43%. Prepn of *m*-, *o*- and *p*-isomers: Roberts *et al.*, *J. Org. Chem.* **24**, 654 (1959). Sepn of *o*- and *p*-isomers: Orton, Bradford, *J. Chem. Soc.* **1927**, 986.

m-**Chloroacetanilide.** *N*-(3-Chlorophenyl)acetamide; 3′-chloroacetanilide. Needles from 50% glacial acetic acid, mp 77-78°. Readily sol in alcohol, benzene, carbon disulfide; very slightly sol in ligroin. uv max (95% ethanol): 245 nm (log ε 4.19).

o-**Chloroacetanilide.** Needles from dil glacial acetic acid, mp 87-88°. Sublimes at about 50-60°. Practically insol in water, alkalies; sol in alc; more sol in benzene than corresponding *p*-isomer. uv max (95% ethanol): 240 nm (log ε 4.02).

p-**Chloroacetanilide.** Orthorhombic crystals from aq glacial acetic acid, alc, or acetone. mp 178-179°. d_4^{22} 1.385. Practically insol in water; readily sol in alc, ether, carbon disulfide; slightly sol in CCl_4, benzene. uv max (95% ethanol): 249 nm (log ε 4.25).

2112. Chloroacetic Acid. [79-11-8] 2-Chloroacetic acid; chloroethanoic acid; monochloroacetic acid; MCA. $C_2H_3ClO_2$; mol wt 94.49. C 25.42%, H 3.20%, Cl 37.52%, O 33.86%. Made by chlorination of glacial acetic acid in presence of small amount of sulfur or iodine; also by hydrolysis of trichlorethylene with 90% H_2SO_4. Lab prepn: Gattermann-Wieland, *Praxis des Organischen Chemikers* (de Gruyter, Berlin, 40th ed., 1961) pp 109, 110. Manuf: *Faith, Keyes & Clark's Industrial Chemicals*, F. A. Lowenheim, M. K. Moran, Eds. (Wiley-Interscience, New York, 4th ed., 1975) pp 254-257. Toxicity: Woodward *et al.*, *J. Ind. Hyg. Toxicol.* **23**, 78 (1941).

Colorless or white, deliquesc crystals. d 1.580. Exists in α, β and γ forms having mp 63°, 55-56° and 50° respectively. mp for acid of commerce is 61-63°. bp for all forms is 189°. Very sol in water; sol in alcohol, benzene, chloroform, ether. *Poisonous, corrosive. Keep well closed and in a cool place.*

Sodium salt. [3926-62-3] Somon. $C_2H_2ClNaO_2$; mol wt 116.48. White crystalline solid. Soly in water at 20°: 85 g/100 ml. LD_{50} in rats, mice, guinea pigs (mg/kg): 76, 255, 80 orally (Woodward).

Caution: Irritating to skin, mucous membranes.

USE: Herbicide. Manuf various dyes and other organic chemicals. Prepn of metal derivatives.

2113. Chloroacetic Anhydride. [541-88-8] Chloroacetic acid 1,1′-anhydride; monochloroacetic acid anhydride; *sym*-dichloroacetic anhydride. $C_4H_4Cl_2O_3$; mol wt 170.97. C 28.10%, H 2.36%, Cl 41.47%, O 28.07%. Prepd by treating monochloroacetic acid in tetrahydrofuran with methoxyacetylene at 15° and distilling the mixture: Eglinton *et al.*, *J. Chem. Soc.* **1954**, 1860; by mixing

monochloroacetic acid and HCN with 1,4-dioxane satd with HCl and fractionating the mixture: Krieble, Smellie, **US 2390106** (1945); by heating chloroacetyl chloride with monochloroacetic acid in the presence of $AlCl_3$: Strosacker, Schwegler, **US 1713104** (1929 to Dow).

Prisms from benzene; d_4^{20} 1.5494; mp 46°. bp_{760} 203°; bp_{116} 163°; bp_{62} 149°; bp_{24} 126°; bp_{10} 109-110°; $bp_{0.05}$ 118-120°. Freely sol in ether, chloroform; slightly sol in benzene, practically insol in cold ligroin.

USE: In *N*-acetylation of amino acids in alkaline soln; prepn of cellulose chloroacetates.

2114. Chloroacetone. [78-95-5] 1-Chloro-2-propanone; chloracetone; monochloroacetone; monochloracetone; acetonyl chloride; chloropropanone; 1-chloro-2-ketopropane; 1-chloro-2-oxopropane. C_3H_5ClO; mol wt 92.52. C 38.95%, H 5.45%, Cl 38.32%, O 17.29%. Prepn by the action of chlorine upon diketene: A. B. Boese, Jr., **US 2209683** (1940 to Carbide and Carbon Chem.); by chlorination of acetone: E. J. Rahrs, **US 2235562** (1941 to Eastman Kodak); G. H. Morey, **US 2243484** (1941 to Commercial Solvents Corp.). Stabilization: *idem*, **US 2229651** (1941 to Commercial Solvents Corp.); E. J. Rahrs, **US 2263010** (1941 to Eastman Kodak). Forms binary azeotropes with many organic liquids: L. H. Horsley, *Azeotropic Data*, Advances in Chemistry Series No. 6 (Washington, 1952) p 73. Reacts with aryl Grignard reagents to form stilbenes: Huang, *J. Chem. Soc.* **1954**, 2539. Toxicology: E. V. Sargent *et al.*, *Am. Ind. Hyg. Assoc. J.* **47**, 375 (1986).

Liquid. Pungent odor. Lacrimator. Turns dark and resinifies on prolonged exposure to light. May be stabilized by addition of 0.1% H_2O or 1.0% $CaCO_3$. d_4^{25} 1.123; d_4^{15} 1.135. mp −44.5°. bp 119.7°, bp_{50} 61°, bp_{12} 20°. Volatile with steam. Surface tension at 20° = 35.27 dyn/cm. *Poisonous, flammable, corrosive.* Dipole moment in hexane, 2.36D. Sol in 10 parts water (w/w). Miscible with alcohol, ether, chloroform. LD_{50} (14 day) in mice, rats (mg/kg): 127, 100 orally; LC_{50} (1 hr) in rats (ppm): 262 by inhalation (Sargent).

2,4-Dinitrophenylhydrazone. Yellow needles from alc, mp 124°.

Caution: Intensely irritating to eyes, skin, mucous membranes.

USE: Has been proposed as tear gas component for police and military use; in the manuf of couplers for color photography; as enzyme inactivator; intermediate in the manuf of perfumes, antioxidants, drugs; in insecticide formulations; in photopolymerization of vinyl compds. Proposed as catalyst in tetraethyllead production, as selective solvent for separating diolefins.

2115. ω-Chloroacetophenone. [532-27-4] 2-Chloro-1-phenylethanone; 2′-chloroacetophenone; α-chloroacetophenone; phenacyl chloride; Chemical Mace; CN. C_8H_7ClO; mol wt 154.59. C 62.16%, H 4.56%, Cl 22.93%, O 10.35%. Chemical warfare agent with lacrimatory properties. Prepn: R. Scholl, H. Korten, *Ber.* **34**, 1902 (1901); H. Rheinboldt, M. Perrier, *J. Am. Chem. Soc.* **69**, 3148 (1947); Schaefer, Sonnenberg, *J. Org. Chem.* **28**, 1128 (1963). Purification: Lofton, **US 2414418** (1947 to Pennsylvania Coal Prod.). Melting point determination and review of prepn: Macy, *J. Chem. Educ.* **24**, 222 (1947). Comparative toxicity with *o*-chlorobenzylidenemalononitrile, *q.v.*: B. Ballantyne, D. W. Swanston, *Arch. Toxicol.* **40**, 75 (1978).

Crystals from dil alcohol, carbon tetrachloride, or light petr. mp 58-59°; also reported as 54° (Macy) and 56.5° (Rheinboldt, Perrier).

bp 244-245°. d^{15} 1.324. Vapor pressure at 20°: 5.4×10^{-3} mm Hg. Practically insol in water. Freely sol in alcohol, ether, benzene. LD_{50} in rats (mg/kg): 41 i.v., 36 i.p., 127 orally; LC_{50} in rats: 8750 mg/min/m³ (Ballantyne, Swanston).

Caution: Potential symptoms of overexposure are irritation of eyes, skin and respiratory system; pulmonary edema. *See NIOSH Pocket Guide to Chemical Hazards* (DHHS/NIOSH 97-140, 1997) p 60.

USE: Riot control agent.

2116.   *p*-Chloroacetophenone. [99-91-2] 1-(4-Chlorophenyl)ethanone; 4'-chloroacetophenone. C_8H_7ClO; mol wt 154.59. C 62.16%, H 4.56%, Cl 22.93%, O 10.35%. Prepd from chlorobenzene and acetic anhydride in the presence of $AlCl_3$: Adams, Noller, *Org. Synth.* **coll. vol. I**, 109 (1941). Physical properties: *eidem, ibid.*; Dreisbach, Martin, *Ind. Eng. Chem.* **41**, 2875 (1949).

Liquid, bp_{24} 124-126°, bp 237°. mp 20-21° (Adams, Noller); also reported as 18.4° (Dreisbach, Martin). d_4^{20} 1.192, d_4^{25} 1.188. n_D^{20} 1.555, n_D^{25} 1.553. Practically insol in water. Miscible with alcohol, ether.

Caution: Highly irritating to eyes, mucous membranes.

2117.   Chloroacetyl Isocyanate. [4461-30-7] 2-Chloroacetyl isocyanate; chloroacetic acid anhydride with isocyanic acid. C_3H_2-$ClNO_2$; mol wt 119.50. C 30.15%, H 1.69%, Cl 29.67%, N 11.72%, O 26.78%. Obtained by the reaction of chloroacetamide with oxalyl chloride: Speziale, Smith, *J. Org. Chem.* **28**, 1808 (1963).

Liquid. *Poisonous.* bp_{20} 50-55°. $n_D^{21.5}$ 1.4580.
USE: Reagent in synthesis.

2118.   Chloroaniline. C_6H_6ClN; mol wt 127.57. C 56.49%, H 4.74%, Cl 27.79%, N 10.98%. Prepn of *m-, o-* and *p*-isomers: Sidgwick, Rubie, *J. Chem. Soc.* **119**, 1013 (1921). Manufacture of *m-* and *p*-isomers by catalytic hydrogenation of chloronitrobenzene: Pray, Trager, *US 2791613* (1957 to Columbia-Southern Chem.); by reduction of chloronitrobenzene with NaHS: Latourette *et al., US 2894035* (1959 to Food Machinery & Chem.). Properties of *o*-isomer: Dreisbach, Martin, *Ind. Eng. Chem.* **41**, 2875 (1949). Toxicity data: H. F. Smyth *et al., Am. Ind. Hyg. Assoc. J.* **23**, 95 (1962).

m-Chloroaniline. [108-42-9] 3-Chlorobenzenamine. Liquid, bp 230.5°. mp −10.4°. d_4^{22} 1.2150. n_D^{20} 1.5931. *Poisonous.* Practically insol in water. Sol in most common organic solvents.

o-Chloroaniline. [95-51-2] Liquid. At 99.61 mol-% purity, bp 208.84°; mp −1.94°. d_4^{22} 1.2114. n_D^{20} 1.5895. *Poisonous.* Practically insol in water. Sol in acid and in most organic solvents.

p-Chloroaniline. [106-47-8] Orthorhombic crystals from alcohol or petr ether, mp 72.5°. bp 232°. d_4^{77} 1.169. Sol in hot water; freely sol in alcohol, ether, acetone, carbon disulfide. LD_{50} orally in rats: 0.31 g/kg (Smyth).

2119.   Chloroarsenol. [151-07-5] (2-Chloro-1-heptenyl)arsonic acid; 2-chloro-1-heptene-1-arsonic acid. $C_7H_{14}AsClO_3$; mol wt 256.56. C 32.77%, H 5.50%, As 29.20%, Cl 13.82%, O 18.71%. Prepn: *DE 296915* (1917 to Bayer).

Needles from water, mp 115°.
Ammonium salt. Ammonium chloroheptenearsonate. C_7H_{17}-$AsClNO_3$. Freely sol in water.

2120.   Chloroauric Acid. [16903-35-8] Hydrogen (*SP*-4-1)-tetrachloroaurate(1−) (1:1); hydrochloroauric acid; aurochlorohydric acid; tetrachloroauric(III) acid; gold trichloride hydrochloride. $AuCl_4H$; mol wt 339.77. Au 57.97%, Cl 41.73%, H 0.30%. $HAuCl_4$. Prepd according to the equation $2Au + 3Cl_2 + 2HCl \rightarrow 2HAuCl_4$: Thomsen, *Ber.* **16**, 1585 (1883); Biltz, Wein, *Z. Anorg. Allg. Chem.* **148**, 192 (1925).

Trihydrate. [16961-25-4] $HAuCl_4 \cdot 3H_2O$; mol wt 393.82.
Tetrahydrate. [1303-50-0] $HAuCl_4 \cdot 4H_2O$; mol wt 411.83. Golden-yellow to reddish-yellow, very hygroscopic and deliquesc monoclinic crystals; readily affected by sunlight. (Also available as brown crystals or cryst masses, contg 50-51% gold, and having the same properties as the yellow crystals.) Has caustic action on the skin (blisters) and then on exposure to light leaves violet-brown spots. Dec on strong heating to Cl_2, HCl, and metallic gold. d about 3.9. Very sol in water, alcohol; sol in ether. *Keep tightly closed and protected from light.*

USE: Photography, gold-plating, gilding glass and porcelain, manuf ruby glass; as a reagent for alkaloids. Graphite-furnace atomic absorption spectroscopy.

2121.   Chlorozodin. [502-98-7] N^1,N^2-Dichloro-1,2-diazenedicarboximidamide; α,α'-azobis[chloroformamidine]; chlorazodin; dichloroazodicarbonamidine; azochloramide. $C_2H_4Cl_2N_6$; mol wt 183.00. C 13.13%, H 2.20%, Cl 38.74%, N 45.92%. Prepn by chlorination of *azo-bis*-formamidine with sodium hypochlorite: F. C. Schmelkes, H. C. Marks, *J. Am. Chem. Soc.* **56**, 1610 (1934); by treatment of an AcOH-NaOAc soln of guanidine nitrate with sodium hypochlorite: Braz *et al., Appl. Chem. USSR* **17**, 565 (1944), *C.A.* **40**, 2267 (1946). Structure and uv spectra: Kumler, *J. Am. Chem. Soc.* **75**, 3092 (1953). Crystal structure: J. H. Bryden, *Acta Crystallogr.* **11**, 158 (1958).

Bright yellow needles, plates, flakes. Faint chlorine odor, burning taste. Dec explosively at ~155°. Decompn is accelerated by contact with metals. Calculated density: 1.764. Very slightly sol in water; sparingly sol in alcohol; slightly sol in glycerol, glyceryl triacetate (1:100), vegetable oils, ether, chloroform. Practically insol in carbon tetrachloride and liquid petrolatum. Oil solns are generally made by diluting a triacetin (glyceryl triacetate) soln of chloroazodin with oil. Solns of chloroazodin in glycerol and in alcohol dec rapidly on warming; all solns dec on exposure to light.

THERAP CAT: Antiseptic, disinfectant.
THERAP CAT (VET): Antiseptic, disinfectant.

2122.   Chlorobenzene. [108-90-7] Monochlorobenzene; benzene chloride. C_6H_5Cl; mol wt 112.56. C 64.02%, H 4.48%, Cl 31.49%. Produced by the chlorination of benzene in the presence of a catalyst. Review of manuf, properties, and use: *Faith, Keyes & Clark's Industrial Chemicals*, F. A. Lowenheim, M. K. Moran, Eds. (Wiley-Interscience, New York, 4th ed., 1975) pp 258-265. Review of toxicology and human exposure: *Toxicological Profile for Chlorobenzene* (PB91-180505, 1990) 100 pp.

Colorless, very refractive liquid; faint, not unpleasant odor. d_4^{20} 1.107. bp 131-132°. Solidif −55°. mp −45°. Flash pt 28°C. n_D^{20} 1.5248. *Flammable.* Insol in water. Freely sol in alcohol, benzene, chloroform, ether.

Caution: Potential symptoms of overexposure are irritation of skin, eyes and nose; drowsiness, incoordination; CNS depression.

See NIOSH Pocket Guide to Chemical Hazards (DHHS/NIOSH 97-140, 1997) p 62.

USE: Manuf phenol, aniline, DDT; organic solvent for paints; heat transfer medium.

2123. p-Chlorobenzenesulfonic Acid. [98-66-8] 4-Chlorobenzenesulfonic acid; closylate. $C_6H_5ClO_3S$; mol wt 192.61. C 37.42%, H 2.62%, Cl 18.41%, O 24.92%, S 16.65%. Prepd by heating chlorobenzene with concd sulfuric acid or oleum under dehydrating conditions: Meyer, Ann. **433**, 327, 333 (1923).

Cl—⟨benzene ring⟩—SO₃H

Monohydrate. Crystals from water, mp 67°. The anhydr material is usually obtained as a syrup, bp_{25} 148°. Sol in water, alc. Practically insol in ether, benzene.

Sodium salt monohydrate. $C_6H_4ClNaO_3S.H_2O$. Cubic crystals from water. Sol in water.

2124. m-Chlorobenzoic Acid. [535-80-8] 3-Chlorobenzoic acid; mol wt 156.57. C 53.70%, H 3.22%, Cl 22.64%, O 20.44%. Prepn by catalytic oxidation of 1-chloro-3-ethylbenzene: Emerson et al., J. Am. Chem. Soc. **71**, 1742 (1949); by chlorination of benzoic acid: Gorvin, Chem. Ind. (London) **1951**, 910; by the von Richter reaction from 1-chloro-4-nitrobenzene and alcoholic KCN: Samuel, J. Chem. Soc. **1960**, 1318.

⟨structure: m-chlorobenzoic acid, COOH and Cl⟩

Crystals, mp 158°. d_4^{25} 1.496. Sol in 2850 parts cold water; more sol in hot water; freely sol in alcohol, ether.

Methyl ester. Methyl m-chlorobenzoate. $C_8H_7ClO_2$. mp 21°, bp 231°.

2125. o-Chlorobenzoic Acid. [118-91-2] 2-Chlorobenzoic acid; mol wt 156.57. C 53.70%, H 3.22%, Cl 22.64%, O 20.44%. Prepd by KMnO₄ oxidation of o-chlorotoluene: Clarke, Taylor, Org. Synth. **coll. vol. II**, 135 (1943). Crystal structure: Ferguson, Sim, Acta Crystallogr. **12**, 941 (1959), C.A. **55**, 24188i (1961).

⟨structure: o-chlorobenzoic acid, COOH and Cl⟩

Monoclinic crystals, mp 142°. d_4^{20} 1.544. Sol in 900 parts cold water; more sol in hot water, freely in alc, ether.

USE: Preservative for glues, paints. Intermediate in the manufacture of fungicides and dyes.

2126. p-Chlorobenzoic Acid. [74-11-3] 4-Chlorobenzoic acid; chlorodracylic acid. $C_7H_5ClO_2$; mol wt 156.57. C 53.70%, H 3.22%, Cl 22.64%, O 20.44%. Prepn by catalytic oxidation of 1-chloro-4-ethylbenzene: Emerson et al., J. Am. Chem. Soc. **71**, 1742 (1949). Manuf by oxidation of p-chlorobenzaldehyde: Shipman, **US 3124611** (1964 to ICI). Crystal structure: Toussaint, Acta Crystallogr. **4**, 71 (1951), C.A. **45**, 6604e (1951); Pollock, Woodward, ibid. **7**, 605 (1954), C.A. **48**, 13331e (1954).

⟨structure: p-chlorobenzoic acid, COOH and Cl⟩

Triclinic crystals, mp 243°. Sol in 5290 parts water; freely sol in alcohol, ether.

Methyl ester. Methyl p-chlorobenzoate. $C_8H_7ClO_2$. mp 44°.

Sodium salt. Sodium p-chlorobenzoate. $C_7H_4ClNaO_2$. White, odorless, crystalline powder, freely sol in water.

USE: Sodium salt as a preservative.

2127. p-Chlorobenzotrifluoride. [98-56-6] 1-Chloro-4-(trifluoromethyl)benzene; (4-chlorophenyl)trifluoromethane; 4-chloro-α,α,α-trifluorotoluene; PCBTF; Oxsol 100. $C_7H_4ClF_3$; mol wt 180.55. C 46.57%, H 2.23%, Cl 19.63%, F 31.57%. Environmentally friendly solvent. Prepn: H. S. Booth et al., J. Am. Chem. Soc. **57**, 2066 (1935). Mass spectrum: F. Belsito et al., Ann. Chim. **69**, 259 (1979). Photochemical reactivity: M. Khan et al., Atmos. Environ. **33**, 1085 (1999). Use as chemical intermediate: E. R. Lavagnino et al., Org. Prep. Proced. Int. **11**, 23 (1979); M. Beller, A. Zapf, Synlett **1998**, 792. GLC/MS determn in fish: M. P. Yurawecz, J. Assoc. Off. Anal. Chem. **62**, 36 (1979). Toxicology: P. E. Newton et al., Inhalation Toxicol. **10**, 33 (1998). Review: C. H. Hare, Mod. Paint Coat. **88**, 30-36 (1998).

Cl—⟨benzene ring⟩—CF₃

Liquid, strong though not unpleasant aromatic odor. Flammable. bp 139°. Freezing pt: −36°. mp −34°. Flash point (closed cup): 109°F (43°C). d^{20} 1.34. d_4^{25} 1.334. n_D^{21} 1.4469. Vapor pressure (20°): 5.3 mm Hg. Soly in water (25°): 29 ppm.

USE: Chemical intermediate for dinitroaniline herbicides, dyes. As dielectric fluid and as solvent.

2128. o-Chlorobenzylidenemalononitrile. [2698-41-1] 2-[(2-Chlorophenyl)methylene]propanedinitrile; o-chlorobenzalmalononitrile; β,β-dicyano-o-chlorostyrene; CS. $C_{10}H_5ClN_2$; mol wt 188.61. C 63.68%, H 2.67%, Cl 18.80%, N 14.85%. Lacrimatory chemical warfare agent. Prepn: B.B. Corson, R. W. Stoughton, J. Am. Chem. Soc. **50**, 2825 (1928). Pharmacology: R. W. Brimblecombe et al., Br. J. Pharmacol. **44**, 561 (1972). Toxicology: C. L. Punte et al., Toxicol. Appl. Pharmacol. **4**, 656 (1962); J. R. Gaskins et al., Arch. Environ. Health **24**, 449 (1972); B. Ballantyne, S. Callaway, Med. Sci. Law **12**, 43 (1972). Comparative toxicity of CS and ω-chloroacetophenone, q.v.: B. Ballantyne, D. W. Swanston, Arch. Toxicol. **40**, 75 (1978).

⟨structure: o-chlorobenzylidenemalononitrile with Cl and two CN groups⟩

White crystalline solid, mp 95-96°, bp 310-315°. Vapor pressure at 20°: 3.4×10^{-5} mm Hg. Sparingly sol in water; sol in acetone, dioxane, methylene chloride, ethyl acetate, benzene. LD_{50} in rats (mg/kg): 28 i.v., 48 i.p.; LC_{50} in rats: 88,480 mg/min/m³ (Ballantyne, Swanston).

Caution: Potential symptoms of overexposure are pain, burning of eyes, lacrimation and conjunctivitis; erythemic eyelids, blepharospasm; irritation of throat, coughing and chest constriction; headache; erythema and vesiculation of skin. See NIOSH Pocket Guide to Chemical Hazards (DHHS/NIOSH 97-140, 1997) p 62.

USE: Riot control agent.

2129. Chlorobutanol. [57-15-8] 1,1,1-Trichloro-2-methyl-2-propanol; β,β,β-trichloro-tert-butyl alcohol; acetone chloroform; chlorbutol; Chloretone; Coliquifilm; Methaform; Sedaform. $C_4H_7Cl_3O$; mol wt 177.45. C 27.07%, H 3.98%, Cl 59.93%, O 9.02%. Chlorobutanol may be anhydrous or it may contain up to about one-half molecule of water. Prepn: Fishburn, Watson, J. Am. Pharm. Assoc. **28**, 491 (1939); E. Bergman, **US 2446453** (1948 to Polymerisable Products Ltd.); P. Riegger, H. Richtzenbain, **DE 1271697** (1968 to Dynamit Nobel A.-G.), C.A. **69**, 95967g (1968); M. Saljoughian, Monatsh. Chem. **114**, 813 (1983). Toxicity studies: E. Impens, Arch. Int. Pharmacodyn. Ther. **8**, 77 (1901).

⟨structure: chlorobutanol, (CH₃)(H₃C)C(OH)CCl₃⟩

Crystals. Camphor odor and taste. A crystal of chlorobutanol placed on a water surface "dances" like a camphor crystal. Sublimes easily. Anhydrous form, mp 97°; hemihydrate, mp 78°. bp_{760} 167°; bp_{246} 135°. One gram dissolves in 1 ml alcohol, in 10 ml glycerol. Freely sol in ether, chloroform, and in volatile oils; slightly sol in water. MLD orally in dogs, rabbits (mg/kg): 238, 213 (Impens).

Compd with chloral hydrate. Ref: **DE 151188** (1904). mp 65°.

USE: Plasticizer for cellulose esters and ethers. Preservative for biological fluids, hypodermic solns, and solns of alkaloids. Pharmaceutic aid (antimicrobial).

THERAP CAT: Analgesic (dental).

THERAP CAT (VET): Has been used as a mild sedative, antipruritic, and antiseptic.

2130.   1-Chloro-2-butene. [591-97-9] α-Chloro-β-butylene; crotyl chloride; γ-methallyl chloride; γ-methylallyl chloride. C_4H_7-Cl; mol wt 90.55. C 53.06%, H 7.79%, Cl 39.15%. Prepn of (E) and (Z)-forms: Hatch, Nesbitt, J. Am. Chem. Soc. **72**, 727 (1950). Manuf: Carlson, US 2494034 (1950 to Shell); Montagna, Hess, US 3055954 (1962 to Union Carbide). Equilibrium constant for isomerization of 3-chloro-1-butene to 1-chloro-2-butene: Dittmer, Marcantonio, J. Org. Chem. **29**, 3473 (1964).

(E)-form

(E)-Form. [4894-61-5] Liquid, bp_{752} 84.8°. n_D^{20} 1.4350, n_D^{25} 1.4327. d_4^{20} 0.9295.

(Z)-Form. [4628-21-1] Liquid, bp_{758} 84.1°. n_D^{20} 1.4390. d_4^{20} 0.9426.

Caution: Irritates eyes, respiratory passages.

2131.   3-Chloro-1-butene. [563-52-0] γ-Chloro-α-butylene; α-methallyl chloride; α-methylallyl chloride. C_4H_7Cl; mol wt 90.55. C 53.06%, H 7.79%, Cl 39.15%. Prepn of DL-form: Böhme, Ber. **71**, 2372 (1938); D. Y. Curtin, S. M. Gerber, J. Am. Chem. Soc. **74**, 4052 (1952); of D(−)-form: Böhme, loc. cit.; of L(+)-form: W. G. Young, F. F. Caserio, Jr., J. Org. Chem. **26**, 245 (1961). Absolute configuration: Young, Caserio, loc. cit. Equilibrium constant for isomerization to 1-chloro-2-butene, q.v.: D. C. Dittmer, A. F. Marcantonio, J. Org. Chem. **29**, 3473 (1964).

DL-Form. Liquid, bp 63.9-64.2°, bp_{26} −5°. d_4^{20} 0.9001. n_D^{20} 1.4150.

D(−)-Form. [130404-07-8] (3R)-3-Chloro-1-butene. Liquid, bp_{26} −5°. $α_D^{20}$ −2.52° (neat).

L(+)-Form. [35729-37-4] (3S)-3-Chloro-1-butene. Liquid, $α_D^{25}$ +5.87° (neat).

2132.   3-Chloro-d-camphor. [508-29-2] 3-Chloro-1,7,7-trimethylbicyclo[2.2.1]heptan-2-one; 3-chloro-d-2-bornanone. C_{10}-$H_{15}ClO$; mol wt 186.68. C 64.34%, H 8.10%, Cl 18.99%, O 8.57%. Of the two isomers, the endo-form is more stable than the exo-form: Cookson, J. Chem. Soc. **1954**, 282. Prepn of endo-form: Kipping, Pope, ibid. **63**, 548 (1893). Prepn of exo-form by isomerization of endo-form: Lowry, Steele, ibid. **107**, 1382 (1915). uv spectra: Lowry, Owen, J. Chem. Soc. **129**, 606 (1926). Crystal structure of endo-form: Wiebenga, Krom, Rec. Trav. Chim. **65**, 663 (1946). Configuration: Cookson, loc. cit.; Kumler et al., J. Am. Chem. Soc. **83**, 2711 (1961). Review: The Terpenes vol. **II**, J. L. Simonsen, Ed. (University Press, Cambridge, 2nd ed., 1949) p 397.

endo-form   exo-form

endo-**Form.** α-Chloro-d-camphor; 3α-chloro-d-camphor; camphor monochlorated. Monochlorinic prisms from alc, mp 94°. Volatile with steam. $[α]_D$ +96.2° (c = 5 in alc). Practically insol in water. Sol in alcohol, chloroform, ether. uv max (cyclohexane): 305 nm (log ε 1.72).

exo-**Form.** α'-Chloro-d-camphor; 3β-chloro-d-camphor. Crystals from alc, mp 117°. $[α]_D$ +35° (c = 5 in alc). Readily sol in all ordinary solvents, except water and formamide. Much more sol in 96% alc than endo-form. On standing, loses HCl. uv max (cyclohexane): 306 nm (log ε 1.75).

2133.   Chlorocresol. [59-50-7] 4-Chloro-3-methylphenol; 3-methyl-4-chlorophenol; 4-chloro-m-cresol; 6-chloro-m-cresol; 6-chloro-3-hydroxytoluene; 2-chloro-5-hydroxytoluene; parachlorometacresol. C_7H_7ClO; mol wt 142.58. C 58.97%, H 4.95%, Cl 24.86%, O 11.22%. Prepd by chlorination of m-cresol: **DE 90847** (1897 to Kalle), Frdl. **4**, 94; Laschinger, US 1847566 (1932), C.A. **26**, 2471 (1932); Sah, Anderson, J. Am. Chem. Soc. **63**, 3164 (1941). HPLC determn in pharmaceutical formulations: R. Gatti et al., J. Pharm. Biomed. Anal. **16**, 405 (1997).

Dimorphous crystals, mp 55.5° and mp 66° (ligroin). Said to be odorless when very pure, but usually a phenolic odor persists. Poisonous. bp 235°. Very sol in alcohol; sol in ether, petr ether, benzene, chloroform, acetone, terpenes, fixed oils, solutions of alkali hydroxides. Volatile with steam. One gram dissolves in 260 ml water at 20°, more sol in hot water. Aq solns turn yellow on exposure to light and air.

USE: Disinfectant; pharmaceutic aid (preservative).

THERAP CAT: Antiseptic, disinfectant.

2134.   Chlorocyanohydrin. [513-96-2] 3,3,3-Trichloro-2-hydroxypropanenitrile; 3,3,3-trichlorolactonitrile; 3,3,3-trichloro-2-hydroxypropionitrile; chloral hydrocyanide. $C_3H_2Cl_3NO$; mol wt 174.41. C 20.66%, H 1.16%, Cl 60.98%, N 8.03%, O 9.17%. Prepd from chloral hydrate and HCN: Chwala, Wassmuth, Monatsh. Chem. **81**, 843 (1950).

Crystals, mp 61°. Odor of HCN and chloral. Sublimes easily. bp 220°. Freely sol in water, alcohol, ether, and most other common organic solvents; moderately sol in carbon tetrachloride. Practically insol in petr ether.

2135.   Chlorodibromomethane. [124-48-1] Dibromochloromethane; chlorobromoform; monochlorodibromomethane; DBCM; NCI-C55254. $CHBr_2Cl$; mol wt 208.28. C 5.77%, H 0.48%, Br 76.73%, Cl 17.02%. Trihalomethane contaminant found in chlorinated water. Prepn: O. Jacobsen, R. Neumeister, Ber. **15**, 599 (1882); M. Fedorynski et al., Synth. Commun. **7**, 287 (1977). Toxicity studies: F. J. Bowman et al., Toxicol. Appl. Pharmacol. **44**, 213 (1978); Y. Aida et al., J. Toxicol. Sci. **17**, 119 (1992). Review of toxicology and human exposure: Toxicological Profile for Bromoform and Dibromochloromethane (PB2006-100001, 2005) 273 pp. Review of carcinogenic risk: IARC Monographs **52**, 243-268 (1991).

Liquid. bp 121.3-121.8° (Fedorynski); also reported as bp 123-125° (Jacobsen). d^{15} 2.4450. n_D^{20} 1.5471. Sol in ethanol, ether, acetone. LD_{50} in male, female rats (mg/kg): 370, 760 orally (Aida); in male, female mice (mg/kg): 800, 1200 by gavage (Bowman).

USE: Chemical reagent/intermediate in organic synthesis.

2136.   Chlorodiisopropylsilane. [2227-29-4] Chlorobis(1-methylethyl)silane; diisopropylchlorosilane. $C_6H_{15}ClSi$; mol wt 150.72. C 47.81%, H 10.03%, Cl 23.52%, Si 18.63%. Silylating reagent used to prepare O-diisopropylsilyl derivatives as intermediates for further reactions. Prepn: F. Metras, J. Valade, Bull. Soc. Chim. Fr. **1965** 1423; and synthetic utility: S. Anwar, A. P. Davis,

Tetrahedron **44**, 3761 (1988); S. H. Bergens *et al., J. Am. Chem. Soc.* **114**, 2121 (1992). Additional synthetic application: S. E. Denmark, T. Kobayashi, *J. Org. Chem.* **68**, 5153 (2003). *Review:* A. P. Davis in *Encyclopedia of Reagents for Organic Synthesis* **2**, L. A. Paquette, Ed. (Wiley, New York, 1995) pp 1116-1118.

Colorless liquid. *Flammable. Corrosive.* bp_{760} 143°; bp_{80} 81-85°; bp_{45} 54-55°. d_4^{20} 0.8720. n_D^{20} 1.4278. Flash pt, closed cup: 100.4°F (38°C). Sol in most aprotic solvents. Moisture sensitive.
USE: Reagent in synthetic organic chemistry.

2137. 1-Chloro-2,4-dinitrobenzene. [97-00-7] 2,4-Dinitro-1-chlorobenzene; 4-chloro-1,3-dinitrobenzene; 6-chloro-1,3-dinitrobenzene. $C_6H_3ClN_2O_4$; mol wt 202.55. C 35.58%, H 1.49%, Cl 17.50%, N 13.83%, O 31.60%. Obtained by nitrating *o*-nitrochlorobenzene: Borsche, Rantscheff, *Ann.* **379**, 152 (1911); Hodgson, Dodgson, *J. Chem. Soc.* **1948**, 1006. The mixture of 2,4- and 2,6-dinitrochlorobenzenes (of which the 2,4-compd is present in largest amount) can be separated with an alkaline ethanol-water soln: Molard, Vaganay, *Mem. Poudres* **39**, 111 (1957). Toxicity study: H. F. Smyth *et al., Am. Ind. Hyg. Assoc. J.* **23**, 95 (1962).

Yellow crystals. *Skin irritant.* d ~1.7. mp 52-54°. bp 315°. Practically insol in water. Sparingly sol in cold, freely sol in hot alc; sol in ether, benzene, CS_2. LD_{50} orally in rats: 1.07 g/kg (Smyth).
Caution: May cause dermatitis of both primary and allergic types.
USE: As a reagent for the detection and determination of nicotinic acid, nicotinamide, and other pyridine compds.

2138. 2-Chloro-1,3-dinitrobenzene. [606-21-3] 1-Chloro-2,6-dinitrobenzene; 2,6-dinitro-1-chlorobenzene. $C_6H_3ClN_2O_4$; mol wt 202.55. C 35.58%, H 1.49%, Cl 17.50%, N 13.83%, O 31.60%. Prepd from 2,6-dinitroaniline by the Sandmeyer reaction: Gunstone, Tucker, *Org. Synth.* **coll. vol. IV**, 160 (1963).

Yellow crystals from benzene + petr ether or from acetic acid. *Skin irritant.* mp 86-87°. $d_4^{16.5}$ 1.6867. bp 315°. Moderately sol in ether, benzene. Sol in alc. Practically insol in water.
Caution: See 1-Chloro-2,4-dinitrobenzene.

2139. Chlorodiphenylphosphine. [1079-66-9] *P,P*-Diphenylphosphinous chloride; diphenylchlorophosphine; diphenylphosphorus chloride; Ph_2PCl. $C_{12}H_{10}ClP$; mol wt 220.64. C 65.32%, H 4.57%, Cl 16.07%, P 14.04%. Chlorophosphine used in a variety of synthetic transformations. Prepn: A. Michaelis, *Ber.* **10**, 627 (1877); A. Michaelis, A. Link, *Ann.* **207**, 193 (1881). Improved prepn and synthetic utility: C. Stuebe *et al., J. Am. Chem. Soc.* **77**, 3526 (1955). Synthetic applications for the prepn of α-alkoxyphosphine oxides: M. Maleki *et al., Tetrahedron Lett.* **22**, 365 (1981); of alkyl bromides and iodides: B. Classon *et al., J. Org. Chem.* **53**, 6126 (1988); of olefins: Z. Liu *et al., ibid.* **55**, 4273 (1990); of triarylphosphines: E. Le Gall *et al., Tetrahedron* **59**, 7497 (2003); of *gem*-dichlorides: G. Aghapour, A. Afzali, *Synth. Commun.* **38**, 4023 (2008).

Yellow liquid. *Corrosive, irritant, lachrymator, reacts violently with water.* mp 15-16°. bp 320°; bp_{57} 210-215°; $bp_{0.3}$ 111-112°. d^{15} 1.2293. n_D^{20} 1.6361. Flash pt, closed cup: 233.6°F (112°C).
USE: Reagent in synthetic organic chemistry.

2140. 2-Chloroethyl Vinyl Ether. [110-75-8] (2-Chloroethoxy)ethene. C_4H_7ClO; mol wt 106.55. C 45.09%, H 6.62%, Cl 33.27%, O 15.02%. Prepd by the action of solid NaOH + triethanolamine upon β,β'-dichlorodiethyl ether: Chitwood, Perkins, **US 2104717** (1938 to Carbide and Carbon Chem.); Cretcher *et al., J. Am. Chem. Soc.* **47**, 1175 (1926). Toxicity study: Smyth *et al., J. Ind. Hyg. Toxicol.* **31**, 60 (1949).

Liquid. d_{15}^{15} 1.0525. bp_{740} 109°. Quite stable to NaOH solns. Even dil acids produce hydrolysis to acetaldehyde and ethylene chlorohydrin [2-chloroethanol]. LD_{50} orally in rats: 250 mg/kg (Smyth).
USE: Mfg anesthetics, sedatives, cellulose ethers.

2141. Chlorofluorocarbons. CFCs; FCCs. Chemically stable series of chlorinated and fluorinated compounds usually with methane or ethane skeleton, marketed under general names such as *Arcton, Freon, Frigen, Genetron.* Known collectively as CFCs, individually identified by a CFC code based on the "rule of 90". To derive the chemical formula for CFC-12, e.g., add "90" to 12; the resulting number "102" indicates 1 carbon, 0 hydrogen, 2 fluorine yielding the formula CCl_2F_2. Initial report on suitability as refrigerants: T. Midgley, Jr., A. L. Henne, *Ind. Eng. Chem.* **22**, 542 (1930). Review including physical and chemical properties of various CFCs: E. Heiskel, *Aerosol Rep.* **22**, 403-415 (1983). CFCs do not decompose in the lower atmosphere. Photodecomposition occurs in the stratosphere via absorption of uv radiation and subsequent release of atomic chlorine which can catalyze ozone breakdown. CFC-ozone depletion hypothesis: M. J. Molina, F. S. Rowland, *Nature* **249**, 810 (1974). Reviews focusing on atmospheric chemistry of CFCs, uses, potential hydrogen-substituted replacements, environmental and regulatory issues: J. P. Glenn, *BioScience* **37**, 647-650 (1987); R. Pool, *Science* **242**, 666-668 (1988); F. S. Rowland, *Environ. Conserv.* **15**, 101-115 (1988); L. B. Weisfeld, *Plast. Compd.* **1988**, 15-22, 40-43; F. S. Rowland, *Am. Sci.* **77**, 36-45 (1989). For prepn information *see* dichlorodifluoromethane (CFC-12), trichlorofluoromethane (CFC-11), cryofluorane (CFC-114).

Colorless, essentially odorless, nonflammable, noncorrosive. Miscible with aliphatic, alicyclic and aromatic hydrocarbons, halogenated hydrocarbons, monovalent low molecular alcohols.
Note: Consult latest Government regulations on use as aerosol propellant.
USE: In aerosol propellants (CFC-11, 12, 113); air conditioning; refrigeration (CFC-12); blowing agents for making foam (CFC-11, 12); cleaning fluids (CFC-113); solvents for the electronics industry, bedding and packaging.

2142. Chloroform. [67-66-3] Trichloromethane. $CHCl_3$; mol wt 119.37. C 10.06%, H 0.84%, Cl 89.09%. Improperly called "formyl trichloride". From the addition of sulfuric acid to acetone and bleaching powder: $2CH_3COCH_3 + 6CaOCl_2 \cdot H_2O \rightarrow 2CHCl_3 + (CH_3COO)_2Ca + 2Ca(OH)_2 + 3CaCl_2 + 6H_2O$. May also be prepd by carefully controlled chlorination of methane: *Faith, Keyes & Clark's Industrial Chemicals,* F. A. Lowenheim, M. K. Moran, Eds. (Wiley-Interscience, New York, 4th ed., 1975) pp 266-269. Has been used as an anesthetic and in pharmaceutical preparations. Toxicity data: H. F. Smyth *et al., Am. Ind. Hyg. Assoc. J.* **23**, 95 (1962); E. T. Kimura *et al., Toxicol. Appl. Pharmacol.* **19**, 699 (1971). Review of toxicology: L. R. Pohl, *Rev. Biochem. Toxicol.* **1**, 79-108 (1979); of carcinogenic risk: *IARC Monographs* **20**, 401-427 (1979); of toxicology and human exposure: *Toxicological Profile for Chloroform* (PB98-101140, 1997) 343 pp. *Review:* M. T. Holbrook in *Kirk-Othmer Encyclopedia of Chemical Technology* **vol. 5** (John Wiley & Sons, New York, 4th ed., 1993) pp 1051-1062.

Highly refractive, nonflammable, heavy, very volatile, sweet-tasting liquid; characteristic odor. d_{20}^{20} 1.484. bp 61-62°. mp −63.5°. n_D^{20} 1.4476. Forms a constant boiling mixture with 7% alc, boiling at 59°. d 1.474-1.478 for U.S.P. chloroform contg 0.5-1% ethanol as stabilizer. One ml dissolves in about 200 ml water at 25°. Misc with alc, benzene, ether, petr ether, carbon tetrachloride, carbon disulfide, oils. Pure chloroform is light sensitive and reagent grade chloroform usually contains 0.75% ethanol as stabilizer. *Poisonous. Protect from light and keep cool.* LD$_{50}$ (14 day) orally in rats: 2.18 ml/kg (Smyth); 0.9 ml/kg (Kimura).

Spirit of Chloroform. An alcoholic soln of chloroform contg 6% by vol of chloroform, corresponding to 10.5% by wt, and ~89% abs alcohol by vol. Has been used as a carminative. Colorless, clear liquid; chloroform odor. d ~0.85.

Caution: Potential symptoms of overexposure are dizziness, mental dullness, nausea and confusion; headache, fatigue; anesthesia; enlarged liver; direct contact may cause irritation to eyes and skin. *See NIOSH Pocket Guide to Chemical Hazards* (DHHS/NIOSH 97-140, 1997) p 64. This substance is reasonably anticipated to be a human carcinogen: *Report on Carcinogens, Twelfth Edition* (PB2011-111646, 2011) p 97.

USE: In the manuf of fluorocarbon-22. As a solvent for fats, oils, rubber, alkaloids, waxes, gutta-percha, resins; as cleansing agent; in fire extinguishers to lower the freezing temp of carbon tetrachloride; in the rubber industry.

2143. Chlorogenic Acid. [327-97-9]; [15016-60-1] ((*E*)-form). (1*S*,3*R*,4*R*,5*R*)-3-[[3-(3,4-Dihydroxyphenyl)-1-oxo-2-propen-1-yl]oxy]-1,4,5-trihydroxycyclohexanecarboxylic acid; 1,3,4,5-tetrahydroxycyclohexanecarboxylic acid 3-(3,4-dihydroxycinnamate); 3-caffeoylquinic acid; 3-(3,4-dihydroxycinnamoyl)quinic acid. C$_{16}$H$_{18}$O$_9$; mol wt 354.31. C 54.24%, H 5.12%, O 40.64%. Important factor in plant metabolism. Isoln from green coffee beans: Freudenberg, *Ber.* **53**, 237 (1920). Chlorogenic acid and its isomers, isochlorogenic acid and neochlorogenic acid, occur also in fruit, leaves, and other tissues of dicotyledenous plants: Sondheimer, *Arch. Pharm.* **293**, 721 (1960). Forms caffeic acid on hydrolysis: Fiedler, *Arzneim.-Forsch.* **4**, 41 (1954). Structure: Fischer, Dangschat, *Ber.* **65**, 1037 (1932); Barnes *et al.*, *J. Am. Chem. Soc.* **72**, 4178 (1950); Corse *et al.*, *Tetrahedron* **18**, 1207 (1962). Synthesis: Panizzi *et al.*, *Gazz. Chim. Ital.* **86**, 913 (1956).

(*E*)-form

Hemihydrate. Needles from water. Becomes anhydr at 110°. mp 208°. [α]$_D^{26}$ −35.2° (c = 2.8). pKa (27°) 2.66. R$_f$ values: Fiedler, *loc. cit.* Soly in water at 25° about 4%, much more sol in hot water. Alkaline solns acquire an orange color. Freely sol in alcohol, acetone. Very slightly sol in ethyl acetate. Heating with dil HCl yields caffeic acid. Forms a black compd with iron, said to be responsible for the blackening of cut and cooked potatoes: *Chem. Ind. (London)* **1958**, 627.

3′-Methyl ether. 3-Feruloylquinic acid. C$_{17}$H$_{20}$O$_9$. Crystals from ethyl acetate + petr ether, mp 196-197°. [α]$_D^{25}$ −42.8° (ethanol). uv max (ethanol): 325 nm (ε 19,200).

2144. Chlorogenin. [562-34-5] (3β,5α,6α,25R)-Spirostan-3,6-diol. C$_{27}$H$_{44}$O$_4$; mol wt 432.65. C 74.96%, H 10.25%, O 14.79%. Isoln from bulbs of the California soap plant, amole *Chlorogalum pomeridianum* (DC.) Kunth, Liliaceae: Liang, Noller, *J. Am. Chem. Soc.* **57**, 525 (1935). Chlorogenin occurs in amole as a saponin which kills or stuns fish without rendering them inedible. Structure: Marker, Rohrmann, *ibid.* **61**, 947, 3479 (1939); Marker *et al.*, *ibid.* **62**, 2537, 3006 (1940). On hydrogenation the 3β,6β-isomer (β-chlorogenin) is produced.

Needles from methanol, mp 273-276°. [α]$_{546}^{24}$ −52° (chloroform or isopropanol).

2145. 1-Chlorohexane. [544-10-5] *n*-Hexyl chloride. C$_6$H$_{13}$Cl; mol wt 120.62. C 59.75%, H 10.86%, Cl 29.39%. Prepd from 1-hexanol by treatment with fuming HCl: Henry, *Chem. Zentralbl.* **1905**, II, 214; with excess SOCl$_2$ or with PCl$_5$ + ZnCl$_2$: Clark, Streight, *Trans. R. Soc. Can. Sect. 3* **23**, III, 77 (1929).

Mobile liquid. d_4^{20} 0.8780. bp$_{760}$ 134°. n_D^{20} 1.4236 (Clark, Streight, *loc. cit.*); n_D^{20} 1.4195 (Mumford, Phillips, *J. Chem. Soc.* **1950**, 75). Insol in water. Refluxing with 10% aq NaOH decomposes 1-chlorohexane to 1-hexanol.

2146. α-Chlorohydrin. [96-24-2] 3-Chloro-1,2-propanediol; 3-chloro-1,2-dihydroxypropane; α-monochlorohydrin; β,β′-dihydroxyisopropyl chloride; glycerol α-monochlorohydrin; 3-chloropropylene glycol; Epibloc. C$_3$H$_7$ClO$_2$; mol wt 110.54. C 32.60%, H 6.38%, Cl 32.07%, O 28.95%. Prepd from glycerol and HCl gas: Conant, Quayle, *Org. Synth.* **coll. vol. I**, 294 (1941). Toxicity study: C. H. Hine *et al.*, *Arch. Ind. Health* **14**, 250 (1956).

Liquid. Sweetish taste. Tendency to turn straw color. d_4^{20} 1.3218. n_D^{20} 1.4831. bp$_{760}$ 213° (dec); bp$_{14}$ 114-120°; bp$_{11}$ 115-117°. Sol in water, alcohol, ether. LD$_{50}$ in mice, rats (g/kg): 0.16, 0.15 orally (Hine).

USE: To lower the freezing point of dynamite; in the manuf of dye intermediates. As rodent chemosterilant.

2147. Chloromethyl Chlorosulfate. [49715-04-0] Chlorosulfuric acid chloromethyl ester; chloromethoxysulfonyl chloride; CMCS. CH$_2$Cl$_2$O$_3$S; mol wt 164.98. C 7.28%, H 1.22%, Cl 42.97%, O 29.09%, S 19.43%. Reagent for chloromethylation reactions. Prepn: K. Fuchs, E. Katscher, *Ber.* **60**, 2288 (1927). Improved prepn and reactivity studies: N. P. Power *et al.*, *Org. Biomol. Chem.* **2**, 1554 (2004). Chloromethylation synthetic applications: E. Binderup, E. T. Hansen, *Synth. Commun.* **14**, 857 (1984); N. Harada *et al.*, *ibid.* **24**, 767 (1994); A. Mäntylä *et al.*, *Tetrahedron Lett.* **43**, 3793 (2002).

Colorless liquid. *Corrosive. Flammable.* bp$_{750}$ 153-155° (dec); bp$_{76}$ 85-89°; bp$_{14}$ 49-50°. d 1.63. Flash pt: 176°F (80.0°C).

USE: Reagent in synthetic organic chemistry.

2148. Chloromethyl Methyl Ether. [107-30-2] Chloromethoxymethane; methyl chloromethyl ether; monochloromethyl ether; chlorodimethyl ether; CMME. C$_2$H$_5$ClO; mol wt 80.51. C 29.84%, H 6.26%, Cl 44.03%, O 19.87%. Prepd by passing HCl through a mixture of formalin and methanol: C. S. Marvel, P. K. Porter, *Org. Synth.* **coll. vol. I**, 377 (1941). *See also: Beilstein* **1**, 580 (1918) and supplements. Commercial product usually contaminated by *sym*-

dichloromethyl ether, *q.v.* Review of carcinogenic risk: *IARC Monographs* **4**, 239-245 (1974).

$$H_3C-O-Cl$$

Colorless liquid, bp 59°. d_4^{20} 1.0605. n_D^{20} 1.39737.

Caution: Potential symptoms of overexposure are irritation of eyes, skin and mucous membranes; pulmonary edema, pulmonary congestion and pneumonia; skin burns, necrosis, coughing, wheezing; blood stained sputum; weight loss; bronchial secretions. *See NIOSH Pocket Guide to Chemical Hazards* (DHHS/NIOSH 97-140, 1997) p 66. The technical grade is listed as a known human carcinogen: *Report on Carcinogens, Twelfth Edition* (PB2011-111646, 2011) p 71.

USE: In synthesis of chloromethylated compounds, plastics and ion-exchange resins.

2149. 1-Chloro-2-methyl-1-propene. [513-37-1] α-Chloro-isobutylene; β,β-dimethylvinyl chloride; isocrotyl chloride; 2-methyl-1-propenyl chloride. C_4H_7Cl; mol wt 90.55. C 53.06%, H 7.79%, Cl 39.15%. Prepn from isobutyraldehyde: Kirrmann, *Bull. Soc. Chim. Fr.* **1948**, 163; by isomerization of 3-chloro-2-methylpropene with 80% H_2SO_4: Backhurst *et al., J. Chem. Soc.* **1959**, 2742.

$$H_3C-C(CH_3)=CH-Cl$$

Liquid, bp 68.1°. d_4^{20} 0.9186. n_D^{20} 1.4221.

Caution: This substance is reasonably anticipated to be a human carcinogen: *Report on Carcinogens, Twelfth Edition* (PB2011-111646, 2011) p 175.

USE: In organic syntheses; research chemical.

2150. 3-Chloro-2-methyl-1-propene. [563-47-3] γ-Chloro-isobutylene; methallylchloride; β-methallyl chloride; β-methylallylchloride. C_4H_7Cl; mol wt 90.55. C 53.06%, H 7.79%, Cl 39.15%. Prepn by chlorination of isobutylene: Burgin *et al., Ind. Eng. Chem.* **31**, 1413 (1939). Manuf: Carlson, US 2494034; Cherniavsky, Brown, US 2612530; Cheney *et al.,* US 2642464 (1950, 1952, 1953 all to Shell).

$$H_2C=C(CH_3)-CH_2-Cl$$

Liquid, bp 71-72°. d_4^{15} 0.9210, d_4^{20} 0.9165 (commercial grade d_{20}^{20} 0.926-0.930). n_D^{15} 1.4318, n_D^{20} 1.4274.

Caution: This substance is reasonably anticipated to be a human carcinogen: *Report on Carcinogens, Twelfth Edition* (PB2011-111646, 2011) p 100.

USE: Insecticide, fumigant. Chemical intermediate in organic syntheses.

2151. 1-Chloronaphthalene. [90-13-1] α-Chloronaphthalene; 1-naphthalenyl chloride; 1-naphthyl chloride. $C_{10}H_7Cl$; mol wt 162.62. C 73.86%, H 4.34%, Cl 21.80%. Prepd by passing chlorine into boiling naphthalene with or without solvent such as chlorobenzene and with or without catalyst such as I_2: DeWitt, Ekeley, *Univ. Colo. Stud.* **18**, 119 (1931), *C.A.* **26**, 2974 (1932). From α-naphthylamine by diazotization and Sandmeyer CuCl reaction: von Auwers, Frühling, *Ann.* **422**, 194, 200, 202 (1921); Hampson, Weissberger, *J. Chem. Soc.* **1936**, 394.

Oily liquid. Volatile with steam. d_4^{20} 1.19382. mp −2.5°. bp_{760} 259.3°; bp_{400} 230.8°; bp_{200} 204.2°; bp_{100} 180.4°; bp_{60} 165.6°; bp_{40} 153.2°; bp_{20} 134.4°; bp_{10} 118.6°; bp_5 104.8°; $bp_{1.0}$ 80.6°. n_D^{20} 1.63321. Sol in benzene, petr ether, alcohol.

Compound with 2,4,6-trinitro-*m*-cresol. $C_{17}H_{12}ClN_3O_7$. mp 78°.

Styphnate. $C_{16}H_{10}ClN_3O_8$. Yellow needles from alcohol, mp 112°.

USE: As immersion liquid in the (microscopic) determn of the refractive index of crystals. Solvent for oils, fat, DDT.

2152. 2-Chloronaphthalene. [91-58-7] β-Chloronaphthalene. $C_{10}H_7Cl$; mol wt 162.62. C 73.86%, H 4.34%, Cl 21.80%. Prepd from β-naphthylamine by diazotization and Sandmeyer CuCl reaction: Scheid, *Ber.* **34**, 1813 (1901); van der Kam, *Rec. Trav. Chim.* **45**, 568 (1926); Hampson, Weissberger, *J. Chem. Soc.* **1936**, 394. By chlorination of gaseous naphthalene at 530° in the presence of iodine, ~50% of the product being 1-chloronaphthalene which is sepd by fractional freezing: Britton, Reed, US 1917822 (1933).

Monoclinic plates, leaflets from dil alcohol. mp 59.5°. bp_{760} 256°; bp_{60} 161°; bp_{20} 132.6°; bp_{11} 119.7°. Volatile with steam. $n_D^{70.7}$ 1.60787. Sol in alcohol, ether, benzene, chloroform, carbon disulfide.

2153. Chloronitrobenzene. Nitrochlorobenzene. C_6H_4-ClNO$_2$; mol wt 157.55. C 45.74%, H 2.56%, Cl 22.50%, N 8.89%, O 20.31%. Prepn of *m*-isomer: W. W. Hartman, M. R. Brethen, *Org. Synth.* **coll. vol. I** (2nd ed., 1964) p 162.

m-**Chloronitrobenzene.** [121-73-3] 1-Chloro-3-nitrobenzene; *m*-nitrochlorobenzene. Pale-yellow orthorhombic prisms from alcohol. d_4^{20} 1.534. mp 46°. bp_{760} 236°; bp_{12} 117°. *Poisonous.* Insol in water. Sparingly sol in cold alcohol; freely sol in hot alcohol, chloroform, ether, carbon disulfide, glacial acetic acid.

o-**Chloronitrobenzene.** [88-73-3] Yellow crystals. d 1.305. mp 32-33°. bp 245-246°. *Poisonous.* Insol in water. Sol in alcohol, benzene, ether.

p-**Chloronitrobenzene.** [100-00-5] Yellow crystals. d 1.520. mp 82-84°. bp 242°. Flash pt 127°C. *Poisonous.* Insol in water. Sparingly sol in cold alcohol, freely in boiling alcohol, ether, carbon disulfide.

Caution: Absorbed by inhalation, skin penetration and ingestion. Potential symptoms of overexposure to *p*-isomer are anoxia; unpleasant taste; anemia; methemoglobinemia. Direct contact may cause allergic dermatitis. Potential occupational carcinogen. *See NIOSH Pocket Guide to Chemical Hazards* (DHHS/NIOSH 97-140, 1997) p 226; *Clinical Toxicology of Commercial Products,* R. E. Gosselin *et al.,* Eds. (Williams & Wilkins, Baltimore, 5th ed., 1984) Section II, p 214, Section III, p 31-36.

USE: In dye chemistry.

2154. 3-Chloroperoxybenzoic Acid. [937-14-4] 3-Chlorobenzenecarboperoxoic acid; 3-chloroperbenzoic acid; *m*-chloroperoxybenzoic acid; *m*-chloroperbenzoic acid; *meta*-chloroperoxybenzoic acid; MCPBA. $C_7H_5ClO_3$; mol wt 172.56. C 48.72%, H 2.92%, Cl 20.54%, O 27.81%. Versatile peracid oxidizing agent; utilized for a variety of transformations including the epoxidation of olefins. Prepn: B. M. Lynch, K. H. Pausacker, *J. Chem. Soc.* **1955**, 1525; R. N. McDonald *et al., Org. Synth.* **coll. vol. VI**, 276 (1988). Synthetic applications in alkene epoxidations: N. N. Schwartz, J. H. Blumbergs, *J. Org. Chem.* **29**, 1976 (1964); F. Fringuelli *et al., Tetrahedron Lett.* **30**, 1427 (1989); in silyl enol ether oxidations: G. M. Rubottom *et al., ibid.* **15**, 4319 (1974); in thiol oxidations: W. G. Filby *et al., J. Org. Chem.* **38**, 4070 (1973); in amine oxidations: K. E. Gilbert, W. T. Borden, *ibid.* **44**, 659 (1979); in Baeyer-Villiger ketone oxidations: K. Kaneda, T. Yamashita, *Tetrahedron Lett.* **37**, 4555 (1996).

White flaky powder, mp 88°. *Organic peroxide, skin sensitizer, irritant.* pKa 7.53. Sol in dichloromethane, chloroform, 1,2-dichloromethane, ethyl acetate, benzene, diethyl ether; slightly sol in hexane. Insol in water. *Shock sensitive and potentially explosive in the condensed phase.* Stable over long periods of time when refrigerated at 2-8°C in polyethylene containers. Commercial forms contain *meta*-chlorobenzoic acid as a contaminant and are less hazardous than the purified form.

USE: Reagent in synthetic organic chemistry.

2155. Chlorophacinone. [3691-35-8] 2-[2-(4-Chlorophenyl)-2-phenylacetyl]-1*H*-indene-1,3(2*H*)-dione; 2-[(*p*-chlorophenyl)phenylacetyl]-1,3-indandione; LM-91; Caid; Drat; Liphadione; Quick; Raviac; Rozol. $C_{23}H_{15}ClO_3$; mol wt 374.82. C 73.70%, H 4.03%, Cl 9.46%, O 12.81%. Prepn: D. Molho *et al.*, **US 3153612** (1964 to Lipha). HPLC determn in serum: M. G. Palazoglu *et al.*, *J. Agric. Food Chem.* **46**, 4260 (1998).

Light yellow silky needles from ethanol or acetone, mp 138°. Absorption max (acetone): 325 nm. Sol in organic solvents; very sparingly sol in water.

USE: Anticoagulant rodenticide.

2156. *p*-Chlorophenacyl Bromide. [536-38-9] 2-Bromo-1-(4-chlorophenyl)ethanone; 2-bromo-4′-chloroacetophenone; 4-chloro-ω-bromoacetophenone; α-bromo-*p*-chloroacetophenone. C_8H_6BrClO; mol wt 233.49. C 41.15%, H 2.59%, Br 34.22%, Cl 15.18%, O 6.85%. Prepn from bromoacetyl chloride, chlorobenzene, and AlCl₃: Collet, *Compt. Rend.* **125**, 718 (1897); by bromination of *p*-chloroacetophenone: *idem, Bull. Soc. Chim. Fr.* [3] **21**, 69 (1899). Toxicity study: Dat-Xuong *et al.*, *Med. Exp.* **11**(3), 137 (1964).

Needles, mp 96-96.5°. LD_{50} orally in mice: >2000 mg/kg (Dat-Xuong).

USE: In the prepn of quaternary salts of methenamine and of chlorophenol glyoximes.

2157. Chlorophenols. C_6H_5ClO; mol wt 128.56. C 56.06%, H 3.92%, Cl 27.57%, O 12.44%. Prepn of *o*-form: Holleman, Rinkes, *Rec. Trav. Chim.* **30**, 48 (1911); Huston, Neely, *J. Am. Chem. Soc.* **57**, 2176 (1935). *cis-trans* equilibria in *o*-halophenols: Baker, *ibid.* **80**, 3598 (1958). Prepn of *p*-form: Crawford, Willson, **US 1910679** (1927 to Monsanto); Neu, *Ber.* **72**, 1505 (1939); Hodgson, Norris, *J. Chem. Soc.* **1949**, Suppl. 1, S181. Manuf: Britton, Keil, **US 2725402** (1955 to Dow). Prepn of *m*-form: Acheson, Taylor, *J. Chem. Soc.* **1956**, 4727. Manuf: Stoesser, Gentry; Barnard, Meyer, **US 2835707** and **US 2852567** (both 1958 to Dow). Toxicity data: Deichmann, *Fed. Proc.* **2**, 76 (1943). Review of toxicology and human exposure: *Toxicological Profile for Chlorophenols* (PB99-166639, 1999) 260 pp.

m-**Chlorophenol.** [108-43-0] 3-Chlorophenol. Needles, mp 33.5°. bp 214°. d_4^{45} 1.245, d_4^{78} 1.214. n_D^{40} 1.5565. Strong medicinal taste and odor. Dipole moment in benzene, 2.10D. *Poisonous.*

Slightly sol in cold water; sol in alcohol, ether, caustic alkali solns. LD_{50} orally in rats: 0.57 g/kg (Deichmann).

o-**Chlorophenol.** [95-57-8] 2-Chlorophenol. Liquid, bp 175°. mp 9.3°. d_4^{23} 1.2573. n_D^{25} 1.5565; *n* 1.5473. Strong medicinal taste and odor. Dipole moment in benzene, 1.33D. *Poisonous.* Freely sol in alcohol, ether, caustic alkali solns; slightly sol in water. LD_{50} orally in rats: 0.67 g/kg (Deichmann).

p-**Chlorophenol.** [106-48-9] 4-Chlorophenol. Crystals with characteristic phenolic odor, mp 43.2-43.7°. bp 220°. d_4^{78} 1.2238. n_D^{55} 1.5419; n_D^{40} 1.5579. Dipole moment in benzene, 2.22D. *Poisonous.* Sparingly sol in water, liquid petrolatum; very sol in alcohol, glycerin, ether, chloroform, fixed and volatile oils: *USP* **XXI**, 1467. LD_{50} orally in rats: 0.67 g/kg (Deichmann).

Caution: Potential symptoms of overexposure are tremors, convulsions, dyspnea, coma. Direct contact may cause skin irritation. *See: Patty's Industrial Hygiene and Toxicology* Vol. 2A, G. D. Clayton, F. E. Clayton, Eds. (Wiley-Interscience, New York, 3rd ed., 1981) pp 2612-2615.

USE: Biocide; disinfectant for home, hospital and farm.

THERAP CAT: Antiseptic.

2158. Chlorophyll. [1406-65-1] The green, photosynthetic pigment of plants. Higher plants and green algae contain chlorophyll a and chlorophyll b in the approx ratio of 3:1. Chlorophyll c is found together with chlorophyll a in many types of marine algae: Jeffrey, *Biochem. J.* **86**, 313 (1963). Red algae *(Rhodophyta)* contain principally chlorophyll a and also chlorophyll d: Manning, Strain, *J. Biol. Chem.* **151**, 1 (1943). Isoln by chromatography: Tswett, *Ber. Dtsch. Bot. Ges.* **24**, 316, 385 (1906); Willstätter, Stoll, *Investigations on Chlorophyll* (transl by Schertz and Merz: Lancaster, 1928); Schertz, *Ind. Eng. Chem.* **30**, 1073 (1938); Fischer-Orth-Stern, *Die Chemie des Pyrrols* **Vol. II**, part 2 (Leipzig, 1940); Zechmeister, Cholnoky, *Principles and Practice of Chromatography* (New York, 1943). Industrial large-scale isoln processes: Judah *et al.*, *Ind. Eng. Chem.* **46**, 2262 (1954). Structure: Fischer-Orth-Stern, *loc. cit.*; Ficken *et al.*, *J. Chem. Soc.* **1956**, 2273. Total synthesis of chlorophyll a: R. B. Woodward *et al.*, *J. Am. Chem. Soc.* **82**, 3800 (1960); *idem, Angew. Chem.* **72**, 651 (1960); Strell *et al., ibid.* 169; R. B. Woodward, *Pure Appl. Chem.* **2**, 383 (1963); *idem et al., Tetrahedron* **46**, 7599 (1990). Abs config of chlorophyll a: I. Fleming, *Nature* **216**, 151 (1967); of chlorophylls a and b: Brockmann, *Ann.* **754**, 139 (1971); Brockmann, Bode, *Ann.* **1974** (7), 1017. ^{13}C-NMR study of chlorophyll a: S. Lötjönen, P. H. Hynninen, *Org. Magn. Reson.* **16**, 304 (1981); of chlorophyll b: N. Risch, H. Brockmann, *Tetrahedron Lett.* **24**, 173 (1983). Review of syntheses: Johnson, *Sci. Prog.* **49**, 77 (1961). Comprehensive reviews with bibliography: Stoll, Wiedemann, *Fortschr. Chem. Org. Naturst.* **1**, 159-254 (1938); *Fortschr. Chem. Forsch.* **2**, 538 (1952); *The Chlorophylls*, L. P. Vernon, G. R. Seely, Eds. (Academic Press, New York, 1966) 679 pp; Inhoffen *et al., Fortschr. Chem. Org. Naturst.* **26**, 284-298 (1968); Inhoffen, *Pure Appl. Chem.* **17**, 443-460 (1968).

	R_1	R_2	R_3
Chlorophyll a	CH₃	CH₂CH₃	X
Chlorophyll b	CHO	CH₂CH₃	X

Chlorophyll a. [479-61-8] $C_{55}H_{72}MgN_4O_5$; mol wt 893.51. Sepn and purification: Anderson, Calvin, *Nature* **194**, 285 (1962).

Waxy blue-black microcrystals, usually aggregates of thin, lancet-like leaflets, mp 117-120°. $[\alpha]_D^{20}$ −262° (acetone). Absorption max (ether): 660, 613, 577, 531, 498, 429, 409 nm. Freely sol in ether, ethanol, acetone, chloroform, carbon disulfide, benzene. Sparingly sol in cold methanol. Practically insol in petr ether. The alcoholic soln is blue-green with a deep-red fluorescence.

Chlorophyll b. [519-62-0] $C_{55}H_{70}MgN_4O_6$; mol wt 907.49. Waxy blue-black microcrystals. Sinters between 86° and 92°, becomes a viscous liquid at 120-130° and then begins to puff up in large bubbles. $[\alpha]_D^{20}$ −267° (acetone-methanol). Absorption max (ether): 642, 593, 565, 545, 453, 427 nm. Sparingly sol in petr ether, ligroin, cold methanol. Freely sol in abs alcohol, ether. The ether soln has a brilliant green color. Solns with other organic solvents are usually green to yellowish-green with red fluorescence.

Chlorophyll c. Structure: Dougherty *et al.*, *J. Am. Chem. Soc.* **88**, 5037 (1966). Reddish-black hexagonal bipyramids or four-sided plates from dil ethanol. Absorption max (acetone): 628, 580, 442 nm ($E_{1cm}^{0.1\%}$ 15.8, 10.7, 115.9). Sol in methanol, ethanol, ethyl acetate; practically insol in ether, acetone.

Chlorophyll d. $C_{54}H_{70}MgN_4O_6$; mol wt 895.48. Chlorophyll a with the 9-vinyl group replaced by a formyl group: Holt, Morley, *Can. J. Chem.* **37**, 507 (1959); Holt, *Can. J. Bot.* **39**, 327 (1961). Absorption max (ether): 686, 445 nm. The chlorophyll of commerce is an intensely dark-green, aq, alc, or oil soln. Careful alkaline hydrolysis of chlorophyll opens the cyclopentanone ring and replaces the methyl and phytyl ester groups with Na or K; the resulting salts are called *chlorophyllins* and are water sol, e.g. sodium magnesium chlorophyllin, $C_{34}H_{31}N_4Na_3MgO_6$. Acid treatment of chlorophyll removes the Mg replacing it with H_2 which can be replaced with other metals, e.g. iron pheophytin, $C_{55}H_{72}FeN_4O_5$, sol in oil. Comprehensive review: "Chlorophyll Derivatives, Their Chemistry, Commercial Preparation and Uses" by J. C. Kephart in *Econ. Bot.* **9**, 3-38 (1955). Review of commercial chlorophyll prepns: Strell *et al.*, *Arzneim.-Forsch.* **5**, 640 (1955); **6**, 8 (1956). Trademarks: *Chloresium, Chlorofolin, Darotol, Ennds, Exodor-Grun, Nullo Chlorophyll, Stozzon-Chlorophyll.*

USE: To color soaps, oils, fats, waxes, confectionery, preserves, liquors, cosmetics, perfumes. Source of phytol. For dyeing leather. As sensitizer for color film. Has been used as antiknock agent in gasoline; as accelerator in the vulcanizing of rubber; in deodorizers.

THERAP CAT: Deodorant.

THERAP CAT (VET): Has been used orally to reduce odors, and topically to promote healing of skin lesions.

2159. Chloropicrin. [76-06-2] Trichloronitromethane; acquinite; nitrochloroform; Larvacide 100; Picfume. CCl_3NO_2; mol wt 164.37. C 7.31%, Cl 64.70%, N 8.52%, O 19.47%. First prepd in 1848 by Stenhouse from picric acid and bleach powder. Review of prepn, properties, physiological action, and uses: Jackson, *Chem. Rev.* **14**, 251 (1934). Manuf from nitromethane and alkaline hypochlorite: Wilhelm, US 3106588 (1963).

Slightly oily liquid, intense odor; bp_{757} 112°. mp −64°; mp −69.2° (corr). d_4^{20} 1.6558, d_4^{25} 1.6483. n_D^{20} 1.4611, n_D^{25} 1.4596. *Poisonous.* Dipole moment in heptane or benzene, 1.80D. Practically insol in water. Soly in water (g/100 ml): 0.2272 (0°), 0.1621 (25°). Miscible with benzene, abs alc, carbon disulfide; sol in ether.

Caution: Potential symptoms of overexposure are irritation of eyes, skin, respiratory system; lacrimation; cough, pulmonary edema; nausea, vomiting. See *NIOSH Pocket Guide to Chemical Hazards* (DHHS/NIOSH 97-140, 1997) p 66.

USE: Disinfecting cereals and grains; in synthesis, esp in manuf of methyl violet; fumigant; soil insecticide; war gas.

2160. Chloroprocaine. [133-16-4] 4-Amino-2-chlorobenzoic acid 2-(diethylamino)ethyl ester; 2-chloro-4-aminobenzoic acid diethylaminoethyl ester; 2-diethylaminoethyl 4-amino-2-chlorobenzoate. $C_{13}H_{19}ClN_2O_2$; mol wt 270.76. C 57.67%, H 7.07%, Cl 13.09%, N 10.35%, O 11.82%. Rapidly hydrolyzed local anesthetic of the *para*-aminobenzoic ester class. Prepn: M. Rubin *et al.*, *J. Am. Chem. Soc.* **68**, 623 (1946). Description of hydrochloride: M. Hädicke, *Pharm. Zentralhalle* **94**, 384 (1955). HPLC determn in pharmaceutical formulations: G. Menon *et al.*, *J. Pharm. Sci.* **73**, 251 (1984); in plasma: P. K. Janicki *et al.*, *J. Chromatogr. B* **675**, 336 (1996). Review of clinical results as spinal anesthetic in general surgery: J. R. Yoos, D. J. Kopacz, *Anesth. Analg.* **100**, 553-558 (2005).

Hydrochloride. [3858-89-7] Nesacaine. $C_{13}H_{19}ClN_2O_2\cdot HCl$; mol wt 307.22. Crystals, mp 176-178° (microstage, Hädicke); also reported as solid from ethanol, mp 171-172° (Rubin). Bitter taste. Practically insol in ether. Very slightly sol in chloroform. Soly in 95% ethanol about one gram in 100 ml. Soly in water at 20° about one gram in 22 ml. Aq solns are just acid to litmus and turn yellow on standing.

THERAP CAT: Anesthetic (local).

2161. β-Chloropropionic Acid. [107-94-8] 3-Chloropropanoic acid. $C_3H_5ClO_2$; mol wt 108.52. C 33.20%, H 4.64%, Cl 32.67%, O 29.49%. CH_2ClCH_2COOH. Prepd by the hydrolysis of ethylene cyanohydrin with hydrochloric acid; by the oxidation of β-chloropropionaldehyde or of trimethylene chlorohydrin by nitric acid: Moureu, Chaux, *Org. Synth.* **8**, 54 (1928); Powell, *ibid.* 58; Paal, Lobeck, *Ber.* **64**, 2142 (1931).

Leaflets from ligroin. Somewhat hygroscopic. mp 41°. bp_{765} 200°; bp_{35} 127°; bp_{25} 124°; bp_{12} 108°. pK (25°) 4.00. Freely sol in water, alc, chloroform, slightly less in ether.

Methyl ester. [6001-87-2] $C_4H_7ClO_2$; mol wt 122.55. From β-chloropropionic acid with methanol and HCl or from acrylic acid chloride and methanol. d_4^0 1.198. bp 155-157°.

2162. β-Chloropropionitrile. [542-76-7] 3-Chloropropanenitrile; 3-chloropropanonitrile. C_3H_4ClN; mol wt 89.52. C 40.25%, H 4.50%, Cl 39.60%, N 15.65%. Prepd from acrylonitrile and hydrogen chloride or bromide: Moureu, Clarke, *Bull. Soc. Chim. Fr.* [4] **27**, 905 (1920); Stewart, Clarke, *J. Am. Chem. Soc.* **69**, 714 (1947); Shirley, *Preparation of Organic Intermediates* (Wiley, New York, 1951) p 82.

Liquid. Acrid, characteristic odor. *Poisonous.* mp −51°. d_4^{25} 1.1363. bp_{760} 176° (dec); bp_{200} 132° (dec); bp_{50} 95.2°; bp_5 46.0°. Flash pt 168°F (75.5°C). n_D^{25} 1.4341. Begins to dec when heated above 130° evolving HCl. Absorbs strongly in the infrared. Transparent to uv above 220 nm. Soly in water at 25°: 4.5 g/100 ml. Soly of water in β-chloropropionitrile at 25°: 2.2 ml/100 g. Miscible with ethanol, ether, acetone, benzene, carbon tetrachloride. LD_{50} orally in mice, rats: 9, 100 mg/kg, Fassett in *Industrial Hygiene and Toxicology* vol. **2**, F. A. Patty, Ed. (Interscience, New York, 2nd ed., 1962) pp 2025-2026.

Caution: Exposure by any route should be avoided. Somewhat less hazardous than acrylonitrile because of lower vapor pressure. Readily penetrates skin to produce systemic cyanide poisoning, death.

USE: In pharmaceutical and polymer synthesis. Combines the reactivity of a nitrile and an alkyl halide. Because of the cyano group the chlorine atom is more reactive than in ordinary alkyl halides.

2163. 6-Chloropurine. [87-42-3] 6-Chloro-9H-purine. $C_5H_3ClN_4$; mol wt 154.56. C 38.86%, H 1.96%, Cl 22.94%, N 36.25%. Prepd by the action of phosphorus oxychloride on hypoxanthine in *N,N*-dimethylaniline: Bendich *et al.*, *J. Am. Chem. Soc.* **76**, 6073 (1954). Antineoplastic activity *in vivo*: A. C. Sartorelli, B. A. Booth, *Experientia* **21**, 457 (1965).

Blunt needles from water, dec 175-177° (hot stage preheated to 170°). Also reported as not melted at 290°. uv max (pH 5.2): 265 nm (ε 9120); (pH 13): 274 nm (ε 8790). Soly in water: 0.5% at 20°, also reported as 1 g/182 ml at 24°. Sol in ether, dimethylformamide.

2164. Chloropyramine. [59-32-5] N^1-[(4-Chlorophenyl)-methyl]-N^2,N^2-dimethyl-N^1-2-pyridinyl-1,2-ethanediamine; 2-[(p-chlorobenzyl)[2-(dimethylamino)ethyl]amino]pyridine; N-dimethyl-aminoethyl-N-p-chlorobenzyl-α-aminopyridine; N,N-dimethyl-N'-(p-chlorobenzyl)-N'-(2-pyridyl)ethylenediamine; halopyramine. $C_{16}H_{20}ClN_3$; mol wt 289.81. C 66.31%, H 6.96%, Cl 12.23%, N 14.50%. Prepn: Phillips, Cates, **US 2607778** (1952 to Merck & Co.); Howard, **US 2569314** (1951 to Am. Cyanamid); **CH 264754**; **CH 266234**; **CH 266235** (1950); **GB 651596** (1951) (all to Geigy).

Light yellow, viscous, oily liquid. Pungent odor. $bp_{0.2}$ 154-155°.
Hydrochloride. [6170-42-9] Synopen; Synpen. $C_{16}H_{20}ClN_3$.HCl; mol wt 326.27. Crystals from acetone, mp 172-174°.
THERAP CAT: Antihistaminic.

2165. Chloroquine. [54-05-7] N^4-(7-Chloro-4-quinolinyl)-N^1,N^1-diethyl-1,4-pentanediamine; 7-chloro-4-(4-diethylamino-1-methylbutylamino)quinoline; SN-7618; RP-3377. $C_{18}H_{26}ClN_3$; mol wt 319.88. C 67.59%, H 8.19%, Cl 11.08%, N 13.14%. Prepd by the condensation of 4,7-dichloroquinoline with 1-diethylamino-4-aminopentane: **DE 683692** (1939); H. Andersag et al., **US 2233970** (1941 to Winthrop); Surrey, Hammer, J. Am. Chem. Soc. **68**, 113 (1946). Review: Hahn in Antibiotics vol. 3, J. W. Corcoran, F. E. Hahn, Eds. (Springer-Verlag, New York, 1975) pp 58-78. Comprehensive description: D. D. Hong, Anal. Profiles Drug Subs. **5**, 61-85 (1976). Comparative clinical trial with dapsone in rheumatoid arthritis: P. D. Fowler et al., Ann. Rheum. Dis. **43**, 200 (1984); with penicillamine: T. Gibson et al., Br. J. Rheumatol. **26**, 279 (1987).

White crystalline powder. mp 87°. Sol in dilute acids, chloroform, ether; very slightly sol in water.
Diphosphate. [50-63-5] Arechin; Avloclor; Malaquin; Resochin. $C_{18}H_{26}ClN_3$.2H_3PO_4; mol wt 515.86. Bitter, colorless crystals. Dimorphic. One modification, mp 193-195°; the other, mp 215-218°. Freely sol in water; pH of 1% soln about 4.5; less sol at neutral and alkaline pH. Stable to heat in solns of pH 4.0 to 6.5. Practically insol in alcohol, benzene, chloroform, ether.
Sulfate. [132-73-0] Aralen; Nivaquine. $C_{18}H_{26}ClN_3$.H_2SO_4; mol wt 417.95.
THERAP CAT: Antimalarial; antiamebic; antirheumatic. Lupus erythematosus suppressant.

2166. N-Chlorosaccharin. [14070-51-0] 2-Chloro-1,2-benzisothiazol-3(2H)-one 1,1-dioxide; o-benzoic N-chlorosulphinide; NCSA; NCSac. $C_7H_4ClNO_3S$; mol wt 217.62. C 38.63%, H 1.85%, Cl 16.29%, N 6.44%, O 22.06%, S 14.73%. Reagent that is a source of electrophilic chlorine; utilized for a variety of chlorination and oxidation reactions. Prepn: F. D. Chattaway, J. Chem. Soc.,

Trans. **87**, 1882 (1905). Green prepn: S. P. L. de Souza et al., Synth. Commun. **33**, 935 (2003). Stability and chlorine potential studies: H.-S. Dawn et al., J. Pharm. Sci. **59**, 955 (1970). Synthetic applications: J. M. Bachhawat et al., Indian J. Chem. **11**, 609 (1973); D. Dolenc, B. Sket, Synlett **1995**, 327; N. Iranpoor et al., Can. J. Chem. **84**, 69 (2006). Oxidimetric titration studies: N. Jayasree, P. Indrasenan, Indian J. Chem. **26A**, 714 (1987). Review: P. E. Gama, Synlett **2008**, 1742-1743.

White crystalline powder, mp 152°. Crystallizes into long, flattened prisms from chloroform, acetic acid. Peculiar odor resembles chloral hydrate. Dec slowly with evolution of gas when heated above 260°. Soly at 25° (g/l): 1,4-dioxane 287.0; acetone 173.0; chloroform 112.0; ethyl acetate 81.1; carbon tetrachloride 4.3; water ~0.1; at 30° (g/l): acetic acid 44.83.
USE: Reagent in synthetic organic chemistry. Titrant in analytical chemistry.

2167. N-Chlorosuccinimide. [128-09-6] 1-Chloro-2,5-pyrrolidinedione; succinchlorimide. $C_4H_4ClNO_2$; mol wt 133.53. C 35.98%, H 3.02%, Cl 26.55%, N 10.49%, O 23.96%. Prepn from succinimide: Hirst, Macbeth, J. Chem. Soc. **121**, 2169 (1922); Zimmer, Audrieth, J. Am. Chem. Soc. **76**, 3856 (1954). Crystal structure: Brown, Acta Crystallogr. **9**, 193 (1956); **14**, 711 (1961). Toxicity data: Stohlman, Smith, Public Health Rep. **59**, 541 (1944).

Orthorhombic crystals, mp 150-151°. Odor of chlorine. Acid to litmus (1:50 aq soln). One gram dissolves in about 70 ml water, 150 ml alcohol, 50 ml benzene. Sparingly sol in ether, chloroform, carbon tetrachloride. Liberates iodine from potassium iodide solns, and bromine from sodium bromide solns. MLD orally in rats: 2.7 g/kg (Stohlman, Smith).
USE: Chlorinating agent.

2168. Chlorosulfonic Acid. [7790-94-5] Chlorosulfuric acid; sulfuric chlorohydrin. $ClHO_3S$; mol wt 116.52. Cl 30.42%, H 0.87%, O 41.19%, S 27.51%. $SO_2(OH)Cl$. Prepd by the reaction of HCl gas with SO_3: Simon, Kratsch, Z. Anorg. Allg. Chem. **242**, 369 (1939); Briggs, **US 1442335** (1922 to General Chem.). Purification: Kaplar, Shechter, Inorg. Synth. **4**, 52 (1953). Review: H. O. Burrus in Kirk-Othmer Encyclopedia of Chemical Technology vol. **5** (Wiley-Interscience, New York, 3rd ed., 1979) pp 873-880.
Colorless or slightly yellow, very corrosive liquid, causes severe burns; fumes in air; pungent odor. d_{20}^{20} 1.76-1.77; d_4^0 1.784; d_4^{20} 1.753. mp −80°. bp_{755} 151-152°; bp_{19} 74-75°; $bp_{2.4}$ 60-64°. n_D^{14} 1.437. On dropping into water dec with explosive violence. Corrosive, poisonous. Keep tightly closed. When used for the prepn of sulfate esters, the common solvent is pyridine. Other solvents are liquid sulfur dioxide and dichloroethane.
Caution: Highly irritating and corrosive to eyes, skin, mucous membranes.
USE: Manuf sulfone compds, saccharin. As chlorosulfonating and condensing agent in organic syntheses.

2169. Chlorothalonil. [1897-45-6] 2,4,5,6-Tetrachloro-1,3-benzenedicarbonitrile; tetrachloroisophthalonitrile; m-tetrachlorophthalodinitrile; 2,4,5,6-tetrachloro-1,3-dicyanobenzene; 1,3-dicyano-2,4,5,6-tetrachlorobenzene; chlorthalonil; DAC-2787; Daconil 2787; Bravo. $C_8Cl_4N_2$; mol wt 265.90. C 36.14%, Cl 53.33%, N 10.54%. Fungicidal properties: N. J. Turner et al., Contrib. Boyce Thompson Inst. **22**, 303 (1964). Prepn: R. D. Battershell, H. Bluestone, **US 3290353** (1966 to Diamond Alkali); R. M. Bimber, **US**

3652637 (1972 to Diamond Shamrock). Toxicity studies in mice: H. Yoshikawa, K. Kawai, *Ind. Health* **4**, 11 (1966). Review of carcinogenic risk: *IARC Monographs* **30**, 319-328 (1983)

Crystals. d_4^{25} 1.7. mp 250-251°. bp_{760} 350°. Vapor press <0.01 at 40°. Practically insol in water (soly at room temp reported as 0.6 ppm). Soly in organic solvents at 25° (w/w): xylene 8%, cyclohexanone 3%, acetone 2%, kerosine <1.0%. LD_{50} orally in rats: >10.0 g/kg (Turner).

USE: Fungicide, bactericide, nematocide. Agricultural and horticultural fungicide.

2170. Chlorothen. [148-65-2] N^1-[(5-Chloro-2-thienyl)-methyl]-N^2,N^2-dimethyl-N^1-2-pyridinyl-1,2-ethanediamine; 2-[(5-chloro-2-thenyl)(2-dimethylaminoethyl)amino]pyridine; *N,N*-dimethyl-N'-(2-pyridyl)-N'-(5-chloro-2-thenyl)ethylenediamine; *N,N*-dimethyl-N'-(α-pyridyl)-N'-(2-methyl-5-chlorothienyl)ethylenediamine; *N*-5-chloro-2-thienylmethyl-N',N'-dimethyl-*N*-2-pyridylethylenediamine; chloropyrilene; chloromethapyrilene; chlorothenylpyramine. $C_{14}H_{18}ClN_3S$; mol wt 295.83. C 56.84%, H 6.13%, Cl 11.98%, N 14.20%, S 10.84%. Prepd by the condensation of 5-chloro-2-thenyl chloride and *N,N*-dimethyl-N'-(2-pyridyl)ethylenediamine in the presence of sodium or potassium amide: R. C. Clapp *et al.*, *J. Am. Chem. Soc.* **69**, 1549 (1947); L. P. Kyrides, US **2581869** (1952 to Monsanto). Toxicity data: J. C. Castillo *et al.*, *J. Pharmacol. Exp. Ther.* **96**, 388 (1949).

Liquid. $bp_{1.0}$ 155-156°. Strong base.
Hydrochloride. Thenclor. $C_{14}H_{18}ClN_3S.HCl$; mol wt 332.29. Crystals, mp 106-108°. Freely sol in water.
Citrate. [148-64-1] Tagathen. $C_{14}H_{18}ClN_3S.C_6H_8O_7$; mol wt 487.95. Crystals, mp 112-116°. On further heating gradually solidifies and remelts 125-140° (dec). uv max: 240 nm ($E_{1cm}^{1\%}$ 390-410). One gram dissolves in 35 ml water, in about 65 ml alc. Practically insol in ether, chloroform, benzene. pH of 1% aq soln 3.9 to 4.1. LD_{50} i.p. in mice: 105 mg/kg (Castillo).
THERAP CAT: Antihistaminic.

2171. Chlorothiazide. [58-94-6] 6-Chloro-2*H*-1,2,4-benzothiadiazine-7-sulfonamide 1,1-dioxide; 6-chloro-7-sulfamoyl-2*H*-1,-2,4-benzothiadiazine 1,1-dioxide; 6-chloro-7-sulfamyl-1,2,4-benzothiadiazine 1,1-dioxide; Chlotride; Diuril; Saluric. $C_7H_6ClN_3O_4S_2$; mol wt 295.71. C 28.43%, H 2.05%, Cl 11.99%, N 14.21%, O 21.64%, S 21.68%. Prepn: F. C. Novello, US **2809194** (1957 to Merck & Co.); F. C. Novello, J. M. Sprague, *J. Am. Chem. Soc.* **79**, 2028 (1957). HPLC determn in urine: R. O. Fullinfaw *et al.*, *J. Chromatogr.* **415**, 347 (1987). Comprehensive description: H. G. Brittain, *Anal. Profiles Drug Subs.* **18**, 33-56 (1989).

Colorless needles from dil alc, mp 342.5-343°. pKa_1 6.85, pKa_2 9.45. Freely sol in DMSO, DMF; readily sol in dilute NaOH;

slightly sol in methanol, pyridine; very slightly sol in water. Practically insol in diethyl ether, chloroform, benzene. Soly (g/L): water (pH 4) 0; (pH 7) 0.65. Sol in alkaline aq solns with decompn upon standing or heating.
Sodium salt. [7085-44-1] $C_7H_5ClN_3NaO_4S_2$; mol wt 317.69.
THERAP CAT: Diuretic; antihypertensive.
THERAP CAT (VET): Diuretic.

2172. Chlorothricin. [34707-92-1] (4*S*,4a*S*,6a*R*,11*E*,12a*R*,-15*R*,16a*S*,21a*R*,21b*R*)-4-[[4-*O*-[3-*O*-(3-Chloro-6-methoxy-2-methylbenzoyl)-2,6-dideoxy-β-D-*arabino*-hexopyranosyl)-2,6-dideoxy-β-D-*arabino*-hexopyranosyl]oxy]-1,2,3,4,4a,6a,7,8,9,10,12a,15,16,-21,21a,21b-hexadecahydro-22-hydroxy-15,21a-dimethyl-18,21-dioxo-18*H*-16a,19-metheno-16a*H*-benzo[*e*]naphtho[2,1-*m*][1,4]dioxacyclopentadecin-14-carboxylic acid. $C_{50}H_{63}ClO_{16}$; mol wt 955.49. C 62.85%, H 6.65%, Cl 3.71%, O 26.79%. Macrolide antibiotic active against gram-positive bacteria. Isolated together with its dechloro analog from Tü 99, a strain of *Streptomyces antibioticus*: Keller-Schierlein *et al.*, *Helv. Chim. Acta* **52**, 127 (1969). Degradation products studies: Muntwyler *et al.*, *ibid.* **53**, 1544 (1970). Structure: Muntwyler, Keller-Schierlein, *ibid.* **55**, 2071 (1972). Non-competitive inhibitor of pyruvate carboxylase: P. W. Schindler, H. Zaehner, *Arch. Mikrobiol.* **82**, 66, (1972); P. W. Schindler, M. C. Scrutton, *ibid.* **55**, 543 (1975). Biosynthetic studies: O. Mascaretti *et al.*, *J. Nat. Prod.* **42**, 455 (1979); *eidem, Biochemistry* **20**, 919 (1981). Partial synthesis of the aglycone, *chlorothricolide*: R. E. Ireland, W. J. Thompson, *J. Org. Chem.* **44**, 3041 (1979); R. E. Ireland *et al.*, *ibid.* **46**, 4863 (1981); W. R. Roush, S. E. Hall, *J. Am. Chem. Soc.* **103**, 5200 (1981). Alternate synthetic approach: R. E. Ireland, M. D. Varney, *J. Org. Chem.* **51**, 635 (1986). *Review:* Keller-Schierlein, *Fortschr. Chem. Org. Naturst.* **30**, 394-396 (1973); H. G. Floss, C.-J. Chang, *Antibiotics* **vol. IV**, J. W. Corcoran, Ed. (Springer-Verlag, New York, 1981) pp 193-214.

Colorless crystals from methylene chloride-methyl acetate, mp 206-207°. uv max (alc): 222, 260 nm (log ε 4.20, 3.81); (0.01*N* alc KOH): 221, 259 nm (log ε 4.09, 3.95). Dibasic acid; pK: 5.01, 7.91. (The above data applies to a 5:1 mixture with dechlorothricin, Muntwyler, Keller-Schierlein, *loc. cit.*). Slightly sol in water; readily sol in organic solvents of medium polarity.

2173. Chlorothymol. [89-68-9] 4-Chloro-5-methyl-2-(1-methylethyl)phenol; 6-chlorothymol; 4-chlorothymol; 1-methyl-3-hydroxy-4-isopropyl-6-chlorobenzene; 6-chloro-4-isopropyl-1-methyl-3-phenol. $C_{10}H_{13}ClO$; mol wt 184.66. C 65.04%, H 7.10%, Cl 19.20%, O 8.66%. Made by the action of sulfuryl chloride on thymol in CCl_4: Satriana *et al.*, *J. Am. Pharm. Assoc.* **39**, 135 (1950);

H. Pahlicke, **DE 905738** (1954 to Diwag), *C.A.* **52**, 16294i (1958). Antifungal properties: M. Iannarone, *Drug Stand.* **25**, 190 (1957).

Crystals. mp 62-64° (59-61° *N.F.* **XII**). One gram dissolves in about 1000 ml water, 0.5 ml alcohol, 2 ml benzene, 2 ml chloroform, 1.5 ml ether, about 10 ml petr ether; also sol in dil aq NaOH.

2174. Chlorotitanium Triisopropoxide. [20717-86-6] (*T*-4)-Chlorotris(2-propanolato)titanium; chlorotriisopropoxytitanium; titanium monochloride triisopropoxide; triisopropoxytitanium chloride; ClTi(O*i*-Pr)$_3$. C$_9$H$_{21}$ClO$_3$Ti; mol wt 260.58. C 41.48%, H 8.12%, Cl 13.60%, O 18.42%, Ti 18.37%. Organotitanium Lewis acid used as an additive to promote enhanced stereocontrol in carbon-carbon bond forming reactions. Prepn: H. Holloway, *Chem. Ind. (London)* **1962**, 214; H. Bürger, *Monatsh. Chem.* **94**, 574 (1963). Synthetic applications: M. T. Reetz *et al.*, *Ber.* **118**, 1421 (1985); M. Nerz-Stormes, E. R. Thornton, *J. Org. Chem.* **56**, 2489 (1991); E. J. Corey *et al.*, **116**, 9345 (1994); G. W. O'Neil, A. J. Philips, *Tetrahedron Lett.* **45** 4253 (2004). *Review:* A. P. G. Macabeo, *Synlett* **2008**, 3247-3248.

Viscous, faintly yellow liquid. Solidifies upon standing at room temp. *Flammable. Corrosive.* bp$_1$ 90°; bp$_{0.1}$ 61-65°. Flash pt, closed cup: 72°F (22°C). Sol in pentane, hexane, toluene, diethyl ether, tetrahydrofuran, dichloromethane. Hygroscopic. May decompose upon exposure to air or moisture. Store under nitrogen.
USE: Reagent in synthetic organic chemistry.

2175. Chlorotoluene. C$_7$H$_7$Cl; mol wt 126.58. C 66.42%, H 5.57%, Cl 28.01%. Prepn from diazotized toluidine: Neogi, Mitra, *J. Chem. Soc.* **1928**, 1332; Marvel, McElvain, *Org. Synth.* **coll. vol. I**, 170 (1941). Manuf of *o*- and *p*-isomers by catalytic chlorination of toluene: Di Bella, **US 3000975** (1959 to Heyden Newport Chem.). Absorption spectrum of *m*- and *p*-isomers: Baly, *J. Chem. Soc.* **99**, 1704 (1911); of *m*- and *o*-isomers: Purvis, *ibid.* 1704. Metabolism of *o*-isomer by bacteria: P. A. Vandenbergh *et al.*, *Appl. Environ. Microbiol.* **42**, 737 (1981); by rats: G. B. Quistad *et al.*, *J. Agric. Food Chem.* **31**, 1158 (1983). Brief review: S. Gelfand in *Kirk-Othmer Encyclopedia of Chemical Technology* **vol. 5** (Wiley-Interscience, New York, 3rd ed., 1979) pp 819-827.

m-**Chlorotoluene.** [108-41-8] 1-Chloro-3-methylbenzene. Liquid, bp 161.75°. d$_4^{18.7}$ 1.0760. mp −47.8°. n$_D^{20}$ 1.5218.
o-**Chlorotoluene.** [95-49-8] 1-Chloro-2-methylbenzene. Liquid, bp 158.97°. Vapor harmful. d$_4^{20}$ 1.0826. mp −35.59°. n$_D^{20}$ 1.5258. Volatile with steam. Slightly sol in water; freely sol in alcohol, benzene, chloroform, ether.
p-**Chlorotoluene.** [106-43-4] 1-Chloro-4-methylbenzene. Liquid, bp 162.4°. d$_4^{20}$ 1.0697. mp 7.5°. n$_D^{20}$ 1.5211. Slightly sol in water; sol in alcohol, benzene, chloroform, ether.
Caution: Potential symptoms of overexposure to *o*-chlorotoluene are irritation of eyes, skin, mucous membranes; dermatitis; drowsiness, incoordination, anesthesia; cough; liver and kidney injury. *See NIOSH Pocket Guide to Chemical Hazards* (DHHS/NIOSH 97-140, 1997) p 68.
USE: Solvent; dyestuff intermediate; in organic syntheses.

2176. Chlorotoluron. [15545-48-9] *N'*-(3-Chloro-4-methylphenyl)-*N*,*N*-dimethylurea; 3-(3-chloro-*p*-tolyl)-1,1-dimethylurea; Dicuran; Tolurex. C$_{10}$H$_{13}$ClN$_2$O; mol wt 212.68. C 56.47%, H 6.16%, Cl 16.67%, N 13.17%, O 7.52%. Photosynthesis inhibitor used to control grass weeds in cereals. Prepn: C. W. Todd, **US 2655445** (1952 to Du Pont); T. Mizuno *et al.*, *Synth. Commun.* **9**, 1675 (2000). Degradation in soil and identification of metabolites: D. Gross *et al.*, *Pestic. Biochem. Physiol.* **10**, 49 (1979). HPLC determn in technical products and formulations: W. Y. Yan *et al.*, *Pestic. Sci.* **49**, 400 (1997). Movement in sandy soil: C. Zander *et al.*, *J. Environ. Qual.* **28**, 1817 (1999). Field trial on grassy weeds in wheat: S. Singh, R. K. Malik *Indian J. Agron.* **39**, 23 (1994).

Crystals from toluene, mp 148.3°.
USE: Herbicide.

2177. Chlorotoxin. [163515-35-3] L-Methionyl-L-cysteinyl-L-methionyl-L-prolyl-L-cysteinyl-L-phenylalanyl-L-threonyl-L-threonyl-L-α-aspartyl-L-histidyl-L-glutaminyl-L-methionyl-L-alanyl-L-arginyl-L-lysyl-L-cysteinyl-L-α-aspartyl-L-cysteinyl-L-cysteinylglycylglycyl-L-lysylglycyl-L-arginylglycyl-L-lysyl-L-cysteinyl-L-tyrosylglycyl-L-prolyl-L-glutaminyl-L-cysteinyl-L-leucyl-L-cysteinyl-L-argininamide cyclic (2 → 19),(5 → 28),-(16 → 33),(20 → 35)-tetrakis(disulfide); ClTx; Ctx. Small neurotoxic peptide (36 amino acids) isolated from the venom of the scorpion, *Leiurus quinquestriatus* characterized by 4 disulfide linkages. Specific blocker of small conductance chloride channels. Isoln, purification, and characterization: J. A. DeBin *et al.*, *Am. J. Physiol.* **264**, C361 (1993). Solid phase synthesis: J. Najib *et al.*, *3rd Int. Symp. 1993 Innovation Perspect. Solid Phase Synth. Collect. Pap.* **1994**, 615. NMR and secondary structure: G. Lippens *et al.*, *ibid.* 583; and solution structure: *eidem*, *Biochemistry* **34**, 13 (1995). Identification of chloride channel in gliomas: N. Ullrich *et al.*, *Neuroscience* **83**, 1161 (1998); inhibition of glioma cell invasion: J. Deshane *et al.*, *J. Biol. Chem.* **278**, 4135 (2003). Use in cell topography: G. Tobasnick, A. S. G. Curtis, *Eur. Cell. Mater.* **2**, 49 (2001).

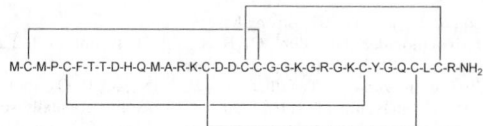

M-C-M-P-C-F-T-T-D-H-Q-M-A-R-K-C-D-D-C-C-G-G-K-G-R-G-K-C-Y-G-Q-C-L-C-R-NH$_2$

USE: Biochemical probe for voltage-gated chloride channels.

2178. Chlorotrianisene. [569-57-3] 1,1',1''-(1-Chloro-1-ethenyl-2-ylidene)tris[4-methoxybenzene]; chlorotris(*p*-methoxyphenyl)ethylene; tri-*p*-anisylchloroethylene; tris(*p*-methoxyphenyl)chloroethylene; Hormonisene; Merbentul; Tace. C$_{23}$H$_{21}$ClO$_3$; mol wt 380.87. C 72.53%, H 5.56%, Cl 9.31%, O 12.60%. Prepd by reacting tri-*p*-anisylethylene or tri-*p*-anisylethanol with Cl in an inert solvent: Basford and I.C.I., **GB 561508** (1944). Synthesis from *p*-(*p*-anisoyl)anisole: Shelton, Van Campen, Jr., **US 2430891** (1947 to Merrell).

Crystals from methanol, mp 114-116°. Softens at 108°. uv max (chloroform): 310 nm, (E$_{1cm}^{1\%}$ 423); min 278 nm. Practically insol in water. Soly in alcohol: 0.28 g/100 ml; in ether: 3.6 g/100 ml. Also sol in glacial acetic acid, acetone, chloroform, carbon tetrachloride, benzene, vegetable oils.

THERAP CAT: Estrogen.

2179. Chlorotriethylsilane. [994-30-9] Triethylchlorosilane; triethylsilyl chloride; TESCl; Et$_3$SiCl. C$_6$H$_{15}$ClSi; mol wt 150.72. C 47.81%, H 10.03%, Cl 23.52%, Si 18.63%. Silylating reagent used in organic synthesis to introduce the triethylsilyl (TES) protecting group. Prepn from triethylethoxysilane: A. Ladenburg, *Ann.* **164**, 300 (1872); from hexaethyldisiloxane: P. A. Di Giorgio *et al.*, *J. Am. Chem. Soc.* **68**, 1380 (1946); from β-chloroethyltriethylsilane: L. H. Sommer, *ibid.* **70**, 2869 (1948). Use in protection of hydroxyl groups: W. Oppolzer *et al.*, *Helv. Chim. Acta* **64**, 2002 (1981); W. R. Roush, S. Russo-Rodriguez, *J. Org. Chem.* **52**, 598 (1987). Additional synthetic applications: S. Danishefsky *et al.*, *J. Am. Chem. Soc.* **107**, 1285 (1985); Y. Fujii *et al.*, *J. Organomet. Chem.* **692**, 375 (2007). *Review:* E. Turos in *Encyclopedia of Reagents for Organic Synthesis* **2**, L. A. Paquette, Ed. (Wiley, New York, 1995) pp 1225-1227.

Liquid. bp 144-147°. d^{20} 0.8967. n$_D^{20}$ 1.4314. *Flammable. Corrosive.* Reacts violently with water. Flash point, closed cup: 86°F (30°C). Sol in most aprotic solvents.

USE: Reagent in synthetic organic chemistry.

2180. Chlorotrimethylsilane. [75-77-4] Trimethylchlorosilane; trimethylsilane chloride; trimethylsilicon chloride; trimethylsilyl chloride. C$_3$H$_9$ClSi; mol wt 108.64. C 33.17%, H 8.35%, Cl 32.63%, Si 25.85%. Silylating reagent and Lewis acid catalyst in synthetic organic chemistry. Prepn from trimethylsilane and chlorine: A. G. Taylor, B. V. De G. Walden, *J. Am. Chem. Soc.* **66**, 842 (1944); from silicon chloride: H. S. Booth, J. F. Suttle, *ibid.* **68**, 2658 (1946); from hexamethyldisiloxane and ammonium chloride: B. O. Pray *et al.*, *ibid.* **70**, 433 (1948). Crystal structure: J. Buschmann *et al.*, *Acta Crystallogr.* **C56**, 121 (2000). Thermodynamic studies: A. E. Beezer, C. T. Mortimer, *J. Chem. Soc. A* **1966**, 514. Synthetic applications: G. A. Olah *et al.*, *J. Org. Chem.* **44**, 4272 (1979); J. G. Lee *et al.*, *Tetrahedron Lett.* **31**, 6677 (1990); J.-M. Lin, B.-S. Liu, *Synth. Commun.* **27**, 739 (1997); J. Eras *et al.*, *J. Org. Chem.* **67**, 8631 (2002); in catalysis: P. Verma, S. Ray, *Indian J. Chem.* **29B**, 652 (1990); L.-W. Xu *et al.*, *Synth. Commun.* **37**, 3095 (2007). Derivitization and gas chromatography: J. Eras *et al.*, *J. Chromatogr. A* **1047**, 157 (2004).

Colorless liquid. Strong camphor-like odor. Fumes slightly in air. *Flammable. Corrosive. Reacts violently with water.* bp 57.3°. fp −57.7° (Booth, Suttle); also reported as fp −40° (Taylor, Walden); mp −97.15° (Buschmann). d$_4^{25}$ 0.846; d^{20} 0.8581. n$_D^{20}$ 1.3884. Flash point, closed cup: −0.4°F (−18°C). Heat of formation: −91.9±0.8 kcal/mol.

USE: Reagent to introduce the trimethylsilyl group in organic synthesis. Catalyst. In compd derivitization to increase volatility for analysis by gas chromatography. In prepn of anhydrous solns of hydrochloric acid.

2181. Chloroxine. [773-76-2] 5,7-Dichloro-8-quinolinol; 5,7-dichloro-8-hydroxyquinoline; Capitrol. C$_9$H$_5$Cl$_2$NO; mol wt 214.05. C 50.50%, H 2.35%, Cl 33.12%, N 6.54%, O 7.47%. Prepd by chlorinating 8-quinolinol: Hebebrand, *Ber.* **21**, 2977 (1888); F. J. Welcher, *Organic Analytical Reagents* **vol. I** (Van Nostrand, 1947) pp 328-329.

Crystals from alc, mp 179-180°. Soluble in benzene, acetone; slightly sol in cold alcohol, acetic acid; readily sol in sodium and potassium hydroxides and in acids, forming yellow solns.

USE: Analytical reagent.

THERAP CAT: Antiseborrheic.

2182. Chloroxylenol. [88-04-0] 4-Chloro-3,5-dimethylphenol; *p*-chloro-*m*-xylenol; 4-chloro-3,5-xylenol; parachlorometaxylenol; 2-chloro-*m*-xylenol; 2-chloro-5-hydroxy-*m*-xylene; 2-chloro-5-hydroxy-1,3-dimethylbenzene; Benzytol; Dettol. C$_8$H$_9$ClO; mol wt 156.61. C 61.35%, H 5.79%, Cl 22.64%, O 10.22%. Prepd by treating 3,5-dimethylphenol with Cl$_2$ or SO$_2$Cl$_2$: Lesser, Gad, *Ber.* **56**, 974, 976 (1923); von Auwers *et al.*, *Chem. Zentralbl.* **1924**, II, 2267; *C.A.* **19**, 2339 (1925); Gladden, Cocker, US 2350677 (1944).

Crystals from benzene, mp 115.5°. Phenolic odor. Volatile with steam. bp 246°. One gram dissolves in 3 liters of water at 20°. More stable in hot water. Soluble in 1 part of 95% alcohol, ether, benzene, terpenes, fixed oils, in solns of alkali hydroxides.

USE: Antiseptic and germicide; for mildew prevention. Claimed to be about 60 times as potent as phenol.

THERAP CAT: Antibacterial; antiseptic (topical and urinary).

THERAP CAT (VET): Antiseptic (topical).

2183. Chlorozotocin. [54749-90-5] 2-[[[(2-Chloroethyl)nitrosoamino]carbonyl]amino]-2-deoxy-D-glucose; 2-[3-(2-chloroethyl)-3-nitrosoureido]-2-deoxy-D-glucopyranose; 1-(2-chloroethyl)-1-nitroso-3-(D-glucos-2-yl)urea; DCNU; NSC-178248. C$_9$H$_{16}$ClN$_3$O$_7$; mol wt 313.69. C 34.46%, H 5.14%, Cl 11.30%, N 13.40%, O 35.70%. Chloroethylnitrosourea derivative with antitumor activity. Similar to carmustine, lomustine, nimustine, ranimustine, *q.q.v;* 2-chloroethyl analog of streptozotocin, *q.v.* Synthesis: H. D. Burns *et al.*, *Org. Prep. Proced. Int.* **6**, 259 (1974); T. P. Johnston *et al.*, *J. Med. Chem.* **18**, 104 (1975). Pharmacology: T. Anderson *et al.*, *Cancer Res.* **35**, 761 (1975); P. S. Schein *et al.*, *Cancer Treat. Rep.* **60**, 801 (1976). Decomposition in aqueous media: J. A. Montgomery *et al.*, *J. Med. Chem.* **18**, 568 (1975).

Ivory colored crystals, mp 147-148° (dec with the evolution of gas), (Burns, Heindel). Also reported as mp 140-141° (dec), (Johnston). Sol in water.

Caution: This substance is reasonably anticipated to be a human carcinogen: *Report on Carcinogens, Twelfth Edition* (PB2011-111646, 2011) p 328.

THERAP CAT: Antineoplastic.

2184. Chlorphenesin. [104-29-0] 3-(4-Chlorophenoxy)-1,2-propanediol; *p*-chlorophenyl α-glyceryl ether; Adermykon; Mycil. C$_9$H$_{11}$ClO$_3$; mol wt 202.63. C 53.35%, H 5.47%, Cl 17.49%, O 23.69%. Prepd by condensing equimol amts of *p*-chlorophenol and glycidol in the presence of a tertiary amine or a quaternary ammonium salt as catalyst: Bradley, Forrest, **GB 628497** (1949 to British Drug Houses).

Crystals, mp 77-79°. Soly in water is less than 1%, may be increased by the addition of solubilizers such as ethylurea or propylene glycol: Berger et al., **US 2468423** (1949 to British Drug Houses).

THERAP CAT: Antifungal (topical).

2185.　Chlorphenesin Carbamate. [886-74-8] 3-(4-Chlorophenoxy)-1,2-propanediol-1-carbamate; carbamic acid 3-(p-chlorophenoxy)-2-hydroxypropyl ester; 3-(p-chlorophenoxy)-2-hydroxypropyl carbamate; 1,2-propanediol-3-(p-chlorophenoxy)-1-carbamate; Maolate; Rinlaxer. $C_{10}H_{12}ClNO_4$; mol wt 245.66. C 48.89%, H 4.92%, Cl 14.43%, N 5.70%, O 26.05%. Prepn: Collins, Matthews, **US 3161567**; Parker, **US 3214336** (1964, 1965 both to Upjohn). Clinical evaluation of analgesic activity: L. J. Cass, W. S. Frederick, *J. New Drugs* **2**, 366 (1962).

Crystals from benzene + toluene, mp 89-91°. Readily sol in 95% ethanol, acetone, ethyl acetate; fairly readily sol in dioxane. Almost insol in cold water, benzene, cyclohexane. LD_{50} orally in rats: 748 mg/kg; i.v. in mice: 239 mg/kg (Cass, Frederick).

THERAP CAT: Muscle relaxant (skeletal).

THERAP CAT (VET): Muscle relaxant (skeletal).

2186.　Chlorpheniramine. [132-22-9] γ-(4-Chlorophenyl)-N,N-dimethyl-2-pyridinepropanamine; 2-[p-chloro-α-(2-dimethylaminoethyl)benzyl]pyridine; 1-(p-chlorophenyl)-1-(2-pyridyl)-3-dimethylaminopropane; 1-(p-chlorophenyl)-1-(2-pyridyl)-3-N,N-dimethylpropylamine; 3-(p-chlorophenyl)-3-(2-pyridyl)-N,N-dimethylpropylamine; chlorprophenpyridamine; chlorphenamine; Haynon. $C_{16}H_{19}ClN_2$; mol wt 274.79. C 69.94%, H 6.97%, Cl 12.90%, N 10.19%. Synthesis: Sperber et al., **US 2567245**, **US 2676964** (1951, 1954, both to Schering). Prepn of d-form: L. A. Walter, **US 3061517** (1962 to Schering). Solutions: Foley, Ilavsky, **US 2766174** (1956 to Schering). Pharmacology: F. E. Roth, W. M. Govier, *J. Pharmacol. Exp. Ther.* **124**, 347 (1958). Toxicity data: R. B. Smith et al., *Toxicol. Appl. Pharmacol.* **28**, 240 (1974). Comprehensive description: C. G. Eckhart, T. McCorkle, *Anal. Profiles Drug Subs.* **7**, 43-80 (1978).

Oily liquid, $bp_{1.0}$ 142°.

Maleate. [113-92-8] Allergisan; Chlor-Trimeton; Piriton; Teldrin. $C_{16}H_{19}ClN_2 \cdot C_4H_4O_4$; mol wt 390.86. Crystals, mp 130-135°. uv max (water): 261 nm (ε 5760). Soly in mg/ml at 25°: ethanol 330; chloroform 240; water 160; methanol 130. Slightly sol in benzene, ether. pH of a 2% aq soln about 5. LD_{50} orally in mice: 162 mg/kg (Smith).

d-Form. [25523-97-1] Dexchlorpheniramine; d-chlorpheniramine. Oily liquid. $[\alpha]_D^{25}$ +49.8° (c = 1 in DMF).

d-Form maleate. [2438-32-6] Phenamin; Polamin; Polaramine; Polaronil. Crystals from ethyl acetate, mp 113-115°. $[\alpha]_D^{25}$ +44.3° (c = 1 in dimethylformamide). pH of 1% soln 4-5.

THERAP CAT: Antihistaminic.

THERAP CAT (VET): Antihistaminic.

2187.　Chlorphenoxamine. [77-38-3] 2-[1-(4-Chlorophenyl)-1-phenylethoxy]-N,N-dimethylethanamine; 2-[(p-chloro-α-methyl-α-phenylbenzyl)oxy]-N,N-dimethylethylamine; β-dimethylaminoethyl (p-chloro-α-methylbenzhydryl) ether; [1-(p-chlorophenyl)-1-phenyl]ethyl (β-dimethylaminoethyl) ether. $C_{18}H_{22}ClNO$; mol wt 303.83. C 71.16%, H 7.30%, Cl 11.67%, N 4.61%, O 5.27%. Prepn: Arnold et al., **US 2785202** (1957 to Asta-Werke). Synthesis: G.

Cahiez et al., *Tetrahedron Lett.* **29**, 3659 (1988). Pharmacology: *eidem, Arzneim.-Forsch.* **4**, 189 (1954); Brock et al., *ibid.* 262. Toxicity studies: Kerley et al., *Toxicol. Appl. Pharmacol.* **4**, 638 (1962). Metabolism: C. Koppel et al., *Arzneim.-Forsch.* **37**, 1062 (1987). GC-MS determn in urine: H. Maurer, K. Pfleger, *J. Chromatogr.* **428**, 43 (1988).

$bp_{0.05}$ 150-155°.

Hydrochloride. [562-09-4] Clorevan; Systral. $C_{18}H_{22}ClNO \cdot HCl$; mol wt 340.29. Needles, mp 128°. Soluble in water; aq solns are stable.

THERAP CAT: Antihistaminic.

2188.　Chlorphentermine. [461-78-9] 4-Chloro-α,α-dimethylbenzeneethanamine; 4-chloro-α,α-dimethylphenethylamine; α,α-dimethyl-p-chlorophenethylamine; 1-(p-chlorophenyl)-2-methyl-2-propylamine; 1-(p-chlorophenyl)-2-methyl-2-aminopropane; clorfentermina; Lucofen; Teramine. $C_{10}H_{14}ClN$; mol wt 183.68. C 65.39%, H 7.68%, Cl 19.30%, N 7.63%. Prepn: Bachman et al., *J. Am. Chem. Soc.* **76**, 3972 (1954); Ferrari, *Farmaco Ed. Sci.* **15**, 337 (1960); **FR M1299** and **FR 1296132** (1962 to Simes), *C.A.* **58**, 4467e, 3352g (1963); **GB 906331** (1962 to Kefalas), *C.A.* **58**, 6654c (1963). Pharmacology: Jun et al., *Can. J. Pharm. Sci.* **4**, 27 (1969); Ciborska et al., *Acta Pol. Pharm.* **26**, 595 (1969); Moeller-Nielsen, Dubnick, *Proc. Int. Symp. Amphetamines Relat. Compounds 1969*, E. Costa, Ed. (Raven Press, New York, 1970) pp 63-73.

Liquid. bp_2 100-102°.

Hydrochloride. [151-06-4] Pre-Sate. $C_{10}H_{14}ClN \cdot HCl$; mol wt 220.14. Crystals from alcohol + ether, mp 234°. Soly in water: >20%. pH of 1% aq soln ~5.5. LD_{50} in mice (mg/kg): 270 orally; 267 s.c. (Ciborska).

Note: This is a controlled substance (stimulant): **21 CFR, 1308.13.**

THERAP CAT: Anorexic.

2189.　Chlorproethazine. [84-01-5] 2-Chloro-N,N-diethyl-10H-phenothiazine-10-propanamine; 2-chloro-10-(3-diethylaminopropyl)phenothiazine; 3-chloro-10-(3-diethylaminopropyl)phenothiazine; RP-4909. $C_{19}H_{23}ClN_2S$; mol wt 346.92. C 65.78%, H 6.68%, Cl 10.22%, N 8.08%, S 9.24%. Prepn: Buisson et al., **US 2769002** (1956 to Rhône-Poulenc).

Hydrochloride. [4611-02-3] Neuriplege. $C_{19}H_{23}ClN_2S \cdot HCl$; mol wt 383.38. Crystals, mp 178°. (Free base bp_1 225-240°). Sensitive to light. Soly in water about 1.0 g/60 ml, ethanol about 1.0 g/300 ml, chloroform 1.0 g/5 ml. Practically insol in acetone, ether, benzene. pH of 1% aq soln 4.8.

THERAP CAT: Topical analgesic.

2190.　Chlorproguanil. [537-21-3] N-(3,4-Dichlorophenyl)-N'-(1-methylethyl)imidodicarbonimidic diamide; 1-(3,4-dichlorophenyl)-5-isopropylbiguanide; N^1-3,4-dichlorophenyl-N^5-isopropylbiguanide; N^1-3,4-dichlorophenyl-N^5-isopropyldiguanide;

M–5943. $C_{11}H_{15}Cl_2N_5$; mol wt 288.18. C 45.85%, H 5.25%, Cl 24.60%, N 24.30%. Method of prepn: Crowther *et al., J. Chem. Soc.* **1951**, 1780; Curd *et al.,* **US 2544827** (1951); Crowther *et al.,* **GB 667116** (1952) (both to ICI). Clinical trial with dapsone, *q.v.,* in drug-resistant malaria: T. Mutabingwa *et al., Lancet* **358**, 1218 (2001).

Hydrochloride. [15537-76-5] Lapudrine. $C_{11}H_{15}Cl_2N_5$.HCl; mol wt 324.63. Crystals, mp 246-247°. Soly in water: 1 g/100 ml. Solns may be boiled without dec.

THERAP CAT: Antimalarial.

2191. Chlorpromazine. [50-53-3] 2-Chloro-*N,N*-dimethyl-10*H*-phenothiazine-10-propanamine; 2-chloro-10-(3-dimethylaminopropyl)phenothiazine; 3-chloro-10-(3-dimethylaminopropyl)phenothiazine; *N*-(3-dimethylaminopropyl)-3-chlorophenothiazine; 2601-A; HL-5746; RP-4560; SKF-2601-A. $C_{17}H_{19}ClN_2S$; mol wt 318.86. C 64.04%, H 6.01%, Cl 11.12%, N 8.79%, S 10.05%. Prepn: Charpentier *et al., Compt. Rend.* **235**, 59 (1952); Charpentier, **US 2645640** (1953 to Rhône-Poulenc). Effects of neuroleptics on dopamine receptors: N.-E. Anden *et al., Eur. J. Pharmacol.* **11**, 303 (1970). Toxicity study: E. I. Goldenthal, *Toxicol. Appl. Pharmacol.* **18**, 185 (1971). Review of analytical methods for determn in pharmaceutical prepns: L. F. S. Chagonda, J. S. Millership, *J. Pharm. Biomed. Anal.* **7**, 271-278 (1989). Brief historical review: G. Curzon, *Trends Pharmacol. Sci.* **11**, 61-63 (1990).

White, crystalline solid. Amine odor. Alkaline reaction. $bp_{0.8}$ 200-205°. *Protect from light.* Freely sol in alc, benzene, chloroform, ether, dilute mineral acids. Practically insol in water, in dilute alkali hydroxides.

Hydrochloride. [69-09-0] Ampliactil; Amplictil; Chloractil; Chlorazin; Contomin; Fenactil; Hibernal; Largactil; Plegomazin; Promacid; Promactil; Propaphenin; Prozin; Taroctyl; Thorazine; Wintermin. $C_{17}H_{19}ClN_2S$.HCl; mol wt 355.32. Crystals, dec 179-180° (capillary); 194-196° (microblock). uv curve: Neuhoff, Auterhoff, *Arch. Pharm.* **288**, 400 (1955). *Protect from light.* pH of 5% aq soln 4.0-5.5. One gram dissolves in 2.5 ml water. Sol in methanol, ethanol, chloroform. Practically insol in ether, benzene. Slightly acid to litmus. LD_{50} orally in rats: 225 mg/kg (Goldenthal).

THERAP CAT: Antiemetic; antipsychotic.

THERAP CAT (VET): Antiemetic; tranquilizer.

2192. Chlorpropamide. [94-20-2] 4-Chloro-*N*-[(propylamino)carbonyl]benzenesulfonamide; 1-(*p*-chlorophenylsulfonyl)-3-propylurea; 1-(*p*-chlorobenzenesulfonyl)-3-propylurea; *N'*-(*p*-chlorobenzenesulfonyl)urea; P-607; Diabinese; Insogen. C_{10}-$H_{13}ClN_2O_3S$; mol wt 276.74. C 43.40%, H 4.74%, Cl 12.81%, N 10.12%, O 17.34%, S 11.58%. Oral sulfonylurea hypoglycemic agent. Prepn: F. J. Marshall, M. V. Sigal, Jr., *J. Org. Chem.* **23**, 927 (1958); **GB 853555**; W. M. McLamore, **US 3349124** (1960, 1967 both to Pfizer); V. J. Bauer *et al., J. Org. Chem.* **31**, 3440 (1960). Pharmacology and metabolism: Khurana *et al., Indian J. Med. Res.* **55**, 1084 (1967); Brotherton *et al., Clin. Pharmacol. Ther.* **10**, 505 (1969); Madsen *et al., Eur. J. Pharmacol.* **13**, 374 (1971). Toxicity study: E. I. Goldenthal, *Toxicol. Appl. Pharmacol.* **18**, 185 (1971). HPLC and GC-MS determn in horse plasma as a screen for horse racing: H. C. Chau *et al, J. Chromatogr. B* **712**, 243 (1998).

Crystals from dil ethanol, mp 127-129°. uv max (0.01*N* HCl): 232.5 nm (ε 16500). Soly in water at pH 6: 2.2 mg/ml. Practically insol at pH 7.3. Sol in alc; moderately sol in chloroform; sparingly sol in ether, benzene. LD_{50} i.p. in rats: 580 mg/kg (Goldenthal).

THERAP CAT: Antidiabetic.

2193. Chlorpropham. [101-21-3] *N*-(3-Chlorophenyl)carbamic acid 1-methylethyl ester; *m*-chlorocarbanilic acid isopropyl ester; isopropyl-*m*-chlorocarbanilate; isopropyl *N*-(3-chlorophenyl)-carbamate; chloro-IPC; chloropropham; CIPC; Chlor-IFC; Furloe; Sprout-Nip. $C_{10}H_{12}ClNO_2$; mol wt 213.66. C 56.22%, H 5.66%, N 6.56%, O 14.98%. Prepn: E. D. Witman, **US 2695225**; Strain, **US 2734911** (1954, 1956 both to Columbia-Southern Chem.); Brockway, **US 2806051** (1957 to B. F. Goodrich). Toxicology: E. M. Boyd, E. Carsky, *Arch. Environ. Health* **19**, 621 (1969).

Solid, mp 40.7-41.1°. bp_2 149°. n_D^{20} 1.5388. Commercial product is a liquid. Slightly sol in water; miscible with most oils and organic solvents. LD_{50} orally in rats: 1.2 g/kg (Boyd, Carsky).

USE: Herbicide; plant growth regulator.

2194. Chlorprothixene. [113-59-7] (3*Z*)-3-(2-Chloro-9*H*-thioxanthen-9-ylidene)-*N,N*-dimethyl-1-propanamine; (*Z*)-2-chloro-*N,N*-dimethylthioxanthene-$\Delta^{9,\gamma}$-propylamine; *cis*-2-chloro-*N,N*-dimethyl-3-thioxanthen-9-ylidenepropylamine; α-2-chloro-10-(3-dimethylaminopropylidene)thiaxanthene; N-714; Taractan; Tarasan; Truxal; Truxaletten. $C_{18}H_{18}ClNS$; mol wt 315.86. C 68.45%, H 5.74%, Cl 11.22%, N 4.43%, S 10.15%. Prepn: **GB 829763**; E. L. Engelhardt, **US 3046283** (1960, 1962 both to Merck & Co.). Structure-activity study: G. E. Bonvicino *et al., J. Org. Chem.* **26**, 2383 (1961). Comprehensive description: B. C. Rudy, B. Z. Senkowski, *Anal. Profiles Drug Subs.* **2**, 63-84 (1973). HPLC determn in serum: M. Bagli *et al., J. Chromatogr. B* **657**, 141 (1994). Clinical pharmacokinetics: *idem et al., Arzneim.-Forsch.* **46**, 247 (1996).

Pale yellow crystals, mp 97-98°. Practically insol in water; sol in alcohol, ether, chloroform. Incompatible with acids, alkalies, phenobarbital, thiopental sodium, mepazine.

Hydrochloride. [6469-93-8] $C_{18}H_{18}ClNS$.HCl. Crystals, mp 221°. Freely sol in water at pH 6 to 6.5.

THERAP CAT: Antipsychotic.

2195. Chlorpyrifos. [2921-88-2] Phosphorothioic acid *O,O*-diethyl *O*-(3,5,6-trichloro-2-pyridinyl) ester; *O,O*-diethyl *O*-3,5,6-trichloro-2-pyridyl phosphorothioate; chlorpyrifos-ethyl; Dowco 179; ENT-27311; Affront; Dursban; Empire; Lock-On; Lorsban; Pyrinex. $C_9H_{11}Cl_3NO_3PS$; mol wt 350.57. C 30.84%, H 3.16%, Cl 30.34%, N 4.00%, O 13.69%, P 8.84%, S 9.15%. Organophosphate insecticide. Prepn: R. H. Rigterink, **FR 1360901**; *idem*, **US 3244586** (1964, 1966 both to Dow); R. H. Rigterink, E. E. Kenaga, *J. Agric. Food Chem.* **14**, 394 (1966). Activity: E. E. Kenaga *et al., J. Econ. Entomol.* **58**, 1043 (1965). Metabolism: G. N. Smith *et al., J. Agric. Food Chem.* **15**, 132 (1967); W. H. Gutenmann *et al., ibid.*

16, 45 (1968). Toxicity study: Schafer, *Toxicol. Appl. Pharmacol.* **21**, 315 (1972). LC-MS/MS determn in human cord blood: H.-T. Liao *et al.*, *J. Chromatogr. B* **879**, 1961 (2011). Review of environmental fate: K. D. Racke, *Rev. Environ. Contam. Toxicol.* **131**, 1-154 (1993); of toxicology and human exposure: *Toxicological Profile for Chlorpyrifos* (PB98-103088, 1997) 217 pp.

White granular crystals, mp 41-42°. Vapor pressure at 25°: 1.87 × 10⁻⁵ mm Hg. Soly at 25°: water 2 ppm; isooctane 79% w/w; methanol 43% w/w. Soly (g/100g): acetone 650, benzene 790, carbon disulfide 590, carbon tetrachloride 310, chloroform 630, diethyl ether 510, ethanol 63, ethyl acetate >200, isooctane 79, kerosene 60, methanol 45, methylene chloride 400, propylene glycol 4, toluene 150, 1,1,1-trichloroethane 400, triethylene glycol 5, xylene 400. Readily sol in other org solvents. uv max: 208, 230, 290 nm. LD_{50} orally in rats: 145 mg/kg (Schafer).

O,O-**Dimethyl analog.** [5598-13-0] Chlorpyrifos-methyl; Dowco 214; ENT-27520; Reldan. $C_7H_7Cl_3NO_3PS$; mol wt 322.52. Crystals, mp 45.5-46.5°. Vapor pressure at 25°: 4.22 × 10⁻⁵ mm Hg. Soly in water at 25°: 5 mg/l.

Caution: Potential symptoms of overexposure are wheezing, laryngeal spasms, salivation; bluish lips and skin; miosis, blurred vision; nausea, vomiting, abdominal cramps, diarrhea. *See NIOSH Pocket Guide to Chemical Hazards* (DHHS/NIOSH 97-140, 1997) p 70.

USE: Insecticide; acaricide.

THERAP CAT (VET): Ectoparasiticide.

2196. Chlorquinaldol. [72-80-0] 5,7-Dichloro-2-methyl-8-quinolinol; 5,7-dichloro-8-quinaldinol; 5,7-dichloro-8-hydroxyquinaldine; 5,7-dichloro-2-methyl-8-hydroxyquinoline; hydroxydichloroquinaldine; chloroquinaldol; Gynotherax; Sterosan; Steroxin. $C_{10}H_7Cl_2NO$; mol wt 228.07. C 52.66%, H 3.09%, Cl 31.09%, N 6.14%, O 7.01%. Prepd by chlorination of 8-hydroxyquinaldine with or without formic acid as solvent: Senn, **US 2411670** (1946 to Geigy).

Yellow needles from alc, mp 114-115° (slight decompn). Medicinal odor. uv max (ethanol): 316 nm ($A_{1cm}^{1\%}$ 170); min 280 nm. Practically insol in water. Soly (25°) in ethanol 1.0 g/100 ml of soln; chloroform 5.0 g; acetone 4.0 g; ether 3.0 g; 0.1N NaOH 1.4 g. Also sol in benzene, glacial acetic acid.

THERAP CAT: Antibacterial.

THERAP CAT (VET): Antibacterial; antifungal.

2197. Chlorsulfuron. [64902-72-3] 2-Chloro-*N*-[[(4-methoxy-6-methyl-1,3,5-triazin-2-yl)amino]carbonyl]benzenesulfonamide; DPX-4189; Glean; Telar. $C_{12}H_{12}ClN_5O_4S$; mol wt 357.77. C 40.29%, H 3.38%, Cl 9.91%, N 19.58%, O 17.89%, S 8.96%. Selective pre- and post-emergence herbicide. Prepn: G. Levitt, **DE 2715786**; *idem*, **US 4127405** (1977, 1978 both to du Pont). Activity: G. Levitt *et al.*, *J. Agric. Food Chem.* **29**, 416 (1981). Mode of action: T. B. Ray, *Pestic. Biochem. Physiol.* **17**, 10 (1982).

Crystals from ether, mp 174-178°. Soly in water: 125 ppm. Moderately sol in methylene chloride; less sol in acetone, acetonitrile. Low soly in hydrocarbon solvents. LD_{50} in male, female rats (mg/kg): 5545, 6293 orally (Levitt).

USE: Herbicide.

2198. Chlortetracycline. [57-62-5] (4S,4aS,5aS,6S,12aS)-7-Chloro-4-dimethylamino-1,4,4a,5,5a,6,11,12a-octahydro-3,6,10,-12,12a-pentahydroxy-6-methyl-1,11-dioxo-2-naphthacenecarboxamide; 7-chlorotetracycline; Acronize; Aureocina; Aureomycin; Biomitsin; Centraureo; Chrysomykine; Orospray. $C_{22}H_{23}ClN_2O_8$; mol wt 478.88. C 55.18%, H 4.84%, Cl 7.40%, N 5.85%, O 26.73%. Antibiotic substance isolated from the substrate of *Streptomyces aureofaciens:* Duggan, *Ann. N.Y. Acad. Sci.* **51**, 177 (1948); **US 2482055** (1949 to Am. Cyanamid); Broschard *et al.*, *Science* **109**, 199 (1949). Structure: Stephens *et al.*, *J. Am. Chem. Soc.* **74**, 4976 (1952); *eidem, ibid.* **76**, 3568 (1954). Crystal structure: Donohue *et al., ibid.* **85**, 851 (1963). Absolute configuration: Dobrynin *et al., Tetrahedron Lett.* **1962**, 901. Purification: Winterbottom *et al.*, **US 2899422** (1959 to Am. Cyanamid). Improved process: Miller *et al.*, **US 2987449**; Goodman, **US 3050446** (1961, 1962 both to Am. Cyanamid). Toxicity: E. I. Goldenthal, *Toxicol. Appl. Pharmacol.* **18**, 185 (1971). Comprehensive description: G. Schwartzman *et al.*, *Anal. Profiles Drug Subs.* **8**, 101-137 (1979).

Golden-yellow crystals, mp 168-169°. $[\alpha]_D^{23}$ −275.0° (methanol). uv max (0.1N HCl): 230, 262.5, 367.5 nm; (0.1N NaOH): 255, 285, 345 nm. Soly in water: 0.5-0.6 mg/ml. Very sol in aq solns above pH 8.5. Freely sol in the Cellosolves, dioxane, and Carbitol. Slightly sol in methanol, ethanol, butanol, acetone, ethyl acetate, benzene. Practically insol in ether, petr ether.

Hydrochloride. [64-72-2] Aureociclina; Isphamycin. $C_{22}H_{23}ClN_2O_8$·HCl; mol wt 515.34. Bitter, yellow rhomboid crystals. Dec above 210°. $[\alpha]_D^{23}$ −240°. Stable in air; slowly affected by light. Soly at about 28° (mg/ml): water 8.6; methanol 17.4; ethanol 1.7. *See:* Weiss *et al.*, *Antibiot. Chemother.* **7**, 374 (1957). Sol in solns of alkali hydroxides and carbonates. Practically insol in acetone, ether, chloroform, dioxane. pH of satd aq soln 2.8-2.9. LD_{50} orally in rats: 10300 mg/kg (Goldenthal).

THERAP CAT: Antibacterial; antiamebic.

THERAP CAT (VET): Antimicrobial.

2199. Chlorthal-dimethyl. [1861-32-1] 2,3,5,6-Tetrachloro-1,4-benzenedicarboxylic acid 1,4-dimethyl ester; 2,3,5,6-tetrachloroterephthalic acid dimethyl ester; dimethyl tetrachloroterephthalate; DCPA; Dacthal. $C_{10}H_6Cl_4O_4$; mol wt 331.95. C 36.18%, H 1.82%, Cl 42.72%, O 19.28%. Prepn: Lindemann, **US 2923634** (1960 to Diamond Alkali). Toxicity data: G. W. Bailey, J. L. White, *Residue Rev.* **10**, 97 (1965). GC-MS determn in water samples: J. E. Picker *et al.*, *Bull. Environ. Contam. Toxicol.* **21**, 612 (1979); of metabolites in drinking water: R. A. Carpenter *et al.*, *Anal. Chem.* **69**, 3314 (1997).

Crystals from methanol, mp 155-156°. Solubility: <5% in water; >5% in acetone, cyclohexanone, xylene. LD_{50} orally in rats: >3000 mg/kg (Bailey, White).

USE: Pre-emergent herbicide.

2200. Chlorthalidone. [77-36-1] 2-Chloro-5-(2,3-dihydro-1-hydroxy-3-oxo-1*H*-isoindol-1-yl)benzenesulfonamide; 2-chloro-5-

(1-hydroxy-3-oxo-1-isoindolinyl)benzenesulfonamide; 3-hydroxy-3-(4-chloro-3-sulfamylphenyl)phthalimidine; 2-chloro-5-(3-hydroxy-1-oxoisoindolin-3-yl)benzenesulfonamide; 1-oxo-3-(3-sulfamyl-4-chlorophenyl)-3-hydroxyisoindoline; 3-(4'-chloro-3'-sulfamoylphenyl)-3-hydroxyphthalimidine; 1-keto-3-(3'-sulfamyl-4'-chlorophenyl)-3-hydroxyisoindoline; chlorphthalidolone; phthalamudine; phthalamodine; G-33182; Higroton; Hydro-long; Hygroton; Thalitone. $C_{14}H_{11}ClN_2O_4S$; mol wt 338.76. C 49.64%, H 3.27%, Cl 10.46%, N 8.27%, O 18.89%, S 9.46%. Prepn: Graf et al., Helv. Chim. Acta 42, 1085 (1959); US 3055904 (1962 to Geigy). Activity and side effects: Holtmeier et al., Med. Welt 1967, 1384; Zsoter et al., J. Pharmacol. Exp. Ther. 180, 723 (1972). Metabolism: Beisenherz et al., Arch. Int. Pharmacodyn. Ther. 161, 76 (1966). Comprehensive description: J. M. Singer et al., Anal. Profiles Drug Subs. 14, 1-36 (1985).

Crystals from 50% acetic acid, dec 224-226°. mp range may extend from 218 to 264° on slow heating. Can form a monohydrate. uv max (methanol): <220 nm. Soly in water: 12 mg/100 ml (20°); 27 mg/100 ml (37°); in 0.1N Na_2CO_3: 577 mg/100 ml (20°); 990 mg/100 ml (37°). More sol in aq solns of NaOH. Soluble in warm ethanol, methanol. Practically insol in ether, chloroform.

THERAP CAT: Diuretic; antihypertensive.

THERAP CAT (VET): Diuretic.

2201. Chlorzoxazone. [95-25-0] 5-Chloro-2(3H)-benzoxazolone; 5-chloro-2-benzoxazolol; 5-chloro-2-hydroxybenzoxazole; 2-hydroxy-5-chlorobenzoxazole; 5-chlorbenzoxazolin-2-one; 5-chlorobenzoxazolidone; Paraflex; Biomioran; Solaxin. $C_7H_4ClNO_2$; mol wt 169.56. C 49.59%, H 2.38%, Cl 20.91%, N 8.26%, O 18.87%. Prepn from 2-amino-5-chlorobenzoxazole: Marsh, US 2895877 (1959 to McNeil Labs.). LC determn in plasma: I. L. Honigsberg et al., J. Pharm. Sci. 68, 253 (1979). Fluorometric determn in tablets and biological fluids: J. T. Stewart, C. W. Chan, ibid. 910. Clinical studies: R. Herman, Curr. Ther. Res. 9, 537 (1967); J. J. Scheiner, ibid. 19, 51 (1976). Metabolism study: R. Twele, G. Spiteller, Arzneim.-Forsch. 32, 759 (1982). Toxicity: G. Hofrichter et al., Arzneim.-Forsch. 17, 242 (1967). Review of clinical studies: J. K. Elenbaas, Am. J. Hosp. Pharm. 37, 1313 (1980). Comprehensive description: J. T. Stewart, C. A. Janicki, Anal. Profiles Drug Subs. 16, 119-144 (1987).

Crystals from acetone, mp 191-191.5°. Sol in solutions of alkali hydroxides, ammonia; sparingly sol in methanol, ethanol, isopropanol; slightly sol in water. LD_{50} in mice (mg/kg): 3650 orally, 380 i.p. (suspensions); 440 orally, 183 i.p. (solns of Na salt) (Hofrichter).

THERAP CAT: Muscle relaxant (skeletal).

2202. Cholane. [548-98-1] $C_{24}H_{42}$; mol wt 330.60. C 87.19%, H 12.81%. Prepn from bisnorcholyl methyl ketone: Wieland et al., Z. Physiol. Chem. 161, 80, 109 (1926); from potassium cholanate: Kazuno et al., Proc. Jpn. Acad. 28, 416 (1952).

Stout prisms from alc, mp 90°; $bp_{0.001}$ 190°. Sparingly sol in methanol.

2203. Cholanic Acid. [25312-65-6] Cholan-24-oic acid; ursocholanic acid; 17β-(1-methyl-3-carboxypropyl)etiocholane. C_{24}-$H_{40}O_2$; mol wt 360.58. C 79.94%, H 11.18%, O 8.87%. Steroidal acid probably formed by the dehydration and hydrogenation of certain bile acids commonly found in animals. Considered to be a chemical trademark certifying the prehistoric presence of some type of animal. See: Seifert, Pure Appl. Chem. 34, 633 (1973). This status as a natural product of exclusive animal origins now questioned by its isolation from the embryo of the jequirity bean, Abrus precatorius, Leguminosae: Mandava et al., Steroids 23, 357 (1974). Prepn from coprostane, cholic acid, lithocholic acid, desoxy- and chenodesoxycholic acid: Wieland, Weil, Z. Physiol. Chem. 80, 287 (1912); Wieland, Boersch, ibid. 106, 193 (1919); Windaus, Neukirchen, Ber. 52, 1915 (1919); Wieland, Vocke, Z. Physiol. Chem. 191, 69 (1930). From 3-ketocholanic acid diethyl thioacetal by hydrogenolysis: Bernstein, Dorfman, US 2440660 (1948 to Am. Cyanamid). 5β-Cholanic acid differs from the thermodynamically more stable 5α-isomer, by being cis at the A/B steroid ring function rather than trans.

Needles from alc, mp 163-164°. $[\alpha]_D^{20}$ +21.7° (chloroform). Sol in chloroform, alcohol, acetic acid. Forms a molecular compd with allocholanic acid, mp 163.5°.

Methyl ester. $C_{25}H_{42}O_2$. mp 86-87°.

Ethyl ester. $C_{26}H_{44}O_2$. Needles from 80% alc, mp 93-94°; bp_{12} 273°. $[\alpha]_D^{19}$ +21° (chloroform).

Note: The name cholanic acid was formerly applied to desoxybilianic acid, $C_{24}H_{36}O_7$.

USE: In chemotaxonomical classification.

2204. Cholanthrene. [479-23-2] 1,2-Dihydrobenz[j]aceanthrylene. $C_{20}H_{14}$; mol wt 254.33. C 94.45%, H 5.55%. Synthesis from 1-β-naphthylhydrindene: Cook et al., J. Chem. Soc. 1935, 667; from 4-indanyl-MgBr and 1-naphthoyl chloride: Fieser, Seligman, J. Am. Chem. Soc. 57, 2174 (1935).

Faintly yellow plates from benzene + ether, mp 173°. Sublimes 210-215° at 0.2 mm Hg. Absorption max: Fieser, Hershberg, J. Am. Chem. Soc. 60, 940 (1938). Sol in benzene, xylene, toluene. Slightly sol in methanol. Insol in water. May be solubilized by aq solns of Na desoxycholate.

Molecular complex with 2,4,7-trinitrofluorenone. Olive green crystals, mp 245-246°: Orchin, Woolfolk, J. Am. Chem. Soc. 68, 1727 (1946).

2205. Cholecystokinin. [9011-97-6] Pancreozymin; cholecystokinin-pancreozymin; CCK-PZ. Polypeptide hormone found in the mammalian gastrointestinal tract and brain. Stimulates pancreatic exocrine secretion and growth. May also play a role in appetite satiation, pain perception, and neuronal transmission. First shown to cause gallbladder contraction: Ivy, Oldberg, Am. J. Physiol. 86, 599 (1928). Discovery of a substance, designated as pancreozymin, which promotes secretion of digestive enzymes by the pancreas: Harper, Raper, J. Physiol. (London) 102, 115 (1943). Identity with pancreozymin: Jorpes et al., Acta Chem. Scand. 18, 2408 (1964). The C-terminal pentapeptide has been shown to be identical to that of gastrin and caerulein: V. Mutt, J. E. Jorpes, Eur. J. Biochem. 6, 156 (1968); eidem, Biochem. J. 125, 57P (1971). Various biologically active, amino-truncated forms have been identified. Cholecystokinin consisting of 33 amino acids (CCK-33) is the predominant gastrointestinal form; CCK-39 and CCK-58 have also been identified. CCK-8 is the predominant CNS form. Identification of CCK in brain: J. J. Vanderhaeghen *et al., Nature* **257**, 604 (1975); G. J. Dockray, *ibid.* **264**, 568 (1976). Distribution and molecular heterogeneity: J. F. Rehfeld, *J. Biol. Chem.* **253**, 4022 (1978). Synthesis of the *C*-terminal dodecapeptide: M. A. Ondetti *et al., J. Am. Chem. Soc.* **92**, 195 (1970). Synthesis of the *N*-terminal hexapeptide of porcine CCK-33: Bodanszky *et al., J. Org. Chem.* **37**, 2303 (1972). Cloning and nucleotide sequence of the human cholecystokinin gene: Y. Takahashi *et al., Proc. Natl. Acad. Sci. USA* **82**, 1931 (1985). Total synthesis of porcine CCK-33: Y. Kurano, *Chem. Commun.* **1987**, 323; of human CCK-33: N. Fujii *et al., ibid.* **1988**, 324. Proposed role in suppression of food intake: M. A. Della-Fera, C. A. Baile, *Science* **206**, 471 (1979); C. J. Savory, M. J. Gentle, *Experientia* **36**, 1191 (1980); M. A. Della-Fera *et al., Science* **212**, 687 (1981); of role in regulation of hypothalamic peptides: S. Itoh *et al., Life Sci.* **25**, 1725 (1979); in modulation of catecholaminergic activity: K. Fuxe *et al., Eur. J. Pharmacol.* **67**, 329 (1980). There is also evidence that CCK acts as a specific antagonist of opiate analgesia: P. L. Faris *et al., Science* **219**, 310 (1983). *Reviews:* E. Straus, R. S. Yalow, *Fed. Proc.* **38**, 2320-2324 (1979); V. Mutt, *Biochem. Soc. Trans.* **8**, 11-14 (1980); *idem, Vitam. Horm.* **39**, 231-426 (1982). Review of physiology: G. J. Dockray, *Br. Med. Bull.* **38**, 253-258 (1982); of role in appetite satiation and pain perception: G. Stacher, *Psychoneuroendocrinology* **11**, 39-48 (1986). Symposium on neuronal CCK: *Ann. N.Y. Acad. Sci.* **448**, 1-697 (1985).

Lys–Ala–Pro–Ser–Gly–Arg–Met–Ser–Ile–Val–Lys–Asn–Leu–Gln–Asn–Leu–Asp
|
H₂NPhe–Asp–Met–Trp–Gly–Met–Tyr–Asp–Arg–Asp–Ser–Ile–Arg–His–Ser–Pro
|
SO₃H

Human CCK - 33

C-Terminal octapeptide *see* Sincalide.

2206. (5α)-Cholestane. [481-21-0] 28,29,30-Trinorlanostane. C₂₇H₄₈; mol wt 372.68. C 87.02%, H 12.98%. The *trans*-decalin homolog of coprostane, *q.v.* Prepd from cholesteryl chloride: Diels, Linn, *Ber.* **41**, 548 (1908); Windaus, *Ber.* **50**, 136 (1917); Ruzicka *et al., Helv. Chim. Acta* **16**, 327 (1933). Crystal structure: Haner, Norton, *Acta Crystallogr.* **20**, 930 (1966).

Scales from ether + alcohol. mp 80°. $[\alpha]_D^{20}$ +24.4° or +30.2° (c = 2 in chloroform). n_D^{88} 1.4887. Freely sol in chloroform, ether, benzene, slightly in abs alcohol.

2207. Cholestanol. [80-97-7] (3β,5α)-Cholestan-3-ol; Dihydrocholesterol; 3β-hydroxycholestane; β-cholestanol. C₂₇H₄₈O; mol wt 388.68. C 83.44%, H 12.45%, O 4.12%. Occurs in human feces, in gallstones, in eggs. Prepn by reduction of cholesterol: Willstätter, Mayer, *Ber.* **41**, 2199 (1908); Ellis, Gardner, *Biochem. J.* **12**, 72 (1918). From coprostenone: Diels, Abderhalden, *Ber.* **39**, 884 (1906). *See also*: Bruce, Ralls, *Org. Synth.* **coll. vol. II**, 191 (1943).

Monohydrate, scales from alc, mp 141.5-142°. $[\alpha]_D^{22}$ +24.2° (c = 1.3 in chloroform). One gram dissolves in about 100 ml alcohol, in 200 ml dry methanol. Freely sol in hot alc, ether, chloroform. Pptd by digitonin.

Methyl ether. C₂₈H₅₀O. Needles from acetone, mp 82.5-83°. $[\alpha]_D^{20}$ +20.0°.

Acetate. C₂₉H₅₀O₂. Prisms from ethyl acetate + methanol, mp 111°. $[\alpha]_D^{20}$ +13.3° (c = 2 in chloroform).

2208. Cholesterol. [57-88-5] (3β)-Cholest-5-en-3-ol; cholesterin. C₂₇H₄₆O; mol wt 386.66. C 83.87%, H 11.99%, O 4.14%. Principal sterol of the higher animals. Found in all body tissues, esp in the brain, spinal cord, and in animal fats or oils. Main constituent of gallstones. Prepd commercially from the spinal cord of cattle by petr ether extraction of the nonsaponifiable matter. Also produced from wool grease. Cholesterol from animal organs always contains cholestanol (dihydrocholesterol) and other satd sterols. Purification by repeated bromination: Schoenheimer, *J. Biol. Chem.* **105**, 355 (1934); Fieser, *Org. Synth.* **coll. vol. IV**, 195 (1963). Laboratory procedure for isoln from gallstones: L. F. Fieser, *Organic Experiments* (Heath, Boston, 3rd ed., 1964) p 70. Total synthesis: Keana, Johnson, *Steroids* **4**, 457 (1964). Reviews and bibliographies: Fieser, Fieser, *Steroids* (Reinhold, New York, Chapman & Hall, London, 1959); Lettré *et al., Ueber Sterine, Gallensäuren und verwandte Naturstoffe* (Stuttgart, 2nd ed., 1955); R. P. Cook, *Cholesterol (Chemistry, Biochemistry and Pathology)* (Academic Press, New York, 1958) 542 pp; J. T. Gwynne, J. F. Strauss, *Endocr. Rev.* **3**, 299-329 (1982).

Monohydrate, pearly leaflets or plates from dil alcohol. Becomes anhydr at 70-80°. When anhydr mp 148.5°. Has been sublimed as orthorhombic needles. bp₀.₅ 233°; bp₇₆₀ 360° (some decompn). d 1.03 (monohydrate); d₁₉¹⁹ 1.052 (anhydr). $[\alpha]_D^{20}$ −31.5° (c = 2 in ether); $[\alpha]_D^{20}$ −39.5° (c = 2 in chloroform). Absorption spectrum: Heilbron *et al., J. Chem. Soc.* **1928**, 47. Practically insol in water (about 0.2 mg/100 ml H₂O). Slightly sol in alc (1.29% w/w at 20°), more sol in hot alc (100 g of satd 96% alcoholic soln contains 28 g at 80°). One gram dissolves in 2.8 ml ether, in 4.5 ml chloroform, in 1.5 ml pyridine. Also sol in acetone, dioxane, ethyl acetate, solvent hexane, benzene, petr ether, oils, fats. Soly in aq solns of bile salts: Rosin, *Z. Physiol. Chem.* **124**, 282 (1923). Solubilization: Gemant, *Life Sci.* **1**, 233 (June 1962). Is pptd by digitonin. Gives intense red color with rosaniline in chloroform soln.

Methyl ether. C₂₈H₄₈O. Crystals from acetone, mp 84°. $[\alpha]_D^{20}$ −45.8° (c = 1.2 in chloroform).

Acetate. C₂₉H₄₈O₂. Needles from acetone, mp 115-116°. $[\alpha]_D^{20}$ −47.4° (c = 2 in chloroform).

Benzoate. C₃₄H₅₀O₂. mp 145.5° (the melt becomes clear at 180°). $[\alpha]_D^{25}$ −13.7° (c = 0.9 in chloroform).

USE: Pharmaceutic aid (emulsifying agent).

2209. Cholestyramine. [11041-12-6] Cholestyramine resin; colestyramin; Dowex 1-X2-Cl; MK-135; Cholybar; Cuemid; Quantalan; Questran. A synthetic, strongly basic anion exchange resin contg quaternary ammonium functional groups which are attached to a styrene-divinylbenzene copolymer. Main constituent: Polystyrene trimethylbenzylammonium as Cl⁻ anion, also contains divinylbenzene. Avg polymeric mol wt >10⁶; Cl⁻ content 14-17%. Effect of particle size on bile salt binding capacity: L. M. Hagerman *et al., Proc. Soc. Exp. Biol. Med.* **139**, 248 (1972). Review of pharmacology: H. R. Casdorph in *Lipid Pharmacology* **Vol. 2**(2), R. Paoletti, C. J. Glueck, Eds. (Academic Press, New York, 1976) pp 222-256. Clinical evaluation in chlordecone detoxification: W. J. Cohn *et al., N. Engl. J. Med.* **298**, 243 (1978). Review of pharmacology and therapeutic efficacy: M. Ast, W. H. Frishman, *J. Clin. Pharmacol.* **30**, 99-106 (1990).

typified structure of main polymeric groups

White to buff-colored, hygroscopic, fine powder. Odorless or has not more than a slight amine-like odor. Insol in water, alcohol, benzene, chloroform, ether.

THERAP CAT: Antilipemic; ion-exchange resin (bile salts).

2210. Cholic Acid. [81-25-4] (3α,5β,7α,12α)-3,7,12-Trihydroxycholan-24-oic acid; cholalic acid; 17β-(1-methyl-3-carboxypropyl)etiocholane-3α,7α,12α-triol; Colalin. $C_{24}H_{40}O_5$; mol wt 408.58. C 70.55%, H 9.87%, O 19.58%. Occurs in conjugation with glycine or taurine in bile of most vertebrates. Extraction from beef bile: Wieland, Siebert, Z. Physiol. Chem. **262**, 1 (1939); laboratory procedure: Gattermann-Wieland, Praxis des Organischen Chemikers (de Gruyter, Berlin, 40th ed., 1961) p 360. Structure and synthesis: Fieser, Fieser, Steroids (Reinhold, New York, 1959).

Monohydrate. Plates from dil acetic acid. Bitter taste with sweetish aftertaste. When anhydr, mp 198°. $[\alpha]_D^{20}$ +37° (c = 0.6 in alcohol). pK = 6.4. Not precipitated by digitonin. Soly at 15° in water 0.28 g/l; in alcohol 30.56 g/l; in ether 1.22 g/l; in chloroform 5.08 g/l; in benzene 0.36 g/l; in acetone 28.24 g/l; in glacial acetic acid 152.12 g/l. Sol in solns of alkali hydroxides or carbonates.

Methyl ester. $C_{25}H_{42}O_5$. Crystals from 95% alcohol + water, mp 155-156°.

Ethyl ester. $C_{26}H_{44}O_5$. Crystals from ethyl acetate + petr ether, mp 162-163°.

Sodium salt. Sodium cholate. $C_{24}H_{39}NaO_5$. Crystals. Soly in water at 15°: >568.9 g/l.

Caution: The names "sodium cholate" and "sodium choleate" are sometimes used for mixtures of bile salts. The term "sodium choleate" is to be preferred for bile salts, see Ox Bile Extract.

Note: The property of forming molecular compds is common to bile acids. For instance a blue molecular compd $(C_{24}H_{40}O_5$-I)$_4$.KI.H$_2$O, may be prepd by mixing an alcoholic soln of cholic acid and a soln of iodine in aq potassium iodide: Barger, Field, J. Chem. Soc. **101**, 1404 (1912).

THERAP CAT: Choleretic.

2211. Choline. [62-49-7] 2-Hydroxy-N,N,N-trimethylethanaminium; (β-hydroxyethyl)trimethylammonium; bilineurine. [C$_5$H$_{14}$NO]$^+$. Quaternary amine that is widely distributed in plants and animals. Essential nutrient required for the biosynthesis of the cell membrane components, phosphatidylcholine (lecithin) and sphingomyelin, q.q.v., and other phospholipids. Also serves as a methyl donor in biomethylation reactions and is a precursor for the neurotransmitter, acetylcholine, q.v. Dietary sources include eggs, meats, peanuts, and wheat germ, where it occurs in both free and esterified forms. Isoln from bile: A. Strecker, Ann. **123**, 353 (1862). Synthesis from trimethylamine with ethylene chlorohydrin: H. C. Klein, R. Kapp, US 2623901 (1952 to Nopco); with ethylene oxide: E. G. Blackett, A. J. Soliday, US 2774759 (1956 to Am. Cyanamid). Prepn of salts: H. C. Klein et al., US 2870198 (1959 to Nopco). Solvent properties of mixture with urea: A. P. Abbott et al., Chem.

Commun. **2003**, 70; of mixture with carboxylic acids: eidem, J. Am. Chem. Soc. **126**, 9142 (2004). LC-MS/MS determn in plasma and urine: S. H. Kirsch et al., J. Chromatogr. B **878**, 3338 (2010). Review of nutritional requirement and dietary sources: E. P. Shronts, J. Am. Dietetic Assoc. **97**, 639-649 (1997); S. H. Zeisel, K.-A. da Costa, Nutr. Rev. **67**, 615-623 (2009); of role in perinatal brain development: S. H. Zeisel, J. Pediatr. **149**, Suppl 5, S131-S136 (2006). Review of metabolism and transport: V. Michel et al., Exp. Biol. Med. **231**, 490-504 (2006); of biosynthesis and homeostasis: Z. Li, D. E. Vance, J. Lipid Res. **49**, 1187-1194 (2008).

Hydroxide. [123-41-1] Bursine; fagine; gossypine; luridine; sincaline; vidine. $C_5H_{15}NO_2$. Viscid, strongly alkaline liq; absorbs CO_2 from the air. Corrosive. Very sol in water, alcohol. Insol in ether. Keep tightly closed.

Chloride. [67-48-1] $C_5H_{14}ClNO$; mol wt 139.62. Colorless crystals or crystalline powder. mp 302°. Hygroscopic. Irritant. Very sol in water, alcohol. The aq soln is practically neutral. Keep tightly closed. LD$_{50}$ in rats (g/kg): 0.400 i.p.; 6.64 orally, see: R. Hartung, H. H. Cornish, Toxicol. Appl. Pharmacol. **12**, 486 (1968).

Dihydrogen citrate. [77-91-8] $C_{11}H_{21}NO_8$; mol wt 295.29. Granules, mp 105-107.5°. Acrid taste. Irritant. Freely sol in water; very slightly sol in alcohol. Practically insol in benzene, chloroform, ether. The pH of a 25% aq soln is 4.25. Keep tightly closed.

Bitartrate. [87-67-2] $C_9H_{19}NO_7$; mol wt 253.25. White, hygroscopic, crystalline powder. Irritant. Freely sol in water; slightly sol in alc. Insol in ether, chloroform. Keep tightly closed.

USE: Nutrient; dietary supplement. In eutectic mixtures as ionic liquid solvent.

2212. Choline Alfoscerate. [28319-77-9] 2-[[[(2R)-2,3-Dihydroxypropoxy]hydroxyphosphinyl]oxy]-N,N,N-trimethylethanaminium inner salt; D-choline hydroxide 2,3-dihydroxypropyl hydrogen phosphate inner salt; L-α-glycerylphosphorylcholine; sn-glycero-3-phosphorylcholine; L-α-GPC; Brezal; Delecit; Gliatilin. $C_8H_{20}NO_6P$; mol wt 257.22. C 37.36%, H 7.84%, N 5.45%, O 37.32%, P 12.04%. Phospholipid; precursor in choline biosynthesis; intermediate in catabolic pathway of phosphatidylcholine. Isoln from bovine pancreas: G. Schmidt et al., J. Biol. Chem. **161**, 523 (1945). Synthesis: E. Baer, M. Kates, J. Am. Chem. Soc. **70**, 1394 (1948); and properties: N. H. Tattrie et al., Biochem. Prep. **6**, 16 (1958). HPLC determn: G. Abbiati et al., J. Chromatogr. **566**, 445 (1991). Metabolism and biodistribution: idem et al., Eur. J. Drug Metab. Pharmacokinet. **18**, 173 (1993). Clinical pharmacology: N. Canal et al., Int. J. Clin. Pharmacol. Ther. Toxicol. **29**, 103 (1991). Clinical pharmacokinetics: G. Gatti et al., ibid. **30**, 331 (1992). Clinical study in multi-infarct dementia: L. Frattola et al., Curr. Ther. Res. **49**, 683 (1991); in Alzheimer's type dementia: L. Parnetti et al., Drugs Aging **3**, 159 (1993).

White crystals, mp 142.5-143°, sinters at 141°. Extremely hygroscopic. Sol in water. $[\alpha]_D^{25}$ −2.7° (c = 2.7 in water, pH 2.5). $[\alpha]_D^{25}$ −2.8° (c = 2.6 in water, pH 5.8).

THERAP CAT: Nootropic.

2213. Choline Salicylate. [2016-36-6] 2-Hydroxy-N,N,N-trimethylethanaminium 2-hydroxybenzoate (1:1); (2-hydroxyethyl)-trimethylammonium salicylate; 2-hydroxy-N,N,N-trimethylethanaminium salt with 2-hydroxybenzoic acid (1:1); choline salicylic acid salt; salicylic acid choline salt; Actasal; Arthropan; Artrobione; Audax; Mundisal. $C_{12}H_{19}NO_4$; mol wt 241.29. C 59.73%, H 7.94%, N 5.81%, O 26.52%. Prepn from choline chloride and sodium salicylate: Broh-Kahn, Int. Rec. Med. **173**, 219 (Apr. 1960); cf. Johnson, GB 8031 (1919); Broh-Kahn et al., US 3069321 (1962 to Labs. for Pharmaceut. Dev.). Variations of process: **BE 583513** (1960 to Mundipharma, AG).

Extremely hygroscopic solid, mp 49.5-50.0°. Very freely sol in water. Also sol in alcohol, acetone, other hydrophilic solvents. Practically insol in ether, petr ether, benzene, oils. Aq solns are stable, they contain the compd in the form of its dissociated choline and salicylate ions. pH of 10% aq soln 6.5. Aq solns are easily discolored by minute traces of iron. The addition of acid to aq solns immediately precipitates free salicylic acid, while choline base, readily recognized by its fishy odor, is liberated upon the addition of alkali.

THERAP CAT: Analgesic; antipyretic.

2214. Cholinesterase. Family of enzymes that catalyze the hydrolysis of choline esters; categorized into 2 major groups based on substrate specificity. Acetylcholinesterase preferentially hydrolyzes acetylcholine, is found primarily in brain, nerve and red blood cells, and is a key modulator of synaptic transmission. Butyrylcholinesterase, found in serum, pancreas and liver, is a nonspecific choline esterase, hydrolyzing butyrylcholine as well as acetyl-, succinyl- and other choline esters. Inhibited by phosphorus-containing insecticides and nerve gases. Isoln from horse serum: E. Stedman *et al.,* *Biochem. J.* **26**, 2056 (1932). Isoln of specific cholinesterase from brain tissue and red blood cells: B. Mendel, H. Rudney, *ibid.* **37**, 59 (1943). Characterization of 2 forms: B. Mendel *et al., ibid.,* 473. Purification of acetylcholinesterase from electric tissue: M. A. Rothenberg, D. Nachmansohn, *J. Biol. Chem.* **168**, 223 (1947). Crystallization and properties: W. Leuzinger *et al., Proc. Natl. Acad. Sci. USA* **59**, 620 (1968). Review of structure, tissue localization, and neurobiology of butrylcholinesterase: S. Darvesh *et al., Nat. Rev. Neurosci.* **4**, 131-138 (2003); of acetylcholinesterase: G. Zimmerman, H. Soreq, *Cell Tissue Res.* **326**, 655-669 (2006). Review of cholinesterase biosensors for environmental monitoring: S. Andreescu, J.-L. Marty, *Biomolec. Eng.* **23**, 1-15 (2006); of assays for cholinesterase activity and inhibition: Y. Miao *et al., Chem. Rev.* **110**, 5216-5234 (2010).

Acetylcholinesterase. [9000-81-1] Acetyl choline esterase; acetylcholine acetylhydrolase; acetylthiocholinesterase; AChE; EC 3.1.1.7.

Butyrylcholinesterase. [9001-08-5] Choline esterase; acylcholine acylhydrolase; pseudocholinesterase; benzoylcholinesterase; propionylcholinesterase; BChE; EC 3.1.1.8.

2215. Choline Theophyllinate. [4499-40-5] 2-Hydroxy-*N,-N,N*-trimethylethanaminium salt with 3,9-dihydro-1,3-dimethyl-1*H*-purine-2,6-dione (1:1); theophylline cholinate; oxtriphylline; oxytrimethylline; Choledyl; Sabidal. $C_{12}H_{21}N_5O_3$; mol wt 283.33. C 50.87%, H 7.47%, N 24.72%, O 16.94%. Prepn: Ladenburg *et al.,* **US 2776287** (1957 to Nepera Chem.).

White granules (contains about 60% of anhydr theophylline). One gram dissolves in 1 ml water. Freely sol in alc; very slightly sol in chloroform.

THERAP CAT: Bronchodilator.

2216. Chondrillasterol. [481-17-4] (3β,5α,22E,24R)-Stigmasta-7,22-dien-3-ol. $C_{29}H_{48}O$; mol wt 412.70. C 84.40%, H 11.72%, O 3.88%. Stereoisomeric with α-spinasterol, *q.v.* Isolated

from the green alga *Scenedesmus obliquus* (Turpin) Kuetz., *Scenedesmaceae:* Bergmann, Feeney, *J. Org. Chem.* **15**, 812 (1950). Synthesis: W. Sucrow *et al., Phytochemistry* **15**, 1533 (1976); M. Anastasia *et al., J. Chem. Soc. Perkin Trans. 1* **1981**, 2561.

Crystals from chloroform-methanol, mp 168-169°. $[\alpha]_D^{24}$ −2° (chloroform).

Acetate. $C_{31}H_{50}O_2$. Crystals from chloroform-methanol, mp 174.5-175.5°. $[\alpha]_D^{24}$ −0.7° (chloroform).

Benzoate. $C_{36}H_{52}O_2$. Crystals from dioxane, mp 194-195°. $[\alpha]_D^{24}$ +3.9° (chloroform).

2217. Chondrocurine. [477-58-7] (13a*R*,25a*S*)-2,3,13a,14,-15,16,25,25a-Octahydro-18,29-dimethoxy-1,14-dimethyl-13*H*-4,-6:21,24-dietheno-8,12-metheno-1*H*-pyrido[3',2':14,15][1,11]dioxacycloeicosino[2,3,4-*ij*]isoquinoline-9,19-diol; *d*-chondocurine; 6,6'-dimethoxy-2,2'-dimethyltubocuraran-7',12'-diol; *d*-tubocurine. $C_{36}H_{38}N_2O_6$; mol wt 594.71. C 72.71%, H 6.44%, N 4.71%, O 16.14%. Isoln from *Chondodendron tomentosum* Ruiz & Pav., *Menispermaceae:* Wintersteiner, Dutcher, *Science* **97**, 467 (1943); King, *J. Chem. Soc.* **1948**, 1945; Bick, Clezy, *ibid.* **1960**, 2402; Boissier *et al., Lloydia* **28**, 191 (1965), *C.A.* **64**, 948b (1966). Structure: Dutcher, *J. Am. Chem. Soc.* **68**, 419 (1946); Hultin, *Acta Chem. Scand.* **15**, 1130 (1961). Configuration: Bick, Clezy, *J. Chem. Soc.* **1953**, 3893; Hultin, *Acta Chem. Scand.* **17**, 753 (1963). Revised structure: Everett *et al., Chem. Commun.* **1970**, 1020. ^{13}C-NMR: L. Koike *et al., J. Org. Chem.* **46**, 2385 (1981).

Slender needles from methanol, mp 232-234°; also reported as mp 218-220° (Boissier). $[\alpha]_D^{24}$ +200° (c = 0.5 in 0.1*N* HCl). $[\alpha]_D^{24}$ +105° (c = 0.9 in pyridine).

Sulfate tetrahydrate. $C_{36}H_{38}N_2O_6 \cdot H_2SO_4 \cdot 4H_2O$. Rectangular plates from water. Contains 9.43% H_2O. After drying, mp 263-265° (dec). $[\alpha]_D^{24}$ +184° (c = 0.375 in methanol).

Methiodide. mp 270-275°.

Dimethiodide. $C_{38}H_{44}I_2N_2O_6$. Pale amorphous precipitate, mp 275° (dec). $[\alpha]_D^{24}$ +184° (c = 0.375 in methanol). uv max (water): 225, 280 nm (ε 62000, 7030).

USE: Methiodides have curarizing power.

2218. Chondrofoline. [31944-97-5] (1β)-6,6',7'-Trimethoxy-2,2'-dimethyltubocuraran-12'-ol; (*R,R*)-7-*O*-methylbebeerine; 7-*O*-methylcurine. $C_{37}H_{40}N_2O_6$; mol wt 608.74. C 73.00%, H 6.62%, N 4.60%, O 15.77%. Alkaloid isolated from the leaves of *Chondrodendron platyphyllum* (A. St. Hil.) Miers, *Menispermaceae:* King, *J. Chem. Soc.* **1940**, 737. Structure, stereochemistry, NMR and mass spectrum: Baldas *et al., Chem. Commun.* **1971**, 132. ^{13}C-NMR: L. Koike *et al., J. Org. Chem.* **46**, 2385 (1981).

Solvated plates from methanol, mp about 135°. $[\alpha]_{5461}^{20}$ $-281°$ (0.1N HCl).

Nitrate hydrate. $C_{37}H_{40}N_2O_6.2HNO_3.H_2O$. Needles from water, dec 225°.

2219. Chondroitin Sulfate. [9007-28-7] Chondroitin hydrogen sulfate; chondroitinsulfuric acid; Chonsurid; Structum. Mol wt estimated at 50,000 depending on source and method of prepn: Schubert, *Fed. Proc.* **17**, 1099 (1958). High viscosity mucopolysaccharides (glycosaminoglycans) with N-acetylchondrosine as a repeating unit and with one sulfate group per disaccharide unit. These biological polymers act as the flexible connecting matrix between the tough protein filaments in cartilage to form a polymeric system similar to reinforced rubber. Chondroitin 4-sulfate and chondroitin 6-sulfate are the most abundant mucopolysaccharides in the body and occur both in skeletal and soft connective tissue. Isoln: Bray *et al.*, *Biochem. J.* **38**, 142 (1944); Patat, Elias, *Z. Physiol. Chem.* **316**, 1 (1959); Kasavina *et al.*, *SU* **157466** (1962); Wheat, Davidson, *Biochem. Prep.* **10**, 52 (1963); Haneno, *JP* **64 7650** (1964 to Yasushi Hano). Structure: Davidson, Meyer, *J. Am. Chem. Soc.* **77**, 4796 (1955). Absorption spectrum of A: Orr, *Biochim. Biophys. Acta* **14**, 173 (1954); of B + C: Mathews, *Nature* **181**, 421 (1958). Clinical trials in atherosclerosis: K. Nakazawa, K. Murata, *J. Int. Med. Res.* **6**, 217 (1978); *eidem*, *Z. Alternsforsch.* **34**, 153 (1979). *Reviews:* K. Meyer, "Chondroitin Sulfates" in *Polysaccharides in Biology*, Trans. 4th Conf. 1958, G. F. Springer, Ed. (Josiah Macy Jr. Foundn., New York, 1959) p 11; Muir, *Am. J. Med.* **47**, 673-690 (1969); Rodén, *Pure Appl. Chem.* **35**, 181-193 (1973). Review of clinical use in osteoarthritis: T. E. McAlindon *et al.*, *J. Am. Med. Assoc.* **283**, 1469-1475 (2000). *See also* Chondrosine.

Chondroitin sulfate A   R = SO₃H   R' = H
Chondroitin sulfate C   R = H   R' = SO₃H

Chondroitin 4-sulfate. [24967-93-9] Chondroitin sulfate A; CSA; Atheroitin. $[\alpha]_D$ -28 to $-32°$.
Chondroitin 4-sulfate disodium salt. [39455-18-0] Condrosulf; Lacrypos.
Chondroitin 6-sulfate. [25322-46-7] Chondroitin sulfate C. $[\alpha]_D$ -12 to $-18°$.
Dermatan sulfate. [24967-94-0] Chondroitin sulfate B; β-heparin. Present in soft connective tissue and abundant in skin, arterial walls and heart valves. Differs from chondroitin sulfate A by containing iduronic acid in place of glucuronic acid, its epimer, at carbon atom 5. Pharmacodynamics: A. M. Traini *et al.*, *J. Int. Med. Res.* **22**, 323 (1994). Clinical evaluation in deep vein thrombosis: B. P. Imbimbo *et al.*, *Thromb. Haemostasis* **71**, 553 (1994). $[\alpha]_D$ -60 to $-70°$.

THERAP CAT: Chondroprotectant; in treatment of osteoarthritis.

2220. Chondrosine. [499-14-9] 2-Amino-2-deoxy-3-*O*-β-D-glucopyranurosyl-D-galactose; β-D-glucopyranosyluronic acid 2-deoxy-2-amino-D-galactose. $C_{12}H_{21}NO_{11}$; mol wt 355.30. C 40.57%, H 5.96%, N 3.94%, O 49.53%. Disaccharide unit of chondroitins 4-

sulfate and 6-sulfate. Isoln: Hebting, *Biochem. Z.* **63**, 353 (1914); Levene, *J. Biol. Chem.* **140**, 267 (1941); Wolfrom *et al.*, *J. Am. Chem. Soc.* **74**, 1491 (1952); Davidson, Meyer, *ibid.* **76**, 5686 (1954). Structure: *eidem*, *ibid.* **77**, 4796 (1955). Synthesis: Takanashi *et al.*, *ibid.* **84**, 3029 (1962). Chondrosine yields on acid hydrolysis chondrosamine (D-galactosamine, *q.v.*). Prepn and derivatives: Stacey, *J. Chem. Soc.* **1944**, 272; Wolfrom, Onodera, *J. Am. Chem. Soc.* **79**, 4737 (1957).

D-Glucuronic acid   Chondrosamine

Crystals from aq ethanol. $[\alpha]_D^{24}$ $+40°$ (0.05 HCl). $[\alpha]_D^{20}$ $+39°$ (water). Dec on heating.
Methyl ester hydrochloride. $C_{13}H_{24}ClNO_{11}$. Crystals from hot ethanol, mp 159-161°. $[\alpha]_D^{23}$ $+42°$ (c = 2 in water).

2221. Chorionic Gonadotropin. Choriogonadotropin; CG. Glycoprotein hormone synthesized by chorionic tissue of the placenta; found in blood and urine during pregnancy. Maintains the function of the corpus luteum and stimulates ovarian steroid secretion in the early stages of gestation. Bioactivity closely resembles that of luteinizing hormone. Structurally similar to LH, FSH, and TSH, *q.q.v.*, all are dimers consisting of noncovalently linked α and β subunits. Within a species, the α-subunits of the four hormones are essentially identical, while the β-subunits, although exhibiting varying degrees of homology, are unique and believed to be responsible for the biological specificity of the individual hormones. Produced by trophoblastic cell neoplasms and used in the diagnosis and management of trophoblastic malignancies. Also produced by cancer cells of various histological types. Discovery in human pregnancy urine: S. Aschheim, B. Zondek, *Klin. Wochenschr.* **6**, 1322 (1927); in serum of pregnant mares: H. H. Cole, G. H. Hart, *Am. J. Physiol.* **93**, 57 (1930). Isoln of human CG: B. Zondek, S. Aschheim, *Klin. Wochenschr.* **7**, 831 (1928); S. Gurin *et al.*, *J. Biol. Chem.* **128**, 525 (1939); P. A. Katzman *et al.*, *ibid.* **148**, 501 (1943). Review of early literature: J. A. Loraine, *J. Reprod. Fertil.* **12**, 23-31 (1966). Purification and physicochemical properties: O. P. Bahl, *J. Biol. Chem.* **244**, 567 (1969). Isoln of α and β subunits: N. Swaminathan, O. P. Bahl, *Biochem. Biophys. Res. Commun.* **40**, 422 (1970); F. J. Morgan, R. E. Canfield, *Endocrinology* **88**, 1045 (1971). Amino acid sequence of α-subunit: R. Bellesario *et al.*, *J. Biol. Chem.* **248**, 6796 (1973); of the β-subunit: R. B. Carlsen *et al.*, *ibid.* **6810**; F. J. Morgan *et al.*, *ibid.* **250**, 5247 (1975). Crystal structure: A. J. Lapthorn *et al.*, *Nature* **369**, 438 (1994). Purification using HPLC: M. A. Chlenov *et al.*, *J. Chromatogr.* **631**, 261 (1993). Comparison of gonadotropins of various species: J. G. Pierce, T. F. Parsons, *Annu. Rev. Biochem.* **50**, 465-495 (1981). Review of biosynthesis: R. O. Hussa, *Endocr. Rev.* **1**, 268-294 (1980). Diagnostic use as a tumor marker: L. A. Cole *et al.*, *Cancer Res.* **48**, 1356 (1988); D. L. Blithe *et al.*, *Trends Endocrinol. Metab.* **1**, 394 (1990). Secretion in normal, nonpregnant humans: W. D. Odell *et al.*, *ibid.* **418**. Synthesis and expression by tumor cells: H. F. Acevedo *et al.*, *Cancer* **76**, 1467 (1995). Review of clinical assay methods: R. J. Norman *et al.*, *Ann. Clin. Biochem.* **27**, 183-194 (1990). Review of structure, secretion and bioactivity of equine CG: B. D. Murphy, S. D. Martinuk, *Endocr. Rev.* **12**, 27-44 (1991).

White or practically white, amorphous powder. Freely sol in water.

Human Chorionic Gonadotropin. [9002-61-3] HCG; pregnancy urine extract; urinary hCG; Choragon; Coriantin; Corulon; Endocorion; Physex; Predalon; Pregnesin; Pregnyl; Primogonyl; Profasi. Stable in dry form. Freely sol in water; sol in aq glycerol and glycols. Insol in alcohol, acetone, ether. pI 2.95. mol wt ~39500 Da. α subunit consists of 92 amino acids; β subunit, 145 amino acids. Total carbohydrate content: ~30% by wt.

Equine Chorionic Gonadotropin. [9002-70-4] Pregnant mare serum gonadotropin; PMSG; serum gonadotropin; Seragon; Serotropin. Sol in water, dil glycerol and glycols. pI 2.60-2.65. α subunit consists of 96 amino acids; β subunit, 149 amino acids. Total carbohydrate content: ~45% by wt.

Choriogonadotropin alfa. [177073-44-8] Chorionic gonadotropin (human α-subunit protein moiety reduced) complex with chorionic gonadotropin (human β-subunit protein moiety reduced); human chorionic gonadotropin, glycoform α; rhCG; Ovidrel; Ovitrelle. Recombinant human CG derived from genetically engineered Chinese hamster ovary cells. Clinical trial in assisted reproductive technology: P. Chang *et al.*, *Fertil. Steril.* **76**, 67 (2001).

THERAP CAT: Gonad-stimulating principle.

THERAP CAT (VET): Gonad-stimulating principle.

2222. Chorismic Acid. [617-12-9] (3*R*,4*R*)-3-[(1-Carboxyethenyl)oxy]-4-hydroxy-1,5-cyclohexadiene-1-carboxylic acid; 3-enolpyruvic ether of *trans*-3,4-dihydroxycyclohexa-1,5-diene carboxylic acid; α-(5-carboxy-1,2-dihydro-2-hydroxyphenoxy)acrylic acid. $C_{10}H_{10}O_6$; mol wt 226.18. C 53.10%, H 4.46%, O 42.44%. The first branch point intermediate in the biosynthesis of aromatic amino acids via the shikimate pathway in bacteria, fungi, and higher plants; naturally occurring as the (−)-form. Its existence was predicted, then discovered during development of a mutant of *A. aerogenes:* M. I. Gibson *et al.*, *Nature* **195**, 1173 (1962); M. I. Gibson, F. Gibson, *Biochim. Biophys. Acta* **65**, 160 (1962). NMR and preliminary structure study: F. Gibson, M. Jackman, *Nature* **198**, 388 (1963). Isoln and metabolism study: M. Gibson, F. Gibson, *Biochem. J.* **90**, 248 (1964). Prepn and characterization of the barium salt: F. Gibson, *ibid.* 256. Structure, relative and abs config, prepn of the monohydrate: J. M. Edwards, L. M. Jackman, *Aust. J. Chem.* **18**, 1227 (1965). Total synthesis of racemic chorismic acid: D. A. McGowan, G. A. Berchtold, *J. Am. Chem. Soc.* **104**, 1153 (1982); B. Ganem *et al.*, *ibid.* 6787; improved synthesis: G. A. Berchtold *et al.*, *ibid.* **105**, 6265 (1983); short formal synthesis: G. H. Posner *et al.*, *J. Org. Chem.* **52**, 4836 (1987). Total synthesis of (−)-form: J. L. Pawlak, G. A. Berchtold, *ibid.* 1765. Potential use in development of herbicides: S. Stinson, *Chem. Eng. News* **60**, 31 (Dec. 6, 1982). Reviews of chorismic acid in biosynthesis of aromatic amino acids: F. Gibson, J. Pittard, *Bacteriol. Rev.* **32**, 465-492 (1968); R. J. Ife *et al.*, *J. Chem. Soc. Perkin Trans. 1* **1976**, 1776-1783; U. Weiss, J. M. Edwards, *The Biosynthesis of Organic Compounds* (Wiley, New York, 1980) pp 134-143. *See also* shikimic acid.

Crystals may be obtained but show marked tendency toward solvent retention, mp 105-108° (dec). $[\alpha]_D^{21}$ −274° (c = 0.16 in water).

Barium salt trihydrate. $C_{10}H_8BaO_6\cdot3H_2O$. Unstable. uv max (aq soln): 272 nm (ε 2700).

Monohydrate. Colorless crystals from ethyl acetate + light petroleum, mp 148-149° (dec); from ether + light petroleum, mp 112° (dec); from ethyl acetate + carbon tetrachloride, mp 115° (dec). $[\alpha]_{5890}^{25}$ −295.5° (c = 0.2 in water). uv max (water): 275 nm (ε 2630).

(±)-Form. Crystals from ethyl acetate/hexane, mp 139.4-141° (dec).

2223. Chromafenozide. [143807-66-3] 3,4-Dihydro-5-methyl-2*H*-1-benzopyran-6-carboxylic acid 2-(3,5-dimethylbenzoyl)-2-(1,1-dimethylethyl)hydrazide; 2′-*tert*-butyl-5-methyl-2′-(3,5-xyloyl)chromane-6-carbohydrazide; ANS-118; Matric. $C_{24}H_{30}N_2O_3$; mol wt 394.52. C 73.07%, H 7.67%, N 7.10%, O 12.17%. Diacylhydrazine ecdysone mimetic for use in food crops. Prepn: M. Yanagi *et al.*, *EP 496342*; *eidem*, *US 5378726* (1992, 1995 both to Nippon Kayaku; Sankyo). Properties and biological activity: M. Yanagi *et al.*, *BCPC Conf. - Pests Dis.* **2000**, 27. Field trial against apple tortix: R. Ichinose *et al.*, *Annu. Rep. Sankyo Res. Lab.* **52**, 59 (2000). Mode of action: T. Toya *et al.*, *Biochem. Biophys. Res. Commun.* **292**, 1087 (2002). Series of articles on design, synthesis, and activ-

ity: Y. Sawada *et al.*, *Pest Manage. Sci.* **59**, 25-57 (2003). Review of development, environmental chemistry, safety, and toxicity: K. Tanaka *et al.*, *Annu. Rep. Sankyo Res. Lab.* **53**, 1-49 (2001).

White crystalline powder, mp 186.4°. Log P (octanol/water): 2.7 (22°). Vapor pressure at 25°: ≤4 × 10⁻⁹ Pa. Soly in water at 20°: 1.12 mg/l. Moderately sol in polar organic solvents. LD_{50} (mg/kg) orally in rats, mice: >5000; dermally in rabbits: >2000; LC_{50} (mg/l) in rats: >4.68 by inhalation; in carp, rainbow trout (96 hr): >47.25, >18.9 (Yanagi).

USE: Insecticide.

2224. Chromic Acetate. [1066-30-4] Acetic acid chromium-(3+) salt (3:1). $C_6H_9CrO_6$; mol wt 229.13. C 31.45%, H 3.96%, Cr 22.69%, O 41.89%. $Cr(CH_3COO)_3$. The commercial material, usually sold as a concd soln of the basic acetate, $Cr(OH)(C_2H_3O_2)_2$, contains Na acetate or Na_2SO_4 impurities. Industrial prepn: Stover, Drew, **US 2650239**; **US 2678328** (1953, 1954 both to Socony-Vacuum Oil). Prepn of hexahydrate: Hein, Herzog in *Handbook of Preparative Inorganic Chemistry* **vol. 2**, G. Brauer, Ed. (Academic Press, New York, 2nd ed., 1965) p 1371. Toxicity data: Cavalli, *Arch. Int. Pharmacodyn.* **62**, 330 (1939). *Review: ACS Monograph Series* **no. 132**, entitled "Chromium," vol. 1, M. J. Udy, Ed. (Reinhold, New York, 1956) pp 229-233.

Hydrate. Approx $Cr(C_2H_3O_2)_3\cdot H_2O$, gray-green powder or violet plates. Slightly sol in water. Practically insol in alc. MLD in frogs, mice, rabbits (mg/kg): 6185, 2290, 1604 i.v. (Cavalli).

Hexahydrate. Hexaaquochromium triacetate. Blue-violet needles. Readily sol in water, with partial hydrolysis, giving a soln which is blue under the incident light and red under transmitted light; solvolyzed by alc.

Basic. $Cr(OH)(C_2H_3O_2)_2$. Violet cryst powder. Freely sol in water.

USE: As a mordant in dyeing; in tanning; in hardening photographic emulsions; to improve light stability and dye affinity of textiles and polymers; in catalyst for polymerization of olefins; as an oxidation catalyst.

2225. Chromic Bromide. [10031-25-1] Chromium bromide (CrB_3). Br_3Cr; mol wt 291.71. Br 82.17%, Cr 17.82%. $CrBr_3$. Prepd by passing Br_2 vapor over Cr powder at 1000°: Hein, Herzog, in *Handbook of Preparative Inorganic Chemistry* **vol. 2**, G. Brauer, Ed. (Academic Press, New York, 2nd ed., 1965) p 1341.

Black, lustrous, hexagonal crystals; green in transmitted and reddish in reflected light. Sol in cold water only upon addition of chromous salts; sol in boiling water.

Hexahydrate. At least two isomeric forms exist. *Dibromotetraaquochromium bromide dihydrate*, $[CrBr_2(H_2O)_4]Br\cdot2H_2O$; green, deliquesc crystals; sol in water, alc. *Hexaaquochromium tribromide*, $[Cr(H_2O)_6]Br_3$; violet, deliquesc crystals; sol in water. Insol in alc. Both forms practically insol in ether.

USE: In catalysts for polymerization of olefins.

2226. Chromic Chloride. [10025-73-7] Chromium chloride ($CrCl_3$). Cl_3Cr; mol wt 158.35. Cl 67.16%, Cr 32.84%. $CrCl_3$. Prepn: Heisig *et al.*, *Inorg. Synth.* **2**, 193 (1946); Pray, *ibid.* **5**, 153 (1953); Vavoulis *et al.*, *ibid.* **6**, 129 (1960); Hein, Herzog in *Handbook of Preparative Inorganic Chemistry* **vol. 2**, G. Brauer, Ed. (Academic Press, New York, 2nd ed., 1965) pp 1338-1340; prepn of hexahydrates: *eidem*, *ibid.* pp 1348-1350. Toxicity data: Cavalli, *Arch. Int. Pharmacodyn.* **62**, 330 (1939).

Violet, lustrous, hexagonal, cryst scales. Greasy feel. mp 1152°; dissociates above 1300°. d^{25} 2.87. The rate of soln in water, acids, organic solvents is extremely slow. Addition of a trace of $CrCl_2$ or wetting agent aid in rapid soln in water, alcohol. Also exists in hygroscopic, sol, peach-blossom colored form. *Keep tightly closed.* MLD in frogs, mice, rabbits (mg/kg): 187, 801, 288 i.v. (Cavalli).

Hexahydrates. Several known isomers. *Dichlorotetraaquochromium chloride dihydrate*, [*trans*-[CrCl$_2$(H$_2$O)$_4$]Cl.2H$_2$O]; dark green salt, monoclinic crystals; d 1.849: Freeman, Dance, *Inorg. Chem.* **4**, 1555 (1965); Morosin, *Acta Crystallogr.* **21**, 280 (1966). *Hexaaquochromium trichloride*, [[Cr(H$_2$O)$_6$]Cl$_3$]; violet, rhombohedral hydrate. *Chloropentaaquochromium dichloride monohydrate*, [[CrCl(H$_2$O)$_5$]Cl$_2$.H$_2$O]; light green isomer. All are deliquesc in air. Sol in water; dil aq solns are violet, concd aq solns are green. pH of 0.2 molar aq soln 2.4. Sol in alcohol; slightly sol in acetone. Practically insol in ether. *Keep well closed.*

USE: In chromizing; manuf of Cr metal and compds; as catalyst for polymerization of olefins and other organic reactions; as textile mordant; in tanning; in corrosion inhibitors; as waterproofing agent.

2227. Chromic Fluoride. [7788-97-8] Chromium fluoride (CrF$_3$). CrF$_3$; mol wt 108.99. Cr 47.71%, F 52.29%. Prepd by heating CrCl$_3$ in a current of HF: Kwasnik in *Handbook of Preparative Inorganic Chemistry* **vol. 1**, G. Brauer, Ed. (Academic Press, New York, 2nd ed., 1963) p. 257. Review of chromium halides: Fergusson in *Halogen Chemistry* **vol. 3**, V. Gutmann, Ed. (Academic Press, New York, 1967) pp 227-302.

Dark green needles, d 3.8. mp 1100°. Sublimes at 1100-1200°. *Corrosive.* Practically insol in water, alcohol. Sol in HCl with violet color.

Trihydrate. Triaquochromium trifluoride. Discussion of other hydrates or Werner complexes: *ACS Monograph Series* **no. 132**, entitled "Chromium," vol. 1, M. J. Udy, Ed. (Reinhold, New York, 1956) p 183. Green crystals from solns of Cr or Cr(OH)$_3$ in hydrofluoric acid. Sparingly sol in water.

USE: The hydrates in printing and dyeing woolens; coloring and hardening marble; mothproofing woolen fabrics; treating silk; polishing metals; halogenation catalyst.

2228. Chromic Formate. [27115-36-2] Formic acid chromium(3+) salt (3:1). C$_3$H$_3$CrO$_6$; mol wt 187.05. C 19.26%, H 1.62%, Cr 27.80%, O 51.32%. Cr(HCOO)$_3$. Prepd from Cr(OH)$_3$ and formic acid: Akhmedli, Negretov, *J. Gen. Chem. USSR* **20**, 2045 (1950).

Hexahydrate. Fine green crystals or gray-green powder. Dec above 300°, evolving CO and CO$_2$. Sol in water; concd aq soln is blue under incident light, red under transmitted light; dil aq soln is green.

USE: In printing cotton in skeins; in leather tanning and waterproofing.

2229. Chromic Hydroxide. [1308-14-1] Chromium hydroxide (Cr(OH)$_3$); chromic oxide gel; chromic oxide, hydrous; chromium trihydroxide. CrH$_3$O$_3$; mol wt 103.02. Cr 50.47%, H 2.94%, O 46.59%. Cr(OH)$_3$. Occurs only as hydrates. Prepn: Ruthroff, *Inorg. Synth.* **2**, 190 (1946); Hein, Herzog in *Handbook of Preparative Inorganic Chemistry* **vol. 2**, G. Brauer, Ed. (Academic Press, New York, 2nd ed., 1965) pp 1345-1346.

Trihydrate. Cr(OH)$_3$.3H$_2$O. Or Cr$_2$O$_3$.9H$_2$O, blue-green powder. Practically insol in water. Sol in dil mineral acids when freshly prepd, giving blue or green soln; becomes insol in acids on aging. Cr(OH)$_3$.*n*H$_2$O, shining, vitrous, jet-black particles. Useful as catalyst in dehydrogenation of alcohols and paraffins, hydrogenation of olefins.

USE: As pigment; in tanning industry; as mordant; as catalyst for organic reactions.

2230. Chromic Nitrate. [13548-38-4] Nitric acid chromium-(3+) salt (3:1); chromium(III) nitrate. CrN$_3$O$_9$; mol wt 238.01. Cr 21.85%, N 17.66%, O 60.50%. Cr(NO$_3$)$_3$. Prepn of anhydr salt from N$_2$O$_5$ and Cr(CO)$_6$: Addison, Chapma, *J. Chem. Soc.* **1964**, 539. Prepn of nonahydrate from Cr(OH)$_3$ and dil HNO$_3$ or by reducing CrO$_3$ in the presence of HNO$_3$: *ACS Monograph Series* **no. 132**, entitled "Chromium," vol. 1, M. J. Udy, Ed. (Reinhold, New York, 1956) pp 203-204. Toxicity study: Smyth *et al.*, *Am. Ind. Hyg. Assoc. J.* **30**, 470 (1969).

Pale green, extremely deliquesc powder. Non-volatile. Dec above 60°. Sol in water, ethyl acetate, DMSO. Practically insol in benzene, CCl$_4$, CHCl$_3$. Reacts with ether, sometimes vigorously.

Nonahydrate. Deep violet, rhombic, monoclinic crystals. mp about 60°; dec above 100°. Sol in water, alcohol. Aq soln slowly becomes green on heating and rapidly recovers the reddish-violet color on cooling. LD$_{50}$ orally in rats: 3.25 g/kg (Smyth).

USE: Prepn of Cr catalyst; in textile printing; corrosion inhibitor.

2231. Chromic Phosphate. [7789-04-0] Phosphoric acid chromium(3+) salt (1:1). CrO$_4$P; mol wt 146.97. Cr 35.38%, O 43.54%, P 21.07%. CrPO$_4$. Prepn: Ness *et al.*, *J. Am. Chem. Soc.* **74**, 4685 (1952); Eickhoff, Kebrich, US 2749214 (1956 to National Lead); Wegenknecht, DE 1046597 (1958); DE 1056104 (1959).

Gray-brown to black crystals of amorphous solid. Does not melt by 1800°. d$^{32.5}$ 2.94. Partially oxidizes to CrO$_3$ on heating in air. Practically insol in water, acetic acid, HCl, aqua regia.

Hemiheptahydrate. Arnaudon's green; Plessy's green. Blue-green powder. d 2.15. Practically insol in water. Sol in acids.

Hexahydrate. Violet crystals. Loses water gradually on heating, becoming anhydr after one hour at 800° or 3-4 hrs at 500°. d^{14} 2.121. Practically insol in water. Slightly sol in acetic acid solns; readily sol in mineral acids, alkalis, oxalic acid solns.

Chromic phosphate P 32. [24381-60-0] Phosphocol P 32. Cr^{32}PO$_4$. Clinical trial for refractory solid tumors: W. Gao *et al.*, *Ann. Nucl. Med.* **22**, 653 (2008); for occult lymphatic metastasis: W. Gao *et al.*, *Nucl. Med. Commun.* **30**, 420 (2009). Contains ^{32}P which is a beta emitter with a half life of 14.3 days.

USE: Green pigment; in wash primers; in catalysts for dehydrogenation of hydrocarbons or polymerization of olefins.

THERAP CAT: Chromic phosphate P 32 as antineoplastic (radiation source).

2232. Chromic Potassium Oxalate. [14217-01-7] Potassium (*OC*-6-11)-tris[ethanedioato(2−)-κO^1,κO^2]chromate(3−) (3:1); tripotassium tris(ethanedioato)chromate(3−); tripotassium tris(oxalato)chromate(3−); potassium trioxalatochromate(III); potassium chromic oxalate. C$_6$CrK$_3$O$_{12}$; mol wt 433.35. C 16.63%, Cr 12.00%, K 27.07%, O 44.30%. K$_3$[Cr(C$_2$O$_4$)$_3$]. Prepd by treatment of oxalic acid and K$_2$C$_2$O$_4$ with K$_2$Cr$_2$O$_7$: Bailar, Jr., Jones, *Inorg. Synth.* **1**, 37 (1939); Hein, Herzog in *Handbook of Preparative Inorganic Chemistry* **vol. 2**, G. Brauer, Ed. (Academic Press, New York, 2nd ed., 1965) p 1372.

Trihydrate. [15275-09-9] Potassium trioxalatotriaquochromate-(III). C$_6$CrK$_3$O$_{12}$.3H$_2$O; mol wt 487.39. Black-green, monoclinic scales with transparent blue edges. Freely sol in water; practically insol in alcohol.

USE: In tanning industry; dyeing chromate colors on wool.

2233. Chromic Potassium Sulfate. [10141-00-1] Sulfuric acid chromium(3+) potassium salt (2:1:1); potassium chromic sulfate; potassium disulfatochromate(III); chromium(III) potassium sulfate. CrKO$_8$S$_2$; mol wt 283.21. Cr 18.36%, K 13.81%, O 45.19%, S 22.64%. KCr(SO$_4$)$_2$. Dodecahydrate produced by reduction of K$_2$Cr$_2$O$_7$ with SO$_2$: Copson in *ACS Monograph Series* **no. 132**, entitled "Chromium," vol. 1, M. J. Udy, Ed. (Reinhold, New York, 1956) p 281; electrolytic manuf: Nishihara *et al.*, JP 60 2164 *C.A.* **55**, 5200e (1961).

Dodecahydrate. [7788-99-0] Chrome alum. CrKO$_8$S$_2$.12H$_2$O; mol wt 499.39. K[Cr(SO$_6$H$_4$)$_2$(H$_2$O)$_2$].6H$_2$O: Duval, *Chim. Anal. (Paris)* **44**, 102 (1962), *C.A.* **57**, 9479d (1962). Large, violet-red to black, octahedral, cubic crystals; ruby-red under transmitted light. d^{25} 1.83. mp 89°; at 400° loses all its H$_2$O. Sol in 4 parts cold, 2 parts boiling water. Practically insol in alcohol. The aq soln is violet when cold, green when hot. The violet color returns after a few weeks at room temp.

USE: Mordant for dyeing fabrics uniformly; tanning leather; printing calico; rendering glue and gum insol; manuf ink, other chromium salts; waterproofing fabrics; hardening photographic emulsions.

2234. Chromic Sulfate. [10101-53-8] Sulfuric acid chromium(3+) salt (3:2); chromium(III) sulfate. Cr$_2$O$_{12}$S$_3$; mol wt 392.16. Cr 26.52%, O 48.96%, S 24.53%. Cr$_2$(SO$_4$)$_3$. Prepn of anhydr salt by dehydration of hydrated forms: Rollinson, Bailar, Jr., *Inorg. Synth.* **2**, 197 (1946). Toxicity data: Cavalli, *Arch. Int. Pharmacodyn.* **62**, 330 (1939).

Peach-colored solid. d 3.012. Practically insol in water and acids. MLD in frogs, mice, rabbits (mg/kg): 37, 246.8, 215 i.v. (Cavalli).

Hydrates. Are known in both green and violet modifications, and have several degrees of hydration up to $18H_2O$: Lukaszewski, Redfern, *Nature* **190**, 805 (1961); Udy in *Chromium* **vol. 1**, M. J. Udy, Ed., A.C.S. Monograph Series no. **132** (Reinhold, New York, 1956) pp 213-217, 288. The technical product comes in the form of a finely granular, dark-green flake or powder approximating the formula $Cr_2(SO_4)_3 \cdot 10H_2O$. Readily sol in water. Almost insol in alc.

Basic chromic sulfates. Of the type $Cr(OH)SO_4 \cdot nH_2O$ are of importance in the tanning industry: Udy, *loc. cit.* and pp 278-280, 305-308. Technical grades are available in two degrees of basicity, one-third and one-half, as finely granular dark-green flakes or powder contg about 25% Cr_2O_3. Readily sol in water.

USE: Insolubilization of gelatin; in catalyst prepn; as mordant in textile industry; in tanning of leather; in chrome plating; in manuf of Cr, CrO_3, and Cr alloys; to improve dispersibility of vinyl polymers in water; in manuf of green varnishes, paints, inks, glazes for porcelain.

2235. Chromium. [7440-47-3] Cr; at. wt 51.9961; at. no. 24; valences 1-6. Group VIB (6). Four naturally occurring isotopes: 50 (4.31%); 52 (83.76%); 53 (9.55%); 54 (2.38%). Artificial radioactive isotopes: 45-49; 51; 55-57; longest-lived isotope is ^{51}Cr (T$_{1/2}$ 27.704 days) prepd by (n,γ) reaction from ^{50}Cr. Abundance in earth's crust: 122 ppm. Isoln from *crocoite* ($PbCrO_4$): L. N. Vauquelin, *J. Mines* **6**, ser. 1, 737 (1787); *idem, Ann.* **70**, 70 (1809). Commercial sources obtained from chrome ore, *chromite* ($FeO \cdot Cr_2O_3$). Reviews of chromium, its alloys and compds: *ACS Monograph Series* **no. 132**, entitled "Chromium," M. J. Udy, Ed. (Reinhold, New York, 1956) **vol. 1**, 433 pp; **vol. 2**, 402 pp; C. L. Rollinson, "Chromium, Molybdenum and Tungsten" in *Comprehensive Inorganic Chemistry* **vol. 3**, J. C. Bailar, Jr. *et al.*, Eds. (Pergamon Press, Oxford, 1973) pp 623-700; *Chemistry of the Elements,* N. N. Greenwood, A. Earnshaw, Eds. (Pergamon Press, New York, 1984) pp 1167-1210; J. H. Westbrook in *Kirk-Othmer Encyclopedia of Chemical Technology* **vol. 6** (Wiley-Interscience, New York, 4th ed., 1993) pp 228-263; B. J. Page, G. W. Loar, *ibid.* pp 263-311. Review of biological function of the Cr(III) ion as essential trace element: Mertz, *Physiol. Rev.* **49**, 163-239 (1969). Review of carcinogenic risk: *IARC Monographs* **2**, 100-125 (1973); *ibid.* **23**, 205-323 (1980); of metabolism and toxicity: M. D. Cohen *et al., Crit. Rev. Toxicol.* **23**, 255-281 (1993); of toxicology and human exposure: *Toxicological Profile for Chromium* (PB2000-108022, 2000) 461 pp. Books: *Chromium: Metabolism and Toxicity,* D. Burrows, Ed. (CRC Press, Boca Raton, 1983) 172 pp; *Chromium in the Natural and Human Environments,* J. O. Nriagu, E. Nieboer, Eds. (John Wiley & Sons, New York, 1988) 571 pp.

Steel-gray, lustrous metal; body-centered cubic structure; hard as corundum and less fusible than platinum. Takes a high polish. mp 1903 ±10°. bp 2642°. d^{20} 7.14. Heat capacity (25°): 5.58 cal/mol/deg C. Heat of fusion: 3.5 kcal/mol; heat of vaporization: 81.7 kcal/mol (at bp); heat of sublimation (25°): 94.8 kcal/mol (Rollinson). Also reported as mp 1875°. bp 2680°. d^{20} 7.19. Specific heat (25° C): 23.9 J/mol/deg K. Heat of fusion: 14.6 kJ/mol; heat of vaporization: 305 kJ/mol (at bp) (Westbrook). Resistant to common corroding agents, highly acid resistant. Reacts with dil HCl, H_2SO_4; not with HNO_3. Resists atmospheric attack at ambient temperatures.

Caution: Potential symptom of overexposure to chromium metal by inhalation is histologic fibrosis of the lungs. *See NIOSH Pocket Guide to Chemical Hazards* (DHHS/NIOSH 97-140, 1997) p 72. Chromic acid and chromate salts are irritating to exposed tissues. Toxic effects due to overexposure may include dermatitis, skin ulcers; nasal inflammation, perforation of nasal septum; cancer of the lungs, nasal cavity and paranasal sinus. *See Patty's Industrial Hygiene and Toxicology* **vol. 2C**, G. D. Clayton, F. E. Clayton, Eds. (John Wiley & Sons, New York, 4th ed., 1994) p 1973-1985. Chromium hexavalent compounds are listed as known human carcinogens: *Report on Carcinogens, Twelfth Edition* (PB2011-111646, 2011) p 106.

USE: In manuf of chrome-steel or chrome-nickel-steel alloys (stainless steel), nonferrous alloys, heat resistant bricks for refractory furnaces. To greatly increase strength, hardness and resistance of metals to abrasion, corrosion and oxidation. For chrome plating of other metals; leather tanning; as pigment and mordant; wood preservative. Use of ^{51}Cr as diagnostic aid *see* sodium chromate(VI).

2236. Chromium Carbonyl. [13007-92-6] (*OC*-6-11)-Chromium carbonyl ($Cr(CO)_6$); chromium hexacarbonyl. C_6CrO_6; mol wt 220.06. C 32.75%, Cr 23.63%, O 43.62%. $Cr(CO)_6$. Prepn from Cr salts and CO in the presence of a Grignard reagent: Owen *et al., Inorg. Synth.* **3**, 156 (1950); in the presence of Mg and ether: Wender, **US 3012858** (1961 to Diamond Alkali); in the presence of Na and diglyme: Podall *et al., J. Am. Chem. Soc.* **83**, 2057 (1961); in the presence of Na and an aromatic hydrocarbon: Pruett, Wyman, **US 3053629** (1962 to Union Carbide); in the presence of I_2 and a nitrile: Wotiz, **US 3100687** (1963 to Diamond Alkali). Toxicity study: Strohmeier, *Z. Naturforsch.* **19b**, 540 (1964).

Orthorhombic, highly refractive crystals. Sublimes at room temp. Sinters at 90°; dec at 130°; explodes at 210°. Burns with a luminous flame. d^{18} 1.77. Vapor pressure (mm): 0.04 (0°); 1.0 (48°); 66.5 (100°). Almost insol in water, ethanol, methanol. Sol in ether, $CHCl_3$, other organic solvents. Solns or impure solid dec by light. LD_{50} i.v. in mice: 100 mg/kg (Strohmeier).

USE: In catalysts for olefin polymerization and isomerization; gasoline additive to increase octane number; prepn of chromous oxide, CrO.

2237. Chromium Dioxide. [12018-01-8] Chromium oxide (CrO_2). CrO_2; mol wt 83.99. Cr 61.91%, O 38.10%. Prepn: Wöhler, *Ann.* **111**, 117 (1859); Thamer *et al., J. Am. Chem. Soc.* **79**, 547 (1957); Swoboda *et al., J. Appl. Phys.* **32**, Suppl. no. 3, 374 (1961); Arthur; Arthur, Ingraham, **US 2959955**; **US 3117093** (1960, 1964 both to du Pont). *Reviews:* Hund, *Farbe Lack* **78**, 11-16 (1972); Rollinson in *Comprehensive Inorganic Chemistry* **vol 3**, J. C. Bailar, Jr. *et al.*, Eds. (Pergamon Press, Oxford, 1973) pp 689-690.

Black, ferromagnetic crystals; rutile structure. d 4.89. Metastable in air; various temperatures (250-500°) reported for decompn to Cr_2O_3.

USE: In magnetic recording tapes; as catalyst.

2238. Chromium(III) Oxide. [1308-38-9] Chromium oxide (Cr_2O_3); anadonis green; chrome green; chrome ocher; chrome oxide green; chromia; chromic oxide; chromium sesquioxide; green cinnabar; green oxide of chromium; green rouge; leaf green; oil green; ultramarine green; C.I. Pigment Green 17; C.I. 77288. Cr_2O_3; mol wt 151.99. Cr 68.42%, O 31.58%. Prepd by reaction of sodium dichromate or chromate with sulfur: Copson in *ACS Monograph Series* **no. 132**, entitled "Chromium," vol. 1, M. J. Udy, Ed. (Reinhold, New York, 1956) pp 277-278. *Review:* Wiesburg, *Paint Ind. Mag.* **71**(2), 11 (1956). *See also: Colour Index* **vol. 4** (3rd ed., 1971) p 4662. Use as catalyst: R. Uma, J. C. Kuriacose, *Indian Chem. Manuf.* **8**, 11 (1970). Cytotoxic effects: V. Bianchi *et al., Toxicology* **17**, 219 (1980).

Light to dark green, fine, hexagonal crystals. mp about 2435°. bp about 3000°. d^{25} 5.22. Turns brown on heating but reverts to green color on cooling. Cryst Cr_2O_3 is extremely hard; will scratch quartz, topaz, zircon. Practically insol in water, alc, acetone. Slightly sol in acids, alkalies.

Caution: Trivalent chromium may cause skin irritation: S. Fregert, H. Rorsman, *Arch. Dermatol.* **90**, 4 (1964).

USE: In abrasives, refractory materials, electric semiconductors; as pigment, particularly in coloring glass; in alloys; printing fabrics and banknotes; as catalyst for organic and inorganic reactions.

2239. Chromium(VI) Oxide. [1333-82-0] Chromium oxide (CrO_3); chromic anhydride; chromium trioxide. CrO_3; mol wt 99.99. Cr 52.00%, O 48.00%. Produced commercially by the action of concd H_2SO_4 on a soln of chromate or dichromate: Faith, Keyes & Clark's Industrial Chemicals, F. A. Lowenheim, M. K. Moran, Eds. (Wiley-Interscience, New York, 4th ed., 1975) pp 270-274.

Dark red, deliquesc bipyramidal prismatic crystals, flakes or granular power. d 2.70. mp 197°. Dec at 250° to Cr_2O_3 and O_2. Very sol in water; sol in H_2SO_4. Powerful oxidizer; oxidizes alcohol and most other organic substances, sometimes with dangerous violence. *Contact with combustible material may cause fire.*

Caution: Potential symptoms of overexposure are respiratory system irritation, nasal septum perforation; liver and kidney damage; leukocytosis, leukopenia, monocytosis, eosinophilia; eye injury, conjunctivitis; skin ulcers, sensitization dermatitis. *See NIOSH Pocket Guide to Chemical Hazards* (DHHS/NIOSH 97-140, 1997) p 70. Chronic overexposure may lead to severe liver damage, CNS involvement and lung cancer. *See Clinical Toxicology of Commercial Products*, R. E. Gosselin *et al.*, Eds. (Williams & Wilkins, 5th ed., 1984) Section II, p 108. *See also Patty's Industrial Hygiene and Toxicology* vol. 2C, G. D. Clayton, F. E. Clayton, Eds. (Wiley-Interscience, New York, 4th ed., 1994) p 1973-1985. Chromium hexavalent compounds are listed as known human carcinogens: *Report on Carcinogens, Twelfth Edition* (PB2011-111646, 2011) p 106.

USE: Chromium plating; copper stripping; aluminum anodizing; corrosion inhibitor; photography; purifying oil and acetylene; hardening microscopical prepns; oxidizing agent in organic chemistry.

THERAP CAT (VET): Has been used in solution as a topical antiseptic and astringent.

2240. Chromium Picolinate. [14639-25-9] Tris(2-pyridinecarboxylato-N^1,O^2)chromium; chromium tripicolinate; chromium-(III) trispicolinate; Chromax. $C_{18}H_{12}CrN_3O_6$; mol wt 418.31. C 51.68%, H 2.89%, Cr 12.43%, N 10.05%, O 22.95%. Biologically active form of chromium. General prepn, chemical and biological properties: G. W. Evans, D. J. Pouchnik, *J. Inorg. Biochem.* **49,** 177 (1993). Clinical effect on insulin metabolism: G. W. Evans, *Int. J. Biosocial Med. Res.* **11,** 163 (1989). Clinical evaluation of lipid-lowering activity: R. I. Press *et al.*, *West. J. Med.* **152,** 41 (1990). Clinical pharmacokinetics: M. L. Gargas *et al.*, *Drug Metab. Dispos.* **22,** 522 (1994). Use as veterinary diet supplement: M. D. Lindemann *et al.*, *J. Anim. Sci.* **73,** 457 (1995). Review of clinical pharmacology: M. F. McCarty, *J. Appl. Nutr.* **43,** 58-66 (1991); *idem, J. Optim. Nutr.* **2,** 36-53 (1993); of safety studies: T. O. Berner *et al.*, *Food Chem. Toxicol.* **42,** 1029-1042 (2004).

Lipophilic. Soly: water (pH 7.0) 0.6 m*M*, chloroform 2.0 m*M*. uv max (water): 264 nm (a_M 15546 L mol^{-1}cm^{-1}).

THERAP CAT: Chromium supplement.

2241. Chromocarb. [4940-39-0] 4-Oxo-4*H*-1-benzopyran-2-carboxylic acid; 2-chromonecarboxylic acid; 4-oxo-4*H*-chromene-2-carboxylic acid; benzo-γ-pyronecarboxylic acid. $C_{10}H_6O_4$; mol wt 190.15. C 63.17%, H 3.18%, O 33.66%. Prepn: S. Ruhemann, H. E. Stapleton, *J. Chem. Soc.* **77,** 1179 (1900); J. Schmutz *et al.*, *Helv. Chim. Acta* **34,** 767 (1951); G. Pifferi *et al.*, *J. Heterocycl. Chem.* **14,** 1257 (1977). Prepn of diethylamine salt: P. A. Tronche, **ZA 6807352**; *eidem*, **US 3816470** (1969, 1974 both to Ferlux). Pharmacology of diethylamine salt in animals: J. Couquelet *et al.*, *C. R. Seances Soc. Biol. Ses Fil.* **164,** 329 (1970); P. Conquet *et al.*, *ibid.* 800. Clinical comparison with dipyridamole of effect on platelet function: A. Vittoria *et al.*, *Curr. Ther. Res.* **35,** 1033 (1984). Clinical trial in diabetes with vascular disease: N. Ciavarella *et al.*, *ibid.* **36,** 293 (1984). Bioavailability in humans: J.-M. Aiache *et al.*, *Biopharm. Drug Dispos.* **7,** 301 (1986).

Colorless needles from alcohol, mp 250-251° (dec) (Ruhemann, Stapleton); also reported as mp 255-256° (Pifferi). uv max: 230, 305 nm (ε 20220, 8075). Sol in alcohol, ammonia. Sparingly sol in water.

Diethylamine. [23915-80-2] Angiophtal; Campel; Fludarene. $C_{14}H_{17}NO_4$; mol wt 263.29. Microcrystalline powder from alcohol

+ acetone, mp 138°. Sol in water. LD$_{50}$ in mice: ~800 mg/kg i.v.; >5 g/kg orally (Tronche, 1974).

THERAP CAT: Diethylamine salt as capillary protectant.

2242. Chromomycin A₃. [7059-24-7] (1*S*)-1-*C*-[(2*S*,3*S*)-7-[[4-*O*-Acetyl-2,6-dideoxy-3-*O*-(2,6-dideoxy-4-*O*-methyl-α-D-*lyxo*-hexopyranosyl)-β-D-*lyxo*-hexopyranosyl]oxy]-3-[[*O*-4-*O*-acetyl-2,6-dideoxy-3-*C*-methyl-α-L-*arabino*-hexopyranosyl-(1 → 3)-*O*-2,6-dideoxy-β-D-*arabino*-hexopyranosyl-(1 → 3)-2,6-dideoxy-β-D-*arabino*-hexopyranosyl]oxy]-1,2,3,4-tetrahydro-5,10-dihydroxy-6-methyl-4-oxo-2-anthracenyl]-5-deoxy-1-*O*-methyl-D-*threo*-2-pentulose; 3^D-*O*-(4-*O*-acetyl-2,6-dideoxy-3-*C*-methyl-α-L-*arabino*-hexopyranosyl)-7-methylolivomycin D; aburamycin B; toyomycin. $C_{57}H_{82}O_{26}$; mol wt 1183.26. C 57.86%, H 6.99%, O 35.15%. Major component of an antitumor antibiotic complex produced by *Streptomyces griseus*. Binds specifically to guanine-cytosine base pairs in DNA; does not intercalate. The aglycone, **chromomycinone**, is identical to that of plicamycin, *q.v.* Isoln of complex and antibacterial activity: M. Shibata *et al.*, *J. Antibiot.* **13B,** 1 (1960), *C.A.* **54,** 22835g (1960). Characterization: K. Mizuno, *J. Antibiot.* **16A,** 22 (1963). Structure: M. Miyamoto *et al.*, *Tetrahedron* **23,** 421 (1967). Abs config: N. Harada *et al.*, *J. Am. Chem. Soc.* **91,** 5896 (1969). Revised structure: J. Thiem, B. Meyer, *J. Chem. Soc. Perkin Trans. 2* **1979,** 1331. Pharmacology and toxicity: M. Slavik, S. K. Carter, *Adv. Pharmacol. Chemother.* **12,** 1 (1975). Fluorescence characteristics: R. H. Jensen, *J. Histochem. Cytochem.* **25,** 573 (1977). Use for analysis and identification of chromosomes: J. H. van de Sande *et al.*, *Science* **195,** 400 (1977); in high-speed chromosome sorting: J. W. Gray *et al.*, *ibid..* **238,** 323 (1987); in flow karyotyping: M. F. Bartholdi, *Pathobiology* **58,** 118 (1990). Review of staining methods: H. A. Crissman, R. A. Tobey, *Methods Cell Biol.* **33,** 97-103 (1990).

Yellow powder, dec. 185°. [α]$_D^{23}$ −57° (ethanol). uv max (ethanol): 230, 281, 304, 318, 330, 412 nm (log ε 4.39, 4.72, 3.85, 3.92, 3.84, 4.07). Excitation max: ~445 nm. Emission max: ~575 nm. Sol in ethanol, ethyl acetate, DMSO, methanol. LD$_{50}$ in mice (mg/kg): 1.85 i.v. (Slavik, Carter). *Protect from light.*

USE: Fluorescent DNA stain in flow cytometry and karyotype analysis of chromosomes.

2243. Chromonar. [804-10-4] 2-[[3-[2-(Diethylamino)ethyl]-4-methyl-2-oxo-2*H*-1-benzopyran-7-yl]oxy]acetic acid ethyl ester; 3-(β-diethylaminoethyl)-4-methyl-7-(carbethoxymethoxy)coumarin; 3-(β-diethylaminoethyl)-4-methyl-7-(carbethoxymethoxy)-2*H*-1-benzopyran-2-one; ethyl [[3-[2-(diethylamino)ethyl]-4-methyl-2-oxo-2*H*-1-benzopyran-7-yl]oxy]acetate; carbochromen; carbocromen. $C_{20}H_{27}NO_5$; mol wt 361.44. C 66.46%, H 7.53%, N 3.88%, O 22.13%. Prepn: **BE 621327**; Ritter *et al.*, **US 3282938** (1963, 1966 both to Casella Farbwerke Mainkur). Toxicity study: R. E. Nitz, E. Potzch, *Arzneim.-Forsch.* **13,** 243 (1963). Metabolic studies: Schraven *et al.*, *ibid.* **20,** 1905 (1970). Series of articles on pharmacology: *ibid.* **13**, 243-268 (1963); on distribution: *ibid.* **22**, 479-511 (1972).

Practically insol in water.

Hydrochloride. [655-35-6] Cassella 4489; Antiangor; Intenkordin; Intensain. $C_{20}H_{27}NO_5 \cdot HCl$; mol wt 397.90. Crystalline powder, mp 159-160°. Freely sol in water, alc, methylene chloride, chloroform. Sparingly sol in acetone, methyl ethyl ketone, benzene, ether. Aq solns show blue fluorescence. LD_{50} in mice (g/kg): 6.3 orally; 0.528 i.p.; 0.0355 i.v. (Nitz, Potzsch).

THERAP CAT: Vasodilator (coronary).

2244. Chromotrope 2B. [548-80-1] 4,5-Dihydroxy-3-[2-(4-nitrophenyl)diazenyl]-2,7-naphthalenedisulfonic acid sodium salt (1:2); C.I. Acid Red 176; *p*-nitrobenzeneazochromotropic acid sodium salt; C.I. 16575. $C_{16}H_9N_3Na_2O_{10}S_2$; mol wt 513.36. C 37.43%, H 1.77%, N 8.19%, Na 8.96%, O 31.17%, S 12.49%. Prepn from diazotized *p*-nitroaniline and chromotropic acid, *see: Colour Index* vol. **4** (3rd ed., 1971) p 4097.

Reddish-brown powder. Sol in water with a yellowish-red color. Insol in alcohol.

USE: As a dye; as a reagent for boric acid or borates.

2245. Chromotropic Acid. [148-25-4] 4,5-Dihydroxy-2,7-naphthalenedisulfonic acid; 1,8-dihydroxynaphthalene-3,6-disulfonic acid. $C_{10}H_8O_8S_2$; mol wt 320.29. C 37.50%, H 2.52%, O 39.96%, S 20.02%. Prepn from 4-chloro-5-hydroxy-2,7-naphthalenedisulfonic acid: **DE 147852** (1904 to BASF); alternate methods: *Frdl.* **3**, 460-466.

Dihydrate. [5808-44-6] $C_{10}H_8O_8S_2 \cdot 2H_2O$; mol wt 356.32. White needles, soluble in water.

Disodium salt dihydrate. [5808-22-0] $C_{10}H_6Na_2O_8S_2 \cdot 2H_2O$; mol wt 400.28. White needles or leaflets. Very sol in water.

USE: Analytical reagent for the determn of nitrate and formaldehyde. Intermediate for azo dyes.

2246. Chromous Acetate. [628-52-4] Acetic acid chromium-(2+) salt (2:1). $C_4H_6CrO_4$; mol wt 170.08. C 28.25%, H 3.56%, Cr 30.57%, O 37.63%. $Cr(C_2H_3O_2)_2$. Prepn of monohydrate: Balthis, Jr., Bailar, Jr., *Inorg. Synth.* **1**, 122 (1939); Hatfield, *ibid.* **3**, 148 (1950); Kranz, Witkowska, *ibid.* **6**, 144 (1960); Ocone, Block, *ibid.* **8**, 125 (1966). Toxicity study: H. F. Smyth *et al., Am. Ind. Hyg. Assoc. J.* **30**, 470 (1969).

Monohydrate. Deep red powder, or monoclinic crystals; composed of dimeric units. Easily oxidized, especially when moist, to chromic acetate. Loses H_2O when dried over P_2O_5 at 100°, changing color to brown. d 1.79. Slightly sol in cold water; readily sol in hot water; sol in and reacts with most acids; slightly sol in alcohol. Practically insoluble in ether. *Keep tightly closed.* LD_{50} orally in rats: 11.26 g/kg (Smyth).

USE: Prepn of other chromous salts; as O_2 absorber in gas analysis.

2247. Chromous Bromide. [10049-25-9] Br_2Cr; mol wt 211.80. Br 75.45%, Cr 24.55%. $CrBr_2$. Prepn: Hein, Herzog in *Handbook of Preparative Inorganic Chemistry* **Vol. 2**, G. Brauer, Ed. (Academic Press, New York, 2nd ed. 1965) p 1340. Review of chromium halides: Fergusson in *Halogen Chemistry* **Vol. 3**, V. Gutmann, Ed. (Academic Press, New York, 1967) pp 227-302.

White monoclinic crystals, becoming yellow when heated. mp 842°; d_4^{25} 4.236. Stable in dry air, oxidizes in moist air. Sol in water (exothermic) giving blue soln; sol in alcohol. *Keep well closed.*

USE: In chromizing.

2248. Chromous Chloride. [10049-05-5] Chromium chloride (CrCl₂). Cl_2Cr; mol wt 122.90. Cl 57.69%, Cr 42.31%. $CrCl_2$. Prepn: Burg, *Inorg. Synth.* **3**, 150 (1950); Hein, Herzog, in *Handbook of Preparative Inorganic Chemistry* **Vol. 2**, G. Brauer, Ed. (Academic Press, New York, 2nd ed. 1965) pp 1336-1338. Prepn of tetrahydrate: Balthis, Jr., Bailar, Jr., *Inorg. Synth.* **1**, 125 (1939); Holah, Fackler, *ibid.* **10**, 26 (1967). Toxicity study: H. F. Smyth *et al., Am. Ind. Hyg. Assoc. J.* **30**, 470 (1969).

Lustrous needles or fused, fibrous mass. mp 824°; d_4^{14} 2.751. Very hygroscopic; stable in dry air but oxidizes rapidly if moist. Powerful reducing agent. Soluble in water giving a blue soln. *Keep well closed.* LD_{50} orally in rats: 1.87 g/kg (Smyth).

Tetrahydrate. Tetraaquochromium dichloride. Bright blue, hygroscopic crystals. Above 38° changes to isomeric green modification. Loses $1H_2O$, forming trihydrate, at 51°. Absorbs O_2 even when dry, forming a greenish-black oxychloride. Sol in water. Almost insol in concd HCl. On standing in soln it is oxidized by the water with liberation of H_2. *Keep well closed.*

USE: In chromizing; in prepn of Cr metal; in catalysts for organic reactions; as O_2 absorbent; in analysis.

2249. Chromous Fluoride. [10049-10-2] CrF_2; mol wt 89.99. Cr 57.78%, F 42.22%. Prepd from $CrCl_2$ + HF: Poulenc, *Compt. Rend.* **116**, 254 (1893); Kwasnik in *Handbook of Preparative Inorganic Chemistry* **Vol. 1**, G. Brauer, Ed. (Academic Press, New York, 2nd ed., 1963) pp 256-257. Review of other preparative methods: Sturm, *Inorg. Chem.* **1**, 665-672 (1962).

Blue-green, monoclinic crystals with iridescent sheen. d 3.79. mp 894°. Sparingly sol in water; practically insol in alcohol; sol in boiling HCl; not attacked by hot dil H_2SO_4 or HNO_3. Changes to Cr_2O_3 when heated in air.

USE: In chromizing; in catalytic cracking of hydrocarbons; alkylation catalyst; in nuclear reactor fuels.

2250. Chromous Formate. [4493-37-2] $C_2H_2CrO_4$; mol wt 142.03. C 16.91%, H 1.42%, Cr 36.61%, O 45.06%. $Cr(HCOO)_2$. Prepd from $CrCl_2$ and Na formate: Lux, Illman, *Ber.* **91**, 2143 (1958); Earnshaw *et al., Proc. Chem. Soc. London* **1963**, 281. Prepn of blue and red anhydrous forms from hydrated formates: Herzog, Kalies, *Z. Chem.* **5**, 273 (1965).

Hemihydrate. Violet to blue crystals. Sol in water.

Monohydrate. Red needles. Sol in water to give blue soln.

Hemipentahydrate. Large, dark-red cubic crystals. Sol in water, alcohol, ether, acetone.

USE: In baths for Cr electroplating; in catalysts for organic reactions.

2251. Chromous Sulfate. [13825-86-0] CrO_4S; mol wt 148.05. Cr 35.12%, O 43.23%, S 21.65%. $CrSO_4$. Prepn of pentahydrate: Lux, Illman, *Ber.* **91**, 2143 (1958); Holah, Fackler, *Inorg. Synth.* **10**, 26 (1967). The common hydrated form, long considered to be the heptahydrate, has been reported to be the pentahydrate: Lux, Illman, *loc. cit.;* Lux *et al., Ber.* **97**, 503 (1964). Prepn of standard soln for use as an analytical reagent: Lingane, Pecsok, *Anal. Chem.* **20**, 425 (1948).

Pentahydrate. Blue crystals. Stable in air if dry. Sol in water; slightly sol in alcohol; practically insol in acetone. Soluble in dil, dec by concd H_2SO_4. Solns are rapidly oxidized by atmospheric oxygen.

USE: Analytical reagent; absorption of O_2 from gas mixtures; dehydrohalogenating and reducing agent.

2252. Chromyl Chloride. [14977-61-8] (*T*-4)-Dichlorodioxochromium; chromium dioxychloride. Cl_2CrO_2; mol wt 154.89. Cl 45.77%, Cr 33.57%, O 20.66%. CrO_2Cl_2. Prepn from CrO_3 +

HCl: Sisler, *Inorg. Synth.* **2**, 205 (1946); Hein, Herzog in *Handbook of Preparative Inorganic Chemistry* **Vol. 2**, G. Brauer, Ed. (Academic Press, New York, 2nd ed., 1965) p 1384; and AlCl₃: Flesch, Svec, *J. Am. Chem. Soc.* **80**, 3189 (1958); from Cr_2O_3 and $TiCl_4$: Braos, Cohen, **US 3111380** (1963 to Natl. Distillers & Chem.). Review of chromium halides: Fergusson in *Halogen Chemistry* **Vol. 3**, V. Gutmann, Ed. (Academic Press, New York, 1967) pp 227-302.

Deep red liquid; appears black under reflected light. Fumes in moist air. *Handle only in well-ventilated hood.* mp $-96.5°$; bp 117°; d_4^{25} 1.91. Slightly less viscous than water. Nonconductor of electricity. Strong oxidizing agent; *can react explosively with combustible organic and inorganic substances.* Hydrolyzes vigorously on contact with water. Reacts vigorously with liquid or gaseous ammonia. Sol in CCl_4, CS_2, benzene, nitrobenzene, $CHCl_3$, $POCl_3$. Its soln in CCl_4 is fairly stable. Liquid CrO_2Cl_2 is stable indefinitely in glass, aluminum, stainless steel containers if protected from light and moisture. CrO_2Cl_2 dissolves CrO_3, yielding a powerful oxidant.

Caution: Potential symptoms of overexposure are irritation of eyes, skin, upper respiratory system; eye, skin burns. *See NIOSH Pocket Guide to Chemical Hazards* (DHHS/NIOSH 97-140, 1997) p 72.

USE: Catalyst for polymerization of olefins; oxidation of hydrocarbons; in the Etard reaction for production of aldehydes and ketones; in the prepn of various coordination complexes of Cr.

2253. Chromyl Fluoride. [7788-96-7] *(T-4)*-Difluorodioxochromium; chromium oxyfluoride. CrF_2O_2; mol wt 121.99. Cr 42.62%, F 31.15%, O 26.23%. CrO_2F_2. Prepn of impure product from CrO_2Cl_2 and F_2: von Wartenberg, *Z. Anorg. Allg. Chem.* **247**, 140 (1941). Prepn from CrO_3 and SeF_4: Bartlett, Robinson, *J. Chem. Soc.* **1961**, 3549; from $K_2Cr_2O_7$ and HF or from Cr, KNO_3 and HF: Wiechert, *Z. Anorg. Allg. Chem.* **261**, 315 (1950). Review of earlier prepns and prepn from CrO_3 and ClF, COF_2, MoF_6 or WF_6: P. J. Green, G. L. Gard, *Inorg. Chem.* **16**, 1243 (1977). Review of chromium halides: Fergusson in *Halogen Chemistry* **vol. 3**, V. Gutmann, Ed. (Academic Press, New York, 1967) pp 227-302.

Red-violet to black rhombic or monoclinic crystals; described as a gas or a liquid in earlier work with impure samples. Sublimation temp 29.6°; mp 31.6° (885 mm). *Extremely reactive;* etches glass, quartz. Ignites hydrocarbons such as methane or butane at elevated temps; mixtures burn with a bright flame producing fumes of Cr_2O_3 and CrF_3. Can be stored in a sealed aluminum phosphate glass tube or in a Kel-F tube. Stable indefinitely at room temp in the dark. Polymerizes slowly if exposed to visible, ultraviolet or infrared light to a gray-white solid, mp 200°.

USE: Fluorination catalyst; to increase olefin-polymer receptivity for dyes.

2254. Chrysanthemaxanthin. [27780-11-6] (3*S*,3'*R*,5*R*,-6'*R*,8*S*)-5,8-Epoxy-5,8-dihydro-β,ε-carotene-3,3'-diol. $C_{40}H_{56}O_3$; mol wt 584.89. C 82.14%, H 9.65%, O 8.21%. Carotenoid pigment. Occurs together with flavoxanthin, *q.v.* Has same structural formula, mp, rotation, and absorption as flavoxanthin, but differs sterically. Can be separated chromatographically: in a zinc carbonate column the chrysanthemaxanthin zone locates itself below flavoxanthin and above lutein epoxide. Isoln from asters: Karrer, Jucker, *Helv. Chim. Acta* **26**, 626 (1943). Partial synthesis from lutein, *eidem, ibid.* **28**, 300 (1945). For absolute configuration see: H. Cadosch *et al., ibid.* **61**, 783 (1978).

Golden yellow leaflets, mp 184-185°, also reported as mp 178.5-180° (Codosch). $[\alpha]_D^{23}$ +104° ±5% (c = 1.833 mg/ml in chloroform). Absorption max ($CHCl_3$): 459, 430 nm. uv max (ethanol): 204, 251, 298, 311, 399, 422, 448 nm (ε 27300, 28800, 10000, 13100, 97700, 150400, 148500). Freely sol in chloroform, benzene, acetone. Less sol in methanol, ethanol. Almost insol in petr ether.

Note: Chrysanthemaxanthin has no vitamin A activity.

2255. Chrysanthemic Acid. [10453-89-1] 2,2-Dimethyl-3-(2-methyl-1-propen-1-yl)cyclopropanecarboxylic acid; chrysanthemummonocarboxylic acid; chrysanthemumic acid. $C_{10}H_{16}O_2$; mol wt 168.24. C 71.39%, H 9.59%, O 19.02%. Occurs as esters in pyrethrum flowers, *see* Pyrethrin I and Cinerin I, *also* Allethrin (a synthetic product). Isoln and structure: Staudinger, Ruzicka, *Helv. Chim. Acta* **7**, 177, 201 (1924). Synthesis: Staudinger *et al., ibid.* 390; Campbell, Harper, *J. Chem. Soc.* **1945**, 283; M. J. Devos, A. Krief, *Tetrahedron Lett.* **1978**, 1845; **1979**, 1891; O. A. Nesmeyanova *et al., Synthesis* **1982**, 296. Asymmetric synthesis: T. Aratani *et al., ibid.* **1977**, 2599. Stereoselective synthesis of *dl-trans*-chrysanthemic acid: Mills *et al., Chem. Commun.* **1971**, 555; *eidem, J. Chem. Soc. Perkin Trans. 1* **1973**, 133; S. C. Welch, T. A. Valdes, *J. Org. Chem.* **42**, 2108 (1977); J. P. Genet *et al., Tetrahedron Lett.* **21**, 3183 (1980); of *dl-cis*-form: J. Mann, A. Thomas, *ibid.* **27**, 3533 (1986). Synthesis of (−)-*cis*- and (+)-*trans*-forms: Gopichand *et al., Indian J. Chem.* **13**, 433 (1975).

d-trans form

dl-cis-Form. [2935-23-1] Cubic prisms from ethyl acetate, mp 115-116°.

dl-trans-Form. [705-16-8] Long prisms, mp 54°. Very sol in ethyl acetate. Exhibits a marked negative heat of soln in ethyl acetate or in methanol.

l-trans-Form. [2259-14-5] Elongated prisms, mp 17-21°. $[\alpha]_D^{25}$ −14.01° (c = 1.535 in abs alc).

d-trans-Form. [4638-92-0] Elongated prisms, mp 17-21°. $[\alpha]_D^{25}$ +14.16° (c = 1.554 in abs alc).

2256. Chrysanthenone. [473-06-3] 2,7,7-Trimethylbicyclo-[3.1.1]hept-2-en-6-one; 2-pinen-7-one. $C_{10}H_{14}O$; mol wt 150.22. C 79.96%, H 9.39%, O 10.65%. A constituent of the essential oil of *Chrysanthemum indicium* L., also extracted from *C. sinense* Sabin, *Compositae*. Prepared by photochemical rearrangement of *l*- or *d*-verbenone, *q.v.*: Hurst, Whitham, *J. Chem. Soc.* **1960**, 2864; Erman, *J. Am. Chem. Soc.* **89**, 3828 (1967); Schuster, Widman, *Tetrahedron Lett.* **1971**, 3571. Structure: Kotake, Nonaka, *Ann.* **607**, 153 (1957); Blanchard, *Chem. Ind. (London)* **1958**, 293. Chemistry: Erman, *J. Am. Chem. Soc.* **91**, 779 (1969).

d-Form. Herbaceous odor. bp₁₂ 88-89°. n_D^{22} 1.4720. $[\alpha]_D$ +37° (c = 2.1 in chloroform). uv max: 290 nm (ε 120). Insoluble in water; sol in alcohol.

2257. Chrysarobin. [491-58-7] 1,8-Dihydroxy-3-methyl-9(10*H*)-anthracenone; chrysophanolanthrone. $C_{15}H_{12}O_3$; mol wt 240.26. C 74.99%, H 5.03%, O 19.98%. Active principle of araroba, *q.v.*, from the South American medicinal tree, *Andira araroba* Aguiar, *Leguminosae*. Reduction product of chrysophanic acid, *q.v.*; prototype of the anthrone class of antipsoriatic agents. Term is also used for the partially purified extract of araroba powder which has been used as a vermifuge, purgative, and dermatologic. Isoln from araroba powder: C. Liebermann, P. Seidler, *Ber. Dtsch. Pharm. Ges.* **11**, 1603 (1878); H. A. D. Jowett, C. E. Potter, *J. Chem. Soc. Trans.* **81**, 1575 (1902). Confirmation of structure: R. Eder, F. Hauser, *Arch. Pharm.* **203**, 321 (1925). Synthesis by reduction of chrysophanic acid: C. A. Naylor, J. H. Gardner, *J. Am. Chem. Soc.* **53**, 4114 (1931). Biogenetic-type synthesis: T. M. Harris *et al., ibid.* **98**, 6065 (1976). Skin tumor promoting activity: J. DiGiovani, R. K. Boutwell, *Carcinogenesis* **4**, 281 (1983); F. H. Kruszewski *et al., Cancer Res.* **47**, 3783 (1987). Histologic alterations in mouse skin: *idem et al., J. Invest. Dermatol.* **92**, 64 (1989).

Yellow needles from glacial acetic acid, mp 203.4-204°.
9-Hydroxy analog. [491-59-8] 3-Methyl-1,8,9-anthracenetriol; 1,8,9-trihydroxy-3-methylanthracene.

2258.   6-Chrysenamine. [2642-98-0] 6-Chrysenylamine; 6-aminochrysene; 6-chrysylamine; CP-1001; Chrysenex. $C_{18}H_{13}N$; mol wt 243.31. C 88.86%, H 5.39%, N 5.76%. Prepd by reduction of 6-nitrochrysene: Newman, Cathcart, *J. Org. Chem.* **5**, 620 (1940).

Leaflets from alc, mp 210-211°. Slightly sol in alcohol, benzene, ethyl acetate.
USE: In biochemical research to produce leukopenia.

2259.   Chrysene. [218-01-9] 1,2-Benzphenanthrene. $C_{18}H_{12}$; mol wt 228.29. C 94.70%, H 5.30%. Occurs in coal tar, *q.v.* Is formed during distillation of coal, in very small amount during distillation or pyrolysis of many fats and oils. Isoln from coal tar: Liebermann, *Ann.* **158**, 299, 307 (1871). Purification by chromatography: Winterstein, Schön, *Z. Physiol. Chem.* **230**, 146 (1934); Winterstein *et al.*, *ibid.* 158. Synthesis by heating H_2 and acetylene to 800°: Meyer, *Ber.* **45**, 1633 (1912). From cholesterol on heating with palladium charcoal or activated charcoal: Schmid, Zentner, *Monatsh. Chem.* **49**, 96 (1928). Review of toxicology and human exposure: *Toxicological Profile for Polycyclic Aromatic Hydrocarbons* (PB95-264370, 1995) 487 pp.

Orthorhombic bipyramidal plates from benzene. d_4^{20} 1.274. mp 254°. Sublimes easily *in vacuo.* bp 448°. Strong fluorescence under ultraviolet light. Absorption spectrum: Marchlewski, Moroz, *Bull. Soc. Chim. Fr.* [4] **33**, 1406 (1923). Slightly sol in alc, ether, carbon bisulfide, glacial acetic acid. At 25° one gram dissolves in 1300 ml abs alc, 480 ml toluene. About 5% is sol in toluene at 100°. Moderately sol in boiling benzene. Insol in water. Slightly sol in cold organic solvents, but fairly sol in these solvents when hot, including glacial acetic acid.

2260.   Chrysin. [480-40-0] 5,7-Dihydroxy-2-phenyl-$4H$-1-benzopyran-4-one; 5,7-dihydroxyflavone; chrysidenon 1438. $C_{15}H_{10}O_4$; mol wt 254.24. C 70.86%, H 3.96%, O 25.17%. From heartwood of *Pinus monticola* Dougl., *P. excelsa* Wall., and *P. aristata* Engelm., *Pinaceae:* Linstedt, *Acta Chem. Scand.* **3**, 1147, 1375 (1949); **4**, 55 (1950); from bark of *Dolichandrone falcata* Seem., *Bisnomiaceae:* Kincl, *Naturwissenschaften* **42**, 646 (1955). Synthesis: Seka, Prosche, *Monatsh. Chem.* **69**, 284 (1936); Hutchins, Wheeler, *J. Chem. Soc.* **1939**, 91; Teoule *et al.*, *Bull. Soc. Chim. Fr.* **1961**, 546.

Light yellow prisms from methanol, mp 285°. uv max: 270, 329 nm (log ε 4.40, 3.90). Practically insol in water; sol in alkali hydroxide solns; slightly sol in alcohol, chloroform, ether.
Diacetoxychrysin. $C_{19}H_{14}O_6$. Crystals from ethanol, mp 194-195°.
Methylchrysin. Tectochrysin. $C_{16}H_{12}O_4$. It is present as such or in the form of a glucoside in buds of *Populus* spp., *Salicaceae.* Use of tectochrysin as diuretic: Perrault, **US 3155579** (1964 to Laroche Navarron). Yellow needles, mp 163°. Sol in alcohol, benzene, chloroform.

2261.   Chrysoidine. [532-82-1] 4-(2-Phenyldiazenyl)-1,3-benzenediamine hydrochloride (1:1); 4-phenylazo-*m*-phenylenediamine hydrochloride; 2,4-diaminoazobenzene hydrochloride; chrysoidine orange; chrysoidine Y; C.I. 11270; C.I. Basic Orange 2. $C_{12}H_{13}ClN_4$; mol wt 248.71. C 57.95%, H 5.27%, Cl 14.25%, N 22.53%. Prepn: Hofmann, *Ber.* **10**, 213 (1877); Maximoff, **US 2053095** (1935 to Azodal Co.). Prepn of salt: **DE 562392** (1933 to Chem.-Pharm. Fabrik Hubold & Bartsch). *See also Colour Index* **vol. 4** (3rd ed., 1971) p 4019; *H. J. Conn's Biological Stains*, R. D. Lillie, Ed. (Williams & Wilkins, Baltimore, 9th ed., 1977) p 87.

Reddish-brown crystalline powder, mp 118-118.5°. Solubility at 15°: water 5.5%; abs ethanol 4.75%; cellosolve 6.0%; anhydr ethylene glycol 9.5%; xylene 0.005%; slightly sol in acetone. Practically insol in benzene. Gives yellow soln in concd H_2SO_4, orange soln in dil H_2SO_4 and in HNO_3.
Citrate. [5909-04-6] 4-Phenylazo-*m*-phenylenediamine hydrochloride citrate; 2,4-diaminoazobenzene hydrochloride citrate; Azoangin; Azohel. $C_{12}H_{13}ClN_4 \cdot C_6H_8O_7$; mol wt 440.84. Sol in water (up to 4%), alcohol.
Free base. [495-54-5] C.I. Solvent Orange 3. $C_{12}H_{12}N_4$.
USE: Dyeing silk and cotton. Biological stain.
THERAP CAT: Citrate as antiseptic.

2262.   Chrysophanic Acid. [481-74-3] 1,8-Dihydroxy-3-methyl-9,10-anthracenedione; 1,8-dihydroxy-3-methylanthraquinone; 3-methylchrysazin; chrysophanol. $C_{15}H_{10}O_4$; mol wt 254.24. C 70.86%, H 3.96%, O 25.17%. Occurs in the free state and as glucoside in cascara sagrada, senna and various species of *Rumex* and *Rheum* (rhubarb). Isoln from rhubarb root: Tutin, Clewer, *J. Chem. Soc.* **99**, 946 (1911); Siesto, Bartoli, *Farmaco Ed. Prat.* **12**, 517 (1957); Carelli, Giuliano, *ibid.* 184; from *Penicillium islandicum* Sopp.: Howard, Raistrick, *Biochem. J.* **46**, 49 (1950); from *Chaetonium affine* Corda: Arkley *et al.*, *Croat. Chem. Acta* **29**, 141 (1957), *C.A.* **53**, 1287h (1959). Synthesis: Eder, Widmer, *Helv. Chim. Acta* **5**, 3 (1922); **6**, 419 (1923); Ayyangar *et al.*, *J. Sci. Ind. Res.* **20B**, 493 (1961). Total synthesis: M. E. Jung, J. A. Lowe, *Chem. Commun.* **1978**, 95.

Hexagonal or monoclinic crystals from alcohol or benzene, mp 196°. Sublimes. Absorption max: 226, 256, 278, 288, 436 nm ($\varepsilon \times 10^{-3}$ 41, 28, 14, 14, 11.8). Practically insol in water. Slightly sol in cold, freely in boiling alc; sol in benzene, chloroform, ether, glacial acetic acid, acetone, solns of alkali hydrides, and in hot solns of alkali carbonates; very slightly sol in petr ether.
Glucoside. Chrysophanein; chrysophaniin. $C_{21}H_{20}O_9$. Fine yellow needles from alc, mp 248-249°. Slightly sol in hot water; sol in pyridine. Practically insol in cold water, chloroform, ether.

2263.   Chymopapain. [9001-09-6] EC 3.4.22.6; BAX-1526; Chymodiactin; Discase. Proteolytic enzyme which is the major component of the crude latex of *Carica papaya*, *Caricaceae.* A sulfhydryl enzyme similar to papain, *q.v.*, with respect to substrate specificities, but differing in electrophoretic mobility, stability and solubility. Original crystallization and partial characterization: Jansen, Balls, *J. Biol. Chem.* **137**, 459 (1941); *eidem*, US 2313875 (1943). Consists of four components, two of which have molecular wts of about 35,-000 and have been isolated and studied. Chymopapain A: Erbata, Yasunobu, *J. Biol. Chem.* **237**, 1086 (1962); chymopapain B: Kunimitsu, Yasunobu, *Biochim. Biophys. Acta* **139**, 405 (1966); Isunoda, Yasunobu, *J. Biol. Chem.* **241**, 4610 (1966). Purification: Stern, US 3558433 (1971 to Baxter Labs.). Specificity studies: Ebata, Takahashi, *Biochim. Biophys. Acta* **118**, 201 (1966). *Reviews:* Kunimitsu, Yasunobu, *Methods Enzymol.* **19**, 244-252 (1970); Glazer, Smith, *The Enzymes* vol. III, P. D. Boyer, Ed. (Academic Press, New York, 3rd ed., 1971) pp 537-538. Review of use in herniated disk treatment: M. J. David, *J. Am. Med. Assoc.* **243**, 2043 (1980).

Powder, more sol in aq soln than papain. Extremely stable at pHs as low as 1.8 and sol in satd NaCl above pH 3. Most active at pHs 2.5 to 4.0. uv max (pH 7): 280 nm ($A_{1cm}^{1\%}$ 18.7 (A); 18.4 (B)). The ratio of milk-clotting to proteolytic activity is twice that of papain although the clotting capacities are the same.

USE: In meat tenderizer.

THERAP CAT: Proteolytic enzyme (used in chemonucleolysis).

2264. Chymotrypsins. Avazyme; Catarase; Chymetin; Chymolase; Kymo-trypure; Quimoral; Quimotrase; Zolyse. A group of major proteolytic enzymes in the pancreatic juice. Produced in the form of inactive chymotrypsinogen by the acinous cells of pancreas, and carried as such by the pancreatic juice into the duodenum where they are activated by trypsin. They split secondary amide or peptide bonds, carboxylic or phenolic ester bonds and even carbon-carbon bonds. Their main function is to hydrolyze peptide bonds during the intestinal digestion of proteins. Cattle pancreatic juice contains nearly equal quantities of two chymotrypsinogens: A which is cationic at pH 8 and B which is anionic. Chymotrypsinogen A is converted by trypsin into π-chymotrypsin. In the presence of large amounts of trypsin, π-chymotrypsin is degraded to δ-chymotrypsin. In the presence of small amounts of trypsin, π-chymotrypsin is autolyzed to α-chymotrypsin. After long standing α-chymotrypsin gives rise to β- and γ-chymotrypsin which differ from it in their crystal habits. Prepn of π- and δ-chymotrypsins: Bettelheim, Neurath, *J. Biol. Chem.* **212**, 241 (1955). Prepn of α-, β-, γ- and δ-chymotrypsins (and chymotrypsin B): Laskowski, *Methods Enzymol.* **2**, 8 (1955). Amino acid sequence of chymotrypsinogen A: Hartley, *Nature* **201**, 1284 (1964); Melorin *et al.*, *Biochim. Biophys. Acta* **130**, 543 (1966); of chymotrypsinogen B: Hartley *et al.*, *Nature* **207**, 1157 (1965); Smillie *et al.*, *ibid.* **218**, 343 (1968). *Reviews:* Bender *et al.*, *J. Polym. Sci.* **49**, 75 (1961); Desnuelle, *The Enzymes* vol. **4**, P. D. Boyer *et al.*, Eds. (Academic Press, New York, 2nd ed., 1960) pp 93-118; Kraut; Blow; Hess, *ibid.* vol. **3** (3rd ed., 1971) 165, 185, 213.

α-**Chymotrypsin.** [9004-07-3] Alpha-Chymocutan; Chymotase (tabl.); Chymozym; Impral; Kimopsin; Kimoral. Arrangement of molecules in monoclinic crystal form: Blow *et al.*, *J. Mol. Biol.* **8**, 65 (1964). Structure: Matthews *et al.*, *Nature* **214**, 652 (1967); Blow *et al.*, *ibid.* **221**, 337 (1969). *Review:* Niemann, *Science* **143**, 1287 (1964).

THERAP CAT: Enzyme (proteolytic).

THERAP CAT (VET): Topically for enzymatic debridement.

2265. Cichoriin. [531-58-8] 7-(β-D-Glucopyranosyloxy)-6-hydroxy-2*H*-1-benzopyran-2-one; 6,7-dihydroxycoumarin-7-glucoside. $C_{15}H_{16}O_9$; mol wt 340.28. C 52.95%, H 4.74%, O 42.32%. Isomeric with esculin. In flowers of the chicory plant *(Cichorium intybus* L., *Compositae)*. Extraction procedure: Merz, *Arch. Pharm.* **270**, 476 (1932). Synthesis: Head, Robertson, *J. Chem. Soc.* **1939**, 1266; also Merz, *loc. cit.* *Review:* Sethna, Shah, *Chem. Rev.* **36**, 1 (1945).

Needles with $2H_2O$, mp 213-215° (after drying in desiccator). $[\alpha]_D^{18}$ −105° (c = 3 in 50% dioxane). Sol in hot water, alc, glacial acetic acid. Insol in ether, petr ether. Sol in dil alkalies with yellow color, but no fluorescence (difference from esculin). Absorption spectrum of soln: Merz, *loc. cit.*

2266. Ciclesonide. [126544-47-6] (11β,16α)-16,17-[[(*R*)-Cyclohexylmethylene]bis(oxy)]-11-hydroxy-21-(2-methyl-1-oxopropoxy)pregna-1,4-diene-3,20-dione; BY-9010; Alvesco; Omnaris; Osonide. $C_{32}H_{44}O_7$; mol wt 540.70. C 71.08%, H 8.20%, O 20.71%. Inhaled corticosteroid prodrug; de-esterified in the lung to the active metabolite, desisobutyryl-ciclesonide. Prepn: J. Calatayud *et al.*, DE 4129535; *eidem*, US 5482934 (1992, 1996 both to Elmu). Clinical pharmacokinetics and pharmacodynamics: H. Derendorf, *J. Clin. Pharmacol.* **47**, 782 (2007). Review of pharmacology and clinical experience in asthma: G. L. Colice, *Expert Opin. Pharmacother.* **7**, 2107-2117 (2006). Clinical trial in pediatric asthma: E. W. Gelfand *et al.*, *J. Pediatr.* **148**, 377 (2006); in allergic rhinitis: P. Chervinsky *et al.*, *Ann. Allergy Asthma Immunol.* **99**, 69 (2007). Review of pharmacology and clinical experience in asthma: P. E. Korenblat, *Expert Opin. Pharmacother.* **11**, 463-479 (2010).

White to yellow-white powder. Freely sol in ethanol, acetone. Practically insol in water.

THERAP CAT: Anti-asthmatic; glucocorticoid.

2267. Cicletanine. [89943-82-5] 3-(4-Chlorophenyl)-1,3-dihydro-6-methylfuro[3,4-*c*]pyridin-7-ol; 1,3-dihydro-3-(4-chlorophenyl)-7-hydroxy-6-methylfuro[3,4-*c*]pyridine; cicletanide; cycletanide. $C_{14}H_{12}ClNO_2$; mol wt 261.71. C 64.25%, H 4.62%, Cl 13.55%, N 5.35%, O 12.23%. Furopyridine derivative with antihypertensive and diuretic properties. Prepn: A. Esanu, BE 891797; *idem*, US 4383998 (1982, 1983 both to Soc. Conseils Recher. Sci.). Studies on mechanism of action: P. Braquet *et al.*, *Lancet* **1**, 1218 (1983); G. R. Elliott *et al.*, *Thromb. Res.* **33**, 549 (1984). Cardiovascular pharmacology in dogs: R. Jouve *et al.*, *J. Cardiovasc. Pharmacol.* **8**, 208 (1986). *In vitro* comparison of antihistaminic activity of isomers: P. Schoeffter *et al.*, *Eur. J. Pharmacol.* **136**, 235 (1987). HPLC determn in biological fluids: G. Cuisinaud *et al.*, *J. Chromatogr.* **341**, 97 (1985). Effect on human prostaglandin metabolism: P. Guinot, J. C. Frölich, *Arzneim.-Forsch.* **35**, 1714 (1985). Effect on potassium ion transport in hypertensive patients: R. Garay *et al.*, *J. Hypertens.* **4**, Suppl. 5, S208 (1986). Pharmacokinetics in humans: J. M. Lize *et al.*, *Therapie* **42**, 399 (1987).

Hydrochloride. [82747-56-6] BN-1270; Coverine; Justar; Secletan; Tenstaten. $C_{14}H_{12}ClNO_2 \cdot HCl$; mol wt 298.16. White crystals, mp 219-228°. Insol in water.

THERAP CAT: Antihypertensive.

2268. Ciclonicate. [53449-58-4] 3-Pyridinecarboxylic acid *rel*-(1*R*,5*S*)-3,3,5-trimethylcyclohexyl ester; *trans*-3,3,5-trimethylcyclohexyl nicotinate; cyclonicate; P-350; Bled. $C_{15}H_{21}NO_2$; mol wt 247.34. C 72.84%, H 8.56%, N 5.66%, O 12.94%. Deriv of nicotinic acid, *q.v.* Prepn: K. Matsuda *et al.*, **DE 1910481** (1969 to Takeda), *C.A.* **72**, 12769g (1970); G. Massaroli, **DE 2406849** (1974 to Poli), *C.A.* **82**, 4129 (1975). Chromatographic study: G. Sekules *et al.*, *Boll. Chim. Farm.* **119**, 521 (1980). Antilipolytic activity in rats: D. Faini *et al.*, *Farmaco Ed. Prat.* **36**, 478 (1981). *In vitro* effect on lipolysis: F. Tessari *et al.*, *Farmaco Ed. Sci.* **36**, 1029 (1981). Clinical study: C. Fossati, *Gazz. Med. Ital.* **139**, 43 (1980).

Relative stereochemistry

Liq, bp$_{0.6}$ 127-128°.
THERAP CAT: Vasodilator.

2269. Ciclopirox. [29342-05-0] 6-Cyclohexyl-1-hydroxy-4-methyl-2(1*H*)-pyridinone. C$_{12}$H$_{17}$NO$_2$; mol wt 207.27. C 69.54%, H 8.27%, N 6.76%, O 15.44%. Broad spectrum antimycotic agent with some antibacterial activity. Prepn: G. Lohaus, W. Dittmar, **ZA 6906039**; *eidem*, **US 3883545** (1970, 1975 both to Hoechst). *In vitro* study: *eidem*, *Arzneim.-Forsch.* **23**, 670 (1973). Series of articles on pharmacokinetics, pharmacology, teratology, toxicity studies: *Oyo Yakuri* **9**, 57-115 (1975), *C.A.* **83**, 53159d, 53538b, 53539c, 71844c, 90833q (1975). Series of articles on chemistry, mechanism of action, toxicology, clinical trials: *Arzneim.-Forsch.* **31**, 1309-1386 (1981). Toxicity data: H. G. Alpermann, E. Schutz, *ibid.* 1328. *Review:* S. G. Jue *et al.*, *Drugs* **29**, 330-341 (1985). Review of clinical experience in seborrheic dermatitis: A. Starova, R. Aly, *Expert Opin. Drug Saf.* **4**, 235-239 (2005).

White to slightly yellowish-white, crystalline powder. mp 144°. Freely sol in ethanol, methylene chloride; sol in ether; slightly sol in water.
Ethanolamine salt (1:1). [41621-49-2] Ciclopirox olamine; HOE-296; Batrafen; Brumixol; Ciclochem; Dafnegin; Loprox; Micoxolamina; Mycoster. C$_{14}$H$_{24}$N$_2$O$_3$; mol wt 268.36. White to slightly yellowish-white, crystalline powder. Very sol in ethanol, methylene chloride; slightly sol in water. Practically insol in cyclohexane. LD$_{50}$ in mice, rats (mg/kg): 2898, 3290 orally (Alpermann, Schutz).
THERAP CAT: Antifungal.

2270. Cicutoxin. [505-75-9] (8*E*,10*E*,12*E*,14*R*)-8,10,12-Heptadecatriene-4,6-diyne-1,14-diol. C$_{17}$H$_{22}$O$_2$; mol wt 258.36. C 79.03%, H 8.58%, O 12.39%. Poisonous principle of the water hemlock, *Cicuta virosa* L., *Umbelliferae*, thought to be one of the most poisonous plants in North America. Naturally occurring as the (−)-form; isomeric with enanthotoxin, *q.v.* Isoln: C. A. Jacobson, *J. Am. Chem. Soc.* **37**, 916 (1915). Structure: E. F. L. J. Anet *et al.*, *J. Chem. Soc.* **1953**, 309. Synthesis of (±)-form: B. E. Hill *et al.*, *ibid.* **1955**, 1770. Synthesis of (−)-form: B. W. Gung, A. O. Omollo, *Eur. J. Org. Chem.* **2009**, 1136. Isoln, absolute configuration, and properties: T. Ohta *et al.*, *Tetrahedron* **55**, 12087 (1999). Review of botanical characterization, pharmacology and symptoms of poisoning: L. J. Schep *et al.*, *Clin. Toxicol.* **47**, 270-278 (2009).

Colorless oil, [α]$_D^{26}$ −14.9° (c = 1.12 in methanol) (Ohta). Also reported as prisms from ether + petr ether, mp 54° (Anet). *Poisonous.* Transformed in air and light to yellow oily resin. [α]$_D^{15}$ −14.5° (c = 1.7 in ethanol). Sol in alc, chloroform, ether, hot water, alkali hydroxides. Practically insol in petr ether. uv max (ether): 335.4, 317.8, 251.6, 241.8 nm (log ε 4.64, 4.66, 4.22, 4.07). LD$_{50}$ i.p. in mice: 2.8 mg/kg (Ohta).
(±)-Form. Crystals from ether + petr ether, mp 67°. uv max (alc): 242, 252, 318.5, 335.5 nm (ε × 10^{-3} 14.6, 21.6, 50.6, 60.3).
Caution: Highly toxic. Symptoms of overexposure include seizures, nausea, vomiting, diarrhea, tachycardia, mydriasis, rhabomyolysis, renal failure, coma, death (Schep).

2271. Cidofovir. [113852-37-2] *P*-[[(1*S*)-2-(4-Amino-2-oxo-1(2*H*)-pyrimidinyl)-1-(hydroxymethyl)ethoxy]methyl]phosphonic acid; (*S*)-1-[3-hydroxy-2-(phosphonylmethoxy)propyl]cytosine; (*S*)-HPMPC; GS-504; Vistide. C$_8$H$_{14}$N$_3$O$_6$P; mol wt 279.19. C 34.42%, H 5.05%, N 15.05%, O 34.38%, P 11.09%. DNA synthesis inhibitor. Prepn: A. Holy *et al.*, **EP 253412**; *eidem*, **US 5142051** (1988, 1992 both to Ceskoslov. Akad. Ved; Rega Inst.); and activity vs cytomegalovirus: R. Snoeck *et al.*, *Antimicrob. Agents Chemother.* **32**, 1839 (1988). Syntheses: J. J. Bronson *et al.*, *Nucleosides Nucleotides* **9**, 745 (1990); P. R. Brodfuehrer *et al.*, *Tetrahedron Lett.* **35**, 3243 (1994). Activity vs herpes simplex virus: G. Andrei *et al.*, *Eur. J. Clin. Microbiol. Infect. Dis.* **11**, 143 (1992). Review of pharmacology and clinical studies: M. J. M. Hitchcock *et al.*, *Antivir. Chem. Chemother.* **7**, 115-127 (1996). Review of clinical potential in poxvirus infections: E. De Clercq, *Trends Pharmacol. Sci.* **23**, 456-458 (2002).

Fluffy white powder, mp 260° (dec). [α]$_D^{20}$ −97.3° (c = 0.80 in water). Monohydrate, uv max (pH 2): 279 nm (ε 13000).
THERAP CAT: Antiviral.

2272. Cifenline. [53267-01-9] 2-(2,2-Diphenylcyclopropyl)-4,5-dihydro-1*H*-imidazole; (±)-2-(2,2-diphenylcyclopropyl)-2-imidazoline; 1-(2-Δ^2-imidazolinyl)-2,2-diphenylcyclopropane; cibenzoline; Ro-22-7796; UP-33-901. C$_{18}$H$_{18}$N$_2$; mol wt 262.36. C 82.41%, H 6.92%, N 10.68%. Prepn: **BE 807630**; J. C. Cognaco, **US 3903104** (1974, 1975 both to Hexachemie). Activity and tolerance: D. Herpin *et al.*, *Acta Cardiol.* **36**, 131 (1981). Electrophysiological effects in man: J. F. Thizy *et al.*, *Lyon Med.* **245**, 119 (1981). Clinical trial: D. Herpin *et al.*, *Curr. Ther. Res.* **30**, 742 (1981).

Crystals from petr ether, mp 103-104°. LD$_{100}$ in rats (mg/kg): 64 i.v. (Cognaco).
Succinate. [100678-32-8] Cibenol; Cipralan; Exacor. C$_{18}$H$_{18}$-N$_2$.C$_4$H$_6$O$_4$; mol wt 380.44. Crystals from ethanol + ether, mp 165°.
THERAP CAT: Antiarrhythmic.

2273. Ciguatoxins. CTX. Potent sodium channel activators found in a wide variety of fish, the toxins were ultimately traced to a dinoflagellate *Gambierdiscus spp.* Family of lipid soluble polyether toxins responsible for ciguatera food poisoning; structural variations are associated with the oceanic region from which the dinoflagellate originates. Isoln from eels: P. J. Scheuer *et al., Science* **155**, 1267 (1967). Purification from dinoflagellate: R. Bagnis *et al., Rev. Int. Oceanogr. Med.* **45-46**, 29 (1977). Structure: M. Murata *et al., J. Am. Chem. Soc.* **112**, 4380 (1990). Purification and characterization of 3 major forms CTX 1-3: R. J. Lewis *et al., Toxicon* **29**, 1115 (1991); absolute configuration: M. Satake *et al., J. Am. Chem. Soc.* **119**, 11325 (1997). Isoln and characterization of Pacific CTX: R. J. Lewis, A. Jones, *Toxicon* **35**, 159 (1997); of Caribbean CTX: J.-P. Vernoux, R. J. Lewis, *ibid.* 889; of Indian CTX: B. Hamilton *et al., ibid.* **40**, 685 (2002). Effect on sodium channels: J.-N. Bidard *et al., J. Biol. Chem.* **259**, 8353 (1984); E. Benoit *et al., Neuroscience* **71**, 1121 (1996). Review of use as research tool: J. Molgo *et al., Methods Neurosci.* **8**, 149-164 (1992); of mechanism of action: C. Frelin *et al., ACS Symp. Ser.* **418**, 192-199 (1996); of neurobiological actions: C. Mattei *et al., J. Soc. Biol.* **193**, 329-344 (1999).

CIGUATOXIN 1

CTX-1. [11050-21-8] $C_{60}H_{86}O_{19}$; mol wt 1111.33. White solid. LD_{50} i.p. in mice: 0.25 μg/kg (Lewis 1991).

CTX-2. [142185-85-1] $C_{60}H_{86}O_{18}$; mol wt 1095.33. White amorphous solid. LD_{50} i.p. in mice: 2.3 μg/kg (Lewis 1991).

CTX-3. [139341-09-6] $C_{60}H_{86}O_{18}$; mol wt 1095.33. Total synthesis: M. Inoue, M. Hirama, *Synlett* **2004**, 577. White amorphous solid. LD_{50} i.p. in mice: 0.9 μg/kg (Lewis 1991).

2274. Cilastatin. [82009-34-5] (2*Z*)-7-[[[(2*R*)-2-Amino-2-carboxyethyl]thio]-2-[[[(1*S*)-2,2-dimethylcyclopropyl]carbonyl]amino]-2-heptenoic acid; MK-791. $C_{16}H_{26}N_2O_5S$; mol wt 358.45. C 53.61%, H 7.31%, N 7.82%, O 22.32%, S 8.94%. Reversible dehydropeptidase I inhibitor; prevents renal metabolism of penem and carbapenem antibiotics. Prepn: D. W. Graham *et al., EP 48301-* (1982 to Merck & Co.); and activity: *idem et al., J. Med. Chem.* **30**, 1074 (1987). Combination with thienamycins: H. Kropp *et al., EP 48025*; F. M. Kahan, H. Kropp, **US 4539208** (1982, 1985 both to Merck & Co.). HPLC determn in serum: C. M. Myers, J. L. Blumer, *Antimicrob. Agents Chemother.* **26**, 78 (1984). Effect on imipenem pharmacokinetics: S. R. Norrby *et al., ibid.* **23**, 300 (1983); on intrathecal penetration of imipenem: A. W. Chow *et al., ibid.* **23**, 634 (1983). Series of articles on pharmacokinetics, tolerance, and efficacy of combination with imipenem: *J. Antimicrob. Chemother.* **12**, Suppl. D, 1-155 (1983); *Infection* **14**, Suppl. 2, S111-S180 (1986). Review of clinical experience in serious infections: J. A. Balfour *et al., Drugs* **51**, 99-136 (1996); of nephroprotective effect with cyclosporin: A. Tejedor *et al., Curr. Med. Res. Opin.* **23**, 505-513 (2007).

Amorphous solid. $[\alpha]_D^{25}$ +17.6° (c = 0.5 in methanol). $[\alpha]_D^{25}$ +14.2° (c = 0.5 in 0.1*N* HCl).

Sodium salt. [81129-83-1] Cilastatin sodium. $C_{16}H_{25}N_2$-NaO_5S. Off-white to yellowish-white hygroscopic, amorphous solid. pKa_1 2.0; pKa_2 4.4; pKa_3 9.2. Very sol in water, methanol.

Mixture of sodium salt with imipenem *see* Imipenem.

THERAP CAT: Antibacterial adjunct (dipeptidase inhibitor).

2275. Cilazapril. [88768-40-5] (1*S*,9*S*)-9-[[(1*S*)-1-(Ethoxycarbonyl)-3-phenylpropyl]amino]octahydro-10-oxo-6*H*-pyridazino[1,2-*a*][1,2]diazepine-1-carboxylic acid; (1*S*,9*S*)-9-[[(*S*)-1-carboxy-3-phenylpropyl]amino]octahydro-10-oxo-6*H*-pyridazino[1,2-*a*][1,2]diazepine-1-carboxylic acid 9-ethyl ester; Ro-31-2848; Dynorm. $C_{22}H_{31}N_3O_5$; mol wt 417.51. C 63.29%, H 7.48%, N 10.06%, O 19.16%. Angiotensin-converting enzyme (ACE) inhibitor; hydrolyzed *in vivo* to the active diacid metabolite. Prepn: M. R. Attwood *et al.,* **DE 3317290**; *eidem,* **US 4512924** (1983, 1985 both to Hoffmann-La Roche); *eidem, J. Chem. Soc. Perkin Trans. 1* **1986**, 1011. HPLC determn in urine and pharmaceutical formulations: J. A. Prieto *et al., J. Chromatogr. B* **714**, 285 (1998). Pharmacology: I. L. Natoff *et al., J. Cardiovasc. Pharmacol.* **7**, 569 (1985). Review of clinical pharmacology: C. H. Kleinbloesem *et al., Am. J. Med.* **87**, Suppl. 6B, 45S-49S (1989). Review of clinical experience in hypertension: T. Szucs, *Drugs* **41**, Suppl. 1, 18-24 (1991); in chronic heart failure: L. Dössegger *et al., J. Cardiovasc. Pharmacol.* **24**, Suppl. 3, S38-S41 (1994); in combination with hydrochlorothiazide: R. C. Pordy, *Cardiology* **86**, 41-48 (1995).

Monohydrate. [92077-78-6] Inhibace; Initiss; Justor; Vascace. $C_{22}H_{31}N_3O_5.H_2O$; mol wt 435.52. White to off-white crystalline powder, mp 98° (dec). $[\alpha]_D^{20}$ −62.51° (c = 1% in ethanol). Soly in water (25°): 0.5 g/100 ml. pKa_1 3.3; pKa_2 6.4. Partition coefficient (octanol/pH 7.4 buffer, 22°): 0.8.

Diacid. [90139-06-3] Cilazaprilat; Ro-31-3113. $C_{20}H_{27}N_3O_5$; mol wt 389.45. Crystals from water, mp 242°. $[\alpha]_D^{20}$ −74.7° (c = 0.5 in 1*M* NaOH).

THERAP CAT: Antihypertensive.

2276. Cilengitide. [188968-51-6] Cyclo(L-arginylglycyl-L-α-aspartyl-D-phenylalanyl-*N*-methyl-L-valyl); cyclo(RGDf-N(Me)-V-); EMD-121974; Impetreve. $C_{27}H_{40}N_8O_7$; mol wt 588.67. C 55.09%, H 6.85%, N 19.04%, O 19.02%. Cyclic RGD-containing pentapeptide with antiangiogenic activity. Inhibits binding of $\alpha_v\beta_3$ and $\alpha_v\beta_5$ integrins to the extracellular matrix. Prepn: A. Jonczyk *et al.,* **EP 770622**; *eidem,* **US 6001961** (1997, 1999 both to Merck KGaA); and antagonist activity: M. A. Dechantsreiter *et al., J. Med. Chem.* **42**, 3033 (1999). Crystal structure in complex with $\alpha_v\beta_3$ integrin: J.-P. Xiong *et al., Science* **296**, 151 (2002). Clinical pharmacokinetics: F. A. L. Eskens *et al., Eur. J. Cancer* **39**, 917 (2003); and pharmacology: S. Hariharan *et al., Ann. Oncol.* **18**, 1400 (2007). Clinical evaluation in glioblastoma multiforme: D. A. Reardon *et al., J. Clin. Oncol.* **26**, 5610 (2008). Review of development and therapeutic potential: D. A. Reardon *et al., Expert Opin. Invest. Drugs* **17**, 1225 (2008).

Hydrochloride. [188969-00-8] EMD-85189. $C_{27}H_{40}N_8O_7 \cdot$ HCl; mol wt 625.12.

THERAP CAT: Antineoplastic.

2277. Ciliary Neurotrophic Factor. CNTF. Neurotrophic factor unrelated to the neurotrophins, *q.v.* Cytosolic peptide produced by Schwann cells in the peripheral nervous system and by astrocytes in the CNS; not expressed by target tissues. Promotes survival of a broad range of neurons and enhances differentiation of neurons and astrocytes. Initial identification as *in vitro* survival factor for chick ciliary neurons: R. Adler *et al., Science* **204**, 1434 (1979). Purification from chick eye: G. Barbin *et al., J. Neurochem.* **43**, 1468 (1984); from rat sciatic nerve: M. Manthorpe *et al., Brain Res.* **367**, 282 (1986). Molecular cloning of rat CNTF: K. A. Stöckli *et al., Nature* **342**, 920 (1989); of rabbit CNTF: L.-F. H. Lin *et al., Science* **246**, 1023 (1989); of human CNTF: A. Lam *et al., Gene* **102**, 271 (1991); A. Negro *et al., Eur. J. Biochem.* **201**, 289 (1991). Effect on motoneuron survival and proposed role as a "lesion factor" released in pathophysiological conditions: M. Sendtner *et al., J. Cell Sci.* **Suppl. 15**, 103 (1991). Review of bioactivity and structural relationship with hematopoietic cytokines: N. Y. Ip, G. D. Yancopoulos, *Prog. Growth Factor Res.* **4**, 139-155 (1992). Review of CNTF receptor: S. Davis, G. D. Yancopoulos, *Curr. Opin. Neurobiol.* **5**, 281-285 (1993). *Review:* M. Manthorpe *et al.* in *Neurotrophic Factors*, S. E. Loughlin, J. H. Fallon, Eds. (Academic Press, San Diego, 1993) pp 443-473.

Human CNTF. [133423-39-9] 200 Amino acid peptide. mol wt ~23,000 Da. pI ~6.0. Exhibits ~83% amino acid homology with rat CNTF; ~87% with rabbit CNTF.

2278. Cilnidipine. [132203-70-4] 1,4-Dihydro-2,6-dimethyl-4-(3-nitrophenyl)-3,5-pyridinedicarboxylic acid 3-(2-methoxyethyl) 5-[(2*E*)-3-phenyl-2-propen-1-yl] ester; (±)-(*E*)-cinnamyl 2-methoxyethyl 1,4-dihydro-2,6-dimethyl-4-(*m*-nitrophenyl)-3,5-pyridinedicarboxylate; FRC-8653; Atelec; Cinalong; Siscard. $C_{27}H_{28}N_2O_7$; mol wt 492.53. C 65.84%, H 5.73%, N 5.69%, O 22.74%. Dihydropyridine calcium channel blocker. Prepn: T. Kutsuma *et al.,* **EP 161877**; *eidem,* **US 4672068** (1985, 1987 both to Fujirebio). Pharmacology: K. Ikeda *et al., Oyo Yakuri* **44**, 433 (1992). Mechanism of action study: M. Hosono *et al., J. Pharmacobio-Dyn.* **15**, 547 (1992). LC-MS determn in plasma: K. Hatada *et al., J. Chromatogr.* **583**, 116 (1992). Clinical study: M. Ishii, *Jpn. Pharmacol. Ther.* **21**, 59 (1993). Acute toxicity study: S. Wada *et al., Yakuri to Chiryo* **20**, Suppl. 7, S1683 (1992), *C.A.* **118**, 32711 (1992).

Crystals from methanol, mp 115.5-116.6°. LD_{50} in male, female mice, rats (mg/kg): ≥5000, ≥5000, ≥5000, 4412 orally; ≥5000 all species s.c.; 1845, 2353, 441, 426 i.p. (Wada).

THERAP CAT: Antihypertensive.

2279. Cilomilast. [153259-65-5] *cis*-4-Cyano-4-[3-(cyclopentyloxy)-4-methoxyphenyl]cyclohexanecarboxylic acid; SB-207499; Ariflo. $C_{20}H_{25}NO_4$; mol wt 343.42. C 69.95%, H 7.34%, N 4.08%, O 18.63%. Selective phosphodiesterase 4 (PDE4) inhibitor. Prepn: S. B. Christensen, **WO 9319749**; *idem,* **US 5552438** (1993, 1996 both to SmithKline Beecham); *idem et al., J. Med. Chem.* **41**, 821 (1998). Anti-inflammatory activity: M. S. Barnette *et al., J. Pharmacol. Exp. Ther.* **284**, 420 (1998); D. E. Griswold *et al., ibid.* 705. Pharmacology and therapeutic potential in asthma and chronic obstructive pulmonary disease (COPD): T. J. Torphy *et al., Pulm. Pharmacol. Ther.* **12**, 131-135 (1999). Clinical pharmacokinetics and bioavailability: B. D. Zussman *et al., Pharmacotherapy* **21**, 653 (2001). Clinical trial in COPD: C. H. Compton *et al., Lancet* **358**, 265 (2001). Review of pharmacology and clinical experience: M. A. Giembycz, *Expert Opin. Invest. Drugs* **10**, 1361-1379 (2001).

Relative stereochemistry

White solid, mp 157°.

THERAP CAT: Antiasthmatic; in treatment of chronic obstructive pulmonary disease.

2280. Cilostazol. [73963-72-1] 6-[4-(1-Cyclohexyl-1*H*-tetrazol-5-yl)butoxy]-3,4-dihydro-2(1*H*)-quinolinone; 6-[4-(1-cyclohexyl-1*H*-tetrazol-5-yl)butoxy]-3,4-dihydrocarbostyril; 6-[4-(1-cyclohexyl-5-tetrazolyl)butoxy]-1,2,3,4-tetrahydro-2-oxoquinoline; OPC-13013; Pletal. $C_{20}H_{27}N_5O_2$; mol wt 369.47. C 65.02%, H 7.37%, N 18.96%, O 8.66%. Phosphodiesterase (PDE III) inhibitor; induces vasodilation and inhibits platelet aggregation. Prepn: **BE 878548**; T. Nishi, K. Nakagawa, **US 4277479** (1979, 1981 both to Otsuka); T. Nishi *et al., Chem. Pharm. Bull.* **31**, 1151 (1983). Physical properties: T. Shimizu *et al., Arzneim.-Forsch.* **35**, 1117 (1985). HPLC determn in plasma: H. Akiyama *et al., J. Chromatogr.* **338**, 456 (1985). Acute toxicity study: G. Nomura *et al., Iyakuhin Kenkyu* **16**, 1200 (1985), *C.A.* **104**, 102207f (1986). Clinical pharmacokinetics: A. Suri *et al., J. Clin. Pharmacol.* **38**, 144 (1998). Clinical evaluation to prevent stent thrombosis: S.-W. Park *et al., Am. J. Cardiol.* **84**, 511 (1999); to prevent restenosis after angioplasty: E. Tsuchikane *et al., Circulation* **100**, 21 (1999). Review of pharmacology and clinical efficacy in intermittent claudication: E. M. Sorkin, A. Markham, *Drugs Aging* **14**, 63-71 (1999); in secondary stroke prevention: Z. A. Al-Qudah *et al., Expert Opin. Pharmacother.* **12**, 1305-1315 (2011).

Colorless needle-like crystals from methanol, mp 159.4-160.3°. uv max (methanol): 257 nm (ε 15200). Freely sol in acetic acid, chloroform, *n*-methyl-2-pyrrolidone, DMSO; slightly sol in methanol, ethanol. Practically insol in ether, water, 0.1*N* HCl, 0.1*N* NaOH. LD_{50} in mice, rats (mg/kg): >2000, >2000 i.p.; >5000, >5000 orally (Nomura).

THERAP CAT: Antithrombotic; in treatment of intermittent claudication.

2281. Cimaterol. [54239-37-1] 2-Amino-5-[1-hydroxy-2-[(1-methylethyl)amino]ethyl]benzonitrile; (±)-5-[1-hydroxy-2-(isopropylamino)ethyl]anthranilonitrile; 1-(4-amino-3-cyanophenyl)-2-isopropylaminoethanol; AB-A-663; CL-263780; AC-263780. $C_{12}H_{17}N_3O$; mol wt 219.29. C 65.73%, H 7.81%, N 19.16%, O 7.30%. β-Adrenergic agonist, related to clenbuterol and mabuterol, *q.q.v.* Prepn: G. Engelhardt *et al.* to Thomae). Pharmacology: G. Engelhardt, *Arzneim.-Forsch.* **34**, 1625 (1984). Pharmacokinetics: T. M. Byrem *et al., J. Anim. Sci.* **70**, 3812 (1992). Effect on lipid metabolism in sheep: C. Y. Hu *et al., ibid.* **66**, 1393 (1988). Efficacy as a repartitioning agent in livestock: R. W. Jones *et al., ibid.* **61**, 905 (1985); D. H. Beermann *et al., ibid.* **62**, 370 (1986); A. P. Molony *et al., Livestock Prod. Sci.* **42**, 23, 35 (1995).

Crystals, mp 159-161°.

THERAP CAT (VET): Repartitioning agent.

2282. Cimetidine. [51481-61-9] *N*-Cyano-*N'*-methyl-*N''*-[2-[[(4-methyl-1*H*-imidazol-5-yl)methyl]thio]ethyl]guanidine; SKF-92334; Aciloc; Acinil; Brumetidina; Cimal; Dyspamet; Stomédine; Tagamet; Ulcedin; Ulcimet. $C_{10}H_{16}N_6S$; mol wt 252.34. C 47.60%, H 6.39%, N 33.31%, S 12.71%. Competitive histamine H_2-receptor antagonist which inhibits gastric acid secretion and reduces pepsin output. Prepn: G. J. Durant *et al.*, **BE 804144**; *eidem*, **US 3950333** (1974, 1976 both to SK&F); P. Kairisalo, E. Honkanen, *Arch. Pharm.* **316**, 688 (1983). Chemistry, pharmacology and toxicology: R. W. Brimblecombe *et al.*, *J. Int. Med. Res.* **3**, 86 (1975); *eidem, Br. J. Pharmacol.* **53**, 435p (1975). Use in combination with other drugs for cancer treatment: R. D. Thornes *et al.*, *Lancet* **2**, 328 (1982); S. Borgström *et al.*, *N. Engl. J. Med.* **307**, 1080 (1982); *ibid.* **308**, 591 (1983). Controlled clinical study in treatment of acute upper gastrointestinal tract bleeding: D. Barer *et al.*, *ibid.* 1571; for prevention of recurrent duodenal ulcer: S. Sontag *et al.*, *ibid.* **311**, 689 (1984). Immunoregulatory effect: J. L. Jorizzo *et al.*, *Ann. Intern. Med.* **92**, 192 (1980); W. B. White, M. Ballow, *N. Engl. J. Med.* **312**, 198 (1985). *In vivo* and *in vitro* effects on angiogenesis and tumor growth: T. Natori *et al.*, *Biomed. Pharmacother.* **59**, 56 (2005). Review of effect on endocrine secretion: C. Scarpignato, G. Bertaccini, *Drugs Exp. Clin. Res.* **5**(4-5), 129-140 (1979). Symposium on clinical efficacy, cytoprotection and antifibrinolytic effects: *Scand. J. Gastroenterol.* **21**, Suppl. 121, 1-62 (1986). Review of safety profile and drug interactions: A. F. Shinn, *Drug Saf.* **7**, 245-267 (1992).

Crystals, mp 141-143°. Bitter taste and characteristic odor. Freely sol in methanol; sol in ethanol, polyethylene glycol 400; very slightly sol in chloroform; sparingly sol in isopropanol. Practically insol in ether. Soly in water at 37°: 1.14%. Soly increased by dil HCl. pKa 6.8. LD_{50} in mice, rats (mg/kg): 2600, 5000 orally; 150, 106 i.v.; 470, 650 i.p. (Brimblecombe).

Hydrochloride. [70059-30-2] $C_{10}H_{16}N_6S$.HCl; mol wt 288.80. White, crystalline powder. Very sol in water; sol in alcohol. pKa 7.11.

THERAP CAT: Antiulcerative.

THERAP CAT (VET): Antiulcerative.

2283. Cimetropium Bromide. [51598-60-8] (1α,2β,4β,5α,-7β)-9-(Cyclopropylmethyl)-7-[(2S)-3-hydroxy-1-oxo-2-phenylpropoxy]-9-methyl-3-oxa-9-azoniatricyclo[3.3.1.0^{2,4}]nonane bromide (1:1); 8-(cyclopropylmethyl)-6β,7β-epoxy-3α-hydroxy-1αH,5αH-tropanium bromide ($-$)-(*S*)-tropate; *N*-cyclopropylmethylscopolamine bromide; DA-3177; Alginor. $C_{21}H_{28}BrNO_4$; mol wt 438.36. C 57.54%, H 6.44%, Br 18.23%, N 3.20%, O 14.60%. Anticholinergic agent with affinity for intestinal muscarinic receptors. Prepn: S. Casadio, A. Donetti, **DE 2316728**; *eidem*, **US 3853886** (1973, 1974 both to De Angeli). Crystal and molecular structure: G. Giuseppetti *et al.*, *Farmaco Ed. Sci.* **35**, 231 (1980). Molecular conformation in solution: A. Gallazzi *et al.*, *ibid.* 913. HPLC-MS/MS determn in plasma: H.-W. Lee *et al.*, *J. Mass Spectrom.* **41**, 855 (2006). Pharmacology: A. Schiavone *et al.*, *Arzneim.-Forsch.* **35**, 796 (1985); C. Scarpignato, G. Bianchi Porro, *Int. J. Clin. Pharmacol. Res.* **5**, 467 (1985). Clinical trial in treatment of irritable bowel syndrome: G. Dobrilla *et al.*, *Gut* **31**, 355 (1990); of infantile colic: F. Savino *et al.*, *J. Pediatr. Gastroenterol. Nutr.* **34**, 417 (2002).

Crystals from acetonitrile, mp 174°. $[\alpha]_D^{20}$ $-18.3°$ (c = 3).

THERAP CAT: Antispasmodic.

2284. Cimigenol. [3779-59-7] (3β,15α,16α,23R,24S)-16,-23:16,24-Diepoxy-9,19-cyclolanostane-3,15,25-triol; cimicifugol. $C_{30}H_{48}O_5$; mol wt 488.71. C 73.73%, H 9.90%, O 16.37%. From the resin of *Cimicifuga racemosa* Nutt. *(Actaea racemosa* L.), *Ranunculaceae:* Corsano, Spano, *Atti Accad. Naz. Lincei Cl. Sci. Fis. Mat. Nat. Rend.* **32**, 674 (1962), *C.A.* **58**, 11408e (1963). Structure: Corsano *et al.*, *Chem. Commun.* **1965**, 185; Corsano, Piancatelli, *Ric. Sci.* **37**, 366 (1967); *eidem, Gazz. Chim. Ital.* **99**, 1140 (1969). Identity with cimicifugol: *J. Pharm. Soc. Jpn.* **87**, 1569 (1967).

Needles from methanol, mp 227.5-228.5°. $[\alpha]_D$ +38° (c = 0.86 in chloroform).

Diacetate. $C_{34}H_{52}O_7$. Needles from hexane, mp 202-204°. $[\alpha]_D$ +46.3° (c = 1.07 in chloroform).

Triacetate. $C_{36}H_{54}O_8$. Needles from hexane, mp 149.5-150.5°. $[\alpha]_D$ +25.4° (chloroform).

2285. Cinacalcet. [226256-56-0] (αR)-α-Methyl-*N*-[3-[3-(trifluoromethyl)phenyl]propyl]-1-napthalenemethanamine; (*R*)-*N*-(3-(3-(trifluoromethyl)phenyl)propyl)-1-(1-napthyl)ethylamine; AMG-073; KRN-1493. $C_{22}H_{22}F_3N$; mol wt 357.42. C 73.93%, H 6.20%, F 15.95%, N 3.92%. Calcimimetic. Prepn: B. C. Van Wagenen *et al.*, **US 6211244** (2001 to NPS Pharmaceuticals). Review of pharmacology and mechanism of action: M. Wada, N. Nagano, *Nephrol. Dial. Transplant.* **18**, Suppl. 3, iii13-iii17 (2003); of clinical development: N. Franceschini *et al.*, *Expert Opin. Invest. Drugs* **12**, 1413-1421 (2003). Clinical trial in secondary hyperparathyroidism: G. A. Block *et al.*, *N. Engl. J. Med.* **350**, 1516 (2004).

Hydrochloride. [364782-34-3] Mimpara; Sensipar. $C_{22}H_{22}F_3$-N.HCl; mol wt 393.88. White to off-white, crystalline solid. Sol in methanol, 95% ethanol; slightly sol in water.

THERAP CAT: Antihyperparathyroid.

2286. Cinchomeronic Acid. [490-11-9] 3,4-Pyridinedicarboxylic acid. $C_7H_5NO_4$; mol wt 167.12. C 50.31%, H 3.02%, N 8.38%, O 38.29%. Prepd in 42% yield by boiling quinine with nitric and fuming nitric acid: Ternájgo, *Monatsh. Chem.* **21**, 448 (1900); Kirpal, *ibid.* **23**, 248 (1902); Kaas, *ibid.* 252; by selenium dioxide oxidation of isoquinoline: Mueller, **US 2436660** (1948 to Allied Chem.).

Crystals from acidulated water, mp 256°. On heating, a small portion sublimes without decompn. Strong acid. Sparingly sol in hot alc, ether, benzene. Insol in chloroform.

Disodium salt dihydrate. $C_7H_3NNa_2O_4.2H_2O$. Plates, moderately sol in water, alcohol.
Dimethyl ester. $C_9H_9NO_4$. Crystals, mp 141°.

2287. Cinchona. Calisaya bark; Peruvian bark; Cinchona bark; Jesuit's bark. Dried bark of stem or root of various species of *Cinchona*, fam. *Rubiaceae*, largely of *C. officinalis* L. (*C. ledgeriana* Moens) cultivated mostly in Java, of *C. officinalis* L. (*C. calisaya* Wedd.) from Bolivia, of *C. officinalis* L. and *C. micrantha* R. & P. from Peru, of *C. pubescens* Vahl. (*C. succirubra* Pav.) from Ecuador and *C. pitayensis* Wedd. from Colombia. *Habit.* South America, cultivated mostly in Java, also in India. *Constit.* About 35 alkaloids; cinchotannic, quinic and quinovic acids; cinchona red; volatile oils. The alkaloid content varies according to the source of the bark. The cultivated bark contains 7-10% total alkaloids of which about 70% is quinine. *Cinchona succirubra* Pav. and some other varieties contain more cinchonidine and cinchonine and sometimes quinidine than the cultivated. Standardized method for the assay of cinchona, named "Brussels 1949": *De Belgische Chemische Industrie (Ind. Chim. Belge)* **15**, 328-338 (1950).

On heating a small portion of the bark in a test tube a characteristic purple vapor is evolved. Dil H_2SO_4 extracts have a blue fluorescence.

THERAP CAT: Antimalarial.

2288. Cinchonamine. [482-28-0] 2-[(1*S*,2*S*,4*S*,5*R*)-5-Ethenyl-1-azabicyclo[2.2.2]oct-2-yl]-1*H*-indole-3-ethanol; 3-(β-hydroxyethyl)-2-(5-vinyl-2-quinuclidyl)indole. $C_{19}H_{24}N_2O$; mol wt 296.41. C 76.99%, H 8.16%, N 9.45%, O 5.40%. From the bark of *Remijia purdieana* Wedd., *Rubiaceae*: Arnaud, *Compt. Rend.* **93**, 593 (1881); Hesse, *Ann.* **225**, 211 (1884). Prepn by reduction of quinamine with lithium aluminum hydride and structure: Goutarel *et al.,Helv. Chim. Acta* **33**, 150 (1950). Synthesis: Chen *et al.,C.A.* **53**, 7219e (1959). Total synthesis: G. Grethe *et al., Helv. Chim. Acta* **59**, 2271 (1976). Stereochemistry: Wenkert, Bringi, *J. Am. Chem. Soc.* **80**, 3484 (1958); Augustine, *Chem. Ind. (London)* **1959**, 1071; Sawa, Matsumura, *Tetrahedron* **26**, 2923 (1970). Biosynthetic studies: Battersby, Parry, *Chem. Commun.* **1971**, 31. Conversion of chinchona alkaloids of the quinoline series to those of the indole series: Ochiai *et al.,C.A.* **59**, 14040h (1963).

Triboluminescent, orthorhombic prisms from methanol, mp 186°, also reported as mp 194°. $[\alpha]_D^{20}$ +123° (c = 0.66 in ethanol). pK in 80% methyl Cellosolve: 8.28. uv max (methanol): 223, 292 nm (log ε 4.60, 3.88). One gram dissolves in about 35 ml alcohol, 105 ml ether. Sol in benzene, chloroform, petr ether, CS_2; practically insol in water.
Hydrochloride monohydrate. $C_{19}H_{24}N_2O.HCl.H_2O$. Cubic crystals. One gram dissolves in 200 ml water. Sol in alc.

2289. Cinchonidine. [485-71-2] (8α,9*R*)-Cinchonan-9-ol; cinchovatine; α-quinidine. $C_{19}H_{22}N_2O$; mol wt 294.40. C 77.52%, H 7.53%, N 9.52%, O 5.43%. Occurs in most varieties of cinchona bark, especially in bark of *Cinchona pubescens* Vahl. (*C. succirubra* Pav.) and *C. pitayensis* Wedd., *Rubiaceae*. Isoln: Leers, *Ann.* **82**, 147 (1852). Structure: Rabe, *ibid.* **365**, 359 (1909). Stereoisomeric with cinchonine: Koenigs, *ibid.* **347**, 182 (1906). Configuration: Prelog, Zalán, *Helv. Chim. Acta* **27**, 535 (1944); Prelog, Häfliger, *ibid.* **33**, 2021 (1950); Roth, *Pharmazie* **16**, 257 (1961); Lyle, Keefer, *Tetrahedron* **23**, 3253 (1967). Biosynthetic studies: Battersby, Parry, *Chem. Commun.* **1971**, 31. Toxicity studies: C. C. Johnson, C. F. Poe, *Acta Pharmacol. Toxicol.* **4**, 265 (1948); E. W. Schafer, Jr. *et al., Ecotoxicol. Environ. Saf.* **6**, 149 (1982).

Orthorhombic plates, prisms from alcohol, mp 210°. $[\alpha]_D^{20}$ −109.2° (alc). uv absorption data: Kamath *et al., Indian J. Chem.* **6**, 510 (1968). Sol in alcohol and chloroform; moderately sol in ether. Practically insol in water. pK_1 5.80, pK_2 10.03. *Protect from light.* LD_{50} i.p. in rats: 206 mg/kg (Johnson, Poe). LD_{50} orally in quail: >316 mg/kg (Schafer).
Dihydrochloride. $C_{19}H_{22}N_2O.2HCl$. White or slightly yellow crystals or powder. Freely sol in water or alcohol. *Protect from light.*
Hydrochloride dihydrate. $C_{19}H_{22}N_2O.HCl.2H_2O$. Cryst powder; loses all of its H_2O at 120°. $[\alpha]_D^{20}$ −117.5° (water). Sol in 25 parts cold water, more sol in boiling water; sol in alcohol, chloroform, slightly in ether. The aq soln is practically neutral. *Protect from light.*
Sulfate trihydrate. $(C_{19}H_{22}N_2O)_2.H_2SO_4.3H_2O$. Silky, acicular crystals; effloresce on exposure to air and darkens in light. mp when anhydr ∼240° with decompn. One gram dissolves in 70 ml water, 20 ml hot water, 90 ml alcohol, 40 ml hot alc, 620 ml chloroform. Practically insol in ether. The aq soln is practically neutral. *Protect from light.*
Epicinchonidine. [550-54-9] (8α,9*S*)-Cinchonan-9-ol. mp 104°, $[\alpha]_D^{20}$ +63° (c = 0.804 in alc): Rabe *et al., Ann.* **492**, 253 (1932).

THERAP CAT: Antimalarial.

2290. Cinchonine. [118-10-5] (9*S*)-Cinchonan-9-ol. $C_{19}H_{22}N_2O$; mol wt 294.40. C 77.52%, H 7.53%, N 9.52%, O 5.43%. Occurs in most varieties of cinchona bark, especially in bark of *Cinchona micrantha* R. & P., *Rubiaceae*. Isoln and structure: Rabe, *Ber.* **41**, 63 (1908). Stereoisomeric with cinchonidine: Koenigs, *Ann.* **347**, 182 (1906). Configuration: Prelog, Zalán, *Helv. Chim. Acta* **27**, 535 (1944); Prelog, Häfliger, *ibid.* **33**, 2021 (1950); Roth, *Pharmazie* **16**, 257 (1961); Lyle, Keefer, *Tetrahedron* **23**, 3253 (1967). Crystal and molecular structure: B. Oleksyn *et al., Acta Crystallogr.* **B35**, 440 (1979). Biosynthetic studies: Battersby, Parry, *Chem. Commun.* **1971**, 31. Toxicity study: C. C. Johnson, C. F. Poe, *Acta Pharmacol. Toxicol.* **4**, 265 (1949).

Prisms, needles from alcohol or ether. mp about 265°; begins to sublime at 220°. $[\alpha]_D$ +229° (alcohol). uv absorption data: Kamath *et al., Indian J. Chem.* **6**, 510 (1968). One gram dissolves in 60 ml alcohol, 25 ml boiling alcohol, 110 ml chloroform, 500 ml ether. Practically insol in water. pK_1 5.85, pK_2 9.92. *Protect from light.* LD_{50} i.p. in rats: 152 mg/kg (Johnson, Poe).
Dihydrochloride. $C_{19}H_{22}N_2O.2HCl$. White or faintly yellow crystals or crystalline powder. Freely sol in water or alcohol. *Protect from light.*
Hydrochloride dihydrate. $C_{19}H_{22}N_2O.HCl.2H_2O$. Fine crystals. mp when anhydr about 215° with dec. One gram dissolves in 20 ml water, 3.5 ml boiling water, 1.5 ml alcohol, 20 ml chloroform; slightly sol in ether. The aq soln is practically neutral. *Protect from light.*

Sulfate dihydrate. $(C_{19}H_{22}N_2O_2)_2.H_2SO_4.2H_2O$. Lustrous, very bitter crystals. mp when anhydr about 198°. One gram dissolves in 65 ml water, 30 ml hot water, 12.5 ml alc, 7 ml hot alc, 47 ml chloroform; slightly sol in ether. The aq soln is practically neutral.

Epicinchonine. [485-70-1] (9R)-Cinchonan-9-ol. mp 83°, $[\alpha]_D^{22}$ +120.3° (c = 0.806 in alc): Rabe et al., Ann. **492**, 253 (1932).

THERAP CAT: Antimalarial.

2291. Cinchophen. [132-60-5] 2-Phenyl-4-quinolinecarboxylic acid; 2-phenylcinchoninic acid. $C_{16}H_{11}NO_2$; mol wt 249.27. C 77.10%, H 4.45%, N 5.62%, O 12.84%. Quinoline derivative formerly used in treatment of chronic gout: A. B. Gutman, Arthritis Rheum. **16**, 431 (1973). Prepd by heating pyruvic acid with aniline and benzaldehyde or with benzylidene aniline in abs alcohol: Doebner, Gieseke, Ann. **242**, 290 (1887). Also from acetophenone and isatinic acid in alcoholic KOH: Pfitzinger, J. Prakt. Chem. **38**, 582 (1882); **56**, 293 (1897). Cinchophen as ulcerogenic agent: T. P. Churchill, F. H. van Wagoner, Arch. Pathol. **13**, 850 (1932); N. Umeda et al., Toxicol. Appl. Pharmacol. **18**, 102 (1971); T. H. Stewart et al., J. Pathol. **131**, 363 (1980). Mitochondrial toxicity of cinchophen and derivatives: H. Vainio et al., Biochem. Pharmacol. **20**, 1589 (1971).

Needles, mp 213-216°. Stable to air, but turns yellow under the influence of light. Slightly bitter taste. Practically insol in water. One gram dissolves in about 400 ml chloroform, in about 100 ml ether and in about 120 ml alcohol.

Hydrochloride. [132-58-1] $C_{16}H_{11}NO_2.HCl$; mol wt 285.73. Yellow cryst powder, slightly bitter, astringent taste; mp about 223°. Practically insol in water. Slightly sol in alcohol; sol in chloroform, ether.

Allyl ester. [524-34-5] Allyl 2-phenylcinchoninate. $C_{19}H_{15}NO_2$; mol wt 289.33. Prepn: Gams, US **1336952** (1920). Long needles from alc, mp 30°. bp_{15} 260°. Practically insol in water. Readily sol in alcohol, ether, acetone, oils. Tasteless.

USE: Experimentally to induce ulcers.

2292. Cinchotoxine. [69-24-9] 3-[(3R,4R)-3-Ethenyl-4-piperidinyl]-1-(4-quinolinyl)-1-propanone; 9-deoxy-9-oxo-1,8-seco-cinchonine; cinchonicine. $C_{19}H_{22}N_2O$; mol wt 294.40. C 77.52%, H 7.53%, N 9.52%, O 5.43%. A rearrangement product of cinchonine or cinchonidine obtained by boiling with acetic acid: Pasteur, Compt. Rend. **37**, 110 (1853); Miller, Rohde, Ber. **33**, 3214 (1900). Conversion to cinchonidinone and cinchonine: Rabe, Ber. **44**, 2088 (1911); **41**, 62 (1908). Prepn from cinchonine: Prostenik, Prelog, Helv. Chim. Acta **26**, 1965 (1943).

Bitter needles. mp 58-60°. $[\alpha]_D^{15}$ +47° in alcohol. Sparingly sol in water; sol in alcohol, ether, chloroform.

2293. Cinerins. Active insecticidal constiuents of pyrethrum flowers. Isoln: Ward, Chem. Ind. (London) **1953**, 586; Stephenson, Pyrethrum Post **5**, no. 4, 22 (1960); C.A. **55**, 13753g (1961). Structure: Schechter et al., J. Am. Chem. Soc. **71**, 3165 (1949); Godin et al., J. Chem. Soc. C **1966**, 332. Stereochemistry: Begley et al.,

Chem. Commun. **1972**, 1276. Review: Crombie, Elliott, Fortschr. Chem. Org. Naturst. **19**, 120-164 (1961). Review of toxicology and human exposure: Toxicological Profile for Pyrethrins and Pyrethroids (PB2004-100004, 2003) 332 pp.

Cinerin I   R = CH_3
Cinerin II   R = $COOCH_3$

Cinerin I. [25402-06-6] (1R,3R)-2,2-Dimethyl-3-(2-methyl-1-propenyl)cyclopropanecarboxylic acid (1S)-3-(2Z)-(2-butenyl)-2-methyl-4-oxo-2-cyclopenten-1-yl-ester. $C_{20}H_{28}O_3$; mol wt 316.44. Viscous liquid. Oxidizes rapidly and becomes inactive in presence of air. $bp_{0.008}$ 136-138°. n_D^{20} 1.5064. $[\alpha]_D^{20}$ $-22°$ (hexane). uv max: 222 nm (ε 21400). Practically insol in water. Sol in alcohol, petr ether, kerosene, carbon tetrachloride, ethylene dichloride, nitromethane.

Cinerin II. [121-20-0] (1R,3R)-3-[(1E)-3-Methoxy-2-methyl-3-oxo-1-propenyl]-2,2-dimethylcyclopropanecarboxylic acid (1S)-3-(2Z)-(2-butenyl)-2-methyl-4-oxo-2-cyclopenten-1-yl ester. $C_{21}H_{28}O_5$; mol wt 360.45. Viscous liquid. Oxidizes rapidly and becomes inactive in presence of air. $bp_{0.001}$ 182-184°. n_D^{20} 1.5183. $[\alpha]_D^{16}$ +16° (isooctane). uv max: 229 nm (ε 28700). Practically insol in water. Sol in alcohol, petr ether (less sol than Cinerin I), kerosene, carbon tetrachoride, ethylene dichloride, nitromethane.

Caution: Direct skin contact and inhalation may cause allergic attacks in sensitive people. Potential symptoms of overexposure are severe dermatitis, asthma, vasomotor rhinitis, anaphylactic reactions; numbness of lips and tongue, sneezing, vomiting, diarrhea; headache, restlessness, tinnitis, incoordination, clonic convulsions, stupor, prostration; death due to respiratory paralysis. See: Clinical Toxicology of Commercial Products, R. E. Gosselin et al., Eds. (Williams & Wilkins, Baltimore, 5th ed., 1984) Section III, pp 352-355.

USE: Insecticide.

2294. Cinidon-ethyl. [142891-20-1] (2Z)-2-Chloro-3-[2-chloro-5-(1,3,4,5,6,7-hexahydro-1,3-dioxo-2H-isoindol-2-yl)phenyl]-2-propenoic acid ethyl ester; ethyl (Z)-2-chloro-3-[2-chloro-5-(1,3-dioxo-4,5,6,7-tetrahydroisoindol-2-yl)phenyl]acrylate; BAS-615H; Bingo; Lotus; Orbit; Solar; Vega. $C_{19}H_{17}Cl_2NO_4$; mol wt 394.25. C 57.88%, H 4.35%, Cl 17.98%, N 3.55%, O 16.23%. Post-emergence herbicide which inhibits protoporphyrinogen oxidase. Prepn: P. Plath et al., DE **3603789**; eidem, US **5062884** (1987, 1991 both to BASF). Biochemical mode of action: K. Grossman, H. Schiffer, Pestic. Sci. **55**, 687 (1999). Field trials in cereals: W. Nuyken et al., Brighton Crop Prot. Conf. - Weeds **1999**, 81.

USE: Herbicide.

2295. Cinitapride. [66564-14-5] 4-Amino-N-[1-(3-cyclohexen-1-ylmethyl)-4-piperidinyl]-2-ethoxy-5-nitrobenzamide. $C_{21}H_{30}N_4O_4$; mol wt 402.50. C 62.67%, H 7.51%, N 13.92%, O 15.90%. Prepn: A. Vega-Noverola et al., DE **2751129** (1978 to Anphar); eidem, US **5026858** (1991 to Walton). Pharmacology: R. Massingham et al., J. Auton. Pharmacol. **5**, 41 (1985). Spectral determn in pharmaceutical prepns: M. Blanco et al., J. Pharm. Sci. **82**, 834 (1993). Clinical evaluation in functional dyspepsia, in comparison with metoclopramide, *q.v.*: F. Mora *et al.*, *Ann. Med. Interne* **10**, 323 (1993).

Tartrate. [96623-56-2] LAS-17177; Cidine. THERAP CAT: Gastroprokinetic.

2296. Cinmethylin. [87818-31-3] *rel*-(1*R*,2*S*,4*S*)-1-Methyl-4-(1-methylethyl)-2-[(2-methylphenyl)methoxy]-7-oxabicyclo-[2.2.1]heptane; (±)-2-*exo*-(2-methylbenzyloxy)-1-methyl-4-isopropyl-7-oxabicyclo[2.2.1]heptane; SD-95481; Cinch. $C_{18}H_{26}O_2$; mol wt 274.40. C 78.79%, H 9.55%, O 11.66%. Pre-emergence grass herbicide. Member of the cineole eucalyptol, *q.v.*, family. Prepn: G. B. Payne *et al.*, **EP 81893** (1983 to Shell); *eidem*, **US 4670041** (1987 to Du Pont). Physical and chemical properties: B. T. Grayson *et al.*, *Pestic. Sci.* **21**, 143 (1987). Mechanism of action study: M. H. El-Deek, F. D. Hess, *Weed Sci.* **34**, 684 (1986). Metabolism in rats: P. W. Lee *et al.*, *J. Agric. Food Chem.* **34**, 162 (1986). Mobility in soil: D. R. Wendt *et al.*, *Proc. South. Weed Sci. Soc.* **1987**, 391; and herbicidal activity: P. C. Lolas, A. Galopoulos, *Zizaniology* **1**, 221 (1985). Field studies in tobacco: P. C. Lolas, *Proc. Br. Crop Prot. Conf. - Weeds* **1985**, 841; in soybeans: P. C. Bhowmik, *Weed Sci.* **36**, 678 (1988).

Relative stereochemistry

Colorless liquid, bp 313 ±2°. d^{20} 1015 kg/m³. Viscosity (20°): 70-90 mPa·s. Vapor pressure (20°): 7.6×10^{-5} mm Hg. Moderately volatile. Partition coefficient (*n*-octanol/water): 6850. Miscible with most organic solvents. Slightly sol in water. Soly in water (20°): 63 ± 2 mg/l. LD$_{50}$ orally in rats: 4.5 g/kg; LD$_{50}$ dermally in rabbits: >2 g/kg (Lee).

USE: Herbicide.

2297. Cinnabarine. [146-90-7] 2-Amino-9-(hydroxymethyl)-3-oxo-3*H*-phenoxazine-1-carboxylic acid; 1-carboxy-2-amino-9-hydroxymethylphenoxazin-3-one; polystictine. $C_{14}H_{10}N_2O_5$; mol wt 286.24. C 58.75%, H 3.52%, N 9.79%, O 27.95%. Antibiotic substance produced by the fungus *Trametes cinnabarina* Jacq.: Gripenberg, *Acta Chem. Scand.* **5**, 590 (1951); from *Coriolus sanguineus* (Fr.) (*Polystictus cinnabarinus* Jacq.): Cavill *et al.*, *J. Chem. Soc.* **1953**, 525. Identity with polystictine: Gripenberg *et al.*, *Acta Chem. Scand.* **11**, 1485 (1957). Structure: Gripenberg, *ibid.* **12**, 603 (1958); **13**, 1305 (1959); Cavill *et al.*, *Tetrahedron* **5**, 275 (1959). Syntheses: Weinstein, Brattesani, *J. Heterocycl. Chem.* **4**, 151 (1967); Schäfer, Schlude, *Tetrahedron Lett.* **1968**, 2161.

Red needles, dec 320°. Absorption max: 234, 430, 455 nm (log ε 4.2, 4.0, 4.0). Soly in acetone: 0.04% (20°); in dioxane: 0.1% (100°). Practically insol in ethanol, chloroform and dimethyl sulfoxide. Sol in cold concd HCl.

O-Methyl ether. $C_{15}H_{12}N_2O_5$. Orange red prisms from ethyl acetate + petr ether, mp 200-202°.

2298. Cinnamaldehyde. [104-55-2] 3-Phenyl-2-propenal; cinnamic aldehyde; phenylacrolein; cinnamal. C_9H_8O; mol wt 132.16. C 81.79%, H 6.10%, O 12.11%. Found in Ceylon and Chinese cinnamon oils as the *trans*-form. Prepn by condensation of benzaldehyde and acetaldehyde: Peine, *Ber.* **17**, 2117 (1884); **JP 163097** (1944 to Ogawa Chem. Ind.); from 2-chloroallylbenzene: Bert, Dorier, *Compt. Rend.* **191**, 332 (1930); Bert, Annequin, *ibid.* **192**, 1315 (1931); by condensation of styrene with formylmethylaniline: **GB 504125** (1939 to I. G. Farben); by oxidation of cinnamyl alc: Holum, *J. Org. Chem.* **26**, 4814 (1961); Traynelis, Hergenrother, *J. Am. Chem. Soc.* **86**, 298 (1964). Isoln from woodrotting fungus, *Stereum subpileatum* Berk. & Curt.: Birkinshaw *et al.*, *Biochem. J.* **66**, 188 (1957). Toxicity data: P. M. Jenner *et al.*, *Food Cosmet. Toxicol.* **2**, 327 (1964). Review of risk assessment as fragrance ingredient: D. Bickers *et al.*, *Food Chem. Toxicol.* **43**, 799-836 (2005); of toxicology: J. Cocchiara *et al.*, *ibid.* 867-923.

Yellowish oily liquid, strong odor of cinnamon. d_{25}^{25} 1.048-1.052. Volatile with steam. mp −7.5°. bp$_{1.0}$ 76.1°; bp$_5$ 105.8°; bp$_{10}$ 120.0°; bp$_{20}$ 135.7°; bp$_{40}$ 152.2°; bp$_{60}$ 163.7°; bp$_{100}$ 177.7°; bp$_{200}$ 199.3°; bp$_{400}$ 222.4°; bp$_{760}$ 246.0° (some dec). n_D^{20} 1.618-1.623. Flash point, closed cup: >100°C. Dissolves in about 700 parts water, in about 7 vols of 60% alc. Misc with alcohol, ether, chloroform, oils. LD$_{50}$ in rats (mg/kg): 2220 orally (Jenner).

USE: In the flavor and perfume industry.

2299. Cinnamic Acid. [621-82-9] 3-Phenyl-2-propenoic acid; *β*-phenylacrylic acid. $C_9H_8O_2$; mol wt 148.16. C 72.96%, H 5.44%, O 21.60%. Occurs in the *trans*-form in storax, balsam Peru or Tolu, oil of cinnamon, coca leaves. Isoln: Beilstein, Kuhlberg, *Ann.* **163**, 123 (1872); von Miller, *Ann.* **188**, 196 (1877). Synthesis (Perkin reaction) from benzaldehyde, acetic anhydride, and potassium acetate: *Org. React.* **I**, 248 (1942); from oxalyl bromide + styrene: Treibs *et al.*, *Naturwissenschaften* **45**, 85 (1958); from acetylene + benzaldehyde: Herbetz, *Ber.* **92**, 541 (1959). Prepn of *cis*- and *trans*-isomers: Comte *et al.*, *Compt. Rend.* **245**, 1144 (1957). Isoln from wood-rotting fungus, *Stereum subpileatum* Berk. & Curt.: Birkinshaw *et al.*, *Biochem. J.* **66**, 188 (1957). Review of risk assessment as fragrance ingredient: D. Bickers *et al.*, *Food Chem. Toxicol.* **43**, 799-836 (2005); of toxicology: C. S. Letizia *et al.*, *ibid.* 925-943.

Monoclinic crystals, honey, floral odor, d_4^4 1.2475. mp 134°; bp$_3$ 147°. Flash point, closed cup: >93.3°C. Distilling at 146° causes decarboxylation to styrene. pK (25°) 4.46. uv max (alc): 273 nm. One gram dissolves in about 2000 ml water at 25° (more sol in hot water), in 6 ml alc, 5 ml methanol, 12 ml chloroform. Freely sol in benzene, ether, acetone, glacial acetic acid, carbon disulfide, oils. The alkali salts are sol in water. LD$_{50}$ (g/kg): 3.57 orally in rats; >5.0 dermally in rabbits (Letizia).

Methyl ester. [103-26-4] $C_{10}H_{10}O_2$. Crystals, odor fruity and balsamic, reminiscent of strawberries. mp 36°. d_0^{36} 1.042. bp 261.9°; bp$_{15}$ 132.5-134°. n_D^{21} 1.5766. Freely sol in alcohol, ether. Practically insol in water. Clearly sol in 80% alc.

Ethyl ester. [103-36-6] Ethyl cinnamate; ethyl phenylacrylate. $C_{11}H_{12}O_2$. Almost colorless, oily liquid, fruity and balsamic odor, reminiscent of cinnamon with an amber note. d_{25}^{25} 1.045-1.048. d_4^{20} 1.049. bp 271°. mp 6-10°. n_D^{20} 1.559-1.561. Miscible with alcohol, ether. Insol in water. Soluble in 3 vols of 70% alc.

n-Butyl ester. [538-65-8] Butyl cinnamate; Eliminoxy. $C_{13}H_{16}O_2$; mol wt 204.27. Liquid, agreeable ethereal odor when pure. bp$_{13}$ 145°. d_4^{18} 1.012. Very sparingly sol in water (<0.5%). Sol in

95% alc, ether, acetone, chloroform, benzene. Incompatible with alkalies; stable to light, air, and storage temps.

Anhydride. [538-56-7] Cinnamic anhydride. $C_{18}H_{14}O_3$. Crystals, mp 136°. Practically insol in water. Freely sol in warm benzene; slightly sol in alc.

USE: Fragrance and flavoring ingredient.

2300. Cinnamon. Ceylon cinnamon. Tropical evergreen tree, *Cinnamomum zeylanicum* Bl. (syn. *C. verum* Presl.), *Lauraceae*; cultivated for the aromatic bark which is widely used in cooking and in traditional medicine. Leaves and root bark are also used for their essential oil. *Habit.* Sri Lanka and southeast Asia. *Saigon cinnamon* is obtained from the bark of *C. loureirii* Nees. *Cassia*, also known as *Chinese cinnamon*, is often used as a substitute; obtained from the related species, *C. cassia* Presl. (syn. *C. aromaticum* Nees). *Constit.* Volatile oil (up to 4%), tannins, catechins, pre-anthocyanidins, resins, mucilage, gum, sugars, calcium oxalate, coumarins, and the insecticidal compounds, cinnzeylanin and cinnzeylanol. Comprehensive description: J. Thomas, P. P. Duethi in *Handbook of Herbs and Spices*, **Vol. 1**, K. V. Peter, Ed., (Woodhead Publishing, Cambridge, 2001) pp 143-153. Analysis of volatile constituents: U. M. Senanayake *et al.*, *J. Agric. Food Chem.* **26**, 822 (1978). Authentication method: M. J. Cikalo *et al.*, *J. Planar Chromatogr.* **5**, 135 (1992). Review of cinnamon: J. Barnes *et al.*, *Herbal Medicines* (Pharmaceutical Press, London, 2nd Ed., 2002) pp 135-136; of cassia: *idem, ibid.* pp 112-113. *Review:* A. Y. Leung, S. Foster, *Encyclopedia of Common Natural Ingredients*, (Wiley-Interscience, Hoboken, 2nd Ed., 2003) pp 167-170.

Cinnamon bark oil. [8015-91-6] Oil of cinnamon, Ceylon. Volatile oil obtained by steam distillation from the dried inner bark of *C. zeylanicum. Constit.* Complex mixture of components chiefly, cinnamaldehyde (60-75%), cinnamyl acetate, eugenol, caryophyllene, linalool. Yellow liquid with a spicy burning taste and the characteristic odor of cinnamon. d_{25}^{25} 1.010-1.030. n_D^{20} 1.573-1.591. Rotation between −2° and 0°. Sol in most fixed oils, propylene glycol, in 3 vols 70% alcohol. Insol in glycerin, mineral oil.

Cinnamon leaf oil. Oil of cinnamon leaf. Volatile oil obtained by steam distillation of the leaves and twigs of *C. zeylanicum. Constit.* Eugenol (80-88%), other components similar to bark oil. Light to dark brown liquid with spicy cinnamon-clove odor and taste. d_{25}^{25} 1.030-1.050. n_D^{20} 1.529-1.537. Rotation between −2° and +1°. Sol in most fixed oils, propylene glycol, in 1.5 vols 70% alcohol. Sol with cloudiness in mineral oil. Insol in glycerin.

Cassia oil. [8007-80-5] Oil of cassia; oil of Chinese cinnamon. Volatile oil obtained by steam distillation from the leaves and twigs of *C. cassia. Constit.* Cinnamaldehyde (75-90%), cinnamyl acetate, salicylaldehyde, methyleugenol, coumarin. Yellow or brown liquid with characteristic odor and taste. Darkens and thickens upon aging or exposure to air. d_{25}^{25} 1.045-1.063. n_D^{20} 1.602-1.614. Rotation between −1° and +1°. Sol in glacial acetic acid, in 2 vols 70% alcohol.

USE: Used extensively as spice and flavoring in foods, beverages. Pharmaceutic aid (flavor). Ingredient in spice blends such as garam masala.

THERAP CAT: Carminative, antiseptic, astringent.

THERAP CAT (VET): Carminative.

2301. Cinnamoyl Chloride. [102-92-1] 3-Phenyl-2-propenoyl chloride; benzylideneacetyl chloride; cinnamic acid chloride; 3-phenylacryloyl chloride. C_9H_7ClO; mol wt 166.60. C 64.89%, H 4.24%, H 21.28%, O 9.60%. Prepn from cinnamic acid: L. Claisen, P. J. Antweiler, *Ber.* **13**, 2123 (1880); from sodium cinnamate: R. Adams, L. H. Ulich, *J. Am. Chem. Soc.* **42**, 599 (1920). Commercial product is predominately the (2E)-form.

(2E)-form

(2E)-Form. [17082-09-6] *trans*-Cinnamoyl chloride. Yellowish crystals. mp 35-36°; bp$_{58}$ 170-171°; bp$_{25}$ 154°; bp$_{16}$ 147°; bp$_2$ 101°; $d_4^{45.3}$ 1.1617; $n_D^{42.5}$ 1.614. *Corrosive.* Moisture sensitive. Hydrolyzes upon contact with water.

USE: Titrimetric determination of small amts of water. Reagent in organic synthesis.

2302. Cinnamyl Acetate. [103-54-8] 3-Phenyl-2-propen-1-ol 1-acetate; 3-acetoxy-1-phenylpropene. $C_{11}H_{12}O_2$; mol wt 176.22. C 74.98%, H 6.86%, O 18.16%. The naturally occurring *trans*-form is a constituent of the essential oils of various species of cinnamon, *q.v.* Isoln from oil of cassia, *Cinnamomum cassia* Blume: F. D. Dodge, A. E. Sherndal, *Ind. Eng. Chem.* **7**, 1055 (1915); F. D. Dodge, *ibid.* **10**, 1005 (1918); R. ter Heide, *J. Agric. Food Chem.* **20**, 747 (1972). Prepn: K. Hess, W. Wustrow, *Ann.* **437**, 256 (1924). Synthesis of stereoisomers from cinnamyl alcohol and acetic anhydride: G. H. Schmid *et al.*, *J. Org. Chem.* **42**, 871 (1977); H. L. Goering, S. S. Kantner, *ibid.* **48**, 721 (1983); H. Eshghi, P. Shafieyoon, *J. Chem. Res.* **2004**, 802. Review of toxicological profile: S. P. Bhatia *et al.*, *Food Chem. Toxicol.* **45**, Suppl. 1, S53-S57 (2007).

trans-Form

Colorless to slightly yellow, oily liquid; sweet, balsamic, floral odor. bp$_{18}$ 140-141°. d_4^{12} 1.0567. n_D^{12} 1.54415. Sol in alcohol, fixed oils. Insol in water, glycerin. LD$_{50}$ orally in rats: 3.3 g/kg; dermally in rabbits: >5 g/kg (Bhatia).

***trans*-Form.** [21040-45-9] (*E*)-Cinnamyl acetate. Colorless liquid, sweet-flowery-fruity, slightly balsamic odor. bp$_{760}$ 263°. bp$_{10}$ 137-139°. bp$_5$ 95°.

***cis*-Form.** [77134-01-1] (*Z*)-Cinnamyl acetate. bp$_{6.5}$ 116°.

USE: Flavor and fragrance ingredient. Modifier for cinnamyl alcohol.

2303. Cinnamyl Alcohol. [104-54-1] 3-Phenyl-2-propen-1-ol; cinnamic alcohol; styryl carbinol; γ-phenylallyl alcohol. C_9H_{10}-O; mol wt 134.18. C 80.56%, H 7.51%, O 11.92%. The *trans*-form occurs (in the esterified form) in storax and in balsam Peru, cinnamon leaves, hyacinth oil. Obtained by the alkaline hydrolysis of storax. Prepd synthetically by reducing cinnamal diacetate with iron filings and acetic acid; from cinnamaldehyde by Meerwein-Ponndorf reduction with aluminum isopropoxide: Meerwein, Schmidt, *Ann.* **444**, 221 (1925). Review of risk assessment as fragrance ingredient: D. Bickers *et al.*, *Food Chem. Toxicol.* **43**, 799-836 (2005); of toxicology: C. S. Letizia *et al.*, *ibid.* 837-866.

Needles or cryst mass. Odor of hyacinth. mp 30°. d_{35}^{35} 1.0397. bp 258°. n_D^{20} 1.58190; n_D^{33} 1.57580. Flash point, closed cup: >93.3°C. When small amounts of impurities are present as in the natural article (cinnamyl alcohol from storax), it remains fluid at lower temps than the melting point. Minimum congealing points specified by the Essential Oil Assn are: 33.0° for cinnamic alcohol pure; 28.0° for cinnamic alcohol prime; 20.0° for cinnamic alcohol from storax. Is oxidized slowly on exposure to heat, light and air. Sol in water, glycerol. Clearly sol in 3 vols 50% alc. Freely sol in alc, ether, other common organic solvents. LD$_{50}$ (g/kg): 2.0 orally in rats; >5.0 dermally in rabbits (Letizia).

USE: In perfumery; as deodorant in 12.5% soln in glycerol.

2304. Cinnamyl Anthranilate. [87-29-6] 1-(2-Aminobenzoate) 3-phenyl-2-propen-1-ol; anthranilic acid cinnamyl ester; 3-phenyl-2-propenyl anthranilate; cinnamyl *o*-aminobenzoate. $C_{16}H_{15}$-NO$_2$; mol wt 253.30. C 75.87%, H 5.97%, N 5.53%, O 12.63%. Synthetic imitation grape or cherry flavoring agent. Prepn: A. Seldner, *Am. Perfum. Essent. Oil Rev.* **54**, 295 (1949); R. P. Staiger, E. B. Miller, *J. Org. Chem.* **24**, 1214 (1959). Carcinogenicity studies: G. D. Stoner *et al.*, *Cancer Res.* **33**, 3069 (1973); *Fed. Regist.* **45**, 85832 (1980). *Review:* D. L. Opdyke, *Food Cosmet. Toxicol.* **13**, Suppl., 751-752 (1975).

Crystals, mp 61-61.5°.

USE: Flavoring agent in food; fragrance in soaps and perfumes.

2305. Cinnamyl Cinnamate. [122-69-0] 3-Phenyl-2-propenoic acid 3-phenyl-2-propen-1-yl ester; cinnamic acid cinnamyl ester; cinnyl cinnamate; styracin. $C_{18}H_{16}O_2$; mol wt 264.32. C 81.79%, H 6.10%, O 12.11%. From buds of *Populus balsamifer* L., *Salicaceae*: Goris, Canal, *Bull. Soc. Chim. Fr.* **3**, 1982 (1936); from *Lavanga scandens* Buch.-Ham., *Lavangalata*: Baslas, Deshapande, *J. Indian Chem. Soc.* **27**, 379 (1950). Synthesis of (*E,E*)-form: Klemm *et al., Tetrahedron* **20**, 871 (1964).

(*E,E*)-Form. [40918-97-6] Needles from abs ethanol, mp 44°. uv max (95% ethanol): 216, 223 nm (log ε 3.45, 3.25). Practically insol in water. 1 g dissolves in 22 ml cold, 3 ml boiling alcohol, 3 ml ether.

2306. Cinnarizine. [298-57-7] 1-(Diphenylmethyl)-4-(3-phenyl-2-propen-1-yl)piperazine; 1-cinnamyl-4-diphenylmethyl-piperazine; *N*-benzhydryl-*N'-trans*-cinnamylpiperazine; 1-*trans*-cinnamyl-4-diphenylmethylpiperazine; 1-cinnamyl-4-benzhydryl-piperazine; 1-diphenylmethyl-4-*trans*-cinnamylpiperazine; cinnipirine; 516-MD; Aplactan; Aplexal; Apotomin; Artate; Carecin; Cerebolan; Cerepar; Cinaperazine; Cinazyn; Cinnacet; Cinnageron; Corathiem; Denapol; Dimitron; Eglen; Folcodal; Giganten; Glanil; Hilactan; Ixertol; Katoseran; Labyrin; Midronal; Mitronal; Olamin; Processine; Sedatromin; Sepan; Siptazin; Spaderizine; Stugeron; Stutgeron; Stutgin; Toliman. $C_{26}H_{28}N_2$; mol wt 368.52. C 84.74%, H 7.66%, N 7.60%. Calcium channel blocker with anti-allergic and anti-vasoconstricting activity. Prepn: Janssen, **US 2882271** (1959, Janssen). Pharmacology: Van Nueten, Janssen, *Arch. Int. Pharmacodyn. Ther.* **204**, 37 (1973). Metabolism: Soudijn, van Wijngaarden, *Life Sci.* **7**, 231 (1968). Clinical study in treatment of intermittent claudication: J. H. Barber *et al., Pharmatherapeutica* **2**, 401 (1980). Assessment of calcium antagonist effects: M. Spedding, *Arch. Pharmacol.* **318**, 234 (1982). HPLC determn in blood and plasma: M. Puttemans *et al., J. Liq. Chromatogr.* **7**, 2237 (1984). Comparison of clinical pharmacokinetics with other antihistamines: D. M. Paton, D. R. Webster, *Clin. Pharmacokinet.* **10**, 477-497 (1985).

Hydrochloride. $C_{26}H_{28}N_2$.HCl. mp 192° (dec). Soly in water: 2 mg/100 ml.

Mixture with vitamin B$_6$. Emesazine.

THERAP CAT: Antihistaminic. Vasodilator (peripheral, cerebral).

2307. Cinnoline. [253-66-7] 1,2-Benzodiazine; benzo[*c*]pyridazine; 1,2-diazanaphthalene; α-phenodiazine. $C_8H_6N_2$; mol wt 130.15. C 73.83%, H 4.65%, N 21.52%. First prepd by reduction of 4-chlorocinnoline: Busch, Rast, *Ber.* **30**, 521, 524 (1897). Synthesis from methyl anthranilate: Jacobs *et al., J. Am. Chem. Soc.* **68**, 1310

(1946); from cinnoline-4-carboxylic acid: Morley, *J. Chem. Soc.* **1951**, 1971. Crystal and molecular structure: C. Huiszoon *et al., Acta Crystallogr.* **33B**, 1867 (1977).

Pale yellow clusters of crystals from ligroin, mp 38°; also reported as mp 40-41° (Morley). Solvated needles from ether (contg 1 mol ether), mp 24-25°. bp$_{0.35}$ 114°. uv spectrum: Hearn *et al., J. Chem. Soc.* **1951**, 3318. Geranium-like odor. *Poisonous*. Bitter taste resembling the taste of quinine. Must be kept under nitrogen. On exposure to air it has a tendency to liquefy and turn green. Strong base: *ibid.* **1949**, 1356. Freely sol in the usual solvents.

Hydrochloride. $C_8H_6N_2$.HCl. Brownish needles from alcohol + ether, mp 156-160°. Volatilizes around 100° on slow heating. Freely sol in water, alcohol.

2308. Cinobufotalin. [1108-68-5] (3β,5β,15β,16β)-16-(Acetyloxy)-14,15-epoxy-3,5-dihydroxybufa-20,22-dienolide; 14,-15β-epoxy-3β,5,16β-trihydroxy-5β-bufa-20,22-dienolide 16-acetate. $C_{26}H_{34}O_7$; mol wt 458.55. C 68.10%, H 7.47%, O 24.42%. Isolated from the Chinese drug Ch'an Su which is prepd from Chinese toads (*Bufo asiaticus* = *bufo gargarizans* Cantor). Isoln: Kotake, Kuwada, *C.A.* **31**, 7065² (1937); Ruckstubl, Meyer, *Helv. Chim. Acta* **40**, 1270 (1957). Structure: Bernoulli *et al., ibid.* **45**, 240 (1962).

Octahedrons from acetone, mp 259-262°. $[\alpha]_D^{20}$ +11°. uv max: 295 nm (log ε 3.72).

2309. Cinoxacin. [28657-80-9] 1-Ethyl-1,4-dihydro-4-oxo-[1,3]dioxolo[4,5-*g*]cinnoline-3-carboxylic acid; 1-ethyl-6,7-methylenedioxy-4(1*H*)-oxocinnoline-3-carboxylic acid; compd 64716; Cinobac; Noxigram; Uronorm. $C_{12}H_{10}N_2O_5$; mol wt 262.22. C 54.97%, H 3.84%, N 10.68%, O 30.51%. Quinolone antibacterial; analog of oxolinic acid, *q.v.* Prepn: W. A. White, **DE 2005104**; *idem*, **US 3669965** (1970, 1972 both to Lilly). *In vitro* and *in vivo* study: W. E. Wick *et al., Antimicrob. Agents Chemother.* **4**, 415 (1973). Pharmacology: J. J. Szwed *et al., J. Antimicrob. Chemother.* **4**, 451 (1978); K. S. Israel *et al., J. Clin. Pharmacol.* **18**, 491 (1978). Metabolism: R. L. Wolen *et al.* in *Stable Isotopes*, T. A. Baillie, Ed. (Univ. Park Press, Baltimore, 1978) pp 113-125. Clinical studies: S. Colleen *et al., J. Antimicrob. Chemother.* **3**, 579 (1977); S. N. Rous, *J. Urol.* **120**, 196 (1978). Toxicity: I. Narama *et al., Chemotherapy (Tokyo)* **28**, Suppl. 4, 406 (1980), *C.A.* **94**, 25007 (1981). Pharmacokinetics: M. Ohkawa *et al., J. Antimicrob. Chemother.* **8**, 447 (1981); R. Barbhaiya *et al., Antimicrob. Agents Chemother.* **21**, 472 (1982). *Review:* J. M. Scavone *et al., Pharmacotherapy* **2**, 266-271 (1982). Review of pharmacology and therapeutic efficacy: T. S. Sisca *et al., Drugs* **25**, 544-569 (1983).

Light tan crystals, mp 261-262° (dec). Sol in most polar organic solvents, alkaline solutions. Insol in water and most common organic solvents. LD$_{50}$ in rats (mg/kg): 4160 orally; 900 i.v. (Narama).

Sodium salt. C$_{12}$H$_9$N$_2$NaO$_5$. White crystalline solid. Sol in aqueous solvents.

THERAP CAT: Antibacterial.

2310. Cinoxate. [104-28-9] 3-(4-Methoxyphenyl)-2-propenoic acid 2-ethoxyethyl ester; p-methoxycinnamic acid 2-ethoxyethyl ester; 2-ethoxyethyl p-methoxycinnamate; Give-Tan; SunDare. C$_{14}$H$_{18}$O$_4$; mol wt 250.29. C 67.18%, H 7.25%, O 25.57%. Prepn: **GB 856411** (1960 to Givaudan).

(E)-form

Viscous liq; bp$_2$ 184-187°. May have a slightly yellow tinge. Practically odorless. Solidifies below −25°. d$_{25}^{25}$ 1.102. n$_D^{20}$ 1.5670. Sapon no. 225.5. Practically insol in water (soly ~0.05%). Soly in glycerol 0.5%, in propylene glycol 5.0%. Miscible with alcohols, esters, vegetable oils.

THERAP CAT: Ultraviolet screen.

2311. Cintredekin Besudotox. [372075-36-0] Toxin hIL13-PE38QQR (plasmid phuIL13-Tx); IL-13PE38. Recombinant chimeric protein consisting of human interleukin 13 (IL-13) fused to a truncated form of *Pseudomonas* exotoxin A, PE38QQR; designed to target human epithelial carcinomas bearing the IL-13 receptor. Prepn: W. Debinski *et al.*, *J. Biol. Chem.* **270**, 16775 (1995); R. K. Puri *et al.*, **WO 9629417**; *eidem*, **US 5919456** (1996, 1999 both to U.S. Dept. Health Human Serv.). Review of pharmacology: S. R. Husain, R. K. Puri, *J. Neuro-Oncol.* **65**, 37-48 (2003). Molecular mode of action study and gene expression profile: J. Han *et al.*, *ibid.* **72**, 35 (2005). Review of clinical experience in malignant glioma: N. G. Rainov, A. Söling, *Curr. Opin. Mol. Ther.* **7**, 170-181 (2005).

THERAP CAT: Antineoplastic.

2312. Ciprofibrate. [52214-84-3] 2-[4-(2,2-Dichlorocyclopropyl)phenoxy]-2-methylpropanoic acid; 2-[4-(2,2-dichlorocyclopropyl)phenoxy]isobutyric acid; Win-35833; Ciprol; Lipanor; Modalim. C$_{13}$H$_{14}$Cl$_2$O$_3$; mol wt 289.15. C 54.00%, H 4.88%, Cl 24.52%, O 16.60%. Hypolipemic agent, related structurally to clofibrate, *q.v.* Prepn: D. K. Phillips, **DE 2343606**; *idem*, **US 3948973** (1974, 1976 both to Sterling). Pharmacokinetics in animals and man: C. Davison *et al.*, *Drug Metab. Dispos.* **3**, 520 (1975). Inhibition of cholesterol biosynthesis in rats: A. Arnold *et al.*, *Atherosclerosis* **32**, 155 (1979). Blood levels, distribution, duration of action: J. Edelson *et al.*, *ibid.* **33**, 351 (1979). Metabolic effects: A. Arnold *et al.*, *J. Pharm. Sci.* **68**, 1557 (1979). Teratology study: H. Tuchman-Duplessis *et al.*, *Toxicology* **12**, 1 (1979).

Pale cream solid from hexane, mp 114-116°.

THERAP CAT: Antilipemic.

2313. Ciprofloxacin. [85721-33-1] 1-Cyclopropyl-6-fluoro-1,4-dihydro-4-oxo-7-(1-piperazinyl)-3-quinolinecarboxylic acid; Bay q 3939. C$_{17}$H$_{18}$FN$_3$O$_3$; mol wt 331.35. C 61.62%, H 5.48%, F 5.73%, N 12.68%, O 14.49%. Fluorinated quinolone antibacterial. Prepn: K. Grohe *et al.*, **DE 3142854**; *eidem*, **US 4670444** (1983, 1987 both to Bayer AG); K. Grohe, H. Heitzer, *Ann.* **1987**, 29. Antibacterial spectrum *in vitro*: B. Watt, F. V. Brown, *J. Antimicrob. Chemother.* **17**, 605 (1986); C. M. Bassey *et al.*, *ibid.* 623. HPLC

determn in biological fluids: W. Gau *et al.*, *J. Liq. Chromatogr.* **8**, 485 (1985). Pharmacokinetics: G. Hoffken *et al.*, *Antimicrob. Agents Chemother.* **27**, 375 (1985); of extended release formulation: C. B. Washington *et al.*, *J. Clin. Pharmacol.* **45**, 1236 (2005). Clinical trials: C. A. Ramirez *et al.*, *ibid.* **28**, 128 (1985); B. E. Scully *et al.*, *Lancet* **1**, 819 (1986). Symposia on antibacterial spectrum and clinical use: *Am. J. Med.* **82**, Suppl. 4A, 1-404 (1987); *J. Antimicrob. Chemother.* **26**, Suppl. F, 3-193 (1990). Review of clinical safety and efficacy in children: R. Kubin, *Infection* **21**, 413-421 (1993); of US FDA approval in postexposure inhalational anthrax: A. Myerhoff *et al.*, *Clin. Infect. Dis.* **29**, 303-308 (2004); of clinical experience in urinary tract infections: A. D. Hickerson, C. C. Carson, *Expert Opin. Invest. Drugs* **15**, 519-532 (2006); of therapeutic potential in a broad spectrum of infections: H. Koch *et al.*, *Clin. Drug Invest.* **26**, 645-654 (2006).

Faintly yellowish to light yellow crystalline powder. Dec 255-257°. Sol in dilute (0.1*N*) HCl. Practically insol in water and ethanol.

Monohydrochloride monohydrate. [86393-32-0] Bay o 9867; Ciflox; Ciloxan; Cipro; Ciprobay; Ciproxan; Ciproxin; Flociprin; Septicide; Uniflox. C$_{17}$H$_{18}$FN$_3$O$_3$·HCl.H$_2$O; mol wt 385.82. Light yellow crystalline powder. mp 318-320°. Slightly sol in water.

THERAP CAT: Antibacterial.

THERAP CAT (VET): Antibacterial.

2314. Circulins. Polypeptide antibiotics related to polymyxins. Produced by *Bacillus circulans*, a microorganism found in soil and dust: Murray, Tetrault, *Proc. Soc. Am. Bact.* **1**, 20 (1948); McLeod, *J. Bacteriol.* **56**, 749 (1948); Murray *et al.*, *ibid.* **57**, 305 (1949). Isoln yields circulins A and B, circulin A being the major component: Tetrault, **US 2676133** (1954 to Purdue Res. Found.). *Review:* Vogner, Studer, *Experientia* **22**, 345 (1966).

DAB = α,γ-diaminobutyric acid

Circulin A   R = (+)-6-methyloctanoyl
Circulin B   R = isooctanoyl

Circulin A. C$_{53}$H$_{100}$N$_{16}$O$_{13}$; mol wt 1169.48. Structure: Fujikawa *et al.*, *Experientia* **21**, 307 (1965). Synthesis: Studer *et al.*, *Helv. Chim. Acta* **49**, 974 (1966).

Sulfate. C$_{53}$H$_{100}$N$_{16}$O$_{13}$·2½H$_2$SO$_4$; mol wt 1414.66. Crystals or amorphous solid, dec 226-228°. [α]$_D^{25}$ −61.6° (c = 1.25). Sol in water; less sol in the lower alcs. Practically insol in acetone and water-immiscible solvents.

Circulin B. C$_{52}$H$_{98}$N$_{16}$O$_{13}$; mol wt 1155.46. Structure: Hayashi *et al.*, *Experientia* **24**, 656 (1968).

2315. Cisapride. [81098-60-4] *rel*-4-Amino-5-chloro-*N*-[1-[(3*R*,4*S*)-3-(4-fluorophenoxy)propyl]-3-methoxy-4-piperidinyl]-2-methoxybenzamide; *cis*-4-amino-5-chloro-*N*-[1-[3-(*p*-fluorophenoxy)propyl]-3-methoxy-4-piperidinyl]-*o*-anisamide. C$_{23}$H$_{29}$-ClFN$_3$O$_4$; mol wt 465.95. C 59.29%, H 6.27%, Cl 7.61%, F 4.08%,

N 9.02%, O 13.73%. Prokinetic 5-HT$_4$ receptor agonist; facilitates gastric release of acetylcholine. Prepn: G. Van Daele, **EP 76530**; *idem*, **US 4962115** (1983, 1990 both to Janssen); and pharmacology: *idem et al.*, *Drug Dev. Res.* **8**, 225 (1986). Spectrophotometric and HPLC determn in pharmaceutical prepns: E. M. Hassan *et al.*, *J. Pharm. Biomed. Anal.* **24**, 659 (2001). HPLC determn in plasma and bioavailability: J. Emami *et al.*, *ibid.* **33**, 513 (2003). Pharmacokinetics: J. A. Barone *et al.*, *Clin. Pharm.* **6**, 640 (1987). Review of pharmacology and therapeutic efficacy in humans: L. R. Wiseman, D. Faulds, *Drugs* **47**, 116-152 (1994); in animals: R. J. Washabau, J. A. Hall, *J. Am. Vet. Med. Assoc.* **207**, 1285-1288 (1995).

Relative stereochemistry

Monohydrate. [260779-88-2] R-51619; Acenalin; Alimix; Prepulsid; Risamol. C$_{23}$H$_{29}$ClFN$_3$O$_4$.H$_2$O; mol wt 483.97. Crystals from 2-propanol, mp 109.8°. Sol in acetone. Sparingly sol in methanol. Practically insol in water.

THERAP CAT: Gastroprokinetic.

THERAP CAT (VET): Gastroprokinetic.

2316. Cisplatin. [15663-27-1] (*SP*-4-2)-Diamminedichloroplatinum; *cis*-diamminedichloroplatinum; *cis*-platinum II; *cis*-DDP; CACP; CPDC; DDP; NSC-119875; Blastolem; Briplatin; Cisplatyl; Neoplatin; Platamine; Platinex; Platiblastin; Platinol; Platosin; Randa. Cl$_2$H$_6$N$_2$Pt; mol wt 300.05. Cl 23.63%, H 2.02%, N 9.34%, Pt 65.02%. Antitumor platinum coordination complex. Originally known as *Peyrone's salt* or *Peyrone's chloride*; of interest in the development of coordination theory. Prepn: M. Peyrone, *Ann.* **51**, 1 (1845); G. B. Kauffman, D. O. Cowan, *Inorg. Synth.* **7**, 239 (1963); S. C. Dhara, *Indian J. Chem.* **8**, 193 (1970). Early structural studies: R. Werner, *Z. Anorg. Chem.* **3**, 267 (1893); H. D. K. Drew *et al.*, *J. Chem. Soc.* **1932**, 988. Discovery of anti-tumor activity: B. Rosenberg *et al.*, *Nature* **205**, 698 (1965); *ibid.* **222**, 385 (1972). Use as neoplasm inhibitor: M. L. Tobe *et al.*, **DE 2318020** (1972 to Rustenburg Platinum Mines Ltd.), *C.A.* **80**, 55897e (1974); M. J. Cleare *et al.*, **DE 2329485** (1972 to Research Corp.), *C.A.* **81**, 21172v (1974). X-ray structure of cisplatin-DNA adduct: S. E. Sherman *et al.*, *Science* **230**, 412 (1985). Inhibition of *in vitro* DNA synthesis: A. L. Pinto, S. J. Lippard, *Proc. Natl. Acad. Sci. USA* **82**, 4616 (1985). Pharmacology: A. Sirica *et al.*, *Proc. Am. Assoc. Cancer Res.* **12**, 4 (1971); C. L. Litterst *et al.*, *Cancer Res.* **36**, 2340 (1976); N. P. Johnson *et al.*, *Chem. Biol. Interact.* **23**, 267 (1978). Metabolism: R. C. Lange *et al.*, *J. Nucl. Med.* **14**, 191 (1973). Clinical studies: J. J. Ochs *et al.*, *Cancer Treat. Rep.* **62**, 239 (1978); H. M. Pinedo *et al.*, *Eur. J. Cancer* **14**, 1149 (1978). Toxicology: R. L. Dixon, *Proc. 7th Int. Congr. Chemother.* **Vol. 2** (University Park Press, Baltimore, 1972) pp 241-243; R. W. Fleishman *et al.*, *Toxicol. Appl. Pharmacol.* **33**, 320 (1975). Review of carcinogenicity studies: *IARC Monographs* **26**, 154-161 (1981); of neurotoxicity: R. J. Cersosimo, *Cancer Treat. Rev.* **16**, 195-211 (1989). Comprehensive description: C. M. Riley, L. M. Sternson, *Anal. Profiles Drug Subs.* **14**, 77-105 (1985). Book: *Cisplatin, Current Status and New Developments*, A. W. Prestayko *et al.*, Eds. (Academic Press, New York, 1980) 527 pp. Review of mechanism of action: M. A. Fuertes *et al.*, *Curr. Med. Chem.* **10**, 257-266 (2003); Z. H. Siddik, *Oncogene* **22**, 7265-7279 (2003).

Yellow to orange crystalline powder. Soly in water 0.253 g/100 g at 25°; slowly changes to *trans*-form in aq soln. Insol in most com-

mon solvents. Sol in DMF. LD$_{50}$ in guinea pigs: 9.7 mg/kg i.p. (Fleishman).

Caution: This substance is reasonably anticipated to be a human carcinogen: *Report on Carcinogens, Twelfth Edition* (PB2011-111646, 2011) p 110.

THERAP CAT: Antineoplastic.

THERAP CAT (VET): Antineoplastic.

2317. Citalopram. [59729-33-8] 1-[3-(Dimethylamino)propyl]-1-(4-fluorophenyl)-1,3-dihydro-5-isobenzofurancarbonitrile; 1-[3-(dimethylamino)propyl]-1-(4-fluorophenyl)-5-phthalancarbonitrile; nitalapram; Lu-10-171. C$_{20}$H$_{21}$FN$_2$O; mol wt 324.40. C 74.05%, H 6.53%, F 5.86%, N 8.64%, O 4.93%. Selective serotonin reuptake inhibitor (SSRI). Prepn: K. P. Boegesoe, A. S. Toft, **DE 2657013**; *eidem*, **US 4136193** (1977, 1979 both to Kefalas); A. J. Bigler *et al.*, *Eur. J. Med. Chem. - Chim. Ther.* **12**, 289 (1977). Prepn of enantiomers: K. P. Boegesoe, J. Perregaard, **EP 347066**; *eidem*, **US 4943590**, reissued as **US RE 34712** (1989, 1990, 1994 all to Lundbeck). Pharmacology: A. V. Christensen *et al.*, *Eur. J. Pharmacol.* **41**, 153 (1977). HPLC determn in plasma and urine: E. Oyehaug *et al.*, *J. Chromatogr.* **308**, 199 (1984). Comparative biotransformation of enantiomers: L. L. Von Moltke *et al.*, *Drug Metab. Dispos.* **29**, 1102 (2001). Evaluation of suicidality risk: G. Laje *et al.*, *Am. J. Psychiatry* **164**, 1530 (2007). Review of clinical pharmacokinetics: K. Brosen, C. A. Naranjo, *Eur. Neuropsychopharmacol.* **11**, 275-283 (2001). Review of clinical experience in depression: M. B. Keller, *J. Clin. Psychiatry* **61**, 896-908 (2000); of escitalopram in depression: P. Malin *et al.*, *Expert Rev. Neurother.* **4**, 769-779 (2004); in anxiety disorders: D. S. Baldwin, R. V. Nair, *ibid.* **5**, 443-449 (2005); S. R. Bareggi *et al.*, *Expert Opin. Drug Metab. Toxicol.* **3**, 741-753 (2007).

bp$_{0.03}$ 175-181°.

Hydrobromide. [59729-32-7] Celexa; Cipramil; Elopram; Seropram. C$_{20}$H$_{21}$FN$_2$O.HBr; mol wt 405.31. Fine, white to off-white powder. Crystals from isopropanol, mp 182-183°. Freely sol in water, ethanol, chloroform.

Escitalopram. [128196-01-0] *S*-(+)-Citalopram. [α]$_D$ +12.33° (c = 1 in methanol).

Escitalopram oxalate. [219861-08-2] Lu-26-054-0; Cipralex; Gaudium; Lexapro. C$_{20}$H$_{21}$FN$_2$O.C$_2$H$_2$O$_4$; mol wt 414.43. Fine white to slightly yellow powder. Crystals from acetone, mp 147-148°. [α]$_D$ +12.31° (c = 1 in methanol). Freely sol in methanol, DMSO; sol in isotonic saline; sparingly sol in water, ethanol; slightly sol in ethyl acetate. Insol in heptane.

THERAP CAT: Antidepressant. Escitalopram as antidepressant, anxiolytic.

2318. Citicoline. [987-78-0] Cytidine 5'-(trihydrogen diphosphate) *P'*-[2-(trimethylammonio)ethyl] ester inner salt; choline cytidine 5'-pyrophosphate (ester); cytidine diphosphate choline ester; CDP-choline; Difosfocin; Nicholin; Recognan; Rexort; Somazina. C$_{14}$H$_{26}$N$_4$O$_{11}$P$_2$; mol wt 488.33. C 34.43%, H 5.37%, N 11.47%, O 36.04%, P 12.69%. Naturally occurring nucleotide; intermediate in the major pathway of lecithin biosynthesis. Identification: E. P. Kennedy, S. B. Weiss, *J. Am. Chem. Soc.* **77**, 250 (1955). Crystallization from yeast extract: I. Lieberman *et al.*, *Science* **124**, 81 (1956). Synthesis: E. P. Kennedy, *J. Biol. Chem.* **222**, 185 (1956); K. Kikugawa *et al.*, *Chem. Pharm. Bull.* **19**, 1011, 2466 (1971). Molecular structure: M. A. Viswamitra *et al.*, *Nature* **258**, 497 (1975). Series of articles on pharmacology and toxicology: *Arzneim.-Forsch.* **33**, 1009-1080 (1983). Acute toxicity: T. Grau *et al.*, *ibid.* 1033. Clinical trial in ischemic stroke: W. M. Clark *et al.*, *Neurology* **49**, 671 (1997). Review of biosynthesis, metabolism, pharmacology: G. B. Weiss, *Life Sci.* **56**, 637-660 (1995); and clinical experience: J. J. Secades, G. Frontera, *Methods Find. Exp. Clin. Pharmacol.* **17**, Suppl. B, 1-54 (1995).

Amorphous, somewhat hygroscopic powder. $[\alpha]_D^{25}$ +19.0° (c = 0.86 in H_2O). uv max (pH 1): 280 nm (ε 12800). Dissolves readily in water to form acidic soln. Practically insol in most organic solvents. pKa 4.4. LD_{50} in mice (mg/kg): 4600 ±335, 4150 ±370 i.v.; both species 8 g/kg orally (Grau).

Sodium salt. [33818-15-4] Acticolin; Brassel; Ceraxon; Neuroton; Sintoclar. $C_{14}H_{25}N_4NaO_{11}P_2$; mol wt 510.31. White, crystalline, spongy, hygroscopic powder, dec 250°. $[\alpha]_D^{20}$ +12.5° (c = 1.0 in H_2O). Sol in water. Practically insol in alcohol.

THERAP CAT: Neuroprotective. In treatment of ischemic stroke and head trauma.

2319. Citiolone. [1195-16-0] N-(Tetrahydro-2-oxo-3-thienyl)acetamide; 2-acetamido-4-mercaptobutyric acid γ-thiolactone; N-acetylhomocysteinethiolactone; α-acetamido-γ-thiobutyrolactone; AHCTL; BO-714; Citiolase. $C_6H_9NO_2S$; mol wt 159.20. C 45.27%, H 5.70%, N 8.80%, O 20.10%, S 20.14%. Prepn: Wagner *et al.*, **DE 1134683** (1962 to Degussa, vorm Roessler), *C.A.* **58**, 4648c (1963); Nakanishi *et al.*, **JP 62 16712** (1962 to Yoshitomi), *C.A.* **59**, 11660h (1963); Takayananagi *et al.*, **JP 64 3420**, *C.A.* **61**, 3193h (1964); Ogino, Yasui, **JP 65 1376** (1964, 1965 both to Sumitomo), *C.A.* **62**, 14822c (1965). Chemistry: Benesch, Benesch, *J. Am. Chem. Soc.* **78**, 1597 (1956); Laliberté *et al.*, *J. Chem. Soc.* **1963**, 2756. Pharmacology: E. -J. Kirnberger *et al.*, *Arzneim.-Forsch.* **8**, 72 (1958); Varga *et al.*, *ibid.* **13**, 1867 (1963).

Needles from toluene, mp 111.5-112.5°. uv max: 238 nm (ε 4400). LD_{50} i.v. in mice: 1200 mg/kg (Kirnberger).

USE: Photographic antifogging agent: Dersch, **US 3068100** (1962); Weber, **DE 1164828** (1964 to Adox Fotowerke Schleussner GmbH), *C.A.* **60**, 14050f (1964); protector against radiation: Langendorff, Koch, *Strahlentherapie* **106**, 451 (1958); Braun *et al.*, *ibid.* **108**, 262 (1959), *C.A.* **52**, 18841h (1958); **53**, 17325e (1959).

THERAP CAT: In treatment of hepatic disorders.

2320. Citraconic Acid. [498-23-7] (2Z)-2-Methyl-2-butenedioic acid; methylmaleic acid. $C_5H_6O_4$; mol wt 130.10. C 46.16%, H 4.65%, O 49.19%. Obtained by carefully heating citric acid at about 175°. Production of citraconic anhydride from itaconic acid: Humphrey, **US 2966498** (1960 to Pfizer). Toxicity study: P. M. Jenner *et al.*, *Food Cosmet. Toxicol.* **2**, 327 (1964).

Hygroscopic monoclinic crystals; characteristic odor. d 1.62. mp ~90° (dec). Freely sol in water, alc, ether, slightly in chloroform. Practically insol in benzene, petr ether. LD_{50} in rats, mice (mg/kg): 1320, 2260 orally (Jenner).

2321. Citral. [5392-40-5] 3,7-Dimethyl-2,6-octadienal. $C_{10}H_{16}O$; mol wt 152.24. C 78.90%, H 10.59%, O 10.51%. Constituent of many commercial oils such as lemon grass, verbena, lemon,

and orange. Citral from natural sources is a 2:1 mixture of two geometric isomers geranial and neral. Isoln from lemongrass: F. W. Semmler, *Ber.* **23**, 2965 (1890); F. D. Dodge, *Am. Chem. J.* **12**, 553 (1890). Separation of isomers: Y. R. Naves, *Bull. Soc. Chim. Fr.* **1952**, 521. IR determn in lemon and orange oils: P. L. Mahia *et al.*, *Food Chem.* **46**, 193 (1993). Stability in food emulsions and beverages: E. J. Freeburg *et al.*, *Perfum. Flavor.* **19**(4), 23 (1994). Review of toxicity: D. L. J. Opdyke, *Food Cosmet. Toxicol.* **17**, 259-266 (1979); of reaction chemistry: R. K. Baslas, B. Gupta, *Indian Perfum.* **32**, 266-272 (1988). Comprehensive reviews: J. L. Simonsen, *The Terpenes* **vol. I**, 83-100 (1947); P. Z. Bedoukian, *Perfumery and Flavoring Synthetics* (Allured Publishing Corporation, Wheaton, IL, 3rd ed., 1986) pp 106-117.

Geranial   Neral

Mobile pale yellow liquid having a strong lemon-like odor. d_{25}^{25} 0.885-0.891. n_D^{20} 1.4860-1.4900. Flash point: 99.5°C (208°F). LD_{50} orally in rats: 4.96 g/kg (Opdyke).

Geranial. [141-27-5] Trans-citral; citral a. Light oily liquid with strong lemon odor. $bp_{2.6}$ 92-93°. d_4^{20} 0.8888. n_D^{20} 1.4898. Practically insol in water. Miscible with alc, ether, benzyl benzoate, diethyl phthalate, glycerol, propylene glycol, mineral oil, essential oils.

Neral. [106-26-3] cis-Citral; citral b. Light oily liquid with a lemon odor not as intense but sweeter than gerianal. $bp_{2.6}$ 91-92°. d_4^{20} 0.8860. n_D^{20} 1.4869. Solubilities same as gerianal.

USE: In the synthesis of vitamin A, ionone and methylionone. As a flavor and in perfumery for its citrus effect.

2322. Citramalic Acid. [597-44-4] 2-Hydroxy-2-methylbutanedioic acid; 2-methylmalic acid; 2-hydroxy-2-methylsuccinic acid; α-hydroxypyrotartaric acid. $C_5H_8O_5$; mol wt 148.11. C 40.55%, H 5.44%, O 54.01%. Enzymatic synthesis: Barker, Blair, *Biochem. Prep.* **9**, 21 (1962). Chemical synthesis: Barker, *ibid.* 25; J. B. Wilkes, R. G. Wall, *J. Org. Chem.* **45**, 247 (1980). Stereoselective synthesis: E. G. J. Staring *et al.*, *Rec. Trav. Chim.* **105**, 374 (1986).

Deliquescent monoclinic prisms from ethyl acetate + petr ether, mp 117°. Sublimes. Freely sol in water, acetone. Sol in ethyl acetate, ether. Practically insol in petr ether, benzene.

d-Form. Crystals, mp 112.2-112.8°. $[\alpha]_D^{22}$ +23.6° (c = 3 in H_2O).
l-Form. Crystals, mp 112-113°. $[\alpha]_D^{20}$ -23.4° (c = 3 in H_2O).

2323. β-Citraurin. [650-69-1] (2E,4E,6E,8E,10E,12E,14E,16E)-17-[(4R)-4-Hydroxy-2,6,6-trimethyl-1-cyclohexen-1-yl]-2,6,11,15-tetramethyl-2,4,6,8,10,12,14,16-heptadecaoctaenal; (3R)-hydroxy-8'-apo-β,ψ-carotenal; citraurin; $C_{30}H_{40}O_2$; mol wt 432.65. C 83.28%, H 9.32%, O 7.40%. Carotenoid pigment found only in orange peel. Isoln by chromatography: Zechmeister, Tuzson, *Ber.* **69**, 1878 (1936); **70**, 1966 (1937). The peels from 100 kilos of oranges yield about 35 mg. Structure: Zechmeister, Tuzson, *loc. cit.*; Karrer, Solmssen, *Helv. Chim. Acta* **20**, 682 (1937); Karrer *et al.*, *ibid.* 1020; Zechmeister, v. Cholnoky, *Ann.* **530**, 291 (1937); Karrer *et al.*, *Helv. Chim. Acta* **21**, 445 (1948). Abs config: Bartlett *et al.*, *J. Chem. Soc. C* **1969**, 2527. Synthesis: H. Pfander *et al.*, *Helv. Chim. Acta* **63**, 1377 (1980).

Thin orange or yellow-colored plates from benzene + petr ether, mp 147°. Absorption max (benzene): 497, 467 nm. Freely sol in acetone, ethanol, ether, benzene, and carbon disulfide. Sparingly sol in petr ether.

2324. Citrazinic Acid. [99-11-6] 1,2-Dihydro-6-hydroxy-2-oxo-4-pyridinecarboxylic acid; 2,6-dihydroxyisonicotinic acid; 2,6-dihydroxy-4-pyridinecarboxylic acid. $C_6H_5NO_4$; mol wt 155.11. C 46.46%, H 3.25%, N 9.03%, O 41.26%. Prepn from citric acid with aq NH_3 at 140-160° under pressure: Bavley, Hamilton, **US 2729647** (1956 to Pfizer). Purification: Bavley *et al.*, **US 2738352** (1956 to Pfizer).

Yellowish powder with a greenish tinge; carbonizes above 300° without melting. Ultrapure material which is white or colorless, has been prepared. Almost insol in water; slightly sol in hot HCl; sol in alkali hydroxide or carbonate solns. Alkaline solns turn blue on standing.

2325. Citric Acid. [77-92-9] 2-Hydroxy-1,2,3-propanetricarboxylic acid; β-hydroxytricarballylic acid. $C_6H_8O_7$; mol wt 192.12. C 37.51%, H 4.20%, O 58.29%. Widely distributed in plants and in animal tissues and fluids. Produced by mycological fermentation on an industrial scale using crude sugar solns, such as molasses and strains of *Aspergillus niger*: Von Loesecke, *Chem. Eng. News* **23**, 1952 (1945); Schweiger, **US 2970084** (1961 to Miles Labs.); *Faith, Keyes & Clark's Industrial Chemicals*, F. A. Lowenheim, M. K. Moran, Eds. (Wiley-Interscience, New York, 4th ed., 1975) pp 275-279. Also extracted from citrus fruits (lemon juice contains 5 to 8%) and from pineapple waste. *Reviews:* Wilson, *Chem. Metall. Eng.* **29**, 787 (1923); Browne, *Ind. Eng. Chem.* **13**, 81 (1921); Warneford, Hardy, *ibid.* **17**, 1283 (1925); E. F. Bouchard, E. G. Merritt in *Kirk-Othmer Encyclopedia of Chemical Technology* vol. 6 (Wiley-Interscience, New York, 3rd ed., 1979) pp 150-179. Toxicity: C. M. Gruber, Jr., W. A. Halbeisen, *J. Pharmacol. Exp. Ther.* **94**, 65 (1948).

Anhydr form, mp 153°. Crystals are monoclinic holohedra and crystallize from hot concd aq soln. d 1.665. At 25°, pK_1 3.128; pK_2 4.761; pK_3 6.396, Bates, Pinching, *J. Am. Chem. Soc.* **71**, 1274 (1949). Soly in water: 54.0% w/w at 10°; 59.2% at 20°; 64.3% at 30°; 68.6% at 40°; 70.9% at 50°; 73.5% at 60°; 76.2% at 70°; 78.8% at 80°; 81.4% at 90°; 84.0% at 100°. Freely sol in alcohol. Very slightly sol in ether. LD_{50} in mice, rats (mmol/kg): 5.0, 4.6 i.p. (Gruber, Halbeisen).

Monohydrate. [5949-29-1] $C_6H_8O_7 \cdot H_2O$; mol wt 210.14. Orthorhombic crystals from cold aq solns. Pleasant, sour taste. d 1.542. Monohydrate crystals lose water of crystn in dry air or when heated at about 40 to 50°, slightly deliquescent in moist air. Softens at 75°. mp ~100°. pH of 0.1*N* soln = 2.2. Densities of aqueous soln (15°/15°): 10% = 1.0392; 20% = 1.0805; 30% = 1.1244; 40% = 1.1709; 50% = 1.2204; 60% = 1.2738. Soly in g/100 g satd soln: ether 2.17; chloroform 0.007; amyl alcohol 15.43; amyl acetate 5.98; ethyl acetate 5.28. Soly at 19° in g/100 g solvent: methanol 197; propanol 62.8.

Barium salt heptahydrate. [6487-29-2] Barium citrate. $C_{12}H_{10}Ba_3O_{14} \cdot 7H_2O$; mol wt 916.28. Powder. Loses all H_2O at 150°. Sol in 1750 parts water; freely sol in dil HCl or HNO_3; practically insol in alcohol.

Ethyl ester. [77-93-0] Ethyl citrate; triethyl citrate. $C_{12}H_{20}O_7$; mol wt 276.29. Bitter, oily liq. d^{20} 1.137. bp_{760} 294°; $bp_{1.0}$ 127°. Viscosity at 25°: 35.2 cP. Pour pt ~10°. n_D^{20} 1.4455. Soly: water ~6.9%; peanut oil 0.8%. Misc with alc, ether.

Disodium salt. [144-33-2] Disodium hydrogen citrate; sodium acid citrate; Alkacitron. $C_6H_6Na_2O_7$; mol wt 236.09. White pow-

der, saline taste. One gram of sesquihydrate dissolves in slightly less than 2 ml water; pH of a 3% w/v soln in water: 4.9 to 5.2. LD_{50} in mice, rats (mmol/kg): 7.5, 7.3 i.p. (Gruber, Halbeisen).

Trisodium salt. *See* Sodium Citrate.

USE: Anticoagulant, generally in solution with glucose, to prevent the clotting of blood intended for transfusion. Acidulant in beverages, confectionery, effervescent salts, in pharmaceutical syrups, elixirs, in effervescent powders and tablets, to adjust the pH of foods and as synergistic antioxidant, in processing cheese. Used in beverages, jellies, jams, preserves and candy to provide tartness. In the manuf of alkyd resins; in esterified form as plasticizer, foam inhibitor. In the manuf of citric acid salts. As sequestering agent to remove trace metals. As mordant to brighten colors; in electroplating; in special inks; in analytical chemistry for determining citrate-soluble P_2O_5; as reagent for albumin, mucin, glucose, bile pigments.

2326. Citrinin. [518-75-2] (3*R*,4*S*)-4,6-Dihydro-8-hydroxy-3,4,5-trimethyl-6-oxo-3*H*-2-benzopyran-7-carboxylic acid; Antimycin. $C_{13}H_{14}O_5$; mol wt 250.25. C 62.39%, H 5.64%, O 31.97%. Antibiotic substance produced by a white spore aspergillus which has been placed under the species name *Aspergillus niveus* (Thorn and Raper). Also produced in small quantities by *Penicillium citrinum*: Hetherington, Raistrick, *Trans. R. Soc. London* **B220**, 269 (1931); Raistrick, Smith, *Chem. Ind. (London)* **60**, 828 (1941); Timonin, *Science* **96**, 494 (1942); Timonin, Rovatt, *Can. J. Public Health* **35**, 80 (1944). Identity with antimycin: Haese, *Arch. Pharm.* **296**, 227 (1963). Structure: Brown *et al.*, *J. Chem. Soc.* **1949**, 867; Warren *et al.*, *J. Am. Chem. Soc.* **79**, 3812 (1957); Kovac *et al.*, *Nature* **190**, 1104 (1961). Synthesis: Cartwright *et al.*, *J. Chem. Soc.* **1949**, 1563; J. A. Barber *et al.*, *J. Chem. Soc. Perkin Trans. 1* **1986**, 2101. Stereochemistry: Cram, *J. Am. Chem. Soc.* **72**, 1001 (1950); Mehta, Whalley, *J. Chem. Soc.* **1963**, 3777; Mathieson, Whalley, *ibid.* **1964**, 4640. Physical characteristics and toxicity: Nagai *et al.*, *Chem. Zentralbl.* **1958**, 8088; *C.A.* **55**, 1914 (1961). Crystal and molecular structure: Rodig, *Chem. Commun.* **1971**, 1553. Biosynthesis: J. Barber *et al.*, *J. Chem. Soc. Perkin Trans. 1* **1981**, 2577; L. Colombo *et al.*, *ibid.* 2594. Physicochemical data: A. E. Pohland *et al.*, *Pure Appl. Chem.* **54**, 2219 (1982). Toxicology: A. M. Ambrose, F. De Eds, *J. Pharmacol. Exp. Ther.* **88**, 173 (1946). *Review:* Saito *et al.*, "Yellowed Rice Toxins" in *Microbial Toxins* vol. VI, A. Ciegler, S. Kadis, A. Ajl, Eds. (Academic Press, New York, 1971) pp 357-367.

Lemon-yellow needles from alcohol, dec 175°. $[\alpha]_D^{18}$ −37.4° (c = 1.15 in alc.). uv max: 250, 331 nm ($E_{1cm}^{1\%}$ 370, 418). Strong acid. Practically insol in water. Sol in alcohol, dioxane, dilute alkali. Solns change color with changes in pH, from lemon-yellow at pH 4.6 to cherry-red at pH 9.9. *Poisonous.* LD_{50} in mice, rats (mg/kg): 35, 67 i.p. (Ambrose, De Eds).

Methyl citrinin. $C_{14}H_{16}O_5$. Plates from benzene, dec 139°. $[\alpha]_D^{18}$ +217.1° (c = 0.38 in acetone). uv max: 260, 334 nm ($E_{1cm}^{1\%}$ 520, 151.6). Sol in hot alcohol; moderately sol in chloroform. Practically insol in petr ether.

Dihydrocitrinin. $C_{13}H_{16}O_5$. Prisms from benzene, dec 171°. $[\alpha]_D^{18}$ −18.8° (c = 4.148 in chloroform). uv max: 260, 330 nm ($E_{1cm}^{1\%}$ 400, 100). Sol in alcohol, acetone, chloroform; sparingly sol in benzene, petr ether.

2327. Citromycetin. [478-60-4] 8,9-Dihydroxy-2-methyl-4-oxo-4*H*,5*H*-pyrano[3,2-*c*][1]benzopyran-10-carboxylic acid; frequentic acid. $C_{14}H_{10}O_7$; mol wt 290.23. C 57.94%, H 3.47%, O 38.59%. Antibiotic substance produced by *Penicillium frequentans* Westling and *P. vesiculosum* Bainier and by *Citromyces* spp: Hetherington, Raistrick, *Philos. Trans. R. Soc. London Ser. B* **220**, 209 (1931); Grove, Brian, *Nature* **167**, 995 (1951). Structure: Robertson *et al.*, *J. Chem. Soc.* **1951**, 2013. Biosynthesis: Birch *et al.*, *ibid.* **1958**, 4576; Money, *Nature* **199**, 592 (1963). Total synthesis: M. Yamauchi *et al.*, *J. Chem. Soc. Perkin Trans. 1* **1987**, 395.

Dihydrate. Yellow crystals, effervescence at 155°, dec 290-300° (considerable antecedent blackening). Freely sol in ethanol; readily sol in aq sodium carbonate soln; sparingly sol in water, chloroform. Insol in benzene, hexane. Stable to acid and alkali at 100°.

2328. Citronellal. [106-23-0] 3,7-Dimethyl-6-octenal. $C_{10}H_{18}O$; mol wt 154.25. C 77.87%, H 11.76%, O 10.37%. Chief constituent of citronella oil; also found in many other volatile oils, such as lemon, lemon grass, melissa: Tiemann, *Ber.* **32**, 834 (1899); Spoon, *Chem. Weekbl.* **54**, 236 (1958). Structure: Naves, *Bull. Soc. Chim. Fr.* **1951**, 505; Eschinazi, *J. Org. Chem.* **26**, 3072 (1961).

Liquid. bp_1 47°. n_D^{20} 1.4460. $[\alpha]_D^{25}$ +11.50°. d 0.848-0.856. Soluble in alcohols; very slightly sol in water.

α-Citronellal. [141-26-4] 3,7-Dimethyl-7-octenal; rhodinal. Liquid. $bp_{1.4}$ 51°. n_D^{20} 1.4410. $[\alpha]_D^{20}$ +9.75°.

USE: In soap perfumes; insect repellent.

2329. β-Citronellol. [106-22-9] 3,7-Dimethyl-6-octen-1-ol; 2,6-dimethyl-2-octen-8-ol; citronellol; cephrol; dihydrogeraniol. $C_{10}H_{20}O$; mol wt 156.27. C 76.86%, H 12.90%, O 10.24%. *l*-Form is a constituent of rose and geranium oils. *d*-Form occurs in Ceylon and Java citronella oils. History: J. L. Simonsen, L. N. Owen, *The Terpenes* vol. I (University Press, Cambridge, 2nd ed, 1947). Prepn of (±)-form: Adams, Garvey, *J. Am. Chem. Soc.* **48**, 477 (1926); Ofner *et al.*, *Helv. Chim. Acta* **42**, 2577 (1959). Prepn of (+)-form: Rienäcker, Ohloff, *Angew. Chem.* **73**, 240 (1961); Naves, Tullen, *Helv. Chim. Acta* **44**, 1867 (1961); Eschinazi, *J. Org. Chem.* **26**, 3072 (1961); Rienäcker, *Chimia* **27**, 97 (1973); C. G. Overberger, J. L. Weise, *J. Am. Chem. Soc.* **90**, 3525 (1968); T. Sato *et al.*, *Tetrahedron Lett.* **1980**, 3377. Prepn of (−)-form: Ohloff, *loc. cit.*; Rienäcker, *loc. cit.*; Shono *et al.*, *Tetrahedron Lett.* **1974**, 1295; K. Mori, T. Sugai, *Synthesis* **1982**, 752. Synthesis of (+) or (−)-form from isoprene: Hidai *et al.*, *Chem. Commun.* **1975**, 170. Stereospecific prepn via microbiological (*Saccharomyces cerevisiae*) reduction: P. Gramatica *et al.*, *Experientia* **38**, 775 (1982). Manuf: Woroch *et al.*; Bain; Webb, US 2990422; US 3005845; US 3028431 (1961, 1961, 1962, all to Glidden); Eschinasi, US 3052730 (1962 to Givaudan). Abs config of the (+)-form: Freudenberg, Hohmann, *Ann.* **584**, 54 (1953); Freudenberg, Lwowski, *ibid.* **587**, 213 (1954). NMR, HPLC determn of *R/S* enantiomer ratios: D. Valentine *et al.*, *J. Org. Chem.* **41**, 62 (1976). *See also* Rhodinol.

R-(+)-β-Citronellol

$d_4^{23.5}$ 0.851. $n_D^{23.5}$ 1.454.
(+)-Form. Oily liquid, bp 224.5°, bp_{10} 108.4°, d_4^{20} 0.8550. n_D^{20} 1.4559. $[\alpha]_D^{20}$ +5.22°. Very slightly sol in water, miscible with alcohol, ether.

(−)-Form. β-Rhodinol; Levocitrol. bp_{10} 108-109°. d_4^{18} 1.4576. $[\alpha]_D^{20}$ −4.76°.

USE: In perfumery.

2330. Citrulline. [372-75-8] N^5-(Aminocarbonyl)-L-ornithine; δ-ureidonorvaline; α-amino-δ-ureidovaleric acid; $N^δ$-carbamylornithine. $C_6H_{13}N_3O_3$; mol wt 175.19. C 41.14%, H 7.48%, N 23.99%, O 27.40%. Ubiquitous amino acid in mammals. First isolated from the juice of watermelon, *Citrullus vulgaris* Schrad., *Cucurbitaceae*: Wada, *Biochem. Z.* **224**, 420 (1930); isoln from casein: Wada, *ibid.* **257**, 1 (1933). Synthesis from ornithine through copper complexes: Kurtz, *J. Biol. Chem.* **122**, 477 (1938); by alkaline hydrolysis of arginine: Fox, *ibid.* **123**, 687 (1938); from cyclopentanone oxime: Fox *et al.*, *J. Org. Chem.* **6**, 410 (1941). Crystallization: Matsuda *et al.*, *JP 71 174* (1971 to Ajinomoto), *C.A.* **74**, 126056u (1971). Crystal and molecular structure: Naganathan, Venkatesan, *Acta Crystallogr. B* **27**, 1079 (1971); Ashida *et al.*, *ibid.* **28**, 1367 (1972). Use in asthenia and hepatic insufficiency: FR 2198739 (1974 to Hublot & Vallet), *C.A.* **82**, 144952c (1975). Clinical trial in treatment of lysinuric protein intolerance: J. Rajantie *et al.*, *J. Pediatr.* **97**, 927 (1980); T. O. Carpenter *et al.*, *N. Engl. J. Med.* **312**, 290 (1985). Use as biomarker to assess intestinal failure: A. H. Herbers *et al.*, *Ann. Oncol.* **21**, 1706 (2010). Review of chemistry, metabolism, and therapeutic use: E. Curis *et al.*, *Amino Acids* **29**, 177-205 (2005).

Prisms from methanol + water, mp 222°. $[\alpha]_D^{20}$ +3.7° (c = 2). pK_1 2.43; pK_2 9.41. Sol in water. Insol in methanol, ethanol.

Hydrochloride. [34312-10-2] $C_6H_{13}N_3O_3.HCl$. Crystals, dec 185°. $[\alpha]_D^{22}$ +17.9° (c = 2).

Malate (salt). [54940-97-5] Stimol. $C_6H_{13}N_3O_3.C_4H_6O_5$; mol wt 309.28.

USE: Biomarker for clinical assessment of intestinal and renal function.

THERAP CAT: Treatment of asthenia.

2331. Citrullol. [1390-93-8] $C_{22}H_{38}O_4$; mol wt 366.54. C 72.09%, H 10.45%, O 17.46%. From fruit pulp of *Citrullus colocynthis* Schrad., *Cucurbitaceae*: Power, Moore, *J. Chem. Soc.* **97**, 99 (1910); Power, Salway, *ibid.* **103**, 399, 1022 (1913); Khadem, Rahman, *Tetrahedron Lett.* **1962**, 1137.

Crystals, mp 282-283°. uv max: 242, 272, 282 nm (log ε 2.85, 2.68, 2.68). Sol in pyridine; practically insol in usual organic solvents.

Diacetate. $C_{26}H_{42}O_6$. Crystals, mp 162°.

2332. Citrus Red 2. [6358-53-8] 1-[2-(2,5-Dimethoxyphenyl)diazenyl]-2-naphthalenol; C.I. Solvent Red 80; C.I. 12156. $C_{18}H_{16}N_2O_3$; mol wt 308.34. C 70.12%, H 5.23%, N 9.09%, O 15.57%. Prepn: H. W. Elley, H. W. Daudt, US 2224904 (1940 to Du Pont). Metabolism: J. L. Radomski, *J. Pharmacol. Exp. Ther.* **134**, 100 (1961); **136**, 378 (1962). Toxicology: M. Sharratt *et al.*, *Food Cosmet. Toxicol.* **4**, 493 (1966). Review of carcinogenicity studies: *IARC Monographs* **8**, 101-106. *See also Colour Index* vol. 4 (3rd ed., 1971) p 4033.

Crystals, mp 155-157°. Slightly sol in water; partially sol in ethanol and vegetable oils.

USE: To color orange skins.

2333. Civet. Zibeth. Unctuous secretion from receptacles between the anus and genitalia of both male and female civet cat. *Constit.* Civetone and similar compds.

Semi-solid, yellowish to brown unctuous substance; unpleasant, subacrid, bitter taste; fusible and burns without leaving much residue. Insol in water; partly sol in hot alcohol or in ether.

USE: As a fixative in perfumery.

2334. Civetone. [542-46-1] (9Z)-9-Cycloheptadecen-1-one. $C_{17}H_{30}O$; mol wt 250.43. C 81.53%, H 12.08%, O 6.39%. Seventeen-membered macrocyclic musk, constituent of civet: Ruzicka, *Helv. Chim. Acta* **9**, 230 (1926); Ruzicka *et al.*, *ibid.* **10**, 695 (1927). Occurs in nature as *cis*-form. Synthesis of *cis*-civetone: Stoll *et al.*, *ibid.* **31**, 543 (1948); J. Tsuji, T. Mondai, *Tetrahedron Lett.* **1977**, 3285; E. Seoane *et al.*, *Chem. Ind. (London)* **1978**, 165. Synthesis of *trans*-form: H. Hunsdiecker, *Ber.* **77**, 185 (1944); H. H. Mathur, S. C. Bhattacharyya, *J. Chem. Soc.* **1968**, 114. Crystal and molecular structure of *cis*-civetone: G. Bernardinelli, R. Gerdil, *Helv. Chim. Acta* **65**, 558 (1982).

(cis)-form

Crystals, mp 31-32°. Musky odor becoming pleasant in extreme dilns. d_4^{33} 0.917. bp_{742} 342°; bp_2 59°. $n_D^{33.4}$ 1.4830.

USE: In perfumery.

2335. Cizolirtine. [142155-43-9] *N,N*-Dimethyl-2-[(1-methyl-1*H*-pyrazol-5-yl)phenylmethoxy]ethanamine; 1-methyl-α-phenyl-*O*-(2-dimethylaminoethyl)-1*H*-pyrazole-5-methanol; E-3710. $C_{15}H_{21}N_3O$; mol wt 259.35. C 69.47%, H 8.16%, N 16.20%, O 6.17%. Antinociceptive agent that inhibits release of calcitonin gene related peptide and substance P, *q.v.*, from the spinal cord. Prepn: A. Colombo *et al.*, **EP 289380**; *eidem*, **US 5017596** (1988, 1991 both to Esteve). Prepn of isomers: J. A. Hueso-Rodriguez *et al.*, *Bioorg. Med. Chem. Lett.* **3**, 269 (1993). CE determn of chemical and enantiomeric purity: A. Gómez-Gomar *et al.*, *J. Chromatogr. A* **950**, 257 (2002). Pharmacology: I. Alvarez *et al.*, *Methods Find. Exp. Clin. Pharmacol.* **22**, 211 (2000). Mechanism of action study: S. Ballet *et al.*, *Neuropharmacology* **40**, 578 (2001). Clinical evaluation in chronic neuropathic pain: P. Shembalkar *et al.*, *Curr. Med. Res. Opin.* **17**, 262 (2001); in renal colic: I. Pavlik *et al.*, *Curr. Ther.* **26**, 1061 (2004). *Review:* N. Monck, *Curr. Opin. Investig. Drugs* **2**, 1269-1272 (2001).

Citrate. [142155-44-0] E-4018. $C_{15}H_{21}N_3O.C_6H_8O_7$; mol wt 451.48.

THERAP CAT: Analgesic.

2336. Cladribine. [4291-63-8] 2-Chloro-2'-deoxyadenosine; 2-chloro-6-amino-9-(2-deoxy-β-D-*erythro*-pentofuranosyl)purine; 2-chlorodeoxyadenosine; 2-CdA; CldAdo; NSC-105014-F; Leustatin; Movectro. $C_{10}H_{12}ClN_5O_3$; mol wt 285.69. C 42.04%, H 4.23%, Cl 12.41%, N 24.51%, O 16.80%. Substituted purine nucleoside with lymphocytotoxic and immunosuppressive activity. Prepn as intermediate in synthesis of 2-deoxynucleosides: H. Venner, *Ber.* **93**, 140 (1960); M. Ikehara, H. Tada, *J. Am. Chem. Soc.* **85**, 2344 (1963); *eidem, ibid.* **87**, 606 (1965). Synthesis and biological activity: L. F. Christensen *et al.*, *J. Med. Chem.* **15**, 735 (1972). Stereospecific synthesis: Z. Kazimierczuk *et al.*, *J. Am. Chem. Soc.* **106**,

6379 (1984); R. K. Robins, G. R. Revankar, **EP 173059**; *eidem*, **US 4760137** (1986, 1988 both to Brigham Young Univ.). Specific toxicity to lymphocytes: D. A. Carson *et al.*, *Proc. Natl. Acad. Sci. USA* **77**, 6865 (1980); *eidem*, *Blood* **62**, 737 (1983). Mechanism of action: S. Seto *et al.*, *J. Clin. Invest.* **75**, 377 (1985). Review of pharmacology and clinical trials in multiple sclerosis: T. P. Leist, P. Vermersch, *Curr. Med. Res. Opin.* **23**, 2667-2676 (2007); of clinical experience in hematologic malignancies: S. Spurgeon *et al.*, *Expert Opin. Invest. Drugs* **18**, 1169-1181 (2009).

Crystals from water, softens at 210-215°, solidifies and turns brown (Christensen). Also reported as crystals from ethanol, mp 220° (softens), resolidifies, turns brown and does not melt below 300° (Kazimierczuk). $[\alpha]_D^{25}$ −18.8° (c = 1 in DMF). uv max in 0.1*N* NaOH: 265 nm; in 0.1*N* HCl: 265 nm.

THERAP CAT: Antineoplastic.

2337. Clanobutin. [30544-61-7] 4-[(4-Chlorobenzoyl)(4-methoxyphenyl)amino]butanoic acid; 4-[*p*-chloro-*N*-(*p*-methoxyphenyl)benzamido]butyric acid; *N*-(*p*-chlorobenzoyl)-γ-(*p*-anisidino)butyric acid; Bykahepar. $C_{18}H_{18}ClNO_4$; mol wt 347.80. C 62.16%, H 5.22%, Cl 10.19%, N 4.03%, O 18.40%. Prepn: K. Klemm *et al.*, **DE 1917036**; *eidem*, **US 3780095** (1971, 1973 both to Byk-Gulden). Series of articles on synthesis, physical and pharmacological properties: *Arzneim.-Forsch.* **29**, 1-15 (1979). *In vitro* biochemical study: H. Wolf *et al.*, *Biochem. Pharmacol.* **29**, 1649 (1980). Effect on bile excretion in rats, dogs: P. Berchtold *et al.*, *Arzneim.-Forsch.* **30**, 1878 (1980).

Cryst from ethyl acetate, mp 115-116°. pKa 5.04. Soly in water at 37°: 4.02×10^{-2} mol/l at pH 7. LD_{50} in rats (mg/kg): >2000 orally; 570 i.v. (Klemm).

THERAP CAT: Choleretic.

THERAP CAT (VET): Choleretic; in treatment of piroplasmosis and anaplasmosis.

2338. Clarithromycin. [81103-11-9] 6-*O*-Methylerythromycin; A-56268; TE-031; Biaxin; Clarosip; Cyllind; Klacid; Klaricid; Macladin; Mavid; Naxy; Veclam; Zeclar. $C_{38}H_{69}NO_{13}$; mol wt 747.96. C 61.02%, H 9.30%, N 1.87%, O 27.81%. Semisynthetic macrolide antibiotic; derivative of erythromycin, *q.v.* Prepn: Y. Watanabe *et al.*, **EP 41355**; *eidem*, **US 4331803** (1981, 1982 both to Taisho); and *in vitro* antibacterial activity: S. Morimoto *et al.*, *J. Antibiot.* **37**, 187 (1984). *In vitro* and *in vivo* antibacterial activity: P. B. Fernandes *et al.*, *Antimicrob. Agents Chemother.* **30**, 865 (1986). Comparative antibacterial spectrum *in vitro*: C. Benson *et al.*, *Eur. J. Clin. Microbiol.* **6**, 173 (1987); H. M. Wexler, S. M. Finegold, *ibid.* 492. HPLC determn in biological fluids: D. Croteau *et al.*, *J. Chromatogr.* **419**, 205 (1987); in plasma: H. Amini, A. Ahmadiani, *J. Chromatogr. B* **817**, 193 (2005). Acute toxicity study: S. Abe *et al.*, *Chemotherapy (Tokyo)* **36**, Suppl. 3, 274 (1988). Symposium on pharmacology and comparative clinical studies: *J. Antimicrob. Chemother.* **27**, Suppl. A, 1-124 (1991). Comprehensive description: I. I. Salem, *Anal. Profiles Drug Subs.*

Excip. **24**, 45-85, (1996). Review of pharmacology and clinical applications: K. N. Williams, W. R. Bishai, *Expert Opin. Pharmacother.* **6**, 2867-2876 (2005).

Colorless needles from chloroform + diisopropyl ether (1:2), mp 217-220° (dec). Also reported as crystals from ethanol, mp 222-225° (Morimoto). uv max (CHCl$_3$): 288 nm (ε 27.9). uv max (CHCl$_3$): 240, 288 nm; (methanol): 211, 288 nm. $[\alpha]_D^{24}$ $-90.4°$ (c = 1 in CHCl$_3$). Stable at acidic pH. Sol in acetone; slightly sol in methanol, ethanol, acetonitrile, and in phosphate buffer at pH values of 2 to 5. Practically insol in water. LD$_{50}$ in male, female mice, male, female rats (mg/kg): 2740, 2700, 3470, 2700 orally, 1030, 850, 669, 753 i.p., >5000 all s.c. (Abe).

THERAP CAT: Antibacterial.

THERAP CAT (VET): Antibacterial.

2339. Clavulanic Acid. [58001-44-8] (2*R*,3*Z*,5*R*)-3-(2-Hydroxyethylidene)-7-oxo-4-oxa-1-azabicyclo[3.2.0]heptane-2-carboxylic acid; MM 14151. C$_8$H$_9$NO$_5$; mol wt 199.16. C 48.25%, H 4.56%, N 7.03%, O 40.17%. β-Lactamase inhibitor. Antibiotic produced by *Streptomyces clavuligerus;* first reported naturally occurring fused β-lactam containing oxygen. Isoln: M. Cole *et al.*, **DE 2517316** (1975 to Beecham), *C.A.* **84**, 72635t (1976); A. G. Brown *et al.*, *J. Antibiot.* **29**, 668 (1976). Structure, x-ray crystallography: T. T. Howarth *et al.*, *Chem. Commun.* **1976**, 266. Total synthesis of (±)-form: P. H. Bentley *et al.*, *ibid.* **1977**, 748, 905; *eidem, Tetrahedron Lett.* **1979**, 1889. β-Lactamase inhibition and antibacterial spectrum: C. Reading, M. Cole, *Antimicrob. Agents Chemother.* **11**, 852 (1977). Mechanism of action: B. G. Spratt *et al.*, *ibid.* **12**, 406 (1977). Antibacterial activity, pharmacology and clinical efficacy of combination with amoxicillin: A. P. Ball *et al.*, *Lancet* **1**, 620 (1980); R. N. Brogden *et al.*, *Drugs* **22**, 337-362 (1981). *In vitro* and *in vivo* synergism with ticarcillin: R. Sutherland *et al.*, *Am. J. Med.* **79**, Suppl. 5B, 13 (1985).

Combination of potassium salt with amoxicillin trihydrate. Co-amoxiclav; Amoksiklav; Augmentin; Ciblor; Clavamox; Clavulin; Klavocin; Neo-Duplamox; Synulox.

Combination of potassium salt with ticarcillin disodium. [116876-37-0] Betabactyl; Timentin.

Methyl ester. C$_9$H$_{11}$NO$_5$. Oil. $[\alpha]_D^{22}$ +38°.

p-Nitrobenzyl ester. C$_{15}$H$_{14}$N$_2$O$_7$. Monoclinic crystals, mp 117.5-118°.

THERAP CAT: Combination with β-lactam antibiotics as antibacterial.

THERAP CAT (VET): Combination with β-lactam antibiotics as antibacterial.

2340. Clazosentan. [180384-56-9] *N*-[6-(2-Hydroxyethoxy)-5-(2-methoxyphenoxy)-2-[2-(2*H*-tetrazol-5-yl)-4-pyridinyl]-4-pyrimidinyl]-5-methyl-2-pyridinesulfonamide; 5-methylpyridine-2-

sulfonic acid 6-(2-hydroxyethoxy)-5-(2-methoxyphenoxy)-2-(2-1*H*-tetrazol-5-ylpyridin-4-yl)pyrimidin-4-ylamide. C$_{25}$H$_{23}$N$_9$O$_6$S; mol wt 577.58. C 51.99%, H 4.01%, N 21.83%, O 16.62%, S 5.55%. Nonpeptide selective endothelin ET$_A$ receptor antagonist. Prepn: V. Breu *et al.*, **WO 9619459**; *eidem,* **US 6004965** (1996, 1999 both to Hoffmann-La Roche). Pharmacology and receptor binding study: S. Roux *et al., J. Pharmacol. Exp. Ther.* **283**, 1110 (1997). Clinical pharmacology: T. J. L. Vuurmans *et al., Hypertension* **41**, 1253 (2003). Cerebrovascular characterization: H. Vatter *et al., J. Neurosurg.* **102**, 1101, 1108 (2005). Clinical effect on vasospasm following subarachnoid hemorrhage: P. Vajkoczy *et al., ibid.* **103**, 9 (2005). Review of development and therapeutic potential: D. Uhlmann, *Curr. Opin. Investig. Drugs* **7**, 272-281 (2006).

White substance from acetonitrile, mp 239-241°.

Disodium salt. [503271-02-1] AXV-034343; Ro-61-1790; VML-588. C$_{25}$H$_{21}$N$_9$Na$_2$O$_6$S; mol wt 621.54. pKa$_1$ 4.5; pKa$_2$ 3.3. Soly in water: 25% at physiol pH.

THERAP CAT: In treatment of cerebral vasospasm.

2341. Clazuril. [101831-36-1] 2-Chloro-α-(4-chlorophenyl)-4-(4,5-dihydro-3,5-dioxo-1,2,4-triazin-2(3*H*)-yl)benzeneacetonitrile; (±)-[2-chloro-4-(4,5-dihydro-3,5-dioxo-*as*-triazin-2(3*H*)-yl)phenyl](*p*-chlorophenyl)acetonitrile; Appertex. C$_{17}$H$_{10}$Cl$_2$N$_4$O$_2$; mol wt 373.19. C 54.71%, H 2.70%, Cl 19.00%, N 15.01%, O 8.57%. Prepn: G. M. Boeckx *et al.*, **EP 170316**; *eidem,* **US 4631278** (both 1986 to Janssen). Anticoccidial effect in pigeons: W. Coussement *et al., Res. Vet. Sci.* **45**, 117 (1988). HPLC determn in eggs and plasma from laying hens: M. Giorgi, G. Soldani, *Br. Poultry Sci.* **49**, 609 (2008).

mp 196.8°.

THERAP CAT (VET): Coccidiostat.

2342. Clebopride. [55905-53-8] 4-Amino-5-chloro-2-methoxy-*N*-[1-(phenylmethyl)-4-piperidinyl]benzamide; 4-amino-*N*-(1-benzyl-4-piperidyl)-5-chloro-*o*-anisamide; *N*-(1'-benzyl-4'-piperidyl)-2-methoxy-4-amino-5-chlorobenzamide. C$_{20}$H$_{24}$ClN$_3$O$_2$; mol wt 373.88. C 64.25%, H 6.47%, Cl 9.48%, N 11.24%, O 8.56%. Dopamine receptor antagonist related to metoclopramide, *q.v.* Prepn: R. G. Spickett *et al.*, **DE 2513136**; A. V. Noverola *et al.*, **US 4138492** (1975, 1979 both to Anphar); J. Prieto *et al., J. Pharm. Pharmacol.* **29**, 147 (1977). Pharmacological study: J. L. Masso, D. J. Roberts, *ibid.* **32**, 727 (1980). *In vitro* metabolism studies: G. Huizing *et al., Xenobiotica* **10**, 211 (1980); G. Huizing, A. H. Beckett, *ibid.* 593. Pharmacokinetics: J. Segura *et al., J. Pharm. Pharmacol.* **33**, 214 (1981). Comparative study of anti-emetic and gastrointestinal effects: P. Berga *et al., Arch. Farmacol. Toxicol.* **7**, 189 (1981). Efficacy in healing exptl ulcers: Y. Matsuo *et al., Oyo Yakuri* **24**, 251 (1982); S. Okabe *et al., ibid.* 261, *C.A.* **97**, 208081n,

208082p (1982). Determn in urine and plasma by radioimmunoassay: M. Yano et al., Chem. Pharm. Bull. **32**, 1491 (1984). Clinical effect in postoperative nausea: D. F. Duarte et al., Clin. Ther. **7**, 365 (1985).

Cryst from methanol, mp 194-195°.

Hydrochloride monohydrate. $C_{20}H_{24}ClN_3O_2.HCl.H_2O$. Cryst, mp 217-219°. Approx oral LD_{50} in male Swiss mice: >1000 mg/kg (Prieto).

Malate. [57645-91-7] Amicos; Clanzol; Clast; Cleboril; Cleprid; Motilex. $C_{20}H_{24}ClN_3O_2.C_4H_6O_5$; mol wt 507.97.

THERAP CAT: Antiemetic; antispasmodic.

2343. Clemastine. [15686-51-8] (2R)-2-[2-[(1R)-1-(4-Chlorophenyl)-1-phenylethoxy]ethyl]-1-methylpyrrolidine; (+)-2-[2-[(p-chloro-α-methyl-α-phenylbenzyl)oxy]ethyl]-1-methylpyrrolidine; 1-methyl-2-[2-(α-methyl-p-chlorobenzhydryloxy)ethyl]pyrrolidine; 1-methyl-2-[2-(methyl-p-chlorodiphenylmethyloxy)ethyl]pyrrolidine; meclastine. $C_{21}H_{26}ClNO$; mol wt 343.90. C 73.34%, H 7.62%, Cl 10.31%, N 4.07%, O 4.65%. Ethanolamine antihistamine. Prepn: **GB 942152** (1963 to Sandoz). Synthesis and abs config: A. Ebnöther, H. P. Weber, Helv. Chim. Acta **59**, 2462 (1976). Pharmacology and toxicology of the hydrogen fumarate: Weidmann et al., Boll. Chim. Farm. **106**, 467 (1967). HPLC determn in plasma: V. Horváth et al., J. Chromatogr. B **816**, 153 (2005). Pharmacology in dogs: H. Hansson et al., Vet. Dermatol. **15**, 152 (2004). Clinical trials in the common cold: J. M. Gwaltney, Jr. et al., Clin. Infect. Dis. **22**, 656 (1996); R. B. Turner et al., ibid. **25**, 824 (1997).

Free base, $bp_{0.02}$ 154°. n_D^{22} 1.5582. $[\alpha]_D^{20}$ +33.6° (ethanol).

Hydrogen fumarate. [14976-57-9] HS-592; Alphamin; Anhistan; Kinotomin; Lecasol; Marsthine; Masletine; Piloral; Tavegil; Tavegyl; Tavist. $C_{25}H_{30}ClNO_5$; mol wt 459.97. Faintly yellow crystalline solid. mp 177-178°. Very slightly sol in water; sparingly sol in alcohol; slightly sol in methanol; very slightly sol in chloroform. $[\alpha]_D^{21}$ +16.9° (methanol). LD_{50} in mice, rats (mg/kg): 730, 3550 orally; 43, 82 i.v. (Weimann).

THERAP CAT: Antihistaminic.

THERAP CAT (VET): Antihistaminic.

2344. Clemizole. [442-52-4] 1-[(4-Chlorophenyl)methyl]-2-(1-pyrrolidinylmethyl)-1H-benzimidazole; 1-(p-chlorobenzyl)-2-(1-pyrrolidinylmethyl)benzimidazole; 1-(p-chlorobenzyl)-2-pyrrolidylmethylenebenzimidazole. $C_{19}H_{20}ClN_3$; mol wt 325.84. C 70.04%, H 6.19%, Cl 10.88%, N 12.90%. Prepn: D. Jerchel et al., Ann. **575**, 162 (1952); **GB 703272**; M. Schenck, W. Heinz, **US 2689853** (both 1954 to Schering A.G.).

Crystals from dil alc, mp 167°.

Hydrochloride. [1163-36-6] Allercur. $C_{19}H_{20}ClN_3.HCl$; mol wt 362.30. White microscopic rods from butanol, mp 239-241°. Sol in water.

THERAP CAT: Antihistaminic.

2345. Clenbuterol. [37148-27-9] 4-Amino-3,5-dichloro-α-[[(1,1-dimethylethyl)amino]methyl]benzenemethanol; 4-amino-α-[(tert-butylamino)methyl]-3,5-dichlorobenzyl alcohol; NAB-365. $C_{12}H_{18}Cl_2N_2O$; mol wt 277.19. C 52.00%, H 6.55%, Cl 25.58%, N 10.11%, O 5.77%. Substituted phenylethanolamine with β_2 adrenergic activity. Prepn: J. Keck et al., **ZA 6705692**; eidem, **US 3536712** (1968, 1970 both to Thomae); eidem, Arzneim.-Forsch. **22**, 861 (1972). Series of articles on pharmacology, toxicology, pharmacokinetics and metabolism: ibid. **26**, 1403-1460 (1976). Toxicity: M. Ueberberg et al., ibid. 1420. Repartitioning effects in veal calves: G. Re et al., Vet. J. **153**, 63 (1997). GC-MS determn in urine: L. Amendola et al., J. Chromatogr. B **773**, 7 (2002). Clinical trial in asthma: D. Wheatley, Curr. Med. Res. Opin. **8**, 113 (1982). Veterinary trial as tocolytic: M. R. Putnam et al., Theriogenology **24**, 385 (1985). Review of clinical pharmacology, detection, and abuse potential: A. Prezelj et al., Curr. Med. Chem. **10**, 281-290 (2003); of bronchodilator use in horses: C. F. Kearns, K. H. McKeever, Vet. J. **182**, 384-391 (2009).

Hydrochloride. [21898-19-1] NAB-365Cl; Monores; Spiropent; Planipart; Ventipulmin. $C_{12}H_{18}Cl_2N_2O.HCl$; mol wt 313.65. Colorless microcrystalline powder from isopropyl alc, mp 174-175.5°. Very sol in water, methanol, ethanol, slightly sol in chloroform. Insol in benzene. LD_{50} in mice, rats, guinea pigs (mg/kg): 176, 315, 67.1 orally; 27.6, 35.3, 12.6 i.v. (Ueberberg).

THERAP CAT: Bronchodilator.

THERAP CAT (VET): Bronchodilator; tocolytic.

2346. Clentiazem. [96125-53-0] (2S,3S)-3-(Acetyloxy)-8-chloro-5-[2-(dimethylamino)ethyl]-2,3-dihydro-2-(4-methoxyphenyl)-1,5-benzothiazepin-4(5H)-one; (+)-(2S,3S)-8-chloro-5-[2-(dimethylamino)ethyl]-2,3-dihydro-3-hydroxy-2-(p-methoxyphenyl)-1,5-benzothiazepin-4(5H)-one acetate (ester). $C_{22}H_{25}ClN_2O_4S$; mol wt 448.96. C 58.86%, H 5.61%, Cl 7.90%, N 6.24%, O 14.25%, S 7.14%. Calcium channel blocker; 8-chloro derivative of diltiazem. Prepn: M. Takeda et al., **EP 127882**; eidem, **US 4567175** (1984, 1986 both to Tanabe); H. Inoue et al., J. Med. Chem. **34**, 675 (1991). Cardiovascular pharmacology: H. Narita et al., Arzneim.-Forsch. **38**, 515 (1988). Clinical pharmacokinetics: A. K. Shah et al., J. Clin. Pharmacol. **33**, 354 (1993). Clinical evaluation in hypertension: S. Kawakita et al., Clin. Cardiol. **14**, 53 (1991).

Maleate. [96128-92-6] TA-3090; Logna. $C_{22}H_{25}ClN_2O_4S.C_4H_4O_4$; mol wt 565.03. Crystals from ethanol, mp 160.5-161.5°. $[\alpha]_D^{20}$ +76.5° (c = 1 in methanol).

THERAP CAT: Antihypertensive.

2347. Clethodim. [99129-21-2] 2-[1-[[[(2E)-3-Chloro-2-propen-1-yl]oxy]imino]propyl]-5-[2-(ethylthio)propyl]-3-hydroxy-2-

cyclohexen-1-one; (±)-2-[(*E*)-1-[(*E*)-3-chloroallyloxyimino]pro-pyl]-5-[2-(ethylthio)propyl]-3-hydroxycyclohex-2-enone; RE-45601; Prism; Select. $C_{17}H_{26}ClNO_3S$; mol wt 359.91. C 56.73%, H 7.28%, Cl 9.85%, N 3.89%, O 13.34%, S 8.91%. Systemic gra-minicide. Prepn: T. Luo, **BE 897413**; *idem*, **US 4440566** (1983, 1984 both to Chevron). Properties and biological profile: R. T. Kin-cade *et al.*, *Proc. Br. Crop Prot. Conf. - Weeds* **1987**, 49. Degrada-tion studies: L. N. Falb *et al.*, *J. Agric. Food Chem.* **38**, 875 (1990). Persistence in soil and efficacy against grasses: A. Rahman *et al.*, *Proc. 41st N. Z. Weed Pest Control Conf.* **1988**, 39. Field trial against grasses: J. H. Hunter, *Weed Technol.* **9**, 432 (1995).

Clear amber liquid. Sol in most organic solvents. LD_{50} in male, female rats (mg/kg): 1630, 1360 orally; LC_{50} in trout: 56 mg/l; 8 day feeding in quail: > 6000 ppm (Kincade).

USE: Post-emergent herbicide.

2348. 1,6-Cleve's Acid. [119-79-9] 5-Amino-2-naphthalene-sulfonic acid; 1-naphthylamine-6-sulfonic acid. $C_{10}H_9NO_3S$; mol wt 223.25. C 53.80%, H 4.06%, N 6.27%, O 21.50%, S 14.36%. Prepn: Erdmann, *Ann.* **275**, 262 (1893).

Crystals. Sol in 1000 parts water. Practically insol in alc, ether.
USE: In manuf of azo dyes.

2349. 1,7-Cleve's Acid. [119-28-8] 8-Amino-2-naphthalene-sulfonic acid; 1-naphthylamine-7-sulfonic acid. $C_{10}H_9NO_3S$; mol wt 223.25. C 53.80%, H 4.06%, N 6.27%, O 21.50%, S 14.36%. Prepn: Erdmann, *Ann.* **275**, 262 (1893); Roos *et al.*, **US 2875243** (1959 to Bayer). Industrial synthesis: Nakahara *et al.*, *J. Synth. Org. Chem. Jpn.* **29**, 1129 (1971).

Needles or prisms from water contg $1H_2O$. Sol in 220 parts water; very slightly sol in alcohol, ether.
USE: In manuf of azo dyes.

2350. Clevidipine. [167221-71-8] 4-(2,3-Dichlorophenyl)-1,4-dihydro-2,6-dimethyl-3,5-pyridinedicarboxylic acid 3-methyl 5-[(1-oxobutoxy)methyl] ester; butyroxymethyl methyl 4-(2′,3′-di-chlorophenyl)-2,6-dimethyl-1,4-dihydropyridine-3,5-dicarboxylate; clevidipine butyrate; Clevelox; Cleviprex. $C_{21}H_{23}Cl_2NO_6$; mol wt 456.32. C 55.28%, H 5.08%, Cl 15.54%, N 3.07%, O 21.04%. Ul-trashort-acting dihydropyridine calcium channel antagonist. Prepn: K. H. Andersson *et al.*, **WO 9512578**; *eidem*, **US 5856346** (1995, 1999 both to Astra). Improved prepn: A. Mattson *et al.*, **WO 0031035** (2000 to AstraZeneca). Capillary GC/MS determn in whole blood: C. Fakt, H. Stenhoff, *J. Chromatogr. B* **723**, 211 (1999). Clinical pharmacokinetics: H. Ericsson *et al.*, *Br. J. Clin. Pharmacol.* **47**, 531 (1999); and pulmonary extraction during cardio-pulmonary bypass: A. Vuylsteke *et al.*, *Br. J. Anaesth.* **85**, 683 (2000). Clinical evaluation in hypertension in postoperative cardiac surgical patients: N. Kieler-Jensen *et al.*, *Acta Anaesthesiol. Scand.* **44**, 186 (2000); J. M. Bailey *et al.*, *Anesthesiology* **96**, 1086 (2002). Review of pharmacology and clinical development: M. Nordlander *et al.*, *Cardiovasc. Drug Rev.* **22**, 227-250 (2004); A. H. Gradman, Y. Vivas, *Expert Opin. Invest. Drugs* **16**, 1449-1457 (2007).

Crystals from diisopropylether, mp 136.2-138.5°. Soly in water: 0.1 mg/ml.
THERAP CAT: Antihypertensive.

2351. Clevudine. [163252-36-6] 1-(2-Deoxy-2-fluoro-β-L-arabinofuranosyl)-5-methyl-2,4-(1*H*,3*H*)-pyrimidinedione; 1-(2-de-oxy-2-fluoro-β-L-arabinofuranosyl)thymine; 2′-fluoro-5-methyl-β-L-arabinofuranosyluridine; L-FMAU. $C_{10}H_{13}FN_2O_5$; mol wt 260.22. C 46.16%, H 5.04%, F 7.30%, N 10.77%, O 30.74%. Nu-cleoside analog; specific inhibitor of viral DNA synthesis. Prepn: C. K. Chu *et al.*, **WO 9520595**; *eidem*, **US 5587362** (1995, 1996 both to Univ. Georgia Res. Found.; Yale Univ.); T. Ma *et al.*, *J. Med. Chem.* **39**, 2835 (1996). Synthesis: M. L. Sznaidman *et al.*, *Nucle-osides Nucleotides Nucleic Acids* **21**, 155 (2002). *In vitro* antiviral activity: C. K. Chu *et al.*, *Antimicrob. Agents Chemother.* **39**, 979 (1995). Pharmacokinetics: J. D. Wright *et al.*, *Pharm. Res.* **12**, 1350 (1995). Mechanism of action study: Y. Chong, C. K. Chu, *Bioorg. Med. Chem. Lett.* **12**, 3459 (2002). Clinical evaluation in chronic hepatitis B: P. Marcellin *et al.*, *Hepatology* **40**, 140 (2004). Review of preclinical pharmacology: C. K. Chu *et al.* in *Therapies for Viral Hepatitis*, R. F. Schinazi *et al.*, Eds. (Int. Med. Press, London, 1998) pp 303-312; of development and clinical experience: G. R. Painter *et al.* in *Frontiers in Viral Hepatitis*, R. F. Schinazi *et al.*, Eds. (El-sevier, New York, 2003) pp 281-300.

Crystals from methanol/chloroform, mp 184-185°. $[\alpha]_D^{25}$ −111.77° (c = 0.23 in methanol). uv max in water: 265.0 (pH 2), 265.5 (pH 7), 265.5 nm (pH 11) (ε 9695, 9647, 7153). LD_{50} in mice, rats (mg/kg): >5000, >3000 orally (Painter).
Triphosphate. [174625-00-4] L-FMAU-TP. $C_{10}H_{16}FN_2O_{14}P_3$; mol wt 500.16.
THERAP CAT: Antiviral.

2352. Clidinium Bromide. [3485-62-9] 3-[(2-Hydroxy-2,2-diphenylacetyl)oxy]-1-methyl-1-azoniabicyclo[2.2.2]octane bro-mide (1:1); 3-hydroxy-1-methylquinuclidinium bromide benzilate; 1-methyl-3-benziloyloxyquinuclidinium bromide; 3-benziloyloxy-1-azabicyclo[2.2.2]octane methobromide; Ro-2-3773; Quarzan. $C_{22}H_{26}BrNO_3$; mol wt 432.36. C 61.12%, H 6.06%, Br 18.48%, N 3.24%, O 11.10%. Anticholinergic. Prepn: Sternbach, **US 2648667** (1953 to Hoffmann-La Roche). Comprehensive description: B. C. Rudy, B. Z. Senkowski, *Anal. Profiles Drug Subs.* **2**, 145-161 (1973).

White to nearly white, practically odorless, crystalline powder. Crystals from methanol + acetone + ether. mp 240-241°. Sol in water, alcohol; slightly sol in benzene, ether.

THERAP CAT: Antispasmodic.

2353. Clinafloxacin. [105956-97-6] 7-(3-Amino-1-pyrrolidinyl)-8-chloro-1-cyclopropyl-6-fluoro-1,4-dihydro-4-oxo-3-quinolinecarboxylic acid. $C_{17}H_{17}ClFN_3O_3$; mol wt 365.79. C 55.82%, H 4.68%, Cl 9.69%, F 5.19%, N 11.49%, O 13.12%. Fluorinated quinolone antibacterial. Prepn: S. Suzue et al., **EP 195316** (1986 to Kyorin). Antibacterial spectrum: M. S. Barrett et al., Diagn. Microbiol. Infect. Dis. **14**, 389 (1991); S. M. H. Qadri et al., Chemotherapy (Basel) **38**, 92 (1992).

Milk-white powder from chloroform-methanol-conc. aq. ammonia, mp 253-258° (dec).
Hydrochloride. [105956-99-8] AM-1091; CI-960; PD-127391. $C_{17}H_{17}ClFN_3O_3 \cdot HCl$; mol wt 402.25. Light yellow prisms from methanol, mp 263-265° (dec).

THERAP CAT: Antibacterial.

2354. Clindamycin. [18323-44-9] Methyl 7-chloro-6,7,8-trideoxy-6-[[[(2S,4R)-1-methyl-4-propyl-2-pyrrolidinyl]carbonyl]-amino]-1-thio-L-threo-α-D-galacto-octopyranoside; 7(S)-chloro-7-deoxylincomycin; 7-deoxy-7(S)-chlorolincomycin; clinimycin (rescinded); U-21251. $C_{18}H_{33}ClN_2O_5S$; mol wt 424.98. C 50.87%, H 7.83%, Cl 8.34%, N 6.59%, O 18.82%, S 7.54%. Semi-synthetic antibiotic prepd from lincomycin, q.v. Prepn: Magerlein et al., Antimicrob. Agents Chemother. **1966**, 727. Manuf process: R. D. Birkenmeyer, **US 3475407** (1969 to Upjohn). Synthesis and structure: R. D. Birkenmeyer, F. Kagan, J. Med. Chem. **13**, 616 (1970). Stability studies: Oesterling, J. Pharm. Sci. **59**, 63 (1970). Activity studies: McGehee et al., Am. J. Med. Sci. **256**, 279 (1968); D. A. Leigh, J. Antimicrob. Chemother. **7**, Suppl. A, 3 (1981). Toxicology: J. E. Gray et al., Toxicol. Appl. Pharmacol. **21**, 516 (1972). Comprehensive description of the hydrochloride: L. W. Brown, W. F. Beyer, Anal. Profiles Drug Subs. **10**, 75-91 (1981).

Yellow, amorphous solid. $[\alpha]_D$ +214° (chloroform).
Hydrochloride. [21462-39-5]; [58207-19-5] (monohydrate). Antirobe; Clinsol; Clintabs; Dalacin C; Sobelin. $C_{18}H_{33}ClN_2O_5S \cdot HCl$; mol wt 461.44. Various hydrated forms exist. White crystals from ethanol-ethyl acetate, mp 141-143° (anhydrous). $[\alpha]_D$ +144° (H_2O). pKa 7.6. Freely sol in water, DMF, methanol; sol in ethanol, pyridine. Practically insol in acetone. LD_{50} in mice (mg/kg): 245 i.v.; 361 i.p.; 2618 orally (Gray).
2-Dihydrogen phosphate. [24729-96-2] U-28508; Cleocin; Dalacin T; Klimicin; Zindaclin. $C_{18}H_{34}ClN_2O_8PS$; mol wt 504.96. White to off-white, hygroscopic, crystalline powder. Sol in water; slightly sol in dehydrated alc; very slightly sol in acetone. Practically insol in chloroform, benzene, ether.

THERAP CAT: Antibacterial.
THERAP CAT (VET): Antibacterial.

2355. Clinofibrate. [30299-08-2] 2,2'-[Cyclohexylidenebis(4,1-phenyleneoxy)]bis[2-methylbutanoic acid]; 2,2'-(4,4'-cyclohexylidenediphenoxy)-2,2'-dimethyldibutyric acid; S-8527; Lipoclin. $C_{28}H_{36}O_6$; mol wt 468.59. C 71.77%, H 7.74%, O 20.49%. Anti-atherosclerosis agent, related structurally to clofibrate, q.v. Prepn: Y. Nakamura et al., **DE 2017331**; eidem, **US 3716583** (1970, 1973 both to Sumitomo). Hypolipidemic activity: K. Toki et al., Atherosclerosis **18**, 101 (1973); K. Suzuki et al., Jpn. J. Pharmacol. **24**, 407 (1974), C.A. **82**, 25821z (1975). Effect on cholesterol and lipoprotein metabolism in rats: K. Suzuki, Biochem. Pharmacol. **25**, 325 (1976). General pharmacology: H. Nakatani et al., Oyo Yakuri **16**, 687 (1978), C.A. **90**, 145808 (1979).

Off-white powder, mp 143-146° (dec). Sol in methanol, ethanol, acetone, chloroform, glacial acetic acid. Slightly sol in CCl_4. Practically insol in water.

THERAP CAT: Antilipemic.

2356. Clobazam. [22316-47-8] 7-Chloro-1-methyl-5-phenyl-1H-1,5-benzodiazepine-2,4(3H,5H)-dione; 1-phenyl-5-methyl-8-chloro-1,2,4,5-tetrahydro-2,4-dioxo-3H-1,5-benzodiazepine; H-4723; HR-376; LM-2717; Frisium; Urbadan; Urbanyl. $C_{16}H_{13}ClN_2O_2$; mol wt 300.74. C 63.90%, H 4.36%, Cl 11.79%, N 9.32%, O 10.64%. Benzodiazepine psychotherapeutic agent in which the nitrogens in the heterocyclic ring are in the 1,5- rather than in the more common 1,4-positions. Prepn: **BE 707667**; S. Rossi, **US 3984398** (1968, 1976 both to Roussel-UCLAF); S. Rossi et al., Chim. Ind. (Milan) **51**, 479 (1969); K. H. Weber et al., Ann. **756**, 128 (1972). Toxicology: E. Schütz, Br. J. Clin. Pharmacol. **7**, 33S (1979). Review of pharmacology and therapeutic use in anxiety: R. N. Brogden et al., Drugs **20**, 161-178 (1980). Symposium on pharmacology: Drug Dev. Res. **1982**, Suppl. 1, 1-186. Metabolism: A. G. Borel, F. S. Abbott, Drug Metab. Dispos. **21**, 415 (1993). GC determn in serum: D. F. LeGatt, D. P. McIntosh, Clin. Biochem. **26**, 159 (1993). Clinical experience in refractory epilepsy: Canadian Clobazam Cooperative Group, Epilepsia **32**, 407 (1991). Review of pharmacology and clinical studies in epilepsy: S. D. Shorvon, "Clobazam" in Antiepileptic Drugs, R. H. Levy et al., Eds. (Raven Press, New York, 3rd ed., 1989) pp 821-840.

Crystals from 50% ethanol, mp 166-168°.
Note: This is a controlled substance (depressant): **21 CFR**, 1308.14.

THERAP CAT: Anxiolytic; anticonvulsant.

2357. Clobenzorex. [13364-32-4] (αS)-N-[(2-Chlorophenyl)methyl]-α-methylbenzeneethanamine; (+)-N-(o-chlorobenzyl)-α-methylphenethylamine; d-N-(1-phenyl-2-propyl)-2-chlorobenzylamine. $C_{16}H_{18}ClN$; mol wt 259.78. C 73.98%, H 6.98%, Cl 13.65%, N 5.39%. Prepn and pharmacology: J. R. Boissier, R. Ratouis, **FR 1429306** (1966 to Soc. Ind. Fabric. Antibiot.), C.A. **66**, 46195h (1967); J. R. Boissier et al., Ann. Pharm. Fr. **24**, 57 (1966). Synthesis of labelled compound: J. Lintermans et al., J. Labelled Compd. **6**, 289 (1970). Metabolic studies: M. Thomasset et al., Experientia **26**, 692 (1970); B. Glasson et al., Arzneim.-Forsch. **21**, 1985 (1971).

bp$_{0.1}$ 132-134°.

Hydrochloride. [5843-53-8] Ba-7205; SD-271-12; Dinintel; Rexigen. $C_{16}H_{18}ClN.HCl$; mol wt 296.24. Crystals, mp 182-183° (isopropanol). $[\alpha]_D^{20}$ +26.3° (water). Sol in water, ethanol. Slightly sol in methanol, chloroform. LD_{50} i.p. in mice: 103 mg/kg (Boissier et al.).

THERAP CAT: Anorexic.

2358. Clobetasol. [25122-41-2] (11β,16β)-21-Chloro-9-fluoro-11,17-dihydroxy-16-methylpregna-1,4-diene-3,20-dione. $C_{22}H_{28}ClFO_4$; mol wt 410.91. C 64.31%, H 6.87%, Cl 8.63%, F 4.62%, O 15.57%. Topical corticosteroid. Prepn: Elks et al., **DE 1902340**; eidem, **US 3721687** (1969, 1973 both to Glaxo). Clinical evaluation of shampoo formulation in severe scalp psoriasis: C. E. M. Griffiths et al., J. Dermatolog. Treat. **17**, 90 (2006). Clinical trial of foam formulation in delayed pressure urticaria: G. A. Vena et al., Br. J. Dermatol. **154**, 353 (2006). Review of pharmacology and clinical efficacy in skin disorders: E. A. Olsen, R. C. Cornell, J. Am. Acad. Dermatol. **15**, 246-255 (1986); of clinical experience of foam formulation in psoriasis: D. C. Reid, A. B. Kimball, Expert Opin. Pharmacother. **6**, 1735-1740 (2005).

17-Propionate. [25122-46-7] GR-2/925; Clobesol; Clobex; Dermovate; Olux; Psorex; Temovate. $C_{25}H_{32}ClFO_5$; mol wt 466.97. White or almost white crystalline powder, mp 195.5-197°. $[\alpha]_D$ +103.8° (c = 1.04 in dioxane). uv max (ethanol): 237 nm (ε 15000). Soluble in acetone, dimethyl sulfoxide, chloroform, methanol, dioxane; sparingly sol in ethanol; slightly sol in benzene, diethyl ether. Insol in water.

THERAP CAT: Glucocorticoid; anti-inflammatory; antipsoriatic.

2359. Clobetasone. [54063-32-0] (16β)-21-Chloro-9-fluoro-17-hydroxy-16-methylpregna-1,4-diene-3,11,20-trione; 21-chloro-11-dehydrobetamethasone. $C_{22}H_{26}ClFO_4$; mol wt 408.89. C 64.62%, H 6.41%, Cl 8.67%, F 4.65%, O 15.65%. Prepn: Elks et al., **DE 1902340**; eidem, **US 3721687** (1969, 1973, both to Glaxo). Activity: Munro, Wilson, Br. Med. J. **3**, 626 (1975). Radioimmunoassay: W. A. Webb, R. V. Brooks, Acta Endocrinol. Suppl. **212**, 203 (1977). Toxicology: J. Tamura et al., J. Toxicol. Sci. **5**, 45, 177 (1980). Comparative clinical studies: A. Lassus, Curr. Med. Res. Opin. **6**, 165 (1979); T. Fredriksson, K. Nordin, ibid. 322. Clinical evaluation of ophthalmic form: D. Lloyd-Jones et al., Br. J. Ophthalmol. **65**, 641 (1981); L. A. Eilon, S. R. Walker, ibid. 644; K. R. Wilhelmus et al., ibid. 699.

17-Butyrate. [25122-57-0] GR-2/1214; Emovate; Eumovate. $C_{26}H_{32}ClFO_5$; mol wt 478.99. Crystals from methanol, mp 90-100°.
THERAP CAT: Glucocorticoid; anti-inflammatory.

2360. Clobutinol. [14860-49-2] 4-Chloro-α-[2-(dimethylamino)-1-methylethyl]-α-methylbenzeneethanol; p-chloro-α-[2-(dimethylamino)-1-methylethyl]-α-methylphenethyl alcohol; 2-(p-chlorobenzyl)-3-dimethylaminomethyl-2-butanol; 1-p-chlorophenyl-2,3-dimethyl-4-dimethylamino-2-butanol. $C_{14}H_{22}ClNO$; mol wt 255.79. C 65.74%, H 8.67%, Cl 13.86%, N 5.48%, O 6.25%. Prepn: **GB 898010**; Berg, **US 3121087** (1962, 1964 both to Thomae). Pharmacology: Engelhorn, Arzneim.-Forsch. **10**, 785 (1960). Metabolism: Beisenherz et al., ibid. **19**, 79 (1969).

bp$_{12}$ 179-180°.

Hydrochloride. [1215-83-4] KAT-256; Biotertussin; Silomat. $C_{14}H_{22}ClNO.HCl$; mol wt 292.24. Minute crystals, mp 169-170°. Slightly bitter, acidic taste. Numbs the tongue. Sol in water. LD_{50} in mice (mg/kg): 600 orally; 130 i.p. (Engelhorn).

THERAP CAT: Antitussive.

2361. Clocapramine. [47739-98-0] 1'-[3-(3-Chloro-10,11-dihydro-5H-dibenz[b,f]azepin-5-yl)propyl][1,4'-bipiperidine]-4'-carboxamide; 3-chloro-5-[3-(4-piperidino-4-carbamoylpiperidino)propyl]-10,11-dihydro-5H-dibenz[b,f]azepine; 3-chlorocarpipramine; clocarpramine; CCP. $C_{28}H_{37}ClN_4O$; mol wt 481.08. C 69.91%, H 7.75%, Cl 7.37%, N 11.65%, O 3.33%. Prepn: M. Nakanishi, C. Tashiro, **DE 1905765**; eidem, **US 3668210** (1969, 1972, both to Yoshitomi); M. Nakanishi et al., J. Med. Chem. **13**, 644 (1970). Pharmacology and toxicity: M. Nakanishi et al., Arzneim.-Forsch. **21**, 391 (1971). Effect on metabolism of serotonin and catecholamines: M. Nakanishi, M. Setoguchi, ibid. **23**, 806 (1973). Metabolism: M. Nakanishi et al., J. Pharm. Soc. Jpn. **91**, 1042 (1971). HPLC determn in plasma: K. Hikida et al., Anal. Sci. **6**, 367 (1990). Review: Jpn. Med. Gaz. **10**(7), 5 (1973).

Dihydrochloride monohydrate. [60789-62-0] Y-4153; Clofekton. $C_{28}H_{37}ClN_4O.2HCl.H_2O$; mol wt 572.01. White, bitter tasting, crystalline powder, mp 267°. Soly in water: <3.5 g/100 ml. Freely sol in glacial acetic acid. Practically insol in ether, acetone. LD_{50} in mice, rats (mg/kg): 160, 125 i.p.; 2550, 6800 orally (Nakanishi).

THERAP CAT: Antipsychotic.

2362. Clocinizine. [298-55-5] 1-[(4-Chlorophenyl)phenylmethyl]-4-(3-phenyl-2-propen-1-yl)piperazine; 1-(p-chloro-α-phenylbenzyl)-4-cinnamylpiperazine; 1-cinnamyl-4-(4-chlorobenzhydryl)piperazine; cliocinizine. $C_{26}H_{27}ClN_2$; mol wt 402.97. C 77.50%, H 6.75%, Cl 8.80%, N 6.95%. Prepn: **GB 809760** (1959 to Janssen). HPLC determn in cough-cold preparations: G. Cavazzutti et al., J. Liq. Chromatogr. **18**, 227 (1995).

Hydrochloride. Crystals, mp 200-201°.
THERAP CAT: Antihistaminic.

2363. Clocortolone. [4828-27-7] ($6\alpha,11\beta,16\alpha$)-9-Chloro-6-fluoro-11,21-dihydroxy-16-methylpregna-1,4-diene-3,20-dione; 9-chloro-6α-fluoro-16α-methyl-1,4-pregnadiene-$11\beta,21$-diol-3,20-dione. $C_{22}H_{28}ClFO_4$; mol wt 410.91. C 64.31%, H 6.87%, Cl 8.63%, F 4.62%, O 15.57%. Prepn: **NL 6412708**; E. K. Kaspar, R. Philippson, **DE 2011559**; *eidem*, **US 3729495** (1965, 1971, 1973 all to Schering AG). Clinical trials: Wortmann, *Wien. Med. Wochenschr.* **122**, 701 (1972); Baumann, *Praxis* **61**, 536 (1972).

Crystals from methanol-methylene chloride, mp 254° (dec).
21-Acetate. [4258-85-9] SH-818. $C_{24}H_{30}ClFO_5$; mol wt 452.95. Crystals from methanol-methylene chloride, mp 252° (dec).
21-Pivalate. [34097-16-0] CL-68; SH-863; Cloderm; Purantix. $C_{27}H_{36}ClFO_5$; mol wt 495.03. White to yellowish-white, odorless powder. Freely sol in chloroform, dioxane; sol in acetone; sparingly sol in alcohol; slightly sol in benzene, ether.
THERAP CAT: Glucocorticoid.

2364. Clodinafop-propargyl. [105512-06-9] (2*R*)-2-[4-[(5-Chloro-3-fluoro-2-pyridinyl)oxy]phenoxy]propanoic acid 2-propynyl-1-yl ester; 2-propynyl (*R*)-2-[4-(5-chloro-3-fluoro-2-pyridyloxy)phenoxy]propionate; CGA-184927; Topik. $C_{17}H_{13}ClFNO_4$; mol wt 349.74. C 58.38%, H 3.75%, Cl 10.14%, F 5.43%, N 4.00%, O 18.30%. Pyridyloxyphenoxy herbicide for use in cereal crops. Prepn: R. Schurter, **EP 248968**; R. Schurter, H. Rempfler, **US 4713109** (both 1987 to Ciba-Geigy). Description: J. Amrein *et al.*, *Brighton Crop Prot. Conf. - Weeds* **1989**, 71-76. HPLC determn in soil and plant samples: S. Roy, S. B. Singh, *J. Chromatogr. A* **1065**, 199 (2005). Review of field trials with the crop safener, cloquintocet, *q.v.*: D. W. Cornes, *ibid.* 729-734; R. D. Welsh, *Proc. 47th N. Z. Plant Prot. Conf.* **1994**, 22-26.

Odorless, crystalline solid, mp 59.3°. $[\alpha]_D^{20}$ +45.4° (c = 2 in acetone). Vapor pressure (20°): 5.3×10^{-6} Pa. Soly in water (20°): 2.5 ppm. Sol in most organic solvents. LD_{50} in rats (mg/kg): 1829 orally; >2000 dermally. LC_{50} in rats (4 hr): >2325 mg/m³ by inhalation (Amrein).
USE: Herbicide.

2365. Clodronic Acid. [10596-23-3] *P,P'*-(Dichloromethylene)bisphosphonic acid; (dichloromethylene)diphosphonic acid; dichloromethanediphosphonic acid; Cl_2MDP; DMDP. CH_4Cl_2 O_6P_2; mol wt 244.88. C 4.90%, H 1.65%, Cl 28.95%, O 39.20%, P 25.30%. Bisphosponate antiresorptive agent. Prepn and use as detergent additive: **BE 672205** (1966 to Procter and Gamble), *C.A.* **67**, 4040u (1967). Prepn: O. T. Quimby *et al.*, *J. Org. Chem.* **32**, 4111 (1967). Effect on hydroxyapatite crystal aggregation: N. M. Hansen *et al.*, *Biochim. Biophys. Acta* **451**, 549 (1976); S. Bisaz *et al.*, *ibid.* 560. HPLC determn in urine: V. Virtanen, L. H. J. Lajunen, *J. Chromatogr.* **617**, 291 (1993). Symposium on therapeutic efficacy in neoplastic bone disease: *Bone* **8**, Suppl. 1, S1-S86 (1987). Review of pharmacology and clinical efficacy in resorptive bone disease: G. L. Plosker, K. L. Goa, *Drugs* **47**, 945-982 (1994). Clinical reduction of new metastases in breast cancer: I. J. Diel *et al.*, *N. Engl. J. Med.* **339**, 357 (1998).

mp 249-251°.
Disodium salt. [22560-50-5]; [88416-50-6] (tetrahydrate). Clodronate disodium; Bonefos; Clasteon; Clastoban; Difosfonal; Loron; Lytos; Mebonat; Ossiten; Ostac. $CH_2Cl_2Na_2O_6P_2$; mol wt 288.85. Occurs as the tetrahydrate.
THERAP CAT: Bone resorption inhibitor.

2366. Clofarabine. [123318-82-1] 2-Chloro-9-(2-deoxy-2-fluoro-β-D-arabinofuranosyl)-9*H*-purin-6-amine; 2-chloro-9-(2-deoxy-2-fluoro-β-D-arabinofuranosyl)adenine; 2-chloro-2'-arabino-fluoro-2'-deoxyadenosine; Cl-F-ara-A; Clolar. $C_{10}H_{11}ClFN_5O_3$; mol wt 303.68. C 39.55%, H 3.65%, Cl 11.67%, F 6.26%, N 23.06%, O 15.81%. Second generation purine nucleoside analog; antimetabolite that inhibits DNA synthesis and resists deamination by adenosine deaminase. Prepn: K. A. Watanabe *et al.*, **EP 219829**; *eidem*, **US 4918179** (1987, 1990 both to Sloan-Kettering Inst. Cancer Res.); J. A. Montgomery *et al.*, *J. Med. Chem.* **35**, 397 (1992). Improved synthesis: W. E. Bauta *et al.*, *Org. Process Res. Dev.* **8**, 889 (2004). Antineoplastic activity: W. R. Waud *et al.*, *Nucleosides Nucleotides Nucleic Acids* **19**, 447 (2000). Clinical pharmacology: H. M. Kantarjian *et al,. J. Clin. Oncol.* **21**, 1167 (2003). Clinical pharmacokinetics: V. Gandhi *et al.*, *Clin. Cancer Res.* **9**, 6335 (2003). Clinical evaluation in acute leukemia: H. Kantarjian *et al.*, *Blood* **102**, 2379 (2003); in pediatric patients: S. Jeha *et al.*, *ibid.* **103**, 784 (2004). Review of development: A. Sternberg, *Curr. Opin. Investig. Drugs* **4**, 1479-1487 (2003); of mechanism of action and clinical experience: S. Faderl *et al.*, *Cancer* **103**, 1985-1995 (2005).

Crystals from water, mp 225-227° (Montgomery). Also reported as crystals from methanol, mp 237° (Bauta). uv max (water): 212, 263 nm (ε 22500, 15989).
THERAP CAT: Antineoplastic.

2367. Clofazimine. [2030-63-9] *N*,5-Bis(4-chlorophenyl)-3,5-dihydro-3-[(1-methylethyl)imino]-2-phenazinamine; 3-(*p*-chloroanilino)-10-(*p*-chlorophenyl)-2,10-dihydro-2-(isopropylimino)phenazine; 2-(4-chloroanilino)-3-isopropylimino-5-(4-chlorophenyl)-3,5-dihydrophenazine; 2-*p*-chloroanilino-5-*p*-chlorophenyl-3,5-dihydro-3-isopropyliminophenazine; G-30320; B-663; Lampren(e). $C_{27}H_{22}Cl_2N_4$; mol wt 473.40. C 68.50%, H 4.68%, Cl 14.98%, N 11.84%. Prepn: V. C. Barry *et al.*, **US 2948726** (1960 to Geigy); Barry *et al.*, *Nature* **179**, 1013 (1957); Barry *et al.*, *J. Chem. Soc.* **1958**, 859; Belton *et al.*, *Proc. R. Irish Acad. Sect. B* **62**, 9 (1961), *C.A.* **58**, 4556d (1963). Activity studies: Vischer, *Arzneim.-Forsch.* **18**, 1529 (1968). HPLC determn in plasma: J. H. Peters *et al.*, *J. Chromatogr.* **229**, 503 (1982). Toxicity data: Stenger *et al.*, *Arzneim.-Forsch.* **20**, 794 (1970). Clinical studies: Mathies, *ibid.* 1838; Karat *et al.*, *Br. Med. J.* **1**, 198 (1970). Review of clinical pharmacokinetics: M. R. Holdiness, *Clin. Pharmacokinet.* **16**, 74-85 (1989); of clinical use: J. C. Garrelts, *DICP Ann. Pharmacother.* **25**, 525-531 (1991). Comprehensive description: V. K. Kapoor, *Anal. Profiles Drug Subs.* **18**, 91-120 (1989); C. M. O'Driscoll, O. I. Corrigan, *Anal. Profiles Drug Subs. Excip.* **21**, 75-108 (1992).

Dark-red crystals, mp 210-212°. uv max (0.01M methanolic HCl): 284, 486 nm (abs. ~1.30, ~0.64). pKa 8.37; also reported as 8.51. Sol in dil acetic acid, DMF, benzene. Sol in 15 parts of chloroform, 700 parts of ethanol, 1000 parts of ether. Sparingly sol in acetone, ethyl acetate. Practically insol in water. Log P (octanol/water): 7.48; (isooctane/pH 5.15 buffer) at 20°: 5.01; (n-octanol/pH 5.15 buffer) at 20°: 4.30; (n-octanol/pH 5.15 buffer) at 37°: 4.40; (n-octanol/pH 5.15 buffer) at 45°: 4.48; (n-octanol/pH 5.15 buffer) at 55°: 4.54. LD$_{50}$ orally in mice, rats, and guinea pigs: >4 g/kg (Stenger).

THERAP CAT: Antibacterial (tuberculostatic, leprostatic).

2368. Clofencet. [129025-54-3] 2-(4-Chlorophenyl)-3-ethyl-2,5-dihydro-5-oxo-4-pyridazinecarboxylic acid; MON-21200. $C_{13}H_{11}ClN_2O_3$; mol wt 278.69. C 56.03%, H 3.98%, Cl 12.72%, N 10.05%, O 17.22%. Hybridizing agent for wheat. Prepn: D. R. Patterson, **EP 49971**; *idem*, **US 4732603** (1982, 1988 both to Rohm & Haas). Improved prepn: J. D. Clark *et al.*, *Org. Process Res. Dev.* **8**, 176 (2004). Sorption behavior in soils: I. G. Dubus *et al.*, *Chemosphere* **45**, 767 (2001). Comprehensive description: *Clofencet Pesticide Fact Sheet* (U.S. EPA, Feb. 1997) 12 pp.

Technical grade: tan/yellow solid, mp 269° (dec). d^{20} 1.44. Soly (%w:v): distilled water >55.2, pH 5 >65.5, pH 7 >69.6, pH 9 >65.8; methanol 1.6; acetone <0.05; dichloromethane <0.04; toluene <0.04; ethyl acetate <0.05; n-hexane <0.06. Vapor pressure (25°): <10^{-7} mm Hg. pKa 2.83 (20°). Log P (octanol/water): −2.2 (25°). LD$_{50}$ in rats (mg/kg/day): 3306 orally; >5000 dermally (U.S. EPA).
Potassium salt. [82697-71-0] MON-21233; Genesis. $C_{13}H_{10}$ ClKN$_2$O$_3$; mol wt 316.78. LC$_{50}$ in rats (mg/l): >3.8 (U.S. EPA).
Sodium salt. [82697-16-3] $C_{13}H_{10}ClN_2NaO_3$; mol wt 300.67. mp 190-191°.
USE: Plant growth regulator.

2369. Clofentezine. [74115-24-5] 3,6-Bis(2-chlorophenyl)-1,2,4,5-tetrazine; bisclofentezin; NC-21314; Apollo. $C_{14}H_8Cl_2N_4$; mol wt 303.15. C 55.47%, H 2.66%, Cl 23.39%, N 18.48%. Tetrazine insecticide active against eggs and early motile stages of phytophagous mites. Prepn: J. H. Parsons, **EP 5912**; *idem*, **US 4237127** (1979, 1980 both to Fisons). Physical properties, toxicity and bioactivity: K. M. G. Bryan *et al.*, *Proc. Br. Crop Prot. Conf. - Pests Dis.* **1981**, 67. LC determn in fruit samples: S. Navickiene, M. L. Ribeiro, *J. AOAC Int.* **87**, 435 (2004). Field trial on apple trees: A. J. Read, *Proc. 36th N.Z. Weed Pest Control Conf.*, 261 (1983). *Review:* F. Rauch, *Def. Veg.* **39**, 11-17 (1985).

Magenta solid. Crystals from ethyl acetate, mp 179-182°. Soly: 50 g/l in chloroform; 2.5 g/l in benzene; <1 g/l in hexane; <1 mg/l in

water. LD$_{50}$ in rats, mice (mg/kg): >3200 orally; LC$_{50}$ (96 hr) in rainbow trout: 100 mg/l (Bryan).
USE: Contact acaricide.

2370. Clofibrate. [637-07-0] 2-(4-Chlorophenoxy)-2-methylpropanoic acid ethyl ester; ethyl 2-(p-chlorophenoxy)-2-methylpropionate; ethyl p-chlorophenoxyisobutyrate; Amotril; Anparton; Apolan; Artevil; Ateculon; Ateriosan; Atheropront; Atromidin; Atromid-S; Bioscleran; Claripex; Clobren-SF; Clofinit; CPIB; Hyclorate; Liprinal; Neo-Atromid; Normet; Normolipol; Recolip; Regelan; Serotinex; Sklerolip; Skleromexe; Sklero-Tablinen; Xyduril. $C_{12}H_{15}ClO_3$; mol wt 242.70. C 59.39%, H 6.23%, Cl 14.61%, O 19.78%. Prepn: W. G. M. Jones *et al.*, **GB 860303**; *eidem*, **US 3262850** (1961, 1966 both to I.C.I.). Clinical trials in the primary prevention of ischaemic heart disease: *Br. Heart J.* **40**, 1069 (1978); *Lancet* **2**, 379 (1980). Toxicity data: G. Metz *et al.*, *Arzneim.-Forsch.* **27**, 1173 (1977). Review of pharmacology: M. Chevais, *Therapie* **35**, 5-22 (1980); of metabolism: M. N. Cayen, *Rev. Drug Metab. Drug Interact.* **3**, 77-103 (1980); of toxicity studies: *IARC Monographs* **24**, 39-58 (1980). Comprehensive description: M. M. A. Hassan, A. A. Elazzouny, *Anal. Profiles Drug Subs.* **11**, 197-224 (1982).

Oil, bp$_{20}$ 148-150°. Practically insol in water. Misc with ethanol, acetone, chloroform, benzene, ether. LD$_{50}$ in mice, rats (g/kg): 1.28, 1.65 orally (Metz).
Mixture with androsterone. [8075-95-4] Atromid.
THERAP CAT: Antilipemic.

2371. Clofibric Acid. [882-09-7] 2-(4-Chlorophenoxy)-2-methylpropanoic acid; 2-(p-chlorophenoxy)-2-methylpropionic acid; α-(p-chlorophenoxy)isobutyric acid; chlorophibrinic acid; Arteriohom; Regulipid. $C_{10}H_{11}ClO_3$; mol wt 214.65. C 55.96%, H 5.17%, Cl 16.52%, O 22.36%. Prepn: Galimberti, Defranceschi, *Gazz. Chim. Ital.* **77**, 431 (1947); Gilman, Wilder, *J. Am. Chem. Soc.* **77**, 6644 (1955); Jones *et al.*, **GB 860303** (1961 to I.C.I.). Prepn and resolution of isomers: Witiak *et al.*, *J. Med. Chem.* **11**, 1086 (1968). Activity due to displacement of thyroxin and inhibition of cholesterol biosynthesis: Chang *et al.*, *Biochem. Pharmacol.* **16**, 2053 (1967); Walsh *et al.*, *Arch. Biochem. Biophys.* **130**, 7 (1969); Witiak *et al.*, *J. Med. Chem.* **14**, 754 (1971). Metabolism: Almirante *et al.*, *Boll. Chim. Farm.* **108**, 292 (1969).

Crystals from water or methanol, mp 118-119°.
Basic aluminum salt. [24818-79-9] Alufibrate; Atherolip; Atherolipin. $C_{20}H_{21}AlCl_2O_7$; mol wt 471.26.
Magnesium salt. [14613-30-0] Magnesium clofibrate; UR-112; Clomag. $C_{20}H_{20}Cl_2MgO_6$; mol wt 451.58. White powder, mp 326-328°. Soly in g/100 mg: water 4.5; abs ethanol 0.7; chloroform 0.02.
Pyridoxine salt. [29952-87-2] Pyridoxine p-chlorophenoxyisobutyrate; Claresan. $C_{10}H_{11}ClO_3 \cdot C_8H_{11}NO_3$; mol wt 383.83. Prepn: Sarbach *et al.*, **DE 1915497** (1970 to Inst. Rech. Sci.), *C.A.* **74**, 13011g (1971).
Ethyl ester *see* Clofibrate.
Etofylline ester *see* Theofibrate.
THERAP CAT: Antilipemic.

2372. Cloflucarban. [369-77-7] N-(4-Chlorophenyl)-N'-[4-chloro-3-(trifluoromethyl)phenyl]urea; 4,4′-dichloro-3-(trifluoromethyl)carbanilide; Irgasan CF3. $C_{14}H_9Cl_2F_3N_2O$; mol wt 349.13. C 48.16%, H 2.60%, Cl 20.31%, F 16.32%, N 8.02%, O 4.58%. Prepn: Schetty *et al.*, **US 2745874** (1956 to Geigy); Sydor, **BE**

617117 (1962 to Am. Cyanamid), compd included but not specifically described in abstract, *C.A.* **59**, 1542a (1963).

White crystalline solid, mp 214-215°. Insol in water; dissolves well in organic solvents.

THERAP CAT: Disinfectant.

2373. Clofoctol. [37693-01-9] 2-[(2,4-Dichlorophenyl)methyl]-4-(1,1,3,3-tetramethylbutyl)phenol; α-(2,4-dichlorophenyl)-4-(1,1,3,3-tetramethylbutyl)-*o*-cresol; Gramplus; Octofene. $C_{21}H_{26}$- Cl_2O; mol wt 365.34. C 69.04%, H 7.17%, Cl 19.41%, O 4.38%. Prepn: J. Debat, US **3830852** (1974 to Inst. Recherches Chim. Biol. Appl.). Activity against gram positive and gram negative bacteria: A. Buogo, *Drugs Exp. Clin. Res.* **10**, 321 (1984). Clinical pharmacokinetics: M. Del Tacca *et al., J. Antimicrob. Chemother.* **19**, 679 (1987). Mechanism of action: J. Combe *et al., J. Pharmacol.* **11**, 411 (1980); F. Yablonsky, G. Simonet, ibid. **13**, 515 (1982). Clinical trials in respiratory infections: J. Vialatte, *Diagnostics* **207**, 65 (1978); in infectious bronchopulmonary diseases: R. Danesi, M. Del Tacca, *Int. J. Clin. Pharmacol. Res.* **5**, 175 (1985); in ear, nose and throat infections: P. L. Ghilardi, A. Casani, *Drugs Exp. Clin. Res.* **11**, 815 (1985).

Crystals from petr ether, mp 78°. LD_{50} orally in male rats: >4 g/kg (Debat).

THERAP CAT: Antibacterial.

2374. Clomazone. [81777-89-1] 2-[(2-Chlorophenyl)methyl]-4,4-dimethyl-3-isoxazolidinone; dimethazone; FMC-57020; Command. $C_{12}H_{14}ClNO_2$; mol wt 239.70. C 60.13%, H 5.89%, Cl 14.79%, N 5.84%, O 13.35%. Pre-emergence herbicide; inhibits biosynthesis of photosynthetic pigments in susceptible species. Prepn: J. H. Chang, BE **889040**; idem, US **4405357** (1982, 1983 both to FMC). Use on soybeans: M. P. Mascianica, *Proc. Annu. Meet. Northeast. Weed Sci. Soc.* **39**, 25 (1985); on white potatoes: C. C. Kupatt *et al., ibid.* 166. Fate in soil: T. L. Mervosh *et al., J. Agric. Food Chem.* **43**, 537 (1995). Review of properties and analytical methods: A. W. Chen in *Comprehensive Analytical Profiles of Important Pesticides*, J. Sherma, T. Cairns, Eds. (CRC Press, Boca Raton, 1993) pp 131-148.

Light brown viscous liquid, mp 25°. Vapor pressure at 25°: 1.4 × 10^{-6} mm Hg. d^{20} 1.19. Soly in water: 1100 ppm. Readily sol in chloroform, methanol, methylene chloride, heptane, acetonitrile, toluene, acetone, dioxane, xylene, hexane. LD_{50} in male rats, female rats, mallard duck-bobwhite quail (mg/kg): 2077, 1369, >2510 orally; in rabbits (mg/kg): >2000 dermally (Chen).

USE: Herbicide.

2375. Clomethiazole. [533-45-9] 5-(2-Chloroethyl)-4-methylthiazole; 4-methyl-5-(β-chloroethyl)thiazole; chlorethiazol; chlormethiazole; S.C.T.Z.. C_6H_8ClNS; mol wt 161.65. C 44.58%, H 4.99%, Cl 21.93%, N 8.67%, S 19.83%. Prepd by condensation of

thioformamide with chloroacetopropyl alcohol: Buchman, *J. Am. Chem. Soc.* **58**, 1803 (1936); by chlorination of 4-methyl-5-(β-ethoxyethyl)thiazole: Clarke, Gurin, ibid. **57**, 1876 (1935); by treating 5-(2-hydroxyethyl)-4-methylthiazole with thionyl chloride: Sawa, Ishida, *J. Pharm. Soc. Jpn.* **76**, 337 (1956); by H_2O_2 oxidation of 2-mercapto-4-methyl-5-(β-chloroethyl)thiazole: **CH 200248** (1938 to Hoffmann-La Roche). Prepn of methane and ethanedisulfonate: Charonnat *et al.,* US **3031457** (1962).

Oily, viscous liquid. d_4^{25} 1.233. Characteristic disagreeable odor of thiazole compds. bp_7 92°.

Hydrochloride. [6001-74-7] $C_6H_8ClNS.HCl$. Crystals from abs ethanol + ether, mp 130°. Sol in water, alcohol.

Methanedisulfonate. $C_{13}H_{20}Cl_2N_2O_6S_4$. Crystals from methanol + ether, mp 120°.

Ethanedisulfonate. [1867-58-9] Distraneurin; Hemineurin; Heminevrin. $C_{14}H_{22}Cl_2N_2O_6S_4$; mol wt 513.48. Crystals from methanol + ether, mp 124°.

USE: Intermediate in some processes of vitamin B_1 manufacture.

THERAP CAT: Sedative, hypnotic, anticonvulsant.

2376. Clometocillin. [1926-49-4] (2S,5R,6R)-6-[[2-(3,4-Dichlorophenyl)-2-methoxyacetyl]amino]-3,3-dimethyl-7-oxo-4-thia-1-azabicyclo[3.2.0]heptane-2-carboxylic acid; 3,4-dichloro-α-methoxybenzylpenicillin; 6-(α-methoxy-3,4-dichlorophenylacetamido)-penicillanic acid; clometacillin; clomethacillin; no. 356; penicillin 356. $C_{17}H_{18}Cl_2N_2O_5S$; mol wt 433.30. C 47.12%, H 4.19%, Cl 16.36%, N 6.47%, O 18.46%, S 7.40%. Semi-synthetic antibiotic related to penicillin. Prepn: Vanderhaeghe *et al., Antimicrob. Agents Chemother.* **1961**, 581; US **3007920** (1961 to Recherche et Ind. Therap.). Activity studies: van Dijck *et al., Antibiot. Chemother.* **12**, 192 (1962).

Sodium salt. [54530-86-8] Rixapen. $C_{17}H_{17}Cl_2N_2NaO_5S$; mol wt 455.28. $[\alpha]_D$ +210-220°. Pure epimers were prepared by fractional crystn. First crop: Crystals, dec 232-235°. $[\alpha]_D^{20}$ +182° (water). Second crop: $[\alpha]_D^{20}$ +207°. Third crop: Crystals, dec 180-183°. $[\alpha]_D^{20}$ +261° (water). The free acid yields two diastereoisomers: $[\alpha]_D$ +177° and $[\alpha]_D$ +257°.

THERAP CAT: Antibacterial.

2377. Clomiphene. [911-45-5] 2-[4-(2-Chloro-1,2-diphenylethenyl)phenoxy]-*N,N*-diethylethanamine; 2-[*p*-(2-chloro-1,2-diphenylvinyl)phenoxy]triethylamine; 2-[*p*-(β-chloro-α-phenylstyryl)phenoxy]triethylamine; 1-[*p*-(β-diethylaminoethoxy)phenyl]-1,2-diphenylchloroethylene; clomifene; chloramiphene; MRL-41; WSM-5008. $C_{26}H_{28}ClNO$; mol wt 405.97. C 76.92%, H 6.95%, Cl 8.73%, N 3.45%, O 3.94%. Synthetic estrogen agonist-antagonist. Prepn: R. E. Allen *et al.,* US **2914563** (1959 to Merrell); F. P. Palopoli *et al., J. Med. Chem.* **10**, 84 (1967). Stereochemistry of the geometric isomers: Ernst *et al., J. Pharm. Sci.* **65**, 148 (1976). Induction of ovulation: R. B. Greenblatt, *Fertil. Steril.* **12**, 402 (1961). Clinical trial in anovulatory women: J. Garcia *et al., ibid.* **28**, 707 (1977). Use in male infertility: P. J. Sorbie, R. Perez-Marrero, *J. Urol.* **131**, 425 (1984). HPLC determn of isomers in human plasma: C. L. Baustian, T. J. Mikkelson, *J. Pharm. Biomed. Anal.* **4**, 237 (1986). Review of development, pharmacology and clinical experience: R. P. Dickey, D. E. Holtkamp, *Hum. Reprod. Update* **2**, 483-506 (1996).

Enclomiphene

Citrate. [50-41-9] Clomid; Clostilbegyt; Pergotime; Serofene; Serophene. $C_{26}H_{28}ClNO.C_6H_8O_7$; mol wt 598.09. White to pale yellow, odorless powder. mp 116.5-118°. Slightly sol in water, chloroform; freely sol in methanol; sparingly sol in alcohol. Insol in ether.

cis-**Form.** [15690-55-8] Zuclomiphene. Hydrochloride salt mp 156.5-158.0°. uv max (methanol): 230, 291 nm (ε 20500, 12700).

trans-**Form.** [15690-57-0] Enclomiphene. Hydrochloride salt mp 149.0-150.5°. uv max (methanol): 239, 297 nm (ε 22100, 11600).

THERAP CAT: Gonad-stimulating principle.

2378. Clomipramine. [303-49-1] 3-Chloro-10,11-dihydro-*N,N*-dimethyl-5*H*-dibenz[*b,f*]azepine-5-propanamine; 3-chloro-5-[3-(dimethylamino)propyl]-10,11-dihydro-5*H*-dibenz[*b,f*]azepine; 5-(γ-dimethylaminopropyl)-3-chloroiminodibenzyl; chlorimipramine; G-34586. $C_{19}H_{23}ClN_2$; mol wt 314.86. C 72.48%, H 7.36%, Cl 11.26%, N 8.90%. Serotonin reuptake inhibitor. Prepn: P. N. Craig *et al.*, *J. Org. Chem.* **26**, 135 (1961); W. Schindler, H. Dietrich, **CH 371799**; *eidem*, **US 3467650** (1963, 1969 both to Geigy). Crystal and molecular structure of the hydrochloride: M. L. Post, A. S. Horn, *Acta Crystallogr.* **33B**, 2590 (1977). HPLC determn in serum: H. Weigmann *et al.*, *J. Chromatogr. B* **710**, 227 (1998). Veterinary trial in canine compulsive disorder: C. J. Hewson *et al.*, *J. Am. Vet. Med. Assoc.* **213**, 1760 (1998); in canine separation anxiety: S. Petit *et al.*, *Rev. Med. Vet.* **150**, 133 (1999). Review of pharmacology and clinical experience in obsessive-compulsive disorder and depression: M. D. Peters *et al.*, *Clin. Pharm.* **9**, 165-178 (1990).

bp$_{0.3}$ 160-170°.

Hydrochloride. [17321-77-6] Anafranil; Clomicalm. $C_{19}H_{23}ClN_2.HCl$; mol wt 351.32. Crystals from acetone-ether/methanol-ether, mp 189-190° (Craig); also reported as crystals from acetone, mp 191.5-192° (Schindler, Dietrich). Freely sol in water, methanol, methylene chloride. Practically insol in ethyl ether, hexane.

THERAP CAT: Antidepressant; antiobsessional agent.

THERAP CAT (VET): In treatment of canine separation anxiety.

2379. Clonazepam. [1622-61-3] 5-(2-Chlorophenyl)-1,3-di-hydro-7-nitro-2*H*-1,4-benzodiazepin-2-one; 7-nitro-5-(2-chloro-phenyl)-3*H*-1,4-benzodiazepin-2(1*H*)-one; Ro-5-4023; Iktorivil; Klonopin; Rivotril. $C_{15}H_{10}ClN_3O_3$; mol wt 315.71. C 57.07%, H 3.19%, Cl 11.23%, N 13.31%, O 15.20%. Benzodiazepine antiepi-leptic agent with anxiolytic and antimanic properties. Prepn: L. H. Sternbach *et al.*, *J. Med. Chem.* **6**, 261 (1963); O. Keller *et al.*, **US 3121076** (1964 to Hoffmann-La Roche). Prepd not claimed: J. Kariss, H. L. Newmark, **US 3116203** (1963 to Hoffmann-La Roche). Alternate process: A. Focella, A. I. Rachlin, **US 3335181** (1967 to Hoffmann-La Roche). Pharmacology: Guerrero-Figueroa *et al.*, *Curr. Ther. Res.* **11**, 40 (1969); Lechat *et al.*, *Therapie* **25**, 893 (1970); D'Armagnac *et al.*, *ibid.* **26**, 439 (1971). Toxicology: Blum

et al., *Arzneim.-Forsch.* **23**, 377 (1973). HPLC determn in plasma: I. F. Bares *et al.*, *J. Pharm. Biomed. Anal.* **36**, 865 (2004). Compre-hensive description: W. C. Winslow, *Anal. Profiles Drug Subs.* **6**, 61-81 (1977). Review of pharmacokinetics: D. J. Greenblatt *et al*, *J. Clin. Psychiatry* **48**, 4-9 (1987); of clinical efficacy in panic at-tacks: M. H. Pollack, *ibid.* 12-14; in acute mania: G. Chouinard, *ibid.* 29-37. Clinical trial in panic disorders: G. Moroz, J. F. Rosen-baum, *ibid.* **60**, 604 (1999).

White crystals from ethanol-methylene chloride, mp 236.5-238.5°. uv max (7.5% methanol in isopropanol): 248, 310 nm (ε 14500, 11600). Soly in mg/ml at 25°: acetone 31; chloroform 15; methanol 8.6; ether 0.7; benzene 0.5; water <0.1. pK$_1$ 1.5, pK$_2$ 10.5. LD$_{50}$ orally in mice: >4000 mg/kg (Blum).

Note: This is a controlled substance (depressant): **21 CFR, 1308.14.**

THERAP CAT: Anticonvulsant.

THERAP CAT (VET): Anticonvulsant.

2380. Clonidine. [4205-90-7] *N*-(2,6-Dichlorophenyl)-4,5-dihydro-1*H*-imidazol-2-amine; 2-(2,6-dichloroanilino)-2-imidazo-line; 2,6-dichloro-*N*-2-imidazolidinylidenebenzenamine; 2-(2,6-di-chloroanilino)-1,3-diazacyclopentene-(2); 2-[(2,6-dichlorophenyl)-imino]-2-imidazoline. $C_9H_9Cl_2N_3$; mol wt 230.09. C 46.98%, H 3.94%, Cl 30.81%, N 18.26%. α$_2$-Adrenergic agonist. Prepn: Zeile *et al.*, **US 3202660** (1965 to Boehringer, Ing.); use in shaving soap formulations: *eidem*, **US 3190802** (1965 to Boehringer, Ing.). Phar-macology: Bolme, Fuxe, *Eur. J. Pharmacol.* **13**, 168 (1971). Re-vised structure: L. M. Jackman, T. Jen, *J. Am. Chem. Soc.* **97**, 2811 (1975). GC determn in plasma: P. O. Edlund, *J. Chromatogr.* **187**, 161 (1980). Activity as α-adrenoceptor agonist: A. G. Roach *et al.*, *J. Pharmacol. Exp. Ther.* **227**, 421 (1983). Symposium on cardio-vascular and psychotropic pharmacology and clinical experience: *Central Blood Pressure Regulation: The Role of α$_2$-Receptor Stim-ulation*, K. Hayduk, K. B. Bock, Eds. (Steinkopff Verlag, Darmstadt, 1983) 284 pp. Effects in acute smoking withdrawal syndrome: A. H. Glassman *et al.*, *Science* **226**, 864 (1984); in alcoholism with-drawal: Z. Jraidi *et al.*, *Therapie* **42**, 21 (1987). Clinical trial in Tourette's syndrome: J. F. Leckman *et al.*, *Neurology* **35**, 343 (1985); in intractable cancer pain: J. C. Eisenach *et al.*, *Pain* **61**, 391 (1995). *Reviews:* A. Walland in *Pharmacological and Biochemical Properties of Drug Substances* vol. 1, M. E. Goldberg, Ed. (Am. Pharm. Assoc., Washington, D.C., 1977) pp 67-107; H. Schmitt, *Handb. Exp. Pharmacol.* **39**, 299-396 (1977); M. C. Houston, *Prog. Cardiovasc. Dis.* **23**, 337-350 (1981). Comprehensive description: M. A. Abounassif *et al.*, *Anal. Profiles Drug Subs. Excip.* **21**, 109-147 (1992). Review of toxicity: D. L. Seger, *Clin. Toxicol.* **40**, 145-155 (2002); of therapeutic potential in perioperative cardiac risk re-duction: A. W. Wallace, *Curr. Opin. Anaesthesiol.* **19**, 411-417 (2006).

White to almost white crystals, mp 130°. Freely sol in methanol, alcohol.

Hydrochloride. [4205-91-8] ST-155; Catapres; Catapressan; Clonistada; Dixarit; Duraclon; Isoglaucon; Paracefan. $C_9H_9Cl_2N_3.HCl$; mol wt 266.55. Crystals, mp 305°. Sol in abs. ethanol; slightly sol in chloroform. Practically insol in ether. One gram is sol in 6 mls water (60°C); about 13 mls water (20°C); about 5.8 mls methanol; about 25 mls ethanol; about 5000 mls chloroform. uv max (water): 213, 271, 302 nm (ε 8290.327, 713.074, 339.876). LD₅₀ in mice, rats (mg/kg): 328, 270 orally; 18, 29 i.v. (Walland).

USE: In shaving soaps.

THERAP CAT: Antihypertensive; analgesic for neuropathic pain.

THERAP CAT (VET): Diagnostic aid (growth hormone deficiency).

2381. Clonixin. [17737-65-4] 2-[(3-Chloro-2-methylphenyl)-amino]-3-pyridinecarboxylic acid; 2-(3-chloro-*o*-toluidino)nicotinic acid; 2-(2-methyl-3-chloroanilino)nicotinic acid; 2-(3-chloro-2-methylanilino)nicotinic acid; clonixic acid; CBA-93626; Sch-10304. $C_{13}H_{11}ClN_2O_2$; mol wt 262.69. C 59.44%, H 4.22%, Cl 13.49%, N 10.66%, O 12.18%. Nonsteroidal anti-inflammatory agent. Prepn: M. H. Sherlock, N. Sperber, **BE 679271**; *eidem*, **US 3337570** (1965, 1967 both to Schering). Pharmacology: A. S. Watnick *et al.*, *Arch. Int. Pharmacodyn.* **190**, 78 (1971). Metabolism in humans: B. Katchen *et al.*, *J. Pharmacol. Exp. Ther.* **187**, 152 (1973). Clinical evaluation of analgesic activity: J. S. Finch, T. J. DeKornfeld, *J. Clin. Pharmacol.* **15**, 371 (1971). GC determn in plasma: C. Giachetti *et al.*, *J. High Resolut. Chromatogr.* **13**, 789 (1990). Toxicology studies: H. Tanaka *et al.*, *Oyo Yakuri* **7**, 655 (1973), *C.A.* **80**, 44091m (1974).

Crystals from isopropyl acetate, mp 233-235°. Extremely bitter taste. LD₅₀ in male mice, rats (mg/kg): 415, 335 orally; 198, 148 i.p.; 296, 325 s.c. (Tanaka).

Lysine salt. [55837-30-4] L-104; Clonix; Deltar; Dolalgial; Dorixina. $C_{13}H_{11}ClN_2O_2 \cdot C_6H_{14}N_2O_2$; mol wt 408.88.

THERAP CAT: Analgesic.

2382. Clopamide. [636-54-4] *rel*-3-(Aminosulfonyl)-4-chloro-*N*-[(2*R*,6*S*)-2,6-dimethyl-1-piperidinyl]benzamide; *cis*-1-(4-chloro-3-sulfamoylbenzamido)-2,6-dimethylpiperidine; *cis*-*N*-(2′,6′-dimethyl-1′-piperidyl)-3-sulfamoyl-4-chlorobenzamide; *cis*-4-chloro-*N*-(2′,6′-dimethyl-1′-piperidyl)-3-sulfamoylbenzamide; chlosudimeprimyl; DT-327; Adurix; Brinaldix. $C_{14}H_{20}ClN_3O_3S$; mol wt 345.84. C 48.62%, H 5.83%, Cl 10.25%, N 12.15%, O 13.88%, S 9.27%. Thiazide-type diuretic. Prepn (stereochemistry unspecified): **BE 610039**; E. Jucker, A. J. Lindenmann, **US 3459756** (1962, 1969 both to Sandoz). Prepn of *cis*- and *trans*-forms: *eidem, Helv. Chim. Acta* **45**, 2316 (1962). HPLC determn: R. T. Sane *et al., J. Chromatogr.* **356**, 468 (1986). Pharmacokinetics and diuretic effect in humans: J. J. McNeil *et al., Clin. Pharmacol. Ther.* **42**, 299 (1987). Clinical trial of combination with pindolol, *q.v.*, in essential hypertension: D. Crowder, E. G. M. Cameron, *Curr. Med. Res. Opin.* **6**, 342 (1979).

Relative stereochemistry

White or almost white crystalline powder. Hygroscopic. Slightly sol in water, anhydrous alc; sparingly sol in methanol.

Mixture with pindolol. [71789-16-7] Viskaldix.

Hydrazine deriv. Crystals from methanol + diisopropyl ether, mp 244-246°.

THERAP CAT: Antihypertensive; diuretic.

2383. Clopenthixol. [982-24-1] 4-[3-(2-Chloro-9*H*-thioxanthen-9-ylidene)propyl]-1-piperazineethanol; 2-chloro-9-[3-[4-(2-hydroxyethyl)-1-piperazinyl]propylidene]thiaxanthene. $C_{22}H_{25}ClN_2$-

OS; mol wt 400.97. C 65.90%, H 6.28%, Cl 8.84%, N 6.99%, O 3.99%, S 8.00%. Thioxanthene neuroleptic. Prepn (configuration not specified): **BE 585338**; P. V. Petersen *et al.*, **US 3116291** (1960, 1963 both to Kefalas A/S). Prepn of the pharmacologically active *cis*-isomer: **BE 816855**; N. Lassen, **US 3996211** (1974, 1976 both to Kefalas A/S). Pharmacology: Cazzullo, Andreola, *Acta Neurol.* **20**, 162 (1965); Weissman, *Mod. Probl. Pharmacopsychiatry* **2**, 15 (1969); Moeller Nielsen, *ibid.* **23**. Metabolism: Khan, *Acta Pharmacol. Toxicol.* **27**, 202 (1969). HPLC determn of isomers in serum: T. Aaes-Jorgensen, *J. Chromatogr.* **188**, 239 (1980). Series of articles on pharmacology and clinical studies: *Acta Psychiatr. Scand.* **64**, Suppl. 294, 1-77 (1981).

cis-isomer

Colorless syrup. Sparingly sol in ether. Readily sol in methanol.

Dihydrochloride. [633-59-0] AY-62021; N-746; Ciatyl; Sordenac; Sordinol. $C_{22}H_{25}ClN_2OS \cdot 2HCl$; mol wt 473.88. Crystals from ethanol, mp 250-260° (dec). Freely sol in water; sparingly sol in alcohol. Practically insol in other organic solvents. LD₅₀ in male mice (mg/kg): 111 i.v. (Lassen).

***cis*(Z)-Form.** [53772-83-1] α-Clopenthixol; zuclopenthixol. Crystals, mp 84-85°.

***cis*(Z)-Form dihydrochloride.** [58045-23-1] Cisordinol; Clopixol. Crystals, mp 250-260° (dec). LD₅₀ in male mice (mg/kg): 105 i.v. (Lassen).

THERAP CAT: Antipsychotic.

2384. Cloperastine. [3703-76-2] 1-[2-[(4-Chlorophenyl)-phenylmethoxy]ethyl]piperidine; 1-[2-[(*p*-chloro-α-phenylbenzyl)-oxy]ethyl]piperidine; *p*-chlorobenzhydryl 2-(1-piperidyl)ethyl ether. $C_{20}H_{24}ClNO$; mol wt 329.87. C 72.82%, H 7.33%, Cl 10.75%, N 4.25%, O 4.85%. Prepn: **GB 670622** (1952 to Parke, Davis); Fujie, *Yakugaku Zasshi* **81**, 693 (1961), *C.A.* **55**, 25848f (1961). Prepn of salts: **GB 1179945** (1970 to Yoshitomi Pharm.), *C.A.* **72**, 132549g (1970). Pharmacology: Takagi *et al., Yakugaku Zasshi* **85**, 550 (1965); *ibid.* **87**, 907 (1967), *C.A.* **63**, 8932f (1965); **67**, 107101u (1967). Metabolism: Kato, Furuta, *Oyo Yakuri* **5**, 735 (1972).

bp₀.₀₆ 172-174°, bp₀.₁₅ 178-180°.

Hydrochloride. [14984-68-0] Hustazol; Nitossil; Novotusil; Seki. $C_{20}H_{24}ClNO \cdot HCl$; mol wt 366.33. Crystals, mp 147.9°.

Methiodide. $C_{20}H_{24}ClNO \cdot CH_3I$. Crystals, mp 140°.

THERAP CAT: Antitussive.

2385. Clopidogrel. [113665-84-2] (α*S*)-α-(2-Chlorophenyl)-6,7-dihydrothieno[3,2-*c*]pyridine-5(4*H*)-acetic acid methyl ester; methyl (+)-(*S*)-α-(*o*-chlorophenyl)-6,7-dihydrothieno[3,2-*c*]pyridine-5(4*H*)-acetate; (+)-methyl-α-5-[4,5,6,7-tetrahydro[3,2-*c*]thienopyridyl]-(2-chlorophenyl)acetate; SR-25990. $C_{16}H_{16}ClNO_2S$; mol wt 321.82. C 59.72%, H 5.01%, Cl 11.02%, N 4.35%, O 9.94%, S 9.96%. Analog of ticlopidine, *q.v.*; inhibits ADP-induced platelet aggregation. Prepn (unspec. stereochem.): D. Aubert *et al.*, **EP 99802**; *eidem*, **US 4529596** (1984, 1985 both to Sanofi). Prepn of (+)-form: A. Badorc, D. Fréhel, **EP 281459**; *eidem*, **US 4847265**

　　Consult the Name Index before using this section.

(1988, 1989 both to Sanofi). Clinical pharmacology and mechanism of action: D. C. B. Mills *et al.*, *Arterioscler. Thromb.* **12**, 430 (1992). Clinical trial vs aspirin for prevention of atherothrombotic events: CAPRIE Steering Committee, *Lancet* **348**, 1329 (1996). Review of pharmacology and therapeutic potential: G. Feuerstein *et al.*, *Expert Opin. Invest. Drugs* **4**, 425-430 (1995); of use in secondary prevention in stroke: H.-C. Diener *et al.*, *Expert Opin. Pharmacother.* **6**, 755-764 (2005).

Colorless oil. $[\alpha]_D^{20}$ +51.52° (c = 1.61 in methanol).

Hydrogen sulfate. [135046-48-9] SR-25990C; Plavix. $C_{16}H_{16}$-$ClNO_2S.H_2SO_4$; mol wt 419.89. White crystals, mp 184°. $[\alpha]_D^{20}$ +55.10° (c = 1.891 in methanol). Sol in water, methanol.

THERAP CAT: Antithrombotic.

2386. Clopidol. [2971-90-6] 3,5-Dichloro-2,6-dimethyl-4-pyridinol; meticlorpindol; clopindol; Coyden. $C_7H_7Cl_2NO$; mol wt 192.04. C 43.78%, H 3.67%, Cl 36.92%, N 7.29%, O 8.33%. Prepn: Stevenson, **US 3206358** (1965 to Dow). Metabolism studies in chickens: Smith, *Poult. Sci.* **48**, 420 (1969). Toxicity study: Plisek *et al.*, *Vet. Spofa* **12**, 111 (1970), *C.A.* **74**, 138780p (1971).

Solid, mp >320°. Practically insol in water. LD_{50} orally in rats: 18 g/kg (Plisek).

Caution: Potential symptoms of overexposure are irritation of eyes, skin, nose, throat; cough. *See NIOSH Pocket Guide to Chemical Hazards* (DHHS/NIOSH 97-140, 1997) p 72.

THERAP CAT (VET): Coccidiostat.

2387. Cloprednol. [5251-34-3] (11β)-6-Chloro-11,17,21-trihydroxypregna-1,4,6-triene-3,20-dione; 6-chloro-Δ^{1,4,6}-pregnatriene-11β,17α,21-triol-3,20-dione; RS-4691; Syntestan. $C_{21}H_{25}$-ClO_5; mol wt 392.88. C 64.20%, H 6.41%, Cl 9.02%, O 20.36%. Prepn: H. J. Ringold, G. Rosenkranz, **FR 1271981** *C.A.* **58**, 1148a (1963); *eidem*, **US 3232965** (1962, 1966 both to Syntex). Metabolic effects: E. Ortega *et al.*, *J. Clin. Pharmacol.* **16**, 122 (1976). Effects on hypothalamic-pituitary-adrenal axis function: *eidem*, *J. Int. Med. Res.* **4**, 326 (1976). Bioavailability in humans: E. Mroszczak *et al.*, *J. Pharm. Sci.* **67**, 920 (1978). HPLC determn in plasma: C. Lee *et al.*, *ibid.* **70**, 669 (1981).

Crystals from acetone/hexane.

THERAP CAT: Glucocorticoid.

2388. Cloprostenol. [40665-92-7] *rel*-(5Z)-7-[(1R,2R,3R,-5S)-2-[(1E,3R)-4-(3-Chlorophenoxy)-3-hydroxy-1-buten-1-yl]-3,5-dihydroxycyclopentyl]-5-heptenoic acid. $C_{22}H_{29}ClO_6$; mol wt 424.92. C 62.19%, H 6.88%, Cl 8.34%, O 22.59%. Aryloxymethyl analog of prostaglandin $F_{2\alpha}$, *q.v.* Prepn: J. Bowler, **DE 2223365**

(1972 to ICI), *C.A.* **78**, 110692v (1973); D. Binder *et al.*, *Prostaglandins* **6**, 87 (1974). Synthesis and biological activity: *eidem, ibid.* **15**, 773 (1978). Short synthesis: N. R. A. Beeley *et al.*, *Tetrahedron* **37**, Suppl. 9, 411 (1981). Disposition in the rat and marmoset: G. R. Bourne *et al.*, *Xenobiotica* **9**, 623 (1979). Effect on fertility in cows: N. Baishya *et al.*, *Br. Vet. J.* **136**, 227 (1980); on superovulation, fertilization, egg transport in ewes: D. Whyman, R. W. Moore, *J. Reprod. Fertil.* **60**, 267 (1980). HPLC separation of enantiomers: K. Kalikova *et al.*, *J. Pharm. Biomed. Anal.* **46**, 892 (2008).

Relative stereochemistry

Sodium salt. [55028-72-3] ICI-80996; Estrumate; Planate. C_{22}-$H_{28}ClNaO_6$; mol wt 446.90.

THERAP CAT (VET): In treatment of infertility in sows, gilts. In synchronization of estrus.

2389. Clopyralid. [1702-17-6] 3,6-Dichloro-2-pyridinecarboxylic acid; 3,6-dichloropicolinic acid; 3,6-DCP; Dowco 290. C_6-$H_3Cl_2NO_2$; mol wt 192.00. C 37.53%, H 1.58%, Cl 36.93%, N 7.30%, O 16.67%. Systemic post-emergence herbicide for use in food crops and mesquite. Prepn: H. Johnston, **BE 644105**; *idem*, **US 3317549** (1964, 1967 both to Dow). Alternate process: S. D. McGregor, **US 4087431** (1978 to Dow). Physicochemical properties, toxicity and herbicidal activity: J. G. Brown, S. D. Uprichard, *Proc. Br. Crop Prot. Conf. - Weeds* **1976**, 119. GLC determn: A. J. Pik, G. W. Hodgson, *J. Assoc. Off. Anal. Chem.* **59**, 264 (1976). Persistence in soil: A. J. Pik *et al.*, *J. Agric. Food Chem.* **25**, 1054 (1977). Field trial in mesquite control: R. W. Bovey, R. E. Meyer, *Weed Sci.* **33**, 349 (1985).

White, odorless crystalline solid, mp 151-152°. Vapor pressure at 25°: 1.2×10^{-5} mm Hg. Soly at 25°: ~1000 ppm in water, >25% w/w in methanol, acetone, xylene. LD_{50} in male, female rats (mg/kg): >5000, 4300 orally; LC_{50} (96 hr) to rainbow trout: 103.5 mg/l (Brown, Uprichard).

Monoethanolamine salt. [57754-85-5] Dow Shield; Lontrel; Reclaim; Stinger; Transline. $C_6H_3Cl_2NO_2.C_2H_7NO$; mol wt 253.08.

USE: Herbicide.

2390. Cloquintocet-mexyl. [99607-70-2] 2-[(5-Chloro-8-quinolinyl)oxy]acetic acid 1-methylhexyl ester; CGA-185072. C_{18}-$H_{22}ClNO_3$; mol wt 335.83. C 64.38%, H 6.60%, Cl 10.56%, N 4.17%, O 14.29%. Safening agent to protect crops from herbicidal injury. Prepn: A. Hubele, **EP 94349**; *idem*, **US 4902340** (1983, 1990 both to Ciba-Geigy). Description: J. Amrein *et al.*, *Brighton Crop Prot. Conf. - Weeds* **1989**, 71-76. Effects on herbicide metabolism in cereals: K. Kreuz *et al.*, *Z. Naturforsch.* **46C**, 901 (1991).

Odorless crystals, mp 69°. Vapor pressure (20°): 2.5×10^{-6} Pa. Soly in water (20°): 0.8 ppm. Sol in most organic solvents. LD_{50}

in rats (mg/kg): >2000 orally; >2000 dermally. LC_{50} (4 hr) in rats: >935 mg/m³ by inhalation (Amrein).

Mixture with clodinafop-propargyl. Célio; Horizon.

USE: Herbicide safener.

2391. Cloransulam-methyl. [147150-35-4] 3-Chloro-2-[[(5-ethoxy-7-fluoro[1,2,4]triazolo[1,5-c]pyrimidin-2-yl)sulfonyl]amino]benzoic acid methyl ester; methyl 3-chloro-2-(5-ethoxy-7-fluoro-[1,2,4]triazolo[1,5-c]pyrimidin-2-ylsulfonamido)benzoate; methyl 3-chloro-N-(5-ethoxy-7-fluoro[1,2,4]triazolo[1,5-c]pyrimidin-2-ylsulfonyl)anthranilate; N-(2-carbomethoxy-6-chlorophenyl)-5-ethoxy-7-fluoro[1,2,4]triazolo[1,5-c]pyrimidine-2-sulfonamide; DE-565; XDE-565; First Rate. $C_{15}H_{13}ClFN_5O_5S$; mol wt 429.81. C 41.92%, H 3.05%, Cl 8.25%, F 4.42%, N 16.29%, O 18.61%, S 7.46%. Herbicide for control of broadleaf weeds in soybeans; inhibits acetolactate synthase (ALS) used by plants to synthesize branched-chain amino acids. Prepn: J. C. Van Heertum et al., **EP 343752** (1989 to Dow Chemical); eidem, **US 5163995** (1992 to DowElanco). GC/MS determn in soybean matrices: D. D. Shackelford et al., J. Agric. Food Chem. **44**, 3570 (1996). HPLC determn in water and soil: M. A. Rodríguez-Delgado, J. Hernández-Borges, J. Sep. Sci. **30**, 8 (2007). Soil metabolism study: J. D. Wolt et al., J. Agric. Food Chem. **44**, 324 (1996). Environmental fate: I. J. van Wesenbeeck et al., ibid. **45**, 3299 (1997). Field studies in soybeans: W. A. Pline et al., Weed Technol. **16**, 737 (2002).

Off-white powder with slight mint odor; mp 216-218°. d^{20} 1.538. pKa 4.81. Vapor pressure (25°): 4×10^{-14} Pa. Log P (octanol/water): 1.12 (pH 5), −0.365 (pH 7), −1.24 (pH 8.5). Soly (mg/l): distilled water 16, water (pH 5) 3, water (pH 7) 184, acetone 4360, acetonitrile 5500, dichloromethane 6980, ethyl acetate 980, methanol 470.

Free acid. [159518-97-5] Cloransulam. $C_{14}H_{11}ClFN_5O_5S$; mol wt 415.78.

USE: Agricultural herbicide.

2392. Clorazepic Acid. [23887-31-2] 7-Chloro-2,3-dihydro-2-oxo-5-phenyl-1H-1,4-benzodiazepine-3-carboxylic acid. $C_{16}H_{11}ClN_2O_3$; mol wt 314.73. C 61.06%, H 3.52%, Cl 11.26%, N 8.90%, O 15.25%. Prepn: **NL 6507637**; J. Schmitt, **US 3516988**; reissued as **US RE 28315** (1965, 1970, 1975 all to Clin-Byla). Synthesis and activity of the dipotassium salt: J. Schmitt et al., Chim. Ther. **4**, 239 (1969). Solution chemistry: R. Raveux, M. Briot, ibid. 303. Metabolism: P. Gros, R. Raveux, ibid. 312. Toxicity data: M. Brunaud et al., Arzneim.-Forsch. **20**, 123 (1970). Series of articles on pharmacology and clinical use: ibid. 123-137. HPLC determn in plasma: P. Colin, G. Sirois, J. Chromatogr. **273**, 367 (1983). Clinical trial in anxiety: W. W. K. Zung, J. Clin. Psychiatry **48**, 13 (1987); in comparison with buspirone, q.v.: K. Rickels et al., Arch. Gen. Psychiatry **45**, 444 (1988). Comprehensive description: J. A. Raihle, V. E. Papendick, Anal. Profiles Drug Subs. **4**, 91-112 (1975).

Dipotassium salt. [57109-90-7] 7-Chloro-2,3-dihydro-2-oxo-5-phenyl-1H-1,4-benzodiazepine-3-carboxylic acid monopotassium salt compd with potassium hydroxide; clorazepate dipotassium; Ab-

bott 35616; CB-4306; Belseren; Mendon; Tranxilène; Tranxilium; Transene; Tranxene. $C_{16}H_{11}ClK_2N_2O_4$; mol wt 408.92. White powder, freely sol in ethanol. Very poorly sol in water. Practically insol in ether, chloroform. Aq solns are alkaline to phenolphthalein. uv max (anhydrous product in water): 231, 311 nm (ε 33500, 2450). LD_{50} in mice (mg/kg): 700 orally; 290 i.p. LD_{50} orally in rats: >1000 mg/kg (Brunaud).

Monopotassium salt. [5991-71-9] CB-4311; Azene. $C_{16}H_{10}ClKN_2O_4$; mol wt 368.81.

Note: This is a controlled substance (depressant): **21 CFR**, 1308.14.

THERAP CAT: Anxiolytic.

2393. Cloricromen. [68206-94-0] 2-[[8-Chloro-3-[2-(diethylamino)ethyl]-4-methyl-2-oxo-2H-1-benzopyran-7-yl]oxy]acetic acid ethyl ester; ethyl [[8-chloro-3-[2-(diethylamino)ethyl]-4-methyl-2-oxo-2H-1-benzopyran-7-yl]oxy]acetate; 8-monochloro-3-(β-diethylaminoethyl)-4-methyl-7-ethoxycarbonylmethoxycoumarin; 8-chlorocarbochromen; AD_6. $C_{20}H_{26}ClNO_5$; mol wt 395.88. C 60.68%, H 6.62%, Cl 8.95%, N 3.54%, O 20.21%. Derivative of coumarin, q.v. Prepn: F. della Valle, **BE 871315**; idem, **US 4452811** (1979, 1984 both to Fidia). In vivo studies of activity as coronary vasodilator: F. Aporti et al., Pharmacol. Res. Commun. **10**, 469 (1978); as inhibitor of platelet aggregation: M. Prosdocimi et al., Thromb. Res. **39**, 399 (1985). Pharmacology in human platelets: R. A. Travagli et al., ibid. **54**, 327 (1989). Mechanism of action: S. Porcellati et al., Agents Actions **29**, 364 (1990).

Crystals from ethyl acetate, mp 147-148°.

Hydrochloride. [74697-28-2] Cromocap; Proendotel. $C_{20}H_{26}ClNO_5$·HCl; mol wt 432.34. mp 219-220°.

THERAP CAT: Antithrombotic; vasodilator (coronary).

2394. Clorophene. [120-32-1] 4-Chloro-2-(phenylmethyl)-phenol; 4-chloro-α-phenyl-o-cresol; 2-benzyl-4-chlorophenol; 5-chloro-2-hydroxydiphenylmethane; o-benzyl-p-chlorophenol; Septiphene; Santophen 1. $C_{13}H_{11}ClO$; mol wt 218.68. C 71.40%, H 5.07%, Cl 16.21%, O 7.32%. Prepn: Klarmann et al., J. Am. Chem. Soc. **54**, 3315 (1932); Klarmann, Gate, **US 1967825** (to Lehn & Fink); from Na or K phenoxide and benzyl chloride followed by chlorination of the resulting o-benzylphenol: Kaiser, **DE 703955** (1941 to Deutsche Hydrierwerke).

Crystals, mp 48.5°. $bp_{3.5}$ 160-162°. $d_{15.5}^{55}$ 1.186-1.190.

USE: In household, hospital and veterinary disinfectant preparations.

2395. Clorsulon. [60200-06-8] 4-Amino-6-(1,2,2-trichloroethenyl)-1,3-benzenedisulfonamide; 4-amino-6-(trichlorovinyl)-m-benzenedisulfonamide; MK-401; Curatrem. $C_8H_8Cl_3N_3O_4S_2$; mol wt 380.64. C 25.24%, H 2.12%, Cl 27.94%, N 11.04%, O 16.81%, S 16.85%. Benzenedisulfonamide deriv with fasciolicidal activity. Prepn: H. H. Mrozik, **DE 2556122** (1976 to Merck & Co.), C.A. **85**, 94100n (1976); H. Mrozik et al., J. Med. Chem. **20**, 1225 (1977). GLC determn in biological fluids: W. J. A. Vanden Heuvel et al., J. Agric. Food Chem. **25**, 389 (1977). Efficacy against liver fluke, Fasciola hepatica: D. A. Ostlind et al., Br. Vet. J. **133**, 211 (1977); M. D. Schulman et al., J. Parasitol. **65**, 555 (1979). In vitro inhibition of glycolysis: M. D. Schulman, D. Valentino, Exp. Parasitol. **49**, 206 (1980). Mechanism of action: M. D. Schulman et al., Mol. Biochem. Parasitol. **5**, 133 (1982). Pharmacokinetics: eidem, J. Parasitol. **68**, 603 (1982).

Crystals from ether, mp 194-203°. Another crystal form from water, mp 203-205°. uv max (CH$_3$OH): 325, 267, 227 nm (ε 4530, 17395, 36310). Freely sol in acetonitrile, methanol; slightly sol in water; very slightly sol in methylene chloride. LD$_{50}$ in mice (mg/kg): 761 i.p.; >10,000 orally (Ostlind).

THERAP CAT (VET): Anthelmintic (Trematodes).

2396. Closantel. [57808-65-8] *N*-[5-Chloro-4-[(4-chlorophenyl)cyanomethyl]-2-methylphenyl]-2-hydroxy-3,5-diiodobenzamide; R-31520; Flukiver; Seponver. C$_{22}$H$_{14}$Cl$_2$I$_2$N$_2$O$_2$; mol wt 663.07. C 39.85%, H 2.13%, Cl 10.69%, I 38.28%, N 4.22%, O 4.83%. Salicylanilide derivative. Prepn: M. A. C. Janssen, V. K. Sipido, BE 839481; eidem, US 4005218 (1976, 1977 both to Janssen). Effectiveness against *Taenia pisiformis* in rabbits: R. A. F. Chevis *et al., Vet. Parasitol.* **7**, 333 (1980); against *Ancylostoma caninum*: J. Guerrero *et al., J. Parasitol.* **68**, 616 (1983); against *Fasciola hepatica* in sheep: B. E. Stromberg *et al., ibid.* **70**, 446 (1984). Prolonged effect on *Haemonchus contortus* in sheep: C. A. Hall *et al., Res. Vet. Sci.* **31**, 104 (1981). Acts by uncoupling oxidative phosphorylation: H. Van den Bossche *et al., Arch. Int. Physiol. Biochim.* **87**, 851 (1979); H. J. Kane *et al., Mol. Biochem. Parasitol.* **1**, 347 (1980).

Crystals from methanol, mp 217.8°.
THERAP CAT (VET): Anthelmintic.

2397. Clostebol. [1093-58-9] (17β)-4-Chloro-17-hydroxyandrost-4-en-3-one; 4-chlorotestosterone. C$_{19}$H$_{27}$ClO$_2$; mol wt 322.87. C 70.68%, H 8.43%, Cl 10.98%, O 9.91%. Prepn: Camerino *et al., Farmaco Ed. Sci.* **11**, 586 (1956); eidem, *J. Am. Chem. Soc.* **78**, 3540 (1956); Ringold *et al., J. Org. Chem.* **21**, 1432 (1956); Camerino, US 2953582 (1960 to Farmitalia); Julian, Printy, US 2933510 (1960 to Julian Labs.). Clinical pharmacology of the acetate: Molinatti *et al., Folia Endocrinol.* **14**, 528 (1961); Krueskemper, Morgner, *Int. Z. Klin. Pharmakol. Ther. Toxikol.* **1**, 455 (1968).

Crystals from acetone + hexane, mp 188-190°. [α]$_D^{20}$ +148° (CHCl$_3$). uv max (95% ethanol): 256 nm (log ε 4.13).

Acetate. [855-19-6] Macrobin; Steranabol. C$_{21}$H$_{29}$ClO$_3$; mol wt 364.91. Crystals from methanol, mp 228-230°. [α]$_D$ +118 ±4° (CHCl$_3$). uv max: 255 nm (ε 13,300). Sol in ethanol.

Note: This is a controlled substance (anabolic steroid): **21 CFR**, 1308.13, as defined in 1300.01.

THERAP CAT: Anabolic.

2398. Clothianidin. [210880-92-5] [*C*(*E*)]-*N*-[(2-Chloro-5-thiazolyl)methyl]-*N*′-methyl-*N*″-nitroguanidine; (*E*)-1-(2-chloro-

1,3-thiazol-5-ylmethyl)-3-methyl-2-nitroguanidine; TI-435; Arena; Poncho. C$_6$H$_8$ClN$_5$O$_2$S; mol wt 249.67. C 28.86%, H 3.23%, Cl 14.20%, N 28.05%, O 12.82%, S 12.84%. Neonicotinoid insecticide for use in food crops. Prepn: H. Uneme *et al.*, EP 376279; eidem, US 5034404 (1990, 1991 both to Takeda); eidem, *Pestic. Sci.* **55**, 202 (1999). Synthesis: P. Maienfisch *et al., Tetrahedron Lett.* **41**, 7187 (2000). Insecticidal activity and physical properties: Y. Ohkawara *et al., Brighton Crop Prot. Conf. - Pests Dis.* **1**, 51 (2002). Field trials in sugar beets: R. H. Meredith *et al., ibid.* **2**, 691 (2002); and cereals: A. M. Dewar *et al., ibid.* 647. Ecotoxicological profile: R. Schmuck, J. Keppler, *Pflanzenschutz-Nachr. Bayer* **56**, 26 (2003). Metabolism studies: O. Klein, *ibid.* 75. Review of use in corn: W. Andersch, M. Schwarz, *ibid.* 147-172.

White crystalline powder, mp 176.8°. Vapor pressure (Pa): 1.3 × 10^{-10} (25°); 3.8 × 10^{-7} (20°). Soly (g/l) at 20°: water 0.327; at 25°: acetone 15.2, methanol 6.26, ethyl acetate 2.03, xylene 0.013. Log P (*n*-octanol/water) at 25°: 0.7. LD$_{50}$ in rats (mg/kg): >5000 orally; >2000 dermally. LC$_{50}$ in rats (mg/l): >6.1 inhalation. LC$_{50}$ in bobwhite quail, mallard duck (ppm): >5200, >5200 (dietary). LC$_{50}$ (96 hr) in rainbow trout, bluegill (mg/l): >100, >120 (Ohkawara).

USE: Insecticide.

2399. Clothiapine. [2058-52-8] 2-Chloro-11-(4-methyl-1-piperazinyl)dibenzo[*b*,*f*][1,4]thiazepine; Entumine; Etumine. C$_{18}$H$_{18}$ClN$_3$S; mol wt 343.87. C 62.87%, H 5.28%, Cl 10.31%, N 12.22%, S 9.32%. Prepn: FR CAM 51 (1964 to Dr. A. Wander), *C.A.* **61**, 8328h (1964). HPLC determn in plasma and tissue samples: F. Sporkert *et al., Forensic Sci. Int.* **170**, 193 (2007). Clinical evaluation in chronic schizophrenia: V. Geller *et al., Schizophr. Res.* **80**, 343 (2005).

Crystals from ether + petr ether, mp 118-120°.
THERAP CAT: Antipsychotic.

2400. Clotiazepam. [33671-46-4] 5-(2-Chlorophenyl)-7-ethyl-1,3-dihydro-1-methyl-2*H*-thieno[2,3-*e*]-1,4-diazepin-2-one; Y-6047; Clozan; Rise; Rize; Rizen; Tienor; Trecalmo; Veratran. C$_{16}$H$_{15}$ClN$_2$OS; mol wt 318.82. C 60.28%, H 4.74%, Cl 11.12%, N 8.79%, O 5.02%, S 10.06%. A thienodiazepine tranquilizer with biological activity similar to the benzodiazepines. Prepn: M. Nakanishi *et al.*, DE 2107356; eidem, US 3849405 (1971, 1974 both to Yoshitomi); eidem, *J. Med. Chem.* **16**, 214 (1973). Pharmacology: eidem, *Arzneim.-Forsch.* **22**, 1905 (1972). Effects on biogenic amine metabolism: eidem, *ibid.* 1914. Metabolism: eidem, *Yakugaku Zasshi* **93**, 311 (1973), *C.A.* **79**, 13397r (1973). Clinical studies: S. Sieberns, *Fortschr. Med.* **97**, 1705 (1979).

Crystals from hexane, mp 105-106°. LD$_{50}$ in mice (mg/kg): 440 i.p., 636 orally (Nakanishi).
Note: This is a controlled substance (depressant): **21 CFR,** 1308.14.

THERAP CAT: Anxiolytic.

2401. Clotrimazole. [23593-75-1] 1-[(2-Chlorophenyl)di-phenylmethyl]-1H-imidazole; 1-(o-chloro-α,α-diphenylbenzyl)-imidazole; 1-[α-(2-chlorophenyl)benzhydryl]imidazole; 1-[(o-chlorophenyl)diphenylmethyl]imidazole; diphenyl-(2-chlorophenyl)-1-imidazolylmethane; 1-(o-chlorotrityl)imidazole; FB-5097; Bay b 5097; Canesten; Canifug; Empecid; Gyne-Lotrimin; Lotrimin; Mono-Baycuten; Mycelex-G; Mycofug; Mycosporin; Pedisafe; Rimazole; Tibatin; Trimysten. C$_{22}$H$_{17}$ClN$_2$; mol wt 344.84. C 76.63%, H 4.97%, Cl 10.28%, N 8.12%. Prepn: K. H. Buechel *et al.*, *ZA 6805392*; *eidem*, *US 3705172* (1969, 1972 both to Bayer). Pharmacology: Plempel *et al.*, *Antimicrob. Agents Chemother.* **1969**, 271; *eidem*, *Dtsch. Med. Wochenschr.* **94**, 1356 (1969). Clinical findings: Oberste-Lehn *et al.*, *ibid.* 1365. Series of articles on prepn, toxicology, pharmacokinetics, clinical studies: *Arzneim.-Forsch.* **22**, 1260-1272, 1276-1299 (1972). Toxicity: D. Tettenborn, *ibid.* 1276. Comprehensive description: J. G. Hoogerheide, B. E. Wyka, *Anal. Profiles Drug Subs.* **11**, 225-255 (1982).

White to pale yellow crystals, mp 147-149°. A weak base. Freely sol in ethanol, methanol, acetone, chloroform, DMF, benzene; sol in ethyl acetate, DMSO, methylene chloride. Practically insol in water. Hydrolyzes rapidly upon heating in aq acids. LD$_{50}$ in male mice, rats (mg/kg): 923, 708 orally (Tettenborn).
Hydrochloride. C$_{22}$H$_{17}$ClN$_2$.HCl. mp 159°.

THERAP CAT: Antifungal.
THERAP CAT (VET): Antifungal.

2402. Clove. Caryophyllus. Evergreen tree, *Syzygium aromaticum* (L.) Merr. et Perry, also known as *Eugenia caryophyllata* Thunb. and *Caryophyllus aromaticus* L., Myrtaceae. *Habit.* Southeast Asia; cultivated in tropical regions worldwide. Parts used are the dried buds (cloves) and the essential oil produced from them. *Constit.* 15-21% volatile oil, sterols, *e.g.* sitosterol, ~6% protein, ~61% carbohydrate, ~20% lipids. *Review:* A. Y. Leung, *Encyclopedia of Common Natural Ingredients* (Wiley-Interscience, New York, 1980) pp 130-132.
Clove oil. Oil of clove. Volatile oil from dried flower buds. *Constit.* 60-90% eugenol, 2-27% eugenyl acetate, 5-12% β-caryophyllene, minor constituents such as methyl amyl ketone, methyl salicylate, benzaldehyde. Extraction procedures: A. A. Clifford *et al.*, *J. Anal. Chem.* **364**, 635 (1999). Colorless to pale yellow liq, becoming darker and thicker with age. d$_{25}^{25}$ 1.038-1.060. n_D^{20} 1.530. Insol in water. Sol in 2 vols 70% alcohol. *Keep well closed, cool and protected from light.*

USE: Flavoring agent in foods; fragrance component in dentifrices, soaps, lotions, perfumes; commercial source of eugenol. Pharmaceutic aid (flavor).

THERAP CAT: Carminative, counterirritant. Clove oil as analgesic (dental).

2403. Cloxacillin. [61-72-3] (2S,5R,6R)-6-[[[3-(2-Chlorophenyl)-5-methyl-4-isoxazolyl]carbonyl]amino]-3,3-dimethyl-7-oxo-4-thia-1-azabicyclo[3.2.0]heptane-2-carboxylic acid; [3-(o-chlorophenyl)-5-methyl-4-isoxazolyl]penicillin; [5-methyl-3-(o-chlorophenyl)-4-isoxazolyl]penicillin; 6-[3-(o-chlorophenyl)-5-methyl-4-isoxazolecarboxamido]penicillanic acid. C$_{19}$H$_{18}$ClN$_3$O$_5$S; mol wt 435.88. C 52.36%, H 4.16%, Cl 8.13%, N 9.64%, O 18.35%, S 7.36%. Semi-synthetic antibiotic related to penicillin; chlorinated deriv of oxacillin, *q.v.* Prepn: F. P. Doyle, J. H. C. Nayler, *US 2996501* (1961); F. P. Doyle *et al.*, *J. Chem. Soc.* **1963**,

5838. Manuf: *Ind. Chem.* **39**, 513 (1963), *C.A.* **60**, 1543a (1964). Properties and pharmacology: J. H. C. Nayler *et al.*, *Nature* **195**, 1264 (1962). Toxicity data: E. I. Goldenthal, *Toxicol. Appl. Pharmacol.* **18**, 185 (1971). Comprehensive description: D. L. Mays, *Anal. Profiles Drug Subs.* **4**, 113-136 (1975).

Sodium salt monohydrate. [7081-44-9]; [642-78-4] (anhydrous). Sodium cloxacillin; BRL-1621; Cloxapen; Cloxypen; Ekvacillin; Latocillin; Orbenin; Methocillin-S; Tegopen. C$_{19}$H$_{17}$ClN$_3$NaO$_5$S.H$_2$O; mol wt 475.88. Microcryst powder, dec 170°. $[\alpha]_D^{20}$ +163° (c = 1 in water). pH of 1% aq soln = 6.0-7.5. Sol in water, methanol, ethanol, pyridine, ethylene glycol; slightly sol in chloroform. LD$_{50}$ in rats, mice (mg/kg): 1630 ±112, 1280 ±50 i.p. (Goldenthal).
Benzathine salt. [23736-58-5] Dry Clox; Noroclox. C$_{54}$H$_{56}$Cl$_2$N$_8$O$_{10}$S$_2$; mol wt 1112.11. White crystals or crystalline powder. Sol in chloroform, methanol; sparingly sol in acetone; slightly sol in water, ethanol, isopropyl alcohol.

THERAP CAT: Antibacterial.
THERAP CAT (VET): Antibacterial.

2404. Cloxazolam. [24166-13-0] 10-Chloro-11b-(2-chlorophenyl)-2,3,7,11b-tetrahydrooxazolo[3,2-d][1,4]benzodiazepin-6(5H)-one; 7-chloro-5-(2-chlorophenyl)tetrahydrooxazolo[5,4-b]-2,3,4,5-tetrahydro-1H-1,4-benzodiazepin-2-one; 10-chloro-11b-(2-chlorophenyl)-2,3,5,6,7,11b-hexahydrobenzo[6,7]-1,4-diazepino[5,4-b]oxazol-6-one; 10-chloro-11b-(2-chlorophenyl)-6-oxo-2,-3,5,6,7,11b-hexahydrooxazolo[3,2-d][1,4]benzodiazepine; CS-370; Betavel; Enadel; Lubalix; Olcadil; Sepazon; Tolestan. C$_{17}$H$_{14}$Cl$_2$N$_2$O$_2$; mol wt 349.21. C 58.47%, H 4.04%, Cl 20.30%, N 8.02%, O 9.16%. Prepn: R. Tachikawa *et al.*, **DE 1812252** and **DE 1952201**; *eidem*, **US 3772371** and **US 3696094** (1969, 1970, 1973, 1972 all to Sankyo); T. Miyadera *et al.*, *J. Med. Chem.* **14**, 520 (1971). Pharmacology: T. Kamioka *et al.*, *Arzneim.-Forsch.* **22**, 884 (1972). Metabolism: Murata *et al.*, *Chem. Pharm. Bull.* **21**, 404 (1973). Multicenter trials and complementary studies: K. A. Fischer-Cornelssen, *Arzneim.-Forsch.* **31**, 1757 (1981).

Crystals, mp 202-204° (dec). Freely sol in glacial acetic acid; sparingly sol in chloroform; slightly sol in acetone, dehydrated ethanol, ethyl acetate, benzene. Practically insol in water. LD$_{50}$ in mice (g/kg): 3.3 orally; >2.0 i.p. (Kamioka).
Note: This is a controlled substance (depressant): **21 CFR,** 1308.14.

THERAP CAT: Anxiolytic.

2405. Cloxyquin. [130-16-5] 5-Chloro-8-quinolinol; 5-chloro-8-hydroxyquinoline; 5-chloro-8-oxychinolin; cloxiquine; Chlorisept. C$_9$H$_6$ClNO; mol wt 179.60. C 60.19%, H 3.37%, Cl 19.74%, N 7.80%, O 8.91%. Prepn: Bratz, Niementowski, *Ber.* **52**, 189 (1919); Weizmann, Bograchov, *J. Am. Chem. Soc.* **69**, 1222 (1947); Manske *et al.*, *Can. J. Res.* **27F**, 359 (1949). Crystal structure: T. Banerjee, N. N. Saha, *Acta Crystallogr.* **C42**, 1408 (1986).

Crystals from ethanol, mp 130°. Sparingly sol in cold, dil HCl.
Hydrochloride. $C_9H_6ClNO.HCl$. Yellow needles. mp 256-258°.
Acetate (ester). 5-Chloro-8-acetoxyquinoline. $C_{11}H_8ClNO_2$.
THERAP CAT: Antibacterial; antifungal.

2406. Clozapine. [5786-21-0] 8-Chloro-11-(4-methyl-1-piperazinyl)-5H-dibenzo[b,e][1,4]diazepine; HF-1854; Clozaril; Leponex. $C_{18}H_{19}ClN_4$; mol wt 326.83. C 66.15%, H 5.86%, Cl 10.85%, N 17.14%. Combined serotonin ($5HT_2$) and dopamine (D_2) receptor antagonist. Prepn: **FR 1334944** (1963 to Wander); Schmutz, Hunziker, **US 3539573** (1970); **NL 293201** (1965 to Wander), *C.A.* **64**, 8221a (1966); Hunziker et al., *Helv. Chim. Acta* **50**, 1588 (1967). Structure-activity studies: Schmutz et al., *Chim. Ther.* **2**, 424 (1967). Pharmacology: Stille et al., *Farmaco Ed. Prat.* **26**, 603 (1971). Metabolism: Gauch, Michaelis, *ibid.* 667. Toxicology: Lindt et al., *ibid.* 585. Review: A. C. Sayers, H. A. Amsler, in *Pharmacological and Biochemical Properties of Drug Substances* vol. **1**, M. E. Goldberg, Ed. (Am. Pharm. Assoc., Washington, DC, 1977) pp 1-31. Review of pharmacology and clinical use: R. J. Baldessarini, F. R. Frankenburg, *N. Engl. J. Med.* **324**, 746-754 (1991). Comprehensive description: M. J. McLeish et al., *Anal. Profiles Drug Subs. Excip.* **22**, 145-184 (1993). Overview of mechanism of action: H. Y. Meltzer, *J. Clin. Psychiatry* **55**, Suppl. B, 47-52 (1994). Clinical trial in suicide prevention in schizophrenia: H. Y. Meltzer et al., *Arch. Gen. Psychiatry* **60**, 82 (2003). Review of safety and tolerability: J. Fitzsimons et al., *Expert Opin. Drug Saf.* **4**, 731-744 (2005).

Yellow crystals from acetone-petr ether, mp 183-184°. Soly w/w at 25° (%): acetone >5; acetonitrile 1.9; chloroform >20; ethyl acetate >5; abs ethanol 4.0; water <0.01. pKa_1 3.70; pKa_2 7.60. Partition coefficient (octanol/water): 0.4 (pH 2); 600 (pH 7); 1000 (pH 7.4); 1500 (pH 8). uv max (ethanol): 215, 230, 261, 297 nm (ε 27400, 25800, 16800, 10500). LD_{50} in mice, rats (mg/kg): 61, 58 i.v.; 199, 260 orally (Lindt).
THERAP CAT: Antipsychotic.

2407. Clupeine. [9007-31-2] Protamine found in herring (*Clupea pallasii*) sperm. Isoln from herring testes contg ripe sperm: Kossel, *The Protamines and Histones* (London, 1928); Rasmussen, *Z. Physiol. Chem.* **224**, 97 (1934); Felix, Mager, *ibid.* **249**, 111 (1937); Block et al., *Proc. Soc. Exp. Biol. Med.* **70**, 494 (1949). Separated into two main fractions, Y and Z, and fraction Y separated into Y_I and Y_{II}: Ando, Sawada, *J. Biochem. (Tokyo)* **49**, 252 (1961). Chemical structure of fraction Z: Ando et al., *Biochim. Biophys. Acta* **56**, 628 (1962); Felix, Hashimoto, *Z. Physiol. Chem.* **330**, 205 (1963). Complete amino acid sequence of Z component: Iwai et al., *J. Biochem. (Tokyo)* **69**, 493 (1971); of Y_{II} component: Suzuki, Ando, *ibid.* **72**, 1419 (1972); of Y_I component: eidem, *ibid.* 1433. Solid-phase synthesis of clupeine Z: Yonezawa et al., *C.A.* **79**, 19093k (1973).

White powder, strongly alkaline reaction. pKa 7.4-8.0; pKb 2.9-3.3.
Disulfate. White powder, $[\alpha]_D^{22}$ -85.49° (satd aq soln). One gram dissolves in 80 ml water at room temp. Freely sol in hot water, separates from the supersatd soln on cooling as a clear, colorless oil contg 50% H_2O, n_D^{20} 1.4435. Clupeine is split by protaminase, active trypsin and by chymotrypsin. Compds of clupeine with nucleic acids are described by Kossel, *loc. cit.*

2408. Cnicin. [24394-09-0] (3R)-3,4-Dihydroxy-2-methylenebutanoic acid (3aR,4S,6E,10Z,11aR)-2,3,3a,4,5,8,9,11a-octahydro-10-(hydroxymethyl)-6-methyl-3-methylene-2-oxocyclodeca[b]-furan-4-yl ester; (Z,E)-6α,8α,15-trihydroxygermacra-1(10),4,11-(13)-trien-12-oic acid 12,6-lactone 8-(3,4-dihydroxy-2-methylenebutyrate); cynisin; centaurin. $C_{20}H_{26}O_7$; mol wt 378.42. C 63.48%, H 6.93%, O 29.59%. Bitter principle of *Cnicus benedictus* L., *Compositae*. Isolation and review: Korte, Bechmann, *Naturwissenschaften* **45**, 390 (1958). Structure: M. Suchy et al., *Tetrahedron Lett.* **1**, 5 (1959); *Ber.* **93**, 2449 (1960). Revised structure: Z. Samek et al., *Tetrahedron Lett.* **1969**, 2931. Stereochemistry: K. Tori et al., *J. Chem. Soc. B* **1971**, 1084.

Crystals, mp 143°. $[\alpha]_D^{20}$ +158°(c = 2.3 in ethanol). uv max: 220 nm (log ε 4.34). Slightly sol in water; freely sol in alcohol. Practically insol in ether.

2409. Coal Tar. Clinitar; Psorigel; T/Gel. A by-product of the carbonization of coal to produce coke or natural gas. *Constit.* Benzene, toluene, naphthalene, anthracene, xylene, and other aromatic hydrocarbons; phenol, cresol, and other phenol bodies; ammonia, pyridine, and some other organic bases; thiophene. Monographs: *Coal Tar Data Book*, 2nd ed., 1965, compiled and published by The Coal Tar Res. Assoc., Gomersal (Leeds), England; H. G. Franck, G. Collin, *Steinkohlenteer* (Springer Verlag, Heidelberg-New York, 1968) 255 pp. Review of carcinogenic risk: *IARC Monographs* **35**, 83-159 (1985); of toxicology and human exposure: *Toxicological Profile for Wood Creosote, Coal Tar Creosote, Coal Tar, Coal Tar Pitch, and Coal Tar Pitch Volatiles* (PB2003-100136, 2002) 394 pp.
Almost black, thick liquid or semisolid; characteristic odor. *Flammable.* A small portion of coal tar dissolves in water; all or almost all dissolves in benzene or nitrobenzene; partly dissolves in alcohol, ether, chloroform, carbon disulfide, methanol, acetone, petr ether, or sodium hydroxide soln. Practically insol in water. Sol in 20 parts alcohol; miscible with abs alcohol, acetone, petrolatum, oils and fats.
Purified coal tar. Pixalbol. Light-yellow, thin, oily liq.
Caution: Coal tar and coal tar pitches are listed as known human carcinogens: *Report on Carcinogens, Twelfth Edition* (PB2011-111646, 2011) p 111.
USE: Prodn of creosote, coal tar pitch, crude naphthalene and anthracene oils; as fuel for furnaces in steel industry.
THERAP CAT: Antieczematic (topical).
THERAP CAT (VET): Topically in skin disorders.

2410. Cob(II)alamin. [14463-33-3] Vitamin B_{12}-Co(II); 5,6-dimethyl-1-(3-O-phosphono-α-D-ribofuranosyl)-1H-benzimidazole monoester with cobinamide-Co(II) inner salt; vitamin B_{12r}; reduced vitamin B_{12}. $C_{62}H_{88}CoN_{13}O_{14}P$; mol wt 1329.37. C 56.02%, H 6.67%, Co 4.43%, N 13.70%, O 16.85%, P 2.33%. A one-electron reduction product of vitamin B_{12} containing a 5-coordinate 2^+ cobalt atom. Undergoes a further one-electron reduction to cob(I)alamin. Intermediate in biosynthesis of cobamamide and in cobalamin-dependent reactions. Prepn: H. Diehl, R. Murie, *Iowa State Coll. J.*

Sci. **26**, 555 (1952); R. O. Brady, H. A. Barker, *Biochem. Biophys. Res. Commun.* **4**, 373 (1961); H.-U. Blaser, J. Halpern, *J. Am. Chem. Soc.* **102**, 1684 (1980). Valence of reduced forms of B$_{12}$: H. A. O. Hill *et al.*, *Chem. Ind. (London)* **1964**, 197. Crystal structure: B. Kräutler *et al.*, *J. Am. Chem. Soc.* **111**, 8936 (1989); of biosynthetic intermediates: M. St. Maurice *et al.*, *Biochemistry* **47**, 5755 (2008). Review of prepns: D. Dolphin, *Methods Enzymol.* **18**, (Pt. C), 34-52 (1971); of role in enzymatic reactions: J. M. Pratt, *Pure Appl. Chem.* **65**, 1513 (1993); G. H. Reed, S. O. Mansoorabadi, *Curr. Opin. Struct. Biol.* **13**, 716-721 (2003); M. Yamanishi *et al.*, *Trends Biochem. Sci.* **30**, 304-308 (2005).

Dark brown amorphous solid. Oxidized by air to aquocobalamin. Absorption max (H$_2$O): 474, 404, 312, 289, 260 nm (log ε 3.99, 3.88, 4.46, 4.29, 4.25).

Cob(I)alamin. [18534-66-2] Cobinamide-*Co*(I) dihydrogen phosphate (ester) 3'-ester with 5,6-dimethyl-1-α-D-ribofuranosyl-1*H*-benzimidazole; hydridocobalamin; vitamin B$_{12s}$. C$_{62}$H$_{89}$CoN$_{13}$-O$_{14}$P; mol wt 1330.38. Obtained as a gray-green product in solution.

2411. Cobalt. [7440-48-4] Co; at. wt 58.933195; at. no. 27; valences 1, 2, 3; rarely 4, 5. Group VIII (9). One naturally occurring isotope: ^{59}Co; artificial, radioactive isotopes: 54-58; 60-64. Widely distributed in nature; abundance in earth's crust 0.001-0.002%. Principle ores include *cobaltite* (CoS$_2$.CoAs$_2$), *linnaeite* (Co$_3$S$_4$), *smaltite* (CoAs$_2$) and erythrite (3CoO.As$_2$O$_5$.8H$_2$O). Metal first isolated in 1735 by Brandt. Reviews of prepn: Whittemore in *Rare Metals Handbook*, C. A. Hampel, Ed. (Reinhold, New York, 1956) pp 105-146; Houot, *Ann. Mines* **1969** (April), 9-36. Prepn of high purity metal: Ware in *Ultrapurification of Semiconductor Materials*, M. S. Brooks, J. K. Kennedy, Eds. (Macmillan, New York, 1962) pp 192-204. Cobalt appears to be essential to life. Plays an important part in animal nutrition; the absence of cobalt-contg vitamin B$_{12}$ causes pernicious anemia. The reactor-produced ^{60}Co (T$_{1/2}$ 5.263 years; β^- 0.314 Mev; γ 1.173, 1.332 Mev) is a widely used source of radioactivity: Centre d'Information du Cobalt, *Cobalt Monograph* (Brussels, 1960) 515 pp. Comprehensive reviews of cobalt and its compds: *ACS Monograph Series* **no. 149**, entitled "Cobalt, Its Chemistry, Metallurgy and Uses," R. S. Young, Ed. (Reinhold, New York, 1960) 424 pp; Nicholls in *Comprehensive Inorganic Chemistry* vol. 3, J. C. Bailar, Jr. *et al*, Eds. (Pergamon Press, Oxford, 1973) pp 1053-1107; F. Planinsek, J. B. Newkirk in *Kirk-Othmer Encyclopedia of Chemical Technology* vol. 6 (Wiley-Interscience, New York, 3rd ed., 1979) pp 481-494. Review of toxicology and human exposure: *Toxicological Profile for Cobalt* (PB2004-104398, 2004) 486 pp.

Gray, hard, magnetic, ductile, somewhat malleable metal. Exists in two allotropic forms. At room temp the hexagonal form is more stable than the cubic form; both forms can exist at room temperature. Stable in air or toward water at ordinary temp. d 8.92. mp 1493°. bp about 3100°. Brinell hardness: 125. Latent heat of fusion 62 cal/g, latent heat of vaporization 1500 cal/g. Specific heat (15-100°): 0.1056 cal/g/°C. Readily sol in dil HNO$_3$; very slowly attacked by HCl or cold H$_2$SO$_4$. The hydrated salts of cobalt are red, and the sol salts form red solns which become blue on adding concd HCl.

^{60}Co-Labeled form. Produced by irradiation of ^{59}Co: ^{59}Co(n, $\gamma)^{60}$Co.

Caution: Potential symptoms of overexposure to metal, dust or fumes are coughing, dyspnea; wheezing and decreased pulmonary function; weight loss; dermatitis; diffuse nodular fibrosis; respiratory hypersensitivity; asthma. *See NIOSH Pocket Guide to Chemical Hazards* (DHHS/NIOSH 97-140, 1997) p 74.

USE: For superalloys in aircraft engines; magnetic alloys and alloys requiring hardness, wear resistance and corrosion resistance. Binder for tungsten carbide cutting tools. In lithium ion cell (LiCoO$_2$); as addition to Ni/Cd battery. ^{60}Co to sterilize medical instruments and devices; as x-ray source for food irradiation; in industrial radiography for non-destructive testing of high stress alloy parts; to crosslink, graft and degrade plastics. For manuf of cobalt salts; in nuclear technology. In the cobalt bomb, a hydrogen bomb surrounded by a cobalt metal shell; considered a "dirty bomb" because of long half-life and intense β^- and γ radiation.

THERAP CAT: Trace mineral; ^{60}Co as antineoplastic (radiation source).

2412. Cobaltic Acetate. [917-69-1] Acetic acid cobalt(3+) salt (3:1); cobalt triacetate. C$_6$H$_9$CoO$_6$; mol wt 236.07. C 30.53%, H 3.84%, Co 24.96%, O 40.66%. Co(C$_2$H$_3$O$_2$)$_3$. Prepd by electrolytic oxidation of Co(C$_2$H$_3$O$_2$)$_2$.4H$_2$O in glacial acetic acid contg 2% (v/v) water: Sharp, White, *J. Chem. Soc.* **1952**, 110; by oxidation of solns of cobaltous salts by alkaline persulfates in the presence of acetic acid: Peschanski, Wormser, *Compt. Rend.* **252**, 1607 (1961). Generally considered to be a hydroxo-bridged binuclear complex.

Dark-green, very hygroscopic powder or green octahedral crystals. Becomes black and dec on heating to 100°. Sol in water, acetic acid, alcohol, *n*-butanol; dec by mineral acids. Aq solns hydrolyze slowly at room temp, rapidly at 60-70°; reduced by light or FeSO$_4$.

USE: Catalyst for cumene hydroperoxide decompn.

2413. Cobaltic-Cobaltous Oxide. [1308-06-1] Cobalt oxide (Co$_3$O$_4$); cobalt tetraoxide; cobalto-cobaltic oxide; cobaltosic oxide; tricobalt tetroxide. Co$_3$O$_4$; mol wt 240.80. Co 73.42%, O 26.58%. Prepn: *Gmelins, Cobalt* (8th ed.) **58** (Part A), 231-235 (1932) and supplement, 202, 491-496 (1961). *Review:* de Bie, Doyen, *Cobalt* **15**, 3-13; **16**, 3-15 (1962).

Black or grey octahedral crystals of cubic system. d 6.11. Commercial material is black and contains about 71% Co. Above 900° loses O$_2$ to form CoO; absorbs O$_2$ at lower temps but cryst structure is unchanged. Absorbs water but no definite hydrate has been identified. Readily reduced to Co metal by C, CO, or H$_2$. Practically insol in water. Sol in acids, alkalies.

USE: In enamels; in semiconductors; in grinding wheels.

2414. Cobaltic Fluoride. [10026-18-3] Cobalt fluoride (CoF$_3$); cobalt trifluoride. CoF$_3$; mol wt 115.93. Co 50.84%, F 49.16%. Prepn from F$_2$, and CoF$_2$, CoCl$_2$, or Co$_2$O$_3$: Priest, *Inorg. Synth.* **3**, 175 (1950); Kwasnik in *Handbook of Preparative Inorganic Chemistry* vol. 1, G. Brauer, Ed. (Academic Press, New York, 2nd ed., 1963) p 268; from ClF$_3$ and CoCl$_2$: Rochow, Kukin, *J. Am. Chem. Soc.* **74**, 1615 (1952).

Minute, light brown, hexagonal crystals, d 3.88. Discolors rapidly on exposure to moist air. Reacts with water giving off O$_2$. Comparatively stable to heat, at 600° the fluorine pressure over the solid is less than 0.1 atm. Volatilizes in an F$_2$ stream at 600-700°. May be stored in hermetically sealed glass, quartz, or metal containers.

USE: Important fluorinating agent, particularly for complete fluorination of hydrocarbons by the Fowler process.

2415. Cobaltic Oxide Monohydrate. [12016-80-7] Cobalt hydroxide oxide (Co(OH)O); cobalt oxyhydroxide; cobaltic hydroxide. CoHO$_2$; mol wt 91.94. Co 64.10%, H 1.10%, O 34.80%. CoO-(OH). Alternate formula Co$_2$O$_3$.H$_2$O. Prepn: Glemser in *Handbook of Preparative Inorganic Chemistry* vol 2, G. Brauer, Ed. (Academic Press, New York, 2nd ed., 1965) pp 1520-1521; Schrader, Petzold, *Z. Anorg. Allg. Chem.* **353**, 186 (1967). Existence of anhydrous cobaltic oxide (Co$_2$O$_3$) has not been established: Pagel *et al.*, *J. Am. Chem. Soc.* **57**, 2552 (1935); Nicholls in *Comprehensive Inorganic Chemistry* vol 3, J. C. Bailar, Jr. *et al.*, Eds. (Pergamon Press, Oxford, 1973) p 1096. *See also*: de Bie, Doyen, *Cobalt* **15**, 3 (1962). Structure: Schrader, Petzold, *loc. cit.*; Delaplane *et al.*, *J. Chem. Phys.* **50**, 1920 (1969).

Dark-brown to black powder; hexagonal crystal structure. Converted to Co$_3$O$_4$ on heating to 148-150° in a vacuum. Practically insol in water. Sol in HCl, evolving Cl$_2$; sol in HNO$_3$, H$_2$SO$_4$.

USE: Oxidation catalysts, in separation of cobalt from nickel.

2416. Cobaltic Potassium Nitrite. [13782-01-9] (*OC*-6-11)-Hexakis(nitrito-κ*N*)cobaltate(3−) potassium (1:3); hexakis(nitrito-*N*)cobaltate(3−) tripotassium; potassium hexanitrocobaltate(III); potassium cobaltinitrite; potassium nitrocobaltate(III); cobalt yellow; Fischer's yellow; C.I. Pigment Yellow 40; C.I. 77357. CoK$_3$N$_6$O$_{12}$; mol wt 452.26. Co 13.03%, K 25.94%, N 18.58%, O 42.45%. K$_3$-Co(NO$_2$)$_6$. Incorrectly called *"Indian Yellow"*. Prepd by addition of KNO$_2$ to a solution of a Co salt: Salyer, Sweet, *Anal. Chem.* **32**, 548 (1962).

Sesquihydrate, yellow, octahedral, cubic crystals. Very slightly sol in water, dil acetic acid. Practically insol in alcohol; dec by mineral acids.

USE: As an oil- and water-color pigment; in painting on glass and porcelain; in coloring rubber; in separation of Co from Ni; in Co analysis.

2417. Cobaltous Acetate. [71-48-7] Acetic acid cobalt(2+) salt (2:1); cobalt(II) acetate. C$_4$H$_6$CoO$_4$; mol wt 177.02. C 27.14%, H 3.42%, Co 33.29%, O 36.15%. Co(C$_2$H$_3$O$_2$)$_2$. Prepd commercially from cobaltous hydroxide or carbonate and an excess of dil acetic acid: Morral in *Kirk-Othmer Encyclopedia of Chemical Technology* vol. 5 (Interscience, New York, 2nd ed., 1964) p 737; prepn from powdered Co and acetic acid: Hahl, *US 3133942* (1964 to BASF); prepn by oxidation of Co carbonyls in the presence of acetic acid: Gwynn *et al.*, *US 3246024* (1966 to Gulf). *Review:* de Bie, Doyen, *Cobalt* **15**, 3-13; **16**, 3-15 (1962).

Light-pink crystals. Readily sol in water.

Tetrahydrate. [6147-53-1] Bis(acetato)tetraaquocobalt. Co(C$_2$-H$_3$O$_2$)$_2$.4H$_2$O; mol wt 249.08. Intense red, monoclinic, prismatic crystals. d 1.705. On heating becomes anhydrous by 140°. Sol in water, alcohols, dil acids, pentyl acetate. pH of 0.2 molar aq soln 6.8.

USE: Bleaching agent and drier for lacquers, varnishes; in anodizing; catalyst for oxidation and esterification; foam stabilizer for malt beverages.

2418. Cobaltous Arsenate. [24719-19-5] Arsenic acid (H$_3$-AsO$_4$) cobalt(2+) salt (2:3); cobalt arsenate (Co$_3$(AsO$_4$)$_2$). As$_2$-Co$_3$O$_8$; mol wt 454.63. As 32.96%, Co 38.89%, O 28.15%. Co$_3$-(AsO$_4$)$_2$. Octahydrate occurs in nature as *erythrite* or *cobalt bloom*. Prepn: *Gmelins, Cobalt* (8th ed.) **58**, (part A), 305 (1932) and supplement, 752 (1961); Charles-Messance *et al.*, *Bull. Soc. Chim. Fr.* **1962**, 574. *See Colour Index* vol. 4 (3rd ed., 1971) p 4664.

Octahydrate. [7785-24-2] C.I. 77350. As$_2$Co$_3$O$_8$.8H$_2$O; mol wt 598.75. Pink to blood-red, monoclinic, fine needles. On heating becomes anhydr by 400°. Dec by 1000° to Co$_6$As$_2$O$_{11}$. d 2.9-3.1. Practically insol in water. Sol in dil mineral acids, in NH$_4$OH.

USE: Painting on glass and porcelain.

2419. Cobaltous Bromide. [7789-43-7] Cobalt bromide (CoBr$_2$); cobalt dibromide. Br$_2$Co; mol wt 218.74. Br 73.06%, Co 26.94%. CoBr$_2$. Prepn of hexahydrate from CoCO$_3$ and HBr: Clark, Buchner, *J. Am. Chem. Soc.* **44**, 230 (1922). Prepn of anhydr: *eidem, loc. cit.*; Watt *et al.*, *ibid.* **77**, 2752 (1955); Wydeven, Gregory, *J. Phys. Chem.* **68**, 3249 (1964).

Bright green solid or lustrous green cryst leaflets. mp 678° (under HBr and N$_2$); d$_4^{25}$ 4.909. Hygroscopic, forms hexahydrate in air. Readily sol in water, methanol, ethanol, acetone, methyl acetate.

Hexahydrate. Red to reddish-purple, deliquesc, prismatic crystals. mp 47-48°. d$_4^{25}$ 2.46. Loses 4H$_2$O at 100° giving the purple dihydrate, and all H$_2$O by 130°. Sol in water giving red or blue soln depending on concn and temp, in methanol giving red soln, in ethanol, acetone, ether, methyl acetate giving blue solns. *Keep well closed.*

USE: Chiefly in hygrometers; also in catalysts for organic reactions.

2420. Cobaltous Carbonate. [513-79-1] Carbonic acid cobalt(2+) salt (1:1). CCoO$_3$; mol wt 118.94. C 10.10%, Co 49.55%, O 40.35%. CoCO$_3$. Occurs in nature as the mineral *cobalt spar* or *sphaerocobaltite*. Prepd by heating a soln of a cobaltous salt with Na$_2$CO$_3$: Schlessinger, *Inorg. Synth.* **6**, 189 (1963) where it is the starting material for the prepn of trinitrotriamminecobalt. *Review:* de Bie, Doyen, *Cobalt* **15**, 3-13; **16**, 3-15 (1962).

Red powder or rhombohedral crystals. d 4.13. Almost insol in water, alcohol, methyl acetate. Does not react with cold concd HNO$_3$ or HCl; when heated, dissolves with evolution of CO$_2$. Oxidized by air or weak oxidizing agents to cobaltic carbonate.

Cobaltous carbonate basic. [12602-23-2] Cobalt carbonate hydroxide. C$_2$H$_6$Co$_5$O$_{12}$; mol wt 516.72. Co$_5$(OH)$_6$(CO$_3$)$_2$. Pale-red powder, usually containing some water. Practically insol in water; sol in dilute acids and ammonia.

USE: In ceramics; manuf of Co pigments; prepn of Co compds.

THERAP CAT (VET): Nutritional factor. Used in cobalt deficiency in ruminants.

2421. Cobaltous Chloride. [7646-79-9] Cobalt chloride (CoCl$_2$); cobalt dichloride; cobalt(II) chloride. Cl$_2$Co; mol wt 129.83. Cl 54.61%, Co 45.39%. CoCl$_2$. Prepn of anhydr from Co powder and Cl$_2$: Osthoff, West, *J. Am. Chem. Soc.* **76**, 4732 (1954); from the acetate and acetyl chloride: Watt *et al.*, *ibid.* **77**, 2752 (1955); by dehydration of the hexahydrate with SOCl$_2$: Hecht, *Z. Anorg. Chem.* **254**, 51 (1947). Prepn of the hexahydrate by treating an aqueous soln of a cobaltous salt with HCl: *ACS Monograph Series* **no. 149**, entitled "Cobalt - Its Chemistry, Metallurgy, and Uses," R. S. Young, Ed. (Reinhold, New York, 1960) p 76. *Review:* de Bie, Doyen, *Cobalt* **15**, 3-13; **16**, 3-15 (1962). Toxicity studies: G. J. A. Speijers *et al.*, *Food Chem. Toxicol.* **20**, 311 (1982); P. P. Singh, A. Y. Junnarkar, *Indian J. Pharmacol.* **23**, 153 (1991). Review of toxicology: B. Venugopal, T. D. Luckey, *Environ. Qual. Safety* Suppl. 1, 4-73 (1975).

Pale-blue hygroscopic leaflets; colorless in very thin layers; turns pink on exposure to moist air. mp 735°; bp 1049°; d$_4^{25}$ 3.367. Dec 400° on long heating in air. Sublimes at 500° in HCl gas, forming iridescent, fluffy, colorless cryst. Sol in water, alcohols, acetone, ether, glycerol, pyridine. LD$_{50}$ in mice, rats (mg/kg): 360.0, 171.0 orally; 92.6, 36.9 i.p.; 23.3, 4.3 i.v. (Singh, Junnarkar).

Hexahydrate. [7791-13-1] CoCl$_2$.6H$_2$O; mol wt 237.92. Structure reported to be [CoCl$_2$(H$_2$O)$_4$].2H$_2$O: Mizuno *et al.*, *J. Phys. Soc. Jpn.* **14**, 383 (1959), *C.A.* **53**, 14630i (1959). Pink to red, slightly deliquesc, monoclinic, prismatic crystals. mp 87°; d^{20} 1.924. On heating loses 4H$_2$O at 52-56° forming the dihydrate, violet or blue crystals, d$_4^{25}$ 2.477, stable unless exposed directly to moisture. Loses another H$_2$O by 100°, giving monohydrate, violet, hygroscopic, amorphous solid or needles. Remaining H$_2$O lost at 120-140°. Sol in water, alcohols, acetone, ether, glycerol. pH of 0.2 molar aq soln 4.6. The aq soln is pink to red, but turns blue when heated or when HCl or H$_2$SO$_4$ is added. *Keep well closed.* LD$_{50}$ orally in rats: 766 mg/kg (Speijers).

Caution: Large amounts of CoCl$_2$ depress erythrocyte production. May lead to death in children. Other effects include cutaneous flushing, chest pains, dermatitides, tinnitus, nausea and vomiting, nerve deafness, thyroid hyperplasia, myxedema, congestive heart failure. *See*: E. Beutler *et al.*, *Clinical Disorders of Iron Metabolism* (Grune & Stratton, New York, 1963) pp 175-178.

USE: Invisible ink; humidity and water indicator; color standard; in hygrometers; temp indicator in grinding; in electroplating; for painting on glass and porcelain; prepn of catalysts; fertilizer and feed additive; foam stabilizer in beer; as absorbent for military poison gas and ammonia; in manuf of vitamin B$_{12}$. Radioactive cobalt chloride, ^{57}CoCl$_2$ (half-life 271.79 days, pure gamma emitter) used in Mössbauer effect (nuclear clock).

THERAP CAT: Hematinic.

THERAP CAT (VET): Nutritional factor. Used in cobalt deficiency in ruminants.

2422.   Cobaltous Chromate(III). [12016-69-2] Chromium cobalt oxide (Cr_2CoO_4); cobalt chromite. $CoCr_2O_4$; mol wt 226.92. Co 25.97%, Cr 45.83%, O 28.20%. Prepn: *Gmelins, Cobalt* (8th ed.) **58**, (part A), 479 (1932) and supplement, 874-876 (1961).

Brilliant greenish-blue powder having a cubic spinel structure. Almost insol in concd HCl and HNO_3.

USE: Green pigment for ceramics.

2423.   Cobaltous Cyanide. [542-84-7] Cobalt cyanide (Co-$(CN)_2$); cobalt dicyanide. C_2CoN_2; mol wt 110.97. C 21.65%, Co 53.11%, N 25.24%. Prepn: Ray, Sahu, *J. Indian Chem. Soc.* **23**, 161 (1946); *Gmelins, Cobalt* (8th ed.) **58**, (part A), 364 (1932) and supplement, 712 (1961). Structure reported as $Co_3[Co(CN)_6]_2$: P. S. Poskozim *et al.*, *J. Inorg. Nucl. Chem.* **35**, 687 (1973). Prepn and structure as $Co(CN)_2$: D. M. S. Mosha, D. Nicholls, *Inorg. Chim. Acta* **38**, 127 (1980).

Deep-blue, very hygroscopic powder. d_4^{25} 1.872.

Di- to trihydrate.   Pink to reddish-brown powder or needles. Practically insol in water, acids, methyl acetate; sol in alkali cyanide solns.

USE: In cobalt catalysts.

2424.   Cobaltous Fluoride. [10026-17-2] Cobalt fluoride (CoF_2); cobalt(II) fluoride; cobalt difluoride. CoF_2; mol wt 96.93. Co 60.80%, F 39.20%. Prepd by the action of HF on anhyd $CoCl_2$: Kwasnik in *Handbook of Preparative Inorganic Chemistry* vol. **1**, G. Brauer, Ed. (Academic Press, New York, 2nd ed., 1963) p 267; on $CoCO_3$: Clark, Buchner, *J. Am. Chem. Soc.* **44**, 230 (1922); on Co: Muetterties, Castle, *J. Inorg. Nucl. Chem.* **18**, 148 (1961).

Rosy-red tetragonal crystals. mp 1100-1200°, forming a red liq. Volatilizes at about 1400°. d 4.43. Sparingly sol in water; readily sol in warm mineral acids. Forms di-, tri-, and tetrahydrates, all sol in water; their aq solns are dec by boiling, forming the oxyfluoride $CoF_2.CoO.H_2O$.

USE: Catalyst for organic reactions.

2425.   Cobaltous Formate. [544-18-3] Formic acid cobalt-(2+) salt (2:1). $C_2H_2CoO_4$; mol wt 148.97. C 16.13%, H 1.35%, Co 39.56%, O 42.96%. Co(HCOO)₂. Prepn: *Gmelins, Cobalt* (8th ed.) **58** (part A), 350 (1932) and supplement, 702 (1961).

Dihydrate.   Red, cryst powder. d_4^{22} 2.13. Sol in water; almost insol in alcohol. Becomes anhyd at 140°.

USE: In prepn of Co catalysts.

2426.   Cobaltous Hydroxide. [21041-93-0] Cobalt hydroxide ($Co(OH)_2$). CoH_2O_2; mol wt 92.95. Co 63.40%, H 2.17%, O 34.42%. $Co(OH)_2$. Prepd from a solution of a cobaltous salt and an alkali hydroxide: Glemser in *Handbook of Preparative Inorganic Chemistry* vol. **2**, G. Brauer, Ed. (Academic Press, New York, 2nd ed., 1965) p 1521; Weiser, Milligan, *J. Phys. Chem.* **36**, 722 (1932). *Review:* de Bie, Doyen, *Cobalt* **15**, 3-13 (1962); *ibid.* **16**, 3-15.

Blue-green or rose-red powder or microscopic rhombohedral crystals; red form is the more stable of the two. d_4^{15} 3.597. Easily oxidized by air or weak oxidizing agents to $Co(OH)_3$. Amphoteric. Loses water on heating, forming CoO at 168° *in vacuo*. Very slightly sol in water; readily sol in acids; practically insol in dil alkalies; sol in ammonia.

USE: Manuf of Co compds; drier for paints; in enhancing drying properties of lithographic printing inks; in storage battery electrode impregnating solns.

2427.   Cobaltous Iodide. [15238-00-3] Cobalt iodide (CoI_2); cobalt diiodide. CoI_2; mol wt 312.74. Co 18.84%, I 81.16%. Prepn: Clark, Buchner, *J. Am. Chem. Soc.* **44**, 230 (1922); Chaigneau, *Bull. Soc. Chim. Fr.* **1957**, 886; Chaigneau, Chastagnier, *ibid.* **1958**, 1192; Glemser in *Handbook of Preparative Inorganic Chemistry* vol. **2**, G. Brauer, Ed. (Academic Press, New York, 2nd ed., 1965) p 1518.

The anhyd salt exists in two isomorphous forms. α-CoI_2: black, graphite-like solid. mp 515-520° (in high vacuum). d_4^{25} 5.584. Very hygroscopic, becomes blackish-green in air. Sol in water to give pink to red soln. β-CoI_2: ochre-yellow powder. Blackens at 400°

and converts to α-form. d_4^{25} 5.45. Very hygroscopic; deliquesc in moist air forming green droplets. Sol in water to give colorless soln which becomes pink on heating.

Hexahydrate.   Dark red hexagonal prisms. Loses H_2O on heating becoming anhyd by 130°. d 2.90. Loses I_2 on exposure to air and light. Sol in water to give soln which is red below 20°, olive green at 20 to 40°, and green at higher temps. Sol in ethanol (blue soln), ether (blue to green soln), chloroform (blue soln), acetone.

USE: Indicator for moisture and humidity; determination of water in organic solvents; catalyst for organic reactions.

2428.   Cobaltous Nitrate. [10141-05-6] Nitric acid cobalt-(2+) salt (2:1); cobalt(II) nitrate. CoN_2O_6; mol wt 182.94. Co 32.21%, N 15.31%, O 52.47%. $Co(NO_3)_2$. Prepn: *Gmelins, Cobalt* (8th ed.) **58** (part A), 252-262 (1932) and supplement, 515-521 (1961); Weigel *et al.*, *Bull. Soc. Chim. Fr.* **1964**, 836; Addison, Sutton, *J. Chem. Soc.* **1964**, 5553. Toxicity study: G. J. A. Speijers *et al.*, *Food Chem. Toxicol.* **20**, 311 (1982). Review of toxicology: B. Venugopal, T. D. Luckey, *Environ. Qual. Safety* Suppl. 1, 4-73 (1975). *Review:* de Bie, Doyen, *Cobalt* **15**, 3-13; **16**, 3-15 (1962).

Pale red powder. Dec at 100-105°. d 2.49. Sol in water. LD in rabbits (mg/kg): 250 orally, 75 s.c. (Venugopal, Luckey).

Hexahydrate.   [10026-22-9] $CoN_2O_6.6H_2O$; mol wt 291.03. Red, deliquesc, monoclinic crystals. mp ~55°. Red liquid becomes green and dec to the oxide above 74°. d 1.88. Very sol in water, alcohol, most organic solvents. *Keep well closed in a cool place.* LD_{50} orally in rats: 691 mg/kg (Speijers).

USE: Manuf of cobalt pigments and invisible inks; decorating stoneware and porcelain; prepn of catalysts; production of vitamin B_{12} supplements.

2429.   Cobaltous Oxalate. [814-89-1] Ethanedioic acid co-balt(2+) salt (1:1). C_2CoO_4; mol wt 146.95. C 16.35%, Co 40.10%, O 43.55%. CoC_2O_4. Prepn: Robin, *Bull. Soc. Chim. Fr.* **1953**, 1078; *Gmelins, Cobalt* (8th ed.) **58** (part A) p 355 (1932) and supplement, pp 704-706 (1961). *Review:* de Bie, Doyen, *Cobalt* **15**, 3-13; **16**, 3-15 (1962).

d_4^{25} 3.021. Readily absorbs moisture from air to form hydrates.

Dihydrate.   Light pink microcryst powder or needles. Almost insol in water; slightly sol in acids; almost insol in oxalic acid; freely sol in aq ammonia. Dec on heating with aq KOH or Na_2CO_3 soln.

Tetrahydrate.   Yellowish-pink amorphous powder. Effloresces on exposure to air. Loses water on heating to 100° giving the dihydrate. Very slightly sol in water; slightly sol in acids; readily sol in aq ammonia.

USE: Prepn of Co catalysts, Co metal powder for powder-metallurgical applications; stabilizer for HCN; temperature indicator.

2430.   Cobaltous Oxide. [1307-96-6] Cobalt oxide (CoO). CoO; mol wt 74.93. Co 78.65%, O 21.35%. Prepn: Amiel *et al.*, *Compt. Rend.* **259**, 3512 (1964); Wilke, *Z. Anorg. Allg. Chem.* **330**, 164 (1964). Toxicity study: Smyth *et al.*, *Am. Ind. Hyg. Assoc. J.* **30**, 470 (1969). *Review:* de Bie, Doyen, *Cobalt* **15**, 3-13; **16**, 3-15 (1962).

Powder, or cubic or hexagonal crystals. Color varies from olive green to red, depending on the particle size, but the commercial material is usually dark grey and contains about 76% Co. mp ~1935°. d 5.7 to 6.7, depending on method of prepn. Readily absorbs O_2 even at room temp. Practically insol in water. Sol in acids or alkalies. Easily reduced to Co by C or CO. Reacts at high temperatures with silica, alumina, zinc oxide to form pigments. LD_{50} orally in rats: 1.70 g/kg (Smyth).

Note: The commercial oxides are usually not definite chemical compds but mixtures of the cobalt oxides.

USE: In pigments for ceramics; glass coloring and decolorization; oxidation catalyst for drying oils, fast-drying paints and varnishes; prepn of cobalt-metal catalysts, Co powder for binder in sintered tungsten carbide; in semiconductors.

2431.   Cobaltous Phosphate. [13455-36-2] Phosphoric acid cobalt(2+) salt (2:3); C.I. Pigment Violet 14; C.I. 77360. $Co_3O_8P_2$; mol wt 366.74. Co 48.21%, O 34.90%, P 16.89%. $Co_3(PO_4)_2$. Prepn from $CoCl_2$ and $(NH_4)_2HPO_4$: Klement, Haselbeck, *Z. Anorg.*

Allg. Chem. **334**, 27 (1964); from Ca(H$_2$PO$_4$)$_2$: Vickery, **US 2914380** (1959 to Horizons). *Review:* de Bie, Doyen, *Cobalt* **15**, 3-13; **16**, 3-15 (1962). *See also Colour Index* vol. 4 (3rd ed., 1971) p 4665.

Octahydrate. Pink to lavender amorph powder. d 2.77. Practically insol in water. Sol in mineral acids.

USE: In ceramic pigments; in artists' colors, plastic resins.

2432. Cobaltous Sulfate. [10124-43-3] Sulfuric acid cobalt-(2+) salt (1:1). CoO$_4$S; mol wt 154.99. Co 38.02%, O 41.29%, S 20.69%. CoSO$_4$. Hexahydrate occurs in nature as the mineral *bieberite*. Prepn: Clark *et al.*, *J. Am. Chem. Soc.* **42**, 2483 (1920); Hammel, *Ann. Chim.* **11**, 247 (1939); *Gmelins, Cobalt* (8th ed.) **58**, (part A) 324-336 (1932) and supplement, 628-647 (1961). *Review:* de Bie, Doyen, *Cobalt* **15**, 3-13; **16**, 3-15 (1962).

Red to lavender dimorphic, orthorhombic crystals. d$_4^{25}$ 3.71. Stable to 708°. Dissolves slowly in boiling water.

Monohydrate. Rose-colored, monoclinic crystals. Structure reported to be Co(H$_2$SO$_5$). d$_4^{25}$ 3.08. Dissolves slowly in boiling water.

Heptahydrate. [10026-24-1] Pink to red monoclinic, prismatic crystals. On heating dehydrates to the hexahydrate (monoclinic, prismatic crystals) at 41.5°, and to the monohydrate at 71°. d$_4^{25}$ 2.03. Sol in water; slightly sol in methanol, ethanol.

Caution: Cobaltous sulfate is reasonably anticipated to be a human carcinogen: *Report on Carcinogens, Twelfth Edition* (PB2011-111646, 2011) p 113.

USE: Usual source of water-soluble cobalt since it is the most economical and it shows less tendency to deliquesc or dehydrate than the chloride or nitrate. Used in storage batteries; in Co-electroplating baths; as drier for lithographic inks, varnishes; in ceramics, enamels, glazes to prevent discoloring; in Co pigments for decorating porcelain.

2433. Cobaltous Sulfide. [1317-42-6] Cobalt sulfide (CoS). CoS; mol wt 90.99. Co 64.77%, S 35.23%. Prepn: Glemser in *Handbook of Preparative Inorganic Chemistry* vol. **2**, G. Brauer, Ed. (Academic Press, New York, 2nd ed., 1965) p 1523.

Exists in two forms. α-CoS: black, amorphous powder. Forms Co(OH)S in air. Sol in HCl. β-CoS: grey powder or reddish-silver octahedral crystals. mp >1100°; d 5.45. Practically insol in water; sol in acids.

USE: Catalyst for hydrogenation or hydrodesulfurization.

2434. Cobaltous Thiocyanate. [3017-60-5] Thiocyanic acid cobalt(2+) salt (2:1); cobaltous rhodanide; cobaltous sulfocyanate. C$_2$CoN$_2$S$_2$; mol wt 175.09. C 13.72%, Co 33.66%, N 16.00%, S 36.62%. Co(SCN)$_2$. Prepn: *Gmelins, Cobalt* (8th ed.) **58** (part A), 380 (1932) and supplement, 720-722 (1961); Schlessinger, *Inorganic Laboratory Preparations* (Chemical Publishing, New York, 1962) p 44.

Yellow-brown powder. Soluble in water to give a rose-colored soln; sol in ethanol, methanol, ether, acetone, CHCl$_3$ to give blue solns.

Trihydrate. Violet to violet-brown rhombic crystals; red in transmitted light. Sol in water to give blue soln which becomes pink on dilution; sol in ethanol, ether, acetone to give blue solns.

USE: As humidity indicator.

2435. Cobamamide. [13870-90-1] *Co*-(5′-Deoxyadenosin-5′-yl)cobinamide *f*-(dihydrogen phosphate) inner salt 3′-ester with (5,6-dimethyl-1-α-D-ribofuranosyl-1*H*-benzimidazole-κN^3); adenosylcobalamin; adenosyl-B$_{12}$; AdoCbl; cobamamidum; coenzyme B$_{12}$; DBC; 5′-deoxyadenosyl-B$_{12}$; 5′-deoxyadenosylcobalamin; dibencozide; dibenzcozamide; dimebenzcozamide; 5,6-dimethylbenzimidazolylcobamide 5′-deoxyadenosine; vitamin B$_{12}$ coenzyme; Ademide; Calomide; Cobaltamin S; Cobanzyme; Coenzile; Enzicoba; Heraclene; Indusil. C$_{72}$H$_{100}$CoN$_{18}$O$_{17}$P; mol wt 1579.61. C 54.75%, H 6.38%, Co 3.73%, N 15.96%, O 17.22%, P 1.96%. One of the metabolically active forms of vitamin B$_{12}$; characterized by a 5′-deoxyadenosyl nucleoside linked by a C to Co bond. Cofactor for several bacterial enzymes that catalyze carbon-skeleton rearrangements, heteroatom eliminations, and amino group migrations. Required by mammals as coenzyme for methylmalonyl-CoA mutase in

the catabolism of certain amino acids and fatty acids. Discovery: H. A. Barker *et al.*, *Proc. Natl. Acad. Sci. USA* **44**, 1093 (1958). Isoln from *Clostridium tetanomorphum: idem et al., J. Biol. Chem.* **235**, 181 (1960); from *Propionibacterium shermanii: eidem, ibid.* 480. Prepn from a cobalamin-thiol complex: M. Murakami *et al.*, **US 3461114** (1969 to Yamanouchi). Crystal structure: P. G. Lenhert, D. C. Hodgkin, *Nature* **192**, 937 (1961). Review of uptake and metabolism by humans and role in deficiency diseases: D. Watkins, D. S. Rosenblatt, *Endocrinologist* **11**, 98-104 (2001). Review of biosynthesis by bacteria: M. J. Warren *et al.*, *Nat. Prod. Rep.* **19**, 390-412 (2002); of mechanisms of B$_{12}$-dependent reactions: R. Banerjee, S. W. Ragsdale, *Annu. Rev. Biochem.* **72**, 209-247 (2003); R. Banerjee, *Chem. Rev.* **103**, 2083-2094 (2003); T. Toraya, *ibid.* 2095-2127 (2003).

Yellow-orange six-faced crystals which become deep red upon exposure to air. Absorption max (H$_2$O): 260, 375, 522 nm (A $\times$ 10^{-6} 34.7, 10.9, 8.0 cm^2/mole). Soly in water (24°): 16.4 mmol. Solns of pH <3.5 are yellow, >3.5, red. Sol in ethanol, phenol. Practically insol in acetone, ether, dichloroethylene, dioxane. pKa 3.5. Stability studies: Collado, Nieto, *Ann. Pharm. Fr.* **27**, 427 (1969). Highly sensitive to light, to cyanide and moderately sensitive to acid. Solns are most stable at pH 6-7 stored in the dark. Heating of acid or alkaline solns produces slow inactivation.

THERAP CAT: Enzyme cofactor; vitamin (hematopoietic).

THERAP CAT (VET): Vitamin (hematopoietic).

2436. Cobicistat. [1004316-88-4] (3*R*,6*R*,9*S*)-12-Methyl-13-[2-(1-methylethyl)-4-thiazolyl]-9-[2-(4-morpholinyl)ethyl]-8,11-dioxo-3,6-bis(phenylmethyl)-2,7,10,12-tetraazatridecanoic acid 5-thiazolymethyl ester; thiazol-5-ylmethyl [(1*R*,4*R*)-1-benzyl-4-[[(2*S*)-2-[[methyl-[[2-(1-methylethyl)thiazol-4-yl]methyl]carbamoyl]amino]-4-(morpholin-4-yl)butanoyl]amino]-5-phenylpentyl]carbamate; GS-9350. C$_{40}$H$_{53}$N$_7$O$_5$S$_2$; mol wt 776.03. C 61.91%, H 6.88%, N 12.63%, O 10.31%, S 8.26%. Selective inibitor of human cytochrome P450 3A (CYP3A); structurally similar to ritonavir, *q.v.*, with no inherent antiviral activity. Increases systemic exposure of coadministered antiretroviral drugs by blocking their metabolism by CYP3A enzymes. Prepn: M. C. Desai *et al.*, **WO 08010921**; *eidem*, **US 080207620** (both 2008 to Gilead Sci.); and structure-activity study: L. Xu *et al.*, *ACS Med. Chem. Lett.* **1**, 209-213 (2010). Clinical pharmacokinetics and pharmacodynamics: A. A. Mathias *et al.*, *Clin. Pharmacol. Ther.* **87**, 322 (2010). Clinical comparison with ritonavir as pharmacoenhancer in treatment of HIV infection: R. Elion *et al.*, *AIDS* **25**, 1881 (2011).

Soly in water (μg/ml): 75 at pH 7.4; >6500 at pH 2.2.
THERAP CAT: Pharmacokinetic enhancer for antiretroviral agents.

2437. Cobrotoxin. [12584-83-7] The main toxic protein in cobra venom. Crystallization and properties: Yang, *J. Biol. Chem.* **240**, 1616 (1965). Composed of a single peptide chain having 62 amino acid residues intramolecularly cross-linked by four disulfide bonds. Amino acid sequence: Yang *et al.*, *Biochim. Biophys. Acta* **188**, 65 (1969). Complete structure: *eidem, ibid.* **214**, 355 (1970). Synthetic study: H. Aoyagi *et al.*, *ibid.* **263**, 823 (1972). *Review:* C. C. Yang, "Biochemical Studies on the Toxic Nature of Snake Venom Cobrotoxin from Formosan Cobra Venom" in *Toxins of Animal and Plant Origin* I, A. de Vries, E. Kochva, Eds. (Gordon and Breach Science Publishers, New York, 1971) pp 205-236. *See also* A. T. Tu, *Venoms: Chemistry and Molecular Biology* (John Wiley and Sons, New York, 1977) pp 174-189.

2438. Coca. Erythroxylon; cuca; hayo; ipado. Dried leaves of *Erythroxylon coca* Lam., *Erythroxylaceae*. *Habit.* Bolivia, Brazil, Peru, cultivated in Java. *Constit.* 0.5-1% alkaloids in the South American, 1.5-2.5% in the Javanese leaves. In the former, the major alkaloid is cocaine; in the latter, there is very little cocaine, the alkaloids consisting chiefly of ecgonine derivatives such as benzoyl ecgonine, methyl ecgonine, etc. The Java leaves also contain small quantities of tropococaine which is apparently absent in the South American leaves. Other "cocaine" alkaloids present are truxillococaine, isatropylcocaine or cocamine, cocaicine.
Note: This is a controlled substance: **21 CFR,** 1308.12.
THERAP CAT: CNS stimulant.

2439. Cocaethylene. [529-38-4] (1*R*,2*R*,3*S*,5*S*)-3-(Benzoyloxy)-8-methyl-8-azabicyclo[3.2.1]octane-2-carboxylic acid ethyl ester; ecgonine ethyl ester benzoate (ester); *O*-benzoyl-*l*-ecgonine ethyl ester; ethylbenzoylecgonine; Homocaine. $C_{18}H_{23}NO_4$; mol wt 317.39. C 68.12%, H 7.30%, N 4.41%, O 20.16%. Homolog of cocaine. Prepn: Merck, *Ber.* **18**, 2952 (1885); Einhorn, *ibid.* **21**, 47 (1888).

Prisms from alcohol, mp 109°. Almost insol in water. Sol in alcohol, ether.
THERAP CAT: Anesthetic (local).

2440. Cocaine. [50-36-2] (1*R*,2*R*,3*S*,5*S*)-3-(Benzoyloxy)-8-methyl-8-azabicyclo[3.2.1]octane-2-carboxylic acid methyl ester; 3β-hydroxy-1α*H*,5α*H*-tropane-2β-carboxylic acid methyl ester benzoate; 2β-carbomethoxy-3β-benzoxytropane; ecgonine methyl ester benzoate; *l*-cocaine; β-cocaine; benzoylmethylecgonine. $C_{17}H_{21}NO_4$; mol wt 303.36. C 67.31%, H 6.98%, N 4.62%, O 21.10%. From the leaves of *Erythroxylon coca* Lam. and other species of *Erythroxylon, Erythroxylaceae* or by synthesis. Extraction proce-

dure: Squibb, *Pharm. J.* [3] **15**, 775, 796; **16**, 67 (1885); Emde in *Ullmanns Encyklopädie der technischen Chemie* Schwyzer, *Die Fabrikation pharmazeutischer und chemisch-technischer Produkte* (Berlin, 1931). Configuration: Findlay, *J. Am. Chem. Soc.* **76**, 2855 (1954); O. Kovacs *et al.*, *Helv. Chim. Acta* **37**, 892 (1954). Synthesis: R. Willstätter *et al.*, *Ann.* **434**, 111 (1923). Stereospecific synthesis of *dl*-form: J. J. Tufariello *et al.*, *Tetrahedron Lett.* **1978**, 1733; *eidem, J. Am. Chem. Soc.* **101**, 2435 (1979). Biosynthesis: E. Leete, S. H. Kim, *J. Am. Chem. Soc.* **110**, 2976 (1988). Absorption spectrum: J. J. Dobbie, J. J. Fox, *J. Chem. Soc.* **103**, 1193 (1913); Fischer, *Arch. Exp. Pathol. Pharmakol.* **170**, 610 (1933). Toxicity: C. L. Rose *et al.*, *J. Lab. Clin. Med.* **15**, 731 (1930). Vapor pressure studies: A. H. Lawrence *et al.*, *Can. J. Chem.* **62**, 1886 (1984). Comprehensive description: F. J. Muhtadi, A. A. Al-Badr, *Anal. Profiles Drug Subs.* **15**, 151-231 (1986). Review of pharmacology and pharmacotherapies for abuse: M. Rocío *et al.*, *Bioorg. Med. Chem.* **12**, 5019-5030 (2004).

Monoclinic tablets from alcohol, mp 98°. Volatile, esp above 90°, but the sublimate is not crystalline. bp$_{0.1}$ 187-188°. $[\alpha]_D^{18}$ −35° (50% alcohol); $[\alpha]_D^{20}$ −16° (c = 4 in chloroform). Aq solns are alkaline to litmus. pKa (15°) 8.61. pKb (15°) 5.59. One gram dissolves in 600 ml water, 270 ml water at 80°, 6.5 ml alcohol, 0.7 ml chloroform, 3.5 ml ether, 12 ml oil turpentine, 12 ml olive oil, 30-50 ml liquid petrolatum. Also sol in acetone, ethyl acetate, carbon disulfide; sparingly sol in mineral oil. LD$_{50}$ i.v. in rats: 17.5 mg/kg (Rose).
Hydrochloride. [53-21-4] Cocaine muriate. $C_{17}H_{21}NO_4 \cdot HCl$. Colorless to white crystals or white crystalline powder. Saline, slightly bitter taste; numbs tongue and lips. mp ~195°. $[\alpha]_D$ −72° (c = 2 in aq soln pH 4.5). One gram dissolves in 0.4 ml water; 3.2 ml cold, 2 ml hot water; 12.5 ml chloroform. Also sol in glycerol, acetone. Insol in ether or oils. Avoid heat in preparing soln as it decomposes. Preserve in well-closed, light-resistant containers.
Note: This is a controlled substance: **21 CFR,** 1308.12.
THERAP CAT: Anesthetic (local).
THERAP CAT (VET): Topical anesthetic (ophthalmic).

2441. Cocculus. Fish-berry; Indian berry; *Cocculus indicus*; oriental berry. Dried fruit of *Anamirta cocculus* (L.) Wight & Arn., *Menispermaceae*. *Habit.* East Indies, Malay Archipelago. *Constit.* Menispermine, paramenispermine, about 1% picrotoxin, picrotoxic acid, cocculine alkaloid, about 50% fat. *Poisonous!*
THERAP CAT: CNS and respiratory stimulant.

2442. Cochineal. Scale insect, *Dactylopius coccus* Costa, formerly known as *Coccus cacti* L. *Habit.* Mexico, Central America; cultivated in West Indies, Canary Islands, Algiers, and Southern Spain. Parasitic to certain species of prickly pear cactus. Females produce a glycosidic red pigment, carminic acid, *q.v.* Extraction methods and determn of constituents by HPLC: M. Gonzalez *et al.*, *J. Agric. Food Chem.* **50**, 6968 (2002). Review of prepn and uses in foods: J. Schul, *IFT Basic Symp. Ser.* **14**, 1-10 (2000); of chemistry and use as histological stain: R. W. Dapson, *Biotech. Histochem.* **82**, 173-187 (2007).
Cochineal extract. [1343-78-8] Prepd from dried, gravid females using water or aq alcohol as the extraction solvent. *Constit.* 2-4% carminic acid; soluble proteins; carbohydrates; ionic salts. Magenta red, water-soluble color. Becomes orange at pH 4.
Carmine. [1390-65-4] Prepd from cochineal extract using aluminum hydroxide to form an insoluble aluminum lake. Contains 50-65% carminic acid; likely as a dimer or tetramer complexed with aluminum. Structural studies: M. Harris *et al.*, *J. Chem. Res.* **2009**, 407. Stable to light and heat. Color ranges from bluish-red depending on the metal ions present in the formulation. May be converted to a water soluble dye by treatment with alkali.
USE: Food colorant; biological stain; source of carminic acid.

2443. Cocillana. Dried bark of *Guarea rusbyi* (Britt.) Rusby, *Meliaceae. Habit.* Bolivia. *Constit.* Rusbyine, about 2.5% resins, about 2.5% fat, tannin.

THERAP CAT: Expectorant.

THERAP CAT (VET): Has been used as an expectorant.

2444. Coclaurine. [486-39-5] (1*S*)-1,2,3,4-Tetrahydro-1-[(4-hydroxyphenyl)methyl]-6-methoxy-7-isoquinolinol; 1-(*p*-hydroxybenzyl)-6-methoxy-7-hydroxy-1,2,3,4-tetrahydroisoquinoline; machiline. $C_{17}H_{19}NO_3$; mol wt 285.34. C 71.56%, H 6.71%, N 4.91%, O 16.82%. Isolated as the racemate from species of *Machilus (Lauraceae)* and *Cocculus (Menispermaceae)*. First isoln from *C. laurifolius* D.C. believed to be of the *d*-form: Kondo, Kondo, *J. Pharm. Soc. Jpn.* **No. 524**, 876 (1925), *C.A.* **20**, 604[7] (1926); *see also*: Johns *et al.*, *Aust. J. Chem.* **20**, 1729 (1967). Structure: Kondo, Kondo, *J. Pharm. Soc. Jpn.* **48**, 1156 (1928); Tomita, Kusada, *ibid.* **72**, 280 (1952). Synthesis: Kratzl, Billek, *Monatsh. Chem.* **82**, 568 (1951); Finkelstein, *J. Am. Chem. Soc.* **73**, 550 (1951). Identity with machiline: Tomita *et al.*, *J. Pharm. Soc. Jpn.* **83**, 218 (1963), *C.A.* **59**, 2874a (1963). Crystal structure and absolute configuration: Fridrichsons, Mathieson, *Tetrahedron* **24**, 5785 (1968).

Plates, tablets from alc, mp 220-221°. Sol in hot alc, hot acetone; slightly sol in water, alc, chloroform, ether, acetone; practically insol in benzene, petr ether.

Hydrochloride. $C_{17}H_{19}NO_3 \cdot HCl$. Crystals, mp 263-264°.

2445. Cocoa. A powder prepd from the roasted and cured kernels of ripe seeds of *Theobroma cacao* L. and other species of *Theobroma, Sterculiaceae.* For bibliography *see* Cacao Shell.

Brownish powder of chocolate odor and taste.

USE: In nutrient beverages; as flavoring.

2446. Coconut Oil. Copra oil. Expressed oil from kernels of *Cocos nucifera* L., *Palmae. Constit.* Trimyristin, trilaurin, tripalmitin, tristearin; also various other glycerides.

White, semisolid, lard-like fat; stable to air. Remains bland and edible for several years under ordinary storage conditions. d_4^0 0.903. mp 21-25°. n_D^{40} 1.4485-1.4495. Sapon. no. 255-258. Iodine no. 8-9.5. Acid no. not over 6. Surface tension (20°): 33.4 dyn/cm; (80°): 28.4 dyn/cm. Practically insol in water, 95% alc, more sol in abs alc; very sol in chloroform, ether, carbon disulfide. Soly data: Rao, Arnold, *J. Am. Oil Chem. Soc.* **33**, 389 (1956).

USE: Manuf soap, edible fats, chocolate, candies; in baking instead of lard; in candles and night lights; in dyeing cotton; as an ointment base; in hair dressing; in massage oil.

2447. Codamine. [21040-59-5] (1*S*)-1-[(3,4-Dimethoxyphenyl)methyl]-1,2,3,4-tetrahydro-6-methoxy-2-methyl-7-isoquinolinol; 1,2,3,4-tetrahydro-6-methoxy-2-methyl-1-veratryl-7-isoquinolinol. $C_{20}H_{25}NO_4$; mol wt 343.42. C 69.95%, H 7.34%, N 4.08%, O 18.63%. Minor opium alkaloid. Constitutes about 0.003% of Turkish opium. Isoln: Hesse, *Ann.* **282**, 213 (1894). Structure: Späth, Epstein, *Ber.* **59B**, 2791 (1926). Synthesis: Schöpf, Thierfelder, *Ann.* **537**, 143 (1939); Billek, *Monatsh. Chem.* **87**, 106 (1956).

dl-**Form.** Large, six-sided prisms from ether, mp 127°. Very sol in alcohol, chloroform. Somewhat sol in boiling water. In soln codamine reacts strongly basic. The salts are bitter in taste, whereas the base is said to be tasteless.

2448. Codeine. [76-57-3] (5α,6α)-7,8-Didehydro-4,5-epoxy-3-methoxy-17-methylmorphinan-6-ol; methylmorphine; morphine monomethyl ether; morphine 3-methyl ether; Codicept. $C_{18}H_{21}NO_3$; mol wt 299.37. C 72.22%, H 7.07%, N 4.68%, O 16.03%. Present in opium from 0.7 to 2.5%, depending on the source, but mostly prepd by methylation of morphine, *q.v.* Discussion of structure and bibliography: Small, Lutz, "Chemistry of the Opium Alkaloids," in *U.S. Public Health Reports* **Suppl. No. 103** (Washington, 1932). Prepn of (+)-codeine and racemate: Goto, Yamamoto, *Proc. Jpn. Acad.* **30**, 769 (1954), *C.A.* **50**, 1052h (1956); of (−)-form: E. J. Bijsterveld, H. J. Sinnige, *Rec. Trav. Chim.* **95**, 24 (1976); H. C. Beyerman *et al.*, *ibid.* **97**, 127 (1978). Manuf from morphine: W. R. Heumann, *Bull. Narc.* **X**, 15 (1958). Facile synthesis from thebaine, *q.v.*: W. G. Dauben *et al.*, *J. Org. Chem.* **44**, 1567 (1979). Toxicity of the hydrochloride: Eddy, Sumwalt, *J. Pharmacol. Exp. Ther.* **67**, 127 (1939). Comprehensive description of codeine and codeine phosphate: F. J. Muhtadi, M. M. A. Hassan, *Anal. Profiles Drug Subs.* **10**, 93-138 (1981).

Monohydrate. Orthorhombic sphenoidal rods or tablets (octahedra) from water or dil alcohol, mp 154-156° (after drying at 80°). Sublimes (when anhydr) at 140-145° under 1.5 mm pressure. Melts to oily drops when heated in an amount of water insufficient for complete soln, crystallizes on cooling. d_4^{20} 1.32. $[\alpha]_D^{15}$ −136° (c = 2 in alcohol), $[\alpha]_D^{15}$ −112° (c = 2 in chloroform). pK (15°) 6.05; pH of satd aq soln 9.8. One gram dissolves in 120 ml water, 60 ml water at 80°, 2 ml alcohol, 1.2 ml hot alcohol, 13 ml benzene, 18 ml ether, 0.5 ml chloroform; freely sol in amyl alcohol, methanol, dil acids. Almost insol in petr ether or in solns of alkali hydroxides.

Hydrochloride. [1422-07-7] $C_{18}H_{21}NO_3 \cdot HCl$. Dihydrate, small needles, mp ~280° with some decompn. $[\alpha]_D^{22}$ −108°. One gram dissolves in 20 ml water, 1 ml boiling water, 180 ml alcohol. pH about 5. LD_{50} s.c. in mice: 300 mg/kg (Eddy, Sumwalt).

Sulfate. [1420-53-7] $(C_{18}H_{21}NO_3)_2 \cdot H_2SO_4$; mol wt 696.81. Trihydrate, white crystals or cryst powder. One gram dissolves in 30 ml water, 6.5 ml water at 80°, 1300 ml alc. Insol in chloroform or ether. pH: 5.0. Store in airtight containers; protect from light.

Phosphate. [52-28-8] Galcodine. $C_{18}H_{21}NO_3 \cdot H_3PO_4$; mol wt 397.36. Fine, white, needle-shaped crystals or cryst powder. Odorless; affected by light. Solns acid to litmus. Freely sol in water; very sol in hot water; slightly sol in alcohol; more sol in boiling alcohol.

This is a controlled substance (opiate): **21 CFR**, 1308.12.

THERAP CAT: Analgesic; antitussive.

THERAP CAT (VET): Analgesic; antitussive.

2449. Cod Liver Oil. Gaduol; Tunol. The partially destearinated fixed oil expressed from fresh livers of *Gadus morrhua* L., and other species of *Gadidae. Constit.* Most important are vitamins A and D, each gram containing at least 850 U.S.P. units vitamin A (255 μg) and at least 85 U.S.P. units vitamin D (2.125 μg); glycerides of palmitic, stearic, etc. acids (≈19% saturated fatty acids; remainder unsaturated); cholesterol, batyl alcohol esters. Source of omega 3 fatty acid: N. Haagsma *et al.*, *J. Am. Oil Chem. Soc.* **59**, 117 (1982). Brief description: D. Hilditch, P. Williams, *The Chemical Constitution of Natural Fats* (Wiley-Interscience, New York, 4th ed., 1964) p 43; *Bailey's Industrial Oil and Fat Products* **Vol. 1**, D. Swern (Wiley-Interscience, New York, 4th ed., 1979) pp 451-453.

Pale-yellow, thin liq; bland, slightly fishy taste and odor. Becomes yellow, acquires a somewhat disagreeable odor on exposure to air and light. d 0.918-0.927. n_D^{40} 1.4705-1.4745. Sapon no. 180-190. Iodine no. 145-180. Acid no. not over 1.2. Slightly sol in alcohol; freely sol in chloroform, ether, carbon disulfide, ethyl acetate, petr ether.

THERAP CAT: Vitamins A and D source.

THERAP CAT (VET): Source of vitamins A and D. Locally to promote healing.

2450. Coenzyme A. [85-61-0] CoA; Adenosine 5'-(trihydrogen diphosphate) 3'-(dihydrogen phosphate) P'-[(R)-3-hydroxy-4-[[3-[(2-mercaptoethyl)amino]-3-oxopropyl]amino]-2,2-dimethyl-4-oxobutyl]ester. $C_{21}H_{36}N_7O_{16}P_3S$; mol wt 767.53. C 32.86%, H 4.73%, N 12.77%, O 33.35%, P 12.11%, S 4.18%. An essential cofactor in enzymatic acetyl transfer reactions. Synthesized in cells from pantothenate, ATP and cysteine. Found ubiquitously in mammalian cells. Isoln from animal sources: Lipmann et al., J. Biol. Chem. **167**, 869 (1947); **186**, 235 (1950). Many microorganisms contain large amounts of the coenzyme. Isoln from Streptomyces fradiae: Kaplan, Lipmann, ibid. **174**, 37 (1948). Purifications: De Vries et al., J. Am. Chem. Soc. **72**, 4838 (1950); Gregory et al., ibid. **74**, 854 (1952). Structure: Baddiley et al., Nature **171**, 76 (1953). Total synthesis: Moffatt, Khorana, J. Am. Chem. Soc. **81**, 1265 (1959); **83**, 663 (1961); Shimizu et al., Chem. Pharm. Bull. **13**, 1142 (1965). Reviews: Lipmann, Bacteriol. Rev. **17**, 1-16 (1953); Baddiley, Adv. Enzymol. **16**, 1 (1955); Jaenicke, Lynen in The Enzymes vol. **3**, P. D. Boyer et al., Eds. (Academic Press, New York, 2nd ed., 1960) pp 3-103. Review of metabolism: J. D. Robishaw, J. R. Neely, Am. J. Physiol. **248**, E1-E9 (1985); of clinical evaluations in hyperlipoproteinemia: A. Perin, G. Fraticelli, Int. J. Tissue React. **13**, 111-114 (1991); of biochemical role in cellular toxicity: E. P. Brass, Chem. Biol. Interact. **90**, 203-214 (1994).

White powder; characteristic thiol odor. May be dried in vacuo over phosphorus pentoxide at 34°. uv max: 259.5 nm (ε 16800). Fairly strong acid. pK 9.6 (thiol); pK 6.4 (secondary phosphate); pK 4.0 (adenine NH_3^+). Soluble in water. Practically insol in ethanol, ether, acetone. Pure dry coenzyme is best stored in evacuated ampuls at room temp. Readily oxidized by air to the catalytically inactive disulfide.

2451. Coffee, Green. Coffee beans. Dry, unroasted seeds of Coffea arabica L., Rubiaceae. Habit. Tropical Africa, especially Abyssinia; cultivated in many tropical countries, e.g., Java, West Indies, Brazil, etc. Constit. Caffeine 1-2%, coffee oil 10-15%, sucrose and other sugars about 8%, proteins about 11%, ash about 5%, chlorogenic and caffeic acids about 6%. Other constituents include cellulose, hemicelluloses, trigonelline, tannic acid, volatile oils. Monograph: M. Sivetz, H. E. Foote, Coffee Processing Technology, 2 vols (Avi Publ., 1963).

2452. Coffee Oil. Coffee bean oil. A fatty oil found in coffee beans to the extent of about 15% (dry basis). Byproduct in the manufacture of instant coffees. Obtained in 2.7% yield by petr ether extraction of green Santos beans after previous dewaxing with tetrachloroethane: Schuette et al., J. Am. Chem. Soc. **56**, 2085 (1934). Coffee oil seems devoid of linolenic acid, however positive tests for linoleic (25.66%) and oleic (12.36%) acids were obtained.

Greenish-brown oil. Odor characteristic of green coffee beans. d_{25}^{25} 0.9653. n_D^{25} 1.4790. Iodine no. (Wijs) 100.72. Saponification no. 195.53; Reichert-Meissl no. 0.36. Saturated acids 33.60%. Unsaturated fatty acids 38.02%. Unsaponifiable matter 12.63%.

2453. Coherin. [9044-70-6] A peptide factor present in the bovine neurohypophysis which may be involved in the normal physiological regulation of intestinal motility in mammals. Appears to stimulate the coordinated or "coherent" contraction of the intestine needed to keep food moving down the tract. Mol wt ~4000. Amino acid analysis (residue/mole): lysine 0.59; aspartic acid 3.9; threonine 0.11; serine 0.22; glutamic acid 4.0; proline 5.1; glycine 4.7; alanine 0.90; half-cystine 4.9; isoleucine 3.3; leucine 3.1; tyrosine 2.9; phenylalanine 0.88. Isoln from bovine posterior pituitary and preliminary data: Goodman, Hiatt, Science **178**, 419 (1972). Localization of formation site: I. Galasinska-Pomykol, M. Chilimoniuk, Endokrynol. Pol. **24**, 429 (1973), C.A. **81**, 102499j (1974). Effect on electrical rhythm of dog ileum in vivo: R. B. Hiatt et al., Am. J. Dig. Dis. **22**, 108 (1977). Effects on idiopathic delayed gastric emptying in humans: E. D. Davidson et al., Am. J. Gastroenterol. **74**, 419 (1980).

Relatively thermostable, losing only 10% of its activity when heated in a boiling water bath at pH 3 for 1 hour. Isoelectric point: pH 6.0.

2454. Colchiceine. [477-27-0] N-[(7S)-5,6,7,9-Tetrahydro-10-hydroxy-1,2,3-trimethoxy-9-oxobenzo[a]heptalen-7-yl]acetamide; 7-acetamido-10-hydroxy-1,2,3-trimethoxy-6,7-dihydrobenzo-[a]heptalen-9(5H)-one. $C_{21}H_{23}NO_6$; mol wt 385.42. C 65.44%, H 6.02%, N 3.63%, O 24.91%. Deriv of colchicine, q.v., isolated from Colchicum autumnale L., Liliaceae: Santavy, Macák, Collect. Czech. Chem. Commun. **19**, 805 (1954). Toxicity data: B. Goldberg et al., Cancer **3**, 124 (1950). Structure and synthesis: Nakamura, Chem. Pharm. Bull. **10**, 299 (1962). Circular dichroism: J. Hrbek et al., Collect. Czech. Chem. Commun. **47**, 2258 (1982). Effect on antibody response induced in vitro: J. Sterzl et al., Folia Microbiol. **27**, 256 (1982).

Yellow crystals from dioxane + ether, mp 178-179°. $[\alpha]_D^{25}$ −255.1° (c = 1 in chloroform). uv max (95% ethanol): 351, 244 nm (log ε 4.28, 4.51). Slightly sol in water; freely sol in alcohol, chloroform. Almost insol in benzene, ether.

Ethyl ether. $C_{23}H_{27}NO_6$. Needles from ethyl acetate + ether, mp 135-139°. $[\alpha]_D^{25}$ −129.4° (c = 1 in chloroform). uv max (95% ethanol): 351, 243.5 nm (log ε 4.21, 4.45). LD_{50} i.p. in mice: 84 mg/kg (Goldberg).

2455. Colchicine. [64-86-8] N-[(7S)-5,6,7,9-Tetrahydro-1,2,-3,10-tetramethoxy-9-oxobenzo[a]heptalen-7-yl]acetamide; Colcrys. $C_{22}H_{25}NO_6$; mol wt 399.44. C 66.15%, H 6.31%, N 3.51%, O 24.03%. Principal alkaloid of the meadow saffron, Colchicum autumnale L., Liliaceae. Binds to tubulin and inhibits microtubule polymerization. Isoln: P. J. Pelletier, J. B. Caventou, Ann. Chim. Phys. [2] **14**, 69 (1820). Extraction procedure from seeds: F. Chemnitius, J. Prakt. Chem. [II] **118**, 29 (1928); F. Santavy, T. Reichstein, Helv. Chim. Acta **33**, 1606 (1950); by supercritical CO_2: E. Ellington et al., Phytochem. Anal. **14**, 164 (2003). Proposed structure: A. Windaus, Ann. **439**, 59 (1924); M. J. S. Dewar, Nature **155**, 141 (1945). Confirmation of structure: M. V. King et al., Acta Crystallogr. **5**, 437 (1952). Abs config: H. Corrodi, E. Hardegger, Helv. Chim. Acta **38**, 2030 (1955). Crystal structure: L. Lessinger, T. N. Margulis, Acta Crystallogr. **B34**, 578 (1978). Total synthesis of (±)-form: J. Schreiber et al., Helv. Chim. Acta **44**, 540 (1961); E. E. van Tamelen et al., Tetrahedron **14**, 8 (1961). Biosynthesis: E. Leete, Tetrahedron Lett. **6**, 333 (1965); A. R. Battersby et al., J. Chem. Soc. **1964**, 4257. Toxicity: S. J. Rosenbloom, F. C. Ferguson, Toxicol. Appl. Pharmacol. **13**, 50 (1968); R. P. Beliles, ibid. **23**, 537 (1972). Comprehensive description: D. K. Wyatt et al., Anal. Profiles Drug Subs. **10**, 139-182 (1981). Review of syntheses: T. Graening, H.-G. Schmalz, Angew. Chem. Int. Ed. **43**, 3230-3256 (2004). Review of interaction with tubulin: B. Bhattacharyya et al., Med. Res. Rev.

28, 155-183 (2008); of pharmacology and clinical experience: C. Cerquaglia *et al.*, *Curr. Drug Targets Inflamm. Allergy* 4, 117-124 (2005); G. Nuki, *Curr. Rheumatol. Rep.* 10, 218-227 (2008).

Nearly colorless to pale yellow needles from ethyl acetate + ether, mp 154-156°. Darkens on exposure to light. $[\alpha]_D^{22}$ −121.6° (c = 1.001 in chloroform). pK at 20°: 12.35. pH of 0.5% soln: 5.9. uv max (ethanol): 247, 350 nm (log ε 4.45; 4.20). One gram dissolves in 22 ml water, 220 ml ether, 100 ml benzene; freely sol in alcohol or chloroform. Practically insol in petr ether. LD_{50} in rats (mg/kg): 1.6 i.v. (Rosenbloom, Ferguson); in mice (mg/kg): 4.13 i.v. (Beliles).

USE: In plant genetics research (for doubling chromosomes).

THERAP CAT: Gout suppressant. In treatment of familial Mediterranean fever.

2456. Colchicum. Meadow saffron; autumn crocus; wild saffron; meadow crocus. Poisonous, flowering plant, *Colchicum autumnale* L., *Liliaceae*; used since ancient times to treat inflammatory conditions. Medicinal parts are the seeds, flowers, or corm. *Habit.* Central and Southern Europe, North Africa. *Constit.* Colchicine, colchicoside, *N*-deacetyl-*N*-formyl-colchicine, demecolcine, colchiceine. Determn of colchicine in medicinal formulations: P. A. W. Self, C. E. Corfield, *Q. J. Pharm. Pharmacol.* 5, 347 (1932). Stability and QC testing of extracts by HPLC: A. Körner, S. Kohn, *J. Chromatogr. A* 1089, 148 (2005). Description and medicinal uses: K. R. Fell, D. Ramsden, *Lloydia*, 30, 123-140 (1967); J. Gruenwald *et al.*, *PDR for Herbal Medicines* (Thomson PDR, Montvale, 3rd Ed., 2004) pp 217-218.

Caution: All parts of the plant are poisonous. Potential symptoms of poisoning include stomach pain, nausea, vomiting, diarrhea, stomach and intestinal hemorrhage, progressive paralysis (Gruenwald).

THERAP CAT: Gout suppressant.

2457. Colesevelam Hydrochloride. [182815-44-7] *N,N,N*-Trimethyl-6-(2-propen-1-ylamino)-1-hexanaminium chloride (1:1) polymer with 2-(chloromethyl)oxirane, 2-propen-1-amine and *N*-2-propenyl-1-decanamine hydrochloride; GT-31-104; CholestaGel; Welchol. Bile acid sequestrant composed of a polyallylamine cross-linked with epichlorohydrin and alkylated with 1-bromodecane and 6-bromohexyltrimethylammonium bromide. Prepn: H. W. Mandeville, III, S. R. Holmes-Farley, WO 9534585; *eidem*, US 5693675 (1995, 1997 both to GelTex); and bile acid binding: W. H. Mandeville *et al.*, *Mater. Res. Soc. Symp. Proc.* 550, 3 (1999). Absorption, distribution and excretion: D. P. Rosenbaum *et al.*, *J. Pharm. Sci.* 86, 591 (1997). Clinical trial in hypercholesterolemia: W. Insull, Jr. *et al.*, *Mayo Clin. Proc.* 76, 971 (2001). Review of pharmacology and clinical experience: M. A. Aldridge, M. K. Ito, *Ann. Pharmacother.* 35, 898-907 (2001).

THERAP CAT: Antilipemic.

2458. Colestilan. [95522-45-5] 2-Methyl-1*H*-imidazole polymer with 2-(chloromethyl)oxirane; 2-methylimidazole polymer with 1-chloro-2,3-epoxypropane; colestimide; MCI-196; Cholebine. $(C_4H_6N_2 \cdot C_3H_5ClO)n$. Bile acid sequestering anion exchange resin. Prepn: H. Toda, K. Kihara, JP Kokai 59138228 (1984 to Mitsubishi Petrochem.); and bile acid binding activity: H. Toda *et al.*, *J. Pharm. Sci.* 77, 531 (1988). Series of articles on pharmacology and toxicology: *Jpn. Pharmacol. Ther.* 24, Suppl. 4, 1-181 (1996). Clinical evaluation in type II hyperlipoproteinemia: Y. Homma *et al.*, *Nutr. Metab. Cardiovasc. Dis.* 6, 211 (1996); in hyperphosphatemia: S. Kurihara *et al.*, *Nephrol. Dial. Transplant.* 20, 424 (2005).

Odorless, tasteless, white powder. Ion exchange capacity: 4.9 mEq/g. Particle size: 44-149 μm.

THERAP CAT: Antilipemic.

2459. Colestipol. [26658-42-4] N^1-(2-Aminoethyl)-N^2-[2-[(2-aminoethyl)amino]ethyl]-1,2-ethanediamine polymer with 2-(chloromethyl)oxirane. A basic anion exchange resin described as a high molecular wt copolymer of diethylenetriamine and 1-chloro-2,3-epoxypropane (hydrochloride), with approx 1 out of 5 amine nitrogens protonated, for which no structural formula has been assigned and for which no specific mol wt information is available, due to the highly cross-linked and insoluble nature of the material. Prepn: Nelson, Van den Berg, DE 1927336; *eidem*, US 3692895 (1969, 1971 to Upjohn); Ledicer, Peery, DE 2053585; *eidem*, US 3803237 (1971, 1974 to Upjohn). Activity studies: Parkinson *et al.*, *Atherosclerosis* 11, 531 (1970); 17, 167 (1973); Marmo *et al.*, *G. Arterioscler.* 8, 229 (1970); Goodman *et al.*, *J. Clin. Invest.* 52, 2646 (1973). Toxicology: Webster, Bollert, *Toxicol. Appl. Pharmacol.* 28, 57 (1974). Review of pharmacology and therapeutic efficacy: R. C. Heel *et al.*, *Drugs* 19, 161-180 (1980).

Hydrochloride. [37296-80-3] U-26597A; Cholestabyl; Colestid; Lestid. Light yellow, hygroscopic resin. Swells when suspended in water or aqueous solutions of acid or alkali. Insol in water and in the common organic solvents. LD_{50} in rats (mg/kg): >4000 i.p.; >1000 orally (Webster, Bollert).

THERAP CAT: Antilipemic.

2460. Colforsin. [66575-29-9] (3*R*,4a*R*,5*S*,6*S*,6a*S*,10*S*,10a*R*,-10b*S*)-5-(Acetyloxy)-3-ethenyldodecahydro-6,10,10b-trihydroxy-3,4a,7,7,10a-pentamethyl-1*H*-naphtho[2,1-*b*]pyran-1-one; 7β-acetoxy-8,13-epoxy-1α,6β,9α-trihydroxylabd-14-en-11-one; forskolin; boforsin (obsolete); HL-362. $C_{22}H_{34}O_7$; mol wt 410.51. C 64.37%, H 8.35%, O 27.28%. Diterpene isolated from *Coleus forskohlii*, Briq. *Labiatae*, possessing vasodilating and cardiostimulatory properties. Isoln and characterization: S. V. Bhat *et al.*, *Tetrahedron Lett.* 1977, 1669; *eidem*, DE 2557784; *eidem*, US 4088659 (1977, 1978 both to Hoechst). Synthesis by hydroxylation of 9-deoxyforskolin: N. J. Hrib, *Tetrahedron Lett.* 28, 19 (1987); *see also* F. E. Ziegler *et al.*, *J. Am. Chem. Soc.* 109, 8115 (1987). Stereocontrolled synthesis of the 3-ring system: S. Hashimoto *et al.*, *Chem. Commun.* 1987, 24. Total synthesis of (±)-forskolin: S. Hashimoto *et al.*, *J. Am. Chem. Soc.* 110, 3670 (1988); E. J. Corey *et al.*, *ibid.* 3672. Positive inotropic and blood-pressure lowering activity: E. Lindner *et al.*, *Arzneim.-Forsch.* 28, 284 (1978). Activates adenylate cyclase: H. Metzger, E. Lindner, *ibid.* 31, 1248 (1981). Binds to catalytic site of adenylate cyclase: T. Pfeuffer, H. Metzger, *FEBS Lett.* 146, 369 (1982). Lowers intraocular pressure: J. Caprioli, M. Sears, *Lancet* 1, 958 (1983). In treatment of glaucoma: *eidem*, US 4476140 (1984 to Yale University). *Reviews:* K. B. Seaman, J. W. Daly, *J. Cyclic Nucleotide Res.* 1981, 201; K. B. Seaman, "Forskolin and Adenylate Cyclase: New Opportunities in Drug Design" in *Annu. Rep. Med. Chem.* 19, D. M. Bailey, Ed. (Academic Press, New York, 1984) pp 293-302.

Colorless crystals from ethyl acetate/petr ether, mp 230-232°. uv max 210, 305 nm (ε 1000, 50). $[\alpha]_D^{25}$ −26.19° (c = 1.68 in CHCl₃).

USE: In purification of adenylate cyclase.

2461. Colfosceril Palmitate. [63-89-8] (7R)-4-Hydroxy-N,-N,N-trimethyl-10-oxo-7-[(1-oxohexadecyl)oxy]-3,5,9-trioxa-4-phosphapentacosan-1-aminium inner salt 4-oxide; L-α-dipalmitoyl lecithin; dipalmitoyl-L-α-glycerylphosphorylcholine; dipalmitoyl phosphatidylcholine; DPPC; Exosurf Neonatal. $C_{40}H_{80}NO_8P$; mol wt 734.05. C 65.45%, H 10.99%, N 1.91%, O 17.44%, P 4.22%. Major phospholipid constituent and primary surface tension lowering component of natural lung surfactant. Isoln from lung extracts: S. J. Thannhauser *et al.*, *J. Biol. Chem.* **166**, 669 (1946); E. S. Brown, *Am. J. Physiol.* **207**, 402 (1964). Enantioselective synthesis: E. Baer, M. Kates, *J. Am. Chem. Soc.* **72**, 942 (1950). Clinical study in combination therapy for neonatal respiratory distress syndrome (RDS): C. J. Morley *et al.*, *Lancet* **1**, 64 (1981). Diagnostic use as predictor of fetal lung maturity: A. Lohninger *et al.*, *Clin. Chem.* **29**, 650 (1983). HPTLC determn in amniotic fluid: J. G. Alvarez *et al.*, *J. Chromatogr. B* **665**, 79 (1995). Review of early literature: D. F. Tierney, *Am. J. Physiol.* **257**, L1-L12 (1989); of role in pulmonary function and clinical use: L. M. G. van Golde *et al.*, *Physiol. Rev.* **68**, 374-455 (1988); of structure and function as surface active agent: B. A. Hills, *Br. J. Anaesth.* **65**, 13-29 (1990); of biosynthesis: H. P. Haagsman, L. M. G. van Golde, *Annu. Rev. Physiol.* **53**, 441-464 (1991).

Occurs as monohydrate, crystals from hot diisobutyl ketone, softens at 75-79°, mp 234-235°. $[\alpha]_D^{23}$ +6.6° (c = 4.2 in 1:1 chloroform-methanol). Readily sol in chloroform, hot diisobutyl ketone, hot dioxane. Readily emulsified in water. Soly at 22-23° (g/100 ml): ethanol 1.5, ether 0.02, acetone 0.02, pyridine 1.1, acetic acid 4.0, methanol 1.4.

THERAP CAT: Pulmonary surfactant. Diagnostic aid (fetal lung maturity).

2462. Colicins. Colicine. Antibiotic substances, or complexes of antibiotic substances, which are highly specific, are produced by certain strains of intestinal bacteria, and act upon other related strains: Fredericq, *Annu. Rev. Microbiol.* **11**, 7 (1957). Colicins produced by various strains may differ in many characteristics, the most conspicuous of which are activity spectra and the specificity of resistant mutants. They give the general reactions of proteins, are antigenic, and their activity is destroyed by proteolytic enzymes: Fredericq, *J. Theor. Biol.* **4**, 159 (1963). Production of colicin A: Barry *et al.*, *Nature* **198**, 211 (1963); of colicin V: Hutton, Goebel, *Proc. Natl. Acad. Sci. USA* **47**, 1498 (1961); of colicins E_4, N, P,V_2, V_3, V_4 and V_5: Hamon, Péron, *Ann. Inst. Pasteur* **107**, 44 (1964). Mechanism of action of colicins: Nomura, *Proc. Natl. Acad. Sci. USA* **52**, 1514 (1964). *Reviews: idem* in *Antibiotics* **vol. 1**, D. Gottlieb, P. D. Shaw, Eds. (Springer-Verlag, New York, 1967) pp 696-704; Wendt, *ibid.* **vol. 3**, J. W. Corcoran, F. E. Hahn, Eds. (1975) pp 588-605.

2463. Colistin. [1066-17-7] Polymyxin E. Cyclic polypeptide antibiotic produced by *Bacillus polymyxa* subsp. *colistinus*. Complex mixture of at least 30 components, primarily colistins A and B. Isoln from Japanese soil: Y. Koyama, *JP 52 1546* (1952). Identification of colistins A, B and C: T. Suzuki *et al.*, *J. Biochem. (Tokyo)* **54**, 25 (1963). Structure of colistin A: *eidem, ibid.* 173, 412. Identity of colistin A with polymyxin E_1: S. Wilkinson, L. A. Lowe, *J. Chem. Soc.* **1964**, 4107. Synthesis of colistin A: R. O. Studer *et al.*, *Helv. Chim. Acta* **48**, 1371 (1965). Prepn of sodium sulfomethyl derivs and toxicity study: M. Barnett *et al.*, *Br. J. Pharmacol. Chemother.* **23**, 552 (1964). Comparative pharmacology of sulfate and methansulfonate salts: A. A. Al-Khayyat, A. L. Aronson, *Chemotherapy (Basel)* **19**, 82 (1973). HPLC determn in medicated animal feeds: B. Cancho-Grande *et al.*, *Chromatographia* **54**, 481 (2001); in human plasma: J. Li *et al.*, *J. Chromatogr. B* **761**, 167 (2001). Review and comparison with polymyxin B, *q.v.*: J. Horton, G. A. Pankey, *Med. Clin. North Am.* **66**, 135-142 (1982); M. E.

Evans *et al*, *Ann. Pharmacother.* **33**, 960-967 (1999). Review of clinical experience in cystic fibrosis: J. M. Littlewood *et al.*, *Respir. Med.* **94**, 632 (2000); in treatment of multi-resistant bacterial infections: J. Li *et al.*, *Int. J. Antimicrob. Agents* **25**, 11-25 (2005).

DAB = α,γ-diaminobutyric acid

Colistin A   R = (+)-6-methyloctanoyl
Colistin B   R = 6-methylheptanoyl

Colistin sodium methanesulfonate. [8068-28-8] Colistimethate sodium; Alficetin; Colimicina; Colimicyne; Colomycin; Coly-Mycin M; Methacolimycin; Promixin. $C_{58}H_{105}N_{16}Na_5O_{28}S_5$; mol wt 1749.81. Prepd by treating colistin complex with formaldehyde and sodium bisulfite. Sol in water and stable in the dry form. LD_{50} i.v. in mice: >550 mg/kg (Barnett).

Sulfate. [1264-72-8] Belcomycine; Colomycin (tabl.). White to slightly yellow powder. Freely sol in water; slightly sol in methanol. Insol in acetone, ether.

Colistin A. [7722-44-3] Polymyxin E_1. $C_{53}H_{100}N_{16}O_{13}$; mol wt 1169.48. $[\alpha]_{5461}^{22}$ −93.3° (2% acetic acid).

Colistin B. [7239-48-7] Polymyxin E_2. $C_{52}H_{98}N_{16}O_{13}$; mol wt 1155.46. $[\alpha]_{5461}^{22}$ −94.5° (2% acetic acid).

THERAP CAT: Antibacterial.

THERAP CAT (VET): Antibacterial.

2464. Collagen. Ossein. Mol wt about 130,000. Polypeptide substance comprising one third of the total protein in mammalian organisms; main constituent of skin, connective tissue, and the organic substance of bones and teeth. Prepd from bones by dissolving the mineral part of the bones with phosphoric acid: Sciallano, **FR 688104** (1929), *C.A.* **25**, 786 (1931). Isoln of sol collagens from corium tissue: Reizo, Sakata, *Kumamoto Med. J.* **13**, 27 (1960). Purification: Bloch, Oneson, **US 2973302** (1961 to Ethicon). Commercial extraction process: Highberger, **US 2979438** (1961 to United Shoe Machinery). Collagen production in the body is preceded by the formation of a much larger molecule, the biosynthetic precursor *procollagen*, which is degraded by specific enzymes to make collagen. Different types of collagens exist. They are all composed of molecules containing three polypeptide chains, α-chains, arranged in a triple helical conformation. The amino acid sequence of the α-chain is mostly a repeating structure with glycine in every third position and proline or 4-hydroxyproline frequently preceding the glycine residues. Slight differences in the primary structure (amino acid sequence) establish differences between types. Review of structural studies: Tanzer, *Science* **180**, 561-566 (1973); P. Bornstein, H. Sage, *Annu. Rev. Biochem.* **49**, 959 (1980). Collagen is differentiated from the accompanying fibrous proteins, elastin, and reticulin, *q.q.v.*, by (1) its content of proline, hydroxyproline and hydroxylysine, by (2) the absence of tryptophan and its low tyrosine and sulfur content, but particularly by (3) its high content of polar groups originating from the difunctional amino acids. The polar groups are responsible for the swelling properties leading eventually to dispersion of collagen in dil acid. Denaturation of collagen is the conversion of the rigidly coiled helix to a random coil called gelatin, *q.v.* Description of the native and denatured states of sol collagen: Boedtker, Doty, *J. Am. Chem. Soc.* **78**, 4267 (1956). Biosynthesized in fibroblastic cells. Review of biosynthesis: Bornstein, *Annu. Rev. Biochem.* **43**, 567-603 (1974). *Reviews:* Gross, *Sci. Am.* **204**, 121 (May 1961); Harrington, von Hippel, *Adv. Protein Chem.* **16**, 1-138 (1961); Grassmann *et al.*, *Fortschr. Chem. Org. Naturst.* **23**, 195-314 (1965); P. Bornstein, W. Traub, "The Chemistry and Biology of Collagen" in *The Proteins* **vol. IV**, H. Neurath, R. L. Hill, Eds. (Academic Press, New York, 1979) pp 411-632; D. R. Eyre, *Science* **207**, 1315-1322 (1980); *Methods Enzymol.* **82**, 3-555 (1982). Summary of recent studies of the role of collagen in disease: *Chem. Eng.*

News **60**, 32 (Jan. 25, 1982). Books: A. Veis, *The Macromolecular Chemistry of Gelatin* (Academic Press, New York, 1964) 433 pp; *Biochemistry of Collagen*, G. N. Ramachandran *et al.*, Eds. (Plenum, New York, 1976) 536 pp; P. P. Fietzek in *Collagen in Health and Disease*, M. I. V. Jayson, J. B. Weiss, Eds. (Churchill-Livingstone, New York, 1982).

USE: As fibers in sutures, in leather substitutes; as a gel in photographic emulsions, in coatings; in food casings.

2465. Collagenase. [9001-12-1] Clostridiopeptidase A; Iruxol; Santyl. Rare proteolytic enzyme capable of digesting native undenatured collagen, *q.v.*, found in certain *Clostridia* bacteria culture filtrates. Crude prepn from *C. histolyticum:* Mandl *et al.*, *J. Clin. Invest.* **32**, 1323 (1953). Purified prepn: Keller, Mandl, *Arch. Biochem. Biophys.* **101**, 81 (1963). Also isolated from culture media of human wound tissue, skin, bone, leukocytes, gingiva, cornea, rheumatoid synovium; from involuting rat uterus and tadpole tailfin tissue. Sepn of fractions A and B having the respective mol wts 105,-000 and 57,400: Harper *et al.*, *Biochem. Biophys. Res. Commun.* **18**, 627 (1965). Amino acid composition: Mandl *et al.*, *Biochemistry* **3**, 1737 (1964); Yoshida, Noda, *Biochim. Biophys. Acta* **105**, 562 (1965). Potential use in treatment of herniated discs: Sussman, **FR 2008611** (1970 to Worthington Biochemical). Review of early studies: Mandl, *Adv. Enzymol.* **23**, 163-264 (1961). *Reviews:* idem, *Science* **169**, 1234-1238 (1970); Nordwig, *Adv. Enzymol. Relat. Areas Mol. Biol.* **34**, 155-205 (1971); Lazarus, *Br. J. Dermatol.* **86**, 193-199 (1972). Book: *Collagenase*, I. Mandl, Ed. (Gordon and Breach, N.Y., 1972) 215 pp.

USE: In investigation of the structure and biosynthesis of collagen; in dispersion of cells for tissue culture studies.

THERAP CAT: Debriding agent.

2466. Collinomycin. [27267-69-2] 6-[2-(4,9-Dihydro-8-hydroxy-5,7-dimethoxy-4,9-dioxonaphtho[2,3-*b*]furan-2-yl)ethyl]-7,8-dihydroxy-1-oxo-1*H*-2-benzopyran-3-carboxylic acid methyl ester; α-rubromycin. $C_{27}H_{20}O_{12}$; mol wt 536.45. C 60.45%, H 3.76%, O 35.79%. Antibiotic substance produced by *Streptomyces collinus:* Brockmann, Renneberg, *Naturwissenschaften* **40**, 166 (1953); **DE 918162** (1954 to Bayer). Identity with α-rubromycin: Brockmann *et al.*, *Tetrahedron Lett.* **1966**, 3525. Structure: Brockmann, Zeeck, *Ber.* **103**, 1709 (1970). Shows considerable *in vitro* activity against *Staph. aureus.*

Orange prisms from chloroform-methanol, mp 280-282°. Moderately sol in chloroform, acetone, dioxane; sparingly sol in ether, low-molecular alcohols. Practically insol in petr ether, water, aq sodium bicarbonate soln.

2467. Collinsonia. Stone-root; knob root; horse balm; richweed. Root of *Collinsonia canadensis* L., *Labiatae. Habit.* North America, from Ontario to Florida and west to Kansas. *Constit.* Resin, saponin, tannin, mucilage.

2468. Collodion. A soln of 4 g pyroxylin (chiefly nitrocellulose) in 100 ml of a mixture of 25 ml alcohol and 75 ml ether. Contains about 70% ether and 24% abs alc by vol.

Colorless, or slightly yellow, clear or slightly opalescent, syrupy liquid. Has the odor of ether. d_{25}^{25} 0.765-0.775. Exposed in thin layers, it evaporates leaving a tough, colorless film. On the addition of water the pyroxylin ppts.

Flexible collodion. Mixture of simple collodion with 2% camphor and 3% castor oil (by wt). Yellow syrupy liquid; contains about 67% ether and about 22% abs alcohol by volume.

Styptic collodion. Mixture of flexible collodion with 18% tannic acid by wt. Contains about 61% ether and about 21% abs alcohol by volume.

Caution: Highly flammable! Keep tightly closed in a cool place and away from flame!

USE: Collodion in photography; manuf lacquers, patent and artificial leathers, artificial pearls; process engraving; in cements.

THERAP CAT: Collodion and flexible collodion as topical protectant; styptic collodion as hemostatic.

THERAP CAT (VET): Skin protectant.

2469. Colloidal Bismuth Subcitrate. [57644-54-9] 2-Hydroxy-1,2,3-propanetricarboxylic acid bismuth(3+) potassium salt (2:1:3); tripotassium dicitrato bismuthate; CBS; De-Nol; De-Noltab; Duosol; Ulcerone. $C_{12}H_{16}BiK_3O_{14}$; mol wt 710.52. C 20.29%, H 2.27%, Bi 29.41%, K 16.51%, O 31.52%. $(C_6H_8O_7)_2BiK_3$. Cytoprotective, polymeric bismuth citrato complex. General prepn of colloidal bismuth solution: L. Vanino, cited in *Mellor's* **vol. IX**, 598 (1929). Manufacturing processes: C. J. McLoughlin, R. B. Himstedt, **DE 2501787**; P. J. H. Bos *et al.*, **EP 75992**; *eidem*, **US 4801608** (1975, 1985, 1989 all to Gist-Brocades). Analytical study: D. R. Williams, *J. Inorg. Nucl. Chem.* **39**, 711 (1977). Crystal structure: E. Asato *et al.*, *Chem. Lett.* **1992**, 1967. Pharmacology: J. Wieriks *et al.*, *Scand. J. Gastroenterol.* **17**, Suppl. 80, 11 (1982). Modes of action: D. W. R. Hall, *ibid.* **24**, Suppl. 157, 3 (1989); S. P. Lee, *ibid.* **26**, Suppl. 185, 1 (1991). Pharmacokinetics and safety: L. Z. Benet *ibid.* 29. Symposia on pharmacology and clinical efficacy: *ibid.* 21, Suppl. 122, 1-54 (1986); *Digestion* **37**, Suppl. 2, 1-64 (1987). Clinical trials in *Helicobacter pylori* associated gastritis and ulcer: T. Rokkas *et al.*, *Gut* **29**, 1386 (1988); W. A. de Boer *et al.*, *Am. J. Gastroenterol.* **89**, 1993 (1994).

White, amorphous powder. Sol in water, dil alkali, ammonia. Forms precipitate of bismuth oxychloride and bismuth citrate at pH <5.

THERAP CAT: Antiulcerative.

2470. Colocynth. Bitter apple; bitter cucumber; bitter gourd. Perennial trailing vine, *Citrullus colocynthis* Schrad., *Cucurbitaceae*, bearing round, yellow, bitter, poisonous fruit with numerous white or brownish seeds. The dried fruit pulp has been used in traditional medicine as a purgative. *Habit.* Semi-arid, tropical and subtropical regions of Asia and N. Africa. *Constit.* Colocynthin, *q.v.*, colocynthidin, cucurbitacins, citrullol, pectin, albuminoids, 1-3% fixed oil. Seeds contain fixed oil, protein, crude fiber. Constituents of seeds: W. N. Sawaya *et al.*, *J. Agric. Food Chem.* **34**, 285 (1986). Review of constituents and medicinal uses: C. P. Khare, *Indian Herbal Remedies* (Springer-Verlag, Berlin, 2004) pp 148-149; J. Gruenwald *et al.*, *PDR for Herbal Medicines* (Thomson PDR, Montvale, 3rd Ed., 2004) pp 85-86.

Caution: Potential symptoms of overexposure by ingestion are vomiting, bloody diarrhea, colic and kidney irritation, convulsions, paralysis, death (Gruenwald).

2471. Colocynthin. [1398-78-3] (9β,10α,16α,23E)-25-(Acetyloxy)-2-(β-D-glucopyranosyloxy)-16,20-dihydroxy-9-methyl-19-norlanosta-1,5,23-triene-3,11,22-trione; 2-*O*-β-D-glucopyranosylcucurbitacin E. $C_{38}H_{54}O_{13}$; mol wt 718.84. C 63.49%, H 7.57%, O 28.93%. Glucoside from fruit of *Citrullus colocynthis* Schrad., *Cucurbitaceae:* Walz, *Neues Jahrb. Pharm.* **9**, 16, 225 (1858); 16, 10 (1861); Henke, *Arch. Pharm.* **221**, 200 (1883); Naylas, Chappel, *Pharm. J.* **25**, 117 (1907); Lavie *et al.*, *Acta Univ. Int. Contra Cancerum* **15**, 177 (1959); Khadem, Rahman, *Tetrahedron Lett.* **1962**, 1137. Other isolns: H. Ripperger, K. Seifert, *Tetrahedron* **31**, 1561 (1975); H. Ripperger, *ibid.* **32**, 1567 (1976). Structure: Khadem, Rahman, *J. Chem. Soc.* **1963**, 4991.

Yellow crystals, mp 158-160°. $[\alpha]_D$ +50° (c = 0.4 in ethanol). uv max: 234-236 nm (log ε 4.11). Sol in ethanol, acetone, chloroform; slightly sol in ether, water.

THERAP CAT: Cathartic.

2472. Colony Stimulating Factors. CSFs. Family of hematopoietic growth factors that regulate the production of myeloid cells. Glycoprotein hormones that stimulate cell proliferation, survival, differentiation commitment, and end-cell functional activity; important to the inflammatory and immune response. Cell types capable of producing one or more CSFs include monocytes, T-lymphocytes, endothelial cells, fibroblasts and skin epithelial cells. Originally identified by their ability to stimulate the *in vitro* growth of hematopoietic progenitor cells into colonies in semisolid media: D. H. Pluznick, L. Sachs, *J. Cell. Comp. Physiol.* **66**, 319 (1965); T. R. Bradley, D. Metcalf, *Aust. J. Exp. Biol. Med. Sci.* **44**, 287 (1966). Distinguished by the type(s) of progeny cells which result from stimulation by the specific factor, CSFs include: granulocyte-CSF (G-CSF), granulocyte-macrophage CSF (GM-CSF), macrophage-CSF (M-CSF) and multi-CSF or interleukin-3 (IL-3), *q.q.v.* G-CSF and M-CSF are thought to act on late progenitor cells already committed to their respective lineages. GM-CSF and IL-3 act on earlier, pluripotential progenitors. Potential clinical applications include the reduction of myelosuppression during cancer chemotherapy; augmentation of host defense in infection; and differentiation induction in hematologic malignancies. Reviews: D. Metcalf, *Science* **229**, 16-22 (1985); S. C. Clark, R. Kamen, *ibid.* **236**, 1229-1237 (1987); J. D. Griffin, *Oncology* **2**, 15-21 (1988). Role in infectious disease: R. M. Rose, *Semin. Oncol.* **19**, 415-421 (1992). Reviews of biological effects and clinical potential: J. L. Gabrilove, *Cancer Chemother. Biol. Response Modif.* **10**, 492-506 (1988); J. E. Groopman *et al.*, *N. Engl. J. Med.* **321**, 1449-1459 (1989); J. A. Hamilton, *Immunol. Today* **14**, 18-24 (1993).

2473. Coltsfoot. Coughwort. Dried leaves of *Tussilago farfara* L., *Compositae*. *Habit.* Northern Europe and Asia; naturalized in Northeastern U.S. *Constit.* Pectin, bitter glucoside, tannin, volatile oil, resin, saponin, caoutchouc.

2474. Colubrines. $C_{22}H_{24}N_2O_3$; mol wt 364.45. C 72.50%, H 6.64%, N 7.69%, O 13.17%. Minor alkaloids found in *Strychnos nux vomica* L.; naturally occurring in two chemically distinct forms, α and β. Isolation from mother liquor of commercial strychnine extraction: K. Warnat, *Helv. Chim. Acta* **14**, 997 (1931). Isolation and characterization from *S. nux vomica* L.: G. B. Marini-Bettolo *et al.*, *J. Assoc. Off. Anal. Chem.* **51**, 185 (1968). Structure of α- and β-colubrine: S. P. Findlay, *J. Am. Chem. Soc.* **73**, 3008 (1951); of β-colubrine: P. Rosenmund, *Ber.* **95**, 2639 (1962); of α-colubrine: J. M. Culver, M. Sainsbury, *J. Chem. Res. Synop.* **1987**, 304. Synthesis from strychnine: P. Rosenmund, H. Franke, *Ber.* **97**, 1677 (1964). Determn by GLC: N. G. Bisset, P. Fouché, *J. Chromatogr.* **37**, 172 (1968); by HPLC: G. M. Iskander *et al.*, *J. Liq. Chromatogr.* **5**, 1481 (1982). Chiroptical properties of β-colubrine: J. W. Snow, T. M. Hooker, Jr., *Can. J. Chem.* **56**, 1222 (1978).

α-Colubrine: R_1 = OCH$_3$ R_2 = H
β-Colubrine: R_1 = H R_2 = OCH$_3$

α-Colubrine. [509-44-4] 3-Methoxystrychnidin-10-one; 11-methoxystrychnine. Pyramids from ethyl acetate, mp 189-193°. $[\alpha]_D^{20}$ −72.4° (c = 0.9 in alc). uv max (ethanol): 255, 297 nm (log ε 4.03, 3.77). Sol in alc, benzene, chloroform.

β-Colubrine. [509-36-4] 2-Methoxystrychnidin-10-one; 10-methoxystrychnine. Crystals from dil alcohol, mp 222°. Very bitter. $[\alpha]_D^{19}$ −107.7° (c = 2.5 in 80% alcohol). $[\alpha]_D^{21}$ −156° (c = 0.042 in chloroform). uv max (ethanol): 262, 297 nm (log ε 4.40, 3.80). Sol in alcohol, benzene, chloroform.

2475. Columbamine. [3621-36-1] 5,6-Dihydro-2-hydroxy-3,9,10-trimethoxydibenzo[*a,g*]quinolizinium; 7,8,13,13aα-tetradehydro-2-hydroxy-3,9,10-trimethoxyberbinium. $[C_{20}H_{20}NO_4]^+$.

Found in *Jatrorrhiza palmata* (DC.) Miers (*J. columba* Miers), *Menispermaceae:* Breslau, *Arch. Pharm.* **245**, 586 (1908); Späth, Burger, *Ber.* **59**, 1486 (1926); Reed, *Diss. Abstr.* **23**, 3638 (1963); from *Berberis lambertii* R. N. Parker, *Berberidaceae:* Chatterjee, Banerjee, *J. Indian Chem. Soc.* **30**, 705 (1953). Synthesis from berberine: Cava, Reed, *J. Org. Chem.* **32**, 1640 (1967).

The free base has not been described, but its quaternary salts are known.

Chloride hemipentahydrate. $C_{20}H_{20}ClNO_4 \cdot 2\frac{1}{2}H_2O$. Orange-yellow needles, mp 194°. Sol in water, alcohol.

Chloride tetrahydrate. $C_{20}H_{20}ClNO_4 \cdot 4H_2O$. Orange-yellow prisms, mp 184°.

Iodide. Orange-yellow needles, mp 224°. Sol in water, alcohol.

2476. Columbin. [546-97-4] (1R,4R,4aR,6aR,9S,10aS,10bS)-9-(3-Furanyl)-1,4,4a,5,6,6a,9,10,10a,10b-decahydro-4-hydroxy-4a,10a-dimethyl-1,4-etheno-3H,7H-benzo[1,2-*c*:3,4-*c'*]dipyran-3,7-dione; 15,16-epoxy-1β,4,12-trihydroxy-5,9-dimethyl-17,18-dinor-8βH,9βH,10α-labda-2,13(16),14-triene-19,20-dioic acid 19,1:-20,12-dilactone. $C_{20}H_{22}O_6$; mol wt 358.39. C 67.03%, H 6.19%, O 26.78%. Major bitter principle from the root of *Jatrorrhiza palmata* (DC.) Miers (*J. columba* Miers), *Menispermaceae*. Isoln: Wittstock, *Poggendorff's Ann.* **19**, 298 (1830). Structure: Barton, Elad, *J. Chem. Soc.* **1956**, 2090. Stereochemistry: Cava *et al.*, *Tetrahedron Lett.* **1959**, 1; Overton *et al.*, *J. Chem. Soc. C* **1966**, 1482; Cheung *et al.*, *J. Chem. Soc. B* **1966**, 853.

Very bitter needles from methanol, mp 195°. $[\alpha]_D$ +52.5° (pyridine). uv max (ethanol): 209 nm (log ε 3.78). Practically insol in water. One gram dissolves in 30 ml acetone, 50 ml ethyl acetate, 75 ml methanol; sol also in chloroform, methylene chloride. Bitterness threshold 1:60,000. Easily isomerized to isocolumbin.

Isocolumbin. $C_{20}H_{22}O_6$; mol wt 358.39. Needles from methanol, dec 190°. $[\alpha]_D^{17}$ +0.17° (alk alc). uv max (alc): 209 nm (log ε 3.80).

2477. Combretastatins. Family of tubulin polymerization inhibitors isolated from the bark and stem wood of the South African medicinal plant, *Combretum caffrum* (Eckl. Zeyh.) Kuntze, *Combretaceae*. Combretastatin A-4 is the most bioactive. Isoln of combretastatin: G. R. Pettit *et al.*, *Can. J. Chem.* **60**, 1374 (1982). Antimitotic activity: E. Hamel, C. M. Lin, *Biochem. Pharmacol.* **32**, 3864 (1983). Isoln of A-1 and B-1: G. R. Pettit *et al.*, *J. Nat. Prod.* **50**, 119 (1987); of A-2, A-3, B-2: G. R. Pettit, S. B. Singh, *Can. J. Chem.* **65**, 2390 (1987). Isoln and structure of A-4: G. R. Pettit *et al.*, *Experientia* **45**, 209 (1989); and synthesis: *idem et al.*, *J. Med. Chem.* **38**, 1666 (1995). Enantioselective synthesis of naturally occuring (−)-combretastatin: A. Ramacciotti *et al.*, *Tetrahedron: Asymmetry* **7**, 1101 (1996). Tubulin binding study: C. M. Lin *et al.*, *Biochemistry* **28**, 6984 (1989). Review of A-4 and A-1 as tumor vascular disrupting agents: G. M. Tozer *et al.*, *Int. J. Exp. Pathol.*

83, 21-38 (2002). Review of isoln, structures and syntheses: A. Cirla, J. Mann, *Nat. Prod. Rep.* **20**, 558-564 (2003).

Combretastatin

Combretastatin A-4

Combretastatin. [82855-09-2] (αR)-3-Hydroxy-4-methoxy-α-(3,4,5-trimethoxyphenyl)benzeneethanol; (−)-combretastatin. $C_{18}H_{22}O_6$; mol wt 334.37. Needles from acetone-hexane, mp 130-131°. $[\alpha]_D^{26}$ −8.51° (c = 1.41 in chloroform). uv max (methanol): 212, 228, 279, 287 nm (log ε 4.33, 4.11, 3.47, 3.36). d 1.33.

Combretastatin A-1. [109971-63-3] 3-Methoxy-6-[(1Z)-2-(3,-4,5,-trimethoxyphenyl)ethenyl]-1,2-benzenediol; CA1. $C_{18}H_{20}O_6$; mol wt 332.35. Small, very thin plates from hexane + chloroform, mp 113-115°. d 1.38. uv max (methanol): 233, 255, 298 nm (ε 7145, 7766, 7848).

Combretastatin A-4. [117048-59-6] 2-Methoxy-5-[(1Z)-2-(3,-4,5,-trimethoxyphenyl)ethenyl]phenol; 3,4,5-trimethoxy-3'-hy-droxy-4'-methoxy-(Z)-stilbene; CA4. $C_{18}H_{20}O_5$; mol wt 316.35. Fine crystals from ethyl acetate + hexane, mp 116°.

Combretastatin A-4 phosphate *see* Fosbretabulin.

2478. Concanavalin A. [11028-71-0] ConA. The most extensively investigated member of the lectin family of plant proteins. Unlike most lectins, it lacks covalently bound carbohydrate and therefore is not a glycoprotein. Isolated from jack bean, *Canavalia ensiformis, Papilionatae*: J. B. Sumner, S. F. Howell, *J. Bacteriol.* **32**, 227 (1936). Its function in *C. ensiformis* is unknown, but it agglutinates a variety of somatic and germ line cells through specific interaction with saccharide-containing cell surface receptors and restores the growth pattern of virus-transformed fibroblasts in tissue culture to that of normal cells: M. M. Burger, K. D. Noonan, *Nature* **228**, 512 (1970); G. M. Edelman, C. F. Millette, *Proc. Natl. Acad. Sci. USA* **68**, 2436 (1971). Differential toxicity on normal and transformed cells *in vitro* and inhibition of tumor development *in vivo* have also been reported: J. Shoham *et al., Nature* **227**, 1244 (1970). The molecule consists of identical polypeptide subunits of mol wt about 27,000, existing as dimers in soln at pH <6 and as tetramers at physiologic pH. The proposed amino acid sequence contains 238 residues; con A has also been shown to have binding site for transition metal ions and calcium ions in addition to saccharide binding sites: G. M. Edelman *et al., Proc. Natl. Acad. Sci. USA* **69**, 2580 (1972). The transition ion, usually Mn^{+2} or Ca^{+2}, apparently stabilizes the formation of the specific saccharide binding site: M. Shoham *et al., Biochemistry* **12**, 1914 (1973). Circular dichroism-NMR study of metal binding sites: A. R. Palmer *et al., ibid.* **19**, 5063 (1980). Oligosaccharide binding study: A. Vanlands *et al., Eur. J. Biochem.* **103**, 307 (1980). Use of con A to study immunoregulation of human T cells: D. M. Dwyer, C. Johnson, *Clin. Exp. Immunol.* **46**, 237 (1981); E. L. Larson *et al., Immunobiology* **161**, 5 (1982). For general refs, *see* Lectins.

USE: As a reagent in analytical and preparative biochemistry; as a probe in studies of cell surface membrane dynamics and cell division.

2479. Condurangin. [1401-98-5] Kondurangin. Bitter principle from the bark of the Condurango vine, *Marsdenia condurango* Reichb. f., *Asclepiadaceae*, also known as **eagle vine, mataperro, condor vine.** Habit. Ecuador, Peru. Isoln: Korte, Korte, *Z. Naturforsch.* **10b**, 223 (1955); Zechner, Zölss, *Sci. Pharm.* **24**, 107 (1956). Separation of fractions: Cellarius, Zechner, *ibid.* **34**, 10 (1966). Proposed structure of the aglycone, **condurangogenin** A: Tschesche *et al., Tetrahedron* **21**, 1777, 1797 (1965); **23**, 1461 (1967).

Crystals from dil methanol, mp 186-188°. uv max (methanol): 277 nm. Soluble in chloroform, methanol; slightly sol in water. Practically insol in ether, petr ether. Bitterness threshold 1:20,000.

THERAP CAT: Astringent.

2480. Conessine. [546-06-5] (3β)-N,N-Dimethylcon-5-enin-3-amine; 3β-(dimethylamino)con-5-enine; neriine; roquessine; wrightine. $C_{24}H_{40}N_2$; mol wt 356.60. C 80.84%, H 11.31%, N 7.86%. Antiamebic, antifungal principle in Kurchi bark, the bark of *Holarrhena anti-dysenterica* Wall., *Apocynaceae*, native to India; also found in African species, such as *H. africana* A.DC.; *H. congolensis* Stapf; *H. wulfsbergii* Stapf, and *H. febrifuga* Klotzsch, *Apocynaceae*. Isoln: Haines, *Trans. Med. Soc. Bombay* **4**, 28 (1858); Warnecke, *Ber.* **19**, 60 (1886); Bertho, *Arch. Pharm.* **277**, 237 (1939); Siddiqui, Pillay, *J. Indian Chem. Soc.* **9**, 553 (1932). Structure: Haworth *et al., J. Chem. Soc.* **1953**, 1102; Haworth, McKenna, *Chem. Ind. (London)* **1957**, 1510; Bertho, Götz, *Ann.* **619**, 96 (1958). Discussion of structure in Fieser, Fieser, *Steroids* (New York, 1959), p 858. Synthesis from Δ^5-pregnene-3β,20β-diol: Barton, Morgan, *Proc. Chem. Soc. London* **1961**, 206; *J. Chem. Soc.* **1962**, 622. Total synthesis: Marshall, Johnson, *J. Am. Chem. Soc.* **84**, 1485 (1962); Stork *et al., ibid.* 2018; Nagata *et al., Tetrahedron Lett.* **1963**, 869.

Leaflets or broad plates from acetone, mp 127-128.5°. $[\alpha]_D^{20}$ +25.3° (c = 0.7 in abs alc). Sparingly sol in water.

Dihydrochloride monohydrate. $C_{24}H_{40}N_2 \cdot 2HCl \cdot H_2O$. Crystals, sol in water, mp >340°. $[\alpha]_D^{20}$ +9.3° (c = 0.9 in H_2O).

Dihydrobromide. [5913-82-6] $C_{24}H_{40}N_2 \cdot 2HBr$. Crystals, dec 340°. Very bitter taste. $[\alpha]_D^{20}$ +7.0° (c = 5 in H_2O). Sol in water. Slightly sol in 95% ethanol. Very slightly sol in ether. Practically insol in petr ether.

2481. Conestat Alfa. [80295-38-1] Rhucin; Ruconest. Recombinant human C1 esterase inhibitor (rhC1INH) expressed in the milk of transgenic rabbits. Prepn: J. H. Nuijens *et al.*, **WO 0157059**; *eidem*, **US 7067713** (2001, 2006 both to Pharming). Analysis of glycosylation pattern: K. Koles *et al., Glycobiology* **14**, 51 (2004). Clinical pharmacokinetics: M. B. A. van Doorn *et al., J. Allergy Clin. Immunol.* **116**, 876 (2005). Overview of development and therapeutic potential in hereditary angioedema (HAE) and cerebral ischemia: H. Longhurst, *Curr. Opin. Investig. Drugs* **9**, 310-323 (2008). Clinical trial in treatment of HAE: B. Zuraw *et al., J. Allergy Clin. Immunol.* **126**, 821 (2010).

THERAP CAT: In treatment of hereditary angioedema.

2482. Congo Red. [573-58-0] 3,3'-[[1,1'-Biphenyl]-4,4'-diylbis(2,1-diazenediyl)]bis[4-amino-1-naphthalenesulfonic acid] sodium salt (1:2); C.I. Direct Red 28; 3,3'-[[1,1'-biphenyl]-4,4'-diylbis(azo)]bis[4-amino-1-naphthalenesulfonic acid] disodium salt; sodium diphenyldiazo-bis-α-naphthylaminesulfonate; C.I. 22120. $C_{32}H_{22}N_6Na_2O_6S_2$; mol wt 696.66. C 55.17%, H 3.18%, N 12.06%, Na 6.60%, O 13.78%, S 9.20%. Prepn: Böttiger, **DRP 28753** (1884); *Frdl.* **1**, 470; Whitehead, *Chem. Trade J.* **77**, 386 (1925); Kline, *J. Chem. Educ.* **15**, 129 (1938). Purification: Bedaux *et al., Pharm. Weekbl.* **98**, 189 (1963). *See also Colour Index* **vol. 4** (3rd ed., 1971) p 4166; H. J. Conn's *Biological Stains*, R. W. Horobin, J. A. Kiernan, Eds. (BIOS Scientific, Oxford, 10th ed., 2002) 132-134. Toxicity: Richardson, Dillon, *Am. J. Med. Sci.* **198**, 73 (1939). Clinical evaluation as diagnostic aid: E. Ouchi *et al., Tohoku J. Exp. Med.* **118**, Suppl., 191 (1976). Use in combination with Sirius red as histological stain: I. R. Hinds, *Lab. Med.* **16**, 366 (1985).

Brownish-red powder. Absorption max (pH 7.3): 488 nm ($E_{1cm}^{1\%}$ 595). Sol in water (yellowish-red) and in ethanol (orange); very slightly sol in acetone. Practically insol in ether. LD_{50} i.v. in rats: 190 mg/kg (Richardson, Dillon).

Congo Red Paper. Riegel's paper; Herzberg's paper. Paper charged with a 0.1% aq soln of Congo Red.

USE: As indicator, usually in 0.1% aq soln for estimating free mineral acids, particularly in presence of organic acids. pH: 3.0 blue-violet, 5.0 red. Detecting and estimating free HCl in gastric contents; detecting acidity of papers. Reagent for bitter-almond water. Dye. As addition to culture media. Biological stain for cytoplasm, elastic fibers, axons, embryonic material, amyloid, cellulose, etc.

THERAP CAT: Diagnostic aid (amyloidosis).

2483. Congressane. [2292-79-7] Decahydro-3,5,1,7-[1,2,3,-4]butanetetraylnaphthalene; pentacyclo[7.3.1.1^{4,12}.0^{2,7}.0^{6,11}]-tetradecane; diamantane. $C_{14}H_{20}$; mol wt 188.31. C 89.30%, H 10.71%. Isoln from petroleum: Hala *et al.*, *Angew. Chem. Int. Ed.* **5**, 1045 (1966). Synthesis: Cupas *et al.*, *J. Am. Chem. Soc.* **87**, 917 (1965); Gund *et al.*, *J. Org. Chem.* **39**, 2979 (1974). Structure: Karle, Karle, *J. Am. Chem. Soc.* **87**, 918 (1965).

Crystals, mp 236-237°.

Note: Structure is congress emblem of the International Congress of Pure and Applied Chemistry.

2484. Conhydrine. [495-20-5] (αR,2S)-α-Ethyl-2-piperidinemethanol; 2-(α-hydroxypropyl)piperidine. $C_8H_{17}NO$; mol wt 143.23. C 67.09%, H 11.96%, N 9.78%, O 11.17%. From seeds of *Conium maculatum* L., *Umbelliferae*. Isoln: Späth, Adler, *Monatsh. Chem.* **63**, 127 (1933). Stereochemistry: Hill, *J. Am. Chem. Soc.* **80**, 1609 (1958); Sicher, Ticky, *Collect. Czech. Chem. Commun.* **23**, 2081 (1958).

Crystals from ether, mp 121°. bp 226°. $[\alpha]_D$ +10°. Slightly sol in water; sol in alc, chloroform, ether.

2485. β-Coniceine. [538-90-9] 2-(1-Propen-1-yl)piperidine; α-allylpiperidine. $C_8H_{15}N$; mol wt 125.22. C 76.74%, H 12.07%, N 11.19%. Prepn from conhydrine: Löffler, Friederick, *Ber.* **42**, 107 (1909); Löffler, Tschunke, *ibid.* 929.

***dl*-Form.** Liquid, solidifies around 8°. bp 168°. d_4^{15} 0.8716. Sol in alcohol, ether; very slightly sol in water.

***dl*-Form hydrochloride.** $C_8H_{15}N.HCl$. mp 206°. Sol in water.

***d*-Form.** Needles, mp 39°. bp 168°. $[\alpha]_D^{45}$ +50°. Sol in alcohol, ether; very slightly sol in water.

***d*-Form hydrochloride.** mp 182°. Sol in water.

***l*-Form.** Crystals, mp 40°. bp 168°. $[\alpha]_D^{45}$ −50°. Sol in alcohol, ether; very slightly sol in water.

***l*-Form hydrochloride.** mp 183°. Sol in water.

2486. γ-Coniceine. [1604-01-9] 2,3,4,5-Tetrahydro-6-propylpyridine; 2-*n*-propyl-3,4,5,6-tetrahydropyridine; 2-*n*-propyl-Δ^1-piperidine. $C_8H_{15}N$; mol wt 125.22. C 76.74%, H 12.07%, N 11.19%. Easily reduced to *dl*-coniine, *q.v.* From seeds of *Conium maculatum* L., *Umbelliferae*: Cromwell, *Biochem. J.* **64**, 259 (1956); Fairbairn, Challen, *ibid.* **72**, 556 (1959); Fairbairn, Sewal,

Phytochemistry **1**, 38 (1961). Structure: Beyerman *et al.*, *Rec. Trav. Chim.* **80**, 513 (1961). Synthesis: Lukes *et al.*, *Collect. Czech. Chem. Commun.* **12**, 641 (1947); Cervinka, *Chem. Listy* **52**, 1145 (1958).

Alkaline liquid, mousy odor. bp 171°; bp$_{15}$ 63°. Volatile with steam. d_4^{15} 0.8753. n_D^{16} 1.4661. Slightly sol in water; freely sol in alcohol, chloroform, ether.

Hydrochloride. $C_8H_{15}N.HCl$. Hygroscopic crystals from ether, mp 143°.

2487. Coniferin. [531-29-3] 4-(3-Hydroxy-1-propen-1-yl)-2-methoxyphenyl β-glucopyranoside; 4-hydroxy-3-methoxy-1-(γ-hydroxypropenyl)benzene-4-D-glucoside; abietin; laricin. $C_{16}H_{22}O_8$; mol wt 342.34. C 56.14%, H 6.48%, O 37.39%. Principal glucoside of the conifers. Also in comfrey root, sugar beet, and asparagus. Extraction from the cambium layer of fir: Solntsev, *C.A.* **38**, 3780 (1944). Synthesis: Pauley, Feuerstein, *Ber.* **60**, 1031 (1927). Hydrolysis by emulsin yields coniferyl alcohol and D-glucose. Yields lignin-like material by enzymatic dehydrogenation and polymerization: Freudenberg *et al.*, *Ber.* **85**, 641 (1952); Freudenberg, Rasenack, *Ber.* **86**, 756 (1953).

Dihydrate. Crystals from water. Anhydr after 4 hrs at 100°. Anhydr coniferin, mp 186°. $[\alpha]_D^{20}$ −68° (c = 0.5). Absorption spectrum: Patterson, Hibbert, *J. Am. Chem. Soc.* **65**, 1862 (1943). One gram dissolves in 200 ml water, freely sol in boiling water; sol in pyridine; sparingly sol in alc. Practically insol in ether.

2488. Coniferyl Alcohol. [458-35-5] 4-(3-Hydroxy-1-propen-1-yl)-2-methoxyphenol; 3-(4-hydroxy-3-methoxyphenyl)-2-propen-1-ol; 4-hydroxy-3-methoxycinnamic alcohol; γ-hydroxyisoeugenol. $C_{10}H_{12}O_3$; mol wt 180.20. C 66.65%, H 6.71%, O 26.64%. In benzoin, especially Siam benzoin, as coniferyl benzoate (up to 78%): Reinitzer, *Arch. Pharm.* **264**, 131 (1926). From coniferin by hydrolysis with emulsin. From ferula aldehyde (coniferyl aldehyde) by the action of fermenting yeast. Small amounts of coniferyl alcohol are found in brandy from beet molasses.

Prisms, mp 74°. Freely sol in ether; moderately sol in alc; almost insol in water; sol in alkalies. Polymerized by dil acids; converted to an amorphous gum. Absorption spectrum: Herzog, Hillmer, *Ber.* **64B**, 1288 (1931).

2489. Coniine. [458-88-8] (2S)-2-Propylpiperidine; cicutine; conicine. $C_8H_{17}N$; mol wt 127.23. C 75.52%, H 13.47%, N 11.01%. Toxic principle of poison hemlock, *Conium maculatum* L., *Umbelliferae*: Ladenburg, *Ber.* **19**, 439 (1886); Koller, *Monatsh. Chem.* **47**, 393 (1926). Occurs naturally as the (S)-(+)-isomer. Review of early literature: Marion in *The Alkaloids* vol. I, R. H. F. Manske, H. L. Holmes, Eds. (Academic Press, New York, 1950) pp 211-217. Isoln from the pitcher plant, *Sarracenia flava* and insect paralyzing properties: N. V. Mody *et al.* *Experientia* **32**, 829 (1976). Absolute configuration: J. C. Craig, S. K. Roy, *Tetrahedron* **21**, 401 (1965). Resolution of (±)-form: J. C. Craig, A. R. Pinder, *J. Org. Chem.* **36**, 3648 (1971). Synthesis of (S)-(+)-form: K. Aketa *et al.*, *Chem. Pharm. Bull.* **24**, 621 (1976); of (R)-(−)-form: D. Lathbury,

T. Gallagher, *Chem. Commun.* **1986**, 114; of (±)-form: T. Nagasaka *et al.*, *Heterocycles* **27**, 1685 (1988). Enantiospecific synthesis of (+)- and (−)-forms: L. Guerrier *et al.*, *J. Am. Chem. Soc.* **105**, 7754 (1983).

Colorless alkaline liquid, darkens and polymerizes on exposure to light and air. Mousy odor. mp ~ −2°. bp 166-166.5°; bp$_{20}$ 65-66°. Volatile with steam. d$_4^{20}$ 0.844-0.848. n$_D^{23}$ 1.4505. [α]$_D^{25}$ +8.4° (c = 4.0 in CHCl$_3$); [α]$_D^{23}$ +14.6° (neat). pK 3.1. One ml dissolves in 90 ml water, less sol in hot water. The base dissolves about 25% water at room temp. Sol in alcohol, ether, acetone, benzene, amyl alcohol; slightly sol in chloroform.

Hydrobromide. [637-49-0] C$_8$H$_{17}$N.HBr. Prisms, mp 211°. One gram dissolves in 2 ml water, 3 ml alcohol; sol in chloroform, ether.

Hydrochloride. [555-92-0] C$_8$H$_{17}$N.HCl. Rhomboids, mp 221°. Freely sol in water, alcohol, chloroform.

(R)-(−)-Form. Liquid, bp$_{756}$ 165°. [α]$_D^{25}$ −8.1° (c = 4.0 in CHCl$_3$); [α]$_D^{23}$ −14.2° (neat).

(±)-Form. bp 200-210°.

Caution: Potential symptoms of overexposure include drowsiness, paresthesias, weakness, ataxia, nausea, profuse salivation, and bradycardia followed by tachycardia. *See: Clinical Toxicology of Commercial Products*, R. E. Gosselin *et al.*, Eds. (Williams & Wilkins, Baltimore, 5th ed., 1984) Section II, pp 249-250.

2490. Conium Fruit. Hemlock; poison hemlock; spotted hemlock; poison parsley; spotted cowbane. Full-grown, but unripe, carefully dried fruit of *Conium maculatum* L., Umbelliferae. *Habit.* Europe, Asia, naturalized in U.S. *Constit.* 0.5-1.5% coniine, conhydrine, pseudoconhydrine, methylconiine, ethylpiperidine, coniic acid, volatile and fixed oil. Has antispasmodic activity.

2491. Conivaptan. [210101-16-9] *N*-[4-[(4,5-Dihydro-2-methylimidazo[4,5-*d*][1]benzazepin-6(1*H*)-yl)carbonyl]phenyl]-[1,1′-biphenyl]-2-carboxamide; 4′-[(2-methyl-1,4,5,6-tetrahydroimidazo[4,5-*d*][1]benzazepin-6-yl)carbonyl]-2-phenylbenzanilide. C$_{32}$H$_{26}$N$_4$O$_2$; mol wt 498.59. C 77.09%, H 5.26%, N 11.24%, O 6.42%. Nonpeptide, dual vasopressin V$_{1a}$/V$_2$ receptor antagonist. Prepn: A. Tanaka *et al.*, **WO 9503305**; *eidem*, **US 5723606** (1995, 1998 both to Yamanouchi); A. Matsuhisa *et al.*, *Chem. Pharm. Bull.* **48**, 21 (2000). Pharmacology: Y. Yatsu *et al.*, *Eur. J. Pharmacol.* **321**, 225 (1997). Binding specificity study: A. Tahara *et al.*, *Peptides* **19**, 691 (1998). Clinical pharmacokinetics: M. Burnier *et al.*, *Eur. J. Clin. Pharmacol..* **55**, 633 (1999). Clinical evaluation in advanced heart failure: J. E. Udelson *et al.*, *Circulation* **104**, 2417 (2001). Review of clinical development: S. A. Doggrell, *Curr. Opin. Investig. Drugs* **6**, 317-326 (2005).

Hydrochloride. [168626-94-6] CI-1025; YM-087; YM-35087; Vaprisol. C$_{32}$H$_{26}$N$_4$O$_2$.HCl; mol wt 535.04. Crystals, mp >250°.
THERAP CAT: In treatment of congestive heart failure.

2492. Conjugated Estrogens. [12126-59-9] Conjugated equine estrogens; CEE; Premarin. Complex mixture of natural estrogens obtained from equine urine or synthetically from estrone and equilin. Principal components are sodium estrone sulfate and sodium equilin sulfate; also known to contain 17α-dihydroequilin, 17α-estradiol, and 17β-dihydroequilin as sodium sulfate conjugates. Purification from urine of pregnant mares: R. W. Bates, H. Cohen, **US 2565115** (1951 to Squibb); E. T. Stiller, E. O'Keefe, **US 2720483**

(1955 to Olin Mathieson). HPLC determn of major components in serum: S. Y. Su *et al.*, *Biomed. Chromatogr.* **6**, 265 (1992). Review of pharmacokinetics and metabolism: B. R. Bhavnani, *Proc. Soc. Exp. Biol. Med.* **217**, 6-16 (1998); of pharmacology and neuroprotective effects: *idem, J. Steroid Biochem. Mol. Biol.* **85**, 473-482 (2003). Clinical assessment of efficacy and safety in combination with progestin in hormone replacement therapy: J. E. Rossouw *et al.*, *J. Am. Med. Assoc.* **288**, 321 (2002); of conjugated estrogens alone: G. L. Anderson *et al.*, *ibid.* **291**, 1701 (2004).
Caution: These substances are listed as known human carcinogens: *Report on Carcinogens, Twelfth Edition* (PB2011-111646, 2011) p 184.
THERAP CAT: Estrogen.

2493. Connexins. Family of gap-junction structural proteins found in virtually all metazoans; the assembly of which permits rapid intercellular communication by ions, second messengers and small metabolites. Named as the species of origin Cx, where x is the predicted molecular mass in kDa, e.g. rat C43. These transmembrane proteins are characterized by 2 extracellular loops, 4 hydrophobic membrane regions and a cytoplasmic loop. Oligomerization into a hexamer forms a **connexon** or hemichannel which interacts across the extracellular gap with a corresponding connexon to complete the intercellular channel. Description of connexon structure and composition: D. L. D. Caspar *et al.*, *J. Cell Biol.* **74**, 605, 629 (1977). Isolation of rat C32 via molecular cloning: D. L. Paul, *ibid.* **103**, 123 (1986). Tissue distribution: E. C. Beyer *et al.*, *ibid.* **108**, 595 (1987). Review of early works on gap junction structure: P. N. T. Unwin, G. Zampighi, *Nature* **283**, 545-549 (1980); of molecular structure and biochemical characterization: E. C. Beyer *et al., J. Membr. Biol.* **116**, 187-194 (1990). Review of biosynthesis, function and turnover: D. W. Laird, *J. Bioenerg. Biomembr.* **28**, 311-318 (1996); of structure/activity: C. Peracchia, X. G. Wang, *Braz. J. Med. Biol. Res.* **30**, 577-590 (1997); of mutations in human genetic diseases: V. Krutovskikh, H. Yamasaki, *Mutat. Res.* **462**, 197-207 (2000). *Review:* D. A. Goodenough *et al., Annu. Rev. Biochem.* **65**, 475-502 (1996).

2494. Conotoxins. Ctx. Family of related peptide neurotoxins isolated from the venom of genus *Conus*, the marine cone snail. Wide variety of venoms are found, almost all of which are under 30 amino acids in length. The most common classes are ω-and δ-conotoxins which target voltage-gated calcium and sodium channels respectively; α-conotoxins which inhibit acetylcholine receptor; μ-conotoxins which block sodium channels. Review characterizing binding of ω-Ctx: E. Sher, F. Clementi, *Neuroscience* **42**, 301-307 (1991); and use: K. Takeda, J. J. Nordmann, *Methods Neurosci.* **8**, 202-222 (1992). Review of α-CTX isolation and binding: J. M. McIntosh, "α-*Conotoxins and Nicotinic Acetylcholine Receptors*", in *Biochemical Aspects of Marine Pharmacology*, P. Lazarovici *et al.*, Eds. (Alaken Inc., Fort Collins, CO, 1996) pp 28-34. *Review:* B. M. Olivera *et al., ACS Symp. Ser.* **418**, 256-278 (1990).
USE: Biochemical probes for ion channels.

2495. Contusugene Ladenovec. [600735-73-7] Ad5CMV-p53; INGN-201; Advexin. Replication-impaired adenoviral vector that carries the human, wild-type *p53* tumor suppressor gene under the control of a cytomegalovirus (CMV) promoter. Derived from adenovirus serotype 5 (Ad5) rendered replication defective by deletion of the envelope glycoprotein E1 region. Used as gene therapy to compensate for mutations in the *p53* gene that are characteristic of tumors. Prepn: W.-W. Wang, J. A. Roth, **WO 9512660**; *eidem*, **US 6410010** (1995, 2002 both to Board of Regents, Univ. Texas). Determn in plasma by PCR-based assay: P. Saulnier *et al., J. Virol. Methods* **114**, 55 (2003). Pharmacokinetics in patients with advanced cancer: A. W. Tolcher *et al., J. Clin. Oncol.* **24**, 2052 (2006). Clinical evaluation as gene replacement therapy with radiation therapy in lung cancer: S. G. Swisher *et al., Clin. Cancer Res.* **9** 93 (2003). Review of development and clinical experience: D. I. Gabrilovich, *Expert Opin. Biol. Ther.* **6**, 823-832 (2006); N. Senzer, J. Nemunaitis, *Curr. Opin. Mol. Ther.* **11**, 54-61 (2009). Review of therapeutic potential in Li-Fraumeni syndrome: J. M. Nemunaitis, J. Nemunaitis, *Future Oncol.* **4**, 759-738 (2008).
THERAP CAT: Antineoplastic.

2496. Convallaria. Lily of the valley; May lily; Park lily; May blossom. Dried flowers of *Convallaria majalis* L., *Liliaceae. Habit.*

U.S., Europe, Northern Asia; cultivated in gardens. *Constit.* of dried flowers: Convallatoxin, convallarin, volatile oil; of dried rhizome and roots: Glycosides of convallamarin, convallatoxin, convallarin. Extraction and identification of glycosides: Tschesche, Seehofer, *Ber.* **87**, 1108 (1954); Erbring, Patt, *Arzneim.-Forsch.* **8**, 554 (1958).

Convallaria glycosides. Convacard; Convallan; Convallen; Convalyt; Convasid. An aqueous extract containing the total glycosides of Convallaria.

THERAP CAT: Cardiotonic.

THERAP CAT (VET): Has been used as a cardiotonic and diuretic.

2497. Convallatoxin. [508-75-8] (3β,5β)-3-[(6-Deoxy-α-L-mannopyranosyl)oxy]-5,14-dihydroxy-19-oxocard-20(22)-enolide; strophanthidin α-L-rhamnoside; Convallaton; Corglykon; Korglykon. C$_{29}$H$_{42}$O$_{10}$; mol wt 550.65. C 63.26%, H 7.69%, O 29.05%. From blossoms of lily of the valley (*Convallaria majalis* L., *Liliaceae*): Karrer, *Helv. Chim. Acta* **12**, 506 (1929); from *Ornithogalum umbellatum* L., *Liliaceae*: Mrozik *et al.*, *ibid.* **42**, 683 (1959); from *Antiaris toxicaria* Lesch, *Moraceae*: Juslen *et al.*, *ibid.* **46**, 117 (1963). Structure: Tschesche, Haupt, *Ber.* **69**, 459 (1936); Fieser, Jacobsen, *J. Am. Chem. Soc.* **59**, 2335 (1937). Cleavage into L-rhamnose and strophanthidin: Reichstein, Katz, *Pharm. Acta Helv.* **18**, 521 (1943); *see also C.A.* **38**, 2046. Synthesis from strophanthidin and acetobromrhamnose: Reyle *et al.*, *Helv. Chim. Acta* **33**, 1541 (1950); Haede *et al.*, **DE 1933090** (1971 to Hoechst). Toxicity study: W. Förster *et al.*, *Arch. Int. Pharmacodyn.* **155**, 165 (1965).

Prisms from methanol + ether. mp 235-242°. [α]$_D^{22}$ −1.7 ± 3° (c = 0.65 in methanol); [α]$_D^{25}$ −9.4 ± 3° (c = 0.72 in dioxane). Sol in alcohol, acetone; slightly sol in chloroform, ethyl acetate, and water (1:2000). Practically insol in ether, petr ether. Gives Legal's reaction. LD$_{50}$ in mice, rats (mg/kg): 10.0 i.p., 16.0 i.v. (Förster).

Tri-O-acetyl-convallatoxin. C$_{35}$H$_{48}$O$_{13}$. Needles from acetone + ether, mp 215-238°. [α]$_D^{25}$ −5.5 ± 2° (c = 0.962 in chloroform).

THERAP CAT: Cardiotonic.

2498. Convicine. [19286-37-4] 6-Amino-5-(β-D-glucopyranosyloxy)-2,4(1H,3H)-pyrimidinedione. C$_{10}$H$_{15}$N$_3$O$_8$; mol wt 305.24. C 39.35%, H 4.95%, N 13.77%, O 41.93%. From seeds of *Vicia sativa* L., *Leguminosae*: Ritthausen, *J. Prakt. Chem.* [2] **24**, 202 (1881); *Ber.* **29**, 894 (1896). Structure: Johnson, *J. Am. Chem. Soc.* **36**, 337 (1914); Fisher, Johnson, *ibid.* **54**, 2038 (1932); Bendick, Clements, *Biochim. Biophys. Acta* **12**, 462 (1953); Bien *et al.*, *J. Chem. Soc. C* **1968**, 496. Synthesis: *eidem, J. Chem. Soc. Perkin Trans. 1* **1973**, 1089.

Leaflets from boiling water, dec without melting at 287°. uv max (pH 7): 245, 271 nm (log ε 3.43, 4.16). Sol in hot water, dil NaOH; slightly sol in cold water, alcohol. Practically insol in chloroform, glacial acetic acid.

2499. COP. [26591-04-8] 2-Methyl-2-propenoic acid 2-oxiranylmethyl ester polymer with ethyl 2-propenoate; poly(glycidyl methacrylate-co-ethyl acrylate); [P(GMA-co-EA)]; glycidyl methacrylate-ethyl acrylate polymer; COP Resist. (C$_7$H$_{10}$O$_3$.C$_5$H$_8$O$_2$)$_x$. Negative electron resist. Prepn and characterization: L. F. Thompson *et al.*, *Tech. Pap. Reg. Tech Conf. Soc. Plast. Eng.* **1973**, (Oct.) 127. Molecular parameters and lithographic performance: L. F. Thompson *et al.*, *J. Vac. Sci. Technol.* **12**, 1280 (1975). Radiation chemistry: E. D. Feit *et al.*, *ACS Org. Coat. Plast. Chem.* **35**, 287 (1975). Use as matrix for deep UV resist materials: C. Kutal, S. K. Weit, *J. Coat. Technol.* **62**, 63 (1990); S. K. Weit *et al.*, *Chem. Mater.* **4**, 453 (1992).

USE: In photosensitive materials as polymer matrix. Photoresist.

2500. Copaene. [3856-25-5] (1R,2S,6S,7S,8S)-1,3-Dimethyl-8-(1-methylethyl)tricyclo[4.4.0.0^{2,7}]dec-3-ene; (1R,2S,6S,7S,8S)-(−)-8-isopropyl-1,3-dimethyltricyclo[4.4.0.0^{2,7}]dec-3-ene; α-copaene. C$_{15}$H$_{24}$; mol wt 204.36. C 88.16%, H 11.84%. Tricyclic sesquiterpene occurring in African copaiba balsam oil; in the so-called supa oil from *Sindora supa* Merr. (*S. wallichii* F.-Vill.), *Leguminosae*: Henderson *et al.*, *J. Chem. Soc.* **1926**, 3077. Also in the essential oil of *Phyllocladus trichomanoides* D. Don, *Podocarpaceae*: Briggs, Sutherland, *J. Org. Chem.* **13**, 1 (1948); from *Cyperus articulatus* Michx., *Cyperaceae*: Couchman *et al.*, *Tetrahedron* **20**, 2037 (1964). Structural studies: Kapadia *et al.*, *ibid.* **21**, 607 (1965). Absolute stereostructure: deMao *et al.*, *ibid.* 619. Synthesis: Heathcock, *J. Am. Chem. Soc.* **88**, 4110 (1966); Heathcock *et al.*, *ibid.* **89**, 4133 (1967); Corey, Watt, *ibid.* **95**, 2303 (1973). *Review:* J. Simonsen, D. H. R. Barton, *The Terpenes* vol. III (University Press, Cambridge, 1952) pp 88-91. *See also* Ylangene.

Oily liquid, bp 246-251°; bp$_{10}$ 119-120°; d$_{15}^{15}$ 0.9077; n$_D^{20}$ 1.4894. [α]$_D^{22}$ −6.3° (c = 1.20 in chloroform).

2501. Copaiba. [8001-61-4] Balsams copaiba; balsam capivi; Jesuit's balsam. Oleoresin from trees of the genus, *Copaifera*, *Leguminosae*, particularly *C. reticulata* Ducke (Para balsam), *C. langsdorffii* Desf. (Maranhao balsam) and *C. guianensis* (Maracaibo balsam). Used in traditional medicine as an antiseptic, diuretic and antiinflammatory. *Habit.* Tropical regions of South America and southern Africa. *Constit.* Varies among the species; volatile oil (20-90%); diterpene acids, such as kaurenoic, copalic, hardwickiic acids. Brief description: D. L. J. Opdyke, *Food Cosmet. Toxicol.* **14**, 687 (1976); of copaiba oil: *idem, ibid.* **11**, 1075 (1973). Chemical characterization of 3 species: V. Cascon, B. Gilbert, *Phytochemistry* **55**, 773 (2000). Volatile constituents of *C. langsdorffii*: N. V. Gramosa, E. R. Silveira, *J. Essent. Oil Res.* **17**, 130 (2005). Review of ethnobotany and uses: C. Plowden, *Econ. Bot.* **58**, 729-739 (2004).

Color, scent and viscosity varies widely ranging from pale yellow fluid to thick, brownish-yellow liquid; characteristic odor; unpleasant taste. Insol in water. Sol in benzene, chloroform, ether, oils, carbon disulfide, abs alcohol, petr ether, partly in 95% alcohol.

Volatile oil. [8013-97-6] Oil of copaiba. Obtained by steam distillation of the oleoresin. *Constit.* Highly variable, chiefly β-caryophyllene, β-bisabolene, α-copaene, α-humulene. Colorless or pale yellow liquid; slightly bitter and pungent taste. d$_{25}^{25}$ 0.880-0.907. bp 250-275°. Rotation −7 to −33°. n$_D^{20}$ 1.493-1.500. Sol in alcohol, most fixed oils, mineral oil. Insol in glycerin, propylene glycol. *Keep well closed, cool and protected from light.*

USE: In varnishes, paints, lacquer; for restoration of oil paintings. As frangrance or odor fixative in soaps, shampoo, lotions, perfumes.

2502. Copal. [9000-14-0] Resin copal; gum copal; anime (soft copal); kaurie; cowrie. Name used for aromatic resins exuding from various species of *Trachylobium*, *Hymenaea courbaril* L., etc., *Leguminosae*. Used by native Mexican and South American cultures as ceremonial incense. Occurs as hard or soft copals. *Habit*. Central and South America, Zanzibar, Mozambique, Australia, Philippine Islands and West Indies. Composed of a complex mixture of diterpenes, hardens upon ageing due to polymerization of communic acids. Analysis of constituents and ageing behavior: D. Scalarone *et al.*, *J. Anal. Appl. Pyrolysis* **68-69**, 115 (2003). Analysis by ATR-FIR spectroscopy: S. Prati *et al.*, *Anal. Bioanal. Chem.* **399**, 3081 (2011).

Yellowish to yellowish-brown pieces of various sizes; conchoidal fracture. Hard copals are almost insol in usual solvents; soft copals are partly sol in alcohol, chloroform or glacial acetic acid; both copals after having been fused are sol in oil turpentine and linseed oil.

USE: In varnishes and cements; as substitute for amber. In dentistry, for modeling compds and cavity varnishes.

2503. Copernicium. [54084-26-3] Ununbium. Cn; Uub; at. no. 112. Group IIb (12). Transuranium element. No stable nuclides. Prepn and decay of isotope $^{277}112$ ($T_{1/2} \sim 240$ μsec, α decay) by $^{208}Pb(^{70}Zn,n)$: S. Hoffmann *et al.*, *Z. Phys. A* **354**, 229 (1996); of isotope $^{283}112$ ($T_{1/2} \sim 81$ sec, decay by SF) by $^{238}U(^{48}Ca,3n)$: Y. T. Oganessian *et al.*, *Eur. Phys. J. A* **5**, 63 (1999). Observation of $^{281}112$ as decay product of isotope $^{285}114$ ($T_{1/2} \sim 890^{+1300}_{-450}$ μsec, α decay): V. Ninov *et al.*, *Phys. Rev. Lett.* **83**, 1104 (1999). Synthesis and decay chains: S. Hofmann *et al*, *Eur. Phys. J. A* **14**, 147 (2002); Y. T. Oganessian *et al.*, *ibid.* **19**, 3 (2004). Review of production and properties: S. Hofmann, *Rep. Prog. Phys.* **61**, 639-689 (1998).

2504. Copper. [7440-50-8] Cu; at. wt 63.546; at. no. 29; valences 1, 2. Group IB (11). Occurrence in the earth's crust: 70 ppm; also present in seawater: 0.001-0.02 ppm. Two naturally occurring isotopes: 63 (69.09%), 65 (30.91%); nine artificial isotopes: 58-62, 64, 66-68. One of the earliest known metals. Found in nature in its native state; also in combined form in several minerals including chalcopyrite, *q.v.*, **chalcocite**, **bornite**, (Cu_5FeS_4), **tetrahedrite** ($Cu_{12}Sb_4S_{13}$), **enargite** (Cu_3AsS_4), **antlerite** ($Cu_3H_4O_8S$). Extraction from ores: Clark-Hawley, *Encyclopedia of Chemistry* (Reinhold, New York, 2nd ed., 1966) p 288. Metallurgy of copper and its alloys: *Metals Reference Book* vols. **1 & 2**, C. J. Smithells, Ed. (Butterworth's, London, 3rd ed., 1962). A trace element essential to many plants and animals. Occurs in biological complexes such as **pheophytin** (analog of chlorophyll), hemocyanin, tyrosinase and ceruloplasmin, *q.q.v.* Reviews of copper and copper compounds: *ACS Monograph Series* **no. 122**, entitled "Copper," A. Butts, Ed. (Reinhold, New York, 1954) 936 pp; Massey, "Copper" in *Comprehensive Inorganic Chemistry* vol. **3**, J. C. Bailar, Jr. *et al.*, Eds. (Pergamon Press, Oxford, 1973) pp 1-78; W. M. Tuddenham, P. A. Dougall, *Kirk-Othmer Encyclopedia of Chemical Technology* vol. **6** (Wiley-Interscience, New York, 3rd ed., 1979) pp 819-869. Book: *Inflammatory Diseases and Copper*, J. R. J. Sorenson, Ed. (Humana Press, Clifton, NJ, 1982) 622 pp. Review of role in human diseases: G. J. Brewer, *Curr. Opin. Chem. Biol.* **7**, 207-212 (2003); of toxicology and human exposure: *Toxicological Profile for Copper* (PB2004-107333, 2004) 318 pp.

Reddish, lustrous, ductile, malleable metal; face-centered cubic structure; commercially available in the form of ingots, sheets, wire or powder. Becomes dull when exposed to air. In moist air gradually becomes coated with green basic carbonate. d 8.94. mp 1083°. bp 2595°. Mohs' hardness 3.0. Resistivity 1.673 microohm-cm. Heat of fusion 48.9 cal/g; heat of vaporization 1150 cal/g. Heat capacity at constant pressure (solid) 0.092 cal/g/°C (20°), (liq) 0.112 cal/g/°C. E^0 (aq) Cu^+/Cu +0.521 V; E^0 (aq) Cu^{2+}/Cu +0.337 V. Very slowly attacked by cold hydrochloric or dil sulfuric acid; readily by dil nitric acid, and by both hot concd H_2SO_4 and HBr. It is also attacked by acetic and other organic acids. Slowly sol in ammonia water. Water-soluble cupric salts yield with sodium hydroxide a bluish-green precipitate of cupric hydroxide which is changed to black cupric oxide on warming. Potassium ferrocyanide produces a brownish-red precipitate of copper ferrocyanide. Hydrogen sulfide produces in acid solns a black precipitate of cupric sulfide which is sol in soln of sodium cyanide. Aluminum, iron or zinc precipitate metallic copper from its solns.

Caution: Potential symptoms of overexposure to dusts and mists are irritation of eyes, nose, pharynx; nasal perforation; metallic taste; dermatitis. Potential symptoms of overexposure to fumes are irritation of eyes, upper respiratory system; metal fume fever (chills, muscle aches, nausea, fever; dry throat, cough, weakness, lassitude); metallic or sweet taste; discoloration of skin and hair. *See NIOSH Pocket Guide to Chemical Hazards* (DHHS/NIOSH 97-140, 1997) p 76.

USE: Manuf bronzes, brass, other copper alloys, electrical conductors, ammunition, copper salts, works of art. Catalyst. Absorption of oxygen.

2505. Copper Phthalocyanine. [147-14-8] (*SP*-4-1)-[29*H*,-31*H*-Phthalocyaninato(2−)-κN^{29},κN^{30},κN^{31},κN^{32}]copper; C.I. Pigment Blue 15; C.I. 74160. $C_{32}H_{16}CuN_8$; mol wt 576.08. C 66.72%, H 2.80%, Cu 11.03%, N 19.45%. Discoverers: Dandridge *et al.*, **GB 322169** (1928 to Scottish Dyes); Prepn: Baumann *et al.*, *Angew. Chem.* **68**, 133 (1956); Sanielevici *et al.*, *Rec. Chim. (Bucharest)* **12**, 281 (1961); Raab, Hoernle, **GB 930150** (1963 to Bayer). Purification: Heinle, Mau, **DE 1167307** (1964 to G. Siegle); Sanielevici *et al.*, **BE 657307** (1965 to Romania, Ministry of Petroleum and Chemical Industry); H. Tomoda *et al.*, *Chem. Lett.* **1980**, 1277. Properties: Dent, Linstead, *J. Chem. Soc.* **1934**, 1027; Easton, Smith, *J. Oil Colour Chem. Assoc.* **49**, 614 (1966). *See also Colour Index* vol. **4** (3rd ed., 1971) p 4618.

Bright blue microcrystals with purple lustre; sol in 98% H_2SO_4 from which it can be almost quantitatively pptd by dilution with water. Practically insol in water, alcohol, and hydrocarbons. Stable toward heat (crystalline and analytically pure sublimate obtained at about 580° in low pressure atmosphere of nitrogen or carbon dioxide), alkalies, dilute acids. Dec by hot nitric acid, or dilute acid permanganate to yield phthalimide. Exists in two forms: the thermodynamically less stable, redder α-form is a better pigment than the more stable greener β-form. In the presence of aromatic solvents, heat, high shear, etc., α-form is converted to β-form.

USE: In inks and paints. In polypropylene sutures.

2506. Copper(I) Thiophene-2-carboxylate. [68986-76-5] (2-Thiophenecarboxylato-κO^2,κS^1)copper; CuTC. $C_5H_3CuO_2S$; mol wt 190.68. C 31.50%, H 1.59%, Cu 33.33%, O 16.78%, S 16.81%. Copper(I) reagent; promotes modified Stille cross-coupling reactions and Ullman-type couplings. Prepn: G. D. Allred, L. S. Liebeskind, *J. Am. Chem. Soc.* **118**, 2748 (1996). Synthetic applications: S. Zhang *et al.*, *J. Org. Chem.* **62**, 2312 (1997); I. Paterson, J. Man, *Tetrahedron Lett.* **38**, 695 (1997); C. Savarin *et al.*, *Org. Lett.* **3**, 91, 2149 (2001); G. Borsato *et al.*, *J. Org. Chem.* **67**, 7894 (2002); M. d'Augustin *et al.*, *Angew. Chem. Int. Ed.* **44**, 1376 (2005). *Brief review:* A. Innitzer, *Synlett* **2005**, 2405-2406.

Tan, air-stable, non-hygroscopic powder.

USE: Mediator in intermolecular cross-coupling reactions, asymmetric 1,4-additions, and allylation reactions.

2507. Coprogen. [31418-71-0] [(3E)-5-[(Hydroxy-κO)[3-[(2S,5S)-5-[3-[(hydroxy-κO)[(2E)-5-hydroxy-3-methyl-1-(oxo-κO)-2-penten-1-yl]amino]propyl]-3,6-dioxo-2-piperazinyl]propyl]amino]-3-methyl-5-(oxo-κO)-3-penten-1-yl N^2-acetyl-N^5-(hydroxy-κO)-N^5-[(2E)-5-hydroxy-3-methyl-1-(oxo-κO)-2-penten-1-yl]-L-or-nithinato(3−)] iron. $C_{35}H_{53}FeN_6O_{13}$; mol wt 821.68. C 51.16%, H 6.50%, Fe 6.80%, N 10.23%, O 25.31%. Iron-contg, growth-promoting complex. Isoln from *Penicillium* spp: Hesseltine *et al.*, *J. Am. Chem. Soc.* **74**, 1362 (1952); Pidacks *et al.*, ibid. **75**, 6064 (1953); Gaeumann, Vischer, US 3297526 (1967 to Ciba). Isoln from *Aspergillaceae:* Zaehner *et al.*, *Arch. Mikrobiol.* **45**, 119 (1963). Structure: Keller-Schierlein, Diekmann, *Helv. Chim. Acta* **53**, 2035 (1970).

Clusters of brick-red needles from ethanol. Absorption max (ethanol): 217, 250, 440 nm (log ε 4.46, 4.22, 3.47). Sol in water, methanol, ethanol (amorphous coprogen only), propanol, benzyl alcohol, Cellosolves. Practically insol in ether, ethyl acetate, chloroform, benzene, Cellosolve esters.

Desferricoprogen. Iron-free coprogen, may be obtained by treating an aq solution with 8-quinolinol yielding a light brown powder.

USE: Coprogen as growth-promoting agent in various organisms. Desferricoprogen to bring about excretion of iron.

2508. Coprostane. [481-20-9] 5β-Cholestane; pseudocholestane. $C_{27}H_{48}$; mol wt 372.68. C 87.02%, H 12.98%. The *cis*-decalin homolog of cholestane. Prepn: Wieland, Jacobi, *Ber.* **59**, 2064 (1926); Young *et al.*, *J. Am. Chem. Soc.* **81**, 1452 (1959); Dart, Henbest, *J. Chem. Soc.* **1960**, 3563; Nickon, Bagli, *J. Am. Chem. Soc.* **83**, 1498 (1961); Caglioti, Grasselli, *Chem. Ind. (London)* **1964**, 153.

Needles from alcohol, mp 72°. $[\alpha]_D^{11}$ +25.1° (c = 2 in CHCl₃). n_D^{88} 1.4884. Freely soluble in ether, chloroform; slightly in abs alcohol.

2509. Coprosterol. [360-68-9] (3β,5β)-Cholestan-3-ol; 3β-coprostanol; stercorin. $C_{27}H_{48}O$; mol wt 388.68. C 83.44%, H 12.45%, O 4.12%. Found in feces of man and of carnivorous animals: Bondzynski, Humnicki, *Z. Physiol. Chem.* **22**, 396 (1896); Wells *et al.*, *Arch. Biochem. Biophys.* **57**, 437 (1955); Samuel *et al.*, *J. Chromatogr.* **14**, 508 (1964). Prepn from coprostenone: Ruzicka *et al.*, *Helv. Chim. Acta* **17**, 1407 (1934); Shoppee, Summers, *J. Chem. Soc.* **1950**, 687. From coprostenol: Schönheimer, Evans, *J. Biol. Chem.* **114**, 567 (1936); Ruzicka *et al.*, *Helv. Chim. Acta* **21**, 498 (1938); Agashe, Summers, *J. Chem. Soc.* **1957**, 3107; Dart, Henbest, *ibid.* **1960**, 3563. Prepn from cholesterol: Rosenfeld, Gallagher, *Steroids* **4**, 515 (1964).

Needles from methanol, mp 101°. $[\alpha]_D^{18}$ +28° (c = 1.8 in CHCl₃). Freely sol in ether, chloroform, benzene; slightly sol in methanol (one gram dissolves in 145 ml MeOH). Insol in water.

2510. Coptine. [1391-14-6] From root of *Coptis trifolia* Salisb., *Ranunculaceae:* Gross, *Am. J. Pharm.* **45**[4], 193 (1873); Mollett, Christensen, *J. Am. Pharm. Assoc.* **23**, 310 (1934); from *C. teeta* Wall.: Chatterjee *et al.*, *J. Indian Chem. Soc.* **28**, 97 (1951); from *C. chinensis* Wils.: Schramm, Tang, *Pharmazie* **14**, 405 (1959).

Crystals. Practically insol in water, ammonia, alkali hydroxides; sol in hot alcohol, dil acids.

2511. Coptis. Goldthread. Dried plant of *Coptis trifolia* Salisb., *Ranunculaceae*. *Habit.* U.S., northern and middle Atlantic States; Canada. *Constit.* Berberine, coptine.

THERAP CAT: Bitter tonic.

2512. Coptisine. [3486-66-6] 6,7-Dihydrobis[1,3]benzodioxolo[5,6-*a*:4′,5′-*g*]quinolizinium; 7,8,13,13a-tetrahydro-2,-3:9,10-bismethylenedioxyberbinium; bis[methylenedioxy]protoberberine. $[C_{19}H_{14}NO_4]^+$. From root of *Coptis japonica* Makino, *Ranunculaceae*. Extraction procedure: Kitasato, *C.A.* **21**, 2700 (1927). Structure: Späth, Posega, *Ber.* **62**, 1029 (1929); *cf.* Huang-Minlon, *Ber.* **69**, 1744 (1936). Prepn from rhoeageninediol: Klásek *et al.*, *Tetrahedron Lett.* **1968**, 4549.

Hydroxide. $C_{19}H_{15}NO_5$. Yellowish needles from alc, mp 218°. Absorption spectrum: Kitasato, *Acta Phytochim.* **3**, 175 (1927). Very slightly sol in water; sparingly in alcohol; sol in alkalies.

Chloride. Orange prisms, not melted at 300°.

Iodide. Yellow needles, dec above 280°.

Sulfate. Yellow crystals, insol in water and alcohol (enables separation from berberine and worenine).

2513. Cord Factors. [61512-20-7] Trehalose 6,6′-dimycolate; 6,6′-di-*O*-mycolyl-α,α-trehalose; (6-*O*-mycolyl-α-D-glucopyranosyl)-6-*O*-mycolyl-α-D-glucopyranoside. Naturally occurring 6,6′-diesters of trehalose and mycolic acids, β-hydroxy α-branched long chain acids produced by *Mycobacteria, Nocardia,* and *Corynebacteria.* 3 major classes identified which are based on the size of the acid and the producing strain: corynomycolic acids (C_{32}), nocardomycolic acids (C_{48}-C_{60}); mycobacteria mycolic acids (C_{60}-C_{90}). The formation of "cords" by tubercle bacilli was first described in 1884 by R. Koch. Isolation of the factor from *Mycobacterium tuberculosis:* H. Bloch, *J. Exp. Med.* **91**, 197 (1950). Structure: J. Asselineau, E. Lederer, *Biochim. Biophys. Acta* **17**, 161 (1955). Synthesis: J. Polonsky *et al.*, *Carbohydr. Res.* **65**, 295 (1978); Y. Gama, *Yakugaku* **44**, 671 (1995). Antitumor activity: A. Bekierkunst *et al.*, *Science* **174**, 1240 (1971); M. V. Pimm *et al.*, *Int. J. Cancer* **24**, 780 (1979). Molecular packing and surface properties: R. Almog, C. A. Mannella, *Biophys. J.* **71**, 3311 (1996). Use as antigen in serodiagnosis of *Mycobacterium avium-intracellulare* complex: K. Enomoto *et al.*, *Microbiol. Immunol.* **42**, 689 (1998); in tuberculosis: J. Pan *et* *al., ibid.* **43**, 863 (1999). Review of early literature: E. Lederer, *Springer Semin. Immunopathol.* **2**, 133-148 (1979). Review of classes of cord factors and their chemistry: *idem, Microbiol. Ser.* **15**, 361-378 (1984).

R = Mycolic acids

2514. Cordycepin. [73-03-0] 3'-Deoxyadenosine; 9-cordyceposidoadenine. $C_{10}H_{13}N_5O_3$; mol wt 251.25. C 47.80%, H 5.22%, N 27.87%, O 19.10%. First reported nucleoside antibiotic. Isoln from culture fluids of *Cordyceps militaris* (Linn.) Link: K. G. Cunningham *et al., J. Chem. Soc.* **1951**, 2299; N. M. Kredich, A. J. Guarino, *Biochim. Biophys. Acta* **41**, 363 (1960). Proposed structure: H. R. Bentley *et al., J. Chem. Soc.* **1951**, 2301. Identity with 3'-deoxyadenosine and revised structure: E. A. Kaczka *et al., Biochem. Biophys. Res. Commun.* **14**, 456 (1964). Biosynthesis: R. Suhadolnik *et al., J. Am. Chem. Soc.* **86**, 948 (1964). Synthesis: A. R. Todd, T. L. Ulbricht, *J. Chem. Soc.* **1960**, 3275; W. W. Lee *et al., J. Am. Chem. Soc.* **83**, 1906 (1961); E. Walton *et al., ibid.* **86**, 2952 (1964); Y. Ito *et al., ibid.* **103**, 6739 (1981). Cordycepin and cordycepin triphosphate have been used extensively in the study of messenger RNA transcription, *see:* H. T. Shigeura, G. E. Boxer, *Biochem. Biophys. Res. Commun.* **17**, 758 (1964); S. Penman *et al., Proc. Natl. Acad. Sci. USA* **67**, 1878 (1970). *Reviews:* J. J. Fox. *et al.,* "Nucleoside Antibiotics" in *Prog. Nucleic Acid Res. Mol. Biol.* **5**, 258-262 (1966); A. J. Guarino, "Cordycepin" in *Antibiotics* **I**, D. Gottlieb, P. Shaw, Eds. (Springer-Verlag, New York, 1967) pp 468-480.

Needles from ethanol, *n*-butanol, *n*-propanol or water. mp 225-226°. $[\alpha]_D^{20}$ −47°. $[\alpha]_D^{27}$ −42°. uv max (ethanol): 260 nm (ε 14600). pH aq soln: 7.1.

Triphosphate. Cordycepin-5'-triphosphate; 3'-deoxy ATP; 3'-deoxyadenosine-5'-(tetrahydro triphosphate). $C_{10}H_{13}N_5O_{12}P_3$. Formation by conversion of 3'-deoxyadenosine in Ehrlich ascites tumor: H. Klenow, *Biochim. Biophys. Acta* **76**, 347 (1963). Metabolism in KB cells: H. Shigeura, S. Sampson, *ibid.* **138**, 26 (1967). Synthesis: J. J. Novak, F. Sorm, *Collect. Czech. Chem. Commun.* **38**, 113 (1973); M. Blandin, *J. Carbohydr. Nucleosides Nucleotides* **3**(5/6), 341 (1976). Inhibits the final step of RNA biosynthesis by termination of the ribonucleotide chain due to the absence of the 3'-hydroxyl group.

2515. Cordyceps. DongChongXiaCao; tochukaso; yartsa gunbu; Chinese caterpillar fungus. Traditional Chinese medicine

used as an invigorant or tonic. Consists of a parasitic fungus, *Cordyceps sinensis* (Berk.) Sacc., *Clavicipitaceae* and its host insect, the larva of *Hepialus armoricanus* Oberthur, *Hepialidae*. Known as "winter worm-summer grass" because of its appearance during different seasons. *Habit*. Higher elevations of the Tibetan plateau in southwest China, Nepal, Bhutan. *Constit.* Cordycepin and other nucleosides, myriocin, mannitol (cordycepic acid), ergosterol, polysaccharides. Overview of habitat, harvest, and economic importance: D. Winkler, *Econ. Bot.* **62**, 291-305 (2008). HPLC determn of constituents: S. Wang *et al., J. Sep. Sci.* **32**, 4069 (2009). Nutritional requirements for mycelial growth of *C. sinensis* in submerged culture: C.-H. Dong, Y.-J. Yao, *J. Appl. Microbiol.* **99**, 483 (2005). Comparison of bioactive constituents of fermented mycelia with natural crude drug: T.-H. Hsu *et al., Food Chem.* **78**, 463 (2002). Review of authentication and quality control methods: S. P. Li *et al., J. Pharm. Biomed. Anal.* **41**, 1571-1584 (2006). Review of bioactive constituents and pharmacology: T. B. Ng, H. X. Wang, *J. Pharm. Pharmacol.* **57**, 1509-1519 (2005); X. Zhou *et al., ibid.* **61**, 279-291 (2009).

USE: In traditional medicine; as health food.

2516. Coriamyrtin. [2571-86-0] (1a*S*,1b*R*,2*S*,2'*R*,5*S*,6a*R*,-7a*R*,8*R*)-Hexahydro-1b-hydroxy-6a-methyl-8-(1-methylethenyl)-spiro[2,5-methano-7*H*-oxireno[3,4]cyclopent[1,2-*d*]oxepin-7,2'-oxiran]-3(2*H*)-one. $C_{15}H_{18}O_5$; mol wt 278.30. C 64.74%, H 6.52%, O 28.74%. Sesquiterpene of the picrotoxane group. Main toxic principle from leaves and fruit of *Coraria myrtifolia* L., *Coriariaceae*: M. J. Riban, *C. R. Hebd. Seances Acad. Sci.* **57**, 798 (1863); and *C. japonica* A. Gray, *Coriariaceae*: Kariyone, Sato, *J. Pharm. Soc. Jpn.* **50**, 106 (1930); Okuda, *Pharm. Bull.* **2**, 185 (1954). Series of articles on structure, absolute configuration, stereochemistry and derivatives: T. Okuda, T. Yoshida, *Chem. Pharm. Bull.* **9**, 379, 404 (1961); *ibid.* **15**, 1687, 1697 (1967). Structural relationship to tutin, *q.v.: eidem, Tetrahedron Lett.* **1965**, 439. Biosynthetic studies: M. Biollaz, D. Arigoni, *Chem. Commun.* **1969**, 633. Synthetic studies: K. Tanaka *et al., Chem. Pharm. Bull.* **31**, 1943, 1958, 1972 (1983). Pharmacology: E. E. Swanson, K. K. Chen, *J. Pharmacol. Exp. Ther.* **57**, 410 (1936). Structure-activity relationship: C. H. Jarboe *et al., J. Med. Chem.* **11**, 729 (1968).

Bitter, monoclinic prisms, mp 229-230°. $[\alpha]_D^{14}$ +79°. Slightly sol in water, cold alc; freely sol in hot alc, in ether. LD_{50} i.p. in mice: 3 mg/kg (Jarboe).

2517. Coriander. Annual plant, *Coriandrum sativum* L., *Umbelliferae*, with characteristic "bug-like" odor. *Habit.* Asia, Europe, North and South America. *Constit.* About 1% volatile oil; fixed oils, malic acid, tannin, mucilage. Medicinal parts are the dried ripe fruit and volatile oil. The leaf, known as *cilantro*, and whole or ground seeds are used in cooking. Description and medicinal uses: M. Wichtl, N. G. Bisset, *Herbal Drugs and Phytopharmaceuticals*, English Ed. (CRC Press, Boca Raton, 1994) pp 159-160. Production and constituents of seed oil: B. M. Smallfield *et al., J. Agric. Food Chem.* **49**, 118 (2001).

Volatile oil. [8008-52-4] Obtained by steam distillation of dried ripe fruit. *Constit.* Chiefly 60-70% linalool, α-pinene, γ-terpinene, limonene, camphor, geraniol and geranyl acetate. Colorless or pale yellow liquid. d_{25}^{25} 0.863-0.875. α_D^{25} +8 to +15°. n_D^{20} 1.4620-1.4720. Almost insol in water; sol in 3 vols 70% alcohol; more sol in stronger alcohol; very sol in chloroform, ether, glacial acetic acid.

USE: Flavoring in foods; component of spice blends such as curry powder.

THERAP CAT: Carminative.

2518. Corifollitropin Alfa. [195962-23-3] Follicle-stimulating hormone (human α-subunit reduced) complex with follicle-stimulating hormone (human β-subunit reduced) fusion protein with 118-chorionic gonadotropin (human β-subunit); FSH-CTP; Org-36286; SCH-900962; Elonva. Long-acting follicle stimulating hormone (FSH) agonist used to stimulate ovulation; recombinant fusion protein consisting of an α-subunit identical to FSH and a β-subunit produced by fusing the C-terminal peptide of the β-subunit of human chorionic gonadotropin (hCG) with the β-subunit of FSH. Prepn: I. Boime *et al.*, **WO 9009800**; *eidem*, **US 5338835** (1990, 1994 both to Washington University). ADME profile: A. van Schanke *et al.*, *Pharmacology* **85**, 77 (2010). Clinical pharmacokinetics: A. H. Balen *et al.*, *J. Clin. Endocrinol. Metab.* **89**, 6297 (2004). Clinical trial for controlled ovarian stimulation before *in vitro* fertilization (IVF): P. Devroey *et al.*, *ibid.* 2062; for IVF in lower-body weight women: Corifollitropin alfa Ensure Study Group, *Reprod. Biomed. Online* **21**, 66 (2010). Clinical comparison with recombinant FSH on ovarian stimulation: P. Devroey *et al.*, *Hum. Reprod.* **24**, 3063 (2009). Review of development and clinical experience: D. Loutradis *et al.*, *Curr. Opin. Investig. Drugs* **10**, 372-380 (2009).

THERAP CAT: Treatment of infertility.

2519. Cork. The bark of the cork oak tree, *Quercus suber* L., *Fagaceae* and to a lesser degree of *Quercus occidentalis* Gray, *Fagaceae*, indigenous to the African and European shores of the western Mediterranean. *Constit.* 35-60% suberin, 30-33% cellulose, 27-32% lignin; small amounts of cerin (a wax), fats, inorganic manganese compds, decacrylic acid. Suberin contains phellonic acid, suberic acid, phloionic acid, phloionolic acid, suberolcarboxylic acid, eicosadicarboxylic acid, stearic acid, cortic acid. Used to seal wine casks since antiquity: Theophrastus, *Historia Plantarum* 3, 16 and 17. Plinius (major): *Historia Naturalis* 16, 8, 13. *Reviews and monographs:* A. Klauber, *Monographie des Korkes* (Weber, Berlin, 1920); W. Herrmann, *Kork* in *Ullmanns Encyklopädie der technischen Chemie* **vol. 10**, (Urban & Schwarzenberg, Munich, 3rd ed., 1958) pp 634-637; G. B. Cooke, *Cork and the Cork Tree* (Pergamon Press, New York, 1961).

Pale tan, elastic mass. Floats on water. Specific gravity about 0.10 to 0.25 g/ml. Excellent insulator against heat and sound. Stable to heat up to about 100°. Unaffected by water, brine, dil acids, alcohols, bland oils. Attacked by concd mineral acids, strong alkali and ammonia, free halogens, hydrogen peroxide, ozone.

USE: Bottle stoppers, sound deadeners, heat insulators, life preservers, gaskets, linoleum manufacture, cork tile, bulletin boards, dart boards, inlays for shoes.

2520. Corn Oil. Maize oil; Maydol; Mazola. Obtained as a byproduct by wet milling the grain of *Zea mays* L., *Gramineae* for the manuf of corn starch, corn syrup, glucose, dextrins, etc.: E. W. Eckey, *Vegetable Fats and Oils* (Reinhold, New York, 1954). *Constit.* Glycerides of the following fatty acids: Myristic 0.1-1.7%, palmitic 8-12%, stearic 2.5-4.5%, hexadecenoic 0.2-1.6%, oleic 19-49%, linoleic 34-62%. Unsaponifiable fraction: 1-3% (γ-tocopherol 0.1%, the rest is mostly isomeric sitosterols and wax such as myricyl and ceryl alcohols). The crude oil may contain up to 2% phospholipids (vegetable lecithin, inositol esters). The following constants are for the refined product.

Yellow oil. Faint characteristic odor and taste. d_{25}^{25} 0.916-0.921. mp −18 to −10°. Titer 14-20°. Flash pt 610°F (321°C). Ignition pt 740°F (393°C). n_D^{25} 1.470-1.474; n_D^{40} 1.464-1.468. Acid value 2-6. Saponification value 187-196. Iodine value 109-133. Thiocyanogen value 71-77. Hydroxyl value 8-12. Reichert-Meissl value <0.5. Polenske value <0.5. Hehner value 92-96. Classed as a semidrying oil. On prolonged exposure to air it thickens and becomes rancid. Miscible with chloroform, ether, benzene, petr ether, solvent hexane. Slightly sol in alc.

USE: As salad and cooking oil; in prepn of margarine. As pharmaceutic aid (solvent). Some use in the preparation of non-yellowing enamel paint.

2521. Corn Steep Liquor. Corn steep water. Cleaned corn grain is washed thoroughly and then steeped for 36 to 40 hours in approx twice its volume of water contg 0.2% SO_2 at a temp of 46-50°C. As the steep water is drawn from the corn it contains between 6 and 9 lbs of solids per 100 lbs of water. The corn should be soft to the touch, but must not be slimy or smeary which occurs when steeped either too hot or too long. The steep liquor is then evaporated under 25 mm Hg pressure in a "Swenson" evaporator to what is called heavy steep liquor with a gravity of 16 to 20° Bé and contg 50-60% of solids. In this form it is sold to manufacturers of antibiotics who may have their own specifications as to age, gravity and SO_2 content. See: Bowden, Peterson, *Arch. Biochem.* **9**, 387-399 (April 1946); Graefe, *Staerke* **4**, 275-282 (1952).

USE: In production of penicillin, *meso*-inositol; source of phytin.

2522. Cornus. Dogwood; flowering dogwood. Dried root bark of *Cornus florida* L., *Cornaceae*. *Habit.* Eastern U.S. and Ontario. *Constit.* Cornin Verbenalin, *q.v.*

THERAP CAT: Astringent bitter.

2523. Coronene. [191-07-1] $C_{24}H_{12}$; mol wt 300.36. C 95.97%, H 4.03%. Planar polycyclic aromatic hydrocarbon; found in almost pure form as the mineral *pendletonite*. Environmental pollutant produced as a byproduct of hydrocarbon combustion; impurity in natural carbon deposits. Possible source of infrared emission from interstellar matter. Prepn: R. Scholl, K. Meyer, *Ber.* **65**, 902 (1932); M. S. Newman, *J. Am. Chem. Soc.* **62**, 1683 (1940). Isoln from coal-hydrogenation oil: M. Orchin, J. Feldman, *J. Org. Chem.* **18**, 609 (1953). UV absorption spectrum: J. W. Patterson, *J. Am. Chem. Soc.* **64**, 1485 (1942). Photophysical properties: R. Kataro *et al.*, *Chem. Phys.* **42**, 121 (1979); N. Nijegorodov *et al.*, *Spectrochim. Acta A* **57**, 2673 (2001). Calorimetry study: Y. Nagano, *J. Chem. Thermodyn.* **32**, 973 (2000). Quantitative structure-property relationships for polycyclic aromatic hydrocarbons: M. M. C. Ferreira, *Chemosphere* **44**, 125 (2001). Thermodynamic properties: J. S. Chickos *et al.*, *J. Chem. Thermodyn.* **34**, 1195 (2002).

mp 429-430° (Scholl, Meyer). Also reported as pale yellow needles from dilute benzene, mp 437-440° (Newman). bp 590.0°. d 1.39. Soly in water at 25°, log *S*: −3.85 g/m³.

USE: Ultraviolet phosphor for charge-coupled devices.

2524. Corticosterone. [50-22-6] (11β)-11,21-Dihydroxypregn-4-ene-3,20-dione; 4-pregnene-11β,21-diol-3,20-dione; Kendall's compound B; Reichstein's substance H; compd B. $C_{21}H_{30}O_4$; mol wt 346.47. C 72.80%, H 8.73%, O 18.47%. Endogenous glucocorticoid; intermediate in the biosynthesis of aldosterone from progesterone. Isoln from adrenal cortex: Mason *et al.*, *J. Biol. Chem.* **114**, 613 (1936); Reichstein, von Euw, *Helv. Chim. Acta* **21**, 1197 (1938); Jeanloz *et al.*, **US 2676904** (1954 to Searle); from adrenal extract: Reichstein, **US 2166877** (1939 to Roche-Organon). Prepn from desoxycholic acid: Wallis, Chakravorty, **US 2341250** (1944 to Research Corp.); from 11-deoxycorticosterone: Zaffaroni, **US 2671752** (1954 to Syntex); from cortisone: Oliveto *et al.*, **US 2927108** (1960 to Schering).

Trigonal plates from acetone, mp 180-182°. $[\alpha]_D^{15}$ +223° (c = 1.1 in alc). uv max: 240 nm. Insol in water. Sol in usual organic solvents. Upon moistening with concd H_2SO_4 it gives an orange-yellow soln which has a strong fluorescence.

21-Acetate. $C_{23}H_{32}O_5$. Clusters of needles from acetone + ether, mp 145°. $[\alpha]_D^{20}$ +195° (c = 0.62 in acetone).

2525. Cortisone. [53-06-5] 17,21-Dihydroxypregn-4-ene-3,-11,20-trione; 17-hydroxy-11-dehydrocorticosterone; 11-dehydro-

17-hydroxycorticosterone; Δ^4-pregnene-17α,21-diol-3,11,20-trione; Kendall's compound E; Wintersteiner's compound F; Reichstein's substance Fa. $C_{21}H_{28}O_5$; mol wt 360.45. C 69.98%, H 7.83%, O 22.19%. Isoln from suprarenal glands: Pfiffner *et al.*, *J. Biol. Chem.* **111**, 585 (1935); *ibid.* **116**, 291 (1936); Mason *et al.*, *ibid.* **114**, 613 (1936); Reichstein, *Helv. Chim. Acta* **19**, 1107 (1936); Kuizenga, Cartland, *Endocrinology* **24**, 526 (1939). Synthesis of the monoacetate from desoxycholic acid: Sarett, *J. Biol. Chem.* **162**, 601 (1946). Further development of prepn methods: Meystre, Wettstein, *Experientia* **3**, 185 (1947); *Helv. Chim. Acta* **30**, 1037, 1256 (1947); Reichstein *et al.*, *ibid.* **26**, 562, 705, 721 (1943); **27**, 821 (1944); Reichstein, **US 2403683** (1946); Kendall *et al.*, *J. Biol. Chem.* **166**, 345 (1946); Gallagher, **US 2447325** (1948); Peterson, Murray, *J. Am. Chem. Soc.* **74**, 1871 (1952); Perlman, *ibid.* 2126; Sarett, *ibid.* **70**, 1454 (1948); **71**, 2443 (1949); Mattox, Kendall, *ibid.* **70**, 882 (1948). Stereospecific total synthesis: Sarett *et al.*, *ibid.* **74**, 4974 (1952).

Rhombohedral platelets from 95% alcohol, mp 220-224° (some decompn) when heated in evac capillary. $[\alpha]_D^{25}$ +209° (c = 1.2 in 95% alcohol); $[\alpha]_{546}^{25}$ +269° (c = 0.125 in benzene); $[\alpha]_{546}^{25}$ +248° (c = 0.1 to 0.2 in alcohol). uv max: 237 nm (ε 1.4×10⁴) *See:* Mason *et al.*, *J. Biol. Chem.* **116**, 267 (1936); Wintersteiner, Pfiffner, *ibid.* 291. Fairly sol in cold methanol, ethanol, acetone; much less sol in ether, benzene, chloroform; slightly sol in water (28 mg/100 ml at 25°). The water soln is neutral. Gives orange-red soln with intense green fluorescence in concd H_2SO_4. Reduces Benedict's soln on heating.

21-Acetate. [50-04-4] Cortisone acetate; Cortistab; Cortisyl; Cortogen; Cortone. $C_{23}H_{30}O_6$; mol wt 402.49. Flat needles from acetone; clusters of radiating rods from chloroform. Becomes opaque at 70-100°, mp 235-238° with slight sintering at 230°. $[\alpha]_D^{25}$ +164° (c = 0.5 in acetone), $[\alpha]_D^{20}$ +208 to +217° (dioxane). uv max: 238 nm (ε 1.58×10⁴), *see* Sarett, *J. Biol. Chem.* **162**, 630 (1946). Soly at 25° in water: 2.2 mg/100 ml; in propylene glycol 44 mg/100 ml; in chloroform 182 mg/g. Reduces ammoniacal silver nitrate soln at room temp. Sol in sulfuric acid giving a yellow soln without fluorescence (difference from hydrocortisone acetate). Sparingly sol in acetone; slightly sol in alcohol; sol in dioxane.

21-Cyclopentanepropionate. [509-00-2] $C_{29}H_{40}O_6$. Needles from diisopropyl ether, mp 158-161°. $[\alpha]_D^{20}$ +190° (chloroform). uv max (ethanol): 239 nm (ε 16350). Sol in ether, glycols, vegetable oils, especially sesame, peanut, and corn oils.

THERAP CAT: Glucocorticoid.

THERAP CAT (VET): Glucocorticoid, anti-inflammatory agent.

2526. Cortistatins. Antiangiogenic steroidal alkaloids produced by the marine sponge, *Corticium simplex*. Cortistatins A and J have the most potent activity. Isoln of cortistatins A-D: S. Aoki *et al.*, *J. Am. Chem. Soc.* **128**, 3148 (2006); of E-H: Y. Watanabe *et al.*, *Tetrahedron* **63**, 4074 (2007); of J-L: S. Aoki *et al.*, *Tetrahedron Lett.* **48**, 4485 (2007). Structure-activity study: S. Aoki *et al.*, *Bioorg. Med. Chem.* **15**, 6758 (2007). Synthesis of A from prednisone: R. A. Shenvi *et al.*, *J. Am. Chem. Soc.* **130**, 7241 (2008). Enantioselective synthesis: H. M. Lee *et al.*, *ibid.* 16864. Total synthesis of A and J: K. C. Nicolaou *et al.*, *ibid.* **131**, 10587 (2009). Protein kinase binding affinity of A: V. J. Cee *et al.*, *Angew. Chem. Int. Ed.* **48**, 8952 (2009).

Cortistatin A

Cortistatin A. [882976-95-6] (3S,3aR,7R,8R,9S,10aR,12aS,-12bR)-9-(Dimethylamino)-1,2,3,3a,4,7,8,9,10,11,12,12b-dodecahydro-3-(7-isoquinolinyl)-3a-methyl-10a,12a-epoxybenzo[4,5]cyclohept[1,2-*e*]indene-7,8-diol. $C_{30}H_{36}N_2O_3$; mol wt 472.63. C 76.24%, H 7.68%, N 5.93%, O 10.16%. Colorless powder. $[\alpha]_D^{20}$ +30.1° (c = 0.56 in methanol). uv max (methanol): 219 nm (ε 45600).

Cortistatin J. [944804-62-0] (3S,3aR,9S,10aR,12aS,12bR)-1,2,-3,3a,4,9,10,11,12,12b-Decahydro-3-(7-isoquinolinyl)-N,N,3a-trimethyl-10a,12a-epoxybenzo[4,5]cyclohept[1,2-*e*]inden-9-amine. $C_{30}H_{34}N_2O$; mol wt 438.62. C 82.15%, H 7.81%, N 6.39%, O 3.65%. Colorless powder. $[\alpha]_D^{20}$ −54.0° (c = 0.26 in chloroform). uv max (methanol): 293, 280, 269, 223 nm (ε 21200, 8500, 3000 46200).

2527. Cortivazol. [1110-40-3] (11β,16α)-21-(Acetyloxy)-11,17-dihydroxy-6,16-dimethyl-2'-phenyl-2'H-pregna-2,4,6-trieno-[3,2-c]pyrazol-20-one; 1,2,3,3a,3b,7,10,10a,10b,11,12,12a-dodecahydro-1,11-dihydroxy-2,5,10a,12a-tetramethyl-7-phenylcyclopenta-[7,8]phenanthro[2,3-c]pyrazol-1-yl hydroxymethyl ketone acetate; 6,16α-dimethyl-11β,17α,21-trihydroxy-2'-phenyl[3,2-c]pyrazolo-4,6-pregnadien-20-one 21-acetate; MK-650; H-3625; Altim; Diaster; Dilaster. $C_{32}H_{38}N_2O_5$; mol wt 530.67. C 72.43%, H 7.22%, N 5.28%, O 15.07%. Prepn: Fried *et al.*, *J. Am. Chem. Soc.* **85**, 236 (1963); Tishler *et al.*, **US 3067194** and **US 3300483** (1962, 1967 to Merck & Co.).

Double mp 160-165° and 229-230°; $[\alpha]_D^{23}$ +14° (CHCl$_3$). uv max (methanol): 283, 315 nm (ε 15700, 19000).

THERAP CAT: Glucocorticoid.

2528. Cortol. [516-38-1] (3α,5β,11β,20S)-Pregnane-3,11,-17,20,21-pentol. $C_{21}H_{36}O_5$; mol wt 368.51. C 68.45%, H 9.85%, O 21.71%. Isolated from human urine after administration of hydrocortisone or ACTH: Fukushima *et al.*, *J. Biol. Chem.* **212**, 449 (1955).

α-Cortol. 5β-Pregnane-3α,11β,17,20α,21-pentol. Crystals from ethyl acetate, mp 250.5-254°. $[\alpha]_D^{26}$ +23.7° (ethanol).

α-Cortol 3,20,21-triacetate. $C_{27}H_{42}O_8$. Crystals from benzene + cyclohexane, mp 168-170°.

β-Cortol. 5β-Pregnane-3α,11β,17,20β,21-pentol. Crystals from methanol, mp 262-264.5°. $[\alpha]_D^{25}$ +33.3° (ethanol).

2529. Cortolone. [516-42-7] (3α,5β,20S)-3,17,20,21-Tetrahydroxypregnan-11-one; 3α,17α,20α,21-pregnanetetrol-11-one. $C_{21}H_{34}O_5$; mol wt 366.50. C 68.82%, H 9.35%, O 21.83%. Isolated from human urine after administration of cortisone, hydrocortisone or ACTH: D. K. Fukushima *et al.*, *J. Biol. Chem.* **212**, 449 (1955).

Synthesis: L. H. Sarett, *J. Am. Chem. Soc.* **71**, 1169 (1949); A. H. Soloway *et al.*, *ibid.* **76**, 2941 (1954).

Crystals from acetone, mp 208-209°. $[\alpha]_D^{25}$ +44° (c = 0.5 in ethanol); also reported as $[\alpha]_D^{28}$ +34.2° (ethanol).

3,20,21-Triacetate. [2638-46-2] $C_{27}H_{40}O_8$. Crystals from methanol, mp 214-216° (Fukushima); also reported as mp 213-214° (Sarett). $[\alpha]_D^{28}$ +28° (ethanol); also reported as $[\alpha]_D^{25}$ +18° (c = 1 in acetone).

2530. Corydaldine. [493-49-2] 3,4-Dihydro-6,7-dimethoxy-1(2H)-isoquinolinone; 3,4-dihydro-6,7-dimethoxyisocarbostyril; 1,2,3,4-tetrahydro-6,7-dimethoxy-1(2H)-1-isoquinolone. $C_{11}H_{13}NO_3$; mol wt 207.23. C 63.76%, H 6.32%, N 6.76%, O 23.16%. Synthesis: Späth, Dobrowsky, *Ber.* **58**, 1274 (1925); Mohunta, Ray, *J. Chem. Soc.* **1934**, 1263; Wiesner *et al.*, *J. Am. Chem. Soc.* **77**, 675 (1955); Brossi *et al.*, *Helv. Chim. Acta* **43**, 1459 (1960); Mahuzier, Hamon, *Bull. Soc. Chim. Fr.* **1969**, 684.

Monoclinic prisms from water or alcohol, mp 175°. Sol in water, alcohol, ether, benzene, chloroform. Practically insol in petr ether.

2531. Corydaline. [518-69-4] (13S,13aR)-5,8,13,13a-Tetrahydro-2,3,9,10-tetramethoxy-13-methyl-6H-dibenzo[a,g]quinolizine; 2,3,9,10-tetramethoxy-13α-methyl-13aβ-berbine; d-corydaline. $C_{22}H_{27}NO_4$; mol wt 369.46. C 71.52%, H 7.37%, N 3.79%, O 17.32%. Alkaloid isolated from many species of *Corydalis*. Discovery in *C. tuberosa*: Wackenroder, *Kustner's Arch.* **8**, 423 (1826), as cited in: C. Wehmer, *Die Pflanzenstoffe* I (G. Fischer, Jena, 1929) p 389. Isoln from *Corydalis aurea* Willd. and *C. solida* (L.) Swartz, *Fumariaceae*: Manske, *Can. J. Res.* **16B**, 81 (1938); *Can. J. Chem.* **34**, 1 (1956). Structure: von Bruchhausen, Stippler, *Arch. Pharm.* **265**, 152 (1927). Synthesis: Späth, Kruta, *Ber.* **62**, 1024 (1929); T. Kametani *et al.*, *J. Chem. Soc. Perkin Trans. 1* **1977**, 1151; M. Cushman, F. W. Dekon, *Tetrahedron* **1978**, 1435; Z. Kiparissides *et al.*, *Can. J. Chem.* **58**, 2770 (1980). Stereochemistry: Bersch, *Arch. Pharm.* **291**, 595 (1958); Jeffs, *Experientia* **21**, 690 (1965). Pharmacology: Berezhinskaya, *C.A.* **78**, 119087j (1973). Toxicity study: R. C. Anderson, K. K. Chen, *Fed. Proc.* **5**, 163 (1946).

Prisms from alc, mp 135°. $[\alpha]_D^{20}$ +311° (c = 0.8 in alc). uv max: 396 nm. Sol in chloroform; moderately sol in ether; sparingly sol in methanol, ethanol. Practically insol in water. LD$_{50}$ in mice (mg/kg): 135.5 ± 12.8 i.v. (Anderson, Chen).

Hydrochloride hydrate. $C_{22}H_{27}NO_4 \cdot HCl \cdot H_2O$. Four-sided columns, mp 230-240°.

dl-Form. mp 135°. Slightly sol in water.
dl-Mesocorydaline. mp 158°. The (13R-cis)-analog of corydaline. Slightly sol in water.
d-Mesocorydaline. Rhombic crystals, mp 152°. $[\alpha]_D^{20}$ +82° (c = 1.4); +180° (c = 3 in chloroform).
l-Mesocorydaline. Rhombic crystals, mp 152°. $[\alpha]_D^{20}$ −85° (c = 1.4); −181° (c = 3 in chloroform).

2532. Corydalis. Squirrel corn; turkey corn. Dried tuber of *Dicentra (Bicuculla) cucullaria* (L.) Bernh., or of *Dicentra canadensis* (DC.) Walp., *Fumariaceae*. *Habit.* Ontario to Kentucky and Missouri. *Constit.* Corydaline, bulbocapnine, isocorydine, corytuberine, corycavine, corybulbine, corydine, fumaric acid, acrid resin, protopine.

2533. Corydine. [476-69-7] (6aS)-5,6,6a,7-Tetrahydro-2,10,11-trimethoxy-6-methyl-4H-dibenzo[de,g]quinolin-1-ol; 2,10,11-trimethoxy-6aα-aporphin-1-ol; 1-hydroxy-2,10,11-trimethoxyaporphine. $C_{20}H_{23}NO_4$; mol wt 341.41. C 70.36%, H 6.79%, N 4.10%, O 18.74%. The methyl ether of corytuberine. Occurs in *Corydalis cava* (L.) Schweigg & Körte (*C. tuberosa* DC), *Fumariaceae*, in the dextrorotatory form. Isoln: Gadamer, Ziegenbein, *Arch. Pharm.* **240**, 94 (1902). Structure: Späth, Berger, *Ber.* **64**, 2038 (1931). Synthesis: Hey, Palluel, *J. Chem. Soc.* **1957**, 2926; Arumugam *et al.*, *Ber.* **91**, 40 (1958); Jackson, Martin, *Chem. Commun.* **1965**, 142, 420; *eidem*, *J. Chem. Soc. C* **1966**, 2222. uv spectrum: Shamma, Yao, *J. Org. Chem.* **36**, 3253 (1971).

Tetragonal prisms from ether, mp 149°. $[\alpha]_D^{20}$ +204° (c = 1.6 in chloroform). Freely sol in chloroform, alcohol, ethyl acetate; moderately sol in ether.

dl-Hydrochloride monohydrate. $C_{20}H_{23}NO_4 \cdot HCl \cdot H_2O$. Crystals from ethanol + ether, sinters at 205°, dec 228°. Sparingly sol in water.

2534. Corynantheine. [18904-54-6] (αE,2S,3R,12bS)-3-Ethenyl-1,2,3,4,6,7,12,12b-octahydro-α-(methoxymethylene)indolo-[2,3-a]quinolizine-2-acetic acid methyl ester; (E)-16,17,18,19-tetradehydro-17-methoxy-17,18-secoyohimban-16-carboxylic acid methyl ester; (16E)-16,17,18,19-tetradehydro-17-methoxycorynan-16-carboxylic acid methyl ester. $C_{22}H_{26}N_2O_3$; mol wt 366.46. C 72.11%, H 7.15%, N 7.64%, O 13.10%. Isolated from *Corynanthe johimbe* K. Schum., and *Pseudocinchona africana* A. Cheval., *Rubiaceae*: Karrer *et al.*, *Helv. Chim. Acta* **9**, 1059 (1926); **35**, 851 (1952). Structure: Janot *et al.*, *ibid.* **34**, 1207 (1951). Stereochemistry: Van Tamelen *et al.*, *J. Am. Chem. Soc.* **79**, 6426 (1957). Total synthesis of dl- form: Van Tamelen, Wright, *Tetrahedron Lett.* **1964**, 295; *eidem*, *J. Am. Chem. Soc.* **91**, 7349 (1969). Total synthesis of d-form: Autrey, Scullard, *ibid.* **90**, 4917 (1968).

Crystals: α-form mp 103-107°; β-form mp 165-166°. $[\alpha]_D$ +28° (c = 1 in methanol). uv max (methanol): 227, 280, 291 nm (log ε 4.64, 3.82, 3.80).

Hydrochloride. $C_{22}H_{26}N_2O_3 \cdot HCl$. Needles from alcohol + ether, mp 194-206°. Also reported as mp 176-179°. $[\alpha]_D^{20}$ +12° (water). Sol in alcohol, sparingly sol in water.

Dihydrocorynantheine. $C_{22}H_{28}N_2O_3$. Total synthesis of *dl*-form: Van Tamelen, Hester, *J. Am. Chem. Soc.* **81**, 3805 (1959); *eidem, ibid.* **91**, 7342 (1969). Plates from alcohol + water, mp 177-177.5°. $[\alpha]_D$ +28°.

Hydrochloride. $C_{22}H_{28}N_2O_3 \cdot HCl$. mp 212-213°. Also reported as crystals from 95% ethanol-ethyl acetate, mp 242.2-243.3° (dec) (sealed tube), Van Tamelen, Hester, *loc. cit.* (1969).

2535. Corynanthine. [483-10-3] (16β,17α)-17-Hydroxyyohimban-16-carboxylic acid methyl ester; rauhimbine. $C_{21}H_{26}N_2O_3$; mol wt 354.45. C 71.16%, H 7.39%, N 7.90%, O 13.54%. From bark of *Pseudocinchona africana* Chev., *Corynanthe johimbe* K. Schum., *Rubiaceae* and *Rauwolfia serpentina* (L.) Benth., *Apocynaceae:* Raymond-Hamet, *Compt. Rend.* **212**, 305 (1941); Jorio, *Ann. Chim. Farm.* **1939**, 50, *C.A.* **33**, 9306[9] (1939); Le Hir *et al., Ann. Pharm. Fr.* **11**, 546 (1953); Hofmann, *Helv. Chim. Acta* **37**, 314 (1954). Identity with rauhimbine: *idem, ibid.* 849. Structure and stereochemistry: Janot *et al., Bull. Soc. Chim. Fr.* **1952**, 1085; *ibid.* **1961**, 637.

Stout prisms from acetone, dec 225-226°. $[\alpha]_D^{19}$ −85° (c = 0.5 in pyridine). uv max (methanol): 226, 283, 290 nm (log ε 4.56, 3.87, 3.79). Practically insol in water or petr ether. Sol in 40 parts of boiling chloroform, in 60 parts of boiling benzene, in 20 parts of boiling ethyl acetate, in 5 parts of boiling alcohol.

O,N-Dibutyrylcorynanthine hydrochloride. Prepn: Reiser *et al.*, US 2975183 (1961 to Chemische Werke Albert). Crystals from benzene, mp 208-210°.

O,N-Dipropionylcorynanthine hydrochloride. Prepn: Reiser *et al., loc. cit.* Crystals from isopropanol, mp 236-237°.

2536. Corypalmine. [6018-40-2] (13aS)-5,8,13,13a-Tetrahydro-2,9,10-trimethoxy-6H-dibenzo[a,g]quinolizin-3-ol; 2,9,10-trimethoxyberbin-3-ol; 3-hydroxy-2,9,10-trimethoxyberbine; tetrahydrojatrorrhizine. $C_{20}H_{23}NO_4$; mol wt 341.41. C 70.36%, H 6.79%, N 4.10%, O 18.74%. In *Corydalis cava* (L.) Schweigg & Körte *(C. tuberosa* D.C.) and other varieties of *Corydalis, Fumariaceae.* Synthesis of *dl*-corypalmine: Govindachari *et al.*, *Ber.* **92**, 1654 (1959).

d-Form. Small crystals, mp 236°. $[\alpha]_D^{16}$ +280° (chloroform). Insol in water. Sol in alcohol and chloroform: Späth *et al., Ber.* **56**, 878 (1923); **58**, 2133 (1925); **60**, 383 (1927).

l-Form. Crystals from chloroform-methanol, mp 246° (in vac): Manske, *Can. J. Res.* **20B**, 57 (1942); **21B**, 111 (1943).

dl-Form. Crystals from methanol, mp 207°.

2537. Corytuberine. [517-56-6] (6aS)-5,6,6a,7-Tetrahydro-2,10-dimethoxy-6-methyl-4H-dibenzo[de,g]quinoline-1,11-diol; 2,10-dimethoxy-6aα-aporphine-1,11-diol; 1,11-dihydroxy-2,10-dimethoxyaporphine. $C_{19}H_{21}NO_4$; mol wt 327.38. C 69.71%, H

6.47%, N 4.28%, O 19.55%. In *Corydalis cava* (L.) Schweigg & Körte *(C. tuberosa* DC), in *Dicentra formosa* (Andr.) Walp., *Fumariaceae.* Isoln: Späth, Berger, *Ber.* **64**, 2038 (1931); Manske, *Can. J. Res.* **10**, 521 (1934). Synthesis of (±)-form: Tomita, Kikkawa, **JP 58 6466** (1958 to Shionogi), *C.A.* **54**, 1584i (1960); T. Kametani, M. Ihara, *J. Chem. Soc. Perkin Trans. 1* **1980**, 629. Biosynthesis studies: Blaschke, *Biochem. Physiol. Alkaloide, Int. Symp., 4th*, K. Mothes, Ed. (Akad.-Verlag, Berlin, 1972) pp 283-286. Biomimetic total synthesis: T. Kametani *et al., Heterocycles* **5**, 175 (1976).

Pentahydrate. Leaflets, plates, turning gray on exposure to light. mp 240° dec (dry). uv max (methanol): 227, 272, 311.5 nm. Sol in alc, hot water; slightly sol in ether, chloroform, ethyl acetate. $[\alpha]_D^{20}$ +283° (alc). The dried crystals are hygroscopic and quickly attract 1 mol H_2O from the air.

Hydrochloride. $C_{19}H_{21}NO_4 \cdot HCl$. Crystals from alcohol-ether, dec above 250°; $[\alpha]_D^{20}$ +168°; sparingly sol in water.

Methyl ether *see* Corydine.

2538. Cositecan. [203923-89-1] (4S)-4-Ethyl-4-hydroxy-11-[2-(trimethylsilyl)ethyl]-1H-pyrano[3′,4′:6,7]indolizino[1,2-b]quinoline-3,14(4H,12H)-dione; 7-[2-(trimethylsilyl)ethyl]camptothecin; BNP-1350; DB-172; Karenitecin. $C_{25}H_{28}N_2O_4Si$; mol wt 448.59. C 66.94%, H 6.29%, N 6.24%, O 14.27%, Si 6.26%. DNA topoisomerase I inhibitor; derivative of camptothecin, *q.v.* Prepn: F. H. Hausheer *et al.*, **WO 9807727**; *eidem*, **US 5910491** (1998, 1999 both to BioNumerik). Synthesis: D. P. Curran *et al.*, **US 6136978** (2000 to Univ. Pittsburgh). HPLC determn in plasma and urine: J. A. Smith *et al., J. Chromatogr. B* **759**, 117 (2001). Pharmacology: A. H. Van Hattum *et al., Int. J. Cancer* **88**, 260 (2000). Clinical evaluation in non-small cell lung cancer: A. A. Miller *et al., Lung Cancer* **48**, 399 (2005); in metastatic melanoma: A. Daud *et al., Clin. Cancer Res.* **11**, 3009 (2005). Review of preclinical and clinical activity: P. N. Munster, A. I. Daud, *Expert Opin. Invest. Drugs* **20**, 1565-1574 (2011).

Tan solid. Highly lipophilic. $[\alpha]_D^{20}$ +29.8° (c = 0.2 in methylene chloride). Sol in acetonitrile, DMSO.

THERAP CAT: Antineoplastic.

2539. Cosyntropin. [16960-16-0] α^{1-24}-Corticotropin; β^{1-24}-corticotropin; tetracosactide; Cortrosyn. $C_{136}H_{210}N_{40}O_{31}S$; mol wt 2933.49. C 55.68%, H 7.22%, N 19.10%, O 16.91%, S 1.09%. Synthetic polypeptide corresponding to the first 24 amino acids of ACTH, *q.v.* Identification of activity: R. G. Shepherd *et al., J. Am. Chem. Soc.* **78**, 5067 (1956). Synthesis: H. Kappeler, R. Schwyzer, *Helv. Chim. Acta* **44**, 1136 (1961); *eidem, ibid.* **46**, 1550 (1963); solid-state synthesis: R. Matsueda *et al., J. Am. Chem. Soc.* **97**, 2573 (1975). Assessment of hormone levels for adrenal function in cats: M. E. Peterson *et al., Res. Vet. Sci.* **37**, 331 (1984); in dogs: L. A. Frank *et al., Domest. Anim. Endocrinol.* **24**, 43 (2003). Review as diagnostic aid in assessment of secondary adrenal insufficiency: T. A. M. Abdu, R. N. Clayton, *Curr. Opin. Endocrinol. Diabetes* **7**, 116-121 (2000); in primary or secondary adrenal insufficiency: R. I. Dorin *et al.*, *Ann. Intern. Med.* **139**, 194-204, E205-E206 (2003).

Ser-Tyr-Ser-Met-Glu-His-Phe-Arg-Trp-Gly-Lys-Pro-Val-Gly
|
Lys
|
Pro-Tyr-Val-Lys-Val-Pro-Arg-Arg-Lys

Hexaacetate. [22633-88-1] α^{1-24}-Corticotropin hexaacetate (salt); Synacthen. $[\alpha]_D^{22}$ −88 ±2° (c = 0.511 in 1% acetic acid). uv max (0.1 *N* NaOH): 283, 289 nm (ε 8550, 8810).
THERAP CAT: Diagnostic aid (adrenal function).
THERAP CAT (VET): Diagnostic aid (adrenal function).

2540. Cotarnine. [82-54-2] 5,6,7,8-Tetrahydro-4-methoxy-6-methyl-1,3-dioxolo[4,5-*g*]isoquinolin-5-ol. $C_{12}H_{15}NO_4$; mol wt 237.26. C 60.75%, H 6.37%, N 5.90%, O 26.97%. Prepd by the oxidation of narcotine with dil nitric acid. Synthesis: Salway, *J. Chem. Soc.* **97**, 1208 (1910). Is tautomeric. Structure studies: Small, Lutz, "Chemistry of the Opium Alkaloids," Supplement No. 103, *U.S. Public Health Reports* (Washington, 1932) p 49; Schneider, Müller, *Ann.* **615**, 34 (1958). Absorption spectrum: Csokán, *Z. Anal. Chem.* **124**, 344 (1942).

Small needles from benzene, dec 132-133°. Sol in alcohol, chloroform, ether, benzene; slightly sol in water; sol in dil acids, in ammonia or sodium carbonate soln, but only slightly in potassium hydroxide soln. Aq or alcoholic solns are yellow.
Hydrochloride. [36647-02-6] Secalysat. $C_{12}H_{15}NO_4 \cdot HCl$; mol wt 273.71.
Chloride. [10018-19-6] 7,8-Dihydro-4-methoxy-6-methyl-1,3-dioxolo[4,5-*g*]isoquinolinium chloride; cotarnine chloride; cotarninium chloride; Stypticin. $C_{12}H_{14}ClNO_3$; mol wt 255.70. Dihydrate, light-yellow powder; deliquesc in moist air. Sol in about 1 part water, 4 parts alc. *Keep well closed.*
Phthalate. [6190-36-9] Styptol. $C_{32}H_{32}N_2O_{10}$; mol wt 604.61.
THERAP CAT: Hemostatic.

2541. Cotinine. [486-56-6] (5*S*)-1-Methyl-5-(3-pyridinyl)-2-pyrrolidinone; *N*-methyl-2-(3-pyridyl)-5-pyrrolidone. $C_{10}H_{12}N_2O$; mol wt 176.22. C 68.16%, H 6.86%, N 15.90%, O 9.08%. Nicotine metabolite first described by: Pinner, *Arch. Pharm.* **231**, 378 (1893). Isolated from autoxidized nicotine, nicotine treated with hydrogen peroxide, from nicotine irradiated with ultraviolet light: Frankenburg, Vaitekunas, *J. Am. Chem. Soc.* **79**, 149 (1957).

Viscous oil, bp_6 210-211°. Absorption spectra: Frankenburg, Vaitekunas, *loc. cit.*
Fumarate. [5695-98-7] Scotine. $(C_{10}H_{12}N_2O)_2 \cdot C_4H_4O_4$; mol wt 468.51.
THERAP CAT: Antidepressant.

2542. Cotton-root Bark. Dried root-bark of one or more of the cultivated species of *Gossypium herbaceum* L. and of other species of *Gossypium*, *Malvaceae*. *Habit.* Asia (India, China, Arabia), Egypt, U.S., West Indies, S. America, Australia, Spain, etc. *Constit.* Yellow chromogen about 8% of a pale yellow resin, fixed oil, sugar.
THERAP CAT: Oxytocic.

2543. Cottonseed Oil. Fixed oil from seeds of cultivated varieties of *Gossypium herbaceum* L. or of other species of *Gossypium*.

Pale yellow, oily, practically odorless liq. d_{25}^{25} 0.915-0.921. Solidif 0° to −5°. n_D^{40} 1.4645-1.4655. Sapon no.: 190-198. Iodine no.: 105-114. Surface tension (20°) = 35.4 dyn/cm; (80°) = 31.3 dyn/cm. Slightly sol in alcohol; miscible with chloroform, ether, solvent hexane, carbon disulfide, petr ether.
USE: Manuf soaps, oleomargarine, hydrogenated fats, lard substitute, glycerol, leather dressings, lubricants, cosmetics; emollient; also as salad and cooking oil; packing fish. Pharmaceutic aid (solvent).
THERAP CAT (VET): Pediculicide, acaricide, laxative, emollient.

2544. Coumachlor. [81-82-3] 3-[1-(4-Chlorophenyl)-3-oxobutyl]-4-hydroxy-2*H*-1-benzopyran-2-one; 3-(α-acetonyl-*p*-chlorobenzyl)-4-hydroxycoumarin; 3-(α-*p*-chlorophenyl-β-acetylethyl)-4-hydroxycoumarin; G-21133. $C_{19}H_{15}ClO_4$; mol wt 342.78. C 66.58%, H 4.41%, Cl 10.34%, O 18.67%. Structural analog of warfarin. Prepn: F. Litvan, W. Stoll, US 2648682 (1953 to J. R. Geigy). Anticoagulant activity: M. Reiff, R. Weismann, *Acta Trop.* **8**, 97 (1951), *C.A.* **50**, 10976d (1956). Review of comparative toxicology: H. Wanntorp, *Acta Pharmacol. Toxicol.* **16**, Suppl. 2, 123 pp (1959).

Crystals, mp 169-171°. Sol in alc, acetone, chloroform. Slightly sol in benzene, ether. Practically insol in water. MLD in dog, swine (mg/kg): <5, <5 orally (Wanntorp).
Caution: Similar to warfarin, with delayed actions on prothrombin level and blood clotting, resulting in death by hemorrhage. *See Clinical Toxicology of Commercial Products*, R.E. Gosselin *et al.*, Eds. (Williams & Wilkins, Baltimore, 5th ed., 1984) Section II, pp 348.
USE: Rodenticide.

2545. Coumalic Acid. [500-05-0] 2-Oxo-2*H*-pyran-5-carboxylic acid; α-pyrone-5-carboxylic acid. $C_6H_4O_4$; mol wt 140.09. C 51.44%, H 2.88%, O 45.68%. Prepd by heating anhydr malic acid with oleum: v. Pechmann, *Ann.* **264**, 262, 272 (1891); Wiley, Smith, *Org. Synth.* **coll. vol. IV**, 201 (1963).

Bright yellow prisms from methanol, mp 205-210° (partial decompn). bp_{120} 218°; sublimes partially. Sparingly sol in cold water. Dec by boiling water. Sol in alcohol, glacial acetic acid. Slightly sol in ether, acetone, ethyl acetate. Insol in chloroform, benzene, ligroin.
Methyl ester. $C_7H_6O_4$. Leaflets from ligroin, mp 73-74°. bp_{60} 178-180°. bp_{760} 250-260°.
Ethyl ester. $C_8H_8O_4$. Crystals, mp 36°. bp_{760} 262-265°. Distills without decomp.

2546. Coumaphos. [56-72-4] Phosphorothioic acid *O*-(3-chloro-4-methyl-2-oxo-2*H*-1-benzopyran-7-yl) *O,O*-diethyl ester; *O,O*-diethyl *O*-(3-chloro-4-methyl-2-oxo-2*H*-1-benzopyran-7-yl) phosphorothioate; *O*-3-chloro-4-methyl-2-oxo-2*H*-chromen-7-yl *O,O*-diethyl phosphorothioate; 3-chloro-7-diethoxyphosphinothioyloxy-4-methylcoumarin; 3-chloro-4-methylumbelliferone *O,O*-diethyl phosphorothioate; *O,O*-diethyl *O*-(3-chloro-4-methyl-7-coumarinyl) phosphorothioate; coumafos; Bayer 21/199; Asuntol; Baymix; Co-ral; Muscatox; Perizin; Resitox. $C_{14}H_{16}ClO_5PS$; mol wt 362.76. C 46.35%, H 4.45%, Cl 9.77%, O 22.05%, P 8.54%, S 8.84%. Organophosphate insectide; cholinesterase inhibitor. Prepn: Schrader, DE 881194; US 2748146 (1951, 1956 both to Bayer). Metabolism: H. R. Krueger *et al.*, *J. Agric. Food Chem.* **7**, 182 (1959). GC determn in honey: U. Menkissoglu-Spiroudi *et al.*, *J. AOAC Int.* **83**, 178 (2000). Toxicity: T. B. Gaines, *Toxicol. Appl.*

Pharmacol. **14**, 515 (1969). Efficacy for nematodes in cattle: M. F. Hansen, S. M. Zeakes, *Trans. Am. Micros. Soc.* **88**, 159 (1969); for cattle ticks: R. B. Davey *et al.*, *Prev. Vet. Med.* **23**, 1 (1995). Use in eradication of bee mites and distribution in hives: P. Tremolada *et al.*, *Ecotoxicology* **13**, 589 (2004).

Crystals, mp 91°. *Poisonous.* Vapor pressure (20°): 1.3×10^{-5} Pa. Log P (octanol/water): 4.13. Practically insol in water. Somewhat sol in acetone, chloroform, corn oil. LD_{50} in female, male rats (mg/kg): 16, 41 orally (Gaines).

Potential symptoms of overexposure by ingestion, inhalation or absorption through skin are dizziness, nausea, vomiting, diarrhea, abdominal cramps; excessive sweating, salivation; constricted pupils; labored breathing, muscle cramps; unconsciousness; convulsions; severe respiratory depression.

USE: Insecticide; acaricide.

THERAP CAT (VET): Anthelmintic; ectoparasiticide.

2547. Coumaran. [496-16-2] 2,3-Dihydrobenzofuran; cumaran; dihydrocoumarone. C_8H_8O; mol wt 120.15. C 79.97%, H 6.71%, O 13.32%. Synthesis: Bennett, Mahmoud Hafez, *J. Chem. Soc.* **1941**, 287; *cf.* Hurd, Hoffmann, *J. Org. Chem.* **5**, 212 (1940); Rindfusz, *J. Am. Chem. Soc.* **41**, 669 (1919); J. M. Bakke, H. M. Roholdt, *Acta Chem. Scand.* **B34**, 73 (1980).

Oily liq, bp 188-189°, bp_{13} 74-75°. d_4^{25} 1.058, n_D^{20} 1.5426. Sol in alcohol, ether, chloroform, carbon disulfide.

2548. *p*-Coumaric Acid. [7400-08-0] 3-(4-Hydroxyphenyl)-2-propenoic acid; *p*-hydroxycinnamic acid; β-[4-hydroxyphenyl]-acrylic acid. $C_9H_8O_3$; mol wt 164.16. C 65.85%, H 4.91%, O 29.24%. Isoln: Hlasiwetz, *Ann.* **136**, 31 (1865); Bamberger, *Monatsh. Chem.* **12**, 459 (1891). Prepn: Eigel, *Ber.* **20**, 2528 (1887); Konek, Pacsu, *Ber.* **51**, 856 (1918). uv data: Wheeler, Covarrubias, *J. Org. Chem.* **28**, 2015 (1963).

Needles, mp 210-213° (reported in early lit as 206°). Crystallizes in anhydrous form from conc hot aq soln, but as monohydrate from dil aq soln on slow cooling. Slightly sol in cold water; sol in hot water and alc, ether. Practically insol in benzene and ligroin. uv max (95% ethanol): 223, 286 nm (ε 14450, 19000).

2549. Coumarilic Acid. [496-41-3] 2-Benzofurancarboxylic acid; coumarone-2-carboxylic acid. $C_9H_6O_3$; mol wt 162.14. C 66.67%, H 3.73%, O 29.60%. Synthesis from coumarin: Fuson *et al.*, *Org. Synth.* **coll. vol. III**, 209 (1955). *Review:* Sethna, Shah, *Chem. Rev.* **36**, 1 (1945).

Needles from water, bitter taste. mp 192-193°. bp 310-315° with slight decompn. Sol in boiling water, in alcohol; slightly sol in chloroform, carbon disulfide.

2550. Coumarin. [91-64-5] 2*H*-1-Benzopyran-2-one; 1,2-benzopyrone; *cis-o*-coumarinic acid lactone; cumarin; coumarinic anhydride; tonka bean camphor. $C_9H_6O_2$; mol wt 146.15. C 73.96%, H 4.14%, O 21.89%. In tonka beans, lavender oil, woodruff

(Asperula species), in sweet clover *(Melilotus).* Crystal and molecular structure: *Chem. Commun. (Univ. Stockholm)* **1976**, 21. Toxicity: Jenner *et al.*, *Food Cosmet. Toxicol.* **2**, 327 (1964). *Review:* Sethna, Shah, *Chem. Rev.* **36**, 1 (1945); W. C. Meuly in *Kirk-Othmer Encyclopedia of Chemical Technology* **vol. 7** (Wiley-Interscience, New York, 3rd ed., 1979) pp 196-206.

Orthorhombic, rectangular plates. Pleasant, fragrant odor resembling that of vanilla beans, burning taste. mp 68-70°. bp 297-299°. bp_5 139°. One gram dissolves in 400 ml cold, 50 ml boiling water. Freely sol in alc, chloroform, ether, oils; also sol in alkali hydroxide solns. LD_{50} orally in rats, guinea pigs: 680, 202 mg/kg (Jenner).

USE: Pharmaceutic aid (flavor).

2551. Coumarin-3-carboxylic Acid. [531-81-7] 2-Oxo-2*H*-1-benzopyran-3-carboxylic acid. $C_{10}H_6O_4$; mol wt 190.15. C 63.17%, H 3.18%, O 33.66%. By heating salicylaldehyde with malonic acid in glacial acetic acid. *Review:* Sethna, Shah, *Chem. Rev.* **36**, 1 (1945).

Needles from water, mp 188° (dec). Slightly sol in water; sol in alc, alkalies. Insol in ether, benzene, petr ether.

2552. Coumestrol. [479-13-0] 3,9-Dihydroxy-6*H*-benzofuro[3,2-*c*][1]benzopyran-6-one; 2-(2,4-dihydroxyphenyl)-6-hydroxy-3-benzofurancarboxylic acid δ-lactone; 7′,6-dihydroxycoumarino(3′,4′,3,2)coumarone. $C_{15}H_8O_5$; mol wt 268.22. C 67.17%, H 3.01%, O 29.82%. An estrogenic factor occurring naturally in forage crops, esp in ladino clover *(Trifolium repens* L.), strawberry clover *(T. fragiferum* L.) and alfalfa *(Medicago sativa* L., *Leguminosae*). Isoln: Bickoff *et al.*, *J. Agric. Food Chem.* **6**, 536 (1958); Bickoff, Booth, **US 2890116** (1959 to U.S.A.). Structure: Bickoff *et al.*, *J. Am. Chem. Soc.* **80**, 3969 (1958). Synthesis: Emerson, Bickoff, *ibid.* 4381; **US 2884427** (1959 to U.S.A.); Jurd, *Tetrahedron Lett.* **1963**, 1151; Kappe, Brandner, *Z. Naturforsch.* **29B**, 292 (1974). Biosynthesis: Grisebach, Barz, *Chem. Ind. (London)* **1963**, 690.

Crystals, mp 385°. Sublimes at 325°; sublimes in high vacuum at about 175°. uv max (methanol): 208, 243, 343 nm. Exhibits bright blue fluorescence in neutral or acid soln, greenish-yellow fluorescence in strong alkali. Absorption and fluorescence spectra: O. S. Wolfbeis, K. Schaffner, *Photochem. Photobiol.* **32**, 143 (1980). Practically insol in water at acid and neutral pH, in petr ether. Sparingly sol in water at alkaline pH (pH 11-12); slightly sol in methanol, chloroform, ether; very slightly sol in carbon tetrachloride, benzene.

Diacetate. $C_{19}H_{12}O_7$. Crystals from acetic acid, mp 237°.

Dimethyl ether. $C_{17}H_{12}O_5$. Crystals from methanol, mp 198°.

2553. Coumingine. [26241-81-6] 3-Hydroxy-3-methylbutanoic acid (2*S*,4a*R*,4b*S*,7*E*,8*R*,8a*S*,10a*R*)-7-[2-[2-(dimethylamino)-ethyl]-2-oxoethylidene]tetradecahydro-1,1,4a,8-tetramethyl-9-oxo-2-phenanthrenyl ester. $C_{29}H_{47}NO_6$; mol wt 505.70. C 68.88%, H 9.37%, N 2.77%, O 18.98%. Cardiotonic principle from the bark of *Erythrophleum couminga* Baillon, *Leguminosae*. Isoln: Ruzicka *et al.*, *Helv. Chim. Acta* **24**, 63 (1941). Structure: Ruzicka *et al.*, *ibid.* 1449.

Thin shiny needles from ether, mp 142°. $[\alpha]_D^{20}$ −70°. Sol in methanol, ethanol, acetone, acetic acid, benzene, chloroform. Practically insol in water, hexane, ether.

Hydrochloride. $C_{29}H_{47}NO_6 \cdot HCl$. Crystals from ethanol + ether, mp 195°. Sol in ethanol, water.

2554. Crabtree's Catalyst. [64536-78-3] [$(1,2,5,6-\eta)$-1,5-Cyclooctadiene](pyridine)(tricyclohexylphosphine)iridium(1+) hexafluorophosphate(1−) (1:1); Felkin-Crabtree catalyst. $C_{31}H_{50}$-IrNP.F_6P; mol wt 804.90. C 46.26%, H 6.26%, Ir 23.88%, N 1.74%, P 7.70%, F 14.16%. Stereoselective organoiridium based catalyst used primarily for hydrogenation of the general formula [Ir(cod)L-(py)]metal complex where cod is 1,5-cyclooctadiene, L is tertiary phosphine, py is pyridine. Prepn: R. H. Crabtree, G. E. Morris, *J. Organomet. Chem.* **135**, 395 (1977); R. H. Crabtree *et al.*, *ibid.* **141**, 205 (1977); R. H. Crabtree, S. M. Morehouse, *Inorg. Synth.* **24**, 173 (1986). Hydrogenation study: J. W. Suggs *et al.*, *Tetrahedron Lett.* **22**, 303 (1981); directing effects: J. M. Brown, S. A. Hall, *Tetrahedron* **41**, 4639 (1985); R. H. Crabtree, M. W. Davis, *J. Org. Chem.* **51**, 2655 (1986). Structure: M. S. Abbassioun *et al.*, *Acta Crystallogr.* **C45**, 331 (1989). Synthetic applications: M. Weck *et al.*, *J. Org. Chem.* **64**, 5463 (1999); J. M. Bueno *et al.*, *Tetrahedron Lett.* **41**, 4379 (2000).

Orange crystalline solid. d 1.67 g/cm³. Air stable. Sol in CH_2Cl_2, $CHCl_3$ and Me_2CO. Insol in alcohols, water, benzene, diethyl ether and hexane.

Tetrafluoroborate. [80409-82-1] $C_{31}H_{50}IrNP.BF_4$; mol wt 746.74.

USE: Catalyst for selective reduction of hindered carbon-carbon double bonds.

2555. Crataegus. Hawthorn; English hawthorn; haws; haw apple; aubépine. Berries, flowers, and leaves of *Crataegus oxyacantha* L., *Rosaceae*. *Habit.* All temperate zones, abundant in Europe, a garden escape in North America. *Constit.* Triterpene acids (oleanolic, ursolic, crataegolic); purines; anthocyanin type pigments and flavone deriv (pelargonin, quercitrin); choline; acetylcholine; trimethylamine; chlorogenic acid; caffeic acid; ascorbic acid; a growth hormone for caterpillars; a substance named RN 30/9: Hahn *et al.*, *Arzneim.-Forsch.* **10**, 825 (1960). Some commercial extracts are: *Curtacrat*; *Crataegus-Kreussler*; *Esbericard*. Review and bibliography: F. Berger, *Handbuch der Drogenkunde* **vol. 3** (Vienna, 1952) pp 202-211.

THERAP CAT: Cardiotonic, vasodilator (coronary).

2556. Creatine. [57-00-1] *N*-(Aminoiminomethyl)-*N*-methylglycine; *N*-amidinosarcosine; (α-methylguanido)acetic acid; *N*-methyl-*N*-guanylglycine; methylglycocyamine. $C_4H_9N_3O_2$; mol wt 131.14. C 36.64%, H 6.92%, N 32.04%, O 24.40%. Amino acid that participates in the transfer of high energy phosphate in muscle cells. Predominantly found in skeletal and cardiac muscle; occurring also in its phosphorylated form, *see* phosphocreatine. Produced by liver, pancreas and kidneys by the transfer of the guanidine moiety of arginine to glycine which is then methylated to give creatine. First identified in meat extracts by Chevreul in 1835; name derived from the Greek "kreas," meaning flesh. Synthesis by heating cyanamide with sarcosine: Strecker, *Jahresber. Chem.* **1868**, 686; *cf.* Volhard, *Z. Chem.* **5**, 318 (1869); Paulmann, *Arch. Pharm.* **232**, 638 (1894); Bergmann, Zervas, *Z. Physiol. Chem.* **173**, 80 (1928); King *J. Chem. Soc.* **1930**, 2374. Use in diagnosis of myocardial infarction: J. Delanghe *et al.*, *Ann. Clin. Biochem.* **25**, 383 (1988). HPLC determn in cardiac muscle: T. Teerlink *et al.*, *Anal. Biochem.* **214**, 278 (1993). Review of role in energy metabolism: S. P. Bessman, C. L. Carpenter, *Annu. Rev. Biochem.* **54**, 831-862 (1985). Review of efficacy and safety as nutritional supplement: A. S. Graham, R. C. Hatton, *J. Am. Pharm. Assoc.* **39**, 803-810 (1999); of effects of dietary supplementation on exercise performance: T. W. Demant, E. C. Rhodes, *Sports Med.* **28**, 49-60 (1999); A. Casey, P. L. Greenhaff, *Am. J. Clin. Nutr.* **72**, Suppl., 607S-617S (2000).

Monoclinic prisms from water as the monohydrate. Becomes anhydr at 100°; dec 303°. Neutral reaction to litmus. pKb (20°) 11.02. Adsorption on various chromatographic agents: Grettie, Williams, *J. Am. Chem. Soc.* **50**, 671 (1928). Absorption spectrum: Abderhalden, Haas, *Z. Physiol. Chem.* **164**, 7 (1927). One gram of the monohydrate dissolves in 75 ml water, in about 9 liters alcohol. Insol in ether.

2557. Creatinine. [60-27-5] 2-Amino-1,5-dihydro-1-methyl-4*H*-imidazol-4-one; 2-amino-1-methyl-4-imidazolidinone; 1-methylhydantoin-2-imide; 1-methylglycocyamidine. $C_4H_7N_3O$; mol wt 113.12. C 42.47%, H 6.24%, N 37.15%, O 14.14%. Metabolic end product of creatine. *q.v.* Found in all bodily fluids, such as blood, urine, sweat, and bile. Formed by the nonenzymatic dehydration of muscle creatine; filtered from the blood by the kidney with little or no tubular reabsorption and excreted in the urine. Concentration in serum is approx 0.55-1.18 mg/dL in healthy adults, is relatively constant, and is dependent upon muscle mass. Creatinine clearance is clinically used as a measure of the glomerular filtration rate. Isoln from urine: Maly, *Ann.* **159**, 279 (1871); Folin, *J. Biol. Chem.* **17**, 463 (1914); Benedict, *ibid.* **18**, 183 (1914). Prepn from commercial creatine by treatment with HCl: Edgar, Hinegardner, *ibid.* **56**, 881 (1923); *Org. Synth.* **4**, 15 (1925). Review of determn methods in urine: T. Smith-Palmer, *J. Chromatogr. B* **781**, 93-106 (2002); of diagnostic tests for renal function: S. Narayanan, H. D. Appleton, *Clin. Chem.* **26**, 1119-1126 (1980); M. Peake, M. Whiting, *Clin. Biochem. Rev.* **27**, 173-184 (2006). Review of creatinine metabolism and excretion: R. D. Perrone *et al.*, *Clin. Chem.* **38**, 1933-1953 (1992); M. Wyss, R. Kaddurah-Dauok, *Physiol. Rev.* **80**, 1107-1213 (2000); of serum concentrations: F. Ceriotti *et al.*, *Clin. Chem.* **54**, 559-566 (2008).

Monoclinic plates. Leaflets from water. Dec about 300°. pKa 4.8; 9.2. pKb (40°) 10.45. Sol in 12 parts water; slightly sol in alcohol. Practically insol in acetone, ether, chloroform.

USE: Biomarker for clinical assessment of renal function.

2558. Creolin. [12751-04-1] Pearson's creolin. Prepd from refined coal-tar oils. Approx composition: Tar acids and oils 75-77%; emulsifying soaps 15-17%; water 8-10%.

Dark brown liquid, characteristic odor resembling that of phenol. Phenol coefficient (against *B. typhosus*) about 10. d 1.02-1.04. Forms stable milky emulsions when dil with much water. Miscible with a small amount of water; also miscible with alcohol, ether, chloroform.

USE: As a general industrial and household disinfectant and deodorant in 1-3% emulsion in water.

THERAP CAT (VET): Antiseptic, parasiticide. Do not use on cats.

2559. Creosol. [93-51-6] 2-Methoxy-4-methylphenol; 2-methoxy-*p*-cresol; 4-methylguaiacol; 3-methoxy-4-hydroxytoluene; 4-hydroxy-3-methoxy-1-methylbenzene. $C_8H_{10}O_2$; mol wt 138.17. C 69.54%, H 7.30%, O 23.16%. Occurs in beechwood tar. It is one of the active constituents of creosote. Obtained by the Clemmensen reduction of vanillin using amalgamated zinc and toluene as auxiliary solvent: Fletcher, Tarbell, *J. Am. Chem. Soc.* **65**, 1431 (1943); Schwarz, Hering, *Org. Synth.* **coll. vol. IV**, 203 (1963).

Colorless to yellowish, strongly refractive, aromatic liq. d_4^{25} 1.092. bp$_{760}$ 220°; bp$_{15}$ 105°; bp$_4$ 79°. mp 5.5°. n_D^{25} 1.5353. Slightly sol in water; miscible with alcohol, benzene, chloroform, ether, glacial acetic acid.

Note: Not to be confused with cresol.

2560. Creosote, Coal Tar. [8001-58-9] Coal tar creosote. A distillate of coal tar produced by high temp carbonization of bituminous coal. *Constit.* Liq and solid, polycyclic aromatic hydrocarbons, tar acids (up to 3%) and tar bases. History and composition: Roche, *J. For. Prod. Res. Soc.* **2**, 75 (1952). Characterization by GLC: F. H. M. Nestler, *Anal. Chem.* **46**, 46 (1974). GC-MS analysis in treated railroad ties: W. Rotard, W. Mailahn, *ibid.* **59**, 65 (1987). Review of constituents, uses and carcinogenic risk: *IARC Monographs* **35**, 83-159 (1985); of toxicology and human exposure: *Toxicological Profile for Wood Creosote, Coal Tar Creosote, Coal Tar, Coal Tar Pitch, and Coal Tar Pitch Volatiles* (PB2003-100136, 2002) 394 pp.

Translucent brown to black, oily liq. *Flammable.* Characteristic sharp, penetrating smoky odor; burning caustic taste. Heavier than water. *Typical specification:* $d_{15.5}^{38.0}$ 1.06. Distillation ranges: Up to 210° not >5%; up to 235° not >25% nor <5%; up to 270° not <20%; up to 355° not >85% nor <60%. Flash pt 165°F (75°C); ignition temp 637°F (335°C). Miscible with alc, ether, fixed or volatile oils. Practically insol in water.

Caution: Readily absorbed through GI tract and skin. Potential symptoms following oral overexposure are intense GI irritation and congestion. Symptoms of systemic poisoning include salivation, vomiting, respiratory difficulties, thready pulse, vertigo, headache, loss of pupillary reflexes, hypothermia, cyanosis and mild convulsions. Direct contact may cause skin, eye or mucous membrane irritation, burning and itching, erythema, papular and vesicular eruptions, keratoconjunctivitis. *See Patty's Industrial Hygiene and Toxicology* Vol. **2B**, G. D. Clayton, F. E. Clayton, Eds. (John Wiley & Sons, Inc., New York, 4th ed., 1994) pp 1602-1605. Coal tar and coal tar pitches are listed as known human carcinogens: *Report on Carcinogens, Twelfth Edition* (PB2011-111646, 2011) p 111.

USE: Wood preservative, disinfectant, insecticide.

THERAP CAT (VET): Has been used as an anthelmintic.

2561. Creosote, Wood. [8021-39-4] Wood creosote; beechwood creosote; Creasote. Liquid obtained from wood tars by distillation; composed chiefly of phenol, guaiacol and creosol, *q.q.v.* Analysis of constituents by GC-MS and HPLC: N. Ogata, T. Baba, *Res. Commun. Chem. Pathol. Pharmacol.* **66**, 411 (1989). Acute toxicity: T. Miyazato *et al.*, *Oyo Yakuri* **21**, 899 (1981), *C.A.* **96**, 28311h (1982). Chronic toxicity and carcinogenicity studies: *eidem, ibid.* **28**, 909, 925 (1984), *C.A.* **102**, 41238b, 57380c (1985). *See also: Clinical Toxicology of Commercial Products*, R. E. Gosselin *et al.*, Eds. (Williams & Wilkins, Baltimore, 5th ed., 1984)

Section II, p 192. Review of toxicology and human exposure: *Toxicological Profile for Wood Creosote, Coal Tar Creosote, Coal Tar, Coal Tar Pitch, and Coal Tar Pitch Volatiles* (PB2003-100136, 2002) 394 pp.

Almost colorless or yellowish, very refractive, oily liq; characteristic smoky odor; caustic, burning taste. d_{25}^{25} 1.076. Begins to boil at about 203° and at least 90% by vol distills between 203-220°. Does not solidify at −20°. Sol in 150-200 parts water, in glycerol, glacial acetic acid, fixed alkali hydroxide solns; miscible with alc, chloroform, ether, oils.

Benzoate. Yellowish, oily liquid. Characteristic smoky odor and sharp, burned taste. Insol in water. Freely sol in alc or ether.

Carbonate. Colorless to yellowish, clear, viscid, oily liquid; odorless and tasteless or slight odor and taste of creosote. d_{25}^{25} not below 1.145. Insol in water. Freely sol in alc; sol in petr ether, fixed oils. Miscible with benzene, chloroform.

Oleate. Oleocreosote. Yellowish, oily liquid. d 0.950. Insol in water. Sol in benzene, chloroform, ether.

Phosphate. Phosote. Mixture of the phosphoric acid esters of creosote. Yellowish, almost odorless, viscid mass. d_{15}^{15} 1.19. bp 230-235° with decompn. Insol in water. Sol in alc.

Valerate. Eosote. Colorless to yellowish liquid. bp about 240°. Insol in water. Sol in alc, ether.

THERAP CAT: Antiseptic; expectorant.

THERAP CAT (VET): Antiseptic, parasiticide, deodorant; has been used as expectorant, gastric sedative and gastrointestinal antiseptic. Do not use in cats.

2562. Creosotic Acid. Mixture of the isomeric cresotic acids. White or reddish-white powder. Slightly sol in water; sol in alcohol, ether, alkali hydroxide solns.

USE: Disinfectant generally and in veterinary medicine in 1:500 aq or soap soln; manuf dyes and artificial tannin.

2563. Creslan®. Fiber X-54; Exlan. A copolymer of acrylonitrile possibly with acrylamide or a substituted acrylamide. *Ref:* R. W. Moncrieff, *Man-Made Fibres* (John Wiley, New York, 1963) p 483.

Off-white fiber. Specific gravity 1.17. Sticks at 210°. Shrinkage in boiling water, 1%. Sunlight has negligible effect. Resistance to acids: good except to mineral acids. Resistance to alkalies: fair to dilute, poor to concd. Good resistance to cleaning solvents. Superior to its forerunner, *fiber X-51*, in the ease with which it can be dyed.

USE: Recommended for those uses that acrylics generally fill. It is claimed that it can be durably pleated.

2564. Cresol. [1319-77-3] Cresylic acid; cresylol; tricresol. C_7H_8O; mol wt 108.14. C 77.75%, H 7.46%, O 14.79%. Mixture of the three isomeric cresols, in which the *m*-isomer predominates. Obtained from coal tar: Paulsen, **US 2998457** (1962 to Ashland Oil & Ref.). Usually contains a few per cent phenol. Prepn by sulfonation of toluene: Englund *et al.*, *Ind. Eng. Chem.* **45**, 189 (1953); by oxidation of toluene: Braunwarth, Winsted, **US 2994722** (1961 to Pure Oil). Toxicity study: W. B. Deichmann, S. Witherup, *J. Pharmacol. Exp. Ther.* **80**, 233 (1944). Review of manuf processes: *Faith, Keyes & Clark's Industrial Chemicals*, F. A. Lowenheim, M. K. Moran, Eds. (Wiley-Interscience, New York, 4th ed., 1975) pp 285-293; of toxicology and human exposure: *Toxicological Profile for Cresols* (PB2009-100002, 2008) 283 pp.

m-Cresol

Colorless, yellowish, brownish-yellow or pinkish liq; phenolic odor; becomes darker with age and on exposure to light. *Poisonous.* d_{25}^{25} 1.030-1.038. Not less than 90% by vol distills between 195-205°. Soluble in about 50 parts water. A soln in water is neutral to bromocresol purple. Also sol in solns of fixed alkali hydroxides. Miscible with alc, benzene, ether, glycerol, petr ether. *Protect from light.*

***m*-Cresol.** [108-39-4] 3-Methylphenol. Obtained from coal tar: Maesawa, Kurakano, **JP 55 8929** (1955 to Osaka Gas), *C.A.* **52**,

1231d (1958); Macak, Rehak, *Brennst.-Chem.* **43**, 80 (1962). Prepn from toluene: Toland, **US 2760991** (1956 to California Res. Corp.); by oxidation of *o*- or *p*-toluic acid: Kaeding *et al.*, *Ind. Eng. Chem.* **53**, 805 (1961). Colorless or yellowish liquid; phenolic odor. d_4^{20} 1.034. bp 202°. mp 11-12°. Flash pt, closed cup: 187°F (86°C). n_D^{20} 1.5398. Sol in about 40 parts water, in solns of fixed alkali hydroxides; miscible with alc, chloroform, ether. LD_{50} orally in rats: 2.02 g/kg (Deichmann, Witherup).

o-**Cresol.** [95-48-7] 2-Methylphenol; *o*-cresylic acid; *o*-hydroxytoluene. Prepn from *m*-toluic acid: Toland, **US 2766294** (1956 to California Res. Corp.); Barnard, Meyer, **US 2852567** (1958 to Dow). Crystals or liq, becoming dark with age and exposure to air and light; phenolic odor. d_4^{20} 1.047. bp 191-192°. mp 30°. Flash pt 81-83°C. n_D^{20} 1.553. Sol in about 40 parts water, in solns of the fixed alkali hydroxides. Miscible with alc, chloroform, ether. *Protect from light.* LD_{50} orally in rats: 1.35 g/kg (Deichmann, Witherup).

p-**Cresol.** [106-44-5] 4-Methylphenol. Obtained from coal tar. Laboratory prepn from *p*-toluenesulfonic acid by fusion with potassium hydroxide: W. W. Hartman, *Org. Synth.* **coll. vol. I**, 175 (2nd ed., 1941); from toluene: Braunwarth, **US 3046305** (1962 to Pure Oil). Crystals. Phenolic odor. d_4^{20} 1.0341. mp 35.5°. bp_{760} 201.8°; bp_{200} 179.4°; bp_{100} 140.0°; bp_{40} 117.7°; bp_{20} 102.3°; bp_{10} 88.6°; bp_5 76.5°; $bp_{1.0}$ 53.0°. Flash pt (closed cup) 86°C (187°F). Volatile in steam. n_D^{20} 1.5395. 100 ml water dissolves about 2.5 g at 50°, about 5 g at 100°. Sol in aq solns of alkali hydroxides; in the usual organic solvents. LD_{50} orally in rats: 1.8 g/kg (Deichmann, Witherup).

Caution: Potential symptoms of overexposure to *o*-, *m*- or *p*-isomers are irritation of eyes, skin and mucous membranes; CNS effects; confusion, depression and respiratory failure; dyspnea, irregular rapid respiration and weak pulse; skin and eye burns; dermatitis; lung, liver, kidney and pancreas damage. *See NIOSH Pocket Guide to Chemical Hazards* (DHHS/NIOSH 97-140, 1997) p 78. Symptoms of intoxication are similar to those produced by phenol, *q.v.* Acute exposure may cause muscular weakness, gastroenteric disturbances, severe depression, collapse and death. Chronic exposure may cause digestive disturbances, liver and kidney damage, and skin eruptions. Cresol has a marked corrosive action on tissues, producing burns and dermatitis. *See Patty's Industrial Hygiene and Toxicology* vol. **2A**, G. D. Clayton, F. E. Clayton, Eds. (Wiley-Interscience, New York, 3rd ed., 1981) pp 2597-2601.

USE: For making synthetic resins; in disinfectants and fumigants; as industrial solvent; *m*-cresol in photographic developers, explosives.

THERAP CAT: Disinfectant.

THERAP CAT (VET): Local antiseptic, parasiticide, disinfectant; has been used as an intestinal antiseptic.

2565. *o*-**Cresolphthalein.** [596-27-0] 3,3-Bis(4-hydroxy-3-methylphenyl)-1(3*H*)-isobenzofuranone; 3,3′-dimethylphenolphthalein. $C_{22}H_{18}O_4$; mol wt 346.38. C 76.29%, H 5.24%, O 18.48%. Prepn: F. J. Welcher, *Organic Analytical Reagents* **vol. 4** (Van Nostrand, New York, 1948) p 482; Hubacher *et al.*, *J. Am. Pharm. Assoc.* **42**, 23 (1953).

Crystals from alc, mp 223°. Slightly sol in water; sol in alcohol. pK 9.4.

USE: As indicator: pH range, 8.2 colorless, 9.8 red. Analytical reagent.

2566. *o*-**Cresolphthalein Complexone.** [2411-89-4] *N,N*′-[(3-Oxo-1(3*H*)-isobenzofuranylidene)bis[(6-hydroxy-5-methyl-3,1-phenylene)methylene]]bis[*N*-(carboxymethyl)glycine]; *o*-cresolphthalein complexon; *o*-cresolphthalexone; 3,3′-dimethyl-5,5′-[(*N*,*N*-bis(carboxymethyl)amino)methyl]phenolphthalein complexon; metalphthalein; phthalein complexone; phthalein purple. $C_{32}H_{32}$-

N_2O_{12}; mol wt 636.61. C 60.37%, H 5.07%, N 4.40%, O 30.16%. Cation chelating dye; forms 1:1 and 2:1 complexes with metal ions. Prepn from *o*-cresolphthalein, *q.v.*: G. Anderegg *et al.*, *Helv. Chim. Acta* **37**, 113 (1954); J. Körbl, R. Pribil, *Chem. Ind. (London)* **1957**, 233. Spectrophotometric ionization studies: F. Gaizer *et al.*, *Talanta* **28**, 127 (1981); S. M. Abu-El-Wafa *et al.*, *Pak. J. Sci. Ind. Res.* **28**, 238 (1985). Polarography studies: X. Gao, M. Zhang, *Anal. Chem.* **56**, 1912 (1984). Metal chelate studies: H. P. Srivastava, D. Tiwari, *J. Chem. Eng. Data* **41**, 821 (1996); S. M. Abu-El-Wafa *et al.*, *J. Coord. Chem.* **57**, 813 (2004); T. M. Ismail *et al.*, *ibid.* 1179; R. M. Issa *et al.*, *Synth. React. Inorg. Met.-Org. Chem.* **34**, 1087 (2005); K. Y. El-Baradie *et al.*, *Monatsh. Chem.* **136**, 1157 (2005). Determn of calcium in biological samples: G. Kessler, M. Wolfman, *Clin. Chem.* **10**, 686 (1964); B. C. Ray Sarkar, U. P. S. Chauhan, *Anal. Biochem.* **20**, 155 (1967); K. Lorentz, *Clin. Chim. Acta* **126**, 327 (1982); C. M. Corns, C. J. Ludman, *Ann. Clin. Biochem.* **24**, 345 (1987).

Solid from ethanol, mp 186° (as monohydrate). *Irritant.* pKa_1 3.95; pKa_2 6.85; pKa_3 7.64; pKa_4 8.50; pKa_5 9.10; pKa_6 10.30. Absorption max (0.1 *M* NH_3-NH_4Cl, pH 9): 580 nm (ε 820). Aq solns colorless at pH < 8; red at pH ≥ 8. Slightly sol in water, ethanol.

USE: Chromogenic indicator for complexometric and spectrophotometric determn of metal ions. In chelation ion chromatography. In determn of calcium in biological samples.

2567. **Cresol Purple.** [2303-01-7] 4,4′-(1,1-Dioxido-3*H*-2,1-benzoxathiol-3-ylidene)bis[3-methylphenol]; *m*-cresolsulfonphthalein. $C_{21}H_{18}O_5S$; mol wt 382.43. C 65.95%, H 4.74%, O 20.92%, S 8.38%. pH sensitive dye. Prepn: W. R. Orndorff, A. C. Purdy, *J. Am. Chem. Soc.* **48**, 2212 (1926). Solvent effects: J. Barbosa *et al.*, *Mikrochim. Acta* **106**, 327 (1992). Ionization and spectral properties: R. Casula *et al.*, *Talanta* **40**, 1781 (1993); M. C. Aragoni *et al.*, *ibid.* **42**, 1157 (1995). Titrimetric determn: K. Vytras *et al.*, *Analyst* **114**, 479 (1989). Use as pH indicator for emission of NO_2 in museum collections: J. Fenn, *Polym. Prepr.* **37**, 170 (1996); in wine: E. N. Gaiao *et al.*, *Analyst* **124**, 1727 (1999); in buffered solns: S. Paula *et al.*, *Biochemistry* **38**, 3025 (1999).

Slightly sol in ethanol; sol in ethanol denatured with 10% methanol. pH 1.2 red; pH 2.8 yellow; pH 7.4 yellow; pH 9.0 purple.

USE: Acid-base indicator.

2568. **Cresol Red.** [1733-12-6] 4,4′-(1,1-Dioxido-3*H*-2,1-benzoxathiol-3-ylidene)bis[2-methylphenol]; α-hydroxy-α,α-bis(4-hydroxy-*m*-tolyl)-*o*-toluenesulfonic acid γ-sultone; *o*-cresolsulfonphthalein. $C_{21}H_{18}O_5S$; mol wt 382.43. C 65.95%, H 4.74%, O 20.92%, S 8.38%. pH sensitive dye. Prepn: M. D. Sohon, *Am. Chem. J.* **20**, 257 (1898); and properties: H. A. Lubs, W. M. Clark, *J. Wash. Acad. Sci.* **6**, 481 (1916). Raman spectrum: K. Machida *et al.*, *J. Raman Spectrosc.* **8**, 172 (1979). Ionization and spectral properties: R. Casula *et al.*, *Talanta* **40**, 1781 (1993); M. C. Aragoni *et al.*, *ibid.* **42**, 1157 (1995). Protein dye binding stoichiometry: I. Molnár-Perl *et al.*, *Food Chem.* **35**, 69 (1990). Use in detection of alkaloids via ion-associates: Z. Skalican *et al.*, *Anal. Lett.* **28**, 1223

(1995); as pH indicator for emission of NO_2 in museum collections: J. Fenn, *Polym. Prepr.* **37**, 170 (1996); for archived photographic films: J. C. Harthan *et al.*, *Imaging Sci. J.* **45**, 81 (1997); for microbiological media: L. A. Actis *et al.*, *J. Microbiol. Methods* **39**, 79 (1999).

Reddish-brown cryst powder. Sol in alcohol. pH 0.2: red; pH 1.8 yellow; pH 7.2 yellow; pH 8.8 reddish-purple.

USE: Acid-base indicator.

2569. m-Cresotic Acid. [50-85-1] 2-Hydroxy-4-methylbenzoic acid; 2,4-cresotic acid; *m*-homosalicylic acid; *m*-cresotinic acid; 2-hydroxy-*p*-toluic acid; γ-cresotic acid. $C_8H_8O_3$; mol wt 152.15. C 63.15%, H 5.30%, O 31.55%. Prepn: Prelog *et al.*, *Helv. Chim. Acta* **30**, 675 (1947); Baine *et al.*, *J. Org. Chem.* **19**, 510 (1954); Hauptschein *et al.*, *J. Am. Chem. Soc.* **77**, 2284 (1955).

Crystals or leaflets, mp 177°; volatile with steam. Soly as of the *o*-acid.

Caution: Toxicity is similar to salicylic acid, *q.v.*

USE: In manuf of dyes.

2570. o-Cresotic Acid. [83-40-9] 2-Hydroxy-3-methylbenzoic acid; 2,3-cresotic acid; *o*-homosalicylic acid; *o*-cresotinic acid; 2-hydroxy-*m*-toluic acid; 3-methylsalicylic acid. $C_8H_8O_3$; mol wt 152.15. C 63.15%, H 5.30%, O 31.55%. Prepn: Baine *et al.*, *J. Org. Chem.* **19**, 510 (1954); Jones, *Chem. Ind. (London)* **1958**, 228; Blicke, McCarty, *J. Org. Chem.* **24**, 1061 (1959); Wessely *et al.*, *Ber.* **93**, 2840 (1960).

White to slightly reddish, odorless cystals, mp 165-166°; volatile with steam. Slightly sol in cold, more sol in hot water; sol in chloroform, alcohol, ether, alkali hydroxides.

Caution: Toxicity is similar to salicylic acid.

USE: In manuf of dyes.

2571. p-Cresotic Acid. [89-56-5] 2-Hydroxy-5-methylbenzoic acid; 2,5-cresotic acid; *p*-homosalicylic acid; *p*-cresotinic acid; 6-hydroxy-*m*-toluic acid. $C_8H_8O_3$; mol wt 152.15. C 63.15%, H 5.30%, O 31.55%. Prepn: Cameron *et al.*, *J. Org. Chem.* **15**, 233 (1950); Baine *et al.*, ibid. **19**, 510 (1954); Thomas *et al.*, *J. Am. Chem. Soc.* **80**, 5864 (1958); Blicke, McCarty, *J. Org. Chem.* **24**, 1061 (1959).

White or slightly reddish, almost odorless crystals, mp 151°. Volatile with steam. Sublimes with partial decompn. Soly as of the *o*-acid.

Caution: Toxicity is similar to salicylic acid.

2572. m-Cresyl Acetate. [122-46-3] Acetic acid 3-methylphenyl ester; *m*-tolyl acetate; *m*-cresol acetic acid ester; acetic acid *m*-cresol ester; acetylmetacresol; metacresol acetate; Cresatin. $C_9H_{10}O_2$; mol wt 150.18. C 71.98%, H 6.71%, O 21.31%. Germicide with pain-relieving properties. Prepn: Claus, Hirsch, *J. Prakt. Chem.* [2] **39**, 62 (1889); Eijkman, *Chem. Weekbl.* **1**, 453 (1904); *see also* US **1031971** (1912 to N. Sulzberger). Purification: Lebeau-Janot, *Traité Pharm. Chim.* **II**, 882 (Paris, 1955). Use in dentistry: J. M. Gergely, P. M. DiFiore, *Gen. Dent.* **41**, 328 (1993).

Oily liquid. Volatile with steam. Characteristic phenolic odor with a reminiscence of acetone. d_4^{26} 1.048. bp 212°; bp_{13} 99°. uv max (methanol): 262.5, 269.5 nm. Practically insol in water, glycerol. Miscible with alc, ether, chloroform, petr ether, benzene. Soly in liquid petrolatum about 5%. Also sol in cottonseed oil.

THERAP CAT: Antiseptic (topical).

THERAP CAT (VET): Antiseptic (topical).

2573. o-Cresyl Acetate. [533-18-6] Acetic acid 2-methylphenyl ester; acetic acid *o*-tolyl ester; *o*-cresol acetate; acetyl-*o*-cresol; *o*-cresylic acetate; *o*-tolyl acetate. $C_9H_{10}O_2$; mol wt 150.18. C 71.98%, H 6.71%, O 21.31%.

Liquid. bp ~208°. Almost insol in cold, sol in hot water; freely sol in usual organic solvents, in oils.

2574. p-Cresyl Phenylacetate. [101-94-0] Benzeneacetic acid 4-methylphenyl ester; *p*-tolyl phenylacetate. $C_{15}H_{14}O_2$; mol wt 226.28. C 79.62%, H 6.24%, O 14.14%. Synthetic organic ester in perfumery. Prepn: W. Autenrieth, G. Thomae, *Ber.* **57**, 423 (1924). Fragrance monograph: *Food Cosmet. Toxicol.* **13**, 775 (1975). Photochemistry study: C.-H. Tung, Y.-M. Ying, *J. Chem. Soc. Perkin Trans.* 2 **1997**, 1319. Synthesis using montmorillonite catalysts: C. R. Reddy *et al.*, *J. Mol. Catal. A* **223**, 117 (2004). Environmentally benign synthesis of perfumery grade product: G. D. Yadav, S. V. Lande, *Org. Process Res. Dev.* **9**, 288 (2005).

Prisms from alcohol, mp 75-76°. Organoleptic character similar to honey, nuts, and butter. LD_{50} (g/kg): >5 orally in rats; >5 dermally in rabbits (*Food Cosmet. Toxicol.*).

USE: In jasmine, narcissus, hyacinth, lily, and jonquille perfume compositions; fixative; fragrance ingredient in soaps and household products.

2575. CRF. [9015-71-8] Corticotropin-releasing factor; CRH; corticoliberin(e); corticotropin-releasing hormone. Hypothalamic substance that stimulates secretion of ACTH, *q.v.* and β-endorphin from the pituitary. First direct evidence for its presence in hypothalami: R. Guillemin, B. Rosenberg, *Endocrinology* **57**, 599 (1955); M. Saffran, A. V. Schally, *Can. J. Biochem. Physiol.* **33**, 408 (1955). Purification and characterization of a 41-residue ovine hypothalamic CRF that is highly potent in stimulating secretion of corticotropin and β-endorphin-like immunoactivities: W. Vale *et al.*,

Science **213**, 1394 (1981). Primary structure: J. Spiess *et al., Proc. Natl. Acad. Sci. USA* **78**, 6517 (1981). When centrally administered, CRF activates the sympathetic nervous system and may therefore function as a key hormone in stress mobilization of the organism: W. Vale *et al., loc. cit.*; M. Brown *et al., Life Sci.* **30**, 207 (1982). CRF also stimulates α-MSH secretion and cyclic AMP accumulation in rat pars intermedia cells: H. Meunier *et al., ibid.* **31**, 2129 (1982). A functional relationship between CRF and dynorphin-related opioid peptides has also been suggested: K. A. Roth *et al., Science* **219**, 189 (1983). *Reviews:* A. V. Schally *et al., Recent Prog. Horm. Res.* **24**, 497-588 (1968); R. Burgus, R. Guillemin, *Annu. Rev. Biochem.* **39**, 499-526 (1970); M. Saffran in *Hypothalamic Peptide Hormones and Pituitary Regulation*, J. C. Porter, Ed. (Plenum Press, New York, 1977) pp 225-235; W. Vale *et al.* in *The Role of Peptides in Neuronal Function*, J. L. Barker, J. G. Smith, Eds. (Dekker, New York, 1980) pp 432-454; several authors in *Polypeptide Hormones*, R. F. Beers, E. G. Bassett, Eds. (Raven Press, New York, 1980) pp 165-271; B. Lutz-Bucher *et al., J. Physiol.* **77**, 939-950 (1981); N. Yasuda *et al., Endocr. Rev.* **3**, 123-140 (1982); E. Stark, G. B. Makara in *Hormonally Active Brain Peptides: Structure and Function*, K. W. McKerns, V. Pantic, Eds. (Plenum Press, New York, 1982) pp 157-179.

Activity is destroyed by trypsin; unaffected by thioglycolate, pepsin, chymotrypsin.

2576. Crilanomer. [37291-07-9] Starch polymer with 2-propenenitrile; Super-Slurper; Clinagel; Intrasite. Hydrolyzed starch-polyacrylonitile graft copolymers capable of absorbing 2000 times their initial weight of liquid. Prepn: L. A. Gugliemelli *et al., J. Appl. Polym. Sci.* **13**, 2007 (1969). Continuous production: Z. Reyes *et al., Ind. Eng. Chem. Process Des. Dev.* **12**, 62 (1973); manufacturing process: M. O. Weaver *et al., Staerke* **29**, 413 (1977). Analysis: R. J. Dennenberg, T.P. Abbott, *J. Polym. Sci. Polym. Lett. Ed.* **14**, 693 (1976). Use in wound dressings: W. R. Spence, *US* **4226232** (1980 to Spenco Med.). Clinical evaluation in leg ulceration: M. Hornemann, G. Raptis, *Z. Hautkrankh.* **62**, 41 (1987); B. Esch, G. Raptis, *ibid.* **64**, 1135 (1989). Brief description: H.-G. Elias, *Polym. News* **4**, 26 (1977).

Dry powder.

USE: In dressings for wounds. Absorption of liquids in products ranging from disposable diapers to seed coatings.

2577. Crizotinib. [877399-52-5] 3-[(1*R*)-1-(2,6-Dichloro-3-fluorophenyl)ethoxy]-5-[1-(4-piperidinyl)-1*H*-pyrazol-4-yl]-2-pyridinamine; PF-2341066; Xalkori. $C_{21}H_{22}Cl_2FN_5O$; mol wt 450.34. C 56.01%, H 4.92%, Cl 15.74%, F 4.22%, N 15.55%, O 3.55%. ATP-competitive inhibitor of anaplastic lymphoma kinase (ALK) and c-Met tyrosine kinases. Prepn: J. J. Cui *et al., WO* **06021884** (2006 to Pfizer); *eidem, US* **7858643** (2010 to Agouron). Mechanism of action and pharmacology: H. Y. Zou *et al., Cancer Res.* **67**, 4408 (2007). Antitumor effect in exptl anaplastic large-cell lymphoma (ALCL): J. G. Christensen *et al., Mol. Cancer Ther.* **6**, 3314 (2007). Clinical evaluation of ALK inhibition in non-small-cell lung cancer: E. L. Kwak *et al., N. Engl. J. Med.* **363**, 1693 (2010); in ALK-rearranged inflammatory myofibroblastic tumor: J. E. Butrynski *et al., ibid.*, 1727. Review of discovery and development: D. E. Gerber, J. D. Minna, *Cancer Cell* **18**, 548-551 (2010).

White to pale-yellow powder. pKa 9.4; 5.6. Log P (octanol/water) at pH 7.4: 1.65. Soly in water (mg/ml): >10 at pH 1.6; <0.1 at pH 8.2.

THERAP CAT: Antineoplastic.

2578. Croceic Acid. [132-57-0] 7-Hydroxy-1-naphthalenesulfonic acid; 2-naphthol-8-sulfonic acid; Bayer's acid. $C_{10}H_8O_4S$; mol wt 224.23. C 53.57%, H 3.60%, O 28.54%, S 14.30%. Prepn: Nietzki, Zübelen, *Ber.* **22**, 453 (1889); Forster, Keyworth, *J. Soc. Chem. Ind. London* **46**, 26T (1927).

On evaporation of an aq soln, the free acid dec into sulfuric acid and β-naphthol. However, very concd solns of the acid can be obtained. The normal sodium salt is sol in water, and readily sol in alcohol, crystallizing with 2 mols of alcohol. *Ref:* Forster, Keyworth, *loc. cit.*

2579. Crocetin. [27876-94-4] (2*E*,4*E*,6*E*,8*E*,10*E*,12*E*,14*E*)-2,6,11,15-Tetramethyl-2,4,6,8,10,12,14-hexadecaheptaenedioic acid; 8,8'-diapo-ψ,ψ-carotenedioic acid; *trans*-crocetin. $C_{20}H_{24}O_4$; mol wt 328.41. C 73.15%, H 7.37%, O 19.49%. Carotenoid-dicarboxylic acid isolated from *Crocus sativus* L.; *C. albiflorus* var. *neapolitanus* Hort.; *C. luteus* Lam., *Iridaceae*. Extraction procedure and structure: Jucker, Karrer, *Carotinoide* (Basel, 1948) p 282.

Brick-red rhombs from acetic anhydride, mp 285°. Absorption max (pyridine): 464, 436, 411 nm. Sol in pyridine, in very dil NaOH solns. Very sparingly sol in water, and in the usual organic solvents except pyridine and similar organic bases. Forms a solid dipyridyl salt.

Dimethyl ester. $C_{22}H_{28}O_4$. Brick-red elongated leaflets, mp 222.5°. Total synthesis: Buchta, Andree, *Naturwissenschaften* **46**, 74 (1959).

Di-gentiobiose ester. [42553-65-1] Crocin. $C_{44}H_{64}O_{24}$; mol wt 976.97. One of the yellow-red pigments of saffron. Isoln and structure: Karrer, Salomon, *Helv. Chim. Acta* **10**, 397 (1927); **11**, 513, 711 (1928); Karrer *et al., ibid.* **12**, 985 (1929); **13**, 392 (1930); Kuhn, Winterstein, *Ber.* **67**, 344 (1934); Reichstein, *Angew. Chem.* **74**, 887 (1962). Hydrated brownish-red needles from methanol, mp 186° (effervescence). Absorption max (methanol): 464, 434 nm. Freely sol in hot water giving an orange-colored soln. Sparingly sol in abs alcohol, ether, other organic solvents.

2580. Croconazole. [77175-51-0] 1-[1-[2-[(3-Chlorophenyl)methoxy]phenyl]ethenyl]-1*H*-imidazole; 1-[1-[2-[(3-chlorobenzyl)oxy]phenyl]vinyl]-1*H*-imidazole; cloconazole. $C_{18}H_{15}ClN_2O$; mol wt 310.78. C 69.57%, H 4.87%, Cl 11.41%, N 9.01%, O 5.15%. Antimycotic vinylimidazole. Prepn and antifungal properties: M. Ogata *et al., BE* **883665**; *eidem, US* **4328348** (1980, 1982 both to Shionogi); *eidem, J. Med. Chem.* **26**, 768 (1983). Pharmacology: K. Yamamoto *et al., Oyo Yakuri* **27**, 533 (1984), *C.A.* **101**, 16881c (1984). Series of articles on metabolism in rats and rabbits: K. Mizojiri *et al., Iyakuhin Kenkyu* **16**, 177-206 (1985), *C.A.* **103**, 47788y, 47789z, 47790t (1985). Mechanism of action study: T. Hiratani, H. Yamaguchi, *Chemotherapy (Tokyo)* **33**, 579 (1985). Clinical trial in treatment of tinea pedis: H. Beierdörffer *et al., Mycoses* **38**, 501 (1995).

mp 72-73°. Sol in ethyl acetate.

Monohydrochloride. [77174-66-4] 710674-S; Pilzcin. $C_{18}H_{15}ClN_2O\cdot HCl$; mol wt 347.24. Crystals from ethyl acetate + acetonitrile, mp 148.5-150°. LD_{50} in rats (mg/kg): 7000 s.c.; 2500 orally (Ogata, 1983).

THERAP CAT: Antifungal (topical).

2581. Cromolyn. [16110-51-3] 5,5'-[(2-Hydroxy-1,3-pro-panediyl)bis(oxy)]bis[4-oxo-4*H*-1-benzopyran-2-carboxylic acid]; 5,5'-[(2-hydroxytrimethylene)dioxy]bis(4-oxo-4*H*-1-benzopyran-2-carboxylic acid); 5,5'-(2-hydroxytrimethylenedioxy)bis(4-oxochro-mene-2-carboxylic acid); 1,3-bis(2-carboxychromon-5-yloxy)-2-hy-droxypropane; 1,3-di(2-carboxy-4-oxochromen-5-yloxy)propan-2-ol; cromoglycic acid. C$_{23}$H$_{16}$O$_{11}$; mol wt 468.37. C 58.98%, H 3.44%, O 37.57%. Chromone complex which blocks mast cell de-granulation. Prepn: **NL 6603997**; C. Fitzmaurice, T. B. Lee, **US 3419578** (1966, 1968 both to Fisons). Metabolism: M. J. Ashton *et al.*, *Toxicol. Appl. Pharmacol.* **26**, 319 (1973). Mechanism of action: T. C. Theoharides *et al.*, *Science* **207**, 80 (1980); R. G. Alvarez *et al.*, *Agents Actions* **11**, 94 (1981). Toxicology in primates: J. E. Beach *et al.*, *Toxicol. Appl. Pharmacol.* **57**, 367 (1981). Clinical study in allergic conjunctivitis: G. A. Friday *et al.*, *Am. J. Ophthal-mol.* **95**, 169 (1983). Review of pharmacology and clinical use: G. G. Shapiro, P. Konig, *Pharmacotherapy* **5**, 156-170 (1985). *Review:* J. S. G. Cox in *Pharmacological and Biochemical Properties of Drug Substances* vol. 1, M. E. Goldberg, Ed. (Am. Pharm. Assoc., Washington, DC, 1977) pp 277-310; W. Storms, M. A. Kaliner, *J. Asthma* **42**, 79-89 (2005).

mp 241-242° (dec).
Monohydrate. Colorless crystals from ethanol + ether, mp 216-217° (dec).
Disodium salt. [15826-37-6] Cromolyn sodium; disodium cro-moglycate; DSCG; FPL-670; Alloptrex; Allergocrom; Colimune; Cromabak; Cromadoses; Cromedil; Cromogen; Cromoptic; Dura-croman; Gastrofrenal; Hay-Crom; Intal; Intercron; Lomudal; Lomu-pren; Lomusol; Multicrom; Nalcrom; Nalcron; Nasalcrom; Ophta-calm; Opticrom; Rynacrom; Vividrin. C$_{23}$H$_{14}$Na$_2$O$_{11}$; mol wt 512.33. Hygroscopic. Freely sol in water (100 mg/ml at 20°). Prac-tically insol in chloroform and alcohol. LD$_{50}$ in mice, rats (mg/kg): >8000 orally (Cox).
THERAP CAT: Antiasthmatic; antiallergic.
THERAP CAT (VET): Antiallergic.

2582. Cropropamide. [633-47-6] *N*-[1-[(Dimethylamino)-carbonyl]propyl]-*N*-propyl-2-butenamide; *N*-(1-dimethylcarbamoyl-propyl)-*N*-propylcrotonamide; α-(*N'*-crotonyl-*N'*-propyl)amino-*N,N*-dimethylbutyramide. C$_{13}$H$_{24}$N$_2$O$_2$; mol wt 240.35. C 64.96%, H 10.07%, N 11.66%, O 13.31%. Prepn: Martin, Gysin, **US 2447587** (1948 to Geigy).

Liquid, bp$_{0.25}$ 128-130°. Easily sol in water, ether.
Combination with crotethamide. [8015-51-8] Prethcamide; Micoren; Respirot.
THERAP CAT: Combination as respiratory stimulant.

2583. Croscarmellose Sodium. [74811-65-7] Ac-Di-Sol. In-ternally cross-linked form of carboxymethylcellulose sodium (Na-CMC), *q.v.* Prepd by lowering pH of NaCMC in solution, then heat-ing: *Technical Literature*, FMC Corp. Use as tablet disintegrant: D. Gissinger, A. Stamm, *Pharm. Ind.* **42**, 189 (1980); M. E. Sangalli *et al.*, *Boll. Chim. Farm.* **128**, 242 (1989). Evaluation of drug-excip-ient interaction: R. G. Hollenbeck *et al.*, *J. Pharm. Sci.* **72**, 325 (1983). Water sorption and thermal behavior: D. Faroongsarng, G. E. Peck, *Drug Dev. Ind. Pharm.* **20**, 779 (1994). Mechanism of action study: R. Thibert, B. C. Hancock, *J. Pharm. Sci.* **85**, 1255 (1996).

Free flowing, odorless, white powder. Bulk density (g/cc) 0.48. pH 5.0-7.0. Hydrophillic. Partially sol in water. Insol in alcohol, ether, other organic solvents.
USE: Pharmaceutic aid (tablet disintegrant).

2584. Crotamine. [37196-57-9] One of the poisonous prin-ciples found in the venom of Brazilian rattlesnakes, *Crotalus terrifi-cus crotaminicus* Goncalves, *Crotalidae*. A polypeptide having mol wt 10,000-15,000 and isoelectric point at pH 10.3. Purification and properties: Goncalves in *Venoms*, E. E. Buckley, N. Porges, Eds. (Publ. no. 44 of the Am. Assoc. Adv. Science, Washington, D.C., 1956) pp 261-274. Pharmacology: Moussatché *et al.*, *ibid.* pp 275-279.

2585. Crotamiton. [483-63-6] *N*-Ethyl-*N*-(2-methylphenyl)-2-butenamide; *N*-ethyl-*o*-crotonotoluidide; crotonyl-*N*-ethyl-*o*-tolu-idine; Crotamitex; Eurax; Euraxil; Veteusan. C$_{13}$H$_{17}$NO; mol wt 203.29. C 76.81%, H 8.43%, N 6.89%, O 7.87%. Scabicide with antipruritic activity. Prepd by treating crotonic acid *o*-toluidide with diethyl sulfate: **CH 253472**; **CH 253473**; **GB 615137** (1949 to Gei-gy). HPLC is plasma and urine: A. Sioufi *et al.*, *J. Chromatogr.* **494**, 361 (1989); in pharmaceutical formulations: S. Izumoto *et al.*, *J. Pharm. Biomed. Anal.* **15**, 1457 (1997). Clinical trial in pediatric scabies: D. Konstantinov *et al.*, *J. Int. Med. Res.* **7**, 443 (1979).

Yellowish oil, bp$_{13}$ 153-155°. Sol in methanol, ethanol.
THERAP CAT: Ectoparasiticide.
THERAP CAT (VET): Scabicide; antipruritic.

2586. Crotethamide. [6168-76-9] *N*-[1-[(Dimethylamino)-carbonyl]propyl]-*N*-ethyl-2-butenamide; *N*-(1-dimethylcarbamoyl-propyl)-*N*-ethylcrotonamide; α-(*N'*-crotonyl-*N'*-ethyl)amino-*N,N*-dimethylbutyramide. C$_{12}$H$_{22}$N$_2$O$_2$; mol wt 226.32. C 63.69%, H 9.80%, N 12.38%, O 14.14%. Prepn: Martin, Gysin, **US 2447587** (1948 to Geigy).

Liquid, bp$_{0.03}$ 132-134°. Easily sol in water, ether.
Combination with cropropamide *see* Cropropamide.
THERAP CAT: Combination as respiratory stimulant.

2587. Crotonaldehyde. [4170-30-3]; [123-73-9] ((*E*)-form); [15798-64-8] ((*Z*)-form). 2-Butenal; crotonic aldehyde; β-methyl-acrolein. C$_4$H$_6$O; mol wt 70.09. C 68.55%, H 8.63%, O 22.83%. Commercial product is a 95:5 mixture of *trans:cis* isomers. Prepn: v. Auwers, Eisenlohr, *J. Prakt. Chem.* **82**, 115 (1910); Hibbert, *J. Am. Chem. Soc.* **37**, 1759 (1915). Alternate synthesis: J. Smidt *et al.*, *Angew. Chem.* **71**, 176 (1959). Toxicity studies: H. F. Smyth, C. P. Carpenter, *J. Ind. Hyg. Toxicol.* **26**, 269 (1944). Review of use in commercial processes: O. Horn, *Ind. Eng. Chem.* **51**, 655-658 (1959). *Reviews:* J. E. Fernandez, T. W. G. Solomons, *Chem. Rev.* **62**, 485-502 (1962); W. Blau *et al.*, in *Ullmann's Encyclopedia of Industrial Chemistry* vol. A8 (VCH, Weinheim, 5th ed., 1987) pp 83-89. Review of carcinogenic potential: V. J. Feron *et al.*, *Mutat. Res.* **259**, 363-385 (1991); of role in DNA adduct formation: S. S. Hecht *et al.*, *Toxicology* **166**, 31-36 (2001).

(*E*)-form

Flammable liq. *Lacrimator.* mp −69°; bp 102.2°; d_{20}^{20} 0.853; $n_D^{17.3}$ 1.4384. Flash pt when anhydr: 13°C (55°F) (open cup). Explosive limits in air 2.95-15.5% v/v. Vapor density 2.41 (air = 1). Heat capacity: 0.7 cal/g/°C; heat of vaporization: 123 cal/g. Dimerizes under strong acid conditions, slowly oxidizes to crotonic acid. Soly in water (g/100 g) at 20°: 18.1; at 5°: 19.2. Soly of water in crotonaldehyde (g/100 g) at 20°: 9.5; at 5°: 8.0. LD$_{50}$ orally in rats: 0.3 g/kg (Smyth, Carpenter).

Caution: Potential symptoms of overexposure are irritation of eyes and respiratory system. *See NIOSH Pocket Guide to Chemical Hazards* (DHHS/NIOSH 97-140, 1997) p 80.

USE: Manuf of butyl alcohol, butyraldehyde, methoxybutyraldehyde, sorbic acid, maleic acid, crotonic acid, crotyl alcohol. In polymer chemistry: manuf of resins and polyvinyl acetals, solvent for polyvinyl chloride, rubber antioxidant, increases rubber strength with ketones. In prepn of insecticides and fertilizers. In production of flavors.

2588. Crotonic Acid. [107-93-7] (2*E*)-2-Butenoic acid; *β*-methylacrylic acid; *α*-crotonic acid; solid crotonic acid; *trans*-crotonic acid. $C_4H_6O_2$; mol wt 86.09. C 55.81%, H 7.03%, O 37.17%. Has been found in clay soil in Texas; formed during the dry distillation of wood. Obtained on a commercial scale exclusively by oxygen- or air-oxidation of crotonaldehyde: Kennedy, **US 2413235** (1946 to Shawinigan Chem.); Leupold, PB report 70249 (1942); Matthews, B.I.O.S. report 758 (1946). Laboratory procedure using alkaline silver oxide: Young, *J. Am. Chem. Soc.* **54**, 2498 (1932); from acetaldehyde and malonic acid in pyridine: v. Auwers, *Ann.* **432**, 46 (1923); Backer, Bloemen, *Rec. Trav. Chim.* **45**, 102 (1926); Florence, *Bull. Soc. Chim. Fr.* [4] **41**, 440 (1927); Letch, Linstead, *J. Chem. Soc.* **1932**, 454. Toxicity study: Smyth, Carpenter, *J. Ind. Hyg. Toxicol.* **26**, 269 (1944). *Review:* W. Blau *et al.,* in *Ullmann's Encyclopedia of Industrial Chemistry* vol. A8 (VCH, Weinheim, 5th ed., 1987) pp 83-89.

Monoclinic needles, prisms from water or ligroin. Corrosive and combustible solid. d_4^{15} 1.018; d_4^{80} 0.964. mp 71.6°. bp$_{10}$ 80.0°; bp$_{20}$ 93.0°; bp$_{40}$ 107.8°; bp$_{60}$ 116.7°; bp$_{100}$ 128.0°; bp$_{200}$ 146.0°; bp$_{400}$ 165.5°; bp$_{760}$ 185.0°. n_D^{80} 1.4228. pKa (25°) 4.817. Heat of combustion: 2.00 MJ/mol. Heat of fusion: 150.9 J/g. Soly in water (g/l) at 0°: 41.5; at 10°: 84.7; at 20°: 76.1; at 25°: 94; at 30°: 122; at 40°: 656. Soly in ethanol at 25°: 52.5% w/w; acetone: 53.0% w/w; toluene: 37.5% w/w. LD$_{50}$ orally in rats: 1.0 g/kg (Smyth, Carpenter).

Methyl ester. $C_5H_8O_2$. Liq, bp 121°; d_4^{20} 0.9444; n_D^{20} 1.4242.
Ethyl ester. $C_6H_{10}O_2$. Liq, bp 138°; d_4^{20} 0.9175; n_D^{20} 1.4245.
Vinyl ester. $C_6H_8O_2$. Liq, bp 133°; d_4^{20} 0.9410; n_D^{20} 1.450.

USE: In the manuf of copolymers with vinyl acetate used in lacquers and paper sizing; in the manuf of softening agents for synthetic rubber. In medicinal chemistry, e.g., in the manuf of DL-threonine, vitamin A.

2589. Croton Oil. Fixed oil expressed from seeds of *Croton tiglium* L., *Euphorbiaceae. Constit.* Croton resin; glycerides of stearic, palmitic, myristic, lauric, tiglic, etc. acids; crotin; phorbol esters, *q.v. Review:* Hecker, Schmidt, *Fortschr. Chem. Org. Naturst.* **31**, 377 (1974).

Pale-yellow or brownish-yellow, rather viscid liq; slight disagreeable odor. *Poisonous.* d_{25}^{25} 0.935-0.950. n_D^{40} 1.470-1.473. Sapon no. 205-220. Iodine no. 102-118. Slightly sol in alcohol, increasing with age. Freely sol in chloroform, ether, oils, petr ether, carbon disulfide, glacial acetic acid. *Keep well closed and protected from light.*

Caution: Potential symptom of overexposure is hemorrhagic gastroenteritis. Direct contact may cause vesiculation, necrosis and sloughing of skin. *See Clinical Toxicology of Commercial Products,* R. E. Gosselin *et al.,* Eds. (William & Wilkins, Baltimore, 5th ed., 1984) Section II, p. 228.

USE: Has been used as an irritant and cocarcinogen in cancer research.

THERAP CAT: Cathartic, counterirritant.

THERAP CAT (VET): Formerly used as a counterirritant and a purgative.

2590. Crotoxin. [9007-40-3] Neurotoxin. One of the toxic principles isolated from rattlesnake venom: Slotta, Fraenkel-Conrat, *Ber.* **71**, 1076 (1938). A polypeptide complex of probably two components, one acidic and one basic, having mol wts of ~9000 and 12,000 resp. The subunits do not retain the toxicity of crotoxin but show synergistic action in combination: Hendon, Fraenkel-Conrat, *Proc. Natl. Acad. Sci. USA* **68**, 1560 (1971). Amino acid analyses and fractionation studies: Fraenkel-Conrat, Singer, *Arch. Biochem. Biophys.* **60**, 64 (1956); Rübsamen *et al.,* *Arch. Pharmacol.* **270**, 274 (1971); Horst *et al.,* *Biochem. Biophys. Res. Commun.* **46**, 1042 (1972). Review of early chemical studies: Fraenkel-Conrat, Singer, *Publ. Am. Assoc. Advan. Sci.* **44**, 259 (1956). *Monograph:* Behringwerk-Mitteilungen Sonderband, *Die Giftschlangen der Erde* (N. G. Elwert, Marburg, 1963) 464 pp. Pharmacology: Brazil, Excell, *J. Physiol. (London)* **212**, 34P (1970); Brazil *et al.,* *ibid.* **234**, 63P (1973); Habermann *et al.,* *Arch. Pharmacol.* **273**, 313 (1972). *See also* Crotamine.

Caution: When injected by snake bite, local pain, redness, hemorrhage and necrosis result. Systemic effects are dizziness, sensory and motor depression, collapse, shock. May be fatal. *Antidote:* Rattlesnake antivenin.

2591. Crotyl Alcohol. [6117-91-5] 2-Buten-1-ol; crotonyl alcohol. C_4H_8O; mol wt 72.11. C 66.63%, H 11.18%, O 22.19%. Prepn by reduction of crotonaldehyde: Nystrom, Brown, *J. Am. Chem. Soc.* **69**, 1197 (1947); Olivier, Young, *ibid.* **81**, 5811 (1959); Smith, Holm, **US 2761883**; Finch, Furman, **US 2763696** (both 1956 to Shell); Foreman, **US 3109865** (1963 to Standard Oil Co., Ohio). Prepn of (*E*) and (*Z*)-forms: Hatch, Nesbitt, *J. Am. Chem. Soc.* **72**, 727 (1950); Hiskey *et al.,* *J. Org. Chem.* **21**, 429 (1956); of (*Z*)-form only: Clarke, Crombie, *Chem. Ind. (London)* **1957**, 143. Toxicity study: H. F. Smyth *et al.,* *Am. Ind. Hyg. Assoc. J.* **23**, 95 (1962).

(*E*)-form

Liquid, bp 122°. d_4^{20} 0.8532. n_D^{20} 1.4285. Sol in 6 parts water; misc with alc. LD$_{50}$ orally in rats: 0.93 ml/kg (Smyth).
(*E*)-Form. [504-61-0] Liquid, bp 121.2°. d_4^{25} 0.8454. n_D^{20} 1.4289, n_D^{20} 1.4270.
(*Z*)-Form. [4088-60-2] Liquid, bp 123.6°. mp −90.15°. d_4^{20} 0.8662. n_D^{20} 1.4342.

2592. Crown Ethers. Crown compounds. Macrocyclic polyethers with the repeating unit (—CH$_2$—CH$_2$—O—)$_n$, where *n* is greater than 2. Crown compounds with other heteroatoms (N, S, P) are known. Prepn: C. J. Pedersen, *J. Am. Chem. Soc.* **89**, 2495, 7017 (1967). Described as "crown" ethers due to appearance of space-filling models and ability to "crown" cations. Proposed nomenclature lists non-ring substituents, number of atoms in ring, the class (crown), and the number of heteroatoms in the ring; e.g. dibenzo-18-crown-6. Bicyclic crowns are called *cryptates.* Crowns complex with cations and solubilize inorganic reagents in organic solvents. Selectivity results from the definite size of the crown cavity which admits only cations of corresponding ionic radii. The stability of complexes depends on the size of the cation and the size of the polyether ring. Chiral crowns are used in optical resolution of enantiomers. *Reviews:* D. J. Cram, J. M. Cram, *Science* **183**, 803-809 (1974); *eidem, Acc. Chem. Res.* **11**, 8-14 (1978); J. J. Christensen *et al., Chem. Rev.* **74**, 351-384 (1974); G. W. Gokel, H. D. Durst, *Synthesis* **1976**, 168-184; A. C. Knipe, *J. Chem. Educ.* **53**, 618-622 (1976); V. Prelog, *Pure Appl. Chem.* **50**, 893-904 (1978). Historical overview: C. J. Pedersen, *Science* **241**, 536-540 (1988).

Dibenzo-18-crown-6

18-Crown-6. 1,4,7,10,13,16-Hexaoxacyclooctadecane. C_{12}-$H_{24}O_6$; mol wt 264.32. Prepn: G. W. Gokel et al., Org. Synth. **57**, 30 (1977). Crystals from acetonitrile, mp 38-39.5°.

Dibenzo-18-crown-6. 2,3,11,12-Dibenzo-1,4,7,10,13,16-hexaoxacyclooctadeca-2,11-diene. $C_{20}H_{24}O_6$; mol wt 360.41. The first crown ether discovered. Prepn: C. J. Pedersen, US 3687978 (1972 to Du Pont). White crystals from acetone, mp 164°, bp$_{679}$ 380-384°.

Dicyclohexano-18-crown-6. Dicyclohexyl-18-crown-6. C_{20}-$H_{36}O_6$; mol wt 372.50. Approx LD in rats (mg/kg): 300 orally (Pedersen, 1967).

Caution: Direct contact may cause eye and skin irritation (Pedersen, 1967).

USE: In organic synthesis as phase transfer reagents, catalysts and in sepn of enantiomers.

2593. CRP. [9007-41-8] C-Reactive Protein. Homogeneous acute phase protein, found in man and most other animals. Discovered as a result of its precipitation with C-polysaccharide (CPS) in sera of patients with infections and inflammatory disease: W. S. Tillet, T. Francis, J. Exp. Med. **52**, 561 (1930). Characterization as a protein and identification of its calcium ion requirement for interaction with CPS: T. J. Abernathy, O. T. Avery, ibid. **73**, 173 (1941). CRP is a trace serum protein consisting of single subunits of mol wt about 21,000. In serum and in the purified state, the subunits aggregate as cyclic pentamers having mol wts of 110,000-144,000. The concentrations of C-reactive protein increase up to 1000-fold in inflammatory conditions and in tissue necrosis. In addition, it can initiate reactions of agglutination, precipitation, and opsonization for phagocytosis and can activate the complement system. Other biological activities with platelets and lymphocytes have also been described, but the full biological functions of CRP have not yet been completely elucidated. Review of CRP and the acute phase response: H. Gewurz et al., Adv. Intern. Med. **27**, 345-372 (1982). Review of structure and function: M. B. Pepys, Eur. J. Rheumatol. Inflammation **5**, 386-397 (1982). Book: Ann. N.Y. Acad. Sci. **389**, entitled "C-Reactive Protein and the Plasma Protein Response to Tissue Injury", I. Kushner et al., Eds. (1982) 482 pp.

2594. Cryofluorane. [76-14-2] 1,2-Dichloro-1,1,2,2-tetrafluoroethane; Freon 114; Frigen 114; Arcton 114. $C_2Cl_2F_4$; mol wt 170.92. C 14.05%, Cl 41.48%, F 44.46%. $ClCF_2CClF_2$. Prepn: Henne, Org. React. **2**, 49 (1944).

Colorless, practically odorless, noncorrosive, nonirritating, nonflammable gas. Faint, ether-like odor in high concentrations. d$_{liq}^0$ 1.5312. mp −94°. bp$_{760}$ 4.1°. n_D^0 1.3092. Critical temp 145.7°; crit pressure 474 lb/sq in. abs. Practically insol in water. Sol in alcohol, ether. Absorbs less than 0.0025% water. Has little, if any, anesthetic or toxic effect on humans, but toxic substances may be formed on contact with a flame or hot metal surface.

Caution: Potential symptoms of overexposure are irritation of respiratory system; asphyxia; cardiac arrhythmias, cardiac arrest; direct contact with liquid may cause frostbite. See NIOSH Pocket Guide to Chemical Hazards (DHHS/NIOSH 97-140, 1997) p 102. *Note:* Consult latest Government regulations on use as aerosol propellant.

USE: Refrigerant, aerosol propellant.

2595. Cryolite. [15096-52-3] Kryolith; ice spar; sodium aluminum fluoride. AlF_6Na_3; mol wt 209.94. Al 12.85%, F 54.30%, Na 32.85%. $3Na.AlF_6$. A mineral named for its resemblance to ice; large deposits existed in Greenland and in the Urals. Synthetic cryolite is usually made from $NaAlO_2$, $NaHCO_3$ and NaF. See the review by Vogel, Chem. Ztg. **75**, 603 (1951).

Snow-white, semi-opaque masses, vitreous fracture. Monoclinic crystals with cubic habit, perfect cleavage (001) (110) (101). The natural form may be colored reddish or brown or even black, but loses this discoloration on heating. d 2.95. Mohs' hardness 2.5-3. Fuses fairly easily, mp 1000°. Sol in concd H_2SO_4. Dec by boiling with aq alkali hydroxides or aq calcium hydroxide.

Caution: Potential symptoms of overexposure are irritation of eyes and respiratory system; nausea, abdominal pain, diarrhea; salivation, thirst and sweating; stiff spine; dermatitis; calcification of ligaments of ribs and pelvis. See NIOSH Pocket Guide to Chemical Hazards (DHHS/NIOSH 97-140, 1997) p 280.

2596. Cryptand 222. [23978-09-8] 4,7,13,16,21,24-Hexaoxa-1,10-diazabicyclo[8.8.8]hexacosane; Kryptofix 222. $C_{18}H_{36}$-

N_2O_6; mol wt 376.49. C 57.42%, H 9.64%, N 7.44%, O 25.50%. Macrocyclic ligand; sequestering agent for metal ions in solution. Prepn: B. Dietrich et al., Tetrahedron **29**, 1629 (1973). Synthesis of complexes with metal ions: A. Knöchel et al., Inorg. Nucl. Chem. Lett. **11**, 787 (1975). Crystal structure: B. Metz et al., J. Chem. Soc. Perkin Trans. 2 **1976**, 423. Thermodynamic properties in water and methanol: M. H. Abraham et al., J. Chem. Soc. Faraday Trans. 1 **76**, 869 (1980). Adsorption properties: M. Carlà et al., J. Phys. Chem. **100**, 11067 (1996). Solution properties: P. Jost et al., Phys. Chem. Chem. Phys. **4**, 335 (2002). Review of thermodynamic properties: Y. Marcus, Rev. Anal. Chem. **23**, 269-302 (2004).

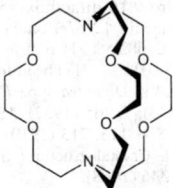

Crystals from hexane, mp 68-69°. d 1.07 ±0.02. Very sol in water. Soly at 25° in cyclohexane: 0.238 M; in tetradecane: 0.086 M. Partition coefficient (cyclohexane/water): 0.0249. Partition coefficient (tetradecane/water): 0.0129. *Irritant.*

USE: In selective complexing and membrane transport of cations, metal purification procedures, phase transfer catalysis, anionic polymerizations, solubilization of insoluble salts in low polarity solvents.

2597. Cryptophycins. Cytotoxic, cyclic depsipeptides isolated from blue-green algae Nostoc sp. Named for their antifungal activity against the genus Cryptococcus sp. The parent compound is cryptophycin-1; at least 21 other naturally-occurring cryptophycins have been isolated; many other analogs have been synthesized. Interacts with the Vinca alkaloid binding domain of tubulin. Isoln: C. F. Hirsch et al., GB 2220657; eidem, US 4946835 (both 1990 to Merck & Co.); R. E. Schwartz et al., J. Ind. Microbiol. **5**, 113 (1990). Prepn: R. E. Moore et al., WO 9640184; eidem, US 6013626 (1996, 2000 both to Univ. of Hawaii and Wayne State Univ.). Structure determn: G. Trimurtulu et al., J. Am. Chem. Soc. **116**, 4729 (1994); and stability and antitumor evaluation: idem et al., ibid. **117**, 12030 (1995). Mechanism of action: C. D. Smith, X. Zhang, J. Biol. Chem. **271**, 6192 (1996); D. Panda et al., Proc. Natl. Acad. Sci. USA **95**, 9313 (1998). Effect on tubulin ring structures: N. R. Watts, et al., Biochemistry **41**, 12662 (2002). Review of syntheses: M. A. Tius, Tetrahedron **58**, 4343-4367 (2002); and anticancer activity: M. Eggen, G. I. Georg, Med. Res. Rev. **22**, 85-101 (2002).

Cryptophycin-1   R = H
Cryptophycin-52   R = CH$_3$

Cryptophycin-52. [186256-67-7] Cyclo[2,2-dimethyl-β-alanyl-(2S)-2-hydroxy-4-methylpentanoyl-(2E,5S,6S)-5-hydroxy-6-[(2R,-3R)-3-phenyloxiranyl]-2-heptenoyl-3-chloro-o-methyl-D-tyrosyl]; LY-355703. $C_{36}H_{45}ClN_2O_8$; mol wt 669.21. C 64.61%, H 6.78%, Cl 5.30%, N 4.19%, O 19.13%. Synthetic cryptophycin analog under investigation for treatment of solid tumors. Enantioselective synthesis: A. K. Ghosh, L. Swanson, J. Org. Chem. **68**, 9823 (2003). Clinical evaluation in lung cancer: M. J. Edelman et al., Lung Cancer **39**, 197 (2003); in platinum-resistant ovarian cancer: G. D'Agostino et al., Int. J. Gynecol. Cancer **16**, 71 (2006). White solid. $[\alpha]_D$ +19.9° (c = 0.5 in $CHCl_3$). uv max (methanol): 204, 218 nm (ε 35100, 20900).

Cryptophycin-1. [124689-65-2] Cyclo[(2R)-2-methyl-β-alanyl-(2S)-2-hydroxy-4-methylpentanoyl-(2E,5S,6S)-5-hydroxy-6-[(2R,3R)-3-phenyloxiranyl]-2-heptenoyl-3-chloro-o-methyl-D-tyrosyl]; cryptophycin A. $C_{35}H_{43}ClN_2O_8$; mol wt 655.19. C 64.16%, H 6.62%, Cl 5.41%, N 4.28%, O 19.54%. White or light colored glassy solid. $[\alpha]_D$ +33.8° (c = 1.83 in methanol). uv max: 208, 218, 228, 280 nm (ε 42400, 33700, 23800, 2210). Sol in organic solvents.

2598. Cryptopine. [482-74-6] 4,6,7,13-Tetrahydro-9,10-dimethoxy-5-methylbenzo[g]-1,3-benzodioxolo[4,5-c]azecin-12(5H)-one; 4,6,7,13-tetrahydro-9,10-dimethoxy-5-methylbenzo[e]-1,3-dioxolo[4,5-l][2]benzazecin-12(5H)-one; cryptocavine. $C_{21}H_{23}$-NO_5; mol wt 369.42. C 68.28%, H 6.28%, N 3.79%, O 21.65%. Occurs in opium (0.003-0.03%). Has been found in *Corydalis sempervirens* (L.) Pers., and in *Dicentra* spp., *Fumariaceae:* Manske, *Can. J. Res.* **8**, 407 (1933); **7**, 265 (1932); **15B**, 274 (1937). Structure: Perkin, *J. Chem. Soc.* **115**, 713 (1919). Synthesis: Haworth, Perkin, *ibid.* **1926**, 1769. Crystal structure: S. R. Hall, F. R. Ahmed, *Acta Crystallogr.* **B24**, 346 (1968).

Crystallizes as six-sided prisms or plates from benzene, mp 220-221°. d 1.35. Almost insol in water and ether. Soluble in chloroform, acetic acid. One part dissolves in 455 parts alc or 80 parts boiling alc. Sparingly sol in most other organic solvents. May be crystallized from hot alc, benzene, petr ether, methyl ethyl ketone, isoamyl alcohol, acetophenone, pyridine, or an alcohol-pyridine mixture. The salts tend to oil out, but can be crystallized by warming, redissolving, and slow cooling.

2599. Cryptoxanthin. [472-70-8] (3R)-β,β-Caroten-3-ol; (R)-all-trans-β-caroten-3-ol; all-trans-cryptoxanthin; β-cryptoxanthin; cryptoxanthol; caricaxanthin. $C_{40}H_{56}O$; mol wt 552.89. C 86.90%, H 10.21%, O 2.89%. Naturally occurring carotenoid pigment with provitamin A activity. Occurs as free alcohol or as monoacyl esters in orange-colored fruits and vegetables, especially papaya, mango, orange, persimmon, and in petals and berries of *Physalis* species. Isoln from *Physalis franchetti*: R. Kuhn, C. Grundmann, *Ber.* **66**, 1746 (1933). Identity with caricaxanthin from papaya: P. Karrer, W. Schlientz, *Helv. Chim. Acta* **17**, 55 (1934). Total synthesis: O. Isler *et al.*, *ibid.* **40**, 456 (1957); D. E. Loeber *et al.*, *J. Chem. Soc. C* **1971**, 404. HPLC determn in serum: A. L. Sowell *et al.*, *Clin. Chem.* **40**, 411 (1994). Distribution of esterified forms in vegetables and fruits: D. E. Breithaupt, A. Bamedi, *J. Agric. Food Chem.* **49**, 2064 (2001).

Red plates with metallic luster from ether + methanol, mp 169°. Absorption max: 452, 480 nm ($E_{1cm}^{1\%}$ 2380, 2080). Freely sol in chloroform, benzene, pyridine, carbon disulfide. Less sol in ligroin, petr ether, alcohol, methanol.

Monoacetate. [52209-35-5] $C_{42}H_{58}O_2$. Deep red leaflets, mp 117.5°.

($\pm$)-Form. [31272-51-2] Crystals from petr ether, mp 158-159°.

2600. Crystallins. Major water-soluble structural proteins found in fiber cells of vertebrate eye lenses; account for the transparency of the lens. Heterogeneous family composed of four groups α, β, γ, δ which have been separated on the basis of size, charge, immunological properties and source. α, β, δ crystallins occur in avian and reptilian lenses while α, β, γ crystallins occur in all other lenses. Several other minor forms have also been identified. Identification and isolation of α, β, γ: C. T. Mörner, *Z. Physiol. Chem.* **18**, 61

(1894); of δ: M. Rabaey, *Exp. Eye Res.* **1**, 310 (1962). Review of early isolation procedures: A. Spector, *Invest. Ophthalmol.* **3**, 182-193 (1964). Structural studies of α: A. F. Van Dam, G. Ten Cates, *Biochim. Biophys. Acta* **121**, 183 (1966); H. A. Kramps *et al.*, *Eur. J. Biochem.* **50**, 503 (1975); of β: H. P. C. Driessen *et al.*, *ibid.* **121**, 83 (1981); G. Wistow *et al.*, *FEBS Lett.* **133**, 9 (1981); of γ: T. Blundell *et al.*, *Nature* **289**, 771 (1981); L. Summers *et al.*, *Pept. Protein Rev.* **3**, 147 (1984); of δ: J. Horwitz, J. Piatigorsky, *Biochim. Biophys. Acta* **624**, 21 (1980); L. A. Williams *et al.*, *ibid.* **708**, 49 (1982). Short-range spatial ordering is essential for lens transparency: M. Delaye, A. Tardieu, *Nature* **302**, 415 (1983). Photoinduced aggregation; implications for cataract formation: U. P. Andley *et al.*, *Photochem. Photobiol.* **40**, 343 (1984); K. Mandal *et al.*, *ibid.* **47**, 583 (1988); M. Kono *et al.*, *ibid.* **593**. Role of crystallin aging in cataract formation: R. J. Siezen *et al.*, *Proc. Natl. Acad. Sci. USA* **82**, 1701 (1985); H. J. Hoenders, G. J. H. Bessems, *Lens Res.* **3**, 281 (1986). Evolution and expression of crystallin genes: J. Piatigorsky, *Cell* **38**, 620 (1984); W. W. de Jong, W. Hendriks, *J. Mol. Evol.* **24**, 121 (1986). *Reviews:* A. Spector, *Invest. Ophthalmol.* **4**, 579-591 (1965); W. W. de Jong, *Co-ord. Regul. Gene Expression - Proc. 2nd Int. Workshop*, R. M. Clayton, Ed. (Plenum Press, New York, 1986) pp 281-291; G. J. Wistow, J. Piatigorsky, *Annu. Rev. Biochem.* **57**, 479-504 (1988). Review of δ-crystallin: J. Piatigorsky, *Mol. Cell. Biochem.* **59**, 33-56 (1984). Book: *Molecular and Cellular Biology of the Eye Lens*, H. Bloemendal, Ed. (John Wiley & Sons, New York, 1981).

α-Crystallin. High molecular wt multimeric (35-50 subunits) polypeptide for which two major subunit forms exist αA_1, αA_2; M_r unit 20 kDa. Most acidic crystallin, pI 4.8-5.0.

β-Crystallin. Dioctamer for which several (6-7) subunit forms exist; M_r unit 23-25 kDa. pI 5.7-7.0.

γ-Crystallin. Monomer, M_r 20 kDa. pI 7.1-8.1.

δ-Crystallin. Tetramer for which two possible subunit forms exist; M_r (monomer) 48-50 kDa. pI 4.9-5.3.

2601. Cubane. [277-10-1] Pentacyclo[4.2.0.0^{2,5}.0^{3,8}.0^{4,7}]octane. C_8H_8; mol wt 104.15. C 92.26%, H 7.74%. Prepn from bromocyclopentadienone dimer: Eaton, Cole, *J. Am. Chem. Soc.* **86**, 3157 (1964). Structure: Fleischer, *ibid.* 3889.

Rhombic crystals from methanol followed by sublimation at just above room temperature and atmospheric pressure, mp 130-131° (sealed capillary). Thermally unstable. Dec at 200°.

2602. Cubeb. Tailed pepper; Java pepper. Dried, unripe, nearly full-grown fruit of *Piper cubeba* L.f., *Piperaceae*. Habit. Southern Asia, Java, Borneo, Sumatra; cultivated in W. Indies and Ceylon. Berries are globular, 4-5 mm in diameter; blackish-gray; intern. whitish and hollow; strong, spicy odor; aromatic, pungent taste. *Constit.* 10-18% volatile oil, cubebin, about 1% cubebic acid, resin, fat, wax. Review of constituents: N. B. Shankaracharya *et al.*, *PAFAI J.* **17**, 33-39 (1995). Botanical description and medicinal uses: J. Gruenwald *et al.*, *PDR for Herbal Medicines* (Medical Economics, Montvale, 2nd Ed., 2000) p 235.

Volatile oil. [8007-87-2] Obtained by steam distillation of the mature, unripe, sun-dried fruit. *Constit.* α- and β-cubebenes, copaene, cubebol, δ-cadinene, humulenes. Description: D. L. J. Opdyke, *Food Cosmet. Toxicol.* **14**, Suppl., 729 (1976). Colorless or pale green to blue-green liquid. d_{25}^{25} 0.898-0.928. n_D^{20} 1.492-1.502. Sol in 10 vols 90% alcohol. Sol in most fixed oils, mineral oil. Insol in glycerin, propylene glycol.

USE: Fragrance and flavoring agent; as a spice in cooking.

THERAP CAT: Urinary antiseptic; expectorant.

2603. Cubebin. [18423-69-3] (2S,3R,4R)-3,4-Bis(1,3-benzodioxol-5-ylmethyl)tetrahydro-2-furanol; ($-$)-(8R,8'R,9S)-cubebin; β-cubebin. $C_{20}H_{20}O_6$; mol wt 356.37. C 67.41%, H 5.66%, O 26.94%. Lignan constituent of the Indonesian medicinal plant, *Piper cubeba*, known as cubeb; also found in *Aristolochia* species. Isoln: Capitaine, Soubeiran, *Ann.* **31**, 190 (1839). Structure: R. D. Haworth, W. Kelly, *J. Chem. Soc.* **1937**, 384; R. Haensel *et al.*, *Arch. Pharm.* **300**, 559 (1967). Synthesis: J. E. Batterbee *et al.*, *J. Chem.*

Soc. C **1969**, 2470. Configuration of hydroxyl group: W. Chen *et al.*, *Arch. Pharm.* **320**, 374 (1987).

Slender prisms from methanol, mp 131-132°. $[\alpha]_D^{14}$ −17° (c = 2.757 in acetone). $[\alpha]_D^{20.4}$ −45.9° (c = 2 in chloroform). uv max: 287, 235 nm (Log ε 3.89, 3.90). Practically insol in water. Sol in alcohol, chloroform, ether.

(2R,3R,4R)-Isomer. [112458-74-9] α-Cubebin; (−)-(8R,8′R,9R)-cubebin.

Note: "Cubebine" is the French designation for ethereal extract of cubeb.

2604. Cucurbitacins. A group of tetracyclic triterpenes, commonly referred to as "bitter principles of cucurbits", which have antineoplastic and anti-gibberellin activity. They are isolated from various spp of cucurbitaceous plants known since antiquity for their beneficial and toxic properties. The plants have been used as vermifuges, emetics, narcotics, and antimalarials and have been implicated in sporadic livestock poisoning in S. Africa. Seventeen cucurbitacins have been isolated, most from plants of the *Cucurbitaceae* family, but also from *Begoniaceae*, *Cruciferae*, *Datisceae*, *Euphorbiaceae*, and *Scrophulariaceae*. Cucurbitacins B and E are the most commonly identified. Isoln of cucurbitacins A, B, C, D, F: P. R. Enslin, *J. Sci. Food Agric.* **5**, 410 (1954); of G, H, J, K, L: *idem, ibid.* **8**, 673 (1957); of E and I: D. Lavie, S. J. Szinai, *J. Am. Chem. Soc.* **80**, 707 (1958); of O, P, Q and antitumor activity: S. Kupchan *et al.*, *J. Org. Chem.* **35**, 2891 (1970). Structures of A, B, C, D, E, I: W. T. DeKock *et al.*, *J. Chem. Soc.* **1963**, 3828; of B, D, F: D. Lavie *et al.*, *Chem. Ind. (London)* **1959**, 951; of G, H: C. W. Holzapfel, P. R. Enslin, *J. S. Afr. Chem. Inst.* **17**, 142 (1964), *C.A.* **62**, 10467c (1965); of J, K, L: P. R. Enslin, K. B. Norton, *J. Chem. Soc.* **1964**, 529. Stereochemistry of B, D, E, F, I: D. Lavie *et al.*, *J. Org. Chem.* **28**, 1790 (1963). ¹³C-NMR study of cucurbitacins: V. V. Velde, D. Lavie, *Tetrahedron* **39**, 317 (1983). Use as plant growth regulators: J. Guha, S. P. Sen, *Nature New Biol.* **244**, 273 (1973). Toxicity studies: J. LeMen *et al.*, *Chim. Ther.* **4**, 459 (1969); D. Albert *et al.*, *Chim. Ther.* **5**, 205 (1970). *Reviews:* D. Lavie, E. Glotter, *Fortschr. Chem. Org. Naturst.* **29**, 308-357 (1971); J. Guha, S. P. Sen, *Plant Biochem. J.* **2**, 12-28 (1975); A. Shrotria, *Botanica* **26**, 28-31 (1976).

Cucurbitacin B

Cucurbitacin B. [6199-67-3] (2β,9β,10α,16α,23E)-25-(Acetyloxy)-2,16,20-trihydroxy-9-methyl-19-norlanosta-5,23-diene-3,-11,22-trione; 1,2-dihydro-α-elaterin. $C_{32}H_{46}O_8$; mol wt 558.71. Crystals from abs ethanol, mp 184-186°. $[\alpha]_D^{25}$ +88° (c = 1.55 in ethanol). LD_{10} orally in mice: 5 mg/kg (LeMen).

Cucurbitacin E. [18444-66-1] (9β,10α,16α,23E)-25-(Acetyloxy)-2,16,20-trihydroxy-9-methyl-19-norlanosta-1,5,23-triene-3,-11,22-trione; α-elaterin. $C_{32}H_{44}O_8$; mol wt 556.70. White hexagonal plates from methanol, mp 232-233° (dec). $[\alpha]_D$ −59° (c = 0.7 in chloroform). uv max (chloroform): 234, 267 nm (ε 11700, 8350). LD_{50} orally in mice: 340 mg/kg (Albert).

2605. Cudbear. Crottle. Common names for the lichen *Ochrolechia tartarea* L., *Lecanoraceae* and for the coloring matter from this and other lichens, especially *Lecanoraceae* and *Roccellaceae*. A source of litmus, *q.v.* Acids in lichens (e.g. lecanoric acid) hydrolyze to orcinol, *q.v.*, which, in the presence of ammonia can be oxi-

dized to the dye orcein, *q.v.* The dyes *French Purple, Persio, Orchil*, and *Orseilles* derive from the salts of orcein: *Colour Index* **Vol. 3** (3rd ed., 1971) p 3241.

Purplish-red powder. Imparts fine purplish-red color to alcohol, acid and neutral liquids. The purple tinge may be covered by adding a few drops of burnt-sugar coloring.

USE: Color for syrups, elixirs etc.

2606. Cuelure. [3572-06-3] 4-[4-(Acetyloxy)phenyl]-2-butanone; 4-(p-hydroxyphenyl)-2-butanone acetate. $C_{12}H_{14}O_3$; mol wt 206.24. C 69.89%, H 6.84%, O 23.27%. Synthetic sex pheromone developed as a lure for melon flies, *Dacus cucurbitae* (Coq.); analog of raspberry ketone. Prepn and biological activity: M. Beroza *et al.*, *Science* **131**, 1044 (1960); B. H. Alexander *et al.*, *J. Agric. Food Chem.* **10**, 270 (1962). Extraction and GC-MS determn in soil: J. P. Alcantara-Licudine *et al.*, *J. Chromatogr. Sci.* **34**, 238 (1996). Field studies in fruit fly eradication: R. T. Cunningham, L. F. Steiner, *J. Econ. Entomol.* **65**, 505 (1972). Acute toxicity study: M. Beroza *et al.*, *Toxicol. Appl. Pharmacol.* **31**, 421 (1975).

Colorless to pale yellow liquid with a raspberry-like odor, $bp_{0.2}$ 123-124°. n_D^{25} 1.5061. d 1.099. Sol in alcohol, hydrocarbons, ether. Insol in water. LD_{50} in rats (mg/kg): 3038 ± 1266 orally; in rabbits (mg/kg): >2025 dermally; LC_{50} (24 hr) in rainbow trout, bluegill sunfish (ppm): ≈21, 18 (Beroza).

USE: As attractant in melon fly traps.

2607. Cumene. [98-82-8] (1-Methylethyl)benzene; isopropylbenzene; 2-phenylpropane; cumol. C_9H_{12}; mol wt 120.20. C 89.93%, H 10.06%. Manuf: *Faith, Keyes & Clark's Industrial Chemicals*, F. A. Lowenheim, M. K. Moran, Eds. (Wiley-Interscience, New York, 1975) pp 294-297. Toxicity study: Smyth *et al.*, *Arch. Ind. Hyg. Occup. Med.* **4**, 119 (1951). *Review:* D. J. Ward in *Kirk-Othmer Encyclopedia of Chemical Technology* vol. **7** (Wiley-Interscience, New York, 3rd ed., 1979) pp 286-290.

Colorless liquid. d_4^{20} 0.862. bp 152-153°. Flash pt, closed cup: 102°F (39°C). n_D^{20} 1.4914: Hirschler, Faulconer, *J. Am. Chem. Soc.* **68**, 210 (1946). Insol in water. Sol in alcohol and many other organic solvents. LD_{50} orally in rats: 2.91 g/kg (Smyth).

Caution: Potential symptoms of overexposure are irritation of eyes, skin and mucous membranes; dermatitis; headache, narcosis, coma. *See NIOSH Pocket Guide to Chemical Hazards* (DHHS/NIOSH 97-140, 1997) p 80.

USE: In manuf of phenol, acetone, acetophenone, α-methylstyrene.

2608. Cumic Acid. [536-66-3] 4-(1-Methylethyl)benzoic acid; p-isopropylbenzoic acid; cuminic acid. $C_{10}H_{12}O_2$; mol wt 164.20. C 73.15%, H 7.37%, O 19.49%.

Triclinic crystals. d^4 1.162. mp 115-117°. Sparingly sol in water; sol in alcohol, ether; also sol in concd H_2SO_4 without decomp, and pptd unchanged by adding water.

2609. Cumic Alcohol. [536-60-7] 4-(1-Methylethyl)benzenemethanol; cuminol; cuminyl alcohol; p-isopropylbenzyl alcohol.

$C_{10}H_{14}O$; mol wt 150.22. C 79.96%, H 9.39%, O 10.65%. Found in caraway seed. Prepd by the reduction of cuminal; also by treating a suspension of cuminylmagnesium chloride in ether with dry oxygen.

H₃C—[benzene ring]—CH₂OH with CH₃ substituent

Liquid; intense, persistent, caraway-like odor and burning, aromatic taste. d^{15} 0.981. bp 248.4°. n_D^{24} 1.522. Insol in water; miscible with alcohol, ether.

2610. Cumidine. [99-88-7] 4-(1-Methylethyl)benzenamine; 4-aminocumene; p-isopropylaniline; 4-amino-1-isopropylbenzene; β-(4-aminophenyl)propane. $C_9H_{13}N$; mol wt 135.21. C 79.95%, H 9.69%, N 10.36%. Prepn from aniline and isopropyl alcohol: Louis, *Ber.* **16**, 111 (1883); by nitration and subsequent reduction of cumene: Constam, Goldschmidt, *Ber.* **21**, 1157 (1898); Vavon, Callier, *Bull. Soc. Chim. Fr.* [4] **41**, 677 (1927).

H₃C—CH—[benzene ring]—NH₂ with CH₃

Liquid. bp_{745} 226-227°. d 0.9526. Insol in water.
Hydrochloride. $C_9H_{13}N.HCl$. Prisms from water or alc.
Sulfate. $C_9H_{13}N.\frac{1}{2}H_2SO_4$. Leaflets, mp 205°. Sol in hot water or alcohol, less sol in cold water, ether.
USE: In the determination of tungsten.

2611. Cuminaldehyde. [122-03-2] 4-(1-Methylethyl)benzaldehyde; p-isopropylbenzaldehyde; cuminal; cumaldehyde. $C_{10}H_{12}$-O; mol wt 148.21. C 81.04%, H 8.16%, O 10.79%. Constituent of eucalyptus, myrrh, cassia, cumin, and other essential oils; prepared synthetically from p-isopropylbenzoyl chloride. Alternate synthesis by formylation of isopropylbenzene under pressure: Crounse, *J. Am. Chem. Soc.* **71**, 1263 (1949). Toxicity study: P. M. Jenner *et al.*, *Food Cosmet. Toxicol.* **2**, 327 (1964).

H₃C—CH—[benzene ring]—CHO with CH₃

Colorless to yellowish, oily liq; strong persistent odor; acrid, burning taste. d^{20} 0.978. bp_{760} 235-236°; bp_{35} 131-135°. n_D^{20} 1.5301. Practically insol in water. Sol in alc, ether. LD_{50} orally in rats: 1390 mg/kg (Jenner).
Thiosemicarbazone. [3811-20-9] SHCH-58. $C_{11}H_{15}N_3S$; mol wt 221.32. Prepn: P. P. T. Sah, T. C. Daniels, *Rec. Trav. Chim.* **69**, 1545 (1950); J. Bernstein *et al.*, *J. Am. Chem. Soc.* **73**, 906 (1951). Stout platelets from methanol, mp 147°. Practically insol in water. Moderately sol in alc.
USE: In perfumery.

2612. Cupferron. [135-20-6] N-Hydroxy-N-nitrosobenzenamine ammonium salt (1:1); N-nitrosophenylhydroxylamine ammonium salt. $C_6H_9N_3O_2$; mol wt 155.16. C 46.45%, H 5.85%, N 27.08%, O 20.62%. Prepd from phenylhydroxylamine by treating with $NaNO_2$ at 0° in presence of HCl, filtering, dissolving in ether and treating with NH_3, or by treating an ether soln of phenylhydroxylamine with butyl or amyl nitrite and NH_3 in the cold: Marvel, Kamm, *Org. Synth.* **4**, 19 (1925).

[benzene ring]—N(—O⁻ NH₄⁺)—N=O

Crystals, mp 163-164°. Freely sol in water or alc. *Keep in well-closed containers to which a piece of ammonium carbonate has been added.*
Caution: This substance is reasonably anticipated to be a human carcinogen: *Report on Carcinogens, Twelfth Edition* (PB2011-111646, 2011) p 123.
USE: As a reagent for separating Sn from Zn, and Cu and Fe from other metals. Ppts iron quantitatively from strongly acid soln; as a quantitative reagent for vanadates with which it gives a dark-red ppt sol in alkali soln, and for Ti with which it forms a yellow ppt; also suitable for the colorimetric estimation of Al.

2613. Cupreine. [524-63-0] (8α,9R)-Cinchonan-6′,9-diol; hydroxycinchonidine; ultraquinine. $C_{19}H_{22}N_2O_2$; mol wt 310.40. C 73.52%, H 7.14%, N 9.03%, O 10.31%. In bark and seeds of *Remijia pedunculata* Flueck., *Rubiaceae.* Isoln: Howard, Hodgkin, *J. Chem. Soc.* **41**, 16 (1882). Separated from homoquinine by treatment with NaOH: Hesse, *Ann.* **230**, 55 (1885). Reduction to *hydrocupreine*: Giemsa, Halberkann, *Ber.* **51**, 1325 (1918). Conversion into cinchonidine: King, *J. Chem. Soc.* **1946**, 523.

[cupreine structure diagram]

Monoclinic plates from alcohol, mp 202°. $[\alpha]_D^{17}$ −176°, (methanol). pK_1 6.57. Sol in alcohol, NaOH soln, but not in NH_4OH. Sparingly sol in water, ether, chloroform, benzene, petr ether. The neutral salts give yellow aq solns; the acid salts remain colorless in aq soln.
Monomethyl ether *see* Quinine.

2614. Cupric Acetate. [142-71-2] Acetic acid copper(2+) salt (2:1); copper(II) acetate. $C_4H_6CuO_4$; mol wt 181.63. C 26.45%, H 3.33%, Cu 34.99%, O 35.23%. $Cu(CH_3COO)_2$. Prepd by the action of acetic acid on CuO or $CuCO_3$: Winter *et al.*, in *Kirk-Othmer Encyclopedia of Chemical Technology* **vol. 6** (Interscience, New York, 2nd ed., 1965) p 267. Toxicity study: H. F. Smyth *et al.*, *Am. Ind. Hyg. Assoc. J.* **30**, 470 (1969).
Monohydrate. [6046-93-1] Neutral verdigris. $C_4H_6CuO_4.H_2$-O; mol wt 199.65. Dark green, monoclinic crystals; exists in dimeric units. mp 115°; dec at 240°; d 1.882. Sol in water, alcohol; slightly sol in ether, glycerol. LD_{50} orally in rats: 0.71 g/kg (Smyth).
Verdigris. [8007-61-2] C.I. Pigment Green 20; C.I. 77408. Prepn: *Colour Index* **vol. 4** (3rd ed, 1971) p 4666. Commercial product is a mixture of *basic cupric acetate* salts with variable waters of hydration, resulting in blue to green powders. Slightly sol in water, ethanol; sol in dil acids, ammonia.
USE: Intermediate in manuf of Paris green, *q.v.*; as fungicide; as catalyst for organic reactions, including rubber aging; in textile dyeing; as pigment for ceramics; in prepn of Fehling's reagent for sugars.

2615. Cupric Acetoarsenite. [12002-03-8] C.I. Pigment Green 21; Paris green; copper acetate arsenite; emerald green; French green; imperial green; mineral green; Mitis green; parrot green; Schweinfurt green; Vienna green; C.I. 77410. $C_4H_6As_6Cu_4O_{16}$; mol wt 1013.79. C 4.74%, H 0.60%, As 44.34%, Cu 25.07%, O 25.25%. $Cu(C_2H_3O_2)_2.3Cu(AsO_2)_2$. Usually contains some water. Prepn: Serciron, US 2159864 (1939); Krefft, US 2268123 (1944 to Chemical Marketing); Glemser, Sauer in *Handbook of Preparative Inorganic Chemistry* **vol. 2**, G. Brauer, Ed. (Academic Press, New York, 2nd ed., 1965) p 1027. *See also Colour Index* **vol. 4** (3rd ed., 1971) p 4667. Toxicity study: T. B. Gaines, *Toxicol. Appl. Pharmacol.* **14**, 515 (1969).
Emerald green, crystalline powder. *Poisonous.* Stable to air, light. Practically insol in water. Dec on prolonged heating in water. Unstable in acids, bases, and toward H_2S. LD_{50} orally in female rats: 100 mg/kg (Gaines).

Caution: Toxicity primarily result of arsenic content. Gastric disturbance, tremors or muscular cramps, and peripheral neuritis. Local effects on the skin, mucous membrances and conjunctiva.

USE: Insecticide; wood preservative; as pigment, particularly for ships and submarines.

2616. Cupric Arsenite. [10290-12-7] Arsonic acid copper-(2+) salt (1:1); arsenious acid copper(2+) salt (1:1); Scheele's green. Compound of variable composition. Usually considered to be Cu-HAsO₃; mol wt 187.46. Also reported as Cu(AsO₂)₂: Bhadraver, *Bull. Chem. Soc. Jpn.* **35**, 1770 (1962). Prepd by reaction of a cupric salt with an alkaline or ammoniacal arsenite: R. N. Kust in *Kirk-Othmer Encyclopedia of Chemical Technology* vol. 7 (Interscience, New York, 3rd ed., 1979) p 102. *See also: Colour Index* vol. 4 (3rd ed., 1971) p 4667.

Yellowish-green powder. *Poisonous.* Practically insol in water, alcohol. Sol in acids, ammonia.

USE: As pigment, wood preservative, insecticide, fungicide, rodenticide.

2617. Cupric Bisglycinate. [13479-54-4] Bis(glycinato-κN,-κO)copper; cupric aminoacetate; glycocoll-copper; Copper Chelazome. $C_4H_8CuN_2O_4$; mol wt 211.66. C 22.70%, H 3.81%, Cu 30.02%, N 13.24%, O 30.24%. Prepn: J. Mauthner, W. Suida, *Monatsh. Chem.* **11**, 373 (1890); and characterization: H. Ley, *Z. Elektrochem.* **10**, 954 (1904); and isomerization study: K. Tomita, *Bull. Chem. Soc. Jpn.* **34**, 280 (1961). Thermodynamics of complex formation: K. P. Anderson *et al.*, *Inorg. Chem.* **5**, 2106 (1966). Kinetics of complex formation: A. F. Pearlmutter, J. Stuehr, *J. Am. Chem. Soc.* **90**, 858 (1968). Clinical evaluation in maintenance of bone mineral density: J. Eaton-Evans *et al.*, *J. Trace Elem. Exp. Med.* **9**, 87 (1996).

Monohydrate. Long, deep-blue, rhombic needles. Loses H₂O at 123°, chars at 213°, and dec with gas evolution at 228°. Sol in water; slightly sol in alcohol.

Dihydrate. Light blue, powdery crystals. Loses one H₂O at 103°, remaining H₂O at about 140°. Dec with gas evolution at about 225°. Sol in water.

THERAP CAT: Copper supplement.

THERAP CAT (VET): Nutritional factor. Given parenterally for copper deficiency in ruminants.

2618. Cupric Borate. [39290-85-2] Boric acid copper salt. A material of indefinite composition prepd by the reaction of a borax soln with CuSO₄: Barnard, Jr. in *ACS Monograph Series* **no. 122**, entitled "Copper," A. Butts, Ed. (Reinhold, New York, 1954) p 794. Blue to blue-green solid. Extremely hard material.

Caution: Symptoms of acute toxicity include vomiting, diarrhea, tremors, convulsions, shock, dermatitis; renal, hepatic and CNS degeneration. Symptoms of chronic ingestion include skin disorders, kidney lesions.

USE: As ceramic color, pigment in oil paint; in dehydrogenation catalysis; as insulating and sealing medium for lead wires in lamps and mercury rectifiers, as fireproofing agent and preservative for wood; as fungicide, insecticide, parasiticide.

2619. Cupric Bromide. [7789-45-9] Copper bromide (CuBr₂); copper dibromide. Br₂Cu; mol wt 223.35. Br 71.55%, Cu 28.45%. Prepn: Watt *et al.*, *J. Am. Chem. Soc.* **77**, 2752 (1955); Glemser, Sauer in *Handbook of Preparative Inorganic Chemistry* vol. 2, G. Brauer, Ed. (Academic Press, New York, 2nd ed., 1965) p 1009.

Almost black, iodine-like, deliquesc monoclinic crystals or cryst powder. mp 498°; bp 900°; d_4^{20} 4.710. Very sol in water; sol in alcohol, acetone, ammonia; practically insol in benzene, ether, concd H₂SO₄. *Keep well closed.*

USE: As intensifier in photography; as brominating agent in organic synthesis; as humidity indicator; as wood preservative; in solid-electrolyte battery; as stabilizer for acetylated polyformaldehyde.

2620. Cupric Butyrate. [540-16-9] Butanoic acid copper(2+) salt (2:1); copper dibutyrate. $C_8H_{14}CuO_4$; mol wt 237.74. C 40.42%, H 5.94%, Cu 26.73%, O 26.92%. Cu(CH₃CH₂CH₂COO)₂. Prepn from a cupric salt and butyric acid or Na butyrate: Graddon, *J. Inorg. Nucl. Chem.* **17**, 222 (1955); Martin, Waterman, *J. Chem. Soc.* **1957**, 2545; from powdered Cu and butyric acid: Hahl, *US 3133942* (1964 to BASF).

Monohydrate. Large, dark green, monoclinic, hexagonal plates. Becomes dull and disintegrates after several days exposure to air. Sol in water, dioxane, benzene; slightly sol in chloroform, alcohol.

USE: In catalysts; in lubricants.

2621. Cupric Carbonate, Basic. [12069-69-1] μ-[Carbonato(2−)-κO:$\kappa O'$]dihydroxydicopper; copper carbonate hydroxide; cupric subcarbonate; Bremen blue; Bremen green. $CH_2Cu_2O_5$; mol wt 221.11. C 5.43%, H 0.91%, Cu 57.48%, O 36.18%. CuCO₃.Cu(OH)₂. Occurs in nature as the mineral *malachite* (dark green monoclinic crystals). Prepd by adding CuSO₄ soln to Na₂CO₃ soln: Winter *et al.*, in *Kirk-Othmer Encyclopedia of Chemical Technology* vol. 6 (Interscience, New York, 2nd ed., 1965) p 270. The commercial product usually contains 50-55% Cu.

Green to blue amorphous powder or dark-green monoclinic crystals. Practically insol in water, alcohol; sol in dil acids, ammonia.

Note: Another form of basic copper carbonate, 2CuCO₃.Cu(OH)₂, occurs in nature as the mineral *azurite* or *chessylite. See also* Burgundy Mixture.

USE: As seed treatment fungicide; in pyrotechnics; as paint and varnish pigment; in animal and poultry feeds; in sweetening of petrol sour crude stock; in manuf of other Cu salts.

THERAP CAT (VET): Nutritional factor. Used in copper deficiency in ruminants.

2622. Cupric Chlorate. [14721-21-2] Chloric acid copper-(2+) salt. Cl₂CuO₆; mol wt 230.44. Cl 30.77%, Cu 27.58%, O 41.66%. Cu(ClO₃)₂. Prepn: *Gmelins, Copper* (8th ed.) **60B**, 332-334 (1958).

Hexahydrate. [13478-36-9] Blue to green, deliquesc, octahedral crystals. mp 65°. Dec at 100°. Very sol in water, alcohol. *Keep well closed and out of contact with organic matter.*

USE: Mordant in dyeing and printing of textiles.

2623. Cupric Chloride. [7447-39-4] Copper chloride (CuCl₂); copper(II) chloride. Cl₂Cu; mol wt 134.45. Cl 52.73%, Cu 47.26%. CuCl₂. Prepn: Pray, *Inorg. Synth.* **5**, 153 (1957); *Gmelins, Copper* (8th ed.) **60B**, 253-295 (1958); Glemser, Sauer in *Handbook of Preparative Inorganic Chemistry* vol. 2, G. Brauer, Ed. (Academic Press, New York, 2nd ed., 1965) p 1008.

Yellow to brown, deliquesc, hygroscopic microcryst powder. Partially dec above 300° to CuCl + Cl₂. mp (extrapolated) 630°. mp of 498° usually reported for this compd is the melting pt of a mixture of CuCl₂ and CuCl. d_4^{25} 3.39. *Corrosive.* Forms the dihydrate in moist air. Sol in water, alc, acetone. *Keep well closed.*

Dihydrate. [10125-13-0] CuCl₂.2H₂O; mol wt 170.48. Green to blue powder or orthorhombic, bipyramidal crystals. Deliquesc in moist air, efflorescent in dry air. Water loss occurs from 70-200°: Bell, Coultard, *J. Chem. Soc. A* **1968**, 1417. mp ~100°; d 2.51. Freely sol in water, methanol, ethanol; moderately sol in acetone, ethyl acetate; slightly sol in ether. The aq soln is acid to litmus; pH of 0.2 molar soln 3.6.

Caution: Potential symptoms of overexposure may include irritation of skin, eyes, and respiratory tract.

USE: As catalyst for organic and inorganic reactions; in petroleum industry as deodorizing, desulfurizing, and purifying agent; as mordant for dyeing and printing textiles; as oxidizing agent for aniline dyestuffs; in indelible, invisible and laundry-marking inks; in metallurgy in wet process for recovering mercury from ores, in refining Cu, Ag, Au; in tinting-baths for Fe, Sn; in electrotyping baths for plating Cu on Al; in analytical chemistry for testing for Mo; in photography as a fixer, desensitizer and reagent; in producing color in pyrotechnic compositions; in manuf of acrylonitrile, fast black (melanin); in pigments for glass, ceramics; as feed-additive, wood preservative, disinfectant.

2624. Cupric Chromate(VI). [13548-42-0] Chromic acid (H₂CrO₄) copper(2+) salt (1:1); neutral cupric chromate; copper chromate. CrCuO₄; mol wt 179.54. Cr 28.96%, Cu 35.39%, O

35.64%. $CuCrO_4$. Prepn from $CuCO_3$, Na_2CrO_4 and CrO_3: Briggs, *J. Chem. Soc.* **1929**, 242; from CuO and CrO_3: Campbell, Lemaire, *Can. J. Res.* **25B**, 243 (1947); from $Cu(OH)_2$ and $K_2Cr_2O_7$: Winter *et al.*, in *Kirk-Othmer Encyclopedia of Chemical Technology* **vol. 6** (Interscience, New York, 2nd ed., 1965) p 270.

Reddish-brown crystals. Gradually dec to $CuCr_2O_4$ above 400°. Sol in acids. Practically insol in water.

Basic cupric chromates. Several cupric chromate-cupric hydroxide compounds are known. $CuCrO_4 \cdot Cu(OH)_2$ forms yellow, copper-red or chocolate-brown to lilac crystals; $CuCrO_4 \cdot 2Cu(OH)_2$, a light brown powder; $2CuCrO_4 \cdot 3Cu(OH)_2$, yellow to yellowish-brown crystals. Prepn: Campbell, Lemaire, *loc. cit.;* Winter *et al.*, *loc. cit.* Practically insol in water.

Caution: Chromium hexavalent (VI) compounds are listed as known human carcinogens: *Report on Carcinogens, Twelfth Edition* (PB2011-111646, 2011) p 106.

USE: In fungicides, seed protectants, and wood preservatives; as mordant in dyeing textiles; in protecting textiles against insects and microorganisms.

2625. Cupric Chromite. [12018-10-9] Chromium copper oxide (Cr_2CuO_4); copper(2+) chromite; cupric chromate(III). $Cr_2\text{-}CuO_4$; mol wt 231.53. Cr 44.92%, Cu 27.45%, O 27.64%. $CuCr_2\text{-}O_4$. Prepd by heating $CuCrO_4$ at 400°: Stroupe, *J. Am. Chem. Soc.* **71**, 569 (1949); or by decompn of $(NH_4)_2Cu(CrO_4)_2 \cdot 2NH_3$ at 700°: Whipple, Wold, *J. Inorg. Nucl. Chem.* **24**, 23 (1962).

Grayish-black to black tetragonal crystals. Dec to $CuCrO_2$ and CrO_3 above 900°. Practically insol in water, dil acids, concd HCl.

Copper-chromium oxide. A mixture of $CuCr_2O_4$ and CuO formed on decompn at temps below 400°, of orange copper ammonium chromate complex prepd from $Cu(NO_3)_2$, $Na_2Cr_2O_7$ and NH_4-OH: Adkins *et al.*, *J. Am. Chem. Soc.* **72**, 2626 (1950); Wagner in *Handbook of Preparative Inorganic Chemistry* **vol. 2**, G. Brauer, Ed. (Academic Press, New York, 2nd ed., 1965) p 1672. Fine, brownish-black to black powder. Stable in atmospheric O_2 and moisture. Practically insol in water, dil acids.

USE: Selective hydrogenation catalysts.

2626. Cupric Citrate. [866-82-0] 2-Hydroxy-1,2,3-propane-tricarboxylic acid copper(2+) salt (1:2); Cuprocitrol. $C_6H_4Cu_2O_7$; mol wt 315.18. C 22.87%, H 1.28%, Cu 40.32%, O 35.53%. Prepd by the interaction of hot aq solns of $CuSO_4$ and Na citrate: *U.S.D.* 25th ed., p 1653.

Hemipentahydrate. Green or bluish-green, odorless, cryst powder. Loses its water at about 100°. Slightly sol in water; sol in ammonia, dil acids; slightly sol in cold, freely in hot solns of alkali citrates.

THERAP CAT: Astringent, antiseptic.

2627. Cupric Ferrocyanide. [13601-13-3] *(OC-6-11)*-Hexakis(cyano-κC)ferrate(4−) copper(2+) (1:2); hexacyanoferrate(4−) dicopper(2+); cupric hexacyanoferrate(II); Hatchett's brown; C.I. Pigment Brown 9. $C_6Cu_2FeN_6$; mol wt 339.05. C 21.26%, Cu 37.48%, Fe 16.47%, N 24.79%. $Cu_2Fe(CN)_6$. Prepd by addition of $K_4Fe(CN)_6$ to an aqueous soln of a sol cupric salt: Weiser *et al.*, *J. Phys. Chem.* **42**, 945 (1938); **46**, 99 (1942).

Reddish-brown powder or cubic crystals; ppts as a colloid or gel. Practically insol in water, dil acids, most organic solvents. Sol in NH_4OH, solns of alkali cyanides.

USE: As pigment; in photographic toning baths; to lower electric resistance of soil and electrode-to-soil contacts.

2628. Cupric Fluoride. [7789-19-7] Copper fluoride (CuF_2); copper difluoride; CuF_2; mol wt 101.54. Cu 62.58%, F 37.42%. Prepd by fluorinating Cu or a Cu salt with F_2: von Wartenberg, *Z. Anorg. Allg. Chem.* **241**, 381 (1939); Jache, Cady, *J. Phys. Chem.* **56**, 1106 (1952); Haendler *et al.*, *J. Am. Chem. Soc.* **76**, 2178 (1954); Ritter, Smith, *J. Phys. Chem.* **70**, 805 (1966); **71**, 2384 (1971).

Monoclinic crystals; turns blue in moist air due to formation of dihydrate. mp $\sim$785° (N_2 atm), 950° (HF atm); d 4.85. Soly in water (20°) 4.7 g/100 ml; hydrolyzed to CuFOH in hot water. May be stored in sealed glass ampuls.

Dihydrate. Blue monoclinic crystals. Dec above 130°. d_4^{25} 2.934. Slightly sol in cold water, hydrolyzed to oxyfluoride in hot water.

USE: Anhydr in cathodes in nonaq galvanic cells; high temperature fluorinating agent; dihydrate as flux in casting gray iron.

2629. Cupric Formate. [544-19-4] Formic acid copper(2+) salt (2:1); copper diformate; tubercuprose. $C_2H_2CuO_4$; mol wt 153.58. C 15.64%, H 1.31%, Cu 41.38%, O 41.67%. $Cu(HCOO)_2$. Prepd from cupric carbonate and formic acid: Martin, Waterman, *J. Chem. Soc.* **1959**, 1359.

Three forms of anhydrous compd exist: powder-blue, turquoise, or royal blue crystals. Sol in water. Practically insol in most organic solvents.

Dihydrate. Very pale blue, monoclinic needles. Loses $2H_2O$ on standing in air. Turquoise, anhydrous modification formed by dehydration at 100° *in vacuo* over phosphoric oxide. Sol in water.

Tetrahydrate. Large, light-blue, monoclinic, holohedral prisms. Powder-blue modification formed by dehydration over $CaCl_2$ under reduced press. Sol in water; very slightly sol in alc. Practically insol in most organic solvents.

USE: As antibacterial agent in the treatment of cellulose.

2630. Cupric Gluconate. [527-09-3] Bis(D-gluconato-κO^1,-κO^2)copper; copper(II) gluconate. $C_{12}H_{22}CuO_{14}$; mol wt 453.84. C 31.76%, H 4.89%, Cu 14.00%, O 49.35%. $Cu[CH_2OH(CHOH)_4\text{-}CO_2]_2$. Prepd from gluconic acid and basic cupric carbonate: Suzuki *et al.*, *JP 63 2889* (1963 to Dainippon), *C.A.* **59**, 11264c (1963).

Hydrate. Light blue to bluish-green crystals or cryst powder. Astringent taste. Soly in water 30 g/100 ml at 25°; slightly sol in alc. Practically insol in most other organic solvents.

USE: Oral deodorant.

THERAP CAT: Copper supplement.

2631. Cupric Hexafluorosilicate. [12062-24-7] Hexafluorosilicate(2−) copper(2+) (1:1); cupric fluosilicate; cupric silicofluoride. CuF_6Si; mol wt 205.62. Cu 30.90%, F 55.44%, Si 13.66%. $CuSiF_6$. Prepn: Worthington, Haring, *Ind. Eng. Chem. Anal. Ed.* **3**, 7 (1931).

Tetrahydrate. Blue, monoclinic, efflorescent crystals. d 2.56: Clark *et al.*, *Can. J. Chem.* **47**, 3859 (1969). Readily sol in water. *Keep well closed.*

USE: Dyeing and hardening white marble; treating plant diseases.

2632. Cupric Hydroxide. [20427-59-2] Copper hydroxide ($Cu(OH)_2$); copper hydrate; hydrated cupric oxide; copper dihydroxide. CuH_2O_2; mol wt 97.56. Cu 65.14%, H 2.07%, O 32.80%. $Cu(OH)_2$. Commercial prepn: Furness, **US 1800828** (1931 to Cellosilk); *idem*, **US RE 24324** (1957 to Copper Research); Rowe, **US 2536096** (1951 to Mountain Copper); laboratory prepn: Weiser *et al.*, *J. Am. Chem. Soc.* **64**, 503 (1942); Gauthier, *Bull. Soc. Chim. Fr.* **1960**, 353.

Blue to blue-green gel or light blue cryst powder. Stability is dependent on the method of prepn; may dec to black CuO on standing a few days or on heating. d 3.37. Practically insol in water. Sol in concd alkali when freshly pptd; sol in acids, NH_4OH.

USE: In manuf of rayon, battery electrodes, other Cu salts; as mordant in dyeing; as pigment; in fungicides, insecticides; as feed additive; in treating and staining paper; in prepn of Schweitzer's reagent; in catalysts.

2633. Cupric Nitrate. [3251-23-8] Nitric acid copper(2+) salt (2:1); copper(II) nitrate. CuN_2O_6; mol wt 187.55. Cu 33.88%, N 14.94%, O 51.18%. $Cu(NO_3)_2$. Prepn: *Gmelins, Copper* (8th ed.) **60B**, 164-179 (1958); Addison, Hathaway, *J. Chem. Soc.* **1958**, 3099. Toxicity: H. F. Smyth *et al.*, *Am. Ind. Hyg. Assoc. J.* **30**, 470 (1969).

Large, blue-green, deliquesc, orthorhombic crystals. Sublimes at 150-225°. mp 255-256°. Sol in water, ethyl acetate, dioxane; dissolves in and reacts vigorously with ether. *Keep well closed.*

Trihydrate. [10031-43-3] Gerhardite. $Cu(NO_3)_2 \cdot 3H_2O$; mol wt 241.60. Blue, deliquesc, rhombic plates. mp 114.5°; d 2.05. Freely sol in water, alcohol. Practically insol in ethyl acetate. pH of 0.2 molar aq soln 4.0. *Keep well closed.* LD_{50} orally in rats: 0.94 g/kg (Smyth).

Hexahydrate. [13478-38-1] $Cu(NO_3)_2 \cdot 6H_2O$; mol wt 295.64. Blue, deliquesc prismatic crystals. Decomposes to trihydrate at 26.4°. d 2.07. Freely sol in water; sol in alc. *Keep well closed.*

Caution: Potential symptoms of overexposure may include eye and skin irritation.

USE: In light-sensitive reproductive papers; as ceramic color; as mordant and oxidizing agent in textile dyeing and printing; as reagent for burnishing iron, for giving a black "antique" finish to copper, for coloring zinc brown; in nickel-plating baths; in aluminum brighteners; in wood-preservatives, fungicides, herbicides; in pyrotechnic compositions; as catalyst component in solid rocket fuel; as nitrating agent for aromatic organosilicon compds; as catalyst for organic reactions.

2634. Cupric Oleate. [1120-44-1] (9Z)-9-Octadecenoic acid copper(2+) salt (2:1); oleic acid copper(2+) salt; copper dioleate. $C_{36}H_{66}CuO_4$; mol wt 626.47. C 69.02%, H 10.62%, Cu 10.14%, O 10.22%. $(C_{17}H_{33}COO)_2Cu$. Prepd from $CuSO_4$ and K oleate: Nelson, Pink, *J. Chem. Soc.* **1954**, 4412.

Blue to green solid. Practically insol in water. Slightly sol in alcohol; sol in ether.

Note: Ingredient of *Cuprex*, a hydrocarbon mixture used as a topical pediculicide.

USE: In antifouling compositions; as emulsifier and dispersing agent; as antioxidant in lubricating oils; as combustion-improver in fuel oils; as stabilizer for amide polymers; as catalyst.

2635. Cupric Oxalate. [814-91-5] [Ethanedioato(2-)-κO^1,-κO^2]copper; ethanedioic acid copper salt. C_2CuO_4; mol wt 151.56. C 15.85%, Cu 41.93%, O 42.22%. CuC_2O_4. Prepd by reaction of $CuSO_4$ with oxalic acid: David, *Bull. Soc. Chim. Fr.* **1960**, 719. Usually contains some water.

Blue-white powder. Loses any hydrated water by 200°; dec in air at 310° to CuO. Practically insol in water, alcohol, ether, acetic acid; sol in NH_4OH.

USE: As catalyst for organic reactions; as stabilizer for acetylated polyformaldehyde; in anticaries compositions; in seed treatments to repel birds and rodents.

2636. Cupric Oxide. [1317-38-0] Copper oxide (CuO); black copper oxide; C.I. 77403; C.I. Pigment Black 15; copper(II) oxide. CuO; mol wt 79.55. Cu 79.88%, O 20.11%. Occurs in nature as the mineral *tenorite*. Prepn: R. Ruer, J. Kuschmann, *Z. Anorg. Allg. Chem.* **154**, 69 (1926); O. Glemser, H. Sauer in *Handbook of Preparative Inorganic Chemistry* vol. 2, G. Brauer, Ed. (Academic Press, New York, 2nd ed., 1965) p 1012. Soly study: A. Hayward *et al., Nature* **215**, 730 (1967). Bioavailability study: S. Aoyagi, D. H. Baker, *Poult. Sci.* **72**, 1075 (1993). Review of physical properties: E. Gmelin, *Indian J. Pure Appl. Phys.* **30**, 596-608 (1992); of use as a nutritional supplement: D. H. Baker, *J. Nutr.* **129**, 2278-2279 (1999).

Pure CuO (tenorite) is a steel-grey to black *p*-type semiconductor, mp ≃1330°. d ≃6.4. Insol in water. Slightly sol in alkalis; sol in acids and ammonia. Heat of formation (25°): −37.1 kcal/mol.

USE: As pigment in glass, ceramics, enamels, porcelain glazes, artificial gems; in mfr of rayon, other Cu compds; in sweetening petr gases; in galvanic electrodes; as flux in metallurgy; in correcting Cu deficiencies in soil; as optical-glass polishing agent; in antifouling paints, pyrotechnic compositions; as exciter in phosphor mixtures; as catalyst for organic reactions; in high temp superconductors; as source of oxygen; in determn of nitrogen.

THERAP CAT: Copper supplement.

THERAP CAT (VET): Copper supplement in ruminants.

2637. Cupric Perchlorate. [13770-18-8] Perchloric acid copper(2+) salt (2:1); copper bis(perchlorate); copper diperchlorate. Cl_2CuO_8; mol wt 262.44. Cl 27.02%, Cu 24.21%, O 48.77%. Cu($ClO_4)_2$. Prepd from $Cu(NO_3)_2$ and perchloric acid or nitrosyl perchlorate: Caven, Bryce, *J. Chem. Soc.* **1934**, 514; Hathaway, Underhill, *ibid.* **1960**, 648; **1961**, 3091; Hathaway, *Proc. Chem. Soc. London* **1958**, 344.

Very pale green, hygroscopic crystals. Volatilizes on heating. Thermally stable to 130°; above 130° dec to a basic perchlorate. mp about 230-240°. Soluble in water, ether, dioxane, ethyl acetate. Practically insol in benzene, CCl_4, hexane.

Hexahydrate. Deep blue, monoclinic crystals. Freely sol in water, methanol, ethanol, acetic acid, acetic anhydride, acetone; slightly sol in ether, ethyl acetate.

USE: Analytical reagent: Kolb, *Ind. Eng. Chem. Anal. Ed.* **11**, 197 (1939); **16**, 38 (1944); Serjeant, *Nature* **186**, 963 (1960). Also in copper electrodeposition; in catalysts for combustion and propellants.

2638. Cupric Phosphate. [7798-23-4] Phosphoric acid copper(2+) salt (2:3); copper phosphate (3:2). $Cu_3O_8P_2$; mol wt 380.58.

Cu 50.09%, O 33.63%, P 16.28%. $Cu_3(PO_4)_2$. Basic salts occur in nature as the minerals: *cornetite, dihydrite, libethenite, phosphorochalcite, pseudolibethenite, pseudomalachite, tagilite*. Prepn from $CuSO_4$ and $(NH_4)_2HPO_4$: Klement, Haselbeck, *Z. Anorg. Allg. Chem.* **334**, 27 (1964).

Trihydrate. Blue or olive orthorhombic crystals or blue to bluish-green powder. Dec on heating. Practically insol in cold water. Slightly sol in hot water; sol in acids, NH_4OH.

USE: Fungicide; catalyst for organic reactions, fertilizer; emulsifier; corrosion inhibitor for H_3PO_4; protectant for metal surfaces against oxidation.

2639. Cupric Salicylate. [16048-96-7] Bis[2-(hydroxy-κO)-benzoato-κO]copper; 2-hydroxybenzoic acid copper salt; copper disalicylate. $C_{14}H_{10}CuO_6$; mol wt 337.77. C 49.78%, H 2.98%, Cu 18.81%, O 28.42%. $Cu[C_6H_4(OH)COO]_2$. Prepd from reaction of $CuSO_4$ with Na salicylate: Hanic, Michalov, *Acta Crystallogr.* **13**, 299 (1960); Inoue *et al., Inorg. Chem.* **3**, 239 (1964).

Brown, blue-green or green powder. Absorbs water in moist air to form the tetrahydrate.

Tetrahydrate. Pale-blue needles or blue-green plates. Blue-green modification loses water on standing in air. Sol in water, alcohol, NH_4OH.

USE: In the separation of monoolefinic from di- or polyolefinic hydrocarbons.

2640. Cupric Selenate. [15123-69-0] Selenic acid copper-(2+) salt (1:1). CuO_4Se; mol wt 206.50. Cu 30.77%, O 30.99%, Se 38.24%. $CuSeO_4$. Prepd by the action of selenic acid on cupric carbonate: Klein, *Ann. Chim.* **14**, 263 (1940).

Pentahydrate. Light-blue triclinic crystals. Loses water above 80°, forming the monohydrate at 150-220° and becoming anhydr by 265°; dec to the selenite and basic selenite at ~480° and to CuO at ~700°. d 2.56. Sol in water; very slightly sol in acetone. Practically insol in alcohol.

USE: In coloring Cu or Cu alloys black.

2641. Cupric Selenide. [1317-41-5] Copper selenide (CuSe); copper(2+) selenide; copper monoselenide. CuSe; mol wt 142.51. Cu 44.59%, Se 55.41%. Occurs in nature as the mineral *klockmannite*. Prepd by reduction of cupric selenite with hydrazine: Benzing *et al., J. Am. Chem. Soc.* **80**, 2657 (1958); Kulifay, *J. Inorg. Nucl. Chem.* **25**, 75 (1965).

Blue-black to greenish-black prismatic needles or hexagonal plates. Dec at a dull red heat. d 6.0-6.6. Sol in HCl with H_2Se evolved, in H_2SO_4 with SO_2 evolved; oxidized to $CuSeO_3$ by HNO_3.

USE: Catalyst in Kjeldahl digestions; in semiconductors.

2642. Cupric Stearate. [660-60-6] Octadecanoic acid copper(2+) salt (2:1); copper bis(octadecanoate); copper distearate. $C_{36}H_{70}CuO_4$; mol wt 630.50. C 68.58%, H 11.19%, Cu 10.08%, O 10.15%. $(C_{17}H_{35}COO)_2Cu$. Prepd by metathesis of alcohol cupric acetate with an alcohol soln of stearic acid: Martin, Waterman, *J. Chem. Soc.* **1957**, 2545; Rai, Mehrotra, *J. Inorg. Nucl. Chem.* **21**, 311 (1961).

Pale-blue to blue-green amorphous powder. Considered to be dimeric. mp ~250°: Grant, *Can. J. Chem.* **42**, 951 (1964). Practically insol in water, methanol, ethanol, acetone, ether. Sol in pyridine, dioxane, acetic anhydride; sol in hot but practically insol in cold benzene, toluene, CCl_4, $CHCl_3$, ethyl acetate.

USE: In antifouling paints; in textile and wood preservatives; in rubber aging; as catalyst.

2643. Cupric Sulfate. [7758-98-7] Sulfuric acid copper(2+) salt (1:1); copper(II) sulfate. CuO_4S; mol wt 159.60. Cu 39.82%, O 40.10%, S 20.09%. $CuSO_4$. Occurs in nature as the mineral *hydrocyanite*. Commercial preparation of pentahydrate: *Faith, Keyes & Clark's Industrial Chemicals*, F. A. Lowenheim, M. K. Moran, Eds. (Wiley-Interscience, New York, 4th ed., 1975) pp 280-284; other prepns: *Gmelins, Copper* (8th ed.) **60B**, 491-560 (1958). Toxicity: Smyth *et al., Am. Ind. Hyg. Assoc. J.* **30**, 470 (1969). Review of toxicology and human exposure: *Toxicological Profile for Copper* (PB2004-107333, 2004) 318 pp.

Grayish-white to greenish-white rhombic crystals or amorphous powder. On heating dec above 560°. d 3.6. Hygroscopic. Sol in water. Practically insol in alc. *Keep tightly closed.*

Monohydrate. [10257-54-2] Dried cupric sulfate. $CuSO_4.H_2O$; mol wt 177.62. Hygroscopic, almost white powder. Sol in water. Practically insol in alc. *Keep tightly closed.*

Pentahydrate. [7758-99-8] Bluestone; blue vitriol; Roman vitriol; Salzburg vitriol. $CuSO_4.5H_2O$; mol wt 249.68. Occurs in nature as the mineral *chalcanthite*. Large, blue or ultramarine, triclinic crystals or blue granules or light-blue powder. Slowly efflorescent in air. Loses $2H_2O$ at 30°, 2 more H_2O at 110°; becomes anhydr by 250°. $d_4^{15.6}$ 2.286. Very sol in water; sol in methanol, glycerol; slightly sol in ethanol. pH of 0.2 molar aq soln 4.0. LD_{50} orally in rats: 960 mg/kg (Smyth).

Caution: Potential symptoms of overexposure are respiratory irritation and vineyard sprayer's lung disease.

USE: Anhydr salt for detecting and removing trace amounts of water from alcohols and other organic compds; as fungicide. Pentahydrate as agricultural fungicide, algicide, bactericide, herbicide, food and fertilizer additive; in insecticide mixtures; in manuf of other Cu salts; as mordant in textile dyeing; in prepn of azo dyes; in preserving hides; in tanning leather; in preserving wood; in electroplating soln; as battery electrolyte; in laundry and metal-marking inks; in petroleum refining; as flotation agent; pigment in paints, varnishes and other materials; in mordant baths for intensifying photographic negatives; in pyrotechnic compositions; in water-resistant adhesives for wood; in metal coloring and tinting baths; in antirusting compositions for radiator and heating systems; as reagent toner in photography and photoengraving; etc.

THERAP CAT: Antidote to phosphorus; antifungal (topical); copper supplement.

THERAP CAT (VET): Nutritional factor. Used in copper deficiency of ruminants. Has also been used as an anthelmintic, emetic, fungicide.

2644. Cupric Sulfate, Basic. [1332-14-5] Sulfuric acid copper(2+) salt, basic; basic cupric sulfate; copper hydroxide sulfate; cupric subsulfate; Basi-Cop; Cuproxat. Salts of variable compositions of cupric sulfate and cupric hydroxide or oxide. Occurs in nature as the mineral *dolerophane*, $CuSO_4.CuO$. *Reviews:* Frear, *Chemistry of the Pesticides* (Van Nostrand, New York, 3rd ed., 1955) pp 316-323; *idem, Agricultural Chemistry* vol. 2 (Van Nostrand, New York, 1951) pp 524-525; *Gmelins, Copper* (8th ed.) **60B**, 579-589 (1958). *See also* Burgundy Mixture.

Copper sulfate tribasic. [1333-22-8] Copper hydroxide sulfate ($Cu_4(OH)_6(SO_4)$); copper oxysulfate. $Cu_4H_6O_{10}S$; mol wt 452.28. Occurs in nature as the mineral *brochantite*; monohydrate as the mineral *langite*. Initial product formed during preparation of Bordeaux mixture, *q.v.*, from $CuSO_4$ and $Ca(OH)_2$: Frear, *loc. cit.* Also prepd from $CuSO_4$ and $Cu(OH)_2$: Denk, Leschhorn, *Z. Anorg. Allg. Chem.* **336**, 58 (1965). Very finely divided, light blue, gelatinous particles. Practically insol in water. Sol in plant and mineral acids, NH_4OH.

Copper sulfate dibasic. [12013-15-9] Copper hydroxide sulfate ($Cu_3(OH)_4(SO_4)$). $Cu_3H_4O_8S$; mol wt 354.72. Occurs in nature as the mineral *antlerite*. Blue-green, rhombic, bipyramidal crystals. Practically insol in water.

USE: Fungicide for plants, seed treatment.

2645. Cupric Sulfide. [1317-40-4] Copper sulfide (CuS); C.I. 77450; C.I. Pigment Blue 34; copper monosulfide. CuS; mol wt 95.61. Cu 66.46%, S 33.53%. Occurs in nature as the mineral *covellite*, or *indigo copper* (blue, hexagonal or monoclinic crystals). Prepn: Glemser, Sauer in *Handbook of Preparative Inorganic Chemistry* vol. 2, G. Brauer, Ed. (Academic Press, New York, 2nd ed., 1965) p 1017.

Black powder. Stable in air when dry, oxidized to $CuSO_4$ by moist air. Practically insol in water, alcohol, dil acids, alkalies. Sol in KCN soln, NH_4OH, hot HNO_3.

USE: In antifouling paints; in prepn of mixed catalysts; in development of aniline black dye in textile printing.

2646. Cupric Tartrate. [815-82-7] (2R,3R)-2,3-Dihydroxybutanedioic acid copper (2+) salt (1:1). $C_4H_4CuO_6$; mol wt 211.62. C 22.70%, H 1.91%, Cu 30.03%, O 45.36%. Prepd by reaction of K tartrate with $Cu(NO_3)_2$: Kirschner, Kiesling, *J. Am. Chem. Soc.* **82**, 4174 (1960).

Trihydrate. Green to blue, odorless powder. Slightly sol in water; sol in acids, alkali solns.

USE: In baths for Cu electroplating.

2647. Cupric Tungstate(VI). [13587-35-4] Copper tungsten oxide ($CuWO_4$); tungstic acid (H_2WO_4) copper(2+) salt (1:1); copper tungstate; cupric wolframate. CuO_4W; mol wt 311.38. Cu 20.41%, O 20.55%, W 59.04%. $CuWO_4$. Prepn: Kosek *et al., Collect. Czech. Chem. Commun.* **24**, 2034 (1959).

Dihydrate. Light green powder. Becomes brown to greyish-yellow on heating, with loss of H_2O. Practically insol in water. Slightly sol in acetic acid; sol in NH_3, H_3PO_4; dec by mineral acids.

USE: In semiconductors, nuclear reactors; as catalyst for polyester formation.

2648. Cuprous Acetate. [598-54-9] Acetic acid copper(1+) salt (1:1); copper monoacetate. $C_2H_3CuO_2$; mol wt 122.59. C 19.60%, H 2.47%, Cu 51.84%, O 26.10%. CH_3COOCu. Obtained as a sublimate by heating cupric acetate *in vacuo* to temps above 220°: Angel, Harcourt, *J. Chem. Soc.* **81**, 1385 (1902); prepn from cuprous oxide and acetic acid-acetic anhydride: Shimizu, Weller, *J. Am. Chem. Soc.* **74**, 4469 (1952). Crystal structure: T. Ogura *et al., J. Am. Chem. Soc.* **95**, 949 (1973).

Transparent, leafy crystals. Volatilizes on heating; dec on strong heating. Rapidly hydrolyzed by water with the formation of yellow Cu_2O.

2649. Cuprous Bromide. [7787-70-4] Copper bromide (CuBr); copper monobromide. BrCu; mol wt 143.45. Br 55.70%, Cu 44.30%. CuBr. Prepn: Briggs, *J. Chem. Soc.* **127**, 496 (1925); Keller, Wycoff, *Inorg. Synth.* **2**, 3 (1946); Wagner, Wagner, *J. Chem. Phys.* **26**, 1597 (1957).

White powder or cubic crystals, (zinc blende structure); turns green to dark-blue on exposure to sunlight. mp 504°; bp 1345°; d_4^{25} 4.72. Slightly sol in cold water, dec in hot water, HNO_3. Sol in HCl, HBr, NH_4OH with formation of complexes. Practically insol in acetone, concd H_2SO_4. *Keep tightly closed in a dark place.*

USE: As catalyst for organic reactions.

2650. Cuprous Chloride. [7758-89-6] Copper chloride (CuCl); copper(I) chloride; Cu-Iyt. ClCu; mol wt 99.00. Cl 35.81%, Cu 64.19%. CuCl. Occurs in nature as the mineral *nantokite* (colorless to grey cubic crystals). Prepn: Keller, Wycoff, *Inorg. Synth.* **2**, 1 (1946); Glemser, Sauer in *Handbook of Preparative Inorganic Chemistry* vol. 2, G. Brauer, Ed. (Academic Press, New York, 2nd ed., 1965) p 1005.

White cryst powder or cubic crystals (zinc-blende structure); stable to air and light if dry, but in presence of moisture turns green on exposure to air and blue to brown on exposure to light. mp 430°; d_4^{25} 4.14. Sparingly sol in water with partial decompn; practically insol in alcohol, acetone; sol in concd HCl, concd NH_4OH with formation of complexes. Solns oxidize rapidly in air. *Keep tightly closed and protected from light.*

USE: As catalyst for organic reactions; catalyst, decolorizer and desulfuring agent in petr industry; in denitration of cellulose; as condensing agent for soaps, fats and oils; in gas analysis to absorb carbon monoxide.

2651. Cuprous Cyanide. [544-92-3] Copper cyanide (Cu-(CN)); cupricin; copper monocyanide. CCuN; mol wt 89.56. C 13.41%, Cu 70.95%, N 15.64%. CuCN. Commercial prepns: Winter *et al., Kirk-Othmer Encyclopedia of Chemical Technology* vol. 6 (Interscience, New York, 2nd ed., 1965) p 271; laboratory prepns: Barber, *J. Chem. Soc.* **1943**, 79; Norberg, Jacobson, *Acta Chem. Scand.* **3**, 174 (1949); Vaughan, McCane, *J. Am. Chem. Soc.* **76**, 2504 (1954). Review of toxicology and human exposure: *Toxicological Profile for Cyanide* (PB2007-100674, 2006) 341 pp.

White to cream-colored powder, colorless or dark green orthorhombic crystals, or dark red monoclinic crystals. *Poisonous.* mp 474°. Sol in NH_4OH and in alkali cyanide solns because of formation of stable cyanocuprate(I) ions. Practically insol in water, alc, cold dil acids.

USE: In electroplating Cu or Fe; as insecticide, fungicide; as antifouling agent in marine paints; as polymerization catalyst.

2652. Cuprous Iodide. [7681-65-4] Copper iodide (CuI); copper monoiodide; Hydro-Giene. CuI; mol wt 190.45. Cu 33.37%, I 66.63%. Occurs in nature as the mineral *marshite* (red-brown crystals). Prepn: Kaufman, Pinnel, *Inorg. Synth.* **6**, 3 (1960); Glemser, Sauer in *Handbook of Preparative Inorganic Chemistry* vol 2, G. Brauer, Ed. (Academic Press, New York, 2nd ed., 1965) p 1007.

Dense powder or cubic crystals (zinc-blende structure). Less photosensitive than CuBr or CuCl. mp 588-606°; bp ~1290°; d_4^{25} 5.63. Extremely insol in water; practically insol in dil acids, alcohol; sol in aq solns of NH_3, alkali cyanides, thiosulfates, iodides; dec by concd H_2SO_4 and HNO_3.

USE: As catalyst in organic reactions; as ice-nucleating chemical; as coating in cathode-ray tubes; as source of iodine in animal feeds; with HgI_2 as temp indicator; bactericide.

2653. Cuprous Mercuric Iodide. [13876-85-2] (*T*-4)-Tetraiodomercurate(2−) copper(1+) (1:2); dicopper(1+) tetraiodomercurate(2−); cuprous tetraiodomercurate(2−); mercuric cuprous iodide. Cu_2HgI_4; mol wt 835.30. Cu 15.22%, Hg 24.01%, I 60.77%. Superionic, thermochromic material; exists in two temperature dependent forms; a structurally ordered low temperature β-form and the more random high temperature α-form. Prepn and structural studies: J. A. A. Ketelaar, *Z. Kristallogr.* **80**, 190 (1931). Prepn and phase diagram: M. Friesel *et al.*, *Phys. Scr.* **35**, 34 (1987); J. Nölting *et al.*, *Ber. Bunsen-Ges. Phys. Chem.* **93**, 1335 (1989). Optical properties: H.-R. C. Jaw *et al.*, *Inorg. Chem.* **26**, 1387 (1987). Crystal structure and phase transition studies: L. Eriksson *et al.*, *Z. Kristallogr.* **197**, 235 (1991). High temperature powder neutron diffraction studies: S. Hull, D. A. Keen, *J. Phys. Condens. Matter* **13**, 5597 (2001). Use in holography: J. M. Yang, D. W. Sweeney, *Appl. Opt.* **18**, 2398 (1979); in titrimetric applications: O. E. S. Godinho *et al.*, *Anal. Lett.* **17**, 135 (1984).

Deep-red, thermochromic cryst powder. Changes from scarlet to chocolate-brown at 67°. *Poisonous.* Practically insol in water, alcohol.

USE: For detecting overheating of machine bearings, etc. In pigments to prepare paints. Recording material for infrared holograms. Analytical reagent in titrimetric determn.

2654. Cuprous Oxide. [1317-39-1] Copper oxide (Cu_2O); dicopper monoxide; red copper oxide; C.I. 77402; yellow cuprocide. Cu_2O; mol wt 143.09. Cu 88.82%, O 11.18%. Occurs in nature as the mineral *cuprite* (red to reddish-brown octahedral or cubic crystals). Prepd commercially by furnace reduction of mixtures of copper oxides with Cu: Drapeau, Johnson, US 2758014; US 2891842 (1956, 1959 both to Glidden); by decompn of copper ammonium carbonate: Rowe, US 2474497; Klein, US 2474533 (both 1949 to Lake Chemical); Rowe, US 2536096; Munn, US 2670273 (1951, 1954 both to Mountain Copper); by treatment of $Cu(OH)_2$ with SO_2: Rowe, US 2665192 (1954 to Mountain Chemical); or by electrolysis of an aq soln of NaCl between Cu electrodes: Arend, *Paint Technol.* **13**, 265 (1948). Laboratory prepns: Glemser, Sauer in *Handbook of Preparative Inorganic Chemistry* vol. 2, G. Brauer, Ed. (Academic Press, New York, 2nd ed., 1965) p 1011. Toxicity study: H. F. Smyth *et al.*, *Am. Ind. Hyg. Assoc. J.* **30**, 470 (1969).

Cubic crystals or microcrystalline powder. Color may be yellow, red, or brown depending on the method of prepn and the particle size. Stable in dry air; gradually oxidizes in moist air to CuO. mp 1232°; d_4^{25} 6.0. Practically insol in water. Sol in NH_4OH; in HCl forming CuCl which dissolves in excess HCl. With dil H_2SO_4 or dil HNO_3, the cupric salt is formed and half the copper is pptd as the metal. LD_{50} orally in rats: 0.47 g/kg (Smyth).

USE: Fungicide; antiseptic for fishnets; in antifouling paints for marine use; in photoelectric cells; as red pigment for glass, ceramic glazes; in brazing pastes; in rectifiers; as catalyst.

2655. Cuprous Potassium Cyanide. [13682-73-0] Bis(cyano-κC)cuprate(1−) potassium (1:1); potassium dicyanocuprate(I); potassio-cuprous cyanide; potassium cyanocuprate(I). C_2CuKN_2; mol wt 154.68. C 15.53%, Cu 41.08%, K 25.28%, N 18.11%. KCu(CN)$_2$. Prepd by evaporation of an aq soln of CuCN and KCN: Staritzky, Walker, *Anal. Chem.* **28**, 419 (1956).

Monoclinic, prismatic crystals. d 2.38. Practically insol in water; dec to CuCN by cold dil H_2SO_4 or heating in water; sol in DMSO.

USE: In electroplating copper and brass.

2656. Cuprous Selenide. [20405-64-5] Copper selenide (Cu_2Se); dicopper monoselenide; berzeline; selenkupfer. Cu_2Se; mol wt 206.05. Cu 61.68%, Se 38.32%. Occurs in nature as the mineral *berzelianite* which in addition to copper and selenium contains 4.73-8.50% silver, 0.35-0.54% iron, up to 0.38% thallium and traces of lead and magnesium. Prepd by reduction of an ammoniacal soln of a cupric salt and H_2SeO_3 with hydrazine: Benzing *et al.*, *J. Am. Chem. Soc.* **80**, 2657 (1958); Kulifay, *J. Inorg. Nucl. Chem.* **25**, 75 (1963); alternate methods: Glemser, Sauer in *Handbook of Preparative Inorganic Chemistry* vol. 2, G. Brauer, Ed. (Academic Press, New York, 2nd ed., 1965) p 1019.

Blue-black to black, tetragonal or cubic crystals with a metallic luster. mp 1113°. d_4^{21} 6.84. Sol in HCl with evolution of H_2Se, in H_2SO_4 with evolution of SO_2, in KCN soln; oxidized by HNO_3 to $CuSeO_3$.

USE: In semiconductors.

2657. Cuprous Sulfide. [22205-45-4] Copper sulfide (Cu_2S); dicopper monosulfide. Cu_2S; mol wt 159.15. Cu 79.86%, S 20.14%. Occurs in nature as the mineral *chalcocite* also called *copper glance* (grey, black, green, blue, or violet rhombic crystals). Prepn: Glemser, Sauer in *Handbook of Preparative Inorganic Chemistry* vol. 2, G. Brauer, Ed. (Academic Press, New York, 2nd ed., 1965) p 1016.

Blue to grayish-black lustrous powder, granules, or orthorhombic crystals. mp ~1100°. d_4^{20} 5.6. On heating in the absence of air it forms Cu and CuS; in the presence of air it forms CuO, $CuSO_4$ and SO_2. Practically insol in water, acetic acid; very sparingly sol in HCl; dec by HNO_3, concd H_2SO_4. Partially sol in NH_4OH, readily in cyanide solns due to complex ion formation.

USE: In luminous paints; in electrodes for thermoelements; in preparation of $CuSO_4$; in solid-lubricant compositions; as catalyst.

2658. Cuprous Sulfite. [35788-00-2] Sulfurous acid dicopper(1+) salt. Cu_2O_3S; mol wt 207.15. Cu 61.35%, O 23.17%, S 15.48%. Cu_2SO_3. Prepn and review: Dasent, Morrison, *J. Inorg. Nucl. Chem.* **26**, 1122 (1964).

Hemihydrate. Etard's salt. White to pale-yellow hexagonal crystals. Slightly sol in water; sol in HCl, NH_4OH, alkali solns. Practically insol in ether, alcohol.

Cupro-cupric sulfate. Chevreul's salt. $Cu_3O_6S_2$; mol wt 350.75. Dihydrate, red microcryst powder or prismatic crystals. Practically insol in water, alcohol. Sol in NH_4OH, HCl.

Rogojski's salt. Formerly considered to be $Cu_2SO_3.H_2O$; shown to be an equimolar mixture of Chevreul's salt and metallic copper (Dasent, Morrison, *loc. cit.*). Brick-red solid.

USE: As fungicide for grape vines; in dyeing polyacrylic fibers; polymerization catalyst. Chevreul's salt is a selective molluscicide.

2659. Cuprous Thiocyanate. [1111-67-7] Thiocyanic acid copper(1+) salt (1:1); cuprous sulfocyanate; copper monothiocyanate. CCuNS; mol wt 121.62. C 9.88%, Cu 52.25%, N 11.52%, S 26.36%. CuSCN. Prepn: Demmerle *et al.*, *Ind. Eng. Chem.* **42**, 2 (1950); Newman, *Analyst* **88**, 500 (1963).

White to yellow amorphous powder. d 2.85. Practically insol in water, dil acids, alc, acetone; sol in NH_4OH, ether, solns of alkali thiocyanates; dec by concd mineral acids.

USE: In marine antifouling paints; in primer compositions for explosives industry.

2660. Cuproxoline. [13007-93-7] Tetrahydrogen bis[8-hydroxy-5,7-quinolinedisulfonato(3−)-$\kappa N^1,\kappa O^8$]cuprate(4−) compd with *N*-ethylethanamine (1:4); 8-hydroxy-5,7-quinolinedisulfonic acid copper derivative compound with diethylamine; cupric bis[8-hydroxyquinoline di(diethylammonium sulfonate)]; copper DOS; Dicuprene; Cujec; Cuprimyl. $C_{34}H_{56}CuN_6O_{14}S_4$; mol wt 964.64. C 42.33%, H 5.85%, Cu 6.59%, N 8.71%, O 23.22%, S 13.29%. Clinical efficacy in treatment of arthritis: W. C. Kuzell *et al.*, *Ann. Rheum. Dis.* **10**, 328 (1951). Veterinary use as parenteral copper supplement: I. J. Cunningham, *N. Z. Vet. J.* **7**, 15 (1959); N. F. Suttle, *Vet. Rec.* **109**, 304 (1981); G. J. Judson *et al.*, *Aust. Vet. J.* **61**, 40 (1984). Review of use of organic copper complexes in treatment of rheumatoid and degenerative diseases: J. R. J. Sorenson, W. Hangarter, *Inflammation* **2**, 217-238 (1977). Review of veterinary use: W. M. Allen, C. B. Mallinson, *Vet. Rec.* **114**, 451-454 (1984).

Consult the Name Index before using this section.

Green platelets. Sol in water, giving a dark-green soln. A 10% aq soln is almost neutral and can be sterilized by autoclaving. LD_{50} i.m. in rats: 126 mg/kg (Kuzell).

THERAP CAT: Antirheumatic.

THERAP CAT (VET): Copper supplement.

2661. Curare. [8063-06-7] Ourari; urari; woorari; woorali; wourara. General term given to the botanical preparations used as arrow poisons by indigenous people of tropical South America. Most frequently derived from the bark of *Chondodendron tomentosum* R. & P., *Menispermaceae* or *Strychos toxifera*, *Loganiaceae*, and closely related species. Acts as a nondepolarizing neuromuscular blocker. The primary physiologically active alkaloid is (+)-tubocurarine chloride, *q.v.* Three kinds of curare have been described and are distinguished by the kind of containers in which they were packed: **Tube curare** or **bamboo curare**, **gourd curare** or **calabash curare**, and **pot curare**. Listing of botanical sources from the family *Menispermaceae* and constituents of tube and pot curares: Krukoff, Moldenke, *Studies of American Menispermaceae, with special reference to species used in the preparation of arrow poisons* in *Brittonia* **3**, 1-74 (1938), also suppl. no. 1-5. Listing of *Strychnos* spp. and constituents of calabash and pot curares: Krukoff, Monachino, *The American Species of Strychnos* in *Brittonia* **4**, 248-322 (1942), also suppl. no. 1-6. Alkaloids from calabash curare and bark of *Strychnos* spp.: P. Karrer, *J. Pharm. Pharmacol.* **8**, 161-164 (1956); Schmid, Karrer, *Helv. Chim. Acta* **29**, 1853 (1946); **30**, 1162 (1947); Marino-Bettolo, *Festschrift Arthur Stoll* (Birkhäuser-Verlag, Basel 1957) pp 257-280. Review of history, chemistry, and use of arrow-poison curare: McIntyre, *Curare* (Chicago, 1947); Bovet *et al.*, *Curare and Curare-like Agents* (Van Nostrand, Princeton, 1959).

Curare is sol in water and in dil alcohol.

2662. *C*-Curarine I. [7168-64-1] $[C_{40}H_{44}N_4O]^{2+}$. Isoln from calabash-curare: Wieland, Pistor, *Ann.* **536**, 68 (1938); Karrer, Schmid, *Helv. Chim. Acta* **29**, 1853 (1946); Zürcher *et al.*, *J. Am. Chem. Soc.* **80**, 1500 (1958). Structure: Nagyvàry *et al.*, *Tetrahedron* **14**, 138 (1961); Grdinic *et al.*, *J. Am. Chem. Soc.* **86**, 3357 (1964); Jones, Nowacki, *Chem. Commun.* **1972**, 805. Prepn of *C*-dihydrotoxiferine chloride: Bernauer *et al.*, *Helv. Chim. Acta* **40**, 1999 (1957).

Dichloride. $C_{40}H_{44}Cl_2N_4O$. Needles from methanol + ether, mp >350°. $[\alpha]_D^{20}$ +73.6° (c = 1 in water). uv max (95% alc): 260, 296

nm (log ε 4.41, 4.07). Sol in water, alc; practically insol in ether, acetone.

2663. Curcumin. [458-37-7] (1*E*,6*E*)-1,7-Bis(4-hydroxy-3-methoxyphenyl)-1,6-heptadiene-3,5-dione; turmeric yellow; diferuloylmethane; C.I. 75300; C.I. Natural Yellow 3. $C_{21}H_{20}O_6$; mol wt 368.39. C 68.47%, H 5.47%, O 26.06%. Natural dyestuff from root of *Curcuma longa* L., *Zingiberaceae*. Isoln: Vogel, *Ann.* **44**, 297 (1842); Perkin, Phipps, *J. Chem. Soc., Trans.* **85**, I, 64 (1904); Rao, Shintre, *J. Soc. Chem. Ind.* **47**, 54T (1928). Synthesis: Lampe, *Ber.* **51**, 1347 (1918). Production: Stieglitz, Horn, DE 859145 (1952 to Hoechst). Biosynthesis studies: Roughley, Whiting, *Tetrahedron Lett.* **1971**, 3741. Chromatography: Srinivasan, *J. Pharm. Pharmacol.* **5**, 448 (1953). *See also:* H. J. Conn's *Biological Stains*, R. D. Lillie, Ed. (Williams & Wilkins, Baltimore, 9th ed., 1977) pp 474-476. Pharmacology and anti-inflammatory activity: Srimal, Dhawan, *ibid.* **25**, 447 (1973).

Orange-yellow, cryst powder, mp 183°. Insol in water, ether. Sol in alcohol, glacial acetic acid. Gives a brownish-red color with alkali; a light-yellow color with acids.

USE: For preparing curcuma paper, pH range 8-9. In the detection of boron.

2664. Curine. [436-05-5] (13a*R*,25a*R*)-2,3,13a,14,15,16,25,-25a-Octahydro-18,29-dimethoxy-1,14-dimethyl-13*H*-4,6:21,24-dietheno-8,12-metheno-1*H*-pyrido[3′,2′:14,15][1,11]dioxacycloeicosino[2,3,4-*ij*]isoquinoline-9,19-diol; (1β)-6,6′-dimethoxy-2,2′-dimethyltubocuraran-7′,12′-diol; *l*-bebeerine. $C_{36}H_{38}N_2O_6$; mol wt 594.71. C 72.71%, H 6.44%, N 4.71%, O 16.14%. From tubocurare *(Chondodendron tomentosum* R. & P., *Menispermaceae)*: Boehm, *Arch. Pharm.* **235**, 660 (1897); Späth *et al.*, *Ber.* **61**, 1698 (1928). Structure: Späth, Kuffner, *Ber.* **67**, 55 (1934); Faltis *et al.*, *Ber.* **69**, 1269 (1936); King, *J. Chem. Soc.* **1939**, 1157. Configuration: Bick, Clezy, *ibid.* **1953**, 3893; Hultin, *Acta Chem. Scand.* **17**, 753 (1963). Biosynthetic study: D. S. Bhakuni *et al.*, *Tetrahedron* **43**, 3975 (1987). Ca^{2+} channel interactions: J. P. Felix *et al.*, *Biochemistry* **31**, 11793 (1992).

Efflorescent crystals from methanol. mp 213°; mp 221° *in vacuo*; four sided prisms from benzene containing one mol benzene, mp 161°. $[\alpha]_D^{20}$ −328° (pyridine). Sol in benzene, chloroform, pyridine.

2665. Curium. [7440-51-9] Cm; at. no. 96; valence 3, 4. Man-made radioactive element. No stable nuclides; known isotopes (mass numbers): 238-251. Longest-lived known isotope: 247 ($T_{1/2}$ 1.56 × 10⁷ years, rel. at. mass 247.0703, α-decay). Prepn of first isotope, ^{242}Cm ($T_{1/2}$ 162.9 days, α-decay), in 1944 by G. T. Seaborg *et al.* in *The Transuranium Elements*, G. T. Seaborg *et al.*, Eds. (McGraw-Hill, New York, 1949) pp 1554-1571. Prepn of metal: J. C. Wolfmann *et al.*, *J. Am. Chem. Soc.* **73**, 493 (1951). Properties: Cunningham, Wallman, *J. Inorg. Nucl. Chem.* **26**, 271 (1964); Smith *et al.*, *J. Chem. Phys.* **50**, 5066 (1969). *Reviews:* M. Haissinsky, J.-P. Adloff, *Radiochemical Survey of the Elements* (Elsevier, New York, 1965) pp 41-43; C. Keller, *The Chemistry of the Transuranium Elements* (Verlag Chemie, Weinheim, English Ed., 1971) pp 529-551; *Comprehensive Inorganic Chemistry* vol 5, J. C. Bailar, Jr. *et al.*, Eds. (Pergamon Press, Oxford, 1973) *passim; Handb. Exp. Pharmakol.* **36**, 689-928 (1973); P. G. Eller, R. A. Penneman in *The*

Chemistry of the Actinide Elements vol. 2, J. J. Katz et al., Eds. (Chapman and Hall, New York, 1986) pp 962-988.

Silvery, lusterous, malleable metal. Two allotropic modifications: double hexagonal close-packed α-form, d 13.51, transforms to β-form at ~1277°; face-centered cubic β-form, d 12.9, exists from ~1277° to mp. mp 1345°. bp (calc) 3110°. Metal dissolves rapidly in dilute acid solns; oxidizes rapidly in the presence of traces of oxygen. Pyrophoric when finely divided. Chemistry of trivalent state similar to that of trivalent lanthanides.

Caution: Radiation hazard; handling requires special equipment and shielding facilities (Katz et al., loc. cit. vol. 2, p 1128).

USE: ^{242}Cm and ^{244}Cm as power sources in radionuclide batteries for space and medical applications. ^{242}Cm as radioactive heat source. ^{248}Cm in accelerator studies to form superheavy elements.

2666. Curvularin. [10140-70-2] (4S)-4,5,6,7,8,9-Hexahydro-11,13-dihydroxy-4-methyl-2H-3-benzoxacyclododecin-2,10-(1H)-dione. $C_{16}H_{20}O_5$; mol wt 292.33. C 65.74%, H 6.90%, O 27.36%. Mold metabolite from a species of Convalaria: Musgrave, J. Chem. Soc. 1956, 4301; from Penicillium steckii: Fennell et al., Chem. Ind. (London) 1959, 1382. Structure: Birch et al., J. Chem. Soc. 1959, 3146. Synthesis: H. Gerlach, Helv. Chim. Acta 60, 3039 (1977); of dl-O,O-dimethyl ether: Baker et al., J. Chem. Soc. C 1967, 1913; T. Takahashi et al., Tetrahedron Lett. 21, 3885 (1980); H. H. Wasserman, R. J. Gambale, ibid. 22, 4849 (1981).

Plates from hot benzene + methanol, mp 206-206.5°. $[\alpha]_D^{18}$ −36.3° (c = 3.8 in ethanol). uv max (ethanol): 223, 272, 304.5 nm (ε 11300; 6350; 5100). Sol in ethanol, methanol, dioxane, acetone, pyridine, concd H_2SO_4; moderately sol in acetic acid, ether; sparingly sol in benzene, petr ether, chloroform, water. Gives yellow solns with aq ammonia, sodium carbonate, and sodium hydroxide. Alkaline solns darken rapidly in air, eventually becoming purple.

Dibenzoate. $C_{30}H_{28}O_7$. Needles from benzene + petr ether, mp 133-134°. $[\alpha]_D^{18}$ −10.8° (c = 1.9 in chloroform). uv max (ethanol): 236 nm (ε 36700).

O,O-Dimethyl ether. $C_{18}H_{24}O_5$. Short rods from aq ethanol, mp 72°. $[\alpha]_D^{18}$ −2.9° (c = 2.7 in chloroform). uv max (ethanol): 223, 267.5 nm (ε 10500; 5100).

2667. Cuscohygrine. [454-14-8] rel-1-[(2R)-1-Methyl-2-pyrrolidinyl]-3-[(2S)-1-methyl-2-pyrrolidinyl]-2-propanone; (R*,S*)-1,3-bis(1-methyl-2-pyrrolidinyl)-2-propanone; meso-cuscohygrine; cuskhygrine; bellaradine. $C_{13}H_{24}N_2O$; mol wt 224.35. C 69.60%, H 10.78%, N 12.49%, O 7.13%. In coca leaves of various origin. Found in crude hygrine, q.v. Readily converted to hygrine by acids and bases. Isoln: Liebermann, Ber. 22, 679 (1898); Liebermann, Cybulski, ibid. 28, 578 (1895). Identity with bellaradine: Steinegger, Phokas, Pharm. Acta Helv. 30, 441 (1955). Structure: Hess, Fink, Ber. 53, 794 (1920); Sohl, Shriner, J. Am. Chem. Soc. 55, 3828 (1933); Rapoport, Jorgensen, J. Org. Chem. 14, 664 (1949). Synthesis: Späth, Tuppy, Monatsh. Chem. 79, 119 (1948); Galinovsky et al., ibid. 82, 551 (1951). Enzymatic synthesis: Tuppy, Faltaous, ibid. 91, 167 (1960). Stereochemistry: Galinovsky, Zuber, ibid. 84, 798 (1953). Biosynthesis: E. Leete, Chem. Commun. 1980, 1170.

Relative stereochemistry

Oily liquid, bp$_{23}$ 169-170°; bp$_{14}$ 152°; bp$_2$ 118-125°. d$_4^{20}$ 0.9733. n$_D^{20}$ 1.4832. Miscible with water. Sol in alcohol, ether, benzene.
Hemiheptahydrate. Needles, mp 40°.
Hydrobromide. $C_{13}H_{24}N_2O.2HBr$. Prisms from alc, mp 234°.

2668. Cy 3. [146397-20-8] 1-[6-[(2,5-Dioxo-1-pyrrolidinyl)oxy]-6-oxohexyl]-2-[3-[1-[6-[(2,5-dioxo-1-pyrrolidinyl)oxy]-6-oxohexyl]-1,3-dihydro-3,3-dimethyl-5-sulfo-2H-indol-2-ylidene]-1-propen-1-yl]-3,3-dimethyl-5-sulfo-3H-indolium inner salt. $C_{43}H_{50}N_4O_{14}S_2$; mol wt 911.01. C 56.69%, H 5.53%, N 6.15%, O 24.59%, S 7.04%. One of a family of indocyanine dyes with high fluorescence. Conjugates easily to proteins and oligonucleotides for labelling. Prepn: A. S. Waggoner et al., DE 3912046; eidem, US 5268486 (1990, 1993 both to Carnegie-Mellon). Use with fiber optic biosensors: L. A. Tempelman et al., Proc. SPIE Int. Soc. Opt. Eng. 2293, 139 (1994). Use with neuropeptides for detection of receptors: N. W. Bunnett et al., Histochem. J. 28, 811 (1996); for detection of synaptic vesicle mobility: K. Kraszewski et al., J. Neurosci. 16, 5905 (1996).

Potassium salt. Water soluble. Absorbance max: 552 nm; emission max: 565 nm; extinction coefficient >130,000 $M^{-1}cm^{-1}$.
USE: Fluorescent label.

2669. Cy 5. [146368-15-2] 1-[6-[-(2,5-Dioxo-1-pyrrolidinyl)oxy]-6-oxohexyl]-2-[5-[1-[6-[(2,5-dioxo-1-pyrrolidinyl)oxy]-6-oxohexyl]-1,3-dihydro-3,3-dimethyl-5-sulfo-2H-indol-2-ylidene]-1,3-pentadien-1-yl]-3,3-dimethyl-5-sulfo-3H-indolium inner salt. $C_{45}H_{52}N_4O_{14}S_2$; mol wt 937.05. C 57.68%, H 5.59%, N 5.98%, O 23.90%, S 6.84%. Indocyanine dye with high fluorescence which is easily coupled to proteins. Named CyX.Y where X is the number of bridge carbons and Y represents a unique dye structure. Prepn: A. S. Waggoner et al., DE 3912046; eidem, US 5268486 (1990, 1993 both to Carnegie-Mellon). Use with fiber optic biosensors: L. A. Tempelman et al., Proc. SPIE Int. Soc. Opt. Eng. 2293, 139 (1994); for antibody/antigen detection: R. M. Wadkins et al., ibid. 2676, 148 (1996). Use in resonance energy transfer: P. R. Selvin et al, J. Am. Chem. Soc. 116, 6029 (1994).

Water soluble.
Potassium salt. Water soluble. Absorbance max 650 nm; emission max 667 nm (ε >200,000). Quantum yield 0.28.
Cy 5.5. [172777-84-3] Water soluble. Absorbance max 678 nm; emission max 703 nm (ε 250,000). Quantum yield 0.28.
USE: Fluorescent label.

2670. Cyacetacide. [140-87-4] 2-Cyanoacetic acid hydrazide; malononitrile hydrazide; cyanoacetohydrazide; cyanazide; cyanizide; cyanacethydrazine; cyanacethydrazide; cyanoethydrazide; cyanacetylhydrazide; Dictyzide; Mackreazid; Armazal; Reacid; Reazide; Hidacian; Leandin; Neohydrazid; Dictycide. $C_3H_5N_3O$; mol wt 99.09. C 36.36%, H 5.09%, N 42.41%, O 16.15%. Prepd by boiling ethyl or methyl cyanoacetate with hydrazine hydrate in alc: Rothenburg, Ber. 27, 687 (1894); Darapsky, Hillers, J. Prakt. Chem. 92[2], 313 (1915); Klosa, Arch. Pharm. 288, 453 (1955). Prepn of dosage forms: Muset, US 2849369 (1958 to Labs OM).

Stout prisms from alc, mp 114.5-115° (evac tube). Rapidly dec by heat. Very slightly acid to litmus. Very sol in water. Sol in alcohol. Practically insol in ether. Aq solns become discolored after a few days and the pH becomes alkaline.

Hydrochloride. [5897-13-2] $C_3H_5N_3O.HCl$; mol wt 135.55. Crystals, mp 145°. Freely sol in water. Acid reaction. Aq solns are more stable than those of the free hydrazide.

THERAP CAT: Antibacterial (tuberculostatic).

THERAP CAT (VET): Anthelmintic (lungworms).

2671. Cyadox. [65884-46-0] 2-Cyanoacetic acid-2-[(1,4-dioxido-2-quinoxalinyl)methylene]hydrazide; 2-formylquinoxaline-1,4-dioxide cyanoacetylhydrazone. $C_{12}H_9N_5O_3$; mol wt 271.24. C 53.14%, H 3.34%, N 25.82%, O 17.70%. Antimicrobial growth promoting feed additive for poultry, calves and pigs. Growth promoting effects in chickens: J. Broz, B. Sevcik, *Biol. Chem. Vyz. Zvirat* **13**, 357, 375 (1977). Prepn: J. Hebky *et al.*, **DE 2829580**; *eidem*, **US 4225604** (1979, 1980 both to Spofa). Adsorptive voltammetry determn in plasma: I. Sestáková, M. Kopanica, *Talanta* **35**, 816 (1988). LC determn in animal feeds and gastrointestinal samples: G. J. De Graaf, T. J. Spierenburg, *J. Chromatogr.* **447**, 244 (1988); in goat tissue: Y. Zhang *et al.*, *Anal. Sci.* **21**, 1495 (2005). Heat stability study: *idem et al.*, *J. Agric. Food Chem.* **53**, 9737 (2005). Subchronic toxicity study: G. Fang *et al.*, *Food Chem. Toxicol.* **44**, 36 (2006).

Yellow solid, mp 255-260°. *Light sensitive.*

THERAP CAT (VET): Growth promotant.

2672. Cyamelide. [462-02-8] 1,3,5-Trioxane-2,4,6-triimine; insoluble cyanuric acid. $C_3H_3N_3O_3$; mol wt 129.08. C 27.92%, H 2.34%, N 32.55%, O 37.18%. Formed by polymerization of cyanic acid (gaseous or liquid). Prepd by digesting equal parts of potassium cyanate and oxalic acid: Liebig, Wöhler, *Berzelius Jahresber.* **11**, 85. Crude cyamelide contains cyanuric acid which is removed by washing with water: Senier, Walsh, *J. Chem. Soc.* **81**, 290 (1902). Mechanism of polymerization: Werner, Fearon, *ibid.* **117**, 1358 (1920).

White amorphous powder. Practically insol in water. Insol in organic solvents. Slightly sol in ammonia water, in concd acids, in concd NaOH with salt formation.

2673. Cyamemazine. [3546-03-0] 10-[3-(Dimethylamino)-2-methylpropyl]-10*H*-phenothiazine-2-carbonitrile; 2-cyano-10-(3-dimethylamino-2-methylpropyl)phenothiazine; 10-(3-dimethylamino-2-methylpropyl)-3-cyanophenothiazine; cyamepromazine; RP-7204; TH-2602; Kyamepromazine; Cianatil; Tercian. $C_{19}H_{21}N_3S$; mol wt 323.46. C 70.55%, H 6.54%, N 12.99%, S 9.91%. Prepn: Jacob, Robert, **US 2877224** (1959 to Rhône-Poulenc). Synthesis: Craig *et al.*, *J. Org. Chem.* **26**, 1138 (1961). Biotransformation studies: Robinson, *J. Pharm. Pharmacol.* **18**, 19 (1966). Chemistry: Kiger, Kiger, *Ann. Pharm. Fr.* **23**, 489 (1965).

Yellow oil, $bp_{0.2-0.5}$ 205-220°; also reported as yellow powder, mp 89-96° (Kiger). Practically insol in water. Sol in ethanol and in organic solvents.

Maleate. $C_{23}H_{25}N_3O_4S$. Pale yellow plates from methanol-ethanol, mp 196-197° (Craig).

THERAP CAT: Antipsychotic.

2674. Cyanamide. [420-04-2] Carbodiimide; hydrogen cyanamide; carbimide; cyanogenamide; amidocyanogen. CH_2N_2; mol wt 42.04. C 28.57%, H 4.80%, N 66.64%. $H_2NC{\equiv}N$. Prepd commercially by continuous carbonation of calcium cyanamide in water. Ref. "Cyanamide," Process Chemicals Dept., Am. Cyanamid (New York, 1959) p 19. Acute toxicity data: J. Doull *et al.*, *Survey of Compounds for Radiation Protection* (AD277689, USAF Radiation Lab., 1962) 124 p.

Deliquescent, orthorhombic, elongated, six-sided tablets from dimethyl phthalate. d_4^{20} 1.282. mp 45-46°. $bp_{0.5}$ 83°. Formn of dicyandiamide begins at 122°. Dipole moment in benzene at 20°: 3.8. Cryoscopic constant (water) 26.8-28.4. Sp ht 0.547 cal/g/°C between 0° and 39°. Heat of formation 14.05 kcal/mole (25°); heat of combustion −176.4 kcal/mole (25°); heat of fusion 2.1 kcal/mole. Heat of vaporization 16.4 kcal/mole. Soly (g/100 g soln) in water at 15°: 77.5, at 43°: 100; in butanol at 20°: 28.8, in methyl ethyl ketone: 50.5, in ethyl acetate: 42.4. Sol in alcohols, phenols, amines, ethers, ketones. Very sparingly sol in benzene, halogenated hydrocarbons. Practically insol in cyclohexane. Solid cyanamide should be stored in a cool, dry place. Polymerizes at 122°. Optimum pH for storage of solns is ~4. Attacks various metals. Solns can be stored in glass provided they are stabilized with phosphoric, acetic, sulfuric, or boric acid. LD_{50} i.p. in male mice: 200-300 mg/kg (Doull).

Note: Term "cyanamide" is also used to designate calcium cyanamide.

Caution: Potential symptoms of overexposure are irritation of eyes, skin, respiratory system; eye, skin burns; miosis, salivation, lacrimation, twitching; Antabuse-like effects. *See NIOSH Pocket Guide to Chemical Hazards* (DHHS/NIOSH 97-140, 1997) p 80.

2675. Cyanazine. [21725-46-2] 2-[[4-Chloro-6-(ethylamino)-1,3,5-triazin-2-yl]amino]-2-methylpropanenitrile; 2-[[4-chloro-6-(ethylamino)-*s*-triazin-2-yl]amino]-2-methylpropionitrile; SD-15418; WL-19805; DW-3418; Bladex; Fortrol. $C_9H_{13}ClN_6$; mol wt 240.70. C 44.91%, H 5.44%, Cl 14.73%, N 34.92%. Selective pre-emergence herbicide. Prepn: **GB 1132306**; W. Schwarze, **US 3505325** (1968, 1970 both to Degussa). Activity: T. Chapman *et al.*, *Proc. 9th Br. Weed Control Conf.* **2**, 1018 (1968). Metabolism: J. V. Crayford, D. H. Hutson, *Pestic. Biochem. Physiol.* **2**, 295 (1975). Persistence in soil: J. T. Majka, T. L. Lavy, *Weed Sci.* **25**, 401 (1977).

White crystals, mp 167.5-169°. Vapor pressure at 20°: 1.6×10^{-9} mmHg. Soly in water at 25°: 171 mg/l. Soly at 25° (g/l): benzene 15, chloroform 210, ethanol 45, hexane 15. LD_{50} in rats, mice (mg/kg): 182, 380 orally (Chapman).

USE: Herbicide.

2676. Cyanic Acid. [420-05-3] Hydrogen cyanate. CHNO; mol wt 43.03. C 27.91%, H 2.34%, N 32.55%, O 37.18%. Best prepd in the laboratory by dry distillation of cyanuric acid: Linhard, *Z. Anorg. Allg. Chem.* **236**, 200 (1938). *See also* Isocyanic Acid.

HO—C≡N

Liquid. *Acrid odor, strong lacrimator and vesicant.* d_4^{20} 1.140; d_4^{-20} 1.156. mp $-86°$. bp$_{760}$ 23.5°. Heat of formation -36.5 kcal/mol. Dipole moment 1.592. pK (20°) 3.66. Rapid heating of the liquid may result in an explosion. Polymerizes on standing forming cyamelide and cyanuric acid. Sol in water with decompn to carbon dioxide and ammonia. Dil solns in ice water may be kept for several hours. Dil solns in ether, benzene, toluene can be kept for several weeks.

Caution: Highly irritating to eyes, skin, and mucous membranes. See *Clinical Toxicology of Commercial Products*, R. E. Gosselin *et al.*, Eds. (Williams & Wilkins, Baltimore, 5th ed., 1984) Section II, p 110.

USE: In formation of some cyanates.

2677. Cyanidin Chloride. [528-58-5] 2-(3,4-Dihydroxy-phenyl)-3,5,7-trihydroxy-1-benzopyrylium chloride (1:1); 3,3′,4′,-5,7-pentahydroxyflavylium chloride; 3,3′,4′,5,7-pentahydroxy-2-phenylbenzopyrylium chloride. $C_{15}H_{11}ClO_6$; mol wt 322.70. C 55.83%, H 3.44%, Cl 10.99%, O 29.75%. Prepd by acid hydrolysis of cyanin chloride: Willstätter, Everest, *Ann.* **401**, 189 (1913). Isoln from bananas: Simmonds, *Nature* **173**, 402 (1954). Prepn by reduction of quercetin: Bauer *et al.*, *Chem. Ind. (London)* **1954**, 433; King, White, *J. Chem. Soc.* **1957**, 3901. Structure: Willstätter, Mallison, *Ann.* **408**, 147 (1915). Synthesis: Willstätter *et al.*, *Ber.* **57**, 1938 (1924); Robertson, Robinson, *J. Chem. Soc.* **1928**, 1528. Biosynthesis: Fritsch *et al.*, *Z. Naturforsch.* **26b**, 581 (1971). *See also* Bioflavonoids.

Metallic brownish-red needles (monohydrate) from dil alcoholic HCl. The anhydr compd does not melt below 300°. Absorption max (methanolic HCl): 535 nm. Freely sol in alcohol and in amyl alcohol giving a violet soln. Sol in sodium carbonate soln with blue color. Sparingly sol in dil HCl or H_2SO_4.

3-Glucoside. Chrysanthemin; asterin; kuromanin. $C_{21}H_{21}ClO_{11}$. From winter aster *(Chrysanthemum indicum* L., *Compositae):* Willstätter, Bolton, *Ann.* **412**, 136 (1917); from wild strawberries *(Fragaria vesca* L., *Rosaceae):* Sondheimer, Karash, *Nature* **178**, 648 (1956); from sweet cherries *(Prunus avium* L., *Rosaceae):* Li, Wagenknecht *ibid.* **182**, 657 (1958). Identity with asterin: Robinson, Willstätter, *Ber.* **61**, 2503 (1928). Structure and synthesis: Murakami *et al.*, *J. Chem. Soc.* **1931**, 2665. Reddish-brown plates or prisms with a metallic shine from dil alcoholic HCl, dec 205° without melting. Absorption max (methanolic HCl): 525 nm. Sol in alc with strong fluorescence and in sodium carbonate with violet-blue color.

3-Rhamnoglucoside. Keracyanin; antirrhinin; sambucin; Meralop; Meralops. $C_{27}H_{31}ClO_{15}$; mol wt 630.98. From skin of sweet cherries *(Prunus avium* L., *Rosaceae):* Willstätter, Zollinger, *Ann.* **412**, 164 (1917); from sour cherries *(Prunus cerasus* L., *Rosaceae):* Li, Wagenknecht, *J. Am. Chem. Soc.* **78**, 979 (1956). Structure: Robertson, Robinson, *J. Chem. Soc.* **1927**, 2196. Prepn by reduction of quercetin-3-rhamnoglucoside: Bauer *et al.*, *Chem. Ind. (London)* **1954**, 433. Reduces time to adjust to darkness: F. Trimarchi *et al.*, *Minerva Oftalmol.* **18**, 143 (1977). Mechanism of action: F. Trimarchi *et al.*, *Ann. Ottalmol. Clin. Ocul.* **105**, 111 (1979). Red needles from dil HCl or dark prisms from dil methanolic HCl. Absorption max (ethanolic HCl): 532, 333, 282 nm. Sol in hot water, alcohol.

3-Galactoside. Idaein; idein. $C_{21}H_{21}ClO_{11}$. From cranberries *(Vaccinium vitis idaea* L., *Ericaceae):* Willstätter, Mallison, *Ann.*

408, 15 (1915); from Winesap apple *(Pyrus malus* Linn., *Rosaceae):* Duncan, Dustman, *J. Am. Chem. Soc.* **58**, 1511 (1936). Structure and synthesis: Grove, Robinson, *J. Chem. Soc.* **1931**, 2722. Red needles with bronze metallic luster from dil ethanolic HCl, dec 210-212°. Sol in water, ethanol, methanol, dil HCl. Practically insol in 7% HCl.

3,5-Diglucoside. Cyanin; shisonin A. $C_{27}H_{31}ClO_{16}$. From cornflower *(Centaurea cyanus* L., *Compositae):* Willstätter, Everest, *loc. cit.* Structure: Léon, Robinson, *J. Chem. Soc.* **1932**, 221. Synthesis: Robinson, Todd, *ibid.* 2488. Plates with a metallic luster from dil alcoholic HCl, mp 203-204°. $[α]_D$ $-258°$ (in 0.05% HCl). Absorption max (methanolic HCl): 522 nm.

3-Sophoroside. Mecocyanin. $C_{27}H_{31}ClO_{16}$. From the flowers of *Papaver rhoeas* L., *Papaveraceae:* Willstätter, Weil, *Ann.* **412**, 231 (1917); from sour cherries: Li, Wagenknecht, *J. Am. Chem. Soc., loc. cit.* Structure: Harborne, *Experientia* **19**, 7 (1963); *Phytochemistry* **2**, 85 (1963). Alternate structure: Grove *et al.*, *J. Chem. Soc.* **1934**, 1608. Dark-red crystals from dil HCl + HOAc or dark-red needles from 2% alcoholic HCl. Absorption max (methanolic HCl): 523 nm. Sol in hot water. Pptd by glacial acetic acid or acetone.

THERAP CAT: 3-Rhamnoglucoside in treatment of night blindness.

2678. Cyanoacetamide. [107-91-5] 2-Cyanoacetamide; malonamide nitrile. $C_3H_4N_2O$; mol wt 84.08. C 42.86%, H 4.80%, N 33.32%, O 19.03%. Prepd by the action of aq or alcoholic ammonia on cyanoacetic ester: Hesse, *Am. Chem. J.* **18**, 724 (1896); Thole, Thorpe, *J. Chem. Soc.* **99**, 429 (1911); Ott, Löpmann, *Ber.* **55**, 1258 (1922); Corson *et al.*, *Org. Synth.* coll. vol. I (2nd ed., 1941) p 179.

Needles from alc, mp 119.5°. One gram dissolves in 6.5 ml of cold water. Soly in 100 ml of 95% alc: 1.3 g at 0°; 3.1 g at 26°; 7.0 g at 44°; 14.0 g at 62°; 21.5 g at 71°.

USE: In organic syntheses (*e.g.*, starting material for vitamin B_6 synthesis).

2679. Cyanoacetic Acid. [372-09-8] 2-Cyanoacetic acid; malonic mononitrile. $C_3H_3NO_2$; mol wt 85.06. C 42.36%, H 3.56%, N 16.47%, O 37.62%. Prepd from chloroacetic acid and NaCN: Ruggli, Businger, *Helv. Chim. Acta* **25**, 35 (1942); Lapworth, Baker, *Org. Synth.* coll. vol. I, 181 (1941); **GB 824640** (1959); Eaker, **US 2539238** (1951 to Monsanto).

Hygroscopic crystals, mp 66°; dec at 160° into CO_2 and acetonitrile. bp$_{15}$ 108°. Sol in water, alcohol, ether, slightly in benzene, chloroform. *Keep well closed.*

USE: Synthesis of intermediates; manuf barbital.

2680. 1-Cyanobenzotriazole. [15328-32-2] 1*H*-Benzotriazole-1-carbonitrile. $C_7H_4N_4$; mol wt 144.14. C 58.33%, H 2.80%, N 38.87%. Cyanating agent. Prepn: M. E. Hermes, F. D. Marsh, *J. Am. Chem. Soc.* **89**, 4760 (1967); T. V. Hughes *et al.*, *J. Org. Chem.* **63**, 401 (1998). Electronic structure: P. Rademacher *et al.*, *J. Mol. Struct.* **513**, 47 (1999). Use in cyanation: A. R. Katritzky *et al.*, *Rev. Roum. Chim.* **36**, 573 (1991); T. V. Hughes, M. P. Cava, *J. Org. Chem.* **64**, 313 (1999); corrected: *ibid.* 2599.

Colorless needles, mp 73-75°. uv max (ethanol): 253, 293 nm ($ε$ 6620, 3300).

USE: In chemical synthesis as a cyanating reagent.

2681.   **Cyanogen.** [460-19-5] Ethanedinitrile; dicyan; oxalic acid dinitrile. C_2N_2; mol wt 52.04. C 46.16%, N 53.83%. N≡C—C≡N. Prepn by adding an aq soln of sodium or potassium cyanide to an aq soln of copper(II) sulfate or chloride: Janz, *Inorg. Synth.* **5**, 43 (1957); from HCN by the use of CuO: Fierce, Millikan, **US 2841472** (1958 to Pure Oil); from HCN and NO_2: Fierce, Sander, *Ind. Eng. Chem.* **53**, 985 (1961). *Review:* Brotherton, Lynn, *Chem. Rev.* **59**, 841-883 (1959). Review of toxicology and human exposure: *Toxicological Profile for Cyanide* (PB2007-100674, 2006) 341 pp.

Gas with almond-like odor, acrid and pungent when in lethal concns. Burns with pink flame having a bluish border. mp $-27.9°$ (also reported as $-34.4°$). bp $-21.17°$. $d_4^{-21.17}$ 0.9537. Log P (octanol/water): 0.07. *Poisonous, flammable.* Heat of vaporization (liquid) 5.778 kcal/mole. Above 500° polymerizes to insol paracyanogen $(CN)_n$. One vol of water dissolves about 4 vols of cyanogen gas. Also sol in alc, ether. Slowly hydrolyzed in aq soln giving oxalic acid and ammonia.

Caution: Potential symptoms of overexposure are irritation of eyes, nose, upper respiratory system; lacrimation; cherry red lips; tachypnea, hyperpnea, bradycardia; headache, vertigo, convulsions; dizziness, loss of appetite, weight loss. Direct contact with liquid may cause frostbite. *See NIOSH Pocket Guide to Chemical Hazards* (DHHS/NIOSH 97-140, 1997) p 82.

2682.   **Cyanogen Azide.** [764-05-6] Carbon pernitride. CN_4; mol wt 68.04. C 17.65%, N 82.35%. Prepd by suspending NaN_3 in dry acetonitrile and distilling cyanogen chloride into the cooled suspension: Marsh, Hermes, *J. Am. Chem. Soc.* **86**, 4506 (1964); *see also Chem. Eng. News* **42**, 51 (Oct. 26, 1964); Marsh, **US 3410658** (1968); Marsh, *J. Org. Chem.* **37**, 2966 (1972). Spectrum, structure, dipole moment: Costain, Kroto, *Can. J. Phys.* **50**, 1453 (1972); Almenningen *et al.*, *Acta Chem. Scand.* **27**, 1531 (1973).

$$\bar{N}{\equiv}N{=}^+{N}{-}C{\equiv}N$$

Clear, colorless, oily liquid. *The pure azide detonates violently upon thermal, electrical or mechanical shock.* Can be handled relatively safely in solvents. Half-life of a 27% soln in acetonitrile is 15 days at room temp, more stable at lower temps.

USE: In organic synthesis. Has a versatility and scope of chemical reactivity that is very broad and useful, *e.g.* reacts with alkanes to give primary alkylcyanamides: Anastassiou *et al.*, *J. Am. Chem. Soc.* **87**, 2296 (1965).

2683.   **Cyanogen Bromide.** [506-68-3] Cyanogen bromide ((CN)Br); bromine cyanide; bromocyanogen; cyanobromide. CBrN; mol wt 105.92. C 11.34%, Br 75.44%, N 13.22%. BrCN. Prepd from potassium or sodium cyanide and bromine: Hartman, Dreger, *Org. Synth.* **11**, 30 (1931). Industrial prepn from sodium bromide, sodium chlorate, and sodium cyanide in 30% H_2SO_4: Ewan, *Chem. Zentralbl.* **1907**, I, 591; Göpner, *ibid.* **1908**, I, 1807; Grignard, Crouzier, *Bull. Soc. Chim. Fr.* [4] **29**, 214 (1921). Brief review of synthetic applications: V. Kumar, *Synlett* **2005**, 1638-1639.

Crystalline solid; dec in the presence of moisture. d_4^{20} 2.015. mp 52°. bp 61-62°. *Poisonous, corrosive.* Freely sol in water, alcohol, ether. Aq solns of alkalies dec it to alkali cyanide and alkali bromide.

Caution: Toxic effects similar to those of hydrogen cyanide, *q.v.*

USE: Reagent for the synthesis of cyanamides.

2684.   **Cyanogen Chloride.** [506-77-4] Cyanogen chloride ((CN)Cl); chlorine cyanide; chlorocyanide; chlorocyanogen. CClN; mol wt 61.47. C 19.54%, Cl 57.67%, N 22.79%. CNCl. Prepd by action of Cl on HCN. Review of toxicology and human exposure: *Toxicological Profile for Cyanide* (PB2007-100674, 2006) 341 pp.

Liquid. *Poisonous, corrosive.* d 1.186. bp 13.8°. mp $-6°$. Sol in water, alc, ether.

Caution: Potential symptoms of overexposure are irritation of eyes, upper respiratory system; cough, delayed pulmonary edema; weakness, headache, giddiness, dizziness, confusion, nausea, vomiting; irregular heart beat; direct contact with liquid may cause skin irritation. *See NIOSH Pocket Guide to Chemical Hazards* (DHHS/NIOSH 97-140, 1997) p 82.

USE: In chemical synthesis. Military poison gas.

2685.   **Cyanogen Iodide.** [506-78-5] Iodine cyanide (I(CN)); iodine monocyanide. CIN; mol wt 152.92. C 7.85%, I 82.99%, N 9.16%. CNI. Prepd by the action of iodine on sodium cyanide: Bak, Hillebert, *Org. Synth.* **coll. vol. IV**, 207 (1963).

White needles; very pungent odor; acrid taste. *Poisonous.* mp 146-147°. Sol in water, alcohol, ether, volatile oils.

Caution: Causes convulsions, paralysis, death from respiratory failure. Lacks inhalation hazard.

USE: Generally for destroying all lower forms of life. In taxidermy for preserving insects, butterflies, etc.

2686.   **1-Cyanoimidazole.** [36289-36-8] 1*H*-Imidazole-1-carbonitrile; *N*-cyanoimidazole. $C_4H_3N_3$; mol wt 93.09. C 51.61%, H 3.25%, N 45.14%. Coupling and cyanating agent. Prepn: H. Giesemann, *J. Prakt. Chem.* **4**(1), 345 (1955); by thermolysis of tetrazolopyrazine: C. Wentrup, *Helv. Chim. Acta* **55**, 565 (1972); P. B. W. McCallum *et al.*, *Aust. J. Chem.* **52**, 159 (1999). Use as a condensing agent: J. P. Ferris *et al.*, *Nucleosides Nucleotides* **8**, 407 (1989); K. J. Luebke, P. B. Dervan, *J. Am. Chem. Soc.* **113**, 7447 (1991); as a cyanating agent: Y. Wu *et al.*, *Org. Lett.* **2**, 795 (2000).

White hygroscopic needles, mp 59.5-60.5°. bp_{760} 185-189°; bp_{27} 95°.

USE: In chemical synthesis as a cyanating agent; as condensing agent in DNA ligation.

2687.   **Cyanophos.** [2636-26-2] Phosphorothioic acid *O*-(4-cyanophenyl) *O,O*-dimethyl ester; phosphorothioic acid *O,O*-dimethyl ester, *O*-ester with *p*-hydroxybenzonitrile; *O*-(4-cyanophenyl) *O,O*-dimethyl phosphorothioate; *O,O*-dimethyl *O*-(4-cyanophenyl) thionophosphate; ciafos; Bay 34727; S-4084; Cyanox. $C_9H_{10}NO_3PS$; mol wt 243.22. C 44.44%, H 4.14%, N 5.76%, O 19.73%, P 12.73%, S 13.18%. Organophosphate insecticide; cholinesterase inhibitor. Prepn: S. Kuramoto *et al.*, **US 3150040** (1964 to Sumitomo); Y. Nishizawa *et al.*, *Agric. Biol. Chem.* **25**, 597 (1961). Insecticidal activity: S. Tamura *et al.*, *ibid.* 773. Toxicology: H. Yamamoto *et al.*, *Oyo Yakuri* **5**, 75 (1971), *C.A.* **76**, 68907a (1972). LC-MS determn in serum: S. Inoue *et al.*, *J. Pharm. Biomed. Anal.* **44**, 258 (2007).

Pale yellow oily liquid. $bp_{0.09}$ 119-120° (dec). mp 14-15°. $n_D^{21.2}$ 1.5457. Very sol in methanol, ethanol, acetone, chloroform. Sparingly sol in *n*-hexane, kerosene; slightly sol in water. LD_{50} in mice (mg/kg): 1000 orally; 880 i.p. (Yamamoto).

USE: Insecticide.

2688.   **Cyanopsin.** Photoreceptor protein found in the retinal cone cells of fresh water and migratory fish, lampreys, and certain amphibians; corresponds to the rod pigment, porphyropsin, *q.v.* Absorption maximum ~620 nm. Composed of the chromophore 11-*cis*-3-dehydroretinal, *q.v.*, bound to a photopsin, the specific protein component of cone cells (*see* Opsins). Proposed existence as a visual pigment and *in vitro* synthesis from 3-dehydroretinal and chicken photopsin: G. Wald, *Science* **118**, 505 (1953). Demonstration of natural occurrence by microspectrophotometry: P. Liebman, G. Entine, *Nature* **216**, 501 (1967). Methods for prepn and assay: R. Hubbard *et al.*, *Methods Enzymol.* **18**, 615-653 (1971). Photochemistry and biological activity: G. Wald, *Science* **162**, 230 (1968). A trichromatic cone system utilizing 3-dehydroretinal as chromophore has been noted to occur in certain amphibians and fish. Three photochemically distinct pigments, corresponding to those of retinal-based color vision, have been identified: P. A. Liebman in *Handbook of Sensory Physiology* **Vol. VII**(1), H. J. A. Dartnall, Ed. (Springer-Verlag, New York, 1972) pp 481-528. A visual system is generally based on one type of chromophore combined with various opsins. However, pigments utilizing both types of chromophore have been found to co-exist in the retina of some of these species. Demonstration of paired-pigment cone systems: E. R. Loew, H. J. A. Dartnall, *Vision Res.* **11**, 551 (1976); A. T. C. Tsin *et al., ibid.* **21**, 943 (1981). Absorption spectrum: J. E. M. Mooij, T. J. T. P. van den Berg, *ibid.* 23 (1983).

2689. Cyantraniliprole. [736994-63-1] 3-Bromo-1-(3-chloro-2-pyridinyl)-*N*-[4-cyano-2-methyl-6-[(methylamino)carbonyl]-phenyl]-1*H*-pyrazole-5-carboxamide; 3-bromo-1-(3-chloro-2-pyridyl)-4'-cyano-2'-methyl-6'-(methylcarbamoyl)pyrazole-5-carboxanilide; DPX-HGW86; Cyazypyr. $C_{19}H_{14}BrClN_6O_2$; mol wt 473.72. C 48.17%, H 2.98%, Br 16.87%, Cl 7.48%, N 17.74%, O 6.75%. Anthranilic diamide pesticide; 4-cyano analog of chlorantraniliprole, *q.v.* Insect-specific ryanodine receptor activator. Prepn: K. A. Hughes *et al.*, **WO 04067528**; *eidem*, **US 7247647** (2004, 2007 both to DuPont). Prepn of crystalline polymorphs: M. R. Oberholzer, **WO 10056720** (2010 to DuPont). LC-MS/MS determn in crop samples: T. Schwarz *et al., J. Agric. Food Chem.* **59**, 814 (2011). Greenhouse trials with *Capsicum annuum* seedlings: A. L. Jacobson, G. G. Kennedy, *Crop Prot.* **30**, 512 (2011).

Polymorphic. Anhydrous crystalline form, mp 218-220°. Practically insol in water.

USE: Insecticide.

2690. Cyanuric Acid. [108-80-5] 1,3,5-Triazine-2,4,6(1*H*,-3*H*,5*H*)-trione; *sym*-triazinetriol; 2,4,6-trihydroxy-1,3,5-triazine; tricyanic acid; trihydroxycyanidine. $C_3H_3N_3O_3$; mol wt 129.08. C 27.92%, H 2.34%, N 32.55%, O 37.18%. Prepn: Schmidt, *J. Prakt. Chem.* [2] **5**, 41-52 (1872); Venable, Moore, *J. Am. Chem. Soc.* **39**, 1752 (1917). HPLC determn: P. Beilstein *et al., J. Agric. Food Chem.* **29**, 1132 (1981); in urine and pool water: T. V. Briggle *et al., J. Assoc. Off. Anal. Chem.* **64**, 1222 (1981). Review of prepn, properties and uses: E. M. Smolin, L. Rapoport, "*s*-Triazines and Derivatives" in *The Chemistry of Heterocyclic Compounds* vol. 13, A. Weissberger, Ed.(Interscience, New York, 1959) pp 17-48; of chemistry, toxicity and antibacterial activity as applied to swimming pools: E. Canelli, *Am. J. Public Health* **64**, 155-162 (1974); F. W. Linda, R. C. Hollenbach, *J. Environ. Health* **40**, 324-329 (1978); of toxicology: B. G. Hammond *et al., Environ. Health Perspect.* **69**, 287-292 (1986). *Review:* J. A. Wojtowicz in *Kirk-Othmer Encyclopedia of Chemical Technology* vol. 7 (John Wiley & Sons, 4th ed., 1993) pp 834-851.

White crystalline solid; does not melt up to 330°, sublimes and dissociates to isocyanic acid at higher temps. pKa_1 6.88; pKa_2 11.40; pKa_3 13.5. uv max (0.1*M* HCl): 214 nm (log ε 3.38). uv max (0.1*M* phosphate buffer pH 7): 214 nm (log ε 4.00). uv max (0.1*M* NaOH) 213 nm (log ε 4.64). Slightly sol in common organic solvents such as acetone, benzene, diethyl ether, ethanol, hexane. Soly (%): DMF 7.2; DMSO 17.4. Soly in water (%): 0.2 at 25°; 2.6 at 90°; 10.0 at 150°. Soly in 96% H_2SO_4 (25°): 14.1%. Sol in hot alcohols, pyridine, concd HCl without decompn; in aq solns of NaOH and KOH. Insol in cold methanol, ether, acetone, benzene, chloroform. Forms dihydrate in aq soln; crystallizes as colorless monoclinic prisms that effloresce in dry air. d^{25} 1.75 (anhydr); d^{25} 1.66 (dihydrate). Predominant tautomeric form in solid state and soln is keto form, *isocyanuric acid*; in basic soln the enol form is more stable. LD_{50} orally in rats: >5.00 g/kg (Canelli).

USE: Convenient lab source of cyanic acid gas. In prepn of melamine, sponge rubber, herbicides, dyes, resin, antimicrobial agents. As stabilizer and disinfectant in swimming pool water.

2691. Cyanuric Chloride. [108-77-0] 2,4,6-Trichloro-1,3,5-triazine; trichloro-*s*-triazine. $C_3Cl_3N_3$; mol wt 184.40. C 19.54%, Cl 57.67%, N 22.79%. Versatile coupling agent for proteins and nucleic acids; reagent in synthetic organic chemistry. Prepn: A. Gautier, *Ann.* **141**, 122 (1867); H. Herlinger, *Angew. Chem.* **76**, 437 (1964). Structural studies: A. W. Hofmann, *Ber.* **19**, 2061 (1886); R. A. Pascal, Jr., D. M. Ho, *Tetrahedron Lett.* **33**, 4707 (1992); S. J. Maginn *et al., ibid.* **34**, 4349 (1993); V. Krishnakumar, R. Ramasamy, *Spectrochim. Acta A* **61**, 3112 (2005). Polarographic determn in air: V. Stará *et al., Anal. Chim. Acta* **147**, 371 (1983). Synthetic applications as a chlorinating agent for alcohols: S. R. Sandler *et al., J. Org. Chem.* **35**, 3967 (1970); for carboxylic acids: K. Venkataraman, D. R. Wagle, *Tetrahedron Lett.* **20**, 3037 (1979); in macrocyclic lactonization: *eidem, ibid.* **21**, 1893 (1980); as a Beckmann rearrangement catalyst: Y. Furuya *et al., J. Am. Chem. Soc.* **127**, 11240 (2005); as a dehydrating agent: K. P. Haval *et al., Tetrahedron* **62**, 937 (2006). Use as a linking agent in molecular biology applications: H.-D. Hunger *et al., Anal. Biochem.* **156**, 286 (1986); *idem et al., ibid.* **165**, 45 (1987); R. A. Abuknesha *et al., J. Immunol. Methods* **306**, 211 (2005).

Crystals from ether, mp 146°. bp 198°. Crystal density: 1.93 g/cm^3.

Caution: Eye irritant; avoid skin contact and inhalation (Hunger, 1986).

USE: Intermediate in the synthesis of active dyes, agricultural products, and drug substances. Reagent in organic synthesis. Coupling agent for nucleic acids and proteins; cyanuric chloride-activated paper is used in capillary and electroblotting applications, dot tests, and hybridization protocols.

2692. Cyanuric Fluoride. [675-14-9] 2,4,6-Trifluoro-1,3,5-triazine; cyanuric trifluoride; trifluoro-1,3,5-triazine. $C_3F_3N_3$; mol wt 135.05. C 26.68%, F 42.20%, N 31.12%. Fluorinating agent. Prepn: A. F. Maxwell *et al., J. Am. Chem. Soc.* **80**, 548 (1958); A. Dorlars **DE 1044091** (1958 to Bayer). Improved prepn: S. Gross *et al., J. Prakt. Chem.* **342**, 711 (2000). Synthetic applications: G. A. Olah *et al., Synthesis* **1973**, 487; *idem et al., ibid.* **1980**, 221; L. A. Carpino *et al., J. Am. Chem. Soc.* **112**, 9651 (1990); G. Kokotos, C. Noula, *J. Org. Chem.* **61**, 6994 (1996); S. Hara *et al., Synlett* **1996**, 993.

Colorless liquid. *Poisonous. Corrosive. Lachrymator.* mp −38°. bp 74°. d_4^{20} 1.5858. uv max (CCl$_4$): 289 nm (log ε 0.727).

USE: Reagent in synthetic organic chemistry.

2693. Cyazofamid. [120116-88-3] 4-Chloro-2-cyano-*N,N*-dimethyl-5-(4-methylphenyl)-1*H*-imidazole-1-sulfonamide; 4-chloro-2-cyano-*N,N*-dimethyl-5-*p*-tolylimidazole-1-sulfonamide; IKF-916; Ranman. $C_{13}H_{13}ClN_4O_2S$; mol wt 324.78. C 48.08%, H 4.03%, Cl 10.92%, N 17.25%, O 9.85%, S 9.87%. Imidazole fungicide for use on food crops; inhibits mitochondrial complex III at the Qi (ubiquinone-reducing) site. Prepn: R. Nasu *et al.*, **AU 8812883**; *eidem*, **US 4995898** (1988, 1991 both to Ishihara Sangyo Kaisha). Mode of action study: S. Mitani *et al., Pestic. Biochem. Physiol.* **71**, 107 (2001). Review of properties and field performance: S. Mitani *et al., Brighton Crop Prot. Conf. - Pests Dis.* **1998**, 351-358; S. Mitani, *Agrochem. Jpn.* **78**, 17-20 (2001).

Consult the Name Index before using this section. **Page 479**

Odorless ivory powder, mp 152.7°. Log P (*n*-octanol/water): 3.2 (25°). Vapor pressure: <1.33 × 10^{-5} Pa. Soly in water at 20°: 0.121 μg/ml (pH 5). LD_{50} orally in mice, rats: >5000, >5000 mg/kg; dermally in rats: >2000 mg/kg. LC_{50} in rats: >5.5 mg/l by inhalation. LC_{50} (48 hr) in carp, rainbow trout: >69.6, >100 mg/l (Mitani).

USE: Agricultural fungicide.

2694. Cycasin. [14901-08-7] [(1*Z*)-2-Methyl-2-oxidodiazenyl]methyl β-D-glucopyranoside; (methyl-*ONN*-azoxy)methyl β-D-glucopyranoside; methylazoxymethanol β-D-glucoside; β-D-glucosyloxyazoxymethane. $C_8H_{16}N_2O_7$; mol wt 252.22. C 38.10%, H 6.39%, N 11.11%, O 44.40%. Toxic substance from seeds of *Cycas revoluta* Thumb. and *C. circinalis* L., *Cycadaceae* from Guam: Nishida *et al.*, *Bull. Agric. Chem. Soc. Jpn.* **19**, 77 (1955), *C.A.* **50**, 13756g (1956); Riggs, *Chem. Ind. (London)* **1956**, 926; Matsumoto, Strong, *Arch. Biochem. Biophys.* **101**, 299 (1963). Biological effects, metabolism and mechanism of action of cycasin and its aglycone, **methylazoxymethanol**: Spatz, *Ann. N.Y. Acad. Sci.* **163**, 848 (1969). Structure: Korsch, Riggs, *Tetrahedron Lett.* **1964**, 523. Toxicology: Laqueur, Spatz, *Cancer Res.* **28**, 2262 (1968). Carcinogenicity and neurotoxicity studies: I. Hirono, *Fed. Proc.* **31**, 1493 (1972). Review of carcinogenicity studies: *IARC Monographs* **10**, 121-138 (1976).

Needles from water + acetone + ether, dec 154°. $[\alpha]_D^{18}$ −44° (c = 0.62). uv max (0.4*M* H_2SO_4): 217 nm. LD_{50} in rats, mice, hamsters, rabbits, guinea pigs (mg/kg): 562, 500, <250, ~30, <20 orally (Hirono).

Tetraacetate. $C_{16}H_{24}N_2O_{11}$. Plates from acetone + petr ether, mp 137°. $[\alpha]_D^{18}$ −27° (c = 0.98 in $CHCl_3$).

Caution: Cycasin has been listed as reasonably anticipated to be a carcinogen: *Fourth Annual Report on Carcinogens* (NTP 85-002, 1985) p 64; delisted because no U.S. residents exposed: *Fifth Annual Report on Carcinogens* (NTP 89-239, 1989) p 340.

2695. Cyclacillin. [3485-14-1] (2*S*,5*R*,6*R*)-6-[[(1-Aminocyclohexyl)carbonyl]amino]-3,3-dimethyl-7-oxo-4-thia-1-azabicyclo-[3.2.0]heptane-2-carboxylic acid; 6-(1-aminocyclohexanecarboxamido)penicillanic acid; (1-aminocyclohexyl)penicillin; ciclacillin; Wy-4508; Calthor; Citosarin; Cyclapen; Syngacillin; Ultracillin; Vastcillin; Vatracin; Vipicil; Wyvital. $C_{15}H_{23}N_3O_4S$; mol wt 341.43. C 52.77%, H 6.79%, N 12.31%, O 18.74%, S 9.39%. Semisynthetic antibiotic related to penicillin. Prepn: Alburn *et al.*, **US 3194802**; C. A. Robinson, J. J. Nescio, **DE 1800698**; *eidem*, **US 3478018** (1965, 1969, 1969, all to Am. Home Prods.). Activity studies: Rosenman *et al.*, *Antimicrob. Agents Chemother.* **1967**, 590; Hopper *et al.*, *ibid.* 597; Yurchenco *et al.*, *ibid.* 602; *Chemotherapy* **15**, 209 (1970). Metabolic studies: Poole, *J. Pharm. Sci.* **59**, 1255 (1970); Tucker *et al.*, *Toxicol. Appl. Pharmacol.* **19**, 361 (1971). Physicochemical data: Alburn *et al.*, *Antimicrob. Agents Chemother.* **1967**, 586; Hou, Poole, *J. Pharm. Sci.* **58**, 1510 (1969); Poole, Bahal, *ibid.* **59**, 1265 (1970).

Crystals, mp 182-183° (anhydrate) (Hou, Poole); also reported as mp 156-158° (dec) (Alburn *et al.*). $[\alpha]_D^{25}$ +268° (water). pK_1, pK_2 in water: 2.68, 7.50; in 50% dioxane: 4.16, 7.04. Stable in acid. Soly at 38°: about 29 mg/ml.

THERAP CAT: Antibacterial.

2696. Cyclamic Acid. [100-88-9] *N*-Cyclohexylsulfamic acid; cyclohexanesulfamic acid; Hexamic Acid. $C_6H_{13}NO_3S$; mol wt 179.23. C 40.21%, H 7.31%, N 7.82%, O 26.78%, S 17.89%. Prepn: Audrieth, Sveda, **US 2275125** (1942 to du Pont); *J. Org. Chem.* **9**, 89 (1944); Thompson, **US 2800501** (1957 to du Pont); Shah, Bernsen, **US 3361799** (1968 to Abbott). Prepn of the Na salt: Robinson, **US 2383617** (1945 to du Pont). Other prepns and metabolism: *See* Calcium Cyclamate. Sweetness of the sodium salt discovered by Michael Sveda at the University of Illinois in 1937. Toxicity: Taylor *et al.*, *Food Cosmet. Toxicol.* **6**, 313 (1968). Review of long-term toxicity and carcinogenicity of sodium cyclamate in mice: Brantom *et al.*, *Food Cosmet. Toxicol.* **11**, 735 (1973).

Sweet-sour crystals. mp 169-170°. Fairly strong acid. Very sparingly soluble in water. Slowly hydrolyzed by hot water.

Sodium salt. Sodium cyclohexylsulfamate; cyclamate sodium; sodium cyclamate; Assugrin; Sucaryl Sodium; Sucrosa. C_6H_{12}-$NNaO_3S$; mol wt 201.22. Crystals. Pleasantly sweet to the taste. Freely sol in water. About 30 times as sweet as refined cane sugar. Sweetness is still easily perceptible at a dilution of 1:10,000 (sugar 1:140; saccharin 1:50,000). pH of 10% aq soln between 5.5 and 7.5. Practically insol in alcohol, ether, benzene, chloroform. LD_{50} orally in mice, rats: 17.0, 15.25 g/kg (Taylor).

Note: Consult latest Government regulations on use in food.

USE: Non-nutritive sweetener.

2697. Cyclandelate. [456-59-7] α-Hydroxybenzeneacetic acid 3,3,5-trimethylcyclohexyl ester; mandelic acid 3,3,5-trimethylcyclohexyl ester; 3,3,5-trimethylcyclohexyl mandelate; 3,5,5-trimethylcyclohexyl amygdalate; 3,3,5-trimethylcyclohexanol α-phenyl-α-hydroxyacetate; BS-572; Cyclergine; Cyclobral; Cyclolyt; Cyclomandol; Cyclospasmol; Natil; Novodil; Perebral; Spasmocyclon. $C_{17}H_{24}O_3$; mol wt 276.38. C 73.88%, H 8.75%, O 17.37%. Prepn: N. Brock *et al.*, *Arzneim.-Forsch.* **2**, 165 (1952); A. B. H. Funcke *et al.*, *ibid.* **3**, 503 (1953); van Sluis, *Chem. Prod.* **17**, 375 (1954); **GB 707227**; W. T. Nauta, **US 2707193** (1954, 1955 both to Brocades-Stheeman). Purification: D. Flitter, **US 3663597** (1972 to Am. Home Prods.). Pharmacokinetics: A. Orr, J. R. Whittier, *Int. J. Nucl. Med. Biol.* **1**, 205 (1974). GLC determn in human plasma: G. Andermann, M. Dietz, *J. Chromatogr.* **223**, 365 (1981). Clinical trial in chronic cerebrovascular disease: S. Bassi *et al.*, *Br. J. Clin. Pract.* **38**, 344 (1984). Review of efficacy as cerebral and peripheral vasodilator: C. B. Blakemore, *ibid.* **34**, Suppl., 3-9 (1984). Symposium on calcium modulation and clinical effects: *Drugs* **33**, Suppl. 2, 1-141 (1987). Comprehensive description: C. M. Shearer, *Anal. Profiles Drug Subs. Excip.* **21**, 149-168 (1992).

White crystals, mp 55.0-56.5°. bp_{14} 192-194°. uv max (ethanol): 269, 258, 251 nm (ε 1575, 2020, 1630). Very sol in chloroform, methanol, alc, ether; freely sol in acetonitrile, ethyl acetate, DMF, toluene. Insol in water.

THERAP CAT: Vasodilator (peripheral, cerebral).

2698. Cyclazocine. [3572-80-3] 3-(Cyclopropylmethyl)-1,-2,3,4,5,6-hexahydro-6,11-dimethyl-2,6-methano-3-benzazocin-8-ol; Win-20740; NSC-107429. $C_{18}H_{25}NO$; mol wt 271.40. C 79.66%, H 9.29%, N 5.16%, O 5.89%. Analgesic with mixed narcotic agonist-antagonist properties. Prepn: Archer, **BE 611000** (1962 to

Sterling Drug), *C.A.* **58**, 2439c (1963). Crystal structure: I. L. Karle *et al.*, *Acta Crystallogr.* **B25**, 1469 (1969). Supraspinal analgesic effects: S. Sasson, C. Kornetsky, *Life Sci.* **38**, 21 (1986). Evaluation in anorexia and motor disruption in rats: J. W. Henck *et al.*, *Pharmacol. Biochem. Behav.* **22**, 671 (1985). Determn by HPLC: I. Jane, A. McKinnon, *J. Chromatogr.* **323**, 191 (1985). Review of use in treatment of opiate addiction: M. J. Goldstein, *Int. J. Addict.* **15**, 939 (1980).

Crystals from methanol, mp 201-204°.
THERAP CAT: Narcotic antagonist.

2699. Cyclen. [294-90-6] 1,4,7,10-Tetraazacyclododecane; tetraaza-12-crown-4. $C_8H_{20}N_4$; mol wt 172.28. C 55.77%, H 11.70%, N 32.52%. Macrocyclic tetraamine ligand. Intermediate for the synthesis of azacrowns; derivatives form stable metal complexes used in biomedical applications. Prepn: H. Stetter, K.-H. Mayer, *Ber.* **94**, 1410 (1961). Modified prepn: G. R. Weisman, D. P. Reed, *J. Org. Chem.* **61**, 5186 (1996); *eidem, ibid.* **62**, 4548 (1997); P. S. Athey, G. E. Kiefer, *ibid.* **67**, 4081 (2002). Synthetic functionalization strategies: F. Denat *et al.*, *Synlett* **2000**, 561.

White crystals, mp 103-107°. bp$_{0.0001}$ 110°. *Irritant.* Sol in chloroform. Purified by sublimation.

Tetrahydrochloride. [10045-25-7] $C_8H_{20}N_4 \cdot 4HCl$; mol wt 318.11. Prepn: J. P. Collman, P. W. Schneider, *Inorg. Chem.* **5**, 1380 (1966); J. E. Richman, T. J. Atkins, *J. Am. Chem. Soc.* **96**, 2268 (1974). White crystals, mp 255-259°. *Irritant.*
USE: Reagent in organic synthesis.

2700. Cyclethrin. [97-11-0] 2,2-Dimethyl-3-(2-methyl-1-propen-1-yl)cyclopropanecarboxylic acid 3-(2-cyclopenten-1-yl)-2-methyl-4-oxo-2-cyclopenten-1-yl ester; 3-(2-cyclopentenyl)-2-methyl-4-oxo-2-cyclopentenyl ester of chrysanthemummonocarboxylic acid; chrysanthemummonocarboxylic acid ester with 3-(2-cyclopenten-1-yl)-2-methyl-4-oxo-2-cyclopenten-1-ol. $C_{21}H_{28}O_3$; mol wt 328.45. C 76.79%, H 8.59%, O 14.61%. Prepn and insecticidal activity: H. L. Haynes *et al.*, *Contrib. Boyce Thompson Inst.* **18**, 1 (1954); H. R. Guest, H. A. Stansbury, **US 2891888** (1959 to Union Carbide). Product is a mixture of isomers including four racemic forms or eight optical and geometric isomers which have not yet been isolated and evaluated. Toxicology: C. P. Carpenter *et al.*, *Arch. Ind. Hyg. Occup. Med.* **10**, 162 (1954).

Liquid. n_D^{30} 1.5120. d_{20}^{20} 1.033. LD$_{50}$ orally in rats: 1.4-2.8 g/kg (Carpenter).
USE: Insecticide for flies, roaches, and grain pests.

2701. Cyclic AMP. [60-92-4] Adenosine cyclic 3′,5′-(hydrogen phosphate); adenosine 3′,5′-cyclic monophosphate; adenosine 3′,5′-cyclic phosphate; adenosine 3′,5′-monophosphate; adenosine 3′,5′-phosphate; cyclic adenosine 3′,5′-monophosphate; acrasin; 3′,5′-AMP; cAMP. $C_{10}H_{12}N_5O_6P$; mol wt 329.21. C 36.48%, H 3.67%, N 21.27%, O 29.16%, P 9.41%. Key intracellular regulator

of a number of cellular processes; found in most animal cells, in bacteria, and in some higher plants. First isoln and identification: Rall *et al.*, *J. Biol. Chem.* **224**, 463 (1957); Sutherland, Rall, *ibid.* **232**, 1077 (1958). Molecular structure and conformation: Cook *et al.*, *J. Am. Chem. Soc.* **79**, 3607 (1957); Lipkin *et al.*, *ibid.* **81**, 6198 (1959); Watenpaugh *et al.*, *Science* **159**, 206 (1968). Syntheses: Smith *et al.*, *J. Am. Chem. Soc.* **83**, 698 (1961); Borden, Smith, *J. Org. Chem.* **31**, 3247 (1966). Physical data: D. Lipkin *et al.*, *J. Am. Chem. Soc.* **81**, 6075 (1959). Functions as a mediator of hormone-action for a variety of hormones such as epinephrine, glucagon and ACTH, *q.q.v.* Activates phosphorylation of proteins by protein kinases. Defined as a "second messenger" because of its response to hormones ("first messengers"). Converted from adenosine triphosphate (ATP) by the enzyme *adenylate cyclase*. Deactivated by *cyclic nucleotide phosphodiesterases* which convert it to 5′-adenylic acid. Reviews of biochemical model: Sutherland *et al.*, *J. Biol. Chem.* **237**, 1220-1243 (1962); Robison *et al.*, *Annu. Rev. Biochem.* **37**, 149 (1968); Jost, Rickenberg, *ibid.* **40**, 741 (1971); Sutherland, *J. Am. Med. Assoc.* **214**, 1281 (1970); Pastan, Perlman, *Nature New Biol.* **229**, 5 (1971); G. A. Robison *et al.*, Eds., *Ann. N.Y. Acad. Sci.* **185**, (1971); Losert, *Pharmazie* **28**, 351 (1973). *See also* Cyclic GMP. Identity with acrasin from cellular slime molds (*Dictyostelium* species), where it acts as a "first" rather than a "second messenger": Konijn *et al.*, *Proc. Natl. Acad. Sci. USA* **58**, 1152 (1967). Review of quantitative methods: *Methods in Molecular Biology* **vol. 3**, "Methods in Cyclic Nucleotide Research", Mark Chasin, Ed. (Marcel Dekker, New York, 1972). Books: G. A. Robison *et al.*, *Cyclic AMP* (Academic Press, New York, 1971); *Advan. Cyclic Nucleotide Res.* **vols. 1 2, 3,**, P. Greengard, G. A. Robison, Eds. (Raven Press, New York, 1972, 1973); *Handb. Exp. Pharmacol.* **58**, "Cyclic Nucleotides", Pt I: Biochemistry and Pt II: Physiology & Pharmacology, J. A. Nathanson, J. W. Kobabian, Eds. (Springer-Verlag, New York, 1982) 736 and 1000 pp resp.

Hydrate crystallizes from water as platelets with pearly luster, mp 219-220° (with gas evolution). $[\alpha]_D$ −51.3° (c = 0.67). uv max (pH 2): 256 nm (ε 14500); (pH 7): 258 nm (ε 14650). Heat stable; resistant to inactivation by acid or alkali. Found in concns of 10^{-7} to 10^{-6} moles/kg in living cells.

2702. Cyclic GMP. [7665-99-8] Guanosine cyclic 3′,5′-(hydrogen phosphate); cyclic guanosine 3′,5′-monophosphate; guanosine 3′,5′-cyclic monophosphate; guanosine 3′,5′-cyclic phosphate; guanosine 3′,5′-monophosphate; 3′,5′-GMP; cGMP. $C_{10}H_{12}N_5O_7P$; mol wt 345.21. C 34.79%, H 3.50%, N 20.29%, O 32.44%, P 8.97%. A cellular regulatory agent which has been described as a "second messenger". *See* Cyclic AMP. First synthesized because of the interest in cyclic nucleotides generated by cyclic AMP: Smith *et al.*, *J. Am. Chem. Soc.* **83**, 698 (1961); Borden, Smith, *J. Org. Chem.* **31**, 3247 (1966). Has subsequently been found in animal and bacterial cells in concentrations of 10^{-8} to 10^{-6} moles/kg. First isoln from rat urine: Ashman *et al.*, *Biochem. Biophys. Res. Commun.* **11**, 330 (1963). Structure and conformation: Chwang, Sundaralingam, *Nature New Biol.* **244**, 136 (1973). Proposed as an antagonist of cyclic AMP in bidirectional systems such as contraction-relaxation or glycogen synthesis-glycogen breakdown. Cyclic GMP levels increase in response to a variety of hormones including acetylcholine, insulin and oxytocin. Formed by conversion of guanosine triphosphate by the enzyme *guanylate cyclase* and hydrolyzed by cyclic nucleotide phosphodiesterases. Found to activate specific protein kinases. Reviews of biochemical model: Goldberg *et al.*, in *Pharmacology and* *the Future of Man* **vol. 5**, R. A. Maxwell, G. H. Acheson, Eds. (Karger, New York, 1973) pp 146-169; *eidem* in *Advan. Cyclic Nucleotide Res.* **vol. 3**, P. Greengard, G. A. Robison, Eds. (Raven Press, New York, 1973) pp 155-223; Kolata, *Science* **182**, 149-151 (1973); *Nature* **246**, 186 (1973); N. D. Goldberg, M. K. Haddox, *Annu. Rev. Biochem.* **46**, 823-896 (1977). Book: *Handb. Exp. Pharmacol.* **58**, entitled "Cyclic Nucleotides", Pt. I: Biochemistry and Pt. II: Physiology & Pharmacology, J. A. Nathanson, J. W. Kobabian, Eds. (Springer-Verlag, New York, 1982) 736 and 1000 pp resp.

Calcium salt. $C_{20}H_{22}CaN_{10}O_{14}P_2$. Decahydrate. uv max (pH 1): 256.5 nm (ε 11350); (pH 7): 254 nm (ε 12950); (pH 12): 262 nm (ε 12400).

2703. Cyclizine. [82-92-8] 1-(Diphenylmethyl)-4-methylpiperazine; *N*-benzhydryl-*N'*-methylpiperazine; *N*-methyl-*N'*-benzhydrylpiperazine; (*N*-benzhydryl)(*N'*-methyl)diethylenediamine; Compd 47-83; Wellcome prepn 47-83; Marzine; Marezine. $C_{18}H_{22}N_2$; mol wt 266.39. C 81.16%, H 8.32%, N 10.52%. Prepn: R. Baltzly *et al., J. Org. Chem.* **14**, 775 (1949); *eidem*, **US 2630435** (1953 to Burroughs Wellcome). Metabolism: R. Kuntzman *et al., Ann. N.Y. Acad. Sci.* **226**, 131 (1973). Physical properties and analysis: S. A. Benezra, *Anal. Profiles Drug Subs.* **6**, 83 (1977). Crystallographic study: V. Bertolasi *et al., Acta Crystallogr.* **B36**, 1975 (1980). HPLC determn: R. B. Walker, I. Kanfer, *J. Chromatogr. B* **672**, 172 (1995). Clinical antiemetic efficacy: G. Rowlands, W. J. Currie, *Br. J. Clin. Pract.* **30**, 197 (1976); W. N. Chestnutt *et al., Eur. J. Anaesthesiol.* **3**, 27 (1986).

Light-sensitive crystals from petr ether, mp 105.5-107.5°. uv max (0.1*N* HCl): 269, 263, 258, 225 (ε 540, 742, 694, 11300). Soly in g/ml at 25°: chloroform 1.1; ether 0.17; ethanol 0.17; in mg/ml: water <0.1.

Hydrochloride. [303-25-3] Valoid. $C_{18}H_{22}N_2 \cdot HCl$; mol wt 302.85. White crystalline powder or colorless crystals. Odorless; slightly bitter taste. Sparingly sol in chloroform; slightly sol in water, alcohol. Insol in ether. *Protect from light.*

THERAP CAT: Antiemetic.

THERAP CAT (VET): Antiemetic.

2704. Cyclobarbital. [52-31-3] 5-(1-Cyclohexen-1-yl)-5-ethyl-2,4,6(1*H*,3*H*,5*H*)-pyrimidinetrione; 5-(1-cyclohexen-1-yl)-5-ethylbarbituric acid; cyclobarbitone; hexemal; tetrahydrophenobarbital; Cavonyl; Cyclodorm; Cyklodorm; Fanodormo; Irifan; Namuron; Palinum; Phanodorm; Phanodorn; Philodorm; Prälumin; Prosonil; Sonaform. $C_{12}H_{16}N_2O_3$; mol wt 236.27. C 61.00%, H 6.83%, N 11.86%, O 20.31%. Prepn: **GB 231150** (1924 to Bayer); Schulemann, Meisenburg, **US 1690796** (1929 to Winthrop); Eckstein, *Przem. Chem.* **9**, 390 (1953), *C.A.* **49**, 11668c (1955). Toxicity data: G. Hofrichter *et al., Arzneim.-Forsch.* **17**, 242 (1967).

Shiny crystals, insipid bitter taste, mp 171-174°. Very slightly sol in cold water, appreciably sol in boiling water. One gram dissolves in 5 ml alcohol, 20 ml ether. LD_{50} in mice, rats (mg/kg): 350, 290 i.p. (Hofrichter).

Calcium salt. [143-76-0] Hexemal calcium; Itridal; Kollerdormfix; Pronox. $C_{24}H_{30}CaN_4O_6$; mol wt 510.60. Minute crystals. Appreciably bitter taste. Slightly basic reaction. Soly in water about 1 part in 70 parts (w/w), in 95% alc about 1 part in 500 parts. Practically insol in ether, chloroform.

Note: This is a controlled substance (depressant): **21 CFR**, 1308.13.

THERAP CAT: Sedative, hypnotic.

THERAP CAT (VET): Has been used as an anesthetic and sedative.

2705. Cyclobendazole. [31431-43-3] *N*-[6-(Cyclopropylcarbonyl)-1*H*-benzimidazol-2-yl]carbamic acid methyl ester; cicloben dazole; CC-2481; R-17147; Haptocil. $C_{13}H_{13}N_3O_3$; mol wt 259.27. C 60.22%, H 5.05%, N 16.21%, O 18.51%. Prepn: J. L. Van Gelder *et al.*, **DE 2029637**; *eidem*, **US 3657267** (1971, 1972 both to Janssen). Synthesis and anthelmintic activity: A. H. M. Raeymaekers *et al., Arzneim.-Forsch.* **28**, 586 (1978). Embryotoxic and antimitotic properties: P. Delatour, Y. Richard, *Therapie* **31**, 505 (1976). Pharmacokinetics: R. R. Brodie *et al., Arzneim.-Forsch.* **27**, 593 (1977). Biotransformation: B. C. Mayo *et al., Drug Metab. Dispos.* **6**, 518 (1978). Clinical evaluation: A. Degrémont, E. Stahel, *Schweiz. Med. Wochenschr.* **108**, 1430 (1978).

Crystals from acetic acid, mp 250.5°.

THERAP CAT: Anthelmintic (Nematodes).

2706. Cyclobenzaprine. [303-53-7] 3-(5*H*-Dibenzo[*a,d*]cyclohepten-5-ylidene)-*N,N*-dimethyl-1-propanamine; *N,N*-dimethyl-5*H*-dibenzo[*a,d*]cyclohepten-$\Delta^{5,\gamma}$-propylamine; 5-(3-dimethylaminopropylidene)dibenzo[*a,e*]cycloheptatriene; 1-(3-dimethylaminopropylidene)-2,3:6,7-dibenzo-4-suberene; proheptatriene; MK-130; Ro-4-1577; RP-9715. $C_{20}H_{21}N$; mol wt 275.40. C 87.23%, H 7.69%, N 5.09%. Prepn: **GB 858187** (1961 to Hoffmann-La Roche); Villani *et al., J. Med. Pharm. Chem.* **5**, 373 (1962); Winthrop *et al., J. Org. Chem.* **27**, 230 (1962). Use as muscle relaxant: N. N. Share, **FR 2100873** (1972 to Frosst), *C.A.* **78**, 47801n (1973). Pharmacology: C. D. Barnes, W. L. Adams, *Neuropharmacology* **17**, 445 (1978); N. N. Share, *ibid.* 721; and toxicology: J. Metysova *et al., Arch. Int. Pharmacodyn. Ther.* **144**, 481 (1963). Metabolism: G. Belvedere *et al., Biomed. Mass Spectrom.* **1**, 329 (1974); H. B. Hucker *et al., Drug Metab. Dispos.* **6**, 184 (1978). Bioavailability: *eidem, J. Clin. Pharmacol.* **17**, 719 (1977). Clinical studies: J. V. Basmajian, *Arch. Phys. Med. Rehabil.* **5**, 58 (1978); B. R. Brown, J. Womble, *J. Am. Med. Assoc.* **240**, 1151 (1978). Comprehensive description: M. L. Cotton, G. R. B. Down, *Anal. Profiles Drug Subs.* **17**, 41-72 (1988).

bp₁ 175-180°. uv max: 224, 289 nm (log ε 4.57, 4.02), (Villani *et al.*)

Hydrochloride. [6202-23-9] Flexeril; Flexiban. $C_{20}H_{21}N.HCl$; mol wt 311.85. White to off-white crystalline powder from isopropanol, mp 216-218°. Soly in water: >20 g/100 ml. Freely sol in methanol, ethanol; sparingly sol in isopropanol; slightly sol in chloroform, methylene chloride. Practically insol in hydrocarbons. uv max: 226, 295 nm (ε 52300, 12000). LD₅₀ in mice (mg/kg): 35 i.v., 250 orally (Metysova).

THERAP CAT: Muscle relaxant (skeletal).

2707. Cyclobutadiene. [1120-53-2] 1,3-Cyclobutadiene; [4]-annulene; diacetylene; CBD; CB. C_4H_4; mol wt 52.08. C 92.25%, H 7.74%. Smallest cyclic hydrocarbon bearing conjugated double bonds. Highly reactive annulene that has been isolated only in a low temperature matrix or molecular container; rapidly undergoes dimerization and other intermolecular reactions. Preliminary synthetic attempts: A. Kekulé, *Ann.* **162**, 77 (1872); R. Willstätter, W. von Schmaedel, *Ber.* **38**, 1992 (1905). Initial prepn: L. Watts *et al.*, *J. Am. Chem. Soc.* **87**, 3253 (1965). Matrix prepn: C. Y. Lin, A. Krantz, *J. Chem. Soc. Chem. Commun.* **1972**, 1111. Prepn and reactions in a hemicarcerand cavity: D. J. Cram *et al.*, *Angew. Chem. Int. Ed. Engl.* **30**, 1024 (1991). Theoretical structural studies: H. Kollmar, V. Staemmler, *J. Am. Chem. Soc.* **99**, 3583 (1977). NMR and IR spectroscopy studies: A. M. Orendt *et al.*, *ibid.* **110**, 2648 (1988). Photoelectron spectrum: D. W. Kohn, P. Chen, *ibid.* **115**, 2844 (1993). Heat of formation determmn: A. Fattahi *et al.*, *Angew. Chem. Int. Ed.* **45**, 4984 (2006). Review of antiaromaticity considerations: T. Bally, *ibid.* 6616-6618. Review of cyclobutadienemetal complexes: A. Efraty, *Chem. Rev.* **77**, 691-744 (1977). *Reviews*: G. Maier, *Angew. Chem. Int. Ed. Engl.* **13**, 425-438 (1974); T. Bally, S. Masumune, *Tetrahedron* **36**, 343-370 (1980).

Heat of formation: 428 ± 16 kJ/mol. Ionization potential: 8.16 ± 0.03 eV. Lifetime in gas phase (0.1 Torr): 2 ms.

2708. Cyclobutane. [287-23-0] Tetramethylene. C_4H_8; mol wt 56.11. C 85.62%, H 14.37%. Prepd by hydrogenaton of cyclobutene in the presence of nickel at 100°: Willstätter, Bruce, *Ber.* **40**, 3988 (1907); Heisig, *J. Am. Chem. Soc.* **63**, 1698 (1941); by pyrolysis of ether, along with other products: Peytral, *Bull. Soc. Chim. Fr.* [4] **35**, 964 (1924). In 39% yield from cyclobutanecarboxylic acid: Cason, Way, *J. Org. Chem.* **14**, 31 (1949); prepn of this acid: Heisig, Stodola, *Org. Synth.* **coll. vol. III**, 213 (1955).

Gas. Burns with a luminous flame. mp −80°. bp₇₄₁ 13.08°. d⁰ 0.7038; d⁻⁵ 0.7185. n_D^0 1.37520. *Flammable*. Insol in water. Freely sol in alcohol, acetone.

2709. Cyclobuxine D. [2241-90-9] (3β,5α,16α,20S)-14-Methyl-3,20-bis(methylamino)-4-methylene-9,19-cyclopregnan-16-ol; cyclobuxine. $C_{25}H_{42}N_2O$; mol wt 386.62. C 77.67%, H 10.95%, N 7.25%, O 4.14%. From *Buxus sempervirens* L., *Buxaceae*: E. Schlittler *et al.*, *Helv. Chim. Acta* **32**, 2209 (1949). Probably identical with Alkaloid A: Heusler, Schlittler, *ibid.* 2226. Structure: Brown, Kupchan, *J. Am. Chem. Soc.* **84**, 4590 (1962). Stereochemistry: *eidem, ibid.* **86**, 4424 (1964).

Crystals, dec 245-247°. $[\alpha]_D^{23}$ +98° (chloroform).

Dihydrobromide. $C_{25}H_{42}N_2O.2HBr$. Crystals, dec 288-292°.

N,N'-Dimethylcyclobuxine D. [4282-16-0] $C_{27}H_{46}N_2O$; mol wt 414.68. Crystals, dec 204-205°. $[\alpha]_D^{25}$ +99° (chloroform).

2710. Cyclodextrins. Cycloamyloses; cycloglucans; Schardinger dextrins. Naturally occurring, cyclic α-(1,4)-linked oligosaccharides of α-D-glucopyranose having a conical structure with a hollow, hydrophobic internal cavity and a hydrophilic outer surface. Form noncovalent inclusion complexes with a wide range of polar and apolar compounds. Originally obtained by the action of *Bacillus macerans* amylase on starch; α-, β- and γ-cyclodextrins are composed of six, seven and eight units, resp. Chemically modified derivatives have also been prepared with expanded inclusion capacities. Isoln of β-cyclodextrin: A. Villiers, *Compt. Rend.* **112**, 536 (1891); of α- and β-cyclodextrin: F. Schardinger, *Z. Untersuch. Nahr. Genussm.* **6**, 865 (1903); of γ-cyclodextrin: K. Freudenberg, R. Jacobi, *Ann.* **518**, 102 (1935). Overview of chemistry: J. Szejtli, *Chem. Rev.* **98**, 1743-1753 (1998); of production using cyclodextrin glycosyl transferase: A. Biwer *et al.*, *Appl. Microbiol. Biotechnol.* **59**, 609-617 (2002). Review of use in separation methods: E. Schneiderman, A. M. Stalcup, *J. Chromatogr. B* **745**, 83-102 (2000); in cosmetics: H. Buschmann, E. Schollmeyer, *J. Cosmet. Sci.* **53**, 185-191 (2002). Review of applications in drug delivery: R. Challa *et al.*, *AAPS PharmSciTech* **6**, E329-E357 (2005); T. Loftsson *et al.*, *Expert Opin. Drug Deliv.* **2**, 335-351 (2005).

α-Cyclodextrin

α-Cyclodextrin. [10016-20-3] Cyclohexaamylose; cyclohexadextrin; cyclomaltohexaose. $C_{36}H_{60}O_{30}$; mol wt 972.85. Total synthesis: Y. Takahashi, T. Ogawa, *Carbohydr. Res.* **164**, 277 (1987). Hexagonal plates from water. $[\alpha]_D$ +150.5°. pK (25°): 12.332. Cavity diameter: 4.7-5.3 Å. Cavity volume: 174 Å³. Soly in water (25°): 145 mg/ml.

β-Cyclodextrin. [7585-39-9] Cycloheptaamylose; cycloheptaglucan; cyclomaltoheptaose. $C_{42}H_{70}O_{35}$; mol wt 1134.99. Monoclinic parallelograms from water. mp 280°. $[\alpha]_D$ +162.0°. pK (25°): 12.202. Cavity diameter: 6.0-6.5 Å. Cavity volume: 262 Å³. Soly in water (25°): 18.5 mg/ml.

γ-Cyclodextrin. [17465-86-0] Cyclooctaamylose; cyclomaltooctaose. $C_{48}H_{80}O_{40}$; mol wt 1297.13. Quadratic prisms from water. $[\alpha]_D$ +177.4°. pK (25°): 12.081. Cavity diameter: 7.5-8.3 Å. Cavity volume: 427 Å³. Soly in water (25°): 232 mg/ml.

USE: Pharmaceutic aid in drug delivery, stabilization and solubilization. Complexing agent in foods, cosmetics, agriculturals. In chiral separation methods.

2711. β-Cyclodextrin Sulfobutyl Ether. [165133-56-2] Heptakis-O-(4-sulfobutyl)-β-cyclodextrin. $C_{70}H_{126}O_{56}S_7$; mol wt 2088.14. C 40.26%, H 6.08%, O 42.91%, S 10.75%. Polyanionic sulfobutyl ether deriv of β-cyclodextrin, *q.v.* Excipient that complexes with otherwise insoluble drugs and enhances their solubility and stability. Prepn and pharmaceutical applications: V. Stella, R. Rajewski, **US 5134127**; *eidem*, **US 5376645** (1992, 1994 both to University of Kansas); and pharmaceutical formulations: V. Stella *et al.*, **US 5874418** (1999 to CyDex). Effect on drug solubility and stability: H. Ueda *et al., Drug Dev. Ind. Pharm.* **24**, 863 (1998). Osmotic properties: E. A. Zannou *et al., Pharm. Res.* **18**, 1226 (2001); V. J. Stella *et al., J. Inclusion Phenom. Macrocyclic Chem.* **44**, 29 (2002). Characterization by capillary electrophoresis: R. J. Tait *et al., J. Pharm. Biomed. Anal.* **10**, 615 (1992). Analysis by chromatography and mass spectrometry: S. Grard *et al., J. Chromatogr. A* **925**, 79 (2001). Review of pharmacokinetics, safety, and toxicology: V. J. Stella, Q. He, *Toxicol. Pathol.* **36**, 30-42 (2008).

R = (CH₂)₄SO₃H or H

Sulfobutyl ether β-cyclodextrin sodium. [182410-00-0] SBE-β-CD; Captisol. Mixture of variably substituted isomers of the general formula $C_{42}H_{70-n}O_{35} \cdot (C_4H_8SO_3Na)_n$ where n, the average degree of substitution, ranges between 6 and 7; mol wt ~2163. White amorphous powder, dec ~275°. Hygroscopic. Very sol in water; sparingly sol in methanol. Practically insol in hexane, acetonitrile, ethyl acetate. d^{25} 1.041 (8.5% w/w soln); d^{25} 1.149 (29.8 % w/w soln); d^{25} 1.29 (49.4% w/w soln). Viscosity at 25°: 1.8 cP (8.5% w/w soln); 5.9 cP (29.8 % w/w soln); 51.9 cP (49.4% w/w soln).

USE: Pharmaceutic aid (excipient).

2712. Cyclodrine. [52109-93-0] α-(1-Hydroxycyclopentyl)-benzeneacetic acid 2-(diethylamino)ethyl ester; 1-hydroxy-α-phenylcyclopentaneacetic acid 2-diethylaminoethyl ester; β-diethylaminoethyl (1-hydroxycyclopentyl)phenylacetate; 2-phenyl-2-(1-hydroxycyclopentyl)acetic acid β-(diethylamino)ethyl ester. $C_{19}H_{29}NO_3$; mol wt 319.45. C 71.44%, H 9.15%, N 4.38%, O 15.02%. Anticholinergic. Prepn: Blicke, **US 2922795** (1960 to Univ. of Michigan). Activity studies: Lands, Luduena, *J. Pharmacol. Exp. Ther.* **117**, 331 (1956); Karczman, Long, *ibid.* **123**, 230 (1958).

Hydrochloride. [78853-39-1] GT-92; Cyclopent. $C_{19}H_{29}NO_3 \cdot$ HCl; mol wt 355.90. Crystals from isopropanol-isopropyl ether, mp 133-135°.

THERAP CAT: Mydriatic.

2713. Cyclofenil. [2624-43-3] 4,4′-(Cyclohexylidenemethylene)bisphenol 1,1′-diacetate; 4-[[4-(acetyloxy)phenyl]cyclohexylidenemethyl]phenol acetate; α-cyclohexylidene-α-(p-hydroxyphenyl)-p-cresol diacetate; bis(p-acetoxyphenyl)cyclohexylidenemethane; 4,4′-diacetoxybenzhydrylidenecyclohexane; F-6066; H-3452; ICI-48213; Fertodur; Neoclym; Ondogyne; Rehibin; Sanocrisin; Sexovid. $C_{23}H_{24}O_4$; mol wt 364.44. C 75.80%, H 6.64%, O

17.56%. Prepn: Miquel *et al., J. Med. Chem.* **6**, 774 (1963); Olsson *et al.*, **US 3287397** (1966). Pharmacology: Hiramatsu *et al., Oyo Yakuri* **6**, 1045 (1972), *C.A.* **78**, 131963z (1973).

Crystals from ethanol, mp 135-136°. uv max (ethanol): 247 nm (ε 17000).

Free diol. F-6060. $C_{19}H_{20}O_2$; mol wt 280.37. mp 235-236°.

THERAP CAT: Gonad-stimulating principle.

2714. Cycloguanil. [516-21-2] 1-(4-Chlorophenyl)-1,6-dihydro-6,6-dimethyl-1,3,5-triazine-2,4-diamine; 4,6-diamino-1-(p-chlorophenyl)-1,2-dihydro-2,2-dimethyl-s-triazine; 2,4-diamino-1-p-chlorophenyl-1,6-dihydro-6,6-dimethyl-1,3,5-triazine; 1-p-chlorophenyl-2,4-diamino-6,6-dimethyl-1,6-dihydro-1,3,5-triazine; chlorazine (Russian); chlorguanide triazine; TCI; M-10580; D-20. $C_{11}H_{14}ClN_5$; mol wt 251.72. C 52.49%, H 5.61%, Cl 14.08%, N 27.82%. Metabolic product formed from the antimalarial drug chlorguanide: Carrington *et al., Nature* **168**, 1080 (1951). Prepn: Modest *et al., J. Am. Chem. Soc.* **74**, 855 (1952); Loo, *ibid.* **76**, 5096 (1954); Carrington *et al., J. Chem. Soc.* **1954**, 1017; Bami, *J. Sci. Ind. Res.* **14C**, 231 (1955); Modest, *J. Org. Chem.* **21**, 1 (1956); **US 2900385** (1959 to Children's Cancer Res. Found.); Elslager, Worth, **US 3074947** (1963 to Parke, Davis).

Prisms from chloroform + ether, mp 146°. uv max (water): 241 nm (log ε 4.11).

Hydrochloride. $C_{11}H_{14}ClN_5 \cdot HCl$. Prisms from water, mp 210-215°. uv max (water): 241 nm (log ε 4.12).

Dihydrochloride. $C_{11}H_{14}ClN_5 \cdot 2HCl$. Crystals from acetone, mp 190-196°.

Pamoate. [609-78-9] Cycloguanil embonate; CI-501; Camolar. $(C_{11}H_{14}ClN_5)_2 \cdot C_{23}H_{16}O_6$; mol wt 891.81. Yellow crystals, mp 231-234°. Soly in water: 0.003%.

THERAP CAT: Antimalarial.

2715. Cycloheptanone. [502-42-1] Suberone; ketoheptamethylene; ketocycloheptane. $C_7H_{12}O$; mol wt 112.17. C 74.95%, H 10.78%, O 14.26%. Prepn from cyclohexanol with 1-(aminomethyl)cyclohexanol as the intermediate: Blicke *et al., J. Am. Chem. Soc.* **74**, 2924 (1952). Other methods: *Org. Synth.* **coll. vol. IV**, 221-228 (1963).

Liquid. d_4^{20} 0.9490. bp_{760} 179-181°. bp_{16} 66-70°. n_D^{20} 1.4608. Dipole moment 2.98. Practically insol in water. Freely sol in alcohol. Sol in ether.

Oxime. Suberoxime. $C_7H_{13}NO$. mp 23°, bp_{20} 152°.

2716. Cyclohexane. [110-82-7] Hexahydrobenzene; hexamethylene; hexanaphthene. C_6H_{12}; mol wt 84.16. C 85.63%, H 14.37%. Occurs in petr (0.5-1.0%). Obtained in the distillation of petr or by hydrogenation of benzene. In the distillation of petr the C_4-400°F boiling range naphthas are fractionated to obtain a C_5-

200°F naphtha contg 10-14% cyclohexane which on superfractionation yields an 85% concentrate (which is sold as such); further purification necessitates isomerization of pentanes to cyclohexane, heat cracking for removing open chain hydrocarbons and sulfuric acid treatment to remove aromatic compds. The hydrogenation of benzene is done in the liq phase at 150° using Raney nickel catalyst and at least 10 atm H_2 pressure: Sabatier, *Ind. Eng. Chem.* **18**, 1005 (1926). Review and bibliography: Sachanen, *Chemical Constituents of Petroleum* (New York, 1945). Prepn of high purity cyclohexane: Seyer *et al., Ind. Eng. Chem.* **31**, 759 (1939). Cyclohexane can exist in two interconvertible conformations, the boat and the chair. In the chair form its 12 extracyclic bonds fall into two classes: six lie parallel to the main axis of symmetry and are designated "axial", while six extend radially outward at ±109.5° angles to the axis and are designated as "equatorial", Barton *et al., Nature* **172**, 1096 (1954); *Science* **119**, 49 (1954). Solubility: F. P. Schwarz, *Anal. Chem.* **52**, 10 (1980). Toxicity: Lazarew, *Arch. Exp. Pathol. Pharmakol.* **143**, 223 (1929). Physical properties and methods of purification: L. Scheflan, M. B. Jacobs, *The Handbook of Organic Solvents* (Van Nostrand, 1953) p 233; *Techniques of Chemistry* vol. **II**, entitled "Organic Solvents", J. D. Riddick, W. B. Bunger, Eds. (Wiley-Interscience, New York, 3rd ed., 1970) p 592. *Review:* M. L. Campbell in *Kirk-Othmer Encyclopedia of Chemical Technology* vol. **12** (Wiley-Interscience, New York, 3rd ed., 1980) pp 931-937.

boat   chair

Liquid with solvent odor, pungent when impure. *Flammable.* d_4^{20} 0.7781; d_4^{0} 0.7206. mp 6.47°. bp_{760} 80.7°; bp_{400} 60.8°; bp_{200} 42.0°; bp_{100} 25.5°; bp_{60} 14.7°; bp_{40} 6.7°. n_D^{20} 1.4264. Flash pt, closed cup: 1°F (−18°C). Flammability limits in air 1.3-8.4% v/v. Soly in water at 23.5°C (w/w): 0.0052%. 100 ml of methanol dissolves 57 grams at 20°C; miscible with ethanol, ethyl ether, acetone, benzene, carbon tetrachloride. LC in mice: ~60-70 mg/l air (Lazarew).

Caution: Potential symptoms of overexposure are irritation of eyes, skin and respiratory system; drowsiness; dermatitis; narcosis, coma. *See NIOSH Pocket Guide to Chemical Hazards* (DHHS/NIOSH 97-140, 1997) p 82.

USE: Organic solvent for lacquers and resins. Paint and varnish remover. In the extraction of essential oils. In analytical chemistry for mol wt determinations (cryoscopic constant 20.3). In the manuf of adipic acid, benzene, cyclohexyl chloride, nitrocyclohexane, cyclohexanol and cyclohexanone. In the manuf of solid fuel for camp stoves. In fungicidal formulations (possesses slight fungicidal action). In the industrial recrystn of steroids.

2717. Cyclohexanecarboxylic Acid. [98-89-5] Hexahydrobenzoic acid. $C_7H_{12}O_2$; mol wt 128.17. C 65.60%, H 9.44%, O 24.97%. Prepn from anisic acid: Lumsden, *J. Chem. Soc.* **87**, 90 (1905); from 2-chlorocycloheptanone: Gutsche, *J. Am. Chem. Soc.* **71**, 3513 (1949); by carbonation of cyclohexylmagnesium chloride: Wagner, Moore, *ibid.* **72**, 974 (1950); from cyclohexane + KI + active Ni: Reppe *et al., Ann.* **582**, 38 (1953); by oxidation of cycloheptanone: Payne, Smith, *J. Org. Chem.* **22**, 1680 (1957); from cyclohexane + HCOOH or CO_2: McKursick *et al., J. Am. Chem. Soc.* **82**, 723 (1960); McKursick, **US 2940913** (1960 to du Pont).

COOH

Liquid. bp 232.5°; bp_{20} 131°; bp_8 110°; $bp_{<1}$ 63-67°. Crystallizes, on cooling, in monoclinic prisms, mp 29°. Odorless but when liquid or in soln has a valerian odor. d_4^{15} 1.0480. n_D^{15} 1.4530. Soly in 100 g water at 15°: 0.201 g. Sol in most organic solvents.

Methyl ester. $C_8H_{14}O_2$. Fragrant liq, bp 183°. d_4^{15} 0.9954.

USE: Solubilizer for vulcanized rubber; clarifier for mineral oil; in insecticide formulations.

2718. Cyclohexanol. [108-93-0] Hexalin; hexahydrophenol. $C_6H_{12}O$; mol wt 100.16. C 71.95%, H 12.08%, O 15.97%. Ob-

tained by oxidation of cyclohexane or hydrogenation of phenol: *Faith, Keyes & Clark's Industrial Chemicals*, F. A. Lowenheim, M. K. Moran, Eds. (Wiley-Interscience, New York, 4th ed., 1975) pp 304-309. Toxicity study: H. F. Smyth *et al., Am. Ind. Hyg. Assoc. J.* **23**, 95 (1962).

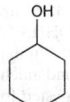

Hygroscopic crystals; camphor odor. d^{20} 0.962. mp 23-25°. bp 161°. Flash pt, closed cup: 154°F (68°C). n_D^{22} 1.465. At 20° soly in water: 3.6% (w/w); soly of water in cyclohexanol: 11% (w/w). Miscible with ethanol, ethyl acetate, linseed oil, petr solvents, aromatic hydrocarbons. LD_{50} orally in rats: 2.06 g/kg (Smyth).

Caution: Potential symptoms of overexposure are irritation of eyes, nose, throat and skin; narcosis. *See NIOSH Pocket Guide to Chemical Hazards* (DHHS/NIOSH 97-140, 1997) p 84.

USE: Solvent for alkyd resins, alcohol-sol phenolic resins, ethyl cellulose. Manuf celluloid; finishing textiles; insecticides.

2719. Cyclohexanone. [108-94-1] Ketohexamethylene; pimelic ketone; Hytrol O; Anone; Nadone. $C_6H_{10}O$; mol wt 98.15. C 73.42%, H 10.27%, O 16.30%. Obtained from cyclohexanol by catalytic dehydrogenation or by oxidation (which yields cyclohexanone and adipic acid) or from cyclohexane by oxidation (yielding cyclohexanone and cyclohexanol): **GB 310055** (1928 to Schering-Kahlbaum); **US 2223493**; **US 2223494**; **US 2285914** (1940, 1940, 1942 to Du Pont). Toxicity data: H. F. Smyth *et al., Am. Ind. Hyg. Assoc. J.* **30**, 470 (1969). Review of manuf processes: *Faith, Keyes & Clark's Industrial Chemicals*, F. A. Lowenheim, M. K. Moran, Eds. (Wiley-Interscience, New York, 4th ed., 1975) pp 304-309.

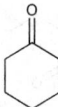

Oily liquid with characteristic odor reminiscent of peppermint and acetone. *Flammable.* d_4^{20} 0.9478; d_4^{25} 0.9421. mp −32.1°. bp_{760} 155.6°; bp_{400} 132.5°; bp_{200} 110.3°; bp_{100} 90.4°; bp_{60} 77.5°; bp_{40} 67.8°; bp_{20} 52.5°; bp_{10} 38.7°; bp_5 26.4°; $bp_{1.0}$ 1.4°. n_D^{20} 1.4507. Flash pt (closed cup): 111°F (44°C). Soly in water: 150 g/l at 10°; 50 g/l at 30°. Soly of water in cyclohexanone: 87 g/l at 20°. Sol in alcohol, ether and other common organic solvents. LD_{50} orally in rats: 1.62 ml/kg (Smyth).

2,4-Dinitrophenylhydrazone. [1589-62-4] 2-(2,4-Dinitrophenyl)hydrazone cyclohexanone. $C_{12}H_{14}N_4O_4$; mol wt 278.27. Golden yellow needles, mp 160°. uv max ($CHCl_3$): 262 nm (ε 22500); (0.01N NaOH): 435 nm (ε 19000).

Caution: Potential symptoms of overexposure are irritation of eyes, skin and mucous membranes; headache; narcosis, coma; dermatitis. *See NIOSH Pocket Guide to Chemical Hazards* (DHHS/NIOSH 97-140, 1997) p 84.

USE: Organic solvent for cellulose acetate, nitrocellulose, natural resins, vinyl resins, crude rubber, waxes, fats, shellac, DDT. In the production of adipic acid for nylon. In the prepn of cyclohexanone resins. 2,4-Dinitrophenylhydrazone deriv as standard in elemental analysis.

2720. Cyclohexene. [110-83-8] 1,2,3,4-Tetrahydrobenzene. C_6H_{10}; mol wt 82.15. C 87.72%, H 12.27%. Occurs in coal tar. Prepd by dehydration of cyclohexanol at high temps over various catalysts. Lab prepn using H_2SO_4 as dehydrating agent: Coleman, Johnstone, *Org. Synth.* **5**, 33 (1925); Wagner, *J. Chem. Educ.* **10**, 113 (1933); by distn of cyclohexanol over silica gel or alumina: Hershberg, Ruhoff, *Org. Synth.* **17**, 25 (1937).

Liquid. d_4^{20} 0.8098; d^{50} 0.7823; d^{100} 0.7355. mp −103.5°. bp$_{760}$ 83°. n_D^{20} 1.4465; n_D^{27} 1.4428. *Flammable.* Absorption spectrum: Purvis, *Proc. Cambridge Philos. Soc.* **23**, 588 (1927); *Chem. Zentralbl.* **1927**, II, 379; *cf.* Hartley, Dobbie, *J. Chem. Soc.* **77**, 846 (1900).

Caution: Potential symptoms of overexposure are irritation of eyes, respiratory system and skin; drowsiness. *See NIOSH Pocket Guide to Chemical Hazards* (DHHS/NIOSH 97-140, 1997) p 84.

USE: Alkylation component. In the manuf of adipic acid, maleic acid, hexahydrobenzoic acid and aldehyde. To prepare butadiene in the laboratory. Has been suggested for the synthesis of maleic acid and as stabilizer for high octane gasoline.

2721. Cycloheximide. [66-81-9] 4-[(2*R*)-2-[(1*S*,3*S*,5*S*)-3,5-Dimethyl-2-oxocyclohexyl]-2-hydroxyethyl]-2,6-piperidinedione; 3-[2-(3,5-dimethyl-2-oxocyclohexyl)-2-hydroxyethyl]glutarimide; actidione; naramycin A; NSC-185; U-4527. $C_{15}H_{23}NO_4$; mol wt 281.35. C 64.04%, H 8.24%, N 4.98%, O 22.75%. Glutarimide antibiotic produced by *Streptomyces griseus*; inhibits protein synthesis in eukaryotic cells. Isoln: B. E. Leach *et al.*, *J. Am. Chem. Soc.* **69**, 474 (1947); J. H. Ford, B. E. Leach, *ibid.* **70**, 1223 (1948); A. J. Whiffen *et al.*, US 2574519 (1951 to Upjohn). Production, assay and activity: A. J. Whiffen, *J. Bacteriol.* **56**, 283 (1948). Structure: E. C. Kornfeld *et al.*, *J. Am. Chem. Soc.* **71**, 150 (1949). Absolute configuration: F. Johnson *et al.*, *J. Am. Chem. Soc.* **87**, 4612 (1965). Total synthesis of *dl-* and *l*-forms: F. Johnson *et al.*, *ibid.* **88**, 149 (1966). Effect on protein synthesis: T. G. Obrig *et al.*, *J. Biol. Chem.* **246**, 174 (1971). Use in selective recovery of *Campylobacter sp.* from contaminated foods: M. P. Doyle, D. J. Roman, *Appl. Environ. Microbiol.* **43**, 1343 (1982); in detection of *Legionella sp.* in water systems: H. Inoue *et al.*, *Biocontrol Sci.* **11**, 69 (2006).

Crystals from water + methanol, mp 119.5-121° (Kornfield); also reported as colorless plates from amyl acetate, mp 115-116.5° (Leach). *Poisonous; skin and eye irritant.* $[\alpha]_D^{29}$ −3.38° (c = 9.47 in methanol); $[\alpha]_D^{25}$ +6.8° (c = 2 in H$_2$O). pKa 11.2. Soly at 2°: water 2.1 g/100 ml, amyl acetate 7 g/100 ml. Also sol in chloroform, ether, acetone, methanol, ethanol, other common organic solvents except satd hydrocarbons. Relatively heat-stable, acid-stable, destroyed by boiling in aq soln at pH 7 for 1 hr, but shows no loss of activity after 15 min boiling. At pH 2 it is not destroyed by boiling for 1 hr. Rapidly inactivated at room temp by dil alkali with the formation of a volatile, fragrant ketone, 2,4-dimethylcyclohexanone. Bitter taste, extremely repellent to rats. LD$_{50}$ i.v. in mice: 150 mg/kg (Leach).

Acetate. $C_{17}H_{25}NO_5$. Glistening plates from methanol, mp 148-149°. $[\alpha]_D^{25}$ +22° (c = 2.3 in methanol).

USE: Fungicide. In selective microbiological media. As protein synthesis inhibitor in biomedical research.

2722. Cyclohexylamine. [108-91-8] Cyclohexanamine; aminocyclohexane; hexahydroaniline. $C_6H_{13}N$; mol wt 99.18. C 72.66%, H 13.21%, N 14.12%. Prepd by the catalytic hydrogenation of aniline at elevated temps and pressures. Fractionation of the crude reaction product yields cyclohexylamine, unchanged aniline, and a high-boiling residue contg *N*-phenylcyclohexylamine (cyclohexylaniline) and dicyclohexylamine. Review and bibliography: Carswell, Morrill, *Ind. Eng. Chem.* **29**, 1247 (1937). Toxicity data: H. F. Smyth *et al.*, *Am. Ind. Hyg. Assoc. J.* **30**, 470 (1969).

Liquid. Strong, fishy, amine odor. d_{25}^{25} 0.8647. mp −17.7°. bp$_{760}$ 134.5°; bp$_{500}$ 118.9°; bp$_{300}$ 102.5°; bp$_{100}$ 72.0°; bp$_{50}$ 56.0°; bp$_{30}$ 45.1°; bp$_{25}$ 41.3°; bp$_{20}$ 36.4°; bp$_{15}$ 30.5°. n_D^{25} 1.4565. *Corrosive, flammable.* Strong base. Completely misc with water and with common organic solvents, including alcohols, ethers, ketones, esters, aliphatic hydrocarbons, aromatic hydrocarbons and their chlorinated derivatives. On distillation with water, cyclohexylamine forms an azeotropic mixture contg 44.2% cyclohexylamine by weight, bp$_{760}$ 96.4°. Reacts with excess ammonia and zinc chloride at 350° to produce α-picoline: Nordt, *Conversion of Hexahydroaniline into Pyridine Bases*, Off. Pub. Bd., Report PB 704 (1941). LD$_{50}$ orally in rats: 0.71 ml/kg (Smyth).

Caution: Potential symptoms of overexposure are irritation of eyes, skin, mucous membranes, respiratory system; eye, skin burns; skin sensitization; cough, pulmonary edema; drowsiness, lightheadedness, dizziness; diarrhea, nausea, vomiting. *See NIOSH Pocket Guide to Chemical Hazards* (DHHS/NIOSH 97-140, 1997) p 84.

USE: In organic synthesis, manuf insecticides, plasticizers, corrosion inhibitors, rubber chemicals, dyestuffs, emulsifying agents, dry-cleaning soaps, acid gas absorbents.

2723. Cyclohexyl Bromide. [108-85-0] Bromocyclohexane. $C_6H_{11}Br$; mol wt 163.06. C 44.20%, H 6.80%, Br 49.00%.

Liquid; penetrating odor. d_{15}^{15} 1.329. bp 163-165°. n_D^{15} 1.4956. Insol in water. Sol in alcohol, etc.

2724. Cyclohexylcarbinol. [100-49-2] Cyclohexanemethanol; cyclohexanecarbinol; hydroxymethylcyclohexane; hexahydrobenzyl alcohol. $C_7H_{14}O$; mol wt 114.19. C 73.63%, H 12.36%, O 14.01%. Prepd from cyclohexylmagnesium bromide and formaldehyde in ether: Gilman, Catlin, *Org. Synth.* **coll. vol. I** (2nd ed., 1941) p 188; Hiers, Adams, *J. Am. Chem. Soc.* **48**, 2385 (1926); Marvel *et al.*, *ibid.* **50**, 2810 (1928). Prepn of cyclohexylmagnesium bromide: Gilman, Caitlin, *loc. cit.;* Gilman, Zoellner, *J. Am. Chem. Soc.* **53**, 1945 (1931).

Liquid. Slight odor of camphor. d_4^{25} 0.9215. bp$_{784}$ 184-186°; bp$_{23}$ 91-92°. n_D^{25} 1.4640. Sol in alc, ether.
Acetate. bp$_{740}$ 199-201°.

2725. Cyclohexyl Chloride. [542-18-7] Chlorocyclohexane. $C_6H_{11}Cl$; mol wt 118.60. C 60.76%, H 9.35%, Cl 29.89%.

Colorless liquid; suffocating odor. d_4^{20} 1.000. mp −44°. bp 142°. n_D^{20} 1.4626. Insol in water. Sol in alcohol, etc.

2726. 2-Cyclohexyl-4,6-dinitrophenol. [131-89-5] 2,4-Dinitro-6-cyclohexylphenol; dinitro-*o*-cyclohexylphenol; DNOCHP; SN-46. $C_{12}H_{14}N_2O_5$; mol wt 266.25. C 54.13%, H 5.30%, N 10.52%, O 30.05%. Prepd from phenol, H$_2$SO$_4$, and cyclohexene or cyclohexanol: GB 620026 (1949 to Pest Control); from *o*-cyclohexylphenol by nitration in chloroform: Baroni, Kleinan, *Monatsh. Chem.* **68**, 251 (1936).

Crystals, mp 106.5-107.5°. Soly in petr oils 2.5 to 6%. Sol in benzene, DMF; very slightly sol in water.

Sodium salt. 2-Cyclohexyl-4,6-dinitrophenate; Anobesina. $C_{12}H_{13}N_2NaO_5$; mol wt 288.23. Crystals, sol in water, alcohol.

USE: Insecticide, esp in control of citrus red mite.

2727. Cycloleucine. [52-52-8] 1-Aminocyclopentanecarboxylic acid; ACPC; CB-1639; NSC-1026. $C_6H_{11}NO_2$; mol wt 129.16. C 55.80%, H 8.58%, N 10.84%, O 24.77%. Synthetic amino acid thought to act as a valine antagonist. Prepn: Zelinsky, Stadnikoff, *Z. Physiol. Chem.* **75**, 350 (1911); Connors, Ross, *J. Chem. Soc.* **1960**, 2119; Cremlyn, *ibid.* **1962**, 3977; Sudo, Ichihara, *Bull. Chem. Soc. Jpn.* **36**, 34 (1963). Manuf: **NL 6607754** (1966 to Rohm & Haas), *C.A.* **67**, 73159b (1967). *Review:* R. B. Ross *et al., J. Med. Pharm. Chem.* **3**, 1 (1961). Immunopharmacology: Rosenthale *et al., J. Pharmacol. Exp. Ther.* **180**, 501 (1972); Brambilla *et al., Cancer Chemother. Rep. Part 1* **56**, 579 (1972).

Crystals from ethanol-water, mp 330° (dec). Soly in water approx 5 g/100 ml. Forms stable metal salts.

Hydrochloride. [92398-48-6]; [66146-62-1] (monohydrate). Prismic crystals from water, mp 274° (dec). LD_{50} in mice, rats (mg/kg): 309, 290 orally; in rats (mg/kg): 340 i.v. (Ross).

USE: Biological testing material for immunosuppressive properties in mice; for amino acid transport studies.

2728. Cyclonite. [121-82-4] Hexahydro-1,3,5-trinitro-1,3,5-triazine; cyclotrimethylenetrinitramine; *sym*-trimethylenetrinitramine; 1,3,5-trinitrohexahydro-*s*-triazine; hexogen; royal demolition explosive; RDX; T_4. $C_3H_6N_6O_6$; mol wt 222.12. C 16.22%, H 2.72%, N 37.84%, O 43.22%. Prepd by treating methenamine with fuming nitric acid: Henning, **DE 104280** (1899); Hale, *J. Am. Chem. Soc.* **47**, 2754 (1925). Two moles of cyclonite can be obtained from one mole of methenamine, if ammonium nitrate and acetic anhydride are added: Bachmann, Sheehan, *ibid.* **71**, 1842 (1949). Manuf by nitrolysis of hexamethylenetetramine: Ruth, **US 3049543** (1962 to Olin Mathieson). Physical data relating to detonation, explosion, and blasting action: *ACS Monograph Series* **no. 139**, entitled "Science of High Explosives," M. A. Cook, Ed. (Reinhold, New York, 1958). Structure: Wood, *Proc. 4th Int. Meet. Mol. Spectrosc.* **2**, 955 (1962). Review of toxicology and human exposure: *Toxicological Profile for RDX* (PB95-264255, 1995) 145 pp.

Orthorhombic crystals from acetone, mp 205-206°. d_4^{20} 1.82. *Very explosive.* One gram dissolves in 25 ml acetone. Slightly sol in methanol, ether, ethyl acetate, glacial acetic acid. Practically insol in water, carbon tetrachloride, carbon disulfide.

Mixture with pentaerythritol tetranitrate. [52441-47-1] Semtex. Plastic explosive manufactured in two forms, **Semtex H** and **Semtex A**, which differ in the proportion of the two components RDX and pentaerythritol tetranitrate, *q.v.* Used in the destruction of PanAm Flight 103 in Dec. 1988 over Lockerbie, Scotland. Identification by MS: J. Yinon, *Can. Soc. Forensic Sci. J.* **21**, 46 (1988);

by time of flight-MS: A. Marshall *et al., Rapid Commun. Mass Spectrom.* **8**, 521 (1994). Description of incidents of illicit use: A. W. Feraday in *Adv. Anal. Detect. Explosives*, J. Yinon, Ed. (Kluwer Academic Publishers, Netherlands, 1993) pp 67-72. Analysis of Semtex H and A: J. R. Hobbs, *ibid.* pp. 409-427.

Caution: Potential symptoms of overexposure are irritation of eyes and skin; headache, irritability, fatigue, weakness, tremor, nausea, dizziness, vomiting, insomnia, convulsions. *See NIOSH Pocket Guide to Chemical Hazards* (DHHS/NIOSH 97-140, 1997) p 86.

USE: High explosive; rat poison.

2729. Cyclonium Iodide. [6577-41-9] 1-[(2-Cyclohexyl-2-phenyl-1,3-dioxolan-4-yl)methyl]-1-methylpiperidinium iodide (1:1); *N*-methyl-*N*-[(2-cyclohexyl-2-phenyl-1,3-dioxolan-4-yl)methyl]piperidinium iodide; 1-(2-phenyl-2-cyclohexyl-1,3-dioxolan-4-yl)methyl-1-methylpiperidinium iodide; ciclonium iodide; oxapium iodide; SH-100; Esperan. $C_{22}H_{34}INO_2$; mol wt 471.42. C 56.05%, H 7.27%, I 26.92%, N 2.97%, O 6.79%. Prepn: M. Kimura *et al.,* **JP 66 6585** (1966 to Toyama), *C.A.* **65**, 7191 (1966); and separation of isomers: I. Saikawa, **JP 70 16475** (1970 to Toyama), *C.A.* **73**, 45355t. Stability studies: Shiuchi, Miyazaki, *Bunseki Kagaku* **18**, 455 (1969), *C.A.* **71**, 74097z (1969). Incompatibility studies: Kunita *et al., Yakuzaigaku* **28**, 84 (1968), *C.A.* **69**, 99335x (1968). Anticholinergic activity and acute toxicity: J. Faff *et al., Pol. J. Pharmacol. Pharm.* **30**, 493 (1978). *Review: Jpn. Med. Gaz.* **8**(9), 11 (1971).

α-Form. White crystals from monochlorobenzene or isopropyl alcohol, mp 195-197°. Insol in trichloroethylene.

β-Form. White crystals from monochlorobenzene, mp 150-152°. Sol in hot trichloroethylene. Both sol in methanol, ethanol, chloroform and tetrachlorethane. Hardly sol in benzene, toluene, xylene and water. LD_{50} i.p. in mice: 106 mg/kg (Faff).

THERAP CAT: Antispasmodic.

2730. 1,5-Cyclooctadiene. [111-78-4] 1,5-COD. C_8H_{12}; mol wt 108.18. C 88.82%, H 11.18%. Cyclic olefin; serves as a reagent when part of a metal complex as well as in its free state. Prepn: R. Willstätter, H. Veraguth, *Ber.* **38**, 1975 (1905); A. C. Cope *et al., J. Am. Chem. Soc.* **72**, 2510 (1950). Ligand and catalyst screening for prepn via butadiene cyclooligomerization: D. E. Bergbreiter, R. Chandran, *J. Org. Chem.* **51**, 4754 (1986). Cycloisomerization and cyclodimerization studies: V. V. Saraev *et al., J. Mol. Catal. A* **315**, 231 (2010). Reactivity in ring opening cross metathesis: J. P. Morgan *et al., Org. Lett.* **4**, 67 (2002). Utility in prepn of the 9-BBN dimer: J. A. Soderquist, A. Negron, *Org. Synth.* **coll. vol. IX**, 95 (1998). Incorporation into organometallic complexes: R. H. Crabtree, G. E. Morris, *J. Organomet. Chem.* **135**, 395 (1977); L. E. Crascall, J. L. Spencer, *Inorg. Synth.* **28**, 126 (1990); J. W. Wielandt, D. Ruckerbauer, *ibid.* **35**, 120 (2010).

Colorless liquid with a stench. *Flammable, irritant, skin sensitizer.* bp_{760} 151°; bp_{25} 51-52°; $bp_{16.5}$ 39.5°. mp −70 to −69°. d_4^{25} 0.8760. n_D^{25} 1.4910; n_D^{20} 1.4942. Flash pt, closed cup: 100°F (38°C). Heat and air sensitive. Protect from light.

USE: Reagent in synthetic organic chemistry; ligand in organometallic chemistry.

2731. Cyclooctatetraene. [629-20-9] 1,3,5,7-Cyclooctatetraene; [8]annulene; COT. C_8H_8; mol wt 104.15. C 92.26%, H 7.74%.

Cyclic hydrocarbon with alternating and clearly localized single and double bonds; adopts a nonplanar, tub-like conformation. Exhibits chemical behavior characteristic of polyenes. Prepn: R. Willstätter, E. Waser, *Ber.* **44**, 3423 (1911). Improved prepn: W. Reppe *et al.*, *Ann.* **560**, 1 (1948). Purification and physical properties: B. H. Eccleston *et al.*, *J. Am. Chem. Soc.* **72**, 3866 (1950). X-ray studies: H. S. Kaufman *et al.*, *Nature* **161**, 165 (1948); K. H. Claus, C. Krüger, *Acta Crystallogr. C* **44**, 1632 (1988). Structure, energetics, and reactions of anion derivatives: S. Kato *et al.*, *J. Am. Chem. Soc.* **120**, 5033 (1998). Review of chemistry: L. A. Paquette, *Tetrahedron* **31**, 2855-2883 (1975); of nonaromatic and antiaromatic considerations: F.-G. Klärner, *Angew. Chem. Int. Ed.* **40**, 3977-3981 (2001); T. Nishinaga *et al.*, *Symmetry* **2**, 76-97 (2010).

Golden yellow liquid, bp$_{760}$ 142-143°; bp$_{17}$ 42.0-42.5°. bp$_{14}$ 36° Colorless orthorhombic crystals when solidified, mp 4°. fp −7.0°. *Flammable, irritant.* d$_4^0$ 0.9382; d$_4^{20}$ 0.9206; d^{20} 0.92094; d^{25} 0.91963; d^{30} 0.91172. n$_D^{20}$ 1.53790; n$_D^{25}$ 1.53501; n$_D^{30}$ 1.53229. Flash pt, closed cup: 72°F (22°C). Store at −20°C.

USE: Reagent in synthetic organic chemistry; ligand in organometallic chemistry.

2732. Cyclooctyne. [1781-78-8] C$_8$H$_{12}$; mol wt 108.18. C 88.82%, H 11.18%. Highly strained, highly reactive cycloalkyne. Prepn: A. T. Blomquist, L. H. Liu, *J. Am. Chem. Soc.* **75**, 2153 (1953); G. Wittig, A. Krebs, *Ber.* **94**, 3260 (1961). Improved synthesis: L. Brandsma, H. D. Verkruijsse, *Synthesis* **1978**, 290. Molecular structure: M. Traetteberg *et al.*, *J. Mol. Struct.* **128**, 217 (1985). Review of synthetic applications: D. Heber *et al.*, *Eur. J. Org. Chem.* **2005**, 4231-4247.

bp$_{740}$ 157.5-158°. bp$_{20}$ 50-55°. mp −34° to−32°. n$_D^{20}$ 1.4876. d$_4^{20}$ 0.868.

USE: Building block in cycloaddition reactions.

2733. Cyclopamine. [4449-51-8] (2′*R*,3*S*,3′*R*,3′a*S*,6′*S*,6a*S*,-6b*S*,7′a*R*,11a*S*,11b*R*)-1,2,3,3′a,4,4′,5′,6,6′,6a,6b,7,7′,7′a,8,11,11a,-11b-Octadecahydro-3′,6′,10,11b-tetramethylspiro[9*H*-benzo[a]fluorene-9,2′(3′*H*)-furo[3,2-*b*]pyridin]-3-ol; 11-deoxojervine. C$_{27}$H$_{41}$-NO$_2$; mol wt 411.63. C 78.78%, H 10.04%, N 3.40%, O 7.77%. Teratogenic *Veratrum* alkaloid; causes cyclopia and other birth defects when ingested by grazing animals during gestation. Inhibits Sonic hedgehog signal transduction. Isoln from *Veratrum album* L. var. *grandiflorum* Maxima, and characterization: T. Masamune *et al.*, *Bull. Chem. Soc. Jpn.* **38**, 1374 (1965). Determn of cyclopian teratogenic effects in sheep: R. F. Keeler, W. Binns, *Teratology* **1**, 5 (1968); R. F. Keeler, *Lipids* **13**, 708 (1978). Structure studies and identification as 11-deoxojervine: *idem*, *Phytochemistry* **8**, 223 (1969). ELISA determn in sheep blood: S. T. Lee *et al.*, *J. Agric. Food Chem.* **51**, 582 (2003). Mechanism of action study: J. K. Chen *et al.*, *Genes Dev.* **16**, 2743 (2002). Inhibition of Sonic hedgehog signaling in pancreas development: S. K. Kim, D. A. Melton, *Proc. Natl. Acad. Sci. USA* **95**, 13036 (1998); in colorectal tumor cells: D. Qualtrough *et al.*, *Int. J. Cancer* **110**, 831 (2004). Review of biological effects and medicinal potential: W. Gaffield *et al.*, *Cell. Mol. Biol.* **45**, 579-588 (1999).

Crystals from methanol, mp 237-238°. [α]$_D^{23}$ −44.2°.

2734. Cyclopentadiene. [542-92-7] 1,3-Cyclopentadiene. C$_5$H$_6$; mol wt 66.10. C 90.85%, H 9.15%. Obtained from the distillates produced in carbonization of coal, esp from the foreruns of coke-oven light oil: Ward, *US 2211038* (1940); process involving the liquefaction of coke-oven gas: Horclois, *Chim. Ind. (Paris)*, special issue April 1934, pp 357-363. Also obtained during the cracking of petr hydrocarbons: Tropsch *et al.*, *Ind. Eng. Chem.* **30**, 169 (1938); *cf.* Dedussenko, *J. Gen. Chem. USSR* **7**, 1467 (1937); *Chem. Zentralbl.* **109**, I, 793 (1938). Synthesis by passing vaporized cyclopentane over activated alumina and oxides of molybdenum, chromium, or vanadium: Grosse *et al.*, *Ind. Eng. Chem.* **32**, 309 (1940); *US 2157202; US 2157203; US 2157939*. Lab prepn by depolymerization of *dicyclopentadiene*: Moffett, *Org. Synth.* **coll. vol. IV**, 238 (1963). Toxicity study: Smyth *et al.*, *Arch. Ind. Hyg. Occup. Med.* **10**, 61 (1954). *Review* and discussion of structure: Wilson, Wells, *Chem. Rev.* **34**, 1 (1944). *Review:* M. Fefer, A. B. Small in *Kirk-Othmer Encyclopedia of Chemical Technology* **vol. 7** (Wiley-Interscience, New York, 3rd ed., 1979) pp 417-429.

Liquid. d$_4^0$ 0.8235; d$_4^{10}$ 0.8131; d$_4^{20}$ 0.8021; d$_4^{25}$ 0.7966; d$_4^{30}$ 0.7914. mp −85°. bp$_{760}$ 41.5-42.0°. n$_D^{16}$ 1.44632. Absorption spectum: Pickett *et al.*, *J. Am. Chem. Soc.* **63**, 1073 (1941). Insol in water. Miscible with alc, ether, benzene, carbon tetrachloride. Sol in carbon disulfide, aniline, acetic acid, liquid petrolatum. Cyclopentadiene polymerizes to dicyclopentadiene on standing. Polymerization is accelerated by the presence of peroxides or trichloroacetic acid. The dimer is cryst, mp 32.5°, with a camphor-like odor, having the structure of a partially hydrogenated indene contg a bridged methylene group. It is a more convenient form in which to handle cyclopentadiene, and is easily depolymerized by distilling at atmospheric pressure. LD$_{50}$ of dimer orally in rats: 0.82 g/kg (Smyth).

Caution: Potential symptoms of overexposure are irritation of eyes and nose. *See NIOSH Pocket Guide to Chemical Hazards* (DHHS/NIOSH 97-140, 1997) p 86.

USE: Manuf resins; in organic synthesis as the diene in the Diels-Alder reaction producing sesquiterpenes, synthetic alkaloids, camphors.

2735. Cyclopentamine. [102-45-4] *N*,α-Dimethylcyclopentaneethanamine; *N*,α-dimethylcyclopentaneethylamine; 1-cyclopentyl-2-methylaminopropane; β-(methylamino)propylcyclopentane; cyclopentadrine; Sinos; Cyklosal; Cyclonarol; Clopane. C$_9$H$_{19}$N; mol wt 141.26. C 76.52%, H 13.56%, N 9.92%. Prepn: Rohrmann, *US 2520015* (1950 to Lilly).

Hydrochloride. [3459-06-1] C$_9$H$_{19}$N.HCl. Crystals, mp 113-115° (base bp$_{30}$ 83-86°, n$_D^{25}$ 1.4500). Bitter taste. Freely sol in water.

THERAP CAT: Adrenergic (vasoconstrictor); nasal decongestant.

2736. Cyclopentane. [287-92-3] Pentamethylene. C$_5$H$_{10}$; mol wt 70.14. C 85.62%, H 14.37%. Occurs in petroleum. Found in petr ether fractions. Prepd by cracking cyclohexane in the presence of alumina at high temps and pressure: Haensel, Ipatieff, *Ind. Eng. Chem.* **35**, 632 (1943); by reduction of cyclopentadiene: David *et al.*, *Bull. Soc. Chim. Fr.* [5] **11**, 561 (1944). Toxicity data: Lazarew, *Arch. Exp. Pathol. Pharmakol.* **149**, 116 (1930).

Mobile, mp −94.4°. bp 49.3°. d$_4^{20}$ 0.7460. n$_D^{20}$ 1.4068. *Flammable.* Insol in water. Miscible with other hydrocarbon solvents, alcohol, ether. LC (2 hr in air) in mice: 110 mg/l (Lazarew).

Caution: Potential symptoms of overexposure are irritation of eyes, skin, nose and throat; lightheadedness, dizziness, euphoria, incoordination, nausea, vomiting, stupor; dry, cracking skin. *See NIOSH Pocket Guide to Chemical Hazards* (DHHS/NIOSH 97-140, 1997) p 86.

2737. Cyclopentanol. [96-41-3] Cyclopentyl alcohol; hydroxycyclopentane. $C_5H_{10}O$; mol wt 86.13. C 69.73%, H 11.70%, O 18.58%. Prepd by modified Meerwein-Pondorf-Verley reduction of cyclopentanone in the presence of aluminum isopropoxide and sodium hydroxide: Truett, Moulton, *J. Am. Chem. Soc.* **73**, 5913 (1951); by catalytic hydrogenation of pure cyclopentanone with copper chromite at 150° and 150 atm: Kögl, Ultée, *Rec. Trav. Chim.* **69**, 1576 (1950); by catalytic hydrogenation of cyclopentanone with platinum oxide and platinum black at 2-3 atm: Noller, Adams, *J. Am. Chem. Soc.* **48**, 1084 (1926); by hydration of cyclopentene in aq H_2SO_4: Hepp, **US 2414646** (1947 to Phillips Petr.); by reduction of cyclopentanone with $LiAlH_4$ in ether at room temp: Nystrom, Brown, *J. Am. Chem. Soc.* **69**, 1197 (1947).

Liquid, odor of amyl alcohol. d_4^0 0.96253; d_4^{15} 0.95078; d_4^{20} 0.9488; d_4^{30} 0.93908. mp −19°. bp 140.85°. Flash pt 124°F. n_D^{15} 1.45512; n_D^{20} 1.4520. *Flammable.* Sparingly sol in water. Sol in ethanol.

2738. Cyclopentanone. [120-92-3] Ketocyclopentane; ketopentamethylene; adipic ketone. C_5H_8O; mol wt 84.12. C 71.39%, H 9.59%, O 19.02%. Prepd by heating adipic acid to 285-295° in the presence of barium hydroxide, distilling, extracting with ether and fractionating: Thorpe, Kon, *Org. Synth.* **coll. vol. I**, 192 (2nd ed., 1941).

Liquid. Agreeable odor, somewhat like peppermint. d_4^{18} 0.9509. mp −58.2°. bp_{760} 130.6°; bp_{10} 23-24°. Flash pt 85°F. n_D^{20} 1.4366. *Flammable.* Slightly sol in water. Miscible with alc, ether. Polymerizes easily, esp in presence of acids.

2739. 1,2-Cyclopentenophenanthrene. [482-66-6] 16,17-Dihydro-15*H*-cyclopenta[*a*]phenanthrene; gona-1,3,5,7,9,11,13-heptaene. $C_{17}H_{14}$; mol wt 218.30. C 93.54%, H 6.46%. Prepn: Ruzicka *et al.*, *Helv. Chim. Acta* **16**, 838 (1933); Kon, *J. Chem. Soc.* **1933**, 1081; Cook, Hewett, *Chem. Ind. (London)* **52**, 451 (1933); Hawthorne, Robinson, *J. Chem. Soc.* **1936**, 763; Bachmann, Kloetzel, *J. Am. Chem. Soc.* **59**, 2207 (1937).

Needles from alc or petr ether, mp 135-136°. Absorption spectrum: Mayneord, Roe, *Proc. R. Soc. London* **A152**, 299 (1935).

2740. Cyclopenthiazide. [742-20-1] 6-Chloro-3-(cyclopentylmethyl)-3,4-dihydro-2*H*-1,2,4-benzothiadiazine-7-sulfonamide 1,1-dioxide; 6-chloro-3-cyclopentylmethyl-3,4-dihydro-7-sulfamoyl-2*H*-1,2,4-benzothiadiazine 1,1-dioxide; 3-cyclopentylmethyl-6-chloro-7-sulfamyl-3,4-dihydro-1,2,4-benzothiadiazine 1,1-dioxide; cyclomethiazide; tsiklometiazid; Su-8341; Navidrex; Salimid. $C_{13}H_{18}ClN_3O_4S_2$; mol wt 379.87. C 41.10%, H 4.78%, Cl 9.33%, N 11.06%, O 16.85%, S 16.88%. Prepn: Whitehead *et al.*, *J. Org. Chem.* **26**, 2814 (1961); **BE 587225** (1960 to Ciba). Pharmacology: W. E. Barrett *et al.*, *Arch. Int. Pharmacodyn.* **131**, 325 (1961).

Crystals from dil alc, mp 230°. LD_{50} in rats, mice (mg/kg): 141.8 ±24, 232.4 ±25 i.v. (Barrett).

THERAP CAT: Antihypertensive.

2741. Cyclopentolate. [512-15-2] α-(1-Hydroxycyclopentyl)benzeneacetic acid 2-(dimethylamino)ethyl ester; 1-hydroxy-α-phenylcyclopentaneacetic acid 2-(dimethylamino)ethyl ester; 2-dimethylaminoethyl 1-hydroxy-α-phenylcyclopentaneacetate; β-dimethylaminoethyl (1-hydroxycyclopentyl)phenylacetate; 2-phenyl-2-(1-hydroxycyclopentyl)ethanoic acid β-(dimethylamino)ethyl ester. $C_{17}H_{25}NO_3$; mol wt 291.39. C 70.07%, H 8.65%, N 4.81%, O 16.47%. Ophthalmic anticholinergic. Prepn: Treves, **US 2554511** (1951 to Schieffelin).

Hydrochloride. [5870-29-1] Ak-Pentolate; Alnide; Mydplegic; Cyclogyl; Mydrilate; Zyklolat. $C_{17}H_{25}NO_3$·HCl; mol wt 327.85. Crystals from ethyl acetate, mp 137-141°. pH of 1% aq soln: 5.0-5.4. Freely sol in water, alcohol. Practically insol in ether.

THERAP CAT: Mydriatic.

2742. Cyclopentyl Methyl Ether. [5614-37-9] Methoxycyclopentane; CPME. $C_6H_{12}O$; mol wt 100.16. C 71.95%, H 12.08%, O 15.97%. Ethereal solvent with high resistance to peroxide formation, high boiling point, and narrow explosion range; alternative to THF, MTBE, and 1,4-dioxane, *q.q.v.*, for attaining green, sustainable processes. Prepn: J. Loevenich *et al.*, *Ber.* **62**, 3084 (1929); W. T. Olson *et al.*, *J. Am. Chem. Soc.* **69**, 2451 (1947). Improved prepn: H. C. Brown *et al.*, *J. Org. Chem.* **50**, 1171 (1985). Synthetic applications: Y. Sawada *et al.*, *Org. Lett.* **6**, 2277 (2004); T. Ooi *et al.*, *J. Am. Chem. Soc.* **127**, 5073 (2005); H. Tokimoto *et al.*, *Tetrahedron: Asymmetry* **16**, 441 (2005). Review of properties as a process solvent in organic synthesis: K. Watanabe *et al.*, *Org. Process Res. Dev.* **11**, 251-258 (2007).

Colorless liquid. bp 106°. mp −134.75°. d^{20} 0.86. *Flammable; irritant.* Vapor specific gravity (air = 1): 3.45. Viscosity (20°): 0.55 cP. Surface tension (20°): 25.17 mN/m. n_D^{20} 1.4189. Dielectric constant (25°): 4.76. Soly in water (23°): 1.1 g/100 g. Flash pt: −1°C (30°F). Ignition pt: 180°. Log P (octanol/water): 1.59. Explosion range: 1.1-9.9 vol%. Vaporization energy: 69.2 kcal/kg. Forms azeotrope with water (16.3%), bp 83°. Relatively stable to acidic and basic conditions.

Caution: Irritating to eyes and skin.

USE: Solvent in organic synthesis.

2743. Cyclophosphamide. [6055-19-2]; [50-18-0] (anhydrous). *N,N*-Bis(2-chloroethyl)tetrahydro-2*H*-1,3,2-oxazaphosphorin-2-amine 2-oxide hydrate (1:1); 1-bis(2-chloroethyl)amino-1-oxo-2-aza-5-oxaphosphoridin monohydrate; bis(2-chloroethyl)phosphamide cyclic propanolamide ester monohydrate; *N,N*-bis(β-chloroethyl)-*N′,O*-propylenephosphoric acid ester diamide monohydrate; cyclophosphane; cytophosphane; B-518; Cycloblastin; Cyclostin; Cytoxan; Endoxan; Procytox; Sendoxan. $C_7H_{15}Cl_2N_2O_2P·H_2O$; mol wt 279.10. C 30.12%, H 6.14%, Cl 25.40%, N 10.04%, O 17.20%, P 11.10%. Oxazaphosphine cytostatic agent; supresses B-cell activity and antibody formation. Prepn: H. Arnold, F. Bourseaux, *Angew. Chem.* **70**, 539 (1958). Optical resolution study: T. Kawashima *et al.*, *J. Org. Chem.* **43**, 1111 (1978). Toxicology: A. G. Wheeler *et al.*, *Toxicol. Appl. Pharmacol.* **4**, 324 (1962). Book: D. L. Hill, *A Review of Cyclophosphamide* (Charles C Thomas, Springfield, 1975) 340 pp. Reviews of carcinogenicity studies: *IARC Monographs* **9**, 135-156 (1975); of field trials as defleecing agent for sheep: M. H. Fahmy, Y. Moride, *Anim. Breed. Abstr.* **52**,

7-19 (1984); of clinical toxicology: L. H. Fraiser *et al.*, *Drugs* **42**, 781-795 (1991); of mechanism of action and clinical use: O. M. Colvin, *Curr. Pharm. Des.* **5**, 555-560 (1999); of analytical methods: M. Malet-Martino *et al.*, *ibid.* 561-586; and metabolism: F. Baumann, R. Preiss, *J. Chromatogr. B* **764**, 173-192 (2001); of clinical pharmacokinetics: M. E. de Jonge *et al.*, *Clin. Pharmacokinet.* **44**, 1135-1164 (2005).

mp 41-45° (Arnold); also reported as mp 47-49° (Hill). Soly in water: 40 g/l. Sol in alcohol, benzene, ethylene glycol, chloroform, dioxane; sparingly sol in ether and acetone. LD$_{50}$ orally in mice, rats: 350, 94 mg/kg (Wheeler). Also marketed as the anhydrous form: crystals from ether, mp 51.5-52.5°.

Caution: This substance is listed as a known human carcinogen: *Report on Carcinogens, Twelfth Edition* (PB2011-111646, 2011) p 124.

USE: Defleecing agent for sheep.

THERAP CAT: Antineoplastic; immunosuppressant

THERAP CAT (VET): Antineoplastic; immunosuppressant.

2744. Cyclopropane. [75-19-4] Trimethylene. C$_3$H$_6$; mol wt 42.08. C 85.63%, H 14.37%. Prepd by reduction of 1,2-dibromocyclopropane with zinc and alcohol: Lott, Christiansen, *J. Am. Pharm. Assoc.* **19**, 341 (1930); Schlatter, *J. Am. Chem. Soc.* **63**, 1733 (1941); from 1,3-dibromopropane with zinc and alcohol in absence of water: Ashdown *et al.*, *ibid.* **58**, 850 (1936); by the action of sodium vapor on 1,3-dibromopropane: Bawn, Hunter, *Trans. Faraday Soc.* **34**, 608 (1938); from 1,3-dichloropropane heated with an excess of zinc dust, iodine and ethanol in 80% yield: Hass *et al.*, *Ind. Eng. Chem.* **28**, 1178 (1936); from 1,3-dichloropropane with zinc in presence of sodium iodide: **US 2102556** (1937); *cf.* **US 2098239; GB 498225; US 2211787; US 2235679; US 2235762; US 2240513; US 2240514; US 2242235.** From ethylene by the reaction with methylene iodide and a zinc-copper couple in 29% yield: Simmons, Smith, *J. Am. Chem. Soc.* **80**, 5323 (1958); *Chem. Eng. News* **36**, 40 (Dec 8, 1958).

Gas with characteristic odor resembling that of petr ether. mp −127°. bp −33°. *Flammable.* Liquefies at 4-6 atms. One liter of cyclopropane (at 1 atm, 0°) weighs 1.879 g. One vol of cyclopropane dissolves in about 2.7 vols of water at 15°. Freely sol in alcohol, ether. Sol in fixed oils. Concd H$_2$SO$_4$ absorbs the gas readily.

Caution: Mixture of cyclopropane with oxygen or air may explode when brought in contact with a flame or other causes of ignition. Explosive limits (% by vol in air), lower: 2.41; upper: 10.3. The explosibility is greater than that of other anesthetic-oxygen mixtures because of the comparatively larger amounts of oxygen that are compatible with cyclopropane anesthesia. Rich oxygen mixtures are therefore to be avoided.

THERAP CAT: Anesthetic (inhalation).

THERAP CAT (VET): Anesthetic (inhalation).

2745. Cyclopropene. [2781-85-3] C$_3$H$_4$; mol wt 40.07. C 89.93%, H 10.06%. Smallest cyclic alkene. The hydrocarbon's extreme reactivity is due to its short double bond length and high strain energy. Preliminary synthetic attempts: P. Freundler, *Bull. Soc. Chim. Fr.* **17**, 615 (1897); *idem*, *Compt. Rend.* **124**, 1157 (1897). Prepn: N. J. Demjanow, M. Dojarenko, *Ber.* **56**, 2200 (1923). Improved prepn: M. J. Schlatter, *J. Am. Chem. Soc.* **63**, 1733 (1941); and reaction in Diels-Alder cycloadditions: G. L. Closs, K. D. Krantz, *J. Org. Chem.* **31**, 638 (1966); P. Binger *et al.*, *ibid.* **61**, 6462 (1996). Thermal dimerization: P. Dowd, A. Gold, *Tetrahedron Lett.* **10**, 85 (1969). Ring strain studies: M. S. Gordon, *J. Am. Chem. Soc.* **102**, 7419 (1980). Review of thermal decomposition: R. Walsh, *Chem. Soc. Rev.* **34**, 714-732 (2005); of chemistry of cyclopropenes:

F. L. Carter, V. L. Frampton, *Chem. Rev.* **64**, 497-525 (1964); M. Rubin *et al.*, *Synthesis* **2006**, 1221-1245.

Gas that condenses into a colorless liquid when collected at −80°. bp$_{744}$ −36 to −35°. d 0.7. Stable in toluene soln at −78° for at least one week. Has the potential to polymerize explosively at room temp.

2746. Cyclopropyl Methyl Ether. [540-47-6] Methoxycyclopropane; cyprome ether. C$_4$H$_8$O; mol wt 72.11. C 66.63%, H 11.18%, O 22.19%. Prepd from 1,3-dibromo-2-methoxypropane by reduction with Zn: Olson *et al.*, *J. Am. Chem. Soc.* **69**, 2454 (1947); *see also* Krantz, Jr. *et al.*, *J. Pharmacol.* **69**, 207 (1940); Krantz, Jr., Drake, **US 2330979** (1944); Haller, **US 2424029** (1947).

Mobile, flammable liq. Odor similar to that of cyclopropane. d$_4^{25}$ 0.786. mp −119°; bp 44.7°; n$_D^{20}$ 1.3802. 100 ml of water dissolve 5.5 g of cyprome ether at 25°. Oil-water coefficient: 6.7 (ethyl ether = 4.5). Vapor press. at 26° (measured in nitrometer): 414 mm (ethyl ether = 555 mm). Explosive mixture, lower limit for air or oxygen: 2.0% v/v (ethyl ether vapor = 2.5 v/v).

Caution: Narcotic in high concns.

2747. Cyclopropyl Phenyl Sulfide. [14633-54-6] (Cyclopropylthio)benzene; phenyl cyclopropyl sulfide; phenylthiocyclopropane. C$_9$H$_{10}$S; mol wt 150.24. C 71.95%, H 6.71%, S 21.34%. Reagent utilized in the prepn of cyclopropane and cyclobutanone derivatives: W. E. Truce *et al.*, *J. Org. Chem.* **33**, 43 (1968); K. Tanaka *et al.*, *Bull. Chem. Soc. Jpn.* **55**, 2965 (1982). Synthetic applications involving the lithiated species: B. M. Trost *et al.*, *J. Am. Chem. Soc.* **95**, 3068 (1972); *eidem*, *ibid.* **99**, 3080 (1977); *eidem*, *ibid.*, 3088; B. M. Trost, W. C. Vladuchick, *Synthesis* **1978**, 821; F. Zutterman, A. Krief, *J. Org. Chem.* **48**, 1135 (1983). *Review*: A. Krief, A.-M. Laval, *Acros Org. Acta* **1**, 49-55 (1995).

Colorless oil. *Poisonous, combustible.* bp$_{4.7}$ 89-90°; bp$_3$ 70-72°; bp$_{1.0}$ 62-63°; bp$_{0.2}$ 58-60°; bp$_{0.10}$ 51-52°. d^{20} 1.052. n$_D^{20}$ 1.5817; n$_D^{25}$ 1.5801. Flash pt, closed cup: 198°F (92°C).

USE: Reagent in synthetic organic chemistry.

2748. Cyclorphan. [4163-15-9] 17-(Cyclopropylmethyl)-morphinan-3-ol; (−)-3-hydroxy-*N*-cyclopropylmethylmorphinan. C$_{20}$H$_{27}$NO; mol wt 297.44. C 80.76%, H 9.15%, N 4.71%, O 5.38%. Prepn: Gates, Montzka, *J. Med. Chem.* **7**, 127 (1964); Gates, **US 3285922** (1966 to Res. Corp.). Pharmacology: Harris *et al.*, *Arch. Int. Pharmacodyn. Ther.* **165**, 112 (1967).

Crystals from ethyl acetate, mp 187.5-189°. [α]$_D^{30}$ −120° (c = 2.26). LD$_{50}$ in mice, rats (mg/kg): 24, 23 i.v. (Gates, 1966).

2749. Cycloserine. [68-41-7] (4*R*)-4-Amino-3-isoxazolidinone; D-4-amino-3-isoxazolidone; orientomycin; PA-94; 106-7; Closina; Farmiserina; Micoserina; Oxamycin; Seromycin. C$_3$H$_6$-N$_2$O$_2$; mol wt 102.09. C 35.30%, H 5.92%, N 27.44%, O 31.34%. Antibiotic substance produced by *Streptomyces garyphalus* sive *orchidaceus*: Kuehl, Jr., *et al.*, *J. Am. Chem. Soc.* **77**, 2344 (1955); Hidy *et al.*, *ibid.* 2345; Shull, Sardinas, *Antibiot. Chemother.* **5**, 398

(1955); Shull et al., **US 2773878** (1956 to Pfizer); Harned, **US 2789983** (1957 to Commercial Solvents); **GB 768007** (1957 to Commercial Solvents), *C.A.* **51**, 10847e (1957); **US 3124590** (1964 to Commercial Solvents); Howe, **US 2845433** (1958 to Merck & Co.). Synthesis: Stammer et al., *J. Am. Chem. Soc.* **77**, 2346 (1955); Peck, **US 2772280** (1956 to Merck & Co.); Plattner et al., *Helv. Chim. Acta* **40**, 1531 (1957); Holly, Stammer, **US 2840565** (1958 to Merck & Co.). Prepn of crystalline calcium and magnesium salts: Harris et al., **US 2832788** (1958 to Merck & Co.). HPLC determn in plasma and urine: D. G. Musson et al., *J. Chromatogr.* **414**, 121 (1987). Comprehensive description: J. W. Lamb, *Anal. Profiles Drug Subs.* **1**, 53-64 (1972); H. A. El-Obeid, A. A. Al-Badr, *ibid.* **18**, 567-597 (1989).

White to pale yellow crystalline powder, decomp 155-156°. $[\alpha]_D^{23}$ +116° (c = 1.17); $[\alpha]_{546}^{25}$ +137° (c = 5 in 2N NaOH). uv max: 226 nm ($E_{1cm}^{1\%}$ 402). Hygroscopic. Sol in water, slightly sol in methanol, propylene glycol. Aq solns have a pH around 6. Forms salts with acids and bases. Neutral or acid solns are unstable. Aq solns buffered to pH 10 with sodium carbonate can be stored without loss for one week at refrigerator temps.

THERAP CAT: Antibacterial (tuberculostatic).

2750. Cyclosporins. Group of nonpolar cyclic oligopeptides with immunosuppressant activity; produced by *Tolypocladium inflatum* Gams (formerly designated as *Trichoderma polysporum* [Link ex Pers.] Rifai) and other fungi imperfecti. Cyclosporins A through Z have been identified with A being the major component. Characterized by a C_9-amino acid, MeBmt, *q.v.*, cyclosporins bind to the cytosolic protein, cyclophilin, to form an active complex which inhibits the enzyme, calcineurin, a key component of T-cell activation. Isoln of A and C, structure of A: A. Rüegger et al., *Helv. Chim. Acta* **59**, 1075 (1976); structure of C: R. Traber et al., *ibid.* **60**, 1247 (1977). Production of A and C: E. Härri et al., **US 4117118** (1978 to Sandoz). Review of isoln, structure elucidation, synthesis and bioactivity: A. von Wartburg, R. Traber, *Prog. Med. Chem.* **25**, 1-33 (1988); of biosynthesis and mode of action: A. Lawen, *ibid.* **33**, 53-97 (1996). Review of role of calcineurin on bioactivity of cyclosporins: K. A. Jorgensen et al., *Scand. J. Immunol.* **57**, 93-98 (2003). HPLC separation of components from fungal fermentation samples: J. Tuominen et al., *Rapid Commun. Mass Spectrom.* **12**, 1085 (1998). HPLC-MS determn in blood: B. Vollenbroeker et al., *Transplant. Proc.* **37**, 1741 (2005).

Cyclosporin A

Cyclosporin A. [59865-13-3] Cyclo[L-alanyl-D-alanyl-N-methyl-L-leucyl-N-methyl-L-leucyl-N-methyl-L-valyl-(3R,4R,6E)-6,7-didehydro-3-hydroxy-N,4-dimethyl-L-2-aminooctanoyl-L-2-aminobutanoyl-N-methylglycyl-N-methyl-L-leucyl-L-valyl-N-methyl-L-leucyl]; cyclosporine; ciclosporin; 27-400; Atopica; Sandimmun(e); Neoral; Optimmune; Restasis. $C_{62}H_{111}N_{11}O_{12}$; mol wt 1202.64. Comprehensive description: M. M. Hassan, M. A. Al-Yahya, *Anal. Profiles Drug Subs.* **16**, 145-206 (1987). Toxicological evaluation: B. Ryffel et al., *Arch. Toxicol.* **53**, 107 (1983). Symposium on therapeutic use in dermatology: *J. Am. Acad. Dermatol.* **23**, part 2, 1241-1334 (1990). Review of use in canine skin diseases: E. Gua-

guere et al., *Vet. Dermatol.* **15**, 61-74 (2004). Review of clinical experience in organ transplantation: C. J. Dunn et al., *Drugs* **61**, 1957-2016 (2001); in treatment of dry eye: H. D. Perry, E. D. Donnenfeld, *Expert Opin. Pharmacother.* **5**, 2099-2107 (2004). White prismatic needles from acetone at −15°. mp 148-151°. $[\alpha]_D^{20}$ −244° (c = 0.6 in chloroform); $[\alpha]_D^{20}$ −189° (c = 0.5 in methanol). Sol in methanol, ethanol, acetone, ether, chloroform, methylene chloride; slightly sol in saturated hydrocarbons. Practically insol in water. LD_{50} in mice, rats, rabbits (mg/kg): 107, 25, >10 i.v.; 2329, 1480, >1000 orally (Ryffel).

Cyclosporin B. [63775-95-1] Ala²-cyclosporine. $C_{61}H_{109}N_{11}O_{12}$; mol wt 1188.61. Amorphous white powder, mp 149-152°. $[\alpha]_D^{20}$ −238° (c = 0.62 in chloroform). $[\alpha]_D^{20}$ −168° (c = 0.56 in methanol).

Cyclosporin C. [59787-61-0] Thr²-cyclosporine. $C_{62}H_{111}N_{11}O_{13}$; mol wt 1218.63. Colorless prismatic needles from acetone at −15°. mp 152-155°. $[\alpha]_D^{20}$ −255° (c = 0.5 in chloroform). $[\alpha]_D^{20}$ −182° (c = 0.5 in methanol). Solubility similar to cyclosporin A.

Cyclosporin D. [63775-96-2] Val²-cyclosporine. $C_{63}H_{113}N_{11}O_{12}$; mol wt 1216.66. Colorless prisms from acetone at −15°. mp 148-151°. $[\alpha]_D^{20}$ −245° (c = 0.52 in CHCl₃). $[\alpha]_D^{20}$ −211° (c = 0.51 in methanol).

Cyclosporin G. [74436-00-3] 7-L-Norvaline cyclosporin A; Nva²-cyclosporine. $C_{63}H_{113}N_{11}O_{12}$; mol wt 1216.66. Immunosuppressive profile: P. C. Hiestand et al., *Transplant. Proc.* **17**, 1362 (1985); eidem, *Immunology* **55**, 249 (1985). Colorless, polyhedric crystals from ether + petroleum ether, mp 196-197°. $[\alpha]_D^{20}$ −245° (c = 1.0 in chloroform). $[\alpha]_D^{20}$ −191° (c = 1.04 in methanol).

Caution: Cyclosporin A is listed as a known human carcinogen: *Report on Carcinogens, Twelfth Edition* (PB2011-111646, 2011) p 125.

THERAP CAT: Cyclosporin A as immunosuppressant.

THERAP CAT (VET): Cyclosporin A as immunosuppressant.

2751. Cyclotheonamides. Naturally occurring family of macrocyclic peptides that inhibit serine proteases such as thrombin, *q.v.* Isolated from the Japanese marine sponge *Theonella swinhoei*; the two major forms, A and B, contain a vinylogous tyrosine (V-Tyr) and an α-ketoarginine residues − amino acids which were previously unknown in nature. Isoln and initial structure of A and B: N. Fusetani et al., *J. Am. Chem. Soc.* **112**, 7053 (1990). Reassignment of stereochemistry and total synthesis of B: M. Hagihara, S. L. Schreiber, *ibid.* **114**, 6570 (1992). Facile synthesis of B: H. M. M. Bastiaans et al., *Tetrahedron Lett.* **36**, 5963 (1995). Isoln of C, D, E: Y. Nakao et al., *Bioorg. Med. Chem.* **3**, 1115 (1995). Inhibition of thrombin and other serine proteases: S. D. Lewis et al., *Thromb. Res.* **70**, 173 (1993). Molecular basis of inhibition and synthesis of A: B. E. Maryanoff et al., *Proc. Natl. Acad. Sci. USA* **90**, 8048 (1993).

Cyclotheonamide A   R = H
Cyclotheonamide B   R = CH₃

Cyclotheonamide A. [129033-04-1] [S(E)]-3-[[4-Amino-5-(4-hydroxyphenyl)-1-oxo-2-pentenyl]amino]-N-formyl-L-alanyl-L-prolyl-N^6-(aminoiminomethyl)-2-oxo-L-3,6-diaminohexanoyl-D-phenylalanine cyclic (4 → 1)-peptide; MS-1144. $C_{36}H_{45}N_9O_8$; mol wt

731.81. uv max (methanol): 278 nm (ε 1940). $[\alpha]_D^{23}$ $-13°$ (c = 0.2 in methanol).

Cyclotheonamide B. [129033-05-2] $C_{37}H_{47}N_9O_8$; mol wt 745.84. $[\alpha]_D^{23}$ $-13.5°$ (c = 0.2 in methanol).

2752. Cyclothiazide. [2259-96-3] 3-Bicyclo[2.2.1]hept-5-en-2-yl-6-chloro-3,4-dihydro-2H-1,2,4-benzothiadiazine-7-sulfonamide 1,1-dioxide; 6-chloro-3,4-dihydro-3-(2-norbornen-5-yl)-2H-1,2,4-benzothiadiazine-7-sulfonamide 1,1-dioxide; 6-chloro-3,4-dihydro-3-(2-norbornen-5-yl)-7-sulfamoyl-1,2,4-benzothiadiazine 1,1-dioxide; 6-chloro-3-(2-norbornen-5-yl)-7-sulfamyl-3,4-dihydro-1,2,4-benzothiadiazine 1,1-dioxide; Lilly 35483; Aquirel; Anhydron; Doburil; Fluidil. $C_{14}H_{16}ClN_3O_4S_2$; mol wt 389.87. C 43.13%, H 4.14%, Cl 9.09%, N 10.78%, O 16.41%, S 16.45%. Prepn: C. W. Whitehead *et al., J. Org. Chem.* **26**, 2814 (1961); E. Müller, K. Hasspacher, **US 3275625** (1966 to Boehringer, Ing.). Comprehensive description: C. D. Wentling, *Anal. Profiles Drug Subs.* **1**, 65-77 (1972).

Crystals from dil alc, mp 234°.

THERAP CAT: Diuretic; antihypertensive.

THERAP CAT (VET): Diuretic.

2753. Cyflufenamid. [180409-60-3] [N(Z)]-N-[[(Cyclopropylmethoxy)amino][2,3-difluoro-6-(trifluoromethyl)phenyl]methylene]benzeneacetamide; (Z)-N-[α-(cyclopropylmethoxyimino)-2,3-difluoro-6-(trifluoromethyl)-2-phenylacetamide; NF-149; Pancho. $C_{20}H_{17}F_5N_2O_2$; mol wt 412.36. C 58.25%, H 4.16%, F 23.04%, N 6.79%, O 7.76%. Amidoxime fungicide active against powdery mildew in food crops. Prepn: I. Kasahara *et al.,* **WO 9619442**; *eidem,* **US 5847005** (1996, 1998 both to Nippon Soda). Mode of action and fungitoxic spectrum: M. Haramoto *et al., J. Pestic. Sci.* **31**, 95 (2006). Field trial for control of powdery mildew: *eidem, ibid.* 116. Field and greenhouse studies on selection pressure, resistance risk, and sensitivity monitoring: M. Haramoto *et al., ibid.* 397. Comprehensive description: A. R. Horgan, *BCPC Int. Congr. - Crop Sci. Technol.* **2005**, 73-78.

White crystals, mp 61.5-62.5°. Soly in water (20°): 0.52 mg/l. Vapor pressure at 20°: 3.54×10^{-5} Pa. LD_{50} rats (g/kg): >5 orally; >2 dermally. LC_{50} in rats (4 hr): >4.41 mg/l by inhalation; in rainbow trout (96 hr): 9.84 mg/l (Horgan). Log P (octanol/water): 4.70 (25°).

USE: Agricultural fungicide.

2754. Cyfluthrin. [68359-37-5] 3-(2,2-Dichloroethenyl)-2,2-dimethylcyclopropanecarboxylic acid cyano(4-fluoro-3-phenoxyphenyl)methyl ester; (R,S)-α-cyano-4-fluoro-3-phenoxybenzyl-(1R,S)-cis,trans-3-(2,2-dichlorovinyl)-2,2-dimethylcyclopropanecarboxylate; cyfoxylate; FCR-1272; BAY FCR 1272; Baythroid; Renounce; Tempo. $C_{22}H_{18}Cl_2FNO_3$; mol wt 434.29. C 60.84%, H 4.18%, Cl 16.33%, F 4.37%, N 3.23%, O 11.05%. Synthetic pyrethroid insecticide. Commercial product is mixture of 8 isomers, the (1R)-isomers primarily responsible for the bioactivity. Prepn of racemic mixture: R. Fuchs *et al.,* **DE 2709264**; *eidem,* **US 4218469** (1978, 1980 both to Bayer AG). Prepn of stereoisomers: *eidem,* **EP 22970**; *eidem,* **US 4287208** (both 1981 to Bayer AG). Chiral phase

HPLC separation of enantiomers: R. A. Chapman, *J. Chromatogr.* **258**, 175 (1983). Metabolism by cell suspension cultures: U. Preiss *et al., Chemosphere* **13**, 861 (1984). Field evaluation against cotton insect pests: J. A. Durant, *J. Agric. Entomol.* **1**, 201 (1984). Review of chemistry, bioactivity and field studies: I. Hammann, R. Fuchs, *Pflanzenschutz-Nachr.* **34**, 121-151 (1981); of formulations and potential uses: W. Behrenz *et al., ibid.* **36**, 127-176 (1983). Review of toxicology and human exposure: *Toxicological Profile for Pyrethrins and Pyrethroids* (PB2004-100004, 2003) 332 pp.

(1R,3R,αR)-form

Yellowish-brown oil, n_D^{23} 1.5511. Soly in water (20°): $1-2 \times 10^{-6}$ g/l. LD_{50} in male, female rats, male, female mice (mg/kg): 500-800, 1200, 300, 600 orally in Lutrol (Behrenz). LC_{50} (96 hr) in rainbow trout: 0.0006 mg/l (Hammann).

(1R,3R,αR)-**Form.** [85649-12-3] Colorless oil. $[\alpha]_D^{20}$ $-15.0°$ (c = 1.0 in $CHCl_3$).

(1R,3R,αS)-**Form.** [85649-15-6] Crystals, mp 50-52°. $[\alpha]_D^{20}$ +24.5° (c = 1.0 in $CHCl_3$).

(1R,3S,αS)-**Form.** [85649-19-0] Crystals from m-hexane mp 68-69°. $[\alpha]_D^{20}$ $-2.1°$ (c = 1.0 in $CHCl_3$).

USE: Agricultural insecticide.

2755. Cyhalofop-butyl. [122008-85-9] (2R)-2-[4-(4-Cyano-2-fluorophenoxy)phenoxy]propanoic acid butyl ester; XDE-537; Clincher. $C_{20}H_{20}FNO_4$; mol wt 357.38. C 67.22%, H 5.64%, F 5.32%, N 3.92%, O 17.91%. Post-emergent herbicide for use in rice crops. Prepn: G. R. Pews *et al.,* **EP 302203**; *eidem,* **US 4894085** (1989, 1990 both to Dow). Multi-residue analysis method in water samples: M. Sakamoto, T. Tsutsumi, *J. Chromatogr. A* **1028**, 63 (2004). Ecotoxicology and aquatic risk assessment: R. Jackson, M. Douglas, in *Human and Environmental Exposure to Xenobiotics,* A. A. M. Del Re *et al.,* Eds. (La Goliardica Pavese, Pavia, 1999) pp 345-354. Field trial for barnyardgrass control in rice: D. A. Ntanos *et al., Weed Technol.* **14**, 383 (2000).

Off-white to buff, waxy solid with faint almond odor. mp 45.5-49.5°. d^{20} 1.172. Log P (octanol/water) at 25°: 3.3158. Vapor pressure: 5.3×10^{-8} kPa. Soly in water (20°): 0.44 mg/l (pH 7). LC_{50} (96 hr) in rainbow trout, bluefill sunfish (mg/l): 1.65, 0.64 (Jackson, Douglas).

USE: Herbicide.

2756. Cyhalothrin. [68085-85-8] 3-(2-Chloro-3,3,3-trifluoro-1-propen-1-yl)-2,2-dimethylcyclopropanecarboxylic acid cyano(3-phenoxyphenyl)methyl ester; α-cyano-3-phenoxybenzyl 3-(2-chloro-3,3,3-trifluoroprop-1-en-1-yl)-2,2-dimethylcyclopropanecarboxylate; Grenade. $C_{23}H_{19}ClF_3NO_3$; mol wt 449.85. C 61.41%, H 4.26%, Cl 7.88%, F 12.67%, N 3.11%, O 10.67%. Synthetic pyrethroid insecticide. Commercial product is a variable mixture of isomers primarily consisting of the pairs of enantiomers having the Z-cis-configurations. The more bioactive pair is known as lambda-cyhalothrin, *q.v.* Prepn: R. K. Huff, **DE 2802962**; *idem,* **US 4183948** (1977, 1980 both to ICI). Prepn and activity of isomers: P. D. Bently *et al., Pestic. Sci.* **11**, 156 (1980). GC determn in apples: N. J. Bostanian *et al., J. Agric. Food Chem.* **41**, 292 (1993). ELISA determn in water samples: H. Gao *et al., ibid.* **54**, 5284 (2006). Veterinary trial as ectoparasiticide: V. K. Stubbs *et al., Aust. Vet. J.* **59**, 152 (1982). Review of toxicology and human exposure: *Toxicological Profile for Pyrethrins and Pyrethroids* (PB2004-100004, 2003) 332 pp.

Viscous liquid.

USE: Insecticide; acaricide.

2757. Cyhexatin. [13121-70-5] Tricyclohexylhydroxystannane; TCTH; tricyclohexylstannol; tricyclohexyltin hydroxide; ENT-27395; Dowco 213; Plictran. $C_{18}H_{34}OSn$; mol wt 385.18. C 56.13%, H 8.90%, O 4.15%, Sn 30.82%. Prepn: E. E. Kenaga, **US 3264177** (1966 to Dow). Activity studies: W. E. Allison *et al.*, *J. Econ. Entomol.* **61**, 1254 (1968); L. R. Jeppson *et al.*, *ibid.* 1502. Biochemical mode of action: S. Ahmad, C. O. Knowles, *Comp. Gen. Pharmacol.* **3**, 125 (1972). Biodegradation: E. H. Blair, *Environ. Qual. Saf. Suppl.* **3**, 406 (1975). Toxicity data: T. B. Gaines, R. E. Linder, *Fundam. Appl. Toxicol.* **7**, 299 (1986). Acute toxicity studies in livestock: J. H. Johnson *et al.*, *Toxicol. Appl. Pharmacol.* **31**, 66 (1975).

Whitish crystalline powder, mp 195-198°. Insol in water. Slightly sol in most organic solvents. Degraded when exposed in thin layers to uv light. LD_{50} in adult male, female rats (mg/kg): 779, 826 orally (Gaines, Linder).

Caution: Potential symptoms of overexposure are irritation of eyes, skin, respiratory system; headache, vertigo; sore throat, cough; abdominal pain, vomiting; skin burns, pruritis. *See NIOSH Pocket Guide to Chemical Hazards* (DHHS/NIOSH 97-140, 1997) p 86.

USE: Acaricide.

2758. Cymarin. [508-77-0] $(3\beta,5\beta)$-3-[(2,6-Dideoxy-3-O-methyl-β-D-*ribo*-hexopyranosyl)oxy]-5,14-dihydroxy-19-oxocard-20(22)-enolide; K-strophanthin-α; Alvonal MR. $C_{30}H_{44}O_9$; mol wt 548.67. C 65.67%, H 8.08%, O 26.24%. A glycoside of cymarose, the aglucon being strophanthidin. Isoln from *Strophanthus kombé* Oliv., *Apocynaceae:* Jacobs, Hoffmann, *J. Biol. Chem.* **67**, 609 (1926); Stoll *et al.*, *Helv. Chim. Acta* **20**, 1484 (1937); from *Adonis vernalis* L., and *A. chrysocyathus* Hook., *Ranunculaceae:* Reichstein, Rosenmund, *Pharm. Acta Helv.* **15**, 150 (1940); Pitra, Cekan, *Collect. Czech. Chem. Commun.* **26**, 1551 (1961); Abubakirov, Yamatova, *Dt. Obshch. Khim.* **31**, 2424 (1961); from *Pentopetia androsaemifolia* Decne., *Asclepiadaceae:* Wyss *et al.*, *Helv. Chim. Acta* **43**, 664 (1960); from *Castilloa elastica* Cerv., *Moraceae:* Adams, Wilkinson, *J. Pharm. Pharmacol.* **13**, 279 (1961). Structure: Kochetkov *et al.*, *Dokl. Akad. Nauk SSSR* **136**, 613 (1961). Toxicity data: V. G. Vogel, E. Kluge, *Arzneim.-Forsch.* **11**, 848 (1961).

Needles from methanol, mp 148°. $[\alpha]_D^{20}$ +39.2° (methanol); $[\alpha]_D^{22}$ +39.0° (c = 1.7 in chloroform). Sol in methanol, chloroform. Practically insol in water. LD_{50} i.v. in rats: 24.8±1.8 mg/kg (Vogel, Kluge).

Sesquihydrate. Hexagonal plates from dil ethanol, mp 184-185°.

Monoacetylcymarin. $C_{32}H_{46}O_{10}$. Needles from dil methanol, mp 175-176°. $[\alpha]_D^{22}$ +45.1° (ethanol). Sol in chloroform. Practically insol in water.

THERAP CAT: Cardiotonic.

2759. Cymarose. [579-04-4] 2,6-Dideoxy-3-O-methyl-*ribo*-hexose; 3-methyldigitoxose. $C_7H_{14}O_4$; mol wt 162.19. C 51.84%, H 8.70%, O 39.46%. By hydrolysis of a number of cardiac glycosides from the *Apocynaceae* family. From *Apocynum cannabinum* L., *A. androsaemifolium* L., *A. venetum* L.: Windaus, Hermanns, *Ber.* **48**, 979 (1915); Trabert, *Arzneim.-Forsch.* **10**, 197 (1960); from *Strophanthus kombé* Oliv., *Periploca graeca* L.: Jacobs, Hoffmann, *J. Biol. Chem.* **67**, 609 (1926); **79**, 519 (1928); from *S. emini* Aschers & Pax: Lamb, Smith, *J. Chem. Soc.* **1936** 442; also from *Castilla elastica* Cerv., *Moraceae:* Adams, Wilkinson, *J. Pharm. Pharmacol.* **13**, 279 (1961). Structure: S. F. Dyke, *The Carbohydrates* (Interscience, New York, 1960) p 104. Synthesis: Bollinger, Ulrich, *Helv. Chim. Acta* **35**, 93 (1952). *Review:* R. C. Elderfield, "The Carbohydrate Components of Cardiac Glycosides" in W. W. Pigman, M. L. Wolfrom, *Advances in Carbohydrate Chemistry* **vol. I** (Academic Press, New York, 1945) pp 147-173.

Prisms from ether + petr ether, mp 93-94°. $[\alpha]_D^{20}$ +54.7° (c = 3.2 in water after 24 hrs). Sol in water, alcohol, acetone; sparingly sol in ether, petr ether, chloroform, benzene. Reduces Fehling soln. Does not form an osazone.

2760. Cymene. [25155-15-1] Methyl (1-methylethyl)benzene; isopropyltoluene; methylisopropylbenzene. $C_{10}H_{14}$; mol wt 134.22. C 89.49%, H 10.51%. Usually prepd by alkylation of toluene; *m*-, *o*- and *p*-isomers obtained: Allen, Yats, *J. Am. Chem. Soc.* **83**, 2799 (1961). Purification and properties of the three isomers: Streiff *et al.*, *Anal. Chem.* **27**, 411 (1955). Separation of the three isomers by gas chromatography: Rihani, Froment, *J. Chromatogr.* **18**, 150 (1965). Toxicity study: P. M. Jenner *et al.*, *Food Cosmet. Toxicol.* **2**, 327 (1964).

m-**Cymene.** [535-77-3] 1-Methyl-3-(1-methylethyl)benzene. Liquid, bp 175.14°. mp −63.75°. d_4^{20} 0.8610, d_4^{25} 0.8570. n_D^{20} 1.4930, n_D^{25} 1.4906. Practically insol in water. Miscible with alcohol, ether.

o-**Cymene.** [527-84-4] 1-Methyl-2-(1-methylethyl)benzene. Liquid, bp 178.15°. mp −71.54°; also reported as −75.24° and −81.53° for two unstable solid forms (Streiff). d_4^{20} 0.8766, d_4^{25} 0.8726. n_D^{20} 1.5006, n_D^{25} 1.4982. Practically insol in water. Miscible with organic solvents.

p-**Cymene.** [99-87-6] 1-Methyl-4-(1-methylethyl)benzene; Dolcymene. Occurs in a number of essential oils. Liquid, bp 177.10°. mp −67.94°. d_4^{20} 0.8573, d_4^{25} 0.8533. n_D^{20} 1.4909, n_D^{25} 1.4885. Flash pt, closed cup: 117°F (47°C). Practically insol in water. Misc with alcohol, ether. LD_{50} orally in rats: 4750 mg/kg (Jenner).

2761. Cymiazole. [61676-87-7] 2,4-Dimethyl-*N*-(3-methyl-2(3*H*)-thiazolylidene)benzenamine; 2-(2′,4′-dimethylphenylimino)-3-methylthiazoline; *N*-(3-methyl-4-thiazolin-2-ylidene)-2,4-xylidene; xymiazole; CGA-50439. $C_{12}H_{14}N_2S$; mol wt 218.32. C 66.02%, H 6.46%, N 12.83%, S 14.68%. Iminophenyl thiazolidine derivative active against mites and ticks. Prepn: D. Duerr, W. D. Traber, **DE 2619724** (1976 to Ciba-Geigy), *C.A.* **86**, 72627h (1977). Activity: R. M. Immler *et al., Proc. Br. Crop Prot. Conf. - Pests Dis.* **1977**, 383. LC determn in honey: P. Cabras *et al., J. AOAC Int.* **76**, 92 (1993). Genotoxicity study: Z. Stanimirovic *et al., Mutat. Res.* **588**, 152 (2005).

Hydrochloride. [121034-85-3] Apitol. $C_{12}H_{14}N_2S \cdot HCl$; mol wt 254.78. Sol in water.

USE: Acaricide.

THERAP CAT (VET): Ectoparasiticide.

2762. Cymoxanil. [57966-95-7] 2-Cyano-*N*-[(ethylamino)-carbonyl]-2-(methoxyimino)acetamide; 2-cyano-*N*-ethylcarbamoyl-2-methoxyiminoacetamide; 1-(2-cyano-2-methoxyiminoacetyl)-3-ethylurea; DPX-3217; Biozate; Curzate. $C_7H_{10}N_4O_3$; mol wt 198.18. C 42.42%, H 5.09%, N 28.27%, O 24.22%. Systemic fungicide primarily for potato blight. Prepn: S. H. Davidson, **US 3957847** (1976 to DuPont). Metabolism in plants: Y. Cohen, U. Gisi, *Crop Prot.* **12**, 284 (1993). *In vitro* sensitivity study: R. A. Hamlen, R. J. Power, *Pestic. Sci.* **53**, 101 (1998). GC determn in plants and food products: R. F. Holt, *ibid.* **10**, 455 (1979). Field trials in blight: M. Absi, *Proc. Br. Crop Prot. Conf. - Pests Dis.* **1979**, 311. Brief description: H. L. Klopping, C. J. Delp, *J. Agric. Food Chem.* **28**, 467 (1980).

Crystals from acetone, mp 160-161°. Soly in water approx. 1000 ppm. For 80% formulations, LD_{50} orally in male rats: 1425 mg/kg; dermally in male rabbits: >3000 mg/kg. LC_{50} orally in bobwhite quails, mallard ducks: 2847, >10000 ppm. LC_{50} (96 hr) in bluegill sunfish, rainbow trout (ppm): 13.5, 18.7 (Klopping, Delp).

USE: Agricultural fungicide.

2763. Cynarine. [30964-13-7] (1*R*,3*R*,4*S*,5*R*)-1,3-Bis[[3-(3,4-dihydroxyphenyl)-1-oxo-2-propen-1-yl]oxy]-4,5-dihydroxycy-clohexanecarboxylic acid; 1,3-dicaffeoylquinic acid; 1,5-dicaffe-ylquinic acid; 3,4-dihydroxycinnamic acid 1-carboxy-4,5-dihy-droxy-1,3-cyclohexylene ester; caffeic acid 1-carboxy-4,5-dihy-droxy-1,3-cyclohexylene ester; 1-carboxy-4,5-dihydroxy-1,3-cyclo-hexylenebis-(3,4-dihydroxycinnamate); quinic acid 1,5-dicaffeic ester; Cinarine; Listrocol; Plemocil. $C_{25}H_{24}O_{12}$; mol wt 516.46. C 58.14%, H 4.68%, O 37.17%. Active principle of artichoke, *Cynara scolymus* L., *Compositae:* L. Panizzi, M. L. Scarpati, *Gazz. Chim. Ital.* **84**, 792 (1954). Synthesis of originally proposed structure: L. Panizzi *et al., ibid.* 806; L. Panizzi *et al.,* **US 2863909; US 3100224** (1958, 1963 both to Farmitalia); Alberti *et al.,* **US 2918477** (1959). Revised structure: L. Panizzi, M. L. Scarpati, *Gazz. Chim. Ital.* **95**, 71 (1965). NMR confirmation of structure: I. Horman *et al., J. Agric. Food Chem.* **32**, 538 (1984). HPLC determn in pharmaceu-tics: A. Bettero, *Boll. Chim. Farm.* **120**, 49 (1981). Inhibition of fatty acid mobilization: P. Dorigo, G. Fassina, *Pharmacol. Res. Commun.* **2**, 109 (1984). Reduction of cholesterol levels in rats: J. Wojcicki, *Drug Alcohol Depend.* **3**, 143 (1978). Clinical trial in hyperlipemic syndrome: M. Montini *et al., Arzneim.-Forsch.* **25**, 1311 (1975).

Crystals from dil acetic acid. Sweet taste. mp 225-227°. $[\alpha]_D^{25}$ −59° (c = 2 in methanol). uv max (methanol): 326 nm ($E_{1cm}^{1\%}$ 616). Sparingly sol in cold, more in boiling water; sol in glacial acetic acid, alcohols.

THERAP CAT: Choleretic.

2764. Cypermethrin. [52315-07-8] 3-(2,2-Dichloroethenyl)-2,2-dimethylcyclopropanecarboxylic acid cyano(3-phenoxyphenyl)-methyl ester; (±)-α-cyano-3-phenoxybenzyl-(±)-*cis,trans*-3-(2,2-dichlorovinyl)-2,2-dimethylcyclopropane carboxylate; NRDC-149; FMC-30980; PP-383; Ammo; Arrivo; Barricade; Basathrin; Cym-bush; Cynoff; Demon; Flectron; Ripcord. $C_{22}H_{19}Cl_2NO_3$; mol wt 416.30. C 63.47%, H 4.60%, Cl 17.03%, N 3.36%, O 11.53%. Syn-thetic pyrethroid insecticide, usually exists as a mixture of *cis* and *trans* isomers. Prepn of racemic mixture: M. Elliot *et al.,* **DE 2326077**; *eidem,* **US 4024163** (1974, 1977 to NRDC). Activity: *eidem, Pestic. Sci.* **6**, 537 (1975); M. H. Breese, *ibid.* **8**, 264 (1977). Soil degradation: T. R. Roberts, M. E. Standen, *ibid.* 305; D. D. Kaufman *et al., J. Agric. Food Chem.* **29**, 239 (1981). Metabolism: T. Shono *et al., ibid.* **27**, 316 (1979); M. J. Crawford *et al., ibid.* **29**, 130 (1981). Analysis: A. Sapiets *et al., Anal. Methods Pestic. Plant Growth Regul.* **13**, 33-51 (1984). Repellent activity against flies on cattle: J. A. Shemanchuk, *Pestic. Sci.* **12**, 412 (1981); J. E. Hillerton *et al., Br. Vet. J.* **141**, 160 (1985). Acute toxicity study: F. Cantala-messa, *Arch. Toxicol.* **67**, 510 (1993). Review of toxicology and human exposure: *Toxicological Profile for Pyrethrins and Pyre-throids* (PB2004-100004, 2003) 332 pp.

Commercial product is a mixture of eight isomers. Viscous semi-solid, mp 60-80°. Insol in water. Sol in methanol, acetone, xylene, methylene dichloride. LD_{50} in 8 day old rats, adult male rats (mg/kg): 14.9, 250.0 orally (Cantalamessa).

USE: Insecticide.

THERAP CAT (VET): Ectoparasiticide.

2765. Cyphenothrin. [39515-40-7] 2,2-Dimethyl-3-(2-meth-yl-1-propen-1-yl)cyclopropanecarboxylic acid cyano(3-phenoxy-phenyl)methyl ester; (*RS*)-α-cyano-3-phenoxybenzyl (1*R*)-*cis,trans*-chrysanthemate; α-cyano-*m*-phenoxybenzyl 2,2-dimethyl-3-(2-methylpropenyl)cyclopropanecarboxylate; 3-phenoxy-α-cyanoben-zylchrysanthemate; S-2703; S-2703 Forte; Gokilaht. $C_{24}H_{25}NO_3$; mol wt 375.47. C 76.77%, H 6.71%, N 3.73%, O 12.78%. Synthetic pyrethroid. Prepn: T. Matsuo *et al.,* **DE 2231312**; *eidem,* **US 3835176** (1973, 1974 to Sumitomo). Insecticidal activity: M. Elliot *et al., Pestic. Sci.* **13**, 407 (1982); T. Matsunaga *et al., Pesticide Chemistry: Human Welfare and the Environment* **2**, J. Miyamoto *et al.,* Eds (Pergamon Press, Oxford, 1983) pp 231-238; T. Itoh *et al., J. Am. Mosq. Control Assoc.* **2**, 503 (1986). Mechanism of action: T. Narahashi, *Neurobehav. Toxicol. Teratol.* **4**, 753 (1982). Series of articles on stereoselective syntheses of labelled *cis* and *trans* forms: H. Kanamaru *et al., Radioisotopes* **35**, 103, 109, 242 (1986).

Yellowish viscous liquid. d_{25}^{25} 1.083. Vapor pressure at 30°: 3.11 $\times 10^{-6}$ mm Hg. Viscosity at 30°: 808.8 cP.

Note: Commercial product is a mixture (35:65) of *cis* and *trans* forms.

USE: Insecticide.

2766. Cyprenorphine. [4406-22-8] (5α,7α)-17-(Cyclopropylmethyl)-4,5-epoxy-3-hydroxy-6-methoxy-α,α-dimethyl-6,14-ethenomorphinan-7-methanol; *N*-(cyclopropylmethyl)-7,8-dihydro-7α-(1-hydroxy-1-methylethyl)-*O*⁶-methyl-6,14-*endo*-ethenonormorphine; *N*-(cyclopropylmethyl)-6,14-*endo*-etheno-7α-(2-hydroxy-2-propyl)tetrahydronororipavine; *N*-(cyclopropylmethyl)-6,14-*endo*-ethenotetrahydronororipavine. $C_{26}H_{33}NO_4$; mol wt 423.55. C 73.73%, H 7.85%, N 3.31%, O 15.11%. Narcotic antagonist closely related to diprenorphine, *q.v.* Prepn (7α- or 7β-linkage unspecified): **BE 629070** (1963 to J. F. Macfarlan), *C.A.* **61**, 13364g (1964); K. W. Bentley, **US 3474101** (1969 to Reckitt & Sons); of 7α: K. W. Bentley, D. G. Hardy, *J. Am. Chem. Soc.* **89**, 3281 (1967). Activity in rats: K. W. Bentley *et al.*, *Nature* **206**, 102 (1965). Use in large animals: M. R. Jainudeen, *Vet. Rec.* **89**, 686 (1971); K. R. Presnell *et al.*, *J. Wildl. Dis.* **9**, 336 (1973). Binding to neuroleptic receptors: A. Czlonkowski *et al.*, *Life Sci.* **22**, 953 (1978). Determn by HPLC: I. Jane, A. McKinnon, *J. Chromatogr.* **323**, 191 (1985).

Prisms from methanol, mp 234° (7α- or 7β-linkage unspecified). **Hydrochloride.** [16550-22-4] M-285. $C_{26}H_{33}NO_4 \cdot HCl$. Crystals from methanol, mp 248°.

Note: This is a controlled substance (opium derivative): **21 CFR**, 1308.11.

THERAP CAT (VET): Etorphine antagonist.

2767. Cypripedium. Lady's slipper; American valerian; nerve root; yellow moccasin flower; Noah's ark. Dried rhizome and roots of *Calypso bulbosa* (L.) Oakes (*Cypripedium bulbosum* L.), *Cypripedium pubescens* Willd., or of *Cypripedium parviflorum* Salisb., *Orchidaceae. Habit.* Nova Scotia south to Alabama, west to Nebraska and Missouri. *Constit.* Volatile oil, volatile acid, tannin, resins.

THERAP CAT: Sedative.

2768. Cyproconazole. [94361-06-5] α-(4-Chlorophenyl)-α-(1-cyclopropylethyl)-1*H*-1,2,4-triazole-1-ethanol; (2RS,3RS)-2-(4-chlorophenyl)-3-cyclopropyl-1-(1*H*-1,2,4-triazol-1-yl)butan-2-ol; SAN-619F; SN-108266; Alto; Atemi; Bonanza; Paindor; Sentinel. $C_{15}H_{18}ClN_3O$; mol wt 291.78. C 61.75%, H 6.22%, Cl 12.15%, N 14.40%, O 5.48%. Broad spectrum triazole antifungal; sterol demethylase inhibitor. Prepn: F. Schaub, **DE 3406993**; *idem*, **US 4664696** (1984, 1987 both to Sandoz). Mode of action: M. S. Fuller *et al.*, *Pestic. Biochem. Physiol.* **36**, 115 (1990). Synergism with herbicides: H. R. Kataria, U. Gisi, *Crop Prot.* **9**, 403 (1990). Field trials in winter wheat: J. P. Devey *et al.*, *Brighton Crop Prot. Conf. - Pests Dis.* **1990**, 801; in cereals in combination with prochloraz, *q.v.*: R. I. Harris *et al.*, *ibid.* 813. *Review:* U. Gisi *et al.*, *Proc. Br. Crop Prot. Conf. - Pests Dis.* **1986**, 33-40.

Colorless crystals from hexane/CH_2Cl_2, mp 103-105°. Soly at 25° (%w/w): water 0.0140 ±0.0004; acetone >23; ethanol >23; xylene 12; DMSO >18. Vapor pressure at 20°: 2.6×10^{-7} torr. LD_{50} in male and female rats (mg/kg): 1020, 1330 orally; in rats: 2000 dermally. LD_{50} in carp: 18.9 mg/l; in trout 7.2 mg/l; in bluegill sunfish 6.0 mg/l in water. LD_{50} in bobwhite quail (mg/kg): 150 orally; LD_{50} (8 day dietary) in bobwhite quail, mallard duck: 816, 1197 mg/kg (Gisi, 1986).

Combination with prochloraz. Sportak Delta; Tiptor; Epicure.

USE: Agricultural fungicide.

2769. Cyprodinil. [121552-61-2] 4-Cyclopropyl-6-methyl-*N*-phenyl-2-pyrimidinamine; *N*-(4-cyclopropyl-6-methylpyrimidin-2-yl)aniline; 2-phenylamino-4-methyl-6-cyclopropylpyrimidine; CGA-219417; Chorus; Unix; Vangard. $C_{14}H_{15}N_3$; mol wt 225.30. C 74.64%, H 6.71%, N 18.65%. Methionine biosynthesis inhibitor. Prepn: A. Hubele, **EP 310550**; *idem*, **US 5153200** (1989, 1992 both to Ciba-Geigy). Comprehensive description: U. J. Heye *et al.*, *Crop Prot.* **13**, 541-549 (1994). Mechanism of action: P. Masner *et al.*, *Pestic. Sci.* **42**, 163 (1994). Review of field trials: N. J. Leadbitter *et al.*, *Br. Crop Prot. Counc. Monogr.* **57**, 73-78 (1994).

White crystalline solid, mp 75.9°. vapor pressure (25°): 5.1×10^{-4} Pa (crystal modif. A); 4.7×10^{-4} Pa (crystal modif. B). Log P (*n*-octanol/water): 3.9 (pH 5.0); 4.0 (pH 7.0); 4.0 (pH 9.0). Soly in water at 25° (mg/l): 20 (pH 5.0); 13 (pH 7.0); 15 (pH 9.0). LD_{50} orally or dermally in rats: >2000 mg/kg (Heye).

USE: Agricultural fungicide.

2770. Cyproheptadine. [129-03-3] 4-(5*H*-Dibenzo[*a,d*]cyclohepten-5-ylidene)-1-methylpiperidine; 1-methyl-4-(5*H*-dibenzo[*a,d*]cycloheptenylidene)piperidine; 5-(1-methylpiperidylidene-4)-5*H*-dibenzo[*a,d*]cycloheptene; 1-methyl-4-(5-dibenzo-[*a,e*]cycloheptatrienylidene)piperidine. $C_{21}H_{21}N$; mol wt 287.41. C 87.76%, H 7.37%, N 4.87%. Serotonin and histamine antagonist. Prepn: E. L. Engelhardt, **US 3014911** (1961 to Merck & Co.); E. L. Engelhardt *et al.*, *J. Med. Chem.* **8**, 829 (1965). Improved prepn: M. M. Cid *et al.*, *Tetrahedron* **44**, 6197 (1988). Metabolism: C. C. Porter *et al.*, *Drug Metab. Dispos.* **3**, 189 (1975). HPLC determn in serum and plasma: E. A. Novak *et al.*, *J. Chromatogr.* **339**, 457 (1985). Toxicity study: J. J. Loux *et al.*, *Arzneim.-Forsch.* **28**, 1644 (1978). Comprehensive description: H. Y. Aboul-Enein, A. A. Al-Badr, *Anal. Profiles Drug Subs.* **9**, 155-179 (1980). Clinical trial as an appetite stimulant in cystic fibrosis: D. N. Homnick *et al.*, *Pediatr. Pulmonol.* **40**, 251 (2005).

Crystals from dil ethanol, mp 112.3-113.3°.

Hydrochloride sesquihydrate. [41354-29-4] Periactin; Periactine; Peritol. $C_{21}H_{22}ClN.1\frac{1}{2}H_2O$; mol wt 350.89. Crystals from abs ethanol + ether, dec 252.6-253.6°. uv max ($0.1N$ H_2SO_4): 224, 285 nm ($E_{1cm}^{1\%}$ 1656, 355). 1.0 g is sol in 1.5 ml methanol, 16 ml chloroform, 35 ml alc, 275 ml H_2O. Practically insol in ether. LD_{50} orally in mice: 74.2 mg/kg (Loux).

Hydrochloride monohydrate. [6032-06-0] Crystals, mp 214-216°. Soly in water: ~5 mg/ml.

THERAP CAT: Antihistaminic; antipruritic; appetite stimulant.

THERAP CAT (VET): Antihistaminic; antipruritic; appetite stimulant.

2771. Cyproterone. [2098-66-0] ($1\beta,2\beta$)-6-Chloro-1,2-dihydro-17-hydroxy-$3'H$-cyclopropa[1,2]pregna-1,4,6-triene-3,20-dione; 6-chloro-17-hydroxy-$1\alpha,2\alpha$-methylenepregna-4,6-diene-3,20-dione; 6-chloro-6-dehydro-17α-hydroxy-1,2α-methyleneprogesterone; 6-chloro-1,2α-methylene-4,6-pregnadien-17α-ol-3,20-dione; SH-881; SH-80881. $C_{22}H_{27}ClO_3$; mol wt 374.91. C 70.48%, H 7.26%, Cl 9.46%, O 12.80%. Prepn of free alcohol: Wiechert, Neumann, **DE 1189991** *C.A.* **63**, 1842e (1965); Wiechert, **US 3234093** (1965, 1966 both to Schering AG). Biodynamics in man: Gerhards *et al., Arzneim.-Forsch.* **23**, 1550 (1973). The free alcohol is an antiandrogen; the acetate is both an anti-androgen and a progestogen. Effect on hormone secretion and on spermatogenesis in man: L. Moltz *et al., Contraception* **21**, 393 (1980). Review of pharmacology and clinical studies (acetate) on acne and hirsutism in women: J. Hammerstein *et al., J. Steroid Biochem.* **19**, 591 (1983).

Crystals from ethyl acetate, mp 237.5-240°.

Acetate. [427-51-0] CPA; SH-714; Androcur; Cyprostat. $C_{24}H_{29}ClO_4$; mol wt 416.94. Crystals from diisopropyl ether, mp 200-201°. uv max (methanol): 281 nm (ε 17280).

Mixture of acetate with ethinyl estradiol. Diane 35; Dianette.

THERAP CAT: The acetate as antiandrogen; combination with estrogen in treatment of acne.

2772. Cyromazine. [66215-27-8] N^2-Cyclopropyl-1,3,5-triazine-2,4,6-triamine; 2-cyclopropylamino-4,6-diamino-*s*-triazine; CGA-72662; Larvadex; Neporex; Trigard; Vetrazin. $C_6H_{10}N_6$; mol wt 166.19. C 43.36%, H 6.07%, N 50.57%. Insect growth regulator. Prepn: H. U. Brechbuhler *et al.*, **DE 2736876**; *eidem*, **US 4225598** (1978, 1980 to Ciba-Geigy). Activity: R. D. Hall, M. C. Foehse, *J. Econ. Entomol.* **73**, 564 (1980); R. J. Hart *et al., Aust. Vet. J.* **59**, 104 (1982).

White or off-white crystalline powder, mp 219-222°. Slightly sol in methanol, water.

USE: Insecticide.

THERAP CAT (VET): Ectoparasiticide.

2773. Cystamine. [51-85-4] 2,2′-Dithiobisethanamine; 2,2′-dithiobis[ethylamine]; β,β'-diaminodiethyl disulfide; bis[β-aminoethyl]disulfide; decarboxycystine. $C_4H_{12}N_2S_2$; mol wt 152.27. C 31.55%, H 7.94%, N 18.40%, S 42.11%. Forms when cystine is distilled: Neuberg, Ascher, *Chem. Zentralbl.* **1907**, II, 1156. Prepd

by H_2O_2 oxidation of $H_2NCH_2CH_2SH$: Mills, Jr., Bogert, *J. Am. Chem. Soc.* **62**, 1173 (1940); Barnett, *J. Chem. Soc.* **1944**, 5.

Viscous oil. Cannot be distilled without decompn (even in high vacuum). Freely sol in water. Sol in alc.

Dihydrochloride. [56-17-7] $C_4H_{12}N_2S_2.2HCl$; mol wt 225.19. Needles from methanol, dec 203-214°. Sol in water.

2774. L-Cystathionine. [56-88-2] S-[(2R)-2-Amino-2-carboxyethyl]-L-homocysteine; L-2-amino-4-[(2-amino-2-carboxyethyl)thio]butyric acid. $C_7H_{14}N_2O_4S$; mol wt 222.26. C 37.83%, H 6.35%, N 12.60%, O 28.79%, S 14.42%. An amino acid whose homocysteine and serine moieties are both L in configuration. Intermediate in transsulfuration whereby the mammal transfers the sulfur of methionine via homocysteine to cysteine: Hope, *Proc. 4th Int. Congr. Biochem.* **13**, 63 (1960). Preparation: du Vigneaud *et al., J. Biol. Chem.* **143**, 59 (1942); Weiss, Stekol, *J. Am. Chem. Soc.* **73**, 2497 (1951); Schöberl, Täuber, *Ann.* **599**, 23 (1956). Isolation from human brain: Tallau *et al., J. Biol. Chem.* **230**, 707 (1958). Abs config: G. Jung *et al., Eur. J. Biochem.* **35**, 436 (1973); C.-S. Chen *et al., Biochim. Biophys. Acta* **538**, 534 (1978). *Review:* J. P. Greenstein, M. Winitz, *Chemistry of the Amino Acids* **vol. 3** (John Wiley, N. Y., 1961) p 2682.

Crystals (rectangular parallelepipeds), dec 312° (darkening at 270°). $[\alpha]_D^{20}$ +23.7° ($1N$ HCl).

2775. Cystatins. Protein superfamily of endogenous cysteine proteinase inhibitors. All inhibit papain-type peptidases, including serveral cathepsins. Widely distributed in animals, plants, and protozoa; at least 12 cystatins have been identified in humans. Type 1 (cystatins A and B), also known as *stefins*, are intracellular, single-chain peptides of ~100 residues with no disulfide bonds or carbohydrate. Type 2, comprising the largest group, are extracellular, secreted proteins, typically 120-125 residues with 2 disulfide bonds; broadly distributed and found in most body fluids. Type 3, known as kininogens, *q.v.*, are intravascular, multi-functional, glycosylated proteins of mol wt 60-120 kDa that also serve as kinin precursors. Discovery of a papain inhibitor in chicken egg white: K. Fossum, J. R. Whitaker, *Arch. Biochem. Biophys.* **125**, 367 (1968). Sequence homologies and evolutionary relationship: W. Müller-Esterl *et al., FEBS Lett.* **191**, 221 (1985). Review of physiological role in malignant progression and prognosis: J. Kos, T. T. Lah, *Oncol. Rep.* **5**, 1349-1361 (1998). *Reviews:* A. J. Barrett, *Biomed. Biochim. Acta* **45**, 1363-1374 (1986); M. Abrahamson *et al., Biochem. Soc. Symp.* **70**, 179-199 (2003).

Cystatin A. [107544-29-6] Stefin A; CPI-A; acidic cysteine proteinase inhibitor. Found primarily in epithelial cells and polymorphonuclear leukocytes; also occurs in amniotic fluid and tears. Purification from human epidermis: M. Järvinen, *J. Invest. Dermatol.* **72**, 114 (1978); from human granulocytes: J. Brzin *et al., Z. Phys. Chem.* **364**, 1475 (1983). Mature human form is a single chain, non-glycosylated peptide containing 98 amino acid residues; mol wt 11.0 kDa. Isoelectric point: 4.5-4.7.

Cystatin B. [99194-04-4] Stefin B; CPI-B; NCPI; neutral cysteine proteinase inhibitor. Broadly distributed in human cells and tissues; general cytosolic inhibitor to protect against leakage of lysosomal enzymes. Mutations in the cystatin B gene have been associated with progressive myoclonus epilepsy. Isoln from human spleen: M. Järvinen, A. Rinne, *Biochim. Biophys. Acta* **708**, 210

(1982); from human liver: G. D. J. Green *et al.*, *Biochem. J.* **218**, 939 (1984). Review of role in Unverricht-Lundborg disease: A.-E. Lehesjoki, *EMBO J.* **22**, 3473-3478 (2003). Mature human form is a single chain, non-glycosylated peptide containing 98 amino acid residues; mol wt 11.2 kDa. Isoelectric point: 5.6-6.3.

Cystatin C. [91448-99-6] Post-γ-globulin; γ-CSF; γ-trace. Found ubiquitously in vertebrates; major extracellular cysteine peptidase inhibitor in mammals. Isoln from human CSF: J. Clausen, *Proc. Soc. Exp. Biol. Med.* **107**, 170 (1961); from urine of patients with renal dysfunction: E. A. Butler, F. V. Flynn, *J. Clin. Pathol.* **14**, 172 (1961). Identification as a cystatin: A. J. Barrett *et al.*, *Biochem. Biophys. Res. Commun.* **120**, 631 (1984). Review of biochemistry and clinical role: M. Mussap, M. Plebani, *Crit. Rev. Clin. Lab. Sci.* **41**, 467-550 (2004); of efficacy as biomarker for glomerular filtration rate: G. Filler *et al.*, *Clin. Biochem.* **38**, 1-8 (2005). Clinical evaluation to predict risk of cardiovascular events in elderly patients: M. G. Shlipak *et al.*, *N. Engl. J. Med.* **352**, 2049 (2005). Mature human form is a single chain, non-glycosylated peptide containing 120 amino acid residues; mol wt 13.3 kDa. Isoelectric point: 9.3. Electrophoretic mobility: γ_3 (agarose gel electrophoresis at pH 8.6). $E_{1cm}^{1\%}$ 9.1 (280 nm). Conc in plasma of healthy adults: 0.8 to 1.2 mg/l.

THERAP CAT: Cystatin C as diagnostic aid (renal function).

2776. Cysteamine. [60-23-1] 2-Aminoethanethiol; mercaptamine; β-mercaptoethylamine; 2-aminoethyl mercaptan; thioethanolamine; decarboxycysteine; MEA; mercamine; L-1573. C_2H_7NS; mol wt 77.15. C 31.14%, H 9.15%, N 18.16%, S 41.56%. Naturally occuring sulfhydryl antioxidant with a variety of biological effects. Precursor in the biosynthesis of taurine and constituent of coenzyme A. Protects against ionizing radiation, depletes prolactin in the pituitary, inhibits release of somatostatin, and is a potent ulcerogen. Prepn: S. Gabriel, E. Leupold, *Ber.* **31**, 2832 (1898); L. Knorr, P. Rössler, *ibid.* **36**, 1278 (1903); E. J. Mills, Jr., M. T. Bogert, *J. Am. Chem. Soc.* **62**, 1173 (1940). Identification as constituent of coenzyme A: J. D. Gregory *et al.*, *ibid.* **74**, 854 (1952). Radioprotective effects: Z. M. Bacq *et al.*, *Science* **117**, 633 (1953). Pharmacology and toxicity: E. Beccari *et al.*, *Arzneim.-Forsch.* **5**, 421 (1955). Metabolism: R. A. Salvador *et al.*, *J. Pharmacol. Exp. Ther.* **121**, 258 (1957). Ulcerogenic effect and use as animal model of duodenal ulcer disease: H. Selye, S. Szabo, *Nature* **244**, 458 (1973); S. Szabo, *Am. J. Pathol.* **93**, 273 (1978). Effect on somatostatin concentration: S. Szabo, S. Reichlein, *Endocrinology* **109**, 2255 (1981); S. M. Sagar *et al.*, *J. Neurosci.* **2**, 225 (1982). Depletion of prolactin concentration in pituitary tissue: W. J. Millard *et al.*, *Science* **217**, 452 (1982). Determn in plasma and tissues: M. W. Duffel *et al.*, *Methods Enzymol.* **143**, 149 (1987); in plasma by capillary electrophoresis: P. Kubalczyk, E. Bald, *Electrophoresis* **29**, 3636 (2008). Bioavailability: L. Tennezé *et al.*, *Br. J. Clin. Pharmacol.* **47**, 49 (1999). Clinical trial in treatment of acetaminophen poisoning: L. F. Prescott *et al.*, *Lancet* **2**, 109 (1976); A. L. Harris, *Br. Med. J.* **284**, 825 (1982); in nephropathic cystinosis: M. Yudkoff *et al.*, *N. Engl. J. Med.* **304**, 141 (1981). Review of clinical experience in treatment of nephropathic cystinosis: R. Kleta, W. A. Gahl, *Expert Opin. Pharmacother.* **5**, 2255-2262 (2004); of potential in treatment of neurodegenerative diseases: C. Gobrat, F. Cicchetti, *Prog. Neuro-Psychopharmacol. Biol. Psychiatry* **35**, 380-389 (2011).

$$H_2N \diagup\diagdown SH$$

Crystals by sublimation *in vacuo*. Disagreeable odor. mp 97-98.5°. Very hygroscopic. Oxidizes to cystamine on standing in air. Freely sol in water, alkaline reaction.

Hydrochloride. [156-57-0] $C_2H_7NS.HCl$; mol wt 113.60. Crystals from alc, mp 70.2-70.7°. Sol in water, alcohol. LD_{50} (cg/kg): 23.19 i.p. in rats; 14.95 i.v. in rabbits (Beccari).

Bitartrate. [27761-19-9] RP-103; Cystagon. $C_2H_7NS.C_4H_6O_6$; mol wt 227.23. Clinical pharmacokinetics: E. B. Belldina *et al.*, *Br. J. Clin. Pharmacol.* **56**, 520 (2003). Clinical trial in nephropathic cystinosis: R. Dohil *et al.*, *J. Pediatr.* **156**, 71 (2010); in non-alcoholic fatty liver disease: *idem et al.*, *Aliment. Pharmacol. Ther.* **33**, 1036 (2011). White powder. Highly sol in water.

USE: Experimentally to produce acute and chronic duodenal ulcers in rats.

THERAP CAT: In treatment of nephropathic cystinosis. Antidote to acetaminophen poisoning.

2777. Cysteic Acid. [13100-82-8] 3-Sulfoalanine; α-amino-β-sulfopropionic acid; DL-cysteic acid. $C_3H_7NO_5S$; mol wt 169.15. C 21.30%, H 4.17%, N 8.28%, O 47.29%, S 18.95%. Has been isolated from human hair oxidized with permanganate: Lissizin, *Z. Physiol. Chem.* **173**, 309 (1928). Occurs normally in the outer part of the sheep's fleece, where the wool is exposed to light and weather: Martin, Synge, *Adv. Protein Chem.* **2**, 3 (1945). Prepd from cystine or cysteine by oxidation with bromine in water: Friedmann, *Beitr. Chem. Physiol. Pathol.* **3**, 25, 38; Gortner, Hoffman, *J. Biol. Chem.* **72**, 435 (1927).

$$\overset{\displaystyle COOH}{\underset{\displaystyle H_2N}{\diagup}}\diagdown SO_3H$$

Crystals, dec 245°.

L-Form. [498-40-8] Octahedra or needles from dil alc (also forms a monohydrate, prismatic needles). When anhydr, dec 260°. $[\alpha]_D^{20}$ +8.66° (1.85 g in 25 ml). pKa_1 (25°): 1.89; pKa_2 8.7; pKb about 12.7. Soluble in water. Insol in alcohol.

2778. Cysteine. [52-90-4] L-Cysteine; Cys; C; β-mercapto-alanine; (*R*)-2-amino-3-mercaptopropanoic acid; 2-amino-3-mercaptopropionic acid; α-amino-β-thiolpropionic acid; half-cystine; thioserine. $C_3H_7NO_2S$; mol wt 121.15. C 29.74%, H 5.82%, N 11.56%, O 26.41%, S 26.46%. A non-essential amino acid in human development. Readily oxides to form a dimeric amino acid, cystine, *q.v.*, in which the two Cys are linked via a disulfide bridge, a common structural feature in proteins. Early chemistry and biochemistry: *Amino Acids and Proteins*, D. M. Greenberg, Ed. (Charles C. Thomas, Springfield, IL, 1951) 950 pp., *passim*; J. P. Greenstein, M. Winitz, *Chemistry of the Amino Acids* vol 1-3 (John Wiley and Sons, Inc., New York, 1961) pp. 1879-1928, *passim*. Simple synthesis of racemic cysteine: V. J. Martens *et al.*, *Angew. Chem. Int. Ed.* **20**, 668 (1981). Determn in proteins: J. G. Hoogerheide, C. M. Campbell, *Anal. Biochem.* **201**, 146 (1992); D. Atherton *et al.*, *ibid.* **212**, 98 (1993). Review of biosynthesis: N. M. Kredich *et al.*, *Ciba Found. Symp.* (Netherlands) **72**, 87-99 (1980). Review of transport in mammalian cells: S. Bannai, *Biochim. Biophys. Acta* **779**, 289-306 (1984). Review of effects on acrylonitrile toxicity: D. E. Nerland *et al.*, *Drug Metab. Rev.* **20**, 233-246 (1989). Review of thermodynamics and kinetics: T. R. Ralph *et al.*, *J. Electroanal. Chem.* **375**, 1-15 (1994); of electrosynthesis: *eidem, ibid.* 17-27. Review of role in chemo- and radioprotectant strategies: J. C. Roberts, *Amino Acids* **8**, 113-124 (1995).

$$HS\diagup\diagdown\overset{\displaystyle O}{\underset{\displaystyle NH_2}{\diagdown}}OH$$

Crystals. $[\alpha]_D^{25}$ +6.5° (5N HCl); $[\alpha]_D^{25}$ +13.0° (glacial acetic acid). pK_1 1.71; pK_2 8.33; pK_3 10.78. Absorption spectrum: Abderhalden, Rossner, *Z. Physiol. Chem.* **178**, 160 (1928). Freely sol in water, alcohol, acetic acid, ammonia water. Insol in ether, acetone, ethyl acetate, benzene, carbon disulfide, carbon tetrachloride. In neutral or slightly alkaline aq solns it is oxidized to cystine by air. More stable in acidic solns.

Hydrochloride. [52-89-1] $C_3H_7NO_2S.HCl$. White crystals or crystalline powder, dec 175-178°. $[\alpha]_D^{25}$ +5.0° (5N HCl); $[\alpha]_D^{25}$ +10.0° (glacial acetic acid). Sol in water, alcohol, acetone; the aq soln is acid. *Keep tightly closed*. Decomposes and oxidizes slowly; hygroscopic.

USE: As dough conditioner.

THERAP CAT (VET): Has been used as a detoxicant.

2779. Cystine. [56-89-3] L-Cystine; [*R*-(*R**,*R**)]-3,3'-dithiobis[2-aminopropanoic acid]; dicysteine; β,β'-dithiodialanine; α-diamino-β-dithiolactic acid; β,β'-diamino-β,β'-dicarboxydiethyl disulfide; bis(β-amino-β-carboxyethyl) disulfide; Gelucystine. $C_6H_{12}N_2O_4S_2$; mol wt 240.29. C 29.99%, H 5.03%, N 11.66%, O

26.63%, S 26.68%. Non-essential amino acid for human development. Formed by the dimerization of two cysteines, *q.v.* through the sulfur. These disulfide bridges occur both within and between polypeptides; often found in extracellular proteins. First amino acid described in 1810 by Wollaston. Isoln from horn hydrolysate: K. A. H. Mörner, *Z. Physiol. Chem.* **28**, 595 (1899). Early chemistry and biochemistry: *Amino Acids and Proteins,* D. M. Greenberg, Ed. (Charles C. Thomas, Springfield, IL, 1951) 950 pp. *passim;* J. P. Greenstein, M. Winitz, *Chemistry of the Amino Acids* vols 1-3 (John Wiley and Sons, Inc., New York, 1961) pp. 1879-1928, *passim.* Distribution in protein: R. C. Fahey *et al., J. Mol. Evol.* **10**, 155 (1977). Review of lysosomal transport including pathophysiology: W. Gahl in *Pathophysiol. Lysosomal Transp.,* J. G. Thoene, Ed. (CRC Press, Boca Raton, FL, 1992) pp 45-71. Review of thermodynamics and kinetics: T. R. Ralph *et al., J. Electroanal. Chem.* **375**, 1-15 (1994); of electrosynthesis: *eidem, ibid.* 17-27.

Hexagonal tablets from water, dec 260-261° (sealed tube). $[\alpha]_D^{20}$ −223.4° (1.0*N* HCl). pK_1 1; pK_2 2.1; pK_3 8.02; pK_4 8.71 at 35°. Soly in water (g/l) at 25°: 0.112; at 50°: 0.239; at 75°: 0.523; at 100°: 1.142. Quite sol in aq solns below pH 2 or above pH 8. Soly curves: Sano, *Biochem. Z.* **168**, 14 (1926). Insol in alc. Absorption spectrum: Marchlewski, Nowotonowna, *Bull. Soc. Chim. Fr.* [4] **39**, 163, 166 (1926).

D-Form. [349-46-2] Crystals. $[\alpha]_D^{20}$ +223° (1.0*N* HCl). Soly in water at 25°: 0.057 g/l.

DL-Form. [923-32-0] Crystals. Soly in water at 25°: 0.057 g/l.

***meso*-Form.** [6020-39-9] Crystals. Soly in water: 0.056 g/l.

2780. L-Cystine S,S-Dioxide. [30452-69-8] 2-Amino-2-carboxyethyl 2-amino-2-carboxyethanethiosulfonate; L-cystine thiosulfonate; sacysyl-cysteine. $C_6H_{12}N_2O_6S_2$; mol wt 272.29. C 26.47%, H 4.44%, N 10.29%, O 35.25%, S 23.55%. Oxidation product of cystine. Prepn: G. Toennies, T. F. Lavine, *J. Biol. Chem.* **113**, 571 (1936); R. Emilozzi, L. Pichat, *Bull. Soc. Chim. Fr.* **1959**, 1887. Orignally thought to be a mixture of two isomers, cystine *S,S*-dioxide and ***cystine S,S'-dioxide***: G. E. Utzinger, *Experientia* **17**, 374 (1961). Elucidaton of structures: G. Axelson *et al., Spectrochim. Acta* **23A**, 2015 (1967); L. D. Setiawan *et al., Surf. Interface Anal.* **7**, 188 (1985). Use in determn of sulfite: T. Ubuka *et al., Anal. Biochem.* **126**, 273 (1982); *idem et al., ibid.* **140**, 449 (1984).

Solid, relatively unstable in aqueous solns; disproportionates to cystine and cysteine sulfinic acid.

USE: In detection of sulfites and thiosulfate.

2781. Cytarabine. [147-94-4] 4-Amino-1-β-D-arabinofuranosyl-2(1*H*)-pyrimidinone; 1-β-D-arabinofuranosylcytosine; Ara-C; β-cytosine arabinoside; aracytidine; CHX-3311; U-19920; Alexan; Aracytine; Cytosar; Depocyte; Udicil. $C_9H_{13}N_3O_5$; mol wt 243.22. C 44.44%, H 5.39%, N 17.28%, O 32.89%. Nucleoside analog; converted by cellular kinases into the active metabolite, *AraCTP*. Prepn: J. H. Hunter, US 3316282 (1963 to Upjohn); T. Y. Shen *et al., J. Org. Chem.* **30**, 835 (1965). NMR soln structure of Ara-C within a DNA dodecamer: B. I. Schweitzer *et al., Biochemistry* **33**, 11460 (1994). Crystal structure of complex with human topoisomerase I: J. E. Chrencik *et al., J. Biol. Chem.* **278**, 12461 (2003). Clinical pharmacology and toxicology: R. C. Donehower *et al., Cancer Treat. Rep.* **70**, 1059 (1986). Symposium on clinical pharmacology, pharmacokinetics and efficacy in leukemia: *Scand. J. Haematol.* **36**, Suppl. 44, 1-74 (1986). Clinical pharmacokinetics of liposomal formulations: A. Hamada *et al., Clin. Pharmacokinet.* **41**, 705 (2002). Review of development of a high dose treatment for acute myeloid

leukemia: R. L. Capizzi, *Invest. New Drugs* **14**, 249-256 (1996); of cellular metabolism and mechanism of action: S. Grant, *Adv. Cancer Res.* **72**, 197-233 (1998).

Crystals from ethanol, mp 212-213°. $[\alpha]_D^{24}$ +153° (c = 0.5 in water). uv max at pH 2: 281.0, 212.5 nm (ε 13171, 10230); at pH 12: 272.5 nm (ε 9259). Freely sol in water; slightly sol in alcohol, chloroform.

THERAP CAT: Antineoplastic.

THERAP CAT (VET): Antineoplastic.

2782. Cythioate. [115-93-5] Phosphorothioic acid *O*-[4-(aminosulfonyl)phenyl] *O,O*-dimethyl ester; phosphorothioic acid *O,O*-dimethyl ester *O*-ester with *p*-hydroxybenzenesulfonamide; *O,O*-dimethyl *O*-*p*-sulfamoylphenylphosphorothioate; CL-26691; ENT-25640; Proban. $C_8H_{12}NO_5PS_2$; mol wt 297.28. C 32.32%, H 4.07%, N 4.71%, O 26.91%, P 10.42%, S 21.57%. Organophosphate insecticide; cholinesterase inhibitor. Prepn: G. Berkelhammer, US 3005004 and R. I. Hewitt, G. Berkelhammer, US 3179560 (1961, 1965 to Am. Cyanamid). Pharmacodynamics: H. G. Smith, R. L. Goulding, *J. Econ. Entomol.* **63**, 1640 (1970). Efficacy as ectoparasiticide: C. P. Doval, I. Gupta, *Indian Vet. J.* **55**, 890 (1978); P. M. Bowen, N. J. Caldwell, *Vet. Med. Small Anim. Clin.* **77**, 79 (1982). Toxicity data: E. E. Kenaga, W. E. Allison, *Bull. Entomol. Soc. Am.* **15**, 85 (1969).

Crystals, mp 70-71°. n_D^{25} 1.5346. LD_{50} orally in rats: 160 mg/kg (Kenaga, Allison).

THERAP CAT (VET): Ectoparasiticide.

2783. Cytidine. [65-46-3] 4-Amino-1-β-D-ribofuranosyl-2-(1*H*)-pyrimidinone; cytosine riboside; 1-β-D-ribofuranosylcytosine. $C_9H_{13}N_3O_5$; mol wt 243.22. C 44.44%, H 5.39%, N 17.28%, O 32.89%. Constituent of nucleic acids. Isoln from yeast nucleic acid: Levene, Jacobs, *Ber.* **43**, 3154 (1910); Levene, La Forge, *ibid.* **45**, 608 (1912). Sepn from other nucleosides by ion-exchange chromatography: Cohn in Chargaff-Davidson, *The Nucleic Acids* vol. I (New York, 1955) p 211. Synthesis: Howard *et al., J. Chem. Soc.* **1947**, 1052. Crystal structure: Furberg *et al., Acta Crystallogr.* **18**, 313 (1965). *Review: Basic Principles in Nucleic Acid Chemistry* vol. 1, P. O. P. Ts'o, Ed. (Academic Press, New York, 1974) *passim.*

Long needles from 90% ethanol, dec 220-230°. $[\alpha]_D^{25}$ +31° (c = 0.7 in water). Freely sol in water, less sol in alcohol. pK (amino, cationic) 4.22; pK (sugar, anionic) 12.5. uv max (pH 8.2): 271 nm

(ε 9100); (pH 2.2): 280 nm (ε 13400), Voet *et al.*, *Biopolymers* **1**, 193 (1963).

Sulfate. $(C_9H_{13}N_3O_5)_2 \cdot H_2SO_4$. Long prismatic needles, mp 224-225° (dec with effervescence). $[\alpha]_{589}^{25}$ +34°; $[\alpha]_{546}^{25}$ +43°.

2784. 2′-Cytidylic Acid. [85-94-9] Cytidine-2′-monophosphate; cytidylic acid *a*; 2′-cytidinephosphoric acid; cytidine-2′-phosphate; 2′-cytosylic acid; 2′-CMP. $C_9H_{14}N_3O_8P$; mol wt 323.20. C 33.45%, H 4.37%, N 13.00%, O 39.60%, P 9.58%. Ribonuclease inhibitor. Prepn from yeast ribonucleic acid: Cohn, Carter, *J. Am. Chem. Soc.* **72**, 2606 (1950); Loring *et al.*, *J. Biol. Chem.* **196**, 807 (1952); by phosphorylation of $N^6,O^{3'},O^{5'}$-tribenzoylcytidine: Rammler, Khorana, *J. Am. Chem. Soc.* **84**, 3112 (1962). Crystal structure and conformation of the trihydrate: Kartha *et al.*, *Science* **179**, 495 (1973). *Reviews: See* Cytidine.

Crystals, dec 238-240°. $[\alpha]_D^{20}$ +20.7° (water); $[\alpha]_D^{25}$ −3° (aq NaOH adjusted to pH 10). uv max at pH 2: 278 nm (ε 12.7); at pH 12: 272 nm (ε 8.6). pKa_1 0.8; pKa_2 4.36; pKa_3 6.17. Sparingly sol in water. Less sol than 3′-cytidylic acid.

2785. 3′-Cytidylic Acid. [84-52-6] Cytidine-3′-monophosphate; cytidylic acid *b*; 3′-cytidinephosphoric acid; cytidine-3′-phosphate; 3′-cytosylic acid; 3′-CMP. $C_9H_{14}N_3O_8P$; mol wt 323.20. C 33.45%, H 4.37%, N 13.00%, O 39.60%, P 9.58%. Ribonuclease inhibitor. Prepn from yeast ribonucleic acid: *see* refs under 2′-Cytidylic Acid. Prepn by pancreatic-ribonuclease-catalyzed ring opening of dicyclohexylguanidinium cytidine 2′,3′-cyclic phosphate: Lohrmann, Khorana, *J. Am. Chem. Soc.* **86**, 4188 (1964). X-ray analysis studies of orthorhombic form: Alver, Furberg, *Acta Chem. Scand.* **13**, 910 (1959); Sundaralingam, Jensen, *J. Mol. Biol.* **13**, 914 (1965); of monoclinic form: Bugg, Marsh, *ibid.* **25**, 67 (1967). *Reviews: See* Cytidine.

Crystals, dec 232-234°. $[\alpha]_D^{20}$ +49.4° (c = 1 in water). $[\alpha]_D^{25}$ +50.0° (c = 1 in aq NaOH adjusted to pH 10). uv max at pH 2: 279 nm (ε 13.0); at pH 12: 272 nm (ε 8.90). pKa_1 0.8; pKa_2 4.28; pKa_3 6.0. Moderately sol in water; more sol than 2′-cytidylic acid.

2786. Cytisine. [485-35-8] (1*R*,5*S*)-1,2,3,4,5,6-Hexahydro-1,5-methano-8*H*-pyrido-[1,2-*a*][1,5]diazocin-8-one; baptitoxine; sophorine; ulexine; Cytiton. $C_{11}H_{14}N_2O$; mol wt 190.25. C 69.45%, H 7.42%, N 14.72%, O 8.41%. Toxic principle in seed of *Laburnum anagyroides* Medik. and other *Leguminosae*. Extraction: Ing, *J. Chem. Soc.* **1931**, 2200; Späth, Galinovsky, *Ber.* **65**, 1526 (1932); **66**, 1338 (1933); Lecoq, *Bull. Soc. Chim. Fr.* **10**, 153 (1943). Structure: Ing, *J. Chem. Soc.* **1932**, 2778. Synthesis: Bohlmann *et al.*, *Angew. Chem.* **67**, 708 (1955); Van Tamelen, Baran, *J. Am. Chem.*

Soc. **77**, 4944 (1955). Absolute configuration: Okuda *et al.*, *Chem. Ind. (London)* **1961**, 1751. Pharmacological properties: R. B. Barlow, L. J. McLeod, *Br. J. Pharmacol.* **35**, 161 (1969).

Orthorhombic prisms from acetone, mp 152-153°. Sublimes. bp_2 218°. $[\alpha]_D^{17}$ −120°. pK_1 6.11; pK_2 13.08. Sol in 1.3 parts water, 13 parts acetone, 1.3 parts methanol, 3.5 parts alcohol, 30 parts benzene, 10 parts ethyl acetate, 2.0 parts chloroform. Practically insol in petr ether. LD_{50} in mice (mg/kg): 1.73 i.v.; 9.4 i.p.; 101 orally (Barlow, McLeod).

Hydrochloride. $C_{11}H_{14}N_2O \cdot HCl$. Deliquescent crystals, sol in water and alcohol. pH of 0.1 molar aq soln 4.3.

Nitrate monohydrate. $C_{11}H_{14}N_2O \cdot HNO_3 \cdot H_2O$. Needles or leaflets, $[\alpha]_D$ −81.5°. Sol in water, slightly sol in alcohol. Practically insol in ether.

2787. Cytochalasins. A class of mold metabolites exhibiting a number of unusual and varied effects on animal cells. More than twenty cytochalasins are known, isolated from several different mold spp. All are characterized by a highly substituted hydrogenated isoindole ring of known configuration to which is fused a macrocyclic ring. This ring may vary from 11 to 14 atoms in size and may be either a carbocycle or a lactone. Isoln and structure of cytochalasins A, B, C, D: Aldridge *et al.*, *J. Chem. Soc. C* **1967**, 1667; *eidem*, *Chem. Commun.* **1967**, 26; of E and F: Aldridge *et al.*, *ibid.* **1972**, 148. Revised structures of cytochalasins E and F: Aldridge *et al.*, *ibid.* **1973**, 551; Büchi *et al.*, *J. Am. Chem. Soc.* **95**, 5423 (1973). Isoln of H, also known as *paspaline-P* or *kodocytochalasin-1*: G. S. Pendse, *Experientia* **30**, 107 (1974). Structure of H: S. A Patwardhan *et al.*, *Phytochemistry* **13**, 1985 (1974); M. A. Beno *et al.*, *J. Am. Chem. Soc.* **99**, 4123 (1977); x-ray crystal and molecular structure: J. A. McMillan *et al.*, *Chem. Commun.* **1977**, 105. Isoln of K, L, M from *Chalara microspora* and proposed structures: T. Fex, *Tetrahedron Lett.* **22**, 2703 (1981). Isoln of E and K from *Aspergillus clavatus* and proposed alternate structure of K: P. S. Steyn *et al.*, *J. Chem. Soc. Perkin Trans. 1* **1982**, 541. Partial synthesis of A and B: S. Masamune *et al.*, *J. Am. Chem. Soc.* **99**, 6756 (1977). Total synthesis of B: G. Stork *et al.*, *ibid.* **100**, 7775 (1978). Major biological effects are the blockage of cytoplasmic cleavage by blocking formation of contractile microfilament structures, resulting in multinucleate cell formation, the reversible inhibition of cell movement, and the induction of nuclear extrusion: Carter, *Nature* **213**, 261 (1967); Krishan, *J. Cell Biol.* **54**, 657 (1972); E. D. Korn, *Physiol. Rev.* **62**, 703 (1982). Correlation between effects of cytochalasins on cellular structures and cellular events and those on actin *in vitro*: I. Yahara *et al.*, *J. Cell Biol.* **92**, 69 (1982). Other reported effects include the inhibition of glucose transport, of thyroid secretion, of growth hormone release, of phagocytosis, and of platelet aggregation and clot contraction. *See*: D. A. Hume *et al.*, *Nature* **272**, 359 (1978). Nomenclature: M. Binder *et al.*, *J. Chem. Soc. Perkin Trans. 1* **1973**, 1146. *Reviews:* M. Binder, C. Tamm, *Angew. Chem. Int. Ed.* **12**, 370 (1973); R. B. Herbert in *The Alkaloids* vol. 7, J. E. Saxton, Ed. (The Chemical Society, London, 1977) pp 29-30; W. G. Thilly *et al.*, *Front. Biol.* **46**, 53-64 (1978); L. V. Domnina *et al.*, *Proc. Natl. Acad. Sci. USA* **79**, 7754-7757 (1982); W. Siess *et al.*, *ibid.* 7709-7713.

Cytochalasin B

Cytochalasin B. [14930-96-2] (7*S*,13*E*,16*R*,20*R*,21*E*)-7,20-Dihydroxy-16-methyl-10-phenyl-24-oxo[14]cytochalasa-6(12),13,21-triene-1,23-dione; (*E*,*E*)-16-benzyl-6,7,8,9,10,12a,13,14,15,15a,-16,17-dodecahydro-5,13-dihydroxy-9,15-dimethyl-14-methylene-2*H*-oxacyclotetradec[2,3-*d*]isoindole-2,18(5*H*)-dione; phomin. $C_{29}H_{37}NO_5$; mol wt 479.62. The most important and biologically studied of the cytochalasins. Formerly isolated from cultures of a *Phoma* sp. and called phomin: Rothweiler, Tamm, *Experientia* **22**, 750 (1966). Physicochemical data: A. E. Pohland *et al.*, *Pure Appl. Chem.* **54**, 2219 (1982). Felted needles from acetone, mp 218-221°. Completely stable under normal conditions. Solutions in DMSO show no decrease in potency when stored at 4° for three years (Aldrich data sheet). Solubility (mg/ml at 24°): acetone 10.3; ethanol 35.4; DMSO 371; DMF 492. Insol in water.

USE: As tools in cytological research and in characterization of polymerization properties of actin, *q.v.*

2788. Cytochrome *c*. [9007-43-6] Myohematin; hematin-protein; Cytorest. Electron transport protein found in the mitochondrial intermembrane space of all eukaryotes. Transfers electrons between complex III and complex IV of the respiratory chain. Synthesized in the cytosol as *apocytochrome c*, then transported across the mitochondrial outer membrane. Heme lyase covalently attaches the heme *c* prosthetic group, a derivative of iron protoporphyrin IX, *q.v.* During apoptosis, cytochrome *c* is released from the mitochondia and acts as a cofactor for caspase, *q.v.* activation. Human form consists of a single polypeptide chain of 104 amino acid residues with the heme group covalently attached through thioether linkages at cysteine residues 14 and 17. Isoln from muscle tissue: C. A. MacMunn, *Proc. R. Soc. London* **39**, 248 (1886). Identification and absorption spectra of cellular pigments called cytochromes: D. Keilin, *Proc. R. Soc. London Ser. B* **98**, 312 (1925). Redox function and isoln from baker's yeast: *idem, ibid.* **106**, 418 (1930). Absorption spectrum: M. Dixon *et al., ibid.* **109**, 29 (1931). Improved prepn from vertebrate and invertebrate sources: E. Margoliash, O. F. Walasek, *Methods Enzymol.* **10**, 339 (1967). Isoln and purification from plant sources: M. Richardson *et al., Phytochemistry* **9**, 2271 (1970). Oxidation-reduction potentials: F. L. Rodkey, E. G. Ball, *J. Biol. Chem.* **182**, 17 (1950). First crystn from King penguin: G. Bodo, *Nature* **176**, 829 (1955). Electron transport function in respiration: E. C. Slater, *Adv. Enzymol. Relat. Subj. Biochem.* **20**, 147 (1958). First elucidation of amino acid sequence was horse heart: E. Margoliash *et al., J. Biol. Chem.* **237**, 2148, 2151, 2161 (1962); of human heart: H. Matsubara, E. L. Smith, *ibid.* **238**, 2732 (1963). Crystal structure of horse heart and bonito cytochrome *c*: R. E. Dickerson *et al., ibid.* **246**, 1511 (1971). Structural properties of apo-, iron-free porphyrin, and native cytochrome *c*: W. R. Fisher *et al., ibid.* **248**, 3188 (1973). Discovery and characterization of apoptotic function: X. Liu *et al., Cell* **86**, 147 (1996); H. Zou *et al., J. Biol. Chem.* **274**, 11549 (1999). Review of role in apoptosis: X. Jiang, X. Wang, *Annu. Rev. Biochem.* **73**, 87-106 (2004). *Reviews:* R. Lemberg, J. W. Legge, *Hematin Compounds and Bile Pigments* (Wiley, New York, 1949); E. Margoliash, A. Schejter, *Adv. Protein Chem.* **21**, 113-286 (1966); H. A. Harbury, R. H. L. Marks, in *Inorganic Biochemistry* vol. 2, G. L. Eichhorn, Ed. (Elsevier, New York, 1973) pp 902-954. *Books: Structure and Function of Cytochromes*, K. Okunuki *et al.*, Eds. (University Park Press, Baltimore, 1968) 742 pp; G. W. Pettigrew, G. R. Moore, *Cytochromes c: Biological Aspects*, (Springer-Verlag, Berlin, 1987) 282 pp.; G. R. Moore, G. W. Pettigrew, *Cytochromes c: Evolutionary, Structural and Physicochemical Aspects*, (Springer-Verlag, Berlin, 1990) 478 pp.; *Cytochrome c: A Multidisciplinary Approach*, R. A. Scott, A. G. Mauk, Eds. (University Science, Sausalito, 1996) 738 pp.

Red, basic, water soluble protein. Mol wt about 12500. pI: 10.65 (0°); 10.05 (20°). $E_0' = +0.254$ v (pH 1.75-7.8). Extinction ratio for sequences containing 1 tryptophan: $E_{550\ red}/E_{280\ ox} = 1.25$; for sequences containing two tryptophans: $E_{550\ red}/E_{280\ ox} = 1.04$-1.06. Reduced form, **Ferrocytochrome *c*:** absorption max 416, 520, 550 nm (ε_{416} 129.1, ε_{550} 29.5 mM); oxidized form, **Ferricytochrome *c*:** 279, 360, 410, 530 nm (ε_{410} 106.1 mM). Contains one heme per mole weight of 12,500.

2789. Cytochromes P$_{450}$. P$_{450}$. Hemoproteins involved in enzyme hydroxylation, demethylation, and *N*-oxidation; present in a number of mammalian tissues but in highest concentrations in adrenal and liver microsomes. Has also been found in insects, yeast,

plants, and bacteria. Existence of mitochondrial P$_{450}$ first reported: Harding *et al., Biochim. Biophys. Acta* **92**, 415 (1964). Extraction and purification: Mitani, Horie, *J. Biochem.* **65**, 269 (1969). Acts as a monooxygenase in catalyzing metabolic functions; influences the rate of drug metabolism. Name derived from the characteristic absorption at 450 nm. *Reviews:* several authors in *Oxidases and Redox Systems* **vol. 2**, King *et al.*, Eds. (University Park Press, Baltimore, 1971); J. T. Groves, *Advances in Inorganic Biochemistry* **vol. 1**, G. L. Eichhorn, L. G. Marzilla, Eds. (Elsevier/North-Holland, New York, 1979) pp 119-145; several authors in *Xenobiotica* **12**, 671-800 (1982). *Books: Cytochromes P-450 & B5: Structure, Function & Interaction*, D. Y. Cooper *et al.*, Eds. (Plenum, New York, 1975) 554 pp; *Cytochrome P-450*, R. Sato, T. Omura, Eds. (Academic Press, New York, 1978) 233 pp; *Biochemistry, Biophysics and Regulation of Cytochrome P-450*, J. A. Gustafsson *et al.*, Eds. (Elsevier/North-Holland, New York, 1980) 626 pp.

2790. Cytohemin. [19554-22-4] (*SP*-5-13)-Chloro[(4*E*,8*E*)-7-ethenyl-17-formyl-12-(1-hydroxy-5,9,13-trimethyl-4,8,12-tetra-decatrienyl)-3,8,13-trimethyl-21*H*,23*H*-porphine-2,18-dipropanoato(4−)-κN^{21},κN^{22},κN^{23},κN^{24}]ferrate(2−) dihydrogen. $C_{49}H_{56}ClFeN_4O_6$; mol wt 888.30. C 66.25%, H 6.35%, Cl 3.99%, Fe 6.29%, N 6.31%, O 10.81%. Isoln from heart muscle: Warburg, Gewitz, *Z. Physiol. Chem.* **288**, 1 (1951); Lemberg, *Biochem. Z.* **338**, 97 (1963). Structure: M. Grassl *et al., ibid.* **337**, 35 (1963); *ibid.* **338**, 771 (1963). Side chain may be a mixture of saturated and unsaturated forms: R. Sayffert *et al., 3rd Proc. Colloq. Hemes Hemoproteins, 1966,* 45-59.

Amorphous powder.

2791. Cytolipin H. [4682-48-8] 1-*O*-(4-*O*-β-D-Galactopyranosyl-β-D-glucopyranosyl)ceramide; lactosylceramide; ceramide-β-lactoside; cerebronylsphingosylglucosidogalactoside; Galβ1 → 4Glcβ1 → 1cer. A glycosphingolipid containing lactose and a ceramide (*N*-acyl fatty acid deriv of a sphingosine). Fatty acid composition is variable but is primarily palmitic, behenic and lignoceric acids. First isolated from human epidermoid carcinoma cells, later found to be one of the major neutrophil glycolipids serving as a neutrophil-differentiation marker. Isoln: M. M. Rapport *et al., Cancer* **12**, 438 (1959). Simplified prepn from ox spleen: M. M. Rapport *et al., J. Biol. Chem.* **237**, 1056 (1962). Structure elucidation: A. C. Schram *et al., Nature* **197**, 1074 (1963). Synthesis: D. Shapiro, E. S. Rachaman, *ibid.* **201**, 878 (1964); K. C. Nicolaou *et al., J. Am. Chem. Soc.* **110**, 7910 (1988). Serological activity: M. M. Rapport, L. Graf, *Nature* **201**, 879 (1964); and effect of fatty acid chain length: L. Graf, M. M. Rapport, *Chem. Phys. Lipids* **13**, 367 (1974). Characterization and distribution of glycolipids in leukemic and nonleukemic blood cells: J. Hildebrand *et al., J. Lipid Res.* **12**, 361 (1971); in neutrophils: B. A. Macher, J. C. Klock, *J. Biol. Chem.* **255**, 2092 (1980). Role as neutrophil cell surface marker: F. W. Symington *et al., J. Immunol.* **134**, 2498 (1985). Intracellular localization: *eidem, J. Biol. Chem.* **262**, 11356 (1987). TLC determn: K. Ogawa *et al., J. Chromatogr.* **426**, 188 (1988). Clinical implications in inflammatory bowel disease: C. R. Stevens *et al., Gut* **29**, 580 (1988). Review of structure, organization and role of glycolipids in the cell surface membrane: S. Hakomori, *Annu. Rev. Biochem.* **50**, 733-764 (1981).

Coral-like clusters from pyridine + acetone, mp 230-240° with sintering at 180-190°. $[\alpha]_D^{24}$ −10.8° (c = 2 in pyridine). Practically insol in water, cold methanol, acetone, cold acetic acid, ether, acetonitrile. Soluble in pyridine, chloroform, hot methanol, hot acetic acid. Dissolves readily in chloroform + methanol mixtures of high chloroform content.

2792. Cytosine. [71-30-7] 6-Amino-2(1*H*)-pyrimidinone; 4-amino-2-oxo-1,2-dihydropyrimidine; 4-amino-2-pyrimidinol; 4-amino-2-hydroxypyrimidine. $C_4H_5N_3O$; mol wt 111.10. C 43.24%, H 4.54%, N 37.82%, O 14.40%. Widely distributed in nature; constituent of nucleic acids. Isoln by hydrolysis of thymus nucleic acids: Levene, Bass, *Nucleic Acids* (New York, 1931) p 57. Syntheses: Hilbert, Johnson, *J. Am. Chem. Soc.* **52**, 1152 (1930); Hilbert *et al.*, *ibid.* **57**, 552 (1935); Hunter, Hlynka, *Biochem. J.* **31**, 486 (1937); Ballweg, *Tetrahedron Lett.* **1968**, 2171; David, Lubineau, *Bull. Soc. Chim. Fr.* **1969**, 816. Synthesis from carbon dioxide and ammonia on kaolinite: G. R. Harvey *et al.*, *Naturwissenschaften* **58**, 624 (1971). Exists in keto form. Structure: Barker, Marsh, *Acta Crystallogr.* **17**, 1581 (1964); of monohydrate: Jeffrey, Kinoshita, *ibid.* **16**, 20 (1963); R. J. McClure, B. M. Craven, *ibid.* **B29**, 1234

(1973). Tautomerism: Y. P. Wong, *J. Am. Chem. Soc.* **95**, 3511 (1973). *Review:* Ts'o, "Bases, Nucleosides and Nucleotides" in *Basic Principles in Nucleic Acid Chemistry* **vol. 1**, P. O. P. Ts'o, Ed. (Academic Press, New York, 1974) pp 453-584.

Monohydrate. Lustrous monoclinic or triclinic platelets from water. Anhydr at 100°; brown around 300°; dec 320-325°. uv max (pH 8.8): 196.5, 267 nm ($\varepsilon \times 10^{-3}$ 22.5, 6.1). pK_1 4.60; pK_2 12.16. One gram dissolves in 130 ml water. Slightly sol in alcohol, insol in ether. Gives red color when dissolved in soln of sodium hypochlorite to which a drop of NH_4OH is added. Forms salts with acids.

D

2793. 2,4-D. [94-75-7] 2-(2,4-Dichlorophenoxy)acetic acid; Trinoxol. $C_8H_6Cl_2O_3$; mol wt 221.03. C 43.47%, H 2.74%, Cl 32.08%, O 21.72%. Prepd from 2,4-dichlorophenol and monochloroacetic acid in aq NaOH: Pokorny, *J. Am. Chem. Soc.* **63**, 1768 (1941); Foster, **GB 573476** (1945); by chlorination of molten phenoxyacetic acid: Manske, **US 2471575** (1949 to U.S. Rubber); from 2,4-dichlorophenol, sodium, and ethyl chloroacetate followed by hydrolysis of the ester: Haskelberg, *J. Org. Chem.* **12**, 426 (1947). Toxicology: V. K. Rowe, T. A. Hymas, *Am. J. Vet. Res.* **15**, 622 (1954); J. M. Way, *Residue Rev.* **26**, 37 (1969).

Crystals from benzene, mp 138°. $bp_{0.4}$ 160°. Almost insol in water. Sol in organic solvents. LD_{50} in mice, rats (mg/kg): 368, 375 orally (Rowe, Hymas).

Sodium salt. $C_8H_5Cl_2NaO_3$. Needles from alc, dec 215°. Soly in water: 3.5%.

Isopropyl ester. Esteron 44. $C_{11}H_{12}Cl_2O_3$; mol wt 263.11. Liquid, d_{25}^{25} 1.255-1.270. Solidifies at +5°. Sol in oil.

Butyl ester. Lironox; Weedone. $C_{12}H_{14}Cl_2O_3$; mol wt 277.14. Prepn: Nagel, **DE 1144288** (1963 to Bayer), *C.A.* **59**, 6315e (1963).

Amine salts. Sol in water. The following amines form useful salts: Dimethylamine, isopropylamine, triethylamine, diethanolamine, triethanolamine. These salts are much more sol than the sodium salt. A small amount of sequestering agent, such as ethylenediaminetetraacetic acid, is generally added to prevent complex formation in hard water.

Note: The name Hedonal was formerly a synonym for carbamic acid 1-methylbutyl ester.

Caution: Potential symptoms of overexposure to 2,4-D are weakness, stupor, hyporeflexia and muscle twitch; convulsions; dermatitis. *See NIOSH Pocket Guide to Chemical Hazards* (DHHS/NIOSH 97-140, 1997) p 88.

USE: Herbicide. To increase latex output of old rubber trees.

2794. Dabigatran. [211914-51-1] *N*-[[2-[[[4-(Aminoiminomethyl)phenyl]amino]methyl]-1-methyl-1*H*-benzimidazol-5-yl]carbonyl]-*N*-2-pyridinyl-β-alanine; 3-[[2-[[4-(carbamimidoylphenyl)amino)methyl]-1-methyl-1*H*-benzoimidazole-5-carbonyl]pyridin-2-ylamino]propionic acid; BIBR-953. $C_{25}H_{25}N_7O_3$; mol wt 471.52. C 63.68%, H 5.34%, N 20.79%, O 10.18%. Nonpeptide, direct thrombin inhibitor. Prepn and crystal structure of complex with thrombin: N. Hauel *et al.*, **WO 9837075**; *eidem*, **US 6087380** (1998, 2000 both to Boehringer, Ing.); *eidem*, *J. Med. Chem.* **45**, 1757 (2002). Review of pharmacology: D. Mungall, *Curr. Opin. Investig. Drugs* **3**, 905-907 (2002). Clinical pharmacokinetics: J. Stangier *et al.*, *J. Clin. Pharmacol.* **45**, 555 (2005); and effect on coagulation assays: J. van Ryn *et al.*, *Thromb. Haemostasis* **103**, 1116 (2010). Clinical trial in acute venous thromboembolism: S. Schulman *et al.*, *N. Engl. J. Med.* **361**, 2342 (2009); for prophylaxis of post-surgical thromboembolism: R. J. Friedman *et al.*, *Thromb. Res.* **126**, 175 (2010).

White crystals, mp 276-277°. Log P (*n*-octanol/buffer, pH 7.4): -2.4.

Etexilate. [211915-06-9] *N*-[[2-[[[4-[[(Hexyloxy)carbonyl]amino]iminomethyl]phenyl]amino]methyl]-1-methyl-1*H*-benzimidazol-5-yl]carbonyl]-*N*-2-pyridinyl-β-alanine ethyl ester; ethyl 3-[[[2-[[4-[[[(hexyloxy)carbonyl]amino]iminomethyl]phenyl]amino]methyl]-1-methyl-1*H*-benzimidazol-5-yl]carbonyl](pyridin-2-yl)amino]pro-

panoate; BIBR-1048; Pradaxa; Rendix. $C_{34}H_{41}N_7O_5$; mol wt 627.75. Colorless crystals, mp 128-129°. Log P (*n*-octanol/buffer, pH 7.4): 2.6.

THERAP CAT: Antithrombotic.

2795. Dacarbazine. [4342-03-4] 5-(3,3-Dimethyl-1-triazen-1-yl)-1*H*-imidazole-4-carboxamide; 5(or 4)-(dimethyltriazeno)imidazole-4(or 5)-carboxamide; DIC; DTIC; NSC-45388; Dacatic; Dacin; DTIC-Dome; Deticene. $C_6H_{10}N_6O$; mol wt 182.19. C 39.56%, H 5.53%, N 46.13%, O 8.78%. Alkylating agent; first synthesized in 1959 at the Southern Research Institute. Prepn: Y. F. Shealy *et al.*, *J. Org. Chem.* **27**, 2150 (1962); Hano *et al.*, *Gann* **59**, 207 (1968), *C.A.* **69**, 42527g (1968). Antitumor activity: Shealy *et al.*, *Biochem. Pharmacol.* **11**, 674 (1962); Hano *et al.*, *Gann* **56**, 417 (1965), *C.A.* **63**, 18856g (1965). Mechanism of action studies: Saunders, Schultz, *Biochem. Pharmacol.* **19**, 911 (1970). Metabolism: Skibba *et al.*, *ibid.* 2043; Mizuno, Humphrey, *Cancer Chemother. Rep. Part 1* **56**, 465 (1972). Electroanalytical behavior: A. J. M. Ordieres *et al.*, *Anal. Chim. Acta* **202**, 141 (1987). Degradation in aqueous soln: B. V. Shetty *et al.*, *J. Pharm. Biomed. Anal.* **10**, 675 (1992). MS study of transition metal coordination: C. G. Hartinger *et al.*, *Polyhedron* **25**, 1971 (2006). Series of articles on history, activity, mechanism of action and clinical studies: *Cancer Treat. Rep.* **60**, 123-214 (1976). Clinical evaluation in pancreatic islet cell cancer: R. K. Ramanathan *et al.*, *Ann. Oncol.* **12**, 1139 (2001). Review of clinical use in metastatic melanoma: A. M. M. Eggermont, J. M. Kirkwood, *Eur. J. Cancer* **40**, 1825-1836 (2004).

Ivory microcrystalline substance; explosive decomp 250-255° (Shealy, *J. Org. Chem.*). Also reported as mp 205° (Hano, 1965). uv max (0.1*N* HCl): 223 nm (7500); (pH 7): 237 nm (11200); both solns protected from light. Stable in neutral soln in absence of light.

Caution: This substance is reasonably anticipated to be a human carcinogen: *Report on Carcinogens, Twelfth Edition* (PB2011-111646, 2011) p 127.

THERAP CAT: Antineoplastic.

THERAP CAT (VET): Antineoplastic.

2796. Daclizumab. [152923-56-3] Anti-(human interleukin 2 receptor) immunoglobulin G1 (human-mouse monoclonal clone 1H4 γ1-chain) disulfide with human-mouse monoclonal clone 1H4 light chain, dimer; humanized anti-Tac; HAT; dacliximab; Ro-24-7375; Zenapax. Humanized monoclonal antibody directed against the α-subunit (p55 alpha, CD25, or Tac subunit) of the human interleukin-2 receptor (IL-2Rα) that is expressed on the surface of activated T lymphocytes. Glycoprotein, mol wt ~150 kDa. DNA sequence is ~90% human origin and ~10% mouse origin. Prepn: C. Queen *et al.*, *Proc. Natl. Acad. Sci. USA* **86**, 10029 (1989). *See also:* C. L. Queen, H. E. Selick, **US 5530101** (1996 to Protein Design Labs). ELISA determn in serum: B. E. Fayer *et al.*, *J. Immunol. Methods* **186**, 47 (1995). Clinical trial in renal transplantation: B. Nashan *et al.*, *Transplantation* **67**, 110 (1999); in cardiac transplantation: A. Beniaminovitz *et al.*, *N. Engl. J. Med.* **342**, 613 (2000). Review of development: J. Hakimi *et al.* in *Antibody Therapeutics*, W. J. Harris, J. R. Adair, Eds. (CRC Press, Boca Raton, 1997) pp 277-300; of clinical efficacy in solid organ transplantation: A. M. Wiland, B. Philosophe, *Expert Opin. Biol. Ther.* **4**, 729-740 (2004).

THERAP CAT: Immunosuppressant.

2797. Dactinomycin. [50-76-0] Actinomycin D; actinomycin-[thr-val-pro-sar-meval]; meractinomycin; actinomycin A_{IV}; actinomycin IV; actinomycin C_1; actinomycin I_1; actinomycin X_1; NSC-3053; Cosmegen. $C_{62}H_{86}N_{12}O_{16}$; mol wt 1255.44. C 59.32%, H 6.90%, N 13.39%, O 20.39%. Antibiotic substance belonging to the actinomycin complex, produced by several *Streptomyces* spp. Historical background of actinomycins and chemistry, toxicology, pharmacology of actinomycin D: *Ann. N.Y. Acad. Sci.* **89**, 285-485 (1960). Isoln from broth cultures of *S. parvulus*: Manaker *et al.*, *Antibiot. Ann.* **1954/55**, 853. Structure: Bullock, Johnson, *J. Chem.*

Soc. **1957**, 3280. Synthesis: Brockmann, Manegold, *Naturwissenschaften* **51**, 383 (1964); Brockmann, Lackner, *ibid.* 384, 435 (1964); Brockmann *et al.*, **DE 1172680** (1964 to Bayer); *eidem*, *Ber.* **101**, 1312 (1968); Meienhofer, *J. Am. Chem. Soc.* **92**, 3771 (1970); T. Tanaka *et al.*, *Bull. Chem. Soc. Jpn.* **53**, 1352 (1980); K. Nakajima *et al.*, *Pept. Chem.* **19**, 143-148 (1981). Conformation from NMR: Conti, DeSantis, *Nature* **227**, 1239 (1970); Lackner, *Ber.* **104**, 3653 (1971). Toxicity data: F. S. Philips *et al.*, *Ann. N.Y. Acad. Sci.* **89**, 348 (1960). Mechanism of action study: P. Delay-Goyet, J. M. Lundberg, *Biochem. Biophys. Res. Commun.* **180**, 1342 (1991). Clinical evaluation in Wilms' tumor: B. de Camargo, E. L. Franco, *Cancer* **73**, 3081 (1993). Review and evaluation of studies of carcinogenic action in laboratory animals: *IARC Monographs* **12**, 29-41 (1976).

H3C—[benzopyran-phenoxazine structure]
Thr–D-Val–Pro
MeVal—Sar

H3C—
Thr–D-Val–Pro
NH2 MeVal—Sar

Bright red crystalline powder. Hygroscopic. Sensitive to light and heat. Freely sol in alcohol; very slightly sol in ether. Sol in water at 10°; slightly sol in water at 37°.

Trihydrate. Bright red, rhomboid prisms from abs alc, dec 241.5-243°. $[\alpha]_D^{28}$ −315° (c = 0.25 in methanol). Abs max (methanol): 244, 441 nm ($A_{1cm}^{1\%}$ 281, 206). Dil solns are very sensitive to light. Sol in alc, propylene glycol, water + glycol mixtures. LD50 orally in mice, rats: 13.0, 7.2 mg/kg (Philips).

THERAP CAT: Antineoplastic.
THERAP CAT (VET): Antineoplastic.

2798. Daidzein. [486-66-8] 7-Hydroxy-3-(4-hydroxyphenyl)-4*H*-1-benzopyran-4-one; 4′,7-dihydroxyisoflavone. $C_{15}H_{10}O_4$; mol wt 254.24. C 70.86%, H 3.96%, O 25.17%. Phytoestrogen found in soybean, trifolium (red clover), and certain other plants; metabolite of formononetin, *q.v.* Converted by intestinal bacteria to equol, *q.v.* Isoln from soybean meal: E. Walz, *Ann.* **489**, 118 (1931); from red clover: E. Wong, *J. Sci. Food Agric.* **13**, 304 (1962). Isoln from soybeans by supercritical fluid extraction: J. M. A. Araujo *et al.*, *Food Chem.* **105**, 266 (2007). Synthesis: W. Baker *et al.*, *J. Chem. Soc.* **1933**, 274; F. Wessely *et al.*, *Ber.* **66**, 685 (1933); H. S. Mahal *et al.*, *J. Chem. Soc.* **1934**, 1769; W. Baker *et al.*, *ibid.* **1953**, 1852. Clinical evaluation to reduce menopausal symptoms: L. Khoadhiar *et al.*, *Menopause* **15**, 125 (2008). Review of determn methods in foods and biological fluids: A. P. Wilkinson *et al.*, *J. Chromatogr. B* **777**, 93-109 (2002); of metabolism and clinical relevance: C. Atkinson *et al.*, *Exp. Biol. Med.* **230**, 155-170 (2005).

HO—[isoflavone structure]—OH

Pale-yellow prisms from dil alc, dec 315-323°. uv max: 250 nm (log ε 4.44). Sol in alcohol, ether.

Daidzin. [552-66-9] 7-(β-D-Glucopyranosyloxy)-3-(4-hydroxyphenyl)-4*H*-1-benzopyran-4-one; daidzein 7-glucoside; daidzoside. $C_{21}H_{20}O_9$; Synthesis: L. Farkas, J. Várady, *Ber.* **92**, 819 (1959). Monohydrate, needles from dec 234-236°. $[\alpha]_D^{20}$ −36.4° (0.02*N* NaOH). Sol in water and aq alcohol.

USE: Dietary supplement.

2799. Dalapon. [75-99-0] 2,2-Dichloropropanoic acid; α,α-dichloropropionic acid. $C_3H_4Cl_2O_2$; mol wt 142.96. C 25.20%, H 2.82%, Cl 49.59%, O 22.38%. Selective systemic herbicide for control of annual and perennial grasses. Prepn: **GB 752761** (1956 to Dow); E. J. Kowolik, J. W. Fisher, **GB 892584** (1962 to British

Celanese); of free acid and sodium salt: Smeykal, Pallutz, *Chem. Tech. (Leipzig)* **15** (11), 654 (1963), *C.A.* **60**, 6741c (1964). Acute toxicity: T. B. Gaines, R. E. Linder, *Fundam. Appl. Toxicol.* **7**, 299 (1986). Toxicity data for the sodium salt: G. W. Bailey, J. L. White, *Residue Rev.* **10**, 97 (1965). Quantitative NMR purity determn: R. J. Wells *et al.*, *J. Agric. Food Chem.* **50**, 3366 (2002).

[chemical structure]

Liquid, bp20 98-99°. d^{20} 1.4014. n_D^{20} 1.4551. LD50 in male, female rats (mg/kg): 7126, 6936 orally (Gaines, Linder).

Sodium salt. [127-20-8] Basfapon; Dowpon; Radapon. $C_3H_3Cl_2NaO_2$; mol wt 164.94. Crystals, dec 174-176°. Soly in water at 25°, 45 g/100 ml; aq solns hydrolyze above 70°. Corrosive to iron.

2-(2,4,5-Trichlorophenoxy)ethyl Ester. [136-25-4] Erbon; Baron; Novon. $C_{11}H_9Cl_5O_3$; mol wt 366.44. Prepn: H. F. Brust, H. O. Senkbeil, **US 2754324** (1956 to Dow). Crystals, mp 49-50°. bp0.5 161-164°. d_4^{50} 1.55. Sol in acetone, alc, kerosene, xylene. Insol in water. LD50 orally in rats: 1120 mg/kg (Bailey, White).

Caution: Potential symptoms of dalapon overexposure are irritation of eyes, skin, upper respiratory system; eye and skin burns; lassitude, loss of appetite, diarrhea, vomiting, slowing of pulse; CNS depression. *See* NIOSH *Pocket Guide to Chemical Hazards* (DHHS/NIOSH 97-140, 1997) p 100.

USE: Herbicide.

2800. Dalbavancin. [171500-79-1] 5,31-Dichloro-38-de-(methoxycarbonyl)-7-demethyl-19-deoxy-56-*O*-[2-deoxy-2-[(10-methyl-1-oxoundecyl)amino]-β-D-glucopyranuronosyl]-38-[[[3-(dimethylamino)propyl]amino]carbonyl]-42-*O*-α-D-mannopyranosyl-N^{15}-methylristomycin A aglycone; BI-397; MDL-63397. $C_{88}H_{100}$-$Cl_2N_{10}O_{28}$; mol wt 1816.71. C 58.18%, H 5.55%, Cl 3.90%, N 7.71%, O 24.66%. Semisynthetic lipoglycopeptide antibiotic derived from the natural glycopeptide, *A-40926*. Structurally related to teicoplanin, *q.v.* Prepn: A. Malabarba *et al.*, *J. Antibiot.* **48**, 869 (1995). *In vitro* antibacterial activity: J. M. Streit *et al.*, *Diagn. Microbiol. Infect. Dis.* **48**, 137 (2004). Clinical pharmacology: A. Leighton *et al.*, *Antimicrob. Agents Chemother.* **48**, 940 (2004). Clinical evaluation in skin and soft-tissue infections: E. Seltzer *et al.*, *Clin. Infect. Dis.* **37**, 1298 (2003); in catheter-related bloodstream infections: I. Raad *et al.*, *ibid.* **40**, 374 (2005). Review of antibacterial spectrum and clinical experience: J. Bailey, K. M. Summers, *Am. J. Health Syst. Pharm.* **65**, 599-610 (2008).

[large glycopeptide structure]

THERAP CAT: Antibacterial.

2801. Dalcetrapib. [211513-37-0] 2-Methylpropanethioic acid *S*-[2-[[[1-(2-ethylbutyl)cyclohexyl]carbonyl]amino]phenyl] ester; *S*-[2-[[1-(2-ethylbutyl)cyclohexane]carbonylamino]phenyl]-2-methylpropanethioate; JTT-705; RO-4607381. $C_{23}H_{35}NO_2S$; mol wt 389.60. C 70.91%, H 9.06%, N 3.60%, O 8.21%, S 8.23%. Cholesteryl ester transfer protein (CETP) inhibitor; increases serum levels of HDL-cholesterol. Prepn: H. Shinkai *et al.*, **WO 9835937**; *eidem*, **US 6426365** (1998, 2002 both to Japan Tobacco); H. Shinkai *et al.*, *J. Med. Chem.* **43**, 3566 (2000). Pharmacology and effect on plasma CETP: H. Okamoto *et al.*, *Eur. J. Pharmacol.* **466**, 147

(2003). Clinical evaluation in combination with atorvastatin: E. A. Stein *et al.*, *Eur. Heart J.* **31**, 480 (2010). Review of pharmacology, safety, and clinical experience: J. G. Robinson, *Expert Opin. Invest. Drugs* **19**, 795-805 (2010).

Colorless solid, mp 63-63.5°.

THERAP CAT: Antilipemic; antiatherosclerotic.

2802. Dalfampridine. [504-24-5] 4-Pyridinamine; fampridine; 4-aminopyridine; γ-aminopyridine; 4-pyridylamine; 4-AP; EL-970; Ampyra. $C_5H_6N_2$; mol wt 94.12. C 63.81%, H 6.43%, N 29.76%. Potassium channel blocker. Prepn: A. Kirpal, *Monatsh. Chem.* **23**, 244 (1902); R. Camps, *Arch. Pharm.* **240**, 345 (1902); C. R. Hauser, G. A. Reynolds, *J. Org. Chem.* **15**, 1224 (1950). Prolongs action potential in demyelinated nerve fibers: R. M. Sherratt *et al.*, *Nature* **283**, 570 (1980). HPLC determn in serum: A. van der Horst *et al.*, *J. Chromatogr.* **574**, 166 (1992). Clinical pharmacokinetics of sustained-release formulation in chronic, incomplete spinal cord injury: K. C. Hayes *et al.*, *Arch. Phys. Med. Rehabil.* **85**, 29 (2004). Neurophysiology and clinical efficacy in multiple sclerosis (MS): H. A. M. van Diemen *et al.*, *J. Neurol. Sci.* **116**, 220 (1993). Clinical evaluation: C. H. Polman *et al.*, *Arch. Neurol.* **51**, 1136 (1994). Clinical trial of sustained-release formulation in MS: A. D. Goodman *et al.*, *Lancet* **373**, 732 (2009). Review of pharmacology and efficacy of aminopyridines in MS: C. T. Bever, Jr., *Ann. Neurol.* **36**, S118-S121 (1994); of clinical experience in MS and spinal cord injury: K. C. Hayes, *Expert Rev. Neurother.* **7**, 453-461 (2007).

White powder. Sol in water, methanol, acetone, THF, isopropanol, acetonitrile, DMF, DMSO, ethanol.

THERAP CAT: In treatment of multiple sclerosis.

2803. Dalfopristin. [112362-50-2] (3*R*,4*R*,5*E*,10*E*,12*E*,14*S*,-26*R*,26a*S*)-26-[[2-(Diethylamino)ethyl]sulfonyl]-8,9,14,15,24,25,-26,26a-octahydro-14-hydroxy-4,12-dimethyl-3-(1-methylethyl)-3*H*-21,18-nitrilo-1*H*,22*H*-pyrrolo[2,1-*c*][1,8,4,19]dioxadiazacyclotetracosine-1,7,16,22(4*H*,17*H*)-tetrone; (26*R*,27*S*)-26-[[2-(diethylamino)ethyl]sulfonyl]-26,27-dihydrovirginiamycin M1; 26-(2-diethylaminoethyl)sulfonylpristinamycin IIB; RP-54476. $C_{34}H_{50}N_4O_9S$; mol wt 690.85. C 59.11%, H 7.30%, N 8.11%, O 20.84%, S 4.64%. Semisynthetic, polyunsaturated macrolactone, streptogramin antibiotic. Prepn: J.-C. Barriere *et al.*, **EP 191662**; *eidem*, **US 4668669** (1986, 1987 both to Rhone-Poulenc). *In vitro* activity: H. C. Neu *et al.*, *J. Antimicrob. Chemother.* **30**, Suppl. A, 83 (1992). HPLC determn in plasma: A. Le Liboux *et al.*, *J. Chromatogr. B* **708**, 161 (1998).

White solid, mp ~150°.

Mixture with quinupristin. [126602-89-9] RP-59500; Synercid. Formulation comprised of two synergistic components in a defined 70:30 mixture of dalfopristin and quinupristin, *q.v.*, as the mesylate salts. HPLC determn for quality control: B. Vasselle *et al.*, *J. Pharm. Biomed. Anal.* **19**, 641 (1999). *In vitro* activity in comparison with pristinamycin, *q.v.*: A. Lozniewski *et al.*, *Pathol. Biol.* **48**, 463 (2000). Clinical trial in vancomycin resistant *Enterococcus faecium* (VREF) infection: R. C. Moellering *et al.*, *J. Antimicrob. Chemother.* **44**, 251 (1999); in skin infections: R. L. Nichols *et al.*, *ibid.* 263. Review: B. Pavan, *Curr. Opin. Investig. Drugs* **1**, 173-180 (2000).

THERAP CAT: Antibacterial.

2804. Dalteparin. Tedelparin. Low molecular weight fragment of heparin, *q.v.*, prepd by nitrous acid depolymerization of porcine mucosal heparin. Mean mol wt 4000-6000 daltons. Prepn: U. Lindahl *et al.*, **WO 8001383**; *eidem*, **US 4303651** (1980, 1981 both to Kabi AB). *See also*: *eidem*, *Proc. Natl. Acad. Sci. USA* **76**, 3198 (1979). Comparison of bioactivity with standard heparin: E. Holmer *et al.*, *Thromb. Res.* **18**, 861 (1980). Clinical pharmacokinetics: G. Bratt *et al.*, *ibid.* **42**, 613 (1986). Symposia on pharmacology and clinical efficacy: *Haemostasis* **16**, Suppl. 2, 1-71 (1986); *Semin. Thromb. Hemostasis* **16**, Suppl., 1-76 (1990). Clinical trial in unstable angina: W. Klein *et al.*, *Circulation* **96**, 61 (1997); in unstable coronary artery disease: FRISC II Investigators, *Lancet* **354**, 701 (1999). *Review:* C. J. Dunn, E. M. Sorkin, *Drugs* **52**, 276-305 (1996).

Sodium salt. Kabi 2165; FR-860; Boxol; Fragmin.

THERAP CAT: Antithrombotic.

2805. Daltroban. [79094-20-5] 4-[2-[[(4-Chlorophenyl)sulfonyl]amino]ethyl]benzeneacetic acid; [*p*-[2-(*p*-chlorobenzenesulfonamido)ethyl]phenyl]acetic acid; BM-13505; SKF-96148. $C_{16}H_{16}ClNO_4S$; mol wt 353.82. C 54.31%, H 4.56%, Cl 10.02%, N 3.96%, O 18.09%, S 9.06%. Thromboxane A_2-receptor antagonist. Prepn: E. C. Witte *et al.*, **DE 3000377**; *eidem*, **US 4443477** (1981, 1984 both to Boehringer, Ingelh.). Receptor binding study: A. Yanagisawa *et al.*, *Eur. J. Pharmacol.* **133**, 89 (1987). Pharmacological evaluations in animals: D. J. Lefer *et al.*, *Arch. Int. Pharmacodyn.* **287**, 89 (1987); E. F. Smith, III, J. McDonald, *Pharmacology* **36**, 340 (1988); P. Löbel *et al.*, *Biomed. Biochim. Acta* **47**, S 86 (1988). GC determn in biological fluids: V. Uebis, *J. Chromatogr.* **419**, 345 (1987).

THERAP CAT: Antithrombotic.

2806. Damar. [9000-16-2] Dammar; gum Damar; resin Damar. Resinous exudate from a species of *Shorea, Dipterocarpaceae. Habit.* East Indies, Philippines. *Constit.* Volatile oil, resins, bitter substance.

Yellowish-white, roundish, or stalactite shaped, friable masses; semi-transparent, conchoidal fracture; varying degrees of hardness. d 1.04-1.12. mp ~120°. Insol in water. Sol in alcohol, chloroform, ether, carbon disulfide, oil rosemary; partly sol in oil turpentine.

USE: In plasters, varnishes, lacquers, etc. A soln of the purified material in chloroform or xylene is used for preserving animal and vegetable specimens for microscopy.

2807. Damascenine. [483-64-7] 3-Methoxy-2-(methylamino)benzoic acid methyl ester; 2-(methylamino)-*m*-anisic acid methyl ester; methyl 2-(methylamino)-3-methoxybenzoate; methyldamascenine; nigelline. $C_{10}H_{13}NO_3$; mol wt 195.22. C 61.53%, H 6.71%, N 7.17%, O 24.59%. Odoriferous principle of oil of *Nigella* from seeds of *Nigella damascena* L. and *Nigella arvensis* L., *Ranunculaceae.* Early syntheses: Ewins, *J. Chem. Soc.* **101**, 544 (1912); Kaufman, Rothlin, *Ber.* **49**, 578 (1916); Sornet, *Manuf. Chem.* **5**, 87 (1924); Keller, Schulze, *Arch. Pharm.* **263**, 481 (1925). Improved synthesis: Mutschler, *ibid.* **298**, 861 (1965); M. Thoinet *et al.*, *Ann.*

Pharm. Fr. **36**, 337 (1978). Toxicity study: Bekemeier *et al.*, *Arch. Int. Pharmacodyn. Ther.* **168**, 199 (1967).

Prisms from abs. alc. Nutmeg-like odor, mp 27-29°. bp 270° (slight dec), bp$_{10}$ 147°. Volatile with steam. Insol in water. Freely sol in alc, ether, chloroform, petr ether, oils.

Hydrochloride. Deliquesc prisms, mp 156°. (Hydrate, mp 122°). LD$_{50}$ orally in mice: 1800 mg/kg (Bekemeier).

2808. β-Damascenone. [23726-93-4]; [23696-85-7] (unspecified stereo). (2*E*)-1-(2,6,6-Trimethyl-1,3-cyclohexadien-1-yl)-2-buten-1-one; *trans*-2,6,6-trimethyl-1-crotonylcyclohexa-1,3-diene; Doricenone. C$_{13}$H$_{18}$O; mol wt 190.29. C 82.06%, H 9.53%, O 8.41%. Terpenic ketone; odorant in fruits, vegetables, honey, wine and beer. Isoln from Bulgarian rose oil, *Rosa damascena* Mill., and synthesis: E. Demole *et al.*, *Helv. Chim. Acta* **53**, 541 (1970). Synthetic studies: G. Büchi, H. Wüest, *ibid.* **54**, 1767 (1971); K. S. Ayyar *et al.*, *J. Chem. Soc. Perkin Trans. 1* **1975**, 1727. Identification in grapes: T. E. Acree *et al.*, *J. Agric. Food Chem.* **29**, 688 (1981); in tomato volatiles: R. G. Buttery *et al.*, *Chem. Ind. (London)* **1988**, 238. Quantitative determn in foods: A. Sen *et al.*, *J. Agric. Food Chem.* **39**, 757 (1991); in beers: F. Chevance *et al.*, *ibid.* **50**, 3818 (2002). Reactivity in wine: M. A. Daniel *et al.*, *ibid.* **52**, 8127 (2004).

Pale yellow to yellow liquid; powerful floral, fruity odor. bp$_{0.001}$ 57°. bp$_{0.08}$ 51°. bp$_{13}$ 116-118°. d$_4^{20}$ 0.942. n$_D^{20}$ 1.5123. Flash pt (closed cup): >100°C. uv max (cyclohexane): 220, 253, 302 nm (ε 10000, 4120, 2270). uv max (ethanol): 228, 255, 310 nm (ε 12000, 500, 2000). Sol in ethanol (95°). Odor threshold in water: 0.02-0.09 ng/g.

USE: In perfumery.

2809. Daminozide. [1596-84-5] 1-(2,2-Dimethylhydrazide)butanedioic acid; butanedioic acid mono(2,2-dimethylhydrazide); *N*-(dimethylamino)succinamic acid; succinic acid 2,2-dimethylhydrazide; B-9; B-995; Alar; B-Nine; Kylar. C$_6$H$_{12}$N$_2$O$_3$; mol wt 160.17. C 44.99%, H 7.55%, N 17.49%, O 29.97%. Prepn: **BE 613799**, *C.A.* **57**, 10281d (1962); H. A. Hageman, W. L. Hubbard, **US 3257414** (1962, 1966 both to U.S. Rubber). Effect in controlling growth in apples: F. W. Southwick *et al.*, *Proc. Am. Soc. Hortic. Sci.* **92**, 71 (1968); D. W. Greene *et al.*, *Fruit Var. J.* **40**, 41 (1986). *In vitro* and *in vivo* mechanism of action studies: K. Ryugo, R. M. Sachs, *J. Am. Soc. Hortic. Sci.* **94**, 529 (1969). Hydrolysis to 1,1-dimethylhydrazine, *q.v.*, and determn in food products: W. H. Newsome, *J. Agric. Food Chem.* **28**, 319 (1980); M. K. Conditt, J. R. Baumgardner, *J. Assoc. Off. Anal. Chem.* **71**, 735 (1988). Discussion of toxicological evaluation: D. Campt, *EPA Journal* **13**, 32 (1987).

Crystals, mp 154-155°. Soly 10% in water, 2.5% in acetone, 5% in methanol.

USE: Plant growth regulator.

2810. Danaparoid. Mucoglucoronan. Low molecular weight heparinoid derived from hog intestinal mucosa. Mixture of glycosa-

minoglycans with a mean mol wt of 6000 Da (range 4000-10000 Da). Consists of heparan sulfate (~83%), dermatan sulfate (~12%) and chondroitin sulfate (~5%). Prepn: A. L. M. Saunders *et al.*, **EP 66908**; *eidem*, **US 4438108** (1982, 1984 both to Akzo). Pharmacology: D. G. Meuleman *et al.*, *Thromb. Res.* **27**, 353 (1982). Determn in plasma: H. ten Cate *et al.*, *Clin. Chem.* **30**, 860 (1984). Series of articles on pharmacology and clinical efficacy: *Haemostasis* **22**, 55-112 (1992). Clinical trial in deep venous thromboembolism: H. W. de Valk *et al.*, *Ann. Intern. Med.* **123**, 1 (1995).

Sodium salt. Org-10172; Orgaran.

THERAP CAT: Antithrombotic.

2811. Danazol. [17230-88-5] (17α)-Pregna-2,4-dien-20-yno-[2,3-*d*]isoxazol-17-ol; 1-ethynyl-2,3,3a,3b,4,5,10,10a,10b,11,12,-12a-dodecahydro-10a,12a-dimethyl-1*H*-cyclopenta[7,8]phenanthro[3,2-*d*]isoxazol-1-ol; 17α-ethynyl-17β-hydroxy-4-androsteno[2,3-*d*]isoxazole; Win-17757; Bonzol; Cyclomen; Danatrol; Danocrine; Danol; Danoval; Ladogal; Winobanin. C$_{22}$H$_{27}$NO$_2$; mol wt 337.46. C 78.30%, H 8.06%, N 4.15%, O 9.48%. Anterior pituitary supressant. Anabolic steroid deriv of ethisterone, *q.v.*, with mild androgenic side effects (an impeded androgen). Prepn: **GB 905844** (1962 to Sterling Drug), *C.A.* **58**, 6895c (1963); Manson *et al.*, *J. Med. Chem.* **6**, 1 (1963); Clinton, Manson, **US 3135743** (1964 to Sterling Drug). Activity studies: Sherins *et al.*, *J. Clin. Endocrinol. Metab.* **32**, 522 (1971); Dmowski *et al.*, *Fertil. Steril.* **22**, 9 (1971). Clinical studies in endometriosis and other endocrine disorders: R. B. Greenblatt *et al.*, *ibid.* 102. Series of articles on pharmacology, pharmacokinetics and clinical use: *Drugs* **19**, 321-372 (1980). Use in idiopathic thrombocytopenic purpura: Y. S. Ahn *et al.*, *N. Engl. J. Med.* **308**, 1396 (1983); in hemophilia: H. R. Gralnick *et al.*, *ibid.* 1393.

White to pale yellow crystals from acetone, mp 224.4-226.8°. [α]$_D^{25}$ +7.5° (ethanol); [α]$_D^{25}$ +21.9° (chloroform). uv max (ethanol): 286 nm (ε 11300). Freely sol in chloroform; sol in acetone; sparingly sol in alcohol, benzene; slightly sol in ether. Practically insol in water, hexane.

THERAP CAT: Antigonadotropin.

2812. Danofloxacin. [112398-08-0] 1-Cyclopropyl-6-fluoro-1,4-dihydro-7-[(1*S*,4*S*)-5-methyl-2,5-diazabicyclo[2.2.1]hept-2-yl]-4-oxo-3-quinolinecarboxylic acid. C$_{19}$H$_{20}$FN$_3$O$_3$; mol wt 357.39. C 63.85%, H 5.64%, F 5.32%, N 11.76%, O 13.43%. Fluorinated quinolone antibacterial. Prepn: M. R. Jefson, P. R. McGuirk, **EP 215650**; *eidem*, **US 4861779** (1987, 1989 both to Pfizer). *See also:* F. R. Busch *et al.*, **EP 342849** (1989 to Pfizer), *C.A.* **113**, 6177m (1990). Evaluation in treatment of pneumonia in cattle: W. T. R. Grimshaw *et al.*, *Dtsch. Tieraerztl. Wochenschr.* **97**, 529 (1990); C. J. Giles *et al.*, *Vet. Rec.* **128**, 296 (1991). Efficacy in treatment of *E. coli* diarrhoea in calves: S. J. Sunderland *et al.*, *Res. Vet. Sci.* **74**, 171 (2003).

mp 268-272°.

Methanesulfonate. [119478-55-6] CP-76136-27; Advocin. C$_{19}$H$_{20}$FN$_3$O$_3$.CH$_3$SO$_3$H; mol wt 453.49.

THERAP CAT (VET): Antibacterial.

2813. Dansyl Chloride. [605-65-2] 5-(Dimethylamino)-1-naphthalenesulfonyl chloride. C$_{12}$H$_{12}$ClNO$_2$S; mol wt 269.74. C

53.43%, H 4.48%, Cl 13.14%, N 5.19%, O 11.86%, S 11.89%. Reagent for fluorescent labelling of amines, amino acids and proteins. Synthesis: G. Weber, *Biochem. J.* **51**, 155 (1952); A. Mendel, *J. Chem. Eng. Data* **15**, 340 (1970). In amino acid analysis: J. Airhart *et al.*, *Anal. Biochem.* **53**, 132 (1973). In HPLC analysis of phenols: R. M. Cassidy, D. S. LeGay, *J. Chromatogr. Sci.* **12**, 85 (1974). In protein microassays: T. Kinoshita *et al.*, *Anal. Biochem.* **66**, 104 (1975). In measurement of desquamation: R. Marks *et al.*, *Br. J. Dermatol.* **111**, 265 (1984).

Yellow-orange crystals from hexane, mp 66.5-68°. Sol in acetone, pyridine, benzene, dioxane. Insol in water.
USE: In fluorescent labelling, column chromatography, HPLC.

2814. Danthron. [117-10-2] 1,8-Dihydroxy-9,10-anthracenedione; 1,8-dihydroxyanthraquinone; chrysazin; dantron; Altan; Antrapurol; Diaquone; Istizin; Modane. $C_{14}H_8O_4$; mol wt 240.21. C 70.00%, H 3.36%, O 26.64%. Prepn. Fierz-David, Blangey, *Farbenchemie* (Vienna, 5th ed., 1943) pp 224-225; Kozlov, *Dokl. Akad. Nauk SSSR* **61**, 281 (1948). Clinical trial with poloxalkol in constipation: L. Mundow, *Br. J. Clin. Pract.* **29**, 95 (1975). Mutation study: J. P. Brown, R. J. Brown, *Mutat. Res.* **40**, 203 (1976). Toxicology: M. T. Case *et al.*, *Drug Chem. Toxicol.* **1**, 89 (1977). Review of toxicity and carcinogenic risk: *IARC Monographs* **50**, 265-275 (1990).

Orange needles from alc, mp 193-197°. Sublimes. Absorption max: 430, 250 nm (log ε 4.35, 4.60). Almost insol in water (6.5 × 10^{-6} mols/l at 25°), in alcohol (1:2000). Moderately sol in ether (1:500), in chloroform; sol in 10 parts hot glacial acetic acid. Very slightly sol in aq solns of alkali hydroxides: about 0.8 g dissolves in 100 ml 0.5N NaOH. (Disodium salt is reported to have a water soly of 0.05%.) LD$_{50}$ orally in mice: <7 g/kg (Case).
Caution: This substance is reasonably anticipated to be a human carcinogen: *Report on Carcinogens, Twelfth Edition* (PB2011-111646, 2011) p 128.
USE: Important intermediate in the manuf of alizarin and indanthrene dyestuffs; forms insol Ca, Ba, Pb lakes. Antioxidant in synthetic lubricants; fungicide.
THERAP CAT: Cathartic.
THERAP CAT (VET): Purgative.

2815. Dantrolene. [7261-97-4] 1-[[[5-(4-Nitrophenyl)-2-furanyl]methylene]amino]-2,4-imidazolidinedione; 1-[[5-(p-nitrophenyl)furfurylidene]amino]hydantoin. $C_{14}H_{10}N_4O_5$; mol wt 314.26. C 53.51%, H 3.21%, N 17.83%, O 25.46%. Direct acting skeletal muscle relaxant. Binds to the ryanodine receptor which modulates the release of calcium from the sarcoplasmic reticulum of skeletal muscle. Prepn: **NL 6612588**; Davis, Snyder, **US 3415821** (1967, 1968 both to Norwich Pharmacal); Snyder *et al.*, *J. Med. Chem.* **10**, 807 (1967); Frimm *et al.*, *Chem. Zvesti* **23**, 916 (1969). HPLC determn in plasma: M. Lalande *et al.*, *J. Chromatogr.* **430**, 187 (1988). Toxicity data: K. O. Ellis *et al.*, *J. Pharm. Sci.* **69**, 327 (1980). Review of pharmacology: G. G. Harrison, *Br. J. Anaesth.* **60**, 279 (1988); and therapeutic use: A. Ward *et al.*, *Drugs* **32**, 130-168 (1986). Review of efficacy in malignant hyperthermia: R. Ben

Abraham *et al.*, *Q. J. Med.* **90**, 13-18 (1997); of pharmacology and clinical indications: T. Krause *et al.*, *Anaesthesia* **59**, 364-373 (2004).

Crystals from aqueous DMF, mp 279-280°.
Sodium salt hemiheptahydrate. [24868-20-0]; [14663-23-1] (anhydrous). F-440; Dantamacrin; Dantrium; Dantrolen. $C_{14}H_9N_4$-NaO$_5$.3½H$_2$O; mol wt 399.29. Orange powder. Sparingly sol in acetone, DMF, glycerine. Slightly soluble in water; more sol in alkaline soln. pKa 7.5. LD$_{50}$ (calculated as base) orally in mice: 1110 mg/kg (Ellis).
THERAP CAT: Muscle relaxant (skeletal); in treatment of malignant hyperthermia.
THERAP CAT (VET): Muscle relaxant (skeletal).

2816. Dapagliflozin. [461432-26-8] (1S)-1,5-Anhydro-1-C-[4-chloro-3-[(4-ethoxyphenyl)methyl]phenyl]-D-glucitol; (2S,3R,-4R,5S,6R)-2-(3-(4-ethoxybenzyl)-4-chlorophenyl)-6-hydroxymethyltetrahydro-2H-pyran-3,4,5-triol. $C_{21}H_{25}ClO_6$; mol wt 408.88. C 61.69%, H 6.16%, Cl 8.67%, O 23.48%. Selective sodium-glucose cotransporter-2 (SGLT-2) inhibitor; blocks renal glucose readsorption and promotes glucosuria. Prepn: B. Ellsworth *et al.*, **US 6515117** (2003 to Bristol-Myers Squibb); W. Meng *et al.*, *J. Med. Chem.* **51**, 1145 (2008). Clinical evaluation of SGLT-2 inhibition: J. F. List *et al.*, *Diabetes Care* **32**, 650 (2009). Clinical pharmacokinetics and pharmacodynamics: B. Komoroski *et al.*, *Clin. Pharmacol. Ther.* **85**, 513 (2009). Clinical effect on glycosuria and glycemic control in type 2 diabetes: J. P. H. Wilding *et al.*, *Diabetes Care* **32**, 1656 (2009). Review of development and clinical experience: M. Kipnes, *Expert Opin. Invest. Drugs* **18**, 327-334 (2009); A. McCord Brooks, S. M. Thacker, *Ann. Pharmacother.* **43**, 1286-1293 (2009).

Glassy, off-white amorphous solid.
Compound with (2S)-1,2-propanediol hydrate. [960404-48-2] dapagliflozin propylene glycol hydrate; BMS-512148-05. $C_{21}H_{25}$-ClO$_6$.C$_3$H$_8$O$_2$.H$_2$O; mol wt 502.99. Prepn: J. Z. Gougoutas *et al.*, **WO 08002824**; *eidem*, **US 080004336** (both 2008 to Bristol-Myers Squibb). Snow white solid.
THERAP CAT: Antidiabetic.

2817. Daphnetin. [486-35-1] 7,8-Dihydroxy-2H-1-benzopyran-2-one; 7,8-dihydroxycoumarin. $C_9H_6O_4$; mol wt 178.14. C 60.68%, H 3.40%, O 35.92%. The aglucon of daphnin. By boiling daphnin with dil mineral acids; by enzymatic hydrolysis of daphnin; by sublimation from daphnin; by synthesis: v. Pechmann, *Ber.* **17**, 934 (1884); Gatterman, Köbner, *Ber.* **32**, 287 (1899); Baker, Savage, *J. Chem. Soc.* **1938**, 1602; Späth, Galinovsky, *Ber.* **70B**, 235 (1937); Sethna, Shah, *Chem. Rev.* **36**, 1 (1945).

Crystals from dil alc, mp 256° (dec). Sublimes. Sol in boiling water, hot dil alcohol, hot glacial acetic acid; very sparingly sol in ether, in carbon bisulfide, in chloroform, in benzene. Sol in alkali and alkali carbonate solns giving a yellow color. Aq solns give a green color with ferric chloride which turns red on addition of sodium carbonate.

2818. Daphnin. [486-55-5] 7-β-(D-Glucopyranosyloxy)-8-hydroxy-2*H*-1-benzopyran-2-one; 7,8-dihydroxycoumarin 7-β-D-glucoside; daphnetin 7-β-D-glucoside; 7-glucosido-8-hydroxycoumarin. $C_{15}H_{16}O_9$; mol wt 340.28. C 52.95%, H 4.74%, O 42.32%. In bark and flower of *Daphne* spp. Extraction from fresh bark of *D. mezereum* L., Thymelaeaceae: *Beilstein* **XXXI**, 248. Structure: Wessely, Sturm, *Ber.* **63**, 1299 (1930); Hattori, *Chem. Zentralbl.* **1930**, II, 2138. Synthesis: Gandini, *Gazz. Chim. Ital.* **70**, 611 (1940), *C.A.* **35**, 1394 (1941).

Monohydrate, prisms from water. Anhydr after drying at 80-90° at 2 mm Hg. When anhydrous, dec 215° (Wessely, Sturm); dec 224° (Hattori); dec 229° (older data). Optical rotations of monohydrate: $[\alpha]_D^{22}$ −124° (15 mg in 4 ml MeOH); after 2 recrystns from water: $[\alpha]_D^{22}$ −115° (abs MeOH) (Wessely, Sturm); $[\alpha]_D^{15}$ −84.2° (MeOH) (Hattori). Sol in hot water, alcohol; sol in aq solns of alkalies and alkali carbonates with yellow color. Insol in ether. Hydrolyzed by dilute acid yielding 7,8-dihydroxycoumarin and D-glucose. Also split by emulsin.

2819. Daphnoline. [479-36-7] (4a*R*,16a*S*)-3,4,4a,5,16a,17,-18,19-Octahydro-21,26-dimethoxy-17-methyl-2*H*-1,24:12,15-dietheno-6,10-metheno-16*H*-pyrido[2′,3′:17,18][1,10]dioxacycloeicosino[2,3,4-*ij*]isoquinoline-9,22-diol; 6,6′-dimethoxy-2-methyloxyacanthan-7,12′-diol; trilobamine. $C_{35}H_{36}N_2O_6$; mol wt 580.68. C 72.40%, H 6.25%, N 4.82%, O 16.53%. From bark of *Daphnandra micrantha* Benth., *Monimiaceae*. Isoln: Pyman, *J. Chem. Soc.* **105**, 1679 (1914). Identity with trilobamine: Bick *et al.*, *ibid.* **1949**, 2767. Structure: Bick *et al.*, *ibid.* **1960**, 4928.

Crystals from chloroform, mp 194-196°. $[\alpha]_D$ +459° (c = 0.3 in chloroform). uv max (methanol): 285 nm (log ε 3.9). Practically insol in water, ethyl acetate, acetone, ether, petr ether. Sparingly sol in methanol, ethanol, xylene, hot chloroform sol in dil acids, cold aq 5% sodium hydroxide.

Daphnandrine. [1183-76-2] 6,6′,12′-Trimethoxy-2-methyloxyacanthan-7-ol. $C_{36}H_{38}N_2O_6$. Needles from methanol, dec 270°. $[\alpha]_D^{16}$ +480° (c = 1.2 in chloroform). uv max (methanol): 284 nm (log ε 3.92). Practically insol in hot water, ethyl acetate, acetone, ether, petr ether. Sparingly sol in hot methanol, ethanol, xylene, cold chloroform; sol in hot chloroform.

2820. Dapiprazole. [72822-12-9] 5,6,7,8-Tetrahydro-3-[2-[4-(2-methylphenyl)-1-piperazinyl]ethyl]-1,2,4-triazolo[4,3-*a*]pyridine; 5,6,7,8-tetrahydro-3-[2-(4-*o*-tolyl-1-piperazinyl)ethyl]-*s*-triazolo[4,3-*a*]pyridine. $C_{19}H_{27}N_5$; mol wt 325.46. C 70.12%, H 8.36%, N 21.52%. α-Adrenergic blocker. Prepn: B. Silvestrini, L. Baiocchi, **DE 2915318**; *eidem*, **US 4252721** (1979, 1981 both to Angelini Francesco). Series of articles on general pharmacology: *Arzneim.-Forsch.* **32**, 674-681 (1982); and toxicity: B. Silvestrini *et al.*, *ibid.* 668. Ocular pharmacokinetics in rabbits: P. Valeri *et al.*, *Pharmacol. Res. Commun.* **18**, 1093 (1986). Topical effects in human eye: A. Reibaldi *et al.*, *Acta Ther.* **10**, 381 (1984); C. Malpassi *et al.*, *ibid.* **12**, 55 (1986). Clinical trial in reversal of mydriasis: R. W. Allinson *et al.*, *Ann. Ophthalmol.* **22**, 131 (1990).

mp 158-160°.
Monohydrochloride. [72822-13-0] AF-2139; Glamidolo; Rev-Eyes. $C_{19}H_{27}N_5$·HCl; mol wt 361.92. Crystals from absolute ethanol, mp 206-207°. LD_{50} in mice (mg/kg): 260 i.p. (Silvestrini).
THERAP CAT: Antiglaucoma; miotic.

2821. Dapoxetine. [119356-77-3] (α*S*)-*N*,*N*-Dimethyl-α-[2-(1-naphthalenyloxy)ethyl]benzenemethanamine; (+)-*N*,*N*-dimethyl-1-phenyl-3-(1-naphthalenyloxy)propanamine; LY-210448. $C_{21}H_{23}$-NO; mol wt 305.42. C 82.58%, H 7.59%, N 4.59%, O 5.24%. Selective serotonin reuptake inhibitor (SSRI). Prepn: D. W. Robertson *et al.*, **EP 288188**; *eidem*, **US 5135947** (1988, 1992 both to Lilly). Synthesis: W. J. Wheeler, D. D. O'Bannon, *J. Labelled Compd. Radiopharm.* **31**, 305 (1992); S. Kang, H.-K. Lee, *J. Org. Chem.* **75**, 237 (2010). HPLC determn in plasma: C. L. Hamilton, J. D. Cornpropst, *J. Chromatogr.* **612**, 253 (1993). Clinical pharmacokinetics: N. B. Modi *et al.*, *J. Clin. Pharmacol.* **46**, 301 (2006). Clinical trial: J. L. Pryor *et al.*, *Lancet* **368**, 929 (2006); C. G. McMahon *et al.*, *J. Sex. Med.* **8**, 524 (2011). Review of pharmacology and clinical experience: B. Feret, *Formulary* **40**, 227-230 (2005); A. M. Feige *et al.*, *Clin. Pharmacol. Ther.* **89**, 125 (2011).

Colorless oil. $[\alpha]_D^{28}$ + 63.2° (c=0.3 in chloroform).
Hydrochloride. [129938-20-1] Priligy. $C_{21}H_{23}NO$·HCl; mol wt 341.88. White crystalline solid from hydrochloric acid and ethyl acetate. $[\alpha]_D$ +135.78° (c = 2.18 in methanol).
THERAP CAT: In treatment of premature ejaculation.

2822. Dapsone. [80-08-0] 4,4′-Sulfonylbisbenzeneamine; 4,4′-sulfonyldianiline; bis(4-aminophenyl)sulfone; 4,4′-diaminodiphenyl sulfone; DDS; diaphenylsulfone; DADPS; 1358F; Avlosulfon; Croysulfone; Diphenasone; Disulone; Dumitone; Eporal; Novophone; Sulfona-Mae; Sulphadione; Udolac. $C_{12}H_{12}N_2O_2S$; mol wt 248.30. C 58.05%, H 4.87%, N 11.28%, O 12.89%, S 12.91%. Prepn: **FR 829926** (1938 to I. G. Farbenind.), *C.A.* **33**, 1761 (1939); Buckles, *J. Chem. Educ.* **31**, 36 (1954); Ferry *et al.*, *Org. Synth.* **coll. vol. III**, 239 (1955). Antibacterial action and metabolism: J. Francis, A. Spinks, *Br. J. Pharmacol.* **5**, 565 (1950). Mechanism of toxic action: Wu, DuBois, *Arch. Int. Pharmacodyn. Ther.* **183**, 36 (1970). Comprehensive description: C. E. Orzech *et al.*, *Anal. Profiles Drug Subs.* **5**, 87-114 (1976). Review of pharmacology and therapeutic use: J. Uetrecht, *Clin. Dermatol.* **7**, 111-120 (1989). Clinical trial with chlorproguanil, *q.v.*, in drug-resistant malaria: T. Mutabingwa *et al.*, *Lancet* **358**, 1218 (2001).

Crystals from 95% ethanol. mp 175-176° (also a higher melting form, mp 180.5°). pKb 13.0. Soluble in alcohol, methanol, acetone, dil hydrochloric acid. Practically insol in water.
USE: Hardening agent in the curing of epoxy resins.
THERAP CAT: Antibacterial (leprostatic); in treatment of dermatitis herpetiformis.
THERAP CAT (VET): Antibacterial; coccidostat.

2823. Daptomycin. [103060-53-3] LY-146032; Cidecin; Cubicin. $C_{72}H_{101}N_{17}O_{26}$; mol wt 1620.69. C 53.36%, H 6.28%, N 14.69%, O 25.67%. Cyclic lipopeptide antibiotic derived from a fermentation product of *Streptomyces roseosporus*; disrupts plasma membrane function in gram-positive bacteria. Prepn: B. J. Abbott *et al.*, US 4537717 (1985 to Eli Lilly); M. Debono *et al.*, *J. Antibiot.* **41**, 1093 (1988). Mechanism of action study: N. E. Allen *et al.*, *Antimicrob. Agents Chemother.* **31**, 1093 (1987); and structure determn: D. Jung *et al.*, *Chem. Biol.* **11**, 949 (2004). Comparative *in vitro* antibacterial spectrum: I. A. Critchley *et al.*, *J. Antimicrob. Chemother.* **51**, 639 (2003). Clinical pharmacokinetics: J. R. Woodworth *et al.*, *Antimicrob. Agents Chemother.* **36**, 318 (1992). Clinical trial in skin infections: R. D. Arbeit *et al.*, *Clin. Infect. Dis.* **38**, 1673 (2004). Review of pharmacology and clinical evaluations: F. P. Tully *et al.*, *Expert Opin. Invest. Drugs* **8**, 1223-1238 (1999); L. P. Kotra *Curr. Opin. Anti-Infect. Invest. Drugs* **2**, 185-205 (2000).

uv max (ethanol): 220, 260 nm (ε 46500, 10000). LD_{50} i.v. in mice: 600 mg/kg (Debono).
THERAP CAT: Antibacterial.

2824. Darapladib. [356057-34-6] *N*-[2-(Diethylamino)ethyl]-2-[[(4-fluorophenyl)methyl]thio]-4,5,6,7-tetrahydro-4-oxo-*N*-[[4′-(trifluoromethyl)[1,1′-biphenyl]-4-yl]methyl]-1*H*-cyclopentapyrimidine-1-acetamide; 1-(*N*-(2-(diethylamino)ethyl)-*N*-(4-(4-trifluoromethylphenyl)benzyl)-aminocarbonylmethyl)-2-(4-fluorobenzyl)thio-5,6-trimethylenepyrimidin-4-one; SB-480848; Tyrisa. $C_{36}H_{38}F_4N_4O_2S$; mol wt 666.78. C 64.85%, H 5.74%, F 11.40%, N 8.40%, O 4.80%, S 4.81%. Selective inhibitor of lipoprotein-associated phospholipase A_2 (Lp-PLA_2). Prepn: D. M. B. Hickey *et al.*, WO 0160805; *eidem*, US 6649619 (2001, 2003 both to SKB). Synthesis and pharmacology: J. A. Blackie *et al.*, *Bioorg. Med. Chem. Lett.* **13**, 1067 (2003). Inhibition of Lp-PLA_2 in swine with diabetes and hypercholesterolemia: R. L. Wilensky *et al.*, *Nat. Med.* **14**, 1059 (2008). Clinical effect on biomarkers of cardiovascular risk: E. R. Mohler *et al.*, *J. Am. Coll. Cardiol.* **51**, 1632 (2008); on coronary plaque stability: P. W. Serruys *et al.*, *Circulation* **118**, 1172 (2008). Review of development and therapeutic potential in atherosclerosis: Q. T. Bui, R. L. Wilensky, *Expert Opin. Invest. Drugs* **19**, 161-168 (2010).

Crystals from isopropyl acetate, mp 125°.
THERAP CAT: Antiatherosclerotic.

2825. Darbepoetin Alfa. [209810-58-2] [30-Asparagine,32-threonine,87-valine,88-asparagine,90-threonine]erythropoietin (human); novel erythropoiesis stimulating protein; NESP; Aranesp. Hyperglycosylated analog of human erythropoietin, *q.v.*, produced by recombinant DNA technology. mol wt ~38 kDa. Engineered to contain 2 extra carbohydrate addition sites, leading to increased circulating half-life. Prepn: S. G. Elliott, T. E. Byrne, WO 9505465 (1995 to Amgen). Review of development and pharmacology: I. C. Macdougall, *Semin. Nephrol.* **20**, 375-381 (2000); of clinical experience in treatment of anemia of kidney disease: A. R. Nissenson, *Am. J. Kidney Dis.* **38**, 1390-1397 (2001); of chemotherapy induced anemia: J. Vansteenkiste, I. Wauters, *Expert Opin. Pharmacother.* **6**, 429-440 (2005). Series of articles on clinical experience in cancer-related anemia: *Br. J. Cancer* **84**, Suppl. 1, 1-38 (2001).
THERAP CAT: Hematopoietic.

2826. Darifenacin. [133099-04-4] (3*S*)-1-[2-(2,3-Dihydro-5-benzofuranyl)ethyl]-α,α-diphenyl-3-pyrrolidineacetamide; 3-(*S*)-(−)-(1-carbamoyl-1,1-diphenylmethyl)-1-[2-(2,3-dihydrobenzofuran-5-yl)ethyl]pyrrolidine; (*S*)-2-[1-[2-(2,3,-dihydrobenzfuran-5-yl)-ethyl]-3-pyrrolidinyl]-2,2-diphenylacetamide; UK-88525. $C_{28}H_{30}N_2O_2$; mol wt 426.56. C 78.84%, H 7.09%, N 6.57%, O 7.50%. Selective muscarinic M_3-receptor antagonist. Prepn: P. E. Cross, A. R. MacKenzie, EP 388054; *eidem*, US 5096890 (1990, 1992 both to Pfizer). HPLC/MS determn in plasma: B. Kaye *et al.*, *Anal. Chem.* **68**, 1658 (1996). Binding profile for receptor subtypes: C. M. Smith, R. W. Wallis, *J. Recept. Signal Transduct. Res.* **17**, 177 (1997); and pharmacology: R. M. Wallis, C. M. Napier, *Life Sci.* **64**, 395 (1999). Pharmacokinetics and metabolism: K. C. Beaumont *et al.*, *Xenobiotica* **28**, 63 (1998). Clinical trial in overactive bladder: F. Haab *et al.*, *Eur. Urol.* **45**, 420 (2004). Review of clinical experience: C. R. Chapple, *Expert Opin. Invest. Drugs* **13**, 1493-1500 (2004).

Foam or colorless glass. $[\alpha]_D^{25}$ −20.6° (c = 1.0 in methylene chloride). pKa (25°): 9.2.
Hydrobromide. [133099-07-7] Emselex; Enablex. $C_{28}H_{31}N_2O_2Br$. mp 229°. $[\alpha]_D^{25}$ −30.3° (c = 1.0 in methylene chloride). Soly at 37° (mg/ml): water 6.03.
THERAP CAT: Antispasmodic; in treatment of urinary incontinence.

2827. Darinaparsin. [69819-86-9] L-γ-Glutamyl-*S*-(dimethylarsino)-L-cysteinylglycine; *S*-(dimethylarsino)glutathione; *N*-[*S*-(dimethylarsino)-*N*-L-γ-glutamyl-L-cysteinyl]-glycine; ZIO-101; Zinapar. $C_{12}H_{22}AsN_3O_6S$; mol wt 411.30. C 35.04%, H 5.39%, As 18.22%, N 10.22%, O 23.34%, S 7.79%. Cytotoxic, organic arsenical compound. Induces apoptosis and cell cycle arrest; inhibits microtubule polymerization and disrupts sonic hedgehog (Shh) signaling. Prepn from glutathione and dimethylchloroarsine: C. H. Banks *et al.*, *J. Med. Chem.* **22**, 572 (1979); from glutathione and dimethylarsinic acid: N. Scott *et al.*, *Chem. Res. Toxicol.* **6**, 102 (1993). Use in cancer treatment: R. A. Zingaro *et al.*, WO 03057012; *eidem*, US 6911471 (2003, 2005 both to Board of Regents, Univ. Texas). Cytotoxic effect: S. Hirano, Y. Kobayashi, *Toxicology* **227**, 45 (2006). Activity vs arsenic trioxide-resistant tumor cell lines: Z. Diaz *et al.*, *Leukemia* **22**, 1853 (2008); S. M. Matulis *et al.*, *Mol. Cancer Ther.* **8**, 1197 (2009). Mechanism of action: T. A. Mason *et al.*, *PLoS ONE* **6**, e27699 (2011). Clinical pharmacology: A. M. Tsimberidou *et al.*, *Clin. Cancer Res.* **15**, 4769 (2009). Clinical evaluation in advanced hepatocellular carcinoma: J. Wu *et al.*, *Invest. New Drugs* **28**, 670 (2010). Review of discovery and clinical pharmacology: A. Quintás-Cardama *et al.*, *Anti-Cancer Agents Med. Chem.* **8**, 904-909 (2008); of cytotoxic activity and clinical trials: K. K. Mann *et al.*, *Expert Opin. Invest. Drugs* **18**, 1727-1734 (2009).

White to off-white powder with slight odor of sulfur. Sol in aq solns. Insol in ethanol.

Pyridinium hydrochloride. [69819-87-0] $C_{12}H_{22}AsN_3O_6S$.-$C_5H_5N.HCl$; mol wt 526.86. Crystals from ethanol, mp 115-118°.

THERAP CAT: Antineoplastic.

2828. Darmstadtium. [54083-77-1] Element 110; ununnilium. Ds, Uun; at. no. 110. Group VIII (10). Transuranium element. No stable nuclides. Prepn and decay of isotopes ^{269}110 ($T_{1/2}$ ~170 μsec, α decay) by ^{208}Pb(^{62}Ni,n) and isotopes ^{271}110 ($T_{1/2}$ ~1.1 msec, α decay) and ^{271m}110 ($T_{1/2}$ ~56 msec, α decay) by ^{208}Pb (^{64}Ni,n): S. Hofmann et al., Z. Phys. A **350**, 277 (1995). Production of isotope ^{273}110 ($T_{1/2}$ ~76 μsec, α decay) as a decay product of isotope ^{277}112: Y. A. Lazarev, Phys. Rev. C **54**, 620 (1996); of isotope ^{273}110 ($T_{1/2}$ ~118 msec, α decay): S. Hofmann et al., Z. Phys. A **354**, 229 (1996). Review of production and properties: idem, Rep. Prog. Phys. **61**, 639-689 (1998).

2829. Darunavir. [206361-99-1] N-[(1S,2R)-3-[[(4-Aminophenyl)sulfonyl] (2-methylpropyl)amino]-2-hydroxy-1-(phenylmethyl)propyl]carbamic acid (3R,3aS,6aR)-hexahydrofuro[2,3-b]furan-3-yl ester; (1S,2R,3'R,3'aS,6'aR)-[3'-hexahydrofuro[2,3-b]-furanyl-[3-(4-aminobenzenesulfonyl)isobutylamino]-1-benzyl-2-hydroxypropyl]carbamate; TMC-114; UIC-94017. $C_{27}H_{37}N_3O_7S$; mol wt 547.67. C 59.21%, H 6.81%, N 7.67%, O 20.45%, S 5.85%. Peptidomimetic HIV-1 protease inhibitor. Prepn: A. K. Ghosh et al., Bioorg. Med. Chem. Lett. **8**, 687 (1998); idem et al., J. Org. Chem. **69**, 7822 (2004). Prepn of solvated crystalline forms: H. W. P. Vermeersch et al., WO 03106461, eidem, US 050250845 (2003, 2005 both to Tibotec). In vitro activity vs multidrug resistant HIV isolates: Y. Koh et al., Antimicrob. Agents Chemother. **47**, 3123 (2003). Crystal structures of complex with protease variants: Y. Tie et al., J. Mol. Biol. **338**, 341 (2004). Clinical trial of combination with ritonavir: B. Clotet et al., Lancet **369**, 1169 (2007). Reviews: J.-M. Molina, A. Hill, Expert Opin. Pharmacother. **8**, 1951-1964 (2007); K. H. S. Busse, S. R. Penzak, Am. J. Health Syst. Pharm. **64**, 1593-1602 (2007).

White amorphous solid, mp 74° (dec). Sol in DMSO.

Monoethanolate. [635728-49-3] Prezista. $C_{27}H_{37}N_3O_7S.C_2$-H_5OH; mol wt 593.74. White to off-white powder. d 1.248. Soly in water (20°): 0.15 mg/ml. Sol in acetone, dicloromethane, ethyl acetate, THF.

THERAP CAT: Antiretroviral.

2830. Darusentan. [171714-84-4] (αS)-α-[(4,6-Dimethoxy-2-pyrimidinyl)oxy]-β-methoxy-β-phenylbenzenepropanoic acid; (+)-(S)-2-(4,6-dimethoxy-pyrimidin-2-yloxy)-3-methoxy-3,3-diphenylpropionic acid; LU-135252. $C_{22}H_{22}N_2O_6$; mol wt 410.43. C 64.38%, H 5.40%, N 6.83%, O 23.39%. Selective endothelin ET_A receptor antagonist. Prepn: H. Reichers et al., EP 785926; eidem, US 5932730 (1995, 1999 both to BASF); and resolution of isomers: H. Riechers et al., J. Med. Chem. **39**, 2123 (1996). Large scale synthesis of active (S)-isomer: R. Jansen et al., Org. Process Res. Dev. **5**, 16 (2001). Radioreceptor assay in plasma and urine: P. Cernacek et al., Clin. Chem. **44**, 1666 (1998). Clinical evaluation in hypertension: R. Nakov et al., Am. J. Hypertens. **15**, 583 (2002); in

chronic heart failure: I. Anand et al., Lancet **364**, 347 (2004). Review of pharmacology: P. Rohmeiss et al., Cardiovasc. Drug Rev. **16**, 391-412 (1998); of development and clinical experience: F. Enseleit et al., Expert Opin. Invest. Drugs **17**, 1255-1263 (2008).

Colorless solid from n-heptane.

(±)-Form. [178306-46-2] LU-127043. White solid from diethyl ether, mp 165-168°.

THERAP CAT: Antihypertensive.

2831. Dasatinib. [302962-49-8] N-(2-Chloro-6-methylphenyl)-2-[[6-[4-(2-hydroxyethyl)-1-piperazinyl]-2-methyl-4-pyrimidinyl]amino]-5-thiazolecarboxamide; BMS-354825; Sprycel. $C_{22}H_{26}$-ClN_7O_2S; mol wt 488.01. C 54.15%, H 5.37%, Cl 7.26%, N 20.09%, O 6.56%, S 6.57%. Orally bioactive, ATP-competitive dual inhibitor of Src and Abl kinases. Prepn: J. Das et al., WO 0062778; eidem US 6596746 (2000, 2003 both to Bristol-Myers Squibb); and Src/Abl kinase inhibition study: L. J. Lombardo et al., J. Med. Chem. **47**, 6658 (2004). In vivo evaluation vs imatinib-resistant mutants in chronic myeloid leukemia: N. P. Shah et al., Science **305**, 399 (2004). Mechanism of action study: M. R. Burgess et al., Proc. Natl. Acad. Sci. USA **102**, 3395 (2005). HPLC-MS determn in plasma: S. De Francia et al., J. Chromatogr. B **877**, 1721 (2009). Clinical trial in imatinib-resistant leukemias: G. Saglio et al., Cancer **116**, 3852 (2010); as first-line treatment of chronic myeloid leukemia: H. Kantarjian et al., N. Engl. J. Med. **362**, 2260 (2010). Review of therapeutic potential for solid tumors: J. Araujo, C. Logothetis, Cancer Treat. Rev. **36**, 492-500 (2010).

White to off-white powder; occurs as the monohydrate. mp 279-280°. Slightly sol in ethanol, methanol. Insol in water.

THERAP CAT: Antineoplastic.

2832. Datiscetin. [480-15-9] 3,5,7-Trihydroxy-2-(2-hydroxyphenyl)-4H-1-benzopyran-4-one; 2',3,5,7-tetrahydroxyflavone; 2'-hydroxycrysidenolon 1493. $C_{15}H_{10}O_6$; mol wt 286.24. C 62.94%, H 3.52%, O 33.54%. Cleavage product of datiscin: Stenhouse, Ann. **98**, 166 (1856). From roots of Datisca cannabina L., Datiscaceae: Marchlewski, Biochem. Z. **3**, 286 (1907). Structure: Leskiewicz, Marchlewski, Ber. **47**, 1599 (1914). Synthesis: Kalf, Robinson, J. Chem. Soc. **127**, 1968 (1925); Simpson, Wally, ibid. **1955**, 166.

Pale yellow needles from alc, mp 271°. uv max (96% alc): 264.0, 375.0 nm (log ε 4.265, 4.005). Sol in alcohol, ether, other organic solvents; practically insol in water. Sol in alkaline solns with yellow color.

2833. Daucol. [887-08-1] (3R,3aS,6S,7S,8aR)-Octahydro-6,8a-dimethyl-3-(1-methylethyl)-1H-3a,6-epoxyazulen-7-ol; 2,3,4,-

5,6,7a,8,8a-octahydro-3-isopropyl-6,8a-dimethyl-1*H*-3a,6-epoxy-azulen-7-ol. $C_{15}H_{26}O_2$; mol wt 238.37. C 75.58%, H 10.99%, O 13.42%. From oil of carrot seeds, *Daucus carota* L., *Umbelliferae:* Richter, *Arch. Pharm.* **247**, 391 (1909). Structure: Sykora *et al.*, *Collect. Czech. Chem. Commun.* **26**, 788 (1961); Zalkow *et al.*, *J. Org. Chem.* **26**, 981 (1961). Stereochemistry: Levisalles, Rudler, *Bull. Soc. Chim. Fr.* **1964**, 2020; **1967**, 2059. Synthesis of (−)-form: DeBroissia *et al.*, *Chem. Commun.* **1972**, 855; *eidem*, *Bull. Soc. Chim. Fr.* **1972**, 4314.

(−)-Daucol

Crystals (from petr ether at −30°), mp 113-115°. bp_2 124-132°. $[\alpha]_D^{20}$ −16.9° (c = 2.76 in ethanol).

2834. Daunorubicin. [20830-81-3] (8*S*,10*S*)-8-Acetyl-10-[(3-amino-2,3,6-trideoxy-α-L-*lyxo*-hexopyranosyl)oxy]-7,8,9,10-tetrahydro-6,8,11-trihydroxy-1-methoxy-5,12-naphthacenedione; daunomycin; leukaemomycin C; rubidomycin; RP-13057; Cerubidin. $C_{27}H_{29}NO_{10}$; mol wt 527.53. C 61.47%, H 5.54%, N 2.66%, O 30.33%. Anthracycline antibiotic related to the rhodomycins, *q.v.* Isolated from fermentation broths of *Streptomyces peucetius:* G. Cassinelli, P. Orezzi, *G. Microbiol.* **11**, 167 (1963), *C.A.* **62**, 9482b (1965); A. Di Marco *et al.*, *Nature* **201**, 706 (1964); *eidem*, **BE 639897**; *eidem*, **US 4012284** (1964, 1977 both to Soc. Farmaceut. Italia); S. Pinnert *et al.*, **US 3997662** (1976 to Rhone-Poulenc). Daunorubicin is a glycoside formed by a tetracyclic aglycone, *daunomycinone*, ($C_{21}H_{18}O_8$) and an amino sugar, *daunosamine*, ($C_6H_{13}NO_3$), 3-amino-2,3,6-trideoxy-L-*lyxo*-hexose: F. Arcamone *et al.*, *J. Am. Chem. Soc.* **86**, 5334, 5335 (1964); R. H. Iwamoto *et al.*, *Tetrahedron Lett.* **1968**, 3891. Absolute stereochemistry: F. Arcamone *et al.*, *Gazz. Chim. Ital.* **100**, 949-989 (1970). Identity with rubidomycin: G. L. Tong *et al.*, *J. Pharm. Sci.* **56**, 1691 (1967). Synthesis of daunosamine: J. P. Marsh *et al.*, *Chem. Commun.* **1967**, 973; T. Yamaguchi, M. Kojimo, *Carbohydr. Res.* **59**, 343 (1977); P. M. Wovkulich, M. R. Uskokovic, *J. Am. Chem. Soc.* **103**, 3956 (1981); of daunomycinone: C. M. Wong *et al.*, *Can. J. Chem.* **51**, 466 (1973); J. S. Swenton, P. W. Reynolds, *J. Am. Chem. Soc.* **100**, 6188 (1978); K. Krohn, K. Tolkiehn, *Ber.* **112**, 3453 (1979); F. M. Hauser, S. Prasanna, *J. Am. Chem. Soc.* **103**, 6378 (1981). Total synthesis of daunorubicin: E. M. Acton *et al.*, *J. Med. Chem.* **17**, 659 (1974). Purification: E. Oppici *et al.*, **BE 898506**; *eidem*, **GB 2133005** (both 1984 to Farmitalia). Toxicity data: A. Di Marco *et al.*, *Cancer Chemother. Rep. Part 1* **53**, 33 (1969). Review of properties, biosynthesis, fermentation: R. J. White, R. M. Stroshane, *Drugs Pharm. Sci.* **22**, 569-594 (1984); of carcinogenic action in laboratory animals: *IARC Monographs* **10**, 145-152 (1976); of toxicology: R. J. Maral *et al.*, *Cancer Treat. Rep.* **65**, Suppl. 4, 9-18 (1981); of use in treatment of solid tumors: R. B. Weiss *et al.*, *ibid.* 25-28; of interactions with nucleic acids: S. Neidle, M. R. Sanderson, in *Molecular Aspects of Anti-cancer Drug Action*, S. Neidle, M. J. Waring, Eds. (Verlag-Chemie, Florida, 1983) pp 35-55; of mechanism of cytotoxicity: H. S. Schwartz, *ibid.* pp 93-125; of metabolism and clinical pharmacokinetics: C. E. Riggs, Jr., *Semin. Oncol.* **11**, Suppl. 3, 2-11 (1984). *Review:* A. DiMarco *et al.*, *Antibiotics* **vol. 3**, J. W. Corcoran, F. E. Hahn, Eds. (Springer Verlag, New York, 1975) pp 101-128.

Daunomycinone

Daunosamine

mp 208-209°. LD_{50} in mice, rats (mg/kg): 20, 13 i.v.; 5, 8 i.p. (DiMarco, 1977).

Hydrochloride. [23541-50-6] Cérubidine; Daunoblastina; Ondena. $C_{27}H_{29}NO_{10}$·HCl; mol wt 563.98. Thin red needles, dec 188-190°. $[\alpha]_D^{20}$ +248 ±5° (c = 0.05-0.1 in methanol). Sol in water, methanol, aq alcohols. Practically insol in chloroform, ether, benzene. Color of aq soln changes from pink at acid pH to blue at alkaline pH. Absorption max (methanol): 234, 252, 290, 480, 495, and 532 nm ($E_{1cm}^{1\%}$ 665, 462, 153, 214, 218, and 112). LD_{50} in mice (mg/kg): 26 i.v. (DiMarco, 1969).

THERAP CAT: Antineoplastic.

2835. Dauricine. [524-17-4] 4-[[(1*R*)-1,2,3,4-Tetrahydro-6,7-dimethoxy-2-methyl-1-isoquinolinyl]methyl]-2-[4-[[(1*R*)-1,2,-3,4-tetrahydro-6,7-dimethoxy-2-methyl-1-isoquinolinyl]methyl]-phenoxy]phenol. $C_{38}H_{44}N_2O_6$; mol wt 624.78. C 73.05%, H 7.10%, N 4.48%, O 15.36%. From *Menispermum dauricum* DC., and *M. canadense* L., *Menispermaceae:* Kondo, Narita, *J. Pharm. Soc. Japan* no. 542, 279 (1927), *C.A.* **21**, 2700 (1927); Manske, *Can. J. Res.* **21B**, 17 (1943). Structure: Kondo *et al.*, *Ber.* **68**, 519 (1935); Kondo, Tomita, *Arch. Pharm.* **274**, 65 (1936); Inubushi, Niwa, *J. Pharm. Soc. Jpn.* **72**, 762 (1952), *C.A.* **47**, 6430e (1953). Total synthesis: Kametani, Fukumoto, *J. Chem. Soc.* **1965**, Suppl. 2, 6141.

Slightly yellow amorphous base, mp 115°. $[\alpha]_D^{11}$ −139° in methanol. Sol in alc, acetone, benzene; slightly sol in ether.

Dimethiodide. $C_{38}H_{44}N_2O_6$·2CH$_3$I. Needles, mp 204°. $[\alpha]_D^{20}$ −114°.

2836. Davunetide. [211439-12-2] L-Asparaginyl-L-alanyl-L-prolyl-L-valyl-L-seryl-L-isoleucyl-L-prolyl-L-glutamine; NAPVSIPQ; AL-108; AL-208. $C_{36}H_{60}N_{10}O_{12}$; mol wt 824.93. C 52.42%, H 7.33%, N 16.98%, O 23.27%. Bioactive fragment of activity-dependent neuroprotective protein (ADNP) derived from mouse neuroglial cells. Binds to brain tubulin and promotes proper microtubule assembly. Protects against toxicity associated with amyloid β peptide, NMDA, and tetrodotoxin. Prepn: I. Gozes *et al.*, **WO 9835042** (1998 to US Dept. Health Human Serv.); *idem et al.*, **US 6613740** (2003 to Ramot Univ. Auth. Appl. Res. Ind. Dev.; US Dept. Health Human Serv.); M. Bassan *et al.*, *J. Neurochem.* **72**, 1283 (1999). Protective effect in neuronal cultures: I. Zemlyak *et al.*, *Regul. Pept.* **96**, 39 (2000). Pharmacology and bioavailability: I. Gozes *et al.*, *CNS Drug Rev.* **11**, 353 (2005). Effect on tau pathology in mouse model of Alzheimer's disease: Y. Matsuoka *et al.*, *J. Pharmacol. Exp. Ther.* **325**, 146 (2008). Review of pharmacology and clinical experience: H. Geerts, *Curr. Opin. Investig. Drugs* **9**, 800-811 (2008).

Asn−Ala−Pro−Val−Ser−Ile−Pro−Gln

THERAP CAT: Neuroprotective.

2837. Dazomet. [533-74-4] Tetrahydro-3,5-dimethyl-2*H*-1,-3,5-thiadiazine-2-thione; 2-thio-3,5-dimethyltetrahydro-1,3,5-thiadiazine; 3,5-dimethyl-2-thionotetrahydro-1,3,5-thiadiazine; dimethylformocarbothialdine; DMTT; UCC 974; Basamid. $C_5H_{10}N_2S_2$; mol wt 162.27. C 37.01%, H 6.21%, N 17.26%, S 39.51%. Prepd by the action of carbon disulfide on trimethyltrimethylenetriimine: Delépine, *Bull. Soc. Chim.* **15**, 891 (1897); from formaldehyde and methylammonium methyldithiocarbamate: Bodendorf, *J. Prakt. Chem.* **126**, 233 (1930); Ainley *et al.*, *J. Chem. Soc.* **1944**, 147.

Prepd commercially by the reaction of carbon disulfide, methylamine and caustic soda: *Chem. Week* **79**, no. 18, 82-83 (1956). Toxicology study: H. F. Smyth, Jr. *et al.*, *Toxicol. Appl. Pharmacol.* **9**, 521 (1966). HPLC determn in commercial formulations: B. Petanovska-Ilievska, *J. Liq. Chromatogr. Relat. Technol.* **24**, 2209 (2001). Degradation in soil: E. Ercag *et al.*, *Anal. Chim. Acta* **505**, 95 (2004). Field trial as soil fumigant: B. S. Park, P. J. Landschoot, *Crop Sci.* **43**, 1387 (2003).

Needles from benzene, mp 106-107°. Soluble in alc with decompn. Dec by water, dil acids. uv max (cyclohexane): 242, 289 nm (ε 7150, 9900). LD$_{50}$ in rats, mice, guinea pigs, rabbits (g/kg): 0.32-0.62, 0.18, 0.16, 0.12 orally; in rats, rabbits, dogs (g/kg): 0.087, 0.127, 0.047-0.063 i.p.; in rabbits (g/kg): 7.1 dermally (Smyth).

USE: Soil fungicide, nematocide, weed killer; slimicide in paper making.

2838. 2,4-DB. [94-82-6] 4-(2,4-Dichlorophenoxy)butanoic acid; 4-(2,4-dichlorophenoxy)butyric acid; Butyrac; Legumex D. C$_{10}$H$_{10}$Cl$_2$O$_3$; mol wt 249.09. C 48.22%, H 4.05%, Cl 28.46%, O 19.27%. Selective post-emergence, translocated herbicide. First described as a plant growth regulator: M. E. Synerholm and P. W. Zimmerman, *Contrib. Boyce Thompson Inst.* **14**, 369 (1947). Prepn: G. W. Kitchingman, A. C. Tucker, **GB 804565** and **GB 883255** (1958, 1961, both to ICI). Similar in activity to 2,4-D, *q.v.*, but more selective. Mechanism of selectivity studies: D. L. King, D. E. Bayer, *Proc. West. Soc. Weed Sci.* **25**, 37 (1972).

White crystals, mp 117-119°. Soly in water at 25°: 46 ppm. Sol in acetone, ethanol, diethyl ether. Slightly sol in benzene, toluene, kerosene.

Sodium salt. Embutox. C$_{10}$H$_9$Cl$_2$NaO$_3$; mol wt 271.07. Water soluble; hard water will show precipitate as the calcium and magnesium salts.

Dimethylamine salt. Butoxone. C$_{12}$H$_{17}$Cl$_2$NO$_3$; mol wt 294.17. Water sol; calcium and magnesium salts precipitate in hard water.

USE: Herbicide.

2839. DBHBT. [63147-28-4] 2-Mercaptoacetic acid [3,5-bis-(1,1-dimethylethyl)-4-hydroxyphenyl]methyl ester; 3,5-di-*tert*-butyl-4-hydroxybenzylthioglycollate. C$_{17}$H$_{26}$O$_3$S; mol wt 310.45. C 65.77%, H 8.44%, O 15.46%, S 10.33%. Thermal stabilizer. Prepn: K. S. Cottman, **US 4020042** (1977 to Goodyear); and stabilization of acrylonitrile-butadiene-styrene (ABS) terpolymer: E. G. Kolawole, J. B. Adeniyi, *Eur. Polym. J.* **18**, 469 (1982). Stabilization of polystyrene and poly(methyl methacylate): E. G. Kolawole, *J. Appl. Polym. Sci.* **27**, 3437 (1982); of Nigerian natural rubber: G. C. Ebi, E. G. Kolawole, *ibid.* **45**, 2239 (1992).

White crystals from benzene, mp 107.5-108.5° (Cottman); also reported as mp 110° (Kolawole, Adeniyi).

USE: Antioxidant.

2840. DBMC. [497-39-2] 2,4-Bis(1,1-dimethylethyl)-5-methylphenol; 4,6-di-*tert*-butyl-*m*-cresol; 3-methyl-4,6-di-*tert*-bu-

tylphenol. C$_{15}$H$_{24}$O; mol wt 220.36. C 81.76%, H 10.98%, O 7.26%. Prepn: Weinrich, *Ind. Eng. Chem.* **35**, 264 (1943).

Crystals, mp 62.1°. bp 282°, bp$_{100}$ 211°, bp$_{20}$ 167°. d$_4^{80}$ 0.912. Practically insol in water, ethylene glycol; sol in alcohol, benzene, carbon tetrachloride, ether, acetone.

USE: Intermediate in the production of rubber chemicals modified phenolic resins, synthetic musks of the ambrette type.

2841. DCM. [51325-91-8] 2-[2-[2-[4-(Dimethylamino)phenyl]ethenyl]-6-methyl-4*H*-pyran-4-ylidene]propanedinitrile; 4-dicyanomethylene-2-methyl-6-*p*-dimethylaminostyryl-4*H*-pyran. C$_{19}$H$_{17}$N$_3$O; mol wt 303.37. C 75.22%, H 5.65%, N 13.85%, O 5.27%. Laser dye characterized by a broad tuning range. Prepn: F. G. Webster, W. C. McColgin, **FR 2156723**; *eidem*, **US 3852683** (1973, 1974 both to Eastman Kodak); P. R. Hammond, *Opt. Commun.* **29**, 331 (1979). Fluorescence lifetime measurements of cis/trans isomers: M. Meyer *et al.*, *Chem. Phys. Lett.* **150**, 484 (1988). Photophysical properties and consequences for use: J.-C. Mialocq, M. Meyer, *Laser Chem.* **10**, 277 (1990). Transient spectra: S. A. Kovalenko *et al.*, *Chem. Phys. Lett.* **258**, 445 (1996). Use in polyimide systems for electro-optic devices: S. Ermer *et al.*, *Appl. Phys. Lett.* **61**, 2272 (1992); in chemical actinometers: J.-C. Mialocq *et al.*, *J. Photochem. Photobiol. A* **69**, 351 (1993).

Dark brown plates from benzene, mp 210-212°.

USE: Laser dye.

2842. DDD (Analytical). [6088-51-3] 6,6'-Dithiobis-2-naphthalenol; 6,6'-dithiodi-2-naphthol; bis(6-hydroxy-2-naphthyl) disulfide; 2,2'-dihydroxy-6,6'-dinaphthyl disulfide; 6,6'-dithiobis(2-naphthol). C$_{20}$H$_{14}$O$_2$S$_2$; mol wt 350.45. C 68.55%, H 4.03%, O 9.13%, S 18.30%. Prepn: Zincke, Dereser, *Ber.* **51**, 352 (1918).

Leaflets, mp 221-222°.

Note: Not to be confused with the insecticide, *p,p'*-DDD [1,1-dichloro-2,2-bis(*p*-chlorophenyl)ethane].

USE: For determination of protein-bound sulfhydryl groups.

2843. DDT. [50-29-3] 1,1'-(2,2,2-Trichloroethylidene)bis[4-chlorobenzene]; 1,1,1-trichloro-2,2-bis(*p*-chlorophenyl)ethane; α,α-bis(*p*-chlorophenyl)-β,β,β-trichlorethane; dichlorodiphenyltrichloroethane; chlorophenothane; clofenotane; dicophane; pentachlorin; *p,p'*-DDT. C$_{14}$H$_9$Cl$_5$; mol wt 354.48. C 47.44%, H 2.56%, Cl 50.00%. Polychlorinated nondegradable pesticide; has been used as an ectoparasiticide. Prepn: Zeidler, *Ber.* **7**, 1180 (1874); P. Müller, **US 2329074** (1944 to Geigy); Rueggeberg, Torrans, *Ind. Eng. Chem.* **38**, 211 (1946); Cook *et al.*, *ibid.* **39**, 868, 1683 (1947). Convenient lab procedures: Bailes, *J. Chem. Educ.* **22**, 122 (1945); Ginsburg, *Science* **108**, 339 (1948). Large scale production: Mosher *et al.*, *Ind. Eng. Chem.* **38**, 916 (1946). Chemical composition of technical grade: H. L Haller *et al.*, *J. Am. Chem. Soc.* **67**, 1591 (1945). Activity: P. Müller, *Helv. Chim. Acta* **29**, 1560 (1946). Comprehensive monograph (in English and German): *DDT*, P. Müller, Ed., 3 vols (Birkhäuser Verlag, Basel and Stuttgart, 1955). GC determn in serum: M. M. Frías *et al.*, *J. Chromatogr. B* **760**, 1 (2001); in sediment: M.-S. Kim *et al., J. Chromatogr. A* **1208**, 25 (2008). Toxicity data: Gaines, *Toxicol. Appl. Pharmacol.* **2**, 88 (1960); **14**, 515 (1969). Review of toxicology and human exposure: *Toxicological Profile for DDT, DDE, DDD* (PB2003-100137, 2002) 497 pp; of toxicology and environmental impact: V. Turusov *et al., Environ. Health Perspect.* **110**, 125-128 (2002).

Biaxial elongated tablets, needles from 95% alc. mp 108.5-109°. uv max (95% alc): 236 nm. Vapor pressure at 20° = 1.5×10^{-7} mm Hg. Practically insol in water, dil acids, alkalies. Soly (g/100 ml): acetone 58; benzene 78; benzyl benzoate 42; carbon tetrachloride 45; chlorobenzene 74; cyclohexanone 116; 95% alc 2; ethyl ether 28; gasoline 10; isopropanol 3; kerosene 8-10; morpholine 75; peanut oil 11; pine oil 10-16; tetralin 61; tributyl phosphate 50; freely sol in pyridine, dioxane. The soly in organic solvents increases sharply with a rise in temp. Resistant to destruction by light and oxidation. Its unusual stability has resulted in difficulties with residue removal from water, soil and foodstuffs. Should not be kept in iron containers and should not be mixed with iron and aluminum salts nor with alkaline substances. High storage temps should also be avoided. Setting point of technical grade: 88.6-91.4°. LD_{50} in male, female rats (mg/kg): 113, 118 orally (Gaines, 1960).

Caution: Poisoning may occur by ingestion, inhalation, or by absorption through skin or respiratory tract. Potential symptoms of overexposure are paresthesia of tongue, lips, face; tremor; anxiety, dizziness, confusion, malaise, headache, lassitude; convulsions; hand paresis; vomitting. Direct contact may cause eye and skin irritation. *See NIOSH Pocket Guide to Chemical Hazards* (DHHS/NIOSH 2005-149, 2007) p 88. This substance is reasonably anticipated to be a human carcinogen: *Report on Carcinogens, Twelfth Edition* (PB2011-111646, 2011) p 143.

Note: DDT is listed as a persistent organic pollutant (POP) in Annex B of the *Stockholm Convention on Persistent Organic Pollutants* (United Nations, Stockholm, 2001) 43 pp; amended (Geneva, 2009) 63 pp.

USE: Contact insecticide.

2844. Deaminooxytocin. [113-78-0] *N*-(3-Mercapto-1-oxopropyl)-L-tyrosyl-L-isoleucyl-L-glutaminyl-L-asparaginyl-L-cysteinyl-L-prolyl-L-leucylglycinamide cyclic(1 → 5) disulfide ; 1-(3-mercaptopropanoic acid)oxytocin; demoxytocin; desaminooxytocin; ODA-914; Sandopart. $C_{43}H_{65}N_{11}O_{12}S_2$; mol wt 992.18. C 52.05%, H 6.60%, N 15.53%, O 19.35%, S 6.46%. Highly potent analog of the posterior pituitary hormone oxytocin, *q.v.*, differing in structure at the 1 position where the free amino group in the half-cystine residue is replaced by hydrogen. Synthesis and biological activity: du Vigneaud *et al., J. Biol. Chem.* **235**, P.C. 64 (1960); Hope *et al., ibid.* **237**, 1563 (1962); Takashima *et al., J. Am. Chem. Soc.* **90**, 1323 (1968). Isoln and purifn: Ferrier *et al., J. Biol. Chem.* **240**, 4264 (1965). Structure-activity correlation: Jarvis, du Vigneaud, *Science* **143**, 545 (1964). Synthesis of D-isomer: G. Flouret, V. du Vigneaud, *J. Med. Chem.* **14**, 556 (1971). Crystal structure analysis: S. P. Wood *et al., Science* **232**, 633 (1986). Pharmacology and comparison with oxytocin: Chan, du Vigneaud, *Endocrinology* **71**, 977 (1962). *Review: Adv. Exp. Med. Biol.* **vol. 2**, 53-104 (1968).

CH₂—CH₂—C—Tyr—Ile—Gln—Asn—Cys—Pro—Leu—GlyNH₂

L-Isomer. White plate-like crystals from water, mp 179° (Ferrier); 182-183° (Takashima). For amorphous powder; $[\alpha]_D^{20}$ −88.3° (Ferrier); $[\alpha]_D^{21}$ −107° (Hope); $[\alpha]_D^{25}$ −95.1° (Takashima); all (c = 0.5 in 1*N* acetic acid). Anhydrous form is very hygroscopic.

D-Isomer. White, fluffy powder, $[\alpha]_D^{20}$ +104° (c = 0.5 in 1*N* acetic acid).

THERAP CAT: Oxytocic.

2845. Deanol. [108-01-0] 2-(Dimethylamino)ethanol; *N,N*-dimethylethanolamine; β-dimethylaminoethyl alcohol; *N,N*-dimethyl-2-hydroxyethylamine; DMAE. $C_4H_{11}NO$; mol wt 89.14. C 53.90%, H 12.44%, N 15.71%, O 17.95%. Naturally occurring precursor of choline, *q.v.*; found in high concentrations in salmon, sardines, and anchovies. Prepn from equimolar amounts of ethylene oxide and dimethylamine: Knorr, *Ber.* **37**, 3508 (1904); Hanhart, Ingold, *J. Chem. Soc.* **1927**, 1012. Determn in animal tissues: C. G. Honegger, R. Honegger, *Nature* **184**, 550 (1959). GCMS determn with choline: D. J. Jenden *et al., Life Sci.* **23**, 291 (1978). Pharmacology and stimulant effect: C. C. Pfeiffer *et al., Science* **126**, 610 (1959). Clinical trial in hyperkinetic children: N. Coleman *et al., Psychosomatics* **17**, 68 (1976); in involuntary movement disorders: D. E. Casey, *Dis. Nerv. Syst.* **38**, 7 (1977). Use in organic synthesis: Z. Lu, R. J. Twieg, *Tetrahedron* **61**, 903 (2005); as corrosion inhibitor: G. Batis *et al., American Concrete Inst. Spec. Publ.* **SP-217**, 469 (2003). Review of role in cosmetic dermatology: R. Grossman, *Am. J. Clin. Dermatol.* **6**, 39-47 (2005).

Very faintly yellow liquid with fishy odor. *Flammable and corrosive.* d_4^{20} 0.8866. bp_{758} 135°. n_D^{20} 1.43. Vapor pressure (20°): 6.12 mm Hg. Vapor density: 3.1 g/l. Flash point (closed cup): 39°C (102°F). Autoignition temp: 245°C. Miscible with water, alcohol, ether. *Keep tightly closed, away from heat and open flame.*

Aceglumate. [3342-61-8] Cleregil; Risatarun. $C_4H_{11}NO.C_7H_{11}NO_5$; mol wt 278.31. Prepn: **FR M2487** (1964 to Interco Fribourg), *C.A.* **61**, 12085c (1964). Sol in water.

Acetamidobenzoate. [3635-74-3] Deaner. $C_4H_{11}NO.C_9H_9NO_3$; mol wt 268.31. Prepn: **GB 879259** (1957 to Riker Labs); A. Lasslo *et al., J. Am. Pharm. Assoc. Sci. Ed.* **48**, 345 (1959). Crystals from absolute ethanol + ethyl acetate, mp 159-161.5°. Sol in water.

Bitartrate. [5988-51-2] $C_4H_{11}NO.C_4H_6O_6$; mol wt 239.22. White powder, mp 109-113°. $[\alpha]_D^{20}$ +18° (c = 3% in water). Sol in water. LD_{50} in mice, rats (g/kg): 3.1, 2.6 orally (Pfeiffer).

Hemisuccinate. Tonibral; Rischiaril. $C_4H_{11}NO.C_4H_6O_4$; mol wt 207.23. White, hygroscopic crystalline powder. Very soluble in water, ethanol; sparingly sol in chloroform, benzene.

USE: Solvent and ligand in organic synthesis; accelerator in epoxy resin formulations. As corrosion inhibitor in steel-reinforced concrete. Salts as dietary supplement, tonic. In cosmetics as antiwrinkle and skin firming agent.

2846. Debrisoquin. [1131-64-2] 3,4-Dihydro-2(1*H*)-isoquinolinecarboximidamide; 3,4-dihydro-2(1*H*)-isoquinolinecarboxamidine; 2-amidino-1,2,3,4-tetrahydroisoquinoline; isocaramidine. $C_{10}H_{13}N_3$; mol wt 175.24. C 68.54%, H 7.48%, N 23.98%. Prepn of derivs: W. Wenner, **BE 629007**; *idem*, **US 3157573** (1963, 1964 both to Hoffmann-La Roche); *idem, J. Med. Chem.* **8**, 125 (1965). Pharmacology: Moe *et al., Curr. Ther. Res.* **6**, 299 (1964); Abrams *et al., J. New Drugs* **4**, 268 (1964). Possible mechanism of action: Medina *et al., Biochem. Pharmacol.* **18**, 891 (1969); Pocelinko *et al., Int. Z. Klin. Pharmakol. Ther. Toxikol.* **2**, 13 (1969). Toxicity study: E. I. Goldenthal, *Toxicol. Appl. Pharmacol.* **18**, 185 (1971).

Sulfate. [581-88-4] Ro-5-3307/1; Declinax; Tendor. $2(C_{10}H_{13}N_3).H_2SO_4$; mol wt 448.54. Crystals, mp 278-280°, 284-285° or 266-268° (H_2O), depending on humidity. Freely sol in water. LD_{50} in neonate, adult rats (mg/kg): 88 ±18, 1580 ±163 orally (Goldenthal).

THERAP CAT: Antihypertensive.

2847. Decaborane(14). [17702-41-9] Decaboron tetradecahydride. $B_{10}H_{14}$; mol wt 122.21. B 88.45%, H 11.55%. Prepd by the pyrolysis of diborane(6). Reviews of chemistry: Griffo, *Diss. Abstr.* **22**, 2976 (1962); Stanko *et al., Usp. Khim.* **34**(6), 1011-1039

(1965). Review of structure and properties: Campbell, Jr., in *Progress in Boron Chemistry* **vol. 1**, Steinberg, McCloskey, Eds. (Macmillan, New York, 1964) pp 173-188. Reviews of toxicity: Levinskas, "Toxicology of Boron Compounds" in *Boron, Metallo-Boron Compounds and Boranes*, R. M. Adams, Ed. (Interscience, New York, 1964) pp 693-737; E. Browning, *Toxicity of Industrial Metals* (Appleton-Century-Crofts, New York, 2nd ed., 1969) pp 92-97. General reviews: Stock, *Hydrides of Boron and Silicon* (Cornell Univ. Press, Ithaca, 1933) *passim*; Siegel, Mack, *J. Chem. Educ.* **34**, 314-317 (1957); Major, *Chem. Eng. Prog.* **54**(3), 49-54 (1958); Lipscomb, *Boron Hydrides* (Benjamin, New York, 1963) *passim;* Adams in *Boron, Metallo-Boron Compounds and Boranes loc. cit.*, pp 647-663; Hawthorn in *Adv. Inorg. Chem. Radiochem.* **5**, 307-345 (1963); Greenwood in *Comprehensive Inorganic Chemistry* **vol. 1**, J. C. Bailar, Jr. *et al.*, Eds. (Pergamon Press, Oxford, 1973) pp 818-837.

Orthorhombic crystals. mp 99.6-99.7°; bp 213°; bp_{19} 100°. d_4^{25} 0.94; d_4^{100} (liq) 0.78. Stable indefinitely at room temp; decomp slowly into B + H_2 at 300°. Heat of fusion: 7.8 kcal/mole; of sublimation: 18.33 kcal/mole; of vaporization: 11.6 kcal/mole. Slightly sol in cold water; hydrolyzes in hot water. Sol in ethyl acetate, 1-bromopropane, ethyl silicate, carbon disulfide, benzene, alcohol, acetic anhydride, acetic acid, ethyl borate, carbon tetrachloride. *Highly reactive.* Reacts with amides, acetone, butyraldehyde, acetonitrile at room temp. Decaborane mixtures with carbon tetrachloride are dangerously shock sensitive.

Caution: Potential symptoms of overexposure are dizziness, headache, nausea, lightheadedness and drowsiness; incoordination, local muscle spasms, tremor and convulsions; fatigue. *See NIOSH Pocket Guide to Chemical Hazards* (DHHS/NIOSH 97-140, 1997) p 88. See also *Patty's Industrial Hygiene and Toxicology* **vol. 2B**, G. D. Clayton, F. E. Clayton, Eds. (Wiley-Interscience, New York, 3rd ed., 1981) 2987-2990.

USE: In rocket propellants; as catalyst in olefin polymerization.

2848. Decabromodiphenyl Ether. [1163-19-5] 1,1'-Oxybis-[2,3,4,5,6-pentabromobenzene]; bis(pentabromophenyl) ether; decabromodiphenyl oxide; deca-BDE; DBDPO; BDE-209. $C_{12}Br_{10}O$; mol wt 959.17. C 15.03%, Br 83.31%, O 1.67%. Polybrominated diphenyl ether (PBDE) flame retardant. Prepn: A. Bonneaud, *Bull. Soc. Chim. Fr.* **7**, 776 (1910). Manufacturing process: **GB 1411524** (1975 to Ugine Kuhlmann). Crystal structure: J. Eriksson *et al.*, *Acta Crystallogr. C* **55**, 2169 (1999). Extraction procedure from recyclable polystyrene materials: A. Altwaiq *et al.*, *Anal. Chim. Acta* **491**, 111 (2003). GC determn in styrenic polymers: M. Pöhlein *et al.*, *J. Chromatogr. A* **1203**, 217 (2008). Photodegradation study: Q. Xie *et al.*, *Chemosphere* **76**, 1486 (2009). Dissipation in soil: H. Huang *et al.*, *Environ. Sci. Technol.* **44**, 663 (2010). Review of human health risk: M. L. Hardy *et al.*, *Crit. Rev. Toxicol.* **39**, Suppl 3, 1-44 (2009).

White powder, mp 300°.

Technical formulations. DE-83R; FR-1210; Saytex 102E. Typically contain small amounts of nonabromodiphenyl ether or other congeners as impurities. Physical properties: M. L. Hardy, *Chemosphere* **46**, 717 (2002); and toxicology: *idem, ibid.*, 757. Melting range 300-310°. Dec 362°. d^{25} 3.4. Vapor pressure (21°): 4.3 × 10^{-6} Pa. Log P (octanol/water): 6.265. Soly (wt %): 0.05 acetone; 0.2 toluene; 0.48 benzene; 0.42 methylene bromide; 0.87 xylene. Soly in water: <0.1 μg/l. Henry's law constant: 1.93 × 10^{-8}. LD_{50} orally in rats: >2000 mg/kg; dermally in rabbits: >2000 mg/kg; LC_{50} in rats by inhalation: >48.2 mg/l (Hardy).

USE: Additive flame retardant for polymers used in electronics, electrical equipment, and upholstery fabrics.

2849. δ-Decalactone. [705-86-2] Tetrahydro-6-pentyl-2*H*-pyran-2-one; 5-decanolide. $C_{10}H_{18}O_2$; mol wt 170.25. C 70.55%, H 10.66%, O 18.79%. Naturally occuring aroma constituent of dairy

products, coconuts, rum, and fruits including raspberries, peaches, and strawberries. Prepn of R-(+)-form by hydrogenation of *Massoi lactone (R-(−)-2-decen-5-olide)*: Th. M. Meijer, *Recl. Trav. Chim. Pays-Bas* **59**, 191 (1940); P. H. van der Schaft *et al.*, *Appl. Microbiol. Biotechnol.* **36**, 712 (1992); of S-(−)-form by hydrogenation of lactones from jasmine oil: M. Winter *et al.*, *Helv. Chim. Acta* **45**, 1250 (1962). Synthesis of racemic mixture: K.-W. Rosenmund, H. Bach, *Ber.* **94**, 2401, 2406 (1961). Synthesis and sensory characteristics of enantiomers: A. Mosandl, M. Gessner, *Z. Lebensm.-Unters. Forsch.* **187**, 40 (1988). Partition behavior and odor characteristics: C. Guyot *et al.*, *J. Agric. Food Chem.* **44**, 2341 (1996). Fragrance monograph: D. L. J. Opdyke, *Food Cosmet. Toxicol.* **14**, Suppl. 1, 739 (1976).

Colorless liquid; creamy-coconut and peach-like flavor and odor. bp_{11} 157°. Very sol in alcohol, propylene glycol. Insol in water. Log P (oil/water): 3.4. LD_{50} orally in rats: >5 g/kg; dermally in rabbits: >5 g/kg (Opdyke).

R-**Form.** [2825-91-4] Fruity-sweet, milky odor. Creamy, fruity peach, apricot taste. $bp_{0.02}$ 117-120°. $d^{27.5}$ 0.9540. n_D^{26} 1.4537. $[\alpha]_D$ +39.68° (neat). $[\alpha]_D^{20}$ +60° (c = 1.0-1.3 in CH_3OH).

S-**Form.** [59285-67-5] Odor and taste similar to the R(+)-form, but more intense. Odor is fruity-sweet, creamy peach, with a fatty, buttery tonality. bp_9 142-144°. d_4^{25} 0.9690. n_D^{25} 1.4544. $[\alpha]_D^{20}$ −49.7° (neat). $[\alpha]_D^{20}$ −61° (c = 1.0-1.3 in CH_3OH).

USE: Ingredient in perfumes, and commercial cream and butter flavorings.

2850. Decalin®. [91-17-8] Decahydronaphthalene; perhydronaphthalene; bicyclo[4.4.0]decane; naphthalane; naphthane; De-Kalin. $C_{10}H_{18}$; mol wt 138.25. C 86.88%, H 13.12%. Occurs in two forms: *cis* and *trans*. Hydrogenation of naphthalene in glacial acetic acid in the presence of platinum catalyst at 25° and 130 atm yields a mixture of 77% *cis*-Decalin and 23% *trans*-Decalin. Hydrogenation of Tetralin under the same conditions yields almost only *cis*-Decalin: Baker, Schuetz, *J. Am. Chem. Soc.* **69**, 1250 (1947). Hydrogenation of $\Delta^{4a(8a)}$-octalin in ethanol in the presence of platinum yields *trans*-Decalin as the main component: Linstead *et al.*, *J. Chem. Soc.* **1937**, 1136. Separation of *trans*- and *cis*-forms: Seyer, Walker, *J. Am. Chem. Soc.* **60**, 2125 (1938). Prepn of *trans*-Decalin from *cis*-Decalin: Zelinsky, *Ber.* **65**, 1299 (1932). Conformation studies: Barton, *J. Chem. Soc.* **1948**, 340; Moniz, Dixon, *J. Am. Chem. Soc.* **83**, 1671 (1961). Toxicity data: Smyth *et al.*, *Arch. Ind. Hyg. Occup. Med.* **4**, 119 (1951).

trans   cis

Liquid. Slight odor resembling menthol. Pure Decalin does not smell of naphthalene. Volatile with steam. The commercial product may be practically all *trans*-Decalin, or a mixture contg up to 60% *cis*-Decalin. The commercial mixture has a flash pt (closed cup) of about 136°F (58°C). Autoignition temp 504°F. Insol in water. Very sol in alcohol, methanol, ether, chloroform. Miscible with propyl and isopropyl alcohol; miscible with most ketones and esters. LD_{50} orally in rats: 4.2 g/kg; LC (in air) in rats: 500 ppm (Smyth).

cis-**Form.** mp −43.26°. bp 195.7°; bp_9 67.0°. d_4^{20} 0.8963. n_D^{20} 1.48113.

trans-**Form.** mp −30.4°. bp 187.25°; bp_9 62.0°. d_4^{20} 0.8700. n_D^{20} 1.46968.

USE: Solvent for naphthalene, fats, resins, oils, waxes; used instead of turpentine in lacquers, shoe polishes, floor waxes. In motor fuel and lubricants. As patent fuel in stoves.

2851. Decamethonium Bromide. [541-22-0] $N^1,N^1,N^1,$-N^{10},N^{10},N^{10}-Hexamethyl-1,10-decanediaminium bromide (1:2); $N,$-N,N,N',N',N'-hexamethyl-1,10-decanediaminium dibromide; deca-methylenebis[trimethylammonium bromide]; C-10; Syncurine. C_{16}-$H_{38}Br_2N_2$; mol wt 418.30. C 45.94%, H 9.16%, Br 38.20%, N 6.70%. Prepn: Blomquist *et al.*, *J. Am. Chem. Soc.* **81**, 678 (1959).

Crystals from methanol + acetone, dec 268-270°. Freely sol in water, alcohol; very slightly sol in chloroform. Practically insol in ether. Aq solns are stable and may be sterilized by autoclaving.

THERAP CAT: Neuromuscular blocking agent.

2852. Decamethylcyclopentasiloxane. [541-02-6] 2,2,4,4,-6,6,8,8,10,10-Decamethylcyclopentasiloxane; D_5. $C_{10}H_{30}O_5Si_5$; mol wt 370.77. C 32.39%, H 8.16%, O 21.58%, Si 37.87%. Cyclic siloxane. Isolated from the hydrolysis product of dimethyldichloro-silane: W. Patnode, D. F. Wilcock, *J. Am. Chem. Soc.* **68**, 358 (1946). HPLC determn and GC-MS characterization of metabolites in urine: S. Varaprath *et al.*, *Drug Metab. Dispos.* **31**, 206 (2003). Physicochemical properties: M. Palczewska-Tulinska, P. Oracz, *J. Chem. Eng. Data* **50**, 1711 (2005).

Colorless volatile liquid. mp −38°. bp$_{760}$ 210°; bp$_{20}$ 101°. d 0.9593. n_D^{20} 1.3982. Soly in water: 17 ppb. Vapor pressure (50°): 2 mm Hg.

USE: Intermediate in the mfr of high mol wt siloxane polymers. Carrier ingredient in personal care products; dry cleaning solvent.

2853. Decamethylene Glycol. [112-47-0] 1,10-Decanediol. $C_{10}H_{22}O_2$; mol wt 174.28. C 68.92%, H 12.72%, O 18.36%. Prepd by the reduction of dimethyl or diethyl sebacate with sodium and ethyl alcohol: Bouveault, Blanc, *Compt. Rend.* **137**, 329 (1903); *Bull. Soc. Chim.* [3] **31**, 1205 (1904); **DE 164294**; *Frdl.* **8**, 1260 (1905); Chuit, *Helv. Chim. Acta* **9**, 264 (1926); Manske, *Org. Synth.* **14**, 20 (1934). By catalytic hydrogenation of sebacic esters: Folkers, Adkins, *J. Am. Chem. Soc.* **54**, 1146 (1932).

Needles from water or dil alcohol. mp 74°. bp$_{20}$ 192°; bp$_{11}$ 179°; bp$_8$ 170°. Freely sol in alcohol, warm ether; almost insol in petr ether and water.

Diethyl ether. [5895-59-0] 1,10-Diethoxydecane. $C_{14}H_{30}O_2$; mol wt 230.39. d_0^0 0.850; bp 260°.

2854. Decamethyltetrasiloxane. [141-62-8] 1,1,1,3,3,5,5,7,-7,7-Decamethyltetrasiloxane. $C_{10}H_{30}O_3Si_4$; mol wt 310.69. C 38.66%, H 9.73%, O 15.45%, Si 36.16%. Prepn: Patnode, Wilcock, *J. Am. Chem. Soc.* **68**, 358 (1946).

Liquid. bp 194°. d 0.8536. n_D^{20} 1.3895. mp ∼ −70°. Stable. Inert to most chemical reagents and rubber. Maintains about the same viscosity over a wide temp range. Sol in benzene and the lighter hydrocarbons; slightly sol in alc and heavy hydrocarbons.

USE: As a basis for silicone oils or fluids designed to withstand extremes of temp; as a foam suppressant in petr lubricating oil.

2855. Dechlorane® Plus. [13560-89-9] 1,2,3,4,7,8,9,10,13,-13,14,14-Dodecachloro-1,4,4a,5,6,6a,7,10,10a,11,12,12a-dodecahy-dro-1,4:7,10-dimethanodibenzo[a,e]cyclooctene; bis(hexachlorocy-clopentadieno)cyclooctane. $C_{18}H_{12}Cl_{12}$; mol wt 653.69. C 33.07%, H 1.85%, Cl 65.08%. Polychlorinated hydrocarbon. Prepn by Diels-Alder reaction between cyclooctadiene and hexachlorocyclopenta-diene: K. Ziegler, H. Froitzheim-Kühlhorn, *Ann.* **589**, 157 (1954). Use as fire-retardant for plastics: E. V. Gouinlock, Jr., **US 3382204** (1965 to Hooker). Accumulation in juvenile Atlantic salmon: V. Zitko, *Chemosphere* **9**, 73 (1980). Brief review of fire retardant props: T. J. Machmer, *Life Prop. Protect., Fire Retard. Chem. Assoc., Semi-Annu. Meet.* **1977**, 156-174.

Colorless crystals, mp >325°. Soluble in *o*-dichlorobenzene.

USE: Fire retardant for plastics.

2856. Decitabine. [2353-33-5] 4-Amino-1-(2-deoxy-β-D-*erythro*-pentofuranosyl)-1,3,5-triazin-2(1*H*)-one; 5-aza-2'-deoxycy-tidine; NSC-127716; Dacogen. $C_8H_{12}N_4O_4$; mol wt 228.21. C 42.11%, H 5.30%, N 24.55%, O 28.04%. Pyrimidine analog that inhibits DNA methylation and induces differentiation of leukemic cells. Prepn: J. Pliml, F. Sorm, *Collect. Czech. Chem. Commun.* **29**, 2576 (1964); A. Piskala, F. Sorm, in *Nucleic Acid Chemistry* Part 1 (Wiley, New York, 1978) pp 443-449. Improved synthesis: J. Ben-Hattar, J. Jiricny, *Nucleosides Nucleotides* **6**, 393 (1987). Chemical stability in soln: K.-T. Lin *et al.*, *J. Pharm. Sci.* **70**, 1228 (1981). HPLC determn in plasma: *eidem*, *J. Chromatogr.* **345**, 162 (1985). Review of pharmacology and mechanism of action: R. L. Momparler, *Pharmacol. Ther.* **30**, 287-299 (1985); *idem et al.*, *Leukemia* **11**, Suppl. 1, 1-6 (1997); of clinical experience in myelodysplastic syndromes: E. Atallah *et al.*, *Expert Opin. Pharmacother.* **8**, 65-73 (2007).

Crystals from methanol, mp 201-202° (dec). $[\alpha]_D^{22}$ +68.5° (30 min) → +57.8° (6 hr) (c = 0.5 in water). uv max (pH 7): 244 nm (log ε 3.86). Sol in DMSO; sparingly sol in water; slightly sol in methanol, methanol/water (50/50), ethanol/water (50/50). LD$_{50}$ in mice (mg/kg): 190 i.p. (Momparler, 1985).

THERAP CAT: Antineoplastic.

2857. Decoquinate. [18507-89-6] 6-(Decyloxy)-7-ethoxy-4-hydroxy-3-quinolinecarboxylic acid ethyl ester; ethyl 6-(*n*-decyl-oxy)-7-ethoxy-4-hydroxyquinoline-3-carboxylate; M & B 15497; Deccox. $C_{24}H_{35}NO_5$; mol wt 417.55. C 69.04%, H 8.45%, N 3.35%, O 19.16%. Anticoccidial quinolone; feed additive for livestock. Description of compd: Ball *et al.*, *Chem. Ind. (London)* **1968**, 56. Prepn: **BE 698305**; M. Davis *et al.*, **US 3485845** (1966, 1969 both to May & Baker). Metabolism and residue studies: E. M. Craine *et al.*, *J. Agric. Food Chem.* **19**, 1228 (1971); R. F. Kouba *et al.*, *ibid.* 1234. Evaluation in experimental ruminant cryptospori-diosis: R. Mancassola *et al.*, *Vet. Parasitol.* **69**, 31 (1997). Mode of action study: E. Del Cacho *et al.*, *Int. J. Parasitol.* **36**, 1515 (2006).

Cream to buff-colored fine amorphous powder. mp 244-246°. Insol in water.

THERAP CAT (VET): Coccidiostat.

2858. *n*-**Decyl Alcohol.** [112-30-1] 1-Decanol; nonylcarbinol. $C_{10}H_{22}O$; mol wt 158.29. C 75.88%, H 14.01%, O 10.11%. Prepn from caprinaldehyde: Krafft, *Ber.* **16**, 1717 (1883); from capric acid methyl ester: Bouveault, Blanc, *Bull. Soc. Chim.* [3] **31**, 674 (1904); from 1-chlorodecane: Schultz, *Ber.* **42**, 3611 (1909); from nonylmagnesium bromide and formaldehyde: Yohe, Adams, *J. Am. Chem. Soc.* **50**, 1507 (1928). Manuf: *Faith, Keyes & Clark's Industrial Chemicals*, F. A. Lowenheim, M. K. Moran, Eds. (Wiley-Interscience, New York, 4th ed., 1975) pp 310-317.

Moderately viscous, strongly refractive liq. Solidifies forming rectangular plates or leaflets; mp 6.4°. bp_{760} 232.9°; bp_{15} 115-120°; bp_8 109.5°. d_4^{20} 0.8297. n_D^{20} 1.43587. Insol in water. Sol in alcohol, ether.

USE: In the manuf of plasticizers, synthetic lubricants, petroleum additives, herbicides, surface active agents, solvents. Has moderate antifoaming capacity.

2859. **Deet.** [134-62-3] *N,N*-Diethyl-3-methylbenzamide; *N,N*-diethyl-*m*-toluamide; M-Det; *m*-DETA; ENT-20218; *m*-Delphene; Detamide; Dieltamid; Flypel; Metadelphene; OFF!; Repel. $C_{12}H_{17}NO$; mol wt 191.27. C 75.36%, H 8.96%, N 7.32%, O 8.36%. Technical deet contains a minimum of 95% meta isomer; commercial formulations contain 11.27-100% deet by weight. Prepn from *m*-toluoyl chloride and diethylamine in benzene or ether: Maxim, *Bull. Soc. Chim. Romania* **11**, 29 (1929); *Chem. Zentralbl.* **1929**, II, 2324, *C.A.* **24**, 94 (1930); *Beilstein* **9**, II, 325; E. T. McCabe *et al.*, *J. Org. Chem.* **19**, 493 (1954). Repellent activity vs mosquitoes: I. H. Gilbert *et al.*, *J. Econ. Entomol.* **48**, 741 (1955). Toxicity study: A. M. Ambrose *et al.*, *Toxicol. Appl. Pharmacol.* **1**, 97 (1959). Metabolism: L. Blomquist, W. Thorsell, *Acta Pharmacol. Toxicol.* **41**, 235 (1977). Physical, chemical and toxicological properties and insect repellent efficacy: *Pesticide Registration Standard - Deet* (U.S. EPA 504/RS-81-004, Washington, D.C., 1980) 82 pp. HPLC determn including metabolites: J. M. Yeung, W. G. Taylor, *Drug Metab. Dispos.* **16**, 600 (1988). Comparative study of efficacy vs mosquitoes: M. S. Fradin, J. F. Day, *N. Engl. J. Med.* **347**, 13 (2002). Review of safety: T. G. Osimitz, R. H. Grothaus, *J. Am. Mosq. Control Assoc.* **11**, 274-278 (1995); and pharmacokinetics and formulation: H. Qiu *et al.*, *ibid.* **14**, 12-27 (1998).

Nearly colorless to amberlike liquid, faint aromatic odor. $bp_{1.0}$ 111°. d^{25} 0.990-1.000. n_D^{25} 1.5206. Flash pt, open cup: 311°F (155°C). Partition coefficient (octane/water): 105. Vapor pressure at 25°: 1.67×10^{-3} mm Hg. Viscosity at 30°: 13.3 cP. Soly in water at 25°: 9.9 mg/ml. Practically insol in glycerin. Miscibile with ethanol, ether, isopropanol, chloroform, benzene, and carbon disulfide. Hygroscopic. Light sensitive. LD_{50} orally in male rats (ml/kg): 2.43; in female rats: 1.78; dermally in rabbits: ≥3.18. LC_{50} by inhalation in rats (mg/l): 5.95 (U.S. EPA).

Caution: Direct contact may cause eye irritation (U.S. EPA). *See also: Clinical Toxicology of Commercial Products*, R. E. Gosselin *et al.*, Eds. (Williams & Wilkins, Baltimore, 5th ed., 1984) Section II, p 346.

USE: Insect repellent.

2860. **Defensins.** [103220-14-0] A class of low molecular weight cationic peptides which have *in vitro* microbiocidal activity against various bacteria, fungi, and viruses, originally isolated from mammalian neutrophils. Possess similar arginine-rich, cysteine-rich structures of 29-35 amino acids with a conserved core of 8-11 residues. Characterized by six cysteines comprising 3 disulfide linkages. Abbreviated NP (neutrophil peptide) where NP is used for rabbit, HNP for human, GPNP for guinea pig etc.; defensins represent a major protein component of granulocytes. Initial identification studies: H. I. Zeya, J. K. Spitznagel, *Science* **142**, 1085 (1961); *eidem*, *J. Bacteriol.* **91**, 750 (1966). Isoln, antimicrobial activities and amino acid sequence of six NPs: M. E. Selsted *et al.*, *Infect. Immun.* **45**, 150 (1984); M. E. Selsted *et al.*, *J. Biol. Chem.* **260**, 4579 (1985); of three HNPs: T. Ganz *et al.*, *J. Clin. Invest.* **76**, 1427 (1985); M. E. Selsted *et al.*, *ibid.* 1436. Crystal structure of HNP-3: C. P. Hill *et al.*, *Science* **251**, 1481 (1991). Direct inactivation of enveloped viruses: K. A. Daher *et al.*, *J. Virol.* **60**, 1068 (1986). *In vitro* cytolysis of tumor cells: A. Lichtenstein *et al.*, *Blood* **68**, 1407 (1986). Mechanism of cytolysis: *eidem*, *J. Immunol.* **140**, 2686 (1988). Identification and isoln of insect defensins from haemolymph of dipteran: J. A. Hoffmann, D. Hoffmann, *Res. Immunol.* **141**, 910 (1990). Structure determn of insect forms: P. Lepage *et al.*, *Eur. J. Biochem.* **196**, 735 (1991). Review of insect forms: J. A. Hoffmann, C. Hetru, *Immunol. Today* **13**, 411-415 (1992). *Review:* B. L. Kagan *et al.*, *Toxicology* **87**, 131-149 (1994); including insect forms: T. Ganz, R. I. Lehrer, *Pharmacol. Ther.* **66**, 191-205 (1995).

2861. **Deferasirox.** [201530-41-8] 4-[3,5-Bis(2-hydroxyphenyl)-1*H*-1,2,4-triazol-1-yl]benzoic acid; ICL-670; ICL-670A; CGP-72670; Exjade. $C_{21}H_{15}N_3O_4$; mol wt 373.37. C 67.56%, H 4.05%, N 11.25%, O 17.14%. Orally active tridentate iron chelator. Prepn: R. Lattmann, P. Acklin, WO 9749395; *eidem*, US 6465504 (1997, 2002 both to Novartis); U. Heinz *et al.*, *Angew. Chem. Int. Ed.* **38**, 2568 (1999). HPLC determn in plasma: M. C. Rouan *et al.*, *J. Chromatogr. B* **755**, 203 (2001). Iron complex formation study: S. Steinhauser *et al.*, *Eur. J. Inorg. Chem.* **2004**, 4177. Pharmacology: C. Hershko *et al.*, *Blood* **97**, 1115 (2001). Clinical pharmacokinetics: R. Galanello *et al.*, *J. Clin. Pharmacol.* **43**, 565 (2003). Clinical trial in thalassemia: E. Nisbet-Brown *et al.*, *Lancet* **361**, 1597 (2003). Review of structure activity: H. Nick *et al.*, *Curr. Med. Chem.* **10**, 1065-1076 (2003); of preclinical and clinical experience: M. D. Cappellini, *Best Pract. Res. Clin. Haematol.* **18**, 289-298 (2005).

Crystals from ethanol, mp 264-265°. Soly in water at 25° (mg/ml): 0.4 (pH 7.40). pKa_1 4.57, pKa_2 8.71, pKa_3 10.56 (Nick). Also reported in $H_2O/DMSO$ ($X_{DMSO} = 0.20$) as pKa_1 4.61, pKa_2 10.12, pKa_3 12.08 (Steinhauser). Partition coefficient (octanol/water): 6.3 (pH 7.40). uv max ($H_2O/DMSO$, pH < 2): 284 nm (ε 1.5×10^4).

THERAP CAT: Chelating agent (iron).

2862. **Deferiprone.** [30652-11-0] 3-Hydroxy-1,2-dimethyl-4(1*H*)-pyridinone; 1,2-dimethyl-3-hydroxypyrid-4-one; DMHP; L1; CP-20; Ferriprox; Kelfer. $C_7H_9NO_2$; mol wt 139.15. C 60.42%, H 6.52%, N 10.07%, O 23.00%. Orally active chelator that promotes urinary iron excretion. Prepn: M. Yasue *et al.*, *Yakugaku Zasshi* **90**, 1222 (1970), *C.A.* **73**, 23102b (1971); G. J. Kontoghiorghes *et al.*, *Arzneim.-Forsch.* **37**, 1099 (1987). Iron binding studies: *idem, Inorg. Chim. Acta* **135**, 145 (1987). HPLC determn in serum and urine: B. Dresow *et al.*, *J. Anal. Chem.* **352**, 562 (1995). Clinical pharmacokinetics: F. N. Al-Refaie *et al.*, *Br. J. Haematol.* **89**, 403 (1995). Clinical studies in transfusional iron overload: *idem et al.*, *ibid.* **91**, 224 (1995); N. F. Olivieri *et al.*, *N. Engl. J. Med.* **332**, 918 (1995); in thalassemia major: A. Maggio *et al.*, *Blood Cells Mol.*

Dis. **42**, 247 (2009). Review of pharmacology and clinical experience: G. J. Kontoghiorghes, *Ann. N.Y. Acad. Sci.* **612**, 339-350 (1990); F. N. Al-Refaie, A. V. Hoffbrand, *Baill. Clin. Haematol.* **7**, 941-963 (1994); of toxicology: G. J. Kontoghiorghes, *Toxicol. Lett.* **80**, 1-18 (1995).

Needles from water, mp 266-268° (Yasue); also reported as mp 263-266° (Kontoghiorghes *et al.*, 1987). Very bitter taste. Soly at 37° (mg/ml): water ~20 (pH 7.4). pKa_1 3.3, pKa_2 9.7. uv max (phosphate buffer saline, pH 7.3): 460 nm (ε 3600). Hydrophilic chelator; forms red colored complexes with iron. LD_{50} i.p. in rats, mice: 650 mg/kg, 0.8-1.0 g/kg; i.g. in rats: 2.0-3.0 g/kg (Kontoghiorghes, 1995).

THERAP CAT: Chelating agent (iron and aluminum).

2863. Deferoxamine. [70-51-9] N^4-[5-[[4-[[5-(Acetylhydroxyamino)pentyl]amino]-1,4-dioxobutyl]hydroxyamino]pentyl]-N^1-(5-aminopentyl)-N^1-hydroxybutanediamide; *N*-[5-[3-[(5-aminopentyl)hydroxycarbamoyl]propionamido]pentyl]-3-[[5-(*N*-hydroxyacetamido)pentyl]carbamoyl]propionohydroxamic acid; 1-amino-6,17-dihydroxy-7,10,18,21-tetraoxo-27-(*N*-acetylhydroxylamino)-6,11,17,22-tetraazaheptaeicosane; desferrioxamine B. $C_{25}H_{48}$-N_6O_8; mol wt 560.69. C 53.55%, H 8.63%, N 14.99%, O 22.83%. Natural product forming iron complexes, isolated from *Streptomyces pilosus* Ettlinger *et al.* Prepn from ferrioxamine B and structure: H. Bickel *et al.*, *Helv. Chim. Acta* **43**, 2129 (1960). Syntheses: V. Prelog, A. Walser, *ibid.* **45**, 631 (1962); **BE 609053**; E. Gaeumann *et al.*, **US 3471476** (1962, 1969 both to Ciba). Derivatives: H. Bickel *et al.*, *Helv. Chim. Acta* **46**, 1385 (1963). Interactions with cellular and plasma iron pools: J. B. Porter *et al.*, *Ann. N.Y. Acad. Sci.* **1054**, 155 (2005). Clinical efficacy in thallasemia major (Cooley's anemia): B. Modell *et al.*, *Br. Med. J.* **284**, 1081 (1982). Treatment of hemodialysis-induced aluminum accumulation in brain: P. Ackril *et al*, *Lancet* **2**, 692 (1980); in bone: H. H. Malluche *et al.*, *N. Engl. J. Med.* **311**, 140 (1984). Neurotoxicity study: N. F. Olivieri *et al.*, *ibid.* **314**, 869 (1986). Review of clinical experience: R. Propper, D. Nathan, *Annu. Rev. Med.* **33**, 509-519 (1982).

Monohydrate. [25442-95-9] Crystals from dil alcohol, mp 138-140°. Sol in water at 20°: 1.2%.

Hydrochloride. [1950-39-6] Ba-29837. $C_{25}H_{48}N_6O_8 \cdot HCl$; mol wt 597.15. Crystals from slightly acidic methanol, mp 172-175°.

Methanesulfonate. [138-14-7] Desferrioxamine mesylate; DFOM; Desferal. $C_{25}H_{48}N_6O_8 \cdot CH_3SO_3H$; mol wt 656.79. Crystals from dil alcohol, mp 148-149°. Sol in water at 20°: >20%. Slightly sol in methanol.

N-Acetyl derivative. [5722-48-5] *N*-Acetyldeferoxamine; *N*-acetyldesferrioxamine B. $C_{27}H_{50}N_6O_9$. Crystals from *n*-propanol, mp 180-182°.

Iron complex. [14836-73-8] Ferrioxamine B. $C_{25}H_{45}FeN_6O_8$. A natural microbial growth factor. Isoln: H. Bickel *et al.*, *Helv. Chim. Acta* **43**, 2118 (1960); E. Gaeumann *et al.*, **US 3118823**; **US 3153621** (both 1964 to Ciba).

Note: Reviews on iron-containing metabolites with growth stimulating properties, **sideramines**, or with antibiotic properties, **sideromycins, ferrimycins**: H. Bickel *et al.*, *Experientia* **16**, 129-133 (1960); V. Prelog, *Pure Appl. Chem.* **6**, 327-338 (1963); F. Knüsel, J. Nüesch, *Nature* **206**, 674-676 (1965); T. F. Emery, *Adv. Enzymol. Relat. Areas Mol. Biol.* **35**, 135-185 (1971); Knüsel, Zimmermann, in *Antibiotics* vol. 3, J. W. Corcoran, F. E. Hahn, Eds. (Springer-Verlag, New York, 1975) pp 653-667.

THERAP CAT: Parenteral chelating agent (iron and aluminum). Methanesulfonate as antidote to iron poisoning.

2864. Defibrotide. [83712-60-1] Fraction P; defibrinotide; Dasovas; Noravid; Prociclide. Sodium salt of polydeoxyribonucleotide extract from mammalian lung which stimulates fibrinolysis. mol wt 45,000-50,000 daltons. Prepn: A. Butti *et al.*, **DE 2154278**; *eidem*, **US 3899481** (1972, 1975 both to Crinos). Antithrombotic activity: R. Niada *et al.*, *Thromb. Res.* **23**, 233 (1981). Pharmacokinetics and profibrinolytic activity in rabbits: R. Pescador *et al.*, *ibid.* **30**, 1 (1983). Clinical trial as antithrombotic: S. Coccheri *et al.*, *Int. J. Clin. Pharmacol. Res.* **2**, 227 (1982). Clinical evaluation in thrombotic microangiopathy: V. Bonomini *et al.*, *Nephron* **40**, 195 (1985). Symposium on pharmacology and clinical efficacy: O. N. Ulutin, Ed., *Haemostasis* **16**, Suppl. 1, 1-68 (1986). Review of pharmacology and therapeutic use: K. J. Palmer, K. L. Goa, *Drugs* **45**, 259-294 (1993).

THERAP CAT: Antithrombotic.

2865. Deflazacort. [14484-47-0] (11β,16β)-21-(Acetyloxy)-11-hydroxy-2'-methyl-5'*H*-pregna-1,4-dieno[17,16-*d*]oxazole-3,20-dione; 11β,21-dihydroxy-2'-methyl-5'βH-pregna-1,4-dieno[17,16-*d*]oxazole-3,20-dione 21-acetate; pregna-1,4-diene-11β,21-diol-3,20-dione[17α,16α-*d*]-2'-methyloxazoline 21-acetate; oxazacort; azacort; DL-458-IT; L-5458; Calcort; Deflan; Dezacor; Flantadin; Lantadin. $C_{25}H_{31}NO_6$; mol wt 441.52. C 68.01%, H 7.08%, N 3.17%, O 21.74%. Systemic corticosteroid; oxazoline derivative of prednisolone, *q.v.* Prepn: G. Nathansohn, G. Winters, **BE 679820**; *eidem*, **GB 1077393**; *eidem*, **US 3436389** (1966, 1967, 1969 all to Lepetit); G. Nathansohn *et al.*, *J. Med. Chem.* **10**, 799 (1967). Pharmacology: P. Schiatti *et al.*, *Arzneim.-Forsch.* **30**, 1543 (1980). Pharmacokinetics: A. Assandri *et al.*, *Eur. J. Drug Metab. Pharmacokinet.* **5**, 207 (1980); *eidem*, *Adv. Exp. Med. Biol.* **171**, 9 (1984). Immunosuppressive activity in humans: B. H. Hahn *et al.*, *J. Rheumatol.* **8**, 783 (1981). Effect on human mineral metabolism: T. J. Hahn *et al.*, *Calcif. Tissue Int.* **31**, 109 (1980); on human glucose metabolism: P. Cavallo-Perin *et al.*, *Eur. J. Clin. Pharmacol.* **26**, 357 (1984). Clinical comparison with prednisone, *q.v.*, in rheumatoid arthritis and lupus: B. Imbimbo *et al.*, *Adv. Exp. Med. Biol.* **171**, 241 (1984).

Crystals from acetone-hexane, mp 255-256.5°. $[\alpha]_D$ +62.3° (c = 0.5 in chloroform). uv max (methanol): 241-242 nm ($E_{1cm}^{1\%}$ 352.5). LD_{50} orally in mice: 5200 mg/kg (Schiatti).

THERAP CAT: Anti-inflammatory; glucocorticoid.

2866. Degarelix. [214766-78-6] *N*-Acetyl-3-(2-naphthalenyl)-D-alanyl-4-chloro-D-phenylalanyl-3-(3-pyridinyl)-D-alanyl-L-seryl-4-[[[(4*S*)-hexahydro-2,6-dioxo-4-pyrimidinyl]carbonyl]amino]-L-phenylalanyl-4-[(aminocarbonyl)amino]-D-phenylalanyl-L-leucyl-N^6-(1-methylethyl)-L-lysyl-L-prolyl-D-alaninamide; Ac-D-2Nal-D-4Cpa-D-3Pal-Ser-4Aph(L-hydroorotyl)-D-4Aph(carbamoyl)-Leu-ILys-Pro-D-Ala-NH₂; [Ac-DNal¹,DCpa²,DPal³,Aph-(Hor)⁵,DAph(Cbm)⁶,ILys⁸,DAla¹⁰] GnRH; Fe-200486. $C_{82}H_{103}$-$ClN_{18}O_{16}$; mol wt 1632.29. C 60.34%, H 6.36%, Cl 2.17%, N 15.45%, O 15.68%. Gonadotropin-releasing hormone (GnRH) antagonist; decapeptide amide with seven unnatural amino acids, five of which are D-amino acids. Prepn: G. Semple *et al.*, **WO 9846634**; *eidem*, **US 5925730** (1998, 1999 both to Ferring); G. Jiang *et al.*, *J. Med. Chem.* **44**, 453 (2001). Microparticle formulation for sustained release: G. Schwach *et al.*, *Eur. J. Pharm. Biopharm.* **56**, 327 (2003). GnRH antagonist activity in rat prostate adenocarcinoma: M. Princivalle *et al.*, *J. Pharmacol. Exp. Ther.* **320**, 1113 (2007). Clinical pharmacokinetics and pharmacodynamics: P. R. Jadhav *et al.*, *J. Pharmacokinet. Pharmacodyn.* **33**, 609 (2006). Clinical evaluation in prostate cancer: M. Gittleman *et al.*, *J. Urol.* **180**, 1986 (2008); and comparison with leuprolide: L. Klotz *et al.*, *Br. J. Urol.*

102, 1531 (2008). Review of pharmacology and clinical experience: C. Doehn *et al.*, *Expert Opin. Invest. Drugs* **18**, 851-860 (2009).

Acetate hydrate. [934246-14-7] Firmagon. $C_{82}H_{103}ClN_{18}$-$O_{16}\cdot xC_2H_4O_2\cdot nH_2O$ White to off-white amorphous powder.
THERAP CAT: Antineoplastic (hormonal).

2867. Deguelin. [522-17-8] (7a*S*,13a*S*)-13,13a-Dihydro-9,10-dimethoxy-3,3-dimethyl-3*H*-[1]benzopyrano[3,4-*b*]pyrano[2,3-*h*][1]benzopyran-7(7a*H*)-one. $C_{23}H_{22}O_6$; mol wt 394.42. C 70.04%, H 5.62%, O 24.34%. From roots of *Tephrosia toxicaria* Pers., leaves of *Tephrosia vogelii* Hook f., *Leguminosae*, and from derris and cubé roots: H. L. Haller, F. B. LaForge, *J. Am. Chem. Soc.* **56**, 2415 (1934); E. Brierly, H. J. Smith, *J. Pharm. Pharmacol.* **20**, 840 (1968). Structure: Clark, *J. Am. Chem. Soc.* **54**, 3000 (1932). Stereoselective synthesis: P. B. Anzeveno, *J. Org. Chem.* **44**,2578 (1979). Total synthesis of racemate: Fukami *et al.*, *Agric. Biol. Chem.* **25**, 252 (1961); Omokawa, Yamashita, *ibid.* **38**, 1731 (1974). Stereochemical studies: Djerassi *et al.*, *J. Chem. Soc.* **1961**, 1448.

Yellow oil, $[\alpha]_D^{27}$ −97.2° (c = 0.2 in benzene).
(±)-**Form.** Pale green crystals, mp 171°. Practically insol in water. Sol in alc.
Caution: Irritates skin. Inhalation may cause fatal pulmonary damage.
USE: Insecticide.

2868. DEHP. [117-81-7] 1,2-Benzenedicarboxylic acid 1,2-bis(2-ethylhexyl) ester; bis(2-ethylhexyl) phthalate; di(2-ethylhexyl) phthalate; DOP; dioctyl phthalate; Octoil. $C_{24}H_{38}O_4$; mol wt 390.56. C 73.81%, H 9.81%, O 16.39%. Dialkyl phthalate ester plasticizer used to impart softness and flexiblilty to PVC products. Prepn: P. J. Garner, G. Watson, US 2508911 (1950 to Shell). Effect on dynamic mechanical properties: D. L. Hertz, Jr., *Elastomerics* **116**, 20 (1984); interaction with PVC polymer: B. Garnaik, S. Sivaram, *Macromolecules* **29**, 185 (1996). HPLC determn in PVC packaging: M. F. Aignasse *et al.*, *Int. J. Pharm.* **113**, 241 (1995). GC/MS determn in toys: R. Stringer *et al.*, *Environ. Sci. Pollut. Res. Int.* **7**, 27 (2000). Evaluation of carcinogenic risk: J. Doull *et al.*, *Regul. Toxicol. Pharmacol.* **29**, 327 (1999); of health effects: M. Fay *et al.*, *Toxicol. Ind. Health* **15**, 651 (1999). *Review:* L. G. Krauskopf, *Plast. Compd.* **6**, 28-38 (1983). Review of environmental fate: C. A. Staples *et al.*, *Chemosphere* **35**, 667-749 (1997); of aquatic toxicity: idem *et al.*, *Environ. Toxicol. Chem.* **16**, 875 (1997); of toxicology and human exposure: *Toxicological Profile for Di(2-ethylhexyl)phthalate* (PB2003-100138, 2002) 337 pp.

Liquid, mp −47°. bp$_5$ 231°. Flash point (closed cup): 206°C (403°F). d_{20}^{20} 0.986. n_D^{20} 1.486. Viscosity (cSt): 386 (0°); 22 (20°); 31 (37.8°); 5 (100°). Vapor pressure (mm Hg): (100°) 0.003; (200°) 1.20; (300°) 96.
Caution: Potential symptoms of overexposure are irritation of eyes and mucous membranes. *See NIOSH Pocket Guide to Chemical Hazards* (DHHS/NIOSH 97-140, 1997) p 118. This substance is reasonably anticipated to be a human carcinogen: *Report on Carcinogens, Twelfth Edition* (PB2011-111646, 2011) p 156.
USE: As plasticizer in PVC applications and flexible vinyls.

2869. Dehydroacetic Acid. [520-45-6] 3-Acetyl-6-methyl-2*H*-pyran-2,4(3*H*)-dione; 2-acetyl-5-hydroxy-3-oxo-4-hexenoic acid δ-lactone; methylacetopyronone; DHA. $C_8H_8O_4$; mol wt 168.15. C 57.14%, H 4.80%, O 38.06%. Polymerization product of ketene: Steele *et al.*,*J. Org. Chem.* **14**, 460 (1949). From ethyl acetoacetate: Arndt, *Org. Synth.* **coll. vol. III**, 231 (1955). Toxicity study: H. C. Spencer *et al.*, *J. Pharmacol. Exp. Ther.* **99**, 57 (1950). Series of articles on toxicity studies, pharmacology, mechanism of action, absorption, distribution, renal action: *ibid.* 57-111.

White to cream crystalline powder, mp 109-111° (sublimes). bp 269.9°. Soly (w/w at 25°): in acetone 22%; benzene 18%; methanol 5%; carbon tetrachloride 3%; U.S.P. ethanol 3%; ether 5%; glycerol <0.1%, *n*-heptane 0.7%; olive oil 1.6%; propylene glycol 1.7%; water <0.1%. LD$_{50}$ orally in rats: 1000 mg/kg (Spencer).
Sodium salt hydrate. DHA-S. $C_8H_7NaO_4\cdot H_2O$; mol wt 208.14. Tasteless white powder. Soly (w/w at 25°): in water 33%; propylene glycol 48%; olive oil <0.1%; methanol 14%; U.S.P. ethanol 1%; *n*-heptane <0.1%; glycerol 15%; ether <0.1%; carbon tetrachloride <0.1%; benzene <0.1%; acetone 0.2%. LD$_{50}$ orally in rats: 570 mg/kg (Spencer).
USE: In organic syntheses; as plasticizer, compatible with nitrocellulose, polystyrene, methacrylate, vinylite resins; as fungicide and bactericide; in antienzyme toothpastes; to reduce pickle bloating.

2870. Dehydroascorbic Acid. [490-83-5] L-*threo*-2,3-Hexodiulosonic acid γ-lactone. $C_6H_6O_6$; mol wt 174.11. C 41.39%, H 3.47%, O 55.13%. The reversibly oxidized form of ascorbic acid. Prepd by the action of benzoquinone on ascorbic acid: Ohle, Erlbach, *Ber.* **67**, 555 (1934); Moll, Wieters, *E. Merck's Jahresber.* **50**, 65 (1936); by the action of iodine: Herbert *et al.*, *J. Chem. Soc.* **1933**, 1270; Kenyon, Munro, *ibid.* **1948**, 158; by oxidn with *peri*-naphthindan-2,3,4-trione hydrate: Moubasher, *J. Biol. Chem.* **176**, 533 (1948); Müller-Mulot, *Z. Physiol. Chem.* **351**, 52 (1970). Isomerization and formn of derivs: Egge, *Tetrahedron Lett.* **1969**, 801. Structure studies: Teichmann, Ziebarth, *Z. Prakt. Chem.* **33**, 124 (1966). Toxicity studies: Gaudiano *et al.*, *Boll. Soc. Ital. Biol. Sper.* **43**, 674 (1967).

Fine needles, dec 225°. Sol in water at 60°. In soln the two carbonyl groups (in position 2 and 3) assume the hydrated form —C(OH)$_2$—C(OH)$_2$—. Practically neutral reaction. pKa: 3.90. $[\alpha]_D^{20}$ +56°. Aq solns are much less stable than those of ascorbic acid. Detailed stability data: Bogdanski, Bogdanska, *Bull. Acad. Pol. Sci. Cl. 2* **3**, 41 (1955). *See also:* Velisek *et al.*, *Collect. Czech. Chem. Commun.* **37**, 1465 (1972). Undecomposed dehydroascorbic acid in soln is easily converted to ascorbic acid by reduction with sulfurous acid. Has same antiscorbutic activity in humans as ascorbic acid (upon oral ingestion).

2871. (3β)-7-Dehydrocholesterol. [434-16-2] (3β)-Cholesta-5,7-dien-3-ol; provitamin D₃. $C_{27}H_{44}O$; mol wt 384.65. C 84.31%, H 11.53%, O 4.16%. Occurs in higher animals and in man. Isoln from the horned snail, *Buccinum undatum:* Windaus *et al., Z. Physiol. Chem.* **241**, 100 (1936); from pigskin: Windaus, Bock, *ibid.* **245**, 168 (1937); Boer *et al., Proc. K. Ned. Akad. Wet.* **39**, 622 (1936). Synthesis: Windaus *et al., Ann.* **520**, 98 (1935); Rosenberg, Tinker, **US 2215727**; Buisman *et al., Rec. Trav. Chim.* **66**, 83 (1947); P. N. Confalone *et al., J. Org. Chem.* **46**, 1030 (1981). Irradiation with ultraviolet light produces vitamin D₃, lumisterol₃ and tachysterol₃: Windaus *et al., Ann.* **533**, 118 (1938).

HO

Hydrated plates from ether-methanol. The water of crystn is held tenaciously: Schenck *et al., Ber.* **69**, 2705 (1936). When anhydr mp 150-151°. $[\alpha]_D^{20}$ −113.6° (c = 1 in chloroform), Koch, Koch, *J. Biol. Chem.* **116**, 757 (1936); $[\alpha]_D^{20}$ −127.1° in benzene (Boer). All provitamins D have the same uv maxima at 260, 270, 281, 293.5 nm. Insol in water, sol in the usual organic solvents.
   Acetate. $C_{29}H_{46}O_2$. Crystals from methanol, mp 129-130°; $[\alpha]_D^{20}$ −85.3° (c = 1.2 in benzene).
   Benzoate. $C_{34}H_{48}O_2$. Plates from chloroform-acetone, mp 139-140°; $[\alpha]_D^{20}$ −53.2° in chloroform.

2872. Dehydrocholic Acid. [81-23-2] (5β)-3,7,12-Trioxocholan-24-oic acid; 3,7,12-triketocholanic acid; Acolen; Bilidren; Cholagon; Chologon; Decholin; Dehychol; Deidrocolico Vita; Didrocolo; Erebile; Felacrinos; Procholon. $C_{24}H_{34}O_5$; mol wt 402.53. C 71.61%, H 8.51%, O 19.87%. Prepd from cholic acid by oxidation with CrO₃ in glacial acetic acid: Hammarsten in *Abderhalden's Handbuch der Biol. Arbeitsmethoden,* Abt. **I**, Teil 6, p 238 (1925); by using chlorine instead of CrO₃: Hinkley, Singleton, **US 2966499** (1960 to Merck & Co.).

Bitter crystals from acetone, mp 237°. $[\alpha]_D^{20}$ +26° (c = 1.4 in alc). Soly at 15° (g/l): water 0.18; alcohol 3.3; ether 0.46; chloroform 9.04; benzene 1.04; acetone 7.76; ethyl acetate 7.4; glacial acetic acid 7.42. Sol in solutions of alkali hydroxides, carbonates.
   Sodium salt. [145-41-5] Sodium dehydrocholate; Carachol; Dycholium; Suprachol. $C_{24}H_{33}NaO_5$; mol wt 424.51. Very sol in water. pH of aq soln about 9.0.
   THERAP CAT: Choleretic.

2873. 11-Dehydrocorticosterone. [72-23-1] 21-Hydroxypregn-4-ene-3,11,20-trione; Δ⁴-pregnen-21-ol-3,11,20-trione; 17-(1-keto-2-hydroxyethyl)-Δ⁴-androsten-3,11-dione; Kendall's compound A. $C_{21}H_{28}O_4$; mol wt 344.45. C 73.23%, H 8.19%, O 18.58%. Found in adrenal cortex. 1000 lbs of beef glands yield 333 mg: Kendall, *Cold Spring Harbor Symp. Quant. Biol.* **5**, 299 (1937). Isoln procedure: Mason *et al., J. Biol. Chem.* **114**, 613 (1936); *see also* Kendall *et al., Proc. Staff Meet. Mayo Clin.* **12**, 136 (1937). Prepn from corticosterone 21-acetate: Reichstein, *Helv. Chim. Acta* **20**, 953 (1937); from desoxycholic acid: Lardon, Reichstein, *ibid.* **26**, 747 (1943); Gallagher, *Recent Progress in Hormone Research* (New York, 1946) p 83; from 3α-acetoxy-11-ketobisnorcholanic

acid: Sarett, *J. Biol. Chem.* **162**, 601 (1946); *J. Am. Chem. Soc.* **68**, 2478 (1946); from 3α-acetyl-11-ketolithocholic acid methyl ester: Wettstein, Meystre, *Helv. Chim. Acta* **30**, 1262 (1947). Manuf: Wettstein, **US 2778776** (1957 to Ciba).

Large prisms from aq acetone, mp 178-180°. Can be distilled in high vacuum. $[\alpha]_{546}^{25}$ +299°, also given as +347° (c = 0.23 in benzene); $[\alpha]_D^{25}$ +258° (alc). Relatively good soly in benzene.
   Acetate. $C_{23}H_{30}O_5$. Needles from alc, mp 179-181°. $[\alpha]_{546}^{18}$ +285° (dioxane); $[\alpha]_D^{18}$ +233.7° (dioxane).

2874. Dehydroemetine. [4914-30-1] (11bS)-3-Ethyl-1,6,7,-11b-tetrahydro-9,10-dimethoxy-2-[[(1R)-1,2,3,4-tetrahydro-6,7-dimethoxy-1-isoquinolinyl]methyl]-4H-benzo[a]quinolizine; 2,3-dehydroemetine; 2,3-didehydro-6',7',10,11-tetramethoxyemetan; 2-dehydroemetine. $C_{29}H_{38}N_2O_4$; mol wt 478.63. C 72.77%, H 8.00%, N 5.85%, O 13.37%. Synthetic analog of emetine, *q.v.;* the (−)-form is therapeutically active. Synthesis of racemic 2,3-dehydroemetine: Brossi *et al., Helv. Chim. Acta* **42**, 772 (1959). Stereospecific synthesis of (±)-form and (±)-*2,3-dehydroisoemetine* (the (1'β)-isomer: Clark *et al., J. Chem. Soc.* **1962**, 2479. Separation of the four possible isomers: Brossi, Burkhardt, *Experientia* **18**, 211 (1962). Absolute configuration and bioactivity of isomers: A. Brossi *et al., Helv. Chim. Acta* **45**, 2219 (1962). Prepn of pure (−)-form: A. Brossi, **BE 629898**; *eidem,* **US 3311633** (1963, 1967 both to Hoffmann-La Roche); N. Whittaker *J. Chem. Soc. C* **1969**, 94. Antiprotozoal activity *in vitro:* G. H. Al-Khateeb *et al., Chemotherapy (Basel)* **23**, 267 (1977). Review of use in amebiasis: G. Woolfe, *Prog. Drug Res.* **8**, 12 (1965).

Crystals from isopropyl ether, mp 94-96°. $[\alpha]_D$ −183°.
   (−)-Form dihydrobromide. $C_{29}H_{38}N_2O_4 \cdot 2HBr$. Crystals, mp 243-245°. $[\alpha]_D$ −97° (methanol). uv max (alc): 282 nm (ε 7300).
   (+)-Form dihydrobromide. Crystals, mp 241-243°. $[\alpha]_D$ +95° (methanol). uv max (alc): 282 nm (ε 7350).
   (±)-Form dihydrochloride. [14358-43-1] Mebadin. $C_{29}H_{38}N_2O_4 \cdot 2HCl$. mol wt 551.55. Crystals from ethanol + ether, mp 235°.
   (±)-2,3-Dehydroisoemetine hydrochloride. Crystals from ethanol + ether, mp 220-225°.
   (−)-2,3-Dehydroisoemetine dihydrobromide. $C_{29}H_{38}N_2O_4 \cdot$ 2HBr. Crystals, mp 257-260°. $[\alpha]_D$ −107° (methanol). uv max (alc): 285 nm (ε 7400).
   (+)-2,3-Dehydroisoemetine dihydrobromide. Crystals, mp 257-258°. $[\alpha]_D$ +109° (methanol). uv max (alc): 285 nm (ε 7450).
   THERAP CAT: Antiamebic.

2875. Dehydroepiandrosterone. [53-43-0] (3β)-3-Hydroxyandrost-5-en-17-one; prasterone; dehydroisoandrosterone; *trans*-dehydroandrosterone; Δ⁵-androsten-3β-ol-17-one; DHEA; Prestara. $C_{19}H_{28}O_2$; mol wt 288.43. C 79.12%, H 9.79%, O 11.09%. Major secretory steroidal product of the adrenal gland; secretion progressively declines with aging. May have estrogen- or androgen-like effects depending on the hormonal milieu. Intracellularly converted to androstenedione, *q.v.* Isoln from male urine: Butenandt, Tscherning, *Z. Physiol. Chem.* **229**, 167 (1934); Butenandt, Dannenbaum, *ibid.* 192. Prepn from cholesterol: Butenandt *et al., ibid.* **237**, 57 (1935); Ruzicka, Wettstein, *Helv. Chim. Acta* **18**, 986 (1935); Wallis, Fernholz, *J. Am. Chem. Soc.* **57**, 1379, 1504 (1935); Schoeller *et al., Naturwissenschaften* **23**, 337 (1935); from sitosterol: Oppenauer, *Nature* **135**, 1039 (1935). High yield prepn: H. Hosoda *et al., J. Org. Chem.* **38**, 4209 (1973). Metabolism study: P. Knapstein *et al. Acta Endocrinol.* **58**, 261 (1968). Toxicity study of the sulfate: M. Yahara *et al., J. Toxicol. Sci.* **2**, 161 (1977). Review of physiological importance: P. Ebeling, V. A. Koivisto, *Lancet* **343**, 1479-1481 (1994). Symposium on role in aging: *Ann. N.Y. Acad. Sci.* **774**, 1-350 (1995). Review of therapeutic potential in aging: B. Allolio, W. Arlt, *Trends Endocrinol. Metab.* **13**, 288-294 (2002); of clinical experience in lupus: P. Kocis, *Am. J. Health Syst. Pharm.* **63**, 2201-2210 (2006).

Dimorphous, needles, mp 140-141°; leaflets, mp 152-153°. $[\alpha]_D^{18}$ +10.9° (c = 0.4 in alc). Pptd by digitonin. Sol in benzene, alcohol, ether; sparingly sol in chloroform, petr ether.

Sulfate. [651-48-9] DHEAS. $C_{19}H_{28}O_5S$; mol wt 368.49. Hydrophilic storage form that circulates in the blood.

Sodium sulfate. [1099-87-2] Sodium dehydroepiandrosterone sulfate; Astenile; Mylis. $C_{19}H_{27}NaO_5S$; mol wt 390.47. White cryst powder, mp 154° (dec). Sol in methanol, slightly sol in water, abs ethanol. Practically insol in acetone, chloroform, benzene.

THERAP CAT: In treatment of menopausal syndrome and systemic lupus erythematosus (SLE).

2876. Dehydroequol. [81267-65-4] 3-(4-Hydroxyphenyl)-2*H*-1-benzopyran-7-ol; 7-hydroxy-3-(4-hydroxyphenyl)-2*H*-1-benzopyran; isoflav-3-ene-4',7-diol; haginin E; idronoxil; phenoxodiol; NV-06. $C_{15}H_{12}O_3$; mol wt 240.26. C 74.99%, H 5.03%, O 19.98%. Naturally occurring isoflavonoid derivative with pro-apoptotic, antitumor activity; proposed metabolite of daidzein, *q.v.* Prepn: A. J. Liepa, *Aust. J. Chem.* **34**, 2647 (1981). Isoln from the legume, *Lespedeza homoloba* Nakai, *Leguminosae*: T. Miyase *et al., Phytochemistry* **52**, 303 (1999). Preparative method from daidzein: A. Heaton, N. Kumar, **WO 0049009** (2000 to Novogen). Synthesis: S.-R. Li *et al., Tetrahedron Lett.* **50**, 2121 (2009). Vasodilatory activity: J. P. F. Chin-Dusting *et al., Br. J. Pharmacol.* **133**, 595 (2001). Inhibition of DNA topoisomerase II: A. I. Constantinou, A. Husband, *Anticancer Res.* **22**, 2581 (2002). Mechanism of apoptotic action: A. B. Alvero *et al., Cancer* **106**, 599 (2006). Clinical pharmacokinetics: P. L. de Souza *et al., Cancer Chemother. Pharmacol.* **58**, 427 (2006). Review of pharmacology and clinical development: G. Mor *et al., Curr. Opin. Investig. Drugs* **7**, 542-548 (2006).

Needles from methanol/benzene, mp 231° (dec). uv max: 245.5, 333 nm (log ε 4.20, 4.39)

THERAP CAT: Antineoplastic.

2877. Dehydroergosterol. [516-85-8] (3β,22*E*)-Ergosta-5,-7,9(11),22-tetraen-3-ol. $C_{28}H_{42}O$; mol wt 394.64. C 85.22%, H 10.73%, O 4.05%. Prepd from ergosterol: Windhaus, Linsert, *Ann.*

465, 148 (1928); Callow, Rosenheim, *J. Chem. Soc.* **1933**, 387. From isopyrocalciferol: Windhaus, Dimroth, *Ber.* **70**, 376 (1937).

Solvated plates from alcohol, needles from ether. When dry, mp 146°. $bp_{0.5}$ 230°. $[\alpha]_D^{15}$ +149.2° (c = 1.9 in chloroform). Absorption spectrum: Morton, de Gouveia, *J. Chem. Soc.* **1934**, 916. One gram dissolves in 800 ml methanol. Freely sol in ether, chloroform, benzene.

Acetate. $C_{30}H_{44}O_2$. mp 146°. $[\alpha]_D^{16}$ +204° (c = 1.1 in chloroform).

Methyl ether. $C_{29}H_{44}O$. mp 106°. $[\alpha]_D^{20}$ +166° (c = 2 in chloroform).

2878. 3-Dehydroretinal. [472-87-1] 3,4-Didehydroretinal; (*all-E*)-3,7-dimethyl-9-(2,6,6-trimethyl-1,3-cyclohexadien-1-yl)-2,-4,6,8-nonatetraenal; *all-trans*-3,4-dehydroretinal; retinal 2; retinene 2; vitamin A₂ aldehyde. $C_{20}H_{26}O$; mol wt 282.43. C 85.05%, H 9.28%, O 5.66%. Carotenoid chromophore of porphyropsin and cyanopsin, *q.q.v.*, which are visual pigments of certain fish, crustaceans and amphibians. May also occur in combination with retinal, *q.v.*, in mixed visual systems. Isoln from retinas of fresh-water fish: G. Wald, *Nature* **139**, 1017 (1937). Recognition as vitamin A₂ aldehyde: R. A. Morton, *ibid.* **153**, 69 (1944). Prepn by the oxidation of vitamin A₂: G. Wald, *Biochim. Biophys. Acta* **4**, 215 (1950). Synthesis: K. R. Farrar *et al., J. Chem. Soc.* **1952**, 1414; of stereoisomers: U. Schwieter *et al., Helv. Chim. Acta* **45**, 517, 528, 541 (1962); R. S. H. Liu *et al., J. Am. Chem. Soc.* **99**, 8095 (1977). HPLC separation: K. Tsukida *et al., J. Chromatogr.* **192**, 395 (1980); T. Suzuki, M. Makino-Tasaka, *Anal. Biochem.* **129**, 111 (1983). Biological activity: G. Wald, *J. Gen. Physiol.* **22**, 775 (1939); *idem, Fed. Proc.* **12**, 606 (1953). Reviews: G. Wald, *Science* **162**, 230-239 (1968); R. Hubbard *et al., Methods Enzymol.* **18**, 615-653 (1971). Book: *The Retinoids* Vol. 1-2, M. B. Sporn *et al.*, Eds. (Academic Press, New York, 1984).

Orange-red prisms from pentane, mp 77-78°. uv max (ethanol): 401 nm ($E_{1cm}^{1\%}$ 1470).

11-*cis*-Isomer. [41470-05-7] Oil from petr ether-ether. uv max (ethanol): 393 nm ($E_{1cm}^{1\%}$ 882).

USE: As tool in biological energy transduction research.

2879. 3-Dehydroretinol. [79-80-1] 3,4-Didehydroretinol; (*all-E*)-3,7-dimethyl-9-(2,6,6-trimethyl-1,3-cyclohexadien-1-yl)-2,-4,6,8-nonatetraen-1-ol; *all-trans*-3-dehydroretinol; retinol₂; vitamin A₂. $C_{20}H_{28}O$; mol wt 284.44. C 84.45%, H 9.92%, O 5.62%. Naturally occurring retinoid originally discovered in livers of fresh-water fish. Constitutes 20-25% of total retinoid content in human epidermis; levels are markedly elevated in patients with hyperproliferative dermatoses such as psoriasis. Discovery in freshwater fish: E. Lederer et al., Nature **140**, 233 (1937); J. R. Edisbury et al., ibid. 234. Isoln from pike liver oils: E. M. Shantz, Science **108**, 417 (1948). Structure and synthesis: K. R. Farrar et al., J. Chem. Soc. **1952**, 2657. Synthesis and characteristics of stereoisomers: U. Schwieter et al., Helv. Chim. Acta **45**, 517, 528, 541, 548 (1962). HPLC determn in fish oils: B. Stancher, F. Zonta, J. Chromatogr. **312**, 423 (1984). Identification in human skin: A. Vahlquist, Experientia **36**, 317 (1980). Comparative levels in psoriasis patients: O. Rollman, A. Vahlquist, Arch. Dermatol. Res. **278**, 17 (1985). Biosynthesis from retinol, q.v., in cultured keratinocytes: O. Rollman et al., Biochem. J. **293**, 675 (1993).

Light yellow needles from petr ether, mp 63-65°. uv max (ethanol): 350 nm (E$_{1cm}^{1\%}$ 1455).

2880.　7-Dehydrositosterol. [521-04-0] (3β)-Stigmasta-5,7-dien-3-ol. C$_{29}$H$_{48}$O; mol wt 412.70. C 84.40%, H 11.72%, O 3.88%. From soy bean oil β-sitosterol: Wunderlich, Z. Physiol. Chem. **241**, 116 (1936). Starting with cholesterol: US 2386635 (1945).

Platelets from alc, browns on contact with air, mp 144-145°. [α]$_D^{21}$ −116° (c = 2 in chloroform). Sparingly sol in methanol, somewhat more in alcohol, freely sol in the other usual organic solvents. Insol in water.
Acetate. C$_{31}$H$_{50}$O$_2$. mp 151-152°. [α]$_D^{21}$ −71° (c = 2 in chloroform).

2881.　Delapril. [83435-66-9] N-[(1S)-1-(Ethoxycarbonyl)-3-phenylpropyl]-L-alanyl-N-(2,3-dihydro-1H-inden-2-yl)glycine; N-[N-[(S)-1-(ethoxycarbonyl)-3-phenylpropyl]-L-alanyl]-N-(indan-2-yl)glycine; ethyl (S)-2-[[(S)-1-[(carboxymethyl)-2-indanylcarbamoyl]ethyl]amino]-4-phenylbutyrate; alindapril; indalapril. C$_{26}$H$_{32}$-N$_2$O$_5$; mol wt 452.55. C 69.01%, H 7.13%, N 6.19%, O 17.68%. Angiotensin-converting enzyme (ACE) inhibitor. Prepn: Y. Oka et al., EP 51391; eidem, US 4385051 (1982, 1983 both to Takeda); J. T. Suh et al., Eur. J. Med. Chem. - Chim. Ther. **20**, 563 (1985); A. Miyake et al., Chem. Pharm. Bull. **34**, 2852 (1986). Pharmacology in animals: Y. Inada et al., Jpn. J. Pharmacol. **42**, 1 (1986); eidem, ibid. 99. Pharmacology and metabolism in humans: T. Ogihara et al., Curr. Ther. Res. **41**, 809 (1987). Pharmacokinetics and evaluation in hypertension: H. Shionoiri et al., Clin. Pharmacol. Ther. **41**, 74 (1987).

Hydrochloride. [83435-67-0] CV-3317; REV-6000A; Adecut; Cupressin. C$_{26}$H$_{32}$N$_2$O$_5$.HCl; mol wt 489.01. Colorless plates

from acetone + hydrochloric acid, mp 166-170° (dec). [α]$_D^{22}$ +18.5° (c = 1 in methanol).
THERAP CAT: Antihypertensive.

2882.　Delavirdine. [136817-59-9] N-[2-[[4-[3-[(1-Methylethyl)amino]-2-pyridinyl]-1-piperazinyl]carbonyl]-1H-indol-5-yl]-methanesulfonamide; 1-(5-methanesulfonamido-1H-indol-2-ylcarbonyl)-4-[3-(1-methylethylamino)pyridinyl]piperazine; 1-[3-[(1-methylethyl)amino]-2-pyridinyl]-4-[[5-[(methylsulfonyl)amino]-1H-indol-2-yl]carbonyl]piperazine; 1-[3-(isopropylamino)-2-pyridyl]-4-[(5-methanesulfonamidoindol-2-yl)carbonyl]piperazine; U-90152. C$_{22}$H$_{28}$N$_6$O$_3$S; mol wt 456.57. C 57.88%, H 6.18%, N 18.41%, O 10.51%, S 7.02%. Bisheteroarylpiperazine non-nucleoside reverse transcriptase inhibitor (NNRTI). Prepn: D. L. Romero et al., WO 9109849; J. R. Palmer et al., US 5563142 (1991, 1996 both to Upjohn). Structure-activity studies: D. L. Romero et al., J. Med. Chem. **36**, 1505 (1993). In vitro activity vs HIV-1: T. J. Dueweke et al., Antimicrob. Agents Chemother. **37**, 1127 (1993); P. J. Pagano, K.-T. Chong, J. Infect. Dis. **171**, 61 (1995). Enzyme kinetic studies: I. W. Althaus et al., Biochem. Pharmacol. **47**, 2017 (1994). NMR characterization of crystalline forms: P. Gao, Pharm. Res. **15**, 1425 (1998). HPLC determn in plasma: C.-L. Cheng et al., J. Chromatogr. B **769**, 297 (2002). Review of discovery and development: W. J. Adams et al., in Pharm. Biotechnol., R. T. Borchardt et al., Eds. (Plenum Press, New York, 1998) pp 285-312; of clinical use in HIV infection: L. J. Scott, C. M. Perry, Drugs **60**, 1411-1444 (2000).

Crystals from ethyl acetate + hexane, mp 226-228°. Aqueous soly at 23° (μg/ml): 2942 (pH 1.0); 295 (pH 2.0); 0.81 (pH 7.4). pKa$_1$ 4.56. pKa$_2$ 8.9. Log P (n-octanol/water): 2.84.
Methanesulfonate. [147221-93-0] Delavirdine mesylate; U-90152S; Rescriptor. C$_{22}$H$_{28}$N$_6$O$_3$S.CH$_3$SO$_3$H; mol wt 552.67.
THERAP CAT: Antiretroviral.

2883.　Delmadinone Acetate. [13698-49-2] 17-(Acetyloxy)-6-chloropregna-1,4,6-triene-3,20-dione; 6-chloro-17-hydroxypregna-1,4,6-triene-3,20-dione acetate; 1,6-bisdehydro-6-chloro-17α-acetoxyprogesterone; Δ^1-chlormadinone acetate; RS-1301; Tardak. C$_{23}$H$_{27}$ClO$_4$; mol wt 402.92. C 68.56%, H 6.75%, Cl 8.80%, O 15.88%. Prepn: Ringold et al., J. Am. Chem. Soc. **81**, 3485 (1959); GB 890315 (1962 to Upjohn), C.A. **57**, 5996c (1962); **76**, 14813y (1972); Campbell, Babcock, DE 1243682 (1967 to Upjohn), C.A. **67**, 100343r (1967). Activity studies: Dorfman, Kincl, Steroids **1**, 185 (1963); Weichert et al., Arzneim.-Forsch. **17**, 1103 (1967); Gerber et al., J. Small Anim. Pract. **14**, 151 (1973).

Crystals, mp 168-170°. [α]$_D$ −83° (chloroform). uv max (ethanol): 229, 258, 297 nm (log ε 4.00, 4.00, 4.03).
THERAP CAT (VET): Progestogen.

2884.　Delmopinol. [79874-76-3] 3-(4-Propylheptyl)-4-morpholineethanol. C$_{16}$H$_{33}$NO$_2$; mol wt 271.45. C 70.80%, H 12.25%,

N 5.16%, O 11.79%. Surface-active agent that reduces dental plaque formation. Prepn: **BE 888052**; S. E. H. Hernestam *et al.*, **US 4636382** (1981, 1987 both to Ferrosan). Stereoselective HPLC determn in plasma: G. Egginger *et al.*, *J. Chromatogr. A* **666**, 275 (1994). *In vitro* effect on bacterial cell walls: J. Rundegren *et al.*, *Oral Microbiol. Immunol.* **10**, 102 (1995). Clinical pharmacokinetics: B. Eriksson *et al.*, *Xenobiotica* **28**, 1075 (1998). Home usage trials of mouthwash: A. J. Elworthy *et al.*, *J. Clin. Periodontol.* **22**, 527 (1995); N. Claydon *et al.*, *ibid.* **23**, 220 (1996).

Hydrochloride. [98092-92-3] M-1650; Decapinol. $C_{16}H_{33}$-NO_2.HCl; mol wt 307.90. mp 70-72°. pKa 7.1. Soly in water: >40%.

THERAP CAT: In treatment of gingivitis.

2885. Delphinidin. [528-53-0] 3,5,7-Trihydroxy-2-(3,4,5-tri-hydroxyphenyl)-1-benzopyrylium chloride (1:1); 3,3′,4′,5,5′,7-hex-ahydroxyflavylium chloride; 3,3′,4′,5,5′,7-hexahydroxy-2-phenyl-benzopyrylium chloride; delphinidol. $C_{15}H_{11}ClO_7$; mol wt 338.70. C 53.19%, H 3.27%, Cl 10.47%, O 33.07%. The aglucone of del-phinin: Willstätter, Mieg, *Ann.* **408**, 61 (1915); the aglucone of vio-lanin: Willstätter, Weil, *ibid.* **412**, 178 (1917). Occurrence in plants: Acheson, *Can. J. Biochem. Physiol.* **37**, 1 (1959). Synthesis: Pratt, Robinson, *J. Chem. Soc.* **127**, 166 (1925); Bradley *et al.*, *ibid.* **1930**, 793.

Chocolate-brown prisms or needles with metallic luster from 5% HCl. Crystallizes from water with 1, 2 or 4 mols H_2O. When anhydrous, does not melt below 350°. Absorption max (methanolic HCl): 544 nm. Sol in methanol, ethanol, ethyl acetate.

3-Glucoside. Myrtillin-a. $C_{21}H_{21}ClO_{12}$. From the whortleberry *(Vaccinium myrtillus* L., *Ericaceae):* Willstätter, Zollinger, *Ann.* **408**, 83 (1915); **412**, 195 (1916); from the pansy *(Viola tricolor* L., *Violaceae):* Karrer, de Meuron, *Helv. Chim. Acta* **16**, 292 (1933). Structure and synthesis: Reynolds, Robinson, *J. Chem. Soc.* **1934**, 1039. Deep purple crystals from dil HCl. Absorption max (meth-anolic HCl): 535 nm.

3,5-Diglucoside. 3,5-Bis(glucosyloxy)-3′,4′,5′,7-tetrahydroxy-flavylium chloride; delphin; delphoside; hyacin. $C_{27}H_{31}ClO_{17}$. From flowers of *Salvia patens* Cav., *Labiatae*, structure and synthesis: Reynolds *et al.*, *J. Chem. Soc.* **1934**, 1235; from flowers of *Verbena hybrida* Hart., *Verbenaceae:* Scott-Moncrieff, Sturgess, *Biochem. J.* **34**, 268 (1940); from grape skins: Bockian *et al.*, *J. Agric. Food Chem.* **3**, 695 (1955). Identity with hyacin: Saito *et al.*, *Proc. Jpn. Acad.* **36**, 340 (1960). Leaflets with bronze luster from dil HCl. Absorption max (methanolic HCl): 534 nm. Practically insol in cold water, alcohol, dil acids; sol in hot dil HCl.

Compound with glucose + hydroxybenzoic acid. Delphinin. $C_{41}H_{39}ClO_{21}$. From *Delphinium consolida* L., *Ranunculaceae:* Willstätter, Mieg, *loc. cit.* Dark red-brown plates or prisms from 3% HCl. dec 200-203°. Sol in water with decompn, dil alcohol, acetone, boiling alcohol.

Compound with glucose + rhamnose + *p*-hydroxycinnamic acid. Violanin. $C_{36}H_{37}ClO_{18}$. From flowers of *Viola tricolor* L., *Violaceae:* Willstätter, Weil, *loc. cit.* Structure: Karrer, de Meuron, *loc. cit.* Bluish-violet plates with green metallic luster from meth-

anolic HCl + ether. Sol in amyl alcohol, in 0.15-0.5% HCl; less sol in 1% HCl; slightly sol in water; practically insol in 5-12% HCl.

2886. Delphinine. [561-07-9] (1α,6α,14α,16β)-1,6,16-Tri-methoxy-4-(methoxymethyl)-20-methylaconitine-8,13,14-triol 8-acetate 14-benzoate. $C_{33}H_{45}NO_9$; mol wt 599.72. C 66.09%, H 7.56%, N 2.34%, O 24.01%. A toxic alkaloid from seeds of *Del-phinium staphisagria* L., *Ranunculaceae*. Isoln: Markwood, *J. Am. Pharm. Assoc.* **16**, 928 (1927). Purification: Schneider, *Arch. Pharm.* **283**, 283 (1950). Structure: Jacobs, Pelletier, *J. Am. Chem. Soc.* **78**, 3542 (1956); Wiesner *et al.*, *Can. J. Chem.* **47**, 2734 (1969). Configuration: Wiesner *et al.*, *Tetrahedron Lett.* **1**, no. 3, 11 (1959); *ibid.* **1960**, no. 15, 23; Gilman, Marion, *ibid.* **1962**, 923; Birnbaum *et al.*, *ibid.* **1971**, 867. Synthesis: Wiesner *et al.*, *Can. J. Chem.* **50**, 1925 (1972).

Orthorhombic, six-sided plates from alc, mp 197.5-199°. $[\alpha]_D^{20}$ +25° (ethanol). Practically insol in water. Sol in 25 parts alcohol, 20 parts chloroform, 10 parts ether.

Hydrochloride. $C_{33}H_{45}NO_9$.HCl. Crystals, dec 214°.

2887. Delsoline. [509-18-2] (1α,6β,14α,16β)-20-Ethyl-6,-14,16-trimethoxy-4-(methoxymethyl)aconitane-1,7,8-triol. $C_{25}H_{41}$-NO_7; mol wt 467.60. C 64.22%, H 8.84%, N 3.00%, O 23.95%. From seeds of *Dephinium consolida* L., *Ranunculaceae:* Mark-wood, *J. Am. Pharm. Assoc.* **13**, 696 (1924); Cionga, Iliescu, *Ber.* **74**, 1031 (1941); Marion, Edwards, *J. Am. Chem. Soc.* **69**, 2010 (1947). Structure: Anet *et al.*, *Can. J. Chem.* **35**, 397 (1957); Skaric, Marion, *J. Am. Chem. Soc.* **80**, 4434 (1958); *eidem*, *Can. J. Chem.* **38**, 2433 (1960); **39**, 1579 (1961).

Delsoline R = CH_3
Delcosine R = H

Prisms from methanol, mp 213-216.5° (when immersed at 205°). $[\alpha]_D^{22}$ +53.4° (c = 2.04 in chloroform). Slightly sol in water, sol in alcohol or chloroform.

Delcosine. [545-56-2] $C_{24}H_{39}NO_7$; mol wt 453.58. Alkaloid also found in *D. consolida.* mp 203-204°, $[\alpha]_D^{25}$ +56.8° (c = 2.01 in chloroform). Quite sol in water.

USE: Glycosidal dyestuff.

2888. Deltamethrin. [52918-63-5] (1*R*,3*R*)-3-(2,2-Dibromo-ethenyl)-2,2-dimethylcyclopropanecarboxylic acid (*S*)-cyano(3-phenoxyphenyl)methyl ester; (*S*)-α-cyano-3-phenoxybenzyl-(1*R*)-*cis*-3-(2,2-dibromovinyl)-2,2-dimethylcyclopropane carboxylate; decamethrin; esbecythrin; FMC-45498; NRDC-161; RU-22974; Bu-tox; Decis; K-Othrine; Scalibor. $C_{22}H_{19}Br_2NO_3$; mol wt 505.21. C 52.30%, H 3.79%, Br 31.63%, N 2.77%, O 9.50%. Synthetic pyreth-roid insecticide. Prepn of racemic mixture: M. Elliot *et al.*, **DE 2439177** (1975 to NRDC), *C.A.* **83**, 73519z (1975); of isomers: *ei-dem*, *Pestic. Sci.* **9**, 105 (1978). Activity: *eidem*, *Nature* **248**, 710 (1974); *eidem*, *Pestic. Sci.* **9**, 112 (1978). Absolute configuration: J. D. Owen, *J. Chem. Soc. Perkin Trans. 1* **1975**, 1865. Photochemistry: L. O. Ruzo *et al.*, *J. Agric. Food Chem.* **25**, 1385 (1977). Metabolism: *eidem, ibid.* **26**, 918 (1978). Toxicology: R. Kavlock *et al.*, *J. Environ. Pathol. Toxicol.* **2**, 751 (1979). Pharmacological effects on central nervous system: P. H. Chanh *et al.*, *Arzneim.-Forsch.* **34**, 175 (1984). Review of toxicology and human exposure: *Toxicological Profile for Pyrethrins and Pyrethroids* (PB2004-100004, 2003) 332 pp.

Crystals, mp 98-101°. Sol in ethanol, acetone, dioxane. Insol in water. LD_{50} in female rats (mg/kg): 31 orally; 4 i.v. (Kavlock).
USE: Insecticide.
THERAP CAT (VET): Ectoparasiticide.

2889. Demecarium Bromide. [56-94-0] 3,3′-[1,10-Decanediylbis[(methylimino)carbonyloxy]]bis[*N,N,N*-trimethylbenzenaminium]bromide (1:2); (*m*-hydroxyphenyl)trimethylammonium bromide, decamethylenebis(methylcarbamate); *N,N′*-bis[3-trimethylammoniumphenoxycarbonyl]-*N,N′*-dimethyldecamethylenediamine dibromide; decamethylenebis[*m*-dimethylaminophenyl *N*-methylcarbamate] dimethobromide; decamethylenebis[*N*-methylcarbamic acid *m*-dimethylaminophenyl ester] bromomethylate; BC-48; Tosmilen; Humorsol. $C_{32}H_{52}Br_2N_4O_4$; mol wt 716.60. C 53.64%, H 7.31%, Br 22.30%, N 7.82%, O 8.93%. Prepn: Schmid, **US 2789-981** (1957 to Oesterreichische Stickstoffwerke).

White or pale yellow, slightly hygroscopic crystalline powder, dec 162-167°. Freely sol in water, alc; sol in ether; sparingly sol in acetone. Aq solns are neutral, stable, and may be sterilized by heat.
THERAP CAT: Cholinergic (ophthalmic).

2890. Demeclocycline. [127-33-3] (4*S*,4a*S*,5a*S*,6*S*,12a*S*)-7-Chloro-4-(dimethylamino)-1,4,4a,5,5a,6,11,12a-octahydro-3,6,10,-12,12a-pentahydroxy-1,11-dioxo-2-naphthacenecarboxamide; 7-chloro-6-demethyltetracycline; demethylchlortetracycline; RP-10192; Bioterciclin; Declomycin; Deganol; Ledermycin; Periciclina. $C_{21}H_{21}ClN_2O_8$; mol wt 464.86. C 54.26%, H 4.55%, Cl 7.63%, N 6.03%, O 27.53%. Antibiotic related to tetracycline produced by *Streptomyces aureofaciens*. Prepn: McCormick *et al.*, *J. Am. Chem. Soc.* **79**, 4561 (1957); **US 2878289** (1959 to Am. Cyanamid). Improved fermentation processes: Szumski, **US 3012946**; Goodman, Matrishin, **US 3019172**; Goodman, **US 3050446** (1961, 1962, 1962 all to Am. Cyanamid); **FR 1344645** (1963 to Merck & Co.); Neidleman, **US 3154476** (1964 to Olin Mathieson). Abs config: Dobrynin *et al.*, *Tetrahedron Lett.* **1962**, 901. Toxicity data: E. I. Goldenthal, *Toxicol. Appl. Pharmacol.* **18**, 185 (1971).

Sesquihydrate. [13215-10-6] mp 174-178° (dec). $[\alpha]_D^{25}$ −258° (c = 0.5 in 0.1*N* H_2SO_4). Solubilities: Marsh, Weiss, *J. Assoc. Off. Anal. Chem.* **50**, 457 (1967).
Hydrochloride. [64-73-3] Clortetrin; Demetraciclina; Detravis; Meciclin; Mexocine. Yellow crystalline powder. Sparingly sol in

water and in solutions of alkali hydroxides and carbonates; slightly sol in ethanol. Practically insol in acetone, chloroform. LD_{50} orally in rats: 2372 mg/kg (Goldenthal).
THERAP CAT: Antibacterial.
THERAP CAT (VET): Antibacterial.

2891. Demecolcine. [477-30-5] (7*S*)-6,7-Dihydro-1,2,3,10-tetramethoxy-7-(methylamino)benzo[*a*]heptalen-9(5*H*)-one; *N*-deacetyl-*N*-methylcolchicine; *N*-methyl-*N*-desacetylcolchicine; colchamine; Santavy's substance F; colcemid. $C_{21}H_{25}NO_5$; mol wt 371.43. C 67.91%, H 6.78%, N 3.77%, O 21.54%. Antitumor alkaloid; depolymerizes microtubules and inhibits spindle formation during metaphase. Isoln from *Colchicum autumnale* L., *Liliaceae*: Santavy, *Pharm. Acta Helv.* **25**, 248 (1950); Santavy, Reichstein, *Helv. Chim. Acta* **33**, 1606 (1950); Schlittler, Uffer, **DE 936268** (1955 to Ciba), *C.A.* **53**, 1396 (1959). Structure: Santavy *et al.*, *Helv. Chim. Acta* **36**, 1319 (1953). Synthesis: Uffer *et al.*, *ibid.* **37**, 18 (1954); H. G. Capraro, A. Brossi, *ibid.* **62**, 965 (1979). Effect on the mitotic cycle: C. L. Rieder, R. E. Palazzo, *J. Cell Sci.* **102**, 387 (1992). Use in oocyte enucleation: E. Ibáñez *et al.*, *Biol. Reprod.* **68**, 1249 (2003).

Pale yellow prisms from ethyl acetate + ether, mp 186°. $[\alpha]_D^{20}$ −129.0° (c = 1 in chloroform). uv max (alc): 245, 355 nm (log ε 4.55, 4.24). Basic reaction. Soluble in acidified water, in alcohol, ether, chloroform, benzene.
USE: Cell synchronization agent; for chromosome visualization; to induce oocyte enucleation for somatic cell cloning.

2892. Demeton. [8065-48-3] Phosphorothioic acid *O,O*-diethyl *O*-[2-(ethylthio)ethyl] ester mixture with *O,O*-diethyl *S*-[2-(ethylthio)ethyl]phosphorothioate; mercaptophos; Bayer 8169; E-1059; Systox. $C_8H_{19}O_3PS_2$; mol wt 258.33. C 37.20%, H 7.41%, O 18.58%, P 11.99%, S 24.82%. Cholinesterase inhibitor. Isomeric mixture consisting of demeton-O and demeton-S (*O,O*-diethyl *O*(and *S*)-ethylmercaptoethyl thiophosphates). Another name for demeton-S is *Isosystox*. Prepn: Schrader, **US 2571989** and **US 2597534** (1951, 1952 both to Bayer); Gardner, Heath, *Anal. Chem.* **25**, 1849 (1953). Toxicity study: T. B. Gaines, *Toxicol. Appl. Pharmacol.* **14**, 515 (1969).

Oily liq. Faint odor. bp_2 134°. Virtually insol in water. Sol in ethanol, propylene glycol, toluene and similar hydrocarbons. Absorbed by plants, rendering the foliage and plant fluids toxic to insects. LD_{50} in female, male rats (mg/kg): 2.5, 6.2 orally; 8.2, 14 dermally (Gaines).
Caution: Potential symptoms of overexposure are miosis, aching eyes, rhinorrhea and headache; chest tightening, wheezing, laryngeal spasms, salivation and cyanosis; anorexia, nausea, vomiting, abdominal cramps and diarrhea; local sweating; muscle fasciculation, weakness and paralysis; giddiness, confusion and ataxia; convul-

Consult the Name Index before using this section.

sions, coma; low blood pressure; cardiac irregularities; irritation of eyes and skin. *See NIOSH Pocket Guide to Chemical Hazards* (DHHS/NIOSH 97-140, 1997) p 90.

USE: Insecticide.

2893. Denatonium Benzoate. [3734-33-6] *N*-[2-[(2,6-Dimethylphenyl)amino]-2-oxoethyl]-*N,N*-diethylbenzenemethaminium benzoate (1:1); benzyldiethyl[(2,6-xylylcarbamoyl)methyl]-ammonium benzoate; lignocaine benzyl benzoate; Bitrex. $C_{28}H_{34}$-N_2O_3; mol wt 446.59. C 75.31%, H 7.67%, N 6.27%, O 10.75%. Prepn: J. E. Hay, **US 3080327** (1963 to T. & H. Smith Ltd.); *idem*, **US 3268575** (1966 to Edinburgh Pharm. Ind.). HPLC determn in rapeseed oil: C. E. Damon, B. C. Pettitt, Jr., *J. Chromatogr.* **195**, 243 (1980). Brief review of substance history and uses: H. A. S. Payne, *Chem. Ind. (London)* **22**, 721-723 (1988).

Crystals from isopropyl alcohol + ethyl acetate, mp 166-170°. Among the most bitter substances known to man. Very sol in chloroform, methanol; soluble in water, alcohol; sparingly sol in acetone. Practically insol in ether.

USE: Pharmaceutic aid (alcohol denaturant; flavor). Added to toxic substances as a deterrent to accidental ingestion. Can replace brucine or quassin as denaturant for ethyl alcohol.

2894. Dendrotoxins. [74811-93-1] DTX. Family of low molecular weight (~7K) proteins isolated from the venom of mamba snakes. Inhibits voltage-dependent potassium (K⁺) channels at nanomolar ranges. Characterized by 3 conserved disulfide linkages in a 57-60 amino acid sequence. Isoln from *Dendroaspis angusticeps*: A. L. Harvey, E. Karlsson, *Arch. Pharmacol.* **312**, 1 (1980). K⁺-channel blockade: J. V. Halliwell *et al.*, *Proc. Natl. Acad. Sci. USA* **83**, 493 (1986); relative potencies as blockers: D. G. Owen *et al.*, *Br. J. Pharmacol.* **120**, 1029 (1997). NMR and 2° structure of I: M.-F. Foray *et al.*, *Eur. J. Biochem.* **211**, 813 (1993). Syntheses and sequences: M. E. Byrnes *et al.*, *Protein Pept. Lett.* **1**, 239 (1995); H. Nishio *et al.*, *Pept. Chem.* **34**, 105 (1996). Structure-activity study of K: L. A. Smith *et al.*, *Biochemistry* **36**, 7690 (1997). Comparison of neurotoxic effects with apamin, *q.v.*: G. Lallement *et al.*, *Toxicology* **104**, 47 (1995). *Review:* A. L. Harvey, *Gen. Pharmacol.* **28**, 7-12 (1997).

USE: Biochemical tool for K⁺ channels.

2895. Denileukin Diftitox. [173146-27-5] *N*-L-Methionyl-387-L-histidine-388-L-alanine-1-388-toxin (*Corynebacterium diphtheriae* strain C7) (388 → 2')-protein with 2-133-interleukin 2 (human clone pTIL2-21a); $DAB_{389}IL2$; LY-335348; Ontak. Recombinant fusion protein consisting of the catalytic and translocation domains of diphtheria toxin fused to interleukin 2, *q.v.* Immunotoxin that selectively targets eukaryotic cells expressing the interleukin 2 receptor. Prepn: D. Williams *et al.*, *J. Biol. Chem.* **265**, 11885 (1990). Clinical evaluation in cutaneous T cell lymphoma: M. Duvic *et al.*, *Am. J. Hematol.* **58**, 87 (1998). Review of development and clinical experience: F. M. Foss *et al.*, *Curr. Top. Microbiol. Immunol.* **234**, 63-81 (1998).

THERAP CAT: Antineoplastic.

2896. Denopamine. [71771-90-9] (*αR*)-α-[[[2-(3,4-Dimethoxyphenyl)ethyl]amino]methyl]-4-hydroxybenzenemethanol; (−)-(*R*)-α-[[(3,4-dimethoxyphenethyl)amino]methyl]-*p*-hydroxybenzyl alcohol; (−)-(*R*)-1-(4-hydroxyphenyl)-2-(3,4-dimethoxyphenethyl-amino)ethanol; TA-064; Carguto; Kalgut. $C_{18}H_{23}NO_4$; mol wt 317.39. C 68.12%, H 7.30%, N 4.41%, O 20.16%. Selective β_1-adrenoceptor agonist with positive inotropic activity. Prepn: M. Ikezaki *et al.*, **BE 833731**; *idem*, **US 4032575** (1976, 1977 both to Tanabe Seiyaku); N. Umino *et al.*, *Chem. Pharm. Bull.* **27**, 1479 (1979). Effect on carbohydrate and lipid metabolism in rats: M. Inamasu *et al.*, *Biochem. Pharmacol.* **33**, 2171 (1984). Cardiovascular pharmacology in animals: T. Nagao *et al.*, *Jpn. J. Pharmacol.* **35**, 415 (1984); in humans: M. Kino *et al.*, *Am. J. Cardiol.* **51**, 802

(1983). Metabolism in humans: T. Suzuki *et al.*, *Drug Metab. Dispos.* **11**, 377 (1983). GC-MS determn in human urine: *eidem*, *Chem. Pharm. Bull.* **33**, 2549 (1985). Binding affinity and selectivity: K. Naito *et al.*, *Jpn. J. Pharmacol.* **38**, 235 (1985). Clinical efficacy in congestive cardiomyopathy: J. Thormann *et al.*, *Am. Heart J.* **110**, 426 (1985). Series of articles on pharmacokinetics, metabolism and disposition in animals: *Arzneim.-Forsch.* **36**, 643-667 (1986).

l-**Form hydrochloride.** $C_{18}H_{23}NO_4 \cdot HCl$. Crystals from isopropanol, mp 138-139.5°. $[\alpha]_D^{25}$ −38.0° (c = 1 in methanol).

dl-**Form hydrochloride.** Crystals from isopropanol + isopropyl ether, mp 164-167°.

THERAP CAT: Cardiotonic.

2897. Denosumab. [615258-40-7] Anti-(human osteoclast differentiation factor) immunoglobulin G2 (human monoclonal AMG162 heavy chain) disulfide with human monoclonal AMG162 light chain, dimer; Anti-(human receptor activator of NF-κB ligand) immunoglobulin G2 (human monoclonal AMG162 heavy chain) disulfide with human monoclonal AMG162 light chain, dimer; AMG-162; Prolia; Xgeva. Human monoclonal antibody directed against receptor activator of NF-κB ligand (RANKL) also known as osteoprotegerin ligand (OPGL). Inhibits the effects of RANKL on osteoclast differentiation and subsequent bone loss. Prepn: W. J. Boyle *et al.*, **WO 03002713** (2003 to Abgenix; Amgen); *eidem*, **US 7364736** (2008 to Amgen). Clinical effect on bone mineral density in postmenopausal women: M. R. McClung *et al.*, *N. Engl. J. Med.* **354**, 821 (2006); H. G. Bone *et al.*, *J. Clin. Endocrinol. Metab.* **93**, 2149 (2008). Clinical evaluation in breast cancer-related bone metastases: A. Lipton *et al.*, *J. Clin. Oncol.* **25**, 4431 (2007); in rheumatoid arthritis: S. B. Cohen *et al.*, *Arthritis Rheum.* **58**, 1299 (2008). Review of clinical experience in osteoporosis: I. Charopoulos *et al.*, *Expert Opin. Drug Saf.* **10**, 205-217 (2011).

THERAP CAT: Bone resorption inhibitor; antiosteoporotic.

2898. Denufosol. [211448-85-0] Uridine 5'-(pentahydrogen tetraphosphate) $P''' \rightarrow 5'$-ester with 2'-deoxycytidine; P^1-(uridine 5')-P^4-(2'-deoxycytidine 5')tetraphosphate; UP_4dC. $C_{18}H_{27}N_5$-$O_{21}P_4$; mol wt 773.32. C 27.96%, H 3.52%, N 9.06%, O 43.45%, P 16.02%. Nucleotide P2Y₂ purinoceptor agonist. Prepn: W. Pendergast *et al.*, **WO 9834942**; *eidem*, **US 6348589** (1998, 2002 both to Inspire); and structure-activity study: S. R. Shaver *et al.*, *Purinergic Signalling* **1**, 183 (2005). Pharmacology: B. R. Yerxa *et al.*, *J. Pharmacol. Exp. Ther.* **302**, 871 (2002). Clinical trial in cystic fibrosis: F. J. Accurso *et al.*, *Am. J. Respir. Crit. Care Med.* **183**, 627 (2011).

Tetrasodium salt. [318250-11-2] INS-37217. $C_{18}H_{23}N_5Na_4$-$O_{21}P_4$; mol wt 861.25.

THERAP CAT: In treatment of cystic fibrosis.

2899. Deoxo-Fluor®. [202289-38-1] (*T*-4)-Trifluoro[2-methoxy-*N*-(2-methoxyethyl)ethanaminato-*κN*]sulfur; bis(2-methoxyethyl)aminosulfur trifluoride. $C_6H_{14}F_3NO_2S$; mol wt 221.24. C 32.57%, H 6.38%, F 25.76%, N 6.33%, O 14.46%, S 14.49%. Broad-spectrum deoxofluorinating agent. Prepn: G. S. Lal *et al.*, **EP 908448**; *eidem* **US 6207860** (1999, 2001 both to Air Prods.

Chem.); *idem et al.*, *Chem. Commun.* **1999**, 215; *idem et al.*, *J. Org. Chem.* **64**, 7048 (1999). Review of use in fluorination reactions of organic compds: R. P. Singh, J. M. Shreeve, *Synthesis* **2002**, 2561-2578.

Yellow liquid, colorless after distillation. bp$_{10}$ 60°. Dec onset ~140°. d 1.2. Sol in hexane, ethers, halogenated organics, toluene. *Caution:* Reacts violently with water to form HF.
USE: Versatile, thermally stable nucleophilic fluorinating reagent.

2900. 6-Deoxy-L-ascorbic Acid. [528-81-4] $C_6H_8O_5$; mol wt 160.13. C 45.00%, H 5.04%, O 49.96%. Prepn starting with 2,3-monoacetone-L-sorbomethylose: Müller, Reichstein, *Helv. Chim. Acta* **21**, 273 (1938). Alternative procedure: **CH 203549** and **CH 209587** (1939 and 1940, both to Hoffmann-La Roche).

Stout prisms from ethyl acetate, mp 168°. Sublimes at 10^{-3} mm and 160°. [α]$_D^{22}$ +36.7° (0.1*N* HCl). Freely sol in water; sol in acetone, alcohol; sparingly sol in ethyl acetate. Practically insol in ether.

2901. Deoxycholic Acid. [83-44-3] (3α,5β,12α)-3,12-Dihydroxycholan-24-oic acid; 17β-(1-methyl-3-carboxypropyl)etiocholane-3α,12α-diol; desoxycholic acid. $C_{24}H_{40}O_4$; mol wt 392.58. C 73.43%, H 10.27%, O 16.30%. Lacks the C-7 hydroxy group of cholic acid. Occurs in bile of man, ox, sheep, dog, goat, and rabbit. Isoln: Wieland, Siebert, *Z. Physiol. Chem.* **262**, 1 (1939). From cholic acid: Sifferd, **US 2765316** (1956 to Armour). Structure and synthesis: L. F. Fieser, M. Fieser, *Steroids* (Reinhold, New York, 1959). Laboratory prepn from ox bile: Gattermann-Wieland, *Praxis des Organischen Chemikers* (de Gruyter, Berlin, 40th ed., 1961) p 361. Forms molecular coordination compds (so-called *choleic acids*) with many substances. Complexes wth fatty acids have been studies extensively: Sobotka, Goldberg, *Biochem. J.* **26**, 555 (1932); Sobotka, *Chem. Rev.* **15**, 362 (1934). Surface activity: *idem*, *Chemistry of the Steroids*, New York (1938).

Crystals from alc, mp 176-178°. Not precipitated by digitonin. [α]$_D^{20}$ +55° (alc). pK 6.58. Soly at 15° in water 0.24 g/l; in alcohol 220.7 g/l; in ether 1.16 g/l; in chloroform 2.94 g/l; in benzene 0.12 g/l; in acetone 10.46 g/l; in glacial acetic acid 9.06 g/l. Sol in solns of alkali hydroxides or carbonates.
Sodium salt. Sodium deoxycholate. $C_{24}H_{39}NaO_4$. Soly in water at 15°: >333 g/l.
THERAP CAT: Choleretic.

2902. Deoxycorticosterone. [64-85-7] 21-Hydroxypregn-4-ene-3,20-dione; 4-pregnen-21-ol-3,20-dione; 21-hydroxyprogesterone; desoxycorticosterone; 11-deoxycorticosterone; cortexone; desoxycortone; Kendall's desoxy compound B; Reichstein's substance Q. $C_{21}H_{30}O_3$; mol wt 330.47. C 76.32%, H 9.15%, O 14.52%. Occurs in adrenal cortex: Reichstein, von Euw, *Helv. Chim. Acta* **21**, 1197 (1938); Steiger, Reichstein, *ibid.* **20**, 1164 (1937). Numer-

ous prepns from other steroids: Schindler *et al.*, *ibid.* **24**, 371 (1941); Reichstein, **DE 875353** (1953 to Schering); Bockmühl *et al.*, **DE 871153** (1953 to Hoechst); Wettstein *et al.*, **US 2778776** (1957 to Ciba); **NL 89575** (1958 to Organon); Kaspar *et al.*, **DE 1028572** (1958 to Schering). Isoln from the prothoracal glands of the water beetle, *Dytiscus marginalis:* Schildknecht *et al.*, *Angew. Chem.* **78**, 392 (1966).

Plates from ether, mp 141-142°. [α]$_D^{22}$ +178° (alc). uv max: 240 nm. Freely sol in alcohol, acetone.
Acetate *see* Deoxycortiscosterone Acetate.
Tetraacetyl-β-D-glucoside. $C_{35}H_{48}O_{12}$. Clusters of needles from 50% alc, mp 176-176.5°. [α]$_D^{24}$ +80° (c = 0.515 in chloroform). Believed to occur in nature in this glycosidic form: Johnson, *J. Am. Chem. Soc.* **63**, 3238 (1941).
THERAP CAT: Mineralocorticoid.

2903. Deoxycorticosterone Acetate. [56-47-3] 21-Acetyloxypregn-4-ene-3,20-dione; 21-hydroxypregn-4-ene-3,20-dione 21-acetate; 11-deoxycorticosterone acetate; cortexone acetate; desoxycorticosterone acetate; deoxycortone acetate; desoxycortone acetate; DCA; Cortate; Cortiron; Decosteron; Doca; Dorcostrin; Percorten; Syncortyl. $C_{23}H_{32}O_4$; mol wt 372.51. C 74.16%, H 8.66%, O 17.18%.

Orthorhombic needles from alcohol, stable in air, mp 154-160°. Sublimes in high vacuum. [α]$_D^{20}$ +168 to +176° (dioxane). Slightly sol in alcohol, methanol, acetone, ether, dioxane, propylene glycol (10 mg/ml), vegetable oils. Practically insol in water.
THERAP CAT: Mineralocorticoid.
THERAP CAT (VET): Adrenocortical steroid, mineralocorticoid.

2904. Deoxyepinephrine. [501-15-5] 4-[2-(Methylamino)ethyl]-1,2-benzenediol; 4-[2-(methylamino)ethyl]pyrocatechol; 3,4-dihydroxyphenylethylamine; methyl[β-(3,4-dihydroxyphenyl)ethyl]amine; *N*-methyldopamine; epinine. $C_9H_{13}NO_2$; mol wt 167.21. C 64.65%, H 7.84%, N 8.38%, O 19.14%. Adrenergic and dopamine receptor agonist; active metabolite of ibopamine, *q.v.* Prepd by heating HCl and 1-keto-6,7-dimethoxy-2-methyltetrahydroisoquinoline obtained from laudanosine or papaverine: Pyman, *J. Chem. Soc.* **95**, 1266, 1610 (1909). Synthesis from veratrole: Kindler, Peschke, *Arch. Pharm.* **270**, 340 (1932); Kindler, Hesse, *ibid.* **271**, 439 (1933). From methylhomoveratrylamine: Buck, *J. Am. Chem. Soc.* **52**, 4119 (1930); Bretschneider, *Monatsh. Chem.* **76**, 335 (1947). Precursor in biosynthesis of epinephrine: W. F. Bridgers, S. Kaufman, *J. Biol. Chem.* **237**, 526 (1962). Cardiovascular pharmacology: I. Martínez-Mir *et al.*, *Gen. Pharmacol.* **31**, 75 (1998). Comparison with 3,4-dihydroxybenzylamine in catecholamine analysis: H. He *et al.*, *J. Chromatogr. B* **701**, 115 (1997).

Crystals, mp 188-189°.

Hydrochloride. $C_9H_{13}NO_2.HCl$. Crystals. Sol in water, alcohol. *Sensitive to light.*

USE: Internal standard in chromatographic analysis of catecholamines.

2905. 2-Deoxy-D-glucose. [154-17-6] 2-Deoxy-D-*arabino*-hexose; D-*arabino*-2-desoxyhexose; 2-deoxyglucose; 2-DG; Ba-2758. $C_6H_{12}O_5$; mol wt 164.16. C 43.90%, H 7.37%, O 48.73%. Antimetabolite of glucose, *q.v.*, with antiviral activity. Synthesis: M. Bergmann *et al., Ber.* **55**, 158 (1922); **56**, 1052 (1923); J. C. Sowden, H. O. L. Fischer, *J. Am. Chem. Soc.* **69**, 1048 (1947); H. R. Bolliger, *Helv. Chim. Acta* **34**, 989 (1954); H. R. Bolliger, M. D. Schmid, *ibid.* 1597, 1671; H. R. Bolliger, "2-Deoxy-D-*arabino*-hexose (2-Deoxy-D-glucose)" in *Methods in Carbohydrate Chemistry* vol. **I**, R. L. Whistler, M. L. Wolfrom, Eds. (Academic Press, New York, 1962) pp 186-189. Inhibition of influenza virus multiplication: E. D. Kilbourne, *Nature* **183**, 271 (1959). Effects on herpes simplex virus: R. J. Courtney *et al., Virology* **52**, 447 (1973). Mechanism of action studies: M. R. Steiner *et al., Biochem. Biophys. Res. Commun.* **61**, 745 (1974); E. K. Ray *et al., Virology* **58**, 118 (1978). Use in human genital herpes infections: H. A. Blough, R. L. Giuntoli, *J. Am. Med. Assoc.* **241**, 2798 (1979); L. Corey, K. K. Holmes, *ibid.* **243**, 29 (1980). Effect vs respiratory syncytial viral infections in calves: S. B. Mohanty *et al., Am. J. Vet. Res.* **42**, 336 (1981).

Cryst from acetone or butanone, mp 142-144°. $[\alpha]_D^{17.5}$ +38.3° (35 min) → +45.9° (c = 0.52 in water); +22.8° (24 hrs) → +80.8° (c = 0.57 in pyridine).

α-Form. Cryst from isopropanol, mp 134-136°. $[\alpha]_D^{26}$ +156° → +103° (c = 0.9 in pyridine).

USE: Exptlly as an antiviral agent.

2906. 1-Deoxynojirimycin. [19130-96-2] (2R,3R,4R,5S)-2-(Hydroxymethyl)-3,4,5-piperidinetriol; D-5-amino-1,5-dideoxyglucopyranose; 1,5-dideoxy-1,5-imino-D-glucitol; (2R,3R,4R,5S)-2-hydroxymethyl-3,4,5-trihydroxypiperidine; moranoline; Bay n 5595; S-GI. $C_6H_{13}NO_4$; mol wt 163.17. C 44.17%, H 8.03%, N 8.58%, O 39.22%. α-Glucosidase inhibitor. Prepd by reduction of nojirimycin: S. Inouye *et al., J. Antibiot.* **19A**, 290 (1966); S. Inouye *et al., Tetrahedron* **23**, 2125 (1968). Isoln from mulberry, *Morus* spp. (*Moraceae*): M. Yagi *et al., Nippon Nogei Kagaku Kaishi* **50**, 571 (1976), *C.A.* **86**, 167851r (1977). Production by *Bacillus subtilis* DSM704: D. C. Stein *et al., Appl. Environ. Microbiol.* **48**, 280 (1984). Total synthesis: H. Setoi *et al., Chem. Pharm. Bull.* **34**, 2642 (1986); H. J. G. Broxterman *et al., Rec. Trav. Chim.* **106**, 571 (1987). HPLC determn: M. D. Cole *et al., J. Chromatogr.* **445**, 295 (1988). Inhibition of α-glucosidase: Y. Yoshikuni, *Agric. Biol. Chem.* **52**, 121 (1988). Antiviral activity: P. S. Sunkara *et al., Biochem. Biophys. Res. Commun.* **148**, 206 (1987). Inhibition of glycoprotein synthesis and secretion: N. Peyrieras *et al., EMBO J.* **2**, 823 (1983); V. Gross *et al., Biochem. J.* **236**, 853 (1986); M. Bollen *et al., Biochem. Pharmacol.* **37**, 905 (1988).

Prisms from water/ethanol, mp 195°. $[\alpha]_D^{21}$ +47° (c = 1.045 in water); $[\alpha]_D^{20}$ +46.7° (c = 0.2 in water); $[\alpha]_D^{20}$ +36° (c = 0.02 in water). pKa 6.6.

2907. Deoxyribonuclease I. [9003-98-9] Deoxyribonuclease; deoxyribonuclease (pancreatic); pancreatic dornase; pancreatic desoxyribonuclease; DNase I. Endonuclease enzyme that hydrolyzes DNA; produced by the mammalian pancreas. Variably glycosylated, single polypeptide chain with two disulfide bridges; multiple isoforms exist. The human form consists of 260 amino acid residues; mol wt ~37 kDa. Prepn from beef pancreas: M. Kunitz, *Science* **108**, 19 (1948); Baumgarten, Johnson, US 2801956; *eidem*, US 3042587 (1957, 1962 both to Merck & Co.). Properties of purified DNase I: Price *et al., J. Biol. Chem.* **244**, 917 (1969). Reviews of purification and characterization: M. Laskowski in *The Enzymes* vol. **4**, P. D. Boyer, Ed. (Academic Press, New York, 3rd ed., 1971) pp 289-311; S. Moore, *ibid.* vol. **14** 281-296 (1981); T.-H. Liao, *Mol. Cell. Biochem.* **34**, 15-22 (1981).

Bovine pancreatic deoxyribonuclease. Deanase; Dinase; Dornavac. Mixture of 4 isoforms. DNAse A is the most predominant. Contains 257 amino acid residues; mol wt 31 kDa. pI ~5.

Dornase alfa. [143831-71-4] Deoxyribonuclease (human clone 18-1 protein moiety); rhDNase I; Pulmozyme. Human DNase I produced in Chinese hamster ovary cells by recombinant DNA technology. Cloning and expression of human form and effect on sputum viscosity: S. Shak *et al., Proc. Natl. Acad. Sci. USA* **87**, 9188 (1990); S. Shak, WO 9007572 (1990 to Genentech). HPLC purification: J. Cacia *et al., J. Chromatogr.* **634**, 229 (1993). Clinical study in cystic fibrosis: H. J. Fuchs *et al., N. Engl. J. Med.* **331**, 637 (1994). Expert report on pharmacology and toxicology: J. D. Green, *Hum. Exp. Toxicol.* **13**, Suppl. 1, S1-S42 (1994).

THERAP CAT: Debriding agent. In treatment of cystic fibrosis and other pulmonary disease to reduce viscosity of sputum.

THERAP CAT (VET): Debriding agent.

2908. Deoxyribonucleic Acid. Desoxyribonucleic acid; DNA; thymus nucleic acid; Desoxiribon; Eucytol. Polynucleotide; essential component of chromosomes in cell nuclei. In its role as the carrier of genetic information, DNA must have two functions: be exactly reproducible in order to transmit its genetic information to future generations; contain information, in chemical code, to direct the development of the cell according to its inheritance. Reviews of biological function: Hotchkiss in *The Nucleic Acids* vol. **2**, E. Chargaff, J. N. Davidson, Eds. (Academic Press, New York, 1955) pp 435-473; Crick, *Nature* **227**, 561 (1970); J. N. Davidson, *The Biochemistry of Nucleic Acids* (Academic Press, New York, 7th ed., 1972) pp 6-28. The purine and pyrimidine bases of the nucleosides are primarily adenine, guanine, cytosine and thymine; the sugar is D-2-deoxyribose, *q.q.v.* The nucleosides are linked together by phosphates in diester linkage from the 3'-hydroxyl of one sugar to the 5'-hydroxyl of the next. The repeating sugar-phosphate linkage forms the backbone of the single polynucleotide strand which is the primary structure of DNA. Chemical analyses of DNA from different species show that the purine content is equal to the pyrimidine content; adenine content equal to thymine; guanine equal to cytosine: Chargaff, *Experientia* **6**, 201 (1950); *idem, Fed. Proc.* **10**, 654 (1951). In the Watson-Crick model of its secondary structure (based on chemical analysis and x-ray studies), DNA consists of two polynucleotide chains forming right-handed helices coiled around the same axis with the sequence of atoms in the two sugar-phosphate backbones running in opposite direction. Two major families of right-handed helix were proposed. *A-DNA* and *B-DNA*, each having its own intrinsic restrictions on chain-folding and structure. B-DNA is believed to be the predominant form in biological systems. The purine and pyrimidine bases are inside the helical structure formed by the sugar phosphate backbones; those on one chain form hydrogen bonds to those on the other. Adenine in one chain is always bonded to thymine in the complementary chain by hydrogen bonds; similarly guanine is bonded to cytosine. The linear sequence of bases in one strand completely determines the sequence in the complementary strand. Thus each strand can serve as a template for the replication of the original DNA molecule: Watson, Crick, *Nature* **171**, 737, 964 (1953). X-ray studies: Wilkins *et al., ibid.* 738; Marvin *et al., J. Mol. Biol.* **3**, 547 (1961); Fuller *et al., ibid.* **12**, 60 (1965). DNA also acts as a template in the formation of ribonucleic acids, *q.v.*, which play a fundamental role in the synthesis of proteins in the cell. Another form of DNA, termed *Z-DNA*, is also known. Its structure is an antiparallel double helix with Watson-Crick base pairing, but it is a left-handed helix with the ribose-phosphate backbone following a zig-zag course. Molecular structure, atomic resolution x-ray crystallographic analysis: A. H.-J. Wang *et al., Nature* **282**, 680 (1979). First identification of Z-DNA in material of biological origin: A. Nordheim *et al., ibid.* **294**, 417 (1981). Studies of B- and Z-DNA: D. J. Patel *et al., Proc. Natl. Acad. Sci. USA* **79**, 1413

(1982). Comparison of A-, B-, and Z-DNA: R. E. Dickerson *et al.*, *Science* **216**, 475 (1982). Demonstration of Z-DNA immunoreactivity in rat tissues: G. Morgenegg *et al.*, *Nature* **309**, 540 (1983).

2909. D-2-Deoxyribose. [533-67-5] 2-Deoxy-D-*erythro*-pentose; desoxyribose; D-2-deoxyarabinose; D-2-ribodesose; D-*erythro*-2-deoxypentose; thyminose. $C_5H_{10}O_4$; mol wt 134.13. C 44.77%, H 7.52%, O 47.71%. Isoln from deoxyribonucleic acid by acidic hydrolysis of purine deoxyribonucleosides which have been isolated by ion-exchange resin chromatography: Laland, Overend, *Acta Chem. Scand.* **8**, 192 (1954). Synthesis: Felton, Freudenberg, *J. Am. Chem. Soc.* **57**, 1637 (1935); Deriaz *et al.*, *J. Chem. Soc.* **1949**, 1879, 2836; Hough, *Chem. Ind. (London)* **1951**, 406; Sowden, *Biochem. Prep.* **5**, 75 (1957); I. Ziderman, E. Dimant, *J. Org. Chem.* **32**, 1267 (1967); J. R. Hauske, H. Rapoport, *ibid.* **44**, 2472 (1979); T. Harada, T. Mukaiyama, *Chem. Lett.* **1981**, 1109. *Review:* Overend, Stacey, in Chargaff-Davidson, *Nucleic Acids* vol. **1**, E. Chargaff, N. J. Davidson, Eds. (Academic Press, New York, 1955) pp 1-80.

Crystals from isopropanol, mp 91°. Shows mutarotation. Final $[\alpha]_D^{22}$ −56.2° (H_2O). Sol in water, pyridine. Slightly sol in alc.
1,3,4-Triacetate. $C_{11}H_{16}O_7$. Needles from methanol, mp 98°. $[\alpha]_D^{23}$ −171.8° (c = 0.56 in chloroform): Allerton, Overend, *J. Chem. Soc.* **1951**, 1480.
3,4,5-Triacetate. Oily liq, $bp_{0.001}$ 105°. $[\alpha]_D^{21}$ +3.4° (c = 4.57 in pyridine): Zinner *et al.*, *Ber.* **90**, 2696 (1957).
1,3,4-Tribenzoate. Small white nodules from ethanol, mp 127°. $[\alpha]_D^{23}$ −65° (c = 1.02 in chloroform) (Allerton, Overend). Probably a mixture of the two anomeric 2-deoxy-D-ribopyranose tribenzoates: Pedersen *et al.*, *J. Am. Chem. Soc.* **82**, 3425 (1960).
3,4,5-Tribenzoate. Fine needles from ethyl acetate + petr ether, mp 118-119°. $[\alpha]_D^{18}$ −2.8° (c = 1.44 in pyridine) (Zinner).

2910. 2-Deoxystreptamine. [2037-48-1] 2-Deoxy-1,3-*myo*-inosadiamine; 1,3-diamino-4,5,6-trihydroxycyclohexane. $C_6H_{14}N_2O_3$; mol wt 162.19. C 44.43%, H 8.70%, N 17.27%, O 29.59%. Important component of several aminocyclitol antibiotics. Production by acid hydrolysis of neamine, *q.v.* and structure: F. A. Kuehl *et al.*, *J. Am. Chem. Soc.* **73**, 881 (1952). Also obtained from hydrolysis products of kanamycins, paromomycins, gentamicins, *q.q.v.* and other aminoglycoside antibiotics. *See:* K. L. Rinehart, *J. Infect. Dis.* **119**, 345 (1969); S. Hanessian, T. H. Haskell, *The Carbohydrates* vol. **IIA**, W. Pigman, D. Horton, Eds. (Academic Press, New York, 2nd ed., 1970) pp 159-172. Configuration: J. Daly *et al.*, *J. Am. Chem. Soc.* **82**, 5928 (1960); H. E. Carter *et al.*, *ibid.* **83**, 3723 (1961); R. U. Lemieux, R. J. Cushley, *Can. J. Chem.* **41**, 858 (1963). Synthesis: M. Nakajima *et al.*, *Tetrahedron Lett.* **1964**, 967; T. Suami *et al.*, *ibid.* **1967**, 2671; S. Ogawa *et al.*, *J. Org. Chem.* **39**, 812 (1974); H. Prinzbach *et al.*, *Angew. Chem. Int. Ed.* **14**, 225

(1975); S. Ogawa *et al.*, *J. Org. Chem.* **42**, 3083 (1977). Review of syntheses: G. F. Busscher *et al.*, *Chem. Rev.* **105**, 775-791 (2005).

Colorless crystals from ethanol, mp 225-228°.
Dihydrochloride. [14429-30-2] $C_6H_{14}N_2O_3$.2HCl. Crystals, mp 325° (dec), H. Hitomi *et al.*, *Chem. Pharm. Bull.* **9**, 340 (1961).

2911. Deoxyuridine. [951-78-0] 2′-Deoxyuridine; 1-(2-deoxy-*β*-D-*erythro*-pentofuranosyl)uracil; 1-(2-deoxy-*β*-D-ribofuranosyl)uracil; uracil deoxyriboside. $C_9H_{12}N_2O_5$; mol wt 228.20. C 47.37%, H 5.30%, N 12.28%, O 35.05%. Prepn: Dekker, Todd, *Nature* **166**, 557 (1950); Brown *et al.*, *J. Chem. Soc.* **1958**, 3035; Prystas *et al.*, *Collect. Czech. Chem. Commun.* **28**, 3140 (1963); Smejkal *et al.*, *ibid.* **31**, 291 (1966). Conformation: Rahman, Wilson, *Nature* **232**, 333 (1971).

Needles from abs alc or 95% alc, mp 163° (Dekker, Todd) and 167° (Brown *et al.*). $[\alpha]_D^{22}$ +50° (c = 1.1 in *N* NaOH).
α-Anomer. 1-(2-Deoxy-α-D-*erythro*-pentofuranosyl)uracil; 1-(2-deoxy-α-D-ribofuranosyl)uracil. Prepn: Prystas *et al.*, *loc. cit.*

2912. Depreotide. [161982-62-3] Cyclo(L-homocysteinyl-*N*-methyl-L-phenylalanyl-L-tyrosyl-D-tryptophyl-L-lysyl-L-valyl), (1 → 1′)-thioether with 3-[(mercaptoacetyl)amino]-L-alanyl-L-lysyl-L-cysteinyl-L-lysinamide; cyclo-(*N*-Me)Phe-Tyr-(D-Trp)-Lys-Val-Hcy(CH_2CO-(*β*-Dap)-Lys-Cys-Lys-NH_2); P829. $C_{65}H_{96}N_{16}$-$O_{12}S_2$; mol wt 1357.70. C 57.50%, H 7.13%, N 16.51%, O 14.14%, S 4.72%. Synthetic cyclic peptide of 10 amino acids; somatostatin analog designed to bind to somatostatin receptor-bearing tumors. ^{99m}Tc-Labeled complex functions as scintigraphic imaging agent. Prepn: R. T. Dean *et al.*, *WO 9500553* (1995 to Diatech); *eidem*, *US 6051206* (2000 to Diatide); D. A. Pearson *et al.*, *J. Med. Chem.* **39**, 1361 (1996). Binding studies: I. Virgolini *et al.*, *Cancer Res.* **58**, 1850 (1998). Clinical study in diagnosis of pulmonary nodules: J. Blum *et al.*, *Chest* **117**, 1232 (2000).

Complex with ^{99m}Tc. Technetium Tc 99m depreotide; ^{99m}Tc-P829; NeoTect.
THERAP CAT: ^{99m}Tc complex as diagnostic aid (radioactive imaging agent).

2913. Deptropine. [604-51-3] (3-*endo*)-3-[(10,11-Dihydro-5*H*-dibenzo[*a,d*]cyclohepten-5-yl)oxy]-8-methyl-8-azabicyclo-[3.2.1]octane; 3α-[(10,11-dihydro-5*H*-dibenzo[*a,d*]cyclohepten-5-

yl)oxy]-1αH,5αH-tropane; dibenzheptropine. $C_{23}H_{27}NO$; mol wt 333.48. C 82.84%, H 8.16%, N 4.20%, O 4.80%. Prepn: Van der Stelt *et al.*, *J. Med. Pharm. Chem.* **4**, 335 (1961). Pharmacology: Funcke *et al.*, *Arch. Int. Pharmacodyn.* **148**, 135 (1964); Timmerman *et al.*, *ibid.* **187**, 291 (1970). Metabolism: Hespe *et al.*, *ibid.* **164**, 397 (1966).

Citrate. [2169-75-7] BS-6987; Brontine. $C_{23}H_{27}NO.C_6H_8O_7$; mol wt 525.60. LD_{50} in mice (mg/kg): 32 i.v.; 300 orally (Timmerman).

Maleate. $C_{23}H_{27}NO.C_4H_4O_4$. Crystals from ethanol, acetone, ethyl acetate or chloroform, mp 133-136°.

THERAP CAT: Antihistaminic.

2914. Dequalinium Chloride. [522-51-0] 1,1′-(1,10-Decanediyl)bis[4-amino-2-methylquinolinium chloride (1:2)]; 1,1′-decamethylenebis[4-aminoquinaldinium chloride]; BAQD-10; decamine; dekamin; Decatylen; Dekadin; Dequadin Chloride; Dequafungan; Dequavet; Dequavagyn; Eriosept; Evazol; Grocreme; Labosept; Optipect; Phylletten; Polycidine; Sorot. $C_{30}H_{40}Cl_2N_4$; mol wt 527.58. C 68.30%, H 7.64%, Cl 13.44%, N 10.62%. Prepn: Taylor *et al.*, **GB 745956** (1956 to Allen & Hanburys), *C.A.* **50**, 16878e (1956); Austin *et al.*, *J. Chem. Soc.* **1958**, 1489.

Crystals from ethanol, mp 326° (dec). Soly in water (25°): about one g/200 ml.

Ingredient of *Efisol, Gargilon, Gramipan, Hexalyse, Micrin*.

USE: Bacteriostat.

THERAP CAT: Antiseptic, disinfectant.

THERAP CAT (VET): Antimicrobial.

2915. Deracoxib. [169590-41-4] 4-[3-(Difluoromethyl)-5-(3-fluoro-4-methoxyphenyl)-1H-pyrazol-1-yl]benzenesulfonamide; SC-046; SC-59046; Deramaxx. $C_{17}H_{14}F_3N_3O_3S$; mol wt 397.37. C 51.38%, H 3.55%, F 14.34%, N 10.57%, O 12.08%, S 8.07%. Selective cyclooxygenase-2 (COX-2) inhibitor. Prepn: J. J. Talley *et al.*, **WO 9515316**; M. J. Graneto, **US 5521207** (1995, 1996 both to Searle); T. D. Penning *et al.*, *J. Med. Chem.* **40**, 1347 (1997). Structure activity study: G. R. Desiraju *et al.*, *Molecules* **5**, 945 (2000). Effect in experimental synovitis in dogs: D. L. Millis *et al*, *Vet. Ther.* **3**, 453 (2002); HPLC determn in plasma: S. K. Cox *et al.*, *J. Chromatogr. B* **819**, 181 (2005).

Light tan solid, mp 159-161°.

THERAP CAT (VET): Anti-inflammatory; analgesic.

2916. Deramciclane. [120444-71-5] *N,N*-Dimethyl-2-[[(1R,-2S,4R)-1,7,7-trimethyl-2-phenylbicyclo[2.2.1]hept-2-yl]oxy]ethanamine; (−)-[1R,2S,4R]-2-(2-dimethylaminoethoxy)-2-phenyl-1,7,7-trimethylbicyclo[2.2.1]heptane. $C_{20}H_{31}NO$; mol wt 301.47. C 79.68%, H 10.37%, N 4.65%, O 5.31%. Serotonin (5-HT$_2$)-receptor antagonist. Prepn: **BE 886579**; Z. Budai *et al.*, **US 4342762** (1981, 1982 both to Egyt); L. Ladanyi *et al.*, *Chirality* **11**, 689 (1999). LC/MS/MS determn in plasma: A. Tolokán *et al.*, *J. Chromatogr. A* **896**, 279 (2000). Receptor binding profile and pharmacology: I. Gacsalyi *et al.*, *Drug Dev. Res.* **40**, 333 (1997). Clinical pharmacokinetics and tolerability: H. Kanerva *et al.*, *Int. J. Clin. Pharmacol. Ther.* **37**, 589 (1999). Clinical evaluation in generalized anxiety disorder: H. Naukkarinen *et al.*, *Eur. Neuropsychopharmacol.* **15**, 617 (2005). *Review*: S. Koks, E. Vasar, *Curr. Opin. Investig. Drugs* **3**, 289-294 (2002).

Colorless oil. Log P (octanol/water): 5.9. Soly in water (25°): 0.0088 g/100 ml.

Fumarate. [120444-72-6] EGIS-3886. $C_{20}H_{31}NO.C_4H_4O_4$; mol wt 417.55. Crystals from DMF, mp 210-213°. $[\alpha]_{436}^{20}$ −88.0° ± 1 (c = 0.4 in DMSO). Log P (octanol/water): 1.41. pKa 9.61.

THERAP CAT: Anxiolytic.

2917. Derris Root. Tuba root; Deguelia root. The roots of plants belonging to the genus *Derris*, of which more than 80 species have been described. *Derris elliptica* (Wall.) Benth. and *D. malaccensis* Prain, *Leguminosae* are cultivated in British Malaya and the Dutch East Indies. Toxicity data: Haag *et al.*, *Proc. Soc. Exp. Biol. Med.* **54**, 140 (1943).

LD_{50} orally in mice: 350 mg/kg (as the dry powdered root (Haag)).

USE: As insecticide; as a source of rotenone and rotenoid compds.

2918. Desaspidin BB. [114-43-2] 2-[[2,4-Dihydroxy-6-methoxy-3-(1-oxobutyl)phenyl]methyl]-3,5-dihydroxy-4,4-dimethyl-6-(1-oxobutyl)-2,5-cyclohexadien-1-one; 3′-[(5-butyryl-2,4-dihydroxy-3,3-dimethyl-6-oxo-1,4-cyclohexadien-1-yl)methyl]-2′,6′-dihydroxy-4′-methoxybutyrophenone; desaspidin; rosapin. $C_{24}H_{30}O_8$; mol wt 446.50. C 64.56%, H 6.77%, O 28.67%. Isoln from the rhizome of *Dryopteris austriaca* (Jacq.) Wojnar, *Polypodiaceae*: Aebi *et al.*, *Helv. Chim. Acta* **40**, 266 (1957); from *D. caucasica* (A. Br.) Fraser-Jenkins et Corley: Widén *et al.*, *ibid.* **56**, 831 (1973). Structure: Aebi *et al.*, *ibid.* **40**, 572 (1957). Toxicity study: Airaksinan *et al.*, *Acta Pharmacol. Toxicol.* **25**, 33 (1967).

Crystals from ether + petr ether, mp 150-150.5°. uv max: 230, 274 nm (ε 23000, 16800) in cyclohexane. Freely sol in ether, benzene, acetone. Practically insol in methanol, ethanol, petr ether. LD_{50} orally in mice: 340 mg/kg (Airaksinan).

2919. Deserpidine. [131-01-1] (3β,16β,17α,18β,20α)-17-Methoxy-18-[(3,4,5-trimethoxybenzoyl)oxy]yohimban-16-carboxylic acid methyl ester; 11-desmethoxyreserpine; canescine; recanescine; Harmonyl; Raunormine. $C_{32}H_{38}N_2O_8$; mol wt 578.66. C 66.42%, H 6.62%, N 4.84%, O 22.12%. Isoln from roots of *Rauwolfia canescens* L., *Apocynaceae:* A. Stoll, A. Hofmann, *J. Am. Chem. Soc.* **77**, 820 (1955); M. W. Klohs *et al., ibid.* 4084; N. Neuss *et al., ibid.* 4087; H. B. MacPhillamy *et al., ibid.* 4335; E. Schlittler *et al., Experientia* **11**, 64 (1955); **GB 791241** (1958 to Penick); **GB 809912**; P. R. Ulshafer, **US 2982769** (1959, 1961 both to Ciba). Stereochemistry: C. F. Huebner *et al., Experientia* **11**, 303 (1955); P. E. Aldrich *et al., J. Am. Chem. Soc.* **81**, 2481 (1959). Synthesis: L. Bláha *et al., Collect. Czech. Chem. Commun.* **25**, 237 (1960); **CA 678216** (1964 to Roussel-UCLAF).

Exists in three cryst forms from methanol: α-form, mp 228-232°; β-form, mp 230-232°; and γ-form, mp 138° and 226-232° with resolidification at 175°. $[\alpha]_D^{20} -163°$ (c = 0.5 in pyridine). pKa 6.68 in 40% methanol. uv max (ethanol): 218, 272, 290 nm (log ε 4.79, 4.26, 4.07).

Hydrochloride. [6033-69-8] $C_{32}H_{38}N_2O_8 \cdot HCl$. Thin rectangular plates from acetone, dec 253-256°.

Nitrate. $C_{32}H_{38}N_2O_8 \cdot HNO_3$. Crystals, dec 254-260°.

Oxalate. [119506-01-3] $C_{32}H_{38}N_2O_8 \cdot C_2H_2O_4$. Crystals, dec 239-243°.

10-Methoxydeserpidine. [865-04-3] methoserpidine; Decaserpyl. $C_{33}H_{40}N_2O_9$; mol wt 608.69.

THERAP CAT: Antihypertensive.

2920. Desflurane. [57041-67-5] 2-(Difluoromethoxy)-1,1,1,2-tetrafluoroethane; (±)-2-difluoromethyl 1,2,2,2-tetrafluoroethyl ether; I-653; Suprane. $C_3H_2F_6O$; mol wt 168.04. C 21.44%, H 1.20%, F 67.84%, O 9.52%. Prepd but not claimed: J. P. Russell *et al.,* **US 3897502** (1975 to Airco). Prepn and use as anesthetic: R. C. Terrell, **US 4762856** (1988 to BOC). Series of articles on pharmacology and physical properties: E. I. Eger II *et al., Anesth. Analg.* **66**, 971-985 (1987); and toxicity studies: *eidem, ibid.* 1227, 1230. Series of articles on clinical pharmacology and kinetics: *Anesthesiology* **74**, 412-439; 479-498. *Review:* R. M. Jones, *Br. J. Anaesth.* **65**, 527-536 (1990).

Volatile liquid. Slight non-pungent odor. Nonflammable, soda lime stable. bp 23.5°. d 1.44. Vapor pressure (20°): 88.53 kPa. Vapor pressure (22-23°): ~700mm. Partition coefficient at 37° (blood/gas): 0.424 ±0.024; (saline/gas): 0.225 ±0.002; (oil/gas): 18.7 ±1.1.

THERAP CAT: Anesthetic (inhalation).

2921. Desipramine. [50-47-5] 10,11-Dihydro-*N*-methyl-5*H*-dibenz[*b,f*]azepine-5-propanamine; 10,11-dihydro-5-[3-(methylamino)propyl]-5*H*-dibenz[*b,f*]azepine; 5-(γ-methylaminopropyl)iminodibenzyl; *N*-(3-methylaminopropyl)iminobibenzyl; desmethylimipramine; norimipramine. $C_{18}H_{22}N_2$; mol wt 266.39. C 81.16%, H 8.32%, N 10.52%. Prepn: **GB 908788; BE 614616** (both 1962 to Geigy); J. H. Biel, C. I. Judd, **GB 980231** (1965 to Lakeside); *eidem,* **US 3454554** (1969 to Colgate Palmolive). Chemistry, pharmacology: E. Eriksoo, O. Rohte, *Arzneim.-Forsch.* **20**, 1561 (1970). Clinical response, plasma levels, pharmacokinetics: P. D. Hrdina, Y. D. Lapierre, *Prog. Neuro-Psychopharmacol.* **4**, 591

(1980). Mechanism of action study: S. A. Checkley *et al., Br. J. Psychiatry* **138**, 248 (1981). Efficacy in depression: W. Z. Potter *et al., Psychopharmacol. Bull.* **17**, 26 (1981); J. W. Stewart *et al., ibid.* 136. Teratological study: L. Aeppli, *Arzneim.-Forsch.* **19**, 1617 (1969). Carcinogenicity and mutagenicity study: H. Kubinski *et al., Mutat. Res.* **89**, 95 (1981). Has been found to be one of the most effective cmpds described for *in vitro* and *in vivo* reversal of chloroquine, *q.v.,* resistance in *Plasmodium falciparum:* A. J. Bitonti *et al., Science* **242**, 1301 (1988).

bp$_{0.02}$ 172-174°. uv max: 213, 252 nm (log ε 4.39, 3.93).

Hydrochloride. [58-28-6] G-35020; JB-8181; NSC-114901; Norpramin; Nortimil; Pertofran; Pertofrane; Petylyl. $C_{18}H_{22}N_2 \cdot$ HCl; mol wt 302.85. White to off-white crystals from methanol ether, mp 215-216°. Freely sol in methanol, chloroform; sol in water, alcohol. Insol in ether. LD$_{50}$ in mice, rats (mg/kg): 500, 385 orally; 94, 48 i.p.; 420, 183 s.c. (Eriksoo, Rohte).

THERAP CAT: Antidepressant.

2922. Deslanoside. [17598-65-1] (3β,5β,12β)-3-[(*O*-β-D-Glucopyranosyl-(1 → 4)-*O*-2,6-dideoxy-β-D-*ribo*-hexopyranosyl-(1 → 4)-*O*-2,6-dideoxy-β-D-*ribo*-hexopyranosyl-(1 → 4)-2,6-dideoxy-β-D-*ribo*-hexopyranosyl)oxy]-12,14-dihydroxycard-20(22)-enolide; deacetyllanatoside C; desacetyldigilanide C; Cedilanide; Desace; Desaci. $C_{47}H_{74}O_{19}$; mol wt 943.09. C 59.86%, H 7.91%, O 32.23%. Cardiac glycoside. Isoln from leaves of *Digitalis lanata* Ehrh., *Scrophulariaceae:* Stoll, Kreis, *Helv. Chim. Acta* **16**, 1049, 1390 (1933). Prepn by alkaline degradation of lanatoside C: Kroszczynski *et al., Acta Pol. Pharm.* **21**, 357 (1964), *C.A.* **62**, 10292b (1965). Hemodynamic effects in heart failure: E. Ambrosioni *et al., Int. J. Clin. Pharmacol. Biopharm.* **17**, 416 (1979). Clinical pharmacokinetics: A. Marzo *et al., Farmaco Ed. Prat.* **37**, 28 (1982). Determn in blood and urine by HPLC/MS/MS: F. Guan *et al., Anal. Chem.* **71**, 4034 (1999).

β-D-glucose-(β-D-digitoxose)$_3$-O

Crystals from methanol, dec 265-268°. $[\alpha]_D^{20}$ +12° (c = 1.084 in 75% alc). One part dissolves in 5000 parts water, 200 parts methanol and 2500 parts ethanol; very slightly sol in chloroform. Practically insol in ether.

THERAP CAT: Cardiotonic.

2923. Desloratadine. [100643-71-8] 8-Chloro-6,11-dihydro-11-(4-piperidinylidene)-5*H*-benzo[5,6]cyclohepta[1,2-*b*]pyridine; descarboethoxyloratadine; Sch-34117; Clarinex. $C_{19}H_{19}ClN_2$; mol wt 310.83. C 73.42%, H 6.16%, Cl 11.40%, N 9.01%. Nonsedating-type histamine H$_1$-receptor antagonist; active metabolite of loratadine, *q.v.* Also inhibits generation and release of inflammatory mediators from basophils and mast cells. Prepn: D. P. Schumacher *et al., EP 208855*; F. J. Villani, J. K. Wong, **US 4659716** (both 1987 to Schering). GLC determn in plasma: R. Johnson *et al., J. Chromatogr. B* **657**, 125 (1994). Pharmacology: W. Kreutner *et al., Arzneim.-Forsch.* **50**, 345 (2000). Series of articles on clinical efficacy in rhinitis, asthma and urticaria: *Allergy* **56**, Suppl. 65, 1-32 (2001). Review of pharmacology, pharmacokinetics and clinical efficacy: R.

S. Geha, E. O. Meltzer, *J. Allergy Clin. Immunol.* **107**, 751-762 (2001); L. M. DuBuske, *Expert Opin. Pharmacother.* **6**, 2511-2523 (2005).

Crystals from hexane, mp 150-151°. Slightly sol in water; very sol in ethanol, propylene glycol.

THERAP CAT: Antihistaminic.

2924. Deslorelin. [57773-65-6] 6-D-Tryptophan-9-(*N*-ethyl-L-prolinamide)-1-9-luteinizing hormone-releasing factor (swine); [D-Trp^6,des-Gly^{10}]-LH-RH ethylamide. C$_{64}$H$_{83}$N$_{17}$O$_{12}$; mol wt 1282.48. C 59.94%, H 6.52%, N 18.57%, O 14.97%. Synthetic nonapeptide agonist analog of LH-RH, *q.v.* Prepn: D. H. Coy *et al.*, *J. Med. Chem.* **19**, 423 (1976); W. Kornreich *et al.*, *Int. J. Pept. Protein Res.* **25**, 414 (1985). Pharmacology: K. Sundaram *et al.*, *Life Sci.* **28**, 83 (1981). Pharmacokinetics and metabolism: B. Candas *et al.*, *J. Clin. Endocrinol. Metab.* **70**, 1046 (1990). Clinical inhibition of spermatogenesis: R. Linde *et al.*, *N. Engl. J. Med.* **305**, 663 (1981). Clinical evaluation in precocious puberty: J. A. Yanovski *et al.*, *N. Engl. J. Med.* **348**, 908 (2003). Induction of ovulation in mares: V. J. Farquhar *et al.*, *J. Equine Vet. Sci.* **20**, 722 (2000). Review of veterinary use in control of reproduction: T. E. Trigg *et al.*, *Theriogenology* **66**, 1507-1512 (2006).

5-oxoPro–His–Trp–Ser–Tyr–D-Trp–Leu–Arg–ProNHCH$_2$CH$_3$

$[\alpha]_D^{24}$ −61° (c = 0.37 in 0.1*M* acetic acid).

Acetate. [82318-06-7] Ovuplant; Suprelorin. C$_{64}$H$_{83}$N$_{17}$O$_{12}$·C$_2$H$_4$O$_2$; mol wt 1342.53.

THERAP CAT (VET): For induction of ovulation in mares; canine male contraceptive.

2925. Desmedipham. [13684-56-5] *N*-[3-[[(Phenylamino)-carbonyl]oxy]phenyl]carbamic acid ethyl ester; ethyl *m*-hydroxycarbanilate carbanilate; ethyl 3-phenylcarbamoyloxycarbanilate; EP-475; SN-38107; Betanex. C$_{16}$H$_{16}$N$_2$O$_4$; mol wt 300.31. C 63.99%, H 5.37%, N 9.33%, O 21.31%. Selective, postemergence herbicide for broadleaf weed control in crops. Analog of phenmedipham, *q.v.*; inhibits photosynthetic electron transport at the photosystem II (PS II) receptor site. Prepn: **GB 1127050** (1966 to Schering AG); and determn methods: C.-H. Röder *et al.*, *Anal. Methods Pestic. Plant Growth Regul.* **10**, 293 (1978). Microbial degradation study: C. O. Knowles, H, J, Benezet, *Bull. Environ. Contam. Toxicol.* **27**, 529 (1981). Field studies in sunflowers: M. D. Anderson, W. E. Arnold, *Weed Sci.* **32**, 310 (1984); in sugar beets: F. Abdollahi, H. Ghadiri, *Weed Technol.* **18**, 968 (2004). Efficacy and metabolism: M. Abbaspoor, J. C. Streibig, *Pest Manag. Sci.* **63**, 576 (2007). LC/MS determn in soil: D. Perret *et al.*, *J. AOAC Int.* **84**, 1407 (2001); in foods: J. Wang, D. Leung, *ibid.* **92**, 279 (2009).

Colorless crystals, mp 120°. Vapor pressure (25°): 3 × 10^{-9} Torr. Readily sol in acetone, methanol, chloroform; less sol in benzene. Soly in water: 7 ppm.

USE: Herbicide.

2926. Desmopressin. [16679-58-6] *N*-(3-Mercapto-1-oxopropyl)-L-tyrosyl-L-phenylalanyl-L-glutaminyl-L-asparaginyl-L-cysteinyl-L-prolyl-D-arginylglycinamide cyclic (1 → 5)-disulfide; 1-(3-mercaptopropanoic acid)-8-D-arginine vasopressin; 8-D-arginine-1-(3-mercaptopropanoic acid) vasopressin; 1-desamino-8-D-arginine vasopressin. C$_{46}$H$_{64}$N$_{14}$O$_{12}$S$_2$; mol wt 1069.22. C 51.67%, H

6.03%, N 18.34%, O 17.96%, S 6.00%. Synthetic analog of vasopressin, *q.v.*, with antidiuretic activity and decreased pressor effects. Improves hemostasis by increasing plasma levels of von Willebrand factor, factor VIII and tissue plasminogen activator, *q.q.v.* Prepn: M. Zaoral *et al.*, **FR 1540536**; *eidem*, **US 3497491** (1968, 1970 both to Ceskoslovenska Akad. Ved). Syntheses: R. L. Huguenin, R. A. Boissonnas, *Helv. Chim. Acta* **49**, 695 (1966); M. Zaoral *et al.*, *Collect. Czech. Chem. Commun.* **32**, 1250 (1967). Review of clinical experience in bleeding disorders: P. M. Mannucci, *Blood* **90**, 2515-2521 (1997); in treatment of nocturnal enuresis: P. E. V. van Kerrebroeck, *BJU Int.* **89**, 420-425 (2002); of mechanism of hemostatic effects: J. E. Kaufmann, U. M. Vischer, *J. Thromb. Haemost.* **1**, 682-689 (2003).

$[\alpha]_D^{25}$ +85.5 ± 2° (calculated for the free peptide).

Monoacetate. [62288-83-9] Adiuretin SD; DDAVP; Desmospray; Minirin; Octim; Octostim; Presinex; Stimate. C$_{46}$H$_{64}$N$_{14}$O$_{12}$S$_2$.C$_2$H$_4$O$_2$; mol wt 1129.28. Occurs as the trihydrate. White fluffy powder. Sol in water, alc, acetic acid.

THERAP CAT: Antidiuretic; hemostatic.

THERAP CAT (VET): Antidiuretic; hemostatic.

2927. Desmosterol. [313-04-2] (3β)-Cholesta-5,24-dien-3-ol; 24-dehydrocholesterol; desmesterol. C$_{27}$H$_{44}$O; mol wt 384.65. C 84.31%, H 11.53%, O 4.16%. Has been isolated from chick embryos and the skin of rats: Stokes *et al.*, *J. Biol. Chem.* **220**, 415 (1956). From barnacles: Fagerlund, Idler, *J. Am. Chem. Soc.* **79**, 6473 (1957); from red algae: Idler *et al.*, *Steroids* **11**, 465 (1968). Synthesis: Dasgupta *et al.*, *J. Org. Chem.* **39**, 1658 (1974); M. A. Apfel, *ibid.* **43**, 2284 (1978). Stereospecific synthesis: M. Koreeda *et al.*, *ibid.* **45**, 1172 (1980); S. Takano *et al.*, *Chem. Commun.* **1983**, 760.

Platelets from methanol, mp 121.5°, also reported as 119-119.5°. $[\alpha]_D^{27}$ −41.0° (c = 1 in chloroform).

2928. Desogestrel. [54024-22-5] (17α)-13-Ethyl-11-methylene-18,19-dinorpregn-4-en-20-yn-17-ol; 17α-ethynyl-18-methyl-11-methylene-Δ^4-estren-17β-ol; Org-2969; Cerazette. C$_{22}$H$_{30}$O; mol wt 310.48. C 85.11%, H 9.74%, O 5.15%. Progestogen with low androgenic potency. Prepn: A. J. van den Broek, **DE 2361120**; *idem*, **US 3927046** (1974, 1975 both to Akzo); A. J. van den Broek *et al.*, *Rec. Trav. Chim.* **94**, 36 (1975). Biological effects: L. Viinikka *et al.*, *Acta Endocrinol.* **83**, 429 (1976). Endocrinological studies in animals: J. van der Vies, J. de Visser, *Arzneim.-Forsch.* **33**, 231 (1983). Radioimmunoassay: L. Viinikka, *J. Steroid Biochem.* **9**, 979 (1978). Metabolism, pharmacokinetics in humans: L. Viinikka *et al.*, *Eur. J. Clin. Pharmacol.* **15**, 349 (1979). Receptor binding study: E. W. Bergink *et al.*, *J. Steroid Biochem.* **14**, 175 (1981). Clinical trial in combination with ethinyl estradiol: M. J. Weijers, *Clin. Ther.* **4**, 359 (1982); as progestogen-only contraceptive: T. Korver *et al.*, *Eur. J. Contracept. Reprod. Health Care* **3**, 169 (1998).

Crystals, mp 109-110°. $[\alpha]_D^{20}$ +55° (chloroform).

Mixture with ethinyl estradiol. [71138-35-7] Cyclosa; Desogen; Dicromil; Marvelon 150/30; Mercilon; Ortho-Cept; Oviol; Varnoline.

THERAP CAT: Progestogen. Alone or in combination with estrogen as oral contraceptive.

2929. Desonide. [638-94-8] ($11\beta,16\alpha$)-11,21-Dihydroxy-16,17-[(1-methylethylidene)bis(oxy)]pregna-1,4-diene-3,20-dione; $11\beta,16\alpha,17,21$-tetrahydroxypregna-1,4-diene-3,20-dione cyclic 16,17-acetal with acetone; 16α-hydroxyprednisolone-16α,17-acetonide; $11\beta,21$-dihydroxy-16α,17-isopropylidenedioxy-1,4-pregnadiene-3,20-dione; 16α-hydroxy-Δ^1-hydrocortisone-16α,17α-acetonide; 16α,17α-isopropylidenedioxyprednisolone; prednacinolone; D-2083; Locapred; Locatop; Sterax; Steroderm; Topifug; Tridesilon. $C_{24}H_{32}O_6$; mol wt 416.51. C 69.21%, H 7.74%, O 23.05%. Prepn and activity: Bernstein *et al.*, *J. Am. Chem. Soc.* **81**, 4573 (1959); Bernstein, Allen, US 2990401 (1961 to Am. Cyanamid); Lee *et al.* and Diassi, Principe, US 3536586 and US 3549498 (both 1970 to Squibb). Pharmacological and toxicological studies: Mascitelli-Coriandoli, Fraia, *Arzneim.-Forsch.* **20**, 111 (1970); Phillips *et al.*, *Toxicol. Appl. Pharmacol.* **20**, 522 (1971). Structure-activity studies: Ringler *et al.*, *Proc. Soc. Exp. Biol. Med.* **107**, 451 (1961). Chemical and physical properties: Mantica *et al.*, *Arzneim.-Forsch.* **20**, 109 (1970).

Small plates or white to off-white odorless powder, mp 274-275° from methanol (Mantica); also reported as mp 263-266° from ethyl acetate-petr ether (Bernstein). $[\alpha]_D^{25}$ +123° (c = 0.5 in DMF). uv max: 242 nm ($E_{1cm}^{1\%}$ 356). LD_{50} in rats (mg/kg): 93 s.c. (Phillips).

THERAP CAT: Anti-inflammatory.

2930. Desosamine. [5779-39-5] 3,4,6-Trideoxy-3-(dimethylamino)-D-*xylo*-hexose; 4-dimethylaminotetrahydro-6-methylpyran-2,3-diol; picrocine. $C_8H_{17}NO_3$; mol wt 175.23. C 54.84%, H 9.78%, N 7.99%, O 27.39%. Sugar component of several macrolide antibiotics, such as the erythromycins. Isoln: Clark, *Antibiot. Chemother.* **3**, 663 (1953). Identity with picrocine: Brockmann *et al.*, *Ber.* **87**, 856 (1954). Stereochemistry: Woo *et al.*, *Tetrahedron Lett.* **1962**, 735. Configuration: Bolton *et al.*, *Chem. Ind. (London)* **1962**, 1945; Foster *et al.*, *J. Chem. Soc.* **1965**, 2318. Synthesis of DL-form: Korte *et al.*, *Tetrahedron* **18**, 657 (1962); Newman, *Chem. Ind. (London)* **1963**, 372; *J. Org. Chem.* **29**, 1461 (1964). Stereospecific synthesis of the natural D-form: Richardson, *J. Chem. Soc.* **1964**, 5364; of the L-form: Baer, Chiu, *Can. J. Chem.* **52**, 122 (1974).

Crystals, mp 86-87°.

Hydrochloride. $C_8H_{17}NO_3$·HCl. Needles from ethanol + acetone, dec 191-193°. $[\alpha]_D^{20}$ +49.5° (c = 10.0); $[\alpha]_D^{20}$ +53.4° (c = 2.1 in ethanol).

2931. Desoximetasone. [382-67-2] ($11\beta,16\alpha$)-9-Fluoro-11,21-dihydroxy-16-methylpregna-1,4-diene-3,20-dione; 9α-fluoro-16α-methyl-17-desoxyprednisolone; 9α-fluoro-16α-methyl-Δ^1-corticosterone; desoxymethasone; A-41-304; R-2113; HOE-304; Esperson; Stiedex LP; Topicort; Topisolon. $C_{22}H_{29}FO_4$; mol wt 376.47. C 70.19%, H 7.76%, F 5.05%, O 17.00%. Prepn: R. Joly *et al.*, FR 1296544; *eidem*, US 3099654 (1962, 1963 to Roussel-UCLAF); BE

614196; Kieslich *et al.*, US 3232839 (1962, 1966 to Schering AG); R. Joly *et al.*, *Arzneim.-Forsch.* **24**, 1 (1974). Activity studies: Branceni *et al.*, *Steroids* **6**, 451 (1965); Schröder *et al.*, *Arzneim.-Forsch.* **24**, 3 (1974). NMR data: Lukacs *et al.*, *C. R. Seances Acad. Sci. Ser. C* **274**, 1458 (1972). Review of pharmacology and therapeutic efficacy in dermatoses: R. C. Heel *et al.*, *Drugs* **16**, 302-321 (1978).

Crystals from ethyl acetate, mp 217°. $[\alpha]_D$ +109° (chloroform). uv max: 238 nm (ϵ 15750). Sol in alcohol, acetone, chloroform, hot ethyl acetate; slightly sol in ether, benzene. Insol in water, dil aq acids and alkalies.

THERAP CAT: Anti-inflammatory; glucocorticoid.

2932. 6-Desoxy-D-glucosamine. [6018-53-7] 2-Amino-2,6-dideoxy-D-glucose; 6-deoxy-D-glucosamine; 2,6-didesoxy-2-amino-D-glucose. $C_6H_{13}NO_4$; mol wt 163.17. C 44.17%, H 8.03%, N 8.58%, O 39.22%. May be prepd by tosylation of the primary alcohol group of a suitable derivative of D-glucosamine, followed by acetylation and replacement of the tosyl group by iodine, and catalytic cleavage of the halide with Raney nickel in the presence of triethylamine: Morel, *Helv. Chim. Acta* **41**, 1501 (1958).

Hydrochloride. $C_6H_{13}NO_4$·HCl. Crystals from abs ethanol + ether, dec 172-173°. Shows mutarotation. Final value: $[\alpha]_D^{20}$ +55° (H_2O).

1,3,4-Tri-*O*-acetyl-*N*-acetyl-6-desoxy-β-D-glucosamine. C_{14}-$H_{21}NO_8$; mol wt 331.32. Crystals from alc, mp 209-210°. $[\alpha]_D^{20}$ +17.5° (chloroform).

USE: In the investigation of amino acid metabolism.

2933. 11-Desoxy-17-hydroxycorticosterone. [152-58-9] 17,21-Dihydroxypregn-4-ene-3,20-dione; 17-hydroxydesoxycorticosterone; 4-pregnene-17α,21-diol-3,20-dione; 17-(1-keto-2-hydroxyethyl)-4-androsten-17α-ol-3-one; Reichstein's Substance S; 11-desoxycortisone; cortexolone. $C_{21}H_{30}O_4$; mol wt 346.47. C 72.80%, H 8.73%, O 18.47%. Isoln and partial synthesis: Reichstein, von Euw, *Helv. Chim. Acta* **21**, 1197 (1938); Reichstein, *ibid.* 1490; Reichstein, von Euw, *ibid.* **23**, 1258 (1940). Partial synthesis involving the chromic acid oxidation of 4-pregnene-17α,20,21-triol-3-one 21-monoacetate: Sarett, *J. Biol. Chem.* **162**, 627 (1946). Prepn from 3α-formoxy-17α-hydroxypregnan-20-one: Gallagher *et al.*, *J. Am. Chem. Soc.* **71**, 3262 (1949); from 16,17-oxido-5-pregnen-3β-ol-20-one acetate: Julian *et al.*, *ibid.* 3574; from 5-pregnen-3β-ol-20-one: Julian *et al.*, *ibid.* **72**, 5145 (1950).

Fine, glistening plates from ether. mp 212.8-216.8°; gives no depression of melting point when mixed with cortisone. Very sparingly sol in water, ether. Sol in acetone, methanol, alcohol. Gives a carmine-red fluorescence reaction with concd. H_2SO_4. Reduces ammoniacal silver nitrate soln at room temp. uv max: 242 nm ($E_{1cm}^{1\%}$ 500). Oxidation with chromic acid in glacial acetic acid yields 4-androstene-3,17-dione.

Acetate. $C_{23}H_{32}O_5$. mp 237.2-240.2° (sinters at 230°); $[\alpha]_D^{24}$ +116° (acetone); uv max (methanol): 242 nm ($E_{1cm}^{1\%}$ 448). When dissolved in concd sulfuric acid produces a typical scarlet color.

2934. Dess-Martin Periodinane. [87413-09-0] Acetic acid $1,1',1''$-(3-oxo-1λ^5-1,2-benziodoxol-1(3H)-ylidyne) ester; triacetoxyperiodinane; 1,1,1-tris(acetyloxy)-1,1-dihydro-1,2-benziodoxol-3(1H)-one. $C_{13}H_{13}IO_8$; mol wt 424.14. C 36.81%, H 3.09%, I 29.92%, O 30.18%. Pentacoordinate organoiodine (periodinane); selectively oxidizes primary and secondary alcohols, aldehydes and ketones. Prepn: D. B. Dess, J. C. Martin, *J. Org. Chem.* **48**, 4155 (1983); R. E. Ireland, L. Liu, *ibid.* **58**, 2899 (1993). Structure and reaction conditions: D. B. Dess, J. C. Martin, *J. Am. Chem. Soc.* **113**, 7277 (1991); acceleration by water: S. D. Meyer, S. L. Schreiber, *J. Org. Chem.* **59**, 7549 (1994). Synthetic applications: N. E. Jenkins *et al.*, *Synth. Commun.* **30**, 947 (2000); A. G. Myers *et al.*, *Tetrahedron Lett.* **41**, 1359 (2000).

White crystalline solid, mp 134°. Sparingly sol in hexane or ether. Sol in chloroform, methylene chloride, acetonitrile and acetone.
USE: Oxidizing reagent.

2935. Desthiobiotin. [533-48-2] (4R,5S)-5-Methyl-2-oxo-4-imidazolidinehexanoic acid; 5-methyl-2-oxo-4-imidazolidinecaproic acid; ε-(4-methyl-5-imidazolidone-2)-caproic acid; 4-methyl-5-(ω-carboxyamyl)imidazolidone-2. $C_{10}H_{18}N_2O_3$; mol wt 214.27. C 56.06%, H 8.47%, N 13.07%, O 22.40%. Prepd from biotin by hydrogenolysis of the sulfide linkage: du Vigneaud *et al.*, *J. Biol. Chem.* **146**, 475 (1942). Improved prepn procedure: Melville *et al.*, *Science* **98**, 497 (1943). Synthesis: Melville, *J. Am. Chem. Soc.* **66**, 1422 (1944); Wood, du Vigneaud, *ibid.* **67**, 210 (1945); Duschinsky, Dolan, *ibid.* 2079; Bourquin *et al.*, *Helv. Chim. Acta* **28**, 528 (1954).

Long needles from water. mp 156-158°. $[\alpha]_D^{21}$ +10.7° (c = 2). Sol in water.
Methyl ester. $C_{11}H_{20}N_2O_3$. Crystals from methanol. Can be sublimed at 10^{-5} mm and 100°. mp 69-70°. $[\alpha]_D^{28}$ +2.6° (c = 2 in chloroform). Readily sol in chloroform.

2936. Destomycin A. [14918-35-5] O-6-Amino-6-deoxy-L-*glycero*-D-*galacto*-heptopyranosylidene-(1 → 2-3)-O-β-D-talo-pyranosyl-(1 → 5)-2-deoxy-N^1-methyl-D-streptamine; 5-O-[2,3-O-[6-(1-amino-2-hydroxyethyl)tetrahydro-3,4,5-trihydroxy-2H-pyran-2-ylidene]-β-D-talopyranosyl]-2-deoxy-N^3-methyl-D-streptamine; Destonate. $C_{20}H_{37}N_3O_{13}$; mol wt 527.52. C 45.54%, H 7.07%, N 7.97%, O 39.43%. Aminoglycoside antibiotic, member of the orthosomycin family. Destomycins A, B and C are known; A is the major component. Isoln of A and B from culture broth of *Streptomyces rimofaciens:* S.-I. Kondo *et al.*, *J. Antibiot.* **18A**, 38 (1965); of C: *eidem, ibid.* **28**, 83 (1975). Prepn: *eidem,* **JP 67 7598** (1967 to Meiji Seika Kaisha). Structure of A: *eidem, J. Antibiot.* **19A**, 139 (1966); of A and B: *eidem, ibid.* **28**, 79 (1975). Stereochemical study: S. Horito *et al.*, *Bull. Chem. Soc. Jpn.* **54**, 2147 (1981). Synthesis of destomycin C: J. Yoshimura *et al.*, *Chem. Lett.* **1985**, 1335; J.-I. Tamura *et al.*, *Carbohydr. Res.* **174**, 181 (1988). Anthelmintic activity in poultry: I. Sawada, *Kiseichugaku Zasshi* **21**, 45 (1972), *C.A.* **77**, 135668j (1972). HPLC determn in pork samples: K. Nakaya *et*

al., *Shokuhin Eiseigaku Zasshi* **28**, 487 (1987), *C.A.* **109**, 5359c (1988).

White powder, mp 180-190° (dec). $[\alpha]_D^{22}$ +7° (c = 2). Freely sol in water, lower alcohols. Insol or poorly sol in most organic solvents. LD_{50} in mice (after 2-week observation) (mg/kg): 5-10 i.v.; 50-100 orally (Kondo, 1965).
THERAP CAT (VET): Broad spectrum antimicrobial; anthelmintic.

2937. Desvenlafaxine. [93413-62-8] 4-[2-(Dimethylamino)-1-(1-hydroxycyclohexyl)ethyl]phenol; O-desmethylvenlafaxine; 1-[2-(dimethylamino)-1-(4-hydroxyphenyl)ethyl]cyclohexanol. C_{16}-$H_{25}NO_2$; mol wt 263.38. C 72.97%, H 9.57%, N 5.32%, O 12.15%. Serotonin noradrenaline reuptake inhibitor (SNRI); major active metabolite of venlafaxine, *q.v.* Prepn: G. E. M. Husbands *et al.*, **EP 112669**; *eidem,* **US 4535186** (1984, 1985 both to Am. Home Prod.); J. P. Yardley *et al.*, *J. Med. Chem.* **33**, 2899 (1990). CE determn of enantiomers: S. Rudaz *et al.*, *J. Pharm. Biomed. Anal.* **23**, 107 (2000). LC/MS/MS determn in plasma: J. Bhatt *et al.*, *J. Chromatogr. B* **829**, 75 (2005). Pharmacological profile in animals: E. A. Muth *et al.*, *Drug Dev. Res.* **23**, 191 (1991). Mechanism of action study: D. C. Deecher *et al.*, *J. Pharmacol. Exp. Ther.* **318**, 657 (2006). Preclinical model in treatment of vasomotor symptoms associated with menopause: *idem et al.*, *Endocrinology* **148**, 1376 (2007). Clinical trial in major depressive disorder: N. A. DeMartinis *et al.*, *J. Clin. Psychiatry* **68**, 677 (2007).

Fumarate monohydrate. [313471-75-9]; [135308-72-4] (anhydrous). Wy-45233. $C_{16}H_{25}NO_2.C_4H_4O_4.H_2O$; mol wt 397.47. mp 140-142°.
Succinate monohydrate. [386750-22-7]; [448904-47-0] (anhydrous). DVS-233; Pristiq. $C_{16}H_{25}NO_2.C_4H_6O_4.H_2O$; mol wt 399.48. White to off-white powder. Sol in water. Partition coefficient (octanol/water, pH 7): 0.21.
THERAP CAT: Antidepressant.

2938. Detaxtran. [9015-73-0] Dextran 2-(diethylamino)ethyl ether; diethylaminoethyldextran; DEAE-dextran; basic dextran. Cationic derivative of dextran, *q.v.*, with antilipemic activity. Prepn: W. McKernan, C. R. Ricketts, *Chem. Ind. (London)* **1959**, 1490; E. Antonini *et al.*, *G. Biochem.* **14**, 88 (1965), *C.A.* **63**, 4375h (1965). Effect on platelet aggregation: R. Brossmer, Th. Pfleiderer, *Naturwissenschaften* **53**, 464 (1966); A. Larcon *et al.*, *Experientia* **28**, 1096 (1972). Use to lower hypercholesterolemia: T. M. Parkinson, **US 3627872** (1971 to Upjohn); to reduce blood lipid levels: F. Kuzuya *et al.*, **US 3851057** (1974 to Meito Sangyo). Clinical trial in hyperlipoproteinemia: F. Pupita, A. Barone, *Int. J. Clin. Pharmacol. Res.* **3**, 287 (1983).

Powder, soluble in water, saline.
Hydrochloride. [9064-91-9] Dexide; Pulsar.
THERAP CAT: Antilipemic.

2939. Detomidine. [76631-46-4] 5-[(2,3-Dimethylphenyl)methyl]-1H-imidazole; 4-(2′,3′-dimethylbenzyl)imidazole; 4-[(2,3-

dimethylphenyl)methyl]-1*H*-imidazole. $C_{12}H_{14}N_2$; mol wt 186.26. C 77.38%, H 7.58%, N 15.04%. α_2-Adrenoceptor agonist with sedative and analgesic activity. Prepn: A. J. Karjalayne, K. O. A. Kurkela, **EP 24829**; *eidem*, **US 4443466** (1981, 1984 both to Farmos). Physical studies: E. Laine *et al.*, *Acta Pharm. Suec.* **20**, 451 (1983). Crystal structure: L. H. J. Lajunen *et al.*, *ibid.* **21**, 163 (1984). Pharmacology: R. Virtanen, L. Nyman, *Eur. J. Pharmacol.* **108**, 163 (1985); R. Virtanen, E. MacDonald, *ibid.* **115**, 277 (1985). Mechanism of action: *eidem*, *J. Vet. Pharmacol. Ther.* **8**, 30 (1985). Radioimmunoassay in biological samples: O. Vakkuri *et al.*, *Life Sci.* **40**, 1357 (1987). GC/MS determn in blood, plasma and urine: A. K. Singh *et al.*, *J. Chromatogr.* **404**, 223 (1987). Analgesic effects in horses: K. Sardari *et al.*, *J. Equine Vet. Sci.* **25**, 262 (2005).

Crystals from acetone, mp 114-116°. LD_{50} i.v. in mice: 35 mg/kg (Karjalayne, Kurkela).

Hydrochloride. [90038-01-0] Dormosedan. $C_{12}H_{14}N_2$·HCl; mol wt 222.72. Crystals, mp 160°. Converts reversibly to monohydrate at room temp, 80% humidity. Sol in water.

THERAP CAT (VET): Sedative and analgesic for horses and cattle.

2940. Detoxin Complex. DX.C. A group of selective antagonists of blasticidin S, *q.v.*, produced by *Streptomyces caespitosus* var. *detoxicus* 7072 GC₁. The complex counteracts the inhibitory action of blasticidin S (BS) against *Bacillus cereus* but not against *Piricularia oryzal*, the causative agent of rice blast disease. DX.C also depresses the phytotoxicity of BS to rice plants, and a combination of the two showed less eye irritation in animals than BS alone. Isoln, production, biological properties: H. Yonehara *et al.*, *J. Antibiot.* **21**, 369 (1968); *eidem*, *Agric. Biol. Chem.* **37**, 2771 (1973). Separation of the complex yields eight groups of detoxins designated A through H, the major components being C and D: N. Otake *et al.*, *J. Antibiot.* **21**, 371 (1968); *eidem*, *Agric. Biol. Chem.* **37**, 2777 (1973). Detoxin D₁, the major component of the D group, has been shown to have the highest specific activity. It is a unique depsipeptide containing the amino acid ***detoxinine (β,3-dihydroxy-2-pyrrolidinepropanoic acid)***. Structure of D₁: K. Kakenuma *et al.*, *Tetrahedron Lett.* **1972**, 2509. ¹³C-NMR: T. Ogita *et al.*, *Agric. Biol. Chem.* **42**, 2403 (1978). Structures of the minor components of the D group: N. Otake *et al.*, *J. Antibiot.* **27**, 484 (1974). Structures of detoxins A₁, B₁, B₃, C₁, C₂, C₃, E₁: N. Otake *et al.*, *Experientia* **37**, 926 (1981); T. Ogita *et al.*, *Agric. Biol. Chem.* **45**, 2605 (1981). Effect of D on BS uptake in *B. cereus*: A. Shimazu *et al.*, *Experientia* **37**, 365 (1981).

Detoxin D₁   R = CH(CH₃)CH₂CH₃
Detoxin C₁   R = CH₃

Light colored fine powder.

Detoxin D₁. [37878-19-6] *N*-(2-Methyl-1-oxobutyl)-L-phenylalanine (1*S*)-1-[(2*S*,3*R*)-3-(acetyloxy)-1-[(2*S*)-2-amino-3-methyl-1-oxobutyl]-2-pyrrolidinyl]-2-carboxyethyl ester. $C_{28}H_{41}N_3O_8$; mol wt 547.65. Fine crystalline powder, mp 156-158°. $[\alpha]_D^{25}$ −16° (c = 1 in methanol). uv max (methanol): 253, 258, 265, 268 nm ($E_{1cm}^{1\%}$ 3.1, 3.58, 2.77, 1.85). Amphoteric. pKa 4.0, 8.0.

Detoxin C₁. [74717-53-6] *N*-Acetyl-L-phenylalanine (1*S*)-1-[(2*S*,3*R*)-3-(acetyloxy)-1-[(2*S*)-2-amino-3-methyl-1-oxobutyl]-2-pyrrolidinyl]-2-carboxyethyl ester; detoxin $C_{\alpha1}$. $C_{25}H_{35}N_3O_8$; mol wt 505.57. Microneedles from water, mp 142-144°. Amphoteric. pKa 8.0, 3.9. $[\alpha]_D^{25}$ −23° (c = 1 in methanol). uv max (methanol): 248, 253, 259, 265, 269 nm ($E_{1cm}^{1\%}$ 3.7, 3.18, 2.65, 2.8, 1.85).

USE: In mixtures with blasticidin S, to decrease phytotoxicity to rice plants.

2941. Deuterium. [7782-39-0] Heavy hydrogen. ²H or D. Exists in the diatomic state, D₂; mol wt 4.028. Stable, non-radioactive isotope of hydrogen, *q.v.* Shows *ortho* and *para* isomerism. Prepd by electrolysis of heavy water, D₂O (obtained by the H₂S/H₂O exchange process) or by fractional distillation of liq hydrogen: Urey *et al.*, *Phys. Rev.* **39**, 164 (1932); Spevack, **US 2787526** (1957 to USAEC). The hydrogen bomb contains lithium deuteride (LiD) as explosive and plutonium (Pu) as initiator. After detonation by Pu the following reaction sequence takes place (nuclear fusion): ⁶Li + D = 2⁴He; ⁶Li + n = ⁴He + T; ⁶Li + T = 2⁴He + n (n = neutrons, coming from Pu in the second reaction). Since the explosion does not start by itself, there is no critical mass and no limit to the size of the bomb. Use as chemical tracer: Wiberg, *Chem. Rev.* **55**, 713-743 (1955). Toxicology: J. F. Thomson, *Biological Effects of Deuterium* (Pergamon Press, 1964). *Review:* Mackay, Dove in *Comprehensive Inorganic Chemistry* vol. **1**, J. C. Bailar, Jr. *et al.*, Eds. (Pergamon Press, Oxford, 1973) pp 77-116; *Chemistry of the Elements*, N. N. Greenwood, A. Earnshaw, Eds. (Pergamon Press, New York, 1984) pp 38-74; J. J. Katz in *Kirk-Othmer Encyclopedia of Chemical Technology* vol. **8** (Wiley-Interscience, New York, 4th ed., 1993) pp 1-17.

Colorless, odorless, flammable gas having properties similar to hydrogen. d (liq, 20.4 K) 0.169. bp −249.49° (23.67 K). mp −254.43° (18.73 K) at 128.5 mm (triple point). Crit temp −234.75°. Crit press. 16.432 atm. Flammable limits in air: 5-75%. Calculation of vapor press. from the triple point to the crit point: Friedman *et al.*, *J. Am. Chem. Soc.* **73**, 1310 (1951).

Caution: Can act as an asphyxiant by displacing air. *See: Matheson Gas Data Book* (Matheson, 6th ed., Lyndhurst, NJ, 1980) pp 215-218.

USE: Used extensively in small amts as tracer in the establishment of rates and kinetics of chemical reactions.

2942. Deuterium Oxide. [7789-20-0] Water-*d₂*; heavy water. ²H₂O. D₂O.

More associated than H₂O. mp 3.81°; triple point temp 3.82°. bp 101.42°. Critical temp 371.5°. d²⁵ 1.1044. Temp of max density 11.23°; d¹¹·²³ 1.1059. Sp heat of liquid (4-25°) 1.028 cal/g/°C. Heat of fusion 1.501 kcal/mole: Long, Kemp, *J. Am. Chem. Soc.* **58**, 1829 (1936). Heat of evapn 9.917 kcal/mole. Dielectric const (25°): 78.06, Vidulich *et al.*, *J. Phys. Chem.* **71**, 656 (1966). Dipole moment in benzene (25°) 1.78, in dioxane 1.87. The surface tension is very slightly smaller and the viscosity at 25° is 1.23 times as great as that of water. On mixing with water, heat is evolved. Reacts with hydrogen at 100° in the presence of 0.2 to 1*N* NaOH forming DOH + HD + OD⁻. Ionization const 1.95×10^{-15}: W. F. K. Wynne-Jones, *Trans. Faraday Soc.* **32**, 1397 (1936). pK (25°) 14.955 (molarity scale); 16.653 (mole fraction scale): A. K. Covington *et al.*, *J. Phys. Chem.* **70**, 3820 (1966). *See also* Water.

USE: To study chemical reaction rates and mechanisms. The cross section of deuterium for the capture of thermal neutrons is very low which makes it useful, in the form of heavy water, as a neutron moderator in nuclear reactors. Produces a considerable decrease in neutron energy per collision.

2943. Devarda's Metal. [8049-11-4] Devarda's alloy. Consists of 50 parts Cu, 45 Al and 5 Zn.

Gray powder. Partly sol in HCl with residue of Cu.

USE: Reducing agent for determination of nitrogen in nitrates and nitrites.

2944. Devil's Claw. Herbaceous, perennial plant, *Harpagophytum procumbens* DC, Pedaliaceae, used in African traditional medicine as a bitter tonic, febrifuge and analgesic. Medicinal formulations are prepared from the dried secondary tubers. *Habit.* Kalahari savannas of southern Africa and Namibia. *Constit.* Iridoid glucosides, primarily harpagoside (the pharmacologically active component), harpagide and procumbide; sugars such as stachyose, raffinose, sucrose; caffeic and cinnamic acids. HPLC determn of active component in commercial extracts: L. Guillerault *et al.*, *J. Liq. Chromatogr.* **17**, 2951 (1994). Pharmacology: B. Baghdikian *et al.*, *Planta Med.* **63**, 171 (1997). Clinical comparison with diacerein, *q.v.*, in osteoarthritis: P. Chantre *et al.*, *Phytomedicine* **7**, 177 (2000). Reviews of medicinal uses: S. Chrubasik, E. Eisenberg, *Pain Clinic* **11**, 171-178 (1999); C. Hansen, *Dtsch. Apoth.* **140**, 85-89 (2000).

Dry extract. Arthrosetten; Arthrotabs; Doloteffin; Herbadon; Rivoltan.

THERAP CAT: Analgesic.

2945. Dexamethasone. [50-02-2] (11β,16α)-9-Fluoro-11,-17,21-trihydroxy-16-methylpregna-1,4-diene-3,20-dione; 9α-fluoro-16α-methylprednisolone; 16α-methyl-9α-fluoro-1,4-pregnadiene-11β,17α,21-triol-3,20-dione; 16α-methyl-9α-fluoroprednisolone; 1-dehydro-16α-methyl-9α-fluorohydrocortisone; 16α-methyl-9α-fluoro-Δ¹-hydrocortisone; hexadecadrol; Aeroseb-Dex; Decadron; Dexacortal; Dexacortin; Dexamonozon; Dexapos; Dexa-sine; Dexasone; Fortecortin; Isopto-Dex; Loverine; Luxazone; Maxidex. $C_{22}H_{29}FO_5$; mol wt 392.47. C 67.33%, H 7.45%, F 4.84%, O 20.38%. Synthetic adrenocortical steroid. Prepn and anti-inflammatory activity: G. E. Arth *et al.*, *J. Am. Chem. Soc.* **80**, 3161 (1958); E. P. Oliveto *et al.*, *ibid.* 4431; G. E. Arth *et al.*, *DE 1113690* (1961 to Merck & Co.); *GB 869511* (1961 to Upjohn). Comprehensive description: E. M. Cohen, *Anal. Profiles Drug Subs.* **2**, 163-197 (1973). Review of analytical methods: S. Saeed-Ul-Hassan *et al.*, *Acta Pharm. Turc.* **43**, 33-42 (2001). HPLC determn in plasma: Y.-K. Song *et al.*, *J. Liq. Chromatogr. Relat. Technol.* **27**, 2293 (2004). Clinical trial as anti-emetic in chemotherapy-induced nausea: M. Markman *et al.*, *N. Engl. J. Med.* **311**, 549 (1984); F. Roila *et al.*, *ibid.* **342**, 1554 (2000). Use as diagnostic aid in depression: B. J. Carroll *et al.*, *Arch. Gen. Psychiatry* **38**, 15 (1981); and suicide prediction: W. Coryell, M. Schlesser, *Am. J. Psychiatry* **158**, 748 (2001). Diagnostic use in Cushing's syndrome: L. Crapo, *Metabolism* **28**, 955 (1979); A. M. Isidori *et al.*, *J. Clin. Endocrinol. Metab.* **88**, 5299 (2003).

White to practically white, odorless crystals from ether, mp 262-264°; mp 268-271° (Arth, 1961). $[\alpha]_D^{25}$ +77.5° (dioxane). Soly in water (25°): 10 mg/100 ml. Sparingly sol in acetone, alc, methanol, dioxane; slightly sol in chloroform; very slightly sol in ether.

21-Acetate. [1177-87-3] Decadronal; Decadron-LA; Dectancyl. $C_{24}H_{31}FO_6$; mol wt 434.50. Crystals, mp 215-221° (Arth, 1958); mp 229-231° (Oliveto); mp 238-240° (Arth, 1961). $[\alpha]_D^{25}$ +73° (chloroform) (Arth, 1958); $[\alpha]_D$ +77.6° (Oliveto). uv max: 239 nm (ε 14900). Freely sol in methanol, acetone, dioxane. Practically insol in water.

21-Phosphate disodium salt. [2392-39-4] Dexamethasone 21-(dihydrogen phosphate) disodium salt; dexamethasone sodium phosphate; Ak-Dex; Dalalone; Desocort; Dexabene; Hexadrol; Oradexon; Orgadrone; SoluDecadron; Soldesam; Solupen N; Totocortin. $C_{22}H_{28}FNa_2O_8P$; mol wt 516.41. Used as an injectable form of dexamethasone. Prepn: Chemerda *et al.*, *US 2939873* (1960 to Merck & Co.); Irmscher, *Chem. Ind. (London)* **1961**, 1035. White or slightly yellow crystalline powder, mp 233-235°. Exceedingly hygroscopic. $[\alpha]_D$ +57° (water). Also reported as $[\alpha]_D^{25}$ +74 ±4° (calcd on water-free and alcohol-free basis, concn of 10 mg/ml): *USP XIX*, p 124. uv max (ethanol): 238-239 nm (ε 14000). Soln in water; slightly sol in alc; very slightly sol in dioxane. Insol in chloroform, ether.

21-Isonicotinate. Dexamethasone 21-(4-pyridinecarboxylate); Auxisone. $C_{28}H_{32}FNO_6$; mol wt 497.56. Crystals, mp 250-252°. $[\alpha]_D^{27}$ +183.5° (dioxane).

17,21-Dipropionate. THS-101; Methaderm. $C_{28}H_{37}FO_7$; mol wt 504.60.

THERAP CAT: Glucocorticoid.; antiemetic; diagnostic aid (Cushing's syndrome, depression).

THERAP CAT (VET): Glucocorticoid.

2946. Dexanabinol. [112924-45-5] (6aS,10aS)-3-(1,1-Dimethylheptyl)-6a,7,10,10a-tetrahydro-1-hydroxy-6,6-dimethyl-6*H*-dibenzo[*b,d*]pyran-9-methanol; (+)-(3S,4S)-7-hydroxy-Δ⁶-tetrahydrocannabinol-1,1-dimethylheptyl; (+)-11-OH-Δ⁸-THC-DMH; HU-211. $C_{25}H_{38}O_3$; mol wt 386.58. C 77.67%, H 9.91%, O 12.42%. NMDA antagonist. Nonpsychotropic derivative of tetrahydrocannabinol, *q.v.* Prepn: R. Mechoulam *et al.*, *DE 3735990*; *eidem*, *US 4876276* (1988, 1989 both to Yissum Res. Dev. Co.); R. Mechoulam *et al.*, *Tetrahedron: Asymmetry* **1**, 315 (1990). Cannabinoid receptor binding study: A. C. Howlett *et al.*, *Neuropharmacology* **29**, 161 (1990). Inhibition of nuclear factor-κB (NF-κB): E. Jüttler *et al.*, *Neuropharmacology* **47**, 580 (2004). Pharmacology: N. Eshhar *et al.*, *Eur. J. Pharmacol.* **283**, 19 (1995). Clinical pharmacokinetics: M. E. Brewster *et al.*, *Int. J. Clin. Pharmacol. Ther.* **35**, 361 (1995). Clinical evaluation in severe head injury: N. Knoller *et al.*, *Crit. Care Med.* **30**, 548 (2002).

Crystals from pentane, mp 141-142°. $[\alpha]_D$ +227° (CHCl₃).

THERAP CAT: Neuroprotective.

2947. Dexetimide. [21888-98-2] (3S)-3-Phenyl-1'-(phenylmethyl)-[3,4'-bipiperidine]-2,6-dione; (S)-(+)-2-(1-benzyl-4-piperidyl)-2-phenylglutarimide; (+)-3-(1-benzyl-4-piperidyl)-3-phenylpiperidine-2,6-dione; (+)-1-benzyl-4-(2,6-dioxo-3-phenyl-3-piperidyl)piperidine; dextrobenzetimide; dexbenzetimide. $C_{23}H_{26}N_2O_2$; mol wt 362.47. C 76.21%, H 7.23%, N 7.73%, O 8.83%. Pharmacologically active enantiomer of benzetimide, *q.v.* Resolution of isomers and comparative pharmacology: Janssen *et al.*, *Arzneim.-Forsch.* **21**, 1365 (1971). Abs config studies: van Wijngaarden *et al.*, *Life Sci.* **9**, part 1, 1289 (1970); Spek *et al.*, *Nature* **232**, 575 (1971). Clinical trials: De Smedt *et al.*, *J. Clin. Pharmacol.* **10**, 207 (1970).

Crystals, mp 181-183°. $[\alpha]_D^{20}$ +125° (chloroform).

Hydrochloride. [21888-96-0] R-16470; Tremblex. $C_{23}H_{26}N_2O_2$·HCl; mol wt 398.93. Crystals, mp 270-275°. $[\alpha]_D^{20}$ +125° (methanol). LD₅₀ i.v. in rats: 45 mg/kg (Janssen).

THERAP CAT: Antiparkinsonian.

2948. Dexmedetomidine. [113775-47-6] 5-[(1S)-1-(2,3-Dimethylphenyl)ethyl]-1*H*-imidazole; (S)-4-[1-(2,3-dimethylphenyl)-ethyl]-1*H*-imidazole; *d*-medetomidine; MPV-1440. $C_{13}H_{16}N_2$; mol wt 200.29. C 77.96%, H 8.05%, N 13.99%. $α_2$-Adrenergic agonist; (+)-isomer of medetomidine, *q.v.* Prepn: A. J. Karjalainen *et al.*, *GB 2206880*; *eidem*, *US 4910214* (1989, 1990 both to Farmos). Physical properties: R. Rajala *et al.*, *Eur. J. Pharm. Sci.* **1**, 219 (1994). LC-MS/MS determn in plasma: J. I. Lee *et al.*, *J. Chromatogr. B* **852**, 195 (2007). Clinical pharmacokinetics: P. Talke *et al*, *Anesth. Analg.* **85**, 1114 (1997). Clinical evaluation as anesthetic adjunct: J. Jalonen *et al.*, *Anesthesiology* **86**, 331 (1997). Veterinary trial as preanesthetic medication in cats: G. M. Mendes *et al.*, *J. Feline Med. Surg.* **5**, 265 (2003). Review of pharmacology and clinical experience for sedation of patients in intensive care: N. Bhana et al., *Drugs* **59**, 263-270 (2000). Series of articles on clinical experience: *Semin. Anesthesia* **25**, 41-84 (2006).

Hydrochloride. [145108-58-3] Dexdomitor; Precedex. $C_{13}H_{16}$-CN_2.HCl; mol wt 248.75. White or almost white crystalline powder, mp 156.5-157.5°. d 1.17 g/cm^3. $[\alpha]$ +52.4° (c = 1 in water). pKa 7.1. Log P (octanol/water): 2.89 (pH 7.4). pH of 1% soln in water: 4.3. Freely sol in water.

THERAP CAT: Sedative; analgesic.

THERAP CAT (VET): Sedative; analgesic.

2949. Dexpanthenol. [81-13-0] (2*R*)-2,4-Dihydroxy-*N*-(3-hydroxypropyl)-3,3-dimethylbutanamide; D(+)-α,γ-dihydroxy-*N*-(3-hydroxypropyl)-β,β-dimethylbutyramide; pantothenylol; *N*-pantoyl-3-propanolamine; pantothenol; pantothenyl alcohol; Alcopan-250; Intrapan; Pantenyl; Panthoderm; Motilyn; Bepanthen; Cozyme; Ilopan; Urupan. $C_9H_{19}NO_4$; mol wt 205.25. C 52.67%, H 9.33%, N 6.82%, O 31.18%. Prepd by the addition of propanolamine to optically active α,γ-dihydroxy-β,β-dimethylbutyrolactone: Schnider, *Jubilee Vol. Emil Barell* **1946**, 85; CH **227706** (1943); GB **582156** (1946); US **2413077** (1946 to Hoffmann-La Roche). Only the D(+)-form has vitamin activity.

Viscous, somewhat hygroscopic liq. Slightly bitter taste. d_{20}^{20} 1.2. bp$_{0.02}$ 118-120°. Easily dec on distn. $[\alpha]_D^{20}$ +29.5° (c = 5). n_D^{20} 1.497. Freely sol in water, alcohol, methanol, propylene glycol; sol in chloroform, ether; slightly sol in glycerin. Natural pH about 9.5. Reasonably stable to usual sterilization time and temp in aq solns adjusted to pH 3.0-4.0, but long heating causes racemization. Hydrolyzed by alkali and strong acid. Usually more stable than salts of pantothenic acid if pH can be adjusted between 3 and 5. Stability data: Rubin, *J. Am. Pharm. Assoc. Sci. Ed.* **37**, 502 (1948). Aq solns can be stabilized with pantolactone: US **2898373** (1959).

dl-**Form.** Panthenol.

THERAP CAT: Cholinergic; *dl*-form as vitamin.

THERAP CAT (VET): Nutritional factor. Dietary source of pantothenic acid.

2950. Dextran. [9004-54-0] Gentran; Hemodex; Intradex; Promit. A term applied to polysaccharides produced by bacteria growing on a sucrose substrate, contg a backbone of D-glucose units linked predominantly α-D(1 → 6). Several organisms produce dextrans but only *Leuconostoc mesenteroides* and *L. dextranicum* (*Lactobacteriaceae*) have been used commercially. Chemical and physical properties of the dextrans vary with the methods of production. Native dextrans usually have high mol wt; lower mol wt clinical dextrans usually prepared by depolymerization of native dextrans or by synthesis. All dextrans are composed exclusively of α-D-glucopyranosyl units, differing only in degree of branching and chain length. Prepn: Tarr, Hibbert, *Can. J. Res.* **5**, 414 (1931); Novak, Stoycos, US **2841578** (1958 to Commonwealth Eng. of Ohio). Enzymic synthesis: Sugg, Hehre, *J. Immunol.* **43**, 119 (1942); Behrens, Ringpfeil, US **3044940** (1962 to Serum Werk Bernburg). The crude dextran may be isolated from the culture by precipitation with methanol. Continuous dialysis process: Shurter, US **2717853** (1955 to C.S.C.). Elimination of pyrogens: Levi, Lozinski, US **2762727** (1956 to Frosst). Method of producing clinical dextran: Novak, Witt, US **2972567** (1961). Structure studies: Fowler et al., *Can. J. Res.* **B15**, 486 (1937); Fairhead et al., *ibid.* **B16**, 151 (1938); Peat et al., *J. Chem. Soc.* **1939**, 581; Goldstein, Whelan, *ibid.* **1962**, 170, 176. ^{13}C-NMR structure study: F. R. Seymour et al., *Carbohydr. Res.* **51**, 179 (1976). *Reviews:* Evans, Hibbert, *Adv. Carbohydr.*

Chem. **2**, 204 (1946); Neely, *ibid.* **15**, 341 (1960); Ricketts, *Prog. Org. Chem.* **5**, 73 (1961); Murphy, Whistler, in *Industrial Gums*, R. L. Whistler, Ed. (Academic Press, New York, 2nd ed., 1973) pp 513-542.

Dextran 1. White to off-white powder. Hygroscopic. Very sol in water; sparingly sol in alc.

Dextran 40. LMD; LMWD; LVD; Gentran 40; Rheomacrodex. Produced by action of *L. mesenteroides* on sucrose; average mol wt: 40,000.

Dextran 70. Gentran 70; Hyskon; Macrodex. Average mol wt: 70,000.

USE: In soft center confections, as a partial substitute for barley malt. Mixed ethers and esters of dextran can be used in lacquers.

THERAP CAT: Plasma volume expander; Dextran 40 also as blood flow adjuvant.

THERAP CAT (VET): Plasma extender.

2951. Dextranase. [9025-70-1] α-1,6-Glucan 6-glucanohydrolase; EC 3.2.1.11. Enzyme which hydrolyzes the α-1 → 6 glucosidic linkages of the bacterial polysaccharide dextran. *Endodextranases* (dextranases which preferentially split glucosidic linkages remote from end groups) are secreted by various molds and a few bacteria. *Exodextranases* occur predominantly in mammalian tissues. Prepn from *Penicillium lilacinum, P. funiculosum,* and *Verticillium coccorum*: Nordström, Hultin, *Sven. Kem. Tidskr.* **60**, 283 (1948), *C.A.* **43**, 3050i (1949); from *Aspergillus*: Carlson, Carlson, *Science* **115**, 43 (1952); *eidem*, US **2709150** (1955 to Enzymatic Chemicals); *eidem*, US **2716084**; Novak, Stoycos, US **2841578** (1955 and 1958 both to Commonwealth Eng. of Ohio); from *P. lilacinum, P. funiculosum, P. verruculosum,* and *Spicaria violacea*: Tsuchiya et al., US **2742399**; Corman, Tsuchiya, US **2776925** (1956 and 1957 both to U.S. Secy. Agr.); from *P. lilacinum*: Charles, Farrell, *Can. J. Microbiol.* **3**, 239 (1957); from *Lactobacillus bifidus*: Bailey, Clarke, *Biochem. J.* **72**, 49 (1959). Tested as dental caries-control agent in hamsters: Fitzgerald et al., *J. Am. Dent. Assoc.* **76**, 301 (1968). *Review:* E. H. Fischer, E. A. Stein, "Cleavage of *O*- and *S*-Glycosidic Bonds (Survey)" in *The Enzymes* vol. 4, P. D. Boyer et al., Eds. (Academic Press, New York, 2nd ed., 1960) pp 304-307.

USE: In prepn of dextran for clinical use; in dentifrices.

2952. Dextranomer. [56087-11-7] Dextran 2,3-dihydroxypropyl 2-hydroxy-1,3-propanediyl ethers; Debrisan; Debrisorb. Three-dimensional hydrophilic network of a dextran polymer, linked by cross-chains of epichlorohydrin, *q.v.;* it absorbs moisture and small molecules from suppurating wounds. Chronic tissue response to implantation: J. Falk, G. Tollerz, *Clin. Ther.* **1**(3), 185 (1977). Potential allergic contact sensitization in guinea pigs: G. Jonsson, *ibid.* **1**(4), 260 (1978). Efficacy in treatment of ulcers and wounds: J. Soul, *Br. J. Clin. Pract.* **32**, 172 (1978); P. N. Sawyer et al., *Surgery* **85**, 201 (1979); S. Di Mascio, *Am. J. Nurs.* **79**, 684 (1979). *Review:* R. C. Heel et al., *Drugs* **18**, 89-102 (1979).

Insol in all solvents; stable in water, salt solns and in alkaline and weakly acidic soln.

THERAP CAT: Vulnerary.

2953. Dextran Sulfate Sodium. [9011-18-1] Dextran hydrogen sulfate sodium salt; dextran sulfuric acid ester sodium salt; Asuro; Colyonal; Dexulate; Dextrarine; MDS. Heparin-like polysaccharide containing ~17% sulfur with up to three sulfate groups per glucose molecule. Mol wt ranges from 4-500 kDa; variations in mol wt are associated with differences in biological activity. Prepn, properties and anticoagulant activity: A. Grönwall *et al.*, *Upsala Laekarefoeren. Foerh.* **50**, 397 (1945); C. R. Ricketts, *Biochem. J.* **51**, 129 (1952). Evaluation of toxicity as a function of mol wt: K. W. Walton, *Br. J. Pharmacol.* **9**, 1 (1954); of carcinogenicity: I. Hirono *et al.*, *Cancer Lett.* **18**, 29 (1983). Use in serum HDL cholesterol determn: P. R. Finley *et al.*, *Clin. Chem.* **24**, 931 (1978); G. R. Warnick *et al.*, *ibid.* **28**, 1379 (1982). Antiscrapie effect: B. Ehlers, H. Diringer, *J. Gen. Virol.* **65**, 1325 (1984); R. H. Kimberlin, C. A. Walker, *Antimicrob. Agents Chemother.* **30**, 409 (1986). Anti-HIV-1 activity *in vitro:* H. Mitsuya *et al.*, *Science* **240**, 646 (1988); M. Baba *et al.*, *Proc. Natl. Acad. Sci. USA* **85**, 6132 (1988).

White powder from alcohol + ether. Freely sol in water. Activity about 17 international heparin units/mg. Aq solns must be buffered (*e.g.*, with sodium bicarbonate) to prevent dec during autoclaving.

USE: Clinical reagent (HDL cholesterol determn).

THERAP CAT: Anticoagulant.

2954. Dextri-Maltose®. [8006-91-5] Malt sugar-dextrin. Maltose and dextrins obtained by enzymic action of barley malt on corn flour.

Light, amorphous powder. Readily sol in water or milk. One leveled tablespoonful (8 grams) supplies 27 calories.

USE: As carbohydrate modifier for use with milk and milk products in infants' formulas.

2955. Dextrin. [9004-53-9] Pyrodextrin; torrefaction dextrin. $(C_6H_{10}O_5)_n \cdot xH_2O$. Produced by the dry heating of unmodified starches. The term also includes products resulting from enzyme or acid-catalyzed hydrolysis of wet starch. *Review:* R. W. Satterthwaite, D. J. Iwinski, in *Industrial Gums*, R. L. Whistler, Ed. (Academic Press, New York, 2nd ed., 1973) pp 577-599.

British gum. Starch gum. Produced at high temp in the absence of acid. Dark brown color, odorous. High viscosity; very sol in cold water. Does not reduce Fehling's soln; gives reddish-brown color with iodine.

Canary dextrin. Yellow dextrin. Hydrolyzed at high temp for long period of time in the presence of small amts of acid. Light brown to yellow color, slight odor. Low viscosity; very sol in cold water.

White dextrin. Hydrolyzed at low temp for short period of time in the presence of large amts of acid. White color, odorless. Slightly sol in cold water giving a red color with iodine. Very sol in hot water giving a blue color with iodine.

USE: Excipient for dry extracts and pills; for preparing emulsions and dry bandages; for thickening dye pastes and mordants used in printing fabrics in fast colors; sizing paper and fabrics; printing tapestries; preparing felt; manuf printer's inks, glues and mucilage; polishing cereals; in matches, fireworks, and explosives.

2956. Dextroamphetamine. [51-64-9] (αS)-α-Methylbenzeneethanamine; (+)-α-methylphenethylamine; *d*-amphetamine; (*S*)-1-phenyl-2-aminopropane; *d*-β-phenylisopropylamine; dexamphetamine. $C_9H_{13}N$; mol wt 135.21. C 79.95%, H 9.69%, N 10.36%. Prepn by resolution of amphetamine: Temmler, **GB 508757** (1939); Nabenhauer, **US 2276508** (1942 to SK&F); Magidson, Garkusha, *J. Gen. Chem. USSR* **11**, 339 (1941); from D-phenylalanine: D. B. Repke *et al.*, *J. Pharm. Sci.* **67**, 1167 (1978). Toxicity data: E. J. Warawa *et al.*, *J. Med. Chem.* **18**, 71 (1975). GC-MS determn in urine: V. A Tetlow, J. Merrill, *Ann. Clin. Biochem.* **33**, 50 (1996). Review of pharmacology: S. J. Chee, *Neurosci. Biobehav. Rev.* **16**, 481-496 (1992).

Sulfate. [51-63-8] Dexamin; Dexedrine; Dextrostat. White, odorless, crystalline powder with bitter taste. mp >300°. $[\alpha]_D^{20}$ +21.8° (c = 2). Freely sol in water (about 1:10); slightly sol in alcohol (about 1:800). Insol in ether. pH 5% aq soln: 5.0 to 6.0. LD_{50} orally in mice: 10 mg/kg (Warawa).

Tannate. [1407-85-8] Tanphetamin; Synatan. Prepn: Cavallito, **US 2950309** (1960 to Irwin, Neisler and Co.).

Note: This is a controlled substance (stimulant): **21 CFR,** 1308.12.

THERAP CAT: CNS stimulant; anorexic.

2957. Dextromethorphan. [125-71-3] ($9\alpha,13\alpha,14\alpha$)-3-Methoxy-17-methylmorphinan; (+)-3-methoxy-*N*-methylmorphinan; *d*-methorphan. $C_{18}H_{25}NO$; mol wt 271.40. C 79.66%, H 9.29%, N 5.16%, O 5.89%. Non-opioid, centrally acting cough suppressant; enantiomer of levomethorphan, *q.v.* Sigma-1 receptor agonist and noncompetitive *N*-methyl-D-aspartate (NMDA) receptor antagonist. Prepn: O. Schnider, A. Grüssner, *Helv. Chim. Acta* **34**, 2211 (1951); *eidem*, **US 2676177** (1954 to Hoffmann-La Roche). Absolute configuration and crystal structure of hydrobromide: L. Gylbert, D. Carlstrom, *Acta Crystallogr. B* **33**, 2833 (1977). HPLC-MS/MS determn in urine: M. L. Constanzer *et al.*, *J. Chromatogr. B* **816**, 297 (2005). Population pharmacokinetics: K. Abduljalil *et al.*, *Clin. Pharmacol. Ther.* **88**, 643 (2010). Evaluation as probe for cytochrome P450 2D6 phenotyping: D. Frank *et al.*, *Eur. J. Clin. Pharmacol.* **63**, 321 (2007). Review of safety and abuse potential: J. L. Bem, R. Peck, *Drug Saf.* **7**, 190-199 (1992); of pharmacology, metabolism and therapeutic potential in pain management: A. Siu, R. Drachtman, *CNS Drug Rev.* **13**, 96-106 (2007); of pharmacology, mechanisms of action, and neuroprotective effects: E.-J. Shin *et al.*, *J. Pharmacol. Sci.* **116**, 137-148 (2011). Review of clinical experience in combination with quinidine for treatment of pseudobulbar affect: K. P. Garnock-Jones, *CNS Drugs* **25**, 435-445 (2011).

White to slightly yellow, odorless, crystalline powder. mp 109-111°. $[\alpha]_D^{20}$ +49.6° (c = 1.5 in alc). Practically insol in water. Freely sol in chloroform.

Hydrobromide. [125-69-9] Ro-1-5470/5; Acodin; AquiTos; Balminil DM; Benylin; Dexir; Rofedex. $C_{18}H_{25}NO \cdot HBr$; mol wt 352.32. Occurs as the monohydrate, crystals, mp 122-124°. $[\alpha]_D^{20}$ +27.6° (c = 1.5 in water). Approx soly in water: 1.5% at 25°, 5% at 50°, 25% at 85°. Freely sol in ethanol, chloroform. Practically insol in ether. pH of a 1% aq soln: 5.2-6.5. Reacts with alkalies forming the free base.

Combination of hydrobromide with quinidine sulfate. Nuedexta.

USE: Probe to determine cyctochrome P450 phenotype.

THERAP CAT: Antitussive. In treatment of pseudobulbar affect.

2958. Dextromoramide. [357-56-2] (3*S*)-3-Methyl-4-(4-morpholinyl)-2,2-diphenyl-1-(1-pyrrolidinyl)-1-butanone; (+)-1-(3-methyl-4-morpholino-2,2-diphenylbutyryl)pyrrolidine; 1-[(3*S*)-3-methyl-4-(4-morpholinyl)-1-oxo-2,2-diphenylbutyl]pyrrolidine; 4-[2-methyl-4-oxo-3,3-diphenyl-4-(1-pyrrolidinyl)butyl]morpholine; *d*-2,2-diphenyl-3-methyl-4-morpholinobutyrylpyrrolidine; pyrrolamidol; R-875; SKF-5137. $C_{25}H_{32}N_2O_2$; mol wt 392.54. C 76.50%, H 8.22%, N 7.14%, O 8.15%. Synthetic opioid analgesic. Synthesis: Janssen, *J. Am. Chem. Soc.* **78**, 3862 (1956); **GB 822055** (1959 to Janssen). Clinical pharmacokinetics: J. G. R. Ufkes *et al.*, *Pharm. World Sci.* **20**, 83 (1998). Clinical trial as adjuvant to methadone maintenance treatment: J. W. de Vos *et al.*, *Addict. Behav.* **24**, 707 (1999).

Crystals, mp 180-184°. $[\alpha]_D^{20}$ +25.5° (c = 5 in benzene). uv max (0.01N isopropanol-HCl): 254, 260, 264 nm. Practically insol in water. Soly in 0.1N HCl: 1:25 (w/v). Soly (g/100 ml) in ethanol 50; in methanol 40; in acetone 50; in ethyl acetate 40; in benzene 5; in chloroform 5. Sol in ether.

Bitartrate. [2922-44-3] Palfium. $C_{25}H_{32}N_2O_2.C_4H_6O_6$. Minute crystals, bitter taste. Dec 189-192°. Soly (w/v) at 25°: Water 20%, chloroform 30%, methanol 40%, ethanol 100%, acetone 100%.

Note: This is a controlled substance (opiate): **21 CFR**, 1308.11.

THERAP CAT: Analgesic.

2959. Dezocine. [53648-55-8] (5R,11S,13S)-13-Amino-5,6,-7,8,9,10,11,12-octahydro-5-methyl-5,11-methanobenzocyclodecen-3-ol; Wy-16225; Dalgan. $C_{16}H_{23}NO$; mol wt 245.37. C 78.32%, H 9.45%, N 5.71%, O 6.52%. Synthetic opioid agonist-antagonist. Prepn: M. E. Freed *et al.*, **DE 2159324** (1972 to Am. Home Products), *C.A.* **79**, 53094w (1973); *eidem, J. Med. Chem.* **16**, 595 (1973). Pharmacology: J. L. Malis *et al., J. Pharmacol. Exp. Ther.* **194**, 488 (1975). Metabolism: S. F. Sisenwine, C. O. Tio, *Drug Metab. Dispos.* **9**, 37 (1981). Pharmacokinetics: S. F. Sisenwine *et al., ibid.* **10**, 366 (1982). Clinical studies: W. Oosterlinck, A. Verbaeys, *Curr. Med. Res. Opin.* **6**, 472 (1980); J. W. Downing *et al., Br. J. Anaesth.* **53**, 59 (1981).

Hydrobromide salt. $C_{15}H_{23}NO.HBr$. Crystalline powder, mp 269-270°. pH (2% aq soln): 4.6. Soly in water: >20 mg/ml. LD_{50} in female mice, rats (mg/kg): 313, 232 orally; 129, 270 i.m. (Malis).

THERAP CAT: Analgesic.

2960. DFDT. [475-26-3] 1,1'-(2,2,2-Trichloroethylidene)-bis[4-fluorobenzene]; 1,1,1-trichloro-2,2-bis(p-fluorophenyl)ethane; difluorodiphenyltrichloroethane; fluorogesarol; HO-2474; Gix. $C_{14}H_9Cl_3F_2$; mol wt 321.57. C 52.29%, H 2.82%, Cl 33.07%, F 11.82%. Prepn: Sumerford, *J. Am. Pharm. Assoc.* **36**, 127 (1947); Cross, **US 2581174** (1952). Toxicity: Piekarski, Holz, *Arch. Exp. Pathol. Pharmakol.* **210**, 71 (1950).

Crystals from aq ethanol, mp 44-45°. Practically insol in water; very sol in oils, most organic solvents.

Caution: Effects similar to DDT [1,1,1-trichloro-2,2-bis(p-chlorophenyl)ethane], *q.v.*

USE: Contact insecticide.

2961. Dhurrin. [499-20-7] (αS)-α-(β-D-Glucopyranosyloxy)-4-hydroxybenzeneacetonitrile; p-hydroxymandelonitrile-β-D-glucoside; β-D-glucopyranosyloxy-L-p-hydroxymandelonitrile. $C_{14}H_{17}NO_7$; mol wt 311.29. C 54.02%, H 5.50%, N 4.50%, O 35.98%. From the fruit of young Egyptian plants of *Sorghum vulgare* Pers., *Gramineae:* Dunstan, Henry, *Chem. News* **85**, 301 (1902). Biosynthesis: Koukol *et al., J. Biol. Chem.* **237**, 3223 (1962). Absolute configuration: Towers *et al., Tetrahedron* **20**, 71 (1964). Allied to amygdalin, *q.v.* and the mandelonitrile glucosides.

Leaflets from water, prisms from alcohol, dec 200°. $[\alpha]_D^{20}$ −65° (alc). uv max: 228 nm (ε 13000). Sol in water, alcohol. Alkali labile.

Pentaacetate. $C_{24}H_{27}NO_{12}$. Needles from ethyl acetate + petr ether, mp 132-132.5°. $[\alpha]_D^{20}$ −50.5° (c = 0.24 in alc).

Caution: Yields HCN on hydrolysis.

2962. Diaboline. [509-40-0] (17R)-1-Acetyl-19,20-didehydro-17,18-epoxycuran-17-ol; N-acetyl-Wieland-Gumlich aldehyde. $C_{21}H_{24}N_2O_3$; mol wt 352.43. C 71.57%, H 6.86%, N 7.95%, O 13.62%. From *Strychnos diaboli* Sandw., *Loganiaceae:* King, *J. Chem. Soc.* **1949**, 955; Bader *et al., Helv. Chim. Acta* **36**, 1256 (1953); Casinovi *et al., Nature* **193**, 1178 (1962). Structure: Battersby, Hodson, *Proc. Chem. Soc. London* **1959**, 123. Synthesis: Deyrup *et al., Helv. Chim. Acta* **45**, 2266 (1962). ^{13}C-NMR study: E. Wenkert *et al., J. Org. Chem.* **43**, 1099 (1978).

Needles from ether, mp 187°. $[\alpha]_D^{22}$ +37.8° (c = 1.72 in chloroform). uv max (ethanol): 249 nm (log ε 4.06).

Hydrochloride monohydrate. $C_{21}H_{24}N_2O_3.HCl.H_2O$. Needles from alcohol + ethyl acetate, mp 300°. For anhydr, $[\alpha]_{5461}^{20}$ +184° (c = 0.57 in water).

2963. Diacerein. [13739-02-1] 4,5-Bis(acetyloxy)-9,10-dihydro-9,10-dioxo-2-anthracenecarboxylic acid; 9,10-dihydroxy-4,5-dihydro-9,10-dioxo-2-anthroic acid diacetate; 1,8-diacetoxy-3-carboxyanthraquinone; 4,5-diacetoxyanthraquinone-2-carboxylic acid; diacerhein; diacetylrhein; DAR; SF-277; Artrodar; Fisiodar. $C_{19}H_{12}O_8$; mol wt 368.30. C 61.96%, H 3.28%, O 34.75%. Diacetyl derivative of rhein, *q.v.* Demonstrates anti-arthritic activity without inhibiting prostaglandin synthesis. Prepn: A. Tschirch, K. Heuberger, *Arch. Pharm.* **240**, 596 (1902); M. Anchel, *J. Biol. Chem.* **177**, 169 (1949); V. K. Murty *et al., Tetrahedron* **23**, 515 (1967). Effect on prostaglandin release: P. Pomarelli *et al., Farmaco Ed. Sci.* **35**, 836 (1980). Inhibition of proteases: L. Raimondi *et al., Pharmacol. Res. Commun.* **14**, 103 (1982). Use in arthritis: C. A. Friedmann, **DE 2711493**; *idem*, **US 4244968** (1977, 1981 both to Proter). Preliminary clinical study in osteoarthritis: A. G. Kay *et al., Curr. Med. Res. Opin.* **6**, 548 (1980). Use in multiple sclerosis, amyotrophic lateral sclerosis: C. A. Friedmann, **US 4346103** (1982); *idem*, **JP Kokai 83 225015** (1983 to Proter), *C.A.* **100**, 144991e (1984). Clinical trial in osteoarthritis: S. Adami *et al., Clin. Ter.* **112**, 439 (1985).

Yellow plates from acetic acid, mp 217-218° (Murty); also reported as mp 250-251° (Anchel).

THERAP CAT: Antiarthritic.

2964. Diacetin. [25395-31-7] 1,2,3-Propanetriol diacetate; glycerol diacetate; glyceryl diacetate. $C_7H_{12}O_5$; mol wt 176.17. C 47.72%, H 6.87%, O 45.41%. Commercial material is probably a mixture of glycerol 1,2- and 1,3-diacetates. Prepn of the 1,2- and 1,3-compounds: Wegscheider, Zmerzlikar, *Monatsh. Chem.* **34**, 1061 (1913); of the 1,2-compd: Golendeev, *J. Gen. Chem. USSR* **6**, 1841 (1936); of the 1,3-compd: Lagenbeck, Bollow, *Naturwissenschaften* **42**, 389 (1955).

1,2-Compd   1,3-Compd

Hygroscopic liquid, bp 259°. d_4^{16} 1.184. n_D^{20} 1.44. Sol in water, alc, ether, benzene; practically insol in CS_2.

dl-1,2-Compd. 1,2-Diacetin. bp_{12} 140-142°, bp_{40} 172-173.5°. d_4^{15} 1.1173.

1,3-Compd. 1,3-Diacetin. bp_{40} 172-174°. d^{15} 1.179. n_D^{20} 1.4395.

USE: Technical grade as a plasticizer and softening agent, and as a solvent.

2965. Diacetonamine. [625-04-7] 4-Amino-4-methyl-2-pentanone. $C_6H_{13}NO$; mol wt 115.18. C 62.57%, H 11.38%, N 12.16%, O 13.89%. Prepd from mesityl oxide with aq NH_3: Haeseler, *J. Am. Chem. Soc.* **47**, 1195 (1925); Orthner, *Ann.* **456**, 245 (1927); Haeseler, *Org. Synth.* **coll. vol. I** (2nd ed., 1941) p 196. From mesityl oxide with liquid ammonia: Smith, Adkins, *J. Am. Chem. Soc.* **60**, 408 (1938).

Liq, $bp_{0.2}$ 25°. Amine odor. Lighter than water. Strongly alkaline. Sol in water, but aq solns prepd in the cold become turbid on heating; miscible with alc, ether. Dec easily forming mesityl oxide and NH_3. Decompn is prone to take place in aq solns or during distillation.

Acid oxalate monohydrate. [5895-86-3] $C_6H_{13}NO.C_2H_2O_4.-H_2O$; mol wt 223.23. Spear-shaped crystals, mp 126-127°.

2966. Diacetone Acrylamide. [2873-97-4] *N*-(1,1-Dimethyl-3-oxobutyl)-2-propenamide; *N*-(1,1-dimethyl-3-oxobutyl)acrylamide; *N*-[2-(2-methyl-4-oxopentyl)]acrylamide. $C_9H_{15}NO_2$; mol wt 169.22. C 63.88%, H 8.94%, N 8.28%, O 18.91%. Highly reactive vinyl monomer. Prepd by the reaction of acrylonitrile with acetone in presence of conc sulfuric acid: Coleman *et al.*, *J. Polym. Sci.* **3A**, 1601 (1965); Coleman, *US 3277056* (1966 to Lubrizol). Toxicity data: K. Hashimoto *et al.*, *Arch. Toxicol.* **47**, 179 (1981). *Review:* idem, "*N*-Oxoalkylacrylamides" in *Encyclopedia of Polymer Science and Technology* vol. 15, N. M. Bikales, Exec. Ed. (Wiley-Interscience, New York, 1971) pp 353-364.

White crystalline solid, mp 57-58°. $bp_{0.1-0.3}$ 93-100°; bp_8 120°. Hygroscopic. Has long shelf life; shows no polymerization on prolonged storage. A 50% neutral water soln is stable at pH 7.5-7.7 for up to six months. Similar to acrylic esters in polymerization behavior. Forms high mol wt polymers in bulk, solution, or emulsion. Esp suitable for electropolymerization. LD_{50} orally in mice: 7.7 mmol/kg (Hashimoto).

USE: In the manufacture of coatings, laminates, sealers, adhesives, lubricating oils.

2967. Diacetone Alcohol. [123-42-2] 4-Hydroxy-4-methyl-2-pentanone; Pyranton. $C_6H_{12}O_2$; mol wt 116.16. C 62.04%, H 10.41%, O 27.55%. Prepd by the action of barium hydroxide or of calcium hydroxide on acetone: Conant, Tuttle, *Org. Synth.* **vol. 1**, p 45 (1921); **coll. vol. I**, 193; *cf.* Jacquemain, *Compt. Rend.* **196**, 1622 (1933). Industrial process using KOH and acetone: Schmitt, Disteldorf, *DE 1052970* (1959 to Hibernia). Toxicity study: Smyth, Carpenter, *J. Ind. Hyg. Toxicol.* **30**, 63 (1948).

Liquid with faint, pleasant odor. d_4^{25} 0.9306 (0.940 for tech grades). bp_{760} 167.9°. bp_{100} 108.2°. bp_{20} 72.0°. bp_{10} 58.8°. $bp_{1.0}$ 22.0°. mp −44°; n_D^{20} 1.4232; Lantz, *J. Am. Chem. Soc.* **62**, 3260 (1940). Flash pt reagent grade: 66°C (151°F). Flash pt commercial grades: 8°C (48°F) (closed cup); 13°C (55°F) (open cup). Vapor density 4.00 (air = 1.00). *Flammable.* Miscible with water, alcohol, ether, and other solvents. Dec by prolonged exposure to alkalies and by distillation at atmospheric pressure. LD_{50} orally in rats: 4.0 g/kg (Smyth).

Caution: Potential symptoms of overexposure are irritation of eyes, nose, throat and skin; corneal damage. *See NIOSH Pocket Guide to Chemical Hazards* (DHHS/NIOSH 97-140, 1997) p 90.

USE: Solvent for cellulose acetate, nitrocellulose, celluloid, fats, oils, waxes, resins. As a preservative in pharmaceutical prepns. In some antifreeze solns and in hydraulic fluids.

2968. Diacetoneglucose. [582-52-5] 1,2:5,6-Bis-*O*-(1-methylethylidene)-α-D-glucofuranose; 1,2:5,6-diisopropylidene-D-glucose; 1,2:5,6-diisopropylidene-D-glucofuranose. $C_{12}H_{20}O_6$; mol wt 260.29. C 55.37%, H 7.75%, O 36.88%. Prepd by shaking glucose with acetone contg 1% HCl: Fischer, Rund, *Ber.* **49**, 90, 93 (1916). Improved procedure (90% yield) using 85% H_3PO_4 and anhydr $ZnCl_2$ in a mechanically stirred suspension of glucose in acetone: Glen *et al.*, *US 2715121* (1955 to Am. Home Prod.).

Needles from ether or petr ether. Bitter taste. mp 110-111°. Sublimes. $[\alpha]_D^{20}$ −18.5° (c = 5). Reduces Fehling's soln. One part dissolves in about 7 parts of boiling water and in about 200 parts of boiling petr ether. Freely sol in chloroform, alc, acetone, warm ether. Precipitated from aq soln by NaOH soln. Is not split by yeast or emulsin.

2969. Diacetyl. [431-03-8] 2,3-Butanedione; biacetyl; dimethyl diketone; dimethylglyoxal; 2,3-diketobutane. $C_4H_6O_2$; mol wt 86.09. C 55.81%, H 7.03%, O 37.17%. $CH_3COCOCH_3$. Found in bay and other oils; also in butter. Made from methyl ethyl ketone by converting to the isonitroso compd and then decomposing to diacetyl by hydrolysis with HCl; by special fermentation of glucose *via* methylacetylcarbinol: Suomalainen, Jännes, *Nature* **157**, 336 (1946). Toxicity study: P. M. Jenner *et al.*, *Food Cosmet. Toxicol.* **2**, 327 (1964).

Yellowish-green liq; quinone odor; the vapors have a Cl-like odor. d_{15}^{15} 0.990. bp 88°. n_D^{18} 1.3933. Sol in ~4 parts water; misc with alcohol, ether. LD_{50} orally in rats: 1580 mg/kg (Jenner).

USE: Carrier of aroma of butter, vinegar, coffee, and other foods.

2970. Diacetyldihydromorphine. [509-71-7] (5α,6α)-4,5-Epoxy-17-methylmorphinan-3,6-diol diacetate (ester); dihydroheroin; paralaudin. $C_{21}H_{25}NO_5$; mol wt 371.43. C 67.91%, H 6.78%, N 3.77%, O 21.54%. Prepn from dihydromorphine: Eddy, Howes, *J. Pharmacol.* **53**, 430 (1935); Small, Mallonee, *J. Org. Chem.* **12**, 558-566 (1947); K. W. Bentley, *The Chemistry of the Morphine Alkaloids* (Oxford, 1954).

Needles, mp 165-167°. Slightly sol in water; sol in chloroform, alcohol, ether.

Hydrochloride. $C_{21}H_{25}NO_5 \cdot HCl$. Flakes, scales. mp 215-218°. $[\alpha]_D^{29} -59°$ (c = 1.2). Very sol in water.

2971. Diacetylmorphine. [561-27-3] (5α,6α)-7,8-Didehydro-4,5-epoxy-17-methylmorphinan-3,6-diol 3,6-diacetate; heroin; diamorphine; acetomorphine. $C_{21}H_{23}NO_5$; mol wt 369.42. C 68.28%, H 6.28%, N 3.79%, O 21.65%. Semisynthetic opioid analgesic. Prepn from morphine and acetic anhydride: O. Hesse, *Ann.* **220**, 203 (1883); from morphine and acetyl chloride: Small, Lutz, *Chemistry of the Opium Alkaloids,* Supplement No. 103, Public Health Reports, Washington (1932); K. W. Bentley, *The Chemistry of the Morphine Alkaloids* (Oxford, 1954). Vapor pressure studies: A. H. Lawrence *et al., Can. J. Chem.* **42**, 1886 (1984). Pharmacology: N. B. Eddy, H. A. Howes, *J. Pharmacol. Exp. Ther.* **53**, 430 (1935). Pharmacodynamics: J. G. Umans, C. E. Inturrisi, *Eur. J. Pharmacol.* **85**, 317 (1982). Clinical pharmacokinetics: F. Girardin *et al., Clin. Pharmacol. Ther.* **74**, 341 (2003). Comprehensive description: D. K. Wyatt, L. T. Grady, *Anal. Profiles Drug Subs.* **10**, 357-403 (1981). Review of determn methods in biological fluids: S. Pichini *et al., Mass Spectrom. Rev.* **18**, 119-130 (1999). Use as adjunct to anesthesia for Caesarean section: S. Lane *et al., Anaesthesia* **60**, 453 (2005). Review of use in maintenance treatment of heroin addiction: M. Ferri *et al., J. Subst. Abuse Treat.* **30**, 63-72 (2006).

Orthorhombic plates, tablets from ethyl acetate. mp 173°. bp$_{12}$ 272-274°. $[\alpha]_D^{25} -166°$ (c = 1.49 in methanol). One gram dissolves in 1.5 ml chloroform, 31 ml alcohol, 100 ml ether, 1700 ml water. Slightly sol in ammonia or sodium carbonate soln, sol in alkalies, dec by boiling with water. Turns pink and emits acetic odor on prolonged exposure to air. LD$_{50}$ i.v. in mice: 59 μmol/kg (Umans, Inturrisi).

Hydrochloride monohydrate. $C_{21}H_{23}NO_5 \cdot HCl \cdot H_2O$. Fine crystals, mp 243-244°. $[\alpha]_D^{24} -156°$ (c = 1.044). Sol in 2 parts water, 11 parts alcohol. Insol in ether.

Methyl iodide. $C_{21}H_{23}NO_5 \cdot CH_3I$. Needles, mp 252°, $[\alpha]_D^{15} -107°$ (c = 0.896).

Note: "China White" has been used as a term for very pure Southeast Asian heroin. This term has also been erroneously used to refer to **3-methylfentanyl** and α-methylfentanyl, *q.v.,* which are potent derivs of fentanyl, *q.v. See:* S. Stinson, *Chem. Eng. News* **59**, 71 (Jan. 19, 1981).

Note: This is a controlled substance (opium derivative): **21 CFR,** 1308.11.

THERAP CAT: Analgesic.

2972. Diafenthiuron. [80060-09-9] *N*-[2,6-Bis(1-methylethyl)-4-phenoxyphenyl]-*N*'-(1,1-dimethylethyl)thiourea; 1-*tert*-butyl-3-(2,6-diisopropyl-4-phenoxyphenyl)thiourea; CGA-106630; Pegasus; Polo. $C_{23}H_{32}N_2OS$; mol wt 384.58. C 71.83%, H 8.39%, N 7.28%, O 4.16%, S 8.34%. Thiourea proinsecticide that acts via its carbodiimide metabolite to inhibit mitochondrial ATPase. Prepn: J. Drabek, M. Böger, **FR 2465720**; *eidem,* **US 4939257** (1981, 1990 both to Ciba-Geigy). Properties and insecticidal activity: H. P. Streibert *et al., Brighton Crop Prot. Conf. - Pests Dis.* **1988**, 25. Field trials: H. P. Streibert, D. Kaeding, *ibid.* **1994**, 743. Mechanism of action studies: F. J. Ruder *et al., Pestic. Biochem. Physiol.* **41**, 207 (1991); E. Petroske, J. E. Casida, *ibid.* **53**, 60 (1995).

Colorless crystals, mp 149.6°. Vapor pressure: 1.8×10^{-8} Pa. Soly at 20° (g/100ml): water 0.00005; cyclohexanone 38; xylene 21; hexane 0.8. LD$_{50}$ in rats (mg/kg): 2068 orally; >2000 dermally; LC$_{50}$ (14 hr) in rats: 558 mg/m^3 by inhalation (Streibert, 1988).

USE: Insecticide.

2973. Diallate. [2303-16-4] *N,N*-Bis(1-methylethyl)carbamothioic acid *S*-(2,3-dichloro-2-propen-1-yl) ester; diisopropylthiocarbamic acid *S*-2,3-dichloroallyl ester; *S*-2,3-dichloroallyl diisopropylthiocarbamate; DATC; CP-15336; Avadex. $C_{10}H_{17}Cl_2NOS$; mol wt 270.21. C 44.45%, H 6.34%, Cl 26.24%, N 5.18%, O 5.92%, S 11.86%. Prepn: **GB 882111**; M. W. Harman, J. J. D'Amico, **US 3330823** (1961, 1967 both to Monsanto). Toxicity data: Bailey, White, *Residue Rev.* **10**, 97 (1965). Mutagenicity studies: F. DeLorenzo *et al., Cancer Res.* **38**, 13 (1978); I. Schuphan *et al., Science* **205**, 1013 (1979). Herbicidal activity: R. Grover *et al., Weed Res.* **19**, 363 (1979). GLC determn in milk and plant tissue: L. W. Cook *et al., J. Assoc. Off. Anal. Chem.* **65**, 215 (1982). Review of carcinogenic risk: *IARC Monographs* **30**, 235-244 (1983).

Brown liquid, bp$_9$ 150°. Very slightly sol in water (40 ppm); sol in acetone, benzene, chloroform, ether, heptane. LD$_{50}$ orally in rats: 395 mg/kg (Bailey, White).

USE: Herbicide.

2974. Diallylamine. [124-02-7] *N*-2-Propen-1-yl-2-propen-1-amine; di-2-propenylamine. $C_6H_{11}N$; mol wt 97.16. C 74.17%, H 11.41%, N 14.42%. Prepd from allylamine and allyl bromide: Ladenburg, *Ber.* **14**, 1879 (1881); from allylamine and allyl chloride: Liebermann, Hagen, *Ber.* **16**, 1641 (1883); from diallylcyanamide by refluxing with dil H_2SO_4: Vliet, *J. Am. Chem. Soc.* **46**, 1307 (1924); *Org. Synth.* **coll. vol. I** (2nd ed.,1941) p 201. Toxicity study: H. F. Smyth *et al., Am. Ind. Hyg. Assoc. J.* **23**, 95 (1962).

Liquid, bp 112°. *Flammable, poisonous, corrosive.* LD$_{50}$ orally in rats: 0.65 g/kg (Smyth).

2975. Diallylcyanamide. [538-08-9] *N,N*-Di-2-propen-1-ylcyanamide; di-2-propenylcyanamide; *N*-cyanodiallylamine. $C_7H_{10}N_2$; mol wt 122.17. C 68.82%, H 8.25%, N 22.93%. Prepd by the action of allyl bromide on disodium cyanamide: Vliet, *J. Am. Chem. Soc.* **46**, 1307 (1924); *Org. Synth.* **coll. vol. I**, 203 (1941); North, **US 1659793** (1929).

Liquid. bp$_{90}$ 140-145°; bp$_{57}$ 128-133°; bp$_{18}$ 105-110°. Insol in water, sol in the usual organic solvents.

2976. Diallyl Trisulfide. [2050-87-5] Di-2-propen-1-yl trisulfide; allyl trisulfide; DATS. $C_6H_{10}S_3$; mol wt 178.33. C 40.41%, H 5.65%, S 53.93%. Component of the essential oil of garlic; formed by pyrolysis of diallyl disulfide. Major component of the Chinese traditional medicine, *allitridium.* Exhibits antifungal, antitumor and antioxidant activity. Isoln from garlic oil: F. W. Semmler, *Arch. Pharm.* **230**, 434 (1892); E. Block *et al., J. Am. Chem. Soc.* **110**, 7813 (1988). Synthesis: B. Milligan *et al., J. Chem. Soc.* **1961**, 4850; A. Banerji, G. P. Kalena, *Tetrahedron Lett.* **21**, 3003 (1980). Inhibitory effect on mammary tumor cells *in vitro*: S. G. Sundaram, J. A. Milner, *Cancer Lett.* **74**, 85 (1993). Mechanism of action study: X. Hu *et al., Arch. Biochem. Biophys.* **336**, 199 (1996). Antifungal activity: J. Shen *et al., Planta Med.* **62**, 415 (1996).

Oil. bp_6 92°; $bp_{0.0008}$ 66-67°. n_D^{20} 1.5896.

2977. Diamine Oxidase. [9001-53-0] Histaminase; benzylamine oxidase; histamine deaminase; histamine oxidase; DAO; EC 1.4.3.6. Copper containing enzyme present in various tissues, esp in the intestinal mucosa; catalyzes the oxidative deamination of histamine and other biogenic amines to form the corresponding aldehydes, ammonia and hydrogen peroxide. Rate-limiting enzyme in the terminal catabolism of polyamines that are essential for cell proliferation. Discovery of a histamine-inactivating enzyme in lung tissue: C. H. Best, *J. Physiol.* **67**, 256 (1929). Identification of activity on other diamines: E. A. Zeller, *Helv. Chim. Acta* **21**, 880 (1938). Extraction from hog kidneys: B. Swedin, *Acta Med. Scand.* **114**, 21 (1943). Improved purification: D. Wilflingseder, H. G. Schwelberger, *J. Chromatogr. B* **737**, 161 (2000). HPLC determn in plasma: P. A. Biondi *et al.*, *J. Chromatogr.* **507**, 333 (1990). *Reviews:* F. Buffoni, *Pharmacol. Rev.* **18**, 1163-1199 (1966); M. C. J. Wolvekamp, R. W. F. de Bruin, *Dig. Dis.* **12**, 2-14 (1994). Review of role in polyamine metabolism: A. Sessa, A. Perin, *Agents Actions* **43**, 69-77 (1994); of role in cell growth and differentiation: P. Pietrangeli, B. Mondovi, *Neurotoxicology* **25**, 317-324 (2004).

2978. 2,4-Diaminoanisole. [615-05-4] 4-Methoxy-1,3-benzenediamine; 4-methoxy-*m*-phenylenediamine; 2,4-DAA; 4-MMPD; C.I. 76050; C.I. Oxidation Base 12. $C_7H_{10}N_2O$; mol wt 138.17. C 60.85%, H 7.30%, N 20.28%, O 11.58%. Prepd by reduction of 2,4-dinitroanisole with iron and acetic acid: **DE 258653** (1912 to BASF); *Frdl.* **11**, 392 (1912-14). Alternate prepn: K. Fries, *Ann.* **454**, 147 (1927). Prepn of hydrochloride: F. Kehrmann, *Ber.* **50**, 562 (1917). Mutagenicity studies: B. N. Ames *et al.*, *Proc. Natl. Acad. Sci. USA* **72**, 2423 (1975); D. J. N. Hossack, J. C. Richardson, *Experientia* **33**, 377 (1977); W. G. H. Blijleven, *Mutat. Res.* **48**, 181 (1977). Toxicity studies: C. Burnett *et al.*, *Food Cosmet. Toxicol.* **13**, 353 (1975); *eidem, J. Toxicol. Environ. Health* **1**, 1027 (1976); *eidem, ibid.* **2**, 657 (1977); G. K. Lloyd *et al.*, *Food Cosmet. Toxicol.* **15**, 607 (1977).

Needles from ether, mp 67-68°. Darkens on exposure to light. LD_{50} of an aq soln containing 0.05% Na_2SO_3: 460 mg/kg orally in rats (Lloyd).

Sulfate. [39156-41-7] 4-MMPDS. $C_7H_{10}N_2O.H_2SO_4$. Off white to violet powder. Sol in water and ethanol. When heated to decomposition, it emits very toxic fumes of nitrogen oxides and sulfur oxides. LD_{50} in rats (mg/kg): 372 i.p.; >4000 orally (Burnett, 1977).

Caution: Potential symptoms of overexposure in exptl animals are skin irritation; thyroid and liver changes; teratogenic effects. *See NIOSH Pocket Guide to Chemical Hazards* (DHHS/NIOSH 97-140, 1997) p 90. 2,4-Diaminoanisole sulfate is reasonably anticipated to be a human carcinogen: *Report on Carcinogens, Twelfth Edition* (PB2011-111646, 2011) p 130.

USE: In prepn of dyes, esp hair and fur dyes; intermediates in the production of C. I. Basic Brown 2; as corrosion inhibitor for steel.

2979. *p*-Diaminoazobenzene. [538-41-0] 4,4'-(1,2-Diazenediyl)bisbenzenamine; 4,4'-azodianiline; 4,4'-azobisbenzenamine; *p*-azoaniline; 4,4'-diaminoazobenzene. $C_{12}H_{12}N_4$; mol wt 212.26. C 67.90%, H 5.70%, N 26.40%. Prepn: Witt, Kopetschini, *Ber.* **45**, 1136 (1912); Brode *et al.*, *J. Am. Chem. Soc.* **77**, 2762 (1955). Lab procedure starting with *p*-aminoacetanilide; Santurri *et al.*, *Org. Synth.* **40**, 18 (1960).

Golden-yellow needles, dec 238-241°. Absorption max (ethanol): 400 nm. Slightly soluble in water, benzene, petr ether; freely sol in alcohol.

2980. 3,5-Diaminobenzoic Acid. [535-87-5] $C_7H_8N_2O_2$; mol wt 152.15. C 55.26%, H 5.30%, N 18.41%, O 21.03%. Prepn from 3,5-dinitrobenzoic acid: Griess, *Ann.* **154**, 327 (1870).

Monohydrate. Needles. Loses its water of crystn at about 110°, mp 228°. Slightly sol in water; the aqueous soln dec on standing. Sol in alcohol or ether.

Ethyl ester. $C_9H_{12}N_2O_2$. Prisms from ether, mp 84°. Moderately sol in hot water. Sol in alcohol, benzene, other organic solvents.

USE: Detection and determination of nitrites.

2981. 4,4'-Diaminodiphenylamine. [537-65-5] N^1-(4-Aminophenyl)-1,4-benzenediamine; *p,p'*-diaminodiphenylamine; C.I. 76120; diazol black C. $C_{12}H_{13}N_3$; mol wt 199.26. C 72.33%, H 6.58%, N 21.09%. Prepn from *p*-phenylenediamine and aniline hydrochloride: Nietzki, *Ber.* **16**, 474 (1883); from *p*-aminoazobenzene: Barbier, Sisley, *Bull. Soc. Chim. Fr.* [3] **33**, 1232 (1905); by reducing bis(*p*-nitrophenyl)nitroxide with tin chloride and hydrochloric acid: Wieland, Roth, *Ber.* **53**, 213, 226 (1920).

Leaflets from water, mp 158°. Oxidized to bromanil and bromonitromethane by bromine and concd nitric acid.

USE: Dyeing fur: **DE 367690**; **DE 371231**; manuf dyes. In the detection of hydrogen cyanide.

2982. *p,p'*-Diaminodiphenylmethane. [101-77-9] 4,4'-Methylenebisbenzenamine; 4,4'-methylenedianiline. $C_{13}H_{14}N_2$; mol wt 198.27. C 78.75%, H 7.12%, N 14.13%. Prepn from aniline and formaldehyde: Scanlon, *J. Am. Chem. Soc.* **57**, 887 (1935); by hydrogenolysis of *p,p'*-diaminobenzophenone with $LiAlH_4$: Conover, Tarbell, *ibid.* **72**, 3586 (1950). Identified as causative agent in clinical outbreak of hepatoxicity called Epping Jaundice: H. Kopelman *et al.*, *Br. Med. J.* **1**, 514 (1966); D. B. McGill, J. D. Motto, *N. Engl. J. Med.* **291**, 278 (1974). Review of toxicology and human exposure: *Toxicological Profile for Methylenedianiline* (PB99-102568, 1998) 194 pp.

Crystals from water or benzene. mp 91.5-92°. bp_{768} 398-399°; bp_{18} 257°; bp_{15} 249-253°; bp_9 232°. Very sol in alc, benzene, ether; slightly sol in cold water.

Caution: Potential symptoms of overexposure by ingestion are jaundice, weakness, abdominal pain, nausea, vomiting, anorexia, fever, chills, hepatotoxicity (Kopelman). Direct contact may cause skin irritation (PB99-102568). This substance and its dihydrochloride are reasonably anticipated to be human carcinogens: *Report on Carcinogens, Twelfth Edition* (PB2011-111646, 2011) p 265.

USE: As chemical intermediate in prodn of isocyanates and polyisocyanates for prepn of polyurethane foams, Spandex fibers; as curing agent for epoxy resins and urethane elastomers; in production of polyamides; in the determination of tungsten and sulfates; in prepn of azo dyes; as corrosion inhibitor.

2983. 2,4-Diamino-6-hydroxypyrimidine. [56-06-4] 2,6-Diamino-4(3*H*)-pyrimidinone. $C_4H_6N_4O$; mol wt 126.12. C 38.09%,

H 4.80%, N 44.42%, O 12.69%. Prepn from guanidine hydrochloride and cyanoacetic ester: Traube, *Ber.* **33**, 1371 (1900).

Crystallizes from water as the monohydrate. When anhydrous, dec 286°.

USE: In the detection of nitrites and nitrates.

2984. 2,3-Diaminophenazine. [655-86-7] 2,3-Phenazinediamine. $C_{12}H_{10}N_4$; mol wt 210.24. C 68.56%, H 4.79%, N 26.65%. Prepn from o-phenylenediamine: Ullmann, Mauthner, *Ber.* **35**, 4304 (1902); Knoevenagel, *J. Prakt. Chem.* [2] **89**, 25 (1914); Richter *Ber.* **44**, 3469 (1911).

Brown to yellow needles. On careful heating sublimes forming yellow leaflets. mp 264°. Sol in alcohol, benzene. Yields 2-aminophenazine and small amounts of phenazine when heated with zinc dust.

USE: Detection of bismuth, cadmium, lead, copper, and mercury.

2985. 2,4-Diaminophenol. [95-86-3] $C_6H_8N_2O$; mol wt 124.14. C 58.05%, H 6.50%, N 22.57%, O 12.89%. Prepn by reduction of 2,4-dinitrophenol: Braude *et al.*, *J. Chem. Soc.* **1954**, 3586; Neilson *et al.*, *ibid.* **1962**, 371; from m-nitroaniline: Bean, **US 2525515** (1950 to Eastman Kodak).

Crystals, dec 78-80°. Somewhat sol in alcohol, acetone; sparingly sol in ether, chloroform, petr ether; readily sol in acids, alkalies.

Dihydrochloride. Acrol; Amidol. $C_6H_8N_2O.2HCl$; mol wt 197.06. Crystals, mp 205°. Soly in water at 15°: 27.5 g/100 ml. Slightly sol in alcohol.

USE: Dihydrochloride as photographic developer, in fur and hair dyeing, in test for formaldehyde and ammonia.

2986. 2,3-Diaminopropionic Acid. [515-94-6] 3-Aminoalanine; 2,3-diaminopropanoic acid; α,β-diaminopropionic acid. $C_3H_8N_2O_2$; mol wt 104.11. C 34.61%, H 7.75%, N 26.91%, O 30.73%. Prepd by treating 2,3-dibromopropionic acid with ammonia: Klebs, *Z. Physiol. Chem.* **19**, 314 (1894); Winterstein, Küng, *ibid.* **59**, 146 (1909); Neuberg, Silbermann, *Ber.* **37**, 341 (1904); Frankland, *J. Chem. Soc.* **97**, 1318 (1910); by treating 2-amino-3-chloropropionic acid with ammonia under pressure: Fisher *et al.*, *Ber.* **40**, 3717 (1907); by the reduction of cyanooximoacetic acid esters: Godefroi, **US 2738363** (1956 to Parke, Davis). Configuration: Karrer *et al.*, *Helv. Chim. Acta* **9**, 314 (1926). *See also:* Bergmann, Grafe, *Z. Physiol. Chem.* **187**, 187 (1930).

Radially arranged cryst masses. Hygroscopic, also absorbs CO_2 from air. Begins to melt at 97°, completely liquid at 110-120°. Sol in water, yielding alkaline solns. Practically insol in alcohol, ether.

Hydrochloride. $C_3H_8N_2O_2.HCl$. Needles or leaflets, mp 236-237°. One gram dissolves in 12 ml water, practically insol in alcohol.

Sulfate hemihydrate. $(C_3H_8N_2O_2)_2.H_2SO_4.\frac{1}{2}H_2O$. Six-sided leaflets from water, dec 233-234°. One gram dissolves in 31 ml water at 20°.

Methyl ester dihydrochloride. $C_4H_{10}N_2O_2.2HCl$. Crystals, dec 170°. Freely sol in water.

Ethyl ester dihydrochloride. $C_5H_{12}N_2O_2.2HCl$. Crystals, mp 164.5-165° (some decompn). Freely sol in water.

L-Form hydrochloride. $C_3H_8N_2O_2.HCl$. Needles from water, $[\alpha]_D^{20}$ +25.0° (c = 5 in 1.0N HCl).

D-Form hydrochloride. $C_3H_8N_2O_2.HCl$. Needles from water, $[\alpha]_D^{20}$ −25.3° (c = 5 in 1.0N HCl).

USE: The salts as growth inhibitors for microorganisms.

2987. 2,6-Diaminopurine. [1904-98-9] 9H-Purine-2,6-diamine; 2,6-diamino-9H-purine. $C_5H_6N_6$; mol wt 150.15. C 40.00%, H 4.03%, N 55.97%. Antagonist of naturally occurring purines: Hitchings, Elion in *Metabolic Inhibitors* **vol. I**, R. M. Hochster, J. H. Quastel, Eds. (Academic Press, New York, 1963) pp 216-232. Molecular and electronic structure: Veillard, Pullman, *J. Theor. Biol.* **4**, 37 (1963). First synthesized by cyclization of 2,4,5,6-tetraaminopyrimidine: Traube, *Ber.* **37**, 4547 (1904). Prepn from guanidine carbonate in 70% yield and refs to earlier prepns: Taylor *et al.*, *J. Am. Chem. Soc.* **81**, 2442 (1959).

Crystals from ethanol + water, mp 302°. uv max (pH 1.9): 241, 282 nm (log ε 3.98, 4.00).

2988. Diamond. [7782-40-3] A crystalline form of carbon. Mined as a mineral, principally in South Africa. (Non-commercial) synthesis from other carbon compds (e.g., lignin) by means of elevated temperatures (about 2700°) and pressures (about 800,000 lbs/sq inch): Desch, *Nature* **152**, 148 (1943); Neuhaus, *Angew. Chem.* **66**, 525 (1954); Hall, *Chem. Eng. News* **33**, 718 (1955); Bridgman, *Sci. Am.* **1955**, 46; Hall, *J. Chem. Educ.* **38**, 484 (1961); Bundy, *Ann. N.Y. Acad. Sci.* **105**, 951-982 (1964). Books: S. Tolansky, *History and Use of Diamond* (London, 1962) 166 pp; R. Berman, *Physical Properties of Diamond* (Oxford, 1965) 442 pp.

Face-centered cubic crystal lattice. Burns when heated with a hot enough flame (over 800°, oxygen torch). d_4^{25} 3.513. n_D^{20} 2.4173. Hardness = 10 (Mohs' scale). Sp heat at 100 K: 0.606 cal/g-atom/K. Entropy at 298.16 K: 0.5684 cal/g-atom/K. Band gap energy: 6.7 ev. Dielectric constant 5.7. Electron mobility: ~1800 cm²/v-sec. Hole mobility: 1200 cm²/v-sec. Can be pulverized in a steel mortar. Attacked by laboratory-type cleaning soln (potassium dichromate + concd H_2SO_4). In the jewelry trade the unit of weight for diamonds is one carat = 200 mg. *Ref: Wall Street J.* **164**, no. 36, p 10 (Aug 19, 1964).

USE: Jewelry. Polishing, grinding, cutting glass, bearings for delicate instruments; manuf dies for tungsten wire and similar hard wires; making styli for recorder heads, long-lasting phonograph needles. In semiconductor research.

2989. Diamyl Sodium Sulfosuccinate. [922-80-5] 2-Sulfobutanedioic acid 1,4-dipentyl ester sodium salt (1:1); sulfosuccinic acid dipentyl ester sodium salt; Aerosol AY; Alphasol AY. $C_{14}H_{25}NaO_7S$; mol wt 360.40. C 46.66%, H 6.99%, Na 6.38%, O 31.07%, S 8.90%. The amyl or 1-methylbutyl diester of the monosodium salt of sulfosuccinic acid or a mixture of both. Wetting agent prepd by the action of the appropriate alcohols on maleic anhydride followed by addtion of sodium bisulfite: Jaeger, **US 2028091** and **US 2176423** (1936, 1939 to Am. Cyanamid).

Available as a mixture of white, hard pellets and powder. Soly in water at 25° = 392 g/liter; at 70° = 502 g/liter. Maximum concn of electrolyte soln in which 1% of the wetting agent is sol: 3% NaCl; 2-4% NH_4Cl (turbid); 3-6% $(NH_4)_2HPO_4$ (turbid); 4% $NaNO_3$ (slightly turbid); 15% Na_2SO_4 (very slightly turbid). Also sol in pine oil, oleic acid, acetone, hot kerosene, carbon tetrachloride, glycerol, hot olive oil; insol in liquid petrolatum. Surface tension in water: 0.001% = 69.4 dyn/cm; 0.02% = 68.3 dyn/cm; 0.1% = 50.2 dyn/cm; 0.25% = 41.6 dyn/cm; 1% = 29.2 dyn/cm. Interfacial tension 1% in water *vs.* liquid petrolatum: 5 seconds = 7.55 dyn/cm; 30 seconds = 7.37 dyn/cm; 15 minutes = 7.03 dyn/cm. Interfacial tension 0.1% in water *vs.* liquid petrolatum: 5 seconds = 29.5 dyn/cm; 30 seconds = 28.6 dyn/cm; 15 minutes = 27.5 dyn/cm. Stable in acid and neutral solns, hydrolyzes in alkaline solns.

USE: As emulsifier in emulsion polymerization and as a wetting agent.

2990. 1,2-Dianilinoethane. [150-61-8] N^1,N^2-Diphenyl-1,2-ethanediamine; N,N'-diphenylethylenediamine; N,N'-diphenyl-α,ω-diaminoethane; *sym*-diphenylethylenediamine. $C_{14}H_{16}N_2$; mol wt 212.30. C 79.21%, H 7.60%, N 13.20%. Prepd by heating aniline with dichloroethane: Büchi *et al.*, *Helv. Chim. Acta* **39**, 950 (1956); by heating aniline with dibromoethane: Wanzlick, Löchel, *Ber.* **86**, 1463 (1953).

Crystals from dil ethanol, mp 67.5°; (also reported as the monohydrate). bp_{12} 228-230°. Very sol in alc, ether.

USE: Identification of aldehydes; stabilizer for resins and rubber. Intermediate in the manufacture of antihistamines.

2991. Dianin's Compound. [472-41-3] 4-(3,4-Dihydro-2,-2,4-trimethyl-2*H*-1-benzopyran-4-yl)phenol; 4-*p*-hydroxyphenyl-2,-2,4-trimethylchroman; *p*-(2,2,4-trimethyl-4-chromanyl)phenol. $C_{18}H_{20}O_2$; mol wt 268.36. C 80.56%, H 7.51%, O 11.92%. Organic clathrate; six host molecules form an hourglass-shaped cage to create inclusion complexes with small molecule guest compounds. Prepn from phenol and mesityl oxide: A. P. Dianin, *J. Russ. Phys. Chem. Soc.* **36**, 1310 (1914); and characterization of inclusion compounds: W. Baker *et al.*, *J. Chem. Soc.* **1956**, 2010; K. J. Harrington, C. P. Garland, *Sep. Sci. Technol.* **17**, 1339 (1982); *idem, ibid.* **17**, 1443 (1982-83). Crystal structure of clathrates with ethanol and chloroform, and cage characterization: J. L. Flippen *et al.*, *J. Am. Chem. Soc.* **92**, 3749 (1970). Crystal structure of "guest-free" cage: F. Imashiro *et al.*, *Acta Crystallogr.* **C54**, 1357 (1998).

Ethanol solvate: hexagonal needles from ethanol, mp 165-166°. Unsolvated form: fine needles, mp 156-157°. d 1.160. Sol in hot aq NaOH or KOH. Log P (octanol/water): 4.487.

USE: Model compound in host-guest chemistry studies.

2992. Dianisidine. [119-90-4] 3,3'-Dimethoxy-[1,1'-biphenyl]-4,4'-diamine; 3,3'-dimethoxybenzidine; 3,3'-dimethoxy-4,4'-diaminobiphenyl. $C_{14}H_{16}N_2O_2$; mol wt 244.29. C 68.83%, H 6.60%, N 11.47%, O 13.10%. Prepn: Meier, Böhler, *Ber.* **89**, 2301 (1956); Sogn, US 2794047 (1957 to Allied Chem.). Manuf of stabilized product: Cashion, US 2966519 (1960 to Allied Chem.). Review of carcinogenicity studies: *IARC Monographs* **4**, 41-47 (1974).

Crystals, becoming violet, mp 137-138°. Sol in alc, benzene, ether. Practically insol in water.

Caution: Potential symptom of overexposure by direct contact is skin irritation. *See NIOSH Pocket Guide to Chemical Hazards* (DHHS/NIOSH 97-140, 1997) p 90. 3,3'-Dimethoxybenzidine and dyes metabolized to 3,3'-dimethoxybenzidine are reasonably anticipated to be human carcinogens: *Report on Carcinogens, Twelfth Edition* (PB2011-111646, 2011) p 164.

USE: Manuf of azo dyes and pigments.

2993. Diatrizoic Acid. [117-96-4] 3,5-Bis(acetylamino)-2,-4,6-triiodobenzoic acid; 3,5-diacetamido-2,4,6-triiodobenzoic acid; amidotrizoic acid. $C_{11}H_9I_3N_2O_4$; mol wt 613.92. C 21.52%, H 1.48%, I 62.01%, N 4.56%, O 10.42%. Iodinated radiocontrast agent. Prepn: A. A. Larsen *et al.*, *J. Am. Chem. Soc.* **78**, 3210 (1956). Use as contrast medium: H. Langecker *et al.*, *Arch. Exp. Pathol. Pharmakol.* **222**, 584 (1954). Comprehensive description: H. H. Lerner, *Anal. Profiles Drug Subs.* **4**, 137-167 (1975). LC determn in pharmaceutical prepns: S. A. Farag, *J. AOAC Int.* **78**, 328 (1995). Use in separation of leukocytes from whole blood: R. J. Perper *et al.*, *J. Lab. Clin. Med.* **72**, 842 (1968); A. Ferrante, Y. H. Thong, *J. Immunol. Methods* **24**, 389 (1978); J. R. Kalmar *et al.*, *ibid.* **110**, 275 (1988). Diagnostic use in contrast enhanced computed tomography: J. S. Morrow, *Curr. Ther. Res.* **27**, 229 (1980). Review of use in small bowel obstruction: S. M. Abbas *et al.*, *Br. J. Surg.* **94**, 404-411 (2007).

Crystals from dil DMF, mp >300°. Sol in DMF, alkali hydroxide solutions; very slightly sol in water, alc.

Sodium salt. [737-31-5] Diatrizoate sodium; sodium amidotrizoate; sodium diatrizoate; Hypaque Sodium. $C_{11}H_8I_3N_2NaO_4$; mol wt 635.90. Rhombic needles, slightly salty taste, mp 261-262° (dec). Soly in water at 20°: 60 g/100 ml. pH of a 50% aq soln 7.0 to 7.5. Slightly sol in alc. Practically insol in acetone, ether. *Store at room temperature. Protect from light.* LD_{50} i.v. in rats: 14.7 g/kg (Langecker).

Meglumine salt. [131-49-7] Diatrizoate meglumine; diatrizoate methylglucamine; meglumine amidotrizoate; meglumine diatrizoate; Angiografin; Cystografin; Hypaque Meglumine. $C_{11}H_9I_3N_2O_4 \cdot C_7H_{17}NO_5$; mol wt 809.13. Rhombic needles, slightly sweet taste. mp 189-193° (dec). Soly in water at 20°: 89 g/100 ml. *Protect from strong light.*

Mixture of sodium and meglumine salts. [8064-12-8] Gastrografin; Renografin; Renovist; Urografin.

USE: Sodium salt as density gradient reagent for blood cell separation.

THERAP CAT: Diagnostic aid (radiopaque medium).

THERAP CAT (VET): Diagnostic aid (radiopaque medium).

2994. Diaveridine. [5355-16-8] 5-[(3,4-Dimethoxyphenyl)methyl]-2,4-pyrimidinediamine; 2,4-diamino-5-veratrylpyrimidine; 2,4-diamino-5-(3',4'-dimethoxybenzyl)pyrimidine. $C_{13}H_{16}N_4O_2$; mol wt 260.30. C 59.99%, H 6.20%, N 21.52%, O 12.29%. Prepn: Falco *et al.*, *J. Am. Chem. Soc.* **73**, 3758 (1951). Hitchings, Falco, US 2624732 (1953 to Burroughs Wellcome); Stenbuck *et al.*, *J. Org. Chem.* **28**, 1983 (1963); Hoffer, US 3341541 (1967 to Hoffmann-La Roche); M. Hoffer *et al.*, *J. Med. Chem.* **14**, 462 (1971). Crystal structure: T. F. Koetzle, G. Williams, *Acta Crystallogr.* **B34**, 323 (1978).

Crystals, mp 233°.

Ingredient of **Darvisul** which also contains sulfaquinoxaline.

THERAP CAT (VET): Antiprotozoan (in combination with sulfaquinoxaline).

2995. 1,5-Diazabicyclo[4.3.0]non-5-ene. [3001-72-7] 2,3,4,-6,7,8-Hexahydropyrrolo[1,2-*a*]pyrimidine; DBN. $C_7H_{12}N_2$; mol wt 124.19. C 67.70%, H 9.74%, N 22.56%. Bicyclic amidine utilized as an organic base, primarily for eliminations and isomerizations. Prepn: W. Reppe *et al.*, *Ann.* **596**, 158 (1955); and uses: H. Oediger *et al.*, *Ber.* **99**, 2012 (1966). Nucleophilicity and carbon basicity studies: M. Baidya, H. Mayr, *Chem. Commun.* **2008**, 1792. *Reviews*: H. Oediger *et al.*, *Synthesis* **1972**, 591-598; A. C. Savoca in *Encyclopedia of Reagents for Organic Synthesis* **2**, L. A. Paquette, Ed. (Wiley, New York, 1995) 1491-1494.

Colorless or light yellow liquid. *Corrosive.* bp_{11} 97-99°; $bp_{7.5}$ 95-98°. d 1.04. n_D^{20} 1.5196. Flash point, closed cup: 201°F (94°C). pKa ~13. Sol in water, ethanol, benzene, chloroform, dichloromethane, tetrahydrofuran, dimethyl sulfoxide. Hygroscopic.

USE: Reagent in synthetic organic chemistry.

2996. 1,8-Diazabicyclo[5.4.0]undec-7-ene. [6674-22-2] 2,-3,4,6,7,8,9,10-Octahydropyrimido[1,2-*a*]azepine; DBU. $C_9H_{16}N_2$; mol wt 152.24. C 71.01%, H 10.59%, N 18.40%. Bicyclic amidine that serves as an organic base for eliminations, isomerizations, and a variety of other synthetic transformations. Prepn: H. Oediger, F. Möller, *Angew. Chem. Int. Ed.* **6**, 76 (1967); F. Möller, **GB 1121924** (1968 to Bayer). Nucleophilicity and carbon basicity studies: M. Baidya, H. Mayr, *Chem. Commun.* **2008**, 1792. *Reviews*: H. Oediger *et al.*, *Synthesis* **1972**, 591-598; A. C. Savoca in *Encyclopedia of Reagents for Organic Synthesis* **2**, L. A. Paquette, Ed. (Wiley, New York, 1995) 1497-1503; N. Ghosh, *Synlett* **2004**, 574-575.

Colorless or light yellow liquid; unpleasant odor. *Corrosive.* bp_{36} 143°; bp_{25} 138-140°; bp_{14} 127-128°; $bp_{0.6}$ 80-83°. d 1.018. Flash point, closed cup: 235°F (113°C). pKa ~13. Sol in water, ethanol, acetone, benzene, ethyl acetate, carbon tetrachloride, diethyl ether, dioxane, dimethyl sulfoxide. Hygroscopic.

USE: Reagent in synthetic organic chemstry.

2997. Diazepam. [439-14-5] 7-Chloro-1,3-dihydro-1-methyl-5-phenyl-2*H*-1,4-benzodiazepin-2-one; 7-chloro-1-methyl-5-phenyl-3*H*-1,4-benzodiazepin-2(1*H*)-one; methyl diazepinone; diacepin; LA-III; Ro-5-2807; Wy-3467; NSC-77518; Apaurin; Bialzepam; Calmpose; Ceregulart; Dialar; Diastat; Diazemuls; Eurosan; Faustan; Gewacalm; Lamra; Noan; Novazam; Paceum; Paxate; Relanium; Seduxen; Servizepam; Stesolid; Unisedil; Valiquid; Valium; Vival. $C_{16}H_{13}ClN_2O$; mol wt 284.74. C 67.49%, H 4.60%, Cl 12.45%, N 9.84%, O 5.62%. Benzodiazepine; positive allosteric modulator of $GABA_A$ receptors. Prepn: L. H. Sternbach, E. Reeder, *J. Org. Chem.* **26**, 4936 (1961); E. Reeder, L. H. Sternbach, **US 3371085** (1968 to Hoffmann-La Roche). Purification: G. Chase, **US 3102116** (1963 to Hoffmann-La Roche). Synthesis: M. Gates, *J. Org. Chem.* **45**, 1675 (1980); M. Ishikura *et al.*, *ibid.* **47**, 2456 (1982). Binding study in rat brain: R. F. Squires, C. Braestrup, *Nature* **266**, 732 (1977). GC-MS determn in urine: S. Kinani *et al.*, *J. Chromatogr. A* **1141**, 131 (2007). Comprehensive description: A.

MacDonald *et al.*, *Anal. Profiles Drug Subs.* **1**, 79-99 (1972). Review of clinical pharmacokinetics: M. Mandelli *et al.*, *Clin. Pharmacokinet.* **3**, 72-91 (1978); and metabolism: L. Bertilsson *et al.*, *Pharmacol. Ther.* **45**, 85-91 (1990). Review of clinical experience and dependency potential in anxiety: J. B. Murray, *J. Psychol.* **124**, 655-674 (1990); of use in seizure treatment: J. M. Pellock, *Drug Saf.* **27**, 383-392 (2004); of therapeutic potential in nerve agent poisoning: T. C. Marrs, *Toxicol. Rev.* **23**, 145-157 (2004).

Colorless to light yellow crystalline compound. Prisms from ethanol, mp 131.5-134.5°. pKa 3.4. Freely sol in chloroform; sol in DMF, benzene, acetone, alc. Practically insol in water.

Note: This is a controlled substance (depressant): **21 CFR, 1308.14.**

THERAP CAT: Anxiolytic; muscle relaxant (skeletal); anticonvulsant.

THERAP CAT (VET): Anxiolytic; muscle relaxant (skeletal); anticonvulsant.

2998. Diazinon. [333-41-5] Phosphorothioic acid O,O-diethyl O-[6-methyl-2-(1-methylethyl)-4-pyrimidinyl] ester; O,O-diethyl O-2-isopropyl-6-methylpyrimidin-4-yl phosphorothioate; O,O-diethyl O-[6-methyl-2-(1-methylethyl)-4-pyrimidinyl] phosphorothioate; diethyl 2-isopropyl-4-methyl-6-pyrimidyl thionophosphate; dimpylate; G-24480; Basudin; Diazol; Neocidol. $C_{12}H_{21}N_2O_3PS$; mol wt 304.34. C 47.36%, H 6.96%, N 9.20%, O 15.77%, P 10.18%, S 10.53%. Organophosphate insecticide; cholinesterase inhibitor. Prepn: H. Gysin, A. Margot, **US 2754243** (1956 to Geigy). Insecticidal properties: Gasser, *Z. Naturforsch.* **8b**, 225 (1953). GC-MS determn in soil: M. S. Díaz-Cruz, D. Barceló, *J. Chromatogr. A* **1132**, 21 (2006). Toxicity data: T. B. Gaines, *Toxicol. Appl. Pharmacol.* **14**, 515 (1969). Review of toxicology and human exposure: *Toxicological Profile for Diazinon* (PB2009-100003, 2008) 298 pp.

Liquid. Faint ester-like odor. d_4^{20} 1.116-1.118. $bp_{0.002}$ 83-84°. n_D^{20} 1.4978-1.4981. Lop P (octanol/water): 3.3. Vapor pressure at 20°: 1.4×10^{-4}; at 40°: 1.1×10^{-3} (~5 times vapor pressure of parathion). Volatility at 20°: 2.4 mg/m³; at 40°: 17.6 mg/m³. Decomposes above 120°. Soly in water at 20°: 0.004%. Miscible with alc, ether, petr ether, cyclohexane, benzene, and similar hydrocarbons. More stable in alkaline formulations, than when at neutral or acid pH. LD_{50} in male, female rats (mg/kg): 250, 285 orally (Gaines).

Caution: Potential symptoms of overexposure are eye irritation; miosis, blurred vision; dizziness, confusion, weakness, convulsions; dyspnea; salivation, abdominal cramps, nausea, vomiting. *See NIOSH Pocket Guide to Chemical Hazards* (DHHS/NIOSH 97-140, 1997) p 92.

USE: Insecticide.

THERAP CAT (VET): Ectoparasiticide.

2999. Diaziquone. [57998-68-2] N,N'-[2,5-Bis(1-aziridinyl)-3,6-dioxo-1,4-cyclohexadiene-1,4-diyl]biscarbamic acid C,C'-diethyl ester; 2,5-bis(1-aziridinyl)-3,6-dioxo-1,4-cyclohexadiene-1,4-dicarbamic acid diethyl ester; 2,5-diaziridinyl-3,6-bis(ethoxycarbonylamino)-1,4-benzoquinone; aziridinylbenzoquinone; AZQ; CI-904; NSC-182986. $C_{16}H_{20}N_4O_6$; mol wt 364.36. C 52.74%, H 5.53%, N 15.38%, O 26.35%. Quinone-containing lipophilic alkylating agent. Prepn as intermediate in dyestuffs: S. Petersen *et al.*, **US 2913453** (1959 to Schenley Ind.). Synthesis and antitumor activity: A. H. Khan, J. S. Driscoll, *J. Med. Chem.* **19**, 313 (1976); J. S. Driscoll *et al.*, **US 4146622** (1979 to U.S. Gov't). *In vivo* antitumor studies: J. S. Driscoll *et al.*, *J. Pharm. Sci.* **68**, 185 (1979). Intracerebral penetration and tissue distribution in humans: N. Savaraj *et al.*, *J. Neuro-Oncol.* **1**, 15 (1983). LC determn: B. A. Allen *et al.*, *J. Chromatogr.* **222**, 146 (1981). Pharmacokinetics: S. Zimm *et al.*, *Cancer Res.* **44**, 1698 (1984). Clinical studies in treatment of CNS neoplasms: R. T. Eagan *et al.*, *J. Neuro-Oncol.* **5**, 309 (1987); E. Tapazoglou *et al.*, *Am. J. Clin. Oncol.* **11**, 474 (1988). *Review:* J. F. Bender *et al.*, *Invest. New Drugs* **1**, 71-84 (1983).

Orange needles from ethanol, mp 230° (dec); also reported as yellowish-brown crystals from ethanol decomposing at temperatures above 250° (Petersen). uv max (methanol): 340 nm (log ε 4.17). Soly in water: 0.5 mg/ml. LD_{50} in mice (mg/m^2): 30.9 i.v. (Bender).

THERAP CAT: Antineoplastic.

3000. Diazoacetic Ester. [623-73-4] 2-Diazoacetic acid ethyl ester; ethyl diazoacetate; ethyl diazoethanoate. $C_4H_6N_2O_2$; mol wt 114.10. C 42.11%, H 5.30%, N 24.55%, O 28.04%. Prepd by the action of sodium nitrite on glycine ethyl ester hydrochloride: Curtius *J. Prakt. Chem.* [2] **38**, 396 (1888); Silberrad, *J. Chem. Soc.* **81**, 600 (1902); Womack, Nelson, *Org. Synth.* **24**, 56 (1944).

Yellow oil. Pungent odor. Very volatile. Explosive. *Distillation, even under reduced pressure, is dangerous.* mp −22°. $d_4^{17.6}$ 1.0852. bp$_5$ 42°; bp$_{12}$ 45°; bp$_{88}$ 85-86°; bp$_{720}$ 140-141°. $n_D^{17.6}$ 1.4588. Slightly sol in water. Neutral reaction. Volatile with steam, ether, and benzene vapors. Miscible with alcohol, benzene, ligroin, ether. Explodes on contact with concd H_2SO_4.

3001. Diazoaminobenzene. [136-35-6] 1,3-Diphenyl-1-triazene; anilinoazobenzene; benzeneazoaniline. $C_{12}H_{11}N_3$; mol wt 197.24. C 73.07%, H 5.62%, N 21.30%. Made by diazotizing aniline dissolved in HCl with NaNO$_2$ and then adding a concd soln of sodium acetate: Hartman, Dickey, *Org. Synth.* **coll. vol. II**, 163 (1943).

Golden-yellow, small crystals. mp 98°. *Explodes when heated to 150°.* Freely sol in benzene, ether, hot alc. Insol in water.

Caution: This substance is reasonably anticipated to be a human carcinogen: *Report on Carcinogens, Twelfth Edition* (PB2011-111646, 2011) p 132.

3002. *p*-Diazobenzenesulfonic Acid. [305-80-6] 4-Sulfobenzenediazonium inner salt; DABS; diazotized sulfanilic acid; Pauly's reagent; sulfanilic acid diazonium salt. $C_6H_4N_2O_3S$; mol wt 184.17. C 39.13%, H 2.19%, N 15.21%, O 26.06%, S 17.41%. Prepn by diazotization of sulfanilic acid: E. Fischer, *Ann.* **190**, 67 (1878); K. B. Whetsel *et al.*, *J. Am. Chem. Soc.* **78**, 3360 (1956); G. Kaupp *et al.*, *Chem. Eur. J.* **8**, 1395 (2002). Crystal structure: C. Bugg *et al.*, *Acta Crystallogr.* **17**, 767 (1964). Use as denaturant for membrane

proteins: R. T. Giaquinta *et al.*, *Arch. Biochem. Biophys.* **162**, 200 (1974); M.-J. Bouchet *et al.*, *J. Med. Chem.* **30**, 2222 (1987); of protein components: S. K. Nishimoto *et al.*, *Anal. Biochem.* **216**, 156 (1994). Determn of bilirubin: J. B. Landis, H. L. Pardue, *Clin. Chem.* **24**, 1690 (1978); P. Fossati *et al.*, *ibid.* **35**, 173 (1989); of phenols: C. Baiocchi *et al.*, *Anal. Lett.* **15**, 1539 (1982); M. C. Quintero *et al.*, *Talanta* **36**, 717 (1989); of halides: M. Pandey *et al.*, *Analyst* **123**, 2319 (1998); S. Gosain *et al.*, *J. AOAC Int.* **82**, 167 (1999).

Solid. *Explodes* upon heating to mp 104°; shock sensitive (hammer and anvil test). Insol in water.

USE: Reagent for determination of phenols, bilirubin, halides and amines. Irreversible probe in membrane protein studies.

3003. Diazolidinyl Urea. [78491-02-8] *N*-[1,3-Bis(hydroxymethyl)-2,5-dioxo-4-imidazolidinyl]-*N*,*N*'-bis(hydroxymethyl)urea; *N*-(hydroxymethyl)-*N*-(1,3-dihydroxymethyl-2,5-dioxo-4-imidazolidinyl)-*N*'-(hydroxymethyl)urea; Germall II. $C_8H_{14}N_4O_7$; mol wt 278.22. C 34.54%, H 5.07%, N 20.14%, O 40.25%. Member of a family of heterocyclic substituted urea compounds with formaldehyde releasing activity. Prepn: P. A. Berke, W. E. Rosen, **WO 8100566**; *eidem*, **US 4271176** (both 1981 to Sutton). Chemistry, toxicity and antimicrobial activity: *eidem*, *Cosmet. Toiletries* **97**, 49 (1982). Determn in cosmetic products: R. J. Geise *et al.*, *J. Soc. Cosmet. Chem.* **45**, 173 (1994). Review of use and safety assessment: *J. Am. Coll. Toxicol.* **9**, 229-245 (1990).

Fine, white, free-flowing powder. Sol in water as a 30% soln. Insol in fats. LD_{50} orally in rats: 2.57 g/kg; dermally in rabbits: >2.0 g/kg (Berke, Rosen).

USE: Antimicrobial preservative in cosmetics and toiletries.

3004. Diazomethane. [334-88-3] Azimethylene. CH_2N_2; mol wt 42.04. C 28.57%, H 4.80%, N 66.64%. $CH_2=N^+=N^-$. Prepd from chloroform and hydrazine by reaction with potassium hydroxide: Staudinger, Kupfer, *Ber.* **45**, 501 (1912); McManus *et al.*, *J. Org. Chem.* **33**, 4272 (1968); from KOH and nitrosomethylurea: Dessaux, Durand, *Bull. Soc. Chim. Fr.* **1963**, 41. These methods yield gaseous diazomethane. The following procedures yield ether solns of diazomethane. From *N*-nitroso-β-methylaminoisobutyl methyl ketone in ether and isopropanol by reaction with sodium isopropoxide or from the same ketone in ether by reaction with sodium cyclohexoxide: Redemann *et al.*, *Org. Synth.* **25**, 28 (1945). By KOH saponification of nitrosomethylurea in ether: Arndt, *Org. Synth.* **coll. vol. II**, 165 (1943); or of nitrosomethylurethan in ether: von Pechmann, *Ber.* **27**, 1888 (1894); *ibid.* **28**, 855 (1895); Meerwein, Burneleit, *Ber.* **61**, 1845 (1928). Laboratory prepn by the action of alkali on *N*-methyl-*N*-nitroso-*N*'-nitroguanidine: McKay, *J. Am. Chem. Soc.* **70**, 1974 (1948); McKay *et al.*, *Can. J. Res.* **28**, 683 (1950); Fieser and Fieser, *Organic Chemistry* (Reinhold, New York, 3rd ed., 1956) p 176. Alternate intermediate: *p*-tolylsulfonylmethylnitrosamide, see De Boer, Backer, *Org. Synth.* **coll. vol. IV**, 250 (1963).

Very toxic yellow gas. *Explosive.* Use safety screen. Insidious poison, a well-ventilated hood is absolutely necessary, avoid vapor. mp −145°; bp −23°. Undil liquid and concd solns may explode violently, especially if impurities are present. Gaseous diazomethane may explode on heating to 100° or on rough glass surfaces. Ground glass apparatus and glass stirrers with glass sleeve bearings where grinding may occur, should not be used. Alkali metals also produce explosions with diazomethane. Sol in ether, dioxane. Such solns dec only slowly at low temps. Decompn is more rapid if alcohols or water are present. Copper powder causes active decompn with the evolution of nitrogen and the formation of insol white flakes of polymethylene $(CH_2)_x$. Solid calcium chloride or boiling stones have the same effect. This phenomenon appears to occur always during the action of diazomethane on solid substances.

Caution: Potential symptoms of overexposure are eye irritation; coughing, shortness of breath; headache; fatigue, flushing of skin, fever; chest pain, pulmonary edema, pneumonitis; asthma; direct contact with liquid may cause frostbite. *See NIOSH Pocket Guide to Chemical Hazards* (DHHS/NIOSH 97-140, 1997) p 92. May act as a respiratory sensitizer. *See Patty's Industrial Hygiene and Toxicology* **vol. 2A**, G. D. Clayton, F. E. Clayton, Eds. (Wiley-Interscience, New York, 3rd ed., 1981) pp 2784-2786.

USE: Powerful methylating agent for acidic compds such as carboxylic acids, phenols, enols. For syntheses with diazomethane *see:* Smith, *Chem. Rev.* **23**, 193 (1938); Eistert, *Z. Angew. Chem.* **54**, 99, 124 (1941) translated by Spangler in *Newer Methods of Preparative Organic Chemistry* (New York, 1948) p 513; J. S. Pizey, *Synthetic Reagents* **vol. 2** (John Wiley, New York, 1974) pp 65-142.

3005. 6-Diazo-5-oxo-L-norleucine. [157-03-9] DON. $C_6H_9N_3O_3$; mol wt 171.16. C 42.10%, H 5.30%, N 24.55%, O 28.04%. Antitumor antibiotic produced by an unidentified species of *Streptomyces* from Peruvian soil: Dion *et al., J. Am. Chem. Soc.* **78**, 3075 (1956); Ehrlich *et al., Antibiot. Chemother.* **6**, 487 (1956). Several methods of synthesis: De Wald, Moore, *ibid.* **80**, 3941 (1958); *eidem,* **US 2965634** (1960 to Parke, Davis). *Review:* Pittillo, Hunt, *Antibiotics* **vol 1**, D. Gottlieb, P. D. Shaw, Eds. (Springer-Verlag, New York, 1967) pp 481-493.

Pale yellow crystals from dil ethanol, dec 145-155°. $[\alpha]_D^{26}$ +21° (c = 5.4). uv max (phosphate buffer): 244, 275 nm ($E_{1cm}^{1\%}$ 376, 683). Freely sol in water. The pH of aq solns should be maintained between 4.5 and 6.5. Also sol in aq solns of methanol, ethanol, acetone. Sparingly sol in absolute alcs.

3006. 5-Diazouracil. [2435-76-9] 5-Diazo-2,4(3H,5H)-pyrimidinedione; DU; 2,4-dioxo-5-diazopyrimidine; 5-diazo-2,4-dioxopyrimidine. $C_4H_2N_4O_2$; mol wt 138.09. C 34.79%, H 1.46%, N 40.57%, O 23.17%. Prepd by diazotization of 5-aminouracil: Johnson *et al., Ber.* **64**, 2629 (1931). Revised structure: Thurber, Townsend, *J. Heterocycl. Chem.* **9**, 629 (1972). Has significant activity against gram-positive and gram-negative bacteria *in vivo:* Hunt, Pittillo, *Appl. Microbiol.* **16**, 1792 (1968).

Crystals, conflagrates at 198°. The stout prisms obtained from ice water should be white, not yellow or red. Sensitive to light, temp and air. Best stored in evacuated, refrigerated ampuls. Acid reactions. IR spectrum: Band at 4.57 μ. Forms alcoholates, also a red monohydrate, $C_4H_4N_4O_3$, from which a potassium salt, $KC_4H_3N_4O_3$, is obtained which is slightly sol in water with neutral reaction.

USE: In cancer research.

3007. Diazoxide. [364-98-7] 7-Chloro-3-methyl-2H-1,2,4-benzothiadiazine 1,1-dioxide; 3-methyl-7-chloro-1,2,4-benzothiadiazine 1,1-dioxide; SRG-95213; Eudemine; Hyperstat; Mutabase; Proglicem; Proglycem. $C_8H_7ClN_2O_2S$; mol wt 230.67. C 41.66%, H 3.06%, Cl 15.37%, N 12.14%, O 13.87%, S 13.90%. ATP-dependent potassium channel agonist; inhibits secretion of insulin from pancreatic beta cells. Prepn: A. A. Rubin *et al., Science* **133**, 2067

(1961); J. G. Topliss *et al.,* **US 2986573**; *eidem,* **US 3345365** (1961, 1967 both to Schering); Raffa, Monzani, *Farmaco Ed. Sci.* **17**, 244 (1962). Crystal and molecular structure: G. Bandoli, M. Nicolini, *J. Cryst. Mol. Struct.* **7**, 229 (1978). Mechanism of action and metabolism: B. Raju, P. E. Cryer, *Am. J. Physiol. Endocrinol. Metab.* **288**, 80 (2005). Clinical evaluation as an adjuvant in schizophrenia: S. Akhondzadeh *et al., J. Clin. Pharm. Ther.* **27**, 453 (2002); in ischemic preconditioning in coronary artery bypass grafting: X. Wang *et al., Eur. J. Cardiothorac. Surg.* **24**, 967 (2003). Review of effect on insulin secretion: *Nutr. Rev.* **30**, 194-198 (1972); of pharmacology and efficacy in hypertension: J. Koch-Weser, *N. Engl. J. Med.* **294**, 1271-1274 (1976). .

Crystals from dil alc, mp 330-331°. uv max (methanol): 268 nm (ε 11300). Very sol in strong alkaline solns; freely sol in DMF; sol in alcohol. Practically insol to sparingly sol in water and in most organic solvents.

THERAP CAT: Antihypoglycemic; antihypertensive.

THERAP CAT (VET): In treatment of insulinomas.

3008. Dibekacin. [34493-98-6] O-3-Amino-3-deoxy-α-D-glucopyranosyl-(1 → 6)-O-[2,6-diamino-2,3,4,6-tetradeoxy-α-D-*erythro*-hexopyranosyl-(1 → 4)]-2-deoxy-D-streptamine; DKB; 3',4'-dideoxykanamycin B; debecacin. $C_{18}H_{37}N_5O_8$; mol wt 451.52. C 47.88%, H 8.26%, N 15.51%, O 28.35%. Semisynthetic analog of kanamycin, *q.v.,* effective against kanamycin-resistant bacteria. Prepn: Umezawa *et al., J. Antibiot.* **24**, 485 (1971); *eidem, Bull. Chem. Soc. Jpn.* **45**, 3624 (1972); Umezawa *et al.,* **DE 2135191** (1972 to Microbiochem. Res. Found.). Improved synthesis: T. Yoneta *et al., Bull. Chem. Soc. Jpn.* **52**, 1131 (1979). Antibacterial activity: A. G. Paradelis *et al., Antimicrob. Agents Chemother.* **14**, 514 (1978). Metabolic studies: Shimizu, *Jpn. J. Antibiot.* **26**, 522 (1973). Toxicology: Koeda *et al., ibid.* 221. Pharmacokinetics and acute toxicity: I. Komiya *et al., J. Pharmacobio-Dyn.* **4**, 356 (1981). HPLC determn in serum: H. Kubo *et al., Antimicrob. Agents Chemother.* **28**, 521 (1985). *Review:* P. Noone, *Drugs* **27**, 548-578 (1984).

$[\alpha]_D^{20}$ +132° (c = 0.65). LD$_{50}$ in mice (mg/kg): 61.0-68.0 i.v., 373.0-380.0 i.m. (Komiya).

Sulfate. [58580-55-5] Débékacyl; Icacine; Orbicin; Panamicin; Panimycin; Tokocin. White or yellowish-white powder. Slightly bitter taste. Sol in water. Practically insol in ethanol, acetone, and other organic solvents.

THERAP CAT: Antibacterial.

3009. Dibenamine. [51-50-3] N-(2-Chloroethyl)-N-(phenylmethyl)benzenemethanamine; N-(2-chloroethyl)dibenzylamine; N,N-dibenzyl-β-chloroethylamine; N,N-dibenzylaminoethyl chloride. $C_{16}H_{18}ClN$; mol wt 259.78. C 73.98%, H 6.98%, Cl 13.65%, N 5.39%. α-Adrenergic blocker. Prepn: W. S. Gump, E. J. Nikawitz, *J. Am. Chem. Soc.* **72**, 1309 (1950); T. F. Wood, **US 2540155** (1951 to Burton T. Bush, Inc.). Pharmacology and toxicity: M. Nickerson, G. M. Nomaguchi, *J. Pharmacol. Exp. Ther.* **101**, 379 (1951). Use in adrenergic receptor differentiation: R. F. Furchgott, *ibid.* **111**, 265 (1954); in specific receptor labelling: R. D. Green *et al., ibid.* **187**, 524 (1973). *In vivo* studies of protection against hepatotoxic agents: H. M. Maling *et al.*, *Toxicol. Appl. Pharmacol.* **27**, 380 (1974); E. K. Weisburger *et al.*, *ibid.* **28**, 477 (1974); H. M. Maling *et al.*, *Biochem. Pharmacol.* **23**, 1479 (1974).

Yellow oily liquid. bp$_{12}$ 194-195°; bp$_5$ 182-184°. n_D^{20} 1.5655.

Hydrochloride. [55-43-6] C$_{16}$H$_{18}$ClN.HCl; mol wt 296.24. Crystals, mp 192°, also given as 180-181°. Practically insol in water near neutrality. Sol in dil acids (2% at pH 2.1, 1% at pH 2.4 and 0.5% at pH 2.7), in 95% alcohol and in propylene glycol. Stable in acid soln, but rapidly loses activity in neutral or alkaline solns. LD$_{50}$ s.c. in mice: 800 mg/kg (Nickerson, Nomaguchi).

3010. Dibenzalacetone. [538-58-9] 1,5-Diphenyl-1,4-penta-dien-3-one; dibenzylidene acetone; distyryl ketone. C$_{17}$H$_{14}$O; mol wt 234.30. C 87.15%, H 6.02%, O 6.83%. Prepn from benzaldehyde + acetone: Conrad, Dolliver, *Org. Synth.* **12**, 22 (1939); Haslam, US 2719863 (1955 to du Pont); Tokár *et al.*, *Acta Chim. Acad. Sci. Hung.* **19**, 83 (1959). Prepn of geometrical isomers: Dinwiddie *et al.*, *J. Org. Chem.* **27**, 327 (1962).

(*E,E*)-form

(*E,E*)-Form. [35225-79-7] Crystals from hot ethyl acetate, mp 110-111°. uv max: 330 nm (ε 34,300). Practically insol in water. Slightly sol in alc, ether; sol in acetone, chloroform.

(1*Z*,4*E*)-Form. [115587-57-0] Light yellow needles from ethanol, mp 60°. uv max: 295 nm (ε 20,000).

(*Z,Z*)-Form. [58321-78-1] Yellow oil, bp$_{0.02}$ 130°. uv max: 287 nm (ε 11,000).

USE: In sun protection preparations.

3011. 1,2:5,6-Dibenzanthracene. [53-70-3] Dibenz[*a,h*]-anthracene. C$_{22}$H$_{14}$; mol wt 278.35. C 94.93%, H 5.07%. From methyl dinaphthyl ketone: Clar, *Ber.* **62**, 350, 1378 (1929); Fieser, Dietz, *ibid.* 1827; Bachmann, *J. Org. Chem.* **1**, 347 (1937). Separation by chromatography: Winterstein, Schön, *Z. Physiol. Chem.* **230**, 146 (1934). Absorption spectrum: Mayneord, Roe, *Proc. R. Soc. London* **A152**, 299 (1935). *Review:* E. Clar, *Polycyclic Hydrocarbons* **Vol. 1 & 2** (Academic Press, New York, 1964). Review of carcinogenic risk: *IARC Monographs* **3**, 178-196 (1973); of toxicology and human exposure: *Toxicological Profile for Polycyclic Aromatic Hydrocarbons* (PB95-264370, 1995) 487 pp.

Plates, leaflets from acetic acid. Crystals may be monoclinic or orthorhombic. Sublimes. mp 266°. Sol in petr ether, benzene, toluene, xylene, oils, other organic solvents; slightly sol in alc, ether. Insol in water.

Caution: This substance is reasonably anticipated to be a human carcinogen: *Report on Carcinogens, Twelfth Edition* (PB2011-111646, 2011) p 354.

3012. Dibenzepin. [4498-32-2] 10-[2-(Dimethylamino)eth-yl]-5,10-dihydro-5-methyl-11*H*-dibenzo[*b,e*][1,4]diazepin-11-one; 5-methyl-10β-dimethylaminoethyl-10,11-dihydro-11-oxodiben-zo[*b,e*][1,4]diazepine; HF-1927. C$_{18}$H$_{21}$N$_3$O; mol wt 295.39. C

73.19%, H 7.17%, N 14.23%, O 5.42%. Prepn: F. Hunziker *et al.*, *Arzneim.-Forsch.* **13**, 324 (1963); **GB 961106**; J. Schmutz, F. Hunziker, US 3312689 and US 3419547 (1964, 1967, 1968 all to Wander). Metabolism: H. Lehner *et al.*, *Arzneim.-Forsch.* **17**, 185 (1967). Comprehensive description: A. Egli, W. R. Michaelis, *Anal. Profiles Drug Subs.* **9**, 181-206 (1980).

mp 116-117°. bp$_{0.01}$ 185°.

Hydrochloride. [315-80-0] Neodalit; Noveril. C$_{18}$H$_{21}$N$_3$O.-HCl; mol wt 331.84. Crystals, mp 238°. uv max (0.1*N* HCl): 204, 220 nm (log ε 4.530, 4.458). pKa 8.25. Sol in water, alcohol, chloroform. LD$_{50}$ orally in mice: 215 mg/kg (Hunziker).

THERAP CAT: Antidepressant.

3013. Dibenzofuran. [132-64-9] 2,2′-Biphenylene oxide; 2,2′-biphenylylene oxide; dibenzo[*b,d*]furan; diphenylene oxide; DBF. C$_{12}$H$_8$O; mol wt 168.20. C 85.69%, H 4.79%, O 9.51%. Parent heterocycle of the persistent organic pollutants known as polychlorinated dibenzofurans. Prepn: W. Hoffmeister *et al.*, *Ann.* **159**, 191 (1871); C. Graebe, F. Ullmann, *Ber.* **29**, 1876 (1896). Isoln from coal tar: G. Kraemer, R. Weissgerber, *ibid.* **34**, 1662 (1901). Improved prepn: A. F. Sierakowski, *Aust. J. Chem.* **36**, 1281 (1983); D. E. Ames, A. Opalko, *Synthesis* **1983**, 234. Crystal structure: A. Banerjee, *Acta Crystallogr. B* **29**, 2070 (1973). Determn of thermodynamic properties: R. D. Chirico *et al.*, *J. Chem. Thermodyn.* **22**, 1075 (1990); H. Fujita *et al.*, *ibid.* **27**, 927 (1995). Quantitation in petroleum-base oil by HPLC and GC/MS: M. T. Cheng *et al.*, *J. Chromatogr. Sci.* **30**, 509 (1992). Gas phase atmospheric chemistry: E. S. C. Kwok *et al.*, *Environ. Sci. Technol.* **28**, 528 (1994). Acute toxicity study: G. A. LeBlanc, *Bull. Environ. Contam. Toxicol.* **24**, 684 (1980). Review of degradation via bacterial metabolism: D. C. Bressler, P. M. Fedorak, *Can. J. Microbiol.* **46**, 397-409 (2000).

Colorless plates from ethanol, mp 80-81° (Sierakowski); also reported as colorless orthorhombic crystals, mp 87° (Banerjee). bp 287°. d 1.270; d$_4^{99.3}$ 1.0886. $n_D^{99.3}$ 1.60794. Flash pt, closed cup: 266.0°F (130.0°C). Sublimes at 125-128°C (0.3 Torr). Vapor pressure at 25°C: 0.302±0.005 Pa. *Irritant.* Volatile in steam. Readily sol in diethyl ether, benzene, acetic acid; fairly sol in ethanol. Insol in water. LC$_{50}$ in *Daphnia magna* (mg/l): 7.5 (24 hr), 1.7 (48 hr) (LeBlanc).

USE: Reagent in synthetic organic chemistry; intermediate for dye production; carrier for textile dyeing and printing; antioxidant in plastics; component of heat-transfer oils.

3014. Dibenzoylmethane. [120-46-7] 1,3-Diphenyl-1,3-pro-panedione; phenyl-α-hydroxystyryl ketone; ω-benzoylacetophen-one; phenyl phenacyl ketone; γ-hydroxychalcone. C$_{15}$H$_{12}$O$_2$; mol wt 224.26. C 80.34%, H 5.39%, O 14.27%. Prepd by the action of sodium methylate on benzylideneacetophenone dibromide: Allen *et al.*, *Org. Synth.* **coll. vol. I** (2nd ed., 1941) p 205.

Orthorhombic plates, tablets from petr ether, mp 80°. bp$_{18}$ 219-221°. 100 parts alc dissolve 4.43 parts at 19°. Sol in ether, chloroform, aq NaOH. Equilibrium mixtures in methanol, alcohol, and glacial acetic acid contain 96-98% of the enol form.

Monoxime. [5741-75-3] $C_{15}H_{13}NO_2$; mol wt 239.27. Prepd by adding 4 mols hydroxylamine to 1 mol dibenzoylmethane in alcohol. Prisms from ether, mp 165°. Sparingly sol in cold alcohol.

3015. 2,3:6,7-Dibenzphenanthrene. [222-93-5] Pentaphene; dibenzo[*b,h*]phenanthrene; *β,β'*-dibenzphenanthrene; naphtho-2',3'-1,2-anthracene. $C_{22}H_{14}$; mol wt 278.35. C 94.93%, H 5.07%. Prepn by pyrolysis of 1,2- or 1,4-di-*o*-toluylbenzene: Clar *et al.*, *Ber.* **62**, 947 (1924); Clar, John, *Ber.* **64**, 986 (1931); Winterstein, Schön, *Z. Physiol. Chem.* **230**, 146 (1934); from phthalic anhydride + *o*-tolylmagnesium bromide: Clar, Stewart, *J. Chem. Soc.* **1951**, 3215; from 2-benzoylbenzoic acid: Marsili, Isola, *Tetrahedron Lett.* **1965**, 3023; Franck, Zander, *Ber.* **99**, 396 (1966).

Yellowish-green needles or leaflets from xylene, mp 257°. Absorption max (alc): 423.5, 412, 399, 379, 356, 345, 329, 314.5, 302, 289.5, 257.5, 245 nm. Sol in benzene, xylene; sparingly sol in alcohol, ether. Solns exhibit blue fluorescence in daylight.

Dipicrate. $C_{22}H_{14}\cdot 2C_6H_3N_3O_7$. Orange needles, mp 184°.

3016. Dibenzylamine. [103-49-1] *N*-(Phenylmethyl)benzenemethanamine. $C_{14}H_{15}N$; mol wt 197.28. C 85.24%, H 7.66%, N 7.10%. Prepd by the action of alcoholic ammonia on benzyl chloride: Mason, *J. Chem. Soc.* **63**, 1312 (1893).

Oil; ammoniacal odor. bp 300° with partial decompn. Practically insol in water; sol in alcohol or ether.
USE: Detection of cobalt, cyanate and iron.

3017. Dibenzyl Chlorophosphonate. [538-37-4] Phosphorochloridic acid bis(phenylmethyl) ester; benzyl phosphorochloridate; dibenzylchlorophosphate; DBPCl; dibenzylphosphoryl chloride. $C_{14}H_{14}ClO_3P$; mol wt 296.69. C 56.68%, H 4.76%, Cl 11.95%, O 16.18%, P 10.44%. Prepd by adding sulfuryl chloride to dibenzyl phosphite at 17-19° in CCl_4 diluent: Atherton *et al.*, *J. Chem. Soc.* **1948**, 1106; by adding chlorine to dibenzyl phosphite in CCl_4 at below 0°: Atherton *et al.*, **US 2490573** (1949 to Hoffmann-La Roche).

Thick oil, dec on standing or distillation. Identified by its reaction with ammonia or amines to form corresponding dibenzyl aminophosphonates.
USE: In solns with inert solvents such as chloroform, carbon tetrachloride, benzene and ether for the phosphorylation of nucleosides and amino acids.

3018. Dibenzyl Disulfide. [150-60-7] Bis(phenylmethyl) disulfide; benzyl disulfide; α-(benzyldithio)toluene; di(phenylmethyl)-disulfide. $C_{14}H_{14}S_2$; mol wt 246.39. C 68.25%, H 5.73%, S 26.02%. Prepd by the reaction of benzyl chloride with sodium disulfide or polysulfide: Blanksma, *Rec. Trav. Chim.* **20**, 137 (1901); Moran, Crandall, **US 2113092** (1938); Wojcik, **US 2185007** (1939).

Leaflets from alcohol, mp 71-72°. Another modification mp 69-70°. Dec >270°. Practically insol in water. Sol in ether, benzene, hot methanol, hot ethanol.
USE: Antioxidant in rubber compounding, stabilizer for petr fractions, additive to silicone oils. The soly in oils is increased by the presence of benzyl alcohol.

3019. Dibenzyl Phosphite. [17176-77-1] Phosphonic acid bis(phenylmethyl) ester; benzyl phosphite; dibenzyl hydrogen phosphite; dibenzyl phosphonate. $C_{14}H_{15}O_3P$; mol wt 262.24. C 64.12%, H 5.77%, O 18.30%, P 11.81%. Prepd by adding dropwise a mixture of dimethylaniline and benzyl alc to a soln of phosphorus trichloride in benzene: F. R. Atherton *et al.*, *J. Chem. Soc.* **1945**, 382; **US 2490573** (1949 to Hoffmann-La Roche). ^{1}H, ^{13}C, ^{31}P-NMR data: J. Perruchon *et al.*, *Synthesis* **2007**, 3553.

Liquid. mp 0 to 5°. $bp_{0.1}$ 160-164°; (decompn); $bp_{0.001}$ 110-120°; also reported as $bp_{0.001}$ 120-130°. n_D^{18} 1.5521.
USE: In preparation of *N*-phosphorylated amines.

3020. Diborane(6). [19287-45-7] Boroethane; diboron hexahydride. B_2H_6; mol wt 27.67. B 78.14%, H 21.86%. Review of methods of prepn: Adams in *Borax to Boranes, Advances Chem. Ser.* No. 32 (American Chemical Society, 1961) pp 60-68. Review of structure and properties: Campell, Jr. in *Progress in Boron Chemistry* vol. 1, Steinberg, McCloskey, Eds. (Macmillan, New York, 1964) pp 167-184. Reviews of reaction chemistry: Schenker, *Angew. Chem.* **73**, 81-107 (1961); Long, *Adv. Inorg. Chem. Radiochem.* **16**, 201-296 (1974). Review of toxicity: *see* Decaborane(14). General reviews: Stock, *Hydrides of Boron and Silicon* (Cornell Univ. Press, Ithaca, 1933); *passim*; Siegel, Mack, *J. Chem. Educ.* **34**, 314-317 (1957); Major, *Chem. Eng. Prog.* **54**(3), 49-54 (1958); Mikhailov, *Usp. Khim.* **31**, 417-451 (1962); *Russian Chem. Rev.* (Eng. Ed.) **31**, 224-235 (1962); Adams, *Boron, Metallo-Boron Compounds and Boranes* (Interscience, New York, 1964) pp 555-605; Lipscomb, *Boron Hydrides* (Benjamin, New York, 1963) *passim*; Long, *Prog. Inorg. Chem.* **15**, 1-99 (1972); Greenwood in *Comprehensive Inorganic Chemistry* vol. 1, J. C. Bailar, Jr., *et al.*, Eds. (Pergamon Press, Oxford, 1973) pp 763-783.

Colorless, flammable gas; repulsive, sickly-sweet odor. mp −165°; bp −92.5°. d^{-112} 0.447; $d^{-29.6}$ 0.33; $d^{15.0}$ 0.210. Critical temp 16.7°C; critical press. 39.5 atm; Cp at 25°: 13.60 cal/mole/°C. Dec at red heat to B + H_2, at lower temps to H_2 and other boron hydrides. Spontaneous ignition temp in air about 40-50°; presence of contaminants may lower the temp limit so that ignition or detonation of diborane(6)-air mixtures may occur at or below room temp. Hydrolyzes in water to H_2 + H_3BO_3. Sol in CS_2. Reacts with NH_3 to form diborane diammoniate; reacts slowly with Br_2 and explosively with Cl_2 to form boron halides; reacts with hydrocarbons or organoboron compds to give alkyl- or arylboron compds; reacts with metal alkyls to form metal borohydrides; reacts with strong electron pair donors to form borane addn compds.

Caution: Potential symptoms of overexposure are tightening of chest, precordial pain, shortness of breath, nonproductive cough and nausea; headache, lightheadedness, vertigo, chills, fever, fatigue, weakness, tremor and muscle fasciculation. *See NIOSH Pocket Guide to Chemical Hazards* (DHHS/NIOSH 97-140, 1997) p 92. *See also Patty's Industrial Hygiene and Toxicology* vol. **2B**, G. D. Clayton, F. E. Clayton, Eds. (Wiley-Interscience, New York, 3rd ed., 1981) pp 2987, 2990-2992.

USE: As catalyst for olefin polymerization; as rubber vulcanizer; as reducing agent; as flame-speed accelerator; in rocket propellants;

as intermediate in prepn of the boron hydrides; in conversion of olefins to trialkylboranes and primary alcohols; as a doping gas.

3021. Diboron Tetrachloride. [13701-67-2] 1,1,2,2-Tetra-chlorodiborane(4); tetrachlorodiborane(4); boron chloride. B_2Cl_4; mol wt 163.42. B 13.23%, Cl 86.77%. Prepd by passing gaseous boron trichloride through flow discharge between mercury or copper electrodes: Urry *et al.*, *J. Am. Chem. Soc.* **76**, 5293 (1954); Wartik *et al.*, *Inorg. Synth.* **10**, 118 (1967); by passing boron trichloride vapor over boron monoxide at 200°: McCloskey *et al.*, *J. Am. Chem. Soc.* **83**, 4750 (1961). Infrared and Raman spectra: Linevsky *et al.*, *J. Am. Chem. Soc.* **75**, 3287 (1953). Review of boron halides: Massey, *Adv. Inorg. Chem. Radiochem.* **10**, 1-152 (1967).

Liquid, ignites in air. mp −92.6°. bp$_1$ −63.5°; bp$_{45}$ 0°; bp (calc) 65.5°.

3022. Dibromantin. [77-48-5] 1,3-Dibromo-5,5-dimethyl-2,4-imidazolidinedione; 1,3-dibromo-5,5-dimethylhydantoin; DBH; DBDMH. $C_5H_6Br_2N_2O_2$; mol wt 285.92. C 21.00%, H 2.12%, Br 55.89%, N 9.80%, O 11.19%. Stable, efficient, and economic bromination reagent; alternative to bromine or NBS, *q.q.v.* Prepn: O. O. Orazi, J. Meseri, *An. Asoc. Quim. Argent.* **38**, 5 (1950); I. Markish, O. Arrad, *Ind. Eng. Chem. Res.* **34**, 2125 (1995). Analytical applications in determn of iodide, iodine values, propylthiouracil, and selenium sulfide: M. Hilp, S. Senjuk, *J. Pharm. Biomed. Anal.* **25**, 363 (2001); *idem, ibid.* **28**, 81, 303, 337 (2002). Use in aromatic bromination reactions: H. Eguchi *et al.*, *Bull. Chem. Soc. Jpn.* **67**, 1918 (1994); X. Herault *et al.*, *Org. Prep. Proced. Int.* **27**, 652 (1995); C. Chassaing *et al.*, *Tetrahedron Lett.* **38**, 4415 (1997); in oxidation of thiols: A. Khazaei *et al.*, *Synthesis* **2004**, 2959. Review of bromination chemistry: R. A. Reed, *Chem. Prod.* **23**, 299-302 (1960); A. Alam, *Synlett* **2005**, 2403-2404.

Yellow to buff colored solid, mp ~190° (dec).

USE: Brominating agent. Analytical reagent for determn of iodide and organic iodine, and for identification tests.

3023. 9,10-Dibromoanthracene. [523-27-3] $C_{14}H_8Br_2$; mol wt 336.03. C 50.04%, H 2.40%, Br 47.56%. Made by brominating a suspension of anthracene in CCl_4: Heilbron, Heaton, *Org. Synth.* **3**, 41 (1923).

Yellow needles from xylene, mp 226°. Sublimes. Absorption max: Conrad-Billroth, *Z. Phys. Chem. (Leipzig)* **33B**, 133 (1936). Insol in water; slightly sol in alcohol, ether, cold benzene; sol in hot benzene, hot toluene. Oxidation gives anthraquinone.

Addition compound with *sym*-trinitrobenzene. $C_{20}H_{11}Br_2N_3O_6$. mp 179°.

3024. *p*-Dibromobenzene. [106-37-6] 1,4-Dibromobenzene. $C_6H_4Br_2$; mol wt 235.91. C 30.55%, H 1.71%, Br 67.74%. Prepn from diazotized *p*-bromoaniline: Fry, Grote, *J. Am. Chem. Soc.* **48**, 710 (1926); by catalytic bromination of bromobenzene: Ferguson *et al., ibid.* **76**, 1250 (1954).

Crystals, mp 87.31°. bp 220.40°. d$^{99.6}$ 0.9641. $n_D^{99.3}$ 1.5743. Practically insol in water. Sol in about 70 parts alcohol; sol in benzene, chloroform; very sol in ether.

3025. Dibromochloropropane. [96-12-8] 1,2-Dibromo-3-chloropropane; 3-chloro-1,2-dibromopropane; DBCP; OS-1897; Fumazone; Nemafume; Nemagon. $C_3H_5Br_2Cl$; mol wt 236.33. C 15.25%, H 2.13%, Br 67.62%, Cl 15.00%. Prepn: Darmstädter, *Ann.* **152**, 320 (1869). Activity: C. W. McBeth, G. B. Bergeson, *Plant Dis. Rep.* **39**, 223 (1955). Toxicity study: T. R. Torkelson *et al., Toxicol. Appl. Pharmacol.* **3**, 545 (1961). Carcinogenicity studies: W. A. Olsen *et al., J. Natl. Cancer Inst.* **51**, 1993 (1973); E. K. Weisburger, *Environ. Health Perspect.* **21**, 7 (1977). Mutagenicity study: H. S. Rosenkranz, *Bull. Environ. Contam. Toxicol.* **14**, 8 (1975). Review of toxicology and human exposure: *Toxicological Profile for 1,2-Dibromo-3-chloropropane* (PB93-110906, 1992) 164 pp.

Brown liquid; pungent odor. bp 196°; bp$_{16}$ 78°; bp$_{0.8}$ 21°. d^{14} 2.093. n_D^{14} 1.553. Vapor press at 21°: 0.8 mm Hg. *Poisonous.* Slightly sol in water; misc with oils, dichloropropane, isopropyl alcohol. LD$_{50}$ in rats, mice (g/kg): 0.17, 0.26 orally (Torkelson).

Caution: Potential symptoms of overexposure are drowsiness; nausea, vomiting; irritation of eyes, nose, throat and skin; pulmonary edema; liver and kidney injury; sterility. *See NIOSH Pocket Guide to Chemical Hazards* (DHHS/NIOSH 97-140, 1997) p 92. *See also Clinical Toxicology of Commercial Products*, R. E. Gosselin *et al.*, Eds. (Williams & Wilkins, Baltimore, 5th ed., 1984) Section II, pp 167-168. This substance is reasonably anticipated to be a human carcinogen: *Report on Carcinogens, Twelfth Edition* (PB2011-111646, 2011) p 134.

USE: Soil fumigant; nematocide; intermediate in organic synthesis.

3026. 1,2-Dibromo-2,4-dicyanobutane. [35691-65-7] 2-Bromo-2-(bromomethyl)pentanedinitrile; 2-bromo-2-(bromomethyl)glutaronitrile; Tektamer 38. $C_6H_6Br_2N_2$; mol wt 265.94. C 27.10%, H 2.27%, Br 60.09%, N 10.53%. Prepn: N. Grier, S. J. Lederer, **DE 2164723**; *eidem*, **US 3833731** (1972, 1974 to Merck & Co.).

Crystals from ethanol, mp 51.2-52.5°. Mildly pungent odor. Very sol in DMF, acetone, chloroform, ethyl acetate, benzene. Sol in methanol, ethanol, ether. Insol in water.

USE: Preservative for latex paints, adhesives, latex emulsions, dispersed pigments, joint cements, metal working fluids.

3027. 4′,5′-Dibromofluorescein. [596-03-2] 4′,5′-Dibromo-3′,6′-dihydroxyspiro[isobenzofuran-1(3*H*),9′-[9*H*]xanthen]-3-one; 4,5-dibromo-3,6-fluorandiol; D & C Orange No. 5; C.I. Solvent Red 72; C.I. 45370:1. $C_{20}H_{10}Br_2O_5$; mol wt 490.10. C 49.01%, H 2.06%, Br 32.61%, O 16.32%. Prepn from fluorescein, *q.v.*: A. Baeyer, *Ann.* **183**, 1 (1876); M. A. Phillips, *J. Chem. Soc.* **1932**, 724. Metabolism: J. M. Webb *et al., J. Pharmacol. Exp. Ther.* **137**, 141 (1962). *See also: Colour Index* vol. **4** (3rd ed., 1971) p 4425.

Red plates, mp 285°. Slightly sol in water (orange with faint yellow fluorescence); sol in ethanol (orange with greenish-yellow fluorescence) and in acetone (pink with yellow fluorescence). Red-yellow soln in conc H_2SO_4, turning yellow-brown with orange ppt on dilution. Absorption spectra: R. C. Gibbs, C. V. Shapiro, *J. Am. Chem. Soc.* **51**, 1769 (1929).

Disodium salt. C.I. Acid Orange 11. $C_{20}H_8Br_2Na_2O_5$.

USE: D & C Orange No. 5 permitted for use in lipsticks, mouthwashes and dentifrices: *Fed. Regist.* **47**, 49632 (1982).

3028. 6,6′-Dibromoindigo. [1277170-99-6]; [19201-53-7] (unspecified stereo). (2*E*)-6-Bromo-2-(6-bromo-1,3-dihydro-3-oxo-2*H*-indol-2-ylidene)-1,2-dihydro-3*H*-indol-3-one; 6,6′-dibromoindigotin; punicin; C.I. Natural Violet 1; C.I. 75800. $C_{16}H_8Br_2N_2O_2$; mol wt 420.06. C 45.75%, H 1.92%, Br 38.04%, N 6.67%, O 7.62%. Major constituent of the natural dye, Tyrian purple, originally obtained from Mediterranean sea snails and known since antiquity for its long-lasting purple hue. The colorless precursor, *tyrindoxyl sulfate*, produced in the molluscan hypobrachial gland, oxidizes and dimerizes during the extraction process to form *tyriverdin*, which is converted to the purple dye upon exposure to sunlight. Isoln from *Purpura capillus*: E. Schunck, *J. Chem. Soc. Trans.* **35**, 589 (1879); from *Murex brandaris* and chemical characterization: P. Friedländer, *Ber.* **42**, 765 (1909). Synthesis: F. Sachs, R. Kempf, *ibid.* **36**, 3299 (1903); F. Sachs, E. Sichel, *ibid.* **37**, 1861 (1904). Crystal structure: P. Süsse, C. Krampe, *Naturwissenschaften* **66**, 110 (1979); S. Larsen, F. Wätjen, *Acta Chem. Scand. A* **34**, 171 (1980). Identification of precursor compounds: J. T. Baker, M. D. Sutherland, *Tetrahedron Lett.* **9**, 43 (1968); Y. Fujise *et al.*, *Chem. Lett.* **9**, 631 (1980). Review of syntheses: J. L. Wolk, A. A. Frimer, *Molecules* **15**, 5473-5508 (2010).

Dark violet crystals with copper luster. Sublimes at 290° (2 Torr). Monoclinic crystals, d 1.96. Abs max (tetrachloroethane): 590, 305 nm (ε 17300, 29030). Insol in water, ether, alcohol. Slightly sol in boiling acetic acid; sol in boiling aniline.

Tyrian purple. Shellfish purple; Byzantine purple; imperial purple; royal purple; purple of the ancients. Purple dye derived exclusively from marine mollusks of the family, *Muricidae*. Term has also been used for pure 6,6′-dibromoindigo. The natural dye also contains small amounts of 6,6′-dibromoindirubin, as well as nonbrominated indigo and indirubin depending on the source. HPLC identification of components: W. Nowik *et al.*, *J. Chromatogr. A* **1218**, 1244 (2011). Review of history, preparation, and use: J. T. Baker, *Endeavour* **33**, 11 (1974); of constituents and biosynthesis: C. J. Cooksey, *Molecules* **6**, 736-769 (2001).

USE: Vat dye for textiles.

3029. Dibromopropamidine. [496-00-4] 4,4′-[1,3-Propanediylbis(oxy)]bis(3-bromobenzenecarboximidamide); 4,4′-(trimethylenedioxy)bis(3-bromobenzamidine); 2′,2″-dibromo-4′,4″-diamidino-1,3-diphenoxypropane. $C_{17}H_{18}Br_2N_4O_2$; mol wt 470.17. C 43.43%, H 3.86%, Br 33.99%, N 11.92%, O 6.81%. Prepn: S. S. Berg, G. Newbery, *GB 598911* (1948 to May & Baker); *eidem, J. Chem. Soc.* **1949**, 642. Antibacterial activity: R. Wien *et al.*, *Lancet* **254**, 711 (1948); A. D. Russell, J. R. Furr, *Int. J. Pharm.* **34**, 115 (1986). Mode of action: W. Woodside, *Microbios* **8**, 23 (1973). Clinical use in *Acanthamoeba* keratitis: J. J. Wiens, W. B. Jackson, *Can. J. Ophthalmol.* **23**, 107 (1988). HPLC determn in cosmetics: B. Wyhowski de Bukanski, M. O. Masse, *Int. J. Cosmet. Sci.* **6**, 283 (1984).

Isethionate. [614-87-9] Brolene Ointment; Brulidine. $C_{21}H_{30}Br_2N_4O_{10}S_2$; mol wt 722.42. Prismatic needles from ethanol, mp 226°. Freely sol in water. One gram dissolves in 2 ml of water at 20°, in 60 g of 95% alcohol at 20°. Sol in glycerol. Practically insol

in ether, chloroform, fixed oils, liquid petrolatum. Aq solns are very slightly acidic. Solns may be sterilized by heating at 100° for 30 min, but they should not be stored for more than a few days, and then only in neutral glass containers since some hydrolysis occurs with the formation of sparingly sol urea derivs. *Incompat.* with chlorides, sulfates, and many organic anions, all of which form sparingly sol salts.

USE: Preservative in cosmetics.

THERAP CAT: Antiseptic; antiamebic.

THERAP CAT (VET): Antiseptic, antimicrobial.

3030. 2,3-Dibromopropene. [513-31-5] 2,3-Dibromo-1-propene; 2,3-dibromopropylene; 2-bromoallyl bromide; α-bromoallyl bromide; α-epidibromohydrin. $C_3H_4Br_2$; mol wt 199.87. C 18.03%, H 2.02%, Br 79.96%. Prepd by the action of sodium hydroxide on 1,2,3-tribromopropane: Lespieau, Bourguel, *Org. Synth.* **5**, 49 (1925); *cf.* Tapley, Giesy, *J. Am. Pharm. Assoc.* **15**, 173 (1926).

Liquid. d_4^{20} 1.9336. bp_{760} 140-143°; bp_{75} 75-76°; bp_{18} 42-43°. n_D^{20} 1.5157. May be distilled under atmospheric pressure with very slight decompn, but becomes highly colored on standing in glass-stoppered bottle.

3031. 2,6-Dibromoquinone-4-chlorimide. [537-45-1] 2,6-Dibromo-4-(chloroimino)-2,5-cyclohexadien-1-one; 2,6-dibromo-*N*-chloro-*p*-benzoquinoneimine; 2,6-dibromo-*N*-chloroquinonimine; BQC reagent. $C_6H_2Br_2ClNO$; mol wt 299.35. C 24.07%, H 0.67%, Br 53.39%, Cl 11.84%, N 4.68%, O 5.34%. Prepn: Hartman *et al.*, *Org. Synth.* **coll. vol. II**, 175 (1943).

Yellow prisms from alc or glacial acetic acid, mp 83°. Sol in about 17,000 parts water; moderately sol in hot alc, hot glacial acetic acid. Gives a blue color with alk phenol soln.

USE: As a reagent for phenol and phosphatases: Ljunggren, *Proc. 13th Int. Dairy Congr.* **3**, 1319 (1953).

3032. 3,5-Dibromosalicylaldehyde. [90-59-5] 3,5-Dibromo-2-hydroxybenzaldehyde. $C_7H_4Br_2O_2$; mol wt 279.92. C 30.04%, H 1.44%, Br 57.09%, O 11.43%. Prepd by bromination of salicylaldehyde in glacial acetic acid with cooling: Wentworth, Brady, *J. Chem. Soc.* **117**, 1043 (1920); Brewster, *J. Am. Chem. Soc.* **46**, 2464 (1924); by bromination in the presence of sodium acetate in glacial acetic acid at 50°: Lindemann, Forth, *Ann.* **435**, 223 (1924).

Pale yellow prisms, mp 86°. Sublimes forming leaflets and needles. Volatile with steam. Sparingly sol in water. Aq solns are yellow. Readily sol in ether, benzene, chloroform, hot petr ether, alcohol, glacial acetic acid.

3033. 3,5-Dibromosalicylic Acid. [3147-55-5] 3,5-Dibromo-2-hydroxybenzoic acid. $C_7H_4Br_2O_3$; mol wt 295.91. C 28.41%, H 1.36%, Br 54.01%, O 16.22%. Prepd from salicylic acid and Br_2: Earle, Jackson, *J. Am. Chem. Soc.* **28**, 111 (1906); Robertson, *J.*

Chem. Soc. **82**, 1481 (1902); and HOBr: Leulier, Pinet, *Bull. Soc. Chim. Fr.* [4] **41**, 1362 (1927).

Needles, mp 223°. Sparingly sol in H_2O; sol in alc, ether.

3034.   2,3-Dibromosuccinic Acid. [526-78-3] 2,3-Dibromobutanedioic acid; α,α'-dibromosuccinic acid; *sym*-dibromosuccinic acid. $C_4H_4Br_2O_4$; mol wt 275.88. C 17.41%, H 1.46%, Br 57.93%, O 23.20%. Prepn of *dl*-form by bromination of maleic acid: McKenzie, *J. Chem. Soc.* **101**, 1196 (1912); Young *et al., J. Am. Chem. Soc.* **61**, 1640 (1939); of *meso*-form by bromination of fumaric acid: *eidem, ibid.*; Rhinesmith, *Org. Synth.* **coll. vol. II**, 177 (1943). Resolution of *dl*-form: McKenzie, *loc. cit.*

dl-**Form.** *threo*-2,3-Dibromosuccinic acid. Crystals, mp 167°. Very sol in cold water.

d-**Form.** Crystalline powder from ethyl acetate + CCl_4, mp 157-158°. $[\alpha]_D^{18}$ +64.4° (c = 5 in water), +135.8° (c = 5 in alc), +147.8° (c = 5 in ethyl acetate).

l-**Form.** Needles from benzene, decomp 157-158°. $[\alpha]_D^{13}$ −148.0° (c = 5.8 in ethyl acetate). Very sol in cold water; sol in ethyl acetate, acetone, methanol, ethanol; sparingly sol in chloroform, petr ether, CCl_4.

meso-**Form.** *erythro*-2,3-Dibromosuccinic acid. Cryst from water, dec 255-256°, also reported as mp 270-273°. Sol in 50 parts cold water, more sol in hot water; sol in alcohol, ether; sparingly sol in chloroform.

3035.   3,5-Dibromo-L-tyrosine. [300-38-9] β-(3,5-Dibromo-4-hydroxyphenyl)alanine; Biotiren; Bromotiren. $C_9H_9Br_2NO_3$; mol wt 338.98. C 31.89%, H 2.68%, Br 47.14%, N 4.13%, O 14.16%. Obtained upon saponification of gorgonin, a substance isolated from coral *(Primnoa lepadifera)* stems: Mörner, *Z. Physiol. Chem.* **88**, 139, 152 (1913). Prepd by the action of bromine vapor on tyrosine: Gorup-Besanez, *Ann.* **125**, 281 (1863). By treating tyrosine in aq HBr with Br: Aloy, Rabaut, *Bull. Soc. Chim. Fr.* [4] **3**, 392 (1908); Zeynek, *Z. Physiol. Chem.* **114**, 275 (1921). Clinical evaluation in thyroid function: A. Isidori, *Int. J. Clin. Pharmacol. Biopharm.* **16**, 180 (1978).

L-Form dihydrate. Efflorescent needles, plates (orthorhombic) from water. When anhyd, dec 245°. $[\alpha]_D^{20}$ +1.3° (c = 5 in 4% HCl). pK_1 2.17; pK_2 6.45; pK_3 7.60. One gram dissolves in 250 ml water at 25°, in 30 ml at boiling temp. Slightly sol in alcohol; insol in ether. Freely sol in alkalies and in dil mineral acids. Forms HBr and HCl salts; stable in boiling water.

DL-Form. Efflorescent prisms, platelets from water probably contg 1 mol water of crystn. Dec at about 245°. Aq solns are acid to litmus. Behaves as a monobasic acid when titrated with NaOH and phenolphthalein. One gram dissolves in about 590 ml water at 20°.

THERAP CAT: Thyroid inhibitor.

3036.   Dibucaine. [85-79-0] 2-Butoxy-*N*-[2-(diethylamino)-ethyl]-4-quinolinecarboxamide; 2-butoxy-*N*-(2-diethylaminoethyl)-

cinchoninamide; cinchocaine. $C_{20}H_{29}N_3O_2$; mol wt 343.47. C 69.94%, H 8.51%, N 12.23%, O 9.32%. Prepn: Miescher, **US 1825623** (1931 to Ciba). Comprehensive description: G. R. Padmanabhan, *Anal. Profiles Drug Subs.* **12**, 105-134 (1983). Spectrometric determn in pharmaceutical formulations: N. T. Abdel-Ghani *et al., Farmaco* **60**, 419 (2005).

White to off-white powder; slight characteristic odor. Darkens on exposure to light. mp 64°. Sol in 1*N* hydrochloric acid, ether; slightly sol in water.

Hydrochloride. [61-12-1] Nupercainal. $C_{20}H_{29}N_3O_2 \cdot HCl$; mol wt 379.93. Hygroscopic crystals. Darkens on exposure to light. Dec 90-98°. uv max (1*N* HCl): 247, 320 nm (ε 24700, 8810). Sol in 0.5 part water; freely sol in alcohol, acetone, chloroform; slightly sol in benzene, ethyl acetate, toluene (on warming). Insol in ether, oils. The aq soln, about 1 in 20, is faintly alkaline to litmus.

THERAP CAT: Anesthetic (local).

3037.   *n*-Dibutylamine. [111-92-2] *N*-Butyl-1-butanamine. $C_8H_{19}N$; mol wt 129.25. C 74.34%, H 14.82%, N 10.84%. Prepn from butyl bromide and ammonia with separation of the mono-, di-, and tributylamines formed: Werner, *J. Chem. Soc.* **115**, 1010 (1919). Manuf: Engel, Hoog, **US 2574693** (1951 to Shell); Davies *et al.*, **US 2609394** (1952 to I.C.I.); Lemon, Myerly, **US 3022349** (1962 to Union Carbide); Brois, Rutkowski, **US 3147310** (1964 to Esso). Toxicity study: H. F. Smyth *et al., Arch. Ind. Hyg. Occup. Med.* **10**, 61 (1954).

Liquid, bp 159-160°. mp −60° to −59°. d_4^{20} 0.7601. n_D^{20} 1.4177. Flash pt, open cup: 135°F (57°C). Sol in water, alc. LD_{50} orally in rats: 550 mg/kg (Smyth).

3038.   Di-*tert*-butyl Dicarbonate. [24424-99-5] Dicarbonic acid C,C'-bis(1,1-dimethylethyl) ester; di-*tert*-butyl oxydiformate; di-*tert*-butyl pyrocarbonate; Boc anhydride; Boc_2O. $C_{10}H_{18}O_5$; mol wt 218.25. C 55.03%, H 8.31%, O 36.65%. Reagent used to introduce the *tert*-butoxycarbonyl (Boc) group to protect amines, often in peptide synthesis. Also used in the protection of alcohols and other functionalities. Prepn: J. H. Howe, L. R. Morris, *J. Org. Chem.* **27**, 1901 (1962); B. M. Pope *et al., Org. Synth.* **coll. vol. VI**, 418 (1988). Functional group protection: D. S. Tarbell *et al., Proc. Natl. Acad. Sci. USA* **69**, 730 (1972); E. Ponnusamy *et al., Synthesis* **1986**, 48; Y. Basel, A. Hassner, *J. Org. Chem.* **65**, 6368 (2000). *Review*: M. Wakselman in *Encyclopedia of Reagents for Organic Synthesis* **3**, L. A. Paquette, Ed. (Wiley, New York, 1995) pp 1602-1608.

Colorless liquid. $bp_{0.5}$ 56-57°; $bp_{0.3}$ 50-51°. mp 21-22°. n_D^{25} 1.4078. *Poisonous. Flammable. Irritant.* Flash pt, closed cup: 98.6°F (37°C). Sol in most organic solvents. Insol in cold water. Store at 2-8°C in absence of moisture.

USE: Reagent in synthetic organic chemistry.

3039.   Di-*tert*-butyldichlorosilane. [18395-90-9] Dichlorobis(1,1-dimethylethyl)silane. $C_8H_{18}Cl_2Si$; mol wt 213.22. C 45.07%, H 8.51%, Cl 33.25%, Si 13.17%. Reagent used to introduce the di-*tert*-butylsilylene (DTBS) group to protect diols. Prepn: L. J. Tyler *et al., J. Am. Chem. Soc.* **70**, 2876 (1948); and use as diol protecting group: B. M. Trost *et al., J. Org. Chem.* **48**, 3252 (1983). Synthetic utility in protection of nucleosides: K. Furusawa, T. Katsura, *Tetrahedron Lett.* **26**, 887 (1985); of glycoconjugates: D. Kumagai *et al.*, *ibid.* **42**, 1953 (2001). *Review*: S. T. Sigurdsson, P. B. Hopkins in *Encyclopedia of Reagents for Organic Synthesis* **3**, L. A. Paquette, Ed. (Wiley, New York, 1995) pp 1608-1610.

Colorless liquid. *Corrosive. Flammable. Irritant. Lachrymator.* bp_{729} 190°; bp_{20} 85-90°; $bp_{0.01}$ 44-48°. mp −15°. d^{20} 1.009. n_D^{20} 1.4561. Flash pt, closed cup: 179.6°F (82°C). Sol in most aprotic solvents. Moisture sensitive.
USE: Reagent in synthetic organic chemistry.

3040. Di-*tert*-butyl Ether. [6163-66-2] 2-(1,1-Dimethylethoxy)-2-methylpropane; 2,2′-oxybis[2-methylpropane]. $C_8H_{18}O$; mol wt 130.23. C 73.78%, H 13.93%, O 12.29%. Prepn: R. J. Moore, G. J. O'Donnell, **GB 652809** (1951 to Bataafsche Petrol.); *C.A.* **46**, 1023d (1952). Synthesis from *t*-butyl chloride and silver carbonate: J. Erickson, W. M. Ashton, *J. Am. Chem. Soc.* **63**, 1769 (1941); from *t*-butyl perbenzoates and Grignard reagents: S. O. Lawesson, N. C. Yang, *ibid.* **81**, 4230 (1959); by alkylation of *t*-butyl alcohol with trimethylcarbenium fluoroantimonate: G. A. Olah *et al.*, *Synthesis* **1975**, 315; from lithium *tert*-butoxide with aromatic sulfonyl chlorides: H. Masada *et al.*, *Tetrahedron Lett.* **1979**, 1315.

Clear liquid with camphorous odor. bp 106.5-107.0°. n_D^{20} 1.3949; d_4^{20} 0.7658. Decomposes in the presence of acids.
USE: Gasoline additive.

3041. Di-*tert*-butyl Malonate. [541-16-2] Propanedioic acid 1,3-bis(1,1-dimethylethyl) ester; propanedioic acid bis(1,1-dimethylethyl) ester; malonic acid di-*tert*-butyl ester. $C_{11}H_{20}O_4$; mol wt 216.28. C 61.09%, H 9.32%, O 29.59%. Prepd from malonic acid, liq isobutylene and sulfuric acid: McCloskey *et al.*, *Org. Synth.* **34**, 26 (1954).

Liquid. mp −6.1 to −5.9°. bp_{31} 112-115°; bp_{10} 93°; bp_1 65-67°. n_D^{20} 1.4184; $n_D^{24.2}$ 1.4161; n_D^{25} 1.4158-1.4161.
USE: In the prepn of ketones.

3042. 2,6-Di-*tert*-butyl-4-methylpyridine. [38222-83-2] 2,6-Bis(1,1-dimethylethyl)-4-methylpyridine; 2,6-di-*tert*-butyl-4-picoline; 4-methyl-2,6-di-*tert*-butylpyridine; DTBMP. $C_{14}H_{23}N$; mol wt 205.35. C 81.89%, H 11.29%, N 6.82%. Non-nucleophilic, sterically hindered organic base that can distinguish between Lewis and Bronsted acids. Serves as an acid scavenger in chemical reactions. Prepn: H. C. Brown, B. Kanner, **US 2780626** (1957 to Research Corp.); A. G. Anderson, P. J. Stang, *J. Org. Chem.* **41**, 3034 (1976); A. T. Balaban, *Org. Prep. Proced. Int.* **9**, 125 (1977). Synthetic applications: P. J. Stang, W. Treptow, *Synthesis* **1980**, 283; M. E. Wright, S. R. Pulley, *J. Org. Chem.* **52**, 1623 (1987); T. R. Forbus, Jr. *et al.*, *ibid.* 4156; D. Kahne *et al.*, *J. Am. Chem. Soc.* **111**, 6881 (1989).

Colorless oil, solidifies and darkens upon standing. *Irritant.* Slowly sublimes at room temp and atmospheric press to form long, thin needles, mp 31-32°. bp_{760} 226°; bp_{95} 148-153°. Flash pt, closed cup: 183°F (84°C). pKa (50% ethanol) 4.41. Sol in ethanol, acetic acid, diethyl ether, water (pH 1); slightly sol in water (pH 4); sparingly sol in water (pH >7). Light and air sensitive.
USE: Reagent in synthetic organic chemistry.

3043. Di-*tert*-butyl Peroxyoxalate. [1876-22-8] Ethanediperoxoic acid 1,2-bis(1,1-dimethylethyl) ester; di-*tert*-butyl peroxalate; *tert*-butyl peroxalate; DBPO; DTBPOX. $C_{10}H_{18}O_6$; mol wt 234.25. C 51.27%, H 7.75%, O 40.98%. Source of the *tert*-butoxyl radical in organic synthesis. Prepn and thermal decompn: P. D. Bartlett *et al.*, *J. Am. Chem. Soc.* **82**, 1762 (1960). Radical formation studies: R. Hiatt, T. G. Traylor, *ibid.* **87**, 3766 (1965). Synthetic applications: N. A. Porter *et al.*, *ibid.* **98**, 6000 (1976); J. L. Courtneidge, M. Bush, *J. Chem. Soc. Perkin Trans. 1* **1992**, 1531. *Review*: J. Boukouvalas in *Encyclopedia of Reagents for Organic Synthesis* **3**, L. A. Paquette, Ed. (Wiley, New York, 1995) pp 1621-1623.

Long clear crystals from pentane, mp 50.5-51.0° (dec). *Explosive. Scratch and shock sensitive.* Sol in most organic solvents.
USE: Reagent in synthetic organic chemistry.

3044. Dibutyl Phthalate. [84-74-2] 1,2-Benzenedicarboxylic acid 1,2-dibutyl ester; *n*-butyl phthalate; phthalic acid dibutyl ester; DBP. $C_{16}H_{22}O_4$; mol wt 278.35. C 69.04%, H 7.97%, O 22.99%. Prepn: Farrar, Wienkauff, *Chem. Ind. (London)* **1962**, 2144. Manuf: Bruno, **US 2628249** (1953 to Pittsburgh Coke & Chemical). Toxicity study: C. C. Smith, *Arch. Ind. Hyg. Occup. Med.* **7**, 310 (1953). Review of toxicology and human exposure: *Toxicological Profile for Di-n-butyl phthalate* (PB2001-109104, 2001) 225 pp.

Oily liquid. bp 340°. d^{20} 1.0459 and 1.0465, Kemppinen, Gokeen, *J. Phys. Chem.* **60**, 126 (1956). n_D^{20} 1.4900. Flash pt, open cup: 340°F (171°C). Miscible with alc, ether. Sol in about 2500 parts water; very sol in acetone, benzene.
Caution: Potential symptoms of overexposure are irritation of eyes, upper respiratory system and stomach. *See NIOSH Pocket Guide to Chemical Hazards* (DHHS/NIOSH 97-140, 1997) p 94.
USE: Plasticizer; solvent for oil-soluble dyes, insecticides and other organics; antifoam agent; textile fiber lubricant; fragrance fixative; insect repellent.

3045. 2,6-Di-*tert*-butylpyridine. [585-48-8] 2,6-Bis(1,1-dimethylethyl)pyridine. $C_{13}H_{21}N$; mol wt 191.32. C 81.61%, H 11.06%, N 7.32%. Prepd by reacting *tert*-butyllithium with 2-*tert*-butylpyridine: Brown, Kanner, *J. Am. Chem. Soc.* **75**, 3865 (1953); **US 2780626** (1957 to Research Corp.).

Liquid, mp 2.2°. bp_{23} 100-101°. n_D^{20} 1.5733. pKa 3.58. The base neutralizes HCl but does not react with alkyl halides or boron trifluoride. Undergoes nuclear sulfonation with sulfur trioxide forming a sulfonic acid, $C_{13}H_{21}NSO_3$, mp 310° (decompn).
Chloroaurate. $C_{13}H_{22}NAuCl_4$. mp 184.2-184.5°.

USE: Has been proposed as an additive for lubricating oil, gasoline and for stabilizing Cl-containing polymers.

3046. Di-*tert*-butyl Succinate. [926-26-1] Butanedioic acid 1,4-bis(1,1-dimethylethyl) ester; succinic acid di-*tert*-butyl ester. $C_{12}H_{22}O_4$; mol wt 230.30. C 62.58%, H 9.63%, O 27.79%. Prepd from succinic acid, liq isobutylene, and sulfuric acid in dioxane: McCloskey *et al., Org. Synth.* coll. vol. IV, 263 (1963).

Liquid, mp 36-37°. bp$_9$ 109-110°; bp$_7$ 105-107°.
USE: In Stobbe condensation of ketones.

3047. Dibutyltin Dilaurate. [77-58-7] Dodecanoic acid 1,1'-(dibutylstannylene) ester; dibutylbis[(1-oxododecyl)oxy]stannane; dibutylbis(lauroyloxy)tin; butynorate; tinostat. $C_{32}H_{64}O_4Sn$; mol wt 631.57. C 60.86%, H 10.21%, O 10.13%, Sn 18.80%. Prepn: Sheverdina *et al., Khim. Prom.* 1962 (10), 707, *C.A.* 59, 8776f (1963); FR 1320473 (1963 to Noury & van der Lande), *C.A.* 59, 8789b (1963), corresp to GB 975369. Compd claimed but not described: Eberly, US 2560034 (1951 to Firestone). Determn in poultry feeds: G. M. George *et al., J. AOAC Int.* 60, 1054 (1977).

Soft crystals or yellow liq, mp 22-24° (Sheverdina). n_D^{20} 1.4683, (FR 1320473). Practically insol in water, methanol. Sol in petr ether, benzene, acetone, carbon tetrachloride, ether, organic esters.
USE: Stabilizer for polyvinyl chloride resins. Catalyst for curing certain silicones.
THERAP CAT (VET): Anthelmintic (vs. chicken tapeworms).

3048. Dibutyltin Oxide. [818-08-6] Dibutyloxostannane; dibutyloxotin; dibutylstannane oxide; Bu$_2$SnO; DBTO. $C_8H_{18}OSn$; mol wt 248.94. C 38.60%, H 7.29%, O 6.43%, Sn 47.69%. Organotin reagent used for numerous synthetic applications, including the manipulation of hydroxyl groups. Prepn: L. I. Zakharkin, O. Y. Okhlobustin, *Izv. Akad. Nauk Ser. Khim.* 1963, 2202; H. Shapiro, P. Kobetz, US 3493592 (1970 to Ethyl Corp.); R. F. Grossman, US 6215010 (2001 to Hammond Group). Spectroscopic studies: W. F. Howard, Jr., W. H. Nelson, *J. Mol. Struct.* 53, 165 (1979). Toxicity in red killifish: H. Nagase *et al., Appl. Organomet. Chem.* 5, 91 (1991). Synthetic applications involving stannylene formation: D. Wagner *et al., J. Org. Chem.* 39, 24 (1974); involving catalysis of macrolide cyclization: K. Steliou *et al.,J. Am. Chem. Soc.* 102, 7578 (1980). Utility in carbohydrate chemistry: G. Hodosi, P. Kovác, *ibid.* 119, 2335 (1997); H.-M. Liu *et al., Carbohydr. Res.* 337, 1763 (2002). *Review:* Q. Wan, *Synlett* 2004, 1847-1848.

White, amorphous, polymeric powder, mp >300°. *Poisonous, irritant.* LC$_{50}$ (48 hr) in red killifish (mmol/dm^3): 0.00337 (Nagase).
USE: Reagent and catalyst in synthetic organic chemistry.

3049. Di-*tert*-butyl Tricarbonate. [24424-95-1] Tricarbonic acid *C,C'*-bis(1,1-dimethylethyl)ester. $C_{11}H_{18}O_7$; mol wt 262.26. C 50.38%, H 6.92%, O 42.70%. Mild reagent which converts most primary amines quantitatively into their corresponding isocyanate. Prepn: C. S. Dean, D. S. Tarbell, *Chem. Commun.* 1969, 728; B. M. Pope *et al., Org. Synth.* 57, 45 (1977). Kinetic study: *eidem,J. Org. Chem.* 43, 2410 (1978). Use in prepn of isocyanates: H. W. I. Peer-

lings, E. W. Meijer, *Tetrahedron Lett.* 40, 1021 (1999); of polyurethanes: R. V. Versteegen *et al., Angew. Chem. Int. Ed.* 38, 2917 (1999).

Crystalline solid, mp 64-65° (Dean); also reported as colorless prisms, mp 62-63° (Pope).
USE: In synthesis of polyurethanes.

3050. Dicamba. [1918-00-9] 3,6-Dichloro-2-methoxybenzoic acid; 3,6-dichloro-*o*-anisic acid; 2-methoxy-3,6-dichlorobenzoic acid; dianat; Velsicol 58-CS-11; Banvel; Clarity. $C_8H_6Cl_2O_3$; mol wt 221.03. C 43.47%, H 2.74%, Cl 32.08%, O 21.72%. Sometimes misnamed *methoxydichlorobenzoate.* Prepn: S. B. Richter, US 3013054 (1961 to Velsicol). Metabolism: D. D. Oehler, G. W. Ivie, *J. Agric. Food Chem.* 28, 685 (1980). Toxicity: Bailey, White, *Residue Rev.* 10, 97 (1965).

Crystals from pentane, mp 114-116°. Vapor pressure at 100°: 3.75 × 10^{-3} mm Hg. Sol in ethanol and acetone. Very slightly sol in water. LD$_{50}$ orally in rats: 1040 mg/kg (Bailey, White).
USE: Herbicide.

3051. Dicentrine. [517-66-8] (7aS)-6,7,7a,8-Tetrahydro-10,11-dimethoxy-7-methyl-5H-benzo[g]-1,3-benzodioxolo[6,5,4-de]quinoline; 9,10-dimethoxy-1,2-(methylenedioxy)aporphine; 1,2-methylenedioxy-9,10-dimethoxyaporphine. $C_{20}H_{21}NO_4$; mol wt 339.39. C 70.78%, H 6.24%, N 4.13%, O 18.86%. *d*-Form prevalent in nature. Found in *Dicentra pusilla* Sieb. & Zucc., *Fumariaceae* and several other *Dicentra* species. Related to actinodaphnine, and laurotetanine, *q.q.v.* Isoln: Y. Asahina, *Arch. Pharm.* 247, 201 (1909). Isoln of *l*-form from *Duguetia* A. St. Hil., *Annonaceae:* Casagrande, Ferrari, *Farmaco Ed. Sci.* 25, 442 (1970). Synthesis of *dl*-form: Haworth *et al., J. Chem. Soc.* 127, 2018 (1925); Cava *et al., Tetrahedron* 29, 2245 (1973). Resolution of isomers: Haworth *et al., J. Chem. Soc.* 129, 29 (1926).

d-Dicentrine

dl-Form. Prisms from methanol, mp 178-179°. Freely sol in chloroform, ethyl acetate, acetone, benzene, hot alcohol. Moderately sol in ether, cold alcohol. Absorption max: Girardet, *J. Chem. Soc.* 1931, 2630.
d-Form. Long prisms from ether. mp 169°. $[\alpha]_D^{17}$ +64.1° (c = 1.433 in chloroform). Freely sol in alcohol, ethyl acetate, benzene.
l-Form. Long prisms from ether. mp 169°. $[\alpha]_D^{17}$ −63.5° (c = 1.70 in chloroform).
dl-Hydrochloride. $C_{20}H_{21}NO_4$.HCl. Small needles from water, dec 263-265°.
dl-Methiodide monohydrate. $C_{20}H_{21}NO_4$.CH$_3$I.H$_2$O. Plates, mp 228-229°.

3052. Dichlobenil. [1194-65-6] 2,6-Dichlorobenzonitrile; H-133; Niagara 5006; Casoron. $C_7H_3Cl_2N$; mol wt 172.01. C 48.88%, H 1.76%, Cl 41.22%, N 8.14%. Prepn: Reich, *Bull. Soc.*

Chim. Fr. [4] **21**, 217 (1917); Norris, Klemka, *J. Am. Chem. Soc.* **62**, 1432 (1940); Chang *et al.*, *C.A.* **53**, 6134d (1959); Koopman, *Rec. Trav. Chim.* **80**, 1075 (1961); Hackmann, ten Haken **GB 861899**; Higson, **GB 862937** (both 1961 to Shell). Use as herbicide: Koopman, Daams, **US 3027248** (1962 to N. A. Phillips). Fate in crops, soil and animals: Beynon, Wright, *Residue Rev.* **43**, 23 (1972); Veroop, *ibid.* 55. Toxicity data: G. W. Bailey, J. L. White, *ibid.* **10**, 97 (1965).

CN ... Cl ... Cl

Crystals from petr ether, mp 144-145°. Soly in water at 25°: 25 ppm; at 20°: 18 ppm. Vapor pressure at 20°: 3×10^{-6} mm Hg; at 25°: 5×10^{-4} mm. Absorption spectrum see Koopman, *loc. cit.* LD_{50} in rats, mice (mg/kg): 2710, 6800 orally (Bailey, White).

USE: Herbicide.

3053. Dichlofluanid. [1085-98-9] 1,1-Dichloro-*N*-[(dimethylamino)sulfonyl]-1-fluoro-*N*-phenylmethanesulfenamide; *N*-[(dichlorofluoromethyl)thio]-*N'*,*N'*-dimethyl-*N*-phenylsulfamide; Bayer 47531; KUE 13032c; Elvaron; Euparen(e). $C_9H_{11}Cl_2FN_2O_2S_2$; mol wt 333.22. C 32.44%, H 3.33%, Cl 21.28%, F 5.70%, N 8.41%, O 9.60%, S 19.24%. Prepn: E. Klauke *et al.*, **BE 609868**; *eidem*, **US 3285929** and **US 3341403** (1962, 1966, 1967 to Bayer). Antifungal activity: E. Kuhle *et al.*, *Angew. Chem.* **76**, 807 (1964); F. Grewe, *Phytiatr.-Phytopharm.* **17**, 47 (1968). UV degradation: T. Clark, D. Watkins, *Pestic. Sci.* **9**, 225 (1978). GC determn in plants and beverages: R. Brennecke, *Pflanzenschutz-Nachr.* **41**, 137 (1988).

White powder, mp 105.0-105.6°. Vapor press at 20°: 1×10^{-6} mm Hg. Insol in water. Sol in acetone, methanol, xylene. Soly at 20° (g/1000 ml): water 0.0013; dichloromethane >200; *n*-hexane 2-5; 2-propanol 10-20; toluene 100-200. Light sensitive; decomp by strong alkaline media. LD_{50} in rats, guinea pigs, rabbits (mg/kg): 1000 orally in all species (Grewe).

USE: Fungicide.

3054. Dichlone. [117-80-6] 2,3-Dichloro-1,4-naphthalenedione; 2,3-dichloro-1,4-naphthoquinone; USR-604; Phygon; Phygon Paste; Phygon XL. $C_{10}H_4Cl_2O_2$; mol wt 227.04. C 52.90%, H 1.78%, Cl 31.23%, O 14.09%. Prepd by chlorination of 1,4-naphthoquinone: Bertheim, *Ber.* **34**, 1554 (1901); Brass, Köhler, *Ber.* **55**, 2554 (1922); Sjöstrand, **US 2975196** (1961 to Svenska Oljeslageri AB); by oxidation of 2,3-dichloro-5,8-dihydro-1,4-naphthohydroquinone: Gaertner, **US 2750427** (1956 to Monsanto); by oxidation of 2,3-dichloro-*p*-benzoquinone: Grinev *et al.*, *Zh. Obshch. Khim.* **29**, 90 (1959). Toxicity data: G. W. Bailey, J. L. White, *Residue Rev.* **10**, 97 (1965). *Review:* Ter Horst, *Ind. Eng. Chem.* **35**, 1255 (1943).

O ... Cl ... Cl ... O

Golden yellow needles or leaflets from alc, mp 193°. Sublimes. Practically insol in water (soly about 1 part in 10 million parts H_2O). Soly in xylene and *o*-dichlorobenzene about 4%. Moderately sol in

acetone, ether, benzene, dioxane. LD_{50} orally in rats: 1300 mg/kg (Bailey, White).

Caution: Irritating to skin, mucous membranes; CNS depressant: *Clinical Toxicology of Commercial Products*, R. E. Gosselin *et al.*, Eds. (Williams & Wilkins, Baltimore, 5th ed., 1984) Section II, p 318.

USE: Fungicide for agriculture and textiles; herbicide.

3055. Dichloralphenazone. [480-30-8] 1,2-Dihydro-1,5-dimethyl-2-phenyl-3*H*-pyrazol-3-one compd with 2,2,2-trichloro-1,1-ethanediol(1:2); dichloralantipyrine; Bonadorm; Dormwell; Welldorm. $C_{15}H_{18}Cl_6N_2O_5$; mol wt 519.02. C 34.71%, H 3.50%, Cl 40.98%, N 5.40%, O 15.41%. $[CCl_3CH(OH)_2]_2.C_{11}H_{12}N_2O$. Prepn: Pfeiffer, Seidel, *Z. Physiol. Chem.* **178**, 97 (1928). GLC determn in pharmaceutical formulation: F. M. Plakogiannis, A. M. Saad, *J. Pharm. Sci.* **66**, 604 (1977). Clinical trial of combination with isometheptene and acetaminophen, *q.q.v.*, in treatment of migraine: S. Diamond, *Headache* **15**, 282 (1976).

Small prismatic needles from water, mp 68°. Freely sol in water, alc, chloroform; sol in dilute acids.

Note: This is a controlled substance (depressant): **21 CFR,** 1308.14.

THERAP CAT: Sedative, hypnotic.

3056. Dichloramine T. [473-34-7] *N*,*N*-Dichloro-4-methylbenzenesulfonamide; *N*,*N*-dichloro-*p*-toluenesulfonamide; *p*-toluenesulfonic acid dichloramide. $C_7H_7Cl_2NO_2S$; mol wt 240.10. C 35.02%, H 2.94%, Cl 29.53%, N 5.83%, O 13.33%, S 13.35%. Prepn from *p*-toluenesulfonic acid: Soper, *J. Chem. Soc.* **125**, 1899 (1924); van Andel, **US 2495489** (1950 to Shell).

Cl ... Cl ... N ... S ... O ... O ... CH3

Prisms from chloroform + petr ether, mp 83°. Strong chlorine odor; dec on exposure to air with loss of chlorine. mp about 80°. Almost insol in water, dec by alcohol when warmed; one gram dissolves in about 1 ml benzene, 1 ml chloroform, 2.5 ml carbon tetrachloride; sol in eucalyptol, chlorinated paraffin hydrocarbons, glacial acetic acid; slightly sol in petr ether. *Keep well closed and protected from light.*

USE: Germicide.

THERAP CAT: Antibacterial.

3057. Dichlorisone. [7008-26-6] (11β)-9,11-Dichloro-17,21-dihydroxypregna-1,4-diene-3,20-dione; 9α,11β-dichloro-1,4-pregnadiene-17α,21-diol-3,20-dione; 9α,11β-dichloro analog of prednisolone; Diloderm; Disoderm. $C_{21}H_{26}Cl_2O_4$; mol wt 413.34. C 61.02%, H 6.34%, Cl 17.15%, O 15.48%. Prepn: Robinson *et al.*, *J. Am. Chem. Soc.* **81**, 2191 (1959); Gould *et al.*, **US 2894963** (1959 to Schering).

O ... OH ... CH3 ... OH ... Cl ... CH3 ... H ... Cl ... H ... O

Crystals from acetone, dec 238-241°. $[\alpha]_D^{20}$ +134° (pyridine). uv max (methanol): 237 nm (ε 15400).

21-Acetate. [79-61-8] Astroderm. $C_{23}H_{28}Cl_2O_5$; mol wt 455.37. Crystals from acetone, dec 246-253°. $[\alpha]_D^{25}$ +162° (dioxane). uv max (methanol): 237 nm (ε 15000).

THERAP CAT: Antipruritic (topical).

3058. Dichlormid. [37764-25-3] 2,2-Dichloro-*N*,*N*-di-2-propen-1-ylacetamide; *N*,*N*-diallyl-2,2-dichloroacetamide; R-25788. $C_8H_{11}Cl_2NO$; mol wt 208.08. C 46.18%, H 5.33%, Cl 34.07%, N 6.73%, O 7.69%. Increases corn tolerance to thiocarbamate and chloroacetanilide herbicides. Field and growth room trials in protection of corn from herbicide injury: F. Y. Chang *et al.*, *Can. J. Plant Sci.* **52**, 707 (1972); *eidem*, *Weed Res.* **13**, 399 (1973). Effects on uptake and metabolism of EPTC, *q.v.*: *eidem*, *J. Agric. Food Chem.* **22**, 245 (1974). Prepn and use in combination with thiocarbamate herbicides: F. M. Pallos *et al.*, **US 4021224** (1977 to Stauffer). Synergistic activity with herbicides that affect photosynthesis: C. Fedtke, R. H. Strang, *Z. Naturforsch.* **45c**, 565 (1990). Effects on lipid biosynthesis in maize: I. Cs. Barta, *ibid.* **46c**, 926 (1991); in barley and wild oats: A. Baldwin *et al.*, *J. Exp. Bot.* **54**, 1289 (2003). Environmental tolerances: *Fed. Regist.* **65**, 16143 (2000). Photodegradation study: A. W. Abu-Qare, H. J. Duncan, *Chemosphere* **46**, 1183 (2002).

bp$_{0.4 kPa}$ 96°. n_D^{30} 1.4990. LD$_{50}$ orally in rats: 2146 mg/kg (*Fed. Regist.*).

Combination with acetochlor. [144115-48-0] Surpass; Top-Notch.

USE: Herbicide safener.

3059. Dichloroacetic Acid. [79-43-6] 2,2-Dichloroacetic acid; bichloracetic acid; dichlorethanoic acid; DCA. $C_2H_2Cl_2O_2$; mol wt 128.94. C 18.63%, H 1.56%, Cl 54.99%, O 24.82%. Transformation of chloral to dichloroacetic acid: Wallach, *Ann.* **173**, 288 (1874); Frantzen, Fikentscher, *ibid.* **623**, 68 (1959); Rosenblum *et al.*, *Chem. Ind. (London)* **1960**, 718. Toxicity: Smyth *et al.*, *Arch. Ind. Hyg. Occup. Med.* **4**, 119 (1951); P. W. Stacpoole *et al.*, *N. Engl. J. Med.* **300**, 372 (1979); J. L. Cicmanec *et al.*, *Fundam. Appl. Toxicol.* **17**, 376 (1991). Mutagenicity studies: V. Herbert *et al.*, *Am. J. Clin. Nutr.* **33**, 1179 (1980). Use in treatment of lactic acidosis: P. W. Stacpoole *et al.*, *N. Engl. J. Med.* **309**, 390 (1983).

Liquid; pungent odor. d$_4^{20}$ 1.563. bp 193-194°. Apparently occurs in two cryst forms, mp 9.7° and −4°. n_D^{22} 1.4659. *Corrosive.* Miscible with water, alc, ether. LD$_{50}$ orally in rats: 2.82 g/kg (Smyth).

Ethyl ester. $C_4H_6Cl_2O_2$. Liquid, bp 158.3-158.7°. d$_4^{20}$ 1.282. n_D^{20} 1.4386. Slightly sol in water; miscible with alcohol, ether.

THERAP CAT: Caustic; keratolytic; topical astringent.

3060. 1,1-Dichloroacetone. [513-88-2] 1,1-Dichloro-2-propanone; *α,α*-dichloroacetone; *uns*-dichloroacetone; dichloromethyl methyl ketone. $C_3H_4Cl_2O$; mol wt 126.96. C 28.38%, H 3.18%, Cl 55.84%, O 12.60%. Prepn: Borsche, Fittig, *Ann.* **133**, 113 (1865).

Oily liq. d$_{15}^{18}$ 1.305. bp 120°. Slightly sol in water; sol in alcohol; miscible with ether. Boiling with Na$_2$CO$_3$ gives acrylic acid.

3061. 1,3-Dichloroacetone. [534-07-6] 1,3-Dichloro-2-propanone; *α,γ*-dichloroacetone; *sym*-dichloroacetone; bis(chloromethyl) ketone. $C_3H_4Cl_2O$; mol wt 126.96. C 28.38%, H 3.18%, Cl 55.84%, O 12.60%. Prepd by the oxidation of dichlorohydrin with sodium dichromate: Conant, Quayle, *Org. Synth.* **2**, 13 (1922).

Plates, needles on distillation. mp 45°. bp 173°. d$_4^{46}$ 1.3826. n_D^{46} 1.47144. Sol in water, very sol in alc, ether.

Caution: Lacrimator, vesicant.

3062. 2,2-Dichloroacetyl Chloride. [79-36-7] Dichloroethanoyl chloride. C_2HCl_3O; mol wt 147.38. C 16.30%, H 0.68%, Cl 72.16%, O 10.86%. Prepd from pentachloroethane: Ott, **DE 362748** (1922 to Weiler ter Meer); from chloroform and carbon monoxide in presence of aluminum chloride at high pressure (21% yield): Frank *et al.*, *Ind. Eng. Chem.* **41**, 2061 (1949). Toxicity study: Smyth *et al.*, *Arch. Ind. Hyg. Occup. Med.* **4**, 119 (1951).

Liquid. Fumes in air; acrid, penetrating odor. d$_4^{16}$ 1.5315. bp$_{760}$ 107-108°; bp$_{739}$ 106-107°. n_D^{16} 1.4638. Dec upon contact with water, alcohol. Miscible with ether. LD$_{50}$ orally in rats: 2.46 g/kg (Smyth).

Caution: Irritating to eyes, mucous membranes.

3063. 3,4-Dichloroaniline. [95-76-1] 3,4-Dichlorobenzeneamine. $C_6H_5Cl_2N$; mol wt 162.01. C 44.48%, H 3.11%, Cl 43.76%, N 8.65%. Prepd by chlorination of *p*-chloroaniline with AlCl$_3$ — HCl: Suthers *et al.*, *J. Org. Chem.* **27**, 447 (1962). Manuf by catalytic reduction of 3,4-dichloro-1-nitrobenzene: Dietzler, Keil, **US 3067253** (1962 to Dow).

Crystals, mp 71-72°. bp 272°. Practically insol in water. Very sol in alcohol, ether; slightly sol in benzene.

3064. *m*-Dichlorobenzene. [541-73-1] 1,3-Dichlorobenzene. $C_6H_4Cl_2$; mol wt 147.00. C 49.02%, H 2.74%, Cl 48.23%. Prepn by Sandmeyer procedure from the appropriate chloroaniline, and, along with *o*- and *p*-dichlorobenzenes, by chlorination of chlorobenzene: Engelsma *et al.*, *Rec. Trav. Chim.* **76**, 325 (1957). Separation of mixture contg *m*-, *o*-, and *p*-dichlorobenzenes by distillation and crystn: Mueller, Wolz, **FR 1374863** (1964 to Bayer), *C.A.* **62**, 4936e (1965); **GB 999845**. Physical properties: *Rev. Environ. Contam. Toxicol.* **106**, 51 (1988). Review of toxicology and human exposure: *Toxicological Profile for Dichlorobenzenes* (PB2007-100672, 2006) 493 pp.

Liquid, bp 173°. mp −24.76°. d$_4^{20}$ 1.2884, d$_4^{25}$ 1.2828. n_D^{20} 1.5459. Log P (octanol/water): 3.53. Log P (olive oil/water) 3.69. Vapor pressure (39°) 5 mm Hg. Soly in water 1.23 mg/L. Sol in alcohol, ether.

3065. *o*-Dichlorobenzene. [95-50-1] 1,2-Dichlorobenzene; orthodichlorobenzene. $C_6H_4Cl_2$; mol wt 147.00. C 49.02%, H 2.74%, Cl 48.23%. Prepn and separation from *m*- and *p*-dichlorobenzenes, *see m*-isomer above. Manuf: *Faith, Keyes & Clark's Industrial Chemicals*, F. A. Lowenheim, M. K. Moran, Eds. (Wiley-Interscience, New York, 4th ed., 1975) pp 258-265. Review of toxicology: *Rev. Environ. Contam. Toxicol.* **106**, 51-68 (1988); and human exposure: *Toxicological Profile for Dichlorobenzenes* (PB2007-100672, 2006) 493 pp.

Liquid, bp 180.5°. mp −17.03°. d$_4^{20}$ 1.3059, d$_4^{25}$ 1.3003. n_D^{20} 1.5515, n_D^{25} 1.5491. Flash pt, closed cup: 151°F (66°C). Vapor pressure (25°) 1.56 mm Hg. Log P (octanol/water): 3.43. Log P (olive oil/water) 3.65. Soly in water 145 mg/L. Miscible with alc, ether, benzene.

Caution: Potential symptoms of overexposure are irritation of nose and eyes; liver and kidney damage; skin blisters. *See NIOSH Pocket Guide to Chemical Hazards* (DHHS/NIOSH 97-140, 1997) p 96. Exposure to high concentrations may result in CNS depression. *See Patty's Industrial Hygiene and Toxicology* **vol. 2B**, G. D. Clayton, F. E. Clayton, Eds. (Wiley-Interscience, New York, 3rd ed., 1981) pp 3611-3617.

USE: Solvent for waxes, gums, resins, tars, rubbers, oils, asphalts; insecticide for termites and locust borers; fumigant; deodorizer; removing sulfur from illuminating gas; as degreasing agent for metals, leather, wool; as ingredient of metal polishes; as heat transfer medium; as intermediate in the manuf of dyes; as precursor for manuf of herbicides.

3066. *p*-Dichlorobenzene. [106-46-7] 1,4-Dichlorobenzene; paracide; PDB; paradichlorobenzene; Para-zene; Di-chloricide; Paramoth. $C_6H_4Cl_2$; mol wt 147.00. C 49.02%, H 2.74%, Cl 48.23%. Prepn and separation from *m*- and *o*-dichlorobenzene, *see m*-dichlorobenzene. Crystal structure of triclinic form (β-modification): Housty, Clastre, *Acta Crystallogr.* **10**, 695 (1957); of monoclinic form (α-modification) and transformation to triclinic form: Panatoni *et al., Gazz. Chim. Ital.* **93**, 813 (1963). Manuf: *Faith, Keyes & Clark's Industrial Chemicals,* F. A. Lowenheim, M. K. Moran, Eds. (Wiley-Interscience, New Yok, 4th ed., 1975) pp 258-265. Toxicity study: T. B. Gaines, R. E. Linder, *Fundam. Appl. Toxicol.* **7**, 299 (1986). Review of toxicology: *Rev. Environ. Contam. Toxicol.* **106**, 51-68 (1988); and human exposure: *Toxicological Profile for Dichlorobenzenes* (PB2007-100672, 2006) 493 pp.

Volatile crystals with the characteristic penetrating odor of mothballs. Sublimes at ordinary temps. mp 53.5° (α-modification), 54° (β-modification). bp 174.12°. n_D^{60} 1.5285. Flash pt (closed cup) 150°F. Log P (octanol/water): 3.44. Log P (olive oil/water) 3.65. Vapor pressure (25°) = 0.4 mm Hg. d^{20} 1.46. Sol in alc, ether, benzene, chloroform, carbon disulfide. Practically insol in water. Non-corrosive; non-staining. LD$_{50}$ in male, female rats (mg/kg): 3863, 3790 orally; >6000, >6000 dermally (Gaines, Linder).

Caution: Potential symptoms of overexposure are eye irritation, periorbital swelling; profuse rhinitis; headache; anorexia, nausea and vomiting; weight loss, jaundice and cirrhosis. *See NIOSH Pocket Guide to Chemical Hazards* (DHHS/NIOSH 97-140, 1997) p 96. Direct contact with fumes or solid may cause skin irritation. *See Patty's Industrial Hygiene and Toxicology* **vol. 2B**, G. D. Clayton, F. E. Clayton, Eds. (Wiley-Interscience, New York, 3rd ed., 1981) pp 3617-3626. This substance is reasonably anticipated to be a human carcinogen: *Report on Carcinogens, Twelfth Edition* (PB2011-111646, 2011) p 139.

USE: Insecticidal fumigant. For domestic use against clothes moths; as space deodorant in room deodorizers, toilet bowl blocks and diaper pail deodorizers. Intermediate in production of plastics for electronic components.

3067. 2,2'-Dichlorobenzidine. [84-68-4] 2,2'-Dichloro[1,1'-biphenyl]-4,4'-diamine. $C_{12}H_{10}Cl_2N_2$; mol wt 253.13. C 56.94%, H 3.98%, Cl 28.01%, N 11.07%. Prepn from *m*-chloronitrobenzene: Laubenheimer, *Ber.* **8**, 1625 (1875); Cain, May, *J. Chem. Soc.* **97**, 723 (1910).

Needles from water or prisms from alc, mp 165°. Almost insol in water; moderately sol in alc; readily sol in ether.

Hydrochloride. $C_{12}H_{10}Cl_2N_2$.2HCl. Leaflets from water, moderately sol in water.

USE: Manuf azo dyes.

3068. 3,3'-Dichlorobenzidine. [91-94-1] 3,3'-Dichloro-[1,1'-biphenyl]-4,4'-diamine; 3,3'-dichloro-4,4'-biphenyldiamine; DCB. $C_{12}H_{10}Cl_2N_2$; mol wt 253.13. C 56.94%, H 3.98%, Cl 28.01%, N 11.07%. Prepn from *o*-chloronitrobenzene: Cohn, *Ber.* **33**, 3552 (1900). Review of carcinogenicity studies: *IARC Monographs* **4**, 49-55 (1974); of bioactivation and toxicology: M. M. Iba, *Drug Metab. Rev.* **21**, 377-400 (1990).

Needles from alc or benzene, mp 132-133°. Readily sol in alc, benzene, glacial acetic acid. Almost insol in water.

Dihydrochloride. [612-83-9] $C_{12}H_{10}Cl_2N_2$.2HCl. Needles. Readily in alc; slightly sol in water.

Caution: Potential symptoms of overexposure are skin sensitization; dermatitis; headache, dizziness; caustic burns; frequent urination, dysuria; hematuria; GI upsets; upper respiratory infection. *See NIOSH Pocket Guide to Chemical Hazards* (DHHS/NIOSH 97-140, 1997) p 96. 3,3'-Dichlorobenzidine and its dihydrochloride are reasonably anticipated to be human carcinogens: *Report on Carcinogens, Twelfth Edition* (PB2011-111646, 2011) p 141.

USE: Manuf azo dyes and pigments; as an intermediate for the Benzidine Yellow pigments. As curing agent for isocyanate polymers and solid urethane plastics.

3069. Dichlorobenzyl Alcohol. [1777-82-8] 2,4-Dichlorobenzenemethanol; Dybenal. $C_7H_6Cl_2O$; mol wt 177.02. C 47.50%, H 3.42%, Cl 40.05%, O 9.04%. Prepn: Van de Lande, *Rec. Trav. Chim.* **51**, 98 (1932); Metayer, Dat-Xuong, *Bull. Soc. Chim. Fr.* **1954**, 615.

Crystals, mp 59.5°.

THERAP CAT: Antiseptic.

3070. 1,1-Dichloro-2,2-bis(*p*-chlorophenyl)ethane. [72-54-8] 1,1'-(2,2-Dichloroethylidene)bis[4-chlorobenzene]; tetrachlorodiphenylethane; TDE; dichlorodiphenyldichloroethane; DDD; *p,p'*-DDD; *p,p'*-TDE; Rhothane. $C_{14}H_{10}Cl_4$; mol wt 320.03. C 52.54%, H 3.15%, Cl 44.31%. Non-degradable pesticide; a component of technical grade DDT, *q.v.* Prepd by condensing dichloroacetaldehyde with chlorobenzene: Haller *et al., J. Am. Chem. Soc.* **67**, 1596, 1600 (1945). Toxicity study: T. B. Gaines, *Toxicol. Appl. Pharmacol.* **14**, 515 (1969). Review of toxicology and human exposure: *Toxicological Profile for DDT, DDE, DDD* (PB2003-100137, 2002) 497 pp.

Crystals, mp 109-110°. The chemical properties and solys are similar to those of DDT. LD$_{50}$ orally in rats: >4000 mg/kg (Gaines).

Caution: Potential symptoms of overexposure are similar to DDT.

USE: Insecticide.

3071. Dichloro(2-chlorovinyl)arsine. [541-25-3]; [50361-05-2] ((*E*)-form); [34461-56-8] ((*Z*)-form). *As*-(2-Chloroethenyl)arsonous dichloride; 2-chlorovinyldichloroarsine; chlorovinylarsine dichloride; Lewisite. $C_2H_2AsCl_3$; mol wt 207.31. C 11.59%, H 0.97%, As 36.14%, Cl 51.30%. Vesicant used as chemical weapon. Prepn: S. J. Green, T. S. Price, *J. Chem. Soc.* **119**, 448 (1921); W. L. Lewis, G. A. Perkins, *Ind. Eng. Chem.* **15**, 290 (1923). Review of military experience: G. N. Jarman in *Adv. Chem. Ser.* **23**, entitled "Metal-Organic Compounds," M. Sittig, Ed. (ACS, Washington DC, 1959) pp 328-337. Determn of degradation products in soil: B. A.

Tomkins *et al., J. Chromatogr. A* **909**, 13 (2001); in urine: J. V. Wooten *et al., J. Chromatogr. B* **772**, 147 (2002). Review of chemistry, toxicology and biological effects: M. Goldman, J. C. Dacre, *Rev. Environ. Contam. Toxicol.* **110**, 75-115 (1989).

(*E*)-form

Colorless, slightly oily liquid; darkens with time becoming violet-black or green. Faint odor of geranium. *Vesicant.* $bp_{12.5}$ 76-77°; bp_{26} 93°; bp_{760} 190°. d_4^{20} 1.888. Sol in the common organic solvents. Insol in water, dil mineral acids. Absorbed by rubber, paint, varnish and porous materials. Rapidly hydrolyzed in aqueous medium. LD_{50} in rats (mg/kg): 50 orally; in rats, rabbits (mg/kg): 24, 6 dermally; 1, 2 s.c. (Goldman, Dacre).

Caution: Vapors are extremely toxic. Contact with skin may produce immediate, persistent stinging, followed by erythema and blistering; irritating to eyes; inhalation of 0.50 mg/l for 5 min is considered lethal in humans (Goldman, Dacre). *Antidote:* Dimercaprol, *q.v.*

USE: Chemical warfare agent.

3072. 2,3-Dichloro-5,6-dicyanobenzoquinone. [84-58-2] 4,5-Dichloro-3,6-dioxo-1,4-cyclohexadiene-1,2-dicarbonitrile; DDQ. $C_8Cl_2N_2O_2$; mol wt 227.00. C 42.33%, Cl 31.23%, N 12.34%, O 14.10%. Prepn: Thiele, Günther, *Ann.* **349**, 45 (1906); Walker, Waugh, *J. Org. Chem.* **30**, 3240 (1965).

Yellow to orange crystals, mp 213.5-215°. Dec in water. Sol in benzene, dioxane, acetic acid. Slightly sol in chloroform, methylene chloride.

USE: Oxidizing agent, especially in steroid synthesis.

3073. Dichlorodifluoromethane. [75-71-8] Difluorodichloromethane; Arcton 12; Freon 12; Frigen 12; Genetron 12; Halon; Isotron 2. CCl_2F_2; mol wt 120.91. C 9.93%, Cl 58.64%, F 31.43%. Prepn: Henne, *Org. React.* **2**, 49 (1944); Swarts, *Chem. Zentralbl.* **1907**, II, 581. Manuf: *Faith, Keyes & Clark's Industrial Chemicals,* F. A. Lowenheim, M. K. Moran, Eds. (Wiley-Interscience, New York, 4th ed., 1975) pp 325-330.

Colorless, practically odorless gas. Faint, ether-like odor in high concentrations. Stable up to 550°. $d_{liq}^{-29.8}$ 1.486. mp −158°. bp_{760} −29.8°; bp_{2atm} −12.2°; bp_{5atm} 16.1°; bp_{10atm} 42.4°; bp_{20atm} 74.0°; bp_{30atm} 95.6°. Critical temp 111.5°; critical pressure 39.6 atm (582 lb/sq inch). Dipole moment 0.51. Non-flammable. Insol in water. Sol in alcohol, ether. Absorbs less than 0.0025% water. Has little, if any, anesthetic or toxic action, but toxic substances may be formed on contact with a flame or hot metal surface.

Caution: Potential symptoms of overexposure are dizziness, tremors, asphyxia, unconsciousness, cardiac arrhythmias and cardiac arrest; direct contact with liquid may cause frostbite. *See NIOSH Pocket Guide to Chemical Hazards* (DHHS/NIOSH 97-140, 1997) p 96. *Note:* Consult latest Government regulations on use as aerosol propellant.

USE: Refrigerant, aerosol propellant.

3074. 1,3-Dichloro-5,5-dimethylhydantoin. [118-52-5] 1,3-Dichloro-5,5-dimethyl-2,4-imidazolidinedione; Dactin; Halane. $C_5H_6Cl_2N_2O_2$; mol wt 197.02. C 30.48%, H 3.07%, Cl 35.99%, N 14.22%, O 16.24%. Prepd by passing chlorine through an aq soln of 5,5-dimethylhydantoin: Biltz, Slotta, *J. Prakt. Chem.* **113**, 248 (1926); Orazi, Orio, *An. Asoc. Quim. Argent.* **41**, 153 (1953), *C.A.* **48**, 13634 (1954). Structure: Bogaert-Verhoogen, Martin, *Bull. Soc. Chim. Belg.* **58**, 567 (1949). Purification by dissolving in concd

H_2SO_4 and diluting with ice-water: Lorenz, **US 2828308** (1958 to Purex).

Four-sided, pointed prisms from chloroform, d_{20}^{20} 1.5. mp 132°. Sublimes at 100°. Turns brown and conflagrates at 212° (after melting at 132°). The dry crystals [combined available chlorine 77.6% (theory)] may be stored without much loss of available chlorine. After 14 weeks at 60° the loss was 1.5% Cl compared with a loss of 37.5% suffered by 70% calcium hypochlorite. On contact with water and esp hot water hypochlorous acid is liberated. At pH 9 nitrogen chloride is formed. Soly in water at 25° 0.21%; at 60° 0.60%. pH of aq soln about 4.4. Freely sol in chlorinated and highly polar solvents at 25°: Chloroform 14%, methylene chloride 30.0%, carbon tetrachloride 12.5%, ethylene dichloride 32.0%, *sym*-tetrachlorethane 17.0%, benzene 9.2%.

Caution: Potential symptoms of overexposure are irritation of eyes, mucous membranes and respiratory system. *See NIOSH Pocket Guide to Chemical Hazards* (DHHS/NIOSH 97-140, 1997) p 98.

USE: Chlorinating agent, disinfectant, industrial deodorant. In water treatment. Active ingredient of powder laundry bleaches such as Sage's Dry Bleach, Colgate's Pruf. Intermediate for amino acids, drugs, insecticides. Stabilizer for vinyl chloride polymers. Polymerization catalyst.

3075. *sym*-Dichloroethyl Ether. [111-44-4] 1,1'-Oxybis[2-chloroethane]; bis(2-chloroethyl) ether; β,β'-dichloroethyl ether; DCEE; Chlorex. $C_4H_8Cl_2O$; mol wt 143.01. C 33.59%, H 5.64%, Cl 49.58%, O 11.19%. Prepn: O. Kamm, J. H. Waldo, *J. Am. Chem. Soc.* **43**, 2223 (1921). Toxicity: H. F. Smyth, Jr., C. P. Carpenter, *J. Ind. Hyg. Toxicol.* **30**, 66 (1948). Evaluation of carcinogenic risk: *IARC Monographs* **9**, 117 (1975). Review of toxicology and human exposure: *Toxicological Profile for Bis(2-chloroethyl)ether* (PB90-168683, 1989) 83 pp.

Colorless, clear liq; pungent odor. Vapor harmful. d_{20}^{20} 1.22. bp 178°. mp −50°. n_D^{20} 1.457. Flash pt, closed cup: 145°F (63°C). Buckman Laboratories, IC literature; also reported as 131°F (55°C), *Ind. Eng. Chem.* **32**, 880 (1940). Insol in water. Sol in most organic solvents. Dissolves oils, fats, greases, etc. LD_{50} orally in rats: 75 mg/kg (Smyth, Carpenter).

Caution: Potential symptoms of overexposure are lacrimation; irritation of nose, throat and respiratory system; coughing; nausea, vomiting. Potential occupational carcinogen. *See NIOSH Pocket Guide to Chemical Hazards* (DHHS/NIOSH 97-140, 1997) p 98.

USE: Reagent for organic synthesis; solvent. Has been used as a scouring agent for textiles; as soil fumigant.

3076. 2',7'-Dichlorofluorescein. [76-54-0] 2',7'-Dichloro-3',6'-dihydroxyspiro[isobenzofuran-1(3*H*),9'-[9*H*]xanthen]-3-one; 2',7'-dichloro-3,6-fluorandiol; dichlorofluorescein. $C_{20}H_{10}Cl_2O_5$; mol wt 401.20. C 59.88%, H 2.51%, Cl 17.67%, O 19.94%. Fluorescent dye. Prepn and use as argentometric titration indicator: I. M. Kolthoff *et al., J. Am. Chem. Soc.* **51**, 3273 (1929). Dissociation and tautomerism: N. O. Mchedlov-Petrossyan *et al., J. Phys. Org. Chem.* **16**, 380 (2003); and NMR spectroscopy: M. Afri *et al., Chem. Phys. Lipids* **131**, 123 (2004). Acid-base equilibrium studies: H. Leonhardt *et al., J. Phys. Chem.* **75**, 245 (1971). Chromatographic purification: W. A. Peeples II, J. R. Heitz, *J. Liq. Chromatogr.* **4**, 51 (1981). Laser dye studies: R. W. Chambers, D. R. Kearns, *Photochem. Photobiol.* **10**, 215 (1969); D. A. Jennings, A. J. Varga, *J. Appl. Phys.* **42**, 5171 (1971); M. Yamashita *et al., J. Chem. Phys.* **66**, 986 (1977). Prepn and applications of fluorescent derivatives: S. C. Burdette *et al., J. Am. Chem. Soc.* **123**, 7831 (2001).

Ochre solid. *Irritant.* Exists as pH dependent tautomers. Unconjugated lactonic form predominates at acidic, neutral pH; lacks typical uv absorption and fluorescence. Lactonic form: Absorption max (DMF): 470 nm (ε 380). Open form is fully conjugated across tricyclic system at basic pH. Open form: Absorption max (ethanol): 509 nm; (DMF): 526 nm (ε 89600). Fluorescence excitation max: 395 nm; emission max: 450 nm. pKa_0 (DMF) < -1; pKa_1 (DMF) 10.4 ± 0.2; pKa_2 (DMF) 13.2 ± 0.1. Sol in ethanol and aq alkali solns; slightly sol in methanol and ether. Practically insol in water.

USE: Fluorescent indicator dye; also used in lasers. Indicator for argentometric titration of chloride and bromide ions. Fluorimetric dip reagent. Fluorescent derivatives as probes for protein labeling and metal ion detection.

3077. **4',5'-Dichlorofluorescein.** [2320-96-9] 4',5'-Dichloro-3',6'-dihydroxyspiro[isobenzofuran-1(3H),9'-[9H]xanthen]-3-one; D & C Orange no. 8; 4,5-dichloro-3,6-fluorandiol. $C_{20}H_{10}Cl_2O_5$; mol wt 401.20. C 59.88%, H 2.51%, Cl 17.67%, O 19.94%. Prepd by condensation of 2-chlororesorcinol with phthalic anhydride: *Colour Index* vol. 4 (3rd ed., 1971) p 4424.

Orange powder, insol in water and in dil acids. Sol in alcohol and in dil alkali yielding solns which are orange with a greenish-yellow fluorescence. Slightly sol in glycerol and the glycols. Practically insol in oils, fats and waxes. If a soln of the dye in strong alkali is heated, a violet color is produced.

Disodium salt. C.I. Solvent Orange 32; C.I. 45365. $C_{20}H_8Cl_2$-Na_2O_5.

Note: Structure depicted as the open tautomer.

3078. **2,6-Dichloroindophenol Sodium.** [620-45-1] 2,6-Dichloro-4-[(4-hydroxyphenyl)imino]-2,5-cyclohexadien-1-one sodium salt (1:1); sodium 2,6-dichloroindophenol; sodium 2,6-dichloro-N-(p-hydroxyphenyl)-p-benzoquinone imine; 2,6-dichloro-1,4-benzoquinone-4-(4-hydroxyanil) sodium; 2,6-dichlorophenol-indophenol sodium; Tillman's reagent. $C_{12}H_6Cl_2NNaO_2$; mol wt 290.07. C 49.69%, H 2.09%, Cl 24.44%, N 4.83%, Na 7.93%, O 11.03%. Prepd from 2,6-dichloro-p-quinone-4-chlorimide and phenol in alkaline soln: Gibbs *et al., Public Health Rep.* **40**, 650 (1925); Tillmans *et al., Z. Unters. Lebensm.* **56**, 273 (1932).

Dark green powder. May contain up to 2 mols H_2O. Freely sol in water, alc. The aq soln is deep blue, changed to red by acids. It liberates iodine from KI in acid solns.

USE: Analytical reagent in the determn of ascorbic acid (vitamin C) which reduces the dye to a colorless hydroxy compd. Indicator.

3079. **sym-Dichloromethyl Ether.** [542-88-1] 1,1'-Oxybis[1-chloromethane]; oxybis[chloromethane]; bis(chloromethyl) ether; BCME. $C_2H_4Cl_2O$; mol wt 114.95. C 20.90%, H 3.51%, Cl 61.68%, O 13.92%. Review of carcinogenic risk: *IARC Mono-*

graphs **4**, 231-238 (1974); of toxicology and human exposure: *Toxicological Profile for Bis (Chloromethyl) Ether* (PB90-168691, 1989) 77 pp.

Colorless liquid; suffocating odor. d_4^{20} 1.315. bp 106°. n_D^{20} 1.4346. Unstable in moist air; dec by water into HCl and formaldehyde.

Caution: Potential symptoms of overexposure are irritation of eyes, skin, mucous membranes and respiratory system; pulmonary congestion, edema; corneal damage, necrosis; decreased pulmonary function, coughing, dyspnea and wheezing; blood stained sputum; bronchial secretions. *See NIOSH Pocket Guide to Chemical Hazards* (DHHS/NIOSH 97-140, 1997) p 64. This substance is listed as a known human carcinogen: *Report on Carcinogens, Twelfth Edition* (PB2011-111646, 2011) p 71.

USE: Manuf of plastics and ion exchange resins.

3080. **4,5-Dichloro-2-octyl-3-isothiazolone.** [64359-81-5] 4,5-Dichloro-2-octyl-3(2H)-isothiazolone; 2-n-octyl-4,5-dichloro-1-isothiazolin-3-one; 4,5-dichloro-2-n-octyl-4-isothiazolin-3-one; DCOI; C-9211; RH-5287; Sea-Nine. $C_{11}H_{17}Cl_2NOS$; mol wt 282.22. C 46.81%, H 6.07%, Cl 25.12%, N 4.96%, O 5.67%, S 11.36%. Environmentally friendly marine antifouling agent. Prepn: E. D. Weiler *et al., J. Heterocycl. Chem.* **14**, 627 (1977). Physical properties and ecological risk assessment: W. D. Shade *et al., ASTM Spec. Tech. Publ.* **STP 1216**, 381 (1993). HPLC analysis and persistence in seawater and sediment: A. Jacobson *et al., ACS Symp. Ser.* **522**, 127 (1993). *Review:* G. L. Willingham, A. H. Jacobson, *ibid.* **640**, 224-233 (1996).

Crystals from hexane, mp 44-46°. Technical product is reported as tan-to-brown waxy solid with pungent aromatic odor, mp 40-41°. d^{25} 1.28 g/ml. Vapor pressure (25°): 7.4×10^{-6} mm Hg. Log P (octanol/water): 2.85 (Willingham, Jacobson); also reported as 4.5 (Shade). Miscible in most organic solvents. Soly (ppm): 6.5 in deionized water; 4.7 in synthetic seawater. LC_{50} (96 hr) in bluegill sunfish, adult bay mussel, fiddler crab: 14, 850, 1312 ppb (Willingham, Jacobson).

USE: Marine antifoulant.

3081. **Dichlorophen.** [97-23-4] 2,2'-Methylenebis[4-chlorophenol]; 2,2'-dihydroxy-5,5'-dichlorodiphenylmethane; 5,5'-dichloro-2,2'-dihydroxydiphenylmethane; bis[5-chloro-2-hydroxyphenyl]methane; di[5-chloro-2-hydroxyphenyl]methane; dichlorophene; G-4; Anthiphen; Dicestal; Didroxane; Di-phenthane-70; Parabis; Plath-Lyse; Preventol G-D; Wespuril. $C_{13}H_{10}Cl_2O_2$; mol wt 269.12. C 58.02%, H 3.75%, Cl 26.35%, O 11.89%. Prepn by adding a CH_2O yielding reagent, e.g. aq HCHO, to p-chlorophenol: Gump, Luthy, **US 2334408** (1944 to Burton T. Bush). Acute toxicity: T. B. Gaines, R. E. Linder, *Fundam. Appl. Toxicol.* **7**, 299 (1986).

Crystals from toluene, mp 177-178°. Practically insol in water. Sparingly sol in toluene. One gram dissolves in 1 g of 95% ethanol, in less than 1 g of ether. Also sol in methanol, isopropyl ether, petr ether. Sol (with decompn) in alkaline aq solns. LD_{50} in adult male, female rats (mg/kg): 1506, 1683 orally (Gaines, Linder).

USE: Agricultural fungicide; antimicrobial; germicide in soaps, shampoos, etc.

THERAP CAT: Anthelmintic (Cestodes).

THERAP CAT (VET): Anthelmintic, antiprotozoan, fungicide.

3082. 2,4-Dichlorophenol. [120-83-2] $C_6H_4Cl_2O$; mol wt 163.00. C 44.21%, H 2.47%, Cl 43.50%, O 9.82%. Key intermediate in synthesis of the herbicide 2,4-D, *q.v.* Prepn by chlorination of phenol: M. Kohn, S. Sussmann, *Monatsh. Chem.* **46**, 575 (1926); B. O. Pray, D. N. Sukow, US 2759981 (1956 to Columbia-Southern Chem. Corp.); by chlorination of *o*- or *p*-chlorophenol: L. G. Groves *et al., J. Chem. Soc.* **1929**, 512; from 2,4-dinitrophenol: A. Ghosh, *J. Indian Chem. Soc.* **28**, 155 (1951). ^{1}H NMR study: J. B. Rowbotham, T. Schaefer, *Can. J. Chem.* **52**, 3037 (1974). Review of toxicology and human exposure: *Toxicological Profile for Chlorophenols* (PB99-166639, 1999) 260 pp.

Needle-like crystals, with strong medicinal taste and odor. mp 45°. bp 209-211°. Volatile with steam. Sol in CCl_4.

USE: Biocide; intermediate in production of herbicides.

3083. 2,6-Dichlorophenol. [87-65-0] $C_6H_4Cl_2O$; mol wt 163.00. C 44.21%, H 2.47%, Cl 43.50%, O 9.82%. Synthesis from ethyl 3,5-dichloro-4-hydroxybenzoate: D. S. Tarbell *et al., Org. Synth.* **coll. vol. III**, 267. Crystal structure: C. Bavoux, P. Michel, *Acta Crystallogr.* **30B**, 2043 (1974). Isoln and identification as sex pheromone of lone star tick: R. S. Berger, *Science* **177**, 704 (1972).

White crystals from petr ether, mp 64.5-65.5°.

3084. 1,3-Dichloro-2-propanol. [96-23-1] α-Dichlorohydrin; *sym*-glycerol dichlorohydrin; glycerol α,γ-dichlorohydrin; *sym*-dichloroisopropyl alcohol. $C_3H_6Cl_2O$; mol wt 128.98. C 27.94%, H 4.69%, Cl 54.97%, O 12.40%. Prepd from glycerol, acetic acid and HCl gas: Conant, Quayle, *Org. Synth.* **2**, 29 (1922); **coll. vol. I** (2nd ed.) p 292. Toxicity study: Smyth *et al., Am. Ind. Hyg. Assoc. J.* **23**, 95 (1962).

Liquid. Ethereal odor. d_4^{17} 1.3506. mp $-4°$. bp_{760} 174.3°; bp_{100} 114.8°; bp_{40} 93°; bp_{20} 78°; bp_5 52°; $bp_{1.0}$ 28°. n_D^{17} 1.480245. Sol in 10 parts water; miscible with alcohol, ether. LD_{50} orally in rats: 110 mg/kg (Smyth).

USE: Solvent for hard resins and nitrocellulose; manuf photographic and Zapon lacquer; cement for celluloid; binder for water colors.

3085. 1,3-Dichloropropene. [542-75-6] 1,3-Dichloro-1-propene; α,γ-dichloropropylene; 1,3-dichloropropylene; γ-chloroallyl chloride; 1,3-D; Curfew; Telone II. $C_3H_4Cl_2$; mol wt 110.97. C 32.47%, H 3.63%, Cl 63.89%. Prepd from 1,2,3-trichloropropane with NaOH: Friedel, Silva, *Compt. Rend.* **75**, 81 (1872); Reboul, *Ann. Chim.* [3] **60**, 37 (1860); from 3-chloro-2-propen-1-ol with PCl_3: Kirrmann *et al., Bull. Soc. Chim. Fr.* [5] **1**, 860 (1934); from acrolein with PCl_5: van Romburgh, *ibid.* [2] **36**, 549 (1881). The usual industrial prepn is from 1,3-dichloro-2-propanol by dehydration with $POCl_3$ or with P_2O_5 in benzene: Hurd, Webb, *J. Am. Chem. Soc.* **58**, 2191 (1936). Prepn of stereoisomers: L. F. Hatch, A. C. Moore, *ibid.* **66**, 285 (1944). Configuration: H. A. Smith, W. H. King, *ibid.* **70**, 3528 (1948); L. F. Hatch, R. H. Perry, *ibid.* **71**, 3262 (1949). Persistence in soil: I. J. Thomason, M. V. McKenry, *Hilgardia* **42**, 393 (1974). GC determn of isomers in rat blood: P. E. Kastl, E. A. Hermann, *J. Chromatogr.* **265**, 277 (1983). Toxicity study: T. R. Torkelson, F. Oyen, *Am. Ind. Hyg. Assoc. J.* **38**, 217 (1977). Review of toxicology and human exposure: *Toxicological Profile for Dichloropropenes* (PB2009-100004, 2008) 361 pp.

(E)-form

Liquid with chloroform odor. d^{25} 1.220. bp 108°. n_D^{22} 1.4735. Technical product is mixture of isomers. Miscible with hydrocarbons, halogenated solvents, esters and ketones. Log P (octanol/water): 1.82. LD_{50} (92% technical product) in male, female rats (mg/kg): 713, 470 orally (10% soln in corn oil); in rabbits (mg/kg): 504 dermally (Torkelson, Oyen).

(E)-Form. [10061-02-6] *trans*-1,3-Dichloro-1-propene. bp 112.0°. d_4^{20} 1.217. n_4^{20} 1.4730. Log P (octanol/water): 2.06. Sol in ether, benzene, chloroform.

(Z)-Form. [10061-01-5] *cis*-1,3-Dichloro-1-propene. bp 104.3°, d_4^{20} 1.224. n_4^{20} 1.4682. Log P (octanol/water): 2.03. Sol in ether, benzene, chloroform.

Caution: Potential symptoms of overexposure are irritation of skin, eyes, and respiratory system; eye and skin burns; lacrimation; headache, dizziness, coma; gasping, refusal to breathe, cough, substernal pain; possible liver, kidney and heart injury. *See NIOSH Pocket Guide to Chemical Hazards* (DHHS/NIOSH 97-140, 1997) p 100; *Clinical Toxicology of Commercial Products*, R. E. Gosselin *et al.*, Eds. (Williams & Wilkins, Baltimore, 5th Ed., 1984) Section III, pp 141-142. The technical grade is reasonably anticipated to be a human carcinogen: *Report on Carcinogens, Twelfth Edition* (PB2011-111646, 2011) p 150.

USE: Soil fumigant; nematocide. In manuf of pesticides.

3086. Dichloroxylenol. [133-53-9] 2,4-Dichloro-3,5-dimethylphenol; 2,4-dichloro-3,5-xylenol; DCMX. $C_8H_8Cl_2O$; mol wt 191.05. C 50.29%, H 4.22%, Cl 37.11%, O 8.37%. Prepn: Lesser, Gad, *Ber.* **56**, 963 (1923); Jones, *J. Chem. Soc.* **1941**, 267; Gemmell, *Manuf. Chem.* **23**, 63 (1952).

Crystals from petr ether, mp 95-96°. Sublimes. Volatile with steam. Soly in water: one part in 5000. Soly at 15° in 100 parts of solvent: benzene, 14; toluene, 15; acetone, 73; diethyl ketone, 59; petr ether, 4; chloroform, 25; CCl_4, 10.

USE: Bacteriostat in soaps; mold inhibitor, preservative.

3087. Dichlorphenamide. [120-97-8] 4,5-Dichloro-1,3-benzenedisulfonamide; 1,3-disulfamyl-4,5-dichlorobenzene; 4,5-dichloro-1,3-disulfamoylbenzene; Antidrasi; Daranide; Oratrol. $C_6H_6Cl_2N_2O_4S_2$; mol wt 305.14. C 23.62%, H 1.98%, Cl 23.24%, N 9.18%, O 20.97%, S 21.01%. Prepn: Schultz, US 2835702 (1958 to Merck & Co.).

Needles from DMSO + water, mp 239-241° (patent, mp 228.5-229°). Practically insol in water. Sol in alkaline solns.

THERAP CAT: Carbonic anhydrase inhibitor.

3088. Dichlorprop. [120-36-5] 2-(2,4-Dichlorophenoxy)propanoic acid; (±)-2-(2,4-dichlorophenoxy)propionic acid; 2,4-DP; dichloroprop; Dicopur DP. $C_9H_8Cl_2O_3$; mol wt 235.06. C 45.99%, H 3.43%, Cl 30.16%, O 20.42%. Selective pre- and post-emergent herbicide. Exists in two optically active forms; only the (+)-isomer shows biological activity. First prepd and described as a plant growth regulator: M. E. Synerholm, P. W. Zimmerman, *Contrib. Boyce Thompson Inst.* **14**, 91 (1945). Commercial prepn: H. A. Stevenson, R. F. Brookes, GB 822199 (1959 to Boots). Resoln of enantiomers and biological activity: S. T. Collins, F. E. Smith: *J. Sci. Food Agric.* **3**, 248 (1952). Field studies and toxicology: W. O.

G. Nuyken *et al.*, *Meded. Fac. Landbouwwet. Rijksuniv. Gent* **52**, 1139 (1987). Enantiomeric effects on degradation: A.W. Garrison *et al.*, *Environ. Sci. Technol.* **30**, 2449 (1996). GC/MS determn in soil: T. Heberer, H.-J. Stan, *J. AOAC Int.* **79**, 1428 (1996).

Colorless crystals, mp 117-118°. pKa 3.2. Soly in water at 20°: 350 ppm. Readily sol in organic solvents. Corrosive to metals in the presence of water.

    (*R*)-(+)-**Form.** [15165-67-0] Dichlorprop-P; Corasil; Optica DP. Colorless needles from benzene, mp 124°. $[\alpha]_D^{21}$ +26.6° (c = 1.23 in ethanol).

    (*R*)-(+)-**Form Potassium salt.** [113963-87-4] Dichlorprop-P-potassium; Duplosan DP. $C_9H_7Cl_2KO_3$; mol wt 273.15.

    USE: Herbicide.

    3089. Dichlorvos. [62-73-7] Phosphoric acid 2,2-dichloroethenyl dimethyl ester; phosphoric acid 2,2-dichlorovinyl dimethyl ester; *O,O*-dimethyl *O*-(2,2-dichlorovinyl) phosphate; dichlorophos; dichlorovos; DDVP; SD-1750; Atgard; Dedevap; Divipan; Doom; Nuvan; Task; Vapona. $C_4H_7Cl_2O_4P$; mol wt 220.97. C 21.74%, H 3.19%, Cl 32.09%, O 28.96%, P 14.02%. Organophosphate insecticide; cholinesterase inhibitor. Prepn: Whetstone, Harman, **US 2956073** (1960 to Shell). Toxicity: T. B. Gaines, *Toxicol. Appl. Pharmacol.* **14**, 515 (1969). Determn in wheat: F. Longobardi *et al.*, *J. Agric. Food Chem.* **53**, 9389 (2005). Review of carcinogenicity studies: M. D. Reuber, *Clin. Toxicol.* **18**, 47-84 (1981); of toxicology and human exposure: *Toxicological Profile for Dichlorvos* (PB98-101124, 1997) 248 pp.

Liquid. *Poisonous.* Practically non-flammable. d_4^{25} 1.415. bp_{20} 140°; $bp_{1.0}$ 84°; $bp_{0.5}$ 72°; $bp_{0.01}$ 30°. n_D^{25} 1.451. Vapor pressure at 20°: 1.2×10^{-2} mm Hg. Misc with alcohol and most non-polar solvents. Soly in water: about 1 g/100 ml; in glycerol: about 0.5 g/100 ml. LD_{50} orally in male, female rats: 80, 56 mg/kg (Gaines).

    Caution: Potential symptoms of overexposure are miosis, aching eyes; rhinorrhea; headache; chest tightening, wheezing, laryngeal spasms and salivation; cyanosis; anorexia, nausea, vomiting and diarrhea; sweating; muscle fasciculation, paralysis, giddiness and ataxia; convulsions; low blood pressure, cardiac irregularities; irritation of eyes and skin. *See NIOSH Pocket Guide to Chemical Hazards* (DHHS/NIOSH 97-140, 1997) p 102.

    USE: Insecticide; fumigant.

    THERAP CAT (VET): Anthelmintic; ectoparasiticide.

    3090. Diclazuril. [101831-37-2] 2,6-Dichloro-α-(4-chlorophenyl)-4-(4,5-dihydro-3,5-dioxo-1,2,4-triazin-2(3*H*)-yl)benzeneacetonitrile; (*p*-chlorophenyl)[2,6-dichloro-4-(4,5-dihydro-3,5-dioxo-*as*-triazin-2(3*H*)-yl)phenyl]acetonitrile; R-64433; Clinacox. $C_{17}H_9Cl_3N_4O_2$; mol wt 407.64. C 50.09%, H 2.23%, Cl 26.09%, N 13.74%, O 7.85%. Nucleotide analog with broad-spectrum anticoccidial activity. Prepn: G. M. Boeckx *et al.*, **EP 170316**; *eidem*, **US 4631278** (both 1986 to Janssen). Mechanism of action studies: L. Maes *et al.*, *J. Parasitol.* **74**, 931 (1988); A. Verheyen *et al.*, *ibid.* 939. Efficacy against *Eimeria* species in turkeys: O. Vanparijs *et al.*, *Avian Dis.* **33**, 479 (1989); in chickens: O. Vanparijs *et al.*, *Poult. Sci.* **69**, 60 (1990). Preliminary evaluation in coccidial enteritis in AIDS patients: K. Kayembe *et al.*, *Lancet* **1**, 1397 (1989).

mp 290.5°.

THERAP CAT (VET): Coccidiostat.

    3091. Diclofenac. [15307-86-5] 2-[(2,6-Dichlorophenyl)amino]benzeneacetic acid; [*o*-(2,6-dichloroanilino)phenyl]acetic acid; Motifene. $C_{14}H_{11}Cl_2NO_2$; mol wt 296.15. C 56.78%, H 3.74%, Cl 23.94%, N 4.73%, O 10.80%. Prepn: **NL 6604752**; A. Sallmann, R. Pfister, **US 3558690** (1966, 1971 both to Geigy). Pharmacology: Renaud, Lecompte, *Thromb. Diath. Haemorrh.* **24**, 577 (1970), *C.A.* **74**, 86215m (1971); Krupp *et al.*, *Experientia* **29**, 450 (1973). HPLC determn in plasma and urine: J. Godbillon *et al.*, *J. Chromatogr.* **338**, 151 (1985). Symposium on pharmacology and clinical experience: *Semin. Arthritis Rheum.* **15**, Suppl. 1, 57-110 (1985); on pharmacology, efficacy and safety: *Am. J. Med.* **80**, Suppl. 4B, 1-87 (1986). Comprehensive description: C. M. Adeyeye, P-K. Li, *Anal. Profiles Drug Subs.* **19**, 123-144 (1990). Review of clinical trials in actinic keratosis: D. C. Peters, R. H. Foster, *Drugs Aging* **14**, 313-319 (1999).

Crystals from ether-petr ether, mp 156-158°.

    Diethylammonium salt. [78213-16-8] Voltarol. $C_{14}H_{11}Cl_2$-$NO_2 \cdot C_4H_{11}N$; mol wt 369.29.

    Sodium salt. [15307-79-6] GP-45840; Allvoran; Benfofen; Dealgic; Deflamat; Delphinac; Dicloflex; Diclomax; Diclophlogont; Dicloreum; Duravolten; Ecofenac; Effekton; Lexobene; Neriodin; Novapirina; Primofenac; Prophenatin; Rewodina; Rhumalgan; Voldal; Voltaren; Xenid. $C_{14}H_{10}Cl_2NNaO_2$; mol wt 318.13. White crystals from water, mp 283-285°. uv max (methanol) 283 nm (ε 1.05×10^5); (phosphate buffer, pH 7.2) 276 nm (ε 1.01×10^5). Soly at 25°C (mg/ml): deionized water (pH 5.2) >9; methanol >24; acetone 6; acetonitrile <1; cyclohexane <1; HCl (pH 1.1) <1; phosphate buffer (pH 7.2) 6. Sol in ethanol. Practically insol in chloroform, ether. pKa 4. Partition coefficient (*N*-octanol/aq. buffer): 13.4. LD_{50} in mice, rats (mg/kg): ~390, 150 orally (Krupp).

    Potassium salt. [15307-81-0] CGP-45840B; Cataflam. $C_{14}H_{10}$-Cl_2KNO_2; mol wt 334.24.

    THERAP CAT: Anti-inflammatory.

    3092. Diclofop-methyl. [51338-27-3] 2-[4-(2,4-Dichlorophenoxy)phenoxy]propanoic acid methyl ester; methyl 2-[4-(2,4-dichlorophenoxy)phenoxy]propionate; HOE-23408; Hoelon; Hoegrass; Illoxan. $C_{16}H_{14}Cl_2O_4$; mol wt 341.18. C 56.33%, H 4.14%, Cl 20.78%, O 18.76%. Selective post-emergence herbicide. Prepn: W. Becker *et al.*, **DE 2223894**; *eidem*, **US 3954442** (1973, 1976 to Hoechst). Soil degradn: A. E. Smith, *J. Agric. Food Chem.* **25**, 893 (1977); **27**, 1145 (1979).

mp 39-41°. $bp_{0.1}$ 175-177°. Soly in water: 0.3 mg/100 ml. Soly (g/100 ml): acetone 249; ethanol 11; xylene 253.

    USE: Herbicide.

    3093. Diclosulam. [145701-21-9] *N*-(2,6-Dichlorophenyl)-5-ethoxy-7-fluoro[1,2,4]triazolo[1,5-*c*]pyrimidine-2-sulfonamide; XDE-564; Strongarm. $C_{13}H_{10}Cl_2FN_5O_3S$; mol wt 406.21. C 38.44%, H 2.48%, Cl 17.45%, F 4.68%, N 17.24%, O 11.82%, S 7.89%. Pre-emergent, soil-applied herbicide for broadleaf weed control in peanuts and soybeans. Prepn: J. C. Van Heertum *et al.*, **EP 343752**; *eidem*, **US 5163995** (1989, 1992 both to DowElanco). Field study in peanuts: W. A. Bailey *et al.*, *Weed Technol.* **13**, 450, 771

(1999); W. J. Grichar *et al.*, *Peanut Sci.* **26**, 23 (1999). Field trial in soybeans: D. R. Shaw *et al.*, *Weed Technol.* **13**, 791 (1999).

mp 234-237°.
USE: Herbicide.

3094. Dicloxacillin. [3116-76-5] (2*S*,5*R*,6*R*)-6-[[[3-(2,6-Dichlorophenyl)-5-methyl-4-isoxazolyl]carbonyl]amino]-3,3-dimethyl-7-oxo-4-thia-1-azabicyclo[3.2.0]heptane-2-carboxylic acid; 6-[3-(2,6-dichlorophenyl)-5-methyl-4-isoxazolecarboxamido]penicillanic acid; 3-(2,6-dichlorophenyl)-5-methyl-4-isoxazolylpenicillin; BRL-1702; Maclicine. $C_{19}H_{17}Cl_2N_3O_5S$; mol wt 470.32. C 48.52%, H 3.64%, Cl 15.07%, N 8.93%, O 17.01%, S 6.82%. Semisynthetic antibiotic related to penicillin. Prepn: J. H. C. Naylor, *GB 978299*; *eidem*, *US 3239507* (1965, 1969 to Beecham). Toxicity data: C. Gloxhuber *et al.*, *Arzneim.-Forsch.* **15**, 322 (1965). Series of articles on chemistry, pharmacology and toxicology: *ibid.* 322-348; Matsuzaki, *Jpn. J. Antibiot.* **21**, 274, 284 (1968), *C.A.* **70**, 95267z, 95265x (1969).

Sodium salt monohydrate. [13412-64-1] Sodium dicloxacillin monohydrate; P-1011; Brispen; Dichlor-Stapenor; Diclocil; Dycill; Dynapen; Noxaben; Pathocil; Pen-Sint; Syntarpen; Veracillin. C_{19}-$H_{16}Cl_2N_3NaO_5S.H_2O$; mol wt 510.32. White crystals, dec 222-225°. $[\alpha]_D^{20}$ +127.2° (water). Freely sol in water; sol in methanol; less sol in butanol; slightly sol in acetone and the usual organic solvents. LD_{50} in mice (g/kg): 0.9 i.v.; in rats (g/kg): 0.63 i.p.; >5 orally (Gloxhuber).

THERAP CAT: Antibacterial.
THERAP CAT (VET): Antibacterial.

3095. Dicobalt Octacarbonyl. [10210-68-1] Di-μ-carbonylhexacarbonyldicobalt (Co-Co); octacarbonyldicobalt; cobalt tetracarbonyl; cobalt octacarbonyl. $C_8Co_2O_8$; mol wt 341.95. C 28.10%, Co 34.47%, O 37.43%. $Co_2(CO)_8$. Prepn: Mond *et al.*, *J. Chem. Soc.* **97**, 798 (1910); Gilmont, Blanchard, *Inorg. Synth.* **2**, 238 (1946); Wender *et al.*, *ibid.* **5**, 190 (1957); Chini *et al.*, *Chim. Ind. (Milan)* **55**, 120 (1973). Crystal structure: Sumner *et al.*, *Acta Crystallogr.* **17**, 732 (1964). Two isomeric forms exist in soln: Noack, *Spectrochim. Acta* **19**, 1925 (1963); Bor, *ibid.* 2065; Noack, *Helv. Chim. Acta* **47**, 1064, 1555 (1964). Toxicity studies: Kincaid *et al.*, *Arch. Ind. Hyg. Occup. Med.* **10**, 210 (1954); Brief *et al.*, *Arch. Environ. Health* **23**, 373 (1971); Spiridonova, Shabalina, *Gig. Sanit.* **1973**, 73, *C.A.* **78**, 119835b (1973). Reviews of prepn, properties and chemistry of cobalt carbonyls: I. Wender *et al.*, *Bull. U.S. Bur. Mines* **600**, 83 pp (1960); *Organic Syntheses via Metal Carbonyls* **vol. 1**, I. Wender, P. Pino, Eds. (Interscience, New York, 1968) *passim*,; Chalk, Harrod, "Catalysis by Cobalt Carbonyls" in *Adv. Organomet. Chem.* **6**, 119-170 (1968); Nicholls in *Comprehensive Inorganic Chemistry* **vol. 3**, J. C. Bailar, Jr. *et al.*, Eds. (Pergamon Press, Oxford, 1973) pp 1059-1064.

Orange platelets obtained by vacuum sublimation. d 1.87. mp 51°; dec above 52°. Stable in an atm of H_2 and CO; dec on exposure to air. Insol in water. Sol in organic solvents such as ether, alcohol, CS_2, naphtha. Slowly attacked by HCl, H_2SO_4, more rapidly by

HNO_3, Br_2. LD_{50} in mice, rats (mg/kg): 377.7, 753.8 by gavage (Spiridonova, Shabalina).

Cobalt carbonyl hydride. Cobalt hydrocarbonyl; tetracarbonyl-hydridocobalt; tetracarbonylhydrocobalt. C_4HCoO_4; mol wt 171.98. Unstable catalytic intermediate; active species in some reactions catalyzed by $Co_2(CO)_8$. Yellow, toxic, foul-smelling liquid. mp −26°. Dec at room temp to $Co_2(CO)_8$. Sparingly sol in water (3 × 10^{-3} moles/l); behaves as a strong acid. Readily oxidized in air.

Caution: Potential symptoms of overexposure are irritation of eyes, skin, mucous membranes; cough, decreased pulmonary function, wheezing, dyspnea. *See NIOSH Pocket Guide to Chemical Hazards* (DHHS/NIOSH 97-140, 1997) p 74.

USE: Catalyst for hydroformylation, hydrogenation, hydrosilation, isomerization, carboxylation, carbonylation and polymerization reactions.

3096. Dicofol. [115-32-2] 4-Chloro-α-(4-chlorophenyl)-α-(trichloromethyl)benzenemethanol; 4,4'-dichloro-α-(trichloromethyl)benzhydrol; 1,1-bis(*p*-chlorophenyl)-2,2,2-trichloroethanol; di(*p*-chlorophenyl)trichloromethylcarbinol; DTMC; ENT-23648; FW-293; Acarin; Kelthane; Mitigan. $C_{14}H_9Cl_5O$; mol wt 370.48. C 45.39%, H 2.45%, Cl 47.84%, O 4.32%. Prepn: Wilson, Craig, **US 2812280**; Wilson, Wolffe, **US 2812362** (both 1957 to Rohm & Haas); Bergmann, Kaluszyner, *J. Org. Chem.* **23**, 1306 (1958); **GB 831421** (1960 to Metal & Thermit Corp.). Miticidal activity: J. S. Baker, F. B. Maughan, *J. Econ. Entomol.* **49**, 458 (1956). Toxicology: R. B. Smith *et al.*, *Toxicol. Appl. Pharmacol.* **1**, 119 (1959). Metabolism: J. R. Brown *et al.*, *ibid.* **15**, 30 (1969); K. Tabata *et al.*, *Appl. Entomol. Zool.* **14**, 490 (1979). Review of carcinogenic risk: *IARC Monographs* **30**, 87-101 (1983).

Crystals from petr ether, mp 77-78°. uv max (ethanol): 226, 258, 266, 276 nm (4.43, 2.82, 2.85, 2.60). Practically insol in water. Sol in most aliphatic and aromatic solvents. LD_{50} in rats (mg/kg): 1495 orally; 1150 i.p. (Brown).

USE: Acaricide.

3097. Dicrotophos. [141-66-2] Phosphoric acid (1*E*)-3-(dimethylamino)-1-methyl-3-oxo-1-propen-1-yl dimethyl ester; phosphoric acid dimethyl ester, ester with *cis*-3-hydroxy-*N,N*-dimethyl-crotonamide; 3-(dimethoxyphosphinyloxy)-*N,N*-dimethyl-*cis*-crotonamide; dimethyl 2-dimethylcarbamoyl-1-methylvinyl phosphate; dimethyl 1-dimethylcarbamoyl-1-propen-1-yl phosphate; C-709; ENT-24482; SD-3562; Bidrin. $C_8H_{16}NO_5P$; mol wt 237.19. C 40.51%, H 6.80%, N 5.91%, O 33.73%, P 13.06%. Cholinesterase inhibitor. Prepn of diethyl ester: Whetstone, Stiles, **US 2802855** (1957 to Shell). Activity: R. A. Corey, *J. Econ. Entomol.* **58**, 112 (1965). Metabolism and degradn: R. I. Beynon *et al.*, *Residue Rev.* **47**, 55 (1973). Toxicity data: T. B. Gaines, *Toxicol. Appl. Pharmacol.* **14**, 515 (1969).

Liquid, commercial grade is brown. bp_{760} 400°. d_{15}^{15} 1.216. n_D^{23} 1.468. Misc with water, ethanol, xylene; somewhat sol in kerosene. Dec at 90° after 7 days; at 75° after 31 days. LD_{50} in female, male rats (mg/kg): 16, 21 orally; 42, 43 dermally (Gaines).

Caution: Potential symptoms of overexposure are headache, nausea, dizziness, anxiety, restlessness, muscle twitching, weakness, tremor, incoordination, vomiting, abdominal cramps, diarrhea; salivation, sweating, lacrimation, rhinitis; anorexia, malaise. *See*

NIOSH Pocket Guide to Chemical Hazards (DHHS/NIOSH 97-140, 1997) p 102.

USE: Insecticide.

3098. Dicryl. [2164-09-2] *N*-(3,4-Dichlorophenyl)-2-methyl-2-propenamide; 3',4'-dichloro-2-methylacrylanilide; *N*-(3,4-dichlorophenyl)methacrylamide; chloranocryl; Niagara 4556. $C_{10}H_9Cl_2$-NO; mol wt 230.09. C 52.20%, H 3.94%, Cl 30.81%, N 6.09%, O 6.95%. Post-emergence herbicide. Prepn from 3,4-dichloroaniline with methacryloyl chloride in the presence of triethylamine. Prepd but not claimed: Thompson, US 3169850 (1965). Toxicity study: G. W. Bailey, J. L. White, *Residue Rev.* **10**, 97 (1965).

Crystals from ethanol + petr ether, mp 128°. Practically insol in water. Soluble in acetone, alcohol, isophorone, dimethyl sulfoxide. LD_{50} orally in rats: 3160 mg/kg (Bailey, White).

USE: Herbicide.

3099. Dictamnine. [484-29-7] 4-Methoxyfuro[2,3-*b*]quinoline; dictamine. $C_{12}H_9NO_2$; mol wt 199.21. C 72.35%, H 4.55%, N 7.03%, O 16.06%. Alkaloid from root of *Dictamnus albus* Linn., *Skimmia repens* Nakai, *Aegle marmelos* Correa, *Zanthoxylum alatum* Roxb., *Rutaceae*: Thoms, *Ber.* **33**, 68 (1923); Asahina *et al.*, *ibid.* **63**, 2045 (1930); Chatterjee, Roy, *J. Indian Chem. Soc.* **36**, 267 (1959); Deb *et al.*, *ibid.* **39**, 493 (1962). Structure: M. F. Brundon, N. J. McCorkindale, *Chem. Ind. (London)* **1956**, 1091. Synthesis: Tuppy, Bohm, *Angew. Chem.* **68**, 388 (1956); M. Sato *et al.*, *Chem. Pharm. Bull.* **35**, 1319 (1987).

Prisms from alcohol, ethyl acetate, benzene + ethyl acetate, mp 133°. Sol in hot alcohol, chloroform; slightly sol in ether. Practically insol in water.

Hydrochloride. $C_{12}H_9NO_2 \cdot HCl$. Needles from alcohol, dec 170°.

3100. Dicumarol. [66-76-2] 3,3'-Methylenebis[4-hydroxy-2*H*-1-benzopyran-2-one]; 3,3'-methylenebis[4-hydroxycoumarin]; bishydroxycoumarin (rescinded); Dicoumarol; Dicoumarin; Dicumol; Melitoxin. $C_{19}H_{12}O_6$; mol wt 336.30. C 67.86%, H 3.60%, O 28.54%. Originally isolated from spoiled sweet clover (improperly cured Melilotus hay): Link *et al.*, *J. Biol. Chem.* **138**, 21, 513, 529 (1941); **142**, 941 (1942); now prepd synthetically by the action of formaldehyde on 4-hydroxycoumarin (obtained as the Na derivative by the action of sodium metal on methyl acetylsalicylate): Link, *Fed. Proc.* **4**, 176 (1945), and patents (The Wisconsin Alumni Res. Found.). Toxicity data: Rose *et al.*, *Proc. Soc. Exp. Biol. Med.* **50**, 228 (1942).

Minute crystals. Slight pleasant odor. Has a slightly bitter taste. mp 287-293°. Sol in aq alkaline solns, in pyridine and similar organic bases. Slightly sol in benzene and chloroform. Practically insol in water, alcohol, ether. LD_{50} orally in rats: 541.6 mg/kg (Rose).

THERAP CAT: Anticoagulant.

3101. Dicyanine. [52260-69-2] 1-Ethyl-2-[3-(1-ethyl-2-methyl-4(1*H*)-quinolinylidene)-1-propen-1-yl]-4-methylquinolinium iodide (1:1); 1-ethyl-2-[3-(1-ethyl-4(1*H*)-quinolylidene)propenyl]lepidinium iodide; 1-ethyl-2-[3-(1-ethyl-2-methyl-4(1*H*)-quinolylidene)propenyl]-4-methylquinolinium iodide; 2',4-dimethyl-1,1'-diethyl-2,4'-carbocyanine iodide. $C_{27}H_{29}IN_2$; mol wt 508.45. C 63.78%, H 5.75%, I 24.96%, N 5.51%. Prepn by J. A. Aeschlimann as reported by: Mills, Odams, *J. Chem. Soc.* **125**, 1913 (1924). See also *Beilstein* 2nd suppl. **vol. 23**, p 284. Has photoconductive properties that increase sharply with a rise in temp and is classed as an O_2-photoconductor: Meier, *Z. Phys. Chem. (Leipzig)* **208**, 325 (1958). Gives a color reaction with magnesium ions: Babenko, *Zh. Anal. Khim.* **13**, 496 (1958).

Olive-green crystals with metallic sheen from methanol. Crystals are usually solvated. Drying at 110° expels methanol of crystn. Decomp 244-252°. Absorption max (methanol): 603.5, 655.5 nm ($A_{1cm}^{1\%}$ 63, 218). Soly in methanol: about 2%. Methanol solns appear deep blue when viewed under incident daylight. Soly in water about 0.2%. Aq solns are dichroic and may appear brownish green. Also sol in abs ethanol (0.5%), in ethylene glycol (1.5%), and in Cellosolve (2.0%). Practically insol in benzene, xylene.

USE: In color photography (sensitizer for extreme reds). Diagnostic stain for cytodiagnosis of ruptured fetal membranes.

3102. Dicyanocobinamide. [27792-36-5] *Co,Co*-Bis(cyano-κ*C*)-cobinamide; dicyanocobyrinic acid *a,b,c,d,e,g*-hexamide *f*-D-2-hydroxypropylamide; cobinamide dicyanide; etiocobalamin; vitamin B_{12p}; factor B. $C_{50}H_{72}CoN_{13}O_8$; mol wt 1042.14. C 57.63%, H 6.96%, Co 5.66%, N 17.47%, O 12.28%. Precursor in the bacterial biosynthesis of vitamin B_{12}. Isoln from calf feces: J. E. Ford, J. W. G. Porter, *Biochem. J.* **51**, v (1952). Prepn by removal of the nucleotide from cyanocobalamin by acid hydrolysis: D. E. Gant *et al.*, *ibid.* **56**, xxxiv (1954); W. Friedrich, K. Bernhauer, *Angew. Chem.* **65**, 627 (1953). Prepn from cobyrinic acid *a,b,c,d,e,g*-hexamide (factor V_{1a}) and D(−)-1-amino-2-propanol: K. Bernhauer *et al.*, *Helv. Chim. Acta* **43**, 696 (1960); *idem*, US 3072674 (1963 to Hoffmann-La Roche). Purification by HPLC: E. Stupperich *et al.*, *Anal. Biochem.* **155**, 365 (1986). Biosynthetic study: J. E. Ford *et al.*, *Biochem. J.* **59**, 86 (1955). Alternate pathway: J. D. Woodson, J. C. Escalante-Semerena, *Proc. Natl. Acad. Sci. USA* **101**, 3591 (2004).

Purple solid. Abs max: 348, 367, 530, 580 nm. Loses cyanide in aqueous solution.

3103. Dicyanodiamide. [461-58-5] *N*-Cyanoguanidine. $C_2H_4N_4$; mol wt 84.08. C 28.57%, H 4.80%, N 66.64%. Prepared by controlled polymerization of cyanamide in water in the presence of ammonia, alkaline earth hydroxides, or other suitable bases: Grube, Nitsche, *Z. Angew. Chem.* **27**, 368 (1914); Pinck, Hetherington, *Ind. Eng. Chem.* **27**, 834 (1935); Pinck, *Inorg. Synth.* **3**, 43 (1950).

Monoclinic prismatic crystals from water or alcohol, mp 209.5°. d_4^{25} 1.400. Eutectic with cyanamide at 35.6° (15% dicyanodiamide). Specific heat 0.456 at 0-204°. Neutral reaction. Soly in water at 13°: 2.26%, more sol in hot water. Solns above 80° dec slowly, yielding ammonia. Soly in abs ethanol at 13°: 1.26%; in ether: 0.01%. Insol in benzene, chloroform. Sol in liquid ammonia.

USE: In the plastics industry (manuf of melamine). In the pharmaceutical industry (barbiturates, guanidine derivs).

3104. Dicyanodiamidine Sulfate. [591-01-5] *N*-(Aminoiminomethyl)urea sulfate (2:1); biuretamidine sulfate; carbamylguanidine sulfate; guanylurea sulfate. $C_4H_{14}N_8O_6S$; mol wt 302.27. C 15.89%, H 4.67%, N 37.07%, O 31.76%, S 10.61%. Prepd by heating dicyanodiamide with H_2SO_4 and H_2O.

Dihydrate. White needles. At 110° loses its water. Sol in about 20 parts cold, 3 parts boiling water; slightly sol in alc.

USE: For detecting and determining Ni and its separation from Co and other metals.

3105. 9-Dicyanomethylene-2,4,7-trinitrofluorene. [1172-02-7] 2-(2,4,7-Trinitro-9*H*-fluoren-9-ylidene)propanedinitrile; DTF. $C_{16}H_5N_5O_6$; mol wt 363.25. C 52.90%, H 1.39%, N 19.28%, O 26.43%. Prepn from 2,4,7-trinitrofluorenone: Mukherjee, Levasseur, *J. Org. Chem.* **30**, 644 (1965).

Yellow crystals from acetonitrile, mp 266-268°. uv max (dichloromethane): 365 nm (log ε 4.38).

USE: Forms charge-transfer complexes with aromatic hydrocarbons and amines.

3106. Dicyclohexylamine. [101-83-7] *N*-Cyclohexylcyclohexanamine; dodecahydrodiphenylamine. $C_{12}H_{23}N$; mol wt 181.32. C 79.49%, H 12.79%, N 7.73%. Prepn by hydrogenation of equimolar amounts of cyclohexanone and cyclohexylamine: C. F. Winans, H. Adkins, *J. Org. Chem.* **54**, 306 (1932); R. M. Robinson, W. C. Braaten, US 3154580 (1964 to Abbott). Forms crystalline derivatives with *N*-protected amino acids: E. Klieger *et al.*, *Ann.* **640**, 157 (1961). *Review:* T. S. Carswell, H. L. Morrill, *Ind. Eng. Chem.* **29**, 1265 (1937).

Liquid. Faint fishy odor. d_{25}^{25} 0.9104. mp −0.1°. pKa 10.4. bp_{760} 255.8°; bp_{300} 214.5°; bp_{200} 199.0°; bp_{100} 174.4°; bp_{50} 154.3°; bp_{25} 135.4°; bp_{11} 121°; bp_4 99.3°; $bp_{1.0}$ 83°. n_D^{25} 1.4823. Flash pt: 110°C. *Corrosive*. Strong base. Sparingly sol in water; sol in the usual organic solvents. Miscible with cyclohexylamine. Readily forms adducts with solvents.

Caution: A skin irritant and possible sensitizer.

USE: Industrial solvent; corrosion inhibitor.

3107. Dicyclohexylcarbodiimide. [538-75-0] *N,N*′-Methanetetraylbiscyclohexanamine; carbodicyclohexylimide; DCC; DCCI. $C_{13}H_{22}N_2$; mol wt 206.33. C 75.68%, H 10.75%, N 13.58%. Coupling agent in peptide synthesis: Sheehan, Hess, *J. Am. Chem. Soc.* **77**, 1067 (1955); Khorana, *Chem. Ind. (London)* **1955**, 1087. Prepn: Schmidt *et al.*, *Ber.* **71**, 1933 (1938); Schmidt, Schnegg, US 2656383 (1953 to Bayer); Stevens *et al.*, *J. Org. Chem.* **32**, 2895 (1967); Bestmann *et al.*, *Ann.* **718**, 24 (1968).

Cryst mass, mp 35-36°. bp_{11} 154-156°; $bp_{0.5}$ 98-100°.

Caution: Contact allergen.

USE: In the synthesis of peptides.

3108. Dicyclomine. [77-19-0] [1,1′-Bicyclohexyl]-1-carboxylic acid 2-(diethylamino)ethyl ester; β-diethylaminoethyl 1-cyclohexylcyclohexanecarboxylate; bis(cyclohexyl)carboxylic acid diethylaminoethyl ester; dicycloverin. $C_{19}H_{35}NO_2$; mol wt 309.49. C 73.74%, H 11.40%, N 4.53%, O 10.34%. Anticholinergic; binds to muscarinic M_1 receptors. Prepn: C. H. Tilford *et al.*, *J. Am. Chem. Soc.* **69**, 2903 (1947); M. G. Van Campen, C. H. Tilford, US 2474796 (1949 to Merrell). Receptor binding study: H. Kilbinger, A. Stein, *Br. J. Pharmacol.* **94**, 1270 (1988). Capillary GC determn in plasma: B. J. Walker *et al.*, *J. Chromatogr.* **416**, 150 (1987). Clinical trial in functional abdominal pain: M. G. Grillage *et al.*, *Br. J. Clin. Pract.* **44**, 176 (1990).

Hydrochloride. [67-92-5] Bentyl; Bentylol; Merbentyl; Wyovin. $C_{19}H_{35}NO_2$·HCl; mol wt 345.95. Crystals from butanone, mp 164-166°. Bitter taste. Sol in water; freely sol in alcohol, chloroform; very slightly sol in ether.

THERAP CAT: Antispasmodic.

3109. Didanosine. [69655-05-6] 2′,3′-Dideoxyinosine; dideoxyinosine; ddI; ddIno; BMY-40900; NSC-612049; Videx. $C_{10}H_{12}N_4O_3$; mol wt 236.23. C 50.84%, H 5.12%, N 23.72%, O 20.32%. Purine nucleoside reverse transcriptase inhibitor (NRTI); converted *in vivo* to the active metabolite, dideoxyadenosine 5′-triphosphate, *q.v.* Enzymatic prepn from dideoxyadenosine: W. Plunkett, S. S. Cohen, *Cancer Res.* **35**, 1547 (1975). Synthesis: R. R. Webb *et al.*, *Nucleosides Nucleotides* **7**, 147 (1988). Antiviral activity vs HIV-1: H. Mitsuya, S. Broder, *Proc. Natl. Acad. Sci. USA* **83**, 1911 (1986). Antiretroviral spectrum *in vitro*: J. E. Dahlberg *et al.*, *ibid.* **84**, 2469 (1987). HPLC determn in human plasma: G. Ray, E. Murrill, *Anal. Lett.* **20**, 1815 (1987). Comprehensive description: M. N. Nassar *et al.*, *Anal. Profiles Drug Subs. Excip.* **22**, 185-227 (1993). Review of pharmacokinetics and clinical experience in HIV infection: C. M. Perry, S. Noble, *Drugs* **58**, 1099-1135 (1999); S. Moreno *et al.*, *ibid.* **67**, 1441-1462 (2007).

White crystalline powder. mp 160-163°. pKa 9.12 ±0.02. Partition coefficient (1-octanol/0.05M phosphate buffer pH 7): 0.068 ±0.005. $[\alpha]_D^{25}$ −26.3° (c = 10 in water). uv max: 248 nm (pH 2); 254 nm (pH 12). Very sol in dimethyl sulfoxide. Soly in water at 25° (mg/ml): 27.3 (pH 6); 429 (pH 10). Soly at 23° (mg/ml): <1 in acetone, acetonitrile, t-butanol, chloroform, ethanol, ethyl acetate, hexane, methylene chloride, 1-propanol, 2-propanol; 1 in PEG-300; 6 in methanol; 8 in propylene glycol; 45 in dimethylacetamide; 200 in DMSO. Unstable in acidic solns.

THERAP CAT: Antiretroviral.

3110. Didecyldimethylammonium Chloride. [7173-51-5] N-Decyl-N,N-dimethyl-1-decanaminium chloride (1:1); dimethyldidecylammonium chloride; Bardac 2250/2280; BTC-1010; Dodigen 1881; Querton 210CL. $C_{22}H_{48}ClN$; mol wt 362.08. C 72.98%, H 13.36%, Cl 9.79%, N 3.87%. Quaternary alkyl ammonium compound with fungicidal and bactericidal activity. Prepn: A. W. Ralston *et al.*, *J. Org. Chem.* **13**, 186 (1948). Conductivity of aqueous soln: *eidem*, *J. Am. Chem. Soc.* **70**, 977 (1948). Mothproofing agent: A. M. Schwartz, C. A. Rader, US 3746767 (1971 to USDA). In carpet dyeing: F. F. Bartsch, R. Feigin, US 3758269 (1971 to Sybron). Germicidal activity: R. J. Wright, DE 2810998; *eidem*, US 4272395 (1978, 1981 both to Unilever). Ion-pair extracting agent in carbapenem isoln: J. D. Hood *et al.*, *J. Antibiot.* **32**, 295 (1979). In froth-flotation recovery of metals: V. Petrovich, US 4225428 (1980). Eliminates ice-nucleating effect of bacteria in plants: E. A. Youngman, R. C. Schnell, EP 37593; *eidem*, US 4311517 (1981, 1982 both to Shell). Breaks oil-in-water emulsions created during oil-spill recovery procedures: J. Newcombe, US 4374734 (1983 to Cities Service). Fungidical wood preservative: A. F. Preston: *J. Am. Oil Chem. Soc.* **60**, 567 (1983).

Extremely hygroscopic. Sol in acetone, extremely sol in benzene. Insol in hexane.

Caution: Causes eye damage and skin irritation. Harmful if swallowed.

USE: General purpose disinfectant, sanitizer; mildew preventative in commercial laundries; water treatment in cooling towers and oil field flood waters; wood preservative.

3111. Didemnins. Biologically active depsipeptides extracted from a Caribbean tunicate (sea squirt) family *Didemnidae*, genus *Trididemnum*. First marine natural product to enter clinical trials as antineoplastic. The three components, didemnins A, B and C are weakly basic compds with antiviral, antitumor activity. Didemnin A is the most abundant, whereas didemnin B is generally the most active. Isoln and bioactivity: K. L. Rinehart, Jr. *et al.*, *Science* **212**, 933 (1981). Extraction and purification: K. L. Rinehart, Jr., EP 48149; *idem*, US 4493796 (1982, 1985 both to Univ. Illinois). Structure determn: K. L. Rinehart, Jr. *et al.*, *J. Am. Chem. Soc.* **103**, 1857 (1981). Revised structure and total synthesis: K. L. Rinehart, Jr. *et al.*, *ibid.* **109**, 6846 (1987). *See also:* U. Schmidt *et al.*, *Tetrahedron Lett.* **29**, 3057 (1988); *eidem*, *ibid.* 4407. Crystal structure of B: M. B. Hossain *et al.*, *Proc. Natl. Acad. Sci. USA* **85**, 4118 (1988). Efficient total synthesis of A and B: Y. Hamada *et al.*, *J. Am. Chem. Soc.* **111**, 669 (1989). Mechanism of action study: L. H. Li *et al.*, *Cancer Lett.* **23**, 279 (1984). *In vitro* and *in vivo* immunosuppressive activity of B: D. W. Montgomery, C. F. Zukowski, *Transplantation* **40**, 49 (1985). HPLC determn in biological fluids: J. N. Hartshorn *et al.*, *J. Liq. Chromatogr.* **9**, 1489 (1986). Clinical pharmacology,

pharmacokinetics: F. A. Dorr *et al.*, *Eur. J. Cancer Clin. Oncol.* **24**, 1699 (1988). Review of bioactivity of A and B: K. L. Rinehart, Jr. *et al.*, *Fed. Proc.* **42**, 87-90 (1983); of activity and toxicology of B: H. G. Chun *et al.*, *Invest. New Drugs* **4**, 279-284 (1986); of biological activity, structure and synthesis: W.-R. Li, M. M. Joullie, *Stud. Nat. Prod. Chem.* **10**, 241-302 (1992).

Didemnin A. [77327-04-9] $C_{49}H_{78}N_6O_{12}$; mol wt 943.19. Greenish-white solid. Sol in methanol, ethanol, dioxane, chloroform. Insol in water.

Didemnin B. [77327-05-0] NSC-325319. $C_{57}H_{89}N_7O_{15}$; mol wt 1112.37. Yellow-white amorphous solid. Sol in methanol, ethanol, dioxane, ethyl acetate, chloroform. Insol in water.

Didemnin C. [77327-06-1] $C_{52}H_{82}N_6O_{14}$; mol wt 1015.26. Oil. Sol in methanol, ethanol, isopropanol, dioxane, ethyl acetate and chloroform. Sparingly sol in toluene. Insol in water.

3112. Dideoxyadenosine. [4097-22-7] 2′,3′-Dideoxyadenosine; 6-amino-9-(2′,3′-dideoxy-β-D-*glycero*-pentofuranosyl)purine; ddA; ddAdo. $C_{10}H_{13}N_5O_2$; mol wt 235.25. C 51.06%, H 5.57%, N 29.77%, O 13.60%. Purine nucleoside reverse transciptase inhibitor (NRTI). Prepn: M. J. Robins, R. K. Robins, *J. Am. Chem. Soc.* **86**, 3585 (1964); G. L. Tong *et al.*, *J. Org. Chem.* **30**, 2854 (1965); J. R. McCarthy *et al.*, *J. Am. Chem. Soc.* **88**, 1549 (1966). Chain-terminating inhibition of DNA synthesis: L. Toji, S. S. Cohen, *Proc. Natl. Acad. Sci. USA* **63**, 871 (1969). Antiviral activity vs HIV-1: H. Mitsuya, S. Broder, *ibid.* **83**, 1911 (1986). HPLC determn in biological fluids: P. A. Blau *et al.*, *J. Chromatogr.* **420**, 1 (1987).

Crystals from ethanol, mp 184-186°. $[\alpha]_D^{25}$ −25.2° (c = 1.01 in water). uv max in methanol: 259.5 nm (ε 14800).

5′-Triphosphate. [24027-80-3] ddATP. $C_{10}H_{16}N_5O_{11}P_3$; mol wt 475.18. Active metabolite of didanosine, *q.v.* RIA determn in blood cells: C. Le Saint *et al.*, *Antimicrob. Agents Chemother.* **48**, 589 (2004). uv max: 261 nm (ε 15200).

USE: Triphosphate in DNA sequencing as 3′-end chain terminator.

THERAP CAT: Antiretroviral.

Dienogest

3117

3113. DIDS. [53005-05-3] 2,2'-(1,2-Ethenediyl)bis[5-iso-thiocyanatobenzenesulfonic acid]; 4,4'-diisothiocyanato-2,2'-stil-benedisulfonic acid. $C_{16}H_{10}N_2O_6S_4$; mol wt 454.50. C 42.28%, H 2.22%, N 6.16%, O 21.12%, S 28.22%. Classic inhibitor of ion transport across biological membranes. Prepn: Z. I. Cabantchik, A. Rothstein, *J. Membr. Biol.* **10**, 311 (1972); of tritated form: S. Ship *et al.*, *ibid.* **33**, 311 (1977). Inhibition of chloride transport: T. Janas *et al.*, *Am. J. Physiol.* **257**, C601 (1989); J.-A. Kim *et al.*, *Biochem. Biophys. Res. Commun.* **309**, 291 (2003); of non-selective cation conductance: A. Diakov *et al.*, *Pfluegers Arch.* **442**, 700 (2001). Effects of ionic strength and pH on membrane penetration: V. A. Tverdislov *et al.*, *Colloids Surf. B* **5**, 205 (1995). Mechanism of inhibition in erythrocytes: J. M. Salhany, *Blood Cells Mol. Dis.* **27**, 901 (2001); of cardiac ryanodine receptor: A. P. Hill, R. Sitsapesan, *Biophys. J.* **82**, 3037 (2002).

White crystalline solid. uv max: 340 nm (ε 16000).
USE: Ion channel inhibitor; especially of anion transport.

3114. Dieldrin. [60-57-1] *rel*-(1a*R*,2*R*,2a*S*,3*S*,6*R*,6a*R*,7*S*,-7a*S*)-3,4,5,6,9,9-Hexachloro-1a,2,2a,3,6,6a,7,7a-octahydro-2,7:3,6-dimethanonaphth[2,3-*b*]oxirene; 1,2,3,4,10,10-hexachloro-6,7-ep-oxy-1,4,4a,5,6,7,8,8a-octahydro-*endo*,*exo*-1,4:5,8-dimethanonaph-thalene; insecticide no. 497; HEOD; ENT-16225; compd 497; Octa-lox. $C_{12}H_8Cl_6O$; mol wt 380.90. C 37.84%, H 2.12%, Cl 55.84%, O 4.20%. Organochlorine pesticide. Stereoisomer of endrin, *q.v.* Activity: C. W. Kearns *et al.*, *J. Econ. Entomol.* **42**, 127 (1949). Prepn from aldrin, *q.v.*: Soloway, US 2676131 (1954 to Shell); Payne, Smith, US 2776301 (1957 to Shell). Synthesis: Korte, Rech-meier, *Ann.* **656**, 131 (1962). GC-MS determn in sediment: M.-S. Kim *et al.*, *J. Chromatogr. A* **1208**, 25 (2008). Metabolism: M. K. Baldwin *et al.*, *Chem. Ind. (London)* **1970**, 595; C. T. Bedford, D. H. Hutson, *ibid.* **1976**, 440. Toxicity data: Gaines, *Toxicol. Appl. Pharmacol.* **14**, 515 (1969). Bioaccumulation and toxic effects in sea birds: C. H. Walker, *Aquat. Toxicol.* **17**, 293 (1990). Review of environmental fate and health effects: J. L. Jorgenson, *Environ. Health Perspect.* **109**, 113-139 (2001); of toxicology and human exposure: *Toxicological Profile for Aldrin/Dieldrin* (PB2003-100134, 2002) 354 pp.

Relative stereochemistry

Crystals, mp 176-177°. Vapor press at 20°: 3.1×10^{-6} mm Hg. Practically insol in water. Moderately sol in common organic sol-vents except aliphatic petr solvents and methyl alcohol. Stable in org and inorg alkalies and acids commonly used in agriculture. Af-fected by strong mineral acids. LD_{50} orally in rats: 46 mg/kg (Gaines).

Caution: Absorbed through skin, respiratory mucosa and GI tract. Potential symptoms of overexposure are malaise, headache, nausea, vomiting, dizziness, sweating, tremors, myoclonic limb jerks; clonic and tonic convulsions, coma, respiratory failure; leukocytosis, in-creased blood pressure, tachycardia, arrhythmia, metabolic acidosis, fever. Potential occupational carcinogen. *See NIOSH Pocket Guide to Chemical Hazards* (DHHS/NIOSH 97-140, 1997) p 104; *Clinical Toxicology of Commercial Products*, R. E. Gosselin *et al.*, Eds. (Wil-liams & Wilkins, Baltimore, 5th ed., 1984) Section III, pp 143-146.

Note: Dieldrin is listed as a persistent organic pollutant (POP) in Annex A of the *Stockholm Convention on Persistent Organic Pollut-ants* (United Nations, Stockholm, 2001) 43 pp; amended (Geneva, 2009) 63 pp.

USE: Formerly as insecticide for termites, wood borers, and textile pests.

3115. Dienestrol. [84-17-3] 4,4'-(1,2-Diethylidene-1,2-eth-anediyl)bisphenol; 4,4'-(diethylideneethylene)diphenol; 3,4-bis(*p*-hydroxyphenyl)-2,4-hexadiene; 4,4'-dihydroxy-γ,δ-diphenyl-β,δ-hexadiene; di(*p*-oxyphenyl)-2,4-hexadiene; dienoestrol; estrodienol; Cycladiene; Dienol; Dinovex; DV; Estroral; Gynefollin; Hormofem-in; Oestrasid; Oestrodiene; Oestroral; Restrol; Retalon; Synestrol. $C_{18}H_{18}O_2$; mol wt 266.34. C 81.17%, H 6.81%, O 12.01%. Synthe-sis: Dodds *et al.*, *Proc. Roy. Soc.* **127B**, 162 (1939); Hobday, Short, *J. Chem. Soc.* **1943**, 609; Short, Hobday, US 2464203 (1949 to Boots); Adler, US 2465505 (1949 to Hoffmann-La Roche). Config-uration: Koch, *Nature* **161**, 309 (1948); Lane, Spialter, *J. Am. Chem. Soc.* **73**, 4408 (1951).

(trans-trans)-form

Minute needles from dil alcohol, mp 227-228°. Sublimes at 130° and 1 mm Hg. The sublimate mp 231-234°. Freely sol in alcohol, methanol, ether, acetone, propylene glycol; sol in chloroform, aq solns of alkali hydroxides; sol in vegetable oils after warming, but crystallizes out on standing. Practically insol in water, dil acids.

Diacetate. [84-19-5] Lipamone; Retalon-Oral. $C_{22}H_{22}O_4$; mol wt 350.41. Prisms from alc, mp 119-120°.

THERAP CAT: Estrogen.
THERAP CAT (VET): Estrogenic hormone therapy.

3116. Dienochlor. [2227-17-0] 1,1',2,2',3,3',4,4',5,5'-Deca-chlorobi-2,4-cyclopentadien-1-yl; bis(pentachloro-2,4-cyclopenta-dien-1-yl); decachlor; HRS-16; Pentac. $C_{10}Cl_{10}$; mol wt 474.61. C 25.31%, Cl 74.69%. Acaricide to control mites on ornamental plants. Prepn from hexachlorocyclopentadiene, characterization: E. T. McBee *et al.*, *J. Am. Chem. Soc.* **77**, 4375 (1955). Prepn and rubber vulcanizing props: E. C. Ladd, US 2732409 (1956 to U.S. Rubber). Prepn and insecticidal props: J. T. Rucker, US 2934470 (1960 to Hooker). Effectiveness as miticide: W. W. Allen *et al.*, *J. Econ. Entomol.* **57**, 187 (1964). Crystal structure: G. Smith *et al.*, *J. Chem. Soc. Perkin Trans. 2* **1976**, 796. Photodegradation: G. B. Quistad, K. M. Mulholland, *J. Agric. Food Chem.* **31**, 621 (1983). Effect on oogenesis of Mexican bean beetle: P. R. Hughes, M. A. Penton, *J. Econ. Entomol.* **76**, 1156 (1983).

Yellow prisms from petr ether, mp 121.5-122°. uv max: 330 nm (ε 2950). Stable to alkali.
USE: Miticide.

3117. Dienogest. [65928-58-7] (17α)-17-Hydroxy-3-oxo-19-norpregna-4,9-diene-21-nitrile; 17α-cyanomethyl-17β-hydroxy-13β-methylgona-4,9-dien-3-one; 17α-cyanomethyl-17β-hydroxy-4,9-estradien-3-one; dienogestril; STS-557; Endometrion. $C_{20}H_{25}$-NO_2; mol wt 311.43. C 77.13%, H 8.09%, N 4.50%, O 10.27%. Derivative of 19-nortestosterone. Prepn: K. Ponsold *et al.*, **DE 2718872**; *eidem*, US 4167517 (1977, 1979 both to VEB Jenapharm); M. Hübner *et al.*, *Arzneim.-Forsch.* **30**, 401 (1980). Structure-activ-ity study: M Hübner, K. Ponsold, *Exp. Clin. Endocrinol.* **81**, 109 (1983). Evaluation of combination with estrogen as oral contracep-tive: J. Spona *et al.*, *Contraception* **56**, 185 (1997); as hormone replacement therapy: T. Gräser *et al.*, *Maturitas* **35**, 253 (2000). Clinical comparison with leuprolide in treatment of endometriosis: T. Strowitzki *et al.*, *Hum. Reprod.* **25**, 633 (2010). Review of pharmacology and clinical experience: R. H. Foster, M. I. Wilde, *Drugs* **56**, 825-833 (1998); in endometriosis: A. E. Schindler, *Int. J. Women's Health* **3**, 175-184 (2011).

Needles from ethyl acetate + acetonitrile, mp 209-214°. $[\alpha]_D^{25}$ −290° (c = 0.5 in pyridine).

Mixture with ethinyl estradiol. [170475-05-5] Certostat; Valette.

THERAP CAT: Progestogen. In combination with estrogen as oral contraceptive.

3118. Diethanolamine. [111-42-2] 2,2′-Iminobisethanol; 2,2′-iminodiethanol; diethylolamine; bis(hydroxyethyl)amine; 2,2′-dihydroxydiethylamine. $C_4H_{11}NO_2$; mol wt 105.14. C 45.70%, H 10.55%, N 13.32%, O 30.43%. Produced along with mono- and triethanolamine by ammonolysis of ethylene oxide. *See refs under* Ethanolamine. Toxicity study: H. F. Smyth *et al., J. Ind. Hyg. Toxicol.* **23**, 259 (1941).

Deliquescent prisms, mp 28°. Usually offered as a viscous liquid. Mild ammoniacal odor. d^{25} 1.094019; d_4^{30} 1.0881; d_4^{60} 1.0693. One U.S. gallon weighs 9.09 lbs at 30°. Viscosity at 30°: 351.9 cP; at 60°: 53.85 cP. bp_{760} 268.8°; bp_{150} 217°; $bp_{0.01}$ 20°. Strong base: pH of 0.1*N* aq soln: 11.0. n_D^{30} 1.4753. Dipole moment 2.81. Flash pt 300°F. Miscible with water, methanol, acetone, alc, chloroform, glycerin. Soly at 25° in benzene: 4.2%, in ether: 0.8%, in carbon tetrachloride: <0.1%, in *n*-heptane: <0.1%. Slightly sol to insol in petroleum ether. LD_{50} orally in rats: 12.76 g/kg (Smyth).

Caution: Potential symptoms of overexposure are irritation of eyes, skin, nose and throat; eye and skin burns, corneal necrosis; lacrimation, cough, sneezing. *See NIOSH Pocket Guide to Chemical Hazards* (DHHS/NIOSH 97-140, 1997) p 104.

USE: To scrub gases as indicated under ethanolamine. Diethanolamine can be used with cracking gases and coal or oil gases which contain carbonyl sulfide that would react with monoethanolamine. As rubber chemicals intermediate. In the manuf of surface active agents used in textile specialties, herbicides, petr demulsifiers. As emulsifier and dispersing agent in various agricultural chemicals, cosmetics, and pharmaceuticals. In the production of lubricants for the textile industry. As humectant and softening agent. In buffer formulations. In organic syntheses.

3119. Diethazine. [60-91-3] *N,N*-Diethyl-10*H*-phenothiazine-10-ethanamine; 10-(2-diethylaminoethyl)phenothiazine; *N*-(diethylaminoethyl)thiodiphenylamine; *N*-(2′-diethylaminoethyl)dibenzoparathiazine; RP-2987; Deparkin; Dinezin; Dolisina; Eazaminum; Ethylemin; Parkazin. $C_{18}H_{22}N_2S$; mol wt 298.45. C 72.44%, H 7.43%, N 9.39%, S 10.74%. Anticholinergic. Prepd by reacting 10-phenothiazineethyl chloride with diethylamine in presence of copper powder or by reacting diethylaminoethyl chloride with phenothiazine: Charpentier, *Compt. Rend.* **225**, 306 (1947); Huttrer, *Enzymologia* **12**, 293 (1948); Charpentier, **US 2530451** (1950 to Rhône-Poulenc); Berg, Ashley, **US 2607773** (1952 to Rhône-Poulenc). Toxicity study: Bovet *et al., Therapie* **2**, 115 (1947).

Oily liquid. bp_{4-5} 195-208°; $bp_{0.4-0.5}$ 167-175°.

Hydrochloride. [341-70-8] Antipar; Aparkazin; Diparcol; Latibon; Thiantan; Thiontan. $C_{18}H_{22}N_2S.HCl$; mol wt 334.91. Crystals, mp 184-186°. Burning taste, producing a temporary numbness of the tongue. One part dissolves in about 5 parts water, 6 parts ethanol, 5 parts chloroform. Practically insol in ether. pH of 10% aq soln 5.0-5.3. LD_{50} orally in mice: 450 mg/kg (Bovet).

THERAP CAT: Antiparkinsonian.

3120. Diethofencarb. [87130-20-9] *N*-(3,4-Diethoxyphenyl)carbamic acid 1-methylethyl ester; isopropyl 3,4-diethoxycarbanilate; isopropyl *N*-(3,4-diethoxyphenyl)carbamate; S-165; S-1605; S-32165; Powmil. $C_{14}H_{21}NO_4$; mol wt 267.33. C 62.90%, H 7.92%, N 5.24%, O 23.94%. Prepn: H. Noguchi *et al.*, **EP 78663**; *eidem*, **US 4608385** (1983, 1986 both to Sumitomo). *In vitro* and greenhouse antifungal activities: M. L. Gullino *et al., Tagungsber. Akad. Landwirtschaftswiss.* **253**, 165 (1987); M. L. Gullino, A. Garibaldi, *Meded. Fac. Landbouwwet. Rijksuniv. Gent* **52**, 895 (1987). Structure-activity study: J. Takahashi *et al., J. Chromatogr.* **436**, 316 (1988). Mechanism of action study: M. Fujimura *et al., ACS Symp. Ser.* **421**, 224-236 (1990). Metabolism in rats: K. Shiba *et al., J. Pestic. Sci.* **15**, 395 (1990). *Review:* M. Fujimura *et al., Jpn. Pestic. Inf.* **57**, 7-11 (1990).

Crystalline solid, mp 101.3°. Vapor pressure at 20°: 6.3×10^{-5}, at 25°: 1.1×10^{-4} mm Hg. Soluble in most organic solvents. Soly at 25 ±2° in water: 26.6 ±0.3 mg/l. LD_{50} in male and female rats (mg/kg): >5000, >5000 orally; >5000, >5000 dermally. LC_{50} (48 hrs) in carp: 13.5 mg/l; (3 hrs) in *Daphnia pulex*: >10 mg/l (Fujimura, review).

USE: Agricultural fungicide.

3121. Diethylacetic Acid. [88-09-5] 2-Ethylbutanoic acid. $C_6H_{12}O_2$; mol wt 116.16. C 62.04%, H 10.41%, O 27.55%.

Colorless liq; odor somewhat resembling that of caproic acid. d_4^{18} 0.920. bp 194-195°, also stated as 190°. mp ∼ −15°. n_D^{10} 1.4179. Slightly sol in water; freely sol in alcohol or ether.

3122. Diethylamine. [109-89-7] *N*-Ethylethanamine. C_4H_{11}-N; mol wt 73.14. C 65.69%, H 15.16%, N 19.15%. Prepn from ethyl iodide and NH_3 with separation of mono-, di-, and triethylamines formed: Watt, Otto, *J. Am. Chem. Soc.* **69**, 836 (1947). Manuf from ethanol and NH_3, obtained along with mono- and triethylamines: Davies *et al.*, **US 2609394** (1952 to ICI); Lemon, Myerly, **US 3022349** (1962 to Union Carbide). Toxicity data: Smyth *et al.*, *Arch. Ind. Hyg. Occup. Med.* **4**, 119 (1951).

Strongly alkaline liquid, bp 55.5°. mp −50°. d_4^{20} 0.7074. n_D^{20} 1.3864. Flash pt <20°F. Forms a hydrate, $B_2.H_2O$, mp −19°. Miscible with water, alc. Usually supplied as a soln. *Flammable, corrosive. Keep well closed.* LD_{50} orally in rats: 540 mg/kg (Smyth).

Hydrochloride. [660-68-4] $C_4H_{11}N.HCl$; mol wt 109.60. Crystals from alcohol + ether, mp 226°. Hygroscopic. bp 320-330°. d_4^{21} 1.048. Sol in water, alcohol, chloroform. Practically insol in ether.

Caution: Potential symptoms of overexposure are eye, skin and respiratory system irritation. *See NIOSH Pocket Guide to Chemical Hazards* (DHHS/NIOSH 97-140, 1997) p 106. *See also Patty's Industrial Hygiene and Toxicology* vol. 2B, G. D. Clayton, F. E. Clayton, Eds. (Wiley-Interscience, New York, 3rd ed., 1981) p 3149.

USE: In the rubber and petroleum industry. In flotation agents, resins, dyes, buffer formulations, and pharmaceuticals.

3123. 2-(Diethylamino)ethanol. [100-37-8] β-Diethylaminoethyl alcohol; 2-hydroxytriethylamine. $C_6H_{15}NO$; mol wt 117.19. C 61.50%, H 12.90%, N 11.95%, O 13.65%. Prepd by the action of ethylene chlorohydrin on diethylamine: Ladenburg, *Ber.* **14**, 1878 (1881); Soderman, Johnson, *J. Am. Chem. Soc.* **47**, 1394 (1925); Hartman, *Org. Synth.* **coll. vol. II**, 183 (1943); by the action of ethylene oxide on diethylamine: Horne, Shriner, *J. Am. Chem. Soc.* **54**, 2928 (1932); Headlee *et al.*, *ibid.* **55**, 1066 (1933).

Liquid. d^{25} 0.8800. bp_{760} 163°; bp_{80} 100°; bp_{10} 55°. n_D^{25} 1.4389. Sol in water, alcohol, ether, benzene.

p-Nitrophenylurethan. Crystals, mp 59-60°.

Caution: Potential symptoms of overexposure are nausea, vomiting; irritation of respiratory system, skin and eyes. *See NIOSH Pocket Guide to Chemical Hazards* (DHHS/NIOSH 97-140, 1997) p 106.

3124. Diethylaminosulfur Trifluoride. [38078-09-0] (*T*-4)-(*N*-Ethylethanaminato)trifluorosulfur; DAST. $C_4H_{10}F_3NS$; mol wt 161.19. C 29.81%, H 6.25%, F 35.36%, N 8.69%, S 19.89%. Versatile fluorinating agent. Prepn: S. P. von Halasz, O. Glemser, *Ber.* **103**, 594 (1970); and fluorination of aldehydes and alcohols: W. J. Middleton, *J. Org. Chem.* **40**, 574 (1975). Thermal decompn study: P. A. Messina *et al.*, *J. Fluorine Chem.* **42**, 137 (1989). Brief review of chemistry: K. A. Jolliffe, *Aust. J. Chem.* **54**, 75 (2001). Review of applications in nucleophilic fluorination reactions: M. Hudlicky, *Org. React.* **25**, 513-637 (1988); R. P. Singh, J. M. Shreeve, *Synthesis* **2002**, 2561-2578.

Pale yellow liquid, bp_{10} 46-47°; bp_{12} 43-44°. n_D^{20} 1.4125. d 1.22-1.23. *Reacts violently with water.*

Caution: Thermally unstable. Undergoes explosion or detonation when heated to >90° (Jolliffe).

USE: Nucleophilic fluorinating reagent.

3125. Diethylaniline. [91-66-7] *N,N*-Diethylbenzenamine. $C_{10}H_{15}N$; mol wt 149.24. C 80.48%, H 10.13%, N 9.39%. Prepd by ethylation of aniline: Whitman, **US 2501556** (1950 to du Pont); Voltz, *J. Org. Chem.* **22**, 48 (1957); Closson *et al.*, *ibid.* **22**, 646 (1957); Horyna, Cerny, *Collect. Czech. Chem. Commun.* **21**, 906 (1956); from bromobenzene, sodium amide + diethylamine: Bunnett, Brotherton, *J. Org. Chem.* **22**, 832 (1957).

Colorless to yellow liq. d_4^{25} 0.9302. bp 215-216°; bp_3 62-66°. mp −38°. n_D^{24} 1.5394. Volatile with steam. uv max (isooctane): 303, 259 nm ($\varepsilon \times 10^{-3}$ 2.37, 16.7). Slightly sol in alcohol, chloroform, ether. One gram dissolves in 70 ml water at 12°.

USE: As dyestuff intermediate, in organic syntheses.

3126. Diethyl Azodicarboxylate. [1972-28-7]; [4143-61-7] ((*E*)-form); [4143-60-6] ((*Z*)-form). 1,2-Diazenedicarboxylic acid 1,2-diethyl ester; azodiformic acid diethyl ester; diethyl azodiformate; ethyl azodicarboxylate; DEAD. $C_6H_{10}N_2O_4$; mol wt 174.16. C 41.38%, H 5.79%, N 16.09%, O 36.75%. Reagent utilized in the Mitsunobu reaction; also applied in other synthetic transformations. Prepn: T. Curtius, K. Heidenreich, *Ber.* **27**, 773 (1894); N. Rabjohn, *Org. Synth.* **coll. vol. III**, 375 (1955); J. C. Kauer, *ibid.* **coll. vol. IV**, 411 (1963). Synthetic applications: O. Mitsunobu, M. Yamada,

Bull. Chem. Soc. Jpn. **40**, 2380 (1967); F. Yoneda *et al.*, *J. Org. Chem.* **32**, 727 (1967); A. Shah, M. V. George, *Tetrahedron* **27**, 1291 (1971); E. E. Smissman, A. Makriyannis, *J. Org. Chem.* **38**, 1652 (1973). *Reviews:* E. J. Stoner in *Encyclopedia of Reagents for Organic Synthesis* vol. **3**, L. A. Paquette, Ed. (Wiley, New York, 1995) pp 1790-1793; S. K. Nune, *Synlett* **2003**, 1221-1222; A. E. Kümmerle, *ibid.* **2008**, 2723-2724.

Orange liquid. bp_{15} 107-111°; bp_{13} 106°; bp_{12} 104.5°; bp_5 93-95°. fp 6°. d_4^{20} 1.1129; $d_4^{44.7}$ 1.0876; $d_4^{61.4}$ 1.0703; $d_4^{85.9}$ 1.0450. n_D^{20} 1.42176. Surface tension (dynes/cm): 33.05 at 22.1°; 31.02 at 42.1°. *Explosive. Irritant. Protect from light and heat.* Flash pt, closed cup: 235°F (113°C). Sol in dichloromethane, ether, toluene. Decomposes vigorously above 100°.

Caution: Irritating to eyes, skin, and respiratory system.

USE: Reagent in synthetic organic chemistry.

3127. *N,N*-Diethylbenzhydrylamine. [519-72-2] *N,N*-Diethyl-α-phenylbenzenemethanamine; diethylaminodiphenylmethane. $C_{17}H_{21}N$; mol wt 239.36. C 85.31%, H 8.84%, N 5.85%. Prepd by addn of an excess of phenylmagnesium bromide to *N,N*-diethylformamide and treating the reaction mixture with ammonium chloride: Maxim, Mavrodineanu, *Bull. Soc. Chim. Fr.* [5] **3**, 1084 (1936). Proposed as an antihistaminic: Capraro, *C.A.* **41**, 6989 (1947).

Viscous liquid, bp_{17} 170°. Solidifies on deep cooling, then mp 56° (Maxim, Mavrodineanu); mp 58-59°. *See:* Sommelet, *Compt. Rend.* **175**, 1149 (1922). Also reported as mp 61° (Titov, *C.A.* **43**, 4217 (1949)).

USE: In the detection of nitrates.

3128. Diethylcarbamazine. [90-89-1] *N,N*-Diethyl-4-methyl-1-piperazinecarboxamide; 1-diethylcarbamoyl-4-methylpiperazine; carbamazine; 1-diethylcarbamyl-4-methylpiperazine; 84L; RP-3799; Carbilazine; Caricide; Cypip; Ethodryl; Notézine; Spatonin. $C_{10}H_{21}N_3O$; mol wt 199.30. C 60.27%, H 10.62%, N 21.08%, O 8.03%. Prepn: Kushner *et al.*, *J. Org. Chem.* **13**, 151 (1948); Kushner, Brancone, **US 2467893**; **US 2467895** (1949 to Am. Cyanamid). Pharmacology and toxicology: Harned *et al.*, *Ann. N.Y. Acad. Sci.* **50**, 141 (1948). GC determn in blood: S. Nene *et al.*, *J. Chromatogr.* **308**, 334 (1984). Mode of action: J.-Y. Cesbron *et al.*, *Nature* **325**, 533 (1987). Review of pharmacology, mechanisms of action and clinical uses: C. D. MacKenzie, M. A. Kron, *Trop. Dis. Bull.* **82**(10), R1-R37 (1985).

Crystals, mp 47-49°. bp_3 108.5-111°.

Hydrochloride. $C_{10}H_{21}N_3O.HCl$. Crystals from acetone, mp 156.5-157°. Very sol in water; sol in chloroform, dioxane.

Citrate. [1642-54-2] Banocide; Dec; Dirocide; Filaribits; Filazine; Franocide; Hetrazan; Loxuran; Longicid. $C_{10}H_{21}N_3O.C_6H_8O_7$; mol wt 391.42. Crystals, mp 141-143°. Freely sol in water (>75% at 20°). Sparingly sol in cold alc; freely sol in hot alc. Practically insol in benzene, acetone, ether, chloroform. LD_{50} orally in rats: 1.38 g/kg (Harned).

Phosphate. Ditrazin. Crystals, freely sol in water.

THERAP CAT: Anthelmintic (Nematodes).

THERAP CAT (VET): Citrate as an anthelmintic.

3129. N,N'-Diethylcarbanilide. [85-98-3] *N,N'*-Diethyl-*N,N'*-diphenylurea; *sym*-diethyldiphenylurea; centralite I; ethyl centralite. $C_{17}H_{20}N_2O$; mol wt 268.36. C 76.09%, H 7.51%, N 10.44%, O 5.96%. Prepn from ethylaniline and phosgene: Michler, *Ber.* **9**, 712 (1876). Alternate synthesis: N. R. Ayyangar *et al.*, *Chem. Ind. (London)* **1988**, 599.

Crystals from alc, mp 79°. Insol in water.

USE: As stabilizer of smokeless explosives.

3130. Diethyl Diethylmalonate. [77-25-8] 2,2-Diethylpropanedioic acid 1,3-diethyl ester; diethyl 3,3-pentanedicarboxylate. $C_{11}H_{20}O_4$; mol wt 216.28. C 61.09%, H 9.32%, O 29.59%. Produced by the action of sodium ethylate and ethylbromide on diethyl malonate.

Liquid. d 0.985-0.990. bp 228-230°. n_D^{20} 1.424. Insol in water; miscible with alcohol, ether.

USE: Manuf barbiturates. Reagent in organic synthesis.

3131. Diethylene Glycol. [111-46-6] 2,2'-Oxybisethanol; 2,2'-oxydiethanol. $C_4H_{10}O_3$; mol wt 106.12. C 45.27%, H 9.50%, O 45.23%. Manuf from ethylene oxide and glycol: *Faith, Keyes & Clark's Industrial Chemicals,* F. A. Lowenheim, M. K. Moran, Eds. (Wiley-Interscience, New York, 4th ed., 1975) pp 397-402. Toxicity study: H. F. Smyth *et al., J. Ind. Hyg. Toxicol.* **23**, 259 (1941).

Colorless, hygroscopic, practically odorless liq; sharply sweetish taste. d_{20}^{20} 1.118. Solidifies at −10.45° (when pure). mp −6.5°. bp 244-245°. n_D^{20} 1.4475. Flash pt, open cup: 290°F (143°C). Misc with water, alc, ether, acetone, ethylene glycol. Practically insol in benzene, carbon tetrachloride. LD_{50} in rats, guinea pigs (g/kg): 20.76, 13.21 orally (Smyth).

Caution: Symptoms on ingestion similar to ethylene glycol, *q.v.* Fatal poisoning resulted from its use as a solvent in an elixir. *See* E. Browning, *Toxicity and Metabolism of Industrial Solvents* (Elsevier, New York, 1965) pp 624-628, 686-690; *Patty's Industrial Hygiene and Toxicology* vol. 2C, G. D. Clayton, F. E. Clayton, Eds. (Wiley-Interscience, New York, 3rd ed., 1982) pp 3832-3838.

USE: In antifreeze soln for sprinkler systems, water seals for gas tanks, etc. (water with 40% diethylene glycol freezes at −18°; with 50% at −28°); as lubricating and finishing agent for wool, worsted, cotton, rayon, and silk; as solvent for vat dyes; in composition corks, glues, gelatin, casein, and pastes to prevent drying out.

3132. Diethylene Glycol Diethyl Ether. [112-36-7] 1,1'-Oxybis[2-ethoxyethane]; Diethyl Carbitol. $C_8H_{18}O_3$; mol wt 162.23. C 59.23%, H 11.18%, O 29.59%. Toxicity data: H. F. Smyth *et al., J. Ind. Hyg. Toxicol.* **23**, 259 (1941).

Liquid. d_4^{20} 0.907. bp 188°. Very sol in water, alcohol, and other organic solvents. LD_{50} orally in rats: 4.97 g/kg (Smyth).

USE: Solvent; high boiling reaction medium.

3133. Diethylene Glycol Monobutyl Ether. [112-34-5] 2-(2-Butoxyethoxy)ethanol; butyl digol; diethylene glycol butyl ether; Butyl Carbitol; Butyl Diicinol. $C_8H_{18}O_3$; mol wt 162.23. C 59.23%, H 11.18%, O 29.59%. Prepn: Riemschneider, Gross, *Monatsh. Chem.* **90**, 783 (1959). Purification: K. F. Ridley, J. Ridley, *GB 795866* (1958 to Esso). Toxicity data: Smyth *et al., J. Ind. Hyg. Toxicol.* **23**, 259 (1941).

Clear liquid, mild ether odor, bp 230.4°. mp −68.1°. *Irritant.* d_{25}^{25} 0.951. d^{20} 0.952. d^{25} 0.948. n_D^{27} 1.4258. Flash pt: 210°F (99°C). Viscosity (25°): 4.9 cP. Sol in water. Miscible in oils. LD_{50} orally in rats, guinea pigs: 6.56, 2.00 g/kg (Smyth).

USE: Solvent for paints, coatings, inks, and cleaners. In mfr of diethylene glycol butly acetate. Deactivator and stabilizer for pesticides. Diluent in hydraulic brake fluid applications. Coalescent for latex adhesives.

3134. Diethylene Glycol Monoethyl Ether. [111-90-0] 2-(2-Ethoxyethoxy)ethanol; diethylene glycol ethyl ether; ethyl digol; Carbitol. $C_6H_{14}O_3$; mol wt 134.18. C 53.71%, H 10.52%, O 35.77%. Prepn from ethylene oxide and 2-ethoxyethanol in the presence of SO_2: E. C. Britton, A. R. Sexton, *US 2807651* (1957 to Dow). Toxicity data: Smyth, Carpenter, *J. Ind. Hyg. Toxicol.* **30**, 63 (1948).

Very hygroscopic liquid, mild odor, bp 201°. fp −43°. d_4^{25} 0.9855, d_{20}^{20} 0.990. n_D^{20} 1.4273. Flash pt, closed cup: 216°F (102°C). Partition coefficient (octanol/water): −0.54. Vapor pressure (20°): 0.07 mmHg. Viscosity (25°): 4.5 cP. Miscible with water, acetone, alc, benzene, chloroform, ether, pyridine; partially miscible with vegetable oils. Immiscible with mineral oils. *See also:* W. M. Jackson, J. S. Drury, *Ind. Eng. Chem.* **51**, 1491 (1959). LD_{50} orally in rats: 8.69 g/kg (Smyth, Carpenter).

Acetate. [112-15-2] $C_8H_{16}O_4$; mol wt 176.21. Liquid, bp 218.5°. mp −25°. d_{20}^{20} 1.0114. n_D^{20} 1.4213. Flash pt, open cup: 110°C. Miscible with water, alc, ether, most oils. LD_{50} orally in rats: 11 g/kg (Smyth, Carpenter).

USE: As solvent for cellulose esters, in lacquer and thinner formulations, in quick-drying varnishes and enamels, for dyestuffs and wood stains. Coupling agent and solvent in cleaners. Pharmaceutic aid (ointment base, solvent). Acetate is used as a solvent and plasticizer for cellulose esters, gums, resins, etc.

3135. Diethylene Glycol Monolaurate. [141-20-8] Dodecanoic acid 2-(2-hydroxyethoxy)ethyl ester; diethylene glycol laurate; diglycol laurate; Glaurin. $C_{16}H_{32}O_4$; mol wt 288.43. C 66.63%, H 11.18%, O 22.19%. Prepared by controlled esterification of diethylene glycol with lauric acid.

Oily liq. d_4^{25} 0.9572; d_{20}^{20} 0.963-0.968. mp 17-18°. bp ~270° (some dec). May be heated to 250° without decompn. Practically insol in water. Sol in methanol, ethanol, benzene, toluene, chlorinated hydrocarbons, acetone, ethyl acetate, cottonseed oil. Slightly sol in petr naphtha.

USE: Dispersing agent, emulsifier, plasticizer.

3136. Diethylene Glycol Monomethyl Ether. [111-77-3] 2-(2-Methoxyethoxy)ethanol; diethylene glycol methyl ether; methyl digol; Methyl Carbitol. $C_5H_{12}O_3$; mol wt 120.15. C 49.98%, H 10.07%, O 39.95%. Toxicity data: Smyth *et al., J. Ind. Hyg. Toxicol.* **23**, 259 (1941).

Colorless liquid, mild odor. mp <−84°. bp 194.1°. d_{25}^{25} 1.020. d^{25} 1.017. n_D^{27} 1.4264. Flash pt, closed cup: 197°F (92°C). Viscosity (25°): 3.5 cP. Vapor pressure (20°): 0.19 mmHg. Miscible with water, alc, glycerol, ether, acetone, DMF. LD_{50} orally in rats: 9.21 g/kg (Smyth).

USE: De-icing additive for jet fuel. Solvent for paints, coatings and inks. Deactivator and stabilizer for pesticides.

3137. Diethyl Ketone. [96-22-0] 3-Pentanone; dimethylacetone; propione; methacetone. $C_5H_{10}O$; mol wt 86.13. C 69.73%, H 11.70%, O 18.58%.

Liquid, acetone odor. d_4^{19} 0.816. bp 101.5°. mp −42°. n_D^{25} 1.3905. *Flammable*. Sol in about 25 parts water; miscible with alcohol, ether. LD_{50} orally in rats: 2.1 g/kg; *see:* Smyth *et al., Arch. Ind. Hyg. Occup. Med.* **10**, 61 (1954).

3138. Diethylmagnesium. [557-18-6] Magnesium diethyl. $C_4H_{10}Mg$; mol wt 82.43. C 58.28%, H 12.23%, Mg 29.49%. Mg-$(C_2H_5)_2$. Prepd by the action of magnesium metal on mercury diethyl in ether: Schlenk, Jr., *Ber.* **64**, 734, 736 (1931).

Solvated crystals from ether, MgEt₂.Et₂O, plates, rods. mp 0°. Liquid at room temp. Spontaneously flammable in air. Violent explosion on contact with water. Loses its ether of crystn on heating *in vacuo*. Not volatile in high vac up to 250° when decompn sets in. Sol in ether. Dec by alc and ammonia. Will glow and catch fire even in CO_2. Must be handled in high vac, or under dry nitrogen or hydrogen.

3139. Diethyl Maleate. [141-05-9] (2Z)-2-Butenedioic acid 1,4-diethyl ester; maleic acid diethyl ester; ethyl maleate. $C_8H_{12}O_4$; mol wt 172.18. C 55.81%, H 7.03%, O 37.17%. Prepn from maleic acid and ethanol: V. M. Mitchovitch, *Bull. Soc. Chim. Fr.* **4**, 1667 (1937). Physical properties: G. H. Jeffrey, A. I. Vogel, *J. Chem. Soc.* **1948**, 658. Copolymerization with styrene: F. M. Lewis *et al., J. Am. Chem. Soc.* **70**, 1519, 1529 (1948). In Diels-Alder reactions: N. L. Bauld *et al., Tetrahedron Lett.* **1972**, 2443. In Michael additions: G. Arsenault *et al., Chem. Commun.* **1983**, 437. Conjugation with glutathione, q.v.: E. Boyland, L. F. Chasseaud, *Biochem. J.* **104**, 95 (1967); *eidem, ibid.* **109**, 651 (1968). Toxicity: H. F. Smyth *et al., J. Ind. Hyg. Toxicol.* **31**, 60 (1949).

Oil, bp₇₅₈ 219.5°. d_4^{20} 1.0674. n_D^{20} 1.4402. Insol in water. Soluble in ethanol, ether. LD_{50} orally in rats: 0.30 g/kg (Smyth).

USE: In organic synthesis.

3140. Diethylmalonic Acid. [510-20-3] 2,2-Diethylpropanedioic acid; 3,3-pentanedicarboxylic acid. $C_7H_{12}O_4$; mol wt 160.17. C 52.49%, H 7.55%, O 39.96%. Prepn: Daimler, *Ann.* **249**, 173 (1888); Speck, *J. Am. Chem. Soc.* **74**, 2876 (1952). Spectral studies: J. L. Delarbre *et al., J. Raman Spectrosc.* **16**, 11 (1985). Crystal structure: A. Dubourg *et al., Acta Crystallogr.* **C44**, 1987 (1988).

Crystals, mp 127°. Dec at 170-180° with liberation of CO_2 and formation of diethylacetic acid. Very sol in water; freely sol in alcohol, ether, slightly in chloroform.

3141. Diethyl Oxalacetate. [108-56-5] 2-Oxobutanedioic acid 1,4-diethyl ester; ethyl oxalacetate; oxaloacetic ester. $C_8H_{12}O_5$; mol wt 188.18. C 51.06%, H 6.43%, O 42.51%. Obtained by the action of sodium on a mixture of ethyl oxalate and ethyl acetate.

Oily liquid. d_4^{20} 1.131. bp₂₄ 132°. $n_D^{16.6}$ 1.45614. Insol in water. Miscible with alc, benzene, ether. Dec when boiled at ordinary pressure. With water it gradually dec into alc and acetic and oxalic acids.

Semicarbazone. [4556-55-2] $C_9H_{15}N_3O_5$; mol wt 245.24. Crystals, mp 162°.

3142. Diethyl Oxalate. [95-92-1] Ethanedioic acid 1,2-diethyl ester; ethyl oxalate; diethyl ethanedioate; oxalic acid diethyl ester. $C_6H_{10}O_4$; mol wt 146.14. C 49.31%, H 6.90%, O 43.79%. Prepn: Clarke, Davis, *Org. Synth.* **coll. vol. I**, 256 (2nd ed., 1941); Kenyon, *ibid.* 261.

Liquid. d_4^{20} 1.0785. mp −40.6°; bp₇₆₀ 185.7°; bp₁₀₀ 130.8°; bp₂₀ 96.8°; bp₁.₀ 47°. n_D^{20} 1.41011. Flash pt, open cup: 168°F (75°C). Sparingly sol in water which dec it gradually; miscible with alcohol, ether, and other usual org solvents.

USE: Manuf phenobarbital, ethylbenzyl malonate, triethylamine, and similar chemicals, plastics, dyestuff intermediates. Solvent for cellulose esters, perfumes.

3143. N,N-Diethylphenylacetamide. [2431-96-1] N,N-Diethylbenzeneacetamide; DEPA. $C_{12}H_{17}NO$; mol wt 191.27. C 75.36%, H 8.96%, N 7.32%, O 8.36%. Substituted amide with broad spectrum repellent activity. Prepn: G. Hausknecht, *Ber.* **22**, 324 (1889); R. Mukherjee, *J. Chem. Soc. D* **1971**, 1113; and structure-activity relationship studies: M. V. S. Suryanarayana *et al., J. Pharm. Sci.* **80**, 1055 (1991). Comparative studies with deet and dimethyl phthalate vs mosquitoes, black flies and land leeches: S. Kumar *et al., Indian J. Med. Res.* **80**, 541 (1984); vs bedbugs: *idem et al., ibid.* **102**, 20 (1995); vs rat flea: K. K. Mathur *et al., ibid.* **83**, 466 (1986); vs sandflies: M. Kalyanasundaram *et al., Med. Vet. Entomol.* **8**, 68 (1994). Comparison with deet and dibutyl phthalate in prevention of scrub typhus: R. Tilak *et al., Indian J. Med. Res.* **113**, 98 (2001). GC determn in urine: S. S. Rao *et al., J. Chromatogr. B* **493**, 210 (1989). Toxicity and metabolism studies: *idem. et al., Toxicology* **58**, 81 (1989); *idem et al., Toxicol. Lett.* **45**, 67 (1989). Development of polysiloxane based slow release formulations and efficacy vs mosquitoes : D. C. Gupta *et al., J. Polym. Mater.* **17**, 215 (2000). *Review*: S. S. Rao, K. M. Rao, *J. Med. Entomol.* **28**, 303-306 (1991).

Liquid, bp₀.₃ 120-122°. Log P (methanol/water): 2.40. Vapor pressure at 30°: 0.1043 cm Hg. LD_{50} in guinea pigs, rats, rabbits, mice (g/kg): 7, 3-4, 3.5, ~2 dermally (Rao, 1991).

USE: Insect repellent.

3144. Diethyl Phosphorochloridite. [589-57-1] Phosphorochloridous acid diethyl ester; chlorodiethoxyphosphine; diethylchlorophosphonite; ethyl phosphorochloridite; ethyl chlorophosphite. $C_4H_{10}ClO_2P$; mol wt 156.55. C 30.69%, H 6.44%, Cl 22.64%, O 20.44%, P 19.79%. Prepd by passing chlorine into diethylphosphorous acid in ligroin contg Na wire: Arbusow, Arbusow, *Ber.* **65**, 195 (1932); by heating triethyl phosphite and phosphorus trichloride; or by adding phosphorus trichloride to a soln of ethanol and diethylaniline in ether: Cook *et al., J. Chem. Soc.* **1949**, 2021.

Liquid. Characteristic odor of acid chlorides. Fumes in moist air; reacts violently with water. d_4^{20} 1.0816; d_0^0 1.0962; d_0^{20} 1.0747. bp_{760} 153-155°; (also reported as bp_{760} 143-148°); bp_{30} 63-65°; bp_{15} 45-53°; bp_{10} 37-38°. n_D^{20} 1.4350.

USE: In peptide synthesis.

3145. Diethylpropion. [90-84-6] 2-(Diethylamino)-1-phenyl-1-propanone; 2-diethylaminopropiophenone; α-benzoyltriethylamine; amfepramone. $C_{13}H_{19}NO$; mol wt 205.30. C 76.06%, H 9.33%, N 6.82%, O 7.79%. Prepn: Hyde *et al., J. Am. Chem. Soc.* **50**, 2287 (1928); Schutte, US 3001910 (1961 to Temmler-Werke). Pharmacology: Cahen *et al., Therapie* **17**, 373-412 (1962). Clinical trial in obesity: R. Abramson *et al., J. Clin. Psychiatry* **41**, 234 (1980).

Hydrochloride. [134-80-5] Anorex; Dobesin; Moderatan; Prefamone; Regenon; Tenuate; Tepanil. $C_{13}H_{19}NO.HCl$; mol wt 241.76. White crystals, dec 168°. Freely sol in water, chloroform, alc. Practically insol in ether.
Note: This is a controlled substance (stimulant): **21 CFR, 1308.14**.

THERAP CAT: Anorexic.

3146. Diethyl Pyrocarbonate. [1609-47-8] Dicarbonic acid C,C'-diethyl ester; oxydiformic acid diethyl ester; diethyl dicarbonate; diethyl oxydiformate; pyrocarbonic acid diethyl ester; DEPC; Ue-5908; Baycovin. $C_6H_{10}O_5$; mol wt 162.14. C 44.45%, H 6.22%, O 49.34%. Prepd according to the equation R_1OOCCl + NaOCOOR$_2$ → $R_1OOCOCOOR_2$ + NaCl: Kovalenko, *Zh. Obshch. Khim.* **22**, 1546 (1952); Thoma, Rinke, *Ann.* **624**, 30 (1959); alternate method: Degering *et al., J. Am. Pharm. Assoc. Sci. Ed.* **39**, 626 (1950). *Review: Handbook of Food Additives*, T. A. Furia, Ed. (Chemical Rubber Co., Cleveland, 1968) pp 181-185.

Viscous liquid. Fruity odor. *Flammable. Irritant.* d_4^{20} 1.12. Flash pt, closed cup: 156°F (69°C). Viscosity at 20° = 1.97 cP. Soluble in alcohols, esters, ketones, hydrocarbons. A 50% w/w soln in ethanol may be prepd. Slowly sol in water with hydrolytic decompn yielding ethanol and CO_2.
Caution: Concd DEPC is irritating to eyes, mucous membranes and skin.
USE: Gentle esterifying agent. Preservative for wines, soft drinks, fruit juices.

3147. Diethylsilane. [542-91-6] 3-Silapentane. $C_4H_{12}Si$; mol wt 88.23. C 54.45%, H 13.71%, Si 31.83%. Prepd from ethylene and silane: Hurd, US 2537763 (1951); Fritz, *Z. Naturforsch.* **7b**, 207 (1952); White, Rochow, *J. Am. Chem. Soc.* **76**, 3897 (1954); Clasen, *Angew. Chem.* **70**, 180 (1958).

Liquid. Stable to air when pure. bp_{741} 56°; d_4^{20} 0.6843; n_D^{20} 1.3921. Quickly oxidized by Ag_2O and HgO. Oxidation can be slowed by soln in petr ether or by cooling to −80°.

3148. Diethylstilbestrol. [56-53-1] 4,4′-[(1E)-1,2-Diethyl-1,2-ethenediyl]bisphenol; (E)-α,α′-diethyl-4,4′-stilbenediol; (E)-3,4-bis(4-hydroxyphenyl)-3-hexene; 4,4′-dihydroxy-α,β-diethylstilbene; stilbestrol; stilboestrol; DES; Apstil; Cyren A; Distilbène; Stilbetin. $C_{18}H_{20}O_2$; mol wt 268.36. C 80.56%, H 7.51%, O 11.92%. Synthetic, nonsteroidal estrogen. Prepn: E. C. Dodds *et al., Nature*

141, 247 (1938); *eidem, Proc. R. Soc. London Ser. B* **127**, 140 (1939). Prepn of ethers: E. E. Reid, E. Wilson, *J. Am. Chem. Soc.* **64**, 1625 (1942). Review of early literature: U. V. Solmssen, *Chem. Rev.* **37**, 481-598 (1945). HPLC determn in plasma and urine: A. T. Rhys Williams *et al., J. Chromatogr.* **235**, 461 (1982). Clinical trial in breast cancer: A. C. Carter *et al., J. Am. Med. Assoc.* **237**, 2079 (1977); in prostate cancer: C. J. Nickel. A. Morales, *Can. J. Surg.* **26**, 434 (1983). Review of metabolism: M. Metzler, *Crit. Rev. Biochem.* **10**, 171-212 (1981). Review of toxicology and use as a growth promotant in livestock: K. E. McMartin *et al., J. Environ. Pathol. Toxicol.* **1**, 279-313 (1978). Book: *Developmental Effects of Diethylstilbestrol (DES) in Pregnancy*, A. L. Herbst, H. A. Bern, Eds. (Thieme-Verlag, New York, 1981) 203 pp. Comprehensive description: A. A. Al-Badr, A. G. Mekkawi, *Anal. Profiles Drug Subs.* **19**, 145-192 (1990).

Small plates from benzene, mp 169-172°. uv max (0.1N NaOH): 259 nm ($E_{1cm}^{1\%}$ 764). Sol in alc, ether, chloroform, fatty oils, dil hydroxides. Almost insol in water.
Diphosphate see Fosfestrol.
Dipropionate. [130-80-3] Cyren B; Dibestil. $C_{24}H_{28}O_4$; mol wt 380.48. Plates from methanol, mp 104°. Sol in organic solvents and in vegetable oils.
Dimethyl ether see Dimestrol.
Monomethyl ether. [18839-90-2] 4-[(1E)-1-Ethyl-2-(4-methoxyphenyl)-1-butenyl]phenol; mestilbol. $C_{19}H_{22}O_2$; mol wt 282.38. Crystals from benzene + petr ether, mp 116-117.5°.
Caution: Diethylstilbestrol is listed as a known human carcinogen: *Report on Carcinogens, Twelfth Edition* (PB2011-111646, 2011) p 159.
THERAP CAT: Antineoplastic (hormonal).

3149. Diethyl Sulfate. [64-67-5] Sulfuric acid diethyl ester. $C_4H_{10}O_4S$; mol wt 154.18. C 31.16%, H 6.54%, O 41.51%, S 20.79%. Prepd from ethanol + sulfuric acid; by absorption of ethylene in sulfuric acid; from diethyl ether and fuming sulfuric acid. Review of prepn and uses: R. Page, J. A. John in *Ethylene and its Industrial Derivatives* S. A. Miller, Ed. (Benn, London, 1969) pp 774-787. Toxicity data: H. F. Smyth *et al., J. Ind. Hyg. Toxicol.* **31**, 60 (1949). Review of carcinogenicity studies: *IARC Monographs* **4**, 277-281 (1974). *Reviews:* C. M. Suter, *The Organic Chemistry of Sulfur* (John Wiley, 1944) pp 62-65; E. E. Gilbert, *Sulfonation and Related Reactions* (Interscience, New York, 1965).

Colorless, oily liquid; peppermint odor. *Poisonous.* Darkens with age. bp 209.5° with decompn; bp_{15} 96°; bp_5 75°. mp −25°. d_4^{23} 1.1774. n_D^{20} 1.40037, A. I. Vogel, D. M. Cowan, *J. Chem. Soc.* **1943**, 16. Flash pt: 220°F. Vapor density (air = 1) 5.31. Vapor pressure at 47° = 1 mm. Practically insol in water, gradually dec by it. Rapidly dec by hot water into monoethyl sulfate and alc. Miscible with alc, ether. LD_{50} orally in rats: 0.88 g/kg (Smyth).
Caution: This substance is reasonably anticipated to be a human carcinogen: *Report on Carcinogens, Twelfth Edition* (PB2011-111646, 2011) p 161.
USE: Chiefly as an ethylating agent; as an accelerator in the sulfation of ethylene; in some sulfonations.

3150. Diethyl Tartrate. [87-91-2] (2R,3R)-2,3-Dihydroxybutanedioic acid diethyl ester; ethyl tartrate. $C_8H_{14}O_6$; mol wt 206.19. C 46.60%, H 6.84%, O 46.56%. Prepn: J. W. Rodger, J. S. S. Brame, *J. Chem. Soc.* **73**, 301 (1898); T. M. Lowry, J. O. Cutter, *ibid.* **121**, 532 (1922).

Colorless, thick, oily liquid. d_4^{20} 1.204. bp 280°; bp_{11} 150°. mp 17°. $[\alpha]_D^{20}$ +7.5°. n_D^{20} 1.4476. Slightly sol in water; miscible with alcohol or ether.

3151. Diethylzinc. [557-20-0] Zinc diethyl. $C_4H_{10}Zn$; mol wt 123.53. C 38.89%, H 8.16%, Zn 52.95%. Prepd by the interaction of zinc and ethyl iodide: Simonowitch, *Chem. Zentralbl.* **1899**, I, 1066; Lachmann, *Am. Chem. J.* **24**, 32 (1900); Dennis, Hance, *J. Am. Chem. Soc.* **47**, 370 (1925); from zinc and ethyl bromide, ethyl iodide: C. R. Noller, *Org. Synth.* **coll. vol. II**, 184. Used in synthesis of cyclopropanes: J. Furukawa *et al.*, *Tetrahedron* **24**, 53 (1968); *eidem*, *Tetrahedron Lett.* **1968**, 3495; in synthesis of ketocarbenes: L. T. Scott, W. D. Cotton, *J. Am. Chem. Soc.* **95**, 2708 (1973); in ring expansion of arenes: S. Miyano, H. Hashimoto, *Chem. Commun.* **1973**, 216; in preservation of papers: J. C. Williams, G. B. Kelly Jr., **US 3969549** (1976 to U.S.A.); *eidem* in *Adv. Chem. Ser.* **193**, entitled "Preservation of Paper and Textiles of Historic and Artistic Value 2," J. C. Williams, Ed. (ACS, Washington DC, 1981) pp 109-117.

Mobile liq. Stable in sealed tube and carbon dioxide. d_4^{20} 1.2065; d_4^8 1.245; mp $-28°$; bp_{760} 118°; bp_{30} 27°; $n_{H\alpha}^{20}$ 1.4936. *Spontaneously combustible, dangerous when wet.* Burns with a blue flame, giving off a peculiar, garlic-like odor. Miscible with ether, petr ether, benzene, other hydrocarbons.

USE: In organic synthesis; in preservation on archival papers.

3152. Difemerine. [80387-96-8] α-Hydroxy-α-phenylbenzeneacetic acid 2-(dimethylamino)-1,1-dimethylethyl ester; 2-(dimethylamino)-1,1-dimethylethyl benzilate. $C_{20}H_{25}NO_3$; mol wt 327.42. C 73.37%, H 7.70%, N 4.28%, O 14.66%. Anticholinergic. Prepn: S. G. Kuznetsov, A. G. El'tsov, *Zh. Obshch. Khim.* **32**, 511 (1962), *C.A.* **58**, 470d (1963); C. Hoffmann, **FR M3406** (1965 to Labs. U.P.S.A.), *C.A.* **63**, 13163ab (1965). Toxicity: S. N. Golikov, S. G. Kuznetsov, *C.A.* **61**, 8790b (1964). Comparison of side effects with those of atropine: L. Moser, P. V. Lundt, *Med. Welt* **31**, 1795 (1980). Crystal structure: A. Carpy *et al.*, *Acta Crystallogr. B* **35**, 882 (1979).

Solid, mp 78-78.3°.
Hydrochloride. [70280-88-5] Luostyl. $C_{20}H_{25}NO_3 \cdot HCl$; mol wt 363.88. Crystals from isopropanol, mp 182°.
THERAP CAT: Antispasmodic.

3153. Difenacoum. [56073-07-5] 3-(3-[1,1'-Biphenyl]-4-yl-1,2,3,4-tetrahydro-1-naphthalenyl)-4-hydroxy-2H-1-benzopyran-2-one; 3-[3-(biphenyl-4-yl)-1,2,3,4-tetrahydro-1-naphthyl]-4-hydroxycoumarin; 3-(3-p-biphenyl-1,2,3,4-tetrahydronaphth-1-yl)-4-hydroxycoumarin; Neosorexa; Ratak. $C_{31}H_{24}O_3$; mol wt 444.53. C 83.76%, H 5.44%, O 10.80%. Second generation anticoagulant rodenticide. Prepn: M. R. Hadler, R. S. Shadbolt, **DE 2424806**; *idem*, **US 3957824** (1975, 1976 both to Ward Blenkinsop); R. S. Shadbolt *et al.*, *J. Chem. Soc. Perkin Trans. 1* **1976**, 1190. Comprehensive description: J. O. Bull, *Proc. Vertebr. Pest Conf.* **7**, 72 (1976). HPLC determn of rat blood and liver: M. J. Kelly *et al.*, *J. Chroma-*

togr. **620**, 105 (1993); of residue levels in rats: H. Atterby *et al.*, *Environ. Toxicol. Chem.* **24**, 318 (2005). Field and laboratory studies against the house mouse: F. P. Rowe *et al.*, *J. Hyg. Camb.* **87**, 171 (1981). Pharmacodynamics and pharmacodynamics in rabbits: A. M. Breckenridge *et al.*, *Br. J. Pharmacol.* **84**, 81 (1985). Pharmacological variation between rodent sexes: M. J. Winn *et al.*, *J. Pharm. Pharmacol.* **39**, 219 (1987).

Crystals from ethyl acetate, mp 215-217°. LD_{50} orally in Wistar male, female rats (mg/kg): 1.8, 2.5; in male rabbits: 2.0 mg/kg; in dogs: 50 mg/kg; in cats: 100 mg/kg (Bull).
USE: Pesticide.

3154. Difenoconazole. [119446-68-3] 1-[[2-[2-Chloro-4-(4-chlorophenoxy)phenyl]-4-methyl-1,3-dioxolan-2-yl]methyl]-1H-1,-2,4-triazole; *cis,trans*-3-chloro-4-[4-methyl-2-(1H-1,2,4-triazol-1-ylmethyl)-1,3-dioxolan-2-yl]phenyl 4-chlorophenyl ether; CGA-169374; Dividend; Plandom; Score; Spectro. $C_{19}H_{17}Cl_2N_3O_3$; mol wt 406.26. C 56.17%, H 4.22%, Cl 17.45%, N 10.34%, O 11.81%. Ergosterol biosynthesis inhibitor. Prepn: A. Hubele, P. Riebli, **GB 2098607** (1982 to Ciba-Geigy). Properties and antifungal activity: W. Ruess *et al.*, *Brighton Crop Prot. Conf. - Pests Dis.* **1988**, 543. Field studies: A. J. Leadbeater *et al.*, *ibid.* 917; A. B. Bassi, Jr. *et al.*, *Br. Crop Prot. Counc. Monogr.* **57**, 91 (1994). Uptake and translocation in crops: H. Dahmen, T. Staub, *Plant Dis.* **76**, 523 (1992).

White crystalline solid, mp 76°. Vapor pressure (20°): 1.2×10^{-7} Pa. Soly (20°): 5 mg/l in water. Very soluble in most organic solvents. LD_{50} in rats (mg/kg): 1453 orally; in rabbits: >2010 dermally (Ruess).
USE: Agricultural fungicide.

3155. Difenoxin. [28782-42-5] 1-(3-Cyano-3,3-diphenylpropyl)-4-phenyl-4-piperidinecarboxylic acid; 1-(3-cyano-3,3-diphenylpropyl)-4-phenylisonipecotic acid; difenoxilic acid; difenoxylic acid; McN-JR-15403-11; Lyspafen. $C_{28}H_{28}N_2O_2$; mol wt 424.54. C 79.22%, H 6.65%, N 6.60%, O 7.54%. Active metabolite of diphenoxylate, *q.v.*, from which it may be prepared by hydrolysis of the ethyl ester: W. Soudijn, I. van Wijngaarden, **DE 1953342**; *eidem*, **US 3646207** (1970, 1972 to Janssen). Series of articles on pharmacology, toxicology, metabolism: *Arzneim.-Forsch.* **22**, 513-531 (1972). Toxicity data: Niemegeers *et al.*, *ibid.* 516.

Hydrochloride. R-15403. $C_{28}H_{28}N_2O_2 \cdot HCl$; mol wt 461.00. White amorphous power, mp 290°. Very slightly sol in water (0.023%); sparingly sol in chloroform, tetrahydrofuran, dimethylacetamide, DMSO. Stable, can be stored for several years under normal conditions; not hygroscopic; not affected by light. LD_{50} orally in rats: 149 mg/kg (Niemegeers).

Note: This is a controlled substance (opiate): **21 CFR**, 1308.11.

THERAP CAT: Antiperistaltic; antidiarrheal.

3156. Difenpiramide. [51484-40-3] *N*-2-Pyridinyl-[1,1'-biphenyl]-4-acetamide; diphenpyramide; Z-876; Difenax. $C_{19}H_{16}N_2$-O; mol wt 288.35. C 79.14%, H 5.59%, N 9.72%, O 5.55%. Prepn: F. Tenconi, R. M. Tagliabue, **DE 2325309** (1973 to Zambeletti); L. Molteni *et al.*, **US 3868380** (1975). Pharmacological properties: S. Caliari *et al.*, *Arzneim.-Forsch.* **27**, 2086 (1977). Physicochemical profile: A. Trebbi, G. Filippi, *Boll. Chim. Farm.* **118**, 729 (1979). Therapeutic use: C. Ortolani *et al.*, *Clin. Ter.* **89**, 391 (1979).

Crystals, mp 122-124°. LD_{50} in male mice, rats (mg/kg): 2590, 2075 orally; 1421, 1396 i.p. (Caliari).

THERAP CAT: Anti-inflammatory.

3157. Difenzoquat. [49866-87-7] 1,2-Dimethyl-3,5-diphenyl-1*H*-pyrazolium. $[C_{17}H_{17}N_2]^+$. Post-emergence herbicide for food crops; also effective vs mildew. Prepn of salts: B. Walworth, E. Klingsberg, **DE 2260485**; *eidem*, **US 3882142** (1973, 1975 both to Am. Cyanamid); B. Cross *et al.*, *J. Heterocycl. Chem.* **17**, 905 (1980). Physical properties and herbicidal activity: N. E. Shafer, *Proc. 12th Br. Weed Control Conf.* **1974**, 831. Effect vs mildew: R. J. Winfield *et al.*, *Proc. Br. Crop Prot. Conf. - Pests Dis.* **1977**, 57. HPLC determn in commercial formulations: C. Barry, R. K. Pike, *J. Assoc. Off. Anal. Chem.* **63**, 64 (1980); in water: M. C. Carneiro *et al.*, *J. Chromatogr. A* **669**, 217 (1994).

Methyl sulfate. [43222-48-6] AC-84777; CL-84777; Avenge; Finaven. $C_{17}H_{17}N_2 \cdot CH_3O_4S$; mol wt 360.43. Crystals from acetone, mp 146-148° (Cross). Soly in water (25°): ~70%. Relatively insol in non-polar solvents. LD_{50} (technical grade) orally in rats: 470 mg/kg; dermally in rabbits: 3540 mg/kg (Shafer).

USE: Herbicide; agricultural fungicide.

3158. Difethialone. [104653-34-1] 3-[3-(4'-Bromo[1,1'-biphenyl]-4-yl)-1,2,3,4-tetrahydro-1-naphthalenyl]-4-hydroxy-2*H*-1-benzothiopyran-2-one; LM-2219; Generation; Hombre. $C_{31}H_{23}$-BrO$_2$S; mol wt 539.49. C 69.02%, H 4.30%, Br 14.81%, O 5.93%, S 5.94%. Second generation anticoagulant rodenticide; member of the hydroxy-4-benzothiopyranone family. Prepn: J.-J. Berthelon, **FR 2562893**; *idem*, **US 4585784** (1985, 1986 both to Lipha). HPLC determn of residue levels in rats: D. A. Goldade *et al.*, *J. Agric. Food Chem.* **46**, 504 (1998). Use against warfarin-resistant rodents: K. Nahas *et al.*, *Anal. Rech. Vét.* **20**, 159 (1989). Toxicity in rodents and non-target species: J. C. Lechevin, R. M. Poche, *Proc. Vertebr. Pest Conf.* **13**, 59 (1988); S. Moran, *Crop Prot.* **12**, 501 (1993). Laboratory and field studies: C. Sheikher, P. Sood, *Indian J. Agric. Sci.* **70**, 312 (2000). Use in conjunction with compounds to improve

bait acceptance: H. Kaur, V. R. Parshad, *Int. Pest Control* **46**, 88 (2004).

Prepd as a variable proportion of the (1*RS*,3*RS*) and (1*RS*,3*SR*) racemates, mp 203-227°. LD_{50} orally in male, female rats, mice (mg/kg): 0.62, 0.42, 0.52, 0.43 (non-resistant); 0.27, 0.39, 0.46, 0.52 (warfarin resistant). LD_{50} in hare: 0.75 mg/kg. LC_{50} in mallard duck, bobwhite quail (ppm): 1.94, 0.56. LD_{50} at 48 hrs in *Daphnia magna*, bluegill sunfish, rainbow trout (μg/l): 4.4, 110, 67 (Lechevin).

USE: Pesticide.

3159. Diflorasone. [2557-49-5] (6α,11β,16β)-6,9-Difluoro-11,17,21-trihydroxy-16-methylpregna-1,4-diene-3,20-dione; 6α,9α-difluoro-16β-methyl-$\Delta^{1,4}$-pregnadiene-11β,17α,21-triol-3,20-dione; 6α,9α-difluoro-16β-methylprednisolone. $C_{22}H_{28}F_2O_5$; mol wt 410.46. C 64.38%, H 6.88%, F 9.26%, O 19.49%. The 16β-analog of flumethasone, *q.v.* Prepn of free alcohol and 21-acetate: **GB 881334** (1961 to Pfizer), *C.A.* **56**, 15586c (1962); F. H. Lincoln *et al.*, **US 3557158** (1962, 1971 both to Upjohn); **GB 912015** (1962 to Merck & Co.). Prepn of the 17,21-diacetate: D. E. Ayer *et al.*, **DE 2308731**; *eidem*, **US 3980778** (1973, 1976 both to Upjohn). Proposed mechanism of action: S. Hammarstrom *et al.*, *Science* **197**, 994 (1977). Pharmacology: S. Wickrema *et al.*, *J. Invest. Dermatol.* **71**, 372 (1978). Clinical study: S. M. Bluefarb *et al.*, *J. Int. Med. Res.* **4**, 454 (1976).

Diacetate. [33564-31-7] U-34865; Dermaflor; Diacort; Florone; Maxiflor; Psorcon; Soriflor. $C_{26}H_{32}F_2O_7$; mol wt 494.53. Crystals from ethyl acetate-Skellysolve C and acetone-methanol, mp 221-223° (dec). uv max (alc): 238 nm (ε 17250). $[\alpha]_D$ +61° (chloroform). Sol in methanol, acetone; sparingly sol in ethyl acetate; slightly sol in toluene; very slightly sol in ether. Insol in water.

THERAP CAT: Anti-inflammatory (topical); glucocorticoid.

3160. Difloxacin. [98106-17-3] 6-Fluoro-1-(4-fluorophenyl)-1,4-dihydro-7-(4-methyl-1-piperazinyl)-4-oxo-3-quinolinecarboxylic acid. $C_{21}H_{19}F_2N_3O_3$; mol wt 399.40. C 63.15%, H 4.80%, F 9.51%, N 10.52%, O 12.02%. Fluoroquinolone antibacterial structurally related to norfloxacin, *q.v.* Prepn: D. T. W. Chu, **EP 131839**; *idem*, **US 4730000** (1985, 1988 both to Abbott); D. T. W. Chu *et al.*, *J. Med. Chem.* **28**, 1558 (1985); H. Narita *et al.*, **JP Kokai 85 237069** (1985 to Toyama). Antibacterial spectrum *in vitro*: J. M. Stamm *et al.*, *Antimicrob. Agents Chemother.* **29**, 193 (1986); *in vivo*: P. B. Fernandes *et al.*, *ibid.* 201. Activity vs anaerobic bacteria: *eidem*, *J. Antimicrob. Chemother.* **18**, 693 (1986). HPLC determn in biological fluids: G. R. Granneman, L. T. Sennello, *J. Chromatogr.* **413**, 199 (1987). Pharmacokinetics in humans: G. R. Granneman *et al.*, *Antimicrob. Agents Chemother.* **30**, 689 (1986).

Crystals from ethyl acetate-ether, mp 240-244°. uv max: 237 nm (ε 16600) (Kieslich, 1969). Also reported as mp 248-249°. $[\alpha]_D^{22}$ +111° (methanol) (Kieslich, 1976).

21-Valerate. [59198-70-8] Neribas; Neriforte; Nerisona; Nerisone; Temetex; Texmeten. $C_{27}H_{36}F_2O_5$; mol wt 478.58. Crystals, mp 195-195.5°. $[\alpha]_D^{22}$ +100.8° (dioxane). Approx LD_{50} in mice, rats: >4, 3.1 g/kg orally; 180, 13 mg/kg s.c.; 450, 98 mg/kg i.p. (Gunzel).

THERAP CAT: Anti-inflammatory.

3163. Diflufenican. [83164-33-4] *N*-(2,4-Difluorophenyl)-2-[3-(trifluoromethyl)phenoxy]-3-pyridinecarboxamide; *N*-(2,4-difluorophenyl)-2-(3-trifluoromethylphenoxy)nicotinamide; 2',4'-difluoro-2-(α,α,α-trifluoro-*m*-tolyloxy)nicotinanilide; M & B 38544; Fenican. $C_{19}H_{11}F_5N_2O_2$; mol wt 394.30. C 57.88%, H 2.81%, F 24.09%, N 7.10%, O 8.12%. Pre and post emergence foliar absorbed herbicide for winter weed control in cereal crops; carotenoid biosynthesis inhibitor. Prepn: M. C. Cramp *et al.*, **EP 53011**; *eidem*, **US 4618366** (1981, 1986 both to May & Baker). Description of physical properties, toxicity and activity: M. C. Cramp *et al.*, *Proc. Br. Crop Prot. Conf. - Weeds* **1985**, 23. Mode of action: G. Britton *et al.*, *ibid.* **1987**, 1015. GC determn in runoff and soil: L. Patty, C. Guyot, *Bull. Environ. Contam. Toxicol.* **55**, 802 (1995). Degradation and persistence in soil: E. Conte *et al.*, *J. Agric. Food Chem.* **46**, 4766 (1998).

White crystalline solid, without odor, from toluene, mp 161-162°. Soly at 20°: acetone 10%; acetophenone 5%; aromasol H 1%; cyclohexane 1%; DMF 10%; ethylcellosolve 1%; isophorone 3.5%; kerosene 1%; xylene 2%; water 0.05 mg/l. Vapor pressure at 25°: 4.25 $\times$ 10^{-6} Pa. LD_{50} in mice, rats (mg/kg): >1000, >2000 orally; in rats (mg/kg): >2000 dermally (Cramp).

USE: Herbicide.

3164. Diflufenzopyr. [109293-97-2] 2-[1-[2-[[(3,5-Difluorophenyl)amino]carbonyl]hydrazinylidene]ethyl]-3-pyridinecarboxylic acid; 2-[1-[4-(3,5-difluorophenyl)semicarbazono]ethyl]nicotinic acid; SAN-835H. $C_{15}H_{12}F_2N_4O_3$; mol wt 334.28. C 53.90%, H 3.62%, F 11.37%, N 16.76%, O 14.36%. Auxin transport inhibitor. Prepn: R. J. Anderson, M. M. Leippe, **JP Kokai 62 45570**; *eidem et al.*, **US 5098466** (1987, 1992 both to Sandoz). Synergy with auxin type herbicides: R. G. Lym, K. M. Christianson, *Proc. West. Soc. Weed Sci.* **51**, 59 (1998). Field trial of combination with dicamba in corn: P. H. Sikkema *et al.*, *Weed Technol.* **13**, 283 (1999). Review of physical properties, mode of action, and activity: S. Bowe *et al.*, *Brighton Crop Prot. Conf. - Weeds* **1999**, 35-40.

Off-white solid, mp 155° (Bowe); also reported as 186° (Anderson). Soly in water: 63 mg/l. Vapor pressure at 20°: <1.3 $\times$ 10^{-5} Pa. Log P (octanol/water, pH 7): 0.3. LD_{50} in rats (mg/kg): >5000 orally; >5000 dermally; LC_{50} in rats (mg/l): 2.93 by inhalation; LC_{50} in rainbow trout, bluegill sunfish (mg/l): 106, 135 (Bowe).

Mixture with dicamba. [211050-90-7] BAS-662; SAN-1269H; Distinct.

USE: Herbicide.

Monohydrochloride. [91296-86-5] Abbott 56619; A-56619; Dicural. $C_{21}H_{19}F_2N_3O_3$·HCl; mol wt 435.86. Crystals, mp >275°.

THERAP CAT (VET): Antibacterial.

3161. Diflubenzuron. [35367-38-5] *N*-[[(4-Chlorophenyl)amino]carbonyl]-2,6-difluorobenzamide; 1-(4-chlorophenyl)-3-(2,6-difluorobenzoyl)urea; difluron; DU-112307; PH-60-40; TH-6040; ENT-29054; OMS-1804; Dimilin; Duphacid; Micromite. $C_{14}H_9$-$ClF_2N_2O_2$; mol wt 310.68. C 54.12%, H 2.92%, Cl 11.41%, F 12.23%, N 9.02%, O 10.30%. Substituted benzoylphenylurea, inhibits chitin deposition in insect cuticle. Prepn: K. Wellinga, R. Mulder, **DE 2123236**; *eidem*, **US 3748356** (1971, 1973 both to Philips). Insecticidal activity: K. Wellinga *et al.*, *J. Agric. Food Chem.* **21**, 993 (1973); R. Mulder, M. J. Gijswijt, *Pestic. Sci.* **4**, 737 (1973). Mode of action: E. Hunter, J. F. V. Vincent, *Experientia* **30**, 1432 (1974); S. M. Meola, R. T. Mayer, *Science* **207**, 985 (1980). Field trials against locusts and grasshoppers: A. Bouaichi *et al.*, *Crop Prot.* **13**, 53, 60 (1994). Brief review: J. L. Marx, *Science* **197**, 1170-1172 (1977). Review of activity, environmental fate, mode of action: A. Verloop, C. D. Ferrell, *ACS Symp. Ser.* **37**, 237-270 (1977); of chemistry and analyses: B. Rabenort *et al.*, *Anal. Methods Pestic. Plant Growth Regul.* **10**, 57-72 (1978).

White to yellowish brown crystalline solid, mp 210-230° (technical grade); mp 239° (pure compd). Soly in water $\approx$0.2 ppm at 20°. Moderate to good soly in polar to very polar solvents; e.g. acetone 0.65 gm/100 ml at 20°. Partition coefficient (dichloromethane): 50 (pH 5.6); (1-octanol/water): ~5000. LD_{50} in mice, rats (formulation with 50% kaolin) (g/kg): 4.64, >10 orally (Mulder, Gijswijt).

USE: Insecticide.

3162. Diflucortolone. [2607-06-9] ($6\alpha,11\beta,16\alpha$)-6,9-Difluoro-11,21-dihydroxy-16-methylpregna-1,4-diene-3,20-dione; $6\alpha,9\alpha$-difluoro-16α-methyl-1,4-pregnadiene-11β,21-diol-3,20-dione; 6α,-9α-difluoro-16α-methyl-1-dehydrocorticosterone. $C_{22}H_{28}F_2O_4$; mol wt 394.46. C 66.99%, H 7.16%, F 9.63%, O 16.22%. The 9α-fluoro deriv of fluocortolone, *q.v.* Prepn: **BE 639708**; K. Kieslich *et al.*, **US 3426128** (1964, 1969, both to Schering AG); *eidem*, *Arzneim.-Forsch.* **26**, 1462 (1976). Toxicity: P. Gunzel *et al.*, *ibid.* 1476. Series of articles on pharmacology, toxicity, metabolism and clinical studies: *ibid.* **26**, 1463-1513 (1976).

3165. Diflunisal. [22494-42-4] 2′,4′-Difluoro-4-hydroxy-[1,1′-biphenyl]-3-carboxylic acid; 2′,4′-difluoro-4-hydroxy-3-biphenylcarboxylic acid; 2′,4′-difluoro-4-hydroxy-[1′,1-diphenyl]-3-carboxylic acid; 2-(hydroxy)-5-(2,4-difluorophenyl)benzoic acid; 5-(2,4-difluorophenyl)salicylic acid; MK-647; Dolobid; Dolobis; Fluniget. $C_{13}H_8F_2O_3$; mol wt 250.20. C 62.41%, H 3.22%, F 15.19%, O 19.18%. Non-steroidal anti-inflammatory drug (NSAID). Prepn: W. V. Ruyle *et al.*, ZA **6701021**; *eidem*, FR **1522570**; *eidem*, US **3714226** (1968, 1968, 1973, all to Merck & Co.); J. Hannah *et al.*, *J. Med. Chem.* **21**, 1093 (1978). HPLC determn in plasma and urine: M. Schwartz *et al.*, *J. Chromatogr.* **380**, 420 (1986). Metabolism: P. J. DeSchepper *et al.*, *Br. J. Clin. Pharmacol.* **4**, 645P (1977). Clinical studies: J. A. Wojtulewski *et al.*, *Curr. Med. Res. Opin.* **5**, 562 (1978); J. A. Hicklin, *ibid.* 572; in sickle cell anemia: E. A. Oyewo, *Clin. Trials J.* **24**, 249 (1987). Review of chemistry, pharmacology, clinical pharmacology: S. L. Steelman *et al.*, *ibid.* 506-514. Comprehensive description: M. Cotton, R. A. Hux, *Anal. Profiles Drug Subs.* **14**, 491-526 (1985). *Review:* Br. J. Clin. Pharmacol. **4**, Suppl. 1, 1S-52S (1977); C. A. Winter *et al.*, in *Pharmacological and Biochemical Properties of Drug Substances* **vol. 3**, M. E. Goldberg, Ed. (Am. Pharm. Assoc., Washington, DC, 1981) pp 291-323. Comprehensive review of pharmacological properties and therapeutic use in pain: R. N. Brogden *et al.*, *Drugs* **19**, 84-106 (1980); *Pharmacotherapy* **3**, no. 2, pt. 2, 1S-82S (1983).

White, crystals from toluene, mp 211-213°. Freely sol in alc, methanol; sol in most organic solvents, ethyl acetate, acetone; slightly sol in chloroform, carbon tetrachloride, methylene chloride. Practically insol in water at neutral or acidic pH. Insol in hexane. LD$_{50}$ orally in female mice: 439 mg/kg (Stone).

THERAP CAT: Analgesic; anti-inflammatory.

3166. 2,4-Difluoroaniline. [367-25-9] 2,4-Difluorobenzenamine. $C_6H_5F_2N$; mol wt 129.11. C 55.82%, H 3.90%, F 29.43%, N 10.85%. Prepn: Swarts, *Rec. Trav. Chim.* **35**, 164 (1916); Kutepov, Rozanova, *Zh. Obshch. Khim.* **27**, 2848 (1957), *C.A.* **52**, 8067 (1958); Nad *et al.*, *C.A.* **53**, 14976i (1959).

mp −7.5°. bp$_{753}$ 169.5°. n_D^{25} 1.5043. Specific gravity at 25°: 10.72 lb/gal. Flash pt 158°F (70°C).

USE: In organic syntheses.

3167. p-Difluorobenzene. [540-36-3] 1,4-Difluorobenzene. $C_6H_4F_2$; mol wt 114.09. C 63.17%, H 3.53%, F 33.30%. Prepd by diazotizing p-fluoroaniline in hydrofluoric acid and heating the diazonium salt to 120°: Swarts, *Chem. Zentralbl.* **1913**, II, 760.

Liquid, pungent, aromatic odor. mp −23.7°. bp 88.82°. d^{20} 1.17006. $n_D^{18.9}$ 1.44219.

3168. 4,4′-Difluorodiphenyl. [398-23-2] 4,4′-Difluoro-1,1′-biphenyl. $C_{12}H_8F_2$; mol wt 190.19. C 75.78%, H 4.24%, F 19.98%. NMR spectrum: T. K. Halstead *et al.*, *J. Chem. Soc. Faraday Trans.* 2 **1981**, 1817.

Crystalline powder; aromatic odor. d 1.04. mp 92-95°. bp 254-255°. Insol in water. Freely sol in alc, chloroform, ether, oils.

3169. Difluoroiodotoluene. [371-11-9] 1-(Difluoriodo)-4-methylbenzene; difluoro(4-methylphenyl)iodine; 4-iodotoluene difluoride; p-tolyliododifluoride. $C_7H_7F_2I$; mol wt 256.03. C 32.84%, H 2.76%, F 14.84%, I 49.57%. Hypervalent iodine(III) difluoride reagent used to transfer of fluorine to organic substrates. Prepn: R. F. Weinland, W. Stille, *Ann.* **328**, 132 (1903); W. Bockemüller, *Ber.* **64**, 522 (1931); and spectral studies: D. Naumann, G. Rüther, *J. Fluorine Chem.* **15**, 213 (1980). Improved prepn: M. Sawaguchi *et al.*, *Synthesis* **2002**, 1802. Synthetic applications in fluorination reactions: W. B. Motherwell *et al.*, *J. Chem. Soc. Perkin Trans. 1* **2002**, 2809; M. A. Arrica, T. Wirth, *Eur. J. Org. Chem.* **2005**, 395.

White needles from hexane, mp 98° (Sawaguchi). Also reported as white solid, mp 110° (dec) (Motherwell). Easily sol in chloroform; sparingly sol in petroleum ether, 40% hydrofluoric acid.

USE: Fluorinating agent.

3170. Difluoromethane. [75-10-5] HFC-32; methylene difluoride; R-32; Genetron 32. CH_2F_2; mol wt 52.02. C 23.09%, H 3.88%, F 73.04%. Hydrofluorocarbon refrigerant with zero ozone-depletion potential. Prepn: A. L. Henne, *J. Am. Chem. Soc.* **59**, 1400 (1937). Structural studies: L. O. Brockway, *J. Phys. Chem.* **41**, 185, 747 (1937); W. C. Hamilton, K. Hedberg, *J. Am. Chem. Soc.* **75**, 5529 (1952). Microwave spectrum: D. R. Lide, Jr., *ibid.* **74**, 3548 (1952). UV absorption spectrum: P. Wagner, A. B. F. Duncan, *ibid.* **77**, 2609 (1955). Thermodynamic properties: P. F. Malbrunot *et al.*, *J. Chem. Eng. Data* **13**, 16 (1968); D. R. Defibaugh *et al.*, *ibid.* **39**, 333 (1994). Viscosity study: M. Takahashi *et al.*, *ibid.* **40**, 900 (1995). Toxicology and metabolism in rats: M. K. Ellis *et al.*, *Fundam. Appl. Toxicol.* **31**, 243 (1996).

Colorless gas, bp −51.6°. mp −136°. d^0 (liq) 1.052. d^{25} (liq) 0.961. Critical temp 78.4°; critical press 57.54 atm; critical d 0.430. Dipole moment 1.98 D. *Flammable.*

USE: Refrigerant.

3171. Difluprednate. [23674-86-4] (6α,11β)-21-(Acetyloxy)-6,9-difluoro-11-hydroxy-17-(1-oxobutoxy)pregna-1,4-diene-3,20-dione; 6α,9-difluoro-11β,17,21-trihydroxypregna-1,4-diene-3,20-dione 21-acetate 17-butyrate; 6α,9α-difluoroprednisolone-21-acetate-17-butyrate; CM-9155; W-6309; Durezol; Epitopic; Myser. $C_{27}H_{34}F_2O_7$; mol wt 508.56. C 63.77%, H 6.74%, F 7.47%, O 22.02%. Topical corticosteroid. Prepn: A. Ercoli, R. Gardi, ZA **6803686**; *eidem*, US **3780177** (1968, 1973 both to Warner-Lambert); R. Gardi *et al.*, *J. Med. Chem.* **15**, 556 (1972). Prepn of ophthalmic formulation: M. Yamaguchi *et al.*, *Int. J. Pharm.* **301**, 121 (2005). Anti-inflammatory properties: G. Di Pasquale *et al.*, *Steroids* **16**, 663, 679 (1970). HPLC determn in aqueous humor: S.-Yasueda *et al.*, *J. Pharm. Biomed. Anal.* **30**, 1735 (2003). Skin penetration and bioavailability study: C. Lafille *et al.*, *Dermatologica* **159**, 277 (1979).

Crystals from methylene chloride/ether/petr ether, mp 191-194°. $[\alpha]_D^{22}$ +31.7° (c = 0.5 in dioxane). Insol in water. uv max (ethanol): 237-238 nm ($E_{1cm}^{1\%}$ 320).

THERAP CAT: Anti-inflammatory.

3172. Digallic Acid. [536-08-3] 3,4-Dihydroxy-5-[(3,4,5-tri-hydroxybenzoyl)oxy]benzoic acid; gallic acid 5,6-dihydroxy-3-car-boxyphenyl ester; 4,5-dihydroxybenzoic acid monogallate; gallic acid 3-monogallate; *m*-digallic acid; *m*-galloylgallic acid. C_{14}-$H_{10}O_9$; mol wt 322.23. C 52.18%, H 3.13%, O 44.69%. Isoln from Aleppo gallotannin and Chinese gallotannin: Nierensten, *Ber*. **43**, 628 (1910). Synthesis: Fischer, Freudenberg, *Ber*. **46**, 1128 (1913). Structure: Nierenstein *et al.*, *J. Am. Chem. Soc*. **47**, 846 (1925). Determn by HPLC: P. Delahaye, M. Verzele, *J. Chromatogr*. **265**, 363 (1983); in tannic acids: M. Verzele *et al.*, *Bull. Soc. Chim. Belg*. **92**, 181 (1983). Role in corrosion inhibition: A. M. Beccaria, E. D. Mor, *Br. Corrosion J*. **11**, 156 (1976).

Hydrated needles from alc + water, anhydr at 110°; dec 280°. Sol in 1900 parts water at 25°; in 50-60 parts boiling water; sol in meth-anol, ethanol, acetone; sparingly sol in ether, glacial acetic acid. Upon addn of KCN to an aq soln a transitory pink color appears, which returns on shaking.

Note: In pharmaceutical literature the name digallic acid is fre-quently confused with tannic acid.

3173. Digalogenin. [6877-35-6] (3β,5α,15β,25R)-Spirostan-3,15-diol. $C_{27}H_{44}O_4$; mol wt 432.65. C 74.96%, H 10.25%, O 14.79%. Aglycon of digalonin. Isoln from saponin mixture and par-tial synthesis from digitogenin: Tschesche, Wulff, *Ber*. **94**, 2019 (1961).

Crystals from methanol, mp 218.5-220.5°. $[\alpha]_D^{21}$ −75° (CHCl₃). IR spectrum: Tschesche, Wulff *loc. cit*. Sol in chloroform.

3174. Diginatigenin. [559-57-9] (3β,5β,12β,16β)-3,12,-14,16-Tetrahydroxycard-20(22)-enolide; 12-hydroxygitoxigenin; 16-hydroxydigoxigenin. $C_{23}H_{34}O_6$; mol wt 406.52. C 67.96%, H 8.43%, O 23.61%. Aglycon of diginatin from *Digitalis lanata* Ehrk., *Scrophulariaceae*: Murphy, *J. Am. Pharm. Assoc. Sci. Ed*. **44**, 719 (1955). Aglycone of lanatoside D: Angliker *et al.*, *Ann*. **607**, 131 (1957). Obtained by microbiological conversion of gitoxi-genin: Tamm, Gubler, *Helv. Chim. Acta* **41**, 1762 (1958); Okado *et al.*, *Chem. Pharm. Bull*. **8**, 530 (1960). Structure: Linde *et al.*, *Helv. Chim. Acta* **42**, 2040 (1959); Okado *et al.*, *Chem. Pharm. Bull*. **8**, 535 (1960).

Needles from water, mp 157°. $[\alpha]_D^{20}$ +34° (methanol). Absorption max (ethanol): 318 nm (log ε 4.18); (98% w/w H_2SO_4): 230, 310, 390, 490 nm ($E_{1cm}^{1\%}$ 160, 130, 210, 85).

3175. Diginatin. [52589-12-5] (3β,5β,12β,16β)-3-[(*O*-2,6-Dideoxy-β-D-*ribo*-hexopyranosyl-(1 → 4)-*O*-2,6-dideoxy-β-D-*ribo*-hexopyranosyl-(1 → 4)-2,6-dideoxy-β-D-*ribo*-hexopyranosyl)-oxy]-12,14,16-trihydroxy-card-20(22)-enolide. $C_{41}H_{64}O_{15}$; mol wt 796.95. C 61.79%, H 8.09%, O 30.11%. Cardiac glycoside from *Digitalis lanata* Ehrk., *Scrophulariaceae*: Murphy, *J. Am. Pharm. Assoc. Sci. Ed*. **44**, 719 (1955); Angliker *et al.*, *Ann*. **607**, 131 (1957); von Wartburg *et al.*, *Experientia* **14**, 439 (1958). Acid hydrolysis yields 1 mol diginatigenin + 3 mols D-digitoxose. The sugar residue is attached to the hydroxyl group at C-3 of the aglycon.

Prisms from acetone, mp 251-253°. Somewhat hygroscopic. $[\alpha]_D^{20}$ +20.5 ±2°; $[\alpha]_{546}^{20}$ +26.3 ±2° (c = 0.6 in methanol). Absorption max (98% w/w H_2SO_4): 230, 310, 390, 480, 560 nm ($E_{1cm}^{1\%}$ 185, 175, 315, 230, 120). Sol in alcohol, dil alcohol, dioxane; slightly sol in acetone. One part sol in 1000 parts water and in 2000 parts chloro-form at 25°.

Acetyldiginatin-α. $C_{43}H_{66}O_{16}$. Flat plates from methanol + chloroform+ ether, mp 199-202°. $[\alpha]_D^{20}$ +30.1° (c = 0.6 in pyridine). uv max: 218 nm (log ε 4.1).

3176. Diginin. [467-53-8] (3β,12α,14β,17α,20S)-3-[(2,6-Di-deoxy-3-*O*-methyl-D-lyxohexopyranosyl)oxy]-12,20-epoxypregn-5-ene-11,15-dione; 3β-(diginosyloxy)-12α,20α-epoxy-14β,17α-pregn-5-ene-11,15-dione. $C_{28}H_{40}O_7$; mol wt 488.62. C 68.83%, H 8.25%, O 22.92%. Isoln from leaves of *Digitalis purpurea* L., *Scro-phulariaceae*: Karrer in *E. Barell Festschrift* (1936) p 238; *Chem. Zentralbl*. **1936**, II, 2727. Isoln: Shoppee, Reichstein, *Helv. Chim. Acta* **23**, 975 (1940); Shoppee, *ibid*. **27**, 426 (1944). Structure: Shoppee *et al.*, *J. Chem. Soc*. **1962**, 3610; Tschesche, Brügmann, *Tetrahedron* **20**, 1469 (1964). Mild hydrolysis yields *diginigenin* and *diginose*. Attempted partial synthesis of diginigenin: Tsches-che, Schwinum, *Ber*. **100**, 464 (1967); Tschesche, Müller-Albrecht, *Ber*. **103**, 350 (1970).

Stout prisms from dil alc. Indistinct melting range: 155-183°. $[\alpha]_D^{14}$ −223° (c = 2.3 in chloroform). uv max (ethanol): 309 nm (log ε 1.94). Freely sol in chloroform; slightly sol in ether, acetone, ethyl acetate, carbon tetrachloride. Practically insol in water. Positive Legal's test. The Keller reaction produces a brilliant yellow band with diginigenin.

Digitalin

3177. Digitalin. [752-61-4] (3β,5β,16β)-3-[(6-Deoxy-4-O-β-D-glucopyranosyl-3-O-methyl-β-D-galactopyranosyl)oxy]-14,16-dihydroxycard-20(22)-enolide; digitalinum verum; digitalinum true; Schmiedeberg's digitalin; Diginorgin. $C_{36}H_{56}O_{14}$; mol wt 712.83. C 60.66%, H 7.92%, O 31.42%. Obtained from seeds of *Digitalis purpurea* L., *Scrophulariaceae*, and from roots of *Adenium honghel* A. DC., *Apocynaceae:* Schmiedeberg, *Arch. Exp. Pathol. Pharmakol.* **3**, 16 (1874); Windaus, Haack, *Ber.* **62**, 475 (1929); Hunger, Reichstein, *Helv. Chim. Acta* **33**, 76 (1950); Sasakawa, *J. Pharm. Soc. Jpn.* **74**, 474 (1954); Miyatake *et al., Pharm. Bull.* **5**, 157 (1957).

Crystals from methanol + ether and methanol + water, mp 240-243°. $[\alpha]_D^{20}$ −1.1° (c = 0.894 in methanol). Slightly sol in water, chloroform or ether; sol in alcohol.

16-Acetyldigitalinum verum. $C_{38}H_{58}O_{15}$. Prepn: Miyatake *et al.*, **US 3023147** (1962). Amorphous powder. Freely sol in water, alcohol, acetone. Practically insol in benzene, ether. uv max: 217 nm (log ε 4.16). $[\alpha]_D^{26}$ −21.1° (methanol).

THERAP CAT: Cardiotonic.

3178. Digitalis. Foxglove; fairy gloves; purple foxglove. Dried leaves of *Digitalis purpurea* L., *Scrophulariaceae. Habit.* Southern and central Europe, cultivated in the U.S. *Constit.* Digitoxin (0.2-0.4%), digitonin, digitalin, antirhinic acid, digitalosmin, digitoflavone, inositol, pectin. One U.S.P. digitalis unit represents the potency of 0.1 g of the U.S.P. Digitalis Reference Standard. Toxicology: A. P. Somlyo, *Am. J. Cardiol.* **5**, 523 (1960). Review of mechanism of action and clinical use: T. W. Smith, *N. Engl. J. Med.* **318**, 358-365 (1988). Book: C. Fisch, B. Surawicz, *Digitalis* (Grune & Stratton, New York, 1969) 244 pp.

Caution: Potential symptoms of overexposure are anorexia, nausea, salivation, vomiting, diarrhea, headache, drowsiness, disorientation, delirium, hallucinations; death may result. *See Clinical Toxicology of Commercial Products,* R. E. Gosselin *et al.*, Eds. (Williams & Wilkins, Baltimore, 4th ed., 1976) Section III, pp 124-133.

THERAP CAT: Cardiotonic.

THERAP CAT (VET): Cardiotonic.

3179. Digitogenin. [511-34-2] (2α,3β,5α,15β,25R)-Spirostan-2,3,15-triol. $C_{27}H_{44}O_5$; mol wt 448.64. C 72.28%, H 9.89%, O 17.83%. The aglycon of digitonin: Tschesche, *Ber.* **68**, 1090 (1935); Tschesche, Hagedorn, *Ber.* **69**, 797 (1936). Structure and stereochemistry: Marker *et al., J. Am. Chem. Soc.* **64**, 1843 (1942); Meystre *et al., Helv. Chim. Acta* **38**, 381 (1955); Klass *et al., J. Am. Chem. Soc.* **77**, 3829 (1955); and IR spectra: Djerassi *et al., ibid.* **78**, 3166 (1956).

Needles from alc, dec 296°. $[\alpha]_D^{19}$ −81° (c = 1.4 in chloroform). Practically insol in water. Sol in 30 parts chloroform, 35 parts boiling alc, 100 parts alc at 20°.

2,3-Diacetate. Crystals from methanol-chloroform, mp 241.5-242°. $[\alpha]_D$ −104°.

Triacetate. Minute needles from ether, mp 190°.

3180. Digitonin. [11024-24-1] (2α,3β,5α,15β,25R)-2,15-Dihydroxyspirostan-3-yl O-β-D-glycopyranosyl-(1 → 3)-O-β-D-galactopyranosyl-(1 → 2)-O-[β-D-xylopyranosyl-(1 → 3)]-O-β-D-glucopyranosyl-(1 → 4)-β-D-galactopyranoside; Digitin. $C_{56}H_{92}O_{29}$; mol wt 1229.32. C 54.71%, H 7.54%, O 37.74%. Obtained from the seeds of *Digitalis purpurea* L., *Scrophulariaceae.* Extraction procedure: Gisvold, *J. Am. Pharm. Assoc.* **23**, 664 (1934). Purification of commercial digitonin and its separation into two fractions: G. Ruhenstroth-Bauer, P. M. Breitenfeld, *Z. Physiol. Chem.* **302**, 111 (1955). Structure: Tschesche, Wulff, *Tetrahedron* **19**, 621 (1963). Use as clinical reagent: H. H. Leffler, *Am. J. Clin. Pathol.* **31**, 310 (1959).

Crystals from alc, sinters 225°. Indistinct mp 235-240°. $[\alpha]_D^{20}$ −54° (0.45 g in 15.8 ml methanol). One gram dissolves in 57 ml abs alc, in 220 ml 95% alc. Practically insol in water, forming a soapy suspension. Also practically insol in chloroform, ether.

USE: Clinical reagent (cholesterol determination).

3181. Digitoxigenin. [143-62-4] (3β,5β)-3,14-Dihydroxy-card-20(22)-enolide; Δ$^{20:22}$-3,14,21-trihydroxynorcholenic acid lactone; cerberigenin; echujetin; evonogenin; Thevetigenin. $C_{23}H_{34}O_4$; mol wt 374.52. C 73.76%, H 9.15%, O 17.09%. The aglycon of digitoxin, thevetin, cerberin, echujin, evomonosid. Prepn by refluxing digitoxin in a mixture of water + alcohol + HCl: Cloetta, *Arch. Exp. Pathol. Pharmakol.* **88**, 113 (1920); Stoll, Kreiss, *Helv. Chim. Acta* **17**, 592 (1934). Structure: Jacobs, Hoffmann, *J. Biol. Chem.* **67**, 333 (1926); Jacobs, Gustus, *ibid.* **78**, 573; **79**, 533 (1928); **82**, 402 (1929); Jacobs, Elderfield, *ibid.* **108**, 497 (1935); Elderfield, *Chem. Rev.* **17**, 187 (1935). Stereochemistry: Meyer, *Helv. Chim. Acta* **30**, 1976 (1947). Synthesis: Danieli *et al., Tetrahedron* **22**, 3189 (1966); W. Fritsch *et al., Ann.* **1974**, 621; S. F. Donovan *et al., Tetrahedron Lett.* **1979**, 3287; R. Marinibettolo *et al., Can. J. Chem.* **59**, 1403 (1981); T. Milkova *et al., Tetrahedron Lett.* **23**, 413 (1982). Biosynthesis from neriifolin, *q.v.:* A. Cruz *et al., J. Org. Chem.* **42**, 3580 (1977).

Stout prisms from 40% alc, mp 253°. $[\alpha]_D^{17}$ +19.1° (c = 1.36 in methanol). Sol in alc, chloroform, acetone; slightly sol in ethyl acetate; very sparingly sol in ether, water.

3-Acetyldigitoxigenin. $C_{25}H_{36}O_5$. Hexagonal plates from acetone + ether, mp 222-227°. $[\alpha]_D^{17}$ +21.4° (c = 1.02 in chloroform).

3182. Digitoxin. [71-63-6] $(3\beta,5\beta)$-3-[(O-2,6-Dideoxy-β-D-ribo-hexopyranosyl-(1 → 4)-O-2,6-dideoxy-β-D-ribo-hexopyranosyl-(1 → 4)-2,6-dideoxy-β-D-ribo-hexopyranosyl)oxy]-14-hydroxycard-20(22)-enolide; digitalin, crystalline; digitophyllin; Carditoxin; Coramedan; Cristapurat; Crystodigin; Digicor; Digilong; Digimerck; Digimed; Digipural; Ditaven; Lanatoxin; Myodigin; Purodigin; Purpurid; Tardigal. $C_{41}H_{64}O_{13}$; mol wt 764.95. C 64.38%, H 8.43%, O 27.19%. Secondary glycoside from *Digitalis purpurea* L., *Scrophulariaceae*. Extracted from the dried leaves with 50% alc. Extraction procedure: Cloetta, *Arch. Exp. Pathol. Pharmakol.* **112**, 261 (1926). Purification: Windaus, Freese, *Ber.* **58**, 2503 (1925). Ten kilo leaves yield about 6 grams pure digitoxin. Identity with digitophyllin: Cloetta, *Arch. Exp. Pathol. Pharmakol.* **88**, 113 (1920). Structure: Elderfield, *Chem. Rev.* **17**, 187 (1935); Stoll, *The Cardiac Glycosides* (London, 1937); Shoppee, *Annu. Rev. Biochem.* **11**, 103 (1942). Acid hydrolysis yields 1 mol digitoxigenin + 3 mol digitoxose. The sugar residue is attached to the hydroxyl group at C-3 of the aglycon. Structure of the sugar moiety: Lichti *et al.*, *Helv. Chim. Acta* **45**, 868 (1962). Digitoxin U.S.P. is either pure digitoxin or a mixture of cardioactive glycosides obtained from *Digitalis purpurea* and consisting chiefly of digitoxin. The potency of digitoxin U.S.P. corresp to the potency of an equal wt of U.S.P. Digitoxin Reference Standard. A deviation of 20% is permitted. The physical characteristics given below are those of the pure compd. Toxicity data: Foerster *et al.*, *Arch. Int. Pharmacodyn. Ther.* **159**, 1 (1966). Comprehensive description: I. M. Jakovljevic, *Anal. Profiles Drug Subs.* **3**, 149-172 (1974).

Very small elongated, rectangular plates from dil alc. May contain ½ or 1 mol H_2O or EtOH which is given up at 118° *in vacuo*. When anhydrous, mp 256-257°. $[\alpha]_D^{20}$ +4.8° (c = 1.2 in dioxane). One gram dissolves in about 40 ml chloroform, about 60 ml alcohol, about 400 ml ethyl acetate. Sol in acetone, amyl alcohol, pyridine; sparingly sol in petr ether. Very slightly sol in ether. Practically insol in water (1 g/100 liter at 20°). LD_{50} in guinea pigs, cats (mg/kg): 60.0, 0.18 orally (Foerster).

Acetyl derivatives see Acetyldigitoxins.

THERAP CAT: Cardiotonic.

THERAP CAT (VET): Cardiotonic.

3183. Digitoxose. [527-52-6] 2,6-Dideoxy-D-ribo-hexose; 2-desoxy-D-altromethylose; 2,6-didesoxy-D-allose. $C_6H_{12}O_4$; mol wt 148.16. C 48.64%, H 8.16%, O 43.19%. Obtained by mild acid hydrolysis of the glycosides digitoxin, gitoxin and digoxin: Cloetta, *Arch. Exp. Pathol. Pharmakol.* **88**, 113 (1920); **112**, 261 (1926); Windaus, Stein, *Ber.* **61**, 2436 (1928); Kraft, *Arch. Pharm.* **250**, 118 (1912); Mannich *et al.*, *ibid.* **268**, 453 (1930); Smith, *J. Chem. Soc.* **1930**, 508; **1931**, 23. Configuration: Micheel, *Ber.* **63**, 347 (1930). Structure: S. F. Dyke, *The Carbohydrates* (Interscience, New York, 1960) p 104. Synthesis: Gut, Prins, *Helv. Chim. Acta* **30**, 1223 (1947); Bolliger, Ulrich, *ibid.* **35**, 93 (1952). Stereochemical study: S. Tsukamoto *et al.*, *J. Chem. Soc. Perkin Trans. 1* **1988**, 2621. *Review:* R. C. Elderfield in W. W. Pigman, M. L. Wolfrom, *Advances in Carbohydrate Chemistry* vol. **I** (Academic Press, New York, 1945) pp 159-164.

Crystals from methanol + ether, from ethyl acetate or from acetone + ether, mp 112°. $[\alpha]_D^{17}$ +46.3° (in water); $[\alpha]_D^{20}$ +39.1° (in methanol); $[\alpha]_D^{18}$ +27.9° → +43.3° (after 24 hrs in pyridine). Freely sol in water; sol in acetone, ethanol. Practically insol in ether.

3184. Diglyme. [111-96-6] 1,1'-Oxybis[2-methoxyethane]; bis(2-methoxyethyl) ether; diethylene glycol dimethyl ether. $C_6H_{14}O_3$; mol wt 134.18. C 53.71%, H 10.52%, O 35.77%. Prepn: Cretcher, Pittenger, *J. Am. Chem. Soc.* **47**, 163 (1925); Gallaugher, Hibbert, *ibid.* **58**, 813 (1936).

Liquid, bp_{760} 162°; bp_{200} 116°; bp_{35} 75°; bp_3 20°. d_{20}^{20} 0.9451. mp −68°. Flash pt (open cup) 158°F (70°C). n_D^{20} 1.4097. Miscible with water, alcohol, ether, hydrocarbon solvents.

USE: Solvent; reaction medium for Grignard and similar syntheses.

3185. Digoxigenin. [1672-46-4] $(3\beta,5\beta,12\beta)$-3,12,14-Trihydroxycard-20(22)-enolide; $\Delta^{20:22}$-3β,12β,14,21-tetrahydroxynorcholenic acid lactone; lanadigenin. $C_{23}H_{34}O_5$; mol wt 390.52. C 70.74%, H 8.78%, O 20.48%. The aglycone of digoxin. By hydrolysis of digoxin: Smith, *J. Chem. Soc.* **1930**, 508. From *Digitalis orientalis* L. and *D. lanata* Ehrh., *Scrophulariaceae:* Mannick, Schneider, *Arch. Pharm.* **279**, 223 (1941); Pataki *et al.*, *Helv. Chim. Acta* **36**, 1295 (1953). Structure: Meyer, Reichstein, *Experientia* **9**, 253 (1953); Cardwell, Smith, *ibid.* 367; *eidem*, *J. Chem. Soc.* **1954**, 2012. Synthesis: P. Welzel, H. Stein, *Tetrahedron Lett.* **22**, 3385 (1981).

Dihydrate. Prismatic rods from dil alc. Anhyd as stout prisms from ethyl acetate, mp 222°. $[\alpha]_{546}^{20}$ +27.0° (c = 1.77 in methanol). Although a 3β-alcohol, it is not precipitated by digitonin (Pataki).

3,12-Diacetyldigoxigenin. Prisms from dil methanol, mp 222-223°. $[\alpha]_{546}^{20}$ +61.3° (c = 2 in methanol).

3186. Digoxin. [20830-75-5] $(3\beta,5\beta,12\beta)$-3-[(O-2,6-Dideoxy-β-D-ribo-hexopyranosyl-(1 → 4)-O-2,6-dideoxy-β-D-ribo-hexopyranosyl-(1 → 4)-2,6-dideoxy-β-D-ribo-hexopyranosyl)oxy]-12,14-dihydroxycard-20(22)-enolide; Digacin; Dilanacin; Eudigox; Lanicor; Lanoxin; Lanoxicaps; Lenoxin; Neo-Dioxanin; Rougoxin. $C_{41}H_{64}O_{14}$; mol wt 780.95. C 63.06%, H 8.26%, O 28.68%. Secondary glycoside from *Digitalis lanata* Ehrh., or *D. orientalis* Lam., *Scrophulariaceae:* S. Smith, *J. Chem. Soc.* **1930**, 508; Stoll, *The Cardiac Glycosides* (London, 1937); M. M. Dhar *et al.*, **IN 62497** (1958 to Council Sci. Indust. Res.), *C.A.* **53**, 653b (1959). *See also* ref under Digoxigenin. Acid hydrolysis of digoxin yields 1 mol digoxigenin + 3 mols digitoxose. The sugar residue is attached to the hydroxyl group at C-3 of the aglycon. Clinical pharmacokinetics: J. K. Aronson, *Clin. Pharmacokinet.* **5**, 137 (1980). Comprehensive description: P. R. B. Foss, S. A. Benezra, *Anal. Profiles Drug Subs.* **9**, 207-243 (1980).

Radially arranged, four- and five-sided triclinic plates from dil alcohol or dil pyridine, decomp 230-265°. $[\alpha]_{Hg}^{25}$ +13.4 to 13.8° (c = 10 in pyridine). uv max (ethanol): 220 nm (ε 12800). Freely sol in pyridine; sol in mixt of chloroform and alcohol. More sol in hot 80% alcohol than gitoxin. Slightly sol in dil alc, chloroform. Practically insol in ether, acetone, ethyl acetate, chloroform, water.

β-Methyldigoxin. [30685-43-9] Medigoxin; metildigoxin; 4'''-O-methyldigoxin; 3β,12β,14β-trihydroxy-5β-card-20(22)-enolide-3-(4'''-O-methyltridigitoxoside); Cardiolan; Lanirapid; Lanitop. $C_{42}H_{66}O_{14}$; mol wt 794.98. Obtained by the O-methylation of digoxin: F. Kaiser *et al.*, **ZA 6806079**; *eidem*, **US 3538078** (1969, 1970 both to Boehringer, Mann.). Pharmacology and toxicity studies: W. Schaumann, R. Wegerle, *Arzneim.-Forsch.* **21**, 225 (1971); H. Czerwek *et al.*, *ibid.* 231. Crystals, mp 227-231°. LD_{50} in rats, mice (mg/kg): 4.8, 4.9 i.v.; 6.2, 4.8 i.p.; 8.3, 7.8 orally (Czerwek).

α-Acetyldigoxin. [5511-98-8] Lanatilin; Sandolanid. $C_{43}H_{66}O_{15}$; mol wt 822.99. Obtained by enzymatic hydrolysis of digilanide. Prisms from methanol + chloroform, dec 225°. $[\alpha]_D^{20}$ +18.9° (pyridine). Very sparingly sol in ethyl acetate.

β-Acetyldigoxin. [5355-48-6] Kardiamed; Longdigox; Novodigal; Stillacor. Needles from alcohol + chloroform, dec 240°. $[\alpha]_D^{20}$ +30.4° (c = 1.2 in alc). More sol in ethyl acetate than the α-form.
THERAP CAT: Cardiotonic.
THERAP CAT (VET): Cardiotonic.

3187. Dihydralazine. [484-23-1] 1,4-Dihydrazinylphthalazine; 2,3-dihydro-1,4-phthalazinedione dihydrazone; 1,4-dihydrazinophthalazine. $C_8H_{10}N_6$; mol wt 190.21. C 50.52%, H 5.30%, N 44.18%. Vasodilator. Prepn: J. Druey, US 2484785 (1949 to Ciba); J. Druey, B. H. Ringier, *Helv. Chim. Acta* **34**, 195 (1951); Zerweck, Kunze, US 2786839 (1957 to Cassella Farbw.). Pharmacology: P. A. Van Zwieten, *Arzneim.-Forsch.* **18**, 79 (1968). Mechanism of action: I. W. Reimann *et al.*, *Clin. Sci.* **61**, 319s (1981); *eidem*, *Clin. Exp. Pharmacol. Physiol.* **12**, 79 (1985). Cerebrovascular effects in rats: L. M. Auer *et al.*, *Acta Med. Scand.* **Suppl. 678**, 73 (1983); D. I. Barry *et al.*, *Stroke* **15**, 102 (1984). Hemodynamic responses in man: G. G. Belz, *Clin. Pharmacol. Ther.* **37**, 48 (1985). Clinical studies: A. Salvadeo *et al.*, *Int. J. Clin. Pharmacol. Ther. Toxicol.* **19**, 372 (1981); P. I. Salmela *et al.*, *Ann. Clin. Res.* **13**, 433 (1981). Toxicity data: G. Steiner *et al.*, *J. Med. Chem.* **24**, 59 (1981).

Orange needles from water, dec about 180°. LD_{50} i.p. in rats: 1084 μmol/kg (Steiner).
Sulfate. [7327-87-9] Pressunic. $C_8H_{10}N_6 \cdot H_2SO_4$; mol wt 288.28. Needles, dec 233°.
Sulfate hemipentahydrate. Depressan; Dihyzin; Nepresol; Népressol.
Methanesulfonate. Dihydralazine mesylate; Nepresol Inject. $C_8H_{10}N_6 \cdot CH_3SO_3H$; mol wt 286.31.
THERAP CAT: Antihypertensive.

3188. Dihydroactinidiolide. [15356-74-8] 5,6,7,7a-Tetrahydro-4,4,7a-trimethyl-2(4H)-benzofuranone; 2-oxo-4,4,7a-trimethyl-2,4,5,6,6,7,7a-hexahydrobenzofuran; (2,6,6-trimethyl-2-hydroxycyclohexylidene)acetic acid lactone. $C_{11}H_{16}O_2$; mol wt 180.25. C 73.30%, H 8.95%, O 17.75%. Found in many plant sources including tea and tobacco. (−)-Form is naturally occurring. Produced by photo-oxidation of carotenoid flavor compounds. Pheromone component for queen recognition in red fire ants. Isoln from black tea aroma: J. Bricout *et al.*, *Helv. Chim. Acta* **50**, 1517 (1967); from *Actinidia polygama* Miq. and structure determn: T. Sakan *et al.*, *Tetrahedron Lett.* **8**, 1623 (1967); from tobacco and synthesis: W. C. Bailey, Jr. *et al.*, *J. Org. Chem.* **33**, 2819 (1968). Synthesis: S. Isoe *et al.*, *Tetrahedron Lett.* **9**, 5561 (1968); by photo-oxidation of β-carotene: *eidem*, *ibid.* **10**, 279 (1969). Abs config study: *eidem*, *ibid.* **13**, 2517 (1972). Prepn of enantiomers: K. Mori, Y. Nakazono, *Tetrahedron* **42**, 283 (1986). Improved prepns: G. V. Subbaraju *et al.*, *Tetrahedron Lett.* **32**, 4871 (1991); A. Bosser *et al.*, *Biotechnol. Prog.* **11**, 689 (1995). Prepn of (R)-form: S. Yao *et al.*, *J. Org. Chem.* **63**, 118 (1998). Germination inhibition in wheat grains: T. Kato *et al.*, *J. Agric. Food Chem.* **51**, 2161 (2003).

(R)-Form

Tan oil, bp_{800} 108-109°. mp 42-43°. Fruity aroma. uv max: 241 nm (ε 10000).
(R)-Form. [17092-92-1] mp 70-71°. $[\alpha]_D^{24}$ −121.0° (c = 1.05 in $CHCl_3$).

(S)-Form. [81800-41-1] mp 67-68°. $[\alpha]_D^{23}$ +120.9° (c = 1.00 in $CHCl_3$).
USE: Flavor ingredient; inhibitor of seed germination.

3189. Dihydrocodeine. [125-28-0] (5α,6α)-4,5-Epoxy-3-methoxy-17-methylmorphinan-6-ol; 6-methoxy-3-methoxy-N-methyl-4,5-epoxymorphinan; dihydroneopine; drocode. $C_{18}H_{23}NO_3$; mol wt 301.39. C 71.73%, H 7.69%, N 4.65%, O 15.93%. Semisynthetic opioid analgesic; metabolized to dihydromorphine, *q.v.*, by cytochrome P450 2D6. Prepd by reduction of codeine or neopine: Skita, Franck, *Ber.* **44**, 2862 (1911); Wieland, Koralek, *Ann.* **433**, 269 (1923); Stein, *Pharmazie* **10**, 180 (1955). GC-MS determn in urine: M. Balikova *et al.*, *J. Chromatogr. B* **752**, 179 (2001). Metabolism in humans: M. F. Fromm *et al.*, *Clin. Pharmacol. Ther.* **58**, 374 (1995). Clinical pharmacokinetics: S. Ammon *et al.*, *Br. J. Clin. Pharmacol.* **48**, 317 (1999). Opioid receptor binding profile: H. Schmidt *et al.*, *Pharmacol. Toxicol.* **91**, 57 (2002). Clinical trial in chronic pain: C. H. Wilder-Smith *et al.*, *Pain* **91**, 23 (2001); maintenance treatment of opioid dependence: J. R. Robertson *et al.*, *Addiction* **101**, 1752 (2006).

Crystals from methanol + water, mp 112-113°; bp_{15} 248°.
Tartrate. [5965-13-9] Dihydrocodeine bitartrate; Codicontin; DF 118; DHC; Dicodin; Paracodin; Parzone; Tiamon. $C_{18}H_{23}NO_3 \cdot C_4H_6O_6$; mol wt 451.47. Crystals from methanol, contains 66.8% dihydrocodeine when completely anhydr. mp 192-193° (Stein). $[\alpha]_D^{25}$ −72 to −75° (c = 1 in H_2O). One gram dissolves in 4.5 ml water. Sparingly sol in alcohol. Insol in ether.
Note: This is a controlled substance (opiate): **21 CFR, 1308.12.**
THERAP CAT: Analgesic; antitussive.

3190. Dihydroequilin. [3563-27-7] (17β)-Estra-1,3,5(10),7-tetraene-3,17-diol. $C_{18}H_{22}O_2$; mol wt 270.37. C 79.96%, H 8.20%, O 11.83%. Occurs in the 17α- and 17β-forms. Prepn of β-form by reduction of equilin with sodium in alc: David, *Acta Brev. Nederland* **4**, 63 (1934); Serini *et al.*, US 2221340 (1940 to Schering). Isoln of β-form from pregnant mares' urine: Gaudry, Glen, *Ind. Chim. Belge* **Suppl. 2**, 435 (1959). Prepn of α-form by reduction of equilin with aluminum isopropoxide: Carol *et al.*, *J. Biol. Chem.* **185**, 267 (1950). Isoln of α-form from pregnant mares' urine: Glen *et al.*, *Nature* **177**, 753 (1956). α-Form was formerly called β-dihydroequilin and vice versa: Banes *et al.*, *J. Biol. Chem.* **187**, 557 (1950).

β-Form

Crystals from actone, mp 174.5-174.6°. $[\alpha]_D^{20}$ +220° (dioxane).
17α-Form. [651-55-8] Plates from 30% ethanol, mp 205.5-205.6°. $[\alpha]_D^{20}$ +213° (ethanol).

3191. Dihydroergotamine. [511-12-6] (5′α,10α)-9,10-Dihydro-12′-hydroxy-2′-methyl-5′-(phenylmethyl)ergotaman-3′,-6′,18-trione; DHE. $C_{33}H_{37}N_5O_5$; mol wt 583.69. C 67.91%, H 6.39%, N 12.00%, O 13.71%. α-Adrenergic blocker with selective venoconstrictor properties. Also binds to serotonin 5HT$_1$-receptors. Prepn from ergotamine: Stoll, Hofmann, *Helv. Chim. Acta* **26**, 2070 (1943). Clinical pharmacology: H. de Marées *et al.*, *Eur. J. Clin. Pharmacol.* **30**, 685 (1986). Neurotransmitter receptor binding study: B. G. McCarthy, S. J. Peroutka, *Headache* **29**, 420 (1989). Clinical trials in migraine: P. Winner *et al.*, *ibid.* **33**, 471 (1993); D. Ziegler *et al.*, *Neurology* **44**, 447 (1994); of inhaled formulation in acute migraine: S. K. Aurora *et al.*, *Headache* **49**, 826 (2009).

Strongly refractive prisms from dil acetone, contg 2 mols acetone and 2 mols water of crystn. mp 239°. $[\alpha]_D^{20}$ −64°; $[\alpha]_{546}^{20}$ −79° (c = 0.5 in pyridine). Insol in water. Sparingly sol in methanol, ethanol, chloroform, benzene.

Tartrate. $(C_{33}H_{37}N_5O_5)_2.C_4H_6O_6$; mol wt 1317.46. Six-sided plates from methanol, dec 210-215°.

Methanesulfonate. [6190-39-2] Dihydroergotamine mesylate; MAP-0004; Agit; Angionorm; Diergo; Diergospray; Dihydergot; Er-gomimet; Ergont; Ergotonin; Ikaran; Levadex; Migranal; Orstanorm; Séglor; Tamik; Tonopres; Verladyn. $C_{33}H_{37}N_5O_5.CH_3SO_3H$; mol wt 679.79. Large prisms from 95% alc. mp 230-235°. Sol in alc; slightly sol in chloroform, water.

THERAP CAT: Antimigraine.

3192. Dihydro-β-erythroidine. [23255-54-1] (2S,13bS)-2,-3,5,6,8,9,10,13-Octahydro-2-methoxy-1H,12H-benzo[i]pyrano[3,4-g]indolizin-12-one; (3β)-1,6-didehydro-14,17-dihydro-3-methoxy-16(15H)-oxaerythrinan-15-one; 12,13-didehydro-2,7,13,14-tetrahydro-α-erythroidine. $C_{16}H_{21}NO_3$; mol wt 275.35. C 69.79%, H 7.69%, N 5.09%, O 17.43%. The more active of the two isomers that constitute the principal alkaloidal fraction of seeds from several *Erythrina* spp. In contrast to curare, *q.v.*, it retains its ability to block neuromuscular transmission even when administered orally: K. Unna, J. G. Greslin, *J. Pharmacol.* **80**, 53 (1944); K. Unna *et al.*, *ibid.* 39. Prepd by catalytic hydrogenation of β-erythroidine or a salt of β-erythroidine: Folkers, Koniuszy, US 2370651 (1945 to Merck & Co.). Structure: Boekelheide *et al.*, *J. Am. Chem. Soc.* **75**, 2550 (1953). Configuration: Hanson, *Proc. Chem. Soc. London* **1963**, 52.

Crystals from anhyd ethyl ether, dec 85-86°. $[\alpha]_D^{25}$ +102.5°. Soluble in ethanol. Alkaline hydrolysis yields sodium dihydro-β-erythroidinate.

Hydrochloride. Crystals from abs ethanol, mp 238°. $[\alpha]_D^{25}$ +124.7°.

Hydrobromide. Crystals from abs ethanol or abs ethanol + abs ether, dec 242°. Bitter taste. $[\alpha]_D^{25}$ +106-107.5°. Soly in water 0.8 g/ml; in ethanol 0.5 g/100 ml.

3193. Dihydroisocodeine. [795-38-0] (5α,6β)-4,5-Epoxy-3-methoxy-17-methylmorphinan-6-ol; DHIC. $C_{18}H_{23}NO_3$; mol wt 301.39. C 71.73%, H 7.69%, N 4.65%, O 15.93%. Differs from dihydrocodeine only by the spatial arrangement of the —CHOH— group. Prepn by catalytic reduction of isocodeine: Speyer, Krauss, *Ann.* **432**, 233 (1923); Lutz, Small, *J. Am. Chem. Soc.* **54**, 4724 (1932); Rapoport, Payne, *J. Org. Chem.* **15**, 1097 (1950); by epimerization of dihydrocodeine: Baizer, US 2774762 (1956 to N.Y. Quinine and Chem. Works).

Prisms from ethanol, mp 199-200°.

Acid tartrate. $C_{18}H_{23}NO_3.C_4H_6O_6$. Crystals, mp 192°, $[\alpha]_D^{29}$ −65.3°.

Acid tartrate trihydrate. $C_{18}H_{23}NO_3.C_4H_6O_6.3H_2O$. Crystals from water, mp about 180°. $[\alpha]_D^{26}$ −62.4° (c = 1.94). Soly at 24° = 4.5 g/100 ml H_2O.

Methiodide. $C_{18}H_{23}NO_3.CH_3I$. Crystals from ethanol, dec 269-272°.

3194. Dihydromorphine. [509-60-4] (5α,6α)-4,5-Epoxy-17-methylmorphinan-3,6-diol. $C_{17}H_{21}NO_3$; mol wt 287.36. C 71.06%, H 7.37%, N 4.87%, O 16.70%. Opioid analgesic; active metabolite of dihydrocodeine, *q.v.*, and precursor in the manuf of hydromorphone, *q.v.* Prepd by hydrogenation of morphine: L. Oldenberg, *Ber.* **44**, 1829 (1911); N. B. Eddy, J. G. Reid *J. Pharmacol.* **52**, 468 (1934). Prepn from neopine: L. Small, *J. Org. Chem.* **12**, 359 (1947). High yield synthesis from tetrahydrothebaine: A. K. Przybyl *et al.*, *ibid.* **68**, 2010 (2003). Comparative pharmacology: A.-K. Gilbert *et al.*, *Eur. J. Pharmacol.* **492**, 123 (2004).

Monohydrate. White crystals from alcohol, mp 155-157°. Insoluble in water. Sol in acetone, alcohol, chloroform, dil acids.

Hydrochloride. [1421-28-9] Paramorphan; paramorfan. $C_{17}H_{21}NO_3.HCl$. Prismatic crystals, unmelted at 280° *in vacuo*. $[\alpha]_D^{25}$ −112° (c = 1.6). Very sol in water; sparingly sol in abs alcohol.

Note: This is a controlled substance (opium derivative): **21 CFR, 1308.11.**

3195. Dihydropyran. [110-87-2] 3,4-Dihydro-2H-pyran; Δ^2-dihydropyran; DHP. C_5H_8O; mol wt 84.12. C 71.39%, H 9.59%, O 19.02%. Reagent for the introduction of the tetrahydropyran group to protect alcohols. Prepn: R. Paul, *Bull. Soc. Chim. Fr.* **1933**, 1489; R. L. Sawyer, D. W. Andrus, *Org. Synth.* **coll. vol. III**, 276 (1955). Synthetic utility in the tetrahydropyranylation of hydroxyl groups: J. H. van Boom, J. D. M. Herschied, *Synthesis* **1973**, 169; N. Miyashita *et al.*, *J. Org. Chem.* **42**, 3772 (1977); Y. Morizawa *et al.*, *Synthesis* **1981**, 899; N. Ravindranath *et al.*, *Synlett* **2001**, 1777.

Colorless liquid. *Flammable, irritant.* bp_{760} 85.4-85.6°; bp_{742} 84.4°. d_4^{20} 0.9261; d_{15}^{19} 0.922. n_D^{20} 1.4420; n_D^{19} 1.4402. Flash pt, closed cup: 3°F (−16°C). Sol in water, ethanol. Difficult to dry; often contains trace quantities of water.

USE: Reagent in synthetic organic chemistry.

3196. Dihydroresorcinol. [504-02-9] 1,3-Cyclohexanedione; hydroresorcinol. $C_6H_8O_2$; mol wt 112.13. C 64.27%, H 7.19%, O 28.54%.

Crystals, mp about 105° with decompn. Sol in water, alcohol, chloroform, acetone, boiling benzene; slightly sol in ether, carbon disulfide, petr ether.

3197. Dihydrostreptomycin. [128-46-1] O-2-Deoxy-2-(methylamino)-α-L-glucopyranosyl-(1 → 2)-O-5-deoxy-3-C-(hydroxymethyl)-α-L-lyxofuranosyl-(1 → 4)-N^1,N^3-bis(aminoimino-methyl)-D-streptamine; DHSM; DST; Abiocine; Vibriomycin. C_{21}-

$H_{41}N_7O_{12}$; mol wt 583.60. C 43.22%, H 7.08%, N 16.80%, O 32.90%. Semi-synthetic antibiotic prepd by reduction of streptomycin: Bartz *et al.*, *J. Am. Chem. Soc.* **68**, 2163 (1946); Fried, Wintersteiner, *ibid.* **69**, 79 (1947); Peck, US 2498574 (1950 to Merck & Co.); Carboni, Regna, US 2522858 (1950 to Pfizer); Levy, US 2663685 (1953 to Schenley); Dolliver, Semenoff, US 2717236 (1955 to Olin Mathieson); Kaplan, US 2790792 (1957 to Bristol); Sokol, Popino, US 2784181 (1957 to Cyanamid); Jurist, US 2945850 (1960 to Olin Mathieson). Isoln from fermentation broth of *Streptomyces humidus:* Nakazawa *et al.*, and Tatsuoka *et al.*, US 2931756 and US 2950277 (both 1960 to Takeda). Crystn of free base (obtained from the sulfate): Rhodehamel *et al.*, *Science* **111**, 233 (1950). Crystn of the hydrochloride: Wolf *et al.*, *ibid.* **109**, 515 (1949); Wolf, US 2590139; US 2594245 (both 1952 to Merck & Co.). Crystn of the sulfate: Solomons, Regna, *Science* **109**, 515 (1949); Wolf *et al.*, *ibid.* 515; Wolf, US 2590140; US 2590141 (both 1952 to Merck & Co.); R. B. Peet, US 2640054 (1953 to Heyden Chem.); Katz, US 2744892 (1956 to Schenley). Total synthesis: Umezawa *et al.*, *J. Am. Chem. Soc.* **96**, 920 (1974); *eidem, Bull. Chem. Soc. Jpn.* **48**, 563 (1975); T. Yamasaki *et al.*, *J. Antibiot.* **31**, 1233 (1978).

Hydrate. Crystals from aq acetone. Chars at 240°, turning black up to 300° without melting.

Trihydrochloride. [6533-54-6] $C_{21}H_{41}N_7O_{12}\cdot3HCl$. Amorphous solid, dec 190-195°, or crystals from methanol. Soly in methanol at 25°: 45 mg/ml for crystalline form; >1 g/ml for amorphous form. Crystals contain methanol of crystn which is lost on heating at 100°C. $[\alpha]_D^{25}$ −95° (1% soln).

Sesquisulfate. [5490-27-7] Didromycine; Double-mycin; Sol-Mycin; Streptomagma. $(C_{21}H_{41}N_7O_{12})_2\cdot3H_2SO_4$; mol wt 1461.41. Amorphous solid; crystalline form from water + (methanol, methyl ethyl ketone, or other low boiling solvent). Crystals, dec 255-265°, also reported as dec 250°. $[\alpha]_D^{25}$ −88.5° (1% soln). Crystals show very little hygroscopicity in contrast to amorphous form. Both forms are very sol in water. Solubility in 50% methanol + water: 0.8 mg/ml for crystals; 100 mg/ml for amorphous form. Soly at 28° (mg/ml): >20 in water; 0.35 in methanol; 0.10 in ethanol. Practically insol in acetone, chloroform. *See also:* Weiss *et al.*, *Antibiot. Chemother.* **7**, 374 (1957). Each milligram contains 800 micrograms of dihydrostreptomycin base on a potency basis.

Pantothenate. [3563-84-6] Didrothenat; Pantostrep. $C_{21}H_{41}\text{-}N_7O_{12}\cdot C_9H_{17}NO_5$; mol wt 802.83.

THERAP CAT: Antibacterial (tuberculostatic).

THERAP CAT (VET): Antibacterial.

3198.   Dihydrotachysterol. [67-96-9] (1*S*,3*E*,4*S*)-4-Methyl-3-[(2*E*)-2-[(1*R*,3a*S*,7a*R*)-octahydro-7a-methyl-1-[(1*R*,2*E*,4*R*)-1,4,5-trimethyl-2-hexen-1-yl]-4*H*-onden-4-ylidene]ethylidene]cyclohexanol; (3β,5*E*,7*E*,10α,22*E*)-9,10-secoergosta-5,7,22-trien-3-ol; dichystrolum; anti-tetany substance 10; AT 10; Antitanil; Calcamine; Dygratyl; Hytakerol; Parterol; Tachyrol. $C_{28}H_{46}O$; mol wt 398.68. C 84.36%, H 11.63%, O 4.01%. Prepn by reduction of tachysterol: Windaus *et al.*, *Ann.* **499**, 198 (1932); v. Werder, US

2228491 (1941 to Winthrop); *Z. Physiol. Chem.* **260**, 119 (1939). Activity similar to parathyroid extract.

Needles from 90% methanol, mp 125-127°. $[\alpha]_D^{22}$ +97.5° (chloroform). uv max: 242, 251, 261 nm ($E_{1cm}^{1\%}$ 870, 1010, 650). Freely sol in ether, chloroform; easily sol in organic solvents; sol in alc; sparingly sol in vegetable oils. Insol in water.

THERAP CAT: Calcium regulator.

3199.   Dihydrothebaine. [561-25-1] (5α)-6,7-Didehydro-4,5-epoxy-3,6-dimethoxy-17-methylmorphinan. $C_{19}H_{23}NO_3$; mol wt 313.40. C 72.82%, H 7.40%, N 4.47%, O 15.31%. Prepd by reduction of thebaine: Small *et al.*, *J. Am. Chem. Soc.* **58**, 1457 (1936); K. W. Bentley, *The Chemistry of the Morphine Alkaloids* (Oxford, 1954) p 204.

Prisms from ethyl acetate, mp 162-163°. $[\alpha]_D^{20}$ −267° (c = 1.02 in benzene). Sol in alcohol, benzene, ethyl acetate. Insol in water, alkalies.

Methyl iodide. $C_{20}H_{23}NO_4\cdot CH_3I$. Rods from alc, mp 257°.

3200.   Dihydroxyacetone. [96-26-4] 1,3-Dihydroxy-2-propanone; 1,3-dihydroxydimethyl ketone; Protosol; Ketochromin. $C_3\text{-}H_6O_3$; mol wt 90.08. C 40.00%, H 6.71%, O 53.28%. Produced from glycerol by *Acetobacter* sp. under aerobic conditions: Bernhauer, Schoen, *Z. Physiol. Chem.* **177**, 107 (1928); Rutten, *Rec. Trav. Chim.* **70**, 449 (1951); Bousfield *et al.*, *J. Inst. Brew.* **53**, 258 (1947); US 2948658 (1960 to Baxter). *Review: Chem. Eng. News* **38**, 62 (Feb. 22, 1960), 54 (Aug. 15, 1960).

White crystalline powder; fairly hygroscopic; characteristic odor; sweet, cooling taste. mp about 75-80°. The normal form is a dimer and is freely sol in water; sol in alc; sparingly sol in ether. When freshly prepd reverts rapidly to monomer in soln. The monomer is freely sol in water, alcohol, ether, acetone.

USE: Artificial tanning agent; active ingredient in *Man-Tan*; *Oxatone*; *Tanorama*; *Q.T.* (*Quick Tan*); *Tan Tone*; *Magic Tan*. Formulations: Barmak, *Am. Perfum.*, March 1960, p 57; Andreadis, Miklean, US 2949403 (1960).

3201.   Dihydroxyaluminum Aminoacetate. [13682-92-3] (*T*-4)-(Glycinato-κ*N*,κ*O*)dihydroxyaluminum; (glycinato-*N*,*O*)dihydroxyaluminum; dihydroxy(glycinato)aluminum; aluminum aminoacetate (basic); aluminum dihydroxyaminoacetate; aluminum glycinate; glycine, aluminum salt; Ada; Alamine; Alubasine; Alzinox;

Aspogen; Dimothyn; Doraxamin; Elcosal; Robalate. $C_2H_6AlNO_4$; mol wt 135.05. C 17.79%, H 4.48%, Al 19.98%, N 10.37%, O 47.39%. $NH_2CH_2COOAl(OH)_2$. Prepd by adding a soln of aluminum isopropoxide in isopropanol to an aq soln of glycine: Krantz et al., *J. Pharmacol.* **82**, 247 (1944).

Very fine powder. Faintly sweet taste. Sol in dilute mineral acids, sols of fixed alkalies. Insol in organic solvents, water. Mixes easily with water forming suspensions which do not separate readily. At 25° the pH of the aq suspension (1 in 25) is 7.4. One gram consumes 200 ml of 0.1N HCl under conditions of the pharmacopoeial test for aluminum hydroxide gel.

THERAP CAT: Antacid.

3202. Dihydroxyaluminum Sodium Carbonate. [12011-77-7] [Carbonato(2−)-κO]dihydroxyaluminate(1−) sodium (1:1); aluminum sodium carbonate hydroxide; Kompensan; Minicid. CH_2Al-NaO_5; mol wt 143.99. C 8.34%, H 1.40%, Al 18.74%, Na 15.97%, O 55.56%. $(HO)_2AlOCO_2Na$. Prepd by the reaction between an aluminum alkoxide and $NaHCO_3$ in water: Grote, US 2783179 (1957 to Chattanooga Medicine).

Amorphous powder or poorly formed crystals. May contain 10 to 11% H_2O. d 2.144. pH of water suspension: 9.7. Acid consuming power: 230 ml 0.1N HCl/gram. Sol in dilute mineral acids with the evolution of carbon dioxide. Practically insol in water, organic solvents.

THERAP CAT: Antacid.

3203. Dihydroxymaleic Acid. [526-84-1] (2Z)-2,3-Dihydroxy-2-butenedioic acid; 1,2-dihydroxyethylenedicarboxylic acid. $C_4H_4O_6$; mol wt 148.07. C 32.45%, H 2.72%, O 64.83%. Prepn from tartaric acid: Fenton, *J. Chem. Soc.* **87**, 811 (1905); Nef, *Ann.* **357**, 291 (1907).

Plates from water, usually crystallizes as the dihydrate. The anhydr acid dec 155° without melting. Slightly sol in cold water, ether, acetic acid; more sol in alcohol. Aq solns are unstable.

USE: In the detection of fluorides and of titanium. Has been proposed as antioxidant for frozen foods.

3204. 9,10-Dihydroxystearic Acid. [120-87-6] 9,10-Dihydroxyoctadecanoic acid. $C_{18}H_{36}O_4$; mol wt 316.48. C 68.31%, H 11.47%, O 20.22%.

White, odorless, tasteless, lustrous crystals; fatty feel. mp 132-136°. Insol in water. Sol in hot alc or acetone, slightly in ether.

USE: Manuf cosmetic and toilet preparations.

3205. Dihydroxytartaric Acid. [76-30-2] 2,2,3,3-Tetrahydroxybutanedioic acid; diketosuccinic acid. $C_4H_6O_8$; mol wt 182.08. C 26.39%, H 3.32%, O 70.29%.

White, cryst powder. mp ~114-115° with decompn. Very sol in water; the aq soln dec on heating.

USE: As a reagent for the determination of sodium.

3206. 2,4-Diiodoaniline. [533-70-0] 2,4-Diiodobenzenamine. $C_6H_5I_2N$; mol wt 344.92. C 20.89%, H 1.46%, I 73.58%, N 4.06%. Prepn: C. Rudolph, *Ber.* **11**, 78 (1878); S. Z. Vatsadze et al., *Russ. Chem. Bull. Int. Ed.* **53**, 471 (2004). Crystal structure: G. Smith, U. D. Wermuth, *Acta Crystallogr. E* **65**, o2108 (2009).

Colorless prisms or needles. d 2.75. mp 95-96°. Slightly sol in cold water, moderately in hot water or in alcohol; freely sol in chloroform, ether, acetone, carbon disulfide, boiling alc.

3207. 4′,5′-Diiodofluorescein. [38577-97-8] 3′,6′-Dihydroxy-4′,5′-diiodospiro[isobenzofuran-1(3H),9′-[9H]xanthen]-3-one; hydroxydiiodo-*o*-carboxyphenylfluorone; D & C Orange No. 10; C.I. Solvent Red 73; C.I. 45425:1. $C_{20}H_{10}I_2O_5$; mol wt 584.10. C 41.13%, H 1.73%, I 43.45%, O 13.70%. Prepd by reaction of fluorescein with I_2 and HIO_4 or ICl in alkali: *Colour Index* **vol. 4** (3rd ed., 1971) p 4427.

Orange-red powder. Slightly sol in water; sol in alcohol, in alkali hydroxide solns.

Sodium salt. [33239-19-9] Erythrosine Extra Yellowish; D & C Orange No. 11; C.I. Acid Red 95; C.I. 45425. $C_{20}H_8I_2Na_2O_5$; mol wt 628.07.

USE: As an adsorption indicator in the determn of iodides in the presence of chlorides and bromides. The salts in dyeing and printing cotton, in printing half-silk, in dyeing jute, straw, etc.

3208. 3,5-Diiodosalicylic Acid. [133-91-5] 2-Hydroxy-3,5-diiodobenzoic acid. $C_7H_4I_2O_3$; mol wt 389.91. C 21.56%, H 1.03%, I 65.09%, O 12.31%. Prepn from salicylic acid and ICl: Woollett, Johnson, *Org. Synth.* **coll. vol. II**, 343 (1943); and iodine + H_2O_2: Jurd, *Aust. J. Sci. Res.* **3A**, 587 (1950).

Colorless, odorless needles or slightly yellow cryst powder. Sweetish, bitter taste. mp 235-236° dec. Sol in 5200 parts water at 25°; freely sol in alcohol, ether and most other organic solvents. Practically insol in $CHCl_3$, benzene.

Bismuth salt. Bijosal. Pollano, *Minerva Med.* **1930** I, 786, *C.A.* **24**, 4854⁷ (1930).

Ethyl ester. Ethyl 3,5-diiodosalicylate. $C_9H_8I_2O_3$; mol wt 404.93. Crystals, mp 133°. Practically insol in H_2O. Sol in alc, ether, benzene.

USE: As I_2 source in foods; in animal feeds as growth promotant.

3209. 3,5-Diiodothyronine. [534-51-0] O-(4-Hydroxyphenyl)-3,5-diiodotyrosine; 3-[4-(p-hydroxyphenoxy)-3,5-diiodophenyl]alanine. $C_{15}H_{13}I_2NO_4$; mol wt 525.08. C 34.31%, H 2.50%, I 48.34%, N 2.67%, O 12.19%. Prepn of DL-form: Harington, McCartney, *Biochem. J.* **21**, 852 (1927); Borrows et al., *J. Chem. Soc.* **1949**, Suppl. Issue No. 1, S199, S204; Siedel et al., US 2894977 (1959 to Hoechst). Prepn of L(+)-form: Harington, *Biochem. J.* **22**, 1429 (1928); Chambers et al., *J. Chem. Soc.* **1949**, 3424; Hillmann, US 2886592 (1959); as the hydrochloride: Anthony, US 2950315 (1960 to Baxter); Meltzer, US 3102136 (1963 to Warner-Lambert). Prepn of D(−)-form: Harington, *loc. cit.*; Elks, Waller, *J. Chem. Soc.* **1952**, 2366.

DL-Form. Plates, dec 256-257°.

L(+)-Form. Crystals, dec 256°. $[\alpha]_D^{22}$ +26° (c = 1.06 in a 1:2 mixture of 1*N* HCl + alcohol).

L(+)-Form hydrochloride. Crystals, mp 235-240°.

D(−)-Form. Crystals, dec 256° (Harington), 265° (Elks, Waller). $[\alpha]_D^{20}$ −27.1° (c = 1 in 1*N* alc HCl:alc, 1:2 by vol).

USE: An intermediate in manuf of thyroxine.

3210. 3,5-Diiodotyrosine. [66-02-4] 3,5-Diiodo-4-hydroxy-β-phenylalanine; iodogorgoic acid; Agontan. $C_9H_9I_2NO_3$; mol wt 432.98. C 24.97%, H 2.10%, I 58.62%, N 3.24%, O 11.09%. Found in the skeletal proteins of corals, sponges, and other marine organisms. Isoln from *Gorgonia cavollini* as the DL acid: Drechsel, *Z. Biol.* **33**, 99; *Jahresber. Tierchemie* **1896**, 574; Henze, *Z. Physiol. Chem.* **38**, 71 (1903); isoln from the common sponge: Wheeler, Mendel, *J. Biol. Chem.* **7**, 1 (1909-10); Low, *J. Mar. Res.* **10**, 239 (1951); from marine algae: Coulson, *Chem. Ind. (London)* **1953**, 997; prepn from L- or DL-tyrosine: Henze, *Z. Physiol. Chem.* **51**, 67 (1907); Borrows *et al.*, *J. Chem. Soc.* **1949**, Suppl. Issue no. 1, S185; Jurd, *J. Am. Chem. Soc.* **77**, 5747 (1955); Boyle, Zlatkis, **US 2835700** (1958 to Basic, Inc.).

L-Form. Bunches of needles from water or 70% alcohol, dec 213°. $[\alpha]_D^{20}$ +2.89° (0.246 g in 5 g 4% HCl); $[\alpha]_D^{20}$ +2.27° (0.227 g in 5 g 25% NH₃). pK₁ 2.12; pK₂ 6.48; pK₃ 7.82. Soly in water (g/l): at 0° = 0.204; at 25° = 0.617; at 50° = 1.862; at 75° = 5.62; at 100° = 17.00. On boiling with dil alc the crystals swell, and after prolonged boiling a gelatinous precipitate is formed.

DL-Form. Double wedge crystals from 70% alc, rectangular plates from water. Dec about 200°. Soly in water (g/l): at 0° = 0.149; at 25° = 0.340; at 50° = 0.773. On cooling a boiled hydroalcoholic soln, no gelatinous precip is formed.

THERAP CAT: Thyroid inhibitor.

3211. Diisoamylamine. [544-00-3] 3-Methyl-*N*-(3-methylbutyl)-1-butanamine; isodiamylamine. $C_{10}H_{23}N$; mol wt 157.30. C 76.36%, H 14.74%, N 8.90%.

Liquid. d_4^{20} 0.767. bp 188°. mp −44°. n_D^{21} 1.4229. Slightly sol in water; sol in alcohol, chloroform, ether.

Hydrochloride. [543-99-7] $C_{10}H_{23}N.HCl$; mol wt 193.76. Vitreous mass, mp 276°. Freely sol in water, alcohol, ether.

Caution: Irritating to skin, mucous membranes. Has pressor effect.

3212. Diisobutylaluminum Hydride. [1191-15-7] Hydrobis(2-methylpropyl)aluminum; hydrodiisobutylaluminum; DIBAH; DIBAL; DIBAL-H; *i*-Bu₂AlH. $C_8H_{19}Al$; mol wt 142.22. C 67.56%, H 13.47%, Al 18.97%. Organoaluminum reducing agent. Prepn: K. Ziegler, **GB 778098** (1957); L. I. Zakharkin *et al.*, *Izv. Akad. Nauk SSSR Ser. Khim.* **1958**, 100. Initial synthetic uses: K. Ziegler *et al.*, *Angew. Chem.* **67**, 425 (1955); G. Wilke, H. Müller, *Ber.* **89**, 444 (1956). Functional group reduction studies: N. M. Yoon, Y. S. Gyoung, *J. Org. Chem.* **50**, 2443 (1985). Stereoselective epoxide reduction: R. S. Lenox, J. A. Katzenellenbogen, *J. Am. Chem. Soc.* **95**, 957 (1973). *Reviews:* E. Winterfeldt, *Synthesis* **1975**, 617-630; P. Galatsis in *Encyclopedia of Reagents for Organic*

Synthesis **3**, L. A. Paquette, Ed. (Wiley, New York, 1995) pp 1908-1912.

Colorless liquid. *Flammable. Corrosive.* bp₁ 116-118°; bp₀.₂ 95.5-96.5°. Flash point, closed cup: −1°F (−18°C). Sol in aprotic solvents. Reacts violently with water.

USE: Reagent in synthetic organic chemistry.

3213. Diisobutyl Sodium Sulfosuccinate. [127-39-9] 2-Sulfobutanedioic acid 1,4-bis(2-methylpropyl) ester sodium salt (1:1); sulfosuccinic acid diisobutyl ester *S*-sodium salt; Aerosol IB; Alphasol IB. $C_{12}H_{21}NaO_7S$; mol wt 332.34. C 43.37%, H 6.37%, Na 6.92%, O 33.70%, S 9.65%. The isobutyl or butyl or 1-methylpropyl diester of the monosodium salt of sulfosuccinic acid or a mixture of all three esters. Wetting agent prepd by the action of the appropriate alcohols on maleic anhydride followed by addition of sodium bisulfite: Jaeger, **US 2028091**; **US 2176423**; **GB 446568**; **FR 776495**.

The mixture of the three esters is available as a white, powder-like, easily grindable material. Soly in water at 25°: 760 g/l; at 60°: 804 g/l. Maximum concn of electrolyte soln in which 1% of the wetting agent is sol: 20% NaCl; 20% NH₄Cl; 30% NaNO₃; 20% Na₂SO₄. Also sol in glycerol, pine oil, oleic acid. Insol in acetone, kerosene, liquid petrolatum, carbon tetrachloride, 2B ethanol, benzene, olive oil. Surface tension (dynes/cm) in water: 0.001% = 72.2; 0.1% = 67.5; 0.5% = 53.4; 1% = 49.1. Interfacial tension 1% in water *vs.* liquid petrolatum: 10 seconds = 32.5 dynes/cm; 30 seconds = 32.3 dynes/cm; 15 minutes = 31.2 dynes/cm. Stable in acid and neutral solns; hydrolyzes in alkaline solns.

Caution: Irritating to eyes, mucous membranes.

USE: Wetting agent.

3214. Diisopropanolamine. [110-97-4] 1,1'-Iminobis-2-propanol; bis(2-hydroxypropyl)amine; DIPA. $C_6H_{15}NO_2$; mol wt 133.19. C 54.11%, H 11.35%, N 10.52%, O 24.02%. Secondary alkanolamine. Prepn: J. N. Wickert, **US 1988225** (1935 to Carbide & Carbon Chem.). Densities, viscosities, diffusivities: H. Hikita *et al.*, *J. Chem. Eng. Jpn.* **14**, 411 (1981). Absorption and mass transfer study: P. M. M. Blauwhoff, W. P. M. Van Swaaij, *ACS Symp. Ser.* **196**, 377 (1982). Safety assessment and use: K. H. Beyer *et al.*, *J. Am. Coll. Toxicol.* **6**, 53 (1987). Review of use in 'Sulfinol' process: K. E. Zarker, *Fert. Sci. Technol. Ser.* **2**, 219-232 (1974); in gas purification: K. F. Butwell *et al.*, *Hydrocarbon Process. Int. Ed.* **61**, 109-116 (1982).

White, waxy solid, mp 32-42°. bp 249°. Freezing pt 107°F. Flash point (open cup) 255°F (127°C). Critical density 0.31 g/cc. d_{20}^{45} 0.9890. n^{60} 1.4450-1.4550. pH of 5% aq soln: 11.5. Completely miscible with water and alcohol, slightly sol in toluene. Insol in hydrocarbons.

USE: Emulsifying and neutralizing agent in cosmetics. Removal of H_2S and CO_2 from natural and industrial gases. Antimicrobial agent in cutting fluids.

3215. Diisopropylamine. [108-18-9] *N*-(1-Methylethyl)-2-propanamine. $C_6H_{15}N$; mol wt 101.19. C 71.22%, H 14.94%, N 13.84%. Prepn: A. Siersch, *Ann.* **148**, 263 (1868); M. Van der Zande, *Rec. Trav. Chim.* **8**, 202 (1889). Enthalpy of formation: M. A. V. Ribeiro da Silva *et al.*, *J. Chem. Thermodyn.* **29**, 1025 (1997). Toxicity study: H. F. Smyth *et al.*, *Arch. Ind. Hyg. Occup. Med.* **10**,

61 (1954). Safety assessment: *J. Am. Coll. Toxicol.* **14**, 182-192 (1995).

Liquid; characteristic odor; strongly alkaline. d^{22} 0.722. bp 84°. Flash pt, open cup: 21°F (-6°C). Sol in water, alc. LD_{50} orally in rats: 0.77 g/kg (Smyth).

Lithium salt. [4111-54-0] LDA; lithiodiisopropylamine; lithium diisopropylamide. Strong nucleophile. Crystal structure: N. D. R. Barnett *et al. J. Am. Chem. Soc.* **113**, 8187 (1991). Use as nucleophile: Y. Tanaka *et al., Bull. Chem. Soc. Jpn.* **60**, 788 (1987); S. Saito *et al., J. Am. Chem. Soc.* **119**, 611 (1997). Review of chemistry: *Reagents for Organic Synthesis* vol. 8, M. Fieser, Ed. (Wiley-Interscience, New York, 1980) pp 292-299. Powder, melts with decompn. Air and moisture sensitive.

Caution: Potential symptoms of overexposure are irritation of eyes, skin and respiratory system; nausea, vomiting; headache; visual disturbance. *See NIOSH Pocket Guide to Chemical Hazards* (DHHS/NIOSH 97-140, 1997) p 110.

USE: pH adjuster in colognes and toilet waters. In organic synthesis, particularly, the lithium salt.

3216. Diisopropylamine Dichloroacetate. [660-27-5] 2,2-Dichloroacetic acid compd with *N*-(1-methylethyl)-2-propanamine (1:1); dichloroacetic acid diisopropylammonium salt; diisopropyl-ammonium dichloroacetate; diisopropylamine dichloroethanoate; DADA; DIPA-DCA; DIEDI; IS-401; Disotat; Kalodil; Oxypangam. $C_8H_{17}Cl_2NO_2$; mol wt 230.13. C 41.75%, H 7.45%, Cl 30.81%, N 6.09%, O 13.90%. Prepn: **GB 862248** (1961 to Italseber). Pharmacology: V. A. E. Kraushaar *et al., Arzneim.-Forsch.* **13**, 109 (1963). Ames mutagenicity test: M. D. Gelernt, V. Herbert, *Nutr. Cancer* **3**, 129 (1982). Pharmacokinetics and TLC determn in racehorses: J.-M. Yang *et al., Gen. Pharmacol.* **19**, 683 (1988). Review of pharmacology: P. W. Stacpoole, *J. Clin. Pharmacol. J. New Drugs* **9**, 282-291 (1969).

Crystals, mp 119-121°. Soly in water: >50%. LD_{50} orally in mice: 1700 mg/kg (Kraushaar).

THERAP CAT: Vasodilator, hypotensive.

3217. Diisopropyl Azodicarboxylate. [2446-83-5] 1,2-Di-azenedicarboxylic acid 1,2-bis(1-methylethyl) ester; diisopropyl azodiformate; DIAD. $C_8H_{14}N_2O_4$; mol wt 202.21. C 47.52%, H 6.98%, N 13.85%, O 31.65%. Alternative reagent to diethyl azodicarboxylate, *q.v.,* for the Mitsunobu reaction; also applied in other synthetic transformations. Prepn: **NL 6408826**; **GB 1012264** (both 1965 to Wallace & Tiernan Inc.); S. A. Rodkin *et al., J. Org. Chem. USSR (Engl. Transl.)* **7**, 2361 (1971). Synthetic applications: D. L. Hughes, R. A. Reamer, *J. Org. Chem.* **61**, 2967 (1996); M. Shi, G.-L. Zhao, *Tetrahedron* **60**, 2083 (2004); L.-X. Shao, M. Shi, *Eur. J. Org. Chem.* **2004**, 426; J. Kroutil *et al., Synthesis* **2004**, 446. Reviews: S. K. Nune, *Synlett* **2003**, 1221-1222; A. E. Kümmerle, *ibid.* **2008**, 2723-2724.

Orange liquid. bp_{10} 109.5°; $bp_{0.25}$ 75.5°. d_4^{20} 1.0400. n_D^{25} 1.4180. *Irritant.* Flash pt, closed cup: 222.8°F (106°C). Decomposes vigorously above 100°.

USE: Reagent in organic chemistry.

3218. *N,N'*-Diisopropylcarbodiimide. [693-13-0] *N,N'*-Methanetetraylbis-2-propanamine; 1,3-diisopropylcarbodiimide;

DIPC; DIC. $C_7H_{14}N_2$; mol wt 126.20. C 66.62%, H 11.18%, N 22.20%. Coupling reagent used to form amide bonds in peptide synthesis; also utilized in the prepn of various heterocycles. Prepn: E. Schmidt, W. Striewsky, *Ber.* **74**, 1285 (1941). Improved prepn: W. R. Ruby, **US 3201463** (1965 to General Aniline and Film). Synthetic utility in formation of peptide bonds: J. Izdebski *et al., Pol. J. Chem.* **54**, 413 (1980); L. A. Carpino, A. El-Faham, *Tetrahedron* **55**, 6813 (1999); in heterocyclic chemistry: B. J. Brown *et al., Synlett* **2000**, 131; S. Crosignani *et al., Tetrahedron Lett.* **45**, 9611 (2004).

Colorless liquid. *Poisonous, flammable, irritant, skin and respiratory sensitizer.* bp 144-144.8°; bp_{15} 44°; bp_{10} 36-37°. d^{25} 0.909. Flash pt, closed cup: 88°F (31°C). Moisture sensitive. Store in cool place under inert gas.

USE: Reagent in synthetic organic chemistry.

3219. Diisopropylethylamine. [7087-68-5] *N*-Ethyl-*N*-(1-methylethyl)-2-propanamine; *N,N*-diisopropylethylamine; ethyldi-isopropylamine; 1,1'-dimethyltriethylamine; Hünig's base; DIPEA; DIEA; *i*Pr$_2$NEt. $C_8H_{19}N$; mol wt 129.25. C 74.34%, H 14.82%, N 10.84%. Sterically hindered, non-nucleophilic organic base. Prepn: R. A. Robinson, *J. Org. Chem.* **16**, 1911 (1951); and use in alkylation and dehydration reactions: S. Hünig, M. Kiessel, *Ber.* **91**, 380 (1958). Synthetic applications: D. A. Evans *et al., J. Am. Chem. Soc.* **103**, 2127 (1981); M. Beyermann *et al., Int. J. Pept. Protein Res.* **37**, 252 (1991); S.-I. Murahashi *et al., J. Org. Chem.* **58**, 1538 (1993); T. Bach, H. Brummerhop, *J. Prakt. Chem.* **341**, 410 (1999); S. A. Reed *et al., J. Am. Chem. Soc.* **131**, 11701 (2009). *Review*: K. L. Sorgi in *Encyclopedia of Reagents for Organic Synthesis* **3**, L. A. Paquette, Ed. (Wiley, New York, 1995) pp 1933-1936.

Colorless liquid. *Flammable. Corrosive. Poisonous.* bp_{760} 126.5°; bp_{731} 119°. d_4^{27} 0.751. n_D^{25} 1.4121. Flash pt, closed cup: 51°F (10.6°C). Sol in most organic solvents.

USE: Reagent and catalyst in synthetic organic chemistry.

3220. Dikegulac. [18467-77-1] 2,3:4,6-Bis-*O*-(1-methyleth-ylidene)-α-L-*xylo*-2-hexulofuranosonic acid; α-2,3:4,6-di-*O*-isopro-pylidene-L-*xylo*-hexulofuranosonic acid; di-*O*-isopropylidene-2-keto-L-gulonic acid; diacetone-2-ketogulonic acid; diacetone-2-oxo-L-gulonic acid; oxogulonic acid diacetonide. $C_{12}H_{18}O_7$; mol wt 274.27. C 52.55%, H 6.62%, O 40.83%. Intermediate in the manuf of ascorbic acid, *q.v.* Manufacturing process: G. M. Jaffe, E. J. Pleven, **DE 2123621**; *eidem,* **US 3832355** (1970, 1974 both to Hoffmann-La Roche). Use as a plant growth regulator: W. Szkrybalo, **DE 2339239**; *eidem,* **US 4337080** (1974, 1982 both to Hoffmann-La Roche); P. Bocian *et al., Nature* **258**, 142 (1975). Physicochemical properties, toxicity and growth retardant effect: W. H. de Silva *et al., Proc. Br. Crop Prot. Conf. - Weeds* **1976**, 349. Activity: S. S. Purohit, *Comp. Physiol. Ecol.* **4**, 264 (1979); **6**, 261 (1981).

Sodium salt. Dikegulac sodium; Ro-7-6145; Atrinal. $C_{12}H_{17}$-NaO_7; mol wt 296.25. Powder, mp >300°. Vapor pressure at 25°: <10^{-10} mm Hg. Soly at 20° (g/l): water 590; methanol 390; ethanol 230; chloroform 60; acetone <10; hexane <10; cyclohexanone <10. LD_{50} in mice, male, female rats (mg/kg): 19500, 31000, 18000 orally; LC_{50} (96 hr) in bluegill sunfish, rainbow trout (ppm): >10000, >5000 (de Silva).

USE: Plant growth regulator; herbicide.

3221. Diketene. [674-82-8] 4-Methylene-2-oxetanone; 3-buteno-β-lactone. $C_4H_4O_2$; mol wt 84.07. C 57.15%, H 4.80%, O 38.06%. Reactive acetoacetylation reagent; also used to introduce functionalized carbon units onto organic compds. Prepn from dimerization of ketene, *q.v.*: F. Chick, N. T. M. Wilsmore, *J. Chem. Soc., Trans.* **93**, 946 (1908). Crystal structure and confirmation of molecular structure: L. Katz, W. N. Lipscomb, *Acta Crystallogr.* **5**, 313 (1952). Process of cracking into ketene: S. Andreades, H. D. Carlson, *Org. Synth.* **coll. vol. V**, 679 (1973). Synthetic applications: S. I. Zavialov *et al., Tetrahedron* **22**, 2003 (1966); A. Nudelman *et al., Synthesis* **1989**, 387. *Reviews*: R. J. Clemens, *Chem. Rev.* **86**, 241-318 (1986); N. Zohreh, *Synlett* **2008**, 1913-1914.

Colorless liquid with pungent odor; turns brownish-yellow upon standing at room temp. bp_{760} 126-127° with slight decompn; bp_{100} 69-71°. mp −7.5°. d_4^{18} 1.0939; d_4^{23} 1.0905. n_D^{23} 1.4342. *Poisonous. Flammable. Severe lachrymator.* Flash pt, closed cup: 93.2°F (34°C). Miscible with most organic solvents. Immiscible with hexane.

USE: Reagent in synthetic organic chemistry.

3222. Dilazep. [35898-87-4] 3,4,5-Trimethoxybenzoic acid 1,1'-[(tetrahydro-1*H*-1,4-diazepine-1,4(5*H*)-diyl)di-3,1-propanediyl] ester; 3,4,5-trimethoxybenzoic acid diester with tetrahydro-1*H*-1,4-diazepine-1,4(5*H*)-dipropanol; 1,4-bis[3-(3,4,5-trimethoxybenzoyloxy)propyl]perhydro-1,4-diazepine; *N,N'*-bis[3-(3,4,5-trimethoxybenzoyloxy)propyl]homopiperazine; *N,N'*-(bis-ω-hydroxypropyl)homopiperazine 3,4,5-trimethoxybenzoate (diester). $C_{31}H_{44}N_2$-O_{10}; mol wt 604.70. C 61.57%, H 7.33%, N 4.63%, O 26.46%. Prepn: **GB 1107470**; H. Arnold *et al., US 3532685** (1968, 1970 both to Asta-Werke). Series of articles on pharmacology and metabolism: *Arzneim.-Forsch.* **22**, 639-666 (1972). Toxicology: H. H. Abel *et al., ibid.* 667. Clinical results: Messerich, *Med. Welt* **1972**, 563.

Dihydrochloride. [20153-98-4] Asta C 4898; Comelian; Cormelian; Labitan. $C_{31}H_{44}N_2O_{10}$·2HCl; mol wt 677.61. Crystals from ethanol, mp 194-198° (monohydrate). LD_{50} in male mice, male rats (mg/kg): 26.6, 19.1 i.v.; 161, 90.1 i.p.; 3740, >2150 orally (Abel).

THERAP CAT: Vasodilator (coronary).

3223. Diloxanide. [579-38-4] 2,2-Dichloro-*N*-(4-hydroxyphenyl)-*N*-methylacetamide; 2,2-dichloro-4'-hydroxy-*N*-methylacetanilide; *N*-dichloroacet-4-hydroxy-*N*-methylanilide; 4-hydroxy-*N*-methyldichloroacetanilide. $C_9H_9Cl_2NO_2$; mol wt 234.08. C 46.18%, H 3.88%, Cl 30.29%, N 5.98%, O 13.67%. Prepn: P. Oxley *et al., US 2912438** (1959 to Boots Pure Drug). Spectrophotometric determn in 2-component tablet formulation: S. M. Galal *et al., J. Pharm. Belg.* **46**, 315 (1991). HPLC determn in bulk drug and formulation with metronidazole: A. Mishal, D. Sober, *J. Pharm. Biomed. Anal.* **39**, 819 (2005). Clinical trial of combination with metronidazole, *q.v.*, in amoebiasis and giardiasis: H. Qureshi *et al.,*

J. Int. Med. Res. **25**, 167 (1997). Review of clinical experience: J. B. McAuley *et al., Clin. Infect. Dis.* **15**, 464-468 (1992).

Crystals from ethyl acetate, mp 175°.

Furoate. [3736-81-0] Diloxanide 2-furoic acid ester; Furamide. White or almost white crystalline powder. Freely sol in chloroform; slightly sol in alc, ether; very slightly sol in water. *Protect from light.*

THERAP CAT: Antiamebic

3224. Diltiazem. [42399-41-7] (2*S*,3*S*)-3-(Acetyloxy)-5-[2-(dimethylamino)ethyl]-2,3-dihydro-2-(4-methoxyphenyl)-1,5-benzothiazepin-4(5*H*)-one; (+)-*cis*-5-[2-(dimethylamino)ethyl]-2,3-dihydro-3-hydroxy-2-(*p*-methoxyphenyl)-1,5-benzothiazepin-4(5*H*)-one acetate (ester). $C_{22}H_{26}N_2O_4S$; mol wt 414.52. C 63.75%, H 6.32%, N 6.76%, O 15.44%, S 7.73%. Calcium channel blocker with vasodilating activity. Prepn (unspec stereochem): H. Kugita *et al., DE 1805714; eidem, US 3562257* (1969, 1971 both to Tanabe Seiyaku); *eidem, Chem. Pharm. Bull.* **19**, 595 (1971). Resolution of optical isomers: H. Inoue *et al., Yakugaku Zasshi* **93**, 729 (1973), *C.A.* **79**, 66331w (1974). Stereospecific synthesis: K. Igarashi, T. Honma, **DE 3415035**; *eidem, US 4552695* (1984, 1985 both to Shionogi). Structure-activity studies: Sato *et al., Arzneim.-Forsch.* **21**, 1338 (1971); T. Nagao *et al., Chem. Pharm. Bull.* **21**, 92 (1973). Pharmacology and toxicity: *eidem, Jpn. J. Pharmacol.* **22**, 467 (1972). Metabolism: Meshi *et al., Chem. Pharm. Bull.* **19**, 1546 (1971). Review of synthesis and pharmacology: H. Inoue, T. Nagao, *Chron. Drug Discovery* **3**, 207-238 (1993). Comprehensive description: D. J. Mazzo *et al., Anal. Profiles Drug Subs. Excip.* **23**, 53-98 (1994). Review of pharmacology and efficacy in angina: M. Chaffman, R. N. Brogden, *Drugs* **29**, 387-454 (1985); in hypertension: M. R. Weir, *J. Clin. Pharmacol.* **35**, 220-232 (1995). Comparative clinical trial in prevention of complications of hypertension: L. Hansson *et al., Lancet* **356**, 359 (2000).

Hydrochloride. [33286-22-5] CRD-401; Adizem; Altiazem; Angizem; Cardizem; Deltazen; Dilacor XR; Diladel; Dilpral; Dilrene; Dilzem; Dilzene; Masdil; Tiazac; Tildiem; Zilden. $C_{22}H_{26}N_2$-O_4S·HCl; mol wt 450.98. Fine needles from ethanol-isopropanol, mp 207.5-212°. Odorless, with a bitter taste. $[\alpha]_D^{24}$ +98.3 ± 1.4° (c = 1.002 in methanol). Freely sol in water, methanol, chloroform, formic acid; sparingly sol in dehydrated alc. Practically insol in benzene; insol in ether. LD_{50} in male, female mice, male, female rats (mg/kg): 61, 58, 38, 39 i.v.; 260, 280, 520, 550 s.c.; 740, 640, 560, 610 orally (Nagao, 1972).

THERAP CAT: Antianginal; antihypertensive; antiarrhythmic (class IV).

THERAP CAT (VET): Antianginal; antihypertensive; antiarrhythmic (class IV).

3225. Dimebolin. [3613-73-8] 2,3,4,5-Tetrahydro-2,8-dimethyl-5-[2-(6-methyl-3-pyridinyl)ethyl]-1*H*-pyrido[4,3-*b*]indole; 9-[2-(2-methyl-5-pyridinyl)ethyl]-3,6-dimethyl-1,2,3,4-tetrahydro-γ-carboline; dimebon. $C_{21}H_{25}N_3$; mol wt 319.45. C 78.96%, H 7.89%, N 13.15%. γ-Carboline derivative with antihistaminic and cognition enhancing activity. Prepn: A. N. Kost *et al.*, *Zh. Obshch. Khim.* **32**, 2050 (1962), *C.A.* **58**, 7917d. Pharmacology: S. K. Shadurskaya *et al.*, *Bull. Exp. Biol. Med.* **101**, 780 (1986); S. Bachurin *et al.*, *Ann. N.Y. Acad. Sci.* **939**, 425 (2001). Clinical evaluation in Alzheimer's disease: R. S. Doody *et al.*, *Lancet* **372**, 207 (2008).

Dihydrochloride. [97657-92-6] $C_{21}H_{25}N_3 \cdot 2HCl$; mol wt 392.37. mp 220-222°.

THERAP CAT: Antihistaminic; nootropic.

3226. Dimecrotic Acid. [7706-67-4] 3-(2,4-Dimethoxyphenyl)-2-butenoic acid; 2,4-dimethoxy-β-methylcinnamic acid; 3-(2,4-dimethoxyphenyl)crotonic acid. $C_{12}H_{14}O_4$; mol wt 222.24. C 64.85%, H 6.35%, O 28.80%. Prepn: **DE 1915023** (1969 to Unicler), *C.A.* **72**, 66630y (1970). TLC determn in pharmaceutics: M.-F. Etcheverry *et al.*, *Sci. Tech. Pharm.* **7**, 51 (1978).

mp 149°.
Magnesium salt. [54283-65-7] Magnesium dimecrotate; Fisiobil; Hepadial. $C_{24}H_{26}MgO_8$; mol wt 466.77. mp 135°. LD_{50} in mice, rats (mg/kg): 1.3, 1.0 i.p. (**DE 1915023**).

THERAP CAT: Choleretic.

3227. Dimefline. [1165-48-6] 8-[(Dimethylamino)methyl]-7-methoxy-3-methyl-2-phenyl-4*H*-1-benzopyran-4-one; 8-[(dimethylamino)methyl]-7-methoxy-3-methylflavone; 8-dimethylaminomethyl-7-methoxy-3-methyl-2-phenylchromone. $C_{20}H_{21}NO_3$; mol wt 323.39. C 74.28%, H 6.55%, N 4.33%, O 14.84%. Prepn: Da Re *et al.*, *Arzneim.-Forsch.* **10**, 800 (1960); **GB 882537**; P. Da Re, **US 3147258** (1961, 1964 both to Recordati). Pharmacology: Setnikar *et al.*, *J. Pharmacol. Exp. Ther.* **128**, 176 (1960); *eidem*, *J. Med. Pharm. Chem.* **3**, 471 (1961).

Hydrochloride. [2740-04-7] Rec-7-0267; Remeflin. $C_{20}H_{21}NO_3 \cdot HCl$; mol wt 359.85. Crystals from alcohol + ether, dec 213-214°.

THERAP CAT: Respiratory stimulant.

3228. Dimefox. [115-26-4] *N,N,N',N'*-Tetramethylphosphorodiamidic fluoride; tetramethyldiamidophosphoric fluoride; bis-(dimethylamido)phosphoryl fluoride; fluophosphoric acid di(dimethylamide); bisdimethylaminofluorophosphine oxide; bis(dimethylamido)fluorophosphate; Pestox XIV. $C_4H_{12}FN_2OP$; mol wt 154.13. C 31.17%, H 7.85%, F 12.33%, N 18.18%, O 10.38%, P 20.10%. Prepd by fluorination of bis(dimethylamido)phosphoryl chloride: Schrader, *Angew. Chem. Suppl.* **62**, 18 (1951); *BIOS Final Report* **No. 714** (1947); Holmsted, *Acta Physiol. Scand. Suppl.* **90**, 33 (1951); Heap, Saunders, *J. Chem. Soc.* **1948**, 1313; **GB 692446** (1953); McCombie *et al.*, **US 2489917** (1949). Toxicity data: A. J. Okinaka *et al.*, *J. Pharmacol. Exp. Ther.* **112**, 231 (1954). *Review:* Kilby, *Chem. Ind. (London)* **1953**, 856.

Liquid. Fishy odor. d_4^{20} 1.1151. $bp_{4.0}$ 67°; bp_{15} 86°. n_D^{20} 1.4267. Freely sol in water, ether, benzene. Aq solns are stable. LD_{50} in rats (mg/kg): 5.0 i.p.; 7.5 orally (Okinaka).

Caution: A highly toxic cholinesterase inhibitor. Symptoms similar to parathion, *q.v.*: *Clinical Toxicology of Commercial Products*, R. E. Gosselin *et al.*, Eds. (Williams & Wilkins, Baltimore, 5th ed., 1984) Section II, p 299.

USE: Insecticide; acaricide.

3229. Dimemorfan. [36309-01-0] (9α,13α,14α)-3,17-Dimethylmorphinan; *d*-3-methyl-*N*-methylmorphinan; AT-17. $C_{18}H_{25}N$; mol wt 255.41. C 84.65%, H 9.87%, N 5.48%. Centrally acting cough suppressant; analog of dextromethorphan. Prepn: Murakami *et al.*, **DE 2128607**; *eidem*, **US 3786054** (1971, 1974 both to Yamanouchi); *eidem*, *Chem. Pharm. Bull.* **20**, 1706 (1972). Synthesis from 3-hydroxymorphinan: J. Y. Kim *et al.*, *Chem. Pharm. Bull.* **56**, 985 (2008). Pharmacology: Ida, Fujii, *Oyo Yakuri* **6**, 1207 (1972); Y. Kasé *et al.*, *Arzneim.-Forsch.* **26**, 353 (1976). Review of clinical efficacy as antitussive: H. Ida, *Clin. Ther.* **19**, 215-231 (1997); of neuroprotective effects: E.-J. Shin *et al.*, *J. Pharmacol. Sci.* **116**, 137-148 (2011).

Pale yellow oil, $bp_{0.3}$ 130-136°; or white crystals from acetone, mp 90-93°.
Phosphate. [36304-84-4] Astomin; Dastosin. $C_{18}H_{25}N \cdot H_3PO_4$; mol wt 353.40. White to yellowish-white, odorless, bitter crystals, mp 267-269°. $[\alpha]_D^{23}$ +25.7° (c = 0.5 in methanol). Freely sol in glacial acetic acid; sparingly sol in water, methanol. Practically insol in ethanol, acetone, chloroform, benzene, ether. LD_{50} in mice (mg/kg): 223 s.c.; 475 orally (Kasé).

THERAP CAT: Antitussive.

3230. Dimenhydrinate. [523-87-5] 8-Chloro-3,7-dihydro-1,3-dimethyl-1*H*-purine-2,6-dione compd with 2-(diphenylmethoxy)-*N,N*-dimethylethanamine (1:1); β-dimethylaminoethyl benzhydryl ether 8-chlorotheophyllinate; 2-(benzhydryloxy)-*N,N*-dimethylethylamine 8-chlorotheophyllinate; *O*-benzhydryldimethylaminoethanol 8-chlorotheophyllinate; 2-(diphenylmethoxy)-*N,N*-dimethylethylamine 8-chlorotheophyllinate; diphenhydramine teoclate; chloranautine; Amosyt; Antemin; Dramamine; Gravol; Nausicalm; Travel-Gum; Vomex A; Xamamina. $C_{24}H_{28}ClN_5O_3$; mol wt 469.97. C 61.34%, H 6.01%, Cl 7.54%, N 14.90%, O 10.21%. Antihistamininic salt of diphenhydramine with 8-chlorotheophylline. Prepn: J. W. Cusic, **US 2499058** (1950 to Searle); *idem*, *Science* **109**, 574 (1949). Efficacy in motion sickness: L. N. Gay, P. E. Carliner, *ibid.* 359. HPLC-MS/MS determn in plasma: V. Tavares *et al.*, *J. Chromatogr. B* **853**, 127 (2007). Clinical trial in treatment of vertigo: K. A. Marill *et al., Ann. Emerg. Med.* **36**, 310 (2000). Review of clinical trials in postoperative nausea and vomiting: P. Kranke *et al., Acta Anaesthesiol. Scand.* **46**, 238-244 (2002); of abuse potential: A. G. Halpert *et al., Neurosci. Biobehav. Rev.* **26**, 61-67 (2002).

Crystals from hot alcohol, mp 103-104°. Freely sol in alcohol, chloroform; sol in benzene. Soly in water: about 3 mg/ml. Almost insol in ether. A satd aq soln has a pH between 6.8 and 7.3.

THERAP CAT: Antiemetic; antivertigo agent; in treatment of motion sickness.

THERAP CAT (VET): Antiemetic.

3231. Dimercaprol. [59-52-9] 2,3-Dimercapto-1-propanol; 1,2-dithioglycerol; British anti-lewisite; BAL; Dicaptol; Sulfactin. $C_3H_8OS_2$; mol wt 124.22. C 29.01%, H 6.49%, O 12.88%, S 51.62%. Developed as an antidote to the vesicant warfare agent, dichloro(2-chlorovinyl)arsine, *q.v.*, also known as Lewisite. Inhibits the toxic action of arsenic on pyruvate dehydrogenase. Discovery and development: R. A. Peters *et al., Nature* **156**, 616 (1945); L. L. Waters, C. Stock, *Science* **102**, 601 (1945). Prepn by hydrogenation of hydroxypropylene trisulfide: W. J. Peppel, F. K. Signaigo, US 2402665 (1946 to DuPont). Stereospecific synthesis: A. K. M. Anisuzzaman, L. N. Owen, *J. Chem. Soc. C* **1967**, 1021. GC/MS determn in plasma: C. E. Byers *et al., J. Anal. Toxicol.* **28**, 384 (2004). Toxicity study: P. Zvirblis, R. I. Ellin, *Toxicol. Appl. Pharmacol.* **36**, 297 (1976). Review of discovery, biochemistry and clinical applications: J. A. Vilensky, K. Redman, *Ann. Emerg. Med.* **41**, 378-383 (2003).

Viscous oily liq. Pungent offensive odor of mercaptans. d_4^{25} 1.2385. n_D^{25} 1.5720. $bp_{0.0001}$ 68-69°. Vapor pressure at 80°: 1.9 mm Hg; at 140°: 40 mm Hg. Soly in water: 8.7 g/100 g. Sol in alc, benzyl benzoate, methanol, vegetable oils. LD_{50} i.m. in rats: 86.7 mg/kg (Zvirblis, Ellin).

THERAP CAT: Antidote (heavy metal poisoning).

THERAP CAT (VET): Antidote (heavy metal poisoning).

3232. 2,3-Dimercapto-1-propanesulfonic Acid. [74-61-3] 2,3-Dithiolpropanesulfonic acid. $C_3H_8O_3S_3$; mol wt 188.27. C 19.14%, H 4.28%, O 25.49%, S 51.09%. Chelating agent related to BAL dimercaprol, *q.v.* Prepn: N. S. Johary, L. N. Owen, *J. Chem. Soc.* **1955**, 1307; V. E. Petrun'kin, *Ukr. Khim. Zh.* **22**, 603 (1956), *C.A.* **51**, 5692h (1957). Distribution and excretion in rats: B. Gabard, *Arch. Toxicol.* **39**, 289 (1978). Metabolism study: B. Gabard, R. Walser, *J. Toxicol. Environ. Health* **5**, 759 (1979). Use in removal of internally deposited gold: B. Gabard, *Br. J. Pharmacol.* **68**, 607 (1980). Protection of mice vs lethal effects of sodium arsenite: H. V. Aposhian *et al., Toxicol. Appl. Pharmacol.* **61**, 385 (1981). Pharmacokinetics in dogs: P. Wiedemann *et al., Biopharm. Drug Dispos.* **3**, 267 (1982). Anti-lewisite activity and stability: H. V. Aposhian *et al., Life Sci.* **31**, 2149 (1982). Toxicological studies: F. Planas-Bohne *et al., Arzneim.-Forsch.* **30**, 1291 (1980).

Sodium salt. [4076-02-2] DMPS; unitiol; Dimaval. C_3H_7-NaO_3S_3; mol wt 210.26. Leaflets, mp 235°. LD_{50} i.p. in mice: 5.22 mmol/kg (Aposhian, 1981).

THERAP CAT: Antidote (heavy metal poisoning).

3233. Dimesna. [16208-51-8] 2,2'-Dithiobisethanesulfonic acid sodium salt (1:2); disodium 2,2'-dithiobis(ethanesulfonate); BNP-7787; Tavocept. $C_4H_8Na_2O_6S_4$; mol wt 326.32. C 14.72%, H 2.47%, Na 14.09%, O 29.42%, S 39.30%. Disulfide form and physiological auto-oxidation product of mesna, *q.v.* Reduced to the active thiol in the kidney; protects against toxic metabolites of chemotherapeutic agents such as cisplatin. Prepn: H. Lemaire, M. Rieger, *J. Org. Chem.* **26**, 1330 (1961); E. Brzezinska, A. L. Ternay, Jr., *ibid.* **59**, 8239 (1994); K. Haridas, US 5808140 (1998 to BioNumerik). HPLC determn in plasma and urine: M. Verschraagen *et al., J. Chromatogr. B* **753**, 293 (2001). Vibrational spectroscopic studies: Y.-S. Li *et al., Spectrochim. Acta A* **59**, 1791 (2003). Mechanism of uroprotection: M. Verschraagen *et al., Br. J. Cancer* **90**, 1654 (2004). Clinical pharmacokinetics in combination with cisplatin: *idem et al., Clin. Pharmacol. Ther.* **74**, 157 (2003). *Review:* R. T. Reilly, *IDrugs* **7**, 64-69 (2004).

Crystals from water. mp 302-303°.

THERAP CAT: Antineoplastic adjunct (uroprotective).

3234. Dimestrol. [130-79-0] 1,1'-[(1E)-1,2-Diethyl-1,2-ethenediyl]bis[4-methoxybenzene]; (E)-α,α'-diethyl-4,4'-dimethoxystilbene; stilbestrol dimethyl ether; (E)-3,4-bis(p-methoxyphenyl)-3-hexene; (E)-3,4-dianisyl-3-hexene. $C_{20}H_{24}O_2$; mol wt 296.41. C 81.04%, H 8.16%, O 10.80%. Nonsteroidal synthetic estrogen; dimethyl ether of diethylstilbestrol, *q.v.* Prepn: Dodds *et al., Nature* **142**, 211, 247 (1938); Reid, Wilson, *J. Am. Chem. Soc.* **64**, 1625 (1942); Sisido, Nozaki, *J. Am. Chem. Soc.* **70**, 777 (1948). Improved stereospecific prepn: T. Hiyama, H. Nozaki, *Bull. Chem. Soc. Jpn.* **46**, 2248 (1973). Crystal structure: G. Ruban, P. Luger, *Acta Crystallogr.* **B31**, 2658 (1975). Biological activities: Y. Inamori *et al., Chem. Pharm. Bull.* **33**, 4478 (1985). Review and bibliography: Solmssen, *Chem. Rev.* **37**, 481 (1945).

Crystals from petr ether, mp 124°. Less sol than the monomethyl ether, mestilbol. Practically insol in water. Sol in alcohol; freely sol in acetone, ether; also sol in dil aq or alcoholic solns of alkali hydroxides and in vegetable oils.

3235. Dimethadione. [695-53-4] 5,5-Dimethyl-2,4-oxazolidinedione; DMO; AC-1198; BAX-1400Z; NSC-30152; Eupractone. $C_5H_7NO_3$; mol wt 129.12. C 46.51%, H 5.46%, N 10.85%, O 37.17%. Active metabolite of trimethadione, *q.v.* Prepn: F. Urech, *Ber.* **13**, 485 (1880); R. W. Stoughton, *J. Am. Chem. Soc.* **63**, 2376 (1941). Anticonvulsant activity: C. D. Withrow *et al., J. Pharmacol. Exp. Ther.* **161**, 335 (1968); in comparison with trimethadione: H. Ferngren, *Acta Pharmacol. Toxicol.* **26**, 177 (1968). Use in measurement of intracellular pH and cellular pH gradients: S. Addanki *et al., J. Biol. Chem.* **243**, 2337 (1968); V. Ehrhardt, *Biochim. Biophys. Acta* **775**, 182 (1984); J. B. Arnold *et al., J. Cereb. Blood Flow Metab.* **5**, 369 (1985). GLC determn: W. Gazdzik, W. Kmiotek, *J. Chromatogr.* **378**, 482 (1986); LC determn in serum: E. Tanaka, S. Misawa, *J. Chromatogr.* **413**, 376 (1987).

Crystals, mp 76-77°. Weak organic acid, pKa (37°) 6.13. LD_{50} i.v. in mice: 450 mg/kg (Stoughton).

USE: *In vivo* measurement of intracellular pH.

THERAP CAT: Anticonvulsant.

3236. Dimethenamid. [87674-68-8] 2-Chloro-*N*-(2,4-dimethyl-3-thienyl)-*N*-(2-methoxy-1-methylethyl)acetamide; SAN-582H; Frontier; Guardsman. $C_{12}H_{18}ClNO_2S$; mol wt 275.79. C 52.26%, H 6.58%, Cl 12.85%, N 5.08%, O 11.60%, S 11.62%. Pre-emergence herbicide for use in food crops. Prepn: K. Seckinger *et al.*, **GB 2114566**; eidem, **US 4666502** (1983, 1987 both to Sandoz). Bioactivity of stereoisomers: M. Couderchet *et al.*, *Pestic. Sci.* **50**, 221 (1997). Comprehensive description: J. Harr *et al.*, *Proc. Br. Crop Prot. Conf. - Weeds* **1991**, 87-92. Review of field trials and persistence in soil: A. Rahman, T. K. James, *Proc. 45th N. Z. Plant Prot. Conf.* **1992**, 84-88.

Yellowish-brown, viscous liquid. Odorless to weak tar-like odor. Mixture of 4 stereoisomers (1*RS*,a*RS*). $bp_{26.7 Pa}$ 127°. Vapor pressure (25°): 36.7 mPa. d^{25} 1.187. Soly (25°): water 1174 mg/l; heptane 28.2 g/100 g; isooctane 22.0 g/100 g; ether, kerosene, ethanol >50%. LD_{50} in rats (mg/kg): 1570 orally; >2000 dermally; LC_{50} in bluegill sunfish, rainbow trout (mg/l): 6.4, 2.6 (Harr).
Dimethenamid P. [163515-14-8] DMTA-P; BAS-656H; Outlook. Consists of the 1*S*,a*RS* isomers. Prepn: K. Seckinger *et al.*, **US 5457085** (1995 to Sandoz). Review of properties and activity: T. Guillet *et al.*, *Phytoma* **546**, 50-53 (2002). Faintly yellow viscous liquid. $[\alpha]_D^{23}$ +3.5° (c = 1.102 in CH_3OH). $bp_{9.3 Pa}$ 122.6°. d^{20} 1.195. Vapor pressure (25°): 2.5×10^{-3} Pa. Soly in water (25°): 1449 mg/l. LD_{50} in rats (mg/kg): 429 orally; >2000 dermally. LC_{50} (96 hr) in rainbow trout (mg/l): 6.3 (Guillet).
USE: Herbicide.

3237. Dimethicone. [9006-65-9] Dimethyl polysiloxane; dimeticon; polydimethylsiloxane; α-(trimethylsilyl)-ω-methylpoly-[oxy(dimethylsilylene)]. Silicone oil consisting of a mixture of fully methylated linear siloxane polymers end-blocked with trimethylsiloxy units. Prepn of homologous liquid methyl siloxane polymers: J. F. Hyde, **US 2441098** (1948 to Corning Glass). Dimethicone mixed with silicon dioxide is known as simethicone. Clinical evaluation of simethicone in functional upper GI disease: J. E. Bernstein, A. M. Kasich, *J. Clin. Pharmacol.* **14**, 617 (1974). Stability of simethicone + antacid mixtures in pharmaceutical preparations: J. A. Rider, *Curr. Ther. Res.* **52**, 681 (1992). Review of use of dimethicones in skin and hair products: A. Disapio, P. Fridd, *Int. J. Cosmet. Sci.* **10**, 75-89 (1988). Review of medical use in soft-tissue augmentation: V. J. Selmanowitz, N. Orentreich, *J. Dermatol. Surg. Oncol.* **3**, 597-611 (1977); D. M. Duffy, *Adv. Dermatol.* **5**, 93-109 (1990). *See also:* Silicones.

Clear colorless liquids. Viscosity increases with degree of polymerization. Miscible with chloroform. Immiscible with water, alcohol. Sol in chlorinated hydrocarbons, benzene, toluene, xylene, *n*-hexane, petroleum spirits, ether, amyl acetate; very slightly sol in isopropyl alc. Insol in methanol, acetone.
Mixture with silicon dioxide. [8050-81-5] Simethicone; activated dimethicone; Antifoam A; Baros; Colicon; Endo-Paractol; Gas-X; Infacol; Lefax; Mylicon; Phasil; Phazyme; sab simplex; Silain. Average number of dimethylsiloxane units is 200 to 350. mol wt 14000-21000. Gray, translucent, viscous liquid. Insol in water, alcohol.
Dimethicone 350. Specific polydimethylsiloxane characterized by the following properties. Viscosity (25°): 350 cSt. d 0.965-0.973. n_D^{25} 1.4013-1.4053.
USE: Oleaginous ointment base, skin protectant, antifoaming agent; prosthetic aid (soft tissue).

THERAP CAT: Antiflatulent.
THERAP CAT (VET): Antibloating agent.

3238. Dimethindene. [5636-83-9] *N,N*-Dimethyl-3-[1-(2-pyridinyl)ethyl]-1*H*-indene-2-ethanamine; 2-[1-[2-[2-(dimethylamino)ethyl]inden-3-yl]ethyl]pyridine; 3-[α-(2′-pyridyl)ethyl]-2-(β-dimethylaminoethyl)indene; dimethpyrindene. $C_{20}H_{24}N_2$; mol wt 292.43. C 82.15%, H 8.27%, N 9.58%. Synthesis: Huebner *et al.*, *J. Am. Chem. Soc.* **82**, 2077 (1960); Huebner, **US 2970149** (1961 to Ciba). Pharmacology: W. E. Barrett *et al.*, *Toxicol. Appl. Pharmacol.* **3**, 534 (1961).

Maleate. [3614-69-5] Fenistil. $C_{20}H_{24}N_2 \cdot C_4H_4O_4$; mol wt 408.50. Crystals, mp 159-161°. LD_{50} in rats (mg/kg): 26.8 i.v.; 618.2 orally (Barrett).
THERAP CAT: Antihistaminic.

3239. Dimethirimol. [5221-53-4] 5-Butyl-2-(dimethylamino)-6-methyl-4(1*H*)-pyrimidinone; 5-butyl-2-(dimethylamino)-6-methyl-4-pyrimidinol; 2-dimethylamino-4-methyl-5-*n*-butyl-6-hydroxypyrimidine; PP-675; Milcurb. $C_{11}H_{19}N_3O$; mol wt 209.29. C 63.13%, H 9.15%, N 20.08%, O 7.64%. Prepn: B. K. Snell *et al.*, **ZA 6701373**; eidem, **US 3980781** (1968, 1976 both to ICI). Systemic fungicidal activity: R. S. Elias, *Nature* **219**, 1160 (1968); K. J. Bent, *Ann. Appl. Biol.* **66**, 103 (1970). Metabolism and mode of action: A. Calderbank, *Acta Phytopathol.* **6**, 355 (1971). Metabolism: H. Bratt *et al.*, *Food Cosmet. Toxicol.* **10**, 489 (1972).

Needles from ethanol, mp 102°. Vapor pressure at 30°: 1.1×10^{-5} mm Hg. uv max (methanol): 229, 304 nm (ε 15500, 7700). Soly in (g/l) at 25°: water 1.2; acetone 45; chloroform 1200; ethanol 65; xylene 360. LD_{50} in female rats (mg/kg): 200-400 i.p., >4000 orally (Elias).
USE: Fungicide.

3240. Dimethisterone. [79-64-1] (6α,17β)-17-Hydroxy-6-methyl-17-(1-propynyl)androst-4-en-3-one; 6α-methyl-17-(1-propynyl)testosterone; 6α,21-dimethyl-17β-hydroxy-17α-pregn-4-en-20-yn-3-one; 6α,21-dimethylethisterone; 17α-ethynyl-6α,21-dimethyltestosterone; 17α-ethynyl-17-hydroxy-6α,21-dimethylandrost-4-en-3-one; Secrosteron (obsolete). $C_{23}H_{32}O_2$; mol wt 340.51. C 81.13%, H 9.47%, O 9.40%. Orally active progestogen; formerly used in combinations as oral contraceptive. Prepn: S. P. Barton *et al.*, *J. Chem. Soc.* **1959**, 1957; eidem, **US 2939819** (1960 to Brit. Drug Houses). Structure-activity study: G. K. Suchowsky, G. Baldratti, *J. Endocrinol.* **30**, 159 (1964). Evaluation of risk for endometrial cancer: N. S. Weiss *et al.*, *N. Engl. J. Med.* **302**, 551 (1980). Review of carcinogenicity studies: *IARC Monographs* **21**, 377-385 (1979).

Crystals, mp 102°. $[\alpha]_D^{20}$ +10° (c = 1 in chloroform). uv max (isopropanol): 240 nm ($E_{1cm}^{1\%}$ 450). Practically insol in water. Sol in ethanol; slightly sol in acetone, chloroform.

Mixture with ethinyl estradiol. [8015-19-8] Oracon (obsolete).

THERAP CAT: Progestogen.

3241. Dimethoate. [60-51-5] Phosphorodithioic acid O,O-dimethyl S-[2-(methylamino)-2-oxoethyl] ester; phosphorodithioic acid O,O-dimethyl ester, ester with 2-mercapto-N-methylacetamide; O,O-dimethyl S-methylcarbamoylmethyl phosphorodithioate; American Cyanamid 12880; Danadim; Perfekthion; Rogor. C_5H_{12}-NO_3PS_2; mol wt 229.25. C 26.20%, H 5.28%, N 6.11%, O 20.94%, P 13.51%, S 27.97%. Organophosphate insecticide; cholinesterase inhibitor. Prepn: Cassaday et al., Young, **US 2494283** and **US 2996531** (1950, 1961, both to Am. Cyanamid); **GB 791824** (1958 to Montecatini), C.A. **52**, 18222 (1958). LC-MS/MS determn in blood: P. Salm et al., J. Chromatogr. B **877**, 568 (2009). Degradation study: L. Zhang et al., J. Hazard. Mater. **149**, 675 (2007). Toxicity data: E. W. Schafer, Toxicol. Appl. Pharmacol. **21**, 315 (1972). Teratogenicity study: K. D. Courtney et al., J. Environ. Sci. Health **B20**, 373 (1985).

Crystals, mp 52-52.5°. d^{65} 1.277. n_D^{65} 1.5334. Burns readily on contact with flame. Very slightly sol in water. Freely sol in most organic solvents, except saturated hydrocarbons. Stable in aq soln; hydrolyzed by aq alkali. LD_{50} orally in rats: 250 mg/kg (Schafer).

USE: Acaricide; insecticide.

3242. Dimethocaine. [94-15-5] 2-[(Diethylamino)methyl]-2-methyl-1-propanol 1-(4-aminobenzoate); 3-(diethylamino)-2,2-dimethyl-1-propanol p-aminobenzoate; 3-diethylamino-2,2-dimethyl-propyl p-aminobenzoate; 1-aminobenzoyl-2,2-dimethyl-3-diethyl-aminopropanol; p-aminobenzoate of diethylaminoneopentyl alcohol; Larocaine. $C_{16}H_{26}N_2O_2$; mol wt 278.40. C 69.03%, H 9.41%, N 10.06%, O 11.49%. Prepn: Mannich, Wilder, Ber. **65**, 378 (1932); C. Mannich, **US 1889678** (1932). Pharmacology: O. Gebner et al., Arch. Exp. Pathol. Pharmakol. **168**, 447 (1932); and toxicology review: A. D. Hirschfelder, R. N. Bieter, Physiol. Rev. **12**, 190-282 (1932).

Hydrochloride. [553-63-9] $C_{16}H_{26}N_2O_2$·HCl. Minute crystals or powder, sometimes fine leaflets. Bitter taste. mp 196-197°. One part dissolves in 3 parts water at 20°, more sol in hot water. One part dissolves also in 10 parts of cold alcohol, in 5 parts of boiling alcohol. Practically insol in ether, oils, fats. Aq solns are slightly acid to litmus, and may be sterilized by heating at 105° for 10 minutes. MLD in mice, rabbits (g/kg): 0.30, 0.15 s.c.; 0.04, 0.015 i.v. (Hirschfelder, Bieter).

THERAP CAT: Anesthetic (local).

3243. Dimethomorph. [110488-70-5] 3-(4-Chlorophenyl)-3-(3,4-dimethoxyphenyl)-1-(4-morpholinyl)-2-propen-1-one; 4-[3-(4-chlorophenyl)-3-(3,4-dimethoxyphenyl)-1-oxo-2-propenyl]morpholine; 3-(4-chlorophenyl)-3-(3,4-dimethoxyphenyl) acrylic acid morpholide; CME-151; Acrobat; Forum. $C_{21}H_{22}ClNO_4$; mol wt 387.86. C 65.03%, H 5.72%, Cl 9.14%, N 3.61%, O 16.50%. Antifungal cinnamic acid derivative; comprised of a mixture of E- and Z- isomers. Process for prepn: J. Curtze, **EP 294907**; idem, **US 4933449** (1988, 1990 both to Shell Int.). Comprehensive description: G. Albert et al., Brighton Crop Prot. Conf. - Pests Dis. **1988**, 17-24. Mechanism of action study: P. J. Kuhn et al., Mycol. Res. **95**, 333 (1991). Efficacy vs downy mildew of grapevines: T. Wicks, B.

Hall, Plant Dis. **74**, 114 (1990). Toxicological evaluation: G. E. Veenstra, D. E. Owen, Arch. Toxicol. **1992**, Suppl 15, 113.

Colorless, odorless, crystalline solid. mp 127-148°. Soly (g/l) at 25°: water <0.05, acetone 50, toluene 20, methanol 20, dichloromethane 500. LD_{50} in rats (mg/kg): 321 i.p.; 3900 orally; >5000 dermally; LC_{50} (4-hr inhalation): >4.2 mg/ml (Veenstra, Owen).

USE: Agricultural fungicide.

3244. Dimethoxane. [828-00-2] 2,6-Dimethyl-1,3-dioxan-4-ol acetate; 6-acetoxy-2,4-dimethyl-m-dioxane; 2,6-dimethyl-m-diox-an-4-yl acetate; acetomethoxane; Dioxin (obsolete); Giv-Gard DXN. $C_8H_{14}O_4$; mol wt 174.20. C 55.16%, H 8.10%, O 36.74%. Prepn: C. S. Marvel et al., J. Org. Chem. **4**, 252 (1939); Späth et al., Ber. **76**, 57 (1943). Description: Am. Perfum. Cosmet. **77**, no. 12, 32-34 (1962), C.A. **58**, 8848c (1963).

Liquid, bp_6 74-75°. n_D^{20} 1.4310; d_4^{20} 1.0655. Mustard-like odor. Miscible with water, many organic solvents.

USE: Preservative for cutting oils, resin emulsions, water-based paints, cosmetics, inks. Effective range of concn: 0.03-0.1%. Gasoline additive: Chafetz et al., **US 3036904** (1962 to Texaco).

3245. 1,2-Dimethoxyethane. [110-71-4] Ethylene glycol dimethyl ether; monoglyme; α,β-dimethoxyethane; glyme; Dimethyl Cellosolve. $C_4H_{10}O_2$; mol wt 90.12. C 53.31%, H 11.19%, O 35.51%. Prepd from ethylene glycol monomethyl ether, methyl sulfate and metallic sodium: Kranzfelder, Vogt, J. Am. Chem. Soc. **60**, 1714 (1938); from ethylene glycol monomethyl ether, methyl chloride and sodium: Capinjola, ibid. **67**, 1615 (1945); from chloromethyl methyl ether in the presence of sodium by Wurtz reaction: Geist, Mason, ibid. **76**, 3728 (1954). Series of articles on toxicology: Environ. Health Perspect. **57**, 1-275 (1984).

Liquid. Sharp ethereal odor. d_4^{15} 0.86877; d_4^{20} 0.86285; d_4^{33} 0.8602; d_{20}^{20} 0.8692. mp −58° (also reported as mp −71°). bp_{760} 82-83°; $bp_{61.2}$ 20°; bp_{50} 16°; bp_{10} −14°. n_D^{24} 1.3739; n_D^{20} 1.3813. Flash pt 4.5°C (40°F). Miscible with water, alcohol. Sol in hydrocarbon solvents.

USE: Solvent, to facilitate the formation of alkali metal-hydrocarbon adducts; in the Reformatsky reaction with methyl γ-bromocrotonate.

3246. 2,6-Dimethoxyquinone. [530-55-2] 2,6-Dimethoxy-2,5-cyclohexadiene-1,4-dione; 2,6-dimethoxybenzoquinone. C_8-H_8O_4; mol wt 168.15. C 57.14%, H 4.80%, O 38.06%. Obtained from pyrogallol-1,3-dimethyl ether-2-acetate upon oxidation with potassium dichromate: Hofmann, Ber. **11**, 337 (1878); or by heating with nitric acid in alc soln: Graebe, Hess, Ann. **340**, 237 (1905); by treating pyrogallol trimethyl ether with nitric acid: Will, Ber. **21**, 608 (1888); from syringa acid: Graebe, Martz, ibid. **221**; from sinapic acid: Gadamer, *Ber.* **30**, 2333 (1897); from antiarole by oxidation with ferric chloride: Baker, *J. Chem. Soc.* **1928**, 1029.

Monoclinic golden-yellow prisms from acetic acid, mp 256°. Sublimes easily. Volatile with steam. Slightly sol in hot water, in ether, in alcohol; freely sol in hot glacial acetic acid and in alkaline solns.

3247. Dimethylacetal. [534-15-6] 1,1-Dimethoxyethane; acetaldehyde dimethyl acetal; ethylidene dimethyl ether. $C_4H_{10}O_2$; mol wt 90.12. C 53.31%, H 11.19%, O 35.51%. Prepn from acetaldehyde and methanol: Meadows, Darwent, *Can. J. Chem.* **30**, 501 (1952); Frevel, Hedelund, **US 2691684** (1954 to Dow). Toxicity study: H. F. Smyth *et al.*, *J. Ind. Hyg. Toxicol.* **31**, 60 (1949).

Liquid, bp 64.5°. d_4^{20} 0.8516. n_D^{20} 1.3665. Miscible with water, alcohol, chloroform, ether. LD_{50} orally in rats: 6.5 g/kg; LC (in air) in rats: 16000 ppm (Smyth).

USE: As Mering's mixture which is 2 vol dimethylacetal and 1 vol chloroform.

3248. *N,N*-Dimethylacetamide. [127-19-5] Acetic acid dimethylamide; DMAC. C_4H_9NO; mol wt 87.12. C 55.15%, H 10.41%, N 16.08%, O 18.36%. Prepn from tris(dimethylamido)-phosphate and acetic anhydride: Dye, **US 2667510** (1954 to Chemstrand); from acetic anhydride and dimethylformamide: Coppinger, *J. Am. Chem. Soc.* **76**, 1372 (1954). Toxicity study: W. Bartsch *et al.*, *Arzneim.-Forsch.* **26**, 1581 (1976).

Liquid. d_4^{25} 0.9366; d_4^{20} 0.9429; d_4^0 0.9599. bp_{760} 163-165°; bp_{80} 96°; bp_{33} 85-87°; bp_{26} 74-74.5°; bp_{15} 66-67°; bp_{12} 62-63°. n_D^{20} 1.4373; also reported as n_D^{20} 1.4230; n_D^{25} 1.4358. Flash pt 66°C (151°F). Miscible with water and most organic solvents. LD_{50} orally in rats: 5.4 ml/kg (Bartsch).

Caution: Potential symptoms of overexposure are jaundice, liver damage; depression, lethargy, hallucinations and delusions; skin irritation. *See NIOSH Pocket Guide to Chemical Hazards* (DHHS/NIOSH 97-140, 1997) p 110.

USE: Solvent for many organic reactions and in industrial applications.

3249. Dimethylaluminum Chloride. [1184-58-3] Chlorodimethylaluminum; dimethylchloroaluminum; Me₂AlCl. C_2H_6AlCl; mol wt 92.50. C 25.97%, H 6.54%, Al 29.17%, Cl 38.32%. Alkylaluminum halide that acts as a Lewis acid catalyst and proton scavenger. Prepn: W. O. Walker, K. S. Wilson, *Refr. Eng.* **34**, 89 (1937); V. F. Hnizda, C. A. Kraus, *J. Am. Chem. Soc.* **60**, 2276 (1938). Synthetic applications in ene reactions: B. B. Snider *et al.*, *ibid.* **104**, 555 (1982); in Diels-Alder reactions: D. A. Evans *et al.*, *Tetrahedron Lett.* **25**, 4071 (1984); in polymerizations: M. Bahadur *et al.*, *Macromolecules* **33**, 9548 (2000); in carbonyl additions: D. A. Evans *et al.*, *J. Am. Chem. Soc.* **123**, 10840 (2001).

White liquid. *Flammable. Corrosive.* bp_{760} 126°; bp_{200} 83-84°. Solid is dimorphic, mp ~ −45°; mp −21.0°. Condenses at −80°C as a viscous liquid. d^{20} 1.0070; d^{30} 0.9915; d^{40} 0.9755; d^{50} 0.9570. Flash pt, closed cup: −1.0°F (−18.0°C). Sol in most organic solvents. *Reacts violently with water. Spontaneously flammable in air.*

USE: Reagent and catalyst in synthetic organic chemistry.

3250. Dimethylamine. [124-40-3] *N*-Methylmethanamine. C_2H_7N; mol wt 45.09. C 53.28%, H 15.65%, N 31.06%. (CH₃)₂-NH. Prepn from methanol + ammonia: Smith, **US 2456599** (1948 to Commercial Solvents); Serban, *Rev. Chim. (Bucharest)* **14**, 451 (1963), *C.A.* **60**, 5097b (1964); by catalytic hydrogenation of nitroso-dimethylamine: Livering, Maury, **GB 797483** (1958 to Hercules Powder). Toxicity studies: R. Hazard *et al.*, *Arch. Int. Pharmacodyn.* **112**, 36 (1957); W. H. Steinhagen *et al.*, *Am. Ind. Hyg. Assoc. J.* **43**, 411 (1982).

Gas at ordinary temp; characteristic odor. d_4^0 (liq) 0.680. bp 7°, mp −96°. *Flammable.* Very sol in water forming a very strong alkaline soln; sol in alcohol or ether. Has been marketed in compressed liquid form in tubes or as a 33% aq soln.

Hydrochloride. [506-59-2] $C_2H_7N.HCl$. Deliquesc leaflets, mp 171°. Very sol in water; sol in alcohol, chloroform. Practically insol in ether. *Keep well closed.* LD_{50} in mice (g/kg): 1.21 i.v., 2.00 s.c. (Hazard). LC_{50} in mice (ppm): 7650 (48 hr), 4725 (14 day); in rats (ppm): 4540 (6 hr) (Steinhagen).

Caution: Potential symptoms of overexposure are irritation of nose and throat; sneezing, coughing and dyspnea; pulmonary edema; conjunctivitis; dermatitis; direct contact with liquid may cause frostbite. *See NIOSH Pocket Guide to Chemical Hazards* (DHHS/NIOSH 97-140, 1997) p 110.

USE: As accelerator in vulcanizing rubber, tanning, manuf detergent soaps, or attracting boll weevils to exterminate them. As reagent for Mg.

3251. *p*-Dimethylaminoazobenzene. [60-11-7] *N,N*-Dimethyl-4-(2-phenyldiazenyl)benzenamine; *N,N*-dimethyl-4-(phenylazo)benzenamine; butter yellow; methyl yellow; C.I. Solvent Yellow 2; C.I. 11020. $C_{14}H_{15}N_3$; mol wt 225.30. C 74.64%, H 6.71%, N 18.65%. Prepn: *Colour Index* vol. **4** (3rd ed., 1971) p 4014. Review of carcinogenic risk: *IARC Monographs* **8**, 125-146 (1975).

Yellow cryst leaflets, mp 114-117°. Insol in water. Sol in alc, benzene, CHCl₃, ether, petr ether, mineral acids, oils.

Caution: Potential symptoms of overexposure are enlarged liver; hepatic and renal dysfunction; contact dermatitis; coughing, wheezing, difficulty breathing; bloody sputum; bronchial secretions; frequent urination, hematuria and dysuria. *See NIOSH Pocket Guide to Chemical Hazards* (DHHS/NIOSH 97-140, 1997) p 110. This substance is reasonably anticipated to be a human carcinogen: *Report on Carcinogens, Twelfth Edition* (PB2011-111646, 2011) p 167.

USE: For determination of free HCl in gastric juice; spot test identification of peroxidized fats; pH indicator (red 2.9, yellow 4.0).

3252. *p*-Dimethylaminobenzaldehyde. [100-10-7] 4-(Dimethylamino)benzaldehyde; 4-dimethylaminobenzenecarbonal; Ehrlich's reagent. $C_9H_{11}NO$; mol wt 149.19. C 72.46%, H 7.43%, N 9.39%, O 10.72%. Prepd by formylation of dimethylaniline with dimethylformamide: Campaigne, Archer, *Org. Synth.* **coll. vol. IV**, 331 (1963).

Small granular crystals or leaflets from alcohol + water. Unless extremely pure, the crystals are lemon-colored and may turn pink on exposure to light. mp 74°. bp_{17} 176-177°. Slightly sol in water; sol in alcohol, ether, chloroform, acetic acid, many other organic solvents.

Hydrochloride. [5988-39-6] $C_9H_{11}NO.HCl$; mol wt 185.65. Crystals, mp 109°.

USE: Manuf dyes. Reagent for arsphenamine, anthranilic acid, antipyrine, indole, skatole, indican, tryptophan, albumin, ergot alkaloids, colon bacteria, typhoid coli. For differentiating between serum eruptions and true scarlet fever. Derivatizing agent.

3253. *p*-Dimethylaminobenzalrhodanine. [536-17-4] 5-[[4-(Dimethylamino)phenyl]methylene]-2-thioxo-4-thiazolidinone; 5-[*p*-(dimethylamino)benzylidene]rhodanine. $C_{12}H_{12}N_2OS_2$; mol wt 264.36. C 54.52%, H 4.58%, N 10.60%, O 6.05%, S 24.25%. Prepn: Mackie, Misra, *J. Chem. Soc.* **1954**, 3919.

Deep red needles from xylene, dec 270°, also reported as dec 246°. Sintering ~260° and melting ~280°. Practically insol in water. Very slightly sol in chloroform, ether, benzene; slightly sol in boiling alc; moderately sol in acetone; sol in strong acids with a yellow color.

USE: As 0.03% soln in acetone for detection of silver, mercury, copper, gold, platinum and palladium ions.

3254. 4-(Dimethylamino)benzoic Acid. [619-84-1] $C_9H_{11}NO_2$; mol wt 165.19. C 65.44%, H 6.71%, N 8.48%, O 19.37%. Prepn: Willstätter, Kahn, *Ber.* **37**, 411 (1904); Morton, Stevens, *J. Am. Chem. Soc.* **53**, 4028, 4031 (1931); Gilman, Banner, *ibid.* **62**, 344 (1940); Bowman, Stroud, *J. Chem. Soc.* **1950**, 1342. Use of esters as sunscreening agents: Kreps, Ohlsson, US 3403207 (1968); Worthington, GB 1162337 and FR 1566396 (both 1969 to Van Dyk).

Crystals from water, mp 242.5-243.5°. Soluble in alcohol, HCl and KOH solns, sparingly sol in ether. Practically insol in acetic acid. pKa 6.027; pKb 11.488.

3-Methylbutyl ester. [21245-01-2] Padimate A; Escalol 506. $C_{14}H_{21}NO_2$; mol wt 235.33.

2-Ethylhexyl ester. [21245-02-3] Padimate O; Escalol 507; Sundown. $C_{17}H_{27}NO_2$; mol wt 277.41. Light yellow, mobile liquid. Sol in alc, isopropyl alc, mineral oil. Practically insol in water, glycerin, propylene glycol.

Glyceryl ester *see* Glyceryl *p*-Aminobenzoate.

THERAP CAT: Esters as ultraviolet screen.

3255. *p*-Dimethylaminobenzophenone. [530-44-9] [4-(Dimethylamino)phenyl]phenylmethanone. $C_{15}H_{15}NO$; mol wt 225.29. C 79.97%, H 6.71%, N 6.22%, O 7.10%. Prepd from benzanilide, dimethylaniline, and phosphorus oxychloride: DE 41751; *Frdl.* **1**, 44 (1887); Meisenheimer *et al., Ann.* **423**, 84 (1921); Hurd, Webb, *Org. Synth.* **7**, 24 (1927).

Leaflets from alc, mp 92-93°. Insol in water. Slightly sol in cold alcohol; freely sol in hot alcohol and in ether. Very weak base, sol in concd mineral acids from which it is precipitated by water.

syn-Oxime. [2998-95-0] (Z)-[4-(Dimethylamino)phenyl]phenylmethanone oxime. $C_{15}H_{16}N_2O$; mol wt 240.31. Prisms from H_2O + alc, mp 163°.

anti-Oxime. [2998-94-9] (E)-[4-(Dimethylamino)phenyl]phenylmethanone oxime. $C_{15}H_{16}N_2O$; mol wt 240.31. mp 176°.

3256. *N*,*N*-Dimethylaniline. [121-69-7] *N*,*N*-Dimethylbenzenamine; dimethylphenylamine. $C_8H_{11}N$; mol wt 121.18. C 79.29%, H 9.15%, N 11.56%. Made by heating aniline, methyl alcohol, and H_2SO_4 under pressure, the sulfate formed being converted by NaOH to the free base. Toxicity data: H. F. Smyth *et al., Am. Ind. Hyg. Assoc. J.* **23**, 95 (1962).

Oily liq. *Poisonous.* d_4^{20} 0.956. bp 192-194°. mp 2°. Flash pt 61°C. n_D^{20} 1.5582. Insol in water. Freely sol in alcohol, chloroform, ether. LD_{50} orally in rats: 1.41 ml/kg (Smyth).

Caution: Potential symptoms of overexposure are anoxia, cyanosis, weakness, dizziness and ataxia; methemoglobinemia. *See NIOSH Pocket Guide to Chemical Hazards* (DHHS/NIOSH 97-140, 1997) p 112.

USE: Solvent; manuf vanillin, Michler's ketone, methyl violet and other dyes. As reagent for methanol, methyl furfural, H_2O_2, nitrate, alcohol, formaldehyde.

3257. 9,10-Dimethyl-1,2-benzanthracene. [57-97-6] 7,12-Dimethylbenz[*a*]anthracene; 1,4-dimethyl-2,3-benzphenanthrene. $C_{20}H_{16}$; mol wt 256.35. C 93.71%, H 6.29%. Synthesis from *o*-(*α*-methyl-*α*-1-naphthyl)toluic acid: Newman, *J. Am. Chem. Soc.* **60**, 1141 (1938). From 9,10-dimethyl-9,10-dihydroxy-9,10-dihydro-1,2-benzanthracene: Bachmann, Chemerda, *J. Am. Chem. Soc.* **60**, 1023 (1938). From 9,10-dimethyl-9,10-dimethoxy-9,10-dihydro-1,2-benzanthracene: *eidem, ibid.* **61**, 2358 (1939). From 9-methyl-1,2-benzanthracen-10-one by a Reformatsky reaction: Mikhailov, Chernova *J. Gen. Chem. USSR* **9**, 2171 (1939), *C.A.* **34**, 4068 (1940).

Plates, leaflets from acetone-alc, faint greenish-yellow tinge. mp 122-123°. Maximum fluorescence at 440 nm: *C.A.* **38**, 1276 (1944). Freely sol in benzene; moderately sol in acetone; slightly sol in alcohol. Insol in water. May be solubilized in water by purines such as caffeine, tetramethyluric acid. The nucleosides, adenosine, and guanosine also show a solvent action.

3258. 5,6-Dimethylbenzimidazole. [582-60-5] 5,6-Dimethyl-1*H*-benzimidazole. $C_9H_{10}N_2$; mol wt 146.19. C 73.94%, H 6.90%, N 19.16%. Obtained by acid hydrolysis of vitamin B_{12}: Brink, Folkers, *J. Am. Chem. Soc.* **71**, 2951 (1949); **72**, 4442 (1950). Synthesis by refluxing 4,5-diamino-1,2-dimethylbenzene and formic acid in 4*N* HCl: Hobrecker, *Ber.* **5**, 921 (1872); Phillips, *J. Chem. Soc.* **1928**, 2393.

Crystals from ether, mp 205-206°. Sublimes at 140° and 3 mm. uv max (95% ethanol acidified with HCl to be 0.01*N*): 274.5, 284 nm. Alkaline to litmus. Soluble in water, chloroform, ether; freely sol in dil acids.

3259. *p*,*α*-Dimethylbenzyl Alcohol. [536-50-5] *α*,4-Dimethylbenzenemethanol; *p*-tolylmethylcarbinol; methyl-*p*-tolylcarbinol;

4-(α-hydroxyethyl)toluene; 4-methyl-α-phenethyl alcohol; 1-*p*-tol-yl-1-ethanol. $C_9H_{12}O$; mol wt 136.19. C 79.37%, H 8.88%, O 11.75%. Constituent of the essential oil from *Curcuma longa* L., *Zingiberaceae* and related plants: Dieterle, Kaiser, *Arch. Pharm.* **271**, 337 (1933). Prepd from *p*-tolylmagnesium bromide and acet-aldehyde in ether: v. Auwers, Kolligs, *Ber.* **55**, 42 (1922); Eisenlohr, Schulz, *Ber.* **57**, 1816 (1924); from 4-methylacetophenone: Gas-taldi, Cherchi, *Gazz. Chim. Ital.* **45**, II, 274 (1915).

Viscous liquid. Odor somewhat like that of menthol. The natural product is probably levorotatory. Constants for the synthetic *dl*-form: $d_4^{15.5}$ 0.9668. bp_{756} 219°; bp_{14} 134°; bp_{11} 115-116°. Very sparingly sol in water. Miscible with abs alcohol, ether. Also sol in isopropanol, liquid petrolatum.

Phenylurethan. $C_{16}H_{17}NO_2$. Crystals, mp 97.5°.

3260. 2,3-Dimethyl-1,3-butadiene. [513-81-5] Diisopropen-yl; β,γ-dimethyl-Δ-α,γ-butadiene. C_6H_{10}; mol wt 82.15. C 87.72%, H 12.27%. Occurs in the 66-70° fraction of Puertollano shale oil. Prepd by slow distillation of anhydr pinacol in the presence of hydrobromic acid: Allen, Bell, *Org. Synth.* **coll. vol. III**, 312; by rapid distillation of anhydr pinacol over alumina at 420-470°: New-ton, Coburn, *ibid.* 313.

Liquid. d_{20}^{20} 0.7273; d_4^{20} 0.7267; d_4^{25} 0.7222. mp −76°. bp_{769} 69.2°; bp_{760} 68.8°. n_D^{25} 1.4362.

USE: In the manuf of synthetic rubber and polymers.

3261. Dimethylcadmium. [506-82-1] Cadmium dimethyl. C_2H_6Cd; mol wt 142.48. C 16.86%, H 4.24%, Cd 78.90%. Cd-$(CH_3)_2$. Prepn from a Grignard reagent and a cadmium halide: Krause, *Ber.* **50**, 1813 (1917); Gilman, Nelson, *Rec. Trav. Chim.* **55**, 518 (1936); Anderson, Taylor, *J. Phys. Chem.* **56**, 161 (1952).

Liquid. Disagreeable odor. May be kept in a sealed tube. mp −4.5°; bp_{758} 105.5°; $d_4^{17.9}$ 1.9846; n_D 1.5488. Dec with explosive violence when heated above 150°. Catches fire when dropped on filter paper and produces dense clouds of first white, then brown cadmium oxide smoke. When thrown into water, it sinks to the bottom in large drops, which dec in a series of sudden explosive jerks, with crack-ling sounds. Sol in hydrocarbons.

USE: In organic synthesis; as polymerization catalyst.

3262. Dimethyl Carbate. [39589-98-5] *rel*-(1*R*,2*S*,3*R*,4*S*)-Bicyclo[2.2.1]hept-5-ene-2,3-dicarboxylic acid dimethyl ester; *cis*-3,6-*endo*-methylene-Δ⁴-tetrahydrophthalic acid dimethyl ester; 5-norbornene-2,3-dicarboxylic acid dimethyl ester; dimalone. C_{11}-$H_{14}O_4$; mol wt 210.23. C 62.85%, H 6.71%, O 30.44%. Prepd by a Diels-Alder condensation of cyclopentadiene and maleic anhydride, followed by esterification of resulting anhydride: Bode, *Ber.* **70**, 1167 (1937); Morgan *et al.*, *J. Am. Chem. Soc.* **66**, 404 (1944). Tox-icology studies: J. H. Draize *et al.*, *J. Pharmacol. Exp. Ther.* **93**, 26 (1948).

Relative stereochemistry

Crystals, mp 38° when very pure. The usual product is a viscous, syrupy liquid. $bp_{12.5}$ 137°; bp_9 130°; $bp_{1.5}$ 115°. d_4^{21} 1.164. n_D^{20} 1.4852. Practically insol in water. Soluble in the usual organic sol-

vents. Solubility in kerosene and mineral oils ~6%. LD_{50} in mice, rats (ml/kg): 1.4, 1.0 orally; in rats (ml/kg): 10.0 dermally (Draize).

Caution: Can cause CNS excitation followed by depression.

USE: Insect repellent.

3263. Dimethyl Carbonate. [616-38-6] Carbonic acid di-methyl ester; methyl carbonate. $C_3H_6O_3$; mol wt 90.08. C 40.00%, H 6.71%, O 53.28%. Prepn: B. Röse, *Ann.* **205**, 227 (1880); V. Grignard *et al.*, *Ann. Chim. (Paris)* **9**, 229 (1920). Review of mfr and characteristics as a fuel additive: M. A. Pacheco, C. L. Marshall, *Energy Fuels* **11**, 2-29 (1997); of synthesis and chemical reactions: Y. Ono, *Appl. Catal. A* **155**, 133-166 (1997); of production and eco-friendly applications: D. Delledonne *et al.*, *ibid.* **221**, 241-251 (2001); and chemistry: P. Tundo, M. Selva, *Acc. Chem. Res.* **35**, 706-716 (2002).

Liquid. *Flammable.* d_4^{20} 1.07. mp 4.6°. bp 90.3°. Viscosity (20°): 0.625 cP. Flash pt, open cup: 71.1°F (21.7°C). Heat of va-porization: 88.2 kcal/kg. Soly in H_2O: 13.9 g/100g. Miscible with alc, ethers, esters, ketones. Dielectric constant at 25°: 3.087. Dipole moment: 0.91. LD_{50} in rats (g/kg): 13.8 orally; 2.5 dermally. LC_{50} (4 hr) in rats (mg/l): 140 by inhalation (Ono).

USE: Environmentally benign substitute for dimethyl sulfate, *q.v.*, and methyl halides in methylation reactions and for phosgene, *q.v.*, in methylcarbonylation reactions.

3264. 5,5-Dimethyl-1,3-cyclohexanedione. [126-81-8] 1,1-Dimethyl-3,5-diketocyclohexane; 1,1-dimethyl-3,5-cyclohexanedi-one; dimethyldihydroresorcinol; dimedone; methone. $C_8H_{12}O_2$; mol wt 140.18. C 68.55%, H 8.63%, O 22.83%. Prepd from mesityl oxide and diethyl malonate: Shriner, Todd, *Org. Synth.* **coll. vol. II**, 200 (1943).

Needles from water, prisms from alcohol + ether. Dec 148-150°. Slightly volatile with steam (50 ml of a distillate contained 0.016 g). The dry crystals may be stored in a brown bottle for several years at room temp. Aq solns oxidize on exposure to air and light or upon prolonged storage in the dark. 100 ml of a satd aq soln contains 0.401 g at 19°; 0.416 g at 25°; 1.185 g at 50°; 3.020 g at 80°; 3.837 g at 90°. Monobasic acid in water. pK (25°): 5.15. Dipole moment 3.46. Also sol in methanol, ethanol, chloroform, benzene, acetic acid, and in 50% ethanol-water mixture.

USE: For the separation and identification of aldehydes. Gives insol condensation products with aldehydes, but not with ketones. The use of piperidine as a catalyst is recommended.

3265. Dimethyldioxirane. [74087-85-7] 3,3-Dimethyl-1,2-dioxirane; DMDO; DMD. $C_3H_6O_2$; mol wt 74.08. C 48.64%, H 8.16%, O 43.19%. Cyclic peroxide that is a versatile oxidizing agent for a variety of functional groups. Effective reagent in epoxidation reactions of alkenes; also used for the oxidation of alkanes, alcohols, amines, and sulfides. Preliminary studies suggesting the existence of dimethyldioxirane as a reactive species: R. E. Montgomery, *J. Am. Chem. Soc.* **96**, 7820 (1974); J. O. Edwards *et al.*, *Photochem. Photobiol.* **30**, 63 (1979). Generation *in situ* and use in epoxidations: R. Curci *et al.*, *J. Org. Chem.* **45**, 4758 (1980); R. Jeyaraman, R. W. Murray, *J. Am. Chem. Soc.* **106**, 2462 (1984). Prepn from the oxi-dation of acetone with potassium peroxymonosulfate and use of the resulting soln in oxidation reactions: R. W. Murray, R. Jeyaraman, *J. Org. Chem.* **50**, 2847 (1985). Synthetic applications as an oxidizing reagent: A. L. Baumstark, P. C. Vasquez, *ibid.* **53**, 3437 (1988); R. W. Murray *et al.*, *ibid.* **54**, 5783 (1989); V. J. Colandrea *et al.*, *Org. Lett.* **5**, 785 (2003). *Reviews:* G. Dyker, *J. Prakt. Chem.* **337**, 162-163 (1995); V. P. Srivastava, *Synlett* **2008**, 626-627. Review of dioxirane chemistry: R. W. Murray, *Chem. Rev.* **89**, 1187-1201 (1989).

H3C   CH3
 \   /
  O — O

Obtained as a dilute yellow soln in acetone with a concn range of 0.07-0.09*M*; can be stored in soln over sodium sulfate at −25°C for up to one week. Volatile. uv max: 335 nm (ε 263). Sol in most organic solvents, but slowly reacts with them. Keep away from light and traces of heavy metals.

USE: Reagent in synthetic organic chemistry.

3266. Dimethyl Ether. [115-10-6] 1,1′-Oxybismethane; methoxymethane; methyl ether. C_2H_6O; mol wt 46.07. C 52.14%, H 13.13%, O 34.73%. Prepn: J. Dumas, E. Peligot, *Ann.* **15**, 1 (1835); and thermal properties: R. M. Kennedy *et al.*, *J. Am. Chem. Soc.* **63**, 2267 (1941). Atmospheric chemistry: S. M. Japar *et al.*, *Int. J. Chem. Kinet.* **22**, 1257 (1990). Review: H. Höver in *Ullmann's Encyclopedia of Industrial Chemistry* **vol. A8** (VCH, Weinheim, 1987) pp 541-544; of use in diesel engines: A. M. Rouhi, *Chem. Eng. News* **73**, 37-39 (May 29, 1995).

H3C   O   CH3
    \ | /

Colorless, flammable gas with a slight ethereal odor. mp −141.50°. bp −24.82°. d^{25} (1 atm) 1.91855 g/l. Heat capacity at 210.51 K: 23.88 cal/deg/mole. Heat of vaporization at bp: 5141.0 cal/mole. Heat of fusion: 1179.8 cal/mole. Flash pt −41°C. One vol water takes up 37 vols gas.

USE: Aerosol propellant; alternative diesel fuel; chemical intermediate.

3267. N,N-Dimethylformamide. [68-12-2] DMF; DMFA. C_3H_7NO; mol wt 73.10. C 49.29%, H 9.65%, N 19.16%, O 21.89%. Prepd from dimethylamine and formic acid: Mitchell, Reid, *J. Am. Chem. Soc.* **53**, 1879 (1931); Brown, *J. Appl. Chem.* **1**, Suppl. Issue no. 2, S159 (1951); Campbell, US 3015674 (1962 to Commercial Solvents); Surman, US 3072725 (1963 to du Pont); from dimethylamine + HCN: Benneville *et al.*, *J. Org. Chem.* **21**, 772 (1956); from HCN + methanol: Fukuoka, Kominami, *Chem. Tech.* **1972** (Nov.), 640. Toxicity study: W. Bartsch *et al.*, *Arzneim.-Forsch.* **26**, 1581 (1976). Reviews of chemical uses: R. S. Kittila, *Dimethylformamide Chemical Uses* (du Pont, Wilmington, 1967) 264 pp and Suppl. (1973) 148 pp; J. S. Pizey, *Synthetic Reagents* **Vol. 1** (John Wiley, New York, 1974) pp 4-99; C. L. Eberling in *Kirk-Othmer Encyclopedia of Chemical Technology* **vol. 11** (Wiley-Interscience, New York, 3rd ed., 1980) pp 263-268. Review of carcinogenic risk: *IARC Monographs* **47**, 171-197 (1989).

     O
     ||
 HC   CH3
    \ /
     N
     |
     CH3

Colorless to very slightly yellow liquid. Faint amine odor. mp −61°. bp$_{760}$ 153°; bp$_{39}$ 76°; bp$_{3.7}$ 25°. d_4^{25} 0.9445. n_D^{25} 1.42803. Flash pt, open cup: 153°F (67°C). Misc with water and most common organic solvents. pH of 0.5 molar soln in H_2O = 6.7. LD$_{50}$ in mice, rats (ml/kg): 6.8, 7.6 orally; 6.2, 4.7 i.p. (Bartsch).

Caution: Potential symptoms of overexposure are irritation of eyes, skin and respiratory system; nausea, vomiting and colic; liver damage, hepatomegaly; high blood pressure; facial flush; dermatitis. *See NIOSH Pocket Guide to Chemical Hazards* (DHHS/NIOSH 97-140, 1997) p 114.

USE: Solvent for liqs and gases. In the synthesis of organic compounds. Solvent for Orlon and similar polyacrylic fibers. Wherever a solvent with a slow rate of evaporation is required. Has been termed the universal organic solvent.

3268. 2,5-Dimethylfuran. [625-86-5] C_6H_8O; mol wt 96.13. C 74.97%, H 8.39%, O 16.64%. Volatile organic compound found in tobacco smoke and in roasted coffee aroma. Prepn: F. Dietrich,

C. Paal, *Ber.* **20**, 1077 (1887); I. M. Heilbron *et al.*, *J. Chem. Soc.* **1946**, 54. Identification in coffee aroma: M. A. Gianturco *et al.*, *Tetrahedron* **20**, 2951 (1964). GC-MS determn in breath of tobacco smokers: S. M. Gordon, *J. Chromatogr.* **511**, 291 (1990); in cigarette smoke: S. M. Charles *et al.*, *Environ. Sci. Technol.* **42**, 1324 (2008). Use in detection of environmental tobacco smoke: M. Alonso *et al.*, *ibid.* **44**, 8289 (2010). Production from biomass-derived fructose and use as biofuel: Y. Roman-Leshkov *et al.*, *Nature* **447**, 982 (2007); J. B. Binder, R. T. Raines, *J. Am. Chem. Soc.* **131**, 1979 (2009). Evaluation of combustion performance in direct-injection engine: S. Zhong *et al.*, *Energy Fuels* **24**, 2891 (2010).

H3C   O   CH3
    \ /   \ /

Dark yellow, clear liquid. *Flammable.* bp 93-94°. d^{20} 0.8897. n_D^{19} 1.4424. uv max (alcohol): 219.5 nm (ε 8000). Flash point, closed cup: 45°F (7°C). Research octane number (RON): 119. Calorific value: 30 kJ/ml. Latent heat of vaporization (20°): 31.91 kJ/mol. Soly in water (25°): ≤1.47 mg/ml.

USE: Liquid biofuel. Marker for detection of tobacco smoke inhalation.

3269. N,N-Dimethylglycine. [1118-68-9] 2-(Dimethylamino)acetic acid; *N*-methylsarcosine; DMG. $C_4H_9NO_2$; mol wt 103.12. C 46.59%, H 8.80%, N 13.58%, O 31.03%. Prepn: H. T. Clarke *et al.*, *J. Am. Chem. Soc.* **55**, 4571 (1933); D. E. Pearson, J. D. Bruton, *ibid.* **73**, 864 (1951). Mutagenicity study: N. Colman *et al.*, *Proc. Soc. Exp. Biol. Med.* **164**, 9 (1980). Constituent of some products sold in the U.S. as "pangamic acid".

        CH3   O
         | ||
 H3C   N   OH
    \   /   \ /
           CH2

Hygroscopic crystals, mp 157-160°.

Hydrochloride. [2491-06-7] $C_4H_9NO_2$·HCl; mol wt 139.58. mp 189-190°. Sol in alc, water. Insol in chloroform, acetone.

3270. Dimethylglyoxime. [95-45-4]; [17117-97-4] (2*E*,3*E*-form). 2,3-Butanedione 2,3-dioxime; diacetyldioxime; Chugaev's reagent. $C_4H_8N_2O_2$; mol wt 116.12. C 41.37%, H 6.94%, N 24.13%, O 27.56%. Prepn: Semon, Damerell, *Org. Synth.* **coll. vol. II**, 204 (1943). Manuf: Kamlet, US 2732404 (1956 to National Distillers Product Corp.). Crystal structure: *Anal. Chem.* **21**, 1428 (1949); Merritt, Lanterman, *Acta Crystallogr.* **5**, 811 (1952).

            CH3
             |
 HO   N   N   OH
    \ //   \   // \ /
    N   C
             |
            CH3

2E, 3E-form

Triclinic crystals from alc + water, mp 238-240°, also reported as melting 242° and 246°. Practically insol in water. Sol in alc, ether, pyridine, acetone.

USE: Ni-specific complexing reagent for its detection, determn, and separation from Co and many other metals. Forms a scarlet red ppt with Ni even in dil solns. Determn and separation of Pd from Sn, Au, Rh, and Ir. Detection of Bi with which it forms a bright yellow color and ppt.

3271. 1,1-Dimethylhydrazine. [57-14-7] *unsym*-Dimethylhydrazine; *asym*-dimethylhydrazine; *N,N*-dimethylhydrazine; UDMH; Dimazine. $C_2H_8N_2$; mol wt 60.10. C 39.97%, H 13.42%, N 46.61%. $(CH_3)_2NNH_2$. Prepd industrially by the reaction of dimethylamine and chloramine; by reduction of nitrosodimethylamine (obtained by treating a dimethylamine salt with sodium nitrite): Hatt, *Org. Synth.* **coll. vol. II**, 211 (1943). Toxicity studies: Witkin, *Arch. Ind. Health* **13**, 34 (1956); Cornish, Hartung, *Toxicol. Appl. Pharmacol.* **15**, 62 (1969). Review of carcinogenic risk: *IARC Monographs* **4**, 137-143 (1974); of toxicity and human exposure: *Toxicological Profile for Hydrazines* (PB98-101025, 1997) 224 pp.

Flammable. Hygroscopic, mobile liquid. Fumes in air and gradually turns yellow. Characteristic ammonia-like fishy odor of aliphatic hydrazines. d_4^{22} 0.791; d_{25}^{25} 0.782. mp $-58°$. bp_{760} 63.9°. $n_D^{22.3}$ 1.40753. Miscible with water with evolution of heat. Also miscible with alc, ether, dimethylformamide, hydrocarbons. LD_{50} in mice, rats (mg/kg): 265, 122 orally; 250, 119 i.v. (Witkin).

Hydrochloride. $C_2H_8N_2.HCl$. Hygroscopic crystals from abs ethanol, mp 83°. Sol in water, ethanol. Practically insol in ether.

Caution: Potential symptoms of overexposure are irritation of eyes and skin; choking, chest pain and dyspnea; lethargy; nausea; anoxia; convulsions; liver injury. *See NIOSH Pocket Guide to Chemical Hazards* (DHHS/NIOSH 97-140, 1997) p 114. *See also Patty's Industrial Hygiene and Toxicology* vol. 2A, G. D. Clayton, F. E. Clayton, Eds. (Wiley-Interscience, New York, 3rd ed., 1981) pp 2801-2803. 1,1-Dimethylhydrazine is reasonably anticipated to be a human carcinogen: *Report on Carcinogens, Twelfth Edition* (PB2011-111646, 2011) p 172.

USE: Base in rocket fuel formulations.

3272. 1,2-Dimethylhydrazine. [540-73-8] *N,N'*-Dimethylhydrazine; *sym*-dimethylhydrazine; SDMH. $C_2H_8N_2$; mol wt 60.10. C 39.97%, H 13.42%, N 46.61%. $CH_3NHNHCH_3$. Prepn from dibenzoylhydrazine: Folpmers, *Rec. Trav. Chim.* **34**, 34 (1915); Hatt, *Org. Synth.* **coll. vol. II**, 208 (1943). Electrosynthesis from nitromethane: Iversen, *Ber.* **104**, 2195 (1971). Toxicity study: Witkin, *Arch. Ind. Health* **13**, 34, (1956). Review of carcinogenic risk: *IARC Monographs* **4**, 145-150 (1974); and of toxicology and human exposure: *Toxicological Profile for Hydrazines* (PB98-101025, 1997) 224 pp.

Flammable, hygroscopic, mobile liquid. Fumes in air and gradually turns yellow. *Corrosive.* Characteristic ammoniacal odor of aliphatic hydrazines. d_4^{20} 0.8274. bp_{753} 81°. n_D^{20} 1.4209. Miscible with water with much evolution of heat. Also miscible with alcohol, ether, dimethylformamide, hydrocarbons. LD_{50} in mice, rats (mg/kg): 36, 160 orally; 29, 175 i.v. (Witkin).

Dihydrochloride. $C_2H_8N_2.2HCl$. Extremely hygroscopic prisms, dec 168°. Freely sol in water, ethanol.

Caution: For toxic symptoms *see* 1,1-Dimethylhydrazine.

3273. 1,3-Dimethyl-2-imidazolidinone. [80-73-9] *N,N'*-Dimethylethyleneurea; DMI. $C_5H_{10}N_2O$; mol wt 114.15. C 52.61%, H 8.83%, N 24.54%, O 14.02%. Aprotic basic solvent. Prepn: W. R. Boon, *J. Chem. Soc.* **1947**, 307; M. W. Farlow, US 2422400 (1947 to du Pont); E. V. Dehmlow, Y. R. Rao, *Synth. Commun.* **18**, 487 (1988). Prepn and pharmacological activity: E. J. Lien, W. D. Kumler, *J. Med. Chem.* **11**, 214 (1968). Prepn and spectral properties: H. Kohn *et al., J. Org. Chem.* **42**, 941 (1977). Evaluation as replacement solvent for HMPA, *q.v.*: C.-C. Lo, P.-M. Chao, *J. Chem. Ecol.* **16**, 3245 (1990).

mp 8.2° (Dehmlow, Rao); also reported as mp below −4° (Lo, Chao). bp_{754} 220°. bp_{17} 106-108°. bp_2 67-68°. n_D^{20} 1.4720. d^{25} 1.0519 g/ml. Dipole moment (dioxane, benzene): 4.09 ±0.01 D, 4.05 ±0.01 D. Dielectric constant (25°): 37.6. Viscosity (25°): 1.944 cP. Soly in toluene: >5%. LD_{50} i.p. in mice: 2840 mg/kg (Lien).

USE: As substitute solvent for HMPA in organic synthesis.

3274. Dimethylmercury. [593-74-8] Methyl mercury. C_2H_6Hg; mol wt 230.66. C 10.41%, H 2.62%, Hg 86.96%. $(CH_3)_2Hg$. Environmental contaminant found together with **monomethyl mercury** compounds in fish and birds. Prepns: Buckton, *Ann.* **108**, 103 (1858); **109**, 219 (1859); Jones, Werner, *J. Am. Chem. Soc.* **40**, 1257 (1918); Gilman, Brown, *ibid.* **52**, 3314 (1930); Wade, US 3636020 (1972 to Ventron); H. H. Sisler *et al., J. Org. Chem.* **45**, 1329 (1980). Synthesized in bottom sediments: Jensen, Jernelöv, *Nature* **223**, 753 (1969). Physical properties: Wilde, *J. Chem. Soc.* **1949**, 72. Formation and degradation studies: Bertilsson, Neujahr, *Biochemistry* **10**, 2805 (1971); DeSimone *et al., Biochim. Biophys. Acta* **304**, 851 (1973); Spangler *et al., Science* **180**, 192 (1973); *eidem, Appl. Micro-*

biol. **25**, 488 (1973). Neurotoxicity studies: Dales, *Am. J. Med.* **53**, 219 (1972); Skerfving, *Food Cosmet. Toxicol.* **10**, 545 (1972); Kojima, Fujita, *Toxicology* **1**, 43 (1973). Review of toxicity and human exposure: *Toxicological Profile for Mercury* (PB99-142416, 1999) 676 pp.

Colorless, volatile, toxic liquid; easily inflammable. bp_{740} 92°. d^{20} 3.1874. n_D^{20} 1.5452. Log P (octanol/water): 2.28. Easily sol in ether, alc. Insol in water.

USE: As inorganic reagent.

3275. Dimethyl(methylene)ammonium Iodide. [33797-51-2] *N*-Methyl-*N*-methylenemethanaminium iodide; Eschenmoser's salt. C_3H_8IN; mol wt 185.01. C 19.48%, H 4.36%, I 68.59%, N 7.57%. Mannich type intermediate originally developed to introduce methyl groups into corrin chromophore. Prepn from trimethylamine and diiodomethane: J. Schreiber *et al., Angew. Chem. Int. Ed.* **10**, 330 (1971); from *N,N,N',N'*-tetramethylmethylenediamine: T. A. Bryson *et al., J. Org. Chem.* **45**, 524 (1980). Used in prepn of Mannich bases: J. Hooz, J. N. Bridson, *J. Am. Chem. Soc.* **95**, 602 (1973). In functionalization of indoles: A. P. Kozikowski, H. Isida, *Heterocycles* **14**, 55 (1980). In prepn of α-methylene carbonyls: J. L. Roberts *et al., Tetrahedron Lett.* **1977**, 1621; of terminal olefins: *eidem, ibid.* 1299.

Colorless crystals from tetrahydrothiophene dioxide, dec ~240°. Sublimes at 120° at 0.05 torr.

USE: In organic synthesis.

3276. *N,N*-Dimethyl-1-naphthylamine. [86-56-6] *N,N*-Dimethyl-1-naphthalenamine; 1-dimethylaminonaphthalene. $C_{12}H_{13}N$; mol wt 171.24. C 84.17%, H 7.65%, N 8.18%. Prepn by heating in a closed tube 1-naphthylamine and methyl iodide in methyl alcohol: Landshoff, *Ber.* **11**, 643 (1878); by heating 1-naphthylamine with dimethyl sulfate in sodium hydroxide in presence of pyridine: Germuth, *J. Am. Chem. Soc.* **51**, 1556 (1929). Use: F. G. Germuth, *Ind. Eng. Chem. Anal. Ed.* **1**, 28 (1929).

Oil; aromatic odor. bp_{711} 274.5°; bp_{90} 193°; bp_{69} 184.5°; bp_{13} 139-140°; d_4^4 1.0522, d_{15}^{15} 1.0446, d_{25}^{25} 1.0391; n_D^{20} 1.622. Readily sol in alcohol or ether.

USE: Detection and determination of nitrites.

3277. Dimethylolpropionic Acid. [4767-03-7] 3-Hydroxy-2-(hydroxymethyl)-2-methylpropanoic acid; 2,2-bis(hydroxymethyl)propionic acid; dihydroxypivalic acid; DMPA. $C_5H_{10}O_4$; mol wt 134.13. C 44.77%, H 7.52%, O 47.71%. Prepn: H. Koch, T. Zerner, *Monatsh. Chem.* **22**, 447, 450 (1891); K. E. Wilzbach *et al., J. Am. Chem. Soc.* **70**, 4069 (1948); R. Riemschneider *et al., Monatsh. Chem.* **88**, 1099 (1957); H. Vieregge, J. F. Arens, *Rec. Trav. Chim.* **78**, 921 (1959).

Free-flowing granular powder, mp 181-185°. Sol in water, methanol. Slightly sol in acetone. Insol in benzene.

USE: In prepn of water-sol alkyl resins.

3278. Dimethyl-*p*-phenylenediamine. [99-98-9] N^1,N^1-Dimethyl-1,4-benzenediamine; *p*-aminodimethylaniline. $C_8H_{12}N_2$; mol wt 136.20. C 70.55%, H 8.88%, N 20.57%. Prepn from *p*-nitrosodimethylaniline which is obtained from dimethylaniline: Gattermann-Wieland, *Praxis des Organischen Chemikers* (de Gruyter,

Berlin, 40th ed., 1961) p 273. By reduction of methyl orange: A. I. Vogel, *Practical Organic Chemistry* (Longmans, London, 3rd ed., 1959) p 624.

Reddish-violet crystals. mp 53°, also stated as 41°. bp 262°. Sol in water, alcohol, chloroform, ether. *Keep tightly closed and protected from light.*

Dihydrochloride. $C_8H_{12}N_2.2HCl$. White to grayish-white, hygroscopic crystalline powder. Freely sol in water; sol in alcohol.

USE: Dihydrochloride in microscopy and in tests for acetone, uric acid, thallic salts, oxydases, lignin, ozone, H_2O_2, H_2S, Br.

3279. Dimethyl Phthalate. [131-11-3] 1,2-Benzenedicarboxylic acid 1,2-dimethyl ester; phthalic acid dimethyl ester; methyl phthalate; dimethyl 1,2-benzenedicarboxylate; DMP; Palatinol M; Fermine; Avolin; Mipax. $C_{10}H_{10}O_4$; mol wt 194.19. C 61.85%, H 5.19%, O 32.96%. Prepd industrially from phthalic anhydride and methanol: *Faith, Keyes & Clark's Industrial Chemicals*, F. A. Lowenheim, M. K. Moran, Eds. (Wiley-Interscience, New York, 4th ed., 1975) pp 318-324. Metabolite of *Gibberella fujikuroi:* Cross *et al., J. Chem. Soc.* **1963**, 2937. Toxicity data: J. H. Draize *et al., J. Pharmacol. Exp. Ther.* **93**, 26 (1948).

Oily liq. Slight aromatic odor. $d_{15.6}^{15.6}$ 1.196; d_{20}^{20} 1.1940; d_{25}^{25} 1.189. One gallon weighs 9.93 lbs. Flash pt 295°F (146°C). mp 5.5° (the commercial product freezes around 0°). bp$_{760}$ 283.7°; bp$_{400}$ 257.8°; bp$_{200}$ 232.7°; bp$_{100}$ 210.0°; bp$_{60}$ 194.0°; bp$_{40}$ 182.8°; bp$_{20}$ 164.0°; bp$_{10}$ 147.6°; bp$_5$ 131.8°; bp$_{1.0}$ 100.3°. Vapor pressure at 20° <0.01 mm. The vapor is heavy, d = 6.69 (air = 1). n_D^{20} 1.5168. Viscosity at 25° = 17.2 cP. Heat of vaporization: 93.1 g-cal/g; heat of combustion: 119.7 kg-cal/mole. uv max (ethanol): 277 nm ($E_{1cm}^{1\%}$ 57.7). Miscible with alcohol, ether, chloroform. Practically insol in water (0.43 g/100 ml), petr ether, and other paraffin hydrocarbons. Soly in mineral oil at 20°: 0.34 g/100 g. LD_{50} in mice, rats, guinea pigs (ml/kg): 7.2, 6.9, 2.4 orally (Draize).

Caution: Potential symptoms of overexposure are irritation of eyes, upper respiratory system; stomach pain. *See NIOSH Pocket Guide to Chemical Hazards* (DHHS/NIOSH 97-140, 1997) p 114.

USE: Solvent and plasticizer for cellulose acetate and cellulose acetate-butyrate compositions. Insect repellent for personal protection against biting insects.

3280. Dimethyl Sulfate. [77-78-1] Sulfuric acid dimethyl ester; DMS. $C_2H_6O_4S$; mol wt 126.13. C 19.05%, H 4.80%, O 50.74%, S 25.42%. $(CH_3)_2SO_4$. Prepn by distillation of a mixture of oleum and methanol: Guyot, Simon, *Compt. Rend.* **169**, 796 (1919). Technical production from dimethyl ether and SO$_3$: *BIOS Final Report* No. 986, p 176. *Review:* C. M. Suter, *The Organic Chemistry of Sulfur* (John Wiley, New York, 1944) p 49-61. Use as methylating agent: L. Fieser, M. Fieser, *Reagents for Organic Synthesis* (John Wiley, New York, 1967) pp 293-295. Toxicity data: Smyth *et al., Arch. Ind. Hyg. Occup. Med.* **4**, 119 (1951). Review of toxicology: E. Browning, *Toxicity and Metabolism of Industrial Solvents* (Elsevier, New York, 1965) pp 713-721; of carcinogenicity studies: *IARC Monographs* **4**, 271-276 (1974); of mutagenicity studies: G. R. Hoffmann, *Mutat. Res.* **75**, 63-129 (1980).

Colorless oily liq. *Poisonous. Corrosive.* bp ~188° (with dec); bp$_{15}$ 76°. mp −27°. d_4^{20} 1.3322. n_D^{20} 1.3874. Flash pt 182°F (83°C). Soly in water 2.8 g/100 ml at 18°. Hydrolysis is rapid at or above this temp. Vapor density 4.35. Sol in ether, dioxane, acetone, aromatic hydrocarbons. Sparingly sol in carbon disulfide, aliphatic hydrocarbons. LD_{50} orally in rats: 440 mg/kg (Smyth).

Caution: Vapors and liquid can be absorbed through the skin or respiratory tract, causing local and systemic toxicity. Potential symptoms of overexposure are irritation of eyes and nose; conjunctivitis; eye and skin burns; headache; giddiness; delerium; photophobia; periorbital edema; dysphonia, aphonia, dysphagia; cough, chest pain; dyspnea; cyanosis; vomiting; diarrhea, dysuria; analgesia; fever; albuminuria, hematuria. *See NIOSH Pocket Guide to Chemical Hazards* (DHHS/NIOSH 97-140, 1997) p 116; P. Grandjean, "Skin Penetration: Hazardous Chemicals at Work" (Taylor & Francis, New York, 1990) pp 173-174. This substance is reasonably anticipated to be a human carcinogen: *Report on Carcinogens, Twelfth Edition* (PB2011-111646, 2011) p 174.

USE: Methylating agent in the manuf of many organic chemicals. War gas.

3281. Dimethyl Sulfide. [75-18-3] 1,1′-Thiobismethane; dimethyl thioether; methyl sulfide. C_2H_6S; mol wt 62.13. C 38.66%, H 9.73%, S 51.60%. Metabolic product of many biosystems. Important flavor constituent in lager beer and in many cooked vegetables such as tomatoes and corn. Principal volatile sulfur compound in seawater; abundantly produced by marine algae. Prepn: V. Regnault, *Ann.* **34**, 24 (1840); and properties: D. T. McAllan *et al., J. Am. Chem. Soc.* **73**, 3627 (1951). Review of role in malting and brewing: C. J. Dickenson, *Chem. Ind.* **24**, 896-898 (1979); B. J. Anness, C. W. Bamforth, *J. Inst. Brew.* **88**, 244-252 (1982). Review of role in global sulfur cycle: P. S. Liss *et al., Philos. Trans. R. Soc. London Ser. B* **352**, 159-169 (1997). Review of gas-phase chemistry: I. Barnes *et al., Chem. Rev.* **106**, 940-975 (2006).

Flammable liquid; characteristic odor and flavor. mp −98.25°. bp 37.3°. d^{20} 0.8483 g/ml; d^{25} 0.8424 g/ml. n_D^{20} 1.4353; n_D^{25} 1.4319. Sol in alcohol, ether; sol in water at concentrations below 300 mM.

USE: Sulfiding agent.

3282. 2,4-Dimethylsulfolane. [1003-78-7] Tetrahydro-2,4-dimethylthiophene 1,1-dioxide; 2,4-dimethylthiacyclopentane 1,1-dioxide; 2,4-dimethyltetramethylene sulfone; 2,4-dimethylcyclotetramethylene sulfone. $C_6H_{12}O_2S$; mol wt 148.22. C 48.62%, H 8.16%, O 21.59%, S 21.63%. Prepd by catalytic hydrogenation of 2,4-dimethyl-3-sulfolene: Morris, Melchior, **US 2451298** (1948 to Shell).

Colorless to yellow liquid. d_4^{20} 1.1362. n_D^{20} 1.4733. bp$_5$ 123.3°; bp 280-281° (some decompn). Miscible with lower aromatic hydrocarbons; partially miscible with naphthenes, olefins, paraffins. Limited miscibility with water.

USE: Solvent for liquid-liquid and vapor-liquid extraction processes.

3283. Dimethyl Sulfone. [67-71-0] 1,1′-Sulfonylbismethane; DMSO$_2$; methyl sulfone; methylsulfonylmethane. $C_2H_6O_2S$; mol wt 94.13. C 25.52%, H 6.43%, O 33.99%, S 34.06%. $CH_3SO_2CH_3$. Has been found in primitive plants such as *Equisetum arvense* L., *Equisetaceae* and in the adrenal cortex of cattle: Pfiffner, North, *J. Biol. Chem.* **134**, 781 (1940). Easily prepd by oxidation of dimethyl sulfide: Douglas, *J. Am. Chem. Soc.* **68**, 1072 (1946); McAllan *et al., ibid.* **73**, 3627 (1951).

Crystals, mp 109°. bp$_{760}$ 238°. Sublimes at 13 mm and 90° to 100°. Infrared absorption (solid) 7600-8700 nm. Dipole moment 4.44 (vapor). Freely sol in water, methanol, ethanol, acetone. Sparingly sol in ether.

USE: High temp solvent for many inorganic and organic substances.

3284. Dimethylsulfoniopropionate. [7314-30-9] (2-Carboxyethyl)dimethylsulfonium inner salt; dimethyl-2-carboxyethylsulfonium hydroxide; dimethyl-β-propiothetin; DMSP. C$_5$H$_{10}$O$_2$S; mol wt 134.19. C 44.75%, H 7.51%, O 23.85%, S 23.89%. Organosulfur zwitterion produced by marine phytoplankton and certain halophytic plants. Maintained at high intracellular concentrations; functions as an osmolyte, cryoprotectant, and antioxidant. Released into the surrounding water during grazing by zooplankton, DSMP acts as a chemoattractant and foraging cue. Key compound in global sulfur cycle; major biological precursor of dimethyl sulfide, *q.v.*, which has a role in geoclimate regulation. Isoln from the marine alga, *Polysiphonia fastigiata*, and identification as dimethyl sulfide precursor: F. Challenger, M. I. Simpson, *J. Chem. Soc.* **1948**, 1591. Activity as biological methyl donor: G. A. Maw, V. du Vigneaud, *J. Biol. Chem.* **174**, 381 (1948). Synthesis: N. F. Blau, C. G. Stuckwisch, *J. Am. Chem. Soc.* **73**, 2355 (1951). LC/MS determn in water samples and algal extracts: A. Spielmeyer, G. Pohnert, *J. Chromatogr. B* **878**, 3238 (2010). Review of biosynthesis, physiological role, and degradation: J. Stefels, *J. Sea Res.* **43**, 183-197 (2000); of uptake and use by marine bacteria: R. P. Kiene *et al., ibid.*, 209-224; and role in global sulfur cycle: D. C. Yoch, *Appl. Environ. Microbiol.* **68**, 5804-5815 (2002). Role as foraging cue for fish: J. L. DeBose *et al., Science* **319**, 1356 (2008); as chemoattractant for marine microorganisms: J. R. Seymour *et al., ibid.* **329**, 342 (2010).

Chloride. [4337-33-1] Dimethyl-β-propiothetin hydrochloride. C$_5$H$_{11}$ClO$_2$S; mol wt 170.65. Colorless needles from abs alcohol, mp 134° (dec) (Challenger, Simpson). Also reported as crystals from hot ethanol + acetone, mp 129° (dec) (Blau, Stuckwisch).

3285. Dimethyl Sulfoxide. [67-68-5] 1,1'-Sulfinylbismethane; methyl sulfoxide; DMSO; SQ-9453; Domoso; Kemsol; Rimso-50. C$_2$H$_6$OS; mol wt 78.13. C 30.75%, H 7.74%, O 20.48%, S 41.03%. Dipolar, aprotic solvent; *methylsulfinyl carbanion* as nucleophilic reagent in organic synthesis. Attributed with a wide range of bioactivities including analgesic, anti-inflammatory and cryoprotective properties. Prepn: A. Saytzeff, *Ann.* **144**, 148 (1867). Obtained as a by-product of paper manufacture: M. D. Robbins, *Chem. Eng.* **68**, 100 (June 26, 1961). Prepn and reactivity of anion: E. J. Corey, M. Chaykovsky, *J. Am. Chem. Soc.* **87**, 1345 (1965); T. J. Broxton *et al., Aust. J. Chem.* **19**, 521 (1966). Toxicity data: W. Bartsch *et al., Arzneim.-Forsch.* **26**, 1581 (1976). Review of properties and use in organic chemistry: D. Martin *et al., Angew. Chem. Int. Ed.* **6**, 318-334 (1967). Reviews of pharmacology and toxicology: S. W. Jacob, D. C. Wood, *Arzneim.-Forsch.* **17**, 1553-1560 (1967); N. A. David, *Annu. Rev. Pharmacol.* **12**, 353-374 (1972). Review of clinical experience: B. N. Swanson, *Rev. Clin. Basic Pharmacol.* **5**, 1-33 (1985); of use in rheumatic disorders: J. M. Trice, R. S. Pinals, *Semin. Arthritis Rheum.* **15**, 45-60 (1985). Review of veterinary uses: C. F. Brayton, *Cornell Vet.* **76**, 61-90 (1986). Brief review of synthetic utility of dimsyl sodium: M. Mondal, *Synlett* **2005**, 2697-2698.

Very hygroscopic liquid. Practically no odor or color. Slightly bitter with sweet after-taste. d$_4^{20}$ 1.100. mp 18.55°. bp$_{760}$ 189°; bp$_{16}$ 79.83°; bp$_{12}$ 72.5°. Flash pt, open cup: 203°F (95°C). n$_D^{20}$ 1.4783. Viscosity (20°): 2.47 cP. Specific heat (25°): 0.4698 cal/g. Dielectric constant: 45. Sol in water. Practically insol in acetone, alcohol, ether, benzene, chloroform. LD$_{50}$ orally in rats: 17.9 ml/kg (Bartsch).

Dimsyl sodium. [15590-23-5] Sulfinylbismethane ion(1−) sodium; methylsulfinyl sodium; sodium dimethylsulfoxylate; sodium methylsulfinylmethylide; sodium DMSO. C$_2$H$_5$NaOS; mol wt 100.11. Prepd by heating sodium hydride with DMSO. Limited stability with exothermic decomposition beginning at 40°. Reacts rapidly with water, carbon dioxide, and oxygen.

Caution: Rapidly absorbed through skin and mucous membranes; enhances dermal absorption of many other chemicals. Repeated topical application may result in mild, erythamatous, scaling dermatitis. Corneal opacities have been produced in exptl animals. *See Clinical Toxicology of Commercial Products*, R. E. Gosselin *et al.*, Eds. (Williams & Wilkins, Baltimore, 5th ed., 1984) Section II, p 407.

USE: Reagent in organic synthesis. Industrial solvent; as antifreeze or hydraulic fluid when mixed with water. To cryopreserve and store cultured cells.

THERAP CAT: In treatment of interstitial cystitis and scleroderma.

THERAP CAT (VET): Topically to reduce swelling due to trauma.

3286. Dimethylsulfoxonium Methylide. [5367-24-8] Corey's reagent; Corey-Chaykovsky reagent; dimethyloxosulfonium methylide; DMSY. C$_3$H$_8$OS; mol wt 92.16. C 39.10%, H 8.75%, O 17.36%, S 34.79%. Methylene transfer reagent. Prepn from trimethylsulfoxonium iodide or chloride and NaH: E. J. Corey, M. Chaykovsky, *J. Am. Chem. Soc.* **84**, 867 (1962); and synthetic applications: *eidem, ibid.* **87**, 1353 (1965). Alternate large scale prepn: J. S. Ng, *Synth. Commun.* **20**, 1193 (1990). NMR characterization and physical properties: H. Schmidbaur, W. Tronich, *Tetrahedron Lett.* **9**, 5335 (1968). Review of chemistry: Y. G. Gololobov *et al., Tetrahedron* **43**, 2609-2651 (1987); J. S. Ng, C. Liu in *Encyclopedia of Reagents for Organic Synthesis*, **vol. 3**, L. A. Paquette, Ed. (John Wiley & Sons, Chichester, 1995) pp 2159-2165; C. Awasthi, *Synlett* **2008**, 1423-1424.

mp 9-10°. bp$_{0.1}$ 41-43°. Sol in THF, DMSO, 1,4-dioxane, DMF. Typically generated in situ. Stable as soln in THF for several months at 0°; decomposes after one week at room temp.

USE: Reagent in organic synthesis for epoxidation of ketones and aldehydes; cyclopropanation of α,β-unsaturated carbonyl compounds; methylation reactions.

3287. *N,N'*-Dimethylthiourea. [534-13-4] Dimethylthiocarbamide. C$_3$H$_8$N$_2$S; mol wt 104.17. C 34.59%, H 7.74%, N 26.89%, S 30.78%.

Colorless, exceedingly deliquesc crystals. mp 60-62°. Very sol in water, alcohol, acetone; sparingly sol in benzene, ether, carbon disulfide, very slightly in petr ether. *Keep tightly closed.*

3288. *N,N*-Dimethyltryptamine. [61-50-7] *N,N*-Dimethyl-1*H*-indole-3-ethanamine; 3-[2-(dimethylamino)ethyl]indole; DMT. C$_{12}$H$_{16}$N$_2$; mol wt 188.27. C 76.56%, H 8.57%, N 14.88%. Occurs naturally in plants with hallucinogenic properties. Isoln from the leaves of *Prestonia amazonica* (Benth.) Macbride (*Haemadictyon amazonicum* Spruce & Benth.), *Apocynaceae*: Hockstein, Paradies, *J. Am. Chem. Soc.* **79**, 5735 (1957). Metabolism: St. Szára, *Experientia* **12**, 441 (1956); B. R. Sitaram *et al., Biochem. Pharmacol.* **36**, 1509 (1987). Relationship between hallucinogenic activity and electronic configuration: Snyder, Merril, *Proc. Natl. Acad. Sci. USA* **54**, 258 (1965). Synthesis: I. Fleming, M. Woolias, *J. Chem. Soc. Perkin Trans. 1* **1979**, 829. Differential interaction with serotonin receptor subtypes: A. V. Deliganis *et al., Biochem. Pharmacol.* **41**, 1739 (1991). Fluorometric detection: O. Jules *et al., Anal. Chim. Acta* **169**, 355 (1985). Review of biosynthesis, metabolism and bioactivity: S. A. Barker *et al., Int. Rev. Neurobiol.* **22**, 83-110 (1981).

Crystals, mp 44.6-46.8°; also reported as plates from ethanol and light petroleum, mp 46° (Fleming, Woolias). bp 60-80°. pKa 8.68 (ethanol-water). Freely sol in dil acetic and dil mineral acids.

Methiodide. mp 216-217°.

Note: This is a controlled substance (hallucinogen): **21 CFR,** 1308.11.

3289. Dimethylzinc. [544-97-8] Zinc dimethyl. C_2H_6Zn; mol wt 95.48. C 25.16%, H 6.33%, Zn 68.51%. $Zn(CH_3)_2$. Prepd by the interaction of zinc and methyl iodide: Simonowitch, *Chem. Zentralbl.* **1899,** I, 1066; Bamford *et al., J. Chem. Soc.* **1946,** 468.

Mobile liquid. Stable in sealed tube and under CO_2. $d_4^{10.5}$ 1.386; mp −40°; bp 46°. *Spontaneously combustible, dangerous when wet.* Burns with a blue flame, giving off a peculiar, garlic-like odor. Very slow oxidation with traces of air produces methylzinc methylate, CH_3ZnOCH_3. Sol in ether, miscible with hydrocarbons.

3290. Dimetridazole. [551-92-8] 1,2-Dimethyl-5-nitro-1*H*-imidazole; RP-8595; Emtryl; Unizole. $C_5H_7N_3O_2$; mol wt 141.13. C 42.55%, H 5.00%, N 29.77%, O 22.67%. Prepn: Bhagwat, Pyman, *J. Chem. Soc.* **127,** 1832 (1925).

Needles from water, mp 138-139°. Freely sol in alcohol; sparingly sol in cold water, ether.

Hydrochloride. $C_5H_7N_3O_2$.HCl. Prisms from dil HCl, mp 195°. Freely sol in water, alcohol; sparingly sol in acetone. Also used as the dihydrogen phosphate, freely sol in water.

THERAP CAT (VET): Antiprotozoal (Histomonas).

3291. Diminazene Aceturate. [908-54-3] *N*-Acetylglycine compd with 4,4'-(1-triazene-1,3-diyl)bis(benzenecarboximidamide) (2:1); *N*-acetylglycine compd with 4,4'-(diazoamino)dibenzamidine; diminazene diaceturate; 4,4'-(diazoamino)dibenzamidine diaceturate; 1,3-bis(*p*-amidinophenyl)triazene bis(*N*-acetylglycinate); 1,3-bis[4-guanylphenyl]triazene diaceturate; 4,4'-diamidinodiazoaminobenzene diaceturate; *p,p'*-diguanyldiazoaminobenzene diaceturate; Berenil. $C_{22}H_{29}N_9O_6$; mol wt 515.53. C 51.26%, H 5.67%, N 24.45%, O 18.62%. Prepn: Brodersen *et al.,* **US 2838485** (1958 to Hoechst); Stavrovskaya, Drusvyatskaya, *C.A.* **59,** 15199a (1963). Mode of action: Festy *et al., C. R. Seances Acad. Sci. Ser. D* **271,** 730 (1970). Chemotherapeutic characteristics: Schmulevich *et al., Veterinariya (Moscow)* **38,** 23 (1961), *C.A.* **56,** 13512b (1962). Review: Newton in *Antibiotics* vol. 3, J. W. Corcoran, F. E. Hahn, Eds. (Springer-Verlag, New York, 1975) pp 34-47.

Yellow solid, dec 217°. Sol in 14 parts water (20°); slightly sol in alcohol; very slightly sol in ether, chloroform.

THERAP CAT (VET): Antiprotozoal (Trypanosoma, Babesia).

3292. Dimoxystrobin. [149961-52-4] (*αE*)-2-[(2,5-Dimethylphenoxy)methyl]-*α*-(methoxyimino)-*N*-methylbenzeneacetamide; (*E*)-2-[*α*-(2,5-dimethylphenoxy)-*o*-tolyl]-2-(methoxyimino)-*N*-methylphenylacetamide; (*E*)-2-methoxyimino-*N*-methyl-2-[2-(2,5-dimethylphenoxymethyl)phenyl]acetamide; BAS-505; SSF-129. $C_{19}H_{22}N_2O_3$; mol wt 326.40. C 69.92%, H 6.79%, N 8.58%, O 14.70%. Broad-spectrum, strobilurin fungicide for control of cereal and fruit diseases. Prepd (not claimed): A. Takase *et al.,* **EP 535928;** *eidem,* **US 5387714** (1993, 1995 both to Shionogi). Structure activity study: M. Ichinari *et al., Pestic. Sci.* **55,** 347 (1999). LC/MS/MS determn of pesticide residues: C. L. Hetherton *et al., Rapid Commun. Mass Spectrom.* **18,** 2443 (2004). Study for control of rice blast fungus and wheat blast disease: A. S. Urashima, H. Kato, *Summa Phytopathol.* **20,** 107 (1994). Aerobic degradation

rate: J. A. McDonald *et al., Soil Sci.* **171,** 239 (2006). Soil adsorption and bioactivity: E. Jastrzebska, J. Kucharski, *Plant Soil Environ.* **53,** 51 (2007).

Crystals from ethanol, mp 154-155.5°.

Mixture with boscalid. Pictor.

Mixture with epoxiconazole. Swing Gold.

USE: Agricultural fungicide.

3293. *β,β'*-Dinaphthylamine. [532-18-3] *N*-2-Naphthalenyl-2-naphthalenamine. $C_{20}H_{15}N$; mol wt 269.35. C 89.19%, H 5.61%, N 5.20%. Prepn from *β*-naphthol: Merz, Weith, *Ber.* **13,** 1300 (1880); Benz, *Ber.* **16,** 15 (1883); from *β*-naphthylamine: Klopsch, *Ber.* **18,** 1586 (1885); Liebermann, Jacobson, *Ann.* **211,** 43 (1882).

Glistening silvery leaves from benzene, mp 170.5°. Slightly sol in boiling alc; readily sol in boiling glacial acetic acid.

USE: Detection of nitrites, nitrates, and chlorates: Sa, *An. Farm. Bioquim.* **5,** 111 (1934), *C.A.* **30,** 6672 (1936).

3294. Diniconazole. [83657-24-3] (*βE*)-*β*-[(2,4-Dichlorophenyl)methylene]-*α*-(1,1-dimethylethyl)-1*H*-1,2,4-triazole-1-ethanol; (*E*)-1-(2,4-dichlorophenyl)-4,4-dimethyl-2-(1,2,4-triazol-1-yl)-1-penten-3-ol; S-3308-10; S-3308L; XE-779L; Ortho Spotless; Spotless; Sumi-8. $C_{15}H_{17}Cl_2N_3O$; mol wt 326.22. C 55.23%, H 5.25%, Cl 21.73%, N 12.88%, O 4.90%. Prepn: Y. Funaki *et al.,* **DE 3010560;** *eidem,* **US 4554007;** resolution of enantiomers: Y. Funaki *et al.,* **EP 54431;** *eidem,* **US 4435203** (1980, 1985, 1982, 1984 all to Sumitomo). Metabolism of racemate in rats: N. Isobe *et al., Nippon Noyaku Gakkaishi* **10,** 475 (1985), *C.A.* **104,** 103703b (1986); of enantiomers: *eidem, ibid.* **12,** 421 (1987), *C.A.* **108,** 1715x (1988). Mechanism of sterol inhibition: Y. Yoshida *et al., Biochem. Biophys. Res. Commun.* **137,** 513 (1986); T. Katagi, *J. Agric. Food Chem.* **36,** 344 (1988). Fungicidal activity of the (−)-form and plant growth regulating activity of the (+)-form: C. S. Kvien *et al., J. Plant Growth Regul.* **6,** 233 (1987); D. J. Daigle *et al., Proc. 14th Ann. Mtg. Plant Growth Regul. Soc. Am. 1987,* p 108. Field trials as foliar fungicide: A. R. Biggs, J. Warner, *Can. J. Plant Pathol.* **9,** 41 (1987); A. K. Hagan *et al., J. Environ. Hort.* **6,** 67 (1988); against soil-borne diseases: A. S. Csinos *et al., Appl. Agric. Res.* **2,** 113 (1987). Comprehensive description: H. Takano, *Jpn. Pestic. Inf.* **49,** 18 (1986).

Crystals from isopropanol, mp 148-149°. Vapor pressure at 20°: 2.2×10^{-5} mm Hg. Soly at 25° (% v/w): water 4.01; at 23°: acetone 9.5; methanol 9.5; xylene 1.4. Stable to heat, sunlight and moisture. LD$_{50}$ in male, female rats (mg/kg): 639, 474 orally; >5000, >5000 dermally (Takano).

(−)-**Form.** [83657-18-5] Crystals from carbon tetrachloride + *n*-hexane, mp 160-161°. [α]$_D^{24}$ −31.7° (c = 1 in CHCl$_3$).

(+)-**Form.** [83657-19-6] Crystals from carbon tetrachloride + *n*-hexane, mp 160-161°. [α]$_D^{24}$ +26° (c = 1 in CHCl$_3$).

USE: Agricultural fungicide.

3295. Dinitolmide. [148-01-6] 2-Methyl-3,5-dinitrobenz-amide; 3,5-dinitro-*o*-toluamide; 3,5-dinitro-2-methylbenzamide; Zo-alene; Zoamix. C$_8$H$_7$N$_3$O$_5$; mol wt 225.16. C 42.68%, H 3.13%, N 18.66%, O 35.53%. Prepn: McGookin *et al.*, *J. Soc. Chem. Ind.* **59**, 92 (1940); Harris *et al.*, **US 2937204** (1960 to Dow). Activity studies: Peterson, *Poult. Sci.* **39**, 739 (1960); Joyner, *Res. Vet. Sci.* **1**, 363 (1960). Metabolism: Smith *et al.*, *J. Agric. Food Chem.* **11**, 253 (1963); Smith, *Anal. Biochem.* **7**, 461 (1964).

Crystals from dil alc, mp 181°.

Caution: Potential symptom of overexposure by direct contact is eczema. *See NIOSH Pocket Guide to Chemical Hazards* (DHHS/NIOSH 97-140, 1997) p 116.

THERAP CAT (VET): Coccidiostat.

3296. 2,4-Dinitroaniline. [97-02-9] 2,4-Dinitrobenzenamine. C$_6$H$_5$N$_3$O$_4$; mol wt 183.12. C 39.35%, H 2.75%, N 22.95%, O 34.95%. Prepn: Wells, Allen, *Org. Synth. coll. vol.* **II**, 221 (1943). Liquid chromatographic determn: A. L. Scher, *J. Assoc. Off. Anal. Chem.* **68**, 474 (1985). Metabolism and disposition in rats: H. B. Matthews *et al.*, *Xenobiotica* **16**, 1 (1986). Toxicity study: E. G. Feldmann, W. O. Foye, *J. Am. Pharm. Assoc. Sci. Ed.* **48**, 419 (1959).

Yellow needles from dil acetone, greenish-yellow plates from alcohol. mp 187.5-188°. pKa 18.46. Practically insol in cold water. Very sparingly sol in boiling water. 5.8 parts dissolve in 1000 parts of 88% alc at 18°; one part dissolves in 132.6 parts of 95% alc at 21°.

Caution: May be irritating to skin, mucous membranes.

USE: In prepn of azo dyes.

3297. 2,6-Dinitroaniline. [606-22-4] 2,6-Dinitrobenzenam-ine. C$_6$H$_5$N$_3$O$_4$; mol wt 183.12. C 39.35%, H 2.75%, N 22.95%, O 34.95%. Synthesis starting with chlorobenzene, fuming sulfuric acid, and potassium nitrate: Schultz, *Org. Synth. coll. vol.* **IV**, 364 (1963).

Light orange needles from alc, mp 139-140°. Soly in 95% ethanol about 0.4 g/100 ml. Also sol in ether, hot benzene. Practically insol in water, petr ether. Dipole moment: 1.9.

3298. 2,4-Dinitrobenzaldehyde. [528-75-6] C$_7$H$_4$N$_2$O$_5$; mol wt 196.12. C 42.87%, H 2.06%, N 14.28%, O 40.79%. Prepn: *p*-Nitrosodimethylaniline hydrochloride, obtained by reacting di-methylaniline and sodium nitrite, is condensed with 2,4-dinitrotol-uene in the presence of sodium carbonate. The product, dinitro-

benzylidene-*p*-aminodimethylaniline is split under steam agitation with HCl: Bennett, Bell, *Org. Synth.* **coll. vol. II**, 223 (1943).

Yellow to light brown crystals, mp 72°. bp$_{10-20}$ 190-210°. Freely sol in alc, ether, benzene, slightly sol in petr ether, water.

USE: In prepn of Schiff bases.

3299. Dinitrobenzene. C$_6$H$_4$N$_2$O$_4$; mol wt 168.11. C 42.87%, H 2.40%, N 16.66%, O 38.07%. Commercial product usu-ally consists of a mixture of the *m*, *o* and *p* isomers. Toxicity study: T. E. Cody *et al.*, *J. Toxicol. Environ. Health* **7**, 829 (1981). Review of toxicology and human exposure: *Toxicological Profile for 1,3-Dinitrobenzene and 1,3,5-Trinitrobenzene* (PB95-264289, 1995) 169 pp.

m-**Dinitrobenzene.** [99-65-0] 1,3-Dinitrobenzene. Yellowish crystals. mp 89-90°. bp 300-303°. Volatile with steam. One gram dissolves in 2000 ml cold water, 320 ml boiling water, 37 ml alc, 20 ml boiling alcohol; freely sol in benzene, chloroform, ethyl acetate. LD$_{50}$ in male, female rats (mg/kg): 91, 81 orally (Cody).

o-**Dinitrobenzene.** [528-29-0] 1,2-Dinitrobenzene. White crys-tals. d 1.57. mp 118°. bp 319°. Volatile with steam. One gram dissolves in 6600 ml cold water, 2700 ml boiling water, about 60 ml alc, 3 ml boiling alc, 20 ml benzene; freely sol in chloroform, ethyl acetate.

p-**Dinitrobenzene.** [100-25-4] 1,4-Dinitrobenzene. White crys-tals; sublimable. d 1.63. mp 173-174°. bp 299°. Volatile with steam. One gram dissolves in 12,500 ml cold water, 555 ml boiling water, 300 ml alcohol; sparingly soluble in benzene, chloroform, ethyl acetate.

Caution: Potential symptoms of overexposure include methemo-globinemia, headache, nausea, dizziness, general malaise (*See Toxi-cological Profile*); anoxia, cyanosis; visual disturbances, central sco-tomas; bad taste, burning mouth, dry throat, thirst; yellowing hair, eyes, skin; anemia; liver damage. *See NIOSH Pocket Guide to Chemical Hazards* (DHHS/NIOSH 97-140, 1997) p. 116-119.

USE: Manuf of dyes, dye intermediates, explosives, plastics.

3300. 2,4-Dinitrobenzenesulfenyl Chloride. [528-76-7] 2,4-Dinitrophenylsulfenyl chloride. C$_6$H$_3$ClN$_2$O$_4$S; mol wt 234.61. C 30.72%, H 1.29%, Cl 15.11%, N 11.94%, O 27.28%, S 13.67%. Prepd from bis(2,4-dinitrophenyl)disulfide and chlorine in ethylene bromide: Kharasch *et al.*, *J. Am. Chem. Soc.* **69**, 1612 (1947); *eidem, ibid.* **72**, 1796 (1950); *eidem, J. Chem. Educ.* **33**, 585 (1956); from 2,4-dinitrothiophenol and chlorine in benzene: Perold, Snyman, *J. Am. Chem. Soc.* **73**, 2379 (1951); from 2,4-dinitrophenylbenzyl sul-fide and sulfuryl chloride in ethylene chloride: Kharasch, Langford, *Org. Synth.* **44**, 47 (1964). *Review:* Kharasch, *Organic Sulfur Com-pounds* vol. 1 (Pergamon Press, New York, 1961) pp 375-396.

Yellow crystals from carbon tetrachloride, mp 96°. Sol in glacial acetic acid, methylene chloride, ethylene chloride, trichlorethylene, benzene, xylene. Somewhat less sol in carbon tetrachloride. Insol in ether. Reacts with alc (even in the cold) to produce ethyl 2,4-dinitrobenzenesulfenate.

USE: Analytical reagent for the characterization of organic compds: Kharasch, *J. Chem. Educ.* **33**, 585 (1956); Langford, Lawson, *ibid.* **34**, 510 (1957).

3301. 3,4-Dinitrobenzoic Acid. [528-45-0] $C_7H_4N_2O_6$; mol wt 212.12. C 39.64%, H 1.90%, N 13.21%, O 45.25%. Prepd by treating 3-nitro-4-aminotoluene with Caro's acid and oxidizing the 3-nitro-4-nitrosotoluene with potassium dichromate: Langley, *Org. Synth.* **22**, 47 (1942); by oxidation of 3,4-dinitrotoluene: Goldstein, Voegeli, *Helv. Chim. Acta* **26**, 475 (1943).

Crystals from water + alc. Bitter taste. mp 166°; (mp 165.5-166.5°, Goldstein). 0.673 g dissolves in 100 parts water at 25°. Freely sol in alc, ether and hot water. Sublimes.

USE: In quantitative sugar analysis.

3302. 3,5-Dinitrobenzoic Acid. [99-34-3] 3-Carboxy-1,5-dinitrobenzene. $C_7H_4N_2O_6$; mol wt 212.12. C 39.64%, H 1.90%, N 13.21%, O 45.25%. Prepd by nitration of benzoic acid with fuming nitric acid: Brewster *et al.*, *Org. Synth.* **22**, 48 (1942); Saunders *et al.*, *Biochem. J.* **36**, 368 (1942). Purification: Jensen *et al.*, *Chem. Weekbl.* **43**, 731 (1947), *C.A.* **42**, 2307d (1948).

Monoclinic prismatic crystals from alc. mp 205-207°, (mp 206.5-207.2° purified). One gram dissolves in 53 parts of boiling water; much less sol in cold water. Very sol in alc and glacial acetic acid; sparingly sol in ether, carbon disulfide and benzene. Sublimes.

USE: Identification of alcohols, alkyl halides; chromatographic determination of the essential oil constituents.

3303. 3,5-Dinitrobenzoyl Chloride. [99-33-2] 3,5-Dinitrobenzoic acid chloride. $C_7H_3ClN_2O_5$; mol wt 230.56. C 36.47%, H 1.31%, Cl 15.38%, N 12.15%, O 34.70%. Prepd from 3,5-dinitrobenzoic acid and PCl_5 at 120-130°: Saunders *et al.*, *Biochem. J.* **36**, 368 (1942).

Needles from light petr (bp 40-60°), mp 69.5°. bp_{10-12} 196°.
Caution: A strong irritant.

USE: Identification of amino acids; characterization of the alcohol part of acetals and ketals without prior hydrolysis and separation of the alcohol.

3304. 4,4′-Dinitrocarbanilide. [587-90-6] *N,N′*-Bis(4-nitrophenyl)urea; DNC; 4,4′-dinitrodiphenylurea. $C_{13}H_{10}N_4O_5$; mol wt 302.25. C 51.66%, H 3.33%, N 18.54%, O 26.47%. Prepd by heating 4-nitroaniline and diphenyl carbonate for 3 hrs at 200°: Vittenet, *Bull. Soc. Chim. Fr.* [3] **21**, 149 (1899); by nitration of diphenylurea

with 68.28% HNO_3 in concd H_2SO_4 at 0°: Kogan, Kutepov, *J. Gen. Chem. USSR* **21**, 1297 (1951), *C.A.* **46**, 2003d (1952); by the action of dil HNO_3 on diphenylurea at up to 100°: *eidem*, *SU 78379* (1949), *C.A.* **48**, 7056i (1954).

Yellow needles from alc, mp 312° (dec). Sublimes above 310°. Moderately sol in boiling nitrobenzene; sparingly sol in boiling alc, somewhat more sol in a mixture of glacial acetic acid and nitric acid. Practically insol in acetone, chloroform, benzene, dioxane, acetic acid, ether and linseed oil. Solubilities in g/100 ml solvent at 25°: water 2×10^{-6}; ethanol 0.007; ethyl acetate 0.015; petr ether 0.0; xylene <0.01; methyl Cellosolve 0.1; dimethyl acetamide 0.14; dimethyl sulfoxide 0.47.

3305. Dinitrocresol. [534-52-1] 2-Methyl-4,6-dinitrophenol; 4,6-dinitro-*o*-cresol; 3,5-dinitro-2-hydroxytoluene; 3,5-dinitro-*o*-cresol; DN; DNC; DNOC; Antinonnin; Detal; Dinitrol; Elgetol; K III; K IV; Ditrosol; Prokarbol; Effusan; Lipan; Selinon; Sinox; Dekrysil. $C_7H_6N_2O_5$; mol wt 198.13. C 42.44%, H 3.05%, N 14.14%, O 40.38%. Prepd by sulfonation of *o*-cresol followed by controlled nitration: Noelting, de Salis, *Ber.* **14**, 987 (1881); Bovini, *Chem. Zentralbl.* **1928**, II, 112; Bures, *C.A.* **22**, 63 (1928); Datta, Varma, *J. Indian Chem. Soc.* **4**, 321 (1927); Monti, Cianetti, *Gazz. Chim. Ital.* **67**, 628 (1937). Acute toxicity data: R. Ben-Dyke *et al.*, *World Rev. Pest Control* **9**, 119-127 (1970). Review of toxicology and human exposure: *Toxicological Profile for Dinitrocresols* (PB95-264321, 1995) 204 pp.

Yellow prisms from alc, mp 87.5°. *Poisonous.* Moderately volatile with steam. Sparingly sol in water; readily sol in alkaline aq solns, in ether, acetone, alcohol (about 10%); sparingly sol in petr ether. LD_{50} in rats (mg/kg): 25-40 orally, 200-600 dermally (Ben-Dyke).

Sodium salt. $C_7H_5N_2O_5Na$. Red powder. Hydrate, yellow needles. Very freely sol in water.

Caution: Potential symptoms of overexposure to dinitrocresol are a sense of well being; headache, fever, lassitude, profuse sweating, excessive thirst, tachycardia, hyperpnea, coughing, shortness of breath and coma. *See NIOSH Pocket Guide to Chemical Hazards* (DHHS/NIOSH 97-140, 1997) p 118. *See also Clinical Toxicology of Commercial Products,* R. E. Gosselin *et al.*, Eds. (Williams & Wilkins, Baltimore, 5th ed., 1984) Section II, p 196.

USE: Selective herbicide, insecticide (ovicidal spray for dormant fruit trees).

3306. Dinitrogen Tetroxide. [10544-72-6] Nitrogen oxide (N_2O_4); dinitrogen tetraoxide. N_2O_4; mol wt 92.01. N 30.45%, O 69.55%. Weakly bound dimer of nitrogen dioxide, NO_2, *q.v.* Prepn from lead nitrate: C. C. Addison, R. Thompson, *J. Chem. Soc.* **1949**, S218; and chemical properties: C. C. Addison *et al.*, *ibid.* **1951**, 1289, 1294. Viscosity study: C. C. Addison, B. C. Smith, *ibid.* **1960**, 1783. Structure and bonding studies: R. Ahlrichs, F. Keil, *J. Am. Chem. Soc.* **96**, 7615 (1974); C. W. Bauschlicher, Jr. *et al.*, *ibid.* **105**, 745 (1983). Absorption spectra: A. P. Altshuller *et al.*, *J. Phys. Chem.* **62**, 607 (1958). Infrared study of equilibrium with NO_2: R. J. Nordstrom, W. H. Chan, *ibid.* **80**, 847 (1976). EPR dissociation study: K. Miaskiewicz, Z. Kecki, *J. Solution Chem.* **14**, 665 (1985). Review of properties and inorganic reactions in N_2O_4/HNO_3 mixtures: C. C. Addison, *Chem. Rev.* **80**, 21-39 (1980); of applications in organic synthesis: M. Shiri, *Synlett* **2006**, 1789-1790.

Colorless diamagnetic gas. bp 21.15°. mp −11.2°. d^{-40} (solid) 1.95. d^0 1.4905. d^{20} 1.447. n$_D^{20}$ 1.420. *Poisonous; oxidizer; corrosive.* Dielectric constant at 18°: 2.42; at −40°: 2.6. Specific conductivity (17°): 1.3×10^{-12} ohm^{-1} cm^{-1}. Surface tension (20°): 26.5 dyn/cm.

Caution: Potential symptoms of overexposure are eye, nose and throat irritation, cough, mucoid frothy sputum, decreased pulmonary function, chronic bronchitis, dyspnea, chest pain, pulmonary edema, cyanosis, tachypnea and tachycardia. *See NIOSH Pocket Guide to Chemical Hazards* (DHHS/NIOSH 2005-149) p 228.

USE: Space shuttle rocket propellant. Reagent in oxidation and nitration reactions in organic synthesis.

3307. Dinitrogen Trioxide. [10544-73-7] Nitrogen oxide (N$_2$O$_3$); nitrogen trioxide; nitrous anhydride. N$_2$O$_3$; mol wt 76.01. N 36.86%, O 63.15%. Weakly bound oxide of nitrogen formed from an equilibrium reaction between nitric oxide and either nitrogen dioxide or dinitrogen tetroxide, *q.q.v.* Occurs in the solid, aqueous, and gaseous phases. The asymmetric isomer is more stable and predominates over the symmetric form. Initial observations: L. J. Gay-Lussac, *Ann. Chim. Phys.* **1**, 394 (1816); P. L. Dulong, *ibid.* **2**, 317 (1816). Prepn: A. W. Shaw, A. J. Vosper, *J. Chem. Soc. A* **1970**, 2193. Crystal structure: T. B. Reed, W. N. Lipscomb, *Acta Crystallogr.* **6**, 781 (1953). Ab initio spectral and structural studies: X. Wang *et al., J. Mol. Struct.* **403**, 245 (1997). Electronic spectrum: A. W. Shaw, A. J. Vosper, *J. Chem. Soc. Dalton Trans.* **1972**, 961. Interconversion studies and spectral characterization: E. M. Nour *et al., J. Phys. Chem.* **87**, 1113 (1983). Hydration and dissociation equilibrium studies: A. J. Vosper, *J. Chem. Soc. Dalton Trans.* **1976**, 135. Role in endogenous nitrosation: S. Basu *et al., Nat. Chem. Biol.* **3**, 785 (2007). *Review:* I. R. Beattie, *Prog. Inorg. Chem.* **5**, 1-26 (1963).

Pale blue solid, fp −100.7. d$^{-67.0}$ 1.5184; d$^{-57.4}$ 1.5022; d$^{-47.5}$ 1.4836; d$^{-37.0}$ 1.4644. *Poisonous.* Decomposes upon warming to greenish liquid.

USE: Nitrosating reagent.

3308. 2,4-Dinitrophenol. [51-28-5] α-Dinitrophenol; Aldifen. C$_6$H$_4$N$_2$O$_5$; mol wt 184.11. C 39.14%, H 2.19%, N 15.22%, O 43.45%. Prepd by the action of NaOH on 1-chloro-2,4-dinitrobenzene. Crystal and molecular structure: F. Iwasaki, Y. Kawano, *Acta Crystallogr.* **33B**, 2455 (1977). Toxicity data: E. W. Schafer, *Toxicol. Appl. Pharmacol.* **21**, 315 (1972). Review of toxicology and human exposure: *Toxicological Profile for Dinitrophenols* (PB95-264339, 1995) 262 pp.

Yellowish to yellow orthorhombic crystals. d 1.683. mp 112-114°. Sublimes when carefully heated. Volatile with steam. Very sparingly sol in cold water; soly in water (g/100 g of satd soln): at 54.5° = 0.137; at 75.8° = 0.301; at 87.4° = 0.587; at 96.2° = 1.22. Soly at 15° (g/100 g soln): Ethyl acetate 15.55; acetone 35.90; chloroform 5.39; pyridine 20.08; carbon tetrachloride 0.423; toluene 6.36. Sol in alcohol, benzene, aq alkaline solns. Forms crystalline sodium salt which is sol in water. LD$_{50}$ orally in rats: 30 mg/kg (Schafer).

Caution: Rapidly absorbed through GI tract, respiratory tract and intact skin. Potential symptoms of overexposure are marked fatigue, tremendous thirst, profuse sweating, flushing of face; nausea, vomiting, abdominal pain, diarrhea; restlessness, anxiety, excitement; rise in body temperature; tachycardia, hyperpnea, dyspnea, cyanosis, muscle cramps; kidney and liver injury. *See Clinical Toxicology of Commercial Products*, R. E. Gosselin *et al.*, Eds. (Williams & Wilkins, Baltimore, 5th ed., 1984) Section III, pp 156-159.

USE: Manuf dyes, diaminophenol, wood preservative, insecticide; also as indicator. pH range: 2.6 colorless, 4.4 yellow. Reagent for the detection of potassium and ammonium ions.

3309. 2,5-Dinitrophenol. [329-71-5] γ-Dinitrophenol. C$_6$H$_4$N$_2$O$_5$; mol wt 184.11. C 39.14%, H 2.19%, N 15.22%, O 43.45%. Review of toxicology and human exposure: *Toxicological Profile for Dinitrophenols* (PB95-264339, 1995) 262 pp.

Yellow crystals. mp 108°; also stated as 104°. Slightly sol in water or cold alcohol; sol in hot alcohol, ether, fixed alkali hydroxides.

USE: Manuf dyes and organic chemicals, as indicator. pH range: 4.0 colorless, 5.4 yellow.

3310. 2,6-Dinitrophenol. [573-56-8] β-Dinitrophenol. C$_6$H$_4$N$_2$O$_5$; mol wt 184.11. C 39.14%, H 2.19%, N 15.22%, O 43.45%. Review of toxicology and human exposure: *Toxicological Profile for Dinitrophenols* (PB95-264339, 1995) 262 pp.

Light yellow crystals. mp 63-64°. Slightly sol in cold water or cold alc; freely sol in chloroform, ether, boiling alc, fixed alkali hydroxide solns.

USE: Manuf dyes and organic chemicals; as indicator. pH range: 2.0 colorless, 4.0 yellow.

3311. (2,4-Dinitrophenyl)hydrazine. [119-26-6] C$_6$H$_6$N$_4$O$_4$; mol wt 198.14. C 36.37%, H 3.05%, N 28.28%, O 32.30%.

Red, cryst powder. mp ~200°. Slightly sol in water or alcohol; sol in moderately dil inorganic acids; readily sol in diglyme.

USE: For the determination of aldehydes and ketones.

3312. 2,4-Dinitroresorcinol. [519-44-8] 2,4-Dinitro-1,3-benzenediol. C$_6$H$_4$N$_2$O$_6$; mol wt 200.11. C 36.01%, H 2.01%, N 14.00%, O 47.97%.

Yellow crystals, mp 146-148°. Explodes when strongly heated. Sublimes partially undec. Very slightly sol in water or cold alcohol; sol in solns of fixed alkali hydroxides.

USE: For dyeing fabrics mordanted with iron a green color. As a reagent for Co (brown-red ppt) and for Fe (olive-green color).

3313. Dinobuton. [973-21-7] Carbonic acid 1-methylethyl 2-(1-methylpropyl)-4,6-dinitrophenyl ester; carbonic acid 2-*sec*-butyl-4,6-dinitrophenyl isopropyl ester; 2-*sec*-butyl-4,6-dinitrophenyl isopropyl carbonate; dinitro-*sec*-butylphenyl isopropyl carbonate; isopropyl 2,4-dinitro-6-*sec*-butylphenyl carbonate; Acrex; Dessin; Sytasol. C$_{14}$H$_{18}$N$_2$O$_7$; mol wt 326.31. C 51.53%, H 5.56%, N 8.59%, O 34.32%. Prepn: Pianka, Polton, US 3234082; US 3234260 (both 1966 to Murphy Chem.). Metabolism studies: Bandal, Casida, *J.*

Agric. Food Chem. **20**, 1235 (1972). Toxicity study: T. B. Gaines, *Toxicol. Appl. Pharmacol.* **14**, 515 (1969).

Crystals from methanol or petr ether, mp 56-57°. LD_{50} in male, female rats (mg/kg): 59, 71 orally (Gaines).

USE: Miticide.

3314. Dinocap. [39300-45-3] 2-Butenoic acid 2(or 4)-isooctyl-4,6(or 2,6)-dinitrophenyl ester; DNOCP; CR-1639; ENT-24727; Karathane. $C_{18}H_{24}N_2O_6$; mol wt 364.40. C 59.33%, H 6.64%, N 7.69%, O 26.34%. Originally thought to be 2-(1-methylheptyl)-4,6-dinitrophenyl butenoate, it is actually a mixture of 2(or 4)-octyl-4,6(or 2,6)-dinitrophenyl butenoate in which the octyl is a mixture of 1-methylheptyl, 1-ethylhexyl, and 1-propylpentyl. Prepn: W. F. Hester *et al.*, US 2526660 (1950 to Rohm & Haas). Fungicidal activity and chemical constitution: A. H. M. Kirby *et al.*, *Ann. Appl. Biol.* **57**, 211 (1966); R. J. W. Byrde *et al.*, *ibid.* 223. Mitocidal activity: N. D. Rishi, A. Q. Rather, *J. Entomol. Res.* **7**, 39 (1983). Degradation in soil: W. Mittelstaedt, F. Fuhr, *J. Agric. Food Chem.* **32**, 1151 (1984). Persistence on crops: B. D. Ripley *et al.*, *Can. J. Plant Sci.* **65**, 229 (1985). HPLC residue analysis in crops: D. Liang *et al.*, *J. Chromatogr.* **387**, 385 (1987). Acute and chronic toxicity: P. S. Larson *et al.*, *Arch. Int. Pharmacodyn. Ther.* **119**, 31 (1959). Evaluation of developmental toxicity in rodents of technical grade and selected isomers: L. E. Gray, Jr. *et al.*, *Teratog. Carcinog. Mutagen.* **6**, 33 (1986); J. M. Rogers *et al.*, *ibid.* **7**, 341 (1987).

R = —CH(CH$_2$)$_5$CH$_3$ or —CH(CH$_2$)$_4$CH$_3$ or —CH(CH$_2$)$_3$CH$_3$
       |CH$_3$ |CH$_2$CH$_3$ |(CH$_2$)$_2$CH$_3$

Dark brown liquid. $bp_{0.05}$ 138-140°. Incompatible with oil, oil based sprays, and lime-sulfur mixtures. LD_{50} in male rats (mg/kg): 23 i.v.; 980 orally (Larson).

Caution: Skin irritant. *See Clinical Toxicology of Commercial Products*, R. E. Gosselin *et al.*, Eds. (Williams & Wilkins, Baltimore, 5th ed., 1984) Section II, p 197.

USE: Acaricide; fungicide.

3315. Dinoseb. [88-85-7] 2-(1-Methylpropyl)-4,6-dinitrophenol; 2-*sec*-butyl-4,6-dinitrophenol; DNBP; ENT-1122; WSX-8365; Chemox PE; Dow General; Premerge; Subitex; Caldon; Basanite. $C_{10}H_{12}N_2O_5$; mol wt 240.22. C 50.00%, H 5.04%, N 11.66%, O 33.30%. Prepn: L. E. Mills, B. L. Fayerweather, US 2192197 (1940 to Dow). Activity: A. S. Crafts, *Science* **101**, 417 (1945). Metabolism: S. K. Bandal, J. E. Casida, *J. Agric. Food Chem.* **20**, 1235 (1972); K. Ingebrigtsen, A. Froeslie, *Acta Pharmacol. Toxicol.* **46**, 326 (1980). Toxicology: R. G. Bough *et al.*, *Toxicol. Appl. Pharmacol.* **7**, 353 (1965); T. B. Gaines, R. E. Linder, *Fundam. Appl. Toxicol.* **7**, 299 (1986).

Orange-brown viscous liquid, mp 38-42°. pKa 4.62. Sol in most organic solvents. LD_{50} in adult male, female rats (mg/kg): 27, 28 orally (Gaines, Linder).
  Acetate. [2813-95-8] HOE-2904; Aretit; Ivosit. $C_{12}H_{14}N_2O_6$; mol wt 282.25. Brown oil, vinegar-like odor, mp 26-27°. Vapor press at 20°: 6×10^{-4} mm Hg. Sol in aromatics.
  Ammonium salt. [6365-83-9] Chemox Selective; Dow Selective; Sinox W. $C_{10}H_{15}N_3O_5$; mol wt 257.25.
  Triethanolamine salt. [6420-47-9] DN-289; Chemox DN; Gebutox. $C_{16}H_{27}N_3O_8$; mol wt 389.41.
  USE: Herbicide; insecticide; miticide.

3316. Dinosterol. [58670-63-6] (3β,4α,5α,22E)-4,23-Dimethylergost-22-en-3-ol; 4α-methyl-5α(H)-Δ^{22}-23,24-dimethylcholesten-3β-ol; Black Sea sterol. $C_{30}H_{52}O$; mol wt 428.75. C 84.04%, H 12.23%, O 3.73%. Biogenetically important marine sterol isolated from the toxic dinoflagellate, *Gonyaulax tamarensis*. Isoln and structure determn: Y. Shimizu *et al.*, *J. Am. Chem. Soc.* **98**, 1059 (1976). Stereochemistry: J. Finer *et al.*, *J. Org. Chem.* **43**, 1990 (1978). Identity with the Black Sea sterol: J. J. Boon *et al.*, *Nature* **277**, 125 (1979). Stereospecific synthesis: A. Y. L. Shu, C. Djerassi, *Tetrahedron Lett.* **22**, 4627 (1981).

Needles from methanol-chloroform, mp 220-222° (Shimizu). Also reported as mp 211-214° (Shu, Djerassi). $[\alpha]_D^{20}$ −2.2° (in $CHCl_3$).

3317. Dinotefuran. [165252-70-0] N″-Methyl-N-nitro-N′-[(tetrahydro-3-furanyl)methyl]guanidine; 1-methyl-2-nitro-3-[(3-tetrahydrofuryl)methyl]guanidine; MTI-446; Starkle. $C_7H_{14}N_4O_3$; mol wt 202.21. C 41.58%, H 6.98%, N 27.71%, O 23.74%. Neonicotinoid insecticide. Prepn: K. Kodaka *et al.*, EP 649845; *eidem*, US 5434181 (both 1995 to Mitsui Toatsu); T. Wakita *et al.*, *Pest Manage. Sci.* **59**, 1016 (2003). Comprehensive description: *eidem*, *Brighton Crop Prot. Conf. - Pests Dis.* **1998**, 21-26. Efficacy vs mosquitoes: V. Corbel *et al.*, *J. Med. Entomol.* **41**, 712 (2004).

White solid from isopropyl ether, mp 94.5-101.5°. Relative density 1.33. Soly in purified water (20°): 54.3 ± 1.3 g/l. Log P (octanol/water): −0.664 (pH 7). LD_{50} in male, female mice, male, female rats (mg/kg): 2450, 2275, 2804, 2000 orally; in rats (mg/kg): >2000 dermally. LC_{50} in carp, rainbow trout, crayfish, daphnia (ppm): >1000 (96 hr), >40 (48 hr), 5-10 (48 hr), 1000 (48 hr) (Kodaka). USE: Insecticide.

3318. DINP. [28553-12-0] 1,2-Benzenedicarboxylic acid 1,2-diisononyl ester; diisononyl phthalate. $C_{26}H_{42}O_4$; mol wt 418.62. C 74.60%, H 10.11%, O 15.29%. Dialkyl phthalate ester plasticizer used to impart softness and flexiblilty to PVC products. Prepn: P. J. Garner, G. Watson, US 2508911 (1950 to Shell). Effects on behavior and properties of PVC: S. V. Patel, M. Gilbert, *Plast. Rubber Process. Appl.* **6**, 321 (1986); *eidem*, *ibid.* **8**, 215 (1987). Effects of temperature on viscosity and density: L. D. Lorenzi *et al.*, *J. Chem. Eng. Data* **43**, 183 (1998). GC/MS determn in toys: R. Stringer *et al.*, *Environ. Sci. Pollut. Res. Int.* **7**, 27 (2000). *Review:* L. G. Krauskopf, *Plast. Compd.* **6**, 28-38 (1983). Review of environmental fate: C. A. Staples *et al.*, *Chemosphere* **35**, 667-749 (1997). Review of health effects and risk assessment: C. F. Wilkinson, J. C. Lamb, *Regul. Toxicol. Pharmacol.* **30**, 140-155 (1999); and environmental fate and toxicology: U. S. Gill *et al.*, *Rev. Environ. Contam. Toxicol.* **172**, 87-127 (2001).

Liquid mp $-48°$. bp$_5$ 252°. Flash point (closed cup): 213°C (415°F). d_{20}^{20} 0.972. n_D^{20} 1.486. Viscosity (cSt): 500 (0°); 102 (20°); 37 (37.8°); 6 (100°). Vapor pressure (mm Hg): (100°) 0.0018; (200°) 0.50; (300°) 40. Partition coefficient (octanol/water): 9.37.

USE: General purpose plasticizers for PVC applications and flexible vinyls.

3319. (R,R)-DIOP. [32305-98-9] 1,1'-[[(4R,5R)-2,2-Dimethyl-1,3-dioxolane-4,5-diyl]bis(methylene)]bis[1,1-diphenylphosphine]; (4R,5R)-4,5-bis(diphenylphosphinomethyl)-2,2-dimethyl-1,3-dioxolane; (−)-2,3-*O*-isopropylidene-2,3-dihydroxy-1,4-bis(diphenylphosphino)butane. $C_{31}H_{32}O_2P_2$; mol wt 498.54. C 74.69%, H 6.47%, O 6.42%, P 12.43%. Biphosphine chelating ligand joined by a chiral backbone. One of the first ligands applied to catalytic asymmetric hydrogenation. Prepn and use of rhodium complexes in enantioselective catalysis: T. P. Dang, H. B. Kagan, *J. Chem. Soc. Chem. Commun.* **1971**, 481; H. B. Kagan, T. P. Dang, *J. Am. Chem. Soc.* **94**, 6429 (1972). Improved prepn: B. A. Murrer *et al.*, *Synthesis* **1979**, 350; S. Zhang *et al.*, *Tetrahedron: Asymmetry* **2**, 173 (1991). Prepn and evaluation of analogues: Y.-Y. Yan, T. V. RajanBabu, *Org. Lett.* **2**, 4137 (2000); D. Liu *et al.*, *Tetrahedron: Asymmetry* **15**, 2181 (2004). Use of DIOP isomer complexes in enantioselective catalysis of hydroformylations: C. Botteghi, C. Salomon, *Tetrahedron Lett.* **15**, 4285 (1974); of keto ester reductions: I. Ojima *et al.*, *J. Org. Chem.* **42**, 1671 (1977); of hydrogenations: A. S. C. Chan, C. R. Landis, *J. Mol. Catal.* **49**, 165 (1989). Mechanistic study of asymmetric hydrogenations: J. M. Brown, P. A. Chaloner, *J. Chem. Soc. Chem. Commun.* **1978**, 321.

Colorless needles from methanol, mp 89-90°. *Irritant.* $[\alpha]_D^{22}$ $-12.3°$ (c = 4.57 in benzene); $[\alpha]_D^{20}$ $-12.4°$ (c = 0.71 in benzene); $[\alpha]_D^{20}$ $-26.5 \pm 0.1°$ (c = 0.015 in dichloromethane). Store at 2-8°C under inert gas.

(S,S)-Form. [37002-48-5] (S,S)-DIOP. White powder. *Irritant.* $[\alpha]_D^{20}$ +27.7 ± 0.2° (c = 0.015 in dichloromethane). Store at 2-8°C under inert gas.

USE: Ligand in organic synthesis.

3320. Dioscin. [19057-60-4] (3β,25R)-Spirost-5-en-3-yl *O*-6-deoxy-α-L-mannopyranosyl-(1 → 2)-*O*-[6-deoxy-α-L-mannopyranosyl-(1 → 4)]-β-D-glucopyranoside; diosgenin bis-α-L-rhamnopyranosyl-(1 → 2 and 1 → 4)-β-D-glucopyranoside. $C_{45}H_{72}O_{16}$; mol wt 869.06. C 62.19%, H 8.35%, O 29.46%. From *Dioscorea tokoro* Mal., *Dioscoreaceae*. Isoln: Tsukamoto *et al.*, *Pharm. Bull.* **4**, 35 (1956); Heitz, *Compt. Rend.* **248**, 283 (1958). Structure: Kawasaki, Yamauchi, *Chem. Pharm. Bull.* **10**, 703 (1962).

Crystals, decomp 275-277°. $[\alpha]_D^{13}$ $-115°$ (c = 0.373 in ethanol).

3321. Dioscorea. Wild yam; colic root; rheumatism root. Dried rhizome of *Dioscorea villosa* L., *Dioscoreaceae*. *Habit.* North America. *Constit.* Saponin, acrid resin. Account of the nature, origins, cultivation and utilization of the useful members of the *Dioscoreaceae:* D. G. Coursey, *Yams* (London, Longmans, 1967) 230 pp. *See also* Yam, Mexican.

3322. Dioscorine. [3329-91-7] (1R,2'S,4R)-2,4'-Dimethylspiro[2-azabicyclo[2.2.2]octane-5,2'-[2H]pyran]-6'(3'H)-one; [1R-(1α,4α,5α)]-2,4'-dimethylspiro[2-azabicyclo[2.2.2]octane-5,2'-[2H]pyran]-6'(3'H)-one. $C_{13}H_{19}NO_2$; mol wt 221.30. C 70.56%, H 8.65%, N 6.33%, O 14.46%. Found in the tubers of *Dioscorea hirsuta* Blume and *D. hispida* Dennst., *Dioscoreaceae:* H. W. Schutte, *Chem. Zentralbl.* **68 II**, 130 (1897); M. K. Gorter, *Rec. Trav. Chim.* **30**, 161 (1911); A. R. Pinder, *Nature* **168**, 1090 (1951); *idem, J. Chem. Soc.* **1952**, 2236. Structure: W. A. M. Davies *et al.*, *Chem. Ind. (London)* **1961**, 1410; *eidem, Tetrahedron* **18**, 405 (1962); Morris, A. R. Pinder, *J. Chem. Soc.* **1963**, 1841. Synthesis: Page, A. R. Pinder, *ibid.* **1964**, 4811. Absolute configuration: A. F. Beecham *et al.*, *Tetrahedron Lett.* **1969**, 3745. Biosynthetic studies: E. Leete, A. R. Pinder, *Chem. Commun.* **1971**, 1499; E. Leete, *J. Am. Chem. Soc.* **99**, 648 (1977). Pharmacology: A. R. Pinder, *J. Chem. Soc.* **1953**, 1826. Evaluation of toxic components: J. Webster *et al.*, *J. Agric. Food Chem.* **32**, 1087 (1984).

Greenish-yellow prisms from ether, mp 54-55°. $[\alpha]_D^{18}$ $-35.0°$ (c = 3.4 in chloroform). uv max (methanol): 215 nm (ε 10160). Distills unchanged *in vacuo*. Sol in water, alcohol, acetone, chloroform; slightly sol in ether, benzene, petr ether.

Hydrochloride. $C_{13}H_{19}NO_2$.HCl. Needles from alcohol + ether, dec 210-211°.

3323. Diosgenin. [512-04-9] (3β,25R)-Spirost-5-en-3-ol; nitogenin. $C_{27}H_{42}O_3$; mol wt 414.63. C 78.21%, H 10.21%, O 11.58%. Aglycone of saponin dioscin. From *Dioscorea tokoro* Makino, *Dioscoreaceae:* Tsukamoto, *J. Pharm. Soc. Jpn.* **56**, 135 (1936); from rhizomes of *Trillium erectum* L., *Liliaceae* and *D. villosa* L., *Dioscoreaceae:* Marker *et al.*, *J. Am. Chem. Soc.* **62**, 2542 (1940). Plant sources for diosgenin: Marker *et al.*, *ibid.* **65**, 1199 (1943); Wall *et al.*, *J. Am. Pharm. Assoc. Sci. Ed.* **46**, 653 (1957). Identity with nitogenin: Marker *et al.*, *J. Am. Chem. Soc.* **65**, 1248 (1943). Structure: Marker *et al.*, *ibid.* **62**, 2525 (1940). Configuration at C$_{25}$: James, *J. Chem. Soc.* **1955**, 637. Synthesis: Mazur *et al.*, *J. Am. Chem. Soc.* **82**, 5889 (1960); Kessar *et al.*, *Tetrahedron Lett.* **1966**, 4319. Obtained commercially from *Dioscorea composita* Hemsl. and *D. terpinapensis* Uline (see under Barbasco). Isoln from barbasco varieties: Julian, *US 3019220* (1962 to Julian Labs.).

Crystals from acetone, mp 204-207°. $[\alpha]_D^{25}$ −129° (c = 1.4 in CHCl₃). Sol in the usual organic solvents, in acetic acid.

Acetate. $C_{29}H_{44}O_4$. Crystals from acetic acid, mp 198°. $[\alpha]_D^{20}$ −119° (pyridine).

USE: Can be converted to pregnenolone and progesterone: Marker et al., J. Am. Chem. Soc. **69**, 2167 (1947).

3324. Diosmetin. [520-34-3] 5,7-Dihydroxy-2-(3-hydroxy-4-methoxyphenyl)-4H-1-benzopyran-4-one; 3′,5,7-trihydroxy-4′-methoxyflavone; cyanidenon-4′-methyl ether 1479; luteolin-4′-methyl ether. $C_{16}H_{12}O_6$; mol wt 300.27. C 64.00%, H 4.03%, O 31.97%. Aglycone of diosmin, q.v. Prepn from diosmin isolated from various plant sources: O. A. Oesterle, G. Wander, Helv. Chim. Acta **8**, 519 (1925); isoln from lemons (Citrus limon Linn., Rutaceae): R. M. Horowitz, J. Org. Chem. **21**, 1184 (1956). Synthesis and structural elucidation: G. Zemplén, R. Bognár, Ber. **76**, 452 (1943). Synthesis: A. Lovecy et al., J. Chem. Soc. **1930**, 817; N. B. Lorette et al., J. Org. Chem. **16**, 930 (1951); J. H. Looker, M. J. Holm, ibid. **24**, 1019 (1959). HPLC determn in biological fluids: D. Baylocq et al., Ann. Pharm. Fr. **41**, 115 (1983).

Hemimethanolate. $C_{16}H_{12}O_6 \cdot \frac{1}{2}CH_3OH$. Yellow needles from alcohol/ethyl acetate, sinters at 248°. mp 253-254°. Also reported as small yellow needles from methanol, mp 258-259° (Horowitz). uv max: 345, 268, 253 nm (log ε 4.32, 4.25, 4.28).

Triacetate. $C_{22}H_{18}O_9$. Colorless needles from methanol, mp 195-196°.

3325. Diosmin. [520-27-4] 7-[[6-O-(6-Deoxy-α-L-manno-pyranosyl)-β-D-glucopyranosyl]oxy]-5-hydroxy-2-(3-hydroxy-4-methoxyphenyl)-4H-1-benzopyran-4-one; 3′,5,7-trihydroxy-4′-methoxyflavone-7-rutinoside; 5-hydroxy-2-(3-hydroxy-4-methoxy-phenyl)-7-(O⁶-α-L-rhamnopyranosyl-β-D-glucopyranosyloxy)chro-men-4-one; 5-hydroxy-2-(3-hydroxy-4-methoxyphenyl)-7-β-rutino-syloxy-4H-chromen-4-one; diosmetin 7-β-rutinoside; barosmin; bu-chu resin; Diosmil; Diosven; Diovenor; Flebosmil; Fleboisten; Hem-erven; Insuven; Litosmil; Tovene; Varinon; Ven-Detrex; Venos-mine. $C_{28}H_{32}O_{15}$; mol wt 608.55. C 55.26%, H 5.30%, O 39.44%. Naturally occurring flavonic glycoside; rhamnoglycoside of dios-metin, q.v. Isolation from various plant sources: O. A. Oesterle, G. Wander, Helv. Chim. Acta **8**, 519 (1925). Elucidation of structure: G. Zemplén, R. Bognár, Ber. **76**, 452 (1943). Prepn from hesperidin, q.v.: eidem, ibid.; N. B. Lorette et al., J. Org. Chem. **16**, 930 (1951). Isoln from lemon peel (Citrus limon Linn. Rutaceae): R. M. Horo-witz, J. Org. Chem. **21**, 1184 (1956); from Zanthoxylum avicennae, Rutaceae: H. R. Arthur et al., J. Chem. Soc. **1956**, 632; H. R. Arthur et al., ibid. **1959**, 4007; from flowers of Sophora microphylla Ait. Leguminosae: L. H. Briggs et al., ibid. **1960**, 1955. Toxicology studies: H. Heusser, W. Osswald, Arch. Farmacol. Toxicol. **3**, 33 (1977). NMR spectrum: J. L. Nieto, A. M. Gutierrez, Spectrosc. Lett. **19**, 427 (1986). Mechanism of action: C. Boudet, L. Peyrin, Arch. Int. Pharmacodyn. **283**, 312 (1986). Pharmacology: J. R. Caseley-Smith, J. R. Caseley-Smith, Agents Actions **17**, 1 (1985); M. Damon et al., Arzneim.-Forsch. **37**, 1149 (1987). HPLC determn in

biological fluids: D. Baylocq et al., Ann. Pharm. Fr. **41**, 115 (1983). Clinical study in post-phlebitic ulcers: M. C. Nguyen, K. Morere, Gaz. Med. **92**, 71 (1985); in acute hemorrhoids: A. Tajana et al., Minerva Med. **79**, 387 (1988). Clinical trial in chronic venous insuf-ficiency: R. Laurent et al., Int. Angiol. **7**, Suppl. 2, 39 (1988).

Monohydrate. $C_{28}H_{32}O_{15} \cdot H_2O$. mp 275-277° (dec) (Zemplén). Also reported as fine needles from aq pyridine or aq DMF, mp 283° (dec) (Briggs). uv max (ethanol): 255, 268, 345 nm (log ε 4.28, 4.25, 4.30). Practically insol in water, alcohol.

Flavonoid extract. Daflon; Flebopex; Flebotropin.

THERAP CAT: Capillary protectant.

3326. Diosphenol. [490-03-9] 2-Hydroxy-3-methyl-6-(1-methylethyl)-2-cyclohexen-1-one; 1-methyl-4-isopropyl-1-cyclo-hexen-2-ol-3-one; 2-hydroxypiperitone; 1-p-menthen-2-ol-3-one; Buchu camphor; Barosma camphor. $C_{10}H_{16}O_2$; mol wt 168.24. C 71.39%, H 9.59%, O 19.02%. The crystalline portion of an oil ob-tained from Buchu leaves which come from various species of Bar-osma, such as B. betulina Bartl. & Wendl., B. serratifolia (Curt.) Willd. and B. crenulata (L.) Hook., Rutaceae: Flückiger, Pharm. J. **11**, 174, 219 (1880); Spica, Gazz. Chim. Ital. **15**, 195 (1885); Shi-moyama, Arch. Pharm. **226**, 403 (1888); Kondakov et al., J. Prakt. Chem. [II] **54**, 433 (1896); [II] **63**, 49 (1901). Structure: Semmler, McKenzie, Ber. **39**, 1160 (1906).

Crystals, mp 83°. Sublimes. bp_{760} 233° (dec); bp_{10} 109°. $d_4^{99.2}$ 0.9542. $n_D^{99.8}$ 1.4607. Absorption spectrum: Lowry, Lishmund, J. Chem. Soc. **1935**, 1313; Gillan et al., ibid. **1941**, 62. Sparingly sol in water; moderately sol in alcohol; sol in ether, chloroform, carbon disulfide.

3327. Dioxadrol. [6495-46-1] 2-(2,2-Diphenyl-1,3-dioxolan-4-yl)piperidine; 2,2-diphenyl-4-(2-piperidyl)-1,3-dioxolane. $C_{20}H_{23}NO_2$; mol wt 309.41. C 77.64%, H 7.49%, N 4.53%, O 10.34%. Prepn of dl-forms and resolution of α-racemates: Hardie, Halver-stadt, BE 613262; US 3262938 (1962, 1966 both to Cutter Labs.); W. R. Hardie et al., J. Med. Chem. **9**, 127 (1966).

Oily liquid.

Hydrochloride. [3666-69-1] Rydar. $C_{20}H_{23}NO_2 \cdot HCl$; mol wt 345.87. Crystals from methanol, mp 256-260°. LD_{50} orally in mice: 240 mg/kg (Hardie).

d-Form hydrochloride. [1162-15-8] Dexoxadrol hydrochloride; Relane. Crystals, dec 254°. $[\alpha]_D^{20}$ +34° (c = 2 in methanol). LD_{50} orally in mice: 340 mg/kg (Hardie).

l-Form hydrochloride. Levoxadrol hydrochloride; Levoxan. Crystals, mp 248-254°. $[\alpha]_D^{20}$ $-34.5°$ (c = 2 in methanol). LD_{50} orally in mice: 230 mg/kg (Hardie).

3328. Dioxane. [123-91-1] 1,4-Dioxane; 1,4-diethylene dioxide. $C_4H_8O_2$; mol wt 88.11. C 54.53%, H 9.15%, O 36.32%. Prepd by distilling ethylene glycol with dil H_2SO_4. Monograph: W. Stumpf, *Chemie und Anwendungen des 1,4-Dioxans* (Verlag Chemie, 1956). Toxicity data: E. P. Laug *et al.*, *J. Ind. Hyg. Toxicol.* **21**, 173 (1939). Carcinogenicity studies: M. F. Argus *et al.*, *J. Natl. Cancer Inst.* **35**, 949 (1965); R. J. Kociba *et al.*, *Toxicol. Appl. Pharmacol.* **30**, 275 (1974).

Faint pleasant odor. d_4^{20} 1.0329. mp 11.80°. bp_{760} 101.1°; bp_{400} 81.8°; bp_{200} 62.3°; bp_{100} 45.1°; bp_{60} 33.8°; bp_{40} 25.2°; bp_{20} 12°. *Flammable.* Cryoscopic constant 4.83. Trouton constant 21.90. Heat of combustion: 581 kcal/mol. Heat of fusion: 2.98 kcal/mol. Specific heat at 20° = 0.0370 kcal/mol/°C. Viscosity (25°): 0.0120 P. Crit temp 312°. Crit press 50.7 atm. Flash pt 5-18°C. n_D^{20} 1.4175. Dipole moment: zero. Azeotropic mixture with water: 81.6% dioxane, bp 87.8°. Azeotropic mixture with ethanol: 9.3% dioxane, bp 78.1°. Sol in water and the usual organic solvents. Tends to form explosive peroxides if anhydr, especially when evaporn to dryness is attempted. LD_{50} in mice, rats (ml/kg): 5.7, 5.2 orally (Laug).

Caution: Potential symptoms of overexposure are drowsiness, headache; nausea, vomiting; irritation of eyes, skin, nose and throat; liver damage; kidney failure. *See NIOSH Pocket Guide to Chemical Hazards* (DHHS/NIOSH 97-140, 1997) p 120. This substance is reasonably anticipated to be a human carcinogen: *Report on Carcinogens, Twelfth Edition* (PB2011-111646, 2011) p 176.

USE: Stabilizer in chlorinated solvents. Organic solvent for cellulose acetate, ethyl cellulose, benzyl cellulose, resins, oils, waxes, oil and spirit-sol dyes, and many other organic as well as some inorganic compds.

3329. 3,6-Dioxaoctane-1,8-dithiol. [14970-87-7] 2,2'-[1,2-Ethanediylbis(oxy)]bisethanethiol; 1,2-bis(2-mercaptoethoxy)-ethane; 1,8-dimercapto-3,6-dioxaoctane; DODT; 2,2'-(ethylenedioxy)diethanethiol; triethylene glycol dimercaptan; triglycol dimercaptan. $C_6H_{14}O_2S_2$; mol wt 182.30. C 39.53%, H 7.74%, O 17.55%, S 35.17%. Dithiol used in organic synthesis and polymer applications. Prepn: J. R. Dann *et al.*, *J. Org. Chem.* **26**, 1991 (1961); D. J. Martin, C. C. Greco, *ibid.* **33**, 1275 (1968); A. W. Snow, E. E. Foos, *Synthesis* **2003**, 509. Mechanism of polyaddition with isocyanates: H. J. Flammersheim *et al.*, *Thermochim. Acta* **229**, 281 (1993). Use as scavenger in protein synthesis: A. Teixeira *et al.*, *Protein Pept. Lett.* **9**, 379 (2002). Use in synthesis of cured epoxy resins: K. Strzelec, *Int. J. Adhes. Adhes.* **27**, 92 (2007); in prepn of luminescent CdS quantum dots: T.-L. Zhang *et al.*, *J. Nanopart. Res.* **10**, 59 (2008).

Clear, very faintly yellow liquid; characteristic stinking odor. bp 225°. bp_5 125-131°; $bp_{0.4}$ 86°. *Poisonous.* Flash pt, closed cup: 129°C (264°F). Slightly sol in water.

USE: In synthesis of crown ethers and polymers; in formulating polymer curing agents. In prepn of self-assembly monolayers and organic electronics. As a scavenger in protein synthesis.

3330. Dioxathion. [78-34-2] Phosphorodithioic acid *S,S'*-1,4-dioxane-2,3-diyl *O,O,O',O'*-tetraethyl ester; 2,3-*p*-dioxanedithiol *S,S*-bis(*O,O*-diethyl phosphorodithioate); AC-528; ENT-22879; Hercules 528; Delnav; Navadel. $C_{12}H_{26}O_6P_2S_4$; mol wt 456.52. C 31.57%, H 5.74%, O 21.03%, P 13.57%, S 28.09%. Organophosphate insecticide; cholinesterase inhibitor. Prepn: Diveley, Lohr, **US 2725328**; Speck, **US 2815350** (1955, 1957 both to Hercules); Diveley *et al.*, *J. Am. Chem. Soc.* **81**, 139 (1959). Toxicity study: T. B. Gaines, *Toxicol. Appl. Pharmacol.* **14**, 515 (1969).

Tan liquid, d_4^{26} 1.257. mp $-20°$. n_D^{20} 1.5420. Practically insol in water. Partly sol in hexane. Hydrolyzed by alkali and by heating. *Poisonous.* LD_{50} in female, male rats (mg/kg): 23, 43 orally; 63, 235 dermally (Gaines).

Caution: Potential symptoms of overexposure are eye, skin irritation; headache, giddiness, vertigo, weakness; rhinorrhea, chest tightness; miosis; nausea, vomiting, abdominal cramps, diarrhea, salivation; muscle fasciculations; confusion, drowsiness. *See NIOSH Pocket Guide to Chemical Hazards* (DHHS/NIOSH 97-140, 1997) p 120.

USE: Insecticide; acaricide.

3331. Dioxethedrine. [497-75-6] 4-[2-(Ethylamino)-1-hydroxypropyl]-1,2-benzenediol; *N*-ethyl-3,4-dihydroxynorephedrine; α-(1-ethylaminoethyl)protocatechuyl alcohol; 2-ethylamino-1-(3',4'-dihydroxyphenyl)-1-propanol; 1-(3',4'-dihydroxyphenyl)-2-ethylamino-1-propanol; C-247. $C_{11}H_{17}NO_3$; mol wt 211.26. C 62.54%, H 8.11%, N 6.63%, O 22.72%. β-Adrenergic agonist. Prepd by catalytic hydrogenation of the corresponding aminoketone: Lespagnol, Cuingnet, *Ann. Pharm. Fr.* **18**, 445 (1960). Only one of the two possible isomers, believed to be the "erythro" form, was isolated.

Hydrochloride. $C_{11}H_{17}NO_3$.HCl. Crystals from methanol + ether, mp 212-214°.

One of the ingredients of *Bexol*.

THERAP CAT: Bronchodilator.

3332. Dioxybenzone. [131-53-3] (2-Hydroxy-4-methoxyphenyl)(2-hydroxyphenyl)methanone; 2,2'-dihydroxy-4-methoxybenzophenone; 4-methoxy-2,2'-dihydroxybenzophenone; benzophenone-8; Cyasorb UV 24 (obsolete); Spectra-Sorb UV 24. $C_{14}H_{12}O_4$; mol wt 244.25. C 68.85%, H 4.95%, O 26.20%. Prepn: Hardy *et al.*, **US 2853521** (1958 to Am. Cyanamid).

Yellow powder, mp 68°. Soly in g/100 ml at 25°: ethanol 21.8; isopropanol 17; propylene glycol 6.2; ethylene glycol 3.0; *n*-hexane 1.5. Freely sol in alc, toluene. Practically insol in water.

THERAP CAT: Ultraviolet screen.

3333. (R,R)-DIPAMP. [55739-58-7] *rel*-(1R,1'R)-1,1'-(1,2-Ethanediyl)bis[1-(2-methoxyphenyl)-1-phenylphosphine]; 1,2-bis[(*o*-anisyl)(phenyl)phosphino]ethane. $C_{28}H_{28}O_2P_2$; mol wt 458.48. C 73.35%, H 6.16%, O 6.98%, P 13.51%. Chiral diphosphine ligand in enantioselective synthesis. Prepn and use of rhodium complexes as catalysts in asymmetric hydrogenations: W. S. Knowles *et al.*, *J. Am. Chem. Soc.* **97**, 2567 (1975); B. D. Vineyard *et al.*, *ibid.* **99**, 5946 (1977). Improved prepn: E. J. Corey *et al.*, *ibid.* **115**, 11000 (1993); U. Schmidt *et al.*, *Synthesis* **1991**, 655. Prepn and evaluation of analogues: B. Zupancic *et al.*, *Adv. Synth. Catal.* **350**, 2024 (2008). Use of ruthenium complexes in the asymmetric hydrogenation of olefins and ketones: J.-P. Genet *et al.*, *Tetrahedron Lett.* **33**, 5343 (1992). Mechanistic studies of asymmetric hydrogenations: J. M. Brown, P. A. Chaloner, *J. Am. Chem. Soc.* **102**, 3040 (1980). Review of utility in asymmetric hydrogenations: W. S. Knowles, *Acc. Chem. Res.* **16**, 106-112 (1983).

Relative stereochemistry

Crystals from hot methanol, mp 102-104°. $[\alpha]_D^{20}$ −85.0° (c = 1.0 in chloroform). *Irritant.* Handle and store under inert gas.

(S,S)-Form. [97858-62-3] (*S,S*)-DIPAMP. Prepn: T. Imamoto *et al.*, *J. Am. Chem. Soc.* **107**, 5301 (1985). Crystals from hot methanol, mp 102-103°. $[\alpha]_D^{24}$ +87.0° (c = 1.0 in chloroform). Handle and store under inert gas.

USE: Ligand in organic synthesis.

3334. 2,5-Di-*tert*-pentylhydroquinone. [79-74-3] 2,5-Bis-(1,1-dimethylpropyl)-1,4-benzenediol; 2,5-di-*tert*-amylhydroquinone; 2,5-bis(1,1-dimethylpropyl)hydroquinone; Santovar A. $C_{16}H_{26}O_2$; mol wt 250.38. C 76.75%, H 10.47%, O 12.78%. Prepn: Erickson, **GB 596461** (1948 to Mathieson Alkali Works).

Crystals, mp 179.4-180.4°.

USE: As a staining protector in rubber.

3335. Diphacinone. [82-66-6] 2-(2,2-Diphenylacetyl)-1*H*-indene-1,3(2*H*)-dione; 2-diphenylacetyl-1,3-diketohydrindene; 2-diphenylacetyl-1,3-indandione; diphenadione; U-1363; Diphacin; Ditrac; Ramik. $C_{23}H_{16}O_3$; mol wt 340.38. C 81.16%, H 4.74%, O 14.10%. Anticoagulant; inhibits vitamin K-dependent synthesis of factors II, VII and X. Prepn: D. G. Thomas, **US 2672483** (1954 to Upjohn). Anticoagulant effects in rodents: J. T. Correll *et al.*, *Proc. Soc. Exp. Biol. Med.* **80**, 139 (1952). Metabolism in rats and mice: C. C. Yu *et al.*, *Drug Metab. Dispos.* **10**, 645 (1982). LC/MS determn in commercial rodenticides: M. Z. Mesmer, R. A. Flurer, *J. Chromatogr. A* **891**, 249 (2000).

Pale yellow crystals from ethanol, mp 146-147°. Practically insol in water. Slightly sol in benzene, hot ethanol. Sol in acetone, acetic acid. LD$_{50}$ orally (mg/kg): 3 in rats; 340 in mice; 35 in rabbits (Correll).

USE: Rodenticide.

3336. Diphemanil Methylsulfate. [62-97-5] 4-(Diphenylmethylene)-1,1-dimethylpiperidinium methyl sulfate (1:1); *p*-(α-phenylbenzylidene)-1,1-dimethylpiperidinium methylsulfate; *N,N*-dimethyl-4-piperidylidene-1,1-diphenylmethane methylsulfate;

Prantal. $C_{21}H_{27}NO_4S$; mol wt 389.51. C 64.76%, H 6.99%, N 3.60%, O 16.43%, S 8.23%. Anticholinergic. Prepn: N. Sperber *et al.*, *J. Am. Chem. Soc.* **73**, 5010 (1951); N. Sperber *et al.*, **US 2739968** (1956 to Schering). Pharmacological properties and toxicity data: S. Margolin *et al.*, *Proc. Soc. Exp. Biol. Med.* **78**, 576 (1951). Clinical evaluation in treatment of gustatory sweating: O. Laccourreye *et al.*, *Laryngoscope* **100**, 1651 (1990). Clinical pharmacokinetics: A. M. Vidal *et al.*, *Eur. J. Clin. Pharmacol.* **42**, 689 (1992).

Crystals, mp 194-195°. Soluble in water. LD$_{50}$ in rats, mice, guinea pigs (mg/kg): 1107, 64, 404 orally (Margolin).

THERAP CAT: In treatment of hyperhidrosis.

3337. Diphenamid. [957-51-7] *N,N*-Dimethyl-α-phenylbenzeneacetamide; *N,N*-dimethyl-2,2-diphenylacetamide; *N,N*-dimethyl-α,α-diphenyl acetamide; 2,2-diphenyl-*N,N*-dimethylacetamide; L-34314; Dymid; Enide. $C_{16}H_{17}NO$; mol wt 239.32. C 80.30%, H 7.16%, N 5.85%, O 6.69%. Selective pre-emergence herbicide. Prepn: Cheney *et al.*, *J. Org. Chem.* **17**, 770 (1952). Toxicity study: G. W. Bailey, J. L. White, *Residue Rev.* **10**, 97 (1965).

Crystals from ethyl acetate, mp 134.5-135.5°. Soluble in water, acetone, dimethylformamide, xylene, phenyl Cellosolve. LD$_{50}$ orally in rats: 700 mg/kg (Bailey, White).

USE: Herbicide.

3338. Diphencyprone. [886-38-4] 2,3-Diphenyl-2-cyclopropen-1-one; 1,2-diphenylcyclopropenone; DPC. $C_{15}H_{10}O$; mol wt 206.24. C 87.36%, H 4.89%, O 7.76%. Contact allergen, possibly the smallest neutral aromatic specie. Prepn: R. Breslow *et al.*, *J. Am. Chem. Soc.* **81**, 247 (1959). Crystal structure: H. L. Ammon, *ibid.* **95**, 7093 (1973). Strain energy: A. Greenberg *et al.*, *ibid.* **105**, 6855 (1983). Mechanism of photodissociation: Y. Hirata, N. Mataga, *Chem. Phys. Lett.* **193**, 287 (1992). Report as contact sensitizer: B. M. Hausen, J. Stute, *Chem. Ind.* **1980**, 699. Clinical evaluation in alopecia totalis: B. Monk, *Clin. Exp. Dermatol.* **14**, 154 (1989); of children with alopecia areata: S. MacDonald Hull *et al.*, *Br. J. Dermatol.* **125**, 164 (1991). Review of clinical efficacy: E. Hoting, A. Boehm, *ibid.* **127**, 625-629 (1992).

Crystallizes as monohydrate from cyclohexane, mp 87-90°. d 1.202 g/cm³. uv max (CH$_3$CN): 297, 282, 226, 220 nm (log ε 4.3, 4.25, 4.13, 4.16). mp (anhydrous): 118-120°.

THERAP CAT: Antialopecia agent.

3339. Diphenhydramine. [58-73-1] 2-Diphenylmethoxy-*N,N*-dimethylethanamine; 2-(benzhydryloxy)-*N,N*-dimethylethylamine; β-dimethylaminoethyl benzhydryl ether; *O*-benzhydryldimethylaminoethanol; β-dimethylaminoethanol diphenylmethyl ether; α-(2-dimethylaminoethoxy)diphenylmethane; benzhydramine. $C_{17}H_{21}NO$; mol wt 255.36. C 79.96%, H 8.29%, N 5.49%, O 6.27%. Antihistamine with sedating and antiemetic effects. Prepn: G. Rieveschl, Jr., US 2421714 (1947 to Parke, Davis); *see also* H. Martin *et al.*, US 2397799 (1946 to Geigy). Toxicology study: O. M. Gruhzit, R. A. Fisken, *J. Pharmacol. Exp. Ther.* **89**, 227 (1947). Comprehensive description of the hydrochloride: I. J. Holcomb, S. A. Fusari, *Anal. Profiles Drug Subs.* **3**, 173-232 (1974). Clinical pharmacokinetics: J. M. Scavone *et al.*, *J. Clin. Pharmacol.* **38**, 603 (1998). Clinical trial in insomnia: K. Rickels *et al.*, *ibid.* **23**, 235 (1983). Spectrofluorometric determn in pharmaceutical preparations: I. Pascual Reguera *et al.*, *Anal. Sci.* **20**, 799 (2004). CE-MS determn in urine: A. Baldacci *et al.*, *Electrophoresis* **25**, 1607 (2004).

bp$_{2.0}$ 150-165°.

Hydrochloride. [147-24-0] Benadryl; Benocten; Nytol; Sedopretten; Sominex; Unisom Sleepgels. $C_{17}H_{21}NO.HCl$; mol wt 291.82. Crystals from abs alcohol + ether, mp 166-170°. Bitter taste. Slowly darkens on exposure to light. Stable under ordinary conditions. One gram dissolves in 1 ml water, 2 ml alcohol, 2 ml chloroform, 50 ml acetone. Very slightly sol in benzene, ether. pH of 1% aq soln about 5.5. The aq soln forms a pink precipitate with satd Reinecke's salt soln. LD$_{50}$ orally in rats: 500 mg/kg (Gruhzit, Fisken).

Di(acefyllinate). [6888-11-5] 1,2,3,6-Tetrahydro-1,3-dimethyl-2,6-dioxo-7*H*-purine-7-acetic acid compd with 2-(diphenylmethoxy)-*N,N*-dimethylethanamine (2:1); diphenhydramine bis(theophyllin-7-ylacetate); bietanautine; Nautamine. $C_{35}H_{41}N_9O_9$; mol wt 731.77. Prepn: M. Mizier, US 2942000 (1960 to Delagrange). Crystals, mp 168-170°. Sol in alc; sparingly sol in water.

THERAP CAT: Antihistaminic; sedative, hypnotic; antiemetic.

THERAP CAT (VET): Antihistaminic; antiemetic.

3340. Diphenic Acid. [482-05-3] [1,1′-Biphenyl]-2,2′-dicarboxylic acid; *o,o′*-bibenzoic acid. $C_{14}H_{10}O_4$; mol wt 242.23. C 69.42%, H 4.16%, O 26.42%. Prepd from diazotized anthranilic acid by treatment with a cuproammonia-sulfite reducing agent: Vorländer, Meyer, *Ann.* **320**, 122 (1902); Atkinson, Lawler, *Org. Synth.* **coll. vol. I** (2nd ed, 1941) p 222. By chromic acid oxidation of phenanthrenequinone: Roberts, Johnson, *J. Am. Chem. Soc.* **47**, 1399 (1925). By heating potassium *o*-bromobenzoate with copper powder: Hurtley, *J. Chem. Soc.* **1929**, 1870.

Monoclinic prismatic rods upon slow cooling from water, leaflets from hot water, needles by careful sublimation. mp 228-229°. The satd aq soln is 0.0052*N* at 25°; soluble in the usual organic solvents.

Dimethyl ester. $C_{16}H_{14}O_4$. Monoclinic prismatic plates, tablets from methanol, mp 73.5°; bp$_{14}$ 204-206°.

Diethyl ester. $C_{18}H_{18}O_4$. Cubes from alcoholic HCl, mp 42°.

3341. Diphenidol. [972-02-1] α,α-Diphenyl-1-piperidinebutanol; 1,1-diphenyl-4-piperidino-1-butanol; diphenyl-[3-(1-piperidyl)propyl]carbinol; defenidol; SKF-478. $C_{21}H_{27}NO$; mol wt 309.45. C 81.51%, H 8.79%, N 4.53%, O 5.17%. Prepn: Miescher, Marxer, US 2411664 (1946 to Ciba); Barrett, Wilkinson, GB 683950 (1952 to Wellcome Foundation), *C.A.* **48**, 2112e (1954). Structure-activity studies: Gautier *et al.*, *Med. Pharmacol. Exp.* **13**, 325 (1965). Clinical studies: Cutt *et al.*, *Aerosp. Med.* **39**, 682 (1968);

Benson, *ibid.* **40**, 589 (1969). Acute toxicity: E. I. Goldenthal, *Toxicol. Appl. Pharmacol.* **18**, 185 (1971).

Needles from petr ether, mp 104-105°. LD$_{50}$ s.c. in rats: 50 mg/kg (Goldenthal).

Hydrochloride. [3254-89-5] SKF-478-A; Ansmin; Cefadol; Celmidol; Difenidolin; Maniol; Mecalmin; Pineroro; Satanolon; Tenesdol; Vontrol; Wansar. $C_{21}H_{27}NO.HCl$; mol wt 345.91. Crystals from chloroform + ethyl acetate, mp 212-214°. Freely sol in methanol; sol in water, chloroform. Practically insol in ether, benzene, petr ether.

THERAP CAT: Antiemetic.

3342. Diphenolic Acid. [126-00-1] 4-Hydroxy-γ-(4-hydroxyphenyl)-γ-methylbenzenebutanoic acid; 4,4-bis[4′-hydroxyphenyl]pentanoic acid; γ,γ-bis-(*p*-hydroxyphenyl)valeric acid; DPA. $C_{17}H_{18}O_4$; mol wt 286.33. C 71.31%, H 6.34%, O 22.35%. Prepd by condensing 2.25-4.0 moles of phenol with one mole of levulinic acid in the presence of HCl: GB 768206 (1957 to S. C. Johnson & Son); from one mole of phenol, 0.5 mole levulinic acid and HCl, H_2SO_4 or H_3PO_4: Bader, Kantowicz, *J. Am. Chem. Soc.* **76**, 4465 (1954); Bader, US 2933520 (1960 to S. C. Johnson & Son).

Crystals from hot water. Higher melting modification, mp 171-172°. Appreciably sol in hot water; sol in acetone, acetic acid, ethanol, isopropanol, methyl ethyl ketone.

USE: Intermediate for surface coatings, lubricating oil additives, cosmetics, surfactants, plasticizers, textile chemicals.

3343. Diphenoxylate. [915-30-0] 1-(3-Cyano-3,3-diphenylpropyl)-4-phenyl-4-piperidinecarboxylic acid ethyl ester; 1-(3-cyano-3,3-diphenylpropyl)-4-phenylisonipecotic acid ethyl ester; ethyl 1-(3-cyano-3,3-diphenylpropyl)-4-phenylisonipecotate; ethyl 1-(3-cyano-3,3-diphenylpropyl)-4-phenyl-4-piperidinecarboxylate; 2,2-diphenyl-4-(4-carbethoxy-4-phenylpiperidino)butyronitrile; R-1132. $C_{30}H_{32}N_2O_2$; mol wt 452.60. C 79.61%, H 7.13%, N 6.19%, O 7.07%. Prepn: Janssen, US 2898340 (1959). Pharmacokinetics and metabolism: Karim *et al.*, *J. Pharmacol. Exp. Ther.* **177**, 546 (1971); *Clin. Pharmacol. Ther.* **13**, 407 (1972). *See also* Difenoxin, the active metabolite of diphenoxylate. Comprehensive description: D. D. Hung, *Anal. Profiles Drug Subs.* **7**, 149-169 (1978).

Hydrochloride. [3810-80-8] $C_{30}H_{32}N_2O_2.HCl$. White crystals, mp 220.5-222°. uv max (methanol): 252, 258, 264 nm. Soly in mg/ml at 25°: acetic acid 500; DMF 500; chloroform 360; methanol >50; ethanol 3; water 0.8; hexane 0.5. Sparingly sol in acetone; slightly sol in isopropanol. Practically insol in ether.

Mixture of hydrochloride with atropine sulfate. Lomotil; Diarsed; Reasec.

Note: This is a controlled substance (opiate): **21 CFR,** 1308.12.

THERAP CAT: Antiperistaltic; antidiarrheal.

3344. Diphenyl. [92-52-4] 1,1'-Biphenyl; bibenzene; phenylbenzene. $C_{12}H_{10}$; mol wt 154.21. C 93.46%, H 6.54%. Toxicity data: Deichmann *et al., J. Ind. Hyg. Toxicol.* **29,** 1 (1947). *Review:* W. C. Weaver *et al.,* in *Kirk-Othmer Encyclopedia of Chemical Technology* **vol. 7** (Wiley-Interscience, New York, 3rd ed., 1979) pp 782-793.

Colorless leaflets; pleasant, peculiar odor. d 1.041. mp 69-71°. bp 254-255°. n_D^{77} 1.588. Insol in water. Sol in alc, ether. LD_{50} orally in rats: 3280 mg/kg (Deichmann).

Caution: Potential symptoms of overexposure are irritation of throat and eyes; headache, nausea, fatigue and numb limbs; liver damage. *See NIOSH Pocket Guide to Chemical Hazards* (DHHS/NIOSH 97-140, 1997) p 120.

USE: As heat transfer agent; fungistat for oranges (applied to inside of shipping container or wrappers); in organic syntheses.

3345. Diphenylacetamide. [519-87-9] *N,N*-Diphenylacetamide; acetyldiphenylamine. $C_{14}H_{13}NO$; mol wt 211.26. C 79.60%, H 6.20%, N 6.63%, O 7.57%.

White, cryst powder. mp 103°. Sublimes without decomposition. Slightly sol in water; sol in alcohol, ether.

3346. Diphenylacetic Acid. [117-34-0] α-Phenylbenzeneacetic acid; diphenylmethane-α-carboxylic acid. $C_{14}H_{12}O_2$; mol wt 212.25. C 79.22%, H 5.70%, O 15.08%. Prepd by the reduction of benzilic acid with hydriodic acid and red phosphorus: C. S. Marvel *et al., Org. Synth.* **coll. vol. I,** 224 (2nd ed., 1941). Ecologically and economically improved synthesis from benzilic acid: P. Strazzolini *et al., Synth. Commun.* **17,** 1919 (1987). Toxicity study: G. W. Bailey, J. L. White, *Residue Rev.* **10,** 97 (1965).

Small plates from alcohol, needles from water, mp 148°. Sublimes. Sol in hot water, alcohol, ether, chloroform.

Methyl ester. [3469-00-9] $C_{15}H_{14}O_2$; mol wt 226.28. Crystals from dil alcohol, mp 60°.

Ethyl ester. [3468-99-3] $C_{16}H_{16}O_2$; mol wt 240.30. Crystals from alcohol, mp 58°. bp_{15} 178°.

Amide. [4695-13-0] $C_{14}H_{13}NO$; mol wt 211.26. mp 167-168°.

Nitrile. [86-29-3] Diphenylacetonitrile; α-cyanodiphenylmethane; diphenatrile; Dipan. $C_{14}H_{11}N$; mol wt 193.25. mp 76°. bp_{12} 181°. Sol in alcohol, ether, propylene glycol. LD_{50} orally in rats: 3500 mg/kg (Bailey, White).

Diphenylacetic acid chloride. [1871-76-7] $C_{14}H_{11}ClO$; mol wt 230.69. Crystals from petr ether, mp 57°; bp_{15} 178°. Very sol in benzene. Sol in hot ligroin, in hot isopropyl ether.

Diphenylacetic acid anhydride. [1760-46-9] $C_{28}H_{22}O_3$; mol wt 406.48. mp 98°; bp_{15} 220°.

USE: Nitrile as herbicide.

3347. Diphenylamine. [122-39-4] *N*-Phenylbenzeneamine. $C_{12}H_{11}N$; mol wt 169.23. C 85.17%, H 6.55%, N 8.28%. Prepd by heating aniline with aniline hydrochloride.

Crystals; floral odor. d 1.16. mp 53-54°. bp 302°. Flash pt 153°C. Discolors in light. One gram dissolves in 2.2 ml alcohol, 4.5 ml propyl alcohol; freely sol in benzene, ether, glacial acetic acid, carbon disulfide. Insol in water. Forms salts with strong acids. *Keep protected from light.*

Hydrochloride. [537-67-7] $C_{12}H_{11}N.HCl$; mol wt 205.69. Crystals, turn blue in air. Freely sol in water, alcohol.

Sulfate. [587-84-8] $C_{12}H_{11}N.H_2SO_4$; mol wt 267.30. White to yellowish powder, mp 123-125°. Practically insol in water. Sol in alc, in H_2SO_4.

Caution: Potential symptoms of overexposure to diphenylamine are irritation of eyes, skin, mucous membranes; eczema; tachycardia, hypertension; cough, sneezing; methemoglobinemia; increased blood pressure and heart rate; proteinuria, hematuria, bladder injury. *See NIOSH Pocket Guide to Chemical Hazards* (DHHS/NIOSH 97-140, 1997) p 120.

USE: Manuf dyes; stabilizing nitrocellulose explosives and celluloid. In analytical chemistry for the detection of NO_3, ClO_3 and other oxidizing substances with which, in the presence of H_2SO_4, it gives a deep-blue color. Indicator for redox titrations.

THERAP CAT (VET): Topically in antiscrewworm mixtures. In tests for nitrate or nitrite poisoning.

3348. Diphenylamine-2,2'-dicarboxylic Acid. [579-92-0] 2,2'-Iminobis[benzoic acid]. $C_{14}H_{11}NO_4$; mol wt 257.25. C 65.37%, H 4.31%, N 5.44%, O 24.88%. Prepn from the alkali salts of anthranilic acid and of 2-chlorobenzoic acid in presence of copper: Ullmann, *Ann.* **355,** 352 (1907).

Yellow crystals from alcohol, mp 296-297° (dec). Insoluble in water. Very slightly sol in alcohol, ether, chloroform, glacial acetic acid.

USE: Instead of diphenylamine in reactions with oxidizing agents. May be used as an oxidation-reduction indicator in strongly acid solns.

3349. *N,N'*-Diphenylbenzidine. [531-91-9] $N^4,N^{4'}$-Diphenyl-[1,1'-biphenyl]-4,4'-diamine; 4,4'-bis(phenylamino)-1,1'-biphenyl. $C_{24}H_{20}N_2$; mol wt 336.44. C 85.68%, H 5.99%, N 8.33%.

Leaflets or plates, mp 242°. Insol in water. Freely sol in boiling toluene or ethyl acetate, slightly in alc, acetone. Is more sensitive to oxidizing substances than diphenylamine. *Keep protected from light.*

USE: *See* diphenylamine.

3350. 1,4-Diphenyl-1,3-butadiene. [886-65-7] 1,1'-(1,3-Butadiene-1,4-diyl)bisbenzene; distyryl; bistyryl; 1,4-diphenylerythrene. $C_{16}H_{14}$; mol wt 206.29. C 93.16%, H 6.84%. Prepd by the condensation of phenylacetic acid and cinnamic aldehyde: Kuhn, Winterstein, *Helv. Chim. Acta* **11,** 103 (1928); Corson, *Org. Synth.* **16,** 28 (1936). This method yields the (*E,E*)-form.

(*E,E*)-form

(E,E)-Form. [538-81-8] Crystals, mp 149.7°. bp$_{720}$ 350°. Soluble in alcohol; sparingly sol in ether.

(Z,Z)-Form. [5807-76-1] Leaflets, needles, mp 70.5°. Changes to the (E,E)-form under the influence of light. d$_4^{100.6}$ 0.9697. n$_{H\alpha}^{100.6}$ 1.61831; n$_{H\beta}^{100.6}$ 1.66748; n$_{He}^{100.6}$ 1.63473. Soluble in ether, chloroform, benzene, petr ether, hot glacial acetic acid, slightly sol in alcohol.

(1E,3Z)-Form. [5808-05-9] Obtained only in the absence of light. Oily liquid or crystals, mp 88°. bp$_{0.1}$ 133-135°. d$_4^{22}$ 0.9974. n$_{H\alpha}^{22.2}$ 1.59679; n$_{H\beta}^{22.2}$ 1.62830; n$_{He}^{22.2}$ 1.60532.

3351. Diphenylcarbamoyl Chloride. [83-01-2] N,N-Diphenylcarbamic chloride; diphenylcarbamyl chloride; DPCC; DPC-Cl. C$_{13}$H$_{10}$ClNO; mol wt 231.68. C 67.40%, H 4.35%, Cl 15.30%, N 6.05%, O 6.91%. Acylating reagent in organic synthesis. Prepn: W. Michler, Ber. 8, 1664 (1875). Preliminary reactivity studies: E. Lellmann, O. Bonhöffer, ibid. 20, 2118 (1887). Use in acylation reactions: D. E. Rivett, J. F. K. Wilshire, Aust. J. Chem. 18, 1667 (1965); eidem, ibid. 19, 165 (1966); J. F. K. Wilshire, ibid. 20, 575 (1967). Use as a guanine residue protecting group in oligonucleotide chemistry: T. Kamimura et al., Tetrahedron Lett. 24, 2775 (1983). Crystal structure: S. Baggio et al., Cryst. Struct. Commun. 2, 531 (1973). Solvolysis studies: M. J. D'Souza et al., J. Org. Chem. 60, 1632 (1995). Activity as an inhibitor of chymotrypsin and trypsin: B. F. Erlanger, W. Cohen, J. Am. Chem. Soc. 85, 348 (1963); eidem, Biochemistry 5, 190 (1966).

White solid from ethanol, mp 85°. Also reported as thin, rhomboidal shaped plates from acetone. *Corrosive, skin sensitizer.* Sol in most organic solvents. Moisture sensitive. Protect from light.

USE: Reagent in synthetic organic chemistry.

3352. sym-Diphenylcarbazide. [140-22-7] 2,2′-Diphenylcarbonic dihydrazide; 1,5-diphenylcarbohydrazide. C$_{13}$H$_{14}$N$_4$O; mol wt 242.28. C 64.45%, H 5.82%, N 23.13%, O 6.60%.

White, cryst powder, gradually becomes pink. mp 168-171°. Very slightly sol in water; sol in hot alcohol, acetone, glacial acetic acid. *Keep protected from light.*

USE: As indicator for chromate and in titrating Fe; for the colorimetric determination of Cr, detection of Cd, Hg, Mg, aldehydes, emetine.

3353. Diphenylcarbazone. [538-62-5]; [119295-41-9] ((E)-form); [119295-40-8] ((Z)-form). 2-Phenyldiazenecarboxylic acid 2-phenylhydrazide. C$_{13}$H$_{12}$N$_4$O; mol wt 240.27. C 64.99%, H 5.03%, N 23.32%, O 6.66%.

(E)-form

Orange-red needles. mp about 157° with decompn. Insol in water. Sol in alcohol, chloroform, benzene.

USE: As a sensitive reagent for Hg, with which it gives a blue color.

3354. Diphenyl Carbonate. [102-09-0] Carbonic acid diphenyl ester; phenyl carbonate. C$_{13}$H$_{10}$O$_3$; mol wt 214.22. C 72.89%, H 4.71%, O 22.41%. Prepn: Eckenroth, Ber. 27, 3410 (1894); Bischoff, ibid. 35, 3434 (1902); Gomberg, Snow, J. Am. Chem. Soc. 47, 198 (1925).

Lustrous needles, mp 80-81°. bp 302-306°; bp$_{15}$ 168°. Practically insol in water. Sol in hot alcohol, benzene, ether, glacial acetic acid.

USE: In the molten state as solvent for nitrocellulose.

3355. Diphenyl Disulfide. [882-33-7] Phenyl disulfide; Ph-SSPh. C$_{12}$H$_{10}$S$_2$; mol wt 218.33. C 66.02%, H 4.62%, S 29.37%. Reagent chemical used to introduce the phenylthio functional group. Early studies: C. Vogt, Ann. 119, 142 (1861); J. Stenhouse, ibid. 149, 247 (1869). Prepn: F. Krafft, W. Vorster, Ber. 26, 2813 (1893); K. S. Ravikumar et al., Org. Synth. coll. vol. XI, 135 (2009). Crystal structure: J. D. Lee, M. W. R. Bryant, Acta Crystallogr. B 25, 2094 (1969); M. Sacerdoti, G. Gilli, ibid. 31, 327 (1975). Synthetic applications: B. M. Trost et al., J. Am. Chem. Soc. 98, 4887 (1976); L. Benati et al., J. Chem. Soc. Perkin Trans. 1 1991, 2103; W. Munbunjong et al., Tetrahedron 65, 2467 (2009). Review: J. H. Byers in Encyclopedia of Reagents for Organic Synthesis 4, L. A. Paquette, Ed. (Wiley, New York, 1995) pp 2214-2218.

Colorless orthorhombic crystals, mp 61°. bp$_{15}$ 190-192°. d 1.34. *Irritant.* Sol in ethanol, tetrahydrofuran, benzene, carbon disulfide; sparingly sol in trifluoroethanol. Insol in water.

USE: Reagent in synthetic organic chemistry.

3356. 1,1-Diphenylethene. [530-48-3] 1,1′-Ethenylidenebisbenzene; unsym-diphenylethylene; α,α-diphenylethylene; α-methylene-diphenylmethane. C$_{14}$H$_{12}$; mol wt 180.25. C 93.29%, H 6.71%. Prepd by the action of phenylmagnesium bromide on ethyl acetate in ether, followed by treatment of the reaction product with ammonium chloride soln: Allen, Converse, Org. Synth. coll. vol. I (2nd ed., 1941) p 226. Absorption spectrum: Lardy, J. Chim. Phys. 21, 361 (1924).

Liquid, mp 8.2°. d$_4^{20}$ 1.0232. bp$_{760}$ 277.0°; bp$_{400}$ 249.8°; bp$_{200}$ 222.8°; bp$_{100}$ 198.6°; bp$_{60}$ 183.4°; bp$_{40}$ 170.8°; bp$_{20}$ 151.8°; bp$_{10}$ 135.0°; bp$_5$ 119.6°; bp$_{1.0}$ 87.4°. n$_D^{20}$ 1.60849.

3357. Diphenyl Ether. [101-84-8] 1,1′-Oxybisbenzene; diphenyl oxide; phenoxybenzene; phenyl ether. C$_{12}$H$_{10}$O; mol wt 170.21. C 84.68%, H 5.92%, O 9.40%. Prepd by heating sodium phenolate with chlorobenzene.

Liquid; characteristic odor. d^{20} 1.075 (liq); mp 28°, and remains liquid at lower temp; bp 259°. Flash pt 115°C. Insol in water; sol in alcohol, benzene, ether, glacial acetic acid.

Caution: Potential symptoms of overexposure to vapor are nausea; irritation of eyes, nose and skin. *See NIOSH Pocket Guide to Chemical Hazards* (DHHS/NIOSH 97-140, 1997) p 248.

USE: As heat transfer medium; in perfuming soaps; in organic syntheses.

3358. 1,3-Diphenylguanidine. [102-06-7] *N,N'*-Diphenyl-guanidine; *sym*-diphenylguanidine; Melaniline; Vulkazit. $C_{13}H_{13}N_3$; mol wt 211.27. C 73.91%, H 6.20%, N 19.89%. Prepn: Naunton, *J. Soc. Chem. Ind. London* **45**, 376T (1926); Macholdt-Erdniss, *Ber.* **91**, 1992 (1958); Ferris, Schutz, *J. Org. Chem.* **28**, 71 (1963). Pharmacology: P. Valade *et al.*, *C. R. Seances Soc. Biol. Ses Fil.* **143**, 815 (1949).

Crystals from ether, mp 150°. Dec at about 170°. d 1.13. Sparingly sol in water; sol in alcohol, chloroform, hot benzene, hot toluene; readily sol in dil mineral acids. Aq soln is strongly alkaline. MLD s.c. in guinea pigs: 200 mg/kg; i.v. in dogs: 25 mg/kg (Valade).

Phthalate. Guantal. $C_{34}H_{32}N_6O_4$; mol wt 588.67. Obtained as the hemihydrate, blue-white to light gray powder. mp 178°. d_4^{25} 1.20. Sol in alcohol. Practically insol in benzene, gasoline.

USE: Recommended as a primary material for standardizing acids. Free base and phthalate as accelerators for vulcanization of rubber.

3359. 1,1-Diphenylhydrazine. [530-50-7] $C_{12}H_{12}N_2$; mol wt 184.24. C 78.23%, H 6.57%, N 15.21%.

Yellow crystals. d 1.19. mp 34.5°; also reported as mp 44°. bp$_{40}$ 220°. Insol in water. Freely sol in alc, ether.

Hydrochloride. [530-47-2] $C_{12}H_{12}N_2$·HCl; mol wt 220.70. White to grayish white, crystalline powder. Slightly sol in water; freely sol in alc.

USE: Hydrochloride as reagent for arabinose and lactose.

3360. Diphenylketene. [525-06-4] 2,2-Diphenylethenone. $C_{14}H_{10}O$; mol wt 194.23. C 86.57%, H 5.19%, O 8.24%. Prepd from benzil monohydrazone: Smith, Hoehn, *Org. Synth.* **coll. vol. III**, 356 (1955).

Reddish-yellow liquid. Best stored under nitrogen. The addition of hydroquinone helps to retard polymerization. Solidifies in refrigerator. $d_4^{13.7}$ 1.1107. bp$_{760}$ 265-270° (dec); bp$_{12}$ 146°; bp$_{3.5}$ 119-121°. Should be distilled under reduced pressure (3 to 5 mm) in an atm of nitrogen. $n_D^{14.1}$ 1.615. Dipole moment in benzene at 25° = −1.9.

3361. Diphenylmagnesium. [555-54-4] Magnesium diphenyl. $C_{12}H_{10}Mg$; mol wt 178.52. C 80.74%, H 5.65%, Mg 13.61%. Mg(C_6H_5)$_2$. Prepd by the action of magnesium on mercury diphenyl: Schlenk, Jr., *Ber.* **64**, 736 (1931); Gilman, Brown, *Rec. Trav. Chim.* **49**, 202 (1930).

Solvated, feathery crystals from ether. Loses its ether *in vacuo* at around 37°. Dec 280°, dissociating into magnesium and diphenyl. Extremely reactive; catches fire in moist, although not in dry, air. Violently dec by water. The etherate is sol in benzene, but the dry compd is not.

3362. Diphenylmethane. [101-81-5] 1,1'-Methylenebisbenzene; benzylbenzene; ditan. $C_{13}H_{12}$; mol wt 168.24. C 92.81%, H 7.19%. Prepd from methylene chloride and benzene with aluminum

chloride as catalyst: Friedel, Crafts, *Bull. Soc. Chim.* [2] **41**, 324 (1884); from benzyl chloride and benzene: Hartman, Phillips, *Org. Synth.* **14**, 34 (1934); L. F. Fieser, *Experiments in Organic Chemistry* (Boston, 3rd ed., 1955) p 157. Absorption spectrum in hexane: Castille, *Bull. Soc. Chim. Belg.* **36**, 296; *Chem. Zentralbl.* **1927**, I, 1126; in alcohol: Ondorff, *J. Am. Chem. Soc.* **49**, 1541 (1927).

Orthorhombic needles. Odor of oranges. d_4^{10} 1.3421 (solid). mp 25.9°. d_4^{26} 1.0008 (liq). bp$_{760}$ 264.5°; bp$_{400}$ 237.5°; bp$_{200}$ 210.7°; bp$_{100}$ 186.3°; bp$_{60}$ 170.2°; bp$_{40}$ 157.8°; bp$_{20}$ 139.8°; bp$_{10}$ 122.8°; bp$_5$ 107.4°; bp$_{1.0}$ 76.0°. n_D^{20} 1.57683. Freely sol in alcohol, ether, chloroform, hexane, benzene. Insol in liquid ammonia.

3363. *N,N'*-Diphenyl-*p*-phenylenediamine. [74-31-7] N^1,N^4-Diphenyl-1,4-benzenediamine; 1,4-dianilinobenzene; DPPD. $C_{18}H_{16}N_2$; mol wt 260.34. C 83.04%, H 6.19%, N 10.76%. Prepd by condensing hydroquinone or *p*-aminophenol with aniline: Calm, *Ber.* **16**, 2805 (1883); Clemens, Magoffin, US **2503712** (1950 to Kodak). Purification: Pecherer, US **2833824** (1958 to Hoffmann-La Roche).

Colorless leaflets from alcohol. Commercial grades are greenish-brown. d 1.20. mp 150-151° (uncorr., Pecherer). bp$_{0.5}$ 220-225°. Sol in monochlorobenzene, benzene, DMF, ether, chloroform, acetone, ethyl acetate, isopropyl alcohol, glacial acetic acid. Slightly sol in alcohol. Almost insol in petr ether, water.

USE: Polymerization inhibitor. Antioxidant for rubber, petr oils, feedstuffs.

THERAP CAT (VET): Antioxidant for feedstuffs.

3364. Diphenylphosphine. [829-85-6] Ph$_2$PH. $C_{12}H_{11}P$; mol wt 186.19. C 77.41%, H 5.96%, P 16.64%. Precursor for the prepn of a variety of functionalized organophosphorus compds. Prepn: A. Michaelis, L. Gleichmann, *Ber.* **15**, 801 (1882); V. D. Bianco, S. Doronzo, *Inorg. Synth.* **16**, 161 (1976). Molecular structure: V. A. Naumov, O. N. Kataeva, *J. Struct. Chem.* **25**, 642 (1985). Synthetic applications involving addition reactions: J. A. van Doorn *et al.*, *J. Chem. Soc. Perkin Trans. 2* **1990**, 479; K. Suzuki *et al.*, *Synlett* **1992**, 125; involving free radical conditions: J. E. Brumwell *et al.*, *Tetrahedron Lett.* **34**, 1215 (1993); involving alkylation chemistry: K. Hirano *et al.*, *Org. Lett.* **6**, 4873 (2004); K. Damian *et al.*, *Appl. Organometal. Chem.* **23**, 272 (2009).

Clear liquid with strong, unpleasant odor. *Pyrophoric, irritant.* bp 280°; bp$_{11}$ 150-154°; bp$_2$ 115-118°; bp$_1$ 108-110°. fp −14.5. d_4^{16} 1.07. n_D^{25} 1.6240; n_D^{20} 1.6269. Flash pt, closed cup: 230°F (110°C). Sol in ethanol, diethyl ether, benzene, concd HCl. Insol in water. Light and air sensitive.

USE: Reagent in synthetic organic chemistry.

3365. *O*-(Diphenylphosphinyl)hydroxylamine. [72804-96-7] (Aminooxy)diphenylphosphine oxide; ODPH. $C_{12}H_{12}NO_2P$; mol wt 233.21. C 61.80%, H 5.19%, N 6.01%, O 13.72%, P 13.28%. Activated hydroxylamine used in electrophilic aminations. Prepn from diphenylphosphinic chloride and hydroxylamine: N. Kreutzkamp, H. Schindler, *Arch. Pharm.* **293**, 296 (1960). Corrected structural assignment and reactivity studies: M. J. P. Harger, *J. Chem. Soc. Chem. Commun.* **1979**, 768; idem, *J. Chem. Soc. Perkin Trans. 1* **1981**, 3284. Synthetic applications: G. Boche *et al.*, *Tetrahedron*

Lett. **23**, 5399 (1982); G. Sosnovsky, K. Purgstaller, *Z. Naturforsch.* **44b**, 582 (1989); R. Badorrey *et al.*, *Tetrahedron: Asymmetry* **6**, 2787 (1995); A. Armstrong *et al.*, *Org. Lett.* **9**, 351 (2007). Reviews: G. Boche in *Encyclopedia of Reagents for Organic Synthesis*, **4**, L. A. Paquette, Ed. (John Wiley & Sons, New York, 1995) pp 2240-2242; E. Bodio, *Synlett* **2008**, 1744-1745.

Solid from methanol, mp 131° (dec) (Kreutzkamp, Schindler); mp 133-134° (Sosnovsky). Sparingly sol in aprotic solvents incl ether, chloroform, dichloromethane. Dec in solns of acetone, dimethylsulfoxide. Stable for up to one year when stored below 0°.

USE: Reagent in synthetic organic chemistry.

3366. Diphenylphosphoryl Azide. [26386-88-9] Phosphorazidic acid diphenyl ester; diphenyl phosphorazidate; DPPA. $C_{12}H_{10}N_3O_3P$; mol wt 275.20. C 52.37%, H 3.66%, N 15.27%, O 17.44%, P 11.26%. Non-explosive azide transfer reagent; utilized in peptide synthesis, the modified Curtius rearrangement, and other synthetic transformations. Prepn and synthetic utility: T. Shioiri *et al.*, *J. Am. Chem. Soc.* **94**, 6203 (1972). Alternate prepn: E. Shi, C. Pei, *Synthesis* **2004**, 2995. Large scale prepn: O. Wolff, S. R. Waldvogel, *ibid.* 1303. Additional synthetic applications: K. Ninomiya *et al.*, *Tetrahedron* **30**, 2151 (1974); S. Yamada *et al.*, *J. Org. Chem.* **39**, 3302 (1974); A. S. Thompson *et al.*, *ibid.* **58**, 5886 (1993). Reviews: A. V. Thomas in *Encyclopedia of Reagents for Organic Synthesis* vol. 4, L. A. Paquette, Ed. (Wiley, New York, 1995) pp 2242-2245; H. Liang, *Synlett* **2008**, 2554-2555.

Colorless liquid. bp$_{0.17}$ 157°. *Poisonous.* Sol in toluene, THF, DMF, *tert*-butanol. Flash pt, closed cup: 233.6°F (112°C).

USE: Reagent in synthetic organic chemistry.

3367. Diphenyl Phosphoryl Chloride. [2524-64-3] Phosphorochloridic acid diphenyl ester; diphenyl chlorophosphate; diphenyl phosphorochloridate. $C_{12}H_{10}ClO_3P$; mol wt 268.63. C 53.65%, H 3.75%, Cl 13.20%, O 17.87%, P 11.53%. Reagent utilized in preparing phosphorylated intermediates. Prepn: G. Jacobsen, *Ber.* **8**, 1519 (1875); M. Rapp, *Ann.* **224**, 156 (1884); and use in phosphorylation reactions: P. Brigl, H. Müller, *Ber.* **72**, 2121 (1939). Improved phosphorylation procedure: S. Jones *et al.*, *J. Org. Chem.* **68**, 5211 (2003). Synthetic applications in the prepn of azirines: J. M. Villalgordo, H. Heimgartner, *Helv. Chim. Acta* **75**, 1866 (1992); in the prepn of diphenyl phosphorazidate, *q.v.*: T. Shioiri, S. Yamada, *Org. Synth.* **coll. vol. VII**, 206 (1990).

Colorless liquid. *Corrosive.* bp$_{272}$ 314-316°; bp$_{13-14}$ 195°; bp$_5$ 165-168°; bp$_1$ 140-142°; bp$_{0.5}$ 133-135°; bp$_{0.05}$ 102-105°. d$_4^{20}$ 1.29604. n$_D^{25}$ 1.5460. Sol in dichloromethane, THF. Moisture sensitive.

USE: Reagent in synthetic organic chemistry.

3368. Diphenyl Phthalate. [84-62-8] 1,2-Benzenedicarboxylic acid 1,2-diphenyl ester; phenyl phthalate. $C_{20}H_{14}O_4$; mol wt 318.33. C 75.46%, H 4.43%, O 20.10%. Prepd from phenol and phthalic anhydride.

White odorless crystals. d^{74} 1.572; mp 70-73°; bp$_{14}$ ~255°. Insol in water; sol in acetone and other ketones, in liquid esters and chlorinated hydrocarbons.

USE: Plasticizer in nitrocellulose lacquers.

3369. 1,1-Diphenyl-2-picrylhydrazyl (Free Radical). [1898-66-4] 2,2-Diphenyl-1-(2,4,6-trinitrophenyl)hydrazinyl; 2,2-diphenyl-1-(2,4,6-trinitrophenyl)hydrazyl; DPPH. $C_{18}H_{12}N_5O_6$; mol wt 394.32. C 54.83%, H 3.07%, N 17.76%, O 24.34%. Prepn: Goldschmidt, Renn, *Ber.* **55B**, 628 (1922); Lyons, Watson, *J. Polym. Sci.* **18**, 141 (1955); Arbuzov, Valitova, *Zh. Obshch. Khim.* **27**, 2354 (1957). Structure: Poirier *et al.*, *J. Org. Chem.* **17**, 1437 (1952).

Large, dark violet prisms from benzene + petr ether, mp 127-129° (dec). mp also reported as 132-133°.

USE: Analytical reagent for reducing substances: Schenck *et al.*, *Tetrahedron Lett.* **1967**, 193.

3370. Diphenylpyraline. [147-20-6] 4-(Diphenylmethoxy)-1-methylpiperidine; diphenylpyrilene; 4-benzhydryloxy-*N*-methylpiperidine; 1-methyl-4-piperidyl benzhydryl ether; 1-methyl-4-hydroxypiperidine benzhydryl ether; P-253. $C_{19}H_{23}NO$; mol wt 281.40. C 81.10%, H 8.24%, N 4.98%, O 5.69%. Prepn: L. H. Knox, R. Kapp, **US 2479843** (1949 to Nopco). Pharmacokinetics: G. Graham, A. G. Bolt, *J. Pharmacokinet. Biopharm.* **2**, 191 (1974). Clinical trial in allergic rhinitis: H. Puhakka *et al.*, *J. Int. Med. Res.* **5**, 37 (1977). GC determn in blood and urine: H. Hattori *et al.*, *J. Chromatogr.* **581**, 213 (1992).

Hydrochloride. [132-18-3] Dayfen; Diafen; Hispril; Histryl; Histyn; Kolton; Lergoban; Lergobine. $C_{19}H_{23}NO \cdot HCl$; mol wt 317.86. Crystals from isopropanol + ether, mp 206°. Sol in water, ethanol, isopropanol. Practically insol in ether, benzene.

Hydrobromide. [5807-87-4] $C_{19}H_{23}NO \cdot HBr$. Prepn: R. F. Phillips, **US 2595405** (1952 to Merck & Co.). Crystals, mp 201-202°. Soluble in water, ethanol.

8-Chlorotheophyllinate. [606-90-6] Piprinhydrinate; diphenylpyraline theoclate; Mepedyl. $C_{19}H_{23}NO \cdot C_7H_7ClN_4O_2$; mol wt 496.01. Prepn: W. A. Schuler, **DE 934890** (1955 to Promonta), *C.A.* **52**, 20065i (1958). White to off-white crystalline powder, mp 174-176°. Sparingly sol in water. Freely sol in alcohol.

THERAP CAT: Antihistaminic.

3371. Diphenyl Sulfide. [139-66-2] 1,1'-Thiobisbenzene; diphenylmercaptan; diphenyl thioether; phenylsulfide. $C_{12}H_{10}S$; mol wt 186.27. C 77.38%, H 5.41%, S 17.21%. Description: Willard, Hall, *J. Am. Chem. Soc.* **44**, 2219 (1922). Toxicity study: H. F. Smyth *et al.*, *Am. Ind. Hyg. Assoc. J.* **23**, 95 (1962).

Colorless, almost odorless liquid. d_{15}^{15} 1.118; mp about $-40°$; bp 295-297°; n_D^{18} 1.6350. Insol in water. Sol in hot alcohol; misc with benzene, ether, carbon disulfide. LD_{50} orally in rats: 0.49 ml/kg (Smyth).

3372. Diphenyl Sulfone. [127-63-9] 1,1'-Sulfonylbisbenzene; phenyl sulfone; sulfobenzide. $C_{12}H_{10}O_2S$; mol wt 218.27. C 66.03%, H 4.62%, O 14.66%, S 14.69%. Prepd by sulfonation of benzene with sulfuric acid; formed as a by-product on production of benzenesulfonic acid and benzenesulfonyl chloride.

White monoclinic prisms or leaflets, mp 128-129°. bp 378-379°. Insoluble in cold water. Slightly sol in boiling water; sol in hot alcohol, in benzene. LD_{50} orally in rats: >2 g/kg; see: *Residue Rev.* **36**, 240 (1971).

USE: Ovicide; acaricide.

3373. *sym*-Diphenylthiourea. [102-08-9] N,N'-Diphenylthiourea; thiocarbanilide; sulfocarbanilide. $C_{13}H_{12}N_2S$; mol wt 228.31. C 68.39%, H 5.30%, N 12.27%, S 14.04%. Prepn from aniline and carbon disulfide: Fry, *J. Am. Chem. Soc.* **35**, 1539 (1913); Stasse, US 2435295 (1948 to Allied Chem. & Dye); from potassium ethylxanthate and aniline: Aravindakshan et al., *Indian J. Chem.* **1**(9), 395 (1963). Physical properties and toxicity data: P. J. Hanzlik, A. Irvine, *J. Pharmacol. Exp. Ther.* **17**, 349 (1921).

White crystalline solid, mp 153-154°. d 1.32. Distinct bitter taste. Practically insol in water. Sol in alcohol, ether, chloroform. MLD orally in rabbits: 1.5 g/kg (Hanzlik, Irvine).

USE: Vulcanizing accelerator; sulfur dyes.

3374. Diphosgene. [503-38-8] Carbonochloridic acid trichloromethyl ester; chloroformic acid trichloromethyl ester; trichloromethylchloroformate; trichloromethylcarbonochloridate. $C_2Cl_4O_2$; mol wt 197.82. C 12.14%, Cl 71.68%, O 16.18%. Used in synthesis as substitute for phosgene, q.v. Prepn by photochlorination of methyl formate: W. Hentschel, *J. Prakt. Chem.* **36**, 209 (1887); F. Grignard et al., *C. R. Hebd. Seances Acad. Sci.* **169**, 1074 (1919); *eidem, ibid.* 1143; by photochlorination of methyl chloroformate: A. Kling et al., *ibid.* 1046; K. Kurita, Y. Iwakura, *Org. Synth.* **59**, 195 (1980). Effect on air-blood barrier in rabbits, toxicity: E. Klika, A. Mysliveckova, *Folia Morphol. (Prague)* **19**, 5 (1971), *C.A.* **75**, 3433c (1971). Use in synthesis of isocyanides: G. Skorna, J. Ugi, *Angew. Chem. Int. Ed.* **16**, 259 (1977).

Suffocating liquid. bp 128°. bp_{50} 49°. d_{15} 1.664. n_D^{22} 1.45664. Stable at room temp, decomposes to phosgene at $\sim300°C$. LC_{100} in rabbits (10-20 minutes exposure to vapor): 0.9 mg/l air (Klika, Mysliveckova).

USE: In organic synthesis; as war gas.

3375. Dipicrylamine. [131-73-7] 2,4,6-Trinitro-N-(2,4,6-trinitrophenyl)benzenamine; 2,4,6,2',4',6'-hexanitrodiphenylamine.

$C_{12}H_5N_7O_{12}$; mol wt 439.21. C 32.82%, H 1.15%, N 22.32%, O 43.71%. *Ref:* Winkel, Maas, *Angew. Chem.* **49**, 827 (1936).

Yellow prisms. mp about 238° with dec. *Explosive.* Insol in water, acetone, alcohol or ether. Sol in alkalies, glacial acetic acid.

USE: Reagent for gravimetric determination of potassium.

3376. Dipin. [738-99-8] 1,4-Bis[bis(1-aziridinyl)phosphinyl]piperazine; 1,4-piperazinediylbis[bis(1-aziridinyl)phosphine oxide]; piperazine-1,4-bis(N,N'-diethylenephosphonediamide); tetraethyleneimidopiperazine-N,N'-diphosphoric acid. $C_{12}H_{24}N_6O_2P_2$; mol wt 346.31. C 41.62%, H 6.99%, N 24.27%, O 9.24%, P 17.89%. Alkylating agent; exptl antineoplastic. Prepn: Kropacheva et al., *Zh. Obshch. Khim.* **30**, 3584 (1960), *C.A.* **55**, 18695c (1961). Spectrophotometric determn: L. Kh. Kartashova, E. M. Salomatin, *Farmatsiya (Moscow)* **33**, 72 (1983), *C.A.* **102**, 119736j (1985). Metabolism in rats: V. V. Chistyakov et al., *Khim. Farm. Zh.* **21**, 398 (1987), *C.A.* **107**, 146734s (1987).

Crystals, mp 187°.

3377. Dipipanone. [467-83-4] 4,4-Diphenyl-6-(1-piperidinyl)-3-heptanone; dl-4,4-diphenyl-6-piperidinoheptan-3-one; 6-piperidino-4,4-diphenylheptan-3-one; 2-(1-piperidino)-4,4-diphenyl-5-heptanone; phenylpiperone; piperidylamidone. $C_{24}H_{31}NO$; mol wt 349.52. C 82.47%, H 8.94%, N 4.01%, O 4.58%. Opioid analgesic. Prepn: Ofner, Walton, *J. Chem. Soc.* **1950**, 2158; GB 654975 (1951 to Wellcome Foundation). Variations of the synthesis using phenyllithium as condensing agent: Kazuhiko, Kubota, *Kumamoto Med. J.* **12**, 304-307 (1960). GC determn in plasma, urine: S. Paterson, *J. Chromatogr. B* **424**, 152 (1988).

Hydrochloride. [75783-06-1] Hoechst 10805. $C_{24}H_{31}NO.HCl$. Component of *Diconal.* Minute prisms from wet alcohol-ether, mp 123-126°. Also reported as mp 126-127°.

Hydrobromide. [909260-86-2] $C_{24}H_{31}NO.HBr$. Crystals from water, mp 103-106°.

Note: This is a controlled substance (opiate): **21 CFR**, 1308.11.

THERAP CAT: Analgesic.

3378. Dipivefrin. [52365-63-6] 2,2-Dimethylpropanoic acid 1,1'-[4-[1-hydroxy-2-(methylamino)ethyl]-1,2-phenylene] ester; ($\pm$)-3,4-dihydroxy-α-[(methylamino)methyl]benzyl alcohol 3,4-dipivalate; 1-(3',4'-dipivaloyloxyphenyl)-2-methylamino-1-ethanol; dipivalyl epinephrine; DPE. $C_{19}H_{29}NO_5$; mol wt 351.44. C 64.94%, H 8.32%, N 3.99%, O 22.76%. Dipivalyl ester of epinephrine, q.v. Prepn: D. Henschler et al., DE 2152058; *eidem,* US 4085270 (1973, 1978 both to Klinge); A. Hussain, J. E. Truelove, DE 2343657; *eidem,* US 3809714 and US 3839584 (all 1974 to

Interx). *In vitro* study: A. H. Neufeld, E. D. Page, *Invest. Ophthalmol. Visual Sci.* **16**, 1118 (1977). Pharmacology: B. C. Wang *et al.*, *J. Pharmacol. Exp. Ther.* **203**, 442 (1977). Effects on intraocular pressure in dogs: R. M. Gwin *et al.*, *Am. J. Vet. Res.* **39**, 83 (1978). Metabolism: I. Abramovsky, J. S. Mindel, *Arch. Ophthalmol.* **97**, 1937 (1979). Clinical study: M. A. Kass *et al.*, *ibid.* 1865. General pharmacology, toxicology and clinical experience in glaucoma: D. A. McClure, *ACS Symp. Ser.* **14**, 224-235 (1975). Comprehensive description: G. M. Wall, T. Y. Fan, *Anal. Profiles Drug Subs. Excip.* **22**, 229-262 (1993).

Crystals from ether, mp 146-147°.

Hydrochloride. [64019-93-8] Diopine; d Epifrin; Diphemin; Pivalephrine; Propine. $C_{19}H_{29}NO_5$.HCl; mol wt 387.90. Crystals from ethyl acetate, mp 158-159°. Sol in water and ethanol. pKa 8.40.

THERAP CAT: Adrenergic (ophthalmic); antiglaucoma.

3379. Diploicin. [527-93-5] 2,4,7,9-Tetrachloro-3-hydroxy-8-methoxy-1,6-dimethyl-11*H*-dibenzo[*b,e*][1,4]dioxepin-11-one; 3,5-dichloro-6-[(3,5-dichloro-6-hydroxy-4-methoxy-*o*-tolyl)oxy]-4,2-cresotic acid ε-lactone; 5,6′-dimethyl-2′,3-dihydroxy-4′-methoxy-2,3′,4,5′-tetrachloro-4-carboxydiphenyl ether 2′,6-lactone. $C_{16}H_{10}Cl_4O_5$; mol wt 424.05. C 45.32%, H 2.38%, Cl 33.44%, O 18.86%. Antibiotic isolated from the lichen *Buellia canescens* (Dicks.) De Not. [*Diploicia canescens* (Dicks.) Massal.], *Lecideaceae*: Zopf, *Ann.* **336**, 58 (1904); Nolan, *Sci. Proc. R. Dublin Soc.* **21**, 67 (1934); Nolan *et al.*, *Chem. Ind. (London)* **1935**, 577; Barry, *Nature* **158**, 131 (1946). Structure: Nolan *et al.*, *Sci. Proc. R. Dublin Soc.* **24**, 319 (1948). Synthesis: Ollis, *ibid.* **27**, 161 (1956); Brown *et al.*, *Proc. Chem. Soc. London* **1960**, 393; Hendrickson, Ramsay, *Chem. Commun.* **1968**, 1101; Hendrickson *et al.*, *J. Am. Chem. Soc.* **94**, 6834 (1972); P. D. Djura *et al.*, *J. Chem. Soc. Perkin Trans. 1* **1976**, 147.

White needles from methanol, mp 233-234°. uv max: 270 nm (log ε 3.79).

Acetate. $C_{18}H_{12}Cl_4O_6$. mp 234-235°.

3380. Diprenorphine. [14357-78-9] (5α,7α)-17-(Cyclopropylmethyl)-4,5-epoxy-18,19-dihydro-3-hydroxy-6-methoxy-α,α-dimethyl-6,14-ethenomorphinan-7-methanol; 21-cyclopropyl-6,7,-8,14-tetrahydro-7α-(1-hydroxy-1-methylethyl)-6,14-*endo*-ethanooripavine; *N*-(cyclopropylmethyl)-19-methylnororvinol; M-5050; RX-5050M. $C_{26}H_{35}NO_4$; mol wt 425.57. C 73.38%, H 8.29%, N 3.29%, O 15.04%. Closely related to cyprenorphine, *q.v.* Prepn: K. W. Bentley, D. G. Hardy, *J. Am. Chem. Soc.* **89**, 3281 (1967). Activity in rats: G. F. Blane, *J. Pharm. Pharmacol.* **19**, 367 (1967); G. F. Blane, D. Dugdall, *ibid.* **20**, 547 (1968). Use as etorphine antagonist in dogs: M. Grange *et al.*, *Rev. Med. Vet.* **124**, 899 (1973); in large animals: B. T. Alford *et al.*, *J. Am. Vet. Med. Assoc.* **164**, 702 (1974). Binding to opiate receptors: C. B. Pert *et al.*, *Life Sci.* **16**, 1849 (1975); J. Pearson *et al.*, *ibid.* **26**, 1047 (1980); to μ opiate receptors: J. J. Frost *et al.*, *ibid.* **38**, 1597 (1986). Treatment of experimental stroke in cats: D. S. Baskin *et al.*, *Neuropeptides* **5**, 307 (1985);

eidem, J. Neurosurg. **64**, 99 (1986). Determn by HPLC: I. Jane, A. McKinnon, *J. Chromatogr.* **323**, 191 (1985). Toxicity data: N. S. Duggett *et al.*, *Toxicol. Appl. Pharmacol.* **31**, 141 (1977).

Crystals from methanol, mp 185°.

Hydrochloride. Revivon. $C_{26}H_{35}NO_4$.HCl; mol wt 462.03. LD_{50} s.c. in mice: 316.0 ± 20 mg/kg (Duggett).

THERAP CAT (VET): Narcotic antagonist.

3381. Dipropalin. [1918-08-7] 4-Methyl-2,6-dinitro-*N,N*-dipropylbenzenamine; 2,6-dinitro-*N,N*-dipropyl-*p*-toluidine; *N,N*-dipropyl-2,6-dinitro-4-methylaniline; 2,6-dinitro-*N,N*-dipropyl-4-methylaniline; 3,5-dinitro-4-dipropylaminotoluene; L-35355. C_{13}-$H_{19}N_3O_4$; mol wt 281.31. C 55.51%, H 6.81%, N 14.94%, O 22.75%. Prepd from 2,6-dinitro-*p*-cresol, *p*-toluenesulfonate and dipropylamine: Hantzsch, *Ber.* **43**, 1662 (1910).

Yellow crystals, mp 80°.

USE: Herbicide.

3382. Dipropetryn. [4147-51-7] 6-(Ethylthio)-N^2,N^4-bis(1-methylethyl)-1,3,5-triazine-2,4-diamine; 2-(ethylthio)-4,6-bis(isopropylamino)-*s*-triazine; GS-16068. $C_{11}H_{21}N_5S$; mol wt 255.38. C 51.74%, H 8.29%, N 27.42%, S 12.55%. Pre-emergence herbicide. Prepn: **NL 6414460**; H. Yamamoto, T. Namekawa, **US 3326912** (1965, 1967 both to Nippon Kayaku). Absorption and translocation in plants: E. Basler *et al.*, *Weed Sci.* **26**, 358 (1978).

Powder, mp 104-106°. Vapor pressure at 20°: 7.3×10^{-7} mm Hg. Soly in water at 20°: 16 mg/l. Soluble in organic solvents.

USE: Herbicide.

3383. *n*-Dipropylamine. [142-84-7] *N*-Propyl-1-propanamine. $C_6H_{15}N$; mol wt 101.19. C 71.22%, H 14.94%, N 13.84%.

Colorless liq; odor of ammonia. d_4^{20} 0.738. bp 110°. mp −63°. n_D^{20} 1.40455. Freely sol in water or alcohol. Forms a hydrate with H_2O. LD_{50} orally in rats: 0.93 g/kg; *see:* H. F. Smyth *et al.*, *Am. Ind. Hyg. Assoc. J.* **23**, 95 (1962).

3384. Dipropylene Glycol Monomethyl Ether. [34590-94-8] 1(or 2)-(2-Methoxymethylethoxy)propanol; DPGME; DPM; Arcosolv DPM; Dowanol DPM; Poly-Solv DPM. $C_7H_{16}O_3$; mol wt 148.20. C 56.73%, H 10.88%, O 32.39%. Commercial product, prepd from propylene oxide and methanol, may consist of as many as four structural isomers. Prepn of major isomer, *1-(2-methoxy-1-methylethoxy)-2-propanol*, depicted below: A. R. Sexton, E. C. Britton, *J. Am. Chem. Soc.* **75**, 4357 (1953). GC determn in urine: B. Hubner *et al.*, *Fresenius J. Anal. Chem.* **342**, 746 (1992). Toxicity studies: H. F. Smyth, Jr. *et al.*, *Am. Ind. Hyg. Assoc. J.* **23**, 95 (1962); T. D. Landry, B. L. Yano, *Fundam. Appl. Toxicol.* **4**, 612 (1984). Review of toxicology: *Patty's Industrial Hygiene and Toxicology* vol. **2D**, G. D. Clayton, F. E. Clayton, Eds. (Wiley-Interscience, New York, 4th ed., 1994) pp 2882-2886; G. Johansson, *NEG and NIOSH Basis for an Occupational Health Standard: Propylene Glycol Ethers and Their Acetates* (PB91-220749, 1991) 47 pp.

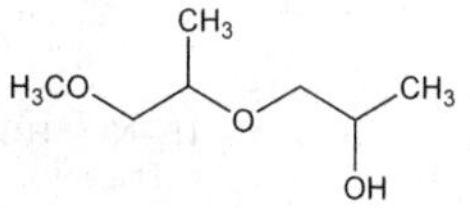

Colorless liquid. bp_{760} 189.6°. fp −80°. d_4^{25} 0.948. n_D^{25} 1.419. Flash pt, open cup: 185°F. Vapor pressure at 25°: 0.41 mm Hg. Misc with water, benzene. LD_{50} orally in rats, dermally in rabbits: 5.66, 10.0 (ml/kg) (Smyth).

Caution: Potential symptoms of overexposure are irritation of eyes, nose, throat; weakness, lightheadedness, headache. *See NIOSH Pocket Guide to Chemical Hazards* (DHHS/NIOSH 97-140, 1997) p 122.

USE: As solvent for automotive fluids, cleaners, dyes, coatings, inks, waxes, adhesives, agricultural products, insect repellents, and cosmetics; chemical intermediate.

3385.   **Dipropyl Ether.** [111-43-3] 1,1'-Oxybispropane; propyl ether. $C_6H_{14}O$; mol wt 102.18. C 70.53%, H 13.81%, O 15.66%. Obtained by heating propyl alcohol with benzenesulfonic acid.

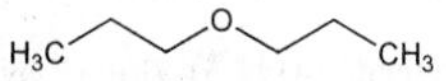

Mobile liquid. *Extremely flammable.* d_4^{20} 0.7360; mp −122°; bp 89-91°; n_D^{20} 1.3807. Flash pt, open cup: −5°F (−20°C). Slightly sol in water; sol in alcohol, ether. Highly volatile. Tends to form explosive peroxides, esp when anhyd. Do not allow to evaporate to near dryness.

3386.   **Dipropyl Ketone.** [123-19-3] 4-Heptanone; butyrone. $C_7H_{14}O$; mol wt 114.19. C 73.63%, H 12.36%, O 14.01%. Made by passing butyric acid over precipitated $CaCO_3$ at 450°.

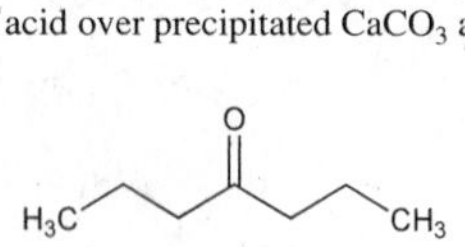

Colorless, very refractive liquid; penetrating odor; burning taste. d_4^{15} 0.821. bp 144°. mp −32.6°. n_D^{22} 1.4073. *Flammable.* Insol in water. Miscible with alcohol, ether.

Caution: Potential symptoms of overexposure are irritation of eyes, skin; CNS depression, dizziness, somnolence, decreased breathing. *See NIOSH Pocket Guide to Chemical Hazards* (DHHS/NIOSH 97-140, 1997) p 122.

3387.   **Dipropyl Sulfide.** [111-47-7] 1,1'-Thiobispropane; dipropyl thioether; propyl sulfide. $C_6H_{14}S$; mol wt 118.24. C 60.95%, H 11.94%, S 27.11%. Prepd from propyl bromide and an alcoholic soln of Na_2S.

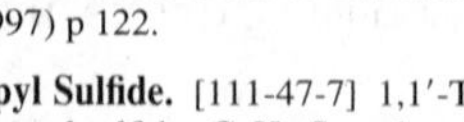

Colorless liquid. d^{17} 0.814; mp ∼ −102°; bp 142°. Insol in water; sol in alcohol, ether.

3388.   **Dipyridamole.** [58-32-2] 2,2',2'',2'''-[(4,8-Di-1-piperidinylpyrimido[5,4-*d*]pyrimidine-2,6-diyl)dinitrilo]tetrakisethanol; 2,6-bis(diethanolamino)-4,8-dipiperidinopyrimido-[5,4-*d*]pyrimidine; NSC-515776; RA-8; Anginal; Cardoxin; Cleridium; Coridil; Coronarine; Curantyl; Dipyridan; Gulliostin; Natyl; Peridamol; Persantine; Piroan; Prandiol; Protangix. $C_{24}H_{40}N_8O_4$; mol wt 504.64. C 57.12%, H 7.99%, N 22.21%, O 12.68%. Phosphodiesterase inhibitor that reduces platelet aggregation; also acts as a coronary vasodilator. Prepn: **GB 807826**; F. G. Fischer, *et al.*, **US 3031450** (1959, 1962 both to Thomae). Activity studies: Saraf, Seth, *Indian J. Physiol. Pharmacol.* **15**, 135 (1971). Toxicological study: F. Takenaka *et al.*, *Arzneim.-Forsch.* **22**, 892 (1972). Symposium on pharmacology and clinical experience as antithrombotic: *Thromb. Res.* **60**, Suppl. 12, 1-99 (1990). Review of use as pharmacological stress agent in echocardiography: M. B. Buchalter *et al.*, *Postgrad. Med. J.* **66**, 531-535 (1990); in ^{201}Tl cardiac imaging: S. G. Beer *et al.*, *Am. J. Cardiol.* **67**, Suppl., 18D-26D (1991).

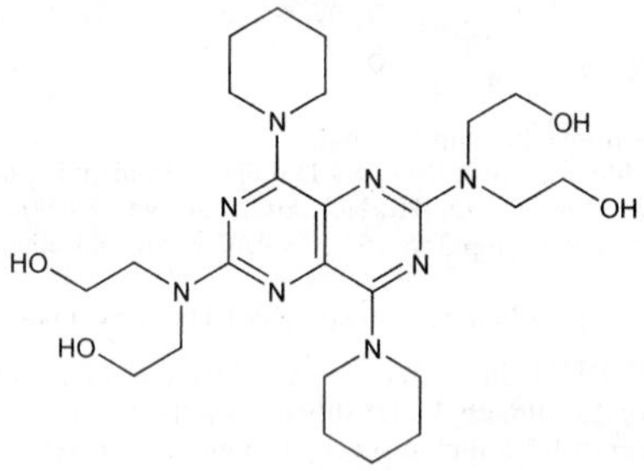

Deep yellow needles from ethyl acetate, mp 163°. Bitter taste. Very sol in methanol, ethanol, chloroform; sol in dil acid having a pH of 3.3 or below; slightly sol in water; very slightly sol in acetone, benzene, ethyl acetate. Solns are yellow and show strong blue-green fluorescence. LD_{50} in rats: 8.4 g/kg orally; 208 mg/kg i.v. (Takenaka).

Combination with aspirin. Aggrenox. Review of pharmacology and clinical efficacy in secondary prevention of stroke: P. S. Hervey, K. L. Goa, *Drugs* **58**, 469-475 (1999).

THERAP CAT: Antithrombotic; diagnostic aid (cardiac stress testing).

3389.   **γ,γ'-Dipyridyl.** [553-26-4] 4,4'-Bipyridine; 4,4'-dipyridyl. $C_{10}H_8N_2$; mol wt 156.19. C 76.90%, H 5.16%, N 17.94%. Prepn: Dimroth, Frister, *Ber.* **55**, 3695 (1922); Smith, *J. Am. Chem. Soc.* **46**, 414 (1924).

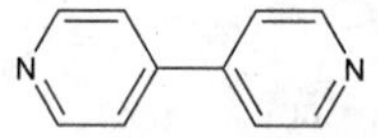

Dihydrate. Bitter needles from water, mp 73°. Anhydr form, mp 111-112°. bp_{760} 304.8°. Freely sol in alc, benzene, chloroform, ether; slightly sol in water. Absorption spectrum: Krumholz, *J. Am. Chem. Soc.* **73**, 3487 (1951).

Dihydrochloride. $C_{10}H_8N_2 \cdot 2HCl$. Prisms from water. Soluble in water; practically insol in ether.

3390.   **Dipyrone.** [5907-38-0] [(2,3-Dihydro-1,5-dimethyl-3-oxo-2-phenyl-1*H*-pyrazol-4-yl)methylamino]methanesulfonic acid sodium salt monohydrate; (antipyrinylmethylamino)methanesulfonic acid sodium salt; 1-phenyl-2,3-dimethyl-5-pyrazolone-4-methylaminomethanesulfonate sodium; noraminopyrine methanesulfonate sodium; 4-methylamino-1,5-dimethyl-2-phenyl-3-pyrazolone sodium methanesulfonate; sodium methylaminoantipyrine methanesulfonate; methylmelubrin; methampyrone; metamizol; analgin; sulpyrin; Alginodia; Algocalmin; Bonpyrin; Conmel; Divarine; Dolazon; D-Pron; Dya-Tron; Espyre; Farmolisina; Feverall; Fevonil; Keypyrone; Metilon; Minalgin; Narone; Nartate; Nevralgina; Nolotil; Novacid; Novaldin; Novalgin; Novemina; Novil; Paralgin; Pyralgin; Pyril; Pyrilgin; Pyrojec; Tega-Pyrone; Unagen. $C_{13}H_{16}N_3$-

NaO$_4$S.H$_2$O; mol wt 351.35. C 44.44%, H 5.16%, N 11.96%, Na 6.54%, O 22.77%, S 9.12%. Prepd by methylating the amino group of sulfamipyrine, then treating with formaldehyde sodium bisulfite soln: **DE 254711**; **DE 259503**; **DE 259577** (all 1911 to Hoechst).

Minute crystals from alc. Sol in water (1 g/1.5 ml), methanol. Less sol in ethanol. Practically insol in ether, acetone, benzene, chloroform. Aq solns are neutral; may acquire a yellow discoloration without apparent potency loss.

Magnesium salt. [6150-97-6] Magnopyrol. C$_{26}$H$_{32}$MgN$_6$O$_8$S$_2$; mol wt 645.00.

THERAP CAT: Analgesic; antipyretic.

THERAP CAT (VET): Analgesic; antispasmodic.

3391. Diquafosol. [59985-21-6] Uridine 5′-(pentahydrogen tetraphosphate) P''' → 5′-ester with uridine; P^1,P^4-diuridine 5′-tetraphosphate; UP$_4$U. C$_{18}$H$_{26}$N$_4$O$_{23}$P$_4$; mol wt 790.31. C 27.36%, H 3.32%, N 7.09%, O 46.56%, P 15.68%. Uridine nucleotide analog. P2Y$_2$ purinoceptor agonist; stimulates mucin secretion from goblet cells. Prepn: M. J. Stutts, III *et al.*, **WO 9640059** (1996 to Univ. North Carolina at Chapel Hill); and receptor activity: W. Pendergast *et al.*, *Bioorg. Med. Chem. Lett.* **11**, 157 (2001). Ocular pharmacology: T. Fujihara *et al.*, *J. Ocul. Pharmacol. Ther.* **18**, 363 (2002). Review of development and therapeutic potential: J. Fischbarg, *Curr. Opin. Investig. Drugs* **4**, 1377-1383 (2003); K. K. Nichols *et al.*, *Expert Opin. Invest. Drugs* **13**, 47-54 (2004). Clinical trial in dry eye disease: J. Tauber *et al.*, *Cornea* **23**, 784 (2004).

Tetrasodium salt. [211427-08-6] INS-365; Prolacria. C$_{18}$H$_{22}$N$_4$Na$_4$O$_{23}$P$_4$; mol wt 878.23.

THERAP CAT: In treatment of dry eye disease.

3392. Diquat Dibromide. [85-00-7] 6,7-Dihydrodipyrido-[1,2-*a*:2′,1′-*c*]pyrazinediium bromide (1:2); 1,1′-ethylene-2,2′-dipyridylium dibromide; FB/2; Reglone. C$_{12}$H$_{12}$Br$_2$N$_2$; mol wt 344.05. C 41.89%, H 3.52%, Br 46.45%, N 8.14%. Prepn: Fielden *et al.*, **US 2823987** (1958 to ICI). Herbicidal properties: Brian *et al.*, *Nature* **181**, 446 (1958). Environmental fate and toxicity: Simsiman *et al.*, *Residue Rev.* **62**, 131-174 (1976). Acute toxicity study: T. B. Gaines, R. E. Linder, *Fundam. Appl. Toxicol.* **7**, 299 (1986). *Review:* Akhavein, Linscott, *ibid.* **23**, 97-145 (1968).

Monohydrate. [6385-62-2] Pale yellow crystals from water, mp <320° (dec). Also reported as mp 335-340°. uv max: 308.31 nm (ε 18000). Soly in water at 20°: 70%. Insoluble in organic solvents. Slightly sol in alcohol. Stable in acid or neutral soln. LD$_{50}$ in male,

female rats (mg/kg): 147, 121 orally; in male rats (mg/kg): 433 dermally (Gaines, Linder).

Caution: Potential symptoms of overexposure are irritation of eyes, skin, mucous membranes, respiratory system; rhinorrhea, epistaxis; skin burns; nausea, vomiting, diarrhea, malaise; kidney and liver injury; cough, chest pain, dyspnea, pulmonary edema; tremor, convulsions; delayed wound healing. See *NIOSH Pocket Guide to Chemical Hazards* (DHHS/NIOSH 97-140, 1997) p 122.

USE: Contact herbicide used also to produce desiccation and defoliation.

3393. 1,3-Di-6-quinolylurea. [532-05-8] *N,N′*-Di-6-quinolinylurea; *sym*-di-(6-quinolyl)urea; bis(6-quinolyl)urea; 6,6′-diquinolylurea; Babesan. C$_{19}$H$_{14}$N$_4$O; mol wt 314.35. C 72.60%, H 4.49%, N 17.82%, O 5.09%. Synthesis starting with 6-aminoquinoline: Schönhöfer, Henecka, **DE 583207** (1933 to I. G. Farben); Haskelberg, *J. Org. Chem.* **12**, 434 (1947); Reuter, *Aust. Chem. Inst. J. Proc.* **16**, 164 (1949).

Crystals from pyridine, mp 262°. Sol in dil acid.

Bismethosulfate. 6,6′-Ureylenebis[1,1′-dimethylquinolinium] sulfate; quinuronium sulfate; 6,6′-ureylenebis(1-methylquinolinium)bis(methosulfate); dimethyl quinolyl methylsulfate urea; SN-5870; Acaprin; Zothelone; Baburan; Pirevan; Pyroplasmin; Atral. C$_{23}$H$_{26}$N$_4$O$_9$S$_2$; mol wt 566.60. Yellow crystals from methanol, dec 237°. Freely sol in water.

THERAP CAT (VET): Antiprotozoal *(Babesia)*.

3394. Dirithromycin. [62013-04-1] (1*R*,2*R*,3*R*,6*R*,7*S*,8*S*,-9*R*,10*R*,12*R*,13*S*,15*R*,17*S*)-7-[(2,6-Dideoxy-3-*C*-methyl-3-*O*-methyl-α-L-*ribo*-hexopyranosyl)oxy]-3-ethyl-2,10-dihydroxy-15-[(2-methoxyethoxy)methyl]-2,6,8,10,12,17-hexamethyl-9-[[3,4,6-trideoxy-3-(dimethylamino)-β-D-*xylo*-hexopyranosyl]oxy]-4,16-dioxa-14-azabicyclo[11.3.1]heptadecan-5-one; [9*S*(*R*)]-9-deoxo-11-deoxy-9,11-[imino[2-(2-methoxyethoxy)ethylidene]oxy]erythromycin; LY-237216; AS-E 136; Dynabac; Noriclan; Nortron; Valodin. C$_{42}$H$_{78}$N$_2$O$_{14}$; mol wt 835.09. C 60.41%, H 9.42%, N 3.35%, O 26.82%. Semi-synthetic derivative of erythromycin, *q.v.* Prepn: **BE 840431** (1976 to Thomae); R. Maier *et al.*, **US 4048306** (1977 to Boehringer, Ing.). Synthesis, ¹H- and ¹³C-NMR, and antimicrobial evaluation: F. T. Counter *et al.*, *Antimicrob. Agents Chemother.* **35**, 1116 (1991). X-ray structure determn: P. Luger, R. Maier, *J. Cryst. Mol. Struct.* **9**, 329 (1979). HPLC determn in plasma: G. W. Whitaker, T. D. Lindstrom, *J. Liq. Chromatogr.* **11**, 3011 (1988). Symposium on antibacterial activity, pharmacology, and clinical experience: *J. Antimicrob. Chemother.* **31**, Suppl. C, 1-185 (1993).

Crystals from ethanol/water, mp 186-189° (dec) (Counter). pKa 9.0 in 66% aq dimethyl fluoride. Very sol in methanol, methylene chloride; very slightly sol in water. LD$_{50}$ in mice (g/kg): >1 s.c.; >1 orally (Maier).

THERAP CAT: Antibacterial.

3395. Dirlotapide. [481658-94-0] 1-Methyl-*N*-[(1*S*)-2-[methyl(phenylmethyl)amino]-2-oxo-1-phenylethyl]-5-[[[4'-(trifluoromethyl)[1,1'-biphenyl]-2-yl]carbonyl]amino]-1*H*-indole-2-carboxamide; (*S*)-*N*-[2-[benzyl(methyl)amino]-2-oxo-1-phenylethyl]-1-methyl-5-[4'-(trifluoromethyl)[1,1'-biphenyl]-2-carboxamido]-1*H*-indole-2-carboxamide; 5-[(4'-trifluoromethyl-biphenyl-2-carbonyl)-amino]-1*H*-indole-2-carboxylic acid benzylmethyl carbamoylamide; CP-742033; Slentrol. $C_{40}H_{33}F_3N_4O_3$; mol wt 674.72. C 71.21%, H 4.93%, F 8.45%, N 8.30%, O 7.11%. Microsomal triglyceride transfer protein inhibitor; blocks the assembly and release of lipoprotein particles into the bloodstream. Prepn: P. Bertinato *et al.,* **WO 03002533**; *eidem,* **US 6720351** (2003, 2004 both to Pfizer); and pharmacology: J. Li *et al., Bioorg. Med. Chem. Lett.* **17,** 1996 (2007).

USE: Canine antiobesity agent.

3396. Dirucotide. [152074-97-0] L-α-Aspartyl-L-α-glutamyl-L-asparaginyl-L-prolyl-L-valyl-L-valyl-L-histidyl-L-phenylalanyl-L-phenylalanyl-L-lysyl-L-asparaginyl-L-isoleucyl-L-valyl-L-threonyl-L-prolyl-L-arginyl-L-threonine; MBP-8298; SF-328. $C_{92}H_{141}N_{25}O_{26}$; mol wt 2013.29. C 54.89%, H 7.06%, N 17.39%, O 20.66%. DENPVVHFFKNIVTPRT. Synthetic peptide corresponding to residues 82 to 98 of human myelin basic protein (MBP); suppresses the autoimmune response directed against MBP in multiple sclerosis. Prepn: K. G. Warren, I. Catz, **WO 9308212** (1993); H. L. Weiner *et al.,* **WO 9321222**; *eidem,* **US 5858980** (1993, 1999 both to Autoimmune). Structure-activity study: K. W. Wucherpfennig *et al., J. Exp. Med.* **179,** 279 (1994). Tolerance induction in MS patients: K. G. Warren *et al., J. Neurol. Sci.* **152,** 31 (1997). Clinical trial in multiple sclerosis: *idem et al., Eur. J. Neurol.* **13,** 887 (2006). *Review:* C. Darlington, *Curr. Opin. Mol. Ther.* **9,** 398-402 (2007).

Asp—Glu—Asn—Pro—Val—Val—His—Phe—Phe—Lys

Thr—Arg—Pro—Thr—Val—Ile—Asn

THERAP CAT: Immunomodulator; in treatment of multiple sclerosis.

3397. Discodermolide. [127943-53-7] (3*R*,4*S*,5*R*,6*S*)-6-[(2*S*,3*Z*,5*S*,6*S*,7*S*,8*Z*,11*S*,12*R*,13*S*,14*S*,15*S*,16*Z*)-14-[(Aminocarbonyl)oxy]-2,6,12-trihydroxy-5,7,9,11,13,15-hexamethyl-3,8,16,18-nonadecatetraen-1-yl]tetrahydro-4-hydroxy-3,5-dimethyl-2*H*-pyran-2-one; YM-19020. $C_{33}H_{55}NO_8$; mol wt 593.80. C 66.75%, H 9.34%, N 2.36%, O 21.55%. Polyhydroxylated lactone isolated from the marine sponge, *Discodermia dissoluta.* Naturally occurring (+)-form stabilizes microtubles and inhibits cell proliferation with the same mechanism as paclitaxel, *q.v.* Isoln and structure determn: S. P. Gunasekera *et al., J. Org. Chem.* **55,** 4912 (1990); correction: *ibid.* **56,** 1346 (1991). Total synthesis: D. T. Hung *et al., J. Am. Chem. Soc.* **118,** 11054 (1996). Review of total syntheses: I. Paterson, G. J. Florence, *Eur. J. Org. Chem.* **2003,** 2193-2208. Series of articles on industrial-scale synthesis: S. J. Mickel *et al., Org. Process Res. Dev.* **8,** 92-130 (2004). NMR soln structure determn: A. B. Smith, III *et al., Org. Lett.* **3,** 695 (2001). Mechanism of action: D. T. Hung *et al., Chem. Biol.* **3,** 287 (1996); E. ter Haar *et al., Biochemistry* **35,** 243 (1996). Synergistic activity with paclitaxel in cancer cells: S. Honore *et al., Cancer Res.* **64,** 4957 (2004).

Monoclinic white crystals, mp 115-116°. uv max (methanol): 210, 226, 235 nm (ε 35400, 19500, 12500). $[\alpha]_D^{20}$ +14.0° (c = 0.6 in methanol); monohydrate: $[\alpha]_D$ +20.1° (c = 1 in methanol).
USE: Reagent for mitotic studies.

3398. Disilane. [1590-87-0] Disilicoethane; disilicon hexahydride; silicoethane; disilicane. H_6Si_2; mol wt 62.22. H 9.72%, Si 90.28%. Si_2H_6. Obtained by separation of mixed silanes prepd from magnesium silicide and hydrochloric acid: Moissan, Smiles, *Compt. Rend.* **134,** 569, 1549 (1902); Stock, Somiesky, *Ber.* **49,** 111 (1916); **54B,** 524 (1921); **56B,** 247 (1923); Culbertson, **US 2551571** (1951 to Union Carbide); prepd by conversion of silane to higher silanes in an ozonizer type electric discharge: Spanier, MacDiarmid, *Inorg. Chem.* **1,** 432 (1962).

Gas; repulsive odor. d_4^{-25} 0.686. mp $-132.5°$. bp $-14.5°$. Dec 300°. Ignites spontaneously in air. Slowly dec in water. Explodes on contact with sulfur hexafluoride; reacts vigorously with carbon tetrachloride and chloroform. Sol in carbon disulfide, ethyl alcohol, benzene, and ethyl silicate. Potassium hydroxide liberates hydrogen.
Caution: A powerful irritant.

3399. Disodium Phenyl Phosphate. [3279-54-7] Phosphoric acid monophenyl ester sodium salt (1:2). $C_6H_5Na_2O_4P$; mol wt 218.06. C 33.05%, H 2.31%, Na 21.09%, O 29.35%, P 14.20%.

White, cryst powder. Freely sol in water, sparingly in alcohol; insol in acetone or ether.
USE: Reagent in the testing of milk for proper pasteurization and for the presence of unpasteurized milk.

3400. Disofenin. [65717-97-7] *N*-[2-[[2,6-Bis(1-methylethyl)phenyl]amino]-2-oxoethyl]-*N*-(carboxymethyl)glycine; [[[(2,6-diisopropylphenyl)carbamoyl]methyl]imino]diacetic acid; (2,6-diisopropylacetanilido)iminodiacetic acid; *N*-(carboxymethyl)-*N*-[2-(2,6-diisopropylphenyl)amino]-2-oxoethylglycine; DISIDA. $C_{18}H_{26}N_2O_5$; mol wt 350.42. C 61.70%, H 7.48%, N 7.99%, O 22.83%. Iminodiacetic acid analog. Prepn of acid and ^{99m}Tc-complex: M. de Schrijver, **DE 2723605**; *idem,* **US 4316883** (1977, 1982 both to Solco Basel). High yield synthesis of acid: M. P. Best *et al., Aust. J. Chem.* **35,** 2371 (1982). Biodistribution: G. Subramanian *et al., Nuklearmedizin Suppl.* **16,** 136 (1978). Chromatographic determn of radiochemical purity: A. M. Zimmer *et al., Eur. J. Nucl. Med.* **7,** 88 (1982). Chemical and biological properties: B. Zmbova *et al., Appl. Radiat. Isot.* **38,** 35 (1987). Pharmacokinetics: I. A. Fraser *et al., Eur. J. Nucl. Med.* **14,** 431 (1988); J. E. Love *et al., ibid.* 436. Clinical evaluation as hepatobiliary imaging agent: R. C. Stadalnik *et al., Radiology* **140,** 797 (1981).

White crystals, mp 196-198° (Zmbova). Also reported as crystals from water, mp 191-192.5° (Best); and as mp 166-175° (de Schrijver). Insol in water.

Complex with ^{99m}Tc. Technetium Tc 99m disofenin; ^{99m}Tc diisopropyl IDA; ^{99m}Tc-DISIDA; Hepatolite. Distribution coefficient (ethylene dichloride/water): 0.0086.

THERAP CAT: ^{99m}Tc complex as diagnostic aid (radioactive imaging agent).

3401. Disophenol. [305-85-1] 2,6-Diiodo-4-nitrophenol; DNP; Ancylol. $C_6H_3I_2NO_3$; mol wt 390.90. C 18.44%, H 0.77%, I 64.93%, N 3.58%, O 12.28%. Prepd by iodination of p-nitrophenol with iodine/potassium iodide soln in aq ammonia: Datta, Prosad, *J. Am. Chem. Soc.* **39**, 446 (1917). As an anthelmintic: Thorson *et al.*, US 3081224 (1963 to Am. Cyanamid). Toxicity studies: J. A. Kaiser, *Toxicol. Appl. Pharmacol.* **6**, 232 (1964); Fowler, *Br. Vet. J.* **127**, 304 (1971). Comparative field trial in sheep: C. A. Hall *et al.*, *Res. Vet. Sci.* **31**, 104 (1981).

Light yellow, feathery crystals from glacial acetic acid, mp 157°. Freely sol in alcohol, very sparingly sol in water. LD_{50} in rats, mice (mg/kg): 170, 212 orally; 105, 88 i.v.; 105, 107 i.p.; 122, 110 s.c. (Kaiser).

THERAP CAT (VET): Anthelmintic (hookworm).

3402. Disopyramide. [3737-09-5] α-[2-[Bis(1-methylethyl)-amino]ethyl]-α-phenyl-2-pyridineacetamide; α-[2-(diisopropylamino)ethyl]-α-phenyl-2-pyridineacetamide; 4-(diisopropylamino)-2-phenyl-2-(2-pyridyl)butyramide; H-3292; SC-7031; Isorythm; Lispine; Ritmodan; Rythmodan. $C_{21}H_{29}N_3O$; mol wt 339.48. C 74.30%, H 8.61%, N 12.38%, O 4.71%. Prepn: J. W. Cusic, H. W. Sause, BE 617730; *idem*, US 3225054 (1962, 1965 both to Searle). Synthesis and biological activity: G. W. Adelstein, *J. Med. Chem.* **16**, 309 (1973). Acute toxicity: P. C. Ruenitz, C. M. Mokler, *ibid.* **22**, 1142 (1979). Conformation-activity study: J. L. Czeisler, R. M. El-Rashidy, *J. Pharm. Sci.* **74**, 750 (1985). Review of analytical methods and pharmacokinetics: E. H. Taylor, A. A. Pappas, *Ann. Clin. Lab. Sci.* **16**, 289-295 (1986); of clinical pharmacology and therapeutic use: P. W. Willis, III, *Angiology* **38**, 165-173 (1987); of effect on left ventricular function: R. Di Bianco *et al.*, *ibid.* 174-183. Clinical trial in obstructive hypertrophic cardiomyopathy: M. V. Sherrid *et al.*, *J. Am. Coll. Cardiol.* **45**, 1251 (2005).

Crystals from hexane, mp 94.5-95.0°. pKa 10.2; also reported as 10.45. Soly in water: 1 mg/ml. Partition coefficient (chloroform/water): 3.1 (pH 7.2). LD_{50} i.p. in mice: 517 μmol/kg (Ruenitz, Mokler).

Phosphate. [22059-60-5] SC-13957; Dirythmin SA; Norpace; Rythmodul. $C_{21}H_{29}N_3O.H_3PO_4$; mol wt 437.48. White odorless powder. mp ~205° (dec). Freely sol in water; slightly sol in alc. Practically insol in chloroform, ether.

THERAP CAT: Antiarrhythmic (class IA).

THERAP CAT (VET): Antiarrhythmic (class IA).

3403. Disparlure. [54910-51-9] (2*S*,3*R*)-2-Decyl-3-(5-methylhexyl)oxirane; (2*S-cis*)-disparlure; (7*R*,8*S*)-disparlure. $C_{19}H_{38}O$; mol wt 282.51. C 80.78%, H 13.56%, O 5.66%. Potent sex pheromone of the gypsy moth *Lymantria dispar* (formerly *Porthetria dispar*); naturally occuring as the (+)-*cis*-enantiomer. Isoln and characterization: B. A. Bierl *et al.*, *Science* **170**, 87 (1970); *eidem*, *J. Econ. Entomol.* **65**, 659 (1972). Synthesis and configuration of natural pheromone: S. Iwaki *et al.*, *J. Am. Chem. Soc.* **96**, 7842 (1974). Stereospecific synthesis: D. G. Farnum *et al.*, *Tetrahedron Lett.* **1977**, 4009. Synthesis of (+)- and (−)- *cis*-forms: K. Mori *et al.*, *Tetrahedron* **35**, 833 (1979); B. E. Rossiter *et al.*, *J. Am. Chem. Soc.* **103**, 464 (1981). Synthesis of (±)-form: K. Eiter, *Angew. Chem. Int. Ed.* **11**, 60 (1972); *idem*, DE 2145454; *idem*, US 3975409 (1973, 1976 both to Bayer); H. J. Bestmann *et al.*, *Ber.* **109**, 3375 (1976). Activity of (+)- and (±)-forms: R. T. Cardé, R. P. Webster, *J. Chem. Ecol.* **5**, 935 (1979). Acute toxicity study: M. Beroza *et al.*, *Toxicol. Appl. Pharmacol.* **31**, 421 (1975).

Viscous colorless oil, $bp_{0.25}$ 146-148°. $[\alpha]_D^{23}$ +0.8 ±0.2° (c = 10 in CCl_4). n_D^{23} 1.4450.

(±)-*cis*-**Form.** [29804-22-6] *cis*-7,8-Epoxy-2-methyloctadecane; ENT-34886; Disrupt II GM. LD_{50} in rats (mg/kg): >34600 orally; in rabbits (mg/kg): >2025 dermally; LC_{50} (24 hr) in rainbow trout, bluegill sunfish (ppm): >100, >100 (Beroza).

USE: Insect attractant and mating disruptant.

3404. Distamycin A. [636-47-5] *N*-[5-[[[(3-Amino-3-iminopropyl)amino]carbonyl]-1-methyl-1*H*-pyrrol-3-yl]-4-[[[4-(formylamino)-1-methyl-1*H*-pyrrol-2-yl]carbonyl]amino]-1-methyl-1*H*-pyrrole-2-carboxamide; β-[1-methyl-4-[1-methyl-4-(1-methyl-4-formylaminopyrrole-2-carboxamido)pyrrole-2-carboxamido]pyrrole-2-carboxamido]propionamidine; stallimycin; F.I. 6426. $C_{22}H_{27}N_9O_4$; mol wt 481.52. C 54.88%, H 5.65%, N 26.18%, O 13.29%. Antiviral antibiotic obtained from *Streptomyces distallicus*. Binds to the minor groove of DNA, preferentially to adenine-thymine rich sequences; inhibits DNA-dependent RNA synthesis. Isoln: F. Arcamone *et al.*, DE 1039198 (1958 to Farmaceutici Italia), *C.A.* **55**, 2012f (1961). Structure and synthesis: *idem et al.*, *Nature* **203**, 1064 (1964); *eidem*, *Gazz. Chim. Ital.* **97**, 1097, 1110 (1967). Total synthesis: M. Bialer *et al.*, *Tetrahedron* **1978**, 2389; L. Grehn, U. Ragnarsson, *J. Org. Chem.* **46**, 3492 (1981). Toxicity and antitumor activity: A. DiMarco *et al.*, *Cancer Chemother. Rep.* **18**, 15 (1962). *Review:* F. E. Hahn in *Antibiotics* vol. 3, J. W. Corcoran, F. E. Hahn, Eds. (Springer-Verlag, New York, 1975) pp 79-100. Review of mechanism of action: H. Grunicke *et al.*, *Rev. Physiol. Biochem. Pharmacol.* **75**, 69-96 (1976). Review of use for targeted delivery of antitumor agents: P. G. Baraldi *et al.*, *Bioorg. Med. Chem.* **15**, 17-35 (2007).

Hydrochloride. [6576-51-8] Herperal. $C_{22}H_{27}N_9O_4.HCl$; mol wt 517.98. White to yellow crystals from dil HCl, mp 184-187° (Arcamone, 1967). Also reported as 189-193° from ethanol-ethyl acetate (Bialer). Hygroscopic. *Protect from light and oxygen.* uv max (96% ethanol): 237, 303 nm (ε 30000, 37000). Sol in water; forms clear yellow to green solution at 20 mg/ml in ethanol. LD_{50} in mice (mg/kg): 75 i.v.; 500 i.p. (DiMarco).

USE: Biochemical tool to inhibit RNA synthesis.

3405. Distigmine Bromide. [15876-67-2] 3,3'-[1,6-Hexane-diylbis[(methylimino)carbonyl]oxy]bis[1-methylpyridinium]bromide (1:2); 3-hydroxy-1-methylpyridinium bromide hexamethylene-bis[methylcarbamate]; hexamethylenebis[methylcarbamic acid] ester of 3-hydroxy-1-methylpyridinium bromide; hexamethylenebis-[N-methylcarbamic acid ester bromomethylate]; hexamethylenebis-[N-methylcarbaminoyl-1-methyl-3-hydroxypyridinium bromide]; hexamarium; BC-51; Ubretid. $C_{22}H_{32}Br_2N_4O_4$; mol wt 576.33. C 45.85%, H 5.60%, Br 27.73%, N 9.72%, O 11.10%. Prepn: Schmid, **US 2789981** (1957 to Oesterr. Stickstoffwerke). Pharmacology: Hertting *et al.*, *Arzneim.-Forsch.* **18**, 479 (1968); Hohenegger, Lindner, *Wien. Klin. Wochenschr.* **81**, 823 (1969). Stability studies: Suzuki, Tanimura, *Yakugaku Zasshi* **90**, 762 (1970).

Crystals, dec 149°.
THERAP CAT: Cholinesterase inhibitor.

3406. Disufenton. [168021-77-0] 4-[[(1,1-Dimethylethyl)-oxidoimino]methyl]-1,3-benzenedisulfonic acid; 2,4-disulfonyl-α-phenyl-*tert*-butylnitrone. $C_{11}H_{15}NO_7S_2$; mol wt 337.36. C 39.16%, H 4.48%, N 4.15%, O 33.20%, S 19.01%. Nitrone-based spin trapping agent; derivative of PBN, *q.v.* Prepn: J. M. Carney, **WO 9517876**; idem, **US 5488145** (1995, 1996 both to Oklahoma Med. Res. Found.). Pharmacology: S. Kuroda *et al.*, *J. Cereb. Blood Flow Metab.* **19**, 778 (1999). Comparative radical trapping study: K. R. Maples *et al.*, *Free Radical Res.* **34**, 417 (2001). Clinical pharmacokinetics: K. R. Lees *et al.*, *Stroke* **32**, 675 (2001). Clinical trial in acute ischemic stroke: idem et al., *N. Engl. J. Med.* **354**, 588 (2006). Reviews of pharmacology and therapeutic potential: P. A. Lapchak, *Curr. Opin. Investig. Drugs* **3**, 1758-1762 (2002); P. A. Lapchak, D. M. Araujo, *CNS Drug Rev.* **9**, 253-262 (2003).

White powder. Soly (g/ml): water >1.
Sodium salt. [168021-79-2] Disodium 4-[(*tert*-butylimino)-methyl]benzene-1,3-disulfonate N-oxide; ARL-16556; CPI-22; CXY-059; NXY-059; Cerovive. $C_{11}H_{13}NNa_2O_7S_2$; mol wt 381.32.
THERAP CAT: Neuroprotective.

3407. Disulfiram. [97-77-8] N,N,N',N'-Tetraethylthioper-oxydicarbonic diamide ([(H$_2$N)C(S)]$_2$S$_2$); tetraethylthioperoxydi-carbonic diamide; bis(diethylthiocarbamoyl) disulfide; tetraethyl-thiuram disulfide; bis(diethylthiocarbamyl) disulfide; teturamin; TTD; Antabuse; Esperal; Etabus. $C_{10}H_{20}N_2S_4$; mol wt 296.52. C 40.51%, H 6.80%, N 9.45%, S 43.25%. Prepn: Bailey, **US 1796977** (1931 to Roessler and Hasslacher); Adams, Newser, **US 1782111** (1931 to Naugatuck); *cf.* Cummings, Simmons, *Ind. Eng. Chem.* **20**, 1173 (1928). Toxicity study: Child, Cramp, *Acta Pharmacol. Toxicol.* **8**, 305 (1952). Comprehensive description: N. G. Nash, R. D. Daley, *Anal. Profiles Drug Subs.* **4**, 168-191 (1975).

White, odorless crystals, mp 70°. d 1.30. Sol in alcohol (3.82 g/100 ml), ether (7.14 g/100 ml), acetone, benzene, chloroform, carbon disulfide. Very slightly sol in water (0.02 g/100 ml). LD$_{50}$ orally in rats: 8.6 g/kg (Child, Cramp).
Caution: Potential symptoms of overexposure are irritation of eyes, skin, respiratory system; sensitization dermatitis; lassitude, fatigue, tremor, restlessness, headache, dizziness; metallic taste, peripheral neuropathy; liver damage. Ingestion of alcohol after disulfiram administration causes intense vasodilation and flushing of face and neck, restlessness, anxiety; tachycardia and tachypnea; headache, nausea, vomiting, hyperpnea, chest pains, sweating, pallor, hypotension. *See NIOSH Pocket Guide to Chemical Hazards* (DHHS/NIOSH 97-140, 1997) p 122; *Clinical Toxicology of Commercial Products*, R. E. Gosselin *et al.*, Eds. (Williams & Wilkins, Baltimore, 5th ed., 1984) Section III, pp 159-163, 383-386.
USE: Rubber accelerator; vulcanizer; seed disinfectant; fungicide.
THERAP CAT: Alcohol deterrent.

3408. Disulfoton. [298-04-4] Phosphorodithioic acid O,O-di-ethyl S-[2-(ethylthio)ethyl] ester; O,O-diethyl-S-ethylmercaptoethyl dithiophosphate; thiodemeton; dithiodemeton; Bay 19639; ENT-23347; Baysiston; Di-Syston. $C_8H_{19}O_2PS_3$; mol wt 274.39. C 35.02%, H 6.98%, O 11.66%, P 11.29%, S 35.05%. Organophosphate insecticide; cholinesterase inhibitor. Homolog of phorate, *q.v.* Prepn: Schrader, Lorenz, **US 2759010**; **GB 797307** (1956, 1958 both to Bayer). GC/MS determn in water samples: P.-S. Chen, S.-D. Huang, *Talanta* **69**, 669 (2006). Toxicity study: T. B. Gaines, *Toxicol. Appl. Pharmacol.* **14**, 515 (1969). Review of toxicology and human exposure: *Toxicological Profile for Disulfoton* (PB95-264347, 1995) 239 pp.

Colorless oil, bp$_{0.01}$ 108°, bp$_{1.5}$ 132-133°. d_4^{20} 1.144. n_D^{20} 1.5348. Vapor pressure at 20°: 1.8×10^{-4} mm Hg. Insol in water. *Poisonous.* LD$_{50}$ in female, male rats (mg/kg): 2.3, 6.8 orally; 6, 15 dermally (Gaines).
Caution: Potential symptoms of overexposure are irritation of eyes, skin; nausea, vomiting, abdominal cramps, diarrhea, salivation; headache, giddiness, vertigo, weakness; rhinorrhea, chest tightness; blurred vision, miosis; cardiac irregularity; muscle fasciculations; dyspnea; eye and skin burns. *See NIOSH Pocket Guide to Chemical Hazards* (DHHS/NIOSH 97-140, 1997) p 124.
USE: Acaricide; insecticide.

3409. Dita Bark. Alstonia cortex; Australian fever bark. The dried bark of *Alstonia scholaris* (L.) R. Br. and of *A. constricta* F. Muell., *Apocynaceae*. *Habit.* India, Ceylon, Philippine Islands, Australia. *Constit.* The alkaloids echitamine, echitenine, alsonine, porphyrine; echicerin—a non-nitrogenous substance; echitin, alstonidine and echitein.

3410. Ditazol. [18471-20-0] 2,2'-[(4,5-Diphenyl-2-oxazol-yl)imino]bisethanol; N-(4,5-diphenyloxazol-2-yl)diethanolamine; 2-[bis(β-hydroxyethyl)amino]-4,5-diphenyloxazole; 2,2'-dihydroxy-N-(4,5-diphenyloxazol-2-yl)diethylamine; diethamphenazol; S-222; Ageroplas. $C_{19}H_{20}N_2O_3$; mol wt 324.38. C 70.35%, H 6.21%, N 8.64%, O 14.80%. Prepn: Marchetti *et al.*, *J. Med. Chem.* **11**, 1092 (1968); **FR 1538009**; E. Marchetti, **US 3557135** (1968, 1971 to Ist. Farmacol. Serono). Series of articles on pharmacology and toxicology: L. Caprino *et al.*, *Arzneim.-Forsch.* **23**, 1272-1291 (1973). Metabolism: Marchetti *et al.*, *ibid.* 1291.

Monohydrate. Crystals from ethyl ether-petr ether, mp 96-98°. LD$_{50}$ in mice, rats (mg/kg): 9621, 11380 orally; 3390, 7770 i.p. (Caprino).
THERAP CAT: Anti-inflammatory.

3411. Dithianon. [3347-22-6] 5,10-Dihydro-5,10-dioxo-naphtho[2,3-*b*]-1,4-dithiin-2,3-dicarbonitrile; 1,4-dithiaanthraquinone-2,3-dicarbonitrile; 2,3-dicyano-1,4-dithiaanthraquinone; IT-931; MV-119A; Delan. C$_{14}$H$_4$N$_2$O$_2$S$_2$; mol wt 296.32. C 56.75%, H 1.36%, N 9.45%, O 10.80%, S 21.64%. Prepn: van Schoor *et al.*, US 2976296 (1961 to E. Merck). Properties and field trials: J. Berker *et al.*, *Proc. 2nd Br. Insectic. Fungic. Conf.* 1963, 351. LC determn in technical formulations: A. R. Hanks, *J. AOAC Int.* 78, 1131 (1995). Environmental fate: M. Ueoka *et al.*, *Chemosphere* 35, 2915 (1997). *Review:* E. Amadori, W. Heupt, *Anal. Methods Pestic. Plant Growth Regul.* 10, 181-187 (1978).

Brown odorless crystals, mp 225°. d^{18} 1.55. Practically insol in water. Sol in dioxane, chlorobenzene, chloroform. LD$_{50}$ orally in rats, guinea pigs: 638, 110 mg/kg (Amadori, Heupt).
USE: Fungicide.

3412. Dithiazanine Iodide. [514-73-8] 3-Ethyl-2-[5-(3-ethyl-2(3*H*)-benzothiazolylidene)-1,3-pentadien-1-yl]benzothiazolium iodide (1:1); 3,3'-diethylthiadicarbocyanine iodide; Abminthic; Anelmid; Anguifugan; Delvex; Dejo; Déselmine; Dilombrin; Dizan; Nectocyd; Partel; Telmicid; Telmid. C$_{23}$H$_{23}$IN$_2$S$_2$; mol wt 518.48. C 53.28%, H 4.47%, I 24.48%, N 5.40%, S 12.37%. Prepn from 1-methylbenzothiazole ethiodide and β-(ethylmercapto)acrolein diethyl acetal: Kendall, Edwards, US 2412815 (1946 to Ilford).

Green needles from methanol, dec 248°. Practically insol in water. Can be solubilized with polyvinylpyrrolidone: CA 676636 (1963 to GAF).
USE: Sensitizer for photographic emulsions.
THERAP CAT: Anthelmintic (Nematodes).
THERAP CAT (VET): Anthelmintic.

3413. 2,2'-Dithiobisbenzothiazole. [120-78-5] MBTS; 2,2'-dibenzothiazyl disulfide; benzothiazyl disulfide; dibenzthiazyl ether; Thiofide. C$_{14}$H$_8$N$_2$S$_4$; mol wt 332.47. C 50.58%, H 2.43%, N 8.43%, S 38.57%. Formed by oxidation of 2-mercaptobenzothiazole.

Pale yellow needles from benzene. d 1.50. mp 180°; also reported as mp 186°; the commercial product, mp 168° min. Insoluble in water. Solubility at 25° (g/100 ml): in alcohol <0.2; in acetone <0.5; in benzene <0.5; in carbon tetrachloride <0.2 (somewhat more in chloroform); in ether <0.2; in naphtha <0.5.
USE: As accelerator in the rubber industry.

3414. 2,4-Dithiobiuret. [541-53-7] Thioimidodicarbonic diamide ([(H$_2$N)C(S)]$_2$NH). C$_2$H$_5$N$_3$S$_2$; mol wt 135.20. C 17.77%, H 3.73%, N 31.08%, S 47.43%. Prepd by the interaction of dicyandiamide and hydrogen sulfide in inert solvents under pressure:

Sperry, US 2371112 (1945 to Am. Cyanamid). *Review:* Kurzer, *Chem. Rev.* 56, 138-144, 179-181 (1956).

Monoclinic or triclinic crystals. *Poisonous.* Apparent bulk density 1.522 g/ml at 30°. Dec 181°. uv max: 225, 280 nm (log ε 2.0, 2.2). Solubility at 27° = 0.27 g/100 ml water; 2.2 g/100 g ethanol; 16 g/100 g acetone; about 34 g/100 g Cellosolve. Solubility in boiling water about 8.0%. pH of satd aq soln at 30° = 5.8. Soluble in alkalies with formation of water-soluble salts. Solubility in 1% NaOH = 3.6 g/100 g soln; in 5% NaOH = 16 g/100 g; in 10% NaOH = 29 g/100 g.
Caution: Limited animal expts suggest high toxicity due to respiratory paralysis.
USE: Plasticizer, rubber accelerator, intermediate in resin manuf, in making insecticides and rodenticides. Can be used to delay the wilting of flowers.

3415. 4,4'-Dithiodimorpholine. [103-34-4] 4,4'-Dithiobis-morpholine; morpholine *N,N'*-disulfide; dimorpholine *N,N'*-disulfide. C$_8$H$_{16}$N$_2$O$_2$S$_2$; mol wt 236.35. C 40.65%, H 6.82%, N 11.85%, O 13.54%, S 27.13%. Prepn: Blake, *J. Am. Chem. Soc.* 65, 1267 (1943); idem, US 2343524 (1944 to Monsanto); Harman, US 2766236 and GB 774570 (1956, 1957, both to Monsanto). Lambrech *et al.*, US 2902402 (1959 to Pfizer).

Crystals, mp 124-125°.
USE: As staining protector in rubber; in vulcanization of rubber. As fungicide: Ladd, US 2429097 (1949 to U.S. Rubber).

3416. 3,3-Dithiodipyridine Dihydrochloride. [538-45-4] 3,3'-Dithiobis[pyridine] dihydrochloride; 3,3-dipyridyl disulfide dihydrochloride. C$_{10}$H$_{10}$Cl$_2$N$_2$S$_2$; mol wt 293.22. C 40.96%, H 3.44%, Cl 24.18%, N 9.55%, S 21.87%. Prepd from 3-pyridinesulfonyl chloride and Na$_2$SO$_3$: Gibbs, Penny, GB 582638 (1946).

Crystals from abs alc, mp 183°. Sol in water.

3417. Dithiopyr. [97886-45-8] 2-(Difluoromethyl)-4-(2-methylpropyl)-6-(trifluoromethyl)-3,5-pyridinedicarbothioic acid S^3,S^5-dimethyl ester; *S,S*-dimethyl 2-(difluoromethyl)-4-isobutyl-6-trifluoromethyl-3,5-pyridinedicarbothioate; MON-15100; MON-7200; Dimension. C$_{15}$H$_{16}$F$_5$NO$_2$S$_2$; mol wt 401.41. C 44.88%, H 4.02%, F 23.66%, N 3.49%, O 7.97%, S 15.97%. Selective herbicide for weed control in turfgrass and selected crops. Prepn: L. F. Lee, EP 133612; idem, US 4692184 (1985, 1987 both to Monsanto). Herbicidal properties: M. Fujiyama, S. Yamane, *Shokubutsu no Kagaku Chosetsu* 24, 49 (1989), *C.A.* 111, 227154n (1989).

Colorless, crystalline solid with faint odor, mp 65°. Vapor pressure at 25°: 4 × 10^{-6} mm Hg. Soly in water (20°): 1.38 ppm. LD$_{50}$ orally in rats: >5000 mg/kg (Fujiyama, Yamane).
USE: Herbicide.

Dithiosalicylic Acid

3418. Dithiosalicylic Acid. [527-89-9] 2-Hydroxybenzene-carbodithioic acid. $C_7H_6OS_2$; mol wt 170.24. C 49.39%, H 3.55%, O 9.40%, S 37.66%. Prepn from salicylaldehyde and ammonium polysulfide: Bruni, Levi, *Atti Accad. Naz. Lincei* **32**, (i), 5 (1923), *C.A.* **18**, 2694 (1924); *Beilstein* 2nd suppl. **vol. 10**, 78.

Orange-yellow needles, mp 48-50°. Moderately sol in water; sol in alcohol, ether, benzene, methanol.

3419. 1,4-Dithiothreitol. [3483-12-3] *rel*-(2R,3R)-1,4-Dimercapto-2,3-butanediol; Cleland's reagent; *threo*-2,3-dihydroxy-1,4-dithiolbutane. $C_4H_{10}O_2S_2$; mol wt 154.24. C 31.15%, H 6.54%, O 20.75%, S 41.57%. Prepn: R. M. Evans *et al.*, *J. Chem. Soc.* **1949**, 253; and use as reagent for protecting SH groups: W. W. Cleland, *Biochemistry* **3**, 480 (1964).

Relative stereochemistry

Slightly hygroscopic needles from ether, mp 42-43°. Can be sublimed at 37° and 0.005 mm press; bp_2 125-130°. Redox potential: −0.33 volts at pH 7. Freely sol in water, ethanol, acetone, ethyl acetate, chloroform, ether.

USE: In organic synthesis as reagent for protection of SH groups.

3420. Dithizone. [60-10-6]; [760132-01-2] (*E*-form). 2-Phenyldiazenecarbothioic acid 2-phenylhydrazide; (phenylazo)thioformic acid 2-phenylhydrazide; diphenylthiocarbazone. $C_{13}H_{12}N_4S$; mol wt 256.33. C 60.91%, H 4.72%, N 21.86%, S 12.51%. Prepn: E. Fischer, E. Besthorn, *Ann.* **212**, 316 (1882). Use as an analytical reagent: A. Stock, E. Pohland, *Angew. Chem.* **39**, 791 (1926); H. Fischer, *ibid.* **46**, 442 (1933); **50**, 919 (1937); *Organic Reagents for Metals* (Hopkins & Williams, 1934). Crystal structure: M. Laing, *J. Chem. Soc. Perkin Trans. 2* **1977**, 1248. Conformation in soln: A. T. Hutton, H. M. Irving, *Chem. Commun.* **1981**, 735. Monograph: *Analytical Sciences Monograph No. 5* entitled "Dithizone", H. M. Irving, Ed. (Chemical Society, London, 1977) 112 pp. *Review: idem, Crit. Rev. Anal. Chem.* **8**, 321-405 (1980).

(*E*)-form

Bluish-black cryst powder, mp 168° (dec). Insol in water. Sparingly sol in alc, freely in carbon tetrachloride, chloroform. Solns are not stable, but may be preserved by a layer of aq SO_2.

Caution: Has caused glycosuria and ocular injury in exptl animals.

USE: Sensitive indicator reagent for several heavy metals, Cd, Co, Cu, Pb, and Hg; esp for the estimation of minute amounts of Pb.

3421. Ditiocarb Sodium. [148-18-5] *N,N*-Diethylcarbamodithioic acid sodium salt (1:1); diethyldithiocarbamic acid sodium salt; sodium diethyldithiocarbamate; dithiocarb; DTC; DDC; DEDC; DDTC; DeDTC; Imuthiol. $C_5H_{10}NNaS_2$; mol wt 171.25. C 35.07%, H 5.89%, N 8.18%, Na 13.42%, S 37.44%. Chelating agent with strong affinity for Hg, Cu, Ni and Zn. Also active as T-cell specific immunostimulant. Prepn: A. M. Clifford, J. G. Lichty, *J. Am. Chem. Soc.* **54**, 1163 (1932); A. L. Klebanskii, L. P. Fomina, *Zh. Obshch. Khim.* **30**, 794 (1960). Absorption spectrum: H. P. Koch, *J. Chem. Soc.* **1949**, 401. Inhibition of superoxide dismutase

in mice: R. E. Heikkila *et al.*, *J. Biol. Chem.* **251**, 2182 (1976); of cisplatin nephrotoxicity in rats: R. F. Borch *et al.*, *Proc. Natl. Acad. Sci. USA* **77**, 5441 (1980). Clinical use in acute nickel carbonyl poisoning: F. W. Sunderman, F. W. Sunderman Jr., *Am. J. Med. Sci.* **236**, 26 (1958); F. W. Sunderman, *Ann. Clin. Lab. Sci.* **9**, 1 (1979); in acute cadmium poisoning: G. R. Gale *et al.*, *ibid.* **11**, 476 (1981) and **13**, 207 (1983). Specific effects on T-cell regulation: A. Pompidou *et al.*, *Int. J. Immunopharmacol.* **7**, 561 (1985). Clinical evaluation in T-cell deficient diseases: E. Lemarie *et al.*, *Methods Find. Exp. Clin. Pharmacol.* **8**, 51 (1986); in AIDS-related complex: A. Pompidou *et al.*, *Comp. Immunol. Microbiol. Infect. Dis.* **9**, 263 (1986). Review of pharmacology, toxicity and clinical uses: G. Renoux, *J. Pharmacol.* **13**, Suppl. 1, 95-134 (1982); of immunopharmacology and use in cancer immunotherapy: G. Renoux *et al.*, *Adv. Exp. Med. Biol.* **166**, 223-239 (1983).

Trihydrate. [20624-25-3] $C_5H_{10}NNaS_2.3H_2O$; mol wt 225.30. Thin, irregular plate-like crystals from acetone, mp 94-102°. Also reported as 90-92° (Sunderman, Sunderman). Freely sol in water; sol in ethanol, methanol, acetone. Insol in ether, benzene. The aq soln is alkaline to litmus and phenolphthalein and slowly dec. (pH of 10% aq soln is 11.6 at room temp). The addition of an acid to the aq soln produces a white turbidity due to the liberation of carbon disulfide. uv max (ethanol): 257, 290 nm (ε 1200, 13000). LD_{50} orally in rats, mice: 2830, 1870 mg/kg; i.v. in mice: >1000 mg/kg (Renoux, 1982).

USE: For colorimetric determination of small quantities of copper and for its separation from other metals.

THERAP CAT: Immunomodulator. Chelating agent (copper); Wilson's Disease treatment. Antidote (nickel, cadmium poisoning).

3422. 1,1-Di-*p*-tolylethane. [530-45-0] 1,1'-Ethylidenebis[4-methylbenzene]; α,α-di-*p*-tolylethane; *asym*-di-*p*-tolylethane; 4,4'-α-trimethyldiphenylmethane. $C_{16}H_{18}$; mol wt 210.32. C 91.37%, H 8.63%. Prepd by passing acetylene into a mixture of toluene and concd H_2SO_4 in presence of mercuric sulfate: Reichert, Nieuwland, *Org. Synth.* **IV**, 23 (1925).

Oily, highly refractive liquid. Aromatic odor. d_4^{20} 0.974. Not solid at −20°. bp 295-300°; bp_{12} 155-157°. Soluble in acetic acid.

3423. 1,2-Di-*p*-tolylethane. [538-39-6] 1,1'-(1,2-Ethanediyl)bis[4-methylbenzene]; α,β-di-*p*-tolylethane; *sym*-di-*p*-tolylethane; 4,4'-dimethyldibenzyl. $C_{16}H_{18}$; mol wt 210.32. C 91.37%, H 8.63%. Prepd by passing *p*-xylene vapor over red hot platinum wires: Meyer, Hofmann, *Monatsh. Chem.* **37**, 690 (1916).

Leaflets from methanol; plates from ligroin. mp 82°; bp_{730} 296-298°; bp_{18} 178°. Sol in benzene, moderately sol in alcohol, petr ether.

3424. *p*-Ditolylmercury. [537-64-4] Bis(4-methylphenyl)-mercury. $C_{14}H_{14}Hg$; mol wt 382.86. C 43.92%, H 3.69%, Hg 52.39%. Prepd by the action of sodium iodide on *p*-tolylmercuric chloride in ethanol: Whitmore *et al.*, *Org. Synth.* **3**, 65 (1923); by

Consult the Name Index before using this section.

boiling *p*-bromotoluene with sodium amalgam in xylene in the presence of ethyl acetate as a catalyst: Dreher, Otto, *Ann.* **154**, 171 (1870); LaCoste, Michaelis, *ibid.* **201**, 246 (1880); *Compt. Rend.* **68**, 1298 (1869).

Needles from xylene. mp 238°. Insol in water. Slightly sol in cold alcohol; sol in hot benzene, xylene, chloroform, carbon disulfide.

3425. Diuron. [330-54-1] *N'*-(3,4-Dichlorophenyl)-*N,N*-dimethylurea; 1,1-dimethyl-3-(3,4-dichlorophenyl)urea; Diurex; Karmex; Urox D. $C_9H_{10}Cl_2N_2O$; mol wt 233.09. C 46.38%, H 4.32%, Cl 30.42%, N 12.02%, O 6.86%. Prepn: Jones, US 2768971 (1956 to I.C.I.). Toxicity data: E. M. Boyd, V. Krupa, *J. Agric. Food Chem.* **18**, 1104 (1970). Review of environmental impact: S. Giacomazzi, N. Cochet, *Chemosphere* **56**, 1021-1032 (2004).

Crystals, mp 158-159°. Soly in water at 25°: 42 ppm. Very low soly in hydrocarbon solvents. Vapor pressure at 50°: 3.1×10^{-6} mm Hg. Partition coefficient (octanol/water): 2.6. LD_{50} orally in rats: 1690 mg/kg (Boyd, Krupa).
Caution: Potential symptoms of overexposure are irritation of eyes, skin, nose, throat. Potential symptoms of overexposure in exptl animals are anemia, methemoglobinemia. *See NIOSH Pocket Guide to Chemical Hazards* (DHHS/NIOSH 97-140, 1997) p 124.
USE: Pre-emergent herbicide.

3426. Divicine. [32267-39-3] 2,6-Diamino-3,6-dihydro-4,5-pyrimidinedione; 2,6-diamino-4,5-pyrimidinediol; 2,6-diamine-5-hydroxy-4(3*H*)-pyrimidinone; 2,4-diamino-5,6-dihydroxypyrimidine. $C_4H_6N_4O_2$; mol wt 142.12. C 33.81%, H 4.26%, N 39.42%, O 22.51%. Hemotoxic aglycone of vicine, *q.v.*, found in fava beans and other *Vicia* spp. Causative agent of favism, a hemolytic anemia induced by ingestion of fava beans in individuals with a genetic deficiency of glucose-6-phosphate dehydrogenase (G6PD). Prepn from vicine by heating with dil H_2SO_4: H. Ritthausen, *Ber.* **29**, 2108 (1896); P. A Levene, J. K. Senior, *J. Biol. Chem.* **25**, 607 (1916). Structure: A. Bendich, G. C. Clements, *Biochim. Biophys. Acta* **12**, 462 (1953). Synthesis: J. Davoll, D. H. Laney, *J. Chem. Soc.* **1956**, 2124; J. H. Chesterfield *et al., ibid.* **1964**, 1001. Extraction and quantification in vetch seeds and fava beans: M. Chevion, T. Navok, *Anal. Biochem.* **128**, 152 (1983). Role in glutathione depletion in G6PD-deficient erythrocytes: J. Mager *et al., Biochem. Biophys. Res. Commun.* **20**, 235 (1965). Hemolytic activity: D. C. McMillan *et al., Chem. Res. Toxicol.* **6**, 439 (1993).

Brownish needles, dec above 280°. One gram dissolves in 100 ml boiling water, in about 350 ml cold water. Soluble in 10% KOH.
Diacetate. $C_8H_{10}N_4O_4$. Crystals from water, dec 309-312°.

3427. Divinyl Ether. [109-93-3] 1,1'-Oxybisethene; divinyl oxide; ethenyloxyethene; vinyl ether; Vinethene; Vinesthene. C_4H_6O; mol wt 70.09. C 68.55%, H 8.63%, O 22.83%. Prepn: Major, Ruigh, US 2021872 (1935 to Merck & Co.); Mittag, Smidt, US 2832807 (1958 to Consortium für electrochem. Ind. GmbH). Toxicity data: Molitar, *J. Pharmacol. Exp. Ther.* **57**, 274 (1936).

Very volatile, flammable liq; characteristic odor. Dec on exposure to light or to acid fumes, forming acetaldehyde, and polymerizes to a solid glass-like mass. d_{20}^{20} 0.774; d_4^{20} 0.773. bp 28.4°. n_D^{20} 1.3989. Crit temp 183° (calc). Critical press. ~30 atm. Specific heat ~0.53 cal/g. Explosives limits (% by vol in air), lower: 1.7, upper: 27.0. Autoignition temp 360° (680°F). 0.53 g dissolves in 100 g water at 37°. Oil-water soly coefficient at 37°: 41. Miscible with alcohol, ether, oils and other organic solvents. Pharmaceutical product contains not less than 96.0% and not more than 97.0% vinyl ether, the remainder consisting of dehydrated alcohol. *Keep protected from light and acid fumes, in a cool place, and away from an open flame.* LC_{50} (3 hr in air) in mice, rats (mg/l): 146.8, 386.5 (Molitor).
THERAP CAT: Anesthetic (inhalation).

3428. Dixanthogen. [502-55-6] Thioperoxydicarbonic acid ([(HO)C(S)]$_2$S$_2$) *OC,OC'*-diethyl ester; dithiobis[thioformic acid] *O,O*-diethyl ester; *O,O*-diethyl dithiobis[thioformate]; bisethylxanthogen; diethyl xanthogenate; ethylxanthic disulfide; preparation K; EXD; Auligen; Aulinogen; Bexide; Herbisan; Lenisarin. $C_6H_{10}O_2S_4$; mol wt 242.38. C 29.73%, H 4.16%, O 13.20%, S 52.91%. Prepn: Losse, Wottgen, *J. Prakt. Chem.* **13**, 260 (1961). Metabolic studies: Gutenmann, Lisk, *J. Agric. Food Chem.* **19**, 200 (1971). Toxicity: G. W. Bailey, J. L. White, *Residue Rev.* **10**, 97 (1965).

Yellow needles, mp 28-32°. Onion-like odor. Soly in alcohol: 2 g/100 ml. Freely sol in benzene, ether, petr ether, oils. Almost insol in water. LD_{50} orally in rats: 480 mg/kg (Bailey, White).
USE: Insecticide formulations; herbicide.
THERAP CAT: Ectoparasiticide.

3429. Dixyrazine. [2470-73-7] 2-[2-[4-[2-Methyl-3-(10*H*-phenothiazin-10-yl)propyl]-1-piperazinyl]ethoxy]ethanol; 10-[2-methyl-3-(4-hydroxyethoxyethyl-1-piperazinyl)propyl]phenothiazine; 1-[2-(2-hydroxyethoxy)ethyl]-4-[2-methyl-3-(10-phenothiazinyl)propyl]piperazine; UCB-3412; Esucos; Roscal. $C_{24}H_{33}N_3O_2S$; mol wt 427.61. C 67.41%, H 7.78%, N 9.83%, O 7.48%, S 7.50%. Prepn: H. Morren, GB 861420 (1961). Partition properties: B.-A. Persson, *Acta Pharm. Suec.* **5**, 335 (1968). GC-MS determn in blood: G. Brante *et al., Eur. J. Clin. Pharmacol.* **20**, 307 (1981). Clinical pharmacokinetics: H. Liedholm *et al., Drug-Nutr. Interact.* **3**, 87 (1985). Clinical trial in postoperative nausea and vomiting: E. Kokinsky *et al., Acta Anaesthesiol. Scand.* **43**, 191 (1999); A. Johansson *et al., ibid.* **44**, 1093 (2000); as topical anesthetic for cataract surgery: E. Monestam *et al., J. Cataract Refract. Surg.* **27**, 445 (2001). Review of clinical chemistry and early clinical studies: T. Fokstuen, *Int. J. Neuropsychiat.* **340**, 294-299 (1965).

pKa_1: 7.8; pKa_2: 3.65.
Dihydrochloride. [60539-20-0] $C_{24}H_{33}N_3O_2S$.2HCl. Crystals from isopropanol, mp 192°.
THERAP CAT: Antipsychotic; antiemetic; anesthetic (local).

3430. Dizocilpine. [77086-21-6] (5*S*,10*R*)-10,11-Dihydro-5-methyl-5*H*-dibenzo[*a,d*]cyclohepten-5,10-imine; (+)-5-methyl-10,11-dihydro-5*H*-dibenzo[*a,d*]cyclohepten-5,10-imine. $C_{16}H_{15}N$; mol wt 221.30. C 86.84%, H 6.83%, N 6.33%. Non-competitive NMDA, *q.v.*, receptor antagonist. Prepn of racemate: M. E. Christy et al., J. Org. Chem. **44**, 3117 (1979); T. R. Lamanec et al., ibid. **53**, 1768 (1988); and of enantiomers: **BE 882361**; P. Anderson et al., **US 4399141** (1980, 1983 both to Merck & Co.). Series of papers on pharmacological activities: B. V. Clineschmidt et al., Drug Dev. Res. **2**, 123-163 (1982). NMDA receptor binding studies: E. H. F. Wong et al., Proc. Natl. Acad. Sci. USA **83**, 7104 (1986); A. C. Foster, E. H. F. Wong, Br. J. Pharmacol. **91**, 403 (1987); J. E. Huettner, B. P. Bean, Proc. Natl. Acad. Sci. USA **85**, 1307 (1988). Metabolism in animals: H. B. Hucker et al., Drug Metab. Dispos. **11**, 54 (1983). Pharmacokinetics: A. S. Troupin et al., Curr. Probl. Epilepsy **4**, 191 (1986). Protective effects against excitotoxic amino acid-induced neuronal degeneration: G. N. Woodruff et al., Neuropharmacology **26**, 903 (1987); A. C. Foster et al., Neurosci. Lett. **76**, 307 (1987); A. C. Foster et al., J. Neurosci. **8**, 4745 (1988); specifically in ischemia: R. Gill et al., Neuroscience **25**, 847 (1988); E. Ozyurt et al., J. Cereb. Blood Flow Metab. **8**, 138 (1988). Anticonvulsant activity in models of epilepsy: K. Sato et al., Brain Res. **463**, 12 (1988); M. E. Gilbert, ibid. **90**. Clinical evaluation in attention deficit disorder: F. W. Reimherr et al., Psychopharmacol. Bull. **22**, 237 (1986).

White solid from cyclohexane, mp 68.5-69°. $[\alpha]_{589}^{20}$ +161.4° (c = 0.038 g/2 ml ethanol).

Maleate. [77086-22-7] MK-801. $C_{16}H_{15}N.C_4H_4O_4$; mol wt 337.38. Crystalline solid, mp 208.5-210°. $[\alpha]_D^{20}$ +114° (c = 0.0128 g/2 ml ethanol).

(−)-Form. [77086-19-2] White solid from cyclohexane, mp 68.5-69.5°. $[\alpha]_{589}$ −160.8° (c = 0.032 g/2 ml ethanol).

USE: Biochemical probe for NMDA receptors.

3431. Djenkolic Acid. [498-59-9] S,S'-Methylenebis-L-cysteine; 3,3'-(methylenedithio)dialanine; 3,3'-methylenedithiobis(2-aminopropanoic acid); L-cysteine thioacetal of formaldehyde. $C_7H_{14}N_2O_4S_2$; mol wt 254.32. C 33.06%, H 5.55%, N 11.02%, O 25.16%, S 25.21%. An amino acid isolated from the djenkol bean (Pithecolobium lobatum Benth., Leguminosae): van Veen, Hyman, Rec. Trav. Chim. **54**, 493 (1935). Synthesis from 1 mol formaldehyde and 2 mols L-cysteine in strong HCl solution: Armstrong, du Vigneaud, J. Biol. Chem. **168**, 373 (1947); cf. du Vigneaud, Patterson, ibid. **114**, 633 (1936); Middlebrook, Phillips, Biochem. J. **41**, 218 (1947). Synthesis of homologs: Frankel, Gertner, J. Chem. Soc. **1960**, 898.

Rosettes of needles of various lengths. Gradually dec between 300 and 350°. $[\alpha]_D^{20.5}$ −65.0° (1.0N HCl); $[\alpha]_D^{25}$ −47.5° (c = 2 in 1.0N HCl). Very sparingly sol in cold water. Soly in boiling water about 1 in 200. Readily sol in aq solns of alkalies or acids.

Monohydrochloride. [4508-48-9] $C_7H_{14}N_2O_4S_2.HCl$; mol wt 290.78. Slender prisms, dec 250-300°. Sol in water.

Dibenzoyldjenkolic acid. $C_{21}H_{22}N_2O_6S_2$; mol wt 462.54. Crystals, mp 87.5-89°.

3432. DMAN. [20734-58-1] N^1,N^1,N^8,N^8-Tetramethyl-1,8-naphthalenediamine; 1,8-bis(dimethylamino)naphthalene; proton sponge. $C_{14}H_{18}N_2$; mol wt 214.31. C 78.46%, H 8.47%, N 13.07%. Prototype of a class of neutral aromatic diamines with exceptionally high basicity known as proton sponges. Prepn: W. G. Brown, N. J. Letang, J. Am. Chem. Soc. **63**, 358 (1941); and basicity study: R. W. Alder et al., J. Chem. Soc. Chem. Commun. **1968**, 723. X-ray crystal structure: H. Einspahr et al., Acta Crystallogr. **B29**, 1611 (1973). Proton transfer studies: F. Hibbert, J. Chem. Soc. Perkin Trans. 2 **1974**, 1862. Ab initio structural studies: J. A. Platts et al., J. Org. Chem. **59**, 4647 (1994); and fluorescence spectroscopy: A. Szemik-Hojniak et al., J. Am. Chem. Soc. **120**, 4840 (1998). Charge distri-

bution: P. R. Mallinson et al., ibid. **121**, 4640 (1999). ^{13}C NMR study of hydrogen bonding: M. Pietrzak et al., ibid. **123**, 4338 (2001). Catalytic activity: I. Rodriguez et al., J. Catal. **183**, 14 (1999). Review of chemistry: H. A. Staab, T. Saupe, Angew. Chem. Int. Ed. **27**, 865-879 (1988); R. W. Alder, Chem. Rev. **89**, 1215-1223 (1989).

bp4 144-145°. mp 47-48°. d 1.12. Absorption max: 335 nm (log ε 3.96). pKa 12.1.

USE: Very strong base in organic synthesis and catalysis.

3433. DMAP. [1122-58-3] N,N-Dimethyl-4-pyridinamine; 4-(dimethylamino)pyridine. $C_7H_{10}N_2$; mol wt 122.17. C 68.82%, H 8.25%, N 22.93%. Basic nucleophilic catalyst. Prepn: E. Koenigs et al., Ber. **58**, 2571 (1925). Fluorescence spectra: S. Mishina et al., J. Photochem. Photobiol. A **141**, 153 (2001). Use in Baylis-Hillman chemistry: K. Y. Lee et al., Bull. Korean Chem. Soc. **23**, 659 (2002); in olefin hydroxylations: Q. Yao, Org. Lett. **4**, 2197 (2002); in regioselective acylations: T. Kurahashi et al., Tetrahedron **58**, 8669 (2002). Review of early literature and use in acylation chemistry: G. Höfle et al., Angew. Chem. Int. Ed. **17**, 569-583 (1978); U. Ragnarsson, L. Grehn, Acc. Chem. Res. **31**, 494-501 (1998); and of alkylation: E. F. V. Scriven, Chem. Soc. Rev. **12**, 129-161 (1983). Brief description and synthetic uses: A. Hassner, "4-(Dimethylamino)pyridine" in Encyclopedia of Reagents for Organic Synthesis **3**, L. A. Paquette, Ed. (John Wiley & Sons, New York, 1995) pp 2022-2023.

Colorless crystalline solid, mp 112-113°. pKa (20°): 9.70. Dipole moment (25°): 4.31 (benzene). Very sol in methanol, ethyl acetate, chloroform, methylene chloride, 1,2-dichloroethane, acetone and acetic acid; less sol in cold hexane, cyclohexane and water.

USE: In a wide variety of organic syntheses as a catalyst.

3434. DMC. [37091-73-9] 2-Chloro-4,5-dihydro-1,3-dimethyl-1H-imidazolium chloride (1:1); 2-chloro-1,3-dimethylimidazolinium chloride. $C_5H_{10}Cl_2N_2$; mol wt 169.05. C 35.52%, H 5.96%, Cl 41.94%, N 16.57%. Dehydrating agent under nearly neutral conditions. For prepn, see: H.-B. König et al., US 3959258 (1976 to Bayer). NMR spectrum: H.-O. Kalinowski, H. Kessler, Org. Magn. Reson. **6**, 305 (1974). Series of articles on properties, dehydration and other reaction types: T. Isobe, T. Ishikawa, J. Org. Chem. **64**, 5832, 6984, 6989 (1999). Use as a dehydrating reagent: M. Node et al., Tetrahedron Lett. **39**, 6331 (1998).

Colorless and odorless prisms, mp 95-100°. Soly at room temperature (mL solvent/g DMC): acetic acid <1; water <1; methanol <1; ethanol <1; isopropanol <2; n-butanol <3; CHCl3 <1; CCl4 >50; n-hexane >50; THF >50; DMF 6.

USE: In organic syntheses for numerous reactions types such as condensations, chlorinations.

3435. DMMP. [756-79-6] P-Methylphosphonic acid dimethyl ester; dimethyl methylphosphonate. $C_3H_9O_3P$; mol wt 124.08. C 29.04%, H 7.31%, O 38.68%, P 24.96%. Testing chemical to simulate nerve agents such as sarin, q.v. Prepn: T. Milobedzki, K. Szulgin, Chem. Zentralbl. **89**, 914 (1918); by photolysis: M. Nakamura et al., J. Chem. Soc. Perkin Trans. 1 **1994**, 141. Photodestruction: L. J. Radziemski, Jr., *J. Environ. Sci. Health* **B16**, 337 (1981). Raman spectroscopic determn: N. Taranenko *et al.*, *J. Raman Spectrosc.* **27**, 379 (1996). Use as simulant in compd destruction study: M. G. Nickelsen *et al.*, *J. Adv. Oxid. Technol.* **3**, 43 (1998). Use as cell volume probe: J. E. Raftos *et al.*, *Biochim. Biophys. Acta* **968**, 160 (1988); J. A. Barry *et al.*, *Biochemistry* **32**, 4665 (1993). Description of use as flame retardant: E. D. Weil in *Kirk-Othmer Encyclopedia of Chemical Technology* vol. **10** (John Wiley, New York, 4th ed., 1993) pp 976-998. Review of toxicology: J. C. Rowland *et al.*, *Health Advisory for Dimethyl Methylphosphonate (DMMP)* (PB 93-117018, 1993) 86 pp.

Colorless liquid, mp <50°. bp$_{754}$ 181°. d 1.145. Vapor pressure: (20°) <0.1 torr; (25°) 2.4 torr; (65°) 20 torr. pKa (20°) in water: 2.37. Sol in water. Miscible in alcohol, ether, benzene, acetone, carbon tetrachloride. Insol in heavy mineral oil. *n* 1.411. Log P (octanol/water): −1.88. LD$_{50}$ in rats, mice (mg/kg): 10190, >6810 orally (Rowland).

USE: NMR probe for cell volume. Flame retardant. Simulant for nerve agents.

3436. DMPA. [299-85-4] *N*-(1-Methylethyl)phosphoramidothioic acid *O*-(2,4-dichlorophenyl) *O*-methyl ester; isopropylphosphoramidothioic acid *O*-2,4-(dichlorophenyl) *O*-methyl ester; *O*-(2,4-dichlorophenyl) *O*-methyl isopropylphosphoramidothioate; K-22023; Dow 1329; ENT-25647; OMS-115; Dowco 118; Zytron. $C_{10}H_{14}Cl_2NO_2PS$; mol wt 314.16. C 38.23%, H 4.49%, Cl 22.57%, N 4.46%, O 10.19%, P 9.86%, S 10.20%. Prepn: Blair *et al.*, *J. Agric. Food Chem.* **11**, 237 (1963). Synthesis of the optical isomers: Seiber, Tolkmith, *Tetrahedron* **25**, 381 (1969). Use as a herbicide: Leasure, US 3074790 (1963 to Dow); as a plant growth regulator: Holmsen, *Weed Sci.* **17**, 187 (1969). Neurotoxicity in chickens: B. M. Francis *et al.*, *J. Environ. Sci. Health* **B15**, 313 (1980). Toxicity study: E. W. Schafer, *Toxicol. Appl. Pharmacol.* **21**, 315 (1972).

Solid, mp 51.4°. Vapor pressure at 150°: 2 mm. Slightly sol in water (5 ppm); freely sol in acetone, benzene, carbon tetrachloride. LD$_{50}$ orally in rats: 270-360 mg/kg (Schafer).

USE: Herbicide; plant growth regulator.

3437. DMPO. [3317-61-1] 3,4-Dihydro-2,2-dimethyl-2*H*-pyrrole 1-oxide; 5,5-dimethyl-1-pyrroline 1-oxide. $C_6H_{11}NO$; mol wt 113.16. C 63.69%, H 9.80%, N 12.38%, O 14.14%. Nitrone spin trap. Prepn: R. Bonnett *et al.*, *J. Chem. Soc.* **1959**, 2094; D. L. Haire *et al.*, *J. Org. Chem.* **51**, 4298 (1986); and purification: E. G. Janzen *et al.*, *Chem. Biol. Interact.* **70**, 167 (1989). uv absorption: L. S. Kaminsky, M. Lamchen, *J. Chem. Soc.* **1968**, 1085. Use as spin trap in biological systems: A. I. Cederbaum, G. Cohen, *Arch. Biochem. Biophys.* **204**, 397 (1980); A. S. W. Li *et al.*, *Biochem. Biophys. Res. Commun.* **146**, 1191 (1987); for neutrophil activation studies: B. E. Britigan, D. R. Hamill, *Arch. Biochem. Biophys.* **275**, 72 (1989); *eidem*, *Free Radical Biol. Med.* **8**, 459 (1990); in reperfusion injury: C. M. Arroyo *et al.*, *FEBS Lett.* **221**, 101 (1987); A. Tosaki, P. Braquet, *Am. Heart J.* **120**, 819 (1990). Use in HPLC quantitation of oxygen free radicals *in vivo*: P. S. Rao *et al.*, *Chromatographia* **30**, 19 (1990).

Nitrone as white crystalline solid. Very hygroscopic. Turns yellow with time even at −20°C sealed under vacuum and protected

from light. uv max in cyclohexane: 246 nm (ε 9000); ethanol: 234 nm (ε 7700); water: 226 nm (ε 8600); *N*-hydrochloric acid: 226 nm (ε 7000); ether: 246 nm.

USE: Spin trap agent for biological systems.

3438. Dobell's Solution. [8021-82-7] Sodium borate solution compound. Antiseptic made from 1.5 g sodium borate, 1.5 g sodium bicarbonate, 0.3 ml liquefied phenol, 3.5 ml glycerol and water to make 100 ml.

Yellowish, clear liquid.

THERAP CAT: Wash for mucous membranes.

THERAP CAT (VET): Has been used as a nonirritant wash for mucous membranes.

3439. Dobesilate Calcium. [20123-80-2] 2,5-Dihydroxybenzenesulfonic acid calcium salt (2:1); calcium dobesilate; hydroquinone calcium sulfonate; Dexium; Doxium. $C_{12}H_{10}CaO_{10}S_2$; mol wt 418.40. C 34.45%, H 2.41%, Ca 9.58%, O 38.24%, S 15.33%. Prepn: ES 335945 (1967 to Labs. Esteve), *C.A.* **69**, 106253z (1968); FR M6163; A. Esteve-Subirana, US 3509207 (1968, 1970 both to Labs. OM). Clinical trials: Berson, *Praxis* **59**, 1305 (1970); *idem*, *ibid*. **61**, 52 (1972). Metabolism: A. Benakis *et al.*, *Therapie* **29**, 211 (1974).

White, powdery crystals from water, mp >300° (dec). Color deepens to pink upon exposure to air. Very soluble in water and alcohol. Practically insol in ether, benzene, chloroform. LD$_{50}$ in mice: 700 mg/kg (Esteve-Subirana).

THERAP CAT: Vasotropic.

3440. Dobutamine. [34368-04-2] 4-[2-[[3-(4-Hydroxyphenyl)-1-methylpropyl]amino]ethyl]-1,2-benzenediol; (±)-4-[2-[[3-(*p*-hydroxyphenyl)-1-methylpropyl]amino]ethyl]pyrocatechol; 3,4-dihydroxy-*N*-[3-(4-hydroxyphenyl)-1-methylpropyl]-β-phenylethylamine; Compound 81929. $C_{18}H_{23}NO_3$; mol wt 301.39. C 71.73%, H 7.69%, N 4.65%, O 15.93%. Synthetic catecholamine with β$_1$-, α$_1$-, and β$_2$-adrenergic agonist activity. Increases myocardial contractility, atrioventricular conduction, and heart rate. Structurally related to dopamine, *q.v.* Prepn: R. R. Tuttle, J. Mills, DE 2317710; *eidem*, US 3987200 (1973, 1976 to Lilly). Pharmacology: R. Weber, R. R. Tuttle in *Pharmacological and Biochemical Properties of Drug Substances* vol. **1**, M. E. Goldberg, Ed. (Am. Pharm. Assoc., Washington, DC, 1977) pp 109-124; E. H. Sonnenblick *et al.*, *N. Engl. J. Med.* **300**, 17 (1979). Comprehensive description: R. H. Bishara, H. B. Long, *Anal. Profiles Drug Subs.* **8**, 139-158 (1979). Review of diagnostic use in stress echocardiography: M. L. Geleijnse *et al.*, *J. Am. Coll. Cardiol.* **30**, 595-606 (1997); in myocardial perfusion imaging: A. Elhendy *et al.*, *J. Nucl. Med.* **43**, 1634-1646 (2002); in cardiovascular magnetic resonance: S. Mandapaka, W. G. Hundley, *J. Magn. Reson. Imaging* **24**, 499-512 (2006).

Hydrochloride. [49745-95-1] Dobuject; Dobutrex. $C_{18}H_{23}$-NO$_3$.HCl; mol wt 337.84. White crystals, mp 184-186°. uv max (methanol): 281, 223 nm (ε 4768, 14400). pKa 9.45. Rapidly oxidized at pH 11-13. Sol in ethanol, pyridine; sparingly sol in water, methanol. LD$_{50}$ i.v. in mice: ~73 mg/kg (Weber, Tuttle).

THERAP CAT: Cardiotonic; diagnostic aid (cardiac stress testing).

THERAP CAT (VET): Cardiotonic.

3441. Docarpamine. [74639-40-0] *C,C'*-[4-[2-[[(2*S*)-2-(Acetylamino)-4-(methylthio)-1-oxobutyl]amino]ethyl]-1,2-phenylene]-

carbonic acid *C,C'*-diethyl ester; (−)-(*S*)-2-acetamido-*N*-(3,4-dihy-droxyphenethyl)-4-(methylthio)butyramide bis(ethylcarbonate) (es-ter); *N*-(*N*-acetyl-L-methionyl)-*O,O*-bis(ethoxycarbonyl)dopamine; *N*-(*N*-acetyl-L-methionyl)-3,4-diethoxycarboxyphenethylamine; TA-870; Tanadopa. $C_{21}H_{30}N_2O_8S$; mol wt 470.54. C 53.60%, H 6.43%, N 5.95%, O 27.20%, S 6.81%. Dopamine prodrug. Prepn: T. Kiguchi *et al.*, **EP 7441**; *eidem*, **US 4228183** (both 1980 to Tanabe Seiyaku). Pharmacology: I. Yamaguchi *et al.*, *J. Cardiovasc. Pharmacol.* **13**, 879 (1989). Mechanism of action: S. Nishiyama *et al.*, *ibid.* **14**, 175 (1989). Pharmacokinetics: M. Yoshikawa *et al.*, *Drug Metab. Dispos.* **16**, 754 (1988). Metabolism in humans: *eidem*, *ibid.* **18**, 212 (1990). HPLC determn in biological fluids: *eidem*, *Biomed. Chromatogr.* **4**, 181 (1990). *Review:* S. Nishiyama *et al.*, *Cardiovasc. Drug Rev.* **10**, 101-116 (1992).

Crystals form ethyl acetate/*n*-hexane, mp 85-90°. $[\alpha]_D^{20}$ −15.6° (c = 2 in methanol). Also reported as crystalline powder, mp 105-108° (Nishiyama, 1992). LD_{50} in male, female rats (mg/kg): 1000-1400, ~1000 s.c.; in rats, dogs (mg/kg): >2000 orally (Nishiyama, 1992).

THERAP CAT: Cardiotonic.

3442. Docetaxel. [114977-28-5] ($\alpha R,\beta S$)-β-[[(1,1-Dimethyl-ethoxy)carbonyl]amino]-α-hydroxybenzenepropanoic acid (2a*R*,-4*S*,4a*S*,6*R*,9*S*,11*S*,12*S*,12a*R*,12b*S*)-12b-(acetyloxy)-12-(benzoyl-oxy)-2a,3,4,4a,5,6,9,10,11,12,12a,12b-dodecahydro-4,6,11-trihy-droxy-4a,8,13,13-tetramethyl-5-oxo-7,11-methano-1*H*-cyclo-deca[3,4]benz[1,2-*b*]oxet-9-yl ester; *N*-debenzoyl-*N*-(*tert*-butoxy-carbonyl)-10-deacetyltaxol; NSC-628503. $C_{43}H_{53}NO_{14}$; mol wt 807.89. C 63.93%, H 6.61%, N 1.73%, O 27.72%. Semisynthetic derivative of paclitaxel, *q.v.*, prepd using a natural precursor, *10-deacetylbaccatin III*, extracted from the needles of the European yew tree, *Taxus baccata* L., *Taxaceae*. Antimitotic agent that promotes the assembly of microtubules and inhibits their depolymeri-zation to free tubulin. Prepn: M. Colin *et al.*, **EP 253738**; *eidem*, **US 4814470** (1988, 1989 both to Rhône-Poulenc Sante); L. Manga-tal *et al.*, *Tetrahedron* **45**, 4177 (1989). Synthesis of the side chain: J.-N. Denis *et al.*, *J. Org. Chem.* **56**, 6939 (1991). Structure-activity study: F. Guéritte-Voegelein *et al.*, *J. Med. Chem.* **34**, 992 (1991). HPLC determn and degradation studies in pharmaceutical formula-tions: D. Kumar *et al.*, *J. Pharm. Biomed. Anal.* **43**, 1228 (2007). LC-MS determn in plasma: R. A. Parise *et al.*, *J. Chromatogr. B* **783**, 231 (2003). Review of clinical pharmacokinetics, metabolism, and analytical methods: S. D. Baker *et al.*, *Clin. Pharmacokinet.* **45**, 235-252 (2006). Review of mechanism of action and synergistic therapies: R. S. Herbst, F. R. Khuri, *Cancer Treat. Rev.* **29**, 407-415 (2003); of clinical experience with solid tumors: A. Montero *et al.*, *Lancet Oncol.* **6**, 229-239 (2003). Review of alternative formula-tions: F. K. Engels *et al.*, *Anti-Cancer Drugs* **18**, 95-103 (2007).

mp 232°. $[\alpha]_D$ −36° (c = 0.74 in ethanol). uv max: 230, 275, 283 nm (ε 14800, 1730, 1670).

Trihydrate. [148408-66-6] RP-56976; Taxotere. $C_{43}H_{53}$-$NO_{14}\cdot 3H_2O$; mol wt 861.94. White to almost-white powder. Highly lipophilic. Practically insol in water.

THERAP CAT: Antineoplastic.

3443. Docosahexaenoic Acid. [6217-54-5] (4*Z*,7*Z*,10*Z*,13*Z*,-16*Z*,19*Z*)-4,7,10,13,16,19-Docosahexaenoic acid; cervonic acid; do-conexent; DHA. $C_{22}H_{32}O_2$; mol wt 328.50. C 80.44%, H 9.82%, O 9.74%. Omega-3 fatty acid found in marine fish oils and in many phospholipids. Major structural component of excitable membranes of the retina and brain; synthesized in the liver from α-linolenic acid, *q.v.* Isoln from oil of *Sardina ocellata* J. and structure: J. M. Whit-cutt, *Biochem. J.* **67**, 60 (1957). Improved isoln from cod liver oil: S. W. Wright *et al.*, *J. Org. Chem.* **52**, 4399 (1987). Effect on brain and behavioral development: P. E. Wainwright, *Neurosci. Biobe-hav. Rev.* **16**, 193 (1992). Review of uptake and metabolism by retinal cells: N. G. Bazan, E. B. Rodriguez de Turco, *J. Ocul. Phar-macol.* **10**, 591-603 (1994). Review of clinical studies in infant for-mula supplementation: M. Makrides *et al.*, *Lipids* **31**, 115-119 (1996).

Clear, faintly yellow oil, mp −44.7 to −44.5°. n_D^{26} 1.5017.

USE: Nutritional supplement.

3444. *n*-Docosanol. [661-19-8] 1-Docosanol; behenic alco-hol; behenyl alcohol; docosyl alcohol; Abreva; Lidakol. $C_{22}H_{46}O$; mol wt 326.61. C 80.90%, H 14.20%, O 4.90%. $H_3C(CH_2)_{21}OH$. Naturally occurring alcohol; active principal of Pygeum africanum extract, *q.v.*, that has been used to treat benign prostatic hypertrophy (BPH). Prepn: R. Willstätter, E. W. Mayer, *Ber.* **41**, 1475 (1908); P. A. Levene, F. A. Taylor, *J. Biol. Chem.* **59**, 905 (1924). Surface thermodynamic properties: S. Pathak, S. S. Katti, *J. Chem. Eng. Data* **14**, 359 (1969). GC-MS determn in Pygeum africanum extract: N. Pierini *et al.*, *Boll. Chim. Farm.* **121**, 27 (1982). Antiviral activ-ity: D. H Katz *et al.*, *Ann. N.Y. Acad. Sci.* **724**, 472 (1994). Study of mechanism of action in BPH: J. Müntzing *et al.*, *Invest. Urol.* **77**, 176 (1979); vs herpes simplex virus: L. E. Pope *et al.*, *Antiviral Res.* **40**, 85 (1998). Clinical evaluation in recurrent herpes labialis: L. Habbema *et al.*, *Acta Derm. Venereol.* **76**, 479 (1996).

mp 70.5-71.5°. bp$_{22mm}$ 180°. d^{75} 0.8063 g/ml; d^{85} 0.7986 g/ml; d^{95} 0.7911 g/ml. Surface tension (dyne/cm): 21.82 (75°); 21.17 (85°); 20.53 (95°). n^{75} 1.4360.

THERAP CAT: Antiviral (topical).

3445. Docusate Calcium. [128-49-4] 2-Sulfobutanedioic acid 1,4-bis(2-ethylhexyl)ester calcium salt (2:1); bis[2-ethylhexyl]-calcium sulfosuccinate; calcium dioctyl sulfosuccinate; dioctyl cal-cium sulfosuccinate; Surfak. $C_{40}H_{74}CaO_{14}S_2$; mol wt 883.22. C 54.40%, H 8.45%, Ca 4.54%, O 25.36%, S 7.26%. Prepd from dioc-tyl sodium sulfosuccinate dissolved in isopropanol and from calcium chloride dissolved in methanol: Klotz, **US 3035973** (1962 to Lloyd Brothers).

White precipitate. Very sol in alcohol, polyethylene glycol 400, corn oil; sol in mineral and vegetable oils; very slightly sol in water. Practically insol in glycerol. Claimed to have greater surface-active wetting properties than the sodium salt.

Ingredient of *Doxidan* which also contains phenolphthalein.

THERAP CAT: Stool softener.

3446. Docusate Sodium. [577-11-7] 2-Sulfobutanedioic acid 1,4-bis(2-ethylhexyl) ester sodium salt (1:1); sulfosuccinic acid 1,4-bis(2-ethylhexyl) ester S-sodium salt; bis(2-ethylhexyl)sodium sulfosuccinate; dioctyl sodium sulfosuccinate; sodium dioctyl sulfosuccinate; DSS; Aerosol OT; Colace; Comfolax; Coprola; Dioctylal; Dioctyl; Diotilan; Disonate; Doxinate; Doxol; Dulcivac; Jamylène; Molatoc; Molcer; Nevax; Regutol; Soliwax; Velmol; Waxsol; Yal. $C_{20}H_{37}NaO_7S$; mol wt 444.56. C 54.04%, H 8.39%, Na 5.17%, O 25.19%, S 7.21%. Prepn: Jaeger, **US 2028091**; **US 2176423** (1936, 1939, both to Am. Cyanamid). Structure and wetting power: Caryl, *Ind. Eng. Chem.* **33**, 731 (1941). Comprehensive description: S. Ahuja, J. Cohen, *Anal. Profiles Drug Subs.* **2**, 199-219 (1973); **12**, 713-720 (1983). For structure see Docusate calcium.

Available as wax-like solid, usually in rolls of tissue-thin material; also as 50-75% solns in various solvents. Soly in water (g/l): 15 (25°), 23 (40°), 30 (50°), 55 (70°). Sol in CCl$_4$, petr ether, naphtha, xylene, dibutyl phthalate, liquid petrolatum, acetone, alcohol, vegetable oils. Very sol in water + alcohol, water + water-miscible organic solvents. Stable in acid and neutral solns; hydrolyzes in alkaline solns.

Docusate potassium. [7491-09-0] Rectalad. $C_{20}H_{37}KO_7S$; mol wt 460.67.

Note: Ingredient of the laxative **Peri-Colace** which also contains casanthranol.

USE: Sodium salt as pharmaceutic aid (surfactant); as wetting agent in industrial, pharmaceutical, cosmetic and food applications; dispersing and solubilizing agent in foods; adjuvant in tablet formation.

THERAP CAT: Stool softener.

THERAP CAT (VET): Stool softener.

3447. Dodecahedrane. [4493-23-6] [5]Fullerane-C$_{20}$-I_h; hexadecahydro-5,2,1,6,3,4-[2,3]butanediyl[1,4]diylidenedipentaleno[2,1,6-cde:2′,1′,6′-gha]pentalene; pentagonal dodecahedrane. $C_{20}H_{20}$; mol wt 260.38. C 92.26%, H 7.74%. Classical uniform convex polyhedrane. Theoretical studies: O. Ermer, *Angew. Chem. Int. Ed.* **16**, 411 (1977); R. L. Disch, J. M. Schulman, *J. Am. Chem. Soc.* **103**, 3297 (1981). Review of synthetic studies: P. E. Eaton, *Tetrahedron* **35**, 2189-2223 (1979). Total synthesis: R. J. Ternansky *et al., J. Am. Chem. Soc.* **104**, 4503 (1982).

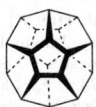

Crystals from benzene, mp >450°.

3448. δ-Dodecalactone. [713-95-1] 6-Heptyltetrahydro-2H-pyran-2-one; 5-dodecanolide. $C_{12}H_{22}O_2$; mol wt 198.31. C 72.68%, H 11.18%, O 16.14%. Naturally occuring aroma constituent of dairy products, pork fat, rum, and fruits including peaches, berries, plums, and mangoes. Prepn: K.-W. Rosenmund, H. Bach, *Ber.* **94**, 2401 (1961); B. M. Dashunin *et al., Zh. Prikl. Khim.* **42**, 1909 (1969). Synthesis of enantiomers: A. Mosandl, M. Gessner, *Z. Lebensm.-Unters. Forsch.* **187**, 40 (1988); H. Nemoto *et al., Synlett* **2007**, 2343. Fragrance monograph: D. L. J. Opdyke, *Food Cosmet. Toxicol.* **17**, Suppl. 1, 773 (1979).

Colorless to yellow liquid; powerful fruity, peach-like and oily odor. On dilution the odor is butter-like. bp$_{12}$ 177°. bp$_{1.5}$ 124-125°. d$_4^{20}$ 0.9555. n$_D^{20}$ 1.4620. Flash pt (closed cup): 113°C (235°F). Very sol in alcohol, propylene glycol, vegetable oil. Insol in water. LD$_{50}$ orally in rats: >5 g/kg; dermally in rabbits: >5 g/kg (Opdyke).

R-Form. [29587-89-1] Fruity-sweet, apricot odor with fatty-green aspects. Fruity-sweet apricot taste. $[\alpha]_D^{20}$ +46° (c = 1.0-1.3 in CH$_3$OH). $[\alpha]_D^{20}$ +32.3° (c = 2.65 in CHCl$_3$). $[\alpha]_D^{20}$ +49.9° (c = 2.00 in THF).

S-Form. [108943-47-1] Similar in odor and taste to the R(+)-form, but more intense, with a distinct aldehyde and sweat odor, and fruity, creamy taste. $[\alpha]_D^{20}$ −46° (c = 1.0-1.3 in CH$_3$OH). $[\alpha]_D^{20}$ −33.2° (c = 1.43 in CHCl$_3$). $[\alpha]_D^{20}$ −48.7° (c = 2.00 in THF).

USE: Ingredient in commercial butter and cream flavors.

3449. Dodecamethylcyclohexasiloxane. [540-97-6] 2,2,4,-4,6,6,8,8,10,10,12,12-Dodecamethylcyclohexasiloxane. $C_{12}H_{36}O_6$-Si$_6$; mol wt 444.92. C 32.40%, H 8.16%, O 21.58%, Si 37.87%. Isolated from the hydrolysis product of dimethyldichlorosilane: Patnode, Wilcock, *J. Am. Chem. Soc.* **68**, 358 (1946).

Oily liquid. mp −3°. bp 245°. bp$_{20}$ 128°. d 0.9762. n$_D^{20}$ 1.4015.

3450. Dodecamethylpentasiloxane. [141-63-9] 1,1,1,3,3,5,-5,7,7,9,9,9-Dodecamethylpentasiloxane. $C_{12}H_{36}O_4Si_5$; mol wt 384.84. C 37.45%, H 9.43%, O 16.63%, Si 36.49%. Prepn: Patnode, Wilcock, *J. Am. Chem. Soc.* **68**, 358 (1946).

Liquid. bp$_{710}$ 229°. d 0.8755. n$_D^{20}$ 1.3925. mp ~−80°. Stable. Inert to most chemical reagents and rubber. Maintains about the same viscosity over a wide temperature range. Sol in benzene and the lighter hydrocarbons; slightly sol in alcohol and the heavy hydrocarbons.

USE: As a basis for silicone oils or fluids designed to withstand extremes of temperature; as a foam suppressant in petr lubricating oil.

3451. 1-Dodecanol. [112-53-8] Dodecyl alcohol; lauryl alcohol; 1-hydroxydodecane. $C_{12}H_{26}O$; mol wt 186.34. C 77.35%, H 14.06%, O 8.59%. Prepd by the reduction of esters of lauric acid with sodium and abs alcohol: Levene, Allen, *J. Biol. Chem.* **27**, 443 (1916); Ford, Marvel, *Org. Synth.* **10**, 62 (1930); or by high pressure hydrogenation of the esters using copper chromite catalyst: Adkins, Folkers, *J. Am. Chem. Soc.* **53**, 1095 (1931).

Leaflets from dil alc, mp 24°. d$_4^{24}$ 0.8309 (liq); d$_4^{40}$ 0.8201; d$_4^{99}$ 0.7781. bp$_{760}$ 259°; bp$_{400}$ 235.7°; bp$_{200}$ 213°; bp$_{100}$ 192°; bp$_{60}$ 177.8°; bp$_{40}$ 167.2°; bp$_{20}$ 150°; bp$_{10}$ 134.7°; bp$_5$ 120.2°; bp$_{1.0}$ 91.0°. Insol in water; sol in alcohol and ether.

Phenylurethan. [5796-07-6] Phenylcarbamic acid dodecyl ester. $C_{19}H_{31}NO_2$; mol wt 305.46. Needles from dilute methanol, mp 84°.

USE: Manuf of sulfuric acid esters which are used as wetting agents.

3452. [S-(all-E)]-3-(1,3,5,7,9-Dodecapentaenyloxy)-1,2-propanediol. [91423-46-0] (2S)-3-[(1E,3E,5E,7E,9E)-1,3,5,7,9-Dodecapentaen-1-yloxy]-1,2-propanediol; Fecapentaene 12. $C_{15}H_{22}O_3$; mol wt 250.34. C 71.97%, H 8.86%, O 19.17%. A potent mutagen detected in human feces: W. R. Bruce *et al., Cold Spring Harbor Conf. Cell Proliferation* **3**, 1641 (1977). Isoln and characterization: T. D. Wilkins *et al., Am. J. Clin. Nutr.* **33**, 2513 (1980). Structure: N. Hirai *et al., J. Am. Chem. Soc.* **104**, 6149 (1982). Implicated in human colon cancer: *Chem. Eng. News* **60**, 22 (Sept. 27, 1982); T. H. Maugh, *Science* **218**, 363 (1982).

Unstable, decomp rapidly in presence of air or acid. uv max: 325, 345, 365 nm. Sol in chloroform, benzene, ether. Insol in water.

3453. Dodecylbetaine. [683-10-3] *N*-(Carboxymethyl)-*N*,*N*-dimethyl-1-dodecanaminium inner salt; *n*-dodecyl-*N*,*N*-dimethyl-glycine; *N*,*N*-dimethyldodecylbetaine; (dodecyldimethylammonio)-acetate; lauryl betaine; Amphitol 20BS; Amphitol 24B. $C_{16}H_{33}$-NO_2; mol wt 271.45. C 70.80%, H 12.25%, N 5.16%, O 11.79%. Amphoteric surfactant. Prepn from dodecyldimethylamine and sodium chloracetate: K. Daimler, C. Platz, **US 2082275** (1937 to GAF); A. H. Beckett, R. J. Woodward, *J. Pharm. Pharmacol.* **15**, 422 (1963); from α-bromomyristic acid and trimethylamine: K. Tori, T. Nakagawa, *Kolloid Z.* **187**, 44 (1963). Physical properties: A. Amin-Alami *et al.*, *Thermochim. Acta* **122**, 171 (1987); Y. Chevalier *et al.*, *Langmuir* **7**, 848 (1991); J. G. Weers *et al.*, *Langmuir* **7**, 854 (1991). Skin penetration and toxicity: G. Ridout *et al.*, *Fundam. Appl. Toxicol.* **16**, 41 (1991). Solubilization of phosphatidylcholine liposomes: A. de la Maza *et al.*, *Chem. Phys. Lipids* **94**, 71 (1998).

Crystals from ethanol:ether (1:20) as monohydrate, mp 183° (Beckett, Woodward); also reported as colorless needles from acetone, mp 185° (dec). Strongly hygroscopic; characteristic bitter taste. Easily sol in water, methanol, ethanol; moderately sol in acetone; sol in benzene. Zwitterionic in slightly acid to strongly alk solns, pI ~6.65. pKa ~1.8. Critical micelle concentration (CMC) at 25°: 2.0 mM/kg. Surface tension (at CMC): 36.7 mN/m. Krafft point: < 0°. LD_{50} in male rats (mg/kg): 53 i.p.; 71 orally; 1300 dermally (Ridout).

USE: High foaming surfactant for industrial and personal care formulations; as shampoo base; as antistatic agent.

3454. Dodemorph. [1593-77-7] 4-Cyclododecyl-2,6-dimethylmorpholine. $C_{18}H_{35}NO$; mol wt 281.48. C 76.81%, H 12.53%, N 4.98%, O 5.68%. Prepn: W. Sanne *et al.*, **BE 614214**; *eidem*, **US 3686399** (1962, 1972 both to BASF); K.-H. König *et al.*, *Angew. Chem. Int. Ed.* **4**, 336 (1965). Activity: J. Kradel, E. H. Pommer, *Proc. 4th Br. Insectic. Fungic. Conf.* **1967**, 170. Toxicity data: D. Marchand, G. Serra, *Def. Veg.* **27**, 144 (1973), *C.A.* **79**, 133569v (1973).

Oil, $bp_{1.5}$ 161-162°. n_D^{25} 1.4907.

Acetate. Cyclomorph; BAS-238F; Meltatox; Milban. $C_{18}H_{35}$-$NO.C_2H_4O_2$; mol wt 341.54. Yellow liquid, d 0.93. Misc with water. LD_{50} i.p. in mice: 100 mg/kg (Marchand, Serra).

USE: Fungicide.

3455. Dodine. [2439-10-3] *N*-Dodecylguanidine acetate (1:1); AC-5223; Carpene; Melprex. $C_{15}H_{33}N_3O_2$; mol wt 287.45. C 62.68%, H 11.57%, N 14.62%, O 11.13%. Prepn: B. Puetzer, **US 2213471** (1940 to Winthrop Chem. Co.). Degradation study: M. C. Goldberg, R. L. Wershaw, *Prepr. Pap. Natl. Meet., Div. Water, Air Waste Chem., Am. Chem. Soc.* **5**, 53 (1965). Mechanism of action: E. Somers, R. J. Pring, *Ann. Appl. Biol.* **58**, 457 (1966). Toxicology: G. J. Levinskas *et al.*, *Toxicol. Appl. Pharmacol.* **3**, 127 (1961); to aquatic organisms: A. Hourdakis *et al.*, *Fresenius Environ. Bull.* **4**, 679 (1995). Determn by titration in formulations: N. R. Pasarela, *J. Assoc. Off. Anal. Chem.* **52**, 1292 (1969); by GLC in fruit: J. Hajslová *et al.*, *J. Chromatogr.* **348**, 437 (1985). Field trial on apple trees: R. G. Ross *et al.*, *Can. J. Plant Sci.* **49**, 655 (1969).

Slightly waxy solid, mp 132-135°. Sol in ethanol, water. Insol in most organic solvents. LC_{50} in harlequin fish: 0.6 mg/l (Hourdakis).

Caution: Strong solns irritating to skin, mucous membranes. Ingestion causes vomiting, diarrhea.

USE: Agricultural fungicide.

3456. Dofetilide. [115256-11-6] *N*-[4-[2-[Methyl[2-[4-[(methylsulfonyl)amino]phenoxy]ethyl]amino]ethyl]phenyl]methanesulfonamide; 1-(4-methanesulfonamidophenoxy)-2-[*N*-(4-methanesulfonamidophenethyl)-*N*-methylamino]ethane; UK-68798; Tikosyn. $C_{19}H_{27}N_3O_5S_2$; mol wt 441.56. C 51.68%, H 6.16%, N 9.52%, O 18.12%, S 14.52%. Potassium channel blocker. Prepn: J. E. Arrowsmith *et al.*, **EP 245997**; P. E. Cross *et al.*, **US 4959366** (1987, 1990 both to Pfizer); *idem et al.*, *J. Med. Chem.* **33**, 1151 (1990). HPLC determn in urine: D. K. Walker *et al.*, *J. Chromatogr.* **568**, 475 (1991). Mechanism of action study: D. Carmeliet, *J. Pharmacol. Exp. Ther.* **262**, 809 (1992). Review of pharmacology and pharmacokinetics: H. S. Rasmussen *et al.*, *ibid.* **20**, Suppl. 2, S96-S105 (1992). Clinical trial in atrial fibrillation and flutter: B. L. Norgaard *et al.*, *Am. Heart J.* **137**, 1062 (1999); in congestive heart failure: C. Torp-Pedersen *et al.*, *N. Engl. J. Med.* **341**, 857 (1999).

Crystals from ethyl acetate/methanol (10:1), mp 147-149° (Cross); from hexane/ethyl acetate, mp 151-152° (Arrowsmith). Also reported as white crystalline solid, mp 161° (Rasmussen). pKa 7.0, 9.0, 9.6. Distribution coefficient (pH 7.4): 0.96. Sol in 0.1*M* NaOH, acetone, 0.1*M* HCl; very slightly sol in water, propan-2-ol.

THERAP CAT: Antiarrhythmic (class III).

3457. Doisynolic Acid. [109784-48-7] 1-Ethyl-1,2,3,4,4a,9,-10,10a-octahydro-7-hydroxy-2-methyl-2-phenanthrenecarboxylic acid; 3-hydroxy-16,17-secoestra-1,3,5(10)-trien-17-oic acid. C_{18}-$H_{24}O_3$; mol wt 288.39. C 74.97%, H 8.39%, O 16.64%. Prepd by alkali fusion of estrone: MacCorquodale *et al.*, *J. Biol. Chem.* **99**, 327 (1933); **101**, 753 (1933); Heer, Miescher, *Helv. Chim. Acta* **28**, 156 (1945). Sixteen isomers are possible. If the hydrogen atom at position 4a is α- and the hydrogen atom at position 10a is β-, the acid belongs to the A series. If the hydrogen at 4a is β- and the hydrogen at 10a is α-, the acid belongs to the B series. If both hydrogens are α-, the acid belongs to the C series. If both hydrogens are β-, the acid belongs to the D series: Miescher, *Chem. Rev.* **43**, 367 (1948). If the hydrogen atom at position 1 and the methyl group at position 2 have either the α,α- or β,β-orientation, the acid is designated *cis;* if they are α,β- the acid is designated *trans:* L. F. Fieser, M. Fieser, *Steroids* (Reinhold, New York, 1959) pp 487-495. Synthesis of *dl-cis* and *dl-trans* of the A series and *dl-cis* of the B series: Anner, Miescher, *Helv. Chim. Acta* **29**, 1889 (1946); **30**, 1422 (1947). Synthesis of *dl-cis* of the C series: Jilck, Protiva, *Collect. Czech. Chem. Commun.* **23**, 692 (1958). Stereochemistry of *d-cis*-form: Iriarte, Crabbe, *Chem. Commun.* **1972**, 1110. *Reviews:* Shoppee, *Annu. Rep. Prog. Chem.* **44**, 190 (1948); Miescher in *Recent Progress in Hormone Research* vol. 3, G. Pincus Ed. (Academic Press, New York, 1948) pp 47-69.

B-type acid

dl-cis **Acid of the A series.** Crystals from methanol, mp 181-182°. Appears to have the highest estrogenic potency.

dl-trans **Acid of the A series.** Plates from methanol, mp 175-177°.

dl-cis **Acid of the B series.** Crystals from methanol, mp 212-214°.

dl-cis **Acid of the C series.** Crystals from methanol, mp 113-117°.

3458. Dolasetron. [115956-12-2] 1*H*-Indole-3-carboxylic acid octahydro-3-oxo-2,6-methano-2*H*-quinolizin-8-yl ester stereoisomer; *endo*-hexahydro-8-(3-indolylcarbonyloxy)-2,6-methano-2*H*-quinolizin-3(4*H*)-one; MDL-73147. $C_{19}H_{20}N_2O_3$; mol wt 324.38. C 70.35%, H 6.21%, N 8.64%, O 14.80%. Bridged pseudopelletierine derivative; specific serotonin (5HT$_3$) receptor antagonist. Prepn: M. W. Gittos, **EP 266730**; *idem et al.*, **US 4906755** (1988, 1990 both to Merrell Dow). Binding study: P. H. Boeijinga *et al.*, *Eur. J. Pharmacol.* **219**, 9 (1992). Pharmacology: R. C. Miller *et al.*, *Drug Dev. Res.* **28**, 87 (1993). Clinical pharmacokinetics: H. Boxenbaum *et al.*, *Biopharm. Drug Dispos.* **13**, 693 (1992); *idem, ibid.* **14**, 131 (1993). GC-MS and LC determn in plasma: T. A. Gillespie *et al.*, *J. Pharm. Biomed. Anal.* **11**, 955 (1993). Clinical trial in emetogenic chemotherapy: A. A. Fauser *et al.*, *Eur. J. Cancer* **32A**, 1523 (1996); for prevention of postoperative nausea: S. G. Graczyk *et al.*, *Anesth. Analg.* **84**, 325 (1997); B. K. Philip *et al.*, *J. Clin. Anesth.* **12**, 1 (2000). Review of veterinary use: G. K. Ogilvie, *J. Am. Anim. Hosp. Assoc.* **36**, 481-483 (2000).

Methanesulfonate. [115956-13-3] Dolasetron mesylate; MDL-73147EF; Anzemet. $C_{19}H_{20}N_2O_3 \cdot CH_3SO_3H$; mol wt 420.48. Crystalline solid, mp 278°. Freely sol in water, propylene glycol; slightly sol in alcohol, saline TS.

THERAP CAT: Antiemetic.

THERAP CAT (VET): Antiemetic.

3459. Dolichodial. [5951-57-5] *rel*-(1*R*,2*S*,3*R*)-2-Formyl-3-methyl-α-methylenecyclopentaneacetaldehyde; (*cis,trans*)-dolichodial. $C_{10}H_{14}O_2$; mol wt 166.22. C 72.26%, H 8.49%, O 19.25%. Lacrimatory, iridoid monoterpene produced by dolichoderine ants; also found in the defensive secretions of the southern walking stick, *Anisomorpha buprestoides*. Occurs naturally as the (−)-isomer, along with the C-2 epimer, **anisomorphal**, and the (*trans,trans*)-diastereomer, **peruphasmal**. Isoln as a mixture of (*trans,cis*)- and (*cis,trans*)-isomers from *Dolichoderus acanthoclinea*: G. W. K. Cavill, H. Hinterberger, *Aust. J. Chem.* **13**, 514 (1960). Isoln of anisomorphal from *A. buprestoides*: J. Meinwald *et al.*, *Tetrahedron Lett.* **3**, 29 (1962). Isoln from cat thyme, *Teucrium marum, Labiatae*, and characterization of stereoisomers: U. M. Pagnoni *et al.*, *Aust. J. Chem.* **29**, 1375 (1976). Characterization of stereoisomers produced by the Argentine ant, *Iridomyrmex humilis*: G. W. K. Cavill *et al.*, *Insect Biochem.* **6**, 483 (1976); by *A. buprestoides*: A. T. Dossey *et al.*, *ACS Chem. Biol.* **1**, 511 (2006). Synthesis and configuration of (−)-dolichodial: T. Yamane *et al.*, *Synthesis* **1995**, 444. Stereoselective synthesis: J. S. Beckett *et al.*, *Org. Lett.* **12**, 1408 (2010).

(−)-Dolichodial

(1S,2R,3S)-Form. [60478-52-6] (−)-Dolichodial. Colorless oil. $[\alpha]_D^{20}$ −72.0° (c = 4.2 in benzene).

(1S,2S,3S)-Form. [3671-76-9] (+)-Anisomorphal. $[\alpha]_D^{20}$ +3.5° (c = 4.3 in benzene).

3460. Dolutegravir. [1051375-16-6] (4*R*,12a*S*)-*N*-[(2,4-Difluorophenyl)methyl]-3,4,6,8,12,12a-hexahydro-7-hydroxy-4-methyl-6,8-dioxo-2*H*-pyrido[1′,2′:4,5]pyrazino[2,1-*b*][1,3]oxazine-9-carboxamide; (4*R*,9a*S*)-5-hydroxy-4-methyl-6,10-dioxo-3,4,6,9,-9a,10-hexahydro-2*H*-1-oxa-4a,8a-diazaanthracene-7-carboxylic acid 2,4-difluorobenzylamide; GSK-1349572. $C_{20}H_{19}F_2N_3O_5$; mol wt 419.38. C 57.28%, H 4.57%, F 9.06%, N 10.02%, O 19.07%. Second generation HIV-1 integrase strand transfer inhibitor (INSTI). Prepn: B. A. Johns *et al.*, **WO 06116764** (2006 to Shionogi); *eidem*, **US 090318421** (2009). *In vitro* antiretroviral properties: M. Kobayashi *et al.*, *Antimicrob. Agents Chemother.* **55**, 813 (2011). Binding and co-crystal structure: S. Hare *et al.*, *Mol. Pharmacol.* **80**, 565 (2011). Clinical pharmacokinetics: S. Min *et al.*, *Antimicrob. Agents Chemother.* **54**, 254 (2010); and evaluation in HIV-1 infection: *eidem*, *AIDS* **25**, 1737 (2011). Review of development and therapeutic potential: J. C. C. Lenz, J. K. Rockstroh, *Expert Opin. Invest. Drugs* **20**, 537-548 (2011).

Sodium salt. [1051375-19-9] GSK-1349572A. $C_{20}H_{18}F_2N_3$-NaO$_5$; mol wt 441.37.

THERAP CAT: Antiviral.

3461. DOM. [15588-95-1] 2,5-Dimethoxy-α,4-dimethylbenzeneethanamine; 2,5-dimethoxy-α,4-dimethylphenethylamine; 1-(2,5-dimethoxy-4-methylphenyl)-2-aminopropane; (±)-2,5-dimethoxy-4-methylamphetamine; STP. $C_{12}H_{19}NO_2$; mol wt 209.29. C 68.87%, H 9.15%, N 6.69%, O 15.29%. Psychedelic compound of "hippie" culture. Prepn: A. T. Shulgin, **US 3547999** (1970 to Dow). Pharmacology: B. T. Ho *et al.*, *J. Med. Chem.* **13**, 26 (1970). Clinical pharmacology: S. H. Snyder *et al.*, *Am. J. Psychiatry* **125**, 357 (1968). Mechanism of action: R. A. Glennon, *Neuropsychopharmacology* **3**, 509 (1990); C. S. Aulakh *et al.*, *J. Pharmacol. Exp. Ther.* **271**, 143 (1994). Abuse potential: P. B. Silverman, B. T. Ho, *Psychopharmacology* **68**, 209 (1980); D. Fiorella *et al., ibid.* **119**, 239 (1995). Activity comparison of enantiomers: F. Benington *et al., Nature* **242**, 185 (1973); sepn and determn of enantiomers by HPLC: J. Goto *et al.*, *J. Liq. Chromatogr.* **2**, 1179 (1979). Analysis by TLC: I. Ojanperä *et al.*, *J. Planar Chromatogr. Mod. TLC* **4**, 373 (1991). Brief description including prepn: A. Shulgin, A. Shulgin, *Pihkal* (Transform Press, Berkeley CA, 1991) pp 637-642.

mp 60.5-61°.

Hydrochloride. [15589-00-1] $C_{12}H_{19}NO_2 \cdot HCl$. White crystals from isopropanol/ether, mp 189-189.5°. LD$_{50}$ i.p. in mice: 89 ±4.2 mg/kg (Ho).

Note: This is a controlled substance (hallucinogen): **21 CFR, 1308.11.**

3462. Domesticine. [476-71-1] (6a*S*)-5,6,6a,7-Tetrahydro-2-methoxy-6-methyl-4*H*-benzo[*de*][1,3]benzodioxolo[5,6-*g*]quinolin-1-ol; 2-methoxy-9,10-(methylenedioxy)-6aα-aporphin-1-ol; 1-hydroxy-2-methoxy-9,10-methylenedioxyaporphine. $C_{19}H_{19}NO_4$; mol wt 325.36. C 70.14%, H 5.89%, N 4.31%, O 19.67%. In *Nandina domestica* Thunb., *Berberidaceae*. Isoln: Kitasato, Shishido,

Ann. **527**, 176 (1937). Syntheses: Govindachari *et al.*, *Indian J. Chem.* **7**, 841 (1969); Kessar *et al.*, *ibid.* **8**, 468 (1970); Kametani *et al.*, *J. Chem. Soc. C* **1971**, 2446, 2712; *eidem*, *Chem. Pharm. Bull.* **21**, 766 (1973); Horii *et al.*, *ibid.* **22**, 583 (1974); Hoshino *et al.*, *ibid.* **23**, 2048 (1975).

From methanol + water, mp 115-116°; from abs methanol or benzene, mp 84-85°; dried at 60° over P_2O_5, mp 152-153°. Easily oxidized in air. uv max (ethanol): 221, 283, 310 nm (log ε 4.56, 4.01, 4.17). Very sol in chloroform, sol in hot alc, ethyl acetate, acetic acid, alkalies; slightly sol in ether. Practically insol in water.

Methyl ether. Nantenine; domestine; epidicentrine. mp 139°. $[\alpha]_D^{18} +102°$ (c = 0.528 in chloroform).

dl-**Form.** Needles from methanol, mp 185-186° (dec).

3463. Domiodol. [61869-07-6] 2-(Iodomethyl)-1,3-dioxolane-4-methanol; 4-hydroxymethyl-2-iodomethyl-1,3-dioxolane; MG-13608; Mucolitico. $C_5H_9IO_3$; mol wt 244.03. C 24.61%, H 3.72%, I 52.00%, O 19.67%. Organic iodide mixture of *cis* and *trans* isomers. Prepn: M. Carissimi *et al.*, **DE 2610704**; *eidem*, **US 4085223** (1976, 1978 both to Maggioni). Prepn, expectorant and mucolytic activities: G. Cantarelli *et al.*, *Farmaco Ed. Prat.* **34**, 393 (1979). Pharmacology, toxicity: K. Kogi *et al.*, *Arzneim.-Forsch.* **33**, 1281 (1983). Separation of isomers and comparison of their pharmacological activity: M. Riva *et al.*, *ibid.* 1091. Metabolism and tissue distribution in rats: T. Ohtsuki *et al.*, *Farmaco Ed. Prat.* **39**, 291 (1984). Comparative study with *S*-carboxymethylcysteine, *q.v.*, in chronic obstructive lung disease: L. Casali *et al.*, *Int. J. Clin. Pharmacol. Ther. Toxicol.* **20**, 554 (1982).

LD_{50} in male, female mice (mg/kg): 79, 89 i.p.; 140, 145 orally (Riva).

cis-**Form.** $bp_{0.2}$ 106-108°. LD_{50} orally in male mice: 135 mg/kg (Riva).

trans-**Form.** $bp_{0.2}$ 114-116°. LD_{50} orally in male mice: 150 mg/kg (Riva).

THERAP CAT: Mucolytic.

3464. Domiphen Bromide. [538-71-6] *N,N*-Dimethyl-*N*-(2-phenoxyethyl)-1-dodecanaminium bromide (1:1); dodecyldimethyl(2-phenoxyethyl)ammonium bromide; (β-phenoxyethyl)dimethyl-dodecylammonium bromide; PDDB; phenododecinium bromide; NSC-39415; Bradosol; Oradol; Modicare; Neo-Bradoral. $C_{22}H_{40}$-BrNO; mol wt 414.47. C 63.75%, H 9.73%, Br 19.28%, N 3.38%, O 3.86%. Prepd by heating phenoxyethyldimethylamine with dodecyl bromide: Hartmann, Bosshard, **US 2581336** (1952 to Ciba).

Crystals, mild characteristic odor, bitter taste, mp 112-113°. Freely sol in water (100 g/100 ml), much less sol at low temps. Sol in ethanol, acetone, ethyl acetate, chloroform; very slightly sol in benzene. Aq solns are clear, colorless, and foam profusely on shaking. pH of 10% soln: 6.42; of 1% commercial product at 25°: 5.5;

of 0.1%: 6.8. Surface tension values at 25° (by the capillary rise method) range from 26.75 dynes/cm (10% soln) to 22.08 dynes/cm (0.1% soln). Incompatible with soap.

THERAP CAT: Anti-infective (topical).

3465. Domitroban. [112966-96-8] (5Z)-7-[(1R,2S,3S,4S)-3-[(Phenylsulfonyl)amino]bicyclo[2.2.1]hept-2-yl]-5-heptenoic acid; (+)-(5Z)-7-[3-*endo*-[(phenylsulfonyl)amino]bicyclo[2.2.1]hept-2-*exo*-yl]-5-heptenoic acid; (+)-S-145. $C_{20}H_{27}NO_4S$; mol wt 377.50. C 63.63%, H 7.21%, N 3.71%, O 16.95%, S 8.49%. Thromboxane A_2-receptor antagonist. Prepn of (±)-form: F. Watanabe *et al.*, **EP 226346**; *eidem*, **US 4861913** (1987, 1989 both to Shionogi); and isomers: M. Narisada *et al.*, *J. Med. Chem.* **31**, 1847 (1988). Prepn of (+)-form: M. J. Martinelli, *J. Org. Chem.* **55**, 5065 (1990); M. Ohtani *et al.*, *ibid.* **56**, 2122 (1991). Pharmacology: A. Arimura *et al.*, *Int. Arch. Allergy Immunol.* **98**, 239 (1992). GC determn in plasma: J. Okamoto *et al.*, *J. Chromatogr.* **583**, 45 (1992); in urine: *eidem*, *ibid.* 53. Clinical pharmacokinetics: A. Fujimura *et al.*, *J. Clin. Pharmacol.* **36**, 409 (1996). Clinical evaluation in asthma: M. Fujimura *et al.*, *Pulm. Pharmacol.* **8**, 251 (1995).

Colorless crystals from toluene + hexane, mp 60-62°. uv max (ethanol): 225 nm (ε 5270). $[\alpha]_{589} +28.7°$ (c = 1.00 in methanol).

Calcium salt. [132747-47-8] S-1452; Anboxan. $C_{40}H_{52}CaN_2$-O_8S_2; mol wt 793.06. Prepd as the dihydrate. Colorless pillars, mp >300° (dec). $[\alpha]_D^{26} +19.0 ±0.6°$ (c = 1.010 in methanol).

(±)-**Form.** [115266-92-7] S-145. Colorless plates from ether + *n*-hexane, mp 85-86°.

THERAP CAT: Antiasthmatic.

3466. Domoic Acid. [14277-97-5] (2S,3S,4S)-2-Carboxy-4-[(1Z,3E,5R)-5-carboxy-1-methyl-1,3-hexadien-1-yl]-3-pyrrolidineacetic acid. $C_{15}H_{21}NO_6$; mol wt 311.33. C 57.87%, H 6.80%, N 4.50%, O 30.83%. Excitatory amino acid isolated from the red alga *Chondria armata* Okamura, *Rhodomelaceae*, known in Japanese as "domoi". Structural analog of kainic acid, *q.v.* Shown to be responsible for amnesic shellfish poisoning associated with ingestion of certain cultured blue mussels. Isoln: T. Takemoto, K. Daigo, *Chem. Pharm. Bull.* **6**, 578 (1958); *eidem*, *Arch. Pharm.* **293**, 627 (1960). Series of articles on isoln, structure, and anthelminthic activity: K. Daigo, *Yakugaku Zasshi* **79**, 350-364 (1959), *C.A.* **53**, 14218 (1959). Structural studies: T. Takemoto *et al.*, *ibid.* **86**, 874 (1966), *C.A.* **66**, 28604m (1967). Total synthesis and revised structure: Y. Ohfune, M. Tomita, *J. Am. Chem. Soc.* **104**, 3511 (1982). Neuroexcitatory activity: T. J. Biscoe *et al.*, *Nature* **255**, 166 (1975); R. Zaczek, J. T. Coyle, *Neuropharmacology* **21**, 15 (1982). Effect on kainate receptor activation: G. Debonnel *et al.*, *Can. J. Physiol. Pharmacol.* **67**, 29, 904 (1989). Identification in toxic mussel extracts: J. L. C. Wright *et al.*, *Can. J. Chem.* **67**, 481 (1989). LC determn in shellfish products: J. F. Lawrence *et al.*, *J. Chromatogr.* **462**, 349, 419 (1989). *Reviews*: M. A. Quilliam, J. L. C. Wright, *Anal. Chem.* **61**, 1053A-1059A (1989); J. Clayden *et al.*, *Tetrahedron* **61**, 5713-5724 (2005).

Dihydrate. mp 217° (dec). $[\alpha]_D^{12} -109.6°$ (c = 1.314 in water). uv max: 242 nm (log ε 4.24). pKa in water: 2.10, 3.72, 4.93, 9.82. Sol in water, acetic acid. Insol in methanol, ethanol, chloroform, acetone, benzene.

3467. Domperidone. [57808-66-9] 5-Chloro-1-[1-[3-(2,3-dihydro-2-oxo-1*H*-benzimidazol-1-yl)propyl]-4-piperidinyl]-1,3-dihydro-2*H*-benzimidazol-2-one; 5-chloro-1-[1-[3-(2-oxo-1-benz-imidazolinyl)propyl]-4-piperidyl]-2-benzimidazolinone; R-33812; Bipéridys; Gastronorm; Motilium; Peridon; Motilyo; Péridys. $C_{22}H_{24}ClN_5O_2$; mol wt 425.92. C 62.04%, H 5.68%, Cl 8.32%, N 16.44%, O 7.51%. Dopamine D_2-receptor antagonist with gastro-prokinetic and antiemetic properties. Prepn: J. Vandenberk *et al.*, **DE 2632870**; *eidem*, **US 4066772** (1977, 1978 both to Janssen). Pharmacology: C. Ennis *et al.*, *J. Pharm. Pharmacol.* **31**, Suppl., 14P (1979). Gastrokinetic properties: J. M. Van Neuten *et al.*, *Life Sci.* **23**, 453 (1978). Receptor binding study: M. Baudry *et al.*, *Arch. Pharmacol.* **308**, 231 (1979). UV spectra and structural study: M. Cignitti *et al.*, *J. Mol. Struct.* **350**, 43 (1995). HPLC determn in serum and breast milk: A. P. Zavitsanos *et al.*, *J. Chromatogr. B* **730**, 9 (1999). Clinical study in diabetic gastroparesis: D. Silvers *et al.*, *Clin. Ther.* **20**, 438 (1998). Review of pharmacology, pharmaco-kinetics and therapeutic efficacy: R. N. Brogden *et al.*, *Drugs* **24**, 360-400 (1982); J. A. Barone, *Ann. Pharmacother.* **33**, 429-440 (1999).

Crystals from DMF/water, mp 242.5°. pKa 7.89. Practically insol in water. Log P (lipid/water): 3.90

THERAP CAT: Antiemetic; gastroprokinetic.

THERAP CAT (VET): Gastroprokinetic.

3468. Donepezil. [120014-06-4] 2,3-Dihydro-5,6-dimeth-oxy-2-[[1-(phenylmethyl)-4-piperidinyl]methyl]-1*H*-inden-1-one; 5,6-dimethoxy-2-[[1-(phenylmethyl)-4-piperidinyl]methyl]-2,3-di-hydro-1*H*-inden-1-one; 1-benzyl-4-[(5,6-dimethoxy-1-indanon-2-yl)methyl]piperidine. $C_{24}H_{29}NO_3$; mol wt 379.50. C 75.96%, H 7.70%, N 3.69%, O 12.65%. Reversible acetylcholinesterase inhib-itor. Prepn: H. Sugimoto *et al.*, **EP 296560**, *eidem*, **US 4895841** (1988, 1990 both to Eisai). Synthesis: Y. Iimura *et al.*, *J. Labelled Compd. Radiopharm.* **27**, 835 (1989). HPLC determn of enantio-mers in plasma: J. Haginaka, C. Seyama, *J. Chromatogr. B* **577**, 95 (1992). Clinical pharmacokinetics: A. Ohnishi *et al.*, *J. Clin. Phar-macol.* **33**, 1086 (1993). Clinical trials in Alzheimer's disease: R. C. Mohs *et al.*, *Neurology* **57**, 481 (2001); B. Winblad *et al.*, *ibid.* 489. Review of pharmacology and clinical experience in cognitive disorders: G. C. Román, S. J. Rogers, *Expert Opin. Pharmacother.* **5**, 161-180 (2004).

Hydrochloride. [120011-70-3] E-2020; Aricept. $C_{24}H_{29}NO_3 \cdot$HCl; mol wt 415.96. White crystalline powder. Crystals from meth-anol/IPE, mp 211-212° (dec). Freely sol in chloroform; sol in water, glacial acetic acid; slightly sol in ethanol, acetonitrile. Practically insol in ethyl acetate, *n*-hexane.

THERAP CAT: Nootropic.

3469. Dopa. 3-Hydroxytyrosine; 3-(3,4-dihydroxyphenyl)ala-nine; β-(3,4-dihydroxyphenyl)-α-alanine; 2-amino-3-(3,4-dihy-droxyphenyl)propanoic acid. $C_9H_{11}NO_4$; mol wt 197.19. C 54.82%, H 5.62%, N 7.10%, O 32.45%. An amino acid found in seedlings, pods and beans of *Vicia faba* L. (broad beans) *Stizolobium deeringianum* L. (velvet beans) *Leguminosae:* Torquati, *Arch. farm. sper.* **15**, 213, 308 (1913), *C.A.* **7**, 2774 (1913); Sealock, *Biochemical Preparations* vol. **O** **1**, 25 (1949). Absolute configuration: Guggen-heim, *Z. Physiol. Chem.* **88**, 276 (1913). Prepn of DL-, D-, and L-dopa: Yamada *et al.*, *Chem. Pharm. Bull.* **10**, 693 (1962).

DL-Form. [63-84-3] Prisms from water or from aq $NaHSO_3$ soln, dec 270-272°. Soly in water: 144 mg/40 ml. Readily sol in dil acids and alkalies; slightly sol in benzene, carbon disulfide. Practically insol in abs alcohol, ether, glacial acetic acid, petr ether, chloroform. Oxidizes readily.

D-Form. [5796-17-8] Needles from water, dec 276-278°. $[\alpha]_D^{11}$ +13.0° (c = 5.27 in 1*N* HCl). Soly in water: 66 mg/40 ml.

L-Form *see* Levodopa.

3470. Dopamine. [51-61-6] 4-(2-Aminoethyl)-1,2-benzene-diol; 4-(2-aminoethyl)pyrocatechol; 3-hydroxytyramine; 3,4-dihy-droxyphenethylamine; α-(3,4-dihydroxyphenyl)-β-aminoethane. $C_8H_{11}NO_2$; mol wt 153.18. C 62.73%, H 7.24%, N 9.14%, O 20.89%. Endogenous catecholamine with α and β-adrenergic activ-ity. Isoln from *Hermidium alipes* (S. Watson) *Nyctaginaceae:* Bue-low, Gisvold, *J. Am. Pharm. Assoc.* **33**, 270 (1944). Prepn from aminotyramine: Waser, Sommer, *Helv. Chim. Acta* **6**, 61 (1923). From homoveratrylamine: Schöpf, Bayerle, *Ann.* **513**, 190-202 (1934); Hahn, Stiehl, *Ber.* **69**, 2627-2654 (1936). Comprehensive description of the hydrochloride: J. E. Carter *et al.*, *Anal. Profiles Drug Subs.* **11**, 257-272 (1982). Review of pharmacology and clin-ical efficacy in oliguria: J. F. Dasta, M. G. Kirby, *Pharmacotherapy* **6**, 304 (1986).

Free base, stout prisms, highly sensitive to oxygen; discolors quickly.

Hydrochloride. [62-31-7] ASL-279; Cardiosteril; Dopastat; Dy-natra; Inovan; Inotropin. $C_8H_{11}NO_2 \cdot$HCl; mol wt 189.64. Rosettes of needles from water, dec 241°; may be recrystallized from meth-anol + ether. Freely sol in water, aq solns of alkali hydroxides; sol in methanol, in hot 95% ethanol. Practically insol in ether, petr ether, chloroform, benzene, toluene.

Hydrobromide. $C_8H_{11}NO_2 \cdot$HBr. Crystals, dec 210-214°.

THERAP CAT: Cardiotonic; antihypotensive.

3471. Dopan. [520-09-2] 5-[Bis(2-chloroethyl)amino]-6-methyl-2,4(1*H*,3*H*)-pyrimidinedione; 5-[bis(2-chloroethyl)amino]-6-methyluracil; 6-methyl-5-[bis(2-chloroethyl)amino]uracil; 4-methyl-5-[bis(β-chloroethyl)amino]uracil; 2,6-dihydroxy-4-methyl-5-bis[2-chloroethyl]aminopyrimidine; Elderfield pyrimidine mus-tard; NSC-23436. $C_9H_{13}Cl_2N_3O_2$; mol wt 266.12. C 40.62%, H 4.92%, Cl 26.64%, N 15.79%, O 12.02%. Description: Larionova, Platonova, *Vopr. Onkol.* **1**, no. 5, 36 (1955); *C.A.* **51**, 6862 (1957). Prepn: Nemets *et al.*, **SU 116912** (1959), *C.A.* **53**, 17438i (1959). Outline of synthesis of the demethyl compd: Petering, Lyttle, *Chem. Eng. News* **36**, 47 (Sept. 22, 1958). Monograph: L. F. Larionov, *Cancer Chemotherapy* (Pergamon Press, New York, 1965).

Snow-white crystals, dec 178-179°. Practically insol in cold wa-ter, acetone and benzene. Slightly sol in alc.

USE: In anticancer research: Louis, *J. Chronic Dis.* **15**, 273-281 (March 1962).

3472. Dopexamine. [86197-47-9] 4-[2-[[6-[(2-Phenylethyl)-amino]hexyl]amino]ethyl]-1,2-benzenediol; 4-[2-[[6-(phenethylami-no)hexyl]amino]ethyl]pyrocatechol; FPL-60278. $C_{22}H_{32}N_2O_2$; mol wt 356.51. C 74.12%, H 9.05%, N 7.86%, O 8.98%. Dopamine receptor and β_2-adrenoreceptor agonist. Prepn: J. B. Farmer *et al.*, **EP 72061** (1983 to Fisons). Unlike dopamine, dopexamine has little or no activity at the α- and β_1-adrenoceptors: R. A. Brown *et al.*, *Br. J. Pharmacol.* **85**, 599 (1985). Cardiovascular activity in dogs: *eidem, ibid.* 609. Hemodynamic effects in chronic congestive heart failure: J. R. Dawson *et al.*, *Br. Heart J.* **54**, 313 (1985); G. Svensson *et al.*, *Eur. Heart J.* **7**, 697 (1986). Human renovascular effects: F. Magrini *et al.*, *Eur. J. Clin. Pharmacol.* **32**, 1 (1987). Symposium on pharmacology and clinical efficacy: *Am. J. Cardiol.* **62**, Suppl., 1C-88C (1988).

Dihydrochloride. [86484-91-5] Dopexamine hydrochloride; FPL-60278AR; Dopacard. $C_{22}H_{32}N_2O_2.2HCl$; mol wt 429.43.
Dihydrobromide. $C_{22}H_{32}N_2O_2.2HBr$. Crystals from ethanol, mp 227-228°.

THERAP CAT: Cardiotonic.

3473. Doramectin. [117704-25-3] 25-Cyclohexyl-5-*O*-de-methyl-25-de(1-methylpropyl)avermectin A_{1a}; 25-cyclohexylaver-mectin B_1; UK-67994; Dectomax; DoraQuest. $C_{50}H_{74}O_{14}$; mol wt 899.13. C 66.79%, H 8.30%, O 24.91%. Mutational biosynthetic antiparasitic antibiotic structurally related to the avermectins, *q.v.* Isoln: S. P. Gibson *et al.*, **EP 214731**; *eidem*, **US 5089480** (1987, 1992 both to Pfizer). Mutational biosynthesis: C. J. Dutton *et al.*, *J. Antibiot.* **44**, 357 (1991). Pharmacokinetics and efficacy in cattle: *Vet. Parasitol.* **49**, 1-119 (1993). Clinical evaluation as prophylactic in canine spirocercosis: E. Lavy *et al.*, *Res. Vet. Sci.* **75**, 217 (2003).

mp 116-119°.

THERAP CAT (VET): Antiparasitic.

3474. Doripenem. [148016-81-3] (4*R*,5*S*,6*S*)-3-[[(3*S*,5*S*)-5-[[(Aminosulfonyl)amino]methyl]-3-pyrrolidinyl]thio]-6-[(1*R*)-1-hy-droxyethyl]-4-methyl-7-oxo-1-azabicyclo[3.2.0]hept-2-ene-2-car-boxylic acid; (1*R*,5*S*,6*S*)-2-[(3*S*,5*S*)-5-sulfamoylaminomethyl-pyrrolidin-3-ylthio]-6-[(1*R*)-1-hydroxyethyl]-1-methylcarbapen-2-em-3-carboxylic acid; (1*R*,5*S*,6*S*)-6-[(1*R*)-1-hydroxyethyl]-2-[(3*S*,5*S*)-5-sulfamidomethylpyrrolidin-3-yl]thio-1-methyl-1-carba-2-pe-nem-3-carboxylic acid; S-4661. $C_{15}H_{24}N_4O_6S_2$; mol wt 420.50. C 42.85%, H 5.75%, N 13.32%, O 22.83%, S 15.25%. Broad spectrum 1β-methyl-carbapenem antibiotic. Prepn: Y. Nishitani *et al.*, **EP 528678**; *eidem*, **US 5317016** (1993, 1994 both to Shionogi). Synthesis and structure-activity relationship: Y. Iso *et al.*, *J. Antibiot.* **49**, 199 (1996). Large-scale synthesis: Y. Nishino *et al.*, *Org. Process Res. Dev.* **7**, 846 (2003). HPLC determn in serum: C. Sutherland, D. P. Nicolau, *J. Chromatogr. B* **853**, 123 (2007). Stability against human recombinant dehydropeptidase-I (DHP-I): M. Mori *et al.*, *J. Antimicrob. Chemother.* **37**, 1034 (1996). Comparative antimicro-bial spectrum *in vitro*: T. R. Fritsche *et al.*, *Clin. Microbiol. Infect.* **11**, 974 (2005). Activity against carbapenem-resistant isolates of *Pseudomonas aeruginosa*: S. Sakyo *et al.*, *J. Antibiot.* **59**, 220 (2006). Pharmacokinetics: T. Hori *et al.*, *In Vivo* **20**, 91 (2006). Clinical comparison with meropenem in intra-abdominal infection: C. Lucasti *et al.*, *Clin. Ther.* **30**, 868 (2008). Review of pharmacol-

ogy and clinical experience: E. B. Chahine *et al.*, *Am. J. Health Syst. Pharm.* **67**, 2015-2024 (2010).

Colorless foam.
Hydrate. [364622-82-2] Finibax. $C_{15}H_{24}N_4O_6S_2.H_2O$; mol wt 438.51. Crystalline powder. Sol in water.

THERAP CAT: Antibacterial.

3475. Dorzolamide. [120279-96-1] (4*S*,6*S*)-4-(Ethylamino)-5,6-dihydro-6-methyl-4*H*-thieno[2,3-*b*]thiopyran-2-sulfonamide 7,7-dioxide. $C_{10}H_{16}N_2O_4S_3$; mol wt 324.43. C 37.02%, H 4.97%, N 8.63%, O 19.73%, S 29.65%. Carbonic anhydrase inhibitor. Prepn: J. J. Baldwin *et al.*, **EP 296879**; *eidem*, **US 4797413** (1988, 1989 both to Merck & Co.). Mechanism of action study: R.-F. Wang *et al.*, *Arch. Ophthalmol.* **109**, 1297 (1991). HPLC determn in plasma and urine: B. K. Matuszewski, M. L. Constanzer, *Chiral-ity* **4**, 515 (1992). Clinical evaluations in glaucoma and ocular hy-pertension: E. A. Lippa *et al.*, *Ophthalmology* **98**, 308 (1991); E. A. Lippa *et al.*, *Arch. Ophthalmol.* **110**, 495 (1992). Review of clinical pharmacokinetics: J. Martens-Lobenhoffer, P. Banditt, *Clin. Phar-macokinet.* **41**, 197-205 (2002).

Hydrochloride. [130693-82-2] MK-507; Trusopt. $C_{10}H_{16}N_2$-$O_4S_3.HCl$; mol wt 360.89. White to off-white crystalline powder, mp 283-285°. $[\alpha]_D^{24}$ $-8.34°$ (c = 1 in methanol). $[\alpha]_{405}^{25}$ $\sim$ $-17°$ (c = 1 in water). Sol in water; slightly sol in methanol and ethanol.
Mixture of hydrochloride with timolol maleate. Cosopt. HPLC determn in commercial formulation: N. Erk, *Pharmazie* **58**, 491 (2003). Review of clinical experience in glaucoma and ocular hypertension: J. E. Frampton, C. M. Perry, *Drugs Aging* **23**, 977-995 (2006).

THERAP CAT: Antiglaucoma agent.

3476. Dosmalfate. [122312-55-4] [μ_7-[7-[[6-*O*-[6-Deoxy-2,-3,4-tri-*O*-(sulfo-κ*O*)-α-L-mannopyranosyl]-2,3,4-tri-*O*-(sulfo-κ*O*)-β-D-glucopyranosyl]oxy]-5-hydroxy-2-[4-methoxy-3-[(sulfo-κ*O*)-oxy]pheny]-4*H*-1-benzopyran-4-onato(7$-$)]]tetradeca-μ-hydroxy-heneicosahydroxytetradecaaluminum; diosmin heptakis (hydrogen sulfate) aluminum complex; flavalfate; F-3616; Diotul. $C_{28}H_{60}Al_{14}$-$O_{71}S_7$; mol wt 2134.88. C 15.75%, H 2.83%, Al 17.69%, O 53.21%, S 10.51%. Cytoprotective derivative of diosmin, *q.v.* Prepn: A. Orjales-Venero, R. Mosquera-Pestana, **EP 558435**; *eidem*, **US 5296469** (1993, 1994 both to FAES). *In vivo* protective effect in exptl colitis: I. Villegas *et al.*, *Eur. J. Pharmacol.* **460**, 209 (2003). Series of articles on pharmacology, toxicity and clinical studies as gastroprotectant and antiulcerative: *Drugs Today* **36**, Suppl. A, 25-85 (2000).

R = SO₃[Al₂(OH₅)]

Yellow solid.
THERAP CAT: Gastroprotectant.

3477. Dotarizine. [84625-59-2] 1-(Diphenylmethyl)-4-[3-(2-phenyl-1,3-dioxolan-2-yl)propyl]piperazine; FI-6026. $C_{29}H_{34}$-N_2O_2; mol wt 442.60. C 78.70%, H 7.74%, N 6.33%, O 7.23%. Mixed calcium channel blocker and serotonin (5-HT$_2$) receptor antagonist. Prepn: R. Foguet et al., **EP 97340**; eidem, **US 4883797** (1984, 1989 both to Ferrer); S. Gubert et al., Arzneim.-Forsch. **37**, 1103 (1987). Clinical pharmacokinetics and tolerability: M. Farré et al., Methods Find. Exp. Clin. Pharmacol. **19**, 343 (1997). Cerebrovascular study in goats: G. Torregrosa et al., Arzneim.-Forsch. **49**, 668 (1999); in rabbits: N. Kuridze et al., J. Neurol. Sci. **175**, 13 (2000). Mechanism of blockade: A. Ruiz-Nuno et al., Eur. J. Pharmacol. **411**, 289 (2001).

White crystalline solid from methanol, mp 100-101° (Gubert); also reported as 93-97° (Foguet).
THERAP CAT: Antimigraine.

3478. Dothiepin. [113-53-1] 3-Dibenzo[b,e]thiepin-11(6H)-ylidene-N,N-dimethyl-1-propanamine; N,N-dimethyldibenzo[b,e]-thiepin-$\Delta^{11(6H),\gamma}$-propylamine; 11-(3-dimethylaminopropylidene)-6,11-dihydrodibenzo[b,e]thiepin; dosulepin. $C_{19}H_{21}NS$; mol wt 295.44. C 77.24%, H 7.16%, N 4.74%, S 10.85%. Tricyclic antidepressant. Prepn: M. Protiva et al., Experientia **18**, 326 (1962); **BE 618591**; M. Protiva et al., **US 3527766** (1962, 1970 both to SPOFA); M. Rajsner, M. Protiva, Cesk. Farm. **11**, 404 (1962), C.A. **59**, 2772g (1963). Synthesis of isomers: M. Rajsner et al., Collect. Czech. Chem. Commun. **34**, 1963 (1969). Pharmacology: J. Metysova et al., Arzneim.-Forsch. **13**, 1039 (1963). Clinical pharmacokinetics and metabolism: K. P. Maguire et al., Br. J. Clin. Pharmacol. **12**, 405 (1981). HPLC determn in plasma: P. J. Taylor et al., J. Chromatogr. **581**, 152 (1992). Review of clinical experience in depression: S. Donovan et al., Prog. Neuro-Psychopharmacol. Biol. Psychiatry **18**, 1143-1162 (1994).

bp$_{0.05}$ 171-172°. mp 55-57°.
Hydrochloride. [897-15-4] Prothiaden. $C_{19}H_{21}NS.HCl$; mol wt 331.90. Crystals from ethanol-ether, mp 218-221°. uv max (methanol): 232, 260, 309 nm (log ε 4.41, 3.97, 3.53).
THERAP CAT: Antidepressant.

3479. Dowicide 9®. [13347-42-7] 4-Chloro-2-cyclopentyl-phenol. $C_{11}H_{13}ClO$; mol wt 196.67. C 67.18%, H 6.66%, Cl 18.03%, O 8.13%. Prepn: **DE 615448** (1935 to Hoffmann-La Roche), C.A. **29**, 6248^1 (1935); Pajeau, Begue, Bull. Soc. Chim. Fr. **1962**, 1923. Bactericidal properties: **NL 6513777**; Lorah, **US 3323988** (1966, 1967 both to Dow).

bp$_{18}$ 181-185°; bp$_{11}$ 160-162°.
USE: Germicide.

3480. Doxacurium Chloride. [106819-53-8] rel-(1R,1'S,2S,-2'R)-2,2'-[(1,4-Dioxo-1,4-butanediyl)bis(oxy-3,1-propane-diyl)]bis[1,2,3,4-tetrahydro-6,7,8-trimethoxy-2-methyl-1-[(3,4,5-trimethoxyphenyl)methyl]isoquinolinium] chloride (1:2); (1R,2S,1S,-2R)-1,2,3,4-tetrahydro-2-(3-hydroxypropyl)-6,7,8-trimethoxy-2-methyl-1-(3,4,5-trimethoxybenzyl)isoquinolinium chloride succinate (2:1); trans,trans-2,2'-[dimethylenebis(carbonyloxytrimethylene)]bis[1,2,3,4-tetrahydro-6,7,8-trimethoxy-2-methyl-1-(3,4,5-trimethoxybenzyl)isoquinolinium] dichloride; BW-A938U; Nuromax. $C_{56}H_{78}Cl_2N_2O_{16}$; mol wt 1106.14. C 60.81%, H 7.11%, Cl 6.41%, N 2.53%, O 23.14%. Benzylisoquinolinium, competitive neuromuscular blocker. Prepn: H. A. El-Sayad et al., **EP 54309** (1982 to Wellcome Found.); eidem, **US 4701460** (1987 to Burroughs Wellcome). Clinical pharmacology: S. J. Basta et al., Anesthesiology **69**, 478 (1988). Clinical evaluation during anesthesia: R. P. F. Scott, J. Norman, Br. J. Anaesth. **62**, 373 (1989).

Relative stereochemistry

Amorphous solid. Sol in water.
THERAP CAT: Neuromuscular blocking agent.

3481. Doxapram. [309-29-5] 1-Ethyl-4-[2-(4-morpholinyl)-ethyl]-3,3-diphenyl-2-pyrrolidinone. $C_{24}H_{30}N_2O_2$; mol wt 378.52. C 76.16%, H 7.99%, N 7.40%, O 8.45%. Respiratory stimulant; acts on both peripheral carotid chemoreceptors and central brainstem respiratory sites. Prepn: C. D. Lunsford, A. D. Cale, **BE 613734**; eidem, **US 3192206** (1962, 1965 both to A. H. Robins); C. D. Lunsford et al., J. Med. Chem. **7**, 302 (1964). GC/MS determn in biological fluids: H. Nichol et al., J. Chromatogr. **182**, 191 (1980). Pharmacokinetics in neonatal apnea: M. A. Beaudry et al., Dev. Pharmacol. Ther. **11**, 65 (1988). Laboratory use in the induction of panic attacks: J. L. Abelson et al., Psychoneuroendocrinology **21**, 375 (1996). Clinical evaluation in bradycardia, hypoxemia and apnea of prematurity: C. F. Poets et al., Biol. Neonate **76**, 207 (1999). Toxicity data: E. I. Goldenthal, Toxicol. Appl. Pharmacol. **18**, 185 (1971). Review of clinical use and mechanism of action: C. S. Yost, CNS Drug Rev. **12**, 236-249 (2006).

Hydrochloride monohydrate. [7081-53-0] AHR-619; Dopram; Respiram. $C_{24}H_{30}N_2O_2.HCl.H_2O$; mol wt 432.99. Crystals from isopropyl ether, mp 217-219°. Bitter taste. Sol in water; sparingly sol in alc; slightly sol in chloroform. Practically insol in ether. LD$_{50}$ orally in rats: 261 mg/kg. (Goldenthal)
USE: In laboratory models to induce panic attack.
THERAP CAT: Respiratory stimulant.
THERAP CAT (VET): Respiratory stimulant.

3482. Doxazosin. [74191-85-8] [4-(4-Amino-6,7-dimethoxy-2-quinazolinyl)-1-piperazinyl](2,3-dihydro-1,4-benzodioxin-2-yl)-methanone; 4-amino-2-[4-(1,4-benzodioxan-2-carbonyl)piperazin-1-yl]-6,7-dimethoxyquinazoline; UK-33274. $C_{23}H_{25}N_5O_5$; mol wt 451.48. C 61.19%, H 5.58%, N 15.51%, O 17.72%. Selective α$_1$-adrenergic blocker related to prazosin, q.v. Prepn: S. F. Campbell, **DE 2847623**; idem, **US 4188390** (1979, 1980 both to Pfizer). Cardiovascular pharmacology: P. B. Timmermans et al., Arch. Int.

Pharmacodyn. Ther. **245**, 218 (1980). HPLC determn in plasma: M. G. Cowlishaw, J. R. Sharman, *J. Chromatogr.* **344**, 403 (1985). Clinical pharmacokinetics: H. L. Elliott *et al.*, *Br. J. Clin. Pharmacol.* **13**, 699 (1982). Symposia on pharmacology and clinical efficacy in hypertension: *Br. J. Clin. Pharmacol.* **21**, Suppl. 1, 1S-92S (1986); *Am. J. Cardiol.* **59**, 1G-104G (1987). Review of clinical trials in benign prostatic hypertrophy: C. G. Roehrborn, R. L. Siegel, *Urology* **48**, 406-415 (1996).

Hydrochloride. $C_{23}H_{25}N_5O_5 \cdot HCl$. Crystals, mp 289-290°.

Methanesulfonate. [77883-43-3] Doxazosin mesylate; UK-33274-27; Alfadil; Alfamedin; Benur; Cardenalin; Cardular; Carduran; Carduran; Diblocin; Doxamax; Zoxan. $C_{23}H_{25}N_5O_5 \cdot CH_3SO_3H$; mol wt 547.58. White to tan-colored powder. Freely sol in formic acid; very slightly sol in water, methanol.

THERAP CAT: Antihypertensive; in treatment of benign prostatic hypertrophy.

3483. Doxepin. [1668-19-5] 3-Dibenz[*b,e*]oxepin-11(6*H*)-ylidene-*N,N*-dimethyl-1-propanamine; *N,N*-dimethyldibenz[*b,e*]-oxepin-Δ$^{11(6H)}$,γ-propylamine; 11-(3-dimethylaminopropylidene)-6,11-dihydrodibenz[*b,e*]oxepin; P-3693A. $C_{19}H_{21}NO$; mol wt 279.38. C 81.68%, H 7.58%, N 5.01%, O 5.73%. Serotonin and noradrenaline reuptake inhibitor with antihistaminic activity; tricyclic antidepressant. Prepn of mixture of *cis*- and *trans*-isomers: K. Stach, F. Bickelhaupt, *Monatsh. Chem.* **93**, 896 (1962); F. Bickelhaupt *et al.*, *ibid.* **95**, 485 (1964); NL **6407758**; K. Stach, US **3438981** (1965, 1969 both to Boehringer Mann.); and separation and activity of isomers: B. M. Bloom, J. R. Tretter, BE **641498**; *eidem*, US **3420851** (1964, 1969 both to Pfizer). Pharmacology: A. Ribbentrop, W. Schaumann, *Arzneim.-Forsch.* **15**, 863 (1965). Stereoselective HPLC determn in plasma and urine: J. Yan *et al.*, *J. Chromatogr. B* **691**, 131 (1997). Review of pharmacology and therapeutic efficacy in depression: R. M. Pinder *et al.*, *Drugs* **13**, 161 (1977). Review of clinical efficacy in histamine mediated diseases and insomnia: H. Singh, P. M. Becker, *Expert Opin. Invest. Drugs* **16**, 1295-1305 (2007).

Oily liquid consisting of a mixture of *cis*- and *trans*-isomers. bp$_{0.03}$ 154-157°, bp$_{0.2}$ 260-270°. LD$_{50}$ in mice, rats (mg/kg): 26, 16 i.v.; 79, 182 i.p.; 135, 147 orally (Ribbentrop, Schaumann).

Hydrochloride. [1229-29-4] Adapin; Aponal; Quitaxon; Silenor; Sinequan. $C_{19}H_{21}NO \cdot HCl$; mol wt 315.84. Crystals, mp 184-186°, 188-189°. Freely sol in alcohol.

trans-**Form hydrochloride.** [3607-18-9] mp 192-193°.

cis-**Form hydrochloride.** [25127-31-5] Cidoxepin hydrochloride; P-4599. Crystals, mp 209-210.5°.

THERAP CAT: Antidepressant.

THERAP CAT (VET): Antipruritic.

3484. Doxercalciferol. [54573-75-0] (1*R*,3*S*,5*Z*)-4-Methylene-5-[(2*E*)-2-[(1*R*,3a*S*,7a*R*)-octahydro-7a-methyl-1-[(1*R*,2*E*)-1,-4,5-trimethyl-2-hexen-1-yl]-4*H*-inden-4-ylidene]ethylidene]-1,3-cyclohexanediol; 1α-hydroxyvitamin D$_2$; (1α,3β,5*Z*,7*E*,22*E*)-9,10-secoergosta-5,7,10(19),22-tetraene-1,3-diol; 1-hydroxyergocalcifer-

ol; Hectorol. $C_{28}H_{44}O_2$; mol wt 412.66. C 81.50%, H 10.75%, O 7.75%. Synthetic vitamin D prohormone. Prepn: H.-Y. P. Lam *et al.*, *Science* **186**, 1038 (1974); *eidem*, *Steroids* **30**, 671 (1977); H. E. Paaren *et al.*, *J. Org. Chem.* **45**, 3253 (1980). Comparative activity and toxicity: G. Sjöden *et al.*, *Proc. Soc. Exp. Biol. Med.* **178**, 432 (1985). Metabolism to bioactive form: J. C. Knutson *et al.*, *Endocrinology* **136**, 4749 (1995). Pharmacology: J. W. Coburn *et al.*, *Nephrol. Dial. Transplant.* **11**, Suppl. 3, 153 (1996). Clinical trial for suppression of secondary hyperparathyroidism in hemodialysis: J. M. Frazao *et al.*, *ibid.* **13**, Suppl. 3, 68 (1998).

Crystals, mp 138-140°. uv max (ethanol): 265 nm (ε 18300). LD$_{50}$ orally in rats: 3.5-6.5 mg/kg (Sjöden).

THERAP CAT: Antihyperparathyroid.

3485. Doxifluridine. [3094-09-5] 5'-Deoxy-5-fluorouridine; 1-(β-D-5'-deoxyribofuranosyl)-5-fluorouracil; 5'-DFUR; 5'-dFUrd; Ro-21-9738; Flutron; Furtulon. $C_9H_{11}FN_2O_5$; mol wt 246.19. C 43.91%, H 4.50%, F 7.72%, N 11.38%, O 32.49%. Fluorinated pyrimidine nucleoside with cytostatic activity. Prepn: A. F. Cook, US **4071680** (1978 to Hoffmann-La Roche); H. Hrebabecky, J. Beranek, *Nucleic Acids Res.* **5**, 1029 (1978); A. F. Cook *et al.*, *J. Med. Chem.* **22**, 1330 (1979). Stereospecific synthesis: J. Kiss *et al.*, *Helv. Chim. Acta* **65**, 1522 (1982). Mechanism of action studies: H.-R. Hartmann, A. Matter, *Cancer Res.* **42**, 2412 (1982); R. D. Armstrong *et al.*, *Cancer Chemother. Pharmacol.* **11**, 102 (1983). Kinetics and metabolism in humans: J.-P. Sommadossi *et al.*, *Cancer Res.* **43**, 930 (1983). Clinical trials in colorectal carcinoma: R. Abele *et al.*, *J. Clin. Oncol.* **1**, 750 (1983); S. D. Fossa *et al.*, *Cancer Chemother. Pharmacol.* **15**, 161 (1985). Series of articles on animal toxicology: *Yakuri to Chiryo* **13**, Suppl. 2, 221-430 (1985); acute toxicity: M. Shimizu *et al.*, *ibid.* 209, *C.A.* **104**, 14673z-14678e (1986). Evaluation of neurotoxicity in humans: M. S. Heier, S. D. Fossa, *Acta Neurol. Scand.* **73**, 449 (1986).

Crystals from ethyl acetate, mp 189-190° (Cook). Also reported as crystals from 2-propanol, mp 186-188° (Hrebabecky, Beranek); needles from methanol + ethyl acetate, mp 192-193° (Kiss). pKa 7.4. [α]$_D^{25}$ +18.4° (c = 0.419 in water). uv max (in methanol): 268-269 nm (ε 8550). LD$_{50}$ (14 day) in mice or rats (mg/kg): >1000 i.v.; >2000 s.c.; in male, female mice, male, female rats (mg/kg): >5000, >5000, 3471, 3390 orally (Shimizu).

THERAP CAT: Antineoplastic.

3486. Doxofylline. [69975-86-6] 7-(1,3-Dioxolan-2-ylmethyl)-3,7-dihydro-1,3-dimethyl-1*H*-purine-2,6-dione; 7-(1,3-dioxolan-2-ylmethyl)theophylline; 2-(7'-theophyllinemethyl)-1,3-dioxolane; doxophylline; dioxyfilline; ABC-12/3; Ansimar; Ventax. $C_{11}H_{14^-}$

N_4O_4; mol wt 266.26. C 49.62%, H 5.30%, N 21.04%, O 24.04%. Prepn: U. Avico *et al.*, *Farmaco Ed. Sci.* **17**, 73 (1962). Use as bronchodilator: **BE 868556**; J. S. Franzone, T. Tamietto, **US 4187308** (1978, 1980 to Istituto Biologico Chemioterapico ABC). Pharmacology: J. S. Franzone *et al.*, *Farmaco Ed. Sci.* **36**, 201 (1981). HPLC determn in plasma: A. Lagana *et al.*, *Biomed. Chromatogr.* **4**, 205 (1990). Clinical pharmacokinetics: E. Bologna *et al.*, *J. Int. Med. Res.* **18**, 282 (1990). Review of pharmacology and clinical efficacy in patients with heart disease: F. L. Dini, R. Cogo, *Curr. Med. Res. Opin.* **16**, 258-268 (2001).

Crystals, mp 144-145.5°. Sol in water, acetone, ethyl acetate, benzene, chloroform, dioxane, hot methanol or hot ethanol. Practically insol in ethyl ether or petr ether. LD_{50} in mice (mg/kg): 841 orally; 215.6 i.v.; in rats: 1022.4 orally, 445 i.p. (Franzone).

THERAP CAT: Bronchodilator.

3487. Doxorubicin. [23214-92-8] (8*S*,10*S*)-10-[(3-Amino-2,-3,6-trideoxy-α-L-*lyxo*-hexopyranosyl)oxy]-7,8,9,10-tetrahydro-6,-8,11-trihydroxy-8-(2-hydroxyacetyl)-1-methoxy-5,12-naphthacenedione; 14-hydroxydaunomycin; NSC-123127; FI-106. $C_{27}H_{29}NO_{11}$; mol wt 543.53. C 59.66%, H 5.38%, N 2.58%, O 32.38%. Anthracycline antibiotic; interferes with topoisomerase II function. Isoln from *Streptomyces peucetius* var *caesius*: F. Arcamone *et al.*, **ZA 6802378**; *eidem*, **US 3590028** (1968, 1971 both to Farmitalia); *eidem*, *Biotechnol. Bioeng.* **11**, 1101 (1969). Structural studies: F. Arcamone *et al.*, *Tetrahedron Lett.* **10**, 1007 (1969). Synthesis from daunomycin, q.v.: *eidem*, *Chim. Ind. (Milan)* **51**, 834 (1969). Biochemical comparison with daunomycin: Wang *et al.*, *Proc. Am. Assoc. Cancer Res.* **12**, No. 62, 77 (1971). In acid environment doxorubicin breaks up into **adriamycinone** and **daunosamine**: A. Di Marco *et al.*, *Cancer Chemother. Rep. Part 1* **53**, 33 (1969). Pharmacokinetic and chemotherapeutic studies: E. Arena *et al.*, *Arzneim.-Forsch.* **21**, 1258 (1971). LC/MS/MS determn of doxorubicin and metabolites: R. D. Arnold *et al.*, *J. Chromatogr. B* **808**, 141 (2004). Toxicity study: C. Bertazzoli *et al.*, *Experientia* **26**, 389 (1970). Comprehensive description: A. Vigevani, M. J. Williamson, *Anal. Profiles Drug Subs.* **9**, 245-274 (1980). Book: *Doxorubicin*, F. Arcamone, Ed. (Academic Press, New York, 1981) 369 pp. Review of clinical development: R. H. Blum, S. K. Carter, *Ann. Intern. Med.* **80**, 249-259 (1974); of efficacy in cancer therapy: H. L. Davis, T. E. Davis, *Cancer Treat. Rep.* **63**, 809-815 (1979); of clinical pharmacokinetics: P. A. J. Speth *et al.*, *Clin. Pharmacokinet.* **15**, 15-31 (1988); of mechanism of cardiotoxicity: R. D. Olson, P. S. Mushlin, *FASEB J.* **4**, 3076-3086 (1990); of mechanism of action: G. Aubel-Sadron, D. Londos-Gagliardi, *Biochimie* **66**, 333-352 (1984); D. A. Gewirtz, *Biochem. Pharmacol.* **57**, 727-741 (1999); of liposomal formulations: D. N. Waterhouse *et al.*, *Drug Saf.* **24**, 903-920 (2001).

amorphous powder or solid. Hygroscopic. mp 205° (dec). $[\alpha]_D^{20}$ +248±2° (c = 0.1 in methanol). Absorption max (methanol): 233, 253, 290, 477, 495, 530 nm (ε 38150, 25500, 8400, 13050, 13000, 7200). Apparent partition coefficient (1-octanol/Tris buffer pH 7.0): 0.52. Sol in acetonitrile, THF, water, methanol, isotonic sodium chloride solution. Practically insol in acetone, benzene, chloroform, ethyl ether and petr ether. Aq solns are yellow-orange at acid pHs, orange-red at neutral pHs and violet-blue at pH >9. LD_{50} i.v. in mice: 21.1 mg/kg (Bertazzoli, 1970).

Liposomal complex of the citrate. [111266-55-8] (doxorubicin citrate). Myocet. Doxorubicin citrate encapsulated in liposome carriers composed of egg phosphatidylcholine and cholesterol. Review of development and clinical efficacy in breast cancer: G. Batist *et al.*, *Expert Opin. Pharmacother.* **3**, 1739-1751 (2002).

Liposomal complex of the hydrochloride. Caelyx; Doxil. Doxorubicin HCl encapsulated in liposome carriers composed of *N*-(carbonyl-methoxypolyethylene glycol 2000)-1,2-distearoyl-*sn*-glycero-3-phosphoethanolamine sodium salt, fully hydrogenated soy phosphatidylcholine, and cholesterol. Review of pharmacology and toxicology: P. K. Working, A. D. Dayan, *Hum. Exp. Toxicol.* **15**, 752-785 (1996); of clinical experience in solid and hematological malignancies and AIDS-related Kaposi's sarcoma: M. Sharpe *et al.*, *Drugs* **62**, 2089-2126 (2002); of pharmacokinetics: A. Gabizon *et al.*, *Clin. Pharmacokinet.* **42**, 419-436 (2003). LD_{50} in mice: 38.3±7.2 mg/kg (Working, Dayan).

Caution: Doxorubicin hydrochloride is reasonably anticipated to be a human carcinogen: *Report on Carcinogens, Twelfth Edition* (PB2011-111646, 2011) p 29.

THERAP CAT: Antineoplastic.

THERAP CAT (VET): Antineoplastic.

3488. Doxycycline. [17086-28-1] (monohydrate); [564-25-0] (anhydrous). (4*S*,4a*R*,5*S*,5a*R*,6*R*,12a*S*)-4-(Dimethylamino)-1,4,4a,-5,5a,6,11,12a-octahydro-3,5,10,12,12a-pentahydroxy-6-methyl-1,11-dioxo-2-naphthacenecarboxamide hydrate (1:1); α-6-deoxy-5-hydroxytetracycline monohydrate; α-6-deoxyoxytetracycline monohydrate; 5-hydroxy-α-6-deoxytetracycline monohydrate; GS-3065; Doxirobe; Supracyclin. $C_{22}H_{24}N_2O_8 \cdot H_2O$; mol wt 462.46. C 57.14%, H 5.67%, N 6.06%, O 31.14%. Prepn of family of 6-deoxytetracyclines: C. R. Stephens *et al.*, *J. Am. Chem. Soc.* **80**, 5324 (1958). *See also:* McCormick, Jensen, **US 3019260** (1962 to Am. Cyanamid). Prepn, separation and configuration of 6α- and 6β-epimers: M. S. von Wittenau *et al.*, *J. Am. Chem. Soc.* **84**, 2645 (1962); C. R. Stephens *et al.*, *ibid.* **85**, 2643 (1963). Prepn of 6α-deoxyoxytetracycline: R. K. Blackwood *et al.*, **US 3200149** (1965 to Pfizer). ^{1}H-NMR study: M. S. von Wittenau, R. K. Blackwood, *J. Org. Chem.* **31**, 613 (1966). Biological properties: English, *Proc. Soc. Exp. Biol. Med.* **122**, 1107 (1966). Pharmacology: Fabre, *Chemotherapia* **11**, 73 (1966); Gibaldi, *ibid.* **12**, 265 (1967). Toxicity of hyclate: Goldenthal, *Toxicol. Appl. Pharmacol.* **18**, 185 (1971). Clinical trial in prophylaxis of leptospirosis: E. T. Takafuji *et al.*, *N. Engl. J. Med.* **310**, 497 (1984). Clinical trial in periodontitis: A. M. Polson *et al.*, *J. Periodontol.* **68**, 110, 119 (1997). *Review:* C. Edwards in *Pharmacological and Biochemical Properties of Drug Substances* vol. 2, M. E. Goldberg, Ed. (Am. Pharm. Assoc., Washington, DC, 1979) pp 305-332.

Adriamycinone

Daunosamine

Hydrochloride. [25316-40-9] Adriacin; Adriblastina; Adriamycin. $C_{27}H_{29}NO_{11} \cdot HCl$; mol wt 579.98. Red-orange crystalline or

Yellow crystalline powder. Freely sol in dilute acid, alkali hydroxide solutions; sparingly sol in alc; very slightly sol in water. Practically insol in chloroform, ether.

Hydrochloride hemiethanolate hemihydrate. [24390-14-5] Doxycycline hyclate; Atridox; Azudoxat; Bassado; Clinofug; Diocimex; Doryx; Doxicrisol; Doxylar; Duradoxal; Granudoxy; Mespafin; Retens; Ronaxan; Spanor; Tetradox; Unacil; Vibramycin; Vibra-Tabs; Vibraveineuse; Vibravenös; Zadorin. $C_{22}H_{25}ClN_2O_8 \cdot \frac{1}{2}C_2-$

$H_6O.\frac{1}{2}H_2O$; mol wt 512.94. Light yellow, crystalline powder from ethanol + HCl. Chars without melting at about 201°. $[\alpha]_D^{25}$ $-110°$ (c = 1 in 0.01N methanolic HCl). uv max (0.01N methanolic HCl): 267, 351 nm (log ε 4.24, 4.12). Sol in water, solutions of alkali hydroxides and carbonates; slightly sol in alc. Practically insol in chloroform, ether. The alcohol and water of crystallization are lost by drying at 100° under reduced pressure. More active biologically than the corresponding 6β-epimer hydrochloride (Wittenau, 1962). LD_{50} i.p. in rats: 262 mg/kg (Goldenthal).

THERAP CAT: Antibacterial.

THERAP CAT (VET): Antibacterial.

3489. Doxylamine. [469-21-6] N,N-Dimethyl-2-[1-phenyl-1-(2-pyridinyl)ethoxy]ethanamine; 2-[α-(2-dimethylaminoethoxy)-α-methylbenzyl]pyridine; phenyl-2-pyridylmethyl-β-N,N-dimethyl-aminoethyl ether; 2-dimethylaminoethoxyphenylmethyl-2-picoline. $C_{17}H_{22}N_2O$; mol wt 270.38. C 75.52%, H 8.20%, N 10.36%, O 5.92%. Prepn: C. H. Tilford *et al., J. Am. Chem. Soc.* **70**, 4001 (1948); N. Sperber *et al., ibid.* **71**, 887 (1949). Pharmacology: B. B. Brown, H. Werner, *J. Lab. Clin. Med.* **33**, 325 (1948). GC determn: H. C. Thompson *et al., J. Chromatogr. Sci.* **20**, 373 (1982). Chronic toxicity study: C. D. Jackson, B. Blackwell, *J. Am. Coll. Toxicol.* **12**, 1 (1993). Review of properties and pharmacology: T. J. Haley, *Dangerous Prop. Ind. Mater. Rep.* **2**, 17-20 (1982). Clinical evaluation as hypnotic: F. Sjöqvist, L. Lasagna, *Clin. Pharmacol. Ther.* **8**, 48 (1967); as antihistaminic for use in colds: R. Eccles *et al., J. Pharm. Pharmacol.* **47**, 990 (1995).

Liquid, $bp_{0.5}$ 137-141°. Sol in acids. Slightly volatile, darkens on exposure to light.

Succinate. [562-10-7] Gittalun; Hoggar; Mereprine; Sedaplus; Unisom. $C_{17}H_{22}N_2O.C_4H_6O_4$; mol wt 388.46. White crystals or powder, mp 100-104°. One gram dissolves in 1 ml water, 2 ml alcohol, 2 ml chloroform. Very slightly sol in benzene and ether. pH (1% aq soln): 4.9 to 5.1. LD_{50} in mice, rabbits (mg/kg): 470, 250 orally; 62, 49 i.v.; in mice, male rats, female rats (mg/kg): 460, 440, 445 s.c. (Brown, Werner).

Combination with pyridoxine hydrochloride. Bendectin; Diclectin. Has been used for nausea of pregnancy. Some formulations also contained dicyclomine, *q.v.* Review of therapeutic use and the issue of teratogenicity: L. J. Sheffield, R. Batagol, *Med. J. Aust.* **143**, 143-147 (1985); R. L. Brent, *Reprod. Toxicol.* **9**, 337-349 (1995).

THERAP CAT: Antihistaminic; sedative, hypnotic.

THERAP CAT (VET): Antihistaminic.

3490. DPPF. [12150-46-8] 1,1'-Bis(diphenylphosphino)ferrocene; 1,1'-ferrocenediylbis(diphenylphosphine). $C_{34}H_{28}FeP_2$; mol wt 554.39. C 73.66%, H 5.09%, Fe 10.07%, P 11.17%. Chelating metalloligand used in conjunction with transition metal catalysts in a variety of reactions, including cross-couplings. Distinguished structurally from other bis(diphenylphosphine) ligands, *q.v.*, by its ferrocene backbone. Prepn: G. Marr, T. Hunt, *J. Chem. Soc. C* **1969**, 1070. Improved prepn: J. J. Bishop *et al., J. Organomet. Chem.* **27**, 241 (1971); R.-J. de Lang *et al., Synth. Commun.* **25**, 2989 (1995). Crystal structure: U. Casellato *et al., J. Crystallogr. Spectrosc. Res.* **18**, 583 (1988). Oxidative electrochemistry: C. Nataro *et al., J. Organomet. Chem.* **673**, 47 (2003). Use in prepn of a palladium complex for cross-coupling catalysis: T. Hayashi *et al., J. Am. Chem. Soc.* **106**, 158 (1984). Utility as a ligand in coupling reactions: C. Carfagna *et al., J. Org. Chem.* **56**, 3924 (1991); G. Mann, J. F. Hartwig, *ibid.* **62**, 5413 (1997). Role in allylation catalysis: Y. S. Wagh *et al., Tetrahedron* **67**, 2414 (2011). Review of complexes: G. Bandoli, A. Dolmella, *Coord. Chem. Rev.* **209**, 161-196 (2000).

Yellow-orange needles from benzene-light petroleum, mp 186-188°. Crystal d 1.32. Very sol in chloroform, dichloromethane, ethanol; sol in pentane. Insol in water. Air stable.

USE: Ligand in organic synthesis.

3491. DPQ. [129075-73-6] 3,4-Dihydro-5-[4-(1-piperidinyl)-butoxy]-1(2H)-isoquinolinone. $C_{18}H_{26}N_2O_2$; mol wt 302.42. C 71.49%, H 8.67%, N 9.26%, O 10.58%. Competitive inhibitor of poly (ADP-ribose)polymerase (PARP). Prepn: M. J. Suto *et al.*, **EP 355720**; *eidem*, **US 5177075** (1990, 1993 both to Warner-Lambert). Use and effects as inhibitor: M. J. L. Eliasson *et al., Nat. Med.* **3**, 1089 (1997); K. Takahashi *et al., Brain Res.* **829**, 46 (1999).

Crystals from water, mp 107-109°. Water soluble.

USE: Reagent for inhibition of PARP.

3492. Dragon's Blood. A resinous secretion found on the fruits of *Daemonorops propinquus* Becc., *D. draco* Blume, and probably other species of *Daemonorops, Palmae* (Rattan palms). *Habit.* Sumatra, Borneo, India. *Constit.* About 55% of a red resin contg about 12-15% of bright-yellow, amorphous dracoresene; 2-3% white amorphous dracoalban. Isoln of the main coloring matter, dracorubin: Brockmann, Haase, *Ber.* **69**, 1950 (1936). Chemical studies of resin pigments: Olaniyi *et al., J. Chem. Soc. Perkin Trans. 1* **1973**, 179.

Red sticks, pieces, or cakes; vitreous fracture; makes a bright-crimson powder; odorless and almost tasteless. mp ~120° with sublimation of some benzoic acid. Insol in water; sol in alcohol.

USE: For coloring lacquers and varnishes; occasionally for coloring plasters; in photoengraving on zinc to protect metal parts against etching.

3493. Drimenin. [2326-89-3] (5aS,9aS,9bR)-5,5a,6,7,8,9,9a,-9b-Octahydro-6,6,9a-trimethylnaphtho[1,2-c]furan-1(3H)-one. $C_{15}H_{22}O_2$; mol wt 234.34. C 76.88%, H 9.46%, O 13.65%. From bark of *Drimys winteri* Forst., *Magnoliaceae:* Appel, Dohr, *Scientia* **25**, 137 (1958); *C.A.* **54**, 4663f (1960). Structure and stereochemistry: Appel *et al., J. Chem. Soc.* **1960**, 4685. Synthesis: Wenkert, Strike, *J. Am. Chem. Soc.* **86**, 2044 (1964); Yamagawa, *et al., Synthesis* **1970**, 257; M. Jallali-Naini *et al., Tetrahedron Lett.* **22**, 2995 (1981).

Crystals from methanol and sublimed at 110°/0.1 mm, mp 133°. $[\alpha]_D$ −42° (c = 0.76 in benzene); $[\alpha]_D^{25}$ −35.8° (chloroform). Soluble in organic solvents; practically insol in water, acids, bases.

3494. Drofenine. [1679-76-1] α-Cyclohexylbenzeneacetic acid 2-(diethylamino)ethyl ester; α-phenylcyclohexaneacetic acid 2-(diethylamino)ethyl ester; 2-diethylaminoethyl α-phenylcyclohexaneacetate; hexahydroadiphenine. $C_{20}H_{31}NO_2$; mol wt 317.47. C 75.67%, H 9.84%, N 4.41%, O 10.08%. Prepn: Miescher, Hoffman, **US 2265184**; **US 2265185** (both 1941 to Ciba). Pharmacology and toxicity: Fleisch et al., Arzneim.-Forsch. **11**, 1119 (1961).

bp$_{0.15}$ 158°.
Hydrochloride. [548-66-3] $C_{20}H_{31}NO_2$·HCl; mol wt 353.93. Crystals from alc + petr ether, mp 145-147°. Freely sol in water; very sparingly sol in alc, ether. A 5% aq soln is neutral to litmus. LD$_{50}$ i.v. in mice: 65.6 mg/kg (Fleisch).

THERAP CAT: Antispasmodic.

3495. Droloxifene. [82413-20-5] 3-[(1E)-1-[4-[2-(Dimethylamino)ethoxy]phenyl]-2-phenyl-1-buten-1-yl]phenol; (E)-α-[p-[2-(dimethylamino)ethoxy]phenyl]-α′-ethyl-3-stilbenol; (E)-1-[4′-(2-dimethylaminoethoxy)phenyl]-1-(3-hydroxyphenyl)-2-phenylbut-1-ene; 3-hydroxytamoxifen. $C_{26}H_{29}NO_2$; mol wt 387.52. C 80.59%, H 7.54%, N 3.61%, O 8.26%. Estrogen-receptor antagonist. Prepn: H. Schickaneder et al., **EP 54168**; eidem, **US 5047431** (1982, 1991 both to Klinge); P. C. Ruenitz et al., J. Med. Chem. **25**, 1056 (1982). Symposium on pharmacology and clinical studies: Am. J. Clin. Oncol. **14**, Suppl. 2, S1-S63 (1991). Clinical trial in breast cancer: P. F. Bruning, Eur. J. Cancer **28A**, 1404 (1992).

Colorless crystals from ether, mp 162-163° (Schickaneder); also reported as white crystals from benzene-hexane, mp 160-162° (Ruenitz).
Citrate. [97752-20-0] K-21060E; FK-435. $C_{26}H_{29}NO_2$·$C_6H_8O_7$; mol wt 579.65. Slightly sol in water, methanol, ethanol. Insol in chloroform.

THERAP CAT: Antineoplastic (hormonal).

3496. Drometrizole. [2440-22-4] 2-(2H-Benzotriazol-2-yl)-4-methylphenol; 2-(2H-benzotriazol-2-yl)-p-cresol; 2-(2′-hydroxy-5′-methylphenyl)benzotriazole; Tinuvin P. $C_{13}H_{11}N_3O$; mol wt 225.25. C 69.32%, H 4.92%, N 18.66%, O 7.10%. Prepn: Heller et al., **US 3004896**; **US 3189615** (1961, 1965 to Geigy). Stabilization mechanism in acrylonitrile-butadiene-styrene (ABS) latex: J. B. Adeniyi, J. Polym. Mater. **3**, 25 (1986). HPLC determn in plastic: B. Niklasson, B. Björkner, Contact Dermatitis **21**, 330 (1989). Review of uses: Dunn, Fogg, J. Appl. Polym. Sci. **2**, 367 (1959); of toxicology and cosmetic use: CTFA, J. Am. Coll. Toxicol. **5**, 455-470 (1986).

Minute crystals, mp 131-133°. bp$_{10}$ 225°. Sol in ethyl acetate, acetone, caprolactam solns, dioctylphthalate, oleyl alcohol, hot petrolatum. Stable to conditions and chemicals used in polymerization or compounding of plastics.

USE: An ultraviolet light absorber for stabilizing plastics and other organic materials against discoloration and deterioration.

THERAP CAT: Ultraviolet screen.

3497. Dromostanolone. [58-19-5] (2α,5α,17β)-17-Hydroxy-2-methyl-androstan-3-one; 17β-hydroxy-2α-methyl-5α-androstan-3-one; 2α-methylandrostan-17β-ol-3-one; 2α-methyldihydrotestosterone; drostanolone. $C_{20}H_{32}O_2$; mol wt 304.47. C 78.90%, H 10.59%, O 10.51%. Synthetic estrogen antagonist. Prepn: H. J. Ringold et al., J. Am. Chem. Soc. **81**, 427 (1959); H. J. Ringold, G. Rosenkranz, **US 3118915** (1964 to Syntex). GC-MS determn of urinary metabolites: D. DeBoer et al., J. Steroid Biochem. Mol. Biol. **42**, 411 (1992).

Crystals from acetone/hexane, mp 149-153°. $[\alpha]_D$ +32° (ethanol).
Propionate. [521-12-0] NSC-12198; Drolban; Emdisterone; Masterid; Masteril; Masterone; Permastril. $C_{23}H_{36}O_3$; mol wt 360.54. Crystals from hexane, mp 126-130°. $[\alpha]_D$ +24°.
Note: This is a controlled substance (anabolic steroid): **21 CFR, 1308.13**, as defined in 1300.01.

THERAP CAT: Antineoplastic.

3498. Dronedarone. [141626-36-0] N-[2-Butyl-3-[4-[3-(dibutylamino)propoxy]benzoyl]-5-benzofuranyl]methanesulfonamide; N,N-dibutyl-3-[4-[(2-butyl-5-methylsulfonamido)benzofuran-3-ylcarbonyl]phenoxy]propylamine; SR-33589. $C_{31}H_{44}N_2O_5S$; mol wt 556.76. C 66.88%, H 7.97%, N 5.03%, O 14.37%, S 5.76%. Multi-channel blocker for atrial fibrillation. Noniodinated analog of amiodarone, q.v. Prepn: J. Gubin et al., **EP 471609**; eidem, **US 5223510** (1992, 1993 both to Sanofi). Electrophysiological effects in canine heart: A. Varró et al., Br. J. Pharmacol. **133**, 625 (2001). Comparison of cardiovascular pharmacology and electrophysiology with amiodarone: S. Kathofer et al., Cardiovasc. Drug Rev. **23**, 217 (2005). Clinical trial in atrial fibrillation: S. H. Hohnloser et al., N. Engl. J. Med. **360**, 668 (2009).

Crystals from hexane, mp 65.3°.
Hydrochloride. [141625-93-6] SR-33589B; Multaq. $C_{31}H_{45}ClN_2O_5S$; mol wt 593.22. Colorless solid from acetone, mp 143°.

THERAP CAT: Antiarrhythmic.

3499. Droperidol. [548-73-2] 1-[1-[4-(4-Fluorophenyl)-4-oxobutyl]-1,2,3,6-tetrahydro-4-pyridinyl]-1,3-dihydro-2H-benzimidazol-2-one; 1-[1-[3-(p-fluorobenzoyl)propyl]-1,2,3,6-tetrahydro-4-pyridyl]-2-benzimidazolinone; 1-[1-[4-(p-fluorophenyl)-4-oxobutyl]-1,2,3,6-tetrahydro-4-pyridyl]-2-benzimidazolinone; R-4749; Dehydrobenzperidol; Dridol; Droleptan; Inapsine. $C_{22}H_{22}FN_3O_2$; mol wt 379.44. C 69.64%, H 5.84%, F 5.01%, N 11.07%, O 8.43%. Prepn: P. A. J. Janssen, J. F. Gardocki, **US 3161645**

(1964 to Janssen). Pharmacology and toxicity: J. Yelnosky *et al.*, *Toxicol. Appl. Pharmacol.* **6**, 37 (1964). Comprehensive description: C. A. Janicki, R. K. Gilpin, *Anal. Profiles Drug Subs.* **7**, 171-192 (1978).

White to light tan, amorphous or microcrystalline powder. Freely sol in chloroform; slightly sol in alc, ether. Practically insol in water.
Hydrate. Crystals, mp 145-146.5°. uv max (9:1, 0.1M HCl:methanol): 245, 280 nm (ε 15600, 7500). Soly at 25° (g/100 ml): chloroform 40; DMF 17; benzene 0.55; methanol 0.41; ethanol 0.34; ether 0.24; 0.1M HCl 0.15; water <0.001. pKa 7.64. Heat and light sensitive. LD$_{50}$ in mice (mg/kg): 125 s.c.; 43 i.v. (Yelnosky).
Note: Ingredient of **Thalamonal**, *Innovar*.
THERAP CAT: Antipsychotic.
THERAP CAT (VET): Tranquilizer.

3500. Dropropizine. [17692-31-8] 3-(4-Phenyl-1-piperazinyl)-1,2-propanediol; 1-phenyl-4-(2,3-dihydroxypropyl)piperazine; 1-(2,3-dihydroxypropyl)-4-phenylpiperazine; 1-phenyl-4-(2,3-dihydroxypropyl)diethylenediamine; UCB-1967; Ribex. C$_{13}$H$_{20}$N$_2$O$_2$; mol wt 236.32. C 66.07%, H 8.53%, N 11.85%, O 13.54%. Cough suppressive phenylpiperazine derivative. Prepn: H. G. Morren, **BE 601394**; *eidem*, **US 3163649** (1961, 1964 to UCB); H. Howell *et al.*, *J. Org. Chem.* **27**, 1709 (1962); J. Bourdais, *Bull. Soc. Chim. Fr.* **1968**, 3246. Prepn of enantiomers: M. Borsa *et al.*, **EP 147847**; *eidem*, **US 4699911** (1985, 1987 both to Dompé). LC-MS-MS determn in plasma: Y. Tang *et al.*, *J. Chromatogr. B* **819**, 185 (2005). Pharmacology: K. Cartwright, J. L. Paterson, *J. Pharm. Pharmacol.* **23**, Suppl., 247S (1971). Controlled clinical trials: G. C. Moreo *et al.*, *Gazz. Med. Ital.* **140**, 409 (1981); A. Ravetta, M. Ravetta, *ibid.* **141**, 531 (1982). Chronic oral toxicity: P. R. B. Noel, *Arzneim.-Forsch.* **19**, 1246 (1969).

Crystals from benzene, mp 105° (Morren); also reported as mp 108° (Bourdais). LD$_{50}$ in rats (mg/kg): 200 i.v., 750 orally (Morren).
S-Form. [99291-24-4] Levodropropizine; Danka; Levotuss; Rapitux. White solid from acetone, mp 98-100°. [α]$_D^{25}$ −10° (c = 1.0 in ethanol). LD$_{50}$ in mice, rats (mg/kg): 1287.2, 886.6 orally; 408.0, 401.3 i.p. (Borsa).
R-Form. Crystals from acetone, mp 104-105°. [α]$_D^{25}$ +9.7° (c = 1.0 in ethanol). LD$_{50}$ in mice, rats (mg/kg): 871.7, 721.3 orally; 319.2, 363.4 i.p. (Borsa).
THERAP CAT: Antitussive.

3501. Drosera. Common sundew; round-leafed sundew; youthwort. Air-dried, flowering plant, *Drosera rotundifolia* L., frequently mixed with closely allied species *D. anglica* Hudson and *D. longifolia* L., Droseraceae. *Habit.* Europe, Asia, North America, south to Florida. *Constit.* Malic and citric acids, resin, tannin.

3502. Drospirenone. [67392-87-4] (2'S,6R,7R,8R,9S,10R,13S,14S,15S,16S)-1,3',4',6,7,8,9,10,11,12,13,14,15,16,20,21-Hexadecahydro-10,13-dimethylspiro[17H-dicyclopropa[6,7:15,16]cyclopenta[a]phenanthrene-17,2'(5'H)-furan]-3,5'(2H)-dione; 6β,7β,15β,16β-dimethylene-3-oxo-4-androstene-[17(β-1')-spiro-5']perhydrofuran-2'-one; 6β,7β,15β,16β-dimethylen-3-oxo-17α-pregn-4-ene-21,17-carbolactone; dihydrospirorenone; ZK-30595.

C$_{24}$H$_{30}$O$_3$; mol wt 366.50. C 78.65%, H 8.25%, O 13.10%. Synthetic progestogen exhibiting antimineralocorticoid and antiandrogenic activity. Prepn: R. Wiechert *et al.*, **DE 2652761**; *eidem*, **US 4129564** (both 1978 to Schering AG); D. Bittler *et al.*, *Angew. Chem.* **94**, 718 (1982). HPLC determn in human plasma: W. Krause, U. Jakobs, *J. Chromatogr.* **230**, 37 (1982). Pharmacological profile: P. Muhn *et al.*, *Contraception* **51**, 99 (1995). Review of synthesis: H. Laurent *et al.*, *J. Steroid Biochem.* **19**, 771-776 (1983); of pharmacology and clinical experience: W. Oelkers, *Mol. Cell. Endocrinol.* **217**, 255-261 (2004).

White powder. mp 201.3°. [α]$_D^{22}$ −182° (c = 0.5 in chloroform). uv (methanol): 265 nm (ε 19000). Freely sol in methylene chloride; sol in acetone, methanol; sparingly sol in ethyl acetate, alc. Practically insol in hexane, water.
Mixture with ethinyl estradiol. Angeliq; Yasmin; Yaz. Clinical trial as oral contraceptive: K. S. Parsey, A. Pong, *Contraception* **61**, 105 (2000); in treatment of menopausal symptoms: R. Schürmann *et al.*, *Climacteric* **7**, 189 (2004).
THERAP CAT: Progestogen. In combination with estrogen as oral contaceptive and in treatment of menopausal symptoms.

3503. Drotaverine. [14009-24-6] 1-[(3,4-Diethoxyphenyl)-methylene]-6,7-diethoxy-1,2,3,4-tetrahydroisoquinoline; 1-(3,4-diethoxybenzylidene)-6,7-diethoxy-1,2,3,4-tetrahydroisoquinoline; isodihydroperparine. C$_{24}$H$_{31}$NO$_4$; mol wt 397.52. C 72.52%, H 7.86%, N 3.52%, O 16.10%. Analogue of papaverine, *q.v.* Prepn: **BE 621917** (1962 to Chinoin), *C.A.* **59**, 8713g (1963). Synthesis: E. Koltai *et al.*, *J. Labelled Compd. Radiopharm.* **16**, 351 (1979). Crystal structure: Z. Böcksei *et al.*, *Acta Crystallogr.* **C51**, 1587 (1995). Absorption study: P. Szentmiklosi, S. Marton, *Pharmazie* **38**, 611 (1983). Platelet aggregation and hemorrheological effects: Z. Kapui *et al.*, *Thromb. Res.* **66**, 693 (1992). Clinical pharmacokinetics and bioavailability: O. O. Bolaji *et al.*, *Eur. J. Drug Metab. Pharmacokinet.* **21**, 217 (1996). HPLC determn in biological fluids: *eidem*, *J. Chromatogr.* **622**, 93 (1993). Clinical evaluation: D. Vecsey, *Ther. Hung.* **32**, 127 (1984). Clinical effects on fetal development: A. E. Czeizel, M. Rockenbauer, *Prenatal Neonat. Med.* **1**, 137 (1996).

Acephyllinate. [61954-97-0] Drotaverine theophylline-7-acetic acid; 6,7,3',4'-tetraethoxy-1-benzyl-3,4-dihydroisoquinoline xanthine-7-acetate; DRA; Depogen. C$_{24}$H$_{31}$NO$_4$.C$_9$H$_{10}$N$_4$O$_4$; mol wt 635.72. Exists as monohydrate.
Hydrochloride. [985-12-6] No-Spa. Pale yellow crystals from ethanol, mp 197-200° (uncorr).
THERAP CAT: Antispasmodic.

3504. Droxidopa. [23651-95-8] (βR)-β,3-Dihydroxy-L-tyrosine; L-*threo*-3-(3,4-dihydroxyphenyl)serine; (−)-(2S,3R)-2-amino-

3-hydroxy-3-(3,4-dihydroxyphenyl)propionic acid; *threo*-dopaserine; L-*threo*-DOPS; L-DOPS; SM-5688; Dops. $C_9H_{11}NO_5$; mol wt 213.19. C 50.71%, H 5.20%, N 6.57%, O 37.52%. Synthetic amino acid precursor of norepinephrine, *q.v.* Prepn of racemate: K. W. Rosenmund, H. Dornsaft, *Ber.* **52B**, 1734 (1919). Separation and resolution of diastereomers: B. Hegedüs *et al.*, *Helv. Chim. Acta* **58**, 147 (1975); B. Hegedüs, A. Krasso, US 3920728 (1975 to Hoffmann-La Roche). Improved process for production: N. Ohashi *et al.*, US 4319040 (1982 to Sumitomo). Pharmacology of stereoisomers: G. Bartholini *et al.*, *J. Pharmacol. Exp. Ther.* **193**, 523 (1975). Clinical pharmacology of L-*threo*-form and clinical evaluation in familial amyloid polyneuropathy (FAP): T. Suzuki *et al.*, *Eur. J. Clin. Pharmacol.* **17**, 429 (1980). Reversed-phase chromatography determn in plasma and urine: F. Boomsma *et al.*, *J. Chromatogr.* **427**, 219 (1988). Pharmacokinetics in FAP: T. Suzuki *et al.*, *Eur. J. Clin. Pharmacol.* **23**, 463 (1982); in parkinsonism: T. Suzuki *et al.*, *Neurology* **34**, 1446 (1984). Metabolism to norepinephrine: T. Suzuki *et al.*, *Life Sci.* **36**, 435 (1985). Clinical studies in Parkinson's disease: N. Ogawa *et al.*, *J. Med.* **16**, 525 (1985); H. Narabayashi *et al.*, *Adv. Neurol.* **45**, 593 (1986).

Crystals from ethanol and ether, mp 232-235° (dec). $[\alpha]_D^{20}$ −39° (c = 1 in 1*N* aq HCl). Also cited as crystals from water and L-ascorbic acid, mp 229-232° (dec) (Ohashi). $[\alpha]_D^{20}$ −42.0° (c = 1 in 1*N* aq HCl).

THERAP CAT: Antiparkinsonian.

3505. DSIP. [62568-57-4] Delta sleep-inducing peptide (rabbit); delta sleep peptide; delta sleep factor. $C_{35}H_{48}N_{10}O_{15}$; mol wt 848.82. C 49.53%, H 5.70%, N 16.50%, O 28.27%. A nonapeptide that shows enhancement and induction of delta (slow-wave) and spindle EEG patterns. Its occurrence was suspected during dialysis of cerebral venous blood of rabbits during sleep induced by electrical stimulation of the thalamus: M. Monnier, L. Hösli, *Science* **146**, 796 (1964). Initial isoln: *eidem*, *Pfluegers Arch.* **282**, 60 (1965). Isoln, characterization: G. A. Schoenenberger *et al.*, *Experientia* **28**, 919 (1972). Amino acid sequence, synthesis of DSIP and analogs: G. A. Schoenenberger, M. Monnier, *Proc. Natl. Acad. Sci. USA* **74**, 1282 (1977). Solid phase synthesis: Y. P. Shvachkin *et al.*, *Zh. Obshch. Khim.* **51**, 719 (1981), *C.A.* **95**, 43644s (1981). Rapid liquid phase synthesis: S. Nozaki, I. Muramatsu, *Bull. Chem. Soc. Jpn.* **55**, 2165 (1982). HPLC separation: M. Dizaroglu *et al.*, *J. Chromatogr.* **237**, 417 (1982). Effect on human sleep: D. Schneider-Helmert *et al.*, *Lancet* **1**, 1256 (1981); *eidem*, *Int. J. Clin. Pharmacol. Ther. Toxicol.* **19**, 341 (1981); D. Schneider-Helmert, G. A. Schoenenberger, *Experientia* **37**, 913 (1981).

Trp–Ala–Gly–Gly–Asp–Ala–Ser–Gly–Glu

3506. DSPA. Vampire bat salivary plasminogen activator; BatPA; v-PA. Plasminogen activator naturally occurring in the saliva of the vampire bat, *Desmodus rotundus*. Four isoforms have been identified. All are single chain peptides that are structurally homologous to human tissue plasminogen activator, *q.v.* DSPA-α1 and -α2 are comprised of 477 amino acid residues containing 4 distinct domains; isoforms β and γ are smaller, truncated versions. DSPA-α1 and -α2 bind to fibrin; all isoforms require fibrin as a cofactor for plasminogen activation. Isoln: C. Hawkey, *Nature* **211**, 434 (1966). Cloning, characterization and structure: S. J. Gardell *et al.*, *J. Biol. Chem.* **264**, 17947 (1989). Identification of isoforms: J. Krätzschmar *et al.*, *Gene* **105**, 229 (1991). Review of purification, properties and activity: S. J. Gardell, P. A. Friedman, *Methods Enzymol.* **223**, 233-249 (1993). Study of structural mechanism of catalytic activity: M. Renatus *et al.*, *Biochemistry* **36**, 13483 (1997).

Desmoteplase. [145137-38-8] Plasminogen activator (Desmodus rotundus isoform α1 protein moiety reduced); rDSPA alpha 1. DSPA-α1 produced by recombinant technology; mol wt 55-58 kDa. Prepn: B. Baldus *et al.*, WO 9009438; *eidem*, US 6008019 (1990, 1999 both to Schering AG); T. Petri *et al.*, *J. Biotechnol.* **39**, 75

(1995). CE/MS method for pharmaceutical formulations: J. A. Chakel *et al.*, *J. Chromatogr. B* **689**, 215 (1997). Pharmacology: W.-D. Schleuning, *Haemostasis* **31**, 118 (2001). Clinical evaluation in acute ischemic stroke: W. Hacke *et al.*, *Stroke* **36**, 66 (2005). *Review*: T. Debens, *Curr. Opin. Mol. Ther.* **6**, 567-575 (2004).

THERAP CAT: Thrombolytic.

3507. DTAF. [51306-35-5] 5-[(4,6-Dichloro-1,3,5-triazin-2-yl)amino]-3′,6′-dihydroxyspiro[isobenzofuran-1(3*H*),9′-[9*H*]xanthen]-3-one; 5-(4,6-dichlorotriazinyl)aminofluorescein. $C_{23}H_{12}Cl_2N_4O_5$; mol wt 495.27. C 55.78%, H 2.44%, Cl 14.32%, N 11.31%, O 16.15%. Fluorochrome labelling agent related to fluorescein, *q.v.* Prepn: V. E. Barskii *et al.*, *Izv. Akad. Nauk Ser. Biol.* **1968**, 744, *C.A.* **70**, 79127 (1969); and use as label: D. Blakeslee, M. G. Baines, *J. Immunol. Methods* **13**, 305 (1976). Practical comparison with FITC, *q.v.*, for labelling: S. P. D. Lalljie, P. Sandra, *Chromatographia* **40**, 519 (1995). Use as label for carbohydrates and proteins: R. Schumann, D. Rentsch, *Mar. Ecol. Prog. Ser.* **163**, 77 (1998); for poloxamers, *q.v.*: F. Ahmed *et al.*, *Langmuir* **17**, 537 (2001). Ultrasensitive chromatographic/fluorescent determn of amino acid herbicides: M. Molina, M. Silva, *Electrophoresis* **23**, 1096 (2002).

Bright yellow powder. Very stable when dry; hydrolyzes in alkaline solutions (pH 9.0). Absorption max: 492 nm; emission max: 513 nm (c = 5 × 10⁻⁷ M in 0.1 *M* sodium borate).

USE: Fluorescent reagent for conjugation of fluorescein to proteins, carbohydrates and selected polymers.

3508. DTBP. [110-05-4] Bis(1,1-Dimethylethyl) peroxide; di-*tert*-butyl peroxide. $C_8H_{18}O_2$; mol wt 146.23. C 65.71%, H 12.41%, O 21.88%.

Flammable liq; d_4^{20} 0.7940; mp −40°; bp_{284} 80°; n_D^{20} 1.3890. Flash pt (Tag open cup) 65°F (18°C). Soluble in organic solvents, in most resin monomers and in partial polymers. Soly in water about 0.01%.

USE: As polymerization catalyst.

3509. Dubnium. [53850-35-4] Element 105; hahnium; nielsbohrium; unnilpentium. Db, Ha, Ns, Unp; at. no. 105. Group VB(5). No stable nuclides. Prepn of α-emitting ²⁶⁰105 isotope (T½ 1.6±0.3 sec) by ²⁴⁹Cf (¹⁵N,4n): A. Ghiorso *et al.*, *Phys. Rev. Lett.* **24**, 1498 (1970). Prepn of ²⁶¹105 isotope (decay by spontaneous fission, T½ 1.8 ±0.6 sec) by ²⁴³Am (²²Ne,4n): G. N. Flerov *et al.*, *At. Energ.* **29**, 243 (1970); *eidem*, *Nucl. Phys. A* **160**, 181 (1971); *eidem*, *Proc. Int. Conf. Heavy Ion Phys.* **1971** pp 125-143, *C.A.* **80**, 139758y (1974). Prepn of α-emitting isotopes ²⁶¹105 (T½ 1.7 sec) by ²⁴⁹Bk (¹⁶O,4n) or by ²⁵⁰Cf (¹⁵N,4n); ²⁶²105 (T½ 40.0 ±10 sec, longest-lived known isotope, rel. at. mass 262.1144) by ²⁴⁹Bk(¹⁸O,5n): A. Ghiorso *et al.*, *Phys. Rev.* **C4**, 1850 (1971); and revised T½ 34.1 ±4.6 sec for ²⁶²105: C. E. Bemis, Jr. *et al.*, *Phys. Rev. Lett.* **39**, 1246 (1977). Discussion of conflicting claims of discovery: Holcomb, *Science* **168**, 810 (1970); G. N. Flerov, *ibid.* **170**, 15 (1970); A. Ghiorso, *ibid.* **171**, 127 (1971). Solution chemistry: K. E. Gregorich *et al.*, *Radiochim. Acta* **43**, 223 (1988); M. K. Gober *et al.*, *ibid.* **57**, 77 (1992); and identification of ²⁶³105 isotope (T½ 27 sec, α-emitter) by ²⁴⁹Bk(¹⁸O,-4n): M. Schädel *et al.*, *ibid.* **85**. Reviews of history, prepn and properties: C. Keller, *The Chemistry of the Transuranium Elements* (Verlag Chemie, Weinheim, English Ed., 1971) pp 619-622; Silva, "Trans-Curium Elements" in *MTP Int. Rev. Sci.: Inorg. Chem., Ser. One* vol. **8**, A. G. Maddock, Ed. (University Park Press, Baltimore, 1972) pp 71-105; R. J. Silva in *The Chemistry of the Actinide Elements* vol. **2**, J. J. Katz *et al.*, Eds. (Chapman and Hall, New York, 1986) pp 1106-1109; E. K. Hyde *et al.*, *Radiochim. Acta* **42**, 57-102 (1987); Transfermium Working Group, *Pure Appl. Chem.* **65**, 1757-1814 (1993). Review of chemistry: D. C. Hoffman, *Proc. Robert A. Welch Found. Conf. on Chem. Res., XXXIV, Fifty Years with Transuranium Elements* (Houston, Texas, 1990) pp 255-276.

3510. Dulcamara. Bittersweet; woody nightshade; scarlet berry. Dried stems of *Solanum dulcamara* L., *Solanaceae*. *Habit.* Europe, Western Asia, Northern Africa, natural in U.S. *Constit.* Solaniceine (about 1%), dulcamarin, dulcamaric and dulcamaretic acids.

3511. Dulcin. [150-69-6] *N*-(4-Ethoxyphenyl)urea; *p*-phenetolcarbamide; *p*-phenetylurea; Sucrol; Valzin. $C_9H_{12}N_2O_2$; mol wt 180.21. C 59.99%, H 6.71%, N 15.55%, O 17.76%. Made by treating *p*-phenetidine with phosgene and then with ammonia: Berlinerblau, *J. Prakt. Chem.* **30**, 103 (1883); from *p*-phenetidine and urea: Kurzer, *Org. Synth.* **coll. vol. IV**, 52 (1963).

Lustrous needles. Very sweet taste, about 250 times as sweet as cane sugar. mp 173-174°. Sol in 800 parts cold water, 50 parts boiling water, 25 parts alcohol.

USE: Non-nutritive sweetener.

3512. Duloxetine. [116539-59-4] (*γS*)-*N*-Methyl-*γ*-(1-naphthalenyloxy)-2-thiophenepropanamine; (+)-(*S*)-*N*-methyl-*γ*-(1-naphthyloxy)-2-thiophenepropylamine; (+)-*N*-methyl-3-(1-naphthalenyloxy)-3-(2-thienyl)propanamine; LY-248686. $C_{18}H_{19}NOS$; mol wt 297.42. C 72.69%, H 6.44%, N 4.71%, O 5.38%, S 10.78%. Selective serotonin and norepinephrine reuptake inhibitor (SNRI). Prepn: D. W. Robertson *et al.*, *EP 273658*; *eidem*, *US 5023269* (1988, 1991 both to Lilly); in abs config: J. Deeter *et al.*, *Tetrahedron Lett.* **31**, 7101 (1990). Improved process: R. A. Berglund, *US 5362886* (1994 to Lilly). Pharmacology: D. T. Wong *et al.*, *Neuropsychopharmacology* **8**, 23 (1993). Neurochemical effects *in vivo*: R. W. Fuller *et al.*, *J. Pharmacol. Exp. Ther.* **269**, 132 (1994). Determn of chiral purity: E. C. Rickard, R. J. Bopp, *J. Chromatogr. A* **680**, 609 (1994). Clinical pharmacokinetics: A. Sharma *et al.*, *J. Clin. Pharmacol.* **40**, 161 (2000). Evaluation of suicidality risk: N. Acharya *et al.*, *J. Clin. Psychopharmacol.* **26**, 587 (2006). Clinical trial in diabetic neuropathic pain: D. K. Kajdasz *et al.*, *Clin. Ther.* **29**, 2536 (2007); in fibromyalgia: L. M. Arnold *et al.*, *J. Womens Health* **16**, 1145 (2007). Review of mechanism of action: K. B. Thor *et al.*, *Int. J. Clin. Pract.* **61**, 1349-1355 (2007). Review of clinical experience in depression: M. Bauer *et al.*, *Expert Opin. Pharmacother.* **7**, 421-427 (2006); in stress urinary incontinence: P. Mariappan *et al.*, *Eur. Urol.* **51**, 67-74 (2007); in generalized anxiety disorder: C. Allgulander *et al.*, *Curr. Med. Res. Opin.* **23**, 1245-1252 (2007). Review of clinical experience in treatment of chronic pain: A. Wright *et al.*, *J. Pain Res.* **4**, 1-10 (2011).

Hydrochloride. [136434-34-9] Ariclaim; Cymbalta; Yentreve. $C_{18}H_{19}NOS.HCl$. White to slightly brownish white solid. pKa in DMF-water (66:34): 9.6. Slightly sol in water.

THERAP CAT: Antidepressant; anxiolytic. In treatment of stress urinary incontinence; in diabetic peripheral neuropathic pain; in fibromyalgia.

3513. DuPHOS Ligands. Series of C_2 symmetric, bidentate bis(phospholane) ligands; rhodium complexes serve as chiral catalysts for asymmetric reactions. Structurally distinct from BPE ligands, *q.v.*; the phenyl ring bridge reduces conformational mobility of the ligand. Prepn and use in enantioselective hydrogenations: M. J. Burk, *J. Am. Chem. Soc.* **113**, 8518 (1991); *idem*, *US 5171892* (1992 to Du Pont); *idem et al.*, *J. Am. Chem. Soc.* **115**, 10125 (1993). Asymmetric hydrogenation applications in the synthesis of warfarin: A. Robinson *et al*, *Tetrahedron Lett.* **37**, 8321 (1996); in the prepn of β-amino acid derivatives: G. Zhu *et al.*, *J. Org. Chem.* **64**, 6907 (1999). Utility in enantioselective copolymerizations: Z. Jiang, A. Sen, *J. Am. Chem. Soc.* **117**, 4455 (1995); in asymmetric cycloisomerizations: S. R. Gilbertson *et al.*, *J. Org. Chem.* **63**, 10077 (1998); in asymmetric cycloadditions: M. Murakami *et al.*, *J. Am. Chem. Soc.* **121**, 4130 (1999). Review of phospholane ligand chemistry: M. J. Burk, *Acc. Chem. Res.* **33**, 363-372 (2000).

(*R,R*)-Me-DuPHOS: R = CH₃
(*R,R*)-Et-DuPHOS: R = CH₂CH₃

(*R,R*)-Me-DuPHOS. [147253-67-6] (2*R*,2'*R*,5*R*,5'*R*)-1,1'-(1,2-Phenylene)bis[2,5-dimethylphospholane]; (−)-1,2-bis[(2*R*,5*R*)-2,5-dimethylphospholano]benzene. $C_{18}H_{28}P_2$; mol wt 306.37. C 70.57%, H 9.21%, P 20.22%. White solid. Air and moisture sensitive. Store under inert gas.

(*S,S*)-Me-DuPHOS. [136735-95-0] Colorless crystals from methanol, mp 79-81°. $[α]_D^{25}$ +476 ± 5° (c = 1 in hexane). Air sensitive. Store under inert gas.

(*R,R*)-Et-DuPHOS. [136705-64-1] (2*R*,2'*R*,5*R*,5'*R*)-1,1'-(1,2-Phenylene)bis[2,5-diethylphospholane]; (−)-1,2-bis[(2*R*,5*R*)-2,5-diethylphospholano]benzene. $C_{22}H_{36}P_2$; mol wt 362.48. C 72.90%, H 10.01%, P 17.09%. Colorless oil, bp$_{0.045}$ 138-145°. $[α]_D^{25}$ −265° (c = 1 in hexane). Air and moisture sensitive. Store under inert gas.

(*S,S*)-Et-DuPHOS. [136779-28-7] Clear, viscous liquid. Sol in most organic solvents. *Irritant*. Air sensitive.

USE: Ligands in organic chemistry.

3514. Durapatite. [1306-06-5] Hydroxylapatite (Ca₅(OH)(PO₄)₃); calcium phosphate hydroxide; calcium orthophosphate basic; hydroxyapatite; Alveograf; Ossopan; Periograf. 3Ca₃(PO₄)₂.Ca(OH)₂ or Ca₁₀(PO₄)₆(OH)₂. Also considered as pentacalcium monohydroxyorthophosphate Ca₅(OH)(PO₄)₃. Calcd as Ca₁₀H₂O₂₆P₆; Ca 39.89%, H 0.20%, O 41.41%, P 18.50%. Occurs as a mineral in phosphate rock. Constitutes the mineral portion of bone. Prepn from Ca(NO₃)₂ and KH₂PO₄: Warington, *J. Chem. Soc.* **26**, 983 (1873); Rathje, *Ber.* **74**, 342 (1941); Hayek in *Handbook of Preparative Inorganic Chemistry*, G. Brauer, Ed. (Academic Press, 2nd ed., 1963) p 545; from calcium phosphate, dibasic: Perloff, Posner, *Inorg. Synth.* **6**, 16 (1960); from Ca(NO₃)₂.4H₂O and (NH₄)₃PO₄ plus NH₄OH: Hayek, Newesely, *ibid.* **7**, 63 (1963). Formation and structure of synthetic bone hydroxyapatites: A. S. Posner *et al.*, *Prog. Cryst. Growth Charact.* **3**, 3 (1980).

Hexagonal needles arranged in rosettes. Dec above 1100°. Practically insol in water, even when freshly prepd. Crystallographic data: a₀ 9.425; C₀ 6.935; C₀/a₀ 0.736.

USE: Prosthetic aid (artificial bone and teeth).

THERAP CAT: Calcium supplement; phosphorus supplement.

3515. Durene. [95-93-2] 1,2,4,5-Tetramethylbenzene; Durol. $C_{10}H_{14}$; mol wt 134.22. C 89.49%, H 10.51%. Occurs in coal tar. Usually prepd from xylene and methyl chloride in the presence of AlCl₃: Smith, *Org. Synth.* **vol. 10**, 32 (1930); *cf.* Smith, Dobrovolny, *J. Am. Chem. Soc.* **48**, 1413 (1926).

Scales with camphor-like odor from alcohol. d_4^{81} 0.84. mp 80°. bp 191-193°. Sublimes and is volatile with steam. Insol in water; freely sol in alcohol, ether, benzene.

3516. Durohydroquinone. [527-18-4] 2,3,5,6-Tetramethyl-1,4-benzenediol; tetramethyl-*p*-hydroquinone; dihydroxydurene. $C_{10}H_{14}O_2$; mol wt 166.22. C 72.26%, H 8.49%, O 19.25%. For prepn *see* refs under Duroquinone.

Needles from alcohol. mp 233°. Begins to sinter at 220°. Sparingly sol in ether. Treatment with ferric chloride yields duroquinone. **Diacetyldurohydroquinone.** Needles from alc, mp 207°.

3517. Duroquinone. [527-17-3] 2,3,5,6-Tetramethyl-2,5-cyclohexadiene-1,4-dione; tetramethyl-*p*-benzoquinone. $C_{10}H_{12}O_2$; mol wt 164.20. C 73.15%, H 7.37%, O 19.49%. Prepn by reduction of dinitrodurene: Smith, *Org. Synth.* **vol. 10**, 40 (1930); Smith, Dobrovolny, *J. Am. Chem. Soc.* **48**, 1420 (1926); by condensation of 2,3-diketopentane with itself in presence of alkalies: von Pechmann, *Ber.* **21**, 1420 (1888); by the action of alkalies on 3,3-dichloropentan-2-one: Faworsky, *J. Prakt. Chem.* [2] **51**, 538 (1895).

Yellow needles from alc, mp 111-112°. Sublimes. Volatile with steam. Insol in water; sol in alcohol, benzene, ether, hot petr ether.

3518. Dutasteride. [164656-23-9] (4a*R*,4b*S*,6a*S*,7*S*,9a*S*,9b*S*,11a*R*)-*N*-[2,5-Bis(trifluoromethyl)phenyl]-2,4a,4b,5,6,6a,7,8,9,-9a,9b,10,11,11a-tetradecahydro-4a,6a-dimethyl-2-oxo-1*H*-indeno-[5,4-*f*]quinoline-7-carboxamide; 17β-*N*-[2,5-Bis(trifluoromethyl)-phenyl]carbamoyl-4-aza-5α-androst-1-en-3-one; GG-745; GI-198745; Avodart; Avolve. $C_{27}H_{30}F_6N_2O_2$; mol wt 528.54. C 61.36%, H 5.72%, F 21.57%, N 5.30%, O 6.05%. Dual inhibitor of 5α-reductase isoenzymes types 1 and 2; structurally related to finasteride, *q.v.* Prepn: K. W. Batchelor, S. V. Frye, **WO 9507927** (1995 to Glaxo). Structure-activity study: R. K. Bakshi *et al.*, *J. Med. Chem.* **38**, 3189 (1995). Clinical pharmacokinetics: P. O. Gisleskog *et al.*, *Br. J. Clin. Pharmacol.* **47**, 53 (1999). Clinical trial in benign prostatic hyperplasia: C. G. Roehrborn *et al.*, *Urology* **60**, 434 (2002). Review of discovery and development: S. V. Frye *et al.*, *Pharm. Biotechnol.* **11**, 393-422 (1998); of clinical experience: B. Djavan *et al.*, *Expert Opin. Pharmacother.* **6**, 311-317 (2005).

White crystalline solid, mp 245-245.5°.
THERAP CAT: In treatment of benign prostatic hyperplasia.

3519. Duthaler-Hafner Reagent. [132068-98-5] Chloro(η^5-2,4-cyclopentadien-1-yl)[(4*R*,5*R*)-2,2-dimethyl-α4,α4,α5,α5-tetraphenyl-1,3-dioxolane-4,5-dimethanolato(2−)-κO^4,κO^5]titanium. $C_{36}H_{33}ClO_4Ti$; mol wt 612.97. C 70.54%, H 5.43%, Cl 5.78%, O 10.44%, Ti 7.81%. Enantioselective allyl-transfer reagent. Prepn: R. O. Duthaler *et al.*, *Pure Appl. Chem.* **62**, 631 (1990). As reagent in prepn of titanium transmetallation complexes: A. Hafner *et al.*, *J. Am. Chem. Soc.* **114**, 2321 (1992). Use in aldol reactions: R. C. Cambie *et al.*, *Aust. J. Chem.* **46**, 583 (1993); in asymmetric allylation: A. Fürstner, K. Langemann, *J. Am. Chem. Soc.* **119**, 9130 (1997). Review of prepn, structure and use: R. O. Duthaler *et al.*, in *Proc. 3rd Symp. Org. Synth. Organomet. 1990*, K. H. Doetz, R. W. Hoffmann Eds. (Vieweg, Braunschweig, Germany, 1991) pp 285-309.

mp 209-213°. $[\alpha]_D^{20}$ −246° (c = 1 in CHCl$_3$). Moisture sensitive, store under nitrogen. *Corrosive*.
(S,S)-Form. [140462-73-3] mp 209-213°. $[\alpha]_D^{20}$ +246° (c = 1 in CHCl$_3$). Moisture sensitive, store under nitrogen. *Corrosive*.
USE: Transmetallation of allyl-Grignard or allyl-Li compounds.

3520. Dyclonine. [586-60-7] 1-(4-Butoxyphenyl)-3-(1-piperidinyl)-1-propanone; 3-piperidino-4′-butoxypropiophenone; β-piperidinoethyl-4-butoxyphenyl ketone; 4-butoxy-β-piperidinopropiophenone; 4-*n*-butoxy-β-(1-piperidyl)propiophenone; 4-butoxyphenyl piperidineethyl ketone; 2-(1-piperidyl)ethyl *p*-butoxyphenyl ketone. $C_{18}H_{27}NO_2$; mol wt 289.42. C 74.70%, H 9.40%, N 4.84%, O 11.06%. Prepd from *p*-butoxyacetophenone by condensation with formaldehyde and piperidine hydrochloride: Pofft, *Chem. Tech. (Berlin)* **4**, 241 (1952), *C.A.* **47**, 10531 (1953).

Hydrochloride. [536-43-6] Dyclone; Tanaclone. $C_{18}H_{27}NO_2$.-HCl; mol wt 325.88. White crystals or crystalline powder. mp 175-176°. Sol in water, alc, acetone, chloroform. Phenol coefficient 3.6.
THERAP CAT: Anesthetic (local).

3521. Dydrogesterone. [152-62-5] (9β,10α)-Pregna-4,6-diene-3,20-dione; 10α-pregna-4,6-diene-3,20-dione; 6-dehydro-*retro*-progesterone; 10α-isopregnenone; Dufaston; Duphaston; Gynorest; Prodel; Retrone. $C_{21}H_{28}O_2$; mol wt 312.45. C 80.73%, H 9.03%, O 10.24%. Prepn: Westerhof, Reerink, *Rec. Trav. Chim.* **79**, 771 (1960); Rappoldt, Westerhof, *ibid.* **80**, 43 (1961).

Consult the Name Index before using this section. Page 635

Crystals from acetone + hexane, mp 169-170°. $[\alpha]_D^{25}$ −484.5° (chloroform). uv max: 286.5 nm (ε 26400).

THERAP CAT: Progestogen.

3522. Dymanthine. [124-28-7] N,N-Dimethyl-1-octadecanamine; N,N-dimethyloctadecylamine; dimantine. $C_{20}H_{43}N$; mol wt 297.57. C 80.73%, H 14.57%, N 4.71%. Prepd from octadecylamine and formaldehyde: Reck et al., J. Org. Chem. **12**, 517 (1947).

mp 22.89°.

Hydrochloride. [1613-17-8] GS-1339; NSC-5547; Thelmesan. $C_{20}H_{43}N.HCl$; mol wt 334.03.

THERAP CAT: Anthelmintic (Nematodes).

THERAP CAT (VET): Anthelmintic (Nematodes).

3523. Dynamin. GTPase mechanoenzyme which plays an essential role in membrane budding and vesicle trafficking. Various isoforms which are tissue or function specific co-exist. Homotetramer which self-assembles into ring and spirals which form collars at the fission point. Isolation from bovine brain: H. S. Shpetner, R. B. Vallee, Cell **59**, 421, (1989). Model for mechanism of membrane fission: M. M. Kozlov, Biophys. J. **77**, 604 (1999). Review of function and nomenclature: R. Urrutia et al., Proc. Natl. Acad. Sci. USA **94**, 377-384 (1997); of functions: M. A. McNiven et al., Trends Biochem. Sci. **25**, 115-120 (2000). Review of structure and assembly: J. E. Hinshaw, Curr. Opin. Struct. Biol. **9**, 260-267 (1999). Review of functional diversity and related dynamin-like proteins: A. M. van der Bliek, Trends Cell Biol. **9**, 96-102 (1999).

3524. Dynel. Staple fiber copolymer of 60% vinyl chloride and 40% acrylonitrile wet spun from acetone, stretched hot as much as 1300% and then annealed by heat treatment: E. E. Stout, Introduction to Textiles (John Wiley, New York, 1960) pp 198-201; R. W. Moncrieff, Man-Made Fibres (John Wiley, New York, 1963) pp 411-420; Kennedy, "Modacrylic Fibers" in Encyclopedia of Polymer Science and Technology vol. 8, N. M. Bikales, Ed. (Interscience, New York, 1968) pp 812-839.

Light cream fiber which can be bleached nearly white. Specific gravity 1.31. Tenacity ranges from 2.0 to 3.5 g/denier. Elongation is 30-40%; hygroscopicity is 0.4% under standard conditions. Has extremely good chemical resistance. Acetone is the best solvent; cyclohexanone and dimethylformamide also have some solvent action. Acetic anhydride, acetaldehyde, aniline, ethylene dichloride, and methyl ethyl ketone all plasticize or swell dynel. Resistant to clothes moths' larvae, to carpet beetles, and to mildew and fungus. Will burn in a flame, but if the flame is removed it is self-extinguishing. Resistant to water and non-felting and non-shrinking below the boil. Hot water delusters dynel. Must be ironed with the lowest iron setting and a dry cotton cover over the fabric. May be heat-set in permanent pleats. Can be dyed readily and may be solution-dyed. Resistant to perspiration and to salt-water deterioration.

USE: In apparel and household furnishings; simulated fur coats; chemically resistant clothing. In making wigs and doll hair; the hair can be washed, combed, set, and in some instances, redyed.

3525. Dynorphin. [74913-18-1] Extremely potent, widely distributed neuropeptide that has 17 amino acid residues and contains leu⁵-enkephalin as its NH_2-terminal sequence. Its name is derived from "dynamis", the Greek word for power, and endorphin, the name applied to the group of opioid peptides to which it belongs. Initially isolated from porcine pituitaries and termed *slow-reversing endorphin*: L. I. Lowney et al., Life Sci. **24**, 2377 (1979). Purification, description of properties and amino acid sequence of the first 13 residues: A. Goldstein et al., Proc. Natl. Acad. Sci. USA **76**, 6666 (1979). Complete amino acid sequence of the heptadecapeptide from porcine pituitary: eidem, ibid. **78**, 7219 (1981). Isoln from porcine duodenum and identity with pituitary dynorphin: S. Tachibana et al., Nature **295**, 339 (1982). Synthesis of porcine dynorphin₁₋₁₃: M. Wakimasu et al., Chem. Pharm. Bull. **29**, 2592 (1981). Soln conformation: R. Maroun, W. L. Mattice, Biochem. Biophys. Res. Commun. **103**, 442 (1981). Radioimmunoassay: V. E. Ghazarossian et al., Life Sci. **27**, 75 (1980). Comparison of distribution of

dynorphin and enkephalin systems in brain: S. J. Watson et al., Science **218**, 1134 (1982). Behavioral effects of dynorphin₁₋₁₃ in mice and rats: J. M. Walker et al., Peptides **1**, 341 (1980); H. Zwiers et al., Life Sci. **28**, 2545 (1981). Opiate activity and receptor selectivity: M. Wuester et al., Neurosci. Lett. **20**, 79 (1980); C. Chavkin, A. Goldstein Nature **291**, 591 (1981). In the guinea pig ileum bioassay it has been shown to be 700 times more potent than leu⁵-enkephalin and its agonist effects are 1/13th as sensitive to naloxone antagonism: eidem, Proc. Natl. Acad. Sci. USA **78**, 6543 (1981). Dynorphin₁₋₁₃ has been proposed as the specific endogenous ligand of the kappa opioid receptor (cf. endorphins): C. Chavkin et al., Science **215**, 413 (1982). It has also been suggested that dynorphin₁₋₈ or dynorphin₁₋₉ may be transmitters or modulators at the kappa binding site and dynorphins 1-13 and 1-17 may act at a distance from the release site: A. D. Corbett et al., Nature **299**, 79 (1982). A possible regulatory role of dynorphin on morphine and β-endorphin-induced analgesia has been proposed: F. Tulunay et al., J. Pharmacol. Exp. Ther. **219**, 296 (1981); E. C. Petrie et al., Peptides **3**, 41 (1982). Several non-opiate effects have also been described: J. M. Walker et al., Science **218**, 1136 (1982); R. Przewlocki et al., ibid. **219**, 71 (1983).

Tyr–Gly–Gly–Phe–Leu–Arg–Arg–Ile–Arg–Pro–Lys–Leu–Lys–Trp–Asp–Asn–Gln

Porcine Dynorphin

Dynorphin₁₋₁₃. 1-13-Dynorphin (pig). $C_{75}H_{126}N_{24}O_{15}$; mol wt 1603.99. $[\alpha]_D^{23}$ −62.9° (c = 0.5 in 1% acetic acid).

3526. Dyphylline. [479-18-5] 7-(2,3-Dihydroxypropyl)-3,7-dihydro-1,3-dimethyl-1H-purine-2,6-dione; 7-(2,3-dihydroxypropyl)theophylline; (1,2-dihydroxy-3-propyl)theophylline; glyphylline; glyfyllin; diprophylline; AFI-phyllin; Astmamasit; Asthmolysin; Astrophyllin; Circair; Coronarin; Cor-Theophyllin; Dilor; Hiphyllin; Hyphyllin; Lufyllin; Neostenovasan; Neothylline; Neotilina; Neo-Vasophylline; Neutrafil; Neutraphylline; Prophyllen; Silbephylline; Solufilin; Solufyllin; Theal ampules; Thefylan. $C_{10}H_{14}N_4O_4$; mol wt 254.25. C 47.24%, H 5.55%, N 22.04%, O 25.17%. Prepn: Jones, Maney, US 2575344 (1951 to State Univ. of Iowa); Roth, Arch. Pharm. **292**, 234 (1959). Toxicity data: R. A. Al'tshuler, Med. Prom. SSSR **16**, 57 (1962), C.A. **61**, 6230g (1964).

White extremely bitter crystals from alcohol, mp 158°. uv max (0.001% in H_2O): 273 nm ($A_{1cm}^{1\%}$ 361). Freely sol in water; one gram dissolves in 3 ml H_2O at 25°. Soly (g/100 ml): alc 2; chloroform 1. pH (1% aq soln) 6.6 to 7.3. Practically insol in ether. LD_{50} in mice (mg/kg): 3400 orally; 1430 s.c. (Al'tshuler).

THERAP CAT: Bronchodilator.

3527. Dypnone. [495-45-4] 1,3-Diphenyl-2-buten-1-one; β-methylchalcone. $C_{16}H_{14}O$; mol wt 222.29. C 86.45%, H 6.35%, O 7.20%. Prepd by self-condensation of acetophenone in the presence of aluminum tert-butoxide: Adkins, Cox, J. Am. Chem. Soc. **60**, 1151 (1938); W. Wayne, H. Adkins, Org. Synth. coll. vol. III, 367 (1955). Use as sunscreen: Ind. Eng. Chem. **46**, 15A (July 1954). Toxicity study: H. F. Smyth et al., J. Ind. Hyg. Toxicol. **31**, 60 (1949).

Liquid. d_4^{15} 1.1080. n_D^{20} 1.6343. bp$_{760}$ 340-345° (partial decompn); bp$_{22}$ 225°; bp$_{1.0}$ 150-155°. Insol in water. Sol in alcohol, ether. LD$_{50}$ orally in rats: 3.6 g/kg (Smyth).

USE: Sunscreen.

3528. Dysidiolide. [182136-94-3] (5R)-5-Hydroxy-4-[(1R)-1-hydroxy-2-[(1R,2R,5S,8aS)-1,2,3,5,6,7,8,8a-octahydro-1,2,5-trimethyl-5-(4-methyl-4-penten-1-yl)-1-naphthalenyl]ethyl]-2(5H)-furanone; (−)-dysidiolide. $C_{25}H_{38}O_4$; mol wt 402.58. C 74.59%, H 9.51%, O 15.90%. Sesterterpene γ-hydroxybutenolide with a unique carbon skeleton isolated from the marine sponge *Dysidea etheria* de Laubenfels. First naturally occurring inhibitor of the protein phosphatase, cdc25; the inhibition of which blocks cell division. Isolation, activity and crystal structure: S. P. Gunasekera *et al.*, *J. Am. Chem. Soc.* **118**, 8759 (1996). Enantioselective synthesis: E. J. Corey, B. E. Roberts, *ibid.* **119**, 12425 (1997). Total synthesis of racemate and preliminary biological activity: S. R. Magnuson *et al.*, *ibid.* **120**, 1615 (1998).

Colorless crystals from CH_3OH/CH_2Cl_2, mp 186-187°. $[\alpha]_D^{24}$ −11.1° (c = 0.6 in CH_2Cl_2/CH_3OH, 1:1).

3529. Dysprosium. [7429-91-6] Dy; at. wt 162.500; at. no. 66; valences 2, 3, 4. A lanthanide; belongs to the yttrium group of rare earth metals. Naturally occurring isotopes (mass numbers): 156 (0.06%); 158 (0.10%); 160 (2.34%); 161 (18.9%); 162 (25.5%); 163 (24.9%); 164 (28.2%); known artificial radioactive isotopes: 145-155; 157; 159; 165-167. Abundance in earth's crust: ~4.5 ppm. Occurs in gadolinite, xenotime, samarskite and other rare earth minerals. Discovered in 1886 by Lecoq de Boisbaudran. Sepn: Urbain, *Compt. Rend.* **139**, 736 (1904); **141**, 521 (1905); **142**, 785 (1906); Spedding *et al.*, *J. Am. Chem. Soc.* **69**, 2812 (1947); **76**, 2557 (1954). Prepn of salt: Engle, Balke, *ibid.* **39**, 53 (1917); prepn of a Dy-Al alloy: Schumacher, Harris, *ibid.* **48**, 3108 (1926). Toxicity studies: Bruce *et al.*, *Toxicol. Appl. Pharmacol.* **5**, 750 (1962); Haley *et al.*, *ibid.* **8**, 37 (1966). Reviews of prepn, properties and compds: *The Rare Earths*, F. H. Spedding, A. H. Daane, Eds. (Krieger, Huntington, N.Y., 1971, reprint of 1961 ed.) 641 pp; Hulet, Bode, "Separation Chemistry of the Lanthanides and Transplutonium Actinides" in *MTP Int. Rev. Sci.: Inorg. Chem., Ser. One* vol. 7, K. W. Bagnall,

Ed. (Univ. Park Press, Baltimore, 1972) pp 1-45; Moeller, "The Lanthanides" in *Comprehensive Inorganic Chemistry* vol. 4, J. C. Bailar, Jr. *et al.*, Eds. (Pergamon Press, Oxford, 1973) pp 1-101; F. H. Spedding in *Kirk-Othmer Encyclopedia of Chemical Technology* vol. 19 (John Wiley & Sons, New York, 3rd ed., 1982) pp 833-854; *Chemistry of the Elements*, N. N. Greenwood, A. Earnshaw, Eds. (Pergamon Press, New York, 1984) pp 1423-1449. Brief review of properties: G. T. Seaborg, *Radiochim. Acta* **61**, 115-122 (1993).

Silver metal; tarnishes in moist air. Hexagonal close-packed crystals. d 8.5500. mp 1412°. bp 2567°. Heat of fusion: 10.782 kJ/mol. Heat of sublimation (25°): 290.4 kJ/mol. Forms greenish-yellow salts.

Oxide. Dysprosia. Dy_2O_3. White substance. Prepd by heating the oxalate or sulfate. d^{27} 7.81.

Hydroxide. $Dy(OH)_3$. A gelatinous precipitate. Prepd by adding ammonia to an aq soln of a dysprosium salt; forms a blue colloidal soln.

Chloride. $DyCl_3$. Obtained by passing S_2Cl_2 over heated dysprosia. A hexahydrate is formed from the aq soln. The anhydr chloride, yellow shining crystals, d 3.67; mp 680°. LD$_{50}$ in mice: 585 mg/kg i.p.; 7.65 g/kg orally (Haley).

Sulfate. $Dy_2(SO_4)_3$. Octahydrate, yellow crystals. Prepd by dissolving the oxide in sulfuric acid and precipitating with alcohol; stable in air at 110°, dehydrated at 360°.

Nitrate. $Dy(NO_3)_3$. Pentahydrate, melts at 88.6° in its water of crystn. Sol in water. LD$_{50}$ (hexahydrate) in rats: 295 mg/kg i.p.; 3.1 g/kg orally (Bruce).

USE: Oxide used in control rods of some nuclear power reactors.

3530. Dystrophin. Protein product of the human Duchenne muscular dystrophy (DMD) gene. Mol wt approx 400,000 daltons. Present in very small amounts in normal muscle (approx 0.002% of total muscle protein) but either absent or abnormal in muscular dystrophy patients. Thought to be an intracellular structural component of the plasma membrane system in normal muscle fibers. Complete cloning of DMD cDNA: M. Koenig *et al.*, *Cell* **50**, 509 (1987). Isoln of dystrophin from mouse and human muscle tissue: E. P. Hoffman *et al.*, *ibid.* **51**, 919 (1987). Subcellular localization ies: E. P. Hoffman *et al.*, *Nature* **330**, 754 (1987); C. M. Knudson *et al.*, *J. Biol. Chem.* **263**, 8480 (1988); E. E. Zubrzycka-Gaarn *et al.*, *Nature* **333**, 466 (1988); S. C. Watkins *et al.*, *ibid.* 863. Amino acid sequence and structural similarity to α-actinin and to the cytoskeletal protein, spectrin: M. Koenig *et al.*, *Cell* **53**, 219 (1988). Characterization of DMD gene expression in normal and diseased human muscle: M. Oronzi Scott *et al.*, *Science* **239**, 1418 (1988); in animal muscle and brain: U. Nudel *et al.*, *Nature* **331**, 635 (1988). Differentiation of muscle and brain DMD mRNA: U. Nudel *et al.*, *ibid.* **337**, 76 (1989). Correlation of clinical phenotype with dystrophin abnormalities: E. P. Hoffman *et al.*, *N. Engl. J. Med.* **318**, 1363 (1988). Use of dystrophin cDNA for prenatal diagnosis and carrier detection in muscular dystrophy: B. T. Darras *et al.*, *Am. J. Med. Genet.* **29**, 713 (1988).

E

3531. Ebastine. [90729-43-4] 1-[4-(1,1-Dimethylethyl)phenyl]-4-[4-(diphenylmethoxy)-1-piperidinyl]-1-butanone; 4'-*tert*-butyl-4-[4-(diphenylmethoxy)piperidino]butyrophenone; 4-diphenylmethoxy-1-[3-(4-*tert*-butylbenzoyl)propyl]piperidine; LAS-W-090; Ebastel; Estivan; Evastel; Kestin; Kestine. $C_{32}H_{39}NO_2$; mol wt 469.67. C 81.83%, H 8.37%, N 2.98%, O 6.81%. Nonsedating type histamine H_1-receptor antagonist. Prepn: J. M. P. Soto *et al.*, **EP 134124**; *eidem*, **US 4550116** (both 1985 to Fordonal). Metabolized *in vivo* to **carebastine**, its active carboxylic acid metabolite. Clinical pharmacology, pharmacokinetics: J. Vincent *et al.*, *Br. J. Clin. Pharmacol.* **26**, 497 (1988). Effect on psychomotor performance: J. Vincent *et al.*, *ibid.* 503.

Fumarate. $C_{32}H_{39}NO_2 \cdot C_4H_4O_4$. Crystals from ethanol, mp 197-198°.

THERAP CAT: Antihistaminic.

3532. Ebrotidine. [100981-43-9] [*N*(*E*)]-*N*-[[[2-[[2-[(Aminoiminomethyl)amino]-4-thiazolyl]methyl]thio]ethyl]amino]methylene]-4-bromobenzenesulfonamide; (*E*)-*N*-*p*-bromobenzenesulfonyl-*N'*-[2-[[[2-[(aminoiminomethyl)amino]-4-thiazolyl]methyl]thio]ethyl]formamidine; *p*-bromo-*N*-[(*E*)-[2-[[[2-[(diaminomethylene)amino]-4-thiazolyl]methyl]thio]ethyl]amino]methylene]benzenesulfonamide; FI-3542; Ebrocit; Ebrodin; Ulsanic. $C_{14}H_{17}BrN_6O_2S_3$; mol wt 477.41. C 35.22%, H 3.59%, Br 16.74%, N 17.60%, O 6.70%, S 20.15%. Histamine H_2-receptor antagonist. Prepn: R. Foguet *et al.*, **EP 159012**; *eidem*, **US 4728655** (1985, 1988 both to Ferrer); L. Anglada *et al.*, *Eur. J. Med. Chem.* **23**, 97 (1988). Clinical pharmacology: S. J. Konturek *et al.*, *Scand. J. Gastroenterol.* **27**, 438 (1992); M. Maczka *et al.*, *J. Physiol. Pharmacol.* **43**, 139 (1992). Clinical evaluation in gastroesophageal reflux disease: E. Sito *et al.*, *ibid.* **44**, 259 (1993). HPLC determn in urine and metabolism study: E. Rozman *et al.*, *J. Pharm. Sci.* **83**, 252 (1994).

Crystals from ethyl acetate. mp 142.5-146°.
THERAP CAT: Antiulcerative.

3533. Ebselen. [60940-34-3] 2-Phenyl-1,2-benzisoselenazol-3(2*H*)-one; PZ-51. $C_{13}H_9NOSe$; mol wt 274.18. C 56.95%, H 3.31%, N 5.11%, O 5.84%, Se 28.80%. Seleno-organic which shows antioxidant effects through glutathione peroxidase-like action. Prepn: R. Lesser, R. Weiss, *Ber.* **57**, 1077 (1924); R. Weber, M. Renson, *Bull. Soc. Chim. Fr.* **1976**, 1124; L. Engman, A. Hallberg, *J. Org. Chem.* **54**, 2964 (1989). One-step synthesis: J. Oppenheimer, L. A. Silks, III, *J. Labelled Compd. Radiopharm.* **38**, 281 (1996). HPLC determn in plasma: R. Terlinden *et al.*, *J. Chromatogr.* **430**, 438 (1988). Clinical trial as neuroprotectant in ischemic stroke: T. Yamaguchi *et al.*, *Stroke* **29**, 12 (1998); A. Ogawa *et al.*, *Cerebrovasc. Dis.* **9**, 112 (1999). Review of pharmacology: M. J. Parnham *et al.*, **32**, 4-9 (1991); and of biochemical interactions: T. Schewe, *Gen. Pharmacol.* **26**, 1153-1169 (1995). Review as glutathione peroxidase mimic: H. Sies, *Free Radical Biol. Med.* **14**, 313-323 (1993).

Crystals from ethanol, mp 180-181°.

3534. Eburnamonine. [474-00-0] Eburnamenin-14(15*H*)-one; (+)-eburnamonine. $C_{19}H_{22}N_2O$; mol wt 294.40. C 77.52%, H 7.53%, N 9.52%, O 5.43%. One of the *Vinca* alkaloids, naturally occurring as the (+)- and (±)-forms. Isoln of the (+)-form from *Hunteria eburnea* Pinchon, *Apocynaceae:* M. F. Bartlett *et al.*, *Compt. Rend.* **249**, 1259 (1959); of the (±)-form from *Vinca minor* L., *Apocynaceae:* J. Mokry *et al.*, *Experientia* **17**, 354 (1961). The (−)-form is obtained by acid hydrolysis of vincamine, *q.v.:* J. Trajanek *et al.*, *Tetrahedron Lett.* **1961**, 702; O. Clauder *et al.*, *ibid.* **1962**, 1147; *eidem*, **HU 151295** (1964 to Gedeon Richter), *C.A.* **60**, 14558e (1964). Structure and synthesis of the (±)-form: M. F. Bartlett, W. I. Taylor, *Tetrahedron Lett.* **20**, 20 (1959); *eidem*, *J. Am. Chem. Soc.* **82**, 5941 (1960); E. Wenkert, B. Wickberg, *ibid.* **87**, 1580 (1965). Short synthesis of the (±)-form: E. Wenkert *et al.*, *ibid.* **100**, 4893 (1978); high yield total synthesis: J. L. Hermann *et al.*, *ibid.* **101**, 1540 (1979); T. Imanishi *et al.*, *Chem. Pharm. Bull.* **30**, 1521 (1982). Synthesis of the (−)-form: D. Cartier *et al.*, *Bull. Soc. Chim. Fr.* **1976**, 1961. Structural and biogenetic relationship to vincamine: J. Mokry *et al.*, *Tetrahedron Lett.* **1962**, 433. Pharmacology of the (−)-form: P. Lacroix *et al.*, *Arzneim.-Forsch.* **29**, 1094 (1979). Series of articles on the pharmacodynamics, metabolism, and therapeutic use of the (−)-form: *Eur. Neurol.* **17**, Suppl. 1, 1-172 (1978). *Reviews:* W. I. Taylor in *The Alkaloids* vol. **8**, R. H. F. Manske, Ed. (Academic Press, New York, 1965) pp 253-259; *idem*, *ibid.* vol. **11**, pp 108-110 (1968).

Cryst, mp 174°. $[\alpha]_D^{25}$ +89° (chloroform). uv max: 241, 268, 296, 302 nm (ε 19800, 10200, 4800, 4800).

(±)-Form. [2580-88-3] (±)-Eburnamenin-14(15*H*)-one; vincanorine. Crystals from methanol, mp 201-202.5°. uv max (ethanol): 227, 287, 294 nm (log ε 4.49, 3.89, 3.87).

(−)-Form. [4880-88-0] (3α,16α)-Eburnamenin-14(15*H*)-one; 16-oxoeburnane; vincamone; CH-846; Cervoxan. Solid, mp 168-170°; also reported as mp 177-178° (Cartier). $[\alpha]_D^{25}$ −102° (chloroform) (Clauder); $[\alpha]_D$ −100° (c = 0.783 in chloroform) (Cartier). uv max: 205, 240, 265, 290, 300 nm (log ε 4.28, 4.16, 3.90, 3.59, 3.57).

(−)-Form phosphate. [94134-60-8] Eburnal.

THERAP CAT: The (−)-form as vasodilator.

3535. Ecabet. [33159-27-2] (1*R*,4a*S*,10a*R*)-1,2,3,4,4a,9,10,-10a-Octahydro-1,4a-dimethyl-7-(1-methylethyl)-6-sulfo-1-phenanthrenecarboxylic acid; dehydro-6-sulfoabietic acid; 12-sulfodehydroabietic acid. $C_{20}H_{28}O_5S$; mol wt 380.50. C 63.13%, H 7.42%, O 21.02%, S 8.43%. Prepn: L. F. Fieser, W. P. Campbell, *J. Am. Chem. Soc.* **60**, 2631 (1938); H. Wada *et al.*, *Chem. Pharm. Bull.* **33**, 1472 (1985). Metabolism in animals: Y. Ito *et al.*, *J. Pharmacobio-Dyn.* **14**, 533 (1991). Efficacy in healing expt ulcers in rats: Y. Onoda *et al.*, *Arzneim.-Forsch.* **41**, 546 (1991). Gastroprotective effects in rats in comparison with sucralfate, *q.v.:* M. Kinoshita *et al.*, *Dig. Dis. Sci.* **40**, 661 (1995). Bactericidal activity against *Helicobacter pylori:* K. Shibata *et al.*, *Antimicrob. Agents Chemother.* **39**, 1295 (1995).

As the hemihydrate, $[\alpha]_D^{25}$ +72.4° (c = 2.5 in alcohol).

Sodium salt. [86408-72-2] TA-2711; Gastron. $C_{20}H_{27}NaO_5S$; mol wt 402.48. Occurs as pentahydrate, mp >300°. $[\alpha]_D^{20}$ +59.4° (c = 0.5).

THERAP CAT: Antiulcerative.

3536. Ecallantide. [460738-38-9] Protein (synthetic human plasma kallikrein-inhibiting); DX-88; Kalbitor. $C_{305}H_{442}N_{88}O_{91}S_8$; mol wt 7053.90. Specific inhibitor of kallikrein, *q.v.* Recombinant, 60 amino acid peptide produced in the yeast, *Pichia pastoris.* Derived from the first Kunitz domain of human tissue factor pathway inhibitor (TFPI), also known as lipoprotein associated coagulation inhibitor (LACI). Identification and synthesis of kallikrein-binding LACI homologs using phage display technique: W. Markland, R. C. Ladner, **WO 9521601** (1995 to Protein Eng. Corp.); *eidem*, **US 5795865**; *eidem*, **US 6057287** (1998, 2000 both to Dyax); W. Markland *et al.*, *Biochemistry* **35**, 8058 (1996). Physiological effects on contact activator induced coagulation: K. A. Tanaka *et al.*, *Thromb. Res.* **113**, 333 (2004). Clinical evaluation in hereditary angioedema (HAE): L. Schneider *et al.*, *J. Allergy Clin. Immunol.* **120**, 416 (2007). Review of development and therapeutic potential: A. Lehmann, *Curr. Opin. Investig. Drugs* **7**, 282 (2006).

THERAP CAT: In treatment of hereditary angioedema.

3537. Ecamsule. [92761-26-7] 3,3′-(1,4-Phenylenedimethylidyne)bis[7,7-dimethyl-2-oxobicyclo[2.2.1]heptane-1-methanesulfonic acid]; terephthalylidene-3,3′-dicamphor-10,10′-disulfonic acid; Mexoryl SX. $C_{28}H_{34}O_8S_2$; mol wt 562.69. C 59.77%, H 6.09%, O 22.75%, S 11.40%. Broad spectrum UVA absorber with some UVB absorption. Prepn: G. Lang *et al.*, **DE 3321679**; *eidem*, **US 4585597** (1983, 1986 both to L'Oreal). Evaluation of photomutagenicity: S.W. Dean *et al.*, *Mutagenesis* **7**, 179 (1992). Evaluation of photoprotection: A. Fourtanier *et al.*, *Photochem. Photobiol.* **55**, 549 (1992); R. D. Ley, A. Fourtanier, *ibid.* **65**, 1007 (1997).

Solid, mp 255° (dec). uv max (ethanol): 345 nm (ε 47000 $M^{-1}cm^{-1}$); uv max (water): 342 nm (ε 42300).

THERAP CAT: Ultraviolet screen.

3538. Ecdysteroids. Polyhydroxylated steroids formerly known as ecdysones. Originally identified as insect molting hormones controlling the pupation of insects. Later shown to be in-

volved in many complex developmental processes in metamorphosis, differentiation and reproduction. Ecdysteroids have been detected in invertebrate species of several phyla belonging to the Protostomia and in some plant species. The terms *zooecdysteroids* and *phytoecdysteroids* are used to distinguish ecdysteroids isolated from animal species from those of plant origin. Nomenclature: T. W. Goodwin *et al.*, *Nature* **272**, 122 (1978). The two major ecdysteroids isolated are ecdysone and 20-hydroxyecdysone. Configuration: M. Koreeda *et al.*, *J. Am. Chem. Soc.* **93**, 4084 (1971). Chromosomal action: M. Ashburner, *Nature* **285**, 435 (1980). GC and HPLC determn: R. P. Evershed *et al.*, *J. Chromatogr.* **390**, 357 (1987). *Reviews:* Kilby, *Discovery* **18**, 13 (1957); P. Karlson, *Angew. Chem. Int. Ed.* **2**, 175-182 (1963); K. Nakanishi, *Pure Appl. Chem.* **25**, 167-195 (1971); M. Koreeda, B. A. Teicher, *Anal. Biochem. Insects 1977*, 207-240; P. Karlson, *Dev. Endocrinol.* **7**, 1-11 (1980). *Book: Ecdysone: From Chemistry to Mode of Action*, J. Koolman, Ed. (Thieme, New York, 1989) 482 pp.

Ecdysone. [3604-87-3] (2β,3β,5β,22R)-2,3,14,22,25-Pentahydroxycholest-7-en-6-one; α-ecdysone; $C_{27}H_{44}O_6$; mol wt 464.64. Secreted by ecdysial tissues, then transformed to more active compound, 20-hydroxyecdysone. First isoln from silkworm moths, *Bombyx mori:* A. Butenandt, P. Karlson, *Z. Naturforsch.* **9b**, 389 (1954). Isoln from rhizomes of *Polypodium vulgare* L.: G. Heinrich, H. Hoffmeister, *Experientia* **23**, 995 (1967); from bracken fern, *Pteridinium aquilinum:* J. N. Kaplanis *et al.*, *Science* **157**, 1436 (1967). Structure: P. Karlson *et al.*, *Ber.* **98**, 2394 (1965). Configuration: R. Huber, W. Hoppe, *ibid.* 2403. Synthesis: U. Kerb *et al.*, *Helv. Chim. Acta* **49**, 1601 (1966); J. B. Siddall *et al.*, *J. Am. Chem. Soc.* **88**, 379, 862 (1966); H. Mori *et al.*, *Chem. Pharm. Bull.* **16**, 563 (1968). mp 238-239°. $[\alpha]_{578}^{20}$ +62°. uv max: 243 nm (ε 11600).

20-Hydroxyecdysone. [5289-74-7] (2β,3β,5β,22R)-2,3,14,20,-22,25-Hexahydroxycholest-7-en-6-one; β-ecdysone; ecdysterone; crustecdysone; isoinokosterone; polypodine A. $C_{27}H_{44}O_7$; mol wt 480.64. Most widely occurring ecdysteroid in both plant and animal species. Isoln from *B. mori:* P. Hocks, R. Wiechert, *Tetrahedron Lett.* **1966**, 2989; from seawater crayfish, *Jasus lalandei:* F. Hampshire, D. H. S. Horn, *Chem. Commun.* **1966**, 37. Isolns from plant sources, *Achyranthes fauriei:* T. Takemoto *et al.*, *Yakugaku Zasshi* **87**, 325 (1967); *P. elatus:* M. N. Galbraith, D. H. S. Horn, *Chem. Commun.* **1966**, 905; *P. vulgare:* J. Jizba *et al.*, *Tetrahedron Lett.* **1967**, 1689. Isoln from parasitic helminths: H. H. Rees, J. G. Mercer, *Adv. Invertebr. Reprod.* **4**, 173 (1986). Configuration: Dammeier, Hoppe, *Ber.* **104**, 1660 (1971). Synthesis: G. Hüppi, J. B. Siddall, *J. Am. Chem. Soc.* **89**, 6790 (1967); U. Kerb *et al.*, *Tetrahedron Lett.* **1968**, 4277; H. Mori, K. Shibata, *Chem. Pharm. Bull.* **17**, 1970 (1969). Total synthesis: T. Kametani *et al.*, *Tetrahedron Lett.* **21**, 4855 (1980). From methanol-ethyl acetate, mp 240-242°. uv max: 243 nm (ε 10300). Unstable in alkaline soln.

3539. Ecgonidine. [484-93-5] (1R,5S)-8-Methyl-8-azabicyclo[3.2.1]oct-2-ene-2-carboxylic acid; anhydroecgonine. C_9H_{13}-NO_2; mol wt 167.21. C 64.65%, H 7.84%, N 8.38%, O 19.14%. Tropane alkaloid related to cocaine. Prepd by phosphorus oxychloride dehydration of ecgonine: A. Einhorn, *Ber.* **20**, 1221 (1887). Isoln from extracts of coca leaves: A. W. K. de Jong, *Rec. Trav. Chim.* **42**, 980 (1923). Prepn and structure: S. P. Findlay, *J. Am. Chem. Soc.* **75**, 1033 (1953). Synthesis of (±)-form: C. Grundmann, G. Ottmann, *Ann.* **605**, 24 (1957). Prepn of methyl ester: H. E. Zaugg *et al.*, *J. Org. Chem.* **23**, 847 (1958); C. R. Holmquist *et* al., Org. Prep. Proced. Int. **29**, 308 (1997). GC-MS determn in forensic tissue: E. T. Shimomura et al., Clin. Chem. **47**, 1040 (2001); in blood and urine: B. D. Paul et al., Biomed. Chromatogr. **19**, 677 (2005).

Crystals from abs alcohol, mp 235°. $[\alpha]_D^{14} -84.6°$ (c = 1.7).

Methyl ester. [43021-26-7] Methyl ecgonidine. $C_{10}H_{15}NO_2$; mol wt 181.24. Pyrolytic byproduct of crack cocaine. Colorless oil. $[\alpha]_D -35.1°$ (c = 1.22 in chloroform). bp_{10} 124-126°. n_D^{25} 1.5006.

(±)-Form. [127379-23-1] 2-Tropidinecarboxylic acid. Crystals from alcohol + ether, dec 235-236°. Soluble in water; sparingly sol in alcohol.

USE: In bioassays as a marker for cocaine exposure.

3540. Ecgonine. [481-37-8] (1R,2R,3S,5S)-3-Hydroxy-8-methyl-8-azabicyclo[3.2.1]octane-2-carboxylic acid; 3β-hydroxy-1αH,5αH-tropane-2β-carboxylic acid. $C_9H_{15}NO_3$; mol wt 185.22. C 58.36%, H 8.16%, N 7.56%, O 25.91%. Tropane alkaloid naturally occuring in coca leaves; precursor and metabolite of cocaine. Isoln from coca leaves: A. W. K. de Jong, Recl. Trav. Chim. Pays-Bas **42**, 980 (1923); idem, ibid. **59**, 687 (1940). Prepn by hydrolysis of cocaine: R. Willstätter et al., Ann. **434**, 111 (1923). Structure: J. Gadamer, C. John, Arch. Pharm. **259**, 227 (1921). Stereochemistry: G. Fodor, Nature **170**, 278 (1952); G. Fodor, O. Kovács, J. Chem. Soc. **1953**, 724; and prepn of esters: S. P. Findlay, J. Am. Chem. Soc. **76**, 2855 (1954). Synthesis of (±)- form: R. Willstätter, M. Bommer, Ann. **422**, 15 (1920). LC/MS/MS determn in blood: E. Jagerdeo et al., J. Chromatogr. B **874**, 15 (2008).

Monohydrate. [5796-30-5] $C_9H_{15}NO_3.H_2O$; mol wt 203.24. Triboluminescent, monoclinic prisms from alc, mp 198° (anhydr, dec 205°). $[\alpha]_D^{15} -45°$ (c = 5). Neutral to litmus. pKa 11.11; pKb 11.22. One gram dissolves in 5 ml water, 67 ml alc, 20 ml methanol, 75 ml ethyl acetate. Sparingly sol in acetone, ether, benzene, chloroform, petr ether.

Hydrochloride. [5796-31-6] $C_9H_{15}NO_3.HCl$; mol wt 221.68. Triclinic plates from water, mp 246°. $[\alpha]_D^{15} -59°$ (c = 10). Sol in water; slightly in alc.

Methyl ester. [7143-09-1] Methylecgonine. $C_{10}H_{17}NO_3$; mol wt 199.25. Nearly colorless oil; hygroscopic. $bp_{0.5}$ 84°; bp_2 104-106°. n_D^{19} 1.4886. d_D^{19} 1.1451. $[\alpha]_D^{20} -12.3°$ (c = 2 in methanol).

(±)-Form. [876657-17-9] Plates from 90% alcohol as the trihydrate, mp 93-118° (anhydr dec 212°).

Note: This is a controlled substance: 21 CFR, 1308.12.

USE: In bioassays as a marker for cocaine exposure.

3541. Echinacea. Purple coneflower. Perennial plant of the family Compositae (Asteraceae). Habit. North American plains and southeast U.S.; cultivated in Europe. Widely used in traditional medicine as an immunostimulant. Medicinal formulations are prepared from the roots and aerial parts of three species: E. angustifolia DC, E. pallida (Nutt.) Nutt. or E. purpurea (L.) Moench. Constit. Alkamides such as **echinacein**, caffeic acid esters such as **echinacoside**, polysaccharides, volatile oil. Isoln of constituents from E. augustifolia DC: A. Stoll et al., Helv. Chim. Acta **33**, 1877 (1950); M. Jacobson, Science **120**, 1028 (1954). Analysis of alkamides: X. He et al., J. Chromatogr. A **815**, 205 (1998); of volatile components: G. Mazza, T. Cottrell, J. Agric. Food Chem. **47**, 3081 (1999). Review of clinical trials in treatment of upper respiratory infection: B. Bar-

rett et al., J. Fam. Pract. **48**, 628-635 (1999); of use in Native American medicine: A. T. Borchers et al., Am. J. Clin. Nutr. **72**, 339-347 (2000).

THERAP CAT: In treatment of respiratory infections.

3542. Echinenone. [432-68-8] β,β-Caroten-4-one; 4-oxo-β-carotene; 4-keto-β-carotene; aphanin; myoxanthin. $C_{40}H_{54}O$; mol wt 550.87. C 87.21%, H 9.88%, O 2.90%. Carotenoid pigment occurring in algae. Isoln from Aphanizomenon flos-aqua: Tischer, Z. Physiol. Chem. **251**, 109 (1938); **260**, 257 (1939); from Paracentrotus lividus: Lederer, Compt. Rend. **201**, 300 (1935); from Oscillatoria rubrescens: Heilbron, Lythgoe, J. Chem. Soc. **1936**, 1376. Structure: Goodwin, Taha, Biochem. J. **47**, 244 (1950); **48**, 513 (1951); Goodwin, ibid. **63**, 481 (1956); Ganguly et al., Arch. Biochem. Biophys. **60**, 345 (1956). Synthesis: Akhtar, Weedon, J. Chem. Soc. **1958**, 3986; **1959**, 4058.

Orange-red crystals from benzene + methanol, mp 178-180°. Absorption max (chloroform): 472-478 nm. Freely sol in carbon disulfide, chloroform, benzene. Slightly sol in pyridine, ether. Practically insol in methanol. Provitamin A activity 54% of that of all-trans-β-carotene.

Oxime. $C_{40}H_{55}NO$. Red crystals from benzene, mp 208°. Absorption max (chloroform): 464-468 nm.

3543. Echinochrome A. [517-82-8] 2-Ethyl-3,5,6,7,8-pentahydroxy-1,4-naphthalenedione; 2-ethyl-3,5,6,7,8-pentahydroxy-1,4-naphthoquinone. $C_{12}H_{10}O_7$; mol wt 266.21. C 54.14%, H 3.79%, O 42.07%. The red pigment of the eggs of the sea urchin (Arbacia pustulosa). Isoln and structure: Kuhn, Wallenfels, Ber. **72**, 1407 (1939). Synthesis from 2-ethyl-1,3,4-trimethoxybenzene and dibenzoyloxymaleic anhydride: Wallenfels, Gauhe, Ber. **76**, 325 (1943).

Deep-red needles from dioxane-water, mp 220° (some decompn). Sublimes at 10^{-4} mm at 120°. Absorption max (chloroform): 533, 497, 462 nm. Very sparingly sol in cold water, yet more than sufficiently sol for chemotaxic effects. Readily sol in carbon disulfide, ether, chloroform, benzene, concd H_2SO_4.

3544. Echinomycin. [512-64-1] N-(2-Quinoxalinylcarbonyl)-O-[N-(2-quinoxalinylcarbonyl)-D-seryl-L-alanyl-3-mercapto-N,S-dimethylcysteinyl-N-methyl-L-valyl]-D-seryl-L-alanyl-N-methylcysteinyl-N-methyl L-valine (8 → 1)-lactone cyclic (3 → 7)-thioester; quinomycin A. $C_{51}H_{64}N_{12}O_{12}S_2$; mol wt 1101.27. C 55.62%, H 5.86%, N 15.26%, O 17.43%, S 5.82%. A **"quinoxaline antibiotic"** similar to the triostins, q.v.: Kuroya, Ishida, J. Antibiot. **14A**, 324 (1961). Powerful, selective inhibitor of nucleic acid synthesis in vitro. Produced by Streptomyces echinatus from soil of Cuanza (Angola): Corbaz et al., Helv. Chim. Acta **40**, 199 (1957). Structure: Keller-Schierlein et al., ibid. **42**, 305 (1959). Revised structure: Dell et al., J. Am. Chem. Soc. **97**, 2497 (1975); D. G. Martin et al., J. Antibiot. **28**, 332 (1975). Identity with quinomycin A: Katagiri, Sugiura, Antimicrob. Agents Chemother. **1961**, 162. Biosynthesis: Arif et al., Indian J. Biochem. **7**, 193 (1970). Isoln and properties of the other quinomycins: Yoshida, Katagiri, J. Antibiot. **14A**, 330 (1961); Otsuka, Shoji, ibid. **19A**, 128 (1966); eidem, Tetrahedron **23**, 1535 (1967). Conformation in soln: H. T. Cheung *et al., J. Am. Chem. Soc.* **100**, 46 (1978). Review of chemistry and biochemistry of echinomycin: M. J. Waring in *Antibiotics* **vol. 5** (pt. 2), F. E. Hahn, Ed. (Springer-Verlag, New York, 1979) pp 173-194.

Slightly hygroscopic crystals, mp 217-218°. $[\alpha]_D^{20}$ −310° (c = 0.86 in chloroform). uv max (methanol): 243, 320 nm ($E_{1cm}^{1\%}$ 622, 100). Has lipophile solubilities. Easily sol in chloroform, dioxane. Insol in water, petr ether, hexane.

3545. Echinopsine. [83-54-5] 1-Methyl-4(1*H*)-quinolinone; 1-methyl-4(1*H*)-quinolone; 1,4-dihydro-1-methyl-4-oxoquinoline; *N*-methyl-4-quinolone. $C_{10}H_9NO$; mol wt 159.19. C 75.45%, H 5.70%, N 8.80%, O 10.05%. Isoln from *Echinops ritro* L. and other spp of *Echinops*, *Compositae*: Greshoff, *Rec. Trav. Chim.* **19**, 360 (1900); Ban'kovskii *et al., Dokl. Akad. Nauk SSSR* **148**, 1073 (1963). Structure: Späth, Kolbe, *Monatsh. Chem.* **43**, 469 (1923). Synthesis: Kondo, Ikawa, *J. Pharm. Soc. Jpn.* **51**, 702 (1931); Allison *et al., J. Chem. Soc.* **1954**, 403; King, Abramo, *J. Org. Chem.* **23**, 1609 (1958); Kamiya, *Chem. Pharm. Bull.* **10**, 669 (1962). Simple synthesis: J. R. Merchant, V. Shankaranarayan, *Chem. Ind. (London)* **1979**, 320. Pharmacology: A. D. Turova *et al., Pharmacol. Toxicol.* **20**, 236 (1957).

Needles from benzene, mp 152°. One gram dissolves in about 60 ml water, 6 ml boiling water. Soluble in alcohol, chloroform, hot benzene; slightly sol in ether. LD$_{100}$ s.c. in mice: 600 mg/kg (Turova).

Hydrochloride. $C_{10}H_9NO.HCl$. Crystals, mp 185-186°.

3546. Echinuline. [1859-87-6] (3*S*,6*S*)-3-[[2-(1,1-Dimethyl-2-propen-1-yl)-5,7-bis(3-methyl-2-buten-1-yl)-1*H*-indol-3-yl]methyl]-6-methyl-2,5-piperazinedione. $C_{29}H_{39}N_3O_2$; mol wt 461.65. C 75.45%, H 8.52%, N 9.10%, O 6.93%. From the dry mycelia of *Aspergillus echinulatus*: Quilico, Panizzi, *Ber.* **76**, 348 (1943). Structure: Casnati *et al., Gazz. Chim. Ital.* **92**, 105 (1962); Romanet *et al., Bull. Soc. Chim. Fr.* **1963**, 1048. Absolute configuration: Nakashima, Slater, *Tetrahedron Lett.* **1967**, 4433; *ibid.* **1971**, 2649. Synthetic studies: Houghton, Saxton, *J. Chem. Soc. C* **1969**, 595, 1003; Takamatsu *et al., Tetrahedron Lett.* **1971**, 4661. Total synthesis of optically active *cis* isomer: eidem, *ibid.* **4665**; S. Inoue *et al., J. Pharm. Soc. Jpn.* **97**, 558 (1977). Epimerization of echinuline

with triethylamine in ethanol gives the *trans* isomer, *epi-echinuline*: Westley *et al., Anal. Chem.* **40**, 1888 (1968).

Needles from butanol, mp 242-243°. $[\alpha]_D^{20}$ −26.0° (chloroform). uv max (ethanol): 230, 279, 286 nm (log ε 4.60, 3.98, 3.96). Sol in glacial acetic acid, chloroform, pyridine, dioxane; less sol in warm alcohol, warm butanol; slightly sol in benzene, ether, petr ether, carbon tetrachloride, acetone, cold alcohol.

3547. Echitamine. [6871-44-9] (1*S*,3*S*,4*E*,8a*S*,13a*R*,14*R*)-4-Ethylidene-2,3,4,5,7,8-hexahydro-1-hydroxy-14-(hydroxymethyl)-14-(methoxycarbonyl)-6-methyl-13*H*-3,8a-methano-1*H*-azepino-[1′,2′:1,2]pyrrolo[2,3-*b*]indolium; (3β,16*R*)-3,17-dihydroxy-16-(methoxycarbonyl)-4-methyl-2,4(1*H*)-cyclo-3,4-secoakuammilanium; ditaine. $[C_{22}H_{29}N_2O_4]^+$. From the bark of *Alstonia scholaris* (L.) R.Br., *A. congensis* Engl., *Apocynaceae; A. neriifolia*: Hesse, *Ann.* **203**, 150 (1880); Goodson, Henry, *J. Chem. Soc.* **127**, 1640 (1925); Goodson, *ibid.* **1932**, 2626; Chakravarti *et al., Bull. Calcutta Sch. Trop. Med.* **16**, 81 (1968), *C.A.* **71**, 3529f (1969); Roy, Chatterjee, *J. Indian Chem. Soc.* **45**, 21 (1968). Structure: Hamilton *et al., Proc. Chem. Soc. London* **1961**, 63; *J. Chem. Soc.* **1962**, 5061. Abs config: Manohar, Ramaseshan, *Tetrahedron Lett.* **2**, 814 (1961). Synthetic studies: Fritz, Fischer, *Tetrahedron* **20**, 1737 (1964); Dolby, Esfandiari, *J. Org. Chem.* **37**, 43 (1972). *Review:* Govindachari, *J. Indian Chem. Soc.* **45**, 945-957 (1968).

Hydroxide. [464-20-0] $C_{22}H_{30}N_2O_5$; mol wt 402.49. White crystals, mp 206°. $[\alpha]_D^{20}$ −29° (alcohol). Sol in water, alcohol, chloroform, ether.

Chloride. [6878-36-0] $C_{22}H_{29}ClN_2O_4$; mol wt 420.93. Long needles from water, mp 295°. $[\alpha]_D^{15}$ −58°. uv max (ethanol): 235, 295 nm (log ε 3.93, 3.55).

3548. Echothiophate Iodide. [513-10-0] 2-[(Diethoxyphosphinyl)thio]-*N,N,N*-trimethylethanaminium iodide (1:1); (2-mercaptoethyl)trimethylammonium iodide *O,O*-diethyl phosphorothioate; diethoxyphosphinylthiocholine iodide; 2-diethoxyphosphinylthioethyltrimethylammonium iodide; *O,O*-diethyl *S*-(2-trimethylammoniumethyl)phosphorothioate iodide; *S*-β-dimethylaminoethyl-*O,O*-diethylthionophosphate methiodide; ecothiopate iodide; 217-MI; Phospholine Iodide. $C_9H_{23}INO_3PS$; mol wt 383.23. C 28.21%, H 6.05%, I 33.11%, N 3.65%, O 12.52%, P 8.08%, S 8.37%. Prepn: Tammelin, *Acta Chem. Scand.* **11**, 1340 (1957); Fitch, **US 2911430** (1959 to Campbell Pharmaceuticals). Comprehensive description: R. D. Daley, *Anal. Profiles Drug Subs.* **3**, 233-251 (1974).

White crystals or hygroscopic solid, mp 138° (Tammelin); mp 124-124.5° (Fitch). Freely sol in water, methanol; sol in dehydrated alcohol, chloroform. Practically insol in organic solvents.

THERAP CAT: Cholinergic (ophthalmic).

3549. Econazole. [27220-47-9] 1-[2-[(4-Chlorophenyl)methoxy]-2-(2,4-dichlorophenyl)ethyl]-1*H*-imidazole; 1-[2,4-dichloro-β-[(*p*-chlorobenzyl)oxy]phenethyl]imidazole; SQ-13050. $C_{18}H_{15}Cl_3$-N_2O; mol wt 381.68. C 56.64%, H 3.96%, Cl 27.86%, N 7.34%, O 4.19%. Prepn: Godefroi *et al.*, *J. Med. Chem.* **12**, 784 (1969); Godefroi, Heeres, **DE 1940388**; *eidem*, **US 3717655** (1970, 1973 both to Janssen). Biological and toxicological properties: Thienpont *et al.*, *Arznein.-Forsch.* **25**, 224 (1975). Review of antifungal activity and therapeutic efficacy: R. C. Heel *et al.*, *Drugs* **16**, 177-201 (1978). Comprehensive description: A. M. Dyas, H. Delargy, *Anal. Profiles Drug Subs. Excip.* **23**, 125-152 (1994).

mp 86.8°.

Nitrate. [24169-02-6] R-14827; Epi-Pevaryl; Gyno-Pevaryl; Ifenec; Micofugal; Micogin; Palavale; Pargin; Pevaryl; Spectazole. $C_{18}H_{15}Cl_3N_2O.HNO_3$; mol wt 444.69. White, odorless crystals from a mixture of 2-propanol, methanol and diisopropyl ether, mp 162°. uv max (methanol): 202, 225 nm. uv max (methanol): 265, 271, 280 nm ($A_{1cm}^{1\%}$ 9.4, 9.7, 4.9). Soly at 20° (g/100 ml): water <0.1; ethanol (96%) 2.0; acetone 1.5. Soluble in methanol; sparingly sol in chloroform; very slightly sol in ether. LD_{50} in mice, rats (mg/kg): 462.7, 667.7 orally (Thienpont).

THERAP CAT: Antifungal.

3550. Ecteinascidins. Family of tetrahydroisoquinoline alkaloids with antitumor activity isolated from the Caribbean tunicate, *Ecteinascidia turbinata.* Ecteinascidin 743 is the most abundant; bends DNA toward the major groove by selectively alkylating guanine from the minor groove. Prepn: K. L. Rinehart, T. G. Holt, **WO 8707610**; *eidem*, **US 5089273** (1987, 1992 both to Univ. Illinois); and structures: K. L. Rinehart *et al.*, *J. Org. Chem.* **55**, 4512 (1990); R. Sakai *et al.*, *Proc. Natl. Acad. Sci. USA* **89**, 11456 (1992). Crystal structures: Y. Guan *et al.*, *J. Biomol. Struct. Dyn.* **10**, 793 (1993). Biosynthetic studies: R. G. Kerr, N. F. Miranda, *J. Nat. Prod.* **58**, 1618 (1995).

Trabectedin

Trabectedin. [114899-77-3] (1'*R*,6*R*,6a*R*,7*R*,13*S*,14*S*,16*R*)-5-(Acetyloxy)-3',4',6,6a,7,13,14,16-octahydro-6',8,14-trihydroxy-

7',9-dimethoxy-4,10,23-trimethylspiro[6,16-(epithiopropanoxy-methano)-7,13-imino-12*H*-1,3-dioxolo[7,8]isoquino[3,2-*b*][3]-benzazocine-20,1'(2'*H*)-isoquinolin]-19-one; ecteinascidin 743; ET-743; Yondelis. $C_{39}H_{43}N_3O_{11}S$; mol wt 761.84. Enantioselective total synthesis: E. J. Corey *et al.*, *J. Am. Chem. Soc.* **118**, 9202 (1996). *In vitro* antitumor activity: E. Isbicka *et al.*, *Ann. Oncol.* **9**, 981 (1998). HPLC determn in plasma: H. Rosing *et al.*, *J. Chromatogr. B* **710**, 183 (1998). Clinical pharmacokinetics: C. van Kesteren *et al.*, *Clin. Cancer Res.* **6**, 4725 (2000). Review of mechanism of action: G. J. Aune *et al.*, *Anti-Cancer Drugs* **13**, 545-555 (2002); of clinical development: C. van Kesteren *et al.*, *ibid.* **14**, 487-502 (2003); of metabolism and hepatotoxicity: J. H. Beumer *et al.*, *Pharmacol. Res.* **51**, 391-398 (2005). $[\alpha]_D^{25}$ +114° (c = 0.1 in methanol).

THERAP CAT: Antineoplastic.

3551. ECTEOLA-Cellulose. [9015-13-8] Cellulose polymer with (chloromethyl)oxirane and 2,2',2''-nitrilotris[ethanol]; cellulose-ECTEOLA. Anion-exchange material prepd from cellulose, epichlorohydrin, and triethanolamine: Peterson, Sober, *J. Am. Chem. Soc.* **78**, 751 (1956); Veder, *J. Chromatogr.* **10**, 507 (1963).

3552. Ectoine. [96702-03-3] (4*S*)-3,4,5,6-Tetrahydro-2-methyl-4-pyrimidinecarboxylic acid; ectoin; RonaCare Ectoin. C_6-$H_{10}N_2O_2$; mol wt 142.16. C 50.69%, H 7.09%, N 19.71%, O 22.51%. Naturally occurring cyclic amino acid. Organic osmolyte; synthesized by microorganisms to maintain osmotic equilibrium in saline environments and to protect enzymes and whole cells from denaturation. Isoln from the halophilic, phototrophic bacterium, *Ectothiorhodospira halochloris*, and characterization as a compatible solute: E. A. Galinski *et al.*, *Eur. J. Biochem.* **149**, 135 (1985). Biosynthesis: P. Peters *et al.*, *FEMS Microbiol. Lett.* **71**, 157 (1990). Calorimetric analysis of biosynthesis efficiency: T. Maskow, W. Babel, *Biochim. Biophys. Acta* **1527**, 4 (2001). Chromatographic determn: V. Riis *et al.*, *Anal. Bioanal. Chem.* **377**, 203 (2003). Osmoprotection in *E. coli*: M. Jebbar *et al.*, *J. Bacteriol.* **174**, 5027 (1992). Mechanism of osmoprotection study: *idem et al.*, *ibid.* **187**, 1293 (2005). Enzyme stabilization study: K. Lippert, E. A. Galinski, *Appl. Microbiol. Biotechnol.* **37**, 61 (1992). *In vitro* prevention of UVA-induced skin damage: J. Buenger, H. Driller, *Skin Pharmacol. Physiol.* **17**, 232 (2004). Inhibition of β-amyloid·peptide aggregation and neurotoxicity: M. Kanapathipillai *et al.*, *FEBS Lett.* **579**, 4775 (2005).

Crystals from water-free methanol. mp ~280°. $[\alpha]_D^{20}$ +140° (c = 1.0 in methanol). Soly at 4°: 6 mol/kg water. Non-ionic character at physiological pH.

USE: Stabilizer in biotechnological processes; protects the taste of food during dehydration; moisturizer in cosmetics; protects skin against photoaging and formation of sunburn cells.

3553. Eculizumab. [219685-50-4] Anti-(human complement C5 α-chain) immunoglobulin (human-mouse monoclonal 5G1.1 heavy chain) disulfide with human-mouse monoclonal 5G1.1 light chain, dimer; h5G1.1 Fab; h5G1.1VHC + h5G1.1VLC; Soliris. Humanized monoclonal antibody directed against complement component C5; designed to prevent activation of complement-mediated inflammation and tissue injury. Prepn: M. J. Evans *et al.*, **WO 9529697**; *eidem*, **US 6355245** (1995, 2002 both to Alexion); and complement inhibition study: T. C. Thomas *et al.*, *Mol. Immunol.* **33**, 1389 (1996). Overview of development and therapeutic potential: M. Kaplan, *Curr. Opin. Investig. Drugs* **3**, 1017-1023 (2002). Clinical trial in paroxysmal nocturnal hemoglobinuria: P. Hillmen *et al.*, *N. Engl. J. Med.* **355**, 1233 (2006).

THERAP CAT: Anti-inflammatory.

3554. Edatrexate. [80576-83-6] *N*-[4-[1-[(2,4-Diamino-6-pteridinyl)methyl]propyl]benzoyl]-L-glutamic acid; 10-ethyl-10-deazaaminopterin; 10-EdAM; 10-EDAAM; CGP-30694. $C_{22}H_{25}$-N_7O_5; mol wt 467.49. C 56.52%, H 5.39%, N 20.97%, O 17.11%. Dihydrofolate reductase inhibitor. Prepn: F. M. Sirotnak, J. I. DeGraw, **FR 2464956** (1981 to SRI International); J. I. DeGraw, Jr., F.

M. Sirotnak, US 4369319 (1983). Synthesis: J. I. DeGraw *et al.*, *J. Med. Chem.* **25**, 1227 (1982); and resolution of diastereomers: *eidem, ibid.* **29**, 1056 (1986). HPLC determn: J. J. Kinahan *et al.*, *Anal. Biochem.* **150**, 203 (1985); O. Van Tellingen *et al.*, *J. Chromatogr.* **529**, 135 (1990). Pharmacology and toxicology: M. P. Fanucchi *et al.*, *Cancer Res.* **47**, 2334 (1987). Clinical evaluation in non-small cell lung cancer: J. S. Lee *et al*, *Invest. New Drugs* **8**, 299 (1990). Review of pharmacology and antitumor activity: F. M. Sirotnak *et al.*, *NCI Monogr.* **5**, 127-131 (1987).

Prepd as 1.75 H_2O, uv max (pH 13): 255, 370 nm (ε 30731, 7582).
THERAP CAT: Antineoplastic.

3555. EDC. [1892-57-5] N^3-(Ethylcarbonimidoyl)-N^1,N^1-dimethyl-1,3-propanediamine; *N*-(3-dimethylaminopropyl)-*N'*-ethylcarbodiimide; 1-ethyl-3-(3'-dimethylaminopropyl)carbodiimide; *N*-ethyl-*N'*-(3-dimethylaminopropyl)carbodiimide; EDAC; EDCI. $C_8H_{17}N_3$; mol wt 155.25. C 61.89%, H 11.04%, N 27.07%. Peptide coupling reagent frequently used to synthesize amide bonds. Exists as a mixture with its cyclic tautomer. Prepn of parent and derivatives: J. C. Sheehan *et al.*, *J. Org. Chem.* **26**, 2525 (1961). Tautomerism studies: T. Tenforde *et al.*, *ibid.* **37**, 3372 (1972); and reaction mechanism studies: A. Williams, I. T. Ibrahim, *J. Am. Chem. Soc.* **103**, 7090 (1981). Spectrophotometric determn: K. Seno *et al.*, *Anal. Sci.* **24**, 505 (2008). Synthetic applications: M. K. Dhaon *et al.*, *J. Org. Chem.* **47**, 1962 (1982); G.-J. Ho *et al.*, *ibid.* **60**, 3569 (1995); D. S. Bose, K. S. Sunder, *Synth. Commun.* **29**, 4235 (1999); H. Sai *et al.*, *Synthesis* **2003**, 201. Cross-linking applications: J. M. Lee *et al.*, *J. Mater. Sci. Mater. Med.* **7**, 531 (1996); H. M. Powell, S. T. Boyce, *Biomaterials* **27**, 5821 (2006). Brief description and review of synthetic uses: R. S. Pottorf, P. Szeto in *Encyclopedia of Reagents for Organic Synthesis* **4**, L. A. Paquette, Ed. (John Wiley & Sons, New York, 1995) pp 2430-2432.

Oil. *Corrosive.* bp$_{0.60}$ 53-54°; bp$_{0.27}$ 47-48°. n_D^{25} 1.4582; n_D^{22} 1.4594. pKa$_1$ 3.1; pKa$_2$ 11.1. Sol in ether.
Hydrochloride. [25952-53-8] $C_8H_{17}N_3$·HCl; mol wt 191.70. White, smooth, powdery crystals from dichloromethane-ether, mp 113.5-114.5°. *Irritant.* Sol in water, dichloromethane, DMF, THF.
Methiodide. [22572-40-3] [3-[(Ethylimidocarbonyl)amino]propyl]trimethylammonium iodide. $C_8H_{17}N_3$·CH$_3$I; mol wt 297.18. White solid from ethyl acetate-chloroform, mp 106.5-107.5°. *Irritant.*
USE: Reagent in synthetic organic chemistry. In the cross-linking of tissue derived biomaterials and of proteins to nucleic acids. In DNA labeling and in the prepn of immunoconjugates.

3556. EDDS. [186459-75-6] *N,N'*-1,2-Ethanediylbisaspartic acid; ethylenediamine *N,N'*-disuccinic acid. $C_{10}H_{16}N_2O_8$; mol wt 292.24. C 41.10%, H 5.52%, N 9.59%, O 43.80%. Hexadentate chelating agent with 2 chiral centers; structural isomer of EDTA, *q.v.* *S,S*-form is naturally occurring and biodegradable. Prepn: M. Barbier *et al.*, *Ann.* **668**, 132 (1963). *See also:* C. Kezerian, W. M. Ramsey, US 3158635 (1964 to Stauffer). Synthesis and metal chelating behavior of *S,S*-form: J. A. Neal, N. J. Rose, *Inorg. Chem.* **7**, 2405 (1968); of isomers: M. Orama *et al.*, *J. Chem. Soc. Dalton Trans.* **2002**, 4644. Bacterial production of *S,S*-form: R. Takahashi *et al.*, *Biosci. Biotechnol. Biochem.* **63**, 1269 (1999). Crystal structure of *S,S*-form: F. E. Scarbrough, D. Voet, *Acta Crystallogr.* **B32**, 2715 (1976). Biodegradation analysis of isomers: R. Takahashi *et*

al., *Biosci. Biotechnol. Biochem.* **61**, 1957 (1997); of *S,S*-form: P. C. Vandevivere *et al.*, *Environ. Sci. Technol.* **35**, 1765 (2001). HPLC determn of *S,S*-form: S. Tandy *et al.*, *J. Chromatogr. A* **1077**, 37 (2005). Study of use in radiochemical decontamination: P. W. Jones, D. R. Williams, *Appl. Radiat. Isot.* **54**, 587 (2001); in heavy metal phytoextraction from soil: H. Grcman *et al.*, *J. Environ. Qual.* **32**, 500 (2003); E. Meers *et al.*, *Chemosphere* **58**, 1011 (2005).

Crystals from methanol, mp 128-132° (Barbier). Also reported as mp 220-222° (Kezerian, Ramsey). Slightly sol in water. Insol in ethanol, acetone, benzene, and most organic solvents. Acid dissociation constants (25°, 0.1 M KNO$_3$): pK$_1$ 2.4; pK$_2$ 3.9; pK$_3$ 6.8; pK$_4$ 9.8.
S,S-Form. [20846-91-7] Crystal density: 1.44 g/cm^3.
USE: Chelating agent in detergents, electroless plating, rust removal, photography, and soil remediation.

3557. Edestin. [9007-57-2] Globular protein originally obtained from hemp seed *(Cannabis sativa):* Osborne, *Am. Chem. J.* **14**, 662 (1892); Stockwell *et al.*, *Proc. Seed Protein Conf.* **1963**, 56. Approx mol wt 300,000. Heat treatment of hemp seed to destroy its narcotic properties also destroys the edestin. Very closely related proteins may be obtained from seeds of *Cucurbita pepo* L. (pumpkin), *C. moschata* Duchesne, and *C. maxima* Duchesne (squash), *Citrullus vulgaris* Schrader (watermelon), *Cucumis melo* L. (cantaloupe), and *Cucumis sativus* L., *Cucurbitaceae* (cucumber). Isoln from *Cucurbita pepo:* Vickery *et al.*, *Biochem. Prep.* **2**, 5 (1952). Extraction from hemp seed, prepn of crystalline form and x-ray data: J. Drenth, E. W. Wiebenga, *Rec. Trav. Chim.* **74**, 813 (1955). The amino acid compositions of these globulins are slightly different from hemp seed globulin and from each other: Smith *et al.*, *J. Biol. Chem.* **164**, 159 (1946); Smith, Greene, *ibid.* **167**, 833 (1947); **172**, 111 (1948). Structure studies: Hall, *ibid.* **185**, 45 (1959); Cleemann, Kratky, *Z. Naturforsch.* **15b**, 526 (1960); Dlouhá *et al.*, *Collect. Czech. Chem. Commun.* **28**, 2779 (1963); **29**, 1835 (1964). Electron microscopy and optical diffraction: A. M. H. Schepman *et al.*, *Biochim. Biophys. Acta* **271**, 279 (1972). Electron spin resonance studies: L. J. Dimmey, W. Gordy, *Proc. Natl. Acad. Sci. USA* **77**, 343 (1980).

Octahedral crystals. Completely sol to a clear soln in 10% salt soln. Sol in dil mineral acids. Forms a water-soluble hydrochloride.

3558. Edifenphos. [17109-49-8] Phosphorodithioic acid *O*-ethyl *S,S*-diphenyl ester; *O*-ethyl *S,S*-diphenyl phosphorodithioate; EDDP; ediphenphos; Bayer 78418; Hinosan. $C_{14}H_{15}O_2PS_2$; mol wt 310.37. C 54.18%, H 4.87%, O 10.31%, P 9.98%, S 20.66%. Prepn: NL 6611860; G. Schrader *et al.*, US 3499951 (1967, 1970 both to Bayer). Degradn in soil: C. Tomizawa *et al.*, *J. Environ. Sci. Health* **B11**, 231 (1976). Metabolism: I. Ueyama *et al.*, *Agric. Biol. Chem.* **42**, 885 (1978). Mechanism of action: O. Kodama *et al.*, *ibid.* **44**, 1015 (1980). Toxicology: T. S. Chen *et al.*, *Toxicol. Appl. Pharmacol.* **23**, 519 (1972).

Clear yellow to light-brown liquid, bp$_{0.01}$ 154°. d$_4^{20}$ 1.23; n_D^{22} 1.61. Practically insol in water. Sol in acetone, xylene. LD$_{50}$ in female, male rats (mg/kg): 25.5, 66.5 i.p. (Chen).
USE: Fungicide.

3559. Edifoligide Sodium. [328538-04-1] DNA d(*P*-thio)(C-T-A-G-A-T-T-T-C-C-C-G-C-G) complex with DNA d(*P*-thio)(G-A-T-C-C-G-C-G-G-G-A-A-A-T) (1:1) hexacosasodium salt; E2F Decoy; E2FD; CGT-003. $C_{272}H_{318}N_{106}Na_{26}O_{138}P_{26}S_{26}$; mol wt 9516.75. C 34.33%, H 3.37%, N 15.60%, Na 6.28%, O 23.20%, P 8.46%, S 8.76%. Double stranded oligodeoxynucleotide that binds and inactivates the cell-cycle transcription factor E2F. Transcription factor decoy designed for *ex vivo* gene suppression therapy to prevent vein graft failure following coronary artery and peripheral artery bypass surgery. Description of sequence and use as decoy: V. J. Dzau *et al.*, **WO 9511687** (1995); *eidem*, **US 6774118** (2004 to Brigham and Women's Hospital). Binding inhibition study in vascular smooth muscle: R. Morishita *et al.*, *Proc. Natl. Acad. Sci. USA* **92**, 5855 (1995); in glomerular mesangial cells: N. Tomita *et al.*, *Am. J. Physiol.* **275**, F278 (1998). Pharmacology: A. Ehsan *et al.*, *J. Thorac. Cardiovasc. Surg.* **121**, 714 (2001). Clinical evaluation in vascular bypass: M. J. Mann *et al.*, *Lancet* **354**, 1493 (1999). Review of mechanism of action: M. J. Mann, *Antisense Nucleic Acid Drug Dev.* **8**, 171-176 (1998); of development and clinical experience: A. W. Hoel, M. S. Conte, *Cardiovasc. Drug Rev.* **25**, 221-234 (2007).

3560. Edotecarin. [174402-32-5] 12-β-D-Glucopyranosyl-12,13-dihydro-2,10-dihydroxy-6-[[2-hydroxy-1-(hydroxymethyl)-ethyl]amino]-5*H*-indolo[2,3-*a*]pyrrolo[3,4-*c*]carbazole-5,7(6*H*)-dione; ED-749; J-107088; PF-804950; PHA-782615. $C_{29}H_{28}N_4O_{11}$; mol wt 608.56. C 57.24%, H 4.64%, N 9.21%, O 28.92%. Topoisomerase I inhibitor that induces single strand DNA cleavage. Prepn: K. Kojiri *et al.*, **EP 760375**; *eidem*, **US 5922860** (1997, 1999 both to Banyu Pharm.); M. Ohkubo *et al.*, *Bioorg. Med. Chem. Lett.* **9**, 3307 (1999). Synthesis: A. Akao *et al.*, *Tetrahedron* **57**, 8917 (2001). Mechanism of action study: T. Yoshinari *et al.*, *Cancer Res.* **59**, 4271 (1999). *In vivo* antitumor activity: H. Arakawa *et al.*, *Jpn. J. Cancer Res.* **90**, 1163 (1999). LC/MS/MS determn in plasma: A. Q. Wang *et al.*, *Rapid Commun. Mass Spectrom.* **16**, 975 (2002). Review of development: W. A. Denny, *IDrugs* **7**, 173-177 (2004); and clinical experience: M. W. Saif, R. B. Diasio, *Clin. Colorectal Cancer* **5**, 27-36 (2005).

Orange solid, mp 330° (dec). $[\alpha]_D^{23}$ +117° (c = 0.8 in 1:1 acetonitrile/water).

THERAP CAT: Antineoplastic.

3561. Edotreotide. [204318-14-9] *N*-[2-[4,7,10-Tris(carboxymethyl)-1,4,7,10-tetraazacyclododec-1-yl]acetyl]-D-phenylalanyl-L-cysteinyl-L-tyrosyl-D-tryptophyl-L-lysyl-L-threonyl-*N*-[(1*R*,2*R*)-2-hydroxy-1-(hydroxymethyl)propyl]-L-cysteinamide cyclic (2 → 7)-disulfide; (DOTA D-Phe^1,Tyr^3)octreotide; DOTATOC; SMT-487. $C_{65}H_{92}N_{14}O_{18}S_2$; mol wt 1421.65. C 54.92%, H 6.52%, N 13.79%, O 20.26%, S 4.51%. Octapeptide analog of somatostatin, *q.v.* Designed for targeted radiotherapy vs somatostatin receptor-expressing tumors. Prepn: R. Albert *et al.*, **EP 714911**; *eidem*, **US 6183721** (1996, 2001 both to Novartis); *idem, et al.*, *Bioorg. Med. Chem. Lett.* **8**, 1207 (1998). Solution-phase synthesis: M. Schottelius *et al.*, *Tetrahedron Lett.* **44**, 2393 (2003). *In vivo* antineoplastic activity: B. Stolz *et al.*, *Eur. J. Nucl. Med.* **25**, 668 (1998). Receptor binding study: J. C. Reubi *et al.*, *ibid.* **27**, 273 (2000). Clinical pharmacokinetics: F. Jamar *et al.*, *Eur. J. Nucl. Med. Mol. Imaging* **30**, 510 (2003). Clinical evaluation in neuroendocrine tumors: C. Waldherr *et al.*, *J. Nucl. Med.* **43**, 610 (2002).

Prepd as acetate. $[\alpha]_{22}^D$ −14.75° (c = 0.52 in 95% acetic acid).

^{90}Y chelate. [322407-70-5] Yttrium Y 90 edotreotide; ^{90}Y-SMT-487; OctreoTher.

THERAP CAT: ^{90}Y chelate as antineoplastic.

3562. Edoxaban. [480449-70-5] N^1-(5-Chloro-2-pyridinyl)-N^2-[(1*S*,2*R*,4*S*)-4-[(dimethylamino)carbonyl]-2-[[(4,5,6,7-tetrahydro-5-methylthiazolo[5,4-*c*]pyridin-2-yl)carbonyl]amino]cyclohexyl]ethanediamide; DU-176. $C_{24}H_{30}ClN_7O_4S$; mol wt 548.06. C 52.60%, H 5.52%, Cl 6.47%, N 17.89%, O 11.68%, S 5.85%. Orally active direct factor Xa inhibitor. Prepn: T. Ohta *et al.*, **WO 03000657**; *eidem*, **US 7365205** (2003, 2008 both to Daiichi). Pharmacology: T. Furugohri *et al.*, *J. Thromb. Haemost.* **6**, 1542 (2008). Clinical pharmacokinetics and pharmacodynamics: K. Ogata *et al.*, *J. Clin. Pharmacol.* **50** 743 (2010). Clinical evaluation for postsurgical thromboprophylaxis: G. Raskob *et al.*, *Thromb. Haemostasis* **104**, 642 (2010). Review of development and therapeutic potential: E. M. Hylek, *Curr. Opin. Investig. Drugs* **8**, 778-783 (2007).

Hydrochloride. [480448-29-1] $C_{24}H_{30}ClN_7O_4S·HCl$; mol wt 584.52.

Tosylate. [480449-71-6] DU-176b; Lixiana. $C_{24}H_{30}ClN_7O_4S·C_7H_8O_3S$; mol wt 720.26. Crystals as monohydrate from ethanol + water, mp 245-248° (dec).

THERAP CAT: Antithrombotic.

3563. Edoxudine. [15176-29-1] 2′-Deoxy-5-ethyluridine; 5-ethyl-2′-deoxyuridine; 5-ethyl-1-(2′-deoxy-β-D-ribofuranosyl)uracil; EDU; EUDR; ORF-15817; Aedurid; Edurid. $C_{11}H_{16}N_2O_5$; mol wt 256.26. C 51.56%, H 6.29%, N 10.93%, O 31.22%. Substituted uracil with anti-herpes activity. Prepn and properties: K. K. Gauri, **GB 1170565**; *eidem*, **US 3553192** (1968, 1971 both to Robugen). Synthesis via organopalladium intermediates: D. E. Bergstrom, J. L. Ruth, *J. Am. Chem. Soc.* **98**, 1587 (1976); D. E. Bergstrom, M. K. Ogawa, *ibid.* **100**, 8106 (1978). Physical properties: M. Swierkowski, D. Shugar, *J. Med. Chem.* **12**, 533 (1969). Antiviral activity: E. De Clercq, D. Shugar, *Biochem. Pharmacol.* **24**, 1073 (1975). Preferential inhibition of herpes simplex virus type 2: C.-Z. Teh, S. L. Sacks, *Antimicrob. Agents Chemother.* **23**, 637 (1983). Synergistic effect with interferon-β, *q.v.*, on herpes simplex virus: C. Janz, R. Wigand, *Arch. Virol.* **73**, 135 (1982). Inhibitory effect on growth of murine leukemia cells: J. Balzarini *et al.*, *Invest. New Drugs* **2**, 35 (1984).

Long clear needles from acetone, mp 152-153°. uv max: 267 nm (ε 9610) at pH 2, 267 nm (ε 7280) at pH 1. pKa: 9.98.

THERAP CAT: Antiviral (herpes simplex).

3564. Edrophonium Chloride. [116-38-1] N-Ethyl-3-hydroxy-N,N-dimethylbenzenaminium chloride (1:1); ethyl(m-hydroxyphenyl)dimethylammonium chloride; (3-hydroxyphenyl)dimethylethylammonium chloride; 3-hydroxy-N,N-dimethyl-N-ethylanilinium chloride; Antirex; Enlon; Tensilon. $C_{10}H_{16}ClNO$; mol wt 201.69. C 59.55%, H 8.00%, Cl 17.58%, N 6.94%, O 7.93%. Cholinesterase inhibitor. Prepn: J. A. Aeschlimann, A. Stempel, US 2647924 (1953 to Hoffmann-La Roche). LC determn in biological fluids: M. G. M. DeRuyter et al., J. Chromatogr. **183**, 193 (1980). Clinical evaluation for reversal of neuromuscular blockade during anesthesia: D. M. Fisher et al., Anesthesiology **61**, 428 (1984); T. Suzuki et al., Can. J. Anaesth. **50**, 879 (2003). Clinical study in supraventricular tachycardia: J. D. Cantwell et al., Arch. Intern. Med. **130**, 221 (1972). Review of diagnostic use in myasthenia gravis: R. M. Pascuzzi, Semin. Neurol. **23**, 83-88 (2003).

Crystals from isopropanol, dec 162-163°. pH of 1% aq soln 4-5. Very sol in water; freely sol in alcohol. Insol in chloroform, ether.

Edrophonium Bromide. [302-83-0] Edrophone Bromide; Ro-2-3198. $C_{10}H_{16}BrNO$; mol wt 246.15. Crystals from ethanol + ether, dec 151-152°. Bitter taste. Sol in water (more than 10%). Moderately sol in alcohol. Practically insol in ether. Solns are stable.

THERAP CAT: Cholinergic; reversal agent for neuromuscular blockade; antidote to curare; diagnostic aid (myasthenia gravis).

THERAP CAT (VET): Cholinergic; reversal agent for neuromuscular blockade; diagnostic aid (myasthenia gravis).

3565. EDTA. [60-00-4] N,N'-1,2-Ethanediylbis[N-(carboxymethyl)glycine]; ethylenebis(iminodiacetic acid); (ethylenedinitrilo)tetraacetic acid; ethylenediaminetetraacetic acid; edathamil; edetic acid; Versene Acid. $C_{10}H_{16}N_2O_8$; mol wt 292.24. C 41.10%, H 5.52%, N 9.59%, O 43.80%. Powerful chelating agent; forms stable complexes with most metal ions. Prepn: F. Münz, US 2130505 (1938 to General Aniline); R. Smith et al., J. Org. Chem. **14**, 355 (1949). Prepn of α- and β-form of crystals: R. B. Le Blanc, H. L. Spell, J. Phys. Chem. **64**, 949 (1960). Efficacy in animal models of cadmium toxicity: L. R. Cantilena, Jr., C. D. Klaassen, Toxicol. Appl. Pharmacol. **53**, 510 (1980). Lead mobilization by calcium and zinc salts: C. F. Brownie, A. L. Aronson, ibid. **75**, 167 (1984). Determn in pharmaceutical formulations and foods: A. A. Krokidis et al., Anal. Chim. Acta **535**, 57 (2005). Clinical evaluation in calcific band keratopathy: D. M. Najjar et al., Am. J. Ophthalmol. **137**, 1056 (2004). Review of laboratory uses: W. C. Broad, Am. Lab. **3**, 47-51 (1971); of toxicology: K. Heindorff et al., Mutat. Res. **115**, 149-173 (1983); of use in personal care products: J. R. Hart, Cosmet. Toiletries **98**, 54-58 (1983); of determn methods: M. Sillanpää, M.-L. Sihvonen, Talanta **44**, 1487-1497 (1997); of properties and analysis methods: F. Belal, A. A. Al-Badr, Anal. Profiles Drug Subs. Excip. **29**, 57-104 (2002); of safety and efficacy in chelation therapy in atherosclerosis: M. T. Grier, D. G. Meyers, Ann. Pharmacother. **27**, 1504-1509 (1993); E. Ernst, Am. Heart J. **140**, 139-141 (2000). Safety assessment: Int. J. Toxicol. Suppl. 2, **21**, 95-142 (2002).

White, crystalline powder, mp 240-241° (dec). Soly in water at 20°: 0.2 g/100 g. Sol in solns of alkali hydroxides. The free acid tends to decarboxylate when heated to temps of 150°. Stable on storage and on boiling in aq soln.

Calcium disodium salt. [62-33-9] Sodium (OC-6-21)-[[N,N'-1,2-ethanediylbis[N-[(carboxy-κO)methyl]glycinato-$\kappa N,\kappa O$]](4$-$)]-calciate(2$-$) (2:1); edetate calcium disodium; Antallin; Calcium Disodium Versenate; Ledclair; Mosatil; Sormetal. $C_{10}H_{12}CaN_2$-Na_2O_8; mol wt 374.27. Prepn: Astakhov, Kiseleva, Zh. Obshch. Khim. **20**, 1780 (1950), C.A. **45**, 2409 (1951). Crystalline granules or powder. Slightly hygroscopic. Freely sol in water: at 30° a 0.1M soln can be prepd (pH $\sim$7). Practically insol in organic solvents.

Cobalt salt. [36499-65-7] Cobalt(2+) (OC-6-21)-[[N,N'-1,2-ethanediylbis[N-[(carboxy-κO)methyl]glycinato-$\kappa N,\kappa O$]]cobaltate(2$-$) (1:1); Dicobalt edetate; Kelocyanor. $C_{10}H_{12}Co_2N_2O_8$; mol wt 406.08.

Disodium salt. [139-33-3]; [6381-92-6] (dihydrate). Edetate disodium; tetracemate disodium; Chelaplex III; Chelatran; Endrate disodium; Titriplex III; Versene Na$_2$. $C_{10}H_{14}N_2Na_2O_8$; mol wt 336.21. Dihydrate: white crystals, mp 252° (dec). pH about 5.3. Soly in water at 20°: 10.8 g/100 ml.

Tetrasodium salt. [64-02-8] Edetate sodium; tetracemate tetrasodium; tetracemin; Endrate Tetrasodium; Versene 100; Sequestrene; Trilon B; Nullapon; Complexone; Distol 8; Calsol; Syntes 12a; Nervanaid B. $C_{10}H_{12}N_2Na_4O_8$; mol wt 380.17. Powder, mp 300° (dec). Apparent density: 6.9 lb/gallon. Sol in water: about 103 g/100 ml. pH of 1% soln 11.3. Less sol in alcohol than the potassium salt.

Trisodium salt. [150-38-9]; [10378-22-0] (monohydrate). Edetate trisodium; Limclair; Versene-9. $C_{10}H_{13}N_2Na_3O_8$; mol wt 358.19. Monohydrate: crystals from water, mp >300°. More sol in water than the corresp disodium salt or free acid. pH of 1% aq soln 9.3.

Magnesium disodium salt. [14402-88-1] Sodium (OC-6-21)-[[N,N'-1,2-ethanediylbis[N-[(carboxy-κO)methyl]glycinato-κN,-κO]](4$-$)]magnesate(2$-$) (2:1). $C_{10}H_{12}MgN_2Na_2O_8$; mol wt 358.50.

USE: Preservative and color retention agent in foods, flavoring agent. As antioxidant in foods. Pharmaceutic aid (chelating agent); anticoagulant for blood collection. Chelating agent in personal care products, in pulp and paper industry, in laboratory titrimetric analysis of metals and solubilization of metal compds.

THERAP CAT: Chelating agent (metal); dicobalt edetate as antidote (cyanide).

THERAP CAT (VET): In treatment of lead and heavy metal poisoning of farm animals.

3566. EEDQ. [16357-59-8] 2-Ethoxy-1(2H)-quinolinecarboxylic acid ethyl ester; N-carbethoxy-2-ethoxy-1,2-dihydroquinoline; N-ethoxycarbonyl-2-ethoxy-1,2-dihydroquinoline; BC-681. $C_{14}H_{17}NO_3$; mol wt 247.29. C 68.00%, H 6.93%, N 5.66%, O 19.41%. Coupling agent used in the synthesis of peptides: Belleau, Malek, J. Am. Chem. Soc. **90**, 1651 (1968); Yajima, Kawatani, Chem. Pharm. Bull. **19**, 1905 (1971); Sipos, Gaston, Synthesis **1971**, 321. Preparation: Weinberg, US 3389142 and US 3452140 (1968, 1969, to Bristol-Myers). Has CNS depressant activity: Belleau, J. Am. Chem. Soc. **90**, 823 (1968). Pharmacological studies: Martel et al., Can. J. Physiol. Pharmacol. **47**, 909 (1969); Chang et al., Pharmacol. Res. Commun. **2**, 63 (1970); Weissman, Muren, J. Med. Chem. **14**, 49 (1971).

mp 56-57°. bp$_{0.1}$ 125-128°.

USE: In the synthesis of peptides.

3567. Efalizumab. [214745-43-4] Anti-(human CD11a (antigen)) immunoglobulin G1 (human-mouse monoclonal hu1124 γ_1-chain) disulfide with human-mouse monoclonal hu1124 light chain, dimer; hu1124; Raptiva; Xanelim. Humanized monoclonal IgG1 antibody directed against the CD11a subunit of lymphocyte function-associated antigen-1 (LFA-1). Inhibits lymphocyte activation, migration and adhesion to endothelial cells. Prepn: P. M. Jardieu, L. G. Presta, **WO 9823761**; eidem, **US 6037454** (1998, 2000 both to

Genentech); W. A. Werther *et al., J. Immunol.* **157**, 4986 (1996). Clinical pharmacokinetics: R. J. Bauer *et al., J. Pharmacokinet. Biopharm.* **27**, 397 (1999). Clinical study in psoriasis: M. Lebwohl *et al., N. Engl. J. Med.* **349**, 2004 (2003). Review of pharmacology and therapeutic potential: R. L. Dedrick *et al., Transplant Immunol.* **9**, 181-186 (2002); J. C. Cather *et al., Expert Opin. Biol. Ther.* **3**, 361-370 (2003); of use in chronic psoriasis: J. C. Cather, A. Menter, *Expert Opin. Biol. Ther.* **5**, 393-403 (2005).

THERAP CAT: Antipsoriatic; immunosuppressant.

3568.   Efaproxiral. [131179-95-8] 2-[4-[2-[(3,5-Dimethyl-phenyl)amino]-2-oxoethyl]phenoxy]-2-methylpropanoic acid; 2-4-[[(3,5-dimethylanilino)carbonyl]methyl]phenoxy-2-methylpropionic acid; RSR-13; Efaproxyn. $C_{20}H_{23}NO_4$; mol wt 341.41. C 70.36%, H 6.79%, N 4.10%, O 18.74%. Allosteric modifier of hemoglobin (Hb). Binds in the central water cavity of the Hb molecule causing a conformational change such that bound oxygen is released more readily. Prepd not claimed: D. J. Abraham *et al.*, **US 5049695** (1991 to Center Innovative Technol.); *eidem, J. Med. Chem.* **34**, 752 (1991). LC/MS/MS determn in urine: M. Thevis *et al., J. Mass Spectrom.* **41**, 332 (2006). Oxygen binding affinities and mechanism of action: D. J. Abraham *et al., Biochemistry* **31**, 9141 (1992). Effect on myocardial oxygen concentration: P. S. Pagel *et al., J. Pharmacol. Exp. Ther.* **285**, 1 (1998). Antimetastatic effects: B. A. Teicher *et al., Cancer Chemother. Pharmacol.* **42**, 24 (1998). Clinical pharmacokinetics and pharmacodynamics: L. Kleinberg *et al., J. Clin. Oncol.* **17**, 2593 (1999). Clinical trial in brain metastases from breast cancer: J. H. Suh *et al., ibid.* **24**, 106 (2006).

Pale yellow, crystalline solid. mp 85°.

THERAP CAT: Antineoplastic adjunct (radiosensitizer).

3569.   Efavirenz. [154598-52-4] (4*S*)-6-Chloro-4-(2-cyclo-propylethynyl)-1,4-dihydro-4-(trifluoromethyl)-2*H*-3,1-benzoxazin-2-one; DMP-266; Stocrin; Sustiva. $C_{14}H_9ClF_3NO_2$; mol wt 315.68. C 53.27%, H 2.87%, Cl 11.23%, F 18.05%, N 4.44%, O 10.14%. Nonnucleoside reverse transcriptase inhibitor (NNRTI). Prepn: S. D. Young *et al.*, **EP 582455**; *eidem*, **US 5519021** (1994, 1996 both to Merck & Co.). Pharmacology: S. D. Young *et al., Antimicrob. Agents Chemother.* **39**, 2602 (1995). Enantioselective synthesis: A. S. Thompson *et al., Tetrahedron Lett.* **36**, 8937 (1995). Total synthesis and resolution of enantiomers: L. A. Radesca *et al., Synth. Commun.* **27**, 4373 (1997). Ionization behavior: S. R. Rabel *et al., Pharm. Dev. Technol.* **1**, 91 (1996). Clinical trial as component of combination therapy in HIV-1 infection in adults: S. Staszewski *et al., N. Engl. J. Med.* **341**, 1865 (1999); in children: S. E. Starr *et al., ibid.* 1874. Review of clinical experience: C. Fortin, V. Joly, *Expert Rev. Anti Infect. Ther.* **2**, 671-684 (2004).

Crystals from toluene:heptane, mp 139-141°. $[\alpha]_D^{20}$ −84.7° (c = 0.005 g/ml in CH_3Cl). $[\alpha]_D^{25}$ −94.1° (c = 0.300 in methanol). pKa 10.2.

THERAP CAT: Antiretroviral.

3570.   Eflornithine. [70052-12-9] 2-(Difluoromethyl)ornithine; DL-α-(difluoromethyl)ornithine; DFMO; RMI-71782. C_6H_{12}-$F_2N_2O_2$; mol wt 182.17. C 39.56%, H 6.64%, F 20.86%, N 15.38%, O 17.56%. Irreversible inhibitor of ornithine decarboxylase, an enzyme involved in polyamine biosynthesis. Prepn: B. W. Metcalf *et*

al., J. Am. Chem. Soc. **100**, 2551 (1978); P. Bey *et al., J. Org. Chem.* **44**, 2732 (1979). Effect on polyamine biosynthesis in trypanosomes: C. J. Bacchi *et al., Science* **210**, 332 (1980). Clinical pharmacokinetics: K. D. Haegele *et al., Clin. Pharmacol. Ther.* **30**, 210 (1981). Reviews of clinical experience in *Pneumocystis carinii* pneumonia: J. Sahai, A. J. Berry, *Pharmacotherapy* **9**, 29-33 (1989); in cancer chemotherapy: F. L. Meyskens, Jr., E. W. Gerner, *Clin. Cancer Res.* **5**, 945-951 (1999); in trypanosomiasis: C. Burri, R. Brun, *Parasitol. Res.* **90**, S49-S52 (2003). Review of topical use in hirsutism: J. A. Barman Balfour, K. McClellen, *Am. J. Clin. Dermatol.* **2**, 197-201 (2001).

Hydrochloride. [68278-23-9]; [96020-91-6] (monohydrate). Ornidyl; Vaniqa. $C_6H_{12}F_2N_2O_2\cdot HCl$; mol wt 218.63. Crystals from ethanol/water as the monohydrate, mp 183°. Freely sol in water; sparingly sol in ethanol.

THERAP CAT: Antineoplastic; antipneumocystic; antiprotozoal (Trypanosoma). In treatment of hirsutism.

3571.   Efonidipine. [111011-63-3] 5-(5,5-Dimethyl-2-oxido-1,3,2-dioxaphosphorinan-2-yl)-1,4-dihydro-2,6-dimethyl-4-(3-nitrophenyl)-3-pyridinecarboxylic acid 2-[phenyl(phenylmethyl)amino]-ethyl ester; 5-(5,5-dimethyl-1,3,2-dioxaphosphorinan-2-yl)-1,4-dihydro-2,6-dimethyl-4-(3-nitrophenyl)-3-pyridinecarboxylic acid 2-[phenyl(phenylmethyl)amino]ethyl ester *P*-oxide; 2-(*N*-benzylanilino)ethyl (±)-1,4-dihydro-2,6-dimethyl-4-(*m*-nitrophenyl)-5-phosphononicotinate, cyclic 2,2-dimethyltrimethylene ester. $C_{34}H_{38}N_3$-O_7P; mol wt 631.67. C 64.65%, H 6.06%, N 6.65%, O 17.73%, P 4.90%. Dihydropyridine calcium channel blocker. Prepn: K. Seto *et al.*, **WO 8704439**; *idem et al.*, **US 4885284** (1987, 1989 both to Nissan); and crystal structure: R. Sakoda *et al., Chem. Pharm. Bull.* **40**, 2362 (1992). Stereoselective synthesis of enantiomers and crystal structure of (*S*)-form: *idem et al., ibid.* 2377. Pharmacology: C. Shudo *et al., J. Pharm. Pharmacol.* **45**, 525 (1993). Mechanism of action study: T. Yamashita *et al., Jpn. J. Pharmacol.* **57**, 337 (1991). Clinical study: T. Saito *et al., Curr. Ther. Res.* **52**, 113 (1992).

Crystals from ethyl acetate, mp 169-170° (Sakoda); also reported as mp 155-156° (Seto).

Hydrochloride. [111011-53-1] $C_{34}H_{38}N_3O_7P\cdot HCl$. LD_{50} in mice (mg/kg): >600 orally (Seto).

Hydrochloride ethanolate. [111011-76-8] NZ-105; Landel. $C_{34}H_{38}N_3O_7P\cdot C_2H_5OH\cdot HCl$; mol wt 714.19. Yellow crystals from aq ethanol, mp 151° (dec).

(*S*)- or (*R*)-Form. Pale yellow crystals from ethanol, mp 190-192°. $[\alpha]_D^{25}$ + or −7.0° resp (c = 0.50 in chloroform).

THERAP CAT: Antihypertensive.

3572.   Efrotomycin. [56592-32-6] ($\alpha S,2R,3R,4R,6S$)-4-[[6-Deoxy-4-*O*-(6-deoxy-2,4-di-*O*-methyl-α-L-mannopyranosyl)-3-*O*-methyl-β-D-allopyranosyl]oxy]-*N*-[(2*E*,4*E*,6*S*,7*R*)-7-[(2*S*,3*S*,4*R*,-5*R*)-5-[(1*E*,3*E*,5*E*)-7-(1,2-dihydro-4-hydroxy-1-methyl-2-oxo-3-pyridinyl)-6-methyl-7-oxo-1,3,5-heptatrien-1-yl]tetrahydro-3,4-di-hydroxy-2-furanyl]-6-methoxy-5-methyl-2,4-octadien-1-yl]-α-eth-yltetrahydro-2,3-dihydroxy-5,5-dimethyl-6-(1*E*,3*Z*)-1,3-pentadien-1-yl-2*H*-pyran-2-acetamide; 31-*O*-[6-deoxy-4-*O*-(6-deoxy-2,4-di-*O*-methyl-α-L-mannopyranosyl)-3-*O*-methyl-β-D-allopyranosyl]-1-methylmocimycin; FR-02A; MK-621. $C_{59}H_{88}N_2O_{20}$; mol wt

1145.35. C 61.87%, H 7.74%, N 2.45%, O 27.94%. Antibiotic produced by *Streptomyces lactamdurans* NRRL 3802: R. G. Wax, W. M. Maiese, **DE 2450813** (1975 to Merck & Co.), *C.A.* **83**, 145755y (1975); R. G. Wax *et al.*, *J. Antibiot.* **29**, 670 (1976). *In vitro* and *in vivo* activity: B. M. Frost *et al.*, *ibid.* 1083; **32**, 626 (1979). Production and growth promoting activity: W. M. Maiese, R. G. Wax, **US 4024251** (1977 to Merck & Co.). Synergism with bottromycin, *q.v.*: B. M. Frost *et al.*, *J. Antibiot.* **32**, 1046 (1979). Structure: R. S. Dewey *et al.*, *ibid.* **38**, 1691 (1985). Stereospecific total synthesis: R. E. Dolle, K. C. Nicolaou, *J. Am. Chem. Soc.* **107**, 1691, 1695 (1985). HPLC determn in feeds: J. D. Strong, *Analyst* **111**, 853 (1986). Effect on gain and feed efficiency in swine: A. G. Foster *et al.*, *J. Anim. Sci.* **65**, 877 (1987).

Pale yellow solid. uv max (pH 7): 232, 327 nm ($E_{1cm}^{1\%}$ 464, 216). LD_{50} in mice (g/kg): >4 orally; >2 s.c. (Frost).

THERAP CAT (VET): Growth stimulant.

3573. EGCG. [989-51-5] 3,4,5-Trihydroxybenzoic acid, (2*R*,3*R*)-3,4-dihydro-5,7-dihydroxy-2-(3,4,5-trihydroxyphenyl)-2*H*-1-benzopyran-3-yl ester; (−)-epigallocatechin 3-*O*-gallate; (−)-epigallocatechol gallate. $C_{22}H_{18}O_{11}$; mol wt 458.38. C 57.65%, H 3.96%, O 38.39%. Polyphenolic constituent of tea; inhibits tumor promotion. Indirectly inhibits AhR (aryl hydrocarbon receptor) transcription activation by directly binding to the C-terminus of hsp90, an AhR chaperone protein. Initial identification and isoln from green tea: M. Tsujimura, *Bull. Agric. Chem. Soc. Jpn.* **6**, 70 (1930), *C.A.* **25**, 3637 (1931); and crystallization: L. Vuataz *et al.*, *J. Chromatogr.* **2**, 173 (1959). Oxidation during tea fermentation: P. Coggon *et al.*, *J. Agric. Food Chem.* **21**, 727 (1973). HPLC/MS extraction from black tea: R. G. Bailey *et al.*, *J. Sci. Food Agric.* **66**, 203 (1994). HPLC determn in plasma and urine: M.-J. Lee *et al.*, *Cancer Epidemiol. Biomarkers Prev.* **4**, 393 (1995). Antitumor promoting activity: S. Yoshizawa *et al.*, *Phytother. Res.* **1**, 44 (1987); T. Yamane *et al.*, *Cancer Res.* **55**, 2081 (1995). Inhibition of metastasis in mice: S. Taniguchi *et al.*, *Cancer Lett.* **65**, 51 (1992). Modulation of endocrine systems and food intake: Y.-H. Kao, *et al.*, *Endocrinology* **141**, 980 (2000). Inhibition of AhR transcription by binding to hsp90: C. M. Palermo *et al.*, *Biochemistry* **44**, 5041 (2005).

White crystals from water, mp 218°. $[\alpha]_D$ −185 ±2°(ethanol). uv max (ethanol): 275 nm (ε 11500).

3574. EGF-Urogastrone. EGF-URO. Related polypeptides that are both potent stimulators of cellular proliferation and inhibitors

of gastric acid secretion. *Urogastrone* was originally detected as an antisecretory agent during experiments on human urine: J. S. Gray *et al.*, *Science* **89**, 489 (1939); M. H. F. Friedman *et al.*, *Proc. Soc. Exp. Biol. Med.* **41**, 509 (1935). Isoln: J. S. Gray *et al.*, *Endocrinology* **30**, 129 (1942); R. A. Gregory, *J. Physiol.* **129**, 528 (1955). Improved procedures led to the isoln and amino acid sequence determn of two polypeptides, *β-urogastrone* and *γ-urogastrone*: H. Gregory, *Nature* **257**, 325 (1975). These two peptides contain 3 disulfide bonds and consist of 53 and 52 amino acid residues, respectively. The only difference between them is the absence of a C-terminal arginine residue in the γ-peptide. *Epidermal growth factor, or EGF*, was isolated from submaxillary glands of male mice after first being detected during purification of a nerve growth promoting protein: S. Cohen, *Proc. Natl. Acad. Sci. USA* **46**, 302 (1960); *idem, J. Biol. Chem.* **237**, 1555 (1962). The primary structure of mouse EGF (mEGF) was found to be a 53 amino acid polypeptide containing 3 disulfide bonds: C. R. Savage *et al.*, *ibid.* **247**, 7612 (1972); **248**, 7669 (1973). Sequence of an mEGF cDNA clone that predicts the synthesis of EGF as a large protein precursor of 1,168 amino acids: A. Gray *et al.*, *Nature* **303**, 722 (1983). mEGF causes premature eye opening and stimulation of epithelial cell tissue growth when injected daily into newborn mice: S. Cohen, *Proc. Natl. Acad. Sci. USA* **46**, 32 (1962); S. Cohen, J. M. Taylor, *Recent Prog. Horm. Res.* **30**, 533 (1974). The structural similarities between EGF and urogastrone led to experiments that established the gastric acid-inhibitory activity of mEGF and the proliferative activity of urogastrone, thus showing that both polypeptides have the same intrinsic biological activities, *cf.* H. Gregory, *loc. cit.* It has also been shown that both mEGF and urogastrone can share the same receptor sites in human tissue with almost equal affinities: M. D. Hollenberg, H. Gregory, *Life Sci.* **20**, 267 (1976). Human EGF has been isolated from urine: S. Cohen, G. Carpenter, *Proc. Natl. Acad. Sci. USA* **72**, 1317 (1975). The similarity of its amino acid composition and physicochemical properties to urogastrone has suggested the identity of these polypeptides. Although the mouse and human polypeptides are chemically distinct (16 differences in amino acid sequence), their identical intrinsic biological activities and their ability to share receptor sites with similar affinities has led to the use of the combined term EGF-urogastrone. *Reviews:* S. Cohen, R. Savage, *Recent Prog. Horm. Res.* **30**, 551-574 (1974); M. D. Hollenberg, *Vitam. Horm.* **37**, 69-110 (1979); D. Gospodarowicz, *Annu. Rev. Physiol.* **43**, 251-263 (1981); P. Walker, *J. Endocrinol. Invest.* **5**, 183-196 (1982); M. Das, *Int. Rev. Cytol.* **78**, 233-256 (1982).

Mouse EGF-URO. Mol wt 6041. uv max: 280 nm ($E_{1cm}^{1\%}$ 30.9). Isoelectric pt 4.60. Heat stable and non-dialyzable. Biological activity stable in boiling water but destroyed by heating in dil acid or alkali. Incubation with chymotrypsin or a bacterial protease also destroys biological activity.

Human EGF-URO. Anthelone; anthelone U; uroanthelone; uroenterone. Mol wt 6201. Isoelectric pt 4.5. Very sol in water. Sol in methanol, ethylene gycol.

3575. Egg Oil. Oil of egg yolk. Obtained from fresh egg yolks by extraction with ethylene dichloride. Contains fatty glycerides, cholesterol, lecithin. The glyceride fraction is a mixture of the glycerides of satd and unsatd fatty acids. Palmitic, stearic, oleic, linoleic and clupanodonic acids have been isolated from both the glyceride and lecithin fractions. Preparation: Levin, Lerman, *J. Am. Oil Chem. Soc.* **28**, 441 (1951); **US 2503312** (to Vio-Bin Corp.).

Dark oil. d 0.95. n_D^{20} 1.4790. Sol in the common organic solvents. Miscible with other oils. Insol in water, but disperses readily when shaken with it to form emulsions.

USE: In the formulation of hydrophilic ointment bases for medicinal ointments and cosmetic creams: Bandelin, Tuschhoff, *J. Am. Pharm. Assoc. Pract. Pharm. Ed.* **14**, 106, 120 (1953); *Schimmel-Briefs*, no. 235 (Oct. 1954).

3576. EGTA. [67-42-5] 3,12-Bis(carboxymethyl)-6,9-dioxa-3,12-diazatetradecanedioic acid; [ethylenebis(oxyethylenenitrilo)]-tetraacetic acid; ethylene glycol bis(β-aminoethyl ether)-*N,N,N′,N′*-tetraacetic acid. $C_{14}H_{24}N_2O_{10}$; mol wt 380.35. C 44.21%, H 6.36%, N 7.37%, O 42.06%. Chelating agent. Prepn: H. Schläpfer, J. Bindler, **US 2709178** (1955 to Geigy). Toxicology: J. E. Wynn *et al.*, *Toxicol. Appl. Pharmacol.* **16**, 807 (1970). Thermal properties: M. F. G. Esteban *et al.*, *Thermochim. Acta* **62**, 267 (1983). Cathodic-stripping determn: M. Ciszkowska, Z. Stojek, *Talanta* **33**,

817 (1986). Use in calcium measurements: J. D. Johnson *et al.*, *Am. J. Physiol.* **272**, C1437 (1997); in heavy metal remobilization from river sediment: K.-C. Yu *et al.*, *Toxicol. Environ. Chem.* **58**, 85 (1997). Review of regulation of Ca^{+2} concentration in biological systems: S. Robertson, J. D. Potter, *Methods Pharmacol.* **5**, 63-75 (1984).

Disodium salt. [26082-78-0] $C_{14}H_{22}N_2Na_2O_{10}$; mol wt 424.31. LD_{50} orally in rats: 3.96 ± 0.50 g/kg (Wynn).

USE: Chelator for heavy metals. Reagent for maintaining calcium concentrations.

3577. Eicosamethylnonasiloxane. [2652-13-3] 1,1,1,3,3,5,-5,7,7,9,9,11,11,13,13,15,15,17,17,17-Eicosamethylnonasiloxane. $C_{20}H_{60}O_8Si_9$; mol wt 681.46. C 35.25%, H 8.88%, O 18.78%, Si 37.09%. Prepd by the reaction of hexamethyldisiloxane with octamethylcyclotetrasiloxane and sulfuric acid; by varying the proportions of these reactants, other methylpolysiloxanes of a desired average molecular weight may be prepd: Patnode, Wilcock, *J. Am. Chem. Soc.* **68**, 362 (1946).

Liquid. $bp_{4.9}$ 173°. d 0.918. n_D^{20} 1.3980. Stable. Inert to most chemical reagents and rubber. Maintains about the same viscosity over a wide temp range. Sol in benzene and the lighter hydrocarbons; slightly sol in alc and the heavy hydrocarbons.

USE: As a basis for silicone oils or fluids designed to withstand extremes of temp; as a foam suppressant in petr lubricating oil.

3578. Eicosapentaenoic Acid. [10417-94-4] (5Z,8Z,11Z,-14Z,17Z)-5,8,11,14,17-Eicosapentaenoic acid; *(all-Z)*-5,8,11,14,17-eicosapentaenoic acid; *all-cis*-fatty acid 20:5 omega-3; EPA; icosapent. $C_{20}H_{30}O_2$; mol wt 302.46. C 79.42%, H 10.00%, O 10.58%. Important polyunsaturated fatty acid of the marine food chain that serves as a precursor for the prostaglandin-3 and thromboxane-3 families. It differs from arachidonic acid, *q.v.* (the eicosatetraenoic acid that is a precursor for the prostaglandin and thromboxane-2 families) by the extra double bond between the third and fourth carbons from the "methyl end" of the molecule. Isoln from cod liver oil: E. Klenk, D. Eberhagen, *Z. Physiol. Chem.* **307**, 42 (1957). Enzymatic conversion to prostaglandin E_3: S. Bergström *et al.*, *J. Biol. Chem.* **239**, PC 4006 (1964). Effects on role of platelets in thrombosis: K. C. Srivastava *et al.*, *Biochem. Exp. Biol.* **16**, 317 (1980). Effects on prostacyclin-like material in human umbilical vasculature: J. Dyerberg, K. A. Jorgensen, *Artery* **8**, 12 (1980). A possible relationship between diets rich in EPA in marine oils and low rates of ischemic heart disease has been proposed: H. O. Bang *et al.*, *Am. J. Clin. Nutr.* **33**, 2657 (1980); J. Dyerberg, *Philos. Trans. R. Soc. London Ser. B* **294**, 373 (1981). Clinical evaluation of lipid lowering effect: Y. Nagakawa *et al.*, *Atherosclerosis* **47**, 71 (1983); of use in rheumatoid arthritis: J. M. Kremer *et al.*, *Ann. Intern. Med.* **106**, 497 (1987); of use in Raynaud's phenomenon: R. A. DiGiacomo *et al.*, *Am. J. Med.* **86**, 158 (1989).

Colorless oil. n_D^{20} 1.49865.
Ethyl ester. [86227-47-6] Epadel. $C_{22}H_{34}O_2$; mol wt 330.51.
THERAP CAT: Antilipemic.

3579. Einsteinium. [7429-92-7] Es; at. no. 99; valence 3, 2. Man-made radioactive element. No stable nuclides; known isotopes (mass numbers): 243-256. Longest-lived known isotope: 252 ($T_{1/2}$ 472 days, α-emitter, rel. at. mass 252.0830). ^{253}Es ($T_{1/2}$ 20.47 days, α-emitter) originally discovered in debris from a thermonuclear test explosion in Nov. 1952. Reviews of discovery: A. Ghiorso *et al.*, *Phys. Rev.* **99**, 1048-1049 (1955); A. Ghiorso, G. T. Seaborg, *Sci. Am.* **195** (6), 67-80 (1956). Prepn of isotopes: A. Chetham-Strode, Jr., L. W. Holm, *Phys. Rev.* **104**, 1314 (1956); B. G. Harvey *et al.*, *ibid.* 1315; G. T. Seaborg, *J. Chem. Educ.* **36**, 38 (1959). *Reviews:* C. Keller, *The Chemistry of the Transuranium Elements* (Verlag Chemie, Weinheim, English Ed., 1971) pp 583-589, Silva, "Trans-Curium Elements" in *MTP Int. Rev. Sci.: Inorg. Chem., Ser. One* vol. 8, A. G. Maddock, Ed. (University Park Press, Baltimore, 1972) pp 71-105; *Comprehensive Inorganic Chemistry* vol. **5**, J. C. Bailar, Jr. *et al.*, Eds. (Pergamon Press, Oxford, 1973) *passim; Handb. Exp. Pharmakol.* **36**, 689-928 (1973); E. K. Hulet in *The Chemistry of the Actinide Elements* vol. 2, J. J. Katz *et al.*, Eds. (Chapman and Hall, New York, 1986) pp 1071-1085; *The Elements Beyond Uranium*, G. T. Seaborg, W. D. Loveland, Eds. (John Wiley & Sons, Inc., New York, 1990) pp 28-38.

Metal; two crystalline forms: double hexagonal close-packed α-form exists below 860°; face-centered cubic β-form, d 8.84. mp 860 ±50°.

Caution: Radiation hazard; handling requires special equipment and shielding facilities (Katz *et al.*, *loc. cit.* vol. 2, p. 1128).

3580. Elagolix. [834153-87-6] 4-[[(1R)-2-[5-(2-Fluoro-3-methoxyphenyl)-3-[[2-fluoro-6-(trifluoromethyl)phenyl]methyl]-3,6-dihydro-4-methyl-2,6-dioxo-1(2H)-pyrimidinyl]-1-phenylethyl]amino]butanoic acid; 3-[2R-(hydroxycarbonylpropylamino)-2-phenylethyl]-5-(2-fluoro-3-methoxyphenyl)-1-[2-fluoro-6-(trifluoromethyl)benzyl]-6-methylpyrimidine-2,4(1H,3H)-dione; NBI-56418. $C_{32}H_{30}F_5N_3O_5$; mol wt 631.60. C 60.85%, H 4.79%, F 15.04%, N 6.65%, O 12.67%. Nonpeptide gonadotropin-releasing hormone (GnRH) antagonist. Prepn: Z. Guo *et al.*, **WO 05007165**; *eidem*, **US 7056927** (2005, 2006 both to Neurocrine Biosci.). Synthesis, structure-activity study, and pharmacology: C. Chen *et al.*, *J. Med. Chem.* **51**, 7478 (2008). Clinical pharmacokinetics and effect on serum gonadotropins: R. S. Struthers *et al.*, *J. Clin. Endocrinol. Metab.* **94**, 545 (2009).

Sodium salt. [832720-36-2] $C_{32}H_{29}F_5N_3NaO_5$; mol wt 653.58. White solid.

THERAP CAT: In treatment of endometriosis.

3581. Elaidic Acid. [112-79-8] (9E)-9-Octadecenoic acid. $C_{18}H_{34}O_2$; mol wt 282.47. C 76.54%, H 12.13%, O 11.33%. Stereoisomer of oleic acid, *q.v.* Prepd by the action of HNO_2 on oleic acid. ^{13}C-NMR: J. G. Batchelor *et al.*, *J. Org. Chem.* **39**, 1698 (1974).

White leaflets. d^{79} 0.851. mp 44-45°; also stated as 51°. bp_{100} 288°; bp_{15} 234°. n_D^{100} 1.4308.

3582. Elaiomycin. [23315-05-1] 4-Methoxy-3-[(1E)-2-[(1Z)-1-octenyl]-2-oxidodiazenyl]-2-butanol; (E,Z)-(2S,3S)-4-methoxy-3-(1-octenyl-*ONN*-azoxy)-2-butanol; (2S,3S)-4-methoxy-3-(1'-*cis*-octenyl-*cis*-azoxy)-2-butanol; D-*threo*-4-methoxy-3-(1-octenylazoxy)-2-butanol. $C_{13}H_{26}N_2O_3$; mol wt 258.36. C 60.44%, H 10.14%, N

10.84%, O 18.58%. Antibiotic substance produced by *Streptomyces hepaticus:* Haskell *et al., Antibiot. Chemother.* **4**, 141 (1954); Ehrlich *et al., ibid.* 338; Anderson *et al., ibid.* **6**, 100 (1956). Inhibits the growth of *Mycobacterium tuberculosis* var. *hominis in vitro:* Karlson, *ibid.* **12**, 446 (1962). Reported to have carcinogenic activity: Schoental, *Nature* **221**, 765 (1969). Structure: Stevens *et al., J. Am. Chem. Soc.* **78**, 3229 (1956); **80**, 6088 (1958). Configuration: Stevens *et al., ibid.* **81**, 1435 (1959); McGahren, Kunstmann, *ibid.* **92**, 1587 (1970). Total synthesis: R. A. Moss, M. Matsuo, *ibid.* **99**, 1643 (1977). Biosynthesis: R. J. Parry *et al., ibid.* **104**, 339 (1982).

Pale yellow oil. Stable to air. $[\alpha]_D^{26}$ +38.4° (c = 2.8 in abs ethanol). n_D^{25} 1.4798. uv max: 237.5 nm ($E_{1cm}^{1\%}$ 428). Sparingly sol in water; sol in practically all of the common organic solvents. Stable in neutral or slightly acidic aq solns. Dec into a yellow product when dissolved in 0.1N NaOH.

Acetate. $C_{15}H_{28}N_2O_4$. Oily liquid, $bp_{0.5}$ 84-90°. $[\alpha]_D^{27}$ +25.3° (c = 3 in alc). uv max (alc): 237.5 nm (ε 11000).

3583. Elastase. Serine proteases produced by many different cell types, named for its ability to break down the connective tissue protein elastin. Active site is a catalytic triad of residues Ser-195, His-57, and Asp-102. Hydrolyzes proteins at N-terminal peptide bonds of small aliphatic residues. First isolated from pancreatic extracts: J. Balo, I. Banga, *Biochem. J.* **46**, 384 (1950). Amino acid sequence comparison of mammalian elastases: T. Tani *et al., J. Biol. Chem.* **263**, 1231 (1988). Review and comparison of porcine pancreatic and human neutrophil elastase: W. Bode *et al., Biochemistry* **28**, 1951-1963 (1989). Review of role in atherosclerosis: L. Robert *et al., Atherosclerosis* **140**, 281-295 (1998); of role in pulmonary hypertension: M. Rabinovitch, *Chest* **114**, Suppl., 213S-224S (1998).

Pancreatic elastase. Elaszym. Stored as an inactive zymogen in the pancreas; secreted into intestines where it is activated by trypsin and participates in digestion. Mol wt about 25,000. Porcine form is a single polypeptide chain of 240 amino acid residues and four disulfide bridges. Purification and characterization: U. J. Lewis *et al., J. Biol. Chem.* **222**, 705 (1956); M. A. Naughton, F. Sanger, *Biochem. J.* **78**, 156 (1961). Primary structure: D. M. Shotton, B. S. Hartley, *Nature* **225**, 802 (1970); *eidem, Biochem. J.* **131**, 643 (1973). Crystal structure: H. C. Watson *et al., Nature* **225**, 806 (1970); D. M. Shotton, H. C. Watson, *ibid.* 811; E. Meyer *et al., Acta Crystallogr.* **B44**, 26 (1988). Review of diagnostic uses: R. Dominici, C. Franzini, *Clin. Chem. Lab. Med.* **40**, 325-332 (2002). White, lyophilized powder of slightly yellowish shade. pI 9.5 ±0.5. Maximum protease activity pH 8.1-8.8.

Neutrophil elastase. Leukocyte elastase; polymorphonuclear elastase; granulocyte elastase. Degrades extracellular matrix proteins including elastin, collagen (types I-IV), fibronectin, laminin, and proteoglycans. Necessary for migration of neutrophils through connective tissue. Defends against infectious bacteria by degrading bacterial structural proteins. Human form is a glycoprotein single peptide chain of 218 amino acid residues and four disulfide bridges. Mol wt about 33,000. Identification of elastolytic activity in polymorphonuclear leukocytes: A. Janoff, J. Scherer, *J. Exp. Med.* **128**, 1137 (1968). Isoln and characterization: A. Janoff, *Lab. Invest.* **29**, 458 (1973); K. Ohlsson, I. Olsson, *Eur. J. Biochem.* **42**, 519 (1974); R. J. Baugh, J. Travis, *Biochemistry* **15**, 836 (1976). Primary structure: S. Sinha *et al., Proc. Natl. Acad. Sci. USA* **84**, 2228 (1987). Crystal structure: H. R. Williams *et al., J. Biol. Chem.* **262**, 17178 (1987). Review of role in emphysema: A. Janoff, *Am. Rev. Respir. Dis.* **132**, 417-433 (1985); of role in acute lung injury: K. Kawabata *et al., Eur. J. Pharmacol.* **451**, 1-10 (2002). Maximum protease activity pH 7.5-9. pI 10-11.

3584. Elastin. Elastic, load-bearing protein fibers of animal connective tissue, particularly the ligaments of the vertebrae and the walls of the large arteries. Elastin is an insoluble, highly cross-linked hydrophobic protein, rich in nonpolar amino acid residues, such as

valine, leucine, isoleucine, and phenylalanine. On the average, about every third residue is glycine and about every ninth residue is a prolyl residue. Unlike collagen, *q.v.*, which is rich in hydroxyproline, elastin contains only modest amounts. At least two types of elastin are distinguishable: type I elastin isolated from *liga-mentum nuchae*, aorta or skin and type II elastin isolated from cartilage. The most ready source of high purity elastin is the *ligamentum nuchae* of the larger ruminants: Partridge *et al., Biochem. J.* **61**, 11 (1955). Amino acid composition studies: Petruska, Sandberg, *Biochem. Biophys. Res. Commun.* **33**, 222 (1968). Its elastic properties are brought about by a cross-linked structure. Identification of new cross-linking amino acids in elastin: Partridge *et al., Biochem. J.* **93**, 30C (1964); Franzblau *et al., Biochem. Biophys. Res. Commun.* **21**, 575 (1965); Starcher *et al., Biochemistry* **6**, 2425 (1967). Molecular model: Gray *et al., Nature* **246**, 461 (1973). *Review:* Partridge, *Adv. Protein Chem.* **17**, 227-302 (1962); C. Franzblau, "Elastin" in *Comprehensive Biochemistry* vol. **26c**, M. Florkin, E. H. Stotz, Eds. (Elsevier, New York, 1971) pp 659-712; Ross, Bornstein, *Sci. Am.* **224**, 44 (June, 1971). Book (in English): I. Banga, *Structure and Function of Elastin and Collagen* (Akademiai Kiado, Budapest, 1967). Series of articles on structure, biosynthesis, and immunology: *Meth-ods Enzymol.* **82**, 559-765 (1982).

Purified elastin has a pale yellow color and a bluish fluorescence in uv light. Resists acid and alkaline hydrolysis. Practically insol in a wide range of hydrogen-bond-breaking solvents at temps up to 100° and swells, but does not dissolve, in phenolic solvents. It appears practically impossible to bring elastin into soln except by hydrolytic reagents capable of rupturing peptide bonds. Elastin is one of only a few polymeric substances which, in the presence of water, exist in a form with rubber-like extensibility and low modulus of elasticity. Enzymes which dissolve fibers of the insol protein elastin undamaged and free from contamination are called elastases.

3585. Elcatonin. [60731-46-6] 1-Butanoic acid 1,7-dicarbacalcitonin (eel); 1,4,7,10,13,16-hexaazacyclotricosane cyclic peptide deriv; [aminosuberic acid 1,7]-eel calcitonin; [ASU^{1,7}]-ECT; carbocalcitonin; HC-58; Calcinil; Elcitonin; Turbocalcin. $C_{148}H_{244}N_{42}$-O_{47}; mol wt 3363.83. C 52.85%, H 7.31%, N 17.49%, O 22.35%. Synthetic analog of eel calcitonin in which the 1 and 7 cystine residues have been modified to deaminosuberic acid. Inhibits osteoclast bone resorption and induces uptake of calcium from body fluids. Prepn: S. Sakakibara *et al., DE 2616399; eidem, US 4086221* (1976, 1978 both to Toyo Jozo); T. Morikawa *et al., Experientia* **32**, 1104 (1976). Structural characterization: P. L. Mauri *et al. Rapid Commun. Mass Spectrom.* **11**, 1292 (1997). Immunoassay determn in plasma: M. Takeyama *et al. Biol. Pharm. Bull.* **18**, 900 (1995); HPLC determn in pharmaceutical prepn: K. Maekawa *et al. Iyakuhin Kenkyu* **32**, 465 (2001). Mechanistic study: M. Ikegame *et al., J. Bone Miner. Res.* **9**, 25 (1994). Clinical pharmacology: T. Ishioka, *Pharmatherapeutica* **1**, 625 (1977); in Paget's disease patients: A. Caniggia *et al., Minerva Med.* **74**, 993 (1983). Clinical trial in osteoporosis: Italian Osteoporosis Network, *Curr. Ther. Res.* **45**, 502 (1989); in combination with alfacalcidol, *q.v.*: T. Fujita *et al. J. Bone Miner. Metab.* **15**, 223 (1997).

THERAP CAT: Calcium regulator. Treatment of Paget's disease.

3586. Elcometrine. [7759-35-5] 17-(Acetyloxy)-16-methylene-19-norpregn-4-ene-3,20-dione; 16-methylene-17α-hydroxy-19-nor-4-pregnene-3,20-dione 17-acetate; 16-methylene-17α-acetoxy-19-norpregn-4-ene-3,20-dione; ST-1435; Nestorone. $C_{23}H_{30}O_4$; mol wt 370.49. C 74.56%, H 8.16%, O 17.27%. Synthetic, non-orally active progestin. Prepn: V. Schwarz *et al., Tetrahedron Lett.* **1967**, 1925; *eidem, Collect. Czech. Chem. Commun.* **33**, 4337 (1968). Pharmacology: N. Kumar *et al., Steroids* **65**, 629 (2000). Clinical pharmacokinetics: G. Noé *et al., Contraception* **48**, 548 (1993). Metabolism: O. Heikinheimo *et al., ibid.* **50**, 275 (1994).

Clinical trial of subdermal implant: S. Diaz *et al., ibid.* **51**, 33 (1995). Clinical trial of vaginal rings delivery system: I. Sivin *et al., ibid.* **69**, 137 (2004); *idem et al., ibid.* **71**, 122 (2005).

Crystals from methanol, mp 178-179°. $[\alpha]_D^{21}$ −105° (c = 1.2 in chloroform). uv max (methanol): 240 nm (log ε 4.21).

THERAP CAT: Progestogen; contraceptive (implantable).

3587. **Eldecalcitol.** [104121-92-8] (1*R*,2*R*,3*R*,5*Z*)-2-(3-Hydroxypropoxy)-4-methylene-5-[(2*E*)-2-[(1*R*,3a*S*,7a*R*)-octahydro-1-[(1*R*)-5-hydroxy-1,5-dimethylhexyl]-7a-methyl-4*H*-inden-4-ylidene]ethylidene]-1,3-cyclohexanediol; (5*Z*,7*E*)-2β-(3-hydroxypropoxy)-9,10-secocholesta-5,7,10(19)-triene-1α,3β,25-triol; 2β-(3-hydroxypropoxy)-1α,25-dihydroxyvitamin D₃; ED-71; Edirol. $C_{30}H_{50}O_5$; mol wt 490.73. C 73.43%, H 10.27%, O 16.30%. Orally active synthetic analog of calcitriol, the bioactive form of vitamin D. Prepn: K. Miyamoto *et al.*, **EP 184206**; *eidem*, **US 4666634** (1986, 1987 both to Chugai); *eidem*, *Chem. Pharm. Bull.* **41**, 1111 (1993). Large scale synthesis: N. Kubodera, S. Hatakeyama, *Heterocycles* **79**, 145 (2009). LC/MS/MS determn in serum: N. Murao *et al., J. Chromatogr. B* **823**, 61 (2005). Pharmacology and anabolic effect on bone metabolism: Y. Uchiyama *et al., Bone* **30**, 582 (2002); N. Okuda *et al., ibid.* **40**, 281 (2007). Clinical study in osteoporosis: T. Matsumoto *et al., J. Clin. Endocrinol. Metab.* **90**, 5031 (2005). Review of development and therapeutic potential: I. Ahmed, *Curr. Opin. Investig. Drugs* **5**, 441-447 (2003); N. Kobudera, *Mini-Rev. Med. Chem.* **9**, 1416-1422 (2009).

Colorless foam. uv max (ethanol): 263 nm.

THERAP CAT: Bone resorption inhibitor. Antiosteoporotic.

3588. **Eledoisin.** [69-25-0] $C_{54}H_{85}N_{13}O_{15}S$; mol wt 1188.41. C 54.58%, H 7.21%, N 15.32%, O 20.19%, S 2.70%. Undecapeptide isolated from the posterior salivary glands of several small octopus species of the genus, *Eledone*: A. Anastasi, V. Erspamer, *Br. J. Pharmacol. Chemother.* **19**, 326 (1962). Amino acid sequence: *eidem, Experientia* **18**, 58 (1962). Synthesis: E. Sandrin, R. A. Boissonnas, *ibid.* 59; *eidem, Helv. Chim. Acta* **47**, 1294 (1964); K. Lubke *et al., Ann.* **679**, 195 (1964); **GB 984810** (1965 to Farmitalia); solidphase synthesis: P. G. Pietta *et al., J. Org. Chem.* **39**, 44 (1974). Its physiologic action resembles that of other tachykinins such as substance P and physalaemin, *q.q.v.* Stimulates extravascular smooth muscle, acts as a potent vasodilator and hypotensive agent; in certain species causes salivation, and increases capillary permeability. Pharmacology: V. Erspamer, G. F. Erspamer, *Br. J. Pharmacol. Chemother.* **19**, 337 (1962); F. Sicuteri *et al., Experientia* **19**, 44 (1963); G. Bertaccini, *Pharmacol. Rev.* **28**, 127 (1976). Effect on human lacri-

mal secretion: M. Impicciatore *et al., Arch. Pharmacol.* **279**, 127 (1973). *Review:* E. G. Erdös, *Adv. Pharmacol.* **4**, 64-72 (1966).

5-oxoPro–Pro–Ser–Lys–Asp–Ala–Phe–Ile–Gly–Leu–MetNH₂

Sesquihydrate. Powder, dec 230°. $[\alpha]_D^{22}$ −44° (c = 1 in 95% acetic acid). Slowly loses activity when incubated in blood.

Trifluoroacetate. [10129-92-7] Eloisin. $C_{54}H_{85}N_{13}O_{15}S.C_2H-F_3O_2$; mol wt 1302.43.

THERAP CAT: Stimulator of lacrimal secretion in dry eye conditions.

3589. **Element 113.** [54084-70-7] Ununtrium. Uut; at. no. 113. Group IIIa (13). Transuranium element. No stable nuclides. Observation of isotope ^{284}113 (T½ 0.48 $^{+0.58}_{-0.17}$ sec, α decay) and isotope ^{283}113 (T½ 100 $^{+490}_{-45}$ msec, α decay) as decay products of ^{288}115 and ^{287}115, respectively: Y. T. Oganessian *et al., Phys. Rev. C* **69**, 021601 (2004); *eidem, ibid.* **72**, 034611 (2005). Prepn and decay of isotope ^{278}113 (T½ 0.344 msec, α decay) by ^{209}Bi(^{70}Zn,n): K. Morita *et al., J. Phys. Soc. Jpn.* **73**, 2593 (2004). Prepn and decay of isotope ^{282}113 (T½ 73 $^{+134}_{-29}$ msec, α decay) by ^{237}Np(^{48}Ca,3n): Y. T. Oganessian *et al., Phys. Rev. C* **76**, 011601 (2007).

3590. **Element 114.** [54085-16-4] Ununquadium; flerovium. Uuq; Fl; at. no. 114. Group IVa (14). Transuranium element. No stable nuclides. Prepn and decay of isotope ^{287}114 (T½ 1.32 sec, α decay) by ^{242}Pu(^{48}Ca,3n): Y. T. Oganessian *et al., Nature* **400**, 242 (1999); of isotope ^{289}114 (T½ 30.4 sec, α decay) by ^{244}Pu(^{48}Ca,3n): *eidem, Phys. Rev. Lett.* **83**, 3154 (1999). Observation of isotope ^{285}114 (T½ 580 $^{+870}_{-280}$ μsec, α decay) as decay product of ^{293}118: V. Ninov *et al., ibid.* 1104. Review of production and properties: G. Münzenberg, *Proc. 7th Int. Conf. Clustering Aspects, Nucl. Struct. Dyn.* **2000**, 377-385. Historical review: Y. T. Oganessian *et al., Sci. Am.* **282**, 63-67 (2000).

3591. **Element 115.** [54085-64-2] Ununpentium. Uup; at. no. 115. Group Va (15). Transuranium element. No stable nuclides. Prepn and decay of isotope ^{288}115 (T½ 87 $^{+105}_{-30}$ msec, α decay) and isotope ^{287}115 (T½ 32 $^{+155}_{-14}$ msec, α decay) by ^{243}Am(^{48}Ca,3n) and ^{243}Am(^{48}Ca,4n), respectively: Y. T. Oganessian *et al., Phys. Rev. C* **69**, 021601 (2004); *eidem, ibid.* **72**, 034611 (2005). Chemical identification of ^{268}Db as the five α-decay descendant of ^{288}115: N. J. Stoyer *et al., Nucl. Phys. A* **787**, 388c (2007).

3592. **Element 116.** [54100-71-9] Ununhexium; livermorium. Uuh; Lv; at. no. 116. Group VIa (16). Transuranium element. No stable nuclides. Preliminary report of the decay of isotope ^{292}116: Y. T. Oganessian *et al., Phys. Rev. C* **63**, 011301 (2000); and synthesis by ^{248}Cm(^{48}Ca,4n): *idem et al., Phys. At. Nucl.* **64**, 1349 (2001). Confirmation of experimental results and mass number change of isotope ^{292}116 to ^{293}116: J. B. Patin *et al., Lawrence Livermore National Laboratory Report*, UCRL-TR-201037 (2003). Prepn and decay of isotope ^{292}116 (T½ 18 $^{+16}_{-6}$ msec, α decay) by ^{248}Cm(^{48}Ca,4n): Y. T. Oganessian *et al., Phys. Rev. C* **70**, 064609 (2004). Initial report of isotope ^{291}116 and isotope ^{290}116: *eidem, ibid.* **69**, 054607 (2004). Prepn and decay of isotope ^{291}116 (T½ 18 $^{+22}_{-8}$ msec, α decay) and isotope ^{290}116 (T½ 7.1 $^{+3.2}_{-1.7}$ msec, α decay) by ^{245}Cm(^{48}Ca,2n) and ^{245}Cm(^{48}Ca,3n), respectively: *eidem, ibid.* **74**, 044602 (2006). Original observation of isotope ^{289}116 as decay product of isotope ^{293}118: V. Ninov *et al., Phys. Rev. Lett.* **83**, 1104 (1999); and subsequent retraction of claim: *eidem, ibid.* **89**, 039901 (2002).

3593. **Element 117.** [54101-14-3] Ununseptium. Uus; at. no. 117. Group VIIa (17). Transuranium element. No stable nuclides. Prepn and decay of isotope ^{294}117 (T½ 78 $^{+370}_{-36}$ msec, α decay) and isotope ^{293}117 (T½ 14 $^{+11}_{-4}$ msec, α decay) by ^{249}Bk(^{48}Ca,3n) and ^{249}Bk(^{48}Ca,4n), respectively: Y. T. Oganessian *et al., Phys. Rev. Lett.* **104**, 142502 (2010).

3594. **Element 118.** [54144-19-3] Ununoctium. Uuo; at. no. 118. Group VIIIa (18). Transuranium element. No stable nuclides. Deduction of the prepn and decay of isotope ^{294}118: Y. Oganessian, *Pure Appl. Chem.* **78**, 889 (2006). Prepn and decay of isotope ^{294}118 (T½ 0.89 $^{+1.07}_{-0.31}$ msec, α decay) by ^{249}Cf(^{48}Ca,3n): Y. T. Oganessian, *et al., Phys. Rev. C* **74**, 044602 (2006). Original account of prepn and decay of isotope ^{293}118: V. Ninov *et al., Phys. Rev. Lett.* **83**,

1104 (1999); and subsequent retraction of claim for synthesis: *eidem, ibid.* **89**, 039901 (2002).

3595. Elenolide. [24582-91-0] (4*S*)-4-[(1*E*)-1-Formyl-1-propen-1-yl]-3,4-dihydro-2-oxo-2*H*-pyran-5-carboxylic acid methyl ester. $C_{11}H_{12}O_5$; mol wt 224.21. C 58.93%, H 5.39%, O 35.68%. Hypotensive lactone from fruits, bark and leaves of the olive tree, *Olea europaea* L., *Oleaceae:* Veer *et al., Rec. Trav. Chim.* **76**, 839 (1957); **GB 789427** (1958 to Organon); **US 3033877** (1962 to Organon). Structure: Panizzi *et al., Gazz. Chim. Ital.* **90**, 1464 (1960); Beyerman *et al., Bull. Soc. Chim. Fr.* **1961**, 1812. Synthesis from secologanin: L. F. Tietze, H. C. Uzar, *Angew. Chem. Int. Ed.* **18**, 539 (1979).

Needles from hot alc, mp 155.5°. $[\alpha]_D^{20}$ +369° (chloroform). uv max (ethanol): 225, 317 nm (log ε 4.29, 1.75).

3596. Elesclomol. [488832-69-5] Propanedioic acid 1,3-bis[2-methyl-2-(phenylthioxomethyl)hydrazide]; N'^1,N'^3-dimethyl-N'^1,N'^3-bis(phenylcarbonothioyl)propanedihydrazide; N'^1,N'^3-dimethyl-N'^1,N'^3-di(phenylcarbonothioyl)malonohydrazide; STA-4783. $C_{19}H_{20}N_4O_2S_2$; mol wt 400.52. C 56.98%, H 5.03%, N 13.99%, O 7.99%, S 16.01%. Cytotoxic agent that selectively induces apoptosis in cancer cells by increasing the generation of reactive oxygen species (ROS) and inducing oxidative stress. Prepn: K. Koya *et al.* **WO 03006430** (2003 to SBR Pharma); *eidem,* **US 6800660** (2004 to Synta). Induction of ROS and pro-apoptotic activity in human tumor cell lines: J. R. Kirshner *et al., Mol. Cancer Ther.* **7**, 2319 (2008). Effect on breast cancer cells in combination with doxirubicin or paclitaxel: Y. Qu *et al., Breast Cancer Res. Treat.* **121**, 311 (2010). Clinical pharmacokinetics in patients with solid tumors: A. Berkenblit *et al., Clin. Cancer Res.* **13**, 584 (2007). Clinical trial in combination with paclitaxel for metastatic melanoma: S. O'Day *et al., J. Clin. Oncol.* **27**, 5452 (2009).

Light yellow crystals from methylene chloride. Sol in DMSO. Poorly sol in water.
THERAP CAT: Antineoplastic.

3597. Eletriptan. [143322-58-1] 3-[[(2*R*)-1-Methyl-2-pyrrolidinyl]methyl]-5-[2-(phenylsulfonyl)ethyl]-1*H*-indole; 5-[2-(benzenesulfonyl)ethyl]-3-(1-methylpyrrolidin-2(*R*)-ylmethyl)-1*H*-indole; UK-116044. $C_{22}H_{26}N_2O_2S$; mol wt 382.52. C 69.08%, H 6.85%, N 7.32%, O 8.37%, S 8.38%. Serotonin 5-HT$_{1B/1D}$ receptor agonist. Prepn: J. E. Macor, M. J. Wythes, **WO 9206973**; *eidem,* **US 5545644** (1992, 1996 both to Pfizer). Pharmacology: E. Willems *et al., Arch. Pharmacol.* **358**, 212 (1998). Affinity and specificity of receptor binding: C. Napier *et al., Eur. J. Pharmacol.* **368**, 259 (1999). Clinical trial in comparison with sumatriptan, *q.v.:* P. J. Goadsby *et al., Neurology* **54**, 156 (2000). Review of pharmacokinetics and clinical efficacy: A. Bardsley-Elliot, S. Noble, *CNS Drugs* **12**, 325-333 (1999); of development and clinical experience: G. Sandrini *et al., Expert Rev. Neurother.* **6**, 1413-1421 (2006).

Hydrobromide. [177834-92-3] Relpax. $C_{22}H_{26}N_2O_2S.HBr$; mol wt 463.43. White to light pale colored powder. Readily sol in water.
THERAP CAT: Antimigraine.

3598. Eleutherobin. [174545-76-7] [(4*R*,4a*S*,5*Z*,7*R*,10*S*,-11*S*,12a*R*)-3,4,4a,7,10,11,12,12a-Octahydro-7-methoxy-1,10-dimethyl-4-(1-methylethyl)-11-[[(2*E*)-3-(1-methyl-1*H*-imidazol-4-yl)-1-oxo-2-propen-1-yl]oxy]-7,10-epoxybenzocyclodecen-6-yl]methyl-β-D-arabinopyranoside 2-acetate. $C_{35}H_{48}N_2O_{10}$; mol wt 656.77. C 64.01%, H 7.37%, N 4.27%, O 24.36%. Cytotoxic glycosylated diterpene isolated from the marine soft coral, *Eleutherobia* sp. Enhances tubulin polymerization and stabilization of microtubules similar to paclitaxel, *q.v.* Isoln and structure determn: W. H. Fenical *et al.,* **US 5473057** (1995 to Univ. Calif.). Induction of tubulin polymerization and mode of action: B. H. Long *et al., Cancer Res.* **58**, 1111 (1998). Total synthesis: K. C. Nicolaou *et al., J. Am. Chem. Soc.* **120**, 8674 (1998); X. T. Chen *et al., ibid.* **121**, 6563 (1999). Solid-state and solution conformations: B. Cinel *et al., Tetrahedron Lett.* **41**, 2811 (2000). Structure-activity: I. Ojima *et al., Proc. Natl. Acad. Sci. USA* **96**, 4256 (1999). Brief review: T. Lindel, *Angew. Chem. Int. Ed.* **37**, 774-776 (1998).

White non-crystalline solid. $[\alpha]_D^{25}$ −49.3° (c = 3 in methanol). uv max (methanol): 290 nm (log ε 3.824).
USE: Biological reagent for tubulin and microtubulin studies.

3599. Eliglustat. [491833-29-5] *N*-[(1*R*,2*R*)-2-(2,3-Dihydro-1,4-benzodioxin-6-yl)-2-hydroxy-1-(1-pyrrolidinylmethyl)ethyl]octanamide; *N*-[(1*R*,2*R*)-1-(2,3-dihydro-1,4-benzodioxin-6-yl)-1-hydroxy-3-(pyrrolidin-1-yl)propan-2-yl]octanamide; Genz-99067. $C_{23}H_{36}N_2O_4$; mol wt 404.55. C 68.29%, H 8.97%, N 6.92%, O 15.82%. Ceramide analog that inhibits glucosylceramide synthase, an enzyme involved in the biosynthesis of glycosphingolipids. Prepn: B. H. Hirth, C. Siegel, **WO 03008399** (2003 to Genzyme); C. Siegel *et al.,* **US 7196205** (2007 to Regents Univ. Mich.; Genzyme). Enzyme specificity and effect on glucosylceramide synthesis and accumulation: K. A. McEachern *et al., Mol. Genet. Metab.* **91**, 259 (2007). Clinical pharmacokinetics: M. J. Peterschmitt *et al., J. Clin. Pharmacol.* **51**, 695 (2011). Clinical trial for substrate reduction therapy of Gaucher disease type 1: E. Lukina *et al., Blood* **116**, 893 (2010); *eidem, ibid.*, 4095. Review of pharmacology and therapeutic potential: T. M. Cox, *Curr. Opin. Investig. Drugs* **11**, 1169-1181 (2010).

White solid, mp 87-88°.
Tartrate. [928659-70-5] Genz-112638. $(C_{23}H_{36}N_2O_4)_2 \cdot C_4H_6O_6$; mol wt 959.19.

THERAP CAT: In treatment of inherited glycosphingolipid lysosomal storage disorders.

3600. Elinogrel. [936500-94-6] 5-Chloro-*N*-[[[4-[6-fluoro-1,4-dihydro-7-(methylamino)-2,4-dioxo-3(2*H*)-quinazolinyl]phenyl]amino]carbonyl]-2-thiophenesulfonamide; [4-(6-fluoro-7-methyl-amino-2,4-dioxo-1,4-dihydro-2*H*-quinazolin-3-yl)-phenyl]-5-chlorothiophen-2-yl-sulfonylurea; PRT-60128. $C_{20}H_{15}ClFN_5O_5S_2$; mol wt 523.94. C 45.85%, H 2.89%, Cl 6.77%, F 3.63%, N 13.37%, O 15.27%, S 12.24%. Direct acting reversible purinoceptor P2Y$_{12}$ antagonist; platelet aggregation inhibitor. Prepn: R. M. Scarborough *et al.* **WO 07056219**; *eidem,* **US 070123547** (both 2007 to Portola Pharm.). Clinical pharmacodynamics and platelet inhibition: P. A. Gurbel *et al., J. Thromb. Haemost.* **8**, 43 (2010). Review of clinical pharmacology and antiplatelet activity: M. Ueno *et al., Future Cardiol.* **6** 445-453 (2010); of development and therapeutic potential: J. H. Oestreich, *Curr. Opin. Investig. Drugs* **11**, 340-348 (2010).

Potassium salt. [936501-01-8] $C_{20}H_{14}ClFKN_5O_5S_2$; mol wt 562.03. Colorless crystals from methanol or dil ethanol. Polymorphic.

THERAP CAT: Antithrombotic.

3601. Elisidepsin. [681272-30-0] *N*-[(4*S*)-4-Methyl-1-oxohexyl]-D-valyl-L-threonyl-L-valyl-D-valyl-D-prolyl-L-ornithyl-D-alloisoleucyl-D-allothreonyl-D-alloisoleucyl-D-valyl-L-phenylalanyl-(2*Z*)-2-amino-2-butenonyl-L-valine (13 → 8)-lactone; 1-[*N*-[(4*S*)-4-methyl-1-oxohexyl]-D-valine]kahalalide F; (4*S*)-MeHex-D-Val-Thr-Val-D-Val-D-Pro-Orn-D-*allo*-Ile-cyclo[D-*allo*-Thr-D-*allo*-Ile-D-Val-Phe-(*Z*)-Dhb-Val]; isokahalalide F. $C_{75}H_{124}N_{14}O_{16}$; mol wt 1477.90. C 60.95%, H 8.46%, N 13.27%, O 17.32%. Synthetic analog of the marine natural product, kahalalide F, *q.v.* Cytotoxic to ErbB3 expressing cells. Prepn: G. T. Faircloth *et al.,* **WO 04035613**; G. T. Faircloth, M. Cuevas Marchante, **US 7507708** (2004, 2009 both to PharmaMar); and structure-activity study: J. C. Jiménez *et al., J. Med. Chem.* **51**, 4920 (2008). LC/MS/MS determn in plasma: J. Yin *et al., Rapid Commun. Mass Spectrom.* **20**, 2735 (2006). Effect on ErbB family proteins and cytotoxicity to human non-small cell lung cancer lines: Y.-H. Ling *et al., Eur. J. Cancer* **45**, 1855 (2009). Review of pharmacology and therapeutic potential: M. Provencio *et al., Clin. Lung Cancer* **10**, 295-300 (2009).

Trifluoroacetate. [915713-02-9] PM-02734; Irvalec. $C_{75}H_{124}N_{14}O_{16}.C_2HF_3O_2$; mol wt 1591.92. Sol in methanol.
THERAP CAT: Antineoplastic.

3602. Ellagic Acid. [476-66-4] 2,3,7,8-Tetrahydroxy[1]-benzopyrano[5,4,3-*cde*][1]benzopyran-5,10-dione; 4,4′,5,5′,6,6′-hexahydrodiphenic acid 2,6,2′,6′-dilactone; benzoaric acid; Lagistase. $C_{14}H_6O_8$; mol wt 302.19. C 55.65%, H 2.00%, O 42.35%. Occurs free or combined in galls. Isoln from the kino of *Eucalyptus maculata* Hook and *E. hemipholia* F. Muell., *Myrtaceae:* Gell *et al., Aust. J. Chem.* **11**, 372 (1958); Hills, Carle, *ibid.* **16**, 147 (1963). Prepd by sodium persulfate oxidation of gallic acid or by acid hydrolysis of crude tannin from walnuts: Perkin, Nierenstein, *J. Chem. Soc.* **87**, 1412 (1905); Jurd, *J. Am. Chem. Soc.* **78**, 3445 (1956); **79**, 6043 (1957). Purification: **FR 1478523** (1967 to Prod. Chim. Celluloses Rey), *C.A.* **68**, 78267r (1968). Physicochemical properties: Press, Hardcastle, *J. Appl. Chem.* **19**, 247 (1969). Pharmacology: Bhargava *et al., J. Pharm. Sci.* **57**, 1728 (1968). Ellagic acid is a potent antagonist of the mutagenicity of bay-region diol epoxides of several aromatic hydrocarbons. Inhibition of the mutagenicity of the ultimate carcinogenic metabolite of benzo[*a*]pyrene: A. W. Wood *et al., Proc. Natl. Acad. Sci. USA* **79**, 5513 (1982). Study of the reaction between this metabolite and ellagic acid: J. M. Sayer *et al., J. Am. Chem. Soc.* **104**, 5562 (1982). Brief discussion of ellagic acid as a prototype of a new class of cancer-preventing drugs: J. Fox, *Chem. Eng. News* **60**, 26 (Oct. 25, 1982).

Cream colored needles from pyridine, mp >360°. uv max (ethanol): 366, 255 nm (log ε 3.93, 4.60). Slightly sol in water or alcohol; sol in alkalies, in pyridine. Practically insol in ether.
Tetraacetate. $C_{22}H_{14}O_{12}$. Needles from acetic anhydride, mp 340°.
THERAP CAT: Hemostatic.

3603. Ellipticine. [519-23-3] 5,11-Dimethyl-6*H*-pyrido[4,3-*b*]carbazole. $C_{17}H_{14}N_2$; mol wt 246.31. C 82.90%, H 5.73%, N 11.37%. Antitumor alkaloid isolated from *Ochrosia elliptica* Labill., *O. sandwicensis* A. DC., *O. viellardii, O. silvatica, Apocynaceae:* Goodwin *et al., J. Am. Chem. Soc.* **81**, 1903 (1959); Kan Fan *et al., Phytochemistry* **9**, 1351 (1970); Cosson, Schmid, *ibid.* 1353. Structure and synthesis: Woodward *et al., J. Am. Chem. Soc.* **81**, 4434 (1959). Syntheses: J. Bergman, R. Carlsson, *Tetrahedron Lett.* **1977**, 4663; D. A. Taylor *et al., J. Chem. Res. Miniprint* **1979**, 4801; S. Kano *et al., J. Org. Chem.* **46**, 2979 (1981); R. Besselievre, H. P. Husson, *Tetrahedron* **37**, 241 (1981). Regioselective synthesis: T.-L. Ho, S.-Y. Hsieh, *Helv. Chim. Acta* **89**, 111 (2006). Toxicity study: Rakieten *et al., U.S. Gov. Res. Dev. Rep.* **67**, 38 (1967). Review of syntheses: M. Sainsbury, *Synthesis* **1977**, 437-448; R. Barone, M. Chanon, *Heterocycles* **16**, 1357-1365 (1981). Review of antineoplastic activity: K. W. Kohn *et al., Antibiotics* **vol. 5** (pt. 2), F. E. Hahn, Ed. (Springer-Verlag, New York, 1979) pp 195-213.

Bright yellow needles from ethyl acetate, mp 311-315° (dec). uv max: 239, 277, 286, 294, 332, 382, 400 nm (log ε 4.23, 4.61, 4.76, 4.74, 3.65, 3.61, 3.53). LD$_{50}$ in mice (mg/kg): 19.5-22.4 i.v.; 178-204 orally (Rakieten).

3604. Elliptinium Acetate. [58337-35-2] 9-Hydroxy-2,5,11-trimethyl-6*H*-pyrido[4,3-*b*]carbazolium acetate (1:1); 9-hydroxy-2-methylellipticinium acetate; HME; NSC-264137; Celiptium. $C_{20}H_{20}N_2O_3$; mol wt 336.39. C 71.41%, H 5.99%, N 8.33%, O 14.27%.

Deriv of ellipticine, *q.v.*, with anticancer activity. Prepn: J. LePecq *et al.*, **DE 2618223**; J. B. LePecq, C. Paoletti, **US 4310667** (1976, 1982 both to Agence Nat. Valorisation Recher.). Activity vs L 1210 mouse leukemia: *eidem, C. R. Seances Acad. Sci. Ser. D* **281**, 1365 (1975). HPLC determn: G. Muzard, J. B. LePecq, *J. Chromatogr.* **169**, 446 (1979). Metabolism, disposition: N. Van-Bac *et al., Cancer Treat. Rep.* **64**, 879 (1980). Antitumor activity, pharmacology, toxicity study: C. Paoletti *et al., Recent Results Cancer Res.* **74**, 107 (1980). Use in treatment of breast cancers: J. Rouesse *et al., Bull. Cancer* **68**, 437 (1981). Cytochemical and autoradiographic study: N. Sales, E. Puvion, *Eur. J. Cancer Clin. Oncol.* **18**, 291 (1982).

THERAP CAT: Antineoplastic.

3605.   Elliptone. [478-10-4] (6a*S*,12a*S*)-12,12a-Dihydro-8,9-dimethoxy[1]benzopyrano[3,4-*b*]furo[2,3-*h*][1]benzopyran-6(6a*H*)-one; derride. $C_{20}H_{16}O_6$; mol wt 352.34. C 68.18%, H 4.58%, O 27.24%. Found in derris resins of low rotenone content. Isoln of naturally occurring (−)-form from *Derris elliptica* (Wall.) Benth., *Leguminosae:* Buckley, *J. Soc. Chem. Ind.* **55**, 285T (1936); Harper, *Chem. Ind. (London)* **57**, 1059 (1938); **58**, 292 (1939); *J. Chem. Soc.* **1939**, 1099; *see also* Meyer, Koolhaas, *Rec. Trav. Chim.* **58**, 207 (1939). Structure: Harper, *J. Chem. Soc.* **1939**, 1424; **1942**, 593. Comparison with similar substances: Seiferle, Frear, *Ind. Eng. Chem.* **40**, 683 (1948). Total synthesis of (±)-form: Fukami *et al., Agric. Biol. Chem.* **29**, 82 (1965), *C.A.* **62**, 13115e (1963). Synthesis of (−)-form from rotenone, *q.v.*: P. B. Anzeveno, *J. Heterocycl. Chem.* **16**, 1643 (1979).

Needles from ethanol, mp 160° (Harper), also reported as mp 177-178° (Anzeveno). $[\alpha]_D^{20}$ −18° (benzene); $[\alpha]_D^{20}$ +55° (acetone).

3606.   Ellman's Reagent. [69-78-3] 3,3′-Dithiobis[6-nitrobenzoic acid]; 5,5′-dithiobis[2-nitrobenzoic acid]; DTNB. $C_{14}H_8$-$N_2O_8S_2$; mol wt 396.34. C 42.43%, H 2.03%, N 7.07%, O 32.29%, S 16.18%. Reacts with thiols to produce the yellow dianion of 5-thio-2-nitrobenzoic acid (TNB). Prepn: G. L. Ellman, *Arch. Biochem. Biophys.* **82**, 70 (1959); and spectral properties: P. W. Riddles *et al., Anal. Biochem.* **94**, 75 (1979). Structure determn: E. Shefter, T. I. Kalman, *J. Chem. Soc. Chem. Commun.* **1969**, 1027. Reaction stoichiometry with protein disulfide groups: R. J. Ackerman, J. F. Robyt, *Anal. Biochem.* **50**, 656 (1972). Use in HPLC determn of alkylthiols: K. Kuwata *et al., Anal. Chem.* **54**, 1082 (1982). Molar absorption coefficients for TNB: P. Eyer *et al., Anal. Biochem.* **312**, 224 (2003).

Yellow crystals from DMF or glacial acetic acid, mp 237-238° (dec) (Ellman); also reported as pale yellow needles from ethanol,

mp 239.7-240.3° (dec) (Riddles). d 1.67. Sol in water (pH 8). Absorption max (pH 7.27, 25°): 324 nm (ε 17780).

Ellman's anion. [77874-90-9] 5-Mercapto-2-nitrobenzoic acid ion(2−); 5-thio-2-nitrobenzoate anion; TNB^{2-}. $C_7H_3NO_4S$; mol wt 197.16. Absorption max (0.1 M phosphate buffer, pH 7.4): 412 nm (ε 14.15×10^3 at 25°, 13.8×10^3 at 37°).

USE: Chromogenic reagent for determn of thiols, covalent modification of protein thiols, kinetic assays of cholinesterase and other enzyme activities.

3607.   Ellman's Sulfinamides. [146374-27-8] 2-Methyl-2-propanesulfinamide; *tert*-butylsulfinamide; 1,1-dimethylethyl sulfinamide. $C_4H_{11}NOS$; mol wt 121.20. C 39.64%, H 9.15%, N 11.56%, O 13.20%, S 26.45%. Versatile chiral auxillary reagents utilized in asymmetric synthesis sequences. Imines formed upon condensation with aldehydes and ketones can be further derivatized with other nucleophiles. Subsequent cleavage of the *tert*-butanesulfinyl group leads to a wide range of enantioenriched amines. Prepn of the racemic mixture (not claimed): M. S. Chambers *et al.*, **EP 514133** (1993 to Merck Sharp & Dohme). Prepn of the (*R*)-form and use in asymmetric synthesis reactions: G. Liu *et al., J. Am. Chem. Soc.* **119**, 9913 (1997); D. A. Cogan *et al., ibid.* **120**, 8011 (1998). Prepn of the (*S*)-form and improved large scale synthesis: D. J. Weix, J. A. Ellman, *Org. Lett.* **5**, 1317 (2003). Large scale prepn of the (*R*)-form: *eidem, Org. Synth.* coll. vol. **XI**, 770 (2009). Procedure for recycling the *tert*-butanesulfinyl group: M. Wakayama, J. A. Ellman, *J. Org. Chem.* **74**, 2646 (2009). Synthetic utility in the prepn of branched amines: D. A. Cogan *et al., Tetrahedron* **55**, 8883 (1999); of amino acid derivatives: D. Naskar *et al., Tetrahedron Lett.* **44**, 8865 (2003); T. P. Tang, J. A. Ellman, *J. Org. Chem.* **67**, 7819 (2002); of amino alcohol derivatives: T. P. Tang *et al., ibid.* **66**, 8772 (2001); T. Kochi *et al., J. Am. Chem. Soc.* **125**, 11276 (2003). *Reviews:* J. Ellman *et al., Acc. Chem. Res.* **35**, 984-995 (2002); X.-Y. Guan, *Synlett* **2010**, 503-504.

White solid, mp 98-102° (Wakayama); also reported as colorless solid, mp 107-109° (Chambers). Store at 2-8°C.

(*R*)-Form. [196929-78-9] Crystals from hexanes, mp 101-102° (Cogan); also reported as white to off-white crystalline solid, mp 102-105° (Weix). $[\alpha]_D^{23}$ +4.9° (c = 1.0 in chloroform); $[\alpha]_D$ +4.6° (c = 0.8 in chloroform). Store at 2-8°C.

(*S*)-Form. [343338-28-3] White to off-white crystalline powder, mp 101-102°. $[\alpha]_D$ −5.1° (c = 0.6 in chloroform). Store at 2-8°C.

USE: Reagents in synthetic organic chemistry.

3608.   Eltoprazine. [98224-03-4] 1-(2,3-Dihydro-1,4-benzodioxin-5-yl)piperazine; 1-(1,4-benzodioxan-5-yl)piperazine. C_{12}-$H_{16}N_2O_2$; mol wt 220.27. C 65.43%, H 7.32%, N 12.72%, O 14.53%. Selective serotonin 5HT$_1$ agonist; psychotropic agent with antiaggressive activity. Prepn: J. Hartog *et al.*, **EP 138280**; *idem et al.*, **EP 189612** (1985, 1986 both to Duphar). Prepd (not claimed): *idem et al.*, **US 4833142** (1989 to Duphar). Crystal structure: M. L. Verdonk *et al., Acta Crystallogr.* **C48**, 2271 (1992). Series of articles on pharmacology, receptor binding and mechanism of action: *Drug Metab. Drug Interact.* **8**, 1-186 (1990). Pharmacokinetics: M. H. de Vries *et al., Eur. J. Clin. Pharmacol.* **41**, 485 (1991). Clinical evaluation in self-injury: W. M. A. Verhoeven *et al., Lancet* **340**, 1037 (1992); in violent psychiatric patients: J. Tiihonen *et al., ibid.* **341**, 307 (1993).

Hydrochloride. [98206-09-8] DU-28853. $C_{12}H_{16}N_2O_2 \cdot HCl$; mol wt 256.73. Crystals from ethanol, mp 256-258°. Soly in water: 19 g/100 ml.

THERAP CAT: Serenic.

3609.   Eltrombopag. [496775-61-2]; [376591-99-0] (unspecified stereo).   3'-[2-[(2Z)-1-(3,4-Dimethylphenyl)-1,5-dihydro-3-methyl-5-oxo-4H-pyrazol-4-ylidene]hydrazinyl]-2'-hydroxy-[1,1'-biphenyl]-3-carboxylic acid; SB-497115. $C_{25}H_{22}N_4O_4$; mol wt 442.48. C 67.86%, H 5.01%, N 12.66%, O 14.46%. Thrombopoietin receptor agonist; stimulates production of platelets from megakaryocytes. Prepn (unspecified stereo): K. J. Duffy *et al.*, **WO 01089457**; *eidem*, **US 7160870** (2001, 2008 both to SmithKline Beecham; Glaxo). Prepn of Z-form bis-monoethanolamine salt: S. Moore, **WO 03098992** (2003 to SmithKline Beecham). Clinical pharmacokinetics and effect on platelet counts: J. M. Jenkins *et al.*, *Blood* **109**, 4739 (2007). Clinical evaluation in hepatitis C related thrombocytopenia: J. G. McHutchison *et al.*, *N. Engl. J. Med.* **357**, 2227 (2007); in idiopathic thrombocytopenic purpura: J. B. Bussel *et al.*, *ibid.* 2237. Review of discovery and pharmacology: K. J. Duffy, C. L. Erickson-Miller in *Target Validaton in Drug Discovery* (Academic Press, Burlington, MA, 2007) pp 241-254.

Orange solid. Soly at 25° (mg/ml): <0.001 in water; 1.9 in methanol. Sol in THF.

Compd with 2-aminoethanol (1:2). [496775-62-3] Eltrombopag olamine; eltrombopag bis-monoethanolamine; Promacta; Revolade. $C_{25}H_{22}N_4O_4 \cdot 2(C_2H_7NO)$; mol wt 564.64. Soly at 25° (mg/ml): 14.2 in water; 6.4 in methanol.

THERAP CAT: Antithrombocytopenic.

3610.   Elvitegravir. [697761-98-1] 6-[(3-Chloro-2-fluorophenyl)methyl]-1,4-dihydro-1-[(1S)-1-(hydroxymethyl)-2-methylpropyl]-7-methoxy-4-oxo-3-quinolinecarboxylic acid; (S)-6-(3-chloro-2-fluorobenzyl)-1-(1-hydroxymethyl-2-methylpropyl)-7-methoxy-4-oxo-1,4-dihydroquinoline-3-carboxylic acid; GS-9137; JTK-303. $C_{23}H_{23}ClFNO_5$; mol wt 447.89. C 61.68%, H 5.18%, Cl 7.91%, F 4.24%, N 3.13%, O 17.86%. HIV-1 integrase strand transfer inhibitor. Prepn: M. Satoh *et al.*, **WO 04046115**; *eidem* **US 7176220** (2004, 2007 both to Japan Tobacco). Discovery and synthesis: M. Sato *et al.*, *J. Med. Chem.* **49**, 1506 (2006). Mechanism of action and resistance profile *in vitro*: K. Shimura *et al.*, *J. Virol.* **82**, 764 (2008). Clinical pharmacokinetics and tolerability: E. DeJesus *et al.*, *J. Acquir. Immune Defic. Syndr.* **43**, 1 (2006). Clinical evaluation in patients with resistant HIV-1: A. R. Zolopa *et al.*, *J. Infect. Dis.* **201**, 814 (2010). Review of development and clinical experience: O. M. Klibanov, *Curr. Opin. Investig. Drugs* **10**, 190-200 (2009); K. Shimura, E. N. Kodama, *Antivir. Chem. Chemother.* **20**, 79-85 (2009).

White crystals from methanol + water, mp 151-152°.

THERAP CAT: Antiretroviral.

3611.   Elymoclavine. [548-43-6] 8,9-Didehydro-6-methylergoline-8-methanol. $C_{16}H_{18}N_2O$; mol wt 254.33. C 75.56%, H 7.13%, N 11.01%, O 6.29%. An ergot alkaloid obtained from cultures of fungi parasitic on *Elymus mollis* Trin.: Abe *et al.*, **US 2835675** (1958). Found in fungi parasitic on *Pennisetum typhoideum* Rich.: Stoll *et al.*, *Helv. Chim. Acta* **37**, 1815 (1954). Structure and stereochemistry: Schreier, *Helv. Chim. Acta* **41**, 1984 (1958). Biosyntheses utilizing *Claviceps* cultures: Naidoo *et al.*, *Chem. Commun.* **1970**, 472; Ogunlana *et al.*, *ibid.* 775; Seiler *et al.*, *ibid.* 1394; Cavender, Anderson, *Biochim. Biophys. Acta* **208**, 345 (1970).

Monoclinic prisms from methanol, 248-252° (dec). $[\alpha]_D^{20}$ −59° (c = 0.1 in ethanol) (Abe); $[\alpha]_D^{20}$ −152° (c = 0.9 in pyridine) (Stoll). uv max: 227, 283, 293 nm (log ε 4.31, 3.84, 3.76). Fairly sol in water with alkaline reaction. Sol in pyridine; very slightly sol in organic solvents.

3612.   Emamectin. [119791-41-2] (4″R)-4″-Deoxy-4″-(methylamino)avermectin B_1; 4″-deoxy-4″-*epi*-methylaminoavermectin B_1. Mixture of semi-synthetic avermectins, *q.v.*, consisting (9:1) of emamectins B_{1a} and B_{1b}. Prepn: H. Mrozik, **US 4874749** (1989 to Merck & Co.); and insecticidal activity: H. Mrozik *et al.*, *Experientia* **45**, 315 (1989). Synthesis: R. J. Cvetovich *et al.*, *J. Org. Chem.* **59**, 7704 (1994). HPLC determn in *Salmo salar* L.: H. Kim-Kang *et al.*, *J. Agric. Food Chem.* **49**, 5294 (2001); in medicated fish feed: L. J. Farer, *J. AOAC Int.* **88**, 462 (2005). Safety and efficacy in *Salmo salar* L.: J. Stone *et al.*, *Aquaculture* **210**, 21 (2002). Avian toxicity studies: A. C. Chukwudebe *et al.*, *Environ. Toxicol. Chem.* **17**, 1118 (1998). *Review:* M. H. Fisher, *ACS Symp. Ser.* **524**, 169-182 (1993); H. Mrozik, *ibid.* **551**, 54-73 (1993).

Component B_{1a} R = CH_2CH_3
Component B_{1b} R = CH_3

LC_{50} (48 hr) in *Daphnia magna*: 2.9 μg/l; LC_{50} in rainbow trout, bluegill sunfish (mg/l) at 24 hrs: >1.0, 0.75; at 48 hrs: >1.0, 0.45; at 96 hrs: 0.67, 0.24 (Mrozik).

Benzoate. [137512-74-4] MK-244; Affirm; Denim; Proclaim; Slice. Exists as the anhydrous and various hydrated forms having different crystal morphologies. Amt of water is nonstoichiometric. 1.1% form is an off white crystalline powder, mp 141-146°. $[\alpha]_D$ −6.9° (c = 0.5% in methanol). uv max (10% methanol in water by volume): 244 nm ($A_1^{1\%}$ = 327 corr). log P (*n*-octanol/aq phosphate): 3.0 (pH 5.1); 5.0 (pH 7.0). pKa 4.2; 7.6. Soly 0.32 mg/ml at pH 5 (aq); 0.024 mg/ml at pH 7 (aq). Freely sol in chloroform and methanol. Insol in hexane. LD_{50} in mallard ducks, northern bobwhite quail (mg/kg): 76, 264 orally (Chukwudebe).

Emamectin B_{1a}. [121124-29-6] (4″R)-5-O-Demethyl-4″-deoxy-4″-(methylamino)avermectin A_{1a}; 4″-*epi*-(methylamino)-4″-deoxyavermectin B_{1a}. $C_{49}H_{75}NO_{13}$; mol wt 886.13.

Emamectin B_{1b}. [121424-52-0] (4″*R*)-5-*O*-Demethyl-25-de(1-methylpropyl)-4″-deoxy-4″-(methylamino)-25-(1-methylethyl)avermectin A_{1a}; 4″-*epi*-(methylamino)-4″-deoxyavermectin B_{1b}. $C_{48}H_{73}NO_{13}$; mol wt 872.11.

USE: Insecticide.

THERAP CAT (VET): Antiparasitic.

3613. Embelin. [550-24-3] 2,5-Dihydroxy-3-undecyl-2,5-cyclohexadiene-1,4-dione; 2,5-dihydroxy-3-undecyl-*p*-benzoquinone; embelic acid. $C_{17}H_{26}O_4$; mol wt 294.39. C 69.36%, H 8.90%, O 21.74%. Bioactive constituent of the Ayurvedic medicinal plant, *Embelia ribes* Burm., *Myrsinaceae*; known for its antitumor, anthelminthic, anti-inflammatory, and wound healing properties. Binds to and inhibits XIAP, the X chromosome-linked inhibitor of apoptosis protein. Isoln from the berries: Heffter, Feuerstein, *Arch. Pharm.* **238**, 15 (1900); by microwave-assisted extraction: C. Latha, *Biotechnol. Lett.* **29**, 319 (2007). Isoln from leaves and wound healing activity: H. M. Kumara Swamy *et al.*, *J. Ethnopharmacol.* **109**, 529 (2007). Structure and synthesis: L. F. Fieser, E. M. Chamberlin, *J. Am. Chem. Soc.* **70**, 71 (1948). Pharmacology: M. Chitra *et al.*, *Chemotherapy (Basel)* **40**, 109 (1994). Pro-apoptotic activity and XIAP binding: Z. Nikolovska-Coleska *et al.*, *J. Med. Chem.* **47**, 2430 (2004). Effect on human prostate cancer cells: M. Danquah *et al.*, *Pharm. Res.* **26**, 2081 (2009).

Glistening orange plates from alcohol, benzene or acetic acid, mp 142-143°. Soluble in the usual hot organic solvents or in alkali hydroxide solns; very slightly sol in petr ether. Practically insol in water. Soly in DMSO: 10 mg/ml.

Disemicarbazone. [5796-42-9] $C_{19}H_{32}N_6O_4$; mol wt 408.50. Pale-brown needles from dil alcohol, dec 236°.

Dioxime. $C_{17}H_{28}N_2O_4$; mol wt 324.42. Pale-yellow needles from acetic acid, mp 278°.

3614. Emedastine. [87233-61-2] 1-(2-Ethoxyethyl)-2-(hexahydro-4-methyl-1*H*-1,4-diazepin-1-yl)-1*H*-benzimidazole; 1-[2-(ethoxy)ethyl]-2-(4-methyl-1-homopiperazinyl)benzimidazole. $C_{17}H_{26}N_4O$; mol wt 302.42. C 67.52%, H 8.67%, N 18.53%, O 5.29%. Histamine H₁-receptor antagonist. Prepn: R. Iemura *et al.*, **EP** 79545; *eidem*, **US 4430343** (1983, 1984 both to Kanebo); *eidem*, *J. Med. Chem.* **29**, 1178 (1986). GC determn in plasma: T. Hamada *et al.*, *Chem. Pharm. Bull.* **34**, 1168 (1986). Series of articles on pharmacology: *Arzneim.-Forsch.* **34**, 801-818 (1984). Toxicity data: T. Fukuda *et al.*, *ibid.* 801.

Difumarate. [87233-62-3] KB-2413; LY-188695; Emadine. $C_{17}H_{26}N_4O.2C_4H_4O_4$; mol wt 534.57. White to faintly yellow crystals from ethanol, mp 148-151°. Soluble in water. LD₅₀ orally in guinea pigs: 744 mg/kg (Fukuda).

THERAP CAT: Antihistaminic.

3615. Emepronium Bromide. [3614-30-0] *N*-Ethyl-*N*,*N*,α-trimethyl-γ-phenylbenzenepropanaminium bromide (1:1); ethyldimethyl(1-methyl-3,3-diphenylpropyl)ammonium bromide; ethyl-(3,3-diphenyl-1-methylpropyl)dimethylammonium bromide; (1-methyl-3,3-diphenylpropyl)dimethylethylammonium bromide; Ceti-

prin; Restenacht; Ripirin; Uro-Ripirin. $C_{20}H_{28}BrN$; mol wt 362.36. C 66.29%, H 7.79%, Br 22.05%, N 3.87%. Prepn: Carlsson, **SE 136606** (1952 to Aktiebolaget Recip.), *C.A.* **48**, 729a (1954).

Crystals, mp 204°.

THERAP CAT: Antispasmodic. In treatment of urinary incontinence.

3616. Emetine. [483-18-1] (2*S*,3*R*,11b*S*)-3-Ethyl-1,3,4,6,7,-11b-hexahydro-9,10-dimethoxy-2-[[(1*R*)-1,2,3,4-tetrahydro-6,7-dimethoxy-1-isoquinolinyl]methyl]-2*H*-benzo[*a*]quinolizine; 6′,7′,-10,11-tetramethoxyemetan; cephaeline methyl ether. $C_{29}H_{40}N_2O_4$; mol wt 480.65. C 72.47%, H 8.39%, N 5.83%, O 13.31%. Principal alkaloid of ipecac, *q.v.*; occurs naturally as (−)-form. Extraction procedures: E. Merck, *BIOS Final Report* **No. 766** (1947); F. E. Hamerslag, *Technology and Chemistry of Alkaloids* (New York, 1950) chapter XI; Stoll, Jucker in *Ullmanns Encyklopädie der technischen Chemie* **3**, Aufl., Bd. III (München-Berlin, 1953) p 231; *Beilstein* 2nd suppl. **vol. 23**, 449. Structure: Robinson, *Nature* **162**, 524 (1948); Janot, *Bull. Soc. Chim. Fr.* **1949**, 185. Stereochemistry: Battersby, *Chem. Ind. (London)* **1958**, 1324; Battersby, Garrett, *J. Chem. Soc.* **1959**, 3512; Van Tamelen *et al.*, *J. Am. Chem. Soc.* **81**, 6214 (1959). Total synthesis: Evstigneeva *et al.*, *Proc. Acad. Sci. USSR* **75**, 539 (1950); Van Tamelen *et al.*, *J. Am. Chem. Soc.* **91**, 7359 (1969); T. Kametani *et al.*, *J. Chem. Soc. Perkin Trans. 1* **1979**, 1211. Alternate syntheses: Burgstahler, Bithos, *J. Am. Chem. Soc.* **82**, 5446 (1960); Openshaw, Whittaker, *J. Chem. Soc.* **1963**, 1461; S. Takano *et al.*, *J. Org. Chem.* **43**, 4169 (1978); T. Fujii, S. Yoshifuji, *Tetrahedron* **36**, 1539 (1980). *See also* Clark *et al.*, **US 3102118** (1963 to Glaxo). Review of syntheses: M. Shamma, *The Isoquinoline Alkaloids* (Academic Press, New York, 1972) pp 426-457. Toxicity: Radomski *et al.*, *J. Pharmacol. Exp. Ther.* **104**, 421 (1952); of dihydrochloride: Child *et al.*, *J. Pharm. Pharmacol.* **16**, 65 (1964). *General review:* Grollman, Jarkovsky, in *Antibiotics* Vol. 3, J. W. Corcoran, F. E. Hahn, Eds. (Springer-Verlag, New York, 1975) pp 420-435. Comprehensive description: L. V. Feyns, L. T. Grady, *Anal. Profiles Drug Subs.* **10**, 289-335 (1981).

White, amorphous powder. mp 74°. Turns yellow on exposure to light and heat. $[\alpha]_D^{20}$ −50° (c = 2 in CHCl₃). Strong alkaline reaction. pK₁ 5.77; pK₂ 6.64. Freely sol in methanol, ethanol, acetone, ethyl acetate, ether, chloroform. Sparingly sol in water, petr ether. Moderately sol in dil ammonium hydroxide, but sparingly in solns of KOH and NaOH. Absorption spectrum: Bayard, *Chem. Zentralbl.* **1943**, II, 305. LD₅₀ i.p. in rats: 12.1 mg/kg (Radomski).

Dihydrochloride. [316-42-7] Emetine hydrochloride; Hemometina. $C_{29}H_{40}N_2O_4.2HCl$; mol wt 553.57. Contains water of crystallization varying from 3 to 8 H_2O. Clusters of needles after drying at 105°, mp 235-255° (dec). $[\alpha]_D$ +11° (c = 1) to $[\alpha]_D$ +21° (c = 8), calculated for the anhydrous salt. One gram of the hydrated salt dissolves in about 7 ml water. pH of aq soln (1 g in 50 ml) 5.6. Sol in alcohol. Solid and solutions turn yellow on exposure to light or heat. LD$_{50}$ (calculated as base) in mice (mg/kg): 32 s.c.; 30 orally (Child).

Emetamine. [483-19-2] 1′,2′,3′,4′-Tetrahydro-6′,7′,10,11-tetramethoxyemetan; tetradehydroemetine. C$_{29}$H$_{36}$N$_2$O$_4$. Occurs in small amounts in ipecac. Isoln: Pyman, *J. Chem. Soc.* **111**, 419 (1917). Structure: Battersby *et al., J. Chem. Soc.* **1959**, 1744. Synthesis: *eidem, J. Chem. Soc.* **1961**, 3899. Crystals from ether, mp 142-143°. uv max (ethanol): 236, 283 nm (log ε 4.85, 3.86).

THERAP CAT: Antiamebic.

THERAP CAT (VET): Hydrochloride has been used as an antiamebic and in lung worm infection.

3617. Emodepside. [155030-63-0] Cyclo[(αR)-α-hydroxy-4-(4-morpholinyl)benzenepropanoyl-*N*-methyl-L-leucyl-(2R)-2-hydroxypropanoyl-*N*-methyl-L-leucyl-(αR)-α-hydroxy-4-(4-morpholinyl)benzenepropanoyl-*N*-methyl-L-leucyl-(2R)-2-hydroxypropanoyl-*N*-methyl-L-leucyl]; Bay-44-4400; PF-1022-221. C$_{60}$H$_{90}$N$_6$O$_{14}$; mol wt 1119.41. C 64.38%, H 8.10%, N 7.51%, O 20.01%. Cyclic octadepsipeptide that binds to presynaptic latrophilin receptors in nematodes. Semisynthetic derivative of PF1022A, a fungal metabolite of *Mycelia sterilia*. Prepn: H. Nishiyama *et al.,* **WO 9319053**; *eidem,* **US 5514773** (1993, 1996 both to Fujisawa). *In vivo* anthelmintic activity: A Harder, G. von Samson-Himmelstjerna, *Parasitol. Res.* **87**, 924 (2001); H. Zahner *et al., Int. J. Parasitol.* **31**, 1515 (2001). Review of mechanism of action studies: A. Harder *et al., Parasitol. Res.* S1-S10 (2005).

White to yellowish powder. Soly in water: 8.1 mg/l (pH 4); 5.2 mg/l (pH 7), 6.1 mg/l (pH 10). Log P (octanol/water): 4.9 (pH 7).

Combination with praziquantel. Profender. Series of articles on therapeutic use in cats, reptiles and rodents: *Parasitol. Res.* **97**, S33-S69 (2005).

THERAP CAT (VET): Anthelmintic (Nematodes and Cestodes).

3618. Emodin. [518-82-1] 1,3,8-Trihydroxy-6-methyl-9,10-anthracenedione; 1,3,8-trihydroxy-6-methylanthraquinone; 4,5,7-trihydroxy-2-methylanthraquinone; frangula emodin; rheum emodin; archin; frangulic acid. C$_{15}$H$_{10}$O$_5$; mol wt 270.24. C 66.67%, H 3.73%, O 29.60%. Occurs mostly as the rhamnoside (*see* Frangulin) in rhubarb root, in alder buckthorn (*Rhamnus frangula* L.), in Cascara sagrada (*Rhamnus purshiana* DC., *Rhamnaceae*), also in *Rumex* and in other *Polygonaceae*. Isoln from rhubarb root: Tutin, Clewer, *J. Chem. Soc.* **99**, 946 (1911); Carelli, Giuliano, *Farmaco Ed. Prat.* **12**, 184 (1957); from bark of alder buckthorn: Bridel, Charaux, *Bull. Soc. Chim. Biol.* **15**, 648 (1933). Identity with archin: Chaudhry *et al., J. Sci. Ind. Res.* **9B**, No. 6, 142 (1950), *C.A.* **44**, 9396h (1950). Synthesis from 3,5-dinitrophthalic anhydride and *m*-cresol: Elder, Widmer, *Helv. Chim. Acta* **6**, 966 (1923); from 2-methylanthraquinone: Ayyangar *et al., J. Sci. Ind. Res.* **20B**, 493 (1961), *C.A.* **57**, 8514b (1962).

Orange needles from alc or by sublimation at 12 mm. mp 256-257°. Absorption max (ethanol): 222, 252, 265, 289, 437 nm (log ε 4.55, 4.26, 4.27, 4.34, 4.10). Practically insol in water; sol in alc, aq alkali hydroxide solns (cherry-red color), Na$_2$CO$_3$ and NH$_3$ solns. Soly at 25° (g/100 ml of satd soln): ether 0.140; chloroform 0.071; carbon tetrachloride 0.010; carbon bisulfide 0.009; benzene 0.041.

3-Methyl ether. 1,8-Dihydroxy-3-methoxy-6-methylanthraquinone; rheochrysidin; physcion. C$_{16}$H$_{12}$O$_5$. Brick-red, monoclinic needles, mp 207°. Occurs naturally as *physcione* or *parietin. See also* Aloe emodin.

Trimethyl ether. C$_{18}$H$_{16}$O$_5$. Pale yellow needles, mp 225°.

THERAP CAT: Cathartic.

3619. Emorfazone. [38957-41-4] 4-Ethoxy-2-methyl-5-(4-morpholinyl)-3(2H)-pyridazinone; M-73101; Nandron; Pentoyl. C$_{11}$H$_{17}$N$_3$O$_3$; mol wt 239.28. C 55.22%, H 7.16%, N 17.56%, O 20.06%. Prepn: K. Satoda *et al.,* **JP 72 24030** (1972 to Morishita), *C.A.* **77**, 164732f (1972); M. Takaya *et al., J. Med. Chem.* **22**, 53 (1979). Metabolism studies: T. Hayashi *et al., Chem. Pharm. Bull.* **26**, 3124 (1978); **27**, 317 (1979). General pharmacological study: M. Sato *et al., Nippon Yakurigaku Zasshi* **75**, 291 (1979), *C.A.* **91**, 117292 (1979). Mechanism of action studies: M. Sato, A. Yamaguchi, *Arzneim.-Forsch.* **32**, 379 (1982). Antigenicity study: M. Sato *et al., Oyo Yakuri* **18**, 65 (1979), *C.A.* **92**, 104391 (1980). Toxicity studies: *eidem, ibid.* **16**, 1011 (1978), *C.A.* **90**, 197741 (1979); C. Onodera *et al., J. Toxicol. Sci.* **4**, 229 (1979); K. Shimpo *et al., ibid.* 255.

Cryst from methanol/isopropyl ether, mp 89-91°. LD$_{50}$ i.p. in mice: 700 mg/kg (Takaya).

THERAP CAT: Anti-inflammatory; analgesic.

3620. EMPA. [1832-53-7] *P*-Methylphosphonic acid monoethyl ester; ethyl methylphosphonate. C$_3$H$_9$O$_3$P; mol wt 124.08. C 29.04%, H 7.31%, O 38.68%, P 24.96%. Degradation product of the chemical warfare agent, VX, *q.v.* Prepn as sodium salt: V. S. Abramov, M. N. Morozova, *J. Gen. Chem. USSR* **22**, 315 (1952). Capillary electrophoretic determn in soils: A.-E. F. Nassar *et al., Anal. Chem.* **71**, 1285 (1999). Determn in human urine samples after exposure to VX: M. Minami *et al., J. Toxicol. Sci.* **23**, Suppl. II 250 (1998).

USE: Marker for the detection of VX.

3621. EMPTA. [18005-40-8] *P*-Methylphosphonothioic acid *O*-ethyl ester. C$_3$H$_9$O$_2$PS; mol wt 140.14. C 25.71%, H 6.47%, O 22.83%, P 22.10%, S 22.88%. Degradation product and precursor of VX, inhibits cholinesterase. Prepn: M. I. Kabachnik *et al., Dokl. Akad. Nauk SSSR* **104**, 861 (1955), *C.A.* **50**, 11240a (1956); F. W. Hoffmann *et al., J. Am. Chem. Soc.* **81**, 148 (1959). Structure: H. Christol *et al., C. R. Seances Acad. Sci. Ser. C* **265**, 1511 (1967). NMR: M. Mikolajczyk *et al., J. Am. Chem. Soc.* **100**, 7003 (1978). Mass spec.: E. R. J. Wils, *Fresenius J. Anal. Chem.* **338**, 22 (1990). Ionspray-MS determn in water: R. Kostiainen *et al., J. Chromatogr.* **634**, 113 (1993).

Yellowish oil, bp$_{0.22mm}$ 63°. n$_D^{25}$ 1.4912. d$_4^{25}$ 1.1735.

USE: Precursor to chemical warfare agents.

3622. Emtricitabine. [143491-57-0] 4-Amino-5-fluoro-1-[(2*R*,5*S*)-2-(hydroxymethyl)-1,3-oxathiolan-5-yl]-2(1*H*)-pyrimidinone; (−)-*cis*-4-amino-5-fluoro-1-(2-hydroxymethyl-1,3-oxathiolan-5-yl)-(1*H*)-pyrimidin-2-one; (−)-(2*R*,5*S*)-5-fluoro-1-[2-(hydroxymethyl)-1,3-oxathiolan-5-yl]cytosine; (−)-β-2′,3′-dideoxy-5-fluoro-3′-thiacytidine; (−)-FTC; 524W91; BW-524W91; Coviracil; Emtriva. $C_8H_{10}FN_3O_3S$; mol wt 247.24. C 38.86%, H 4.08%, F 7.68%, N 17.00%, O 19.41%, S 12.97%. Reverse transcriptase inhibitor; nucleoside analog structurally related to lamivudine, *q.v.* Prepn: D. C. Liotta *et al.*, **WO 9214743** (1992 to Emory University); G. Dionne, **US 5538975** (1996 to BioChem Pharma, Inc.); L. S. Jeong *et al.*, *J. Med. Chem.* **36**, 181 (1993). Absolute configuration: P. van Roey *et al.*, *Antivir. Chem. Chemother.* **4**, 369 (1993). HPLC-NMR determn of urinary metabolites: J. P. Shockcor *et al.*, *Xenobiotica* **26**, 189 (1996). Comparative efficacy of enantiomers vs HIV: R. F. Schinazi *et al.*, *Antimicrob. Agents Chemother.* **36**, 2423 (1992). Pharmacokinetics: L. W. Frick *et al.*, *ibid.* **38**, 2722 (1994). Mechanism of action study: J. Y. Feng *et al.*, *FASEB J.* **13**, 1511 (1999). Clinical study in combination with didanosine and efavirenz: J.-M. Molina *et al.*, *J. Infect. Dis.* **182**, 599 (2000). *Review:* P. Cahn, *Expert Opin. Invest. Drugs* **13**, 55-68 (2004).

White solid from ether and methanol, mp 136-140°. $[\alpha]_D^{25}$ −133.60° (c = 0.23 in MeOH). uv max (water): 287.8 nm (pH 2); 280.0 nm (pH 7); 279.8 nm (pH 11) (ε 14210, 11090, 11810). pKa 2.65. Soly in water (25°): ∼112 mg/ml.

THERAP CAT: Antiretroviral.

3623. Enalapril. [75847-73-3] *N*-[(1*S*)-1-(Ethoxycarbonyl)-3-phenylpropyl]-L-alanyl-L-proline; 1-[*N*-[(*S*)-1-carboxy-3-phenylpropyl]-L-alanyl]-L-proline 1′-ethyl ester. $C_{20}H_{28}N_2O_5$; mol wt 376.45. C 63.81%, H 7.50%, N 7.44%, O 21.25%. Angiotensin-converting enzyme (ACE) inhibitor; de-esterified *in vivo* to its active diacid metabolite. Prepn: A. A. Patchett *et al.*, **EP 12401**; E. E. Harris *et al.*, **US 4374829** (1980, 1983 both to Merck & Co.); A. A. Patchett *et al.*, *Nature* **288**, 280 (1980). Similtaneous determn with enalaprilat in plasma by HPLC-MS/MS: S. Lu *et al.*, *J. Pharm. Biomed. Anal.* **49**, 163 (2009). Comprehensive description: D. P. Ip, G. S. Brenner, *Anal. Profiles Drug Subs.* **16**, 207-243 (1987). Review of pharmacology and clinical experience in congestive heart failure: P. A. Todd, K. L. Goa, *Drugs* **37**, 141-161 (1989); in hypertension: *eidem, ibid.* **43**, 346-381 (1992). Review of clinical pharmacokinetics: R. J. MacFadyen *et al.*, *Clin. Pharmacokinet.* **25**, 274-282 (1993). Review of combination with hydrochlorothiazide: P. L. Malini, *Adv. Ther.* **10**, 253-262 (1993); with lercanidipine: V. Barrios *et al.*, *Vasc. Health Risk Manag.* **4**, 847-853 (2008).

Maleate. [76095-16-4] MK-421; Amprace; Cardiovet; Enacard; Enapren; Glioten; Hipoartel; Innovace; Naprilene; Pres; Renitec; Reniten; Renivace; Vasotec; Xanef. $C_{20}H_{28}N_2O_5.C_4H_4O_4$; mol wt 492.53. White to off-white crystalline powder, mp 143-144.5°. $[\alpha]_D^{25}$ −42.2° (c = 1 in methanol). pH (1% water) 2.6. pKa$_1$ 3.0; pKa$_2$ (25°) 5.4. Soly (g/ml): water 0.025; alcohol 0.08; methanol 0.20. Freely sol in DMF; slightly sol in semipolar organic solvents. Practically insol in nonpolar organic solvents.

Diacid dihydrate. [84680-54-6]; [76420-72-9] (anhydrous). Enalaprilat; enalaprilic acid; MK-422; Vasotec I.V. $C_{18}H_{24}N_2O_5.2H_2O$; mol wt 384.43. Needles from H_2O, mp 148-151°. $[\alpha]_D$ −67.0° (0.1*M* HCl). Sparingly sol in methanol, DMF; slightly sol in water,

isopropyl alc; very slightly sol in acetone, alc, hexane. Practically insol in acetonitrile, chloroform.

THERAP CAT: Antihypertensive.

THERAP CAT (VET): In treatment of heart failure in dogs.

3624. Enanthotoxin. [20311-78-8] (2*E*,8*E*,10*E*,14*R*)-2,8,10-Heptadecatriene-4,6-diyne-1,14-diol; oenanthotoxin. $C_{17}H_{22}O_2$; mol wt 258.36. C 79.03%, H 8.58%, O 12.39%. The poisonous principle of the hemlock water dropwort, *Oenanthe crocata* L., *Umbelliferae*, believed to be the most poisonous plant in Great Britain. Isomeric with cicutoxin, *q.v.*, produced by the closely related North American water hemlock. Isoln: S. G. C. Clarke *et al.*, *J. Pharm. Pharmacol.* **1**, 377 (1949). Purification and structure: E. F. L. J. Anet *et al.*, *J. Chem. Soc.* **1953**, 309. Absolute configuration: G. Appendino *et al.*, *J. Nat. Prod.* **72**, 962 (2009). Synthesis of DL-form: F. Bohlmann, H. G. Viehe, *Ber.* **88**, 1245 (1955). Fluorimetric determn: B. Del Castillo *et al.*, *Ital. J. Biochem.* **29**, 233 (1980). Chromatographic determn in stomach contents: G. C. Kite *et al.*, *J. Chromatogr. B* **838**, 63 (2006). Blocking of sodium current and intramembrane charge movement: J. M. Dubois, M. F. Schneider, *Nature* **289**, 685 (1981); *eidem, Toxicon* **20**, 49 (1982). Toxicity study: M. P. Martinez-Honduvilla *et al.*, *Arch. Farmacol. Toxicol.* **7**, 197 (1981). Review of botanical characterization, pharmacology, and symptoms of poisoning: L. J. Schep *et al.*, *Clin. Toxicol.* **47**, 270-278 (2009).

Large prisms, mp 87°. *Poisonous.* Unstable, dec by light and air to brown insol resin. $[\alpha]_D^{15}$ +30.5° (c = 2.0 in methanol). uv max: 213, 252, 267, 281, 296, 315.5, 337.5 (ε × 10^{-3} 17.5, 33, 29, 17.5, 30.5, 40, 29). Practically insol in water, petr ether, alkalies, dil mineral acids. Readily sol in chloroform, ethanol, methanol, ether, benzene. Average LD i.p. in mice: 0.83 mg/kg (Clarke). LD$_{50}$ i.p. in rats: 2.94 mg/kg (Martinez-Honduvilla).

DL-Form. Star-shaped crystals, mp 68°.

Caution: Highly toxic. Symptoms of overexposure include seizures, nausea, vomiting, diarrhea, tachycardia, mydriasis, rhabomyolysis, renal failure, coma, death (Schep).

3625. Encainide. [66778-36-7] 4-Methoxy-*N*-[2-[2-(1-methyl-2-piperidinyl)ethyl]phenyl]benzamide; (±)-2′-[2-(1-methyl-2-piperidyl)ethyl]-*p*-anisanilide; 4-methoxy-2′-[2-(1-methyl-2-piperidyl)ethyl]benzanilide. $C_{22}H_{28}N_2O_2$; mol wt 352.48. C 74.97%, H 8.01%, N 7.95%, O 9.08%. Anti-arrhythmic benzanilide derivative. Prepn: S. J. Dykstra, J. L. Minielli, **DE 2210154** (1972 to Bristol-Myers); *C.A.* **78**, 4138j (1973); *eidem*, **US 3931195** (1976 to Mead Johnson); S. J. Dykstra *et al.*, *J. Med. Chem.* **16**, 1015 (1973). Anti-arrhythmic pharmacology in animals: J. E. Byrne *et al.*, *J. Pharmacol. Exp. Ther.* **200**, 147 (1977). Clinical pharmacology and efficacy in chronic ventricular arrhythmia: D. M. Roden *et al.*, *N. Engl. J. Med.* **302**, 877 (1980). Electrophysiology and effects on cardiac conduction: M. Sami *et al.*, *Am. J. Cardiol.* **44**, 526 (1979). Hemodynamic effects: M. Sami *et al.*, *ibid.* **52**, 507 (1983). Adverse effects: R. A. Winkle *et al.*, *Am. Heart J.* **102**, 857 (1981). Comparison with other class I anti-arrhythmic agents: A. Pottage, *Am. J. Cardiol.* **52**, 24C (1983). Series of articles on pharmacology, pharmacokinetics, metabolism, clinical safety and efficacy: *ibid.* **58**(5), 1C-116C (1986). *Review:* L. B. Mitchell, R. A. Winkle in *New Drugs Annual: Cardiovascular Drugs* **vol. 1**, A. Scriabine, Ed. (Raven Press, New York, 1983) pp 93-107.

Hydrochloride. [66794-74-9] MJ-9067; Enkaid. $C_{22}H_{28}N_2O_2$.-HCl; mol wt 388.94. Crystals, mp 131.5-132.5°. Freely sol in water; slightly sol in ethanol. Insol in heptane. LD_{50} in mice, dogs (mg/kg): 86, 43 orally; 16, 17 i.v. (Mitchell, Winkle).

THERAP CAT: Antiarrhythmic (class IC).

3626. Endiandric Acids. A group of novel polycyclic compounds isolated from leaves of the 'Dorrigo Plum', *Endiandra introrsa* C. T. White, *Lauraceae*, a large tree occurring in rain forests of Australia. Although they have a number of asymmetric centers, endiandric acids occur in nature in racemic rather than enantiomeric forms. This fact led to a proposed hypothesis for the "biogenesis" of these compounds in nature from achiral precursors by a series of non-enzymatic electrocyclizations called the "endiandric acid cascade". Isoln from *E. introrsa* leaves (obtained in 1958) and structure of endiandric acid A: W. M. Bandaranayake *et al.*, *Chem. Commun.* **1980**, 162. Postulated electrocyclic reactions, prediction of endiandric acid D as a natural product: *eidem, ibid.* 902. X-ray crystallographic structural elucidation of A and detailed report of isoln: *eidem, Aust. J. Chem.* **34**, 1655 (1981). Isoln and structure of endiandric acid B: *eidem, ibid.* **35**, 557 (1982); of C: *eidem, ibid.* 567. Stereocontrolled total synthesis of A and B, description of "endiandric acid cascade": K. C. Nicolaou *et al.*, *J. Am. Chem. Soc.* **104**, 5555 (1982). Stereocontrolled total synthesis of endiandric acids C-G: *eidem, ibid.* 5557. Synthesis of precursors and thermal studies: *eidem, ibid.* 5558, 5560.

Endiandric acid A
R = CH₂COOH

Endiandric acid B

R = ⌒COOH

Endiandric acid C

Endiandric Acid A. [74591-03-0] *rel*-(1*R*,1a*R*,2a*R*,5*S*,5a*S*,7a*S*,-7b*R*,7c*R*)-1a,2,2a,5,5a,7a,7b,7c-Octahydro-5-phenyl-1*H*-cyclobut-[*bc*]acenaphthylene-1-acetic acid; 2-(6'-phenyltetracyclo[5,4,2,-$0^{3,13}$,$0^{10,12}$]trideca-4',8'-dien-11'-yl)acetic acid. $C_{21}H_{22}O_2$; mol wt 306.41. Rods from aq ethanol, mp 147-149°. uv max (95% ethanol): 242, 255, 261, 268, 286 nm (log ε 2.19, 2.36, 2.45, 2.32, 1.45). pKa 5.1, 5.0. Gives a yellow color with tetranitromethane.

Endiandric Acid B. [76060-33-8] *rel*-(2*E*)-4-[(1*R*,1a*S*,3a*S*,4*S*,-6a*R*,7a*R*,7b*R*,7c*R*)-1a,3a,4,6a,7,7a,7b,7c-Octahydro-4-phenyl-1*H*-cyclobut[5,4,2,$0^{3,13}$,$0^{10,12}$]trideca-4',8'-dien-11'-yl)but-2-enoic acid. $C_{23}H_{24}O_2$; mol wt 332.44. Rosettes from aq ethanol and chloroform + petr ether, mp 163-165°. uv max (95% ethanol): 252, 258, 262, 265, 269 nm (log ε 3.11, 3.08, 3.07, 3.01, 2.97).

Endiandric Acid C. [76060-34-9] *rel*-(1*R*,1a*R*,3*S*,3a*R*,6*S*,6a*S*,-6b*R*,7*S*)-1,1a,2,3,3a,6,6a,6b-Octahydro-1-[(2*E*,4*E*)-5-phenyl-2,4-pentadien-1-yl]-3,6-methanocyclobut[*cd*]indene-7-carboxylic acid; 4-[(*E*,*E*)-5'-phenylpenta-2',4'-dien-1'-yl]tetracyclo[5,4,0,0^{2,5},-$0^{3,9}$]undec-10-ene-8-carboxylic acid. $C_{23}H_{24}O_2$; mol wt 332.44. Cryst from ethanol and methanol, mp 125-132° (variable and dec). uv max (95% ethanol): 222, 228, 236, 280, 288 nm (log ε 4.10, 4.05, 3.89, 4.51, 4.53).

3627. Endorphins. Generic name derived from the term "endogenous morphine" and applied to a group of neuropeptides that are endogenous ligands of the opiate receptors. They are found in brain, pituitary gland and peripheral tissues of all vertebrates; the effects of endorphins on cells resemble those of opiates such as morphine. Their existence was postulated as a result of the discovery of stereospecific binding sites for narcotic analgesics in animal brain: A. Goldstein *et al.*, *Proc. Natl. Acad. Sci. USA* **68**, 1742 (1971); C. B. Pert, S. H. Snyder, *Science* **179**, 1011 (1973); E. J. Simon *et al.*,

Proc. Natl. Acad. Sci. USA **70**, 1947 (1973). Three types of stereochemically related opiate receptors have been postulated, *mu, delta, kappa:* W. R. Martin *et al.*, *J. Pharmacol. Exp. Ther.* **197**, 517 (1976); P. E. Gilbert, W. R. Martin, *ibid.* **198**, 66 (1976); J. A. H. Lord *et al.*, *Nature* **267**, 495 (1977). Of the endorphin group of peptides, several important members are known: α-, β- and γ-endorphins (β-endorphin being the most potent), α-neo-endorphin, β-neo-endorphin, dynorphin, *q.v.*, and met-enkephalin and leu-enkephalin, two naturally occurring pentapeptides belonging to the endorphin class. The amino acid sequences of α-, β- and γ-endorphins are contained within the sequence 61-91 of β-lipotropin, and in each, the amino terminal pentapeptide sequence is that of met-enkephalin. Dynorphin and α- and β-neo-endorphins have common terminal leu-enkephalin sequences. Initial isoln of active oligopeptide: L. Terenius, A. Wahlstrom, *Acta Pharmacol. Toxicol.* **35**, Suppl. 1, 55 (1974); J. Hughes, *Brain Res.* **88**, 295 (1975). Isoln, characterization and synthesis of two active pentapeptides (*enkephalins*): J. Hughes *et al.*, *Life Sci.* **16**, 1753 (1975); G. W. Pasternak *et al.*, *ibid.* 1765; J. Hughes *et al.*, *Nature* **258**, 577 (1975); J. D. Bower *et al.*, *J. Chem. Soc. Perkin Trans. 1* **1976**, 2488. Isoln and properties of β-endorphin from bovine pituitary: H. Teschemacher *et al.*, *Life Sci.* **16**, 1771 (1975); B. M. Cox *et al.*, *ibid.* 1777. Identity of β-endorphin with carboxy terminal amino acid sequence (61-91) of β-lipotropin: C. H. Li, D. Chang, *Proc. Natl. Acad. Sci. USA* **73**, 1145 (1976); B. M. Cox *et al.*, *ibid.* 1821. Production of biologically active β-endorphin via expression of cloned gene sequences by *E. coli:* J. Shine *et al.*, *Nature* **285**, 456 (1980). For a review of isoln, amino acid sequences, and synthesis of β-endorphins from various species, *see Hormonal Proteins and Peptides* vol. X, entitled "β-Endorphin", C. H. Li, Ed. (Academic Press, New York, 1981) pp 2-30. Isoln and structure of α-endorphin or *LPH (61-76)*: R. Guillemin *et al.*, *C. R. Seances Acad. Sci. Ser. D* **282**, 783 (1976); isoln, structure, synthesis of α- and γ-endorphin or *LPH (61-77)*: N. Ling *et al.*, *Proc. Natl. Acad. Sci. USA* **73**, 3942 (1976). High yield synthesis of met⁵-enkephalin: B. J. Dhotre *et al.*, *J. Indian Chem. Soc.* **55**, 1128 (1978). Rapid synthesis of leu⁵-enkephalin: E. Vilkas *et al.*, *Int. J. Pept. Protein Res.* **15**, 29 (1980). Isoln of α-neo-endorphin: K. Kagawa *et al.*, *Biochem. Biophys. Res. Commun.* **86**, 153 (1979). Amino acid sequence: *eidem, ibid.* **99**, 871 (1981). Isoln, purification, amino acid sequence of β-neo-endorphin: N. Minamino *et al.*, *ibid.* 864. Metabolism and physiological effects of endorphins: H. W. Kosterlitz, J. Hughes, *Life Sci.* **17**, 91 (1975); L.-F. Tseng *et al.*, *Proc. Natl. Acad. Sci. USA* **73**, 4187 (1976). Although the physiological role of the endorphins has not been completely elucidated, they are known to block inhibitory pathways in the vertebrate CNS and their possible effects on pain perception, addictive states, and psychiatric disorders have been investigated: D. T. Krieger, A. S. Liotta, *Science* **205**, 366 (1979); R. A. Nicoll *et al.*, *Nature* **287**, 22 (1980). Studies have also suggested that endorphins may be involved in gluco-regulation: M. Feldman *et al.*, *N. Engl. J. Med.* **308**, 350 (1983). *Reviews:* A. Goldstein, *Science* **193**, 1081-1086 (1976); *idem, Harvey Lect.* **79**, 291-314 (1979); R. A. North, *Life Sci.* **24**, 1527-1546 (1979); C. R. Beddell *et al.*, *Prog. Med. Chem.* **17**, 1-39 (1980); several authors in *Adv. Biochem. Pharmacol.* **22**, 145-642 (1980); M. S. Gold *et al.*, *Med. Res. Rev.* **2**, 211-246 (1982); S. Zakarian, D. G. Smyth, *Biochem. J.* **202**, 561-571 (1982). Books: *Opiates and Endogenous Opioid Peptides*, H. W. Kosterlitz, Ed. (Elsevier, New York, 1976) 466 pp; *Endorphins: Proceedings*, L. Graf *et al.*, Eds. (Elsevier, New York, 1979) 471 pp; *Endorphins and Opiate Antagonists in Psychiatric Research*, N. Shah, A. G. Donald, Eds. (Plenum, New York, 1982) 425 pp.

Human β-Endorphin

β-Endorphin. β-Lipotropin C-fragment; β-lipotropin (61-91); LPH (61-91).

Met⁵-Enkephalin. Methionine⁵-enkephalin; met⁵-E; L-tyrosyl-glycylglycyl-L-phenylalanyl-L-methionine; LPH (61-55). $C_{27}H_{35}$-

N_5O_7S; mol wt 573.67. Needles from hot methanol, mp 196-198°. $[\alpha]_{589}^{22}$ −21.9° (c = 1 in DMF); +14° (c = 1 in N HCl).

Leu⁵-Enkephalin. Leucine⁵-enkephalin; leu⁵-E; L-tyrosylglycylglycyl-L-glycyl-L-phenylalanyl-L-leucine. $C_{28}H_{37}N_5O_7$; mol wt 555.63. White crystalline solid, mp 206° (dec). $[\alpha]_{589}^{22}$ −23.4° (c = 1 in DMF); +18.3° (c = 1 in N HCl).

3628. Endostatin. [187888-07-9] Endogenous inhibitor of angiogenesis and tumor growth. 20 kDa, C-terminal fragment of collagen XVIII, a collagen-like peptide localized around blood vessels. Released by the action of tumor-produced proteases on the collagen molecule. Isoln from murine hemangioendothelioma: M. S. O'Reilly et al., Cell **88**, 277 (1997); from human plasma: L. Ständker et al., FEBS Lett. **420**, 129 (1997). Crystal structure: E. Hohenester et al., EMBO J. **17**, 1656 (1998). Antitumor activity in mice: T. Boehm et al., Nature **390**, 404 (1997). Review of biological activity: Y. Fu et al., IUBMB Life **61**, 613-626 (2009); of clinical experience in cancer therapy: M. V. Karamouzis, S. J. Moschos, Expert Opin. Biol. Ther. **9**, 641-648 (2009).

3629. Endosulfan. [115-29-7] 6,7,8,9,10,10-Hexachloro-1,-5,5a,6,9,9a-hexahydro-6,9-methano-2,4,3-benzodioxathiepin 3-oxide; 1,4,5,6,7,7-hexachloro-5-norbornene-2,3-dimethanol cyclic sulfite; 1,2,3,4,7,7-hexachlorobicyclo[2.2.1]-2-heptene-5,6-bisoxymethylene sulfite; chlorthiepin; Malix; Thiodan; Thionex. $C_9H_6Cl_6$-O_3S; mol wt 406.90. C 26.57%, H 1.49%, Cl 52.27%, O 11.80%, S 7.88%. Prepd by reaction of hexachlorocyclopentadiene with cis-butene-1,4-diol to form the bicyclic dialcohol, followed by esterification and cyclization with $SOCl_2$: Geering, Nelson, **US 2983732** (1961 to Hooker). Configuration studies: Reimschneider, Wuscherpfenning, Z. Naturforsch. **17b**, 585 (1962); Forman et al., J. Org. Chem. **30**, 169 (1965). Toxicity study: T. B. Gaines, Toxicol. Appl. Pharmacol. **14**, 515 (1969). Review of toxicology and human exposure: Toxicological Profile for Endosulfan (PB2000-108023, 2000) 323 pp.

Commercial product brown crystals. mp 70-100° (pure mp 106°). Practically insol in water. Sol in most organic solvents. Stable toward dil mineral acids; hydrolyzed rapidly by alkalies. Commercial product is a mixture of α-isomer, mp 108-110°, and β-isomer, mp 208-210°. LD_{50} orally in male, female rats: 43, 18 mg/kg (Gaines).

Caution: Potential symptoms of overexposure are skin irritation; nausea, confusion, agitation, flushing, dry mouth, tremors, convulsions, headache. See NIOSH Pocket Guide to Chemical Hazards (DHHS/NIOSH 97-140, 1997) p 126.

USE: Insecticide.

3630. Endothall. [145-73-3] 7-Oxabicyclo[2.2.1]heptane-2,3-dicarboxylic acid; 3,6-endoxohexahydrophthalic acid; endothal. $C_8H_{10}O_5$; mol wt 186.16. C 51.62%, H 5.41%, O 42.97%. Prepn: Olin, **US 2550494** (1951 to Sharples Chemicals). Of the three possible isomers, the exo-cis form has the greatest biological activity. Fate in the environment and toxicity: Simsiman et al., Residue Rev. **62**, 131-174 (1976); T. B. Gaines, R. E. Linder, Fundam. Appl. Toxicol. **7**, 299 (1986).

Solubility of the acid in g/100 g at 20°: water 10; acetone 7; benzene 0.01; methanol 28. Converted to the anhydride at 90°. LD_{50} in adult male, female rats (mg/kg): 57, 46 orally (Gaines, Linder).

Dipotassium salt. Aquathol; Aquathol K. $C_8H_8K_2O_5$; mol wt 262.34.

Caution: May be very irritating to skin, eyes, mucous membranes. Ingestion may cause severe G.I. inflammation with erosion. See Clinical Toxicology of Commercial Products, R. E. Gosselin et al., Eds. (Williams & Wilkins, Baltimore, 5th ed., 1984) Section II, p 342.

USE: Herbicide; defoliant.

3631. Endothelin. [117399-94-7] Endothelin-1; ET-1. Potent vasoconstrictor peptide produced by mammalian vascular endothelial cells; present in low concentrations in normal serum. Composed of 21 amino acid residues; mol wt ~2500 Da. Isoln, characterization and cloning of porcine endothelin: M. Yanagisawa et al., Nature **332**, 411 (1988). Structural identity of porcine and human endothelin: Y. Itoh et al., FEBS Lett. **231**, 440 (1988). Structural similarity with the snake venom toxins, *sarafotoxins S6*: C. Y. Lee, V. A. Chiappinelli, Nature **335**, 303 (1988); C. Takasaki et al., ibid. 303. Identification in serum by radioimmunoassay: K. Ando et al., FEBS Lett. **245**, 164 (1989). Two additional isoforms, known as *endothelin-2* and *endothelin-3*, have been identified. Structures and comparative pharmacology of human endothelin isopeptides: A. Inoue et al., Proc. Natl. Acad. Sci. USA **86**, 2863 (1989). Mechanism of action study: Y. Hirata et al., Biochem. Biophys. Res. Commun. **154**, 868 (1988). Possible role in acute renal failure: J. D. Firth et al., Lancet **2**, 1179 (1988). In vivo effects on pulmonary and systemic hemodynamics: H. L. Lippton et al., J. Appl. Physiol. **66**, 1008 (1989). Review: T. Masaki, J. Cardiovasc. Pharmacol. **13**, Suppl. 5, S1-S4 (1989).

C–S–C–S–S–L–M–D–K–E–C–V–Y–F–C–H–L–D–I–I–W

3632. Endralazine. [39715-02-1] (3-Hydrazinyl-7,8-dihydropyrido[4,3-c]pyridazin-6(5H)-yl)phenylmethanone; 6-benzoyl-5,6,7,8-tetrahydropyrido[4,3-c]pyridazin-3(2H)-one hydrazone; 6-benzoyl-3-hydrazino-5,6,7,8-tetrahydropyrido[4,3-c]pyridazine. $C_{14}H_{15}N_5O$; mol wt 269.31. C 62.44%, H 5.61%, N 26.01%, O 5.94%. Prepn: E. Schenker, **DE 2221808**; idem, **US 3838125** (1972, 1974 both to Sandoz); E. Schenker, R. Salzmann, Arzneim.-Forsch. **29**, 1835 (1979). Pharmacology: R. Salzmann et al., ibid. 1843. Hemodynamic study: H. U. Lehmann et al., Z. Cardiol. **66**, 203 (1977). HPLC determn in human plasma: P. A. Reece et al., J. Chromatogr. **225**, 151 (1981). Clinical pharmacology: F. C. Reubi, Eur. J. Clin. Pharmacol. **13**, 185 (1978). Clinical study in hypertension: H. U. Lehmann et al., Med. Welt **29**, 1007 (1978).

Crystals from acetonitrile/water, mp 220-223° (dec).

Methanesulfonate. Endralazine mesylate; BQ 22-708; Miretilan. $C_{14}H_{15}N_5O.CH_3SO_3H$; mol wt 365.41. Cryst, mp 185-188° (dec).

THERAP CAT: Antihypertensive.

3633. Endrin. [72-20-8] rel-(1aR,2R,2aR,3R,6S,6aS,7S,7aS)-3,4,5,6,9,9-Hexachloro-1a,2,2a,3,6,6a,7,7a-octahydro-2,7:3,6-dimethanonaphth[2,3-b]oxirene; 3,4,5,6,9,9-hexachloro-6,7-epoxy-1,4,4a,5,6,7,8,8a-octahydro-endo,endo-1,4:5,8-dimethanonaphthalene; mendrin; nendrin; hexadrin; experimental insecticide no. 269; compd 269; ENT-17251. $C_{12}H_8Cl_6O$; mol wt 380.90. C 37.84%, H 2.12%, Cl 55.84%, O 4.20%. Organochlorine pesticide; stereoisomer of dieldrin, q.v. Prepn from isodrin: H. Bluestone, **US 2676132** (1954 to Shell); Marks, **US 2899446** (1959 to Velsicol). Metabolism: M. K. Baldwin et al., Pestic. Sci. **7**, 575 (1976). Carcinogenicity studies: M. D. Rueber, Sci. Total Environ. **12**, 101 (1979). Toxicity study: T. B. Gaines, Toxicol. Appl. Pharmacol. **14**, 515 (1969). Soil degradation study: T. K. S. Gowda, N. Sethunathan, Soil Sci. **124**, 5 (1977). GC-MS determn in cereals and animal feed: S. Walorczyk, J. Chromatogr. A **1208**, 202 (2008). Review of toxicology and human exposure: Toxicological Profile for Endrin (PB97-

121040, 1996) 227 pp; of chemistry and environmental fate: V. Zitko in *The Handbook of Environmental Chemistry*, H. Fiedler, Ed. (Springer-Verlag, Berlin, 2003) pp 47-90.

Relative stereochemistry

Crystals, mp 245° (dec). Vapor pressure at 25°: 2×10^{-7} mm Hg. Soly in g/100 ml solvent at 25°: acetone 17, benzene 13.8, carbon tetrachloride 3.3, hexane 7.1, xylene 18.3. *Poisonous.* LD_{50} in female, male rats (mg/kg): 7.5, 18 orally (Gaines).

Caution: Potential symptoms of overexposure are epileptiform convulsions; stupor, headache and dizziness; abdominal discomfort, nausea and vomiting; insomnia; aggressiveness, confusion; lethargy, weakness; anorexia. *See NIOSH Pocket Guide to Chemical Hazards* (DHHS/NIOSH 97-140, 1997) p 126.

Note: Endrin is listed as a persistent organic pollutant (POP) in Annex A of the *Stockholm Convention on Persistent Organic Pollutants* (United Nations, Stockholm, 2001) 43 pp; amended (Geneva, 2009) 63 pp.

USE: Formerly as insecticide, rodenticide.

3634. Enduracidin. [11115-82-5] enramycin. Cyclodepsipeptide antibiotic produced by *Streptomyces fungicidicus*, strain no. B5477, from a soil sample collected in Nishinomiya, Hyogo Prefecture, Japan: Higashide *et al.*, *J. Antibiot.* **21**, 126 (1968). Originally thought to be a single compound. Isoln and characterization: Asai *et al.*, *ibid.* 138; **FR 1514139**; **GB 1163270** (1968, 1969 to Takeda). Manufacturing process: **JP Kokai 82 79896** (1982 to Takeda), *C.A.* **97**, 108525s (1982). Separation into two components, enduracidins A and B: Hori *et al.*, *Chem. Pharm. Bull.* **21**, 1171 (1973). The two components are each composed of seventeen amino acids, sixteen of which form a macrocyclic peptide lactone, and only differ by one methylene group in their fatty acid moieties. Structural studies: Hori *et al.*, *ibid.* 1175. Final structure: Iwasaki *et al.*, *ibid.* 1184. Antibacterial activity: Goto *et al.*, *J. Antibiot.* **21**, 119 (1968); Tsuchiya *et al.*, *ibid.* 147; M. Kawakami *et al.*, *ibid.* **24**, 583 (1971). Toxicology: H. Yokotani *et al.*, *Takeda Kenkyusho Nempo* **28**, 76 (1969), *C.A.* **72**, 77300s (1970).

Monohydrochloride. Enradin. Yellowish powder, dec 234-238°. Slightly sol in water, methanol. Practically insol in acetone, ethanol, chloroform, benzene. LD_{50} in mice and rats (mg/kg): >10000 orally; >5000 i.m.; >5000 s.c. (Yokotani).

Enduracidin A hydrochloride. [33386-20-6] $C_{107}H_{138}Cl_2N_{26}$-O_{31}.HCl; mol wt 2391.79. mp 240-245°. $[\alpha]_D^{23}$ +92° (c = 0.5 in DMF). uv max (0.1N HCl): 231, 272 nm.

Enduracidin B hydrochloride. [34765-98-5] $C_{108}H_{140}Cl_2N_{26}$-O_{31}.HCl; mol wt 2405.82. mp 238-241°. $[\alpha]_D^{23}$ +92° (c = 0.5 in DMF). uv max (0.1N HCl): 231, 272 nm.

THERAP CAT: Antibacterial.

3635. Enflurane. [13838-16-9] 2-Chloro-1-(difluoromethoxy)-1,1,2-trifluoroethane; 2-chloro-1,1,2-trifluoroethyl difluoromethyl ether; methylflurether; compd 347; NSC-115944; Alyrane; Efrane; Ethrane. $C_3H_2ClF_5O$; mol wt 184.49. C 19.53%, H 1.09%, Cl 19.22%, F 51.49%, O 8.67%. Prepn by fluorination of the corresp dichloromethyl ether: R. C. Terrell, **GB 1138406**; *idem*, **US 3469011**; **US 3527813** (1969, 1969, 1970 all to Air Reduction). Synthesis and anesthetic properties: R. C. Terrell *et al.*, *J. Med. Chem.* **14**, 517 (1971); **15**, 604 (1972). Enantiomeric resolution: J. Meinwald *et al.*, *Science* **251**, 560 (1991).

Colorless, stable, volatile, non-flammable liq. bp 56.5°. n_D^{20} 1.3025. d_{25}^{25} 1.5167. Does not degrade in the presence of alkali or light. Miscible with other organic solvents incl. fats and oils. Slightly sol in water.

Caution: Potential symptoms of overexposure are eye irritation; CNS depression, analgesia, anesthesia, seizures, respiratory depression. *See NIOSH Pocket Guide to Chemical Hazards* (DHHS/NIOSH 97-140, 1997) p 128.

THERAP CAT: Anesthetic (inhalation).

3636. Enfuvirtide. [159519-65-0] N-Acetyl-L-tyrosyl-L-threonyl-L-seryl-L-leucyl-L-isoleucyl-L-histidyl-L-seryl-L-leucyl-L-isoleucyl-L-α-glutamyl-L-α-glutamyl-L-seryl-L-glutaminyl-L-asparaginyl-L-glutaminyl-L-glutaminyl-L-α-glutamyl-L-lysyl-L-asparaginyl-L-α-glutamyl-L-glutaminyl-L-α-glutamyl-L-leucyl-L-leucyl-L-α-glutamyl-L-leucyl-L-α-aspartyl-L-lysyl-L-tryptophyl-L-alanyl-L-seryl-L-leucyl-L-tryptophyl-L-asparaginyl-L-tryptophyl-L-phenylalaninamide; pentafuside; DP-178; T-20; Fuzeon. $C_{204}H_{301}N_{51}O_{64}$; mol wt 4491.95. C 54.55%, H 6.75%, N 15.90%, O 22.79%. HIV fusion inhibitor. Synthetic, 36 amino acid peptide corresponding to a region of gp41, the transmembrane subunit of the HIV-1 envelope protein. Prepn and biological activity: D. P. Bolognesi *et al.*, **WO 9428920**; *eidem*, **US 5464933** (1994, 1995 both to Duke Univ.); C. T. Wild *et al.*, *Proc. Natl. Acad. Sci. USA* **91**, 9770 (1994). Mechanism of action study: R. A. Furuta *et al.*, *Nat. Struct. Biol.* **5**, 276 (1998). Clinical pharmacokinetics: J. M. Kilby *et al.*, *AIDS Res. Hum. Retroviruses* **18**, 685 (2002). Clinical trial in drug-resistant HIV-1: J. P. Lalezari *et al.*, *N. Engl. J. Med.* **348**, 2175 (2003); A. Lazzarin *et al.*, *ibid.* 2186. Review of development and clinical experience: A. Lazzarin, *Expert Opin. Pharmacother.* **6**, 453-464 (2005).

White to off-white amorphous solid. Negligible soly in pure water. Soly increases in aq buffers (pH 7.5) to 85-142 g/100 ml.

THERAP CAT: Antiretroviral.

3637. Enilconazole. [35554-44-0] 1-[2-(2,4-Dichlorophenyl)-2-(2-propen-1-yloxy)ethyl]-1H-imidazole; (±)-1-[β-(allyloxy)-2,4-dichlorophenethyl]imidazole; imazalil; R-23979; Clinafarm; Imaverol. $C_{14}H_{14}Cl_2N_2O$; mol wt 297.18. C 56.58%, H 4.75%, Cl 23.86%, N 9.43%, O 5.38%. Broad spectrum antimycotic with agricultural and veterinary applications. Prepn: E. F. Godefroi, J. L. Schuermans, **DE 2063857**; *eidem*, **US 3658813** (1971, 1972 both to Janssen). Efficacy vs citrus decay: P. R. Harding, *Plant Dis. Rep.* **60**, 643 (1976); S. Ben-Yehoshua *et al.*, *Pestic. Sci.* **12**, 485 (1981). Biological and toxicological properties: D. Thienpont *et al.*, *Arzneim.-Forsch.* **31**, 309 (1981). Residue analysis in treated grapefruit: E. R. Stein *et al.*, *J. Environ. Sci. Health* **B16**, 427 (1981). LC-MS determn in citrus fruit: N. Yoshioka *et al.*, *Food Control* **21**, 212 (2010).

Solidified oil. Slightly sol in organic solvents; poorly sol in water.

USE: Agricultural fungicide.

THERAP CAT (VET): Topical antifungal.

3638. Eniluracil. [59989-18-3] 5-Ethynyl-2,4(1H,3H)-pyrimidinedione; 5-ethynyluracil; 5-EU; 776C85. $C_6H_4N_2O_2$; mol wt 136.11. C 52.95%, H 2.96%, N 20.58%, O 23.51%. Irreversible inhibitor of dihydropyrimidine dehydrogenase, the rate-determining enzyme in catabolism of 5-fluorouracil, *q.v.* Prepn: J. Perman *et al.*, *Tetrahedron Lett.* **1976**, 2427; P. J. Barr *et al.*, *Nucleic Acids Res.* **3**, 2845 (1976). Large scale synthesis: N. G. Kundu, S. A. Schmitz, *J. Heterocycl. Chem.* **19**, 463 (1982). Crystal structure determn: M. Sacchetti *et al.*, *J. Pharm. Sci.* **90**, 1049 (2001). Mechanism-based inactivation of dihydropyrimidine dehydrogenase: D. J. T. Porter *et al.*, *J. Biol. Chem.* **267**, 5236 (1992). Prolongs plasma concentration of 5-fluorouracil: D. P. Baccanari *et al.*, *Proc. Natl. Acad. Sci. USA*

90, 11064 (1993). Clinical safety and efficacy: R. A. Humerickhouse *et al.*, *Clin. Cancer Res.* **5**, 291 (1999). Clinical study with 5-fluorouracil in advanced pancreatic cancer: M. L. Rothenberg *et al.*, *Ann. Oncol.* **13**, 1576 (2002); in metastatic/advanced colorectal cancer: R. L. Schilsky *et al.*, *J. Clin. Oncol.* **20**, 1519 (2002). Review of clinical development: J. A. Hohneker, *Oncology* **12**, 52-56 (1998).

Cream colored solid from methanol-water, mp 320° (dec). d 1.527. uv max (methanol): 284, 225 nm (ε 9450, 10330).

THERAP CAT: Antineoplastic adjunct (chemomodulator).

3639. Enniatins. Ionophore antibiotics produced by the fungus, *Fusarium orthoceras* var. *enniatum*, and other *Fusaria:* Gaumann *et al.*, *Experientia* **3**, 202 (1947); Plattner, Nager, *ibid.* 325; Plattner *et al.*, *Helv. Chim. Acta* **31**, 594 (1948). They are cyclic depsihexapeptides related structurally to valinomycin, *q.v.* Enniatins A, B, and C are known; A and B are obtained from natural sources and C is a synthetic homolog. Structure and synthesis of enniatin A: Quitt *et al.*, *ibid.* **46**, 1715 (1963); **47**, 166 (1964); of enniatin B: Plattner *et al.*, *ibid.* **46**, 927 (1963); Shemyakin *et al.*, *Tetrahedron Lett.* **1963**, 885. Conformation of enniatin B: Ovchinnikov *et al.*, *Biochem. Biophys. Res. Commun.* **37**, 668 (1969). Synthesis of C: *eidem*, *Izv. Akad. Nauk SSSR Ser. Khim.* **1964**, 1912, *C.A.* **62**, 28246b (1965). Enniatins are able to form "sandwich" complexes with cations, a factor used to explain their ability to induce permeability in biological membranes to the complexed ion. *See:* V. T. Ivanov, *Ann. N.Y. Acad. Sci.* **264**, 221 (1975). *Review:* Y. A. Ovchinnikov, V. T. Ivanov, "The Cyclic Peptides: Structure, Conformation, and Function" in *The Proteins* **vol. V**, H. Neurath, R. L. Hill, Eds. (Academic Press, New York, 3rd ed., 1982) pp 365-373, 516-529.

Enniatin A   R = CH(CH$_3$)CH$_2$CH$_3$
Enniatin B   R = CH(CH$_3$)$_2$
Enniatin C   R = CH$_2$CH(CH$_3$)$_2$

Enniatin A. [2503-13-1] $C_{36}H_{63}N_3O_9$; mol wt 681.91. Long needles from ethanol + water, mp 122-122.5°. Very slowly sublimes at 10^{-4} mm and 127-128° (oil bath temp). $[\alpha]_D^{18}$ −91.9° (c = 0.926 in chloroform). Sol in ether, benzene, ethyl acetate; sparingly sol in water. Solns are stable to heat, inactivated by alkali.

Enniatin B. [917-13-5] $C_{33}H_{57}N_3O_9$; mol wt 639.83. Crystals from petr ether, mp 175-175.5°. $[\alpha]_D^{20}$ −107.9° (c = 0.63 in chloroform). Very sol in organic solvents; slightly sol in petr ether. Practically insol in water.

Enniatin C. [19893-23-3] $C_{36}H_{63}N_3O_9$; mol wt 681.91. Crystals, mp 160-161°. $[\alpha]_D^{23}$ −24°.

3640. Enocitabine. [55726-47-1] *N*-(1-β-D-Arabinofuranosyl-1,2-dihydro-2-oxo-4-pyrimidinyl)docosanamide; N^4-behenoyl-1-β-D-arabinofuranosylcytosine; behenoylcytosine arabinoside; BH-AC; NSC-239336; Sunrabin. $C_{31}H_{55}N_3O_6$; mol wt 565.80. C 65.81%, H 9.80%, N 7.43%, O 16.97%. Deriv of cytarabine, *q.v.* Prepn: T. Ishida *et al.*, **DE 2426304**; *eidem*, **US 3991045** (1975, 1976 both to Asahi); M. Akiyama *et al.*, *Chem. Pharm. Bull.* **26**, 981 (1978). Antitumor activity: M. Aoshima *et al.*, *Cancer Res.* **36**, 2726 (1976); *eidem*, *ibid.* **37**, 2481 (1977). Distribution, excretion: M. Fukama *et al.*, *Gan to Kagaku Ryoho* **7**, 2109 (1980), *C.A.* **95**, 35186 (1981). Effect and mode of action: T. Kataoka, Y. Sakurai, *Recent Results Cancer Res.* **70**, 147 (1980). Pharmacological and clinical studies: K. Yamada *et al.*, *ibid.* 219.

Crystals from DMSO, mp 141-142°. $[\alpha]_D$ +70° (c = 1 in THF, 22°). uv max (isopropyl alcohol): 216, 248, 303 nm (ε 16400, 15200, 8200).

THERAP CAT: Antineoplastic.

3641. Enoxacin. [74011-58-8] 1-Ethyl-6-fluoro-1,4-dihydro-4-oxo-7-(1-piperazinyl)-1,8-naphthyridine-3-carboxylic acid; AT-2266; CI-919; PD-107779; Comprecin; Flumark. $C_{15}H_{17}FN_4O_3$; mol wt 320.32. C 56.25%, H 5.35%, F 5.93%, N 17.49%, O 14.98%. Fluorinated quinolone antibacterial; nalidixic acid analog. Prepn and antibacterial activity: J. Matsumoto *et al.*, **EP 9425**; *eidem*, **US 4352803**; *eidem*, **US 4359578** (1980, 1982, 1982 all to Dainippon and Roger Bellon); J. Matsumoto *et al.*, *J. Med. Chem.* **27**, 292 (1984). HPLC determn in plasma, urine: T. B. Vree *et al.*, *J. Chromatogr.* **343**, 449 (1985). *In vitro* comparison with other quinolones and azaquinolones: S. Selwyn, M. Bakhtiar, *Drugs Exp. Clin. Res.* **10**, 653 (1984). Pharmacokinetics, therapeutic efficacy: K. G. Naber *et al.*, *Infection* **13**, 219 (1985). Efficacy in treatment of urinary tract infections: R. R. Bailey, B. A. Peddie, *N. Z. Med. J.* **98**, 286 (1985). Toxicology studies: H. Senda *et al.*, *Chemotherapy (Tokyo)* **32**, Suppl. 3, 192 (1984). Series of articles on activity, pharmacology, metabolism, early clinical studies: *ibid.* 1-1093; *J. Antimicrob. Chemother.* **14**, Suppl. C, 1-96 (1984); on activity, pharmacokinetics, clinical safety and efficacy: *Infection* **14**, Suppl. 3, S183-S220 (1986).

Crystals from ethanol/methylene chloride, mp 220-224°. LD_{50} in male, female mice, male, female rats (mg/kg): 327, 391, 236, 294 i.v.; 1237, 1320, >2000, >2000 s.c.; all >5000 orally (Senda).

Sesquihydrate. [84294-96-2] Bactidan; Enoxen; Enoxor; Gyramid. $C_{15}H_{17}FN_4O_3 \cdot 1\frac{1}{2}H_2O$; mol wt 347.35.

THERAP CAT: Antibacterial.

3642. Enoxaparin. Low molecular weight fragment of heparin, *q.v.*, prepd by alkaline depolymerization of a benzylic ester of porcine intestinal mucosal heparin. The majority of components present a 2-*O*-sulfo-α-L-*threo*-hex-4-enopyranuronic acid structure at the non-reducing end and a 2-*N*,6-*O*-disulfo-D-glucosamine structure at the reducing end of the chain. Average mol wt ~4500 daltons. Prepn: J. Mardiguian, **EP 40144** (1981 to Pharmindustrie), *C.A.* **96**, 218191s (1982). Determn in human plasma: L. Bara *et al.*, *Haemostasis* **17**, 127 (1987). Symposium on pharmacology and clinical efficacy: *ibid.* **16**, 69-188 (1986). Review of development and clinical experience: J. M. Galla, K. W. Mahaffey, *Expert Opin. Pharmacother.* **6**, 1241-1251 (2004); of pharmacology and use in prevention and treatment of deep venous thrombosis: M. M. Buckley, E. M. Sorkin, *Drugs* **44**, 465-497 (1992); in unstable angina: S. Noble, C. M. Spencer, *ibid.* **56**, 259-272 (1998); in acute coronary syndromes: S. Lee, C. M. Gibson, *Expert Rev. Cardiovasc. Ther.* **5**, 387-399 (2007).

Sodium salt. [679809-58-6] PK-10169; RP-54563; Clexane; Lovenox.

THERAP CAT: Antithrombotic.

3643. Enoximone. [77671-31-9] 1,3-Dihydro-4-methyl-5-[4-(methylthio)benzoyl]-2*H*-imidazol-2-one; 4-methyl-5-[*p*-(methylthio)benzoyl]-4-imidazolin-2-one; fenoximone; MDL-17043; RMI-17043; Perfane; Perfan. $C_{12}H_{12}N_2O_2S$; mol wt 248.30. C 58.05%, H 4.87%, N 11.28%, O 12.89%, S 12.91%. Selective phosphodiesterase inhibitor with vasodilating and positive inotropic activity. Prepn: **BE 883856** (1980 to Richardson-Merrell); R. A. Schnettler *et al.*, **US 4405635** (1983 to Merrell-Dow); *eidem, J. Med. Chem.* **25**, 1477 (1982). Cardiovascular pharmacology in animals: R. C. Dage *et al., J. Cardiovasc. Pharmacol.* **4**, 500 (1982); L. E. Roebel *et al.*, *ibid.* 721. *In vitro* mechanism of action studies: T. Kariya *et al.*, *ibid.* 509; H. S. Ahn *et al.*, *Biochem. Pharmacol.* **35**, 1113 (1986). Acute hemodynamic effects in congestive heart failure: B. F. Uretsky *et al., Circulation* **67**, 823 (1983). HPLC determn in plasma: K. Y. Chan *et al., J. Chromatogr.* **272**, 396 (1983). Pharmacokinetics in humans: R. G. Alken *et al., Clin. Pharmacol. Ther.* **36**, 209 (1984). Symposium on pharmacology and clinical efficacy in congestive heart failure: *Am. J. Cardiol.* **60**, 1C-90C (1987).

Crystals from isopropanol + water, mp 255-258° (dec).
THERAP CAT: Cardiotonic.

3644. Enoxolone. [471-53-4] (3β,20β)-3-Hydroxy-11-oxoolean-12-en-29-oic acid; 3β-hydroxy-11-oxoolean-12-en-30-oic acid; glycyrrhetic acid; 18β-glycyrrhetinic acid; uralenic acid; Arthrodont; Biosone; P.O. 12. $C_{30}H_{46}O_4$; mol wt 470.69. C 76.55%, H 9.85%, O 13.60%. From glycyrrhizic acid, *q.v.* Structure: Ruzicka *et al., Helv. Chim. Acta* **26**, 2143, 2278 (1943). Stereochemistry: Beaton, Spring, *J. Chem. Soc.* **1955**, 3126. Prepn from shredded licorice root: Mer, *Am. Perfum. Aromat.* **74**(6), 39 (1959). Manuf: **GB 833184** (1960 to Carlo Erba). Identity with uralenic acid: Belous *et al., Zh. Obshch. Khim.* **35**, 401 (1965). Metabolism: Parke *et al., J. Pharm. Pharmacol.* **15**, 500 (1963). Mechanism of action: Helbing, Berntsen, *Pharm. Weekbl.* **100**, 1438 (1965).

Needles from alcohol + petr ether, mp 296°. $[\alpha]_D^{21}$ +86° (alc); $[\alpha]_D^{20}$ +145.5° (dioxane); $[\alpha]_D^{20}$ +163° (chloroform). Freely sol in chloroform, dioxane. Soluble in alcohol, pyridine, acetic acid. Practically insol in petr ether.

18α-Hydrogen Form. Platelets from dil alcohol, mp 335°. $[\alpha]_D^{20}$ +140° (alcohol); $[\alpha]_D$ +98° (c = 0.1 in chloroform). Sol in alcohol, dioxane, chloroform.

THERAP CAT: Anti-inflammatory (topical).

3645. Enprostil. [73121-56-9] *rel*-7-[(1*R*,2*R*,3*R*)-3-Hydroxy-2-[(1*E*,3*R*)-3-hydroxy-4-phenoxy-1-buten-1-yl]-5-oxocyclopentyl]-4,5-heptadienoic acid methyl ester; *(dl)*-9-keto-11α,15α-dihydroxy-16-phenoxy-17,18,19,20-tetranorprosta-4,5,13-*trans*-trienoic acid methyl ester; RS-84135; Camleed; Gardrin(e). $C_{23}H_{28}O_6$; mol wt 400.47. C 68.98%, H 7.05%, O 23.97%. Prostaglandin E$_2$ derivative with antisecretory activity. Prepn: A. R. Van Horn *et al.*, **US 4178457** (1979 to Syntex). Pharmacokinetics: D. R. Stanski *et al., Clin. Pharmacol. Ther.* **31**, 273 (1982). Cytoprotective effect against aspirin-induced gastromucosal injury in humans: M. M. Cohen *et al., Gastroenterology* **88**, 382 (1985). Clinical comparison with cimetidine of antisecretory and antigastrin activity in duodenal ulcer: V. Mahachai *et al., ibid.* **89**, 555 (1985); of efficacy in ulcer healing: L. Carling *et al., Scand. J. Gastroenterol.* **22**, 325 (1987). Series of articles on pharmacology, clinical efficacy and safety: *Am. J. Med.* **81**, Suppl. 2A, 1-88 (1986).

White to off-white waxy solid. Softens at 30°, liquid at 46°. Very slightly sol in water. Sol in alcohol, propylene glycol, propylene carbonate. uv max (methanol): 220, 265, 271, 277 nm (log ε 4.01, 3.14, 3.24, 3.16).

THERAP CAT: Antiulcerative.

3646. Enrofloxacin. [93106-60-6] 1-Cyclopropyl-7-(4-ethyl-1-piperazinyl)-6-fluoro-1,4-dihydro-4-oxo-3-quinolinecarboxylic acid; CFPQ; Bay Vp 2674; Baytril. $C_{19}H_{22}FN_3O_3$; mol wt 359.40. C 63.50%, H 6.17%, F 5.29%, N 11.69%, O 13.35%. Fluorinated quinolone antibacterial. Prepn: K. Grohe *et al.*, **DE 3142854**; *eidem*, **US 4670444** (1983, 1987 both to Bayer AG); K. Grohe, H. Heitzer, *Ann.* **1987**, 29. Use as plant fungicide: K. Grohe *et al.*, **US 4563459** (1986 to Bayer AG). Pharmacokinetics in calves: J. N. Davidson *et al., Proc. West. Pharmacol. Soc.* **29**, 129 (1986); in chickens: G. M. Conzelman *et al., ibid.* **30**, 393 (1987). Spectrofluorometric determn of residues in poultry tissues: T. B. Waggoner, M. C. Bowman, *J. Assoc. Off. Anal. Chem.* **70**, 813 (1987). Toxicology and physical properties: P. Altreuther, *Vet. Med. Rev.* **2**, 87 (1987). Series of articles on pharmacology, *in vitro* antibacterial activity and field trials: *ibid.* 90-140.

Pale yellow crystals, mp 219-221°. Slightly sol in water at pH 7. LD$_{50}$ in male, female mice (mg/kg): >5000, 4336 orally; ~200, ~200 i.v.; in male rats, male rabbits (mg/kg): >5000, 500-800 orally (Altreuther).

THERAP CAT (VET): Antibacterial.

3647. Ensulizole. [27503-81-7] 2-Phenyl-1*H*-benzimidazole-6-sulfonic acid; phenylbenzimidazole sulfonic acid; Eusolex 232; Neo Heliopan Hydro; Parsol HS. $C_{13}H_{10}N_2O_3S$; mol wt 274.29. C 56.93%, H 3.67%, N 10.21%, O 17.50%, S 11.69%. UV-B filter in sunscreens and cosmetics. Synergistic increase in SPF occurs in combination with oil-soluble UV absorbers. Prepn: V. G. Sayapin *et al., Khim. Geterotsikl. Soedin.* **5**, 681 (1970), *C.A.* **73**, 45450p (1970). *See also:* U. Heywang *et al.*, **US 5473079** (1995 to Merck Patent GmbH). DNA photosensitization study: C. Stevenson, R. J. H. Davies, *Chem. Res. Toxicol.* **12**, 38 (1999). Photochemical studies: J. J. Inbaraj *et al., Photochem. Photobiol.* **75**, 107 (2002). Fluorometric determn in urine: M. T. Vidal *et al., Talanta* **59**, 591 (2003). Complexation with cyclodextrins: S. Scalia *et al., Eur. J. Pharm. Sci.* **22**, 241 (2004).

mp 410° (dec). Absorption max (phosphate buffer, pH 7.4): 300 nm. Soly in water: ~0.03%.

Sodium salt. [5997-53-5] $C_{13}H_9N_2NaO_3S$; mol wt 296.28. Soly: water >30%; 1,2-propylene glycol >20%.

THERAP CAT: Ultraviolet screen.

3648. Entacapone. [130929-57-6] (2E)-2-Cyano-3-(3,4-di-hydroxy-5-nitrophenyl)-N,N-diethyl-2-propenamide; (E)-α-cyano-N,N-diethyl-3,4-dihydroxy-5-nitrocinnamamide; (E)-2-cyano-N,N-diethyl-3-(3,4-dihydroxy-5-nitrophenyl)acrylamide; OR-611; Comtan; Comtess. $C_{14}H_{15}N_3O_5$; mol wt 305.29. C 55.08%, H 4.95%, N 13.76%, O 26.20%. Peripherally acting inhibitor of catechol-O-methyl transferase (COMT), an enzyme involved in the metabolism of catecholamine neurotransmitters and related drugs. Prepn (stereochemistry unspecified): R. J. Bäckström et al., **DE 3740383**; eidem, **US 5446194** (1988, 1995 both to Orion). Prepn of (E)-isomer: A. K. Pippuri et al., **EP 426468**; eidem, **US 5135950** (1991, 1992 both to Orion). Pharmacology: E. Nissinen et al., Arch. Pharmacol. **346**, 262 (1992). LC determn in plasma and urine: M. Karlsson, T. Wikberg, J. Pharm. Biomed. Anal. **10**, 593 (1992). Identification of metabolites: T. Wikberg et al., Eur. J. Drug Metab. Pharmacokinet. **18**, 359 (1993). Clinical pharmacokinetics: T. Keränen et al., Eur. J. Clin. Pharmacol. **46**, 151 (1994). Clinical effect on levodopa bioavailability in Parkinson's disease: H. M. Routtinen, U. K. Rinne, J. Neurol. Neurosurg. Psychiatry **60**, 36 (1996).

Crystals from acetic acid + HCl, mp 162-163°. pKa ~4.5.

THERAP CAT: Antiparkinsonian.

3649. Entecavir. [142217-69-4] 2-Amino-1,9-dihydro-9-[(1S,3R,4S)-4-hydroxy-3-(hydroxymethyl)-2-methylenecyclopentyl]-6H-purin-6-one. $C_{12}H_{15}N_5O_3$; mol wt 277.28. C 51.98%, H 5.45%, N 25.26%, O 17.31%. Deoxyguanine nucleoside analog; inhibits hepatitis B virus (HBV) DNA polymerase. Prepn: R. Zahler, W. A. Slusarchyk, **EP 481754**; eidem, **US 5206244** (1992, 1993 both to Squibb); G. S. Bisacchi et al., Bioorg. Med. Chem. Lett. **7**, 127 (1997). In vitro antiviral activity: S. F. Innaimo et al, Antimicrob. Agents Chemother. **41**, 1444 (1997). Review of pharmacology and clinical experience: P. Honkoop, R. A. de Man, Expert Opin. Invest. Drugs **12**, 683-688 (2003); T. Shaw, S. Locarnini, Expert Rev. Anti Infect. Ther. **2**, 853-871 (2004). Clinical comparisons with lamivudine in chronic hepatitis B: T.-T. Chang et al., N. Engl. J. Med. **354**, 1001 (2006); C.-L. Lai et al., ibid. 1011.

Monohydrate. [209216-23-9] BMS-200475; SQ-200475; Baraclude. $C_{12}H_{15}N_5O_3 \cdot H_2O$; mol wt 295.30. White to off-white powder, mp >220°. $[\alpha]_D +35.0°$ (c = 0.38 in water). Soly in water: 2.4 mg/ml. pH of saturated soln in water is 7.9 at 25°±0.5°.

THERAP CAT: Antiviral.

3650. Enterobactin. [28384-96-5] N,N',N''-[(3S,7S,11S)-2,-6,10-Trioxo-1,5,9-trioxacyclododecane-3,7,11-triyl]tris[2,3-dihydroxy]benzamide; N,N',N''-(2,6,10-trioxo-1,5,9-trioxacyclododecane-3,7,11-triyl)tris-o-pyrocatechuamide; enterochelin. $C_{30}H_{27}N_3O_{15}$; mol wt 669.55. C 53.82%, H 4.06%, N 6.28%, O 35.84%. Physiologically active macrocyclic iron sequestering agent of the phenolate type, involved in microbial transport and metabolism of iron; it is overproduced by E. coli and related enteric bacteria under low-iron stress. Isoln from culture medium of Salmonella typhimurium LT2 and structure: J. R. Pollack, J. B. Neilands, Biochem. Biophys. Res. Commun. **38**, 989 (1970); also produced by Escherichia coli and Aerobacter aerogenes: I. G. O'Brien, F. Gibson, Biochim. Biophys. Acta **215**, 393 (1970). Conformation: M. Leinás et al., Biochemistry **12**, 3836 (1973). Total synthesis: E. J. Corey, S. Bhattacharyya, Tetrahedron Lett. **1977**, 3919. Synthesis of enterobactin and enantioenterobactin: W. H. Rastetter et al., J. Org. Chem. **45**, 5011 (1980); **46**, 3579 (1981). Biosynthetic study: J. R. Pollack et al., J. Bacteriol. **104**, 635 (1970). Stability constants and electrochemical behavior of ferric enterobactin: K. N. Raymond, J. Am. Chem. Soc. **101**, 6097 (1979). Kinetics and mechanism of iron removal from transferrin: C. J. Carrano et al., ibid. 5401. Review: J. B. Neilands in Inorg. Biochem. **1**, G. Eichorn, Ed. (Elsevier, New York, 1973) pp 167-202; idem, Bioinorg. Chem. **II**, K. N. Raymond, Ed. (A.C.S., Washington, 1977) pp 3-33.

Crystals from ethanol/water, mp 202-203°. $[\alpha]_D^{25} +7.40°$ (ethanol). uv max (ethyl acetate): 316 nm (ε 9390). Sol in acetone, dioxane, dimethyl sulfoxide, methanol. Practically insol in water, soly increases at pH 7 to 8. Neutral on paper electrophoresis at pH 5. Displays bright blue fluorescence under uv light.

Ferric enterobactin. $[C_{30}H_{21}FeN_3O_{15}]^{3-}$. Anionic complex of enterobactin. uv max: 495 nm (ε 5600).

USE: As a growth-promoting agent in various organisms.

3651. Enterogastrone. [9007-67-4] Anthelone E; enteroanthelone; Duosan; Ileogastrone. An inhibitor of gastric secretion, apparently different from urogastrone. Obtained from extracts of the intestinal mucosa of mammals: Gray, Ivy, Cold Spring Harbor Symp. Quant. Biol. **5**, 405 (1937); Ivy, Greengard, **US 2477541** (1949 to Research Corp.). Comparison with urogastrone: Friedman, Vitam. Horm. **9**, 313 (1951); Gregory, J. Physiol. **129**, 528 (1955). Amorphous buff-colored powder. Freely sol in water.

THERAP CAT: Gastric secretion inhibitor.

3652. Enterolactone. [78473-71-9] rel-(3R,4R)-Dihydro-3,4-bis[(3-hydroxyphenyl)methyl]-2(3H)-furanone; trans-(±)-2,3-bis(3'-hydroxybenzyl)-γ-butyrolactone; HPMF; HBBL; compd 180/442. $C_{18}H_{18}O_4$; mol wt 298.34. C 72.47%, H 6.08%, O 21.45%. Phenolic compound having 2,3-dibenzylbutane skeleton. First lignan, q.v., found in animal species. Determn in female urine from humans, vervet monkeys and characterization: S. R. Stitch et al., Nature **287**, 738 (1980); K. D. R. Setchell et al., ibid. 740. Structure determn by GC-MS: eidem, Biochem. J. **197**, 447 (1981); eidem, Biomed. Mass Spectrom. **10**, 227 (1983). ¹H-NMR, X-ray crystal structure: G. Cooley et al., J. Chem. Soc. Perkin Trans. 1 **1984**, 489. Synthesis: M. B. Groen, J. Leemhuis, Tetrahedron Lett. **21**, 5043 (1980); G. Cooley et al., ibid. **22**, 349 (1981); A. Pelter et al., ibid. 1549; P. A. Ganeshpure, R. Stevenson, Chem. Ind. (London) **1981**,

778; A. Pelter *et al.*, *J. Chem. Soc. Perkin Trans. 1* **1983**, 643. Lignan excretion peaks during luteal phase of menstrual cycle: K. D. R. Setchell *et al.*, *J. Steroid Biochem.* **12**, 375 (1980). Excretion as glucuronide conjugate: M. Axelson, K. D. R. Setchell, *FEBS Lett.* **122**, 49 (1980). Biosynthesis by intestinal bacteria: *eidem, ibid.* **123**, 337 (1981). Effect of diet on lignan excretion: H. Aldercreutz *et al.*, *Med. Biol.* **59**, 337 (1981).

Relative stereochemistry

Gum, mp 141-143°. uv max (ethanol): 227, 261 nm (log ε 4.66, 4.64).

3653. Entprol. [102-60-3] 1,1',1'',1'''-(1,2-Ethanediyldinitrilo)tetrakis-2-propanol; 1,1',1'',1'''-(ethylenedinitrilo)tetra-2-propanol; *N,N,N',N'*-tetrakis(2-hydroxypropyl)ethylenediamine; Quadrol. $C_{14}H_{32}N_2O_4$; mol wt 292.42. C 57.50%, H 11.03%, N 9.58%, O 21.88%. Prepd by the action of propylene oxide on ethylenediamine: Lundsted, Schulz, **US 2697118** (1954 to Wyandotte).

Viscous liq. bp$_{1.0}$ 190°. Miscible with water. Sol in methanol, ethanol, toluene, ethylene glycol, perchloroethylene.

USE: Cross-linking agent and catalyst in the manuf of urethan foams; in epoxy resin curing; as complexing agent, humectant, emulsifier, plasticizer.

3654. Enviomycin. [33103-22-9] Cyclo[(2Z)-3-[(aminocarbonyl)amino]-2,3-didehydroalanyl-(2S)-2-[(4R)-2-amino-3,4,5,6-tetrahydro-4-pyrimidinyl]glycyl-(2S)-2-[[(3R,4R)-3,6-diamino-4-hydroxy-1-oxohexyl]amino]-β-alanyl-L-seryl-L-seryl]; tuberactinomycin N; (R)-1-[(3R,4R)-4-hydroxy-3,6-diaminohexanoic acid]-6-[L-2-(2-amino-1,4,5,6-tetrahydro-4-pyrimidinyl)glycine]viomycin. $C_{25}H_{43}N_{13}O_{10}$; mol wt 685.70. C 43.79%, H 6.32%, N 26.56%, O 23.33%. Polypeptide antibiotic produced by a mutant of *Streptomyces griseoverticillatus* var. *tuberacticus*. Isoln and characterization: T. Ando *et al.*, *J. Antibiot.* **24**, 680 (1971); A. Nagata *et al.*, in *Advan. Antimicrob. Antineopl. Chemother.* **vol. 1/2**, M. Hejzlar *et al.*, Eds. (University Park Press, Baltimore, 1972) pp 1039-1041. Prepn: J. Abe *et al.*, **DE 2133181**; *eidem*, **US 3892732** (1972, 1975 both to Toyo Brewing Co., Ltd.). Structural studies: H. Yoshioka *et al.*, *Tetrahedron Lett.* **1971**, 2043; T. Wakamiya, T. Shiba, *J. Antibiot.* **28**, 292 (1975). Pharmacology: H. Hamakawa *et al.*, *Oyo Yakuri* **8**, 817 (1974); *C.A.* **82**, 51530y (1975). Metabolism: T. Shimizu *et al.*, *Jpn. J. Antibiot.* **27**, 279 (1974). Total synthesis: T. Shiba *et al.*, *J. Antibiot.* **32**, 1078 (1979).

Hydrochloride. $C_{25}H_{43}N_{13}O_{10}\cdot 3HCl$. White crystalline powder, mp >245° (dec). $[\alpha]_D^{21}$ −19.1°. uv max (water, 0.1N HCl): 268 nm ($E_{1cm}^{1\%}$ 342); uv max (0.1N NaOH): 288 nm ($E_{1cm}^{1\%}$ 215). Very sol in water, slightly sol in methanol, ethanol. Insol in common organic solvents.

Sulfate. Tuberactin. $C_{50}H_{92}N_{26}O_{32}S_3$; mol wt 1665.62. White crystalline powder. Sol in water, slightly sol in usual organic solvents. LD$_{50}$ in mice, rats (mg/kg): 485, 680 i.v. (Hamakawa).

THERAP CAT: Antibacterial (tuberculostatic).

3655. Enviroxime. [72301-79-2] (1E)-[2-Amino-1-[(1-methylethyl)sulfonyl]-1H-benzimidazol-6-yl]phenylmethanone oxime; 6-[(E)-(hydroxyimino)phenylmethyl]-1-[(1-methylethyl)sulfonyl]-1H-benzimidazol-2-amine; (E)-2-amino-6-benzoyl-1-(isopropylsulfonyl)benzimidazole oxime; *anti*-2-amino-1-(isopropylsulfonyl)-6-(α-hydroxyiminobenzyl)benzimidazole; LY-122772. $C_{17}H_{18}N_4O_3S$; mol wt 358.42. C 56.97%, H 5.06%, N 15.63%, O 13.39%, S 8.94%. Benzimidazole deriv that inhibits rhinovirus multiplication. Prepn: C. J. Paget *et al.*, **DE 2638551**; *eidem*, **US 4118742** (1977, 1978 both to Lilly). Synthesis and sepn of *syn* and *anti* isomers: J. H. Wikel *et al.*, *J. Med. Chem.* **23**, 368 (1980). Inhibition of rhinovirus replication in organ culture: D. C. De Long, S. E. Reed, *J. Infect. Dis.* **141**, 87 (1980). Metabolic studies: J. F. Quay *et al.*, *Fed. Proc.* **39**, 3079 (1980); C. J. Parli *et al.*, *ibid.* 214. Clinical activity vs rhinovirus infection: R. J. Phillpotts *et al.*, *Lancet* **1**, 1342 (1981). Prophylactic activity: F. G. Haden, J. M. Gwaltney, *Antimicrob. Agents Chemother.* **21**, 892 (1982).

Cryst from acetonitrile, mp 198-199°. uv max (methanol): 218, 290 nm (ε 45600, 27100).

(Z)-isomer. Zinviroxime. Cryst from methanol, mp 182-183°. uv max (methanol): 254, 285 nm (ε 20800, 13200).

3656. Enzastaurin. [170364-57-5] 3-(1-Methyl-1H-indol-3-yl)-4-[1-[1-(2-pyridinylmethyl)-4-piperidinyl]-1H-indol-3-yl]-1H-pyrrole-2,5-dione; LY-317615. $C_{32}H_{29}N_5O_2$; mol wt 515.62. C 74.54%, H 5.67%, N 13.58%, O 6.21%. Bisindolylmaleimide inhibitor of protein kinase Cβ (PKCβ), AKT, and GSK3β; structural analog of staurosporine, *q.v.* Prepn: W. F. Heath *et al.*, **WO 9517182**; *eidem*, **US 5668152** (1995, 1997 both to Lilly); M. M. Faul *et al.*, *Tetrahedron* **59**, 7215 (2003). Inhibition of cell growth through the AKT pathway: J. R. Graff *et al.*, *Cancer Res.* **65**, 7462 (2005). *In vivo* and *in vitro* pharmacology in gastric cancer cells: K.-W. Lee *et al.*, *ibid.* **68**, 1916 (2008). Clinical pharmacology: P. A. Welch *et al.*, *J. Clin. Pharmacol.* **47**, 1138 (2007). Clinical evaulation in relapsed or refractory mantle cell lymphoma: F. Morschhauser *et al.*, *Ann. Oncol.* **19**, 247 (2008); in second- or third-line therapy of non-small-cell lung cancer: Y. Oh *et al.*, *J. Clin. Oncol.* **26**, 1135 (2008). Review of development, clinical experience, and therapeutic potential: R. S. Herbst *et al.*, *Clin. Cancer Res.* **13**, 4641s-4646s (2007).

Orange solid.

Hydrochloride. [359017-79-1] $C_{32}H_{29}N_5O_2$.HCl; mol wt 552.08. Prepn: J. K. Bush *et al.*, **WO 04006928** (2004 to Lilly). Crystals, mp ~256°.

THERAP CAT: Antineoplastic.

3657. Eosine I Bluish. [548-24-3] 4′,5′-Dibromo-3′,6′-dihydroxy-2′,7′-dinitrospiro[isobenzofuran-1(3*H*),9′-[9*H*]xanthen]-3-one sodium salt (1:2); 4′,5′-dibromo-2′,7′-dinitrofluorescein disodium salt; hydroxydibromodinitro-*o*-carboxyphenylfluorone sodium; C.I. Acid Red 91; C.I. 45400. $C_{20}H_6Br_2N_2Na_2O_9$; mol wt 624.06. C 38.49%, H 0.97%, Br 25.61%, N 4.49%, Na 7.37%, O 23.07%. Prepd by nitration of 4′,5′-dibromofluorescein: *Colour Index* **vol. 4** (3rd ed., 1971) p 4427.

Red powder. Freely sol in water with green fluorescence; sol in alc.

USE: Dyeing wool, cotton, and paper. In histology as a stain for epithelia, muscular fibers, nuclei, etc.

3658. Eosin Y. [17372-87-1] 2′,4′,5′,7′-Tetrabromo-3′,6′-dihydroxyspiro[isobenzofuran-1(3*H*),9′-[9*H*]xanthen]-3-one sodium salt (1:2); 2′,4′,5′,7′-tetrabromofluorescein; bromoeosine; tetrabromofluorescein sod; bromofluoresceic acid; eosin; eosine; eosin yellowish; D & C Red No. 22; C.I. Acid Red 87; C.I. 45380. $C_{20}H_6Br_4Na_2O_5$; mol wt 691.86. C 34.72%, H 0.87%, Br 46.20%, Na 6.65%, O 11.56%. Xanthene dye capable of photodynamic generation of singlet oxygen. Prepd by bromination of fluorescein: *Colour Index* **vol. 4** (3rd ed., 1971) p 4426. Photodynamic injury to leaf tissue and inhibition of photosynthesis: J. P. Knox, A. D. Dodge, *Planta* **164**, 22; 30 (1985). *Review:* H. J. Conn's Biological Stains, R. W. Horobin, J. A. Kiernan, Eds. (BIOSIS Scientific, Oxford, 10th ed., 2002), pp 231-233.

Red crystals with bluish tinge, or brownish-red powder. Freely sol in water, less in alcohol. Insol in ether. The concd aq soln is deep brownish-red, the dilute (1:500) soln is yellowish-red with greenish fluorescence; the alcoholic soln exhibits a strong green fluorescence.

Free acid. [15086-94-9] Eosine acid; D & C Red No. 21; C.I. Solvent Red 43; C.I. 45380:2. $C_{20}H_8Br_4O_5$. Bright yellowish red. Sol in acetone, ethanol. Insol in mineral oil, toluene.

USE: Fluorescent pH indicator in the range 0.0-3.0. In microbiological differential media. Colorant in drugs and cosmetics such as lipstick and nail-polish; dyeing wool, silk, leather, and paper; in red inks. Biological stain for cellular cytoplasm and associated structures, blood and cytological smears, mineralized bone, and proteins separated by polyacrylamide electrophoretic gels. Diagnostic viability stain of spermatozoa. Tracer for assessing water flow in xylem of plants.

3659. Eotaxin. Chemoattractant protein that is highly selective for eosinophils. Member of the family of proinflammatory substances known as CC (or β-) chemokines. Purification and proposed role in allergic asthma: D. A. Griffiths-Johnson *et al.*, *Biochem. Biophys. Res. Commun.* **197**, 1167 (1993); P. J. Jose *et al.*, *J. Exp. Med.* **179**, 881 (1994). Cloning of guinea pig eotaxin: *idem et al.*, *Biochem. Biophys. Res. Commun.* **205**, 788 (1994); of human eotaxin: P. D. Ponath *et al.*, *J. Clin. Invest.* **97**, 604 (1996). Cloning of the eosinophil eotaxin receptor (CKR-3): B. L. Daugherty *et al.*, *J. Exp. Med.* **183**, 2349 (1996); P. D. Ponath *et al.*, *ibid.* 2437. Discussion of role in allergic inflammation: H. Kita, G. J. Gleich, *ibid.* 2421-2426.

3660. Epalrestat. [82159-09-9] (5*Z*)-5-[(2*E*)-2-Methyl-3-phenyl-2-propen-1-ylidene]-4-oxo-2-thioxo-3-thiazolidineacetic acid; 5-[(*E*,*E*)-β-methylcinnamylidene]-4-oxo-2-thioxo-3-thiazolidineacetic acid; 3-carboxymethyl-5-(2-methylcinnamylidene)rhodanine; ONO-2235; Kinedak; Sorbistat. $C_{15}H_{13}NO_3S_2$; mol wt 319.39. C 56.41%, H 4.10%, N 4.39%, O 15.03%, S 20.08%. Aldose reductase inhibitor. Prepn: T. Tadao *et al.*, **EP 47109**; *eidem*, **US 4464382** (1982, 1984 both to Ono Pharmaceutical). Inhibitory effect on aldose reductase *in vitro:* H. Terashima *et al.*, *J. Pharmacol. Exp. Ther.* **229**, 226 (1983). Effect on motor nerve conduction and sorbitol levels in diabetic rats: R. Kikkawa *et al.*, *Diabetologia* **24**, 290 (1983). Effect in rats on peripheral nerve dysfunction: *eidem*, *Metabolism* **33**, 212 (1984); on retinal microangiopathy: K. Kojima *et al.*, *Jpn. J. Ophthalmol.* **29**, 99 (1985).

Crystals from ethanol-water, mp 210-217°.

N-Methyl-D-glucamine salt. $C_{22}H_{30}N_2O_8S_2$. Crystals from methanol, mp 163-165°.

THERAP CAT: Treatment of diabetic neuropathy.

3661. Eperisone. [64840-90-0] 1-(4-Ethylphenyl)-2-methyl-3-(1-piperidinyl)-1-propanone; 4′-ethyl-2-methyl-3-piperidinopropiophenone. $C_{17}H_{25}NO$; mol wt 259.39. C 78.72%, H 9.72%, N 5.40%, O 6.17%. Spasmolytic agent related structurally to tolperisone, *q.v.* Prepn: E. Morita *et al.*, **DE 2458638**; E. Morita, T. Kanai, **US 4181803** (1975, 1980 both to Eisai). Pharmacological study: K. Tanaka *et al.*, *Nippon Yakurigaku Zasshi* **77**, 511 (1981), *C.A.* **95**, 35471t (1981). Absorption, distribution, excretion in rats and guinea pigs: T. Fujita *et al.*, *Oyo Yakuri* **21**, 835 (1981), *C.A.* **96**, 28193w (1982). Toxicity study: H. Miyagawa *et al.*, *ibid.* 939, *C.A.* **96**, 79722a (1982).

Hydrochloride. [56839-43-1] EMPP; E-646; Mional; Myonal. $C_{17}H_{25}NO$.HCl; mol wt 295.85. Needles from isopropanol, mp 170-172°. LD_{50} in male S.D. rats, Wistar rats, mice (mg/kg): 1300, 1850, 1024 orally (Miyagawa).

THERAP CAT: Muscle relaxant (skeletal).

3662. Ephedra. Ma Huang. Stems and leaves of *Ephedra equisetina* Bunge, *Ephedraceae*, *E. sinica* Stapf. and other *Ephedra* species. *Habit.* China, India, Europe. *Constit.* 0.75-1% ephedrine, pseudoephedrine, norpseudoephedrine, *N*-methylephedrine, *q.q.v.*, (+)-*N*-methylpseudoephedrine, (−)-norephedrine. Used in traditional Chinese medicine as a diaphoretic, stimulant, and antiasthmatic. Brief description: V. E. Tyler in *The Honest Herbal* (Pharmaceutical Products Press, New York, 3rd ed., 1993) pp 119-121. Pharmacology of psychoactive alkaloids: P. Kalix, *J. Ethnopharmacol.* **32**, 201 (1991). GC determn of alkaloids in Ma Huang preparations: J. M. Betz *et al.*, *J. AOAC Int.* **80**, 303 (1997). Review of cardiovascular and CNS effects of weight-loss supplements containing ephedra: C. A. Haller, N. L. Benowitz, *N. Engl. J. Med.* **343**, 1833-1838 (2000). Review of clinical experience: S. R. Mehendale *et al., Am. J. Chin. Med.* **32**, 1-10 (2004).

Morman Tea. Brigham tea; teamsters' tea; popotillo; whorehouse tea. Fresh or dried stems of *Ephedra nevadensis* Wats., *Ephedraceae*, a North American species, generally considered to be alkaloid free. Used by early American settlers as a remedy for venereal disease, colds and kidney disorders. Brief description: V. E. Tyler, *loc. cit.*, 215-216.

3663. Ephedrine. [299-42-3] (αR)-α-[(1S)-1-(Methylamino)ethyl]benzenemethanol; (1R,2S)-2-methylamino-1-phenylpropan-1-ol; L-*erythro*-2-(methylamino)-1-phenylpropan-1-ol; *l*-ephedrine. $C_{10}H_{15}NO$; mol wt 165.24. C 72.69%, H 9.15%, N 8.48%, O 9.68%. α- and β-Adrenergic agonist. Occurs in plants of the genus *Ephedra* (Ephedraceae) known in traditional medicine as Ma Huang. Isoln from *E. vulgaris*: Nagai, *Chem. Zentralbl.* **59**, 130 (1888). Syntheses: E. Späth, R. Göhring, *Monatsh. Chem.* **41**, 319 (1920); R. H. F. Manske, T. B. Johnson, *J. Am. Chem. Soc.* **51**, 580, 1906 (1929). Stereochemistry: K. Freudenberg *et al., ibid.* **54**, 234 (1932); K. Freudenberg, F. Nikolai, *Ann.* **510**, 223 (1934). Properties of crystalline forms: E. E. Moore, D. L. Tabern, *J. Am. Pharm. Assoc.* **24**, 211 (1935). Symposium on sympathomimetic agents: *Ind. Eng. Chem.* **37**, 116-148 (1945). Toxicity: M. D. Fairchild, G. A. Alles, *J. Pharmacol. Exp. Ther.* **158**, 135 (1967). Electrophoretic determn in urine: M. Chicharro *et al., J. Chromatogr.* **622**, 103 (1993). HPLC determn in Ma Huang: B. J. Gurley *et al., J. Pharm. Sci.* **87**, 1547 (1998). Clinical pharmacokinetics of botanical ephedrine: *eidem, Ther. Drug Monit.* **20**, 439 (1998). Review of use as bronchodilator: M. M. Weinberger, *Pediatr. Clin. North Am.* **22**, 121-127 (1975).

Waxy solid, crystals or granules. Gradually dec on exposure to light. Anhydr material is hygroscopic, mp 38.1°. Also occurs as the hemihydrate, mp 40°. bp$_{745}$ 260°. Sol in water, alcohol, chloroform, ether; moderately and slowly sol in mineral oil, the solution becoming turbid if the ephedrine contains more than about 1 percent of water. *Keep well closed in a cool place.*

Hydrochloride. [50-98-6] Caniphedrin. $C_{10}H_{15}NO.HCl$; mol wt 201.69. Fine, white, odorless crystals or powder, affected by light, mp 217-220°. $[\alpha]_D^{25}$ −33 to −35.5° (c = 5). Freely sol in water; sol in alcohol. Practically insol in ether.

Sulfate. [134-72-5] Isofedrol. $(C_{10}H_{15}NO)_2.H_2SO_4$; mol wt 428.54. Fine, white, odorless crystals or powder. Darkens on exposure to light. mp 245° (dec). $[\alpha]_D^{20}$ −30 to −32.5° (c = 5). Freely sol in water; sparingly sol in alcohol.

dl-Form. [90-81-3] Racemic ephedrine; racephedrine. Crystals from chloroform-petr ether, mp 75°. Sol in water, alcohol, ether, chloroform, oils.

dl-Form hydrochloride. [134-71-4] Racephedrine hydrochloride. Crystals from alcohol-acetone, mp 189°. One gram dissolves in 4 ml water, in about 40 ml of 95% alc at 20°. Practically insol in ether. pH about 6.

(+)-*threo*-Form see Pseudoephedrine.

THERAP CAT: Bronchodilator.

THERAP CAT (VET): Bronchodilator; in treatment of urinary incontinence.

3664. Epiandrosterone. [481-29-8] (3β,5α)-3-Hydroxyandrostan-17-one; 3β-hydroxy-17-androstanone; isoandrosterone; 3β-androstanol-17-one; 3β-hydroxyetioallocholan-17-one. $C_{19}H_{30}O_2$; mol wt 290.45. C 78.57%, H 10.41%, O 11.02%. Present in normal human urine as a minor constituent, it is a less active androgen than androsterone. Synthesis: Cardwell *et al., J. Chem. Soc.* **1953**, 361; Johnson *et al., J. Am. Chem. Soc.* **75**, 2275 (1953); Johnson *et al., ibid.* **78**, 6331 (1956). *Review:* R. I. Dorfman, R. A. Shipley, *Androgens* (Wiley, New York, 1956).

dl-Form. Crystals, mp 161-162°. Gives off a musk-like odor when hot.

d-Form. Crystals from ethyl acetate + petr ether, mp 174.5°. $[\alpha]_D^{20}$ +88° (in methanol). Precipitated by digitonin. Practically insol in water. Soluble in organic solvents.

Acetate. $C_{21}H_{32}O_3$. Stout prisms, mp 103-104°. $[\alpha]_D^{18}$ +68.5° (chloroform).

Benzoate. $C_{26}H_{34}O_3$. Crystals, mp 210-212°.

3665. Epibatidine. [140111-52-0] (1R,2R,4S)-2-(6-Chloro-3-pyridinyl)-7-azabicyclo[2.2.1]heptane. $C_{11}H_{13}ClN_2$; mol wt 208.69. C 63.31%, H 6.28%, Cl 16.99%, N 13.42%. First naturally occurring non-opioid, organochlorine, alkaloid analgesic; isolated from skin of the Ecuadoran poison frog, *Epipedobates tricolor*. Isoln and structural determn: T. F. Spande *et al., J. Am. Chem. Soc.* **114**, 3475 (1992). First natural product identified which contains the 7-azabicyclo[2.2.1]heptane ring system. Synthesis of ($\pm$)-form: C. A. Broka, *Tetrahedron Lett.* **1993**, 3251.

Basic, relatively polar, colorless oil. uv max (methanol): 217 nm, shoulder 250-280 nm.

3666. Epichlorohydrin. [106-89-8] 2-(Chloromethyl)oxirane; *dl*-α-epichlorohydrin; 1-chloro-2,3-epoxypropane; γ-chloropropylene oxide. C_3H_5ClO; mol wt 92.52. C 38.95%, H 5.45%, Cl 38.32%, O 17.29%. Prepn: H. T. Clarke, W. W. Hartman, *Org. Synth.* **coll. vol. I**, 233 (2nd ed., 1941); G. Braun, *ibid.* **coll. vol. II**, 256 (1943). Manuf: *Faith, Keyes & Clark's Industrial Chemicals,* F. A. Lowenheim, M. K. Moran, Eds. (Wiley-Interscience, New York, 4th ed., 1975) pp 335-338. Toxicity data: H. F. Smyth, C. P. Carpenter, *J. Ind. Hyg. Toxicol.* **30**, 63 (1948).

Liquid. d_4^{20} 1.1812; d_4^{25} 1.1750; d_4^{50} 1.1436; d_4^{75} 1.1101. bp$_{760}$ 117.9°; bp$_{400}$ 98.0°; bp$_{200}$ 79.3°; bp$_{100}$ 62.0°; bp$_{40}$ 42.0°; bp$_{10}$ 16.6°; bp$_{1.0}$ −16.5°. mp −25.6°. $n_D^{11.6}$ 1.44195; n_D^{16} 1.43969; n_D^{25} 1.43585. Flash pt, open cup: 105°F (40°C). *Poisonous, flammable.* Insol in water. Misc with alcohol, ether, chloroform, trichloroethylene, carbon tetrachloride. Immiscible with petr hydrocarbons. LD$_{50}$ orally in rats: 0.09 g/kg (Smyth, Carpenter).

Caution: Potential symptoms of overexposure are nausea, vomiting; abdominal pain; respiratory distress, coughing; cyanosis; irritation of eyes and skin with deep pain; reproductive effects. See *NIOSH Pocket Guide to Chemical Hazards* (DHHS/NIOSH 97-140, 1997) p 128. See also *Patty's Industrial Hygiene and Toxicology* **vol. 2A**, G. D. Clayton, F. E. Clayton, Eds. (Wiley-Interscience, New York, 3rd ed., 1981) pp 2242-2247. This substance is reasonably anticipated to be a human carcinogen: *Report on Carcinogens, Twelfth Edition* (PB2011-111646, 2011) p 180.

USE: Solvent for natural and synthetic resins, gums, cellulose esters and ethers, paints, varnishes, nail enamels and lacquers, cement for celluloid. As stabilizer.

3667. Epicholestanol. [516-95-0] (3α,5α)-Cholestan-3-ol; 3α-hydroxycholestane; ε-cholestanol. $C_{27}H_{48}O$; mol wt 388.68. C 83.44%, H 12.45%, O 4.12%. The 3α-hydroxy epimer of cholestanol. Prepn from cholestanone: Ruzicka, *Helv. Chim. Acta* **17**, 1407 (1934); *cf.* Marker, *et al., J. Am. Chem. Soc.* **57**, 2359 (1935); Barnett *et al., J. Chem. Soc.* **1940**, 1390.

Needles from alcohol, mp 185-186°. $[\alpha]_D^{20}$ +34° (c = 1.7 in chloroform). Less sol than cholestanol. Not precipitated by digitonin.

Acetate. $C_{29}H_{50}O_2$. mp 95.5-96°. Crystals from methanol.

Benzoate. $C_{34}H_{52}O_2$. mp 102-103°. $[\alpha]_{546}$ +27.2° in chloroform.

3668. Epicholesterol. [474-77-1] (3α)-Cholest-5-en-3-ol. $C_{27}H_{46}O$; mol wt 386.66. C 83.87%, H 11.99%, O 4.14%. The 3α-hydroxy epimer of cholesterol. Prepd from 8-oxocholesteryl chloride or from cholest-5-en-3-one or by passing O_2 into a soln of cholesteryl Mg chloride: Ruzicka, Goldberg, *Helv. Chim. Acta* **19**, 1407 (1936); Marker, *et al., J. Am. Chem. Soc.* **58**, 481 (1936); Marker, **US 2117355** (1938); Barnett *et al., J. Chem. Soc.* **1940**, 1390.

Crystals from alcohol, mp 141.5°. $[\alpha]_D^{30}$ −35° (c = 1 in alcohol). Not precipitated by digitonin.

Acetate. $C_{29}H_{48}O_2$. Crystals from methanol, mp 85°.

3669. Epicillin. [26774-90-3] (2S,5R,6R)-6-[[(2R)-2-Amino-2-(1,4-cyclohexadien-1-yl)acetyl]amino]-3,3-dimethyl-7-oxo-4-thia-1-azabicyclo[3.2.0]heptane-2-carboxylic acid; 6-[D-α-amino-2-(1,4-cyclohexadien-1-yl)acetamido]penicillanic acid; D-α-amino-(1,4-cyclohexadien-1-yl)methylpenicillin; SQ-11302; Dexacilina; Dexacillin; Spectacillin. $C_{16}H_{21}N_3O_4S$; mol wt 351.42. C 54.69%, H 6.02%, N 11.96%, O 18.21%, S 9.12%. Semi-synthetic antibiotic related to penicillin. Prepn: Weisenborn *et al.,* **US 3485819** (1969 to Squibb); Dolfini *et al., J. Med. Chem.* **14**, 117 (1971). Activity studies: Basch *et al., Infect. Immun.* **4**, 44 (1971); Gadebusch *et al., ibid.* 50. Clinical trials: Woodruff *et al., N. Y. State J. Med.* **71**, 1087 (1971); Beck *et al., Curr. Ther. Res.* **13**, 530 (1971); Reyes, *ibid.* 602; Landa, *ibid.* 654.

Crystals, dec 202° (hemihydrate).

THERAP CAT: Antibacterial.

3670. Epicocconone. [371163-96-1] (6S,9aS)-5,6-Dihydro-6-(hydroxymethyl)-3-[(1Z,4E,6E,8E)-1-hydroxy-3-oxo-1,4,6,8-decatetraen-1-yl]-9a-methyl-2H-furo[3,2-g][2]benzopyran-2,9-(9aH)-dione; Deep Purple; Lightning Fast. $C_{23}H_{22}O_7$; mol wt 410.42. C 67.31%, H 5.40%, O 27.29%. Fluorescent natural product isolated from the fungus, *Epicoccum nigrum*. Reacts reversibly with primary amines to produce a highly fluorescent enamine.

Prepn: P. J. L. Bell, P. Karuso, **WO 0181351** (2001 to Maquarie Res. Ltd.); *eidem,* **US 030157518** (2003). Isoln, characterization, and fluorescence applications: *eidem, J. Am. Chem. Soc.* **125**, 9304 (2003). Protein gel staining characteristics: J. A. Mackintosh *et al., Proteomics* **3**, 2273 (2003). Photostability study: G. B. Smejkal *et al., Electrophoresis* **25**, 2511 (2004). Mechanism of protein staining: D. R. Coghlan *et al., Org. Lett.* **7**, 2401 (2005).

Orange solid. $[\alpha]_D^{25}$ +110° (c = 0.009 in acetonitrile). Absorption max (methanol): 432, 555 nm (ε 10000, 4000). Absorption max (acetonitrile): 209, 290(sh), 317, 441 nm (ε 6000, 3000, 3000, 10000).

USE: Biodegradable fluorescent stain for detection of proteins in 1D and 2D gel electrophoresis.

3671. 16-Epiestriol. [547-81-9] (16β,17β)-Estra-1,3,5(10)-triene-3,16,17-triol; $\Delta^{1,3,5}$-estratriene-3,16β,17β-triol; 16-epioestriol; Actriol. $C_{18}H_{24}O_3$; mol wt 288.39. C 74.97%, H 8.39%, O 16.64%. Isoln from urine of pregnant women: Marrian, Bauld, *Biochem. J.* **59**, 136 (1955); Watson, Marrian, *ibid.* **63**, 64 (1956). From human placenta: Diczfalusy, Halla, *Acta Endocrinol.* **27**, 303 (1958). Prepn by reduction of 3,16β-diacetoxy-1,3,5(10)-estratrien-17-one: Biggerstaff, Gallagher, *J. Org. Chem.* **22**, 1220 (1957).

Crystals from methanol + benzene, mp 289-291°. $[\alpha]_D^{15}$ +76° (c = 0.297 in ethanol).

Epiestriol-3-allyl ether. Prepn: Huffman, **US 3002009** (1961 to Lasdon Found). Crystals, mp 156-156.5°.

3672. Epimedii Herba. Yinyanghuo. Chinese herbal medicine made from the dried aerial parts of *Epimedium* sp. (*Berberidaceae*). Used as a tonic and in the treatment of rheumatic disease and hypertension. Principal constituents are prenylflavone glycosides such as icariin, epimedosides, sagittatosides, and baohuosides. Isoln of icariin from *E. macranthum*: S. Akai, *J. Pharm. Soc. Jpn.* **55**, 537 (1935), *C.A.* **29**, 5850 (1935). Revised structure: Y. Tokuoka *et al., Yakugaku Zasshi* **95**, 825 (1975), *C.A.* **83**, 111172r (1975). HPLC determn of constituents: M. Mizuno *et al., Chem. Pharm. Bull.* **36**, 3487 (1988); Y. Ito *et al., J. Chromatogr.* **456**, 392 (1988). Immunoregulatory effects of icariin *in vitro:* W. He *et al., Arzneim.-Forsch.* **45**, 910 (1995). Review: W. Tang, G. Eisenbrand in *Chinese Drugs of Plant Origin* (Springer-Verlag, Berlin, 1992) pp 491-498.

Icariin

Icariin. [489-32-7] 3-[(6-Deoxy-α-L-mannopyranosyl)oxy]-7-(β-D-glucopyranosyloxy)-5-hydroxy-2-(4-methoxyphenyl)-8-(3-methyl-2-butenyl)-4H-1-benzopyran-4-one; 4′-O-methyl-8-γ,γ-dimethylallylkaempferol-3-rhamnoside-7-glucoside. $C_{33}H_{40}O_{15}$; mol wt 676.67. Sesquihydrate, mp 231.5°. $[\alpha]_D^{15}$ −87.09° (in pyridine). Sol in pyridine. Insol in water, alcohol, chloroform, acetone, methanol, ethyl acetate.

3673. Epinastine. [80012-43-7] 9,13b-Dihydro-1H-dibenz[c,f]imidazo[1,5-a]azepin-3-amine; 3-amino-9,13b-dihydro-1H-dibenz[c,f]imidazo[1,5-a]azepine; WAL-801. $C_{16}H_{15}N_3$; mol wt 249.32. C 77.08%, H 6.06%, N 16.85%. Tetracyclic, non-sedating histamine H_1 receptor antagonist. Prepn: G. Walther *et al.*, **GB 2071095**; *eidem*, US 4313931 (1981, 1982 both to Boehringer, Ing.); and receptor binding studies: *idem et al.*, *Arzneim.-Forsch.* **40**, 440 (1990). Pharmacology: A. Fügner *et al.*, *ibid.* **38**, 1446 (1988). Clinical evaluation: W. S. Adamus *et al.*, *ibid.* **37**, 569 (1987). Mechanism of action: K. Tasaka *et al.*, *Oyo Yakuri* **39**, 365 (1990). Toxicology: J. Nishikawa *et al.*, *ibid.* **42**, 151 (1991).

Crystals from acetonitrile, mp 205-208°. log P_a (octanol/aqueous buffer, pH 7.4, 20°) −0.70. pKa 11.2.

Hydrobromide. [127786-29-2] Crystals from methanol/ethyl actate, mp 284-286°.

Hydrochloride. [108929-04-0] WAL-801CL; Alesion; Azusaleon; Elestat; Flurinol; Relestat; Talerc. $C_{16}H_{15}N_3$·HCl; mol wt 285.78. Crystals from methanol/ether, mp 273-275°. Freely sol in water. LD_{50} in male, female rats (mg/kg): 314, 192 orally; 17, 22 i.v. (Nishikawa).

THERAP CAT: Antihistaminic.

3674. Epinephrine. [51-43-4] 4-[(1R)-1-Hydroxy-2-(methylamino)ethyl]-1,2-benzenediol; (−)-3,4-dihydroxy-α-[(methylamino)methyl]benzyl alcohol; l-1-(3,4-dihydroxyphenyl)-2-(methylamino)ethanol; l-3,4-dihydroxy-1-[1-hydroxy-2-(methylamino)ethyl]-benzene; l-methylaminoethanolcatechol; adrenaline; levorenin; Anapen; EpiPen; Primatene Mist; Sus-phrine. $C_9H_{13}NO_3$; mol wt 183.21. C 59.00%, H 7.15%, N 7.65%, O 26.20%. Endogenous catecholamine with combined α- and β-agonist activity. Principal sympathomimetic hormone produced by the adrenal medulla. Isoln from animal adrenal glands: Takamine, *J. Soc. Chem. Ind.* **20**, 746 (1901); Aldrich, *Am. J. Physiol.* **5**, 457 (1901). Synthesis of dl-form: Stolz, *Ber.* **37**, 4149 (1904); Payne, *Ind. Chem.* **37**, 523 (1961). Historic review of syntheses: Loewe, *Arzneim.-Forsch.* **4**, 583 (1954). Resolution of dl-form: Flächer, *Z. Physiol. Chem.* **58**, 189 (1908). Configuration: Pratesi *et al.*, *J. Chem. Soc.* **1958**, 2069. Acute toxicity: A. M. Lands *et al.*, *J. Pharmacol. Exp. Ther.* **90**, 110 (1947). HPLC determn in plasma and urine: C. R. Benedict, *J. Chromatogr.* **385**, 369 (1987). Comprehensive description: D. H. Szulczewski, W.-H. Hong, *Anal. Profiles Drug Subs.* **7**, 193-229 (1978). Physiologic review: Malmejac, *Physiol. Rev.* **44**, 186 (1964). Review of biosynthesis: L. A. Pohorecky, R. J. Wurtman, *Pharmacol. Rev.* **23**, 1-35 (1971); of pharmacology and clinical use in cardiopulmonary resuscitation: N. A. Paradis, E. M. Koscove, *Ann. Emerg. Med.* **19**, 1288-1301 (1990); P. Hebert *et al.*, *J. Emerg. Med.* **9**, 487-495 (1991). Review of use in anaphylaxis: A. P. C. McLean-Tooke *et al.*, *Br. Med. J.* **327**, 1332-1335 (2003).

White microcrystals, gradually darkening on exposure to light and air. mp 211-212°. mp ~215° (dec) when rapidly heated. $[\alpha]_D^{25}$

−50.0° to −53.5° (in 0.6N HCl). Readily sol in aq solns of mineral acids, NaOH, and KOH; very slightly sol in water, alc. Insol in chloroform, ether, acetone, oils, as well as, aq solns of ammonia and of the alkali carbonates. LD_{50} i.p. in mice: 4 mg/kg (Lands).

Hydrochloride. [55-31-2] Adrenalin; Epifrin; Glaucon; Suprarenin. $C_9H_{13}NO_3$·HCl; mol wt 219.67.

d-Bitartrate. [51-42-3] Asthmahaler; Medihaler-Epi. $C_9H_{13}NO_3$·$C_4H_6O_6$; mol wt 333.29. White, grayish-white or light brownish-gray crystals, mp 147-154° (some dec). Darkens slowly on exposure to air and light. One gram dissolves in about 3 ml water. Slightly sol in alc. Practically insol in chloroform, ether.

dl-Form. [329-65-7] Racepinephrine. Sparingly sol in water, alcohol.

dl-Form hydrochloride. [329-63-5] Vaponefrin. Crystals from alcohol, mp 157°. Readily sol in water; sparingly sol in abs alc.

THERAP CAT: Bronchodilator; cardiostimulant; mydriatic; antiglaucoma.

THERAP CAT (VET): Vasoconstrictor; cardiostimulant.

3675. Epiquinidine. [572-59-8] (9R)-6′-Methoxycinchonan-9-ol. $C_{20}H_{24}N_2O_2$; mol wt 324.42. C 74.05%, H 7.46%, N 8.64%, O 9.86%. Occurs in cinchona bark. Isoln from quinoidine: Dirscherl, Thron, *Ann.* **521**, 48 (1938). By epimerization of quinine or quinidine: Rabe, Höter, *J. Prakt. Chem.* **154**, 66 (1940). Stereochemistry: Prelog, Zalán, *Helv. Chim. Acta* **27**, 535, 545 (1944); Prelog, Häfliger, *ibid.* **33**, 2021 (1950); Roth, *Pharmazie* **16**, 257 (1961). Synthesis: Grethe *et al.*, *Helv. Chim. Acta* **55**, 1044 (1972); Gutzwiller, Uskokovic, *ibid.* **56**, 1494 (1973).

Lustrous leaflets from ether, mp 111-113°. $[\alpha]_D^{25}$ +107.8° (c = 1.02 in ethanol). Freely sol in alcohol; moderately sol in ether. Shows more blue fluorescence in H_2SO_4 than quinidine or quinine. Forms a double sulfate with epiquinine.

Dihydrochloride. $C_{20}H_{24}N_2O_2$·2HCl. Crystals from alc, dec 195-196°. $[\alpha]_D^{20}$ +46° (c = 0.8 in 99% alc).

Neutral dibenzoyl-d-tartrate. $(C_{20}H_{24}N_2O_2)_2$·$C_{18}H_{14}O_8$. Crystals from alcohol or acetone, dec 166-167°. $[\alpha]_D^{20}$ +3.7° (4:1 alcohol + chloroform).

3676. Epiquinine. [572-60-1] (8α,9S)-6′-Methoxycinchonan-9-ol. $C_{20}H_{24}N_2O_2$; mol wt 324.42. C 74.05%, H 7.46%, N 8.64%, O 9.86%. For refs *see* Epiquinidine.

Viscous oil, $[\alpha]_D^{22}$ +43° (c = 0.95 in 99% alc). Freely sol in organic solvents. Shows more blue fluorescence in H_2SO_4 than quinine. Forms a double sulfate with epiquinidine.

Dihydrochloride. $C_{20}H_{24}N_2O_2$·2HCl. Crystals from acetone, dec 196°. $[\alpha]_D^{21}$ +33° (c = 0.8 in 99% alc).

Neutral dibenzoyl-d-tartrate. $(C_{20}H_{24}N_2O_2)_2$·$C_{18}H_{14}O_8$. Crystals from acetone, dec 159°, $[\alpha]_D^{21}$ −24.3° (c = 0.93 in ethanol).

3677. Epirizole. [18694-40-1] 4-Methoxy-2-(5-methoxy-3-methyl-1*H*-pyrazol-1-yl)-6-methylpyrimidine; 2-(3-methyl-5-methoxy-1-pyrazolyl)-4-methoxy-6-methylpyrimidine; 2-(3-methoxy-5-methylpyrazol-2-yl)-4-methoxy-6-methylpyrimidine; 1-(4-methoxy-6-methyl-2-pyrimidinyl)-3-methyl-5-methoxypyrazole; mepirizole; DA-398; Mebron. $C_{11}H_{14}N_4O_2$; mol wt 234.26. C 56.40%, H 6.02%, N 23.92%, O 13.66%. Prepn: Naito *et al.*, **ZA 6704936** (1968 to Daiichi Seiyaku), *C.A.* **70**, 57876q (1969); Naito *et al.*, *Chem. Pharm. Bull.* **17**, 1467 (1969). Pharmacology and metabolism: Oshima *et al.*, *ibid.* 1492; Takabatake *et al.*, **18**, 1900 (1970). Toxicity data: H. Ogura *et al.*, *J. Med. Chem.* **15**, 923 (1972).

Minute, white or cream-colored crystals from isopropyl ether, mp 90-92°. Characteristic odor, bitter taste. Sparingly sol in water. Sol in dil acids; freely sol in ethanol, benzene, dichloroethane. Also sol in ether, acetone. LD_{50} orally in mice: 820 mg/kg (Ogura).

THERAP CAT: Anti-inflammatory.

3678. Epirubicin. [56420-45-2] (8*S*,10*S*)-10-[(3-Amino-2,-3,6-trideoxy-α-L-*arabino*-hexopyranosyl)oxy]-7,8,9,10-tetrahydro-6,8,11-trihydroxy-8-(2-hydroxyacetyl)-1-methoxy-5,12-naphthacenedione; 3-glycoloyl-1,2,3,4,6,11-hexahydro-3,5,12-trihydroxy-10-methoxy-6,11-dioxo-1-naphthacenyl-3-amino-2,3,6-tridioxy-α-L-arabino-hexopyranoside; 4′-epidoxorubicin; 4′-epiadriamycin; pidorubicin; 4′-epi-DX; IMI-28. $C_{27}H_{29}NO_{11}$; mol wt 543.53. C 59.66%, H 5.38%, N 2.58%, O 32.38%. Analog of the anthracycline antibiotic doxorubicin, *q.v.*, differing only in the position of the C-4 hydroxy group of the sugar moiety. Prepn: F. Arcamone *et al.*, **DE 2510866**; *eidem*, **US 4058519** (1975, 1977 both to Soc. Farmaceut. Italia); *eidem*, *J. Med. Chem.* **18**, 703 (1975); S. Penco, *Process Biochem.* **15**(5), 12 (1980). Purification: E. Oppici *et al.*, **BE 898506**; *eidem*, **GB 2133005** (both 1984 to Farmitalia Carlo Erba). Comparison with doxorubicin of *in vitro* activity: A. Di Marco *et al.*, *Cancer Res.* **36**, 1962 (1976). Tissue distribution: C. Italia *et al.*, *Br. J. Cancer* **47**, 545 (1983); F. Arcamone *et al.*, *Cancer Chemother. Pharmacol.* **12**, 157 (1984). HPLC determn in biological fluids: P. E. Deesen, B. Leyland-Jones, *Drug Metab. Dispos.* **12**, 9 (1984); G. Cassinelli *et al.*, *ibid.* 506. Clinical trials in solid tumors: K. Kolaric *et al.*, *J. Cancer Res. Clin. Oncol.* **106**, 148 (1983). Review of activity in experimental tumors: A. Goldin *et al.*, *Invest. New Drugs* **3**, 3-21 (1985). Review of pharmacology and clinical efficacy: F. Ganzina, *Cancer Treat. Rev.* **10**, 1-22 (1983); R. J. Cersosimo, W. K. Hong, *J. Clin. Oncol.* **4**, 425-439 (1986); P. Hurteloup, F. Ganzina, *Drugs Exp. Clin. Res.* **13**, 233-246 (1986). Review of epirubicin and other anthracycline antineoplastics: R. B. Weiss *et al.*, *Cancer Chemother. Pharmacol.* **18**, 185-97 (1986).

Hydrochloride. [56390-09-1] Ellence; Farmorubicina. $C_{27}H_{29}NO_{11}.HCl$; mol wt 579.98. Red-orange crystals, mp 185° (dec). $[\alpha]_D^{20}$ +274° (c = 0.01 in methanol). Solution should be protected from sunlight.

THERAP CAT: Antineoplastic.

3679. Epitestosterone. [481-30-1] (17α)-17-Hydroxyandrost-4-en-3-one; *cis*-testosterone. $C_{19}H_{28}O_2$; mol wt 288.43. C 79.12%, H 9.79%, O 11.09%. Naturally occurring, non-anabolic 17α-epimer of testosterone, *q.v.* Isoln from rabbit liver slices: L. C. Clark, Jr., C. D. Kochakian, *J. Biol. Chem.* **170**, 23 (1947). Synthesis: F. Alvarez, *Steroids* **2**, 393 (1963). Isoln from human urine: S. G. Korenman *et al.*, *J. Biol. Chem.* **239**, 1004 (1964). Crystal structure: N. W. Isaacs *et al.*, *J. Chem. Soc. Perkin Trans. 2* **1972**, 2335. HPLC/MS/MS determn of epitestosterone and testosterone conjugates in urine: D. J. Borts, L. D. Bowers, *J. Mass Spectrom.* **35**, 50 (2000). Isotope ratio MS determn in urine: R. Aguilera *et al.*, *Clin. Chem.* **48**, 629 (2002). Review of physiological properties: L. Stárka, *J. Steroid Biochem. Mol. Biol.* **87**, 27-34 (2003).

White needles, mp 218-221° (Clark, Kochakian). Also reported as crystals from acetone, mp 211-215° (Alvarez). uv max (ethanol): 240 nm (ε 15500). d 1.14.

Note: This is a controlled substance (anabolic steroid): **21 CFR,** 1308.13, as defined in 1300.01.

USE: Reference standard in doping control of testosterone abuse.

3680. Epitiostanol. [2363-58-8] (2α,3α,5α,17β)-2,3-Epithioandrostan-17-ol; 10275-S; Thiodrol. $C_{19}H_{30}OS$; mol wt 306.51. C 74.45%, H 9.87%, O 5.22%, S 10.46%. Anabolic steroid. Episulfide deriv of androstane, *q.v.* Prepn: **GB 977599**; T. Komeno, **US 3230215** (1964, 1966 both to Shionogi); K. Takeda *et al.*, *Tetrahedron* **1965**, 329; P. D. Klimstra *et al.*, *J. Med. Chem.* **9**, 693 (1966). Antitumor effect in mice: A. Matsuzawa, T. Yamamoto, *Cancer Res.* **37**, 4408 (1977). Teratogenicity study: T. Minesita *et al.*, *Oyo Yakuri* **7**, 723 (1973), *C.A.* **80**, 116474p (1974). Toxicity study: *eidem*, *ibid.* 805, *C.A.* **80**, 66865u (1974). Use in treatment of breast cancer: M. Fujimoro *et al.*, *Cancer* **31**, 789 (1973).

Crystals from acetone, mp 127-128°. $[\alpha]_D^{27.5}$ +24.4° (c = 1.054 in chloroform). uv max (alcohol): 262 nm. LD_{50} in mice, rats (mg/kg): 1, 5 i.p. (Minesita, p 805).

THERAP CAT: Antineoplastic.

3681. Eplerenone. [107724-20-9] (7α,11α,17α)-9,11-Epoxy-17-hydroxy-3-oxopregn-4-ene-7,21-dicarboxylic acid γ-lactone 7-methyl ester; 9α,11-epoxy-7α-(methoxycarbonyl)-3-oxo-17α-pregn-4-ene-21,17-carbolactone; 9α,11α-epoxy-7α-methoxycarbonyl-20-spirox-4-ene-3,21-dione; epoxymexrenone; CGP-30083; SC-66110; Inspra. $C_{24}H_{30}O_6$; mol wt 414.50. C 69.54%, H 7.30%, O 23.16%. Selective aldosterone receptor antagonist (SARA); structurally similar to spironolactone, *q.v.* Prepn: J. Grob, J. Kalvoda, **EP 122232**; *eidem*, **US 4559332** (1984, 1985 both to Ciba-Geigy); J. Grob *et al.*, *Helv. Chim. Acta* **80**, 566 (1997). Pharmacology and receptor binding studies: M. de Gasparo *et al.*, *J. Pharmacol. Exp. Ther.* **240**, 650 (1987). Clinical pharmacokinetics: W. R. Ravis *et al.*, *J. Clin. Pharmacol.* **45**, 810 (2005). Clinical trial in congestive heart failure after myocardial infarction: B. Pitt *et al.*, *N. Engl. J. Med.* **348**, 1309 (2003); in hypertension vs losartan: M. H. Weinberger *et al.*, *Am. Heart J.* **150**, 426 (2005). Review of pharmacology and clinical experience: J. A. Delyani *et al.*, *Cardiovasc. Drug Rev.*

19, 185-200 (2001); E. Burgess, *Expert Opin. Pharmacother.* **5**, 2573-2581 (2004).

Crystals from methylene chloride + isopropanol, mp 240-242°. $[\alpha]_D$ +5° (c = 0.437 in chloroform). uv max: 240 nm (ε 16800). Partition coefficient (octanol/water): 7.1 (pH 7.0). Sol in dichloromethane, acetonitrile; slightly sol in ethanol; sparingly sol in methyl ethyl ketone, methanol; very slightly sol in water.

THERAP CAT: Antihypertensive. In treatment of congestive heart failure.

3682. EPN. [2104-64-5] *P*-Phenylphosphonothioic acid *O*-ethyl *O*-(4-nitrophenyl) ester; ethyl *p*-nitrophenyl benzenethiophosphonate; *O*-ethyl *O*-4-nitrophenyl phenylphosphonothioate. $C_{14}H_{14}NO_4PS$; mol wt 323.30. C 52.01%, H 4.36%, N 4.33%, O 19.79%, P 9.58%, S 9.92%. Organophosphate insecticide; cholinesterase inhibitor. Prepn: A. G. Jelinek, **US 2503390** (1950 to du Pont). Manufacturing process: N. Shindo *et al.*, **US 3327026** (1967 to Nissan). Insecticidal activity: D. A. Wolfenbarger *et al.*, *J. Econ. Entomol.* **63**, 1568 (1970). Cholinesterase inhibition and toxicology: H. C. Hodge *et al.*, *J. Pharmacol. Exp. Ther.* **112**, 29 (1954). Toxicity data: T. B. Gaines, *Toxicol. Appl. Pharmacol.* **2**, 88 (1960). Metabolism in animals: R. L. Chrzanowski, A. G. Jelinek, *J. Agric. Food Chem.* **29**, 580 (1981). Multiresidue determn in rice by GC-MS: S.-K. Cho *et al.*, *Biomed. Chromatogr.* **21**, 602 (2007).

Light yellow oil, aromatic odor. d^{25} 1.268. n_D^{25} 1.6021. Practically insol in water. Miscible with benzene, toluene, xylene, acetone, isopropanol, methanol. *Poisonous.* LD_{50} in female, male rats (mg/kg): 7.7, 36 orally; 25, 230 dermal (Gaines).

Caution: Potential symptoms of overexposure are miosis, lacrimation; rhinorrhea; headache; tight chest, wheezing and laryngeal spasm; salivation; cyanosis; anorexia, nausea, abdominal cramps and diarrhea; paralysis, convulsions; low blood pressure; cardiac irregularities; irritation of skin and eyes. *See NIOSH Pocket Guide to Chemical Hazards* (DHHS/NIOSH 97-140, 1997) p 128.

USE: Insecticide; acaricide.

3683. Epostane. [80471-63-2] (4α,5α,17β)-4,5-Epoxy-3,17-dihydroxy-4,17-dimethylandrost-2-ene-2-carbonitrile; (2α,4α,5α,-17β)-4,5-epoxy-17-hydroxy-4,17-dimethyl-3-oxoandrostane-2-carbonitrile; Win-32729. $C_{22}H_{31}NO_3$; mol wt 357.49. C 73.92%, H 8.74%, N 3.92%, O 13.43%. 3β-Hydroxysteroid dehydrogenase inhibitor; derivative of trilostane, *q.v.* Prepn as ketone tautomer: R. G. Christiansen, **DE 2855091**; *idem*, **US 4160027** (both 1979 to Sterling). Pharmacology and interceptive activity: J. E. Creange *et al.*, *Contraception* **24**, 289 (1981); R. G. Christiansen *et al.*, *J. Med. Chem.* **27**, 928 (1984). Clinical pharmacology: Z. M. Van der Spuy *et al.*, *Clin. Endocrinol.* **19**, 521 (1983); N. S. Pattison *et al.*, *Fertil. Steril.* **42**, 875 (1984). Clinical evaluation as abortifacient: L. Birgerson *et al.*, *Contraception* **35**, 111 (1987); M. J. Crooij *et al.*, *N. Engl. J. Med.* **319**, 813 (1988). Comparative clinical study with mifepristone, *q.v.*, as abortifacient: L. Birgerson, V. Odlind, *Fertil. Steril.* **48**, 565 (1987). Induction of labor in swine: P. A. Martin *et al.*, *J. Anim. Sci.* **64**, 497 (1987); in sheep: M. Silver, *Vet. Rec.* **120**, 299 (1987). Regulation of ovulation in sheep: R. Webb, *J. Reprod.*

Fertil. **79**, 231 (1987). Abortifacient efficacy in dogs: D. M. Keister *et al.*, *Theriogenology* **30**, 497 (1988).

Crystals, mp 191-194° from DMF/H_2O. $[\alpha]_D^{25}$ +67.4° (c = 1 in pyridine).

3684. Epothilones. Microtubule-stablilizing, 16-membered macrolides which mimic the biological effects of paclitaxel, *q.v.*, but are structurally lacking the taxane ring. Isoln of A and B from the myxobacterium *Sorangium cellulosum*: G. Höfle *et al.*, **DE 4138042**, (1993 to Ges. Biotech. Forsch.), *C.A.* **120**, 52841 (1994). Purification of A and B, and mechanism of action: D. M. Bollag *et al.*, *Cancer Res.* **55**, 2325 (1995). Total synthesis of A: A. Balog *et al.*, *Angew. Chem. Int. Ed.* **35**, 2801 (1996); of B and D: D.-S. Su *et al.*, *ibid.* **36**, 757 (1997). Crystal structure of A and B, and conformation in soln: G. Höfle *et al.*, *ibid.* **35**, 1567 (1996). Isoln and structural studies of 39 natural epothilone variants: I. H. Hardt *et al.*, *J. Nat. Prod.* **64**, 847 (2001). Isoln and crystallization of D: R. L. Arslanian *et al.*, *ibid.* **65**, 570 (2002). Effects of A and B on microtubule proteins in comparison with paclitaxel: R. J. Kowalski *et al.*, *J. Biol. Chem.* **272**, 2534 (1997). Review of properties and clinical potential: D. M. Bollag, *Expert Opin. Invest. Drugs* **6**, 867-873 (1997); of biology, chemistry and therapeutic potential: K. C. Nicolaou *et al.*, *Angew. Chem. Int. Ed.* **37**, 2014-2045 (1998); *idem et al.*, *Chem. Commun.* **2001**, 1523-1535. Review of pharmacology and clinical experience with epothilone B in cancer treatment: B. Bystricky, I. Chau, *Expert Opin. Invest. Drugs* **20**, 107-117 (2011).

Epothilone A   R = H
Epothilone B   R = CH_3

Epothilone D

Epothilone A. [152044-53-6] (1*S*,3*S*,7*S*,10*R*,11*S*,12*S*,16*R*)-7,11-Dihydroxy-8,8,10,12-tetramethyl-3-[(1*E*)-1-methyl-2-(2-methyl-4-thiazolyl)ethenyl]-4,17-dioxabicyclo[14.1.0]heptadecane-5,9-dione. $C_{26}H_{39}NO_6S$; mol wt 493.66. Colorless crystals from ethyl acetate/toluene, mp 95°. Soly at 20° (g/l): water 0.7. uv max (methanol): 211, 249 nm (ε 17800, 12500). $[\alpha]_D^{21}$ −47.1° (c = 1.0 in methanol).

Epothilone B. [152044-54-7] Patupilone; EPO-906. $C_{27}H_{41}NO_6S$; mol wt 507.69. Colorless crystals from ethyl acetate, mp 93-

94°. uv max (methanol): 211, 249 nm (ε 18600, 14100). $[\alpha]_D^{21}$ −35° (c = 0.7 in methanol).

Epothilone D. [189453-10-9] (4S,7R,8S,9S,13Z,16S)-4,8-Dihydroxy-5,5,7,9,13-pentamethyl-16-[(1E)-1-methyl-2-(2-methyl-4-thiazolyl)ethenyl]oxacyclohexadec-13-ene-2,6-dione; desoxyepothilone B; KOS-862. $C_{27}H_{41}NO_5S$; mol wt 491.69. Colorless crystals from ethanol + water, mp 120-121° (Arslanian). Also reported as colorless amorphous solid. $[\alpha]_D^{22}$ −61.3° (c = 2.5 in methanol). uv max (methanol): 210, 248 nm (ε 18400, 13200) (Hardt).

3685. Epoxiconazole. [133855-98-8] rel-1-[[(2R,3S)-3-(2-Chlorophenyl)-2-(4-fluorophenyl)-2-oxiranyl]methyl]-1H-1,2,4-triazole; (2RS,3RS)-1-[3-(2-chlorophenyl)-2,3-epoxy-2-(4-fluorophenyl)-propyl]-1H-1,2,4-triazole; BAS-480-F; Opus. $C_{17}H_{13}ClFN_3O$; mol wt 329.76. C 61.92%, H 3.97%, Cl 10.75%, F 5.76%, N 12.74%, O 4.85%. Ergosterol biosynthesis inhibitor. Prepn: S. Karbach et al., **EP 196038**; eidem, **US 4906652** (1986, 1990 both to BASF). Properties and field trials: E. Ammermann et al., Brighton Crop Prot. Conf. - Pests Dis. **1990**, 407; R. Saur et al., ibid. 831. Mode of action: A. Akers et al., ibid. 837. HPLC determn of enantiomers in water and soil: M. Hutta et al., J. Chromatogr. A **959**, 143 (2002). Use in combination with fenpropimorph, q.v.: R. Saur et al., Gesunde Pflanz. **46**, 61 (1994). Brief review: A. Floquet, N. Martin, Phytoma **449**, 54-57 (1993).

Relative stereochemistry

Crystals from diisopropyl ether, mp 136.2°. Soly at 20° (g/100 ml): water 6.63 × 10⁻⁴, acetone 18, dichloromethane 14, n-heptane <0.1. Partition coefficient (n-octanol/ water at pH 7) 3.44. LD_{50} in rats (mg/kg): >5000 orally, >2000 dermally; LC_{50} in rats (4 hr): >5.3 mg/l (dust aerosol) (Ammermann).

USE: Agricultural fungicide.

3686. Epoxomicin. [134381-21-8] N-Acetyl-N-methyl-L-isoleucyl-L-isoleucyl-N-[(1S)-3-methyl-1-[[(2R)-2-methyl-2-oxiranyl]carbonyl]butyl]-L-threoninamide; BU-4061T. $C_{28}H_{50}N_4O_7$; mol wt 554.73. C 60.63%, H 9.09%, N 10.10%, O 20.19%. α',β'-epoxyketone peptide which specifically inhibits proteasome activity. Isoln from actinomycetes: M. Konishi et al., **EP 411660**; eidem, **US 5071957** (both 1991 to Bristol-Myers); and characterization: M. Hanada et al., J. Antibiot. **45**, 1746 (1992). Total synthesis: N. Sin et al., Bioorg. Med. Chem. Lett. **9**, 2283 (1999). Proteasome selectivity and antiinflammatory activity: L. Meng et al., Proc. Natl. Acad. Sci. USA **96**, 10403 (1999). Crystal structure and mechanism of selectivity: M. Groll et al., J. Am. Chem. Soc. **122**, 1237 (2000). Use as inhibitor to modulate antigen presentation: K. Schwarz et al., J. Immunol. **164**, 6147 (2000).

Colorless powder, mp 107-109°. $[\alpha]_D^{24.5}$ −66.1 ± 0.4° (c = 0.5 in MeOH). Readily sol in methanol, methylene chloride and ethyl acetate. Practically insol in water.

USE: In studies of proteasome biology.

3687. Epratuzumab. [205923-57-5]; [501423-23-0] (⁹⁰Y-labeled form). Anti-(human CD22 (antigen)) immunoglobulin G1

(human-mouse monoclonal IMMU-hLL2 γ-chain) disulfide with human-mouse monoclonal IMMU-hLL2 κ-chain, dimer; hLL2; AMG-412; IMMU-hLL2; LymphoCide. Humanized monoclonal antibody directed against CD22, a transmembrane sialoglycoprotein expressed on the surface of B-cell malignancies. Prepn: S.-O. Leung, H. Hansen, **WO 9604925**; eidem, **US 5789554** (1996, 1998 both to Immunomedics); idem et al., Mol. Immunol. **32**, 1413 (1995). Characterization and binding study: J. Carnahan et al., Clin. Cancer Res. **9**, Suppl., 3982s (2003). Clinical pharmacokinetics and evaluation in non-Hodgkin's lymphoma: J. P. Leonard et al., Clin. Cancer Res. **10**, 5327 (2004). Review of clinical development: A. B. Siegel et al., Semin. Oncol. **30**, 457-464 (2003).

Yttrium Y 90 epratuzumab tetraxetan. [501423-25-2] Epratuzumab, 1,4,7,10-tetraazacyclododecane-1,4,7,10-tetraacetic acid conjugate, yttrium-⁹⁰Y chelate; ⁹⁰Y-DOTA-hLL2. Prepn and biodistribution studies: S. V. Govindan et al., Bioconjugate Chem. **9**, 773 (1998); G. L. Griffiths et al., J. Nucl. Med. **44**, 77 (2003). Clinical pharmacokinetics and efficacy: R. M. Sharkey et al., ibid. 2000. Clinical evaluation in non-Hodgkin's lymphoma: O. Lindén et al., Clin. Cancer Res. **11**, 5215 (2005).

THERAP CAT: Antineoplastic. Yttrium Y 90 epratuzumab tetraxetan in radioimmunotherapy.

3688. Eprazinone. [10402-90-1] 3-[4-(2-Ethoxy-2-phenylethyl)-1-piperazinyl]-2-methyl-1-phenyl-1-propanone; 3-[4-(β-ethoxyphenethyl)-1-piperazinyl]-2-methylpropiophenone. $C_{24}H_{32}$-N_2O_2; mol wt 380.53. C 75.75%, H 8.48%, N 7.36%, O 8.41%. Prepn of the dihydrochloride: R. Y. Mauvernay, **NL 6602581**; idem, **US 3448192** (1966, 1969 both to Riom). Pharmacodynamics: J. Vacher et al., Arch. Int. Pharmacodyn. Ther. **165**, 1 (1967). Pharmacological study: Y. Kase et al., Nippon Yakurigaku Zasshi **73**, 605 (1977), C.A. **88**, 15657 (1978). Efficacy: W. Spitzer, Fortschr. Med. **98**, 871 (1980). Metabolism: P. Toffel-Nadolny, W. Gielsdorf, Arzneim.-Forsch. **31**, 719 (1981).

Dihydrochloride. [10402-53-6] 746-CE; Eftapan; Mucitux. $C_{24}H_{32}N_2O_2 \cdot 2HCl$; mol wt 453.45. White, bitter-tasting crystalline powder, mp 201° (Toffel-Nadolny, Gielsdorf); also reported as crystals from methanol, mp 160° (Mauvernay). LD_{50} in mice (mg/kg): 729 orally, 38 i.v. (Vacher).

THERAP CAT: Antitussive.

3689. Eprinomectin. [123997-26-2] (4″R)-4″-(Acetylamino)-4″-deoxyavermectin B₁; MK-397; Eprinex. Mixture of semisynthetic avermectins, q.v.; containing 90% or more of component B_{1a} and 10% or less of component B_{1b}. Prepn: H. H. Mrozik, **US 4427663** (1984 to Merck & Co.). Synthesis: R. J. Cvetovich et al., J. Org. Chem. **59**, 7704 (1994); and biological activity: T. K. Jones et al., J. Agric. Food Chem. **42**, 1786 (1994). Mechanism of action study: J. P. Arena et al., J. Parasitol. **81**, 286 (1995).

Component B_{1a}, R = C_2H_5
Component B_{1b}, R = CH_3

White powder, mp 163.3-165.7°. Insol in cold water. LD_{50} orally in mice: 24 mg/kg (Jones).

Component B$_{1a}$. [133305-88-1] (4″R)-4″-(Acetylamino)-5-O-demethyl-4″-deoxyavermectin A$_{1a}$. C$_{50}$H$_{75}$NO$_{14}$; mol wt 914.14.

Component B$_{1b}$. [133305-89-2] (4″R)-4″-(Acetylamino)-5-O-demethyl-25-de(1-methylpropyl)-4″-deoxy-25-(1-methylethyl)avermectin A$_{1a}$. C$_{49}$H$_{73}$NO$_{14}$; mol wt 900.12.

THERAP CAT (VET): Antiparasitic.

3690. Epristeride. [119169-78-7] (17β)-17-[[(1,1-Dimethylethyl)amino]carbonyl]androsta-3,5-diene-3-carboxylic acid; 17β-(N-tert-butylcarboxamido)androsta-3,5-diene-3-carboxylic acid; SKF-105657. C$_{25}$H$_{37}$NO$_3$; mol wt 399.58. C 75.15%, H 9.33%, N 3.51%, O 12.01%. 5α-Reductase inhibitor. Prepn and in vitro activity: D. A. Holt et al., EP 289327; eidem, US 4910226 and US 5017568 (1988, 1990, 1991 all to SmithKline Beecham); D. A. Holt et al., J. Med. Chem. 33, 943 (1990). Pharmacology: J. C. Lamb et al., Endocrinology 130, 685 (1992). Antitumor effects in rats: J. C. Lamb et al., Prostate 21, 15 (1992). Mechanism of action study: M. A. Levy et al., Biochemistry 29, 2815 (1990). HPLC determn in human plasma: V. K. Boppana et al., J. Chromatogr. 631, 251 (1993).

White crystals from ethyl acetate, mp 242-249°. pK 4.8.
THERAP CAT: Treatment of benign prostatic hypertrophy.

3691. Eprosartan. [133040-01-4] (αE)-α-[[2-Butyl-1-[(4-carboxyphenyl)methyl]-1H-imidazol-5-yl]methylene]-2-thiophenepropanoic acid; (E)-3-[2-butyl-1-[(4-carboxyphenyl)methyl]-imidazol-5-yl]-2-(2-thienylmethyl)-2-propenoic acid; SKF-108566. C$_{23}$H$_{24}$N$_2$O$_4$S; mol wt 424.52. C 65.07%, H 5.70%, N 6.60%, O 15.07%, S 7.55%. Prototype of the imidazoleacrylic acid angiotensin II receptor antagonists. Prepn: J. A. Finkelstein et al., EP 403159 (1990 to SmithKline Beckman); eidem, US 5185351 (1993 to SmithKline Beecham); J. Weinstock et al., J. Med. Chem. 34, 1514 (1991); R. M. Keenan et al., ibid. 36, 1880 (1993). Pharmacology: R. M. Edwards et al., J. Pharmacol. Exp. Ther. 260, 175 (1992). Receptor binding study: H. T. Schambye et al., Mol. Pharmacol. 47, 425 (1995). Clinical pharmacokinetics: D. Tenero et al., Biopharm. Drug Dispos. 19, 351 (1998). Clinical trial in hypertension: W. J. Elliott et al., J. Hum. Hypertens. 13, 413 (1999). Review of clinical experience: L. Ruilope, B. Jäger, Expert Opin. Pharmacother. 4, 107-114 (2003).

Crystals from methanol, mp 260-261°.
Monomethanesulfonate. [144143-96-4] Eprosartan mesylate; SKF-108566J; Teveten. C$_{23}$H$_{24}$N$_2$O$_4$S.CH$_3$SO$_3$H; mol wt 520.62. White to off-white crystalline powder, mp 248-250°. Insol in water. Freely sol in ethanol.
THERAP CAT: Antihypertensive.

3692. Eprotirome. [355129-15-6] 3-[[3,5-Dibromo-4-[4-hydroxy-3-(1-methylethyl)phenoxy]phenyl]amino]-3-oxopropanoic acid; N-[3,5-dibromo-4-(4-hydroxy-3-isopropylphenoxy)phenyl]-malonamic acid; KB-2115. C$_{18}$H$_{17}$Br$_2$NO$_5$; mol wt 487.14. C 44.38%, H 3.52%, Br 32.81%, N 2.88%, O 16.42%. Thyroid hor-

mone mimetic; stimulates bile acid synthesis and lowers LDL cholesterol. Prepn: T. J. Friends et al., WO 0160784; eidem, US 6800605 (2001, 2004 both to Bristol-Myers Squibb). Improved process: R. Chidambaram et al., US 6806381 (2004 to Bristol-Myers Squibb). Clinical pharmacology and pharmacokinetics: A. Berkenstam et al., Proc. Natl. Acad. Sci. USA 105, 663 (2008). Clinical trial in statin-treated dyslipidemia: P. W. Ladenson et al., N. Engl. J. Med. 362, 906 (2010).

Crystals from methanol + water.
THERAP CAT: Antilipemic.

3693. Eprozinol. [32665-36-4] 4-(2-Methoxy-2-phenylethyl)-α-phenyl-1-piperazinepropanol; 1-(2-methoxy-2-phenylethyl)-4-(3-hydroxy-3-phenylpropyl)piperazine. C$_{22}$H$_{30}$N$_2$O$_2$; mol wt 354.49. C 74.54%, H 8.53%, N 7.90%, O 9.03%. Prepn: Saunders, GB 1188505 (1970); to R. Y. Mauvernay, N. Busch, US 3705244 (1972). Pharmacology: Duchene-Marullaz et al., Therapie 26, 155 (1971). Clinical studies: Sors, Dutarte, ibid. 163.

Dihydrochloride. [27588-43-8] Alecor; Brovel; Eupnéron. C$_{22}$H$_{30}$N$_2$O$_2$.2HCl; mol wt 427.41. White crystalline powder, mp 164°. Sol in water and alcohol. LD$_{50}$ orally in mice: 500 mg/kg (Mauvernay, Busch).
THERAP CAT: Bronchodilator.

3694. Espiprantel. [98123-83-2] 2-(Cyclohexylcarbonyl)-2,-3,6,7,8,12b-hexahydropyrazino[2,1-a][2]benzazepin-4(1H)-one; 2-cyclohexylcarbonyl-4-oxo-1,2,3,4,6,7,12b-octahydropyrazino[2,1-a][2]benzazepine; BRL-38705; Cestex. C$_{20}$H$_{26}$N$_2$O$_2$; mol wt 326.44. C 73.59%, H 8.03%, N 8.58%, O 9.80%. Anthelmintic structurally related to praziquantel, q.v. Prepn: R. J. Dorgan, R. L. Elliott, EP 134984; eidem, US 4661489 (1985, 1987 both to Beecham); M. D. Brewer et al., J. Med. Chem. 32, 2058 (1989). Veterinary trials in dogs: R. M. Corwin et al., Am. J. Vet. Res. 50, 1076 (1989); in cats and dogs: B. R. Manger, M. D. Brewer, Br. Vet. J. 145, 384 (1989).

White crystals from chloroform/petr ether, mp 189-190°.
THERAP CAT (VET): Anthelmintic (Cestodes).

3695. Eptastigmine. [101246-68-8] N-Heptylcarbamic acid (3aS,8aR)-1,2,3,3a,8,8a-hexahydro-1,3a,8-trimethylpyrrolo[2,3-b]-indol-5-yl ester; N-demethyl-N-heptylphysostigmine; heptylphysostigmine; heptylstigmine. C$_{21}$H$_{33}$N$_3$O$_2$; mol wt 359.51. C 70.16%, H 9.25%, N 11.69%, O 8.90%. Cholinesterase inhibitor; analog of physostigmine, q.v. Prepn: M. Pomponi et al., EP 154864; M. Brufani *et al.*, **US 4831155** (1985, 1989 both to Consiglio Naz. Ricerche). Prepn of tartrate: P. G. Pagella *et al.*, **EP 298202** (1989 to Mediolanum Farm. and Consiglio Naz. Ricerche). Prepn and pharmacology: M. Marta *et al.*, *Life Sci.* **43**, 1921 (1988). Anticholinesterase activity: P. De Sarno *et al.*, *Neurobiochem. Res.* **14**, 971 (1989). Clinical cholinesterase inhibition: L. K. Unni *et al.*, *Eur. J. Clin. Pharmacol.* **41**, 83 (1991).

Crystals from heptane, mp 60-64°; also reported as 58-62° (Marta). uv max (methanol): 303, 253 nm (ε 3300, 14200).

Tartrate. [121652-76-4] MF-201. $C_{21}H_{33}N_3O_2 \cdot C_4H_6O_6$; mol wt 509.60. Microcrystalline powder, mp 122-123°. Soly in water 70%. LD_{50} i.p. in mice: 35 mg/kg (De Sarno).

THERAP CAT: Cholinergic.

3696. Eptazocine. [72522-13-5] (1*S*,6*S*)-2,3,4,5,6,7-Hexahydro-1,4-dimethyl-1,6-methano-1*H*-4-benzazonin-10-ol; (1*S*)-1,4-dimethyl-10-hydroxy-2,3,4,5,6,7-hexahydro-1,6-methano-1*H*-4-benzazonine; ST-2121; Sedapain. $C_{15}H_{21}NO$; mol wt 231.34. C 77.88%, H 9.15%, N 6.05%, O 6.92%. Opioid agonist-antagonist analgesic, related to pentazocine, *q.v.* Prepn: M. Ikeda *et al.*, **DE 2422309**; *eidem*, **US 4008219** (1974, 1977 both to Nihon Iyakuhin Kogyo). Chromatographic study: I. Hayashi *et al.*, *Iyakuhin Kenkyu* **12**, 442 (1981), *C.A.* **95**, 220196v (1981). Series of articles on pharmacological activities: *Nippon Yakurigaku Zasshi* **78**, 599-645 (1981), *C.A.* **96**, 62879, 79732, 79733 (1982). Opioid receptor binding study: T. Nabeshima *et al.*, *Res. Commun. Chem. Pathol. Pharmacol.* **48**, 173 (1985).

Hydrobromide. $C_{15}H_{22}BrNO$. Crystals, mp 207-210°.

THERAP CAT: Analgesic.

3697. EPTC. [759-94-4] *N,N*-Dipropylcarbamothioic acid *S*-ethyl ester; dipropylthiocarbamic acid *S*-ethyl ester; *S*-ethyl dipropylthiocarbamate; R-1608; FDA-1541; Eptam. $C_9H_{19}NOS$; mol wt 189.32. C 57.10%, H 10.12%, N 7.40%, O 8.45%, S 16.93%. Selective pre-planting herbicide. Prepn: **GB 808753**; H. Tilles, J. Antognini, **US 2913327** (both 1959 to Stauffer). Activity: J. Antognini *et al.*, *Proc. Northeast. States Weed Control Conf.* **1957**, 3. Metabolism: V. Y. Ong, S. C. Fang, *Toxicol. Appl. Pharmacol.* **17**, 418 (1970); J. P. Hubbell, J. E. Casida, *J. Agric. Food Chem.* **25**, 404 (1977).

Liquid, bp$_{20}$ 127°. d^{30} 0.9546; n$_D^{30}$ 1.4750. Vapor pressure at 35°: 3.4×10^{-2} mm Hg. Soly in water at 20°: 365 mg/l. Miscible with benzene, alc, toluene, xylene. LD_{50} orally in rats: 1631 mg/kg (Antognini).

USE: Herbicide.

3698. Eptifibatide. [188627-80-7] N^6-(Aminoiminomethyl)-N^2-(3-mercapto-1-oxopropyl)-L-lysylglycyl-L-α-aspartyl-L-tryptophyl-L-prolyl-L-cysteinamide cyclic (1 → 6)-disulfide; Integrilin. $C_{35}H_{49}N_{11}O_9S_2$; mol wt 831.97. C 50.53%, H 5.94%, N 18.52%, O 17.31%, S 7.71%. Fibrinogen receptor antagonist. Specifically in-

hibits fibrinogen binding to the platelet integrin GPIIb-IIIa which prevents platelet aggregation and subsequent clot formation. Synthetic, disulfide-linked cyclic heptapeptide based on *barbourin*, a specific GPIIb-IIIa antagonist isolated from venom of the southeastern pigmy rattlesnake, *Sistrurus m. barbouri*. Contains the Lys-Gly-Asp (KGD) amino acid sequence, a modification of the Arg-Gly-Asp (RGD) recognition site which appears to confer specificity: R. M. Scarborough *et al.*, *J. Biol. Chem.* **266**, 9359 (1991). Design and structure-activity study of integrin antagonists: *eidem, ibid.* **268**, 1066 (1993). Prepn: R. M. Scarborough *et al.*, **US 5686570** (1997 to COR Therapeutics). Pharmacology: F. A. Nicolini *et al.*, *Circulation* **89**, 1802 (1994); K. Uthoff *et al.*, *ibid.* **90**, pt 2, II-269 (1994). Clinical trial in patients undergoing percutaneous cardiovascular intervention: IMPACT-II Investigators, *Lancet* **349**, 1422 (1997). Clinical trial in acute coronary syndromes: PURSUIT Trial Investigators, *N. Engl. J. Med.* **339**, 436 (1998).

THERAP CAT: Antithrombotic.

3699. Equilenin. [517-09-9] 3-Hydroxyestra-1,3,5,7,9-pentaen-17-one; 11,12,13,14,15,16-hexahydro-3-hydroxy-13-methyl-17*H*-cyclopenta[*a*]phenanthren-17-one; 1,3,5:10,6,8-estrapentaen-3-ol-17-one. $C_{18}H_{18}O_2$; mol wt 266.34. C 81.17%, H 6.81%, O 12.01%. Estrogenic steroid isolated from urine of pregnant mares: Girard *et al.*, *Compt. Rend.* **195**, 981 (1932). Occurs naturally in the *d*-form. Not found in human urine. Component of Conjugated Estrogenic Hormones, *q.v.* Isoln by chromatography: Duschinsky, Lederer, *Bull. Soc. Chim. Biol.* **17**, 1534 (1935). Total synthesis: Bachmann *et al.*, *J. Am. Chem. Soc.* **61**, 974 (1939); **62**, 824 (1940); Johnson *et al.*, *ibid.* **67**, 2274 (1945); **69**, 2942 (1947); **72**, 505 (1950); Hughes, Smith, *Chem. Ind. (London)* **1960**, 1022; Bailey *et al.*, *Chem. Commun.* **1967**, 1253; Stein *et al.*, *Tetrahedron* **26**, 1917 (1970). Synthesis from estrone: Corbellini *et al.*, *Farmaco Ed. Sci.* **19**, 913 (1964); O. N. Minailova *et al.*, *Zh. Obshch. Khim.* **49**, 2633 (1979); A. R. Daniewski, T. Kowalczyk-Przewloka, *Tetrahedron Lett.* **1982**, 2411. Review of synthetic studies: Taub, "Naturally Occurring Aromatic Steroids" in *The Total Synthesis of Natural Products* vol. 2, J. ApSimon, Ed. (John Wiley & Sons, New York, 1973) pp 642-663.

Needles from dil alc, mp 258-259°. Sublimes at 170-180° at 0.01 mm Hg. [α]$_D^{16}$ +87° (12.8 mg made up to 1.8 ml in dioxane). uv max (ethanol): 231, 270, 282, 292, 325, 340 nm. Soly/100 ml alcohol 0.63 (18°); 2.5 g (boiling).

Acetate. $C_{20}H_{20}O_3$. mp 156-157°.

Benzoate. $C_{25}H_{22}O_3$. mp 222-223° (vac).

Methyl ether. $C_{19}H_{20}O_2$. Needles from alc, mp 197-198°; mp 193-194° (vac).

THERAP CAT: Estrogen.

3700. Equilin. [474-86-2] 3-Hydroxyestra-1,3,5(10) 7-tetraen-17-one; 1,3,5,7-estratetraen-3-ol-17-one. $C_{18}H_{20}O_2$; mol wt 268.36. C 80.56%, H 7.51%, O 11.92%. Steroidal hormone isolated from urine of pregnant mares: Girard *et al.*, *Compt. Rend.* **195**, 981 (1932); Cartland, Meyer, *J. Biol. Chem.* **112**, 9 (1935/6). Structure: Serini, Logemann, *Ber.* **71**, 186 (1938); Pearlman, Wintersteiner, *J. Biol. Chem.* **130**, 35 (1939). Separation from estrone: Serini, Logemann, **US 2221340** (1941 to Schering). Synthesis: Zderic *et al.*, *J. Am. Chem. Soc.* **80**, 2596 (1958); Bagli *et al.*, *Tetrahedron Lett.* **1964**, 387; Stein *et al.*, *ibid.* **1966**, 5015; *eidem, Tetrahedron* **26**, 1917 (1970). Prepn from testosterone by bacterial fermentation: Bowers *et al.*, **US 3067212** (1962 to Syntex). Prepn by conversion of equilenin: Bailey *et al.*, *Chem. Commun.* **1967**, 1253; Marshall, Deghenghi, *Can. J. Chem.* **47**, 3127 (1969). Component of Conjugated Estrogenic Hormones, *q.v.* Review of synthetic studies: Taub, "Naturally Occurring Aromatic Steroids" in *The Total Synthesis of Natural Products* vol. 2, J. ApSimon, Ed. (John Wiley & Sons, New York, 1973) pp 664-670.

Orthorhombic sphenoidal plates from ethyl acetate, mp 238-240°. $[\alpha]_D^{25}$ +308° (c = 2 in dioxane); $[\alpha]_D^{25}$ +325° (c = 2 in alc). uv max: 283-285 nm. Soluble in alcohol, dioxane, acetone, ethyl acetate, in other organic solvents; sparingly sol in water.

Benzoate. $C_{25}H_{24}O_3$. mp 196-197°.

Methyl ether. $C_{19}H_{22}O_2$. Needles from alc, mp 161-162°.

THERAP CAT: Estrogen.

3701. Equol. [531-95-3] (3S)-3,4-Dihydro-3-(4-hydroxy-phenyl)-2H-1-benzopyran-7-ol; (S)-7-hydroxy-3-(4'-hydroxyphen-yl)chroman; (S)-(−)-4',7-isoflavandiol; (S)-4',7-dihydroxyisofla-vane. $C_{15}H_{14}O_3$; mol wt 242.27. C 74.37%, H 5.82%, O 19.81%. Bioactive metabolite of the phytoestrogen, daidzein, q.v. Produced from dietary sources by intestinal bacteria; exhibits estrogenic, va-sorelaxant, and antioxidant activity. Isoln from horse urine: G. F. Marrian, G. A. D. Haslewood, *Biochem. J.* **26**, 1227 (1932); *eidem*, *ibid.* **29**, 1586 (1935). Structure: F. Wessely, F. Prillinger, *Ber.* **72**, 629 (1939); E. L. Anderson, G. F. Marrian, *J. Biol. Chem.* **127**, 649 (1939). Abs config: K. Kurosawa *et al.*, *Chem. Commun.* **1968**, 1265. Prepn of (±)-form: J. A. Lamberton *et al.*, *Aust. J. Chem.* **31**, 455 (1978). Enantioselective synthesis: J. M. Heemstra *et al.*, *Org. Lett.* **8**, 5441 (2006). LC-MS/MS determn in urine and serum: P. B. Grace *et al.*, *J. Chromatogr. B* **853**, 138 (2007). Review of meta-bolic formation from soy isoflavones and estrogenic activity: K. D. R. Setchell *et al.*, *J. Nutr.* **132**, 3577-3584 (2002); and cardiovascular pharmacology: K. A. Jackman *et al.*, *Curr. Med. Chem.* **14**, 2824-2830 (2007).

White solid, mp 192-193°. $[\alpha]_D^{24}$ −23.5° (ethanol). Readily sol in ethanol, methanol, ether, ethyl acetate, acetone.

(R)-Form. [221054-79-1] Isoequol.

3702. Erabutoxins. Neurotoxic principles isolated from venom of the sea-snake, *Laticauda semifasciata* (in Japanese, *erabu-umihebi*). Their action on post-synaptic membrane blocks neuro-muscular transmission. Separation and crystn of erabutoxins A and B: Tamiya, Arai, *Biochem. J.* **99**, 624 (1966). Structures are single polypeptide chains containing 62 amino acid residues with four S-S bridges, *erabutoxin A* differing from *erabutoxin B* only in the amino acid residue at position 26: Sato, Tamiya, *ibid.* **122**, 453 (1971); Endo *et al.*, *ibid.* 463. Isoln and structure of the minor component, *erabutoxin C*: Tamiya, Abe, *ibid.* **130**, 547 (1972). Corrected amino acid sequences: N. Maeda, N. Tamiya, *Biochem. J.* **167**, 289 (1977). Three-dimensional structure of B, description of active site: B. W. Low, *Adv. Cytopharmacol.* **3**, 141 (1979). Molecular confor-mation: M. R. Kimball *et al.*, *Biochem. Biophys. Res. Commun.* **88**, 950 (1979); F. Inagaki *et al.*, *Eur. J. Biochem.* **109**, 129 (1980). Raman spectra: T. Takamatsu *et al.*, *Biochim. Biophys. Acta* **622**, 189 (1980). Immunological studies: A. Tatsuya *et al.*, *Toxicon* **17**, 571 (1979), *C.A.* **92**, 141459t (1980). Effect on transmission in au-tonomic ganglia: V. A. Chiappinelli *et al.*, *Brain Res.* **211**, 107 (1981).

LD_{50} in mice, rats (g/g): 0.15, 0.07 i.m. (Tamiya, Arai).

3703. Erbium. [7440-52-0] Er; at. wt 167.26; at. no. 68; va-lence 3. Rare earth metal of the yttrium group; member of the lan-thanide series. Naturally occurring isotopes (mass numbers): 162 (0.14%); 164 (1.61%); 166 (33.6%); 167 (22.95%); 168 (26.8%); 170 (14.9%). Known artificial radioactive isotopes: 147; 148; 150-161; 163; 165; 169; 171-173. Abundance in earth's crust: 2.47-3.5

ppm. Occurs in small quantities in all the rare earth minerals; main sources: xenotime, fergusonite, gadolinite, euxenite, polycrase, blomstrandine. Discovery: Mosander, *Skand. Naturför. Förh.* **3**, 387 (1842); *Philos. Mag.* [3] **23**, 241 (1843). Separation of oxide: James, *J. Am. Chem. Soc.* **32**, 517 (1910); **34**, 757 (1912); Hofmann, Burger, *Ber.* **41**, 308 (1908); Prandtl, *Z. Anorg. Chem.* **198**, 157 (1931). Sepn by ion-exchange: Spedding *et al.*, *J. Am. Chem. Soc.* **69**, 2812 (1947); **76**, 2557 (1954). Prepn of the metal: Klemm, Bommer, *Z. Anorg. Chem.* **213**, 138 (1937). Prepn of chloride, bro-mide: Jantsch *et al.*, *Z. Anorg. Chem.* **207**, 353 (1932). Radioactiv-ity induced by neutron bombardment: Sugden, *Nature* **135**, 469 (1935); McLennan, Rann, *ibid.* **136**, 831 (1935). Spectrum: Eder, *Ber. Wien. Akad.* [2a] **124**, 790 (1915); deGramont, *Compt. Rend.* **171**, 1106 (1920); Mott, McDonald, *Trans. R. Soc. Can. Sect. 3* **21**, 230 (1927). Natural isotopic composition: Hayden *et al.*, *Phys. Rev.* **77**, 299 (1950). Toxicity study: Haley, *J. Pharm. Sci.* **54**, 663 (1965). Reviews of prepn, properties and compds: *The Rare Earths*, F. H. Spedding, A. H. Daane, Eds. (Krieger, Huntington, N.Y., 1971, reprint of 1961 ed.) 641 pp; Hulet, Bode, "Separation Chemistry of the Lanthanides and Transplutonium Actinides" in *MTP Int. Rev. Sci.: Inorg. Chem., Ser. One* vol. 7, K. W. Bagnall, Ed. (University Park Press, Baltimore, 1972) pp 1-45; Moeller, "The Lanthanides" in *Comprehensive Inorganic Chemistry* vol. 4, J. C. Bailar Jr. *et al.*, Eds. (Pergamon Press, Oxford, 1973) pp 1-101; F. H. Spedding in *Kirk-Othmer Encyclopedia of Chemical Technology* vol. **19** (John Wiley & Sons, New York, 3rd ed., 1982) pp 833-854; *Chemistry of the Elements*, N. N. Greenwood, A. Earnshaw, Eds. (Pergamon Press, New York, 1984) pp 1423-1449. Brief review of properties: G. T. Seaborg, *Radiochim. Acta* **61**, 115-122 (1993).

Dark-gray metallic powder; hexagonal close-packed crystal lat-tice. d 9.066. mp 1529°. bp 2868°. Heat of fusion: 19.90 kJ/mol. Heat of sublimation (25°): 317.10 kJ/mol. E°(aq) Er^{3+}/Er −230 V (calc). Similar to the other rare earth metals, possesses two reduction potentials 1.770 and 1.875 volts (ref to the normal calomel elec-trode), Noddack, Brukl, *Angew. Chem.* **50**, 362 (1937).

Oxide. Erbia. Er_2O_3. Pinkish powder changing into cubic crys-tals on heating at 1300°; d 8.64; sp heat 0.065; prepd by igniting the oxalate or basic nitrate. Readily sol in acids. Soly in water: 1.28×10^{-5} g-mol/l at 29°.

Hydroxide. Er(OH)$_3$. Pale pink gelatinous ppt. Prepd by action of alkali hydroxide on a soln of erbium nitrate.

Chloride. ErCl$_3$. Hexahydrate, deliquesc crystals. Sol in water; slightly sol in alcohol. Dehydrated by heating in a stream of hydro-gen chloride. Anhydr form is pinkish powder, d 4.1. LD_{50} in mice (mg/kg): 535 i.p.; 6.2 orally (Haley).

Bromide. ErBr$_3$. Nonahydrate, deliquesc rose crystals.

Nitrate. Er(NO$_3$)$_3$. Pentahydrate, reddish, deliquesc crystalline solid. Loses 4 mols of water on heating to 130°. LD_{50} of hexahy-drate in female rats (mg/kg): 230 i.p.; 35.8 i.v. (Haley).

3704. Erdosteine. [84611-23-4] 2-[[2-Oxo-2-[(tetrahydro-2-oxo-3-thienyl)amino]ethyl]thio]acetic acid; (±)-[[[(tetrahydro-2-oxo-3-thienyl)carbamoyl]methyl]thio]acetic acid; DL-S-[2-[N-3-(2-oxotetrahydrothienyl)acetamido]]thioglycolic acid; N-(carboxy-methylthioacetyl)homocysteine thiolactone; dithiosteine; RV-144; Erdotin; Vectrine. $C_8H_{11}NO_4S_2$; mol wt 249.30. C 38.54%, H 4.45%, N 5.62%, O 25.67%, S 25.72%. Homocysteine thiolactone derivative with mucomodulating activity. Prepn: J. Gonella, **EP 61386**; *idem*, **US 4411909** (1982, 1983 both to Refarmed); M. Gob-etti *et al.*, *Farmaco Ed. Sci.* **41**, 69 (1986). Review of clinical devel-opment: K. L. Dechant, S. Noble, *Drugs* **52**, 875-881 (1996). Met-abolism: S. Savu *et al.*, *Int. J. Clin. Pharmacol. Ther.* **38**, 415 (2000). Clinical trial in chronic obstructive bronchitis: C. F. Mar-chioni *et al.*, *Int. J. Clin. Pharmacol. Ther.* **33**, 612 (1995); in COPD: M. Moretti *et al.*, *Drugs Exp. Clin. Res.* **30**, 143 (2004).

Crystals from ethanol, mp 156-158°. pKa 3.71. LD_{50} in mice and rats (g/kg): >10 orally; >3.5 i.v. (Gonella).

THERAP CAT: Mucolytic.

3705. Ergocornine. [564-36-3] $(5'\alpha)$-12'-Hydroxy-2',5'-bis(1-methylethyl)ergotaman-3',6',18-trione. $C_{31}H_{39}N_5O_5$; mol wt 561.68. C 66.29%, H 7.00%, N 12.47%, O 14.24%. Natural ergot alkaloid derived from lysergic acid; component of ergotoxine, q.v. Isoln from ergot: Stoll, Hofmann, *Helv. Chim. Acta* **26**, 1570 (1943). Structure: Stoll *et al.*, *ibid.* **34**, 1544 (1951). Separation and purification: Stoll, Hofmann, **US 2447214** (1948 to Sandoz). Synthesis: Stadler *et al.*, *Helv. Chim. Acta* **52**, 1549 (1969). Determn in ergot by capillary electrophoresis: K. Frach, G. Blaschke, *J. Chromatogr. A* **808**, 247 (1998).

Solvated, polyhedra from methanol, dec 181° (contains 1 mole methanol). $[\alpha]_D^{20}$ −110° (pyridine); −175° (chloroform). uv max (methanol): 311 nm (log ε 3.91). Soluble in acetone, chloroform, ethyl acetate; slightly sol in ethyl and methyl alcohol. Nearly insol in water.

8α-Epimer. [564-37-4] Ergocorninine. Prisms from alc, dec 228°. $[\alpha]_D^{20}$ +409° (chloroform). uv max (methanol): 240.5, 312.5 (log ε 4.31, 3.92). Soluble in 15 parts boiling ethanol, 25 parts boiling methanol, 30 parts boiling benzene, 30 parts boiling ethyl acetate; freely sol in acetone, chloroform. Practically insol in water.

3706. Ergocristine. [511-08-0] $(5'\alpha)$-12'-Hydroxy-2'-(1-methylethyl)-5'-(phenylmethyl)ergotaman-3',6',18-trione. $C_{35}H_{39}N_5O_5$; mol wt 609.73. C 68.95%, H 6.45%, N 11.49%, O 13.12%. Natural ergot alkaloid derived from lysergic acid; a member of the ergotoxine group. Isoln from ergot: Stoll, Burckhardt, *Z. Physiol. Chem.* **250**, 1 (1937); Stoll, Hofmann, *Helv. Chim. Acta* **26**, 1570 (1943). Structure: Stoll *et al.*, *ibid.* **34**, 1544 (1951). Separation and purification: Stoll, Hofmann, **US 2447214** (1948 to Sandoz). Synthesis: Stadler *et al.*, *Helv. Chim. Acta* **52**, 1549 (1969).

Orthorhombic crystals with $2C_6H_6$ from benzene. mp 155-157° (dec) (solvent-free base). $[\alpha]_D^{20}$ −183° (chloroform). Very sol in ethyl and methyl alcohol, acetone, chloroform, ethyl acetate. Slightly sol in ether. Practically insol in water, petr ether.

Phosphate. [6424-36-8] Crystals, dec 195°.

Ethanesulfonate. [6055-56-7] Crystals, dec 207°.

8α-isomer. [511-07-9] Ergocristinine. Crystallizes solvent-free, unlike ergocristine which tends to retain the solvent of crystn. Long prisms from abs alc, mp 226° (dec). $[\alpha]_D^{20}$ +366° (c = 0.68 in chloroform); +471° (c = 0.35 in pyridine). uv max (methanol/methylene chloride): 313 nm (log ε 3.96). Much less sol than ergocristine. Does not seem to form salts.

3707. Ergocryptine. Ergokryptine. $C_{32}H_{41}N_5O_5$; mol wt 575.71. C 66.76%, H 7.18%, N 12.16%, O 13.90%. Pair of ergot alkaloid isomers which differ only in the butyl side chain. Upon hydrolysis, α-ergocryptine yields L-leucine, while β-ergocryptine yields L-isoleucine. Epimeric with ergocryptinine, q.v. The ergocryptine discussed in the literature prior to 1967 is now referred to as α-ergocryptine. Isoln from ergot: Stoll, Hofmann, *Helv. Chim. Acta* **26**, 1570 (1943). Structure: Stoll *et al.*, *ibid.* **34**, 1544 (1951). Separation and purification: Stoll, Hofmann, **US 2447214** (1948 to Sandoz). Separation of β-ergocryptine from α-ergocryptine: Schlientz *et al.*, *Experientia* **23**, 991 (1967); *see also eidem*, *Pharm. Acta Helv.* **43**, 497 (1968). Synthesis of α- and β-ergocryptine: Stadler *et al.*, *Helv. Chim. Acta* **52**, 1549 (1969).

α-Ergocryptine   R = $CH_2CH(CH_3)_2$
β-Ergocryptine   R = $CH(CH_3)CH_2CH_3$

α-Ergocryptine. [511-09-1] $(5'\alpha)$-12'-Hydroxy-2'-(1-methylethyl)-5'-(2-methylpropyl)ergotaman-3',6',18-trione. Solvated prisms from acetone, benzene, methanol. With MeOH of crystn, mp 212° (dec). $[\alpha]_D^{20}$ −120° (pyridine); −198° (chloroform). uv max (methanol): 241, 312.5 nm (log ε 4.31, 3.95). Freely sol in alcohol, chlorofom. Almost insol in water.

β-Ergocryptine. [20315-46-2] $(5'\alpha)$-12'-Hydroxy-2'-(1-methylethyl)-5'-[(1S)-1-methylpropyl]ergotaman-3',6',18-trione. Rectangular plates from benzene, mp 173° (dec). $[\alpha]_D^{20}$ −98° (c = 0.5 in pyridine); −179° (c = 0.5 in chloroform). uv max (methanol): 312 (log ε 3.93).

THERAP CAT: *See* Ergot.

3708. Ergocryptinine. $C_{32}H_{41}N_5O_5$; mol wt 575.71. C 66.76%, H 7.18%, N 12.16%, O 13.90%. Alkaloid pair isomeric with α- and β-ergocryptine, resp., but differing by an α-configuration at C-8. The literature prior to 1967 refers to α-ergocryptinine as ergocryptinine. Isolation, structure, separation and purification *see* ergocryptine. Production: Abe *et al.*, **US 2835675** (1958). Prepn of β-ergocryptinine from β-ergocryptine: Schlientz *et al.*, *Experientia* **23**, 991 (1967). Synthesis of α- and β-ergocryptinin: Stadler *et al.*, *Helv. Chim. Acta* **52**, 1549 (1969).

α-Ergocryptinine. [511-10-4] $(5'\alpha,8\alpha)$-12'-Hydroxy-2'-(1-methylethyl)-5'-(2-methylpropyl)ergotaman-3',6',18-trione; α-ergokryptinine. Fine needles from methanol, dec 240-242°. $[\alpha]_D^{20}$ +408° (chloroform); +485° (c = 0.5 in pyridine). uv max (methanol): 241.5, 312.5 nm (log ε 4.30, 3.94). Sol in 20 parts boiling ethanol, 50 parts boiling methanol; freely sol in acetone, chloroform. Almost insol in water.

β-Ergocryptinine. [19467-61-9] $(5'\alpha,10\alpha)$-12'-Hydroxy-2'-(1-methylethyl)-5'-[(1S)-1-methylpropyl]ergotaman-3',6',18-trione. Colorless needles from methylene chloride/methanol, mp 217-218° (dec). $[\alpha]_D^{20}$ +421° (chloroform); +497° (pyridine). uv max (methanol): 240.5, 312 nm (log ε 4.31, 3.94).

3709. Ergoflavin. [3101-51-7] $(1S,1'S,3S,3'S,4S,4'S,4aS,4'aS,9aR,9'aR)$-1,1',3,3',4,4',9a,9'a-Octahydro-4,4',8,8',9a,9'a-hexahydroxy-3,3'-dimethyl-[7,7'-bi-1,4a-(epoxymethano)-4aH-xanthene]-9,9',11,11'($2H,2'H$)-tetrone; ergochrome CC(2,2'). $C_{30}H_{26}O_{14}$; mol wt 610.52. C 59.02%, H 4.29%, O 36.69%. Principal pigment from ergot: Freeborn, *Pharm. J.* **88**, 568 (1912); Eglinton *et al.*, *J. Chem. Soc.* **1958**, 1833. Structure: McPhail *et al.*, *ibid.* **1966**, *Sect. B*, 18. Review of ergoflavin and other ergochromes: Franck, Flasch in *Fortschr. Chem. Org. Naturst.* **30**, 151-206 (1973).

Yellow needles from methanol, decomp 350°. $[\alpha]_D^{21}$ +37.5° (c = 1.236 in acetone). uv max: 240, 260, 381 nm (E$_{1cm}^{1\%}$ 350, 346, 130). Sol in acetone, pyridine; moderately sol in methanol, alcohol, ethyl acetate, dioxane; sparingly sol in ether, benzene. Practically insol in 2N aq NaHCO$_3$.

Hexaacetate. C$_{42}$H$_{38}$O$_{20}$. Prisms from chloroform + petr ether, dec 248-249°. $[\alpha]_D^{20}$ +61.2° (c = 0.62 in dioxane). uv max: 340, 338 nm (E$_{1cm}^{1\%}$ 343, 61).

3710. Ergoloid Mesylates. [8067-24-1] Dihydroergotoxine methanesulfonate (1:1); co-dergocrine mesylate; CCK-179; Circanol; Coristin; Dacoren; DCCK; Decril; Dulcion; Ergodesit; Ergohydrin; Ergoplus; Hydergine; Lysergin; Novofluen; Orphol; Pérénan; Progeril; Redergin; Sponsin; Trigot. Hydrogenated ergot alkaloids, specifically an equiproportional mixture of dihydroergocornine methanesulfonate, dihydroergocristine methanesulfonate and α- and β-dihydroergocryptine methanesulfonate in the ratio of 1.5-2.5:1. Neuropharmacological investigations: D. M. Loew et al., *Postgrad. Med. J.* **52**, Suppl. 1, 40 (1976).

Dihydroergocornine	R = CH(CH$_3$)$_2$
Dihydroergocristine	R = CH$_2$C$_6$H$_5$
Dihydro-α-ergocryptine	R = CH$_2$CH(CH$_3$)$_2$
Dihydro-β-ergocryptine	R = CH(CH$_3$)CH$_2$CH$_3$

White to off-white, microcrystalline or amorphous, practically odorless powder. Soluble in methanol, alc; sparingly sol in acetone; slightly sol in water.

Dihydroergocristine methanesulfonate. [24730-10-7] Decme; Defluina; Enirant; Insibrin; Nehydrin; Simactil; Unergol. C$_{36}$H$_{45}$N$_5$O$_8$S; mol wt 707.84.

THERAP CAT: α-Adrenergic blocker (treatment of impaired mental function in the elderly).

3711. Ergometrinine. [479-00-5] (8α)-9,10-Didehydro-N-[(1S)-2-hydroxy-1-methylethyl]-6-methylergoline-8-carboxamide; D-lysergic acid D-propanolamide; ergonovinine; C$_{19}$H$_{23}$N$_3$O$_2$; mol wt 325.41. C 70.13%, H 7.12%, N 12.91%, O 9.83%. An alkaloid isomeric with ergonovine, q.v. Ergometrinine and ergonovine can be interconverted by simple chemical procedures: Smith, Timmis, *J. Chem. Soc.* **1936**, 1166. Chromatographic determn in cereal grains: G. H. Ware et al., *J. Assoc. Off. Anal. Chem.* **69**, 697 (1986).

Stout prisms from acetone, dec 196°. $[\alpha]_D^{20}$ +416° (c = 0.45 in chloroform). pK 7.3. Freely sol in pyridine, moderately in chloroform; slightly sol in alcohol, acetone. Nearly insol in water.

Hydrobromide monohydrate. C$_{19}$H$_{24}$BrN$_3$O$_2$.H$_2$O. Needles from aq acetone + ether, dec 190°.

Hydrochloride monohydrate. C$_{19}$H$_{24}$ClN$_3$O$_2$.H$_2$O. Needles, dec 175-180°.

Mononitrate. C$_{19}$H$_{24}$N$_4$O$_5$. Stout prisms from aq MeOH + ether, dec 235°. $[\alpha]_D^{20}$ +282° (c = 1).

Perchlorate. C$_{19}$H$_{24}$ClN$_3$O$_6$. Needles, dec 225° (brown at 210°).

3712. Ergonovine. [60-79-7] (8β)-9,10-Didehydro-N-[(1S)-2-hydroxy-1-methylethyl]-6-methylergoline-8-carboxamide; N-[α-(hydroxymethyl)ethyl]-D-lysergamide; D-lysergic acid L-2-propanolamide; ergometrine; Ergobasine; Ergotocine; Ergostetrine; Ergotrate; Ergoklinine; Syntometrine. C$_{19}$H$_{23}$N$_3$O$_2$; mol wt 325.41. C 70.13%, H 7.12%, N 12.91%, O 9.83%. From some ergots: A. Stoll et al., **US 2809920** (1957 to Saul & Co.). Prepn from D-lysergic acid and L(+)-2-amino-1-propanol: A. Stoll, A. Hofmann, *Helv. Chim. Acta* **26**, 944 (1943); Pioch, **US 2736728** (1956 to Lilly); Patelli, Bernardi, **US 3141887** (1964 to Farmitalia). Structure: A. Stoll et al., *Helv. Chim. Acta* **34**, 1544 (1951). Total synthesis: E. C. Kornfeld et al., *J. Am. Chem. Soc.* **78**, 3087 (1956). Metabolism: Slaytor, Wright, *J. Med. Pharm. Chem.* **5**, 483 (1962). Comprehensive description of the maleate: V. D. Reif, *Anal. Profiles Drug Subs.* **11**, 273-312 (1982). Toxicity data: R. P. Beliles, *Toxicol. Appl. Pharmacol.* **23**, 537 (1972). Clinical use in diagnosis of angina: L. A. DiCarlo, Jr. et al., *Am. J. Cardiol.* **54**, 744 (1984); R. Nordlander, R. Orinius, *Acta Med. Scand.* **221**, 47 (1987).

Tetrahedra from ethyl acetate, fine needles from benzene. Tends to form solvated crystals, mp 162°. $[\alpha]_D^{20}$ +90° (in water). pK 6.8. Freely sol in lower alcohols, ethyl acetate, acetone; more sol in water than the other principal alkaloids of ergot; slightly sol in chloroform.

Hydrochloride. C$_{19}$H$_{23}$N$_3$O$_2$.HCl. Needles from ethyl alcohol, dec 246°. $[\alpha]_D^{25}$ +63° (c = 0.9). More sol in water than the hydrobromide.

Maleate. [129-51-1] Cornocentin; Ergotrate Maleate; Ermetrine. C$_{19}$H$_{23}$N$_3$O$_2$.C$_4$H$_4$O$_4$; mol wt 441.48. Crystals, dec 167°. $[\alpha]_D^{25}$ +48 to +57°. One gram dissolves in 36 ml water, 120 ml alcohol. Nearly insol in ether and chloroform. LD$_{50}$ i.v. in mice: 8.26 mg/kg (Beliles).

Tartrate hydrate. Basergin; Neofemergen. (C$_{19}$H$_{23}$N$_3$O$_2$)$_2$.C$_4$-H$_6$O$_6$.H$_2$O; mol wt 818.93. Crystals, slightly sol in water.

Hydracrylate. Ergotrate-H. C$_{41}$H$_{52}$N$_6$O$_7$; mol wt 740.90.

THERAP CAT: Oxytocic.

THERAP CAT (VET): Oxytocic.

3713. Ergosine. [561-94-4] (5'α)-12'-Hydroxy-2'-methyl-5'-(2-methylpropyl)ergotaman-3',6',18-trione. C$_{30}$H$_{37}$N$_5$O$_5$; mol wt 547.66. C 65.79%, H 6.81%, N 12.79%, O 14.61%. Isoln of ergosine and ergosinine from ergot: Smith, Timmis, *J. Chem. Soc.* **1937**, 396. Structure: Stoll et al., *Helv. Chim. Acta* **34**, 1544 (1951). Stereochemistry: Stoll et al., *ibid.* **37**, 2039 (1954). Total synthesis of ergosine and ergosinine: Stadler et al., *ibid.* **47**, 1911 (1964).

Prisms from ethyl acetate, dec 228°. $[\alpha]_D^{20}$ −161° (chloroform). Sol in chloroform; fairly sol in methanol, acetone; sparingly sol in ethyl acetate, benzene.

8α-Isomer. Ergosinine; ergoclavinine. C$_{30}$H$_{37}$N$_5$O$_5$. Prisms from 90% alcohol, aq acetone, benzene or ethyl acetate, dec 228°.

Also reported as colorless needles, mp 190-191° (dec), Stadler *et al.*, *loc. cit.* $[\alpha]_D^{20}$ +420° (chloroform); $[\alpha]_D^{20}$ +380° (acetone). Very readily sol in chloroform; readily sol in acetone; less sol in ethyl acetate; sparingly sol in benzene; very sparingly sol in methyl alcohol; almost insol in water.

Note: **Ergoclavine** is an equimolar mixture of ergosine and ergosinine.

3714. Ergostane. [511-20-6] (5α)-Ergostane. $C_{28}H_{50}$; mol wt 386.71. C 86.97%, H 13.03%. Prepn from allocholanic acid: E. Fernholz, *Ber.* **69**, 1792 (1936).

Scales, plates from ether + methanol, mp 85°. $[\alpha]_D^{20}$ +17° (c = 2 in chloroform).

3715. Ergostanol. [6538-02-9] (3β,5α)-Ergostan-3-ol. $C_{28}H_{50}O$; mol wt 402.71. C 83.51%, H 12.52%, O 3.97%. Prepn by hydrogenation of ergosta-14,22-dien-7-one: Chen, *Ber.* **70**, 1432 (1937).

Crystals, mp 144-145°. $[\alpha]_D^{20}$ +15.9° (c = 1.8 in chloroform). Is precipitated by digitonin.

Acetate. $C_{30}H_{52}O_2$. mp 145°. $[\alpha]_D^{20}$ +6.0° (c = 1.8 in chloroform).

Benzoate. $C_{35}H_{54}O_2$. mp 163-165°.

3716. Ergosterol. [57-87-4] (3β,22E)-Ergosta-5,7,22-trien-3-ol; ergosta-5:6,7:8,22:23-trien-3-ol; ergosterin. $C_{28}H_{44}O$; mol wt 396.66. C 84.78%, H 11.18%, O 4.03%. Most important of the provitamins D. Usually obtained from yeast which synthesizes it from simple sugars such as glucose. Damp yeast yields about 2.5% ergosterol, the variety of the yeast being very important. Isoln procedure: Green *et al.*, US 3006932 (1961 to Vitamins Ltd.). When irradiated with uv light, ergosterol develops powerful vitamin D₂, *q.v.*, activity. Askew *et al.*, in England and Windaus and collaborators in Germany isolated the antirachitic vitamin D₂. The main irradiation products of ergsterol are lumisterol → tachysterol → vitamin D₂. Structure: Chuang, *Ann.* **500**, 270 (1933). Oxidation products: Fuerst, *Arch. Pharm.* **300**, 144 (1967).

Small hydrated plates from alcohol, in hydrated needles from ether. The best crystallized form contains 1½ mol H_2O, mp 168°: Bills, Honeywell, *J. Biol. Chem.* **80**, 15 (1928). Complete removal of water is almost impossible and results in an amorphous mass,

melting range 166-183°. bp$_{0.01}$ 250°. $[\alpha]_D^{20}$ −135° (c = 1.2 in $CHCl_3$ calcd as anhydr). $[\alpha]_{546}^{20}$ −171° ($CHCl_3$). uv max (ethanol): 262, 271, 282, 293 nm: Hogness *et al.*, *ibid.* **120**, 239 (1937). Practically insol in water. One gram dissolves in 660 ml alcohol, in 45 ml boiling alcohol, in 70 ml ether, in 39 ml boiling ether, in 31 ml chloroform. Precipitated by digitonin. Affected by light and air, turns yellow. Oxygen forms peroxides and hydrogen may form polyhydro compds.

Acetate. $C_{30}H_{46}O_2$. mp 179°, clear at 181°.

Benzoate. $C_{35}H_{48}O_2$. Plates or needles, mp 169-171°. $[\alpha]_D^{23}$ −71° (c = 1.1 in $CHCl_3$), $[\alpha]_{546}$ −88° ($CHCl_3$).

22,23-Dihydro analog. 22,23-Dihydroergosterol. $C_{28}H_{46}O$. Prepn: A. Windhaus, R. Langer, *Ann.* **508**, 105 (1934); D. H. R. Barton *et al.*, *J. Chem. Soc. Perkin Trans. 1* **1976**, 821; D. J. Curry *et al.*, *ibid.* **1977**, 822. uv irradiation gives vitamin D₄, *q.v.* Solvated needles from ethyl acetate + methanol, mp 152-153° (dried). $[\alpha]_D^{19}$ −109° ($CHCl_3$) (Windhaus); also reported as needles from chloroform-methanol, mp 128-130°. $[\alpha]_D^{19}$ −121° (c = 0.1). uv max: 262, 272, 282, 294 nm (ε 8000, 11200, 11800, 6800) (Barton).

22,23-Dihydro analog acetate. $C_{30}H_{48}O_2$. mp 157-158°. $[\alpha]_D^{17}$ −75° (c = 2.1 in $CHCl_3$).

THERAP CAT: Vitamin (antirachitic).

3717. Ergot. Secale cornutum; spurred rye. Dried sclerotia of the fungus *Claviceps purpurea* (Fries) Tul., *Hypocreaceae*, parasitic on rye plants. *Habit.* Europe; cultivated in Spain, Germany, and France. Four main classes of ergot alkaloids can be distinguished: clavine alkaloids, lysergic acids, lysergic acid amides and ergot peptide alkaloids. There are ten ergot peptide alkaloids which are ergotamine, ergosine, ergocristine, ergocryptine, ergocornine, ergotaminine, ergosinine, ergocristinine, ergocryptinine, and ergocorninine, the last five alkaloids being isomers of the first five. These allklaloids are typified by a structure consisting of lysergic acid, dimethylpyruvic acid, proline, and phenylalanine joined in amide linkages. In 1943 A. Stoll and A. Hofmann, *Helv. Chim. Acta* **26**, 1570 (1943) made it clear that the ergotoxine reported by G. Barger and F. H. Carr, *J. Chem. Soc.* **91**, 377 (1907) and by F. Kraft, *Arch. Pharm.* **244**, 336 (1906) was but a mixture of ergocristine, ergocryptine, and ergocornine, and that the ergotinine, first reported by C. Tanret, *Compt. Rend.* **81**, 891 (1875), was also a mixture consisting of ergocristinine, ergocryptinine, and ergocorninine. Other constituents of ergot are ergonovine, ergometrinine, ergoclavine, elymoclavine, trimethylamine, putrescine, cadaverine, agmatine, histamine, tyramine, histidine, tyrosine, valine, leucine, betaine, choline, acetylcholine, ergothioneine, 15-30% fatty oil, ergosterol, mannitol, lactic acid, and succinic acid. Production of ergocryptinine and elymoclavine by cultures of fungi parasitic on *Elymus mollis* Trin: Abe *et al.*, US 2835675. Production of ergot alkaloids by saprophytic cultures: Adams, US 3117917 (1964 to Miles Labs.). Ergot alkaloid fermentation: W. J. Kelleher, *Adv. Appl. Microbiol.* **vol. 11**, 211 (1969). Biosynthesis of ergot alkaloids: H. G. Floss, *Tetrahedron* **32**, 873 (1976). *Reviews:* Gröger in *Microbial Toxins* **vol. VIII**, S. Kadis *et al.*, Eds. (Academic Press, New York, 1972) pp 321-373; Stadler, Stütz, "The Ergot Alkaloids" in *The Alkaloids* **vol. 15**, R. H. F. Manske, Ed. (Academic Press, New York, 1975) pp 1-40. Books: F. J. Bove, *The Story of Ergot* (Karger, Basel, 1970) 297 pp; *Handb. Exp. Pharmacol.* **49** entitled "Ergot Alkaloids and Related Compounds", B. Berde, H. O. Schild, Eds. (Springer-Verlag, New York, 1978) 1003 pp.

Caution: Potential symptoms of overexposure are CNS stimulation, nausea, vomiting, weakness, tremors, excitement, confusion, convulsions, tachycardia, mydriasis, peripheral vasoconstriction progressing to gangrene. *See Clinical Toxicology of Commercial Products*, R. E. Gosselin *et al.*, Eds. (Williams & Wilkins, Baltimore, 5th ed., 1984) Section II, pp 218-219.

THERAP CAT: Vasoconstrictor (specific in migraine).

THERAP CAT (VET): Has been used as an oxytocic.

3718. Ergotamine. [113-15-5] (5′α)-12′-Hydroxy-2′-methyl-5′-(phenylmethyl)ergotaman-3′,6′,18-trione. $C_{33}H_{35}N_5O_5$; mol wt 581.67. C 68.14%, H 6.07%, N 12.04%, O 13.75%. Vasoconstrictor found in ergot of Central Europe. Extraction procedure: Stoll, *Helv. Chim. Acta* **28**, 1283 (1945). Pharmacology: E. Rothlin, *Schweiz. Med. Wochenschr.* **76**, 1254 (1946). Structure: Stoll *et al.*, *ibid.* **34**, 1544 (1951). Total synthesis: Hofmann *et al.*, *Experientia* **17**, 206 (1961). Stereochemistry: Hofmann *et al.*, *ibid.* **46**, 2306

(1963). Comprehensive description: B. Kreilgard, *Anal. Profiles Drug Subs.* **6**, 113-159 (1977). LC determn in tablets: U. R. Cieri, *J. Assoc. Off. Anal. Chem.* **70**, 538 (1987); GC/MS determn in plasma: N. Feng *et al.*, *J. Chromatogr.* **575**, 289 (1992). Bioavailablity and efficacy in migraine: V. Ala-Hurula, *Headache* **22**, 167 (1982). Review of clinical pharmacokinetics and treatment of headache: V. L. Perrin, *Clin. Pharmacokinet.* **10**, 334-352 (1985). Review of teratogenic risk: G. V. Raymond, *Teratology* **51**, 344-347 (1995).

Elongated prisms from benzene. Very hygroscopic. Darkens and dec on exposure to air, heat and light. Dec 212-214°. $[\alpha]_D^{20}$ −160° (chloroform). Sol in about 70 parts methanol, 150 parts acetone, 300 parts alcohol; freely sol in chloroform, pyridine, glacial acetic acid; moderately sol in ethyl acetate; slightly in benzene. Almost insol in water, petr ether. LD_{50} in mice, rats, rabbits (mg/kg): 62, 80, 3 i.v.; in cats: 11 s.c. (Rothlin).

Hydrochloride. $C_{33}H_{35}N_5O_5 \cdot HCl$. Rectangular plates from 90% alc, mp 212° (dec). Sol in water-alcohol mixtures; sparingly in water or alcohol.

Tartrate. [379-79-3] Ergomar; Ergostat; Gynergen; Lingraine. $(C_{33}H_{35}N_5O_5)_2 \cdot C_4H_6O_6$; mol wt 1313.43. Colorless crystals or white to yellowish-white crystalline powder, mp ~180° (dec). One gram dissolves in 3200 mL water; in the presence of a slight excess of tartaric acid 1 gram dissolves in about 500 mL of water. Slightly sol in alc. *Protect from heat and light.*

THERAP CAT: Antimigraine.

THERAP CAT (VET): Tartrate has been used as an oxytocic.

3719. Ergotaminine. [639-81-6] (5′α,8α)-12′-Hydroxy-2′-methyl-5′-(phenylmethyl)ergotaman-3′,6′,18-trione. $C_{33}H_{35}N_5O_5$; mol wt 581.67. C 68.14%, H 6.07%, N 12.04%, O 13.75%. An alkaloid isomeric with ergotamine, *q.v.* Isoln from ergot: Stoll, *Helv. Chim. Acta* **28**, 1283 (1945); Stoll *et al.*, US 2809920 (1957 to Saul & Co.); Tabor, Vining, *Can. J. Microbiol.* **3**, 55 (1957).

Crystallizes solvent-free, unlike ergotamine which tends to retain the solvent of crystn. Thin rhombic plates from methanol, dec 241-243°. $[\alpha]_D^{20}$ +369° (c = 0.5 in chloroform). Much less sol than ergotamine; sol in about 1000 parts boiling alcohol, 1500 parts boiling methanol; fairly sol in chloroform, pyridine, glacial acetic acid. Does not seem to form salts.

3720. Ergothioneine. [497-30-3] (αS)-α-Carboxy-2,3-dihydro-*N,N,N*-trimethyl-2-thioxo-1*H*-imidazole-4-ethanaminium inner salt; [1-carboxy-2-[2-mercaptoimidazol-4-yl]ethyl]trimethylammonium hydroxide inner salt; L(+)-ergothioneine; thioneine; thiolhistidine-betaine; thiasine; sympectothion. $C_9H_{15}N_3O_2S$; mol wt 229.30. C 47.14%, H 6.59%, N 18.33%, O 13.95%, S 13.98%. First discovered in the sclerotia of the ergot fungus, *Claviceps purpurea,*: Tanret, *J. Pharm. Chim.* **30**, 145 (1909). Has since been found to occur in blood, semen and various mammalian tissues, principally liver and kidneys. Biosynthesis from histidine by *Claviceps purpurea* cultures: Heath, Wildy, *Nature* **179**, 196 (1957); also produced by *Neurospora crassa:* Melville *et al.*, *Fed. Proc.* **15**, 314 (1956). Chemical synthesis: Heath *et al.*, *J. Chem. Soc.* **1951**, 2215. Occurrence in the *Linulus polyphemus* L. (king crab): Ackermann, List, *Z. Physiol. Chem.* **313**, 30 (1958). *Review:* Melville, *Vitam. Horm.* **17**, 155 (1959).

Dihydrate. Needles or leaflets from dil ethanol, dec 256-257°. $[\alpha]_D^{20}$ +116.5°; $[\alpha]_D^{27}$ +115° (H_2O). uv max (water): 258 nm (ε 16000). One gram dissolves in 5 ml at 25°, much more sol in hot water. Slightly sol in hot methanol, hot ethanol, acetone. Practically insol in ether, chloroform, benzene.

3721. Ergotinine. [8006-08-4] Ergotoxinine. A 1:1:1 mixture of ergocornine, ergocristinine, and ergocryptinine. Isomeric with the ergotoxine mixture, *q.v.* First reported: Tanret, *Compt. Rend.* **81**, 891 (1875). Resolution into three alkaloids: Stoll, Hofmann, *Helv. Chim. Acta* **26**, 1570 (1943).

Crystallizes solvent-free, unlike ergotoxine which tends to retain the solvent of crystn. Long prisms from acetone. mp 229° with decompn. $[\alpha]_D^{20}$ +365° (c = 0.35 in chloroform). Very sol in chloroform. Sol in 25 parts acetone, 420 parts alcohol, 1000 parts abs ether.

3722. Ergotoxine. [8006-25-5] Ecboline. A mixture of ergocornine, ergocristine, and ergocryptine, *q.q.v.*, with oxytocic activity. *See also* the isomeric ergotinine mixture. Prepn: Barger, Carr, *J. Chem. Soc.* **91**, 337 (1907); Smith, Timmis, *ibid.* **1930**, 1390; Kofler, *Arch. Pharm.* **275**, 455 (1937); Stoll, Hofmann, *Helv. Chim. Acta* **26**, 1570 (1943); Pitra, Sapara, *Cesk. Farm.* **5**, 585 (1956), *C.A.* **51**, 8369f (1957). Toxicity data: H. Kreitmair, *Arch. Exp. Pathol. Pharmakol.* **176**, 171 (1934).

Orthorhombic crystals with $2C_6H_6$ from benzene. The solvent of crystn is given up only after long drying and heating in high vacuum. mp ~190° (dec) (solvent-free base). $[\alpha]_D^{20}$ −197° (chloroform). Very sol in ethyl and methyl alcohol, acetone, chloroform, ethyl acetate; slightly sol in ether. Almost insol in water, petr ether. LD in mice (mg/kg): 0.107 s.c.; 0.032 i.v. (Kreitmair).

Ergotoxine phosphate. Clusters of needles, dec 187°. Sol in 320 parts water, 15 parts boiling alcohol.

Ergotoxine ethanesufonate. Acicular crystals, dec 209°. Sol in methanol; slightly in alcohol. Almost insol in water. Solid and solns are sensitive to light and air.

3723. Eribulin. [253128-41-5] (2*R*,3*R*,3a*S*,7*R*,8a*S*,9*S*,10a*R*,11*S*,12*R*,13a*R*,13b*S*,15*S*,18*S*,21*S*,24*S*,26*R*,28*R*,29a*S*)-2-[(2*S*)-3-Amino-2-hydroxypropyl]hexacosahydro-3-methoxy-26-methyl-20,27-bis(methylene)-11,15:18,21:24,28-triepoxy-7,9-ethano-12,15-methano-9*H*,15*H*-furo[3,2-*i*]furo[2′,3′:5,6]pyrano[4,3-*b*][1,-4]dioxacyclopentacosin-5(4*H*)-one; ER-086526. $C_{40}H_{59}NO_{11}$; mol wt 729.91. C 65.82%, H 8.15%, N 1.92%, O 24.11%. Synthetic analog of the marine polyether macrolide, halichondrin B. Inhibits tumor cell proliferation by suppressing microtubule growth. Prepn: B. Littlefield *et al.*, WO 9965894 (1999); *eidem*, US 6214865 (2001 to Eisai); W. Zheng *et al.*, *Bioorg. Med. Chem. Lett.* **14**, 5551 (2004). Anticancer activities: M. J. Towle *et al.*, *Cancer Res.* **61**, 1013 (2001). Mechanism of action: M. A. Jordan *et al.*, *Mol. Cancer Ther.* **4**, 1086 (2005). LC/MS/MS determn in plasma and urine: C. DesJardins *et al.*, *J. Chromatogr. B* **875**, 373 (2008). Clinical evaluation in breast cancer: J. Cortes *et al.*, *J. Clin. Oncol.* **28**, 3922 (2010). Review of pharmacology and clinical experience: S. Newman, *Curr. Opin. Investig. Drugs* **8**, 1057-1066 (2007); in treatment of breast cancer: T. Cigler, L. T. Vahdat, *Expert Opin. Pharmacother.* **11**, 1587-1593 (2010).

Monomethanesulfonate. [441045-17-6] Eribulin mesylate; E-7389; Halaven. $C_{40}H_{59}NO_{11}.CH_3SO_3H$; mol wt 826.01.

THERAP CAT: Antineoplastic.

3724. Erigeron. Fleabane; horseweed. Leaves and tops of *Conyza canadensis* (L.) Cron. *(Erigeron canadensis* L.*), Compositae. Habit.* Northern and central U.S. *Constit.* Volatile oil, tannin, gallic acid.

3725. Eriochrome® Black T. [1787-61-7] 3-Hydroxy-4-[2-(1-hydroxy-2-naphthalenyl)diazenyl]-7-nitro-1-naphthalenesulfonic acid sodium salt (1:1); 3-hydroxy-4-[(1-hydroxy-2-naphthalenyl)-azo]-7-nitro-1-naphthalenesulfonic acid monosodium salt; C.I. Mordant Black 11; C.I. 14645. $C_{20}H_{12}N_3NaO_7S$; mol wt 461.38. C 52.07%, H 2.62%, N 9.11%, Na 4.98%, O 24.27%, S 6.95%. Prepn: Hagenbach, **US 790363** (1904 to Geigy); *Colour Index* **vol. 4** (3rd ed., 1971) p 4067.

Brownish-black powder with a faint metallic sheen. Sol in hot water giving a reddish-brown soln when cold. Violet-brown precipitate with excess HCl. Deep blue, then red in aq soln of NaOH. Sol in concd H_2SO_4 giving a blackish-blue soln which yields a brown precipitate on dilution.

USE: To dye wool from an acid bath reddish-black, which can be converted to blue-black by afterchroming. As indicator for complexometric titrations and in the determn of the total calcium and magnesium content of water.

3726. Eriodictyol. [552-58-9] (2S)-2-(3,4-Dihydroxyphenyl)-2,3-dihydro-5,7-dihydroxy-4H-1-benzopyran-4-one; 3',4',5,7-tetrahydroxyflavanone. $C_{15}H_{12}O_6$; mol wt 288.26. C 62.50%, H 4.20%, O 33.30%. Isoln from *Eriodictyon californicum* (H. & A.) Greene, *Hydrophyllaceae*: Geissman, *J. Am. Chem. Soc.* **62**, 3258 (1940); from lemon: Mager, *Z. Physiol. Chem.* **274**, 109 (1942); Horowitz, *J. Am. Chem. Soc.* **79**, 6561 (1957); **US 2857318** (1958 to U.S. Dept. Agr.). Structure and synthesis: Reichel *et al., Ann.* **550**, 146 (1942); Zemplén *et al., Ber.* **76B**, 1112 (1943); Pew, *J. Org. Chem.* **27**, 2935 (1962). Synthesis: G. Wurm, U. Geres, *Arch. Pharm.* **315**, 183 (1982). *See also* Bioflavanoids.

Needles with 1½ H_2O from dil alc, dec 257° (rapid heating). After drying in vacuo at 100° for 6 hours, dec 267°. uv max (alc): 290, 326 nm (log ε 2.54, 2.16). Sparingly sol in boiling water, hot alcohol, ether, glacial acetic acid; sol in dil alkalies.

7-L-Rhamnoside. Eriodictin. $C_{21}H_{22}O_{10}$. Isoln from citrin: Bruckner, Szent-Györgyi, *Nature* **138**, 1057 (1936). Structure: Mager, *loc. cit.* Crystals from ethyl acetate, dec 184-186°. $[\alpha]_D^{20}$ −51.5° (in pyridine). Sol in water, acetone, alcohol; practically insol in ether.

3727. Eriodictyon. Yerba santa; consumptive's weed; bear's weed; mountain balm; gum plant. Dried leaves of *Eriodictyon californicum* (H. & A.) Greene, *Hydrophyllaceae. Habit.* U.S. (California). *Constit.* Volatile oil, eriodictyol, homoeriodictyol, chrysoeriodictyol, xanthoeriodictyol, eriodonol, eriodictyonic acid, ericolin, resin.

USE: Pharmaceutic aid (flavor).

3728. Eritadenine. [23918-98-1] (αR,βR)-6-Amino-α,β-dihydroxy-9H-purine-9-butanoic acid; 2(R),3(R)-dihydroxy-4-(9-adenyl)butyric acid; 4-(6-amino-9H-purin-9-yl)-4-deoxy-D-erythronic acid; lentinacin; lentysine. $C_9H_{11}N_5O_4$; mol wt 253.22. C 42.69%, H 4.38%, N 27.66%, O 25.27%. Hypocholesterolemic principle isolated from the Shiitake mushroom, *Lentinus edodes* Sing. Isoln, activity and structure: I. Chibata *et al., Experientia* **25**, 1237 (1969). Synthesis: T. Kamiya *et al., J. Heterocycl. Chem.* **9**, 359 (1972). GC/MS determn in mushroom extracts: G. Vitányi *et al., Rapid Commun. Mass Spectrom.* **12**, 120 (1998). Effect on phospholipid metabolism: K. Sugiyama *et al., J. Nutr.* **125**, 2134 (1995); on linoleic acid metabolism: *idem et al., Lipids* **32**, 859 (1997).

Crystals from 10% acetic acid, mp 278-279° (dec). $[\alpha]_D$ +51.4° (c = 1.6 in 0.1N NaOH). uv max (water): 261.5 nm (ε 14508).

Sodium salt. $C_9H_{10}N_5NaO_4$. Crystals from 50% ethanol, mp 275° (dec). $[\alpha]_D^{20}$ +45.5° (c = 1 in H_2O). IR, NMR: Chibata, *loc. cit.*

3729. Eritoran. [185955-34-4] 3-O-Decyl-2-deoxy-6-O-[2-deoxy-3-O-[(3R)-3-methoxydecyl]-6-O-methyl-2-[[(11Z)-1-oxo-11-octadecen-1-yl]amino]-4-O-phosphono-β-D-glucopyranosyl]-2-[[(1,3-dioxotetradecyl)amino]-α-D-glucopyranose 1-(dihydrogen phosphate). $C_{66}H_{126}N_2O_{19}P_2$; mol wt 1313.68. C 60.34%, H 9.67%, N 2.13%, O 23.14%, P 4.72%. Toll-like receptor 4 (TLR4) antagonist. Synthetic analog of the lipid A component of endotoxin, a major constituent of the outer membrane of Gram negative bacteria. Prepn: W. J. Christ *et al., WO 9639411; eidem,* **US 5750664** (1996, 1998 both to Eisai). Pharmacology: M. Mullarkey *et al., J. Pharmacol. Exp. Ther.* **304**, 1093 (2003). Clinical pharmacokinetics: Y. N. Wong *et al., J. Clin. Pharmacol.* **43**, 735 (2003); and lipid distribution profile: D. P. Rossignol *et al., Antimicrob. Agents Chemother.* **48**, 3233 (2004). Clinical evaluation in experimental endotoxemia: M. Lynn *et al., J. Pharmacol. Exp. Ther.* **308**, 175 (2004). Review of pharmacology and clinical experience: D. P. Rossignol, M. Lynn, *Curr. Opin. Investig. Drugs* **6**, 496-502 (2005).

Tetrasodium salt. [185954-98-7] E-5564; B-1287. $C_{66}H_{122}N_2Na_4O_{19}P_2$; mol wt 1401.60.

THERAP CAT: In treatment of severe sepsis.

3730. Erlotinib. [183321-74-6] N-(3-Ethynylphenyl)-6,7-bis(2-methoxyethoxy)-4-quinazolinamine. $C_{22}H_{23}N_3O_4$; mol wt 393.44. C 67.16%, H 5.89%, N 10.68%, O 16.27%. Selective epidermal growth factor receptor (EGFR)-tyrosine kinase inhibitor. Prepn: R. C. Schnur, L. D. Arnold, **WO 9630347**; *eidem*, US **5747498** (1996, 1998 both to Pfizer). Mechanism of action study: J. D. Moyer *et al.*, *Cancer Res.* **57**, 4838 (1997). Enzyme inhibition and antitumor activity: V. A. Pollack *et al.*, *J. Pharmacol. Exp. Ther.* **291**, 739 (1999). HPLC determn in plasma: E. R. Lepper *et al.*, *J. Chromatogr. B* **796**, 181 (2003). Clinical pharmacokinetics: M. Hidalgo *et al.*, *J. Clin. Oncol.* **19**, 3267 (2001). Clinical trial in non-small-cell lung cancer: F. Cappuzzo *et al.*, *Lancet* **11**, 521 (2010). Review of pharmacology and clinical development: T. E. Kim, J. R. Murren, *Curr. Opin. Investig. Drugs* **3**, 1385-1395 (2002); and use in non-small cell lung cancer: F. H. Blackhall *et al.*, *Expert Opin. Pharmacother.* **6**, 995-1002 (2005).

Hydrochloride. [183319-69-9] CP-358774; OSI-774; Tarceva. $C_{22}H_{23}N_3O_4 \cdot HCl$; mol wt 429.90. mp 228-230°. pKa (25°): 5.42. Soly in water (pH ~2): ~0.4 mg/ml. Slightly sol in methanol. Practically insol in acetonitrile, acetone, ethyl acetate, hexane.

THERAP CAT: Antineoplastic.

3731. Ertapenem. [153832-46-3] [4R,5S,6S]-3-[[(3S,5S)-5-[[(3-Carboxyphenyl)amino]carbonyl]-3-pyrrolidinyl]thio]-6-[(1R)-1-hydroxyethyl]-4-methyl-7-oxo-1-azabicyclo[3.2.0]hept-2-ene-2-carboxylic acid; (1R,5S,6S,8R,2'S,4'S)-2-(2-(3-carboxyphenylcarbamoyl)pyrrolidin-4-ylthio)-6-(1-hydroxyethyl)-1-methylcarbapenem-3-carboxylic acid; ZD-443. $C_{22}H_{25}N_3O_7S$; mol wt 475.52. C 55.57%, H 5.30%, N 8.84%, O 23.55%, S 6.74%. Group I carbapenem antibiotic. Prepn: M. J. Betts *et al.*, **WO 9315078**; *eidem*, **US 5478820** (1993, 1995 both to Zeneca). Large-scale synthesis: J. M. Williams *et al.*, *J. Org. Chem.* **70**, 7479 (2005). Pharmacokinetics in primates: J. G. Sundelof *et al.*, *Antimicrob. Agents Chemother.* **41**, 1743 (1997). *In vivo* efficacy and pharmacokinetics: C. J. Gill *et al.*, *ibid.* **42**, 1996 (1998). *In vitro* efficacy and β-lactamase stability: J. Kohler *et al.*, *ibid.* **43**, 1170 (1999). Stability in aq solns: M. Zajac *et al.*, *J. Pharm. Biomed. Anal.* **43**, 445 (2007). HPLC determn in plasma and urine: D. G. Musson *et al.*, *J. Chromatogr. B* **720**, 99 (1998); of residual crystallization solvents: T. K. Natishan, Y. Wu, *J. Chromatogr. A* **800**, 275 (1998). Series of articles on pharmacology and clinical experience in complicated community-acquired infections: *J. Antimicrob. Chemother.* **53**, Suppl. S2, ii1-ii86 (2004). Review of clinical experience: O. Burkhardt *et al.*, *Expert Opin. Pharmacother.* **8**, 237-256 (2007).

Sodium salt. [153773-82-1] MK-826; Invanz. $C_{22}H_{24}N_3NaO_7S$; mol wt 497.50. White to off-white hygroscopic, weakly crystalline powder. Sol in water, 0.9% sodium chloride soln. Practically insol in ethanol; insol in isopropyl acetate, THF.

THERAP CAT: Antibacterial.

3732. Erucic Acid. [112-86-7] (13Z)-13-Docosenoic acid; Δ^{13}-*cis*-docosenoic acid. $C_{22}H_{42}O_2$; mol wt 338.58. C 78.04%, H

12.50%, O 9.45%. A monoethenoid acid found in the seed fats of *Cruciferae* and *Tropaeolaceae*. It constitutes 40 to 50% of the total fatty acids of rapeseed, mustard and wallflower seed, and it represents up to 80% of fatty acids of nasturtium seeds, *cf.* K. S. Markley, *Fatty Acids* Part I (Interscience, New York, 2nd ed., 1960) p 138-139. Prepn of a crude product by alkaline hydrolysis of rapeseed oil: Noller, Talbot, *Org. Synth.* **coll. vol. II**, 258 (1943). A purer product is obtained by fractional precipitation and crystallization: Dorée, Pepper, *J. Chem. Soc.* **1942**, 477. Prepn of a pure product by acid soap crystallization: Chobanov *et al.*, *Chem. Ind. (London)* **1965**, 606. Synthesis: Bowman, *J. Chem. Soc.* **1950**, 177, 325; Bounds *et al.*, *ibid.* **1953**, 2393. Treatment with nitric acid yields the *trans* isomer, brassidic acid, *q.v.*: Dorée, Pepper, *loc. cit.* Brief review: E. Lower, *Manuf. Chem.* **56**(6), 61-63 (1985).

Needles from alcohol, mp 33.8°. Iodine value 74.98; neutralization value 165.72. d_4^{55} 0.860. bp_{760} 381.5° (decompn); bp_{400} 358.8°; bp_{100} 314.4°; bp_{60} 300.2°; bp_{20} 270.6°; bp_5 239.7°; $bp_{1.0}$ 206.7°. n_D^{45} 1.4534; n_D^{65} 1.44794. Insol in water. About 175 g dissolve in 100 ml ethanol and about 160 g dissolve in 100 ml methanol. Very sol in ether.

3733. Erythritol. [149-32-6] *rel*-(2R,3S)-1,2,3,4-Butanetetrol; *meso*-erythritol; tetrahydroxybutane; erythrol; erythroglucin; phycite. $C_4H_{10}O_4$; mol wt 122.12. C 39.34%, H 8.25%, O 52.40%. All natural bulk sweetener found in various fruits and fermented foods; 60 to 70% as sweet as sucrose. Isoln from algae, lichens, grasses: Bamberger, Landsiedl, *Monatsh. Chem.* **21**, 571 (1900); Hesse, *J. Prakt. Chem.* **92**, 425 (1915); Hofmann, *Ber.* **7**, 508 (1874). Prepn by *Aspergillus niger*: Yuill, *Nature* **162**, 652 (1948); by *Penicillium herquei*: Galarraga *et al.*, *Biochem. J.* **61**, 456 (1955); from 2-butene-1,4-diol: Reppe, Schnabel, **DE 734025** (1943 to I. G. Farbenind.); from periodate-oxidized starch: Jeanes, Hudson, *J. Org. Chem.* **20**, 1565 (1955). Structure: Shimada, *Acta Crystallogr.* **11**, 748 (1958). Review of metabolism, toxicology and clinical safety studies: I. C. Munro *et al.*, *Food Chem. Toxicol.* **36**, 1139-1174 (1998); of use in beverage industry: P. de Cock, C.-L. Bechert, *Pure Appl. Chem.* **74**, 1281-1289 (2002).

Relative stereochemistry

Tetragonal prisms, mp 121.5°. About twice as sweet as sucrose. bp 329-331°. Very sol in water (satd soln contains about 61% w/w); sol in pyridine (satd soln contains 2.5% w/w); slightly sol in alcohol. Practically insol in ether. pKa (18°): 13.903. LD$_{50}$ in male, female rats (g/kg): 6.6, 9.6 i.v.; >16, >16 s.c.; 13.1, 13.5 orally (Munro).

USE: Non-nutritive sweetener in beverages.

3734. Erythritol Anhydride. [564-00-1] *rel*-(2R,2'S)-2,2'-Bioxirane; *meso*-1,2:3,4-diepoxybutane; *meso*-1,3-butadiene diepoxide; *meso*-diepoxybutane. $C_4H_6O_2$; mol wt 86.09. C 55.81%, H 7.03%, O 37.17%. Prepn from erythrityl chlorohydrin: Przybytek, *Ber.* **17**, 1091 (1884); by refluxing 1,4-dichloro-2,3-butanediol in tetrahydrofuran with NaOH: Reppe, *Ann.* **596**, 141 (1955). Prepn of *meso*-form from 1,4-dihydroxy-2-butene or from 3,4-epoxy-1-butene and *dl*-form from 1,4-dibromo-2-butene: Beech, *J. Chem. Soc.* **1951**, 2483. Use to prevent microbial spoilage: H. D. Michener, J. C. Lewis, **US 2934439** (1960 to U.S. Secy of Agr.). Toxicity data: H. F. Smyth *et al.*, *Arch. Ind. Hyg. Occup. Med.* **10**, 61 (1954).

Relative stereochemistry

Liquid. mp $-19°$. d_4^{18} 1.113. bp$_{761}$ 140-142°; bp$_{23}$ 50-51°. *Flammable.* Misc with water which hydrolyzes it to erythritol. LD$_{50}$ orally in rats: 0.078 g/kg (Smyth).

DL-Form. [298-18-0] *rel*-(2R,2'R)-2,2'-Bioxirane; *dl*-1,2:3,4-diepoxybutane; DL-diepoxybutane. mp 4°. bp$_{35}$ 58-60°.

Caution: This substance is reasonably anticipated to be a human carcinogen: *Report on Carcinogens, Twelfth Edition* (PB2011-111646, 2011) p 152.

USE: Curing polymers; crosslinking textile fibers.

3735. Erythrityl Tetranitrate. [7297-25-8] *rel*-(2R,3S)-1,2,-3,4-Butanetetrol 1,2,3,4-tetranitrate; erythritol tetranitrate; erythrol tetranitrate; tetranitrol; tetranitrin; nitroerythrite; Cardilate; Cardiloid. $C_4H_6N_4O_{12}$; mol wt 302.11. C 15.90%, H 2.00%, N 18.55%, O 63.55%. Made by nitration of erythritol.

Relative stereochemistry

Leaflets from alcohol, mp 61°. Sol in alcohol, ether, glycerol. Insol in water. Reduces Fehling's soln. *Explodes on percussion.* Pharmaceutical preparations are mixed with carbohydrate substances, such as lactose, and formulated into tablets which are nonexplosive.

THERAP CAT: Vasodilator (coronary).

THERAP CAT (VET): Vasodilator.

3736. Erythrocentaurin. [50276-98-7] 3,4-Dihydro-1-oxo-1H-2-benzopyran-5-carboxaldehyde; 5-formyl-3,4-dihydroisocoumarin; 5-formyl-3,4-dihydro-1H-2-benzopyran-1-one. $C_{10}H_8O_3$; mol wt 176.17. C 68.18%, H 4.58%, O 27.24%. From *Centaurium umbellatum* Gilib., *(Erythraea centaurium* Pers.), *Gentianaceae* or *Swertia japonica* (Maxim.) Makino, *Gentianaceae.* By hydrolysis of swertiamarin and erytaurin with emulsin. Isoln: Kariyone, Matsushima, *J. Pharm. Soc. Jpn.* **47**, 25 (1927). Structure: Kubota, Tomita, *Chem. Ind. (London)* **1958**, 230; Kubota *et al., Tetrahedron Lett.* **1961**, 223. Synthesis: Wenkert *et al., J. Org. Chem.* **29**, 2534 (1964).

Long needles, mp 140-141°. Turns red on exposure to sunlight. uv max: 223, 290 nm (log ε 4.30, 3.13).

THERAP CAT: Bitter tonic.

3737. α-Erythroidine. [466-80-8] (2R,9aS,13bS)-2,6,8,9,-9a,10-Hexahydro-2-methoxy-1H,12H-benzo[i]pyrano[3,4-g]indolizin-12-one; (3β,12β)-1,2,6,7-tetradehydro-12,17-dihydro-3-methoxy-16(15H)-oxaerythrinan-15-one. $C_{16}H_{19}NO_3$; mol wt 273.33. C 70.31%, H 7.01%, N 5.12%, O 17.56%. Isoln from *Erythrina* spp, *Leguminosae:* Folkers, Major, US 2373952 (1945 to Merck & Co.); Boekelheide, Grundon, *J. Am. Chem. Soc.* **75**, 2563 (1953). Structure: Godfrey *et al., ibid.* **77**, 3342 (1955). Absolute configuration: Hill, Shearer, *J. Org. Chem.* **27**, 921 (1962); Wenzinger, Boekelheide, *Proc. Chem. Soc. London* **1963**, 53. Conversion of α- to β-erythroidine: Boekelheide, Morrison, *J. Am. Chem. Soc.* **80**, 3905 (1958). Biosynthesis studies: Leete, Ahmad, *ibid.* **88**, 4722 (1966).

Needles from pentane, mp 58-60°. $[\alpha]_D^{27}$ +136° (c = 0.5 in water). Unstable on exposure to air.

Hydrochloride. $C_{16}H_{20}ClNO_3$. Prisms from ethanol, dec 226-228°. $[\alpha]_D^{32}$ +118° (c = 0.5 in water). uv max (ethanol): 224 nm (log ε 4.55).

Methiodide. $C_{17}H_{22}INO_3$. Yellow prisms from ethanol, mp 219-220°.

3738. β-Erythroidine. [466-81-9] (2R,13bS)-2,6,8,9,10,13-Hexahydro-2-methoxy-1H,12H-pyrano[4',3':3,4]pyrido[2,1-i]indol-12-one; (3β)-1,2,6,7-tetradehydro-14,17-dihydro-3-methoxy-16(15H)-oxaerythrinan-15-one; 12,13-didehydro-13,14-dihydro-α-erythroidine. $C_{16}H_{19}NO_3$; mol wt 273.33. C 70.31%, H 7.01%, N 5.12%, O 17.56%. Skeletal muscle relaxant. Isolated from the seeds and other plant parts of *Erythrina* spp., *Leguminosae:* Folkers, Major, *J. Am. Chem. Soc.* **59**, 1580 (1937); *eidem*, US 2373952; US 2385266; US 2407713 (1945 and 1946 to Merck & Co.). Structural studies: Koniuszy, Folkers, *J. Am. Chem. Soc.* **72**, 5519 (1950); Boekelheide *et al., ibid.* **75**, 2550 (1953). Absolute configuration: Wenzinger, Boekelheide, *Proc. Chem. Soc. London* **1963**, 53; *eidem, J. Org. Chem.* **29**, 1307 (1964). Toxicity data: F. M. Berger, R. P. Schwartz, *J. Pharmacol. Exp. Ther.* **93**, 362 (1948). *Review:* Boekelheide, *Rec. Chem. Prog.* **16**, 227-239 (1955).

Crystals from abs ethanol, mp 99.5-100°. $[\alpha]_D^{25}$ +88.8°. Sol in water, benzene, chloroform, methanol, ethanol; moderately sol in diethyl ether. Reacts with NaOH to form sodium erythroidinate. LD$_{50}$ i.p. in mice: 24.0 mg/kg (Berger, Schwartz).

Hydrochloride. $C_{16}H_{19}NO_3$·HCl. Small needles from abs ethanol, mp 232° (dec). $[\alpha]_D^{25}$ +10°. Unstable in air and light. Bitter taste. Sol in water, benzene, chloroform, methanol, ethanol; moderately sol in diethyl ether. *Incompat.* with oxidizing agents, alkaline solns, and akaloidal reagents.

Hydrochloride hemihydrate. $C_{16}H_{19}NO_3$·HCl·½H$_2$O. Small needles from abs ethanol, mp 229.5-230° (dec). $[\alpha]_D$ +95°. Has curare-like action which is antagonized by prostigmine.

Hydrobromide. $C_{16}H_{19}NO_3$·HBr. Crystals from abs ethanol, mp 222.5°. $[\alpha]_D^{25}$ +111.2°.

Hydriodide. $C_{16}H_{19}NO_3$·HI. Crystals from abs ethanol, mp 206°. $[\alpha]_D^{25}$ +108.1°.

Methiodide. $C_{16}H_{19}NO_3$·CH$_3$I. White prisms from alcohol, mp 211°.

3739. Erythromycin. [114-07-8] E-Base; E-Mycin; Erythromycin A; Aknemycin; Aknin; Emgel; Ery-Derm; Erymax; Ery-Tab; Erythromid; ERYC; Erycen; Erycin; Erycinum; Ermysin; Gallimycin; Ilotycin; Inderm; PCE; Retcin; Staticin; Stiemycin. $C_{37}H_{67}NO_{13}$; mol wt 733.94. C 60.55%, H 9.20%, N 1.91%, O 28.34%. Antibiotic substance produced by a strain of *Streptomyces erythreus* (Waksman) Waksman & Henrici, found in a soil sample from the Philippine Archipelago. Isoln: McGuire *et al., Antibiot. Chemother.* **2**, 281 (1952); Bunch, McGuire, US 2653899 (1953 to Lilly); Clark, Jr., US 2823203 (1958 to Abbott). Properties: Flynn *et al., J. Am. Chem. Soc.* **76**, 3121 (1954). Solubility data: Weiss *et al., Antibiot. Chemother.* **7**, 374 (1957). Structure: Wiley *et al., J. Am. Chem. Soc.* **79**, 6062 (1957). Configuration: Hofheinz, Grisebach, *Ber.* **96**, 2867 (1963); Harris *et al., Tetrahedron Lett.* **1965**, 679. There are three erythromycins produced during fermentation, designated A, B, and C; A is the major and most important component. Erythromycins A and B contain the same sugar moieties, desosamine, *q.v.*, and cladinose (3-O-methylmycarose). They differ in position 12 of the aglycone, erythronolide, A having an hydroxyl substituent. Component C contains desosamine and the same aglycone present in A but differs by the presence of mycarose, *q.v.*, instead of cladinose. Structure of B: P. F. Wiley *et al., J. Am. Chem. Soc.* **79**, 6074 (1957); of C: *eidem, ibid.* 6074. Synthesis of the aglycone, erythronolide B: E. J. Corey *et al., ibid.* **100**, 4618, 4620 (1978); of erythronolide A: *eidem, ibid.* **101**, 7131 (1979). Asymmetric total synthesis of erythromycin A: R. B. Woodward *et al.*, *ibid.* **103**, 3215 (1981). NMR spectrum of A: D. J. Ager, C. K. Sood, *Magn. Reson. Chem.* **25**, 948 (1987). HPLC determn in plasma: W. Xiao *et al.*, *J. Chromatogr. B* **817**, 153 (2005). Biosynthesis: Martin, Goldstein, *Prog. Antimicrob. Anticancer Chemother., Proc. 6th Int. Congr. Chemother.* **II**, 1112 (1970); Martin *et al.*, *Tetrahedron* **31**, 1985 (1975). Cloning and expression of clustered biosynthetic genes: R. Stanzak *et al.*, *Biotechnology* **4**, 229 (1986). *Reviews:* T. J. Perun in *Drug Action and Drug Resistance in Bacteria* **1**, S. Mitsuhashi, Ed. (University Park Press, Baltimore, 1977) pp 123-152; Oleinick in *Antibiotics* **vol. 3**, J. W. Corcoran, F. E. Hahn, Eds. (Springer-Verlag, New York, 1975) pp 396-419; *Infection* **10**, Suppl. 2, S61-S118 (1982). Comprehensive description: W. L. Koch, *Anal. Profiles Drug Subs.* **8**, 159-177 (1979).

Erythromycin A

Hydrated crystals from water, mp 135-140°, resolidifies with second mp 190-193°. Melting point taken after drying at 56° and 8 mm. $[\alpha]_D^{25}$ $-78°$ (c = 1.99 in ethanol). uv max (pH 6.3): 280 nm (ε 50). pKa$_1$ 8.8. Basic reaction. Readily forms salts with acids. Soly in water: ~2 mg/ml. Freely sol in acetone, acetonitrile, ethyl acetate. Sol in alc, chloroform, ether. Moderately sol in ethylene dichloride, amyl acetate.

Ethylsuccinate. [41342-53-4] Anamycin; Arpimycin; E.E.S.; Eritrocina; Eryliquid; Eryped; Erythroped; Esinol; Monomycin; Paediathrocin; Pediamycin; Refkas. $C_{43}H_{75}NO_{16}$; mol wt 862.06. Prepn: **GB 830846**; R. K. Clark, **US 2967129** (1960, 1961 both to Abbott). Hydrated crystals from acetone + water, mp 109-110°. $[\alpha]_D$ $-42.5°$. Freely soluble in alc, chloroform, polyethylene glycol 400; very slightly sol in water.

THERAP CAT: Antibacterial.
THERAP CAT (VET): Antibacterial.

3740. Erythromycin Estolate. [3521-62-8] Erythromycin 2′-propanoate dodecyl sulfate (1:1); erythromycin propionate lauryl sulfate; lauryl sulfate salt of the propionic acid ester of erythromycin; propionylerythromycin lauryl sulfate; Eritroger; Eromycin; Ilosone; Lauromicina; Neo-Erycinum; PELS; Roxomicina; Stellamicina; Eriscel; Estomicina; Eupragin; Marcoeritrex; Togiren. $C_{52}H_{97}NO_{18}S$; mol wt 1056.40. C 59.12%, H 9.26%, N 1.33%, O 27.26%, S 3.03%. $C_{40}H_{71}NO_{14} \cdot CH_3(CH_2)_{11}OSO_3H$. Prepn from propionyl erythromycin: Stephens *et al.*, *J. Am. Pharm. Assoc.* **48**, 620 (1959); Bray, Stephens, **US 3000874** (1961 to Lilly). Toxicity: E. I. Goldenthal, *Toxicol. Appl. Pharmacol.* **18**, 185 (1971). Comprehensive description: J. M. Mann, *Anal. Profiles Drug Subs.* **1**, 101-117 (1972). For structure see Erythromycin.

Long needles, mp 135-140° (dec). pKa 6.9. Soly in water: 0.024 mg/ml; sol in alcohol, acetone, chloroform. LD$_{50}$ orally in rats: >5000 mg/kg (Goldenthal).

THERAP CAT: Antibacterial.

3741. Erythromycin Lactobionate. [3847-29-8] Erythromycin 4-*O*-β-D-galactopyranosyl-D-gluconate (1:1); Erythrocin Lactobionate. $C_{49}H_{89}NO_{25}$; mol wt 1092.23. C 53.88%, H 8.21%, N 1.28%, O 36.62%. Semi-synthetic macrolide antibiotic prepd from erythromycin base and lactobiono-δ-lactone in water-acetone: Hoffhine, **US 2761859** (1956 to Abbott); alternate prepn and antibacterial activity: S. K. Dutta, K. S. Basu, **US 4137397** (1979 to Jadavpur

Univ.). Pharmacokinetics: R. L. Parsons *et al.*, *J. Int. Med. Res.* **8**, suppl. 2, 15 (1980).

White, amorphous powder, mp 145-150°. Soly in water about 200 mg/ml. Freely sol in alcohol, methanol; slightly sol in acetone, chloroform. Practically insol in ether. pH of a 2% aq soln 6.0-7.5. A 5% stock soln can be prepd with water or glucose soln but not with NaCl isotonic soln, because NaCl at this concn precipitates erythromycin base. Lower concns of inorganic salts are compatible.

THERAP CAT: Antibacterial.
THERAP CAT (VET): Antibacterial.

3742. Erythromycin Propionate. [134-36-1] Erythromycin 2′-propanoate; propionylerythromycin; monopropionylerythromycin; Propiocine. $C_{40}H_{71}NO_{14}$; mol wt 790.00. C 60.82%, H 9.06%, N 1.77%, O 28.35%. Semi-synthetic macrolide antibiotic. Prepn: V. C. Stephens, **US 2993833** (1961 to Lilly); V. C. Stephens, J. W. Conine, *Antibiot. Annu.* **1958-59**, 346. Pharmacology and toxicology: C. Lee *et al.*, *ibid.* 354. Clinical studies: R. S. Griffith, *ibid.* 364; D. M. Perry *et al.*, *ibid.* 375. Clinical pharmacokinetics: M. Ducci *et al.*, *Int. J. Clin. Pharmacol. Ther. Toxicol.* **19**, 494 (1981). HPLC determn in plasma: W. Xiao *et al.*, *J. Chromatogr. B* **817**, 153 (2005).

Monohydrate. Crystals from acetone + water, mp 122-126°. $[\alpha]_D^{25}$ $-81.6°$ (acetone). pKa: 6.9. Very slightly sol in water. Readily sol in methanol, ethanol, acetone, ethyl acetate, DMF. LD$_{50}$ in mice, rats (g/kg): 2.87, >5.0 orally; >5.0, >5.0 s.c. (Lee).

THERAP CAT: Antibacterial.

3743. Erythromycin Stearate. [643-22-1] Erythromycin octadecanoate (1:1); Abboticine; Bristamycin; Dowmycin E; Erypar; Eryprim; Erythro S; Erythrocin; Ethril; Ethryn; Gallimycin; Meberyt; Pantomicina; Pfizer-E; Wemid. $C_{55}H_{103}NO_{15}$; mol wt 1018.42. C 64.87%, H 10.19%, N 1.38%, O 23.56%. Semi-synthetic macrolide antibiotic. Prepn: O. F. Walasek, **US 2881163** (1959 to Abbott). Bioavailability: A.-S. Malmborg, *J. Antimicrob. Chemother.* **5**, 591 (1979). Pharmacokinetics and tolerance: D. C. Shanson *et al.*, *ibid.* **14**, 157 (1984). Clinical efficacy in respiratory tract infections: J. P. Butler *et al.*, *Chemotherapy (Basel)* **25**, 367 (1979); in acute otitis media: C. Rosen *et al.*, *Acta Otolaryngol. (Stockholm)* **Suppl. 407**, 23 (1984). Prepn of erythromycin stearate (ester): Booth, Murray, **US 2862921** (1958 to Upjohn).

White or slightly yellow crystals or powder. Slightly bitter taste. Soluble in methanol, alc, ether, chloroform. Practically insol in water.

THERAP CAT: Antibacterial.
THERAP CAT (VET): Antibacterial.

3744. Erythropoietin. [11096-26-7] Erythropoiesis stimulating factor; hemopoietine; ESF; Ep; Epo. Glycoprotein hormone which stimulates red blood cell formation in higher organisms: P. Carnot, C. Deflandre, *Compt. Rend.* **143**, 384 (1906). Single chain, 165 amino acid polypeptide containing approx 40% carbohydrate; mol wt ~30 kDa. Produced in response to hypoxia, primarily in the kidneys and to a lesser extent in the liver. Occurs normally in low concentrations in plasma; elevated levels in urine and plasma under conditions of anemic or hypoxic stress. Human erythropoeitin produced by recombinant technology is known as *epoetin* or *rHuEPO*; several glycoforms have been produced. Isoln from sheep plasma: W. F. White *et al.*, *Recent Prog. Horm. Res.* **16**, 219 (1960); from human urine: J. Espada, A. Gutnisky, *Biochem. Med.* **3**, 475 (1970). Purification from human urine: T. Miyake *et al.*, *J. Biol. Chem.* **252**, 5558 (1977). Amino acid sequence: P.-H. Lai *et al.*, *J. Biol. Chem.* **261**, 3116 (1986). Biogenesis and control of production: J. W. Fisher, *Proc. Soc. Exp. Biol. Med.* **173**, 289 (1983). Production by recombinant DNA technology in *E. coli*: S. Lee-Huang, *Proc. Natl. Acad. Sci. USA* **81**, 2708 (1984); in mammalian cells: K. Jacobs *et al.*, *Nature* **313**, 806 (1985); F.-K. Lin *et al.*, *Proc. Natl. Acad. Sci. USA* **82**, 7580 (1985). Review of chemistry, pharmacology and clinical efficacy of recombinant human Epo: K. K. Flaharty *et al.*, *Clin. Pharm.* **8**, 769-782 (1989). Review of clinical screening methods and abuse potential in sports: G. Lippi, G. Guidi, *Clin. Chem. Lab. Med.* **38**, 13-19 (2000). Review of clinical experience: K.-U. Eckardt, *Nephrol. Dial. Transplant.* **16**, 1745-1749 (2001); D. H. Henry, *Expert Opin. Pharmacother.* **6**, 295-310 (2005).

Epoetin alfa. [113427-24-0] 1-165-Erythropoietin (human clone λHEPOFL13 protein moiety), glycoform α; Binocrit; Epogen;

Eprex; Erypo; Procrit. Human Epo produced by recombinant technology in Chinese hamster ovary cells. Review of pharmacology and clinical experience: T. Littlewood, G. Collins, *Expert Rev. Anticancer Ther.* **5**, 947-956 (2005).

Epoetin beta. [122312-54-3] 1-165-Erythropoietin (human clone λHEPOFL13 protein moiety), glycoform β; Epogin; NeoRecormon; Recormon. Human Epo produced by recombinant technology in Chinese hamster ovary cell line DN2-323. Review of safety and efficacy in chemotherapy-induced anemia: D. Spaëth, *Expert Rev. Anticancer Ther.* **8**, 875-885 (2008).

Epoetin theta. [762263-14-9] 1-165 Erythropoietin (human), glycoform θ; Biopoin; Eporatio. Human Epo produced by recombinant technology. Clinical safety and efficacy in anemic hemodialysis patients: B. Gertz *et al.*, *Curr. Med. Res. Opin.* **26**, 2393 (2010).

Epoetin zeta. [604802-70-2] 1-165 Erythropoietin (human clone B03XA01); Retacrit; Silapo. Human Epo produced by recombinant technology. Clinical trial in chemotherapy-induced anemia: V. Tzekova *et al.*, *Curr. Med. Res. Opin.* **25**, 1689 (2009).

THERAP CAT: Hematopoietic.

THERAP CAT (VET): Hematopoietic.

3745. Erythropterin. [7449-03-8] 3-(2-Amino-4,5,6,8-tetrahydro-4,6-dioxo-7(3*H*)-pteridinylidene)-2-oxopropanoic acid; 2-amino-3,4,5,6-tetrahydro-4,6-dioxo-7-pteridinepyruvic acid. C_9H_7-N_5O_5; mol wt 265.19. C 40.76%, H 2.66%, N 26.41%, O 30.17%. Pigment responsible for the red, orange and yellow color spots on the wings of butterflies. Has been detected also in the integument of other insects. Isoln: C. Schöpf, E. Becker, *Ann.* **524**, 49 (1936). Structure: R. Tschesche, H. Ende, *Ber.* **91**, 2074 (1958). Revised structure: W. Pfleiderer, *ibid.* **95**, 2195 (1962). Synthesis: M. Viscontini, H. Stierlin, *Helv. Chim. Acta* **46**, 51 (1963). Correlation to body coloration: E. J. Pfeiler, Jr., *J. Res. Lepid.* **7**, 183 (1968); C. Melber, G. H. Schmidt, *Comp. Biochem. Physiol.* **116A**, 17 (1997). Quantification: *eidem, ibid.* **101B**, 115 (1992); Y. Bel *et al.*, *Arch. Insect Biochem. Physiol.* **34**, 83 (1997).

As the monohydrate, red microcrystals. Abs max (pH 1.0): 450 nm (log ε 4.02). Orange fluorescence from 254-365 nm.

3746. D-Erythrose. [583-50-6] (2*R*,3*R*)-2,3,4-Trihydroxybutanal. $C_4H_8O_4$; mol wt 120.10. C 40.00%, H 6.71%, O 53.29%. From calcium D-arabonate by oxidation with H_2O_2: Ruff, *Ber.* **32**, 3674 (1899); **33**, 1799 (1900). From D-glucose: Perlin, *Methods Carbohydr. Chem.* **1** (Academic Press, New York, 1962) p 64. Synthesis of DL-erythrose: Sonogashira, Nakagawa, *Bull. Chem. Soc. Jpn.* **45**, 2616 (1972).

Syrup. Shows mutarotation. $[\alpha]_D^{20}$ +1° → −14.5° (3 days, c = 11). Sol in water. Slowly reduces cold Fehling's soln. Sodium amalgam reduces it to natural, inactive erythritol. No aldehyde reaction with benzenesulfhydroxamic acid. Not fermented by yeast.

Phenylosazone. $C_{16}H_{18}N_4O_2$. mp 164°.

3747. L-Erythrose. [533-49-3] (2*S*,3*S*)-2,3,4-Trihydroxybutanal. $C_4H_8O_4$; mol wt 120.10. C 40.00%, H 6.71%, O 53.29%. From calcium L-arabonate by oxidation with H_2O_2: Ruff, Meusser, *Ber.* **34**, 1365 (1901). From L-arabinose oxime: Wohl, *Ber.* **32**, 3667 (1899). From L-arabonamide by treatment with alkaline NaOCl: Weerman, *Rec. Trav. Chim.* **37**, 35 (1918).

Syrup. Sweet taste. Shows mutarotation. $[\alpha]_D^{24}$ +11.5° (8 min) → +15.2° (120 min) → +30.5° (final, c = 3): Felton, Freudenberg, *J. Am. Chem. Soc.* **57**, 1640 (1935). Soluble in water. Heating with HCl yields lactic acid. Oxidation with Br converts it to L-erythronic acid. Reduces Fehling's soln slowly in the cold. Not fermented by yeast.

Phenylosazone. $C_{16}H_{18}N_4O_2$. mp 164°.

3748. D-Erythrose 4-Phosphate. [585-18-2] (2*R*,3*R*)-2,3-Dihydroxy-4-(phosphonooxy)butanal; 4-D-erythrosephosphoric acid. $C_4H_9O_7P$; mol wt 200.08. C 24.01%, H 4.53%, O 55.97%, P 15.48%. Occurs in minute amounts in muscle flesh of all animals. Important natural intermediate in the Embden-Meyerhof scheme of alcoholic fermentation and glycolysis. Prepn by chemical synthesis: Ballou *et al.*, *J. Am. Chem. Soc.* **77**, 2658, 5967 (1955). Outline: *Chem. Eng. News* **34**, 2506 (1956).

Obtained in aq soln only. Shows no observable optical rotation. Is condensed with phosphoenolpyruvate by a cell-free extract from an *E. coli* mutant to give a 90% yield. Best stored and transported as the cyclohexylammonium salt of 4-phosphoryl-D-erythrose dimethyl acetal from which it can be prepared readily when needed.

3749. Erythrosine. [16423-68-0] 3',6'-Dihydroxy-2',4',-5',7'-tetraiodospiro[isobenzofuran-1(3*H*),9'-[9*H*]xanthen]-3-one sodium salt (1:2); 2',4',5',7'-tetraiodofluorescein disodium salt; erythrosine BS; erythrosine B; FD & C Red No. 3; C.I. Food Red 14; C.I. Acid Red 51; C.I. 45430. $C_{20}H_6I_4Na_2O_5$; mol wt 879.86. C 27.30%, H 0.69%, I 57.69%, Na 5.23%, O 9.09%. Prepn: Gilliard *et al.*, **DE** 108838 (1899); *Frdl.* **5**, 215; Gomberg, Tabern, *J. Ind. Eng. Chem.* **14**, 1115 (1922); Dolinsky, Jones, *J. Assoc. Off. Agric. Chem.* **34**, 114 (1951); *Colour Index* **vol. 4** (3rd ed., 1971) p 4428. Structure: Holmes, Scanlan, *J. Am. Chem. Soc.* **49**, 1594 (1927). HPLC determn: F. E. Lancaster, J. F. Lawrence, *J. Chromatogr.* **388**, 248 (1987). Acute toxicity: K. R. Butterworth *et al.*, *Food Cosmet. Toxicol.* **14**, 525 (1976); S. L. Yankell *et al.*, *J. Periodontol.* **48**, 228 (1977). Carcinogenicity studies in rodents: J. F. Borzelleca *et al.*, *Food Chem. Toxicol.* **25**, 723, 735 (1987).

Brown powder. Absorption max (water): 524 nm; in 95% alcohol: 531 nm. Soluble in water to cherry-red soln; sol in alcohol. HCl added to an aq soln produces a yellowish-brown ppt. Sodium hydroxide produces a red ppt sol in an excess of the reagent. LD_{50} in male, female mice, male, female rats (g/kg): 0.40, 0.32, 0.34, 0.37 i.p.; 6.7, 6.9, 7.4, 6.8 orally (Butterworth). Also reported as LD_{50} in mice, rats (mg/kg): 2558, 2891 orally (Yankell).

Note: Consult latest governmental regulations on use in food, drugs and cosmetics.

USE: Biological stain; color additive.

3750. L-Erythrulose. [533-50-6] (3*S*)-1,3,4-Trihydroxy-2-butanone; L-*glycero*-tetrulose. $C_4H_8O_4$; mol wt 120.10. C 40.00%, H 6.71%, O 53.29%. Prepn by bacterial oxidation of *meso*-erythritol: Müller *et al.*, *Helv. Chim. Acta* **20**, 1468 (1937); Whistler, Underköfler, *J. Am. Chem. Soc.* **60**, 2507 (1937). Synthesis from D-fructose: Gorin *et al.*, *J. Chem. Soc.* **1955**, 2699.

CH_2OH
|
C=O
|
HO—CH
|
CH_2OH

Syrup. $[\alpha]_D^{18}$ +11.4° (c = 2.4 in water). Very sensitive to alkali. Soluble in water, abs alc.

Phenylosazone. $C_{16}H_{18}N_4O_2$. Crystals from aq ethanol, mp 162°. $[\alpha]_D$ +32° (10 min) $\rightarrow$ 0° (24 hr, constant) (c = 0.75 in pyridine + ethanol, 3:2 v/v).

3751. Escin. [6805-41-0] Aescin; Reparil. A mixture of saponins occurring in the seed of the horse chestnut tree, *Aesculus hippocastanum* L., *Hippocastanaceae:* Winterstein, *Z. Physiol. Chem.* **199**, 25 (1931); Steiner, Holtzem in Paech-Tracey, *Moderne Methoden der Pflanzenanalyse III* (Springer-Verlag, 1955) p 117. Isoln by chromatography and purification: Fiedler, *Arzneim.-Forsch.* **4**, 213 (1953); using ion-exchange resins: Erbring *et al.*, US 3238190 (1966 to Madaus). Previously thought to be built up from the aglycon escigenin, glucuronic acid, glucose and xylose: Jermstadt, Waaler, *Pharm. Acta Helv.* **28**, 265 (1953); Patt, Winkler, *Arzneim.-Forsch.* **10**, 273 (1960); Tschesche *et al.*, *Ann.* **669**, 171 (1963). Structural studies indicate that the two major glycosides in the mixture are built up from the aglycon, *protoescigenin*, which is acylated at C-22 by acetic acid, and from the sugar moiety, glucuronic acid and two D-glucose molecules. The two aglycons differ only at the C-21 position which is acylated by either angelic acid or tiglic acid, *q.q.v.* Structure and stereochemistry: Wulff, Tschesche, *Tetrahedron* **25**, 415 (1969); Wagner *et al.*, *Arzneim.-Forsch.* **20**, 205 (1970); *eidem*, *Z. Physiol. Chem.* **351**, 1133 (1970). Early work identified two forms, *α-escin* and *β-escin*: Wagner, Basse, *ibid.* **320**, 27 (1960). Identity of *prosaponin B* with β-escin: Voigtlander, Rosenberg, *Arzneim.-Forsch.* **13**, 385 (1963). β-Escin is the natural form and can be converted to α-escin: Wagner, Schlemmer, US 3450691 (1969 to Klinge); Wagner *et al.*, *Arzneimittel-Forsch.*, *loc. cit.* Pharmacology: H. Hampel *et al.*, *ibid.* **20**, 209 (1970); Lang, Mennicke, *ibid.* **22**, 1928 (1972). Review: Tschesche, Wulff in *Fortschr. Chem. Org. Naturst.* **30**, 461-606 (1973). Review of pharmacology and clinical experience: C. R. Sirtori, *Pharmacol. Res.* **44**, 183-193 (2001).

R = Tiglic acid or Angelic acid

Major glycosides of Escin

α-Escin. [66795-86-6] Amorphous powder, mp 225-227°. $[\alpha]_D^{25}$ −13.5° (c = 5 in methanol). Very sol in water. Hemolytic index: 1:20,000. LD_{50} in mice, rats, guinea pigs (mg/kg): 3.2, 5.4, 15.2 i.v.; 320, 720, 475 orally (Hampel).

β-Escin. [11072-93-8] Flogencyl. Leaflets from dil ethanol, mp 222-223°. $[\alpha]_D^{27}$ −23.7° (c = 5 in methanol). Practically insol in water. Hemolytic index: 1:40,000. LD_{50} in mice, rats, guinea pigs (mg/kg): 1.4, 2.0, 7.2 i.v.; 134, 400, 188 orally (Hampel).

Sodium salt. White, highly hygroscopic solid, mp 251-252°.

THERAP CAT: In treatment of peripheral vascular disorders.

3752. Esculetin. [305-01-1] 6,7-Dihydroxy-2*H*-1-benzopyran-2-one; 6,7-dihydroxycoumarin; cichorigenin. $C_9H_6O_4$; mol wt 178.14. C 60.68%, H 3.40%, O 35.92%. The aglucon of esculin and of cichoriin. By hydrolysis of esculin or of cichoriin: Merz, *Arch. Pharm.* **270**, 486 (1932). By synthesis: Gattermann, Köbner, *Ber.* **32**, 288 (1899); Bert, *Compt. Rend.* **214**, 230 (1942). *Review:* Sethná, Shah, *Chem. Rev.* **36**, 1 (1945).

Prisms from glacial acetic acid; leaflets by vacuum sublimation. mp 268-270°. Soluble in dil alkalies with blue fluorescence; moderately sol in hot alcohol and in glacial acetic acid; almost insol in ether, in boiling water.

USE: In filters for absorption of ultraviolet light.

3753. Esculin. [531-75-9] 6-(β-D-Glucopyranosyloxy)-7-hydroxy-2*H*-1-benzopyran-2-one; 6,7-dihydroxycoumarin 6-glucoside; esculoside; bicolorin; enallachrome; polychrome; Escosyl. $C_{15}H_{16}O_9$; mol wt 340.28. C 52.95%, H 4.74%, O 42.32%. In leaves and bark of horse chestnut tree *Aesculus hippocastanum* L., *Hippocastanaceae.* Extraction procedure: Tumann, *Chem. Zentralbl.* **1916**, I, 1277. Synthesis: Merz, Hagemann, *Naturwissenschaften* **29**, 650 (1941); *eidem*, *Arch. Pharm.* **282**, 79 (1944); Amiard, Nominé, *Bull. Soc. Chim. Fr.* **1948**, 476. Use in bile-esculin test for identification of group D streptococci: C. Chuard, L. B. Reller, *J. Clin. Microbiol.* **36**, 1135 (1998).

Sesquihydrate. Needles from hot water, mp 204-206°. One gram dissolves in 580 ml water, 13 ml boiling water. Sol in hot alcohol, methanol, pyridine, ethyl acetate, acetic acid. Aq solns show blue fluorescence above pH 5.8. $[\alpha]_D^{18}$ −78.4° (c = 2.5 in 50% dioxane). Absorption spectrum: Merz, *Arch. Pharm.* **270**, 482 (1932); Goodwin, Pollock, *Arch. Biochem. Biophys.* **49**, 1 (1954). Has vitamin P activity.

Pentaacetate dihydrate. Needles from methanol, mp 163-164°.

4-Methylesculin. $C_{16}H_{18}O_9$. Prepn: Velluz, Amiard, *Bull. Soc. Chim. Fr.* **1948**, 1109.

USE: Reagent for microbiological identification test.

3754. Eslicarbazepine. [104746-04-5] (10*S*)-10,11-Dihydro-10-hydroxy-5*H*-dibenz[*b*,*f*]azepine-5-carboxamide; (*S*)-licarbazepine; (*S*)-10-hydroxycarbazepine; BIA 2-194. $C_{15}H_{14}N_2O_2$; mol wt 254.29. C 70.85%, H 5.55%, N 11.02%, O 12.58%. Voltage-gated sodium channel blocker; dibenzazepine antiepileptic similar to carbamazepine *q.v.*, and major metabolite of oxcarbazepine, *q.v.* Identification as active metabolite: M. Theisohn, G. Heimann, *Eur. J. Clin. Pharmacol.* **22**, 545 (1982). Prepn and activity of isomers: J. Benes *et al.*, *J. Med. Chem.* **42**, 2582 (1999). Prepn of esters: J. Benés, P. M. V. A. Soares Da Silva, WO 9702250; *eidem*, US 5753646 (1997, 1998 both to Portela). HPLC determination in plasma: G. Alves *et al.*, *Biomed. Chromatogr.* **21**, 1127 (2007).

Mechanism of action: A. Ambrósio *et al.*, *Neurochem. Res.* **27**, 121 (2002). Pharmacokinetics in pediatric and adolescent epilepsy: L. Almeida *et al.*, *J. Clin. Pharmacol.* **48**, 966 (2008). Clinical trial as adjunctive therapy in refractory partial-onset seizures: C. Elger *et al.*, *Epilepsia* **50**, 454 (2009). Review of pharmacology and clinical experience: T. Mestre, J. J. Ferreira, *Expert Opin. Invest. Drugs* **18**, 221-229 (2009); P. L. McCormack, D. M. Robinson, *CNS Drugs* **23**, 71-79 (2009).

mp 189-191°. $[\alpha]_D^{20}$ +199° (c = 1.10 in pyridine).

Acetate. [236395-14-5] (10*S*)-10-(Acetyloxy)-10,11-dihydro-5*H*-Dibenz[*b*,*f*]azepine-5-carboxamide; BIA 2-093; SEP-0002093; Erelib; Pazzul; Stedesa; Zebinix. $C_{17}H_{16}N_2O_3$; mol wt 296.33. Whitish, odorless, non-hygroscopic, crystalline powder, mp 186-187°. Sol in acetonitrile. Soly in aq buffer: < 1 mg/ml. $[\alpha]_D^{20}$ −20.5° (c = 1.10 in pyridine).

THERAP CAT: Anticonvulsant.

3755. Esmolol. [81147-92-4] 4-[2-Hydroxy-3-[(1-methylethyl)amino]propoxy]benzenepropanoic acid methyl ester; (±)-methyl 3-[4-[2-hydroxy-3-(isopropylamino)propoxy]phenyl]propionate; methyl *p*-[2-hydroxy-3-(isopropylamino)propoxy]hydrocinnamate. $C_{16}H_{25}NO_4$; mol wt 295.38. C 65.06%, H 8.53%, N 4.74%, O 21.67%. Cardioselective β-adrenergic blocker. Prepn: E. I. Carlsson *et al.*, **EP 41491** (1982 to A.B. Hässle), *C.A.* **96**, 122391f (1982); P. W. Erhardt *et al.*, **EP 53435**; *eidem*, **US 4593119** (1982, 1986 both to Am. Hosp. Supply); P. W. Erhardt *et al.*, *J. Med. Chem.* **25**, 1408 (1982). Pharmacology: R. J. Gorczynski *et al.*, *J. Cardiovasc. Pharmacol.* **5**, 668 (1983); *eidem*, *ibid.* **6**, 1548 (1984). Pharmacokinetics: A. Yacobi *et al.*, *J. Pharm. Sci.* **72**, 711 (1983). GC-MS determn in blood: C. Y. Sum, A. Yacobi, *ibid.* **73**, 1177 (1984). Clinical studies of effect in supraventricular tachycardia: G. Klein *et al.*, *Int. J. Clin. Pharmacol. Ther. Toxicol.* **22**, 112 (1984); R. J. Gray *et al.*, *J. Am. Coll. Cardiol.* **5**, 1451 (1985). Symposium on pharmacology and clinical efficacy: *Am. J. Cardiol.* **56**(11), 1F-62F (1985). Review of pharmacology and clinical efficacy: P. Benfield, E. M. Sorkin, *Drugs* **33**, 392-412 (1987).

Oil, gradually forming crystalline rosettes at room temp, mp 48-50°.

Hydrochloride. [81161-17-3] ASL-8052; Brevibloc. $C_{16}H_{25}$-NO_4·HCl; mol wt 331.84. Crystals from methanol-ether, mp 85-86°. Very sol in water; freely sol in acetone. Partition coefficient (octanol/water, pH 7.0): 0.42.

THERAP CAT: Antiarrhythmic.

THERAP CAT (VET): Antiarrhythmic.

3756. Esparto Wax. Spanish grass wax; halfa wax. A wax derived from a tall, tough grass of the Mediterranean region (S. Europe and N. Africa and Libya). The grass is shipped to Scotland, where it is dewaxed and made into fine paper. The wax is a byproduct. Two species of grass are cultivated for their excellent cellulose content: *Stipa tenacissima* L. (*Macrochloa tenacissima* (L.) Kunth.), *Graminaceae* and *Lygeum spartum* L., *Graminaceae. Constit.* Esparto wax consists of 15-17% free wax acids, 20-22% alcohols and hydrocarbons, 63-65% esters. The principal hydrocarbon is hentriacontane ($C_{31}H_{64}$, mp 68°). The acids include cerotic, montanic, myr-

icinic ($C_{30}H_{60}O_2$, mp 68°), lacceric ($C_{32}H_{64}O_2$, mp 70.5°) and hydroxy acids: A. H. Warth, *Chemistry and Technology of Waxes* (Reinhold, New York, 1947) p 139. Extraction of wax from grass: **DD 74703** (1970). TLC identification: H. Schmidt, *Am. Cosmet. Perfum.* **87**, 35 (1972).

Hard, tough wax. d_4^{25} 0.9887. mp 78.1°. Solidifies at 68.8°. Acid value 23.9. Saponification value 69.8. Soly in ethanol (25°): 0.244 g/100 ml; in ethylene chloride (37°): 1.48 g/100 ml.

USE: Substitute or extender for carnauba wax, *q.v.* Blends well with other waxes. Emulsifies easily and imparts smoothness to polishes. Preferred in the manufacture of carbon papers.

3757. Estazolam. [29975-16-4] 8-Chloro-6-phenyl-4*H*-[1,2,-4]triazolo[4,3-*a*][1,4]benzodiazepine; 8-chloro-6-phenyl-4*H*-*s*-triazolo[4,3-*a*][1,4]benzodiazepine; D-40TA; Cannoc; Esilgan; Eurodin; Julodin; Nemurel; Nuctalon; ProSom; Somnatrol. $C_{16}H_{11}ClN_4$; mol wt 294.74. C 65.20%, H 3.76%, Cl 12.03%, N 19.01%. Prepn: J. B. Hester, **DE 2012190**; *idem*, **US 3701782** (1970, 1972 to Upjohn); Meguro, Kuwada, *Tetrahedron Lett.* **1970**, 4039; *eidem*, **DE 1955349** (1971 to Toyama), *C.A.* **74**, 88078t (1971); Tawada *et al.*, **DE 2114441** (1971 to Takeda), *C.A.* **76**, 34320p (1972); Hester *et al.*, *J. Med. Chem.* **14**, 1078 (1971). Structure-activity studies: R. Nakajima *et al.*, *Jpn. J. Pharmacol.* **21**, 489 (1971). Pharmacology: *eidem*, *ibid.* 497; *eidem*, *Takeda Kenkyushoho* **31**, 349 (1972), *C.A.* **78**, 24138m (1972). Toxicity: Yokotani *et al.*, *ibid.* **32**, 152 (1973), *C.A.* **79**, 133006j (1973). Metabolism: Tanayama *et al.*, *Xenobiotica* **4**, 33-64 (1974); *see also: ibid.* 229, 441. Molecular structure studies: Kamiya *et al.*, *Chem. Pharm. Bull.* **21**, 1520 (1973).

Crystals from ethyl acetate-Skellysolve B hexanes, mp 228-229°. LD_{50} orally in male mice, rats and rabbits: 740, 3200, 300 mg/kg (Yokotani).

Note: This is a controlled substance (depressant): **21 CFR, 1308.14.**

THERAP CAT: Sedative, hypnotic.

3758. Estradiol. [50-28-2] (17β)-Estra-1,3,5(10)-triene-3,17-diol; β-estradiol; *cis*-estradiol; 3,17-epidihydroxyestratriene; dihydrofollicular hormone; dihydrofolliculin; dihydroxyestrin; dihydrotheelin; Climara; Estrace; Estraderm; Estradot; Estring; Estrofem; Estrogel; Evorel; Menostar; Systen; Vivelle; Vivelle-Dot; Zumenon. $C_{18}H_{24}O_2$; mol wt 272.39. C 79.37%, H 8.88%, O 11.75%. Potent mammalian estrogenic hormone produced by the ovary. Triggers the production of gonadotropins leading to ovulation. Isoln from follicular fluid of sow ovaries: D. W. MacCorquodale *et al.*, *J. Biol. Chem.* **115**, 435 (1936). Has also been isolated from urine of pregnant women and mares. Prepn from estrone: A. Butenandt, C. Georgens, *Z. Physiol. Chem.* **248**, 129 (1937); F. Hildebrandt *et al.*, **US 2096744** (1938 to Schering); from cholesterol: H. H. Inhoffen, G. Zühlsdorff, *Ber.* **74**, 1914 (1941). Total syntheses: U. Eder *et al.*, *Ber.* **109**, 2948 (1976); W. Oppolzer, D. A. Roberts, *Helv. Chim. Acta* **63**, 1703 (1980). Pharmacology: R. W. Lievertz, *Am. J. Obstet. Gynecol.* **156**, 1289 (1987). HPLC determn in plasma and urine: W. Slikker *et al.*, *J. Chromatogr.* **224**, 205 (1981). Pharmacokinetics: W. Kuhnz *et al.*, *Arzneim.-Forsch.* **43**, 966 (1993). Clinical evaluation in osteoporosis: J. W. W. Studd *et al.*, *Br. J. Obstet. Gynaecol.* **101**, 787 (1994). Comprehensive description: E. G. Salole, *Anal. Profiles Drug Subs.* **15**, 283-318 (1986). Physiologic review: H. G. Burger, *Int. J. Gynaecol. Obstet.* **1989**, Suppl. 1, 5-9. Review of clinical safety and efficacy in estrogen replacement therapy: A. Cheang *et al.*, *Drug Saf.* **9**, 365-379 (1993); H. L. Judd, *J. Reprod. Med.* **39**, 343-352 (1994).

White crystalline powder. Prisms from 80% alc, stable in air, mp 173-179°. $[\alpha]_D^{25}$ +76 to +83° (dioxane). uv max: 225, 280 nm. Precipitated by digitonin. Hygroscopic. 1 mg = 10,000 international units. Sol in alc, acetone, dioxane, chloroform, other organic solvents, solns of fixed alkali hydroxides; sparingly sol in vegetable oils. Practically insol in water.

Hemihydrate. [35380-71-3] Estroclim; Menorest; Vagifem. $C_{18}H_{24}O_2.\frac{1}{2}H_2O$; mol wt 281.40.

3-Acetate. [4245-41-4] Femring; Femtrace. $C_{20}H_{26}O_3$; mol wt 314.43.

3-Benzoate. [50-50-0] Estradiol benzoate; Agofollin; Duralease; Ovahormon; Progynon B. $C_{25}H_{28}O_3$; mol wt 376.50. Crystals from alc, mp 191-196°. Stable in air. $[\alpha]_D^{25}$ +58 to +63° (c = 2 in dioxane). Sol in alc, acetone, dioxane; slightly sol in diethyl ether, vegetable oils.

17β-Cyclopentanepropanoate. [313-06-4] Estradiol cypionate; ECP. $C_{26}H_{36}O_3$; mol wt 396.57. Prepn: Ott, **US 2611773** (1952 to Upjohn). Crystals from benzene + petr ether, mp 151-152°. $[\alpha]_D^{25}$ +45° (chloroform). Soluble in alc, acetone, dioxane, ether, methanol, benzene, chloroform; sparingly sol in peanut oil, cottonseed oil, corn oil, sesame oil. Insol in water.

Dipropionate. [113-38-2] Ovahormon Depot. $C_{24}H_{32}O_4$; mol wt 384.52. Prepn: K. Miescher, C. Scholz, **US 2205627** (1940 to Ciba). mp 104-105°.

17-Heptanoate. [4956-37-0] Estradiol enanthate; SQ-16150. $C_{25}H_{36}O_3$; mol wt 384.56. Prepn: Gauthier et al., Ann. Pharm. Fr. **16**, 757 (1958). Crystals from diisopropyl ether, mp 94-96°.

17-Valerate. [979-32-8] Climaval; Delestrogen; Pelanin Depot; Progynon Depot; Progynova. $C_{23}H_{32}O_3$; mol wt 356.51. Prepn: K. Miescher, C. Scholz, **US 2233025** (1941 to Ciba). Comprehensive description: K. Florey, Anal. Profiles Drug Subs. **4**, 192-208 (1975). Crystals, mp 144-145°.

Caution: These substances are listed as known human carcinogens: *Report on Carcinogens, Twelfth Edition* (PB2011-111646, 2011) p 184.

THERAP CAT: Estrogen.

THERAP CAT (VET): Estrogen.

3759. α-Estradiol. [57-91-0] (17α)-Estra-1,3,5(10)-triene-3,17-diol; 1,3,5-estratriene-3,17α-diol; 3,17-dihydroxyestratriene. $C_{18}H_{24}O_2$; mol wt 272.39. C 79.37%, H 8.88%, O 11.75%. Has been isolated from pregnancy urine of mares. Prepn from β-estradiol by inversion of the hydroxyl group at C-17 after tosylation: Allais, Hoffmann, **US 2835681** (1958).

Needles with $\frac{1}{2}H_2O$ from 80% alcohol, mp 220-223°. $[\alpha]_D^{20}$ +53 to +56° (c = 0.9 in dioxane). Not precipitated by digitonin (in 80% alcoholic soln). Soluble in alcohol, acetone, aq alkalies. One gram dissolves in more than 100 ml of boiling benzene. Slightly sol in ether, chloroform. Insol in water, aq dil acids.

Diacetate. $C_{22}H_{28}O_4$. mp 140-142°.

3-Benzoate. $C_{25}H_{28}O_3$. mp 156-157°; also reported as polymorphous: I, mp 63°; II, mp 153°; III, mp 158°.

3760. Estragole. [140-67-0] 1-Methoxy-4-(2-propen-1-yl)-benzene; p-allylanisole; chavicol methyl ether; esdragol. $C_{10}H_{12}O$; mol wt 148.21. C 81.04%, H 8.16%, O 10.79%. Main constituent of *tarragon oil* (*estragon oil*), the oil from *Artemisia dracunculus* L., Compositae (esdragon) where it occurs to an extent of 60-75%: Grimaux, Bull. Soc. Chim. [3] **11**, 34 (1894); Daufresne, ibid. [4] **3**, 333 (1908). Occurs also in pine oil and in American turpentine oil. Prepn: Tiffeneau, Compt. Rend. **139**, 482 (1904); Verley, **DE 154654**; D. Wigfield, K. Taymaz, Tetrahedron Lett. **1973**, 4841; P. Gramatica et al., Gazz. Chim. Ital. **104**, 629 (1974). Toxicity study: P. M. Jenner et al., Food Cosmet. Toxicol. **2**, 327 (1964).

Liquid. d_4^{21} 0.9645. bp_{764} 216°; bp_{25} 108-114°; bp_{12} 95-96°. $n_D^{17.5}$ 1.5230. Sol in alcohol, chloroform. Forms azeotropic mixtures with water. LD_{50} in rats, mice (mg/kg): 1820, 1250 orally (Jenner).

USE: In perfumes and as flavor in foods and liqueurs.

3761. Estramustine. [2998-57-4] (17β)-Estra-1,3,5(10)-triene-3,17-diol 3-[bis(2-chloroethyl)carbamate]; estradiol 3-bis(2-chloroethyl)carbamate; estra-1,3,5(10)triene-3,17β-diol 3-[N,N-bis(2-chloroethyl)carbamate]; Ro-21-8837. $C_{23}H_{31}Cl_2NO_3$; mol wt 440.41. C 62.73%, H 7.10%, Cl 16.10%, N 3.18%, O 10.90%. Carbamate ester linking estradiol to nor-nitrogen mustard; depolymerizes microtubules by binding to microtubule associated proteins (MAPs). Prepn: **BE 646319**; Fex et al., **US 3299104** (1963, 1967 to AB Leo); Niculescu-Duvaz et al., J. Med. Chem. **10**, 172 (1967). Clinical results: Anderes, Praxis **60**, 1375 (1971); Muntzing, Nilsson, Z. Krebsforsch. Klin. Onkol. **77**, 166 (1972). Review of mechanism of action: K. D. Tew, M. E. Stearns, Pharmacol. Ther. **43**, 299-319 (1989); of pharmacokinetics and pharmacodynamics: A. T. Bergenheim, R. Henriksson, Clin. Pharmacokinet. **34**, 163-172 (1998).

Crystals from benzene-petr ether, mp 104-105°. $[\alpha]_D^{20}$ +50° (in dioxane). uv max (alcohol): 270.7, 276.5 nm.

17-(Dihydrogen phosphate). [4891-15-0] Estramustine phosphate; Estracyt; $C_{23}H_{32}Cl_2NO_6P$; mol wt 520.38. mp 155° (dec). $[\alpha]_D^{20}$ +30° (dioxane). Sol in aqueous and alkali soln.

17-(Dihydrogenphosphate) disodium salt monohydrate. [1227300-83-5]; [52205-73-9] (anhydrous). Estramustine phosphate sodium; Emcyt. $C_{23}H_{30}Cl_2NNa_2O_6P.H_2O$; mol wt 582.36. Off-white powder; readily sol in water.

THERAP CAT: Antineoplastic.

3762. Estriol. [50-27-1] (16α,17β)-Estra-1,3,5(10)-triene-3,16,17-triol; 1,3,5-estratriene-3β,16α,17β-triol; 3,16α,17β-trihydroxy-$\Delta^{1,3,5}$-estratriene; 16α-hydroxyestradiol; follicular hormone hydrate; oestriol; trihydroxyestrin; Aacifemine; Colpogyn; Destriol; Gynäsan; Hormomed; Klimax E; Oekolp; Ortho-Gynest; Ovesterin; Ovestin; Ovo-Vinces; Theelol; Trophicreme; Tridestrin; Triovex. $C_{18}H_{24}O_3$; mol wt 288.39. C 74.97%, H 8.39%, O 16.64%. A metabolite of, and considerably less potent than the hormone estradiol, q.v. It is usually the predominant estrogenic metabolite found in urine. During pregnancy the placenta produces relatively large amounts of estriol. Isoln from human pregnancy urine: Marrian, Biochem. J. **23**, 1090, 1233 (1929); probably occurs as a glycuronide: Cohen, Marrian, ibid. **29**, 1577 (1935). Isoln from human placenta: Collip, Br. Med. J. **II**, 1080 (1930); Collip et al., Endocrinology **18**, 71 (1934). Also obtained from plant sources. Isoln from pussywillows: Skarzynski, Nature **131**, 766 (1933). Structure: Huffman, Lott, J. Am. Chem. Soc. **69**, 1835 (1947). Crystal and molecular structure: Cooper et al., Acta Crystallogr. **25B**, 814

(1969). Soly studies: Ruchelman, Howe, *J. Chromatogr. Sci.* **7**, 340 (1969). Partial synthesis: Huffman, *J. Biol. Chem.* **169**, 167 (1947). Syntheses: Huffman, Lott, *J. Am. Chem. Soc.* **71**, 719 (1949); Leeds *et al., ibid.* **76**, 2943 (1954).

Very small, monoclinic crystals from dil alc. d 1.27. mp 282°. During heating on the microscope heating stage, rearrangement of the crystal structure takes place at 270° and 275° (rate of heating, 4°/min, Kofler microscope heating stage). $[\alpha]_D^{25}$ +58° ±5° (0.04 g in 1 ml dioxane). uv max: 280 nm. Precipitated by digitonin. Freely sol in pyridine, in solns of fixed alkali hydroxides; sol in acetone, dioxane, chloroform, ether, vegetable oils; sparingly sol in alc. Practically insol in water.

Triacetate. $C_{24}H_{30}O_6$. Crystals from 90% alc. mp 126°.

16,17-Bis(sodium hemisuccinate). [113-22-4] Estriol succinate; Orgastyptin; Stiptanon; Synapause. $C_{26}H_{30}Na_2O_9$; mol wt 532.50. Prepn: **GB 879014** (1960 to Organon).

THERAP CAT: Estrogen.

THERAP CAT (VET): Estrogenic hormone therapy.

3763. Estrone. [53-16-7] 3-Hydroxyestra-1,3,5(10)-trien-17-one; 1,3,5-estratrien-3-ol-17-one; oestrone; folliculin; follicular hormone; tokokin; thelykinin; ketohydroxyestrin; Hiestrone; Menformon; Glandubolin; Cristallovar; Destrone; Endofolliculina; Estrol; Femidyn; Folikrin; Kolpon; Crinovaryl; Folisan; Disynformon; Hormovarine; Oestroperos; Wynestron; Thelestrin; Kestrone; Estrusol; Estrugenone; Femestrone Inj; Folipex; Follestrine; Follidrin (tablets); Folliculodis; Hormofollin; Oestrin; Ovifollin; Perlatan; Ketodestrin; Theelin. $C_{18}H_{22}O_2$; mol wt 270.37. C 79.96%, H 8.20%, O 11.83%. A metabolite of 17β-estradiol, *q.v.*, possessing considerably less biological activity; isolated as *d*-form. Occurs in pregnancy urine of women and mares, in follicular liquor of many animals, in human placenta, in urine of bulls and stallions, in palm-kernel oil. Isoln: Butenandt, *Naturwissenschaften* **17**, 879 (1929); Doisy *et al., Am. J. Physiol.* **90**, 329 (1929). Also isolated from moghat roots and date palm pollen grains: Amin *et al., Phytochemistry* **8**, 295 (1969). Manuf: E. A. Doisy *et al.*, **US 1967350, US 1967351** (1934 both to St. Louis University); Joly *et al.*, **FR 1305992** (1962 to Roussel-Uclaf.) Synthesis of a stereoisomer of estrone: W. E. Bachmann *et al., J. Am. Chem. Soc.* **64**, 974 (1942). Total synthesis of natural estrone: G. Anner, K. Miescher, *Helv. Chim. Acta* **31**, 2173 (1948); *ibid.* **32**, 1957 (1949); *ibid.* **33**, 1379 (1950); W. S. Johnson *et al., J. Am. Chem. Soc.* **72**, 1426 (1950); *ibid.* **74**, 2832 (1952); I. V. Torgov *et al., Dokl. Akad. Nauk SSSR* **135**, 73 (1960), *C.A.* **55**, 11462f (1961). Stereochemistry of estrone isomers: W. S. Johnson *et al., J. Am. Chem. Soc.* **80**, 661 (1958). Review of estrone syntheses: L. F. Fieser, M. Fieser, *Steroids* (Reinhold, New York, 1959) pp 495-502; T. B. Windholz, M. Windholz, *Angew. Chem. Int. Ed.* **3**, 353 (1964); Taub, "Naturally Occurring Aromatic Steroids" in *The Total Synthesis of Natural Products* vol. 2, J. ApSimon, Ed. (John Wiley & Sons, New York, 1973) pp 670-725. Recent syntheses: P. A. Bartlett, W. S. Johnson, *J. Am. Chem. Soc.* **95**, 7501 (1973); S. Danishefsky, P. Cain, *ibid.* **98**, 4975 (1976); T. Kametani *et al., ibid.* **99**, 3461 (1977); P. A. Grieco *et al., J. Org. Chem.* **45**, 2247 (1980). Comprehensive description: D. Both, *Anal. Profiles Drug Subs.* **12**, 135-189 (1983).

Crystals from acetone, mp 254.5-256°. $[\alpha]_D^{22}$ +152° (c = 0.995 in CHCl$_3$). uv max (*p*-dioxane): 282, 296 nm (ε 2300, 2130); (conc

H_2SO_4): 300, 450 nm; (0.1*M* NaOH): 239, 293 nm. Exists in three cryst phases, one monoclinic, the other two orthorhombic. Pptd by digitonin. Soly in water (25°): 0.003 g/100 ml. One gram of estrone dissolves in 250 ml of 96% alcohol at 15°, in 50 ml of boiling alcohol; in 50 ml acetone at 15°, in 110 ml chloroform at 15°, in 145 ml boiling benzene. Sol in dioxane, pyridine, vegetable oils; slightly sol in ether, fixed alkali hydroxide solns.

Acetate. [901-93-9] Hogival. $C_{20}H_{24}O_3$; mol wt 312.41. Crystals from alcohol, mp 125-127°.

Propionate. [975-64-4] $C_{21}H_{26}O_3$. mp 134-135°, **US 2156599**.

Sulfate. [481-97-0] Conjugal. Price, **US 2917522** (1959 to Parke, Davis).

Sulfate piperazine salt. [7280-37-7] Estropipate; piperazine estrone sulfate; Harmogen; Ogen; Sulestrex Piperazine. $C_{22}H_{32}N_2$-O_5S; mol wt 436.57. Comprehensive description: Z. L. Chang, *Anal. Profiles Drug Subs.* **5**, 375-402 (1976). White to yellowish-white crystalline powder, mp 190° to form a liquid which resolidifies and then decomp at 245°. $[\alpha]_D^{25}$ +87.8° (c = 1 in 0.4% NaOH). uv max (0.04% NaOH): 275, 268 nm (ε 838, 851). Soluble in warm water; very slightly sol in water, alc, chloroform, ether.

***dl*-Form.** [19973-76-3] Crystals from acetone, mp 251-254°.

Methyl ether. [1091-94-7] $C_{19}H_{24}O_2$. Synthesis: T. A. Bryson, C. J. Reichel, *Tetrahedron Lett.* **21**, 2381 (1980). *dl*-Form, crystals from acetone + methanol, mp 143.2-144.2°. *d*-Form, crystals from methanol, mp 164-165°.

Caution: These substances are listed as known human carcinogens: *Report on Carcinogens, Twelfth Edition* (PB2011-111646, 2011) p 184.

USE: In the prepn of commercial 19-norsteroids.

THERAP CAT: Estrogen.

THERAP CAT (VET): Estrogen.

3764. Etafedrine. [7681-79-0] α-[1-(Ethylmethylamino)ethyl]benzenemethanol; *N*-ethylephedrine; α-[1-(ethylmethylamino)-ethyl]benzyl alcohol; 2-methylethylamino-1-phenyl-1-propanol; Menetryl; Novedrin. $C_{12}H_{19}NO$; mol wt 193.29. C 74.57%, H 9.91%, N 7.25%, O 8.28%. β-Adrenergic agonist. Prepn: Skita, Keil, *Ber.* **63**, 34 (1930); Ueda *et al., Pharm. Bull.* **3**, 465 (1955).

Hydrochloride. [5591-29-7] Nethamine. $C_{12}H_{19}NO.HCl$; mol wt 229.75. Crystals from acetone + alcohol, mp 183-184°. One gram dissolves in 1.5 ml water, 8.0 ml alcohol. Aq solns are very stable.

THERAP CAT: Bronchodilator.

3765. Etafenone. [90-54-0] 1-[2-[2-(Diethylamino)ethoxy]-phenyl]-3-phenyl-1-propanone; 2'-[2-(diethylamino)ethoxy]-3-phenylpropiophenone; *o*-diethylaminoethoxy-β-phenylpropiophenone; 2'-(β-diethylaminoethoxy)-3-phenylpropiophenone; β-phenyl-*o*-(diethylaminoethoxy)propiophenone; LG-11457. $C_{21}H_{27}NO_2$; mol wt 325.45. C 77.50%, H 8.36%, N 4.30%, O 9.83%. Prepn: G. DiPaco, S. C. Tauro, *Ann. Chim. (Rome)* **48**, 1215 (1958); *eidem*, **DE 1265758** (1968 to Guidotti), *C.A.* **69**, 58950a (1968). Toxicity data: H.-J. Hapke, W. Sterner, *Arzneim.-Forsch.* **19**, 1664 (1969). Series of articles on pharmacology, toxicology, metabolism: *ibid.* 1664-1681.

Liquid, bp$_{30}$ 264-268°.

Hydrochloride. Hetaphenone; Asamedol; Baxacor; Corodilan; Dialicor; Pagano-Cor; Relicor. $C_{21}H_{27}NO_2.HCl$; mol wt 361.91.

Crystals, mp 129-130°. LD$_{50}$ in rats (mg/kg): 716 orally; 20.8 i.v. (Hapke, Sterne).

THERAP CAT: Vasodilator (coronary).

3766. Etamiphyllin. [314-35-2] 7-[2-(Diethylamino)ethyl]-3,7-dihydro-1,3-dimethyl-1*H*-purine-2,6-dione; 7-(2-diethylaminoethyl)theophylline; 7-(2-diethylaminoethyl)-1,3-dimethylxanthine; 1,3-dimethyl-7-(2-diethylaminoethyl)xanthine; dietamiphylline; etamiphylline. C$_{13}$H$_{21}$N$_5$O$_2$; mol wt 279.34. C 55.90%, H 7.58%, N 25.07%, O 11.45%. Prepn: Quevauviller *et al.*, *Bull. Soc. Chim. Biol.* **31**, 532 (1946); Moussalli *et al.*, **GB 669070** (1952); Klosa, *Arch. Pharm.* **288**, 301 (1955). Toxicity data on combinations: Baettig, *Arzneim.-Forsch.* **21**, 354 (1971). Pharmacokinetics, metabolism, and urinary determn in camels: M. Elghazali *et al.*, *J. Vet. Pharmacol. Ther.* **25**, 43 (2002); GC/MS and LC/MS determn in greyhound urine: M. C. Dumasia, P. Teale, *J. Pharm. Biomed. Anal.* **36**, 1085 (2005).

Waxy solid, mp 75°. Very sol in water, acetone; slightly sol in ethanol, ether.

Hydrochloride. [17140-68-0] Solufilina. C$_{13}$H$_{21}$N$_5$O$_2$.HCl. Crystals, mp 239-241°. Very sol in water, forms stable neutral soln.

Camphorsulfonate. [19326-29-5] Etamiphylline camsylate; Millophyline. C$_{23}$H$_{37}$N$_5$O$_6$S; mol wt 511.64. Crystals, mp 174°.

THERAP CAT: Bronchodilator.

THERAP CAT (VET): Bronchodilator.

3767. Etanercept. [185243-69-0] 1-235-Tumor necrosis factor receptor (human) fusion protein with 236-467-immunoglobulin G1 (human γ_1-chain Fc fragment); human tumor necrosis factor receptor p75 Fc fusion protein; TNFR:Fc; Enbrel. Recombinant protein consisting of the human soluble TNF receptor p75 linked to the Fc portion of human immunoglobulin G$_1$. Dimerizes via the cysteine residues in the Fc fragment to form an immunoglobulin-like structure. Inhibits the biological effects of tumor necrosis factor, *q.v.* Description of prepn and medicinal use: C. A. Smith, C. A. Jacobs, **WO 9406476**; *eidem*, **US 5605690** (1994, 1997 both to Immunex). Clinical pharmacokinetics: H. Lee *et al.*, *Clin. Pharmacol. Ther.* **73**, 348 (2003). Clinical trial as monotherapy in plaque psoriasis: C. L. Leonardi *et al.*, *N. Engl. J. Med.* **349**, 2014 (2003). Review of clinical experience in rheumatoid and psoriatic arthritis: B. Goffe, J. C. Cather, *J. Am. Acad. Dermatol.* **49**, S105-S111 (2003); S. Nanda, J. M. Bathon, *Expert Opin. Pharmacother.* **5**, 1175-1186 (2004).

THERAP CAT: Anti-inflammatory; antipsoriatic.

3768. Etanidazole. [22668-01-5] *N*-(2-Hydroxyethyl)-2-nitro-1*H*-imidazole-1-acetamide; DUP-453; NSC-301467; SR-2508; Radinyl. C$_7$H$_{10}$N$_4$O$_4$; mol wt 214.18. C 39.26%, H 4.71%, N 26.16%, O 29.88%. Oxygen-mimetic which renders hypoxic cells sensitive to chemo- and radiotherapy. Prepn: A. G. Beaman *et al.*, *Antimicrob. Agents Chemother.* **1967**, 520. *See also:* W. W. Lee *et al.*, **US 4371540** (1983 to U.S. Sec'y Health Human Serv.). Pharmacology: J. M. Brown *et al.*, *Int. J. Radiat. Oncol. Biol. Phys.* **7**, 695 (1981). HPLC determn in biological materials: R. Ward, P. Workman, *J. Chromatogr.* **420**, 223 (1987). Clinical pharmacokinetics and efficacy in radiation therapy: T. H. Wasserman *et al.*, *Radiother. Oncol.* **20**, Suppl. 1, 129 (1991). Evaluation in combination with alkylating agents: L. N. Shulman *et al.*, *Int. J. Radiat. Oncol. Biol. Phys.* **29**, 545 (1994).

Crystals from ethanol, mp 162-163° (Beaman); also reported as white powder from ethyl acetate + methanol, mp 164.5-165° (Lee). Partition coefficient (octanol/water): 0.046. uv max (isopropanol): 313 nm (ε 7800). Soly in isotonic saline: 200 mg/ml.

THERAP CAT: Antineoplastic adjunct (radiosensitizer).

3769. Etaqualone. [7432-25-9] 3-(2-Ethylphenyl)-2-methyl-4(3*H*)-quinazolinone; 2-methyl-3-(*o*-ethylphenyl)-4-quinazolone; ethinazone; Aolan. C$_{17}$H$_{16}$N$_2$O; mol wt 264.33. C 77.25%, H 6.10%, N 10.60%, O 6.05%. Prepd by condensation of *N*-acetylanthranilic acid and 2-ethylaniline with POCl$_3$: **BE 615282** (1962 to P. Beiersdorf AG), *C.A.* **58**, 13971ef (1963). Structure-activity studies: Parmar *et al.*, *J. Med. Chem.* **12**, 138 (1969).

mp 81°.

Hydrochloride. C$_{17}$H$_{16}$N$_2$O.HCl. mp 247°.

THERAP CAT: Sedative, hypnotic.

3770. Eterobarb. [27511-99-5] 5-Ethyl-1,3-bis(methoxymethyl)-5-phenyl-2,4,6(1*H*,3*H*,5*H*)-pyrimidinetrione; 5-ethyl-1,3-bis(methoxymethyl)-5-phenylbarbituric acid; *N*,*N'*-dimethoxymethylphenobarbital; eterobarbital; Ex-12-095; Antilon. C$_{16}$H$_{20}$N$_2$O$_5$; mol wt 320.35. C 59.99%, H 6.29%, N 8.74%, O 24.97%. Alkoxymethyl deriv of phenobarbital, *q.v.*, reported to have little or no hypnotic activity. Prepn: C. M. Samour, J. A. Vita, **DE 1939987**; J. A. Vita, **US 3595862** (1970, 1971 both to Kendall); C. M. Samour *et al.*, *J. Med. Chem.* **14**, 187 (1971); J. Gal, *J. Pharm. Sci.* **68**, 1562 (1979). Spectrophotographic and polarographic analysis: M. Romer *et al.*, *Anal. Chim. Acta* **88**, 261 (1977). Pharmacologic study: J. A. Vida, M. L. Hasker, *J. Med. Chem.* **16**, 602 (1973). Metabolism: M. A. Goldberg *et al.*, *Ann. Neurol.* **5**, 121 (1979). Efficacy in febrile convulsions: R. M. Julien, G. G. Fowler, *Neuropharmacology* **16**, 719 (1977).

Crystals from ethanol, mp 116-118°. LD$_{50}$ in mice: 470 mg/kg orally (Vida, Hooker).

Note: This is a controlled substance (depressant): **21 CFR,** 1308.13.

THERAP CAT: Anticonvulsant.

3771. Ethacridine. [442-16-0] 7-Ethoxy-3,9-acridinediamine; 6,9-diamino-2-ethoxyacridine; 2,5-diamino-7-ethoxyacridine; 2-ethoxy-6,9-diaminoacridine; etakridin. C$_{15}$H$_{15}$N$_3$O; mol wt 253.31. C 71.12%, H 5.97%, N 16.59%, O 6.32%. Prepn: **DE 364033**; **DE 364037**; **DE 393411**; **DE 395683** (1922); *Frdl.* **14**, 804, 807-812; A. Albert, W. Gledhill, *J. Soc. Chem. Ind.* **61**, 159 (1942). Toxicity: S. D. Rubbo, *Br. J. Exp. Pathol.* **28**, 1 (1947); N. J. Joshi *et al.*, *Indian J. Exp. Biol.* **16**, 1038 (1978). Pharmacokinetics: T. J. Rising *et al.*, *Arzneim.-Forsch.* **27**, 872 (1977). Use in protein analysis: R. H. Yue *et al.*, *Thromb. Diath. Haemorrh.* **31**, 439 (1974); F. Franek, *Methods Enzymol.* **121**, 631-638 (1986); in serological detection of *Brucella sp.*: J. D. Huber, P. Nicoletti, *Am. J. Vet. Res.* **47**, 1529 (1986). Clinical evaluation in diarrhea: N. Madanagopalan *et al.*, *Curr. Ther. Res.* **18**, 546 (1975); R. Raedsch *et al.*, *Klin. Wochenschr.* **69**, 863 (1991); as an abortifacient: S. Gardo, M. Nagy, *Arch. Gynecol. Obstet.* **247**, 39 (1990).

Orange-yellow crystals from 50% alc. mp 226°.

Lactate monohydrate. [6402-23-9] Acrinol; Rimaon; Rivanol; Vucine; Acrolactine; Ethodin; Metifex. $C_{15}H_{15}N_3O.C_3H_6O_3.H_2O$; mol wt 361.40. Light yellow crystals from 90% alcohol + ether. Darkens at 200°, mp 235°. Slowly sol in 15 parts water; sol in 9 parts boiling water; in 110 parts alcohol at 22°, in 100 parts boiling alcohol. Solns are yellow, fluorescent and stable to boiling. LD_{50} s.c. in mice: 0.12 g/kg (Rubbo).

USE: Reagent in serological testing and protein chemistry.

THERAP CAT: Antiseptic. Abortifacient.

3772. Ethacrynic Acid. [58-54-8] 2-[2,3-Dichloro-4-(2-methylene-1-oxobutyl)phenoxy]acetic acid; [2,3-dichloro-4-(2-methylenebutyryl)phenoxy]acetic acid; [4-(methylenebutyryl)-2,3-dichlorophenoxy]acetic acid; MK-595; Edecril; Edecrin; Hydromedin. $C_{13}H_{12}Cl_2O_4$; mol wt 303.14. C 51.51%, H 3.99%, Cl 23.39%, O 21.11%. Unsaturated ketone derivative of an aryloxyacetic acid. Prepn: Schultz et al., J. Med. Pharm. Chem. **5**, 660 (1962); E. M. Schultz, J. M. Sprague, BE 612755, C.A. **59**, 12712b (1963); eidem, US 3255241 (1962, 1966 both to Merck & Co.). HPLC determn in plasma: F. P. LaCreta et al., J. Chromatogr. **571**, 271 (1991); in urine: P. Campíns-Falcó et al.,, Anal. Chim. Acta **284**, 67 (1993). Pharmacology: Beyer et al., J. Pharmacol. Exp. Ther. **147**, 1 (1965). Crystal structure: J. Lamotte et al., Acta Crystallogr. **B34**, 2636 (1978). Toxicity study: Peck et al., Fed. Proc. **23**, 438 (1964). Reviews: Kim et al., Am. J. Cardiol. **27**, 407-415 (1971); H. E. Williamson, J. Clin. Pharmacol. **17**, 663-672 (1977).

White, or practically white, crystalline powder, mp 121-122° (corr). pKa' 3.50. Very slightly sol in water; sol in most organic solvents such as alcohols, chloroform, benzene. LD_{50} in mice (mg/kg): 176 i.v.; 627 orally (Peck).

Sodium salt. [6500-81-8] Ethacrynate sodium. $C_{13}H_{11}Cl_2$-NaO_4; mol wt 325.12. uv max (water): 225 nm (ε 15287). Soly in water at 25°: up to 7%. Solns at pH 7 at 25° stable for short periods of time. Stability decreased with an increase in pH, temperature and time.

THERAP CAT: Diuretic.

THERAP CAT (VET): Diuretic.

3773. Ethadione. [520-77-4] 3-Ethyl-5,5-dimethyl-2,4-oxazolidinedione; 3-ethyl-5,5-dimethyl-2,4-diketooxazolidine; Petidiol; Didione; Petidion; Epinyl; Etydion; Petisan; Neo-Absentol. C_7H_{11}-NO_3; mol wt 157.17. C 53.49%, H 7.05%, N 8.91%, O 30.54%. Prepn: Davis, Hook, GB 626971 (1949 to Brit. Schering Res. Labs.).

Flat, vitreous prisms from ether, mp 76-77°.

THERAP CAT: Anticonvulsant.

3774. Ethalfluralin. [55283-68-6] N-Ethyl-N-(2-methyl-2-propen-1-yl)-2,6-dinitro-4-(trifluoromethyl)benzenamine; N-ethyl-N-α,α,α-trifluoro-N-(2-methylallyl)-2,6-dinitro-p-toluidine; N-ethyl-N-methallyl-4-trifluoromethyl-2,6-dinitroaniline; EL-161; Sonalan; Sonalen. $C_{13}H_{14}F_3N_3O_4$; mol wt 333.27. C 46.85%, H 4.23%,

F 17.10%, N 12.61%, O 19.20%. Selective, pre-emergent herbicide. Prepn: H. D. Porter, DE 2511897; idem, GB 1505249 (1975, 1978 both to Eli Lilly). Activity: G. Skylakakis et al., Proc. 12th Br. Weed Control Conf. **2**, 795 (1974). Review of chemistry and analytical methods: E. W. Day, "Ethalfluralin" in Analytical Methods for Pesticides and Plant Growth Regulators Vol. 10, G. Zweig, J. Sherma, Eds. (Academic Press, New York, 1978) pp 341-352. Persistence in soil: B. J. Hayden, A. E. Smith, Bull. Environ. Contam. Toxicol. **25**, 508 (1983).

Yellow crystals from petr ether, mp 54-57°. Readily sol in acetone, acetonitrile, benzene, chloroform, hexane, methanol, xylene. Soly in water (25°): 0.3 ppm. uv max (methanol): 374, 267 nm. Vapor pressure (25°): 8.2×10^{-5} torr. LD_{50} in mice, rats (g/kg): >10 orally; LC_{50} in bluegill sunfish, rainbow trout, goldfish (ppb): 22, 37, 260 (Day).

USE: Pre-emergent herbicide.

3775. Ethambutol. [74-55-5] (2S,2'S)-2,2'-(1,2-Ethanediyldiimino)bis-1-butanol; (+)-2,2'-(ethylenediimino)di-1-butanol; d-N,N'-bis(1-hydroxymethylpropyl)ethylenediamine; EMB. $C_{10}H_{24}$-N_2O_2; mol wt 204.31. C 58.79%, H 11.84%, N 13.71%, O 15.66%. Prepn: Wilkinson et al., J. Am. Chem. Soc. **83**, 2212 (1961); eidem, J. Med. Chem. **5**, 835 (1962). Pharmacology: Kulig et al., Diss. Pharm. Pharmacol. **23**, 463 (1971), C.A. **76**, 81239d (1972). Comprehensive description: C. S. Lee, L. Z. Benet, Anal. Profiles Drug Subs. **7**, 231-249 (1978). Mechanism of action: W. H. Beggs in Antibiotics vol. 5(pt. 1), F. E. Hahn, Ed. (Springer-Verlag, New York, 1979) pp 43-66.

mp 87.5-88.8°. $[\alpha]_D^{25}$ +13.7° (c = 2 in water). Sol in chloroform, methylene chloride; less sol in benzene; sparingly sol in water.

Dihydrochloride. [1070-11-7] Dexambutol; Ebutol; Etibi; Etapiam; Myambutol; Mycobutol; Sural; Tibutol. $C_{10}H_{24}N_2O_2.2HCl$; mol wt 277.23. White crystalline powder. mp 198.5-200.3°; also reported as mp 201.8-202.6°. $[\alpha]_D^{25}$ +7.6° (c = 2 in water). Freely sol in water; sol in alc, methanol, DMSO; sparingly sol in ethanol; slightly sol in ether, acetone, chloroform.

Mixture with isoniazid methanesulfonate. [41663-50-7] (+)-2,2-(Ethylenediimino)dibutanol diisoniazide methanesulfonate; isobutol; Isoetam. Prepn: C. Ferrer-Salat et al., BE 753862; US 3718655 (1971, 1973 both to Lab. Ferrer). White powder, crystallizes as needles, mp 121-122° (dec). Sol in water, methanol, N,N-dimethylformamide. Insol in ethanol, ethyl ether, chloroform. uv max: 265 nm. $[\alpha]_D^{20}$ +3° (c = 5% in water). LD_{50} in mice (g/kg): 2.800 orally; 2.210 i.p. (Ferrer-Salat).

THERAP CAT: Antibacterial (tuberculostatic).

3776. Ethamivan. [304-84-7] N,N-Diethyl-4-hydroxy-3-methoxybenzamide; N,N-diethylvanillamide; vanillic acid diethylamide; vanillic diethylamide; Vandid; Cardiovanil; Emivan. C_{12}-$H_{17}NO_3$; mol wt 223.27. C 64.56%, H 7.68%, N 6.27%, O 21.50%. Prepd by dissolving vanillic acid O-acyl amide in NaOH and neutralizing with CO_2 or acid: Kratzl, Kvasnicka, Monatsh. Chem. **83**, 18 (1952); by refluxing a mixture of vanillic acid, P_2O_5 and powdered glass in xylene and treating with K_2CO_3: eidem, US 2641612 (1953 to Oesterr. Stickstoffwerke). Toxicity study: Caujolle et al., Compt. Rend. **243**, 609 (1956).

Needles from ligroin, mp 95-95.5°. LD$_{50}$ i.p. in rats: 28 mg/kg (Caujolle).

THERAP CAT: CNS stimulant; respiratory stimulant.

3777. Ethamsylate. [2624-44-8] 2,5-Dihydroxybenzene-sulfonic acid compd with N-ethylethanamine (1:1); diethylammonium 2,5-dihydroxybenzenesulfonate; 1-hydroxy-4-oxo-2,5-cyclohexadiene-1-sulfonic acid compd with diethylamine; diethylammonium cyclohexadien-4-ol-1-one-4-sulfonate; cyclonamine; etamsylate; MD-141; E-141; Aglumin; Altodor; Biosinon; Dicynene; Dicynone. C$_{10}$H$_{17}$NO$_5$S; mol wt 263.31. C 45.62%, H 6.51%, N 5.32%, O 30.38%, S 12.18%. Prepn: **GB 895709** (1962 to Labs. OM); **ES 279303** (1962 to Labs. Esteve), *C.A.* **60**, 1652g (1964). Pharmacology: Demars, *Med. Exp.* **4**, 173 (1961); Huguet *et al.*, *Therapie* **24**, 429 (1969). Toxicity: Esteve *et al.*, *Therapie* **15**, 110 (1960).

Crystals from ethanol, mp 125°. LD$_{50}$ i.v. in mice, rats: 800, 1350 mg/kg (Esteve).

THERAP CAT: Hemostatic.

3778. Ethane. [74-84-0] Bimethyl; dimethyl; methylmethane; ethyl hydride. C$_2$H$_6$; mol wt 30.07. C 79.89%, H 20.11%. CH$_3$CH$_3$. Constituent of natural gas (about 9%). Can be recovered from the gases produced during the distillation of crude petroleum. *Review:* Smith, Hanson, *Oil Gas J.* **44**, no. 10, 119 (1945).

Colorless, odorless gas. *Flammable.* Burns with a faintly luminous flame. d$_4^0$ 1.0493 (air = 1) or 1.3562 g/l. d$_4^0$ liq 0.446. mp −172°. bp −88°. Crit temp 32°; crit press. 48.2 atm. Heat of combustion: 1727 Btu/cu ft at 25° (one pound of CH$_3$CH$_3$ yields 20,420 Btu [net] at 15.56°). Flammable limits in air: 3.2 to 12.5% by vol. Ignition temp (in air): 530°. Soly in water at 20°: 4.7 ml/100 ml H$_2$O; in alc at 4°: 46 ml/100 ml alc.

Caution: Narcotic in high concns. A simple asphyxiant.

USE: In the manuf of chlorinated derivs; as refrigerant in some two-stage refrigeration systems where relatively low temps are produced; as fuel gas (so called "bottled gas" or "suburban propane" contains about 90% propane, 5% ethane, and 5% butane).

3779. 1,2-Ethanedisulfonic Acid. [110-04-3] 1,2-Ethylene-disulfonic acid. C$_2$H$_6$O$_6$S$_2$; mol wt 190.18. C 12.63%, H 3.18%, O 50.48%, S 33.72%. Prepn from 1,2-dibromoethane and sodium sulfite via disodium salt: McElvain *et al.*, *J. Am. Chem. Soc.* **67**, 1578 (1945); Ohba *et al.*, **US 3022172** (1962 to Fuji Photo Film). Review of prepns: Suter, *Organic Chemistry of Sulfur* (Wiley, New York, 1944) p 166; Gilbert, *Sulfonation and Related Reactions* (Interscience, New York, 1965).

Dihydrate. [5982-56-9] C$_2$H$_6$O$_6$S$_2$.2H$_2$O; mol wt 226.21. Crystals from acetic acid + acetic anhydride, mp 111-112°. Heating at 145° and 1 mm for 6 hours gives anhydr compd, mp 172-174°. Somewhat soluble in anhydr ether, very sol in dioxane.

Sodium salt dihydrate. [5982-57-0] C$_2$H$_4$Na$_2$O$_6$S$_2$.2H$_2$O; mol wt 270.18. Monoclinic prisms. Not very readily sol in water. Practically insol in alcohol.

3780. 1,2-Ethanedithiol. [540-63-6] Dithioethyleneglycol; ethylenedimercaptan. C$_2$H$_6$S$_2$; mol wt 94.19. C 25.50%, H 6.42%,

S 68.08%. Prepd by reacting ethanol, thiourea and ethylene dibromide and subsequent alka-line hydrolysis of the ethylenediisothiuronium bromide: Speciale, *Org. Synth.* **30**, 35 (1950); modified procedure: Grogan *et al.*, *J. Org. Chem.* **20**, 50 (1955).

Liquid. d$^{23.5}$ 1.123. bp$_{760}$ 146°; bp$_{150}$ 76-81°; bp$_{46}$ 63°; bp$_{24}$ 51-52°. n$_D^{20}$ 1.5589. Freely sol in alcohol and in alkalies.

Caution: Vapors of ethanedithiol may cause severe headache and nausea.

3781. Ethanethiol. [75-08-1] Ethyl mercaptan; mercaptoethane; ethyl sulfhydrate; thioethyl alcohol. C$_2$H$_6$S; mol wt 62.13. C 38.66%, H 9.73%, S 51.60%. Found in urine of rabbits after ingestion of cabbage. Is formed in vinous fermentation. Occurs in illuminating gas, in "sour" natural gas of W. Texas; in petroleum distillates from which it may be separated by chemical or physical methods: Thompson *et al.*, *Anal. Chem.* **27**, 175 (1955). Prepn from sodium ethylsulfate and KSH: Klason, *Ber.* **20**, 3407 (1887); catalytically from ethanol and hydrogen sulfide: Kramer, Reid, *J. Am. Chem. Soc.* **43**, 880 (1921). Review on occurrence, prepn, properties and reactions: E. Emmet Reid, *Organic Chemistry of Bivalent Sulfur* vol. I, 15-261 (Chemical Publishing Co., New York, 1958).

Liquid; penetrating leek-like odor: minimum detectable concn = 1 part in 50 billion parts of air. d$_4^0$ 0.8617; d$_4^{25}$ 0.83147. bp 34.7-35.04°; bp$_{400}$ 17.7°; bp$_{200}$ 1.5°; bp$_{100}$ −13.0°; bp$_{10}$ −50.2°; bp$_1$ −76.1°. mp −147.97 to −144.4°. n$_D^{20}$ 1.431; n$_D^{25}$ 1.420. Crit temp 225.5°. Crit pressure 54.2 atm. Explosive limits 2.8 and 18.2% by vol of vapor. Min ignition temp in air 299°, in oxygen 261°. Entropy at 25°: 70.77 cal/deg/mole. Heat capacity at 25°: 17.37 cal/deg/mole. Surface tension at 2°: 23.63 dynes/cm; at 16.7°: 21.62 dynes/cm. Viscosity at 20°: 0.003155 g/cm sec. Azeotrope with n-pentane (51% ethanethiol) bp 30.46°; with ether (40% ethanethiol) bp 31.50°. Soly in water at 20°: 6.76 g/l or 0.112 moles/l. Sol in alcohol, ether. *Keep tightly closed and in a cool place.*

Octadecahydrate. Needles. Practically insol in water, ethanethiol.

Sodium salt. C$_2$H$_5$NaS. Hydrolyzes instantly in water.

Caution: Potential symptoms of overexposure to ethanethiol are headache, nausea and irritation of mucous membranes. *See NIOSH Pocket Guide to Chemical Hazards* (DHHS/NIOSH 97-140, 1997) p 140.

USE: Odorant for natural gas; intermediate and starting material in manuf of plastics, insecticides, antioxidants.

3782. Ethanolamine. [141-43-5] 2-Aminoethanol; monoethanolamine; β-aminoethyl alcohol; 2-hydroxyethylamine; β-hydroxyethylamine; ethylolamine; colamine. C$_2$H$_7$NO; mol wt 61.08. C 39.33%, H 11.55%, N 22.93%, O 26.19%. Prepd on a large scale by ammonolysis of ethylene oxide: Knorr, *Ber.* **30**, 909 (1897); **FR 650574** (1928 to I. G. Farben); Reid, Lewis, **US 1904013** (1933 to Carbide & Carbon); Schwoegler, Olin, **US 2373199** (1945 to Sharples). Also from nitromethane and formaldehyde: *Ullmanns Encyklopädie der technischen Chemie* **3**, 102 (3rd ed., 1953). Manuf: *Faith, Keyes & Clark's Industrial Chemicals*, F. A. Lowenheim, M. K. Moran, Eds. (Wiley-Interscience, New York, 4th ed., 1975) pp 339-344. Toxicity: H. F. Smyth *et al.*, *J. Ind. Hyg. Toxicol.* **23**, 259 (1941).

Clear, viscous, hygroscopic liq. Ammoniacal odor. Absorbs CO$_2$, d$_4^{25}$ 1.0117; d$_4^{40}$ 0.9998; d$_4^{60}$ 0.9844. One gallon weighs 8.45 lbs in the U.S.A. Viscosity at 25°: 18.95 cP; at 60°: 5.03 cP. mp 10.3°. bp$_{760}$ 170.8°; bp$_{12}$ 70-72°. Strong base. pKa at 25°: 9.4. pH of 25% aq soln: 12.1; of 0.1N aq soln: 12.05. n$_D^{20}$ 1.4539. Dipole moment 2.27. Flash pt 195°F. Misc with water, methanol, acetone, alc, glycerin, chloroform. Immiscible with ether, solvent hexane, fixed oils, although it dissolves many essential oils. Soly at 25°: benzene 1.4%; ether 2.1%; carbon tetrachloride 0.2%; *n*-heptane <0.1%. LD$_{50}$ orally in rats: 10.20 g/kg (Smyth).

Hydrochloride. [2002-24-6] C$_2$H$_7$NO.HCl; mol wt 97.54. Deliquesc crystals from alc, mp 75-77°.

Oleate. [2272-11-9] Antivariz; Esclerosina; Ethamolin. C$_2$H$_7$-NO.C$_{18}$H$_{34}$O$_2$; mol wt 343.55. Use as a sclerosing agent: S. E. Hedberg *et al.*, *Am. J. Surg.* **143**, 426 (1982).

Caution: Potential symptoms of overexposure to ethanolamine are irritation of eyes, skin, respiratory system; lethargy. *See NIOSH Pocket Guide to Chemical Hazards* (DHHS/NIOSH 97-140, 1997) p 128.

USE: To remove CO$_2$ and H$_2$S from natural gas and other gases; in the synthesis of surface active agents; in polishes, hair waving solns, in emulsifiers; as softening agent for hides; dispersing agent for agricultural chemicals. Is reacted with other substances to form an accelerator in the manuf of antibiotics. Pharmaceutic aid (surfactant; emulsifying and solubilizing agent).

THERAP CAT: Oleate as sclerosing agent.

3783. Ethaverine. [486-47-5] 1-[(3,4-Diethoxyphenyl)methyl]-6,7-diethoxyisoquinoline; 6,7-diethoxy-1-(3,4-diethoxybenzyl)-isoquinoline; ethylpapaverine; Dyscural. C$_{24}$H$_{29}$NO$_4$; mol wt 395.50. C 72.89%, H 7.39%, N 3.54%, O 16.18%. Tetraethyl homolog of papaverine. Synthesis: Wolf, **US 1962224** (1934).

Crystals from alcohol + ether, mp 99-101°. Insol in water. Readily sol in hot alc; slightly in ether and chloroform.

Hydrochloride. [985-13-7] Barbonin; Circubid; Diquinol; Ethabid; Isovex; Laverin; Perparin; Perperine. C$_{24}$H$_{29}$NO$_4$.HCl; mol wt 431.96. Prepn of stable solns: Hereld, **US 2971888** (1961). Crystals from abs alc, mp 186-188° (dec). One gram dissolves in about 40 ml water. pH of 1% soln 3.6; of 0.1% soln 4.6.

Acid oxalate. C$_{24}$H$_{29}$NO$_4$.(COOH)$_2$. Crystals from alc, soluble in hot water.

Salt with 7-iodo-8-hydroxyquinoline-5-sulfonic acid. Prepn: **HU 106906** (1931); v. Issekutz *et al.*, *Arch. Exp. Pathol. Pharmakol.* **164**, 158 (1932).

THERAP CAT: Antispasmodic.

3784. Ethchlorvynol. [113-18-8] 1-Chloro-3-ethyl-1-penten-4-yn-3-ol; 5-chloro-3-ethylpent-1-yn-4-en-3-ol; ethyl β-chlorovinyl ethchlorvynol; Placidyl; Arvynol; Serenesil; Roeridorm; Normoson. C$_7$H$_9$ClO; mol wt 144.60. C 58.14%, H 6.27%, Cl 24.52%, O 11.06%. Tertiary carbinol sedative-hypnotic. Prepn: W. M. McLamore *et al.*, *J. Org. Chem.* **20**, 109 (1955); Bavley, McLamore, **US 2746900** (1956 to Pfizer). Pharmacological studies: S. Y. P'an *et al.*, *J. Pharmacol. Exp. Ther.* **114**, 326 (1955). Pharmacodynamics: Y. C. Martin, *Biochem. Pharmacol.* **16**, 2041 (1967). Evaluation as hypnotic: J. W. Middleton *et al.*, *Curr. Ther. Res.* **8**, 391 (1966); J. H. Pattison *et al.*, *J. Am. Geriatr. Soc.* **20**, 398 (1972). Study of metabolites: J. P. Horwitz *et al.*, *Drug Metab. Dispos.* **8**, 77 (1980).

Liquid, pungent aromatic odor. Slowly darkens on exposure to light and air. d$_4^{25}$ 1.065-1.070 (also reported as d$_4^{20}$ 1.070). bp$_{760}$ 173-174° (also reported as bp$_{760}$ 181°); bp$_{0.1}$ 28.5-30°. n$_D^{25}$ 1.4675-1.4800. Immiscible with water; miscible with most organic solvents. LD$_{50}$ in mice (mg/kg): 290 orally; 240 s.c. (P'an).

Note: This is a controlled substance (depressant): **21 CFR**, 1308.14.

THERAP CAT: Sedative, hypnotic.

3785. 3-Ethenylpyridine. [1121-55-7] β-Vinylpyridine; 3-EP. C$_7$H$_7$N; mol wt 105.14. C 79.97%, H 6.71%, N 13.32%. Combustion product of nicotine found in mainstream and sidestream smoke from cigarettes and cigars. Prepn: H. A. Iddles *et al.*, *J. Am. Chem. Soc.* **59**, 1945 (1937). Electronic spectra: R. Abu-Eittah *et al.*, *Int. J. Quantum Chem.* **39**, 211 (1991). Toxicology: K. D. Brunnemann *et al.*, *Cancer Lett.* **65**, 107 (1992). Use as tracer for environmental tobacco smoke exposure: M. W. Ogden, *Anal. Commun.* **33**, 197 (1996); J. M. Daisey *et al.*, *J. Expo. Anal. Environ. Epidemiol.* **8**, 313 (1998). Rapid chromatographic determn in indoor environments: G. Bertoni *et al.*, *Chromatographia* **43**, 296 (1996).

Light-yellow liquid. Polymerizes at room temperatures. Sol in ether, partially sol in alcohol and very slightly sol in water. uv max (cyclohexane): 278, 239 nm (ε 3320, 3920 L/mol cm); uv max (methanol): 279, 238 nm (ε 3322, 3930 L/mol cm).

USE: Tracer compound for environmental tobacco smoke.

3786. Ethenzamide. [938-73-8] 2-Ethoxybenzamide; ethbenzamide; 2-ethoxybenzenecarbonamide; salicylamide *o*-ethyl ether; Lucamide; Protopyrin; Trancalgyl. C$_9$H$_{11}$NO$_2$; mol wt 165.19. C 65.44%, H 6.71%, N 8.48%, O 19.37%. Prepn: **GB 656746** (1951 to H. Lundbeck); Shapiro *et al.*, *J. Am. Chem. Soc.* **81**, 3728 (1959); Bryk, Osowski, **PL 47822** (1961 to Polfa), *C.A.* **61**, 5573f (1964). Metabolism: Davison *et al.*, *J. Pharmacol. Exp. Ther.* **136**, 226 (1962). Toxicity data: Starmer *et al.*, *Toxicol. Appl. Pharmacol.* **19**, 20 (1971).

Crystals from ethyl acetate + hexane, mp 132-134°. LD$_{50}$ orally in mice: 1160 mg/kg (Starmer).

THERAP CAT: Analgesic.

3787. Ethephon. [16672-87-0] *P*-(2-Chloroethyl)phosphonic acid; 2-chloroethanephosphonic acid; CEPA; camposan; Cerone; Ethrel; Prep. C$_2$H$_6$ClO$_3$P; mol wt 144.49. C 16.63%, H 4.19%, Cl 24.53%, O 33.22%, P 21.44%. Ethylene-releasing agent used to control growth and ripening of fruit crops. Prepn: Kabachnik, Rossiiskaya, *Izv. Akad. Nauk SSSR Otd. Khim. Nauk* **1946**, 403-410, *C.A.* **42**, 7242e (1948); *Beilstein* **EIII 4**, 1780. Toxicity data: G. Hennighausen *et al.*, *Pharmazie* **32**, 181 (1977). Review of use in viticulture: E. Szyjewicz *et al.*, *Am. J. Enol. Vitic.* **35**, 117-123 (1984).

Very hygroscopic needles from benzene (must be dried over P$_2$O$_5$). mp 74-75°. Freely sol in water, methanol, acetone, ethylene glycol, propylene glycol. Slightly sol in benzene, toluene. Practically insol in petr ether. Aq solns are stable below pH 3.5. Above pH 3.5 hydrolysis begins with the release of free ethylene. LD$_{50}$ orally in mice: 2850 mg/kg (Hennighausen).

Caution: Spray formulations are quite acidic, about pH 1.0. May be irritating to exposed skin and eyes, or if inhaled.

USE: Plant growth regulator.

3788. Ethinyl Estradiol. [57-63-6] (17α)-19-Norpregna-1,-3,5(10)-trien-20-yne-3,17-diol; 17α-ethynyl-1,3,5(10)-estratriene-

3,17β-diol; 17-ethinylestradiol; ethynylestradiol; Estinyl; Feminone; Lynoral; Orestralyn; Primogyn C; Progynon C. $C_{20}H_{24}O_2$; mol wt 296.41. C 81.04%, H 8.16%, O 10.80%. Synthetic steroid with high oral estrogenic potency: Inhoffen, Hohlweg, *Naturwissenschaften* **26**, 96 (1938). Prepn from estrone: Inhoffen *et al.*, *Ber.* **71**, 1024 (1938). *See also* **DE 702063**; **GB 516444**; **US 2243887**; **US 2251939**; **US 2265976**; **US 2267257**. Properties: Petit, Muller, *Bull. Soc. Chim. Fr.* **1951**, 121; L. Ehmann, A. Wettstein, *Pharm. Acta Helv.* **25**, 297 (1950). NMR: Hampel, Kraemer, *Ber.* **98**, 3255 (1965). Toxicity: E. I. Goldenthal, *Toxicol. Appl. Pharmacol.* **18**, 185 (1971). Randomized double-blind clinical studies: S. Koetsawang *et al.*, *Contraception* **25**, 231 (1982); A. Sheth *et al.*, *ibid.* 243. Clinical evaluation in gonadal dysgenesis: L. Cuttler *et al.*, *J. Clin. Endocrinol. Metab.* **60**, 1087 (1985). General review: K. W. Thompson, *J. Clin. Pharmacol.* **8**, 1088-1098 (1948). Review of metabolism and pharmacokinetics: K. Fotherby, *Methods Find. Exp. Clin. Pharmacol.* **4**, 133-141 (1982); of carcinogenicity studies: *IARC Monographs* **21**, 233-255 (1979); *ibid.* Suppl. 4, 186-188 (1982).

White to creamy white, odorless, crystalline powder. Sol in alc, chloroform, ether, vegetable oils, in solutions of fixed alkali hydroxides. Insol in water.

Hemihydrate. Fine needles from methanol + water, mp 141-146°, $[\alpha]_D^{25}$ 0 ± 1° (dioxane). Dehydrates after melting and further heating, mp 182-184°. $[\alpha]_D^{24}$ +3.5 ± 0.5° (c = 2 in dioxane); −29.5 ± 1° (c = 2 in pyridine). uv max (ethanol): 281 nm (ε 2040 ± 60). Soly: 1 part in 6 of ethanol, 1 in 4 of ether, 1 in 5 of acetone, 1 in 4 of dioxane, and 1 in 20 of chloroform. Sol in vegetable oils, and in solns of fixed alkali hydroxides. Practically insol in water. LD_{50} in rats, mice (mg/kg): 2952, 1737 orally (Goldenthal).

3-Acetate. $C_{22}H_{26}O_3$. Crystals, mp 152-153°. $[\alpha]_D^{20}$ +3° (chloroform).

3-Benzoate. $C_{27}H_{28}O_3$. Needles from methanol, mp 200-202°.

Note: Also used in combination with chlormadinone acetate, desogestrel, ethynodiol, gestodene, lynestrenol, norethindrone or norgestrel, *q.q.v.* Has been used in combination with dimethisterone, medroxyprogesterone, or megestrol acetate, *q.q.v.*

Caution: These substances are listed as known human carcinogens: *Report on Carcinogens, Twelfth Edition* (PB2011-111646, 2011) p 184.

THERAP CAT: Estrogen. In combination with progestogen as oral contraceptive.

THERAP CAT (VET): Estrogen.

3789. Ethiodized Oil. [8008-53-5] Ethyl ester of iodinated fatty acid of poppyseed oil; Ethiodol. Contains 37% organically bound iodine. Ref: Hom *et al.*, *J. Am. Pharm. Assoc. Sci. Ed.* **46**, 254 (1957). Use in lymphography: R. M. Paxton *et al.*, *Br. Med. J.* **I**, 120 (1975).

Amber-colored oil, extremely fluid, sp gr 1.28. Soluble in acetone, chloroform, ether, hexane. Insol in water.

THERAP CAT: Diagnostic aid (radiopaque medium). Antineoplastic when part of the iodine is ^{131}I.

3790. Ethion. [563-12-2] Phosphorodithioic acid $S^P,S^{P'}$-methylene $O^P,O^P,O^{P'},O^{P'}$-tetraethyl ester; ethyl methylene phosphorodithioate; O,O,O',O'-tetraethyl S,S'-methylenediphosphorodithioate; bis[S-(diethoxyphosphinothioyl)mercapto]methane; diethion; ENT-24105; FMC-1240; Niagara 1240; Commando; Rhodocide; Tafethion. $C_9H_{22}O_4P_2S_4$; mol wt 384.46. C 28.12%, H 5.77%, O 16.65%, P 16.11%, S 33.36%. Organophosphate insecticide; cholinesterase inhibitor. Prepn: Willard, Henahan, **US 2873228**; **US 3014058** (1959, 1961 both to FMC). Determn in air using ion mobility spectrometry: M. T. Jafari, *Talanta* **69**, 1054 (2006). Toxicity study: T. B. Gaines, *Toxicol. Appl. Pharmacol.* **14**, 515 (1969). Review of toxicology and human exposure: *Toxicological Profile for*

Ethion (PB2000-108024, 2000) 218 pp. Field study vs horn fly in cattle: O. S. Anziani *et al.*, *Vet. Parasitol.* **91**, 147 (2000).

Liquid, mp −13 to −12°. d_4^{20} 1.220. n_D^{20} 1.5490. Vapor pressure: 1.5×10^{-6} mm Hg. Sol in xylene, methylated naphthalene, chloroform, acetone and kerosene + 1% methyl ethyl ketone or benzene; slightly sol in water. LD_{50} in female, male rats (mg/kg): 27, 65 orally; 62, 245 dermally (Gaines).

Caution: Potential symptoms of overexposure are eye, skin irritation; nausea, vomiting, abdominal cramps, diarrhea, salivation; headache, giddiness, vertigo, weakness; rhinorrhea, chest tightness; blurred vision, miosis; cardiac irregularities; muscle fasciculations; dyspnea. *See NIOSH Pocket Guide to Chemical Hazards* (DHHS/NIOSH 97-140, 1997) p 130.

USE: Insecticide; acaricide.

THERAP CAT (VET): Ectoparasiticide.

3791. Ethionamide. [536-33-4] 2-Ethyl-4-pyridinecarbothioamide; 2-ethylthioisonicotinamide; 3-ethylisothionicotinamide; 2-ethylisothionicotinamide; 2-ethyl-4-thiocarbamoylpyridine; α-ethylisonicotinoylthioamide; amidazine; ethioniamide; Bayer 5312; 1314-Th; Nisotin; Trescatyl; Aetina; Ethimide; Iridocin; Tio-Mid. $C_8H_{10}N_2S$; mol wt 166.24. C 57.80%, H 6.06%, N 16.85%, S 19.29%. Prepn: Libermann *et al.*, *Compt. Rend.* **242**, 2409 (1956); *Bull. Soc. Chim. Fr.* **1958**, 687; **GB 800250** (1958 to Chimie et Atomistique). Mechanism of action study: K. Johnsson *et al.*, *J. Am. Chem. Soc.* **117**, 5009 (1995).

Minute yellow crystals from ethanol, dec 164-166°. Freely sol in pyridine; sol in methanol, hot acetone, dichloroethane; sparingly sol in ethanol, propylene glycol; slightly sol in water, chloroform, ether.

THERAP CAT: Antibacterial (tuberculostatic).

3792. Ethionine. [13073-35-3] S-Ethyl-L-homocysteine; 2-amino-4-(ethylthio)butyric acid; α-amino-γ-(ethylmercapto)butyric acid; homocysteine S-ethyl ether. $C_6H_{13}NO_2S$; mol wt 163.24. C 44.15%, H 8.03%, N 8.58%, O 19.60%, S 19.64%. Prepn from ethanethiol and acrolein, purification via Zn salt: Norton, **US 2840587** (1958 to Dow). Prepn of DL-form and L-form: Armstrong, Lewis, *J. Org. Chem.* **16**, 749 (1951). Resolution of D- and L-isomers: Greenstein *et al.*, *J. Biol. Chem.* **204**, 307 (1953). Review: E. Farber, "Ethionine Carcinogenesis" in A. Haddow, S. Weinhouse, *Advan. Cancer Res.* **vol. 7** (Academic Press, New York, 1963) pp 383-474.

Crystals, dec 272-274°. $[\alpha]_D^{24}$ +25.1°. Also reported as $[\alpha]_D^{23}$ +20.1° (1N HCl) (Armstrong, Lewis).

DL-Form. [67-21-0] Crystals, dec 257-260°. Also reported as dec 272-284° (Armstrong, Lewis).

3793. Ethiprole. [181587-01-9] 5-Amino-1-[2,6-dichloro-4-(trifluoromethyl)phenyl]-4-(ethylsulfinyl)-1H-pyrazole-3-carbonitrile; 5-amino-3-cyano-1-(2,6-dichloro-4-trifluoromethylphenyl)-4-ethylsulfinylpyrazole; RPA-107382; Curbix; Kirappu. $C_{13}H_9Cl_2$-F_3N_4OS; mol wt 397.20. C 39.31%, H 2.28%, Cl 17.85%, F 14.35% N 14.11% O 4.03%, S 8.07%. Insecticidal phenylpyrazole; analog of fipronil, *q.v.* Prepn: C. L. Haas *et al.*, **WO 9722593** (1997 to Rhone-Poulenc Agrochimie). Field trial vs stored-grain insects: F. H. Arthur, *J. Econ. Entomol.* **95**, 1314 (2002). Photochemistry, metabolism and GABAergic action compared with fipronil: P. Caboni *et al.*, *J. Agric. Food Chem.* **51**, 7055 (2003).

mp ~174°. LD$_{50}$ in house flies (24 hr topical): 0.50 ±0.03 μg/g (Caboni).

USE: Insecticide for chewing and sucking insects.

3794. Ethirimol. [23947-60-6] 5-Butyl-2-(ethylamino)-6-methyl-4(3*H*)-pyrimidinone; 2-ethylamino-4-methyl-5-*n*-butyl-6-hydroxypyrimidine; PP-149; Milstem; Milgo; Milcurb Super. C$_{11}$H$_{19}$N$_3$O; mol wt 209.29. C 63.13%, H 9.15%, N 20.08%, O 7.64%. Prepn: B. K. Snell *et al.*, **ZA 6701373**; *eidem*, **US 3980781** (1968, 1976 both to ICI). Systemic fungicidal activity: R. M. Bebbington, *Chem. Ind. (London)* **1969**, 1512; K. J. Bent, *Ann. Appl. Biol.* **66**, 103 (1970). Metabolism and mode of action: A. Calderbank, *Acta Phytopathol.* **6**, 355 (1971).

Crystalline solid, mp 159-160°. Vapor pressure at 25°: 2 × 10^{-6} mm Hg. Soly in water at 25°: 200 mg/l. Sol in chloroform, trichloroethylene, aq solns of strong acids and bases. Slightly sol in ethanol; sparingly sol in acetone. LD$_{50}$ orally in rats: 4000 mg/kg (Bebbington).

USE: Fungicide.

3795. Ethisterone. [434-03-7] 17α-Hydroxypregn-4-en-20-yn-3-one; 17α-ethynyltestosterone; 17α-ethynyl-17β-hydroxy-4-androsten-3-one; 17α-ethinyltestosterone; 17α-ethynyl-4-androsten-17β-ol-3-one; anhydrohydroxyprogesterone; pregneninolone. C$_{21}$H$_{28}$O$_2$; mol wt 312.45. C 80.73%, H 9.03%, O 10.24%. Synthetic progestogen; metabolite of danazol, *q.v.*; intermediate in the synthesis of spironolactone, *q.v.* Prepn: Inhoffen *et al.*, *Ber.* **71**, 1024 (1938). Crystal structure and photostability study: J. Reisch *et al.*, *Monatsh. Chem.* **124**, 1169 (1993). GC/MS characterization as a metabolite of danazol: J. Y. Kim *et al.*, *J. Vet. Pharmacol. Ther.* **24**, 147 (2001). HPLC/CD determn of isomers: D. Szegvári *et al.*, *Anal. Bioanal. Chem.* **375**, 713 (2003).

Crystals from ethyl acetate, mp 269-275°. Sublimes in high vacuum at 190-195°. [α]$_D^{23}$ +23.8° (dioxane); [α]$_D^{25}$ −32.0° (pyridine). Crystal density: 1.220 g/cm^3. uv max (methanol): 241 nm (E$_{1cm}^{1\%}$ 513). Practically insol in water. Slightly sol in alcohol, acetone, ether, chloroform, vegetable oils.

3796. Ethofumesate. [26225-79-6] 2-Ethoxy-2,3-dihydro-3,3-dimethyl-5-benzofuranol 5-methanesulfonate; NC-8438; Nortron; Tramat. C$_{13}$H$_{18}$O$_5$S; mol wt 286.34. C 54.53%, H 6.34%, O 27.94%, S 11.20%. Prepn: P. S. Gates *et al.*, **US 3689507** (1972 to Fisons). Biological properties and use: R. K. Pfeiffer, *3rd Symp. New Herbic.* **1969**, 1. Toxicity data: *Anal. Ref. Standards Suppl.*

Data (PB85-143766, 1984) p 55. Field studies in grass: G. E. Coats, J. V. Krans, *Weed Sci.* **34**, 930 (1986); B. J. Johnson *et al*, *HortScience* **24**, 102 (1989). HPLC determn in plant, soil, water: M. A. Alawi, *Anal. Lett.* **23**, 1695 (1990).

White crystalline solid, mp 70-72°. Vapor pressure at 25°: 8.6×10^{-7} mbar. Soly in water at 25°: 110 ppm. Relatively stable in water. LD$_{50}$ orally in rats: 1130 mg/kg (PB85-143766).

USE: Herbicide.

3797. Ethohexadiol. [94-96-2] 2-Ethyl-1,3-hexanediol; 2-ethyl-3-propyl-1,3-propanediol; octylene glycol; 6-12. C$_8$H$_{18}$O$_2$; mol wt 146.23. C 65.71%, H 12.41%, O 21.88%. Prepn: Kulpinsky, Nord, *J. Org. Chem.* **8**, 256 (1943); B. G. Wilkes, **US 2407205** (1946 to Carbide & Carbon Chemicals). Activity: P. Granett, H. L. Haynes, *J. Econ. Entomol.* **38**, 671 (1945); W. V. King, *ibid.* **44**, 339 (1951); B. V. Travis, C. N. Smith, *ibid.* 428. As an additive in mosquito repellant: W. G. Reifenrath, L. C. Rutledge, *J. Pharm. Sci.* **72**, 169 (1983). Toxicity study: B. Ballantyne *et al.*, *Vet. Hum. Toxicol.* **27**, 491 (1985).

Colorless, odorless, slightly viscous liq. d$_{20}^{20}$ 0.9422. d$_4^{22}$ 0.9325. bp$_{760}$ 244.2°; bp$_{50}$ 163°; bp$_{10}$ 129°; bp$_3$ 102°; bp$_{0.5}$ 94-96°. n$_D^{22}$ 1.4530. Flash pt 260°F. Vapor pressure at 20° <0.01 mm. Abs viscosity: 271 cP. Soly in water 0.6% w/w; soly of water in 2-ethyl-1,3-hexanediol 10.8% w/w; sol in ethanol, isopropanol, propylene glycol, castor oil. LD$_{50}$ in male, female rats (ml/kg): 9.85, 4.92 orally (Ballantyne).

Caution: Little or no skin absorption; ingestion causes CNS depression. *See: Clinical Toxicology of Commercial Products*, R. E. Gosselin *et al.*, Eds. (Williams & Wilkins, Baltimore, 5th ed., 1984) Section II, pp 347.

USE: Insect repellent.

3798. Ethopabate. [59-06-3] 4-(Acetylamino)-2-ethoxybenzoic acid methyl ester; methyl 4-acetamido-2-ethoxybenzoate; 2-ethoxy-4-acetamidobenzoic acid methyl ester; ethyl pabate. C$_{12}$H$_{15}$NO$_4$; mol wt 237.26. C 60.75%, H 6.37%, N 5.90%, O 26.97%. Prepn: Grimme, Schmitz, *Ber.* **87**, 179 (1954); E. F. Rogers, R. L. Clark, **BE 613166**; *eidem*, **US 3211610** (1962, 1965 both to Merck & Co.); **FR 1407055**; M. L. Thominet, **US 3357978** (1965, 1967 both to Soc. d'Etudes Sci. Ind. de l'Ile-de-France). Anticoccidial activity studies: Rogers *et al.*, *Proc. Soc. Exp. Biol. Med.* **117**, 488 (1964). Metabolism in chickens: Buhs *et al.*, *J. Pharmacol. Exp. Ther.* **154**, 357 (1966).

White to pinkish-white, practically odorless crystals from methanol and water, mp 148-149°. uv max (methanol): 298, 267 nm (A$_{1cm}^{1\%}$ 805, 365). Sol in methanol, acetone, acetonitrile, dehydrated alc; sparingly sol in isopropanol, *p*-dioxane, ethyl acetate, methylene chloride; slightly sol in ether; very slightly sol in water, isooctane.

Mixture with amprolium. Amprol Plus.

THERAP CAT (VET): Mixture as coccidiostat.

3799. Ethoprop. [13194-48-4] Phosphorodithioic acid *O*-ethyl *S,S*-dipropyl ester; *O*-ethyl *S,S*-dipropylphosphorodithioate;

ethoprophos; VC9-104; Mocap. $C_8H_{19}O_2PS_2$; mol wt 242.33. C 39.65%, H 7.90%, O 13.20%, P 12.78%, S 26.46%. Prepn: L. E. Goyette, **US 3112244** (1963 to Virginia-Carolina Chem. Corp.); J. H. Wilson, Jr., **US 3268393** (1966 to Mobil). Activity: C. R. Harris, J. L. Hitchon, *J. Econ. Entomol.* **63**, 2 (1970). Metabolism: Z. M. Iqbal, R. E. Menzer, *Biochem. Pharmacol.* **21**, 1569 (1972). Soil degradation: J. H. Smelt *et al., Pestic. Sci.* **8**, 147 (1977).

Pale yellow oil, $bp_{0.2}$ 86-91°. d_4^{20} 1.094. Vapor press. at 26°: 3.5 × 10^{-4} mm Hg. Soly in water: 750 mg/l. Sol in most organic solvents.

USE: Insecticide; nematocide.

3800. Ethopropazine. [522-00-9] *N,N*-Diethyl-α-methyl-10*H*-phenothiazine-10-ethanamine; 10-(2-diethylaminopropyl)phenothiazine; 10-(2-diethylamino-2-methylethyl)phenothiazine; 2-diethylamino-1-propyl-*N*-dibenzoparathiazine; phenopropazine; profenamine; RP-3356; W-483; Isothazine; Isothiazine; Parkin. $C_{19}H_{24}N_2S$; mol wt 312.48. C 73.03%, H 7.74%, N 8.97%, S 10.26%. Anticholinergic. Prepd from Grignard complexes of diethylaminopropyl halide and phenothiazine: S. S. Berg, J. N. Ashley, **US 2607773** (1952 to Rhône-Poulenc). Pharmacology: M. E. Farquharson, R. G. Johnston, *Br. J. Pharmacol.* **14**, 559 (1959). Pharmacokinetics: M. Maboudian-Esfahani, D. R. Brocks, *Biopharm. Drug Dispos.* **20**, 159 (1999). HPLC determn in plasma: *eidem, J. Chromatogr. B* **715**, 417 (1998).

Crystals, mp 53-55°. Usually obtained as an oil because of contamination with 10-(2-diethylamino-1-methylethyl)phenothiazine.

Hydrochloride. [1094-08-2] Dibutil; Lysivane; Pardisol; Parphezein; Parphezin; Parsidol; Parsitan; Parsotil; Rodipal. $C_{19}H_{24}N_2S$.HCl; mol wt 348.93. Crystals from ethylene dichloride, mp 223-225° (some decompn). Lower melting points reported are caused by admixture with 10-(2-diethylamino-1-methylethyl)phenothiazine-HCl which melts at 166-168°. One gram dissolves in 400 ml water at 20°, in 20 ml water at 40°. Sol in ethanol, chloroform. Soly in abs ethanol at 25° = 1.0 g/30 ml. Sparingly sol in acetone. Practically insol in ether, benzene. pH of a 5% aq soln is about 5.8. LD_{50} s.c. in mice: 670 mg/kg (Farquharson, Johnston).

THERAP CAT: Antiparkinsonian.

3801. Ethosuximide. [77-67-8] 3-Ethyl-3-methyl-2,5-pyrrolidinedione; 2-ethyl-2-methylsuccinimide; α-ethyl-α-methylsuccinimide; 3-methyl-3-ethylpyrrolidine-2,5-dione; Atysmal; Capitus; Emeside; Epileo Petitmal; Ethymal; Mesentol; Pemal; Peptinimid; Petinimid; Petnidan; Pyknolepsinum; Simatin; Succimal; Suxilep; Suximal; Suxinutin; Zarontin. $C_7H_{11}NO_2$; mol wt 141.17. C 59.56%, H 7.85%, N 9.92%, O 22.67%. Prepn: Sircar, *J. Chem. Soc.* **1927**, 1252. Acute toxicity data: H. Najer *et al., Bull. Soc. Chim. Fr.* **3**, 1119 (1966). Comparative clinical trial with valproic acid in epilepsy: S. Sato *et al., Neurology* **32**, 157 (1982). Brief review: S. J. Wallace, *Neurol. Clin.* **4**, 601 (1986).

Crystals from acetone + ether, mp 64-65° Very soluble in alc, ether; freely sol in water, chloroform; very slightly sol in solvent hexane. LD_{50} in mice (g/kg): 1.65 i.p.; 1.75 orally (Najer).

THERAP CAT: Anticonvulsant.

3802. Ethotoin. [86-35-1] 3-Ethyl-5-phenyl-2,4-imidazolidinedione; 3-ethyl-5-phenylhydantoin; 1-ethyl-2,5-dioxo-4-phenyl-imidazolidine; Peganone. $C_{11}H_{12}N_2O_2$; mol wt 204.23. C 64.69%, H 5.92%, N 13.72%, O 15.67%. Prepd by heating the potassium salt of 5-phenylhydantoin with ethyl bromide in alc at 100° in a sealed tube: Pinner, *Ber.* **21**, 2320 (1888). Use as anticonvulsant: W. J. Close, **US 2793157** (1957 to Abbott Labs.). Metabolism: Dudley *et al., J. Pharmacol. Exp. Ther.* **175**, 27 (1970).

Stout prisms from water, mp 94°. Freely sol in dehydrated alcohol, chloroform, benzene, dil aq solns of alkali hydroxides; sol in ether. Insol in water.

THERAP CAT: Anticonvulsant.

3803. 2-Ethoxyethanol. [110-80-5] Ethylene glycol monoethyl ether; Cellosolve; Oxitol. $C_4H_{10}O_2$; mol wt 90.12. C 53.31%, H 11.19%, O 35.51%. Manuf from ethylene oxide and ethanol: *Faith, Keyes & Clark's Industrial Chemicals*, F. A. Lowenheim, M. K. Moran, Eds. (Wiley-Interscience, New York, 4th ed., 1975) pp 403-407. Toxicity data: H. F. Smyth *et al., J. Ind. Hyg. Toxicol.* **23**, 259 (1941). Series of articles on toxicology: *Environ. Health Perspect.* **57**, 1-275 (1984).

Colorless, practically odorless liquid. d_{20}^{20} 0.931. bp 135°. mp −70°. Flash pt, closed cup: 112°F (44°C); open cup: 120°F (49°C). n_D^{25} 1.406. Misc with water, alc, ether, acetone, liquid esters. It dissolves many oils, resins, waxes, etc. LD_{50} orally in rats: 3 g/kg (Smyth).

Caution: Potential symptoms of overexposure in exptl animals are irritation of eyes, respiratory system; blood changes; liver, kidney, lung damage; reproductive, teratogenic effects. *See NIOSH Pocket Guide to Chemical Hazards* (DHHS/NIOSH 97-140, 1997) p 130. *See also Patty's Industrial Hygiene and Toxicology* **vol. 2D**, G. D. Clayton, F. E. Clayton, Eds. (Wiley-Interscience, New York, 4th ed., 1994) pp 2765, 2777-2785.

USE: Solvent for nitrocellulose, lacquers and dopes; in varnish removers, cleansing solns, dye baths; finishing leather with water pigments and dye solns; increasing stability of emulsions.

3804. 2-Ethoxyethyl Acetate. [111-15-9] 2-Ethoxyethanol 1-acetate; cellosolve acetate; ethylene glycol monoethyl ether acetate. $C_6H_{12}O_3$; mol wt 132.16. C 54.53%, H 9.15%, O 36.32%. Toxicity data: Smyth *et al., J. Ind. Hyg. Toxicol.* **23**, 259 (1941). Series of articles on toxicology: *Environ. Health Perspect.* **57**, 1-275 (1984).

Colorless liq, pleasant odor. d_{20}^{20} 0.975. bp 156°. Flash pt, open cup: 134°F (56°C). Sol in about 6 parts water. LD_{50} orally in rats: 5.1 g/kg (Smyth).

Caution: Potential symptoms of overexposure are vomiting; kidney damage; paralysis. Direct contact may cause irritation of eyes and nose. *See NIOSH Pocket Guide to Chemical Hazards* (DHHS/NIOSH 97-140, 1997) p 130. *See also Patty's Industrial Hygiene and Toxicology* **vol. 2D**, G. D. Clayton, F. E. Clayton, Eds. (Wiley-Interscience, New York, 4th ed., 1994) pp 2921-2925.

USE: In automobile lacquers to retard evaporation and impart high gloss.

3805. Ethoxymethylfurfural. [1917-65-3] 5-(Ethoxymethyl)-2-furancarboxaldehyde; 5-(ethoxymethyl)-2-furaldehyde; 2-formyl-5-ethoxymethylfuran; ω-ethoxymethylfurfuraldehyde; EMF. $C_8H_{10}O_3$; mol wt 154.17. C 62.33%, H 6.54%, O 31.13%. Naturally occurring aroma component of fortified wines. High energy organic liquid; prototype of the biofuels known as "furanics". Prepn from ω-bromomethylfurfuraldehyde: W. F. Cooper, W. H. Nuttall, *J. Chem. Soc., Trans.* **99**, 1193 (1911). Identification in port wine: R. F. Simpson, *J. Sci. Food Agric.* **31**, 214 (1980). Quantification in sweet fortified wines: I. Cutazch *et al.*, *J. Agric. Food Chem.* **47**, 2837 (1999). Prepn via dehydration of fructose: D. W. Brown *et al.*, *J. Chem. Technol. Biotechnol.* **32**, 920 (1982). Prepn from glucose-containing starting material and use as biofuel: G. J. M. Gruter, F. Dautzenberg, **EP 1834950** (2007 to Avantium). Direct prepn from cellulose: M. Mascal, E. B. Nikitin, *Angew. Chem. Int. Ed.* **47**, 7924 (2008).

Yellow oil. bp 235-240°; bp_{35} 145°. $d^{15.5}_{15.5}$ 1.1096. Energy density: 8.7 kWh/l. Readily sol in usual organic solvents. Sol in hot petroleum; insol in cold. Moderately sol in water.
USE: Alternative fuel.

3806. 2-Ethoxynaphthalene. [93-18-5] Ethyl β-naphthyl ether; ethyl β-naphtholate; nerolin bromelia; β-naphthyl ethyl ether. $C_{12}H_{12}O$; mol wt 172.23. C 83.69%, H 7.02%, O 9.29%. Prepd by heating β-naphthol with potassium ethyl sulfate, or with alcohol and H_2SO_4: Yokoyama *et al.*, *Yakugaku Zasshi* **78**, 123 (1958), *C.A.* **52**, 10986c (1958); from β-naphthol + *p*-toluenesulfonyl chloride + ethyl alcohol: Drakowzal, Klamann, *Monatsh. Chem.* **82**, 588 (1951); from β-naphthol + H_2SO_4 + ethyl acetate: S. Patai, M. Bentov, *J. Am. Chem. Soc.* **74**, 6118 (1952); from β-naphthol + triethylphosphate + *p*-toluenesulfonic acid: A. Bell, **US 2683748** (1954 to Eastman Kodak). Thermochemical properties: M. Colomina *et al.*, *J. Chem. Thermodyn.* **8**, 869 (1976).

Lustrous crystals, mp 37-38°. bp 282°. Floral sweet naphtha, orange blossom or grape odor. Heavy, powdery, floral taste with berry and grape nuance. $n_D^{47.3}$ 1.5932. d_{20}^{20} 1.0640. *Irritant.* Practically insol in water. Sol in alcohol, chloroform, ether, carbon disulfide, toluene, petr ether.
USE: In flavors and perfumery.

3807. Ethoxyquin. [91-53-2] 6-Ethoxy-1,2-dihydro-2,2,4-trimethylquinoline; 1,2-dihydro-6-ethoxy-2,2,4-trimethylquinoline; EMQ; Santoflex; Santoquin. $C_{14}H_{19}NO$; mol wt 217.31. C 77.38%, H 8.81%, N 6.45%, O 7.36%. Prepn: Knoevenagel, *Ber.* **54**, 1722 (1921); Baird *et al.*, **GB 505113** (1939 to ICI); C. C. Tung, *Tetrahedron* **19**, 1685 (1963). Structure: Cliffe, *J. Chem. Soc.* **1933**, 1327. Metabolism: R. H. Wilson *et al.*, *Agric. Food Chem.* **7**, 206 (1959); J. U. Skaare, E. Solheim, *Xenobiotica* **9**, 649 (1979). Toxicity data: V. I. Piul'skaya *et al.*, *C.A.* **90**, 4601c (1979).

Yellow liquid. bp_2 123-125°. n_D^{25} 1.569-1.672. d_{25}^{25} 1.029-1.031. LD_{50} orally in rats, mice: 1920, 1730 mg/kg (Piul'skaya).
USE: Antioxidant in feed and food; antidegradation agent for rubber.

3808. Ethoxysulfuron. [126801-58-9] *N*-[[(4,6-Dimethoxy-2-pyrimidinyl)amino]carbonyl]sulfamic acid 2-ethoxyphenyl ester;

3-(4,6-dimethoxypyrimidin-2-yl)-1-(2-ethoxyphenoxysulfonylurea); HOE-095404; Gladium; Grazie; Hero; Skol; Sunrice; Sunstar. $C_{15}H_{18}N_4O_7S$; mol wt 398.39. C 45.22%, H 4.55%, N 14.06%, O 28.11%, S 8.05%. Post-emergence herbicide for use in grains and sugarcane; inhibits the enzyme, acetohydroxyacid synthase. Prepn: H. Kehne *et al.*, **EP 342569**; *eidem*, **US 5104443** (1989, 1992 both to Hoechst). Comprehensive description: E. Hacker *et al.*, *Brighton Crop Prot. Conf. - Weeds* **1995**, 73. Mode of selective action: H. Köcher, G. Dickerhof, *ibid.* 249. Field study in summer rice: M. K. Bhowmick, R. K. Ghosh, *Adv. Plant Sci.* **15**, 499 (2002).

Crystals from diethyl ether, mp 145-147°. Vapor pressure (20°): 6.6×10^{-5} Pa. Partition coefficient (octanol/water) at 20°: 773 (pH 3); 1.01 (pH 7); 0.06 (pH 9). Soly in water at 20° (ppm): 26 (pH 5); 1353 (pH 7); 9628 (pH 9). LD_{50} in rats (mg/kg): >3270 orally; >4000 dermally (Hacker).
USE: Herbicide.

3809. Ethoxzolamide. [452-35-7] 6-Ethoxy-2-benzothiazolesulfonamide; ethoxyzolamide; Ethamide; Glaucotensil. $C_9H_{10}N_2O_3S_2$; mol wt 258.31. C 41.85%, H 3.90%, N 10.85%, O 18.58%, S 24.82%. Carbonic anhydrase inhibitor. Prepn: **GB 795174**; J. Korman, **US 2868800** (1958, 1959 both to Upjohn).

Crystals from ethyl acetate + Skellysolve B, mp 188-190.5°.
THERAP CAT: Diuretic.
THERAP CAT (VET): Diuretic.

3810. Ethybenztropine. [524-83-4] (3-*endo*-)-3-(Diphenylmethoxy)-8-ethyl-8-azabicyclo[3.2.1]octane; 3α-(diphenylmethoxy)-8-ethyl-1αH,5αH-nortropane; *N*-ethylnortropine benzhydryl ether; tropehydrylin; *N*-ethyl-8-aza-3-bicyclo[3.2.1]octyl benzhydryl ether; *N*-ethylbenztropine; ethylbenzatropine; etybenzatropine. $C_{22}H_{27}NO$; mol wt 321.46. C 82.20%, H 8.47%, N 4.36%, O 4.98%. Anticholinergic. Prepn: **GB 804837** (1958 to Sandoz); Boehringer *et al.*, **GB 824875** (1959 to Boehringer, Ing.).

Hydrochloride. [26598-44-7] Ponalide. $C_{22}H_{27}NO.HCl$; mol wt 357.92. Crystals from acetone, mp 190-191°.
Hydrobromide. [24815-25-6] $C_{22}H_{27}NO.HBr$; mol wt 402.38. Crystals from methanol + ether, mp 226-228°.
THERAP CAT: Antiparkinsonian.

3811. Ethyl Acetate. [141-78-6] Acetic acid ethyl ester; acetoxyethane; ethyl ethanoate; acetic ether; vinegar naphtha. $C_4H_8O_2$; mol wt 88.11. C 54.53%, H 9.15%, O 36.32%. Obtained by the slow distillation of a mixture of acetic acid, ethyl alc, and sulfuric acid: Alheritiere, Mercier, **US 2787636** (1957 to Usines de Melle);

Faith, Keyes & Clark's Industrial Chemicals, F. A. Lowenheim, M. K. Moran, Eds. (Wiley-Interscience, New York, 4th ed., 1975) pp 350-354. Toxicity: H. F. Smyth *et al.*, *Am. Ind. Hyg. Assoc. J.* **23**, 95 (1962).

H₃C — C(=O) — O — CH₃

Clear, volatile, characteristic fruity odor; pleasant taste when diluted. Slowly dec by moisture, then acquires an acid reaction. Absorbs water (up to 3.3% w/w). d_4^{20} 0.902; d_{25}^{25} 0.898. bp 77°. mp −83°. Ignition temp 800°F. Explosive limits (% vol in air): 2.2 to 11.5. n_D^{25} 1.3719. Vapor density 3.04 (air = 1). One ml dissolves in 10 ml water at 25°; more sol at lower and less sol at higher temps. Misc with alc, acetone, chloroform, ether. Azeotropic mixture with water (6.1% w/w) bp 70.4°. Azeotropic mixture with water (7.8% w/w) and alc (9.0% w/w) bp 70.3°. *Flammable. Keep tightly closed in a cool place.* LD₅₀ orally in rats: 11.3 ml/kg (Smyth).

Caution: Potential symptoms of overexposure are irritation of eyes, skin, nose and throat; narcosis; dermatitis. *See NIOSH Pocket Guide to Chemical Hazards* (DHHS/NIOSH 97-140, 1997) p 130.

USE: Pharmaceutic aid (flavor); artificial fruit essences. Extraction medium and chromatography reagent. Organic solvent for nitrocellulose, varnishes, lacquers, and aeroplane dopes; manuf smokeless powder, artificial leather, photographic films and plates, artificial silk, perfumes; cleaning textiles, etc.

3812. Ethyl Acetoacetate. [141-97-9] 3-Oxobutanoic acid ethyl ester; acetoacetic acid ethyl ester; acetoacetic ester; ethyl 3-oxobutanoate. C₆H₁₀O₃; mol wt 130.14. C 55.38%, H 7.75%, O 36.88%. Only the equilibrium mixture of the keto and enol forms is described here. Prepd from ethyl acetate by the action of sodium, sodium ethoxide, sodamide, or calcium: Inglis, Roberts, *Org. Synth.* **coll. vol. I**, 235 (2nd ed., 1941); Hansley, Schott, US 2843623 (1958 to Natl. Distillers); Scheibler, *Ann.* **565**, 176 (1949); Gattermann-Wieland, *Praxis des Organischen Chemikers* (de Gruyter, Berlin, 40th ed., 1961) p 218. Discussion of keto-enol tautomerism: Ward, *J. Chem. Educ.* **39**, 95 (1962). Absorption spectrum: Morton, Rosney, *J. Chem. Soc.* **1926**, 711. Toxicity data: H. F. Smyth *et al.*, *J. Ind. Hyg. Toxicol.* **31**, 60 (1949).

H₃C — C(=O) — CH₂ — C(=O) — O — CH₃

Liquid. Agreeable odor. d_4^{10} 1.0357; d_4^{17} 1.0288; d_4^{25} 1.0213; d_4^{54} 0.9924; d_4^{75} 0.9703. mp −45°. bp₇₆₀ 180.8°; bp₄₀₀ 158.2°; bp₂₀₀ 138.0°; bp₂₀ 106°; bp₆₀ 81.1°; bp₅ 54.0°; bp₁.₀ 28.5°. n_D^{20} 1.41937. Flash pt, closed cup: 184°F. Sol in about 35 parts water; misc with the usual organic solvents. LD₅₀ orally in rats: 3.98 g/kg (Smyth).

Caution: Moderately irritating to skin, mucous membranes.

3813. Ethyl Acrylate. [140-88-5] 2-Propenoic acid ethyl ester; acrylic acid ethyl ester. C₅H₈O₂; mol wt 100.12. C 59.98%, H 8.05%, O 31.96%. Prepd from ethylene chlorohydrin or acrylonitrile, ethanol, and sulfuric acid; also by an oxo reaction from acetylene, carbon monoxide, and ethanol in the presence of suitable catalysts. *See* the refs under Methyl Acrylate.

H₃C — O — C(=O) — CH=CH₂

Monomer. Liquid, acrid, penetrating odor, retained by clothing. *Lacrimator.* d_4^{20} 0.9405. fp below −72°. bp₇₆₀ 99.4°; bp₃₉.₂ 20° (polymerizes on distn). n_D^{20} 1.404. Specific heat at −60°: 0.442 cal/g/°C. Heat of vaporization 8.27 kcal/mol; heat of combustion 655.49 kcal/mol. Flash pt, open cup: 60°F (15°C). Vapor density 3.45 (air = 1). *Flammable.* Soly in water at 20°: 2 g/100 ml. Soly of water in ethyl acrylate at 20°: 1.5 g/100 g. Sol in alcohol, ether. Azeotropes: 45.0% water = bp 81°; 56.8% ethanol = bp 76°. Easily polymerizes on standing; polymerization process speeded up by heat,

light, and peroxides. If pure, the monomer can be stored below +10° without incurring polymerization.

Polymer. Transparent, elastic substance. Practically no odor. Little adhesive power. Resists the usual solvents.

Caution: Potential symptoms of overexposure to the monomer are irritation of eyes, respiratory system and skin. *See NIOSH Pocket Guide to Chemical Hazards* (DHHS/NIOSH 97-140, 1997) p 132. *See also Patty's Industrial Hygiene and Toxicology* **vol. 2A**, G. D. Clayton, F. E. Clayton, Eds. (Wiley-Interscience, New York, 3rd ed., 1981) p 2292-2296. This substance was formerly listed as reasonably anticipated to be a human carcinogen: *Eighth Report on Carcinogens* (PB99-128746, 1998) p III-627; delisted because the relevant data are not sufficient to meet the current criteria for this listing: *Ninth Report on Carcinogens* (PB2000-107509, 2000) p B-2.

USE: The monomer in the manuf of water emulsion paint vehicles; in production of emulsion-based polymers used in textile and paper coatings, leather finish resins and adhesives. Imparts flexibility to hard films.

3814. Ethyl Alcohol. [64-17-5] Ethanol; absolute alcohol; anhydrous alcohol; dehydrated alcohol; ethyl hydrate; ethyl hydroxide. C₂H₆O; mol wt 46.07. C 52.14%, H 13.13%, O 34.73%. CH₃CH₂OH. Manuf: by fermentation of starch, sugar, and other carbohydrates; from ethylene, acetylene, sulfite waste liquors, and synthesis gas (CO + H); by hydrolysis of ethyl sulfate, and oxidation of methane. Toxicity: G. S. Wiberg *et al.*, *Toxicol. Appl. Pharmacol.* **16**, 718 (1970). Embryotoxicity in mammals: N. A. Brown *et al.*, *Science* **206**, 573 (1979). Possible mechanism for actions of ethanol on the brain: G. Aston-Jones *et al.*, *Nature* **296**, 857 (1982). Ethanol-induced chromosomal abnormalities in mice: M. H. Kaufman, *ibid.* **302**, 258 (1983). Disruption of reproductive function in female primates following alcohol self-administration: N. K. Mello *et al.*, *Science* **221**, 677 (1983). Review of metabolism and toxicity: C. S. Lieber in *Reviews in Biochemical Toxicology* **vol. 5**, E. Hodgson *et al.*, Eds. (Elsevier, New York, 1983) pp 267-312; of pharmacology: L. Pohorecky, J. Brick, *Pharmacol. Ther.* **36**, 335-427 (1988); of hepatotoxicity: C. S. Lieber, L. M. DeCarli, *J. Hepatol.* **12**, 394-401 (1991). General reviews: P. Baud, "Ethyl Alcohol Industry" in Grignard, *Traité de Chimie Organique* **vol. 5** (Masson, 1937) pp 841-975; Zabel; *Chem. Ind.* **64**, 212 (1949); Faith, Keyes & Clark's Industrial Chemicals, F. A. Lowenheim, M. K. Moran, Eds. (Wiley-Interscience, New York, 4th ed., 1975) pp 355-364; P. D. Sherman, P. R. Kavasmaneck, "Ethanol" in *Kirk-Othmer Encyclopedia of Chemical Technology* **vol. 9** (Interscience, New York, 3rd ed., 1980) pp 338-380.

Clear, colorless, very mobile, pleasant odor; burning taste. Absorbs water rapidly from air. d_4^{20} 0.789. bp 78.5°. mp −114.1°. n_D^{20} 1.361. Flash pt, closed cup: 13°C. Miscible with water and with many organic liquids. *Flammable. Keep tightly closed, cool, and away from flame.* LD₅₀ in young, old rats (g/kg): 10.6, 7.06 orally (Wiberg). The terms *95% alcohol* and *alcohol* (when used alone) refer to a binary azeotrope having a distillate composition of 95.57% ethyl alcohol (by wt) and bp 78.15°. *Alcohol, USP* is specified as containing not less than 92.3% and not more than 93.8% by weight, corresponding to not less than 94.9% and not more than 96.0% by vol of C₂H₅OH at 15.56°. d_{25}^{25} 0.810; d 0.816 at 15.56° (60°F). *Diluted alcohol,* prepd from equal vols 95% alcohol and water, contains about 41.5% by wt or about 48.9% by vol of C₂H₅OH. d_{25}^{25} 0.931; d 0.936 at 15.56° (60°F). *See: USP* **XXI**, 22, 1530 (1985).

Caution: Potential symptoms of overexposure are irritation of eyes, skin, nose; headache, nausea, vomiting, drowsiness, fatigue, narcosis; cough; flushing, rapid pulse, sweating; mental excitement or depression, impaired perception, incoordination, stupor, coma; liver damage; anemia; reproductive and teratogenic effects. *See NIOSH Pocket Guide to Chemical Hazards* (DHHS/NIOSH 97-140, 1997) p 132; *Clinical Toxicology of Commercial Products*, R. E. Gosselin *et al.*, Eds. (Williams & Wilkins, Baltimore, 5th ed., 1984) Section III, pp 166-171.

USE: Most ethyl alcohol is used in alcoholic beverages in suitable dilutions. Reagent in synthetic organic chemistry and chromatography. Industrial and laboratory organic solvent. Other uses are in manuf of denatured alcohol, pharmaceuticals (rubbing compds, lotions, tonics, colognes), in perfumery. Octane booster in gasoline. Pharmaceutic aid (solvent).

THERAP CAT: Antiseptic.

THERAP CAT (VET): Antiseptic. To destroy nerve tissue. Solvent and dehydrating agent.

3815. Ethyl Alcohol, Denatured. Denatured alcohol. Ethyl alcohol to which has been added some substance or substances which, while allowing the use of the alcohol in the most varied industries and arts, renders it entirely unfit for consumption as a beverage. The most commonly used denaturants, either alone or in combination, are the following: Methanol, camphor, Aldehol, amyl alcohol, gasoline, isopropanol, terpineol, benzene, castor oil, acetone, nicotine, aniline dyes, ether, cadmium iodide, pyridine bases, sulfuric acid, kerosene, diethyl phthalate. *Formula 1* is 5 gallons approved wood alcohol added to 100 gal of 95% ethanol. *Formula 2B* is 0.5 gal benzene added to 100 gal of 95% ethanol. Similarly *formula 3A* contains 5 gal commercial methanol, *formula 6B* contains 0.5 gal pyridine bases, *formula 12A* 5 gal benzene, *formula 13A* 10 gal ethyl ether, *formula 19* 4 gal methyl isobutyl ketone and 1 gal kerosene, *formula 20* 5 gal crude chloroform, *formula 23A* 10 gal acetone, *formula 28* 10 gal benzene, *formula 28A* 1 gal gasoline, *formula 30* 10 gal methanol, *formula 32* 5 gal ethyl ether, *formula 33* 30 lbs methyl violet, *formula 35A* 5 gal ethyl acetate, *formula 39C* 1 gal diethyl phthalate; *formula 44* contains 20 gal *n*-butanol. Additional permissible formulas are given in *Appendix to Regulations No. 3, Formulae for Completely and Specially Denatured Alcohol*, published by the U.S. Treasury Dept., Bureau of Industrial Alcohol.

Caution: Denaturants, particularly methanol, may modify and increase toxic symptoms caused by ingestion and exposure to fumes.

3816. Ethylaluminum Dichloride. [563-43-9] Dichloroethylaluminum; EtAlCl₂. $C_2H_5AlCl_2$; mol wt 126.94. C 18.92%, H 3.97%, Al 21.26%, Cl 55.85%. Organoaluminum Lewis acid that also functions as a proton scavenger. Prepn: F. C. Hall, A. W. Nash, *J. Inst. Petroleum Tech.* **23**, 679 (1937); A. V. Grosse, J. M. Mavity, *J. Org. Chem.* **5**, 106 (1940). Synthetic applications in Friedel-Crafts acylations: B. B. Snider, A. C. Jackson, *ibid.* **47**, 5393 (1982); in ene reactions: *idem et al., ibid.* **48**, 464 (1983); in hydrosilylations: T. Sudo *et al., ibid.* **64**, 2494 (1999); in polymerizations: R. P. F. Guiné, J. A. A. M. Castro, *J. Appl. Polym. Sci.* **82**, 2558 (2001); in cycloadditions: K. Takasu *et al., J. Org. Chem.* **69**, 517 (2004).

White crystals from pentane, mp 32°. bp₅₀ 114.5-115.5°; bp₀.₂₅ 40°. d_4^{40} 1.2461. *Flammable. Corrosive.* Flash pt, closed cup: 0°F (−18°C). Sol in most organic solvents. *Reacts violently with water. Vapors may form explosive mixture with air.*

USE: Reagent and catalyst in synthetic organic chemistry.

3817. Ethylamine. [75-04-7] Ethanamine; monoethylamine; aminoethane. C_2H_7N; mol wt 45.09. C 53.28%, H 15.65%, N 31.06%. Prepn from ethyliodide + liq ammonia: Watt, Otto, *J. Am. Chem. Soc.* **69**, 836 (1947); from ethanol + ammonia: Davies *et al.*, **US 2609394** (1952 to ICI); Lemon, Myerly, **US 3022349** (1962 to Union Carbide). Toxicity study: H. F. Smyth *et al., Arch. Ind. Hyg. Occup. Med.* **10**, 61 (1954).

Liquid; ammonia odor; strong alkaline reaction. d_{15}^{15} 0.689. bp 16.6°. Solidif −80°. Miscible with water, alcohol, ether. *Flammable. Keep tightly closed and in cold place.* LD₅₀ orally in rats: 0.40 g/kg (Smyth).

Hydrochloride. [557-66-4] C_2H_7N·HCl. Crystals from ethanol + water, mp 110°. d 1.22. Soluble in 0.4 part water; freely sol in alcohol; slightly sol in chloroform or acetone. Practically insol in ether. *Keep well closed.*

Hydriodide. [506-58-1] C_2H_7N·HI. Hygroscopic crystals, mp 188°. d 2.10. Freely sol in water or alcohol. Practically insol in chloroform, ether. *Keep well closed and protected from light.*

Oleate. [39664-27-2] Ethanamine (Z)-9-octadecenoate; Etalate. C_2H_7N·$C_{18}H_{34}O_2$; mol wt 327.55. Commercial prepn is a 5% soln with 2% benzyl alcohol as anodyne.

Caution: Potential symptoms of overexposure are irritation of eyes, skin and respiratory system; skin burns; dermatitis. *See NIOSH Pocket Guide to Chemical Hazards* (DHHS/NIOSH 97-140, 1997) p 132.

USE: In resin chemistry; stabilizer for rubber latex; intermediate for dyestuffs, medicinals; in oil refining; in organic syntheses.

THERAP CAT: Oleate as a sclerosing agent.

3818. N-Ethylamphetamine. [457-87-4] N-Ethyl-α-methylbenzeneethanamine; N-ethyl-α-methylphenethylamine; N-ethyl-ω-phenylisopropylamine; 2-ethylamino-1-phenylpropane. $C_{11}H_{17}N$; mol wt 163.26. C 80.93%, H 10.50%, N 8.58%. Prepn: Keil, Dobke, **DE 767263** (1952 to Theodor H. Temmler); Leonard *et al., J. Am. Chem. Soc.* **80**, 4858 (1958). Separation of isomers: **GB 814339** (1959 to Sterling Drug).

bp₁₄ 104.5-106°. n_D^{25} 1.4986.

Hydrochloride. [1858-47-5] MG-19973; Apetinil-Depo. $C_{11}H_{17}N$·HCl; mol wt 199.72.

d-Form hydrochloride. $C_{11}H_{17}N$·HCl. mp 154-156°. $[\alpha]_D^{15}$ +17.2° (c = 2 in water).

l-Form hydrochloride. $C_{11}H_{17}N$·HCl. mp 155-156°. $[\alpha]_D^{25}$ −17.3° (c = 2 in water).

Note: This is a controlled substance (stimulant): **21 CFR**, 1308.11.

THERAP CAT: Anorexic.

3819. Ethylaniline. [103-69-5] N-Ethylbenzenamine; ethylphenylamine. $C_8H_{11}N$; mol wt 121.18. C 79.29%, H 9.15%, N 11.56%. Produced by heating aniline hydrochloride and alcohol at 180°. Toxicology: M. Sziza, L. Podhragyai, *Arch. Gewerbepathol. Gewerbehyg.* **15**, 447 (1957).

Very refractive liquid; rapidly becomes brown on exposure to light and air. Aniline-like odor. d_{25}^{25} 0.958. Solidifies below 80°. bp 204.5°. mp −63.5°. n_D^{20} 1.5559. Insol in water. Miscible with alc, ether and many other organic solvents. *Keep well closed and protected from light.* LD₅₀ in rats (g/kg): 0.18 i.p., 0.28 orally, 4.7 s.c. (Sziza, Podhragyai).

3820. Ethylbenzene. [100-41-4] C_8H_{10}; mol wt 106.17. C 90.50%, H 9.49%. Prepn from acetophenone: Clemmensen, *Ber.* **46**, 1838 (1913); Gattermann-Wieland, *Praxis des Organischen Chemikers* (de Gruyter, Berlin, 40th ed., 1961) p 332; by Huang-Minlon modification of Wolff-Kishner reduction: A. I. Vogel, *Practical Organic Chemistry* (Longmans, 3rd ed., 1959) p 516. Physical properties: L. C. Gibbons *et al., J. Am. Chem. Soc.* **68**, 1130 (1946). Manuf: *Faith, Keyes & Clark's Industrial Chemicals*, F. A. Lowenheim, M. K. Moran, Eds. (Wiley-Interscience, New York, 4th ed., 1975) pp 365-370. Toxicity study: H. F. Smyth *et al., Am. Ind. Hyg. Assoc. J.* **23**, 95 (1962). Review of toxicology and human exposure: *Toxicological Profile for Ethylbenzene* (PB2010-100004, 2010) 319 pp.

Colorless liquid. *Flammable.* d_{25}^{25} 0.866. bp 136.25°. mp −95.01°. n_D^{25} 1.4932. Flash pt, closed cup: 64°F (18°C). Practically insol in water. Misc with the usual organic solvents. Sol in alc, ether. LD₅₀ orally in rats: 5.46 g/kg (Smyth).

Caution: Potential symptoms of overexposure are irritation of eyes, skin and mucous membranes; headache; dermatitis; narcosis, coma. *See NIOSH Pocket Guide to Chemical Hazards* (DHHS/NIOSH 97-140, 1997) p 132.

USE: For conversion to styrene monomer; as resin solvent.

3821. Ethyl Benzoate. [93-89-0] Benzoic acid ethyl ester. $C_9H_{10}O_2$; mol wt 150.18. C 71.98%, H 6.71%, O 21.31%.

Colorless, clear, refractive liq; aromatic odor; vapors cause cough. d_4^{25} 1.050. bp 211-213°. mp −34°. n_D^{20} 1.506. Almost insol in water; miscible with alcohol, chloroform, ether, petr ether. LD_{50} orally in rats: 6.48 g/kg; *see:* Smyth *et al., Arch. Ind. Hyg. Occup. Med.* **10**, 61 (1954).

USE: In perfumery under the name *Essence de Niobe*; in manuf of *Peau d'Espagne*; artificial fruit essence.

3822. Ethyl Benzoylacetate. [94-02-0] β-Oxobenzenepropanoic acid ethyl ester. $C_{11}H_{12}O_3$; mol wt 192.21. C 68.74%, H 6.29%, O 24.97%.

Liquid; pleasant odor; becomes yellow on exposure to air and light. d^{15} 1.122. bp 265-270° with decompn. Volatile with steam. n_D^{20} 1.5338. Insol in water; miscible with alcohol, ether. *Keep protected from air and light.*

3823. α-Ethylbenzyl Alcohol. [93-54-9] α-Ethylbenzenemethanol; ω-ethylbenzyl alcohol; 1-phenyl-1-propanol; 1-phenylpropyl alcohol; ethyl phenyl carbinol; α-hydroxypropylbenzene; SH-261; Ejibil; Livonal; Phenycholon; Phenicol; Phenychol; Felicur; Felitrope. $C_9H_{12}O$; mol wt 136.19. C 79.37%, H 8.88%, O 11.75%. Prepd from benzaldehyde or from ethyl phenyl ketone: Norris, Cortese, *J. Am. Chem. Soc.* **49**, 2640 (1927). Toxicity study: O. Linét *et al., Arzneim.-Forsch.* **12**, 347 (1962).

Oily liquid. Weak, ester-like odor. Sweetish, slightly irritating taste. bp$_{760}$ 219°; bp$_{15}$ 107°; bp$_3$ 78°. d_4^{25} 0.9915. n_D^{23} 1.5169. uv max (methanol): 250, 260 nm (ε 173, 114). Misc with methanol, ethanol, ether, benzene, toluene, olive oil. LD_{50} orally in rats: 1.6 ml/kg (Linét).

USE: As heat transfer medium; in perfumery.

THERAP CAT: Choleretic.

3824. Ethylbenzylaniline. [92-59-1] *N*-Ethyl-*N*-phenylbenzenemethanamine; *N*-ethyl-*N*-phenylbenzylamine. $C_{15}H_{17}N$; mol wt 211.31. C 85.26%, H 8.11%, N 6.63%. Prepn: Martin, MacQueen, **US 1887772** (1933 to Dow); Burgstahler, *J. Am. Chem. Soc.* **73**, 3021 (1951).

Light yellow, oily liq. d_4^{19} 1.034. bp$_{710}$ 287° with slight decompn; bp$_{14}$ 170-180°. n_D^{23} 1.5938. Practically insol in water. One ml dissolves in 5.5 ml alcohol; sol in the usual organic solvents.

USE: Manuf dyes; in organic syntheses.

3825. Ethyl Biscoumacetate. [548-00-5] 4-Hydroxy-α-(4-hydroxy-2-oxo-2*H*-1-benzopyran-3-yl)-2-oxo-2*H*-1-benzopyran-3-acetic acid ethyl ester; bis(4-hydroxy-2-oxo-2*H*-1-benzopyran-3-yl)-acetic acid ethyl ester; ethyl bis(4-hydroxycoumarinyl)acetate; ethyldicoumarol acetate; bis-3,3′-(4-hydroxycumarinyl)acetic acid ethyl ester; 3,3′-carboxymethylene bis(4-hydroxycoumarin)ethyl ester; ethyl 4,4′-dihydroxydicoumarinyl-3,3′-acetate; B.O.E.A.; Dicumacyl; Pelentan; Stabilene; Tromexan. $C_{22}H_{16}O_8$; mol wt 408.36. C 64.71%, H 3.95%, O 31.34%. Prepn: Rosicky, *Cas. Lek. Cesk.* **83**, 1200 (1944); US 2482510; US 2482511; US 2482512 (all 1949 to Spojené farm. zovody); Stahmann *et al., J. Am. Chem. Soc.* **65**, 2285 *sqq* (1943). Toxicity study: C. M. Gruber *et al., Fed. Proc.* **10**, 303 (1951).

Crystals, dimorphous, mp 177-182° and mp 154-157°. Bitter, persistent taste. Practically insol in water. Sol in 20 parts acetone, also sol in benzene. Slightly sol in alcohol, ether. LD_{50} in mice, rats (g/kg): 0.88, 0.26 orally; 0.32, 1.1 i.p. (Gruber).

THERAP CAT: Anticoagulant.

3826. Ethyl Bromide. [74-96-4] Bromoethane; monobromoethane; bromic ether; hydrobromic ether. C_2H_5Br; mol wt 108.97. C 22.04%, H 4.63%, Br 73.33%. Made by distilling from a mixture of HBr, ethyl alcohol and H_2SO_4: Kamm, Marvel, *Org. Synth.* **coll. vol. I**, 29 (1941). By phosphorus and bromine method: Goshorn *et al., ibid.* 36. Absorption spectrum: Hantzsch, *Ber.* **58**, 619 (1925). Physical properties: Mumford, Phillips, *J. Chem. Soc.* **1950**, 75. Toxicity data: E. H. Vernot *et al., Toxicol. Appl. Pharmacol.* **42**, 417 (1977).

Colorless, volatile liq; ethereal odor; burning taste; becomes yellowish on exposure to air and light. Vapor harmful. d_4^{20} 1.4612; d_4^{25} 1.4515. bp 38.2°. mp −119°. n_D^{20} 1.4242. *Poisonous.* Soly in water (g/100 g) at 0°: 1.067; 10°: 0.965; 20°: 0.914; 30°: 0.896; miscible with alcohol, ether, chloroform and with other organic solvents. Explosive limits (% by vol in air), lower 6.75, upper 11.25. Autoignition temp 952°F (511°C). LC_{50} rats, mice (ppm): 27000, 16200 (Vernot).

Caution: Potential symptoms of overexposure are irritation of eyes, respiratory system and skin; CNS depression; pulmonary edema; liver and kidney disease; cardiac arrhythmias; cardiac arrest. *See NIOSH Pocket Guide to Chemical Hazards* (DHHS/NIOSH 97-140, 1997) p 134.

USE: Ethylating agent in organic synthesis; as refrigerant. Formerly used as a topical and inhalation anesthetic.

3827. Ethyl α-Bromopropionate. [535-11-5] 2-Bromopropanoic acid ethyl ester. $C_5H_9BrO_2$; mol wt 181.03. C 33.17%, H 5.01%, Br 44.14%, O 17.68%.

Liquid; sharp, pungent odor; becomes yellow on exposure to light. d_{20}^{20} 1.447. bp 159-160°; also stated as 160-165°. n_D^{20} 1.4469. Insol in water; miscible with alcohol, ether. *Protect from light.*

3828. Ethyl Butylacetylaminopropionate. [52304-36-6] *N*-Acetyl-*N*-butyl-β-alanine ethyl ester; 3-(*N*-butyl-*N*-acetyl)-aminopropionic acid ethyl ester; ethyl-3-(*N*-*n*-butyl-*N*-acetyl)aminopropionate; insect repellent 3535; IR 3535. $C_{11}H_{21}NO_3$; mol wt 215.29. C 61.37%, H 9.83%, N 6.51%, O 22.29%. *N*-Disubstituted derivative of β-alanine, *q.v.* Prepd (not claimed): M. Klier *et al.*, **DE 2246433**; *eidem*, **US 4127672** (1974, 1978 both to Beiersdorf). Prepn, toxicity and insect repellency studies: M. Klier, F. Kuhlow, *J. Soc. Cosmet. Chem.* **27**, 141 (1976). Chemical, physical and toxicological properties: *WHO Interim Specification for IR 3535* (WHO/IS/TC/667/2001) 15 pp. Lab and field tests of efficacy vs mosquitoes: U. Thavara *et al.*, *J. Am. Mosq. Control Assoc.* **17**, 190 (2001); and comparison to other repellents: D. R. Barnard *et al.*, *J. Med. Entomol.* **39**, 895 (2002); M. S. Fradin, J. F. Day, *N. Engl. J. Med.* **347**, 13 (2002). HPLC determn in topical gel formulation: S. C. Marselos, H. A. Archontaki, *J. Chromatogr. A* **946**, 295 (2002). Assessment of efficiency of repellent formulations: R. Milutinovic *et al.*, *SOFW-J.* **128**, 14 (2002). Review of repellency spectrum vs target pests: F. Marchio, *ibid.* **122**, 478-485 (1996).

Colorless to slightly yellowish, almost odorless liquid. mp < −20°. bp$_{0.2}$ 108-110°. bp$_{0.5}$ 126-127°. n_D^{20} 1.452-1.455. Flash point: 318°F (159°C). Log P (octanol/water): 1.7 (23°). Vapor pressure at 20°: 0.15 Pa. Soly at 20°: water 70 ±3 g/l (non-buffered). Sol in *n*-heptane, dichloromethane, ethyl acetate, *p*-xylene, acetone, methanol. LD$_{50}$ in rats (mg/kg): >5000 orally; LC$_{50}$ in rats (mg/l): >5.1 by inhalation (WHO).

USE: Insect repellent.

3829. Ethyl *tert*-Butyl Ether. [637-92-3] 2-Ethoxy-2-methylpropane; *tert*-butyl ethyl ether; ethyl *tert*-butyl oxide; 1,1-dimethylethyl ethyl ether; ethyl 1,1-dimethylethyl ether; ETBE. C$_6$H$_{14}$O; mol wt 102.18. C 70.53%, H 13.81%, O 15.66%. Prepn: J. U. Nef, *Ann.* **309**, 126 (1899). Synthesis: J. F. Norris, G. W. Rigby, *J. Am. Chem. Soc.* **54**, 2088 (1932). Physical properties: T. W. Evans, K. R. Edlund, *Ind. Eng. Chem.* **28**, 1186 (1936). Thermal decomposition: N. J. Daly, C. Wentrup, *Aust. J. Chem.* **21**, 1535 (1968). Brief review focusing on use as gasoline additive: M. Iborra *et al.*, *Chemtech* **18**, 120-122 (1988).

fp −94.0°. bp$_{760}$ 72.8° (Evans, Edlund). Also reported as bp$_{760}$ 73.1° (Norris, Rigby). n_D^{25} 1.3728. n_D^{20} 1.3760. d$_4^{15}$ 0.7456; d$_4^{20}$ 0.7404; d$_4^{25}$ 0.7353, d$_4^{30}$ 0.7300. Vapor pressure at 25°: 130 mm Hg. Heat vaporization: 74.3 cal/g. Specific heat (liquid) at 25°: 0.51 cal/g/°C. Surface tension at 24°: 19.8 dynes/cm. *Flammable.* Soly in water (20°): 1.2 g/100 g soln. Soly of water in compound (20°): 0.5 g/100 g soln.

USE: Gasoline additive.

3830. Ethyl Butyrate. [105-54-4] Butanoic acid ethyl ester; butyric acid ethyl ester; ethyl *n*-butyrate. C$_6$H$_{12}$O$_2$; mol wt 116.16. C 62.04%, H 10.41%, O 27.55%. Toxicity data: P. M. Jenner *et al.*, *Food Cosmet. Toxicol.* **2**, 327 (1964).

Colorless liquid; pineapple odor. d$_4^{20}$ 0.879. bp 120-121°. mp −93°. n_D^{20} 1.400. Flash pt, closed cup: 78°F (25°C); open cup: 85°F (29°C). *Flammable.* Sol in about 150 parts water; misc with alcohol, ether. LD$_{50}$ orally in rats: 13,050 mg/kg (Jenner).

USE: Manuf artificial rum; perfumery; the alcoholic soln constitutes the so-called *"pineapple oil"*.

3831. Ethyl Caprate. [110-38-3] Decanoic acid ethyl ester; ethyl decanoate. C$_{12}$H$_{24}$O$_2$; mol wt 200.32. C 71.95%, H 12.08%, O 15.97%.

Colorless liq. d^{20} 0.862. bp 243-245°. Insoluble in water; miscible with alcohol, chloroform, ether.

USE: Manuf wine bouquets, cognac essence.

3832. Ethyl Caproate. [123-66-0] Hexanoic acid ethyl ester; ethyl hexanoate. C$_8$H$_{16}$O$_2$; mol wt 144.21. C 66.63%, H 11.18%, O 22.19%.

Colorless to yellowish liquid; pleasant odor. d^{20} 0.873. bp 166-167°. Insol in water; miscible with alcohol, ether.

USE: Manuf artificial fruit flavors.

3833. Ethyl Caprylate. [106-32-1] Octanoic acid ethyl ester; ethyl octanoate; ethyl octylate. C$_{10}$H$_{20}$O$_2$; mol wt 172.27. C 69.72%, H 11.70%, O 18.57%.

Colorless, clear, very mobile liquid; pleasant, pineapple odor. d^{17} 0.878. bp 207-209°. Insol in water; misc with alc, ether. LD$_{50}$ orally in rats: 25,960 mg/kg; *see:* P. M. Jenner *et al.*, *Food Cosmet. Toxicol.* **2**, 327 (1964).

USE: Manuf fruit ethers; constit of enanthic, cocoic, and cognac ethers.

3834. Ethyl β-Carboline-3-carboxylate. [74214-62-3] 9*H*-Pyrido[3,4-*b*]indole-3-carboxylic acid ethyl ester; ethyl norharmancarboxylate; β-CCE. C$_{14}$H$_{12}$N$_2$O$_2$; mol wt 240.26. C 69.99%, H 5.03%, N 11.66%, O 13.32%. Deriv of **β-carboline** that is a potent displacer of ^{3}H-diazepam from brain benzodiazepine receptors. Isoln from human urine and brain and binding site study: C. Braestrup *et al.*, *Proc. Natl. Acad. Sci. USA* **77**, 2288 (1980). Initially thought to be an endogenous ligand for benzodiazepine receptors in mammalian CNS, it is now believed to be formed during isoln and extraction procedures: R. F. Squires in *GABA and Benzodiazepine Receptors*, E. Costa *et al.*, Eds. (Raven Press, New York, 1980) pp 129-138; M. Nielson *et al.*, *J. Neurochem.* **36**, 276 (1981). Synthesis and psychotropic activity: **JP Kokai 81 43283** (to Schering AG), *C.A.* **95**, 115508a (1981); U. Eder *et al.*, **EP 30254** (1981 to A/S Ferrosan; Schering AG). β-CCE has been shown to lower seizure threshold and to reverse the sedative effect of flurazepam, *q.v.:* P. J. Cowen *et al.*, *Nature* **290**, 54 (1981). Neurochemical and pharmacological actions of β-CCE and other β-carbolines: M. Cain *et al.*, *J. Med. Chem.* **25**, 1081 (1982). Anxiogenic and convulsant properties: L. Prado de Carvalho *et al.*, *Nature* **301**, 64 (1983).

mp 229-233°. uv max (pH 7): 215, 242, 279 nm.

3-Hydroxymethyl-β-carboline. [65474-79-5] 9*H*-Pyrido[3,4-*b*]indole-3-methanol; 3-HMC. C$_{12}$H$_{10}$N$_2$O; mol wt 198.23. Prepn: F. Hamaguchi, S. Ohki, *Heterocycles* **8**, 383 (1977); M. Cain *et al.*, *loc. cit.* Antagonism of anticonvulsant and anxiolytic actions of diazepam: P. Skolnick *et al.*, *Eur. J. Pharmacol.* **68**, 381 (1980). Crystals, mp 225-228°.

USE: As tools for studying benzodiazepine receptors.

3835. Ethyl Carbonate. [105-58-8] Carbonic acid diethyl ester; diethyl carbonate; Eufin. $C_5H_{10}O_3$; mol wt 118.13. C 50.84%, H 8.53%, O 40.63%. Prepn: Palomaa *et al., Ber.* **72**, 313 (1939). Manuf: Mador, Blackham, US 3114762 (1963 to Natl. Distillers & Chem.).

Liquid, bp 126°. Pleasant ethereal odor, mp −43°. Flash pt, closed cup: 77°F (25°C). d_4^{20} 0.9764. n_D^{20} 1.3843. Practically insol in water; miscible with alcohol, ether.

USE: Solvent for nitrocellulose; manuf radio tubes; fixing rare earths to cathode elements.

3836. Ethyl Cellulose. [9004-57-3] Cellulose ethyl ether; Ethocel. Prepd from wood pulp or chemical cotton by treatment with alkali and ethylation of the alkali cellulose with ethyl chloride. Review and bibliography: E. Ott, *Cellulose and Cellulose Derivatives* (New York, 2nd ed., 1955).

Free-flowing white to light tan powder. Commercial ethyl cellulose has an ethoxy content of 43-50%. Soly is dependent upon the degree of substitution. A product containing <46.5% of ethoxy groups is freely sol in tetrahydrofuran, methyl acetate, chloroform and in mixtures of aromatic hydrocarbons with alcohol. A product containing not less than 46.5% softens at 140° and is freely sol in alc, methanol, toluene, chloroform, ethyl acetate. Sol in ethylene dichloride, benzene, xylene, butyl acetate, acetone, ethanol, butanol, carbon tetrachloride. Insol in water, glycerin, propylene glycol. To avoid brittleness, ethyl cellulose formulations usually include an antioxidant such as hydroquinone monobenzyl ether, 4-hexylpyrocatechol, or diphenylamine.

USE: In the manuf of plastics and lacquers. Pharmaceutic aid (tablet binder).

3837. Ethyl Chloride. [75-00-3] Chloroethane; monochlorethane; chlorethyl; aethylis chloridum; ether chloratus; ether hydrochloric; ether muriatic; Kelene; Chelen; Anodynon; Chloryl Anesthetic; Narcotile. C_2H_5Cl; mol wt 64.51. C 37.24%, H 7.81%, Cl 54.95%. Prepd by the action of chlorine on ethylene in the presence of HCl and light: US 2393509 (1946); by the action of chlorine on ethylene in the presence of the chlorides of copper, iron, antimony, and calcium: DE 298931; Bähr, Zieler, *Angew. Chem.* **1930**, 233, 286; by heating alcohol, HCl and $ZnCl_2$: US 2396639 (1946). Review of mfg processes: *Faith, Keyes & Clark's Industrial Chemicals,* F. A. Lowenheim, M. K. Moran, Eds. (Wiley-Interscience, New York, 4th ed., 1975) pp 371-375; of toxicology and human exposure: *Toxicological Profile for Chloroethane* (PB99-121956, 1998) 186 pp.

Gas at ordinary temp and pressure. Characteristic ethereal odor; burning taste. At low temps or under increased pressure, ethyl chloride is a mobile, very volatile liquid. *Flammable.* Keep away from heat, sparks, and open flame. Keep container closed, out of sun, and away from heat. Use with adequate ventilation. Carbon dioxide is recommended for extinguishing ethyl chloride fires. d_4^0 0.9214. Vapor density: 2.22 (air = 1.00). mp −138.7°. bp_{760} 12.3°. When ethyl chloride is liberated at ordinary room temp from its sealed container (usually a tube with automatic closure) it vaporizes at once. $bp_{2\,atm}$ 32.5°; $bp_{10\,atm}$ 92.6°; $bp_{30\,atm}$ 149.5°; $bp_{50\,atm}$ 180.5°. Crit temp = 187.2°; crit press = 52.0 atm. Soly in water: 0.574 g/100 ml at 20°; in alcohol: 48.3 grams/100 ml. Miscible with ether. Ethyl chloride burns with a smoky, greenish flame, producing hydrogen chloride. Flash pt: −50°C (−58°F) (closed cup); −43°C (−45°F) (open cup). Explosive limits (% by vol in air): lower 3.6; upper 14.8.

Caution: Potential symptoms of overexposure are incoordination; inebriation; abdominal cramps; cardiac arrhythmias, cardiac arrest; liver and kidney damage. *See NIOSH Pocket Guide to Chemical Hazards* (DHHS/NIOSH 97-140, 1997) p 134.

USE: Refrigerant, solvent, alkylating agent, starting point in the manuf of tetraethyl lead: US 1907701 (1933).

THERAP CAT: Anesthetic (topical).
THERAP CAT (VET): Anesthetic (topical).

3838. Ethyl Chloroacetate. [105-39-5] 2-Chloroacetic acid ethyl ester. $C_4H_7ClO_2$; mol wt 122.55. C 39.20%, H 5.76%, Cl 28.93%, O 26.11%. Prepn: Vogel, *J. Chem. Soc.* **1948**, 644.

Liquid; pungent odor. d_4^{20} 1.1498. mp −26°. bp 144-146°. Flash pt 54°C. n_D^{20} 1.4227. *Poisonous, flammable.* Insol in water. Miscible with alcohol, ether.

Caution: Direct contact with vapors may be irritating to eyes.

3839. Ethyl Chloroformate. [541-41-3] Carbonochloridic acid ethyl ester; chloroformic acid ethyl ester; ethyl chlorocarbonate. $C_3H_5ClO_2$; mol wt 108.52. C 33.20%, H 4.64%, Cl 32.67%, O 29.49%. Prepn from phosgene and alc: Dumas, *Ann.* **10**, 277 (1834).

Liquid. Flash pt slightly below 61°F. bp 95°. d_4^{20} 1.1403. n_D^{20} 1.3947. *Poisonous, flammable, corrosive.* Practically insol and gradually decomp by water. Miscible with alcohol, benzene, chloroform, ether.

Caution: Vapors strongly irritate eyes, mucous membranes, and skin.

3840. Ethyl α-Chloropropionate. [535-13-7] 2-Chloropropanoic acid ethyl ester. $C_5H_9ClO_2$; mol wt 136.58. C 43.97%, H 6.64%, Cl 25.96%, O 23.43%.

Liq; pleasant odor. d_4^{20} 1.087. bp 147-148°. n_D^{20} 1.4185. Insol in water; miscible with alcohol, ether.

3841. Ethyl Cyanoacetate. [105-56-6] 2-Cyanoacetic acid ethyl ester; cyanoacetic ester; ethyl cyanoethanoate; malonic acid ethyl ester nitrile. $C_5H_7NO_2$; mol wt 113.12. C 53.09%, H 6.24%, N 12.38%, O 28.29%. Made by the interaction of sodium cyanide and chloroacetic acid and subsequent esterification of the cyanoacetic acid formed: Kohler, Allen, *Org. Synth.* **3**, 53 (1923); Inglis, *ibid.* **8**, 74 (1928).

Liquid. Slight, pleasant odor. d_4^{25} 1.0560; d_4^{50} 1.0306; d_4^{70} 1.0110. mp −22°. bp_{760} 206.0°; bp_{100} 152.8°; bp_{40} 133.8°; bp_{20} 119.8°; bp_{10} 106.0°; bp_5 93.5°; $bp_{1.0}$ 67.8°. $n_D^{20.5}$ 1.41793. Insol in water. Miscible with alcohol, ether; sol in ammonia water, aq solns of alkalies.

3842. Ethyl Cyanoacrylate. [7085-85-0] 2-Cyano-2-propenoic acid ethyl ester; ECA; ethyl α-cyanoacrylate; ethyl 2-cyanoacrylate; Krazy Glue. $C_6H_7NO_2$; mol wt 125.13. C 57.59%, H 5.64%, N 11.19%, O 25.57%. Main constituent of *super glue*; polymerizes rapidly in various media including water. Preparative methods: A. E. Ardis, US 2467927 (1949 to B. F. Goodrich); C. H. McKeever, US 2912454 (1959 to Rohm & Haas). Prepn, polymerization and degradation: F. Leonard *et al., J. Appl. Polym. Sci.* **10**, 259 (1966). Bioadhesive and histotoxic properties: B. J. Zumpano *et al., Surg. Neurol.* **18**, 452 (1982). Prepn and drug delivery applications of polymer nanoparticles: G. Cavallaro *et al., Int. J. Pharm.* **111**, 31 (1994). *In vitro* and *in vivo* studies in sclerotherapy: J.-C. Lin *et al.,*

J. Biomed. Mater. Res. **53**, 799 (2000). Review of chemistry and toxicology: R. Cary, *Concise Int. Chem. Assess. Doc. No. 36* (WHO, Geneva, 2001) 33 pp.

Clear, colorless liquid; strong, acrid odor; bp 54-56° (0.21-0.40 kPa); mp −20 to −25°. d^{20} 1.040. n_D^{20} 1.4391. Surface tension: 34.32 dynes/cm. Vapor pressure at 25°: <0.27 kPa. Sol in methyl ethyl ketone, toluene, acetone, DMF, nitromethane.

Polymer. [25067-30-5] Poly(ethyl 2-cyanoacrylate); PECA.

USE: In adhesives; in mfr of plastics, electronics, scientific instruments, jewelry, sports equip; in cable joining, manicuring, dentistry, mortuaries, fingerprint development. Polymer nanoparticles as pharmaceutic aid for controlled release drug delivery.

3843. **Ethyl Dibunate.** [5560-69-0] 3,6-Bis(1,1-dimethylethyl)-1-naphthalenesulfonic acid ethyl ester; 3,6-di-*tert*-butyl-1-naphthalenesulfonic acid ethyl ester; ethyl 3,6-di-*tert*-butyl-1-naphthalenesulfonate; 2,7-di-*tert*-butylnaphthalene-4-sulfonic acid ester; ethyl 2,7-di-*tert*-butylnaphthalene-4-sulfonate; dibunate ethyl; NDR-304; Neodyne. $C_{20}H_{28}O_3S$; mol wt 348.50. C 68.93%, H 8.10%, O 13.77%, S 9.20%. Prepn: Menard *et al., Can. J. Chem.* **39**, 729 (1961). Pharmacology: Shemano, *Arch. Int. Pharmacodyn. Ther.* **165**, 410 (1967). Clinical trial in chronic cough: H. Sevelius, J. P. Colmore, *Clin. Pharmacol. Ther.* **8**, 381 (1967).

Crystals from ethanol, mp 138-139°.
THERAP CAT: Antitussive.

3844. **Ethyldimethyl-9-octadecenylammonium Bromide.** [6458-13-5] *N*-Ethyl-*N,N*-dimethyl-9-octadecen-1-aminium bromide (1:1); Onyxide. $C_{22}H_{46}BrN$; mol wt 404.52. C 65.32%, H 11.46%, Br 19.75%, N 3.46%. Prepn: Du Bois, **US 2519747** (1950 to Onyx Oil & Chem.).

trans-form

USE: A cationic surface active agent. Used in the control of algae. Sold as a 75% concentrate in isopropanol or propylene glycol.

3845. **Ethylene.** [74-85-1] Ethene; elayl; olefiant gas. C_2H_4; mol wt 28.05. C 85.64%, H 14.37%. $CH_2=CH_2$. Plant hormone; occurs in ripening fruit, in illuminating gas (up to 4%). Prepd by decompn of petr gases or by dehydration of alcohol. Lab prepn from alc: Gattermann-Wieland, *Praxis des Organischen Chemikers* (de Gruyter, Berlin, 40th ed, 1961) p 98; F. A. Wunder, E. I. Leupold, *Angew. Chem. Int. Ed.* **19**, 126 (1980). Biosynthesis from 1-aminocyclopropane-1-carboxylic acid: D. Adams, S. F. Yang, *Proc. Natl. Acad. Sci. USA* **76**, 170 (1979). Proposed mechanism of biosynthesis: M. C. Pirrung, *J. Am. Chem. Soc.* **105**, 7207 (1983). Toxicity data: Flury, *Arch. Exp. Pathol. Pharmakol.* **138**, 65 (1928). Comprehensive monograph: S. A. Miller, *Ethylene and its Industrial Derivatives* (Ernest Benn Ltd., London, 1969) 1321 pp. Review of ethylene in biological systems: Spencer, *Fortschr. Chem. Org. Naturst.* **27**, 31-80 (1969); of manuf: *Faith, Keyes & Clark's Industrial Chemicals,* F. A. Lowenheim, M. K. Moran, Eds. (Wiley-Interscience, New York, 4th ed., 1975) pp 376-384; of carcinogenic risk: *IARC Monographs* **60**, 45-71 (1994). *Review:* K. M. Sundaram *et al.,* in *Kirk-Othmer Encyclopedia of Chemical Technology* **vol. 9** (Wiley-Interscience, New York, 4th ed., 1994) pp 877-915.

Colorless gas with faint sweet odor. Burns with a luminous flame. Solidif −181° forming monoclinic prisms. mp −169.4°, bp$_{700}$ −102.4°. Vapor density 0.978 (air = 1.00). One liter of ethylene gas at 760 mm and 0° weighs 1.260 g. bp$_{2\text{ atm}}$ −90.8°; bp$_{10\text{ atm}}$ −52.8°; bp$_{30\text{ atm}}$ −14.2°; bp$_{50\text{ atm}}$ 8.9°. $d_4^{6.5}$ (liquid): 0.30342. Critical temp +9.6°; crit press. 50.7 atm. *Flammable.* One volume of ethylene gas dissolves in about 4 vols of water at 0°, in about 9 vols of water at 25°, in 0.5 vol alcohol at 25°, in about 0.05 vol ether at 15.5°. Sol in acetone, benzene. Explosive limits (% vol in air), lower: 3.02; upper: 34. Autoignition temp: 1009°F (+543°C). LC for mice in air: 950,000 ppm (Flury).

Caution: Can act as a simple asphyxiant due to oxygen displacement. Exposure to high concns may cause CNS depression, unconsciousness. *See Patty's Industrial Hygiene and Toxicology* **vol. 2B**, G. D. Clayton, F. E. Clayton, Eds. (John Wiley & Sons, New York, 4th ed., 1994) p 1241-1244.

USE: Oxyethylene welding and cutting metals; manuf alcohol, mustard gas, and many other organics. Manuf ethylene oxide (for plastics), polythene, polystyrene and other plastics. Plant growth regulator; used commercially to accelerate the ripening of various fruits.

THERAP CAT: Anesthetic (inhalation).

THERAP CAT (VET): Has been used as inhalation anesthetic.

3846. **Ethylene Bromohydrin.** [540-51-2] 2-Bromoethanol; β-bromoethyl alcohol; glycol bromohydrin. C_2H_5BrO; mol wt 124.97. C 19.22%, H 4.03%, Br 63.94%, O 12.80%. Prepd by the action of hydrobromic acid on ethylene oxide: Thayer *et al., Org. Synth.* **6**, 12 (1926).

Hygroscopic liq. d_4^0 1.7902; d_4^{15} 1.7696; d_4^{20} 1.7629; d_4^{25} 1.7560; d_4^{30} 1.7494. bp$_{750}$ 149-150° (decompn); bp$_{20}$ 56-57°; bp$_{13}$ 48.5°. n_D^{20} 1.49361. Miscible with water; sol in the usual organic solvents, except petr ether. Hydrolysis of aq solns is accelerated by heat, acids, alkalies. Aq solns have a sweet, burning taste.

Caution: Ethylene bromohydrin vapors are an irritant to the eyes and the mucosa.

3847. **Ethylene Chlorohydrin.** [107-07-3] 2-Chloroethanol; 2-chloroethyl alcohol; glycol chlorohydrin. C_2H_5ClO; mol wt 80.51. C 29.84%, H 6.26%, Cl 44.03%, O 19.87%. Made from ethylene by action of a hypochlorite. Toxicity data: H. F. Smyth *et al., J. Ind. Hyg. Toxicol.* **23**, 259 (1941). *Review:* G. H. Riesser in *Kirk-Othmer Encyclopedia of Chemical Technology* **vol. 5** (Wiley-Interscience, New York, 3rd ed., 1979) pp 848-864.

Colorless liq. *Poisonous, flammable.* d_4^{20} 1.197. bp 128-130°. mp −67°. n_D^{20} 1.4419. Flash pt, open cup: 105°F (40°C). Miscible with water, alcohol. LD$_{50}$ orally in rats: 0.095 g/kg (Smyth).

Caution: Potential symptoms of overexposure are irritation of mucous membranes; nausea, vomiting; vertigo, incoordination; numbness; visual disturbance; headache; thirst; delirium; low blood pressure; collapse, shock and coma; liver and kidney damage. *See NIOSH Pocket Guide to Chemical Hazards* (DHHS/NIOSH 97-140, 1997) p 134. *See also Patty's Industrial Hygiene and Toxicology* **vol. 2C**, G. D. Clayton, F. E. Clayton, Eds. (Wiley-Interscience, New York, 4th ed., 1982) pp 4675-4684.

USE: Solvent; manuf insecticides; treating sweet potatoes before planting.

3848. **Ethylene Cyanohydrin.** [109-78-4] 3-Hydroxypropanenitrile; hydracrylonitrile; glycol cyanohydrin; β-hydroxypropionitrile. C_3H_5NO; mol wt 71.08. C 50.69%, H 7.09%, N 19.71%, O 22.51%. Prepd by the interaction of ethylene chlorohydrin and alkali cyanide: E. C. Kendall, B. McKenzie, *Org. Synth.* **coll. vol. I**, 256 (2nd ed., 1941). Toxicity study: H. F. Smyth, C. P. Carpenter, *J. Ind. Hyg. Toxicol.* **26**, 269 (1944).

Liquid. d_4^{25} 1.0404. mp −46°. bp_{760} 228° (slight dec); bp_{200} 178°; bp_{100} 157.7°; bp_{60} 144.7°; bp_{40} 134.1°; bp_{20} 117.9°; bp_{10} 102°; bp_5 87.8°; $bp_{1.0}$ 58.7°. n_D^{25} 1.4241. Miscible with water, acetone, methyl ethyl ketone, ethyl alcohol. Slightly sol in ether (2.3% w/w at 15°). Insol in benzene, petr ether, carbon disulfide, carbon tetrachloride. LD_{50} orally in rats: 10.0 g/kg (Smyth, Carpenter).

USE: Solvent for some cellulose esters and many inorganic salts.

3849. Ethylenediamine. [107-15-3] 1,2-Ethanediamine; 1,2-diaminoethane. $C_2H_8N_2$; mol wt 60.10. C 39.97%, H 13.42%, N 46.61%. Manuf from ethylene dichloride and ammonia: *Faith, Keyes & Clark's Industrial Chemicals,* F. A. Lowenheim, M. K. Moran, Eds. (Wiley-Interscience, New York, 4th ed., 1975) pp 385-388. Toxicity: H. F. Smyth *et al., J. Ind. Hyg. Toxicol.* **23**, 259 (1941).

$H_2N \diagup\!\!\diagdown NH_2$

Colorless, clear, thick liq; ammonia odor. d_4^{25} 0.898. bp 116-117°. mp 8.5°. n_D^{D} 1.4540. Flash pt, closed cup: 110°F (43°C). *Corrosive, flammable.* Volatile with steam. Miscible with water, alc. Sol in benzene unless insufficiently dried; slightly sol in ether. Properties: S. G. Boas-Traube *et al., Nature* **162**, 960 (1948). Strongly alkaline and may readily absorb CO_2 from air to form a nonvolatile carbonate. Protect against undue exposure to the atmosphere. LD_{50} orally in rats: 1.16 g/kg (Smyth).

Monohydrate. [6780-13-8] $C_2H_8N_2 \cdot H_2O$; mol wt 78.12. mp 10°; bp 118°.

Dihydrochloride. [333-18-6] $C_2H_8N_2 \cdot 2HCl$; mol wt 133.02. Crystals. Sublimes without melting. Freely sol in water. Practically insol in alcohol. Aq soln is practically neutral.

Caution: Potential symptoms of overexposure to ethylenediamine are irritation of nose, respiratory system; sensitization dermatitis; asthma; liver and kidney damage. *See NIOSH Pocket Guide to Chemical Hazards* (DHHS/NIOSH 97-140, 1997) p 136.

USE: Solvent for casein, albumin, shellac, and sulfur; emulsifier; stabilizing rubber latex; as inhibitor in antifreeze solns; in textile lubricants. Pharmaceutic aid (aminophylline injection stabilizer).

THERAP CAT (VET): Dihydrochloride has been used as a urinary acidifier.

3850. Ethylene Dibromide. [106-93-4] 1,2-Dibromoethane; *sym*-dibromoethane; ethylene bromide; EDB; Dowfume W 85. $C_2H_4Br_2$; mol wt 187.86. C 12.79%, H 2.15%, Br 85.07%. $BrCH_2$-CH_2Br. Made from ethylene and bromine; also from acetylene and HBr. Manuf: *Faith, Keyes & Clark's Industrial Chemicals,* F. A. Lowenheim, M. K. Moran, Eds. (Wiley-Interscience, New York, 4th ed., 1975) pp 389-391. Toxicity data: G. W. Fischer *et al., J. Prakt. Chem.* **320**, 133 (1978). Carcinogenicity studies: W. A. Olson *et al., J. Natl. Cancer Inst.* **51**, 1993 (1973); *Clin. Toxicol.* **14**, 473 (1979). History of controversial use as a fumigant: J. Walsh, *Science* **215**, 1592 (1982). Review of toxicology and human exposure: *Toxicological Profile for 1,2-Dibromoethane* (PB93-110740, 1992) 173 pp.

$Br \diagup\!\!\diagdown Br$

Heavy liquid; chloroform odor. d_{25}^{25} 2.172. bp 131-132°. mp 9°. n_D^{20} 1.5379. Vapor pressure at 25°: 11 mm Hg. Sol in about 250 parts water; misc with alc, ether. *Poisonous. Protect from light.* LD_{50} i.p. in mice: 220 mg/kg (Fischer).

Caution: Potential symptoms of overexposure are irritation of skin, respiratory system and eyes; dermatitis with vesiculation; liver, heart, spleen, kidney damage; reproductive effects. *See NIOSH Pocket Guide to Chemical Hazards* (DHHS/NIOSH 97-140, 1997) p 136. *See also Patty's Industrial Hygiene and Toxicology* vol. **2B**, G. D. Clayton, F. E. Clayton, Eds. (Wiley-Interscience, New York, 3rd ed., 1981) p 3497-3502. This substance is reasonably anticipated to be a human carcinogen: *Report on Carcinogens, Twelfth Edition* (PB2011-111646, 2011) p 135.

USE: Soil and grain fumigant; as lead scavenger in anti-knock gasolines.

3851. Ethylene Dichloride. [107-06-2] 1,2-Dichloroethane; *sym*-dichloroethane; ethylene chloride; EDC; Dutch liquid; Brocide.

$C_2H_4Cl_2$; mol wt 98.95. C 24.28%, H 4.07%, Cl 71.65%. $ClCH_2$-CH_2Cl. Made from ethylene and chlorine; also from acetylene and HCl: *Beilstein* I, 84 (1918). Manuf: *Faith, Keyes & Clark's Industrial Chemicals,* F. A. Lowenheim, M. K. Moran, Eds. (Wiley-Interscience, New York, 4th ed., 1975) pp 392-396. Toxicity data: Smyth *et al., Am. Ind. Hyg. Assoc. J.* **30**, 470 (1969). Review of carcinogenic risk: *IARC Monographs* **20**, 429-444 (1979); of toxicology and human exposure: *Toxicological Profile for 1,2-Dichloroethane* (PB2001-109103, 2001) 297 pp.

Heavy liq; burns with smoky flame; pleasant odor; sweet taste. d_4^{20} 1.2569. bp 83-84°. mp ~ −40°. Flash pt, closed cup: 56°F (13°C); open cup: 65°F (18°C). n_D^{20} 1.4443. *Flammable, poisonous.* Sol in about 120 parts water; misc with alc, chloroform, ether. LD_{50} orally in rats: 770 mg/kg (Smyth).

Caution: Potential symptoms of overexposure are CNS depression; nausea, vomiting; dermatitis; irritation of eyes, corneal opacity; liver, kidney, cardiovascular system damage. *See NIOSH Pocket Guide to Chemical Hazards* (DHHS/NIOSH 97-140, 1997) p 136. *See also Clinical Toxicology of Commercial Products,* R. E. Gosselin *et al.,* Eds. (Williams & Wilkins, Baltimore, 5th ed., 1984) Section II, pp 163-164. This substance is reasonably anticipated to be a human carcinogen: *Report on Carcinogens, Twelfth Edition* (PB2011-111646, 2011) p 145.

USE: Manuf of vinyl chloride, acetyl cellulose; solvent for fats, oils, waxes, gums, resins, and particularly for rubber. Has been used as insect and soil fumigant. Organic solvent.

3852. Ethylene Glycol. [107-21-1] 1,2-Ethanediol; 1,2-dihydroxyethane; 2-hydroxyethanol. $C_2H_6O_2$; mol wt 62.07. C 38.70%, H 9.74%, O 51.55%. Prepd on a large scale by the hydration of ethylene oxide. Flowsheets: *Ullmanns Encyklopädie der technischen Chemie* vol. **3** (3rd ed., 1953) p 137; M. Sittig, *Organic Chemical Process Encyclopedia* (Noyes Dev. Corp., Park Ridge, N.J., 1967) p 265. Manuf from ethylene and oxygen: *Faith, Keyes & Clark's Industrial Chemicals,* F. A. Lowenheim, M. K. Moran, Eds. (Wiley-Interscience, New York, 4th ed., 1975) pp 397-402. Toxicity studies: Smyth *et al., J. Ind. Hyg. Toxicol.* **23**, 259 (1941); Bornmann, *Arzneim.-Forsch.* **4**, 643 (1954). Monograph: G. O. Curme, F. Johnston, *Glycols* (Reinhold, New York, 1952). GC/MS determn in plasma: C. Giachetti *et al., Biomed. Environ. Mass Spectrom.* **18**, 592 (1989). Review of toxicology and human exposure: *Toxicological Profile for Ethylene Glycol* (PB2010-10005, 2010) 315 pp.

$HO \diagup\!\!\diagdown OH$

Slightly viscous liq. Sweet taste. *Poisonous. Do not swallow.* Considerably hygroscopic: Absorbs twice its weight of water at 100% relative humidity. d_4^0 1.1274. d_4^4 1.1204. d_4^{20} 1.1135. d_4^{30} 1.1065. One gallon weighs 9.3 lbs. Flash pt, open cup: 240°F (115°C). mp −13°. bp_{760} 197.6°; bp_{97} 140°; bp_{18} 100°; $bp_{3.0}$ 70°; $bp_{0.06}$ 20°. n_D^{15} 1.43312; n_D^{20} 1.43063. Log P (octanol/water): −1.36. Viscosity (cP): 26 (15°); 21 (20°); 17.3 (25°). Dielectric constant at 20° and 150 meters wavelength: 38.66 esu. Dipole moment 2.20. Spec heat (20°): 0.561 cal/g/°C. Heat of formation −108.1 kcal/mol. Heat of fusion 44.7 cal/g. Heat of vaporization 191 cal/g. Heat of soln −6.5 cal/g of soln at 17° when 37 parts are mixed with 63 parts H_2O (w/w). Parachor 148.9 (theory 152.2). Surface tension at 20° = 48.4 dynes/cm. Miscible with water and many organic solvents including alcohol, ether, chloroform, hexanes, glycerol, acetic acid, acetone and similar ketones, aldehydes, pyridine and similar coal tar bases. Practically insol in benzene and its homologs, chlorinated hydrocarbons, petr ether, oils. Density and freezing point of ethylene glycol-water mixtures: 10.15% ethylene glycol by wt d_4^{20} 1.013, fp −3.5°; 20.44% ethylene glycol d_{20}^{20} 1.027, fp −8°; 29.88% ethylene glycol d_{20}^{20} 1.040, fp −15°; 40.23% ethylene glycol d_{20}^{20} 1.054, fp −24°; 50.18% ethylene glycol d_{20}^{20} 1.067, fp −36°; 58.37% ethylene glycol d_{20}^{20} 1.0770, fp −48°. LD_{50} in rats, guinea pigs (g/kg): 8.54, 6.61 orally (Smyth); in mice (ml/kg): 13.79 orally (Bornmann).

Caution: Potential symptoms of overexposure by ingestion are nausea, vomiting, abdominal pain, weakness; dizziness, stupor, convulsions, CNS depression. Direct contact may cause irritation of eyes, nose, skin; skin sensitization. *See NIOSH Pocket Guide to Chemical Hazards* (DHHS/NIOSH 97-140, 1997) p 136. *See also*

Patty's Industrial Hygiene and Toxicology vol. 2F, G. D. Clayton, F. E. Clayton, Eds. (Wiley-Interscience, New York, 4th ed., 1994) pp 4645-4657.

USE: Antifreeze in cooling and heating systems. In hydraulic brake fluids and de-icing solutions. Industrial humectant. Ingredient of electrolytic condensers where it serves as solvent for boric acid and borates. Solvent in the paint and plastics industries. In the formulation of printers' inks, stamp pad inks, ball-point pen ink. Softening agent for cellophane. Stabilizer for soybean foam used to extinguish oil and gasoline fires. In the synthesis of safety explosives, glyoxal, unsatd ester type alkyd resins, plasticizers, elastomers, synthetic fibers (Terylene, Dacron), and synthetic waxes. To create artificial smoke and mist for theatrical uses.

3853. Ethylene Glycol Diacetate. [111-55-7] 1,2-Ethanediol 1,2-diacetate; glycol diacetate; ethylene diacetate. $C_6H_{10}O_4$; mol wt 146.14. C 49.31%, H 6.90%, O 43.79%. Prepn from ethylene bromide, glacial acetic acid and potassium acetate: Henry, *Bull. Soc. Chim.* [3] **17**, 207 (1897); Gattermann-Wieland, *Praxis des Organischen Chemikers* (de Gruyter, Berlin, 40th ed., 1961) p 107; from synthesis gas via homogeneous ruthenium catalysis: J. F. Knifton, *Chem. Commun.* **1981**, 188. Toxicity study: Smyth *et al.*, *J. Ind. Hyg. Toxicol.* **23**, 259 (1941).

H₃C structure diagram

Liquid. d 1.104. bp 190-191°. mp −31°. n_D^{20} 1.415. Flash pt, open cup: 205°F (96°C). Sol in 7 parts water; miscible with alc, ether. LD_{50} orally in rats: 6.86 g/kg (Smyth).

USE: Solvent for oils, cellulose esters, explosives, etc.

3854. Ethylene Glycol Dinitrate. [628-96-6] 1,2-Ethanediol 1,2-dinitrate; glycol dinitrate; nitroglycol; EGDN. $C_2H_4N_2O_6$; mol wt 152.06. C 15.80%, H 2.65%, N 18.42%, O 63.13%. Explosive. Prepn: L. Henry, *Ann. Chim. (Paris)* **4**, 253 (1872); P. Golding *et al.*, *Tetrahedron* **49**, 7037 (1993). Determn in blood by GC: M. H. Litchfield, *Analyst* **93**, 653 (1968); in explosion debris by HPLC: R. J. Prime, J. Krebs, *Can. Soc. Forensic Sci. J.* **13**, 27 (1980); in vapor phase by ion mobility spectrometry: A. H. Lawrence, P. Neudorfl, *Anal. Chem.* **60**, 104 (1988). Measurement of vapor pressure: P. A. Pella, *J. Chem. Thermodyn.* **9**, 301 (1977). Adsorption kinetics: D. O. Henderson *et al.*, *Appl. Spectrosc.* **47**, 528 (1993). Brief description of properties: *Am. Ind. Hyg. Assoc. J.* **27**, 574-577 (1966); V. Lindner in *Kirk-Othmer Encyclopedia of Chemical Technology* vol. 10, (John Wiley & Sons, New York, 4th ed., 1993) pp 4, 21-26.

O₂N structure diagram

Colorless, odorless liquid with sweetish taste, mp −22.8°. d 1.490 g/cm³. n_D^{20} 1.395. Insol in water. Very sol in alcohol and ether. *Combustible, explosive.*

Caution: Potential symptoms of overexposure are throbbing headache; dizziness; nausea; vomiting; abdominal pain; hypotension; flushing; palpitations; methemoglobinemia; delirium; CNS depression. Direct contact may cause skin irritation. *See NIOSH Pocket Guide to Chemical Hazards* (DHHS/NIOSH 97-140, 1997) p 138.

USE: Explosive for mining and fuel industries. Additive to dynamite. Detection of hidden bombs by analysis of ambient air for EGDN.

3855. Ethylene Glycol Monoacetate. [542-59-6] 1,2-Ethanediol 1-acetate; 1,2-ethanediol monoacetate; glycol-monoacetin. $C_4H_8O_3$; mol wt 104.11. C 46.15%, H 7.75%, O 46.10%.

H₃C structure diagram

Liquid. d 1.108. bp 182°. Miscible with water, alcohol. LD_{50} orally in rats: 8.25 g/kg; *see:* Smyth *et al.*, *J. Ind. Hyg. Toxicol.* **23**, 259 (1941).

3856. Ethylene Oxide. [75-21-8] Oxirane; Anprolene. C_2H_4O; mol wt 44.05. C 54.53%, H 9.15%, O 36.32%. Prepd from ethylene chlorohydrin and KOH: A. Wurtz, *Ann.* **110**, 125 (1859). Manuf by catalytic oxidation of ethylene: T. E. Lefort, *US* **1998878** (1935 to Société Française de Catalyse Généralisée) reissued as **US RE 20370** (1937 to Carbide and Carbon). Physical props: W. H. Perkin, *J. Chem. Soc.* **63**, 488 (1893). Leukemia in workers exposed to ethylene oxide: C. Hogstedt *et al.*, *J. Am. Med. Assoc.* **241**, 1132 (1979). Hazards and handling: L. G. Hess, V. V. Tilton, *Ind. Eng. Chem.* **42**, 1251 (1950). Genotoxicity: E. Agurell *et al.*, *Mutat. Res.* **250**, 229 (1991). Review of toxicology and human exposure: *Toxicological Profile for Ethylene Oxide* (PB91-180554, 1990) 115 pp; of carcinogenic risk: *IARC Monographs* **60**, 73-159 (1994). *Review:* H. C. Schultze in *Glycols*, G. O. Curme, Jr., Ed. (Reinhold, New York, 1952) pp 74-113; J. P. Dever *et al.*, in *Kirk-Othmer Encyclopedia of Chemical Technology* vol. 9 (John Wiley & Sons, New York, 4th ed., 1994) pp 915-959.

Colorless, flammable gas at ordinary room temp and pressure; liquid below 12°. *Explosive.* bp 10.7°. d_4^4 0.891, d_7^7 0.887, d_{10}^{10} 0.882. mp −111°. n_D^7 1.3597. Sol in water, alc, ether.

Caution: Potential symptoms of overexposure are irritation of eyes, skin, nose and throat; peculiar taste; headache; nausea, vomiting and diarrhea; dyspnea, cyanosis and pulmonary edema; drowsiness, incoordination; EKG abnormalities; reproductive effects; direct contact with liquid or high vapor concentrations may cause burns to eyes and skin; direct contact with liquid may cause frostbite. *See NIOSH Pocket Guide to Chemical Hazards* (DHHS/NIOSH 97-140, 1997) p 138. This substance is listed as a known human carcinogen: *Report on Carcinogens, Twelfth Edition* (PB2011-111646, 2011) p 188.

USE: Fumigant for foodstuffs and textiles; to sterilize surgical instruments; agricultural fungicide. In organic syntheses, esp in the production of ethylene glycol. Starting material for the manuf of acrylonitrile and nonionic surfactants.

3857. Ethylene Sulfide. [420-12-2] Thiirane; ethylene episulfide. C_2H_4S; mol wt 60.11. C 39.96%, H 6.71%, S 53.34%. Parent compd of the class of heterocycles known as thiiranes or episulfides; utilized as a reagent for the thioethylation of amines. Prepn: M. Delépine, *Bull. Soc. Chim. Fr.* **27**, 740 (1920); and reactivity studies: M. Delépine, S. Eschenbrenner, *ibid.* **33**, 703 (1923). Improved prepn: S. Searles *et al.*, *Org. Synth.* coll. vol. V, 562 (1973). Synthetic utility in the prepn of aminothiols: H. R. Snyder *et al.*, *J. Am. Chem. Soc.* **69**, 2672 (1947); R. J. Wineman *et al.*, *J. Org. Chem.* **27**, 4222 (1962); R. Luhowy, F. Meneghini, *ibid.* **38**, 2405 (1973). Toxicity studies: J. R. Brown, E. Mastromatteo, *Am. Ind. Hyg. Assoc. J.* **25**, 560 (1964). Review of thiirane chemistry: W. Chew, D. N. Harpp, *Sulfur Rep.* **15**, 1-39 (1993).

Colorless, malodorous liquid. *Poisonous, flammable, irritant.* bp 55-56°. d_4^{20} 1.0113. n_D^{20} 1.4946. Flash pt, closed cup: 50°F (10°C). Sol in most organic solvents. Insol in water. Air sensitive. Store under inert gas at 2-8°C. LD_{50} in rats (mg/kg): 178 orally; 42 i.p. (Brown).

USE: Reagent in synthetic organic chemistry.

3858. Ethylene Thiourea. [96-45-7] 2-Imidazolidinethione; imidazoline-2-thiol; 2-mercaptoimidazoline; ETU; Akrochem ETU-22; NA-22; Robac 22; Sanceller 22; Vulkacit NPV/C. $C_3H_6N_2S$; mol wt 102.16. C 35.27%, H 5.92%, N 27.42%, S 31.38%. Degradation product of ethylenebisdithiocarbamate fungicides such as mancozeb, maneb, zineb, q.q.v. Prepn: C. F. H. Allen *et al.*, *Org. Synth.* coll. vol. III, 394 (1955); G. Matolcsy, *Ber.* **101**, 522 (1968). Identification as decomposition product: R A. Ludwig *et al.*, *Can. J. Bot.* **32**, 48 (1954); G. Czegledi-Janko, *J. Chromatogr.* **31**, 89 (1967). Determn in various foods by GLC: W. H. Newsome, *J. Agric. Food Chem.* **20**, 967 (1972); by GLC, HPLC: J. H. Onley *et al.*, *J. Assoc. Off. Anal. Chem.* **60**, 1105 (1977); J. H. Onley, *ibid.* 1111. Review of determn methods: P. Bottomley *et al.*, *Residue*

Rev. **95**, 45-89 (1985). Metabolism in mice: K. Savolainen, H. Pyysalo, *J. Agric. Food Chem.* **27**, 1177 (1979); in cats and rats: F. Iverson, *Toxicol. Appl. Pharmacol.* **52**, 16 (1980). Aquatic toxicity data: C. J. Van Leeuwen, *Aquat. Toxicol.* **7**, 145 (1985). Goitrogenic activity: S. L. Graham, W. H. Hansen, *Bull. Environ. Contam. Toxicol.* **7**, 19 (1972); D. M. Smith, *Br. J. Ind. Med.* **41**, 362 (1984). Use in rubber industry as crosslinking agent: J. T. Oetzel, *Rubber World* **172**, 55 (1975); C. Hepburn, M. S. Mahdi, *Kautsch. Gummi Kunstst.* **39**, 629 (1986). Review of toxicology: L. Fishbein, *J. Toxicol. Environ. Health* **1**, 713-756 (1976); of genotoxicity studies: K. L. Dearfield, *Mutat. Res.* **317**, 111-132 (1994).

Needles, prisms from alc or amyl alc. mp 203-204°. Soly in 100 ml water: 2 g at 30°, 9 g at 60°, 44 g at 90°. Moderately sol in methanol, ethanol, ethylene glycol, and pyridine. Insol in acetone, ether, chloroform, benzene, ligroin. LD$_{50}$ orally in rats: 1832 mg/kg (Graham, Hansen).

Caution: Potential symptom of overexposure is eye irritation. *See NIOSH Pocket Guide to Chemical Hazards* (DHHS/NIOSH 97-140, 1997) p 138. This substance is reasonably anticipated to be a human carcinogen: *Report on Carcinogens, Twelfth Edition* (PB2011-111646, 2011) p 191.

USE: Accelerator in synthetic rubber production.

3859. Ethylenimine. [151-56-4] Aziridine; azacyclopropane; dimethylenimine. C$_2$H$_5$N; mol wt 43.07. C 55.77%, H 11.70%, N 32.52%. Prepd by treating 2-chloroethylamine hydrochloride with NaOH: Wystrach *et al.*, *J. Am. Chem. Soc.* **77**, 5915 (1955); **78**, 1263 (1956); from β-aminoethylsulfuric acid by reaction with NaOH: Allen *et al.*, *Org. Synth.* **coll. vol. IV**, 433 (1963). Acute toxicity: H. F. Smyth *et al.*, *J. Ind. Hyg. Toxicol.* **23**, 259 (1941). Toxicology: Weightman, Hoyle, *J. Am. Med. Assoc.* **189**, 543 (1964). Review of carcinogenicity studies: *IARC Monographs* **9**, 37 (1975).

Liquid. Intense odor of ammonia. *Poisonous. Handle in hood only.* Strongly alkaline. Polymerizes easily. d$_4^{24}$ 0.8321. mp −73.96°. bp$_{760}$ 56-57°. n$_D^{25}$ 1.412. Vapor press. (20°) 160 mm Hg. Miscible with water. Sol in alc. Can be stored for some time over a few pellets of sodium hydroxide. LD$_{50}$ orally in rats: 15 mg/kg (Smyth).

Caution: Potential symptoms of overexposure are nausea, vomiting; headache, dizziness; pulmonary edema; liver and kidney damage; eye burns; skin sensitization; irritation of eyes, skin, nose and throat. Potential occupational carcinogen. *See NIOSH Pocket Guide to Chemical Hazards* (DHHS/NIOSH 97-140, 1997) p 138.

USE: In the manuf of triethylenemelamine.

3860. Ethylestrenol. [965-90-2] (17α)-19-Norpregn-4-en-17-ol; 17α-ethylestr-4-en-17β-ol; 17α-ethyl-17β-hydroxy-4-estrene; 17β-hydroxy-17α-ethyl-19-nor-4-androstene; Orabolin; Durabolin-O; Orgaboral; Maxibolin; Orgabolin (obsolete). C$_{20}$H$_{32}$O; mol wt 288.48. C 83.27%, H 11.18%, O 5.55%. Prepn: de Winter *et al.*, *Chem. Ind. (London)* **1959**, 905; Szpilfogel, de Winter, US **2878267** (1959 to Organon).

Crystals, mp 76-78°. [α]$_D$ +31° (chloroform).

Note: This is a controlled substance (anabolic steroid): **21 CFR**, 1308.13, as defined in 1300.01.

THERAP CAT: Anabolic.

3861. Ethyl Ether. [60-29-7] 1,1'-Oxybisethane; ethoxyethane; ether; diethyl ether; ethyl oxide; diethyl oxide; sulfuric ether; anesthetic ether. C$_4$H$_{10}$O; mol wt 74.12. C 64.82%, H 13.60%, O 21.59%. Produced on a large scale by dehydration of ethanol or by hydration of ethylene, both processes being carried out in the presence of sulfuric acid. Review of mfg processes: Himmler in *Ullmanns Encyklopädie der technischen Chemie* **vol. 5** (1954) pp 777-782; D. E. Keeley in *Kirk-Othmer Encyclopedia of Chemical Technology* **vol. 9** (Wiley-Interscience, New York, 3rd ed., 1980) pp 381-393.

Mobile, very volatile, highly flammable liq. *Explosive.* Vapor heavier than air. Characteristic, sweetish, pungent odor, more agreeable than chloroform. Burning taste. Tends to form explosive peroxides under the influence of air and light, esp when evaporation to dryness is attempted. Peroxides may be removed from ether by shaking with 5% aq ferrous sulfate soln. Addition of naphthols, polyphenols, aromatic amines, and aminophenols has been proposed for the stabilization of ethyl ether. d$_4^0$ 0.7364; d$_4^{10}$ 0.7249; d$_4^{20}$ 0.7134; d$_4^{30}$ 0.7019. Vapor density 2.55 (air = 1.0). mp −116.3° (stable crystals); mp −123.3° (metastable crystals). bp$_{760}$ 34.6°; bp$_{400}$ 17.9°; bp$_{200}$ 2.2°; bp$_{100}$ −11.5°; bp$_{10}$ −48.1°; bp$_{1.0}$ −74.3°. Satd vapor press. at 0°: 184.9 mm; at 10°: 290.8 mm; at 20°: 439.8 mm; at 50°: 1276 mm; at 70°: 2304 mm. Critical temp 192.7°; crit press. 35.6 atm. Flash pt, closed cup: −49°F (−45°C). Air-ether mixtures containing more than 1.85 volume-% of ether vapor, are explosive hazards. Autoignition temp 180-190°. n$_D^{15}$ 1.35555. Dielectric constant at 26.9° and 85.8 kilocycles = 4.197; good insulator. When shaken under absolutely dry conditions ether can generate enough static electricity to start a fire. Surface tension at 20°: 17.06 dynes/cm. Viscosity at 20°: 0.2448 cP. Heat of vaporization at 30°: 89.80 cal/g. Produces considerable coldness on quick evaporation. Heat of formation −907 cal/g; heat of combustion −8.807 kcal/g. Sol in water. A satd water soln contains 8.43% (w/w) of ether at 15° and 6.05% (w/w) at 25°. Ether satd with water contains 1.2% H$_2$O at 20°. Soly in water increased by HCl. Sol in concd hydrochloric acid. May explode when brought in contact with anhydr nitric acid. Miscible with lower aliphatic alcohols, benzene, chloroform, petr ether, methylene chloride, solvent hexane, other fat solvents, many fixed and volatile oils. Azeotrope with water (1.3%), bp 34.2°.

Caution: Potential symptoms of overexposure are dizziness; drowsiness; headache, excitedness and narcosis; nausea, vomiting; irritation of eyes, upper respiratory system and skin. *See NIOSH Pocket Guide to Chemical Hazards* (DHHS/NIOSH 97-140, 1997) p 140.

USE: Solvent for waxes, fats, oils, perfumes, alkaloids, gums. Excellent solvent for nitrocellulose when mixed with alc. Important reagent in organic syntheses, esp in Grignard and Wurtz type reactions. Easily removable extractant of active principles (hormones, etc.) from plant and animal tissues. In the manuf of gun powder. As primer for gasoline engines.

THERAP CAT: Anesthetic (inhalation).

THERAP CAT (VET): Anesthetic (inhalation). Has been used orally in colic, subcutaneously as a stimulant.

3862. Ethyl Formate. [109-94-4] Formic acid ethyl ester. C$_3$H$_6$O$_2$; mol wt 74.08. C 48.64%, H 8.16%, O 43.19%. Toxicity data: H. F. Smyth *et al.*, *Arch. Ind. Hyg. Occup. Med.* **10**, 61 (1954).

Mobile, flammable liq. d$_4^{20}$ 0.917. bp 53-54°. mp −80°. Flash pt, closed cup: −4°F (−20°C). n$_D^{20}$ 1.3597. Sol in about 10 parts water with gradual decompn into free acid and alcohol; miscible with alcohol, ether. *Keep tightly closed and preferably in contact with calcium chloride.* LD$_{50}$ orally in rats: 4.29 g/kg (Smyth).

Caution: Potential symptoms of overexposure are irritation of eyes and upper respiratory system. *See NIOSH Pocket Guide to Chemical Hazards* (DHHS/NIOSH 97-140, 1997) p 140.

USE: As flavor for lemonades and essences; for manuf artificial rum and arrac; also as a solvent for nitrocellulose; as fungicide and larvicide for tobacco, cereals, dried fruits, etc.; in organic synthesis.

3863. 2-Ethyl-1-hexanol. [104-76-7] 2-Ethylhexyl alcohol. $C_8H_{18}O$; mol wt 130.23. C 73.78%, H 13.93%, O 12.29%. Manuf from butyraldehyde: *Faith, Keyes & Clark's Industrial Chemicals*, F. A. Lowenheim, M. K. Moran, Eds. (Wiley-Interscience, New York, 4th ed., 1975) pp 413-417. Toxicity study: H. F. Smyth *et al., Am. Ind. Hyg. Assoc. J.* **30**, 470 (1969).

Colorless liq. d_{20}^{20} 0.8344. bp 184-185°. n_D^{20} 1.4300. Flash pt 81°C. Soluble in about 720 parts water, in many organic solvents. It dissolves about 2.5% its weight of water at 25°. LD_{50} orally in rats: 12.46 ml/kg (Smyth).

USE: Mercerizing textiles; as a solvent for dyes, resins, oils; also claimed to possess antifoaming properties.

3864. Ethylhexyl Triazone. [88122-99-0] 4,4′,4″-(1,3,5-Tri-azine-2,4,6-triyltriimino)trisbenzoic acid 1,1′,1″-tris(2-ethylhexyl) ester; octyl triazone; 2,4,6-trianilino-*p*-(carbo-2′-ethylhexyl-1′-oxy)-1,3,5-triazine; Uvinul T 150. $C_{48}H_{66}N_6O_6$; mol wt 823.09. C 70.04%, H 8.08%, N 10.21%, O 11.66%. Substituted *s*-triazine; uv absorber in cosmetic sunscreen products. Prepn: U. Hoppe *et al.*, **EP 87098**; *eidem*, **US 4617390** (1983, 1986 both to BASF). HPLC determn in cosmetic products: L. Gagliardi *et al., Anal. Lett.* **23**, 2123 (1990). Photochemical study: L. Douarre *et al., J. Photochem. Photobiol. A* **96**, 71 (1996).

Crystals from gasoline, mp 128°. uv max (ethanol): 218.7, 313.3 nm ($E_{1cm}^{1\%}$ 456, 1585). Readily sol in polar cosmetic oils.

THERAP CAT: Ultraviolet screen.

3865. Ethylidene Chloride. [75-34-3] 1,1-Dichloroethane; dichloromethylmethane. $C_2H_4Cl_2$; mol wt 98.95. C 24.28%, H 4.07%, Cl 71.65%. Prepd by the action of PCl_5 on acetaldehyde. Review of toxicology and human exposure: *Toxicological Profile for 1,1-Dichloroethane* (PB91-180539, 1990) 115 pp.

Oily liquid; odor and taste as of chloroform. d_4^{20} 1.1757; d_4^{25} 1.1680. bp 57.3°. mp ~−98°. n_D^{20} 1.4167. Sol in about 200 parts water; miscible with alcohol.

Caution: Potential symptoms of overexposure are CNS depression; skin irritation; lung, liver and kidney damage. *See NIOSH*

Pocket Guide to Chemical Hazards (DHHS/NIOSH 97-140, 1997) p 98.

USE: Reagent and chemical intermediate. Solvent for paint, varnishes; degreaser. Has been used as a surgical anesthetic.

3866. Ethylidene Diacetate. [542-10-9] 1,1-Ethanediol 1,1-diacetate; 1,1-diacetoxyethane. $C_6H_{10}O_4$; mol wt 146.14. C 49.31%, H 6.90%, O 43.79%. Made by heating acetaldehyde and acetic anhydride. Control of fungal and bacterial growth in crops and animal feedstuffs: D. L. Kensler *et al.*, **US 3931412** and **US 4012526** (1976 and 1977 to Chevron).

Liq; sharp, fruity odor. d_4^{12} 1.061. bp 167-169°. Slightly sol in water; miscible with alcohol. With NaOH it yields acetaldehyde.

USE: Agricultural fungicide.

3867. Ethylidene Dicoumarol. [1821-16-5] 3,3′-Ethyli-denebis[4-hydroxy-2*H*-1-benzopyran-2-one]; 3,3′-ethylidenebis(4-hydroxycoumarin); 3,3-ethylidenebis(4-oxycoumarin); Pertrombon. $C_{20}H_{14}O_6$; mol wt 350.33. C 68.57%, H 4.03%, O 27.40%. Prepn: Jansen, Jensen, *Z. Phys. Chem.* **277**, 66 (1943); Sullivan *et al., J. Am. Chem. Soc.* **66**, 2288 (1943); Stahmann *et al., ibid.* **67**, 900 (1944); **AT 196874** (1958 to Spofa), *C.A.* **52**, 10212d (1958); Fucik, **CS 85918** (1957), *C.A.* **52**, 2930d (1958).

Crystals from ethanol + dioxane, mp 178°.
Dibenzoate. $C_{34}H_{22}O_8$. mp 209-210°.

THERAP CAT: Anticoagulant.

3868. Ethyl Iodide. [75-03-6] Iodoethane. C_2H_5I; mol wt 155.97. C 15.40%, H 3.23%, I 81.36%. Made by the action of iodine on alcohol in the presence of red phosphorus. Ethyl iodide contains about 98% CH_3CH_2I and about 2% alcohol.

Heavy, clear, very refractive liquid; ethereal odor. When freshly prepared it is colorless, but soon becomes red (on exposure to light and air) due to liberation of iodine. Silver leaf prevents or greatly retards decompn. d_{20}^{20} 1.950. bp 72°. mp −108°. n_D^{15} 1.5168. Sol in 250 parts water with gradual decompn; miscible with alcohol and most organic solvents. *Keep dry and cool, tightly closed, protected from light.*

3869. Ethyl Isobutyrate. [97-62-1] 2-Methylpropanoic acid ethyl ester. $C_6H_{12}O_2$; mol wt 116.16. C 62.04%, H 10.41%, O 27.55%.

Liquid; aromatic, fruity odor. d_{20}^{20} 0.870. bp 110-111°. mp −88°. n_D^{20} 1.3903. *Flammable.* Slightly sol in water; miscible with alcohol, ether.

USE: Manuf flavoring compds and essences.

3870. Ethyl Isothiocyanate. [542-85-8] Isothiocyanatoeth-ane; ethyl mustard oil. C_3H_5NS; mol wt 87.14. C 41.35%, H 5.78%, N 16.07%, S 36.79%. Obtained by the action of mercuric chloride on the product of the reaction of carbon disulfide with eth-ylamine.

Consult the Name Index before using this section.

Colorless to yellowish liq; pungent odor. d_4^{18} 1.003. bp 130-132°. mp −6°. n_D^{18} 1.5142. Insol in water. Miscible with alcohol, ether.
Caution: Vapors irritate the eyes; blister the skin.
USE: Military poison gas.

3871. Ethyl Isovalerate. [108-64-5] 3-Methylbutanoic acid ethyl ester. $C_7H_{14}O_2$; mol wt 130.19. C 64.58%, H 10.84%, O 24.58%.

Colorless, oily liq; apple odor. d_{20}^{20} 0.868. bp 135°. mp −99°. n_D^{20} 1.4009. Sol in about 350 parts water; miscible with alcohol, benzene, ether.
USE: In alc soln for flavoring confectionery and beverages.

3872. Ethyl Lactate. [97-64-3] 2-Hydroxypropanoic acid ethyl ester; ethyl α-hydroxypropionate. $C_5H_{10}O_3$; mol wt 118.13. C 50.84%, H 8.53%, O 40.63%.

Colorless liquid; characteristic odor. d_4^{14} 1.042. bp 154°. $[\alpha]_D^{14}$ −10°. Flash pt, closed cup: 117°F (47°C). Miscible with water (with partial decompn), alcohol, ether. *Flammable. Keep well closed.*
USE: As solvent for nitrocellulose and cellulose acetate.

3873. Ethyl Laurate. [106-33-2] Dodecanoic acid ethyl ester. $C_{14}H_{28}O_2$; mol wt 228.38. C 73.63%, H 12.36%, O 14.01%.

Oil. d_{19}^{19} 0.867. mp −10°. bp 269°; bp_{25} 163°. n_D 1.4321. Insol in water. Very sol in alcohol, ether.

3874. Ethyl Levulinate. [539-88-8] 4-Oxopentanoic acid ethyl ester. $C_7H_{12}O_3$; mol wt 144.17. C 58.32%, H 8.39%, O 33.29%.

Liquid. d_{20}^{20} 1.012. bp 205-206°. n_D^{20} 1.4229. Freely sol in water; miscible with alcohol.

3875. Ethyl Linoleate. [544-35-4] (9Z,12Z)-9,12-Octadeca-dienoic acid ethyl ester; mandenol. $C_{20}H_{36}O_2$; mol wt 308.51. C 77.86%, H 11.76%, O 10.37%. Prepn from sunflower seed oil: Mc-Cutcheon, *Org. Synth.* coll. vol. III, 526 (1955); *see also* Parker *et al.*, *Biochem. Prep.* **4**, 88 (1955).

Colorless oil. More stable to air oxidation than linoleic acid. $d_4^{15.5}$ 0.8846; d^{20} 0.8919. bp_6 193°; $bp_{2.5}$ 175°; $bp_{0.001}$ 133°. n_D^{20}

1.46753; n_D^{48} 1.4489. Iodine value 162.5. Miscible with dimethyl-formamide, fat solvents, oils.
USE: In the vitamin industry.

3876. Ethyl Loflazepate. [29177-84-2] 7-Chloro-5-(2-fluo-rophenyl)-2,3-dihydro-2-oxo-1*H*-1,4-benzodiazepine-3-carboxylic acid ethyl ester; ethyl 7-chloro-5-(*o*-fluorophenyl)-2,3-dihydro-2-oxo-1*H*-1,4-benzodiazepine-3-carboxylate; CM-6912; Meilax; Victan. $C_{18}H_{14}ClFN_2O_3$; mol wt 360.77. C 59.93%, H 3.91%, Cl 9.83%, F 5.27%, N 7.77%, O 13.30%. Benzodiazepine tranquilizer. Prepn: J. Hellerbach *et al.*, **DE 2001276**; *eidem*, **US 3657223** (1970, 1972 both to Hoffmann-La Roche). Analysis in plasma and urine: W. Cantreels, J. R. Jeanniot, *Biomed. Mass Spectrom.* **7**, 565 (1980). GC analysis of metabolites: J. P. Cano, *J. Chromatogr.* **226**, 413 (1981). Toxicologic evaluation: G. Mozue *et al.*, *Int. J. Clin. Pharmacol. Ther. Toxicol.* **19**, 453 (1981).

Crystals from ether, mp 193-194°.
Note: This is a controlled substance (depressant): **21 CFR**, 1308.14.
THERAP CAT: Anxiolytic.

3877. *N*-Ethylmaleimide. [128-53-0] 1-Ethyl-1*H*-pyrrole-2,5-dione. $C_6H_7NO_2$; mol wt 125.13. C 57.59%, H 5.64%, N 11.19%, O 25.57%. Prepd by heating *N*-ethylmaleamic acid in par-affin: Marrian, *J. Chem. Soc.* **1949**, 1515.

Crystals, mp 45°. Lacrimator when liquid.
Caution: Strong irritant.
USE: In cancer research (possible antimitotic activity.)

3878. Ethyl Malonate. [105-53-3] Propanedioic acid 1,3-di-ethyl ester; diethyl malonate; malonic ester. $C_7H_{12}O_4$; mol wt 160.17. C 52.49%, H 7.55%, O 39.96%. Prepd from chloroacetic acid and sodium cyanide followed by esterification with ethanol and H_2SO_4: L. Gattermann, T. Wieland, *Praxis des Organischen Chemikers* (de Gruyter, Berlin, 40th ed., 1961) pp 220-221.

Liquid. Slightly aromatic, pleasant odor. d 1.055. bp 198-199°; bp_{20} 95°. mp −50°. n_D^{20} 1.4143. One gram dissolves in about 50 ml of water. Miscible with alcohol, ether.
USE: Manuf of barbiturates.

3879. Ethyl Maltol. [4940-11-8] 2-Ethyl-3-hydroxy-4*H*-py-ran-4-one; 2-ethylpyromeconic acid; Veltol-Plus. $C_7H_8O_3$; mol wt 140.14. C 60.00%, H 5.75%, O 34.25%. Synthetic analog of maltol, *q.v.* Prepn: C. R. Stephens, Jr. *et al.*, **BE 651427**; C. R. Stephens, Jr., R. P. Allingham, **US 3376317** (1965, 1968 both to Pfizer). Synthesis: I. Ichimoto, C. Tatsumi, *Agric. Biol. Chem.* **34**, 961 (1970). Toxicity studies: E. J. Gralla *et al.*, *Toxicol. Appl. Pharmacol.* **15**, 604 (1969). GC determn in apple juice: S. W. Gunner *et al.*, *J. AOAC Int.* **51**, 959 (1968). Solid phase microextraction and micro-wave-assisted extraction from food products: Y. Wang *et al.*, *J. High Resolut. Chromatogr.* **20**, 213 (1997). Metabolism in dogs: H.

H. Rennhard, *J. Agric. Food Chem.* **19**, 152 (1971). Crystal structures of polymorphs: S. D. Brown *et al.*, *Acta Crystallogr.* **C51**, 1335 (1995). Review of use as a food additive: D. T. LeBlanc, H. A. Akers, *Food Technol.* **43**, 78-84 (1989).

Crystals from chloroform, mp 89-90° (Ichimoto). Also reported as crystals from water, mp 89-92° (Gunner). Cotton candy, caramel-like odor. uv max (methanol): 278 nm (ε 8200). LD$_{50}$ orally in male mice, male rats, female rats, chicks (mg/kg): 780, 1150, 1200, 1270 (Gralla).

USE: Flavor and fragrance enhancer in foods, esp baked goods, beverages, and synthetic berry and citrus flavorings; minimizes undesirable flavors in tobacco products, cough syrup, vitamins, cosmetics, and saccharin-containing products.

3880. Ethyl Menthane Carboxamide. [39711-79-0] *N*-Ethyl-5-methyl-2-(1-methylethyl)cyclohexanecarboxamide; *N*-ethyl-4-menthane-3-carboxamide; WS3. $C_{13}H_{25}NO$; mol wt 211.35. C 73.88%, H 11.92%, N 6.63%, O 7.57%. Prepn: NL 7201523; H. R. Watson *et al.*, US 4193936 (1972, 1980 both to Wilkinson Sword). Structure-activity study: H. R. Watson *et al.*, *J. Soc. Cosmet. Chem.* **29**, 185 (1978). Brief review: M. A. Parrish, *Manuf. Chem.* **58**, 31-32 (February, 1987).

White to nearly white powder, mp 88°. [α]$_D$ −49 to −54°. Insol in water. Sol in most organic solvents. Average oral cooling threshold: 0.2 μg. LD$_{50}$ in mice, rats (g/kg): 5.3, 2.9 orally (Parrish).

USE: Physiological coolant in foods, beverages, toiletries, cosmetics and pharmaceuticals.

3881. Ethylmercuric Chloride. [107-27-7] Chloroethylmercury; Granosan. C_2H_5ClHg; mol wt 265.10. C 9.06%, H 1.90%, Cl 13.37%, Hg 75.67%. CH_3CH_2HgCl. Prepd from ethylmercuric bromide by treating with methanolic KOH, filtering and neutralizing with HCl: Slotta, Jacobi, *J. Prakt. Chem.* [2] **120**, 249 (1929). Prepn from HgCl$_2$ and tetraethyl lead: Whelen in *Adv. Chem. Ser.* **23**, entitled "Metal-Organic Compounds," M. Sittig, Ed. (ACS, Washington DC, 1960) pp 82-86. For the prepn of the bromide from ethylmagnesium bromide and mercuric bromide *see* Slotta, Jacobi, *loc. cit.*, Marvel *et al.*, *J. Am. Chem. Soc.* **47**, 3009 (1925). *Review:* Krause, von Grosse, *Die Chemie der Metallorganischen Verbindungen* (Berlin, 1937).

White, silvery leaflets from ethanol, mp 192°. Sublimes easily. Solubility in water at 18° = 1.4×10^{-4} g/100 ml, at 100° = 2.5×10^{-4} g/100 ml; at 78° = 0.75 g/100 g; at 78° = 3.5 g/100 g; in chloroform at 18° = 2.6 g/100 g. Slightly sol in ether.

Note: A close analog, *chloro-(2-methoxyethyl)mercury, Ceresan Wet*, is also used as seed fungicide. *Caution:* Highly toxic. Causes skin burns; is absorbed through the skin. Chronic exposure has caused permanent injury to brain.

USE: Applied at 2% strength (soln or mixed with solids) as a fungicide for treating seeds.

3882. Ethyl Methanesulfonate. [62-50-0] Methanesulfonic acid ethyl ester; ethyl methanesulfonic acid; ethyl mesylate; EMS; NSC-26805. $C_3H_8O_3S$; mol wt 124.15. C 29.02%, H 6.50%, O 38.66%, S 25.82%. Prepn: O. C. Billeter, *Ber.* **38**, 2015 (1905). Mutagenicity studies: T. Alderson, *Nature* **207**, 164 (1965); J. B. Jenkins, *Mutat. Res.* **4**, 90 (1967); A. P. Schalet, *ibid.* **49**, 313 (1978). Review of carcinogenicity studies: *IARC Monographs* **7**, 245-252 (1974). Review of comparative mutagenicity of EMS and methyl

methanesulfonate, *q.v.*: S. Kondo, *Environ. Sci. Res.* **24**, 743-785 (1981).

Liquid, bp$_{761}$ 213-213.5°; bp$_{10}$ 85-86°. d$_4^{22}$ 1.1452.

Caution: This substance is reasonably anticipated to be a human carcinogen: *Report on Carcinogens, Twelfth Edition* (PB2011-111646, 2011) p 194.

USE: Exptlly as mutagen, teratogen and brain carcinogen.

3883. Ethyl Methyl Ether. [540-67-0] Methoxyethane; methyl ethyl ether. C_3H_8O; mol wt 60.10. C 59.96%, H 13.42%, O 26.62%.

Liquid. d$_0^0$ 0.725. bp 10.8°. *Flammable.* Sol in water; miscible with alcohol, ether.

3884. Ethylmorphine. [76-58-4] (5α,6α)-7,8-Didehydro-4,5-epoxy-3-ethoxy-17-methylmorphinan-6-ol; codéthyline. $C_{19}H_{23}NO_3$; mol wt 313.40. C 72.82%, H 7.40%, N 4.47%, O 15.31%. Semisynthetic opioid analgesic. Prepd by ethylation of morphine: Baizer, Ellner, *J. Am. Pharm. Assoc.* **39**, 581 (1950); Gorecki, *Ann. Pharm. (Poznan)* **7**, 21 (1969). Toxicity data: M. Aurousseau, J. Navarro, *Ann. Pharm. Fr.* **15**, 640 (1957). GC-MS determn in blood: H. Gjerde *et al.*, *Forensic Sci. Int.* **51**, 105 (1991). Biotransformation and pharmacokinetics: T. A. Aasmundstad *et al.*, *Br. J. Clin. Pharmacol.* **39**, 611 (1995).

Crystals from ethanol, mp 199-201°.

Hydrochloride dihydrate. [6746-59-4]; [125-30-4] (anhydrous). $C_{19}H_{23}NO_3 \cdot HCl \cdot 2H_2O$; mol wt 385.89. Component of *Cosylan, Phol-Tux, Sano-Tuss.* White to faintly yellow crystalline powder, mp about 123° (dec); anhydrous form, mp about 170° (dec). One gram dissolves in 10 ml water and in 25 ml alcohol; slightly sol in chloroform, ether. LD$_{50}$ in mice: 264.6 mg/kg s.c.; 0.771 g/kg orally (Aurousseau, Navarro).

Note: This is a controlled substance (opiate): **21 CFR**, 1308.12.

THERAP CAT: Antitussive.

3885. Ethyl Nitrite. [109-95-5] Nitrous acid ethyl ester; nitrous ether. $C_2H_5NO_2$; mol wt 75.07. C 32.00%, H 6.71%, N 18.66%, O 42.62%. Prepd by the action of sodium nitrite on a mixture of alcohol and H$_2$SO$_4$ in the cold. The article of commerce contains 90-95% ethyl nitrite, the remainder is chiefly alcohol as preservative. Laboratory generation of ethyl nitrite gas by placing 20 liters of 90% ethanol in a suitable vessel, diluting with 200 liters of water, and while stirring, adding to the dil alc 18.3 kg of nitrosyl chloride at the rate of 2.25 kg per hour; *see also* Chase, US 2615896 (1952).

Colorless or yellowish, clear, exceedingly volatile liq. Characteristic odor; burning, sweetish taste. d$_{15}^{15}$ 0.90. bp 17°. Slightly sol in water and dec by it; miscible with alcohol, ether. On keeping, it gradually dec, becoming acid and oxides of nitrogen form. Decompn is hastened by air, light, and moisture. *Flammable, poisonous. Keep tightly closed, in a cool place, protected from light.*

Spirit of Ethyl Nitrite. Spirit of nitrous ether; sweet spirit of niter. An alcoholic soln containing 3.5-4.5% ethyl nitrite. Pale yellow or faintly greenish-yellow, clear, mobile, volatile, flammable liquid; ethereal, pungent odor; sharp, burning taste. The ethyl nitrite evaporates rapidly; dec on exposure to air; dec in light. $d^{25} \leq 0.823$. Miscible with water and alcohol. *Flammable, poisonous. Keep tightly closed, protected from light, in a cool place, remote from fire.*

Caution: May cause methemoglobinemia and hypotension and, in high concns, narcosis.

THERAP CAT: Has been used as diaphoretic.

3886. Ethyl Nitrobenzoate. $C_9H_9NO_4$; mol wt 195.17. C 55.39%, H 4.65%, N 7.18%, O 32.79%.

Colorless crystals. Insol in water. Freely sol in alc, ether.

o-Ester. mp 30°. bp$_{18}$ 173°.

m-Ester. mp 47°. bp about 298°.

p-Ester. mp 57°.

3887. N-Ethyl-N-nitrosourea. [759-73-9] *N*-Nitroso-*N*-ethylurea; ENU. $C_3H_7N_3O_2$; mol wt 117.11. C 30.77%, H 6.03%, N 35.88%, O 27.32%. Potent mouse mutagen. Prepn: E. A. Werner, *J. Chem. Soc.* **115**, 1093 (1919); F. Arndt, H. Scholz, *Angew. Chem.* **46**, 47 (1933). Carcinogenicity studies: M. F. Rajewsky, R. Goth, *Mol. Base Malig., Sel. Pap. Int. Symp.* **1976**, 2. Mutagenicity studies: W. L. Russell *et al.*, *Proc. Natl. Acad. Sci. USA* **16**, 5818 (1979). Toxicity studies: H. Druckrey *et al.*, *Nature* **210**, 1378 (1966).

Pale buff-yellow, hexagonal plates, mp 103-104°. LD$_{50}$ i.v. in rats: 240 mg/kg (Druckrey).

Caution: This substance is reasonably anticipated to be a human carcinogen: *Report on Carcinogens, Twelfth Edition* (PB2011-111646, 2011) p 312.

USE: Exptlly as mutagen; ethylating agent.

3888. Ethylnorepinephrine. [536-24-3] 4-(2-Amino-1-hydroxybutyl)-1,2-benzenediol; α-(1-aminopropyl)-3,4-dihydroxybenzyl alcohol; α-(1-aminopropyl)protocatechuyl alcohol; 1-(3,4-dihydroxyphenyl)-2-aminobutanol; 1-(3,4-dihydroxyphenyl)-1-hydroxy-2-aminobutane; ethylnoradrenaline; ethylnorsuprarenin; E.N.E.; E.N.S.. $C_{10}H_{15}NO_3$; mol wt 197.23. C 60.90%, H 7.67%, N 7.10%, O 24.34%. Prepd by catalytic hydrogenolysis of α-benzhydrylamino-3,4-dibenzyloxybutyrophenone-HCl: Suter, Ruddy, *J. Am. Chem. Soc.* **66**, 747 (1944); US 2431285 (1947 to Winthrop-Stearns).

Hydrochloride. [3198-07-0] Bronkephrine. $C_{10}H_{15}NO_3$·HCl; mol wt 233.69. Crystals, dec 199-200°. Soluble in water. Marketed as an 0.2% aq soln.

THERAP CAT: Bronchodilator.

3889. Ethyl Oenanthate. [106-30-9] Heptanoic acid ethyl ester; ethyl heptanoate; ethyl *n*-heptoate; oenanthic ether; cognac oil, synthetic; oil of grapes; aether oenanthicus; oleum vitis viniferae. $C_9H_{18}O_2$; mol wt 158.24. C 68.31%, H 11.47%, O 20.22%. Obtained commercially either by esterification of heptanoic acid, yielding a high-grade product; or by the direct esterification of coconut oil yielding a mixture of the ethyl esters of lauric, myristic, palmitic, and other higher fatty acids plus ethyl and isoamyl alcohol. For some flavoring applications the latter product is more desirable. Prepn: Rogers, *J. Am. Pharm. Assoc. Sci. Ed.* **12**, 503 (1923). Toxicity: P. M. Jenner *et al.*, *Food Cosmet. Toxicol.* **2**, 327 (1964).

Pure ester, liquid. Fruity, wine-like odor and taste with burning aftertaste. d_4^{15} 0.8723; d_4^{25} 0.8630. bp$_{760}$ 189°; bp$_{35}$ 95°; bp$_8$ 68°. Viscosity at 25°: 0.0111 g/cm/sec. Insol in water. Misc with alcohol, ether, chloroform. Forms an azeotrope with water, the azeotropic mixture boils at 98.5° and contains 72% w/w of ethyl oenanthate. LD$_{50}$ orally in rats: >34640 mg/kg (Jenner).

Note: Artificial Ethyl Oenanthate, also known as *artificial cognac essence* or *oil of wine,* is a mixture of amyl and ethyl caprates with ethyl and isoamyl butyrates and caprylates.

USE: In the manuf of liqueurs. Plays an important part in the formulation of raspberry, gooseberry, grape, cherry, apricot, currant, bourbon, and other artificial essences.

3890. Ethylparaben. [120-47-8] 4-Hydroxybenzoic acid ethyl ester; ethyl *p*-hydroxybenzoate; Nipagin A; Ethyl Parasept; Solbrol A. $C_9H_{10}O_3$; mol wt 166.18. C 65.05%, H 6.07%, O 28.88%. Prepd by esterification of *p*-hydroxybenzoic acid: Cavill, Vincent, *J. Soc. Chem. Ind. London* **66**, 175 (1947).

Colorless crystals or white powder, mp 116°. bp 297-298° (decompn). Freely sol in alcohol, ether, acetone, propylene glycol; sol in water at 20°: 0.070% w/w; at 25°: 0.075% w/w. Slightly sol in glycerin.

USE: Antimicrobial preservative for pharmaceuticals.

3891. Ethyl Pelargonate. [123-29-5] Nonanoic acid ethyl ester; ethyl nonanoate; wine ether. $C_{11}H_{22}O_2$; mol wt 186.30. C 70.92%, H 11.90%, O 17.18%. Toxicity data: P. M. Jenner *et al.*, *Food Cosmet. Toxicol.* **2**, 327 (1964).

Liquid. d_4^{18} 0.866. bp ~220°. mp −44°. Insol in water. Misc with alc, ether. LD$_{50}$ orally in rats: >43,000 mg/kg (Jenner). Marketed product is an alcoholic soln of various essences in form of a yellowish liq, d ~0.823. Used in production of cognac-like beverages; 1 part ethyl pelargonate and 20 parts alcohol constitute one kind of artificial "Cognac Essence".

3892. 2-Ethylphenol. [90-00-6] Phlorol; *o*-ethylphenol; 1-ethyl-2-hydroxybenzene. $C_8H_{10}O$; mol wt 122.17. C 78.65%, H 8.25%, O 13.10%. Prepn: W. Suida, S. Plohn, *Monatsh. Chem.* **1**, 175 (1880). Purification and properties: D. P. Biddiscombe *et al.*, *J. Chem. Soc.* **1963**, 5764. Colorimetric determn of isomeric purities in reagent grade chemicals: I. M. Jakovljevic, R. H. Bishara, *J. Chromatogr.* **192**, 425 (1980). Thermolytic reactions: P. Zhou, B. L. Crynes, *Ind. Eng. Chem. Process Des. Dev.* **25**, 898 (1986). Decomposition and oxidation of aqueous solns: C. J. Martino, P. E. Savage, *Ind. Eng. Chem. Res.* **38**, 1775 (1999).

Colorless liquid; phenol odor. Cooling the liquid to −30° gives a metastable cryst form, mp about −28°; when kept for 24-48 hrs, this changes to the stable cryst form, mp −3.4°. fp −3.31 ± 0.01° for repurified sample. bp$_{760}$ 204.52°. d_{20} 1.01885 ± 0.00001; d_{25} 1.01459; d_{30} 1.01033 ± 0.00001 (g/ml). Practically insol in water. Freely sol in alcohol, benzene, glacial acetic acid.

Consult the Name Index before using this section.

USE: Starting material and intermediate for pharmaceutical and agricultural chemicals.

3893. Ethyl Phenylacetate. [101-97-3] Benzeneacetic acid ethyl ester; α-toluic acid ethyl ester. $C_{10}H_{12}O_2$; mol wt 164.20. C 73.15%, H 7.37%, O 19.49%. Prepd by heating benzyl cyanide with alcohol and H_2SO_4: Adams, Thal, *Org. Synth.* **2**, 27 (1922). Absorption spectrum: Baly, Collie, *J. Chem. Soc.* **87**, 1344 (1905); Baly, Tryhorn, *ibid.* **107**, 1063 (1915).

Liquid. Pleasant odor. d_4^{20} 1.0333. bp_{760} 226°; bp_{32} 135°; bp_{20} 121°. $n_D^{18.5}$ 1.49921.
USE: In perfumery.

3894. 3-Ethyl-4-picoline. [529-21-5] 3-Ethyl-4-methylpyridine; β-collidine; 3-ethyl-γ-picoline; β-ethyl-γ-methylpyridine. $C_8H_{11}N$; mol wt 121.18. C 79.29%, H 9.15%, N 11.56%. Prepd from 2,6-dichloro-3-ethyl-4-methylpyridine by reduction with HI and red phosphorus: Ruzicka, Fornasir, *Helv. Chim. Acta* **2**, 338 (1919); from 3-acetyl-1,4-dimethyl-1,2,5,6-tetrahydropyridine: Prelog, Komzak, *Ber.* **74**, 1705 (1941); from γ-methylnicotinic acid: Rabe, Jantzen, *Ber.* **54**, 925 (1921); from β-ethylpyridine: Koenigs, Hoffmann, *Ber.* **58**, 194 (1925); from crotonaldehyde and ammonia: Tschitschibabin, Oparina, *Ber.* **60**, 1877 (1927).

Liquid, aromatic odor. d_4^{17} 0.9286. bp_{753} 195-196°; bp_{23} 88-90°; bp_{12} 76°. Volatile with steam. Sparingly sol in water. Soluble in alcohol, ether, chloroform, dil acids.
Hydrochloride. $C_8H_{11}N.HCl$. Hygroscopic platelets, sol in water.

3895. 4-Ethyl-2-picoline. [536-88-9] 4-Ethyl-2-methylpyridine; α-methyl-γ-ethylpyridine; 4-ethyl-α-picoline; α-collidine. $C_8H_{11}N$; mol wt 121.18. C 79.29%, H 9.15%, N 11.56%. Prepd from 2-picoline and acetic anhydride in the presence of zinc dust: H. Maier-Bode, J. Altpeter, *Das Pyridin und Seine Derivate* (Halle, 1934) p 54; **DE 390333**.

Liquid. d_4^0 0.9291; d_4^{16} 0.9268. bp_{751} 179°. Volatile with steam. Sol in alcohol, ether, benzene, dil acids. Sparingly sol in water.

3896. 5-Ethyl-2-picoline. [104-90-5] 5-Ethyl-2-methylpyridine; aldehyde-collidine; aldehydine; 2-methyl-5-ethylpyridine; 5-ethyl-α-picoline; 3-ethyl-6-methylpyridine. $C_8H_{11}N$; mol wt 121.18. C 79.29%, H 9.15%, N 11.56%. Prepd by heating acetaldehyde ammonia in a double vol of abs alc: Ador, Bayer, *Ann.* **155**, 297 (1870); by heating ammonia water and paraldehyde in presence of ammonium acetate: Dürkopf, Schlaugk, *Ber.* **21**, 294 (1888); **DE 347820**; **DE 349184**; **DE 349267**; H. Maier-Bode, J. Altpeter, *Das Pyridin und Seine Derivate* (Halle, 1934) p 56; Frank *et al.*, *J. Am. Chem. Soc.* **68**, 1368 (1946); Dunn, **US 2717897** (1955 to UCC); R. L. Frank *et al.*, *Org. Synth.* **coll. vol. IV**, 451 (1963). Continuous process: Takeba *et al.*, **US 2935513** (1960 to Takeda).

Liquid. Aromatic odor. d_4^{23} 0.9184. bp_{747} 177.8°; bp_{20} 74-75°; bp_{17} 65-66°. n_D^{20} 1.4971. Volatile with steam. Practically insol in water. Sol in alcohol, ether, benzene, dil acids, concd H_2SO_4.
Hydrochloride. $C_8H_{11}N.HCl$. Hygroscopic needles, freely sol in water.

3897. 1-Ethyl-3-piperidinol. [13444-24-1] 1-Ethyl-3-hydroxypiperidine. $C_7H_{15}NO$; mol wt 129.20. C 65.08%, H 11.70%, N 10.84%, O 12.38%. Prepn: Biel *et al.*, *J. Am. Chem. Soc.* **74**, 1485 (1952); Biel, **US 2802007** (1957 to Lakeside).

Liquid. bp_{15} 93-95°. n_D^{14} 1.4777.

3898. Ethyl Propiolate. [623-47-2] 2-Propynoic acid ethyl ester; ethyl acetylenecarboxylate. $C_5H_6O_2$; mol wt 98.10. C 61.22%, H 6.17%, O 32.62%. Michael acceptor in organic synthesis. Prepn: W. H. Perkin, Jr., J. L. Simonsen, *J. Chem. Soc.* **1907**, 816. IR spectrum and rotational isomerism study: S. W. Charles *et al.*, *J. Mol. Struct.* **26**, 249 (1975). Use in photochemical addition reactions: G. Büchi, S. H. Feairheller, *J. Org. Chem.* **34**, 609 (1969); in Michael addition reactions: A. Hong, J. M. Friedman, *Synth. Commun.* **27**, 2971 (1997); in halo aldol reactions with aldehydes: D. Chen *et al.*, *Eur. J. Org. Chem.* **2004**, 3330. Review of use as a peptide coupling reagent: B. Iorga, J.-M. Campagne, *Synlett* **2004**, 1826-1828.

Distills at 119° (745 mm); also reported as bp 122° (Büchi, Feairheller). d_4^4 0.9788. d_{15}^{15} 0.9676. d_{25}^{25} 0.9583.
USE: In organic synthesis; peptide coupling reagent.

3899. Ethyl Propionate. [105-37-3] Propanoic acid ethyl ester. $C_5H_{10}O_2$; mol wt 102.13. C 58.80%, H 9.87%, O 31.33%.

Colorless liquid; fruity odor. d_4^{20} 0.891. bp 99°. mp −73°. n_D^{20} 1.3844. Flash pt, closed cup: 12°C. *Flammable.* Sol in about 60 parts water; miscible with alcohol, ether.

3900. 3-Ethylpyridine. [536-78-7] β-Lutidine. C_7H_9N; mol wt 107.16. C 78.46%, H 8.47%, N 13.07%.

Colorless to brownish liquid. d^{23} 0.940. bp 162-165°. n_D^{22} 1.5021. Slightly sol in water; freely sol in alcohol, or ether.

3901. 4-Ethylpyridine. [536-75-4] γ-Ethylpyridine. C_7H_9N; mol wt 107.16. C 78.46%, H 8.47%, N 13.07%. Found in California petr: Hackmann *et al.*, *Rec. Trav. Chim.* **62**, 229 (1943). Best prepd from pyridine by treatment with acetic anhydride and zinc: Dohrn, Horsters, **DE 390333** (1924 to Schering); Wibaut, Arens, *Rec. Trav. Chim.* **60**, 119 (1941); Frank, Smith, *Org. Synth.* **27**, 38 (1947). Using iron powder: Tenenbaum, Fand, **US 2712019** (1955 to Nepera Chem.).

Ethyl Salicylate

Liquid. Obnoxious odor. Turns brown if not very pure. d_4^{22} 0.9404. bp_{750} 169.6-170.0°. n_D^{18} 1.5029. Sparingly sol in water; sol in alcohol, ether.

Caution: Symptoms similar to pyridine, *q.v.*

3902. Ethyl Salicylate. [118-61-6] 2-Hydroxybenzoic acid ethyl ester; ethyl *o*-hydroxybenzoate; salicylic acid ethyl ester; salicylic ether; sal ethyl. $C_9H_{10}O_3$; mol wt 166.18. C 65.05%, H 6.07%, O 28.88%. Prepd by esterification of salicylic acid with ethanol. GC/MS determn in biological fluids: T. Kakkar, M. Mayersohn, *J. Chromatogr. B* **718**, 69 (1998). Thermophysical properties: Y.-W. Sheu, C.-H. Tu, *J. Chem. Eng. Data* **50**, 1706 (2005). Brief description: D. L. J. Opdyke, *Food Cosmet. Toxicol.* **16**, Suppl. 1, 751-752 (1978). Review of toxicology as a fragrance ingredient: A. Lapczynski *et al.*, *Food Chem. Toxicol.* **45**, Suppl. 1, S397-S401 (2007).

Clear, colorless to very pale yellow, very refractive liquid; pleasant wintergreen-type odor. d_4^{20} 1.131. d^{25} 1.12500. bp 231-234°. mp 1°. n_D^{25} 1.52022. Flash pt, closed cup: >212°F (>100°C). Slightly sol in water; miscible with alc, ether. Viscosity (25°): 2.831 mPa·s. Surface tension (25°): 36.3 mN/m. LD_{50} (g/kg): 1.32 orally in rats; >5 dermally in rabbits (Opdyke). *Protect from light.*

USE: Fragrance ingredient in cosmetics, toiletries, household cleaners and detergents.

3903. Ethyl Silicate. [78-10-4] Silicic acid (H_4SiO_4) tetraethyl ester; silicon tetraethoxide; tetraethoxysilane; tetraethyl silicate. $C_8H_{20}O_4Si$; mol wt 208.33. C 46.12%, H 9.68%, O 30.72%, Si 13.48%. Prepn from abs alcohol and silicon tetrachloride: J. J. Ebelmen, *Ann.* **57**, 319 (1846); A. W. Dearing, E. E. Reid, *J. Am. Chem. Soc.* **50**, 3058 (1928); G. Sumrell, E. E. Ham, *ibid.* **78**, 5573 (1956). Toxicological studies: H. F. Smyth, J. Seaton, *J. Ind. Hyg. Toxicol.* **22**, 288 (1940); V. H. Rowe *et al.*, *ibid.* **30**, 332 (1948). *Review:* H. D. Cogan, C. A. Setterstrom, *Ind. Eng. Chem.* **39**, 1364 (1947).

Colorless, flammable liquid. mp −77°. bp 165-166°. Flash point 125°F (52°C). d_4^{20} 0.933; n_D^{25} 1.3818. Viscosity: 0.6 cP. Practically insol in water, and slowly dec by it. Miscible with alcohol.

Caution: Potential symptoms of overexposure are irritation of eyes and nose. *See NIOSH Pocket Guide to Chemical Hazards* (DHHS/NIOSH 97-140, 1997) p 142.

USE: In weatherproofing and hardening stone, arresting decay and disintegration; manuf weatherproof and acidproof mortars and cements. In the "lost wax" process for casting of high-melting alloys.

3904. Ethyl Sulfate. [540-82-9] Sulfuric acid monoethyl ester; ethyl hydrogen sulfate; ethylsulfuric acid; sulfethylic acid; sulfovinic acid. $C_2H_6O_4S$; mol wt 126.13. C 19.05%, H 4.80%, O 50.74%, S 25.42%. $C_2H_5OSO_2OH$. Prepn: Hamid *et al.*, *J. Chem. Soc.* **129**, 1098 (1926); Atwood, US 2683731 (1954 to Natl. Petro-Chemicals). *Review:* C. M. Suter, *The Organic Chemistry of Sulfur* (John Wiley, 1944) p 23-27.

Corrosive.

Barium salt. Barium ethyl sulfate. $C_4H_{10}BaO_8S_2$. Dihydrate, lustrous leaflets. Freely soluble in water; slightly sol in alcohol. Not pptd by H_2SO_4 or sulfates.

Calcium salt. Calcium ethyl sulfate; calcium sulfovinate. C_4H_{10}-CaO_8S_2. Dihydrate, crystals. Soluble in water, alcohol.

See also Diethyl Sulfate.

USE: Intermediate in the manuf of ethanol from ethylene.

3905. Ethyl Sulfide. [352-93-2] 1,1′-Thiobisethane; diethyl sulfide; thioethyl ether. $C_4H_{10}S$; mol wt 90.18. C 53.28%, H 11.18%, S 35.55%. Made by distillation of sodium ethyl sulfate with a soln of K_2S.

Liquid; ethereal odor. d_4^{20} 0.837. bp 92°. n_D^{20} 1.44233. Insol in water; miscible with alcohol, ether.

USE: Solvent for anhydr mineral salts; in plating baths for coating metals with gold or silver.

3906. 2-(Ethylthio)ethanol. [110-77-0] β-Hydroxydiethyl sulfide. $C_4H_{10}OS$; mol wt 106.18. C 45.25%, H 9.49%, O 15.07%, S 30.19%. Prepn from 2-chloroethanol and ethyl mercaptan: Demuth, Meyer, *Ann.* **240**, 310 (1887); Steinkopf *et al.*, *Ber.* **53**, 1010 (1920); from 2-thioethanol and diethyl sulfate: Bergmann, *J. Am. Chem. Soc.* **74**, 829 (1952).

Liquid. d_{20}^{20} 1.015-1.025. bp_{760} 184.5°; bp_{28} 99°. Soluble in ether, other organic solvents.

USE: Intermediate for pesticides, lubricating and cutting oil additives, flotation agents, plasticizers.

3907. Ethyl *p*-Toluenesulfonate. [80-40-0] 4-Methylbenzenesulfonic acid ethyl ester. $C_9H_{12}O_3S$; mol wt 200.25. C 53.98%, H 6.04%, O 23.97%, S 16.01%.

Monoclinic crystals. d 1.17. mp 33°. bp_{15} 173°.

USE: For ethylation.

3908. Ethyl Vanillin. [121-32-4] 3-Ethoxy-4-hydroxybenzaldehyde; ethylprotocatechuic aldehyde; bourbonal; ethylprotal; vanillal; Ethavan; Ethovan. $C_9H_{10}O_3$; mol wt 166.18. C 65.05%, H 6.07%, O 28.88%.

Colorless flakes possessing a finer and more intense vanilla odor and taste than vanillin. mp 77-78°. Freely soluble in alcohol, chloroform, ether, sols of alkali hydroxides; sol in glycerol, propylene glycol; sparingly sol in water at 50°. Soly in 95% alc about 1 g/2 ml. LD_{50} orally in rats: >2000 mg/kg, P. M. Jenner *et al.*, *Food Cosmet. Toxicol.* **2**, 327 (1964).

USE: In flavoring and perfumery.

3909. Ethynodiol. [1231-93-2] (3β,17α)-19-Norpregn-4-en-20-yne-3,17-diol; 17α-ethynyl-19-norandrost-4-ene-3β,17β-diol; 17α-ethynyl-4-estrene-3β,17β-diol; ED. $C_{20}H_{28}O_2$; mol wt 300.44. C 79.96%, H 9.39%, O 10.65%. Prepn: F. B. Colton, US 2843609 (1958 to Searle). Prepn of the 3-acetate, 17-acetate, and diacetate: P. D. Klimstra, US 3176013 (1965 to Searle); *see also* F. Sondheimer, Y. Klibansky, *Tetrahedron* **5**, 15 (1959). Pharmacokinetics and metabolism: C. J. Lewis *et al.*, *Xenobiotica* **10**, 705 (1980). Com-

Consult the Name Index before using this section.

parative clinical study in oral contraceptives: M. H. Briggs, *J. Reprod. Med.* **28**, Suppl. 1, 92 (1983). Review of carcinogenicity studies: *IARC Monographs* **21**, 387-398 (1979). Comprehensive description: E. P. K. Lau, J. L. Sutter, *Anal. Profiles Drug Subs.* **3**, 253-279 (1974).

Diacetate. [297-76-7] 3β,17β-Diacetoxy-17α-ethynyl-4-estrene; SC-11800; Femulen; Luteonorm; Luto-Metrodiol; Metrodiol. $C_{24}H_{32}O_4$; mol wt 384.52. Crystals from methanol + water, mp ~126-127°. $[\alpha]_D$ -72.5° (chloroform). Very soluble in chloroform; freely sol in ether; sol in alc; sparingly sol in fixed oils. Insol in water.

Diacetate mixture with mestranol. Luteolas; Metrulen; Ovaras; Ovulen.

Diacetate mixture with ethinyl estradiol. Conova; Demulen; Miniluteolas.

THERAP CAT: Progestogen; in combination with estrogen as oral contraceptive.

3910. Ethynylbenzene. [536-74-3] Phenylacetylene. C_8H_6; mol wt 102.14. C 94.07%, H 5.92%. Prepd by dropping β-bromostyrene on molten potassium hydroxide and distilling: Hessler, *Org. Synth.* **2**, 67 (1922).

Liquid. d_4^{20} 0.9300. mp -44.8°. bp_{760} 142.4°; bp_{90} 75°; bp_{15} 39°. n_D^{20} 1.5489. Insol in water; miscible with alcohol, ether, other organic solvents.

3911. Etidocaine. [36637-18-0] *N*-(2,6-Dimethylphenyl)-2-(ethylpropylamino)butanamide; 2-(ethylpropylamino)-2',6'-butyroxylidide. $C_{17}H_{28}N_2O$; mol wt 276.42. C 73.87%, H 10.21%, N 10.13%, O 5.79%. Prepn: H. J. F. Adams *et al.*, **DE 2162744**; *eidem*, **US 3812147** (1972, 1974 both to Astra). Activity and toxicity studies: *eidem, J. Pharm. Sci.* **61**, 1829 (1972).

Hydrochloride. [36637-19-1] W-19053; Duranest. $C_{17}H_{28}N_2$O.HCl; mol wt 312.88. Crystals from abs ethanol-ether and isopropanol-isopropylether, mp 203-203.5°. LD_{50} in female mice (mg/kg): 6.7 i.v.; 99 s.c. (Adams, 1972).

THERAP CAT: Anesthetic (local).

3912. Etidronic Acid. [2809-21-4] *P,P*'-(1-Hydroxyethylidene)bisphosphonic acid; (1-hydroxyethylidene)diphosphonic acid; ethane-1-hydroxy-1,1-diphosphonic acid; Dequest 2010; Fostex P. $C_2H_8O_7P_2$; mol wt 206.03. C 11.66%, H 3.91%, O 54.36%, P 30.07%. Bisphosphonate antiresorptive agent. Prepn: H. von Baeyer, K. A. Hofmann, *Ber.* **30**, 1973 (1897); and characterization of the acid and disodium salt: F. Kasparek, *Monatsh. Chem.* **99**, 2016 (1968); B. Blaser *et al.*, *Z. Anorg. Allg. Chem.* **381**, 247 (1971). Use as detergent builder: F. L. Diehl, **US 3159581** (1964 to Proctor & Gamble). Determn in pharmaceutical formulations: E. W. Tsai *et al., J. Pharm. Biomed. Anal.* **11**, 513 (1993). Symposium on clinical efficacy in malignancy-related hypercalcemia: *Am. J. Med.* **82**, Suppl 2A, 1-78 (1987). Review of pharmacology and therapeutic

efficacy in resorptive bone disease: C. J. Dunn *et al., Drugs Aging* **5**, 446-474 (1994); of clinical safety and efficacy in osteoporosis: T. P. van Staa *et al., Pharmacotherapy* **18**, 1121-1128 (1998).

Crystallizes from water as the monohydrate. pK_1 1.35 ±0.08; pK_2 2.87 ±0.01; pK_3 7.03 ±0.01; pK_4 11.3. Very sol in water (69% at 20° C). Insol in acetic acid.

Disodium salt. [7414-83-7] Disodium dihydrogen (1-hydroxyethylidene)bis[phosphonate]; etidronate disodium; Didronel; Diphos; Etidron. $C_2H_6Na_2O_7P_2$; mol wt 249.99. White powder. Crystallizes from water as the di- or tetrahydrate. Freely soluble in water. Practically insol in alc.

USE: Sequestering and chelating agent; scale and corrosion inhibitor.

THERAP CAT: Bone resorption inhibitor.

3913. Etifoxine. [21715-46-8] 6-Chloro-*N*-ethyl-4-methyl-4-phenyl-4*H*-3,1-benzoxazin-2-amine; 6-chloro-2-(ethylamino)-4-methyl-4-phenyl-4*H*-3,1-benzoxazine; HOE-36801. $C_{17}H_{17}ClN_2$O; mol wt 300.79. C 67.88%, H 5.70%, Cl 11.79%, N 9.31%, O 5.32%. Psychotropic agent with anxiolytic and anticonvulsant activity. Prepn: **FR M7358**; H. Kuch *et al.*, **US 3725404** (1969, 1973 both to Hoechst); I. Hoffmann *et al., Arzneim.-Forsch.* **20**, 975 (1970). Exptl pharmacological study: J. R. Boissier *et al.*, *Therapie* **27**, 325 (1972). Analysis of EEG effects: D. Bente *et al., Arzneim.-Forsch.* **25**, 944 (1975). Evaluation of psychotropic effect: R. Corsico *et al., Psychopharmacologia* **45**, 301 (1976).

Colorless crystals from petr ether, mp 90-92°. uv max (ethanol): 273 nm (ε 21200). LD_{50} orally in mice: 12 g/kg (Hoffmann).

Hydrochloride. [56776-32-0] Stresam. $C_{17}H_{17}ClN_2$O.HCl; mol wt 337.24. Crystals, mp 150-151°.

THERAP CAT: Anxiolytic.

3914. Etilefrin. [709-55-7] α-[(Ethylamino)methyl]-3-hydroxybenzenemethanol; α-[(ethylamino)methyl]-*m*-hydroxybenzyl alcohol; *m*-hydroxy-α-(ethylaminomethyl)benzyl alcohol; α-(*m*-hydroxyphenyl)-β-(ethylamino)ethanol; ethylphenylephrine; etiladrianol. $C_{10}H_{15}NO_2$; mol wt 181.24. C 66.27%, H 8.34%, N 7.73%, O 17.66%. Sympathomimetic amine. Prepn: T. Goto, *J. Pharm. Soc. Jpn.* **74**, 318 (1954), *C.A.* **49**, 3960 (1955); **ES 273595** (1962 to Labs. Fher S.A.), *C.A.* **60**, 1649e (1964). Clinical pharmacology: A. J. Coleman *et al., Eur. J. Clin. Pharmacol.* **8**, 41 (1975). Clinical pharmacokinetics and disposition: J. H. Hengstmann *et al.*, *ibid.* **9**, 179 (1975). HPLC determn in plasma: K. Kojima *et al., J. Chromatogr.* **525**, 210 (1990). Clinical trial in arterial hypotension: T. Taivainen, *Acta Anaesthesiol. Scand.* **35**, 164 (1991); in neurocardiogenic syncope: F. Ammirati *et al., Am. J. Cardiol.* **86**, 472 (2000).

Crystals, mp 147-148°.

Hydrochloride. [534-87-2] Cardanat; Circupon; Effontil; Effortil; Efortil; Kertasin; Thomasin. $C_{10}H_{15}NO_2$.HCl; mol wt 217.69.

Crystals, mp 121°. Bitter taste. Freely sol in water; sol in alcohol. Practically insol in chloroform.

THERAP CAT: Antihypotensive.

THERAP CAT (VET): Antihypotensive.

3915. Etiocholane. [438-23-3] (5β)-Androstane; 5-epiandrostane. $C_{19}H_{32}$; mol wt 260.47. C 87.61%, H 12.38%. Parent compd of alkyl substituted etiocholanes, such as pregnane, cholane, coprostane. Prepd from etiocholane-17-one semicarbazone: Butenandt, Dannenbaum, *Z. Physiol. Chem.* **229**, 192 (1934).

Needles from acetone, mp 78-80°.

3916. Etiocholanic Acid. [438-08-4] (5β,17β)-Androstane-17-carboxylic acid; aetiocholanic acid; etianic acid; etiocholane-17β-carboxylic acid. $C_{20}H_{32}O_2$; mol wt 304.47. C 78.90%, H 10.59%, O 10.51%. Prepn: Wieland *et al., Z. Physiol. Chem.* **161**, 80 (1926); Jacobs, Elderfield, *J. Biol. Chem.* **108**, 497 (1935); Tschesche, *Ber.* **68**, 9 (1935); Steiger, Reichstein, *Helv. Chim. Acta* **21**, 841 (1938).

Needles from glacial acetic acid; elongated leaflets from acetic acid. Sublimes at 0.002 mm press. and 160° bath temp, mp 228-229°. Insol in water; sol in pentane.

Methyl ester. $C_{21}H_{34}O_2$. Needles from methanol, mp 99-101°.

Note: The stem name "etianic acid" was proposed by the Subcommittee on Steroid Nomenclature of the National Research Council as a replacement for "etiocholanic acid" in order to avoid the use of the same name for parent hydrocarbons of different carbon content: *J. Am. Chem. Soc.* **74**, 2817 (1952).

3917. Etioporphyrin. [26608-34-4] 2,7,12,18-Tetraethyl-3,-8,13,17-tetramethyl-21H,23H-porphine; 1,3,5,8-tetramethyl-2,4,6,7-tetraethylporphine; etioporphyrin III; mesoetioporphyrin. $C_{32}H_{38}N_4$; mol wt 478.68. C 80.29%, H 8.00%, N 11.70%. Occurs in Bavarian oil shale, crude petr, ozokerite, amber, cannel coal and other hard varieties of bituminous coal: Treibs, *Ann.* **510**, 60 (1934); **517**, 184 (1935); *Angew. Chem.* **49**, 682 (1936). Prepd by decarboxylation of mesoporphyrin IX: Fischer, Treibs, *Ann.* **466**, 191, 206 (1928). Synthesis: Fischer, Stangler, *Ann.* **462**, 265 (1928); Johnson *et al., J. Chem. Soc.* **1959**, 3416. Structure: Abraham *et al., ibid.* **1961**, 3468. *Review:* Rimington, Kennedy, in M. Florkin, H. S. Mason, *Comparative Biochemistry* (Academic Press, New York, 1962) pp 557-614.

Long prismatic needles or butterflies from pyridine or from chloroform-petr ether, mp 360-363°. pKa 18. Absorption max: 246, 269, 396, 497, 532, 566, 620, 645 nm (log ε 3.90, 3.89, 5.22, 4.13, 3.99, 3.81, 3.65, 2.62).

Copper salt. $C_{32}H_{36}N_4Cu$. Red needles from pyridine-acetic acid.

Magnesium salt. $C_{32}H_{36}N_4Mg$. Crystals from methanol.

3918. Etiproston. [59619-81-7] (5Z)-7-[(1R,2R,3R,5S)-3,5-Dihydroxy-2-[(1E)-2-[2-(phenoxymethyl)-1,3-dioxolan-2-yl]ethenyl]cyclopentyl]-5-heptenoic acid; (5Z,13E)-(8R,9S,11R,12R)-9,11-dihydroxy-15,15-ethylenedioxy-16-phenoxy-17,18,19,20-tetranorprostadienoic acid; 15-deoxy-15,15-ethylenedioxy-16-phenoxy-17,18,19,20-tetranorprostaglandin $F_{2\alpha}$; Prostavet; Vetiprost. $C_{24}H_{32}O_7$; mol wt 432.51. C 66.65%, H 7.46%, O 25.89%. Prostaglandin $F_{2\alpha}$ analog with estrus cycle synchronizing activity. Prepn: W. Skuballa *et al.,* **DE 2434133**; *eidem,* **US 4088775** (1976, 1978 both to Schering AG); and biological activity: W. Skuballa *et al., J. Med. Chem.* **21**, 443 (1978).

Colorless oil.

THERAP CAT (VET): Luteolytic.

3919. Etizolam. [40054-69-1] 4-(2-Chlorophenyl)-2-ethyl-9-methyl-6H-thieno[3,2-f][1,2,4]triazolo[4,3-a][1,4]diazepine; 1-methyl-6-o-chlorophenyl-8-ethyl-4H-s-triazolo[3,4-c]thieno[2,3-e]-1,4-diazepine; Y-7131; Depas. $C_{17}H_{15}ClN_4S$; mol wt 342.85. C 59.56%, H 4.41%, Cl 10.34%, N 16.34%, S 9.35%. Prepn: M. Nakanishi *et al.,* **DE 2229845**; *eidem,* **US 3904641** (1972, 1973 both to Yoshitomi). Pharmacology and toxicity studies: T. Tsumagari *et al., Arzneim.-Forsch.* **28**, 1158 (1978). Effect on monoamine metabolism in brain: M. Setoguchi *et al., ibid.* 1165; on rage responses in cats: T. Fukuda, T. Tsumagari, *Jpn. J. Pharmacol.* **33**, 885 (1983).

Crystals from toluene, mp 147-148°. LD_{50} in male, female rats, male, female mice (mg/kg): 3619, 3509, 4358, 4258 orally; 865, 825, 830, 783 i.p.; >5000 s.c. (Tsumagari).

THERAP CAT: Anxiolytic.

3920. Etodolac. [41340-25-4] 1,8-Diethyl-1,3,4,9-tetrahydropyrano[3,4-b]indole-1-acetic acid; etodolic acid; AY-24236; Etogesic; Lodine; Tedolan; Ultradol. $C_{17}H_{21}NO_3$; mol wt 287.36. C 71.06%, H 7.37%, N 4.87%, O 16.70%. Prepn: C. A. Demerson *et al.,* **DE 2301525** (1973 to Ayerst); *eidem,* **US 3939178** (1976 to Am. Home Products); *eidem, J. Med. Chem.* **19**, 391 (1976). Review of synthesis, pharmacology, metabolism and efficacy: L. G. Humber, *Med. Res. Rev.* **7**, 1-28 (1987). HPLC determn of enantiomers in plasma and urine: F. Jamali *et al., J. Pharm. Sci.* **77**, 963 (1988). Veterinary trial for osteoarthritis in dogs: S. C. Budsberg *et al., J. Am. Vet. Med. Assoc.* **214**, 206 (1999). Clinical trial for lumbar disc herniation: M. Hatori, S. Kokubun, *Curr. Med. Res. Opin.* **15**, 193 (1999). Review of pharmacology and therapeutic efficacy: J. A. Balfour, M. M.-T. Buckley, *Drugs* **42**, 274-299 (1991); of clinical pharmacokinetics: D. R. Brocks, F. Jamali, *Clin. Pharmacokinet.* **26**, 259-274 (1994).

Crystals from hexane/chloroform, mp 145-148°. pKa 4.65. Insol in water. Sol in alcohols, chloroform, DMSO, aqueous polyethylene glycol.

THERAP CAT: Anti-inflammatory; analgesic.

THERAP CAT (VET): Anti-inflammatory; analgesic.

3921. Etofenamate. [30544-47-9] 2-[[3-(Trifluoromethyl)-phenyl]amino]benzoic acid 2-(2-hydroxyethoxy)ethyl ester; N-(α,-α,α-trifluoro-m-tolyl)anthranilic acid 2-(2-hydroxyethoxy)ethyl ester; B-577; TV-485; Bayrogel; Glasel; Rheumon gel; Traumon Gel. $C_{18}H_{18}F_3NO_4$; mol wt 369.34. C 58.54%, H 4.91%, F 15.43%, N 3.79%, O 17.33%. Percutaneously active antiphlogistic agent. Prepn: K. H. Boltze et al., **DE 1939112**; eidem, **US 3692818** (1971, 1972 both to Troponwerke). Series of articles on chemistry, analysis, biochemistry, pharmacology, toxicology, and clinical studies: Arzneim.-Forsch. **27**, 1300-1363 (1977). Toxicity: H. Jacobi et al., ibid. 1333. Metabolism and GC determn: ibid. **31**, 9-21 (1981).

Pale yellow viscous oil, thermolabile at 180°. bp$_{.001}$ 130-135°. uv max (methanol): 286 (E$_{1cm}^{1\%}$ 423). n_D^{25} 1.564. Sol in lower alcohols, ethyl acetate, acetone, chloroform, ether, benzene. Soly in water at 22°: 0.16 mg/100 ml. LD$_{50}$ in male, female rats (mg/kg): 292, 470 orally; 140, 226 i.v.; 373, 397 i.p.; 643, 568 s.c. (Jacobi).

THERAP CAT: Anti-inflammatory.

3922. Etofenprox. [80844-07-1] 1-[[2-(4-Ethoxyphenyl)-2-methylpropoxy]methyl]-3-phenoxybenzene; 2-(4-ethoxyphenyl)-2-methylpropyl-3-phenoxybenzyl ether; ethofenprox; ethoproxyfen; MTI-500; OMS-3002; Vectron. $C_{25}H_{28}O_3$; mol wt 376.50. C 79.75%, H 7.50%, O 12.75%. Non-ester pyrethroid insecticide. Prepn: K. Nakatani et al., **FR 2481695**; eidem, **US 4397864** (1981, 1983 both to Mitsui Toatsu); T. K. Kim et al., Bull. Korean Chem. Soc. **8**, 128 (1987). Fluorometric determn in potable water sources: M. Nakamura et al., Fresenius J. Anal. Chem. **367**, 658 (2000). Field trial vs leafhoppers and planthoppers: N. V. Krishnaiah, M. B. Kalode, Crop Prot. **12**, 532 (1993); vs malarial vectors: S. Nalim et al., Southeast Asian J. Trop. Med. Public Health **28**, 851 (1997). Review of physical properties and field trials: T. Tanaka, Agrochem. Jpn. **1999**, no. 74, 13-15.

White crystalline solid, mp 36.4-38.0°. Vapor pressure (100°): 2.4×10^{-4}mm Hg. Soly in water (25°): <1 ppb. $n_D^{20.2}$ 1.5732. LD$_{50}$ in rats, mice (mg/kg): >2880, >107200 orally; >2140, >2140 dermally (Tanaka).

USE: Insecticide.

3923. Etofibrate. [31637-97-5] 3-Pyridinecarboxylic acid 2-[2-(4-chlorophenoxy)-2-methyl-1-oxopropoxy]ethyl ester; nicotinic acid 2-hydroxyethyl ester 2-(p-chlorophenoxy)-2-methylpropionate (ester); ethylene glycol 1-[2-(p-chlorophenoxy)-2-methylpropionate]-2-nicotinate; α-[(p-chlorophenoxy)isobutyroyl]-β-(nicotinoyl)-glycol; ethofibrate; Lipo-Merz. $C_{18}H_{18}ClNO_5$; mol wt 363.79. C 59.43%, H 4.99%, Cl 9.74%, N 3.85%, O 21.99%. Glycol diester of nicotinic and clofibric acids, q.q.v. Prepn: A. Scherm, D. Peteri, **DE**

1941217; eidem, **US 3723446** (1971, 1973 both to Merz). Lipid lowering effects in rats: W. Sterner, A. Schultz, Arzneim.-Forsch. **24**, 1990 (1974). Metabolism: H. Oelschläger et al., ibid. **30**, 984 (1980). In vitro effect on human platelet function: M. P. Ortega et al., Thromb. Res. **19**, 409 (1980). HPLC determn in plasma: E. R. Garrett, M. R. Gardner, J. Pharm. Sci. **71**, 14 (1982). Bioavailability and pharmacokinetics in humans: K. I. Johnson et al., Arzneim.-Forsch. **34**, 1785 (1984). Review of pharmacology and comparison with other hypolipidemic agents: R. Paoletti et al., Am. J. Cardiol. **52**, Suppl B, 21B-27B (1983).

Crystalline powder, mp 100°.

THERAP CAT: Antilipemic.

3924. Etofylline. [519-37-9] 3,7-Dihydro-7-(2-hydroxyethyl)-1,3-dimethyl-1H-purine-2,6-dione; 7-(2-hydroxyethyl)theophylline; oxyethyltheophylline; Oxphylline. $C_9H_{12}N_4O_3$; mol wt 224.22. C 48.21%, H 5.39%, N 24.99%, O 21.41%. Obtained by the action of glycol monochlorohydrin upon theophylline in an alkaline medium: Lespagnol et al., Bull. Soc. Pharm. Lille **1948**, no. 2, p 18; Roth, Arch. Pharm. **292**, 234 (1959); Fabbrini, Cencioni, Farmaco Ed. Sci. **17**, 660 (1962). HPLC determn in plasma: R. V. S. Nirogi et al., J. Chromatogr. B **848**, 271 (2007). Clinical pharmacokinetics: J. Zuidema et al., Int. J. Clin. Pharmacol. Res. **19**, 310 (1981).

Crystals from abs ethanol, mp 158°. Freely sol in water. Moderately sol in alcohol. pH of a 5% aq soln 6.5-7.0. Solns may be sterilized by heating.

Nicotinate. [13425-39-3] $C_{15}H_{15}N_5O_4$; mol wt 329.32. Prepn: Pongratz, Zirm, Monatsh. Chem. **88**, 330 (1957). Pharmacology: Fischbach, Haas, Arzneim.-Forsch. **17**, 313 (1967). Fine dendritic needles from abs alcohol, mp 151-152°. Hydrated crystals from water. One gram dissolves in 50 ml boiling water; moderately sol in alcohol.

THERAP CAT: Bronchodilator.

3925. Etoglucid. [1954-28-5] 2,2'-(2,5,8,11-Tetraoxadodecane-1,12-diyl)bisoxirane; 1,2:15,16-diepoxy-4,7,10,13-tetraoxahexadecane; 1,2-bis[2-(2,3-epoxypropoxy)ethoxy]ethane; triethylene glycol diglycidyl ether; TDE; ethoglucid; ICI-32865; Epodyl. $C_{12}H_{22}O_6$; mol wt 262.30. C 54.95%, H 8.45%, O 36.60%. Prepn: Greenshields et al., **GB 901876** (1962 to ICI); Blyakhman, Zh. Org. Khim. **3**, 1423 (1967), C.A. **68**, 59384k (1968); X.-P. Gu et al., Synthesis **1985**, 649. Pharmacology and metabolism: Duncan, Snow, Biochem. J. **82**, 8P (1962); James, Solheim, Xenobiotica **1**, 43 (1971). Clinical evaluation in bladder tumors: M. R. G. Robinson et al., J. Urol. **118**, 972 (1977); J. Flamm, F. Grof, Urologe **B27**, 26 (1987).

Colorless liquid, mp −15 to −11°. bp$_{0.005}$ 140°; bp$_{0.1}$ 133-149°; bp$_2$ 195-197°. d^{20} 1.1312. n_D^{20} 1.4584.

THERAP CAT: Antineoplastic.

3926. Etomidate. [33125-97-2] 1-[(1R)-1-Phenylethyl]-1H-imidazole-5-carboxylic acid ethyl ester; (R)-(+)-1-(α-methylbenzyl)-imidazole-5-carboxylic acid ethyl ester; R-16659; Amidate; Hypnomidate. $C_{14}H_{16}N_2O_2$; mol wt 244.29. C 68.83%, H 6.60%, N 11.47%, O 13.10%. Imidazole hypnotic used in the induction of anesthesia. Prepn of the (±)-form: BE 662474; E. F. Godefroi, C. A. M. van der Eijcken, US 3354173 (1965, 1967 both to Janssen); E. F. Godefroi et al., J. Med. Chem. 8, 220 (1965). Prepn of (R)-(+)-form: L. F. C. Roevens et al., DE 2609573; eidem, US 3991072 (both 1976 to Janssen). Metabolism: E. Goetz, Anaesthesist 23, 331 (1974). Pharmacokinetics: M. J. Van Hamme et al., Anesthesiology 49, 274 (1978). Pharmacology: P. A. J. Janssen et al., Arch. Int. Pharmacodyn. Ther. 214, 92 (1975). HPLC determn in plasma: M. P. McIntosh, R. A. Rajewski, J. Pharm. Biomed. Anal. 24, 689 (2001). Clinical study: R. J. Fragen, N. Caldwell, Anesthesiology 50, 242 (1979). Clinical trial vs. midazolam in pediatric sedation and analgesia: L. Di Liddo et al., Ann. Emerg. Med. 48, 433 (2006). Comprehensive description: Z. L. Chang, J. B. Martin, Anal. Profiles Drug Subs. 12, 191-214 (1983). Veterinary use: S. Robertson, Vet. Clin. North Am. Small Anim. Pract. 22, 277 (1992). Review of use in rapid sequence intubation: A. J. Oglesby, Emerg. Med. J. 21, 655-659 (2004).

Crystals from diisopropyl ether, mp 67°. $[\alpha]_D^{20}$ +66° (c = 1 in ethanol). uv max (isopropanol): 240 nm (ε 12200). Soly in water at 25°: 0.0045 mg/100 ml. Sol in chloroform, methanol, ethanol, propylene glycol, acetone. LD_{50} in mice, rats (mg/kg): 29.5, 14.8-24.3 i.v. (Janssen).

THERAP CAT: Anesthetic (intravenous).
THERAP CAT (VET): Anesthetic (intravenous).

3927. Etonitazene. [911-65-9] 2-[(4-Ethoxyphenyl)methyl]-N,N-diethyl-5-nitro-1H-benzimidazole-1-ethanamine; 1-[(2-diethylamino)ethyl]-2-(p-ethoxybenzyl)-5-nitrobenzimidazole; 2-p-ethoxybenzyl-1-(2-diethylaminoethyl)-5-nitrobenzimidazole. $C_{22}H_{28}$-N_4O_3; mol wt 396.49. C 66.65%, H 7.12%, N 14.13%, O 12.11%. Opioid analgesic. Prepn: Hunger et al., Experientia 13, 400 (1957); Hoffmann et al., US 2935514 (1960 to Ciba); F. I. Carroll, M. C. Coleman, J. Med. Chem. 18, 318 (1975). Crystal structure: C. Humblet et al., Acta Crystallogr. B34, 3828 (1978). Binds selectively to mu-opioid receptors: M. S. Moolten et al., Life Sci. 52, PL199 (1993).

Needles from ether-petroleum ether, mp 77-78°.
Hydrochloride. [2053-25-0] $C_{22}H_{28}N_4O_3$·HCl. Crystals from abs ethanol, mp 163-164.5°.
Note: This is a controlled substance (opiate): 21 CFR, 1308.11.
USE: Reinforcer in animal models of drug addiction.

3928. Etonogestrel. [54048-10-1] (17α)-13-Ethyl-17-hydroxy-11-methylene-18,19-dinorpregn-4-en-20-yn-3-one; 17-ethynyl-17β-hydroxy-18-methyl-11-methyleneestr-4-en-3-one; 3-ketodesogestrel; 3-oxodesogestrel; Org-3236; Implanon; Nexplanon. C_{22}-$H_{28}O_2$; mol wt 324.46. C 81.44%, H 8.70%, O 9.86%. Biologically active metabolite of desogestrel, q.v. Prepn: H. Hofmeister et al., EP 51762; eidem, US 4371529 (1982, 1983 both to Schering AG); H. Gao et al., Steroids 62, 398 (1997). Clinical pharmacokinetics

and bioavailability from contraceptive implant: R. Wenzi et al., Contraception 58, 283 (1998). Review of clinical safety and efficacy of contraceptive vaginal ring: M.-S. Wagner et al., Expert Opin. Pharmacother. 8, 1769-1777 (2007); of implant: P. Darney et al., Fertil. Steril. 91, 1646 (2009).

mp 199-201°. $[\alpha]_D$ +87.6°.
Combination with ethinyl estradiol. NuvaRing.
THERAP CAT: Progestogen; implantable contraceptive.

3929. Etoposide. [33419-42-0] (5R,5aR,8aR,9S)-9-[(4,6-O-(1R)-Ethylidene-β-D-glucopyranosyl)oxy]-5,8,8a,9-tetrahydro-5-(4-hydroxy-3,5-dimethoxyphenyl)furo[3′,4′:6,7]naphtho[2,3-d]-1,3-dioxol-6(5aH)-one; 4′-demethylepipodophyllotoxin 9-[4,6-O-ethylidene-β-D-glucopyranoside]; EPEG; NSC-141540; VP-16-213; Lastet; Vepesid. $C_{29}H_{32}O_{13}$; mol wt 588.56. C 59.18%, H 5.48%, O 35.34%. DNA topoisomerase II inhibitor. Semi-synthetic deriv of podophyllotoxin, related structurally to teniposide, q.q.v. Prepn: C. Keller-Juslén et al., US 3524844 (1970 to Sandoz); idem et al., J. Med. Chem. 14, 936 (1971). Teratogenicity and cytogenicity study: S. M. Sieber et al., Teratology 18, 31 (1978). Review of pharmacokinetics and assay methods: P. I. Clark, M. L. Slevin, Clin. Pharmacokinet. 12, 223-252 (1987). Comprehensive description: J. J. M. Holthuis et al., Anal. Profiles Drug Subs. 18, 121-151 (1989). Review of pharmacology and clinical experience: J. D. Hainsworth, F. A. Greco, Ann. Oncol. 6, 325-341 (1995); S. Joel, Cancer Treat. Rev. 22, 179-221 (1996).

Crystals from methanol, mp 236-251°. $[\alpha]_D^{20}$ −110.5° (c = 0.6 in chloroform). uv max (abs methanol): 283 nm (ε 4245). pKa 9.8. Sparingly sol in methanol; slightly sol in alc, chloroform, ethyl acetate, methylene chloride; very slightly sol in water.
Phosphate. [117091-64-2] BMY-40481; Etopophos. $C_{29}H_{33}$-$O_{16}P$; mol wt 668.54. Review: A. H. I. Witterland et al., Pharm. World Sci. 18, 163-170 (1996). Sol in water. Practically insol in organic solvents.
THERAP CAT: Antineoplastic.

3930. Etoricoxib. [202409-33-4] 5-Chloro-6′-methyl-3-[4-(methylsulfonyl)phenyl]-2,3′-bipyridine; 5-chloro-3-(4-methylsulfonyl)phenyl-2-(2-methyl-5-pyridinyl)pyridine; MK-663; MK-0663; Arcoxia. $C_{18}H_{15}ClN_2O_2S$; mol wt 358.84. C 60.25%, H 4.21%, Cl 9.88%, N 7.81%, O 8.92%, S 8.93%. Selective cyclooxygenase-2 (COX-2) inhibitor. Prepn: D. Dube et al., WO 9803484; eidem, US 5861419 (1998, 1999 both to Merck Frosst); R. W. Friesen et al., Bioorg. Med. Chem. Lett. 8, 2777 (1998). Practical synthesis: I. W. Davies et al., J. Org. Chem. 65, 8415 (2000). HPLC determn in urine and plasma: C. Z. Matthews et al., J. Chromatogr.

B **751**, 237 (2001). Pharmacology: D. Riendeau *et al.*, *J. Pharmacol. Exp. Ther.* **296**, 558 (2001); A. Dallob *et al.*, *J. Clin. Pharmacol.* **43**, 573 (2003). Clinical comparison with indomethacin in acute gouty arthritis: H. R. Schumacher, Jr. *et al.*, *Br. Med. J.* **324**, 1488 (2002). Clinical trial in rheumatoid arthritis: A. K. Matsumoto *et al.*, *J. Rheumatol.* **29**, 1623 (2002); in osteoarthritis: S. P. Curtis *et al.*, *BMC Musculoskelet. Disord.* **6**, 58 (2005); in ankylosing spondylitis: D. van der Heijde *et al.*, *Arthritis Rheum.* **52**, 1205 (2005). Gastrointestinal safety profile in inflammatory bowel diseases: Y. El Miedany *et al.*, *Am. J. Gastroenterol.* **101**, 311 (2006). Review of pharmacology and clinical development: P. Patrignani *et al.*, *Expert Opin. Pharmacother.* **4**, 265-284 (2003).

White solid, mp 127-128°. pKa 4.5. uv max (acetonitrile-phosphate buffer): 238, 280 nm.

THERAP CAT: Anti-inflammatory.

3931. Etorphine. [14521-96-1] (*αR*,5*α*,7*α*)-4,5-Epoxy-3-hydroxy-6-methoxy-*α*,17-dimethyl-*α*-propyl-6,14-ethenomorphinan-7-methanol; tetrahydro-7*α*-(1-hydroxy-1-methylbutyl)-6,14-*endo*-ethenooripavine; 7,8-dihydro-7*α*-[1(*R*)-hydroxy-1-methylbutyl]-*O*⁶-methyl-6,14-*endo*-ethenomorphine; 7*α*-[1(*R*)-hydroxy-1-methylbutyl]-6,14-*endo*-ethenotetrahydrooripavine; 19-propylorvinol; tetrahydro-7*α*-(2-hydroxy-2-pentyl)-6,14-*endo*-ethenooripavine. $C_{25}H_{33}NO_4$; mol wt 411.54. C 72.96%, H 8.08%, N 3.40%, O 15.55%. Narcotic analgesic. Prepn: **BE 618392**; K. W. Bentley, **GB 937214** (1962, 1963 both to J. F. MacFarlan); and structure: K. W. Bentley, D. G. Hardy, *Proc. Chem. Soc. London* **1963**, 220. Improved prepn: W. R. Hydro, **US 3763167** (1973 to U.S. Army). Determn by HPLC: I. Jane, A. McKinnon, *J. Chromatogr.* **323**, 19 (1985). Analgesic activity: G. F. Blane, *J. Pharm. Pharmacol.* **19**, 367 (1967). Pharmacology: Lister, *ibid.* **16**, 364 (1964). Use as tranquilizer in dogs: J. L. Crooks *et al.*, *Vet. Rec.* **87**, 498 (1970); M. Grange *et al.*, *Rev. Med. Vet.* **124**, 899 (1973); in large animals: A. M. Harthoorn, *Am. Vet. Med. Assoc.* **149**, 875 (1966); B. T. Alford *et al.*, *ibid.* **164**, 702 (1974).

White crystals from aqueous ethoxyethanol, mp 214-217°. (7*β*-isomer, melts at 280°).

Hydrochloride. M-99. $C_{25}H_{33}NO_4.HCl$; mol wt 448.00. mp 266-267°. (7*β*-isomer hydrochloride, dec at 290°)°.

Hydrochloride in combination with acepromazine. Immobilon.

3-Acetate. Acetorphine; *O*³-acetyl-7,8-dihydro-7*α*-[1(*R*)-hydroxy-1-methylbutyl]-*O*⁶-methyl-6,14-*endo*-ethenomorphine; 3-*O*-acetyl-7*α*-[1(*R*)-hydroxy-1-methylbutyl]-6,14-*endo*-ethenotetrahydrooripavine; 3-*O*-acetyl-17-propylorvinol. $C_{27}H_{35}NO_5$. Crystals from methanol, mp 195°.

3-Acetate hydrochloride. M-183. $C_{27}H_{35}NO_5.HCl$; mol wt 490.04. mp 206°.

Note: Acetorphine and etorphine are controlled substances (opium derivatives): **21 CFR**, 1308.11. Etorphine hydrochloride is a controlled substance (opiate): **21 CFR**, 1308.12.

Caution: Use with extreme care; very small amounts may bring about respiratory paralysis and death: D. W. Upson, *Upson's Handbook of Clinical Veterinary Pharmacology* (VM Publishing, Bonner Springs, 1980) pp 380-381.

THERAP CAT (VET): Used to immobilize large animals.

3932. Etoxadrol. [28189-85-7] (2*S*)-2-[(2*S*,4*S*)-2-Ethyl-2-phenyl-1,3-dioxolan-4-yl]piperidine; (+)-2-(2-ethyl-2-phenyl-1,3-dioxolan-4-yl)piperidine; 2-ethyl-2-phenyl-4-(2-piperidyl)-1,3-dioxolane. $C_{16}H_{23}NO_2$; mol wt 261.37. C 73.53%, H 8.87%, N 5.36%, O 12.24%. NMDA receptor antagonist. Prepn of *dl*-form: Hardie *et al.*, *J. Med. Chem.* **9**, 127 (1966); Hardie, Halverstadt, **US 3262938** (1966 to Cutter Labs). Resolution of the *α*-racemates: Allen *et al.*, **DE 2001616** (1970 to Cutter Labs), *C.A.* **74**, 13129b (1971). Pharmacological studies: Traber *et al.*, *J. Pharmacol. Exp. Ther.* **175**, 395 (1970); Hidalgo *et al.*, *Anesth. Analg.* **50**, 231 (1971); Kelly *et al.*, *ibid.* 262. Structure-activity and NMDA receptor affinity: M. Sax, B. Wünsch, *Curr. Top. Med. Chem.* **6**, 723-732 (2006).

Hydrochloride. [23239-37-4] CL-1848C. $C_{16}H_{23}NO_2.HCl$; mol wt 297.82. Crystals from isopropanol, mp 221.5-222°. $[\alpha]_D^{25}$ +16.63°.

3933. Etoxazole. [153233-91-1] 2-(2,6-Difluorophenyl)-4-[4-(1,1-dimethylethyl)-2-ethoxyphenyl]-4,5-dihydrooxazole; 4-(4-*tert*-butyl-2-ethoxyphenyl)-2-(2,6-difluorophenyl)-4,5-dihydrooxazole; 2-(2,6-difluorophenyl)-4-(2-ethoxy-4-*tert*-butylphenyl)-2-oxazoline; YI-5301; Baroque; TetraSan; Zeal. $C_{21}H_{23}F_2NO_2$; mol wt 359.42. C 70.18%, H 6.45%, F 10.57%, N 3.90%, O 8.90%. Insect growth regulator; molting inhibitor for control of mites and aphids on fruits, vegetables and ornamentals. Prepn: J. Suzuki *et al.*, **WO 9322297**; *eidem*, **US 5478855** (1993, 1995 both to Yashima); and ovicidal and insecticidal activity: *idem et al.*, *J. Pestic. Sci.* **26**, 215 (2001). Comprehensive description: T. Ishida *et al.*, *Brighton Crop Prot. Conf. - Pests Dis.* **1994**, 37-44. Efficacy as a tick control agent: Y. Tamura *et al.*, *Med. Entomol. Zool.* **55**, 303 (2004).

Colorless crystals from hexane, mp 101-102°. Vapor pressure (25°): 2.18×10^{-6} Pa. Log P (octanol/water): 5.59 at 25°. Soly at 20°(g/l solvent): water 75.4×10^{-6}; methanol 100; acetone 450; THF 850; *n*-hexane 15. LD_{50} in rats (mg/kg): >5000 orally; >2000 dermally. LC_{50} (96 hr) in Japanese carp: 0.89 mg/l (Ishida).

USE: Acaricide.

3934. Etozolin. [73-09-6] 2-[3-Methyl-4-oxo-5-(1-piperidinyl)-2-thiazolidinylidene]acetic acid ethyl ester; 2-carbethoxymethylene-3-methyl-5-piperidino-4-thiazolidone; 3-methyl-4-oxo-5-piperidino-Δ²,*α*-thiazolidineacetic acid ethyl ester; Gö-687; W-2900A; Elkapin. $C_{13}H_{20}N_2O_3S$; mol wt 284.37. C 54.91%, H 7.09%, N 9.85%, O 16.88%, S 11.27%. A member of a new class of heterocyclic compounds, the 2-methylenethiazolidones; a diuretic with choleretic properties. Prepn: Satzinger, *Ann.* **665**, 150 (1963); **US 3072653** (1963 to Warner-Lambert). Pharmacology: O. Heidenreich *et al.*, *Arzneim.-Forsch.* **14**, 1242 (1964). Series of articles on activity, metabolism and toxicology: G. Satzinger *et al.*, *ibid.* **27**, 1742-1817 (1977); M. Herrmann *et al.*, *ibid.* 1745.

Crystals from methanol, mp 140°. uv max (methanol): 283, 243 nm (log ε 4.32, 4.0). LD_{50} in male mice, rats (mg/kg): 1210, 1575 i.p.; in male, female mice, rats (mg/kg): 8670, 9360, 11040, 10250 orally (Herrmann).

Hydrochloride. $C_{13}H_{20}N_2O_3S.HCl$. Crystals, mp 158-159°.

Free acid. Ozolinone; Gödecke 3282. $C_{11}H_{16}N_2O_3S$.

THERAP CAT: Diuretic.

3935.   **Etravirine.** [269055-15-4] 4-[[6-Amino-5-bromo-2-[(4-cyanophenyl)amino]-4-pyrimidinyl]oxy]-3,5-dimethylbenzonitrile; 4-[[4-amino-5-bromo-6-(4-cyano-2,6-dimethylphenyloxy)-2-pyrimidinyl]amino]benzonitrile; R-165335; TMC-125; Intelence. $C_{20}H_{15}BrN_6O$; mol wt 435.29. C 55.19%, H 3.47%, Br 18.36%, N 19.31%, O 3.68%. Diarylpyrimidine, nonnucleoside reverse transcriptase inhibitor (NNRTI). Prepn: B. De Corte *et al.*, **WO 0027825**; *eidem*, **US 7037917** (2000, 2006, both to Janssen); and anti-HIV activity: D. W. Ludovici *et al.*, *Bioorg. Med. Chem. Lett.* **11**, 2235 (2001). Crystal structure of complex with HIV reverse transcriptase: K. Das *et al.*, *J. Med. Chem.* **47**, 2550 (2004). Antiretroviral spectrum and resistance potential: J. Vingerhoets *et al.*, *J. Virol.* **79**, 12773 (2005). Clinical pharmacokinetics and antiretroviral activity in drug-resistant patients: M. Boffito *et al.*, *AIDS* **21**, 1449 (2007). Clinical trials in treatment experienced patients: J. V. Madruga *et al.*, *Lancet* **370**, 29 (2007); A. Lazzarin *et al.*, *ibid.* 39.

Crystals from acetonitrile, mp 255°.

THERAP CAT: Antiretroviral.

3936.   **Etretinate.** [54350-48-0] (2E,4E,6E,8E)-9-(4-Methoxy-2,3,6-trimethylphenyl)-3,7-dimethyl-2,4,6,8-nonatetraenoic acid ethyl ester; Ro-10-9359; Tigason. $C_{23}H_{30}O_3$; mol wt 354.49. C 77.93%, H 8.53%, O 13.54%. Aromatic analog of retinoic acid, *q.v.* Prepn: W. Bollag *et al.*, **DE 2414619**; *eidem*, **US 4105681** and **US 4215215** (1974, 1978, 1980 all to Hoffmann-La Roche). Effects on skin metabolism: W. P. Raab, B. M. Gmeiner, *Arch. Dermatol. Res.* **256**, 247 (1976). Toxicity study in organ culture: D. R. Bard, I. Lasnitzki, *Br. J. Cancer* **35**, 110 (1977). Pharmacokinetics: R. Haenni, *Dermatologica* **157**, Suppl, 5 (1978). Mutation study: H. J. Juhl *et al.*, *Mutat. Res.* **58**, 317 (1978). HPLC determn in plasma: R. Haenni *et al.*, *J. Chromatogr.* **162**, 615 (1979). Toxicology, carcinogenicity and teratogenicity study: J. J. Kamm, *J. Am. Acad. Dermatol.* **6**, 652 (1982). Clinical evaluation in psoriatic arthritis: M. L. Ciompi *et al.*, *Int. J. Tissue React.* **10**, 25 (1988). *Review*: H. Mayer *et al.*, *Experientia* **34**, 1105-1119 (1978). Review of pharmacology and therapeutic efficacy: A. Ward *et al.*, *Drugs* **26**, 9-43 (1983); of clinical pharmacology: A. Vahlquist, O. Rollman, *Dermatologica* **175**, Suppl. 1, 20-27 (1987).

Crystals, mp 104-105°. LD_{50} in mice (1 day): >4000 mg/kg i.p. (Bollag). LD_{50} (20 day) in mice, rats (mg/kg): 1176, >2000 i.p.; >2000, >4000 orally (Kamm).

Free acid *see* Acitretin.

THERAP CAT: Antipsoriatic.

3937.   **β-Eucaine.** [500-34-5] 2,2,6-Trimethyl-4-piperidinol 4-benzoate; α-4-benzoyloxy-2,2,6-trimethylpiperidine; α-vinyldiacetonalkamine benzoate; benzamine; betacaine; eucaine B. $C_{15}H_{21}NO_2$; mol wt 247.34. C 72.84%, H 8.56%, N 5.66%, O 12.94%. Prepn from lower melting form (α-form) of 2,2,6-trimethyl-4-piperidinol and benzoyl chloride: Harries, *Ann.* **417**, 107, *see* p 175 (1918). Stereochemistry: King, *J. Chem. Soc.* **125**, 41 (1924); Stenlake, *J. Pharm. Pharmacol.* **6**, 164 (1954). Configuration: Perks, Russell, *ibid.* **19**, 318 (1966). Toxicity data: Hirschfelder, *Physiol. Rev.* **12**, 262 (1932).

Crystals from petr ether, mp 70-71° (corr). MLD in frogs, rabbits (mg/kg): 1300, 400-500 s.c.; in rats, cats (mg/kg): 15-25, 10.0-12.5 i.v.; in guinea pigs (mg/kg): 310 s.c., 180 i.p., 30 i.v. (Hirschfelder).

Hydrochloride. Crystals from water, mp 277-279° (corr). pKa 9.4. One gram dissolves in 30 ml water, 35 ml alcohol, 6 ml chloroform; more sol in boiling water or boiling alcohol. Aq soln is neutral to litmus.

d-**Form.** Crystals from petr ether, mp 57-58° (corr).

d-**Form hydrochloride.** $[\alpha]_D$ +11.5° (water).

l-**Form.** Crystals from petr ether, mp 57-58° (corr).

l-**Form hydrochloride.** Crystals, mp 244-245° (corr). $[\alpha]_D$ −11.3° (water).

THERAP CAT: Anesthetic (local).

THERAP CAT (VET): Has been used as a local anesthetic.

3938.   **Eucalyptol.** [470-82-6] 1,3,3-Trimethyl-2-oxabicyclo-[2.2.2]octane; 1,8-epoxy-*p*-menthane; 1,8-cineole; cajeputol; cineole. $C_{10}H_{18}O$; mol wt 154.25. C 77.87%, H 11.76%, O 10.37%. The chief constituent of oil of eucalyptus, *q.v.*; also found in essential oils of laurel, rosemary, and many other aromatic plants. Isoln: Cloez, *Ann.* **154**, 372 (1870); Berry, *Australas. J. Pharm.* **1929**, 203; from wormwood oil: Wallach, Brass, *Ann.* **225**, 291 (1884). Identity with cajeputol: Wallach, *ibid.* 314. Identity with cineole: Jahns, *Ber.* **17**, 2941 (1884). Structure: Wallach, *Ann.* **291**, 342 (1896). Biosynthesis in *Eucalyptus*: A. J. Birch *et al.*, *Tetrahedron Letters* **1** (3), 1 (1959). Chemical synthesis: J. Adams, M. Belley, *ibid.* **27**, 2075 (1986). Review of safety and toxicology: D. L. J. Opdyke, *Food Cosmet. Toxicol.* **13**, 105-106 (1975); M. De Vincenzi *et al.*, *Fitoterapia* **73**, 269-275 (2002). Clinical trial as mucolytic in rhinosinusitis: W. Kehrl *et al.*, *Laryngoscope* **114**, 738 (2004).

Colorless liquid; fresh, camphor-like odor; spicy, cooling taste. d_{25}^{25} 0.921-0.923. bp 176-177°. mp 1.5°. n_D^{20} 1.455-1.460. Flash pt, closed cup: 118°F (48°C). Practically insol in water; miscible with alcohol, chloroform, ether, glacial acetic acid, oils. LD_{50} orally in rats: 2480 mg/kg. *See* P. M. Jenner *et al.*, *Food Cosmet. Toxicol.* **2**, 327 (1964).

USE: Fragrance and flavoring agent in foods, candies, cough drops, personal care products. Pharmaceutic aid (flavor).

3939.   **Eucalyptus.** Genus of evergreen trees of the Family *Myrtaceae*, having aromatic blue-grey leaves that yield a characteristic essential oil. *Habit.* Australia, cultivated in subtropical regions of Europe, Africa, Asia and U.S. More than 700 species are known; the Australian blue-gum, *Eucalyptus globulus* Labill., is the most predominant and said to be richest in essential oil. *Constit.* 0.5-3.5% volatile oil, resins, tannins, polyphenolic acids, and flavonoids such as quercetin, rutin, eucalyptin. Review of botany, cultivation and uses: M. Forrest, *Biologist* **47**, 139-142 (2000); of pharmacology and medicinal uses: A. Y. Leung, S. Foster, *Encyclopedia of Common Natural Ingredients*, (Wiley-Interscience, Hoboken, 2nd Ed., 2003) pp 232-234; J. Gruenwald *et al.*, *PDR for Herbal Medicines* (Thomson PDR, Montvale, 3rd Ed., 2004) pp 293-297.

Volatile oil. [8000-48-4] Oil of eucalyptus. Obtained by steam distillation from fresh leaves of *E. globulus*. *Constit.* 70-80% Eucalyptol (1,8-cineole), α-pinene, *d*-limonene, *p*-cymene, α-phellandrene, 1-α-terpineol. Description: E. Guenther, *The Essential Oils* vol. 4 (van Nostrand, New York, 1950) pp 437-525. Colorless to pale yellow liquid; characteristic camphoraceous odor; pungent, spicy, cooling taste. d_{25}^{25} 0.905-0.925. n_D^{20} 1.458-1.470. Almost insol in water; sol in 5 vols 70% alcohol. *Keep well closed, cool and protected from light.*

USE: Wood for timber, pulp, fuel, charcoal; cut foliage in floral arrangements. Oil as fragrance component in soaps, creams, lotions and as flavoring agent in toothpastes, mouthwashes. Pharmaceutic aid (flavor).

THERAP CAT: Oil as expectorant, antiseptic, externally for rheumatism.

3940. Eugenol. [97-53-0] 2-Methoxy-4-(2-propen-1-yl)phenol; 4-allyl-2-methoxyphenol; allylguaiacol; eugenic acid; caryophyllic acid. $C_{10}H_{12}O_2$; mol wt 164.20. C 73.15%, H 7.37%, O 19.49%. Obtained from many natural sources: *Beilstein* **vol. 6**, 961. Prepn: Claisen, *Ann.* **418**, 69 (1919); from oil of cloves: Waterman, Priester, *Rec. Trav. Chim.* **48**, 1272 (1929). Toxicity study: E. C. Hagan *et al.*, *Toxicol. Appl. Pharmacol.* **7**, 18 (1965).

Colorless or pale yellow liquid, bp 255°. Darkens and thickens on exposure to air. Odor of cloves; spicy, pungent taste. mp −9.2 to −9.1°. d_4^{20} 1.0664. n_D^{20} 1.5410. Practically insol in water. Misc with alcohol, chloroform, ether, oils. One ml dissolves in 2 ml 70% alcohol; sol in glacial acetic acid, in aq fixed alkali hydroxide solns. Ferric chloride, potassium permanganate. LD_{50} in rats, mice (mg/kg): 2680, 3000 orally (Hagan).

Benzoate. *O*-Benzoyleugenol. $C_{17}H_{16}O_3$. Crystals, mp 69-70°. bp 360°. Freely sol in benzene, chloroform, ether, hot alcohol. Practically insol in water.

USE: In perfumery instead of oil of cloves; manuf vanillin. As insect attractant.

THERAP CAT: Analgesic (dental).

3941. Euonymus. Wahoo; Indian arrow wood; bitter ash; burning bush; strawberry tree; spindle tree. Dried root bark of *Euonymus atropurpureus* Jacq., *Celastraceae*. *Habit.* U.S., Ontario to Florida east of Mississippi. *Constit.* Euonymol—an alcohol; euonysterol, atropurpurin, asparagine, citrullol, dulcite; malic, citric and tartaric acids.

THERAP CAT: Cathartic, diuretic.

3942. Euparin. [532-48-9] 1-[6-Hydroxy-2-(1-methylethenyl)-5-benzofuranyl]ethanone; 6-hydroxy-2-isopropenyl-5-benzofuranyl methyl ketone; 5-acetyl-6-hydroxy-2-isopropenylbenzofuran. $C_{13}H_{12}O_3$; mol wt 216.24. C 72.21%, H 5.59%, O 22.20%. From roots of *Eupatorium purpureum* L., *E. cannabinum* L., *Compositae*. Kamthong, Robertson, *J. Chem. Soc.* **1939**, 925; Jermanowska, *C.A.* **48**, 5848h (1954). Structure: Kamthong, Robertson, *J. Chem. Soc.* **1939**, 933. Synthesis: Ramachandran *et al.*, *J. Org. Chem.* **28**, 2744 (1963). *Review:* Gizycki, *Pharmazie* **6**, 686 (1951).

Yellow needles from alcohol, mp 121-122°. uv max (ethanol): 263, 358 nm (ε 34400, 5900). Sol in alcohol, benzene, chloroform, ether. Practically insol in water, alkali.

Acetate. $C_{15}H_{14}O_4$. Prisms from petr ether, mp 80°.

***O*-Methyl derivative.** *O*-Methyleuparin; 2-isopropenyl-6-methoxy-5-benzofuranyl methyl ketone. $C_{14}H_{14}O_3$. Needles from dil alcohol, mp 76-77°. Sol in the usual organic solvents.

3943. Eupatorin. [855-96-9] 5-Hydroxy-2-(3-hydroxy-4-methoxyphenyl)-6,7-dimethoxy-4*H*-1-benzopyran-4-one; 3′,5-dihydroxy-4′,6,7-trimethoxyflavone. $C_{18}H_{16}O_7$; mol wt 344.32. C 62.79%, H 4.68%, O 32.53%. Isoln from *Eupatorium semiserratum* DC., *Compositae* and structure: Kupchan *et al.*, *J. Pharm. Sci.* **54**, 929 (1965).

Crystals from dioxane + water, mp 196-198°. uv max (alcohol): 243, 254, 274, 342 nm (ε 17400, 19300, 19800, 27700).

3944. Eupatorium. Boneset; thoroughwort. Flowering tops and dried leaves of *Eupatorium perfoliatum* L., *Compositae*. *Habit.* Canada to Florida and west to Texas and Nebraska. *Constit.* Eupatorin, volatile oil, resin, tannin, sugar, inulin, wax.

3945. Euphorbia. Pill-bearing spurge; snake-weed; cat's hair; Queensland asthma weed; flowery-headed spurge. Dried herb of *Euphorbia hirta* L. or *E. pilulifera* L., *Euphorbiaceae*. *Habit.* Queensland (Australia), India, widely distributed in tropical countries. *Constit.* Several resins, an unstable glucoside.

3946. Europium. [7440-53-1] Eu; at. wt 151.964; at. no. 63; valences 2, 3. A lanthanide; belongs to cerium group of rare earth metals. Naturally occurring isotopes (mass numbers): 151 (47.8%); 153 (52.2%); known artificial radioactive isotopes: 138-150; 152; 154-160. Abundance in earth's crust: 1.06-2.1 ppm. Commercially important sources are the rare earth minerals monazite and bastnaesite; also found in gadolinite. Has been detected spectroscopically in the sun and in certain stars. Discovered and prepd as the oxide: Demarçay, *Compt. Rend.* **122**, 728 (1896); **130**, 1019, 1469 (1900); **132**, 1484 (1901). Separation: Spedding *et al.*, *J. Am. Chem. Soc.* **76**, 2557 (1954); by paper chromatography: Lederer, *Nature* **176**, 462 (1953). Toxicity study: Haley, *J. Pharm. Sci.* **54**, 663 (1965). Reviews of prepn, properties and compds: *The Rare Earths*, F. H. Spedding, A. H. Daane, Eds. (Krieger, Huntington, N.Y., 1971, reprint of 1961 ed.) 641 pp; S. P. Sinha, *Europium* (Springer, New York, 1967) 164 pp; Hulet, Bode, "Separation Chemistry of the Lanthanides and Transplutonium Actinides" in *MTP Int. Rev. Sci.: Inorg. Chem., Ser. One* vol. **7**, K. W. Bagnall, Ed. (University Park Press, Baltimore, 1972) pp 1-45; Moeller, "The Lanthanides" in *Comprehensive Inorganic Chemistry* vol. **4**, J. C. Bailar, Jr. *et al.*, Eds. (Pergamon Press, Oxford, 1973) pp 1-101; F. H. Spedding in *Kirk-Othmer Encyclopedia of Chemical Technology* vol. **19** (John Wiley & Sons, New York, 3rd ed., 1982) pp 833-854; *Chemistry of the Elements*, N. N. Greenwood, A. Earnshaw, Eds. (Pergamon Press, New York, 1984) pp 1423-1449. Brief review of properties: G. T. Seaborg, *Radiochim. Acta* **61**, 115-122 (1993).

Body-centered cubic crystal lattice; d 5.244; mp 826°. bp 1429°. Heat of fusion: 9.221 kJ/mol. Heat of sublimation (25°): 144.7 kJ/mol. Sol in liquid ammonia. Shows two reduction potentials −0.710 and −2.510 v. (referred to a normal calomel electrode): Noddack, Brukl, *Angew. Chem.* **50**, 362 (1937); gives two definite series of salts, in one the metal is divalent, and in the other it is trivalent.

Sesquioxide. Europia. Eu_2O_3. Pink powder, d 7.42, prepd by heating the hydroxide, nitrate, oxalate or sulfate at 1600°. The oxide of the divalent metal is prepd by reduction of the sesquioxide at elevated temp.

Hydroxide. $Eu(OH)_3$. Prepd by adding ammonia or an alkali hydroxide to a soln of an europic salt.

Europic chloride. EuCl$_3$. Greenish-yellow needles; mp 623° in nitrogen (in a closed tube), d^{35} 4.471, prepd by passing sulfur chloride over the heated oxide at 200-500°. LD$_{50}$ of trichloride in mice: 550 mg/kg i.p.; 5 g/kg orally (Haley).

Europous chloride. EuCl$_2$. Prepd by reduction of EuCl$_3$ with hydrogen at 600°. White amorphous powder, sol in water.

Europic sulfate. Eu$_2$(SO$_4$)$_3$. Octahydrate, a pinkish cryst solid, prepd by dissolving the oxide in sulfuric acid. Soly in water: 2.56 parts per 100 parts at 20°, 1.93 parts per 100 parts at 40°. On heating at 375° yields the anhydr sulfate.

Europic nitrate. Eu(NO$_3$)$_3$. Hexahydrate, mp 85° in its water of crystallization (sealed tube). LD$_{50}$ in rats (mg/kg): 210 i.p.; >5000 orally (Haley).

Europous sulfate. EuSO$_4$. Colorless crystals. Insol in water and in dil acids. Prepd by electrolytic reduction of europic salts.

USE: The salts in cathode ray tube coatings for color television receivers. Eu has a very high cross-section for the capture of thermal neutrons which is of value in the construction of electric atomic power stations. Organic derivs as shift reagents in NMR spectroscopy: C. C. Hinckley, *J. Am. Chem. Soc.* **91**, 5160 (1969); R. E. Sievers, *Nuclear Magnetic Resonance Shift Reagents* (Academic Press, New York, 1973).

3947. Evacetrapib. [1186486-62-3] *trans*-4-[[(5S)-5-[[[3,5-Bis(trifluoromethyl)phenyl]methyl](2-methyl-2H-tetrazol-5-yl)amino]-2,3,4,5-tetrahydro-7,9-dimethyl-1H-1-benzazepin-1-yl]methyl]-cyclohexanecarboxylic acid; LY-2484595. C$_{31}$H$_{36}$F$_6$N$_6$O$_2$; mol wt 638.66. C 58.30%, H 5.68%, F 17.85%, N 13.16%, O 5.01%. Cholesteryl ester transfer protein (CETP) inhibitor. Prepn: X. Chen *et al.*, *WO 06002342*; *idem et al.*, *US 7786108*; *idem et al.*, *US 100331309* (2006, 2010, 2010 all to Lilly). Pharmacology: G. Cao *et al.*, *J. Lipid Res.* **52**, 2169 (2011). Clinical effect on serum cholesterol levels in patients with dyslipidemia: S. J. Nicholls *et al.*, *J. Am. Med. Assoc.* **306**, 2099 (2011).

White amorphous solid or hydrated crystals from water.
THERAP CAT: Antilipemic.

3948. Evan's Blue. [314-13-6] 6,6'-[(3,3'-Dimethyl[1,1'-biphenyl]-4,4'-diyl)bis(2,1-diazenediyl)]bis[4-amino-5-hydroxy-1,3-naphthalenedisulfonic acid] sodium salt (1:4); 6,6'-[(3,3'-dimethyl-[1,1'-biphenyl]-4,4'-diyl)bis(azo)]bis[4-amino-5-hydroxy-1,3-naphthalenedisulfonic acid] tetrasodium salt; C.I. Direct Blue 53; 4,4'-bis[7-(1-amino-8-hydroxy-2,4-disulfo)naphthylazo]-3,3'-bitolyl tetrasodium salt; C.I. 23860; T-1824; Azovan Blue. C$_{34}$H$_{24}$N$_6$Na$_4$-O$_{14}$S$_4$; mol wt 960.79. C 42.50%, H 2.52%, N 8.75%, Na 9.57%, O 23.31%, S 13.35%. Prepd by coupling 1 mol of diazotized *o*-tolidine with 2 mols of Chicago acid (1-amino-8-naphthol-2,4-disulfonic acid): **DE 35341**; **DE 38802** *Frdl.* **1**, 469, 488 (1877-1887); **DE 3949**; **DE 57327**; **DE 75469** *Frdl.* **3**, 685, 687, 690 (1890-1894); Hartwell, Fieser, *Org. Synth.* **coll. vol. II**, 145 (1943). Diagnostic use: M. H. Nielsen, N. C. Nielsen, *Scand. J. Clin. Lab. Invest.* **14**, 605 (1962); O. Linderkamp *et al.*, *Eur. J. Pediatr.* **125**, 135 (1977).

Blue crystals with bronze to green luster. Sol in water, alcohol, acids, alkalies. Indicator changing color near pH 10. Destroyed by strong oxidizing and reducing agents and precipitated from soln by strong concns of neutral salts. Rather stable in aq soln, and may be autoclaved at 15 lbs pressure for 30 min. Dye made up in physiological saline should not be autoclaved.

THERAP CAT: Diagnostic aid (blood volume determination).

3949. Evening Primrose Oil. EPO. Seed oil of the evening primrose, *Oenothera biennis* L., *Onagraceae*, which contains approx 72% linoleic acid and approx 9% γ-linolenic acid, *q.q.v.*, as the two main constituents. Unique among vegetable oils because of its high content of γ-linolenic acid. Effect on prostaglandin biosynthesis in rats: B. A. Schölkens *et al.*, *Prostaglandins Leukotrienes Med.* **8**, 273 (1982). Clinical studies in atopic eczema: C. R. Lovell *et al.*, *Lancet* **1**, 278 (1981); S. Wright, J. L. Burton, *ibid.* **2**, 1120 (1982); P. L. Biagi *et al.*, *Drugs Exp. Clin. Res.* **14**, 285 (1988). Ingredient in cosmetics for aging skin: J. P. Marty, **DE 3447618** (1985 to Roussel-UCLAF), *C.A.* **103**, 146984r (1985). Brief review including discussion of uses: A. J. Barber, *Pharm. J.* **240**, 723-725 (1988). Clear, golden yellow oil. d$_{15}$ 0.9283. n$_D^{25}$ 1.4782. Sapon. no. 287.8. Iodine no. 154.8.

Note: Evening primrose oil products include *Efamol*, *Efamast*, *Epogam*.

USE: Dietary supplement.
THERAP CAT: In treatment of atopic eczema and mastaglia.

3950. Everolimus. [159351-69-6] 42-O-(2-Hydroxyethyl)rapamycin; 40-O-(2-hydroxyethyl)rapamycin; RAD-001; SDZ RAD; Afinitor; Certican; Votubia; Zortress. C$_{53}$H$_{83}$NO$_{14}$; mol wt 958.24. C 66.43%, H 8.73%, N 1.46%, O 23.37%. Macrolide immunosuppressant; derivative of rapamycin, *q.v.* Inhibitor of the serine-threonine kinase, mTOR. Suppresses cytokine-mediated lymphocyte proliferation. Prepn: S. Cottens, R. Sedrani, **WO 9409010**; *eidem*, **US 5665772** (1994, 1997 both to Sandoz). Pharmacology: W. Schuler *et al.*, *Transplantation* **64**, 36 (1997). Whole blood determn by LC/MS: N. Brignol *et al.*, *Rapid Commun. Mass Spectrom.* **15**, 898 (2001); by HPLC: S. Baldelli *et al.*, *J. Chromatogr. B* **816**, 99 (2005). Clinical pharmacokinetics in combination with cyclosporine: J. M. Kovarik *et al.*, *Clin. Pharmacol. Ther.* **69**, 48 (2001). Clinical study in prevention of cardiac-allograft vasculopathy: H. J. Eisen *et al.*, *N. Engl. J. Med.* **349**, 847 (2003); in advanced renal cell carcinoma: R. J. Motzer *et al.*, *Lancet* **372**, 449 (2008). Clinical trial for treatment of subependymal giant-cell astrocytomas (SEGA) in tuberous sclerosis: D. A. Krueger *et al.*, *N. Engl. J. Med.* **363**, 1801 (2010). *Review:* F. J. Dumont *et al.*, *Curr. Opin. Investig. Drugs* **2**, 1220-1234 (2001); B. Nashan, *Ther. Drug Monit.* **24**, 53-58 (2002).

THERAP CAT: Immunosuppressant; antineoplastic.

3951. Evodiamine. [518-17-2] (13bS)-8,13,13b,14-Tetrahydro-14-methylindolo[2',3':3,4]pyrido[2,1-*b*]quinazolin-5(7H)-one.

$C_{19}H_{17}N_3O$; mol wt 303.37. C 75.22%, H 5.65%, N 13.85%, O 5.27%. From *Evodia rutaecarpa* Hook. & Thoms and bark of *Zanthoxylum rhetsa* DC., *Rutaceae:* Y. Asahina, K. Kashiwaki, *J. Pharm. Soc. Jpn.* **1915**, 1293, *C.A.* **10**, 607 (1916); Gopinath *et al.*, *Tetrahedron* **8**, 293 (1960). Structure: Y. Asahina *J. Pharm. Soc. Jpn.* **1924**, 1; Ohta, *J. Pharm. Soc. Jpn.* **65**, 15 (1945), *C.A.* **45**, 5697 (1951). Synthesis: Asahina, Ohta, *Ber.* **61B**, 319 (1928); T. Kametani *et al.*, *J. Am. Chem. Soc.* **98**, 6186 (1976); *eidem*, *Heterocycles* **4**, 23 (1976). Biosynthesis: M. Yamazaki *et al.*, *Tetrahedron Lett.* **1966**, 3221; *ibid.* **1967**, 3317. Mass spec.: J. Tamas *et al.*, *Acta Chim. Acad. Sci. Hung.* **89**, 85 (1976).

Yellow plates from alc, mp 278°. $[\alpha]_D^{15}$ +352° (acetone); $[\alpha]_D$ +440° (chloroform). uv max (acetonitrile): 272, 280, 291, 335 nm (log ε 4.06, 4.02, 3.90, 3.30). Sol in acetone; slightly sol in alcohol, ether, chloroform. Practically insol in water, petr ether, benzene. Does not seem to form salts.

3952. Exaltolide®. [106-02-5] Oxacyclohexadecan-2-one; 15-hydroxypentadecanoic acid ε-lactone. $C_{15}H_{28}O_2$; mol wt 240.39. C 74.95%, H 11.74%, O 13.31%. Musk compd responsible for the pleasant odor of angelica root oil. Synthesis: Ruzicka, Stoll, *Helv. Chim. Acta* **11**, 1159 (1928); Carnduff *et al.*, *Chem. Ind. (London)* **1960**, 559; Dhenkne *et al.*, *J. Chem. Soc.* **1962**, 2348; Mathur, Bhattacharya, *ibid.* **1963**, 3505; Becker, Ohloff, *Helv. Chim. Acta* **54**, 2889 (1971). Alternate syntheses: T. Mukaiyama *et al.*, *Chem. Lett.* **1977**, 441; K. Narasaka *et al.*, *ibid.* 763; W. H. Kruizinga, R. M. Kellogg, *Chem. Commun.* **1979**, 286.

Thick oil. Odor of amber and musk. $bp_{0.25}$ 110°; bp_{15} 176°. d_4^{20} 0.9549. n_D^{20} 1.4708. When sublimed in vacuum, needles, mp 32°. Sol in alc.

USE: As a fixative in perfumery.

3953. Exametazime. [105613-48-7] *rel*-(2E,2'E,3R,3'R)-3,3'-[(2,2-Dimethyl-1,3-propanediyl)diimino]bis-2-butanone 2,2'-dioxime; (RR,SS)-4,8-diaza-3,6,6,9-tetramethylundecane-2,10-dione bisoxime; (±)-(3RS,3'RS)-3,3'-[(2,2-dimethyltrimethylene)diimino]di-2-butanone dioxime; *d,l*-hexamethylpropyleneamine oxime; *d,l*-HM-PAO; hexametazine; Ceretec. $C_{13}H_{28}N_4O_2$; mol wt 272.39. C 57.32%, H 10.36%, N 20.57%, O 11.75%. Propyleneamine oxime derivative that forms a neutral, lipophilic complex with ^{99m}Tc which diffuses across cellular membranes, including the blood brain barrier. Used in regional cerebral blood flow imaging. Prepn: L. R. Canning *et al.*, **EP 194843**; *eidem*, **US 4789736** (1986, 1988 both to Amersham). Prepn and radiopharmacology of ^{99m}Tc complex: D. P. Nowotnik *et al.*, *Nucl. Med. Commun.* **6**, 499 (1985); R. D. Neirinckx *et al.*, *J. Nucl. Med.* **28**, 191 (1987). Clinical comparison of isomers: P. F. Sharp *et al.*, *ibid.* **27**, 171 (1986). Biodistribution in humans: K. Nakamura *et al.*, *Eur. J. Nucl. Med.* **15**, 100 (1989). Clinical evaluation in single photon emission tomography (SPECT) of meningiomas: S. Nakano *et al.*, *J. Nucl. Med.* **30**, 1101 (1989). ^{99m}Tc radiolabeling of human platelets: H. J. Danpure, S. Osman, *Nucl. Med. Commun.* **9**, 267 (1988); of human leukocytes: H. J. Danpure *et al.*, *ibid.* 465. Clinical study of labeled leukocytes for imaging inflammatory lesions: J. H. Reynolds, *Clin. Radiol.* **42**, 195 (1990).

Series of articles on basic and clinical studies: *J. Cereb. Blood Flow Metab.* **8**, S1-S126 (1988).

Relative stereochemistry

Crystals from ethyl acetate, mp 128-130°.
Complex with ^{99m}Tc. Technetium Tc 99m exametazime.
THERAP CAT: ^{99m}Tc complex as diagnostic aid (radioactive imaging agent).

3954. Exemestane. [107868-30-4] 6-Methyleneandrosta-1,4-diene-3,17-dione; FCE-24304; Aromasin. $C_{20}H_{24}O_2$; mol wt 296.41. C 81.04%, H 8.16%, O 10.80%. Irreversible aromatase inhibitor. Prepn: E. Di Salle *et al.*, **DE 3622841**; F. Buzzetti *et al.*, **US 4808616** (1987, 1989 both to Farmitalia Carlo Erba). Pharmacology: D. Giudici *et al.*, *J. Steroid Biochem.* **30**, 391 (1988). HPLC determn in plasma: M. Breda *et al.*, *J. Chromatogr.* **620**, 225 (1993). Clinical pharmacology: T. R. J. Evans *et al.*, *Cancer Res.* **52**, 5933 (1992). Review of clinical experience in primary and secondary therapy for breast cancer: J. M. Dixon, *Expert Rev. Anticancer Ther.* **2**, 267-275 (2002). Clinical trial in sequential therapy following tamoxifen in breast cancer: R. C. Coombes *et al.*, *N. Engl. J. Med.* **350**, 1081 (2004).

White to slightly yellow crystalline powder. mp 188-191°. uv max: 247 nm (ε 13750). Freely sol in DMF; sol in methanol. Practically insol in water.

THERAP CAT: Antineoplastic (hormonal).

3955. Exendins. Bioactive peptides isolated from the venom of *Heloderma suspectum* Helodermatidae (Gila monster) and *H. horridum* (Mexican beaded lizard). Carboxy-amidated peptides containing 39 amino acid residues; differ only at positions 2 and 3. Interact with mammalian receptors for glucagon like peptide-1 (GLP-1), a physiological regulator of insulin secretion, and exhibit similar activities. Structurally related peptides known as *helospectin* and *helodermin* have also been isolated from venom of these species and were considered exendins-1 and -2 but differ in bioactivity. Review of isoln, structure and activity: J. Eng, *Mt. Sinai J. Med.* **59**, 147-149 (1992); J.-P. Raufman, *Regul. Pept.* **61**, 1-18 (1996).

Exendin-3. [130391-54-7] Isoln from *H. horridum:* J. Eng *et al.*, *J. Biol. Chem.* **265**, 20259 (1990).

Exendin-4. [141732-76-5] Exenatide; AC-2993; LY-2148568; Byetta. Isoln from *H. suspectum:* J. Eng *et al.*, *J. Biol. Chem.* **267**, 7402 (1992). GLP-1 receptor binding study: R. Göke *et al.*, *J. Biol. Chem.* **268**, 19650 (1993). Use as probe for human GLP-1 receptor: G. G. Chicchi *et al.*, *Peptides* **18**, 319 (1997). Review of pharmacology: L. L. Nielsen *et al.*, *Regul. Pept.* **117**, 77-88 (2004). Clinical trial in type 2 diabetes: J. B. Buse *et al.*, *Diabetes Care* **27**, 2628 (2004). Review of pharmacology and clinical experience: G. M. Bray, *Am. J. Health Syst. Pharm.* **63**, 411-418 (2006).

Exendin (9-39). [133514-43-9] 9-39-Exendin 3 (Heloderma horridum); 9-39-exendin 4; exendin (9-39) amide. Amino truncated fragment; specific antagonist of GLP-1 receptor. Prepn from exendin 3: J.-P. Raufman *et al.*, *J. Biol. Chem.* **266**, 2897 (1991). Use as GLP-1 antagonist for *in vivo* studies: F. Kolligs *et al.*, *Diabetes* **44**, 16 (1995).

USE: Biological probe.

THERAP CAT: Antidiabetic.

3956. Exisulind. [59973-80-7] (1*Z*)-5-Fluoro-2-methyl-1-[[4-(methylsulfonyl)phenyl]methylene]-1*H*-indene-3-acetic acid; *cis*-5-fluoro-2-methyl-1-(*p*-methylsulfonylbenzylidene)-3-acetic acid; sulindac sulfone; FGN-1; Aptosyn. $C_{20}H_{17}FO_4S$; mol wt 372.41. C 64.50%, H 4.60%, F 5.10%, O 17.18%, S 8.61%. Sulfone metabolite of sulindac, *q.v.*; induces apoptosis of cancer cells. Identification as metabolite: H. B. Hucker *et al.*, *Drug Metab. Dispos.* **1**, 721 (1973). Prepn: H. Jones *et al.*, *J. Carbohydr. Nucleosides Nucleotides* **3**, 369 (1974). Antiproliferative effect vs colon cancer cells: L. J. Hixson *et al.*, *Cancer Epidemiol. Biomarkers Prev.* **3**, 433 (1994). HPLC determn in serum: M. Siluveru, J. T. Stewart, *J. Chromatogr. B* **673**, 91 (1995); and clinical pharmacokinetics: G. F. Ray *et al.*, *J. Pharm. Biomed. Anal.* **14**, 213 (1995). Review of mechanism of action studies: D. J. Ahnen, *Eur. J. Surg. Suppl.* **582**, 111-114 (1998).

mp 194-196°. $[\alpha]_D$ −62° (c = 1 in CHCl$_3$).

THERAP CAT: In treatment of colorectal polyps.

3957. Exosurf®. [99732-49-7] (7*R*)-4-Hydroxy-*N,N,N*-trimethyl-10-oxo-7-[(1-oxohexadecyl)oxy]-3,5,9-trioxa-4-phosphapentacosan-1-aminium, inner salt, 4-oxide mixture with formaldehyde polymer with oxirane and 4-(1,1,3,3-tetramethylbutyl)phenol and 1-hexadecanol; Surfexo. $C_{40}H_{80}NO_8P.C_{16}H_{34}O.(C_{14}H_{22}O.C_2H_4O.CH_2O)_x$. Protein-free synthetic lung surfactant composed of a 13.5:1.5:1 mixture of colfosceril palmitate, cetyl alcohol and tyloxapol, *q.q.v.* Prepn: J. A. Clements, **US 5110806** (1992 to Univ. of Calif.). Pharmacology: D. J. Durand *et al.*, *J. Pediatr.* **107**, 775 (1985); W. H. Tooley *et al.*, *Am. Rev. Respir. Dis.* **136**, 651 (1987). Clinical trial in neonatal respiratory distress syndrome (RDS): W. Long *et al.*, *J. Pediatr.* **118**, 595 (1991). Symposium on pulmonary mechanics and clinical use in RDS: *ibid.* **120**, Suppl., S1-S50 (1992). Review of pharmacodynamics, clinical efficacy and tolerance: K. L. Dechant, D. Faulds, *Drugs* **42**, 877-894 (1991); of metabolism: R. L. DeAngelis, J. W. Findlay, *Clin. Perinatol.* **20**, 697-710 (1993).

THERAP CAT: Pulmonary surfactant.

3958. Ezetimibe. [163222-33-1] (3*R*,4*S*)-1-(4-Fluorophenyl)-3-[(3*S*)-3-(4-fluorophenyl)-3-hydroxypropyl]-4-(4-hydroxy-phenyl)-2-azetidinone; Sch-58235; Ezetrol; Zetia. $C_{24}H_{21}F_2NO_3$; mol wt 409.43. C 70.41%, H 5.17%, F 9.28%, N 3.42%, O 11.72%. Cholesterol absorption inhibitor. Prepn: S. B. Rosenblum *et al.*, **WO 9508532**; *eidem*, **US 5767115** (1995, 1998 both to Schering); *idem et al.*, *J. Med. Chem.* **41**, 973 (1998). Enantioselective synthesis: G. Wu *et al.*, *J. Org. Chem.* **64**, 3714 (1999). Activity in animals: M. van Heek *et al.*, *J. Pharmacol. Exp. Ther.* **283**, 157 (1997). Metabolism and distribution: *eidem*, *Br. J. Pharmacol.* **129**, 1748 (2000). Review of pharmacology and clinical studies: H. Bays, *Expert Opin. Invest. Drugs* **11**, 1587-1604 (2002).

White solid, mp 164-166°. $[\alpha]_D^{22}$ −33.9° (c = 3 in methanol).

THERAP CAT: Antilipemic.

3959. Ezogabine. [150812-12-7] *N*-[2-Amino-4-[[(4-fluoro-phenyl)methyl]amino]phenyl]carbamic acid ethyl ester; 2-amino-4-(4-fluorobenzylamino)-1-ethoxycarbonylaminobenzene; *N*-[2-amino-4-(4-fluorobenzylamino)phenyl] carbamic acid ethyl ester; retigabine; D-23129; Potiga; Trobalt. $C_{16}H_{18}FN_3O_2$; mol wt 303.34. C 63.35%, H 5.98%, F 6.26%, N 13.85%, O 10.55%. Neuronal potassium channel opener acting at K_v7.2-7.5 (KCNQ2-5) channels. Prepn: H.-R. Dieter *et al.*, **DE 4200259**; *eidem*, **US 5384330** (1993, 1995 both to Asta Medica). HPLC/MS determn in plasma: N. G. Knebel *et al.*, *J. Chromatogr. B* **748**, 97 (2000). LC/MS/MS determn of the active, *N*-acetyl metabolite in plasma: W. Bu *et al.*, *J. Chromatogr. B* **852**, 465 (2007). Characterization of binding site and channel activation: T. V. Wuttke *et al.*, *Mol. Pharmacol.* **67**, 1009 (2005); A. Schenzer *et al.*, *J. Neurosci.* **25**, 5051 (2005). Clinical pharmacokinetics: G. M. Ferron *et al.*, *J. Clin. Pharmacol.* **42**, 175 (2002). Review of chemistry and pharmacology: G. Blackburn-Munro *et al.*, *CNS Drug Rev.* **11**, 1-20 (2005); R. J. Porter *et al.*, *Neurotherapeutics* **4**, 149-154 (2007). Clinical trials in partial epilepsy: M. J. Brodie *et al.*, *Neurology* **75**, 1817 (2010); J. A. French *et al.*, *ibid.* **76**, 1555 (2011).

White to slightly colored, odorless, tasteless, non-hygroscopic crystalline powder, mp 138-145°. pKa 10.8.

Dihydrochloride. [150812-13-8] D-20443. $C_{16}H_{18}FN_3O_2$.2HCl; mol wt 376.25. Colorless to slightly pink crystals from ethanol ether, mp 182-186°. Hygroscopic.

THERAP CAT: Anticonvulsant.

F

3960. Factor V. [9001-24-5] Blood-coagulation factor V; proaccelerin; labile factor; accelerator globulin; Ac-globulin. Mol wt above 300,000. Participates in the later stages of blood coagulation. Activated by factor X, *q.v.*, and along with Ca^{2+} and phospholipid forms a complex which converts prothrombin to thrombin, *q.q.v.* Accelerates the interaction of prothrombin, thromboplastin and Ca^{2+} ions: Ware, Seegers, *Fed. Proc.* **7**, 131 (1948). Thrombin produces serum Ac-globulin, *accelerin*, a far more potent accelerator, from plasma Ac-globulin, apparently by catalytic action: Ware *et al.*, *Science* **106**, 618 (1947); Rapaport *et al.*, *Blood* **21**, 221 (1963). Purification: Ware, Seegers, *J. Biol. Chem.* **172**, 699 (1948); Blombäck, Blombäck, *Nature* **198**, 886 (1963). Role in the scheme of blood coagulation: Pavlovsky in *Fibrinogen and Fibrin Turnover of Clotting Factors*, R. B. Hunter *et al.*, Eds. (Schattauer-Verlag, Stuttgart, 1963) pp 443-454. Proaccelerin deficiency is characterized by prolonged Quick's prothrombin time, and cephalin time, abnormal thromboplastin generation test, abnormal prothrombin consumption, and prolonged whole blood clotting time: Owren in *Thrombolytic Activity and Related Phenomena*, I. S. Wright *et al.*, Eds. (Schattauer-Verlag, Stuttgart, 1961) pp 387-391. *Review:* R. W. Colman, *Prog. Hemostasis Thromb.* **3**, 109-143 (1976). Purification, stability, amino acid composition, physical properties of bovine and human factor V: R. W. Colman, R. M. Weinberg, *Methods Enzymol.* **45B**, 107-122 (1976).

3961. Factor VII. [9001-25-6] Blood-coagulation factor VII; proconvertin; co-thromboplastin; serum prothrombin conversion accelerator; SPCA. Enzyme which, in conjunction with tissue factor and Ca^{2+} ions initiates the activation of factor X, *q.v.* Isoln from human plasma or serum: Fantt, Osborn, *Thromb. Diath. Haemorrh.* **8**, 286 (1962); from bovine plasma: J. Jesty, Y. Nemerson, *J. Biol. Chem.* **249**, 509 (1974). Separation from factor X by DEAE cellulose: Hougie, Bunting, in *New Blood Clotting Factors*, I. S. Wright *et al.*, Eds. (Schattauer-Verlag, Stuttgart, 1960) pp 40-42. Purification and structure of bovine factor VII: J. Jesty, Y. Nemerson, *J. Biol. Chem.* **249**, 509 (1974); R. Radcliffe, Y. Nemerson, *ibid.* **250**, 388 (1975); *eidem, Methods Enzymol.* **45B**, 49 (1976). Its deficiency results in retarded prothrombin conversion and an elevated one-stage prothrombin time: Alexander in *Thrombolytic Activity and Related Phenomena*, I. S. Wright *et al.*, Eds. (Schattauer-Verlag, Stuttgart, 1961) pp 392-402. Factor VII deficiency is associated with severe liver disease, vitamin K deficiency and broad spectrum antibiotic therapy, or may follow anticoagulant coumarin therapy.

3962. Factor VIII. [9001-27-8] Blood-coagulation factor VIII, complex; antihemophilic globulin; antihemophilic factor A; AHG; AHF; Alphanate; Hemofil-M; Humate-P; Koate-HP; Monoclate-P; Nordiocto; Profilate. Crucial, nonenzymatic cofactor in the intrinsic coagulation pathway leading to the localized generation of thrombin. Deficiency results in the bleeding disorder known as hemophilia A or classical hemophilia. Factor VIII is an X-linked gene product synthesized by hepatocytes. Mature, single chain protein contains 2332 amino acid residues; mol wt 265 kDa. Released into the circulation as a set of heterodimers that rapidly interact with von Willebrand factor to form a stable, noncovalent complex; present at extremely low concn in plasma. Converted by thrombin into a heterotrimer (factor VIIIa) that accelerates the activation of factor X by factor IXa. Inactivated by protein C. Identification in plasma: T. Addis, *J. Pathol. Bacteriol.* **15**, 427 (1911). Role in hemostasis: A. J. Patek, Jr., F. H. L.Taylor, *J. Clin. Invest.* **16**, 113 (1937). Isoln and characterization of human factor VIII: E. J. Hershgold *et al.*, *J. Lab. Clin. Med.* **77**, 185 (1971). Purification and properties of bovine: G. A. Vehar, E. W. Davie, *Biochemistry* **19**, 401 (1980). Characterization of the human factor VIII gene: J. Gitschier *et al.*, *Nature* **312**, 326 (1984). Series of articles on cloning, expression, and structure: *ibid.* 330-347. Review of bioregulation: R. J. Kaufman, *Annu. Rev. Med.* **43**, 325-339 (1992). Review of production methods and clinical use in treatment of factor VIII deficiency: J. C. Gill, *Semin. Thromb. Hemostasis* **19**, 1-12 (1993). Review of biosynthesis, structure and function: P. Lollar, *Adv. Exp. Med. Biol.* **386**, 3-17 (1995); P. J. Lenting *et al.*, *Blood* **92**, 3983-3996 (1998).

Octocog alfa. [139076-62-3] Advate; Kogenate; Recombinate. Human factor VIII produced by recombinant DNA technology. Re-

view of development and clinical experience with octocog preparations: N. Ananyeva *et al.*, *Expert Opin. Pharmacother.* **5**, 1061-1070 (2004); A. D. Shapiro, *Vasc. Health Risk Manag.* **3**, 555-565 (2007).

THERAP CAT: Antihemophilic factor (human).

3963. Factor IX. [9001-28-9] Blood-coagulation factor IX; Christmas factor; PTC; plasma thromboplastin component; antihemophilic factor B; autoprothrombin II. A plasma and serum glycoprotein that participates in the middle phases of blood coagulation. Activated by factor XI and Ca^{2+}. Interacts then with factor VIII, Ca^{2+}, and phospholipid to form the complex which converts factor X into factor X$_a$. Isoln and characterization of bovine factor IX: K. Fujikawa *et al.*, *Biochemistry* **12**, 4938 (1973); K. Fujikawa, E. W. Davie, *Methods Enzymol.* **45B**, 74 (1976). Deficiency results in a congenital bleeding disorder known as Christmas disease or hemophilia B. *Review:* Macfarlane in *Thrombolytic Activity and Related Phenomena*, I. S. Wright *et al.*, Eds. (Schattauer-Verlag, Stuttgart, 1961) pp 408-415; B. N. Bouma, J. A. Van Mourik, *Haemostasis: Biochemistry, Physiology & Pathology*, D. Ogston, B. Bennett, Eds. (Wiley-Interscience, New York, 1977) pp 56-77.

Factor IX complex (human). Konyne; Profilnine; Proplex T. A mixture of human factors II, VII, IX, and X.

Factor IX (human). AlphaNine; Mononine.

THERAP CAT: Hemostatic.

3964. Factor X. [9001-29-0] Blood-coagulation factor X; Stuart-Prower factor; Stuart factor; Prower factor. Enzyme which is activated by a complex of factors IX, VIII, calcium ions and phospholipid, or by factor VII and tissue factor in the intrinsic and extrinsic pathways of blood coagulation resp. Non-physiological activators are trypsin and a specific protein from Russel's viper venom: *see* Fujikawa *et al.*, *Biochemistry* **11**, 4892 (1972). Active factor X apparently acts upon factor V, *q.v.* in the presence of calcium ions and lipid to form a complex which activates the conversion of prothrombin to thrombin. Present in both plasma and serum because neither increases nor decreases during clotting. Presumably synthesized in the liver because level is reduced with coumarin drugs and in liver disease. Isoln and characterization of bovine factor X: *eidem, ibid.* 4882. Factor X$_a$ is a serine protease which, in the presence of factor V, Ca^{2+} ions and phospholipids, activates prothrombin. Severe deficiency causes hemorrhagic state resembling hemophilia. Mild deficiency may result in easy bruising. In severe cases deep muscular and joint hemorrhages occur. Improved purification and bibliography of earlier isoln procedures: Bajaj, Mann, *J. Biol. Chem.* **248**, 7729 (1973). Comparison of the molecular forms generated by the different methods of activation: Radcliffe, Barton, *ibid.* 6788. *Review:* Graham, Hougie in *Thombolytic Activity and Related Phenomena*, I. S. Wright *et al.*, Eds. (Schattauer-Verlag, Stuttgart, 1961) pp 416-420; C. M. Jackson, *Thromb. Diath. Haemorrh. Suppl.* **57**, 197-216 (1974). Reviews on prepn, physical properties and function in the coagulation process: K. Fujikawa, E. W. Davie, *Methods Enzymol.* **45B**, 89 (1976); J. Jesty, Y. Nemerson, *ibid.* p 95.

Stable at room temperature for several days. Adsorbed by BaSO$_4$, Al(OH)$_3$, Ca$_3$(PO$_4$)$_2$, Seitz filters, bentonite and DEAE cellulose. At 56° is destroyed rapidly, in serum, less rapidly as citrate eluate. At 65° is destroyed rapidly in any form. Stable between pH 6.1-9.0. Soluble in water. Precipitates from water between pH 8-4. Isoelec pt pH 4.3.

3965. Factor XI. [9013-55-2] Blood-coagulation factor XI; PTA; plasma thromboplastin antecedent. Participates in the initial phases of blood coagulation. In normal plasma, factor XI is present in a precursor form and is converted to an active form, factor XI$_a$, which in turn activates factor IX to factor IX$_a$ in the presence of Ca^{2+} ions: Ratnoff, Davie, *Biochemistry* **1**, 677 (1962); Schiffman *et al.*, *Blood* **22**, 733 (1963); K. Fujikawa *et al.*, *Biochemistry* **13**, 4508 (1974). Deficiency is characterized by the moderate bleeding symptoms of hemophilia C; by pathological clotting and recalcification times, impaired prothrombin consumption, and delayed thromboplastin formation: Duckert, Soulier in *New Blood Clotting Factors*, I. S. Wright *et al.*, Eds. (Schattauer-Verlag, Stuttgart, 1960) pp 145-147, 123-131. Review on bovine factor XI: T. Koide, *Methods Enzymol.* **45B**, 65-73 (1976).

3966. Factor XII. [9001-30-3] Blood-coagulation factor XII; Hageman factor; HF. Mol wt about 82,000. An enzyme which circulates in zymogen form in blood and when activated initiates the first of a series of events in coagulation of blood plasma. Enzymatic nature of blood coagulation is described by the "cascade" theory in which clotting factors interact with one another in a stepwise manner, one acting as an enzyme, the other as substrate, in a sequence of reactions leading to the formation of thrombin, *q.v.*: Davie, Ratnoff, *Science* **145**, 1310 (1964); MacFarlane, *Nature* **202**, 498 (1964). Factor XII initiates clotting when blood or plasma comes into contact with glass, collagen, or surface-active agents. Once activated, it interacts with factor XI, *q.v.*, to convert it to an enzyme. Factor XII is a sialoglycoprotein with esterase activity. Its activation by glass may be viewed as a dynamic alteration of the tertiary structure of the protein moiety: Schoenmakers *et al.*, *Biochim. Biophys. Acta* **101**, 166 (1965). Isoln of human and rabbit factor XII in zymogen form: C. G. Cochrane, K. D. Wuepper, *J. Exp. Med.* **134**, 986 (1971); H. Saito *et al.*, *Circ. Res.* **34**, 641 (1974). Highly purified Hageman factor undergoes a change in physical properties during activation: Donaldson, Ratnoff, *Science* **150**, 754 (1965). Factor XII deficiency is termed the Hageman trait. Unlike deficiencies in all other clotting factors, persons with Hageman trait do not show significant bleeding tendencies. *Reviews:* Ratnoff in *New Blood Clotting Factors*, I. S. Wright *et al.*, Eds. (Schattauer-Verlag, Stuttgart, 1960) pp 116-122; Ratnoff *et al.*, in *Thrombolytic Activity and Related Phenomena*, I. S. Wright *et al.*, Eds. (Schattauer-Verlag, Stuttgart, 1961) pp 364-378; several authors in *Recent Advances in Blood Coagulation*, L. Poller, Ed. (J. & A. Churchill, London, 1969); J. Spragg, K. F. Austen, *Compr. Immunol.* **3**, 125-143 (1977); J. H. Griffin, C. G. Cochrane, *Methods Enzymol.* **45B**, 56-65 (1976).

Sedimentation coefficient, $s^{\circ}_{20,\omega} = 7.08$. Diffusion constant, $D^{\circ}_{20,\omega} = 7.14 \times 10^{-7}$ cm²/sec. uv max at pH 7: 280 nm ($E^{1\%}_{1cm}$ 12.0). Isoelectric pt pH 8.0.

3967.   Factor XIII. [9013-56-3] Blood-coagulation factor XIII; fibrin-stabilizing factor; FSF; fibrinase; Laki-Lorand factor; LLF; Fibrogammin. Plasma enzyme precursor which, when activated by thrombin in the presence of Ca^{2+}, converts soluble fibrin gel to a tough insoluble clot: Laki, Lorand, *Science* **108**, 280 (1948); Lorand, *Nature* **166**, 694 (1950). During the terminal stage of blood-clotting, factor XIII catalyzes the formation of cross-links in the fibrin gel by eliminating NH_3 in the transamidation reaction in which the ε-amino group of lysine from a fibrin molecule forms a peptide linkage with a glutaminyl residue from a nearby fibrin monomer: Lorand *et al.*, *Biochem. Biophys. Res. Commun.* **25**, 629 (1966); Fuller, Doolittle, *ibid.* 694. Trypsin, papain, and reptilase also activate factor XIII: Lorand *et al.*, *ibid.* **31**, 222 (1968). Structure contains two each of two types of soluble proteinaceous subunits called the α and β chains; mol wt of subunits each about 80,000: Schwartz *et al.*, *J. Biol. Chem.* **248**, 1395 (1973). Amino acid sequence studies: Holbrook *et al.*, *Biochem. J.* **135**, 901 (1973). Clinical studies: Cucuianu *et al.*, *Thromb. Diath. Haemorrh.* **30**, 480 (1973). *Review:* L. Lorand, *Ann. N.Y. Acad. Sci.* **202**, 6-30 (1972); A. G. Loewy, *ibid.* 41-58; C. G. Curtis, L. Lorand, *Haemostasis: Biochemistry, Physiology & Pathology*, D. Ogston, B. Bennett, Eds. (Wiley-Interscience, New York, 1977) pp 186-201; *eidem*, *Methods Enzymol.* **45B**, 177-191 (1976).

uv max: 280 nm ($E^{1\%}$ 13.8). Retains activity in plasma kept at 0°; inactivated in a few days at room temperature. Activity destroyed if serum heated at 60° for 10 min. Does not dialyse out from plasma. Inactivated by *Echis coloratus* venom. Pretreatment with compounds containing an SH-group, such as cysteine, reduced glutathione, DL-penicillamine, or mercapto-1-propanol prevents inactivation as well as restores activity after venom treatment.

THERAP CAT: Antihemorrhagic.

3968.   Fadrozole. [102676-47-1] 4-(5,6,7,8-Tetrahydroimidazo[1,5-*a*]pyridin-5-yl)benzonitrile; 5-*p*-cyanophenyl-5,6,7,8-tetrahydroimidazo[1,5-*a*]pyridine. $C_{14}H_{13}N_3$; mol wt 223.28. C 75.31%, H 5.87%, N 18.82%. Aromatase inhibitor. Prepn: L. J. Browne, **EP 165904**; *idem*, **US 4617307** (1985, 1986 both to Ciba-Geigy); L. J. Browne *et al.*, *J. Med. Chem.* **34**, 725 (1991). Effects on estrogen biosynthesis: R. E. Steele *et al.*, *Steroids* **50**, 147 (1987). Pharmacology: K. Schieweck *et al.*, *Cancer Res.* **48**, 834 (1988). GC/MS determn in plasma and urine: R. Ackermann, G. Kaiser, *Biomed. Environ. Mass Spectrom.* **18**, 558 (1989). Pharmacokinetics: G. M. Kochak *et al.*, *J. Clin. Endocrinol. Metab.* **71**, 1349

(1990). Clinical evaluation in metastatic breast cancer: J. I. Raats *et al.*, *J. Clin. Oncol.* **10**, 111 (1992).

mp 117-118°.

Hydrochloride. [102676-31-3] CGS-16949A; Afema. $C_{14}H_{13}$-N_3.HCl; mol wt 259.74. Crystals from 2-propanol, mp 231-233°. Sol in water.

THERAP CAT: Antineoplastic.

3969.   Fagarine. [524-15-2] 4,8-Dimethoxyfuro[2,3-*b*]quinoline; γ-fagarine; 8-methoxydictamnine. $C_{13}H_{11}NO_3$; mol wt 229.24. C 68.11%, H 4.84%, N 6.11%, O 20.94%. Antiarrhythmic principle in *Fagara coco* (Gill.) Engl., *Rutaceae*. Isoln: Deulofeu *et al.*, *J. Am. Chem. Soc.* **64**, 2326 (1942); Briggs, Cambi, *Tetrahedron* **2**, 256 (1958). Structure: Berinzaghi *et al.*, *J. Org. Chem.* **10**, 181 (1945). Synthesis: Grundon, McCorkindale, *Chem. Ind. (London)* **1956**, 1091.

Prisms from alcohol, mp 142°. uv max: 238, 332, 370 nm (log ε 4.76, 3.88, 3.89). Sol in chloroform, benzene, ether; slightly sol in water, petr ether.

Hydrochloride. $C_{13}H_{12}ClNO_3$. Needles from chloroform + ether, mp 158-159°.

3970.   Famciclovir. [104227-87-4] 2-[2-(2-Amino-9*H*-purin-9-yl)ethyl]-1,3-propanediol 1,3-diacetate; 9-[4-acetoxy-3-(acetoxymethyl)but-1-yl]-2-aminopurine; FCV; BRL-42810; Oravir; Famvir. $C_{14}H_{19}N_5O_4$; mol wt 321.34. C 52.33%, H 5.96%, N 21.79%, O 19.92%. Prodrug of penciclovir, *q.v.* Prepn: M. R. Harnden, R. L. Jarvest, **AU 85 47560**; *eidem*, **US 5246937** (1986, 1993 both to Beecham); M. R. Harnden *et al.*, *J. Med. Chem.* **32**, 1738 (1989). HPLC determn in plasma and urine: J. R. McMeekin *et al.*, *Anal. Proc.* **29**, 178 (1992). Review of metabolism and mode of action: R. A. Vere Hodge, *Antivir. Chem. Chemother.* **4**, 67-84 (1993). Series of articles on pharmacology and pharmacokinetics: *ibid.* Suppl. 1, 37-68 (1993). Review of clinical efficacy in herpes zoster and genital herpes: R. Circelli *et al.*, *Antiviral Res.* **29**, 141-151 (1996).

White shiny plates from ethyl acetate-hexane, mp 102-104°. uv max (methanol): 222, 244, 309 nm (ε 27500, 4890, 7160). Sol in water (25°): >25% w/v initially; rapidly ppts as sparingly sol monohydrate (2-3% w/v). Freely sol in acetone, methanol; sparingly sol in ethanol, isopropanol.

THERAP CAT: Antiviral.

3971.   Famotidine. [76824-35-6] 3-[[[2-[(Aminoiminomethyl)amino]-4-thiazolyl]methyl]thio]-*N*-(aminosulfonyl)propanimidamide; [1-amino-3-[[[2-[(diaminomethylene)amino]-4-thiazolyl]methyl]thio]propylidene]sulfamide; *N*-sulfamoyl-3-[(2-guanidino-thiazol-4-yl)methylthio]propionamide; YM-11170; MK-208; Amfamox; Fadul; Famodil; Famosan; Famoxal; Ganor; Gaster; Gastridin; Gastropen; Lecedil; Motiax; Muclox; Pepcid; Pepcidac; Pepcidine;

Pepdine; Pepdul; Peptan; Ulfamid. $C_8H_{15}N_7O_2S_3$; mol wt 337.44. C 28.48%, H 4.48%, N 29.06%, O 9.48%, S 28.50%. Histamine H_2-receptor antagonist. Prepn, NMR and mass spectral data: H. Yasufumi *et al.*, **BE 882071**; *eidem*, **US 4283408**; **JP Kokai 81 55383**, (1980, 1981, 1981 all to Yamanouchi). Inhibition of gastric acid and pepsin secretion in rats: M. Takeda *et al.*, *Arzneim.-Forsch.* **32**, 734 (1982); in man: M. Miwa *et al.*, *Int. J. Clin. Pharmacol. Ther. Toxicol.* **22**, 214 (1984). Effect on disposition of antipyrine in liver: Ch. Staiger *et al.*, *Arzneim.-Forsch.* **34**, 1041 (1984). Chromatographic determn in plasma and urine: W. C. Vincek *et al.*, *J. Chromatogr.* **338**, 438 (1985). Clinical trial in Zollinger-Ellison syndrome: J. M. Howard *et al.*, *Gastroenterology* **88**, 1026 (1985). Symposia on pharmacology and clinical efficacy: *Am. J. Med.* **81**, Suppl. 4B, 1-64 (1986); *Scand. J. Gastroenterol.* **22**, Suppl. 134, 1-62 (1987). Tolerability and safety profile: C. W. Howden, G. N. J. Tytgat, *Clin. Ther.* **18**, 36-54 (1996). Review of clinical pharmacokinetics: H. Echizen, T. Ishizaki, *Clin. Pharmacokinet.* **21**, 178-194 (1991).

White to pale yellow crystalline powder, mp 163-164°. Soly at 20° (%, w/v): 80 in DMF; 50 in acetic acid; 0.3 in methanol; 0.1 in water; <0.01 in ethanol, chloroform. Practically insol in acetone, ether, ethyl acetate. LD_{50} i.v. in mice: 244.4 mg/kg (Yasufumi).

THERAP CAT: Antiulcerative.

THERAP CAT (VET): Antiulcerative.

3972. Famoxadone. [131807-57-3] 5-Methyl-5-(4-phenoxyphenyl)-3-(phenylamino)-2,4-oxazolidinedione; 3-anilino-5-methyl-5-(4-phenoxyphenyl)-1,3-oxazolidine-2,4-dione; DPX-JE874; Famoxate. $C_{22}H_{18}N_2O_4$; mol wt 374.40. C 70.58%, H 4.85%, N 7.48%, O 17.09%. Inhibits electron transport via blockage of ubiquinol:cytochrome c oxidoreductase. Prepd not claimed: D. Geffken, D. R. Rayner, **US 4957933** (1990 to DuPont). Biological activity and field trials: M. M. Joshi, J. A. Sternberg, *Proc. Br. Crop Prot. Conf. - Pests Dis.* **1996**, 21. Mode of action: D. B. Jordan *et al.*, *Pestic. Sci.* **55**, 105 (1999). Environmental fate: K. M. Jernberg, P. W. Lee, *ibid.* 587. Metabolic fate in plants and animals: P. W. Lee *et al.*, *ibid.* 589. Review of synthesis and structure-activity relationship: J. A. Sternberg *et al.*, *ACS Symp. Ser.* **686**, 216-227 (1998).

mp 140.3-141.8°. Log P (*n*-octanol/water): 4.65 (pH 7). Soly in water: 52 µg/L. LD_{50} in rats (mg/kg): >5000 orally; >2000 dermally (Joshi, Sternberg).

USE: Agricultural fungicide.

3973. α-Farnesene. [502-61-4] (3*E*,6*E*)-3,7,11-Trimethyl-1,-3,6,10-dodecatetraene; 2,6,10-trimethyl-2,6,9,11-dodecatetraene; farnesene. $C_{15}H_{24}$; mol wt 204.36. C 88.16%, H 11.84%. Isoln of (*E*,*E*)-form from natural coating of apples: Huelin, Murray, *Nature* **210**, 1260 (1966); Murray, *Aust. J. Chem.* **22**, 197 (1969); from Dufour's gland in ants: Cavill *et al.*, *Tetrahedron Lett.* **1967**, 2201. Isoln of (*Z*,*E*)-form from oil of perilla: T. Sakai, Y. Hirose, *Bull. Chem. Soc. Jpn.* **42**, 3615 (1969). Oxidation products of farnesene are believed to cause scald, a serious storage disorder of apples. Four possible geometric isomers. Synthetic studies: Ruzicka, *Helv. Chim. Acta* **6**, 490, 501 (1923); Ruzicka, Capato, *ibid.* **8**, 267 (1925). Configuration of (*E*,*E*)-form confirmed by synthesis from *trans*-β-farnesene: Brieger *et al.*, *J. Org. Chem.* **34**, 3789 (1969). Stereospecific synthesis of (*E*,*Z*)- and (*Z*,*Z*)-isomers: Anet, *Aust. J. Chem.* **23**, 2101 (1970). Synthesis of (*E*,*E*)-form: Tanaka *et al.*, *J. Am. Chem. Soc.* **97**, 3252 (1975). (*E*,*E*)- and (*Z*,*E*)-forms are components of aphid alarm pheromones: J. A. Pickett, D. C. Griffiths, *J. Chem. Ecol.* **6**, 349 (1980); of the trail pheromone of red imported fire ants: R. K. Vandermeer *et al.*, *Tetrahedron Lett.* **22**, 1651 (1981).

(E,E)- α-Farnesene

Thin oil. bp$_{12}$ about 125°. d_4^{20} 0.8410. n_D^{20} 1.4836. Practically insol in water. Misc with hydrocarbon solvents. uv max of (*E*,*E*)-form (alc): 233 nm (ε 27,000); of (*Z*,*E*)-form (alc): 238 nm (ε 11300).

3974. β-Farnesene. [18794-84-8] (6*E*)-7,11-Dimethyl-3-methylene-1,6,10-dodecatriene. $C_{15}H_{24}$; mol wt 204.36. C 88.16%, H 11.84%. The naturally occurring *trans* form or (*E*)-isomer is a constituent of various essential oils, *see* F. Sorm *et al.*, *Collect. Czech. Chem. Commun.* **14**, 699 (1949); **15**, 626 (1951); also an alarm pheromone of several aphid species: Bowers *et al.*, *Science* **177**, 1121 (1972); Edwards *et al.*, *Nature* **241**, 126 (1973); W. S. Bowers *et al.*, *J. Insect Physiol.* **23**, 697 (1977). Isoln from leaves of wild potato, *Solanum berthaultii* Hawkes, *Solanaceae*: R. W. Gibson, J. A. Pickett, *Nature* **302**, 608 (1983). Prepd by dehydration of farnesol: Bhati, *Perfum. Essent. Oil Rec.* **54**, 376 (1963); Brieger, *J. Org. Chem.* **32**, 3720 (1967); alternate syntheses: Tanaka *et al.*, *J. Am. Chem. Soc.* **97**, 3252 (1975); S. Akutagawa *et al.*, *Chem. Lett.* **1976**, 485; O. P. Vig *et al.*, *Indian J. Chem. Sect. B* **18**, 33 (1979). Synthesis of (*Z*)-isomer: Anet, *Aust. J. Chem.* **23**, 2101 (1970); O. P. Vig *et al.*, *J. Indian Chem. Soc.* **47**, 999 (1970); *eidem*, *Indian J. Chem.* **13**, 1244 (1975).

(E)- β-Farnesene

(*E*)-**Form.** Oil, bp 124°. d_4^{20} 0.8310. n_D^{20} 1.4870. uv max (hexane): 224 nm (ε 14000).

(*Z*)-**Form.** Oil, bp$_{3-4}$ 95-107°. n_D^{32} 1.4780. uv max (hexane): 224 nm (ε 17300).

3975. Farnesol. [4602-84-0] 3,7,11-Trimethyl-2,6,10-dodecatrien-1-ol; farnesyl alcohol. $C_{15}H_{26}O$; mol wt 222.37. C 81.02%, H 11.79%, O 7.19%. Sesquiterpene alcohol found in oils of citronella, neroli, cyclamen, lemon grass, tuberose, rose, musk, balsam Peru, and tolu. Isoln: Elge, *Chem. Ztg.* **34**, 857 (1910); **37**, 1422 (1913); Kerschbaum, *Ber.* **46**, 1732 (1913); Naves, *Helv. Chim. Acta* **32**, 1798, 2181 (1949); LaFace, *ibid.* **33**, 249 (1950). Synthesis: Ruzicka, *ibid.* **6**, 492 (1923); Ruzicka, Firmenich, *ibid.* **22**, 392 (1939); Nazarov *et al.*, *Zh. Obshch. Khim.* **28**, 1444 (1958); Shvarts, Petrov, *ibid.* **30**, 3598 (1960); Popjak *et al.*, *J. Biol. Chem.* **237**, 56 (1962). Four possible stereoisomers. Stereochemistry: Bates *et al.*, *Chem. Ind. (London)* **1961**, 1907; *J. Org. Chem.* **28**, 1086 (1963). *trans,trans*-Farnesol is the only stereoisomer present in many essential oils but occurs mixed with *cis,trans*-farnesol in petitgrain oil and several other oils: Naves, *Compt. Rend.* **251**, 900 (1960). Stereospecific synthesis of *trans,trans*-farnesol: Corey *et al.*, *J. Am. Chem. Soc.* **92**, 6637 (1970).

trans,trans-Farnesol

trans,trans-**Farnesol.** [106-28-5] (2*E*,6*E*)-3,7,11-Trimethyl-2,-6,10-dodecatrien-1-ol; 2,6-di-*trans*-farnesol; (*E*,*E*)-farnesol. Liquid with characteristic, flowery scent. bp$_{0.35}$ 111°; bp$_{1.3}$ 151-152°. n_D^{25} 1.4872. d^{25} 0.879. Flash point, closed cup: 96°C (204.8°F). uv max: 192-196 nm (ε 28,500).

USE: In perfumery, to emphasize the odor of sweet floral perfumes, such as lilac and cyclamen.

3976. Faropenem. [106560-14-9] (5*R*,6*S*)-6-[(1*R*)-1-Hydroxyethyl]-7-oxo-3-[(2*R*)-tetrahydro-2-furanyl]-4-thia-1-azabicyclo-[3.2.0]hept-2-ene-2-carboxylic acid; fropenem; (5*R*,6*S*,8*R*,2'*R*)-2-(2'-tetrahydrofuryl)-6-hydroxyethylpenem-3-carboxylate. $C_{12}H_{15}$-

NO$_5$S; mol wt 285.31. C 50.52%, H 5.30%, N 4.91%, O 28.04%, S 11.24%. Orally active, β-lactamase stable, penem antibiotic. Prepn: M. Ishiguro *et al.*, **EP 199446**; *eidem*, **US 4997829** (1986, 1991 both to Suntory); *eidem*, *J. Antibiot.* **41**, 1685 (1988). Pharmacokinetics: A. Tsuji *et al.*, *Drug Metab. Dispos.* **18**, 245 (1990). *In vitro* antimicrobial spectrum: J. M. Woodcock *et al.*, *J. Antimicrob. Chemother.* **39**, 35 (1997). β-Lactamase stability: A. Dalhoff *et al.*, *Chemotherapy (Basel)* **49**, 229 (2003). HPLC determn in plasma: R. V. S. Nirogi *et al.*, *Arzneim.-Forsch.* **55**, 762 (2005). Clinical trial in urinary tract infections: S. Arakawa *et al.*, *Nishinihon J. Urol.* **56**, 300 (1994); in bacterial sinusitis: R. Siegert *et al.*, *Eur. Arch. Otorhinolaryngol.* **260**, 186 (2003).

Sodium salt. [122547-49-3] Furopenem; ALP-201; SUN-5555; SY-5555; WY-49605; Farom. C$_{12}$H$_{15}$NNaO$_5$S; mol wt 308.30. $[\alpha]_D^{22}$ +60° (c = 0.10).

Daloxate. [141702-36-5] (5*R*,6*S*)-6-[(1*R*)-1-Hydroxyethyl]-7-oxo-3-[(2*R*)-tetrahydro-2-furanyl]-4-thia-1-azabicyclo[3.2.0]hept-2-ene-2-carboxylic acid (5-methyl-2-oxo-1,3-dioxol-4-yl)methyl ester; faropenem medoxomil; Bay-56-6854; SUN-208; Orapem. C$_{17}$H$_{19}$NO$_8$S; mol wt 397.40. Prepn: H. Iwata *et al.*, **WO 9203442**; *eidem*, **US 5830889** (1992, 1998 both to Suntory). Pale yellow crystals.

THERAP CAT: Antibacterial (antibiotics).

3977. Fasciculins. [86697-68-9] FAS. Group of low molecular wt (~6750 Da) toxic peptides isolated from the venom of green mamba snakes. Allosterically inhibits ($K_i \simeq 10^{-10}M$) most synaptic acetylcholinesterases (AChE). Cationic peptides characterized by 4 disulphide bridges. Isoln from *Dendroaspis angusticeps*: D. Rodriguez-Ithurralde *et al.*, *Neurochem. Int.* **5**, 267 (1983). X-ray crystallography: M. H. le Du *et al.*, *J. Biol. Chem.* **267**, 22122 (1992). Inhibition of AChE: P. Fossier *et al.*, *Cell. Mol. Neurobiol.* **6**, 221 (1986). Allosteric control of inhibition: Z. Radic *et al.*, *J. Biol. Chem.* **270**, 20391 (1995). Amino acid sequence and structure-activity study: C. Cervenansky *et al.*, *Toxicon* **34**, 718 (1996). Review of neuropharmacology: F. Dajas *et al.*, *Methods Neurosci.* **8**, 258-270 (1992).

USE: Research tool.

3978. Fast Green FCF. [2353-45-9] *N*-Ethyl-*N*-[4-[[4-[ethyl[(3-sulfophenyl)methyl]amino]phenyl](4-hydroxy-2-sulfophenyl)methylene]-2,5-cyclohexadien-1-ylidene]-3-sulfobenzenemethanaminium inner salt sodium salt (1:2); FD & C Green No. 3; C.I. Food Green 3; C.I. 42053. C$_{37}$H$_{34}$N$_2$Na$_2$O$_{10}$S$_3$; mol wt 808.84. C 54.94%, H 4.24%, N 3.46%, Na 5.68%, O 19.78%, S 11.89%. Prepn: H. Johnson, P. Staub, *Ind. Eng. Chem.* **19**, 497 (1927). Toxicity studies: F. C. Lu, A. Lavalle, *Can. Pharm. J.* **97**, 30 (1964); W. H. Hansen *et al.*, *Food Cosmet. Toxicol.* **4**, 389 (1966). Review of carcinogenicity studies: *IARC Monographs* **16**, 187-197 (1978). *See also: Colour Index* **vol.** 4 (3rd ed., 1971) p 4383.

Dark green powder or granules with a metallic lustre. Absorption max: 628 nm. Very sol in water; sol in ethanol. Dull orange soln

in conc H$_2$SO$_4$, changing to dull green on dilution. Orange soln in conc HCl or conc HNO$_3$. Bright blue soln in 10% aq NaOH. LD$_{50}$ orally in rats: >2 g/kg (Lu, Lavalle).

USE: Biological stain; color additive in food, drugs, cosmetics.

3979. Fasudil. [103745-39-7] 5-[(Hexahydro-1*H*-1,4-diazepin-1-yl)sulfonyl]isoquinoline; hexahydro-1-(5-isoquinolinylsulfonyl)-1*H*-1,4-diazepine; *N*-(5-isoquinolinesulfonyl)-1,4-perhydrodiazepine; 1-(5-isoquinolinesulfonyl)homopiperazine. C$_{14}$H$_{17}$N$_3$O$_2$-S; mol wt 291.37. C 57.71%, H 5.88%, N 14.42%, O 10.98%, S 11.00%. Intracellular calcium antagonist. Prepn: H. Hidaka, T. Sone, **EP 187371**; *eidem*, **US 4678783** (1986, 1987 both to Asahi); A. Morikawa *et al.*, *Chem. Pharm. Bull.* **40**, 770 (1992). Pharmacology: T. Asano *et al.*, *Br. J. Pharmacol.* **98**, 1091 (1989). Mechanism of action: *idem et al.*, *J. Pharmacol. Exp. Ther.* **241**, 1033 (1987). Clinical trials: M. Shibuya *et al.*, *J. Neurosurg.* **76**, 571 (1992). Acute toxicity: H. Koga *et al.*, *Yakuri to Chiryo* **20**, S1433 (1992), *C.A.* **117**, 226005b (1992). Review of pharmacology and clinical evaluation: M. Shirotani *et al.*, *Cardiovasc. Drug Rev.* **10**, 333-357 (1992).

Hydrochloride. [105628-07-7] AT-877; HA-1077; Eril. C$_{14}$H$_{17}$N$_3$O$_2$S.HCl; mol wt 327.83. Crystals from water, mp 220.5°. Also reported as white crystalline powder, mp 219.3° (Shirotani). Sol in water up to 2 × 10^{-2}M at pH 5.0-7.0. LD$_{50}$ in mice, rats (mg/kg): 67.5, 59.9 i.v.; 124.5, 123.2 s.c.; 273.9, 335.0 orally (Koga).

THERAP CAT: Vasodilator (cerebral).

3980. Fazadinium Bromide. [49564-56-9] 1,1′-(1,2-Diazenediyl)bis[3-methyl-2-phenylimidazo[1,2-*a*]pyridinium] bromide (1:2); 1,1′-azobis[3-methyl-2-phenylimidazo[1,2-*a*]pyridinium] dibromide; AH-8165; Fazadon. C$_{28}$H$_{24}$Br$_2$N$_6$; mol wt 604.35. C 55.65%, H 4.00%, Br 26.44%, N 13.91%. A short-acting curarimimetic with rapid onset. Prepn: D. Jack, E. E. Glover, **DE 2127355**; *eidem*, **US 3773746** (1971, 1973 both to Allen & Hanburys); E. E. Glover, M. Yorke, *J. Chem. Soc. C* **1971**, 3280. Pharmacology: L. Bolger *et al.*, *Nature* **238**, 354 (1972); and toxicity data: P. Bellani *et al.*, *Farmaco Ed. Sci.* **39**, 846 (1984). Clinical pharmacokinetics: T. Busi *et al.*, *Minerva Anestesiol.* **49**, 485 (1983). Clinical use: G. Minutella, G. Bongiovanni, *ibid.* 301; A. Lusini *et al.*, *ibid.* 393.

Dihydrate. Yellow crystals from water, mp 215-219° (softens at 196°). Also reported as yellow solid from methanol/ethyl acetate, mp 218-220°. uv max (H$_2$O): 285, 297 nm (log ε 4.04, 4.34). LD$_{50}$ i.v. in rats: 1.5 μM/k (Bellani).

THERAP CAT: Neuromuscular blocking agent.

3981. Febantel. [58306-30-2] *N*,*N*′-[[2-[(2-Methoxyacetyl)-amino]-4-(phenylthio)phenyl]carbonimidoyl]biscarbamic acid *C*,*C*′-dimethyl ester; dimethyl [[2-(2-methoxyacetamido)-4-(phenylthio)phenyl]imidocarbonyl]dicarbamate; Bay Vh 5757; Bay h 5757; Rintal. C$_{20}$H$_{22}$N$_4$O$_6$S; mol wt 446.48. C 53.80%, H 4.97%, N 12.55%, O 21.50%, S 7.18%. Prepn: H. Koelling *et al.*, **DE 2423679**; *eidem*, **US 3993682** (1975, 1976 both to Bayer). Anthelmintic efficacy: H.

Wollweber *et al.*, *Arzneim.-Forsch.* **28**, 2193 (1977); H. Thomas, *Res. Vet. Sci.* **25**, 290 (1978). Safety evaluation in horses: J. A. Shmidl, *Vet. Med. Small Anim. Clin.* **73**, 775 (1978).

Crystals, mp 129-130°.

THERAP CAT (VET): Anthelmintic.

3982. Febrifugine. [24159-07-7] 3-[3-[(2R,3S)-3-Hydroxy-2-piperidinyl]-2-oxopropyl]-4(3H)-quinazolinone; 3-[3-(3-hydroxy-2-piperidyl)acetonyl]-4(3H)-quinazolinone; 3-[β-keto-γ-(3-hydroxy-2-piperidyl)propyl]-4-quinazolone; β-dichroine. $C_{16}H_{19}$-N_3O_3; mol wt 301.35. C 63.77%, H 6.36%, N 13.94%, O 15.93%. Alkaloid with antimalarial properties; active component in traditional Chinese medicines. Isoln from *Dichroa febrifuga* Lour., *Saxifragaceae*: J. B. Koepfli *et al.*, *J. Am. Chem. Soc.* **69**, 1837 (1947); *eidem, ibid.* **71**, 1048 (1949); from the common hydrangea: F. Ablondi *et al.*, *J. Org. Chem.* **17**, 14 (1952). Identity with β-dichroine: Fu, Yang, *C.A.* **43**, 1530a (1949). Structure: J. B. Koepfli *et al.*, *J. Am. Chem. Soc.* **72**, 3323 (1950). Synthesis: B. R. Baker *et al.*, *J. Org. Chem.* **17**, 132 (1952); **18**, 178 (1953). Config: Hill, Edwards, *Chem. Ind. (London)* **1962**, 858. Revised stereochemistry: D. F. Barringer, Jr. *et al.*, *J. Org. Chem.* **38**, 1937 (1973); and asymmetric synthesis: S. Kobayashi *et al.*, *Tetrahedron Lett.* **40**, 2175 (1999); Y. Takeuchi *et al.*, *Tetrahedron* **59**, 1639 (2003).

Dimorphic crystals: needles from ethanol, mp 139-140°; from chloroform, mp 154-156°. $[\alpha]_D^{25}$ +6° (c = 0.5 in chloroform); $[\alpha]_D^{25}$ +28° (c = 0.5 in ethanol). Freely sol in methanol + chloroform, water + ethanol; slightly sol in water, ethanol, acetone, chloroform. Practically insol in ether, benzene, petr ether. LD$_{50}$ orally in white mice: 2.5-3.0 mg/kg (Koepfli, 1949).

Dihydrochloride. [32434-42-7] $C_{16}H_{19}N_3O_3 \cdot 2HCl$; mol wt 374.26. Crystals from 90% ethanol, mp 220-222° (dec). $[\alpha]_D^{25}$ +12.8° (c = 0.8).

3983. Febuxostat. [144060-53-7] 2-[3-Cyano-4-(2-methylpropoxy)phenyl]-4-methyl-5-thiazolecarboxylic acid; 2-(3-cyano-4-isobutyloxyphenyl)-4-methyl-5-thiazolecarboxylic acid; TEI-6720; TMX-67; Uloric. $C_{16}H_{16}N_2O_3S$; mol wt 316.38. C 60.74%, H 5.10%, N 8.85%, O 15.17%, S 10.13%. Xanthine oxidase/xanthine dehydrogenase inhibitor. Prepn: S. Kondo *et al.*, *EP* **513379**; *eidem, US* **5614520** (1992, 1997 both to Teijin). Synthesis: M. Hasegawa, *Heterocycles* **47**, 857 (1998). Mechanism of action and crystal structure study: K. Okamoto *et al.*, *J. Biol. Chem.* **278**, 1848 (2003). Pharmacology: Y. Osada *et al.*, *Eur. J. Pharmacol.* **241**, 183 (1993). Clinical pharmacokinetics: M. D. Mayer *et al.*, *Am. J. Ther.* **12**, 22 (2005). Clinical evaluation in hyperuricemia and gout: M. A. Becker *et al.*, *N. Engl. J. Med.* **353**, 2450 (2005). Review of pharmacology and clinical experience: H. R. Schumacher Jr. *et al.*, *Rheumatology* **48**, 188-194 (2009).

Crystals from ethanol, mp 238-239° (dec). Also reported as crystals from acetone, mp 201-202° (Hasegawa).

THERAP CAT: Treatment of hyperuricemia and chronic gout.

3984. Felbamate. [25451-15-4] 2-Phenyl-1,3-propanediol 1,3-dicarbamate; carbamic acid 2-phenyltrimethylene ester; ADD-03055; W-554; Felbatol; Taloxa. $C_{11}H_{14}N_2O_4$; mol wt 238.24. C 55.46%, H 5.92%, N 11.76%, O 26.86%. Antiepileptic, structurally similar to meprobamate, *q.v.* Prepn: F. M. Berger, B. J. Ludwig, US **2884444** (1959 to Carter); B. J. Ludwig *et al.*, *J. Med. Chem.* **12**, 462 (1969). Use in treatment of epileptic seizures: R. D. Sofia, US **4978680** (1990 to Carter-Walace). HPLC determn in plasma: R. P. Remmel *et al.*, *Ther. Drug Monit.* **12**, 90 (1990). Pharmacology: E. A. Swinyard *et al.*, *Epilepsia* **27**, 27 (1986). Metabolism in animals: J. T. Yang *et al.*, *Drug Metab. Dispos.* **19**, 1126 (1991). Penetration of blood brain barrier: E. M. Cornford *et al.*, *Epilepsia* **33**, 944 (1992). Clinical pharmacokinetics: A. J. Wilensky *et al.*, *ibid.* **26**, 602 (1985). Clinical trial in refractory partial seizures: E. Faught *et al.*, *Neurology* **43**, 688 (1993). Efficacy in childhood epileptic encephalopathy (Lennox-Gastaut syndrome): F. J. Ritter *et al.*, *N. Engl. J. Med.* **328**, 29 (1993). Review of pharmacology, toxicology and clinical evaluation: R. D. Sofia *et al.*, *Epilepsy Res.* Suppl. 3, 103-108 (1991); of clinical experience: J. M. Pellock *et al.*, *Epilepsy Res.* **71**, 89-101 (2006).

White, odorless powder, mp 151-152°. Sparingly sol in water, methanol, ethanol, acetone and chloroform; freely sol in DMSO, 1-methyl-2-pyrrolidinone and DMF. LD$_{50}$ i.p. in mice: 4000 mg/kg (Ludwig *et al.*).

THERAP CAT: Anticonvulsant.

THERAP CAT (VET): Anticonvulsant.

3985. Felbinac. [5728-52-9] [1,1'-Biphenyl]-4-acetic acid; 4-biphenylacetic acid; 4-carboxymethylbiphenyl; BPAA; LY-61017; L-141; LJC-10141; Napageln; Traxam. $C_{14}H_{12}O_2$; mol wt 212.25. C 79.22%, H 5.70%, O 15.08%. One of the metabolites of fenbufen, *q.v.* Prepn: E. Schwenk, D. Papa, *J. Org. Chem.* **11**, 798 (1946). Improved prepn: G. R. Malone, A. I. Meyers, *ibid.* **39**, 618 (1974). Pharmacology: A. E. Sloboda, A. C. Osterberg, *Inflammation* **1**, 415 (1976). Mode of action study in ocular inflammation: E. L. Tolman *et al.*, *Invest. Ophthalmol.* **15**, 1005 (1976). HPLC determn in serum: J. S. Fleitman *et al.*, *J. Chromatogr.* **228**, 372 (1982).

White needles from ethyl ether, mp 164-165°. LD$_{50}$ orally in rats: 164 mg/kg (Sloboda, Osterberg).

Ethyl ester. [14062-23-8] LM-001; Daitec. $C_{16}H_{16}O_2$; mol wt 240.30.

THERAP CAT: Anti-inflammatory, analgesic.

3986. Felinine. [471-09-0] S-(3-Hydroxy-1,1-dimethylpropyl)-L-cysteine; L-3-[(3-hydroxy-1,1-dimethylpropyl)thio]alanine; (−)-felinine. $C_8H_{17}NO_3S$; mol wt 207.29. C 46.35%, H 8.27%, N 6.76%, O 23.15%, S 15.47%. Sulfur containing amino acid found in cat urine. Isoln: R. G. Westall, *Biochem. J.* **55**, 244 (1953). Synthesis of (±)-felinine: S. Trippett, *J. Chem. Soc.* **1957**, 1929; of (−)-felinine: H. Eggerer, *Ann.* **657**, 212 (1962); A. Schöberl *et al.*, *Ber.* **101**, 373 (1968). Determn in urine by capillary electrophoresis: S. M. Rutherfurd *et al.*, *Amino Acids* **27**, 49 (2004). Review of biological properties and distribution in feline species: W. H. Hendriks *et al.*, *Comp. Biochem. Physiol. B* **112**, 581-588 (1995).

Needles from water + acetone, dec 177°. $[\alpha]_D^{25}$ −11.5° (c = 2.81 in water). Sol in water, ethanol. Practically insol in acetone, ethyl acetate, ether.

(±)-**Felinine.** [168639-57-4] Crystals from aq ethanol, dec 181°.

3987. Felodipine. [72509-76-3] 4-(2,3-Dichlorophenyl)-1,4-dihydro-2,6-dimethyl-3,5-pyridinedicarboxylic acid 3-ethyl 5-methyl ester; H-154/82; Agon; Feloday; Flodil; Hydac; Munobal; Plendil; Prevex; Splendil. $C_{18}H_{19}Cl_2NO_4$; mol wt 384.25. C 56.26%, H 4.98%, Cl 18.45%, N 3.65%, O 16.65%. Dihydropyridine calcium channel blocker marketed as the racemate. Prepn: P. B. Berntsson et al., **EP 7293**; eidem, **US 4264611** (1980, 1981 both to AB Hassle). Conformation: P. B. Berntsson, R. E. Carter, Acta Pharm. Suec. **18**, 221 (1981). Interaction with calmodulin, q.v.: S. L. Boström et al., Nature **292**, 777 (1981). Diuretic-natriuretic properties: G. F. Dibona et al., Clin. Res. **30**, 571A (1982). Hemodynamic effects in coronary patients: P. Decoster et al., Eur. J. Clin. Invest. **12**, 43 (1982). Enantioselective determn in plasma: P. A. Soons et al., J. Chromatogr. **528**, 343 (1990); HPLC-MS/MS determn in plasma: L. H. Miglioranca et al., J. Chromatogr. B **814**, 217 (2005). Pharmacokinetics and pharmacodynamics: B. Edgar et al., Biopharm. Drug Dispos. **8**, 235 (1987). Pharmacokinetics, vascular selectivity, effects on hypertension, hemodynamics and renal function: Drugs **29**, Suppl. 2, 1-212 (1985).

Crystals from isopropyl ether, mp 145°. Freely sol in acetone, methanol; very slightly sol in heptane. Insol in water.

THERAP CAT: Antihypertensive; antianginal.

3988. Felypressin. [56-59-7] 2-(L-Phenylalanine)-8-L-lysinevasopressin; 2-(phenylalanine)-8-lysine vasopressin; Phe²-Lys⁸-vasopressin; Phe²-Phe²-Lys⁸-oxytocin; PLV-2; Octapressin. $C_{46}H_{65}$- $N_{13}O_{11}S_2$; mol wt 1040.23. C 53.11%, H 6.30%, N 17.50%, O 16.92%, S 6.16%. Synthetic analog of the antidiuretic hormone, vasopressin, q.v. Prepn: Boissonnas, Guttmann, Helv. Chim. Acta **43**, 190 (1960); Meienhofer, du Vigneaud, J. Am. Chem. Soc. **82**, 6336 (1960); **GB 928607**; R. Boissonnas, S. Guttmann, **US 3232923** (1963, 1966 both to Sandoz). LC determn in pharmaceutical formulations: M. Svensson, K. Gröningsson, J. Chromatogr. A **521**, 141 (1990). Clinical trial in combination with prilocaine for dental procedures: V. Goldman, H. Evers, Dent. Pract. Dent. Rec. **19**, 225 (1969). Cardiovascular pharmacology: R. Cecanho et al., Anesth. Prog. **53**, 119 (2006).

Cys–Phe–Phe–Gln–Asn–Cys–Pro–Lys–GlyNH₂

White or almost white powder or flakes. Freely sol in water. Practically insol in ethanol, acetone.

THERAP CAT: Vasoconstrictor.

3989. Femoxetine. [59859-58-4] (3R,4S)-3-[(4-Methoxyphenoxy)methyl]-1-methyl-4-phenylpiperidine; (+)-trans-1-methyl-3-[p-(methoxy)phenoxymethyl]-4-phenylpiperidine. $C_{20}H_{25}NO_2$; mol wt 311.43. C 77.13%, H 8.09%, N 4.50%, O 10.27%. Serotonin uptake inhibitor. Prepn: J. A. Christensen, R. F. Squires, **BE 810310**; eidem, **US 3912743** (1974, 1975 both to A/S Ferrosan). Neuropharmacology: J. B. Lassen et al., Psychopharmacologia **42**, 21 (1975). Pharmacology and toxicity: eidem, Eur. J. Pharmacol. **32**, 108 (1975). GC determn in body fluids: E. Bechgaard, J. Lund, J. Chromatogr. **133**, 147 (1977). Metabolism: H. Larsson, J. Lund, Acta Pharmacol. Toxicol. **48**, 424 (1981). Bioavailability and pharmacokinetics: H. Mengel et al., Arzneim.-Forsch. **33**, 462 (1983). Double-blind clinical trials in treatment of migraine: P. G. Andersson, E. N. Petersen, Acta Neurol. Scand. **64**, 280 (1981); in treatment

of endogenous depression: L.-E. Dahl et al., Acta Psychiatr. Scand. **66**, 9 (1982); P. N. Reebye et al., Pharmacopsychiatria **15**, 164 (1982).

Hydrochloride. [56222-04-9] FG-4963; Malexil. $C_{20}H_{25}NO_2$.- HCl; mol wt 347.88. LD_{50} in female, male mice (mg/kg): 48, 45 i.v.; 941, 723 s.c.; 1408, 1687 orally (Lassen).

THERAP CAT: Antidepressant.

3990. Fenamidone. [161326-34-7] (5S)-3,5-Dihydro-5-methyl-2-(methylthio)-5-phenyl-3-(phenylamino)-4H-imidazol-4-one; (S)-1-anilino-4-methyl-2-methylthio-4-phenylimidazolin-5-one; RPA-407213; Reason. $C_{17}H_{17}N_3OS$; mol wt 311.40. C 65.57%, H 5.50%, N 13.49%, O 5.14%, S 10.30%. Systemic, broad spectrum fungicide; mitochondrial electron transport inhibitor. Prepn: G. Lacroix et al., **EP 551048**; eidem, **US 6002016** (1993, 1999 both to Rhone-Poulenc). Physical and biological properties: R. T. Mercer et al., Brighton Crop Prot. Conf. - Pests Dis. **1998**, 319. Greenhouse studies: M. P. Latrose et al., ibid. 863. Field trials: G. DeWever et al., Meded. Fac. Landbouwwet. Univ. Gent **65**, 807 (2000).

White woolly powder, mp 137°. d 1.285. Soly in water (20°): 7.8 mg/l. Log P (octanol/water): 2.8 (20°). Vapor pressure (25°): 3.4 × 10⁻⁷ Pa. LD_{50} in male, female rats (mg/kg): >5000, 2028 orally; in rats (mg/kg): >2000 dermally. LC_{50} in bobwhite quail, mallard duck (mg/kg): >5200, >5200 dietary (Mercer).

USE: Fungicide.

3991. Fenamiphos. [22224-92-6] N-(1-Methylethyl)phosphoramidic acid ethyl 3-methyl-4-(methylthio)phenyl ester; isopropylphosphoramidic acid ethyl 3-methyl-4-(methylthio)-m-tolyl ester; ethyl 3-methyl-4-(methylthio)-m-tolyl isopropylphosphoramidate; ethyl 3-methyl-4-(methylthio)phenyl (1-methylethyl)phosphoramidate; phenamiphos; Bay 68138; SRA-3886; B-68138; Bayer 68138; Nemacur. $C_{13}H_{22}$- NO_3PS; mol wt 303.36. C 51.47%, H 7.31%, N 4.62%, O 15.82%, P 10.21%, S 10.57%. Organophosphate insecticide; cholinesterase inhibitor. Prepn: H. Kayser, G. Schrader, **US 2978479** (1961 to Bayer). Properties and nematocidal activity: B. Homeyer, Pflanzenschutz-Nachr. **24**, 48 (1971). Field trials: W. M. Zeck, ibid. **24**, 114 (1971). Degradation in soil: T. B. Waggoner, A. M. Khasawinah, Residue Rev. **53**, 79 (1974). Microbial degradation study: T. P. Cáceres et al., Curr. Microbiol. **57**, 643 (2008). Multiresidue determn by GC-MS: R. M. González-Rodríguez et al., J. Chromatogr. A **1196-1197**, 100 (2008). Toxicity to birds: E. F. Hill, M. B. Camardese, Ecotoxicol. Environ. Saf. **8**, 551 (1984).

Crystals, mp 49°. d_4^{49} 1.14°. Vapor pressure (30°): ∼1×10⁻⁶ mm Hg. Soly in water (20°): 700 ppm. Sol in most organic solvents.

Poisonous. LD$_{50}$ in male, female rats (mg/kg): 15.3, 19.4 orally (Homeyer).

Caution: Potential symptoms of overexposure are nausea, vomiting, abdominal cramps, diarrhea, salivation; headache, giddiness, vertigo, weakness; rhinorrhea, chest tightness; blurred vision, miosis; cardiac irregularities; muscle fasciculations; dyspnea. *See NIOSH Pocket Guide to Chemical Hazards* (DHHS/NIOSH 97-140, 1997) p 142.

USE: Nematocide.

3992. Fenarimol. [60168-88-9] α-(2-Chlorophenyl)-α-(4-chlorophenyl)-5-pyrimidinemethanol; 2,4'-dichloro-α-(pyrimidin-5-yl)benzhydryl alcohol; EL-222; Rubigan. C$_{17}$H$_{12}$Cl$_2$N$_2$O; mol wt 331.20. C 61.65%, H 3.65%, Cl 21.41%, N 8.46%, O 4.83%. Prophylactic and curative leaf fungicide; inhibits ergosterol biosynthesis. Prepn: **NL 6806106**; H. M. Taylor *et al.*, **US 3818009** (1968, 1974 both to Lilly). Improved prepn: *eidem*, **US 3869456**; use as fungicide: *eidem*, **US 3887708** (both 1975 to Lilly). Mode of action, efficacy against various fungi: H. Buchenauer, *Proc. Br. Crop Prot. Conf. - Pests Dis.* **1977**, 699. Absorption by leaf tissue, local systemic activity: I. F. Brown, H. R. Hall, *ibid.* **1981**(2), 573. In control of apple scab: J. M. Olivier, J. Guillaumes, *Monogr. Br. Crop Prot. Counc.* **31**, 485 (1985). Post-infection efficacy: A. L. O'Leary, T. B. Sutton, *Phytopathology* **76**, 119 (1986). Brief review: J.-M. Beraud *et al.*, *Def. Veg.* **34**, 17 (1980).

White odorless crystals, mp 117-119°. Practically insol in water (13.7 ppm at pH 7). Sol in most organic solvents. Vapor pressure at 25° <10^{-7} millibar. LD$_{50}$ in mice, rats (mg/kg): 4500, 2500 orally (Beraud).

USE: Plant fungicide.

3993. Fenazaquin. [120928-09-8] 4-[2-[4-(1,1-Dimethylethyl)phenyl]ethoxy]quinazoline; 4-[2-[4-(*tert*-butyl)phenyl]ethoxy]-quinazoline; 4-*tert*-butylphenethyl quinazolin-4-yl ether; DE-436; EL-436; Magister. C$_{20}$H$_{22}$N$_2$O; mol wt 306.41. C 78.40%, H 7.24%, N 9.14%, O 5.22%. Acaricidal quinazoline; inhibits Complex I of the mitochondrial electron transport chain. Prepn: B. A. Dreikorn *et al.*, **EP 326329** (1989 to Lilly); *idem et al.*, **US 5411963** (1995 to DowElanco). Mechanism of action study: E. Wood *et al.*, *Pestic. Biochem. Physiol.* **54**, 135 (1996). Photodecomposition study: J. Bhattacharyya *et al.*, *J. Agric. Food Chem.* **51**, 4013 (2003). GC determn in environmental and crop samples: A. R. Gambie, J. M. Perkins, *Brighton Crop Prot. Conf. - Pests Dis.* **1992**, 895. Field trials vs spider mites on ornamentals: R. T. Pollak *et al.*, *ibid.* 1181. Comprehensive description: C. Longhurst *et al.*, *ibid.* 51-58.

White to tan solid, mp 70-71° (Dreikorn); also reported as mp 77.5-80.0° (Gambie). Vapor pressure at 25°: 1.6×10^{-4} Pa. Log P at 25° (octanol/water): 5.51. Soly (g/ml): hexane 0.033-0.05; dichloromethane 70.6; acetone 0.4-0.5; acetonitrile 0.33-0.50; water 0.0001. Molar extinction coefficient at 262.2 nm: 1.24×10^4 (pH 7.83). LD$_{50}$ orally in rats, mice (mg/kg): 134, 1480; dermally in

rabbits (mg/kg): >5000. LC$_{50}$ (96 hr) in bluegill, trout (μg/l): 34.1, 3.8 (Longhurst).

USE: Acaricide and insecticide.

3994. Fenbendazole. [43210-67-9] *N*-[6-(Phenylthio)-1*H*-benzimidazol-2-yl]carbamic acid methyl ester; 5-(phenylthio)-2-benzimidazolecarbamic acid methyl ester; methyl 5-(phenylthio)-2-benzimidazolecarbamate; HOE-881v; Panacur. C$_{15}$H$_{13}$N$_3$O$_2$S; mol wt 299.35. C 60.19%, H 4.38%, N 14.04%, O 10.69%, S 10.71%. Prepn: H. Loewe *et al.*, **DE 2164690**; *eidem*, **US 3954791** (1973, 1976 both to Hoechst); C. Baeder *et al.*, *Experientia* **30**, 753 (1974). Efficacy vs gastrointestinal nematodes in swine: Enigk *et al.*, *Dtsch. Tieraerztl. Wochenschr.* **81**, 177 (1974).

Light brownish-gray, odorless, tasteless crystalline powder, mp 233° (dec). Freely sol in DMSO; sparingly sol in DMF; very slightly sol in methanol. Practically insol in water.

THERAP CAT (VET): Anthelmintic (Nematodes).

3995. Fenbuconazole. [114369-43-6] α-[2-(4-Chlorophenyl)ethyl]-α-phenyl-1*H*-1,2,4-triazole-1-propanenitrile; (*R,S*)-4-(4-chlorophenyl)-2-phenyl-2-[(1*H*-1,2,4-triazol-1-yl)methyl]butyronitrile; fenethanil; RH-7592; Enable; Govern; Indar. C$_{19}$H$_{17}$ClN$_4$; mol wt 336.82. C 67.75%, H 5.09%, Cl 10.52%, N 16.63%. Systemic triazole antifungal for food and ornamental crops. Prepn: S. H. Shaber *et al.*, **DE 3721786**; *idem*, **US 5087635** (1988, 1992 both to Rohm & Haas). Comprehensive description: D. Driant *et al.*, *Brighton Crop Prot. Conf. - Pests Dis.* **1988**, 33-40. Field trials vs *Botrytis* infection in roses: Y. Elad *et al.*, *Crop Prot.* **12**, 69 (1993); vs root rot in barley: P. J. Cotterill, *ibid.* 273.

White crystalline solid, mp 124-126°. vapor pressure (20°): 0.37 × 10^{-7} torr. Soly in water (25°): 0.2 ppm. Sol in common organic solvents. Insol in aliphatic hydrocarbons. LD$_{50}$ in rats (mg/kg): >2000 orally; >5000 dermally (Driant).

USE: Agricultural fungicide.

3996. Fenbufen. [36330-85-5] γ-Oxo-[1,1'-biphenyl]-4-butanoic acid; 3-(4-biphenylylcarbonyl)propionic acid; β-*p*-phenylbenzoylpropionic acid; diphenyl-4-γ-oxo-γ-butyric acid; 4-(4-biphenylyl)-4-oxobutyric acid; CL-82204; Bufemid; Cinopal; Cinopol; Lederfen. C$_{16}$H$_{14}$O$_3$; mol wt 254.29. C 75.57%, H 5.55%, O 18.87%. Prepn: **FR 798941** (1936 to I. G. Farbenind.), *C.A.* **30**, 7729^4 (1936); D. H. Hey, R. Wilkinson, *J. Chem. Soc.* **1940**, 1030; M. Weizmann *et al.*, *Chem. Ind. (London)* **1940**, 402; W. Reppe *et al.*, *Ann.* **596**, 223 (1955); A. S. Tomcufcik *et al.*, **DE 2147111**; *eidem*, **US 3784701** (1972, 1974 to Am. Cyanamid). Series of articles on chemistry, pharmacology, reproductive toxicology and clinical experience: *Arzneim.-Forsch.* **30**, 695-720, 725-746 (1980). Toxicity: H. F. Bolte *et al.*, *ibid.* 721. Review of pharmacology and therapeutic use: R. N. Brogden *et al.*, *Drugs* **21**, 1-22 (1981).

Crystals from ethanol, mp 185-187°. LD_{50} in various strains of mice, rats (mg/kg): 795-1673, 200-720 orally; 506-811, 265-575 i.p. (Bolte).

THERAP CAT: Anti-inflammatory.

3997. Fenbutatin Oxide. [13356-08-6] 1,1,1,3,3,3-Hexakis(2-methyl-2-phenylpropyl)distannoxane; di[tri-(2-methyl-2-phenylpropyl)tin]oxide; hexakis(β,β-dimethylphenethyl)distannoxane; SD-14114; Torque; Vendex. $C_{60}H_{78}OSn_2$; mol wt 1052.70. C 68.46%, H 7.47%, O 1.52%, Sn 22.55%. Selective organotin miticide. Prepn: **DE 2225666**; C. A. Horne Jr., **US 3657451** (1971, 1972 both to Shell Oil Co.). General description and analysis: G. Zweig, J. Sherma, *Anal. Methods Pestic. Plant Growth Regul.* **10**, 139 (1978). HPLC/MS determn in fruits and vegetables: K. A. Barnes *et al.*, *Rapid Commun. Mass Spectrom.* **11**, 159 (1997). Suppression of citrus rust mite: V. French, *J. Rio Grande Val. Hortic. Soc.* **35**, 121 (1982).

White crystalline powder, mp 145°. Soly at 23° (g/l): acetone 6; octanol 33; xylene 53; benzene 143; dichloromethane 377; methylene chloride 380. Soly in water: <5 μg/l. Converts to the hydroxide in the presence of water. Nontoxic to bees, toxic to fish. LD_{50} in rats, rabbits (mg/kg): >2000 orally; >2000 percutaneously (Zweig).

USE: Acaricide.

3998. Fencamfamine. [1209-98-9] *N*-Ethyl-3-phenylbicyclo[2.2.1]heptan-2-amine; *N*-ethyl-3-phenyl-2-norbornanamine; 2-ethylamino-3-phenylnorcamphane; 2-phenyl-3-ethylaminonorbornane; 2-ethylamino-3-phenylnorbornane; 2-ethylamino-3-phenylbicyclo[2.2.1]heptane; Euvitol. $C_{15}H_{21}N$; mol wt 215.34. C 83.67%, H 9.83%, N 6.50%. Prepd from 3-amino-2-phenylnorbornane and acetaldehyde followed by hydrogenation in the presence of PtO_2: Thesing *et al.*, **DE 1110159** (1961 to E. Merck), *C.A.* **56**, 2352g (1962). Pharmacology: R. Hotovy *et al.*, *Arzneim.-Forsch.* **11**, 20 (1961).

$bp_{0.1}$ 128-131°.
Hydrochloride. [2240-14-4] H-610. $C_{15}H_{21}N$.HCl; mol wt 251.80. Principal ingredient of *Reactivan* which also contains vitamins B_1, B_6, B_{12} and C. Crystals from acetone, mp 192°. Freely sol in water, ethanol, methanol, chloroform; slightly sol in benzene. Practically insol in ether. LD_{50} in mice, rats (mg/kg): 135.0, 113.0 orally; 85.5, 68.5 s.c.; 15.7, 23.5 i.v. (Hotovy).
Note: This is a controlled substance (stimulant): **21 CFR**, 1308.14.

THERAP CAT: CNS stimulant.

3999. Fencamine. [28947-50-4] 3,7-Dihydro-1,3,7-trimethyl-8-[[2-[methyl(1-methyl-2-phenylethyl)amino]ethyl]amino]-1*H*-purine-2,6-dione; 8-[[2-[methyl(α-methylphenethyl)amino]ethyl]amino]caffeine; 8-[2-(*N*,α-dimethyl-β-phenyl)ethylamino]-1,3,7-trimethyl-2,6-dioxopurine; N^1-(1,3,7-trimethyl-2,6-dioxopurin-8-yl)-N^2-(1-methyl)phenethyl-N^2-methylethylenediamine; *N*-8-caffeyl-*N'*-methyl-*N'*-(α-methylphenethyl)ethylenediamine; phencamine; ST-374. $C_{20}H_{28}N_6O_2$; mol wt 384.48. C 62.48%, H 7.34%, N 21.86%, O 8.32%. Prepn: **ES 347509**; **ES 352077** (both 1969 to Labs. Miquel); Pitarch *et al.*, *Quim. Ind. (Madrid)* **17**, 71 (1971). Pharmacology: *eidem, ibid.* 76. Colorimetric and TLC determn in urine: J. Mallol *et al.*, *Arzneim.-Forsch.* **24**, 1301 (1974). Toxicity

data: E. Usdin, D. H. Efron, *Psychotropic Drugs and Related Compounds* (National Institute of Mental Health, Rockville, MD, 2nd ed., 1972) p 229. Clinical trial in depression: P. G. Quiros e Isla, *Rev. Clin. Esp.* **119**, 437 (1970).

Crystals from methanol, mp 150-152°.
Hydrochloride. [63918-50-3] Altimina; Sicoclor. $C_{20}H_{28}N_6$-O_2.HCl; mol wt 420.94. Slightly hygroscopic, bitter crystals, decomp 278-279°. uv max (water): 296 nm ($E_{1cm}^{1\%}$ 398). Sol in water; slightly sol in many organic solvents. LD_{50} in rats, mice (mg/kg): 93, 82 i.p.; 508, 418 orally (Usdin, Efron).

THERAP CAT: Analeptic.

4000. *d*-Fenchone. [4695-62-9] (1*S*,4*R*)-1,3,3-Trimethylbicyclo[2.2.1]heptan-2-one; *d*-1,3,3-trimethyl-2-norbornanone; *d*-1,3,3-trimethyl-2-norcamphanone. $C_{10}H_{16}O$; mol wt 152.24. C 78.90%, H 10.59%, O 10.51%. Occurs in fennel oil and in the essential oil of *Lavandula stoechas* L., *Labiatae*. Isoln: Wallach, *Ann.* **263**, 129 (1891); **353**, 209 (1907); **369**, 63 (1909); Shavrygin, *J. Appl. Chem. USSR* **12**, 1201 (1939). Total synthesis: Boyle *et al.*, *Chem. Commun.* **1971**, 395; G. Buchbauer, H. C. Rohner, *Ann.* **1981**, 2093. Toxicity study: P. M. Jenner *et al.*, *Food Cosmet. Toxicol.* **2**, 327 (1964). *Review:* Simonsen, *The Terpenes* vol. II (Cambridge, 2nd ed., 1949) pp 560-580; D. L. J. Opdyke, *Food Cosmet. Toxicol.* **14**, Suppl., 769-771 (1976).

Oily liq. Camphor-like odor. d_4^{18} 0.948. mp 6.1°. bp_{760} 193.5°; bp_{100} 122°; bp_{20} 82°; bp_{15} 66°. $[\alpha]_D^{20}$ +66.9°. n_D^{18} 1.4636. Practically insol in water (pH of satd soln 6.82). Very sol in abs alcohol, ether. LD_{50} orally in rats: 6.16 g/kg (Jenner).

USE: As flavor in foods; in perfumes.

THERAP CAT: Counterirritant.

4001. Fenclozic Acid. [17969-20-9] 2-(4-Chlorophenyl)-4-thiazoleacetic acid; 2-(*p*-chlorophenyl)thiazol-4-ylacetic acid; acidum fenclozicum; ICI-54450; Myalex. $C_{11}H_8ClNO_2S$; mol wt 253.70. C 52.08%, H 3.18%, Cl 13.97%, N 5.52%, O 12.61%, S 12.64%. Prepn: W. Hepworth, G. J. Stacey, **NL 6614130**; *eidem*, **US 3538107** (1967, 1970 both to ICI); Aries, **FR 1561433** (1969), *C.A.* **72**, 43654y (1970). Pharmacology: W. Hepworth *et al.*, *Nature* **221**, 582 (1969); Newbould, *Br. J. Pharmacol.* **35**, 487 (1969). Metabolic studies: Foulkes, *J. Pharmacol. Exp. Ther.* **172**, 115, 449 (1970). Clinical evaluation: Chalmers *et al.*, *Ann. Rheum. Dis.* **28**, 590, 595 (1969).

Colorless crystalline solid from ethyl acetate, mp 155-156°. Soluble in most organic solvents; sparingly sol in water. LD_{50} in rats, mice (mg/kg): 850, 1000 orally; 250, 300 i.v. (Hepworth).

THERAP CAT: Anti-inflammatory.

4002. Fendiline. [13042-18-7] γ-Phenyl-*N*-(1-phenylethyl)-benzenepropanamine; *N*-(3,3-diphenylpropyl)-α-methylbenzylamine; *N*-(1-phenylethyl)-3,3-diphenylpropylamine. $C_{23}H_{25}N$; mol wt 315.46. C 87.57%, H 7.99%, N 4.44%. Calcium blocking agent. Prepn: **BE 621300**; Harsányi *et al.*, **US 3262977** (1962, 1966 both to Chinoin); *eidem*, *J. Med. Chem.* **7**, 623 (1964); Klosa, *J. Prakt. Chem.* **34**, 312 (1966). Prepn and studies of the labelled compound: Volford, Harsányi, *J. Labelled Compd.* **9**, 219 (1973). Structure-activity studies: Leszkovszky *et al.*, *Acta Physiol. Acad. Sci. Hung.* **29**, 283 (1966). Pharmacologic properties: W. R. Kukovetz *et al.*, *Arzneim.-Forsch.* **26**, 1321 (1976); A. Fleckenstein *et al.*, *ibid.* **27**, 562 (1977). Pharmacokinetics and tolerance: R. Weyhenmeyer *et al.*, *ibid.* **37**, 58 (1987). Use in ischemic heart disease: Z. Antaloczy, I. Preda, *Ther. Hung.* **27**, 71 (1979). Assessment of Ca^{2+}-antagonist effects: M. Spedding, *Arch. Pharmacol.* **318**, 234 (1982).

bp$_{0.3}$ 206-210°.

Hydrochloride. [13636-18-5] HK-137; Cordan; Fendilar; Sensit. $C_{23}H_{25}N \cdot HCl$; mol wt 351.92. Almost white or slightly pink powder, mp 204-205°. Very slightly sol in water; easily sol in methanol, ethanol, chloroform. LD$_{50}$ in mice (mg/kg): 14.5 i.v.; 950 orally (Harsányi).

THERAP CAT: Vasodilator (coronary).

4003. Fenethylline. [3736-08-1] 3,7-Dihydro-1,3-dimethyl-7-[2-[(1-methyl-2-phenylethyl)amino]ethyl]-1*H*-purine-2,6-dione; 7-[2-[(α-methylphenethyl)amino]ethyl]theophylline; 7-[β-(α-methyl-β-phenylethylamino)ethyl]theophylline; 7-(phenylisopropylaminoethyl)theophylline; 7-(3-phenyl-2-propylaminoethyl)theophylline; theophyllineethylamphetamine. $C_{18}H_{23}N_5O_2$; mol wt 341.42. C 63.32%, H 6.79%, N 20.51%, O 9.37%. Prepn from 7-(β-chloroethyl)theophylline and amphetamine: Kholstaedt, Klinger, **DE 1123329** (1962 to Degussa), *C.A.* **57**, 5933c (1962). Metabolism in man: T. Ellison *et al.*, *Eur. J. Pharmacol.* **13**, 123 (1970).

Hydrochloride. [1892-80-4] H-814; Captagon. $C_{18}H_{23}N_5O_2 \cdot HCl$; mol wt 377.87. Crystals, two different modifications, mp 227-229° and mp 237-239°.

d-**Form hydrochloride.** Crystals, mp 246-247°.

l-**Form hydrochloride.** Crystals, mp 246-247°.

Note: This is a controlled substance (stimulant): **21 CFR**, 1308.11.

THERAP CAT: CNS stimulant.

4004. Fenfluramine. [458-24-2] *N*-Ethyl-α-methyl-3-(trifluoromethyl)benzeneethanamine; *N*-ethyl-α-methyl-*m*-(trifluoromethyl)phenethylamine; 2-ethylamino-1-(3-trifluoromethylphenyl)-propane; S-768. $C_{12}H_{16}F_3N$; mol wt 231.26. C 62.32%, H 6.97%, F 24.65%, N 6.06%. Prepn: L. G. Beregi *et al.*, **FR M1658**; *eidem*, **US 3198833** (1963, 1965 both to Sci. Union et Cie Soc. Franc. Recherche Méd.). Prepn of optical isomers: *eidem*, **US 3198834** (1965 to Sci. Union et Cie Soc. Franc. Recherche Med.). Pharmacology: *Presse Med.* **71**, 181 (1963). Pharmacology and toxicity of isomers and racemate: J. C. Le Douarec *et al.*, *Arch. Int. Pharmacodyn. Ther.* **161**, 206 (1966). Pharmacokinetics: S. Caccia *et al.*, *Eur. J. Clin. Pharmacol.* **29**, 221 (1985). Clinical trial of dextrofenfluramine in refractory obesity: N. Finer *et al.*, *Curr. Ther. Res.* **38**, 847 (1985). Comprehensive review: Pinder *et al.*, *Drugs* **10**, 241-323 (1975).

bp$_{12}$ 108-112°. LD$_{50}$ i.p. in mice: 144 mg/kg (**US 3198833**).

Hydrochloride. [404-82-0] Acino; Adipomin; Obedrex; Pesos; Ponderal; Ponderax; Ponderex; Pondimin; Rotondin. $C_{12}H_{16}F_3N \cdot HCl$; mol wt 267.72. Crystals from ethanol + ether, mp 166°.

d-**Form.** [3239-44-9] Dexfenfluramine; dextrofenfluramine. $[\alpha]_D^{25}$ +9.5° (c = 8 in ethanol). LD$_{50}$ orally in rats: 114.6 mg/kg (Le Douarec).

d-**Form hydrochloride.** [3239-45-0] Adifax; Glypolix; Isomeride; Redux. Crystals from ethyl acetate, mp 160-161°.

l-**Form.** [37577-24-5] $[\alpha]_D^{25}$ −9.6° (c = 8 in ethanol). LD$_{50}$ orally in rats: 195 mg/kg (Le Douarec).

l-**Form hydrochloride.** [3616-78-2] Crystals from ethyl acetate, mp 160-161°.

Note: This is a controlled substance: **21 CFR**, 1308.14.

THERAP CAT: Anorexic.

4005. Fenhexamid. [126833-17-8] *N*-(2,3-Dichloro-4-hydroxyphenyl)-1-methylcyclohexanecarboxamide; 1-methyl-cyclohexanecarboxylic acid (2,3-dichloro-4-hydroxy-phenyl)-amide; KBR-2738; Decree; Elevate. $C_{14}H_{17}Cl_2NO_2$; mol wt 302.20. C 55.64%, H 5.67%, Cl 23.46%, N 4.64%, O 10.59%. Hydroxyanilide fungicide. Prepn: B.-W. Krüger *et al.*, **EP 339418**; *eidem*, **US 5059623** (1989, 1991 both to Bayer). Review of properties and activities: H.-J. Rosslenbroich *et al.*, *Brighton Crop Prot. Conf. - Pests Dis.* **1998**, 327-334; of field trials on fruit crops: N. M. Adam, P. A. Birch, *ibid.* 849-856.

mp 141°. Soly in water at 20°: 20 mg/l (pH 5-7). Vapor pressure at 20°: 4×10^{-7} Pa (extrapolated). Log P (*n*-octanol/water) at 20°: 3.51 (pH 7). LD$_{50}$ in rats (mg/kg): >5000 orally; >5000 dermally (24 hr). LC$_{50}$ (4hr) in rats (mg/m^3): >5057 by inhalation. (Rosslenbroich).

USE: Fungicide.

4006. Fenipentol. [583-03-9] α-Butylbenzenemethanol; α-butylbenzyl alcohol; phenylbutylcarbinol; 1-phenyl-1-hydroxypentane; phenylpentanol; PC 1; Ph BC; Pancoral. $C_{11}H_{16}O$; mol wt 164.25. C 80.44%, H 9.82%, O 9.74%. Synthesis: Fourneau, Puyal, *An. Soc. Esp. Fis. Quim.* **18**, 323 (1920); Adams, Vander Werf, *J. Am. Chem. Soc.* **72**, 4368 (1950); Protiva *et al.*, *Chem. Listy* **46**, 37 (1952). Prepn of the drug: Scheffler, Engelhorn, **GB 915815**; *eidem*, **US 3084100** (both 1963 to Thomae). Pharmacology: Engelhorn, *Arzneim.-Forsch.* **10**, 255 (1960); Beck, Bierwisch, *ibid.* **20**, 693 (1970). Metabolism: Koss *et al.*, *ibid.* **14**, 195 (1964).

Colorless or slightly yellow liquid. bp$_{12}$ 123-124°. n_D^{20} 1.5112. Miscible with organic liquids. Practically insol in water.

THERAP CAT: Choleretic. Also used in treatment of mild chronic pancreatitis.

4007. Fenitrothion. [122-14-5] Phosphorothioic acid *O,O*-dimethyl *O*-(3-methyl-4-nitrophenyl) ester; *O,O*-dimethyl *O*-4-nitro-

m-tolyl phosphorothioate; *O,O*-dimethyl *O*-(3-methyl-4-nitrophenyl) phosphorothioate; *O,O*-dimethyl *O*-4-nitro-*m*-tolyl thiophosphate; MEP; metathion; Bayer 41831; Bayer S 5660; ENT-25715; OMS-45; AC-47300; Sumithion. $C_9H_{12}NO_5PS$; mol wt 277.23. C 38.99%, H 4.36%, N 5.05%, O 28.86%, P 11.17%, S 11.56%. Organophosphate insecticide; cholinesterase inhibitor. Prepn: **BE 594669** (1960 to Sumitomo); **BE 596091** (1960 to Bayer). Activity: Y. Nishizawa *et al.*, *Agric. Biol. Chem.* **25**, 605 (1961). Toxicity study: G. Schrader, *Angew. Chem.* **73**, 331 (1961). HPLC determn in water: A. Sánchez-Ortega *et al.*, *J. Chromatogr. A* **1094**, 70 (2005). GC determn in stored barley: U. Uygun *et al.*, *Food Chem.* **100**, 1165 (2007).

Yellow oil. bp$_{0.05}$ 118°. n_D^{25} 1.5528. d_4^{25} 1.3227. Log P (octanol/water): 3.43. Vapor press at 20°: 6×10^{-6} mm Hg. uv max: 269.5 nm (ε 6756). Practically insol in water. Low soly in aliphatic hydrocarbons; sol in most organic solvents. LD$_{50}$ orally in rats: 250 mg/kg (Schrader).

USE: Agricultural insecticide.

4008. Fennel. Large fennel; sweet fennel. Dried, ripe fruit of cultivated varieties of *Foeniculum vulgare* Mill., *Umbelliferae. Habit.* Southern Europe, Western Asia, widely cultivated. *Constit.* 2-6% volatile oil, fixed oil, protein, organic acids and flavonoids. Clinical trial of fennel oil emulsion in infantile colic: I. Alexandrovich *et al.*, *Altern. Ther. Health Med.* **9**, 58 (2003). *Review:* F. S. D'Amelio, Sr., *Botanicals* (CRC Press, Boca Raton, 1999) pp 105-106.

Volatile oil. [8006-84-6] Oil of fennel. *Constit.* 50-70% *trans*-anethole, up to 20% (+)-fenchone, methylchavicol, anisaldehyde, terpenoids including α-pinene, limonene, α-phellandrene. Colorless or pale yellow liquid with characteristic odor of fennel. d_{25}^{25} 0.953-0.973. α_D^{25} +12 to +24°. n_D^{20} 1.532-1.543. Slightly sol in water; sol in 1 vol 90% or in 8 vols 80% alcohol; very sol in chloroform, ether. *Keep well closed, cool and protected from light.*

USE: Flavoring agent in foods. Pharmaceutic aid (flavor). Used in eye washes and in facial steams to soothe and clean skin.

THERAP CAT: Carminative.

4009. Fenofibrate. [49562-28-9] 2-[4-(4-Chlorobenzoyl)-phenoxy]-2-methylpropanoic acid 1-methylethyl ester; isopropyl [4'-(*p*-chlorobenzoyl)-2-phenoxy-2-methyl]propionate; fenofibric acid isopropyl ester; procetofen; procetofene; LF-178; Lipanthyl; Lipantil; Lipidil; Lipoclar; Lipofene; Lipsin; Nolipax; Secalip; Supralip; Tricor. $C_{20}H_{21}ClO_4$; mol wt 360.83. C 66.57%, H 5.87%, Cl 9.82%, O 17.74%. Lipid lowering prodrug metabolized *in vivo* to fenofibric acid. Prepn: A. Mieville, **DE 2250327** (1973 to Fournier); *idem*, **US 4058552** (1977 to Orchimed). Series of articles on synthesis, metabolism, pharmacology and clinical trials: *Arzneim.-Forsch.* **26**, 885-909 (1976). Toxicity data: R. Sornay *et al.*, *ibid.* 885. Determn of active metabolite in plasma and urine by GC-MS: L. F. Elsom *et al.*, *J. Chromatogr.* **123**, 463 (1976); in plasma by LC: B. Streel *et al.*, *J. Chromatogr. B* **742**, 391 (2000). Mechanism of action: W. Wülfert *et al.*, *Artery* **9**, 120 (1981). Review of pharmacology and clinical experience in dyslipidemia, metabolic syndrome and diabetes: G. M. Keating, K. F. Croom, *Drugs* **67**, 121-153 (2007).

White crystals from isopropanol, mp 80-81°. Very soluble in methylene chloride; sol in acetone, ether, benzene, chloroform;

slightly sol in methanol, ethanol. Practically insol in water. LD$_{50}$ in mice: 1600 mg/kg orally (Sornay).

Free acid. [42017-89-0] Fenofibric acid; procetofenic acid; LF-153. $C_{17}H_{15}ClO_4$; mol wt 318.75. Crystals from toluene, mp 185°.

Fenofibric acid choline salt. [856676-23-8] 2-Hydroxy-*N,N,N*-trimethylethanaminium 2-[4-(4-chlorobenzoyl)phenoxy]-2-methyl-propanoate (1:1); choline fenofibrate; ABT-335; Trilipix. $C_{17}H_{14}$-$ClO_4 \cdot C_5H_{14}NO$; mol wt 421.92. Prepn: R. D. Cink *et al.*, **US 7259186** (2007 to Abbott). Clinical evaluation in combination with simvastatin: S. M. Mohiuddin *et al.*, *Am. Heart J.* **157**, 195 (2009). Review of clinical trials in patients with mixed dyslipidemia and diabetes: P. H. Jones *et al.*, *Am. J. Cardiovasc. Drugs* **10**, 73-84 (2010). White to yellow powder, mp ~210°. Freely sol in water.

THERAP CAT: Antilipemic.

4010. Fenoldopam. [67227-56-9] 6-Chloro-2,3,4,5-tetrahydro-1-(4-hydroxyphenyl)-1*H*-3-benzazepine-7,8-diol; 6-chloro-7,8-dihydroxy-1-*p*-hydroxyphenyl-2,3,4,5-tetrahydro-1*H*-3-benzaze-pine; SKF-82526. $C_{16}H_{16}ClNO_3$; mol wt 305.76. C 62.85%, H 5.27%, Cl 11.59%, N 4.58%, O 15.70%. Dopamine D$_1$-receptor agonist. Prepn: J. Weinstock, **DE 2751258**; *idem*, **US 4197297** (1978, 1980 both to SmithKline); J. Weinstock *et al.*, *J. Med. Chem.* **23**, 973 (1980). HPLC determn in urine and plasma: V. K. Boppana *et al.*, *J. Chromatogr.* **317**, 463 (1984). Clinical pharmacology: R. M. Stote *et al.*, *Clin. Pharmacol. Ther.* **34**, 309 (1983); R. M. Carey *et al.*, *J. Clin. Invest.* **74**, 2198 (1984). Hemodynamic effects in hypertension: H. O. Ventura *et al.*, *Circulation* **69**, 1142 (1984); M. P. Caruana *et al.*, *Br. J. Clin. Pharmacol.* **24**, 721 (1987). Clinical evaluation in congestive heart failure: G. S. Francis *et al.*, *Am. Heart J.* **116**, 473 (1988).

Hydrobromide. $C_{16}H_{16}ClNO_3 \cdot HBr$. mp 277° (dec).

Methanesulfonate. [67227-57-0] Fenoldopam mesylate; SKF-82526-J; Corlopam. $C_{16}H_{16}ClNO_3 \cdot CH_3SO_3H$; mol wt 401.86. White to off-white powder. mp 274° (dec). Soluble in water.

THERAP CAT: Antihypertensive.

4011. Fenoprofen. [29679-58-1] α-Methyl-3-phenoxybenzeneacetic acid; (±)-*m*-phenoxyhydratropic acid; α-*dl*-2-(3-phenoxyphenyl)propionic acid; Lilly 53838. $C_{15}H_{14}O_3$; mol wt 242.27. C 74.37%, H 5.82%, O 19.81%. Prepn: W. S. Marshall, **FR 2015728**; *idem*, **US 3600437** (1970, 1971 both to Lilly). Pharmacology: Rubin *et al.*, *J. Pharm. Sci.* **60**, 1797 (1971); **61**, 800 (1972); Herrmann, *Proc. Soc. Exp. Biol. Med.* **139**, 548 (1972). Metabolism: Rubin *et al.*, *J. Pharmacol. Exp. Ther.* **183**, 449 (1972). Toxicology: J. L. Emmerson *et al.*, *Toxicol. Appl. Pharmacol.* **25**, 444 (1973). Comprehensive description: C. K. Ward, R. E. Schirmer, *Anal. Profiles Drug Subs.* **6**, 161-182 (1977). *Review:* R. N. Brogden *et al.*, *Drugs* **13**, 241-265 (1977); R. Nickander *et al.*, in *Pharmacological and Biochemical Properties of Drug Substances* **vol. 1**, M. E. Goldberg, Ed. (Am. Pharm. Assoc., Washington, DC, 1977) pp 183-213.

Viscous oil, bp$_{0.11}$ 168-171°. n_D^{25} 1.5742. pKa 7.3.

Calcium salt dihydrate. [53746-45-5] Lilly 69323; Fenopron; Fepron; Feprona; Nalfon; Nalgesic; Progesic. $C_{30}H_{26}CaO_6 \cdot 2H_2O$; mol wt 558.64. White crystalline powder. Soly in mg/ml at 37°: *n*-

hexanol 11; methanol 8; water 2.5; chloroform 0.01. pKa 4.5. Aq solns sensitive to intense uv light. LD_{50} orally in mice: 800 mg/kg (Emmerson).

THERAP CAT: Anti-inflammatory; analgesic.

4012. Fenoterol. [13392-18-2] 5-[1-Hydroxy-2-[[2-(4-hydroxyphenyl)-1-methylethyl]amino]ethyl]-1,3-benzenediol; 3,5-dihydroxy-α-[[(p-hydroxy-α-methylphenethyl)amino]methyl]benzyl alcohol; 1-(3,5-dihydroxyphenyl)-1-hydroxy-2-[(4-hydroxyphenyl)-isopropylamino]ethane; 1-(p-hydroxyphenyl)-2-[[β-hydroxy-β-(3′,5′-dihydroxyphenyl)ethyl]aminopropane; TH-1165. $C_{17}H_{21}NO_4$; mol wt 303.36. C 67.31%, H 6.98%, N 4.62%, O 21.10%. $β_2$-Adrenergic agonist. Prepn and sepn of stereoisomers: **BE 640433**; Zelle *et al.*, **US 3341593** (1962, 1967 to Boehringer, Ing.). Pharmacology: Schuster, Baum, *Arzneim.-Forsch.* **19**, 1905 (1969); O'Donnell, *Eur. J. Pharmacol.* **12**, 35 (1970). Toxicity: E. I. Goldenthal, *Toxicol. Appl. Pharmacol.* **18**, 185 (1971). Metabolism in mice: S. Kojima *et al.*, *Arzneim.-Forsch.* **30**, 959 (1980). HPLC determn in plasma: S. Kramer, G. Blaschke, *J. Chromatogr. B* **751**, 169 (2001). Clinical evaluation: Tweel, *Ann. Allergy* **29**, 142 (1971); Rebuck, Saunders, *Med. J. Aust.* **1**, 225 (1972). Comparison with ritodrine in preterm labor: J. Gerris *et al.*, *Eur. J. Clin. Pharmacol.* **18**, 443 (1980). Review of pharmacology and therapeutic efficacy: R. C. Heel *et al.*, *Drugs* **15**, 3-32 (1978); N. Svedmyr, *Pharmacotherapy* **5**, 109-126 (1985).

Hydrobromide. [1944-12-3] TH-1165a; Airum; Berotec; Dosberotec; Partusisten. $C_{17}H_{21}NO_4.HBr$; mol wt 384.27. Crystals from methanol-ether, mp 222-223°. LD_{50} in mice (mg/kg): 1100 s.c.; 1990 orally (Goldenthal).
Hydrochloride. [1944-10-1] $C_{17}H_{21}NO_4.HCl$. mp 183° (acetonitrile-ether).

THERAP CAT: Bronchodilator; tocolytic.

4013. Fenoverine. [37561-27-6] 2-[4-(1,3-Benzodioxol-5-ylmethyl)-1-piperazinyl]-1-(10*H*-phenothiazin-10-yl)ethanone; 10-[[4-(1,3-benzodioxol-5-ylmethyl)-1-piperazinyl]acetyl]-10*H*-phenothiazine; 10-[4-(3,4-dioxymethylenebenzyl)-1-piperazinylacetyl]-phenothiazine; 10-[(4-piperonyl-1-piperazinyl)acetyl]phenothiazine; Spasmopriv. $C_{26}H_{25}N_3O_3S$; mol wt 459.56. C 67.95%, H 5.48%, N 9.14%, O 10.44%, S 6.98%. Piperonylpiperazine derivative with spasmolytic activity. Prepn: A. Buzas, R. Pierre, **FR 2092639** (1972), *C.A.* **77**, 140143p (1972). Pharmacology, clinical studies in gastrointestinal disorders, dysmenorrhea: R. Pierre, R. Roustan, *Med. Interne* **15**, 49 (1980). In treatment of colon dysfunction: Claudon, *Rev. Fr. Gastro-Enterol.* **198**, 31 (1984).

Crystals from isopropyl ether, mp 141-142°. LD_{50} in mice (g/kg): ~1.50 orally, ~2.50 i.p. (Buzas, Pierre).

THERAP CAT: Antispasmodic.

4014. Fenoxanil. [115852-48-7] *N*-(1-Cyano-1,2-dimethylpropyl)-2-(2,4-dichlorophenoxy)propanamide; AC-382042; NNF-9425; Achieve. $C_{15}H_{18}Cl_2N_2O_2$; mol wt 329.22. C 54.72%, H 5.51%, Cl 21.54%, N 8.51%, O 9.72%. Melanin biosynthesis inhibitor for control of rice blast fungus. Commercial product consists of a mixture of 4 isomers: 85% (1*R*,2*RS*) and 15% (1*S*,2*RS*). Prepn (stereochem unspec): W. Buck, E. Raddatz, **EP 262393** (1987 to

Shell). Review of properties and field trials: E. Sieverding *et al.*, *Brighton Crop Prot. Conf. - Pests Dis.* **1998**, 359-366.

Off white, odorless solid, mp 69.0-71.5°. d^{20} 1.22 g/cm³. Partition coefficient (*n*-octanol/water): 3390 ±133 (25°). Soly in water (20°): 30.7 ±0.3 × 10^{-3} g/l. Sol in most organic solvents. Vapor pressure (25°): 0.21 ±0.021 × 10^{-4} Pa. LD_{50} orally in mice, male rats, female rats (mg/kg): >5000, >5000, 4211 (Sieverding).

USE: Agricultural fungicide.

4015. Fenoxaprop-ethyl. [66441-23-4] 2-[4-[(6-Chloro-2-benzoxazolyl)oxy]phenoxy]propanoic acid ethyl ester; HOE-33171. $C_{18}H_{16}ClNO_5$; mol wt 361.78. C 59.76%, H 4.46%, Cl 9.80%, N 3.87%, O 22.11%. Selective postemergent herbicide to control grassy weeds in broadleaved crops and established turfgrass. Prepn: **BE 858618**; R. Handte *et al.*, **US 4130413** (both 1978 to Hoechst). Metabolism in soybeans: O. Wink *et al.*, *J. Agric. Food Chem.* **32**, 187 (1984). Persistence in soil: A. E. Smith, *ibid.* **33**, 483 (1985). Field trials in agricultural crops: G. W. Kerse *et al.*, *Proc. 36th N. Z. Weed Pest Control Conf.* 265 (1983); in turfgrass: P. H. Dernoeden, J. D. Fry, *Proc. Annu. Meet. Northeast. Weed Sci. Soc.* **39**, 282 (1985). Brief description: J. Bieringer *et al.*, *Proc. Br. Crop Prot. Conf. - Weeds* **1982**, 11-17.

(*R*)-Form

mp 84-85°. bp 200° at 100 Pa. Vapor pressure at 20°: 0.19×10^{-5} Pa. Soly at 20° (%): >0.5 in hexane; >1 in cyclohexane, ethanol, 1-octanol; >20 in ethyl acetate; >30 in toluene; >50 in acetone. Soly in water at 25°: 0.9 mg/l. LD_{50} in male, female rats (mg/kg): 2357, 2500 orally; 739, 864 i.p. (Bieringer).

(*R*)-Form. [71283-80-2] Fenoxaprop-P-ethyl; HOE-46360; Acclaim; Furore Super; Puma; Ricestar; Whip. Comprehensive description: H. P. Huff *et al.*, *Brighton Crop Prot. Conf. - Weeds* **1989**, 717-722. LD_{50} in male, female rats (mg/kg): 3040, 2090 orally; >2000, >2000 dermally (Huff).

USE: Postemergent herbicide.

4016. Fenoxazoline. [4846-91-7] 4,5-Dihydro-2-[[2-(1-methylethyl)phenoxy]methyl]-1*H*-imidazole; 2-[(*o*-cumenyloxy)-methyl]-2-imidazoline; 2-(*o*-isopropylphenoxymethyl)-2-imidazoline; phenoxazoline. $C_{13}H_{18}N_2O$; mol wt 218.30. C 71.53%, H 8.31%, N 12.83%, O 7.33%. Sympathomimetic with vasoconstrictor properties. Preparation: **FR 1365971** (1964 to Lab. Dausse and Soc. BMC). Use as sympathomimetic agent: Giudicelli, **US 3198703** (1965 to Lab. Dausse).

Hydrochloride. [21370-21-8] Aturgyl; Nasofelin; Nebulicina. $C_{13}H_{18}N_2O.HCl$; mol wt 254.76. Crystals, mp 174°. Sol in water, ethanol.

THERAP CAT: Nasal decongestant.

4017. Fenoxycarb. [72490-01-8] *N*-[2-(4-Phenoxyphenoxy)-ethyl]carbamic acid ethyl ester; ethyl [2-(4-phenoxyphenoxy)ethyl]-

carbamate; ABG-6215; Ro-13-5223; Comply; Insegar. $C_{17}H_{19}NO_4$; mol wt 301.34. C 67.76%, H 6.36%, N 4.65%, O 21.24%. Insect growth regulator. Prepn: U. Fischer *et al.*, **EP 4334**; *eidem*, **US 4215139** (1979, 1980 both to Hoffmann-La Roche). Insecticidal properties: S. Dorn *et al.*, *Z. Pflanzenkrankh. Pflanzenschutz* **88**, 269 (1981). Field trials, persistence in water, and bioaccumulation in fish: C. H. Schaefer *et al.*, *J. Econ. Entomol.* **80**, 126 (1987).

Crystals from petr ether, mp 53-54°. Soly in water (23°): 5.76 ppm. LD_{50} in rats (mg/kg): 9220 i.p.; 16800 orally (Dorn).
USE: Insecticide.

4018. Fenpiclonil. [74738-17-3] 4-(2,3-Dichlorophenyl)-1*H*-pyrrole-3-carbonitrile; 3-(2,3-dichlorophenyl)-4-cyanopyrrole; 4-cyano-3-(2,3-dichlorophenyl)pyrrole; CGA-142705; Beret; Galbas; Gambit. $C_{11}H_6Cl_2N_2$; mol wt 237.08. C 55.73%, H 2.55%, Cl 29.91%, N 11.82%. Phenylpyrrole fungicide for seed treatment; structurally related to pyrrolnitrin, *q.v.* Prepd (not claimed): K. Ohkuma *et al.*, **GB 2024824**; *eidem*, **US 4229465** (both 1980 to Nippon Soda). Process: P. Martin, **EP 174910**; *idem*, **US 4812580** (1986, 1989 both to Ciba-Geigy). Comprehensive description: D. Nevill *et al.*, *Brighton Crop Prot. Conf. - Pests Dis.* **1988**, 65-72. Field trials in potatoes: A. J. Leadbeater, W. W. Kirk, *ibid.* **1992**, 657. Mechanism of action study: A. B. K. Jespers, M. A. De Waard, *Pestic. Biochem. Physiol.* **49**, 53 (1994).

Colorless, odorless crystals, mp 152.9°. Soly in water (20°): 2 ppm. Log P (*n*-octanol/water): 4.3. LD_{50} in rats, mice, rabbits (mg/kg): >5000 orally. LD_{50} dermally in rats: >2000 mg/kg. LC_{50} (4 hr) in rats: >1502 mg/m³ by inhalation (Nevill).
USE: Agricultural fungicide.

4019. Fenpropathrin. [39515-41-8] 2,2,3,3-Tetramethylcyclopropanecarboxylic acid cyano(3-phenoxyphenyl)methyl ester; α-cyano-3-phenoxybenzyl 2,2,3,3-tetramethylcyclopropanecarboxylate; fenpropanate; S-3206; SD-41706; WL-41706; Danitol; Meothrin; Rody. $C_{22}H_{23}NO_3$; mol wt 349.43. C 75.62%, H 6.63%, N 4.01%, O 13.74%. Synthetic pyrethroid insecticide with repellant and contact activity. Prepn: T. Matsuo *et al.*, **DE 2231312**; *eidem*, **US 3835176** (1973, 1974 to Sumitomo). Commercial product is mixture of stereoisomers, the (*S*)-isomer primarily responsible for the bioactivity. Separation of enantiomers by chiral phase HPLC: R. A. Chapman, *J. Chromatogr.* **258**, 175 (1983). Pesticidal activity: M. H. Breese, *Pestic. Sci.* **8**, 264 (1977). Metabolism: M. J. Crawford, D. H. Hutson, *ibid.* 579. Degradn in soil: T. R. Roberts, M. E. Standen, *ibid.* 600; R. A. Chapman, *Bull. Environ. Contam. Toxicol.* **26**, 513 (1981). Fish toxicity: J. R. Coats, *ibid.* **23**, 250 (1979). Mammalian toxicity study: R. D. Verschoyle, W. N. Aldridge, *Arch. Toxicol.* **45**, 325 (1980). Review of toxicology and human exposure: *Toxicological Profile for Pyrethrins and Pyrethroids* (PB2004-100004, 2003) 332 pp.

Pale yellow oil, n_D^{26} 1.5283. LC_{50} (24 hr) in rainbow trout: 76.7 ppb (Coats). LD_{50} in rats (mg/kg): 2.5 i.v. (Verschoyle); in male, female rats (mg/kg): 24-36, 18-24 orally (Crawford).
USE: Insecticide, acaricide.

4020. Fenpropidin. [67306-00-7] 1-[3-[4-(1,1-Dimethylethyl)phenyl]-2-methylpropyl]piperidine; (*RS*)-1-[3-(4-*tert*-butylphenyl)-2-methylpropyl]piperidine; Tern. $C_{19}H_{31}N$; mol wt 273.46. C 83.45%, H 11.43%, N 5.12%. Piperidine fungicide active against powdery mildews of grain. Prepn: **BE 861002**; A. Pfiffner, **US 4241058** (1978, 1980 both to Hoffmann-La Roche). Improved process: N. Goetz, L. Hupfer, **EP 17893**; *eidem*, **US 4283534** (1980, 1981 both to BASF). Structure-activity relationships: W. Himmele, E.-H. Pommer, *Angew. Chem. Int. Ed.* **19**, 184 (1980). Inhibition of ergosterol biosynthesis: R. I. Baloch *et al.*, *Phytochemistry* **23**, 2219 (1984).

Oil, $bp_{0.2}$ 117°; $bp_{0.045}$ 125°; also reported as $bp_{0.032}$ 104°.
USE: Agricultural fungicide.

4021. Fenpropimorph. [67564-91-4] *rel*-(2*R*,6*S*)-4-[3-[4-(1,1-Dimethylethyl)phenyl]-2-methylpropyl]-2,6-dimethylmorpholine; BAS-42100F; Ro-14-3169/000; Corbel; Mistral. $C_{20}H_{33}NO$; mol wt 303.49. C 79.15%, H 10.96%, N 4.62%, O 5.27%. Systemic fungicide for control of powdery mildew, rust in cereal crops. Prepn, fungicidal activity: W. Himmele *et al.*, **DE 2656747** (1978 to BASF), *C.A.* **89**, 109522k (1978). Synthesis using an ionic liquid solvent: S. A. Forsyth *et al.*, *Org. Process Res. Dev.* **10**, 94 (2006). Field trials in control of cereal diseases: J. C. Atkin *et al.*, *Proc. Br. Crop Prot. Conf. - Pests Dis.* **1981**, 307. Inhibition of ergosterol biosynthesis: R. I. Baloch *et al.*, *Phytochemistry* **23**, 2219 (1984). Structure-activity relationships of 3-phenylpropylamines: W. Himmele, E.-H. Pommer, *Angew. Chem. Int. Ed.* **19**, 184 (1980); of substituted morpholines: E.-H. Pommer, *Pestic. Sci.* **15**, 285 (1984). Brief review: K. Bohnen *et al.*, *Proc. Br. Crop Prot. Conf. - Pests Dis.* **1979**, 541-548.

Relative stereochemistry

Liquid, $bp_{0.05}$ 120°. Soly in water: 1 gm/l; sol in most organic solvents. Vapor pressure at 20°: 2.5×10^{-5} mm Hg. LD_{50} in male, female rats (mg/kg): 3650, 3420 orally; 4200, 4380 dermally; in male, female mice (mg/kg): 1180, 1270 i.p. (Bohnen).
USE: Systemic fungicide.

4022. Fenproporex. [16397-28-7] 3-[(1-Methyl-2-phenylethyl)amino]propanenitrile; (±)-3-[(α-methylphenethyl)amino]propionitrile; (±)-*N*-2-cyanoethylamphetamine. $C_{12}H_{16}N_2$; mol wt 188.27. C 76.56%, H 8.57%, N 14.88%. Deriv of amphetamine, *q.v.* Prepn: **FR M4364**; P. Pohrbach, J. Blum, **US 3485924** (1966, 1969 both to Bottu). Pharmacological studies: B. M. Beecham *et al.*, *J. Pharm. Pharmacol.* **23**, 140 (1971); A. H. Beckett *et al.*, *ibid.* **24**, 194 (1972). Peripheral effects in human and rat adipose tissue: M. Dubost *et al.*, *Br. J. Pharmacol.* **58**, 436P (1976). Chromatographic identification of amphetamine in urine of patients treated with fenproporex: R. B. Sznelvar, *Eur. J. Toxicol. Environ. Hyg.* **8**, 5 (1975). Clinical trial: G. Hertel, W. Fallot-Burghardt, *Fortschr. Med.* **96**, 2380 (1978).

Liquid, bp_2 126-127°.

Hydrochloride. [18305-29-8] Gacilin; Solvolip. $C_{12}H_{16}N_2$·HCl; mol wt 224.73. White, cryst, odorless powder from abs ethanol, mp 146°. Bitter taste. Sol in water, 95% ethanol.

Note: This is a controlled substance (stimulant): **21 CFR,** 1308.14.

THERAP CAT: Anorexic.

4023. Fenprostalene. [69381-94-8] *rel*-(4*S*)-7-[(1*R*,2*R*,3*R*,- 5*S*)-3,5-Dihydroxy-2-[(1*E*,3*R*)-3-hydroxy-4-phenoxy-1-buten-1-yl]- cyclopentyl]-4,5-heptadienoic acid methyl ester; (±)-9α,11α,15α- trihydroxy-16-phenoxy-17,18,19,20-tetranorprosta-4,5,13-*trans*-tri- enoic acid methyl ester; RS-84043; Bovilene; Porcilene; Synchro- cept B. $C_{23}H_{30}O_6$; mol wt 402.49. C 68.64%, H 7.51%, O 23.85%. Synthetic analog of prostaglandin $F_{2α}$, related structurally to prosta- lene, *q.v.* Prepn: J. M. Muchowski, J. H. Fried, US 3985791; A. R. Van Horn *et al.,* US 4178457 (1976, 1979 both to Syntex). Effect on pregnancy in beagles: B. Vickery, G. McRae, *Biol. Reprod.* **22,** 438 (1980). Duration of action study: B. H. Vickery *et al., Prosta- glandins Med.* **5,** 93 (1980).

Relative stereochemistry

uv max (methanol): 220, 265, 271, 278 nm (log ε 3.99, 3.11, 3.23, 3.16).

THERAP CAT (VET): Luteolytic.

4024. Fenpyroximate. [134098-61-6] 4-[[[(*E*)-[(1,3-Dimeth- yl-5-phenoxy-1*H*-pyrazol-4-yl)methylene]amino]oxy]methyl]ben- zoic acid 1,1-dimethylethyl ester; *tert*-butyl (*E*)-α-(1,3-dimethyl-5- phenoxypyrazol-4-ylmethyleneaminooxy)-*p*-toluate; NNI-850; HOE-555-02A; FujiMite; Sequel. $C_{24}H_{27}N_3O_4$; mol wt 421.50. C 68.39%, H 6.46%, N 9.97%, O 15.18%. Phenoxypyrazole based acaricide; mitochondrial electron-transport inhibitor. Chemical, bi- ological properties and field evaluation: T. Konno *et al., Proc. Br. Crop Prot. Conf. - Pests Dis.* **1990,** 71. Mode of action: K. Motoba *et al., Pestic. Biochem. Physiol.* **43,** 37 (1992). LC determn in ap- ples: B. L. Halvorsen *et al., J. Chromatogr. A* **880,** 121 (2000). Control of spider mites: M. Van de Veire, D. Degheele, *Meded. Fac. Landbouwwet. Univ. Gent* **57,** 925 (1992). Metabolism in rats: H. Nishizawa *et al., J. Pestic. Sci.* **18,** 59 (1993). Soil degradation: Y. Izawa *et al., ibid.* 67.

White crystalline powder, mp 101.1-102.4°. Vapor pressure at 25°: 5.6×10^{-8} mm Hg. Soly at 20° (g/l): water 0.015×10^{-3}; at 25°: methanol 15; *n*-hexane 4; xylene 175. Log P (octanol/water) at 20°: 5.01. LD_{50} in male, female rats (mg/kg): 480, 245 orally; >2000, >2000 dermally. LC_{50} (48hr) in carp: 6.1 μg/l; LC_{50} (3hr) in *Daphnia pulex*: 85 μg/l (Konno).

USE: Acaricide.

4025. Fenretinide. [65646-68-6] *N*-(4-Hydroxyphenyl)retin- amide; 4-HPR; *all-trans-N*-4'-hydroxyretinanilide; 4-(*all-trans*-ret- inoyl)aminophenol; McN-R-1967. $C_{26}H_{33}NO_2$; mol wt 391.56. C 79.75%, H 8.50%, N 3.58%, O 8.17%. Synthetic analog of vitamin A, *q.v.* Prepn: R. J. Gander, J. A. Gurney, BE 847942; *eidem*, US 4190594 (1977, 1980 both to J & J); and pharmacology: R. C. Moon *et al., Cancer Res.* **39,** 1339 (1979). Large scale prepn and physical

properties: Y. F. Shealy *et al., J. Pharm. Sci.* **73,** 745 (1984). Clin- ical efficacy and safety in bladder cancer: A. Decensi *et al., Eur. J. Cancer Prev.* **3,** 377 (1994). Review of pharmacology and clinical evaluation in breast cancer prevention: *idem et al., Oncol. Rep.* **1,** 817-824 (1994); M. A. Cobleigh, *Leukemia* **8,** Suppl. 3, s59-s63 (1994); and oral leukoplakia: A. Costa *et al., Cancer Res.* **54,** Suppl., 2032s-2037s (1994).

Crystallizes in several polymorphic or solvated forms. Crystals from ethanol/water, mp 173-175° (Shealy). Crystals from chloro- form/hexanes, mp 162-163°; solvated form obtained by slow crystal- lization from methanol, mp 178-181° (Moon). uv max (chloroform): 370 nm (ε 44500). uv max (methanol): 362 nm (47900).

THERAP CAT: Antineoplastic.

4026. Fenspiride. [5053-06-5] 8-(2-Phenylethyl)-1-oxa-3,8- diazaspiro[4.5]decan-2-one; decaspiride; DESP. $C_{15}H_{20}N_2O_2$; mol wt 260.34. C 69.20%, H 7.74%, N 10.76%, O 12.29%. α-Adrener- gic blocker. Prepn: NL 6504602; Regnier *et al.,* US 3399192 (1965 and 1968, both to Sci. Union et Cie-Soc. Franc. Recherche Méd.). Pharmacology: LeDouarec *et al., Arzneim.-Forsch.* **19,** 1263 (1969); Duhault *et al., ibid.* **22,** 1947 (1972).

Hydrochloride. [5053-08-7] NAT-333; NDR-5998A; Decaspir; Fluiden; Pneumorel; Respiride; Tegencia; Viarespan. $C_{15}H_{20}N_2$- O_2·HCl; mol wt 296.80. Crystals decomp 232-233°. Soluble in wa- ter. LD_{50} i.v. in mice: 106 mg/kg; orally in rats: 437 mg/kg (Le- Douarec).

THERAP CAT: Bronchodilator.

4027. Fensulfothion. [115-90-2] Phosphorothioic acid *O,O*- diethyl *O*-[4-(methylsulfinyl)phenyl] ester; *O,O*-diethyl *O*-[*p*-(meth- ylsulfinyl)phenyl] phosphorothioate; DMSP; Bay 25141. $C_{11}H_{17}$- O_4PS_2; mol wt 308.35. C 42.85%, H 5.56%, O 20.75%, P 10.05%, S 20.79%. Organophosphate insecticide; cholinesterase inhibitor. Prepn: Homeyer, Schrader, BE 666012 (1965 to Bayer), *C.A.* **64,** 20555f (1966). Acute toxicity and anticholinesterase activity: K. P. DuBois, F. Kinoshita, *Toxicol. Appl. Pharmacol.* **6,** 78 (1964). HPLC determn in natural waters: Y. He, H. K. Lee, *J. Chromatogr. A* **1122,** 7 (2006).

Liquid. *Poisonous.* $bp_{0.01}$ 138-141°. LD_{50} in male, female rats (mg/kg): 5.5, 1.5 i.p.; 10.5, 2.2 orally; 30.0, 3.5 dermally; in male, female mice (mg/kg): 10.5, 7.0 i.p.; in male guinea pigs (mg/kg): 5.4 i.p., 9.0 orally (DuBois, Kinoshita).

Caution: Potential symptoms of overexposure are skin irritation; nausea, vomiting, abdominal cramps, diarrhea, salivation; headache, giddiness, vertigo, weakness; rhinorrhea, chest tightness; blurred vi- sion, miosis; cardiac irregularities; muscle fasciculations; dyspnea.

See NIOSH Pocket Guide to Chemical Hazards (DHHS/NIOSH 97-140, 1997) p 142.

USE: Nematocide; insecticide.

4028. Fentanyl. [437-38-7] *N*-Phenyl-*N*-[1-(2-phenylethyl)-4-piperidinyl]propanamide; *N*-(1-phenethyl-4-piperidyl)propionanilide; *N*-(1-phenethyl-4-piperidinyl)-*N*-phenylpropionamide; phentanyl; R-4263; Matrifen; Duragesic; Durogesic. C$_{22}$H$_{28}$N$_2$O; mol wt 336.48. C 78.53%, H 8.39%, N 8.33%, O 4.75%. Prototype anilidopiperidine opioid analgesic. Prepn: P. A. J. Janssen, **FR 1344366**; *idem*, **US 3164600** (1963, 1965 both to Janssen). Pharmacology: J. F. Gardocki, J. Yelnosky, *Toxicol. Appl. Pharmacol.* **6**, 48 (1964); R. Hess *et al.*, *J. Pharmacol. Exp. Ther.* **179**, 474 (1971). LC-MS/MS determn in human plasma and urine: N.-H. Huynh *et al.*, *J. Pharm. Biomed. Anal.* **37**, 1095 (2005). Effects on cerebral circulation and metabolism in rats: C. Carlsson *et al.*, *Anesthesiology* **57**, 375 (1982). Clinical studies: E. A. Welchew, J. A. Thornton, *Anaesthesia* **37**, 309 (1982); M. J. Stephens *et al.*, *Med. J. Aust.* **1**, 419 (1982). Clinical comparison with morphine for postoperative pain: H. S. Minkowitz *et al.*, *Pain Med.* **8**, 657 (2007). Review of clinical experience with transdermal formulation: R. B. R. Muijsers, A. J. Wagstaff, *Drugs* **61**, 2289-2307 (2001).

Crystals, mp 83-84°. Partition coefficient (*n*-octanol/water): 860. pKa 8.4.

Citrate. [990-73-8] Abstral; Actiq; Fentanest; Leptanal; Sublimaze. C$_{22}$H$_{28}$N$_2$O.C$_6$H$_8$O$_7$; mol wt 528.60. White crystalline powder, mp 149-151°. Bitter taste. One gram dissolves in about 40 ml water. Sol in methanol; slightly sol in chloroform. LD$_{50}$ in mice (mg/kg): 11.2 i.v.; 62 s.c. (Gardocki, Yelnosky).

Note: This is a controlled substance (opiate): **21 CFR**, 1308.12.

THERAP CAT: Analgesic.

THERAP CAT (VET): Analgesic; tranquilizer.

4029. Fenthion. [55-38-9] Phosphorothioic acid *O,O*-dimethyl *O*-[3-methyl-4-(methylthio)phenyl] ester; *O,O*-dimethyl *O*-[3-methyl-4-(methylthio)phenyl] phosphorothioate; *O,O*-dimethyl *O*-4-(methylthio)-*m*-tolyl phosphorothioate; *O,O*-dimethyl *O*-(4-methylmercapto-3-methylphenyl) thiophosphate; Bayer 29493; ENT-25540; S-1752; Baycid; Baytex; Lebaycid; Queletox; Tiguvon. C$_{10}$H$_{15}$O$_3$PS$_2$; mol wt 278.32. C 43.16%, H 5.43%, O 17.25%, P 11.13%, S 23.04%. Organophosphate insecticide; cholinesterase inhibitor. Prepn: E. Schlegk, G. Schrader, **DE 1116656**; *eidem*, **US 3042703** (1961, 1962 both to Bayer). Properties: H. F. Jung, *Bull. WHO* **21**, 215 (1959); Schrader, *Die Entwicklung neuer insektizider Phosphorsäure-Ester* (Verlag Chemie, 3rd ed., 1963) pp 298-313. Toxicity: Gaines, *Toxicol. Appl. Pharmacol.* **2**, 88 (1960). Degradation and persistence in treated oranges: Y. Picó *et al.*, *Anal. Chem.* **79**, 9350 (2007). Determn in olive oil and in river water: T. G. Díaz *et al.*, *Talanta* **76**, 809 (2008).

Pale yellow oil, slight odor of garlic. *Poisonous*. bp$_{0.01}$ 105°. d$_4^{20}$ 1.250. n$_D^{20}$ 1.5698. Log P (octanol/water): 4.8. Vapor pressure at 20°: 3×10^{-5} mm Hg. Thermally stable up to 210°; resistant to alkalies up to pH 9. Readily sol in methanol, ethanol, ether, acetone and many other organic solvents, esp chlorinated hydrocarbons.

Practically insol in water (55 mg/l). LD$_{50}$ in male, female rats (mg/kg): 215, 245 orally (Gaines).

Caution: Potential symptoms of overexposure are nausea, vomiting, abdominal cramps, diarrhea, salivation; headache, giddiness, vertigo, weakness; rhinorrhea, chest tightness; blurred vision, miosis; cardiac irregularities; muscle fasciculations; dyspnea. *See NIOSH Pocket Guide to Chemical Hazards* (DHHS/NIOSH 97-140, 1997) p 144.

USE: Insecticide; acaricide.

THERAP CAT (VET): Ectoparasiticide.

4030. Fentiazac. [18046-21-4] 4-(4-Chlorophenyl)-2-phenyl-5-thiazoleacetic acid; BR-700; CH-800; Donorest; Flogene; Norvedan. C$_{17}$H$_{12}$ClNO$_2$S; mol wt 329.80. C 61.91%, H 3.67%, Cl 10.75%, N 4.25%, O 9.70%, S 9.72%. Prepn: K. Brown, **GB 1145884**; *idem*, **US 3476766** (both 1969 to Wyeth). Anti-inflammatory activity: K. Brown *et al.*, *Nature* **219**, 164 (1968). Crystallographic study: R. Destro, *Acta Crystallogr.* **B34**, 959 (1978). Absorption: G. Zanolo *et al.*, *Boll. Chim. Farm.* **119**, 209 (1980). Metabolism: S. Fumero *et al.*, *Arzneim.-Forsch.* **30**, 1253 (1980). Pharmacokinetics in humans: M. Quattrini *et al.*, *ibid.* **31**, 1046 (1981). Toxicity study: D. A. Shriver *et al.*, *Toxicol. Appl. Pharmacol.* **42**, 75 (1977). Review of exptl and clinical pharmacology: E. Marmo, *Curr. Med. Res. Opin.* **6**, 53-63 (1979).

Colorless needles from benzene, mp 161-162°. LD$_{50}$ in rats, mice (mg/kg): 661, 692 orally (Marmo).

THERAP CAT: Anti-inflammatory.

4031. Fenticlor. [97-24-5] 2,2'-Thiobis[4-chlorophenol]; bis[2-hydroxy-5-chlorophenyl]; 2,2'-dihydroxy-5,5'-dichlorodiphenyl sulfide; S-7; Novex. C$_{12}$H$_8$Cl$_2$O$_2$S; mol wt 287.15. C 50.19%, H 2.81%, Cl 24.69%, O 11.14%, S 11.16%. Prepn by chlorination of bis[2-hydroxyphenyl]sulfide: Muth, **DE 568944** (1931 to I. G. Farben); Dunning *et al.*, *J. Am. Chem. Soc.* **53**, 3466 (1931).

Fine needles from toluene, mp 175°. Sol in aq solns of NaOH, in alcohol, in hot benzene.

USE: Fungicide; esp used against *Monosporium apiospermum*: Reifferscheid, Seeliger, *Dtsch. Med. Wochenschr.* **80**, 1841, 1850 (1955).

THERAP CAT: Anti-infective.

4032. Fenticonazole. [72479-26-6] 1-[2-(2,4-Dichlorophenyl)-2-[[4-(phenylthio)phenyl]methoxy]ethyl]-1*H*-imidazole; 1-[2,4-dichloro-β-[[*p*-(phenylthio)benzyl]oxy]phenethyl]imidazole; 1-(2,4-dichlorophenyl)-2-(*N*-imidazolyl)ethyl-4-phenylthiobenzyl ether; 2,4-dichloro-4'-phenylthio-(*N*-imidazolylmethyl)dibenzyl ether. C$_{24}$H$_{20}$Cl$_2$N$_2$OS; mol wt 455.40. C 63.30%, H 4.43%, Cl 15.57%, N 6.15%, O 3.51%, S 7.04%. Broad spectrum antimycotic, also active as antibacterial. Prepn: D. Nardi *et al.*, **DE 2917244**; *eidem*, **US 4221803** (1979, 1980 both to Recordati); *eidem*, *Arzneim.-Forsch.* **31**, 2123 (1981). Series of articles on physico-chemical properties, antimicrobial specturm and pharmacology: *ibid.* 2127-2154. Toxicity data: G. Graziani *et al.*, *ibid.* 2145. Quality control analysis of pharmaceutical raw material by HPCE: M. G. Quaglia *et al.*, *J. Sep. Sci.* **24**, 392 (2001). Preliminary clinical comparison with miconazole in vaginal candidiasis: A. Gastaldi, *Curr. Ther. Res.* **38**, 489 (1985). Comparative clinical trials in dermatomycoses: A. Finzi *et al.*, *Mykosen* **29**, 41 (1986); E. M. Kokoschka *et al.*, *ibid.* **45**. Review of use in dermatology and gynecology: S. Veraldi, R. Milani, *Drugs* **68**, 2183-2194 (2008).

Mononitrate. [73151-29-8] Rec-15-1476; Falvin; Fenizolan; Fentigyn; Gynoxin; Lomexin. $C_{24}H_{20}Cl_2NS.HNO_3$; mol wt 488.40. Odorless, white crystalline powder, mp 136°. uv max (methanol): 252 nm (ε 13894). Soly at 20° (mg/ml): water <0.1; ethyl ether <0.1; ethanol 30; methanol 100; chloroform 300; DMF 600. pKa 6.54. LD_{50} in mice, male rats, female rats (mg/kg): 1191, 440, 309 i.p.; all >3000 orally (Graziani).

THERAP CAT: Antifungal (topical).

4033. Fentonium Bromide. [5868-06-4] (3-*endo*,8-*anti*)-8-(2-[1,1′-Biphenyl]-4-yl-2-oxoethyl)-3-[(2S)-3-hydroxy-1-oxo-2-phenylpropoxy]-8-methyl-8-azoniabicyclo[3.2.1]octane bromide; 3α-hydroxy-8-(*p*-phenylphenacyl)-1αH,5αH-tropanium bromide (−)-tropate; N-(4-phenylphenacyl)-1-hyoscyaminium bromide; phentonium bromide; FA-402; Z-326; Ketoscilium; Ulcesium. $C_{31}H_{34}BrNO_4$; mol wt 564.52. C 65.96%, H 6.07%, Br 14.15%, N 2.48%, O 11.34%. Anticholinergic. Prepn: U. Teotino, D. Della Bella, **GB 1026640**; *eidem*, **US 3356682** (1966, 1967 both to Whitefin Holding); Teotino *et al.*, *Chim. Ther.* **3**, 453 (1968). Human pharmacology: Azzollini *et al.*, *Curr. Ther. Res.* **12**, 734 (1970). Clinical studies: Alberto *et al.*, *Minerva Med.* **62**, 852 (1971).

Crystals, mp 203-205° (dec). $[\alpha]_D^{23}$ −5.68° (c = 5 in DMF). Also reported as mp 193-194°. $[\alpha]_D^{25}$ −4.7° (c = 5 in DMF). LD_{50} in mice (mg/kg): 12.1 i.v.; >400 s.c. and orally (Teotino, Della Bella).

THERAP CAT: Antispasmodic.

4034. Fentrazamide. [158237-07-1] 4-(2-Chlorophenyl)-N-cyclohexyl-N-ethyl-4,5-dihydro-5-oxo-1H-tetrazole-1-carboxamide; 1-(2-chlorophenyl)-4-(N-cyclohexyl-N-ethylcarbamoyl)-5(4H)-tetrazolinone; BAY-YRC-2388; NBA-061; YRC-2388; Lecs. $C_{16}H_{20}ClN_5O_2$; mol wt 349.82. C 54.94%, H 5.76%, Cl 10.13%, N 20.02%, O 9.15%. Tetrazolinone herbicide for use in rice. Prepn: T. Goto *et al.*, **EP 612735**; *eidem*, **US 5362704** (both 1994 to Nihon Bayer). Mode of action study: C. Fedtke *et al.*, *Pestic. Sci.* **55**, 566 (1999). Field trial in combination with propanil on seeded rice: H. Fürsch, *Brighton Crop Prot. Conf. - Weeds* **1999**, 99. Comprehensive review: Y. Yasui *et al.*, *ibid.* **1997**, 67-72. Series of articles on properties, metabolism, and field trials: *Pflanzenschutz-Nachr. Bayer (Engl. Ed.)* **54**, 5-142 (2001).

Colorless crystals, mp 79°. d^{20} 1.30. Soly at 20° (g/l): pure water 0.0025; tap water (pH 4, 7, 9) 0.0023; *n*-heptane 20; 2-propanol 32; acetone >250; xylene >250. Vapor pressure at 20° (hPa): 5×10^{-10}; at 25°: 1×10^{-9}. Log P (octanol/water): 3.60 at 20°. LD_{50} in rats (mg/kg): >5000 orally, >5000 dermally. LC_{50} in rats (mg/m³): >5000 by inhalation. LC_{50} (48 hr) in carp, rainbow trout (mg/l): 3.2, 3.4 (Yasui).

USE: Herbicide.

4035. Fenugreek. Greek Hay. *Trigonella foenumgraecum* L., *Leguminosae*, an annual herb, cultivated in Southern Europe, Northern Africa and India for its seeds which are used in making curry. The seeds contain the alkaloid trigonelline and the steroidal sapogenin, diosgenin, *q.q.v.*: Fazli, Hardman, *Trop. Sci.* **10**, 66-78 (1968).

USE: Has been used in making imitation maple syrup and for culinary spices other than curry, flavoring agent.

THERAP CAT (VET): Emollient.

4036. Fenuron. [101-42-8] N,N-Dimethyl-N′-phenylurea; N-phenyl-N′,N′-dimethylurea; Dybar. $C_9H_{12}N_2O$; mol wt 164.21. C 65.83%, H 7.37%, N 17.06%, O 9.74%. Prepn: Crosby, Niemann, *J. Am. Chem. Soc.* **76**, 4458 (1954); Gilbert, Sorma, **US 2729677** (1956 to Allied Chem.); Jones, **US 2768971** (1956 to ICI); Applegath *et al.*, **US 2857430** (1958 to Monsanto). Prepn of trichloroacetate: E. E. Gilbert *et al.*, **US 2782112** (1957 to Allied Chem.). Toxicity data: G. W. Bailey, J. L. White, *Residue Rev.* **10**, 97 (1965).

Crystals, mp 131-133°. Sparingly sol in water (0.29% at 24°); in hydrocarbons. LD_{50} orally in rats: 7500 mg/kg (Bailey, White).

Trichloroacetate. [4482-55-7] Fenuron TCA; Urab. $C_{11}H_{13}Cl_3N_2O_3$; mol wt 327.59. mp 65-68°. Esp effective in woody plant control.

Caution: If hydrolyzed to aniline can cause methemoglobinemia: *Clinical Toxicology of Commercial Products*, R. E. Gosselin *et al.*, Eds. (Williams & Wilkins, Baltimore, 5th ed., 1984) Section II, p 331.

USE: Herbicide.

4037. Fenvalerate. [51630-58-1] 4-Chloro-α-(1-methylethyl)benzeneacetic acid cyano(3-phenoxyphenyl)methyl ester; α-cyano-3-phenoxybenzyl α-(4-chlorophenyl)isovalerate; cyano(3-phenoxyphenyl)methyl 4-chloro-α-(1-methylethyl)benzeneacetate; α-cyano-3-phenoxybenzyl-2-(4-chlorophenyl)-3-methylbutyrate; fenvalerate; S-5602; SD-43775; WL-43775; Belmark; Pydrin; Pyridin; Sumicidin; Tirade. $C_{25}H_{22}ClNO_3$; mol wt 419.91. C 71.51%, H 5.28%, Cl 8.44%, N 3.34%, O 11.43%. Synthetic pyrethroid insecticide without the usual cyclopropane ring. Prepn: **DE 2335347**; K. Fujimoto *et al.*, **US 3996244** (1974, 1976, both to Sumitomo); D. A. Wood, **DE 2651341**; *idem*, **US 4061664** (1977, both to Shell). Has two chiral centers giving four possible optical isomers; prepn and comparative insecticidal activity of isomers: M. Hirano *et al.*, **DE 2737297**; *eidem*, **US 4503071** (1978, 1985 both to Sumitomo). Absolute configuration of most active isomer, esfenvalerate: K. Aketa *et al.*, *Agric. Biol. Chem.* **42**, 895 (1978). GC separation of isomers: G. R. Cayley, B. W. Simpson, *J. Chromatogr.* **356**, 123 (1986); determn of esfenvalerate in technical prepn: S. Sakaue *et al.*, *Agric. Biol. Chem.* **51**, 1671 (1987). Comprehensive description of activity and physical properties of esfenvalerate: H. Oo'uchi, *Jpn. Pestic. Inf.* **1985**, 21-24. Comparative soil metabolism of isomers: P.W. Lee *et al.*, *J. Agric. Food Chem.* **35**, 384 (1987). Review of toxicology and human exposure: *Toxicological Profile for Pyrethrins and Pyrethroids* (PB2004-100004, 2003) 332 pp.

Clear yellow viscous liquid at 23°. d^{23} 1.17. n_D^{20} 1.5533. Vapor press at 25° = 1.1×10^{-8} mm Hg. Soly at 20° (g/l): acetone, >450; chloroform, >450; methanol, >450; hexane, 77. Insol in water. More stable in acidic soln than in alk soln. Dec gradually between 150-300°. No significant breakdown after 100 hrs at 75°. LD_{50} orally in rats: 451 mg/kg (DMSO); >3200 mg/kg (aqueous suspension), Shell Technical Data Bulletin. Highly toxic to fish and bees.

(S,S)-Isomer. [66230-04-4] Esfenvalerate; Aα; OMS-3023; S-1844; S-5602α; Sumi-alpha. White crystalline solid, mp 59-60.2°. $[\alpha]_D^{25}$ −15.0° (c = 2.0 in CH_3OH). d_{23}^{23} 1.163. Vapor pressure at 20°: 2.63×10^{-7} mm Hg; at 25°: 5.00×10^{-7} mm Hg. Soly (1%): acetonitrile >60, chloroform >60, DMF >60, DMSO >60, ethyl acetate >60, acetone >60, ethyl cellosolve 40-50, n-hexane 1-5, kerosene <1, methanol 7-10, α-methylnaphthalene 50-60, xylene >60.

Caution: Eye, skin irritant.

USE: Insecticide.

THERAP CAT (VET): Ectoparasiticide.

4038. Fepradinol. [36981-91-6] α-[[(2-Hydroxy-1,1-dimethylethyl)amino]methyl]benzenemethanol; 1-phenyl-2-(α,α-dimethylethanolamino)ethanol. $C_{12}H_{19}NO_2$; mol wt 209.29. C 68.87%, H 9.15%, N 6.69%, O 15.29%. Prepn: E. Biekert, J. Sonnenbichler, *Ber.* **95**, 1451 (1962); G. Staibano, C. Protto, *Farmaco Ed. Sci.* **27**, 929 (1972); G. Staibano, **BE 774629** (1972 to Gentili). Pharmacology: J. M. Masso *et al.*, *J. Pharm. Pharmacol.* **45**, 959 (1993). Mechanism of action study: *eidem, Arzneim.-Forsch.* **44**, 68 (1994). Clinical trial: C. Bestit *et al.*, *Curr. Ther. Res.* **45**, 53 (1989).

White cryst from ethanol, mp 142-143°; also reported as mp 140-142° (Staibano). Cryst from benzene, mp 139° (Biekert). uv max (0.1N HCl): 250.5, 256.4, 262.0 nm (log ε 2.17, 2.28, 2.25).

Hydrochloride. Dalgen; Flexidol. $C_{12}H_{19}NO_2 \cdot HCl$; mol wt 245.75.

THERAP CAT: Anti-inflammatory.

4039. Feprazone. [30748-29-9] 4-(3-Methyl-2-buten-1-yl)-1,2-diphenyl-3,5-pyrazolidinedione; 4-prenyl-1,2-diphenyl-3,5-pyrazolidinedione; 4-prenyl-1,2-diphenyl-3,5-dioxopyrazolidine; 4-(β-isoamylenyl)-1,2-diphenyl-3,5-pyrazolidinedione; 4-(2-isopentenyl)-1,2-diphenyl-3,5-pyrazolidinedione; phenylprenazone; prenazone; DA-2370; Analud; Methrazone; Zepelin. $C_{20}H_{20}N_2O_2$; mol wt 320.39. C 74.98%, H 6.29%, N 8.74%, O 9.99%. Non-steroidal anti-inflammatory drug. Prepn: S. Casadio, G. Pala, **DE 2031238**; *eidem*, **US 3703528** (1971, 1972 both to Ist. De Angeli); Casadio *et al., Arzneim.-Forsch.* **22**, 171 (1972). Series of articles on physical-chemical data, pharmacology, toxicology and clinical studies: *ibid.* 174-281. Toxicity data: C. Bianchi, G. Bonardi, *ibid.* 196.

Fine, white odorless and tasteless crystalline powder, mp 156.5° (ethanol). Very sol in acetone, chloroform, DMF; sparingly sol in acetonitrile, benzene, 10% NaOH. Slightly sol in ether, methanol, ethanol, cyclohexane. Practically insol in 10% HCl, 10% acetic acid, water. uv max (ethanol): 246 nm (log ε 4.19); (pH 9-12 buffer): 264 nm (log ε 4.322-4.326). pKa 5.09 ± 0.07. LD_{50} in male mice, rats (mg/kg): 408.8, 386.4 i.p.; 1067, >2000 orally (Bianchi, Bonardi).

THERAP CAT: Anti-inflammatory.

4040. Ferbam. [14484-64-1] (OC-6-11)-Tris(N,N-dimethylcarbamodithioato-κS,κS')iron; (OC-6-11)-tris(dimethylcarbamodithioato-S,S')iron; tris(dimethyldithiocarbamato)iron; dimethyldithiocarbamic acid iron salt; ferric dimethyldithiocarbamate; Carbamate; Ferbeck; Fermate; Ferradow; Karbam Black. $C_9H_{18}FeN_3S_6$; mol wt 416.47. C 25.96%, H 4.36%, Fe 13.41%, N 10.09%, S 46.19%. $[(CH_3)_2NCS_2]_3Fe$. Prepd from Na dimethyldithiocarbamate and a soln of a ferric salt: Tisdale, Williams, **US 1972961** (1934 to du Pont). Toxicity data: H. C. Hodge *et al., J. Am. Pharm. Assoc.* **41**, 662 (1952).

Black solid. mp >180° (dec). Solubility in water 120 ppm, pH of soln 5.0; sol in acetone, chloroform, pyridine, acetonitrile. Tends to decompose on prolonged storage or exposure to heat and moisture. LD_{50} in male mice, rats (mg/kg): 3000, 2700 i.p. (Hodge).

Caution: Potential symptoms of overexposure are irritation of eyes and respiratory tract; dermatitis; GI disturbance. *See NIOSH Pocket Guide to Chemical Hazards* (DHHS/NIOSH 97-140, 1997) p 144. *See also Clinical Toxicology of Commercial Products,* R. E. Gosselin *et al.*, Eds. (Williams & Wilkins, Baltimore, 5th ed., 1984) Section II, p 311.

USE: Fungicide.

4041. Fermium. [7440-72-4] Fm; at. no. 100; valence 3, 2. Man-made, radioactive element. No stable nuclides; known isotopes (mass numbers): 242-259. Longest-lived known isotope: 257 ($T_{1/2}$ 101 days, α-emitter, rel. at. mass 257.0951), produced by neutron bombardment of plutonium (or other elements with higher at. no.). ^{255}Fm ($T_{1/2}$ 20.07 hrs, α-emitter) originally discovered in debris from a thermonuclear test explosion in Nov. 1952. Reviews of discovery: A. Ghiorso *et al., Phys. Rev.* **99**, 1048-1049 (1955); A. Ghiorso, G. T. Seaborg, *Sci. Am.* **195**(6), 67-80 (1956). *Reviews:* G. T. Seaborg, *J. Chem. Educ.* **36**, 38-44 (1959); C. Keller, *The Chemistry of the Transuranium Elements* (Verlag Chemie, Weinheim, English Ed., 1971) pp 591-594; Silva, "Trans-Curium Elements" in *MTP Int. Rev. Sci.: Inorg. Chem., Ser. One* vol. 8, A. G. Maddock, Ed. (University Park Press, Baltimore, 1972) pp 71-105; *Comprehensive Inorganic Chemistry* vol. 5, J. C. Bailar, Jr. *et al.*, Eds. (Pergamon Press, Oxford, 1973) *passim*; *Handb. Exp. Pharmakol.* **36**, 689-738 (1973); R. J. Silva in *The Chemistry of the Actinide Elements* vol. 2, J. J. Katz *et al.*, Eds. (Chapman and Hall, New York, 1986) pp 1086-1092; *The Elements Beyond Uranium,* G. T. Seaborg, W. D. Loveland, Eds. (John Wiley & Sons, Inc., New York, 1990) pp 28-38.

Caution: Radiation hazard; handling requires special equipment and shielding facilities (Katz *et al., loc. cit.* vol. 2, p. 1128).

4042. Ferredoxins. A group of electron transfer factors found in plants and bacteria, which are non-heme iron-sulfur proteins and which play an important role in photosynthesis, nitrogen and carbon dioxide fixation, and respiration. They are generally classified by the presence of either 2 or 4 iron atom clusters and an equivalent amount of inorganic or "acid-labile" sulfide bonded to the peptide chain through 4 cysteine sulfhydryl groups. The two-iron ferredoxins are found primarily in plants and blue-green algae and are sometimes referred to as chloroplast or "plant type" ferredoxins; the four-iron ferredoxins are predominant in bacteria. Mol wts of chloroplast ferredoxins are about 12,000; those of bacterial ferredoxins range from about 6,000 to 24,000. Isoln from *Clostridium pasteurianum:* L. E. Mortenson *et al., Biochem. Biophys. Res. Commun.* **7**, 448 (1962). Prepn of bacterial ferredoxins: L. E. Mortenson, **US 3344130** (1960 to duPont). The amino acid sequence of several bacterial and chloroplast ferredoxins has been elucidated. Amino acid sequence of ferredoxin from *Clostridium pasteurianum:* Tanaka *et al., Biochem. Biophys. Res. Commun.* **16**, 422 (1964); *eidem, Biochemistry* **5**, 1666 (1966); from spinach: Matsubara, Sasaki, *J. Biol. Chem.* **243**, 1732 (1968). Synthesis of peptide chain of *C. pasteurianum* ferredoxins: Bayer *et al., Tetrahedron* **24**, 4853 (1968); Trakatellis, Schwartz, *Proc. Natl. Acad. Sci. USA* **63**, 436 (1969). *Reviews:* Buchanan, *Struct. Bonding* **1**, 109 (1966); Malkin, Rabinowitz, *Annu. Rev. Biochem.* **36**, 113 (1967); Arnon, *Naturwissenschaften* **56**, 295 (1969); Buchanan, Arnon, *Adv. Enzymol.* **33**, 119 (1970); W. Lovenberg, "Ferredoxin and Rubredoxin" in *Microbial Iron Metabolism,* J. B. Neilands, Ed. (Academic Press, New York, 1974) pp 161-182; D. C. Yoch, R. P. Carithers, *Microbiol. Rev.* **43**, 384-421 (1979).

Absorption max of bacterial ferredoxins: 280, 385-400 nm; of plant ferredoxins: 280, 325, 420, 463 nm. All show negative oxidation-reduction potentials near that of the hydrogen electrode:

−0.390 V *(C. pasteurianum);* −0.420 V (spinach): Tagawa, Arnon, *Biochim. Biophys. Acta* **153**, 602 (1968). Autoxidizable. Acidification, as well as treatment with iron-chelating agents or mercurials, results in evolution of H_2S and loss of the visible absorption.

4043. Ferric Acetate, Basic. [10450-55-2] Bis(acetato-κO)-hydroxyiron; basic iron(III) acetate. $C_4H_7FeO_5$; mol wt 190.94. C 25.16%, H 3.70%, Fe 29.25%, O 41.90%. $Fe(OH)(CH_3COO)_2$. Prepd by heating ferric acetate soln: Casey, Doyle in *Kirk-Othmer Encyclopedia of Chemical Technology* vol. 12 (Interscience, New York, 2nd ed., 1967) p 24.
Brownish-red scales or amorphous powder; faint acetic odor. Practically insol in water; sol in alcohol, acids. *Protect from light.*
USE: In the textile industry as a mordant in dyeing and printing, and for the weighting of silk and felt; as wood preservative; in leather dyes.

4044. Ferric Ammonium Citrate. [1185-57-5] 2-Hydroxy-1,2,3-propanetricarboxylic acid ammonium iron (3+) salt (1:?:?); ammonium ferric citrate; iron(III) ammonium citrate; FerriSeltz. Compounds of NH_3, Fe, and citric acid ($C_6H_8O_7$) of undetermined formula and structure. Prepn: H. J. Kruse, H. C. Mounce, **US 2644828** (1953 to Mallinckrodt). ESR characterization of green and brown forms: R. A. Uphaus, M. I. Blake, *Naturwissenschaften* **68**, 270 (1981). Decomposition study: G. A. M. Hussein, *Powder Technol.* **80**, 265 (1994). Use in prevention of anemia in pigs: B. G. Harmon *et al., J. Anim. Sci.* **26**, 1051 (1967). Bioavailability: N. P. Wong *et al., Nutr. Rep. Int.* **29**, 135 (1984). Clinical evaluation in iron deficiency anemia: M. Taniguchi *et al., J. Nutr. Sci. Vitaminol.* **37**, 161 (1991); of use as an MRI contrast agent: S. Hirohashi *et al., Magn. Reson. Imaging* **12**, 837 (1994).
Two forms exist. The brown hydrated form exists as garnet-red transparent scales or granules, or brownish-yellow powder, and contains ~9% NH_3, 16.5-18.5% Fe, and ~65% hydrated citric acid. The green hydrated form exists as green transparent, deliquesc scales, granules, or powder, and contains ~7.5% NH_3, 14.5-16% Fe, and ~75% hydrated citric acid. Very sol in water. Insol in alcohol. Reduced to ferrous salt by light. *Keep well closed and protected from light.*
USE: For blueprints; in photography.
THERAP CAT: Hematinic.
THERAP CAT (VET): In iron deficiency anemia.

4045. Ferric Bromide. [10031-26-2] Iron bromide ($FeBr_3$). Br_3Fe; mol wt 295.56. Br 81.10%, Fe 18.89%. $FeBr_3$. Prepn from Fe and Br_2: Gregory, Thackrey, *J. Am. Chem. Soc.* **72**, 3176 (1950); Gregory, *ibid.* **73**, 472 (1951); from $Fe_2(SO_4)_3$ and LiBr: Lieser, Elias, *Z. Anorg. Allg. Chem.* **316**, 208 (1962).
Dark-red or black hexagonal, rhombic plates. Very hygroscopic. Loses some Br_2 on exposure to air and light. Sol in water, alcohol, ether, acetic acid; slightly sol in liquid NH_3; aq soln dec to $FeBr_2$ + Br_2 on boiling. *Keep cool, well-closed and protected from light.*
Caution: Irritant.
USE: Catalyst for organic reactions, particularly in bromination of aromatic compds.

4046. Ferric Chloride. [7705-08-0] Iron chloride ($FeCl_3$); iron(III) chloride; flores martis. Cl_3Fe; mol wt 162.20. Cl 65.57%, Fe 34.43%. $FeCl_3$. Occurs in nature as the mineral *molysite.* Prepn: Tarr, *Inorg. Synth.* **3**, 191 (1950); Pray, *ibid.* **5**, 153 (1957); Epperson *et al., ibid.* **7**, 163 (1963); Lieser, Elias, *Z. Anorg. Allg. Chem.* **316**, 208 (1962); Attwood, Shelton, *J. Inorg. Nucl. Chem.* **26**, 1758 (1964); Bardawil *et al., Inorg. Chem.* **3**, 149 (1964). Acute toxicity: C. S. Hosking, *Aust. Paediatr. J.* **6**, 92 (1970). Use as clinical reagent: L. Cassidei *et al., Clin. Chim. Acta* **90**, 121 (1978).
Hexagonal, dark leaflets or plates. Red by transmitted light, green by reflected light; sometimes appears brownish-black. Very hygroscopic. Melts and volatilizes about 300°; bp ~316°; d^{25} 2.90. Vapor density measurements show that it is dimeric at about 400° but monomeric above 750°. Dissociates at high temps to $FeCl_2$ and Cl_2. Readily absorbs water in air to form the hexahydrate. Readily sol in water, alcohol, ether, acetone; slightly sol in CS_2. Practically insol in ethyl acetate. *Corrosive. Keep well closed.*
Hexahydrate. [10025-77-1] $Cl_3Fe.6H_2O$; mol wt 270.29. Brownish-yellow or orange monoclinic crystals. d 1.82. Structure reported as *trans*-$[FeCl_2(H_2O)_4]Cl.2H_2O$: Lind, *J. Chem. Phys.* **47**,

990 (1967). Usually slight odor of HCl; very hygroscopic. mp ~37°. Readily sol in water, alcohol, acetone, ether; pH of 0.1 molar aq soln 2.0. *Keep well closed.* LD_{50} i.v. in mice: 0.049 mg Fe/g (Hosking).
Caution: Potential symptoms of overexposure are irritation of eyes, skin, mucous membranes; abdominal pain, diarrhea, vomiting; possible liver damage. *See NIOSH Pocket Guide to Chemical Hazards* (DHHS/NIOSH 97-140, 1997) p 174.
USE: In photoengraving; photography; manuf of other Fe salts, pigments, ink; as catalyst in organic reactions; purifying factory effluents and deodorizing sewage; chlorination of Ag and Cu ores; as mordant in dyeing and printing textiles; oxidizing agent in dye manuf. Clinical reagent (amino acids in urine, esp in phenylketonuria).
THERAP CAT: Hexahydrate as astringent, styptic.
THERAP CAT (VET): Styptic, astringent.

4047. Ferric Chromate(VI). [10294-52-7] Chromic acid (H_2CrO_4) iron(3+) salt (3:2); C.I. Pigment Yellow 45; C.I. 77505; Siderin yellow. $Cr_3Fe_2O_{12}$; mol wt 459.67. Cr 33.93%, Fe 24.30%, O 41.77%. $Fe_2(CrO_4)_3$. Preparation: Boericke, Bangert, *Bur. Mines Rep. Invest.* **3813**, 19 pp (1945), *C.A.* **39**, 5164[7] (1945).
Yellow powder. Sol in HCl. Practically insol in water.
Caution: Chromium hexavalent (VI) compounds are listed as known human carcinogens: *Report on Carcinogens, Twelfth Edition* (PB2011-111646, 2011) p 106.
USE: Pigment for ceramics, glass and enamels.

4048. Ferric Citrate. [2338-05-8] (unspecified stoichiometry); [28633-45-6] (undefined iron(III) salt); [3522-50-7] (1:1 iron-(III) citrate salt). 2-Hydroxy-1,2,3-propanetricarboxylic acid iron-(3+) salt (1:?). A combination of iron and citric acid of indefinite composition. Prepn: Belloni, *Gazz. Chim. Ital.* **50**, II, 159 (1920). Complex formation study: R. C. Warner, I. Weber, *J. Am. Chem. Soc.* **75**, 5086 (1953). Polymerization studies: T. G. Spiro *et al., ibid.* **89**, 5555, 5559 (1967). Role in iron uptake in plants: H. F. Bienfait, M. R. Scheffers, *Plant Soil* **143**, 141 (1992). Tumorigenicity study in mice: K. Inai *et al., Food Chem. Toxicol.* **32**, 493 (1994). Clinical evaluation in hyperphosphataemia: W.-C. Yang *et al., Nephrol. Dial. Transplant.* **17**, 265 (2002).
Garnet-red, transparent scales or pale-brown powder. Slowly but completely sol in cold water, readily sol in hot water, but diminishing in soly with age. Practically insol in alc.
THERAP CAT: Hematinic.
THERAP CAT (VET): In iron deficiency.

4049. Ferric Fluoride. [7783-50-8] Iron fluoride (FeF_3). F_3Fe; mol wt 112.84. F 50.51%, Fe 49.49%. FeF_3. Prepn from $FeCl_3$ and HF: Kwasnik in *Handbook of Preparative Inorganic Chemistry* vol. 1, G. Brauer, Ed. (Academic Press, New York, 2nd ed., 1963) p 266; Shinn *et al., Inorg. Chem.* **5**, 1927 (1966); from Fe_2O_3 or Fe_2S_3 and SF_4: Oppegard *et al., J. Am. Chem. Soc.* **82**, 3835 (1960); from $FeCl_3$ and CH_3CHF_2: Natta *et al.,* **GB 995186** (1965 to Montecatini).
Green hexagonal crystals. Sublimes >1000°. d 3.87. Very slightly sol in water; freely sol in dil HF; practically insol in alcohol, ether, benzene. Soly in liquid HF: 0.008 g/100 g HF. *See:* Jache, Cady, *J. Phys. Chem.* **56**, 1106 (1952).
USE: As catalyst in organic reactions.

4050. Ferric Formate. [555-76-0] Formic acid iron(3+) salt. $C_3H_3FeO_6$; mol wt 190.90. C 18.88%, H 1.58%, Fe 29.25%, O 50.28%. $Fe(HCOO)_3$. Prepn: Weinland, Reihlin, *Ber.* **46**, 3144 (1913); Starke, *J. Inorg. Nucl. Chem.* **25**, 823 (1963). May contain 1 or 2 H_2O.
Red or yellow microcryst powder. Readily hydrolyzed to basic formates. Sol in water; very slightly sol in alcohol. *Protect from light.*
USE: Preservation of silage.

4051. Ferrichromes. Growth-promoting iron chelates. Ferrichrome and ferrichrome A first isolated from the rust fungus, *Ustilago sphaerogena:* Neilands, *J. Am. Chem. Soc.* **74**, 4846 (1952); Garibaldi, Neilands, *ibid.* **77**, 2429 (1955). Ferrichrome and four related compounds, **ferrichrysin, ferricrocin, ferrirubin** and **ferrirhodin** have also been isolated from *Aspergillaceae* and from *Penicillium resticulosum:* Zähner *et al., Arch. Mikrobiol.* **45**, 119 (1963);

Keller-Schierlein, Deer, *Helv. Chim. Acta* **46**, 1907 (1963); Keller-Schierlein, *ibid.* 1920. Ferrichromes are cyclic hexapeptides containing three small, neutral amino acids and three derivatives of δ-*N*-hydroxy-L-ornithine [HONH(CH$_2$)$_3$CH(NH$_2$)COOH]. Structure: Rogers, Neilands, *Biochemistry* **3**, 1850 (1964). Crystal and molecular structure of ferrichrome A: Zalkin *et al.*, *J. Am. Chem. Soc.* **88**, 1810 (1966). Synthesis of ferrichrome: Keller-Schierlein, Maurer, *Helv. Chim. Acta* **52**, 603 (1969); Isowa *et al.*, *Bull. Chem. Soc. Jpn.* **47**, 215 (1974). Reviews of ferrichromes and other hydroxamic acids: Neilands, *Struct. Bonding* **1**, 59-108 (1966); Emery, *Adv. Enzymol. Relat. Areas Mol. Biol.* **35**, 135-185 (1971); J. B. Neilands, *Bioinorg. Chem.* **II**, K. N. Raymond, Ed. (A.C.S., Washington, 1977), pp 3-32.

| Ferrichrome | R = CH$_3$ | R' = H |
| Ferrichrome A | R = *trans*-HOOCCH$_2$(CH$_3$)C=CH | R' = CH$_2$OH |

Ferrichrome. C$_{27}$H$_{42}$FeN$_9$O$_{12}$; mol wt 740.53. Long, yellow needles from anhyd methanol, shrink and blacken at 240-242° without melting. [α]$_D$ +300° (c = 0.04). Absorption max (methanol): 425 nm (E$_{1cm}^{1\%}$ 39.4). Sol in water and hot methanol; sparingly sol in ethanol, acetone, ether, chloroform.

Ferrichrome A. C$_{41}$H$_{58}$FeN$_9$O$_{20}$; mol wt 1052.80. Crystals from water. Absorption max (0.1*M* phosphate buffer; pH 7): 440 nm (E$_{1cm}^{1\%}$ 33.8). Very sol in methanol, ethanol, propanol; slightly sol in hot water. Practically insol in acetone, petr ether, ether, chloroform.

4052. Ferric Hydroxide. [20344-49-4] Iron hydroxide oxide (Fe(OH)O); ferric hydroxide oxide; hydrated ferric oxide. FeHO$_2$; mol wt 88.85. Fe 62.85%, H 1.13%, O 36.01%. FeO(OH). Occurs in nature as the minerals *goethite* [α-FeO(OH)], *lepidocrocite* [γ-FeO(OH)], and *limonite* [FeO(OH).*n*H$_2$O]. Other known allomorphic forms: β-FeO(OH); δ-FeO(OH). The hydroxide Fe(OH)$_3$ is not known. Prepn: Lux in *Handbook of Preparative Inorganic Chemistry* vol 2, G. Brauer, Ed. (Academic Press, New York, 2nd ed., 1965) p 1499. Crystal structure of α-FeO(OH): Sampson, *Acta Crystallogr.* **25B**, 1683 (1969). *Review:* Bernal *et al.*, *Clay Miner. Bull.* **4**, 15-30 (1959).

Red to brown powder or crystals. Loses H$_2$O to form Fe$_2$O$_3$. d 3.4-3.9. Practically insol in water, alcohol. Sol in mineral acids.

USE: In purifying water; as absorbent in chemical processing; as pigment; as catalyst.

4053. Ferric Hypophosphite. [7783-84-8] Phosphinic acid iron(3+) salt. FeH$_6$O$_6$P$_3$; mol wt 250.81. Fe 22.27%, H 2.41%, O 38.27%, P 37.05%. Fe(H$_2$PO$_2$)$_3$. Prepn: *U.S.D.* 25th ed, p 573.

White or grayish-white powder. Odorless, tasteless. Sol in 2300 parts cold water, 1200 parts boiling water; more sol in water in presence of H$_3$PO$_2$; sol in warm concd solns of alkali citrates. *Protect from light.* Should not be heated or triturated with chlorates, nitrates, or other oxidizing agents.

USE: Formerly as dietary supplement for phosphorus.

4054. Ferric Nitrate. [10421-48-4] Nitric acid iron(3+) salt (3:1); iron(III) nitrate. FeN$_3$O$_9$; mol wt 241.86. Fe 23.09%, N

17.37%, O 59.53%. Fe(NO$_3$)$_3$. Prepn: *Gmelins, Iron* (8th ed.) **59**, part B, 161-172 (1932). Toxicity study: H. F. Smyth *et al.*, *Am. Ind. Hyg. Assoc. J.* **30**, 470 (1969).

Nonahydrate. [7782-61-8] Fe$_3$N$_3$O$_9$.9H$_2$O; mol wt 515.68. Pale-violet to grayish-white, somewhat deliquesc crystals. mp 47°. Dec below 100°. d^{21} 1.68. *Oxidizer.* Freely sol in water, alcohol, acetone; slightly sol in cold concd HNO$_3$. LD$_{50}$ orally in rats: 3.25 g/kg (Smyth).

Caution: Potential symptoms of overexposure are irritation of eyes, skin, mucous membranes; abdominal pain, diarrhea, vomiting; liver damage. *See NIOSH Pocket Guide to Chemical Hazards* (DHHS/NIOSH 97-140, 1997) p 174.

USE: As mordant in dyeing, weighting silks, tanning; as reagent in analytical chemistry for phosphate determn; as corrosion inhibitor.

4055. Ferric Oxide. [1309-37-1] Iron oxide (Fe$_2$O$_3$); ferric sesquioxide; jeweler's rouge. Fe$_2$O$_3$; mol wt 159.69. Fe 69.94%, O 30.06%. α-Form occurs in nature as the mineral *hematite*. γ-Form occurs in nature as the mineral *maghemite*; prepd by dehydration of α-FeO(OH): Giovanoli, Brütsch, *Chimia* **28**, 188 (1974). Prepn of a third allomorphic form, ε-Fe$_2$O$_3$: Schrader, Büttner, *Z. Anorg. Allg. Chem.* **320**, 220 (1963); Trautmann, Forestier, *Compt. Rend.* **261**, 4423 (1965). Color and appearance of Fe$_2$O$_3$ are dependent upon the size and shape of the particles and the amount of combined water. Preparation and properties: *Gmelins, Iron* (8th ed.) **59**, part B, 63-94 (1932); Baudisch, Hartung, *Inorg. Synth.* **1**, 185 (1939); *Ullmanns Encyklopädie der technischen Chemie* vol. 6, 421-423 (1955); Bernal *et al.*, *Clay Miner. Bull.* **4**, 15-30 (1959). Toxicology: L. T. Fairhall, *Industrial Toxicology* (Hafner, New York, 2nd ed., 1969) pp 64-66.

Red or yellow powder or a blending of the two basic colors. Insol in water, organic solvents. Dissolves in HCl upon warming with a small amount of insol residue usually remaining.

Note: The composition of the substance called δ-Fe$_2$O$_3$ is actually FeO(OH) (Bernal *et al.*).

Caution: Potential symptom of overexposure to dust and fumes is benign pneumoconiosis with x-ray shadows indistinguishable from fibrotic pneumoconiosis. *See NIOSH Pocket Guide to Chemical Hazards* (DHHS/NIOSH 97-140, 1997) p 172.

USE: As pigment for rubber, paints, paper, linoleum, ceramics, glass; in paint for ironwork, ship hulls; as polishing agent for glass, precious metals, diamonds; in electrical resistors and semiconductors; in magnets, magnetic tapes; as catalyst; colloidal solns as stain for polysaccharides.

4056. Ferric Phosphate. [10045-86-0] Phosphoric acid iron-(3+) salt (1:1); iron orthophosphate. FeO$_4$P; mol wt 150.81. Fe 37.03%, O 42.43%, P 20.54%. FePO$_4$. Occurs in nature as the minerals *beraunite, cacoxenite, dufrenite, koninckite, phosphosiderite, strengite*. Prepn from Fe(H$_2$PO$_4$)$_3$: Remy, Boulle, *Compt. Rend.* **253**, 2699 (1961); from Fe(CO)$_5$ and H$_3$PO$_4$: Cate *et al.*, *Soil Sci.* **88**(3), 130 (1959); from phosphate rock: Vickery, *US 2914380* (1959 to Horizons Inc.); from mill scale and H$_3$PO$_4$: Alexander, Mathes, *US 3070423* (1962 to Chemetron).

Dihydrate. White, grayish-white, or light pink, orthorhombic or monoclinic crystals or amorphous powder. Loses water above 140°. d 2.87. Practically insol in water. Slowly sol in HNO$_3$; readily sol in HCl.

USE: As food and feed supplement, particularly in bread enrichment; as fertilizer.

4057. Ferric Pyrophosphate. [10058-44-3] Diphosphoric acid iron(3+) salt (3:4); Ferro Angelini. Fe$_4$O$_{21}$P$_6$; mol wt 745.20. Fe 29.98%, O 45.09%, P 24.94%. Fe$_4$(P$_2$O$_7$)$_3$. Prepn: *Gmelins, Iron* (8th ed.) **59**, part B, 777 (1932); H. M. Knight, J. T. Kelly, **US 3014784** (1962 to American Oil). Magnetic structural study: W. M. Reiff, C. C. Torardi, *Hyperfine Interact.* **53**, 403 (1990). Bioavailability: H. Tsuchita *et al.*, *J. Agric. Food Chem.* **39**, 316 (1991). Clinical evaluation in hemodialysis: A. Gupta *et al.*, *Kidney Int.* **55**, 1891 (1999).

Nonahydrate. [10049-18-0] Yellowish-white powder. Practically insol in water or acetic acid. Sol in mineral acids.

USE: As catalyst; in fireproofing of synthetic fibers; in corrosion-preventing pigments.

THERAP CAT: Hematinic.

4058. Ferric Sodium Edetate. [15708-41-5] (*OC*-6-21)-[[*N,N'*-1,2-Ethanediylbis[*N*-[(carboxy-*κO*)methyl]glycinato-*κN*,-*κO*]](4−)]-ferrate(1−) sodium (1:1); sodium [(ethylenedinitrilo)tetraacetato]ferrate(1−); (ethylenedinitrilo)tetraacetic acid sodium salt iron complex; ferric monosodium ethylenediaminetetraacetate; edetic acid sodium iron salt; sodium iron edetate; sodium feredetate; NaFeEDTA; Ferrostrane; Ferrostrene; Sybron; Ferrazone. $C_{10}H_{12}FeN_2NaO_8$; mol wt 367.05. C 32.72%, H 3.30%, Fe 15.21%, N 7.63%, Na 6.26%, O 34.87%. Bioavailable iron source. Prepn from disodium ethylenediaminetetraacetic acid and ferric nitrate: D. T. Sawyer, J. M. McKinnie, *J. Am. Chem. Soc.* **82**, 4191 (1960). Review of pharmacology and use as iron fortificant in foods: J. Heimbach *et al.*, *Food Chem. Toxicol.* **38**, 99-111 (2000); T. H. Bothwell, A. P. MacPhail, *Int. J. Vitam. Nutr. Res.* **74**, 421-434 (2004). Review of clinical trials for treatment of iron deficiency: B. Wang *et al.*, *Br. J. Nutr.* **100**, 1169-1178 (2008).

Crystals from water + ethanol.
USE: Food supplement as source of iron.
THERAP CAT: Hematinic.

4059. Ferric Sulfate. [10028-22-5] Sulfuric acid iron(3+) salt (3:2); diiron trisulfate; ferric persulfate; ferric sesquisulfate. $Fe_2O_{12}S_3$; mol wt 399.86. Fe 27.93%, O 48.01%, S 24.05%. $Fe_2(SO_4)_3$. Prepn: *Gmelins, Iron* (8th ed.) **59**, part B, 439-462 (1932).
Grayish-white powder, or rhombic or rhombohedral crystals. Very hygroscopic. Commercial product usually contains about 20% water and is yellowish in color. d^{18} 3.097. Slowly sol in water, rapidly sol in the presence of a trace of $FeSO_4$; sparingly sol in alcohol; practically insol in acetone, ethyl acetate. Hydrolyzed slowly in aq soln. *Keep well closed and protected from light.*
Caution: Potential symptoms of overexposure are irritation of eyes, skin, mucous membranes; abdominal pain, diarrhea, vomiting; liver damage. *See NIOSH Pocket Guide to Chemical Hazards* (DHHS/NIOSH 97-140, 1997) p 174.
USE: In preparation of iron alums, other iron salts and pigments; as coagulant in water purification and sewage treatment; in etching aluminum; in pickling stainless steel and copper; as mordant in textile dyeing and calico printing; in soil conditioners; as polymerization catalyst.

4060. Ferric Tannate. Ferric gallotannate. Variable composition. Contains 8-10% Fe, 70-80% tannin.
Bluish-black powder. Insol in water; sol in dil mineral acids.
USE: In inks.

4061. Ferric Thiocyanate. [4119-52-2] Thiocyanic acid iron(3+) salt; ferric sulfocyanate; ferric sulfocyanide. $C_3FeN_3S_3$; mol wt 230.08. C 15.66%, Fe 24.27%, N 18.26%, S 41.80%. Fe(SCN)$_3$. Prepn: *Gmelins, Iron* (8th ed.) part B, 747-761 (1932).
Sesquihydrate. Red, deliquesc crystals. Dec on heating. Sol in water, alcohol, ether, acetone, pyridine, ethyl acetate. Practically insol in $CHCl_3$, CCl_4, CS_2, toluene. *Keep well closed.*
USE: Analytical reagent.

4062. Ferrite. [1317-54-0] Ferrospinel. A crystalline, usually man-made material, having a spinel structure and consisting essentially of ferric oxide and at least one other metallic oxide which is usually, although not always, divalent in nature. When molded into compressed bodies, the material is characterized by high magnetic permeability. Typified composition: Fe_2O_3 67-70%; ZnO 10-10.5%; MnO_2 20-22.5%; CuO 0.1-10%; Co_3O_4 0.1%. Ferrites are prepd by ceramic techniques. The oxides or carbonates are milled in steel ball mills, and the mixture of very fine particles is dried and prefired in order to obtain a homogeneous end product: Hilpert, *Ber.* **42**, 2248 (1909). Prepn: J. O. Simpkiss, **US 2723238**; R. L. Harvey, **US 2723239** (both 1955 to RCA). Prepn of single crystals: Rooymans, *Colloq. Int. Cent. Nat. Rech. Sci.* **No. 205**, 151 (1972). Books: Snoek, *New Developments in Ferromagnetic Materials* (Elsevier, New York, 1947); Smit, Wijn, *Ferrites* (John Wiley, New York, 1959); Soohov, *Theory and Applications of Ferrites* (Prentice Hall,

1960); Standley, *Oxide Magnetic Materials* (Clarendon Press, Oxford, 1962); *Ferrites, Proc. Int. Conf.*, Y. Hoshino *et al.*, Eds. (University Park Press, Baltimore, 1971) 671 pp; E. E. Riches, *Ferrites, A Review of Materials and Applications* (Mills and Boon, London, 1972) 88 pp. Reviews with bibliographies: Gorter, *Proc. IRE* **43**, 1945-1973 (1955); Fresh, "Methods of Preparation and Crystal Chemistry of Ferrites," *ibid.* **44**, 1303-1311 (1956); Brailsford, *Magnetic Materials* (John Wiley, New York, 3rd ed., 1960) pp 160-181; Hogen, *Sci. Am.* **202**, 92-104 (1960); Economos in *Kirk-Othmer Encyclopedia of Chemical Technology* vol. 8 (Interscience, New York, 2nd ed., 1965) pp 881-901; Gray, "Oxide Spinels" in *High Temperature Oxides*, Part IV, A. M. Alper, Ed. (Academic Press, New York, 1971) pp 77-107.
Note: The term "ferrites" has been expanded to mean any oxidic magnetic material.
USE: Magnetic cores for inductors and transformers; microwave devices; information storage; electromechanical transducers: E. E. Riches, *loc. cit.*; Brockman, *Ceram. Ind.* **99**, 24 (1972).

4063. Ferritin. [9007-73-2] Ferrofolin; Ferrol; Ferrosprint; Ferrostar; Sanifer; Sideros; Unifer. Major iron storage protein; widely distributed in the plant and animal kingdoms. Consists of a 24-subunit protein shell surrounding a crystalline hydrous ferric oxide core. The core may contain up to 4500 Fe^{3+} ions. The protein shell, *apoferritin*, has a mol wt of ~445,000. Isoln and crystallization of horse spleen ferritin: V. Laufberger, *Bull. Soc. Chim. Biol.* **19**, 1575 (1937); S. Granick, *J. Biol. Chem.* **146**, 451 (1942). X-ray structural study of apoferritin: P. M. Harrison, *J. Mol. Biol.* **6**, 404 (1963). Review of properties and role in iron metabolism: S. Granick, *Chem. Rev.* **38**, 379-403 (1946); of structure, biosynthesis and function: R. R. Crichton, *N. Engl. J. Med.* **284**, 1413-1422 (1971); P. M. Harrison, T. G. Hoy, "Ferritin" in *Inorganic Biochemistry* vol. 1, G. L. Eichhorn, Ed. (Elsevier, New York, 1973) pp 253-279; H. N. Munro, M. C. Linder, *Physiol. Rev.* **58**, 317-396 (1978); of applications in clinical medicine: J. W. Halliday, L. W. Powell, *Prog. Hematol.* **11**, 229-266 (1979); of biology: M. Worwood, *Blood Rev.* **4**, 259-269 (1990); of structure and role in iron mineralization: N. D. Chasteen, P. M. Harrison, *J. Struct. Biol.* **126**, 182-194 (1999); of *mitochondrial ferritin:* J. Drysdale *et al.*, *Blood Cells Mol. Dis.* **29**, 376-383 (2002); of role in dietary iron supplementation: E. C. Theil, *Annu. Rev. Nutr.* **24**, 327-343 (2004).
Red-brown, water-soluble protein. Forms cubic and orthorhombic crystals.
USE: Tool for the study of protein synthesis and regulatory mechanisms.
THERAP CAT: Hematinic.

4064. Ferrocene. [102-54-5] Dicyclopentadienyliron; biscyclopentadienyliron. $C_{10}H_{10}Fe$; mol wt 186.04. C 64.56%, H 5.42%, Fe 30.02%. Prepn: Kealy, Pauson, *Nature* **168**, 1039 (1951); Pauson, **US 2680756** (1954 to Du Pont); Miller *et al.*, *J. Chem. Soc.* **1952**, 632; Anzilotti, Weinmayr, **US 2791597** (1957 to DuPont). Other prepn: Pruett, Morehouse in *Adv. Chem. Ser.* **23**, entitled "Metal-Organic Compounds," M. Sittig, Ed. (ACS, Washington DC, 1959) pp 368-371; Wilkinson, *Org. Synth.* **coll. vol. IV**, 473 (1963); Cordes, **FR 1341880** (1963 to BASF), *C.A.* **60**, 6873a (1964). Structure studies: Wilkinson *et al.*, *J. Am. Chem. Soc.* **74**, 2125 (1952); Seibold, Sutton, *J. Chem. Phys.* **23**, 1967 (1955). Synthesis of a helical ferrocene: T. J. Katz, J. Pesti, *J. Am. Chem. Soc.* **104**, 346 (1982). *Reviews:* Rausch *et al.*, *J. Chem. Educ.* **34**, 268-272 (1957); M. Rosenblum, *Chemistry of the Iron Group Metallocenes* (John Wiley, New York, 1965) 241 pp; Bruce, *Organomet. Chem. Rev. Sect. B* **10**, 75-122 (1972); B. W. Rockett, G. Marr, *J. Organomet. Chem.* **211**, 215-278 (1981).

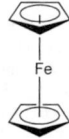

Orange needles from methanol or ethanol; odor of camphor. mp 173-174°. Sublimes above 100°. Volatile in steam. Practically insol in water, 10% NaOH, and concd boiling HCl. Sol in alcohol, ether, benzene. Dissolves in dil nitric and concd sulfuric acids forming a deep red soln with blue fluorescence. The molecule is diamagnetic and the dipole moment is effectively zero.

Caution: Potential symptoms of overexposure are irritation of eyes, skin and respiratory system. *See NIOSH Pocket Guide to Chemical Hazards* (DHHS/NIOSH 97-140, 1997) p 104.

USE: Antiknock additive for gasoline; catalyst.

4065. Ferrocholinate. [1336-80-7] 2-Hydroxy-*N,N,N*-trimethylethanaminium (*OC*-6-44)-triaqua[2-(hydroxy-κO)-1,2,3-propanetricarboxylato(4−)-$\kappa O^1,\kappa O^2$]ferrate(1−) (1:1); [hydrogen citrato(3−)]triaquoiron, choline salt; iron choline citrate complex. $C_{11}H_{24}FeNO_{11}$; mol wt 402.15. C 32.85%, H 6.02%, Fe 13.89%, N 3.48%, O 43.76%. Chelate prepd by interaction of equimolar quantities of choline dihydrogen citrate and freshly prepd $Fe(OH)_3$ or $FeCO_3$: F. J. Bandelin, US 2575611 (1951 to Flint Eaton); by treatment of a freshly prepd solution of ferric citrate with an equimolar amount of choline: J. K. Chakrabarti, S. P. Sen, *Chem. Ind. (London)* **1961**, 1407. Pediatric use in anemia: J. Rose, *Clin. Med.* **69**, 1601 (1962). Clinical evaluation in iron deficient infants: J. C. Lee, *J. Catholic Med. Coll.* **30**, 479 (1977). Pharmaceutical prepn, characteristics and applications: G. N. Bookwalter *et al., J. Food Sci.* **38**, 618 (1973). Bioavailability in pigs: E. R. Miller *et al., J. Anim. Sci.* **52**, 783 (1981). Brief review in anemia: P. P. Gerbino, K. J. Tietze, *Am. J. Pharm.* **155**, 49-62 (1983).

Greenish-brown, reddish-brown or brown amorphous solid with glistening surface upon fracture. Freely sol in water, yielding stable solns; sol in acids, alkalies. One gram of pharmaceutical grade is equivalent to 120 mg of elemental iron and 360 mg of choline base.

Note: Was sold commerically under the names **Chel-Iron, Ferrolip, Chelafer**. Ferrocholinate is also used for other combinations of iron, choline and citric acid.

THERAP CAT: Formerly as a hematinic.

4066. Ferroglycine Sulfate. [17169-60-7] (*T*-4)-(Glycinato-$\kappa N,\kappa O$) [sulfato(2−)-$\kappa O,\kappa O'$]ferrate(1−) hydrogen (1:1); ferrous aminoacetosulfate; iron sulfate-glycine complex; Ferrocontin; Plesmet; Ferrosanol. $C_2H_5FeNO_6S$; mol wt 226.97. C 10.58%, H 2.22%, Fe 24.60%, N 6.17%, O 42.29%, S 14.13%. Chelate of ferrous sulfate and glycine, *q.q.v.* Prepn: W. Rummell, US 2877253 (1959 to Schwarz Arzneimittelfabrik). Tolerability study: J. Pomeranze, R. J. Gadek, *N. Engl. J. Med.* **257**, 73 (1957). Use in iron fortification of salt: S. Ranganathan *et al., Food Chem.* **57**, 311 (1996). Clinical evaluation in iron deficiency anemia vs ferrous fumarate, *q.v.*: A. Aronstam, D. L. Aston, *Pharmatherapeutica* **3**, 263 (1982).

THERAP CAT: Hematinic.

4067. Ferrosoferric Oxide. [1317-61-9] Iron oxide (Fe_3O_4); ferriferrous oxide; triiron tetraoxide; black iron oxide; magnetic iron oxide; ethiops iron. Fe_3O_4; mol wt 231.53. Fe 72.36%, O 27.64%. Occurs in nature as the mineral *magnetite* (red-black lumps). Prepn: *Gmelins, Iron* (8th ed) part B, 36-62 (1932); *Ullmanns Encyklopädie der technischen Chemie* **vol. 6**, 420 (1955). *Review:* Robl, *Angew. Chem.* **70**, 367 (1958).

Black cubes or amorphous powder. mp 1538°; d 5.2. Oxidized to Fe_2O_3 on heating in air. Practically insol in water. Sol in acids.

USE: Pigment in paints, linoleum, ceramic glazes; in coloring glass; as a polishing compd; in the textile industry; in cathodes; as catalyst.

4068. Ferrous Bisglycinate. [20150-34-9] Bis(glycinato-$\kappa N,\kappa O$)iron; bisglycino-iron(II) chelate; Ferrochel. $C_4H_8FeN_2O_4$; mol wt 203.96. C 23.56%, H 3.95%, Fe 27.38%, N 13.74%, O 31.38%. Nutritionally functional, totally reacted, metal amino acid chelate. Prepn: B. W. Fitzsimmons *et al., Inorg. Chim. Acta* **106**, 109 (1985). Clinical evaluation of tolerability vs ferrous sulfate: M. Coplin *et al., Clin. Ther.* **13**, 606 (1991); of iron bioavailability: M. Layrisse *et al., J. Nutr.* **130**, 2195 (2000). Clinical study in iron deficiency anemia in infants: O. Pineda, H. D. Ashmead, *Nutrition* **17**, 381 (2001); in children: L. H. S. Miglioranza *et al., ibid.* **19**, 419 (2003). Toxicity study: R. B. Jeppsen, J. F. Borzelleca, *Food Chem. Toxicol.* **37**, 723 (1999). *Reviews:* M. R. Motyka, *Agro-Food-Ind. Hi-Tech* **12**, 6-8, 14-16 (2001).

Off-white solid. LD_{50} orally in rats: 560 mg iron/kg (Jeppsen).
THERAP CAT: Hematinic.

4069. Ferrous Bromide. [7789-46-0] Iron bromide ($FeBr_2$); iron dibromide. Br_2Fe; mol wt 215.65. Br 74.11%, Fe 25.90%. $FeBr_2$. Prepn: Baxter, *Z. Anorg. Chem.* **38**, 236 (1904); Baxter *et al., ibid.* **70**, 333 (1911); Schimmel, *Ber.* **62**, 963 (1929); Kühnl, Ernst, *Z. Anorg. Allg. Chem.* **317**, 84 (1962).

Light yellow to dark brown hygroscopic crystals. mp 684°. d_4^{25} 4.63. Very sol in water and alcohol. *Keep tightly closed.*

Hexahydrate. Pale green to bluish-green rhombic prisms. Forms tetrahydrate at 49°; dihydrate at 83°. Rapidly oxidized in moist air. *Keep tightly closed.*

USE: Polymerization catalyst.

4070. Ferrous Chloride. [7758-94-3] Iron chloride ($FeCl_2$); iron dichloride. Cl_2Fe; mol wt 126.75. Cl 55.94%, Fe 44.06%. $FeCl_2$. Occurs in nature as the mineral *lawrencite*. Preparation: Kovacic, Brace, *Inorg. Synth.* **6**, 172 (1960); Kühnl, Ernst, *Z. Anorg. Allg. Chem.* **317**, 84 (1962). Prepn of dihydrate: Gayer, Woontner, *Inorg. Synth.* **5**, 179 (1957).

White rhombohedral crystals; may sometimes have a green tint. Very hygroscopic. mp 674°; bp 1023°; d^{25} 3.16. *Corrosive.* Can be sublimed in a stream of HCl at about 700°. Forms $FeCl_3$ and Fe_2O_3 on heating in air. Freely sol in water, alc, acetone; slightly sol in benzene. Practically insol in ether.

Dihydrate. White monoclinic crystals with pale green tint. Loses 1 H_2O at 120°: Gayer, Woontner, *loc. cit.;* also reported to lose 1 H_2O at 150-160°: Schafer, *Z. Anorg. Allg. Chem.* **258**, 69 (1949). Sol in water.

Tetrahydrate. Pale green to blue-green, monoclinic crystals or cryst powder. Loses $2H_2O$ at about 105-115°: Schafer, *loc. cit.* d 1.93. Sol in water, alcohol. The technical product may not be completely sol without the addn of acid. Aq solns are readily oxidized.

Caution: Potential symptoms of overexposure are irritation of eyes, skin, mucous membranes; abdominal pain, diarrhea, vomiting; possible liver damage. *See NIOSH Pocket Guide to Chemical Hazards* (DHHS/NIOSH 97-140, 1997) p 174.

USE: In metallurgy; as reducing agent; in pharmaceutical prepns; as mordant in dyeing.

4071. Ferrous Citrate. [23383-11-1] 2-Hydroxy-1,2,3-propanetricarboxylic acid iron(2+) salt (1:?); citric acid iron(2+) salt. Several forms of this salt are known. Prepn from citric acid and Fe powder: Oroshnik, Haffcke, US 2904573 (1959 to Ortho); and ferrous salts: Carlson, US 3091626 (1963 to Scherer).

Monohydrate. Monoferrous acid citrate monohydrate. Powder. Practically insol in water, alcohol, acetone.

Decahydrate. Triferrous dicitrate decahydrate. Very slightly colored powder or white crystals. Very stable to air oxidation. If H_2O is removed by vacuum desiccation, the dehydrated ferrous salt rapidly oxidizes to a ferric salt. Practically insol in water, acetone.

THERAP CAT: Hematinic.

4072. Ferrous Fluoride. [7789-28-8] Iron fluoride (FeF$_2$); iron difluoride. F$_2$Fe; mol wt 93.84. F 40.49%, Fe 59.51%. FeF$_2$. Prepn from FeCl$_2$ and HF gas: Kwasnik in *Handbook of Preparative Inorganic Chemistry*, G. Brauer, Ed. (Academic Press, New York, 2nd ed., 1963) p 266; from Fe powder and liq HF: Muetterties, Castle, *J. Inorg. Nucl. Chem.* **18**, 148 (1961).

Tetragonal crystals (rutile type) or powder. Sublimes at about 1100°. d 4.09. Sparingly sol in water; more sol in dil HF. Practically insol in alcohol, ether, benzene.

USE: As catalyst in organic reactions.

4073. Ferrous Fumarate. [141-01-5] (2*E*)-2-Butenedioic acid iron(2+) salt (1:1); Erco-Fer; Ferronat; Ferrotemp; Fersaday; Fersamal; Fumafer; Fumiron; Galfer; Heferol; Ircon; Meterfer; Palafer; Toleron; Tolferain. C$_4$H$_2$FeO$_4$; mol wt 169.90. C 28.28%, H 1.19%, Fe 32.87%, O 37.67%. Carboxyl groups in the fumarate act as bidentate ligands; overall structure is polymeric. Prepn: H. C. Bertsch, J. F. Lemp, US 2848366 (1958 to Mallinckrodt); and structural studies: C. P. Prabhakaran, C. C. Patel, *Indian J. Chem.* **7**, 266 (1969). Animal toxicity and human studies in treatment of anemia: M. C. Berenbaum *et al., Blood* **15**, 540 (1960). Pharmacokinetics in rats: F. Ryszka *et al., Boll. Chim. Farm.* **137**, 178 (1998). Clinical evaluation in dietary iron fortification: R. F. Hurrell *et al., Br. J. Nutr.* **65**, 271 (1991); with ascorbic acid, *q.v.*, in anemia in infants: S. Zlotkin *et al., Am. J. Clin. Nutr.* **74**, 791 (2001).

Reddish-orange to reddish-brown, granular powder. d^{25} 2.435. Odorless; almost tasteless. Slightly soluble in water; very slightly sol in alc. Soly in dilute hydrochloric acid is limited by the separation of fumaric acid. Practically insol in common organic solvents including styrene and α-methylstyrene. LD$_{50}$ in mice, rats (mg Fe/kg): 630, 580 orally (Berenbaum).

THERAP CAT: Hematinic.

4074. Ferrous Gluconate. [299-29-6] Bis(D-gluconato-*κO*^{1},-*κO*^{2})iron; Fergon; Ferlucon; Ferronicum; Iromon; Irox; Nionate. C$_{12}$H$_{22}$FeO$_{14}$; mol wt 446.14. C 32.31%, H 4.97%, Fe 12.52%, O 50.21%. Fe[HOCH$_2$(CHOH)$_4$CO$_2$]$_2$. Prepd from Ba gluconate and FeSO$_4$. Prepn of isotonic solns: Hammarlund, *Pharm. Acta Helv.* **35**, 593 (1960). Toxicity study: Hoppe *et al., Am. J. Med. Sci.* **230**, 491 (1955).

Dihydrate. Yellowish-gray or pale greenish-yellow powder. Slight odor of caramel. Acid to litmus. Sol in water. Practically insol in alcohol. Aq solns are stabilized by the addition of glucose. Extensive stability studies on aq solns: Johnson, Thomas, *J. Pharm. Pharmacol.* **6**, 1037 (1954). LD$_{50}$ in mice: 114 mg/kg i.v.; 3.7 g/kg orally (Hoppe).

USE: As coloring and flavoring in foods.

THERAP CAT: Hematinic.

4075. Ferrous Hydroxide. [18624-44-7] Iron hydroxide (Fe-(OH)$_2$); iron dihydroxide. FeH$_2$O$_2$; mol wt 89.86. Fe 62.15%, H 2.24%, O 35.61%. Fe(OH)$_2$. Prepn: Rihl, Fricke, *Z. Anorg. Allg. Chem.* **251**, 406 (1943); Leussing, Kolthoff, *J. Am. Chem. Soc.* **75**, 2476 (1953). Oxidation products: Feitknecht, Keller, *Z. Anorg. Chem.* **262**, 61 (1950); Mayne, *J. Chem. Soc.* **1953**, 129; Shipko, Douglas, *J. Phys. Chem.* **60**, 1519 (1956).

White, amorphous powder or white to pale green hexagonal crystals. Oxidized on exposure to air; may ignite spontaneously on exposure to air if finely divided. Practically insol in water. More sol in solns of NH$_4$ salts; sol in concd NaOH soln.

4076. Ferrous Iodide. [7783-86-0] Iron iodide (FeI$_2$); iron diiodide. FeI$_2$; mol wt 309.65. Fe 18.03%, I 81.97%. Prepn: Chaigneau, *Bull. Soc. Chim. Fr.* **1957**, 886; Lieser, Elias, *Z. Anorg. Allg. Chem.* **316**, 208 (1962); Lux in *Handbook of Preparative Inorganic Chemistry* vol. 2, G. Brauer, Ed. (Academic Press, New York, 2nd ed., 1965) p 1495.

Large, thin, red-violet crystals or black leaflets. Very hygroscopic. Sol in water, alcohol, ether; aq soln is readily oxidized by air. *Keep tightly closed and protected from light.*

USE: As catalyst for organic reactions.

4077. Ferrous Lactate. [5905-52-2] 2-Hydroxypropanoic acid iron(2+) salt (2:1). C$_6$H$_{10}$FeO$_6$; mol wt 233.99. C 30.80%, H 4.31%, Fe 23.87%, O 41.02%. Fe(C$_3$H$_5$O$_3$)$_2$. Prepn: *Gmelins, Iron* (8th ed.) **59**, part B, p 532 (1932). Prepn of isotonic solns: Hammarlund, *Pharm. Acta Helv.* **35**, 593 (1960). Toxicity study: T. H. Eickholt, W. F. White, *J. Pharm. Sci.* **54**, 1211 (1965). Review of toxicology: B. Venugopal, T. D. Luckey, *Environ. Qual. Safety* Suppl. 1, 4-73 (1975).

LD$_{50}$ orally in mice: 147 ±30 mg/kg (Eickholt, White).

Trihydrate. [6047-24-1] Greenish-white powder or cryst masses. Slight characteristic odor; mild, sweet and ferruginous taste. Sol in water; freely sol in alkali citrates forming a green soln. Almost insol in alc. On exposure to air it becomes darker and incompletely sol. *Keep well closed and protected from light.* LD in rabbits (mg/kg): 578 s.c., 287 i.v. (Venugopal, Luckey).

USE: Food coloring.

THERAP CAT: Hematinic.

4078. Ferrous Oxalate. [516-03-0] [Ethanedioato(2−)-*κO*^{1},-*κO*^{2}]iron; ferrox; iron oxalate. C$_2$FeO$_4$; mol wt 143.86. C 16.70%, Fe 38.82%, O 44.48%. FeC$_2$O$_4$. Prepn of dihydrate: *Gmelins, Iron* (8th ed.) **59**, part B, 532-533 (1932); prepn of anhydr salt: Waterman, Vivian, *J. Org. Chem.* **14**, 289 (1949).

Dihydrate. Pale yellow, odorless, cryst powder. d 2.28. Dec at 150-160° on heating in air. Slightly sol in water; sol in dil mineral acids.

USE: As a photographic developer for silver bromide-gelatin plates; to impart a greenish-brown tint to optical glass (sunglasses, windshields, railroad car windows), for decorative glassware; pigment for plastics, paints, lacquers.

4079. Ferrous Oxide. [1345-25-1] Iron oxide (FeO); iron monoxide. FeO; mol wt 71.84. Fe 77.74%, O 22.27%. Prepn: Lux in *Handbook of Preparative Inorganic Chemistry* vol. 2, G. Brauer, Ed. (Academic Press, New York, 2nd ed., 1965) p 1497.

Jet-black powder. mp 1360°; d 5.7. Easily oxidized by air. It is a strong base and readily absorbs CO$_2$. Practically insol in water and alkalies. Readily sol in acids.

USE: In manuf of green, heat-absorbing glass; in steel manuf; in enamels; as catalyst.

4080. Ferrous Phosphate. [14940-41-1] Phosphoric acid iron(2+) salt (2:3); ferrophosphate; iron(2+) phosphate. Fe$_3$O$_8$P$_2$; mol wt 357.47. Fe 46.87%, O 35.80%, P 17.33%. Fe$_3$(PO$_4$)$_2$. Prepn of octahydrate: *Gmelins, Iron* (8th ed.) **59**, part B, 771 (1932). Due to unavoidable oxidation, the article of commerce contains basic ferric phosphate.

Octahydrate. Grayish-blue powder or monoclinic crystals. d 2.58. Practically insol in water. Sol in mineral acids. *Keep well closed and protected from light.*

USE: In ceramics, as catalyst.

4081. Ferrous Phosphide. [1310-43-6] Iron phosphide (Fe$_2$P); iron hemiphosphide; diiron phosphide. Fe$_2$P; mol wt 142.66. Fe 78.29%, P 21.71%. Prepn from Fe and red P: Rundqvist, Jellinek, *Acta Chem. Scand.* **13**, 425 (1959); Cadeville, Meyer, *Compt. Rend.* **252**, 1124 (1961).

Gray, hexagonal needles or blue-gray powder. d 6.85. Ferromagnetic. Insol in water, dil acid, dil alkali. Reacts with hot mineral acids.

4082. Ferrous Selenide. [1310-32-3] Iron selenide (FeSe); iron monoselenide. FeSe; mol wt 134.81. Fe 41.42%, Se 58.57%. Prepn from the elements: Gronvold, Westrum, *Acta Chem. Scand.* **13**, 241 (1959).

Black mass with metallic luster. Stable in air. Dec when heated in O$_2$. d 6.78. Practically insol in water. Sol in HCl with the evolution of H$_2$Se.

USE: In electrical semiconductors.

4083. Ferrous Succinate. [10030-90-7] Butanedioic acid iron(2+) salt (1:1); Cerevon; Ferromyn. C$_4$H$_4$FeO$_4$; mol wt 171.92. C 27.95%, H 2.35%, Fe 32.48%, O 37.22%. Fe^{2+}(OOCCH$_2$CH$_2$COO)$^{2-}$. Prepn: Franke, *Ann.* **491**, 30 (1931).

Tetrahydrate. Sparingly sol in water. Also obtained as a dihydrate.

THERAP CAT: Hematinic.

4084. Ferrous Sulfate. [7720-78-7] Sulfuric acid iron(2+) salt (1:1); iron(2+) sulfate; iron(II) sulfate. FeO_4S; mol wt 151.90. Fe 36.76%, O 42.13%, S 21.11%. $FeSO_4$. Hydrates occur in nature as the minerals: *melanterite, siderotil, szomolnikite, tauriscite*. Heptahydrate prepd commercially by the action of H_2SO_4 on Fe: *Faith, Keyes & Clark's Industrial Chemicals*, F. A. Lowenheim, M. K. Moran, Eds. (Wiley-Interscience, New York, 4th ed., 1975) pp 418-421. Crystal structure of heptahydrate: Baur, *Acta Crystallogr.* **17**, 1167 (1964). Acute toxicity: Hoppe *et al., Am. J. Med. Sci.* **230**, 491 (1955).

Monohydrate. [17375-41-6] Dried ferrous sulfate; exsiccated ferrous sulfate; Feromax; Feroritard; Ferro-Gradumet; Fespan; Tetucur. $FeSO_4 \cdot H_2O$; mol wt 169.92. White to yellow cryst powder. Loses H_2O at about 300°. Dec at higher temps. Slowly sol in water. Insol in alc.

Heptahydrate. [7782-63-0] Copperas; green vitriol; iron vitriol; Feosol; Feospan; Fesofor; Fero-Gradumet; Fer-in-Sol; Haemofort; Ironate; Mol-Iron; Presfersul; Sulferrous. $FeSO_4 \cdot 7H_2O$; mol wt 278.01. Blue-green, monoclinic, odorless crystals or granules. Efflorescent in dry air; oxidizes in moist air forming a brown coating of basic ferric sulfate. Forms tetrahydrate at 56.6° and monohydrate at 65°. d 1.897. Freely sol in water; very sol in boiling water. Practically insol in alcohol. Aq solns are oxidized slowly by air when cold, rapidly when hot; rate of oxidation increased by addn of alkali or exposure to light. LD_{50} in mice: 65 mg/kg i.v.; 1.52 g/kg orally (Hoppe).

Caution: Potential symptoms of overexposure are irritation of eyes, skin, mucous membranes; GI disturbances, abdominal pain, vomiting, diarrhea; dehydration; shock, pallor, cyanosis, coldness; rapid, weak pulse; low blood pressure; rapid, shallow respirations; drowsiness; hyporeflexia; dilated pupils; coma; liver damage. *See NIOSH Pocket Guide to Chemical Hazards* (DHHS/NIOSH 97-140, 1997) p 174; *Clinical Toxicology of Commercial Products*, R. E. Gosselin *et al.*, Eds. (Williams & Wilkins, Baltimore, 5th ed., 1984) Sect. III, pp 179-185.

USE: In manufacture of Fe, Fe compds, other sulfates; in Fe electroplating baths; in fertilizer; as food and feed supplement; in radiation dosimeters; as reducing agent in chemical processes; as wood preservative; as weed-killer; in prevention of chlorosis in plants; in other pesticides; in writing ink; in process engraving and lithography; as dye for leather; in etching aluminum; in water treatment; in qualitative analysis ("brown ring" test for nitrates); as polymerization catalyst.

THERAP CAT: Hematinic.

THERAP CAT (VET): In iron deficiency. Astringent.

4085. Ferrous Sulfide. [1317-37-9] Iron sulfide (FeS); iron sulfuret. FeS; mol wt 87.91. Fe 63.53%, S 36.47%. Occurs in nature as the minerals: *magnetkies, pyrrhotine, troillite*. Usually prepd from the elements: Lux in *Handbook of Preparative Inorganic Chemistry* **vol 2**, G. Brauer, Ed. (Academic Press, New York, 2nd ed., 1965) p 1502. The commercial product is ~75-80% FeS.

Colorless hexagonal crystals when pure; usually gray to brownish-black lumps, rods or granular powder. mp 1194°; d 4.84. Trimorphic with transition points at 135° and 325°. Oxidized by moist air to S and Fe_3O_4. Practically insol in water. Soluble in acids with the evolution of H_2S.

USE: As a laboratory source of H_2S; in the ceramic industry; as a paint pigment; in anodes; in lubricant coatings.

4086. Ferrous Thiocyanate. [6010-09-9] Thiocyanic acid iron(2+) salt; ferrous sulfocyanate; iron dithiocyanate. $C_2FeN_2S_2$; mol wt 172.00. C 13.97%, Fe 32.47%, N 16.29%, S 37.28%. $Fe(SCN)_2$. Prepn: *Gmelins, Iron* (8th ed) **59**, part B, 743-747 (1932); O'Brien, *Chem. Eng. News* **33**, 2008 (1955).

Trihydrate, pale green monoclinic prisms. Rapidly oxidized on exposure to air; dec by heat. Freely sol in water, alcohol, ether. *Keep tightly closed and protected from light.*

USE: Indicator for peroxides in organic solns.

4087. Fertilysin. [1477-57-2] *N,N'*-1,8-Octanediylbis[2,2-dichloroacetamide]; *N,N'*-octamethylenebis(2,2-dichloroacetamide); *N,N'*-bis(dichloroacetyl)-1,8-octamethylenediamine; Win-18446. $C_{12}H_{20}Cl_4N_2O_2$; mol wt 366.10. C 39.37%, H 5.51%, Cl 38.73%, N 7.65%, O 8.74%. Inhibits spermatogenesis without affecting production of gonadotropins in male laboratory animals.

Prepn and biological activity: Surrey, Mayer, *J. Med. Pharm. Chem.* **3**, 419 (1961); Surrey, **US 3143566** (1964 to Sterling Drug). Assessment of antifertility activity in rats: S. Nag, J. J. Ghosh, *J. Steroid Biochem.* **11**(1B), 681 (1979). Teratological study: P. Taleporos *et al., Teratology* **18**, 5 (1978). Toxicology study: F. Coulston *et al., Toxicol. Appl. Pharmacol.* **2**, 715 (1960).

Crystals, mp 122.4-123.6°. LD_{50} orally in mice: >16000 mg/kg (Coulston).

4088. Fertirelin. [38234-21-8] 9-(*N*-Ethyl-L-prolinamide)-1-9-luteinizing hormone-releasing factor (swine); 9-(*N*-ethyl-L-prolinamide)-10-deglycinamide-luteinizing hormone-releasing factor (pig); Des-Gly^{10}-NH$_2$-LH-RH-ethylamide; TAP-031. $C_{55}H_{76}N_{16}O_{12}$; mol wt 1153.31. C 57.28%, H 6.64%, N 19.43%, O 16.65%. Synthetic nonapeptide analog of LH-RH, *q.v.* Prepn: M. Fujino *et al.*, **DE 2321174**; *eidem*, **US 3853837** (1973, 1974 both to Takeda); S. Shinagawa, M. Fujino, *Chem. Pharm. Bull.* **23**, 229 (1975); *see also:* M. Fujino *et al., Biochem. Biophys. Res. Commun.* **49**, 863 (1972). Enzyme immunoassay in bovine plasma: J. Okada, S. Kondo, *ibid.* **33**, 4464 (1985). HPLC determn of bulk drugs and formulations: P. A. Hartman, J. D. Stodola, *J. Chromatogr.* **444**, 177 (1988). Field trial in bovine cystic ovarian disease: T. Nakao *et al., Jpn. J. Vet. Sci.* **45**, 269 (1983); to induce ovulation for insemination of dairy cows: T. Nakao *et al., Theriogenology* **20**, 111 (1983); K. Yamada *et al., Anim. Reprod. Sci.* **74**, 27 (2002).

5-oxoPro–His–Trp–Ser–Tyr–Gly–Leu–Arg–ProNHCH$_2$CH$_3$

Acetate. [66002-66-2] U-69689E; Conceral; Ovalyse. $C_{55}H_{76}N_{16}O_{12} \cdot C_2H_4O_2$; mol wt 1213.37. Pentahydrate as white, fluffy powder, $[\alpha]_D^{25}$ −53.6° (c = 0.5 in 5% acetic acid).

THERAP CAT (VET): Gonad stimulating principle.

4089. Ferulic Acid. [1135-24-6] 3-(4-Hydroxy-3-methoxyphenyl)-2-propenoic acid; 4-hydroxy-3-methoxycinnamic acid; 3-methoxy-4-hydroxycinnamic acid; caffeic acid 3-methyl ether. $C_{10}H_{10}O_4$; mol wt 194.19. C 61.85%, H 5.19%, O 32.96%. Widely distributed in small amounts in plants. Isoln from *Ferula foetida* Reg. *Umbelliferae:* H. Hlasiwetz, L. Barth, *Ann.* **138**, 61 (1866); from *Pinus laricio* Poir. *Abietineae:* M. Bamberger, *Monatsh. Chem.* **12**, 441 (1891); *see also Beilstein* **10**, 436 (1927) and supplements. Prepd by the interaction of vanillin, malonic acid and piperidine in pyridine for 3 weeks, then precipitating with HCl: Vorsatz, *J. Prakt. Chem.* **145**, 265 (1936); Pearl, Beyer, *J. Org. Chem.* **16**, 216 (1951). Sepn of *cis* and *trans* isomers: Comte *et al., Compt. Rend.* **245**, 1144 (1957). ^{13}C NMR study: C. J. Kelley *et al., J. Org. Chem.* **41**, 449 (1976). Discovery as a component of cell walls in wheat and barley: M. G. Smart, T. P. O'Brien, *Aust. J. Plant Physiol.* **6**, 485 (1979). Use as food preservative: T. Tsuchiya, M. Takasawa, **JP Kokai 75 18621** (1975 to Kyokuto Shibosan), *C.A.* **83**, 7602v (1975).

trans - Ferulic Acid

cis-**Form.** Yellow oil. uv max (alcohol): 316 nm.

trans-**Form.** Orthorhombic needles from water, mp 174°. uv max (alcohol): 236, 322 nm. Sol in hot water, alcohol, ethyl aetate. Moderately sol in ether. Sparingly sol in petr ether, benzene. Forms a sodium salt.

USE: Food preservative.

4090. Ferumoxides. [119683-68-0] AMI-25; Feridex; Endorem. Active ingredient of colloidal super paramagnetic iron oxide (SPIO) consisting of nonstoichiometric magnetite cores coated with dextran. Designed for use as hepatic MRI contrast agent. Prepn: E. V. Groman, L. Josephson, **WO 8800060**; *eidem*, **US 4770183** (both 1988 to Advanced Magnetics). Pharmacokinetics: S. Majumdar *et al.*, *Invest. Radiol.* **25**, 771 (1990). Physical and chemical properties: C. W. Jung, P. Jacobs, *Magn. Reson. Imaging* **13**, 661 (1995). Surface properties: C. W. Jung, *ibid.* 675. Clinical studies in hepatic imaging: T. J. Vogl *et al.*, *Radiology* **198**, 881 (1996); E. Senéterre *et al.*, *ibid.* **200**, 785 (1996). Series of articles on toxicology: *Jpn. Pharmacol. Ther.* **22**, 57-141 (1994). Review of physicochemical characteristics and clinical applications: O. Clement *et al.*, *Top. Magn. Reson. Imaging* **9**, 167-182 (1998).

Aqueous suspension, black to reddish-brown liquid. Osmolality: ~340 mOsm/kg. d 1.04.

THERAP CAT: Diagnostic aid (MRI contrast agent).

4091. Ferumoxsil. [171544-35-7] AMI-121; GastroMARK; Lumiren. Active ingredient of colloidal super paramagnetic iron oxide (SPIO) consisting of nonstoichiometric magnetite cores coated with siloxane (FeO$_x$[C$_3$H$_{13}$N$_2$SiO$_2$]$_y$); designed for use as negative MRI contrast agent. Prepn: M. S. Chagnon *et al.*, **EP 125995**, R. A. Whitehead *et al.*, **US 4554088** (1984, 1985 both to Advanced Magnetics). Physical and chemical properties: C. W. Jung, P. Jacobs, *Magn. Reson. Imaging* **13**, 661 (1995). Surface properties: C. W. Jung, *ibid.* 675. Clinical study in lower abdominal imaging: R. C. H. Heusler *et al.*, *J. Magn. Reson. Imaging* **4**, 385 (1995); A. D'Arienzo *et al.*, *Am. J. Gastroenterol.* **95**, 720 (2000).

Aqueous suspension, turbid, slightly viscous, dark brown to orange-brown liquid. Osmolality: 250 mOsm/kg. d 1.01.

THERAP CAT: Diagnostic aid (MRI contrast agent).

4092. Ferumoxtran 10. [189047-99-2] AMI-227; BMS-180549; Code-7227; Combidex; Sinerem. Ultrasmall superparamagnetic iron oxide (USPIO) magnetite nanoparticles covered with low molecular weight dextran T-10; designed for use as an MRI imaging agent to detect metastatic cancer in lymph nodes. Prepn: E. V. Groman *et al.*, **WO 8800060**; S. Palmacci, L. Josephson, **US 5262176** (1988, 1993 both to Advanced Magnetics). Physical and chemical properties: C. W. Jung, P. Jacobs, *Magn. Reson. Imaging* **13**, 661 (1995); surface properties: C. W. Jung, *ibid.* 675. Biodistribution study: H. H. Bengele *et al.*, *ibid.* **12**, 433 (1994). Clinical studies of diagnostic use: Y. Anai *et al.*, *Radiology* **228**, 777 (2003); A. G. Rockall *et al.*, *J. Clin. Oncol.* **23**, 2813 (2005). Review of clinical experience: M. G. Harisinghani *et al.*, *RadioGraphics* **24**, 867-878 (2004); of pharmacology and preclinical studies: P. Bourrinet *et al.*, *Invest. Radiol.* **41**, 313 2006.

Diameter of iron oxide core: 4.3-6.0 nm; of coated particle: 17-21 nm.

THERAP CAT: Diagnostic aid (MRI contrast agent).

4093. Ferumoxytol. [722492-56-0] 7228; Feraheme. Iron oxide nanoparticle coated with polyglucose sorbitol carboxymethylether with superparamagnetic properties; mol wt 731 kDa. Designed for use as an MRI contrast agent; has activity as iron replacement therapy in anemia. Prepn: E. V. Groman *et al.*, **WO 0061191**; *eidem*, **US 6599498** (2000, 2003 both to Advanced Magnetics). Physiochemical properties: V. S. Balakrishnan *et al.*, *Eur. J. Clin. Invest.* **39**, 489 (2009). Clinical potential as a contrast agent in cardiovascular MRI: M. R. Prince *et al.*, *J. Xray Sci. Technol.* **11**, 231 (2003). Clinical pharmacokinetics: R. Landry *et al.*, *Am. J. Nephrol.* **25**, 400 (2005). Clinical trials in anemic patients with chronic kidney disease: B. S. Spinowitz *et al.*, *J. Am. Soc. Nephrol.* **19**, 1599 (2008); R. Provenzano *et al.*, *Clin. J. Am. Soc. Nephrol.* **4**, 386 (2009). Review of clinical experience in iron deficiency anemia: D. W. Coyne, *Expert Opin. Pharmacother.* **10**, 2563-2568 (2009).

Pharmaceutical formulation is a black to reddish brown, aqueous colloidal product; pH 6 to 8. Isotonic; osmolality of 291-330 mOsm/kg. Average particle size 17-31 nm.

THERAP CAT: Hematinic. Diagnostic Aid (MRI Contrast Agent).

4094. Fervenulin. [483-57-8] 6,8-Dimethylpyrimido[5,4-*e*]-1,2,4-triazine-5,7-(6*H*,8*H*)-dione; 6,8-dimethyl-5,7-dioxo-5,6,7,8-tetrahydropyrimido[5,4-*e*]-*as*-triazine; 1,3-dimethylazalumiazine; planomycin. C$_7$H$_7$N$_5$O$_2$; mol wt 193.17. C 43.52%, H 3.65%, N

36.26%, O 16.56%. Antibiotic from culture filtrates of *Streptomyces fervens:* Eble *et al.*, *Antibiot. Annu.* **1959-1960**, 227. Structure: Daves *et al.*, *J. Org. Chem.* **26**, 5256 (1961). Synthesis: Pfleiderer, Schündehütte, *Ann.* **615**, 42 (1958); Daves *et al.*, *J. Am. Chem. Soc.* **84**, 1724 (1962); Yoneda, Nagamatsu, *Bull. Chem. Soc. Jpn.* **48**, 2884 (1975); Taylor, Sowinski, *J. Org. Chem.* **40**, 2321 (1975); S. Senda *et al.*, *J. Am. Chem. Soc.* **99**, 7358 (1977).

Yellow orthorhombic crystals, mp 178-179°. uv max (ethanol): 238, 275, 340 nm (ε 18,500, 1600, 4200). Sol in practically all of the common organic solvents; sol in cold water to about 2 mg/ml; in hot water to about 40 mg/ml. Practically insol in hydrocarbons. Labile to alkali; stable to acid.

4-Oxide. C$_7$H$_7$N$_5$O$_3$. Synthesis: M. Ichiba *et al.*, *J. Heterocycl. Chem.* **14**, 175 (1977); K. Senga *et al.*, *Heterocycles* **6**, 273 (1977); synthesis and conversion to fervenulin: M. Ichiba *et al.*, *J. Org. Chem.* **43**, 469 (1978). Crystals from alc, mp 179-180°. uv max (alc): 240, 304 nm (log ε 4.10, 3.21).

4095. Fesoterodine. [286930-02-7] 2-Methylpropanoic acid 2-[(1*R*)-3-[bis(1-methylethyl)amino]-1-phenylpropyl]-4-(hydroxymethyl)phenyl ester; 2-[(1*R*)-3-[bis(1-methylethyl)amino]-1-phenylpropyl]-4-(hydroxymethyl)phenyl 2-methylpropanoate; (*R*)-(+)-2-(3-diisopropylamino-1-phenylpropyl)-4-hydroxymethylphenylisobutyrate ester. C$_{26}$H$_{37}$NO$_3$; mol wt 411.59. C 75.87%, H 9.06%, N 3.40%, O 11.66%. Muscarinic M$_3$-receptor antagonist. Prepn: B. Sparf, C. O. Meese, **EP 0957073**; *eidem*, **US 6713464** (1999, 2004 both to Schwarz). Prepn of stable salts: C. Meese, **DE 19955190**; *idem*, **US 6858650** (2001, 2005 both to Schwarz). Clinical trial in overactive bladder: C. Chapple *et al.*, *Eur. Urol.* **52**, 1204 (2007).

Colorless, viscous oil.

Fumarate. [286930-03-8] SPM-907; SPM-8272; Toviaz. C$_{26}$H$_{37}$NO$_3$.C$_4$H$_4$O$_4$; mol wt 527.66. Colorless flakes from cyclohexane/2-butanone (90:10), mp 103°. Absorption max: 191, 193, 200, 220 nm (A$_{1cm}^{1\%}$ 1306, 1305, 1143, 456).

THERAP CAT: Antispasmodic; in treatment of overactive bladder.

4096. Feverfew. Featherfew; featherfoil; midsummer daisy. *Tanacetum parthenium* (L.) Sch. Bip., (formerly *Chrysanthemum parthenium* (L.) Bernh.) *Compositae;* a perennial, strongly aromatic herb found in Britain and the Balkan peninsula. Used medicinally since the Middle Ages as a febrifuge. Constituents include sesquiterpene lactones such as parthenolide, *q.v.:* P. J. Hylands, D. M. Hylands in *Development of Drugs and Modern Medicines*, J. W. Gorrod *et al.*, Eds. (Ellis Horwood, Chichester, 1986) pp 100-104. Inhibition of prostaglandin biosynthesis by feverfew extract: H. O. J. Collier *et al.*, *Lancet* **2**, 922 (1980). Effect on human platelet phospholipase: A. N. Makheja, J. M. Bailey, *ibid.* 1054 (1981); *eidem*, *Prostaglandins Leukotrienes Med.* **8**, 653 (1982); J. K. Thakkar *et al.*, *Biochim. Biophys. Acta* **750**, 134 (1983). Inhibition of platelet secretory activity: S. Heptinstall *et al.*, *Lancet* **1**, 1071 (1985); S. Heptinstall *et al.*, *J. Pharm. Pharmacol.* **39**, 459 (1987). Structure and anti-secretory activity study: W. A. Groenewegen *et al.*, *ibid.* **38**, 709 (1986). Clinical trials in migraine using freeze dried feverfew leaves: E. S. Johnson *et al.*, *Br. Med. J.* **291**, 569 (1985). Use of oil extract in migraine: E. S. Johnson *et al.*, *US* **4758433** (1988 to R. P. Scherer). *Review:* M. I. Berry, *Pharm. J.* **232**, 611-614 (1984). Brief review of activity and possible side effects: C. A. Baldwin *et al.*, *ibid.* **239**, 237-238 (1987).

4097. Fexofenadine. [83799-24-0] 4-[1-Hydroxy-4-[4-(hydroxydiphenylmethyl)-1-piperidinyl]butyl]-α,α-dimethylbenzeneacetic acid; carboxyterfenadine; terfenadine carboxylate; MDL-16455. $C_{32}H_{39}NO_4$; mol wt 501.67. C 76.61%, H 7.84%, N 2.79%, O 12.76%. Nonsedating-type histamine H_1-receptor antagonist. Prepn: A.A. Carr *et al.*, **DE 3007498**; *eidem*, **US 4254129** (1980, 1981 both to Richardson-Merrell). Identification as active metabolite of terfenadine, *q.v.*: D. A. Garteiz *et al.*, *Arzneim.-Forsch.* **32**, 1185 (1982). Synthesis: S. H. Kawai *et al.*, *J. Org. Chem.* **59**, 2620 (1994). HPLC separation from terfenadine: K. Y. Chan *et al.*, *J. Chromatogr.* **571**, 291 (1991); determn in biological fluids: A. Terhechte, G. Blaschke, *J. Chromatogr. A* **694**, 219 (1995). Effects on cardiac K^+ channels: D. Rampe *et al.*, *Mol. Pharmacol.* **44**, 1240 (1993). Comparative analysis of cardiotoxic potential: J. A. Hey *et al.*, *Arzneim.-Forsch.* **46**, 153 (1996). Clinical pharmacology in children: F. E. R. Simons *et al.*, *J. Allergy Clin. Immunol.* **98**, 1062 (1996).

Crystals from methanol-butanone, mp 195-197° (Carr); also reported as white crystals from methanol, mp 142-143° (Kawai).

Hydrochloride. [153439-40-8] Allegra; Telfast. $C_{32}H_{39}NO_4$·HCl; mol wt 538.13.

THERAP CAT: Antihistaminic.

4098. Fialuridine. [69123-98-4] 1-(2-Deoxy-2-fluoro-β-D-arabinofuranosyl)-5-iodo-2,4(1*H*,3*H*)-pyrimidinedione; 1-(2-deoxy-2-fluoro-β-D-arabinofuranosyl)-5-iodouracil; 5-iodo-2′-fluoroarauracil; FIAU. $C_9H_{10}FIN_2O_5$; mol wt 372.09. C 29.05%, H 2.71%, F 5.11%, I 34.11%, N 7.53%, O 21.50%. Exptl antiviral agent; nucleoside analog with antihepatitis B activity. Prepn: K. A. Watanabe *et al.*, *J. Med. Chem.* **22**, 21 (1979). Antiviral activity: J. M. Colacino, C. Lopez, *Antimicrob. Agents Chemother.* **24**, 505 (1983); K. A. Staschke *et al.*, *Antiviral Res.* **23**, 45 (1994). Clinical pharmacokinetics: R. R. Bowsher *et al.*, *Antimicrob. Agents Chemother.* **38**, 2134 (1994). Report of trial suspension: S. R. Ahmed, *Lancet* **342**, 166 (1993). Evaluation of mechanism of hepatotoxicity: L. Cui *et al.*, *J. Clin. Invest.* **95**, 555 (1995). Clinical toxicological profile: W. Stevenson *et al.*, *Transplant. Proc.* **27**, 1219 (1995).

Crystals from ethanol, mp 216-217°.

4099. Fibrin. [9001-31-4] Fibrin monomer is fibrinogen from which one or two peptides have been removed by means of thrombin: Laki, Chandrasekhar, *Nature* **197**, 1267 (1963). The term fibrin is usually applied to polymerized fibrin monomer. Terminal clotting

takes place in four steps: (1) fibrinogen hydrolyzes under the influence of thrombin into fibrin and fibrinopeptide fragments; (2) fibrin forms soft clots which can be readily dispersed; (3) thrombin activates fibrin-stabilizing factor, *q.v.*, an enzyme precursor, present in blood plasma; (4) fibrin in the networks cross-links under the influence of the activated FSF to give the final hard clots: *Chem. Eng. News* **43**, no. 32, 38 (1965); O. D. Ratnoff, B. Bennett, *Science* **179**, 1291 (1973). Fibrin occurs in two principal forms, *fibrin-i*, "insoluble" fibrin, differing from *fibrin-s*, "soluble" fibrin, by urea solubility as well as other characteristics. Fibrin-i is formed through the reaction of a fibrinogen-like plasma protein, FSF, which in the presence of Ca^{2+} converts what would otherwise be a "soluble" weakly bonded gel into a covalently bonded, insol clot: Rosenberg, Carman, *Nature* **204**, 994 (1964). Chemical studies of crosslinking segments: Chen, Doolittle, *Proc. Natl. Acad. Sci. USA* **66**, 472 (1970); *eidem*, *Biochemistry* **10**, 4486 (1971); Doolittle *et al.*, *Biochem. Biophys. Res. Commun.* **44**, 94 (1971). *Reviews:* W. H. Seegers, *Prothrombin* (Harvard University Press, 1962) 728 pp; Laki, Gladner, *Physiol. Rev.* **44**, 127 (1964); Lorand, *Fed. Proc.* **24**, no. 4, part 1, 784 (1965); A. L. Copley, *Thromb. Res.* **14**, 249 (1979). Review of chemistry: several authors, *Thromb. Diath. Haemorrh. Suppl.* **39** (1970).

4100. Fibrinogen. [9001-32-5] Factor I; blood-coagulation factor I. Plasma glycoprotein, synthesized and secreted by hepatic parenchymal cells, that is essential to the clotting of blood. Present to the extent of 0.3-0.4 g/100 ml in human plasma. Synthesis is greatly increased during acute inflammatory challenge; raised fibrinogen levels are associated with increased risk of vascular disease. The fibrinogen molecule exists in plasma as a dimer consisting of 2 sets of three polypeptide chains, Aα, Bβ, and γ, stabilized by 29 disulfide bonds; mol wt 340 kDa. Thrombin releases fibrinopeptides A and B from the amino-terminal ends of the α and β chains of fibrinogen in the formation of fibrin during coagulation. Because fibrinogen is less sol than other plasma proteins it is readily separated by precipitation with sodium chloride: M. Florkin, *J. Biol. Chem.* **87**, 629 (1930). Prepn from human plasma: P. R. Morrison *et al.*, *J. Am. Chem. Soc.* **70**, 3103 (1948). Mechanism of fibrin formation by thrombin: K. Bailey *et al.*, *Nature* **167**, 233 (1951). Review of structure and functions: R. F. Doolittle, *Annu. Rev. Biochem.* **53**, 195-229 (1984); M. W. Mosesson, *J. Thromb. Haemost.* **3**, 1894-1904 (2005); of biosynthesis: C. M. Redman, H. Xia, *Ann. N.Y. Acad. Sci.* **936**, 480-495 (2001); as predictor of vascular disease: A. I. Kakafika *et al.*, *Curr. Pharm. Des.* **13**, 1647-1659 (2007). Review of clinical experience as hemostatic agent in treatment of fibrinogen deficiency: C. Fenger-Eriksen *et al.*, *Expert Opin. Biol. Ther.* **9**, 1325-1333 (2009).

Sparingly sol in water. Aq solns are viscous. Isoelectric point 5.5. Readily denatured by heating to 56° or higher, and by chemical agents such as salicylaldehyde, naphthoquinone sulfonates, ninhydrin, and alloxan. Small amounts of papain will clot fibrinogen, but larger amounts will digest the clot.

Fibrinogen concentrate (human). Clottagen; Haemocomplettan P; RiaSTAP. Purified fibrinogen prepared from pooled human plasma.

THERAP CAT: Replacement therapy in congenital fibrinogen deficiency.

4101. Fibroblast Growth Factor. [62031-54-3] FGF. Growth stimulatory factor originally isolated from bovine brain and pituitary and found to stimulate DNA synthesis in cultured fibroblast cells. Isoln: D. Gospodarowicz, *Nature* **249**, 123 (1974). Mitogenic effect on cultured cell lines and induction of amphibian limb regeneration *in vivo*: D. Gospodarowicz *et al.*, *Adv. Metab. Disord.* **8**, 301 (1975). Two closely related forms have been identified, known as basic (*bFGF*) and acidic (*aFGF*) fibroblast growth factors, having a total amino acid sequence homology of 55%. Both induce the proliferation and differentiation of a wide variety of cell types, including corneal and vascular endothelial cells, myoblasts, chondrocytes, osteoblasts and glial cells. FGF has neurotrophic and angiogenic activity and may play an important role in the wound healing process. Purification of bFGF from pituitary: D. Gospodarowicz, *J. Biol. Chem.* **250**, 2515 (1975); from brain: D. Gospodarowicz *et al.*, *ibid.* **253**, 3736 (1978). Identification of aFGF from bovine brain: K. A. Thomas *et al.*, *ibid.* **255**, 5517 (1980). Comparison of fibroblast growth factors: S. K. Lemmon *et al.*, *J. Cell Biol.* **95**, 162 (1982). Purification and characterization of aFGF: K. A. Thomas *et al.*,

Proc. Natl. Acad. Sci. USA **81**, 357 (1984). Identity of bFGF from brain and pituitary: D. Gospodarowicz *et al., ibid.* 6963. Amino acid sequence of bFGF: F. Esch *et al., ibid.* **82**, 6507 (1985); of aFGF: G. Gimenez-Gallego *et al., Science* **230**, 1385 (1985); F. Esch *et al., Biochem. Biophys. Res. Commun.* **133**, 554 (1985). Possible identity of aFGF with *endothelial cell growth factor* (*ECGF*) and *eye-derived growth factor-II* (*EDGF-II*): A. B. Schreiber *et al., J. Cell Biol.* **101**, 1623 (1985); of bFGF with *macrophage-derived growth factor* (*MDGF*): A. Baird *et al., Biochem. Biophys. Res. Commun.* **126**, 358 (1985); of FGFs with *retina-derived endothelial cell growth factor*: A. Baird *et al., Biochemistry* **24**, 7855 (1985). Cloning of cDNA for bovine bFGF: J. A. Abraham *et al., Science* **233**, 545 (1986); for human bFGF: T. Kurokawa *et al., FEBS Lett.* **213**, 189 (1987); M. Iwane *et al., Biochem. Biophys. Res. Commun.* **146**, 470 (1987). Expression of a chemically synthesized gene for bioactive bovine aFGF: D. L. Linemeyer *et al., Biotechnology* **5**, 960 (1987). Receptor binding study: G. Neufeld, D. Gospodarowicz: *J. Biol. Chem.* **261**, 5631 (1986). FGF-like factors have been isolated from several human tumor cell lines: R. R. Lobb *et al., Biochem. Biophys. Res. Commun.* **139**, 861 (1986); M. Klagsbrun *et al., Proc. Natl. Acad. Sci. USA* **83**, 2448 (1986); D. Moscatelli *et al., J. Cell. Physiol.* **129**, 273 (1986). FGF has also been shown to be structurally homologous to the protein products of several oncogenes: C. Dickson, G. Peters, *Nature* **326**, 833 (1987); M. Taira *et al., Proc. Natl. Acad. Sci. USA* **84**, 2980 (1987); P. Delli Bovi *et al., Cell* **50**, 729 (1987). Review of tissue distribution and bioactivity of bFGF: A. Baird *et al., Recent Prog. Horm. Res.* **42**, 143-205 (1986). Review of structural characterization and biological functions: D. Gospodarowicz *et al., Endocr. Rev.* **8**, 95-114 (1987). Potential role in the control of pituitary and gonad development: D. Gospodarowicz, N. Ferrara, *J. Steroid Biochem.* **32**, 183-191 (1989). *Reviews:* K. A. Thomas, G. Gimenez-Gallego, *Trends Biochem. Sci.* **11**, 81-84 (1986); D. Gospodarowicz *et al., Mol. Cell. Endocrinol.* **46**, 187-204 (1986); K. A. Thomas, *FASEB J.* **1**, 434-440 (1987).

Acidic fibroblast growth factor. [106096-92-8] pI 5-7. Exists in 2 microheterogeneous forms: *aFGF-1*, a 140 amino acid peptide, mol wt 15,900 daltons, and *aFGF-2*, an amino truncated form lacking 6 amino terminal residues, mol wt 15,200 daltons.

Basic fibroblast growth factor. [106096-93-9] pI 9.6. 146 amino acid peptide, mol wt ~16,000 daltons. Also exists as an amino truncated form, *des 1-15 bFGF*, lacking the first 15 amino acid residues.

4102. Fibroins. [9007-76-5] Protein filaments produced by members of the phylum *Arthropoda*, particularly by certain species belonging to the classes *Insecta* (insects) and *Arachnida* (spiders, etc.). Fibroin is the main protein of silk and is secreted by the insect in its silk glands together with *sericin*, the second silk protein, in aq soln and converted into silk by a process called "spinning." Organization of amino acid sequences in silk fibroin of *Bombyx mori* and general review of silk proteins: F. Lucas, K. M. Rudall in *Comprehensive Biochemistry* vol. **26B**, M. Florkin, E. M. Stotz, Eds. (Elsevier, New York, 1968) pp 475-558. *See also:* P. M. Lizardi, *Cell* **18**, 581 (1979); L. P. Gage, R. F. Manning, *J. Biol. Chem.* **255**, 9444 (1980). Chemical and crystalline structures: B. Lotz, F. Colonna Cesari, *Biochimie* **61**, 205 (1979).

Pale yellow mass resembling silk. Insol in water, alcohol, ether, dil alkalies. Sol in concd alkalies, concd mineral acids and in ammoniacal nickel oxide soln.

4103. Fibrolase. [116036-70-5] Fibrinolytic enzyme isolated from the venom of the southern copperhead snake, *Agkistrodon contortrix contortrix*. Single chain, zinc metalloproteinase which preferentially cleaves the Aα chain of fibrinogen and fibrin. Several isoforms have been identified; the longest form has 203 amino acid residues with mol wt of 23 kDa. Isoln: N. B. Egen *et al., Toxicon* **25**, 1189 (1987); and effect on human coagulation system: A. D. Retzios, F. S. Markland, Jr., *Thromb. Res.* **52**, 541 (1988). Amino acid sequence: A. Randoph *et al., Protein Sci.* **1**, 590 (1992). Resolution of isoforms: S. L. Loayza *et al., J. Chromatogr. B* **662**, 227 (1994). Three-dimensional structure: M. B. Bolger *et al., AAPS PharmSci.* **3**, E16 (2001). Review and comparison with other snake venom fibrinolytic enzymes: S. Swenson, F. S. Markland, Jr., *Toxicon* **45**, 1021-1039 (2005).

pI ~6.8.

4104. Fibronectins. α_2-SB glycoproteins; α_2-opsonins; CIG; CSP; CAF; GAP A; LETS; Zeta protein. High mol wt multifunctional glycoproteins, found on cell surfaces, in body fluids (especially plasma), in soft connective tissue matrices, and in most basement membranes. Although fibronectins apparently function as adhesive ligand-like molecules, the full range of their biological activities and relationships are still being elucidated. Their importance in cell adhesion, oncogenic transformation, reticuloendothelial system function, embryonic differentiation, phagocytosis, hemostasis, and chemotaxis is being studied. Discovered as a result of isoln of a partially purified fraction of human plasma and initially termed "cold-insoluble globulin" or CIG: P. R. Morrison *et al., J. Am. Chem. Soc.* **70**, 3103 (1948). Subsequent studies described various proteins or factors, named according to sources or biological activities, that are now designated as fibronectins. At least two types are known to exist, termed plasma and cellular fibronectin, respectively. Both forms contain subunits of mol wt >200,000, joined by disulfide bonds. They are similar in amino acid compositions, carbohydrate structures and secondary and tertiary structures; they cannot be distinguished in biological activity in assays of cell interactions with substrates or in opsonic activity for macrophages. They differ in their effects on cell morphology, on alignment of transformed cells and on hemagglutination; they also have differences in solubility and in the number of subunits linked by disulfide bonds. Monoclonal antibody studies have indicated that the two forms are distinct: B. T. Atherton, R. O. Hynes, *Cell* **25**, 133 (1981); K. D. Noonan *et al., J. Supramol. Struct. Cell. Biochem.* **5**, Suppl, 302 (1981). Regulation of fibronectin biosynthesis: D. R. Senger *et al., Am. J. Physiol.* **245**, 144 (1983). Structure-function relationships: T. Vartio, A. Vaheri, *Trends Biochem. Sci.* **8**, 442 (1983). Role in cellular adhesion, spreading and cytoskeletal organization: I. Virtanen *et al., Nature* **298**, 660 (1982); in phagocytosis: L. Van de Water *et al., Science* **220**, 201 (1983); in wound healing: G. R. Martin *et al.*, "Regulation of Tissue Structure and Repair by Collagen and Fibronectin" in *The Surgical Wound*, P. Dineen, C. Hildrick-Smith, Eds. (Lea & Febiger, Philadelphia, 1981) pp 110-122. Use in treatment of corneal trophic ulcer therapy: T. Nishida *et al., Arch. Ophthalmol.* **101**, 1046 (1983). Review of role in cellular adhesion: S. K. Akiyama *et al., J. Supramol. Struct. Cell. Biochem.* **16**, 345-358 (1981); role in inflammation: D. F. Mosher *et al., Adv. Inflammation Res.* **2**, 187-207 (1981); C. Bianco, *Ann. N.Y. Acad. Sci.* **408**, 602-609 (1983); actvity in various disease states: S. K. Akiyama, K. M. Yamada, "Fibronectin in Disease" in *Monographs in Pathology* No. 24, N. Kaufman, Ed., entitled "Connective Tissue Diseases", B. M. Wagner *et al.*, Eds. (Williams & Wilkins, Baltimore, 1983). *General Reviews:* E. Pearlstein *et al., Mol. Cell. Biochem.* **29**, 103-128 (1980); M. W. Mosesson, D. L. Amrani, *Blood* **56**, 145-158 (1980); E. Ruoslahti, *J. Oral Pathol.* **10**, 3-13 (1981); R. O. Hynes, K. M. Yamada, *J. Cell Biol.* **95**, 369-377 (1982); R. O. Hynes, *Sci. Am.* **254**, 42-51 (1986).

4105. Fichtelite. [2221-95-6] (1*S*,4a*S*,4b*S*,7*S*,8a*S*,10a*S*)-Tetradecahydro-1,4a-dimethyl-7-(1-methylethyl)phenanthrene; 18-norabietane. $C_{19}H_{34}$; mol wt 262.48. C 86.94%, H 13.06%. From decayed wood of conifers: Bromeis, *Ann. Pharm.* **37**, 304 (1841). Structure: L. Ruzicka, E. Waldmann, *Helv. Chim. Acta* **18**, 611 (1935); Crowfoot, *J. Chem. Soc.* **1938**, 1241. Stereochemistry: Burgstahler, Marx, *Tetrahedron Lett.* **1964**, 3333. Synthesis from abietic acid: Jensen, Johnson, *J. Org. Chem.* **32**, 2045 (1967); Burgstahler, Marx, *ibid.* **34**, 1562 (1969). Synthesis of *dl*-form: Johnson *et al., J. Am. Chem. Soc.* **88**, 3859 (1966); **90**, 5872 (1968); D. F. Taber, S. A. Saleh, *ibid.* **102**, 5085 (1980).

Crystals from methanol, mp 45-46°. bp$_{43}$ 235-236°. d$_4^{22}$ 0.9380. n$_D^{20}$ 1.5052. [α]$_D$ +19°.

4106. Ficin. [9001-33-6] Ficus proteinase; ficus protease; ficain; EC 3.4.22.3; Debricin; Higueroxyl Delabarre. A proteolytic enzyme of est. mol wt 23,800-25,500 which requires a free sulfhydryl group for activity and as such is a member of a group which includes papain and bromelain, *q.q.v.* Occurs in the latex of tropical trees of the genus *Ficus* subgenus *Pharmacosyce, Moraceae* (Oje trees). The commercial product is a concentrate prepd by filtering and drying the latex of *Ficus glabrata* H. B. & K., *Moraceae.* First crystallized from fresh fig latex: Walti, *J. Am. Chem. Soc.* **60**, 493 (1938). Characterization: Cohen, *Nature* **182**, 659 (1958); Englund *et al., Biochemistry* **7**, 163 (1968). Purification: Gibian, Bratfisch, **US 2950227** (1960 to Schering AG). Amino acid composition: Wong, Liener, *Biochem. Biophys. Res. Commun.* **17**, 470 (1964); Metrione *et al., Arch. Biochem. Biophys.* **122**, 137 (1967); Husain, Lowe, *Biochem. J.* **117**, 333 (1970). Pharmacodynamics: H. Heisto, M. K. Fagerhol, *Transfusion* **19**, 545 (1979); A. Perkash *et al., ibid.* **20**, 301 (1980). Toxicology: H. Molitor *et al., J. Pharmacol. Exp. Ther.* **71**, 20 (1941). *Reviews:* Liener, Friedenson, *Methods Enzymol.* **19**, 260-273 (1970); Glazer, Smith, *The Enzymes* **vol. III**, P. D. Boyer, Ed. (Academic Press, New York, 3rd ed., 1971) pp 538-542.

Buff to cream-colored hygroscopic powder. Acrid odor, growing stronger with age. Bulky, approx 3 ml/g, not free-flowing. Appears dry, even with 15% H$_2$O present. Not completely sol in water, 2-10% insol material. The sol portion of 1 g dissolves in 3 ml H$_2$O. pH (2% soln): 4.1. Insol in usual organic solvents. Loses about 10-20% activity when stored 1-3 yrs at ordinary temp and atm conditions. Aq solns inactivated at 100°, solid partially inactivated within a few hours. Solns relatively stable between pH 4-8.5; incompatible with iron, copper, aluminum. Gelatin, coagulated egg white, casein, meat and most protein-like material hydrolyzed in aq ficin solns. Ficin is 10-20 times as active as papain in regard to milk clotting; 4-10 times as active, in general. LD$_{50}$ orally in rats, mice: ~10 g/kg; in rabbits, guinea pigs: ~5 g/kg (Molitor).

Note: The name ficin is currently used to describe both the crude dried latex from different species of the genus *Ficus,* as well as the enzyme itself.

Caution: Handle ficin carefully because of its tissue-dissolving properties. Can cause irritation to skin, eyes, mucous membranes. Large doses by mouth cause purging.

USE: Protein digestant. In the brewing industry as a chill proofing agent in beer. In the cheese industry as a substitute for rennet in the coagulation of milk. In the meat industry as a meat tenderizer and as an agent for removing casings from formed sausage. In the leather industry for the bating of leather. In the textile industry for shrink-proofing wool, for removing gelatin from sized thread, and mixed with amylases and maltases as a spot remover. In the prepn of peptones. For solubilizing protein material in spent grains. Also used in the determination of the Rh factor. Speeds 10 times the agglutination of human blood cells by the Rh factor when in contact with the anti-Rh serum.

THERAP CAT (VET): Has been used as a trichuricide.

4107. Fidarestat. [136087-85-9]; [105300-43-4] (unspecified stereo). (2S,4S)-6-Fluoro-2,3-dihydro-2′,5′-dioxospiro[4H-1-benzopyran-4,4′-imidazolidine]-2-carboxamide; (2S,4S)-6-fluoro-2′,5′-dioxospiro[chroman-4,4′-imidazolidine]-2-carboxamide; SNK-860. C$_{12}$H$_{10}$FN$_3$O$_4$; mol wt 279.23. C 51.62%, H 3.61%, F 6.80%, N 15.05%, O 22.92%. Aldose reductase inhibitor for treatment of diabetic complications. Prepn (stereo unspec): M. Kurono, *et al.*, **EP 193415**; *eidem,* **US 4740517** (1986, 1988 both to Sanwa). Prepn of isomers: T. Yamaguchi *et al., Arzneim.-Forsch.* **44**, 344 (1994). Pharmacological profile: K. Mizuno *et al.* in *Current Concepts of Aldose Reductase and Its Inhibitions,* N. Sakamoto *et al.*, Eds. (Elsevier, Amsterdam, 1990) pp 89-96. Configuration and crystal structure of complex with aldose reductase: M. Oka *et al., J. Med. Chem.* **43**, 2479 (2000). Clinical efficacy in diabetic peripheral neuropathy: N. Hotta *et al., Diabetes Care* **24**, 1776 (2001). Clinical suppression of sorbitol accumulation in erythrocytes of diabetic patients: T. Asano *et al., J. Diabetes Complications* **16**, 133 (2002); *eidem, ibid.* **18**, 336 (2004). Review of clinical development: N. Giannoukakis, *Curr. Opin. Investig. Drugs* **4**, 1233-1239 (2003).

Crystals from methanol, mp 290-300° (dec). [α]$_D^{27}$ +168° (c = 1.0 in methanol). LD$_{50}$ in rats, dogs (mg/kg): >2000 orally (Yamaguchi).

THERAP CAT: Treatment of diabetic neuropathy.

4108. Fidaxomicin. [873857-62-6] (3E,5E,8S,9E,11S,12R,-13E,15E,18S)-3-[[[6-Deoxy-4-O-(3,5-dichloro-2-ethyl-4,6-dihydroxybenzoyl)-2-O-methyl-β-D-mannopyranosyl]oxy]methyl]-12-[[6-deoxy-5-C-methyl-4-O-(2-methyl-1-oxopropyl)-β-D-*lyxo*-hexopyranosyl]oxy]-11-ethyl-8-hydroxy-18-[(1R)-1-hydroxyethyl]-9,-13,15-trimethyloxacyclooctadeca-3,5,9,13,15-pentaen-2-one; lipiarmycin A3; tiacumicin B; clostomicin B$_1$; OPT-80; PAR-101; Dificid. C$_{52}$H$_{74}$Cl$_2$O$_{18}$; mol wt 1058.05. C 59.03%, H 7.05%, Cl 6.70%, O 27.22%. Macrolide antibiotic with activity against *Clostridium difficile;* inhibits bacterial RNA polymerase. Major component of the antibiotic complex originally isolated as lipiarmycin from a strain of *Actinoplanes deccanensis.* Isoln of complex: C. Coronelli *et al.*, **DE 2455230**; *eidem,* **US 3978211** (1975, 1976 both to Lepetit). Mechanism of action: M. Talpaert *et al., Biochem. Biophys. Res. Commun.* **63**, 328 (1975). Isoln of lipiarmycins A3 and A4 and elucidation of structure: A. Arnone *et al., J. Chem. Soc. Perkin Trans. 1* **1987**, 1353. Isoln as tiacumicin B from *Dactylosporangium aurantiacum* subsp. *hamdenensis:* R. J. Theriault *et al., J. Antibiot.* **40**, 567 (1987); J. E. Hochlowski *et al., ibid.*, 575. Prepn of crystalline polymorphs: Y.-H. Chiu *et al.*, **US 7378508** (2008 to Optimer). Antibacterial spectrum vs anaerobes: K. L. Credito, P. C. Appelbaum, *Antimicrob. Agents Chemother.* **48**, 4430 (2004). Clinical pharmacokinetics: Y. K. Shue *et al., ibid.* **52**, 1391 (2008). Clinical evaluation in *C. difficile* infection: T. Louie *et al., ibid.* **53**, 223 (2009). Review of development and clinical experience: K. M. Sullivan, L. M. Spooner, *Ann. Pharmacother.* **44**, 352-359 (2010); M. Miller, *Expert Opin. Pharmacother.* **11**, 1569-1578 (2010).

White crystals from ethyl acetate + hexane, mp 161-165° (Arnone). Also reported as white needles from isopropanol, mp 166-169° (Chiu). [α]$_D$ −6.9° (c = 2.0 in methanol). Sol in methanol, isopropanol, acetonitrile, ethyl acetate. Practically insol in water.

THERAP CAT: Antibacterial.

4109. Filicinic Acid. [2065-00-1] 3,5-Dihydroxy-4,4-dimethyl-2,5-cyclohexadien-1-one; 1,1-dimethylcyclohexane-2,4,6-trione; *gem*-dimethylphloroglucinol; filicinsäure (German). C$_8$H$_{10}$O$_3$; mol wt 154.17. C 62.33%, H 6.54%, O 31.13%. Isolated from male fern and by hydrolytic decompn of filixic acid or aspidin: Boehm, *Ann.* **302**, 171 (1898); **329**, 321 (1903). Synthesis: Robertson, Sandrock, *J. Chem. Soc.* **1933**, 1617; Angus *et al., Chem. Ind. (London)* **1954**,

546; Inagaki *et al.*, *J. Pharm. Soc. Jpn.* **76**, 1258 (1956); Hoefer, Riedl, *Angew. Chem.* **74**, 501 (1962); *eidem, Ann.* **656**, 127 (1962).

Prisms from water or benzene, mp 214-215° (dec). pK 5.8. Practically insol in water, petr ether. Sol in 70 parts boiling water, 10 parts boiling alcohol; sparingly sol in ether, benzene, glacial acetic acid. Reduces Tollen's reagent.

3,5-Diacetylfilicinic acid. $C_{12}H_{14}O_5$. Crystals from methanol, mp 65-66°. pK 5.0. Sol in cold methanol, benzene, ether; slightly sol in hot water.

4110. Filipin. [11078-21-0] Filimarisin; U-5956; NSC-3364. Polyene antibiotic complex containing at least eight pentaene compounds, which has been resolved into three pure components, filipin II, III (major), and IV, which differ in the number of hydroxyl groups present (8, 9, 9, resp.) and filipin I, a mixture of at least 5 components. Described as a single entity in the earlier literature. Initial isolation from *Streptomyces filipenensis* in Philippine soil: Whitfield *et al.*, *J. Am. Chem. Soc.* **77**, 4799 (1955). Early structural work: Dhar *et al.*, *Proc. Chem. Soc. London* **1960**, 310; Djerassi *et al.*, *Tetrahedron Lett.* **1961**, 383; Golding, Rickards, *ibid.* **1964**, 2615; Dhar *et al.*, *J. Chem. Soc.* **1964**, 842; Ceder, Ryhage, *Acta Chem. Scand.* **18**, 588 (1964). Separation of filipin complex into components: Bergy, Eble, *J. Antibiot.* **23**, 414 (1970); *see also* Rickards *et al.*, *ibid.* 603. Mechanism of action: R. W. Holz in *Antibiotics* vol. **5**, pt. 2, F. E. Hahn, Ed. (Springer-Verlag, New York, 1979) pp 313-340. Filipin also interacts specifically with 3β-hydroxysterols (e.g. cholesterol): P. M. Elias *et al.*, *J. Histochem. Cytochem.* **27**, 1247 (1979); *see also* N. J. Severs, H. J. Simons, *Nature* **303**, 637 (1983).

Filipin III

Yellow, feathery needles from chloroform, mp 195-205°. Sensitive to air. $[\alpha]_D^{22}$ −148.3° (c = 0.89 in methanol). uv max (methanol): 322, 338, 355 nm ($E_{1cm}^{1\%}$ 910, 1360, 1330). Freely sol in DMF, pyridine. Also sol in 95% ethanol, methanol, butanol, isopropanol, glacial acetic acid, ether. Practically insol in water, chloroform.

Filipin III. [480-49-9] (3R,4S,6S,8S,10R,12R,14R,16S,17E,-19E,21E,23E,25E,27S,28R)-4,6,8,10,12,14,16,27-Octahydroxy-3-[(1R)-1-hydroxyhexyl]-17,28-dimethyloxacyclooctacosa-17,19,21,-23,25-pentaen-2-one; 15-deoxylagosin. $C_{35}H_{58}O_{11}$; mol wt 654.84. Isomeric with filipin IV. Crystals from propanol, mp 163-180° (Bergy, Eble). $[\alpha]_D^{25}$ −245° (c = 0.8 in DMF). uv max (methanol): 243, 308, 321, 337, 354 nm ($E_{1cm}^{1\%}$ 62, 413, 851, 1368, 1343).

USE: Produces arthritis rapidly in rabbits: Pras, Weissman, *Drug Trade News*, July 4, 1966, p 40; as a sterol probe in freeze-fracture cytochemistry.

THERAP CAT: Antifungal.

4111. Filixic Acids. Filicin; filicic acid; Filixsäure (German). Natural filixic acid from male fern, *Dryopteris filix-mas*, is a mixture of six homologs; the 3 main components, filixic acid -BBB, -PBB, and -PBP, are obtained by recrystallization from ethyl acetate. When raw filixic acid is purified by treatments with methanol but without recrystns from ethyl acetate, a product is obtained which seems to contain 3 additional homologs with an analogous composition but

with acetyl groups in the side chain. Isoln: Luck, *Ann.* **54**, 119 (1845); Boehm, *ibid.* **318**, 253 (1901). Structure studies: Riedl, *Ann.* **585**, 32 (1954); Chan, Hassal, *Experientia* **13**, 349 (1957). Structure of the three main filixic acids: Penttilä, Sundman, *Acta Chem. Scand.* **17**, 191 (1963).

	R_1	R_2
BBB	$R_1 = R_2 = C_3H_7$	
PBB	$R_1 = C_2H_5$	$R_2 = C_3H_7$
PBP	$R_1 = R_2 = C_2H_5$	

Natural filixic acid, pale-yellow plates from ethyl acetate, mp 184-185°. uv max: 228, 288 nm (ε 41000, 29000). Practically insol in water, methanol, acetone. Sol in chloroform, benzene, warm ethyl acetate, acetic acid; slightly sol in ether.

Filixic acid BBB. [4482-83-1] 2,2′-[[2,4,6-Trihydroxy-5-(1-oxobutyl)-1,3-phenylene]bis(methylene)]bis[3,5-dihydroxy-4,4-dimethyl-6-(1-oxobutyl)-2,5-cyclohexadien-1-one]. $C_{36}H_{44}O_{12}$; mol wt 668.74. Crystals from ethyl acetate, mp 172-174°.

Filixic acid PBB. [49582-09-4] 2-[[3-[[2,4-Dihydroxy-3,3-dimethyl-6-oxo-5-(1-oxobutyl)-1,4-cyclohexadien-1-yl]methyl]-2,-4,6-trihydroxy-5-(1-oxobutyl)phenyl]methyl]-3,5-dihydroxy-4,4-dimethyl-6-(1-oxopropyl)-2,5-cyclohexadien-1-one. $C_{35}H_{42}O_{12}$; mol wt 654.71. Crystals from ethyl acetate, mp 184-186°.

Filixic acid PBP. [51005-85-7] 2,2′-[[2,4,6-Trihydroxy-5-(1-oxobutyl)-1,3-phenylene]bis(methylene)]bis[3,5-dihydroxy-4,4-dimethyl-6-(1-oxopropyl)-2,5-cyclohexadien-1-one]. $C_{34}H_{40}O_{12}$; mol wt 640.68. Crystals from ethyl acetate, mp 192-194°.

4112. Finafloxacin. [209342-40-5] 8-Cyano-1-cyclopropyl-6-fluoro-7-[(4aS,7aS)-hexahydropyrrolo[3,4-b]-1,4-oxazin-6(2H)-yl]-1,4-dihydro-4-oxo-3-quinolinecarboxylic acid; 8-cyano-1-cyclopropyl-6-fluoro-7-[(1S,6S)-2-oxa-5,8-diazabicyclo[4.3.0]non-8-yl]-1,4-dihydro-4-oxo-3-quinolinecarboxylic acid. $C_{20}H_{19}FN_4O_4$; mol wt 398.39. C 60.30%, H 4.81%, F 4.77%, N 14.06%, O 16.06%. Fluorinated quinoline antibacterial that maintains activity under acidic conditions. Prepn: M. Matzke *et al.*, *DE 19652239*; *eidem*, *US 6133260* (1998, 2000 both to Bayer); J. Hong *et al.*, *Tetrahedron Lett.* **50**, 2525 (2009). Antibacterial activity under acidic pH: N.-C. Emrich *et al.*, *J. Antimicrob. Chemother.* **65**, 2530 (2010). Comparative activity vs multidrug resistant isolates of *Acinetobacter baumannii*: P. G. Higgins *et al.*, *Antimicrob. Agents Chemother.* **54**, 1613 (2010).

Crystals from ethanol, mp 294° (dec).

Hydrochloride. [209342-41-6] BAY-35-3377. $C_{20}H_{19}FN_4O_4 \cdot$HCl; mol wt 434.85. mp 314-316° (dec). $[\alpha]_D^{23}$ −112° (c = 0.29 in 1N NaOH).

THERAP CAT: Antibacterial.

4113. Finasteride. [98319-26-7] (4aR,4bS,6aS,7S,9aS,9bS,-11aR)-N-(1,1-Dimethylethyl)-2,4a,4b,5,6,6a,7,8,9,9a,9b,10,11,11a-tetradecahydro-4a,6a-dimethyl-2-oxo-1H-indeno[5,4-f]quinoline-7-carboxamide; 17β-(N-tert-butylcarbamoyl)-4-aza-5α-androst-1-en-3-one; (5α,17β)-N-(1,1-dimethylethyl)-3-oxo-4-azaandrost-1-ene-17-carboxamide; MK-906; Chibro-Proscar; Finastid; Propecia; Proscar; Prostide. $C_{23}H_{36}N_2O_2$; mol wt 372.55. C 74.15%, H 9.74%, N 7.52%, O 8.59%. Inhibitor of 5α-reductase, the enzyme which converts testosterone to the more potent androgen, 5α-dihydrotestosterone. Prepn: G. H. Rasmusson, G. F. Reynolds, **EP 155096**; *eidem*, **US 4760071** (1985, 1988 both to Merck & Co.); G. H. Rasmusson *et al.*, *J. Med. Chem.* **29**, 2298 (1986); A. Bhattacharya *et al.*, *J. Am. Chem. Soc.* **110**, 3318 (1988). HPLC determn in plasma and urine: J. R. Carlin *et al.*, *J. Chromatogr.* **427**, 79 (1988). Review of pharmacology and clinical uses: S. L. Sudduth, M. J. Koronkowski, *Pharmacotherapy* **13**, 309-329 (1993). Clinical trials in benign prostatic hyperplasia: J. L. Tenover *et al.*, *Clin. Ther.* **19**, 243 (1997); J. D. McConnell *et al.*, *N. Engl. J. Med.* **338**, 557 (1998); in male pattern hair loss: K. D. Kaufman *et al.*, *J. Am. Acad. Dermatol.* **39**, 578 (1998). Review of use in benign prostatic hyperplasia: J. E. Edwards, R. A. Moore, *BMC Urol.* **2**, 14 (2002); in alopecia: J. F. Libecco, W. F. Bergfeld, *Expert Opin. Pharmacother.* **5**, 933-940 (2004); in prevention of prostate cancer: E. D. Canby-Hagino *et al.*, *ibid.* **7**, 899-905 (2006).

White to off-white crystalline solid, mp ~257° (dried to anhydrous under N_2). Also reported as mp 252-254° (Rasmusson, 1988). $[\alpha]_{405}$ −59° (c = 1 in methanol). Freely sol in chloroform, DMSO, ethanol, methanol, *n*-propanol; sparingly sol in propylene glycol, polyethylene glycol 400; very slightly sol in 0.1*N* HCl, 0.1*N* NaOH, water.

THERAP CAT: In treatment of benign prostatic hypertrophy. Antialopecia agent.

THERAP CAT (VET): In treatment of benign prostatic hypertrophy.

4114. Fingolimod. [162359-55-9] 2-Amino-2-[2-(4-octylphenyl)ethyl]-1,3-propanediol. $C_{19}H_{33}NO_2$; mol wt 307.48. C 74.22%, H 10.82%, N 4.56%, O 10.41%. Sphingosine-1-phosphate receptor modulator. Regulates the egress of mature lymphocytes from secondary lymphoid tissue into systemic circulation. Structure derived from myriocin, *q.v.*, an immunosuppressant found in the traditional Chinese medicine, cordyceps. Prepn: T. Fujita *et al.*, **WO 9408943**; *eidem*, **US 5604229** (1994, 1997 both to Yoshitomi); K. Adachi *et al.*, *Bioorg. Med. Chem. Lett.* **5**, 853 (1995); M. Kiuchi *et al.*, *J. Med. Chem.* **43**, 2946 (2000). Efficient synthesis: P. Durand *et al.*, *Synthesis* **2000**, 505. Immunopharmacology: P. Troncoso, B. D. Kahan, *Clin. Biochem.* **31**, 369-373 (1998). Clinical pharmacokinetics: B. D. Kahan *et al.*, *Transplantation* **76**, 1079 (2003). Clinical evaluation in relapsing multiple sclerosis (MS): L. Kappos *et al.*, *N. Engl. J. Med.* **355**, 1124 (2006). Review of mechanism of action: V. Brinkmann, K. R. Lynch, *Curr. Opin. Immunol.* **14**, 569-575 (2002). Review of discovery and SAR: K. Adachi, K. Chiba, *Perspect. Med. Chem.* **1**, 11-23 (2007); of pharmacology and clinical experience in multiple sclerosis: A. Horga, X. Montalban, *Expert Rev. Neurother.* **8**, 699-714 (2008).

Crystals from ethyl acetate, mp 103-105°.

Hydrochloride. [162359-56-0] FTY-720; Gilenya. $C_{19}H_{33}$-NO_2·HCl; mol wt 343.94. Crystals from ethanol, mp 118-120°. Also reported as white solid from ethyl alcohol + ethyl acetate, mp 107-108° (Kiuchi). Freely sol in water, ethanol; sol in propylene glycol. LD_{50} orally in rats: 300-600 mg/kg (Troncoso).

THERAP CAT: Immunomodulator; in treatment of multiple sclerosis.

4115. Fipexide. [34161-24-5] 1-[4-(1,3-Benzodioxol-5-yl-methyl)-1-piperazinyl]-2-(4-chlorophenoxy)ethanone; 1-(1,3-benzodioxol-5-ylmethyl)-4-[(4-chlorophenoxy)acetyl]piperazine; 1-[(*p*-chlorophenoxy)acetyl]-4-piperonylpiperazine; 1-[2-(4-chlorophenoxy)acetyl]-4-(3,4-methylenedioxybenzyl)piperazine. $C_{20}H_{21}$-ClN_2O_4; mol wt 388.85. C 61.78%, H 5.44%, Cl 9.12%, N 7.20%, O 16.46%. Prepn of hydrochloride: A. Buzas, R. Pierre, **FR M7524** (1970 to Lab Bouchard), *C.A.* **75**, 112865r (1971). Manufacturing process: G. P. Gardini *et al.*, **US 4225714** (1980 to Farmaceutici Geymon.). Improved prepn: G. P. Gardini *et al.*, *Synth. Commun.* **12**, 887 (1982). Determn by TLC: G. Musumarra *et al.*, *J. Chromatogr.* **350**, 151 (1985). Toxicology and pharmacology: G. David *et al.*, *Acta Ther.* **11**, 387 (1985). Clinical trial in the elderly: B. Bompani, G. Scali, *Curr. Med. Res. Opin.* **10**, 99 (1986).

Hydrochloride. [34161-23-4] BP-662; Attentil; Vigilor. C_{20}-$H_{21}ClN_2O_4$·HCl; mol wt 425.31. Crystals from ethanol, mp 230-232°. LD_{50} in Swiss mice, Sprague-Dawley rats, Wistar rats (mg/kg): 4150, 4482, 7000 orally; 499, 537, 450 i.p. (David).

THERAP CAT: Nootropic.

4116. Fipronil. [120068-37-3] 5-Amino-1-[2,6-dichloro-4-(trifluoromethyl)phenyl]-4-[(trifluoromethyl)sulfinyl]-1*H*-pyrazole-3-carbonitrile; 5-amino-3-cyano-1-(2,6-dichloro-4-trifluoromethyl-phenyl)-4-trifluoromethylsulfinylpyrazole; (±)-5-amino-1-(2,6-dichloro-α,α,α-trifluoro-*p*-tolyl)-4-trifluoromethylsulfinylpyrazole-3-carbonitrile; MB-46030; Frontline; Termidor. $C_{12}H_4Cl_2F_6N_4OS$; mol wt 437.14. C 32.97%, H 0.92%, Cl 16.22%, F 26.08%, N 12.82%, O 3.66%, S 7.33%. GABA-gated chloride channel blocker. Prototype of the phenylpyrazole insecticides known as *fiproles*. Prepn: I. G. Buntain *et al.*, **EP 295117** (1988 to May & Baker); L. R. Hatton *et al.*, **US 5232940** (1993). Mechanism of action study: L. M. Cole *et al.*, *Pestic. Biochem. Physiol.* **46**, 47 (1993). Comprehensive description: F. Colliot *et al.*, *Brighton Crop Prot. Conf. - Pests Dis.* **1992**, 29-34.

White solid, mp 200.5-201°. Vapor pressure (20°): 2.8×10^{-9} mm Hg. Log P (*n*-octanol/water): 4.0. Soly: water 2 mg/l; acetone >50%; corn oil >10,000 mg/l. LD_{50} in rats (mg/kg): 100 orally; >2000 dermally (Colliot); in mice (mg/kg): 32 i.p. (Cole).

USE: Pesticide.

THERAP CAT (VET): Ectoparasiticide.

4117. Firefly Luciferin. [2591-17-5] (4*S*)-4,5-Dihydro-2-(6-hydroxy-2-benzothiazolyl)-4-thiazolecarboxylic acid; 2-(6-hydroxy-benzothiazol-2-yl)-2-thiazoline-4-carboxylic acid; D-(−)-luciferin. $C_{11}H_8N_2O_3S_2$; mol wt 280.32. C 47.13%, H 2.88%, N 9.99%, O 17.12%, S 22.87%. Light emission in the American firefly, *Photinus pyralis*, has been shown to involve the interaction of magnesium ion, oxygen, ATP, the enzyme luciferase, and the oxidizable substrate luciferin. Isoln from fireflies (yield from 15,000 active fireflies about 9 mg): Bitler, McElroy, *Arch. Biochem. Biophys.* **72**, 358 (1957); from Japanese firefly (*Luciola cruciata*): Kishi *et al.*, *Tetrahedron Lett.* **1968**, 2847. Structure and synthesis: E. H. White *et al.*, *J. Am. Chem. Soc.* **83**, 2402 (1961); **85**, 337 (1963); *J. Org. Chem.* **30**, 2344 (1965); S. Seto *et al.*, *Bull. Chem. Soc. Jpn.* **36**, 331 (1963). Configuration: G. E. Blank *et al.*, *Biochem. Biophys. Res. Commun.* **42**, 583 (1971). Measurement of ATP: D. M. Karl, O. Holm-Hansen,

Anal. Biochem. **75**, 100 (1976). *Review:* L. J. Bowie, *Methods Enzymol.* **57**, 15 (1978).

Pale yellow needles from methanol, dec 189.5-190°. Recrystallizes with difficulty and sublimes with decomposition and decarboxylation. $[\alpha]_D^{22}$ −36° (c = 1.2 in DMF). uv max (H_2O): 268, 327 nm (log ε 3.88, 4.27). Slightly sol in water at pH 6.5. Sol in alkaline aq solns, methanol, acetone, ethyl acetate, DMF. Aqueous solns are sensitive to extremes in pH, esp. in presence of light and oxygen. Racemization occurs rapidly in some solvents. Should be stored for extended periods dry, under nitrogen atmosphere in light-tight containers.

USE: In the assay of ATP.

4118. Firocoxib. [189954-96-9] 3-(Cyclopropylmethoxy)-5,5-dimethyl-4-[4-(methylsulfonyl)phenyl]-2(5*H*)-furanone; ML-1785713; Previcox. $C_{17}H_{20}O_5S$; mol wt 336.40. C 60.70%, H 5.99%, O 23.78%, S 9.53%. Selective cyclooxygenase-2 (COX-2) inhibitor. Prepn: M. Belley *et al.,* **WO 9714691**; *eidem* **US 5981576** (1997, 1999 both to Merck Frosst). Pharmacokinetics and therapeutic efficacy in inflammatory synovitis in dogs: M. E. McCann *et al., Am. J. Vet. Res.* **65**, 503 (2004).

THERAP CAT (VET): Anti-inflammatory; analgesic.

4119. Fisetin. [528-48-3] 2-(3,4-Dihydroxyphenyl)-3,7-dihydroxy-4*H*-1-benzopyran-4-one; 3,3′,4′,7-tetrahydroxyflavone; 5-desoxyquercetin; fisidenolon 1521; C.I. 75620; C.I. Natural Brown 1. $C_{15}H_{10}O_6$; mol wt 286.24. C 62.94%, H 3.52%, O 33.54%. Flavanoid present in the bark and stems of a variety of trees. Isolation from *Rhus cotinus* L. *Anacardiaceae*, Venetian Sumach: Chevreul, *Lecons Chim. Appl. a la Teint.* **2**, 169 (1833); *cf.* J. Schmid, *Ber.* **19**, 1734 (1886); from heartwood of *Acacia* spp, *Leguminosae:* Roux *et al., Biochem. J.* **78**, 834 (1961). Structure: Seshadri, *Annu. Rev. Biochem.* **20**, 492 (1951). Synthesis: St. Kostanecki *et al., Ber.* **37**, 784 (1904); J. Allan, R. Robinson, *J. Chem. Soc.* **129**, 2334 (1926). Inhibition of aflatoxin cytotoxicity: A. G. Schwartz, W. R. Rate, *J. Environ. Pathol. Toxicol.* **2**, 1021 (1979). Mutagenicity studies: J. P. Brown *et al., Biochem. Soc. Trans.* **5**, 1489 (1977); *idem*, P. S. Dietrich, *Mutat. Res.* **66**, 223 (1979). TLC determn: D. Heimler, *J. Chromatogr.* **366**, 407 (1986).

Yellow needles from dil alc, dec 330°. uv max (ethanol): 252, 320, 360 nm (log $E_{1cm}^{1\%}$ 2.62, 2.51, 2.73). Sol in alcohol, acetone, acetic acid, solns of fixed alkali hydroxides. Practically insol in water, ether, benzene, chloroform and petr ether.

4120. FITC. [27072-45-3] (isomeric mixture); [3326-32-7] (5-isothiocyanate); [18861-78-4] (6-isothiocyanate). 3′,6′-Dihy-

droxy-5(or 6)-isothiocyanatospiro[isobenzofuran-1(3*H*),9′-[9*H*]-xanthen]-3-one; fluorescein isothiocyanate. $C_{21}H_{11}NO_5S$; mol wt 389.38. C 64.78%, H 2.85%, N 3.60%, O 20.54%, S 8.23%. Isomeric mixture of the 5- and 6-isothiocyanates; structural analog of the fluorochrome, fluorescein, *q.v.* Prepn: A. H. Coons *et al., J. Immunol.* **45**, 159 (1942); J. L. Riggs *et al., J. Am. Pathol.* **34**, 1081 (1958). Determn of purity of commerical product: R. M. McKinney *et al., Anal. Biochem.* **7**, 74 (1964); of properites and binding stoichiometry: A. Jobbágy, G. M. Jobbágy, *J. Immunol. Methods* **2**, 159, 169 (1972). Quantitative determn of bound FITC: H. G. de Bruin *et al., Vox Sang.* **45**, 373 (1983). Use as a pH probe: E. Lanz *et al., J. Fluoresc.* **7**, 317 (1997); and evaluation of protonation kinetics: J. Widengren *et al., Chem. Phys.* **249**, 259 (1999). Immunofluorescent dye for microscopy: A. Entwistle, M. Noble, *J. Microsc.* **168**, 219 (1992). Label for amino acids: K. Muramoto *et al., Meth. Protein Seq. Anal.* **1993**, 29; F. Dang, Y. Chen, *Sci. China Ser. B* **42**, 663 (1999). Detection of anionic detergents: H.-W. Gao, D.-Y. Zhou, *Bull. Korean Chem. Soc.* **23**, 29 (2002).

Orange solid, dec upon heating.

USE: Fluorescent indicator and probe for biologic systems. Dye for sewage monitoring.

4121. Flavaspidic Acid. [114-42-1] 3,5-Dihydroxy-4,4-dimethyl-2-(1-oxobutyl)-6-[[2,4,6-trihydroxy-3-methyl-5-(1-oxobutyl)phenyl]methyl]-2,5-cyclohexadien-1-one; 3′-[(5-butyryl-2,4-dihydroxy-3,3-dimethyl-6-oxo-1,4-cyclohexadien-1-yl)methyl]-5′-methylphlorobutyrophenone; polystichocitrin; Toxifren. $C_{24}H_{30}O_8$; mol wt 446.50. C 64.56%, H 6.77%, O 28.67%. Isolated from the rhizomes of male fern: Boehm, *Ann.* **318**, 253 (1901); **329**, 310 (1903). Structure and synthesis: McGookin *et al., J. Chem. Soc.* **1953**, 1828; Riedl, *Ann.* **585**, 32 (1954); Aebi, *Helv. Chim. Acta* **39**, 153 (1956). Crystal structure of α- and β-forms: Erämetsä, Penttilä, *Acta Chem. Scand.* **24**, 3335 (1970). Toxicity study: Airaksinen *et al., Acta Pharmacol. Toxicol.* **25**, 33 (1967).

α-Form. Orthorhombic crystals from methanol or ethanol, mp 92°; solidifies again at 110° and melts again at 156°. LD$_{50}$ orally in mice: 690 mg/kg (Airaksinen).

β-Form. Monoclinic crystals from benzene, xylene or acetic acid, mp 156°.

4122. Flavin-Adenine Dinucleotide. [146-14-5] Riboflavin 5′-(trihydrogen diphosphate) *P′* → 5′-ester with adenosine; adenosine 5′-(trihydrogen diphosphate) 5′ → 5′-ester with riboflavine; FAD; riboflavin 5′-adenosine diphosphate; isoalloxazine-adenine dinucleotide; Fademin; Flavitan. $C_{27}H_{33}N_9O_{15}P_2$; mol wt 785.56. C 41.28%, H 4.23%, N 16.05%, O 30.55%, P 7.89%. One of the biologically active forms of riboflavin, *q.v. See also:* flavin mononucleotide. Redox cofactor involved in biological energy metabolism; reduced to FADH$_2$ to carry high energy electrons as part of the electron transport chain. Prosthetic group of certain flavoproteins including D-amino acid oxidase, glucose oxidase, glycine oxidase, fumaric hydrogenase, histaminase, and xanthine oxidase. Isoln from yeast: Warburg *et al.*, *Biochem. Z.* **297**, 417 (1938). Structure and isoln from liver, kidneys, hearts, muscles: Warburg, Christian, *ibid.* **298**, 150 (1938); isoln from the mycelium of *Eremothecium ashbyii:* Yagi, Tada, *Biochem. Prep.* **7**, 51 (1959); Masuda *et al.*, *US 2973305* (1961 to Takeda). Synthesis: Christie *et al.*, *J. Chem. Soc.* **1954**, 46; Huennekens, Kilgour, *J. Am. Chem. Soc.* **77**, 6716 (1955); De-Luca, Kaplan, *J. Biol. Chem.* **223**, 569 (1956); Moffatt, Khorana, *J. Am. Chem. Soc.* **80**, 3756 (1958). Review of FAD and other flavin coenzymes: Beinert, *The Enzymes* vol. **2**, P. D. Boyer *et al.*, Eds. (Academic Press, New York, 2nd ed., 1960) pp 339-416; of covalent flavoproteins: M. Mewies *et al.*, *Protein Sci.* **7**, 7-20 (1998). Structure-sequence analysis of FAD-containing proteins: O. Dym, D. Eisenberg, *ibid.* **10**, 1712-1728 (2001). Review of diverse biological roles: S. O. Mansoorabadi *et al.*, *J. Org. Chem.* **72**, 6329-6342 (2007).

Barium salt. $C_{27}H_{31}BaN_9O_{15}P_2$. Small yellow spheres clustered like grapes. Absorption max: 366, 445 nm. The absorption curve is practically identical with that of riboflavin. There is some stronger absorption between 450 and 510 nm resulting in aq solns which are more reddish and less green than those of riboflavin. The appearance of a strong greenish fluorescence indicates decomposition and loss of catalytic activity.

THERAP CAT: Vitamin (enzyme cofactor).

4123. Flavin Mononucleotide. [146-17-8] Riboflavin 5′-(dihydrogen phosphate); riboflavin monophosphate; FMN; vitamin B$_2$ phosphate. $C_{17}H_{21}N_4O_9P$; mol wt 456.35. C 44.74%, H 4.64%, N 12.28%, O 31.55%, P 6.79%. One of the bioactive forms of riboflavin, *q.v. See also:* flavin-adenine dinucleotide. Prosthetic group for several oxidoreductases involved in biological energy metabolism; cofactor for the blue-light photoreceptor, phototropin. Prepd by phosphorylation of riboflavin with chlorophosphoric acid: L. A. Flexser, W. G. Farkas, *US 2610179* (1952 to Hoffmann-La Roche); with pyrophosphoric acid: Breivogel, Ridge, *US 2535385* (1950 to White Labs.); with metaphosphoric acid: Viscontini *et al.*, *Helv. Chim. Acta* **35**, 457 (1952); *US 2740775* (1956 to Hoffmann-La Roche); with pyrocatechol cyclic phosphate: Ukita *et al.*, *US 3118876* (1964 to Takeda). Review of role as phototropin cofactor: W. R. Briggs *et al.*, *Antioxid. Redox Signal.* **3**, 775-788 (2001); in nitric oxide synthase electron transfer: D. J. Stuehr *et al.*, *FEBS J.* **276**, 3959-3974 (2009).

Monosodium salt. [130-40-5] Hyryl; Ribo. $C_{17}H_{20}N_4NaO_9P$; mol wt 478.33. Occurs as the dihydrate. Yellow crystals. Soly in water at pH 6.9 = 112 mg/ml; at pH 5.6 = 68 mg/ml; at pH 3.8 = 43 mg/ml. *Protect from light.*

THERAP CAT: Vitamin (enzyme cofactor).

THERAP CAT (VET): Vitamin (enzyme cofactor).

4124. Flavone. [525-82-6] 2-Phenyl-4*H*-1-benzopyran-4-one; 2-phenylchromone; 2-phenyl-γ-benzopyrone; 2-phenyl-1,4-benzopyrone. $C_{15}H_{10}O_2$; mol wt 222.24. C 81.07%, H 4.54%, O 14.40%. Prepn: Feuerstein, Kostanecki, *Ber.* **31**, 1757 (1898); Kostanecki, Tambor, *Ber.* **33**, 330 (1900); Kostanecki, Szabranski, *Ber.* **37**, 2634 (1904); Ruhemann, *Ber.* **46**, 2192 (1913); Bogert, Marcus, *J. Am. Chem. Soc.* **41**, 89 (1919). Isoln from *Primula malacoides*, Franch., *Primulaceae* and biological properties: Weller *et al.*, *Antibiot. Chemother.* **3**, 603 (1953). Syntheses: Wheeler, *Org. Synth.* **coll. vol. IV**, 478 (1963); Y. Ashihara *et al.*, *Bull. Chem. Soc. Jpn.* **50**, 3298 (1977); A. Banerji, N. C. Goomer, *Synthesis* **1980**, 874.

Crystals from petr ether, mp 99-100°. Practically insol in water. Sol in most organic solvents. Absorption max: 350, 405 nm.

4125. Flavopereirine. [486-18-0] 3-Ethylindolo[2,3-*a*]quin-olizin-5-ium inner salt; melinonine G. $C_{17}H_{14}N_2$; mol wt 246.31. C 82.90%, H 5.73%, N 11.37%. From *Strychnos melinoniana* Baillon., *Loganiaceae:* Bächli *et al.*, *Helv. Chim. Acta* **40**, 1167 (1957); from *Geissospermum vellosii* Baillon., *Apocynaceae:* Rapoport *et al.*, *J. Am. Chem. Soc.* **80**, 1601 (1958); Bertho *et al.*, *Ber.* **91**, 2581 (1958). Structure: Bejar *et al.*, *Compt. Rend.* **244**, 2066 (1957). Synthesis: Le Hir *et al.*, *Bull. Soc. Chim. Fr.* **1958**, 551; Prasad, Swan, *J. Chem. Soc.* **1958**, 2024; G. W. Gribble, D. A. Johnson, *Tetrahedron Lett.* **28**, 5259 (1987); of perchlorate: J. Ninomiyo *et al.*, *Heterocycles* **9**, 1527 (1978).

Orange crystals from acetone, mp 233-235°. uv max (ethanol): 230, 238, 248, 294, 351, 390 nm (log ε 4.40, 4.43, 4.39, 4.14, 4.25, 4.14).

Perchlorate. $C_{17}H_{15}ClN_2O_4$. Crystals from methanol, mp 307-308°. uv max (0.015*N* HCl-ethanol): 238, 294, 350, 389 nm (log ε 4.57, 4.22, 4.31, 4.21).

4126. Flavopiridol. [146426-40-6] 2-(2-Chlorophenyl)-5,7-dihydroxy-8-[(3*S*,4*R*)-3-hydroxy-1-methyl-4-piperidinyl]-4*H*-1-benzopyran-4-one; (−)-*cis*-2-(2-chlorophenyl)-5,7-dihydroxy-8-[4*R*-(3*S*-hydroxy-1-methyl)-piperidinyl]-4*H*-1-benzopyran-4-one; alvocidib. $C_{21}H_{20}ClNO_5$; mol wt 401.84. C 62.77%, H 5.02%, Cl 8.82%, N 3.49%, O 19.91%. Cyclin-dependent kinase (CDK) inhibitor; induces apoptosis in certain tumor cells. Prepn: S. L. Kattige *et al.*, *EP 241003; eidem*, *US 4900727* (1987, 1990 both to Hoechst). Properties and formulation study: R.-M. Dannenfelser *et al.*, *PDA J. Pharm. Sci. Technol.* **50**, 356 (1996). Mode of action study: S. Brüsselbach *et al.*, *Int. J. Cancer* **77**, 146 (1998). Clinical evaluation in refractory cancers: A. M. Senderowicz *et al.*, *J. Clin. Oncol.* **16**, 2986 (1998). Clinical pharmacokinetics: J. P. Thomas *et al.*, *Cancer Chemother. Pharmacol.* **50**, 465 (2002). Review of chemistry, pharmacology, and antitumor spectrum: H. H. Sedlacek *et al.*, *Int. J. Oncol.* **9**, 1143-1168 (1996); of clinical experience in chronic lymphocytic leukemia: B. A. Christian *et al.*, *Clin. Lymphoma Myeloma Leuk.* **9**, Suppl. 3, S179-S185 (2009).

Hydrochloride. [131740-09-5] L-86-8275; HMR-1275; NSC-649890. $C_{21}H_{20}ClNO_5$·HCl; mol wt 438.30. mp 169.5 ± 0.5° (Dannenfelser). Also reported as yellow powder, mp 188° containing 4.73-6.2% water of crystallization (Sedlacek). pKa 5.68 ± 0.06. $[\alpha]_D^{24}$ −1.73 to −3.9°. Soly (mg/ml): water, D5W, and normal saline all <5 at pH 4. Soly (mg/ml): ethanol 10.1; PEG 400 >73.8; propylene glycol >88.1; M-pyrol very much greater than 103.0; τ-butyrolactone very much greater than 105.5.

THERAP CAT: Antineoplastic.

4127. Flavoxanthin. [512-29-8] (3S,3'R,5R,6'R,8R)-5,8-Epoxy-5,8-dihydro-β,ε-carotene-3,3'-diol. $C_{40}H_{56}O_3$; mol wt 584.89. C 82.14%, H 9.65%, O 8.21%. Carotenoid pigment; stereoisomer of chrysanthemaxanthin, q.v. Often found in plants, but in minute amounts only and never as a principal pigment. Isoln from *Ranunculus acris* L., *Ranunculaceae:* Kuhn, Brockmann, *Z. Physiol. Chem.* **213**, 192 (1932). Structure: Karrer, Rutschmann, *Helv. Chim. Acta* **25**, 1144 (1942). Partial synthesis: Karrer, Jucker, *ibid.* **28**, 300 (1945). Absolute configuration: H. Cadosch *et al., ibid.* **61**, 783 (1978).

Golden-yellow aggregates of prisms, mp 184°. Absorption max (chloroform): 430, 459 nm. $[\alpha]_{Cd}^{20}$ +190° (c = 0.04 in benzene). Freely sol in chloroform, benzene, acetone; less sol in methanol, ethanol. Almost insol in petr ether.

4128. Flavoxate. [15301-69-6] 3-Methyl-4-oxo-2-phenyl-4H-1-benzopyran-8-carboxylic acid 2-(1-piperidinyl)ethyl ester; 3-methylflavone-8-carboxylic acid β-piperidinoethyl ester; 2-piperidinoethyl 3-methyl-4-oxo-2-phenyl-4H-1-benzopyran-8-carboxylate; 2-piperidinoethyl 3-methylflavone-8-carboxylate. $C_{24}H_{25}NO_4$; mol wt 391.47. C 73.64%, H 6.44%, N 3.58%, O 16.35%. Smooth muscle relaxant. Prepn of the hydrochloride: P. Da Re *et al.,J. Med. Pharm. Chem.* **2**, 263 (1960); P. Da Re, **US 2921070** and **US 3350411** (1960 to Recordati and 1967 to Seceph). Pharmacological studies: I. Setnikar *et al., J. Pharmacol. Exp. Ther.* **130**, 356 (1960). Mechanism of action: P. Cazzulani *et al., Arch. Int. Pharmacodyn.* **274**, 189 (1985). Pharmacokinetics in rats: I. Setnikar *et al., Arzneim.-Forsch.* **25**, 1916 (1975); in man: M. Bertoli *et al., Pharmacol. Res. Commun.* **8**, 417 (1976). *In vivo* activity: C. Pietra, P. Cazzulani, *Farmaco Ed. Prat.* **41**, 267 (1986). Determn in tissues: A. Cova, I. Setnikar, *Arzneim.-Forsch.* **25**, 1707 (1975). Clinical evaluations: D. V. Bradley, R. J. Cazort, *J. Clin. Pharmacol.* **10**, 65 (1970); A. Zanollo, F. Catanzaro, *Urol. Int.* **35**, 176 (1980). Clinical comparison with phenazopyridine: S. Gould, *Urology* **5**, 612 (1975). Short review: R. Ruffmann, A. Sartini, *Drugs Exp. Clin. Res.* **13**, 57 (1987).

Crystals, pK 7.3. Soly in water at 37°: 0.001% (w/v). Sol in ethanol and chloroform. LD$_{50}$ in rats (mg/kg): 1110 orally; 20.8 i.v. (Setnikar, 1975).

Hydrochloride. [3717-88-2] DW-61; Rec-7-0040; Bladderon; Genurin; Patricin; Spasuret; Urispas. $C_{24}H_{25}NO_4$·HCl; mol wt 427.93. Crystals from ethanol + ether, mp 232-234°. Slightly soluble in water, alc, methylene chloride. LD$_{50}$ i.v. in rats: 27.4 mg/kg (Cazzulani).

Succinate. [28782-19-6] $C_{24}H_{25}NO_4$·$C_4H_6O_4$. Soly in water at 37°: 33.7% (w/v).

THERAP CAT: Antispasmodic; in treatment of urinary incontinence.

4129. Flazasulfuron. [104040-78-0] N-[[(4,6-Dimethoxy-2-pyrimidinyl)amino]carbonyl]-3-(trifluoromethyl)-2-pyridinesulfonamide; 1-(4,6-dimethoxypyrimidin-2-yl)-3-(3-trifluoromethyl-2-pyridylsulfonyl)urea; OK-1166; SL-160; Katana; Shibagen. $C_{13}H_{12}F_3$-N_5O_5S; mol wt 407.32. C 38.33%, H 2.97%, F 13.99%, N 17.19%, O 19.64%, S 7.87%. Sulfonylurea herbicide. Prepn: F. Kimura *et al.*, **EP 184385**; *eidem*, **US 4744814** (1986, 1988 both to Ishihara Sangyo Kaisha); T. Haga *et al., ACS Symp. Ser.* **443**, 107 (1991). Description of properties and biological activity: B. Hashizume, *Jpn. Pestic. Inf.* **57**, 27 (1990). Field trial for control of ragwort: T. K. James *et al., Proc. 50th N. Z. Plant Prot. Conf.* **1997**, 447. Weed control management in grapevines: P. Bourdrez, J.-M. Béraud, *Phytoma* **521**, 66 (1999).

Odorless, white crystalline solid, mp 164-166°. Soly at 20° (w/w%): acetone 1.2%; toluene 0.06%; at 25° (w/v%): acetic acid 0.67%. Soly in water: 4.1 ppm (pH 1). Vapor pressure: 3.1×10^{-8} mm Hg. pKa (24°): 4.6. Log P (octanol/water): 1.08. LD$_{50}$ mice, rats (mg/kg): >5000, >5000 orally; in rats (mg/kg): >2000 dermally (Hashizume).

USE: Herbicide.

4130. Flecainide. [54143-55-4] N-(2-Piperidinylmethyl)-2,5-bis(2,2,2-trifluoroethoxy)benzamide. $C_{17}H_{20}F_6N_2O_3$; mol wt 414.35. C 49.28%, H 4.87%, F 27.51%, N 6.76%, O 11.58%. Prepn: E. H. Banitt, W. R. Brown, **US 3900481** (1975 to Riker); of the acetate: *eidem*, **US 4005209** (1977 to Riker); E. H. Banitt *et al., J. Med. Chem.* **20**, 821 (1977). Preliminary pharmacological study: J. R. Schmid *et al., Fed. Proc.* **34**, 775 (1975). *In vitro* electrophysiological study: A. B. Hodess *et al., J. Cardiovasc. Pharmacol.* **1**, 427 (1979). Antiarrhythmic effects: P. Somani, *Clin. Pharmacol. Ther.* **27**, 464 (1980). Use in acute exptl myocardial infarction: H. Gülker *et al., Z. Cardiol.* **70**, 124 (1981). Clinical study in ventricular arrhythmias: J. L. Anderson *et al., N. Engl. J. Med.* **305**, 473 (1981). Determn of acetate in human plasma by spectrophotofluorometry: S. F. Chang *et al., Arzneim.-Forsch.* **33**, 251 (1983). Review of pharmacology and clinical efficacy: D. M. Roden, R. L. Woosley, *N. Engl. J. Med.* **315**, 36-41 (1986). Symposium on clinical experience: *Am. J. Cardiol.* **62**, Suppl., 1D-67D (1988). Comprehensive description: S. Alessi-Severini *et al., Anal. Profiles Drug Subs. Excip.* **21**, 169-195 (1992).

uv max (ethanol): 205, 230, 300 nm (E$_{1cm}^{1\%}$ 521, 219, 59).

Monoacetate. [54143-56-5] R-818; Almarytm; Apocard; Ecrinal; Flécaine; Tambocor. $C_{17}H_{20}F_6N_2O_3$·$C_2H_4O_2$; mol wt 474.40. White granular solid from isopropyl alcohol/isopropyl ether, mp 145-147°. Soly at 37° (mg/ml): water 48.4, alcohol 300.

THERAP CAT: Antiarrhythmic (class IC).

4131. Fleroxacin. [79660-72-3] 6,8-Difluoro-1-(2-fluoroeth-yl)-1,4-dihydro-7-(4-methyl-1-piperazinyl)-4-oxo-3-quinolinecarboxylic acid; AM-833; Ro-23-6240; Megalocin; Megalone; Quinodis. $C_{17}H_{18}F_3N_3O_3$; mol wt 369.34. C 55.28%, H 4.91%, F 15.43%, N 11.38%, O 13.00%. Fluorinated quinolone antibacterial. Prepn: **BE 887574**; T. Irikura *et al.*, **US 4398029** (1981, 1983 both to Kyorin). Antibacterial spectrum *in vitro* and *in vivo:* K. Hirai *et al.*, *Antimicrob. Agents Chemother.* **29**, 1059 (1986). *In vitro* activity vs anaerobic bacteria: J. Wüst, U. Hardegger, *Eur. J. Clin. Microbiol.* **6**, 688 (1987). HPLC determn in biological fluids: H. Kusajima *et al.*, *J. Chromatogr.* **381**, 137 (1986). Pharmacokinetics and bioavailability: E. Weidekamm *et al.*, *Antimicrob. Agents Chemother.* **31**, 1909 (1987). Preliminary clinical evaluation in gonorrhea: J. B. J. Boerema *et al.*, *J. Antimicrob. Chemother.* **21**, 140 (1988).

Hydrochloride. $C_{17}H_{18}F_3N_3O_3 \cdot HCl$. Crystals from water, mp 269-271° (dec).

THERAP CAT: Antibacterial.

4132. Flibanserin. [167933-07-5] 1,3-Dihydro-1-[2-[4-[3-(trifluoromethyl)phenyl]-1-piperazinyl]ethyl]-2*H*-benzimidazol-2-one; BIMT-17. $C_{20}H_{21}F_3N_4O$; mol wt 390.41. C 61.53%, H 5.42%, F 14.60%, N 14.35%, O 4.10%. Serotonin 5HT$_{1A}$ receptor agonist and 5HT$_{2A}$ receptor antagonist. Prepn: M. Turconi *et al.*, **EP 526434**; G. Bietti *et al.*, **US 5576318** (1993, 1996 both to Boehringer Ingelheim). Receptor binding study: F. Borsini *et al.*, *Arch. Pharmacol.* **352**, 276 (1995). Electrophysiological characterization: *eidem, ibid.* **283**. *In vivo* modulation of dopaminergic and serotonergic activity: B. Ferger *et al.*, *ibid.* **381**, 573 (2010). Clinical efficacy for sexual dysfunction in depression: S. Kennedy, *J. Sex. Med.* **7**, 3449 (2010). Review of pharmacology: F. Borsini *et al.*, *CNS Drug Rev.* **8**, 117-142 (2002); and therapeutic potential for hypoactive sexual desire disorder (HSDD): A. H. Clayton *et al.*, *Womens Health* **6**, 639-653 (2010).

Hydrochloride. [147359-76-0] $C_{20}H_{21}F_3N_4O \cdot HCl$; mol wt 426.87. Crystals from isopropanol, mp 230-231°.

THERAP CAT: In treatment of hypoactive sexual desire disorder.

4133. Flindersine. [523-64-8] 2,6-Dihydro-2,2-dimethyl-5*H*-pyrano[3,2-*c*]quinolin-5-one; 2,2'-dimethyl-α-pyrano(5',6',3,4)-2(1*H*)-quinolone. $C_{14}H_{13}NO_2$; mol wt 227.26. C 73.99%, H 5.77%, N 6.16%, O 14.08%. From wood of *Flindersia australis* R. Br., *Rutaceae:* Matthes, Schreiber, *Ber. Dtsch. Pharm. Ges.* **24**, 385 (1914). Structure: Brown *et al.*, *Aust. J. Chem.* **7**, 348 (1954). Synthesis: *eidem, ibid.* **9**, 277 (1956); Piozzi *et al.*, *Gazz. Chim. Ital.* **99**, 711 (1969); Bowman *et al.*, *Chem. Commun.* **1970**, 666; Huffman, Hsu, *Tetrahedron Lett.* **1972**, 141.

Crystals from methanol, dec 185-186°. uv max (methanol): 235, 333, 350, 365 nm (log ε 4.42, 4.00, 4.10, 3.93). Sol in alc, benzene, chloroform, glacial acetic acid, paraffin, fatty oils, alkali hydroxides, slightly sol in petr ether. Practically insol in water.

N-**Methylflindersine.** $C_{15}H_{15}NO_2$. Crystals from petr ether, mp 84°.

Dihydroflindersine. $C_{14}H_{15}NO_2$. Hexagons from ethyl acetate, mp 229°. uv max (methanol): 225, 272, 283, 312 nm (log ε 4.44, 3.94, 3.96, 3.93).

4134. Flocoumafen. [90035-08-8] 4-Hydroxy-3-[1,2,3,4-tetrahydro-3-[4-[[4-(trifluoromethyl)phenyl]methoxy]phenyl]-1-naphthalenyl]-2*H*-1-benzopyran-2-one; 4-hydroxy-3-[1,2,3,4-tetrahydro-3-[4-(4-trifluoromethylbenzyloxy)phenyl]-1-naphthyl]; WL-108366; Storm. $C_{33}H_{25}F_3O_4$; mol wt 542.55. C 73.06%, H 4.64%, F 10.51%, O 11.80%. Second generation anticoagulant rodenticide. Prepn: I. D. Entwistle, P. Boehm, **EP 98629**; *eidem*, **US 4520007** (1983, 1985 both to Shell). Synthesis and separation of isomers: O. S. Park, B. S. Jang, *Arch. Pharmacal Res.* **18**, 277 (1995). HPLC determn in liver tissue: A. Jones, *Bull. Environ. Contam. Toxicol.* **56**, 8 (1996). Absorption and distribution in rats: K. R. Huckle *et al.*, *Pestic. Sci.* **25**, 297 (1989). Efficacy against seven rodent species: J. E. Gill, *Int. Biodeterior. Biodegrad.* **30**, 65 (1992). Review of physical properties, mode of action, and field trials: D. J. Bowler *et al.*, *Proc. Br. Crop Prot. Conf. - Pests Dis.* **1984**, 397.

White powder. Soly in water: 1.1 mg/l. Soly in acetone, alcohol, chloroform, dichloromethane: >10 g/l. Vapor pressure at 25°: approx 1×10^{-13} mm Hg. LD$_{50}$ in rats, mice, rabbits (mg/kg): 0.25, 0.8, 0.2 orally (Bowler).

USE: Rodenticide.

4135. Floctafenine. [23779-99-9] 2-[[8-(Trifluoromethyl)-4-quinolinyl]amino]benzoic acid 2,3-dihydroxypropyl ester; *N*-[8-(tri-fluoromethyl)-4-quinolyl]anthranilic acid 2,3-dihydroxypropyl ester; 4-[*o*-(2',3'-dihydroxypropyloxycarbonyl)phenyl]amino-8-trifluoro-methylquinoline; 8-trifluoromethyl-7-deschloroglafenine; R-4318; RU-15750; Idalon; Idarac; Novodolan. $C_{20}H_{17}F_3N_2O_4$; mol wt 406.36. C 59.12%, H 4.22%, F 14.03%, N 6.89%, O 15.75%. Prepn: A. Allais, J. Meier, **DE 1815467**; *eidem*, **US 3644368** (1969, 1972, both to Roussel-UCLAF); G. Mouzin *et al.*, *Synthesis* **1980**, 54. Analgesic activity: Allais *et al.*, *Chim. Ther.* **8**, 154 (1973); M. Peterfalvi *et al.*, *Arch. Int. Pharmacodyn. Ther.* **216**, 97 (1975). Clinical investigation: Stenport, *Curr. Ther. Res.* **18**, 303 (1975). Toxicology: R. Glomot *et al.*, *Toxicol. Appl. Pharmacol.* **36**, 173 (1976). Metabolism: R. K. Lynn *et al.*, *J. Clin. Pharmacol.* **19**, 20 (1979).

Crystals from methanol, mp 179-180°. Sol in alcohol, acetone; very slightly sol in ether, chloroform, methylene chloride. Insol in water. LD$_{50}$ in male mice, rats (mg/kg): 3400, 960 orally; 180, 160 i.v. (Glomot).

THERAP CAT: Analgesic.

4136. Flomoxef. [99665-00-6] (6*R*,7*R*)-7-[[2-[(Difluoro-methyl)thio]acetyl]amino]-3-[[[1-(2-hydroxyethyl)-1*H*-tetrazol-5-yl]thio]methyl]-7-methoxy-8-oxo-5-oxa-1-azabicyclo[4.2.0]oct-2-

ene-2-carboxylic acid; 7-β-difluoromethylthioacetamido-7α-methoxy-3-[[1-(2-hydroxyethyl)-1*H*-tetrazol-5-yl]thiomethyl]-1-oxa-3-cephem-4-carboxylic acid. $C_{15}H_{18}F_2N_6O_7S_2$; mol wt 496.46. C 36.29%, H 3.65%, F 7.65%, N 16.93%, O 22.56%, S 12.92%. Semisynthetic oxacephalosporin antibiotic. Prepn: T. Tsuji *et al.*, **BE 898541**; *eidem*, **US 4532233** (1984, 1985 both to Shionogi); *eidem*, *J. Antibiot.* **38**, 466 (1985). *In vitro* activity and β-lactamase stability: H. C. Neu, N.-X. Chin, *Antimicrob. Agents Chemother.* **30**, 638 (1986). Determn in serum by capillary electrophoresis: T. Kitahashi, I. Furuta, *J. Chromatogr. Sci.* **41**, 173 (2003). Clinical trial for prophylaxis of post-surgical infection: S. Togo *et al.*, *J. Antimicrob. Chemother.* **59**, 964 (2007).

Crystals from acetone + methylene chloride, mp 82.5-87.5°.
Sodium salt. [92823-03-5] 6315-S; Flumarin. $C_{15}H_{17}F_2N_6$-NaO_7S_2; mol wt 518.44. White to light yellowish white powder or masses. mp 100-150° (dec). Partition coefficient: (octanol/water): 0.001. Very sol in water; freely sol in methanol; sparingly sol in ethanol.

THERAP CAT: Antibacterial.

4137. Flonicamid. [158062-67-0] *N*-(Cyanomethyl)-4-(trifluoromethyl)-3-pyridinecarboxamide; *N*-(cyanomethyl)-4-(trifluoromethyl)nicotinamide; F-1785; IKI-220; Aria; Beleaf; Carbine; Turbine. $C_9H_6F_3N_3O$; mol wt 229.16. C 47.17%, H 2.64%, F 24.87%, N 18.34%, O 6.98%. Selective systemic aphicide. Prepn: T. Toki *et al.*, **EP 580374**; *eidem*, **US 5360806** (both 1994 to Ishihara Sangyo Kaisha). Comprehensive description: M. Morita *et al.*, *BCPC Conf. - Pests Dis.* **2000**, 59-65. Field performance in cotton: H. G. Hancock, *Proc. Beltwide Cotton Conf.* **2004**, 1629.

Odorless, white crystalline powder, mp 157.5°. Vapor pressure (20°): 9.43×10^{-7} Pa. Log P (octanol/water): 0.30. Soly in water (20°): 5.2 g/l. LD_{50} in male, female rats (mg/kg): 884, 1768 orally; in rats (mg/kg): >5000 dermally; in rats (mg/m³): >4900 by inhalation. LC_{50} in carp, rainbow trout (mg/l): >100, >91.9 (Morita).

USE: Insecticide.

4138. Flopropione. [2295-58-1] 1-(2,4,6-Trihydroxyphenyl)-1-propanone; 2′,4′,6,-trihydroxypropiophenone; phloropropiophenone; RP-13907; Argobyl; Cospanon; Flopion; Gallepronin; Gasstenon; Labroda; Labrodax; Pasmus; Profenon; Spamorin; Spasmoril; Supanate; Supazlun. $C_9H_{10}O_4$; mol wt 182.18. C 59.34%, H 5.53%, O 35.13%. Prepn: Shinoda, *J. Pharm. Soc. Jpn.* **1927**, 111; Canter *et al.*, *J. Chem. Soc.* **1931**, 1245; Howells, Little, *J. Am. Chem. Soc.* **54**, 2451 (1932).

Monohydrate. Needles from water. Anhydr compd, mp 175-176°. Sol in ethanol, ether, ethyl acetate, hot water; very slightly sol in cold water.

THERAP CAT: Antispasmodic.

4139. Florasulam. [145701-23-1] *N*-(2,6-Difluorophenyl)-8-fluoro-5-methoxy-[1,2,4]triazolo[1,5-*c*]pyrimidine-2-sulfonamide; 2′,6′,8-trifluoro-5-methoxy-*s*-triazolo[1,5-*c*]pyrimidine-2-sulfonanilide; Boxer; Nikos; Primus. $C_{12}H_8F_3N_5O_3S$; mol wt 359.28. C 40.12%, H 2.24%, F 15.86%, N 19.49%, O 13.36%, S 8.92%. Triazolopyrimidine herbicide for use in cereals. Acetolactate synthase (ALS) inhibitor. Prepn: J. C. Van Heertum *et al.*, **WO 8911782** (1989 to Dow); *eidem*, **US 5163995** (1992 to DowElanco). Comprehensive description: A. R. Thompson *et al.*, *Brighton Crop Prot. Conf. - Weeds* **1999**, 73-80. Review of field trials: A. D. Bailey *et al.*, *ibid.* 205-209. Photolytic degradation in soil and water: M. S. Krieger *et al.*, *J. Agric. Food Chem.* **48**, 3710 (2000).

mp 220-221° (dec) (Van Heertum); also reported as 193.5-230.5° (Thompson). pKa: 4.54. Partition coefficient (octanol/water): 0.060 (pH 7.0). Vapor pressure (25°): 1×10^{-5} Pa. Soly in water (20°): 6.36 g/l (pH 7.0). LD_{50} in rat, quail (mg/kg): >6000, 1046 orally; percutaneously in rabbit: >2000 mg/kg. LC_{50} in rainbow trout, bluegill (96 hr): >96, >98 mg/l (Thompson).

USE: Herbicide.

4140. Florbetapir F 18. [956103-76-7] 4-[(1*E*)-2-[6-[2-[2-[2-(Fluoro-^{18}F)ethoxy]ethoxy]ethoxy]-3-pyridinyl]ethenyl]-*N*-methylbenzenamine; (*E*)-4-(2-(6-(2-(2-(2-^{18}F-fluoroethxy)ethoxy)ethoxy)-pyridin-3-yl)vinyl)-*N*-methyl benzenamine; (*E*)-2-(2-(2-(2-^{18}F-fluoroethoxy)ethoxy)ethoxy)-5-(4-methylaminostyryl)pyridine; [^{18}F]AV-45; Amyvid. $C_{20}H_{25}{}^{18}FN_2O_3$. Radiolabeled styrylpyridine; positron emission tomography (PET) imaging agent for detection of β-amyloid plaques in the brain. Prepn: H. F. Kung, M.-P. Kung, **WO 07126733**; *eidem*, **US 7687052** (2007, 2010 both to Trustees Univ. Penn.). Automated synthesis: Y. Liu *et al.*, *Nucl. Med. Biol.* **37**, 917 (2010); C.-H. Yao *et al.*, *Appl. Radiat. Isot.* **68**, 2293 (2010). Bioavailability and binding affinity studies: S. R. Choi *et al.*, *J. Nucl. Med.* **50**, 1887 (2009). Clinical biodistribution and binding to β-amyloid plaques: K.-J. Lin *et al.*, *Nucl. Med. Biol.* **37**, 497 (2010). Clinical pharmacokinetics: D. F. Wong *et al.*, *J. Nucl. Med.* **51**, 913 (2010). Clinical trial comparing imaging and autopsy results: C. M. Clark *et al.*, *J. Am. Med. Assoc.* **305**, 275 (2011).

Prepd as a clear, colorless solution in buffered saline; labeled with ^{18}F which has a half-life of 109.3 minutes.

THERAP CAT: Diagnostic aid (radioactive imaging agent).

4141. Florfenicol. [73231-34-2] 2,2-Dichloro-*N*-[(1*S*,2*R*)-1-(fluoromethyl)-2-hydroxy-2-[4-(methylsulfonyl)phenyl]ethyl]-acetamide; D-*(threo)*-1-*p*-methylsulfonylphenyl-2-amino-3-fluoro-1-propanol; fluorothiamphenicol; Sch-25298; Aquafen; Aquaflor; Florocol; Nuflor. $C_{12}H_{14}Cl_2FNO_4S$; mol wt 358.21. C 40.24%, H 3.94%, Cl 19.79%, F 5.30%, N 3.91%, O 17.87%, S 8.95%. Fluorinated derivative of thiamphenicol, *q.v.* Inhibits bacterial protein synthesis by binding to ribosome 50S and 70S subunits. Prepn: T. Nagabhushan, **EP 14437**; **US 4235892** (both 1980 to Schering); D. P. Schumacher *et al.*, *J. Org. Chem.* **55**, 5291 (1990). Improved prepn: G. Wu *et al.*, *ibid.* **62**, 2996 (1997). Determn in fish plasma by HPLC: C. Vue *et al.*, *J. Chromatogr. B* **780**, 111 (2002); in animal tissues by LC-MS/MS: P. Luo *et al.*, *J. Chromatogr. B* **878**, 207 (2010). *In vitro* antibacterial activity: H. C. Neu, K. P. Fu, *Antimicrob. Agents Chemother.* **18**, 311 (1980); V. P. Syriopoulou *et al.*, *ibid.* **19**, 294 (1981); vs bovine and porcine respiratory disease isolates: S. J. Shin *et al.*, *Vet. Microbiol.* **106**, 73 (2005). Pharmacokinetics in veal calves: K. J. Varma *et al.*, *J. Vet. Pharmacol. Ther.* **9**, 412 (1986); in sheep: J. Shen *et al.*, *ibid.* **27**, 163 (2004). Therapeutic effect on pseudotuberculosis in yellowtail tuna: N. Yasunaga, S. Yasumoto, *Fish Pathol.* **23**, 1 (1988).

Crystals from 2-propanol/H_2O, mp 152-154°. $[\alpha]^{26}$ +17.9° (DMF). Sol in water.

THERAP CAT (VET): Antibacterial.

4142. Flosequinan. [76568-02-0] 7-Fluoro-1-methyl-3-(methylsulfinyl)-4(1*H*)-quinolinone; flosequinon; BTS-49465; Manoplax. $C_{11}H_{10}FNO_2S$; mol wt 239.26. C 55.22%, H 4.21%, F 7.94%, N 5.85%, O 13.37%, S 13.40%. Mixed arterial and venous vasodilator. Prepn: R. V. Davies *et al.*, **DE 3011994**; *eidem*, **US 4302460** (1980, 1981 both to Boots). Pharmacology: J. G. Smith, G. T. Kinasewitz, *J. Cardiovasc. Pharmacol.* **8**, 878 (1986); M. F. Sim *et al.*, *Br. J. Pharmacol.* **94**, 371 (1988). HPLC determn in plasma, serum and urine: M. B. Slegowski *et al.*, *J. Chromatogr.* **425**, 227 (1988). Pharmacokinetics and hemodynamics in humans: A. J. Cowley *et al.*, *J. Hypertens.* **2**, Suppl. 3, 547 (1984); R. D. Wynne *et al.*, *Eur. J. Clin. Pharmacol.* **28**, 659 (1985). Preliminary evaluation in hypertension: A. J. Cowley *et al.*, *ibid.* **33**, 203 (1987); in congestive heart failure: P. D. Kessler, M. Packer, *Am. Heart J.* **113**, 137 (1987); P. D. Kessler *et al.*, *J. Cardiovasc. Pharmacol.* **12**, 6 (1988). Clinical study in chronic heart failure: A. J. Cowley *et al.*, *Br. Med. J.* **297**, 169 (1988).

Crystals, mp 226-228°.
THERAP CAT: Antihypertensive.

4143. Floxacillin. [5250-39-5] (2*S*,5*R*,6*R*)-6-[[[3-(2-Chloro-6-fluorophenyl)-5-methyl-4-isoxazolyl]carbonyl]amino]-3,3-dimethyl-7-oxo-4-thia-1-azabicyclo[3.2.0]heptane-2-carboxylic acid; 3-(2-chloro-6-fluorophenyl)-5-methyl-4-isoxazolylpenicillin; 6-[3-(2-chloro-6-fluorophenyl)-5-methyl-4-isoxazolecarboxamido]penicillanic acid; flucloxacillin; BRL-2039. $C_{19}H_{17}ClFN_3O_5S$; mol wt 453.87. C 50.28%, H 3.78%, Cl 7.81%, F 4.19%, N 9.26%, O 17.63%, S 7.06%. Semi-synthetic antibiotic active against penicillin-resistant staphylococci. Halogen-substituted derivative of oxacillin, *q.v.* Prepn: **ZA 6304323** (1964 to Beecham); Nayler, **GB 978299**; *idem*, **US 3239507** (1964, 1966 both to Beecham). Pharmacology and toxicity: R. Sutherland *et al.*, *Br. Med. J.* **4**, 460 (1970). Clinical studies: Harding *et al.*, *Clin. Trials J.* **7**, 368 (1970); Qureshi *et al.*, *ibid.* 375.

Sodium salt monohydrate. [34214-51-2] Abboflox; Culpen; Floxapen; Ladropen; Stafoxil; Staphlipen; Staphylex. $C_{19}H_{16}ClFN_3NaO_5S.H_2O$; mol wt 493.87. Freely sol in water. LD_{50} in mice (g/kg): 2.2 s.c., 3.8 orally (Sutherland).
THERAP CAT: Antibacterial.

4144. Floxuridine. [50-91-9] 2'-Deoxy-5-fluorouridine; 1-(2-deoxy-β-D-ribofuranosyl)-5-fluorouracil; 5-fluoro-2'-deoxy-β-uridine; NSC-27640; FUDR. $C_9H_{11}FN_2O_5$; mol wt 246.19. C 43.91%, H 4.50%, F 7.72%, N 11.38%, O 32.49%. Prepn: Hoffer *et al.*, *J. Am. Chem. Soc.* **81**, 4112 (1959); Heidelberger, Duschinsky, **US 2885396** (1959); Duschinsky *et al.*, **US 2970139** (1961); Hoffer, **US 2949451**; *idem*, **US 3041335** (1960, 1962 both to Hoffmann-La Roche). Structure: Lemieux, Hoffer, *Can. J. Chem.* **39**, 110 (1961). Crystal and molecular structure: Harris, *Diss. Abstr.* **24**, 4425 (1964); Harris, McIntyre, *Biophys. J.* **4**, 203 (1964). Conformation of furanose ring in molecule: Sundaralingam, *J. Am. Chem. Soc.* **87**, 599 (1965).

Crystals from butyl acetate, mp 150-151° (Hoffer, 1959); also reported as mp 145° (**US 3041335**). $[\alpha]_D$ +37° (water), +48.6° (DMF). uv max (pH 7.2): 268 nm (ε 7570); (pH 14): 270 nm (ε 6480).

α-Anomer. 1-(2-Deoxy-α-D-*erythro*-pentofuranosyl)-5-fluorouracil. Crystals from butyl acetate, mp 150-151°. $[\alpha]_D^{25}$ −21° (c = 2 in water).
THERAP CAT: Antiviral; antineoplastic.

4145. Fluacizine. [30223-48-4] 3-(Diethylamino)-1-[2-(trifluoromethyl)-10*H*-phenothiazin-10-yl]-1-propanone; 10-[3-(diethylamino)-1-oxopropyl]-2-(trifluoromethyl)-10*H*-phenothiazine; 10-(*N*,*N*-diethyl-β-alanyl)-2-(trifluoromethyl)phenothiazine; fluoracisine; fluoracizine; ftoracizine; Phtorazisin. $C_{20}H_{21}F_3N_2OS$; mol wt 394.46. C 60.90%, H 5.37%, F 14.45%, N 7.10%, O 4.06%, S 8.13%. Phenothiazine derivative with psychotropic activity. Prepn: Y. I. Vikhlyaev *et al.*, **SU 360342** (1972), *C.A.* **78**, 97683w (1973); of hydrochloride: S. V. Zhuravlev *et al.*, **GB 1191800** (1970), *C.A.* **73**, 25493h (1970); Y. I. Vikhlyaev *et al.*, **FR 2035748** (1971), *C.A.* **75**, 140872j (1971); *eidem*, **DE 1805659** (1979), *C.A.* **75**, 49101w (1971). Use as antidepressant: *eidem*, **SU 356992** (1972) (all to Inst. Pharmacol. Chemother., Acad. Med. Sci., USSR), *C.A.* **78**, 75888q (1973). Mass spectrum: S. Morosawa *et al.*, *Org. Mass Spectrom.* **17**, 309 (1982). Effect on drug metabolism: V. Avakumov *et al.*, *Biochem. Pharmacol.* **27**, 2177 (1978). Pharmacokinetics in animals, man: V. P. Zherdev *et al.*, *Farmakol. Toksikol. (Moscow)* **45**, 83 (1982).

Hydrochloride. Fluoracyzine; toracizin. $C_{20}H_{21}F_3N_2OS.HCl$. White cryst powder, mp 163-165°. Sol in water, warm alcohols.
THERAP CAT: Antidepressant.

4146. Fluanisone. [1480-19-9] 1-(4-Fluorophenyl)-4-[4-(2-methoxyphenyl)-1-piperazinyl]-1-butanone; 4'-fluoro-4-[4-(*o*-methoxyphenyl)-1-piperazinyl]butyrophenone; haloanisone; R-2028; Sedalande. $C_{21}H_{25}FN_2O_2$; mol wt 356.44. C 70.76%, H 7.07%, F 5.33%, N 7.86%, O 8.98%. Prepn: Janssen, **US 2997472** (1961). Determn in human plasma: M. P. Quaglio, A. M. Bellini, *Farmaco Ed. Prat.* **36**, 204 (1981). Comparison between biochemical and behavioral effects: G. B. Fregnan, R. Porta, *Arzneim.-Forsch.* **31**, 70 (1981). Toxicity study: C. Cascio *et al.*, *Farmaco Ed. Sci.* **35**, 605 (1980).

Crystals, mp 67.5-68.5°. Sol in chloroform; sparingly sol in methanol; slightly sol in ether. Practically insol in water. LD$_{50}$ i.p. in mice: 200 mg/kg (Cascio).

Hydrochloride. mp 205-205.5°.

THERAP CAT: Antipsychotic.

4147. Fluazifop-butyl. [69806-50-4] 2-[4-[[5-(Trifluoromethyl)-2-pyridinyl]oxy]phenoxy]propanoic acid butyl ester; butyl 2-[4-(5-trifluoromethyl-2-pyridyloxy)phenoxy]propionate; PP-009; TF-1169; Fusilade. C$_{19}$H$_{20}$F$_3$NO$_4$; mol wt 383.37. C 59.53%, H 5.26%, F 14.87%, N 3.65%, O 16.69%. Selective post-emergence herbicide. Prepn: R. Nishiyama *et al.*, **DE 2812571** (1979 to Ishihara Sangyo Kaisha), *C.A.* **90**, 152017g (1979). Activity: R. E. Plowman *et al.*, *Proc. Br. Crop Prot. Conf. - Weeds* **1980**, 29; J. R. Finney, P. B. Sutton, *ibid.* 429.

bp$_{0.05}$ 167°.

USE: Herbicide.

4148. Fluazinam. [79622-59-6] 3-Chloro-*N*-[3-chloro-2,6-dinitro-4-(trifluoromethyl)phenyl]-5-(trifluoromethyl)-2-pyridinamine; 3-chloro-*N*-(3-chloro-5-trifluoromethyl-2-pyridinyl)-α,α,α-trifluoro-2,6-dinitro-*p*-toluidine; *N*-(3-chloro-5-trifluoromethyl-2-pyridyl)-2,6-dinitro-3-chloro-4-trifluoromethylaniline; IKF-1216; ICI-A-192; PP-192; Frowncide; Shirlan. C$_{13}$H$_4$Cl$_2$F$_6$N$_4$O$_4$; mol wt 465.09. C 33.57%, H 0.87%, Cl 15.24%, F 24.51%, N 12.05%, O 13.76%. Blight fungicide with uncoupling activity on oxidative phosphorylation. Prepn: R. Nishiyama *et al.*, **EP 31257**; *eidem*, **US 4331670** (1981, 1982 both to Ishihara Sangyo Kaisha). Mechanism of action study: Z. Guo *et al.*, *Biochim. Biophys. Acta* **1056**, 89 (1991). Field trials in bulbs: B. P. Anema *et al.*, *Meded. Fac. Landbouwwet. Rijksuniv. Gent* **53**, 635 (1988); in potatoes: B. P. Anema *et al.*, *Brighton Crop Prot. Conf. - Pests Dis.* **1992**, 663.

mp 100-102°. pKa 7.11.

USE: Agricultural fungicide.

4149. Flubendazole. [31430-15-6] *N*-[6-(4-Fluorobenzoyl)-1*H*-benzimidazol-2-yl]carbamic acid methyl ester; 5-(*p*-fluorobenzoyl)-2-benzimidazolecarbamic acid methyl ester; R-17889; Flubenol; Flumoxal; Fluvermal. C$_{16}$H$_{12}$FN$_3$O$_3$; mol wt 313.29. C 61.34%, H 3.86%, F 6.06%, N 13.41%, O 15.32%. Fluoro analog of mebendazole, *q.v.* Prepn: J. L. H. Van Gelder *et al.*, **DE 2029637**; *eidem*, **US 3657267** (1971, 1972 both to Janssen); A. H. M. Raeymaekers *et al.*, *Arzneim.-Forsch.* **28**, 586 (1978). Pharmacological, biological properties: D. Thienpont *et al.*, *ibid.* 605. Anthelmintic activity in rats: O. Vanparijs *et al.*, *Vet. Parasitol.* **5**, 237 (1979); in pigs: E. Telléz-Giron *et al.*, *Am. J. Trop. Med. Hyg.* **30**, 135 (1981).

Crystals, mp 260°. LD$_{50}$ in mice, rats, guinea pigs (mg/kg): >2560 orally (Thienpont).

THERAP CAT (VET): Anthelmintic.

4150. Flubendiamide. [272451-65-7] *N*²-[1,1-Dimethyl-2-(methylsulfonyl)ethyl]-3-iodo-*N*¹-[2-methyl-4-[1,2,2,2-tetrafluoro-1-(trifluoromethyl)ethyl]phenyl]-1,2-benzenedicarboxamide; 3-iodo-*N*¹-(2-mesyl-1,1-dimethylethyl)-*N*-[4-[1,2,2,2-tetrafluoro-1-(trifluoromethyl)ethyl]-*o*-tolyl]phthalamide; NNI-0001; Belt; Fame; Fenos; Synapse. C$_{23}$H$_{22}$F$_7$IN$_2$O$_4$S; mol wt 682.39. C 40.48%, H 3.25%, F 19.49%, I 18.60%, N 4.11%, O 9.38%, S 4.70%. Phthalic acid diamide pesticide; insect-specific ryanodine receptor activator. Prepn: M. Tohnishi *et al.*, **EP 1006107**; *eidem*, **US 6603044** (2000, 2003 both to Nihon Nohyaku); *eidem*, *J. Pestic. Sci.* **30**, 354 (2005). Mechanism of action: T. Masaki *et al.*, *Mol. Pharmacol.* **69**, 1733 (2006). HPLC determn in rice: M. Gopal, E. Mishra, *Bull. Environ. Contam. Toxicol.* **81**, 360 (2008). Comprehensive description: T. Nishimatsu *et al.*, *BCPC Int. Conf. - Crop Sci. Technol.* **2005**, 57-64. Review of field trials: D. Ebbinghaus *et al.*, *Pflanzenschutz-Nachr. Bayer (Engl. Ed.)* **60**, 219-246 (2007).

White crystalline powder, mp 218-221°. Log P (octanol/water): 4.20 (25°). Vapor pressure (25°): <10⁻⁴ Pa. Soly in water (20°): 29.9 μg/l. LD$_{50}$ in rats (mg/kg): >2000 orally; >2000 dermally. LC$_{50}$ (96 hr) in carp: >548 μg/l (Nishimatsu).

USE: Insecticide.

4151. Flucarbazone. [145026-88-6] 4,5-Dihydro-3-methoxy-4-methyl-5-oxo-*N*-[[2-(trifluoromethoxy)phenyl]sulfonyl]-1*H*-1,2,4-triazole-1-carboxamide. C$_{12}$H$_{11}$F$_3$N$_4$O$_6$S; mol wt 396.30. C 36.37%, H 2.80%, F 14.38%, N 14.14%, O 24.22%, S 8.09%. Acetolactate synthase inhibitor for control of weeds in wheat. Prepn: K.-H. Müller *et al.*, **EP 507171**; *eidem*, **US 5534486** (1992, 1996 both to Bayer). Comprehensive description: H. J. Santel *et al.*, *Brighton Crop Prot. Conf. - Weeds* **1999**, 23-28. Field trials in wheat: K. J. Kirkland *et al.*, *Weed Technol.* **15**, 48 (2001).

Sodium salt. [181274-17-9] BAY MKH 6562; MKH 6562; SJO 0498; Everest. C$_{12}$H$_{10}$F$_3$N$_4$NaO$_6$S; mol wt 418.28. Colorless, odorless crystalline powder, mp 200° (dec). Vapor pressure (20°): <1 × 10⁻⁹ Pa. d 1.59. pKa 1.9. Soly in water (20°): 44 g/l (pH 4-9). Log P (octanol/water): −0.89 (pH 4); −1.85 (pH 7); −1.89 (pH 9). LD$_{50}$ in rats (mg/kg): >5000 orally; >5000 dermally. LD$_{50}$ by inhalation in rats: >5.13 mg/l. LC$_{50}$ in bluegill sunfish, rainbow trout: >99.3, >96.7 mg/l.

USE: Herbicide.

4152. Fluchloralin. [33245-39-5] *N*-(2-Chloroethyl)-2,6-dinitro-*N*-propyl-4-(trifluoromethyl)benzenamine; *N*-(2-chloroethyl)-α,α,α-trifluoro-2,6-dinitro-*N*-propyl-*p*-toluidine; BAS-392H; Basalin. C$_{12}$H$_{13}$ClF$_3$N$_3$O$_4$; mol wt 355.70. C 40.52%, H 3.68%, Cl 9.97%, F 16.02%, N 11.81%, O 17.99%. Soil incorporated, pre-planting herbicide. Prepn: K. H. Karl, K. Kiehs, **DE 2161879** (1973 to BASF), *C.A.* **79**, 65990y (1973). Photochemistry: G. P. Nilles, M. J. Zabik, *J. Agric. Food Chem.* **22**, 684 (1974). Degradn and metabolism in soil: S. Otto, *Environ. Qual. Saf. Suppl.* **3**, 277 (1975).

Orange-yellow cryst solid, mp 42-43°. Soly in water: 10 ppm.
USE: Herbicide.

4153. Fluconazole. [86386-73-4] α-(2,4-Difluorophenyl)-α-(1*H*-1,2,4-triazol-1-ylmethyl)-1*H*-1,2,4-triazole-1-ethanol; 2,4-difluoro-α,α-bis(1*H*-1,2,4-triazol-1-ylmethyl)benzyl alcohol; 2-(2,4-difluorophenyl)-1,3-bis(1*H*-1,2,4-triazol-1-yl)propan-2-ol; UK-49858; Diflucan; Elazor; Triflucan. $C_{13}H_{12}F_2N_6O$; mol wt 306.28. C 50.98%, H 3.95%, F 12.41%, N 27.44%, O 5.22%. Orally active bistriazole antifungal agent. Prepn: K. Richardson, **GB 2099818**; *idem*, **US 4404216** (1982, 1983 both to Pfizer). Antifungal activity *in vivo*: K. Richardson *et al.*, *Antimicrob. Agents Chemother.* **27**, 832 (1985); and *in vitro*: T. E. Rogers, J. N. Galgiani, *ibid.* **30**, 418 (1986). Pharmacokinetics: M. J. Humphrey *et al.*, *ibid.* **28**, 648 (1985). GC determn in human plasma and urine: P. R. Wood, M. H. Tarbit, *J. Chromatogr.* **383**, 179 (1986). Evaluation as maintenance therapy for cryptococcal meningitis in AIDS patients: S. A. Bozzette *et al.*, *N. Engl. J. Med.* **324**, 580 (1991). Symposium on mode of action, toxicology, animal models and clinical studies: *International Telesymposium on Recent Trends in the Discovery, Development and Evaluation of Antifungal Agents* **Section 2**, R. A. Fromtling, Ed. (J. R. Prous Science Publ., Barcelona, 1987) pp 77-174. Infection prophylaxis in bone marrow transplantation patients: M. L. MacMillan *et al.*, *Am. J. Med.* **112**, 369 (2002). Review of use in treatment of invasive candidiasis: C. Charlier *et al.*, *J. Antimicrob. Chemother.* **57**, 384-410 (2006).

Crystals from ethyl acetate/hexane, mp 138-140°. Freely soluble in methanol; sol in alc, acetone; sparingly sol in isopropanol, chloroform; slightly sol in water, saline; very slightly sol in toluene.
Dihydrogen phosphate ester. [194798-83-9] Fosfluconazole; 2-(2,4-difluorophenyl)-1,3-bis(1*H*-1,2,4-triazol-1-yl)propyl dihydrogen phosphate; Prodif. $C_{13}H_{13}F_2N_6O_4P$; mol wt 386.26. Prepn: C. W. Murtiashaw *et al.*, **WO 9728169**; S. Green *et al.*, **US 6790957** (1997, 2004 both to Pfizer). Review of discovery and commercial process development: A. Bentley *et al.*, *Org. Process Res. Dev.* **6**, 109-112 (2002). Water-soluble prodrug. White powder, mp 223-224°.
THERAP CAT: Antifungal.
THERAP CAT (VET): Antifungal.

4154. Flucycloxuron. [113036-88-7]; [94050-52-9] (*E*-isomer); [94050-53-0] (*Z*-isomer). *N*-[[[4-[[[[(4-Chlorophenyl)cyclopropylmethylene]amino]oxy]methyl]phenyl]amino]carbonyl]-2,6-difluorobenzamide; 1-[α-(4-chloro-α-cyclopropylbenzylideneaminooxy)-*p*-tolyl]-3-(2,6-difluorobenzoyl)urea; DU-319722; PH-70-23; UBI-A1335; Andalin. $C_{25}H_{20}ClF_2N_3O_3$; mol wt 483.90. C 62.05%, H 4.17%, Cl 7.33%, F 7.85%, N 8.68%, O 9.92%. Insect growth regulator which interferes with chitin biosynthesis. Prepn: M. S. Brouwer, A. C. Grosscurt, **EP 117320**; *eidem*, **US 4550202** (1984, 1985 both to Duphar). Comprehensive description: P. Scheltes *et al.*, *Brighton Crop Prot. Conf. - Pests Dis.* **1988**, 559. Mechanism of action and insecticidal activity: A. C. Grosscurt *et al.*, *Pestic. Sci.* **22**, 51 (1988). Determn in fruit and soil samples: J. van Zijtveld *et al.*, *J. Chromatogr.* **600**, 211 (1992).

E-isomer

Odorless crystalline powder. Product consists of 50-80% *E*-isomer and 50-20% *Z*-isomer. Data given for ~70:30 *E:Z* mixture. mp

143.6°. vapor pressure (20°): <4.4 mPa. Soly (20°): water <1μg/l; xylene 0.2 g/l; cyclohexane 3.3 g/l; ethanol 3.9 g/l. LC$_{50}$ (4 hr) in rats: 3.3 mg/l; LC$_{50}$ (96 hr) in rainbow trout, bluegill sunfish: >100 mg/l; LD$_{50}$ in rats (mg/kg): >5000 orally; >2000 dermally (Scheltes).
USE: Insecticide.

4155. Flucythrinate. [70124-77-5] 4-(Difluoromethoxy)-α-(1-methylethyl)benzeneacetic acid cyano(3-phenoxyphenyl)methyl ester; (±)-cyano-(3-phenoxyphenyl)methyl (+)-4-(difluoromethoxy)-α-(1-methylethyl)benzeneacetate; AC-222705; Cybolt; Payoff. $C_{26}H_{23}F_2NO_4$; mol wt 451.47. C 69.17%, H 5.14%, F 8.42%, N 3.10%, O 14.18%. Synthetic pyrethroid insecticide with contact and stomach poison activity. Prepn: G. Berkelhammer, V. Kameswaran, **DE 2757066**; *eidem*, **US 4199595** (1977, 1980 both to Am. Cyanamid). Activity: A. F. Saad *et al.*, *Proc. Br. Crop Prot. Conf. - Pests Dis.* **1981**, 381; K. Wettstein, *ibid.* 563. Spectrophotometric determn method: R. V. P. Raju, R. R. Naidu, *Talanta* **41**, 761 (1994). Efficacy in cattle fly control: S. M. Taylor *et al.*, *Vet. Rec.* **116**, 566 (1985). Review of toxicology and human exposure: *Toxicological Profile for Pyrethrins and Pyrethroids* (PB2004-100004, 2003) 332 pp.

Viscous liquid. Sol in acetone, xylene, 2-propanol.
USE: Insecticide.
THERAP CAT (VET): Ectoparasiticide.

4156. Flucytosine. [2022-85-7] 6-Amino-5-fluoro-2(1*H*)-pyrimidinone; 5-fluorocytosine; 5-FC; 2-hydroxy-4-amino-5-fluoropyrimidine; Ro-2-9915; Ancobon; Ancotil. $C_4H_4FN_3O$; mol wt 129.09. C 37.22%, H 3.12%, F 14.72%, N 32.55%, O 12.39%. Fluorinated pyrimidine; antifungal activity results from conversion to 5-fluorouracil, *q.v.*, in susceptible fungal cells. Prepn: R. Duschinsky *et al.*, *J. Am. Chem. Soc.* **79**, 4559 (1957); R. Duschinsky, C. Heidelberger, **US 2945038** (1960 to Hoffmann-La Roche); Undheim, Gacek, *Acta Chem. Scand.* **23**, 294 (1969). Patents as a fungicide: J. Berger, R. Duschinsky, **BE 628615**; *eidem*, **US 3368938** (1963, 1968 both to Hoffmann-La Roche). Activity studies: Grunberg *et al.*, *Antimicrob. Agents Chemother.* **1963**, 566; Shadomy *et al.*, *ibid.* **1968**, 452; Grunberg *et al.*, in *5th Int. Congr. Chemother.*, *Proc.* vol. IV, K. Spitzy, Ed. (Verlag Wiener Med. Akad., 1967, Austria) p 69. Metabolic studies: B. A. Koechlin *et al.*, *Biochem. Pharmacol.* **15**, 435 (1966). Clinical results: Utz *et al.*, *Antimicrob. Agents Chemother.* **1968**, 344; Warner *et al.*, *ibid.* **1970**, 473. Comprehensive reviews: E. H. Waysek, J. H. Johnson, *Anal. Profiles Drug Subs.* **5**, 115-138 (1976); A. Vermes *et al.*, *J. Antimicrob. Chemother.* **46**, 171-179 (2000).

Odorless, white crystalline solid, mp 295-297° (dec). uv max (0.1*N* HCl): 285 nm (ε 8900). Soly in water: 1.5 g/100 ml at 25°C; slightly sol in alc. Practically insol in chloroform, ether. pKa$_1$: 2.90 pKa$_2$: 10.71. LD$_{50}$ in mice (mg/kg): >2000 orally and s.c.; 1190 i.p.; 500 i.v. (Grunberg, 1963).
THERAP CAT: Antifungal.
THERAP CAT (VET): Antifungal.

4157. Fludarabine. [21679-14-1] 9-β-D-Arabinofuranosyl-2-fluoro-9*H*-purin-6-amine; 9-β-D-arabinofuranosyl-2-fluoroadenine; 2-fluorovidarabine; 2-fluoro-9-β-D-arabinofuranosyladenine; 2-F-araA; NSC-118218; NSC-118218-H. $C_{10}H_{12}FN_5O_4$; mol wt 285.24. C 42.11%, H 4.24%, F 6.66%, N 24.55%, O 22.44%. Adenosine deaminase-resistant purine nucleoside antimetabolite. Prepn and *in vitro* cytotoxicity: J. A. Montgomery, K. Hewson, *J. Med. Chem.* **12**, 498 (1969). Improved prepn: J. A. Montgomery *et al.*, *J. Heterocycl. Chem.* **16**, 157 (1979); J. A. Montgomery, US **4210745** (1980 to U.S. Dept. Health, Education and Welfare). Inhibition of DNA synthesis and *in vivo* antileukemic activity: R. W. Brockman *et al.*, *Biochem. Pharmacol.* **26**, 2193 (1977). Metabolized to 5'-monophosphate: R. W. Brockman *et al.*, *Cancer Res.* **40**, 3610 (1980). HPLC determn in human leukemia cells: V. Gandhi *et al.*, *J. Chromatogr.* **413**, 293 (1987). Prepn of 5'-monophosphate: J. A. Montgomery, A. T. Shortnacy, US **4357324** (1982 to U.S. Dept. of Health and Human Services). Pharmacokinetics in humans: M. R. Hersh *et al.*, *Cancer Chemother. Pharmacol.* **17**, 277 (1986). Evaluation of therapeutic efficacy and CNS toxicity in acute refractory leukemia: R. P. Warrell, Jr., E. Berman, *J. Clin. Oncol.* **4**, 74 (1986); H. G. Chun *et al.*, *Cancer Treat. Rep.* **70**, 1225 (1986). Series of articles on pharmacology and therapeutic use: *Semin. Oncol.* **17**, Suppl. 8, 1-78 (1990).

Crystals from ethanol + water, mp 260°. $[\alpha]_D^{25}$ +17 ±2.5° (c = 0.1 in ethanol). uv max (pH 1, pH 7, pH 13): 262, 261, 262 nm ($\varepsilon \times 10^{-3}$ 13.2, 14.8, 15.0). Sparingly sol in water, organic solvents.

5'-Monophosphate. [75607-67-9] 2-F-ara-AMP; NSC-328002; NSC-312887; Fludara. $C_{10}H_{13}FN_5O_7P$; mol wt 365.21. White to off-white crystalline, hygroscopic powder. Freely sol in dimethylformamide; slightly sol in water and 0.1*M* HCl. Practically insol in ethanol.

THERAP CAT: Phosphate as antineoplastic.

4158. Fludeoxyglucose F 18. [105851-17-0] 2-Deoxy-2-(fluoro-^{18}F)-α-D-glucopyranose; 2-[^{18}F]fluoro-2-deoxy-D-glucose; ^{18}FDG. $C_6H_{11}^{18}FO_5$. Radioactive glucose analog used clinically to evaluate glucose metabolism by PET. Prepn: T. Ido *et al.*, *J. Labelled Compd. Radiopharm.* **14**, 175 (1978). Review of syntheses: T. J. Tewson, *Nucl. Med. Biol.* **16**, 533-551 (1989). Biodistribution: B. M. Gallagher *et al.*, *J. Nucl. Med.* **19**, 1154 (1978). Cerebral glucose measurement in man: M. Reivich *et al.*, *Circ. Res.* **44**, 127 (1979). Clinical evaluation of metabolic imaging in myocardium: R. J. Gropler *et al.*, *J. Nucl. Med.* **31**, 1749 (1990); in tumors: J. Okada *et al.*, *ibid.* **33**, 325 (1992); N. Y. Tse *et al.*, *Ann. Surg.* **216**, 27 (1992). Reviews of imaging in cerebrovascular disease: K. Herholz, W. D. Heiss, *Semin. Neurol.* **9**, 293-300 (1989); Alzheimer's disease: J. M. Hoffman *et al.*, *Eur. Neurol.* **29**, Suppl. 3, 16-24 (1989); brain tumors: R. E. Coleman *et al.*, *J. Nucl. Med.* **32**, 616-622 (1991); pancreatic cancer: M. Zimny, U. Buell, *Ann. Oncol.* **10**, Suppl. 4, S28-S32 (1999).

THERAP CAT: Diagnostic aid (radioactive imaging agent).

4159. Fludiazepam. [3900-31-0] 7-Chloro-5-(2-fluorophenyl)-1,3-dihydro-1-methyl-2*H*-1,4-benzodiazepin-2-one; ID-540; Ro-5-3438; Erispan. $C_{16}H_{12}ClFN_2O$; mol wt 302.73. C 63.48%, H 4.00%, Cl 11.71%, F 6.28%, N 9.25%, O 5.28%. Fluorinated analog of diazepam, *q.v.* Prepn: E. Reeder *et al.*, **DE 1136709**; E. Reeder, L. H. Sternbach, US **3371085** (1962, 1968 both to Hoffmann-La-

Roche); L. H. Sternbach *et al.*, *J. Org. Chem.* **27**, 3788 (1962). Synthesis and pharmacology: Y. Asami *et al.*, *Arzneim.-Forsch.* **24**, 1563 (1974). Pharmacodynamics: M. Nakamura, H. Fukushima, *J. Pharm. Pharmacol.* **30**, 56, 254 (1978); T. Tsuchiya, H. Fukushima, *Eur. J. Pharmacol.* **48**, 421 (1978). Anxiolytic vs sedative properties: M. Babbini *et al.*, *Life Sci.* **25**, 15 (1979).

Colorless prisms from *n*-hexane/isopropanol, mp 88-92°, (Asami); also reported as mp 69-72° (Sternbach). LD_{50} in mice (mg/kg): 910 orally; 360 i.p.; 1150 s.c. (Asami).
Note: This is a controlled substance (depressant): **21 CFR**, 1308.14.

THERAP CAT: Anxiolytic.

4160. Fludioxonil. [131341-86-1] 4-(2,2-Difluoro-1,3-benzodioxol-4-yl)-1*H*-pyrrole-3-carbonitrile; 3-(2,2-difluorobenzodioxol-4-yl)-4-cyanopyrrole; CGA-173506; Celest; Maxim; Medallion; Scholar. $C_{12}H_6F_2N_2O_2$; mol wt 248.19. C 58.07%, H 2.44%, F 15.31%, N 11.29%, O 12.89%. Nonsystemic phenylpyrrole fungicide structurally related to pyrrolnitrin, *q.v.* Prepn: R. Nyfeler, J. Ehrenfreund, **EP 206999**; *eidem*, US **4705800** (1986, 1987 both to Ciba-Geigy). Field trial for postharvest control of blue mold in apples: D. Errampalli *et al.*, *Postharvest Biol. Technol.* **39**, 101 (2006). Efficacy and residue levels in stone fruit: S. D'Aquino *et al.*, *J. Agric. Food Chem.* **55**, 825 (2007). Comprehensive description: K. Gehmann *et al.*, *Brighton Crop Prot. Conf. - Pests Dis.* **1990**, 399-406. Review of field trials as seed treatment: E. Koch, A. J. Leadbeater, *ibid.* **1992**, 1137-1146.

Colorless, odorless crystals, mp 199.4°. Log P (*n*-octanol/water): 2.6. Vapor pressure (25°): $2.9 \times 10^{.9}$ mmHg. Soly in water (25°): 1.8 mg/ml. LD_{50} in rats (mg/kg): >2000 orally; >2000 dermally; LC_{50} (4 hr) in rats: >2600 mg/m^3 (Gehmann).

USE: Agricultural fungicide.

4161. Fludrocortisone. [127-31-1] (11β)-9-Fluoro-11,17,21-trihydroxypregn-4-ene-3,20-dione; 9α-fluorohydrocortisone; 9α-fluoro-17-hydroxycorticosterone; 9α-fluorocortisol; fluodrocortisone; fluohydrisone; fluohydrocortisone; Astonin H. $C_{21}H_{29}FO_5$; mol wt 380.46. C 66.30%, H 7.68%, F 4.99%, O 21.03%. Prepn: J. Fried, E. F. Sabo, *J. Am. Chem. Soc.* **76**, 1455 (1954); **GB 792224**; J. Fried, US **2852511** (both 1958 to Olin Mathieson). Series of articles on pharmacology and metabolism: *Arzneim.-Forsch.* **21**, 1103-1158 (1971). Comprehensive description of the acetate: K. Florey, *Anal. Profiles Drug Subs.* **3** 281-306 (1974).

Crystals, dec 260-262°. $[\alpha]_D^{23}$ +139° (c = 0.55 in 95% ethanol). uv max (ethanol): 239 nm (ε 17600). Soly in water: 0.14 mg/ml.

21-Acetate. [514-36-3] Florinef. $C_{23}H_{31}FO_6$; mol wt 422.49. White to pale yellow crystals, polymorphic, mp 233-234°. Also reported as mp 205-208°, resolidifying on further heating, then mp 226-228°. Crystallizing procedure: R. P. Graber, C. S. Snoddy, **US 2957013** (1960 to Merck & Co.). $[\alpha]_D^{23}$ +123° (c = 0.64 in chloroform). uv max (ethanol): 238 nm (ε 16800). Hygroscopic. Soly (mg/ml): water 0.04; acetone 56; alc 20; chloroform 20; ether 4.

THERAP CAT: Mineralocorticoid.

THERAP CAT (VET): Mineralocorticoid.

4162. Flufenacet. [142459-58-3] *N*-(4-Fluorophenyl)-*N*-(1-methylethyl)-2-[[5-(trifluoromethyl)-1,3,4-thiadiazol-2-yl]oxy]acetamide; *N*-isopropyl-(5-trifluoromethyl-1,3,4-thiadiazol-2-yl)-(4′-fluorooxyacetanilide); BAY FOE-5043; FOE-5043; Define. $C_{14}H_{13}F_4N_3O_2S$; mol wt 363.33. C 46.28%, H 3.61%, F 20.92%, N 11.57%, O 8.81%, S 8.82%. Oxyacetamide herbicide. Prepn: H. Förster *et al.*, **DE 3821600**; *eidem*, **US 4968342** (1989, 1990 both to Bayer). Ecotoxicological profile: T. Hall, *Pflanzenschutz-Nachr. Bayer (Engl. Ed.)* **50**, 187 (1997). Series of articles on chemical and biological properties, determn and field trials: *ibid.* 105-236. Review of properties and grass control in crops: R. Deege *et al.*, *Brighton Crop Prot. Conf. - Weeds* **1995**, 43.

White to tan solid, mp 75-77°. Does not dissociate. Soly in water at 25° (mg/l): 56 pH 4; 56 pH 7; 54 pH 9. Log P (octanol/water): 3.2. Vapor pressure at 20°: 9.0×10^{-5} Pa. LD$_{50}$ in rats (mg/kg): 589-1617 orally; >2000 dermally. LD$_{50}$ orally in mice, bobwhite quail, mallard duck: 1331-1756, 1608, >2000 mg/kg. LC$_{50}$ in sunfish, rainbow trout, sheepshead minnow: 2.1-2.4, 3.5-5.8, 3.3 mg/l (Hall).

USE: Pre-emergent herbicide.

4163. Flufenamic Acid. [530-78-9] 2-[[3-(Trifluoromethyl)phenyl]amino]benzoic acid; *N*-(α,α,α-trifluoro-*m*-tolyl)anthranilic acid; 3′-trifluoromethyldiphenylamine-2-carboxylic acid; CI-440; INF-1837; Achless; Ansatin; Arlef; Fullsafe; Meralen; Paraflu; Parlef; Ristogen; Sastridex; Surika; Tecramine. $C_{14}H_{10}F_3NO_2$; mol wt 281.23. C 59.79%, H 3.58%, F 20.27%, N 4.98%, O 11.38%. Prepn from *m*-aminobenzotrifluoride and *o*-iodobenzoic acid: Wilkinson, Finar, *J. Chem. Soc.* **1948**, 32. Pharmacology: Winder *et al.*, *Arthritis Rheum.* **6**, 36 (1963); **12**, 472 (1969). Antiviral activity: Inglot, *J. Gen. Virol.* **4**, 203 (1969). Toxicity data: Zoni *et al.*, *Farmaco Ed. Sci.* **26**, 191 (1971). Comprehensive description: E. Abignente, P. deCaprariis, *Anal. Profiles Drug Subs.* **11**, 313-346 (1982).

Pale yellow needles from 50% alc, mp 125°. LD$_{50}$ in mice: 715 mg/kg orally (Zoni).

Aluminum salt. [61891-34-7] Aluminum flufenamate; Alfenamin; Opyrin. $C_{42}H_{27}AlF_9N_3O_6$; mol wt 867.66.

Butyl ester. [67330-25-0] Fenazol; Combec. $C_{18}H_{18}F_3NO_2$; mol wt 337.34.

THERAP CAT: Anti-inflammatory; analgesic.

4164. Flufenoxuron. [101463-69-8] *N*-[[[4-[2-Chloro-4-(trifluoromethyl)phenoxy]-2-fluorophenyl]amino]carbonyl]-2,6-difluorobenzamide; 1-[4-(2-chloro-α,α,α-trifluoro-*p*-tolyloxy)-2-fluorophenyl]-3-(2,6-difluorobenzoyl)urea; *N*-(2,6-difluorobenzoyl)-*N*′-(2-fluoro-4-[2-chloro-4-(trifluoromethyl)phenoxy]phenylurea); WL-115110; DPX EY-059; Cascade. $C_{21}H_{11}ClF_6N_2O_3$; mol wt 488.77.

C 51.61%, H 2.27%, Cl 7.25%, F 23.32%, N 5.73%, O 9.82%. Acylurea insecticide which inhibits chitin synthesis. Prepn: M. Anderson, **AU 8540924** (1985 to Shell); *idem*, **US 4698365** (1987 to Du Pont). HPLC determn in grapes and wine: G. E. Miliadis *et al.*, *J. Chromatogr. A* **835**, 113 (1999). Metabolism in caterpillars: B. S. Clarke, P. J. Jewess, *Pestic. Sci.* **28**, 357 (1990). Inhibition of chitin synthesis: *eidem, ibid.* 377. Field trial against German cockroach: B. L. Reid *et al., J. Econ. Entomol.* **48**, 1194 (1992). Review of physical properties and field trials: M. Anderson *et al., Proc. Br. Crop Prot. Conf. - Pests Dis.* **1986**, 89-96.

Colorless, odorless, crystals, mp 169-172° (dec). Soly (g/l) at 25°: water 4×10^{-6}, acetone 82, xylene 6, dichloromethane 24. LD$_{50}$ in rats (mg/kg): >3000 orally; >2000 percutaneously (Anderson).

USE: Insecticide; acaricide.

4165. Fluindione. [957-56-2] 2-(4-Fluorophenyl)-1*H*-indene-1,3(2*H*)-dione; 2-(*p*-fluorophenyl)-1,3-indandione; LM-123; Previscan. $C_{15}H_9FO_2$; mol wt 240.23. C 75.00%, H 3.78%, F 7.91%, O 13.32%. Prepn: Geiger *et al.*, **DE 1130439** (1962 to USV), *C.A.* **57**, 12403a (1962); Molho, Boschetti, **FR 1369396** and **FR M 6913** (1964, 1969 to LIPHA), *C.A.* **62**, 3988g (1965); **74**, 141378u (1971). Alternate syntheses: Shapiro *et al., J. Org. Chem.* **26**, 3580 (1961); Hrnciar, Kovalcik, *Chem. Zvesti* **16**, 200 (1962). Activity and toxicity data: Fontaine *et al., Med. Pharmacol. Exp.* **17**, 497 (1967).

Crystals from acetic acid, mp 120°. LD$_{50}$ orally in mice: 240 mg/kg (Fontaine).

THERAP CAT: Anticoagulant.

4166. Flumazenil. [78755-81-4] 8-Fluoro-5,6-dihydro-5-methyl-6-oxo-4*H*-imidazo[1,5-*a*][1,4]benzodiazepine-3-carboxylic acid ethyl ester; ethyl-8-fluoro-5,6-dihydro-5-methyl-6-oxo-4*H*-imidazo[1,5-*a*][1,4]benzodiazepine-3-carboxylate; flumazepil; Ro-15-1788; Anexate; Lanexat; Romazicon. $C_{15}H_{14}FN_3O_3$; mol wt 303.29. C 59.40%, H 4.65%, F 6.26%, N 13.86%, O 15.83%. Imidazodiazepine which selectively blocks the central effects of classic benzodiazepines. Prepn: W. Haefely *et al.*, **EP 27214**; M. Gerecke *et al.*, **US 4316839** (1981, 1982 both to Hoffmann-La Roche). Specific inhibition of benzodiazepine-receptor binding: W. Hunkeler *et al., Nature* **290**, 514 (1981). Electrophysiological study in animals: P. Pole *et al., Arch. Pharmacol.* **316**, 317 (1981). Antagonist, agonist and inverse agonist properties: S. E. File *et al., Psychopharmacology* **89**, 113 (1986). HPLC determn in human plasma: U. Timm, M. Zell, *Arzneim.-Forsch.* **33**, 358 (1983). Clinical reversal of benzodiazepine-induced sedation: A. Darragh *et al., Lancet* **2**, 8 (1981); *eidem, ibid.* 1042; B. Ricou *et al., Br. J. Anaesth.* **58**, 1005 (1986). Clinical evaluation in drug overdose: G. F. O'Sullivan *et al., Clin. Pharmacol. Ther.* **42**, 254 (1987). Review of effects in neuropsychiatric disorders: A. L. Malizia, D. J. Nutt, *Clin. Neuropharmacol.* **18**, 215-232 (1995).

Crystals from alcohol, mp 201-203°. pKa 1.7. Soly in water: 0.4 g/L; slightly sol in acidic aqueous solutions. Partition coefficient (*n*-octanol/aq. sulfate buffer, pH 7.4): 14. LD_{50} in mice, rats (mg/kg): 4000, 1360 i.p.; 4300, 6000 orally (Hunkeler).

THERAP CAT: Benzodiazepine antagonist.

THERAP CAT (VET): Benzodiazepine antagonist.

4167. Flumecinol. [56430-99-0] α-Ethyl-α-phenyl-3-(trifluoromethyl)benzenemethanol; α-ethyl-3-(trifluoromethyl)benzhydrol; RGH-3332; Zixoryn; Zyxorin. $C_{16}H_{15}F_3O$; mol wt 280.29. C 68.56%, H 5.39%, F 20.33%, O 5.71%. Hepatic microsomal enzyme inducer. Prepn: E. Toth *et al.*, **DE 2438399**; *eidem*, **US 4039589** (1975, 1977 both to Gedeon-Richter). GC determn in biological fluids: I. Klebovich, L. Vereczkey, *J. Chromatogr.* **221**, 403 (1980). Toxicity study: M. Ledniczky *et al.*, *Arzneim.-Forsch.* **28**, 669 (1978). Series of articles on metabolism, CNS effects, enzyme induction, pharmacological properties: *ibid.* 663-679.

Oil, $bp_{0.03}$ 106-108°. uv max (ethanol): 259, 265, 271 nm. LD_{50} orally in adult rats: 2235 mg/kg (Ledniczky).

THERAP CAT: Enzyme inducer (hepatic).

4168. Flumequine. [42835-25-6] 9-Fluoro-6,7-dihydro-5-methyl-1-oxo-1*H*,5*H*-benzo[*ij*]quinolizine-2-carboxylic acid; R-802; Apurone; Fantacin. $C_{14}H_{12}FNO_3$; mol wt 261.25. C 64.37%, H 4.63%, F 7.27%, N 5.36%, O 18.37%. Fluorinated quinolone antibacterial. Prepn: J. F. Gerster, **DE 2264163**; *idem*, **US 3896131** (1973, 1975 both to Riker). *In vitro* study: G. Stilwell *et al.*, *Antimicrob. Agents Chemother.* **7**, 483 (1975). Bioevaluation: S. R. Rohlfing *et al.*, *J. Antimicrob. Chemother.* **3**, 615 (1977). Activity vs *E. coli:* D. Greenwood, *Antimicrob. Agents Chemother.* **13**, 479 (1978).

White microcrystalline powder, mp 253-255°. Sol in alkaline solns and alcohol. Insol in water.

THERAP CAT: Antibacterial.

THERAP CAT (VET): Antibacterial.

4169. Flumethasone. [2135-17-3] (6α,11β,16α)-6,9-Difluoro-11,17,21-trihydroxy-16-methylpregna-1,4-diene-3,20-dione; 6α,9α-difluoro-16α-methylprednisolone; 6α-fluorodexamethasone; Aniprime; Cortexilar; Flucort; Methagon. $C_{22}H_{28}F_2O_5$; mol wt 410.46. C 64.38%, H 6.88%, F 9.26%, O 19.49%. Prepn: **GB 902292**; F. H. Lincoln *et al.*, **US 3499016** (1962, 1970 both to Upjohn). Prepn of the 21-acetate: J. A. Edwards *et al.*, *J. Am. Chem. Soc.* **81**, 3156 (1959); **82**, 2318 (1960).

21-Acetate. $C_{24}H_{30}F_2O_6$. Crystals from acetone + hexane, mp 260-264°. $[α]_D$ +91°. uv max (ethanol): 237 nm (log ε 4.16).

21-Pivalate. [2002-29-1] Locacorten; Locorten; Lorinden. $C_{27}H_{36}F_2O_6$; mol wt 494.58. White to off-white crystalline powder. Slightly soluble in methanol; very slightly sol in chloroform, methylene chloride. Insol in water.

THERAP CAT: Glucocorticoid; anti-inflammatory.

THERAP CAT (VET): Glucocorticoid.

4170. Flumethiazide. [148-56-1] 6-(Trifluoromethyl)-2*H*-1,-2,4-benzothiadiazine-7-sulfonamide 1,1-dioxide; 6-trifluoromethyl-7-sulfamoyl-4*H*-1,2,4-benzothiadiazine 1,1-dioxide; 6-trifluoromethyl-7-sulfamyl-1,2,4-benzothiadiazine 1,1-dioxide; trifluoromethylthiazide; Ademol; Fludemil. $C_8H_6F_3N_3O_4S_2$; mol wt 329.27. C 29.18%, H 1.84%, F 17.31%, N 12.76%, O 19.44%, S 19.47%. Synthesis: Holdrege *et al.*, *J. Am. Chem. Soc.* **81**, 4807 (1959); Yale *et al.*, *ibid.* **82**, 2042 (1960); **US 3040042** (1962 to Olin Mathieson); **GB 861809** (1961 to SK&F).

Crystals, dec 305.4-307.8°. uv max: 278 nm ($E_{1cm}^{1\%}$ 335) (50% diglyme + 50% 0.1*N* HCl). Sparingly sol in water (50 mg/ml in boiling water with decompn); sol in methanol, ethanol, DMF. Practically insol in ethyl acetate, methyl ethyl ketone, benzene, toluene. Unstable in alkaline soln (conversion to precursor α,α,α-trifluoro-3-amino-4,6-disulfamoyltoluene).

THERAP CAT: Carbonic anhydrase inhibitor.

4171. Flumethrin. [69770-45-2] 3-[2-Chloro-2-(4-chlorophenyl)ethenyl]-2,2-dimethylcyclopropanecarboxylic acid cyano(4-fluoro-3-phenoxyphenyl)methyl ester; 3'-phenoxy-4'-fluoro-α-cyanobenzyl 2,2-dimethyl-3-[2-(4-chlorophenyl)-2-chlorovinyl]-cyclopropane carboxylate; Bayticol; Bayvarol. $C_{28}H_{22}Cl_2FNO_3$; mol wt 510.39. C 65.89%, H 4.34%, Cl 13.89%, F 3.72%, N 2.74%, O 9.40%. Synthetic pyrethroid insecticide. Prepn: R. Fuchs *et al.*, **DE 2730515**; *eidem*, **US 4276306** (1979, 1981 both to Bayer). Series of articles on laboratory evaluation, efficacy and toxicology: *Vet. Med. Rev.* **2**, 115-139, 158-177 (1982). Review of toxicology and human exposure: *Toxicological Profile for Pyrethrins and Pyrethroids* (PB2004-100004, 2003) 332 pp.

Oil, n_D^{25} 1.5831.

USE: Insecticide; acaricide.

THERAP CAT (VET): Ectoparasiticide.

4172. Flumetsulam. [98967-40-9] *N*-(2,6-Difluorophenyl)-5-methyl[1,2,4]triazolo-[1,5-*a*]pyrimidine-2-sulfonamide; DE-498; Broadstrike; Python. $C_{12}H_9F_2N_5O_2S$; mol wt 325.29. C 44.31%, H 2.79%, F 11.68%, N 21.53%, O 9.84%, S 9.86%. Selective inhibitor of acetolactate synthase, a key enzyme in branched chain amino acid biosynthesis. Prepn: W. A. Kleschick *et al.*, **EP 142152** (1985 to Dow). Improved process: L. H. McKendry, **US 4910306** (1990 to Dow). Synthesis, toxicity and herbicidal activity: W. A. Kleschick *et al.*, *J. Agric. Food Chem.* **40**, 1083 (1992). Degradation in soil: R. G. Lehmann *et al.*, *Weed Res.* **33**, 187 (1993). Metabolism in wheat, corn and barley: D. S. Frear *et al., Pestic. Biochem. Physiol.* **45**, 178 (1993).

White powder, mp 250-251°. pKa 4.6. Partition coefficient (octanol/water): 0.21. Vapor pressure (25°): 3.7×10^{-13} Pa. Soly in water (g/l): 0.049 (pH 2.5); 5.65 (pH 7). LD_{50} in rats (mg/kg): >5000 orally; in rabbits (mg/kg): >2000 dermally (Kleschick, 1992).

USE: Herbicide.

4173. Flumiclorac. [87547-04-4] 2-[2-Chloro-4-fluoro-5-(1,-3,4,5,6,7-hexahydro-1,3-dioxo-2*H*-isoindol-2-yl)phenoxy]acetic acid; 2-chloro-4-fluoro-5-[(3,4,5,6-tetrahydro)phthalimido]phenoxyacetic acid; [2-chloro-5-(cyclohex-1-ene-1,2-dicarboximido)-4-fluorophenoxy]acetic acid. $C_{16}H_{13}ClFNO_5$; mol wt 353.73. C 54.33%, H 3.70%, Cl 10.02%, F 5.37%, N 3.96%, O 22.61%. Post-emergent herbicide for use in food crops. Prepn: E. Nagano *et al.*, **EP 83055**; *eidem*, **US 4770695** (1983, 1988 both to Sumitomo). Comprehensive description: K. Kamoshita *et al.*, in *Pest Management in Soybean*, L. G. Copping *et al.*, Eds. (Elsevier, London, 1992) pp 317-324.

Pentyl ester. [87546-18-7] V-23031; S-23031; Resource. $C_{21}H_{23}ClFNO_5$; mol wt 423.87. mp 88.87-90.13°. Vapor pressure (22.4°): $<1.0 \times 10^{-7}$ mm Hg. d^{20} 1.3316. Soly (25°): methanol 47.8 g/l; hexane 3.28 g/l; acetone 590 g/l; acetonitrile 589 g/l; water 0.189 mg/l. LD_{50} in rats (mg/kg): >5000 orally; >2000 dermally (Kamoshita).

USE: Herbicide.

4174. Flumioxazin. [103361-09-7] 2-[7-Fluoro-3,4-dihydro-3-oxo-4-(2-propyn-1-yl)-2*H*-1,4-benzoxazin-6-yl]-4,5,6,7-tetrahydro-1*H*-isoindole-1,3(2*H*)-dione; *N*-(7-fluoro-3,4-dihydro-3-oxo-4-prop-2-ynyl-2*H*-1,4-benzoxazin-6-yl)cyclohex-1-ene-1,2-dicarboximide; S-53482; V-53482; Sumisoya; Valor. $C_{19}H_{15}FN_2O_4$; mol wt 354.34. C 64.40%, H 4.27%, F 5.36%, N 7.91%, O 18.06%. Selective postemergence herbicide that inhibits protoporphyrinogen oxidase. Prepn: E. Nagano *et al.*, **EP 170191**; *eidem*, **US 4640707** (1986, 1987 both to Sumitomo). Field study in peanut: S. D. Askew *et al., Weed Technol.* **13**, 594 (1999). Metabolism in rat: Y. Tomigahara *et al., J. Agric. Food Chem.* **47**, 305, 2429 (1999). Review of physical properties, mode of action, and field trials: R. Yoshida *et al., Brighton Crop Prot. Conf. - Weeds* **1991**, 69-75.

Yellowish brown, odorless, powder, mp 201.83-203.83°. Soly in water at 25°: 1.79 mg/l. Sol in common organic solvents. Vapor pressure at 22°: 2.41×10^{-6} mmHg. d^{20} 1.5132 g/ml. LD_{50} in rats (mg/kg): >5000 orally; >2000 dermally. LC_{50} (4 hr) in rats (mg/m³): >3930 by inhalation. LC_{50} (96 hr) in bluegill, rainbow trout (mg/l): >21, 2.3 (Yoshida).

USE: Herbicide.

4175. Flunarizine. [52468-60-7] 1-[Bis(4-fluorophenyl)-methyl]-4-[(2*E*)-3-phenyl-2-propenyl]piperazine; (*E*)-1-[bis(*p*-fluorophenyl)-methyl]-4-cinnamylpiperazine; 1-cinnamyl-4-(di-*p*-fluorobenzhydryl)piperazine. $C_{26}H_{26}F_2N_2$; mol wt 404.50. C 77.20%, H 6.48%, F 9.39%, N 6.93%. Calcium channel blocker; fluorinated deriv of cinnarizine, *q.v.* Prepn: P. A. J. Janssen, **DE 1929330** and **FR 2014487**; *idem*, **US 3773939** (1970, 1970, 1973, all to Janssen). Pharmacology: L. K. Desmedt *et al., Arzneim.-Forsch.* **25**, 1408 (1975); T. Godfraind, *Eur. J. Pharmacol.* **53**, 273 (1979). Clinical studies: J. Schetz *et al., Curr. Ther. Res.* **23**, 131 (1978); G. Rudofsky *et al., Angiology* **30**, 479 (1979). Rheological effects in humans: J. D. Cree *et al., ibid.* **505**. Mechanism of calcium blocking activity: T. Godfraind, *Fed. Proc.* **40**, 2866 (1981). Clinical study in retinal vasculopathy: P. Nihard, *Angiology* **33**, 37 (1982). Clinical trial in chronic cerebrovascular disorders: A. Agnoli *et al., Int. J. Clin. Pharmacol. Res.* **8**, 189 (1988). Series of articles on absorption, distribution, excretion and metabolism: *Arzneim.-Forsch.* **33**, 1135-1151 (1983). Review of pharmacology, toxicology and clinical studies: B. Holmes *et al., Drugs* **27**, 6 (1984).

Dihydrochloride. [30484-77-6] Dinaplex; Flugeral; Flunagen; Flunarl; Fluxarten; Gradient; Issium; Mondus; Sibelium. $C_{26}H_{26}F_2N_2 \cdot 2HCl$; mol wt 477.42. Crystals from a mixture of 2-propanol/ethanol, mp 251.5°.

THERAP CAT: Vasodilator (cerebral and peripheral).

4176. Flunisolide. [3385-03-3] (6α,11β,16α)-6-Fluoro-11,21-dihydroxy-16,17-[(1-methylethylidene)bis(oxy)]pregna-1,4-diene-3,20-dione; 6α-fluoro-11β,16α,17,21-tetrahydroxypregna-1,4-diene-3,20-dione cyclic 16,17-acetal with acetone; 6α-fluoro-11β,21-dihydroxy-16α,17-isopropylidenedioxy-$\Delta^{1,4}$-pregnadiene-3,20-dione; RS-3999; Aerobid; Bronalide; Lunis; Nasalide; Rhinalar; Synaclyn; Syntaris. $C_{24}H_{31}FO_6$; mol wt 434.50. C 66.34%, H 7.19%, F 4.37%, O 22.09%. Synthetic fluorinated corticosteroid related to prednisolone, *q.v.* Prepn using *S. roseochromogenes*: **GB 933867** (1963 to Am. Cyanamid), *C.A.* **60**, 3070f (1964). Prepn using *Cunninghamella blakesleeana* 8688b: H. J. Ringold, G. Rosenkranz, **US 3124571**; H. J. Ringold *et al.*, **US 3126375** (both 1964 to Syntex). Metabolism: N. I. Chu *et al., Drug Metab. Dispos.* **7**, 81 (1979). Use in rhinitis: J. N. Sahay *et al., Clin. Allergy* **9**, 17 (1979); J. K. Sarsfield, G. E. Thomson, *Br. Med. J.* **2**, 95 (1979). Use in bronchial asthma: D. R. Webb *et al., Ann. Allergy* **42**, 80 (1979). Clinical study in chronic asthma: R. G. Slavin *et al., J. Allergy Clin. Immunol.* **66**, 379 (1980). Review of pharmacology and therapeutic efficacy: G. E. Pakes *et al., Drugs* **19**, 397-411 (1980); of clinical studies with HFA formulation: S. Shafazand, G. Colice, *Expert Opin. Pharmacother.* **5**, 1163-1173 (2004).

White to creamy-white crystalline powder, mp 245°. Soluble in acetone; sparingly sol in chloroform; slightly sol in methanol. Practically insol in water.

21-Acetate. [4533-89-5] RS-1320. $C_{26}H_{33}FO_7$; mol wt 476.54.

THERAP CAT: Glucocorticoid; antiasthmatic.

4177. Flunitrazepam. [1622-62-4] 5-(2-Fluorophenyl)-1,3-dihydro-1-methyl-7-nitro-2H-1,4-benzodiazepin-2-one; 1-methyl-7-nitro-5-(2-fluorophenyl)-3H-1,4-benzodiazepin-2(1H)-one; Ro-5-4200; Narcozep; Rohypnol; Roipnol. $C_{16}H_{12}FN_3O_3$; mol wt 313.29. C 61.34%, H 3.86%, F 6.06%, N 13.41%, O 15.32%. Prepn: L. H. Sternbach et al., J. Med. Chem. **6**, 261 (1963); J. Kariss, H. L. Newmark, US **3116203** and US **3123529**; O. Keller et al., US **3203990** (1963, 1964, 1965 all to Hoffmann-La Roche). Industrial prepn and purifn: J. M. Autin, FR **2529203** (1983 to Fabre), C.A. **100**, 191913r (1984). Pharmacology and clinical studies: E. Eidelberg et al., Neurology **5**, 223 (1965); Schuler et al., Psychopharmacologia **27**, 123 (1972); J. M. Monti, H. Altier, ibid. **32**, 343 (1973); Schallek et al., Arch. Int. Pharmacodyn. Ther. **206**, 161 (1973); Kaplan et al., J. Pharm. Sci. **63**, 527 (1974). Review of pharmacology and therapeutic use: M. A. K. Mattila, H. M. Larni, Drugs **20**, 353-374 (1980).

Pale yellow needles from methylene chloride-hexane, mp 166-167°. Also reported as crystals from acetonitrile and methanol, mp 170-172°.

Note: This is a controlled substance (depressant): **21 CFR,** 1308.14.

THERAP CAT: Hypnotic.

4178. Flunixin. [38677-85-9] 2-[[2-Methyl-3-(trifluoromethyl)phenyl]amino]-3-pyridinecarboxylic acid; 2-(2-methyl-3-trifluoromethylanilino)nicotinic acid; 2-($\alpha^3,\alpha^3,\alpha^3$-trifluoro-2,3-xylidino)-nicotinic acid; Sch-14714. $C_{14}H_{11}F_3N_2O_2$; mol wt 296.25. C 56.76%, H 3.74%, F 19.24%, N 9.46%, O 10.80%. Cyclooxygenase inhibitor. Prepn: M. H. Sherlock, N. Sperber, BE **679271**; eidem, US **3337570** (1966, 1967 both to Schering). Prepn of the acid and meglumine salt: M. H. Sherlock, BE **812772**; idem, US **3839344** (both 1974 to Schering). Pharmacodynamics in horses: M. M. Hardee, J. N. Moore, Res. Vet. Sci. **40**, 152 (1986); and pharmacokinetics: L. R. Soma et al., J. Vet. Pharmacol. Ther. **15**, 292 (1992). GC determn in urine: M. Johansson, E. -L. Anlér, J. Chromatogr. **427**, 55 (1988). Clinical evaluation in horses: J. W. Houdeshell, P. W. Hennessey, J. Equine Med. Surg. **1**, 57 (1977). Experimental use in food-producing animals: M. Kopcha, A. S. Ahl, J. Am. Vet. Med. Assoc. **194**, 45 (1989). Comparative toxicological study in horses: C. G. MacAllister et al., ibid. **202**, 71 (1993).

Crystals from acetone/hexane, mp 226-228°. pKa′ 5.82.

Meglumine salt. [42461-84-7] Banamine; Binixin; Finadyne. $C_{14}H_{11}F_3N_2O_2.C_7H_{17}NO_5$; mol wt 491.46. Colorless crystals from ethanol/ether, mp 135-137°. Crystals from acetonitrile, mp 136-139°. Soluble in water, alc, methanol. Practically insol in ethyl acetate.

THERAP CAT (VET): Anti-inflammatory; analgesic; antipyretic.

4179. Flunoxaprofen. [66934-18-7] (αS)-2-(4-Fluorophenyl)-α-methyl-5-benzoxazoleacetic acid; RV-12424; Priaxim. $C_{16}H_{12}FNO_3$; mol wt 285.27. C 67.37%, H 4.24%, F 6.66%, N 4.91%, O 16.83%. Nonsteroidal anti-inflammatory. Prepn of racemate: D. Evans et al., DE **2324443**; eidem, US **3912748** (1973, 1975 both to Lilly). Prepn of racemate and anti-inflammatory activity: D. W. Dunwell et al., J. Med. Chem. **18**, 53-58 (1975). Improved process

for prepn of racemate and enantiomers: F. Mauri, R. Signorini, BE **877887**; eidem, US **4304918** (1979, 1981 both to Ravizza); R. Signorini, A. Verga, DE **3325672** (1984 to Ravizza). HPLC determn of enantiomers in body fluids: S. Pedrazzini et al., J. Chromatogr. **415**, 214 (1987). Prophylactic and therapeutic treatment of asthma and other immediate hypersensitivity diseases: W. Dawson, US **4416892** (1983 to Lilly).

Crystals from acetone-water or acetic acid, mp 162-164°. $[\alpha]_D^{20}$ +50° (c = 2% in DMF). LD_{50} orally in mice: ~1200 mg/kg (Dunwell).

THERAP CAT: Anti-inflammatory.

4180. Fluoboric Acid. [16872-11-0] Hydrogen tetrafluoroborate(1−) (1:1); borofluoric acid; hydrofluoboric acid. BF_4H; mol wt 87.81. B 12.31%, F 86.54%, H 1.15%. HBF_4. Prepd from H_3BO_3 + HF: Fichter, Thiele, Z. Anorg. Allg. Chem. **67**, 302 (1910); Mathers et al., J. Am. Chem. Soc. **37**, 1516 (1915); Funk, Binder, Z. Anorg. Allg. Chem. **155**, 327; **159**, 121 (1926); Sheintsis, J. Appl. Chem. USSR **13**, 1101 (1940); Wamser, Christian, J. Am. Chem. Soc. **70**, 1209 (1948); Kwasnik in Handbook of Preparative Inorganic Chemistry **vol. 1**, G. Brauer, Ed. (Academic Press, New York, 2nd ed., 1963) pp 221-222. For other methods of prepn see H. S. Booth, D. R. Martin, Boron Trifluoride and Its Derivatives (New York, 1949). Review: Sharp, Adv. Fluorine Chem. **1**, 68-128 (1960).

Liquid. Poisonous. bp 130° (decompn). Miscible with water, alcohol. Strong acid. n_D^{20} of a 20% aq soln 1.3284. Heat of formn 388.5 kcal. Undergoes limited hydrolysis in water to form hydroxyfluoborate ions; major product is BF_3OH^-. Pure HBF_4 may be stored in glass vessels at room temps. Forms cryst salts with metals. The heavy metal salts and $LiBF_4$ and $NaBF_4$ are very sol in water. See Potassium Tetrafluoborate about color phenomena appearing in concd solns of fluoboric acid.

Caution: Strong caustic action on skin, mucous membranes. Irritating to eyes, respiratory tract. Plating solns contg fluoborates are considered toxic.

USE: As catalyst for preparing acetals, esterifying cellulose; to clean metal surfaces before welding; to brighten aluminum; as a solute in electrolytes for plating metals such as chromium, iron, nickel, copper, silver, zinc, cadmium, indium, tin, and lead (has a high throwing power). Reagent for sodium in the presence of magnesium and potassium ions; for making stabilized diazo salts (diazonium and tetrazonium fluoborates). An 0.1 to 0.5% soln retards fermentation: Homeyer, Pharm. Ztg. **34**, 761 (1889).

4181. Fluocinolone Acetonide. [67-73-2] (6α,11β,16α)-6,9-Difluoro-11,21-dihydroxy-16,17-[(1-methylethylidene)bis(oxy)]-pregna-1,4-diene-3,20-dione; 6α,9α-difluoro-11β,16α,17,21-tetrahydroxypregna-1,4-diene-3,20-dione cyclic 16,17-acetal with acetone; 6α,9α-difluoro-16α-hydroxyprednisolone 16,17-acetonide; 6α,9α-difluoro-16α,17α-isopropylidenedioxy-1,4-pregnadiene-3,20-dione; Coriphate; Cortiplastol; Dermalar; Fluonid; Fluovitef; Fluvean; Fluzon; Jellin; Localyn; Synalar; Synamol; Synandone; Synemol; Synotic; Synsac. $C_{24}H_{30}F_2O_6$; mol wt 452.49. C 63.71%, H 6.68%, F 8.40%, O 21.21%. Prepn: J. S. Mills et al., J. Am. Chem. Soc. **82**, 3399 (1960); J. S. Mills, A. Bowers, US **3014938**; H. J. Ringold et al., US **3126375** (1961, 1964 both to Syntex). IR spectrum: Sammul et al., J. Assoc. Off. Agric. Chem. **47**, 952 (1964).

White, odorless crystals from acetone + hexane, mp 265-266°. $[\alpha]_D$ +95° (chloroform). uv max: 238 nm (log ε 4.21). Soluble in methanol; slightly sol in ether, chloroform. Insol in water.

THERAP CAT: Glucocorticoid; anti-inflammatory.

THERAP CAT (VET): Adrenocortical steroid (topical use).

4182. Fluocinonide. [356-12-7] $(6\alpha,11\beta,16\alpha)$-21-(Acetyloxy)-6,9-difluoro-11-hydroxy-16,17-[(1-methylethylidene)bis(oxy)]pregna-1,4-diene-3,20-dione; $6\alpha,9$-difluoro-11β,16α,17,21-tetrahydroxypregna-1,4-diene-3,20-dione, cyclic 16,17-acetal with acetone, 21-acetate; 21-acetoxy-6α,9α-difluoro-11β-hydroxy-16α,17α-isopropylidenedioxy-1,4-pregnadiene-3,20-dione; 16α,17α-isopropylidene-6α-fluorotriamcinolone 21-acetate; fluocinolide; fluocinolone acetonide acetate; Biscosal; Dermaplus; Lidex; Metosyn; Straderm; Topsym; Topsymin; Topsyne; Topsyn. $C_{26}H_{32}F_2O_7$; mol wt 494.53. C 63.15%, H 6.52%, F 7.68%, O 22.65%. Prepn: **GB 916996** (1963 to Olin Mathieson), *C.A.* **59**, 1716b (1963); H. J. Ringold, G. Rosenkranz, **US 3124571**; H. J. Ringold *et al.*, **US 3126375** (both 1964 to Syntex); J. Fried, **US 3197469** (1965 to Pharm. Res. Prod.). NMR: A. D. Cross, P. W. Landis, *J. Am. Chem. Soc.* **86**, 4005 (1964).

White to cream-colored crystals from methanol. mp 308-311°. $[\alpha]_D$ +83° (chloroform). uv max: 237 nm (log ε 4.18). Sparingly soluble in acetone, chloroform; slightly sol in alc, methanol, dioxane; very slightly sol in ether. Practically insol in water.

THERAP CAT: Anti-inflammatory; glucocorticoid.

4183. Fluocortin Butyl. [41767-29-7] $(6\alpha,11\beta,16\alpha)$-6-Fluoro-11-hydroxy-16-methyl-3,20-dioxopregna-1,4-dien-21-oic acid butyl ester; butyl 6α-fluoro-11β-hydroxy-16α-methyl-3,20-dioxopregna-1,4-dien-21-oate; SH K 203; Varlane; Vaspit. $C_{26}H_{35}FO_5$; mol wt 446.56. C 69.93%, H 7.90%, F 4.25%, O 17.91%. The butyl ester deriv of fluocortolone-21-acid, a metabolite of fluocortolone, *q.v.*, in humans. Prepn: H. Laurent *et al.*, **DE 2150268, DE 2150270**; *eidem*, **US 3824260** (1973, 1973, 1974, all to Schering AG); H. Laurent *et al.*, *Angew. Chem.* **87**, 70 (1975). Description of the pregnan-21-oic esters as a novel corticoid structure type, their activity, and methods for synthesis: H. Laurent *et al.*, *J. Steroid Biochem.* **6**, 185 (1975). Toxicity data: P. Günzel *et al.*, *Arzneim.-Forsch.* **27**, 2217 (1977). Series of articles on synthesis, pharmacology and clinical trials: *ibid.* 2185-2246.

Crystals from acetone/hexane, mp 195.1°. $[\alpha]_D^{25}$ +136° (c = 0.5 in chloroform). uv max (methanol): 242 nm (ε 16800). Soluble in chloroform, ethanol; poorly sol in ethyl ether. Insol in water. LD_{50} in mice, rats (g/kg): >4 orally and s.c. (Günzel).

THERAP CAT: Anti-inflammatory.

4184. Fluocortolone. [152-97-6] $(6\alpha,11\beta,16\alpha)$-6-Fluoro-11,21-dihydroxy-16-methylpregna-1,4-diene-3,20-dione; 6α-fluoro-16α-methyl-1-dehydrocorticosterone; 6α-fluoro-16α-methyl-$\Delta^{1,4}$-pregnadiene-11β,21-diol-3,20-dione; Ultralan oral. $C_{22}H_{29}FO_4$; mol wt 376.47. C 70.19%, H 7.76%, F 5.05%, O 17.00%. Prepn: **BE 614196**; Kieslich *et al.*, **US 3232839** (1962, 1966 to Schering AG); Doménico *et al.*, *Arzneim.-Forsch.* **15**, 46 (1965); Kieslich *et al.*, *Ann.* **726**, 168 (1969). Pharmacology: Doménico *et al.*, *loc. cit;* Gillich *et al.*, *Int. Z. Klin. Pharmakol. Ther. Toxikol.* **1**, 197 (1968); Von Schoening *et al.*, *Klin. Wochenschr.* **48**, 1448 (1970). Metabolism: Gerhards *et al.*, *Acta Endocrinol.* **68**, 98 (1971). Clinical studies: Breuer *et al.*, *Arzneim.-Forsch.* **15**, 50 (1965).

Crystals, mp 188-190.5°. Soly (mg/l): 295 in water (37°); 120 in ethanol (20°); 440 in toluene (20°). $[\alpha]_D^{20}$ +100° (dioxane). uv max (methanol): 242 nm (ε 16300).

21-Acetate. $C_{24}H_{31}FO_5$. Crystals, mp 237-239°. uv max (methanol): 240 nm (ε 15860).

21-Hexanoate. [303-40-2] Fluocortolone 21-caproate; SH-770; Ficoid; Ultralanum. $C_{28}H_{39}FO_5$; mol wt 474.61. Crystals, mp 242-245°. $[\alpha]_D^{20}$ +98.5° (dioxane). uv max (methanol): 242 nm (ε 16200). Soly (mg/l): 7.8 in water (37°); 450 in ethanol (20°); 440 in toluene (20°).

21-Pivalate. Fluocortolone trimethylacetate. $C_{27}H_{37}FO_5$. Crystals, mp 187°. Almost insol in water. Sol in chloroform, methanol. Slightly sol in ether.

THERAP CAT: Glucocorticoid.

4185. Fluometuron. [2164-17-2] N,N-Dimethyl-N'-[3-(trifluoromethyl)phenyl]urea; 1,1-dimethyl-3-(α,α,α-trifluoro-*m*-tolyl)urea; N-(3-trifluoromethylphenyl)-N',N'-dimethylurea; Ciba 2059; Cotoran; Cottonex. $C_{10}H_{11}F_3N_2O$; mol wt 232.21. C 51.72%, H 4.77%, F 24.54%, N 12.06%, O 6.89%. Made by the reaction of dimethylamine on 3-trifluoromethylphenyl isocyanate: Abel, *Chem. Ind. (London)* **1957**, 1106. Herbicidal preparations: Martin, Aebi, **US 3134665** (1964 to Ciba). Toxicity data: R. Ben-Dyke *et al.*, *World Rev. Pest Control* **9**, 119 (1970). Review of carcinogenic risk: *IARC Monographs* **30**, 245-253 (1983).

Crystals, mp 163-164.5°. Soly in water at 25°: 80 ppm; sol in acetone, ethanol, isopropanol, DMF and other organic solvents. LD_{50} in rats (mg/kg): 8900 orally; in rabbits (mg/kg): >10,000 dermally (Ben-Dyke).

USE: Herbicide.

4186. Fluopicolide. [239110-15-7] 2,6-Dichloro-N-[[3-chloro-5-(trifluoromethyl)-2-pyridinyl]methyl]benzamide; AE-C638206. $C_{14}H_8Cl_3F_3N_2O$; mol wt 383.58. C 43.84%, H 2.10%, Cl 27.73%, F 14.86%, N 7.30%, O 4.17%. Acyl picolide fungicide for use in food crops; delocalizes a spectrin-like protein in the fungal mycelium and perturbs the plasma membrane. Prepn: B. A. Maloney *et al.*, **WO 9942447** (1999 to AgrEvo); *eidem*, **US 6503933** (2003 to Aventis CropSci.). Discovery and chemistry: G. Briggs *et al.*, *Pflanzenschutz-Nachr. Bayer (Engl. Ed.)* **59**, 141 (2006). Mode of action: V. Toquin *et al.*, *ibid.*, 171. Efficacy vs Phytophthora blight in peppers: J. Shin *et al.*, *Plant Pathol. J.* **26**, 367 (2010); in squash: K. L. Jackson *et al.*, *Crop Prot.* **29**, 1421 (2010). Persistance in soil and grapes: S. Mohapatra *et al.*, *Bull. Environ. Contam. Toxicol.* **86**, 238 (2011).

Beige powder, mp 150°. Dec before boiling at 320°. Vapor pressure (25°): 8.03×10^{-7} Pa. Log P (octanol/water): 2.9. Soly at 20° (g/l): acetone 74.7; dichloromethane 126; DMSO 183; ethanol 19.2; ethyl acetate 37.7; *n*-hexane 0.20; toluene 20.5. Soly in water at 20° (g/l): 0.00280 (pH 4); 0.00286 (pH 9).

USE: Agricultural fungicide.

4187. 9H-Fluorene. [86-73-7] *o*-Biphenylenemethane; diphenylenemethane; 2,2′-methylenebiphenyl. $C_{13}H_{10}$; mol wt 166.22. C 93.94%, H 6.06%. Occurs in coal tar, *q.v.* (about 1.6%). Isoln: Kruber, *Ber.* **70**, 1556 (1937). Also found in coke-oven tar: Weiss, Downs, *Ind. Eng. Chem.* **15**, 1022 (1923). From acetylene and hydrogen in red-hot tube: Meyer, *Ber.* **45**, 1609 (1912); Meyer, Taeger, *Ber.* **53**, 1261 (1921). From charcoal by boiling with fuming HNO_3: Dimroth, Kerkovins, *Ann.* **399**, 120 (1913). From 2,2′-dibromodiphenylmethane on boiling with hydrazine hydrate in presence of Pd: Busch, Weber, *J. Prakt. Chem.* [2] **146**, 47 (1936). Reactions: Rieveschl Jr., Ray, *Chem. Rev.* **23**, 287 (1938). Review of toxicology and human exposure: *Toxicological Profile for Polycyclic Aromatic Hydrocarbons* (PB95-264370, 1995) 487 pp.

Dazzling white leaflets or flakes from alc. d 1.202. Sublimes easily in high vacuum. mp 116-117°. bp 295°. Absorption spectrum: Mayneord, Roe, *Proc. R. Soc. London* **A158**, 634 (1937). Freely sol in glacial acetic acid; sol in carbon disulfide, ether, benzene, hot alcohol. Soly data: Mortimer, *J. Am. Chem. Soc.* **45**, 633 (1923).

4188. 9H-Fluorene-2,7-diamine. [525-64-4] 2,7-Diaminofluorene. $C_{13}H_{12}N_2$; mol wt 196.25. C 79.56%, H 6.16%, N 14.27%. Prepd by nitrating fluorene and reducing the 2,7-dinitrofluorene formed with tin and hydrochloric acid: Schmidt, Hinderer, *Ber.* **64**, 1793 (1931).

Needles from water, mp 165°. Slightly sol in cold water, more sol in hot water. Readily sol in alc.

Hydrochloride. $C_{13}H_{12}N_2$.HCl. Crystals, readily sol in hot water.

USE: Detection of bromide, chloride, nitrate, persulfate, cadmium, copper, cobalt, zinc.

4189. N-2-Fluorenylacetamide. [53-96-3] *N*-9*H*-Fluoren-2-ylacetamide; 2-acetylaminofluorene; AAF; 2-FAA. $C_{15}H_{13}NO$; mol wt 223.28. C 80.69%, H 5.87%, N 6.27%, O 7.17%. Synthesis: Hayashi, Nakayama, *J. Soc. Chem. Ind. Jpn.* [Suppl] **36**, 127B (1933). Carcinogenicity studies: R. H. Wilson *et al.*, *Cancer Res.* **1**, 595 (1941). Toxicity studies: Haley *et al.*, *Proc. Soc. Exp. Biol. Med.* **143**, 1117 (1973); **146**, 648 (1974).

Crystals from alcohol + water, mp 194°. uv max: 285 nm. Insol in water. Sol in alcohols, glycols, fat solvents.

Caution: Potential symptoms of overexposure are reduced function of liver, kidneys, bladder and pancreas. *See NIOSH Pocket Guide to Chemical Hazards* (DHHS/NIOSH 97-140, 1997) p 4. This substance is reasonably anticipated to be a human carcinogen: *Report on Carcinogens, Twelfth Edition* (PB2011-111646, 2011) p 24.

USE: As a positive control to study the carcinogenicity and mutagenicity of aromatic amines.

4190. 9-Fluorenylmethyl Chloroformate. [28920-43-6] Carbonochloridic acid 9*H*-fluoren-9-ylmethyl ester; 9-fluorenylmethoxycarbonyl chloride; 9-fluorenylmethyl chlorocarbonate; Fmoc chloride; Fmoc-Cl. $C_{15}H_{11}ClO_2$; mol wt 258.70. C 69.64%, H 4.29%, Cl 13.70%, O 12.37%. Reagent used to introduce the 9-fluorenylmethoxycarbonyl (Fmoc) group to protect amines and alcohols. Derivatizing agent for HPLC fluorescence detection. Prepn and use in amine protection: L. A. Carpino, G. Y. Han, *J. Am. Chem. Soc.* **92**, 5748 (1970); *idem, J. Org. Chem.* **37**, 3404 (1972). Additional functional group protection: C. Gioeli, J. B. Chattopadhyaya, *J. Chem. Soc. Chem. Commun.* **1982**, 672; T. Johnson *et al., ibid.* **1993**, 369. Amino acid derivatization, separation, and analysis: S. Einarsson *et al., J. Chromatogr.* **282**, 609 (1983). *Review:* R. L. Polt in *Encyclopedia of Reagents for Organic Synthesis* **4**, L. A. Paquette, Ed. (Wiley, New York, 1995) pp 2545-2548.

Colorless crystals from ether, mp 61.5-63°. *Corrosive.* Absorption max: 267 nm (ε 20500, ethyl acetate); 264 nm (ε 19810, 30% pH 3.5 phosphate buffer in methanol). Fluorescence excitation: 206, 259 nm (acetonitrile). Fluorescence emission: 311 nm (acetonitrile). Sol in dichloromethane, tetrahydrofuran, dioxane. Store at 2-8° in absence of moisture.

USE: Reagent in synthetic organic chemistry.

4191. Fluorescamine. [38183-12-9] 4-Phenylspiro[furan-2(3*H*),1′(3′*H*)-isobenzofuran]-3,3′-dione; 4-phenylspiro[furan-2(3*H*),1′-phthalan]-3,3′-dione; Ro-20-7234; Fluram. $C_{17}H_{10}O_4$; mol wt 278.26. C 73.38%, H 3.62%, O 23.00%. Non-fluorescent reagent that reacts readily with primary amines to form highly fluorescent compds: S. Udenfriend *et al., Science* **178**, 871 (1972). Prepn: M. Weigele *et al., J. Am. Chem. Soc.* **94**, 5927 (1972); *idem, J. Org. Chem.* **41**, 388 (1976). Use as fluorometric reagent: W. Leimgruber, M. Weigele, *DE 2350179*; *idem, US 3830629* (both 1974 to Hoffmann-La Roche). Review of analytical uses: C. Y. Lai, *Methods Enzymol.* **47**, 236-243 (1977); S. Stein, *Peptides in Neurobiology*, H. Gainer, Ed. (Plenum, New York, 1977) pp 9-37; S. Udenfriend, *Pharmacology* **19**, 223-227 (1979).

mp 154-155°. uv max (ether): 235, 276, 284, 306 nm (ε 25900, 3950, 4100, 3800).

USE: Analytical reagent.

4192. Fluorescein. [2321-07-5] 3′,6′-Dihydroxyspiro[isobenzofuran-1(3*H*),9′-[9*H*]xanthen]-3-one; 9-(*o*-carboxyphenyl)-6-

hydroxy-3*H*-xanthen-3-one; 3',6'-dihydroxyfluoran; 3',6'-fluorandiol; 9-(*o*-carboxyphenyl)-6-hydroxy-3-isoxanthenone; resorcinolphthalein; D & C Yellow no. 7; C.I. Solvent Yellow 94; C.I. 45350:1. $C_{20}H_{12}O_5$; mol wt 332.31. C 72.29%, H 3.64%, O 24.07%. Prepd by heating phthalic anhydride with resorcinol: Fischer, Bollmann, *J. Prakt. Chem.* **104**, 123 (1922); McKenna, Sowa, *J. Am. Chem. Soc.* **60**, 124 (1938). Structure: Ramart-Lucas, *Compt. Rend.* **205**, 864 (1937); Nagase *et al.*, *J. Pharm. Soc. Jpn.* **73**, 1033, 1039 (1953). Review of synthesis, properties and histological use: R. F. Steiner, H. Edelhoch, *Chem. Rev.* **62**, 457 (1962). Use as label in immunoassays: E. F. Ullman *et al.*, *J. Biol. Chem.* **251**, 4172 (1976); Y. Suzuki *et al.*, *Jpn. J. Exp. Med.* **49**, 179 (1979). Toxicity studies in fish: L. L. Marking, *Prog. Fish Cult.* **31**, 139 (1969). Toxicity data: S. L. Yankell, J. J. Loux, *J. Periodontol.* **48**, 228 (1977). *See also: Colour Index* vol. **4** (3rd ed., 1971) p 4424; *H. J. Conn's Biological Stains*, R. D. Lillie, Ed. (Williams & Wilkins, Baltimore, 9th ed., 1977) p 337.

Yellowish-red to red powder. mp 314-316° in sealed tube, with decompn. Sol in hot alcohol or glacial acetic acid; also sol in alkali hydroxides or carbonates with a bright green fluorescence appearing red by transmitted light. Insol in water, benzene, chloroform, ether. Absorption max: 493.5, 460 nm.

Disodium salt. [518-47-8] Soluble fluorescein; resorcinol phthalein sodium; uranin(e); uranine yellow; D & C yellow No. 8; C.I. Acid Yellow 73; C.I. 45350; Ak-Fluor; Fluorescite; Fluorets; Fluori-strip; Ful-Glo; Funduscein; Irescein. $C_{20}H_{10}Na_2O_5$; mol wt 376.27. Hygroscopic orange-red powder. Freely sol in water with yellowish-red color and intense yellowish-green fluorescence perceptible down to a dil of 0.02 ppm under uv light. The fluorescence disappears when the soln is made acid, and reappears when the soln is again made neutral or alkaline. Absorption max (water): 493.5 nm. Sparingly sol in alc. LD_{50} in mice, rats (mg/kg): 4738, 6721 orally (Yankel, Loux).

USE: In examining subterranean waters. Serves to ascertain source of springs, connections between streams and sea, determining approx vol of water delivered by a spring, detecting source of contamination of drinking water, infiltration of soil with waste waters of factories. In externally applied drugs and cosmetics. Analytical reagent (protein label). Clinical reagent (immunohistological stain, immunofluorescent label).

THERAP CAT: Diagnostic aid (corneal trauma indicator), ophthalmic angiography, contact lens fitting.

THERAP CAT (VET): Diagnostic aid (corneal lesions, intraocular inflammation).

4193. Fluorescin. [518-44-5] 2-(3,6-Dihydroxy-9*H*-xanthen-9-yl)benzoic acid; resorcinolphthalin. $C_{20}H_{14}O_5$; mol wt 334.33. C 71.85%, H 4.22%, O 23.93%. Obtained by heating fluorescein with NaOH and zinc dust. Formation by *Pseudomonas aeruginosa*: King *et al.*, *Can. J. Res.* **26C**, 514 (1948); Totter, Moseley, *J. Bacteriol.* **65**, 45 (1953).

Bright yellow powder, mp 125-127°. Readily oxidizes to fluorescein. Practically insol in water. Sol in alkali hydroxides or carbonates, alcohol, ether. *Keep well closed.*

USE: Reagent for oxidases, peroxidases.

4194. Fluorine. [7782-41-4] F; at. wt 18.9984032; at. no. 9; valence 1. A halogen, Group VIIA (17). Does not exist as elemental state, F, in nature. Occurs as diatomic molecule, F_2. Occurrence in earth's crust 0.065% by wt. Naturally occurring isotope (mass number): 19 (100%). Know artificial radioactive isotopes; 15-17, 18 (longest-lived known isotope, $T_{1/2}$ 109.77 min, β^+ emitter), 20-27. Primary commercial source is fluorite, or fluorspar; other important sources are cryolite, and fluorapatite. Discovered in 1771 by Scheele. Isolated in 1886 by electrolyzing a soln of potassium fluoride in anhyd hydrogen fluoride at $-23°$, using platinum-iridium electrodes: H. Moissan, *Compt. Rend.* **102**, 1543 (1886); **103**, 202, 256. Toxicity data: Keplinger, Suissa, *Am. Ind. Hyg. Assoc. J.* **29**, 10 (1968). *Reviews:* Finger, "Fluorine Resources and Fluorine Utilization" in *Advances in Fluorine Chemistry* vol. **2**, M. Stacey *et al.*, Eds. (Butterworths, London, 1961) pp 35-54; A. J. Rudge, *The Manufacture and Use of Fluorine and its Compounds* (Oxford University Press, 1962); T. A. O'Donnell, "Fluorine" in *Comprehensive Inorganic Chemistry* vol. **2**, J. C. Bailar, Jr. *et al.*, Eds. (Pergamon Press, Oxford, 1973) pp 1009-1106; G. Shia in *Kirk-Othmer Encyclopedia of Chemical Technology* vol. **11** (Wiley-Interscience, New York, 4th ed., 1994) pp 241-267. Review of toxicology and human exposure: *Toxicological Profile for Fluorides, Hydrogen Fluoride and Fluorine* (PB2004-100002, 2003) 404 pp. Book: *Fluorine: The First Hundred Years (1886-1986)*, R. E. Banks *et al.*, Eds. (Elsevier Sequoia, Laussanne, 1986) 404 pp.

Pale yellow, diatomic gas with sharp penetrating odor. mp $-219.61°$ (53.54 K); bp $-188.13°$ (85.02 K); d (liq, $-188.13°$) 1.5127; vapor pressure data: Hu *et al.*, *J. Am. Chem. Soc.* **75**, 5642 (1953); White *et al.*, *ibid.* **76**, 2584 (1954). Crit temp: $-129°$; crit pressure: 55 atm. *Poisonous, oxidizer, corrosive.* Most chemically reactive element; most powerful oxidizing agent, higher oxidation potential than ozone; most electronegative element; $E°$ (calc) ½F/F$^-$ 2.9 V. F-F bond weaker than Cl-Cl and Br-Br bonds; enthalpy of dissociation: 37.7 kcal. Reacts vigorously with most oxidizable substances at room temp, frequently with ignition. Combines directly or indirectly, to form fluorides with all the elements except helium, neon and argon. Dec water, giving hydrofluoric acid, HF, oxygen fluoride, OF_2, hydrogen peroxide, oxygen and ozone. Reacts with nitric acid, forming the explosive gas, fluorine nitrate, NO_3F; with sulfuric acid, giving fluorosulfuric acid, $HFSO_3$. Yields the metal fluorides, water, oxygen and oxygen fluoride when made to react with metal hydroxides in the cold. Reacts violently with organic compds, usually with disintegration of the molecule. Under controlled conditions, however, hydrocarbon vapors may be fluorinated with elemental fluorine. Solid fluorine explodes when brought in contact with liquid hydrogen. Under ordinary conditions it does not react directly with oxygen, nor does it react with oxides of sodium, potassium or calcium. LC_{50} (1 hr) inhalation by rats, mice, guinea pigs: 185, 150, 170 ppm (by vol) (Keplinger, Suissa).

Caution: Potential symptoms of overexposure are irritation of eyes, nose and respiratory system; laryngeal spasms, bronchial spasms; pulmonary edema; eye and skin burns. *See NIOSH Pocket Guide to Chemical Hazards* (DHHS/NIOSH 97-140, 1997) p 146. Chronic ingestion of high concentrations from water supply can cause mottled enamel of teeth and osteosclerosis. *See Clinical Toxicology of Commercial Products*, R. E. Gosselin *et al.*, Eds. (Williams & Wilkins, Baltimore, 5th ed., 1984) Section III, pp 185-193.

USE: In manuf of UF_6 for nuclear power generation, of SF_6 for dielectrics, of fluorinating and metal fluoride compounds.

4195. Fluorine Dioxide. [7783-44-0] Oxygen fluoride ((O_2)-F_2); dioxygen difluoride. F_2O_2; mol wt 69.99. F 54.29%, O 45.72%. Prepd from O_2 and F_2: Ruff, Menzel, *Z. Anorg. Allg. Chem.* **211**, 204 (1933); *ibid.* **217**, 85 (1934); Goetschel *et al.*, *J. Am. Chem. Soc.* **91**, 4702 (1969). Chemical behavior: Streng, *ibid.* **85**, 130 (1963). *Review:* Kemmitt, Sharp, *Adv. Fluorine Chem.* **4**, 214-215 (1965).

Thermally unstable gas at room temp; begins to dec into fluorine and oxygen at $-100°$. Pale yellow solid or yellow liquid at low temp. mp $-154°$C (119 K). Early prepns, described as brown gas, red liq and orange solid (mp $-163.5°$), probably contained other fluorine-oxygen compounds as impurities: Goetschel *et al.*, *loc. cit.* Strong oxidizing and fluorinating agent.

4196. Fluorine Monoxide. [7783-41-7] Oxygen fluoride (OF_2); difluorene monoxide; fluorine oxide; oxydifluoride. F_2O;

mol wt 54.00. F 70.36%, O 29.63%. Prepd by passing fluorine slowly through an aq NaOH soln: Yost, *Inorg. Synth.* **1**, 109 (1939); Schnizlein *et al.*, *J. Phys. Chem.* **56**, 233 (1952). Toxicity data: Darmer *et al.*, *Am. Ind. Hyg. Assoc. J.* **33**, 661 (1972). *Review:* Kemmitt, Sharp, *Adv. Fluorine Chem.* **4**, 213-314 (1965).

Colorless gas. Yellowish-brown when liq. Peculiar smell. Does not attack glass in the cold. Corrodes mercury. Reacts very slowly with water. The gas may be kept over water unchanged for a month. d (liq; −224°) 1.90. mp −228.8°. bp −145.3°. Trouton constant 20.65. Soly in water (0°) 6.8 ml gas/100 ml H_2O. LC_{50} (1 hr) inhalation by rats, mice: 2.6, 1.5 ppm (Darmer).

Caution: Potential symptoms of overexposure are headache; eyes, skin, respiratory system irritation, pulmonary edema; direct contact with gas under pressure may cause eye and skin burns. *See NIOSH Pocket Guide to Chemical Hazards* (DHHS/NIOSH 97-140, 1997) p 238.

4197. Fluorine Nitrate. [7789-26-6] Nitroxy fluoride; nitrogen trioxyfluoride; nitryl hypofluorite. FNO_3; mol wt 81.00. F 23.45%, N 17.29%, O 59.26%. $FONO_2$. Prepd by the action of fluorine on nitric acid: Cady, *J. Am. Chem. Soc.* **56**, 2635 (1934); Ruff, Kwasnik, *Angew. Chem.* **48**, 238 (1935); Kwasnik in *Handbook of Preparative Inorganic Chemistry* vol. **1**, G. Brauer, Ed. (Academic Press, New York, 2nd ed., 1963) pp 187-189. *Reviews:* Kemmitt, Sharp, *Adv. Fluorine Chem.* **4**, 216-218 (1965); Woolf, *ibid.* **5**, 1-30 (1965); Schmutzler, *Angew. Chem. Int. Ed.* **7**, 440-455 (1968).

Colorless gas. Moldy, acrid odor. mp −175°. bp −45.9°. d (liq at bp) 1.507. $d^{-193.2}$ (solid) 1.951. Trouton const 20.8. The liquid explodes on slight percussion. Hydrolyzed by water to OF_2, O_2, HF and HNO_3. Sol in acetone. Conflagrates on contact with alcohol, ether, aniline. May be stored in vacuum-sealed glass ampuls cooled by liquid oxygen. Powerful oxidizing agent.

USE: Oxidizing agent in rocket propellants.

4198. Fluoroacetamide. [640-19-7] 2-Fluoroacetamide; fluoroacetic acid amide; monofluoroacetamide; 1081; Fluorakil 100; Fussol. C_2H_4FNO; mol wt 77.06. C 31.17%, H 5.23%, F 24.65%, N 18.18%, O 20.76%. CH_2FCONH_2. Numerous syntheses, e.g. from fluoroacetyl chloride and NH_3: Truce, *J. Am. Chem. Soc.* **70**, 2828 (1948). Mode of action study and toxicity data: F. Matsumura, R. D. O'Brien, *Biochem. Pharmacol.* **12**, 1201 (1963).

Crystals. Sublimes on heating. Freely sol in water; sol in acetone; sparingly sol in chloroform. LD_{50} i.p. in mice: 85 mg/kg (Matsumura, O'Brien).

USE: Rodenticide. Insecticide proposed mainly for use on fruits to combat scale insects, aphids, and mites.

4199. Fluoroacetic Acid. [144-49-0] 2-Fluoroacetic acid; cymonic acid; fluoroethanoic acid; gifblaar poison; monofluoroacetic acid; FAA. $C_2H_3FO_2$; mol wt 78.04. C 30.78%, H 3.87%, F 24.34%, O 41.00%. Toxic constituent of the poisonous South African plant, *Dichapetalum cymosum* (Hook.) Engl., *Dichapetalaceae*, commonly known as gifblaar ("poison leaf"). Also contained in other poisonous plant species. Metabolically converted to fluorocitrate which inhibits the tricarboxylic acid cycle resulting in cellular hypoxia. Prepn via methyl ester: F. Swarts, *Bull. Soc. Chim.* [3] **15**, 1134 (1896); B. C. Saunders, G. J. Stacey, *J. Chem. Soc.* **1948**, 1773; by fluorination of disodium malonate: V. Grakauskas, *J. Org. Chem.* **34**, 2446 (1969). Isoln from gifblaar: J. S. C. Marais, *Onderstepoort J. Vet. Res.* **20**, 67 (1944). Mechanism of toxicity: R. A. Peters, *Endeavour* **13**, 147 (1954). Determn in biological samples using HS-SPME: F. Sporkert *et al.*, *J. Chromatogr. B* **772**, 45 (2002). Review of properties and use: A. J. Elliott in *Kirk-Othmer Encyclopedia of Chemical Technology* vol. **11** (Wiley-Interscience, New York, 4th ed., 1994) pp 544-550. Review of toxicology: N. V. Goncharov *et al.*, *J. Appl. Toxicol.* **26**, 148-161 (2006).

Colorless needles, mp 31-32°. bp 165°. pKa 2.59. d 1.37. Sol in water. Burns with a green flame.

Sodium salt. [62-74-8] Sodium fluoroacetate; compound 1080. $C_2H_2FNaO_2$; mol wt 100.02. Hygroscopic white solid. mp 200-

202° (dec when heated above mp). Soly at 25° (g/100g): water 111, methanol 5, ethanol 1.4, acetone 0.04, carbon tetrachloride 0.004.

Methyl ester. [453-18-9] Methyl fluoroacetate; MFA. C_3H_5-FO_2. Mobile liquid, fp −32°. Practically odorless or faint, fruit-like odor. d_4^{20} 1.1744; d_4^{15} 1.1613. n_D^{20} 1.3679. bp 104.5°. Sol in water; slightly sol in petr ether. LD_{50} in rabbits, mice (mg/kg): 0.25 i.v.; 7-10 s.c. (Saunders, Stacey).

Ethyl ester. [459-72-3] $C_4H_7FO_2$. Liq, odor of ethyl acetate. $d^{20.5}$ 1.0926. bp_{758} 121.6°. $n_D^{20.5}$ 1.3767. Sol in water.

Caution: Fluoroacetates are highly toxic. Potential symptoms of overexposure are vomiting; apprehension; clonic-tonic convulsions; depression; loss of consciousness; metabolic acidosis; hypotension; cardiac arrhythmias; respiratory failure; hypocalcemia; renal failure; death (Goncharov).

USE: Rodenticide.

4200. 4-Fluoroaniline. [371-40-4] 4-Fluorobenzenamine. C_6H_6FN; mol wt 111.12. C 64.85%, H 5.44%, F 17.10%, N 12.61%. Prepd by reduction of 1-fluoro-4-nitrobenzene by Raney nickel: Benington *et al.*, *J. Org. Chem.* **18**, 1508 (1953); Finger *et al.*, *J. Am. Chem. Soc.* **81**, 98 (1959); by sulfurated sodium borohydride: Lalancette, Brindle, *Can. J. Chem.* **49**, 2990 (1971).

F—⟨benzene ring⟩—NH₂

Liquid. d_4^{20} 1.1725; d_4^{25} 1.1690. mp −1.9°. bp 188°, bp_{20} 86°. n_D^{20} 1.51954. Heat of combustion 780.4 kcal/mol. Very slightly sol in water.

USE: Intermediate in the manuf of herbicides and plant growth regulators.

4201. Fluorobenzene. [462-06-6] Monofluorobenzene; phenyl fluoride. C_6H_5F; mol wt 96.10. C 74.99%, H 5.24%, F 19.77%. Obtained by warming benzenediazonium chloride with concd HF. Solubility value: P. M. Gross *et al.*, *J. Am. Chem. Soc.* **55**, 650 (1933). Pressure, vol, temp constants: Douslin *et al.*, *ibid.* **80**, 2031 (1958).

F—⟨benzene ring⟩

Liquid; benzene odor. d_4^{20} 1.024. bp_{760} 84.73°; $bp_{13\ atm}$ 200°; $bp_{38\ atm}$ 275°. mp −40°. n_D^{20} 1.4677. *Flammable.* Miscible with alcohol, ether. Sol (30°C): 1.54 g/1000 g water.

4202. N-Fluorobenzenesulfonimide. [133745-75-2] *N*-Fluoro-*N*-(phenylsulfonyl)benzenesulfonamide; *N*-fluorobis(phenylsulfonyl)amine; FBS; NFBS; NFSI; $(PhSO_2)_2NF$. $C_{12}H_{10}FNO_4S_2$; mol wt 315.33. C 45.71%, H 3.20%, F 6.02%, N 4.44%, O 20.29%, S 20.33%. Reagent used for electrophilic fluorinations: E. Differding, H. Ofner, *Synlett* **1991**, 187. Use in phosphonate fluorinations: E. Differding *et al.*, *ibid.* 395; S. D. Taylor *et al.*, *Tetrahedron* **54**, 1691 (1998). In regiospecific aromatic fluorinations: V. Snieckus *et al.*, *Tetrahedron Lett.* **35**, 3465 (1994). Synthetic utility as electrophilic aminating reagent: C. F. Rosewall *et al.*, *J. Am. Chem. Soc.* **131**, 9488 (2009). Reactivity studies with nucleophilic reagents: J. M. Antelo *et al.*, *J. Chem. Soc. Perkin Trans. 2* **2000**, 2071.

⟨structure of N-fluorobenzenesulfonimide⟩

White solid, mp 114-116°. Thermally stable up to 180°C. Sol in most organic solvents.

USE: Reagent in synthetic organic chemistry.

4203. 4-Fluorobenzoic Acid. [456-22-4] $C_7H_5FO_2$; mol wt 140.11. C 60.01%, H 3.60%, F 13.56%, O 22.84%. Prepd by treating *p*-carbethoxybenzenediazonium chloride with fluoboric acid, followed by thermal decompn of the resulting *p*-carbethoxybenzenediazonium fluoborate and by hydrolysis of the ensuing ethyl ester of *p*-fluorobenzoic acid: Schiemann, Winkelmüller, *Org. Synth.* **13**, 52 (1933).

Monoclinic prisms from water. Peculiar sweet taste. mp 182.6°. Sparingly sol in cold water; freely sol in hot water. Soly in water at 32°: 1.1 g/l. Water solns evaporate without leaving a residue. Sol in alc, ether. pK (25°): 3.85.

Ethyl ester. $C_9H_9FO_2$. Crystals, mp 26°. bp 210°.

4204.   1-Fluoro-2,4-dinitrobenzene. [70-34-8] 2,4-Dinitro-1-fluorobenzene; 2,4-dinitrophenyl fluoride; DNFB; FDNB; Sanger's reagent. $C_6H_3FN_2O_4$; mol wt 186.10. C 38.72%, H 1.62%, F 10.21%, N 15.05%, O 34.39%. Prepn: A. F. Holleman, J. W. Beekman, *Rec. Trav. Chim.* **23**, 225 (1904); Cook, Saunders, *Biochem. J.* **41**, 558 (1947). Use in peptide analysis: Sanger, *Biochem. J.* **39**, 507 (1945); **40**, 261 (1946); **45**, 563 (1949); Porter, Sanger, *ibid.* **42**, 287 (1948). Tumor promoting activity: F. G. Bock *et al.*, *Cancer Res.* **29**, 179 (1969). Mutagenicity study: D. R. Jagannath *et al.*, *Mutat. Res.* **78**, 91 (1980). Review of uses: *Reagents for Organic Synthesis*, L. F. Fieser, M. Fieser, Eds. (Wiley, New York, 1967) pp 321-322.

Pale yellow crystals from ether, mp 26°. $bp_{2.0}$ 137°. Sol in benzene, ether, propylene glycol.

Caution: Vesicant. For proper handling see: J. S. Thompson, O. P. Edmunds, *Ann. Occup. Hyg.* **23**, 27 (1980).

USE: Reagent for labeling a terminal amino acid group; in modified Wohl degradations of aldoses. As hapten.

4205.   Fluoroform. [75-46-7] Trifluoromethane. CHF_3; mol wt 70.01. C 17.16%, H 1.44%, F 81.41%. Prepd from $CHCl_3$ and HF: Meslans, *Compt. Rend.* **110**, 717 (1890); Valentiner, **US 643835** (1900); Ruff, *Ber.* **69**, 299 (1936); Whallay, *J. Soc. Chem. Ind. London* **66**, 429 (1947); Kwasnik in *Handbook of Preparative Inorganic Chemistry* vol. **1**, G. Brauer, Ed. (Academic Press, New York, 2nd ed., 1963) pp 204-205. Early industrial prepn: Pearlson in *Fluorine Chemistry* vol. **I**, J. H. Simons, Ed. (Academic Press, New York, 1950) p 467.

Colorless, odorless gas. Stable up to 1150°. Chemically very inert. d (solid) 1.935. d (liq; −100°) 1.52. mp −160°. bp −84.4°. Critical temp 33°, critical pressure 47 atm, critical density 0.516. May be stored over water. Practically non-toxic: N. V. Sidgwick, *The Chemical Elements and Their Compounds* vol. **II** (Oxford, 1950) p 1130.

Caution: May be slightly irritating to respiratory tract, and, in high concns, narcotic.

USE: Refrigerant for low temps.

4206.   Fluoromethane. [593-53-3] Methyl fluoride. CH_3F; mol wt 34.03. C 35.30%, H 8.89%, F 55.83%. Prepn in 82% yield by heating fluorosulfonic acid methyl ester with KF: Zappel, Jonas, **DE 1131197** (1962 to Bayer).

Gas. Agreeable ether-like odor. Burns with evolution of HF, the flame being about as colorless as that of alc: Dumas, Peligot, *Ann.* **15**, 59 (1835). d (liq; −78°) 0.8774. d (gas) 1.1951 (air = 1); d (gas) 1.0813 (oxygen = 1). Dipole moment 1.81. Molecular volume: 22.03. Van der Waals constants: 0.00923; 0.002350. Critical temp 44.9°; crit press. 62.0 atm. Dielectric constant (for wavelengths below 10^4 cm) = 1.00948. mp −141.8°. bp_{872} −75.7°; bp_{760} −78.2°. bp_{143} −103.7°. One hundred vols of water dissolve 166 vols of the gas at 15°. Freely sol in alcohol, ether.

Caution: Narcotic in high concns.

4207.   Fluorometholone. [426-13-1] (6α,11β)-9-Fluoro-11,17-dihydroxy-6-methylpregna-1,4-diene-3,20-dione; 21-desoxy-

9α-fluoro-6α-methylprednisolone; 21-desoxy-6α-methyl-9α-fluoro-prednisolone; fluormetholon; Delmeson; Efflumidex; Fluaton; Flumetholon; Fluor-Op; FML. $C_{22}H_{29}FO_4$; mol wt 376.47. C 70.19%, H 7.76%, F 5.05%, O 17.00%. Prepn: F. H. Lincoln *et al.*, **US 2867637** (1959 to Upjohn). Prepn of acetate and other esters: B. J. Magerlein *et al.*, **US 3038914** (1962 to Upjohn). Pharmacology of acetate: A. Kupferman *et al.*, *Arch. Ophthalmol.* **100**, 640 (1982). HPLC determn: H. Tokunaga *et al.*, *Chem. Pharm. Bull.* **32**, 4012 (1984). Clinical evaluation in ocular inflammation: H. M. Leibowitz *et al.*, *Ann. Ophthalmol.* **16**, 1110 (1984).

White to yellowish-white, odorless crystals from acetone, mp 292-303°. Slightly soluble in alc; very slightly sol in chloroform, ether. Practically insol in water.

17-Acetate. [3801-06-7] Eflone; Flarex. $C_{24}H_{31}FO_5$. Crystals from ethyl acetate + hexanes, mp 230-232°. $[\alpha]_D$ +28° (chloroform).

THERAP CAT: Glucocorticoid; anti-inflammatory.

4208.   5-Fluoroorotic Acid. [703-95-7] 5-Fluoro-1,2,3,6-tetrahydro-2,6-dioxo-4-pyrimidinecarboxylic acid; 5-fluoroorotate; 5-fluoro-6-carboxyuracil; 5-FOA; ENT-26398; NSC-31712; Ro-2-9945; WR-152520. $C_5H_3FN_2O_4$; mol wt 174.09. C 34.50%, H 1.74%, F 10.91%, N 16.09%, O 36.76%. Pyrimidine precursor; selectively toxic to yeast cells that synthesize the enzyme orotidine-5′-P decarboxylase. Prepn: R. Duschinsky *et al.*, *J. Am. Chem. Soc.* **79**, 4559 (1957); **GB 806584** (1958 to Hoffmann-La Roche); R. Duschinsky, C. Heidelberger, **US 2948725** (1960). Improved synthesis: D. H. R. Barton *et al.*, *J. Chem. Soc. Perkin Trans. 1* **1974**, 2095; and mass spectrum: S. N. Alam *et al.*, *Acta Pharm. Suec.* **12**, 375 (1975). Antitumor activity: C. Heidelberger *et al.*, *Nature* **179**, 663 (1957). Inhibition of RNA synthesis: C. T. Garrett *et al.*, *Arch. Biochem. Biophys.* **155**, 342 (1973). Clinical evaluation of antimycotic activity: K. Nikolova *et al.*, *Methods Find. Exp. Clin. Pharmacol.* **9**, 85 (1987). Review of *in vitro* use in the positive selection of genetically transformed yeast cells: J. D. Boeke *et al.*, *Methods Enzymol.* **154**, 164-175 (1987).

Monohydrate. Crystals, mp 255° (dec) (Duschinsky, 1957); also reported as 258-259° (Barton). Partially sol in water. uv max (0.1*N* HCl): 284-285 nm (ε 7100). LD_{50} i.p. in male mice: 300 mg/kg (Nikolova).

USE: Research tool in molecular genetics.

4209.   *p*-Fluorophenylacetic Acid. [405-50-5] 4-Fluorobenzeneacetic acid. $C_8H_7FO_2$; mol wt 154.14. C 62.34%, H 4.58%, F 12.33%, O 20.76%. Prepd by treating *p*-fluorobenzyl cyanide with sulfuric acid: Olah *et al.*, *J. Org. Chem.* **22**, 879 (1957).

Crystals from chloroform, mp 86°. bp_2 164°.

β-Dimethylaminoethyl ester HCl. Crystals, mp 103°.

USE: Intermediate in the manuf of fluorinated anesthetics.

4210. Fluorosalan. [4776-06-1] 3,5-Dibromo-2-hydroxy-*N*-[3-(trifluoromethyl)phenyl]benzamide; 3,5-dibromo-α,α,α-trifluoro-*m*-salicylotoluidide; 3,5-dibromo-3′-trifluoromethylsalicylanilide; flusalan. $C_{14}H_8Br_2F_3NO_2$; mol wt 439.03. C 38.30%, H 1.84%, Br 36.40%, F 12.98%, N 3.19%, O 7.29%. Prepn: Stecker, US 3041236 (1962).

USE: Biocide in cosmetics.
THERAP CAT: Disinfectant.

4211. Fluorosulfonic Acid. [7789-21-1] Fluorosulfuric acid; fluosulfonic acid. FHO_3S; mol wt 100.06. F 18.99%, H 1.01%, O 47.97%, S 32.04%. HSO_3F. Prepn: Thorpe, Kirman, *J. Chem. Soc.* **61**, 921 (1892); Meyer, Schramm, *Z. Anorg. Allg. Chem.* **206**, 25 (1932); Kwasnik in *Handbook of Preparative Inorganic Chemistry* vol. 1, G. Brauer, Ed. (Academic Press, New York, 2nd ed., 1963) pp 177-178. Reviews of properties and chemistry: Gillespie, *Acc. Chem. Res.* **1**, 202-209 (1968); Thompson, "Fluorosulfuric Acid" in *Inorganic Sulfur Chemistry*, G. Nickless, Ed. (Elsevier, New York, 1968) pp 587-606; Jache in *Adv. Inorg. Chem. Radiochem.* **16**, 177-200 (1974).

Colorless liquid; fumes in moist air. d_4^{18} 1.740; d_4^{25} 1.726. mp −89°. bp_{760} 163°; bp_{120} 110.0°; bp_{19} 77.0°. Stable to 900°. Considerably more acidic than 100% H_2SO_4. *Corrosive.* Does not attack glass when anhydr and pure. Violent reaction with water although it is incompletely and reversibly hydrolyzed. Reddish-brown color with acetone. Forms stable salts which are little hydrolyzed by water and which may be recrystallized from water.
Methyl ester *see* Methyl Fluorosulfonate.
Caution: May be highly irritating to skin, mucous membranes.
USE: Fluorinating agent. Catalyst in alkylation, acylation, polymerization and condensation reactions; in hydrofluorination of olefins; in production of substituted pyridines. In production of petroleum products. *See* Thompson, *loc. cit.* **Magic Acid**, a 1:1 HSO_3F-SbF_5 soln, is used in the study of stable solns of alkyl- and arylcarbonium ions: Olah, *Science* **168**, 1298 (1970).

4212. Fluorotoluene. Fluoromethylbenzene. C_7H_7F; mol wt 110.13. C 76.34%, H 6.41%, F 17.25%. Prepn of *o*-, *m*-, *p*-isomers: A. F. Holleman, J. W. Beekman, *Rec. Trav. Chim.* **23**, 225 (1904); A. F. Holleman, *ibid.* **25**, 330 (1906); E. D. Bergmann, S. Berkovic, *J. Org. Chem.* **26**, 919 (1961).

Colorless liquids. *Flammable.* Insol in water. Miscible with alc, ether.
o-Fluorotoluene. [95-52-3] 1-Fluoro-2-methylbenzene. d^{13} 1.004. mp ∼ −80°. bp 114°. n_D^{20} 1.4704.
m-Fluorotoluene. [352-70-5] d^{13} 0.997. mp −111°. bp 116°. n_D^{20} 1.4691.
p-Fluorotoluene. [352-32-9] Odor of oil of bitter almonds. d^{16} 1.001. bp 116°. n_D^{20} 1.4702.

4213. Fluorouracil. [51-21-8] 5-Fluoro-2,4(1*H*,3*H*)-pyrimidinedione; 2,4-dioxo-5-fluoropyrimidine; 5-FU; Ro-2-9757; NSC-19893; Adrucil; Arumel; Carac; Efudex; Efudix; Fluoroplex; Fluroblastin. $C_4H_3FN_2O_2$; mol wt 130.08. C 36.93%, H 2.32%, F 14.61%, N 21.54%, O 24.60%. Pyrimidine analog; nucleotide metabolites inhibit DNA synthesis and incorporate into RNA. Prepn: R. Duschinsky *et al.*, *J. Am. Chem. Soc.* **79**, 4559 (1957); C. Heidelberger, R. Duschinsky, US 2802005 (1957); D. H. R. Barton *et al.*, *J. Org. Chem.* **37**, 329 (1972); O. Miyashita *et al.*, *Chem. Pharm.*

Bull. **29**, 3181 (1981). Site of action: F. Maley, G. F. Maley, *FEBS Symp.* **57**, 21 (1979). Clinical trial in actinic keratoses: C. S. Jury *et al.*, *Br. J. Dermatol.* **153**, 808 (2005); as an adjuvant in rectal cancer: S. R. Smalley *et al.*, *J. Clin. Oncol.* **24**, 3542 (2006). Comprehensive description: B. C. Rudy, B. Z. Senkowski, *Anal. Profiles Drug Subs.* **2**, 221-244 (1973); S. M. Bayomi, A. A. Al-Badr, *ibid.* **18**, 599-639 (1989). Review of pharmacokinetics: W. Sadee, C. G. Wong, *Clin. Pharmacokinet.* **2**, 437-450 (1977); of clinical trials: R. M. Hansen, *Cancer Invest.* **9**, 637-642 (1991); in colon cancer: W. Kelder *et al.*, *Expert Rev. Anticancer Ther.* **6**, 785-794 (2006).

Crystals from water or methanol-ether, mp 282-283° (dec). Sublimes (0.1 mm) 190-200°. uv max (0.1*N* HCl): 265-266 nm (ε 7070). Sparingly sol in water; slightly sol in alc. Practically insol in chloroform, ether.
THERAP CAT: Antineoplastic.
THERAP CAT (VET): Antineoplastic.

4214. Fluosilicic Acid. [16961-83-4] Hexafluorosilicate(2−) hydrogen (1:2); hexafluosilicic acid; hydrogen hexafluorosilicate; hydrosilicofluoric acid; hydrofluosilicic acid; silicofluoric acid; fluorosilicic acid. F_6H_2Si; mol wt 144.09. F 79.11%, H 1.40%, Si 19.49%. H_2SiF_6. Prepd from HF + SiO_2; also prepd by the action of water on SiF_4; by the action of H_2SO_4 on $BaSiF_6$: Hempel, *Ber.* **18**, 1434 (1885); Baur, Glaessner, *Ber.* **36**, 4215 (1903); Söll, *FIAT-Review* **23**, 257 (1946); Kwasnik in *Handbook of Preparative Inorganic Chemistry* vol. 1, G. Brauer, Ed. (Academic Press, New York, 2nd ed., 1963) p 214-215; Lange in *Fluorine Chemistry* vol. I, J. H. Simon, Ed. (New York, 1950) p 129. Commercial grades of fluosilicic acid soln are obtained as a by-product in the superphosphate industry. *Review:* Colton, *J. Chem. Educ.* **35**, 562-563 (1958).

Liquid, when anhydr dissociates almost instantly into SiF_4 and HF. Marketed as aq soln only. A 60-70% soln solidifies around 19°, forming a cryst dihydrate. May be distilled without decompn only as a 13.3% aq soln. Fairly strong acid. Sour, pungent odor. $d_{17.5}^{17.5}$ 5% soln: 1.0407; 10%: 1.0834; 15%: 1.1281; 20%: 1.1748; 25%: 1.2235; 30%: 1.2742; 34%: 1.3162. The more concd solns (but not the anhydr liq) can be stored in glass, although some etching will take place around the surface. Usually stored in iron containers.
Caution: Severe corrosive effect on skin, mucous membranes.
USE: A 1-2% soln is used widely for sterilizing equipment in brewing and bottling establishments. Other concns are used in the electrolytic refining of lead, in electroplating, for hardening cement, crumbling lime or brick work, for the removal of lime from hides during the tanning process, to remove molds, as preservative for timber.

4215. Fluosol DA. [75216-20-5] Biologically inert fluorocarbon emulsion of perfluorodecalin and perfluorotripropylamine that performs the oxygen-carrying function of red blood cells. Has a particle size of 0.05 to 0.25μ and is stable at room temperature for years. Can be administered regardless of blood type; does not carry hepatitis or other infectious diseases. Prepn: Y. Yokoyama *et al.*, DE 2630586; *eidem*, US 4252827 (1977, 1981 to Green Cross). Hemodynamic and oxygen transport effects: K. K. Tremper *et al.*, *Crit. Care Med.* **8**, 738 (1980). Clinical studies: H. Ohyanagi *et al.*, *Clin. Ther.* **2**, 306 (1979); T. Suyama *et al.*, *Prog. Clin. Biol. Res.* **55**, 609 (1981); T. Mitsuno *et al.*, *Ann. Surg.* **195**, 60 (1982). Review: *Int. Congr. Ser.* **486**, 1-471 (1979). Review of use as red cell substitute: K. C. Lowe, *Comp. Biochem. Physiol.* **87A**, 825-838 (1987).
20% Emulsion, **FDA-20**. Viscosity at 37°: 2.3 cP. pH 7.4.
Perfluorodecalin. [306-94-5] Octadecafluorodecahydronaphthalene; perflunafene; FDC. $C_{10}F_{18}$; mol wt 462.08. Prepn: E. T. McBee, L. D. Bechtol, *Ind. Eng. Chem.* **39**, 380 (1947). mp −10 to −7°. bp 140°. d_4^{20} 1.9456. n_D^{20} 1.3118. Vapor pressure at 37°: 12.7 mm Hg.
Perfluorotripropylamine. [338-83-0] 1,1,2,2,3,3,3-Heptafluoro-*N*,*N*-bis(heptafluoropropyl)-1-propanamine; heneicosafluorotri-

propylamine; perfluamine; FTPA. $C_9F_{21}N$; mol wt 521.07. Prepn: R. N. Haszeldine, *J. Chem. Soc.* **1951**, 102. bp 129.5-130.5°. d_4^{25} 1.822. n_D^{25} 1.279. Vapor press. at 37°: 20.0 mm Hg.

THERAP CAT: Blood substitute.

4216. Fluoxastrobin. [361377-29-9] (1*E*)-[2-[[6-(2-Chlorophenoxy)-5-fluoro-4-pyrimidinyl]oxy]phenyl](5,6-dihydro-1,4,2-dioxazin-3-yl)methanone *O*-methyloxime; HEC-5725; Disarm; Vigold. $C_{21}H_{16}ClFN_4O_5$; mol wt 458.83. C 54.97%, H 3.52%, Cl 7.73%, F 4.14%, N 12.21%, O 17.43%. Leaf-systemic broad-spectrum fungicide for use in cereal and food crops; member of methoxyiminodihydro-dioxazines. Prepn (stereochem. unspecified): U. Heinemann *et al.*, DE 19602095; *eidem*, US 6103717 (1997, 2000 both to Bayer). Comprehensive description: S. Dutzmann *et al.*, *BCPC Conf. - Pests Dis.* **2002**, 365. Field trial in winter wheat seeds: I. Haeuser-Hahn *et al.*, *BCPC Int. Congr. - Crop Sci. Technol.* **2003**, 801. Series of articles on chemistry, biology, determn, and environmental fate: *Pflanzenschutz-Nachr. Bayer (Engl. Ed.)* **57**, 299-449 (2004). Ecotoxicology: P. Breuer, *ibid.* 319.

White crystals with slight characteristic odor, mp 103-108°. bp 497° (est.). d^{20} 1.422. Log P (octanol/water): 2.86 (20°). Vapor pressure at 20° (extrapolated): 6×10^{-10} Pa. Soly at 20° (g/l): *n*-heptane 0.04; 2-propanol 6.7; xylene 38.1; dichloromethane >250; in water (mg/l): 2.56 (unbuffered); 2.43 (pH 4); 2.29 (pH 7); 2.27 (pH 9). LD_{50} in rats, bobwhite quail (mg/kg): >2500, >2000 orally; LC_{50} (96 hr) rainbow trout, bluegill sunfish, carp (mg/l): 0.44, 0.97, 0.57 (Breuer).

USE: Agricultural fungicide.

4217. Fluoxetine. [54910-89-3] *N*-Methyl-γ-[4-(trifluoromethyl)phenoxy]benzenepropanamine; (±)-*N*-methyl-3-phenyl-3-[(α,α,α-trifluoro-*p*-tolyl)oxy]propylamine; *dl*-*N*-methyl-3-(*p*-trifluoromethylphenoxy)-3-phenylpropylamine. $C_{17}H_{18}F_3NO$; mol wt 309.33. C 66.01%, H 5.87%, F 18.43%, N 4.53%, O 5.17%. Selective serotonin reuptake inhibitor (SSRI). Prepn: B. B. Malloy, K. K. Schmiegel, DE 2500110; *eidem*, US 4314081 (1975, 1982 both to Lilly). Pharmacology: D. T. Wong *et al.*, *Life Sci.* **15**, 471 (1974). GC determn in plasma: J. F. Nash *et al.*, *Clin. Chem.* **28**, 2100 (1982). Series of articles on clinical pharmacology and therapeutic efficacy in depression: *J. Clin. Psychiatry* Suppl. 3 (2), **46**, 2-67 (1985). Toxicity data: P. Stark *et al.*, *ibid.* 7. Clinical trial in bulimia nervosa: D. J. Goldstein *et al.*, *Br. J. Psychiatry* **166**, 660 (1995); in obsessive-compulsive disorder: J. J. Lopez-Ibor, Jr. *et al.*, *Eur. Neuropsychopharmacol.* **6**, 111 (1996); in post-traumatic stress disorder: F. Martenyi, V. Soldatenkova, *ibid.* **16**, 340 (2006). Veterinary trial in canine aggression: N. H. Dodman *et al.*, *J. Am. Vet. Med. Assoc.* **209**, 1585 (1996). Evaluation of suicidality risk: R. H. Perlis *et al.*, *Psychother. Psychosom.* **76**, 40 (2007). Comprehensive description: D. S. Risley, R. J. Bopp, *Anal. Profiles Drug Subs.* **19**, 193-219 (1990). Review of clinical experience: P. E. Stokes, A. Holtz, *Clin. Ther.* **19**, 1135-1250 (1997); in premenstrual dysphoric disorder: T. Pearlstein, K. A. Yonkers, *Expert Opin. Pharmacother.* **3**, 979-991 (2002); in anorexia nervosa: S. S. Kim, *Ann. Pharmacother.* **37**, 890-892 (2003); of safety profile: J. F. Wernicke, *Expert Opin. Drug Saf.* **3**, 495-504 (2004); of discovery and clinical development: D. T. Wong *et al.*, *Nat. Rev. Drug Discovery* **4**, 764-774 (2005).

Hydrochloride. [59333-67-4] LY-110140; Adofen; Fluctin; Fluneurin; Fluoxeren; Fluox-Puren; Flusol; Fluxet; Fontex; Foxetin; Lovan; Prozac; Reconcile; Sarafem. $C_{17}H_{18}F_3NO\cdot HCl$; mol wt 345.79. White to off-white crystalline powder. mp 158.4-158.9°. Soly (mg/ml): methanol, ethanol >100; acetone, acetonitrile, chloroform 33-100; dichloromethane 5-10; ethyl acetate 2-2.5; toluene, cyclohexane, hexane 0.5-0.67. Maximum soly in water: 14 mg/ml. Practically insol in ether. uv max (methanol) 227, 264, 268, 275 nm ($E_{1cm}^{1\%}$ 372.0, 29.2, 29.3, 21.5). LD_{50} in mice, rats (mg/kg): 248, 452 orally (Stark).

Oxalate. $C_{19}H_{18}F_3NO_5$. Crystals from ethyl acetate-methanol, mp 179-182° (dec).

THERAP CAT: Antidepressant; antiobessional; antibulimic.

THERAP CAT (VET): In treatment of behavioral disorders in dogs and cats.

4218. Fluoxymesterone. [76-43-7] (11β,17β)-9-Fluoro-11,17-dihydroxy-17-methylandrost-4-en-3-one; 11β,17β-dihydroxy-9α-fluoro-17α-methyl-4-androsten-3-one; 9α-fluoro-11β-hydroxy-17α-methyltestosterone; Halotestin. $C_{20}H_{29}FO_3$; mol wt 336.45. C 71.40%, H 8.69%, F 5.65%, O 14.27%. Prepn: Herr *et al.*, *J. Am. Chem. Soc.* **78**, 501 (1956). *cf.* Herr, US 2813881 (1957 to Upjohn). Comprehensive description: J. Kirschbaum, *Anal. Profiles Drug Subs.* **7**, 251-275 (1978).

White crystals, dec 270°. $[\alpha]_D$ +109° (ethanol). uv max (ethanol): 240 nm (ε 16700). Sol in pyridine; sparingly sol in methanol, alc; slightly sol in acetone, chloroform. Practically insol in water, ether, benzene, hexanes.

Note: This is a controlled substance (anabolic steroid): **21 CFR**, 1308.13, as defined in 1300.01.

THERAP CAT: Androgen.

4219. Flupentixol. [2709-56-0] 4-[3-[2-(Trifluoromethyl)-9*H*-thioxanthen-9-ylidene]propyl]-1-piperazineethanol; 2-trifluoromethyl-9-[3-[4-(β-hydroxyethyl)-1-piperazinyl]thioxanthene; flupenthixol; N-7009; LC-44. $C_{23}H_{25}F_3N_2OS$; mol wt 434.52. C 63.58%, H 5.80%, F 13.12%, N 6.45%, O 3.68%, S 7.38%. Neuroleptic agent related structurally to thiothixene, *q.v.* Prepn: P. N. Craig, C. L. Zirkle, US 3282930 (1966 to SKF). Prepn of *cis*- and *trans*-forms: GB 925538 (1963 to SKF); P. V. Petersen, T. Ammitzboll, US 3681346 (1972 to Kefalas). Metabolism of decanoate: A. Jorgensen *et al.*, *Acta Pharmacol. Toxicol.* **29**, 339 (1971). α-Flupentixol, the *cis*-isomer, shows greater pharmacological activity than β-flupentixol, the *trans*-isomer: I. Moller Nielsen *et al.*, *Acta Pharmacol. Toxicol.* **33**, 353 (1973). X-ray crystallography of isomers: M. L. Post *et al.*, *Nature* **256**, 342 (1975).

cis-form

Dihydrochloride. [2413-38-9] Emergil; Fluanxol; Siparlol; Metamin. $C_{23}H_{25}F_3N_2OS\cdot2HCl$; mol wt 507.44.

Decanoate. [30909-51-4] Lu-5-110; Depixol; Fluanxol Dépôt; Viscoleo.

THERAP CAT: Antipsychotic.

4220. Fluphenazine. [69-23-8] 4-[3-[2-(Trifluoromethyl)-10*H*-phenothiazin-10-yl]propyl]-1-piperazineethanol; 1-(2-hydroxy-ethyl)-4-[3-(trifluoromethyl)-10-phenothiazinyl)propyl]piperazine; 10-[3'-[4''-(β-hydroxyethyl)-1''-piperazinyl]propyl]-3-trifluoromethylphenothiazine; 2-(trifluoromethyl)-10-[3-[1-(β-hydroxyethyl)-4-piperazinyl]propyl]phenothiazine; S-94; SQ-4918. C$_{22}$H$_{26}$F$_3$N$_3$OS; mol wt 437.53. C 60.39%, H 5.99%, F 13.03%, N 9.60%, O 3.66%, S 7.33%. Prepn: H. L. Yale, F. Sowinski, *J. Am. Chem. Soc.* **82**, 2039 (1960); **GB 829246**; G. E. Ullyot, **US 3058979** (1960, 1962 both to SKF); **GB 833474** (1960 to Scherico), *C.A.* **54**, 21143e (1960); E. L. Anderson *et al.*, *Arzneim.-Forsch.* **12**, 937 (1962); H. L. Yale, R. C. Merrill, **US 3194733** (1965 to Olin Mathieson). Metabolism: J. Dreyfuss, A. J. Cohen, *J. Pharm. Sci.* **60**, 826 (1971). Comprehensive description of the enanthate ester: K. Florey, *Anal. Profiles Drug Subs.* **2**, 245-262 (1973); of the dihydrochloride: *idem, ibid.* 263-294; of the decanoate ester: G. Clarke, *ibid.* **9**, 275-294 (1980).

Dark brown viscous oil, bp$_{0.5}$ 268-274°; bp$_{0.3}$ 250-252°.
Dihydrochloride. [146-56-5] Anatensol; Dapotum; Lyogen; Moditen; Omca; Pacinol; Permitil; Prolixin; Siqualone; Tensofin; Valamina. C$_{22}$H$_{26}$F$_3$N$_3$OS.2HCl; mol wt 510.44. Crystals from abs ethanol, mp 235-237°. Also reported as mp 224.5-226°. Freely sol in water; slightly sol in acetone, alc, chloroform. Practically insol in benzene, ether.
Decanoate. [5002-47-1] SQ-10733; QD-10733; Modecate. C$_{32}$H$_{44}$F$_3$N$_3$O$_2$S; mol wt 591.78. Pale yellow-orange, viscous liquid. Slowly crystallizes at room temp. mp 30-32°. Very sol in chloroform, ether, cyclohexane, methanol, ethanol. Insol in water.
Enanthate. [2746-81-8] SQ-16144. C$_{29}$H$_{38}$F$_3$N$_3$O$_2$S; mol wt 549.70. Pale yellow to yellow-orange viscous liquid or oily solid. Freely sol in alc, chloroform, ether. Insol in water.
THERAP CAT: Antipsychotic.

4221. Flupirtine. [56995-20-1] *N*-[2-Amino-6-[[(4-fluoro-phenyl)methyl]amino]-3-pyridinyl]carbamic acid ethyl ester; 2-amino-6-[(*p*-fluorobenzyl)amino]-3-pyridinecarbamic acid ethyl ester; D-9998. C$_{15}$H$_{17}$FN$_4$O$_2$; mol wt 304.33. C 59.20%, H 5.63%, F 6.24%, N 18.41%, O 10.51%. Substituted pyridine with central analgesic properties. Prepn: K. Thiele, W. von Bebenburg, **ZA 6902364** (1970 to Degussa); W. von Bebenburg *et al.*, *Chem. Ztg.* **103**, 387 (1979); *eidem, ibid.* **105**, 217 (1981). Prepn of maleate: W. von Bebenburg, S. Pauluhn, **BE 890331**; *eidem*, **US 4481205** (1980, 1984 both to Degussa). Comparison of pharmacology with other analgesics: V. Jakovlev *et al.*, *Arzneim.-Forsch.* **35**, 30 (1985). Pharmacokinetic studies: K. Obermeier *et al.*, *ibid.* 60. Effect on driving ability: B. Biehl, *ibid.* 77. Clinical trials in treatment of cancer pain: W. Scheef, D. Wolf-Gruber, *ibid.* 75. Efficacy in treatment of pain after hysterectomy: R. A. Moore *et al.*, *Br. J. Anaesth.* **55**, 429 (1983). Symposium on pharmacology and clinical efficacy: *Postgrad. Med. J.* **63**, Suppl. 3, 1-113 (1987).

Crystals from isopropanol, mp 115-116°. 5% ethanol soln is colorless, turns green on exposure to air for 20 hours. LD$_{50}$ orally in mice, rats: 617, 1660 mg/kg (Jakovlev).

Hydrochloride. C$_{15}$H$_{17}$FN$_4$O$_2$.HCl. Crystals from water, mp 214-215°. When prepd industrially contains intensely blue by-product.
Maleate. [75507-68-5] Katadolon. C$_{15}$H$_{17}$FN$_4$O$_2$.C$_4$H$_4$O$_4$; mol wt 420.40. Colorless crystals from isopropanol, mp 175.5-176°. Formed as mixture of two crystalline forms A and B; mixtures containing 60-90% A are preferred.
THERAP CAT: Analgesic.

4222. Fluprednidene Acetate. [1255-35-2] (11β)-21-(Acetyloxy)-9-fluoro-11,17-dihydroxy-16-methylenepregna-1,4-diene-3,20-dione; 9α-fluoro-11β,17,21-trihydroxy-16-methylenepregna-1,4-diene-3,20-dione 21-acetate; 9α-fluoro-16-methylene-Δ^{1,4}-pregnadiene-11β,17,21-triol-3,20-dione 21-acetate; 16-methylene-9α-fluoroprednisolone 21-acetate; 9α-fluoro-16-methyleneprednisolone 21-acetate; fluprednylidene 21-acetate; StL-1106; Decoderm. C$_{24}$H$_{29}$FO$_6$; mol wt 432.49. C 66.65%, H 6.76%, F 4.39%, O 22.20%. Prepn: Taub *et al., J. Org. Chem.* **25**, 2258 (1960); **29**, 3486 (1964); v. Werder *et al., Arzneim.-Forsch.* **18**, 7 (1968); Wendler, Taub; Taub, Wendler; Taub *et al.;* Wendler *et al.*, **US 3065239; US 3068224; US 3068226** and **US 3136760** (1960 and 1964 to Merck & Co.).

Crystals, mp 231-234°. [α]$_D$ +43° (CHCl$_3$); [α]$_D$ +32° (dioxane) (v. Werder). uv max (methanol): 238 nm (ε 15700).
THERAP CAT: Anti-inflammatory (topical).

4223. Fluprednisolone. [53-34-9] (6α,11β)-6-Fluoro-11,-17,21-trihydroxypregna-1,4-diene-3,20-dione; 6α-fluoroprednisolone; 6α-fluoro-1,4-pregnadiene-11β,17α,21-triol-3,20-dione; 6α-fluoro-1-dehydrohydrocortisone; U-7800; NSC-47439; Etadrol. C$_{21}$H$_{27}$FO$_5$; mol wt 378.44. C 66.65%, H 7.19%, F 5.02%, O 21.14%. Prepn: Hogg, Spero, **US 2841600** (1958 to Upjohn); Batres *et al.*, **DE 1079042** (1960 to Syntex); Lettré, Hotz, **DE 1088953** (1960 to Bayer).

Crystals, mp 208-213°. [α]$_D$ +92°.
21-Acetate. C$_{23}$H$_{29}$FO$_6$. Crystals, mp 235-238°.
THERAP CAT: Glucocorticoid; anti-inflammatory.
THERAP CAT (VET): Anti-inflammatory.

4224. Fluprostenol. [40666-16-8] *rel*-(5Z)-7-[(1R,2R,3R,-5S)-3,5-Dihydroxy-2-[(1E,3R)-3-hydroxy-4-[3-(trifluoromethyl)-phenoxy]-1-buten-1-yl]cyclopentyl]-5-heptenoic acid; ICI-81008; Equimate. C$_{23}$H$_{29}$F$_3$O$_6$; mol wt 458.47. C 60.26%, H 6.38%, F 12.43%, O 20.94%. Potent luteolytic agent related to prostaglandin F$_{2α}$, *q.v.*: Dukes *et al.*, *Nature* **250**, 330 (1974). Prepn: Binder *et al.*, *Prostaglandins* **6**, 87 (1974). Induction of estrus and luteolysis in mares: Cooper, Farr, *Vet. Rec.* **94**, 161 (1974); Berwyn-Jones, Irvine, *N. Z. Vet. J.* **22**, 107 (1974).

Relative stereochemistry

Sodium salt. [55028-71-2] ICI-80008. $C_{23}H_{28}F_3NaO_6$; mol wt 480.46.

THERAP CAT (VET): Treatment of infertility in mares.

4225. Flupyrsulfuron-methyl. [144740-53-4] 2-[[[[(4,6-Dimethoxy-2-pyrimidinyl)amino]carbonyl]amino]sulfonyl]-6-(trifluoromethyl)-3-pyridinecarboxylic acid methyl ester; methyl 2-[(4,6-dimethoxypyrimidin-2-ylcarbamoyl)sulfamoyl]-6-(trifluoromethyl)-nicotinate. $C_{15}H_{14}F_3N_5O_7S$; mol wt 465.36. C 38.72%, H 3.03%, F 12.25%, N 15.05%, O 24.07%, S 6.89%. Sulfonylurea postemergent herbicide which inhibits acetolactate synthetase. Prepn: T. A. Andrea, P. H. T. Liang, **EP 502740**; *eidem,* **US 5393734** (1992, 1995 both to Du Pont). Description of chemical and biological activities: S. R. Teaney *et al., Brighton Crop Prot. Conf. - Weeds* **1995**, 49. Metabolism and selectivity: M. K. Koeppe *et al., Pestic. Biochem. Physiol.* **59**, 105 (1998). LC determn in soil and water: M. E. Powley, P. A. de Bernard, *J. Agric. Food Chem.* **46**, 514 (1998). Environmental fate: S. K. Singles *et al., Pestic. Sci.* **55**, 288 (1999).

Solid, mp 130-137°.
Sodium salt. [144740-54-5] DPX-KE459; Lexus; Millenium. $C_{15}H_{13}F_3N_5NaO_7S$; mol wt 487.34. Solid, mp 172-173°. Most stable in aqueous solutions, pH 5-7; stable in most organic solvents including methanol, acetonitrile, acetone, ethyl acetate and methylene chloride. Soly in water at 25° (ppm): 62.7 (pH 5); 603 (pH 7). pKa: 4.9. LD_{50} in rats, Japanese quail, mallard duck (mg/kg): >5000, >2250, >2250 orally; in rabbits (mg/kg): >2000 dermally. LC_{50} (96hr) in rainbow trout, carp (mg/l): 470, 820; LC_{50} in rats (mg/l): >5.8 by inhalation (Teaney).

USE: Herbicide.

4226. Fluquinconazole. [136426-54-5] 3-(2,4-Dichlorophenyl)-6-fluoro-2-(1H-1,2,4-triazol-1-yl)-4(3H)-quinazolinone; SN-597265; Castellan; Jockey; Vista. $C_{16}H_8Cl_2FN_5O$; mol wt 376.17. C 51.09%, H 2.14%, Cl 18.85%, F 5.05%, N 18.62%, O 4.25%. Prepn: D. E. Green, A. Percival, **EP 183458**; *eidem,* **US 4731106** (1985, 1988 both to Schering Agrochem). Chemical and biological properties: P. E. Russell *et al., Brighton Crop Prot. Conf. - Pests Dis.* **1992**, 411. Fungicidal activity: G. Schnabel, L. Parisi, *Acta Hortic.* **466**, 83 (1998). Efficacy against root disease, Take-all, in seed treatment: M. Wenz *et al., Brighton Crop Prot. Conf. - Pests Dis.* **1998**, 907; in various crop stages: A. M. Löchel *et al., ibid.* 89. Review of residue analysis: R. Hänel *et al., Nachrichtenbl. Dtsch. Pflanzenschutzdienst* **50**, 118 -126 (1998).

Off-white particulate, mp 191.5-193°. Soly at 20° (g/l): water 0.001; acetone 44; xylene 10; ethanol 3; DMSO 150. Log P (octanol/water): 3.4. Vapor pressure (20°): 6.4×10^{-9} Pa. LD_{50} in male, female mice, male, female rats (mg/kg): 325, 180, 112, 112 orally; in male, female rats (mg/kg): 2679, 625 dermally (Russell).

USE: Agricultural fungicide primarily for cereal crops.

4227. Flurandrenolide. [1524-88-5] (6α,11β,16α)-6-Fluoro-11,21-dihydroxy-16,17-[(1-methylethylidene)bis(oxy)]pregn-4-ene-3,20-dione; 6α-fluoro-11β,16α,17,21-tetrahydroxypregn-4-ene-3,20-dione cyclic 16,17-acetal with acetone; 6α-fluoro-16α,17α-isopropylidenedioxy-4-pregnene-11β,21-diol-3,20-dione; 6α-fluoro-11β,16α,17,21-tetrahydroxyprogesterone cyclic 16,17-acetal with acetone; fludroxycortide; fluorandrenolone; flurandrenolone; flurandrenolone acetonide; Cordran; Denison; Drocort; Sermaka. $C_{24}H_{33}FO_6$; mol wt 436.52. C 66.04%, H 7.62%, F 4.35%, O 21.99%. Prepn: H. J. Ringold *et al.,* **DE 1131213**; *eidem,* **US 3126375** (1962, 1964 both to Syntex). HPLC determn in pharmaceutical cream: R. Pearlman *et al., Int. J. Pharm.* **18**, 53 (1984); in urine: E. Neufeld *et al., J. Chromatogr. B* **718**, 273 (1998). Clinical trial in psoriasis: G. G. Krueger *et al., J. Am. Acad. Dermatol.* **38**, 186 (1998).

Crystals from acetone + hexane, mp 247-255°. $[\alpha]_D$ +140-150° (CHCl₃). uv max: 236 nm (log ε 4.17). Freely sol in chloroform; sol in methanol; sparingly sol in alc. Practically insol in water, ether.

THERAP CAT: Glucocorticoid; antipsoriatic.

4228. Flurazepam. [17617-23-1] 7-Chloro-1-[2-(diethylamino)ethyl]-5-(2-fluorophenyl)-1,3-dihydro-2H-1,4-benzodiazepin-2-one; Felmane; Noctosom; Stauroderm. $C_{21}H_{23}ClFN_3O$; mol wt 387.88. C 65.03%, H 5.98%, Cl 9.14%, F 4.90%, N 10.83%, O 4.12%. Prepn: G. A. Archer *et al.,* **BE 629005** *C.A.* **60**, 15896f (1964); **NL 6401335** *C.A.* **62**, 5289a (1965); G. A. Archer *et al.,* **US 3299053** (1963, 1964, 1967 all to Hoffmann-La Roche). Manuf process: R. Fryer, L. H. Sternbach, **US 3567710** (1971 to Hoffmann-La Roche). Synthesis: S. Inaba *et al., Chem. Pharm. Bull.* **19**, 263 (1971). Structure-activity studies: L. H. Sternbach *et al., J. Med. Chem.* **8**, 815 (1965); Armagnac *et al., Therapie* **26**, 439 (1971). Metabolism: Schwartz *et al., J. Med. Chem.* **11**, 770 (1968); *eidem, J. Pharm. Sci.* **59**, 1800 (1970). Pharmacology: L. O. Randall *et al., Arch. Int. Pharmacodyn. Ther.* **178**, 216 (1969). Toxicology: H. C. Rosenberg, *Pharmacol. Biochem. Behav.* **13**, 415 (1980). GC determn in plasma: Z. Salama *et al., Arzneim.-Forsch.* **38**, 400 (1988). Comprehensive description: B. C. Rudy, B. Z. Senkowski, *Anal. Profiles Drug Subs.* **3**, 307-331 (1974).

White rods from ether-petr ether, mp 77-82°.
Dihydrochloride. [1172-18-5] Flurazepam hydrochloride; Ro-5-6901; Benozil; Dalmadorm; Dalmane; Dalmate; Dormodor; Felison; Insumin; Lunipax; Somlan. $C_{21}H_{23}ClFN_3O.2HCl$; mol wt

460.80. Pale yellow crystals from methanol + ether, mp 190-220°. Freely sol in water, alc; slightly sol in isopropyl alc, chloroform. LD_{50} in mice (mg/kg): 290 i.p.; 870 orally; 84 i.v. (Randall).

Note: This is a controlled substance (depressant): **21 CFR,** 1308.14.

THERAP CAT: Sedative, hypnotic.

4229. Flurbiprofen. [5104-49-4] 2-Fluoro-α-methyl[1,1'-biphenyl]-4-acetic acid; 2-(2-fluoro-4-biphenylyl)propionic acid; 3-fluoro-4-phenylhydratropic acid; BTS-18322; U-27182; Adfeed; Ansaid; Antadys; Cebutid; Froben; Flurofen; Ocufen; Stayban; Zepolas. $C_{15}H_{13}FO_2$; mol wt 244.27. C 73.76%, H 5.36%, F 7.78%, O 13.10%. Prepn: **FR M5737**; Adams *et al.*, **US 3755427** (1968, 1973 both to Boots Co., Ltd.). Pharmacology: Chalmers *et al.*, *Ann. Rheum. Dis.* **31**, 319 (1972); *ibid.* **32**, 58 (1973); Glenn *et al.*, *Agents Actions* **3**, 210 (1973); Nishizawa *et al.*, *Thromb. Res.* **3**, 577 (1973). HPLC determn in urine and plasma: J. M. Hutzler *et al.*, *J. Chromatogr. B* **749**, 119 (2000). Symposium on pharmacokinetics and clinical efficacy in pain management: *Am. J. Med.* **80**, Suppl. 3A, 1-157 (1986).

Crystals from petr ether, mp 110-111°. Freely sol in acetone, dehydrated alc, ether, methanol; readily sol in most polar solvents; sol in acetonitrile. Practically insol in water.

THERAP CAT: Anti-inflammatory; analgesic.

4230. Flurogestone Acetate. [2529-45-5] (11β)-17-(Acetyloxy)-9-fluoro-11-hydroxypregn-4-ene-3,20-dione; 9-fluoro-11β,17-dihydroxypregn-4-ene-3,20-dione-17-acetate; 17α-acetoxy-9α-fluoro-11β-hydroxyprogesterone; 9-fluoro-11β,17-dihydroxyprogesterone 17-acetate; flugestone acetate; SC-9880; Chronogest; Cronolone; Synchronate. $C_{23}H_{31}FO_5$; mol wt 406.49. C 67.96%, H 7.69%, F 4.67%, O 19.68%. Prepn: Bergstrom *et al.*, *J. Am. Chem. Soc.* **81**, 4432 (1959); Bergstrom, Dodson; Bergstrom, Nicholson, **US 2892851**; **US 2963498** (1959, 1960 to Searle).

Crystals from benzene + petr ether, or ethyl acetate + petr ether, mp 266-269°. $[\alpha]_D$ +77.6° (in chloroform). uv max (methanol): 238 nm (ε 17500).

Component of *Syncro-Mate.*

THERAP CAT: Progestogen.

THERAP CAT (VET): Progestogen. Estrus regulation.

4231. Flurothyl. [333-36-8] 1,1'-Oxybis[2,2,2-trifluoroethane]; bis(2,2,2-trifluoroethyl) ether; hexafluorodiethyl ether; bis(trifluoroethyl) ether; Indoklon. $C_4H_4F_6O$; mol wt 182.07. C 26.39%, H 2.21%, F 62.61%, O 8.79%. Prepn: **GB 814493** (1959 to Pennsalt Chem.); **GB 889282** (1962 to Air Reduction). Toxicology in animals: A. Cherkin, *Psychopharmacologia* **15**, 404 (1969). Review of pharmacology: L. Arce, *Psychosomatics* **11**, 358-360 (1970); of clinical trials in psychiatric convulsant therapy: J. G. Small, I. F. Small, *Semin. Psychiatry* **4**, 13-26 (1972).

Colorless, mobile liq, mild ethereal, pleasant odor. d_4^{20} 1.41. bp 63.9°. Practically insol in water. Sol in alcohol.

THERAP CAT: CNS stimulant.

4232. Fluroxene. [406-90-6] (2,2,2-Trifluoroethoxy)ethene; 2,2,2-trifluoroethyl vinyl ether; Fluoromar (formerly). $C_4H_5F_3O$; mol wt 126.08. C 38.11%, H 4.00%, F 45.21%, O 12.69%. First clinical fluorinated inhalation anesthetic agent. Prepd (not claimed): J. G. Shukys, **US 2830007**; P. W. Townsend, **US 2870218** (1958, 1959 to Air Reduction). Anesthetic action: J. C. Krantz, Jr. *et al.*, *J. Pharmacol. Exp. Ther.* **108**, 488 (1953). Review of metabolism and toxicity: V. Fiserova-Bergerova, *Environ. Health Perspect.* **21**, 225-230 (1977); L. S. Kaminsky, J. M. Fraser, *Crit. Rev. Toxicol.* **19**, 87-112 (1988).

Liquid. bp_{751} 42.5°. n_D^{20} 1.3192. d_4^{20} 1.135. *Potentially explosive.*

Caution: Potential symptoms of overexposure are eye irritation; CNS depression, analgesia, anesthesia, seizures, respiratory depression. *See NIOSH Pocket Guide to Chemical Hazards* (DHHS/NIOSH 97-140, 1997) p 146.

THERAP CAT: Formerly as anesthetic (inhalation).

4233. Fluroxypyr. [69377-81-7] 2-[(4-Amino-3,5-dichloro-6-fluoro-2-pyridinyl)oxy]acetic acid. $C_7H_5Cl_2FN_2O_3$; mol wt 255.03. C 32.97%, H 1.98%, Cl 27.80%, F 7.45%, N 10.98%, O 18.82%. Auxin-type herbicide for use in cereals. Prepn: S. D. McGregor, **US 3761486** (1973 to Dow). Description: J. A. Paul *et al.*, *Proc. Br. Crop Prot. Conf. - Weeds* **1985**, 939. Field trials in grassland: A. R. Thompson, *ibid.* **1987**, 735. Mode of action: G. E. Sanders, K. E. Pallet, *Ann. Appl. Biol.* **111**, 385 (1987). HPLC determn: R. G. Lehmann, J. R. Miller, *J. Chromatogr.* **485**, 581 (1989). Persistence in soil: L. F. Bergström *et al.*, *Pestic. Sci.* **29**, 405 (1990). Uptake and metabolism in weeds: R. L. MacDonald *et al.*, *Weed Res.* **34**, 333 (1994).

White crystalline solid, mp 232-233°. Vapor pressure (25°): 9.42 × 10^{-7} mm Hg. Soly at 20° (g/l): water 0.091; acetone 41.6. LD_{50} orally in rats: 2405 mg/kg; i.p. in male, female rats: 458, 519 mg/kg; percutaneous in rabbits >5000 mg/kg (Paul).

1-Methylheptyl ester. [81406-37-3] Fluroxypyr-meptyl; Dowco 433; Spotlight; Starane; Vista; WideMatch S. $C_{15}H_{21}Cl_2FN_2O_3$; mol wt 367.24.

USE: Herbicide.

4234. Flurprimidol. [56425-91-3] α-(1-Methylethyl)-α-[4-(trifluoromethoxy)phenyl]-5-pyrimidinemethanol; α-isopropyl-α-[p-(trifluoromethoxy)phenyl]-5-pyrimidinemethanol; Compd 72500; EL-500; Cutless. $C_{15}H_{15}F_3N_2O_2$; mol wt 312.29. C 57.69%, H 4.84%, F 18.25%, N 8.97%, O 10.25%. Growth regulator for use on established turfgrass. Prepn: R. L. Benefiel, **BE 815245** (1974 to Lilly), *C.A.* **83**, 97354t (1975); R. L. Benefiel, E. V. Krumkalns, **US 3967949**, **US 4002628** (1976, 1977 both to Lilly). Growth retardant properties, effect on turfgrass: L. G. Thompson, *Proc. West. Soc. Weed Sci.* **35**, 113 (1982). Effect on Kentucky bluegrass, comparison with other growth retardants: T. L. Watschke, *Proc. Annu. Meet. Northeast. Weed Sci. Soc.* **35**, 322 (1981); N. E. Christians, *J. Am. Soc. Hortic. Sci.* **110**, 765 (1985). Long term study of effect on bluegrass-fescue turf: P. H. Dernoeden, *Agron. J.* **76**, 807 (1984). Efficacy in dwarfing southern pine seedlings: R. C. Hare, *Can. J. For. Res.* **14**, 123 (1984).

Non-volatile white crystals, mp 94-96°. Sol in acetone, ethanol, methanol, DMSO, diethyl ether. LD_{50} dermally in rabbits: >2000 mg/kg (Thompson).

USE: Growth retardant for grasses.

4235. Flurtamone. [96525-23-4] 5-(Methylamino)-2-phenyl-4-[3-(trifluoromethyl)phenyl]-3(2H)-furanone; 5-methylamino-2-phenyl-4-(α,α,α-trifluoro-m-tolyl)furan-3(2H)-one; RE-40885; Baccara. $C_{18}H_{14}F_3NO_2$; mol wt 333.31. C 64.86%, H 4.23%, F 17.10%, N 4.20%, O 9.60%. Bleaching herbicide; inhibits carotenoid biosynthesis. Prepn: C. E. Ward, **DE 3422346**; *idem*, **US 4568376** (1984, 1986 both to Chevron). LC determn in soil: T. C. Mueller *et al.*, *J. Assoc. Off. Anal. Chem.* **73**, 298 (1990). Mode of action study: G. Sandmann *et al.*, *Plant Physiol.* **94**, 476 (1990). Persistence in soil: T. C. Mueller *et al.*, *Weed Sci.* **38**, 411 (1990). Crystal structure: G. Pèpe *et al.*, *Acta Crystallogr.* **C52**, 1514 (1996). Review of physical properties, mode of action, and field trials: D. D. Rogers *et al.*, *Proc. Br. Crop Prot. Conf. - Weeds* **1987**, 69-75.

Ivory powder, mp 152-155°; also reported as colorless monoclinic crystals. d 1.38 mg/m³. Soly in water at 20°: 35 mg/l; in ethanol at 25°: 9.92 g/g solvent. Sol in acetone, methanol, methylene chloride. Slightly sol in isopropanol. LD_{50} in rats (mg/kg): 500 orally; in rabbits (mg/kg): 500 dermally (Rogers).

USE: Herbicide.

4236. Flusilazole. [85509-19-9] 1-[[Bis(4-fluorophenyl)methylsilyl]methyl]-1H-1,2,4-triazole; bis(4-fluorophenyl)methyl-(1H-1,2,4-triazol-1-ylmethyl)silane; fluzilazol; DPX-H6573; Nustar; Olymp; Punch. $C_{16}H_{15}F_2N_3Si$; mol wt 315.40. C 60.93%, H 4.79%, F 12.05%, N 13.32%, Si 8.90%. Sterol-inhibiting, broad spectrum foliar fungicide. Prepn: W. K. Moberg, **EP 68813**; *idem*, **US 4510136** (1983, 1985 both to Du Pont); W. K. Moberg *et al.*, *Pest. Sci. Biotech., Proc. 6th Int. Congr. Pest. Chem.*, 57 (1987). Field performance, physical properties and toxicity: T. M. Fort, W. K. Moberg, *Proc. Br. Crop Prot. Conf. - Pests Dis.* **1984**, 413. Field trials against apple scab: A. R. Biggs, J. Warner, *Can. J. Plant Pathol.* **9**, 41 (1987); and mildew: A. A. J. Swait *et al.*, *Tests Agrochem. Cultiv.* **8**, 42 (1987).

Crystalline solid, mp 55°. Soly: >2 g/ml in many organic solvents. LD_{50} in male, female rats (mg/kg): 1110, 674 orally; in rabbits >2000 dermally (Fort, Moberg).

USE: Agricultural fungicide.

4237. Fluspirilene. [1841-19-6] 8-[4,4-Bis(4-fluorophenyl)butyl]-1-phenyl-1,3,8-triazaspiro[4.5]decan-4-one; R-6128; Imap; Redeptin. $C_{29}H_{31}F_2N_3O$; mol wt 475.58. C 73.24%, H 6.57%, F 7.99%, N 8.84%, O 3.36%. Prepn: **BE 633914**; P. A. J. Janssen, **US 3238216** (1963, 1966 both to Janssen). HPLC determn: A. H. Hikal, H. I. Al-Shoura, *J. Liq. Chromatogr.* **5**, 2205 (1982). Metabolism in rat: J. P. P. Heykants, *Life Sci.* **8**, Part I, 1029 (1969). Pharmacology: P. A. J. Janssen *et al.*, *Arzneim.-Forsch.* **20**, 1689 (1970). Calcium channel blocking activity: R. J. Gould *et al.*, *Proc. Natl. Acad. Sci. USA* **80**, 5122 (1983); J. P. Galizzi *et al.*, *ibid.* **83**, 7513 (1986). Clinical trials in schizophrenia: H. Immich *et al.*, *Arzneim.-Forsch.* **20**, 1699 (1970); G. Chouinard *et al.*, *J. Clin. Psychopharmacol.* **6**, 21 (1986).

White to yellowish amorphous or crystalline solid, mp 187.5-190°. Soly in water: 0.015-0.020 mg/ml. LD_{50} i.m. in rats: 146 ±14 mg/kg (Janssen).

THERAP CAT: Antipsychotic.

4238. Flutamide. [13311-84-7] 2-Methyl-N-[4-nitro-3-(trifluoromethyl)phenyl]propanamide; α,α,α-trifluoro-2-methyl-4′-nitro-m-propionotoluidide; 4′-nitro-3′-trifluoromethylisobutyranilide; niftolid; Sch-13521; Drogenil; Eulexin; Euflex; Flucinom; Flutamin; Fugerel. $C_{11}H_{11}F_3N_2O_3$; mol wt 276.22. C 47.83%, H 4.01%, F 20.63%, N 10.14%, O 17.38%. Prepn: Baker *et al.*, *J. Med. Chem.* **10**, 93 (1967); Neri, Topliss, **DE 2130450** (1972 to Sherico), *C.A.* **78**, 58091g (1973); Gold, **DE 2261293**; *idem*, **US 3847988** (1973, 1974 both to Schering). Activity studies: Neri *et al.*, *Endocrinology* **91**, 427 (1972); Peets *et al.*, *ibid.* **94**, 532 (1974); B. A. Gladue, L. G. Clemens, *Endocrinology* **106**, 1917 (1980). Effect in prostatic cancer: P. C. Sogani, W. F. Whitmore, *J. Urol.* **122**, 640 (1979). Clinical evaluation in pancreatic cancer: B. A. Greenway, *Br. Med. J.* **316**, 1935 (1998).

Crystals from benzene, mp 111.5-112.5°. Freely sol in acetone, ethyl acetate, methanol; sol in chloroform, ether. Practically insol in mineral oil, petroleum ether, water.

THERAP CAT: Antiandrogen; antineoplastic (hormonal).

THERAP CAT (VET): Antiandrogen.

4239. Fluthiacet-methyl. [117337-19-6] 2-[[2-Chloro-4-fluoro-5-[(tetrahydro-3-oxo-1H,3H-[1,3,4]thiadiazolo[3,4-a]pyridazin-1-ylidene)amino]phenyl]thio]acetic acid methyl ester; 9-(4-chloro-2-fluoro-5-methoxycarbonylmethylthiophenylimino)-8-thia-1,6-diazabicyclo-[4.3.0]nonane-7-one; KIH-9201; CGA-248757; Action. $C_{15}H_{15}ClFN_3O_3S_2$; mol wt 403.87. C 44.61%, H 3.74%, Cl 8.78%, F 4.70%, N 10.40%, O 11.88%, S 15.88%. Selective, post-emergence, isourazole herbicide which upon conversion to the urazole inhibits protoporphyrinogen oxidase. Prepn: M. Yamaguchi *et al.*, **EP 273417**; *eidem*, **US 4906279** (1988, 1990 both to Kumiai Chem. Ind.; Ihara Chem. Ind.). Description of physical properties, toxicity, and activity: T. Miyazawa *et al.*, *Brighton Crop Prot. Conf. - Weeds* **1993**, 23. Conversion to active metabolite and mode of action study: T. Shimizu *et al.*, *Plant Cell Physiol.* **36**, 625 (1995). Field study as harvest-aid for cotton: R. G. Lemon *et al.*, *Proc. Beltwide Cotton Conf.* **1999**, 605.

White powder, mp 104.6°. Soly in water at 29°: 0.64 mg/l. Vapor pressure: 9.2×10^{-6} Pa. Log P (octanol/water): 3.55. LD_{50} in rats (mg/kg): >5000 orally; >2000 dermally (Miyazawa).

USE: Herbicide.

4240. Fluticasone Propionate. [80474-14-2] $(6\alpha,11\beta,16\alpha,-17\alpha)$-6,9-Difluoro-11-hydroxy-16-methyl-3-oxo-17-(1-oxopropoxy)androsta-1,4-diene-17-carbothioic acid S-(fluoromethyl) ester; S-fluoromethyl $6\alpha,9\alpha$-difluoro-11β-hydroxy-16α-methyl-17α-propionyloxy-3-oxoandrosta-1,4-diene-17β-carbothioate; CCI-18781; Cutivate; Flixonase; Flixotide; Flonase; Flunase. $C_{25}H_{31}$-F_3O_5S; mol wt 500.57. C 59.99%, H 6.24%, F 11.39%, O 15.98%, S 6.40%. Derivative of flumethasone, $q.v.$ Prepn: **NL 81 00707**; G. H. Phillipps $et\ al.$, **US 4335121** (1981, 1982 both to Glaxo). Series of articles on structure-activity, pharmacology and clinical studies: $Respir.\ Med.$ **84**, Suppl. A, 19-35 (1990). Clinical trials in allergic rhinitis: E. O. Meltzer $et\ al.$, $J.\ Allergy\ Clin.\ Immunol.$ **86**, 221 (1990). Evaluation in untreated celiac disease: H. C. Mitchison $et\ al.$, Gut **32**, 260 (1991). Clinical trials in COPD: P. L. Paggiaro $et\ al.$, $Lancet$ **351**, 773 (1998); P. S. Burge $et\ al.$, $Br.\ Med.\ J.$ **320**, 1297 (2000). Clinical comparison with zafirlukast, in persistant asthma: J. H. Brabson $et\ al.$, $Am.\ J.\ Med.$ **113**, 15 (2002). Clinical trial in asthma in adults: M. Masoli $et\ al.$, $Eur.\ Respir.\ J.$ **28**, 960 (2006); in children: R. L. Wasserman $et\ al.$, $Ann.\ Allergy\ Asthma\ Immunol.$ **96**, 808 (2006).

Crystals, mp 272-273° (dec). $[\alpha]_D$ +30° (c = 0.35). Practically insol in water. Freely sol in DMSO, DMF; slightly sol in methanol, 95% ethanol.

THERAP CAT: Antiallergic; antiasthmatic; anti-inflammatory.

THERAP CAT (VET): Antihistaminic; anti-inflammatory.

4241. Flutolanil. [66332-96-5] N-[3-(1-Methylethoxy)phenyl]-2-(trifluoromethyl)benzamide; o-trifluoromethyl-m'-isopropoxybenzoic anilide; α,α,α-trifluoro-3'-isopropoxy-o-toluanilide; NNF-136; Prostar; Moncut. $C_{17}H_{16}F_3NO_2$; mol wt 323.32. C 63.15%, H 4.99%, F 17.63%, N 4.33%, O 9.90%. Systemic benzanilide fungicide. Prepn: K. Yabutani $et\ al.$, **DE 2731522**; $eidem$, **US 4093743** (both 1978 to Nihon Nohyaku). Properties and biological activity: F. Araki, K. Yabutani, $Proc.\ Br.\ Crop\ Prot.\ Conf.\ -\ Pests\ Dis.$ **1981**, 3; F. Araki, $Jpn.\ Pestic.\ Inf.$ **47**, 23 (1985). Mode of action: K. Motoba $et\ al.$, $Agric.\ Biol.\ Chem.$ **52**, 1445 (1988). Degradation study: M. Uchida $et\ al.$, $J.\ Pestic.\ Sci.$ **8**, 529 (1983). Control of rice sheath blight: H. Mochizuki $et\ al.$, $ibid.$ **12**, 29 (1987); of fungal diversity on peanuts: R. E. Baird $et\ al.$, $Phytoprotection$ **76**, 101 (1995).

Odorless, white, crystalline solid. mp 108° (Araki, Yabutani), also reported as mp 104-105° (Araki, 1985). Stable in acid and alkaline solutions. Soly at 20° (g/l): water 0.0096; chloroform 341; methanol 480; benzene 131; xylene 29. Vapor pressure at 20°: 1.33×10^{-5} mm Hg. LD_{50} in male and female rats, male and female mice (mg/kg): >10,000, >10,000 orally; in male and female rats (mg/kg): >5000 dermally (Araki).

USE: Agricultural fungicide.

4242. Flutriafol. [76674-21-0] α-(2-Fluorophenyl)-α-(4-fluorophenyl)-1H-1,2,4-triazole-1-ethanol; (RS)-2,4'-difluoro-α-

(1H-1,2,4-triazol-1-ylmethyl)benzhydryl alcohol; flutriafen; R-152450; PP-450; Impact; Topguard; Vincit. $C_{16}H_{13}F_2N_3O$; mol wt 301.30. C 63.78%, H 4.35%, F 12.61%, N 13.95%, O 5.31%. Prepn: K. P. Parry $et\ al.$, **EP 15756** (1980 to ICI). Activity, physical properties, and toxicology: A. M. Skidmore $et\ al.$, $Proc.\ 10th\ Int.\ Congr.\ Plant\ Prot.$ **1**, 368 (1983). Mode of action: B. C. Baldwin $et\ al.$, $Meded.\ Fac.\ Landbouwwet.\ Rijksuniv.\ Gent$ **49**, 303 (1984). Field trials: P. J. Northwood $et\ al.$, $Proc.\ 10th\ Int.\ Congr.\ Plant\ Prot.$ **3**, 930 (1983); J. S. Brown, $Crop\ Prot.$ **6**, 157 (1987). Comparative persistence in soil: R. H. Bromilow $et\ al.$, $Pestic.\ Sci.$ **55**, 1135 (1999).

White crystalline solid, mp 130°. Vapor pressure at 20°: 3.00×10^{-9} mm Hg. Soly at 20° (g/l): acetone 190; dichloromethane 150; hexane 0.30; methanol 69; xylene 12; water (pH 7.0) 0.13. LD_{50} male, female rats (mg/kg): 1140, 1480 orally; rats, rabbits: >1000, >2000 percutaneously (Skidmore).

USE: Agricultural fungicide.

4243. Flutrimazole. [119006-77-8] 1-[(2-Fluorophenyl)(4-fluorophenyl)phenylmethyl]-1H-imidazole; 1-[o-fluoro-α-(p-fluorophenyl)-α-phenylbenzyl]imidazole; UR-4056; Micetal. $C_{22}H_{16}$-F_2N_2; mol wt 346.38. C 76.29%, H 4.66%, F 10.97%, N 8.09%. Prepn: J. Bartroli, M. Anguita, **EP 352352**; $eidem$, **US 5149707** (1990, 1992 both to Uriach). Series of articles on synthesis, antifungal activity, pharmacokinetics: $Arzneim.$-$Forsch.$ **42**, 832-863 (1992). Toxicity study: M. L. Vericat $et\ al.$, $ibid.$ 841. Clinical trial in dermatomycoses: O. Binet $et\ al.$, $Mycoses$ **37**, 455 (1994).

Crystals from acetonitrile, mp 164-167°. Also reported as white solid, mp 161-163°. LD_{50} in male, female mice, male, female rats (mg/kg): >1000, >1000, 808, 1214 orally; >2000, >2000, 1079, 1446 i.p. (Vericat).

THERAP CAT: Antifungal (topical).

4244. Fluvalinate. [69409-94-5] N-[2-Chloro-4-(trifluoromethyl)phenyl]valine cyano(3-phenoxyphenyl)methyl ester; ZR-3210; Mavrik. $C_{26}H_{22}ClF_3N_2O_3$; mol wt 502.92. C 62.09%, H 4.41%, Cl 7.05%, F 11.33%, N 5.57%, O 9.54%. Synthetic pyrethroid insecticide without the usual cyclopropane ring. Prepn: C. A. Henrick, B. A. Garcia, **DE 2812169** (1978 to Zoecon), $C.A.$ **90**, 122072d (1970). Activity: C. A. Henrick $et\ al.$, $Pestic.\ Sci.$ **11**, 224 (1980). Review of toxicology and human exposure: $Toxicological\ Profile\ for\ Pyrethrins\ and\ Pyrethroids$ (PB2004-100004, 2003) 332 pp.

Yellow-amber liquid. Vapor press. at 25°: 1×10^{-7} mm Hg. Soly in water: 2.0 ppb. Sol in organic solvents.

USE: Insecticide.

4245. Fluvastatin. [93957-54-1] *rel*-(3*R*,5*S*,6*E*)-7-[3-(4-Fluorophenyl)-1-(1-methylethyl)-1*H*-indol-2-yl]-3,5-dihydroxy-6-heptenoic acid; (±)-(3*R**,5*S**,6*E*)-7-[3-(*p*-fluorophenyl)-1-isopropylindol-2-yl]-3,5-dihydroxy-6-heptenoate. $C_{24}H_{26}FNO_4$; mol wt 411.47. C 70.06%, H 6.37%, F 4.62%, N 3.40%, O 15.55%. Synthetic HMG-CoA reductase inhibitor. Prepn: F. G. Kathawala, **WO 8402131**; *idem*, **US 4739073** (1984, 1988 both to Sandoz). Manufacturing process: P. C. Fuenfschilling *et al.*, *Org. Process Res. Dev.* **11**, 13 (2007). Clinical pharmacology: J. Yuan *et al.*, *Atherosclerosis* **87**, 147 (1991). Clinical pharmacokinetics: F. L. S. Tse *et al.*, *J. Clin. Pharmacol.* **32**, 630 (1992). Metabolism in humans: J. G. Dain *et al.*, *Drug Metab. Dispos.* **21**, 567 (1993). HPLC determn in plasma: G. Kalafsky *et al.*, *J. Chromatogr.* **614**, 307 (1993). Clinical study in familial hypercholesterolaemia: E. Leitersdorf *et al.*, *Eur. J. Clin. Pharmacol.* **45**, 513 (1993); in prevention of cardiac events following percutaneous coronary intervention: P. W. J. Serruys *et al.*, *J. Am. Med. Assoc.* **287**, 3215 (2002). Review of pharmacology and clinical experience: A. Asberg, H. Holdaas, *Expert Rev. Cardiovasc. Ther.* **2**, 641-652 (2004); K. J. McDonald, A. G. Jardine, *Expert Opin. Pharmacother.* **9**, 1407-1414 (2008).

Relative stereochemistry

Sodium salt. [93957-55-2] Fluindostatin; XU 62-320; Lescol; Lipaxan; Primexin. $C_{24}H_{25}FNNaO_4$; mol wt 433.46. White to pale yellow, hygroscopic powder. mp 194-197°. Sol in water, ethanol, methanol.

THERAP CAT: Antilipemic.

4246. Fluvoxamine. [54739-18-3] (1*E*)-5-Methoxy-1-[4-(trifluoromethyl)phenyl]-1-pentanone *O*-(2-aminoethyl)oxime; 5-methoxy-4'-(trifluoromethyl)valerophenone (*E*)-*O*-(2-aminoethyl)oxime. $C_{15}H_{21}F_3N_2O_2$; mol wt 318.34. C 56.60%, H 6.65%, F 17.90%, N 8.80%, O 10.05%. Selective serotonin reuptake inhibitor (SSRI). Prepn: **NL 7503310**; H. B. A. Welle, V. Claassen, **US 4085225** (1975, 1978 both to Philips-Duphar). Inhibition of 5-HT uptake: V. Claassen *et al.*, *Br. J. Pharmacol.* **60**, 505 (1977). HPLC determn in plasma: G. J. De Jong, *J. Chromatogr.* **183**, 203 (1980). Series of articles on pharmacology, pharmacokinetics and clinical trials in depression: *Br. J. Clin. Pharmacol.* **15**, Suppl. 3, 347S-450S (1983). Clinical trial in obsessive-compulsive disorder: L. M. Koran *et al.*, *J. Clin. Psychopharmacol.* **16**, 121 (1996). Review of clinical experience in depression: M. R. Ware, *J. Clin. Psychiatry* **58**, Suppl. 5, 15-23 (1997); in obsessive-compulsive disorder: B. Dell'Osso *et al.*, *Expert Opin. Pharmacother.* **6**, 2727-2740 (2005).

Maleate. [61718-82-9] DU-23000; MK-264; Dumirox; Faverin; Fevarin; Floxyfral; Luvox; Maveral. $C_{15}H_{21}F_3N_2O_2 \cdot C_4H_4O_4$; mol wt 434.41. Crystals from acetonitrile, mp 120-121.5°. Freely sol in ethanol, chloroform; sparingly sol in water. Practically insol in diethyl ether.

THERAP CAT: Antidepressant; antiobsessional.

4247. FM1-43. [149838-22-2] 4-[2-[4-(Dibutylamino)phenyl]ethenyl]-1-[3-(triethylammonio)propyl]pyridinium bromide (1:2); Neurodye GH1-43; SynaptoGreen. $C_{30}H_{49}Br_2N_3$; mol wt 611.55.

C 58.92%, H 8.08%, Br 26.13%, N 6.87%. Green fluorescent styryl dye that binds to but is not translocated through biological membranes. Prepn and optical activity of *p*-aminostyrylpyridinium probes: L. M. Loew *et al.*, *Biochemistry* **17**, 4065 (1978); A. Grinvald *et al.*, *Biophys. J.* **39**, 301 (1982). Optical analysis of synaptic vesicle recycling: W. J. Betz, G. S. Bewick, *Science* **255**, 200 (1992); W. J. Betz *et al.*, *J. Neurosci.* **12**, 363 (1992); *eidem*, *Curr. Opin. Neurobiol.* **6**, 365 (1996). Interaction with lipid membranes: U. Schote, J. Seelig, *Biochim. Biophys. Acta* **1415**, 135 (1998); with background quenching: A. R. Kay *et al.*, *Neuron* **24**, 809 (1999); J. L. Pyle *et al.*, *ibid.* 803. Use in membrane recycling study: T. J. Diefenbach *et al.*, *J. Neurosci.* **19**, 9436 (1999). Review of use and mechanism of action: M. A. Cousin, P. J. Robinson, *J. Neurochem.* **73**, 2227-2239 (1999).

Red-orange solid. Amphiphilic. Sol in water, DMSO. Practically nonfluorescent in aqueous solutions, becoming fluorescent upon membrane binding. Abs max (methanol): 510 nm (ε 56000). Emission max: 626 nm.

USE: Biological probe for membrane dynamics.

4248. Folic Acid. [59-30-3] *N*-[4-[[(2-Amino-3,4-dihydro-4-oxo-6-pteridinyl)methyl]amino]benzoyl]-L-glutamic acid; *N*-[*p*-[[(2-amino-4-hydroxy-6-pteridinyl)methyl]amino]benzoyl]glutamic acid; pteroylglutamic acid; *N*-(*p*-[(2-amino-4-hydroxypyrimido[4,5-*b*]pyrazin-6-yl)methylamino]benzoyl)glutamic acid; PGA; liver *Lactobacillus casei* factor; vitamin Bc; vitamin M; folsäure; Folacin; Foldine; Foliamin; Folicet; Folipac; Folettes; Folsan; Folvite; Incafolic; Millafol. $C_{19}H_{19}N_7O_6$; mol wt 441.40. C 51.70%, H 4.34%, N 22.21%, O 21.75%. Hematopoietic vitamin present, free or combined with one or more additional molecules of L(+)-glutamic acid, in liver, kidney, mushrooms, spinach, yeast, green leaves, grasses: Mitchell *et al.*, *J. Am. Chem. Soc.* **63**, 2284 (1941). Isoln: Pfiffner *et al.*, *ibid.* **69**, 1476 (1947); Stokstad *et al.*, *ibid.* **70**, 3 (1948). Structure: Mowat *et al.*, *ibid.* 14. Crystal structure: D. Mastropaolo *et al.*, *Science* **210**, 334 (1980). History of the different folic acid factors: *Ann. N.Y. Acad. Sci.* **48**, 255-350 (1946). *See also* reviews by Subbarow *et al.* in *Vitam. Horm.* **3**, 237-296 (1945); Pfiffner, Hogan, *ibid.* **4**, 1-13 (1946). Several syntheses, *see* reviews by Gates, *Chem. Rev.* **41**, 63-95 (1947); Hutchings, Mowat, *Vitam. Horm.* **6**, 1-25 (1948); Sletzinger *et al.*, *J. Am. Chem. Soc.* **77**, 6365 (1955); **US 2786056**; **US 2816109**; **US 2821527**; **US 2821528** (all 1958 to Merck & Co.); Sadao Kawanishi, **US 2956057** (1960 to Kongo Kagaku Kabushiki Kaisha). Alternate syntheses: Bieri, Viscontini, *Helv. Chim. Acta* **56**, 2905 (1973); E. Khalifa *et al.*, *ibid.* **59**, 242 (1976). Comprehensive reviews: Jaenicki, Kutzbach, *Fortschr. Chem. Org. Naturst.* **21**, 183-274 (1963); Marchetti, *Acta Vitaminol. Enzymol.* **25**, 41-64 (1971); F. J. Al-Shammary *et al.*, *Anal. Profiles Drug Subs.* **19**, 221-259 (1990). Review of role in prevention of neural tube defects: G. J. Locksmith, P. Duff, *Obstet. Gynecol.* **91**, 1027-1034 (1998).

Yellowish-orange crystals. Extremely thin platelets (elongated at two ends) from hot water, no mp. Darkens and chars from about 250°. $[\alpha]_D^{25}$ +23° (c = 0.5 in 0.1*N* NaOH). uv max (pH 13): 256, 283, 368 nm (log ε 4.43, 4.40, 3.96). Very slightly sol in cold water (0.0016 mg/ml at 25°), sol to about 1% in boiling water. Slightly sol in methanol, appreciably less sol in ethanol and butanol. Insol in acetone, chloroform, ether, benzene. Relatively sol in acetic acid, phenol, pyridine, solns of alkali hydroxides and carbonates. Sol in hot dil HCl and H_2SO_4. *Protect from light.*

Sodium salt. [6484-89-5] Sodium folate; sodium pteroylglutamate. $C_{19}H_{18}N_7NaO_6$; mol wt 463.39. Prepd for injection as sterile soln of folic acid in water with sodium hydroxide or sodium carbonate yielding a clear, yellow to orange-yellow liquid, pH 8.0-11.0.

THERAP CAT: Vitamin (hematopoietic).

THERAP CAT (VET): Nutritional factor (dietary requirement in poultry).

4249. Folinic Acid. [58-05-9] *N*-[4-[[(2-Amino-5-formyl-3,-4,5,6,7,8-hexahydro-4-oxo-6-pteridinyl)methyl]amino]benzoyl]-L-glutamic acid; *N*-[*p*-[[(2-amino-5-formyl-5,6,7,8-tetrahydro-4-hydroxy-6-pteridinyl)methyl]amino]benzoyl]glutamic acid; 5-formyl-5,6,7,8-tetrahydropteroyl-L-glutamic acid; 5-formyl-5,6,7,8-tetrahydrofolic acid; CF; citrovorum factor; leucovorin. $C_{20}H_{23}N_7O_7$; mol wt 473.45. C 50.74%, H 4.90%, N 20.71%, O 23.65%. Intermediate product of the metabolism of folic acid; the active form into which that acid is converted in the body, ascorbic acid being a necessary factor in the conversion process. First reported as the *Leuconostoc citrovorum* 8081 growth factor: H. E. Sauberlich, C. A. Baumann, *J. Biol. Chem.* **176**, 165 (1948). Isoln from houseflies: S. Miller, A. S. Perry, *Life Sci.* **4**, 1573 (1965). Prepn: J. A. Brockman, Jr., *et al.*, *J. Am. Chem. Soc.* **72**, 4325 (1950); E. Khalifa *et al.*, *Helv. Chim. Acta* **63**, 2554 (1980). Isomers: D. B. Cosulich *et al.*, *J. Am. Chem. Soc.* **74**, 4215 (1952). Structure: May *et al.*, *ibid.* **73**, 3067 (1951); Pohland *et al.*, *ibid.* 3247. Manuf: Shive, **US 2741608** (1956 to Res. Corp.). Stereoselective synthesis: J. Owens, *et al.*, *J. Chem. Soc. Perkin Trans. 1* **7**, 871 (1993). Used as an antidote to folic acid antagonists such as methotrexate, *q.v.*, which block the conversion of folic acid into folinic acid. Review of clinical combination therapy with methotrexate: J. R. Bertino *et al.*, *Ann. N.Y. Acad. Sci.* **186**, 486-495 (1971). Pharmacokinetics: P. F. Nixon, *Clin. Exp. Pharmacol. Physiol.* **Suppl. 5**, 35 (1979). Comprehensive description: L. O. Pont *et al.*, *Anal. Profiles Drug Subs.* **8**, 315-350 (1979). Review of clinical synergy with fluorouracil in cancer: R. J. DeLap, *Yale J. Biol. Med.* **61**, 23-34 (1988).

Crystals, dec 240-250°. uv max (0.1*N* NaOH): 282 nm (% T = 27.0 for 10 mg/l). $[\alpha]_D^{20}$ +14.26° (c = 3.42 as anhydr Ca salt). pKa (3 groups): 3.1, 4.8, and 10.4. Sparingly sol in water. pH of satd aq soln 2.8-3.0 at which pH partial decompn takes place. More stable at neutral or mildly alkaline pH.

Calcium salt pentahydrate. [6035-45-6] Calcium folinate; NSC-3590; Folaren; Foliben; Lederfolat; Lederfolin; Leucovorin; Leucosar; Rescufolin; Rescuvolin; Tonofolin; Wellcovorin. C_{20}-$H_{21}CaN_7O_7.5H_2O$; mol wt 601.58. Off-white to light beige amorphous, odorless powder. Very sol in water. Practically insol in alc. $[\alpha]_D^{21}$ +14.9° (c = 1 in water).

l-Form calcium salt. [80433-71-2] Calcium (6*S*)-folinate; calcium levofolinate; levoleucovorin calcium; Elvorine. $[\alpha]_D^{20}$ −15.1° (c = 1.82).

THERAP CAT: Antidote to folic acid antagonists; antianemic (folate deficiency).

4250. Follicle-Stimulating Hormone. [9002-68-0] FSH; follitropin. Gonadotropic hormone secreted by the anterior pituitary under the regulation of gonadotropin-releasing hormone and inhibin, *q.q.v.* In the female, FSH stimulates estrogen production by the granulosa cells of the ovary and induces maturation of the Graafian follicle. In the male, stimulates the Sertoli cells of the testes and promotes spermatogenesis. Glycoprotein with mol wt ~34 kDa; multiple isoforms exist. Structure is a heterodimer of noncovalently linked α and β subunits; analogous with LH, TSH and chorionic gonadotropin, *q.q.v.* Isoln from pituitary: H. L. Fevold *et al.*, *En-*

docrinology **26**, 999 (1940); C. H. Li *et al.*, *Science* **109**, 445 (1949). Prepn of human FSH from pituitaries and from postmenopausal urine: P. Roos, *Acta Endocrinol.* **59**, Suppl. 131, 1-93 (1968). Amino acid sequence of human FSH α-subunit: P. Rathnam, B. B. Saxena, *J. Biol. Chem.* **250**, 6735 (1975); of human β-subunit: B. B. Saxena, P. Rathnam, *ibid.* **251**, 993 (1976). Production in Chinese hamster ovary cells by recombinant DNA technology: J. L. Keene *et al.*, *J. Biol. Chem.* **264**, 4769 (1989). Characterization of isoforms: P. G. Stanton *et al.*, *Mol. Cell. Endocrinol.* **125**, 133 (1996). Review of structure, biosynthesis and receptor binding: J. G. Pierce, T. F. Parsons, *Annu. Rev. Biochem.* **50**, 465-495 (1981). Review of role in spermatogenesis: R. I. McLachlan *et al.*, *J. Endocrinol.* **148**, 1-9 (1996); in ovarian function and clinical use: B. C. Fauser, A. M. Van Heusden, *Endocr. Rev.* **18**, 71-106 (1997).

Solid. Isoelectric point 4.5. Soluble in water, physiological saline soln, 50% alc. uv max: 277 nm (pH 7.0 phosphate buffer); 291 nm (0.1*N* NaOH-8*M* urea).

Follitropin Alfa. Gonal-F. Recombinant human FSH, glycoform α. Clinical pharmacokinetics: J.-Y. le Cotonnec *et al.*, *Fertil. Steril.* **61**, 669, 679 (1994). Clinical trial: T. Strowitzki *et al.*, *Hum. Reprod.* **10**, 3097 (1995).

Follitropin Beta. Org-32489; Follistim; Puregon. Recombinant human FSH, glycoform β. Clinical pharmacokinetics: B. Mannaerts *et al.*, *Fertil. Steril.* **59**, 108 (1993). Clinical trial: B. Hedon *et al.*, *Hum. Reprod.* **10**, 3102 (1995).

Menotropins. Menotrophin; human menopausal gonadotropins; Humegon; Menogon; Pergonal; Repronex. Extract of human postmenopausal urine containing both FSH and LH.

Urofollitropin. [97048-13-0] Urofollitrophin; Fertinorm; Metrodin; Orgafol. Purified extract of human postmenopausal urine containing FSH.

THERAP CAT: Gonad-stimulating principle.

THERAP CAT (VET): Gonad-stimulating principle.

4251. Folpet. [133-07-3] 2-[(Trichloromethyl)thio]-1*H*-isoindole-1,3(2*H*)-dione; *N*-(trichloromethylthio)phthalimide; *N*-(trichloromethylmercapto)phthalimide; Phaltan. $C_9H_4Cl_3NO_2S$; mol wt 296.55. C 36.45%, H 1.36%, Cl 35.86%, N 4.72%, O 10.79%, S 10.81%. *Cf:* Captan. Prepn: Kittleson, **US 2553770** (1951 to Standard Oil). Toxicity: T. B. Gaines, R. E. Linder, *Fundam. Appl. Toxicol.* **7**, 299 (1986).

Crystals from benzene, mp 177°. LD_{50} orally in adult male, female rats: >5000 mg/kg (Gaines, Linder).

Caution: May irritate mucosal surfaces. *See Clinical Toxicology of Commercial Products*, R. E. Gosselin *et al.*, Eds. (Williams & Wilkins, Baltimore, 5th ed., 1984) Section II, p 317.

USE: Agricultural fungicide.

4252. Fomecins. Antibacterial substances produced by several strains of the basidiomycete *Fomes juniperinus* Schrenk. Isoln of fomecin A (major) and fomecin B (minor) from cultures grown in corn steep liquor: M. Anchel *et al.*, *Proc. Natl. Acad. Sci. USA* **38**, 655 (1952). Structures: T. C. McMorris, M. Anchel, *Can. J. Chem.* **42**, 1595 (1964). Synthesis of B: S. M. Al-Mousawi *et al.*, *Bull. Soc. Chim. Belg.* **88**, 883 (1979); of A and B: K. Hayashi *et al.*, *Chem. Pharm. Bull.* **28**, 1971 (1980).

Fomecin A   R = CH_2OH
Fomecin B   R = CHO

Fomecin A. 2,3,4-Trihydroxy-6-(hydroxymethyl)benzaldehyde. $C_8H_8O_5$; mol wt 184.15. Cream-colored to orange crystals from ethanol-water, ethanol-benzene, acetone-benzene, or ethyl acetate. Dec above 160° without melting. Optically inactive in ethanol. uv max (ethanol): 241, 304 nm (ε 10800, 15300). Weakly acidic. Sparingly sol in water (1 mg/ml); slightly more sol in ethanol, acetone, ethyl acetate. Less sol in chloroform, benzene. Aq solns at neutral or acid pH are stable. The activity is lost around pH 8.

Fomecin B. 3,4,5-Trihydroxy-1,2-benzenedicarboxaldehyde. $C_8H_6O_5$; mol wt 182.13. Yellow needles from ethyl acetate. Darkens on heating; mp ~230°. uv max (ethanol): 263, 336 nm (ε 26400, 9200).

4253. Fomepizole. [7554-65-6] 4-Methyl-1H-pyrazole; 4-MP; Antizol. $C_4H_6N_2$; mol wt 82.11. C 58.51%, H 7.37%, N 34.12%. Alcohol dehydrogenase inhibitor. Prepn: H. Pechmann, E. Burkard, *Ber.* **33**, 3590 (1900); D. S. Noyce *et al., J. Org. Chem.* **20**, 1681 (1955); T. Momose *et al., Heterocycles* **30**, 789 (1990). Inhibition of human liver alcohol dehydrogenase: T.-K. Li, H. Theorell, *Acta Chem. Scand.* **23**, 892 (1969). Toxicity study: G. Magnusson *et al., Experientia* **28**, 1198 (1972). GC determn in plasma and urine: R. Achari, M. Mayersohn, *J. Pharm. Sci.* **73**, 690 (1984). Clinical pharmacology: D. Jacobsen *et al., Alcohol. Clin. Exp. Res.* **12**, 516 (1988). Pharmacokinetics: *eidem, Eur. J. Clin. Pharmacol.* **37**, 599 (1989). Clinical trial in ethylene glycol poisoning: J. Brent *et al., N. Engl. J. Med.* **340**, 832 (1999); in methanol poisoning: *idem et al., ibid.* **344**, 424 (2001). *Review:* J. Likforman *et al., J. Toxicol. Clin. Exp.* **7**, 373-382 (1987). Review of use in methanol poisoning: M. B. Mycyk, J. B. Leikin, *Am. J. Ther.* **10**, 68-70 (2003).

mp 15.5-18.5°. bp$_{18mm}$ 98.5-99.5°; bp$_{730}$ 204-205°. n_D^{22} 1.4913. uv max in 95% ethanol: 220 nm (log ε 3.47); in 6N HCl: 226 nm (log ε 3.65). Sol in water, alcohol. LD_{50} (7 days) in mice, rats (mmol/kg): 3.8, 3.8 i.v.; 7.8, 6.5 orally (Magnusson).

THERAP CAT: Antidote to methanol and ethylene glycol poisoning.

THERAP CAT (VET): Antidote to ethylene glycol poisoning in dogs.

4254. Fomesafen. [72178-02-0] 5-[2-Chloro-4-(trifluoromethyl)phenoxy]-N-(methylsulfonyl)-2-nitrobenzamide; Flex; Reflex. $C_{15}H_{10}ClF_3N_2O_6S$; mol wt 438.76. C 41.06%, H 2.30%, Cl 8.08%, F 12.99%, N 6.38%, O 21.88%, S 7.31%. Post-emergence diphenyl ether herbicide used for early control of broad-leaf weeds in soybean and other legume crops. Prepn: D. Cartwright, D. J. Collins, **EP 3416**; *eidem,* **US 4285723** (1979, 1981 both to Imperial Chemical Industries). Properties and field trial in soybeans: S. R. Colby, J. W. Barnes, *Proc. 10th Int. Congr. Plant Prot.* **1**, 295 (1983); in green beans: C. M. Knott, *Brighton Crop Prot. Conf. - Weeds* **1993**, 1053. Mode of selectivity and metabolism in plant crops and weeds: J. D. H. L. Evans *et al., Proc. Br. Crop Prot. Conf. - Weeds* **1987**, 345. Metabolism and hepatoxicity in rodents: J. Krijt *et al., Hum. Exp. Toxicol.* **18**, 338 (1999). Environmental fate in soils: J. B. Weber, *Pestic. Sci.* **39**, 31 (1993); J. Guo *et al., Water Air Soil Pollut.* **148**, 77 (2003). Ecotoxicological effects in aquatic systems: T. Caquet *et al., Environ. Toxicol. Chem.* **24**, 1116 (2005).

White crystalline solid, mp 220-221°. pKa 2.9. Soly in distilled water: ~50 mg/l. LD_{50} orally in male, female rats (mg/kg): 1250-2000, 1600; LC_{50} (48, 96 hr.) in rainbow trout and bluegill (mg/l): 830, 680, 6740, 6030 (Colby).

Sodium salt. [108731-70-0] Flexstar. $C_{15}H_{10}ClF_3N_2NaO_6S$; mol wt 461.75. Percutaneous absorption: J. Hilton *et al., Pharm.*

Res. **11**, 1396 (1994). Soly at 25° (mg/ml): propylene glycol 631; octanol 12; ethyl decanoate 10. LD_{50} orally (mg/kg) in male, female rats, mice: 1860, 1500, 750, 770; in female guinea pigs: 490-980; in male rabbit: 490 (Colby).

USE: Herbicide.

4255. Fominoben. [18053-31-1] N-[3-Chloro-2-[[methyl-[2-(4-morpholinyl)-2-oxoethyl]amino]methyl]phenyl]benzamide; 3'-chloro-α-[methyl[(morpholinocarbonyl)methyl]amino]-o-benzotoluidide; 3'-chloro-β-[N-methyl-N-[(morpholinocarbonyl)methyl]-aminomethyl]benzanilide. $C_{21}H_{24}ClN_3O_3$; mol wt 401.89. C 62.76%, H 6.02%, Cl 8.82%, N 10.46%, O 11.94%. Prepn: **FR 1482547; FR Addn. 92488** (1967, 1968 both to Thomae), *C.A.* **68**, 95494e (1968); **72**, 21938p (1970); Kruger *et al.,* **US 3661903** (1972 to Boehringer, Ing.); *eidem, Arzneim.-Forsch.* **23**, 290 (1973). Series of articles on pharmacology and clinical findings: *ibid.* 296-375. Metabolism: Zimmer, *ibid.* 1798. Toxicity data: S. Püschmann, R. Engelhorn, *ibid.* 296.

Base, mp 122.5-123°.

Hydrochloride. [24600-36-0] PB-89; Finaten; Noleptan; Oleptan; Terion; Tussirama. $C_{21}H_{24}ClN_3O_3 \cdot HCl$; mol wt 438.35. Crystals, mp 206-208° (dec). Soly in water: 0.1 mg/100 ml; in 0.05N aq tartaric acid: 5 g/100 ml. LD_{50} in mice, rats (mg/kg): 630, 1201 i.p.; 2200, 1250 orally (Püschmann, Engelhorn).

THERAP CAT: Antitussive; respiratory stimulant.

4256. Fomivirsen. [144245-52-3] d(P-Thio)(G-C-G-T-T-T-G-C-T-T-C-T-T-C-T-T-C-T-T-G-C-G)DNA. $C_{204}H_{263}N_{63}O_{114}P_{20}S_{20}$; mol wt 6682.35. C 36.67%, H 3.97%, N 13.21%, O 27.29%, P 9.27%, S 9.60%. Synthetic, phosphorothioate antisense oligonucleotide designed to inhibit gene expression in human cytomegalovirus (CMV). Consists of 21 nucleotides in a sequence complementary to the specific portion of CMV messenger RNA which encodes regulatory proteins. Prepn: K. P. Anderson, K. G. Draper, **WO 9203456**; *idem et al.,* **US 5442049** (1992, 1995 both to Isis). Large scale synthesis: M. Andrade *et al., Bioorg. Med. Chem. Lett.* **4**, 2017 (1994). Determn in pharmaceutical formulations by capillary gel electrophoresis: G. S. Srivatsa *et al., J. Chromatogr. A* **680**, 469 (1994). Antiviral activity vs human CMV: R. F. Azad *et al., Antimicrob. Agents Chemother.* **37**, 1945 (1993). Mechanism of action study: K. P. Anderson *et al., ibid.* **40**, 2004 (1996). Pharmacokinetics: J. M. Leeds *et al., Drug Metab. Dispos.* **25**, 921 (1997). Retinal toxicity study: M. Flores-Aguilar *et al., J. Infect. Dis.* **175**, 1308 (1997). Review of pharmacology and therapeutic potential: J. Temsamani *et al., Expert Opin. Invest. Drugs* **6**, 1157-1167 (1997). Series of articles on clinical experience for CMV retinitis in AIDS patients: *Am. J. Ophthalmol.* **133**, 467-498 (2002).

Eicosasodium salt. [160369-77-7] Formivirsen sodium; ISIS-2922; Vitravene. $C_{204}H_{243}N_{63}Na_{20}O_{114}P_{20}S_{20}$; mol wt 7121.99.

THERAP CAT: Antiviral.

4257. Fomocaine. [17692-39-6] 4-[3-[4-(Phenoxymethyl)-phenyl]propyl]morpholine; 4-[3-(α-phenoxy-p-tolyl)propyl]morpholine; P-652; Erbocain. $C_{20}H_{25}NO_2$; mol wt 311.43. C 77.13%, H 8.09%, N 4.50%, O 10.27%. Prepn: **GB 786128** (1957 to Promonta), *C.A.* **52**, 9225h (1958); H. Oelschläger, *Arzneim.-Forsch.* **9**, 313 (1959). Improved synthesis: *eidem, ibid.* **27**, 1625 (1977). Pharmacology: O. Nieschulz *et al., ibid.* **8**, 539 (1958). Metabolism: H. Blume, H. Oelschläger, *ibid.* **28**, 956 (1978). Bioavailability and local anesthetic effect: H. Oelschläger, D. Rothley, *ibid.* **29**, 693 (1979).

Colorless crystals from petr ether, mp 52-53°. bp$_{1.1}$ 238-240°. uv max (ethanol): 220, 269 nm (ε 15820, 1373). LD$_{50}$ i.v. in mice: 175 mg/kg (Nieschulz).

THERAP CAT: Anesthetic (local).

4258. Fonazine. [7456-24-8] 10-[2-(Dimethylamino)propyl]-*N*,*N*-dimethyl-10*H*-phenothiazine-2-sulfonamide; 10-(2-dimethyl-aminopropyl)-3-dimethylsulfamidophenothiazine; 3-dimethylsulfamido-10-(2-dimethylaminopropyl)phenothiazine; dimethothiazine; dimethiotazine; dimetiotazine; RP-8599. C$_{19}$H$_{25}$N$_3$O$_2$S$_2$; mol wt 391.55. C 58.28%, H 6.44%, N 10.73%, O 8.17%, S 16.38%. Serotonin inhibitor. Prepn: **GB 814512** (1959 to Rhône-Poulenc). Pharmacokinetics and distribution in dog and rat: O. R. W. Lewellen, R. Templeton, *Adv. Biochem. Psychopharmacol.* **9**, 213 (1974). Clinical evaluation in spasticity: W. B. Matthews *et al.*, *Acta Neurol. Scand.* **48**, 635 (1972). HPLC determn in plasma: J. Holt *et al.*, *Br. J. Clin. Pharmacol.* **13**, 282 (1982).

Hydrochloride. C$_{19}$H$_{25}$N$_3$O$_2$S$_2$.HCl. Crystals, dec 214°.

Methanesulfonate. [7455-39-2] Dimethothiazine mesylate; fonazine mesylate; Banistyl; Bonpac; Calsekin; Migristène; Neomestine; Promaquid; Yoristen. C$_{19}$H$_{25}$N$_3$O$_2$S$_2$.CH$_3$SO$_3$H; mol wt 487.65.

THERAP CAT: Antimigraine.

4259. Fondaparinux Sodium. [114870-03-0] Methyl *O*-2-deoxy-6-*O*-sulfo-2-(sulfoamino)-α-D-glucopyranosyl-(1 → 4)-*O*-β-D-glucopyranuronosyl-(1 → 4)-*O*-2-deoxy-3,6-di-*O*-sulfo-2-(sulfoamino)-α-D-glucopyranosyl-(1 → 4)-*O*-2-*O*-sulfo-α-L-idopyranuronosyl-(1 → 4)-2-deoxy-2-(sulfoamino)-α-D-glucopyranoside 6-(hydrogen sulfate) sodium salt (1:10); fondaparin sodium; SR-90107A; Org-31540; IC-851589; Arixtra. C$_{31}$H$_{43}$N$_3$Na$_{10}$O$_{49}$S$_8$; mol wt 1728.03. C 21.55%, H 2.51%, N 2.43%, Na 13.30%, O 45.37%, S 14.84%. Synthetic pentasaccharide corresponding to the antithrombin binding site of heparin, *q.v.* Prepn: M. Petitou *et al.*, **EP 84999**; *eidem*, **US 4818816** (1983, 1989 both to Choay); M. Petitou *et al.*, *Carbohydr. Res.* **167**, 67 (1987). Review of pharmacology: J. M. Herbert *et al.*, *Cardiovasc. Drug Rev.* **15**, 1-26 (1997). Clinical pharmacokinetics: B. Boneu *et al.*, *Thromb. Haemostasis* **74**, 1468 (1995). Clinical trial for prevention of post-operative deepvein thrombosis: A. G. G. Turpie *et al.*, *N. Engl. J. Med.* **344**, 619 (2001); for treatment of superficial-vein thrombosis: H. Decousus *et al.*, *ibid.* **363**, 1222 (2010). Review of clinical safety and efficacy for prevention and treatment of venous thromboembolism: A. G. G. Turpie, *Expert Opin. Drug Saf.* **4**, 707-721 (2005).

White powder after lyophilisation. $[\alpha]_D^{23}$ +48° (c = 0.61 in water).

THERAP CAT: Antithrombotic.

4260. Fonofos. [944-22-9] *P*-Ethylphosphonodithioic acid *O*-ethyl *S*-phenyl ester; *O*-ethyl *S*-phenyl ethylphosphonothiolothionate; N-2790; Dyfonate. C$_{10}$H$_{15}$OPS$_2$; mol wt 246.32. C 48.76%, H 6.14%, O 6.50%, P 12.57%, S 26.03%. Organophosphorus insecticide; cholinesterase inhibitor. Prepn: Szabo *et al.*, **US 2988474** (1961 to Stauffer); H. M. Pitt, R. A. Simone, **DE 2002629**; *eidem*, **US 3642960** (1970, 1972 both to Stauffer). Toxicity and biochemistry of racemate and isomers: P. W. Lee *et al.*, *Pestic. Biochem. Physiol.* **8**, 146 (1978). Acute toxicity data: R. Ben-Dyke *et al.*, *World Rev. Pest Control* **9**, 119 (1970).

Light yellow liquid, bp$_{0.1}$ 130°. d$_{25}^{25}$ 1.16. n$_D^{30}$ 1.5883. Practically insol in water. Miscible with organic solvents. LD$_{50}$ in mice (mg/kg): 14.0 orally, 4.8 i.p. (Lee). LD$_{50}$ in rats (mg/kg): 3-17 orally, 147 dermally (Ben-Dyke).

Caution: Potential symptoms of overexposure are nausea, vomiting, abdominal cramps, diarrhea, salivation; headache, giddiness, vertigo, weakness; rhinorrhea, chest tightness; blurred vision, miosis; cardiac irregularities; muscle fasciculations; dyspnea. *See NIOSH Pocket Guide to Chemical Hazards* (DHHS/NIOSH 97-140, 1997) p 146.

USE: Soil insecticide.

4261. Foramsulfuron. [173159-57-4] 2-[[[[(4,6-Dimethoxy-2-pyrimidinyl)amino]carbonyl]amino]sulfonyl]-4-(formylamino)-*N*,*N*-dimethylbenzamide; AE-F130360; Option. C$_{17}$H$_{20}$N$_6$O$_7$S; mol wt 452.44. C 45.13%, H 4.46%, N 18.58%, O 24.75%, S 7.09%. Sulfonylurea herbicide for use in corn; acetolactate synthase inhibitor. Prepn: G. Schnabel *et al.*, **DE 4415049**; *eidem*, **US 5922646** (1995, 1999 both to Hoechst Schering AgrEvo). Review of properties and activity: B. Collins *et al.*, *BCPC Conf. - Weeds* **2001**, 35-42.

Light beige solid, mp 199.5°. Vapor pressure (20°): 4.2 × 10^{-11} Pa. Log P (octanol/water): 4.01 (pH 5); 0.166 (pH 7); 0.0106 (pH 8). Soly in water at 20° (g/l): 0.04 (pH 5); 3.3 (pH 7); 94.6 (pH 8). LD$_{50}$ in rats (mg/kg): >5000 orally; >2000 dermally (Collins).

USE: Herbicide.

4262. Forchlorfenuron. [68157-60-8] *N*-(2-Chloro-4-pyridinyl)-*N*'-phenylurea; 1-(2-chloro-4-pyridyl)-3-phenylurea; CPPU; KT-30; Prestige; Sitofex. C$_{12}$H$_{10}$ClN$_3$O; mol wt 247.68. C 58.19%, H 4.07%, Cl 14.31%, N 16.97%, O 6.46%. Synthetic cytokinin used to improve size and quality of fruit crops. Prepn: K. Shudo *et al.*, **DE 2843722**; *eidem*, **US 4193788**; **US RE31550** (1978, 1980, 1984); and cytokinin activity: S. Takahashi *et al.*, *Phytochemistry* **17**, 1201 (1978). HPLC determn in watermelon: J.-Y. Hu, J.-Z. Li, *J. AOAC Int.* **89**, 1635 (2006). Persistence in soil, water, and grapes: D. Sharma, M. D. Awasthi, *Chemosphere* **50**, 589 (2003). Adsorption in soils: K. Banerjee *et al.*, *Bull. Environ. Contam. Toxicol.* **80**, 201 (2008). Effect on size and quality of grapes: D. H. Diaz, L. A. Maldonado, *Proc. Plant Growth Regul. Soc. Am.* **19**, 123 (1992); of kiwifruit: J. G. Kim *et al.*, *Sci. Hortic.* **110**, 219 (2006).

Crystals from acetone-ethyl ether, mp 173-174°. d^{21} 1.44. vapor pressure (25°): 3.5 × 10^{-5} torr. Log P (octanol/water): 3.2. Soly in water (21°): 39 mg/l. Sol in acetonitrile.

USE: Plant growth regulator.

4263. Formaldehyde. [50-00-0] Methanal; oxomethane; oxymethylene; methylene oxide; formic aldehyde; methyl aldehyde. CH$_2$O; mol wt 30.03. C 40.00%, H 6.71%, O 53.28%. HCHO.

Formed by incomplete combustion of many organic substances. Present in coal and wood smoke, esp in smoke as produced for smoking ham and fish. Found in the atm, esp over large cities. Prepd commercially by catalytic vapor phase oxidation of methanol using air as the oxidizing agent and heated silver, copper, alumina, or coke as catalysts. Process using molybdenum iron oxide catalyst: Allyn *et al.*, **US 2812309** (1957) and **US 2849492** (1958 to Reichhold Chem.). Prepn of stable formaldehyde by heating low molecular polyoxymethylenes with P_2O_5: **DE 1070611** (1959 to BASF). Mfg processes: *Faith, Keyes & Clark's Industrial Chemicals*, F. A. Lowenheim, M. K. Moran, Eds. (Wiley-Interscience, New York, 4th ed., 1975) pp 422-429. Prepn of semicarbazone: M. Pomerantz *et al.*, *J. Org. Chem.* **47**, 2217 (1982). Toxicity data: H. F. Smyth *et al.*, *J. Ind. Hyg. Toxicol.* **23**, 259 (1941). Vapor-pressure relationship studies: R. Spence, W. Wild, *J. Chem. Soc.* **1935**, 506. Carcinogenicity study: J. A. Swenberg *et al.*, *Cancer Res.* **40**, 3398 (1980). Assessment, regulation and evaluation of carcinogenicity: F. Perera, C. Petito, *Science* **216**, 1285 (1982). *Review:* H. R. Gerberich *et al.*, in *Kirk-Othmer Encyclopedia of Chemical Technology* vol. 11 (John Wiley & Sons, New York, 4th ed., 1994) pp 929-951. Review of toxicology: H. d'A. Heck *et al.*, *Crit. Rev. Toxicol.* **20**, 397-426 (1990); of oral toxicity: P. Restani, C. L. Galli, *ibid.* **21**, 315-328 (1991).

Colorless gas at ordinary temp. Pungent suffocating odor. d 1.067 (air = 1.000). d_4^{-20} 0.815. mp $-118.0°$. bp$_{760}$ $-19.5°$; bp$_{400}$ $-33.0°$; bp$_{200}$ $-46.0°$; bp$_{100}$ $-57.3°$; bp$_{60}$ $-65.0°$; bp$_{40}$ $-70.6°$; bp$_{20}$ $-79.6°$; bp$_{10}$ $-88.0°$. Ignition temp about 300° (572°F). *Flammable, corrosive.* Very sol in water, up to 55%; sol in alc, ether. Very reactive, combines readily with many substances, and polymerizes easily. *See also* Paraformaldehyde.

Formaldehyde solution. Formalin; Formol; Morbicid; Veracur. A soln of about 37% by wt of formaldehyde gas in water, usually with 10-15% methanol added to prevent polymerization. This soln is the full strength and also known as Formalin 100% or Formalin 40 which signifies that it contains 40 grams of formaldehyde within 100 ml of the soln. Colorless liq; pungent odor. On standing, esp in the cold, may become cloudy and on exposure to very low temp a ppt of trioxymethylene is formed. When evaporated, some formaldehyde escapes, but most of it is changed to trioxymethylene. It is a powerful reducing agent especially in presence of alkali. In the air it slowly oxidizes to formic acid. d_{25}^{25} 1.081-1.085. One gallon weighs 9.1 lbs. bp$_{760}$ 96°. n_D^{20} 1.3746. Flash pt 60°C (140°F). pH 2.8-4.0. Misc with water, alcohol, acetone. *Keep well closed in a moderately warm place.* LD$_{50}$ orally in rats: 0.80 g/kg (Smyth).

Formaldehyde semicarbazone. [14066-69-4] 2-Methylenehydrazinecarboxamide. $C_2H_5N_3O$; mol wt 87.08. Crystals from ethanol, mp 120-121°. uv max (CH$_3$CN): 227 nm (log ε 3.73).

Caution: Potential symptoms of overexposure to formaldehyde are irritation of eyes, nose, throat and respiratory system; lacrimation; coughing; bronchial spasm. *See NIOSH Pocket Guide to Chemical Hazards* (DHHS/NIOSH 97-140, 1997) p 148. May act as a primary irritant on skin causing erythmatous or eczematous dermatitis. Sensitization can result. *See Patty's Industrial Hygiene and Toxicology* vol. 2A, G. D. Clayton, F. E. Clayton, Eds. (Wiley-Interscience, New York, 3rd ed., 1981) pp 2637-2646. Ingestion may cause immediate intense pain in mouth, pharynx and stomach; nausea, vomiting, hematemesis, abdominal pain, diarrhea; pale, clammy skin and other signs of shock; difficult urination, hematuria, anuria; vertigo, convulsions, stupor, coma; death due to respiratory failure. *See Clinical Toxicology of Commercial Products*, R. E. Gosselin *et al.*, Eds. (Williams & Wilkins, New York, 5th ed., 1984) Section III, pp 196-197. Formaldehyde is listed as a known human carcinogen: *Report on Carcinogens, Twelfth Edition* (PB2011-111646, 2011) p 195.

USE: In the prodn of amino, phenolic, and polyacetal resins, wood products, plastics, fertilizers and foam insulation. As a textile finish, preservative, stabilizer, disinfectant and antibacterial food additive. Reagent in synthetic organic chemistry. Reducing agent. Prevention of polymerization.

THERAP CAT: Solution as disinfectant.

THERAP CAT (VET): Solution as antiseptic; fumigant; has been used in tympany, diarrhea, mastitis, pneumonia, internal bleeding.

4264. Formaldehyde Sodium Bisulfite. [870-72-4] 1-Hydroxymethanesulfonic acid sodium salt (1:1); sodium formaldehyde bisulfite; methylolsulfonic acid sodium salt. CH_3NaO_4S; mol wt 134.08. C 8.96%, H 2.26%, Na 17.15%, O 47.73%, S 23.91%. HOCH$_2$SO$_3$Na. Prepn: Skrabal, Skrabal, *Monatsh. Chem.* **69**, 17 (1936); Chhabria *et al.*, **DE 1173059** (1964 to Schill & Seilacher). Structure: Lauer, Langkammerer, *J. Am. Chem. Soc.* **57**, 2360 (1935); Caughlan, Tartar, *ibid.* **63**, 1265 (1941).

Monohydrate. Needles from water. pK (25°): 6.92.

USE: Fixing agent for keratin-containing fibers. Flotation of lead-zinc ores. Protecting color pictures exposed to high humidities against fading and staining.

4265. Formamide. [75-12-7] Methanamide; carbamaldehyde; formimidic acid. CH_3NO; mol wt 45.04. C 26.67%, H 6.71%, N 31.10%, O 35.52%. HCONH$_2$. Prepd on a large scale from carbon monoxide and ammonia at high press. and temp: Meyer, Orthner, *Ber.* **54**, 1705 (1921); **55**, 857 (1922); Meyer, **DE 390798** (1924); **DE 392409** (1924); **DE 414257** (1925 to BASF); Wietzel, Herbst, **US 1843434** (1932 to I. G. Farben). Toxicity study: Pham-Huu-Chank *et al.*, *Toxicol. Appl. Pharmacol.* **26**, 596 (1973). *Review:* C. L. Eberling in *Kirk-Othmer Encyclopedia of Chemical Technology* vol. 11 (Wiley-Interscience, New York, 3rd ed., 1980) pp 258-263.

Slightly viscous, odorless, colorless liq. Industrial grades may have faint odor of ammonia. mp 2.55°. bp$_{760}$ 210.5° (partial decomp into CO and NH$_3$ at atm pressure beginning at 180°); bp$_{400}$ 193.5°; bp$_{200}$ 175.5°; bp$_{60}$ 147.0°; bp$_{20}$ 122.5°; bp$_{10}$ 109.5°; bp$_{1.0}$ 70.5°. d_4^{15} 1.13756; d_4^{20} 1.13340; d_4^{30} 1.12483. Dielectric constant ε = 84. Viscosity η × 10^5 at 15° = 4320, at 30° = 2926. Surface tension γ at 20° = 58.35. n_D^{15} 1.44911; n_D^{20} 1.44754; n_D^{110} 1.4170; n_D^{130} 1.4095. Flash pt, open cup: 310°F (154°C). pH of 0.5 molar aq soln 7.1. Misc with water, methanol, ethanol, acetone, acetic acid, dioxane, ethylene glycol, U.S.P. glycerol, phenol. Very slightly sol in ether, benzene. Dissolves casein, glucose, zein, tannins, starch, lignin, polyvinyl alcohol, cellulose acetate, nylon, the chlorides of copper, lead, zinc, tin, cobalt, iron, aluminum, nickel, the acetates of the alkali metals, some inorganic sulfates and nitrates. LD$_{50}$ in mice, rats (g/kg): 4.6, 5.7 i.p. (Pham-Huu-Chanh).

Caution: Potential symptoms of overexposure are irritation of eyes, skin, mucous membranes; drowsiness, fatigue; nausea, acidosis; skin eruptions. *See NIOSH Pocket Guide to Chemical Hazards* (DHHS/NIOSH 97-140, 1997) p 148.

USE: As ionizing solvent, manuf formic esters, hydrocyanic acid by catalytic dehydration, as softener for paper, animal glues, water-sol gums.

4266. Formanilide. [103-70-8] *N*-Phenylformamide; formylaniline. C_7H_7NO; mol wt 121.14. C 69.40%, H 5.82%, N 11.56%, O 13.21%. Review on methods of prepn: de Jonge *et al.*, *Rec. Trav. Chim.* **75**, 5 (1956).

Crystals from ether + petr ether, mp 46.6-47.5°. bp 271°; bp$_{120}$ 216°; bp$_{14}$ 166°. uv max (96% ethanol): 242-243 nm (ε 13700); in water: 239-240 nm (ε 11200); in cyclohexane: 240 nm (ε 12600). Soly in water at 20°: 25.4 g/l; at 25°: 28.6 g/l.

4267. Formebolone. [2454-11-7] (11α,17β)-11,17-Dihydroxy-17-methyl-3-oxoandrosta-1,4-diene-2-carboxaldehyde; 2-formyl-17α-methylandrosta-1,4-diene-11α,17β-diol-3-one; 2-formyl-11α-hydroxy-Δ^1-methyltestosterone; formyldienolone; Esiclene. $C_{21}H_{28}O_4$; mol wt 344.45. C 73.23%, H 8.19%, O 18.58%. Prepn: Canonica *et al.*, *Gazz. Chim. Ital.* **95**, 138 (1968); Gomarasca, **GB 1168931**; **FR 1584960** (1969, 1970 both to Lab. Prod. Biol. Braglia), *C.A.* **72**, 90742g (1970); **73**, 131212a (1970). Pharmacology: *idem*, *Minerva Med.* **62**, 842 (1971); Vaccari, Livini, *ibid.* 846; De Marchi *et al.*, *Arzneim.-Forsch.* **23**, 1583 (1973). GC/MS determn of metabolites in urine: R. Massé *et al.*, *Anal. Chim. Acta* **247**, 211 (1991). Toxicological studies: Travella, *Gazz. Int. Med. Chir.* **76**, 400 (1971), *C.A.* **75**, 30792a (1971).

Crystals from ethyl acetate, mp 209-212°. $[\alpha]_D^{25}$ -105° ($CHCl_3$). Water soluble. Approx LD_{50} in rats, mice (mg/kg): 104, 187 i.p.; 270, 293 s.c.; orally in rats: >1000 (Travella).

11-β-Epimer. mp 95-103° from ethyl ether-petr ether.

Note: This is a controlled substance (anabolic steroid): **21 CFR,** 1308.13, as defined in 1300.01.

THERAP CAT: Anabolic.

4268. Formestane. [566-48-3] 4-Hydroxyandrost-4-ene-3,17-dione; 4-OHA; CGP-32349; Lentaron. $C_{19}H_{26}O_3$; mol wt 302.41. C 75.46%, H 8.67%, O 15.87%. Selective aromatase inhibitor. Prepn: R. D. Burnett, D. N. Kirk, *J. Chem. Soc. Perkin Trans. 1* **1973**, 1830; J. Mann, B. Pietrzak, *ibid.* **1983**, 2681; P. G. Ciattini *et al., Synth. Commun.* **22**, 1949 (1992). Aromatase inhibition and effect on estrogen-dependent processes *in vivo:* A. M. H. Brodie *et al., Endocrinology* **100**, 1684 (1977). Radioimmunoassay in plasma: J. Khubieh *et al., J. Steroid Biochem.* **35**, 377 (1990). GC determn in plasma and urine: P. H. Degen, W. Schneider, *J. Chromatogr.* **565**, 67 (1991). Pharmacokinetics and endocrine effects in breast cancer patients: M. Dowsett *et al., Eur. J. Cancer* **28**, 415 (1992). Clinical evaluation in breast cancer: R. C. Coombes, *J. Steroid Biochem. Mol. Biol.* **43**, 145 (1992).

Needles from aq methanol, mp 199-202° (Burnett, Kirk); also reported as crystals from ethyl acetate, mp 203.5-206° (Brodie). uv max (99.5% ethanol): 278 nm (ε 11030). $[\alpha]_D^{20}$ +181° (c = 7.7 in chloroform).

THERAP CAT: Antineoplastic (hormonal).

4269. Formic Acid. [64-18-6] Ameisensäure (German). CH_2O_2; mol wt 46.03. C 26.09%, H 4.38%, O 69.52%. HCOOH. Found in the bites and stings of various insects including bees and ants; metabolite of methanol. Observed by S. Fisher in 1670 in the products resulting from the distillation of ants. Production: P. G. Jessop *et al., Nature* **368**, 231 (1994). Acute and chronic toxicity: G. Malorny, *Z. Ernaehrungswiss.* **9**, 332 (1969). Review of metabolism and toxicology: J. Liesivuori, H. Savolainen, *Pharmacol. Toxicol.* **69**, 157-163 (1991). Clinical trial in treatment of warts: R. M. Bhat *et al., Int. J. Dermatol.* **40**, 415 (2001). *Reviews:* A. Aguilo, T. Horlenko, *Hydrocarbon Process. Int. Ed.* **59**, 120-130 (1980); D. J. Drury in *Kirk-Othmer Encyclopedia of Chemical Technology* **vol.** 11 (Wiley-Interscience, New York, 4th ed., 1994) pp 951-958.

Colorless liquid; pungent odor. bp_{760} 100.8°. mp 8.4°. d^{20} 1.220. pKa (20°) 3.75. n_D^{20} 1.3714. Flash point, open cup: 59°C. Surface tension at 20°: 37.67 dyn/cm. Viscosity at 20°: 1.784 cP. Dielectric constant at 20°: 57.9. Vapor pressure at 20°: 33.55 mm Hg. Heat of fusion: 3031 cal/mole; of vaporization at 100°: 104 cal/g. Azeotropic mixture with water: 77.5 wt% formic acid, bp_{760} 107.3°. *Corrosive.* Strong reducing agent. Misc with water, ether, acetone, ethyl acetate, methanol, ethanol. Partially sol in benzene, toluene, xylenes. LD_{50} in mice (mg/kg): 1100 orally; 145 i.v. (Malorny).

Spirit of Formic Acid. Spirit of ants. Composed of 40 ml of 25% formic acid and 225 ml water in sufficient alcohol to make one liter; corresponds to ~1% formic acid and ~70% abs alcohol by vol. Colorless liquid. Miscible with water, alcohol.

Caution: Potential symptoms of overexposure are eye, skin, throat irritation; skin burns, dermatitis; lacrimation; rhinorrhea; coughing and dyspnea; nausea. *See NIOSH Pocket Guide to Chemical Hazards* (DHHS/NIOSH 97-140, 1997) p 148.

USE: Preservative in foods and silage; acidulant in dyeing of natural and synthetic fibers, leather tanning; coagulating latex in rubber production; reagent in synthetic organic chemistry.

THERAP CAT: Caustic.

4270. Formicin. [625-51-4] *N*-(Hydroxymethyl)acetamide; formaldehyde acetamide; methylal acetamide. $C_3H_7NO_2$; mol wt 89.09. C 40.45%, H 7.92%, N 15.72%, O 35.92%. Obtained by warming acetamide with formaldehyde and a small amount of potash: Fuchs, *Pharm. Ztg.* **1905**, 803; **DE 164610** (1905 to Kalle).

Colorless, very hygroscopic mass. Marketed only as a slightly yellowish, syrupy liq. d 1.14-1.18. Very sol in water, alcohol; sol in chloroform, glycerol. Insol in ether.

USE: Antiseptic, disinfectant esp for surgical instruments.

4271. Formononetin. [485-72-3] 7-Hydroxy-3-(4-methoxyphenyl)-4*H*-1-benzopyran-4-one; 7-hydroxy-4'-methoxyisoflavone; biochanin B; formononetol; neochanin. $C_{16}H_{12}O_4$; mol wt 268.27. C 71.64%, H 4.51%, O 23.86%. Phytoestrogen found predominantly in trifolium (red clover), also in soybean and other legumes; metabolic precursor of daidzein, *q.v.* Prepn from daidzein: E. Walz, *Ann.* **489**, 118 (1931). Proposed structure: W. Baker *et al., J. Chem. Soc.* **1933**, 274. Synthesis: F. Wessely *et al., Ber.* **66**, 685 (1933); W. Baker *et al., J. Chem. Soc.* **1953**, 1852; S.-R. Li *et al., Tetrahedron Lett.* **50**, 2121 (2009). Isoln from clover species *Trifolium subterraneum* L. and *T. pratense* L., *Leguminosae:* R. B. Bradbury, D. E. White, *J. Chem. Soc.* **1951**, 3447; E. C. Bate-Smith *et al., Chem. Ind. (London)* **1953**, 1127. Identity with biochanin B: J. L. Bose, *J. Sci. Ind. Res.* **15B**, 325 (1956). HPLC determn in urine: J. Tekel *et al., J. Agric. Food Chem.* **47**, 3489 (1999). Metabolism in humans: S.-M. Heinonen *et al., ibid.* **52**, 6802 (2004). Antioxidant and estrogenic effects: H. Mu *et al., Phytomedicine* **16**, 314 (2009).

Needles from alcohol, mp 258°. uv max (ethanol): 250, 300 nm (ε 27440, 11240).

Ononin. [486-62-4] 7-(β-D-Glucopyranosyloxy)-3-(4-methoxyphenyl)-4*H*-1-benzopyran-4-one; formononetin 7-glucoside; 4'-methyldaidzin. $C_{22}H_{22}O_9$; mol wt 446.40. Isoln from *Ononis spinosa* L., *Leguminosae:* Hlasiwetz, *J. Prakt. Chem.* **65**, 415 (1855). Synthesis: L. Farkas, J. Varady, *Ber.* **92**, 819 (1959). Needles from water as monohydrate, mp 210-214°. When anhydr, dec 245°. $[\alpha]_D^{25}$ -24.2° (pyridine). Sol in alcohol; slightly sol in water, ether.

USE: Dietary supplement.

4272. Formoterol. [73573-87-2] *rel-N*-[2-Hydroxy-5-[(1*R*)-1-hydroxy-2-[[(1*R*)-2-(4-methoxyphenyl)-1-methylethyl]amino]ethyl]phenyl]formamide; 3-formylamino-4-hydroxy-α-[*N*-[1-methyl-2-(*p*-methoxyphenyl)ethyl]aminomethyl]benzyl alcohol; (±)-2'-hydroxy-5'-[(*RS*)-1-hydroxy-2-[[(*RS*)-*p*-methoxy-α-methylphenethyl]amino]ethyl]formanilide. $C_{19}H_{24}N_2O_4$; mol wt 344.41. C 66.26%, H 7.02%, N 8.13%, O 18.58%. Selective β$_2$-adrenergic receptor agonist. Mixture of *R,R* (-) and *S,S* (+) enantiomers. Prepn: M. Murakami *et al.,* **DE 2305092**; *eidem,* **US 3994974** (1973, 1976 both to Yamanouchi); K. Murase *et al., Chem. Pharm. Bull.* **25**, 1368 (1977). Absolute configuration and activity of isomers: *eidem, ibid.* **26**, 1123 (1978). Toxicity studies: T. Yoshida *et al., Pharmacometrics* **26**, 811 (1983). HPLC determn in plasma: J. Campestrini *et al.,*

J. Chromatogr. B **704**, 221 (1997). Review of pharmacology: G. P. Anderson, *Life Sci.* **52**, 2145-2160 (1993); and clinical efficacy: R. A. Bartow, R. N. Brogden, *Drugs* **55**, 303-322 (1998).

R,R-Form

Fumarate dihydrate. [43229-80-7] BD-40A; Atock; Foradil; Oxeze. $(C_{19}H_{24}N_2O_4)_2 \cdot C_4H_4O_4 \cdot 2H_2O$; mol wt 840.92. Crystals from 95% isopropyl alcohol, mp 138-140°. pKa_1 7.9; pKa_2 9.2. Log P (octanol/water): 0.4 (pH 7.4). Freely sol in dimethyl sulfoxide, acetic acid; sol in methanol; sparingly sol in ethanol, isopropanol; slightly sol in water, 2-propanol. Practically insol in acetonitrile, ethyl acetate, diethyl ether. LD_{50} in male, female, rats, mice (mg/kg): 3130, 5580, 6700, 8310 orally; 98, 100, 72, 71 i.v.; 1000, 1100, 640, 670 s.c.; 170, 210, 240, 210 i.p. (Yoshida).

R,R-**Form.** [67346-49-0] Arformoterol.

R,R-**Form L-tartrate.** [200815-49-2] Arformoterol tartrate; Brovana. $C_{19}H_{24}N_2O_4 \cdot C_4H_6O_6$; mol wt 494.50. Prepn: Y. Gao *et al.*, **WO 9821175**; *eidem*, **US 6040344** (1998, 2000 both to Sepracor). Pharmacology: D. A. Handley *et al.*, *Pulm. Pharmacol. Ther.* **15**, 135 (2002). Off-white powder, mp 184°. Slightly sol in water.

THERAP CAT: Antiasthmatic.

4273. Formothion. [2540-82-1] Phosphorodithioic acid *S*-[2-(formylmethylamino)-2-oxoethyl] *O,O*-dimethyl ester; phosphorodithioic acid *O,O*-dimethyl ester *S*-ester with *N*-formyl-2-mercapto-*N*-methylacetamide; *S*-(*N*-formyl-*N*-methylcarbamoylmethyl) *O,O*-dimethyl phosphorodithioate; J-38; OMS-968; Aflix; Anthio. $C_6H_{12}NO_4PS_2$; mol wt 257.26. C 28.01%, H 4.70%, N 5.44%, O 24.88%, P 12.04%, S 24.92%. Organophosphate insecticide; cholinesterase inhibitor. Prepn: Lutz, Schuler, **GB 900557** (1962 to Sandoz). Multi-residue determn in cereals by GC-MS/MS: S. Walorczyk, *J. Chromatogr. A* **1208**, 202 (2008). Toxicity: K. Liesche, *Dtsch. Apoth.* **108**, 604 (1968), *C.A.* **69**, 26287y (1968).

Yellowish liq, mp ~25°. Practically insol in water. Miscible with alcohols, ether, chloroform, benzene. LD_{50} in rats, rabbits (mg/kg): 350, 410 orally (Liesche).

USE: Acaricide, insecticide.

4274. Formyl Fluoride. [1493-02-3] CHFO; mol wt 48.02. C 25.01%, H 2.10%, F 39.56%, O 33.32%. HCOF. Prepn by the interaction of anhydr formic acid, potassium fluoride and benzoyl chloride (16% yield): Nyesmejanov, Kahn, *Ber.* **67**, 203 (1946); from benzoyl fluoride and formic acid (36% yield): Masentshev, *J. Gen. Chem. USSR* **16**, 203 (1946); from benzoyl chloride, formic acid, and KHF_2 (35% yield): Olah *et al.*, *Ber.* **89**, 862 (1956); Olah, Kuhn, *J. Am. Chem. Soc.* **82**, 2381 (1960); from acetic formic anhydride, $CH_3COOOCH$, and anhydr HF (67% yield): Olah, Kuhn, *ibid.* 2380.

Gas at ordinary temps and pressure: bp_{760} −29°. mp −142°. Trouton constant 20.8. Heat of vaporization at −24°: 5.175 kcal.

Caution: Strong irritant.

USE: Acylating agent in organic synthesis.

4275. Fortimicins. Aminoglycoside antibiotic complex produced by *Micromonospora olivoasterospora*. The main components, fortimicins A and B, both exhibit broad-spectrum antibacterial activity. Isoln of A: T. Nara *et al.*, **JP Kokai 75 29789** (to Kyowa); *eidem*, **US 3976768** (1976 to Abbott). Isoln of B: *eidem*, **DE 2418349** (1974 to Kyowa); *eidem*, **US 3931400** (1976 to Abbott). Isoln and biological properties of A and B: *eidem*, *J. Antibiot.* **30**, 533 (1977); physical and chromatographic properties: R. Okachi *et*

al., *ibid.* 541. Structures of A and B: R. S. Egan *et al.*, *ibid.* 552. *In vitro* activity: R. Girolami, J. S. Stamm, *ibid.* 564. Isoln and properties of minor components, fortimicins C, D, KE: M. Sugimoto *et al.*, *ibid.* **32**, 868 (1979). Synthesis of B: T. Suami, Y. Honda, *Chem. Lett.* **1980**, 641; Y. Honda, T. Suami, *Bull. Chem. Soc. Jpn.* **55**, 1156 (1982).

Fortimicin A   Fortimicin B

Fortimicin A. [55779-06-1] 4-Amino-1-[(aminoacetyl)methylamino]-1,4-dideoxy-3-*O*-(2,6-diamino-2,3,4,6,7-pentadeoxy-β-L-*lyxo*-heptopyranosyl)-6-*O*-methyl-L-*chiro*-inositol; astromicin; Abbott 44747. $C_{17}H_{35}N_5O_6$; mol wt 405.50. White amorphous powder, mp >200° (dec). $[\alpha]_D^{25}$ +87.5° (c = 0.1 in water). Sol in water and lower alcohols. Insol in organic solvents. LD_{50} (of the sulfate salt) in mice (mg/kg): 380 i.v.; 400 s.c. (Nara, 1977).

Fortimicin B. [54783-95-8] 4-Amino-1,4-dideoxy-3-*O*-(2,6-diamino-2,3,4,6,7-pentadeoxy-β-L-*lyxo*-heptopyranosyl)-6-*O*-methyl-1-(methylamino)-L-*chiro*-inositol. $C_{15}H_{32}N_4O_5$; mol wt 348.44. White amorphous powder, mp 101-103°. $[\alpha]_D^{25}$ +22.2° (c = 0.1 in water). Sol in water and lower alcohols. Insol in organic solvents.

THERAP CAT: Antibacterial.

4276. Fosamprenavir. [226700-79-4] *N*-[(1*S*,2*R*)-3-[[(4-Aminophenyl)sulfonyl](2-methylpropyl)amino]-1-(phenylmethyl)-2-(phosphonooxy)propyl]carbamic acid *C*-[(3*S*)-tetrahydro-3-furanyl] ester; (3*S*)-tetrahydro-3-furyl [(α*S*)-α-[(1*R*)-1-hydroxy-2-(*N*[1]-isobutylsulfanilamido)ethyl]phenethyl]carbamate dihydrogen phosphate (ester); VX-175. $C_{25}H_{36}N_3O_9PS$; mol wt 585.61. C 51.28%, H 6.20%, N 7.18%, O 24.59%, P 5.29%, S 5.47%. Peptidomimetic HIV protease inhibitor; water soluble prodrug of amprenavir, *q.v.* Prepn: R. D. Tung *et al.*, **WO 9933815**; *eidem*, **US 6559137** (1999, 2003 both to Vertex). Prepn of crystalline calcium salt: I. G. Armitage *et al.*, **WO 0004033** (2000 to Glaxo); *eidem*, **US 6514953** (2003 to SKB). Clinical pharmacokinetics: C. Falcoz *et al.*, *J. Clin. Pharmacol.* **42**, 887 (2002). Review of pharmacology and clinical experience in HIV: T. M. Chapman *et al.*, *Drugs* **64**, 2101-2124 (2004); C. Arvieux, O. Tribut, *ibid.* **65**, 633-659 (2005).

Calcium salt. [226700-81-8] GW-433908G; Lexiva; Telzir. $C_{25}H_{34}CaN_3O_9PS$; mol wt 623.67. White microcrystalline needles, mp 282-284°. Soly in water (25°): 0.31 mg/ml.

THERAP CAT: Antiretroviral.

4277. Fosbretabulin. [222030-63-9] 2-Methoxy-5-[(1*Z*)-2-(3,4,5-trimethoxyphenyl)ethenyl]phenol 1-(dihydrogen phosphate); combretastatin A-4 phosphate; CA4P. $C_{18}H_{21}O_8P$; mol wt 396.33. C 54.55%, H 5.34%, O 32.29%, P 7.82%. Phosphate prodrug of combretastatin A-4, a naturally occuring stilbenoid isolated from the South African bushwillow. Tubulin-binding agent that disrupts tumor endothelial vasculature and reduces blood flow to the tumor site. Prepn: G. R. Pettit *et al.*, *Anti-Cancer Drug Design* **10**, 299 (1995); *idem*, **US 5561122** (1996 to Arizona Board of Regents); of tromethamine salt: J. J. Venit *et al*, **WO 0222626**; *eidem*, **US 6670344** (2002, 2003 both to Bristol-Myers Squibb). LC-MS/MS determn in plasma: X. Wang *et al., J. Chromatogr. B* **877**, 3813 (2009). Mechanisms of action vs leukemic cells: I. Petit *et al., Blood* **111**, 1951 (2008). Clinical evaluation in anaplastic thyroid carcinoma: C. J. Mooney *et al., Thyroid* **19**, 233 (2009). Review of pharmacology and clinical experience: G. Nahaiah, S. C. Remick, *Future Oncol.* **6**, 1219-1228 (2010).

Oily mass. pKa 1.2; 6.2. Practically insol to very slightly sol in water.

Disodium salt. [168555-66-6] Combretastatin A-4 disodium phosphate; disodium combretastatin A-4 3-*O*-phosphate. $C_{18}H_{19}$-Na_2O_8P; mol wt 440.30. Cream colored solid from methanol, mp 190-195° (dec). uv max (methanol): 218, 292 nm (log ε 4.16, 3.78).

Tromethamine. [404886-32-4] 2-Amino-2-(hydroxymethyl)-1,3-propanediol compd with 2-methoxy-5-[(1*Z*)-2-(3,4,5-trimethoxyphenyl)ethenyl]phenyl dihydrogen phosphate (1:1); CA4P mono-TRIS salt; Zybrestat. $C_{18}H_{21}O_8P \cdot C_4H_{11}NO_3$; mol wt 517.47. White solid from aq isopropanol, mp 196°. Soly in water (mg/ml): 3.37 (pH 4.8); 191 (pH 8.2).

THERAP CAT: Antineoplastic.

4278. Foscarnet Sodium. [63585-09-1] 1,1-Dihydroxyphosphinecarboxylic acid 1-oxide sodium salt (1:3); dihydroxyphosphinecarboxylic acid oxide trisodium salt; trisodium phosphonoformate; trisodium carboxyphosphate; A-29622. CNa_3O_5P; mol wt 191.95. C 6.26%, Na 35.93%, O 41.67%, P 16.14%. Pyrophosphate analog; inhibits viral DNA polymerase and reverse transcriptase. Prepn: P. Nylén, *Ber.* **57B**, 1023 (1924). Crystal structure: R. R. Naqvi *et al., J. Chem. Soc. A* **1971**, 2751; S. C. Abrahams, *Acta Crystallogr.* **B28**, 2886 (1972). Activity against herpes simplex virus: E. Helgstrand *et al., Science* **201**, 819 (1978). Effect on HIV-1 reverse transcriptase: P. S. Sarin *et al., Biochem. Pharmacol.* **34**, 4075 (1985). HPLC determn in plasma: B. B. Ba *et al., J. Chromatogr. B* **724**, 127 (1999); in pharmaceutical formulations: J. Garcia *et al., Biomed. Chromatogr.* **20**, 1024 (2006). Series of articles on pharmacology and clinical experience in cytomegalovirus retinitis: *Am. J. Med.* **92**, Suppl. 1, S1-S35 (1992). Clinical trial in multidrug resistant HIV infection: A. Canestri *et al., Antivir. Ther.* **11**, 561 (2006).

White to almost white, crystalline powder. Sol in water. Practically insol in alc.

Hexahydrate. [34156-56-4] Foscavir. $CNa_3O_5P \cdot 6H_2O$; mol wt 300.04. White or almost white, crystalline powder. mp >250°. Sol in water. Practically insol in ethanol.

THERAP CAT: Antiviral.

4279. Fosetyl Al. [39148-24-8] Phosphonic acid monoethyl ester aluminum salt (3:1); aluminum tris(ethyl phosphite); aluminum tris(*O*-ethylphosphonate); efosite Al; phosethyl Al; LS-74-783; Aliette; Chipco Signature. $C_6H_{18}AlO_9P_3$; mol wt 354.10. C 20.35%, H 5.12%, Al 7.62%, O 40.66%, P 26.24%. Systemic fungicide translocated both upwards and downwards in plants. Prepn: V. V. Orlovskii *et al., J. Gen. Chem. USSR* **42**, 1924 (1972). Alternate prepn, fungicidal properties: J. Ducret *et al.,* **DE 2456627**; *eidem*, **US 4139616** (1975, 1979 both to Rhone-Poulenc); J. Abblard *et al.,* **US 4143059** (1979 to Philagro). LC-MS/MS determn in vegetable samples: F. Hernández *et al., J. AOAC Int.* **86**, 832 (2003). Biological properties and field trials against vine mildew and other *Phycomycetes* pathogens: A. Bertrand *et al., Phytiatr.-Phytopharm.* **26**, 3 (1977). Evaluation of carcinogenicity: J. A. Quest *et al., Regul. Toxicol. Pharmacol.* **14**, 3 (1991). Review of properties and fungi-

cidal action on tropical, temperate crops: D. J. Williams *et al., Proc. Br. Crop Prot. Conf. - Pests Dis.* **1977**, 565.

White odorless crystals, mp >300°. Soly in water at 20°: 120 g/l. Practically insol in acetonitrile, propylene glycol (<80 mg/l). LD_{50} in rats, mice, Japanese quail, rabbits (mg/kg): 5800, 3700, 4997, 2680 orally; LD_{50} i.p. in rats: 550 mg/kg (Williams).

USE: Agricultural fungicide.

4280. Fosfestrol. [522-40-7] 4,4'-[(1*E*)-1,2-Diethyl-1,2-ethenediyl]bisphenol 1,1'-bis(dihydrogen phosphate); α,α'-diethyl-4,4'-stilbenediol diphosphoric acid ester; diethylstilbestrol diphosphate; diethylstilbestryl diphosphate; diethyldihydroxystilbene diphosphate; stilbestrol diphosphate; Stilphostrol. $C_{18}H_{22}O_8P_2$; mol wt 428.31. C 50.48%, H 5.18%, O 29.88%, P 14.46%. Synthetic, nonsteroidal estrogen. Prepd by treating stilbestrol with phosphorus oxychloride in pyridine: Miescher, Heer, **US 2234311** (1941); Arnold, **US 2802854** (1957 to Asta-Werke). Prepn of stable solns at pH 10: Fonner, Collins, **US 2828244** (1958 to Miles Labs.). Prepn of sodium salts: Dawson, **US 2971975** (1961 to Miles Labs.). Clinical pharmacokinetics: H. Oelschläger *et al., Arzneim.-Forsch.* **36**, 1284 (1986). Review of clinical trials in prostate cancer: J-P. Droz *et al, Cancer* **71**, 1123-1130 (1993).

Voluminous white cryst powder from dil HCl. Dec 204-206°. Soluble in alc, dilute alkali; sparingly sol in water.

Disodium salt. [5965-09-3] $C_{18}H_{20}Na_2O_8P_2$; mol wt. Crystals. Softens at 190°, uniform melt at 230°. Sol in water. pH of aq soln 6.5. pH also reported as 4.85 to 5.00.

Tetrasodium salt. [4719-75-9] Honvan; Honvol; ST-52; Stilbostatin. $C_{18}H_{18}Na_4O_8P_2$; mol wt 516.24.

THERAP CAT: Antineoplastic (hormonal).

4281. Fosfomycin. [23155-02-4] *P*-[(2*R*,3*S*)-3-Methyl-2-oxiranyl]phosphonic acid; (−)-(1*R*,2*S*)-(1,2-epoxypropyl)phosphonic acid; fosfonomycin; phosphonomycin; MK-955. $C_3H_7O_4P$; mol wt 138.06. C 26.10%, H 5.11%, O 46.35%, P 22.44%. Antibiotic produced by *Streptomyces* strains: D. Hendlin *et al., Science* **166**, 122 (1969); B. G. Christensen *et al., ibid.* 123; also produced by *Pseudomonas syringae:* J. Shogi *et al., J. Antibiot.* **39**, 1011 (1986). Isoln: D. Hendlin *et al.,* **BE 718507**; *eidem*, **US 3914231** (1969, 1975 both to Merck & Co.). Synthesis and resolution: B. G. Christensen *et al.,* **BE 723072**; **BE 723073** (both 1969 to Merck & Co.). Alternate synthesis: Girotra, Wendler, *Tetrahedron Lett.* **1969**, 4647; Glamkowski *et al., J. Org. Chem.* **35**, 3510 (1970). Series of articles on characterization, activity and clinical testing: *Antimicrob. Agents Chemother.* **1969**, 284-351; H. B. Woodruff *et al., Chemotherapy* **23**, Suppl. 1, 1-22 (1977). Mechanism of action: Kahan *et al., Ann. N.Y. Acad. Sci.* **235**, 354 (1974). Pharmacokinetics in humans: M. Goto *et al., Antimicrob. Agents Chemother.* **20**, 393 (1981). Pharmacology of *cofosfolactamines*, fixed dose combinations of fosfomycin with β-lactam antibiotics: P. Periti, *Drugs Exp. Clin. Res.* **6**, 305 (1980); and clinical trial: T. Barreca *et al., ibid.* **10**, 55 (1984). Electrophoretic determn in biological fluids: D. Levêque *et al., J. Chromatogr. B* **655**, 320 (1994). Clinical trial in cystic fibrosis: A. Mirakhur *et al., J. Cyst. Fibros.* **2**, 19 (2003).

Crystals, mp ~94°. Soluble in water.

Benzylammonium salt. $C_{10}H_{16}NO_4P$. mp 170-174°. $[\alpha]_{405}$ −9.1° (c = 5).

Calcium salt monohydrate. [26016-98-8] (anhydrous). Afos; Biocin; Biofos; Endociclina; Faremicin; Fonofos; Fosfobiotic; Fosfocin; Fosfocina; Fosfogram; Fosforal; Fosfotricina; Fosmicin; Foximin; Francital; Gram-Micina; Ipamicina; Lancetina; Lofoxin; Neofocin; Palmofen; Priomicina; Selemicina; Ultramicina; Valemicina. $C_3H_5CaO_4P.H_2O$; mol wt 194.14.

Tromethamine. [78964-85-9] (2R-cis)-(3-Methyloxiranyl)-phosphonic acid compd with 2-amino-2-(hydroxymethyl)-1,3-propanediol (1:1); fosfomycin trometamol; Monuril; Monurol. $C_7H_{18}NO_7P$; mol wt 259.19. Clinical trial in urinary tract infection: G. Bonfiglio et al., Chemotherapy (Basel) **51**, 162 (2005).

THERAP CAT: Antibacterial.

4282. Fosfosal. [6064-83-1] 2-(Phosphonooxy)benzoic acid; salicylic acid dihydrogen phosphate; o-carboxyphenyl phosphate; salicyl phosphate; UR-1521; Disdolen. $C_7H_7O_6P$; mol wt 218.10. C 38.55%, H 3.24%, O 44.01%, P 14.20%. Salicylic acid deriv. Prepn: A. C. Marin Moga, E. Francia Barra, DE 2641526 (1978 to Uriach), C.A. **88**, 136317h (1978). Pharmacology and comparison with aspirin: J. Garcia Rafanell et al., Arzneim.-Forsch. **30**, 1091 (1980). Toxicity studies: M. S. Sanchez et al., ibid. 1098. Pharmacokinetics: V. Rimbau et al., Arch. Farmacol. Toxicol. **7**, 61 (1981). Double-blind trials in musculoskeletal/articular pain: C. Diaz et al., Clin. Ther. **4**, 121 (1981); L. Madera Cat, J. Garcia Rafanell, Med. Clin. **76**, 18 (1981).

White solid, mp 168-170°. Sol in water, ethanol, acetone. Insol in non-polar organic solvents. Hydrolyzes in aq soln. LD_{50} in male, female mice, male, female rats at pH 1.0, aq soln (mg/kg): 94, 105, 153, 257 i.v.; 352, 253, 338, 360 i.p.; 1455, 1539, 1104, 1213 orally; at pH 3.5 (mg/kg): 117, 118, 207, 215 i.v.; 1592, 1483, 1085, 1128 i.p.; 1702, 2007, 1685, 2225 orally (Sanchez).

THERAP CAT: Analgesic.

4283. Fosinopril. [98048-97-6] (4S)-4-Cyclohexyl-1-[2-[(R)-[(1S)-2-methyl-1-(1-oxopropoxy)propoxy](4-phenylbutyl)phosphinyl]acetyl]-L-proline; (4S)-4-cyclohexyl-1-[[[(RS)-1-hydroxy-2-methylpropoxy](4-phenylbutyl)phosphinyl]acetyl]-L-proline propionate (ester); fosenopril. $C_{30}H_{46}NO_7P$; mol wt 563.67. C 63.93%, H 8.23%, N 2.48%, O 19.87%, P 5.50%. Phosphinic acid containing angiotensin converting enzyme (ACE) inhibitor. Prepn: E. W. Petrillo, Jr., US 4337201 (1982 to Squibb); of active diacid form: J. Krapcho et al., J. Med. Chem. **31**, 1148 (1988). Metabolism and pharmacokinetics: S. M. Singhvi et al., Br. J. Clin. Pharmacol. **25**, 9 (1988). GC determn of diacid: M. Jemal et al., J. Chromatogr. **345**, 299 (1985). Clinical trial in hypertension: P. A. Sullivan et al., Am. J. Hypertens. 1, 280S (1988). Brief description: E. W. Petrillo, Jr., et al., Clin. Exp. Theory Prac. **A9**, 235 (1987). Series of articles on pharmacology, pharmacokinetics, and clinical experience: Drug Invest. **3**, Suppl. 4, 1-53 (1991).

Sodium salt. [88889-14-9] SQ-28555; Dynacil; Elidiur; Fosinorm; Fositen; Fozitec; Monopril; Staril; Tensozide. $C_{30}H_{45}NNaO_7P$; mol wt 585.65.

Diacid. [95399-71-6] SQ-27519. $C_{23}H_{34}NO_5P$; mol wt 435.50. mp 149-153°. $[\alpha]_D$ −24° (c = 1 in methanol).

THERAP CAT: Antihypertensive.

4284. Fosmidomycin. [66508-53-0] P-[3-(Formylhydroxyamino)propyl]phosphonic acid. $C_4H_{10}NO_5P$; mol wt 183.10. C 26.24%, H 5.51%, N 7.65%, O 43.69%, P 16.92%. Inhibitor of microbial nonmevalonate isoprenoid biosynthesis. Prepn: T. Kamiya et al., US 4206156 (1980 to Fujisawa); K. Hemmi et al., Chem. Pharm. Bull. **29**, 646 (1981); eidem, ibid. **30**, 111 (1982). Synthesis: E. Ohler, S. Kanzler, Synthesis **1995**, 539 (1995). CE determn in serum and urine: S. Bronner et al., J. Chromatogr. B **806**, 255 (2004). Mode of action: T. Kuzuyama et al., Tetrahedron Lett. **39**, 7913 (1998); H. Jomaa et al., Science **285**, 1573 (1999). Pharmacokinetics: T. Murakawa et al., Antimicrob. Agents Chemother. **21**, 224 (1982); H.-P. Kuemmerle et al., Int. J. Clin. Pharmacol. Ther. Toxicol. **23**, 515, 521 (1985). Clinical evaluation in malaria: M. A. Missinou et al., Lancet **360**, 1941 (2002); in combination with clindamycin: S. Borrmann et al., J. Infect. Dis. **189**, 901 (2004); idem et al., ibid. **190**, 1534 (2004). Interaction with other antimalarial drugs and synergy with clindamycin: J. Wiesner et al., Antimicrob. Agents Chemother. **46**, 2889 (2002). Review: idem et al., Parasitol. Res. **90**, S71-S76 (2003).

Colorless crystals from methanol, mp 158-160° (dec) (Ohler); also reported as crystals from water-ethanol, mp 160-166° (dec) (Hemmi). LD_{50} in mice and rats (mg/kg): >11000 orally; ~8000 s.c. (Wiesner, 2003).

Monosodium salt. [66508-37-0] FR-31564. $C_4H_{10}NNaO_5P$; mol wt 206.09. Crystals from methanol-ethanol, mp 189-191° (dec).

THERAP CAT: Antimalarial.

4285. Fosphenytoin. [93390-81-9] 5,5-Diphenyl-3-[(phosphonooxy)methyl]-2,4-imidazolidinedione; 3-(hydroxymethyl)-5,5-diphenylhydantoin phosphate ester; (3-phosphoryloxymethyl)phenytoin; prophenytoin. $C_{16}H_{15}N_2O_6P$; mol wt 362.28. C 53.05%, H 4.17%, N 7.73%, O 26.50%, P 8.55%. Prodrug of phenytoin, q.v. Prepn: V. J. Stella, K. B. Sloan, US 4260769 (1981 to INTERx). Series of articles on chemical properties and pharmacokinetics: S. A. Varia et al., J. Pharm. Sci. **73**, 1068-1090 (1984). Pharmacology: R. D. Smith et al., Epilepsia **30**, Suppl. 2, S15 (1989). Clinical bioavailablilty and pharmacokinetics: B. D. Jamerson et al., ibid. **31**, 592 (1990). Series of articles on use in epilepsy: Neurology **46**, Suppl. 1, S1-S28 (1996). Clinical trial in epilepsy and neurosurgery: B. J. Wilder et al., Arch. Neurol. **53**, 764 (1996).

White crystals from acetone, mp 173-176.5°.

Disodium salt. [92134-98-0] ACC-9653; Cerebyx; Pro-Epanutin. $C_{16}H_{13}N_2Na_2O_6P$; mol wt 406.24. White crystals as the dihydrate from water-ethanol-acetone, mp 220° (softens). pH of a saturated aqueous soln: ~9. Soly at 25° (mg/ml): 142 in water. LD_{50} in mice, rats (mg/kg): 234, 363 i.v. (Smith).

THERAP CAT: Anticonvulsant.

4286. Fospropofol. [258516-89-1] 1-[2,6-Bis(1-methylethyl)phenoxy]methanol 1-dihydrogen phosphate; phosphono-O-methyl-2,6-diisopropylphenol. $C_{13}H_{21}O_5P$; mol wt 288.28. C 54.16%, H 7.34%, O 27.75%, P 10.74%. Prodrug of propofol, q.v. Prepn: V. J. Stella et al., WO 0008033; eidem, US 6204257 (2000, 2001 to Univ. Kansas). Clinical pharmacokinetics and pharmacodynamics: J. Fechner et al., Anesthesiology **99**, 303 (2003). Clinical evaluation to induce and maintain sedation: eidem, Anesth. Analg. **100**, 701

(2005). Clinical comparison with propofol: E. Gibiansky *et al.*, *Anesthesiology* **103**, 718 (2005); M. M. R. Struys *et al.*, *ibid.* 730. Clinical trial in moderate sedation in flexible bronchoscopy: G. A. Silvestri *et al.*, *Chest* **135**, 41 (2009). *Review*: M. D. Krasowski, *Curr. Opin. Investig. Drugs* **6**, 90-98 (2005).

Colorless oil, unstable on standing at room temp.
Disodium salt. [258516-87-9] Fospropofol disodium; GPI-15715; Aquavan; Lusedra. $C_{13}H_{19}Na_2O_5P$; mol wt 332.24. White powder. Sol in water.
Note: This is a controlled substance (depressant): **21 CFR**, 1308.14.
THERAP CAT: Anesthetic (intravenous).

4287. Fostamatinib. [901119-35-5] 6-[[5-Fluoro-2-[(3,4,5-trimethoxyphenyl)amino]-4-pyrimidinyl]amino]-2,2-dimethyl-4-[(phosphonooxy)methyl]-2*H*-pyrido[3,2-*b*]-1,4-oxazin-3-(4*H*)-one; N^4-(2,2-dimethyl-4-[(dihydrogen phosphonoxy)methyl]-3-oxo-5-pyrido[1,4]oxazin-6-yl)-5-fluoro-N^2-(3,4,5-trimethoxyphenyl)-2,4-pyrimidinediamine; R-788; R-935788. $C_{23}H_{26}FN_6O_9P$; mol wt 580.47. C 47.59%, H 4.51%, F 3.27%, N 14.48%, O 24.81%, P 5.34%. Spleen tyrosine kinase (Syk) inhibitor; prodrug for the active metabolite, *R-406*. Blocks Fc receptor signaling and reduces immune complex mediated inflammation. Prepn of active compd: R. Singh *et al.*, **WO 05012294**; *eidem*, **US 7122542** (2005, 2006 both to Rigel); of prodrug: *idem et al.*, **WO 06078846**; S. Bhamidipati *et al.*, **US 7449458** (2006, 2008 both to Rigel); of prodrug salts: *eidem*, **WO 08064274** (2008 to Rigel). Pharmacology: S. Braselmann *et al.*, *J. Pharmacol. Exp. Ther.* **319**, 998 (2006); P. R. Pine *et al.*, *Clin. Immunol.* **124**, 244 (2007). Clinical trial in treatment of rheumatoid arthritis: M. E. Weinblatt *et al.*, *Arthritis Rheum.* **58**, 3309 (2008). Review of pharmacology and clinical experience: M. Bajpai, *IDrugs* **12**, 174-185 (2009).

Off-white, fluffy solid. Soly in water: >5 mg/ml.
Disodium salt. [1025687-58-4]; [914295-16-2] (hexahydrate). Tamatinib fosdium. $C_{23}H_{24}FN_6Na_2O_9P$; mol wt 624.43. White, monoclinic crystals as the hexahydrate. Begins to dehydrate at 40°, becoming anhydrous at ~110° with decompn. d 1.485. Stable under ambient conditions.
THERAP CAT: Antiarthritic.

4288. Fosthiazate. [98886-44-3] *P*-(2-Oxo-3-thiazolidinyl)-phosphonothioic acid *O*-ethyl *S*-(1-methylpropyl) ester; *S-sec*-butyl-*O*-ethyl(2-oxo-3-thiazolidinyl)phosphonothioate; IKI-1145; ASC-66824; Nemathorin. $C_9H_{18}NO_3PS_2$; mol wt 283.34. C 38.15%, H 6.40%, N 4.94%, O 16.94%, P 10.93%, S 22.63%. Non-fumigant insecticide and nematicide. Prepn: T. Haga *et al.*, **EP 146748**; *eidem*, **US 4590182** (1985, 1986 both to Ishihara Sangyo Kaisha); and pesticidal activity: T. Koyanagi *et al.*, *J. Pestic. Sci.* **22**, 187 (1997). Activity of isomers: *idem et al.*, *ACS Symp. Ser.* **443**, 387 (1991). Field trial in potatoes: J. Kimpinski *et al.*, *J. Nematol.* **29**, Suppl. 4,

685 (1997). Mode of action: S. R. Woods *et al.*, *Ann. Appl. Biol.* **135**, 409 (1999). Review of physical properties, mode of action, and biological characteristics: T. Toki, O. Imai, *Agrochem. Jpn.* **65**, 13 (1994).

Pale yellow oil. bp$_{0.5}$ 198°. $n_D^{19.6}$ 1.5334. Soly in water at 20°: 9.85 g/l. Vapor pressure at 25°: 4.2×10^{-6} mmHg. LD$_{50}$ in male, female mice, male, female rats (mg/kg): 104, 91, 73, 57 orally; in male, female rats (mg/kg): 2396, 861 dermally. LC$_{50}$ in male, female rats (mg/l): 0.832, 0.558 by inhalation (Toki).
USE: Nematicide.

4289. Fotemustine. [92118-27-0] *P*-[1-[[[(2-Chloroethyl)nitrosoamino]carbonyl]amino]ethyl]phosphonic acid diethyl ester; (±)-diethyl [1-[3-(2-chloroethyl)-3-nitrosoureido]ethyl]phosphonate; 1-[*N*-(2-chloroethyl)-*N*-nitrosoureido]ethylphosphonic acid diethyl ester; S-10036; Muphoran. $C_9H_{19}ClN_3O_5P$; mol wt 315.69. C 34.24%, H 6.07%, Cl 11.23%, N 13.31%, O 25.34%, P 9.81%. Amino acid-linked nitrosourea alkylating agent. Prepn: G. Lavielle, C. Cudennec, **FR 2536075**; *eidem*, **US 4567169** (1984, 1986 both to ADIR). Clinical evaluation in advanced cancers: D. Khayat *et al.*, *Cancer Res.* **47**, 6782 (1987); in disseminated malignant melanoma: D. Khayat *et al.*, *J. Natl. Cancer Inst.* **80**, 1407 (1988).

mp 85°.
THERAP CAT: Antineoplastic.

4290. Francium. [7440-73-5] Eka-cesium. Fr; at. no. 87; valence 1. Group IA (1). No stable nuclides. Naturally occurring isotope: 223, *Actinum K*, (longest lived isotope, T$_{1/2}$ 21.8 min, β^- emitter; rel. at. mass 223.0197) formed by α-decay of actinium ^{227}Ac in ^{235}U series. Known artificial isotopes: 201-222, 224-232. Found in uranium minerals. Terrestrial abundance ~2×10^{18} ppm. Also obtainable by proton bombardment of thorium. First obtained in 1939 from an actinium prepn by Perey, *J. Phys. Radium* **10**, 439 (1939); *J. Chim. Phys.* **43**, 155, 262 (1946); Hyde, Ghiorso, *Phys. Rev.* **90**, 267 (1953); Hyde, *J. Am. Chem. Soc.* **74**, 4181 (1952). Isoln by paper chromatography: Perey, Adloff, *Compt. Rend.* **236**, 1163 (1953). *Reviews:* Hyde, *J. Chem. Educ.* **36**, 15-21 (1959); Whaley, "Sodium, Potassium, Rubidium, Cesium and Francium" in *Comprehensive Inorganic Chemistry* vol. **1**, J. C. Bailar Jr., *et al.*, Eds. (Pergamon Press, Oxford, 1973) pp 369-529; *Chemistry of the Elements* N. N. Greenwood, A. Earnshaw, Eds. (Pergamon Press, New York, 1984) pp 75-116.
Most electropositive element. Chemical behavior similar to that of other alkali metals.

4291. Frangula. Buckthorn bark; alder buckthorn; black dogwood; berry alder; arrow wood; Persian berries. Dried bark of *Rhamnus frangula* L., Rhamnaceae. *Habit.* Europe, Russian Asia, Mediterranean coast of Africa. *Constit.* Frangulin, emodin, chrysophanic acid.

4292. Frangulin. [60529-33-1] Franguloside; avornin; Cascarin. In berries, bark, and rootbark of *Rhamnus* spp., especially in alder buckthorn (*Rhamnus frangula* L.), *Rhamnus carthartica* L., and *Rhamnus purshiana* DC. (*Cascara sagrada*), Rhamnaceae. Prepn from bark of alder buckthorn: Bridel, Charaux, *Bull. Soc. Chim. Biol.* **15**, 642 (1933); M. Kubiak, *Herba Pol.* **23**, 217 (1977), *C.A.* **88**, 166260b (1978). Consists of the two glucosides, frangulins A and B, which were originally thought to be isomeric. Structure and synthesis of frangulin A: Hörhammer, Wagner, *Z. Naturforsch.* **27B**, 959 (1972). Structure of frangulin B: Wagner, Demuth, *Tetrahedron Lett.* **1972**, 5013.

Frangulin A   R =

Frangulin B   R =

Frangulin A. [521-62-0] 1,3,8-Trihydroxy-6-methylanthraquinone-*l*-rhamnoside; emodin-*l*-rhamnoside; rhamnoxanthin. C_{21}-$H_{20}O_9$; mol wt 416.38. Crystals, mp 228°. Absorption max: 225, 264, 282, 300, 430 nm (log ε 4.52, 4.28, 4.15, 3.97, 4.05).

Frangulin B. [14101-04-3] 6-*O*-(D-Apiofuranosyl)-1,6,8-trihydroxy-3-methylanthraquinone. $C_{20}H_{18}O_9$; mol wt 402.36. mp 196°.

THERAP CAT: Cathartic.

4293. Frataxin. Highly conserved, mitochondrial protein involved in cellular iron homeostasis and resistance to oxidative stress. In humans, frataxin deficiency is associated with Friedreich's ataxia, an autosomal recessive, neurodegenerative disease caused by a mutation in the frataxin (*FXN*) gene. Identification of the *FXN* gene and amino acid sequence of human frataxin: V. Campuzano *et al.*, *Science* **271**, 1423 (1996). Identification as mitochondrial protein: T. J. Gibson *et al.*, *Trends Neurosci.* **19**, 465 (1996). Role in iron homeostasis: M. Babcock *et al.*, *Science* **276**, 1709 (1997). Quantification in human lymphocytes by electrochemiluminescence assay (ECLIA): H. Steinkellner *et al*, *Anal. Chim. Acta* **659**, 129 (2010). Review of structure and function: K. Z. Bencze *et al.*, *Crit. Rev. Biochem. Mol. Biol.* **41**, 269-291 (2006); M. Pandolfo, A. Pastore, *J. Neurol.* **256**, Suppl. 1, 9-17 (2009). Review of role in FeS cluster biogenesis: T. L. Stemmler *et al.*, *J. Biol. Chem.* **285**, 26737-26743 (2010). Review of physiological functions and role in Friedreich's ataxia: R. Santos *et al.*, *Antioxid. Redox Signal.* **13**, 651-690 (2010).

Frataxin homologs are small, acidic proteins containing 100-200 amino acids with isoelectric points ~4.5. Human frataxin is expressed as a 210 amino acid precursor protein which is transported to mitochondria and processed to the active 130 residue protein.

4294. Fraxetin. [574-84-5] 7,8-Dihydroxy-6-methoxy-2*H*-1-benzopyran-2-one; 7,8-dihydroxy-6-methoxycoumarin. $C_{10}H_8O_5$; mol wt 208.17. C 57.70%, H 3.87%, O 38.43%. The aglucon of fraxin. Obtained by heating fraxin with dil H_2SO_4. Structure: Wessely, Demmer, *Ber.* **61**, 1279 (1928). Synthesis: Aghoramurthy, Seshadri, *J. Chem. Soc.* **1954**, 3065.

Plates from alc, mp 228°. Turns yellow at 150° and brown at mp. Sol in 10 liters cold water, in 300 ml boiling water; somewhat more sol in alcohol; hardly sol in ether.

Dimethyl ether. 6,7,8-Trimethoxycoumarin. $C_{12}H_{12}O_5$. Orthorhombic bipyramidal crystals, mp 104°. $bp_{0.2}$ 90-100°.

4295. Fraxin. [524-30-1] 8-(β-D-Glucopyranosyloxy)-7-hydroxy-6-methoxy-2*H*-1-benzopyran-2-one; 7,8-dihydroxy-6-methoxycoumarin-8-β-D-glucoside; 8-(glucosyloxy)-6-methoxyumbelli-

ferone; fraxetin-8-glucoside; fraxoside; paviin. $C_{16}H_{18}O_{10}$; mol wt 370.31. C 51.90%, H 4.90%, O 43.20%. From bark of the common European ash (*Fraxinus excelsior* L.): Salm-Horstmar, *Pogg. Ann.* **100**, 607 (1857); from *F. oxyphylla* Mar., *Oleaceae:* Paris, Stambouli, *Compt. Rend.* **253**, 313 (1961). Identity with paviin: Stokes, *J. Chem. Soc.* **12**, 126 (1859). From the horse chestnut tree (*Aesculus hippocastanum* L., *Hippocastanaceae*): Reppel, *Planta Med.* **4**, 199 (1956); from *Diervilla* spp., *Caprifoliaceae:* Charaux, *J. Pharm. Chim.* [7] **4**, 248 (1911). Structure: Wessely, Demmer, *Ber.* **62**, 120 (1929).

Yellow hydrated needles from water or dil alc. Slightly bitter, astringent taste. The water of crystn (about 3 mols) is removed at 130° and 0.2 mm Hg. The anhydr substance, mp 205° (rapid heating). Sparingly sol in cold water; freely sol in hot water, hot alcohol. Practically insol in ether. Alkaline solns are sulfur-yellow and on dilution show characteristic blue-green fluorescence.

4296. Fredericamycin A. [80455-68-1] (2*S*)-6',7'-Dihydroxy-4,9,9'-trihydroxy-6-methoxy-3'-[(1*E*,3*E*)-1,3-pentadien-1-yl]spiro-[2*H*-benz[*f*]indene-2,8'-[8*H*]cyclopent[*g*]isoquinoline]-1,1',3,5,8-(2'*H*)-pentone; FCRC-A48; NSC-305263. $C_{30}H_{21}NO_9$; mol wt 539.50. C 66.79%, H 3.92%, N 2.60%, O 26.69%. Antitumor antibiotic produced by *Streptomyces griseus* (FCRC-48); represents new structural class of antibiotics containing a spiro[4,4]nonane ring system. Isoln together with two biologically inactive components, fredericamycins B and C: R. C. Pandey *et al.*, *J. Antibiot.* **34**, 1389 (1981). Antimicrobial and cytotoxic activity *in vitro* and antitumor activity *in vivo:* D. J. Warnick-Pickle *et al.*, *ibid.* 1402. Prepn and biological activity of water soluble salts: R. Misra, *ibid.* **51**, 976 (1988). Structure determn by x-ray crystallography: R. Misra *et al.*, *J. Am. Chem. Soc.* **104**, 4478 (1982). Spectroscopic and mass spectral characterization: *eidem, J. Antibiot.* **40**, 786 (1987). Synthetic studies: A. V. R. Rao *et al.*, *Chem. Commun.* **1984**, 1119; K. A. Parker *et al.*, *Tetrahedron Lett.* **26**, 2181 (1985). Synthesis of (±)-form: T. R. Kelly *et al.*, *J. Am. Chem. Soc.* **108**, 7100 (1986). Biosynthetic study: K. M. Byrne *et al.*, *Biochemistry* **24**, 478 (1985). Mechanism of action study: B. D. Hilton *et al.*, *ibid.* **25**, 5533 (1986).

Thin, platelet-like crystals from acetonitrile + water, mp >350° (dec). pKa (DMF) 6.80; 8.88. Sol in acetic acid, DMSO, DMF, pyridine; partially sol in acidic methanol, acidic chloroform, acidic ethyl acetate. Insol in water, petr ether, ether. Acts as an indicator: Red in acidic soln, green to blue in basic soln.

Potassium salt. Solubility (mg/ml): water (1.0); DMSO (8/1.5); 1:1 DMSO-H_2O (3.0); also readily sol in DMF, dimethylacetamide, pyridine. Sparingly sol in ethyl acetate, acetonitrile, methanol, chloroform. Insol in hexanes, benzene, acetone, ether.

4297. Fremy's Salt. [14293-70-0] Nitrosodisulfonic acid potassium salt (1:2); potassium nitrosodisulfonate; potassium peroxylaminedisulfonate. $K_2NO_7S_2$; mol wt 268.32. K 29.14%, N 5.22%, O 41.74%, S 23.90%. $K_2[NO(SO_3)_2]$. Radical oxidizing agent. Prepn: E. Fremy, *Ann. Chim. Phys.* **15**, 408 (1845). Structural studies: W. Moser, R. A. Howie, *J. Chem. Soc. A* **1968**, 3039; R. A. Howie *et al.*, *ibid.* 3043. Mechanism of oxidation: B. Giethlen, J. M. Schaus, *Tetrahedron Lett.* **38**, 8483 (1997). Synthetic applications: H. J. Teuber, G. Jellinek, *Naturwissenschaften* **38**, 259 (1951); H. Schlude, *Tetrahedron* **29**, 4007 (1973); P. M. Deya *et al.*, *ibid.* **43**, 3523 (1987); G. A. Kraus, N. Selvakumar, *Tetrahedron Lett.* **40**, 2039 (1999). EPR studies: B. L. Bales, *Chem. Phys. Lett.* **10**, 361 (1971); M. T. Jones *et al.*, *J. Phys. Chem.* **83**, 1327 (1979); B. L. Bales *et al.*, *J. Magn. Reson.* **118**, 227 (1996). Photochemistry study: H. Kunkely *et al.*, *J. Photochem. Photobiol. A* **112**, 9 (1998). Review of oxidation chemistry: H. Zimmer *et al.*, *Chem. Rev.* **71**, 229-246 (1971).

Occurs in two distinct crystalline forms: as monoclinic yellow-orange needles, and as triclinic orange-brown rhombs. Absorption max (water): 248, 545 nm (ε 1690, 21).

USE: Reagent for the oxidation of anilines and phenols to quinones. EPR standard.

4298. Frenolicin. [10023-07-1] (1*R*,3*S*,4a*S*,10a*R*)-3,4,5,10-Tetrahydro-9-hydroxy-5,10-dioxo-1-propyl-4a,10a-epoxy-1*H*-naphtho[2,3-*c*]pyran-3-acetic acid. $C_{18}H_{18}O_7$; mol wt 346.34. C 62.42%, H 5.24%, O 32.34%. Antibiotic produced by *Streptomyces fradiae*: Van Meter *et al.*, *Antimicrob. Agents Annu.* **1960**, 77. Structure: Ellestad *et al.*, *J. Am. Chem. Soc.* **88**, 4109 (1966); **90**, 1325 (1968). Total synthesis: A. Ichihara *et al.*, *Tetrahedron Lett.* **21**, 4469 (1980).

Pale yellow needles from benzene, mp 160-161°. $[\alpha]_D^{25}$ −37.7° (c = 1.5 in methanol). uv max (methanol): 234, 362 nm (ε 18300, 5200). Sol in methanol, ethanol, ethyl acetate, acetone, glacial acetic acid, carbon tetrachloride; slightly sol in benzene. Practically insol in water, cyclohexane, petr ether.

4299. Frequentin. [29119-03-7] 6-(1,3-Heptadien-1-yl)-3,4-dihydroxy-2-oxocyclohexanecarboxaldehyde. $C_{14}H_{20}O_4$; mol wt 252.31. C 66.65%, H 7.99%, O 25.36%. Mold metabolite with antineoplastic and antibiotic activity; isolated from *Penicillium frequentans* Westling and *P. palitans* Westling: Curtis *et al.*, *Nature* **167**, 557 (1951); Birkinshaw, *Biochem. J.* **51**, 271 (1952). Structure: Sigg, *Helv. Chim. Acta* **46**, 1061 (1963). Effect on Ehrlich ascites carcinoma cells: J. Fuska *et al.*, *Int. Congr. Chemother., Proc., 7th, Prague, 1971* vol. 2 (Univ. Park Press, Baltimore, 1972) pp 97-98. Synthetic study: C. G. Unson, *Diss. Abstr. B* **39**, 1296 (1978).

Needles from boiling water, mp 134.5°. When dissolved, exists as the 1-hydroxymethylene tautomer. $[\alpha]_D^{22}$ +65° (c = 0.95 in $CHCl_3$); $[\alpha]_D^{22}$ +61° (c = 0.96 in methanol). uv max (dioxane): 290, 232 nm (log ε 3.74, 4.52). Sol in acetone, dioxane; less sol in chloroform, ethanol, benzene; slightly sol in carbon tetrachloride, water.

4300. Friedelin. [559-74-0] (4β,5β,8α,9β,10α,13α,14β)-5,9,13-Trimethyl-24,25,26-trinoroleanan-3-one; *D:A*-friedooleanan-3-one; friedelan-3-one. $C_{30}H_{50}O$; mol wt 426.73. C 84.44%, H 11.81%, O 3.75%. Major triterpene constituent of cork, obtained by extraction of ground cork with alc. Isoln: Istrati, Ostrogovich, *Compt. Rend.* **128**, 1581 (1899). Isolated also from *Ceratopetalum apetalum* D. Don, *Cunoniaceae*: Jefferies, *J. Chem. Soc.* **1954**, 473. Structure, sterochemistry: Brownlie *et al.*, *Chem. Ind. (London)* **1955**, 1156; Kane, Stevenson, *Tetrahedron* **15**, 223 (1961); Steven-

son, *J. Org. Chem.* **28**, 188 (1963). Total synthesis of (±)-form: R. E. Ireland, D. M. Walba, *Tetrahedron Lett.* **1976**, 1071.

Needles from ethyl acetate or alcohol, mp 263-263.5°. $[\alpha]_D$ −27.8° (chloroform). One gram dissolves in 8.6 ml chloroform, 264 ml 99% alcohol.

4301. Frovatriptan. [158747-02-5] (3*R*)-2,3,4,9-Tetrahydro-3-(methylamino)-1*H*-carbazole-6-carboxamide; 3-methylamino-6-carboxamido-1,2,3,4-tetrahydrocarbazole; SB-209509. $C_{14}H_{17}N_3$-O; mol wt 243.31. C 69.11%, H 7.04%, N 17.27%, O 6.58%. Serotonin 5-HT$_{1B/1D}$ receptor agonist. Prepn: F. D. King *et al.*, **WO 9300086**; *eidem*, **US 5464864** (1993, 1995 both to SmithKline Beecham). Pharmacology: A. A. Parsons *et al.*, *J. Cardiovasc. Pharmacol.* **30**, 136 (1997). Series of articles on pharmacology, safety and clinical studies: *Headache* **42**, Suppl. 2, S45-S99 (2002). Review of clinical experience in cluster headaches: H. C. Siow *et al.*, *Cephalalgia* **24**, 1045-1048 (2004); in acute migraine: N. Poolsup *et al.*, *J. Clin. Pharm. Ther.* **30**, 521-532 (2005).

Succinate monohydrate. [158930-17-7]; [158930-09-7] (anhydrous). SB-209509AX; VML-251; Frova; Frovelan; Migard; Miguard. $C_{14}H_{17}N_3O.C_4H_6O_4.H_2O$; mol wt 379.41. White to off-white powder. Sol in water.

THERAP CAT: Antimigraine.

4302. Fructose. [57-48-7] D-Fructose; β-D-fructose; levulose; fruit sugar; Fructosteril; Laevoral; Levugen; Laevosan. C_6-$H_{12}O_6$; mol wt 180.16. C 40.00%, H 6.71%, O 53.28%. Occurs in a large number of fruits, honey, and as the sole sugar in bull and human semen: Auerbach, Bodlander, *Angew. Chem.* **36**, 602 (1923); Mann, *Nature* **157**, 79 (1946); Pryde, *ibid.* 660. Prepd by adding abs alcohol to the syrup obtained from the acid hydrolysis of inulin: Bates *et al.*, *Natl. Bur. Stand. Circ.* **C440**, 399 (1942). Prepn from dextrose: Cantor, Hobbs, **US 2354664** (1944 to Corn Prod. Refining). From sucrose by enzymatic conversion: Koepsell *et al.*, **US 2729587** (1956 to U.S.A.). Crystal and molecular structure: J. A. Kanters *et al.*, *Acta Crystallogr.* **B33**, 665 (1977). *Review:* Barry, Honeyman, *Adv. Carbohyd. Chem.* **7**, 53-98 (1952); M. Chen, R. L. Whistler, *ibid.* **34**, 285-343 (1977). Occurs in both the furanose and pyranose forms. An aq soln at 20° contains about 20% of the furanose form.

Orthorhombic, bisphenoidal prisms from alc, dec 103-105°. Sweetest of the sugars. Shows mutarotation. $[\alpha]_D^{20}$ −132° to −92° (c = 2). Rapid and anomalous mutarotation involves pyranose-furanose interconversion. The final value is obtained instantly in the presence of hydroxyl ions. pKa (18°): 12.06. Freely sol in water.

One gram dissolves in 15 ml alc, in 14 ml methanol. Slightly sol in cold, freely in hot acetone; sol in pyridine, ethylamine, methylamine.

Ferric Form. [12286-76-9] CB-302; Ferritose. Prepn: Stitt *et al.*, *Proc. Soc. Exp. Biol. Med.* **110**, 70 (1962); Saltman, Charley, US 3074927 and US 3275514 (1966).

USE: To prevent sandiness in ice cream.

THERAP CAT: Fluid and nutrient replenisher, and ferric form as hematinic.

THERAP CAT (VET): For bovine ketosis.

4303. DL-Fructose. [30237-26-4] α-Acrose; methose. $C_6H_{12}O_6$; mol wt 180.16. C 40.00%, H 6.71%, O 53.28%. Component of formose (polymerization product of formaldehyde): Vogel, *Helv. Chim. Acta* **11**, 370 (1928).

Needles from methanol. d_4^{16} 1.665. mp 129-130° (slow heating). Reduces Fehling's soln.

Phenylosazone. $C_{18}H_{22}N_4O_4$. mp 216-217°.

4304. Fructose-1,6-diphosphate. [488-69-7] D-Fructose 1,6-bis(dihydrogen phosphate); 1,6-fructosediphosphoric acid; hexose diphosphate; Harden-Young ester. $C_6H_{14}O_{12}P_2$; mol wt 340.11. C 21.19%, H 4.15%, O 56.45%, P 18.21%. Formed from fructose-6-phosphate in the presence of ATP, Mg^{2+} and the enzyme phosphohexokinase. Prepn from glucose, mannose, fructose, sucrose by the action of yeasts: A. Harden, *Alcoholic Fermentation* (Longmans, Green & Co., New York, 4th ed., 1932); v. Lebedev, *Biochem. Z.* **36**, 254 (1911); *cf* DE 292817; DE 293864; DE 301590. Fructose-1,6-diphosphate is reversibly split in the presence of aldolase forming 1-phosphodihydroxyacetone and 3-phosphoglyceraldehyde: Meyerhof *et al.*, *Biochem. Z.* **286**, 301 (1936). Metabolism regulation study: M. E. Kirtley, M. McKay, *Mol. Cell. Biochem.* **18**, 141 (1977).

Sodium salt monohydrate. Esafosfina.
THERAP CAT: Roborant; tonic.

4305. Fructose-6-phosphate. [643-13-0] D-Fructose 6-(dihydrogen phosphate); D-fructose-6-phosphoric acid; fructose monophosphate; hexose phosphate; hexose monophosphate; Neuberg ester. $C_6H_{13}O_9P$; mol wt 260.13. C 27.70%, H 5.04%, O 55.35%, P 11.91%. Present in animal tissues as an equilibrium mixture with glucose-6-phosphate. The glucose-6-phosphate may be reversibly transformed into fructose-6-phosphate by the enzyme phosphohexose isomerase. Prepn by hydrolysis of 1,6-fructose diphosphate with dil acid: Neuberg, *Biochem. Z.* **88**, 432 (1917); DE 334250 (Bayer); *Chem. Zentralbl.* **1921**, II, 961; *C.A.* **13**, 948. Role in metabolic regulation and heat generation: E. A. Newsholme, *Biochem. Soc. Trans.* **4**, 978 (1976).

Very sol in water. $[\alpha]_D^{21}$ +2.5° (c = 3). *See:* Meyerhof, Lohmann, *Biochem. Z.* **185**, 117 (1927). $[\alpha]_D$ +1.2° (c = 0.9). See: Lohmann, *ibid.* **262**, 145 (1933). The magnesium and zinc salts also are sol in water. The equilibrium mixture of 75-80% glucose-6-phosphate and 20-25% fructose-6-phosphate is called *lactacidogen* or *Embden ester.*

4306. Fucosamine. [24724-90-1] 2-Amino-2,6-dideoxygalactose; DL-fucosamine. $C_6H_{13}NO_4$; mol wt 163.17. C 44.17%, H 8.03%, N 8.58%, O 39.22%. Isoln of D-form from lipopolysaccharide of *Chromobacterium violaceum:* Crumpton, Davies, *Biochem. J.* **70**, 729 (1958); from *Bacillus licheniformis:* Sharon *et al.*, *ibid.*

93, 210 (1964). Isoln of L-form from type V *Pneumococcus* capsular polysaccharide: Barker *et al.*, *Nature* **189**, 303 (1961). Synthesis of L-form: Kuhn *et al.*, *Ann.* **628**, 186 (1959); J. Lehmann *et al.*, *Ber.* **112**, 1470 (1979); of D-form: Zehavi, Sharon, *J. Org. Chem.* **29**, 3654 (1964).

α-D-Fucosamine

D-Form hydrochloride. $C_6H_{13}NO_4 \cdot HCl$. Crystals from aq acetone, dec 170-175°. $[\alpha]_D^{20}$ +91° (water). Absorption max: 400 nm.

N-Acetyl-D-fucosamine. $C_8H_{15}NO_5$. Crystals from ethanol, dec 196-197°. $[\alpha]_D^{22}$ +92° (c = 2 in water).

L-Form hydrochloride. $C_6H_{13}NO_4 \cdot HCl$. Rods from methanol + isopropanol, mp 192-193°. $[\alpha]_D^{27}$ −92° (c = 0.89 in water).

N-Acetyl-L-fucosamine. $C_8H_{15}NO_5$. Crystals from ethanol, mp 197-198°. $[\alpha]_D^{26}$ −82° (c = 1.46 in water).

4307. D-Fucose. [3615-37-0] 6-Deoxy-D-galactose; D-galactomethylose; rhodeose. $C_6H_{12}O_5$; mol wt 164.16. C 43.90%, H 7.37%, O 48.73%. Obtained from glucosides found in various species of *Convolvulaceae, e.g.*, convolvulin, jalapin, β-turpethein. Isoln: Votocek, *Z. Zuckerind. Boehm.* **24**, 249; *Chem. Zentralbl.* **1901**, I, 1042; **1902**, II, 1361; **1905**, II, 1528. Isoln from jalap resin: Votocek, Bulir, *ibid.* **1906**, I, 1818. Prepn by acid hydrolysis of chartreusin: Sternbach *et al.*, *J. Am. Chem. Soc.* **80**, 1639 (1958).

α-Form, needles from alc. Sweet taste. mp 144°. Shows mutarotation, $[\alpha]_D^{19}$ +127.0° (7 min) → +89.4° (31 min) → +77.2° (71 min) → +76.0° (final value 146 min, c = 10). Soluble in water; moderately sol in alcohol.

Pentaacetate. $C_{16}H_{22}O_{10}$. mp 115.5°.
Oxime. $C_6H_{13}NO_5$. mp 188.5°.
Phenylhydrazone. $C_{12}H_{18}N_2O_4$. mp 172°.
Phenylosazone. $C_{18}H_{22}N_4O_3$. mp 176.5°.

4308. L-Fucose. [2438-80-4] 6-Deoxy-L-galactose; L-galactomethylose. $C_6H_{12}O_5$; mol wt 164.16. C 43.90%, H 7.37%, O 48.73%. Occurs in seaweed: *Ascophyllum nodosum* (L.) Ledol. (*Fucus nodosus* L.), *Fucus vesiculosus* L., *F. serratus* L., *F. virsoides* (Don) J. Ag., *Fucaceae*, and in gum tragacanth. Isoln from seaweed: Clark, *J. Biol. Chem.* **54**, 65 (1922); Hockett *et al.*, *J. Am. Chem. Soc.* **61**, 1658 (1939). Manuf from fucoidan: Schweiger, US 3240775 (1966 to Kelco). Synthesis from D-galactose: Dejterjuszynski, Flowers, *Carbohydr. Res.* **28**, 144 (1973); from D-glucose: T. Chiba, S. Tejima, *Chem. Pharm. Bull.* **27**, 2838 (1979); from D-mannose: J. Defaye *et al.*, *Carbohydr. Res.* **94**, 131 (1981). Review of chemistry and biochemistry: H. M. Flowers, *Adv. Carbohydr. Chem. Biochem.* **39**, 279-345 (1981).

α-Form. Minute needles from abs alcohol, mp 140°. Shows mutarotation. $[\alpha]_D^{20}$ −124.1° (10 min) → −108.0° (20 min) → −91.5° (36 min) → −78.6° (70 min) → −75.6° (final value, 24 hrs, c = 9). Soluble in water and alcohol.

4309. Fucosterol. [17605-67-3] $(3\beta,24E)$-Stigmasta-5,24-(28)-dien-3-ol; 24-ethylidenecholest-5-en-3β-ol. $C_{29}H_{48}O$; mol wt 412.70. C 84.40%, H 11.72%, O 3.88%. Isoln from *Fucus vesicu-*

losus L., *Fucaceae:* Heilbron *et al., J. Chem. Soc.* **1934**, 1572; from marine brown algae *(Phaeophyceae):* Tsuda, *Chem. Pharm. Bull.* **6**, 724 (1958). Structure: MacPhillamy, *J. Am. Chem. Soc.* **64**, 1732 (1942). Stereochemistry: C. Brooks *et al., Steroids* **20**, 487 (1972). Synthesis: Hayazu, *Pharm. Bull.* **5**, 452 (1957).

Crystals from methanol, mp 124°. $[\alpha]_D^{20}$ $-38.4°$ (chloroform). Soluble in most organic solvents.

Acetate. $C_{31}H_{50}O_2$. Crystals, mp 120-121°. $[\alpha]_D^{20}$ $-45°$ (chloroform).

4310. Fucoxanthin. [3351-86-8] $(3S,3'S,5R,5'R,6S,6'R)$-6',7'-Didehydro-5,6-epoxy-4',5',6,7-tetrahydro-3,3',5',-trihydroxy-β,β-caroten-8(5H)-one; $(3S,3'S,5R,5'R,6S,6'R)$-3'-(acetyloxy)-6',7'-didehydro-5,6-epoxy-5,5',6,6',7,8-hexahydro-3,5'-dihydroxy-8-oxo-β,β-carotene; 6',7'-didehydro-5,6-epoxy-4',5,5',6,7,8-hexahydro-3,3',5'-trihydroxy-8-oxo-α-carotene 3'-acetate; *all-trans*-fucoxanthin. $C_{42}H_{58}O_6$; mol wt 658.92. C 76.56%, H 8.87%, O 14.57%. Carotenoid pigment found in fresh brown algae: *Fucus virsoides* (Don) J. Ag., *Fucaceae.* Also found in *Zygnema pectinatum* (Vauch.) Ag., *Zygnemaceae* and in *Polysiphonia nigrescens* (Dillw.) Grev., *Rhodomelaceae.* Isoln: Karrer *et al., Helv. Chim. Acta* **14**, 628 (1931); Willstätter, Page, *Ann.* **404**, 237 (1914). Structure: Bonnet *et al., Chem. Commun.* **1966**, 515; *J. Chem. Soc. C* **1969**, 429. Abs config: DeVille *et al., Chem. Commun.* **1969**, 1311; K. Bernhard *et al., Tetrahedron Lett.* **1976**, 115.

Needles from ether + petr ether, mp 160°. $[\alpha]_D^{18}$ +72.5 ±9° (chloroform). Abs max (chloroform): 492, 457 nm. Abs max (ethanol): 450 nm ($E_{1cm}^{1\%}$ 1140): Antia, *Can. J. Chem.* **43**, 302 (1965). Freely sol in ethanol; less sol in carbon disulfide; sparingly sol in ether. Practically insol in petr ether. 1.66 g dissolves in 100 g boiling methanol.

4311. Fucus. Bladder-wrack; sea-wrack; bladder fucus; kelpware; black-tang; cut-weed; sea-oak. Dried thallus of *Fucus vesiculosus* L., *F. serratus* L., or *F. siliquosus*, L. *Fucaceae.* Habit. Atlantic and Pacific Oceans. *Constit.* Algin, about 0.01% iodine and some bromine mannite.

4312. Fuller's Earth. Floridin. A nonplastic variety of kaolin containing an aluminum magnesium silicate. The name is derived from an ancient process of cleaning or fulling wool, to remove the oil and dirt particles, with a water slurry of earth or clay. At the present time, the term fuller's earth is applied to any clay that has adequate decolorizing and purifying capacity to be used commercially in oil refining without chemical treatment. It is sometimes considered to be synonymous with montmorillonite, *q.v.,* **kaolinite** ($Al_2O_3.2SiO_2.2H_2O$) and **Halloysite** ($Al_2O_3.2SiO_2.4H_2O$). List of minerals likely to be found in fuller's earth: Porter, *U.S. Geol. Surv. Bull.* **315**, 268 (1907), *C.A.* **1**, 1684 (1907); and the opinion is expressed that fuller's earth results from the decompn of hornblendes

and augites rather than from feldspars. Fuller's earth has for its base a series of amorphous, hydrous aluminum silicates that have a rather persistent colloidal (used in its widest sense) structure. It is to this colloidal structure, which is not lost at 130° and possibly higher, that the bleaching power is due. The bleaching efficiency of fuller's earth is usually increased by treatment with dilute acids.

USE: Decolorizer for oils and other liquids; filtering medium; filler for rubber; in agricultural formulations; also instead of absorbent charcoal.

4313. Fulvestrant. [129453-61-8] $(7\alpha,17\beta)$-7-[9-[(4,4,5,5,5-Pentafluoropentyl)sulfinyl]nonyl]estra-1,3,5(10)-triene-3,17-diol; ICI-182780; Faslodex. $C_{32}H_{47}F_5O_3S$; mol wt 606.78. C 63.34%, H 7.81%, F 15.66%, O 7.91%, S 5.28%. Steroidal estrogen antagonist reported to lack any partial agonist activity. Prepn: J. Bowler, B. S. Tait, **EP 138504**; *eidem,* **US 4659516** (1985, 1987 both to ICI). Pharmacology: A. E. Wakeling *et al., Cancer Res.* **51**, 3867 (1991). Clinical pharmacokinetics: A. Howell *et al., Br. J. Cancer* **74**, 300 (1996). Clinical trial in advanced breast cancer: J. N. Ingle *et al., J. Clin. Oncol.* **24**, 1052 (2006). Review of pharmacology, mode of action and clinical development: S. J. Johnston, K. L. Cheung, *Curr. Med. Chem.* **17**, 902-914 (2010). Review of clinical experience in estrogen-sensitive breast cancer: P. Kabos, V. F. Borges, *Expert Opin. Pharmacother.* **11**, 807-816 (2010).

THERAP CAT: Antineoplastic (hormonal).

4314. Fulvoplumierin. [20867-01-0] $(7E)$-7-$(2E)$-Buten-1-ylidene-1,7-dihydro-1-oxocyclopenta[c]pyran-4-carboxylic acid methyl ester; 3-(2-butenylidene)-2-carboxy-α-(hydroxymethylene)-1,4-cyclopentadiene-1-acetic acid δ-lactone methyl ester; methyl 7-crotonylidenecyclopenta[c]pyran-1-(7H)-one-4-carboxylate. $C_{14}H_{12}O_4$; mol wt 244.25. C 68.85%, H 4.95%, O 26.20%. Occurs together with plumieride and plumericin. Isoln from *Plumeria acutifolia* Poir., *Apocynaceae,* also from roots of *P. rubra* var *alba:* Grumbach *et al., Experientia* **8**, 224 (1952). Structure: Schmid, Bencze, *Helv. Chim. Acta* **36**, 206, 1468 (1953). Stereochemistry: Albers-Schönberg *et al., ibid.* **45**, 1406 (1962). Synthesis: Büchi, Carlson, *J. Am. Chem. Soc.* **90**, 5336 (1968); **91**, 6470 (1969).

Orange needles from chloroform + petr ether, ethyl acetate, or alcohol, dec 151-152°. Sublimes in high vacuum. uv max (ethanol): 272, 365 nm (ε 7,000; 33,700). Sol in chloroform, hot ethyl acetate, benzene, alcohol; less sol in pyridine, acetone. Practically insol in water, petr ether.

4315. Fumagillin. [23110-15-8] $(2E,4E,6E,8E)$-2,4,6,8-Decatetraenedioic acid 1-[(3R,4S,5S,6R)-5-methoxy-4-[(2R,3R)-2-methyl-3-(3-methyl-2-buten-1-yl)-2-oxiranyl]-1-oxaspiro[2.5]oct-6-yl] ester; 2,4,6,8-decatetraenedioic acid mono[4-(1,2-epoxy-1,5-dimethyl-4-hexenyl)-5-methoxy-1-oxaspiro[2.5]oct-6-yl] ester. $C_{26}H_{34}O_7$; mol wt 458.55. C 68.10%, H 7.47%, O 24.42%. Antibiotic substance produced by *Aspergillus fumigatus:* T. E. Eble, F. R. Hanson, *Antibiot. Chemother.* **1**, 55 (1951); F. R. Hanson, T. E. Eble, **US 2652356** (1953 to Upjohn). Purification: D. S. Tarbell *et al., J. Am. Chem. Soc.* **77**, 5613 (1955). Structure: *eidem, ibid.* **82**, 1005 (1960); **83**, 3096 (1961). Stereochemistry: N. J. McCorkindale, J. G. Sime, *Proc. Chem. Soc. London* **1961**, 331; J. R. Turner, D. S. Tarbell, *Proc. Natl. Acad. Sci. USA* **48**, 733 (1962). Biosynthesis:

A. J. Birch, S. F. Hussain, *J. Chem. Soc. C* **1969**, 1473. Total synthesis of (±)-form: E. J. Corey, B. B. Snider, *J. Am. Chem. Soc.* **94**, 2549 (1972). Anti-amebic activity: M. C. McCowen *et al.*, *Science* **113**, 202 (1951). Acute toxicity: J. A. DiPaolo *et al.*, *Antibiot. Annu.* **1958-1959**, 541. HPLC determn in trout muscle: J. Guyonnet *et al.*, *J. Chromatogr. B* **666**, 354 (1995). Field trial in honey bees: T. C. Webster, *J. Econ. Entomol.* **87**, 601 (1994); in rainbow trout: R. le Gouvello *et al.*, *Aquaculture* **171**, 27 (1999). Clinical evaluation in intestinal microsporidiosis in immunocompromised patients: J.-M. Molina *et al.*, *N. Engl. J. Med.* **346**, 1963 (2002).

Yellow needles from methanol, mp 194-195°. $[\alpha]_D^{25}$ −26.6° (c = 1 in 95% ethanol). Absorptivity: 156.0 at 335 nm and 146.5 at 351 nm (soln of 100 mg in 10 ml chloroform diluted with alcohol to 0.0004% fumagillin and 0.04% chloroform). Practically insol in water, dil acids, satd hydrocarbons. Sol in most other organic solvents, in aq solns of bicarbonates and alkali hydroxides. Best stored in dark, evacuated ampuls at low temps. Stability data: T. E. Eble, E. R. Garrett, *J. Am. Pharm. Assoc.* **43**, 536 (1954); E. R. Garrett, *ibid.* 539. LD$_{50}$ in mice (mg/kg): ~800 s.c. (DiPaolo).

Dicyclohexylamine salt. [41567-78-6] Bicyclohexylammonium fumagillin; Fumidil B.

THERAP CAT: Antiprotozoal.

THERAP CAT (VET): Antiprotozoal. Control of *Nosema apis* in honey bees.

4316. Fumaric Acid. [110-17-8] (2E)-2-Butenedioic acid; *trans*-1,2-ethylenedicarboxylic acid; allomaleic acid; boletic acid. $C_4H_4O_4$; mol wt 116.07. C 41.39%, H 3.47%, O 55.14%. Occurs in many plants, *e.g.*, in *Fumaria officinalis* L., *Fumariaceae*, in *Boletus scaber* Bull., *Boletaceae*, and in *Fomes igniarius* (Fries) Kickx., *Polyporaceae*. Essential to vegetable and animal tissue respiration. Prepd industrially from glucose by the action of fungi such as *Rhizopus nigricans*: Foster, Waksman, *J. Am. Chem. Soc.* **61**, 127 (1939). Laboratory prepn by the oxidation of furfural with sodium chlorate in the presence of vanadium pentoxide: Milas, *Org. Synth.* **coll. vol. II**, 302 (1943). Molecular structure: J. L. Derissen, *J. Mol. Struct.* **38**, 177 (1977). *Review:* W. D. Robinson, R. A. Mount in *Kirk-Othmer Encyclopedia of Chemical Technology* vol. 14 (Wiley-Interscience, New York, 3rd ed., 1981) pp 770-793.

Monoclinic, prismatic needles or leaflets from water. d 1.625. Sublimes at 200°. Sublimes at 165° at 1.7 mm pressure. Partial carbonization and formation of maleic anhydride occur at 230° (open vessel). mp 287° (closed capillary, rapid heating). pK$_1$ (25°): 3.03; pK$_2$: 4.54. Absorption spectrum: Macbeth, Stewart, *J. Chem. Soc.* **111**, 830 (1917). Soly in 100 g water at 25°: 0.63 g; at 40°: 1.07 g; at 60°: 2.4 g; at 100°: 9.8 g; in 100 g 95% alcohol at 30°: 5.76 g; in 100 g acetone at 30°: 1.72 g; in 100 g ether at 25°: 0.72 g. Almost insol in olive oil, chloroform, carbon tetrachloride, benzene, xylene, molten camphor, liquid ammonia.

Monomethyl ester. $C_5H_6O_4$. Prisms from alc, mp 144.5°.

Dimethyl ester. $C_6H_8O_4$. Crystals, mp 102°. bp 192°.

USE: Substitute for tartaric acid in beverages and baking powders; as a replacement or partial replacement for citric acid in fruit drinks. As an antioxidant. Manuf polyhydric alcohols, synthetic resins. As mordant in dyeing.

4317. Fumigatin. [484-89-9] 3-Hydroxy-2-methoxy-5-methyl-2,5-cyclohexadiene-1,4-dione; 3-hydroxy-2-methoxy-5-methyl-

p-benzoquinone; 6-hydroxy-5-methoxy-*p*-toluquinone; 3-hydroxy-4-methoxy-2,5-toluquinone. $C_8H_8O_4$; mol wt 168.15. C 57.14%, H 4.80%, O 38.06%. Fungal toxin with antibiotic properties isolated from metabolism soln of *Aspergillus fumigatus* Fres: Anslow, Raistrick, *Biochem. J.* **32**, 687 (1938); Waksman, Geiger, *J. Bacteriol.* **47**, 391 (1944). Synthesis: Baker, Raistrick, *J. Chem. Soc.* **1941**, 670; Posternak, Ruelius, *Helv. Chim. Acta* **26**, 2045 (1943); Seshadri, Venkatasubramanian, *J. Chem. Soc.* **1959**, 1660. Biosynthesis: Pettersson, *Acta Chem. Scand.* **17**, 1323 (1963); Simonart, Verachtert, *Bull. Soc. Chim. Biol.* **49**, 543 (1967). Formation and polarographic assay: J. Lafond-Grellety *et al.*, *Ann. Microbiol.* **129B**, 3 (1978). *Review:* Wilson, "Miscellaneous Aspergillus Toxins," in *Microbial Toxins* vol. VI, A. Ciegler *et al.*, Eds. (Academic Press, New York, 1971) p 281.

Maroon-colored needles or hexagonal plates from petr ether, mp 116°. Sublimes *in vacuo*. Volatile with steam. Sparingly sol in water, petr ether; freely sol in acetone, ether, chloroform, benzene, ethyl acetate, alcohol.

4318. Fumonisin B$_1$. [116355-83-0] (2R,2'R)-1,2,3-Propanetricarboxylic acid 1,1'-[(1S,2R)-1-[(2S,4R,9R,11S,12S)-12-amino-4,9,11-trihydroxy-2-methyltridecyl]-2-[(1R)-1-methylpentyl]-1,2-ethanediyl] ester; macrofusine; FB$_1$. $C_{34}H_{59}NO_{15}$; mol wt 721.84. C 56.57%, H 8.24%, N 1.94%, O 33.25%. Most prevalent of a family of mycotoxins produced by *Fusarium moniliforme*, a common mold associated with corn; also isolated from other *Fusarium* species. Isolation: W. C. A. Gelderblom *et al.*, *Appl. Environ. Microbiol.* **54**, 1806 (1988). Structure elucidation of family: S. C. Bezuidenhout *et al.*, *Chem. Commun.* **1988**, 743. Causative agent of pulmonary edema in pig: L. R. Harrison *et al.*, *J. Vet. Diagn. Invest.* **2**, 217 (1990). Association of B$_1$, B$_2$ with human esophageal cancer: J. P. Rheeder *et al.*, *Phytopathology* **82**, 353 (1992). Metabolism: G. S. Shephard *et al.*, *Toxicon* **30**, 768 (1992). Toxicity and carcinogenicity in rat: W. C. A. Gelderblom *et al.*, *Carcinogenesis* **12**, 1247 (1991). Toxicology in pig: W. H. Haschek *et al.*, *Mycopathologia* **117**, 83 (1992). LC determn in corn of B series fumonisins: M. E. Stack, R. M. Eppley, *J. Assoc. Off. Anal. Chem.* **75**, 834 (1992); P. A. Murphy *et al.*, *J. Agric. Food Chem.* **41**, 263 (1993). Review of animal toxicoses: P. F. Ross *et al.*, *Mycopathologia* **117**, 109-114 (1992). Review: W. P. Norred, *J. Toxicol. Environ. Health* **38**, 309-328 (1993).

4319. Fungichromin. [6834-98-6] (3R,4S,6S,8S,10R,12R,-14R,15R,16R,17E,19E,21E,23E,25E,27S,28R)-4,6,8,10,12,14,15,-16,27-nonahydroxy-3-[(1R)-1-hydroxyhexyl]-17,28-dimethyloxacyclooctacosa-17,19,21,23,25-pentaen-2-one; Antibiotic A 246; cogomycin; lagosin; pentamycin; FemiFect; Pruri-ex. $C_{35}H_{58}O_{12}$; mol wt 670.84. C 62.67%, H 8.72%, O 28.62%. Antifungal polyene macrolide antibiotic, related structurally to filipin, *q.v.* Isoln from *Streptomyces cellulosae*: A. A. Tytell *et al.*, *Antibiot. Annu.* **1954-1955**, 716. Isoln (as pentamycin) from *S. penticus*: S. Umezawa, Y. Tanaka, *J. Antibiot.* **11A**, 26 (1958). Isoln (as lagosin) from *S. roseoluteus*: C. J. Bessel *et al.*, *US 3013947* (1961 to Glaxo). Isoln from *S. griseus* (FCRC-21): R. C. Pandey *et al.*, *Biomed. Mass Spectrom.* **7**, 93 (1980). Structures: A. C. Cope, H. E. Johnson, *J. Am. Chem. Soc.* **80**, 1504 (1958); A. C. Cope *et al.*, *ibid.* **84**, 2170 (1962); M. L. Dhar *et al.*, *J. Chem. Soc.* **1964**, 842; M. P. Berry, M.

C. Whiting, *ibid.* 862; V. Pozgay *et al., J. Antibiot.* **29**, 472 (1976). Identity of fungichromin with lagosin and cogomycin and comparison of reported physico-chemical constants: R. C. Pandey *et al., J. Antibiot.* **35**, 988 (1982). Biosynthesis and NMR assignment: H. Noguchi *et al., J. Am. Chem. Soc.* **110**, 2938 (1988). Clinical evaluation and systemic absorption of intravaginal formulation: B. Frey Tirri *et al., Chemotherapy (Basel)* **56**, 190 (2010).

Light yellow cryst. mp 157-162° (dec). $[\alpha]_D^{20}$ −227.7° (c = 0.53 in DMF). uv max (methanol): 357, 338, 322 nm ($E_{1cm}^{1\%}$ 1231, 1250, 786). LD_{50} in mice (mg/kg): 1624 orally; 33.3 i.p. (Umezawa, Tanaka).

THERAP CAT: Antifungal (topical).

4320. Fungisterol. [516-78-9] $(3\beta,5\alpha)$-Ergosta-7-en-3-ol. $C_{28}H_{48}O$; mol wt 400.69. C 83.93%, H 12.08%, O 3.99%. Formed from ergosterol by fungi. Isoln from *Penicillium chrysogenum:* A. Saito, *J. Ferment. Technol.* **29**, 457 (1951), *C.A.* **47**, 12507f (1953). Structure: *idem, ibid.* **31**, 141 (1953), *C.A.* **48**, 5276d (1954); *idem, ibid.* **32**, 138, 140 (1954), *C.A.* **49**, 9009i (1955).

Crystals from alcohol + ether + ethyl acetate, mp 147.5°. $[\alpha]_D^{15}$ −21.9° (chloroform).
Acetate. $C_{30}H_{46}O_2$. Crystals, mp 158.5°. $[\alpha]_D^{15}$ −15.7° (chloroform).

4321. Funtumine. [474-45-3] $(3\alpha,5\alpha)$-3-Aminopregnan-20-one; 3α-amino-20-oxo-5α-pregnane. $C_{21}H_{35}NO$; mol wt 317.52. C 79.44%, H 11.11%, N 4.41%, O 5.04%. Steroidal alkaloid isolated from *Funtumia latifolia* Stapf., *Apocynaceae:* Janot *et al., C. R. Hebd. Seances Acad. Sci.* **246**, 3076 (1958); from leaves of *Holorrhena febrifuga* Stapf., *Apocynaceae:* H. Dodoun *et al., Phytochemistry* **12**, 923 (1973). Structure: Janot *et al., Bull. Soc. Chim. Fr.* **1960**, 1640, 1669. Prepn: H. Kapnang *et al., Tetrahedron Lett.* **1977**, 3469. *Review:* R. Goutarel, *Bull. Soc. Chim. Fr.* **1960**, 769. Effect on liver carcinogenesis in rats: A. Lacassagne *et al., C. R. Seances Acad. Sci. Ser. D* **274**, 2830 (1972).

Prisms from ethyl acetate, mp 126°. $[\alpha]_D$ +95° (c = 1.7 in chloroform). pK 9.18. Sol in the usual organic solvents. LD_{50} i.v. in mice: 30 mg/kg (Lacassagne).

4322. Fura-2. [96314-98-6] 2-[6-[Bis(carboxymethyl)amino]-5-[2-[2-[bis(carboxymethyl)amino]-5-methylphenoxy]ethoxy]-2-benzofuranyl]-5-oxazolecarboxylic acid. $C_{29}H_{27}N_3O_{14}$; mol wt 641.54. C 54.29%, H 4.24%, N 6.55%, O 34.91%. Fluorescent calcium imaging agent; structurally related to BAPTA, *q.v.* Ca^{2+} binding causes a spectral shift to shorter wavelengths; ratio of the fluorescence at each wavelength is related to changes in Ca^{2+} binding and is independent of dye concentration. Also used as pentaacetoxymethyl ester, **Fura-2/AM**, to enhance crossing of membranes. Prepn: G. Grynkiewicz *et al., J. Biol. Chem.* **260**, 3440 (1985). Binding kinetics to Ca^{2+}: A. P. Jackson *et al., FEBS Lett.* **216**, 35 (1987). Determn of K_d: D. L. Groden *et al., Cell Calcium* **12**, 279 (1991). Spectra of intracellular forms: C. S. Owen, *ibid.* 385. Photophysics: V. Van den Bergh *et al., Biophys. J.* **68**, 1110 (1995). HPLC determn: M. Castle, E. Neuteboom, *J. Chromatogr. A* **696**, 93 (1995). Determn in tissue: N. N. P. Tran *et al., Cell Calcium* **18**, 420 (1995). Practical aspects of use in intracellular Ca^{2+} measurement: K. R. Sipido, G. Callewaert, *Cardiovasc. Res.* **29**, 717 (1995).

uv max absorption (100 m*M* KCl): 362 nm free dye; 335 nm Ca^{2+} complex. emission max: 512 nm free dye; 505 nm Ca^{2+} complex.

USE: Fluorescent probe for measurement of calcium in biological systems.

4323. Furaltadone. [139-91-3] 5-(4-Morpholinylmethyl)-3-[[(5-nitro-2-furanyl)methylene]amino]-2-oxazolidinone; 5-morpholinomethyl-3-(5-nitrofurfurylideneamino)-2-oxazolidinone; 3-(5-nitro-2-furfurylideneamino)-5-(4-morpholinomethyl)-2-oxazolidone; furmethonol; nitrofurmethone; NF-260; Altafur; Altabactina; Furazolin; Ibifur; Medifuran; Nitraldone; Otifuril; Sepsinol; Ultrafur; Unifur; Valsyn. $C_{13}H_{16}N_4O_6$; mol wt 324.29. C 48.15%, H 4.97%, N 17.28%, O 29.60%. Prepn: Gever, US 2802002 (1957 to Norwich). LC-MS/MS determn of residues in meat: P. Mottier *et al., J. Chromatogr. A* **1067**, 85 (2005).

Yellow crystals from 95% ethanol, dec 206°. Sparingly sol in water: about 75 mg/100 ml at 25°.

THERAP CAT: Antibacterial.

THERAP CAT (VET): Antibacterial.

4324. Furametpyr. [123572-88-3] 5-Chloro-*N*-(1,3-dihydro-1,1,3-trimethyl-4-isobenzofuranyl)-1,3-dimethyl-1*H*-pyrazole-4-carboxamide; *N*-(1,1,3-trimethyl-2-oxa-4-indanyl)-5-chloro-1,3-dimethylpyrazole-4-carboxamide; S-658; Limber. $C_{17}H_{20}ClN_3O_2$; mol wt 333.82. C 61.17%, H 6.04%, Cl 10.62%, N 12.59%, O 9.59%. Fungicide developed to control rice sheath blight. Prepn: T. Mori *et al.,* EP 315502; *eidem,* US 4877441 (1989, 1989 both to Sumitomo). *In vitro* and *in vivo* metabolism studies: H. Nagahori *et*

4325 Furan

al., *J. Agric. Food Chem.* **48**, 5754 (2000); *eidem, ibid.* 5760. Brief description: Y. Oguri, *Agrochem. Jpn.* **1997**, no. 70, 15-16.

Colorless or light brownish crystals, mp 150.2°. Soly in water (25°): 225 mg/l. Sol in most organic solvents. Vapor pressure (25°): 4.7×10^{-6} Pa. LD_{50} in male, female rats, male, female mice (mg/kg): 640, 590, 660, 730 orally. LD_{50} in rats (mg/kg): >2000 dermally (Oguri).

USE: Fungicide.

4325. Furan. [110-00-9] Furfuran; oxole; tetrole; divinylene oxide. C_4H_4O; mol wt 68.08. C 70.57%, H 5.92%, O 23.50%. Occurs in oils obtained by the distillation of rosin contg pine wood. Prepd by decarboxylation of 2-furancarboxylic acid: Wilson, *Org. Synth.* **coll. vol. I** (2nd ed., 1941), p 274. Has been prepd directly from furfural over hot soda-lime or by dropping furfural on a fused mixt of sodium and potassium hydroxides: Hurd *et al., J. Am. Chem. Soc.* **54**, 2532 (1932). Toxicity data: Henderson, *J. Pharmacol. Exp. Ther.* **57**, 394 (1936). Thermodynamic properties: G. B. Guthrie, Jr. *et al., J. Am. Chem. Soc.* **74**, 4662 (1952).

Liquid. $d_4^{19.4}$ 0.9371. bp_{760} 31.36°; bp_{758} 32°. n_D^{20} 1.4216. Flash pt, closed cup: $-32°F (-35°C)$. *Flammable.* Absorption spectrum: Purvis, *J. Chem. Soc.* **97**, 1648, 1655 (1910). Insol in water. Freely sol in alcohol and ether. Stable to alkalies; resinifies on evaporation or when in contact with mineral acids. LC (in air) in rats: 30400 ppm (Henderson).

Caution: The vapors are anesthetic. Can be absorbed through skin. *See: Toxicity of Industrial Metals,* E. Browning (Appleton-Century-Crofts, New York, 2nd ed., 1969) p 698. This substance is reasonably anticipated to be a human carcinogen: *Report on Carcinogens, Twelfth Edition* (PB2011-111646, 2011) p 205.

USE: In organic syntheses.

4326. 2-Furanacrylic Acid. [539-47-9] 3-(2-Furanyl)-2-propenoic acid; β-2-furylacrylic acid; furacrylic acid; 2-furalacetic acid; furfurylidene acetic acid; $C_7H_6O_3$; mol wt 138.12. C 60.87%, H 4.38%, O 34.75%. Prepd by the condensation of furfural with malonic acid in the presence of pyridine. Laboratory procedure: S. Rajagopalan, P. V. A. Raman, *Org. Synth.* **25**, 51 (1945). Large scale procedure: Johnson, *ibid.* **20**, 55 (1940). Acid dissociation determn: C. C. Price, E. A. Dudley, *J. Am. Chem. Soc.* **78**, 68 (1956).

Colorless needles from water, mp 141°. Sublimes in high vacuum at 112°. bp_{760} 286°; also reported as bp_{760} 226°; bp_8 117°. pKa in ethanol (25°): 6.49. Volatile in steam. One gram dissolves in 500 ml water at 15°. 100 ml of a satd benzene soln contains 1.14 g at 19°. Sol in alcohol, ether, glacial acetic acid.

Labile Form. mp 104°. Can be converted to the stable form by exposure to sunlight in benzene soln contg some iodine.

4327. 2-Furanacrylonitrile. [7187-01-1] 3-(2-Furanyl)-2-propenenitrile; 3-(2-furyl)acrylonitrile. C_7H_5NO; mol wt 119.12. C 70.58%, H 4.23%, N 11.76%, O 13.43%. Prepd from furfural and cyanoacetic acid in the presence of ammonium acetate and pyridine: Patterson, *Org. Synth.* **40**, 46 (1960).

Liquid. bp_{760} 95-97°. n_D^{25} 1.5824. Miscible with toluene, dimethylformamide.

4328. 2,5-Furandicarboxaldehyde. [823-82-5] 2,5-Diformylfuran; 5-formylfurfural; furan-2,5-dialdehyde; 2,5-furandicarbaldehyde. $C_6H_4O_3$; mol wt 124.10. C 58.07%, H 3.25%, O 38.68%. Useful synthetic building block for the prepn of complex organic compds. Prepn from chloromethylfurfuraldehyde: W. F. Cooper, W. H. Nuttall, *J. Chem. Soc.* **101**, 1074 (1912); from 2,5-bis(hydroxymethyl)furan: A. F. Oleinik, K. Y. Novitskii, *J. Org. Chem. USSR (Engl. Transl.)* **6**, 2643 (1970); from (hydroxymethyl)furfural: E. L. Clennan, M. E. Mehrsheikh-Mohammadi, *J. Am. Chem. Soc.* **106**, 7112 (1984); from fructose: G. A. Halliday *et al., Org. Lett.* **5**, 2003 (2003). Synthetic applications: C. Domínguez *et al., Synthesis* **1989**, 172; M. Badri *et al., J. Am. Chem. Soc.* **112**, 5618 (1990); Z. Hui, A. Gandini, *Eur. Polym. J.* **28**, 1461 (1992); A. Obrocka *et al., Pol. J. Chem.* **80**, 1915 (2006). Review: J. Lewkowski, *ARKIVOC* **2001**, 17-54.

Feathery, irridescent plates from light petroleum/chloroform, mp 109.5-110°. Forms thin, waxy, transparent plates from both ether and chloroform. Sublimes below 100°, yielding aggregates of glistening plates that resemble cottonwool. Abs max (water): 290 nm (ε 16900). Extremely sol in chloroform; readily sol in alcohol, benzene, ether, ethyl acetate; somwhat sol in water. Insol in light petroleum.

USE: Versatile intermediate in organic chemistry, especially in the synthesis of macrocycles and polymers.

4329. Furazabol. [1239-29-8] (5α,17β)-17-Methylandrostano[2,3-c][1,2,5]oxadiazol-17-ol; 17β-hydroxy-17α-methyl-5α-androstano[2,3-c]furazan; androfurazanol; furazalon; DH-245; Miotolon. $C_{20}H_{30}N_2O_2$; mol wt 330.47. C 72.69%, H 9.15%, N 8.48%, O 9.68%. Anabolic steroid with hypocholesterolemic properties. Prepn: G. Ohta *et al.*, **BE 645743**; *eidem*, **US 3245988** (1964, 1966 both to Daiichi Seiyaku); Shimizu *et al., Chem. Pharm. Bull.* **13**, 895 (1965); Ohta *et al., ibid.* 1445. Pharmacological studies: Kasahara *et al., ibid.* **13**, 1460 (1965); **16**, 1456, 1460 (1968). Toxicity: *eidem, ibid.* **14**, 285 (1966). Metabolic studies: Takegoshi *et al., ibid.* **20**, 1243 (1972). GC-MS determn in urine: T. Kim *et al., J. Chromatogr. B* **687**, 79 (1996).

Needles from methanol, mp 152-153°. $[\alpha]_D$ +39.4° (c = 1.42 in $CHCl_3$). uv max (ethanol): 217 nm (ε 4300). LD_{50} in mice (g/kg): 2.330 orally; >4 s.c.; 0.494 i.p. (Kasahara, 1966).

Note: This is a controlled substance (anabolic steroid): **21 CFR**, 1308.13, as defined in 1300.01.

4330. Furazolidone. [67-45-8] 3-[[(5-Nitro-2-furanyl)methylene]amino]-2-oxazolidinone; 3-(5-nitrofurfurylideneamino)-2-oxazolidinone; *N*-(5-nitro-2-furfurylidene)-3-amino-2-oxazolidone; NF-180; Furovag; Furoxane; Furoxone; Giarlam; Giardil; Medaron; Neftin; Nicolen; Nifulidone; Ortazol; Roptazol; Tikofuran; Topazone. $C_8H_7N_3O_5$; mol wt 225.16. C 42.68%, H 3.13%, N 18.66%, O 35.53%. Prepn: **GB 735136**; G. Gever, **US 2742462**; G. D. Drake *et al.*, **US 2759931**; G. Gever, C. J. O'Keefe, **US 2927110** (1955, 1956, 1956, 1960 all to Norwich); G. Gever *et al., J. Am. Chem. Soc.* **77**, 2277 (1955). LC-MS/MS determn of residues in meat: P. Mottier *et al., J. Chromatogr. A* **1067**, 85 (2005). Antimicrobial spectrum and toxicity: J. A. Yurchenco *et al., Antibiot. Chemother.* **3**, 1035 (1953); G. S. Rogers *et al., ibid.* **6**, 231 (1956). Pharmacology and toxicology: B. H. Ali, *Vet. Res. Commun.* **6**, 1 (1983).

Yellow crystals from DMF, mp 256-257°. Darkens under strong light. Soly in water (pH 6): approx 40 mg/l. Dec by alkali. Practically insol in alc, carbon tetrachloride.

THERAP CAT: Anti-infective (topical); topical antiprotozoal (Trichmonas).

THERAP CAT (VET): Antimicrobial.

4331. Furcellaran. [9000-21-9] Furcellaria gum; Danish agar; Burtonite 44. A gum obtained from a seaweed of the *Rhodophyceae*, the red alga *Furcellaria fastigiata*, fam. *Furcellariaceae*, order *Gigartinales*. The weed is found primarily in Northern European waters, especially in the Kattegat (between Sweden and Denmark). The gum is the potassium salt of the sulfuric acid ester of a high molecular weight polysaccharide. Consists mainly of D-galactose, 3,6-anhydro-D-galactose, and the half-ester sulfates of these sugars; one sulfate group occurs for each three or four monomeric units, which are arranged in an alternating sequence of $(1 \rightarrow 3)$ and $(1 \rightarrow 4)$-linked units. *Review:* Bjerre-Petersen *et al.* in *Industrial Gums*, R. L. Whistler, Ed. (Academic Press, New York, 2nd ed., 1973) pp 123-136.

The processed gum is a white, odorless powder. Sol in hot or warm water. Easily dispersed in cold water to a homogeneous suspension without lumps; the furcellaran particles hydrate, swell and become almost invisible but do not dissolve unless heated. Forms agar-like gels at low concns. The strength of the gel can be increased by adding salts, esp potassium salts. Highly viscous. Solns in neutral medium are not adversely affected by prolonged exposure to high heat. However, exposure to heat in acidic media results in rapid hydrolysis and loss of gelling power.

USE: Natural colloid, gelling agent, viscosity control agent used primarily in food products but also in pharmaceuticals. Also in products for diabetics, proprietaries for reducing excess body wt, toothpastes. As carrier for food preservatives, bactericides. In bacteriological culture media.

4332. Furethidine. [2385-81-1] 4-Phenyl-1-[2-[(tetrahydro-2-furanyl)methoxy]ethyl]-4-piperidinecarboxylic acid ethyl ester; 4-phenyl-1-[2-(tetrahydrofurfuryloxy)ethyl]isonipecotic acid ethyl ester; ethyl 1-(tetrahydrofurfuryloxyethyl)-4-phenylpiperidine-4-carboxylate; 1-(2'-tetrahydrofurfuryloxyethyl)norpethidine. $C_{21}H_{31}NO_4$; mol wt 361.48. C 69.78%, H 8.64%, N 3.87%, O 17.70%. Prepn: Frearson, Stern, GB 797448 (1958 to J. F. Macfarlan & Co.); Frearson *et al.*, *J. Chem. Soc.* **1960**, 2103.

bp$_{0.5}$ 210°; bp$_{0.3}$ 175-183°. mp ~28°. n_D^{20} 1.5219. pKa 7.48.
Methiodide. $C_{22}H_{34}INO_4$. Crystals from ethyl acetate, mp 174°.
Note: This is a controlled substance (opiate): **21 CFR**, 1308.11.

4333. Furfural. [98-01-1] 2-Furancarboxaldehyde; 2-furaldehyde; pyromucic aldehyde; artificial oil of ants; "furfurol". $C_5H_4O_2$; mol wt 96.09. C 62.50%, H 4.20%, O 33.30%. Occurs in some essential oils. Prepd industrially from pentosans which are contained in cereal straws and brans. Laboratory prepn from corncobs: R. Adams, V. Voorhees, *Org. Synth.* **coll. vol. I**, 280 (2nd ed., 1941). May also be prepd from pyridine. Toxicity study: P. M. Jenner *et al.*, *Food Cosmet. Toxicol.* **2**, 327 (1964).

Colorless oily liq. Peculiar odor, somewhat resembling the odor of benzaldehyde. Turns yellow to brown on exposure to air and light and resinifies (the polymerization is greatly accelerated by hot alkali). d_4^{25} 1.1563. bp$_{760}$ 161.8°; bp$_{100}$ 103°; bp$_{20}$ 67.8°; bp$_{1.0}$ 18.5° mp −36.5°. Volatile in steam. n_D^{20} 1.5261. Absorption spectrum: Purvis, *J. Chem. Soc.* **97**, 1655 (1910). Sol in 11 parts water; very sol in alcohol, ether. Flash pt, closed cup, 140°F (60°C); open cup, 155°F (68°C). Lower explosive limit: 2.1% by vol in air. Autoignition temp 797°F (392°C). *Keep in airtight container and protect from light.* LD$_{50}$ orally in rats: 127 mg/kg (Jenner).

Caution: Potential symptoms of overexposure are irritation of eyes, skin and upper respiratory system; headache; dermatitis. See *NIOSH Pocket Guide to Chemical Hazards* (DHHS/NIOSH 97-140, 1997) p 150.

USE: In the manufacture of furfural-phenol plastics such as Durite; in solvent refining of petroleum oils; in the prepn of pyromucic acid. As a solvent for nitrated cotton, cellulose acetate, and gums; in the manuf of varnishes; for accelerating vulcanization; as insecticide, fungicide, germicide. As a catalyst; as a reagent in analytical chemistry. In the synthesis of furan derivatives.

4334. Furfuryl Alcohol. [98-00-0] 2-Furanmethanol; 2-furylcarbinol; 2-furancarbinol; α-furylcarbinol; furfuralcohol; 2-hydroxymethylfuran. $C_5H_6O_2$; mol wt 98.10. C 61.22%, H 6.17%, O 32.62%. Prepd from furfural which is obtained by the processing of corncobs. Has been obtained by yeast reduction of furfural. The oil obtained by steam distillation of roasted coffee bean meal consists of 50% furfuryl alcohol after all organic acids have been removed. Laboratory prepn from furfural by the Cannizzaro reaction: Wilson, *Org. Synth.* **coll. vol. I** (2nd ed., 1941) p 276; *cf.* US 2041184 (to Quaker Oats), *C.A.* **30**, 4515 (1936). Prepd industrially by the catalytic reduction of furfural using nickel and Cu-CrO catalysts: Peters, US 1906873 (1933 to Quaker Oats); Wojcik, *Ind. Eng. Chem.* **40**, 210 (1948). Acute toxicity study: K. H. Jacobson *et al.*, *Am. Ind. Hyg. Assoc. J.* **19**, 91 (1958). Toxicology and carcinogenesis: *NTP Technical Report 482* (NIH PB99-3972, 1999) 248 p.

Liquid. Faint burning odor. Bitter taste. *Poisonous.* d_4^{23} 1.1282. bp$_{760}$ 170°; bp$_{400}$ 151.8°; bp$_{200}$ 133.1°; bp$_{100}$ 115.9°; bp$_{60}$ 104.0°; bp$_{40}$ 95.7°; bp$_{20}$ 81.0°; bp$_{10}$ 68.0°; bp$_5$ 56.0°; bp$_{1.0}$ 31.8°. n_D^{23} 1.48515. Miscible with water, but unstable in water. Sol in alcohol, benzene, chloroform. Very sol in ether. Easily resinified by acids. Spontaneous ignition temp +490° (915°F). Flash pt 75°C (167°F). LC$_{50}$ (4 hr) in rats: 233 ppm (Jacobson).

Caution: Potential symptoms of overexposure are irritation of eyes and mucous membranes; dizziness; nausea, diarrhea; diuresis; respiration, body temperature depression; vomiting; dermatitis. See *NIOSH Pocket Guide to Chemical Hazards* (DHHS/NIOSH 97-140, 1997) p 150.

USE: Solvent; manuf wetting agents, resins.

4335. α-Furildioxime. [522-27-0] 1,2-Di-2-furanyl-1,2-ethanedione 1,2-dioxime. $C_{10}H_8N_2O_4$; mol wt 220.18. C 54.55%, H 3.66%, N 12.72%, O 29.07%.

Monohydrate. Needle-like crystals, mp 166-168°. Very sol in alcohol, ether, slightly sol in benzene, petr ether.
USE: As a reagent, in alcohol soln, for nickel with which it gives an orange-red compd.

4336. 2-Furoic Acid. [88-14-2] 2-Furancarboxylic acid; α-furoic acid; pyromucic acid; Brenzschleimsäure (German). $C_5H_4O_3$; mol wt 112.08. C 53.58%, H 3.60%, O 42.82%. Prepd from furfural by a Cannizzaro reaction: Wilson, *Org. Synth.* **coll. vol. I**

Furonazide

(2nd ed., 1941) p 276; *cf.* **US 2041184** (to Quaker Oats), *C.A.* **30**, 4515 (1936); Wojcik, *Ind. Eng. Chem.* **40**, 210 (1948); Harrisson, Moyle, *Org. Synth.* **coll. vol. IV**, 493 (1963). Formed in man and several animal species as a metabolite of furfural and related compds.

Elongated monoclinic prisms from water or by sublimation at 130-140° and 50-60 mm pressure. mp 133-134°. bp$_{760}$ 230-232°; bp$_{20}$ 141-144°. pK (25°) 3.12. Absorption spectrum: Hartley, Dobbie, *J. Chem. Soc.* **73**, 600 (1898). One gram dissolves in 26 ml water at 15°; in 4 ml boiling water. More sol in alcohol; freely sol in ether.

Ethyl ester. [614-99-3] Ethyl 2-furoate; ethyl pyromucate. C$_7$H$_8$O$_3$. Colorless crystals. d$_4^{20}$ 1.117. mp 34-36°. bp$_{706}$ 195°. Insol in water; sol in alcohol.

Methyl ester. *See* Methyl 2-furoate.

4337. Furonazide. [3460-67-1] 4-Pyridinecarboxylic acid 2-[1-(2-furanyl)ethylidene]hydrazide; isonicotinic acid α-methylfurfurylidenehydrazide; 2-furyl methyl ketone isonicotinoylhydrazone; α-methylfurfurylidenehydrazide of isonicotinic acid; INF; Furilazone; Clitizina; Menazone. C$_{12}$H$_{11}$N$_3$O$_2$; mol wt 229.24. C 62.87%, H 4.84%, N 18.33%, O 13.96%. Prepn: Miyatake *et al., J. Pharm. Soc. Jpn.* **75**, 1066 (1955).

Crystals, mp 199-201.5°.
THERAP CAT: Antibacterial (tuberculostatic).

4338. Furosemide. [54-31-9] 5-(Aminosulfonyl)-4-chloro-2-[(2-furanylmethyl)amino]benzoic acid; 4-chloro-N-furfuryl-5-sulfamoylanthranilic acid; 4-chloro-N-(2-furylmethyl)-5-sulfamoylanthranilic acid; frusemide; fursemide; LB-502; Aisemide; Beronald; Desdemin; Discoid; Diural; Dryptal; Durafurid; Errolon; Eutensin; Frusetic; Frusid; Fulsix; Fuluvamide; Furesis; Furo-Puren; Furosedon; Hydro-rapid; Impugan; Katlex; Lasilix; Lasix; Lowpston; Macasirool; Mirfat; Nicorol; Odemase; Oedemex; Profemin; Rosemide; Rusyde; Trofurit; Urex. C$_{12}$H$_{11}$ClN$_2$O$_5$S; mol wt 330.74. C 43.58%, H 3.35%, Cl 10.72%, N 8.47%, O 24.19%, S 9.69%. Prepn: K. Stürm *et al.*, **DE 1122541**; *eidem,* **US 3058882** (both 1962 to Hoechst). Molecular and crystal structure: J. Lamotte *et al., Acta Crystallogr.* **B34**, 1657 (1978). Review of pharmacokinetics: L. Z. Benet, *J. Pharmacokinet. Biopharm.* **7**, 1-27 (1979); R. E. Cutler, A. D. Blair, *Clin. Pharmacokinet.* **4**, 279-296 (1979). Pharmacodynamics and pharmacokinetics of a slow-release formulation in man: A. Ebihara *et al., Arzneim.-Forsch.* **33**, 163 (1983). Toxicity: E. I. Goldenthal, *Toxicol. Appl. Pharmacol.* **18**, 185 (1971). Comprehensive description: A. M. Al-Obaid *et al., Anal. Profiles Drug Subs.* **18**, 153-193 (1989).

Crystals from aq ethanol, mp 206°. uv max (95% ethanol): 288, 276, 336 nm (E$_{1cm}^{1}$ 945, 588, 144); (0.1N NaOH): 226, 273, 336 nm (E$_{1cm}^{1\%}$ 1147, 557, 133). Slightly sol in water, chloroform, ether. Sol in acetone, methanol, DMF, aq solns above pH 8.0. Less sol in ethanol. *Pharmaceut. Incompat:* Calcium gluconate, ascorbic acid, tetracyclines, urea, epinephrine. LD$_{50}$ orally in female, male rats: 2600, 2820 mg/kg (Goldenthal).

THERAP CAT: Diuretic; antihypertensive.
THERAP CAT (VET): Diuretic.

4339. 2-Furoyl Chloride. [527-69-5] 2-Furancarbonyl chloride; 2-(chlorocarbonyl)furan; α-furoic chloride; pyromucyl chloride. C$_5$H$_3$ClO$_2$; mol wt 130.53. C 46.01%, H 2.32%, Cl 27.16%, O 24.51%. Prepn: P. F. Frankland, F. W. Aston, *J. Chem. Soc. Trans.* **79**, 511 (1901); W. W. Hartman, J. B. Dickey, *Ind. Eng. Chem.* **24**, 151 (1932).

Colorless, strongly refracting liquid. *Combustible. Corrosive.* mp −2°. bp 170°. bp$_{10}$ 66°. Flash point, closed cup: 185°F (85°C). Dec by water or alcohol. Sol in ether.

4340. Fursultiamine. [804-30-8] N-[(4-Amino-2-methyl-5-pyrimidinyl)methyl]-N-[4-hydroxy-1-methyl-2-[[(tetrahydro-2-furanyl)methyl]dithio]-1-buten-1-yl]formamide; thiamine tetrahydrofurfuryl disulfide; Alinamin F; Diteftin; Judolor; TTFD. C$_{17}$H$_{26}$N$_4$O$_3$S$_2$; mol wt 398.54. C 51.23%, H 6.58%, N 14.06%, O 12.04%, S 16.09%. Prepn: S. Yurugi, T. Fushimi, **US 3016380** (1962 to Takeda). Metabolism: S. Kikuchi *et al., Eur. J. Pharmacol.* **9**, 367 (1970); C. Mitoma, *Drug Metab. Dispos.* **1**, 698 (1973). Series of papers on function of tetrahydrofurfuryl mercaptan moiety: *J. Biochem.* **74**, 717 (1973). Crystal structure: W. Shin, Y. C. Kim, *Bull. Korean Chem. Soc.* **7**, 331 (1986).

E - form

Colorless prisms from ethyl acetate, mp 132° (dec). d 1.29. Sparingly sol in water; sol in organic solvents, dil mineral acids. LD$_{50}$ in rats (mg/kg): 2200 orally; 540 i.p. (Yurugi, Fushimi).

THERAP CAT: Vitamin (enzyme cofactor).

4341. Furtrethonium. [7618-86-2] N,N,N-Trimethyl-2-furanmethanaminium; furfuryltrimethylammonium; trimethylfurfurylammonium; furtrimethonium; Furmethide. [C$_8$H$_{14}$NO]$^+$. Prepn of salts: Nabenhauer, **US 2185220** (1938 to SK & F); Weilmuenster, Jordan, *J. Am. Chem. Soc.* **67**, 415 (1945); Khromov-Borisov *et al., J. Gen. Chem. USSR* **24**, 2021 (1954).

Iodide. [541-64-0] Furamon; Furanol. C$_8$H$_{14}$INO; mol wt 267.11. Crystals from ethanol + ethyl acetate, mp 116-117° (Weilmuenster, Jordan); also reported as mp 118-120° (Khromov-Borisov). Soluble in water, alcohol. Practically insol in benzene. pH of 1% aq soln, 5.3-6.0.

Benzenesulfonate. Benzamon. C$_{14}$H$_{19}$NO$_4$S; mol wt 297.37. Crystals from ethanol + butyl acetate, mp 134-135° (Kromov-Borisov). Soluble in water, alcohol.

p-**Toluenesulfonate.** Fisostina. C$_{15}$H$_{21}$NO$_4$S; mol wt 311.40.
THERAP CAT: Cholinergic. Iodide as parasympathomimetic.

4342. Fusafungine. [1393-87-9] S-314; Biofusal; Fusaloyos; Fusarine; Locabiotal. Antibiotic used as an aerosol for local application. Isoln from *Fusarium* species: Couchoud, **FR 1164181** (1958), *C.A.* **54**, 20074b (1960); Servier, **BE 612474** (1962), *C.A.* **57**, 11320b (1962); *idem,* **GB 1018626** (1966 to Biofarma), *C.A.* **64**, 16585b (1966). Bacteriological activity: D. Haler, *J. Int. Med. Res.* **5**, 61 (1977); *idem, Vie Med.* **61**, 507 (1980).

Solid, mp 125-129°. Stable up to 180°. Practically insol in water. Sol in glycols and fats.

THERAP CAT: Antibacterial.

4343. Fusaric Acid. [536-69-6] 5-Butyl-2-pyridinecarboxylic acid; 5-butylpicolinic acid. C$_{10}$H$_{13}$NO$_2$; mol wt 179.22. C

Consult the Name Index before using this section.

67.02%, H 7.31%, N 7.82%, O 17.85%. Antibiotic (wilting agent) first isolated from the fungus *Fusarium heterosporium*, Nees: Yabuta *et al.*, *J. Agric. Chem. Soc. Jpn.* **10**, 1059 (1934). Isoln from other *Fusarium* species and from *Gibberella fujikuroi* and synthesis: Plattner *et al.*, *Helv. Chim. Acta* **37**, 1379 (1954). Prepn: Hardegger, Nikles, *ibid.* **39**, 505 (1956); **40**, 2428 (1957); Schreiber, Adam, *Ber.* **93**, 1848 (1960); Umezawa, Nagatsu, **DE 2005255** (1970 to Microbiochem. Res. Found.); R. Tschesche, W. Führer, *Ber.* **111**, 3502 (1978). Dopamine β-hydroxylase inhibitor and hypotensive activity: Suda *et al.*, *Chem. Pharm. Bull.* **17**, 2377 (1969); Nagatsu *et al.*, *Biochem. Pharmacol.* **19**, 35 (1970). Toxicity study: Ishii *et al.*, *Arzneim.-Forsch.* **25**, 55 (1975).

Colorless crystals, mp 96-98°. LD_{50} orally in mice: 230 mg/kg (Ishii).

Copper salt. Bluish-violet crystals from water, mp 258-259°.

4344. Fusarubin. [1702-77-8] 3,4-Dihydro-3,6,9-trihydroxy-7-methoxy-3-methyl-1*H*-naphtho[2,3-*c*]pyran-5,10-dione; 5,8-dihydroxy-2-(hydroxymethyl)-6-methoxy-3-(2-oxopropyl)-1,4-naphthalenedione; 3-acetonyl-5,8-dihydroxy-2-(hydroxymethyl)-6-methoxy-1,4-naphthoquinone; oxyjavanicin. $C_{15}H_{14}O_7$; mol wt 306.27. C 58.83%, H 4.61%, O 36.57%. Isoln from *Fusarium solani*: Ruelius, Gauhe, *Ann.* **569**, 38 (1950). Structure: Hardegger *et al.*, *Helv. Chim. Acta* **47**, 2027 (1964). Efficient total synthesis: Y. Tanoue *et al.*, *Bull. Chem. Soc. Jpn.* **60**, 2927 (1987).

Red prisms from benzene, dec 218°. Absorption max (ether): 535, 499 nm. Sol in glacial acetic acid, tetrahydrofuran, acetone, dioxane, pyridine; slightly sol in cold chloroform, cold alcohol, ether. Practically insol in carbon disulfide, cyclohexane, cold benzene. Dissolves in dil NaOH with a violet color. Practically insol in bicarbonate soln.

4345. Fusel Oil. [8013-75-0] Fusel alcohols. Mixture of higher alcohols formed as a by-product of carbohydrate fermentation in the production of ethyl alcohol. Flavor constituent of beer, wines and distilled spirits. Varies widely in composition, depending on the fermentation raw material used, but contains chiefly isopentyl alcohol and 2-methyl-1-butanol, as well as isobutyl alcohol (20%), *n*-propyl alcohol (3-5%), and small amounts of other alcohols, esters and aldehydes. Composition of grape brandy fusel oil: A. D. Webb *et al.*, *Anal. Chem.* **24**, 1944 (1952). HPLC determn of constituents: M. E. Neale, *J. Chromatogr.* **447**, 443 (1988). *Review*: A. G. Patil *et al.*, *Int. Sugar J.* **104**, 51-58 (2002).

Refined fusel oil is a colorless to pale yellow liquid with wine or whiskey-like odor. bp 128-130°. n_D^{20} 1.405-1.410. d_{25}^{25} 0.807-0.813. Rotation: −0.5° to −2.0°. Sol in propylene glycol, vegetable oils, in 3 parts 70% ethanol. Insol in water.

USE: Flavoring agent, solvent, source of amyl alcohols.

4346. Fusidic Acid. [6990-06-3] (3α,4α,8α,9β,11α,13α,-14β,16β,17Z)-16-(Acetyloxy)-3,11-dihydroxy-29-nordammara-17-(20),24-dien-21-oic acid; (Z)-3α,11α,16β-trihydroxy-29-nor-8α,-9β,13α,14β-dammara-17(20),24-dien-21-oic acid 16-acetate; 3α,-11α,16β-trihydroxy-4α,8,14-trimethyl-18-nor-5α,8α,9β,13α,14β-cholesta-17(20),24-dien-21-oic acid 16-acetate; 3,11,16-trihydroxy-4,8,10,14-tetramethyl-17-(1'-carboxyisohept-4'-enylidene)cyclopentanoperhydrophenanthrene 16-acetate; ramycin; Fucithalmic. $C_{31}H_{48}O_6$; mol wt 516.72. C 72.06%, H 9.36%, O 18.58%. Antibiotic isolated from the fermentation broth of *Fusidium coccineum*; structurally similar to cephalosporin P_1, *q.v.* Inhibits bacterial protein synthesis by interference with elongation factor G. Isoln and

structure: W. O. Godtfredsen *et al.*, *Nature* **193**, 987 (1962); *Lancet* **I**, 928 (1962); W. O. Godtfredsen, S. Vangedal, *Tetrahedron* **18**, 1029 (1962). Identity with ramycin: H. Vanderhaeghe *et al.*, *Nature* **205**, 710 (1965). Structure: D. Arigoni *et al.*, *Experientia* **19**, 521 (1963). Stereochemistry: W. O. Godtfredsen *et al.*, *Tetrahedron* **21**, 3505 (1965). Synthetic studies: W. G. Dauben *et al.*, *J. Am. Chem. Soc.* **94**, 8593 (1972); R. E. Ireland, U. Hengartner, *ibid.* 3652; M. Tanabe *et al.*, *Tetrahedron Lett.* **1977**, 1481. Total synthesis: W. G. Dauben *et al.*, *J. Am. Chem. Soc.* **104**, 303 (1982). Review of structure-activity relationships: W. von Daehne *et al.*, *Adv. Appl. Microbiol.* **25**, 95-146 (1979). Series of articles on pharmacology and clinical experience: *Int. J. Antimicrob. Agents* **12**, Suppl. 2, S1-S93 (1999). Review of use in *Staphylococcus aureus* infections: D. Dobie, J. Gray, *Arch. Dis. Child.* **89**, 74-77 (2004).

Crystals from ether or benzene, mp 192-193°. $[\alpha]_D^{20}$ −9° (chloroform). uv max: 204 nm (ε 9900). pK: 5.35 in water. Sol in alc, acetone, chloroform, pyridine, dioxane; sparingly sol in water, ether, hexane. LD_{50} in mice (g/kg): 1.2 s.c.; 1.5 orally (Godtfredsen).

Sodium salt. [751-94-0] Sodium fusidate; ZN-6; Fucidin; Fucidine. $C_{31}H_{47}NaO_6$; mol wt 538.70. Crystals, sol in water. LD_{50} in mice (g/kg): 0.2 i.v. (Godtfredsen, *Nature* 1962).

THERAP CAT: Antibacterial.

THERAP CAT (VET): Antibacterial.

4347. Fustin. [20725-03-5] *rel*-(2R,3R)-2-(3,4-Dihydroxyphenyl)-2,3-dihydro-3,7-dihydroxy-4*H*-1-benzopyran-4-one; *trans*-(±)-3,3',4',7-tetrahydroxyflavanone; dihydrofisetin. $C_{15}H_{12}O_6$; mol wt 288.26. C 62.50%, H 4.20%, O 33.30%. Flavanone which occurs naturally as a racemic mixture of *trans*-isomers. From wood of *Rhus cotinus* L. (Venice sumac) and *R. succedanea* L., *Anacardiaceae*: Schmid, *Ber.* **19**, 1734 (1886); from *Gleditsia triacanthos* L., *Leguminosae*: Chadenson *et al.*, *Compt. Rend.* **249**, 1362 (1955). Structure: T. Oyamada, *Ann.* **538**, 44 (1939). Stereochemistry: K. Weinges, *Ann.* **627**, 229 (1959); D. G. Roux, E. Paulus, *Biochem. J.* **77**, 315 (1960); W. Gaffield, *Tetrahedron* **26**, 4093 (1970).

Relative stereochemistry

Crystals, mp 226-228°. Natural racemate shows low negative rotation: $[\alpha]_D^{21}$ −2.4°.

Tetraacetate. $C_{23}H_{20}O_{10}$; mol wt 456.40. Crystals from methanol, mp 147-148°.

(2S,3S)-Form. [17654-28-3] (−)-Fustin. Crystals, mp 216-218°. $[\alpha]_D^{25}$ −26° (c = 2 in 1:1 acetone, water).

(2S,3S)-Form tetraacetate. Tetraacetyl-(−)-fustin. Crystals, mp 117-118°. $[\alpha]_D^{23}$ −25.2° (c = 0.8 in tetrachloroethane).

(2R,3R)-Form. [4382-36-9] (+)-Fustin. Needles from water, mp 228-229°. $[\alpha]_D^{23}$ +28.3° (c = 0.9 in 1:1 acetone, water).

(2R,3R)-Form tetraacetate. Tetraacetyl-(+)-fustin. Crystals from ethanol, mp 116-119°. $[\alpha]_D^{25}$ +24.4° (c = 1.3 in tetrachloroethane).

G

4348. Gabapentin. [60142-96-3] 1-(Aminomethyl)cyclohexaneacetic acid; CI-945; Gö-3450; GOE-3450; Gralise; Neurontin. $C_9H_{17}NO_2$; mol wt 171.24. C 63.13%, H 10.01%, N 8.18%, O 18.69%. Lipophilic analog of γ-aminobutyric acid (GABA), *q.v.*, that crosses the blood brain barrier and exhibits anticonvulsant and antinociceptive properties. Specifically binds to and inhibits voltage gated calcium channels containing the $α_2δ$-1 subunit. Prepn: G. Satzinger *et al.*, **DE 2460891** (1976 to Gödecke); *eidem*, **US 4024175** (1977 to Warner-Lambert). Formal synthesis: J. S. Bryans *et al.*, *Tetrahedron* **59**, 6221 (2003); R. Cagnoli *et al.*, *ibid.*, 9951. Analysis of binding conformation: J. S. Bryans *et al.*, *Bioorg. Med. Chem.* **7**, 715 (1999). GC determn in biological fluids: W. D. Hooper *et al.*, *J. Chromatogr.* **529**, 167 (1990). LC-MS/MS determn in plasma: N. V. S. Ramakrishna *et al.*, *J. Pharm. Biomed. Anal.* **40**, 360 (2006). Review of pharmacokinetics and clinical trials in epilepsy: K. L. Goa, E. M. Sorkin, *Drugs* **46**, 409-427 (1993); of pharmacology and clinical experience in neuropathic pain: M. A. Rose, P. C. A. Kam, *Anaesthesia* **57**, 451-462 (2002); in postoperative pain: K.-Y. Ho *et al.*, *Pain* **126**, 91-101 (2006). Review of mechanisms of action: G. J. Sills, *Curr. Opin. Pharmacol.* **6**, 108-113 (2006); C. P. Taylor, *Pain* **142**, 13-16 (2009).

White to off-white crystalline solid with bitter taste. Crystals from ethanol/ether, mp 162-166° (Satzinger); also reported as mp 165-167° (Schmidt). pKa_1 (25°) 3.68; pKa_2 10.70. Isoelectric point 7.14. Partition coefficient (octanol/buffer): 0.075 (pH 7.4). Freely sol in water and in alkaline and acidic solutions.

THERAP CAT: Anticonvulsant; analgesic.

THERAP CAT (VET): Anticonvulsant; analgesic.

4349. Gabapentin Encarbil. [478296-72-9] 1-[[[[1-(2-Methyl-1-oxopropoxy)ethoxy]carbonyl]amino]methyl]cyclohexaneacetic acid; 1-[[(α-isobutanoyloxyethoxy)carbonyl]aminomethyl]-1-cyclohexaneacetic acid; ASP-8825; GSK-1838262; XP-13512. $C_{16}H_{27}NO_6$; mol wt 329.39. C 58.34%, H 8.26%, N 4.25%, O 29.14%. Prodrug of gabapentin, *q.v.* Prepn: M. A. Gallop *et al.*, **WO 02100347**; *eidem*, **US 6818787** (2002, 2004 both to Xenoport); and pharmacology: K. C. Cundy *et al.*, *J. Pharmacol. Exp. Ther.* **311**, 315 (2004). Bioavailability and absorption: *idem et al.*, *ibid.*, 324. Clinical pharmacokinetics: *idem et al.*, *J. Clin. Pharmacol.* **48**, 1378 (2008). Clinical trial in restless legs syndrome: R. K. Bogan *et al.*, *Mayo Clin. Proc.* **85**, 512 (2010). Review of pharmacology and clinical experience: G. Merlino *et al.*, *Curr. Opin. Investig. Drugs* **10**, 91-102 (2009); P. Agarwal *et al.*, *Neuropsychiatr. Dis. Treat.* **6**, 151-158 (2010).

White crystalline solid from ethyl acetate/heptane, mp 65°. pKa 5.0.

THERAP CAT: In treatment of restless legs syndrome.

4350. Gabexate. [39492-01-8] 4-[[6-[(Aminoiminomethyl)-amino]-1-oxohexyl]oxy]benzoic acid ethyl ester; *p*-hydroxybenzoic acid ethyl ester 6-guanidinohexanoate; *p*-carbethoxyphenyl ε-guanidinocaproate. $C_{16}H_{23}N_3O_4$; mol wt 321.38. C 59.80%, H 7.21%, N 13.08%, O 19.91%. Non-peptide proteolytic enzyme inhibitor which also inhibits the hydrolytic effects of thrombin, plasmin, and kallikrein, trypsin but not chymotrypsin; *cf.* aprotinin. Prepn as the

p-toluenesulfonate salt: S. Fujii, T. Watanabe, **DE 2050484**; *eidem*, **US 3751447** (1971, 1973 both to Ono). Enzyme inhibition: M. Muramatu, S. Fujii, *Biochim. Biophys. Acta* **268**, 221 (1972); S. Tamura *et al.*, *ibid.* **484**, 417 (1977). Pharmacology: T. Okegada *et al.*, *Nippon Yakurigaku Zasshi* **71**, 71 (1975), *C.A.* **84**, 218m (1976). Metabolism: M. Sugiyama *et al.*, *Oyo Yakuri* **9**, 733 (1975), *C.A.* **83**, 188145s (1975). Metabolism of inhibitory effect on platelet aggregation: G. Kosaki *et al.*, *Thromb. Res.* **20**, 587 (1980). Beneficial action in traumatic shock: A. M. Lefer *et al.*, *IRCS Med. Sci. Libr. Compend.* **8**, 278 (1980); in exptl acute pancreatitis: J. R. Wisner *et al.*, *Pancreas* **2**, 181 (1987). Comparative clinical study in acute pancreatitis: N. Tanaka *et al.*, *Adv. Exp. Med. Biol.* **120**, 367 (1979). Teratology and toxicity study: T. Fujita *et al.*, *Oyo Yakuri* **9**, 743 (1975), *C.A.* **83**, 188322x (1975).

Methanesulfonate. [56974-61-9] Gabexate mesylate; FOY; Megacert. $C_{16}H_{23}N_3O_4.CH_3SO_3H$; mol wt 417.48. White crystals. Sol in water, ethanol, chloroform. Slightly sol in acetone. Practically insol in ether. pH of soln (1:100): 4.0-5.0. LD_{50} in mice (mg/kg): 8000 orally; 4700 s.c.; 25 i.v. (Fujita).

THERAP CAT: Enzyme inhibitor (proteinase).

4351. Gaboxadol. [64603-91-4] 4,5,6,7-Tetrahydroisoxazolo[5,4-*c*]pyridin-3(2*H*)-one; 4,5,6,7-tetrahydroisoxazolo[5,4-*c*]-pyridin-3-ol; THIP; Lu-02-030; MK-0928. $C_6H_8N_2O_2$; mol wt 140.14. C 51.42%, H 5.75%, N 19.99%, O 22.83%. Selective extrasynaptic GABA agonist (SEGA); structural analog of muscimol, *q.v.* Prepn: P. Krogsgaard-Larsen, *Acta Chem. Scand. B* **31**, 584 (1977); *idem*, **EP 167**; *idem*, **US 4278676** (1979, 1981 both to H. Lundbeck & Co.). GABA agonist effects: *idem et al.*, *Nature* **268**, 53 (1977). ^{3}H-THIP binding study: E. Falch, P. Krogsgaard-Larsen, *J. Neurochem.* **38**, 1123 (1982). HPLC determn: S. M. Madsen, *J. Chromatogr.* **238**, 509 (1982). Pharmacokinetics: *idem et al.*, *Acta Pharmacol. Toxicol.* **53**, 353 (1983). Pharmacodynamics and potential therapeutic uses: A. V. Christensen *et al.*, *Pharm. Weekbl. Sci. Ed.* **4**, 145 (1982). Clinical effects on nocturnal sleep and hormone secretion: M. Lancel *et al.*, *Am. J. Physiol. Endocrinol. Metab.* **281**, E130 (2001); on night sleep and cognitive performance: S. Mathias *et al.*, *Neuropsychopharmacology* **30**, 833 (2005). Review of clinical development: R. Huckle, *Curr. Opin. Investig. Drugs* **5**, 766-773 (2004).

Colorless cryst, mp 242-244° (dec). uv max (methanol): 212 nm (log ε 3.64). pKa (water, 25°): 4.44 ±0.03; 8.48 ±0.04.

Hydrobromide. [65202-63-3] $C_6H_8N_2O_2.HBr$. Faintly reddish cryst from methanol-ether, mp 162-163° (dec). LD_{50} in mice (mg/kg): 80 i.v., 145 i.p.; >320 orally (U.S. patent).

Hydrochloride. [85118-33-8] $C_6H_8N_2O_2.HCl$; mol wt 176.60.

USE: As a molecular probe to study GABA receptors.

THERAP CAT: Sedative, hypnotic.

4352. Gadobenate Dimeglumine. [127000-20-8] 1-Deoxy-1-(methylamino)-D-glucitol [4-(carboxy-κ*O*)-5,8,11-tris[(carboxy-κ*O*)methyl]-1-phenyl-2-oxa-5,8,11-triazatridecan-13-oato(5−)-κ*N*[5],κ*N*[8],κ*N*[11],κ*O*[13]]gadolinate(2−) (2:1); gadolinium benzyloxypropionictetraacetate dimeglumine; Gd-BOPTA/Dimeg; B-19036/7; MultiHance. $C_{36}H_{62}GdN_5O_{21}$; mol wt 1058.16. C 40.86%, H 5.91%, Gd 14.86%, N 6.62%, O 31.75%. Intravascular paramagnetic MRI contrast agent. Prepn: E. Felder *et al.*, **EP 230893**; *eidem*,

US 4916246 (1987, 1990 both to Bracco); F. Ungerri *et al., Inorg. Chem.* **34**, 633 (1995). HPLC determn in biological samples: T. Arbughi *et al., J. Chromatogr. B* **713**, 415 (1998). Physicochemical properties: C. de Haen *et al., J. Comput. Assist. Tomogr.* **23**, Suppl. 1, S161 (1999). Pharmacology: P. Tirone *et al., ibid.* S195. Pharmacokinetics: V. Lorusso *et al., ibid.* S181. Toxicology: A. Morisetti *et al., ibid.* S207. Clinical study in MRI of liver lesions: J. Petersein *et al. Radiology* **215**, 727 (2000). Review of clinical studies: B. Hamm *et al., J. Comput. Assist. Tomogr.* **23**, Suppl. 1, S53-S60 (1999).

Hygroscopic powder. mp 124°. Freely sol in water, sol in methanol. Practically insol in *n*-butanol, *n*-octanol, chloroform. Abs max 257.8 nm (ε 203). $[\alpha]^{20}_{365}$ −26.9° (c = 1.45 in water). Prepd as 0.5*M* soln, osmolality (37°) 1.97 mol/kg. d^{20} 1.22. Viscosity (mPa·s): 9.2 (20°), 5.3 (37°). LD_{50} i.v. in mice (mmol/kg): 5.7 (at 1 mL/min), 7.9 (at 0.2 mL/min); LD_{50} i.v. in rats (mmol/kg): 6.6 (at 6 mL/min), 9.2 (at 1 mL/min) (Morisetti).

THERAP CAT: Diagnostic aid (MRI contrast agent).

4353. Gadobutrol. [138071-82-6] [10-[2,3-Dihydroxy-1-(hydroxymethyl)propyl]-1,4,7,10-tetraazacyclododecane-1,4,7-triacetato(3−)-$\kappa N^1,\kappa N^4,\kappa N^7,\kappa N^{10},\kappa O^1,\kappa O^4,\kappa O^7$]gadolinium; Gd-DO3A-butrol; Gadovist. $C_{18}H_{31}GdN_4O_9$; mol wt 604.72. C 35.75%, H 5.17%, Gd 26.00%, N 9.27%, O 23.81%. Neutral, macrocyclic gadolinium chelate. Prepn: J. Platzek *et al.,* **EP 448191** (1991 to Schering AG). Physicochemical properties and *in vivo* imaging studies: H. Vogler *et al., Eur. J. Radiol.* **21**, 1 (1995). Clinical pharmacokinetics: T. Staks *et al., Invest. Radiol.* **29**, 709 (1994). Clinical evaluation of diagnostic use for cerebral metastases: T. J. Vogl *et al., Radiologe* **35**, 508 (1995); for glioblastomas: M. Hartmann *et al., Fortschr. Röntgenstr.* **164**, 119 (1996).

Hydrophilic. Osmolality (osmol/kg): 0.57 (0.5 mol/l); 1.39 (1 mol/l). Viscosity (cP): 1.43 (0.5 mol/l); 3.7 (1 mol/l). Partition coefficient (butanol/water): 0.006. LD_{50} i.v. in mice: 23 mmol/kg (Vogler).

THERAP CAT: Diagnostic aid (MRI contrast agent).

4354. Gadodiamide. [131410-48-5] [5,8-Bis[(carboxy-κO)-methyl]-11-[2-(methylamino)-2-(oxo-κO)ethyl]-3-(oxo-κO)-2,5,8,11-tetraazatridecan-13-oato(3−)-$\kappa N^5,\kappa N^8,\kappa N^{11},\kappa O^{13}$]gadolinium; gadolinium diethylenetriamine pentaacetic acid bismethylamide; Gd-DTPA-BMA; S-041. $C_{16}H_{26}GdN_5O_8$; mol wt 573.66. C 33.50%, H 4.57%, Gd 27.41%, N 12.21%, O 22.31%. Nonionic, low-osmolar paramagnetic gadolinium chelate complex. Prepn: S. C. Quay, **WO 8602841** (1986 to New Salutar); *idem,* **US 4687659** (1987 to Salutar). Pharmacokinetics, hemodynamics and toxicology: E. S. Harpur *et al., Invest. Radiol.* **28**, Suppl. 1, S28 (1993). Clinical pharmacokinetics: M. VanWagoner, D. Worah, *ibid.* S44. Clinical study of CNS imaging: G. Sze *et al., ibid.* **28**, Suppl. 1, S49 (1993); J. Valk *et al., Neuroradiology* **35**, 173 (1993); of myocardial imaging: M.-C. Dulce *et al., Am. J. Roentgenol.* **160**, 963 (1993).

Review of physicochemical properties: C. A. Chang, *Invest. Radiol.* **28**, Suppl. 1, S21-S27 (1993).

Osmolality (37°, 0.1*M*): 111 μmol/kg water. Freely sol in water, methanol; sol in ethyl alc; slightly sol in acetone, chloroform. LD_{50} i.v. in mice: 14 mmol/kg (Chang).

Gadodiamide injection. [131410-51-0] Omniscan. $C_{16}H_{28}$-$GdN_5O_9.C_{16}H_{26}CaN_5NaO_8$; mol wt 1071.15. Osmolality (37°, 0.5*M*): 789 mOsM/kg water. Viscosity (cP): 2.0 (20°), 1.4 (37°). d^{20} 1.13. Log P (butanol/water): −2.1. LD_{50} i.v. in mice: 34 mmol/kg (Harpur).

THERAP CAT: Diagnostic aid (MRI contrast agent).

4355. Gadofosveset. [201688-00-8] (*SA*-8-11252634)-[[(8*R*)-8-[Bis[(carboxy-κO)methyl]amino-κN]-3-6-bis[(carboxy-κO)methyl]-11-[(4,4-diphenylcyclohexyl)oxy]-11-hydroxy-10-oxa-3,6-diaza-11-phosphaundecanoic acid-$\kappa N^3,\kappa N^6,\kappa O^1$]-11-oxidato(6−)]gadolinate(3−) hydrogen (1:3); (*SA*-8-11252634)-[[4-[bis-[(carboxy-κO)methyl]amino-κN]-6,9-bis[(carboxy-κO)methyl]-1-[(4,4-diphenylcyclohexyl)oxy]-1-hydroxy-2-oxa-6,9-diaza-1-phosphaundecan-11-oic acid-$\kappa N^6,\kappa N^9,\kappa O^{11}$]-1-oxidato(6−)]gadolinate(3−) trihydrogen; MS-325. $C_{33}H_{41}GdN_3O_{14}P$; mol wt 891.92. C 44.44%, H 4.63%, Gd 17.63%, N 4.71%, O 25.11%, P 3.47%. Intravascular contrast agent exhibiting strong but reversible binding to human serum albumin, *q.v.*, in plasma. Prepn: T. J. McMurray *et al.,* **WO 9623526** (1996 to Metasyn). Pharmacology and image enhancing characteristics: R. B. Lauffer *et al., Radiology* **207**, 529 (1998). Pharmacokinetics: D. J. Parmelee *et al., Invest. Radiol.* **32**, 741 (1997). Physicochemical properties: R. N. Muller *et al., Eur. J. Inorg. Chem.* **1999**, 1949. Clinical evaluation in cornary artery angiography: D. A. Bluemke *et al., Radiology* **219**, 114 (2001); in peripheral vascular angiography: P. Perreault *et al., ibid.* **229**, 811 (2003).

Trisodium salt. [193901-90-5] Vasovist. $C_{33}H_{38}GdN_3Na_3O_{14}$-P; mol wt 957.87.

THERAP CAT: Diagnostic aid (MRI contrast agent).

4356. Gadolinium. [7440-54-2] Gd; at. wt 157.25; at. no. 64; valence 3. A lanthanide; belongs to yttrium group of rare earth metals. Naturally occurring isotopes (mass numbers): 152 (0.20%), radioactive, $T_{1/2}$ 1.08 × 10¹⁴ years, α-emitter; 154 (2.18%); 155 (14.80%); 156 (20.47%); 157 (15.65%); 158 (24.84%); 160 (21.86%); known artificial radioactive isotopes: 137, 139, 142-151, 153, 159, 161, 162. Abundance in earth's crust: 6.1-6.36 ppm. Sources: samarskite, gadolinite (ytterbate), xenotime, and other rare earth minerals. Discovered by J. C. G. de Marignac in 1880. Prepn of metal: Trombe, *Compt. Rend.* **200**, 459 (1935); *idem, Bull. Soc. Chim. Fr.* **2**, 660 (1935); Klemm, Bommer, *Z. Anorg. Chem.* **231**, 138 (1937). Sepn from other rare earths: F. H. Spedding *et al., J. Am. Chem. Soc.* **69**, 2812 (1947); **76**, 2557 (1954). Spectrum: Albertson, *Phys. Rev.* **47**, 370 (1935); Spedding *et al., J. Chem. Phys.* **5**, 33 (1937). Toxicity study: Haley, *J. Pharm. Sci.* **54**, 663 (1965).

Reviews of prepn, properties and compds: *The Rare Earths*, F. H. Spedding, A. H. Daane (Krieger, Huntington, N.Y., 1971, reprint of 1961 ed.) 641 pp; Hulet, Bode, "Separation Chemistry of the Lanthanides and Transplutonium Actinides" in *MTP Int. Rev. Sci.: Inorg. Chem., Ser. One* vol. 7, K. W. Bagnall, Ed. (University Park Press, Baltimore, 1971) pp 1-45; T. Moeller, "The Lanthanides" in *Comprehensive Inorganic Chemistry* vol. 4, J. C. Bailar, Jr. *et al.*, Eds. (Pergamon Press, Oxford, 1973) pp 1-101; F. H. Spedding in *Kirk-Othmer Encyclopedia of Chemical Technology* vol. 19 (John Wiley & Sons, New York, 3rd ed., 1982) pp 833-854; *Chemistry of the Elements*, N. N. Greenwood, A. Earnshaw, Eds. (Pergamon Press, New York, 1984) pp 1423-1449. Brief review of properties: G. T. Seaborg, *Radiochim. Acta* **61**, 115-122 (1993).

Colorless or faintly yellowish metal; tarnishes in moist air. Crystalline forms: hexagonal close-packed α-form, d 7.886, transforms to β-form at 1262°; body-centered cubic β-form exists at >1262°. mp 1312°. bp 3273°. Heat of fusion: 10.05 kJ/mol. Heat of sublimation (25°): 397.5 kJ/mol. E°(aq) Gd^{3+}/Gd −2.4 V (calc). Experimental reduction potentials (referred to a normal calomel electrode): −1.810, −1.955 V: Noddack, Brukl, *Angew. Chem.* **50**, 362 (1937).

Oxide. Gadolinia. Gd_2O_3. Colorless, hygroscopic powder; prepd by igniting the hydroxide, nitrate, carbonate or oxalate, d^{15} 7.407, absorbs CO_2 from the air.

Hydroxide. $Gd(OH)_3$. Gelatinous precipitate, prepd by the action of alkali or ammonium hydroxide on a soln of a gadolinium salt. Absorbs CO_2 from the air.

Chloride. $GdCl_3$. White monoclinic crystals, prepd by heating the oxide with excess of ammonium chloride above 200°. d^0 4.52. mp ~609°. Sol in water; forms double salts with platinic and auric chlorides. A hexahydrate, $GdCl_3.6H_2O$, deliquesc crystals, d 2.424, is obtained from the aq soln. LD_{50} in mice (mg/kg): 550 i.p.; >2000 orally (Haley).

Sulfate. $Gd_2(SO_4)_3$. Octahydrate, colorless monoclinic crystals. Soly in water decreases with rise in temp. On heating at 400° yields the anhydr sulfate, d 4.139; begins to dec at 500°.

Nitrate. $Gd(NO_3)_3$. Hexahydrate, deliquesc triclinic crystals. d 2.332. mp 91°. Sol in water, in alcohol. A pentahydrate, prismatic crystals, mp 92°, d 2.406, very insol, has been prepd. LD_{50} (hexahydrate) in rats (mg/kg): 230 i.p.; >5000 orally (Haley).

USE: Oxide in control rods of some nuclear power reactors.

4357. Gadopentetic Acid. [80529-93-7] [*N,N*-Bis[2-[bis-[(carboxy-κO)methyl]amino-κN]ethyl]glycinato(5−)-$\kappa N,\kappa O$]gadolinate(2−) hydrogen (1:2); *N,N*-bis[2-[bis(carboxymethyl)amino]-ethyl]glycine gadolinium complex; gadolinium diethylenetriaminepentaacetic acid; Gd-DTPA. $C_{14}H_{20}GdN_3O_{10}$; mol wt 547.58. C 30.71%, H 3.68%, Gd 28.72%, N 7.67%, O 29.22%. Evaluation as water-soluble paramagnetic relaxation reagent (PARR) for NMR spectrometry: J. J. Dechter, G. C. Levy, *J. Magn. Reson.* **39**, 207 (1980); and prepn: T. J. Wenzel *et al.*, *Anal. Chem.* **54**, 615 (1982). Use as diagnostic medium for NMR imaging: H. Gries *et al.*, **DE 3129906**; *eidem*, **US 4647447** (1983, 1987 both to Schering AG). Physicochemical properties: H. Gries, H. Miklautz, *Physiol. Chem. Phys. Med. NMR* **16**, 105 (1984); E. Roux, L. De Broe, *J. Belge Radiol.* **71**, 31 (1988). Evaluation as NMR contrast-enhancing agent in animals: R. C. Brasch *et al.*, *Am. J. Roentgenol.* **142**, 625 (1984). Pharmacokinetics and toxicity: H. J. Weinmann *et al.*, *ibid.* 619; in humans: H. J. Weinmann *et al.*, *Physiol. Chem. Phys. Med. NMR* **16**, 167 (1984). HPLC determn: M. M. Vora *et al.*, *J. Chromatogr.* **369**, 187 (1986). Exptl use in imaging reperfused myocardium in dogs: S. Schaefer *et al.*, *J. Am. Coll. Cardiol.* **12**, 1064 (1988). Clinical studies of diagnostic use in brain imaging: R. Felix *et al.*, *Radiology* **156**, 681 (1985); J. P. Stack *et al.*, *Neuroradiology* **30**, 145 (1988). Safety and tolerance in humans: H. A. Goldstein *et al.*, *Radiology* **174**, 17 (1990). Reviews of diagnostic use in intracranial lesions: J. R. Hesselink, G. A. Press, *Radiol. Clin. North Am.* **26**, 873-887 (1988); in spinal disease: G. Sze, *ibid.* 1009-1024.

Dimeglumine salt. [86050-77-3] SHL-451A; ZK-93035; Magnevist. $C_{14}H_{20}GdN_3O_{10}.2(C_7H_{17}NO_5)$; mol wt 938.01. Freely sol in water. LD_{50} i.v. in rats: 10 mmol/kg (Weinmann).

THERAP CAT: Diagnostic aid (MRI contrast agent).

4358. Gadoteridol. [120066-54-8] [10-[2-(Hydroxy-κO)propyl]-1,4,7,10-tetraazacyclododecane-1,4,7-triacetato(3−)-κN^1,-$\kappa N^4,\kappa N^7,\kappa N^{10},\kappa O^1,\kappa O^4,\kappa O^7$]gadolinium; 10-(2-hydroxypropyl)-1,-4,7,10-tetraazacyclododecane-1,4,7-triacetic acid gadolinium complex; gadolinium(III) 1,4,7-tris(carboxymethyl)-10-(2′-hydroxypropyl)-1,4,7,10-tetraazacyclododecane; Gd(HP-DO3A); SQ-32692. $C_{17}H_{29}GdN_4O_7$; mol wt 558.69. C 36.55%, H 5.23%, Gd 28.15%, N 10.03%, O 20.05%. Nonionic, low-osmolar paramagnetic gadolinium (III) chelate. Prepn: M. F. Tweedle *et al.*, **EP 292689**; *eidem*, **US 4885363** (1988, 1989 both to Squibb); D. D. Dischino *et al.*, *Inorg. Chem.* **30**, 1265 (1991). Biodistribution: P. Wedeking *et al.*, *Magn. Reson. Imaging* **10**, 641 (1992). RIA determn in biological fluids: M. D. Ogan *et al.*, *J. Pharm. Sci.* **82**, 475 (1993). Toxicological studies: R. A. Soltys, *Invest. Radiol.* **27**, Suppl. 1, S7 (1992). Symposium on physicochemical properties, pharmacokinetics and clinical use in neurological disease: *ibid.* 1-63. Comprehensive description: K. Kumar *et al.*, *Anal. Profiles Drug Subs. Excip.* **24**, 209-241 (1996).

White solid obtained as an aggregate clump of fine needle-like microcrystals from methanol/acetone, mp >225°. Hydrophilic. At pH 7, log P (octanol/water): −3.68; (butanol/water): −1.98. Soly (mg/ml): water 737, methanol 119, isopropanol 41, dimethylformamide 10.1, acetonitrile 6.1, methylene chloride 5.2, ethyl acetate 0.5, acetone 0.4, hexane 0.2, toluene 0.3. uv max (water): 274 nm (a_m 2.5).

Gadoteridol injection. ProHance. Osmolality (37°): 630 mOsM/kg water. Viscosity (cP): 2.0 (20°), 1.3 (37°). d^{25} 1.140. LD_{50} in mice, rats (mmol/kg): 11-14, >10 i.v. (Soltys).

THERAP CAT: Diagnostic aid (MRI contrast agent).

4359. Gadoversetamide. [131069-91-5] [6,9-Bis(carboxy-κO)methyl]-3-[2-[(2-methoxyethyl)amino]-2-(oxo-κO)-15-oxa-3,-6,9,12-tetraazahexadecanoato(3−)-$\kappa N^3,\kappa N^6,\kappa N^9,\kappa O^1$]gadolinium; [8,11-bis(carboxymethyl)-14-[2-[(2-methoxyethyl)amino]-2-oxoethyl]-6-oxo-2-oxa-5,8,11,14-tetraazahexadecan-16-oato(3−)]gadolinium; [*N,N″*-bis[*N*-(2-methoxyethyl)-carbamoylmethyl]diethylenetriamine-*N,N′,N″*-triaceto]gadolinium(III); gadolinium diethylenetriamine pentaacetic acid bis(methyoxyethylamide); Gd-DTPA-BMEA; MP-1177; OptiMARK. $C_{20}H_{34}GdN_5O_{10}$; mol wt 661.77. C 36.30%, H 5.18%, Gd 23.76%, N 10.58%, O 24.18%. Nonionic paramagnetic contrast agent. Prepn: R. W. Weber, M. P. Periasamy, **WO 9001024**; R. W. Weber, **US 5130120** (1990, 1992 both to Mallinckrodt); M. Periasamy *et al.*, *Invest. Radiol.* **26**, S217 (1991). Paramagentic profile: K. Adzamli *et al.*, *ibid.* **34**, 410 (1999). Clinical pharmacokinetics: K. Sw. Swan *et al.*, *J. Magn. Reson. Imaging* **9**, 317 (1999). Clinical study in liver imaging: D. L. Rubin *et al.*, *ibid.* 240; in CNS imaging: R. I. Grossman *et al.*, *Invest. Radiol.* **35**, 412 (2000).

Prepd as 0.5 mmol/ml soln, osmolality (37°) 1110 mOsmol/kg water. Viscosity (cP): 3.1 (20°), 2.0 (37°). d^{25} 1.160. Log P (butanol/water): −1.93. Freely sol in water. LD_{50} i.v. in mice: 30.3 mmol/kg (Periasamy).

THERAP CAT: Diagnostic aid (MRI contrast agent).

4360. Gadoxetic Acid. [135326-11-3] (*SA*-8-11252634)-[*N*-[(2*S*)-2-[Bis[(carboxy-*κO*)methyl]amino-*κN*]-3-(4-ethoxyphenyl)-propyl]-*N*-[2-[bis[(carboxy-*κO*)methyl]amino-*κN*]ethyl]glycinato-(5−)-*κN*,*κO*]gadolinate(2−) hydrogen (1:2); 3,6,9-triaza-3,6,9-tris-(carboxymethyl)-4-(4-ethoxybenzyl)-undecanedioic acid gadolinium complex; gadolinium ethoxybenzyldiethylenetriaminepentaacetic acid; Gd-EOB-DTPA. $C_{23}H_{30}GdN_3O_{11}$; mol wt 681.75. C 40.52%, H 4.44%, Gd 23.07%, N 6.16%, O 25.81%. Paramagnetic hepatobiliary contrast agent. Prepn: H. Schmitt-Willich *et al.*, **EP 405704**; *eidem*, **US 5695739** (1991, 1997 both to Schering AG). Pharmacology: H.-J. Weinmann *et al.*, *Magn. Reson. Med.* **22**, 233 (1991); and toxicology: A. Mühler *et al.*, *Invest. Radiol.* **28**, 26 (1993). Physicochemical characterization: L. Vander Elst *et al.*, *Acta Radiol.* **38**, Suppl. 412, 135 (1997). Clinical pharmacokinetics: G. Shuhmann-Giampieri *et al.*, *J. Clin. Pharmacol.* **37**, 587 (1997). Clinical evaluation of hepatic imaging: W. Stern *et al.*, *Acta Radiol.* **41**, 255 (2000). Review of preclinical studies: Y. Ni, G. Marchal, *Top. Magn. Reson. Imaging* **9**, 183-195 (1998).

Sodium salt. [135326-22-6] Gadoxetate; SHL-569B; Eovist. $C_{23}H_{28}GdN_3Na_2O_{11}$; mol wt 725.72. Prepd as 0.5 mol/l soln, osmolality 1.53 osmol/kg. Viscosity (37°) 2.32 mPa·s. Amphiphilic. LD_{50} i.v. in mice: 7.5 mmol/kg (Weinmann).

THERAP CAT: Diagnostic aid (MRI contrast agent).

4361. Galactaric Acid. [526-99-8] Mucic acid; galactosaccharic acid; tetrahydroxyadipic acid; saccharolactic acid; Schleimsäure (German). $C_6H_{10}O_8$; mol wt 210.14. C 34.29%, H 4.80%, O 60.91%. Prepd by oxidation of lactose and of galactose: Kent, Tollens, *Ann.* **227**, 221 (1885); Maurer, Drefahl, *Ber.* **75B**, 1489 (1942). Manuf from wood sawdust: Acree, **GB 160777** (1921). *Review:* B. A. Lewis *et al.*, "Galactaric Acid and Its Derivatives" in Whistler, Wolfrom, *Methods in Carbohydrate Chemistry* **vol. II** (Academic Press, New York, 1963) pp 38-46.

Cryst powder, dec ~255° when rapidly heated, also reported as 225°. Soluble in 300 parts cold water, 60 parts boiling water, alkalies; practically insol in alcohol, ether.

Ammonium salt. $(NH_4)_2C_6H_8O_8$. Acicular crystals. Soluble in water.

USE: Has been proposed to replace potassium bitartrate in baking powder and for manuf of granular effervescing salts.

4362. Galactitol. [608-66-2] Dulcitol; dulcite; dulcose; euonymit; melampyrite; melampyrum; melampyrin. $C_6H_{14}O_6$; mol wt 182.17. C 39.56%, H 7.75%, O 52.69%. Found in dulcite or Madagascar manna (*Melampyrum nemorosum* L.) and in other species of *Melampyrum*, *Scrophulariaceae*, and *Evonymus atropurpureus* Jacq., *Celastraceae*. Isoln: Hünefeld, *Ann.* **24**, 241 (1837); Bouchardat, *Ann. Chim. Phys.* **27**, 68 (1872); Fischer, Hertz, *Ber.* **25**, 1261 (1892); Rogerson, *J. Chem. Soc.* **101**, 1040 (1912). Prepn by

catalytic isomerization of D-glucitol: Wright, Hartmann, *J. Org. Chem.* **26**, 1588 (1961). Synthesis: Lespieau, *Bull. Soc. Chim. Fr.* [5] **1**, 1374 (1934); Delepine, Horeau, *ibid.* **4**, 1524 (1937); Wiemann, Gordon, *ibid.* **1958**, 433. Structure: R. L. Lohmar "The Polyols" in W. Pigman, *The Carbohydrates* (Academic Press, New York, 1957) p 247.

Crystals from methanol + water, mp 188-189°. Slightly sweet taste. d^{20} 1.47. bp_1 275-280°. One gram dissolves in 30 ml water, in 2 ml boiling water. Slightly sol in alc. Ka at 18° = 3.5×10^{-14}.

Hexa-*O*-acetylgalactitol. $C_{18}H_{26}O_{12}$. Crystals from ethanol, mp 168-169°.

Hexanitrate. Nitrodulcitol. mp 94-95°. Has explosive properties: Taylor, Rinkenbach, *J. Franklin Inst.* **204**, 374 (1927).

4363. Galactoflavin. [5735-19-3] 1-Deoxy-1-(3,4-dihydro-7,8-dimethyl-2,4-dioxobenzo[*g*]pteridin-10(2*H*)-yl)-D-galactitol; 7,8-dimethyl-10-(D-*galacto*-2,3,4,5,6-pentahydroxyhexyl)benzo[*g*]pteridine-2,4(3*H*,10*H*)-dione; 7,8-dimethyl-10-(D-*galacto*-2,3,4,5,6-pentahydroxyhexyl)isoalloxazine; 7,8-dimethyl-10-(*d*-1′-dulcityl)-isoalloxazine; 6,7-dimethyl-9-(*d*-1′-dulcityl)isoalloxazine; 6,7-dimethyl-9-(1-deoxy-D-galactitol-1-yl)isoalloxazine. $C_{18}H_{22}N_4O_7$; mol wt 406.40. C 53.20%, H 5.46%, N 13.79%, O 27.56%. Prepd from 1-deoxy-1-(3,4-dimethyl-6-phenylazo)anilino-D-galactitol and barbituric acid: Berezovskii, Eremenko, *Zh. Obshch. Khim.* **32**, 4056 (1962), *C.A.* **59**, 736b (1963). Structure: Emerson *et al.*, *J. Biol. Chem.* **160**, 165 (1945). Pharmacology: Lane, Brindley, *Proc. Soc. Exp. Biol. Med.* **116**, 57 (1964). Produces congenital malformations in animals: Nelson *et al.*, *J. Nutr.* **58**, 125 (1956); Miller *et al.*, *J. Biol. Chem.* **237**, 968 (1962); Mackler, *Pediatrics* **43**, 915 (1969).

Yellow crystals, dec 260°. Absorption max: 223, 267, 370, 445 nm (ε 2730, 28100, 9100, 10800). Compd has yellow-green fluorescence in water.

USE: Riboflavine antagonist.

4364. D-Galactosamine. [7535-00-4] 2-Amino-2-deoxy-D-galactose; chondrosamine; GalN. $C_6H_{13}NO_5$; mol wt 179.17. C 40.22%, H 7.31%, N 7.82%, O 44.65%. Amino sugar isolated from chondroitin sulfate, *q.v.*: P. A. Levene, F. B. La Forge, *J. Biol. Chem.* **18**, 123 (1914). Sepn of α- and β-anomers: P. A. Levene, *ibid.* **57**, 337 (1923). Synthesis: S. P. James *et al.*, *Nature* **156**, 308 (1945) *eidem*, *J. Chem. Soc.* **1946**, 625; R. Kuhn, W. Kirschenlohr, *Ann.* **600**, 126 (1956); P. A. Gent *et al.*, *J. Chem. Soc. Perkin Trans. 1* **1972**, 277. Chemistry: D. Horton in *The Amino Sugars* **Vol. 1A**, R. W. Jeanloz, Ed. (Academic, New York, 1969) pp 133-145. Inducer of exptl hepatitis: D. Keppler *et al.*, *Exp. Mol. Pathol.* **9**, 279 (1968); K. Decker, D. Keppler in *Progress in Liver Diseases* **Vol.**

IV, H. Popper, F. Schaffner, Eds. (Grune & Stratton, New York, 1972) p 183. Powerful inhibitor of hepatic RNA synthesis: D. Keppler *et al.*, *J. Biol. Chem.* **249**, 211 (1974); T. Anukarahanonta *et al.*, *Eur. J. Cancer* **16**, 1171 (1980).

α-form

Hydrochloride. $C_6H_{14}ClNO_5$. Crystals, mp 180° (dec). Shows mutarotation. α-Form: $[\alpha]_D^{23}$ +124° → +93° (water). β-Form: $[\alpha]_D^{23}$ +47° → +93° (water).

4365. D-Galactose. [59-23-4] Cerebrose; brain sugar. C_6-$H_{12}O_6$; mol wt 180.16. C 40.00%, H 6.71%, O 53.28%. Constituent of many oligo- and polysaccharides occurring in pectins, gums, and mucilages. Prepn: Kent, Tollens, *Ann.* **227**, 224 (1885); E. P. Clark, *J. Biol. Chem.* **47**, 2 (1921). Mutarotation and purification of β-form: C. S. Hudson, E. Yanosky, *J. Am. Chem. Soc.* **39**, 1021 (1917). Structural configuration: J. Pryde, *J. Chem. Soc.* **123**, 1809 (1923); W. Charlton *et al.*, *ibid.* **1926**, 94; W. N. Haworth *et al.*, *ibid.* **1927**, 2428; E. L. Jackson, C. S. Hudson, *J. Am. Chem. Soc.* **59**, 994 (1937); R. M. Hann *et al.*, *ibid.* **66**, 1912 (1944). Isoln in the processing of the red algae, *Porphyra umbilicalis:* S. Peat *et al.*, *J. Chem. Soc.* **1961**, 1590. *Review:* W. Pigman, *The Carbohydrates* (Academic Press, New York, 1957) pp 88-90. Review of diagnostic use: W. J. Schirmer *et al.*, *J. Surg. Res.* **41**, 543 (1986).

α-form

α-Form. Prisms from water or ethanol, mp 167°. $[\alpha]_D$ +150.7° → +80.2° (water). Soluble in about 0.5 parts water; freely sol in hot water; final soly in water at 25° = 68%; sol in pyridine; slightly sol in alcohol.

β-Form. Crystals, mp 167°. $[\alpha]_D$ +52.8° → +80.2° (water). Sol in 1.7 parts water at 17°.

Monohydrate. Prisms from water, mp 118-120°.

Microparticulate Form. [90881-70-2] SH U 454; Echovist. Suspension of galactose microparticle granules in a galactose solution. Prepn: J. S. Rasor, E. G. Tickner, **EP 131540** (1985 to Schering). Series of articles on *in vivo* use in echocardiography: *Arzneim.-Forsch.* **36**, 1030-1040 (1986). Review of formulations and clinical diagnostic use: R. Schürmann, R. Schlief, *Radiol. Med.* **87**, Suppl. 1, 15-23 (1994).

Transpulmonary Microparticulate Form. [144046-30-0] SH U 508A; Levovist. Suspension of galactose microparticle granules containing 0.1% physiologic palmitic acid in a sterile water solution. THERAP CAT: Diagnostic aid (hepatic function). Microparticulate forms as diagnostic aid (ultrasound contrast agent).

4366. α-Galactosidase. [9025-35-8] α-D-Galactoside galactohydrolase; α-galactosidase A; α-Gal; ceramide trihexosidase; melibiase; E.C. 3.2.1.22. Enzyme that hydrolyzes terminal α-D-galactose residues in oligosaccharides and galactolipids. Widely distributed in microorganisms, plants and animals. Human form is a homodimeric glycoprotein, mol wt ~101 kDa; genetic deficiency of the enzyme results in the glycosphingolipid storage disorder known as Fabry's disease. Production by *E. coli*: J. L. Koppel *et al.*, *J. Gen. Physiol.* **36**, 703 (1953); and chromogenic assay of activity: C. J. Porter *et al.*, *ibid.* **37**, 271 (1953). Isoln in crystalline form from *Mortierella vinacea*: H. Suzuki *et al.*, *J. Biol. Chem.* **245**, 781 (1970). Identification of human enzyme and role in disease: R. O. Brady *et al.*, *N. Engl. J. Med.* **276**, 1163 (1967); J. A. Kint, *Science* **167**, 1268 (1970). Crystal structure of human α-Gal: S. C. Garman, D. N. Garboczi, *J. Mol. Biol.* **337**, 319 (2004); and catalytic mecha-

nism: A. I. Guce *et al.*, *J. Biol. Chem.* **285**, 3625 (2010). Use of purified human enzyme in treatment of Fabry's disease: R. J. Desnick *et al.*, *Proc. Natl. Acad. Sci. USA* **76**, 5326 (1979). Production methods and use in soy food products for oligosaccharide hydrolysis: D. L. Falkoski *et al.*, *J. Agric. Food Chem.* **54**, 10184 (2006). Clinical evaluation as antiflatulant: M. Di Stefano *et al.*, *Dig. Dis. Sci.* **52**, 78 (2007). Review of role in Fabry's disease: R. J. Desnick *et al.* in *The Metabolic and Molecular Bases of Inherited Disease*, C. R. Scriver *et al.*, Eds. (McGraw-Hill, New York, 7th Ed., 1995) pp 2741-2784; and clinical experience as enzyme replacement therapy: O. Lidove *et al.*, *Genet. Med.* **12**, 668-679 (2010).

Agalsidase. [104138-64-9] α-Galactosidase (human clone λAG18 isoenzyme A subunit protein moiety reduced). Protein subunit of recombinant human α-galactosidase having the identical amino acid sequence as the natural human lysosomal enzyme. Monomer consists of 398 amino acid residues; mol wt ~45 kDa.

Agalsidase alfa. Replagal. Agalsidase glycoform α; produced by recombinant DNA technology in cultured human cells. *See*: R. F. Selden *et al.*, **WO 9811206** (1998 to Transkaryotic Therapies). Clinical trial in Fabry's disease: A. Mehta *et al.*, *Lancet* **374**, 1986 (2009). Review of clinical experience: U. Ramaswami, *Drug Des. Devel. Ther.* **5**, 155-173 (2011).

Agalsidase beta. Fabrazyme. Agalsidase glycoform β; produced by recombinant DNA technology in Chinese hamster ovary cells. *See*: R. J. Desnick *et al.*, **US 5356804** (1994 to Mt. Sinai School of Med.). Clinical trial in Fabry's disease: C. M. Eng *et al.*, *N. Engl. J. Med.* **345**, 9 (2001). Review of clinical experience: G. M. Keating, D. Simpson, *Drugs* **67**, 435-455 (2007).

USE: In food processing to reduce the content of raffinose oligosaccharides in soy and bean products.

THERAP CAT: Digestive aid and antiflatulant. Agalsidase as enzyme replacement therapy for Fabry's disease.

4367. D-Galacturonic Acid. [685-73-4] $C_6H_{10}O_7$; mol wt 194.14. C 37.12%, H 5.19%, O 57.69%. Obtained by hydrolysis of pectin where it is present as polygalacturonic acid: Ehrlich, *Chem. Ztg.* **41**, 197 (1917); Ehrlich, Guttmann, *Biochem. Z.* **259**, 100 (1933); *Ber.* **66**, 220 (1933); Niemann, Link, *J. Biol. Chem.* **95**, 203 (1932); **104**, 743 (1934); Morell, Link, *ibid.* **100**, 385 (1933); Anderson, King, *J. Chem. Soc.* **1961**, 5333. Isoln from mustard seeds: Goering, **US 2987448** (1961 to Oil Seed Prod.).

α-form

α-Form. Monohydrate, needles, mp 159°. $[\alpha]_D^{20}$ +98.0° → +50.9° (water). Soluble in water; slightly sol in hot alcohol. Practically insol in ether.

β-Form. mp 166°. $[\alpha]_D$ +27° → +55.6° (water).

Phenylhydrazone. mp 141°.

4368. Galanga. Galangal; colic root; East India root; Chinese ginger. Dried rhizome of *Alpinia officinarum* Hance, *Zingiberaceae*. *Habit.* China. *Constit.* Volatile oil, resin, kaempferid, galangin, dioxyflavanol, galangol.

4369. Galangin. [548-83-4] 3,5,7-Trihydroxy-2-phenyl-4*H*-1-benzopyran-4-one; 3,5,7-trihydroxyflavone; norizalpinin. C_{15}-$H_{10}O_5$; mol wt 270.24. C 66.67%, H 3.73%, O 29.60%. Isoln from galanga root, *Alpininia officinarum*, Hance and characterization: E. Jahns, *Ber.* **14**, 2807 (1881). Prepn: T. Heap, R. Robinson, *J. Chem. Soc.* **129**, 2336 (1926); J. J. Chavan, R. Robinson, *ibid.* **1933**, 368. Mutagenicity studies: J. T. MacGregor, L. Jurd, *Mutat. Res.* **54**, 297 (1978); J. P. Brown, P. S. Dietrich, *ibid.* **66**, 223 (1979).

Yellowish needles from ethanol, mp 214-215°. Moderately sol in ethanol, ether; insol in water. Very sol in chloroform, benzene.

4370. Galantamine. [357-70-0] (4aS,6R,8aS)-4a,5,9,10,-11,12-Hexahydro-3-methoxy-11-methyl-6H-benzofuro[3a,3,2-ef]-[2]benzazepin-6-ol; galanthamine; lycoremine. C$_{17}$H$_{21}$NO$_3$; mol wt 287.36. C 71.06%, H 7.37%, N 4.87%, O 16.70%. Selective acetylcholinesterase inhibitor. Isoln from Caucasian snowdrops, *Galanthus woronowii* Vel., *Amaryllidaceae:* N. F. Proskurnina, A. P. Yakovleva, *J. Gen. Chem. USSR* **22**, 1899 (1952); from *Narcissus* spp: Boit *et al., Ber.* **90**, 725, 2197 (1957). Structure work: S. Kobayashi *et al., Chem. Ind. (London)* **1956**, 177. Synthesis and stereochemistry: D. H. R. Barton, G. W. Kirby, *Proc. Chem. Soc. London* **1960**, 392; *eidem, J. Chem. Soc.* **1962**, 806. Asymmetric synthesis of isomers from L-tyrosine: K. Shimizu *et al., Heterocycles* **8**, 277 (1977). Biosynthesis studies: D. H. R. Barton *et al., J. Chem. Soc.* **1963**, 4545; W. Döbke, *Heterocycles* **6**, 551 (1977). Toxicology study: S. L. Friess *et al., Toxicol. Appl. Pharmacol.* **3**, 347 (1961). Clinical pharmacokinetics: U. Bickel *et al., Clin. Pharmacol. Ther.* **50**, 420 (1991). Review of synthesis and pharmacology: J. Marco-Contelles *et al., Chem. Rev.* **106**, 116-133 (2006); of clinical experience in Alzheimer's disease: G. Razay, G. K. Wilcock, *Expert Rev. Neurother.* **8**, 9-17 (2008).

Crystals from benzene, mp 126-127°. [α]$_D^{20}$ $-118.8°$ (c = 1.378 in ethanol). Monoacidic base. Fairly sol in hot water; freely sol in alcohol, acetone, chloroform. Less sol in benzene, ether.
Hydrochloride. C$_{17}$H$_{21}$NO$_3$.HCl. Crystals from water, dec 256-257°. Sparingly sol in cold, more sol in hot water. Very sparingly sol in alcohol, acetone.
Hydrobromide. [1953-04-4] Nivalin; Razadyne; Reminyl. C$_{17}$-H$_{21}$NO$_3$.HBr; mol wt 368.27. Crystals from water, dec 246-247°. [α]$_D^{20}$ $-93.1°$ (c = 0.1015 in 15 ml H$_2$O). Sol in 0.1N sodium hydroxide; sparingly sol in water; very slightly sol in alc. Insol in n-propanol. LD$_{50}$ i.v. in mice (mg/kg): 5.2 ± 0.2 (Friess).
THERAP CAT: Cholinesterase inhibitor.

4371. Galegine. [543-83-9] N-(3-Methyl-2-buten-1-yl)guanidine; N-3,3-dimethylallylguanidine; isoamyleneguanidine. C$_6$-H$_{13}$N$_3$; mol wt 127.19. C 56.66%, H 10.30%, N 33.04%. Isoprenoid guanidine deriv from seeds of *Galega officinalis* L., *Leguminosae:* Tanret, *Compt. Rend.* **158**, 1182, 1426 (1914); **159**, 108 (1914); Markovic, Dittertová, *Chem. Zvesti* **9**, 576 (1955), *C.A.* **50**, 8137d (1956). Structure: Barger, White, *Biochem. J.* **17**, 827 (1923). Synthesis: Späth, Spitzy, *Ber.* **58**, 2273 (1925); Babor, Jezo, *Chem. Zvesti* **8**, 18 (1954), *C.A.* **49**, 7495f (1955). Metabolic effects: G. Weitzel *et al., Z. Physiol. Chem.* **353**, 535 (1972). Effects on mitochondria: B. Lotina *et al., Arch. Biochem. Biophys.* **159**, 520 (1973). Biosynthetic study: J. Steiniger, G. Reuter, *Biochem. Physiol. Pflanz.* **166**, 275 (1974). Review: Braun, *J. Chem. Educ.* **8**, 2175 (1931).

Hygroscopic, bitter crystals. mp 60-65°. Freely sol in water, alcohol; slightly sol in ether. *Keep well closed.*

4372. Gallacetophenone. [528-21-2] 1-(2,3,4-Trihydroxyphenyl)ethanone; 2',3',4'-trihydroxyacetophenone; Alizarine yellow C; C.I. 57000. C$_8$H$_8$O$_4$; mol wt 168.15. C 57.14%, H 4.80%, O 38.06%. Prepn: Hart, Woodruff, *J. Am. Chem. Soc.* **58**, 1957 (1936); Campbell, Coppinger, **US 2686123** (1954 to U.S. Secy.

Agr.); Knowles, **US 2763691** (1956 to Kodak); Price, Israelstam *J. Org. Chem.* **29**, 2800 (1964).

White to brownish-gray, cryst powder, mp 173°. uv max (methanol): 237, 296 nm (ε 8560, 12,500). Sol in 600 parts cold water, more in hot water; sol in alcohol, ether, soln of sodium acetate.
USE: Antiseptic.

4373. Gallamine Triethiodide. [65-29-2] 2,2',2''-[1,2,3-Benzenetriyltris(oxy)]tris[N,N,N-triethylethanaminium] iodide (1:3); [ν-phenenyltris(oxyethylene)]tris[triethylammonium triiodide]; 1,2,3-tris(2-triethylammonium ethoxy)benzene triiodide; 1,2,3-tris(2-diethylaminoethoxy)benzene tris(ethyl iodide); tri(β-diethylaminoethoxy)-1,2,3-benzene triiodoethylate; pyrogallol 1,2,3-diethylaminoethyl ether) tris(ethyl iodide); benzcurine iodide; RP-3697; F-2559; Relaxan; Flaxedil. C$_{30}$H$_{60}$I$_3$N$_3$O$_3$; mol wt 891.54. C 40.42%, H 6.78%, I 42.70%, N 4.71%, O 5.38%. Curarizing properties: D. Bovet *et al., Compt. Rend.* **225**, 74 (1947); F. Depierre, *ibid.* 956. Prepn: E. Fourneau, **US 2544076** (1951 to Rhone-Poulenc). Comparative clinical pharmacokinetics: W. Buzello, S. Agoston, *Anaesthesist* **27**, 313 (1978). Mode of action: D. Colquhoun, R. E. Sheridan, *Br. J. Pharmacol.* **66**, 78 (1979); *eidem, Proc. R. Soc. London Ser. B* **211**, 181 (1981). Effects in mammalian and amphibian nerve fibers: K. J. Smith, C. L. Schauf, *Science* **212**, 1170 (1981).

White cryst from acetone/water, mp 152-153° (indefinite). Hygroscopic. Very sol in water; freely sol in dil acetone; sparingly sol in anhydr acetone, ether, benzene, alc; very slightly sol in chloroform.
THERAP CAT: Neuromuscular blocking agent.
THERAP CAT (VET): Neuromuscular blocking agent.

4374. Gallein. [2103-64-2] 3',4',5',6'-Tetrahydroxyspiro-[isobenzofuran-1(3H),9'-[9H]xanthen]-3-one; 3',4',5',6'-tetrahydroxyfluoran; 3',4',5',6'-tetrahydroxyspiro[phthalan-1,9'-xanthen]-3-one; pyrogallolphthalein; C.I. 45445; mordant violet 25. C$_{20}$-H$_{12}$O$_7$; mol wt 364.31. C 65.94%, H 3.32%, O 30.74%. Obtained by heating 1 part phthalic anhydride with 2 parts of pyrogallol or gallic acid: Baeyer, *Ber.* **4**, 457 (1871); Buchka, *Ann.* **209**, 261 (1881). Use as a biological stain: R. D. Lillie *et al., Stain Technol.* **49**, 339 (1974); R. Welsh, *ibid.* **52**, 261 (1977). *See also* H. J. Conn's *Biological Stains,* R. D. Lillie, Ed. (Williams & Wilkins, Baltimore, 9th ed., 1977) p 351.

Brownish-red powder or crystals with 1½ H$_2$O, or red crystals with greenish-yellow color when anhydr. Loses the water of crystn at about 180° and blackens above this temp. Does not melt even at

300°. pH 3.8 brownish-yellow; pH 6.6 rose-red. Almost insol in water, benzene, chloroform. Slightly sol in ether; sol in alc, acetone, alkalies.

Disodium salt. Alizarin violet. $C_{20}H_{10}Na_2O_7$. pH 10.6 rose; pH 13.0 violet.

USE: Clinical reagent (phosphates in urine). Monophosphates give a yellow, dibasic a red, tribasic a violet color. Used in soln of 0.5 g in 100 ml 50% alc; 2-3 drops for 100 ml liq. As sensitive indicator for acids, alkali hydroxides, NH_3, but not for carbonates. Biological stain.

4375. Gallic Acid. [149-91-7] 3,4,5-Trihydroxybenzoic acid. $C_7H_6O_5$; mol wt 170.12. C 49.42%, H 3.56%, O 47.02%. Obtained by alkaline or acid hydrolysis of the tannins from nutgalls; also by enzymatic hydrolysis using spent broths from *Penicillium glaucum* or *Aspergillus niger* which contain tannase: A. G. Perkin, O. Gunnell, *J. Chem. Soc.* **69**, 1303 (1896); Hsias, *Huang Hai* No. **7**, 51 (1946), *C.A.* **42**, 3901i (1948); Cochrane, *Econ. Bot.* **2**, 145 (1948); Toth, Henster, *Acta Chim. Acad. Sci. Hung.* **2**, 209 (1952). Prepn from tannin containing materials: Krueger *et al.*, US 2723992 (1955 to Mallinckrodt). Synthesis from aliphatic materials: Shipchandler *et al.*, *J. Chem. Soc. Perkin Trans. 1* **1975**, 1400. Biosynthesis: Haslam *et al.*, *J. Chem. Soc.* **1961**, 1854. Study of polymorphic forms: E. Lindpainter, *Mikrochemie* **27**, 21 (1939). Toxicity studies: J. W. Dollahite *et al.*, *Am. J. Vet. Res.* **23**, 1264 (1962).

Needles from abs methanol or chloroform, formerly reported as dec 235-240° (Perkin, Gunnell). Sublimes at 210° giving a stable form with mp 258-265° (dec) and an unstable form mp 225-230° (Lindpainter). One gram dissolves in 87 ml water, 3 ml boiling water, 6 ml alcohol, 100 ml ether, 10 ml glycerol, 5 ml acetone. Practically insol in benzene, chloroform, petr ether. *Protect from light.* LD_{50} in rabbits (g/kg): 5.0 orally (Dollahite).

Monohydrate. [5995-86-8] $C_7H_6O_5 \cdot H_2O$; mol wt 188.14.

Methyl ester. [99-24-1] Methyl gallate; gallicin. $C_8H_8O_5$; mol wt 184.15. Monoclinic prisms from methanol, often hydrated or solvated. When dry, mp 202°. Sol in hot water, alcohol, methanol, ether.

Propyl ester *see* Propyl Gallate.

USE: Manuf gallic acid esters, pyrogallol, inks; as photographic developer; in tanning; in dyeing; in testing for free mineral acids, dihydroxyacetone, and alkaloids. Esters as antioxidants.

THERAP CAT: Formerly as astringent, styptic.

THERAP CAT (VET): Has been used as intestinal astringent.

4376. Gallium. [7440-55-3] Ga; at. wt 69.723; at. no. 31; valences 3, 2, 1. Group IIIA (13). Natural isotopes: 69 (60.2%); 71 (39.8%); artificial radioactive isotopes: 63-68; 70; 72-76. Best source is the mineral germanite, a copper sulfide ore; occurs in very small quantities in zinc blendes, in aluminum clays, found in ores of iron, chromium, manganese; constitutes 5×10^{-4}% of the crust of the earth. Discovered by L. Boisbaudran, *Compt. Rend.* **81**, 493, 1100 (1875); **82**, 163, 1036 (1876); isolated pure by L. Boisbaudran and E. Jungfleisch, *Bull. Soc. Chim.* [2] **31**, 50 (1879). Isoln from rhenium-rich copper schist: Feit, *Angew. Chem.* **46**, 216 (1933). From bauxite: *Chem. Eng. News* **34**, 4300 (1956). Purification by zone melting: *Chem. Ztg.* **80**, 787 (1956). Alternate methods of purification: Gebauhr, US 2928731; Merkel, US 2927853 (both 1960 to Siemens-Schuckert). Spectra: L. Boisbaudran *et al.*, cited in *Mellor's* vol. **5**, 378 (1929). *Reviews:* Wagner, Gitzen, *J. Chem. Educ.* **29**, 162 (1952); Greenwood, *Inorg. Chem. Radiochem.* **5**, 91-134 (1963); Wade, Banister, *Comprehensive Inorganic Chemistry* vol. **1**, J. C. Bailar, Jr. *et al.*, Eds. (Pergamon Press, Oxford, 1973) pp 997-1000, 1069-1117; P. de la Bretèque in *Kirk-Othmer Encyclopedia of Chemical Technology* vol. **11** (Wiley-Interscience, New York, 3rd ed., 1980) pp 604-620. Review of antineoplastic properties and clinical applications of gallium salts: P. Collery *et al.*, *Crit. Rev. Oncol. Hematol.* **42**, 283-296 (2002).

Grayish metal; possesses a greenish-blue reflection; tin- or silver-like when molten; has a crystalline orthorhombic texture. mp 29.78°. bp approx 2400°: Cochran, Foster, *J. Electrochem. Soc.* **109**, 144 (1962). Shows a tendency to remain in supercooled state. Contracts on melting; $d^{29.65}$ (solid) 5.9037; $d^{29.8}$ (liq) 6.0947: Richards, Boyer, *J. Am. Chem. Soc.* **43**, 274 (1921). Heat capacity: 0.09 cal/g/°C (0-24°, solid). Latent heat of fusion 19.16 cal/g. *Corrosive.* Stable in dry air; tarnishes in moist air or oxygen. Reacts with alkalies with evolution of hydrogen; attacked by cold concd hydrochloric acid; rendered passive by hot nitric acid; readily attacked by halogens.

Suboxide. [12024-20-3] Ga_2O. Brown powder; obtained by heating the sesquioxide and the metal at 700°; stable in dry air; dec above 800°; converted to the trivalent state by nitric acid or bromine.

Hydroxide. [12023-99-3] $Ga(OH)_3$. A gelatinous precipitate; obtained by the action of ammonia or alkali hydroxide on a soln of a gallic salt.

Caution: Administration to humans has caused metallic taste, skin rashes, bone marrow depression: H. E. Stokinger in *Patty's Industrial Hygiene and Toxicology* vol. **2A**, G. D. Clayton, F. E. Clayton, Eds. (Wiley-Interscience, New York, 3rd ed., 1981) pp 1630-1637.

4377. Gallium Arsenide. [1303-00-0] AsGa; mol wt 144.64. As 51.80%, Ga 48.20%. GaAs. Prepd by passing a mixture of hydrogen and arsenic vapor over gallium(III) oxide at 600°: V. M. Goldschmidt, *Skr. Akad. Oslo* **1926**, no. 8, pp 34, 100; *idem Ber.* **60**, 1263 (1927). Simplified procedure: Juza, Schulz, *Z. Anorg. Allg. Chem.* **275**, 65 (1954); Minden, *Sylvania Technol.* **11**, no. 1, p 19 (Jan. 1958). Vapor phase crystal growth: McAleer *et al.*, *J. Electrochem. Soc.* **108**, 1168 (1961). Use in high-speed microcircuits: A. L. Robinson, *Science* **219**, 275 (1983). Review of toxicology and human exposure: *Toxicological Profile for Arsenic* (PB2008-100002, 2007) 559 pp.

Cubic crystals. mp 1238°. Dark gray with metallic sheen. Garlic odor. Hardness 4.5. d_4^{25} 5.31. Thermal expansion coefficient: 5.9×10^{-6}. Thermal conductivity 0.52 watt units. Specific heat 0.086 cal/g/°C. Sol in HCl. Molten gallium arsenide attacks quartz, therefore graphite boats or quartz boats, which are carbon coated by pyrolytic decompn of methane, should be used in zone refining. GaAs single crystals have been grown by the Czochralski technique and by the floating zone method. Extensive twinning occurs. Intrinsic electron concn 10^7. Energy gap at room temp 1.38 electron volts. Electron mobility 8800 cm^2/volt sec. Hole mobility 450 cm^2/volt sec. Effective mass for electrons 0.06 m_0. Lattice constant 5.654Å. Dielectric constant 11.1. Intrinsic resistivity at 300 K = 3.7×10^8 ohm-cm. Electron lattice mobility at 300 K = 10,000 cm^2/volt-sec. Intrinsic charge density at 300 K = 1.4×10^6/cm^3. Electron diffusion constant at 300 K = 310 cm^2/sec. Hole diffusion constant = 11.5 cm^2/sec.

USE: In semiconductor applications (transistors, solar cells, lasers).

4378. Gallium Chloride. [13450-90-3] Gallium chloride ($GaCl_3$); gallium trichloride; trichlorogallium. Cl_3Ga; mol wt 176.07. Cl 60.40%, Ga 39.60%. $GaCl_3$. Occurs as a bridged dimer, Ga_2Cl_6. Prepn: W. C. Johnson, C. A. Haskew, *Inorg. Synth.* **1**, 26 (1939); R. A. Kovar, *ibid.* **17**, 167 (1977). Vapor phase studies: A. W. Laubengayer, F. B. Schirmer, *J. Am. Chem. Soc.* **62**, 1578 (1940). Crystal structure: S. C. Wallwork, I. J. Worrall, *J. Chem. Soc.* **1965**, 1816. Use in synthesis of organogallium compounds: R. A. Kovar *et al.*, *Inorg. Chem.* **14**, 2809 (1975). Review of applications in organic synthesis: R. M. Kellogg, *Chemtracts Org. Chem.* **16**, 79-92 (2003); R. Amemiya, M. Yamaguchi, *Eur. J. Org. Chem.* **2005**, 5145-5150.

Hygroscopic solid, mp 78-79°. bp_{760} 200.0°. d 2.47. Very sol in water; sol in hydrocarbons, ether-type solvents. *Reacts with air.*

USE: Lewis acid catalyst in organic synthesis; reagent for generating organogallium compounds, metallic gallium, and gallium semiconductors.

4379. Gallium Citrate. [27905-02-8]; [30403-03-3] (unspecified stoichiometry). 2-Hydroxy-1,2,3-propanetricarboxylic acid gallium salt (1:1). $C_6H_5GaO_7$; mol wt 258.82. C 27.84%, H 1.95%, Ga 26.94%, O 43.27%. $Ga(C_6H_5O_7)$. Organogallium complex administered to provide an aqueous source of gallium that accumulates in specific biological tissues. Prepd from gallium hydroxide and citric acid: P. Neogi, S. K. Nandi, *J. Indian Chem. Soc.* **13**, 399

(1936); H. C. Dudley, *J. Am. Chem. Soc.* **72**, 3822 (1950). Review of early clinical use as radiolabeled form: J. I. Hirsch *et al.*, *Drug Intell. Clin. Pharm.* **7**, 519-523 (1973). Mechanism studies of tumor localization: S. R. Vallabhajosula *et al.*, *Int. J. Nucl. Med. Biol.* **8**, 363 (1981); and metabolism: H. Kriegel, *Nuklearmedizin* **2**, 53 (1984). Clinical management of lymphoma with SPECT/CT system: B. Palumbo *et al.*, *Eur. J. Nucl. Med. Mol. Imaging* **32**, 1011 (2005). Toxicity studies: H. C. Dudley *et al.*, *J. Pharmacol. Exp. Ther.* **98**, 409 (1950); H. D. Brunner, *et al.*, *Radiology* **61**, 550 (1953). Review of diagnostic use as imaging agent in oncology: H. A. Macapinlac *et al.*, *Nucl. Med. Biol.* **21**, 731-738 (1994); in infection: C. J. Palestro, *Semin. Nucl. Med.* **24**, 128-141 (1994); in lymphoma: E. Even-Sapir, O. Israel, *Eur. J. Nucl. Med. Mol. Imaging* **30**, Suppl. 1, S65-S81 (2003).

Crystalline powder. Soly at 20° (g/100ml): water 92.4, 95% ethanol 0.12, abs ethanol 0.07. Insol in ether, benzene, acetone. LD_{50} i.v. in rats: >220 mg Ga/kg (Brunner).

Gallium citrate Ga 67. [41183-64-6] Neoscan. ^{67}Ga-Labeled compound prepd from radioactive gallium which has a half-life of 78.26 hours.

THERAP CAT: ^{67}Ga-labeled form as diagnostic aid (radioactive imaging agent).

4380. Gallium Nitrate. [13494-90-1] Nitric acid gallium salt (3:1); gallium trinitrate; NSC-15200; Ganite. GaN_3O_9; mol wt 255.74. Ga 27.26%, N 16.43%, O 56.30%. $Ga(NO_3)_3$. Formed by the reaction of elemental gallium with nitric acid; alters the mineral, matrix, and cellular properties of bone. Prepn: A. Dupré, *C. R. Hebd. Seances Acad. Sci.* **86**, 720 (1878); of octahydrate: E. Einecke, *Angew. Chem.* **55**, 40 (1942); of nonahydrate: R. Reinmann, A. Tanner, *Z. Naturforsch.* **20b**, 71 (1965). Toxicity and antitumor activity: M. M. Hart *et al.*, *J. Natl. Cancer Inst.* **47**, 1121 (1971). Clinical pharmacokinetics: D. P. Kelsen *et al.*, *Cancer* **46**, 2009 (1980). Clinical trial in Paget's disease: R. S. Bockman *et al.*, *J. Clin. Endocrinol. Metab.* **80**, 595 (1995). Review of properties and clinical development: B. J. Foster *et al.*, *Cancer Treat. Rep.* **70**, 1311-1319 (1986); of clinical studies in treatment of bone metastases: R. P. Warrell, Jr., *Cancer* **80**, 1680-1685 (1997); of therapeutic uses: G. Apseloff, *Am. J. Ther.* **6**, 327-339 (1999). Series of articles on pharmacology and mechanism of action: *Semin. Oncol.* **18**, Suppl. 5, 1-31 (1991); on clinical experience in multiple myeloma, lymphoma, bladder cancer and cancer-related hypercalcemia: *ibid.* **30**, Suppl. 5, 1-41 (2003).

White crystalline powder. Sol in warm and cold aqueous solvents, absolute alcohol, ether.

Nonahydrate. [135886-70-3] White, slightly hygroscopic, crystalline powder. Sol in water. LD_{50} in male mice, female rats (mg/kg): 80.0, 67.5 i.p. (Hart).

Octahydrate. mp ~65°.

THERAP CAT: Antineoplastic; antipagetic; bone resorption inhibitor.

4381. Gallium Nitride. [25617-97-4] Gallium mononitride. GaN; mol wt 83.73. Ga 83.27%, N 16.73%. Semiconductor material. Prepn: W. C. Johnson *et al.*, *J. Phys. Chem.* **36**, 2651 (1932); and wurtzite structure determn: R. Juza, H. Hahn, *Z. Anorg. Allg. Chem.* **239**, 282 (1938). Prepn of single crystalline GaN by vapor deposition: H. P. Maruska, J. J. Tietjen, *Appl. Phys. Lett.* **15**, 327 (1969). Luminescence studies: H. G. Grimmeiss, H. Koelmans, *Z. Naturforsch.* **14a**, 264 (1959). P-type doping: H. Amano *et al.*, *Jpn. J. Appl. Phys.* **28**, L2112 (1989). Fabrication of blue light emitting diodes (LEDs): S. Nakamura *et al.*, *ibid.* **30**, L1998 (1991). Potential use of GaN LEDs in phototherapy for jaundice: D. S. Seidman *et al.*, *J. Pediatr.* **136**, 771 (2000). Review of properties and crystal growth techniques: S. Strite, H. Morkoç, *J. Vac. Sci. Technol. B* **10**, 1237-1266 (1992); and applications: S. Keller, S. P. Denbaars, *Curr. Opin. Solid State Mater. Sci.* **3**, 45-50 (1997); S. J. Pearton *et al.*, *Materials Today* **5**, 24-31 (June, 2002); of thin film growth techniques and optical properties: R. F. Davis *et al.*, *Proc. IEEE* **90**, 993-1004 (2002).

Equilibrium crystal structure is hexagonal wurtzite; lattice constants at 300 K: $a = 3.189$ Å; $c = 5.185$ Å. Less thermodynamically stable cubic zinc blende structures can be grown on cubic substrates; lattice constant $a = 4.503$ Å. mp 2500° (2800 K). Thermal conductivity: 1.3 W/cm K. Bandgap at 300 K: 3.39 eV. n (1 eV) = 2.33. n (3.38 eV) = 2.67. Exceedingly chemically and thermally stable.

Insol in H_2O, acids, and bases at room temp. Dissolves slowly in hot alkalis.

USE: Blue and UV light emitter with applications in semiconductor devices including: LEDs, laser diodes, lighting, displays, and data storage.

4382. Gallium Oxide. [12024-21-4] Gallia; gallium(III) oxide; gallium sesquioxide. Ga_2O_3; mol wt 187.44. Ga 74.40%, O 25.61%. Wide band gap material; exhibits both conduction and luminescence properties. Prepn: L. de Boisbaudran, *Compt. Rend.* **86**, 941 (1878). Synthesis of nanostructures: S. Sharma, M. K. Sunkara, *J. Am. Chem. Soc.* **124**, 12288 (2002); and photoluminescence properties of nanowires: X. C. Wu *et al.*, *Chem. Phys. Lett.* **328**, 5 (2000). Polymorphism study: R. Roy *et al.*, *J. Am. Chem. Soc.* **74**, 719 (1952). Crystal structure of β-form: S. Geller, *J. Chem. Phys.* **33**, 676 (1960). Electrical properties: M. R. Lorenz *et al.*, *J. Phys. Chem. Solids* **28**, 403 (1967). Toxicity study: R. K. Wolff *et al.*, *J. Appl. Toxicol.* **8**, 191 (1988). Use as a dehydration catalyst: B. H. Davis *et al.*, *J. Org. Chem.* **44**, 2142 (1979).

Multiple crystal forms exist; β-Ga_2O_3 is the only stable form. mp 1725±15°. d 5.94. Band gap, $E_g = 4.9$ eV.

USE: In semiconductors; gas sensing; catalysis. Nanostructures as blue and UV light emitters in optoelectronic device applications.

4383. Gallium Phosphide. [12063-98-8] GaP; mol wt 100.70. Ga 69.24%, P 30.76%. Prepn and description: Folberth, Oswald, *Z. Naturforsch.* **9a**, 1050 (1954); Wolff *et al.*, *Phys. Rev.* **94**, 753 (1954); Antell, Effer, *J. Electrochem. Soc.* **106**, 509 (1959); **107**, 252 (1960); Frosch, Derick, *ibid.* **108**, 251 (1961); Addamiano, *J. Am. Chem. Soc.* **82**, 1537 (1960); Gershenzon, Mikulyak, *ibid.* **108**, 548 (1961); Pizzarello, *ibid.* **109**, 226 (1962). Seguin, Gans, US 2862787 (1958); Chang, US 2921905 (1960 to Westinghouse).

Translucent, amber-colored crystals of the zinc blende type. Greenish-yellow, opaque cryst mass, when unreacted gallium is present. mp 1465°. Dielectric constant 8.4. Energy gap 2.25 ev. Hole mobility 70. Electron mobility 1200. Current density at 4 volts = 25 amps/cm^2 at 25° (in rectifier circuitry).

USE: In semiconductor electronics.

4384. Gallium Trifluoride. [7783-51-9] Gallium fluoride (GaF_3). F_3Ga; mol wt 126.72. F 44.98%, Ga 55.02%. GaF_3. Prepd by thermal decompn of ammonium hexafluogallate $(NH_4)_3(GaF_6)$: Hannebohn, Klemm, *Z. Anorg. Allg. Chem.* **229**, 342 (1936); Kwasnik in *Handbook of Preparative Inorganic Chemistry* vol. **1**, G. Brauer, Ed. (Academic Press, New York, 2nd ed., 1963) pp 227-228; from Ga and HF: Brewer *et al.*, *J. Inorg. Nucl. Chem.* **9**, 56 (1959); Muetterties, Castle, *ibid.* **18**, 148 (1961).

White powder. d 4.47 (after heating to 630° in a current of F_2). mp >1000°. bp ~950°. Can be sublimed in N_2 current at about 800° without decomp. Soly in water (25°) 0.0024 g/100 ml. Soly in hot dil HCl 0.0028 g/100 ml.

Trihydrate. White substance, mp >140°. More sol in water in anhydr form.

4385. Gallocyanine. [1562-85-2] 1-Carboxy-7-(dimethylamino)-3,4-dihydroxyphenoxazin-5-ium chloride (1:1); C.I. Mordant Blue 10; C.I. 51030. $C_{15}H_{13}ClN_2O_5$; mol wt 336.73. C 53.50%, H 3.89%, Cl 10.53%, N 8.32%, O 23.76%. Made by introducing nitrosodimethylaniline hydrochloride into a suspension of gallic acid in boiling methanol: Koechlin, DE 19580 (1881), *C.A.* **1**, 269; Nietzki, Otto, *Ber.* **21**, 1736 (1888); *Colour Index* vol. **4** (3rd ed., 1971) p 4460.

Green crystals. Practically insol in cold water. Slightly sol in hot water; sol in alcohol, glacial acetic acid, in alkali carbonates with reddish color, in concd HCl with a blue color which becomes red upon diluting with water.

USE: As a dye; in alkali carbonate soln as a reagent for lead with which it forms a deep violet color.

4386. Gallopamil. [16662-47-8] α-[3-[[2-(3,4-Dimethoxy-phenyl)ethyl]methylamino]propyl]-3,4,5-trimethoxy-α-(1-methyl-ethyl)benzeneacetonitrile; 5-[(3,4-dimethoxyphenethyl)methylami-no]-2-isopropyl-2-(3,4,5-trimethoxyphenyl)valeronitrile; α-isopro-pyl-α-[(N-methyl-N-homoveratryl)-γ-aminopropyl]-3,4,5-trimeth-oxyphenylacetonitrile; methoxyverapamil; D-600. $C_{28}H_{40}N_2O_5$; mol wt 484.64. C 69.39%, H 8.32%, N 5.78%, O 16.51%. Calcium channel blocking agent; methoxy analog of verapamil, q.v. Prepn: F. Dengel, **BE 615861**; idem, **US 3261859** (1962, 1966 to Knoll). Stereoselective synthesis of enantiomers: L. J. Theodore, W. L. Nelson, J. Org. Chem. **52**, 1309 (1987). Pharmacology: H. Haas, E. Busch, Arzneim.-Forsch. **17**, 257 (1967); eidem, ibid. **18**, 401 (1968); A. Fleckenstein, ibid. **20**, 1317 (1970). Clinical trial in stable angina pectoris: N. S. Khurmi et al., Am. J. Cardiol. **53**, 684 (1984). Symposium on pharmacology and clinical efficacy: J. Cardiovasc. Pharmacol. **20**, Suppl. 7, S1-S95 (1992). Review of pharmacokinetics and therapeutic potential in ischemic heart disease: R. N. Brogden, P. Benfield, Drugs **47**, 93-115 (1994).

Pale yellow viscous oil. n_D^{25} 1.5402.

Hydrochloride. [16662-46-7] Algocor; Procorum. $C_{28}H_{40}N_2O_5 \cdot HCl$; mol wt 521.10. mp 145-148°.

d-Form hydrochloride. [38176-09-9] White crystalline solid from propanol/cyclohexane, mp 160.5-161.5°. $[\alpha]_D^{25}$ +11.7° (c = 5.02 in ethanol).

l-Form hydrochloride. [36222-39-6] White crystalline solid from propanol/cyclohexane, mp 160.5-161.5°. $[\alpha]_D^{23}$ −11.7° (c = 5.04 in ethanol).

THERAP CAT: Antianginal.

4387. Galvinoxyl. [2370-18-5] 4-[[3,5-Bis(1,1-dimethyleth-yl)-4-oxo-2,5-cyclohexadien-1-ylidene]methyl]-2,6-bis(1,1-dimeth-ylethyl)phenoxy; Coppinger's radical; 2,6-di-tert-butyl-α-(3,5-di-tert-butyl-4-oxo-2,5-cyclohexadiene-1-ylidene)-p-tolyloxy radical; 2,6,3′,5′-tetra-tert-butyl-4′-phenoxy-4-methylene-2,5-cyclohexadi-ene-1-one radical. $C_{29}H_{41}O_2$; mol wt 421.65. C 82.61%, H 9.80%, O 7.59%. Stable phenoxyl radical. Prepn: G. M. Coppinger, J. Am. Chem. Soc. **79**, 501 (1957); M. S. Kharasch, B. S. Joshi, J. Org. Chem. **22**, 1435 (1957). Radical scavenging study: P. D. Bartlett, T. Funahashi, J. Am. Chem. Soc. **84**, 2596 (1962). Electrical properties: D. D. Eley et al., Trans. Faraday Soc. **62**, 3192 (1966). Structure determn: D. E. Williams, Mol. Phys. **16**, 145 (1969). Resonance Raman study: G. N. R. Tripathi, Chem. Phys. Lett. **81**, 375 (1981). Theoretical study of magnetic interactions: F. Dietz et al., J. Phys. Chem. B **102**, 3912 (1998). Ab initio studies of ferromagnetic properties: S. J. Luo, K. L. Yao, J. Magn. Magn. Mater. **257**, 11 (2003). Use in detection of phospholipid phase transitions: M. A. Singer et al., Anal. Biochem. **94**, 322 (1979). Use in measurement of hydrogen-donating activity and antioxidant activity in phenols: H. Shi et al., Methods Enzymol. **335**, 157 (2001).

Deep blue needles from absolute ethanol, mp 157.5° (Kharasch, Joshi). Also reported as dark blue needles from dry ethanol, mp 153° (Eley). Absorption max (iso-octane): 280, 289, 400, 423 nm (ε ×10³ 9.39, 9.4, 24.3, 180.0). Absorption max (benzene): 407 nm, 431 nm (ε 30000, 154000). Absorption max (ethanol): 428 nm. Sol in petr ether, benzene, cyclohexane.

USE: Free radical scavenger.

4388. Gamabufotalin. [465-11-2] (3β,5β,11α)-3,11,14-Trihydroxybufa-20,22-dienolide; gamabufogenin; gamabufagin. $C_{24}H_{34}O_5$; mol wt 402.53. C 71.61%, H 8.51%, O 19.87%. One of the genins found in the venom of a Japanese toad (gama; Bufo vulgaris formosus). Isoln and structure: Kotake, Ann. **465**, 11 (1928); Wieland, Vocke, Ann. **481**, 215 (1930); Chen et al., J. Pharmacol. Exp. Ther. **49**, 26 (1933); Kondo, Ohno, J. Pharm. Soc. Jpn. **59**, 186 (1939); Jensen, J. Am. Chem. Soc. **59**, 767 (1937); Kuno Meyer, Helv. Chim. Acta **32**, 1599 (1949). Also found in Ch'an-Su, A Chinese drug prepared from Chinese toads (Bufo asiaticus = Bufo gargarizans Cantor): Ruckstuhl, Meyer, Helv. Chim. Acta **40**, 1270 (1957).

Bitter, solvated prisms from alcohol + ether, mp 254°; when dry dec 262-263°. $[\alpha]_D^{18}$ +1.26° (c = 0.793 in methanol). Produces numbness of the tongue. uv max about 300 nm. Very sparingly sol in chloroform, acetone, water. Somewhat more sol in methanol.

4389. Gambir. Catechu; pale catechu; gambir catechu; terra japonica. Dried aqueous extract of leaves and twigs of Uncaria gambier (Hunter) Roxb. (Ourouparia gambier (Hunter) Baill.), Rubiaceae. Habit. Southern Asia. Constit. 7-30% catechol, 20-50% catechutannic acid; quercetin, catechu-red, gambir-fluorescein, fixed oil, wax. Contains not less than 70% water-sol substance and not less than 60% is sol in alcohol. Incompat: Iron compds, gelatin, limewater, mercuric chloride, zinc sulfate, alkalies.

USE: Tanning, dyeing fabrics brown or black.

THERAP CAT: Astringent.

THERAP CAT (VET): Astringent (intestinal).

4390. Gamboge. [9000-25-3] Cambogia. Gum-resin from the trunk of various Garcinia trees, particularly, Garcinia hanburyi Hook. f., Guttiferae. Habit. India, Southeast Asia. Constit. 25-30% mucilages, 70-75% resin containing polyprenylated xanthones, primarily gambogic acid, q.v. Used in traditional medicine as a cathartic and vermifuge; topically as an antimicrobial and anti-inflammatory. Term also refers to the yellow pigment prepared from the resin. Pharmaceutical identification tests: F. O. Taylor, J. Ind. Eng. Chem. **2**, 208 (1910). Analysis of constituents: H. Autherhoff et al., Arch. Pharm. **295**, 833 (1962); J. Asano et al., Phytochemistry **41**, 815 (1996); J.-Z. Song et al., J. Sep. Sci. **30**, 304 (2007). Anti-inflammatory activity of ethyl acetate extract: A. Panthong et al., J. Ethnopharmacol. **111**, 335 (2007). Description and medicinal uses: J. Gruenwald et al., PDR for Herbal Medicines (Thomson PDR, Montvale, 3rd Ed., 2004) pp 342-343.

Gamboge pigment. [114921-13-0] C.I. Natural Yellow 24; rattan yellow; gummigutt; teng huang. Review of properties and use as artistic pigment: J. Winter in Artists' Pigments: A Handbook of Their History and Characteristics, **Vol. 3**, E. W. FitzHugh, Ed., (National Gallery of Art, Washington DC, 1997) pp 143-155. Spectroscopic detectn in works of art: P. Vandenabeele et al., Anal. Chim. Acta **407**, 261 (2000). Electrochemical determn in archeological fabrics: A. Doménech-Carbó et al., Talanta **66**, 769 (2005). Bright yellow powder. Poisonous. Sol in water. Fades in bright light.

USE: Transparent, yellow pigment in watercolors and oils; in colored laquers and varnishes.

4391. Gambogic Acid. [2752-65-0] (2Z)-2-Methyl-4-[(1R,-3aS,5S,11R,14aS)-3a,4,5,7-tetrahydro-8-hydroxy-3,3,11-trimethyl-13-(3-methyl-2-buten-1-yl)-11-(4-methyl-3-penten-1-yl)-7,15-dioxo-1,5-methano-1H,3H,11H-furo[3,4-g]pyrano[3,2-b]xanthen-1-yl]-2-butenoic acid; β-guttiferin. $C_{38}H_{44}O_8$; mol wt 628.76. C 72.59%, H 7.05%, O 20.36%. Cytotoxic principle of gamboge, q.v., a gum-resin from the latex of garcinia trees, especially *Garcinia hanburyi* Hook. f., *Guttiferae*. Isoln: M. Amorosa, L. Lipparini, *Ann. Chim. (Rome)* **45**, 977 (1955). Structure: W. D. Ollis *et al.*, *Tetrahedron* **21**, 1453 (1965); S. A. Ahmad *et al.*, *J. Chem. Soc. C* **1966**, 772; G. Cardillo, L. Merlini, *Tetrahedron Lett.* **8**, 2529 (1967). HPLC determn in plasma: L. Ding *et al.*, *J. Chromatogr. B* **846**, 112 (2007). Apoptosis inducing activity: H.-Z. Zhang *et al.*, *Bioorg. Med. Chem.* **12**, 309 (2004). Antiangiogenic activity: T. Yi *et al.*, *Cancer Res.* **68**, 1843 (2008). Toxicology: Q. Guo *et al.*, *Basic Clin. Pharmacol. Toxicol.* **99**, 178 (2006).

Yellow-orange solid. $[\alpha]_D^{20}$ −685° (methanol). uv max (ethanol): 217, 280, 291, 362 nm (ε 26000, 16700, 17000, 14900). Sol in DMSO, ethanol. LD_{50} i.p. in mice: 45.9 mg/kg (Guo).

Methyl ester monomethyl ether. [2752-66-1] $C_{40}H_{48}O_8$. Yellow prisms from methanol, mp 130-131°. $[\alpha]_D^{22}$ −560° (c = 0.7 in chloroform). uv max (ethanol): 224, 299 nm (ε 36000, 13600).

Pyridine salt. [2631-91-6] $C_{43}H_{49}NO_8$. Orange needles from ether + petr ether, mp 147-149°. $[\alpha]_D$ −550° (chloroform). uv max (ethanol): 291.5, 359.5 nm (ε 22300, 18100).

Amide. [286935-60-2] (2Z)-2-Methyl-4-[(1R,3aS,5S,11R,14aS)-3a,4,5,7-tetrahydro-8-hydroxy-3,3,11-trimethyl-13-(3-methyl-2-buten-1-yl)-11-(4-methyl-3-penten-1-yl)-7,15-dioxo-1,5-methano-1H,3H,11H-furo[3,4-g]pyrano[3,2-b]xanthen-1-yl]-2-butenamide. $C_{38}H_{45}NO_7$; mol wt 627.78. Prepn: S. X. Cai *et al.*, **WO 00044216**; *eidem*, **US 6462041** (2000, 2002 both to Cytovia). Neurotrophic activity: S.-W. Jang *et al.*, *Proc. Natl. Acad. Sci. USA* **104**, 16329 (2007).

4392. Ganaxolone. [38398-32-2] (3α,5α)-3-Hydroxy-3-methylpregnan-20-one; CCD-1042. $C_{22}H_{36}O_2$; mol wt 332.53. C 79.46%, H 10.91%, O 9.62%. Synthetic analog of the neuroactive steroids known as epalons that allosterically modulate the γ-aminobutyric acid type A (GABA$_A$) receptor. Prepn: M. C. Cook *et al.*, **DE 2162555**; *eidem*, **US 3953429** (1972, 1976 both to Glaxo); D. J. Hogenkamp *et al.*, *J. Med. Chem.* **40**, 61 (1997). Binding study and anticonvulsant activity: R. B. Carter *et al.*, *J. Pharmacol. Exp. Ther.* **280**, 1284 (1997). Clinical pharmacology: E. P. Monaghan *et al.*, *Epilepsia* **38**, 1026 (1997). Determn in serum by HPLC-MS-MS: K. Ramu *et al.*, *J. Chromatogr. B* **751**, 49 (2001). Clinical evaluation in epilepsy: V. A. Pieribone *et al.*, *Epilepsia* **48**, 1870 (2007).

White crystals from methanol, mp 190-192°. $[\alpha]_D$ +103°. Insol in aqueous media.

THERAP CAT: Anticonvulsant.

4393. Ganciclovir. [82410-32-0] 2-Amino-1,9-dihydro-9-[[2-hydroxy-1-(hydroxymethyl)ethoxy]methyl]-6H-purin-6-one; 9-[(1,3-dihydroxy-2-propoxy)methyl]guanine; 2′-nor-2′-deoxyguanosine; DHPG; 2′NDG; BIOLF-62; BW-B759U; BW-759; BW-759U; RS-21592; Virgan; Vitrasert; Zirgan. $C_9H_{13}N_5O_4$; mol wt 255.23. C 42.35%, H 5.13%, N 27.44%, O 25.07%. Nucleoside analog structurally related to acyclovir, q.v. Prepn: J. P. Verheyden, J. C. Martin, **US 4355032** (1982 to Syntex); K. K. Ogilvie *et al.*, *Can. J. Chem.* **60**, 3005 (1982); W. T. Ashton *et al.*, *Biochem. Biophys. Res. Commun.* **108**, 1716 (1982); J. C. Martin *et al.*, *J. Med. Chem.* **26**, 759 (1983). Antiviral spectrum *in vitro*: K. O. Smith *et al.*, *Antimicrob. Agents Chemother.* **22**, 55 (1982). Mode of action study: Y.-C. Cheng *et al.*, *Proc. Natl. Acad. Sci. USA* **80**, 2767 (1983). HPLC dertermn in plasma: Y. Dao *et al.*, *J. Chromatogr. B* **867**, 270 (2008). Review of pharmacology and clinical efficacy in cytomegalovirus (CMV) infection: C. S. Crumpacker, *N. Engl. J. Med.* **335**, 721-729 (1996); in transplant patients: J. K. McGavin, K. L. Goa, *Drugs* **61**, 1153-1183 (2001).

Crystals from methanol, mp 250° (dec) (Verheyden, Martin); also reported as crystalline monohydrate from water, mp 248-249° (dec) (Ashton); crystals from water, mp >300° (Martin). uv max (methanol): 254 nm (ε 12880). Soly in water (25°): 4.3 mg/ml at pH 7. LD_{50} i.p. in mice: 1-2 g/kg (Martin).

Sodium salt. [107910-75-8] Cymevan; Cymevene; Cytovene; Denosine. $C_9H_{12}N_5NaO_4$; mol wt 277.22.

THERAP CAT: Antiviral.

4394. Gangliosides. Animal glycosphingolipids occurring in highest concentration in the central nervous system but widely distributed in other tissues. Primarily located in the outer surface of mammalian cell plasma membranes, they are also found in the synaptic membrane of the central nervous system. More than 60 gangliosides have been characterized. Each consists of a fatty acid (often stearic acid) and an oligosaccharide moiety, both attached to sphingosine. Sialic acid, q.v., is the identifying sugar. The sialic acid residues of gangliosides and those of glycoprotein are the main cause of cell surface negative charge. Gangliosides are also thought to be involved in many different cell functions, e.g. metabolism, growth, malignant transformation. Ganglioside nomenclature: L. Svennerholm, "Ganglioside Metabolism" in *Comprehensive Biochemistry* vol. 18, M. Florkin, E. H. Stotz, Eds. (Elsevier, Amsterdam, 1970) pp 201-204. Isoln from normal brain and from the brain of patients with Tay-Sachs disease: Klenk, *Z. Physiol. Chem.* **268**, 50 (1941); **273**, 76 (1942); Trams, Lauter, *Biochim. Biophys. Acta* **60**, 350 (1962); Svennerholm, *Acta Chem. Scand.* **17**, 239 (1963). Occurrence of different gangliosides: Kuhn, Egge, *Angew. Chem.* **72**, 805 (1960). Chromatographic studies: Kuhn *et al.*, *ibid.* **73**, 580 (1961); Klenk, Gielen, *Z. Physiol. Chem.* **326**, 144 (1961). Structure of ganglioside G_I (or G_{M1}): Kuhn, Wiegandt, *Ber.* **96**, 866 (1963). Structure of gangliosides G_I and G_{II} (or G_{D1a}): Kuhn, Egge, *ibid.* **96**, 3338 (1963). Structure of gangliosides G_{II}, G_{III} (or G_{D1b}) and G_{IV} (or G_{T1b}): Kuhn, Wiegandt, *Z. Naturforsch.* **18b**, 541 (1963). Molecular diversity, biological implications of brain and thymus gangliosides: Y. Nogai, M. Iwamori, *Mol. Cell. Biochem.* **29**, 81 (1980). Biosynthesis in tissues: S. Basu *et al.*, *Adv. Exp. Med. Biol.* **125**, 213 (1980). Action as biotransducers of membrane-mediated information: R. O. Brady, P. H. Fishman, *Adv. Enzymol.* **50**, 303 (1979). Reviews: Wiegandt, *Angew. Chem. Int. Ed.* **7**, 87-96 (1968). Collection of articles on structure and function: *Adv. Exp. Med. Biol.* **125**, 1-555 (1980); on structure analysis, function and properties, regeneration and recovery in the nervous system, clinical trials in amyotropic lateral sclerosis and diabetic neuropathy: *ibid.* **174**, 1-649 (1984). Book: *Gangliosides in Neurological and Neuromuscular Function, Development, and Repair* M. M. Rapport, A. Gorio, Eds. (Raven Press, New York, 1981) 267 pp.

Ganglioside G_{M1}

Gangliosides are colorless crystallizable substances which melt with decompn. Insol in non-polar solvents. Soly in polar solvents increases with the size of the sugar residue and the sialic acid content. Forms micelles in aq soln, having mol wt of about 200,000-250,000. Forms molecular solns in DMF or tetrahydrofuran, having mol wt of 1000-3000.

Ganglioside G_{M1}. [37758-47-7] Siagoside; Sygen. $C_{73}H_{131}N_3$-O_{31}; mol wt 1546.84.

Combination of gangliosides G_{M1}, G_{D1a}, G_{D1b}, and G_{T1b}. Cronassial; Gangliovet.

THERAP CAT: In treatment of peripheral neuropathies.

THERAP CAT (VET): In treatment of peripheral neuropathies.

4395. Ganirelix. [124904-93-4] *N*-Acetyl-3-(2-naphthalenyl)-D-alanyl-4-chloro-D-phenylalanyl-3-(3-pyridinyl)-D-alanyl-L-seryl-L-tyrosyl-N^6-[bis(ethylamino)methylene]-D-lysyl-L-leucyl-N^6-[bis(ethylamino)methylene]-L-lysyl-L-prolyl-D-alaninamide; [*N*-Ac-D-Nal(2)¹-D-*p*Cl-Phe²-D-Pal(3)³-D-hArg(Et₂)⁶-hArg(Et₂)⁸-D-Ala¹⁰]GnRH. $C_{80}H_{113}ClN_{18}O_{13}$; mol wt 1570.35. C 61.19%, H 7.25%, Cl 2.26%, N 16.06%, O 13.24%. Decapeptide GnRH antagonist. Prepn and structure-activity studies: J. J. Nestor, Jr. *et al.* in *Peptides 1988: Proc. 20th Eur. Peptide Symp.*, G. Jung, E. Bayer, Eds. (Walter de Gruyter, Berlin, 1989) pp 592-594; *eidem, J. Med. Chem.* **35**, 3942 (1992). Improved synthesis: H. B. Arzeno *et al., Int. J. Pept. Protein Res.* **41**, 342 (1993). Degradation in aq soln: R. G. Strickley *et al., Pharm. Res.* **7**, 530 (1990). Clinical pharmacology and pharmacokinetics: J. Rabinovici *et al., J. Clin. Endocrinol. Metab.* **75**, 1220 (1992). Clinical trial in assisted fertilization: P. Devroey *et al., Hum. Reprod.* **13**, 3023 (1998); C. B. Lambalk *et al., ibid.* **21**, 632 (2006).

Acetate. [129311-55-3] Org-37462; RS-26306; Orgalutran. $C_{80}H_{113}ClN_{18}O_{13}.2C_2H_4O_2$; mol wt 1690.45. pKa: 4.2 (3-pyridinylalanine); 9.8 (tyrosine). Sol in water.

THERAP CAT: In treatment of infertility.

4396. Ganoderic Acids. Family of bioactive, lanostane-type triterpenes isolated from the medicinal mushroom, *Ganoderma lucidum* (Fr.) Karst, *Ganodermataceae*. Numerous closely related structures have been identified that exhibit hepatoprotectant, antimicrobial, antihypertensive, antilipemic, or antitumor activities. Isoln and structures of ganoderic acids A and B: T. Kubota *et al., Helv. Chim. Acta* **65**, 611 (1982); of ganoderic acids T through Z: J. O. Toth *et al., Tetrahedron Lett.* **24**, 1081 (1983). Prodn by liquid fermentation: Q.-H. Fang, J.-J. Zhong, *Biotechnol. Prog.* **18**, 51 (2002). Structural analysis and bioactivities: M.-S. Shiao *et al., ACS Symp. Ser.* **547**, 342 (1994). Quantitative determn of major components by HPLC: X.-M. Wang *et al., J. Pharm. Biomed. Anal.* **41**, 838 (2006).

Analysis in chloroform extracts of *G. lucidum* by HPLC/ESI-MS: M. Yang *et al., J. Am. Soc. Mass Spectrom.* **18**, 927 (2007). HPLC determn in plasma: X. Wang *et al., Biomed. Chromatogr.* **21**, 389 (2007). Review of structures and properties: R. J. Cole, M. A. Schweikert, *Handbook of Secondary Fungal Metabolites* **Vol. 2** (Academic Press, San Diego, 2003) pp 271-350. *Review*: Y. Zhou *et al., Food Rev. Int.* **22**, 259-273 (2006).

Ganoderic Acid B

Ganoderic Acid A. [81907-62-2] (7β,15α,25R)-7,15-Dihydroxy-3,11,23-trioxolanost-8-en-26-oic acid. $C_{30}H_{44}O_7$; mol wt 516.68. Bitter, amorphous powder. $[\alpha]_D^{27}$ +153.8° (c = 0.156 in chloroform).

Ganoderic Acid B. [81907-61-1] (3β,7β,25R)-3,7-Dihydroxy-11,15,23-trioxolanost-8-en-26-oic acid. $C_{30}H_{44}O_7$; mol wt 516.68. Pale yellow needles from ethyl acetate + benzene, mp 212-213°. $[\alpha]_D^{24}$ +138° (c = 0.2 in methanol). uv max in ethanol: 254 nm (ε 10740).

4397. Ganoderma. Lingzhi; reishi; mannentake. Basidiomycete mushroom, *Ganoderma lucidum* (Fr.) Karst., *Ganodermataceae*, used in Asian traditional medicine as a tonic, immunomodulator, and cancer treatment. Other species have also been used medicinally, such as *G. applanatum. Habit.* Temperate regions of Europe and Asia, particularly China, Japan. Widely cultivated as natural occurrence is rare. *Constit.* Several hundred bioactive components, particularly polysaccharides and triterpenoids, such as lucidenic and ganoderic acids, *q.v.* Botanical description, constituents and uses: A. Y. Leung, S. Foster, *Encyclopedia of Common Natural Ingredients* (Wiley-Interscience, Hoboken, 2nd Ed., 2003) pp 255-260. Cultivation by submerged fermentation: R. Wagner *et al., Food Technol. Biotechnol.* **41**, 371 (2003). Review of medicinal uses and constituents: R. Russell, M. Paterson, *Phytochemistry* **67**, 1985-2001 (2006); and cultivation methods: B. Boh *et al., Biotechnol. Annu. Rev.* **13**, 265-301 (2007).

USE: Dietary supplement. In skin-care creams and lotions as moisturizer and replenisher.

4398. Garcinia. Genus of tropical fruit trees of the family, *Guttiferae*, alternatively known as *Clusiaceae. Habit.* Tropical regions of Asia, Africa, and Polynesia. Several species are used in traditional medicine, in food preparation, for their edible seed oil, or as a source of pigments. The gum-resin from the latex is known as gamboge, *q.v.* Description of medicinal uses: C. P. Khare, *Indian Herbal Remedies* (Springer, Berlin, 2004) pp 229-230. Review of constituents and commercial uses: V. K. Raju, M. Reni in *Handbook of Herbs and Spices*, **Vol. 1**, K. V. Peter, Ed. (Woodhead Publishing, Cambridge, 2001) pp 207-215.

Cambodge. Malabar tamarind. Refers to *Garcinia cambogia* Desr. Fruit rind is used as a condiment for flavoring curries and to cure dried fish. Extracts are used in weight loss preparations as a source of hydroxycitric acid, *q.v.* HPLC determn of acids in fruit rinds: G. K. Jayaprakasha, K. K. Sakariah, *J. Chromatogr. A* **806**, 337 (1998). Clinical trial of fruit extract on visceral fat accumulation: K. Hayamizu *et al., Curr. Ther. Res.* **64**, 551 (2003).

Kokam. Kokum. Refers to *Garcinia indica* Choisy. Edible fruit with agreeable flavor and sweetish, acid taste. *Constit.* Organic acids, primarily hydroxycitric acid, malic acid; carbohydrates; anthocyanin pigments, especially cyanidin glycosides; isoprenylated benzophenones such as *garcinol*, a fat-soluble yellow pigment. Seeds yield an edible oil known as *kokam butter* which is rich in stearic and oleic acids.

USE: Dietary supplement; condiment; oils as emollient and vulnerary.

4399. Gardenins. A group of flavones isolated from the resinous exudate of leaf buds of *Gardenia lucida* Roxb. *Rubiaceae,* a small spiny tree occurring throughout India. Isoln of the main flavonoid pigment, gardenin A (originally thought to be the only component and referred to in early literature as "gardenin"): J. Stenhouse, C. G. Groves, *J. Chem. Soc.* **1877**, 552; *Ann.* **200**, 311 (1880); K. J. Balakrishna, T. R. Seshadri, *Proc. Indian Acad. Sci.* **27A**, 91 (1948). Isoln and structure of gardenins B, C, D, E: A. V. R. Rao *et al., Indian J. Chem.* **8**, 398 (1970). Isoln of B from *Citrus jambhiri* lush. *Rutaceae* and synthesis: B. P. Chaliha *et al., Tetrahedron* **21**, 1441 (1965). Structure of A: P. K. Bose, *J. Indian Chem. Soc.* **22**, 233 (1945). Revised structure: A. V. R. Rao, K. Venkataraman, *Indian J. Chem.* **6**, 677 (1968). Synthesis of A: M. Krishnamurti *et al., ibid.* **8**, 575 (1970); M. Kamalam, A. V. R. Rao, *ibid.* 573; of C: A. J. Kalra *et al., ibid.* **11**, 96 (1973); of E: *eidem, ibid.* 1092. Antineoplastic activity and cytotoxicity: J. Edwards *et al., J. Nat. Prod.* **42**, 85 (1979). ^{13}C-NMR study: M. Iinuma *et al., Chem. Pharm. Bull.* **28**, 708 (1980).

Gardenin A R = R' = OCH$_3$
Gardenin B R = R' = H
Gardenin C R = OH, R' = OCH$_3$
Gardenin D R = OH, R' = H
Gardenin E R = R' = OH

Gardenin A. [21187-73-5] 5-Hydroxy-6,7,8-trimethoxy-2-(3,-4,5-trimethoxyphenyl)-4*H*-1-benzopyran-4-one. C$_{21}$H$_{22}$O$_9$; mol wt 418.40. Golden yellow needles from ethanol, mp 162-163°. Sol in alcohol, chloroform. All the gardenins give a green color with alcoholic ferric chloride.

Gardenin B. [2798-20-1] 5-*O*-Desmethyltangeretin. C$_{19}$H$_{18}$O$_7$; mol wt 358.35. Fine yellow needles from benzene/petr ether, mp 176-177°. uv max (ethanol): 292, 330 nm (log e 4.40, 4.35); (1% AlCl$_3$): 310, 350 nm (log ε 4.48, 4.49).

Gardenin C. [29550-05-8] C$_{20}$H$_{20}$O$_9$; mol wt 404.37. Yellow flakes from ethyl acetate-petr ether, mp 179-180°. uv max (methanol) 304, 323 nm (log ε 4.18, 4.20).

Gardenin D. [29202-00-4] C$_{19}$H$_{18}$O$_8$; mol wt 374.35. Cryst, mp 190-192°. uv max (ethanol): 256, 280, 343 nm.

Gardenin E. [29550-07-0] C$_{19}$H$_{18}$O$_9$; mol wt 390.34. Golden yellow needles from ethyl acetate/petr ether, mp 234-235°. uv max (methanol): 280, 325 nm (log ε 4.37, 4.36).

4400. Gardinol Type Detergents. A mixture of the sodium salts of sulfated fatty alcohols made by reducing the mixed fatty acids of coconut oil or of cottonseed oil, and of fish oils. Sometimes natural waxes such as spermaceti, wool fat, and beeswax are sulfated directly. The mixture of the sulfated alcohols which goes by the commercial name "lauryl alcohol" consists of about 15% mixed C$_8$ and C$_{10}$ (octyl and decyl) alcohols, 40% C$_{12}$ (lauryl or dodecyl) alcohol, 30% C$_{14}$ (myristyl or tetradecyl) alcohol, and 15% mixed C$_{16}$ and C$_{18}$ (cetyl, stearyl, and oleyl) alcohols. Some of the tradenames designating this type of detergent are: *Gardinol; Duponol; Modinal; Aurinol; Maprofix; Tergavon; Sadopan; Cyclopon; Cyclanon; Sapidan; Lissapol; Teepol. See also* Sodium Lauryl Sulfate.

Caution: May cause local sensitivity reactions.

USE: Detergents, wetting, emulsifying, dispersing agents for cosmetics, as dry cleaning aids, in fungicidal sprays, in metal cleaning, leather processing, in textile manuf.

4401. Gardol®. [137-16-6] *N*-Methyl-*N*-(1-oxododecyl)glycine sodium salt (1:1); *N*-lauroylsarcosine sodium salt; sodium *N*-lauroyl sarcosinate; Medialan LL-99. C$_{15}$H$_{28}$NNaO$_3$; mol wt 293.38. C 61.41%, H 9.62%, N 4.77%, Na 7.84%, O 16.36%. Prepn: Jungermann *et al., J. Am. Chem. Soc.* **78**, 172 (1956).

Aq soln, *Medialan LL-33.*

USE: Detergent, foaming agent, antienzyme for dentifrices: King, US 2689170 (1954 to Colgate-Palmolive).

4402. Garenoxacin. [194804-75-6] 1-Cyclopropyl-8-(difluoromethoxy)-7-[(1*R*)-2,3-dihydro-1-methyl-1*H*-isoindol-5-yl]-1,4-dihydro-4-oxo-3-quinolinecarboxylic acid; T-3811. C$_{23}$H$_{20}$F$_2$N$_2$O$_4$; mol wt 426.42. C 64.78%, H 4.73%, F 8.91%, N 6.57%, O 15.01%. Des-F(6)-quinolone antibacterial; topoisomerase II inhibitor. Prepn: Y. Todo *et al.,* WO 9729102; *eidem,* US 6025370 (1997, 2000 both to Toyama); and antibacterial activity: K. Hayashi *et al., Arzneim.-Forsch.* **52**, 903 (2002). HPLC determn in serum: D. Xuan *et al., J. Chromatogr. B* **765**, 37 (2001). Mechanism of action study: D. Ince *et al., Antimicrob. Agents Chemother.* **46**, 3370 (2002). Comparative antimicrobial spectra: M. Takahata *et al., ibid.* **43**, 1077 (1999); M. Bassetti *et al., ibid.* **46**, 234 (2002). Clinical pharmacokinetics: D. A. Gajjar *et al., ibid.* **47**, 2256 (2003); and pharmacodynamics: S. Van Wart *et al., ibid.* **48**, 4766 (2004). Toxicology study: A. Nagai *et al., J. Toxicol. Sci.* **27**, 219 (2002). Review of pharmacology and therapeutic potential: R. Frechette, *Curr. Opin. Investig. Drugs* **2**, 1706-1711 (2001).

Pale yellow plates from ethanol as 0.25 hydrate, mp 226-227°. $[\alpha]_D^{27} -9.0°$ (c = 0.10 in *N,N*-dimethylformamide). Also reported as hydrate, mp 234-235°.

Methanesulfonate. [223652-82-2]; [223652-90-2] (monohydrate). Garenoxacin mesylate; BMS-284756; T-3811ME. C$_{23}$H$_{20}$F$_2$N$_2$O$_4$.CH$_4$O$_3$S; mol wt 522.52.

THERAP CAT: Antibacterial.

4403. Garlic. Allium. Strongly scented perennial herb, *Allium sativum* L., *Liliaceae,* used extensively in foods and in traditional medicine to treat respiratory infections. *Habit.* Central Asia, Southern Europe, U.S., cultivated worldwide. *Constit.* Enzymes, volatile oil (0.1-0.36%), carbohydrates (26-30%), protein and amino acids, lipids, minerals, vitamins, saponins. Upon bruising of the bulb, the enzyme allinase converts alliin to allicin, *q.q.v.,* the source of the garlic odor. Description and uses: A. Y. Leung, *Encyclopedia of Common Natural Ingredients* (John Wiley & Sons, New York, 1980) pp 176-178. Review of chemistry: E. Block, *Sci. Am.* **252**, 114-119 (March, 1985). Series of articles on cardiovascular effects: *Br. J. Clin. Pract.* **44**, Suppl. 69, 1-39 (1990). Review of antioxidant properties and therapeutic potential in aging: K. Rahman, *Ageing Res. Rev.* **2**, 39-56 (2003).

Volatile oil. [8000-78-0] Obtained by steam distillation of the crushed bulb. *Constit.* Ajoene, alliin, allicin, diallyl disulfide, diallyl trisulfide, methyl allyl trisulfide, allyl sulfide, terpenes including citral, geraniol, linalool, α- and β-phellandrene. Brief review: E. Guenther, *The Essential Oils* vol. 6 (Van Nostrand, New York, 1952) pp 67-69. Comprehensive description: N. A. Shaath *et al., Dev. Food Sci.* **37B**, 2025 (1995). Clear yellow to red-orange liquid; strong garlic odor. d$_{25}^{25}$ 1.050-1.095. n$_D^{20}$ 1.550-1.580. Sol in most fixed oils, mineral oil, incompletely sol in alcohol. Insol in glycerine, propylene glycol. *Keep well closed, cool and protected from light.*

USE: As a spice and seasoning in foods.

THERAP CAT: In treatment of hypertension and hyperlipidemia.

4404. Garner's Aldehyde. [127589-93-9] 4-Formyl-2,2-dimethyl-3-oxazolidinecarboxylic acid 1,1-dimethylethyl ester; 1,1-dimethylethyl 4-formyl-2,2-dimethyloxazolidine-3-carboxylate. C_{11}-$H_{19}NO_4$; mol wt 229.28. C 57.62%, H 8.35%, N 6.11%, O 27.91%. Both enantiomers are configurationally stable building blocks for use in asymmetric synthesis. Prepn of *R*-form: P. Garner, *Tetrahedron Lett.* **25**, 5855 (1984); of *S*-form: P. Garner, J. M. Park, *J. Org. Chem.* **52**, 2361 (1987). Improved prepn of *S*-form: A. McKillop *et al.*, *Synthesis* **1994**, 31. Review of chemistry: X. Liang *et al.*, *J. Chem. Soc. Perkin Trans. 1* **2001**, 2136-2157.

(*S*)-Garner Aldehyde

R-Form. [95715-87-0] (*R*)-Garner aldehyde. $[\alpha]_D^{27}$ +103° (c = 1.0 in CHCl$_3$).

S-Form. [102308-32-7] (*S*)-Garner aldehyde. Colorless liquid, bp$_{1.0-1.4}$ 83-88°; bp$_{0.3}$ 72-75°. $[\alpha]_D$ −91.7° (c = 1.34 in CHCl$_3$).

USE: Chiral, non-racemic synthon for asymmetric synthesis of amino sugars and other nitrogen-containing targets.

4405. Gasoline. Petrol (British); Benzin (German). A mixture of C_4 to C_{12} hydrocarbons. Natural gasoline, obtained by fractional distillation of petroleum, contains mostly saturated hydrocarbons; commercial grades of motor gasoline contain paraffins, olefins, naphthenes, and aromatics, all in substantial concns. Motor gasolines are made chiefly by cracking processes, in which heavier petr fractions are converted into more volatile fractions by thermal or catalytic decompn; have also been made commercially by catalytic high-pressure hydrogenation of soft coal and by catalytic synthesis of hydrocarbons from carbon monoxide and hydrogen. Some gasolines sold in the U.S. contain a minor proportion of tetraethyllead added to motor gasoline to increase the octane numbers and thereby prevent "knock" in engines in which the gasoline is used as fuel. Knock is the audible manifestation of an excessive rate of pressure rise when the gasoline vapor is ignited under compression in an engine. The relative knocking tendencies of gasolines are measured in terms of "Octane Number," which is defined as the percentage of iso-octane, having "100 Octane No.," to be blended with *n*-heptane, having "0 Octane No." by definition, in order to obtain the same degree of knock as is obtained with the gasoline being rated, under standard conditions in a standardized test engine. Additives such as ethanol, methanol, benzene, toluene, MTBE and MMT, *q.q.v.*, are replacing tetraethyllead. *Review:* J. C. Lane in *Kirk-Othmer Encyclopedia of Chemical Technology* vol. 11 (Wiley-Interscience, New York, 3rd ed., 1984) pp 652-695. Review of toxicity: N. K. Weaver, *Ann. N.Y. Acad. Sci.* **534**, 441-451 (1988); of carcinogenicity: M. A. Mehlman, *Toxicol. Ind. Health* **7**, 143-152 (1990); of toxicology and human exposure: *Toxicological Profile for Gasoline* (PB95-264206, 1995) 224 pp. Symposium on toxicology, exposure and health effects: *Environ. Health Perspect.* **101**, Suppl. 6, 1-212 (1993).

Mobile liquid with characteristic odor; evaporates quickly. Flash pt 40-70°C. bp 32-210°C. *Flammable.* Explosive limits, vol % in air: lower 1.3, upper 6.0. Insol in water. Freely sol in abs alcohol, ether, chloroform, benzene. Dissolves fats, oils, natural resins. LD$_{50}$ orally in rats: 18.85 ml/kg (Weaver).

Caution: Potential symptoms due to mild overexposure by ingestion may include inebriation, vomiting, vertigo, drowsiness, confusion and fever; severe overexposure by ingestion may result in mild excitation, loss of consciousness, convulsions, cyanosis, congestion, capillary hemorrhaging of the lungs and internal organs, and death due to circulatory failure. Aspiration may cause severe inflammatory reaction and chemical pneumonitis. Direct contact may cause irritation of skin, eye, mucous membranes. Prolonged skin exposure may result in a chemical burn; repeated exposure may cause defatting of the skin. Potential symptoms due to acute overexposure by inhalation of vapor may include irritation and burning in the respiratory system, dizziness, anesthesia, bronchopneumonia, asphyxiation;

inhalation of vapor at high levels may induce CNS depression, coma, convulsions, myocardial irritability. Potential symptoms due to chronic overexposure by inhalation may include vomiting, diarrhea, insomnia, headache, dizziness, anemia, impairment of renal function, muscle and neurological symptoms. *See Patty's Industrial Hygiene and Toxicology* vol. **2B**, G. D. Clayton, F. E. Clayton, Eds. (Wiley-Interscience, New York, 4th ed., 1994) p 1395-1406; Weaver *loc. cit.* Potential occupational carcinogen. *See NIOSH Pocket Guide to Chemical Hazards* (DHHS/NIOSH 97-140, 1997) p 150.

USE: As fuel in internal combustion engines of the spark-ignited, reciprocating type; diluent; finishing agent; industrial solvent.

4406. Gastrins. Gastrointestinal hormones isolated from the mucosal lining of the gastric antrum of various mammalian species. Highly potent gastric secretion stimulants, first discovered by J. S. Edkins: *Proc. R. Soc. London* **76B**, 376 (1905). Several gastrins have been identified; they are referred to as "little gastrin", "big gastrin" and "minigastrin". "Little gastrin" or G-17 exists in two forms, 18-34-gastrin I or G-17-I and 18-34-gastrin II or G-17-II, which are heptadecapeptides that are identical in amino acid sequence, with the latter having sulfated tyrosine residues. There are relatively small species differences in amino acid sequences of G-17, although there are differences in the ratio of the non-sulfated to the sulfated forms. This ratio is about 3:4 in hog compared to about 2:1 in man. "Big gastrin" or 1-34-gastrin, also found in two forms, consists of the heptadecapeptides of "little gastrins" extended from their *N*-termini by additional heptadecapeptides with amino acid compositions different from G-17. For each species, the G-17-I and G-17-II portions of the G-34 gastrin are identical. Pairs of shorter gastrins, or "minigastrins" (22-34-gastrin) are C-terminal tetradecapeptides. Structure of porcine "little gastrins": Gregory *et al.*, *Nature* **204**, 391 (1964); human: Bentley *et al.*, *ibid.* **209**, 583 (1966). Synthesis of porcine "little gastrins": Anderson *et al.*, *ibid.* **204**, 933 (1964); human: Beecham *et al.*, *ibid.* **209**, 585 (1966); human and canine: Agarwal, Kenner, *J. Chem. Soc. C* **1969**, 2213; ovine: Agarwal *et al.*, *ibid.* 954; feline: *eidem*, *Experientia* **25**, 346 (1969). General synthetic method: E. Brown *et al.*, *Chem. Commun.* **1980**, 1093. Isoln of "big gastrins" from Zollinger-Ellison tumor tissue: R. A. Gregory, H. J. Tracy, *Lancet* **2**, 797 (1972). Isoln of "minigastrins" from Zollinger-Ellison tumor tissue: *eidem*, *Gut* **15**, 683 (1974). Amino acid sequences of porcine and human "big gastrins": *eidem*, in *Gastrointestinal Hormones*, J. C. Thompson, Ed. (Univ. Texas Press, Austin, 1975) pp 13-14. Revised sequence of porcine: G. J. Dockray *et al.*, *Bioorg. Chem.* **8**, 465 (1979); of human: A. M. Choudhury *et al.*, *Z. Physiol. Chem.* **361**, 1719 (1980). Synthesis of human: G. Wendlberger *et al.*, *Monatsh. Chem.* **112**, 1297 (1981). Structure and synthesis of "minigastrins": R. A. Gregory *et al.*, *Z. Physiol. Chem.* **360**, 73 (1979). Review of chemical studies: Kenner, Sheppard, *Proc. R. Soc. London* **170B**, 89 (1968). Review of physiological advances: Gregory, *ibid.* 81. General reviews: Sanders, Schimmel, *Am. J. Med.* **49**, 380 (1970); McGuigan, *Vitam. Horm.* **32**, 47-88 (1974); V. Mutt, *ibid.* **39**, 231-426 (1982).

5-oxo-Pro–Gly–Pro–Trp–Leu–Glu–Glu–Glu–Glu–Glu–Ala–Tyr–Gly–Trp–Met–Asp–PheNH$_2$

18-34-Gastrin I (human)

4407. Gastrodia. Chi-jian; red arrow. Parasitic herbaceous plant, *Gastrodia elata* Blume, *Orchidaceae*; used in Asian traditional medicine as an analgesic and anticonvulsant. Medicinal portion is the dried tuber, known as **tianma**. Growth is dependent on 2 fungi, *Armillaria mellea* for nutrients and *Mycena osmundicola* for seed germination. *Habit.* Woodlands of Japan, Korea and the central provinces of China. *Constit.* Gastrodin, vanillin, β-sitosterol, *q.q.v.*, vanillyl alcohol, 4-hydroxybenzyl alcohol, 4-hydroxybenzyl aldehyde, parishin, gastrodioside. Isoln of constituents: H. Taguchi *et al.*, *Chem. Pharm. Bull.* **29**, 55 (1981). HPLC determn of active constituents in tuber extracts: C. L. Liu *et al.*, *Chromatographia* **55**, 317 (2002). Effects on learning and memory in rats: C.-R. Wu *et al.*, *Planta Med.* **62**, 317 (1996); on amyloid β peptide-induced cell death: H.-J. Kim *et al.*, *J. Ethnopharmacol.* **84**, 95 (2003). Review of constituents and pharmacology: W. Tang, *Chinese Drugs of Plant Origin* (Springer-Verlag, Berlin, 1992) pp 545-548. Review of life cycle and cultivation: J. Xu, S. Guo, *Chin. Med. J.* **113**, 686-692 (2000).

4408. Gastrodin. [62499-27-8] 4-(Hydroxymethyl)phenyl-β-D-glucopyranoside; 4-(β-D-glucopyranosyloxy)benzyl alcohol. $C_{13}H_{18}O_7$; mol wt 286.28. C 54.54%, H 6.34%, O 39.12%. Pharmacologically active constituent of the traditional medicinal plant, gastrodia, q.v. Prepn: B. Helferich et al., Ann. **508**, 192 (1934); A. E. Pavlov et al., Russ. J. Gen. Chem. **71**, 1811 (2001). Isoln from Gastrodia elata tuber: H. Taguchi et al., Chem. Pharm. Bull. **29**, 55 (1981); H.-B. Li, F. Chen, J. Chromatogr. A **1052**, 229 (2004). CE determn in medicinal formulations: Y. Zhao et al., ibid. **849**, 277 (1999). Pharmacokinetics in rats: R. Wang et al., J. Liq. Chromatogr. Relat. Technol. **25**, 857 (2002). Pharmacology in gerbils: S.-J. An et al., J. Neurosci. Res. **71**, 534 (2003).

mp 169-170°. $[\alpha]_D^{20}$ −17.5° (c = 1 in pyridine). pKa 9.10. uv max (aq soln): 249 nm (log ε 4.140).

Hemihydrate. Colorless needles from methanol + ethyl acetate, mp 156-157°. $[\alpha]_D^{33}$ −62.1° (in ethanol). uv max (ethanol): 223, 273, 278 (log ε 3.91, 2.88, sh 2.78).

4409. Gatifloxacin. [112811-59-3] 1-Cyclopropyl-6-fluoro-1,4-dihydro-8-methoxy-7-(3-methyl-1-piperazinyl)-4-oxo-3-quinolinecarboxylic acid; Tequin; Zymar. $C_{19}H_{22}FN_3O_4$; mol wt 375.40. C 60.79%, H 5.91%, F 5.06%, N 11.19%, O 17.05%. Fluorinated quinolone antibacterial. Prepn: K. Masuzawa et al., **EP 230295**; eidem, **US 4980470** (1987, 1990 both to Kyorin); J. P. Sanchez et al., J. Med. Chem. **38**, 4478 (1995); of the sesquihydrate: T. Matsumoto et al., **US 5880283** (1999 to Kyorin). In vitro antibacterial activity: A. Bauernfeind, J. Antimicrob. Chemother. **40**, 639 (1997); H. Fukuda et al., Antimicrob. Agents Chemother. **42**, 1917 (1998). Clinical pharmacokinetics: M. Nakashima et al., ibid. **39**, 2635 (1995). Clinical study in urinary tract infection: H. Nito, 10th Mediterranean Congr. Chemother. **1996**, 327; in respiratory tract infection: S. Sethi, Expert Opin. Pharmacother. **4**, 1847 (2003).

Pale yellow prisms from methanol as hemihydrate, mp 162°.
Sesquihydrate. [180200-66-2] AM-1155. $C_{19}H_{22}FN_3O_4$·1½H_2O; mol wt 402.42.
THERAP CAT: Antibacterial.

4410. GDNF. Glial cell line-derived neurotrophic factor. Survival factor for dopaminergic neurons and spinal motoneurons. Prototype of a family of neurotrophic factors known as GDNF family ligands (GFL) which also includes **neurturin, persephin,** and **artemin.** Glycosylated, disulfide-linked homodimer; mol wt 40-45 kDa; structurally related to the TGF-β superfamily. Widely expressed in peripheral tissues and brain during embryonic and early postnatal development. Isoln from rat glial cells and prepn of recombinant human form: L. H. Lin et al., Science **260**, 1130 (1993). Review of pharmacology and therapeutic potential: P. A. Lapchak, Rev. Neurosci. **7**, 165-176 (1996); of neuroprotective and neurorestorative properties: D. M. Gash et al., Ann. Neurol. **44**, Suppl. 1, S121-S125 (1998). Analgesic effect in neuropathic pain models: T. J. Boucher et al., Science **290**, 124 (2000). Reviews: M. C. Bohn et al., Int. J. Dev. Neurosci. **18**, 679-684 (2000); M. S. Airaksinen, M. Saarma, Nat. Rev. Neurosci.. **3**, 383-394 (2002).
Liatermin. [188630-14-0] Methionyl neurotrophic factor (human glial-derived), dimer. Production in E. coli by recombinant

DNA technology: L. H. Lin et al., **US 6093802** (2000 to Amgen). Clinical evaluation in Parkinson's disease: S. S. Gill et al., Nat. Med. **9**, 589 (2003).

4411. Gefarnate. [51-77-4] (4E,8E)-5,9,13-Trimethyl-4,8,12-tetradecatrienoic acid (2E)-3,7-dimethyl-2,6-octadienyl ester; geranyl farnesylacetate; DA-688; Alsanate; Arsanyl; Dixnalate; Gefanil; Gefarnil; Gefarnyl; Gefulcer; Osteol; Salanil; Zackal. $C_{27}H_{44}O_2$; mol wt 400.65. C 80.94%, H 11.07%, O 7.99%. Prepn: Cardani et al., J. Med. Chem. **6**, 457 (1963); **BE 617994** (1962 to Ist. de Angeli). Metabolism: G. Coppi et al., Arzneim.-Forsch. **19**, 1519 (1969). Clinical evaluation in duodenal ulcer: C. R. Newman, D. A. Montgomery, Br. J. Clin. Pract. **27**, 85 (1973).

Slightly yellowish liquid with weak terpenic smell. $bp_{0.05}$ 165-168°. n_D^{20} 1.4900. uv max: 204 nm ($E_{1cm}^{1\%}$ 486). Sol in alc, ether, dimethylformamide, acetone, fatty oils. Practically insol in water, formamide, ethylene glycol, propylene glycol, glycerine.
THERAP CAT: Antiulcerative.

4412. Gefitinib. [184475-35-2] N-(3-Chloro-4-fluorophenyl)-7-methoxy-6-[3-(4-morpholinyl)propoxy]-4-quinazolinamine; 4-(3′-chloro-4′-fluoroanilino)-7-methoxy-6-(3-morpholinopropoxy)-quinazoline; ZD-1839; Iressa. $C_{22}H_{24}ClFN_4O_3$; mol wt 446.91. C 59.13%, H 5.41%, Cl 7.93%, F 4.25%, N 12.54%, O 10.74%. Selective epidermal growth factor receptor (EGFR)-tyrosine kinase inhibitor which blocks signal transduction pathways implicated in the proliferation and survival of cancer cells. Prepn: K. H. Gibson, **WO 9633980**; idem, **US 5770599** (1996, 1998 both to Zeneca). LC/MS/MS determn in biological samples: M. Zhao et al., J. Chromatogr. B **819**, 73 (2005). Clinical pharmacokinetics and tolerability: H. Swaisland et al., Clin. Pharmacokinet. **40**, 297 (2001). In vitro and in vivo inhibition of EGFR-expressing cancer lines: N. G. Anderson et al., Int. J. Cancer **94**, 774 (2001). Proposed mechanism of action: A. Hirata et al., Cancer Res. **62**, 2554 (2002). Clinical study in non-small cell lung cancer: M. G. Kris et al., J. Am. Med. Assoc. **290**, 2149 (2003). Review: C. L. Arteaga, D. H. Johnson, Curr. Opin. Oncol. **13**, 491-498 (2001). Review of clinical evaluations: D. Raben et al., Semin. Oncol. **29**, Suppl. 4, 37-46 (2002); of use in non-small cell lung cancer: K. Tamura, M. Fukuoka, Expert Opin. Pharmacother. **6**, 985-993 (2005).

White powder. Crystals from toluene, mp 119-120°. pKa 5.4, 7.2. Freely sol in glacial acetic acid, DMSO; sol in pyridine; sparingly sol in tetrahydrofuran; slightly sol in methanol, ethanol (99.5%), ethyl acetate, propan-2-ol, and acetonitrile.
THERAP CAT: Antineoplastic.

4413. Geissoschizoline. [18397-07-4] (16α)-Curan-17-ol; pereirine. $C_{19}H_{26}N_2O$; mol wt 298.43. C 76.47%, H 8.78%, N 9.39%, O 5.36%. From bark of Geissospermum vellosii Allem., Apocynaceae: Hesse, Ann. **202**, 141 (1880); Bertho, Moog, ibid. **509**, 241 (1934). Identity with pereirine: Bertho, Koll, Naturwis-

senschaften **48**, 49 (1961). Structure: Janot *et al.*, *Compt. Rend.* **250**, 4383 (1960); Bertho, Koll *Ber.* **94**, 2737 (1961). Synthesis: Hymon, Schmid, *Helv. Chim. Acta* **49**, 2067 (1966); of *dl*-form: Dadson, Harley-Mason, *Chem. Commun.* **1969**, 665; Harley-Mason, Taylor, *ibid.* **1970**, 812.

Coarse crystals from abs methanol, mp 142.5-143°. $[\alpha]_D^{21}$ +32° (ethanol). uv max in (ethanol): 245, 300 nm (log ε 3.93, 3.47). Sol in alcohol, chloroform, ether. Practically insol in water.

Methyl chloride. $C_{19}H_{26}N_2O.CH_3Cl$. Rods from abs methanol, mp 297°.

Methiodide. $C_{19}H_{26}N_2O.CH_3I$. Rods from abs methanol, mp 254°.

4414. Geissospermine. [427-01-0] ($\alpha R,1S,3aR,9S,11aS,$-$11bS,12S,13aS,14S$)-14-Ethyl-α-[(2S,3E,12bS)-3-ethylidene-1,2,3,-4,6,7,12,12b-octahydroindolo[2,3-a]quinolizin-2-yl]-2,3,11a,11b,-13,13a-hexahydro-12H-1,12-ethano-9H-[1,3]oxazino[3,4,5-lm]pyrrolo[2,3-d]carbazole-9-acetic acid methyl ester; (16R,19E)-19,20-didehydro-16-[(10β,13β,21β)-23-deoxy-21,22-dihydro-11-oxa-12,14-secostrychnidin-10-yl]corynan-17-oic acid methyl ester; 19,20-didehydro-16-(15-ethyl-3a,5,5a,7,8,13a-hexahydro-4H-4,6-ethano-1H,3H-[1,3]oxazino[3,4,5-lm]pyrrolo[2,3-d]carbazol-1-yl)-corynan-17-oic acid methyl ester; β,17-epoxy-α-(3-ethylidene-1,2,-3,4,6,7,12,12b-octahydroindole[2,3-a]quinolizin-2-yl)-curan-1-propanoic acid methyl ester. $C_{40}H_{48}N_4O_3$; mol wt 632.85. C 75.92%, H 7.65%, N 8.85%, O 7.58%. From bark of *Geissospermum vellosii* Allem., *Apocynaceae*: Hesse, *Ann.* **202**, 141 (1880). Structure: Puisieux, LeHir, *Compt. Rend.* **252**, 902 (1961); Janot, *Tetrahedron* **14**, 113 (1961). Crystal structure: A. Chiaroni, C. Riche, *Acta Crystallogr.* **35B**, 1820 (1979). Pharmacology: M. M. Aurousseau, *Ann. Pharm. Fr.* **19**, 175 (1961). On hydrolysis with HCl splits into geissoschizine and geissoschizoline, *q.v.*

Anhydr form, crystals from abs acetone, mp 213-214° (dec). $[\alpha]_D^{20}$ −101° (ethanol). uv max (methanol): 251, 285, 293 nm (log ε 4.10, 3.91, 3.90). Slightly sol in water, ether; sol in alcohol.

Dihydrate. mp 207° (sinters at 145°).

4415. Gelatin. [9000-70-8] Gelfilm; Gelfoam. Heterogeneous mixture of water-soluble proteins of high average mol wt. Derived from the denaturation and hydrolysis of collagen, *q.v.* Amino acid composition is variable: glycine or alanine account for one third to half of the amino acid residues; approx 25% is proline or hydroxyproline; close to 25% is basic or acidic; contains no tryptophan. Amino acid composition studies: R. E. Neuman, *Arch. Biochem.* **24**, 289 (1949); J. E. Eastoe, *Biochem. J.* **61**, 589 (1955). Surgical use in hemostasis: R. U. Light, H. R. Prentice, *J. Neurosurg.* **2**, 435 (1945). Review of the chemistry and structure of collagen with emphasis on its transformation to gelatin: A. Veis, *The Macromolecular Chemistry of Gelatin* (Academic Press, New York, 1964) 433 pp.

Review of therapeutic use as an embolic agent in neurologic disease: A. Berenstein, E. Russell, *Radiology* **141**, 105-112 (1981); of prepn, properties, and use in food and pharmaceutical industries: K. B. Djagny *et al.*, *Crit. Rev. Food Sci. Nutr.* **41**, 481-492 (2001).

Colorless or slightly yellow, transparent, brittle, practically odorless, tasteless sheets, flakes, or coarse powder. Amphoteric. Swells up and absorbs 5-10 times its weight of water to form a gel in solutions below 35-40°. Sol in hot water, glycerol, acetic acid. Insol in cold water, alc, chloroform, ether, organic solvents, and in fixed and volatile oils.

USE: As stabilizer, thickener and texturizer in food; manuf rubber substitutes, adhesives, cements, lithographic and printing inks, plastic compds, artificial silk, photographic plates and films, matches, light filters for mercury lamps; clarifying agent; in hectographic masters; sizing paper and textiles; for inhibiting crystn in bacteriology, for preparing cultures. Pharmaceutic aid (suspending agent; encapsulating agent; tablet binder; tablet and coating agent).

THERAP CAT: Hemostatic.

THERAP CAT (VET): Plasma expander; hemostasis (sponge).

4416. Geldanamycin. [30562-34-6] $C_{29}H_{40}N_2O_9$; mol wt 560.64. C 62.13%, H 7.19%, N 5.00%, O 25.68%. Benzoquinone ansamycin antibiotic with antitumor activity; produced by *Streptomyces hygroscopicus* var *geldanus* var *nova*. Binds to the ATP binding site of the molecular chaperone, heat shock protein 90 (Hsp90), inhibiting its folding and leading to cell cycle disruption. Isoln: C. De Boer *et al.*, *J. Antibiot.* **23**, 442 (1970). Structure: K. Sasaki *et al.*, *J. Am. Chem. Soc.* **92**, 7591 (1970). Biosynthesis: R. D. Johnson *et al.*, *ibid.* **96**, 3316 (1974). Total synthesis: M. B. Andrus *et al.*, *Org. Lett.* **4**, 3549 (2002); H.-L. Qin, J. S. Panek, *ibid.* **10**, 2477 (2008). Mechanism of action: L. Whitesell *et al.*, *Proc. Natl. Acad. Sci. USA* **91**, 8324 (1994). Review of Hsp90 binding, antitumor activity, and use in biochemical studies of heat shock proteins: L. Neckers *et al.*, *Invest. New Drugs* **17**, 361-373 (1999); H.-J. Ochel *et al.*, *Cell Stress Chaperones* **6**, 105-112 (2001); and role of derivatives in cancer therapy: J. R. Porter *et al.*, *Curr. Top. Med. Chem.* **9**, 1386-1418 (2009).

Yellow needles from diethyl ether, mp 252-255°; $[\alpha]_D^{25}$ +55° (c = 0.638 in chloroform) (De Boer). Also reported as orange-yellow solid, $[\alpha]_D^{23}$ +29.7° (c = 0.064 in chloroform) (Andrus).

4417. Gellan Gum. [71010-52-1] Native gellan gum; high acyl gellan gum; polysaccharide S-60; PS-60. Extracellular polysaccharide obtained by aerobic fermentation of *Pseudomonas elodea*. Anionic hydrocolloid composed of acetylated linear tetrasaccharide repeat unit β-D-glucose, β-D-glucuronic acid and α-L-rhamnose in a 2:1:1 molar ratio. Prepn of native form: K. S. Kang, G. T. Veeder, US **4326053**; of deacetylated form: K. S. Kang *et al.*, US **4326052** (both 1982 to Merck & Co.); R. Moorhouse *et al.*, *ACS Symp. Ser.* **150**, 111 (1981). Structural studies: M. A. O'Neill *et al.*, *Carbohydr. Res.* **124**, 123 (1983); P.-E. Jansson *et al.*, *ibid.* 135. Gelation properties: H. Grasdalen, O. Smidsrod, *Carbohydr. Polym.* **7**, 371 (1987). Texture profile analysis: G. R. Sanderson *et al.* in *Gums and Stabilizers for the Food Industry* vol. 4, G. O. Phillips *et al.*, Eds. (IRL Press, Oxford, 1988) pp 219-229. Determn in food gels: J. K. Baird, W. W. Smith, *Food Hydrocolloids* **3**, 407 (1989); in food products: H. D. Graham, *ibid.* 435 (1990). Use as agar substitute in culture media: D. Shungu *et al.*, *Appl. Environ. Microbiol.* **46**, 840 (1983). Physicochemical properties: M. Milas *et al.*, *Biopolymers*

30, 451 (1990). Rheological study: P. B. Deasy, K. J. Quigley, *Int. J. Pharm.* **73**, 117 (1991). Comparative review of gelling agents: G. R. Sanderson *et al., Cereal Foods World* **34**, 991-998 (1989). *Reviews:* G. R. Sanderson in *Food Gels*, P. Harris, Ed. (Elsevier Appl. Sci., London, 1990) pp 201-232; W. Gibson in *Thickening and Gelling Agents in Food*, A. Imeson, Ed. (Blackie Academic & Professional, London, 1992) pp 227-249.

Low Acyl form

Sol in hot or cold deionized water. Forms thermoreversible gels.

Low acyl purified gellan gum. Deacylated gellan gum; deacylated PS-60; Gelrite; Kelcogel. Unsubstituted tetrasaccharide. White to tan powder, mol wt ~0.5×10^6. Partially sol in cold water; aqueous dispersions dissolve >70°. Forms thermoreversible gels. Incompat with strong oxidizers. Bulk density ~50 lb/cu ft. LD_{50} orally in rats: >5000 mg/kg (Sanderson, 1990).

USE: In foods, pharmaceuticals, and personal care products as gelling, texturizing, stabilizing, film forming and suspending agent. For microbiological media and plant tissue culture.

4418. Gelsemine. [509-15-9] $C_{20}H_{22}N_2O_2$; mol wt 322.41. C 74.51%, H 6.88%, N 8.69%, O 9.92%. CNS stimulant from roots and rhizome of *Gelsemium sempervirens* (L.) Ait., *Loganiaceae.* Isoln: Gerrard, *Pharm. J.* **13**, 641 (1883); Moore, *J. Chem. Soc.* **97**, 2223 (1910); *ibid.* **99**, 1231 (1911); Sayre, Watson, *J. Am. Pharm. Assoc.* **8**, 708 (1919); Chou, *Chin. J. Physiol.* **5**, 131 (1931), *C.A.* **25**, 4085[6] (1931); Schwarz, Marion, *Can. J. Chem.* **31**, 958 (1953). Structure: Conroy, Chakrabarti, *Tetrahedron Lett.* **1959** [4] 6; Lovell *et al., ibid.* 1; Roe, Gates, *Tetrahedron* **11**, 148 (1960). NMR spectroscopic study: Y. Schun, G. A. Cordell, *J. Nat. Prod.* **48**, 969 (1985). Partial syntheses: W. E. Earley *et al., Tetrahedron Lett.* **29**, 3781, 3785 (1988).

Crystals from acetone, mp 178°. *Poisonous.* $[\alpha]_D^{20}$ +13° (c = 1.2 in chloroform). pKa 7.75 in 80% methylcellosolve. uv max (methanol): 210, 252, 280 nm (log ε 4.50, 3.87, 3.15). Slightly sol in water; sol in alcohol, benzene, chloroform, ether, acetone, dilute acids.

Hydrochloride. $C_{20}H_{23}ClN_2O_2$. Prisms from methanol + ether, mp 326°. $[\alpha]_D$ +5° (c = 1.072 in water). Sol in water, slightly sol in alcohol.

4419. Gelsemium. Yellow jasmine; yellow jessamine; wild woodbine. CNS stimulant from dried rhizome and roots of *Gelsemium sempervirens* (L.) Ait., *Loganiaceae. Habit.* Southern U.S. *Constit.* Gelsemine, gelsemoidine, scopoletin, gelsemic acid, volatile oil, resin.

Poisonous.

4420. Gemcitabine. [95058-81-4] 2'-Deoxy-2',2'-difluorocytidine; 1-(2-oxo-4-amino-1,2-dihydropyrimidin-1-yl)-2-deoxy-2,2-difluororibose; dFdC; dFdCyd; LY-188011; Gemzar. $C_9H_{11}F_2N_3O_4$; mol wt 263.20. C 41.07%, H 4.21%, F 14.44%, N 15.97%, O 24.31%. Prepn: L. W. Hertel, *GB 2136425; idem, US 4808614* (1984, 1989 both to Lilly); L. W. Hertel *et al., J. Org. Chem.* **53**, 2406 (1988); T. S. Chou *et al., Synthesis* **1992**, 565. Antitumor activity: L. W. Hertel *et al., Cancer Res.* **50**, 4417 (1990). Mode of action study: V. W. T. Ruiz *et al., Biochem. Pharmacol.* **46**, 762 (1993). Clinical pharmacokinetics and toxicity: J. L. Abbruzzese *et al., J. Clin. Oncol.* **9**, 491 (1991). Review of clinical studies: B. Lund *et al., Cancer Treat. Rev.* **19**, 45-55 (1993).

Crystals from water, pH 8.5. $[\alpha]_{365}$ +425.36°; $[\alpha]_D$ +71.51° (c = 0.96 in methanol). uv max (ethanol): 234, 268 (ε 7810, 8560). LD_{10} i.v. in rats: 200 mg/m² (Abbruzzese).

Hydrochloride. [122111-03-9] $C_9H_{11}F_2N_3O_4 \cdot HCl$. Crystals from water-acetone, mp 287-292° (dec). $[\alpha]_D$ +48°; $[\alpha]_{365}$ +257.9° (c = 1.0 in deuterated water). uv max (water): 232, 268 nm (ε 7960, 9360). Sol in water; slightly sol in methanol. Practically insol in alc, polar organic solvents.

THERAP CAT: Antineoplastic.

4421. Gemeprost. [64318-79-2] (2E,11α,13E,15R)-11,15-Dihydroxy-16,16-dimethyl-9-oxoprosta-2,13-dien-1-oic acid methyl ester; 16,16-dimethyl-*trans*-Δ^2-PGE_1 methyl ester; ONO-802; Cergem; Cervagem(e); Preglandin. $C_{23}H_{38}O_5$; mol wt 394.55. C 70.02%, H 9.71%, O 20.27%. Analog of prostaglandin E_1, *q.v.* Prepn: M. Hayashi *et al., DE 2700021; eidem, US 4052512* (both 1977 to Ono); H. Suga *et al., Prostaglandins* **15**, 907 (1978). Effects on uterine contractility and steroid hormone plasma levels: K. Oshima *et al., J. Reprod. Fertil.* **55**, 353 (1979). Effects on reproductive function: K. Matsumoto *et al., Nippon Yakurigaku Zasshi* **79**, 15 (1982), *C.A.* **96**, 98392 (1982). Use in termination of first trimester pregnancy: O. Reiertsen *et al., Prostaglandins Leukotrienes Med.* **8**, 31 (1982).

THERAP CAT: Abortifacient; oxytocic.

4422. Gemfibrozil. [25812-30-0] 5-(2,5-Dimethylphenoxy)-2,2-dimethylpentanoic acid; 2,2-dimethyl-5-(2,5-xylyloxy)valeric acid; CI-719; Decrelip; Genlip; Gevilon; Lipozid; Lipur; Lopid. $C_{15}H_{22}O_3$; mol wt 250.34. C 71.97%, H 8.86%, O 19.17%. Serum lipid regulating agent. Prepn: P. L. Creger, *DE 1925423; eidem, US 3674836* (1969, 1972, both to Parke, Davis). Production: O. P. Goel, *US 4126637* (1978 to Warner-Lambert). Pharmacology: A. H. Kissebach *et al., Atherosclerosis* **24**, 199 (1976); M. T. Kahonen *et al., ibid.* **32**, 47 (1979). Series of articles on metabolism, clinical pharmacology, kinetics and toxicology: *Proc. R. Soc. Med.* **69**, Suppl 2, 1-120 (1976). Toxicity data: S. M. Kurtz *et al., ibid.* 15. Clinical trial in hyperlipidemia: J. E. Lewis *et al., Pract. Cardiol.* **9**, 99 (1983). Clinical reduction of cardiovascular risk in patients with low HDL levels: H. B. Rubins *et al., N. Engl. J. Med.* **341**, 410 (1999).

Crystals from hexane, mp 61-63°. $bp_{0.02}$ 158-159°. Sol in alc, methanol, chloroform. Practically insol in water. LD_{50} in mice, rats (mg/kg): 3162, 4786 orally (Kurtz).

THERAP CAT: Antilipemic.

4423. Gemifloxacin. [175463-14-6] 7-[(4Z)-3-(Aminomethyl)-4-(methoxyimino)-1-pyrrolidinyl]-1-cyclopropyl-6-fluoro-1,4-dihydro-4-oxo-1,8-naphthyridine-3-carboxylic acid; SB-265805; LB-20304. $C_{18}H_{20}FN_5O_4$; mol wt 389.39. C 55.52%, H 5.18%, F

4.88%, N 17.99%, O 16.43%. Third generation fluorinated quinolone antibacterial. Prepn: J. H. Kwak *et al.*, **EP 688772**; C. Y. Hong *et al.*, **US 5633262** (1995, 1997 both to LG Chemical); *idem et al.*, *J. Med. Chem.* **40**, 3584 (1997); of methanesulfonate: A. R. Kim *et al.*, **WO 9842705** (1998 to LG Chemical). Antibacterial spectrum *in vitro*: R. Wise, J. M. Andrews, *J. Antimicrob. Chemother.* **44**, 679 (1999). LC/MS/MS determn in human plasma: E. Doyle *et al.*, *J. Chromatogr. B* **746**, 191 (2000). Clinical pharmacokinetics: A. Allen *et al.*, *Antimicrob. Agents Chemother.* **44**, 1604 (2000). Review of antibacterial activity, pharmacology, and clinical trials: J. M. Blondeau, B. Missaghi, *Expert Opin. Pharmacother.* **5**, 1117-1152 (2004).

Off-white, amorphous solid from chloroform-ethanol as the monohydrate, mp 235-237°.
Methanesulfonate. [210353-53-0] Gemifloxacin mesylate; Factive. $C_{18}H_{20}FN_5O_4 \cdot CH_3SO_3H$; mol wt 485.49. White to light brown solid, mp 195° (dec). Soly in water at 25° (mg/ml): 162 (pH 2); 19 (pH 6); 0.5 (pH 7).
THERAP CAT: Antibacterial.

4424. Gemtuzumab Ozogamicin. [220578-59-6] CDP-771; CMA-676; Mylotarg. Immunoconjugate of *N*-acetyl-γ-calicheamicin with the humanized mouse monoclonal antibody IgG4 κ antibody, hP67.6, directed against the human CD33 antigen located on the surface of normal and leukemic myeloid cells, but not on normal hematopoietic stem cells. Designed to deliver antibody-targeted chemotherapy for treatment of acute myelocytic leukemia (AML). Prepn of conjugate: P. R. Hamann *et al.*, **EP 689845**; *eidem*, **US 5773001** (1996, 1998 both to Am. Cyanamid). *In vitro* antineoplastic activity: K. Naito *et al.*, *Leukemia* **14**, 1436 (2000). Clinical evaluation in AML: E. L. Sievers *et al.*, *Blood* **93**, 3678 (1999). Reviews of development and therapeutic potential: I. D. Bernstein, *Leukemia* **14**, 474-475 (2000); I. Niculescu-Duvaz, *Curr. Opin. Mol. Ther.* **2**, 691-696 (2000); of clinical experience: F. J. Giles, *Expert Rev. Anticancer Ther.* **2**, 630-640 (2002).
Sensitive to light; protect from direct and indirect sunlight and unshielded fluorescent light.
THERAP CAT: Antineoplastic.

4425. Geneserine. [25573-43-7] (4a*S*,9a*S*)-2,3,4,4a,9,9a-Hexahydro-2,4a,9-trimethyl-1,2-oxazino[6,5-*b*]indol-6-ol 6-(*N*-methylcarbamate); physostigmine aminoxide; eseridine; eserine aminoxide; eserine oxide. $C_{15}H_{21}N_3O_3$; mol wt 291.35. C 61.84%, H 7.27%, N 14.42%, O 16.47%. From Calabar bean *(Physostigma venenosum* Balf., *Leguminosae):* Polonovski, Nitzberg, *Bull. Soc. Chim. Fr.* [5] **17**, 244 (1915); **21**, 191 (1917). Structure: Stedman, Barger, *J. Chem. Soc.* **127**, 247 (1925). Revised structure: Hootele, *Tetrahedron Lett.* **1969**, 2713. Stereochemistry: Robinson, Moorcroft, *J. Chem. Soc. C* **1970**, 2077. Total synthesis of racemate: K. Shishido *et al.*, *Chem. Commun.* **1986**, 904.

Rectangular prisms from ether, mp 129°. $[\alpha]_D^{15}$ −175° (alc). Weak base, just alkaline to litmus; does not form crystn salts with mineral acids. Almost insol in water. Sol in alc, chloroform, benzene, ether, petr ether, acetone, dil acids.

4426. Genistein. [446-72-0] 5,7-Dihydroxy-3-(4-hydroxyphenyl)-4*H*-1-benzopyran-4-one; 4′,5,7-trihydroxyisoflavone; prunetol; genisteol; sophoricol. $C_{15}H_{10}O_5$; mol wt 270.24. C 66.67%, H 3.73%, O 29.60%. Phytoestrogen with anti-proliferative activity; isoflavone predominantly found in soybean, trifolium (red clover), and other legumes. The aglucon of genistin and of sophoricoside; metabolite of biochanin A, *q.v.* Isoln from dyer's broom, *Genista tinctoria*: A. G. Perkin, F. G. Newbury, *J. Chem. Soc., Trans.* **75**, 830 (1899). Structure and identity with prunetol: W. Baker, R. Robinson, *ibid.* **127** 1981 (1925); *eidem*, *J. Chem. Soc.* **1926**, 2713. Synthesis: *eidem, ibid.* **1928**, 3115; R. L. Shriner, C. J. Hull, *J. Org. Chem.* **10**, 288 (1945). Isoln from soybean: E. Walz, *Ann.* **489**, 118 (1931); E. D. Walter, *J. Am. Chem. Soc.* **63**, 3273 (1941); by supercritical fluid extraction: J. M. A. Araujo *et al.*, *Food Chem.* **105**, 266 (2007). HPLC determn in biological fluids: B. F. Thomas *et al.*, *J. Chromatogr. B* **760**, 191 (2001). Review of synthesis and estrogenic activity: R. A. Dixon, D. Ferreira, *Phytochemistry* **60**, 205-211 (2002); of anticancer activity and therapeutic potential: M. H. Ravindranath *et al.*, *Adv. Exp. Med. Biol.* **546**, 121-165 (2004); of safety and carcinogenic risk: C. K. Taylor *et al.*, *Nutr. Rev.* **67**, 398-415 (2009).

Rectangular or six-sided rods from 60% alcohol. Dendritic needles from ether. mp 297-298° (slight decompn). Sol in the usual organic solvents, in dil alkalies with yellow color. Sparingly sol in cold alcohol, acetic acid. Practically insol in water. uv max: 262.5 nm (ε 138).
Sophoricoside. [152-95-4] Genistein-4′-glucoside. $C_{21}H_{20}O_{10}$. Isoln from the green pods of *Sophora japonica* L., *Leguminosae:* Charaux, Rabate, *Bull. Soc. Chim. Biol.* **20**, 454 (1938). Structure: Zemplén *et al., Ber.* **76**, 267 (1943). Synthesis: Bognár, Szabo, *Acta Chim. Acad. Sci. Hung.* **4**, 383 (1954); *Chem. Ind. (London)* **1954**, 518. Crystals from alcohol, mp 298°. $[\alpha]_D^{20}$ −47° (pyridine). $[\alpha]_D^{20}$ −32° (10% aq pyridine). uv max (abs ethanol): 262 nm. Sparingly sol in water, alc, acetic acid; more sol in hot alc, hot acetic acid; sol in pyridine, dil alkalies. Practically insol in ethyl acetate, acetone.
Genistin. [529-59-9] Genistein-7-*O*-β-D-glucoside. $C_{21}H_{20}O_{10}$. Synthesis: Zemplén, Farkas, *Ber.* **76B**, 1110 (1943). Pale yellow plates from 80% ethanol, mp 256°. $[\alpha]_D^{21}$ −28° (c = 0.6 in 0.02*N* NaOH); $[\alpha]_D^{26}$ −21.4° (pyridine). uv max (85% ethanol): 262.5 nm (a 90.5). Practically insol in cold water. Slightly sol in hot water, hot ethanol, hot methanol; sol in hot 80% ethanol, hot 80% methanol, hot acetone, pyridine.
USE: Dietary supplement. Chemical probe to explore signal transduction pathways.

4427. Gentamicin. [1403-66-3] Gentamycin. Antibiotic complex produced by fermentation of *Micromonospora purpurea* or *M. echinospora* and variants thereof: M. J. Weinstein *et al.*, *Antimicrob. Agents Chemother.* **1963**, 1. Isoln, purification and characterization: J. P. Rosselet *et al., ibid.* 14. Industrial pats.: G. M. Luedemann, M. J. Weinstein; Charney, **US 3091572**; **US 3136704** (1963, 1964, both to Schering). Consists of three closely related components, gentamicins C_1, C_2, C_{1a}, and also gentamicin A which differs from the other members of the complex but is similar to kanamycin C, *q.v.* Separation of gentamicin C components: H. Maehr, C. P. Schaffner, *J. Chromatogr.* **30**, 572 (1967); Wagman *et al., ibid.* **34**, 210 (1968). Structures contain 2-deoxystreptamine, *q.v.*, linked to two saccharide units, these being *garosamine* and a *purpurosamine* in the C series gentamicins. Structure studies: D. J. Cooper *et al.*, *J. Chem. Soc. C* **1971**, 960, 2876, 3126. Structure of gentamicin A: H. Maehr, C. P. Schaffner, *J. Am. Chem. Soc.* **89**, 6787 (1967); **92** 1697 (1969). Sepn and structures of gentamicins A_1 to A_4: Nagabhushan *et al., J. Org. Chem.* **40**, 2830, 2835 (1975). Synthetic studies: W. Meyer zu Reckendorf, Bischof, *Ber.* **105**, 2546 (1972); M. Chmielewski *et al., Carbohydr. Res.* **70**, 275 (1979). Review of activity, toxicity and clinical pharmacology: J.

Black *et al.*, *Antimicrob. Agents Chemother.* **1963**, 138-147. Comprehensive description: B. E. Rosenkrantz *et al.*, *Anal. Profiles Drug Subs.* **9**, 295-340 (1980). Determn in serum by immunoassay: H. A. Holt *et al.*, *J. Antimicrob. Chemother.* **34**, 747 (1994). Review of clinical use: S. B. Shrimpton *et al.*, *ibid.* **31**, 599-606 (1993). Clinical effect on chloride transport in cystic fibrosis: M. Wilschanski *et al.*, *Am. J. Respir. Crit. Care Med.* **161**, 860 (2000).

purpurosamine

2-deoxystreptamine

garosamine

Gentamicin C₁	R₁ = R₂ = CH₃
Gentamicin C₂	R₁ = CH₃, R₂ = H
Gentamicin C₁ₐ	R₁ = R₂ = H

Gentamicin C_1 $R_1 = R_2 = CH_3$
Gentamicin C_2 $R_1 = CH_3, R_2 = H$
Gentamicin C_{1a} $R_1 = R_2 = H$

White amorphous powder, mp 102-108°. $[\alpha]_D^{25}$ +146°. Freely sol in water; sol in pyridine, DMF, in acidic media with salt formation; moderately sol in methanol, ethanol, acetone. Practically insol in benzene, halogenated hydrocarbons.

Gentamicin C₁. [25876-10-2] $C_{21}H_{43}N_5O_7$; mol wt 477.60. mp 94-100°. $[\alpha]_D^{25}$ +158°.

Gentamicin C₂. [25876-11-3] $C_{20}H_{41}N_5O_7$; mol wt 463.58. mp 107-124°. $[\alpha]_D^{25}$ +160°.

Gentamicin C₁ₐ. [26098-04-4] *O*-3-Deoxy-4-*C*-methyl-3-(methylamino)-β-L-arabinopyranosyl-(1 → 6)-*O*-[2,6-diamino-2,3,-4,6-tetradeoxy-α-D-*erythro*-hexopyranosyl-(1 → 4)]-2-deoxy-D-streptamine; gentamicin D. $C_{19}H_{39}N_5O_7$; mol wt 449.55.

Gentamicin A. [13291-74-2] *O*-2-Amino-2-deoxy-α-D-glucopyranosyl-(1 → 4)-*O*-[3-deoxy-3-(methylamino)-α-D-xylopyranosyl-(1 → 6)]-2-deoxy-D-streptamine. $C_{18}H_{36}N_4O_{10}$; mol wt 468.50.

Hydrochloride. mp 194-209°. $[\alpha]_D^{25}$ +113°. Freely sol in water, methanol; slightly in ether. Practically insol in other organic solvents.

C complex sulfate. [1405-41-0] Alcomicin; Cidomycin; Duragentam; Garamycin; Garasol; Genoptic; Gentacin; Gent-Ak; Gentalline; Gentalyn; Gentibioptal; Genticin; Gentogram; Gent-Ophtal; Lugacin; Ophtagram; Pangram; Refobacin; Septopal; Sulmycin. White, hygroscopic powder, mp 218-237°. $[\alpha]_D^{25}$ +102°. Sol in ethylene glycol, formamide. LD₅₀ in mice (mg base/kg): 430 i.p.; 485 s.c.; 75 i.v.; >9050 orally (Black).

THERAP CAT: Antibacterial.

THERAP CAT (VET): Antibacterial.

4428. Gentian. Yellow gentian; pale gentian; bitter root. Dried rhizome and roots of *Gentiana lutea* L., *Gentianaceae*. *Habit.* Central and Southern Europe. *Constit.* Gentiin, gentiamarin, gentisin, gentisic acid, gentiopicrin, gentianose (a sugar), pectin.

THERAP CAT: Bitter tonic.

THERAP CAT (VET): Bitter tonic.

4429. Gentianine. [439-89-4] 5-Ethenyl-3,4-dihydro-1*H*-pyrano[3,4-*c*]pyridin-1-one; 4-(2-hydroxyethyl)-5-vinylnicotinic acid δ-lactone; erythricine. $C_{10}H_9NO_2$; mol wt 175.19. C 68.56%, H 5.18%, N 8.00%, O 18.26%. From *Gentiana kirilowi*, *Gentianaceae*: Proskurnina, *J. Gen. Chem. USSR* **14**, 1148 (1944), *C.A.* **40**, 7213 (1946); from *Anthocleista procera* Afz. and *Fagraea fragrans* Roxb., *Loganiaceae*: Lavie, Taylor-Smith, *Chem. Ind. (London)* **1963**, 781; Wan, Chow, *J. Pharm. Pharmacol.* **16**, 484 (1964). Isoln from *Slevogtia orientalis* Gris., *Gentianaceae*, structure and synthesis: Govindachari *et al.*, *J. Chem. Soc.* **1957**, 551, 2725.

Needles from ether or petr ether, mp 82-83°. uv max: 220 nm (log ε 4.38). Sol in alkali.

Hydrochloride. $C_{10}H_9NO_2$.HCl. Needles from alcohol + ether, dec 169-170°.

Dihydrogentianine. $C_{10}H_{11}NO_2$. Crystals from ether + petr ether, mp 74-76°. uv max: 270 nm (log ε 3.4).

4430. Gentian Violet. [548-62-9] *N*-[4-[Bis[4-(dimethylamino)phenyl]methylene]-2,5-cyclohexadien-1-ylidene]-*N*-methylmethanaminium chloride (1:1); C.I. Basic Violet 3; hexamethylpararosaniline chloride; hexamethyl-*p*-rosaniline chloride; aniline violet; crystal violet; methylrosanilinium chloride; C.I. 42555; Axuris; Badil; Gentiaverm; Pyoktanin. $C_{25}H_{30}ClN_3$; mol wt 407.99. C 73.60%, H 7.41%, Cl 8.69%, N 10.30%. Prepn and properties: *Colour Index* **vol. 4** (3rd ed., 1971) p 4391. Toxicity studies: H. C. Hodge *et al.*, *Toxicol. Appl. Pharmacol.* **22**, 1 (1972). Review of metabolism and mode of action: R. Docampo, S. N. Moreno, *Drug Metab. Rev.* **22**, 161-178 (1990); of use as a biological stain: *Conn's Biological Stains*, R. W. Horobin, J. A. Kiernan, Eds. (BIOS Scientific Publishers Ltd, Oxford, UK, 10th ed., 2002) 193-195.

Dark green powder or greenish, glistening pieces with metallic luster. Soly (%): water 0.2-1.7; ethanol 3-14; acetone 0.4; chloroform 5.1. Sol in glycerin. Insol in ether, xylene. Absorption max (water): 590 nm. Changes from yellow at pH 0.0 to blue-violet at pH 2.0. LD₅₀ orally in mice, rats: 1.2, 1.0 g/kg (Hodge).

Note: Commercial product, which is usually admixed with **pentamethylpararosaniline chloride** and **tetramethylpararosaniline chloride**, contains not less than 96% gentian violet.

USE: As dye for wood, silk, paper; in inks. Biological stain for bacterial components, vascular plant tissues, amyloid, etc. Used in Gram stain. As addition to microbiological culture media. Indicator for copper salts. Used to quench autofluorescence.

THERAP CAT: Anti-infective (topical). Has been used as anthelmintic (Nematodes). Blood additive to prevent transmission of Chagas' disease by blood transfusion.

THERAP CAT (VET): Antimicrobial (topical); mycostatic agent in poultry feed.

4431. Gentiobiose. [554-91-6] 6-*O*-β-D-Glucopyranosyl-D-glucose; 6-(β-D-glucosido)-D-glucose; amygdalose. $C_{12}H_{22}O_{11}$; mol wt 342.30. C 42.11%, H 6.48%, O 51.41%. From gentianose by partial hydrolysis with 0.2% H_2SO_4 or with invertin. From D-glucose by enzymatic synthesis with emulsin: Helferich, Lette, *Org. Synth.* **22**, 53 (1942). Prepn and structure: Haworth, Wylam, *J. Chem. Soc.* **123**, 3120 (1923); Hudson, *J. Am. Chem. Soc.* **51**, 1708 (1930). Structure: Hassid, Ballou in W. Pigman, *The Carbohydrates* (Academic Press, New York, 1957) p 492. Synthesis: Helferich, Klein, *Ann.* **450**, 219 (1926); Reynolds, Evans, *J. Am. Chem. Soc.* **60**, 2559 (1938). Hydrolysis with almond emulsin gives 2 mols D-glucose.

α-Form. Lentil-shaped crystals with 2CH$_3$OH from methanol. Bitter taste. Very hygroscopic. mp 86°. Shows mutarotation. $[\alpha]_D^{22}$ +16° (3 min) → +8.3° (3½ hrs c = 4). Sol in water, hot methanol, hot 90% alcohol.

β-Form. Anhydr crystals from alc, mp 190-195°. Shows mutarotation. $[\alpha]_D^{22}$ −5.9° (6 min) → +9.6° (6 hrs, c = 3). Sol in water, hot methanol, hot 90% alcohol.

4432. Gentiopicrin. [20831-76-9] (5R,6S)-5-Ethenyl-6-(β-D-glucopyranosyloxy)-5,6-dihydro-1H,3H-pyrano[3,4-c]pyran-1-one; gentiopicroside. C$_{16}$H$_{20}$O$_9$; mol wt 356.33. C 53.93%, H 5.66%, O 40.41%. The principal bitter glucoside of common gentians, which was isolated in 1862 from *Gentiana lutea*, L., *Gentianaceae*. Isoln from roots of *G. lutea*: Korte, *Ber.* **87**, 512 (1954); Korte *et al.*, *ibid.* **91**, 759 (1958). Structure: L. Canonica *et al.*, *Tetrahedron Lett.* **1**, no. 45, 7 (1960); *eidem*, *Tetrahedron* **16**, 192 (1961). Revised structure: H. Inouye *et al.*, *Tetrahedron Lett.* **9**, 4429 (1968). Absolute configuration: Manitto, Pagnoni, *Gazz. Chim. Ital.* **94**, 229 (1964).

Crystals from anhydr ethyl acetate or abs alcohol, mp 191°. $[\alpha]_D^{20}$ −199° (ethanol). uv max (c = 0.0285 g/l methanol): 270 nm (log ε 3.96).

Hemihydrate. Crystals from 50% ethanol, mp 121°.

Tetraacetate. C$_{24}$H$_{28}$O$_{13}$. Crystals from methanol. mp 138.5-139.5°. uv max (c = 0.0214 g/l methanol): 272 nm (log ε 3.84).

THERAP CAT: Antimalarial.

4433. Gentisic Acid. [490-79-9] 2,5-Dihydroxybenzoic acid; 5-hydroxysalicylic acid. C$_7$H$_6$O$_4$; mol wt 154.12. C 54.55%, H 3.92%, O 41.52%. Occurs in gentian: Redgrove, *Pharm. J.* **122**, 324 (1929). Found in urine of dogs after ingestion of salicylates: Neuberg, *Berl. Klin. Wochenschr.* **48**, 799 (1911). Prepd from hydroquinone: Senhofer, Sarlay, *Monatsh. Chem.* **2**, 448 (1881); Juch, *ibid.* **26**, 839 (1905); Brunner, *Ann.* **351**, 321 (1907); Zeltner, Landau, *DE 258887*; *Frdl.* **11**, 210; Dyson, *Manual of Organic Chemistry* vol. I (London, 1950) p 614; by oxidation of salicylic acid with potassium persulfate: *DE 81297* (to Schering AG); *Frdl.* **4**, 127; from *p*-hydroxyphenol: Meyer, *US 2588336* (1952 to Monsanto); from 5-bromo-2-hydroxybenzoic acid: Lowenthal, Pepper, *J. Am. Chem. Soc.* **72**, 3292 (1950); by Kolbe synthesis: Clemens, *US 2816137* (1957 to Eastman Kodak). Metabolic product of *Penicillium patulum*: Tanenbaum, Bassett, *Biochim. Biophys. Acta* **28**, 21 (1958); of *Polyporus tumulosus* Cooke: Crowden, Ralph, *Aust. J. Chem.* **14**, 475 (1961). Biosynthesis: Gatenback, Linnroth, *Acta Chem. Scand.* **16**, 2298 (1962).

Needles, monoclinic prisms from water, mp 199-200°. Crystals are dimorphic and undergo phase inversion upon heating. The stable phase begins to sublime at 200° and melts at 205°. pK (25°): 2.93.

Sol in water (about 1 part in 200 parts H$_2$O at 5°, much more sol in hot water), alcohol, ether. Practically insol in carbon disulfide, chloroform, benzene.

Sodium salt. [4955-90-2] Sodium gentisate; Gentinatre; Gentisod; Legential; Gentisine U.C.B. C$_7$H$_5$NaO$_4$; mol wt 176.10. Crystals with 5½H$_2$O. Rapidly loses 3H$_2$O on exposure to air, but holds ½H$_2$O tenaciously even at 100°. Sol in water.

THERAP CAT: Analgesic; anti-inflammatory.

4434. Gentisin. [437-50-3] 1,7-Dihydroxy-3-methoxy-9H-xanthen-9-one; 4,7-dihydroxy-2-methoxyxanthone; gentianic acid; gentianin; gentiin. C$_{14}$H$_{10}$O$_5$; mol wt 258.23. C 65.12%, H 3.90%, O 30.98%. From root of *Gentiana lutea* L., *Gentianaceae*: Henry, Caventou, *J. Pharm. Chim.* **1821**, 178; Canonica, Pelizzoni, *Gazz. Chim. Ital.* **85**, 1007 (1955). Structure: Korte, *Ber.* **87**, 1357 (1954). Synthesis: Anand, Venkataraman, *Proc. Indian Acad. Sci.* **25A**, 438 (1947); Rao, Seshadri, *ibid.* **37A**, 710 (1953).

Yellow needles from alc, mp 266-267°. Absorption max (methanol): 260, 275, 315, 410 nm (log ε 4.35, 4.30, 4.10, 3.70). Very slightly sol in water or organic solvents.

Diacetate. C$_{18}$H$_{14}$O$_7$. Crystals from alc, mp 196-197°. uv max (methanol): 240, 270, 300 nm (log ε 4.58, 4.05, 4.10).

4435. Geosmin. [19700-21-1] (4S,4aS,8aR)-Octahydro-4,8a-dimethyl-4a(2H)-naphthalenol; octahydro-4α,8aβ-dimethyl-4aα-(2H)-naphthol; 1,10-*trans*-dimethyl-*trans*-(9)-decalol. C$_{12}$H$_{22}$O; mol wt 182.31. C 79.06%, H 12.16%, O 8.78%. Major volatile component of beet essence, also determined to be the potent earthy odor contaminant of fish, beans, water: A. C. Thaysen, *Ann. Appl. Biol.* **23**, 99 (1936); J. Silvey, A. W. Roach, *J. Am. Water Works Assoc.* **56**, 60 (1964); A. A. Rosen *et al.*, *Water Treat. Exam.* **19** (Pt. 2), 106 (1970); R. G. Buttery *et al.*, *J. Agric. Food Chem.* **24**, 419, 1246 (1976). Isoln from *Actinomyetes*, algae: N. N. Gerber, H. A. Lechevalier, *Appl. Microbiol.* **13**, 935 (1965); A. A. Rosen *et al.*, *ibid.* **16**, 178 (1968); R. S. Safferman *et al.*, *Environ. Sci. Technol.* **1**, 429 (1967). Isoln from beets: K. E. Murray, P. A. Bannister, *Chem. Ind. (London)* **1975**, 973; L. D. Tyler *et al.*, *J. Agric. Food Chem.* **26**, 1466 (1978). Structure: N. N. Gerber, *Biotechnol. Bioeng.* **9**, 321 (1967); *eidem*, *Tetrahedron Lett.* **1968**, 2971. Biological degradation: L. V. Narayan, W. J. Nunez, *J. Am. Water Works Assoc.* **66** (Pt. 1), 532 (1974).

Colorless neutral oil, bp 270°. $[\alpha]_D^{25}$ −16.5°. Threshold odor conc: 0.1 ppb. Decomp in acid to odorless compds.

4436. Gephyrotoxin. [55893-12-4] (1R,3aR,5aR,6R,9aS)-Dodecahydro-6-(2Z)-2-penten-4-ynylpyrrolo[1,2-a]quinoline-1-ethanol; HTX D; histrionicotoxin D. C$_{19}$H$_{29}$NO; mol wt 287.45. C 79.39%, H 10.17%, N 4.87%, O 5.57%. Parent member of a class of tricyclic perhydrobenzoindolizine neurotoxin alkaloids isolated from the skin secretions of the Colombian poison-dart frogs *Dendrobates histrionicus* and *Dendrobates occultator*, family *Dendrobatidae*. The name gephyrotoxin was coined from the Greek word "gephyra" meaning "bridge", and refers to the bridge presumably formed by addition of the nitrogen function to one side of a 2,6-disubstituted piperidine precursor to form the bicyclic indolizine. Isolation of naturally occurring *l*-form: T. Tokuyama *et al.*, *Helv. Chim. Acta* **57**, 2597 (1974). Structure and absolute configuration: J. W. Daly *et al.*, *ibid.* **60**, 1128 (1977). Revised configuration: R. Fujimoto, Y. Kishi, *Tetrahedron Lett.* **1981**, 4197. Total synthesis of (±)-form: R. Fujimoto *et al.*, *J. Am. Chem. Soc.* **102**, 7154 (1980); D. J. Hart, *J. Org. Chem.* **46**, 3576 (1981); D. J. Hart, K. Kanai, *J. Am. Chem. Soc.* **105**, 1255 (1983); L. E. Overman *et al.*, *ibid.* 5373. Complete proton and ^{13}C NMR assignment: M. W. Edwards, A. Bax, *ibid.* **108**, 918 (1986). Effect on neuromuscular transmission: C. Souccar *et al.*, *Mol. Pharmacol.* **25**, 384, 395 (1984). Receptor binding study: R. S. Aronstam *et al.*, *Neurochem. Res.* **11**, 1227 (1986). Review including discussion of gephyrotoxin congeners: J. W. Daly, *Fortschr. Chem. Org. Naturst.* **41**, 283-300 (*see also* refs pp 326-340) (1982).

mp 231-232° (dec). uv max (ethanol): 225 nm (ε 8400). $[\alpha]_D^{25}$ −51.5° (c = 1 in ethanol).

4437. Geraniol. [106-24-1] (2*E*)-3,7-Dimethyl-2,6-octadien-1-ol; *trans*-3,7-dimethyl-2,6-octadien-8-ol; lemonol. $C_{10}H_{18}O$; mol wt 154.25. C 77.87%, H 11.76%, O 10.37%. An olefinic terpene alcohol constituting the chief part of oil of rose and oil of palmarosa; also found in many other essential oils such as citronela, lemon grass, etc. Isomeric with linalool. Isoln: Jacobsen, *Ann.* **157**, 234 (1871). Structure: Verley, *Bull. Soc. Chim. Fr.* **25**, 68 (1919); J. L. Simonsen, *The Terpenes* **vol. I** (University Press, Cambridge, 2nd ed., 1947) pp 40-52. Stereochemistry: Burrell *et al.*, *Proc. Chem. Soc. London* **1959**, 263; Bates *et al.*, *J. Org. Chem.* **28**, 1086 (1963). Synthesis: Burrell *et al.*, *J. Chem. Soc. C* **1966**, 2144; K. Takabe *et al.*, *Chem. Lett.* **1977**, 1025; K. K. Mathew *et al.*, *Indian J. Chem.* **B20**, 340 (1981).

Oily liq. Sweet rose odor. bp_{757} 229-230°; bp_{12} 114-115°. d_4^{20} 0.8894. n_D^{20} 1.4766. uv max: 190-195 nm (ε 18000). Practically insol in water. Miscible with alcohol, ether.

Acetate. $C_{12}H_{20}O_2$. Sweet, fragrant liq. bp ∼242° with decompn. d_{15}^{15} 0.9174. n_D^{13} 1.4628. Almost insol in water. Very sol in alcohol; miscible with ether.

Butyrate. $C_{14}H_{24}O_2$. Liquid. Characteristic fragrant odor. bp_{18} 152°. d_4^{17} 0.901. Almost insol in water. Sol in alc, ether.

Formate. $C_{11}H_{18}O_2$. Liquid. Odor of roses and of green rose leaves. bp_{15} 113-114°. d_4^{20} 0.927. Almost insol in alc, ether.

USE: In perfumery. As insect attractant. Butyrate for compounding artificial attar of rose. Formate as constituent of artificial neroli oil and of artificial orange blossom oil.

4438. Geranium. Cranesbill; storksbill; alum root. Dried rhizome of *Geranium maculatum* L., *Geraniaceae*. *Habit.* Canada and Eastern U.S., south to Georgia. *Constit.* 10-28% tannin, gallic, acid, resin, sugar, pectin.

THERAP CAT: Astringent.

THERAP CAT (VET): Astringent (intestinal).

4439. Geranylhydroquinone. [10457-66-6] 2-[(2*E*)-3,7-Dimethyl-2,6-octadien-1-yl]-1,4-benzenediol; *trans*-(3,7-dimethyl-2,6-octadienyl)hydroquinone; *trans*-1,4-dihydroxy-2-(3,7-dimethyl-2,6-octadienyl)benzene; geranyl-1,4-benzenediol; geroquinol; Béradia. $C_{16}H_{22}O_2$; mol wt 246.35. C 78.01%, H 9.00%, O 12.99%. Prepn and use: Baranger, **FR M2694** (1964), *C.A.* **61**, 15940e (1964). Radioprotective activity: G. Rudali, *C. R. Seances Soc. Biol. Ses Fil.* **160**, 1365 (1966). Effects on cancers in mice: G. Rudali, L. Menetrier, *Therapie* **22**, 895 (1967). Discovery as a potent contact allergen from trichomes of *Phacelia crenulata* var. *funerea* J. Voss, *Hydrophyllaceae*: G. Reynolds, E. Rodriguez, *Phytochemistry* **18**, 1567 (1979).

Colorless needles from *n*-hexane/ethyl acetate, mp 61-62°. *See:* H. Inouye *et al.*, *Ber.* **101**, 4057 (1968).

USE: Exptly as a radioprotective agent.

4440. Germane. [7782-65-2] Germanium hydride (GeH_4); germanium tetrahydride. GeH_4; mol wt 76.66. Ge 94.74%, H 5.26%. Prepd by the action of lithium aluminum hydride on a germanium halide in ether soln: Finholt *et al.*, *J. Am. Chem. Soc.* **69**, 2692 (1947); by reduction of GeO_2 by sodium hydoborate: Griffiths, *Inorg. Chem.* **2**, 375 (1963).

Colorless gas. mp −165°, bp −90°, d_4^{-142} 1.523. *Poisonous, flammable.* Slightly sol in hot hydrochloric acid, dec in nitric acid.

Caution: Potential symptoms of overexposure are malaise, headache, giddiness, fainting; dyspnea; nausea, vomiting; kidney injury; hemolytic effects. *See NIOSH Pocket Guide to Chemical Hazards* (DHHS/NIOSH 97-140, 1997) p 150.

4441. Germanium. [7440-56-4] Ge; at. wt 72.63; at. no. 32; valences 4, 2. Group IVA (14). Five naturally occurring isotopes: 70 (20.55%); 72 (27.37%); 73 (7.67%); 74 (36.74%); 76 (7.67%); artificial, radioactive isotopes: 65-69; 71; 75; 77; 78. Extent of occurrence in the earth's crust about 0.0007%. Predicted and called ekasilicon by Mendeléeff. Discovered in 1886 by Clemens Winkler: *J. Prakt. Chem.* **34**, 177 (1886). Obtained industrially from the flue dusts of smelters processing zinc-bearing ores: Jaffee *et al.*, *Trans. Electrochem. Soc.* **89**, 277 (1946). Purification by zone refining: Pfann, *J. Met.* **4**, 747 (1952). Physical properties: Hassion *et al.*, *J. Phys. Chem.* **59**, 1076 (1955). Inhalation toxicity studies: J. H. E. Arts *et al.*, *Food Chem. Toxicol.* **28**, 571 (1990). Review and description of modern isolation techniques: Pirest in L. P. Hunter, *Handbook of Semiconductor Electronics* (McGraw-Hill, New York, 1956), section 6. Comprehensive monograph: V. I. Davydov, *Germanium* (Gordon & Breach, New York, 1966) 417 pp. *Reviews:* Rochow in *Comprehensive Inorganic Chemistry* **vol. 2**, J. C. Bailar, Jr. *et al.*, Eds. (Pergamon Press, Oxford, 1973) pp 1-41; J. H. Adams in *Kirk-Othmer Encyclopedia of Chemical Technology* **vol. 11** (Wiley-Interscience, New York, 3rd ed., 1980) pp 791-802.

Grayish-white, lustrous, brittle metalloid. Diamond-cubic structure when cryst. Poor conductor of electricity. d_4^{25} 5.323. Reported melting points range from 925-975°; best value 937.2° (Hassion). Vol smaller by a few % when molten. bp 2700°. Thermal expansion coefficient (at ∼25°): 6.1×10^{-6}/°C. Thermal conductivity (at 25°): 0.14 cal/sec cm/°C. Specific heat (0-100°): 0.074 cal/g/°C. Lattice constant at 25°: 5.657×10^{-8} cm. Atoms/cc = 4.42×10^{22}. Volume compressibility: 1.3×10^{-12} cm²/dyn. Dielectric constant: 16. Covalent bond ionization energy at 0 K = 1.2 ev. Band gap: 0.67 ev. Impurity atom ionization energy: ∼0.01 ev. Intrinsic resistivity at 300 K = 47 ohm-cm. Electron mobility at 300 K = 3900 cm²/v sec. Hole mobility at 300 K = 1900 cm/v sec. Magnetic (mass) susceptibility ($\chi \times 10^6$) = −0.12. Intrinsic charge density at 300 K = 2.4×10^{13}. Electron diffusion constant at 300 K = 100. Hole diffusion constant at 300 K = 49. Insol in water, hydrochloric acid, dil alkali hydroxides. Attacked by aqua regia, concd nitric or sulfuric acids, fused alkalies, alkali peroxides, nitrates, or carbonates. Relatively stable, unaffected by air, becomes oxidized above 600°; is slowly oxidized by hydrogen peroxide at room temp, fairly rapidly at 90°; is attacked by hydrogen above 1000°. When finely divided, burns in chlorine or bromine.

USE: In electronics: Manuf rectifying devices (germanium diodes), transistors, in red-fluorescing phosphors; in dental alloys; in the production of glass capable of transmitting infrared radiation. Review of uses: Aldington, Cumming, *Endeavour* **14**, 200-204 (1955); *New Uses for Germanium*, F. I. Metz, Ed. (Midwest Research Institute, 1974) 120 pp.

THERAP CAT (VET): Astringent (intestinal).

4442. Germanium Dichloride. [10060-11-4] Germanium chloride ($GeCl_2$). Cl_2Ge; mol wt 143.53. Cl 49.40%, Ge 50.60%.

GeCl$_2$. Obtained as residue upon low temp distillation of GeHCl$_3$ (prepd by the reaction of hydrogen chloride with germanium monosulfide): Moulton, Miller, *J. Am. Chem. Soc.* **78**, 2702 (1956).

Unstable substance, dec into polymeric subchlorides even at low temps. Sol in ether, benzene.

4443. Germanium Dioxide. [1310-53-8] Germanium oxide (GeO$_2$); germanium(IV) oxide. GeO$_2$; mol wt 104.63. Ge 69.42%, O 30.58%.

White powder. Sol in about 250 parts cold water, 100 parts boiling water; sol in acids or in solns of the fixed alkalies. LD$_{50}$ i.p. in rats: 750 mg/kg; *see:* Rosenfeld, Wallace, *Arch. Ind. Hyg. Occup. Med.* **8**, 466 (1953).

4444. Germanium Tetrachloride. [10038-98-9] Tetrachlorogermane; germanium(IV) chloride. Cl$_4$Ge; mol wt 214.43. Cl 66.13%, Ge 33.87%. GeCl$_4$. Prepd from Ge and Cl$_2$ or GeO$_2$ and HCl: Bauer, Burschkies, *Ber.* **66**, 277 (1933); Foster *et al.*, *Inorg. Synth.* **2**, 109 (1946).

Mobile liq. Fumes in air. Peculiar, acidic odor, but can be distinguished from that of concd HCl. Appreciably volatile at room temp. d$_{20}^{20}$ 1.879. bp$_{760}$ 83.1°. mp −49.5°. Hydrolyzed by water with a crackling noise. Is stable, but not very sol in 6N HCl. If more concd HCl is used, the soly of GeCl$_4$ is reduced. Sol in benzene, ether, other organic solvents.

Caution: Fumes irritating to eyes, mucous membranes of respiratory tract.

4445. Germanium Tetrafluoride. [7783-58-6] Tetrafluorogermane. F$_4$Ge; mol wt 148.62. F 51.13%, Ge 48.87%. GeF$_4$. Prepd by heating BaGeF$_6$ to around 700°: Dennis, Laubengayer, *Z. Phys. Chem.* **130**, 520 (1927); Dennis, *Z. Anorg. Allg. Chem.* **174**, 119 (1928); Biltz, *ibid.* **207**, 65 (1932); Hoffman, Gutowsky, *Inorg. Synth.* **4**, 147 (1953).

Colorless gas. Odor of garlic. Thermally stable up to about 1000°. d^0 (liq) 2.162; d (solid; −195°) 3.148. mp −15° (under 3032 mm pressure). Sublimes at −36.5°. On contact with water, hydrolyzes to GeO$_2$ and H$_2$GeF$_6$. Does not attack glass when absolutely dry. Corrodes mercury, grease.

Trihydrate. Obtained by slow evaporation of a soln of germanium dioxide in 20% HF. Sol in water.

Caution: Highly irritating to eyes, skin, lungs, mucous membranes.

4446. Germine. [508-65-6] (3β,4α,7α,15α,16β)-4,9-Epoxycevane-3,4,7,14,15,16,20-heptol. C$_{27}$H$_{43}$NO$_8$; mol wt 509.64. C 63.63%, H 8.50%, N 2.75%, O 25.11%. Alkamine present in many polyester alkaloids which occur in *Veratrum* and *Zygadenus* species. Isoln: Poethke, *Pharm. Monatsh.* **18**, 77 (1937); *idem, Arch. Pharm.* **275**, 571 (1937); Seiferle *et al.*, *J. Econ. Entomol.* **35**, 35 (1942); Klohs *et al.*, *J. Am. Chem. Soc.* **75**, 4925 (1953); Kupchan, Deliwala, *ibid.* **76**, 5545 (1954); Myers *et al.*, *ibid.* **77**, 3348 (1955); **78**, 1621 (1956). Structure and configuration: Kupchan, Narayanan, *ibid.* **81**, 1913 (1959). Comparative toxicity: O. Krayer *et al.*, *J. Pharmacol. Exp. Ther.* **82**, 167 (1944).

Crystals from methanol, mp 221.5-223°. [α]$_D^{25}$ +4.5° (95% ethanol); [α]$_D^{16}$ +23.1° (c = 1.13 in 10% acetic acid). Sol in chloroform,

methanol, ethanol, acetone, water; slightly sol in ether. LD$_{50}$ i.v. in mice: 139.0 mg/kg (Krayer).

3-Acetate. C$_{29}$H$_{45}$NO$_9$. Needles from ether, mp 219-221°. [α]$_D^{23}$ +10° (c = 1.05 in pyridine).

16-Acetate. C$_{29}$H$_{45}$NO$_9$. Crystals from chloroform, mp 225-227°. [α]$_D^{23}$ −19° (c = 0.93 in pyridine).

3,4,7,15,16-Pentaacetate. C$_{37}$H$_{53}$NO$_{13}$. Prisms from acetone + petr ether, dec 285-287°. [α]$_D^{23}$ −65° (c = 0.65 in pyridine).

4447. Gestodene. [60282-87-3] (17α)-13-Ethyl-17-hydroxy-18,19-dinorpregna-4,15-dien-20-yn-3-one; 17α-ethynyl-17β-hydroxy-18-methyl-4,15-estradien-3-one; 17α-ethynyl-13-ethyl-17β-hydroxy-4,15-gonadien-3-one; SH B 331. C$_{21}$H$_{26}$O$_2$; mol wt 310.44. C 81.25%, H 8.44%, O 10.31%. Orally active gestogen with progesterone-like profile of activity. Prepn: **BE 847090**; H. Hofmeister *et al.*, **US 4081537** (1977, 1978 both to Schering AG). Pharmacokinetics: B. Duesterberg *et al.*, *Contraception* **24**, 673 (1981); B. Duesterberg, *Steroids* **43**, 43 (1984). Comparative study of anti-aldosterone activity: W. Losert *et al.*, *Arzneim.-Forsch.* **35**, 459 (1985). Clinical trial as oral contraceptive: G. Hoppe, *Adv. Contracep.* **3**, 159 (1987).

Crystals from acetone-hexane, mp 197.9°.
Mixture with ethinyl estradiol. [109852-02-0] Femodene; Femovan; Ginoden; Gynera; Harmonet; Meliane; Milvane; Minulet; Phaeva; Tri-Minulet.

THERAP CAT: Progestogen. In combination with estrogen as oral contraceptive.

4448. Gestonorone Caproate. [1253-28-7] 17-[(1-Oxohexyl)oxy]-19-norpregn-4-ene-3,20-dione; 17-hydroxy-19-norpregn-4-ene-3,20-dione hexanoate; 17α-hydroxy-19-norprogesterone caproate; 17β-acetyl-17-hydroxyestr-4-ene-3-one hexanoate; gestronol caproate; SH-582; Depostat; Primostat. C$_{26}$H$_{38}$O$_4$; mol wt 414.59. C 75.32%, H 9.24%, O 15.44%. Prepn: Popper *et al.*, *Arzneim.-Forsch.* **19**, 352 (1969); Popper *et al.*, **DE 1074582** (1960 to Schering). Pharmacological studies: Junkmann, *Anglo-Ger. Med. Rev.* **1**, 385 (1962). Metabolism: Breuer, Lisboa, *Acta Endocrinol.* **51**, 114 (1966).

Crystals, mp 123-124°. [α]$_D$ +13° (chloroform). uv max: 239 nm (ε 17540).

THERAP CAT: Progestogen; in treatment of prostatic hypertrophy.

4449. Gestrinone. [16320-04-0] (17α)-13-Ethyl-17-hydroxy-18,19-dinorpregna-4,9,11-trien-20-yn-3-one; 13β-ethyl-17α-ethynyl-17β-hydroxy-4,9,11-gonatrien-3-one; 13β-ethyl-17α-ethynyl-Δ^{4,9,11}-gonatriene-17β-ol-3-one; ethylnorgestrienone; A-46745; R-2323; RU-2323; Dimetriose; Dimetrose; Nemestran; Tridomose. C$_{21}$H$_{24}$O$_2$; mol wt 308.42. C 81.78%, H 7.84%, O 10.37%. Steroidal antiestrogen, antiprogestogen, analog of norgestrienone, *q.v.* Prepn: G. Nomine *et al.*, **US 3257278**; **NL 6607609**, *C.A.* **67**, 44029d (1967); D. Bertin, A. Pierdet, **US 3478067** (1966, 1969 all to Roussel-UCLAF). Radioimmunoassay in human plasma: J. Frick *et al.*, *Urol. Res.* **5**, 55 (1977). HPLC-MS-MS determn in plasma: Q. Wang *et al.*, *J. Chromatogr. B* **746**, 151

(2000). Clinical evaluation of contraceptive efficacy: G. Azadian-Boulanger *et al.*, *Am. J. Obstet. Gynecol.* **125**, 1049 (1976); of use in fibrocystic breast disease: E. M. Coutinho, G. Azadian-Boulanger, *Int. J. Gynaecol. Obstet.* **22**, 363 (1984); in endometriosis: E. J. Thomas, I. D. Cooke, *Br. Med. J.* **294**, 272 (1987); E. M. Coutinho, G. Azadian-Boulanger, *Fertil. Steril.* **49**, 418 (1988).

Crystals from ethyl acetate and benzene-cyclohexane (1:1), mp 154°. $[\alpha]_D^{20}$ +84.6° (c = 0.41 in methanol).

THERAP CAT: Antigonadotropin.

4450. Ghatti Gum. [9000-28-6] Gum Ghatti; Indian gum. The gummy exudate from stems of *Anogeissus latifolia* Wall., *Combretaceae*, abundant in India and Ceylon, *cf.* C. L. Mantell, *The Water-Soluble Gums* (New York, 1947). Name derived from the word *ghats*, meaning passes, and given to the gum because of its ancient mountain transportation routes. Structure is a complex, water-soluble, polysaccharide occurring as a calcium-magnesium salt. Composed of L-arabinose, D-galactose, D-mannose, D-xylose, D-glucuronic acid, in a molar ratio of 10:6:2:1:2, and traces of 6-deoxyhexose: Aspinall *et al.*, *J. Chem. Soc.* **1955**, 1160. Early investigation of chemistry and mol wt: Shaw *et al.*, *Proc. S. D. Acad. Sci.* **15**, 46 (1935); **16**, 34 (1936); **17**, 27 (1937); **19**, 130 (1939); **21**, 78 (1941). *Review:* Meer *et al.*, in *Industrial Gums*, R. L. Whistler, Ed. (Academic Press, New York, 2nd ed., 1973) pp 265-271.

Forms a very viscous mucilage, more viscous but less adhesive than acacia. Insol in 90% alcohol. $[\alpha]_D^{25}$ +42° (dil H_2SO_4). Gum ghatti solns may be colored slightly due to traces of pigment remaining in the gum. Does not form a true gel. Autoclaving will make all of the gum water-sol. The *U.S. Dispensatory* (24th ed.) states that gum Ghatti suitable as clinical laboratory reagent is entirely sol in 5 parts of cold water.

USE: As substitute for acacia. As emulsifying agent in pharmaceuticals, oils, waxes.

4451. Ghi. Ghee; samli; clarified butter. Prepd from cream or butter by melting and heating to 122° for about 25 mins. The water evaporates, and lactose, salt, and albuminous substances sink to the bottom. This treatment kills existing microorganisms by heat and deprives incoming microorganisms of moisture and necessary nutrients.

Properly clarified butter is of finely grained consistency, yellow color, mp 30° and retains most of the vitamin A content of the original butter. Keeps much longer than ordinary butter and contains less than 0.7% water. Used in India instead of butter.

4452. GHK. [49557-75-7] Glycyl-L-histidyl-L-lysine; liver cell growth factor; GHL. $C_{14}H_{24}N_6O_4$; mol wt 340.38. C 49.40%, H 7.11%, N 24.69%, O 18.80%. Naturally occuring tripeptide found in plasma. Spontaneously forms complexes with copper and other transition metals; involved in the physiological transport and uptake of copper ions. Stimulates synthesis of collagen, glycosaminoglycans, and extracellular matrix proteins; has growth promoting and wound healing properties. Isoln from human serum: L. Pickart, M. M. Thaler, *Nature New Biol.* **243**, 85 (1973). Confirmation of peptide structure: D. H. Schlesinger *et al.*, *Experientia* **33**, 324 (1977). Characterization of copper complexes: S. Lau, B. Sarkar, *Biochem. J.* **199**, 649 (1981); J. H. Freedman *et al.*, *Biochemistry* **21**, 4540 (1982); C. Conato *et al.*, *Biochim. Biophys. Acta* **1526**, 199 (2001). HPLC determn in plasma: T. Endo *et al.*, *J. Chromatogr. B* **692**, 37 (1997). Overview of use in cell culture systems: L. Pickart, S. Lovejoy, *Methods Enzymol.* **147B**, 314-328 (1987). Effect of copper complex on wound healing: F.-X. Maquart *et al.*, *J. Clin. Invest.* **92**, 2368 (1993); A. Siméon *et al.*, *J. Invest. Dermatol.* **115**, 962 (2000). Veterinary trial in rabbits: I. T. Cangul *et al.*, *Vet. Dermatol.* **17**, 417

(2006). Membrane permeability study: L. Mazurowska, M. Mojski, *Talanta* **72**, 650 (2007).

Hygroscopic. pKa 2.91, 6.53, 7.93, 10.44.

Acetate. [72957-37-0] $C_{14}H_{24}N_6O_4 \cdot C_2H_4O_2$; mol wt 400.44. Crystals from methanol + acetic acid, mp 146-148°.

Complex with copper. [89030-95-5] [Glycyl-κN-L-histidyl-κN,κN^3-L-lysinato(2−)] copper; GHK-Cu. $C_{14}H_{22}CuN_6O_4$; mol wt 401.91. Dark purple-blue, octahedral crystals from water + ethanol.

Bis complex with copper. [130120-56-8] Hydrogen [N^2-(glycyl-L-histidyl-κN^3)-L-lysinato][N^2-(glycyl-κN-L-histidyl-κN^3)-L-lysinato(2−)]cuprate(1−); PC-1020. $C_{28}H_{46}CuN_{12}O_8$; mol wt 742.30.

Bis complex with copper diacetate. [130120-57-9] Hydrogen (glycyl-L-histidyl-κN^1-L-lysinato)[glycyl-κN-L-histidyl-κN,κN^3-L-lysinato(2−)]cuprate(1−) diacetate; prezatide copper acetate; Iamin. $C_{28}H_{46}CuN_{12}O_8 \cdot 2C_2H_4O_2$; mol wt 862.40. Clinical trial for diabetic ulcers: G. D. Mulder *et al.*, *Wound Repair Regen.* **2**, 259 (1994).

USE: Growth factor in cell culture systems. In cosmetics as anti-aging ingredient.

THERAP CAT: Vulnerary.

THERAP CAT (VET): Vulnerary.

4453. Ghrelin. [304853-26-7] Orexigenic hormone secreted primarily by the stomach and duodenum; implicated in meal-time hunger and long-term regulation of body weight. Identified as the endogenous ligand for the growth hormone secretagogue receptor (GHSR). Released in response to fasting, resulting in appetite stimulation by the hypothalamus and growth hormone secretion by the pituitary. Contains 28 amino acid residues of which one is acylated by *n*-octanoic acid. Octanoylation is essential for bioactivity, possibly to facilitate passage across the blood-brain barrier. Isoln from rat stomach and cloning and synthesis of human ghrelin: M. Kojima *et al.*, *Nature* **402**, 656 (1999). Effect on growth hormone release in humans: E. Arvat *et al.*, *J. Endocrinol. Invest.* **23**, 493 (2000). Structure-activity study: M. A. Bednarek *et al.*, *J. Med. Chem.* **43**, 4370 (2000). Identification of stomach as primary source and effect of feeding state on plasma levels: H. Ariyasu *et al.*, *J. Clin. Endocrinol. Metab.* **86**, 4753 (2001). Clinical role in meal initiation: D. E. Cummings *et al.*, *Diabetes* **50**, 1714 (2001); in regulation of body weight: *idem et al.*, *N. Engl. J. Med.* **346**, 1623 (2002). *Reviews:* A. Inui, *Nat. Rev. Neurosci.* **2**, 1-10 (Aug. 2001); G. Wang *et al.*, *Regul. Pept.* **105**, 75-81 (2002); M. Kojima, K. Kangawa, *Physiol. Rev.* **85**, 495-522 (2005).

Human ghrelin. [258279-04-8] Glycyl-L-seryl-*O*-(1-oxooctyl)-L-seryl-L-phenylalanyl-L-leucyl-L-seryl-L-prolyl-L-α-glutamyl-L-histidyl-L-glutaminyl-L-arginyl-L-valyl-L-glutaminyl-L-glutaminyl-L-arginyl-L-lysyl-L-α-glutamyl-L-seryl-L-lysyl-L-lysyl-L-prolyl-L-prolyl-L-alanyl-L-lysyl-L-leucyl-L-glutaminyl-L-prolyl-L-arginine; ghrelin (human clone CTB-187P1 gene GHRELIN). Recombinant form is identical to the naturally occuring human peptide.

4454. Gibberellic Acid. [77-06-5] (1α,2β,4aα,4bβ,10β)-2,-4a,7-Trihydroxy-1-methyl-8-methylenegibb-3-ene-1,10-dicarboxylic acid 1,4a-lactone; gibberellin X; gibberellin A$_3$; Activol; Gibrel. $C_{19}H_{22}O_6$; mol wt 346.38. C 65.88%, H 6.40%, O 27.71%. Plant hormone; most outstanding of the plant-growth promoting metabolites of *Gibberella fujikuroi*. Isoln: P. J. Curtis, B. E. Cross, *Chem. Ind. (London)* **1954**, 1066; B. E. Cross, *J. Chem. Soc.* **1954**, 4670; P. W. Brian *et al.*, US 2842051; C. T. Calam, P. J. Curtis, US 2950288; A. J. Birch *et al.*, US 2977285 (1958, 1960, 1961, all to ICI). Stereochemistry and structure: G. Stork, H. Newman, *J. Am. Chem. Soc.* **81**, 5518 (1959); B. E. Cross *et al.*, *Proc. Chem. Soc. London* **1959**, 302; F. McCapra *et al.*, *ibid.* **1962**, 185; D. C. Aldridge *et al.*, *J.*

Chem. Soc. **1963**, 143; P. M. Bourn *et al.*, *ibid.* **1963**, 154. Partial synthesis: E. J. Corey *et al.*, *J. Am. Chem. Soc.* **93**, 7316 (1971). Stereospecific total synthesis: *eidem, ibid.* **100**, 8034 (1978). Promotes growth of seedlings: M. J. Bukovac, S. H. Wittwer, *Q. Bull. Mich. Agric. Exp. Stn.* **39**, 307 (1956); J. M. Merritt, *J. Agric. Food Chem.* **6**, 184 (1958). *Reviews:* B. E. Cross *et al.* in *Adv. Chem. Ser.* **28**, entitled "Gibberellins," J. M. Merritt, Ed. (ACS, Washington DC, 1961) p 13; series of articles in *Plant Growth Subst., Proc. 7th Int. Conf.*, D. J. Carr, Ed. (Springer-Verlag, Berlin, 1972).

Crystals from ethyl acetate, mp 233-235° (effervescence). $[\alpha]_D^{19}$ +86° (c = 2.12). pK 4.0. Slightly sol in water, ether. Freely sol in methanol, ethanol, acetone. Moderately sol in ethyl acetate. Sol in aq solns of sodium bicarbonate and sodium acetate.

Methyl ester. $C_{20}H_{24}O_6$. Needles from benzene + methanol, mp 209-210°. $[\alpha]_D^{20}$ +75° (c = 0.5).

USE: Plant growth regulator.

4455. Gibberellins. GAs. A class of plant growth hormones first isolated in 1938 from cultures of *Gibberella fujikuroi* (Sawada) Wollenweber *(Fusarium moniliforme* Sheldon) the fungus causing Bakanae disease in rice: Yabuta, Sumiki, *J. Agric. Chem. Soc. Jpn.* **14**, 1526 (1938). Isolated also from higher plants; for source references see review by Lang, *Annu. Rev. Plant Physiol.* **21**, 537 (1970). More than 60 gibberellins are known of which gibberellin A_3, *q.v.*, is the most important. GA_3 and mixtures of GA_4 and GA_7 are available commercially. All gibberellins are diterpenoid acids based on the *gibberellane* skeleton containing the *gibbane* nucleus. Major structural differences lie in the substituents at positions 4a, 7, 8 (gibbane numbering) and the presence or absence of a γ-lactone ring. Total synthesis of racemic gibberellins A_2, A_4, A_9, A_{10}: Mori *et al.*, *Tetrahedron* **25**, 1293 (1969); of gibberellin A_4: A. L. Cossey *et al.*, *Tetrahedron Lett.* **1980**, 4383; of gibberellin A_{15}: Nagata *et al.*, *J. Am. Chem. Soc.* **93**, 5740 (1971); of gibberellins A_{15} and A_{37}: E. Fujita *et al.*, *J. Chem. Soc. Perkin Trans. 1* **1977**, 611. Stereochemistry: Meguro, Fuzimura, *Tetrahedron Lett.* **1968**, 6305. Biosynthesis: Cross *et al.*, *J. Chem. Soc. C* **1968**, 1054; Shechter, West, *J. Biol. Chem.* **244**, 3200 (1969). Nomenclature: MacMillan, Takahasni, *Nature* **217**, 170 (1968). *Reviews:* Brian *et al.*, *Fortschr. Chem. Org. Naturst.* **18**, 350 (1960); Paleg, *Annu. Rev. Plant Physiol.* **16**, 291 (1965); *Outlook Agric.* **7**, 14 (1972); Cleland in *The Physiology of Plant Growth and Development*, M. B. Wilkins, Ed. (McGraw-Hill, New York, 1969) pp 49-81; L. Rappaport, "Applications of Gibberellins in Agriculture", in *Plant Growth Subst., Proc. 10th Int. Conf.*, F. K. Skoog, Ed. (Springer-Verlag, Berlin, 1980) pp 377-391; I. D. Railton, *Cell Biol. Int. Rep.* **6**, 319-337 (1982). Comprehensive synthetic review: E. Fujita, M. Node, *Heterocycles* **7**, 709 (1977).

Gibbane   Gibberellane

USE: Plant growth hormone.

4456. Gibbs Reagent. [101-38-2] 2,6-Dichloro-4-(chloroimino)-2,5-cyclohexadien-1-one; 2,6-dichloro-*p*-benzoquinone-4-chloroimine; *N*,2,6-trichloro-*p*-quinoneimine; 2,6-dichloroquinone chloroimide; *N*,2,6-trichloroquinoneimine; *N*,2,6-trichlorobenzoquinone imine. $C_6H_2Cl_3NO$; mol wt 210.44. C 34.25%, H 0.96%, Cl 50.54%, N 6.66%, O 7.60%. Reagent used to determine the presence of phenols. The reaction with phenols unsubstituted in the *para* position is called the Gibbs Reaction: H. D. Gibbs, *Chem. Rev.* **3**, 291 (1927). Prepn of the reagent: *idem, J. Biol. Chem.* **72**, 649 (1927); G. I. Mikhailov, *Trans. Inst. Pure Chem. Reagents* **16**, 83 (1939). Analysis of the Gibbs color reaction: J. C. Dacre, *Anal. Chem.* **43**, 589 (1971).

Yellow needles from alcohol, mp 65-67°.

USE: In determination of phenols.

4457. Gilsonite. [12002-43-6] Uintahite; uintaite. Naturally occurring, solid hydrocarbon classified as an asphaltite. Coal-like mineral found almost exclusively in the Uinta valley, near Fort Duchesne, Utah: E. S. Dana, *A System of Mineralogy* (John Wiley, New York, 6th ed., 1901) p 1020. Review: K. R. Neel in *Kirk-Othmer Encyclopedia of Chemical Technology* **vol. 11** (John Wiley & Sons, New York, 3rd ed., 1980) pp 803-806. Review of uses and mining methods: D. Jackson, *Eng. Min. J.* **182**, 88-91 (July, 1981); of health effects from occupational exposure: D. G. Keimig *et al.*, *Am. J. Ind. Med.* **11**, 287-296 (1987); G. J. Kullman *et al.*, *Am. Ind. Hyg. Assoc. J.* **50**, 413-418 (1989).

Black, lustrous masses; streak and powder a rich brown. Softening point: 160°. d_{15}^{15} 1.04. Specific heat (J/kg): 2180 (149°); 2550 (260°). Viscosity (cP): 500 (260°); 140 (316°). Nonconductor of electricity. Sol in heavy lubricating petroleum, warm oil of turpentine, and alc.

USE: In automotive body sealers, photogravure inks, anticorrosive paints, oil drilling fluids, and in nuclear-grade graphite.

4458. Ginger. Perennial plant, *Zingiber officinale* Roscoe, Zingiberaceae, with tuberous rhizomes that are widely used in traditional medicine and foods. *Habit.* Tropical Asia; cultivated in Jamaica, Africa, Australia, most tropical countries. *Constit.* Volatile oil (1-3%); oleoresin (4-7.5%) containing the pungent gingerols and shogaols; starch (40-60%); proteins (10%); and fats (10%). Chemistry of oil and oleoresin: D. W. Connell *et al.*, *Flavour Ind.* **1**, 677 (1970). Comprehensive review: V. S. Govindarajan, *Crit. Rev. Food Sci. Nutr.* **17**, 1-96; 189-258 (1982). Review of cultivation, processing and uses: B. M. Lawrence, *Perfum. Flavor.* **9**, 1-40 (1984); of history and use: E. Langner *et al.*, *Adv. Ther.* **15**, 25-44 (1998). Review of medicinal uses: J. Gruenwald *et al.*, *PDR for Herbal Medicines* (Thomson PDR, Montvale, 3rd Ed., 2004) pp 362-367; of clinical trials in nausea and vomiting of pregnancy: F. Borrelli *et al.*, *Obstet. Gynecol.* **105**, 849 (2005).

Volatile oil. [8007-08-7] Oil of ginger. Obtained by steam distillation of the dried ground rhizome. *Constit.* Chiefly, zingiberene (30-70%), β-sesquiphellandrene (15-20%), β-bisabolene (10-15%), α-farnesene, *ar*-curcumene, *d*-camphene, phellandrene, borneol, cineol, citral. Yellowish, viscid liquid. d_{25}^{25} 0.870-0.882. Angular rotation: −28 to −47°. n_D^{20} 1.488-1.494. Sol in most fixed oils, mineral oil, in alcohol with turbidity. Insol in glycerin, propylene glycol.

USE: Flavor in foods and beverages.

THERAP CAT: Antiemetic; carminative.

THERAP CAT (VET): Carminative.

4459. [6]-Gingerol. [23513-14-6] (5*S*)-5-Hydroxy-1-(4-hydroxy-3-methoxyphenyl)-3-decanone. $C_{17}H_{26}O_4$; mol wt 294.39. C 69.36%, H 8.90%, O 21.74%. The major phenol and most important of the pungent principles of ginger oil, isolated from rhizome of *Zingiber officinale* Roscoe, Zingiberaceae: Thresh, *Pharm. J.* [3] **10**, 171 (1879); *ibid.* **12**, 721 (1881); *ibid.* **14**, 798 (1883); *ibid.* **15**, 208 (1884); Garnett, Grier, *ibid.* [4] **25**, 118 (1907); Nelson, *J. Am. Chem. Soc.* **39**, 1466 (1917). Structure studies: Lapworth *et al.*, *J. Chem. Soc.* **111**, 777 (1917). Review of early literature: Redgrove, *Pharm. J.* **125**, 54 (1930); Jacobs, *Am. Perfum.* **48**, no. 7, 60, 62 (1946). Demonstration of the presence of an homologous series of phenolic ketones in the pungent constituents of ginger, [6]-gingerol being the major member: D. W. Connell, M. D. Sutherland, *Aust. J. Chem.* **22**, 1033 (1969). Synthesis of (±)-form: Hirao *et al.*, *Chem. Pharm. Bull.* **20**, 2287 (1972); K. Banno, T. Mukaiyama, *Bull. Chem. Soc. Jpn.* **49**, 1453 (1976); P. Denniff, D. A. Whiting, *Chem. Commun.* **1976**, 712; P. Denniff *et al.*, *J. Chem. Soc. Perkin Trans. 1* **1981**, 82. Stereoselective synthesis of *S*(+)-form: D. Enders *et al.*, *Ber.* **112**, 3703 (1979). Biosynthesis of (±)-form: P. Denniff, D. A. Whiting, *Chem. Commun.* **1976**, 711; I. Macleod, D. A. Whiting, *ibid.* **1979**, 1152; P. Denniff *et al.*, *J. Chem. Soc. Perkin Trans. 1* **1980**, 2637. Physical properties and cardiotonic effects of gingerols: N. Shoji *et al.*, *J. Pharm. Sci.* **71**, 1174 (1982). Mutagenicity studies: H. Nakamura, T. Yamamoto, *Mutat. Res.* **103**, 119 (1982); *eidem*, *ibid.* **122**, 87 (1983).

Normally obtained as pungent, yellow oil. n_D^{25} 1.5224. uv max (ethanol) 282 nm (ε 2560) (Connell, Sutherland). Crystalline form, mp 30-32°. $[\alpha]_D$ +27.8° (c = 1 in CHCl$_3$). uv max (ethanol): 284 nm (ε 2700). Sol in 50% alcohol, ether, chloroform, benzene; moderately sol in hot petr ether.

4460. Ginkgo. Maidenhair tree; kew tree. *Ginkgo biloba* Linn; the only living member of the *Ginkgoaceae*. Valued as a street and park tree, the ripe fruit gives off a foul odor. The seeds and leaves have been used in traditional medicine for asthma and vascular disease. *Habit.* China, Japan, cultivated in Eastern U.S. and Canada. A wide variety of compounds have been identified in the fruit, leaves and bark including: ginkgolides, bilobalide, *q.q.v.*, biflavones, flavonol glycosides. Fruit pulp contains toxic principles: ginkgolic acid, ginkgol, bilobol. Analysis of constituents of leaves: S. Furukawa, *Sci. Papers Inst. Phys. Chem. Res. Jpn.* **19**, 27 (1932); **21**, 273 (1933), *C.A.* **27**, 303, 5745 (1933). Isoln and characterization of biflavonyl constituents of ginkgo leaves: W. Baker *et al.*, *J. Chem. Soc.* **1963**, 1477; of phenolic constituents: K. Weinges *et al.*, *Arzneim.-Forsch.* **18**, 539 (1968). Isoln of ginkgolides from root bark: M. Maruyama *et al.*, *Tetrahedron Lett.* **1967**, 299-326. HPLC determn of biflavones in leaf extracts: F. Briançon-Scheid *et al.*, *Planta Med.* **49**, 204 (1983); of terpenes: P. G. Pietta *et al.*, *Chromatographia* **29**, 251 (1990). HPLC fingerprint analysis of leaf extracts: Y.-B. Ji *et al.*, *J. Chromatogr. A* **1066**, 97 (2005). *Reviews:* W. Tang, G. Eisenbrand, *Chinese Drugs of Plant Origin* (Springer-Verlag, Berlin, 1992) pp 555-565; P. Houghton, *Pharm. J.* **253**, 122-123 (1994).

EGb 761. Extractum *Ginkgo biloba* 761; Rökan; Tanakan; Tebonin. A defined extract obtained from the leaves: P. Kloss, H. Jaggy, **DE 2117429** (1972 to Willmar Schwabe), *C.A.* **78**, 47787 (1973). Clinical evaluation in arterial insufficiency: U. Bauer, *Arzneim.-Forsch.* **34**, 716 (1984). Clinical trial in dementia: P. L. Le Bars *et al.*, *J. Am. Med. Assoc.* **278**, 1327 (1997). *Book:* F. V. DeFeudis, *Ginkgo biloba Extract (EGb 761): Pharmacological Activities and Clinical Applications* (Elsevier, Amsterdam, 1991) 187 pp.

THERAP CAT: Nootropic. In treatment of vascular insufficiency.

4461. Ginkgolides. Family of bioactive terpenes isolated from the root bark and leaves of *Ginkgo biloba* L., Ginkgoaceae. Specific platelet activating factor (PAF) antagonists. Isoln from leaves: S. Furukawa, *Sci. Papers Inst. Phys. Chem. Res. Jpn.* **19**, 27 (1932), *C.A.* **27**, 303 (1933). Isoln from root bark and characterization: M. Maruyama *et al.*, *Tetrahedron Lett.* **1967**, 299; K. Nakanishi, *Pure Appl. Chem.* **14**, 89 (1967); from leaves: N. Sakabe *et al.*, *Chem. Commun.* **1967**, 259; K. Okabe *et al.*, *J. Chem. Soc. C* **1967**, 2201. Crystal structure of B: L. Dupont *et al.*, *Acta Crystallogr.* **C42**, 1759 (1986); of A and C: M. Sbit *et al.*, *ibid.* **C43**, 2377 (1987). Total synthesis of (±)-B: E. J. Corey *et al.*, *J. Am. Chem. Soc.* **110**, 649 (1988); of (±)-A: E. J. Corey, A. K. Ghosh, *Tetrahedron Lett.* **29**,

3205 (1988). Enantioselective route to B: E. J. Corey, A. V. Gavai, *ibid.* 3201. Review of syntheses: E. J. Corey, *Chem. Soc. Rev.* **17**, 111-133 (1988). ^{1}H- and ^{13}C-NMR: C. Roumestand *et al.*, *Tetrahedron* **45**, 1975 (1989). Effect on PAF receptor binding: R. Korth *et al.*, *Eur. J. Pharmacol.* **152**, 101 (1988). *Review:* P. Braquet *et al.*, *Med. Res. Rev.* **11**, 295-355 (1991). Review of physical properties: T. A. van Beek, *Bioorg. Med. Chem.* **13**, 5001-5012 (2005).

Ginkgolide B

Ginkgolides are sol in acetone, ethanol, methanol, ethyl acetate, THF, dioxane, acetic acid, trifluoroacetic acid, acetonitrile, pyridine, DMSO; sparingly sol in ether, water. Insol in hexane, benzene, chloroform, carbon tetrachloride.

Ginkgolide A. [15291-75-5] BN-52020. $C_{20}H_{24}O_9$; mol wt 408.40. Bitter crystals from ethanol, mp ~300° (dec). $[\alpha]_D^{24}$ −53.4° (c = 1 in ethanol). uv max (ethanol): 219 nm (log ε 2.72).

Ginkgolide B. [15291-77-7] (1β)-1-Hydroxyginkgolide A; BN-52021. $C_{20}H_{24}O_{10}$; mol wt 424.40. Review of pharmacology and PAF-antagonism: P. Braquet, R. H. Bourgain, *Adv. Exp. Med. Biol.* **215**, 215-235 (1987). Clinical trial in asthma: K.-H. Hsieh, *Chest* **99**, 877 (1991); in severe sepsis: J. F. Dhainaut *et al.*, *Crit. Care Med.* **22**, 1720 (1994). Bitter crystals from ethanol, mp ~300° (dec). $[\alpha]_D^{24}$ −52.6° (c = 1 in ethanol). uv max (ethanol): 219 nm (log ε 2.37). Log P (water/*n*-octanol): 1.72. pKa$_1$ 7.14; pKa$_2$ 8.60; pKa$_3$ 11.89.

Ginkgolide C. [15291-76-6] (1α,7β)-1,7-Dihydroxyginkgolide A; BN-52022. $C_{20}H_{24}O_{11}$; mol wt 440.40. Bitter crystals from ethanol, mp ~300° (dec). $[\alpha]_D^{24}$ −14.7° (c = 1 in ethanol). uv max (ethanol): 220 nm (log ε 2.30).

USE: Reference compd for PAF receptors in biological systems.

4462. Ginseng. Panax. Root of *Panax ginseng* C.A. Mey., *Araliaceae*, a perennial herb indigenous to Eastern Asia, *P. quinquefolium* L., found in Eastern U.S. and Canada, and *P. pseudoginseng* Wall, found in India, China and Japan. The biologically active constituents are considered to be a series of saponin glycosides known as *ginsenosides*, *panaxosides* or *panaquilins*. Review of constituents: J. P. Hov, *Comp. Med. East West* **5**, 123-145 (1977). Isoln and identification of ginseng saponin glycosides: S. Shibata *et al.*, *Tetrahedron Lett.* **1962** 419; G. B. Elyakov *et al.*, *ibid.* **1964**, 3591; R. Kasai *et al.*, *Chem. Pharm. Bull.* **31**, 2120 (1983). Chemico-pharmacological study: T. Kaku, *Arzneim.-Forsch.* **25**, 539 (1975). Effect on brain biogenic amines: V. Petkov, *ibid.* **28**, 388 (1978). TLC analysis of saponin content of commercial ginseng products: L. E. Liberti, A. D. Marderosian, *J. Pharm. Sci.* **67**, 1487 (1978). Use in oriental medicine as tonic: K. Chimin Wong, Wu Lien-teh, *History of Chinese Medicine* (Shanghai, 2nd ed., 1936) 906 pp. Comprehensive review of morphology, cultivation and uses: Baranov, *Econ. Bot.* **20**, 403-406 (1966). Review of chemical constituents: J. P. Hou, *Comp. Med. East West* **5**, 123-145 (1977). Brief reviews: W. E. Court, *Pharm. J.* **214**, 180-181 (1975); B. J. Spalding, *Chem. Week* **139**, 19-21 (1986).

Sweet, slightly aromatic taste.

THERAP CAT: Tonic.

4463. Giractide. [24870-04-0] 1-Glycine-18-L-argininamide-$\alpha^{(1-18)}$-corticotropin. $C_{100}H_{156}N_{34}O_{22}S$; mol wt 2218.62. C 54.14%, H 7.09%, N 21.47%, O 15.86%, S 1.45%. Polypeptide correspo to the first 18 amino acid residues in corticotropin, in which the 1-serine is replaced by glycine. Prepn: H. Otsuka *et al.*, *Bull. Chem. Soc. Jpn.* **43**, 196 (1970); of the hexaacetate: *eidem*, **JP 70 19061** (1970 to Shionogi), *C.A.* **73**, 110138r (1970). Adrenocorticotropic potency in man: *J. Clin. Endocrinol. Metab.* **33**, 355 (1971). Fluorometric study: T. Muraki *et al.*, *Experientia* **32**, 1605 (1976). Absorption: M. Hirata *et al.*, *Chem. Pharm. Bull.* **26**, 1061

(1978). Chromatographic studies: S. Terabe *et al.*, *J. Chromatogr.* **172**, 163 (1979); M. Schoeneshoefer, A. Fenner, *ibid.* **224**, 472 (1981).

Gly–Tyr–Ser–Met–Glu–His–Phe–Arg–Trp–Gly–Lys–Pro–Val–Gly–Lys–Lys–Arg–ArgNH₂

$[\alpha]_D^{23.5}$ −51.4 ±1.9° (c = 0.472 in 0.1N acetic acid). uv max (0.1N NaOH): 281, 288 nm (ε 6750, 6490).

Hexaacetate salt. [29365-11-5] S-50022; Acthormon. C_{100}-$H_{156}N_{34}O_{22}S.6C_2H_4O_2$; mol wt 2578.94.

THERAP CAT: Adrenocorticotropic hormone.

4464. Girard Reagent D. [539-64-0] *N,N*-Dimethylglycine hydrazide hydrochloride (1:1). $C_4H_{12}ClN_3O$; mol wt 153.61. C 31.28%, H 7.87%, Cl 23.08%, N 27.36%, O 10.42%. Prepd from *N,N*-dimethylglycine ethyl ester and hydrazine hydrate in abs alc: Viscontini, Meier, *Helv. Chim. Acta* **33**, 1773 (1950).

Crystals from alc, mp 181°.

Dihydrochloride. [5787-71-3] $C_4H_{11}N_3O.2HCl$; mol wt 190.07. mp 214.5°.

USE: Reagent for aldehydes and ketones.

4465. Girard Reagents. Quaternary ammonium acetylhydrazine chlorides which form water-soluble hydrazones with carbonyl compounds. The hydrazones formed can subsequently be hydrolyzed in order to regenerate the original carbonyl compounds. Prepn: A. Girard, G. Sandulesco, *GB 6640* (1934); *eidem*, *Helv. Chim. Acta* **19**, 1095 (1936); A. Girard, *Org. Synth.* **coll. vol. II**, 85 (1943). *Review:* Wheeler, *Chem. Rev.* **62**, 205 (1962). Use of Girard reagents in separation of carbonyl compounds: Schubert, Wehrberger, *Endokrinologie* **48**, 70 (1965); R. E. J. Mitchel, H. C. Birnboim, *Anal. Biochem.* **81**, 47 (1977); W. Holstein, D. Severin, *Erdoel Kohle, Erdgas, Petrochem.* **32**, 487 (1979), *C.A.* **92**, 44207b (1980).

Girard reagent T

Girard reagent P

Girard reagent T. [123-46-0] 2-Hydrazinyl-*N,N,N*-trimethyl-2-oxoethanaminium chloride (1:1); (carboxymethyl)trimethylammonium chloride hydrazide; betaine hydrazide hydrochloride. C_5H_{14}-ClN_3O; mol wt 167.64. Highly hygroscopic needles. May be stored in well-stoppered containers. Material that has developed an odor should be recrystallized from abs ethanol. mp 192° (slight decomp). Very freely sol in water; sol in about 150 parts of absolute ethanol; more sol in methanol. Also very sol in acetic acid, glycerol, ethylene glycol. Practically insol in organic solvents devoid of hydroxyl groups.

Girard reagent P. [1126-58-5] 1-(2-Hydrazinyl-2-oxoethyl)-pyridinium chloride (1:1); 1-(carboxymethyl)pyridinium chloride hydrazide. $C_7H_{10}ClN_3O$; mol wt 187.63. Non-hygroscopic crystals from methanol, dec 200°. Less soluble in polar solvents than Girard reagent T.

USE: In the isoln of 17-ketosteroids and other carbonyl compounds.

4466. Gitogenin. [511-96-6] (2α,3β,5α,25R)-Spirostan-2,3-diol; digin. $C_{27}H_{44}O_4$; mol wt 432.65. C 74.96%, H 10.25%, O 14.79%. Obtained from gitonin by heating with dil HCl. Structure: Tschesche, *Ber.* **68**, 1090 (1935); Jacobs, Simpson, *J. Biol. Chem.* **110**, 429 (1935); Marker, Rohrmann, *J. Am. Chem. Soc.* **61**, 2724 (1939); **62**, 647 (1940); Noller, Lieberman, *ibid.* **63**, 2131 (1941); Klass *et al*, *ibid.* **77**, 3829 (1955). Pharmacological study: H. K. Iwamoto *et al.*, *J. Pharmacol. Exp. Ther.* **91**, 130 (1947).

Leaflets from benzene, dec 271.5-275°. $[\alpha]_D^{20}$ −75° (c = 1.02 in CHCl₃). Sol in chloroform, hot alc; sparingly sol in cold ethyl acetate, in ether; practically insol in water. Not precipitated by digitonin.

Diacetate. $C_{31}H_{48}O_6$. Long needles from ether + methanol, mp 251-254°. $[\alpha]_D^{20}$ −96° (c = 1.92 in CHCl₃).

4467. F-Gitonin. [28591-01-7] (2α,3β,5α,25R)-2-Hydroxy-spirostan-3-yl *O*-β-D-glucopyranosyl-(1 → 2)-*O*-[β-D-xylopyranosyl-(1 → 3)]-*O*-β-D-glucopyranosyl-(1 → 4)-β-D-galactopyranoside; gitogenin β-lycotetraoside. $C_{50}H_{82}O_{23}$; mol wt 1051.18. C 57.13%, H 7.86%, O 35.01%. Leaf saponin from *Digitalis purpurea* L., *Scrophulariaceae*: Kawasaki, Nishioka, *Chem. Pharm. Bull.* **12**, 1311 (1964). Structure: Kawasaki *et al.*, *Tetrahedron* **21**, 299 (1965). A gitogenin tetraglycoside of which the sugar composition is 2 moles D-glucose, 1 mole D-galactose, and 1 mole D-xylose; differs from **gitonin** (the gitogenin tetraglycoside in *D. purpurea* seeds), in which the sugar composition is 2 galactose, 1 glucose, and 1 xylose: Tschesche, Wulff, *Ber.* **94**, 2019 (1961).

Dihydrate. Needles from butanol + water, dec 252-255°. $[\alpha]_D^{25}$ −58.5° (c = 0.53 in pyridine).

4468. Gitoxigenin. [545-26-6] (3β,5β,16β)-3,14,16-Trihydroxycard-20(22)-enolide; $\Delta^{20,22}$-3,14,16,21-tetrahydroxynorcholenic acid lactone; $C_{23}H_{34}O_5$; mol wt 390.52. C 70.74%, H 8.78%, O 20.48%. The aglycon of gitoxin. By refluxing gitoxin in a mixture of water + alcohol + HCl: Smith, *J. Chem. Soc.* **1931**, 23. Structure: Jacobs, Elderfield, *J. Biol. Chem.* **100**, 671 (1933); Elderfield, *Chem. Rev.* **17**, 217 (1935); Henderson, Chen, *J. Med. Pharm. Chem.* **5**, 988 (1962). Configuration: Moore, *Helv. Chim. Acta* **37**, 659 (1954); Repke, Klesczewski, *Arch. Exp. Pathol. Pharmakol.* **239**, 131 (1960). *Cf.* ref under Digoxigenin.

Sesquihydrate. Plates from dil alc. After drying at 100° *in vacuo* mp 234°. $[\alpha]_{545}^{20}$ +38.5° (c = 0.68 in methanol). Absorption max

(96% H_2SO_4): 310, 485, 520 nm. Slightly sol in alcohol, acetone, ethyl acetate. Treatment with alcoholic HCl yields digitaligenin with loss of $2H_2O$.

3,16-Diacetylgitoxigenin. mp 249-250°.

3,16-Dibenzoylgitoxigenin. mp 262°.

4469. Gitoxin. [4562-36-1] $(3\beta,5\beta,16\beta)$-3-[(O-2,6-Dideoxy-β-D-$ribo$-hexopyranosyl-$(1 \rightarrow 4)$-O-2,6-dideoxy-β-D-$ribo$-hexopyranosyl-$(1 \rightarrow 4)$-2,6-dideoxy-β-D-$ribo$-hexopyranosyl)oxy]-14,16-dihydroxycard-20(22)-enolide; anhydrogitalin; bigitalin; pseudodigitoxin. $C_{41}H_{64}O_{14}$; mol wt 780.95. C 63.06%, H 8.26%, O 28.68%. Secondary glycoside mainly from $Digitalis\ purpurea$ L., also from $D.\ lanata$ Ehrh., $Scrophulariaceae$. Byproduct of digitoxin manuf. Isoln: Kraft, $Arch.\ Pharm.$ **250**, 118 (1912); Cloetta, $Arch.\ Exp.\ Pathol.\ Pharmakol.$ **112**, 261 (1926); Smith, $J.\ Chem.\ Soc.$ **1931**, 23. Purification: McChesney $et\ al.,J.\ Am.\ Pharm.\ Assoc.$ **37**, 364 (1948). Acid hydrolysis yields 1 mol gitoxigenin + 3 mols digitoxose, $q.q.v.$: Windaus, Schwarte, $Ber.$ **58**, 1515 (1925). $See\ also:$ Satoh, Aoyama, $Chem.\ Pharm.\ Bull.$ **18**, 94 (1970). Pharmacokinetics in humans: M. Lesne, $Int.\ J.\ Clin.\ Pharmacol.\ Biopharm.$ **16**, 456 (1978). Prepn of pentaacetate: **BE 668116**; **NL 6506250** (1965, 1966 both to VEB Arzneimittelwerk Dresden), $C.A.$ **65**, 23366 (1966); **67**, 91092d (1967). Toxicity study: Foerster $et\ al.,$ $Arch.\ Int.\ Pharmacodyn.$ **155**, 165 (1965). $Review:$ R. Megges $et\ al., Pharmazie$ **32**, 665-667 (1977).

Stout prisms from chloroform + methanol, dec 285° (rapid heating). $[\alpha]_{546}^{20}$ +3.5° (c = 1.02 in pyridine). Absorption max (98% H_2SO_4): 315, 415, 495, 530 nm ($E_{1cm}^{1\%}$ 275, 185, 430, 505). Almost insol in chloroform, ethyl acetate and acetone. Dissolves in a mixture of chloroform and alcohol or in pyridine or in dil alcohol. Less sol in hot 80% alcohol than digoxin.

α- and β-**Acetylgitoxin.** $C_{43}H_{66}O_{15}$. Obtained by enzymatic hydrolysis of digilanide B. β-Form: Long, thin, hair-like prisms from dil methanol, dec 220-225°. $[\alpha]_D^{20}$ +15.7° (c = 1.28 in pyridine). Sol in 80-100 parts methanol; sparingly sol in water and ether.

Pentaacetate. [7242-04-8] Pengitoxin; penta-O-acetylgitoxin; Carnacid-Cor; Cordoval; Pentagit. $C_{51}H_{74}O_{19}$; mol wt 991.13. Mass spectral studies: Blessington, Morton, $Org.\ Mass\ Spectrom.$ **3**, 95 (1970). Rhomboid crystals, mp 151-155°. $[\alpha]_D^{20}$ +14.0 ±1.5° (c = 1.6 in pyridine). LD_{50} in mice (mg/kg): 6.4 i.p.; in rats (mg/kg): 21.0 i.v. (Foerster).

THERAP CAT: Cardiotonic.

4470. Glatiramer. [28704-27-0] L-Glutamic acid polymer with L-alanine, L-lysine and L-tyrosine; copolymer 1; COP 1 (polyamide); COP 1. Random basic synthetic copolymer of L-alanine, L-lysine, L-glutamic acid and L-tyrosine in a molar ratio of 6:1.9:4.7:1. Mol wt of 14,000 to 23,000 Da. Immunologically cross reactive with myelin basic protein. Prepn: D. Teitelbaum $et\ al., Eur.\ J.\ Immunol.$ **1**, 242 (1971). Suppression of exptl allergic encephalomyelitis (EAE): $idem, ibid.$ **3**, 273 (1973). Inhibition of T-cell response to myelin basic protein: D. Teitelbaum $et\ al., Proc.\ Natl.\ Acad.\ Sci.\ USA$ **89**, 137 (1992). Review of pharmacology and early clinical evaluation: L. A. Rolak, $Clin.\ Neuropharmacol.$ **10**, 389 (1987). Clinical comparison with interferon beta-1b in multiple sclerosis: P. O'Connor $et\ al., Lancet\ Neurol.$ **8** 889 (2009). Review of clinical experience in multiple sclerosis: K. P. Johnson, D. L. Due, $Expert\ Rev.\ Pharmacoecon.\ Outcomes\ Res.$ **9**, 205-214 (2009); N. J. Carter, G. M. Keating, $Drugs$ **12**, 1545-1577 (2010).

Acetate. [147245-92-9] Copaxone.

THERAP CAT: Immunomodulator; in treatment of multiple sclerosis.

4471. Glaucine. [475-81-0] $(6aS)$-5,6,6a,7-Tetrahydro-1,2,-9,10-tetramethoxy-6-methyl-$4H$-dibenzo[de,g]quinoline; 1,2,9,10-tetramethoxyaporphine; boldine dimethyl ether; Bromcholitin; Glauvent. $C_{21}H_{25}NO_4$; mol wt 355.43. C 70.97%, H 7.09%, N 3.94%, O 18.01%. d-Form prevalent in nature. Found in $Glaucium\ flavum$ Crantz ($G.\ luteum$ Scop.), $Papaveraceae$ and in $Dicentra$ and $Corydalis$ species, $Fumariaceae$. Isoln: Fischer, $Arch.\ Pharm.$ **239**, 421 (1901). Structure and prepn: Gadamer, $ibid.$ **249**, 680 (1911); Späth, Tharrer, $Ber.$ **66**, 904 (1933). Configuration: Faltis, Adler, $Arch.\ Pharm.$ **284**, 281 (1951). Synthesis of dl-form: Chan, Maitland, $J.\ Chem.\ Soc.\ C$ **1966**, 753; Jackson, Martin, $ibid.$ 2061; Cava $et\ al., J.\ Org.\ Chem.$ **35**, 175 (1970). Pharmacology and toxicity of dl-glaucine phosphate: Y. Kasé $et\ al., Arzneim.-Forsch.$ **33**, 936 (1983). Sites of antitussive action: $eidem, ibid.$ 947.

d-Glaucine

Orthorhombic plates, prisms from ethyl acetate or ether, mp 120°. $[\alpha]_D^{20}$ +115° (c = 3 in alc). Sol in acetone, alcohol, chloroform, ethyl acetate. Moderately sol in ether, petr ether. Practically insol in water and benzene.

Hydrochloride trihydrate. $C_{21}H_{25}NO_4 \cdot HCl \cdot 3H_2O$. Needles. mp 232° (anhydrous). Sol in water, alcohol, chloroform.

Hydrobromide. $C_{21}H_{25}NO_4 \cdot HBr$. Crystals, mp 235°. Less sol than the hydrochloride.

dl-**Form phosphate.** DL-832. $(C_{21}H_{25}NO_4)_2 \cdot 3H_3PO_4$; mol wt 1004.85. Crystalline powder. LD_{50} in mice (mg/kg): 98 i.v.; 401 orally (Kasé).

4472. Gliadin. [9007-90-3] A simple protein, one of the prolamins, derived from the gluten of wheat, rye, etc. May contain up to 43% glutamine: $C.A.$ **50**, 15792a (1956). Studies on the physical nature of gliadin: Holme, Briggs, $Cereal\ Chem.$ **36**, 321 (1959). Use of deamidized gliadin in food products: McDonald, **US 3030211** (1962 to USDA).

Practically insol in water, abs alcohol, and other neutral solvents. Sol in 70-80% alcohol, dil acid, dil alkali.

4473. Glibornuride. [26944-48-9] N-[[[(1S,2S,3R,4R)-3-Hydroxy-4,7,7-trimethylbicyclo[2.2.1]hept-2-yl]amino]carbonyl]-4-methylbenzenesulfonamide; (1R,2R,3S,4S)-1-(2-hydroxy-3-bornyl)-3-(p-tolylsulfonyl)urea; D-3-$endo$-p-tosylureidoborneol; 1-(p-tolylsulfonyl)-3-(2-$endo$-hydroxy-3-$endo$-D-bornyl)urea; Ro-6-4563; Gluborid; Glutril. $C_{18}H_{26}N_2O_4S$; mol wt 366.48. C 58.99%, H 7.15%, N 7.64%, O 17.46%, S 8.75%. Prepn: H. Bretschneider $et\ al., Monatsh.\ Chem.$ **100**, 2133 (1969). $See\ also:\ idem,$ **ZA 6706161**; $idem,$ **US 3770761**; $idem,$ **ZA 6902395**; $idem,$ **US 3654357** (1968, 1973, 1969, 1972, all to Hoffmann-La Roche). Comprehensive review of pharmacology: $Arzneim.-Forsch.$ **22**, 2153-2222 (1972).

Crystals, mp 192-195° (ethanol-water); also reported as 195-198°. $[\alpha]_D$ +63.8° (ethanol).

THERAP CAT: Antidiabetic.

4474.   Gliclazide. [21187-98-4] *N*-[[(Hexahydrocyclopenta-[*c*]pyrrol-2(1*H*)-yl)amino]carbonyl]-4-methylbenzenesulfonamide; 1-(hexahydrocyclopenta[*c*]pyrrol-2(1*H*)-yl)-3-(*p*-tolylsulfonyl)urea; *N*-(4-methylbenzenesulfonyl)-*N'*-(3-azabicyclo[3.3.0]oct-3-yl)urea; 1-(3-azabicyclo[3.3.0]oct-3-yl)-3-(*p*-tolylsulfonyl)urea; S-1702; Diamicron; Glimicron; Nordialex. $C_{15}H_{21}N_3O_3S$; mol wt 323.41. C 55.71%, H 6.55%, N 12.99%, O 14.84%, S 9.91%. Prepn: Beregi *et al.*, **FR 1510714** and **US 3501495** (1968 and 1970, both to Sci. Union & Co.-Soc. Franc. Rech. Med.). Series of articles on pharmacology: *Arzneim.-Forsch.* **22**, 1686-1695 (1972). Toxicity data: J. Duhault *et al.*, *ibid.* 1682. Review of pharmacology, efficacy: B. Holmes *et al.*, *Drugs* **27**, 301 (1984).

Crystals from anhydrous ethanol, mp 180-182°. LD_{50} orally in mice: >3 g/kg (Duhault).

THERAP CAT: Antidiabetic.

4475.   Glimepiride. [93479-97-1] 3-Ethyl-2,5-dihydro-4-methyl-*N*-[2-[4-[[[[(*trans*-4-methylcyclohexyl)amino]carbonyl]amino]sulfonyl]phenyl]ethyl]-2-oxo-1*H*-pyrrole-1-carboxamide; *N*-[4-[2-(3-ethyl-4-methyl-2-oxo-3-pyrroline-1-carboxamido)-ethyl]-benzenesulfonyl]-*N'*-4-methylcyclohexylurea; 1-[4-[2-(3-ethyl-4-methyl-2-oxo-3-pyrroline-1-carboxamido)ethyl]phenylsulfonyl]-3-(4-methylcyclohexyl)urea; HOE-490; Amarel; Amaryl; Solosa. $C_{24}H_{34}N_4O_5S$; mol wt 490.62. C 58.76%, H 6.99%, N 11.42%, O 16.30%, S 6.53%. Oral sulfonylurea; promotes insulin release at ATP-sensitive potassium channels on pancreatic β-cells via binding to a 65 kDa subunit of the sulfonylurea receptor. Prepn: R. Weyer *et al.*, **DE 2951135**; *eidem*, **US 4379785** (1981, 1983 both to Hoechst). Synthesis: R. Weyer, V. Hitzel, *Arzneim.-Forsch.* **38**, 1079 (1988). Pharmacology: K. Geisen, *ibid.* 1120. Effects on insulin and glucagon secretion: V. Leclercq-Meyer *et al.*, *Biochem. Pharmacol.* **42**, 1634 (1991). HPLC determn in plasma: Y.-K. Song *et al.*, *J. Chromatogr. B* **810**, 143 (2004). Clinical pharmacokinetics: K. Ratheiser *et al.*, *Arzneim.-Forsch.* **43**, 856 (1993). Toxicity study: U. Schollmeier *et al.*, *ibid.* 1038. Series of articles on pharmacology and clinical efficacy: *Diabetes Res. Clin. Pract.* **28** Suppl., S115-S149 (1995). Review of pharmacology and clinical use in type 2 diabetes: H. D. Langtry, J. A. Balfour, *Drugs* **55**, 563-584 (1998); A. L. McCall, *Expert Opin. Pharmacother.* **2**, 699-713 (2001).

Relative stereochemistry

White to yellowish-white crystalline powder. mp 207°. Sol in DMSO, DMF; sparingly sol in methylene chloride; slightly sol in acetone, methanol; very slightly sol in acetonitrile. Practically insol in water.

THERAP CAT: Antidiabetic.

THERAP CAT (VET): Antidiabetic.

4476.   Gliotoxin. [67-99-2] (3*R*,5a*S*,6*S*,10a*R*)-2,3,5a,6-Tetrahydro-6-hydroxy-3-(hydroxymethyl)-2-methyl-10*H*-3,10a-epidithiopyrazino[1,2-*a*]indole-1,4-dione; $C_{13}H_{14}N_2O_4S_2$; mol wt 326.39. C 47.84%, H 4.32%, N 8.58%, O 19.61%, S 19.65%. Antibiotic substance produced by various spp of *Trichoderma, Gladiocladium fimbriatum, Aspergillus fumigatus,* and *Penicillium* spp: Weindling, Emerson, *Phytopathology* **26**, 1068 (1936); **27**, 1175

(1937); Johnson *et al.*, *J. Am. Chem. Soc.* **65**, 2005 (1943); Menzel *et al.*, *J. Biol. Chem.* **152**, 419 (1944). Structure: Bell *et al.*, *J. Am. Chem. Soc.* **80**, 1001 (1958); Beecham *et al.*, *Tetrahedron Lett.* **1966**, 3131. Crystallographic data: McCrone, *Anal. Chem.* **26**, 1662 (1954). Biosynthesis: Suhadolnik, Chenoweth, *J. Am. Chem. Soc.* **80**, 4391 (1958); Winstead, Suhadolnik, *ibid.* **82**, 1644 (1960); J. D. M. Herscheid *et al.*, *J. Org. Chem.* **45**, 1885 (1980). Synthetic studies: Poisel, Schmidt, *Ber.* **104**, 1714 (1971); *ibid.* **105**, 625 (1972); Oehler *et al.*, *ibid.* 625. Total synthesis of *dl*-form: T. Fukuyama, Y. Kishi, *J. Am. Chem. Soc.* **98**, 6723 (1976); T. Fukuyama *et al.*, *Tetrahedron* **37**, 2045 (1981).

Monoclinic needles from methanol or benzene, dec 221°. $[\alpha]_D^{25}$ −290° (c = 0.08 in ethanol). uv max: 270 nm (ε 4500). Soly in mg/ml at 7°: acetic acid 12; acetone 9.0; acetonitrile 10.2; benzene 5.5; carbon tetrachloride 0.8; chloroform 20; dioxane 73 (decompn); dimethylformamide 17; ethyl acetate 8.5; ethanol 4.7; methanol 1.4; pyridine 77; water (30°): 0.07. Sensitive to oxidation and heat; inactivated by heating for 10 min at 100°.

Monoacetate. $C_{15}H_{16}N_2O_5S_2$. Isolated from cultures of *Penicillium terlikowski* Zaleski: Johnson *et al.*, *J. Am. Chem. Soc.* **75**, 2110 (1953). Orthorhombic crystals from benzene, mp 162-163°.

4477.   Glipizide. [29094-61-9] *N*-[2-[4-[[[(Cyclohexylamino)carbonyl]amino]sulfonyl]phenyl]ethyl]-5-methyl-2-pyrazinecarboxamide; 1-cyclohexyl-3-[[*p*-[2-(5-methylpyrazinecarboxamido)-ethyl]phenyl]sulfonyl]urea; glydiazinamide; K-4024; Glibenese; Glucotrol; Mindiab; Minidiab; Ozidia. $C_{21}H_{27}N_5O_4S$; mol wt 445.54. C 56.61%, H 6.11%, N 15.72%, O 14.36%, S 7.20%. Second generation sulfonylurea with hypoglycemic activity. Prepn: V. Ambrogi, W. Logemann, **DE 2012138**; *eidem*, **US 3669966** (1970, 1972 both to Carlo Erba); Ambrogi *et al.*, *Arzneim.-Forsch.* **21**, 200 (1971). Pharmacology: *eidem, ibid.* 208; Marigo *et al., ibid.* 215. Metabolism: Goldaniga *et al., ibid.* **23**, 242 (1973); Fuccella *et al., J. Clin. Pharmacol.* **13**, 68 (1973). Pharmacokinetics and pharmacodynamics of extended-release formulation: M. Chung *et al.*, *J. Clin. Pharmacol.* **42**, 651 (2002). Evaluation in diabetic cats: E. C. Feldman *et al.*, *J. Am. Vet. Med. Assoc.* **210**, 772 (1997). Clinical evaluation in diabetes: D. C. Simonson *et al.*, *Diabetes Care* **20**, 597 (1997). Toxicity: Ambrogi *et al.*, *Arzneim.-Forsch.* **21**, 208 (1971). Review of pharmacology and therapeutic efficacy: R. N. Brogden *et al.*, *Drugs* **18**, 329-353 (1979); H. E. Lebovitz, *Pharmacotherapy* **5**, 63-77 (1985).

Crystals from ethanol, mp 208-209°. Also reported as mp 200-203°. pKa 5.9 Freely sol in DMF; sol in 0.1 *N* NaOH; slightly sol in methylene chloride. Insol in water, alcohols. LD_{50} in mice, rats (g/kg): >3, 1.2 i.p. (Ambrogi).

THERAP CAT: Antidiabetic.

THERAP CAT (VET): Antidiabetic.

4478.   Gliquidone. [33342-05-1] *N*-[(Cyclohexylamino)carbonyl]-4-[2-(3,4-dihydro-7-methoxy-4,4-dimethyl-1,3-dioxo-2(1*H*)-isoquinolinyl)ethyl]benzenesulfonamide; 1-cyclohexyl-3-[[*p*-[2-(3,4-dihydro-7-methoxy-4,4-dimethyl-1,3-dioxo-2(1*H*)-isoquinolyl)ethyl]phenyl]sulfonyl]urea; 1,2,3,4-tetrahydro-2-[*p*-(*N'*-cyclohexylureido-*N*-sulfonyl)phenethyl]-4,4-dimethyl-7-methoxyisoquinoline-1,3-dione; AR-DF 26; Glurenorm. $C_{27}H_{33}N_3O_6S$; mol wt

527.64. C 61.46%, H 6.30%, N 7.96%, O 18.19%, S 6.08%. Prepn: E. Kutter *et al.*, **DE 2000339** (1971 to Thomae); *eidem*, **US 3708486** (1973 to Boehringer, Ing.); *eidem*, **DE 2011126** (1971 to Thomae), *C.A.* **76**, 14359e (1972). Pharmacokinetics and metabolism: Kopitar, *Arzneim.-Forsch.* **25**, 1455 (1975); Kopitar *et al.*, *ibid.* 1933.

Crystals from boiling methanol, mp 180-182°.
Sodium salt. Sinters at 160°. LD_{50} in mice: >2 g/kg orally; 234 mg/kg i.v. (Kutter, 1973).
THERAP CAT: Antidiabetic.

4479. Glisoxepid. [25046-79-1] *N*-[2-[4-[[[[(Hexahydro-1*H*-azepin-1-yl)amino]carbonyl]amino]sulfonyl]phenyl]ethyl]-5-methyl-3-isoxazolecarboxamide; 1-(hexahydro-1*H*-azepin-1-yl)-3-[[*p*-[2-(5-methyl-3-isoxazolecarboxamido)ethyl]phenyl]sulfonyl]urea; 4-[4-[β-(5-methylisoxazole-3-carboxamido)ethyl]phenylsulfonyl]-1,1-hexamethylenesemicarbazide; BS-4231; RP-22410; Pro-Diaban. $C_{20}H_{27}N_5O_5S$; mol wt 449.53. C 53.44%, H 6.05%, N 15.58%, O 17.80%, S 7.13%. Prepn: H. Plümpe, W. Puls, **ZA 6806886**; *eidem*, **US 3668215** (1969, 1972 both to Bayer). Pharmacology: Loubatieres *et al.*, *C. R. Seances Acad. Sci. Ser. D* **271**, 1446 (1970); *J. Pharmacol.* **3**, 171, 229 (1972). Toxicity study: Tettenborn, *Arzneim.-Forsch.* **24**, 409 (1974). Series of articles: *ibid.* 363-452.

Colorless crystals from ethanol, mp 189°. LD_{50} in mice, rats, cats, dogs (g/kg): >10.0, >10.0, >4.0, >2.0 orally; in mice, rats (mg/kg): 283, 196 i.v. (Tettenborn).
THERAP CAT: Antidiabetic.

4480. Globin. The colorless, basic protein of hemoglobin. Formed from tissue protein in the body, the globin part of catabolized hemoglobin is re-used (unlike heme, the prosthetic group of hemoglobin). Prepn from ox hemoglobin: Anson, Mirsky, *J. Gen. Physiol.* **13**, 469 (1930). Globin from normal adult human hemoglobin consists of four polypeptide chains — two α-chains and two β-chains. The α-chain contains 141, the β-chain 146 amino acids. Thus globin contains 574 amino acids and has an approx mol wt of 62,000. Abnormal globins may contain γ- and δ-chains. Structure: Braunitzer *et al.*, *Z. Physiol. Chem.* **325**, 283 (1961); **331**, 1 (1963); Konigsberg *et al.*, *J. Biol. Chem.* **237**, 1549, 2547 (1962); **238**, 2016, 2028 (1963). Review of prepn and properties: Rossi Fanelli *et al.*, *Adv. Protein Chem.* **19**, 124 (1964). *Review:* Braunitzer *et al.*, *ibid.* 1.

Denatures rapidly above 17°. At pH values near neutrality, combines with ferroprotoporphyrin to yield hemoglobin, or with ferriprotoporphyrin to yield methemoglobin.

4481. Glucagon. [9007-92-5] Glukagon; hyperglycemic-glycogenolytic factor; HG-factor; HGF. $C_{153}H_{225}N_{43}O_{49}S$; mol wt 3482.80. Polypeptide hormone produced in the alpha cells of the islets of Langerhans in the pancreas; stimulates hepatic glucose output. Identification as hyperglycemic factor in pancreatic extracts: J. R. Murlin *et al.*, *J. Biol. Chem.* **56**, 253 (1923); C. P. Kimball, J. R. Murlin, *ibid.* **58**, 337 (1923). Purification and properties: M. Bürger, W. Brandt, *Z. Ges. Exp. Med.* **96**, 375 (1935); A. Staub *et al.*,

Science **117**, 628 (1953); *eidem*, *J. Biol. Chem.* **214**, 619 (1955). Amino acid sequence of porcine: W. W. Bromer *et al.*, *J. Am. Chem. Soc.* **79**, 2807 (1957); of bovine: *idem et al.*, *J. Biol. Chem.* **246**, 2822 (1971); of human: J. Thomsen *et al.*, *FEBS Lett.* **21**, 315 (1972). *Note:* Amino acid sequences of porcine, bovine, and human glucagon are identical. *Review:* O. K. Behrens, W. W. Bromer, *Vitam. Horm.* **16**, 263-301 (1958). *Books:* P. J. Lefèbvre, R. H. Unger, *Glucagon* (Pergamon Press, New York, 1972) 535 pp; *Glucagon* **Pts. I and II**, P. Lefèbvre, Ed. (Springer, New York, 1983) 700 pp. Review of cardiovascular effects and use as antidote to beta-blocker overdose: C. M. White, *J. Clin. Pharmacol.* **39**, 442-447 (1999); of role in glucose homeostasis: G. Jiang, B. B. Zhang, *Am. J. Physiol. Endocrinol. Metab.* **284**, E671-E678 (2003). Review of use to decrease gastrointestinal motility during endoscopy: T. Hashimoto *et al.*, *Aliment. Pharmacol. Ther.* **16**, 111 (2002).

His–Ser–Gln–Gly–Thr–Phe–Thr–Ser–Asp–Tyr–Ser–Lys–Tyr–Leu–Asp–Ser–Arg

Thr–Asn–Met–Leu–Trp–Gln–Val–Phe–Asp–Gln–Ala–Arg

Rhombic dodecahedra. Stable. Sol in dilute alkali and acidic solns, *i.e.* below pH 3 and above pH 9.5. Practically insol in water; insol in most organic solvents.
Glucagon (rDNA origin). GlucaGen. Structurally identical to naturally occurring human glucagon. Administered as hydrochloride.
THERAP CAT: Antihypoglycemic. Diagnostic aid (adjunct for endoscopy).
THERAP CAT (VET): Antihypoglycemic in small animals; in treatment of fatty liver disease in cattle.

4482. Glucagon-Like Peptides. Bioactive peptides (GLP-1 and GLP-2) encoded within the preproglucagon gene; formed by tissue-specific, post-translational processing of proglucagon by intestinal L-cells. Key regulators of energy homeostasis and metabolism. Released into the circulation in response to a meal; rapidly inactivated by dipeptidylpeptidase-IV (DPP-IV). GLP-1 is a potent, glucose-dependent stimulator of insulin secretion (incretin). GLP-2 has growth-promoting and cytoprotective effects on intestinal epithelium. Identification in hamster preproglucagon: G. I. Bell *et al.*, *Nature* **302**, 716 (1983); in human: *idem et al.*, *ibid.* **304**, 368 (1983). Review of structures, tissue distribution, and physiological activites: T. J. Kieffer, J. F. Habener, *Endocr. Rev.* **20**, 876-913 (1999). Review of biosynthesis and regulation of secretion: P. L. Brubaker, Y. Anini, *Can. J. Physiol. Pharmacol.* **81**, 1005-1012 (2003). Review of bioactivities: L. L. Baggio, D. J. Drucker, *Best Pract. Res. Clin. Endocrinol. Metab.* **18**, 531-554 (2004); and therapeutic potential: D. J. Drucker, *Nat. Clin. Pract. Endocrinol. Metab.* **1**, 22-31 (2005).

His–Ala–Glu–Gly–Thr–Phe–Thr–Ser–Asp–Val–Ser–Ser–Tyr–Leu–Glu–Gly–Gln–Ala

Gly–Arg–Gly–Lys–Val–Leu–Trp–Ala–Ile–Phe–Glu–Lys–Ala

hGLP-1 (7-37)

His–Ala–Asp–Gly–Ser–Phe–Ser–Asp–Glu–Met–Asn–Thr–Ile–Leu–Asp–Asn–Leu–Ala

Asp–Thr–Ile–Lys–Thr–Gln–Ile–Leu–Trp–Asn–Ile–Phe–Asp–Arg–Ala

hGLP-2

GLP-1. [89750-14-1]; [87805-34-3] (human). Glucagon-like peptide I; glucagon-related peptide I. Originally identified as a 37 amino acid peptide; bioactive forms are GLP-1 (7-37) and GLP-1 (7-36) amide. Stimulates insulin secretion, suppresses glucagon secretion, and delays gastric emptying. Review of physiology and therapeutic potential in diabetes: M. A. Nauck, *Acta Diabetol.* **35**, 117-129 (1998); M. M. J. Combettes, *Curr. Opin. Pharmacol.* **6**, 598-605 (2006).
GLP-2. [89750-15-2]; [223460-79-5] (human). Glucagon-like peptide II; glucagon-related peptide II. Intestinotrophic peptide containing 33 amino acid residues. Stimulates crypt-cell proliferation

and inhibits apoptosis in the gastrointestinal mucosa. Identification of bioactivity: D. J. Drucker *et al.*, *Proc. Natl. Acad. Sci. USA* **93**, 7911 (1996). Comprehensive review: J. L. Estall, D. J. Drucker, *Annu. Rev. Nutr.* **26**, 391-411 (2006).

4483. Glucametacin. [52443-21-7] 2-[[2-[1-(4-Chlorobenzoyl)-5-methoxy-2-methyl-1*H*-indol-3-yl]acetyl]amino]-2-deoxy-D-glucose; 2-[2-[1-(*p*-chlorobenzoyl)-5-methoxy-2-methylindol-3-yl]acetamido]-2-deoxy-D-glucose; glucametacine; glucamethacin; indomethacin glucosamide. $C_{25}H_{27}ClN_2O_8$; mol wt 518.95. C 57.86%, H 5.24%, Cl 6.83%, N 5.40%, O 24.66%. Deriv of indomethacin, *q.v.* Prepn: A. Demetrio *et al.*, **DE 2223051** (1973 to SIR Lab. Chem. Biol. SpA), *C.A.* **80**, 83529e (1974). Pharmacological study: E. Paroli *et al.*, *Arzneim.-Forsch.* **28**, 819 (1978). Clinical studies: P. Petera *et al.*, *Int. J. Clin. Pharmacol. Biopharm.* **15**, 581 (1977); L. Capelli *et al.*, *Curr. Med. Res. Opin.* **7**, 227 (1981).

Monohydrate. Euminex; Teorema; Teoremac. $C_{25}H_{27}ClN_2O_8 \cdot H_2O$; mol wt 536.96.

THERAP CAT: Anti-inflammatory.

4484. Glucamine. [488-43-7] 1-Amino-1-deoxy-D-glucitol; 1-amino-1-deoxysorbitol; glycamine; D-glucamine. $C_6H_{15}NO_5$; mol wt 181.19. C 39.77%, H 8.34%, N 7.73%, O 44.15%. Prepn from D-glucose: Holly *et al.*, *J. Am. Chem. Soc.* **72**, 5416 (1950); from *N*-benzylglycamine: Kagan *et al.*, *ibid.* **79**, 3541 (1957); by catalytic reduction of glucose in the presence of hydrazine: Lemieux, **US 2830983** (1958 to Natl. Res. Council, Ottawa). Commercial prepn: Flint, Salzberg, **US 2016962** (1935 to du Pont); Groggins, Stirton, *Ind. Eng. Chem.* **29**, 1358 (1937).

Crystals from methanol, mp 127°. Sharp, slightly sweet taste. $[\alpha]_D^{15}$ −7.95° (c = 10 in water). Very sol in water; slightly sol in alcohol. Practically insol in ether.

4485. D-Glucaric Acid. [87-73-0] Saccharic acid; D-glucosaccharic acid; D-tetrahydroxyadipic acid. $C_6H_{10}O_8$; mol wt 210.14. C 34.29%, H 4.80%, O 60.91%. Best prepd by nitric acid oxidation of starch; yields as high as 65% are obtained in contrast to much lower yields from glucose or sucrose: Kiliani, *Ber.* **58**, 2344 (1925); Schmidt *et al.*, *ibid.* **70**, 2402 (1937). Prepn from D-glucose: Mehltretter *et al.*, **US 2472168** (1949 to U.S. Secy. Agr.); Truchan, **US 2809989** (1957 to Cowles Chem.); Phillips *et al.*, *J. Chem. Soc.* **1958**, 3522; **BE 615023** (1962 to Ciba).

Needles from 95% ethanol, mp 125-126°. Shows mutarotation. $[\alpha]_D^{19}$ +6.86° $\rightarrow$ +20.60° (H_2O). $Ka_1 = 1.0 \times 10^{-5}$ at 25°. Soluble in water, ethanol. Sparingly sol in ether.

1,4-Lactone. [389-36-6] Saccharolactone. $C_6H_8O_7$; mol wt 192.12. Strong inhibitor of β-glucuronidase: Levvy, *Biochem. J.* **52**, 464 (1952); Boyland, Williams, *ibid.* **64**, 578 (1956). Monohydrate, crystals, mp 90°.

4486. D-Glucoascorbic Acid. [528-88-1] D-*arabino*-Hept-2-enonic acid γ-lactone; 3-keto-D-glucoheptonofuranolactone. $C_7H_{10}O_7$; mol wt 206.15. C 40.78%, H 4.89%, O 54.33%. A physiologically inactive homolog of ascorbic acid. Prepn: Ault *et al.*, *J. Chem. Soc.* **1933**, 741; Baird *et al.*, *ibid.* **1934**, 62; Reichstein *et al.*, *Helv. Chim. Acta* **17**, 510 (1934); Stacey, Turton, *J. Chem. Soc.* **1946**, 661; Stedehouder, *Rec. Trav. Chim.* **71**, 831 (1952). Review on analogs of ascorbic acid: Smith, *Adv. Carbohydr. Chem.* **2**, 79 (1946).

Monohydrate. Clusters of rod-like crystals with pointed ends from acetone + methanol + petr ether, mp 101-105° (Reichstein); mp 138° (Baird). Becomes anhydr at 70° under high vacuum, mp 191° (Ault). $[\alpha]_D^{20}$ −14° (c = 1 as hydrate); $[\alpha]_D^{20}$ −22° (c = 1 as hydrate in methanol); $[\alpha]_D^{14.5}$ −37.8° (c = 2.41 in 0.01*N* HCl); $[\alpha]_D^{20}$ −80° (c = 0.75 neutralized with NaOH). pK_1 4.26; pK_2 11.58. Soluble in water; moderately sol in alcohol.

4487. Glucocerebrosidase. [37228-64-1] Glucosylceramidase; acid β-glucosidase; glucosylcerebrosidase; β-glucocerebrosidase; glucosylsphingosine β-glucosidase; GlcCer-β-glucosidase; EC 3.2.1.45. Lysosomal enzyme that hydrolyzes the β-glucosyl linkage of glucosylceramide. Human form is a monomeric glycoprotein with 497 amino acid residues and containing ~12% carbohydrate; mol wt ~67 kDa. Genetic deficiency of the enzyme results in the glycolipid storage disorder known as Gaucher's disease. Purification from human spleen: R. O. Brady *et al.*, *J. Biol. Chem.* **240**, 39 (1965); from human placental tissue: P. G. Pentchev *et al.*, *ibid.* **248**, 5256 (1973). Role in Gaucher's disease: R. O. Brady *et al.*, *Biochem. Biophys. Res. Commun.* **18**, 221 (1965). Cloning and sequence determn: J. Sorge *et al.*, *Proc. Natl. Acad. Sci. USA* **82**, 7289 (1985). Structure of oligosaccharide units: S. Takasaki *et al.*, *J. Biol. Chem.* **259**, 10112 (1984). Sequential deglycosylation to expose mannose residues and effect on macrophage uptake: F. S. Furbish *et al.*, *Biochim. Biophys. Acta* **673**, 425 (1981); G. J. Murray, *Methods Enzymol.* **149**, 25 (1987). X-ray crystal structure: H. Dvir *et al.*, *EMBO Rep.* **4**, 704 (2003).

Alglucerase. [143003-46-7] Glucosylceramidase (human placental isoenzyme protein moiety reduced); mannose-terminated human placental glucocerebrosidase; Ceredase. Modified form of human glucocerebrosidase designed to target the tissue macrophages in which the lipid accumulates. Prepd by sequential deglycosylation of the native placental-derived enzyme to expose terminal mannose residues that can be recognized by macrophage cell surface receptors. Mol wt 59.3 kDa; contains ~6% carbohydrate. Clinical evaluation in Gaucher's disease: N. W. Barton *et al.*, *N. Engl. J. Med.* **324**, 1464 (1991); M. L. Figueroa *et al.*, *ibid.* **327**, 1632 (1992). Review of pharmacology and therapeutic efficacy: R. Whittington, K. L. Goa, *Drugs* **44**, 72-93 (1992). Effect of long-term treatment: N. J. Weinreb *et al.*, *Am. J. Med.* **113**, 112-119 (2002).

THERAP CAT: Enzyme replacement therapy in Gaucher's disease.

4488. Glucofrangulin. [52731-38-1] Erbalax-N; Irgalax. $C_{27}H_{30}O_{14}$; mol wt 578.52. C 56.06%, H 5.23%, O 38.72%. Anthraquinone glycoside from bark of *Rhamnus frangula* L., *Rhamnaceae* (alder buckthorn): Casparis, Maeder, *Bull. Soc. Chim. Biol.* **9**, 324 (1927); Cucu, Jarpo, *Pharmazie* **14**, 316 (1959); Knap *et al.*, *CS* **110024** (1964), *C.A.* **61**, 4159e (1964). Consists of two isomers, glucofrangulin A and glucofrangulin B, which differ by the linkage to the sugar moiety at the 3 position of the aglycone. Structure: Schindler, *Helv. Chim. Acta* **29**, 411 (1946); Hörhammer *et al.*, *Naturwissenschaften* **51**, 310 (1964); Longo *et al.*, *Arch. Pharm.* **297**,

248 (1964); Wagner, Hörhammer, *Naturwissenschaften* **53**, 585 (1966). Proof of structure of glucofrangulin A and partial synthesis: *eidem, Z. Naturforsch.* **24B**, 1408 (1969).

Glucofrangulin A. [21133-53-9] 3-[(6-Deoxy-α-L-manno-pyranosyl)oxy]-1-(β-D-glucopyranosyloxy)-8-hydroxy-6-methyl-9,10-anthracenedione.
Glucofrangulin A octaacetate. $C_{43}H_{46}O_{22}$. Needles from methanol, mp 228-230°. $[\alpha]_D^{20} -124°$ (c = 1.16 in acetone). uv max: 212, 264, 360 nm (log ε 4.57, 4.56, 4.26).

THERAP CAT: Cathartic.

4489. α-Glucogallin. [53318-36-8] α-D-Glucopyranose 1-(3,4,5-trihydroxybenzoate); α-D-glucopyranose-1-gallate; 1-galloyl-α-D-glucose. $C_{13}H_{16}O_{10}$; mol wt 332.26. C 46.99%, H 4.85%, O 48.15%. Synthesis: Schmidt, Herok, *Ann.* **587**, 63 (1954); Schmidt, Schmadel, *ibid.* **649**, 149 (1961).

Crystals, dec 179-181°. $[\alpha]_D^{20} +83°$ (c = 3 in methanol).
Dihydrate. Prisms from water, mp 171-173°. $[\alpha]_D^{25} +79.1°$ (water). Freely sol in water, methanol, ethanol, dioxane, acetic acid. Sparingly soluble in ethyl acetate, ether and acetone.

4490. β-Glucogallin. [13405-60-2] β-D-Glucopyranose 1-(3,4,5-trihydroxybenzoate); β-D-glucopyranose-1-gallate; glucogallic acid; 1-galloyl-β-D-glucose. $C_{13}H_{16}O_{10}$; mol wt 332.26. C 46.99%, H 4.85%, O 48.15%. Glucoside or glucotannoid from chinese rhubarb, *Rheum officinale*, Baill., *Polygonaceae:* Gilson, *Compt. Rend.* **136**, 385 (1903). Structure and synthesis: Fischer, Bergmann, *Ber.* **51**, 1760 (1918); Schmidt, Schmadel, *Ann.* **649**, 149 (1961).

Bitter microscopic prisms from water, methanol or 80% ethanol, mp 207°. $[\alpha]_D^{25} -24.5°$ (c = 1.75 in water). Freely sol in hot water. Sparingly sol in cold water, methanol, ethanol, acetone, ethyl acetate. Practically insol in ether, benzene, chloroform, petr ether.

4491. Glucoheptonic Acid. [87-74-1] D-*glycero-*D-*gulo*-Heptonic acid; α-glucoheptonic acid; glucosemonocarboxylic acid; glucomonocarbonic acid. $C_7H_{14}O_8$; mol wt 226.18. C 37.17%, H 6.24%, O 56.59%. Obtained by treating glucose with HCN yielding a cyanohydrin which is saponified to glucoheptonic acid: Kiliani, *Ber.* **19**, 769 (1886); Fischer, *Ann.* **270**, 71 (1892); Armestar, *C.A.*

45, 2865 (1951). Process starting with calcium cyanide and glucose: Clevenot, **US 2735866** (1956 to Lab. Clevenot). Diagnostic use of ^{99m}Tc complexes in renal scintigraphy: R. E. Boyd *et al., Br. J. Radiol.* **46**, 604 (1973); in brain scanning: J. Léveillé *et al., J. Nucl. Med.* **18**, 957 (1977); T. W. Ryerson *et al., Radiology* **127**, 429 (1978). Subacute toxicity study: L. Belbeck *et al., Can. J. Comp. Med.* **45**, 299 (1981).

Lactonizes upon evapn. The lactone forms large sweetish crystals, mp 145-148°. $[\alpha]_D^{20} -56.0°$ (shows mutarotation). Sol in water.
Sodium salt. [13007-85-7] Gluceptate sodium; sodium gluco-heptonate. $C_7H_{13}NaO_8$. Prepn from corn syrup: Behnke, **US 3022343** (1962 to Pfanstiehl Labs). Crystals (α-form), dec 161°. $[\alpha]_D^{20} +6.06°$ (c = 10 in H_2O). Freely sol in water.
Calcium salt. [17140-60-2] Gluceptate calcium; calcium gluco-heptonate; calcium glucosemonocarbonate; calcium glucomonocar-bonate; Calciforte; Calheptose. $C_{14}H_{26}CaO_{16}$; mol wt 490.42. Prepn from Na salt: Holstein, **US 3033900** (1962 to Pfanstiehl Labs.). Hygroscopic crystals, somewhat acrid taste, dec 200°. Sol in water.
Magnesium salt. [74347-32-3] Magnesium glucoheptonate; magnesium glucosemonocarbonate; magnesium glucomonocarbon-ate; Navolin. $C_{14}H_{26}MgO_{16}$; mol wt 474.65. Prepn: Cipelli, **US 3063896** (1962 to Merck & Co.). Water-sol crystals, pleasant taste.
Complex with ^{99m}**Tc.** Technetium Tc 99m gluceptate; techne-tium Tc 99m glucoheptonate; Glucoscan; TechneScan gluceptate.

USE: Pharmaceutic aid.
THERAP CAT: ^{99m}Tc complex as diagnostic aid (radioactive imaging agent).

4492. Gluconic Acid. [526-95-4] D-Gluconic acid; dextronic acid; maltonic acid; glyconic acid; glycogenic acid; pentahydroxy-caproic acid. $C_6H_{12}O_7$; mol wt 196.16. C 36.74%, H 6.17%, O 57.09%. Prepd by oxidation of glucose: H. Hlasiwetz, J. Haber-mann, *Ann.* **155**, 120 (1870); J. Habermann, *ibid.* **162**, 297 (1872). Fermentative prepn using *Aspergillus niger:* K. Bernhauer, L. Schu-lof, **US 1849053** (1932 to Pfizer); A. J. Moyer *et al., Ind. Eng. Chem.* **32**, 1379 (1940). Review of prepns and uses: F. J. Prescott *et al., ibid.* **45**, 338 (1953); M. Roehr *et al.* in *Biotechnology*, **Vol. 6**, H. Rehm, G. Reed, Eds. (VCH, Weinheim, 2nd ed, 1996) pp 347-362. *See also* Gluconolactone.

Crystals, mp 131°. Mild acid taste. $[\alpha]_D^{20} -6.7°$ (c = 1). pK (25°) 3.60. Freely sol in water, slightly sol in alcohol. Insol in ether and most other organic solvents. In aq solns the acid is partially transformed into an equilibrium mixt with gamma and delta gluconolactones.
Magnesium salt. [3632-91-5] Magnesium gluconate; Almora; Ultra-Mg. $C_{12}H_{22}MgO_{14}$; mol wt 414.60. Clinical pharmacokinetics and use as magnesium supplement: J. White *et al., Clin. Ther.* **14**, 678 (1992). Also occurs as the dihydrate. Sol in water, slightly sol in alcohol. Insol in ether.
Potassium salt. [299-27-4] Potassium gluconate; Gluconsan K; Kalimozan; Kaon; Potasoral; Potassuril; Tumil-K. $C_6H_{11}KO_7$; mol wt 234.25. Yellowish-white crystals. Stable in air. Mild, slightly saline taste. Dec 180°. Freely sol in water. Practically insol in abs alcohol, ether, benzene, chloroform. Aq solns are alkaline to litmus and have a pH of 7.5-8.5.
Sodium salt. [527-07-1] Sodium gluconate. $C_6H_{11}NaO_7$; mol wt 218.14. Crystals. The technical grade may have a pleasant odor.

Soly in water at 25°: 59 g/100 ml. Sparingly sol in alcohol. Insol in ether. Aq solns are stable to short boiling periods.

Zinc complex. [4468-02-4] $(T\text{-}4)$-Bis-(D-gluconato-$\kappa O^1,\kappa O^2$)-zinc; zinc gluconate; Rubozinc. $C_{12}H_{22}O_{14}Zn$; mol wt 455.70. Review of clinical use in treatment of colds: M. L. Garland, K. O. Hagmeyer, *Ann. Pharmacother.* **32**, 63-69 (1998). Clinical trial in inflammatory acne: J. Meynadier, *Eur. J. Dermatol.* **10**, 269 (2000).

USE: Chelating agent. In high alkalinity bottle washes and other cleansers; in finish removers; in the tanning and textile industry.

THERAP CAT: Magnesium salt as magnesium replenisher; potassium salt as potassium supplement; zinc complex as zinc supplement.

THERAP CAT (VET): Potassium salt as potassium supplement.

4493. Gluconolactone. [90-80-2] D-Gluconic acid δ-lactone; glucono delta lactone; delta gluconolactone; Fujiglucon. $C_6H_{10}O_6$; mol wt 178.14. C 40.45%, H 5.66%, O 53.89%. Prepn by oxidation of glucose with bromine water: Isbell, Pigman, *J. Res. Natl. Bur. Stand.* **10**, 337 (1933); by oxidation of glucose in *Acetobacter suboxydans*: King, Cheldelin, *Biochem. J.* **68**, 31P (1958). Structure: J. Staněk *et al.*, *The Monosaccharides* (Academic Press, New York, 1963) p 271.

Crystals, dec 153°. Sweet taste (different from gluconic acid). $[\alpha]_D^{20} +61.7°$ (c = 1). Soly in water 59 g/100 ml; in alc about 1 g/100 g. Insol in ether. Hydrolyzed to gluconic acid by water. A freshly prepd 1% aq soln has a pH of 3.6 changing to pH 2.5 within 2 hrs.

USE: Component of many cleaning cmpds because of the sequestering ability of the gluconate radical which remains active in alk solns; in the dairy industry to prevent milkstone; in breweries to prevent beerstone; as latent acid catalyst for acid colloid resins, particularly in textile printing; as a coagulant for tofu.

4494. Glucosamine. [3416-24-8] 2-Amino-2-deoxy-D-glucose; chitosamine. $C_6H_{13}NO_5$; mol wt 179.17. C 40.22%, H 7.31%, N 7.82%, O 44.65%. Found in chitin, in mucoproteins, and in mucopolysaccharides. Isoln from chitin: Ledderhose, *Z. Physiol. Chem.* **2**, 213 (1878); Hackman, *Aust. J. Biol. Sci.* **7**, 168 (1954). Synthesis: Fischer, Leuchs, *Ber.* **35**, 3787 (1902); **36**, 24 (1903). Separation of α- and β-forms: Westphal, Holzmann, *ibid.* **75B**, 1274 (1942). Structure: Haworth *et al.*, *J. Chem. Soc.* **1939**, 271; Cutler, Peat, *ibid.* 782; Cox, Jeffrey, *Nature* **143**, 894 (1939). Pharmacokinetics in dog and man: I. Setnikar *et al.*, *Arzneim.-Forsch.* **36**, 729 (1986). HPTLC/densitometry determn in herbal supplement: V. Esters *et al.*, *J. Chromatogr. A* **1112**, 156 (2006). HPLC determn in plasma and bioequivalence: L. Zhang *et al.*, *J. Chromatogr. B* **842**, 8 (2006). Antioxidant activity and immunostimulating properties: Y. Yan *et al.*, *Int. Immunopharmacol.* **7**, 29 (2007). Clinical trials in arthrosis: Y. Vajarudal, *Clin. Ther.* **3**, 336 (1981); M. J. Tapadinhas *et al.*, *Pharmatherapeutica* **3**, 157 (1982). Review: Foster, Stacey, "The Chemistry of the 2-Amino Sugars" in C. S. Hudson *et al.*, *Advan. Carbohyd. Chem.* vol. 7 (Academic Press, New York, 1952) pp 247-288. Review of pharmacology, toxicology and clinical efficacy: J. W. Anderson *et al.*, *Food Chem. Toxicol.* **43**, 187-201 (2005).

α-Form. [28905-11-5] Crystals, mp 88°. $[\alpha]_D^{20} +100°$ changing to +47.5° after 30 min (water).

β-Form. [28905-10-4] Needles from methanol, dec 110°. $[\alpha]_D^{20} +28°$ changing to +47.5° after 30 min (water). Very sol in water, sol in about 38 parts boiling methanol; sparingly sol in cold methanol or ethanol. Practically insol in ether, chloroform.

N-Acetylglucosamine. [7512-17-6] $C_8H_{15}NO_6$. Needles from methanol + ether, mp 205°. $[\alpha]_D^{18} +64°$ changing to +40.9° (in water).

Sulfate salt. [29031-19-4] Dona; Faximin. $C_6H_{13}NO_5\cdot xH_2SO_4$.

USE: Pharmaceutic aid.

THERAP CAT: Antiarthritic.

THERAP CAT (VET): Antiarthritic.

4495. Glucose. [50-99-7] D-Glucose; dextrose; blood sugar; grape sugar; corn sugar; Dextropur; Dextrosol; Glucolin. $C_6H_{12}O_6$; mol wt 180.16. C 40.00%, H 6.71%, O 53.28%. A main source of energy for living organisms. Occurs naturally and in the free state in fruits and other parts of plants. Combined in glucosides, in di- and oligosaccharides, in the polysaccharides cellulose and starch, and in glycogen. Normal human blood contains 0.08-0.1%. Manuf on a large scale from starch: Dean, Gottfried, *Adv. Carbohydr. Chem.* **5**, 127 (1950). Below 50°, α-D-glucose hydrate is the stable cryst form, above 50° the anhydr form is obtained and at still higher temps β-D-glucose is formed: W. Pigman, *The Carbohydrates* (Academic Press, New York, 1957) p 92. Structure: Kjaer, Lindberg, *Acta Chem. Scand.* **13**, 1713 (1959). Conformation: E. Percival, *Structural Carbohydrate Chemistry* (J. Garnet Miller, London, 1962) pp 51-57. Comprehensive monograph: H. Bartelheimer *et al.*, *D-Glucose und verwandte Verbindungen in Medizin und Biologie* (Enke, Stuttgart, 1966) 1126 pp.

α-Form monohydrate. Crystals from water, mp 83°. $[\alpha]_D +102.0° \rightarrow +47.9°$ (water). 0.74 times as sweet as sucrose. One gram dissolves in about 1 ml water and in about 60 ml alcohol.

α-Form anhydr. Crystals from ethanol or water, mp 146°. $[\alpha]_D +112.2° \rightarrow +52.7°$ (c = 10 in water). The final value is obtained instantly in the presence of hydroxyl ions. Formula for varying concns: $[\alpha]_D^{20} +52.5° + 0.0188p$ (p = g/100 ml). pH of 0.5 molar aq soln 5.9. $d_{17.5}^{17.5}$ of water solns w/v: 5% = 1.019; 10% = 1.038; 20% = 1.076; 30% = 1.113; 40% = 1.149. n_D^{20} 10% soln 1.3479. One gram dissolves in 1.1 ml water at 25°; in 0.8 ml at 30°; in 0.41 ml at 50°; in 0.28 ml at 70°; in 0.18 ml at 90°; in 120 ml methanol at 20°. Very sparingly sol in abs alcohol, ether, acetone; sol in hot glacial acetic acid, pyridine, aniline.

β-Form. Crystals from hot water + ethanol, from dil acetic acid, or from pyridine, mp 148-155°. $[\alpha]_D +18.7° \rightarrow +52.7°$ (c = 10 in water).

THERAP CAT: Antihypoglycemic; fluid and nutrient replenisher.

THERAP CAT (VET): Antihypoglycemic; fluid and nutrient replenisher.

4496. Glucose Oxidase. [9001-37-0] β-D-Glucopyranose aerodehydrogenase; β-D-glucose:oxygen 1-oxidoreductase; EC 1.1.3.4; P-FAD; corylophyline; microcide; mikrotsid; notatin. An enzyme obtained from mycelia of fungi, such as *Aspergilli* and *Penicillia*; a typical aerobic dehydrogenase which catalyzes the oxidation of glucose to gluconic acid (molecular oxygen is reduced to hydrogen peroxide). It is a flavoprotein, the prosthetic group being flavine-adenine dinucleotide (FAD). Commercial prepns frequently contain appreciable amounts of another enzyme, catalase, which is desirable for certain uses since it removes hydrogen peroxide aerobically generated by glucose oxidase. Names of some commercial prepns are: **DeeO, Fermcozyme, OxyBan, Ovazyme**. Isoln from *Penicillia* cultures: Coulthard *et al.*, *Biochem. J.* **39**, 24 (1945). Commercial production from *Aspergilli* and *Penicillia*: Goldsmith *et al.*, US 2926122 (1960); from *Aspergillus niger*: Faucett *et al.*, US 3102081 (1963 to Miles Labs.). Removal of proteolytic enzymes from glucose oxidase (contg catalase) obtained from *Aspergilli* or *Penicillia* cultures: Ohlmeyer, US 2940904 (1960 to Ben L. Sarett).

Separation from catalase: Pazur *et al.*, *Biochim. Biophys. Acta* **65**, 369 (1962). Properties: Muller, *Enzymologia* **10**, 40 (1941); Keilin, Hartree, *Biochem. J.* **42**, 221 (1948), **50**, 331 (1952). *Reviews:* L. A. Underkofler "Glucose Oxidase: Production, Properties, Present and Potential Applications" in *Soc. Chem. Ind. (London) Monograph* no. **11**, 72-86 (1961); R. Bentley, "Glucose Oxidase" in *The Enzymes* vol. **7**, P. D. Boyer *et al.*, Eds. (Academic Press, New York, 1963) pp 567-586. Review of use as analytical reagent: J. Raba, H. A. Mottola, *Crit. Rev. Anal. Chem.* **25**, 1-42 (1995).

Amorphous powder or crystals. Abs max between 270-280, 375-380, and 450-460 nm (aq soln). Freely sol in water giving yellowish-green solns. Most active at pH 5.5-6.0 and 30-35°. Stable between pH 4.5 and 7.0. Stable to pepsin and trypsin. A glucose oxidase unit is defined as that quantity of enzyme which will cause the uptake of 10 mm^3 oxygen per min in a Warburg manometer at 30° in the presence of excess air and excess catalase with a substrate contg 3.3% glucose monohydrate and 0.1M phosphate buffer, pH 5.9 with 0.4% sodium dehydroacetate: Scott, *J. Agric. Food Chem.* **1**, 727 (1953).

USE: Analytical reagent for the selective determn of glucose. Food additive for the removal of glucose during the prepn of dried egg products. Antioxidant in food and food wrappers. Stabilizer for ascorbic acid and vitamin B$_{12}$.

4497. α-Glucose-1-phosphate. [59-56-3] α-D-Glucopyranose 1-(dihydrogen phosphate); α-glucose-1-phosphoric acid; α-D-glucopyranose-1-phosphate; Cori ester. C$_6$H$_{13}$O$_9$P; mol wt 260.13. C 27.70%, H 5.04%, O 55.35%, P 11.91%. Found widely in both plants and animals. In plants it is the immediate precursor of starch, and in animals of glycogen, being also the first product in the breakdown and utilization of these substances. Isoln from muscle and synthesis using trisilver phosphate: Cori *et al.*, *J. Biol. Chem.* **121**, 465 (1937); Krahl, Cori, *Biochem. Prep.* **1**, 33 (1949). Prepn from α-acetobromglucose + silver diphenyl phosphate: Posternak, *J. Am. Chem. Soc.* **72**, 4824 (1950); by phosphorolysis of starch using phosphorylase and orthophosphate: McCready, Hassid, *Biochem. Prep.* **4**, 63 (1955). Structure: Wolfrom, Pletcher, *J. Am. Chem. Soc.* **63**, 1050 (1941). Configuration: Wolfrom *et al.*, *ibid.* **64**, 23 (1942); Harmon, *Diss. Abstr.* **24**, 4400 (1964); Beevers, Maconochie, *Acta Crystallogr.* **18**, 232 (1965).

[α]$_D^{25}$ +120°. pK$_1$ = 1.11; pK$_2$ = 6.13. Stronger acid than H$_3$PO$_4$. Extremely sol in water.
Barium salt trihydrate. [6056-70-8] Nonhygroscopic, stable powder. [α]$_D^{25}$ +75° (c = 1.26). Easily sol in water.
Calcium salt. [119119-85-6] Actigam.
Dipotassium salt dihydrate. [5996-14-5] Crystals from ethanol. [α]$_D^{20}$ +78° (c = 4); [α]$_{549}^{20}$ +90° (c = 4). Freely sol in water.
THERAP CAT: Calcium salt as roborant.

4498. Glucose-6-phosphate. [56-73-5] D-Glucose 6-(dihydrogen phosphate); glucose-6-phosphoric acid; Robison ester. C$_6$H$_{13}$O$_9$P; mol wt 260.13. C 27.70%, H 5.04%, O 55.35%, P 11.91%. A normal constituent of resting muscle, probably always existing in equilibrium with fructose-6-phosphate. For the enzymatic conversion from the 1-phosphate *see* α-Glucose-1-phosphate. Isoln from a crude mixture of hexose phosphates, obtained by yeast fermentation: Robison, King, *Biochem. J.* **25**, 323 (1931). Prepn by the action of phosphoglucomutase on α-glucose-1-phosphate: Colowick, Sutherland, *J. Biol. Chem.* **144**, 423 (1942); from acetone glucose: Levene, Raymond, *ibid.* **92**, 757 (1931); by phosphorylation of 1,2,3,4-tetra-acetylglucose followed by deacetylation: Fischer, Lardy, *ibid.* **164**, 513 (1946); *Biochem. Prep.* **2**, 39 (1952). Prepn from starch: de Chatelperron *et al.*, **FR 1379068** (1964), *C.A.* **62**, 9394b (1965).

Barium salt. C$_6$H$_{11}$O$_5$PO$_4$Ba. Nonhygroscopic, stable powder. [α]$_D^{24}$ +17.9°. Easily sol in water.
Dipotassium salt. C$_6$H$_{11}$O$_5$PO$_4$K$_2$. Precipitate from methanol. [α]$_D^{24}$ +21.2° (c = 1.3). Freely sol in water.

4499. Glucosulfone Sodium. [554-18-7] 1,1'-[Sulfonyl-bis(4,1-phenyleneimino)]bis[1-deoxy-1-sulfo-D-glucitol] sodium salt (1:2); *p,p'*-sulfonyldianiline *N,N'*-diglucoside disodium disulfonate; *p,p'*-sulfonyldianiline-*N,N'*-di-D-glucose sodium bisulfite compd; *p,p'*-diaminodiphenylsulfone-*N,N'*-di(dextrose sodium sulfonate); disodium *p,p'*-diaminodiphenylsulfone-*N,N'*-diglucose sulfonate; 501-P; Protomin; Promin; Promanide. C$_{24}$H$_{34}$N$_2$Na$_2$O$_{18}$S$_3$; mol wt 780.69. C 36.92%, H 4.39%, N 3.59%, Na 5.89%, O 36.89%, S 12.32%. Prepd by refluxing a mixture of 4,4'-diaminodiphenylsulfone, glucose, sodium bisulfite, and 80% ethanol: **CH 234108** (1944 to B. Siegfried), *C.A.* **43**, 4297a (1949); B. C. Jain *et al.*, *Sci. Cult.* **11**, 568 (1946). Structure activity study: W. Logemann, G. P. Miori, *Arzneim.-Forsch.* **5**, 213 (1955). Clinical evaluation in ocular leprosy: A. Bouzas, *Arch. Ophthalmol.* **31**, 629 (1971).

White, amorphous powder. Soluble in water; slightly sol in ethanol. Insol in ether, benzene, methanol, ethyl acetate, pyridine. Aq solns may be sterilized by autoclaving. LD$_{50}$ in rats (g/kg): 3-4 orally, 3-3.5 i.v. (Logemann, Miori).
THERAP CAT: Antibacterial (leprostatic).

4500. Glucovanillin. [494-08-6] 4-(β-D-Glucopyranosyloxy)-3-methoxybenzaldehyde; vanillin-D-glucoside; avenein; vaniloside. C$_{14}$H$_{18}$O$_8$; mol wt 314.29. C 53.50%, H 5.77%, O 40.72%. From green fruit of vanilla: Goris, *Compt. Rend.* **179**, 70 (1924). From coniferin by oxidation with CrO$_3$: Tremann, *Ber.* **18**, 1595 (1885). Structure and synthesis: Fischer, Raske, *Ber.* **42**, 1475 (1909); Thorpe, Williams, *J. Chem. Soc.* **1937**, 494.

Needles from methanol, mp 189-190°. Bitter taste. [α]$_D^{20}$ −89.9° (water). Soluble in hot water, alcohol. Almost insol in ether.
Tetraacetate. C$_{22}$H$_{26}$O$_{12}$. Crystals from dil alcohol, mp 142-143°. [α]$_D$ −48.3° (chloroform).
USE: Pharmaceutic aid (flavor).

4501. D-Glucuronic Acid. [6556-12-3] C$_6$H$_{10}$O$_7$; mol wt 194.14. C 37.12%, H 5.19%, O 57.69%. Widely distributed in the plant and animal kingdoms. Usually occurs in "paired" form, *i.e.* as a glycosidic combination with phenols, alcohols, etc. Such glucuronides form in the liver to detoxify poisonous hydroxyl-containing substances. The glucuronides present in normal urine are those of phenol, cresol, and indoxyl. After the ingestion of poisons such as morphine, chloral hydrate, camphor, or turpentine, glucuronides formed with the poison or its hydroxylated derivatives appear in the urine. Review and bibliography: Stacey, *Adv. Carbohydr. Chem.* **2**, 161 (1946); Jones, Smith, *ibid.* **4**, 243 (1949). Structure: Pryde, Williams, *Nature* **128**, 187 (1931); Levene, Meyer, *J. Biol. Chem.* **92**, 257 (1931); Levene, Kreider, *ibid.* **120**, 597 (1937). Review of syntheses: Mehltretter, *Adv. Carbohydr. Chem.* **8**, 231 (1953). Prepn by irradiation of D-glucose in dil aq soln: Phillips *et al., J. Chem. Soc.* **1958**, 3522; by γ-irradiation of aq sucrose soln: Phillips, Moody, *ibid.* **1960**, 762. Electrophoretic sepn of D-glucuronic acid and its C-5 epimer, *L-iduronic acid*: I. Miyamoto, S. Nagase, *Anal. Biochem.* **115**, 308 (1981). Monographs: N. E. Artz, E. M. Osman, *Biochemistry of Glucuronic Acid* (Academic Press, New York, 1950); G. J. Dutton, Ed., *Glucuronic Acid, Free and Combined* (Academic Press, New York, 1966) 629 pp.

β-Form. Needles from alcohol or ethyl acetate. mp 165°. Shows mutarotation: $[\alpha]_D^{24}$ +11.7° → +36.3° (2 hrs, c = 6). Soluble in water, alcohol. Reduces Fehling's soln.

4502. β-Glucuronidase. [9001-45-0] β-D-Glucuronoside glucuronosohydrolase; EC 3.2.1.31; Glusulase. Glucuronide-splitting enzyme found in liver, spleen, and certain tissues of the endocrine and reproductive systems. Rats seem to have a higher concn than other mammals. Also found in fish liver, snails, mollusks, and some insects. Isoln from rat livers, kidneys, and spleens: Fishman, Talalay, *Science* **105**, 131 (1947). *Review and bibliography:* Fishman, *Adv. Enzymol.* **16**, 361-409 (1955).

White powder. Soluble in water.

USE: In the determination of urinary steroids and of steroid conjugates in blood.

4503. D-Glucuronolactone. [32449-92-6] D-Glucuronic acid γ-lactone; D-glucofuranurono-6,3-lactone; glucurolactone; glucurone; Dicurone; Glucoxy; Guronsan. $C_6H_8O_6$; mol wt 176.12. C 40.92%, H 4.58%, O 54.50%. Found in many plant gums in polymeric combination with other carbohydrates. Important structural constituent of practically all fibrous and connective tissues in the animal organism, *cf.* D-glucuronic acid. Prepd synthetically from many polysaccharides or suitable glucosides where the hydroxyl at carbon 6 may be oxidized while the other sensitive groups are protected. Prepn: Stacey, *J. Chem. Soc.* **1939**, 1529; Hardegger, Spitz, *Helv. Chim. Acta* **33**, 337 (1950); Marsh, *Proc. Biochem. Soc. [Biochem. J.]*, **50**, XI (1951); Mehltretter *et al., J. Am. Chem. Soc.* **73**, 2424 (1951); Phillips, Moody, *J. Chem. Soc.* **1960**, 762. Structure: J. Stanek *et al., The Monosaccharides* (Academic Press, New York, 1963) p 259. For isoln procedures see the ref under glucuronic acid.

Crystals from ethanol, mp 176-178°. (Commercial grades, mp 172°.) d_4^{30} 1.76. $[\alpha]_D^{25}$ +19.8° (c = 5.19). Soluble in water (26.9 g/100 ml of soln); slightly sol in methanol (2.8 g/100 ml). Very slightly sol in abs ethanol (0.7 g/100 ml), in glacial acetic acid (0.3 g/100 ml). The free acid is more sol than the lactone. At room temp an aq soln of glucuronolactone reaches an equilibrium of about 20% lactone and 80% acid within 2 months. At 100° an equilibrium of 60% lactone and 40% free acid is reached within 2 hrs. Initial pH of 10% aq soln 3.5, after 1 week the pH is about 2.5.

THERAP CAT: Detoxicant.

4504. Glufosfamide. [132682-98-5] 1-[N,N'-Bis(2-chloroethyl)phosphorodiamidate]-β-D-glucopyranose; β-D-glucopyrano-

syl N,N'-di(2-chloroethyl) phosphoric acid diamide; β-D-glucosyl-isophosphoramide mustard; D-19575. $C_{10}H_{21}Cl_2N_2O_7P$; mol wt 383.16. C 31.35%, H 5.52%, Cl 18.50%, N 7.31%, O 29.23%, P 8.08%. Glucose coupled to isophosphoramide mustard, the active metabolite of the alkylating agent ifosfamide, *q.v.* Enters cells through upregulated glucose transport proteins. Prepn: M. Wiessler, M. Dickes, **DE 3835772**; *eidem*, **US 5622936** (1990, 1997 both to Deutsches Krebsforschungszentrum Stiftung). Pharmacology and toxicity studies: J. Pohl *et al., Cancer Chemother. Pharmacol.* **35**, 364 (1995). Mechanism of action study: H. Seker *et al., Br. J. Cancer* **82**, 629 (2000). Clinical evaluation in pancreatic cancer: E. Briasoulis *et al., Eur. J. Cancer* **39**, 2334 (2003); in non-small cell lung cancer: G. Giaccone *et al., ibid.* **40**, 667 (2004). Review of pharmacology and clinical development: I. Niculescu-Duvaz, *Curr. Opin. Investig. Drugs* **3**, 1527-1532 (2002).

LD_{50} in rats, mice (mg/kg): 1575, 1575 i.v.; 1470, 1470 orally (Pohl).

THERAP CAT: Antineoplastic.

4505. Glutamic Acid. [56-86-0] L-Glutamic acid; Glu; E; glutaminic acid; (*S*)-2-aminopentanedioic acid; α-aminoglutaric acid; 1-aminopropane-1,3-dicarboxylic acid; Glutacid; Glutaminol. $C_5H_9NO_4$; mol wt 147.13. C 40.82%, H 6.17%, N 9.52%, O 43.50%. Non-essential amino acid for human development; referred to as an excitatory amino acid (EAA) due to its role in neurotransmission. Isoln from wheat gluten: H. Ritthausen, *J. Prakt. Chem.* **99**, 454 (1866). Early chemistry and biochemistry: *Amino Acids and Proteins*, D. M. Greenberg, Ed. (Charles C. Thomas, Springfield, IL, 1951) 950 pp., *passim*; J. P. Greenstein, M. Winitz, *Chemistry of the Amino Acids* vols **1-3** (John Wiley and Sons, Inc., New York, 1961) pp. 1929-1954, *passim*; C. W. Huffman, W. G. Skelly, *Chem. Rev.* **63**, 625-644 (1963). Interconversion from L-proline: S. Yoshifuji *et al., Tetrahedron Lett.* **21**, 2963 (1980). Determn in serum: C. D. Stalikas *et al., Eur. J. Clin. Chem.* **32**, 767 (1994); of naturally occurring levels in food: D. H. Daniels *et al., Food Addit. Contam.* **12**, 21 (1995). Identification as excitatory neurotransmitter: D. R. Curtis *et al., J. Physiol.* **150**, 656 (1960). Review of biosynthesis: A. Hamberger *et al., Adv. Biochem. Psychopharmacol.* **27**, 115-126 (1981); of metabolism and associated disorders: S. B. Prusiner, *Annu. Rev. Med.* **32**, 521-542 (1981). Review of receptor binding: P. A. Briley *et al., Mol. Cell. Biochem.* **39**, 347 (1981). Review as neurotransmitter: B. Engelsen, *Acta Neurol. Scand.* **74**, 337-355 (1986); E. Marmo, *Med. Res. Rev.* **8**, 441-458 (1988). Review of role in learning and memory: W. J. McEntee, T. H. Crook, *Psychopharmacology* **111**, 391-401 (1993); in Parkinson's disease: M. S. Starr, *Synapse* **19**, 264-293 (1995). *Books*: R. Powell, *Monosodium Glutamate and Glutamic Acid* (Noyes Dev. Corp., Park Ridge, N.J., 1968) 256 pp.; *Glutamic Acid: Advances in Biochemistry and Physiology* L. J. Filer, Jr. *et al.*, Eds. (Raven, New York, 1979).

Orthorhombic, bisphenoidal crystals from aq alc. d_4^{20} vac 1.538. Melts 160° with conversion to L-pyrrolidonecarboxylic acid. Sublimes at 200°. $[\alpha]_D^{22.4}$ +31.4° (6N HCl). pK_1 2.19; pK_2 4.25; pK_3 9.67. Soly in water (g/l); 8.64 (25°); 21.86 (50°); 55.32 (75°); 140.00 (100°). Insol in methanol, ethanol, ether, acetone, cold glacial acetic acid and common neutral solvents.

Hydrochloride. [138-15-8] Acidulin; Hypochylin. $C_5H_9NO_4$.HCl; mol wt 183.59. Orthorhombic bisphenoidal plates, dec 214°.

Sodium salt *see* Monosodium Glutamate.

Magnesium salt hydrobromide monohydrate. Magnesium glutamate hydrobromide; magnesium bromoglutamate; Psicosoma; Psychoverlan. $C_{10}H_{17}BrMgN_2O_8.H_2O$; mol wt 415.48.

DL-Form. [617-65-2] Orthorhombic crystals from water, 225-227° (dec). d_D^{20} 1.4601. Soly in water (g/l): 20.54 (25°); 49.34 (50°); 118.6 (75°); 284.9 (100°). Sparingly sol in alcohol, ether, petr ether.

D-Form. [6893-26-1] Shiny leaflets from water. $[\alpha]_D^{20}$ −30.5° (c = 1.00 in 6N HCl).

THERAP CAT: Nutritional supplement. Hydrochloride as gastric acidifier. Magnesium salt hydrobromide has been used as anxiolytic.

4506. L-Glutamic Acid 5-Ethyl Ester. [1119-33-1] L-Glutamic acid γ-ethyl ester; γ-ethyl L-glutamate; Glutestere. $C_7H_{13}NO_4$; mol wt 175.18. C 47.99%, H 7.48%, N 8.00%, O 36.53%. Prepn from glutamic acid, ethanol, and hydrogen chloride: Hegedus, *Helv. Chim. Acta* **31**, 737 (1948); Pravda, *Collect. Czech. Chem. Commun.* **24**, 2083 (1959); CS 88344 (1959), *C.A.* **54**, 8660h (1960).

Hydrochloride. [73270-47-0] $C_7H_{13}NO_4.HCl$; mol wt 211.64. Crystals from ethanol + ether, mp 134-136°.

4507. Glutamine. [56-85-9] L-Glutamine; Gln; Q; 2-aminoglutaramic acid; (S)-2,5-diamino-5-oxopentanoic acid; glutamic acid 5-amide; Cebrogen; Glumin; Levoglutamina; Stimulina. $C_5H_{10}N_2O_3$; mol wt 146.15. C 41.09%, H 6.90%, N 19.17%, O 32.84%. Non-essential amino acid for human development; most abundant free amino acid in plasma and tissue. Isoln from sugarbeet juice: E. Schulze, E. Bosshard, *Ber.* **16**, 312 (1883). Due to ease of conversion to glutamic acid, q.v., it was not isolated from protein until 1932: M. Damodoran *et al.*, *Biochem. J.* **26**, 1704 (1932). Early chemistry and biochemistry: R. M. Archibald, *Chem. Rev.* **37**, 161-208 (1945); *Amino Acids and Proteins*, D. M. Greenberg, Ed. (Charles C. Thomas, Springfield, IL, 1951) 950 pp., *passim*; J. P. Greenstein, M. Winitz, *Chemistry of the Amino Acids* **vol 1-3** (John Wiley and Sons, Inc., New York, 1961) pp. 1929-1954, *passim*. HPLC determn in biotech samples: B. Polanuer *et al.*, *J. Chromatogr.* **594**, 173 (1992). Review of nutritive needs: R. J. Smith, D. W. Wilmore, *J. Parenter. Enteral Nutr.* **14**, (Suppl.) 94S-99S (1990); and metabolism: W. W. Souba, *J. Nutr. Biochem.* **4**, 2-9 (1993); L. M. Castell *et al.*, *Amino Acids* **7**, 231-243 (1994). Review of metabolism and physiologic implications: R. J. Smith, *J. Parenter. Enteral Nutr.* **14**, (Suppl.) 40S-44S (1990); B. Moskovitz *et al.*, *Pharmacol. Res.* **30**, 61-71 (1994).

Fine opaque needles from water or dil ethanol, dec 185-186°. $[\alpha]_D^{23}$ +6.1° (c = 3.6). pK$_1$ 2.17; pK$_2$ 9.13. One gram dissolves in 20.8 ml water at 30°, in 38.5 ml at 18°, in 56.7 ml at 0°. Practically insol in methanol (3.5 mg/100 ml at 25°), ethanol (0.46 mg/100 ml at 25°), ether, benzene, acetone, ethyl acetate, chloroform.

DL-Form. [585-21-7] Synthesis: F. E. King, D. A. A. Kidd, *J. Chem. Soc.* **1949**, 3315; G. B. Kline, S. H. Cox, *J. Org. Chem.* **26**, 1854 (1961). Prisms from dil acetone, mp 185-186° (King); mp 173-174.5° (Kline). One gram dissolves in 38.5 ml water at 18°.

4508. Glutaraldehyde. [111-30-8] Pentanedial; glutaral; glutaric dialdehyde; 1,3-diformylpropane; Cidex; Glutarol; Novaruca; Sonacide; Ucarcide; Verucasep; Verutal. $C_5H_8O_2$; mol wt 100.12. C 59.98%, H 8.05%, O 31.96%. Prepn: C. Harries, L. Tank, *Ber.* **41**, 1701 (1908); A. C. Cope *et al.*, *Org. Synth.* **coll. vol. IV**, 816 (1963). Toxicity data: H. F. Smyth *et al.*, *Am. Ind. Hyg. Assoc. J.* **23**, 95 (1962). Clinical evaluation in treatment of viral warts: R. Hirose *et al.*, *J. Dermatol.* **21**, 248 (1994). Review of role in electron microscopy: M. A. Hayat, *Micron Microsc. Acta* **17**, 115-135

(1986); of toxicology: R. O. Beauchamp, Jr. *et al.*, *Crit. Rev. Toxicol.* **22**, 143-174 (1992); of microbiocidal activity and use as chemosterilizing agent: A. D. Russell, *Infect. Control Hosp. Epidemiol.* **15**, 724-733 (1994).

Oil, fp −14°. bp$_{760}$ 187-189° (dec); bp$_{50}$ 106-108°; bp$_{10}$ 71-72°. n_D^{25} 1.43300. Vapor pressure (20°): 0.0152 torr (50% aq soln); 0.0012 torr (2% aq soln). Sol in ethanol, benzene, ether, water; volatile in steam. Polymerizes in water to a glassy form which regenerates the dialdehyde on vacuum distillation. LD$_{50}$ of 25% soln orally in rats: 2.38 ml/kg; by skin penetration in rabbits: 2.56 ml/kg (Smyth).

Dioxime. $C_5H_{10}N_2O_2$. Crystals from water or pyridine, mp 178°. Sublimes. Treatment with hot mineral acids gives pyridine.

Caution: Potential symptoms of overexposure are irritation of eyes, skin, respiratory system; dermatitis, sensitization of skin; cough, asthma; nausea, vomiting. *See: NIOSH Pocket Guide to Chemical Hazards* (DHHS/NIOSH 97-140, 1997) p 152.

USE: Disinfectant; in sterilization of endoscopic instruments; as a tanning agent for leather; fixative for electron microscopy.

THERAP CAT: Keratolytic.

4509. Glutaric Acid. [110-94-1] Pentanedioic acid; 1,3-propanedicarboxylic acid. $C_5H_8O_4$; mol wt 132.12. C 45.45%, H 6.10%, O 48.44%. Occurs in green sugar beets; is found in water extracts of crude wool. Manuf from cyclopentanone by oxidative ring fission with hot 50% nitric acid in the presence of vanadium pentoxide. Lab prepn by acid hydrolysis of trimethylene cyanide: Marvel, Tuley, *Org. Synth.* **5**, 69 (1925), or of methylenedimalonic ester: Otterbacher, *ibid.* **10**, 58 (1930). Several new methods: Paris *et al.*, *Org. Synth.* **coll. vol. IV**, 496 (1963); English, Dayan, *ibid.* 499. Absorption spectrum: *Compt. Rend.* **189**, 915 (1929)

Large monoclinic prisms, mp 97.5-98°. d_4^{15} 1.429. bp$_{760}$ 302-304° (very slight decompn); bp$_{20}$ 200°; bp$_{10}$ 195-198°, pK$_1$ (25°): 4.34; pK$_2$: 5.22. n_D^{106} 1.41878. Soly in water (g/l): at 0°: 429; at 20°: 639; at 50°: 957; at 65°: 1118. Freely sol in abs alcohol, ether; sol in benzene, chloroform; slightly in petr ether.

Dimethyl ester. [1119-40-0] $C_7H_{12}O_4$. Liquid, faint agreeable odor. d_4^{15} 1.0934. bp$_{752}$ 213.5-214°; bp$_{13}$ 93.5-94.5°. n_D^{20} 1.4246. Very sol in alcohol and ether.

4510. Glutaronitrile. [544-13-8] Pentanedinitrile; glutaric acid dinitrile; trimethylene dicyanide; trimethylene cyanide. $C_5H_6N_2$; mol wt 94.12. C 63.81%, H 6.43%, N 29.76%. Prepd by the action of potassium cyanide on trimethylene bromide: Marvel, McColm, *Org. Synth.* **5**, 103 (1925).

Viscous liquid. Bitter-sweet taste. d^{23} 0.9888. mp −29°. bp$_{760}$ 286°; bp$_{100}$ 206°; bp$_{60}$ 190°; bp$_{40}$ 176°; bp$_{20}$ 157°; bp$_{10}$ 140°; bp$_5$ 124°; bp$_{1.0}$ 91.3°. n_D^{23} 1.4365. Soluble in water, alcohol, chloroform. Insol in ether, carbon disulfide.

4511. Glutathione. [70-18-8] L-γ-Glutamyl-L-cysteinylglycine; L-glutathione; glutathione-SH; GSH; Agifutol S; Copren; Delthathione; Glutamed; Glutasan; Glutathin; Glutathiol; Glutathion; Glutinal; Isethion; Neuthion; Tathiclon; Tathion; Tationil; Triptide. $C_{10}H_{17}N_3O_6S$; mol wt 307.32. C 39.08%, H 5.58%, N 13.67%, O 31.24%, S 10.43%. The major low mol wt thiol compound of the living plant or animal cell. Isoln from yeast: Hopkins, *J. Biol. Chem.* **84**, 269 (1929). Synthesis: du Vigneaud, Miller, *Biochem. Prep.* **2**, 87 (1952); Goldschmidt *et al.*, *Ber.* **97**, 2434 (1964); Y. Ozawa *et al.*, *Bull. Chem. Soc. Jpn.* **53**, 2592 (1980). Review of early syntheses: Jeschkeit *et al.*, *Pharmazie* **18**, 658 (1963). Review of metabolism: A. Meister, M. E. Anderson, *Annu. Rev. Biochem.* **52**, 711-760 (1983); of metabolic role in antineoplastic chemotherapy: B. A. Arrick, C. F. Nathan, *Cancer Res.* **44**, 4224-4232 (1984). *Monographs:* S. Colowick *et al.*, *Glutathione* (Academic Press, New York, 1954); *Glutathione*, E. M. Crook, Ed., Biochem. Soc. Symposium No. 17, London, 1958 (Cambridge University Press, 1959); *Glutathione*, L. Flohe *et al.*, Eds. (Academic Press, New York, 1974); *Glutathione: Metabolism & Function* I. M. Arias, W. B. Jackoby, Eds. (Raven, New York, 1976).

Crystals from 50% ethanol, mp 195°. $[\alpha]_D^{25}$ −18.9° (c = 4.653). $[\alpha]_D^{27}$ −21° (c = 2.74). pK$_1'$ 2.12; pK$_2'$ 3.53; pK$_3'$ 8.66; pK$_4'$ 9.12. Freely sol in water, dil alcohol, liquid ammonia, dimethylformamide.

Disulfide. [27025-41-8] L-γ-Glutamyl-L-cysteinyl-glycine disulfide; *N,N'*-[dithiobis[1-[(carboxymethyl)carbamoyl]ethylene]]diglutamine; GSSG; oxidized glutathione. $C_{20}H_{32}N_6O_{12}S_2$; mol wt 612.63. Crystals, mp 123°. $[\alpha]_D^{20}$ −108° (c = 2 in water).

4512. Gluten. Protein substance of wheat which is intermixed with the starchy endosperm of the grain. Causes the carbon dioxide produced during dough fermentation to be retained by the dough in a manner which provides the porous and spongy structure of bread. Prepn from wheat: Rist, *Sugar J.* **11**, no. 9, 26 (1949), *C.A.* **43**, 9505c (1949); Christensen, **US 2583684** (1952 to Gateway Chemurgic). Amino acid composition: Pence *et al.*, *Cereal Chem.* **27**, 335 (1950). *Reviews:* M. J. Blish "Wheat Gluten" in M. L. Anson, J. T. Edsall, *Advan. Protein Chem.* **vol. II** (Academic Press, New York, 1945) pp 337-359; Meredith, *Cereal Sci. Today* **9**, 33, 54 (1964).

Yellowish-gray powder. Practically insol in water. Partly sol in alcohol, dil acids; sol in alkalies.

USE: As adhesive and as substitute for flour.

4513. Glutethimide. [77-21-4] 3-Ethyl-3-phenyl-2,6-piperidinedione; 2-ethyl-2-phenylglutarimide; α-ethyl-α-phenylglutarimide; 3-ethyl-3-phenyl-2,6-dioxopiperidine; 3-ethyl-3-phenyl-2,6-diketopiperidine; Elrodorm; Doriden. $C_{13}H_{15}NO_2$; mol wt 217.27. C 71.87%, H 6.96%, N 6.45%, O 14.73%. Synthesis: Tagmann *et al.*, *Helv. Chim. Acta* **35**, 1541 (1952); Salmon-Legagneur, Neveu, *Compt. Rend.* **234**, 1060 (1952); *Bull. Soc. Chim. Fr.* **1953**, 70; Hoffmann, Tagmann, **US 2673205** (1954 to Ciba). Resolution: Kukalja *et al.*, *Croat. Chem. Acta* **33**, 41 (1961), *C.A.* **55**, 27193g (1961). Abs config of antipodes: Finch *et al.*, *Experientia* **31**, 1002 (1975). Comprehensive description: H. Y. Aboul-Enein, *Anal. Profiles Drug Subs.* **5**, 139-187 (1976).

dl-Form. Crystals from ether or from ethyl acetate + petr ether, mp 84°. uv max (methanol): 251, 257, 263 nm. Freely sol in ethyl acetate, acetone, ether, chloroform; sol in ethanol, methanol. Practically insol in water.

d-Form. Crystals, mp 102.5-103°. $[\alpha]_D^{20}$ +176° (methanol).

l-Form. Crystals, mp 102-103°. $[\alpha]_D^{20}$ −181° (methanol).

Note: This is a controlled substance (depressant): **21 CFR, 1308.12.**

THERAP CAT: Sedative; hypnotic.

4514. Glyburide. [10238-21-8] 5-Chloro-*N*-[2-[4-[[[(cyclohexylamino)carbonyl]amino]sulfonyl]phenyl]ethyl]-2-methoxybenzamide; 1-[[*p*-[2-(5-chloro-*o*-anisamido)ethyl]phenyl]sulfonyl]-3-cyclohexylurea; *N*-[4-(β-(2-methoxy-5-chlorobenzamido)ethyl)-benzosulfonyl]-*N'*-cyclohexylurea; N^1-[4-[β-(2-methoxy-5-chlorobenzoylamino)ethyl]benzenesulfonyl]-N^2-cyclohexylurea; glybenzcyclamide; glibenclamide; HB-419; U-26452; Azuglucon; Bastiver-

it; Diabasan; Diabeta; Daonil; Duraglucon; Euglucon; Gilemal; Glimidstada; Glycolande; Libanil; Maninil; Micronase; Praeciglucon. $C_{23}H_{28}ClN_3O_5S$; mol wt 494.00. C 55.92%, H 5.71%, Cl 7.18%, N 8.51%, O 16.19%, S 6.49%. Second generation sulfonylurea with hypoglycemic activity. Prepn: Aumuller *et al.*, *Arzneim.-Forsch.* **16**, 1640 (1966); **NL 6603398** (1966 to Boehringer, Mann.), *C.A.* **66**, 65289h (1967); **NL 6610580**; H. Weber *et al.*, **US 3454635** (1967, 1969 both to Hoechst). Pharmacology: Loubatières, Mariani, *C. R. Seances Acad. Sci. Ser. D* **265**, 643 (1967). Toxicity: Mizukami *et al.*, *Arzneim.-Forsch.* **19**, 1413 (1969). Series of articles on synthesis, pharmacology, toxicology and clinical studies: *ibid.* 1323-1494. Effect on release of insulin, glucagon and somatostatin: S. Efendic *et al.*, *Proc. Natl. Acad. Sci. USA* **76**, 5901 (1979). Symposium on pharmacology, mechanism of action and clinical trials: *Ann. Clin. Res.* **15**, Suppl. 37, 1-35 (1983). Comprehensive description: P. G. Takla, *Anal. Profiles Drug Subs.* **10**, 337-355 (1981). Review of pharmacology and clinical efficacy: J. M. Feldman, *Pharmacotherapy* **5**, 43-62 (1985).

Crystals from methanol, mp 169-170° (Weber); also reported as mp 172-174° (Aumüller). pKa 5.3. Sparingly sol in water, sol in the usual organic solvents. LD$_{50}$ in rats and mice (g/kg): >20 orally; >12.5 i.p.; >20 s.c. (Mizukami).

THERAP CAT: Antidiabetic.

4515. Glybuthiazole. [535-65-9] 4-Amino-*N*-[5-(1,1-dimethylethyl)-1,3,4-thiadiazol-2-yl]benzenesulfonamide; N^1-(5-*tert*-butyl-1,3,4-thiadiazol-2-yl)sulfanilamide; sulfatertiobutylthiadiazole; 2-(*p*-aminobenzenesulfamido)-5-tertiobutyl-1,3,4-thiadiazole; glybuthizol; RP-2259; Glipasol. $C_{12}H_{16}N_4O_2S_2$; mol wt 312.41. C 46.14%, H 5.16%, N 17.93%, O 10.24%, S 20.52%. Prepn: **GB 828963** (1960 to Rhône-Poulenc).

Needles from ethanol, mp 221-223°. Sol in ethanol (1.0 g/65 ml); in acetone (1.0 g/15 ml); in DMF (1.0 g/3 ml). Insol in water, ether, benzene. Forms a water-sol Na salt.

THERAP CAT: Antidiabetic.

4516. Glybuzole. [1492-02-0] *N*-[5-(1,1-Dimethylethyl)-1,-3,4-thiadiazol-2-yl]benzenesulfonamide; *N*-(5-*tert*-butyl-1,3,4-thiadiazol-2-yl)benzenesulfonamide; 2-benzenesulfonamido-5-*tert*-butyl-1,3,4-thiadiazole; desaglybuzole; TH-1395; RP-7891; AN-1324; Gludiase. $C_{12}H_{15}N_3O_2S_2$; mol wt 297.39. C 48.47%, H 5.08%, N 14.13%, O 10.76%, S 21.56%. Prepn: Macrae, Drain, **GB 822947** (1959 to Smith & Nephew), *C.A.* **54**, 4622h (1960); **FR M3389** (1965 to Rhône-Poulenc), *C.A.* **64**, 3553a (1966). Pharmacodynamics: Bargeton *et al.*, *Arch. Int. Pharmacodyn.* **153**, 379 (1965).

Needles, mp 163°.

THERAP CAT: Antidiabetic.

4517. Glyceraldehyde. [56-82-6] 2,3-Dihydroxypropanal; DL-glyceraldehyde; glyceric aldehyde; α,β-dihydroxypropionaldehyde. $C_3H_6O_3$; mol wt 90.08. C 40.00%, H 6.71%, O 53.28%. Obtained with its isomer dihydroxyacetone from glycerol by mild oxidation: Witzemann, *J. Am. Chem. Soc.* **36**, 2227 (1914). *See also Org. Synth.* **coll. vol. II**, 305 (1943). The equilibrium mixture of glyceraldehyde and dihydroxyacetone is called **glycerose**. The two isomers are convertible into another through a common enediol resulting from the migration of hydrogen atoms (Lobry de Bruyn-van Eckenstein rearrangement). The equilibrium mixture plays an important role in the fermentation of sugars and in the biogenesis of constituents of the animal organism, *cf.* L.F. Fieser, M. Fieser, *Advanced Organic Chemistry* (Reinhold, New York, 1961) pp 78, 284, 405. Glyceraldehyde is the simplest aldose; the D- and L-forms are the configurational reference standard for carbohydrates. The two forms have been obtained through the action of nitrous acid on the corresponding form of 3-amino-2-hydroxypropanal: Wohl, Momber, *Ber.* **47**, 3346 (1914); Pictet, Barbier, *Helv. Chim. Acta* **4**, 924 (1921); *cf.* Baer, Fischer, *Science* **88**, 108 (1938). Prepn of L-glyceraldehyde from L-sorbose and of D-glyceraldehyde from D-fructose: Perlin, *Methods Carbohydr. Chem.* **1**, 61 (1962).

Tasteless crystals from alcohol + ether, d_{18}^{18} 1.455. mp 145°. Distills at 140-150° (bath temp) and 0.8 mm pressure. 100 ml water dissolve 3 g at 18°. Insol in benzene, petr ether, pentane. Osazone, mp 132°.

L-Form. [497-09-6] (S)-$(-)$-2,3-Dihydroxypropanol. $[\alpha]_D^{25}$ $-8.7°$ (c = 2 in H_2O). Its dimethylacetal, bp_{17-20} 126-129°. $[\alpha]_D^{25}$ $-20.9°$ (p = 9.22).

D-Form. [453-17-8] $[\alpha]_D^{25}$ $+8.7°$ (c = 2 in H_2O). Its dimethylacetal, bp_{17} 127-129°, bp_{10} 123-126°. $[\alpha]_D^{15}$ $+21.2°$ (c = 18).

4518. Glyceraldehyde 3-Phosphate. [591-59-3] 2-Hydroxy-3-(phosphonooxy)propanal; 3-phosphoglyceraldehyde. $C_3H_7O_6P$; mol wt 170.06. C 21.19%, H 4.15%, O 56.45%, P 18.21%. An intermediate product of carbohydrate metabolism. Prepn of DL-form by enzymatic route: Meyerhof, Junowicz-Kocholaty, *J. Biol. Chem.* **149**, 71 (1943); by reductive cleavage of glyceraldehyde 1-benzyl ether 3-phosphoric acid: Fischer, Baer, *Ber.* **65**, 337, 1040 (1932); by hydrolysis of dimeric glyceraldehyde 1,3-diphosphoric acid: Baer, Fischer, *J. Biol. Chem.* **150**, 213 (1943); by hydrolysis of dimeric glyceraldehyde 1-bromide 3-phosphoric acid: *eidem, ibid.* 223. *See also* Baer in *Biochem. Prep.* **1**, 50 (1949). Prepn of D-form by enzymatic route: Meyerhof, Junowicz-Kocholaty, *loc. cit.*; by synthetic route: Ballou, Fischer, *J. Am. Chem. Soc.* **77**, 3329 (1955).

DL-Form calcium salt dihydrate. $C_3H_5CaO_6P.2H_2O$. Crystals, sol in water. Aq solns are not stable, particularly when alkaline. The dioxane addition compd of DL-glyceraldehyde 1-bromide 3-phosphoric acid described by Baer is relatively stable and may be stored in the refrigerator. An aq soln of the Na or K salt of glyceraldehyde-3-phosphoric acid (contg dioxane and bromide ion) is readily obtained by dissolving the dioxane compd in cold water and carefully neutralizing to pH 7.

D-Form calcium salt dihydrate. Amorphous powder. $[\alpha]_D^{25}$ $+14.5°$ (c = 1.2 in 0.1N HCl calcd as the free acid). Prepn of calcium-free aq soln: Ballou, Fischer, *loc. cit.*

4519. Glyceric Acid. [473-81-4] 2,3-Dihydroxypropanoic acid; α,β-dihydroxypropionic acid; DL-glyceric acid. $C_3H_6O_4$; mol wt 106.08. C 33.97%, H 5.70%, O 60.33%. Prepn from α,β-dibromopropionic acid by treatment with silver oxide: Karrer, Klarer, *Helv. Chim. Acta* **7**, 931 (1924); from isoserine and nitrous acid:

Fischer, Jacobs, *Ber.* **40**, 1069 (1907); from glycerol and nitrous acid: Mulder, *Ber.* **9**, 1902 (1876); Beilstein, *Ann.* **120**, 229 (1861). Dextrorotatory glyceric acid is obtained by the action of *Penicillia* or *Aspergilli* on the DL-form: McKenzie, Harden, *J. Chem. Soc.* **83**, 431 (1903). Levorotatory glyceric acid has been obtained by the oxidation of D(+)-glyceric aldehyde: Wohl, Schellenberg, *Ber.* **55**, 1408 (1922).

Syrup, dec on distn. pK (25°): 3.55. Miscible with water, alcohol, acetone. Nearly insol in ether.

L(+)-Form. [28305-26-2] $(2S)$-2,3-Dihydroxypropanoic acid. Syrup. Its esters and salts are levorotatory and its salts are much more sol in water than those of the DL-form.

L-Form calcium salt. [6057-35-8] $Ca(C_3H_5O_4)_2.2H_2O$; mol wt 286.25. Monoclinic sphenoidal crystals, mp 137°. One gram dissolves in 10 ml water. $[\alpha]_D^{20}$ $-14.6°$ (c = 5).

D(−)-Form. [6000-40-4] $(2R)$-2,3-Dihydroxypropanoic acid. Syrup. Its salts are dextrorotatory.

D-Form calcium salt. [6000-41-5] $Ca(C_3H_5O_4)_2.2H_2O$; mol wt 286.25. Prisms, mp 138°. $[\alpha]_D^{20}$ $+14.5°$ (c = 5).

4520. Glycerol. [56-81-5] 1,2,3-Propanetriol; glycerin; glycerine; trihydroxypropane; incorporation factor; IFP; Bulbold; Cristal; Glyceol; Ophthalgan. $C_3H_8O_3$; mol wt 92.09. C 39.13%, H 8.76%, O 52.12%. Obtained from oils and fats as byproduct in the manuf of soaps and fatty acids. During World War I, supplementary quantities were produced by the "Protol" fermentation process from sugar, a process based upon the fixation of acetaldehyde by sodium sulfite. Just prior to World War II, the synthesis of glycerol from propylene was announced. Production from sugars by fermentation: Onishi, US 3012945 (1961 to Noda). Identity with incorporation factor: Kuehl *et al., J. Am. Chem. Soc.* **82**, 2079 (1960). In nucleic acid the incorporation factor may exist as a bound form of glycerol. Acute toxicity: W. Bartsch *et al., Arzneim.-Forsch.* **26**, 1581 (1976). Reviews and bibliographies: J. W. Lawrie, *Glycerol and the Glycols* (New York, 1928); G. Leffingwell, M. Lesser, *Glycerin* (Brooklyn, 1945); C. S. Miner, N. N. Dalton, *Glycerol* (New York, 1953); C. Lüttgen, *Glyzerin und glyzerinähnliche Stoffe* (Heidelberg, 2nd ed., 1955); J. C. Kern in *Kirk-Othmer Encyclopedia of Chemical Technology* vol. **11** (Wiley-Interscience, New York, 3rd ed., 1980) pp 921-932.

Syrupy liquid. Sweet warm taste. About 0.6 times as sweet as cane sugar. Hygroscopic; also absorbs H_2S, HCN, SO_2. *Contact with strong oxidizing agents such as chromium trioxide, potassium chlorate, or potassium permanganate may produce an explosion.* Neutral to litmus. Solidifies after prolonged cooling at 0° forming shiny orthorhombic crystals, mp 17.8°. bp_{760} 290.0° (dec); bp_{400} 263.0°; bp_{200} 240.0°; bp_{100} 220.1°; bp_{60} 208.0°; bp_{20} 182.2°; bp_{10} 167.2°; bp_5 153.8°; $bp_{1.0}$ 125.5°. n_D^{15} 1.4758; n_D^{20} 1.4746; n_D^{25} 1.4730. d_{15}^{15} 1.26557; d_{20}^{20} 1.26362; d_{25}^{25} 1.26201. Flash point, open cup: 350°F (176°C). Specific gravities of 95% aq soln w/w (U.S.P. grade): d_{15}^{15} 1.25270; d_{20}^{20} 1.25075; d_{25}^{25} 1.24910; 90% aq soln w/w: d_{15}^{15} 1.23950; d_{20}^{20} 1.23755; d_{25}^{25} 1.23585; 80% d_{15}^{15} 1.213; 70% d_{15}^{15} 1.185; 60% d_{15}^{15} 1.157; 50% d_{15}^{15} 1.129; 20% d_{15}^{15} 1.049; 5% d_{15}^{15} 1.0122. Viscosity (cP at 20°): 5% soln 1.143; 10% 1.311; 25% 2.095; 50% 6.050; 60% 10.96; 70% 22.94; 83% 111. Freezing points of aq solns w/w: 10% $-1.6°$; 30% $-9.5°$; 50% $-23.0°$; 66.7% $-46.5°$; 80% $-20.3°$; 90% $-1.6°$. Miscible with water, alcohol. One part dissolves in 11 parts ethyl acetate, in about 500 parts ethyl ether. Insol in benzene, chloroform, carbon tetrachloride, carbon disulfide, petr ether and in fixed and volatile oils. LD_{50} in rats (ml/kg): >20 orally; 4.4 i.v. (Bartsch).

Caution: Potential symptoms of overexposure to mist are irritation of eyes, skin, respiratory system; headache, nausea, vomiting; kidney injury. *See NIOSH Pocket Guide to Chemical Hazards* (DHHS/NIOSH 97-140, 1997) p 152.

USE: As solvent, humectant, plasticizer, emollient, sweetener, in the manuf of nitroglycerol (dynamite), cosmetics, liq soaps, liqueurs, confectioneries, blacking, printing and copying inks, lubricants, elastic glues, lead oxide cements; to keep fabrics pliable; to preserve printing on cotton; for printing rollers, hectographs; to keep frost from windshields; as antifreeze in automobiles, gas meters and hydraulic jacks, in shock absorber fluids. In fermentation nutrients in the production of antibiotics. Titration reagent. In tests for heavy metals. Pharmaceutic aid (humectant; solvent, vehicle). Leffingwell and Lesser (*op. cit.*) give 1583 different uses.

THERAP CAT: Diagnostic aid (ophthalmic).

THERAP CAT (VET): Emollient, demulcent. As a source of glucose in bovine ketosis.

4521. Glycerol Formal. Methylidinoglycerol; Glicerinformal; Sericosol N. $C_4H_8O_3$; mol wt 104.11. C 46.15%, H 7.75%, O 46.10%. Mixture of isomeric α,α'- and α,β-forms prepd from glycerin and formaldehyde: M. Schultz, B. Tollens, *Ann.* **289**, 20 (1895). Prepn and separation of isomers: H. Hibbert, N. M. Carter, *J. Am. Chem. Soc.* **50**, 3120 (1928); J. D. van Roon, *Rec. Trav. Chim.* **48**, 173 (1929). Prepn and pharmacology: P. Gimeno, **ES 475962** (1979 to Calipe), *C.A.* **93**, 26445u (1980). Toxicological evaluation for use as pharmaceutical solvent: D. M. Sanderson, *J. Pharm. Pharmacol.* **11**, 150, 446 (1959). Teratogenicity studies: V. Aliverti *et al.*, *Arch. Sci. Biol.* **61**, 89 (1977); E. Giavini, M. Prati, *Acta Anat.* **106**, 203 (1980).

α,α'-form   α,β-form

Liquid, bp$_{760}$ 191-195°, bp$_{20}$ 95-97°. d$_4^{20}$ 1.215; n$_D^{20}$ 1.451. Sol in water, alc, chloroform. pH of 10% soln: 4-6.5. LD$_{50}$ in rats (ml/kg): 8.6 orally, 3.5 i.v. (Gimeno).

α,α'-Form. 1,3-Dioxan-5-ol; α,α'-methylene glycerin; α,α'-formaldehyde glycerin; 5-*m*-dioxanol. bp 193.8°. d$_4^{25}$ 1.2200. n$_D^{25}$ 1.4527.

α,β-Form. 1,3-Dioxolane-4-methanol; α,β-methylene glycerin; α,β-formaldehyde glycerol; 4-(hydroxymethyl)-1,3-dioxolane. bp 192.5°. d$_4^{25}$ 1.2008. n$_D^{25}$ 1.4469.

USE: Pharmaceutic aid (solvent).

4522. Glycerophosphoric Acid. [57-03-4] 1,2,3-Propanetriol 1-(dihydrogen phosphate); DL-α-glycerol phosphate; phosphoric acid glycerol esters. $C_3H_9O_6P$; mol wt 172.07. C 20.94%, H 5.27%, O 55.79%, P 18.00%. Three isomers exist: the L(−)-α-form, the D(+)-α-form, and the β-form. The L-α-form is naturally occurring; the β-form, present in hydrolyzates of lecithins from natural sources, arises from migration of the phosphoryl group from the α-carbon atom. Prepn by phosphorylation of glycerol results in a mixture of the α- and β-forms: Cherbuliez, Weniger, *Helv. Chim. Acta* **29**, 2006 (1946). Prepn and configuration of the L-α-form: Baer, Fischer, *J. Biol. Chem.* **128**, 491 (1939). Prepn of the D-α-form: *eidem, ibid.* **135**, 321 (1940). Separation of α-forms from the β-form and polyglycerophosphoric acids: Carrara, **IT 460219** (1950), *C.A.* **46**, 5077a (1952).

L-α-form

Absolute acid (commercial mixture of α- and β-forms); clear syrupy liquid, mp −25°. d$_4^{14}$ 1.59. Tends to dec during concn; hence, usually marketed as a 25-50% soln. Soluble in water, alcohol.

L-α-form. [5746-57-6] (2*S*)-1,2,3-Propanetriol 1-(dihydrogen phosphate). Syrup. Readily sol in water, methanol, ethanol. Practically insol in ether. [α]$_D$ −1.45° (barium salt, c = 10.3 in 2*N* HCl).

D-α-form. [17989-41-2] (2*R*)-1,2,3-Propanetriol 1-(dihydrogen phosphate).

β-form. [17181-54-3] 1,2,3-Propanetriol 2-(dihydrogen phosphate). White solid. bp 230°. Hygroscopic. Sol in DMSO.

Note: **Phosphatidic acids** are fatty acid diesters of glycerophosphoric acid.

USE: Absolute acid used to manuf certain glycerophosphates or to impart taste to solns of glycerophosphates which are generally used medicinally. *See also:* Calcium Glycerophosphate.

4523. Glyceryl *p*-Aminobenzoate. [136-44-7] 1,2,3-Propanetriol 1-(4-aminobenzoate); *p*-aminobenzoic acid monoglyceryl ester; monoglyceryl *p*-aminobenzoate; Escalol 106. $C_{10}H_{13}NO_4$; mol wt 211.22. C 56.86%, H 6.20%, N 6.63%, O 30.30%. Prepd by controlled esterification of *p*-aminobenzoic acid with glycerol.

Semisolid, waxy mass or syrup. Faint aromatic odor. Liquefies and congeals very slowly. Soluble in methanol, ethanol, isopropanol, glycerol, propylene glycol. Insol in water, oils, fats.

USE: In cosmetic sunscreen prepns (up to 1%).

4524. Glyceryl Monostearate. [31566-31-1] Octadecanoic acid monoester with 1,2,3-propanetriol; Monostearin. The commercial product is a mixture of variable proportions of glyceryl monostearate and glyceryl monopalmitate.

White, wax-like solid or wax-like beads, or flakes, mp 56-58°. Saponification value 164-170. Iodine value not more than 6. Soluble in hot organic solvents such as alcohol, benzene, ether, acetone, mineral or fixed oils. Insol in water, but may be dispersed in hot water with the aid of a small amount of soap or other suitable surface active agent.

USE: In pharmaceutical dispensing, *see* Green, *J. Am. Pharm. Assoc. Pract. Pharm. Ed.* **7**, 299 (1946).

4525. Glycidol. [556-52-5] 2-Oxiranemethanol; 2,3-epoxy-1-propanol; 3-hydroxypropylene oxide. $C_3H_6O_2$; mol wt 74.08. C 48.64%, H 8.16%, O 43.19%. Prepd by the action of perbenzoic acid on allyl alc: Prileshajew, *Ber.* **42**, 4813 (1909); from glycerol-1-monochlorohydrin by the action of KOH in alc, or by the action of sodium in ether: Rider, Hill, *J. Am. Chem. Soc.* **52**, 1521 (1930). L-form prepd from L-1-(*p*-toluenesulfonyl)glycerol: Sowden, Fischer, *ibid.* **64**, 1291 (1942). Acute toxicity data: E. D. Thompson, R. A. Hiles, *Food Cosmet. Toxicol.* **19**, 347 (1981); E. D. Thompson, D. P. Gibson, *Food Chem. Toxicol.* **22**, 665 (1984).

DL-Form. Slightly viscous liq. d$_4^{25}$ 1.1143. bp$_{760}$ 167° (decompn); bp$_{2.5}$ 66°; bp$_{0.9}$ 25°. Miscible with water. LD$_{50}$ in female rats (g/kg): 0.42 orally, 0.20 i.p. (Thompson, Hiles). LD$_{50}$ in male, female rats (g/kg): 0.35, 0.21 i.p.; 0.76, 0.64 orally (Thompson, Gibson).

L-Form. [α]$_D^{20}$ +15° (neat).

Caution: Potential symptoms of overexposure are irritation of eyes, nose, throat and skin; narcosis. *See NIOSH Pocket Guide to Chemical Hazards* (DHHS/NIOSH 97-140, 1997) p 152. *See also Patty's Industrial Hygiene and Toxicology* vol. 2A, G. D. Clayton, F. E. Clayton, Eds. (John Wiley & Sons, New York, 4th ed., 1994) 422-425. This substance is reasonably anticipated to be a human carcinogen: *Report on Carcinogens, Twelfth Edition* (PB2011-111646, 2011) p 215.

USE: Stabilizer in manuf of vinyl polymers; intermediate in synthesis of glycerol, glycidyl ethers, and amines; additive for oil and synthetic hydraulic fluids; epoxy resin diluent.

4526. Glycine. [56-40-6] Gly; G; aminoacetic acid; aminoethanoic acid; glycocoll; Gyn-Hydralin. $C_2H_5NO_2$; mol wt 75.07. C 32.00%, H 6.71%, N 18.66%, O 42.62%. Non-essential amino acid for human development. Only amino acid with no asymmetric carbon. Major inhibitory neurotransmitter. Isoln from gelatin: H.

Braconnot, *Ann. Chim. Phys.* **13**, 113 (1820). Early chemistry and biochemistry: *Amino Acids and Proteins*, D. M. Greenberg, Ed. (Charles C. Thomas, Springfield, IL, 1951) 950 pp., *passim*; J. P. Greenstein, M. Winitz, *Chemistry of the Amino Acids* **vols 1-3**, (John Wiley and Sons, Inc., New York, 1961) pp. 1955-1970, *passim*. Exists in three polymorphic forms, α-, β-, and γ-: Y. Iitaka, *Nature* **183**, 390 (1959). Review of metabolism and radioprotection: S. Capalna, *Rev. Roum. Physiol.* **9**, 17-34 (1972); of metabolism in humans: A. A. Jackson, M. H. Golden, *Clin. Sci.* **58**, 517-522 (1980); in anerobes: J. R. Andreesen, *Antonie van Leeuwenhoek* **66**, 223-227 (1994). Review of role in neurotransmission: S. M. Paul in *Psychopharmacology: The Fourth Generation of Progress*, F. E. Bloom, D. J. Kupfer, Eds. (Raven Press, New York, 1995) pp 87-94. *Book: Glycine Neurotransmission*, O. P. Ottersen, J. Storm-Mathisen, Eds. (John Wiley & Sons, Chichester, U.K., 1990) pp 489.

Sweet, monoclinic prisms from alc, starts to dec at 233°, completely sintered at 290°. d 1.595. pK_1 2.34; pK_2 9.60. pH of 0.2 molar soln in H_2O = 4.0. Adsorption on various chromatographic agents: Grettie, Williams, *J. Am. Chem. Soc.* **50**, 671 (1928). Soly in 100 ml water at 25°: 25.0 g; at 50°: 39.1 g; at 75°: 54.4 g; at 100°: 67.2 g. 100 g of abs alc dissolve about 0.06 g. Sol in 164 parts pyridine; very slightly sol in ether, alc.

Hydrochloride. [6000-43-7] $C_2H_5NO_2 \cdot HCl$; mol wt 111.53. Hygroscopic prisms from HCl, mp 182°.

USE: Buffer. Reagent in synthetic organic chemistry.

4527. Glycine Sulfate. [513-29-1] Triglycine sulfate. $C_6H_{17}N_3O_{10}S$; mol wt 323.27. C 22.29%, H 5.30%, N 13.00%, O 49.49%, S 9.92%. $(NH_2CH_2COOH)_3 \cdot H_2SO_4$. Prepn: Horsford, *Ann.* **60**, 1 (1846); *see also* Matthias, Miller, Remeika, *Phys. Rev.* **104**, 849 (1956). Crystal growing during prepn: Konstantinova *et al.*, *Kristallografiya* **4**, 69-73 and 125-129 (1959).

Orthorhombic crystals. Very freely sol in water. Has ferroelectric properties: Curie point 47°. Spontaneous polarization at room temp: 2.2×10^{-6} coul/cm². Coercive field 220 v/cm.

USE: In electronics research.

4528. Glycinin. [9007-93-6] Chief protein constituent of soybeans: Osborne, Campbell, *J. Am. Chem. Soc.* **20**, 419 (1898); Smith, Circle, *Ind. Eng. Chem.* **31**, 1282 (1939). *Review:* H. Neurath, K. Bailey, Eds., *The Proteins* **vol. I**, Part A (Academic Press, New York, 1953) pp 208-209, 223; *eidem, ibid.* **vol. II**, part A, (1954) pp 503, 506.

Soluble in water in the pH range 1-4, as well as above 7.

4529. Glycocholic Acid. [475-31-0] N-[(3α,5β,7α,12α)-3,-7,12-Trihydroxy-24-oxocholan-24-yl]glycine; N-cholylglycine. $C_{26}H_{43}NO_6$; mol wt 465.63. C 67.07%, H 9.31%, N 3.01%, O 20.62%. The product of conjugation of cholic acid with glycine; chief ingredient of the bile of herbivorous animals. In the weakly alkaline bile fluid glycocholic acid exists as the sodium salt. Prepn from bile: Hammarsten in *Abderhalden's Handbuch der Biol. Arbeitsmethoden*, Abt. I, Teil 6, p 211 (1925). Prepn from cholic acid: Cortese, *J. Am. Chem. Soc.* **59**, 2532 (1937). Synthesis: Cortese, Bauman, *ibid.* **57**, 1393 (1935); Bergstrom, Norman, *Acta Chem. Scand.* **7**, 1126 (1953). Separation: Antonides, *GB* **928635** (1963 to Armour). Metabolism: Norman, *Scand. J. Gastroenterol.* **5**, 231 (1970).

Sesquihydrate. Crystals from 5% alc, mp about 130°. $[\alpha]_D^{23}$ +30.8° (c = 7.5 in 95% ethanol). Anhydr form, mp 165-168°. pK 4.4. Soly in water at 15°: 0.33 g/l; in boiling water: 8.3 g/l. Is hydrolyzed to cholic acid and glycine by acids and alkalies. Forms addition compds with nitrobenzene, aniline, benzyl alcohol, benzaldehyde, triolein.

Sodium salt. $C_{26}H_{42}NNaO_6$. Crystals from 95% alcohol + ether, mp 230-240°. $[\alpha]_D^{24}$ +32° (water). Soly at 15° in water >274 g/l; in alcohol >340 g/l.

4530. Glycocyamine. [352-97-6] N-(Aminoiminomethyl)-glycine; N-amidinoglycine; guanidineacetic acid; guanidoacetic acid. $C_3H_7N_3O_2$; mol wt 117.11. C 30.77%, H 6.03%, N 35.88%, O 27.32%. Prepd from S-ethylthiourea hydrobromide, sodium hydroxide, and glycine: Brand, Brand, *Org. Synth.* **coll. vol. III**, 440 (1955). Crystal and molecular structure: Guha, *Acta Crystallogr.* **B29**, 2163 (1973); J. Berthou *et al., ibid.* **B32**, 1529 (1976).

Crystals, dec 280-284°. Appreciably sol in water.

THERAP CAT: In combination with betaine as cardiotonic.

4531. Glycogen. [9005-79-2] Animal starch; liver starch. $(C_6H_{10}O_5)_n$; mol wt from about 2.7×10^5 to 3.5×10^6. Reserve carbohydrate of the animal organism. High molecular wt polymer having branched-chain structure composed of D-glucopyranose residues. Distributed through the cell protoplasm. Found esp in the liver and in rested muscle. Occurs also in insects and lower plants including fungi and yeasts. Isoln by alkaline destruction of the other cell constituents: Claude Bernard, *Lecons sur le diabete* (Paris, 1877) p 553; by destruction with trichloroacetic acid: Bell, Young, *Biochem. J.* **28**, 882 (1934); by centrifugation: Meyer, Jeanloz, *Adv. Enzymol.* **3**, 112 (1943); by hydraulic pressure: Stockhausen, Silbereisen, *Biochem. Z.* **287**, 276 (1936). For biological synthesis and lysis from the Cori ester (glucose-1-phosphate) *see* the review and bibliography by Meyer, *Adv. Enzymol.* **3**, 109 (1943); *see also* Nord, *Chem. Rev.* **26**, 423 (1940); Kalckar, *ibid.* **28**, 71 (1941). Isoln from the causal agent of cotton root rot, *Phymatotrichum omnivorum* (Shear) Duggar: Ergle, *J. Am. Chem. Soc.* **69**, 2061 (1947). Studies on linkages: Bahl, Smith, *J. Org. Chem.* **31**, 2915 (1966).

White powder. $[\alpha]_D^{25}$ +196 to +197°. Sol in water with opalescence. Insol in alc. Does not reduce Fehling's soln. With iodine, brown to violet colors are produced.

4532. Glycolaldehyde. [141-46-8] Hydroxyacetaldehyde; diose; 2-hydroxyethanal; glycollic aldehyde; methylolformaldehyde. $C_2H_4O_2$; mol wt 60.05. C 40.00%, H 6.71%, O 53.29%. Simplest possible monosaccharide; isomer of methyl formate and acetic acid, *q.q.v.* Detected in interstellar space and thought to be key to the prebiotic formation of sugars and amino acids. Exists as a dimer in solid form, dissociates into the monomer in solution. Prepn from dihydroxymaleic acid: H. J. H. Fenton, H. Jackson, *J. Chem. Soc.* **75**, 575 (1899); from divinyl ether: N. A. Milas *et al., J. Am. Chem. Soc.* **61**, 1844 (1939); from ethylene glycol: H. Ukeda *et al., Biosci. Biotechnol. Biochem.* **62**, 1589 (1998). Dissociation of dimer in soln: N. P. McCleland, *J. Chem. Soc.* **99**, 1827 (1911); R. P. Bell, J. P. H. Hirst, *ibid.* **1939**, 1777; C. I. Stassinopoulou, C. Zioudrou, *Tetrahedron* **28**, 1257 (1972). Autocatalytic role in the production of sugars from formaldehyde (formose reaction): R. Breslow, *Tetrahedron Lett.* **1**, No. 21, 22 (1959). Use in organic synthesis: P. Dinprasert *et al., Tetrahedron Lett.* **30**, 1149 (1989); in polyol synthesis of silver nanostructures: S. E. Skrabalak *et al., Nano Lett.* **8**, 2077 (2008). Role in prebiotic synthesis of amino acids: G. Zubay, *Chemtracts Biochem. Mol. Biol.* **10**, 407 (1997); of carbohydrates: O. P. Pestunova *et al.* in *Biosphere Origin and Evolution* (Springer-Verlag, Heidelberg, 2008) pp 103-117. Detection in interstellar clouds: J. M. Hollis *et al., Astrophys. J.* **540**, L107 (2000); outside the galactic center: M. T. Beltrán *et al., ibid.* **690**, L93 (2009).

Colorless, transparent, oblique plates with sweet taste. mp 95-97°. $bp_{4.5}$ 90-98°; bp_{12} 110-120°. d^{100} 1.366; d^{16} 1.391. n_D^{11} 1.4811. Sol in cold water, hot alcohol; very sparingly sol in ether.

Dimer. [23147-58-2] 1,4-Dioxane-2,5-diol; 2,5-dihydroxy-1,4-dioxane. $C_4H_8O_4$; mol wt 120.10. White crystalline powder, mp 80-90°.

USE: Starting material and reagent in organic synthesis.

4533. Glycol Dilaurate. [624-04-4] Dodecanoic acid 1,1'-(1,2-ethanediyl) ester; ethylene dilaurate. $C_{26}H_{50}O_4$; mol wt 426.68. C 73.19%, H 11.81%, O 15.00%.

Colorless, amorphous mass, mp 50-52°. bp_{20} 188°. Insol in alcohol, ether.

USE: In lacquers and varnishes as a plasticizer.

4534. Glycolic Acid. [79-14-1] 2-Hydroxyacetic acid; hydroxyethanoic acid. $C_2H_4O_3$; mol wt 76.05. C 31.59%, H 5.30%, O 63.11%. Naturally occuring α-hydroxy acid; constituent of sugar cane juice, also found in sugar beets and unripe grapes. Forms biodegradable polymers. Prepn from glycine and nitrous acid: A. Strecker, *Ann.* **68**, 47 (1848); from formaldehyde, carbon monoxide and water: D. J. Loder, US 2152852 (1939 to DuPont). Use as skin peeling agent: H. Murad *et al., Dermatol. Clin.* **13**, 285 (1995). Review of synthesis and applications: E. Lower, *Spec. Chem.* **16**, 54-55 (1996); R. Datta in *Kirk-Othmer Encyclopedia of Chemical Technology* vol. 14 (Wiley-Interscience, Hoboken, 5th Ed., 2005) pp 126-130. Review of use in drug delivery systems: S. Fredenberg *et al., Int. J. Pharm.* **415**, 34-52 (2011).

Odorless, somewhat hygroscopic crystals, mp 80°. *Corrosive.* pK (25°) 3.83. Sol in water, methanol, alcohol, acetone, acetic acid, ether. pH of aq solns: 2.5 (0.5%); 2.33 (1.0%); 2.16 (2.0%); 1.91 (5.0%); 1.73 (10.0%). LD_{50} orally in rats: 1.95 g/kg; *see:* H. F. Smyth *et al., J. Ind. Hyg. Toxicol.* **23**, 259 (1941).

USE: In skin care products as exfollient and keratolytic. In biopolymers for absorbable sutures and drug delivery systems. In the processing of textiles, leather, and metals; in pH control, in the manuf of adhesives, in copper brightening, decontamination cleaning, dyeing, electroplating, in pickling, cleaning and chemical milling of metals.

4535. Glycol Salicylate. [87-28-5] 2-Hydroxybenzoic acid 2-hydroxyethyl ester; monoglycol salicylate; ethylene glycol monosalicylate; 2-hydroxyethyl salicylate; glysal; GL-7; Lumbinon; Phardol; Phlogont. $C_9H_{10}O_4$; mol wt 182.18. C 59.34%, H 5.53%, O 35.13%.

Almost colorless, odorless liq. bp_{12} 169-172°. Soluble in about 110 parts water, 8 parts olive oil; very sol in alcohol, benzene, chloroform, ether.

THERAP CAT: Counterirritant, anti-inflammatory (topical).

4536. Glyconiazide. [3691-74-5] D-Glucuronic acid γ-lactone 1-[(4-pyridinylcarboxyl)hydrazone]; D-glucuronolactone isonicotinoylhydrazone; isonicotinoylhydrazone of D-glucuronic acid lactone; isonicotinic acid hydrazide hydrazone with glucuronic acid lactone; Galatone; Gatalone; Glucazide; Gluconiazide; Gluronazide;

Guidazide; Hydronsan; INH-G; Mycobactyl. $C_{12}H_{13}N_3O_6$; mol wt 295.25. C 48.82%, H 4.44%, N 14.23%, O 32.51%. Prepd by heating isonicotinic acid hydrazide with D-glucuronolactone in methanol: Sah, *J. Am. Chem. Soc.* **75**, 2512 (1953); Sah, US 2940899 (1960 to U. of Calif.).

Plates and rods from methanol, needles from abs ethanol. Dec 150-160°. Freely sol in water. Practically insol in cold alc; 1.2 g dissolve in 100 ml methanol at 66°.

THERAP CAT: Antibacterial (tuberculostatic).

4537. Glycopyrrolate. [596-51-0] 3-[(2-Cyclopentyl-2-hydroxy-2-phenylacetyl)oxy]-1,1-dimethylpyrrolidinium bromide (1:1); 3-hydroxy-1,1-dimethylpyrrolidinium bromide α-cyclopentylmandelate; α-cyclopentylmandelic acid ester with 3-hydroxy-1,1-dimethylpyrrolidinium bromide; 1-methyl-3-pyrrolidyl α-cyclopentylmandelate methobromide; 1-methyl-3-pyrrolidyl α-phenyl-α-cyclopentylglycolate methobromide; 3-(2-phenyl-2-cyclopentylglycoloyloxy)-1,1-dimethylpyrrolidinium bromide; glycopyrronium bromide; AHR-504; Nodapton; Robanul; Robinul; Tarodyl; Tarodyn. $C_{19}H_{28}BrNO_3$; mol wt 398.34. C 57.29%, H 7.09%, Br 20.06%, N 3.52%, O 12.05%. Synthetic, quaternary ammonium anticholinergic. Prepn: Franko, Lunsford, *J. Med. Pharm. Chem.* **2**, 523 (1960); Lunsford, US 2956062 (1960 to A. H. Robins). Pharmacodynamics: E. Kaltiala *et al., J. Pharm. Pharmacol.* **26**, 352 (1974). Toxicology: B. V. Franko *et al., Toxicol. Appl. Pharmacol.* **17**, 361 (1970). Clinical comparison with atropine in anaesthetic practice: F. Kongsrud, S. Sponheim, *Acta Anaesthesiol. Scand.* **26**, 620 (1982); A. I. Webb, R. M. McMurphy, *Am. J. Vet. Res.* **48**, 1733 (1987); B. V. G. Malling *et al., Br. J. Anaesth.* **60**, 426 (1988). Brief review of pharmacology and clinical use: R. K. Mirakhur, J. W. Dundee, *Anaesthesia* **38**, 1195-1204 (1983).

White crystals from butanone, mp 193.2-194.5°. Sol in water, alc. Practically insol in chloroform, ether. LD_{50} (72 hr.) in female mice, female rats (mg/kg): 107, 196 i.p.; in male rats (mg/kg): 1150 orally (Franko).

THERAP CAT: Antispasmodic; preanesthetic medicant.

THERAP CAT (VET): Preanesthetic medicant.

4538. Glycosine. [6873-15-0] 1-Methyl-2-(phenylmethyl)-4(1H)-quinazolinone; 2-benzyl-1-methylquinazol-4-one; arborine. $C_{16}H_{14}N_2O$; mol wt 250.30. C 76.78%, H 5.64%, N 11.19%, O 6.39%. Found in the toothbrush plant, *Glycosmis pentaphylla* (Retz.) Corr., and *G. arborea* Corr., *Rutaceae.* Isoln from dried, powdered leaves: Chatterjee, Majumdar, *J. Am. Chem. Soc.* **76**, 2459 (1954). Identity of arborine and glycosine, structure: Chakravarti *et al., Tetrahedron* **16**, 224 (1961). Synthesis: Pakrashi *et al., Indian J. Chem.* **6**, 472 (1968); Ziegler *et al., Monatsh. Chem.* **100**, 948 (1969); T. Kametani *et al., Heterocycles* **9**, 1585 (1978).

Rhombohedral prisms from chloroform + ethyl acetate, mp 155-156°. uv max (ethanol): 231, 268, 277, 306 nm. Freely sol in chloroform, ethyl acetate, benzene, ethanol. Sparingly sol in ether.

Hydrochloride. $C_{16}H_{14}N_2O\cdot HCl$. Leaflets from 90% ethanol, dec 209-210°.

4539. Glycylglycine. [556-50-3] $C_4H_8N_2O_3$; mol wt 132.12. C 36.36%, H 6.10%, N 21.20%, O 36.33%. The simplest of all peptides. Prepn from 2,5-diketopiperazine: Schott et al., J. Org. Chem. **12**, 490 (1947); Greenstein, Winitz, Chemistry of the Amino Acids vol. **2**, (New York, 1961) p 803. From tritylglycylglycine: Zervas et al., J. Am. Chem. Soc. **78**, 1359 (1956). From phthalylglycylglycine: Sheehan, Frank, ibid. **71**, 1856 (1949). From the dicyclohexylamine salt of trifluoroacetylglycylglycine: Weygand, Reiher, Ber. **88**, 26 (1955).

Crystals from dil alc. Crystal shape described as small tetrahedral leaves with a lustrous ball in center. Dec 262-264°. pK_1' 3.12; pK_2' 8.17. Heat of combustion: 472.4 kcal/mole. Soluble in hot water; slightly sol in ethanol. Practically insol in ether.

Hydrochloride monohydrate. $C_4H_8N_2O_3\cdot HCl\cdot H_2O$. Crystals from water + ethanol.

Ethyl ester hydrochloride. Crystals from abs ethanol, dec 182°.

USE: In the synthesis of more complicated peptides.

4540. Glycyrrhiza. Licorice; liquorice; sweet root. Dried rhizome and roots of Glycyrrhiza glabra L., var. typica Regel & Herder (Spanish licorice), or of G. glabra L., var. glandulifera (Waldst. & Kit.) Regel & Herder (Russian licorice), or of other varieties of G. glabra yielding a yellow and sweet wood, Leguminosae. Habit. Southern Europe to Central Asia. Constit. 6-14% glycyrrhizin (the glucoside of glycyrrhetic acid), asparagine, sugars, resin. Used chiefly in the form of glycyrrhiza syrup. Incompat. Acids, metallic salts.

USE: Extract and syrup as pharmaceutic aids (flavor and flavored vehicles).

4541. Glycyrrhizic Acid. [1405-86-3] $(3\beta,20\beta)$-20-Carboxy-11-oxo-30-norolean-12-en-3-yl 2-O-β-D-glucopyranuronosyl-α-D-glucopyranosiduronic acid; glycyrrhizin; glycyrrhizinic acid; glycyrrhetinic acid glycoside. $C_{42}H_{62}O_{16}$; mol wt 822.94. C 61.30%, H 7.59%, O 31.11%. Triterpene saponin used in traditional Chinese medicinal preparations for its anti-inflammatory, antiulcerous and antiallergic effects. Extraction from Glycyrrhiza glabra L., Leguminosae: Karrer, Chao, Helv. Chim. Acta **4**, 100 (1921); Ruzicka, Louenberger, ibid. **19**, 1402 (1936). From commercial glycyrrhizinum ammoniacale: Tschirch, Cederberg, Arch. Pharm. **245**, 97 (1907); Voss et al., Ber. **70**, 122 (1937). Revised method of isoln: Conn, Conn, J. Lab. Clin. Med. **47**, 20 (1956). Structure: Lythgoe, Trippett, J. Chem. Soc. **1950**, 1983. Revised structure: Marsh, Levvy, Biochem. J. **63**, 9 (1956). See also: I. Kitagawa et al., Chem. Pharm. Bull. **36**, 3710 (1988); T. Hatano et al., ibid. **39**, 1238 (1991). Synthesis of derivatives: Brieskorn, Sax, Arch. Pharm. **303**, 905 (1970). LC-MS/MS determn in plasma: Z. J. Lin et al., J. Chromatogr. B **814**, 201 (2005). Review: Nieman, Chem. Weekbl. **48**, 213 (1952).

glycyrrhetinic acid

Crystals from glacial acetic acid. Intensely sweet taste. $[\alpha]_D^{17}$ +46.2° (c = 1.5 in alc). Freely sol in hot water, alcohol; practically insol in ether.

Ammonium glycyrrhizinate pentahydrate. $C_{42}H_{65}NO_{16}\cdot5H_2O$. Needles from 75% aqueous ethanol, decomp 212-217°. $[\alpha]_D^{20}$ +46.9° (c = 1.5 in 40% ethanol). uv max: 248 nm (ε 11400). Sol in ammonia water, glacial acetic acid.

Dipotassium salt. Rizinsan K2 A2. $C_{42}H_{60}K_2O_{16}$; mol wt 899.12.

4542. Glyhexamide. [451-71-8] N-[(Cyclohexylamino)carbonyl]-2,3-dihydro-1H-indene-5-sulfonamide; 1-cyclohexyl-3-(5-indanylsulfonyl)urea; 1-cyclohexyl-3-(5-hydrindenylsulfonyl)urea; SQ-15860; Subose. $C_{16}H_{22}N_2O_3S$; mol wt 322.42. C 59.60%, H 6.88%, N 8.69%, O 14.89%, S 9.94%. Prepd from hydrindene-5-sulfonamide and cyclohexyl isocyanate: Hoehn, Breuer, US 3097242 (1963 to Olin Mathieson). Clinical pharmacology: Grinnell et al., Am. J. Med. Sci. **253**, 312 (1967).

Crystals from 70% acetone, mp 153-155°.

THERAP CAT: Antidiabetic.

4543. Glymidine. [339-44-6] N-[5-(2-Methoxyethoxy)-2-pyrimidinyl]benzenesulfonamide; 2-benzenesulfonamido-5-(β-methoxyethoxy)pyrimidine; glycodiazine. $C_{13}H_{15}N_3O_4S$; mol wt 309.34. C 50.48%, H 4.89%, N 13.58%, O 20.69%, S 10.36%. Prepn: BE 609270; H. Priewe et al., US 3275635 (1962, 1966 both to Schering, AG); Gutsche et al., Arzneim.-Forsch. **14**, 373 (1964). Series of articles on pharmacology: ibid. 377-412. Activity: Losert et al., ibid. **23**, 1251 (1973). Metabolism: Soyfer et al., Chim. Ther. **5**, 441 (1970). Toxicity data: Kramer et al., Arzneim.-Forsch. **14**, 377 (1964).

Crystals, mp 152-154°. Soly in ethanol: 0.91%; in toluene: 0.67%.

Sodium salt. [3459-20-9] SH-717; Glyconormal; Gondafon; Lycanol; Redul. $C_{13}H_{14}N_3NaO_4S$; mol wt 331.32. Crystals, mp 221-226°. Sparingly sol in alc. Soly in water at 37°: 70.5%. LD_{50} in mice, rats (g/kg): 1.48, 2.00 i.v.; 5.30, 2.85 orally (Kramer).

THERAP CAT: Antidiabetic.

4544. Glyoxal. [107-22-2] Ethanedial; biformyl; diformyl; oxalaldehyde. $C_2H_2O_2$; mol wt 58.04. C 41.39%, H 3.47%, O 55.13%. OHCCHO. Prepd by the oxidation of acetaldehyde by nitric or selenious acid: Lubawin, Ber. **8**, 768 (1875); Wyss, Ber. **10**, 1366 (1877); Kölln, Ann. **416**, 230 (1918); Riley et al., J. Chem. Soc. **1932**, 1881; Ronzio, Waugh, Org. Synth. coll. vol. III, 438 (1955); by hydrolysis of dichlorodioxane: Butler, Cretcher, J. Am. Chem. Soc. **54**, 2988 (1932). Review of commercial development: J. F. Bohmfalk et al., Ind. Eng. Chem. **43**, 786 (1951). Toxicity study: H. F. Smyth et al., J. Ind. Hyg. Toxicol. **23**, 259 (1941). Review: A. B. Boese et al. in Glycols, G. O. Curme, F. Johnston, Eds. (Reinhold, New York, 1952) pp 125-128.

Yellow prisms or irregular pieces turning white on cooling. d^{20} 1.14. Opaque at 10°, mp 15°. bp_{776} 51°. The vapors are green and burn with a purple flame. Explosive: Mixtures with air may explode! $n_D^{20.5}$ 1.3826. Sol in anhyd solvents. pH of a 40% aq soln: 2.1-2.7; d_4^{20} 1.27. Polymerizes quickly on standing, on contact with water (violent reaction), or when dissolved in solvents contg water. The anhyd polymer changes to the monomer on heating. Solns of the monomer are obtained on heating the polymer with anethole, phenetole, safrole, methyl nonyl ketone, or benzaldehyde. LD_{50} in rats, guinea pigs (mg/kg): 2020, 760 orally (Smyth).

Dihydrate. $(OHCCHO)_3\cdot2H_2O$. Crystalline powder, nonhygroscopic. More sol in hot water than in cold water. Commercially available in anhydr form as crystalline dihydrate, or as a 40% aq soln which may contain polymerization inhibitors.

Caution: Moderately irritating to skin, mucous membranes.

USE: In textiles, organic synthesis, glues, biocides.

4545. Glyoxal-Sodium Bisulfite. [517-21-5] 1,2-Dihydroxy-1,2-ethanedisulfonic acid disodium salt; glyoxal compd with sodium bisulfite. $C_2H_4Na_2O_8S_2$; mol wt 266.15. C 9.03%, H 1.51%, Na 17.28%, O 48.09%, S 24.09%. Prepn: Ronzio, Waugh, *Org. Synth.* **coll. vol. III**, 438 (1955).

Monohydrate. Hard crystals. Faint SO_2 odor. Freely sol in water. Practically insol in alcohol.

4546. Glyoxylic Acid. [298-12-4] 2-Oxoacetic acid; formylformic acid; glyoxalic acid; oxoethanoic acid. $C_2H_2O_3$; mol wt 74.04. C 32.44%, H 2.72%, O 64.83%. Occurs in unripe fruit and in young green leaves; has also been found in very young sugar beets. Prepd by heating dibromoacetic acid with some water: Grimaux, *Bull. Soc. Chim.* [2] **26**, 483; Cramer, *Ber.* **25**, 714 (1892); by electrolytic reduction of oxalic acid: Meyer, *Ber.* **37**, 3592 (1904); by the action of *Aspergillus niger* on calcium acetate, malonic or citric acid: Challenger *et al.*, *J. Chem. Soc.* **1927**, 205, 207. Prepn of glyoxylic acid soln for analytical use: *Beilstein* vol. III, 594.

Hemihydrate. Crystals from water, mp 70-75°. Also obtained in anhydr form as monoclinic crystals from water, mp 98°. Obnoxious odor. Strong, corrosive acid. $K = 4.6 \times 10^{-4}$. Deliquesces rapidly and forms a syrup on short exposure to air. Sparingly sol in alc, ether, benzene. Freely sol in water; aq solns tend to acquire a yellowish tint. Attacks most base metals except certain stainless steel alloys.

Monohydrate. Crystals, mp ~50°. Highly hygroscopic.

Caution: Irritant, corrosive.

4547. Glyphosate. [1071-83-6] *N*-(Phosphonomethyl)glycine; MON-0573. $C_3H_8NO_5P$; mol wt 169.07. C 21.31%, H 4.77%, N 8.28%, O 47.31%, P 18.32%. Broad-spectrum post-emergence, translocated herbicide. Prepn: J. E. Franz, *DE 2152826*; *idem, US 3799758* and *US 3853530* (1972, 1974, 1974 all to Monsanto). Metabolism and degradation in soil and water: M. L. Rueppel *et al.*, *J. Agric. Food Chem.* **25**, 517 (1977). Toxicity study: E. A. Bababunmi *et al.*, *Toxicol. Appl. Pharmacol.* **45**, 319 (1978).

White solid, mp 230° (dec). Soly in water at 25°: 12 g/l. Insol in most organic solvents. LD_{50} in rats, mice (mg/kg): 4873, 1568 orally (Bababunmi).

Mono(isopropylamine) salt. [38641-94-0] MON-2139; Accord; Durango; Glyphomax; Glypro; Rodeo; Roundup. $C_3H_8NO_5P.C_3H_9N$; mol wt 228.18. Very sol in water.

Trimethylsulfonium salt. [81591-81-3] Sulfosate; glyphosate-trimesium; Avans-330; SC-0224; ICI-A-0224; Touchdown. $C_3H_7NO_5P.C_3H_9S$; mol wt 245.23. Description of biological and physical properties: B. Trouslard, *Phytoma* **429**, 47 (1991); A. D. Baylis *et al.*, *J. Biosci.* **8**, 173 (1997). bp$_{760}$ 110°. d 1.27. Soly in water (20°): 1050 g/l. LD_{50} in rats, mice (mg/kg): 748, 1383 orally; in rabbits (mg/kg): >2000 dermally. LC_{50} in rainbow trout (48 hrs): 4800 mg/l (Trouslard).

USE: Herbicide.

4548. Goitrin. [500-12-9] (5*S*)-5-Ethenyl-2-oxazolidinethione; 5-vinyl-2-thiooxazolidone; (−)-5-vinyl-2-oxazolidinethione. C_5H_7NOS; mol wt 129.18. C 46.49%, H 5.46%, N 10.84%, O 12.39%, S 24.82%. An antithyroid compd isolated from seeds of different species of *Brassica, Cruciferae:* Astwood *et al.*, *J. Biol. Chem.* **181**, 121 (1949); Greer, *J. Am. Chem. Soc.* **78**, 1260 (1956). Stereochemistry: Kjaer *et al.*, *Acta Chem. Scand.* **13**, 144 (1959). Activity: Langer *et al.*, *Endokrinologie* **57**, 225 (1971).

Large prisms from ether, mp 50°. $[\alpha]_D^{31}$ −70.5° (c = 2 in methanol). Behaves as a weak acid; pKa 10.5. Stable in alkali, but not in acid.

4549. Gold. [7440-57-5] Au; at. wt 196.966569; at. no. 79; valences 1, 3. Group IB (11). Occurrence in the earth's crust: 0.005 ppm. One natural isotope: 197; artificial isotopes (mass numbers): 177-179, 181, 183, 185-196, 198-203. Probably the first pure metal known to man. Occurs in nature in its native form and in minute quantities in almost all rocks and in seawater. Gold ores include *calavarite*, (AuTe₂), *sylvanite*, [(Ag,Au)Te₂], *petzite*, [(Ag,Au)₂Te]. Methods of mining, extracting and refining: Hull, Stent in *Modern Chemical Processes* Vol. 5 (Reinhold, New York, 1958) pp 60-71. Lab prepn of gold powder from gold pieces: Block, *Inorg. Synth.* **4**, 15 (1953). Chemistry of gold drugs in the treatment of rheumatoid arthritis: D. H. Brown, W. E. Smith, *Chem. Soc. Rev.* **9**, 217 (1980). Use as catalyst in oxidation of organic compds by NO₂: R. E. Sievers, S. A. Nyarady, *J. Am. Chem. Soc.* **107**, 3726 (1985). Least reactive metal at interfaces with gas or liquid: B. Hammer, J. K. Norskov, *Nature* **376**, 238 (1995). *Reviews:* Gmelins, Gold (8th ed.) **62**, parts 2, 3 (1954); Johnson, Davis, "Gold" in *Comprehensive Inorganic Chemistry* vol. 3, J. C. Bailar Jr. *et al.*, Eds. (Pergamon Press, Oxford, 1973) pp 129-186; J. G. Cohn, E. W. Stern in *Kirk-Othmer Encyclopedia of Chemical Technology* vol. 11 (Wiley-Interscience, New York, 3rd ed., 1980) pp 972-995.

Yellow, soft metal; face-centered cubic structure; when prepared by volatilization or precipitation methods, deep violet, purple, or ruby powder, mp 1064.76°; bp 2700°. d 19.3. Hardness (Mohs') 2.5-3.0; (Brinell's) 18.5. Extremely inactive; not attacked by acids, air or oxygen. Superficially attacked by aq halogens at room temp. Reacts with aqua regia, with mixtures contg chlorides, bromides, or iodides if they can generate nascent halogens, with many oxidizing mixtures especially those contg halogens. Also with alkali cyanides, solns of thiocyanates and double cyanides.

USE: In manuf jewelry; in gold plating other metals; as a standard of currency; most frequently alloyed with silver and copper. For use in medicine, *see* Gold, Radioactive, Colloidal.

4550. Gold Monochloride. [10294-29-8] Gold chloride (Au-Cl); aurous chloride. AuCl; mol wt 232.42. Au 84.75%, Cl 15.25%. Prepd by thermal decompn of gold trichloride: Biltz, Wein, *Z. Anorg. Allg. Chem.* **148**, 192 (1925); Capella, Schwab, *C. R. Hebd. Seances Acad. Sci.* **260**, 4337 (1965).

Yellowish powder. d 7.57. Dec at about 289° into gold and Cl₂. Practically insol in water, but slowly dec by it, more rapidly on heating, with formation of gold trichloride and separation of metallic gold; sol in alkali cyanides. With solns of alkali bromides, metallic gold and potassium auribromide are formed.

4551. Gold Monocyanide. [506-65-0] Gold cyanide (Au(CN)); aurous cyanide. CAuN; mol wt 222.98. C 5.39%, Au 88.33%, N 6.28%. AuCN. May be prepd by decompn of Na[Au(CN)₂] with HCl: Wogrinz, *Metalloberflaeche* **8**, B162 (1954).

Yellow, odorless powder; iridescent in sunlight; slowly dec in presence of moisture. When warmed with HCl, HCN is evolved. When ignited, dec into metal Au and CN. d₄²⁰ 7.14. Hexagonal when cryst. Practically insol in water, alcohol, ether, dil acids; sol in ammonia, soln of NaCN; dissolved by aqua regia. On warming with concd H₂SO₄ half of the gold separates as metal.

Caution: May liberate HCN.

4552. Gold, Radioactive, Colloidal. [10043-49-9] Gold isotope of mass 198; radioactive colloidal gold; colloidal gold ¹⁹⁸Au;

radio-gold (^{198}Au) colloid; gold colloid ^{198}Au; Aurcoloid-198; Aurcoscan-198; Aureotope. Colloidal dispersion of radioactive gold (^{198}Au) for parenteral administration.

Particle diameter range, 3-7 mμ. Has half-life of 2.7 days and emits beta and gamma radiation. Stable to heat except autoclaving under pressure. Compatible with saline solns, radiopaque media and other agents; flocculated by polyvalent metal ion.

THERAP CAT: Antineoplastic.

4553. Gold Sodium Thiomalate. [12244-57-4] 2-Mercaptobutanedioic acid gold(1+) sodium salt (1:1:?); sodium aurothiomalate; Myochrysine; Shiosol; Tauredon. Disease modifying antirheumatic drug (DMARD). Mixture of the mono- ($C_4H_4AuNaO_4S$) and disodium ($C_4H_3AuNa_2O_4S$) salts of gold thiomalic acid. Prepn as the monohydrate of the disodium salt: M. Delépine, **US 1994213** (1935 to Rhone Poulenc). Detection in blood by atomic absorption spectroscopy: A. I. A. Rodgers *et al.*, *Anal. Proc.* **19**, 87 (1982). Clinical pharmacokinetics: K. L. N. Blocka *et al.*, *Clin. Pharmacokinet.* **11**, 133 (1986). Discussion of mechanisms of action: H. A. Capell, *Agents Actions Suppl.* **24**, 158 (1988). Clinical trial in rheumatoid arthritis: R. Munro *et al.*, *Ann. Rheum. Dis.* **57**, 88 (1998).

White to yellowish-white powder, metallic taste. Very sol in water. Practically insol in alcohol, ether. Aq solns are colorless to pale yellow. pH of a 5% aq soln: 5.8-6.5.

THERAP CAT: Antirheumatic.

4554. Gold Sodium Thiosulfate. [15283-45-1] (*T*-4)-Bis-[monothiosulfato(2−)-$\kappa O,\kappa S$]-aurate(3−) sodium (1:3); hyposulfite of gold and sodium; sodium aurothiosulfate; Aurocidin; Aurolin; Crisalbine; Fosfocrisolo; Sanocrysin. AuNa$_3$O$_6$S$_4$; mol wt 490.11. Au 40.18%, Na 14.07%, O 19.58%, S 26.16%. Na$_3$Au(S$_2$O$_3$)$_2$. Prepn of dihydrate: Fordos, Gélis, *Ann. Chim. Phys.* **13**, 394 (1845); H. Brown, *J. Am. Chem. Soc.* **49**, 958 (1927); *Roger's Inorganic Pharmaceutical Chemistry*, T. O. Soine, C. O. Wilson, Eds. (Lea & Febiger, Philadelphia, 8th ed., 1967) pp 343-345. Crystal structure of dihydrate: H. Ruben *et al.*, *Inorg. Chem.* **13**, 1836 (1974). Efficacy of antidotes: M. A. Basinger *et al.*, *J. Rheumatol.* **12**, 274 (1985). Clinical study of contact allergy patch test reactions: K. E. McKenna *et al.*, *Contact Dermatitis* **32**, 143 (1995). Clinical evaluation in rheumatoid arthritis: M. L. Ciompi *et al.*, *Reumatismo* **54**, 251 (2002).

Dihydrate. [33614-49-2] White, glistening, needle-like or prismatic crystals. d 3.14. Darkens slowly on exposure to light. One gram dissolves in 2 ml water. Insol in alc and most other organic solvents. A 1:20 aq soln is neutral or slightly alkaline to litmus. LD$_{50}$ i.p. in mice: 110 mg/kg (Basinger).

Caution: Potential symptoms of overexposure via therapeutic i.m. administration are dermatitis, nausea, vomiting, diarrhea, nephritis, blood disorders, peripheral neuritis, hepatitis, encephalitis. *See Clinical Toxicology of Commercial Products*, R. E. Gosselin *et al.*, Eds. (Williams & Wilkins, Baltimore, 5th ed., 1984) p II-144.

THERAP CAT: Antirheumatic; diagnostic aid (contact allergen).

4555. Gold Stannate. [1345-24-0] C.I. Pigment Red 109; aurous stannate; C.I. 77482; gold-tin precipitate; gold-tin purple; purple of Cassius. Contains gold, tin and oxygen; composition of commercial products varies and in some cases may be a complex mixture of gold and stannic acid. Prepn: *Colour Index* vol. 4 (3rd ed., 1971) p 4669.

Brown powder. Practically insol in water. Sol in ammonia.

USE: Manuf ruby glass, colored enamels, and painting porcelain.

4556. Gold Trichloride. [13453-07-1] Gold chloride (AuCl$_3$); auric chloride. AuCl$_3$; mol wt 303.32. Au 64.94%, Cl 35.06%. Conveniently prepd in the laboratory from metallic gold and iodine monochloride: Gutmann, *Z. Anorg. Allg. Chem.* **264**, 169 (1951).

Dihydrate. Dark orange-red crystals, deliquesc in moist air. d$_4^{20}$ 3.9. Sublimes at 180° (760 mm). bp 229°. Decomp 254°. Soluble in water, alcohol, ether. *Keep well closed and protected from light.*

4557. Gold Trihydroxide. [1303-52-2] Gold hydroxide (Au(OH)$_3$); auric hydroxide. AuH$_3$O$_3$; mol wt 247.99. Au 79.43%, H 1.22%, O 19.35%. Au(OH)$_3$. Usually contains about 3H$_2$O and hence about 65% gold. May be prepd in lab according to the equation 2KAuCl$_4$ + 3Na$_2$CO$_3$ + 3H$_2$O → 2Au(OH)$_3$ + 6NaCl + 2KCl + 3CO$_2$: Lydén, *Z. Anorg. Allg. Chem.* **240**, 157 (1939).

Brown powder; dec by sunlight to metallic gold; also slowly dec with age or at 100°, and completely at 250°. Practically insol in water. Sol in soln of NaCN, in HCl or concd HNO$_3$. With NH$_3$ yields gold fulminate which *explodes easily in dry form. Protect from light.*

USE: In gold-plating solns; for decorating porcelains.

4558. Gold Trioxide. [1303-58-8] Gold oxide (Au$_2$O$_3$); auric oxide; digold trioxide; gold sesquioxide; gold oxide. Au$_2$O$_3$; mol wt 441.93. Au 89.14%, O 10.86%. Prepd from the hydroxide: Roseveare, Buehner, *J. Am. Chem. Soc.* **49**, 1221 (1927).

Brown powder; begins to evolve oxygen at 110°; at 250° it is entirely dec to metallic gold; also slowly dec by sunlight. Practically insol in water. Sol in HCl, concd HNO$_3$ and in NaCN soln. *Keep protected from light.*

4559. Gold Trisulfide. [1303-61-3] Gold sulfide (Au$_2$S$_3$); auric sulfide. Au$_2$S$_3$; mol wt 490.38%, S 19.62%. Prepd by treating dry lithium aurichloride Li(AuCl$_4$) with hydrogen sulfide at −10°. The reaction product consists of HCl, lithium chloride, and auric sulfide. The lithium chloride can be removed by extraction with alcohol. *Ref:* Antony, Lucchesi, *Gazz. Chim. Ital.* **19**, 552 (1889). Prepd from AuCl$_3$ or HAuCl$_4$ and H$_2$S in abs ether at low temp: Guthier, Dürrwachter, *Z. Anorg. Allg. Chem.* **121**, 266 (1922).

Black powder. Heating to 200° dec it into its elements.

USE: In photography.

4560. Golimumab. [476181-74-5] Anti-(human tumor necrosis factor α) immunoglobulin G1 (human monoclonal CNTO 148 γ_1-chain) disulfide with human monoclonal CNTO 148 κ-chain, dimer; CNTO-148; Simponi. Fully human monoclonal antibody directed against tumor necrosis factor α (TNFα); specifically binds to both soluble and membrane-bound forms of TNFα. Prepn: J. Giles-Komar *et al.*, **WO 02012502**; G. Heavner *et al.*, **US 7250165** (2002, 2007 both to Centocor). Clinical pharmacokinetics and safety: H. Zhuo *et al.*, *J. Clin. Pharmacol.* **47**, 383 (2007). Clinical trial in ankylosing spondylitis: R. D. Inman *et al.*, *Arthritis Rheum.* **58**, 3402 (2008); in rheumatoid arthritis: J. S. Smolen *et al.*, *Lancet* **374**, 210 (2009). Review of pharmacology and clinical experience: G. Hutas, *Curr. Opin. Mol. Ther.* **10**, 393-406 (2008).

THERAP CAT: Anti-inflammatory.

4561. Gonadotropin-Releasing Hormone. [9034-40-6] Luteinizing hormone-releasing factor; LH-RH; luteinizing hormone-releasing hormone; GnRH; gonadotropin-releasing factor; gonadoliberin; luliberin. C$_{55}$H$_{75}$N$_{17}$O$_{13}$; mol wt 1182.31. C 55.87%, H 6.39%, N 20.14%, O 17.59%. Peptide hormone produced in the hypothalamus that regulates the reproductive system by stimulating the secretion of the pituitary hormones, LH (luteinizing hormone) and FSH (follicle-stimulating hormone) *q.q.v.* First discovered in swine and subsequently found in all classes of vertebrates; variants have also been found in invertebrate species. The hypothalamic hormone is known as GnRH-I. A second isoform, GnRH-II, is structurally conserved throughout vertebrate species; is particularly abundant in kidney, bone marrow, and prostate; and has been shown to act as a neuromodulator. Isoln of GnRH-I from porcine hypothalamic extracts: A. V. Schally *et al.*, *Biochem. Biophys. Res. Commun.* **43**, 393 (1971). Amino acid sequence of porcine: H. Matsuo *et al.*, *ibid.* 1334; Y. Baba *et al.*, *ibid.* **44**, 459 (1971); of ovine: R. Burgus *et al.*, *Proc. Natl. Acad. Sci. USA* **69**, 278 (1972). Confirmation of biological activity: A. V. Schally *et al.*, *Science* **173**, 1036 (1971). Physiology and implications in fertility control: A. V. Schally *et al.*, *Fertil. Steril.* **22**, 703 (1971); R. Guillemin, *Contraception* **6**, 1 (1972). Comparison of structural variants: J. A. King, R. P. Millar, *Science* **206**, 67 (1979). Identification of GnRH-II in chicken brain: K. Miyamoto *et al.*, *Proc. Natl. Acad. Sci. USA* **81**, 3874 (1984). Review of structure and function of GnRH-II: R. P. Millar, *Trends Endocrinol. Metab.* **14**, 35-43 (2003). Structural and phylogenetic analysis of GnRH variants: A. Gorbman, S. A. Sower, *Gen. Comp. Endocrinol.* **134**, 207 (2003). *Reviews:* F. Schneider *et* *al., Theriogenology* **66**, 691-709 (2006); L. W. T. Cheung, A. S. T. Wong, *FEBS J.* **275**, 5479-5495 (2008).

5-oxoPro–His–Trp–Ser–Tyr–Gly–Leu–Arg–Pro–GlyNH$_2$

Gonadorelin

Gonadorelin. [33515-09-2] Luteinizing hormone-releasing factor (swine); Fertiral; Kryptocur; Relefact LH-RH. C$_{55}$H$_{75}$N$_{17}$O$_{13}$; mol wt 1182.31. Decapeptide corresponding to the mammalian form of GnRH-I. Solid phase synthesis: M. W. Monahan, J. Rivier, *Biochem. Biophys. Res. Commun.* **48**, 1100 (1972); D. H. Coy *et al.*, *Methods Enzymol.* **37**, 416 (1975). Field trial in lactating dairy cows: A. H. Souza *et al.*, *Theriogenology* **72**, 271 (2009). [α]$_D^{25}$ −50° (1% acetic acid).

Gonadorelin Acetate. [34973-08-5]; [52699-48-6] (hydrate). Cystorelin; Fertagyl; Hypocrine; Lutrelef; Lutrepulse. C$_{55}$H$_{75}$N$_{17}$O$_{13}$.xC$_2$H$_4$O$_2$.yH$_2$O, as the diacetate tetrahydrate or a mixture of monoacetate and diacetate hydrates. White to slightly yellowish powder. Sol in water; sparingly sol in methanol.

Gonadorelin Hydrochloride. [51952-41-1] AY-24031; HRF; Factrel. C$_{55}$H$_{75}$N$_{17}$O$_{13}$.xHCl, where x = 1-2; mol wt 1219-1255. Hygroscopic, white powder. Sol in water.

THERAP CAT: Gonad-stimulating principle.

THERAP CAT (VET): In treatment of ovarian cysts in dairy cattle and for estrus induction.

4562. Goserelin. [65807-02-5] 6-[*O*-(1,1-Dimethylethyl)-D-serine]-1-9-luteinizing hormone-releasing factor (swine) 2-(aminocarbonyl)hydrazide; 6-[*O*-(1,1-dimethylethyl)-D-serine]-10-deglycinamideluteinizing hormone-releasing factor (pig) 2-(aminocarbonyl)hydrazide; D-Ser(Bu^t)^{6}Azgly^{10}-gonadorelin; ICI-118630. C$_{59}$H$_{84}$N$_{18}$O$_{14}$; mol wt 1269.43. C 55.82%, H 6.67%, N 19.86%, O 17.64%. Synthetic peptide agonist analog of gonadotropin-releasing hormone, *q.v.* Prepn: A. S. Dutta *et al.*, *DE 2720245; eidem, US 4100274* (1977, 1978 both to I.C.I.); *eidem, J. Med. Chem.* **21**, 1018 (1978). Radioimmunoassay in serum: R. N. Clayton *et al.*, *Clin. Endocrinol.* **22**, 453 (1985). Endocrine effects in women: C. P. West, D. T. Baird, *ibid.* **26**, 213 (1987). Review of pharmacokinetics and therapeutic efficacy in sex hormone related disorders: P. Chrisp, K. L. Goa, *Drugs* **41**, 254-288 (1991). Clinical trial in treatment of uterine fibroids: J. Gerris *et al.*, *Horm. Res.* **45**, 279 (1996); in prostate cancer: M. Bolla *et al.*, *N. Engl. J. Med.* **337**, 295 (1997). Review of clinical experience in premenopausal breast cancer: A. Rody *et al.*, *Expert Rev. Anticancer Ther.* **5**, 591-604 (2005).

5-oxoPro–His–Trp–Ser–Tyr–D-Ser(*t*-Bu)–Leu–Arg–Pro–NHNHCONH$_2$

Acetate. Zoladex. C$_{59}$H$_{84}$N$_{18}$O$_{14}$.xC$_2$H$_4$O$_2$, where x = 1 to 2.4. Off-white powder. Freely sol in acetic acid; sol in water, 0.1M HCl, 0.1M NaCl, DMF, DMSO. Insol in acetone, chloroform, ether.

THERAP CAT: Antineoplastic (hormonal).

4563. Gossyplure. [50933-33-0] 7,11-Hexadecadien-1-ol 1-acetate. C$_{18}$H$_{32}$O$_2$; mol wt 280.45. C 77.09%, H 11.50%, O 11.41%. Sex pheromone of pink bollworm, *Pectinophora gossypiella* (Saunders): Hummel *et al.*, *Science* **181**, 873 (1973). Isoln and prepn of 1:1 mixture of (Z,Z) and (Z,E) isomers: B. A. Bierl *et al.*, *J. Econ. Entomol.* **67**, 211 (1974). Improved prepn: R. J. Anderson, C. A. Henrick, *US 3919329* (1975 to Zoecon); *eidem, J. Am. Chem. Soc.* **97**, 4327 (1975). Stereoselective synthesis of isomers: K. Mori *et al.*, *Agric. Biol. Chem.* **38**, 1551 (1974); *eidem, Tetrahedron* **31**, 1846 (1975); H. Su, P. G. Mahany, *J. Econ. Entomol.* **67**, 319 (1974); H. J. Bestmann *et al.*, *Tetrahedron Lett.* **1976**, 353; J. M. Muchowski, M. C. Venuti, *J. Org. Chem.* **46**, 459 (1981). Activity of isomers: H. M. Flint *et al.*, *Environ. Entomol.* **6**, 274 (1977). Degradn: R. D. Henson, *ibid.* 821.

(Z,Z) - form

Yellow liquid. Sol in most org solvents. Extremely flammable.

(Z,Z)-Form. [52207-99-5] bp 130-132°. n_D^{21} 1.4592.
(Z,E)-Form. [53042-79-8] bp 132-134°. n_D^{21} 1.4591.
USE: Insect attractant.

4564. Gossypol. [303-45-7] 1,1′,6,6′,7,7′-Hexahydroxy-3,3′-dimethyl-5,5′-bis(1-methylethyl)[2,2′-binaphthalene]-8,8′-dicarboxaldehyde; 2,2′-bis[1,6,7-trihydroxy-3-methyl-5-isopropyl-8-aldehydonaphthalene]; 2,2′-bis[8-formyl-1,6,7-trihydroxy-5-isopropyl-3-methylnaphthalene]. C$_{30}$H$_{30}$O$_8$; mol wt 518.56. C 69.49%, H 5.83%, O 24.68%. Yellow pigment found in cottonseed; functions as a natural insecticide in plants. Name derived from the botanical name of the cotton plant, *Gossypium* L., *Malvaceae*. Isoln: J. Longmore, *J. Soc. Chem. Ind.* **5**, 200 (1886); L. Marchlewski, *J. Prakt. Chem.* **60**, 84 (1899); K. N. Campbell *et al.*, *J. Am. Chem. Soc.* **59**, 1723 (1937). Structural studies: R. Adams *et al.*, *ibid.* **60**, 2193 (1938). Synthesis: Edwards, *J. Am. Oil Chem. Soc.* **47**, 441 (1970). NMR studies: J. W. Jaroszewski *et al.*, *NMR Spectrosc. Drug Res.* **26**, 75 (1988). Vibrational CD structural studies: T. B. Freedman *et al.*, *Chirality* **15**, 196 (2003). Prepn of enantiomers: M. K. Dowd, *ibid.* 486. HPLC determn: G. B. Marcelle *et al.*, *J. Pharm. Sci.* **73**, 396 (1984). Metabolism studies: M. B. Abou-Donia *et al.*, *Lipids* **5**, 938 (1970). Mechanism of action study: C.-Y. G. Lee *et al.*, *Mol. Cell. Biochem.* **47**, 65 (1982). Structure-activity study in tumor cell lines: M. D. Shelley *et al.*, *Anti-Cancer Drugs* **11**, 209 (2000). Clinical evaluation in malignant glioma: P. Bushunow *et al.*, *J. Neuro-Oncol.* **43**, 79 (1999); in male contraception: E. M. Coutinho *et al.*, *Contraception* **61**, 61 (2000). Review of chemistry: R. Adams *et al.*, *Chem. Rev.* **60**, 555-574 (1960); of toxicity in livestock: S. E. Morgan, *Vet. Clin. North Am. Food Anim. Pract.* **5**, 251-262 (1989); of clinical pharmacology and use as a male contraceptive agent: D. Wu, *Drugs* **38**, 333-341 (1989); E. M. Coutinho, *Contraception* **65**, 259-263 (2002).

Exists in 3 tautomeric forms. Yellow crystals from ether, mp 184°; from chloroform, mp 199°; from ligroin, mp 214°. *Light sensitive.* Very sol in methanol, ethanol, cold dioxane, diethylene glycol, ether, ethyl acetate, acetone, carbon tetrachloride, pyridine, chloroform, DMF, lipids. Freely sol (with slow decompn) in dil aq solns of ammonia and sodium carbonate. Slightly sol in glycerol, cyclohexane. Insol in water. Absorption max: 385 nm (ε 18000).

R-Form. [90141-22-5] (−)-Gossypol. Bright yellow crystals from acetone, mp 181-184° (dec). [α]$_D^{25}$ −386 to −390° (c = 0.5 g/100 g soln in CHCl$_3$) (Dowd).

S-Form. [20300-26-9] (+)-Gossypol. Isoln from *Thespesia populnea*: T. J. King, L. B. de Silva, *Tetrahedron Lett.* **9**, 261 (1968); S. C. Datta *et al.*, *Indian J. Chem.* **10**, 263 (1972). Synthesis: A. I. Meyers, J. J. Willemsen, *Tetrahedron* **54**, 10493 (1998). Bright yellow crystals from acetone, mp 181-184° (dec). [α]$_D^{28}$ +386 to +390° (c = 0.5 g/100 g soln in CHCl$_3$) (Dowd). uv max (ethanol): 237.5, 277. 378 nm (log ε 4.8, 4.4, 4.2) (Datta).

Gossypol Acetic Acid. [12542-36-8] Bright yellow plates, mp 187° (Adams *et al.*).

Caution: Potentially toxic to animals overfed cottonseed products; symptoms of overexposure may include cardiac failure, shortness of breath, pulmonary edema, and reproductive effects (Morgan).

4565. Gougerotin. [2096-42-6] 1-(4-Amino-2-oxo-1(2*H*)-pyrimidinyl)-1,4-dideoxy-4-[(*N*-methylglycyl-D-seryl)amino]-β-D-glucopyranuronamide; 1-(4-amino-2-oxo-1(2*H*)-pyrimidinyl)-1,4-dideoxy-4-[D-2-[2-(methylamino)acetamido]hydracrylamido]glucopyranuronamide; 1-[4-deoxy-4-(sarcosyl-D-seryl)amino-β-D-glucopyranuronamide]cytosine; aspiculamycin; asteromycin. $C_{16}H_{25}$-N_7O_8; mol wt 443.42. C 43.34%, H 5.68%, N 22.11%, O 28.86%. Antibiotic substance with antibacterial and antineoplastic activity. Isoln from *Streptomyces gougerotii:* Kanzaki *et al., J. Antibiot.* **15A**, 93 (1962). Identity with asteromycin: Ikeuchi *et al., ibid.* **25**, 548 (1972). Structure: Iwasaki, *Yakugaku Zasshi* **82**, 1358 (1962). Revised structure: Fox *et al., Tetrahedron Lett.* **9**, 6029 (1968); Watanabe *et al., Chem. Pharm. Bull.* **17**, 416 (1969). Total synthesis: *eidem, J. Am. Chem. Soc.* **94**, 3272 (1972); Lichtenthaler *et al., Tetrahedron Lett.* **16**, 3527 (1975). Identity with aspiculamycin: Lichtenthaler *et al., ibid.* 665. Mechanism of action study: J. C. Lacal *et al., J. Antibiot.* **33**, 441 (1980). *Reviews:* Clark in *Antibiotics* **vol. 1**, D. Gottlieb, P. D. Shaw, Eds. (Springer-Verlag, New York, 1967) pp 278-282; Yukioka, *ibid.* **vol. 3**, J. W. Corcoran, F. E. Hahn, Eds. (1975) pp 448-458.

Needles, mp 211-217° (dec). $[\alpha]_D^{27}$ +53° (c = 0.8). uv max (water): 267, 235 nm (ε 9400, 9300); in 0.1N HCl: 275 nm (ε 13600); in 0.1N NaOH: 267 nm (ε 9800). LD_{50} in mice (mg/kg): 57 i.v. (Kanzaki).

4566. G-Proteins. GTP binding proteins. Distinct class of membrane associated *guanine nucleotide binding proteins* characterized by their function as couplers between a wide variety of receptors and their effector molecules in transmembrane signalling pathways. An example is the retinal G-protein, *transducin*, which links the photon receptor, rhodopsin, *q.v.*, to cGMP phosphodiesterase. G-Proteins are heterotrimeric, with apparent mol wt of 100 kDa and composed of α, β, γ subunits. The α subunit contains the guanine nucleotide binding site, possesses GTPase activity, and is specific for each G-protein. β and γ subunits form a noncovalent, membrane attached complex. *Reviews:* A. M. Spiegel, *Mol. Cell. Endocrinol.* **49**, 1-16 (1987); *idem, Annu. Rep. Med. Chem.* **23**, 235-242 (1988); P. J. Casey, A. G. Gilman, *J. Biol. Chem.* **263**, 2577-2580 (1988); H. R. Bourne, *Nature* **337**, 504-505 (1989); L. Birnbaumer, *Annu. Rev. Pharmacol. Toxicol.* **30**, 675-705 (1990); M. E. Linder, A. G. Gilman, *Sci. Am.* **267**, 56-65 (July, 1992). Review of role in disease: A. C. Dolphin, *Trends Neurosci.* **10**, 53-57 (1987).

4567. Gramicidin S. [113-73-5] Gramicidin S (Soviet); gramicidin C (Soviet). $C_{60}H_{92}N_{12}O_{10}$; mol wt 1141.47. C 63.13%, H 8.12%, N 14.73%, O 14.02%. Cyclic decapeptide antibiotic produced by a strain of *Bacillus brevis.* Isoln: Gause *et al., C. R. Acad. Sci. USSR* **43**, 217 (1944); *C.A.* **39**, 1195 (1945); Gause, Brazhnikova, *Lancet* **247**, 715 (1944). More closely related to tyrocidines in biological and chemical properties than to true gramicidins, *q.q.v.* Structure: Synge, *Biochem. J.* **39**, 363 (1945); Consden *et al., ibid.* **41**, 596 (1947); Battersby, Craig, *J. Am. Chem. Soc.* **73**, 1887 (1951); Erlanger, Goode, *Nature* **174**, 840 (1954). Synthesis and absorption spectrum: Schwyzer, Sieber, *Helv. Chim. Acta* **40**, 624 (1957); Waki, Izuniya, *Bull. Chem. Soc. Jpn.* **40**, 1687 (1967). Solid phase synthesis: Losse, Neubert, *Tetrahedron Lett.* **11**, 1267 (1970); M. Ohno *et al., J. Am. Chem. Soc.* **93**, 5251 (1971). Improved synthesis via a linear pentapeptide: Y. Minematsu *et al., Tetrahedron Lett.* **21**, 2179 (1980); via a linear decapeptide: T. Mukaiyama *et al., Chem. Lett.* **1981**, 1367. Industrial procedure: **GB 836725** (1960 to Ciba). *Review:* Y. A. Ovchinnikov, V. T. Ivanov, "The Cyclic Peptides: Structure, Conformation, and Function" in *The Proteins* **vol. V**, H. Neurath, R. L. Hill, Eds. (Academic Press, New York, 3rd ed., 1982) pp 547-555.

Val–Orn–Leu–D-Phe–Pro
Pro–D-Phe–Leu–Orn–Val

Hydrochloride. $C_{60}H_{92}N_{12}O_{10}$·2HCl. Prisms from ethanol + aq HCl, dec 277-278°. $[\alpha]_D^{24}$ −289° (c = 0.43 in 70% ethanol). Freely sol in alcohol; slightly sol in acetone. Practically insol in water, acids, alkalies. LD_{50} i.p. in rats: 17 mg/kg (Gause, Brazhnikova).

THERAP CAT: Topical antibacterial.

4568. Gramicidins. Gramicidin D (Dubos); linear gramicidins; Gramoderm. Polypeptide antibiotic complex first isolated from the antibiotic mixture, tyrothricin, along with tyrocidine, *q.q.v.*, from cultures of *Bacillus brevis:* Dubos, Hotchkiss, *J. Exp. Med.* **73**, 629 (1941); *eidem, J. Biol. Chem.* **141**, 155 (1941). Commercial extraction: Baron, **US 2534541** (1950 to Penick). Commercial preparation is a mixture of the four components, gramicidin A, B, C, and D, comprising about 87.5, 7.1, 5.1, 0.3 percent resp: Gross, Witkop, *Biochemistry* **4**, 2495 (1965). Each of the components A, B, and C consist of 2 chains, one with valine in position 1, comprising 80-95% of the component, and the other with isoleucine in position 1. Structure, characterization, and synthesis of the two isoforms of gramicidin A, N-formylvaline and isoleucine: Sarges, Witkop, *J. Am. Chem. Soc.* **86**, 1862 (1964); **87**, 2011, 2020 (1965); Bauer *et al., Biochemistry* **11**, 3266 (1972). Structure of gramicidin B: Sarges, Witkop, *J. Am. Chem. Soc.* **87**, 2027 (1965); of gramicidin C: *eidem, Biochemistry* **4**, 2491 (1965). Synthesis of valine-gramicidin B and C: K. Noda, E. Gross in *Chemistry and Biology of Peptides, Proc. 3rd Am. Peptide Symp.*, J. Meienhofer Ed. (Ann Arbor Science Publishers, Michigan, 1972) pp 241-250. *Review:* Hunter, Schwartz, "Gramicidins" in *Antibiotics* **I**, S. Gottlieb, P. Shaw, Eds. (Springer-Verlag, New York, 1967) pp 642-648. Comprehensive description: G. A. Brewer, *Anal. Profiles Drug Subs.* **8**, 179-218 (1979).

HC–Val–Gly–Ala–D-Leu–Ala–D-Val–Val–D-Val–Trp–D-Leu–Trp–D-Leu–Trp–D-Leu–Trp–NHCH$_2$CH$_2$OH

Valine-Gramicidin A

Spear-shaped or lenticular platelets, mp 229-230°. Soluble in the lower alcohols, acetic acid, pyridine; moderately sol in dry acetone and dioxane. Practically insol in ether, hydrocarbons; insol in water (0.6 mg/100 ml). Tends to form colloidal suspensions in water.

THERAP CAT: Antibacterial.

THERAP CAT (VET): Antimicrobial.

4569. Gramine. [87-52-5] N,N-Dimethyl-1H-indole-3-methanamine; 3-(dimethylaminomethyl)indole; Donaxine. $C_{11}H_{14}N_2$; mol wt 174.25. C 75.82%, H 8.10%, N 16.08%. In chlorophyll-deficient mutants of barley: Euler *et al., Z. Physiol. Chem.* **217**, 23 (1933). In the Asiatic reed *Arundo donax* L., *Gramineae:* Orechoff, Norkina, *Ber.* **68**, 436 (1935). From *Acer saccharinum* L. (the Silver Maple) and *A. rubrum* L., *Aceraceae:* Pachter *et al., J. Org. Chem.* **24**, 1285 (1959); Pachter, *J. Am. Pharm. Assoc. Sci. Ed.* **48**, 670 (1959). Synthesis: Kühn, Stein, *Ber.* **70**, 567 (1937). Biosynthesis from tryptophan in barley: Bowden, Marion, *Can. J. Chem.* **29**, 1037 (1951); O'Donovan, Leete, *J. Am. Chem. Soc.* **85**, 461 (1963); Gower, Leete, *ibid.* 3683; *see also* Gross *et al., Tetrahedron Lett.* **1971**, 4047.

Shiny, flat needles or plates from acetone, mp 138-139°. Absorption spectrum: Kanakoa *et al., Chem. Pharm. Bull.* **8**, 294 (1960). Sol in alcohol, ether, chloroform; slightly sol in cold acetone. Practically insol in petr ether, water.

Hydrochloride. $C_{11}H_{14}N_2$·HCl. Crystals from ethanol + ether, dec 191°. Sol in water.

Methiodide. $C_{11}H_{14}N_2$·CH$_3$I. Crystals from methanol + benzene, mp 168-169°. Sol in water: Geissman, Armen, *J. Am. Chem. Soc.* **74**, 3916 (1952).

4570.   Granaticin. [19879-06-2] (3a*S*,5*S*,8*S*,9*R*,11*R*,13b*S*,-15*R*)-3,3a,5,8,11,13b-Hexahydro-7,8,12,15-tetrahydroxy-5,9-dimethyl-8,11-ethanofuro[2,3-*e*]naphtho[2,3-*c*:6,7-*c'*]dipyran-2,6,13-(9*H*)-trione; antibiotic WR 141; litmomycin. $C_{22}H_{20}O_{10}$; mol wt 444.39. C 59.46%, H 4.54%, O 36.00%. Antibiotic substance produced by *Streptomyces olivaceus* from soil of Portuguese West Africa. Isoln and antibacterial activity: R. Corbaz *et al.*, *Helv. Chim. Acta* **40**, 1262 (1957). Determn by microbiological diffusion assay: A. Ricicova, M. Podojil, *Folia Microbiol.* **10**, 299 (1965). Isoln of granaticin B, the α-L-rhodinoside of granaticin: S. Barcza *et al.*, *Helv. Chim. Acta* **49**, 1736 (1966); **FR 1525993**; W. Keller, H. Zaehner, **US 3836642** (1968, 1974 both to Ciba-Geigy). Structure of granaticin and granaticin B: W. Keller-Schierlein *et al.*, *Helv. Chim. Acta* **51**, 1257 (1968); M. Brufani, M. Dobler, *ibid.* 1269; *Naturally Occurring Quinones*, R. H. Thomson, Ed. (Academic Press, New York, 2nd ed., 1971) pp 298-302. Identity of granaticin with antibiotic litmomycin: C.-J. Chang *et al.*, *J. Antibiot.* **28**, 156 (1975). Biosynthesis: C. E. Snipes *et al.*, *J. Nat. Prod.* **42**, 627 (1979); *eidem*, *J. Am. Chem. Soc.* **101**, 701 (1979). Total synthesis of (±)-form: K. Nomura *et al.*, *ibid.* **109**, 3402 (1987); of the natural (−)-form: K. Okazaki *et al.*, *Chem. Commun.* **1989**, 354. Cytotoxic action on carcinoma cells: E. Sturdik, L. Drobnica, *Neoplasma* **30**, 3 (1983). Inhibition of RNA synthesis: A. Ogilvie *et al.*, *Biochem. J.* **152**, 517 (1975); P. Heinstein, *J. Pharm. Sci.* **71**, 197 (1982).

Deep red, garnet-like crystals from acetone, dec 204-206°. Also reported as mp 211-213° (dec). Acts as an indicator: red in acids, blue in alkalies. Absorption max (abs ethanol): 223, 286, 532, 576 nm (log ε 4.58, 3.76, 3.87, 3.75).

Tetraacetylgranaticin. [1401-59-8] $C_{30}H_{28}O_{14}$. Yellow crystals from alc, mp 242-243°. $[\alpha]_D^{20}$ −100° (c = 0.818 in chloroform).

Granaticin B. [19879-03-9] $C_{28}H_{30}O_{12}$; mol wt 558.54. Red crystalline solid from methanolic-HCl, mp 117-119°. $[\alpha]_D^{22}$ +17.2° (c = 0.83 in pyridine). Absorption max (methanol): 223, 285, 527, 566 nm (log ε 4.42, 3.68, 3.76, 3.57).

4571.   Grandisol. [26532-22-9] (1*R*,2*S*)-1-Methyl-2-(1-methylethenyl)cyclobutaneethanol; *cis*-(+)-2-isopropenyl-1-methylcyclobutaneethanol; (+)-(1*R*,2*S*)-1-(2′-hydroxyethyl)-1-methyl-2-isopropenylcyclobutane. $C_{10}H_{18}O$; mol wt 154.25. C 77.87%, H 11.76%, O 10.37%. Major component of *grandlure*, the sex pheromone of the boll weevil (*Anthonomus grandis*, Boheman). Isoln and synthesis: J. H. Tumlinson *et al.*, *Science* **166**, 1010 (1969); *eidem*, *J. Org. Chem.* **36**, 2616 (1971). Synthesis of optically active grandisol: P. D. Hobbs, P. D. Magnus, *Chem. Commun.* **1974**, 856; *eidem*, *J. Am. Chem. Soc.* **98**, 4594 (1976); K. Mori, *Tetrahedron* **34**, 915 (1978); of enantiomerically pure grandisol: J. B. Jones *et al.*, *Can. J. Chem.* **60**, 2007 (1982). Synthesis of racemate: B. M. Trost *et al.*, *J. Am. Chem. Soc.* **99**, 3088 (1977). Short stereoselective synthesis of (±)-grandisol: I. Aljancic-Solaja *et al.*, *Helv. Chim. Acta* **70**, 1302 (1987). Review of syntheses: J. A. Katzenellenbogen, *Science* **194**, 139-148 (1976); J. M. Brand *et al.*, *Fortschr. Chem. Org. Naturst.* **37**, 18-29 (1979), *see also* refs pp 157-190; K. Mori, "The Synthesis of Insect Pheromones" in *The Total Synthesis of Natural Products* **vol. 4**, J. ApSimon, Ed. (Wiley-Interscience, New York, 1981) pp 80-85.

Liquid, bp$_{1.0}$ 50-60°. $[\alpha]_D^{21.5}$ +18.5° (c = 1 in hexane). n_D^{20} 1.4748.

4572.   Granisetron. [109889-09-0] 1-Methyl-*N*-[(3-*endo*)-9-methyl-9-azabicyclo[3.3.1]non-3-yl]-1*H*-indazole-3-carboxamide; Sancuso. $C_{18}H_{24}N_4O$; mol wt 312.42. C 69.20%, H 7.74%, N 17.93%, O 5.12%. Selective serotonin 5HT$_3$-receptor antagonist. Prepn: F. D. King, **EP 200444**; *idem*, **US 4886808** (1986, 1989 both to Beecham). 5HT$_3$ receptor binding study: G. J. Kilpatrick *et al.*, *Nature* **330**, 746 (1987). Series of articles on pharmacology and clinical trials to prevent chemotherapy-induced emesis: *Semin. Oncol.* **22**, Suppl. 10, 1-30 (1995). LC-MS/MS determn in plasma: Y. Jiang *et al.*, *J. Pharm. Biomed. Anal.* **42**, 464 (2006). Review of pharmacology and efficacy of transdermal formulation in emetogenic chemotherapy: S. T. Duggan, M. P. Curran, *Drugs* **69**, 2597-2605 (2009).

White to off-white solid. Insol in water.

Hydrochloride. [107007-99-8] BRL-43694A; Kytril. $C_{18}H_{24}$-N_4O.HCl; mol wt 348.88. White to off-white solid, mp 290-292°. Readily sol in water at 20°; sol in methanol.

THERAP CAT: Antiemetic.

4573.   Granulocyte Colony-Stimulating Factor. [143011-72-7] CSF-β; G-CSF; GM-DF; MGI-2; pluripoietin. Hematopoietic growth factor that stimulates the development of committed progenitor cells to neutrophils and enhances the functional activities of the mature end-cell. Glycoprotein of mol wt 18-22 kDa; produced in response to specific stimulation by a variety of cells including monocytes, fibroblasts and endothelial cells. Murine and human molecules exhibit cross species reactivity. Originally identified as a differentiation factor for murine leukemic cells. Characterization in mouse serum: J. Lotem *et al.*, *Int. J. Cancer* **25**, 763 (1980); A. W. Burgess, D. Metcalf, *ibid.* **26**, 647 (1980). Purification of murine G-CSF: N. A. Nicola *et al.*, *J. Biol. Chem.* **258**, 9017 (1983); of human G-CSF: K. Welte *et al.*, *Proc. Natl. Acad. Sci. USA* **82**, 1526 (1985); N. Nicola *et al.*, *Nature* **314**, 625 (1985); H. Nomura *et al.*, *EMBO J.* **5**, 871 (1986). Production of human G-CSF by recombinant DNA technology: L. M. Souza *et al.*, *Science* **232**, 61 (1986); S. Nagata *et al.*, *Nature* **319**, 415 (1986). Review of clinical potential and comparison with GM-CSF: W. P. Steward, *Lancet* **342**, 153-157 (1993). *Reviews:* G. D. Demetri, J. D. Griffin, *Blood* **78**, 2791-2808 (1991); L. S. Tkatch, D. J. Tweardy, *Lymphokine Cytokine Res.* **12**, 477-488 (1993). Review of role in infectious diseases: D. C. Dale *et al.*, *J. Infect. Dis.* **172**, 1061-1075 (1995).

Filgrastim. [121181-53-1] *N*-L-Methionylcolony-stimulating factor (human clone 1034); recombinant methionyl human G-CSF; r-metHuG-CSF; KRN-8601; Neupogen; Zarzio. 175 amino acid peptide produced in *E. coli* by recombinant DNA technology; mol wt 18.8 kDa. Differs from endogenous human G-CSF by the addition of an *N*-terminal methionine and the absence of glycosylation. Prepn: L. M. Souza, **EP 237545** (1987 to Kirin-Amgen); *idem*, **US 5580755** (1996 to Amgen). Clinical trial in severe chronic neutropenia: D. C. Dale *et al.*, *Blood* **81**, 2496 (1993). Book: *Filgrastim in Clinical Practice*, G. Morstyn, T. M. Dexter, Eds. (Marcel Dekker, New York, 1994) 351 pp.

Pegfilgrastim. [208265-92-3] 3-Hydroxypropyl-*N*-methionylcolony-stimulating factor (human) 1-ether with α-methyl-ω-hydroxypoly(oxy-1,2-ethanediyl); Neulasta. Pegylated form of filgrastim; modified by the addition of a 20 kDa linear molecule of polyethylene glycol. Pharmacology: G. Molineux *et al.*, *Exp. Hematol.* **27**, 1724 (1999). Series of articles on chemistry, pharmacokinetics and clinical efficacy: *Pharmacotherapy* **23**, 1S-19S (2003). Review of prepn and development: G. Molineux, *Curr. Pharm. Des.* **10**, 1235-1244 (2004); of clinical experience in chemotherapy-induced neutropenia: A.-R. Waladkhani, *Eur. J. Cancer Care* **13**, 371-379 (2004).

Lenograstim. [135968-09-1] Granocyte; Neutrogin. Glycoprotein with 174 amino acids and 4% sugar chains produced in Chinese hamster ovary cells by recombinant DNA technology; mol wt ~20 kDa. Closely resembles endogenous human G-CSF. Prepn: M.

Ono *et al.*, **WO 8604605**; *eidem*, **EP 215126** (1986, 1987 both to Chugai). Clinical trial in bone-marrow transplantation: C. Gisselbrecht *et al.*, *Lancet* **343**, 696 (1994). Review of pharmacology and clinical experience: C. J. Dunn, K. L. Goa, *Drugs* **59**, 681-717 (2000).

Nartograstim. [134088-74-7] 1-(*N*-L-Methionyl-L-alanine)-3-L-threonine-4-L-tyrosine-5-L-arginine-17-L-serinecolony-stimulating factor (human clone 1034); marograstim; KW-2228. G-CSF mutein of 175 amino acids produced in *E. coli* by recombinant DNA technology; mol wt 18.9 kDa. Prepn: Y. Yokoo *et al.*, **JP 2234692** (1990 to Kyowa), *C.A.* **114**, 183900 (1991). *In vitro* activity: T. Suzuki *et al.*, *Acta Haematol.* **87**, 181 (1992). ELISA determn in plasma: T. Kuwabara *et al.*, *J. Pharmacobio-Dyn.* **25**, 121 (1992).

THERAP CAT: Hematopoietic; antineutropenic.

4574. Granulocyte-Macrophage Colony-Stimulating Factor. [83869-56-1] Colony-stimulating factor-2; CSF-2; CSFα; GM-CSF; NIF-T. Hematopoietic growth factor that stimulates the development of neutrophils and macrophages and promotes the proliferation and development of early erythroid, megakaryocytic and eosinophilic progenitor cells. Produced by endothelial cells, monocytes, fibroblasts and T lymphocytes. Inhibits neutrophil migration and enhances the functional activities of the mature end-cells. Purification of murine GM-CSF: A. W. Burgess *et al.*, *J. Biol. Chem.* **252**, 1998 (1977); of human GM-CSF: N. A. Nicola *et al.*, *Blood* **54**, 614 (1979); and identity with neutrophil migration inhibition factor (NIF-T): J. C. Gasson *et al.*, *Science* **226**, 1339 (1984). Partial amino acid sequence of murine: L. G. Sparrow *et al.*, *Proc. Natl. Acad. Sci. USA* **82**, 292 (1985). Cloning and expression of murine GM-CSF: N. M. Gough *et al.*, *Nature* **309**, 763 (1984); of human: G. G. Wong *et al.*, *Science* **228**, 810 (1985); F. Lee *et al.*, *Proc. Natl. Acad. Sci. USA* **82**, 4360 (1985); M. A. Cantrell *et al.*, *ibid.* 6250. Pharmacokinetics: W. P. Petros, *Pharmacotherapy* **12**, Suppl 2-2, 32S (1992). Clinical comparison with G-CSF: S. Blackwell, J. Crawford, *ibid.* 20S. *Review:* J. C. Gasson, *Blood* **77**, 1131-1145 (1991). Review of clinical and biological effects: S. G. Louie, B. Jung, *Am. J. Hosp. Pharm.* **50**, Suppl 3, S10-S18 (1993).

Molgramostim. [99283-10-0] Colony-stimulating factor 2 (human clone pHG25 protein moiety reduced); Sch-39300; Leucomax. Nonglycosylated peptide of 127 amino acids produced in *E. coli* by recombinant DNA technology. Mol wt 14.5 kDa. Pharmacokinetics in AIDS and ARC: R. G. Hewitt *et al.*, *Antimicrob. Agents Chemother.* **37**, 512 (1993).

Regramostim. [127757-91-9] Colony-stimulating factor 2 (human clone pCSF-1 protein moiety reduced) glycoform GMC 89-107. Glycoprotein of 127 amino acids produced in Chinese hamster ovary cells by recombinant DNA technology. Mol wt 21-34 kDa.

Sargramostim. [123774-72-1] 23-L-Leucinecolony-stimulating factor 2 (human clone pHG25 protein moiety); BI 61.012; Leukine; Prokine. Variably glycosylated glycoprotein of 127 amino acids produced in yeast by recombinant DNA technology. Mol wt 15.5-19.5 kDa. Differs from endogenous human GM-CSF by the substitution of a leucine moiety at position 23 of the amino acid sequence. Clinical evaluation in myelodysplasia: W. J. Gradishar *et al.*, *Blood* **80**, 2463 (1992); in Crohn's disease: J. R. Korzenik *et al.*, *N. Engl. J. Med.* **352**, 2193 (2005).

THERAP CAT: Hematopoietic; antineutropenic.

4575. Graphite. [7782-42-5] Plumbago; black lead; mineral carbon. Obtained by mining, especially in Canada and Ceylon. Monograph: A. R. Ubbelohde, F. A. Lewis, *Graphite and Its Crystal Compounds* (Oxford, 1960). *Review:* Holliday *et al.* in *Comprehensive Inorganic Chemistry* **vol. 1**, J. C. Bailar, Jr. *et al.*, Eds. (Pergamon Press, Oxford, 1973) pp 1250-1294.

Crystallized carbon with traces of Fe, SiO_2, etc. Usually soft, black scales, crystals rare. d 2.09-2.23. Mohs' hardness = 1.0. Commercial varieties usually withstand temps up to 2820°. Sol in molten iron.

Caution: Potential symptoms of overexposure are coughing, dyspnea, black sputum, decreased pulmonary function and lung fibrosis. *See NIOSH Pocket Guide to Chemical Hazards* (DHHS/NIOSH 97-140, 1997) p 154.

USE: For "lead" pencils, refractory crucibles, stove polish; as pigment, lubricant, graphite cement; for matches and explosives, commutator brushes, anodes, arc-lamp carbons, electroplating; polishing

compds, rust and needle-paper; coating for cathode ray tubes; moderator in nuclear piles.

4576. Graphite Fluoride. [11113-63-6] Fluorine compd with graphite (1:?). Non-wettable layered solid, CF_x, (0.5<x<1.3), obtained by direct fluorination of graphite at high temp. Two covalent forms have been produced by controlling the quality of the graphite and the reaction temp: $(CF)_n$ at (300°-600°) and $(C_2F)_n$ at (350°-400°). Composition and color are a function of the fluorination temp; the F/C ratio increases with increasing temp, and color changes from black to gray to white. Prepn: O. Ruff, O. Bretschneider, *Z. Anorg. Allg. Chem.* **217**, 1 (1934); and thermal/electrical characteristics: C.-C. Hung *et al.*, *SAMPE Q.* **19**, 12 (1988). Evaluation of lubricating characteristics of $(CF_x)_n$: R. L. Fusaro, H. E. Sliney, *ASLE Trans.* **13**, 56 (1970); of $(C_2F)_n$: H. Miyake *et al.*, *Proc. JSLE Int. Tribiol. Conf.* **2**, 395 (1985). Lubrication mechanism and wear life: R. L. Fusaro, *Wear* **53**, 303 (1979). Brief review of physical properties: Central Glass Co. in *Proc. BMRA Symp.*, Brussels 1983, A. Kozawa, M. Nagayama, Eds. (Battery Mat. Res. Assoc., 1984) 135-141; of prepn, properties and uses: T. Nakajima, N. Watanabe, *Chemtech* **1990**, 426-430. Book: N. Watanabe *et al.*, *Graphite Fluorides* (Elsevier Press, Amsterdam, 1988), 262 pp.

Poly(carbon monofluoride). Cefbon-CF. $(CF)_n$. Gray-white (commercial). d 2.58 g/cm³; packed bulk density 0.7 g/cm³. Insol in all solvents. Fluorine content: 61-64 wt%.

Poly(dicarbon monofluoride). [144913-72-4] Cefbon-C_2F. $(C_2F)_n$. Dark gray-black (commercial). d 2.79 g/cm³; packed bulk density 0.8 g/cm³. Insol in all solvents. Fluorine content: 49-53 wt%.

USE: Solid lubricant; as cathode material in lithium-fluoride batteries; in controlling wettability of surfaces.

4577. Graphitic Acid. Graphite oxide; graphitic oxide. This material, obtained by oxidation of graphite, was first prepd by Brodie in 1859; Hummers, Offeman, *J. Am. Chem. Soc.* **80**, 1339 (1958). Its composition is not well defined but usually given as $C_4O(OH)$: Aragon de la Cruz, Cowley, *Nature* **196**, 468 (1962), *Acta Crystallogr.* **16**, 531 (1963). Prepn and manufacture: Hummers, US **2798878** (1957 to National Lead); Hummers, Offeman, *loc. cit.*; Ruskin, US **2933381** and US **2944881** (both 1960 to Union Carbide). Crystal structure: Aragon de la Cruz, Cowley, *loc. cit.*

Very light to dark brown, or yellowish-brown solid.

USE: In rocket propellant mixtures.

4578. Grayanotoxins. Toxic diterpenoids present in leaves of the various species of *Rhododendron*, *Kalmia*, and *Leucothoe*, Ericaceae; also found in honey from rhododendron flowers. Eighteen grayanotoxins have been isolated, the first three being the most important. Isoln of grayanotoxins I, II, III: Kakisawa *et al.*, *Tetrahedron* **21**, 3091 (1965); of IV and V: Okuno *et al.*, *ibid.* **26**, 4765 (1970); of V, VI, and VII: Hikino *et al.*, *Chem. Pharm. Bull.* **18**, 2357 (1970); of VIII, IX, X, and XI: Hikino *et al.*, *ibid.* **19**, 1289 (1971); of XII and XIII: Hikino *et al.*, *ibid.* **20**, 422 (1972). Approaches to synthesis of the grayanotoxin skeleton: T. Shiozaki *et al.*, *Tetrahedron Lett.* **1972**, 657; T. Kametani *et al.*, *Chem. Pharm. Bull.* **27**, 152 (1979); *eidem*, *Tetrahedron Lett.* **22**, 2379 (1981); *eidem*, *Tetrahedron* **37**, 3813 (1981). Toxicity study: H. Hikino *et al.*, *Toxicol. Appl. Pharmacol.* **35**, 303 (1976).

	R_1	R_2	R_3
Grayanotoxin I	OH	CH_3	$COCH_3$
Grayanotoxin III	OH	CH_3	H
Grayanotoxin II	$R_1R_2 = =CH_2$		R_3 = H

Grayanotoxin I. [4720-09-6] (3β,6β,14R)-Grayanotoxane-3,5,-6,10,14,16-hexol 14-acetate; G-I; acetylandromedol; andromedotoxin; rhodotoxin; asebotoxin. $C_{22}H_{36}O_7$; mol wt 412.52. From *Leucothoe grayana* Max., *Ericaceae:* Miyajimi, Takei, *J. Agric. Chem. Soc. Jpn.* **10**, 1093 (1934); from *Rhododendron maximum* L., *Ericaceae:* Wood *et al., J. Am. Chem. Soc.* **76**, 5689 (1954). Identity with acetylandromedol, andromedotoxin, rhodotoxin: Tallent *et al., ibid.* **79**, 4548 (1957). Stereochemistry: Iwasa, Nakamura, *Tetrahedron Lett.* **1969**, 3973; Narayanan *et al., ibid.* **1970**, 3943; Hikino *et al., Chem. Pharm. Bull.* **18**, 1071 (1970). Approach to synthesis: Okuno, Matsumoto, *Tetrahedron Lett.* **1969**, 4077. Has hypotensive action: Moran *et al., J. Pharmacol. Exp. Ther.* **110**, 415 (1954). Crystals from ethyl acetate, mp 258-260 to 267-270°, depending on rate of heating. $[\alpha]_D^{25}$ −8.8° (c = 2.3 in ethanol). Sol in hot water, alcohol, acetic acid, hot chloroform; very slightly sol in benzene, ether, petr ether. LD_{50} i.p. in mice: 1.31 mg/kg (Hikino, 1976).

Grayanotoxin II. [4678-44-8] (3β,6β,14R)-Grayanotox-10(20)-ene-3,5,6,14,16-pentol; G-II; deacetylanhydroandromedotoxin. $C_{20}H_{32}O_5$; mol wt 352.47. From *L. grayana* Max., *Ericaceae:* Miyajimi, Takei, *loc. cit.* Identity with deacetylanhydromedotoxin: Meguri, *Yakugaku Zasshi* **79**, 1060 (1959); *C.A.* **54**, 5599g (1960). Stereochemistry: Iwasa, Nakamura, *loc. cit.*; Kumazawa, Iriye, *Tetrahedron Lett.* **1970**, 927; Yasue *et al., Chem. Pharm. Bull.* **18**, 2586 (1970). Synthesis: S. Gasa *et al., Tetrahedron Lett.* **1976**, 553. Columnar crystals, mp 199-200°. $[\alpha]_D^{28}$ −41.88°. LD_{50} i.p. in mice: 26.1 mg/kg (Hikino, 1976).

Grayanotoxin III. [4678-45-9] (3β,6β,14R)-Grayanotoxane-3,5,6,10,14,16-hexol; G-III; deacetylandromedotoxin. $C_{20}H_{34}O_6$; mol wt 370.49. From *L. grayana* Max., *Ericaceae:* Miyajimi, Takei, *J. Agric. Chem. Soc. Jpn.* **12**, 947 (1936), *C.A.* **30**, 6747[9] (1936). Stereochemistry: Hikino *et al., Chem. Pharm. Bull.* **18**, 1071 (1970). LD_{50} i.p. in mice: 0.84 mg/kg (Hikino, 1976).

4579. Green Fluorescent Protein. GFP. Class of autofluorescent proteins found in bioluminescent coelenterates where they function as energy transfer acceptors, emitting a green fluorescent light (λ_{max} = 509 nm). Acidic globular proteins consisting of 238 amino acids, monomeric mol wt ∼30,000. Developed commercially as a biochemical tool to visualize cellular structure and monitor dynamic cellular events via fluorescence resonance energy transfer (FRET) assays. Purifn, characterization and energy transfer studies from *Aequorea victoria:* H. Morise *et al., Biochemistry* **13**, 2656 (1974); from *Renilla reniformis:* W. W. Ward, M. J. Cormier, *J. Biol. Chem.* **254**, 781 (1979). Properties of naturally occurring proteins: M. Chalfie, *Photochem. Photobiol.* **62**, 651 (1995). Structural studies of chromophore fragment: G. N. Phillips, Jr., *Curr. Opin. Struct. Biol.* **7**, 821 (1997); B. R. Branchini *et al., J. Am. Chem. Soc.* **120**, 1 (1998). Fluorescence and spectral properties of variants genetically engineered for enhanced fluorescence: G. H. Patterson *et al., Biophys. J.* **73**, 2782 (1997); R. H. Stauber *et al., BioTechniques* **24**, 462 (1998); for Ca²⁺ visualization: T. Nagai *et al., Proc. Natl. Acad. Sci. USA* **98**, 3197 (2001). Review of expression and detection: S. R. Kain, P. Kitts, *Methods Mol. Biol.* **63**, 305-324 (1997). Review of research applications: T. Misteli, D. L. Spector, *Nat. Biotechnol.* **15**, 961-964 (1997); as reporter gene: S. R. Kain *et al., BioTechniques* **19**, 650-655 (1995); as fluorescent protein tag: H.-H. Gerdes, C. Kaether, *FEBS Lett.* **389**, 44-47 (1996). Bibliography: L. J. Kricka, P. E. Stanley, *J. Biolumin. Chemilumin.* **12**, 113-134 (1997).

USE: Research tool in cell biology.

4580. Grepafloxacin. [119914-60-2] 1-Cyclopropyl-6-fluoro-1,4-dihydro-5-methyl-7-(3-methyl-1-piperazinyl)-4-oxo-3-quinolinecarboxylic acid. $C_{19}H_{22}FN_3O_3$; mol wt 359.40. C 63.50%, H 6.17%, F 5.29%, N 11.69%, O 13.35%. Fluorinated quinolone antibacterial. Prepn: J. M. Domagala *et al.*, **WO 8906649**; *eidem*, **US 4920120** (1989, 1990 both to Warner-Lambert); S. E. Hagen *et al., J. Med. Chem.* **34**, 1155 (1991). Comparative *in vitro* activity: F. Marco *et al., J. Antimicrob. Chemother.* **33**, 647 (1994); R. C. Arduino *et al., ibid.* **34**, 403. HPLC determn in human bronchoalveolar lavage samples: J. M. Woodcock *et al., FEMS Microbiol. Lett.* **119**, 315 (1994). Clinical pharmacokinetics: J. Child *et al., Antimicrob. Agents Chemother.* **39**, 513 (1995). Tissue concentration in lung: P. J. Cook *et al., J. Antimicrob. Chemother.* **35**, 317 (1995). Review of pharmacology and clinical trials: A. J. Wagstaff, J. A. Balfour, *Drugs* **53**, 817-824 (1997).

Dihydrate. mp 190-192°.

Hydrochloride. [161967-81-3] OPC-17116; Raxar. $C_{19}H_{22}FN_3O_3 \cdot HCl$; mol wt 395.86.

THERAP CAT: Antibacterial.

4581. Grindelia. Gum-plant (of California). Dried leaves and flowering tops of *Grindelia camporum* Greene or of *G. humilis* H. & A. (*G. cuneifolia* Auth.), *Compositae.* Habit. North America (California). *Constit.* Volatile oil, over 20% resin, grindelol, saponin, tannin, robustic acid.

THERAP CAT: Expectorant.

4582. Grindelic Acid. [1438-57-9] (1'R,4'aS,5S,8'aS)-4,4'a,-5,5',6',7',8',8'a-Octahydro-2',5,5',5',8'a-pentamethylspiro[furan-2(3H),1'(4'H)-naphthalene]-5-acetic acid; 9,13-epoxylabd-7-en-15-oic acid. $C_{20}H_{32}O_3$; mol wt 320.47. C 74.96%, H 10.07%, O 14.98%. Major grindelane diterpenoid isolated from the resin of *Grindelia robusta* Nutt., *Compositae.* Isoln and structure: L. Panizzi *et al., Gazz. Chim. Ital.* **92**, 522 (1962). Stereochemical studies: L. Mangoni, M. Belardini, *ibid.* **92**, 1379 (1962); **93**, 455, 465 (1963). Synthesis: M. Adinolfi *et al., ibid.* **106**, 625 (1976). Absolute configuration: M. Adinolfi *et al., Phytochemistry* **27**, 1878 (1988); L. A. Paquette, H.-L. Wang, *J. Org. Chem.* **61**, 5352 (1996).

Crystals from acetic acid, mp 100-101°. $[\alpha]_D$ −102.2°.

Methyl ester. $C_{21}H_{34}O_3$. Crystals from methanol, mp 70-70.5°. $[\alpha]_D$ −134.1° (c = 1.46 in methanol).

4583. Grisein. [1391-82-8] Antibiotic substance produced by strains of *Streptomyces griseus.* Isoln: Reynolds *et al., Proc. Soc. Exp. Biol. Med.* **64**, 50 (1947); Reynolds, Waksman, *J. Bacteriol.* **55**, 739 (1948). Improved method of isoln: F. A. Kuehl, L. Chaiet, **US 2505053** (1950 to Merck & Co.); F. A. Kuehl *et al., J. Am. Chem. Soc.* **73**, 1770 (1951). Analysis of composition: $C_{40}H_{61}FeN_{10}O_{20}S$. Degradation of grisein by acid hydrolysis yielded 3-methyluracil and at least two amino acids. One of the acids appears to be glutamic acid: F. A. Kuehl *et al., loc. cit.* Probably is a mixture of components; similar or identical to albomycin, *q.v.:* Stapley, Ormond, *Science* **125**, 587 (1957); Turková *et al., Collect. Czech. Chem. Commun.* **31**, 2444 (1966). Toxicity study: V. I. Aksenov, *Veterinariya (Moscow)* **12**, 93 (1974), *C.A.* **83**, 54127d (1975). Biosynthesis: V. V. Kuklin *et al., Antibiotiki* **25**, 403 (1980), *C.A.* **93**, 146173a (1980).

Amorphous red powder. Sol in water; slightly sol in 95% alcohol. Practically insol in abs alcohol, ether, acetone, chloroform, benzene. The activity remains unchanged when an aq soln is heated to 100° for 10 min. LD_{50} in mice (mg/kg): 600 orally; 34 s.c. (Aksenov).

4584. Griseofulvin. [126-07-8] (1'S,6'R)-7-Chloro-2',4,6-trimethoxy-6'-methylspiro[benzofuran-2(3H),1'-[2]cyclohexene]-3,4'-dione; 7-chloro-4,6-dimethoxycoumaran-3-one-2-spiro-1'-(2'-methoxy-6'-methylcyclohex-2'-en-4'-one); amudane; Curling factor; Fulcin; Fulvicin; Grifulvin; Grisactin; Griséfuline; Grisovin; Gris-PEG; Grysio; Lamoryl; Likuden; Polygris; Poncyl-FP; Spirofulvin; Sporostatin. $C_{17}H_{17}ClO_6$; mol wt 352.77. C 57.88%, H 4.86%, Cl 10.05%, O 27.21%. Antibiotic substance produced by

Penicillium griseofulvum Dierckx and by *P. janczewskii* Zal. [same as *P. nigricans* (Banier)Thom]. Isoln: Oxford *et al., Biochem. J.* **33**, 240 (1939); Brian *et al., Trans. Br. Mycol. Soc.* **29**, 173 (1946); Hockenhull, Dorey *et al.,* **US 3069328, US 3069329** (both 1962 to Glaxo). Structure: Grove *et al., Chem. Ind. (London)* **1951**, 219; *J. Chem. Soc.* **1952**, 3977. Stereochemistry: MacMillan, *ibid.* **1959**, 1823; Brown, Sim, *ibid.* **1963**, 1050. Total synthesis: Brossi *et al., Helv. Chim. Acta* **43**, 1444, 2071 (1960); Taub *et al., Tetrahedron* **19**, 1 (1963); Stork, Tomasz, *J. Am. Chem. Soc.* **86**, 471 (1964); S. Danishefsky, F. J. Walker, *ibid.* **101**, 7018 (1979). Conformation: Levine, Hicks, *Tetrahedron Lett.* **1971**, 311. Crystal structure: G. Malmros *et al., Cryst. Struct. Commun.* **6**, 463 (1977). Review and evaluation of studies of carcinogenic action in laboratory animals: *IARC Monographs* **10**, 153-161 (1976). *Review:* Grove, *Q. Rev. Chem. Soc.* **17**, 1 (1963); Huber in *Antibiotics* **vol. 3**, J. W. Corcoran, F. E. Hahn, Eds. (Springer-Verlag, New York, 1975) pp 606-613. Comprehensive description: E. R. Townley, *Anal. Profiles Drug Subs.* **8**, 219-249 (1979).

Stout octahedra or rhombs from benzene, mp 220°. $[\alpha]_D^{17}$ +370° (satd CHCl$_3$ soln). uv max: 286, 325 nm. Soly in DMF at 25°: 12 to 14 g/100 ml. Sol in acetone, chloroform; sparingly sol in alc; slightly sol in methanol, benzene, CHCl$_3$, ethyl acetate, acetic acid; very slightly sol in water. Practically insol petr ether.

THERAP CAT: Antifungal.

THERAP CAT (VET): Antifungal.

4585. Grubbs' Catalyst. [172222-30-9] (*SP*-5-31)-Dichloro-(phenylmethylene)bis(tricyclohexylphosphine)ruthenium; benzyli-denebis(tricyclohexylphosphine)ruthenium dichloride; Grubbs' first generation catalyst. C$_{43}$H$_{72}$Cl$_2$P$_2$Ru; mol wt 822.97. C 62.76%, H 8.82%, Cl 8.62%, P 7.53%, Ru 12.28%. Transition metal carbene complex used as a catalyst in the synthesis of olefins from ring-closing metathesis and ring-opening metathesis polymerization reactions. Prepn: P. Schwab *et al., Angew. Chem. Int. Ed.* **34**, 2039 (1995). Prepn and evaluation of structurally related catalysts: *eidem, J. Am. Chem. Soc.* **118**, 100 (1996). Ring-closing synthesis of amino acids and peptides containing olefins: S. J. Miller *et al., ibid.* **9606**; of dienes: T. A. Kirkland, R. H. Grubbs, *J. Org. Chem.* **62**, 7310 (1997). Mechanism of ring-opening metathesis: J. A. Tallarico *et al., J. Am. Chem. Soc.* **119**, 7157 (1997). Utility in yne-ene cross metathesis applications: M. Schuster, S. Blechert, *Tetrahedron Lett.* **39**, 2295 (1998). Review of olefin metathesis: A. Fürstner, *Angew. Chem. Int. Ed.* **39**, 3012-3043 (2000); of the design and development of olefin metathesis catalysts: T. M. Trnka, R. H. Grubbs, *Acc. Chem. Res.* **34**, 18-29 (2001).

R =

Purple microcrystalline solid. *Flammable; irritant.* Sol in dichloromethane, toluene, benzene, 1,1,2-trichloroethane, 1,2-dichloroethane, methanol, water. Protect from light. Store under inert gas at 2-8°C.

USE: Catalyst for olefin metathesis reactions.

4586. Grubbs' Second Generation Catalyst. [246047-72-3] (*SP*-5-41)-[1,3-Bis(2,4,6-trimethylphenyl)-2-imidazolidinylidene]-

dichloro(phenylmethylene) (tricyclohexylphosphine)ruthenium; benzylidene(1,3-dimesityl-4-imidazolidin-2-ylidene) (tricyclohexyl-phosphine)ruthenium dichloride; Grubbs II catalyst. C$_{46}$H$_{65}$Cl$_2$N$_2$-PRu; mol wt 848.98. C 65.08%, H 7.72%, Cl 8.35%, N 3.30%, P 3.65%, Ru 11.90%. Ruthenium carbene complex; *N*-heterocyclic analogue of Grubbs' catalyst, *q.v.* Utilized as a catalyst to mediate carbon-carbon bond rearrangements, including cross metathesis, ring-opening metathesis polymerizations, and ring-closing metathesis. Prepn and use as an olefin methathesis catalyst: M. Scholl *et al., Org. Lett.* **1**, 953 (1999). Improved prepn: L. Jafarpour *et al., Organometallics* **21**, 442 (2002). ^{1}H NMR studies: M. M. Gallagher *et al., J. Organomet. Chem.* **693** 1252 (2008). Synthesis of olefins by cross metathesis: A. K. Chatterjee, R. H. Grubbs, *Org. Lett.* **1**, 1751 (1999); by ring-opening metathesis: C. W. Bielawski, R. H. Grubbs, *Angew. Chem. Int. Ed.* **39**, 2903 (2000); J. P. Morgan *et al., Org. Lett.* **4**, 67 (2002). Utility in ring-closing and hydrosilylation reactions: V. Polshettiwar, R. S. Varma, *J. Org. Chem.* **73**, 7417 (2008). Mechanistic studies of olefin metathesis reactions: M. S. Sanford *et al., J. Am. Chem. Soc.* **123**, 749 (2001); *idem et al.,* 6543. Comprehensive review of ruthenium-based heterocyclic carbene catalysts: G. C. Vougioukalakis, R. H. Grubbs, *Chem. Rev.* **110**, 1746-1787 (2010).

R =

Pinkish-brown microcrystalline solid. *Flammable.* Insol in hexanes. Protect from light. Store under inert gas at 2-8°C.

USE: Catalyst for olefin metathesis reactions.

4587. Grundmann's Ketone. [66251-18-1] (1*R*,3a*R*,7a*R*)-1-[(1*R*)-1,5-Dimethylhexyl]octahydro-7a-methyl-4*H*-inden-4-one; (1*R*,3a*R*,7a*R*)-7a-methyl-1-((*R*)-6-methylheptan-2-yl)-octahydroin-den-4-one; Windaus-Grundmann ketone. C$_{18}$H$_{32}$O; mol wt 264.45. C 81.75%, H 12.20%, O 6.05%. Prepd by ozonolysis of vitamin D$_3$, *q.v.* Preparative method: A. Windaus, W. Grundmann, *Ann.* **524**, 295 (1936). Prepn: H. H. Inhoffen *et al., Ber.* **90**, 664 (1957). Synthetic applications: H. Nemoto *et al., J. Org. Chem.* **51**, 5311 (1986); P. Bovicelli *et al., ibid.* **57**, 5052 (1992); M. C. Clasby, D. Craig, *Synth. Commun.* **24**, 481 (1994); R. R. Sicinski, H. F. DeLuca, *Bioorg. Med. Chem. Lett.* **5**, 159 (1995); W. H. Okamura *et al., J. Org. Chem.* **67**, 1637 (2002).

Colorless oil, bp$_{0.001}$ 115-120°. $[\alpha]_D^{20}$ +8.9°.

USE: Intermediate in the synthesis of vitamin D active compounds.

4588. Guaiac. [9000-29-7] Guaiacum (resin); gum guaiac; resin guaiac; guaiacum. Resin from *lignum vitae*, the wood of *Guajacum officinale* L. or *G. sanctum* L., *Zygophyllaceae. Constit.* About 70% α- and β-guaiaconic acids, about 11% guaiacic acid, related compds and guaiaretic acid, 15% vanillin, guaiac yellow, guaiac saponin (guaiacin). Use as clinical reagent for occult blood: R. H. Wilkinson, W. A. F. Penfold, *Lancet* **2**, 847 (1969). Acute toxicity: P. M. Jenner *et al.*, *Food Cosmet. Toxicol.* **2**, 327 (1964).

Brown or greenish-brown, irregular lumps. mp 85-90°. Insol in water. Freely sol in alcohol, chloroform, ether, creosote, soln of chloral hydrate, alkalies; slightly sol in benzene, carbon disulfide. LD_{50} orally in rats: >5000 mg/kg (Jenner).

USE: Clinical reagent (blood or hemoglobin).

4589. Guaiacol. [90-05-1] 2-Methoxyphenol; methylcatechol; *o*-hydroxyanisole; 1-hydroxy-2-methoxybenzene; Anastil. C_7-H_8O_2; mol wt 124.14. C 67.73%, H 6.50%, O 25.78%. Isolated from guaiac resin: Sobrero, *Ann.* **48**, 19 (1843); from hardwood tar: McGinness *et al.*, *Tappi* **43**, 1027 (1960). Prepd by mercuric oxide oxidation of lignin: Lewis, Pearl, US 2433227 (1947 to Sulphite Prod.); by oxidation of anisole with trifluoroperoxyacetic acid: McClure, Williams, *J. Org. Chem.* **27**, 627 (1962); from acetovanillone + $ZnCl_2$: Read, US 3057927 (1962 to Ontario Res. Found.); from the diazonium salt of *o*-anisidine: Herbst, DE 1148236 (1963 to Hoechst). Toxicity data: Taylor *et al.*, *Toxicol. Appl. Pharmacol.* **6**, 378 (1964).

White or slightly yellow cryst mass or colorless to yellowish, very refractive liquid; characteristic odor. Darkens on exposure to air and light. d (crystals) 1.129; d (liq) ~1.112. Solidif 28°, but may remain liquid for a long time even at a much lower temp. bp 204-206°; bp_4 53-55°. One gram dissolves in 60-70 ml water, 1 ml glycerol; miscible with alcohol, chloroform, ether, oils, glacial acetic acid. Slightly sol in petr ether; sol in NaOH soln; with moderately concd KOH it forms a sparingly sol compd. *Protect from light.* LD_{50} orally in rats: 725 mg/kg (Taylor).

Phenylacetate. [4112-89-4] Gujaphenyl; Gunyl. $C_{15}H_{14}O_3$; mol wt 242.27.

THERAP CAT: Expectorant.

THERAP CAT (VET): Expectorant.

4590. Guaiazulene. [489-84-9] 1,4-Dimethyl-7-(1-methylethyl)azulene; 7-isopropyl-1,4-dimethylazulene; *S*-guaiazulene; AZ 8; AZ 8 Beris; Eucazulen; Kessazulen; Vaumigan. $C_{15}H_{18}$; mol wt 198.31. C 90.85%, H 9.15%. Isoln from chamomile oil: Sorm *et al.*, *Collect. Czech. Chem. Commun.* **16**, 626 (1951); from guaiac wood oil: Joos, CH 314487 (1956), *C.A.* **52**, 443b (1958). Total synthesis: Plattner *et al.*, *Helv. Chim. Acta* **32**, 2452 (1949); Sorm *et al.*, *Collect. Czech. Chem. Commun.* **16**, 168 (1951); Jacob *et al.*, *Tetrahedron* **20**, 2821 (1964); J. Mukherjee *et al.*, *J. Am. Chem. Soc.* **101**, 251 (1979). Pharmacokinetics of guaiazulene soluble in animals: H. Mukai *et al.*, *J. Pharmacobio-Dyn.* **8**, 329, 337 (1985). Effect on gastric and duodenal ulcers in rats: S. Okabe *et al.*, *Nippon Yakurigaku Zasshi* **88**, 467 (1986), *C.A.* **106**, 43769 (1987).

Blue oil. bp_{10} 165-170°.

3-Sulfonate sodium salt. [6223-35-4] 5-Isopropyl-3,8-dimethyl-1-azulenesulfonic acid sodium salt; sodium gualenate; guaiazulene soluble; Azulon. $C_{15}H_{17}NaO_3S$; mol wt 300.35.

Trinitrobenzene derivative. [4968-29-0] $C_{15}H_{18}.C_6H_3N_3O_6$. Violet to black needles from ethanol, mp 151°.

THERAP CAT: Anti-inflammatory; antiulcerative.

4591. Guaifenesin. [93-14-1] 3-(2-Methoxyphenoxy)-1,2-propanediol; glycerol mono(2-methoxyphenyl) ether; glycerol α-(2-

methoxyphenyl) ether; guaiacyl glyceryl ether; glyceryl guaiacyl ether; glycerol guaiacolate; α-glyceryl guaiacol ether; *o*-methoxyphenyl glyceryl ether; 1,2-dihydroxy-3-(2-methoxyphenoxy)propane; guaiacol glyceryl ether; guaiphenesin; guaiacuran; MY-301; XL-90; Colrex; Mucinex; Myoscain; Relaxil G; Resyl; Robitussin. $C_{10}H_{14}O_4$; mol wt 198.22. C 60.59%, H 7.12%, O 32.29%. Centrally acting muscle relaxant with expectorant properties. Prepn: Marle, *J. Chem. Soc.* **101**, 305 (1912); Yale *et al.*, *J. Am. Chem. Soc.* **72**, 3710 (1950); Roviralta, Astoul, ES 212920 (1954), *C.A.* **49**, 8332b (1955). Prepn from 2-methoxyphenol and glycidol: W. Merk *et al.*, DE 3106995; *eidem*, US 4390732 (1982, 1983 to Degussa AG). GLC determn in blood: W. R. Maynard, R. B. Bruce, *J. Pharm. Sci.* **59**, 1346 (1970). Clinical use in chronic respiratory disease: D. G. Workman *et al.*, *Curr. Ther. Res.* **7**, 665 (1965). Clinical efficacy as antitussive: J. J. Kuhn *et al.*, *Chest* **82**, 713 (1982). Pharmacokinetics and cardiopulmonary effects in horses: J. A. E. Hubbell *et al.*, *Am. J. Vet. Res.* **41**, 1751 (1980). Use in equine anesthesia: J. L. Grandy, W. N. McDonell, *J. Am. Vet. Med. Assoc.* **176**, 619 (1980); G. J. Brouwer, *Equine Vet. J.* **17**, 133 (1985).

Minute rhombic prisms from ether, mp 78.5-79°. bp_{19} 215°. Slightly bitter aromatic taste. One gram dissolves in 20 ml water at 25°; much more sol in hot water. Freely sol in ethanol; sol in alc, chloroform, glycerol, propylene glycol, DMF; moderately sol in benzene; sparingly sol in glycerin. Practically insol in petr ether.

THERAP CAT: Expectorant.

THERAP CAT (VET): Expectorant; muscle relaxant.

4592. Guaiol. [489-86-1] (3*S*,5*R*,8*S*)-1,2,3,4,5,6,7,8-Octahydro-α,α,3,8-tetramethyl-5-azulenemethanol; 3,8-dimethyl-5-(α-hydroxyisopropyl)-Δ⁹-octahydroazulene; champaca camphor; champacol; guaiac alcohol; Guajol. $C_{15}H_{26}O$; mol wt 222.37. C 81.02%, H 11.79%, O 7.19%. A sesquiterpene alc from guaiac wood: *Michelia champaca* L., *Magnoliaceae*: also from oil of wood of *Bulnesia sarmienti* Lorentz, *Zygophyllaceae*. Isoln: Plattner, Lemay, *Helv. Chim. Acta* **23**, 897 (1940). Structure: Plattner, Magyar, *ibid.* **25**, 581 (1942). Stereochemistry: Takeda, Minato, *Tetrahedron Lett.* **1** (43), 33 (1960); Minato, *Chem. Pharm. Bull.* **9**, 625 (1961); *idem*, *Tetrahedron* **18**, 365 (1962). Total synthesis of *dl*-form: Buchanan, Young, *Chem. Commun.* **1971**, 643; *eidem*, *J. Chem. Soc. Perkin Trans. 1* **1973**, 2404; Marshall *et al.*, *Tetrahedron Lett.* **12**, 855 (1971); Marshall, Greene, *J. Org. Chem.* **37**, 982 (1972); Andersen, Uh, *Tetrahedron Lett.* **14**, 2079 (1973).

Trigonal pyramidal crystals from alc, mp 91°. d_{20}^{100} 0.9074. bp_{760} 288° (slight decompn); bp_{17} 165°; bp_{10} 148°. $[α]_D^{20}$ −30° (c = 4 in alc). n_D^{100} 1.4716. Sol in alc, ether. Insol in water.

Methyl ether. $C_{16}H_{28}O$. Liq, d_4^{25} 0.9332. bp_9 142°. $[α]_D^{20}$ −31.8°. $n_D^{18.5}$ 1.4896.

Note: The name "guaiol" is also applied to 1,2-dimethylacrolein, isolated from guaiacum resin.

4593. Guanabenz. [5051-62-7] 2-[(2,6-Dichlorophenyl)methylene]hydrazinecarboximidamide; [(2,6-dichlorobenzylidene)-amino]guanidine; *N*-(2,6-dichlorobenzylidene)-*N'*-amidinohydrazine; NSC-68982. $C_8H_8Cl_2N_4$; mol wt 231.08. C 41.58%, H 3.49%, Cl 30.68%, N 24.25%. α₂-Adrenergic agonist. Prepn: J. Yates, E. Haddock, GB 1019120 (1966 to Shell), *C.A.* **64**, 11132h (1966). Use as antihypertensive: W. J. Houlihan *et al.*, DE 1804634

(1969 to Sandoz), *C.A.* **71**, 89976j (1969). Pharmacology: T. Baum *et al.*, *J. Pharmacol. Exp. Ther.* **171**, 276 (1970); E. Lampa *et al.*, *Experientia* **36**, 228 (1980). Disposition of ^{14}C-guanabenz in humans: R. H. Meacham *et al.*, *Clin. Pharmacol. Ther.* **27**, 44 (1980). Mechanism of action: G. F. DiBona, *J. Cardiovasc. Pharmacol.* **6**, Suppl. 3, S543 (1984). Radioimmunoassay determn in plasma: H. Tatsumi, *Arzneim.-Forsch.* **34**, 1704 (1984). Clinical studies: A. Reppelli *et al.*, *Boll. Soc. Ital. Cardiol.* **23**, 177 (1978); C. V. Ram *et al.*, *J. Clin. Pharmacol.* **19**, 148 (1979). Clinical studies in opiate withdrawal: F. S. Tennant, R. A. Rawson, *NIDA Res. Monogr. Ser.* **49**, 338 (1984); J. T. Murphy, *Drug Intell. Clin. Pharm.* **19**, 32 (1985). Review of pharmacodynamic properties and therapeutic efficacy: B. Holmes *et al.*, *Drugs* **26**, 212-229 (1983). Comprehensive description: C. M. Shearer, *Anal. Profiles Drug Subs.* **15**, 319-336 (1986).

White solid from acetonitrile, mp 227-229° (dec).

Monoacetate. [23256-50-0] Wy-8678; Rexitene; Tenelid; Wytensin. $C_{10}H_{12}Cl_2N_4O_2$; mol wt 291.13. Solid, mp 192.5° (dec). Soly at 25° (mg/ml): water 11; alcohol 50; propylene glycol 100; chloroform 0.6; ethyl acetate 1; sparingly sol in 0.1 *N* HCl.

THERAP CAT: Antihypertensive.

4594. Guanadrel. [40580-59-4] *N*-(1,4-Dioxaspiro[4.5]dec-2-ylmethyl)guanidine. $C_{10}H_{19}N_3O_2$; mol wt 213.28. C 56.32%, H 8.98%, N 19.70%, O 15.00%. Orally active postganglionic sympathetic inhibitor. Prepn: W. R. Hardie, J. E. Aaron, **ZA 6706328**; *eidem*, **US 3547951** (1968, 1970 both to Cutter). Antihypertensive activity: L. Hansson *et al.*, *Clin. Pharmacol. Ther.* **14**, 204 (1973). *In vitro* adrenergic neuron blocking activity: L. Roller, *Aust. J. Pharm. Sci.* **5**, 35 (1976). Pharmacologic study: E. M. Johnson, F. E. Hunter, *Biochem. Pharmacol.* **28**, 1525 (1979). Effect on patients with thyrotoxicosis: S. Rubenfeld *et al.*, *Arch. Intern. Med.* **138**, 1106 (1978). Review of pharmacology and efficacy in hypertension: F. A. Finnerty Jr., R. N. Brogden, *Drugs* **30**, 22-31 (1985).

Sulfate. [22195-34-2] CL-1388R; U-28288D; Hylorel. $(C_{10}H_{19}N_3O_2)_2 \cdot H_2SO_4$; mol wt 524.63. Cryst from methanol/ethanol, mp 213.5-215°. Sol in water; sparingly sol in methanol; slightly sol in alc, acetone.

THERAP CAT: Antihypertensive.

4595. Guanethidine. [55-65-2] *N*-[2-(Hexahydro-1(2*H*)-azocinyl)ethyl]guanidine; [2-(octahydro-1-azocinyl)ethyl]guanidine; 1-(2-guanidinoethyl)octahydroazocine; 2-(1'-azacyclooctyl)ethylguanidine; *N*-(2-perhydroazocin-1-ylethyl)guanidine; 2-(1-*N,N*-heptamethylenimino)ethylguanidine; oktadin; oktatenzin. $C_{10}H_{22}N_4$; mol wt 198.31. C 60.57%, H 11.18%, N 28.25%. Prepn: R. A. Maxwell *et al.*, *Experientia* **15**, 267 (1959); R. P. Mull, **US 2928829** (1960 to Ciba). Pharmacology: H. J. Sah *et al.*, *Arzneim.-Forsch.* **16**, 53 (1966). Clinical trial in hypertension: E. A. Ramirez *et al.*, *Circulation* **55**, 519 (1977); in glaucoma: R. A. Hitchings, D. Glover, *Br. J. Ophthalmol.* **66**, 247 (1982).

Sulfate. [60-02-6] Su-5864. $(C_{10}H_{22}N_4)_2 \cdot H_2SO_4$; mol wt 494.70. Crystals from dil ethanol, mp 276-281° (dec).

Monosulfate. [645-43-2] Sanotensin; Ismelin. $C_{10}H_{22}N_4 \cdot H_2SO_4$; mol wt 296.39. Colorless, crystalline, almost odorless powder. Sol in 1.5 parts water at 20°. Sparingly sol in alcohol. Practically insol in ether, chloroform.

THERAP CAT: Antihypertensive; antiglaucoma.

4596. Guanfacine. [29110-47-2] *N*-(Aminoiminomethyl)-2,6-dichlorobenzeneacetamide; *N*-amidino-2-(2,6-dichlorophenyl)-acetamide; [(2,6-dichlorophenyl)acetyl]guanidine. $C_9H_9Cl_2N_3O$; mol wt 246.09. C 43.93%, H 3.69%, Cl 28.81%, N 17.08%, O 6.50%. Centrally acting α_2-adrenoceptor agonist. Prepn: J. B. Bream, C. W. Picard, **FR 1584670**; *eidem*, **US 3632645** (1969, 1972 both to Wander). Pharmacology: H. F. Oates *et al.*, *Arch. Int. Pharmacodyn. Ther.* **231**, 148 (1978). Determn in biological fluids: M. Guerrat *et al.*, *J. Pharm. Sci.* **68**, 219 (1979). Symposium on pharmacology and clinical experience in hypertension: *Br. J. Clin. Pharmacol.* **10**, Suppl. 1, 1S-208S (1980). *Review:* L. A. Cornish, *Clin. Pharm.* **7**, 187-197 (1988). Clinical trial in attention deficit hyperactivity disorder (ADHD) in children: L. Scahill *et al.*, *Am. J. Psychiatry* **158**, 1067 (2001).

White grains from methanol/ether, mp 225-227°.

Hydrochloride. [29110-48-3] BS-100-141; LON-798; Estulic; Intuniv; Tenex. $C_9H_9Cl_2N_3O \cdot HCl$; mol wt 282.55. White needles from ethanol, mp 215-217°.

THERAP CAT: Antihypertensive. In treatment of attention deficit hyperactivity disorder.

4597. Guanidine. [113-00-8] Aminomethanamidine; carbamamidine; carbamidine; aminoformamidine; iminourea. CH_5N_3; mol wt 59.07. C 20.33%, H 8.53%, N 71.14%. Strong organic base existing primarily as guanidinium ion at physiological pH. Found in turnip juice, mushrooms, corn germ, rice hulls, mussels, earthworms. Prepn of nitrate from dicyanodiamide + ammonium nitrate: Smith *et al.*, *Ind. Eng. Chem.* **23**, 1124 (1931); Davis, *Org. Synth.* **coll. vol. I** (2nd ed., 1941) p 302; from SO_2, CO_2, + NH_3: Boivin, **US 2762843** (1956); from urea: Mackay, **US 2590257** (1952 to Am. Cyanamid); Craig, Minor, **US 3009949** (1961 to Deere); Shaver, **US 3108999** (1963 to Monsanto); from ammonium thiocyanate or thiourea + ammonia: Watt, Makosky, *Ind. Eng. Chem.* **46**, 2599 (1954). Pharmacology as muscle stimulant: A. I. Podlesnaya, *Bull. Exp. Biol. Med.* **61**, 291 (1966). Clinical evaluation with pyridostigmine in Lambert-Eaton myasthenic syndrome: S. J. Oh *et al.*, *Muscle Nerve* **20**, 1146 (1997). Review of mode of action: D. R. Tershak *et al.*, "Guanidine" in *Handbook of Experimental Pharmacology* Vol. **61**, G. V. R. Born *et al.*, Eds. entitled "Chemotherapy of Viral Infections" P. E. Came, L. A. Caliguiri, Eds. (Springer-Verlag, New York, 1982) pp 343-375. Review of antiviral activity studies: F. Davidoff, *N. Engl. J. Med.* **289**, 141-146 (1973). General reviews: M. Schenck, *Pharmazie* **3**, 5 (1948); G. Schaefer, "Guanidines and Biguanidines" in *International Encyclopedia of Pharmacology and Therapeutics* **107**, M. Erecinska, Ed. (Pergamon, Oxford 1981) pp 165-185.

Deliquescent, cryst mass, mp ~50°. pKa ~12.5. Absorbs CO_2 from air. Very sol in water, alcohol. On heating to 160° it is converted to melamine and NH_3. *Keep well closed.* LD_{50} i.p. in mice: 350 mg/kg (Podlesnaya).

Hydrochloride. [50-01-1] $CH_5N_3 \cdot HCl$. Cryst powder. Freely sol in water, alcohol. Aq soln is neutral.

Nitrate. [506-93-4] $CH_5N_3 \cdot HNO_3$. Cryst powder, mp 214°. Soluble in 10 parts water; in alcohol. Aq soln is neutral.

THERAP CAT: Cholinergic.

Consult the Name Index before using this section.

4598. Guanidinium Aluminum Sulfate Hexahydrate.
[10199-21-0] Sulfuric acid aluminum salt compd with guanidine
hydrate (2:1:1:6); GASH. $CH_{18}AlN_3O_{14}S_2$; mol wt 387.26. C
3.10%, H 4.69%, Al 6.97%, N 10.85%, O 57.84%, S 16.56%. Prepd
from an aq soln of an equimolecular mixture of guanidine sulfate and
aluminum sulfate: Ferraboschi, *Proc. Cambridge Philos. Soc.* **14**,
473 (1908); Holden *et al.*, *Phys. Rev.* **101**, 962 (1956); Wieder, *Proc.
IRE* **45**, 1094 (1957). Crystal structure: Schein *et al.*, *J. Chem. Phys.*
47, 5183 (1967). Crystal growth: T. A. Zarembovskaya, *Zh. Fiz.
Khim.* **45**, 2504 (1971). Optical properties: P. M. Nikolic *et al.*,
Fizika (Zagreb) **12**, Suppl. 1, 165 (1980).

Large hexagonal plates belonging to the trigonal system with perfect basal cleavage. Has ferroelectric properties.

4599. Guanine. [73-40-5] 2-Amino-1,9-dihydro-6*H*-purin-6-
one; 2-aminohypoxanthine. $C_5H_5N_5O$; mol wt 151.13. C 39.74%,
H 3.33%, N 46.34%, O 10.59%. Constituent of nucleic acids; widespread in animal and plant kingdom. First isolated from guano. Syntheses: Fisher, *Ber.* **30**, 2226 (1897); Traube, *ibid.* **33**, 1371 (1900);
DE 134984 (1903); **DE 158591** (1903); **DE 162336** (1904). Prepn
of ^{15}N-isotopic guanine following Traube's synthesis: Plentl,
Schoenheimer, *J. Biol. Chem.* **153**, 203 (1944). Several desmotropic
forms. Crystal structure of hydrochloride monohydrate: Broom-
head, *Acta Crystallogr.* **4**, 92 (1951). *Reviews:* Shapiro, *Prog. Nucleic Acid Res. Mol. Biol.* **8**, 73-112 (1968); Ts'o, "Bases, Nucleosides and Nucleotides" in *Basic Principles in Nucleic Acid Chemistry*
vol. **1**, P. O. P. Ts'o, Ed. (Academic Press, New York, 1974) pp 453-
584.

Usually amorphous. Small rhombic crystals by slow evaporation
of aq soln contg large excess of NH_3. Dec above 360° with partial
sublimation. uv max (pH 6.2): 246, 275 nm ($\varepsilon \times 10^{-3}$ 10.7, 8.1).
Freely sol in ammonia water, aq KOH solns, dil acids; sparingly sol
in alcohol, ether. Almost insol in water. pKb 3.22; pKa 9.92, detd
at 40°. Many compds with acids, bases and metals have been prepared.
Hydrochloride monohydrate. $C_5H_5N_5O.HCl.H_2O$. Cryst
powder. Loses H_2O at 100°, HCl at 200°. Practically insol in water,
alcohol, ether; sol in acidulated water.

4600. Guano. Bird manure. The dried excrements of sea birds
(cormorants) and bats from coastal islands of Peru, Chile, West Indies, and Africa. Usually mixed with feathers and bones. Contains
about 9% nitrogen, 6% phosphorus, 2% potassium, and 15-20%
moisture. Used as fertilizer.

4601. Guanosine. [118-00-3] 2-Amino-9-β-D-ribofuranosyl-
9*H*-purine-6(1*H*)-one; guanine riboside; vernine. $C_{10}H_{13}N_5O_5$; mol
wt 283.24. C 42.41%, H 4.63%, N 24.73%, O 28.24%. Constituent
of nucleic acids. Prepn from yeast nucleic acid: P. A. Levene, L. W.
Bass, *Nucleic Acids* (New York, 1931) p 163. Prepn from polynucleotides: P. A. Levene, E. Jorpes, *J. Biol. Chem.* **86**, 389 (1930).
Prepn from plants: H. Stendel, E. Peiser in G. Klein, *Handbuch der
Pflanzenanalyse* **IV** (Vienna, 1933) p 448. Structure: Levene, Tipson, *J. Biol. Chem.* **97**, 491 (1932); Gulland *et al.*, *J. Chem. Soc.*
1934, 1639; Tsuboi *et al.*, *Biochim. Biophys. Acta* **55**, 1 (1962).
Synthesis: Davoll, *J. Am. Chem. Soc.* **1958**, 1593. Tautomerism in aq
soln: Miles *et al.*, *Science* **142**, 1458 (1963). Crystal structure and
conformation: Bugg *et al.*, *Biochem. Biophys. Res. Commun.* **3**, 436
(1968). *Review: Basic Principles in Nucleic Acid Chemistry* vol. **1**,
P. O. P. Ts'o, Ed. (Academic Press, New York, 1974) *passim*.

Dihydrate. Needles from water. Anhydr at 110°. Dec 240° in
sealed tube (rapid heating). $[\alpha]_D^{20}$ −61° (in water); $[\alpha]_D^{24}$ −72° (c =
0.96 in 0.1*N* NaOH). uv max (pH 5.5): 188.3, 252.5 nm ($\varepsilon \times 10^{-3}$
26.8, 13.7): Voet *et al.*, *Biopolymers* **1**, 193 (1963). One gram dissolves in 1320 ml water at 18°, in 33 ml boiling water. Soluble in
dil mineral acids, in hot acetic acid, and in dil bases. Insol in alcohol,
ether, chloroform, benzene.

4602. Guanoxabenz. [24047-25-4] 2-[(2,6-Dichlorophenyl)-
methylene]-*N*-hydroxyhydrazinecarboximidamide; 1-[(2,6-dichlo-
robenzylidene)amino]-3-hydroxyguanidine; Compd 43-663. C_8H_8-
Cl_2N_4O; mol wt 247.08. C 38.89%, H 3.26%, Cl 28.70%, N
22.68%, O 6.48%. Prepn: W. J. Houlihan, R. E. Manning, **DE
1902449**; *eidem*, **US 3591636** (1969, 1971 both to Sandoz). Pharmacological study: J. C. Doxey, A. S. Hersom, *Br. J. Pharmacol.*
70, 171 (1980).

Hydrochloride. [23256-40-8] Benzerial. $C_8H_8Cl_2N_4O.HCl$;
mol wt 283.54. Crystals from ethanol/ether, mp 173-175°.
THERAP CAT: Antihypertensive.

4603. Guanoxan. [2165-19-7] *N*-[(2,3-Dihydro-1,4-benzo-
dioxin-2-yl)methyl]guanidine; (1,4-benzodioxan-2-ylmethyl)guani-
dine; 2-guanidinomethyl-1,4-benzodioxan. $C_{10}H_{13}N_3O_2$; mol wt
207.23. C 57.96%, H 6.32%, N 20.28%, O 15.44%. Prepn: **BE
632701**; J. Augstein, S. M. Green, **US 3247221** (1963, 1966, both to
Pfizer); Monro, *Chem. Ind. (London)* **1964**, 1806; Gardner, **GB
996708** (1965 to SKF), *C.A.* **63**, 14877d (1965); Yu, Shen, *C.A.* **66**,
94969k (1967). Prepn and properties of (+) and (−)-forms: Stenlake *et al.*, *J. Pharm. Pharmacol.* **20** (Suppl), 82 (1968). Pharmacology: Augstein *et al.*, *J. Med. Chem.* **8**, 446 (1965); Cession-Fossion,
Arch. Int. Pharmacodyn. Ther. **164**, 419 (1966); Vidal-Beretewide *et
al.*, *Arzneim.-Forsch.* **19**, 947 (1969). Metabolic studies: Jack *et al.*,
J. Pharm. Pharmacol. **23**, 2225 (1971); *Xenobiotica* **2**, 35 (1972).

Crystals, mp 164-165°.
Sulfate. [5714-04-5] Envacar. $(C_{10}H_{13}N_3O_2)_2.H_2SO_4$; mol wt
512.54.
THERAP CAT: Antihypertensive.

4604. 3′-Guanylic Acid. [117-68-0] Guanosine 3′-mono-
phosphate; guanine riboside-3-phosphoric acid; guanylic acid b.
$C_{10}H_{14}N_5O_8P$; mol wt 363.22. C 33.07%, H 3.89%, N 19.28%, O
35.24%, P 8.53%. From yeast or pancreas. Prepn: P. A. Levene, L.
W. Bass, *Nucleic Acids* (New York, 1931) pp 224-227. Structure:
Levene, Jorpes, *J. Biol. Chem.* **81**, 579 (1929); Levene, Harris, *ibid.*
95, 755 (1932); **98**, 9 (1932). Early work probably done on a mixture
of 2′- and 3′-guanylic acids; see physical data below. Separation of
two isomers: Cohn, *J. Am. Chem. Soc.* **72**, 1471 (1950); Khym,
Cohn, *ibid.* **76**, 1818 (1954); *eidem*, *Biol. Prepn.* **5**, 40 (1957). Ab-

sorption spectrum: Voet *et al.*, *Biopolymers* **1**, 193 (1963). *Reviews: see* Guanine.

Dihydrate. Long prisms from water. The water of crystn is given up at 118° and is taken up again at room temp. When anhydrous, dec 180° (closed tube). $[\alpha]_D^{25} -8°$ (c = 2); $-65°$ (c = 2 in 5% NaOH). Acid to litmus. Soluble in cold water, freely sol in hot water. Boiling with dil mineral acids yields guanine, H_3PO_4, and D-ribose.

Neutral sodium salt. $Na_2C_{10}H_{12}N_5O_8P$. Flakes from water, contains 21.1% H_2O. Sol in cold, freely sol in hot water.

Brucine salt heptahydrate. $C_{10}H_{14}N_5O_8P.(C_{23}H_{26}N_2O_4)_2.$-$7H_2O$. Rectangular leaflets from alc. When anhydr, dec 233-240°. $[\alpha]_D^{20} -26°$ (35% alc). One gram dissolves in 100 ml water.

4605. 5'-Guanylic Acid. [85-32-5] Guanosine 5'-monophosphate; GMP; guanosine 5'-phosphate; guanine riboside-5-phosphoric acid. $C_{10}H_{14}N_5O_8P$; mol wt 363.22. C 33.07%, H 3.89%, N 19.28%, O 35.24%, P 8.53%. Nucleotide widely distributed in nature; found in hydrolyzates of RNA. Isolated together with inosinic acid from sardines or yeast extract: Kuninaka *et al.*, *New Food Ind.* **3**, no. 1, 21 (1961). Also by direct biosynthesis using microorganisms or enzymes: Abrams, Bentley, *Arch. Biochem. Biophys.* **79**, 91 (1959); Magasanik, Karibian, *J. Biol. Chem.* **235**, 2672 (1960); Okumura *et al.*, **US 3249511** (1966). Chemical synthesis: Michelson, Todd, *J. Chem. Soc.* **1949**, 2483; Chambers *et al.*, *J. Am. Chem. Soc.* **79**, 3747 (1957); Gilham, Tener, *Chem. Ind. (London)* **1959**, 542; Tener, *J. Am. Chem. Soc.* **83**, 159 (1961); Koransky *et al.*, *Z. Naturforsch.* **17B**, 291 (1962). Prepn of Na salt: Ishibashi, Ito, **US 3190877** (1965 to Takeda). Monograph on synthesis of nucleotides: G. R. Pettit, *Synthetic Nucleotides* vol. 1 (Van Nostrand Reinhold, New York, 1972) 252 pp. *Reviews: See* Guanidine.

Microcrystals, dec 190-200°. Sparingly sol in cold water.

Barium salt octahydrate. $C_{10}H_{12}N_5O_8PBa.8H_2O$. White powder. uv max (ε 12400); (pH 12): 256 nm (ε 12400); (pH 12): 260 nm (ε 12100).

Disodium salt monohydrate. $C_{10}H_{12}N_5O_8PNa_2.H_2O$. Hygroscopic crystals, decomp at about 250°. Characteristic meaty taste. a_M (molar absorbancy): 13.7×10^3 at 252.5 nm (pH 7). Soly in water at 25° about 25 g/100 ml. Practically insol in alcohol, acetone, ether.

USE: The disodium salt as flavor intensifier, often in combination with sodium inosinate or sodium glutamate.

4606. Guaran. [9000-30-0] Principal polysaccharide from endosperm of guar seeds, *Cyamopsis tetragonaloba* (L.) Taub., *Leguminosae:* Heyne, Whistler, *J. Am. Chem. Soc.* **70**, 2249 (1948). Structure: Whistler, Durso, *ibid.* **74**, 5140 (1952). Configuration:

Koleske, Kurath, *J. Polym. Sci. A* **2**, 4123 (1964). *Review:* Deuel *et al.*, *Chimia* **8**, 64 (1954).

$[\alpha]_D^{25}$ +53° (1N NaOH). Sol in cold water.

Triacetate. Fibrous material, mp 226-227°. Can be formed into strong films which can be elongated 550%. Becomes birefringent and does not develop crystallinity.

USE: In textile and paper industry.

4607. Guar Gum. [9000-30-0] Guar flour; gum cyamopsis; cyamopsis gum; Burtonite V-7-E; Decorpa; Guarina; Glucotard; Guarem. Mol wt about 220,000. The ground endosperms of *Cyamopsis tetragonolobus* (L.) Taub., *Leguminosae* which is cultivated in India as livestock feed. The water soluble fraction (85%) of guar flour is called guaran which consists of linear chains of $(1 \to 4)$-β-D-mannopyranosyl units with α-D-galactopyranosyl units attached by $(1 \to 6)$ linkages. Ratio of D-galactose to D-mannose is 1:2. Effect on lipid metabolism: D. J. A. Jenkins *et al.*, *Br. Med. J.* **2**, 1555 (1979); on glucose and lipid levels in diabetic and healthy volunteers: U. Smith, G. Holm, *Atherosclerosis* (Shannon, Ire.) **45**, 1 (1982); on renal tumors in diabetic rats: B. C. Chin *et al.*, *Biomed. Res.* **5**, 273 (1984). As source of fiber in patients with non-insulin dependent diabetes: M. E. McIvor *et al.*, *Am. J. Clin. Nutr.* **41**, 891 (1985). Toxicology studies: S. L. Graham *et al.*, *Food Cosmet. Toxicol.* **19**, 287 (1981). Comparative study of commercial guar gums: I. A. Schlakman, A. J. Bartilucci, *Drug Stand.* **25**, 149-154 (1957). Comprehensive monograph: F. Smith, R. Montgomery, *The Chemistry of Plant Gums and Mucilages* (Reinhold, New York, 1959) 627 pp. *Review:* Goldstein *et al.* in *Industrial Gums*, R. L. Whistler, Ed. (Academic Press, New York, 2nd ed., 1973) p 303-321. Comprehensive description: K. Yu *et al.*, *Anal. Profiles Drug Subs. Excip.* **24**, 243-276 (1996).

White to yellowish white, nearly odorless, free flowing powder. Dispersible in hot or cold water, forming a colloidal solution. Slightly sol in water but not in organic solvents. Water solns are tasteless, odorless, nontoxic, of a pale, translucent gray color, and neutral. Stable to heat. Has five to eight times the thickening power of starch. Water solns may be converted to a gel by small amounts of borax. Aq dispersions are neutral. LD_{50} in male, female rats (g/kg): 7.35, 6.77 orally (Graham).

USE: In paper sizing; as a protective colloid, stabilizer, thickening and film forming agent for cheese, salad dressings, ice cream, soups; as a binding and disintegrating agent in tablet formulations; in pharmaceutical jelly formulations; in suspensions, emulsions, lotions, creams, toothpastes; in the mining industry as a flocculant, as a filtering agent; in water treatment as a coagulant aid.

THERAP CAT: Adjunct to diet, insulin or oral hypoglycemics in control of diabetes.

4608. Guggulsterone. [95975-55-6] Pregna-4,17(20)-diene-3,16-dione. $C_{21}H_{28}O_2$; mol wt 312.45. C 80.73%, H 9.03%, O 10.24%. Bioactive component in guggulu, *q.v.*, the gum resin of the Ayurvedic medicinal plant, *Commiphora mukul*, *Burseraceae*, used in treatment of lipid disorders. Both the *E*- and *Z*- isomers are naturally occurring and bioactive. Synthesis and stereochemistry: W. R. Benn, R. M. Dodson, *J. Org. Chem.* **29**, 1142 (1964). Isoln from guggulu: V. D. Patil *et al.*, *Tetrahedron* **28**, 2341 (1972). HPLC determn of isomers in serum: N. Verma *et al.*, *J. Chromatogr. B* **708**, 243 (1998). LC determn of isomers in commercial guggulu formulations: M. Nagarajan *et al.*, *J. AOAC Int.* **84**, 24 (2001). Inhibition of the bile acid receptor FXR and cholesterol lowering activity: N. L. Urizar *et al.*, *Science* **296**, 1703 (2002). Brief review of proposed mechanism of action: C. J. Sinal, F. J. Gonzalez, *Trends Endocrinol. Metab.* **13**, 275-276 (2002).

Z-form

E-Form. [39025-24-6] Prisms from aq methanol, mp 170-171.5°. $[\alpha]_D^{26}$ −30° (c = 1 in chloroform). uv max (methanol): 241 nm (ε 27600).

Z-Form. [39025-23-5] Crystals from aq methanol, mp 188-190°. $[\alpha]_D^{26}$ −61° (c = 1 in chloroform). uv max (methanol): 241 nm (ε 25000).

4609. Guggulu. Guggul; gum guggulu; gum guggul. Oleoresin exudate of the guggul tree, *Commiphora mukul* (Hook, ex Stocks), also known as *C. wightii* (Arnott), *Burseraceae*. Used in Aruyvedic medicine for treatment of lipid disorders, obesity and arthritis. *Constit.* Ethyl acetate-soluble fraction containing guggulsterone, *q.v.*, guggulsterols, diterpenoids, ferulic acid esters of guggultetrols, lignins; essential oil consisting primarily of myrcene, camphorene; insoluble carbohydrate gum. Identification of steroidal constituents: V. D. Patil *et al.*, *Tetrahedron* **28**, 2341 (1972). Review of chemical constituents and bioactivity: S. Dev in *Studies in Natural Products Chemistry* **Vol. 5**, A. Rahman, Ed. (Elsevier, New York, 1989) pp 695-719. Clinical evaluation in hyperlipidemia: S. K. Verma, A. Bordia, *Indian J. Med. Res.* **87**, 356 (1988). Review of medicinal uses: G. V. Satyavati, *ibid.* 327-335 (1988); of pharmacognosy: A. K. Tajuddin *et al.*, *Curr. Res. Med. Aromat. Plants* **16**, 75-86 (1994); of clinical studies in hyperlipidemia: C. Ulbricht *et al.*, *Complement. Ther. Med.* **13**, 279-290 (2005).

Ethyl acetate extract. Gugulipid; Guglip. Standardized to contain 4.0% guggulsterones. Clinical evaluation in hypercholesterolemia: R. B. Singh *et al.*, *Cardiovasc. Drugs Ther.* **8**, 659 (1994); P. O. Szapary *et al.*, *J. Am. Med. Assoc.* **290**, 765 (2003). LD$_{50}$ in mice (mg/kg): 1600 i.p. and orally (Dev).

THERAP CAT: Antilipemic.

4610. Guinea Green B. [4680-78-8] *N*-Ethyl-*N*-[4-[[4-[ethyl[(3-sulfophenyl)methyl]amino]phenyl]phenylmethylene]-2,5-cyclohexadien-1-ylidene]-3-sulfobenzenemethanaminium inner salt sodium salt (1:1); C.I. Acid Green 3; C.I. Food Green 1; FD & C Green 1; C.I. 42085. C$_{37}$H$_{35}$N$_2$NaO$_6$S$_2$; mol wt 690.80. C 64.33%, H 5.11%, N 4.06%, Na 3.33%, O 13.90%, S 9.28%. Prepn: Jones *et al.*, *J. Assoc. Off. Agric. Chem.* **38**, 977 (1955). Toxicity studies: F. C. Lu, A. Lavalle, *Can. Pharm. J.* **97**, 30 (1964); W. H. Hansen *et al.*, *Food Cosmet. Toxicol.* **4**, 389 (1966). *See also: Colour Index* **vol. 4** (3rd ed., 1971) p 4385.

A dull, dark green powder, or a bright, crystalline solid. Sol in water to a green soln which becomes brownish-yellow on addn of HCl and blackish-green with NaOH. An excess of NaOH decolorizes the soln. Sparingly sol in alcohol; it dissolves in concd H$_2$SO$_4$ to a yellow soln which, when diluted with water, turns first yellowish-red, then green. LD$_{50}$ orally in rats: >2 g/kg (Lu, Lavalle).

USE: Limited use as a dye for silk and wool fabrics; as biological stain.

4611. D-Gulonic Acid. [20246-33-7] C$_6$H$_{12}$O$_7$; mol wt 196.16. C 36.74%, H 6.17%, O 57.09%. Prepd as the sodium salt by reduction of sodium glucuronate with sodium amalgam in alkaline medium: Fischer, Piloty, *Ber.* **24**, 525 (1891); from D-gulonic acid γ-lactone: Rehorst, Naumann, *ibid.* **77**, 24 (1944).

$[\alpha]_D^{20}$ −6° (10 min) → −38.6° (15 days). The free acid forms the lactone spontaneously. pK (25°): 3.68.

Sodium salt. C$_6$H$_{11}$NaO$_7$. Crystals. $[\alpha]_D^{20}$ +11.5°. Sol in water.

Calcium salt. Ca(C$_6$H$_{11}$O$_7$)$_2$. $[\alpha]_D^{21}$ −14.45° (c = 1.73). Precipitated from aq soln by alc.

4612. L-Gulonic Acid. [526-97-6] Xylosecarboxylic acid. C$_6$H$_{12}$O$_7$; mol wt 196.16. C 36.74%, H 6.17%, O 57.09%. Prepd from L-xylose and HCN followed by hydrolysis of the nitrile: Fischer, Stahel, *Ber.* **24**, 529 (1891). Prepn from D-glucuronic acid: **DE 618907** (1935 to Hoffmann-La Roche); from L-gulonolactone: Ishidate *et al.*, *Chem. Pharm. Bull.* **13**, 173 (1965).

Crystallizes as the lactone on evapn of an aq soln.

Sodium salt. $[\alpha]_D^{20}$ +12.7° (c = 9). Freely sol in water.

4613. D-Gulose. [4205-23-6] C$_6$H$_{12}$O$_6$; mol wt 180.16. C 40.00%, H 6.71%, O 53.28%. Prepd by sodium amalgam reduction of an acid soln of the γ-lactone of D-gulonic acid: Fischer, Stahel, *Ber.* **24**, 532 (1891); van Ekenstein, Blanksma, *Rec. Trav. Chim.* **27**, 3 (1908). Alternate synthesis: Meyer zu Reckendorf, *Angew. Chem. Int. Ed.* **6**, 177 (1967); *idem, Methods Carbohydr. Chem.* **6**, 129 (1972); R. Köster *et al.*, *Angew. Chem. Int. Ed.* **19**, 547 (1980).

Syrup. Sweet taste. $[\alpha]_D^{20}$ −20.4°. Sol in water, slightly sol in alcohol. Not fermentable by yeast.

4614. L-Gulose. [6027-89-0] C$_6$H$_{12}$O$_6$; mol wt 180.16. C 40.00%, H 6.71%, O 53.28%. Prepd by sodium amalgam reduction of an acid soln of the γ-lactone of L-gulonic acid: Fischer, Piloty, *Ber.* **24**, 526 (1891). *See also* van Ekenstein, Blanksma, *Rec. Trav. Chim.* **27**, 3 (1908); Levene, LaForge, *J. Biol. Chem.* **20**, 430 (1915); Talen, *Rec. Trav. Chim.* **44**, 891 (1925); Isbell, *J. Am. Chem. Soc.* **55**, 2167 (1933). Synthesis from D-mannose: Evans, Parrish, *Carbohydr. Res.* **28**, 359 (1973); from D-glucose: D. K. Minster, S. M. Hecht, *J. Org. Chem.* **43**, 3987 (1978).

Syrup. $[\alpha]_D^{20}$ +61.6°. $[\alpha]_D$ +21.3° (c = 4.58) (Evans, Parrish). Freely sol in water; slightly sol in alcohol. Not fermentable by yeast.

4615. Gum Benzoin. [9000-05-9] Resin benzoin; resin benjamin; gum benjamin. Balsamic resin from *Styrax benzoin* Dryand., known as Sumatra benzoin, or from *S. tonkinensis* (Pierre) Craib, *Styracaceae*, or other species of *Styrax* known as Siam benzoin. *Habit.* Thailand, Cambodia, S. Vietnam, Sumatra, Java, and Sunda Islands. *Constit.* Ethereal oil, free and combined benzoic and cinnamic acids up to 39%, vanillin, coniferyl benzoate, resin (a mixture of benzoresinol and benzoresinotannol) esterified with benzoic acid, styrol, styracin. Not less than 90% of Siam and not less than 75% of Sumatra benzoic is sol in alc (U.S.P.). *Ref:* Reinitzer, *Arch. Pharm.* **264**, 131 (1926); Brans, *Pharm. Weekbl.* **73**, 374 (1936); Freudenberg, Bittner, *Ber.* **83**, 600 (1950).

USE: Preserving ointments; preparing natural benzoic acid; for fumigating pastilles; in perfumery and cosmetics.

THERAP CAT: Topical protectant.

THERAP CAT (VET): Tincture is used topically as an antiseptic and to promote healing; as an inhalant for bronchitis, and orally as an expectorant.

4616. Gum Tragacanth. [9000-65-1] Tragacanth. Mol wt about 840,000. The dried gummy exudation from *Astragalus gummifer* Labill. (white gavan) or other Asiatic species of *Astragalus, Leguminosae*, found largely in Iran, also in Asia Minor and in Syria. When mixed with water gives a soluble fraction, as a hydrosol, called *tragacanthin* which is a complex mixture of polysaccharides containing D-galacturonic acid, other sugars, and traces of starch and cellulose. The insoluble fraction swells to a gel and consists of 60-70% bassorin. Structural studies: Norman, *Biochem. J.* **25**, 200 (1931); James, Smith, *J. Chem. Soc.* **1945**, 739, 749; Aspinall, Baillie, *ibid.* **1963**, 1702, 1714. *Reviews:* D. C. Beach in *Adv. Chem. Ser.* **11**, entitled "Natural Plant Hydrocolloids," (ACS, Washington DC, 1954) pp 38-44; Meer *et al.* in *Industrial Gums*, R. L. Whistler, Ed. (Academic Press, New York, 2nd ed., 1973) pp 289-299. *Book:* F. Smith, R. Montgomery, *The Chemistry of Plant Gums and Mucilages* (Reinhold, New York, 1959) 627 pp.

Odorless. Insipid, mucilaginous taste. Acid reaction. One gram requires 0.9 ml 0.1N NaOH for neutralization to phenolphthalein: Gabel, *J. Am. Pharm. Assoc.* **23**, 341 (1934). Viscosity of tragacanth mucilages is reduced by adding acid, alkali, and NaCl particularly if the mucilage is heated: Mantell, *The Water-Soluble Gums* (New York, 1947). Maximum initial viscosity of solns at pH 8; maximum stable viscosity near pH 5. Forms a deep yellow stringy precipitate when a soln is boiled with a few drops of 10% aqueous ferric chloride soln. A stringy precipitate formed also on heating a soln with Schweitzer reagent. Tragacanth is entirely insol in alcohol.

USE: In pharmaceutical compounding and dispensing, *e.g.*, to suspend heavy insol powders, as an excipient for tablets and to impart consistence to troches; also in making emulsions and emulsifying agents; as stabilizer, thickener, texturizer in food; in adhesives (mucilages, pastes); in textile sizing, textile printing and general printing inks, and in dyeing with insol color lakes.

4617. Gusperimus. [98629-43-7] 7-[(Aminoiminomethyl)-amino]-N-[2-[[4-[(3-aminopropyl)amino]butyl]amino]-1-hydroxy-2-oxoethyl]heptanamide; (±)-N-[[[4-[(3-aminopropyl)amino]butyl]-carbamoyl]hydroxymethyl]-7-guanidinoheptanamide; 1-amino-19-guanidino-11-hydroxy-4,9,12-triazanonadecane-10,13-dione; deoxyspergualin; (±)-15-deoxyspergualin. $C_{17}H_{37}N_7O_3$; mol wt 387.53. C 52.69%, H 9.62%, N 25.30%, O 12.39%. Synthetic derivative of the antitumor antibiotic *spergualin*. Prepn: **BE 894651**; H. Umezawa *et al.*, **US 4518532** (1983, 1985 both to Microbiochem. Res. Found.); Y. Umeda *et al.*, *J. Antibiot.* **38**, 886 (1985). Prepn of the active (−)-isomer: H. Iwasawa *et al.*, *ibid.* **35**, 1665 (1982); H. Umezawa *et al.*, **EP 94632**; *eidem*, **US 4525299** (1983, 1985 both to Microbiochem. Res. Found.). Synthesis and bioactivity of isomers: Y. Umeda *et al.*, *J. Antibiot.* **40**, 1316 (1987). Mechanism of action study: W. E. G. Müller *et al.*, *ibid.* 1028. HPLC determn in plasma: R. Nakanuma *et al.*, *J. Chromatogr.* **527**, 208 (1990). Pharmacokinetics: J. F. Muindi *et al.*, *Cancer Res.* **51**, 3096 (1991). Prelimi-

nary evaluation in renal transplant rejection: H. Amemiya *et al.*, *Transplant. Proc.* **25**, 730 (1993). Series of articles on synthesis and immunomodulating activity: *Ann. N.Y. Acad. Sci.* **685**, 123-206 (1993).

Trihydrochloride. [85468-01-5] BMS-181173; NKT-01; NSC-356894; Spanidin. $C_{17}H_{37}N_7O_3 \cdot 3HCl$; mol wt 496.90. White powder, prepd as the sesquihydrate. Sol in water. pH of 50 mg/ml saline soln: ~4.9. LD_{50} in mice (mg/kg): 25-50 i.p. (Umezawa).

(−)-Form trihydrochloride. [84937-45-1] Prepd as the dihydrate. Colorless syrup, no def mp. $[\alpha]_D^{25}$ −7.3° (c = 1 in H_2O). LD_{50} in mice (mg/kg): 35-40 i.v. or i.p. (Iwasawa).

THERAP CAT: Immunosuppressant.

4618. Gutta-Percha. [9000-32-2] The purified, coagulated, milky exudate of various trees of the genus *Palaquium, Sapotaceae*. *Habit.* Malayan Archipelago. Extensive review: Williams, *Econ. Bot.* **18**, 5-26 (1964). Defined as a *trans* isomer of rubber. Rubber has a repeat period of 8.2 Å, whereas α-gutta-percha has 8.7 Å and β-gutta-percha has 4.8 Å. The short period of β-gutta-percha identifies it almost uniquely as an all-*trans* polyisoprene.

Becomes pliable at 25-30°, plastic at 60°. mp 100° (partial dec). On exposure to air and sunlight, it absorbs oxygen and becomes brittle. 90% or more dissolves in chloroform, petr ether. Partially sol in hot alcohol, benzene, carbon disulfide, oil of turpentine. Insol in water. *Keep under water and protected from light.*

USE: Insulator in electrotechnics, as dental cement; in orthopedics for fracture splints; manuf surgical instruments; covering golf balls.

4619. Guvacine. [498-96-4] 1,2,5,6-Tetrahydro-3-pyridine-carboxylic acid; 1,2,5,6-tetrahydronicotinic acid. $C_6H_9NO_2$; mol wt 127.14. C 56.68%, H 7.14%, N 11.02%, O 25.17%. From betel nuts, the seeds of *Areca catechu* L., *Palmae*. Extraction: Jahns, *Ber.* **24**, 2615 (1891). Synthesis: Freudenberg, *ibid.* **51**, 976, 1669 (1918); Hess, Leibbrandt, *ibid.* **51**, 806; **52**, 206 (1919).

Prisms from water, dec 295°. Neutral to litmus. Sol in water. Almost insol in abs alc, ether, chloroform, benzene.

Hydrochloride. $C_6H_{10}ClNO_2$. Needles from water, dec 318°. Sol in water.

Hydrobromide. $C_6H_{10}BrNO_2$. Needles from abs alcohol, dec 280°. Almost insol in acetone.

USE: Has been proposed as growth factor for *Staphylococcus aureus* and *Proteus vulgaris* instead of nicotinic acid.

4620. Gymnemic Acid. [1399-64-0] Gymnemin. Sweetness-inhibiting principle occurring in the leaves of *Gymnema sylvestre* R. Br. and allied *Asclepiadaceae*: Hooper, *Chem. News* **59**, 159 (1889); Mhaskar, Caius, *Indian J. Med. Res.* **16**, 1 (1930); Warren, Pfaffman, *J. Appl. Physiol.* **14**, 40 (1959); Stöcklin *et al.*, *Helv. Chim. Acta* **50**, 474 (1967). A complex mixture of at least nine closely related acidic glycosides, the major active component being gymnemic acid A_1: Stöcklin, *J. Agric. Food Chem.* **17**, 704 (1969); Dateo, Long, *ibid.* **21**, 899 (1973). Separation of major components: Sinsheimer *et al.*, *J. Pharm. Sci.* **59**, 622, 629 (1970). Completely obtunds taste for several hours for sweet, but not for bitter, sour, or salty substances: G. Hellekant *et al.*, *Acta Physiol. Scand.* **124**, 399 (1985).

The acid is a yellow to brown amorphous, bitter powder. Almost insol in water; sol in alcohol. The potassium salt is a reddish-brown cryst mass, sol in water or alcohol.

H

4621. Hachimycin. [1394-02-1] Trichomycin; Trichonat. Heptaene macrolide antibiotic substance produced by *Streptomyces hachijoensis* from soil of the Pacific Island Hachijo Jima: S. Hosoya *et al.*, *J. Antibiot.* **5**, 564 (1952); T. Yamaguchi, *ibid.* **7A**, 10 (1954). Purification and prepn of insol derivs: S. Hosoya *et al.*, *ibid.* **8A**, 5 (1955). At one time, trichomycin was believed to be identical to candicidin and hamycin, *q.q.v.*: H. Burrows, D. Calam, *J. Chromatogr.* **53**, 566 (1970). Subsequent HPLC studies have shown the three polyene antibiotics to be different entities: P. Helboe *et al.*, *ibid.* **189**, 249 (1980). Active against *Trichomonas vaginalis*, *Treponema pallidum*, *Trichophyton*, *Candida*, and yeast, weak activity against *Aspergillus* and *Penicillium*: T. Yamaguchi, *J. Antibiot.* **7A**, 10 (1954). Toxicity: S. Hosoya *et al.*, *J. Antibiot.* **6A**, 98 (1953). *Review*: S. A. Waksman, H. A. Lechevalier, *The Actinomycetes* vol. **III** (Williams & Wilkins, Baltimore, 1962) pp 397-399.

Yellow crystals. Acid reaction. Forms a water-sol sodium salt. Can be pptd as a water-insol salt by acridine derivs such as acriflavine, by sulfa drugs such as homosulfanilamide, by basic antibiotics such as streptomycin base, by enzymes such as papain and lysozyme, by dyes such as methylene blue, and by metallic salts such as $CaCl_2$. LD_{50} i.p. in mice: 0.05 mg/10-12 g mouse (Hosoya).

THERAP CAT: Antifungal. Antiprotozoal (Trichomonas).

4622. Hadacidin. [689-13-4] *N*-Formyl-*N*-hydroxyglycine; *N*-formyl-*N*-hydroxyaminoacetic acid; *N*-hydroxyformamidoacetic acid. $C_3H_5NO_4$; mol wt 119.08. C 30.26%, H 4.23%, N 11.76%, O 53.74%. Antitumor antibiotic originally isolated from cultures of *Penicillium frequentans* Westling. Synthesis: Kaczka *et al.*, *Biochemistry* **1**, 340 (1962); Kinnel, Schoenewaldt, US 3154578 (1964 to Merck & Co.). Biosynthesis: Stevens, Emery, *Biochemistry* **5**, 74 (1966). *Review*: Shigeura, "Hadacidin" in *Antibiotics* **I**, D. Gottlieb, P. Shaw, Eds. (Springer-Verlag, New York, 1967) pp 451-456.

Unstable crystals, mp 119-120°. Turns brown and liquefies on standing. The decompn products are formic acid and *N*-hydroxyglycine. Dibasic acid, potentiometric titration shows pH peak at 3.5 and 9.1. Soluble in water, methanol, ethanol, acetone, ether.

Monosodium salt. $C_3H_4NNaO_4$. Crystals. Easily forms a hydrate, very freely sol in water.

4623. Hafnium. [7440-58-6] Hf; at. wt 178.49; at. no. 72; valences 4; also 2, 3. Group IVB (4). Six naturally occurring isotopes: 180 (35.22%); 178 (27.1%); 177 (18.56%); 179 (13.75%); 176 (5.21%); 174 (0.163%; α-emitter, $T_{1/2}$ 2 × 10^{15} years); artificial isotopes: 157; 158; 168-173; 175; 181-183. Abundance in earth's crust: 5 ppm. Found in all zirconium-contg minerals. Discovered in 1923 by Coster and Hevesy: *Nature* **111**, 79 (1923). Extraction from the mineral cyrtolite: Larsen *et al.*, *Inorg. Synth.* **3**, 67 (1950). Prepd by thermal decompn of its iodide; by reduction of the tetrachloride or of the hydrofluohafniate with metallic sodium; by reduction of the oxide with a mixture of calcium and sodium: van Arkel, de Boer, *Z. Anorg. Chem.* **148**, 345 (1925); de Boer, Fast, *ibid.* **187**, 193 (1930); US 2741628. Spectra: Coster *et al.*, cited by Mellor, *A Comprehensive Treatise on Inorganic and Theoretical Chemistry* **7**, 170 (1930). A dibromide and tribromide have been prepd: Schumb, Morehouse, *J. Am. Chem. Soc.* **69**, 2696 (1947). Reviews of hafnium and its compds: Larsen, "Zirconium and Hafnium Chemistry" in *Adv. Inorg. Chem. Radiochem.* **13**, 1-133 (1970); Bradley, Thornton, "Zirconium and Hafnium" in *Comprehensive Inorganic Chemistry* vol. **3**, J. C. Bailar, Jr. *et al.*, Eds. (Pergamon Press, Oxford, 1973) pp 419-490; R. H. Nielson in *Kirk-Othmer Encyclopedia of Chemical Technology* vol. **12** (Wiley-Interscience, New York, 3rd ed., 1980) pp 67-80.

Highly lustrous, ductile metal of hexagonal cryst structure. mp 2227°; d 13.3. *Powder spontaneously combustible*. Resembles zirconium and thorium.

Dioxide. HfO_2. Obtained by igniting the hydroxide, the oxalate, or the sulfate. d^{20} 9.68. mp 2774°.

Tetrachloride. $HfCl_4$. White cryst mass, obtained by heating the oxide in the presence of chlorine and a reducing agent. Hydrolyzed to hafnyl chloride, $HfOCl_2$, by water. Volatilizes appreciably at 250°.

Sulfate. $Hf(SO_4)_2$. Prepd by the action of fuming sulfuric acid on the tetrachloride, dec above 500°.

Caution: Potential symptoms of overexposure to hafnium in exptl animals are irritation of eyes, skin, mucous membranes; liver damage. *See NIOSH Pocket Guide to Chemical Hazards* (DHHS/NIOSH 97-140, 1997) p 156.

4624. Halazepam. [23092-17-3] 7-Chloro-1,3-dihydro-5-phenyl-1-(2,2,2-trifluoroethyl)-2*H*-1,4-benzodiazepin-2-one; Sch-12041; Paxipam. $C_{17}H_{12}ClF_3N_2O$; mol wt 352.74. C 57.89%, H 3.43%, Cl 10.05%, F 16.16%, N 7.94%, O 4.54%. Polyfluoroalkyl analog of diazepam, *q.v.* Prepn: J. G. Topliss, US 3429874 (1969 to Schering); M. Steinman *et al.*, *J. Med. Chem.* **16**, 1354 (1973). Pharmacology: J. B. Petel *et al.*, *Psychopharmacology* **61**, 25 (1979). Use in schizophrenia: K. Y. Ota *et al.*, *Curr. Ther. Res.* **15**, 327 (1973); in anxiety: K. Rickels *et al.*, *Int. Pharmacopsychiatry* **13**, 118 (1978).

Crystals from acetone-hexane, mp 164-166°. LD_{50} in mice: >4000 mg/kg orally (Steinman).

Note: This is a controlled substance (depressant): **21 CFR**, 1308.14.

THERAP CAT: Anxiolytic.

4625. Halazone. [80-13-7] 4-[(Dichloroamino)sulfonyl]benzoic acid; *p*-(dichlorosulfamoyl)benzoic acid; *p*-sulfondichloramidobenzoic acid; *p*-carboxybenzenesulfondichloroamide; Pantocid. $C_7H_5Cl_2NO_4S$; mol wt 270.08. C 31.13%, H 1.87%, Cl 26.25%, N 5.19%, O 23.70%, S 11.87%. Prepd by chlorination of *p*-sulfamylbenzoic acid in alkaline medium; product is a mixture of mono- and dichloroamides with the dichloro compd predominating: Proschko, **DE 492249**; *Chem. Zentralbl.* **101**, I, 2630 (1930); US 1697139, *C.A.* **23**, 1138 (1929); Claas, **DE 318899**; *Chem. Zentralbl.* **91**, IV, 14 (1920). Improved synthesis: M. Saljoughian, M. T. Sadeghi, *Monatsh. Chem.* **117**, 553 (1986). Water disinfection study: R. Kfir *et al.*, *Water Sci. Technol.* **21**, 207 (1989). Review of use as antiseptic in water disinfection: J. T. O'Connor, S. K. Kapoor, *J. Am. Water Works Assoc.* **62**, 80-84 (1970).

Crystals or white powder, dec about 195°. Odor of chlorine. Sol in glacial acetic acid; very slightly sol in water, chloroform. Dissolves in solns of alkali hydroxides and carbonates with the formation of a salt.

USE: Water disinfectant.

4626. Halcinonide. [3093-35-4] (11β,16α)-21-Chloro-9-fluoro-11-hydroxy-16,17-[(1-methylethylidene)bis(oxy)]pregn-4-ene-3,20-dione; 21-chloro-9-fluoro-11β,16α,17-trihydroxypregn-4-ene-3,20-dione cyclic 16,17-acetal with acetone; 21-chloro-9α-fluoro-11β-hydroxy-16α,17α-isopropylidenedioxy-4-pregnene-3,20-dione; 9α-fluoro-21-chloro-11β,16α,17α-trihydroxypregn-4-ene-3,20-dione 16,17-acetonide; SQ-18566; Halciderm; Halcimat; Halog. $C_{24}H_{32}ClFO_5$; mol wt 454.96. C 63.36%, H 7.09%, Cl 7.79%, F 4.18%, O 17.58%. Prepn: Bernstein *et al.*, *J. Org. Chem.*

27, 690 (1962); L. T. Difazio, M. A. Augustine, **DE 2355710**; *eidem*, **US 3892857** (1972, 1975 both to Squibb). Pharmacological evaluation: Bagatell, Augustine, *Curr. Ther. Res.* **16**, 748 (1974); R. C. Millonig, E. Yiakas, in *Pharmacological and Biochemical Properties of Drug Substances* vol. 1, M. E. Goldberg, Ed. (Am. Pharmaceut. Assoc., Washington, DC, 1977) pp 215-231. Comprehensive description: J. Kirshbaum, *Anal. Profiles Drug Subs.* **8**, 251-281 (1979).

Crystals from acetone-petr ether, mp 264-265° (dec). uv max (methanol): 238 nm (ε 16400). $[\alpha]_D^{25}$ +155° (CHCl$_3$). Sol in acetone, chloroform, DMSO; slightly sol in benzene, ethanol, ethyl ether, methanol. Insol in water, 0.1M HCl, 0.1M NaOH, hexanes. LD$_{50}$ i.p. in mice: 150 mg/kg (Millonig, Yiakas).

THERAP CAT: Anti-inflammatory (topical).

4627. Halibut Liver Oil. From the liver of the halibut, *Hippoglossus hippoglossus* L. It usually contains not less than 60,000 U.S.P. vitamin A units and not less than 500 U.S.P. vitamin D units per gram.

Pale yellow liq; slight fishy odor and taste. d 0.922-0.925. n_D^{40} 1.470-1.478. Sapon no. 170-180. Iodine no. 120-136. Unsaponifiable matter 8-13%.

THERAP CAT: Source of vitamins A and D.

THERAP CAT (VET): Source of vitamins A and D.

4628. Halichondrins. Group of antimitotic, polyether macrolides produced by several marine sponges. Halichondrin B has the most potent bioactivity and is the prototype for the anticancer agent, eribulin, *q.v.* Isoln from *Halichondria okadai* Kadota: D. Uemura *et al., J. Am. Chem. Soc.* **107**, 4796 (1985); and antitumor activity: Y. Hirata, D. Uemura, *Pure Appl. Chem.* **58**, 701 (1986). Isoln from *Axinella* sp.: G. R. Pettit *et al., J. Med. Chem.* **34**, 3339 (1991). Total synthesis of halichondrin B and norhalichondrin B: T. D. Aicher *et al., J. Am. Chem. Soc.* **114**, 3162 (1992). Effect on bovine brain tubulin: R. F. Luduena *et al., Biochem. Pharmacol.* **45**, 421 (1993). Review of potential production methods: D. Sipkema *et al., Biotechnol. Bioeng.* **90**, 201-222 (2005).

Halichondrin B

Halichondrin A. [102721-98-2] (1″*R*,2*S*,2′*S*,2″*R*,3a*S*,3′a*S*,-3″a*R*,3″b*S*,5*R*,5″*R*,7*S*,7′*S*,7a*S*,7′a*S*,8″*S*,11″,*S*14″*S*,16″*R*,18″*R*,19″a*S*,-20″a*S*,21″a*R*,24″a*R*,25″a*S*,26″*S*,26″a*R*,30″*R*,31″a*S*,32″*S*,33″a*R*,38″*R*)-Tetracontahydro-2″,38″-dihydroxy-7,7′,16″,26″-tetramethyl-10″,17″-bis(methylene)-2-[(1*S*,3*R*)-1,3,4-trihydroxybutyl]dispiro[5*H*-furo-[3,2-*b*]pyran-5,5′-[5*H*]furo[3,2-*b*]pyran-2′(3′*H*),23″(6″*H*)-[1,5:8,-11:14,18]triepoxy[30,32]ethano[2,5]methano[2*H*,5*H*,28*H*]furo[2′,-3′:5,6]pyrano[4,3-*b*]furo[2″,3″:5′,6′]pyrano[2′,3′:5,6]pyrano[3,2-*i*]-[1,4,8]trioxacyclopentacosin]-28″-one. C$_{60}$H$_{86}$O$_{21}$; mol wt 1143.33.

Halichondrin B. [103614-76-2] 12,13-Dideoxyhalichondrin A. C$_{60}$H$_{86}$O$_{19}$; mol wt 1111.33. Crystalline solid, mp 164-166°. $[\alpha]_D$ −58.9° (c = 0.94 in methanol).

Halichondrin C. [101383-38-4] 13-Deoxyhalichondrin A. C$_{60}$-H$_{86}$O$_{20}$; mol wt 1127.33. Crystalline solid, mp 169-172°. $[\alpha]_D$ −41.6° (c = 0.49 in methanol).

4629. Halobetasol Propionate. [66852-54-8] (6α,11β,16β)-21-Chloro-6,9-difluoro-11-hydroxy-16-methyl-17-(1-oxopropoxy)-pregna-1,4-diene-3,20-dione; 21-chloro-6α,9-difluoro-11β,17-dihydroxy-16β-methylpregna-1,4-diene-3,20-dione 17-propionate; ulobetasol propionate; BMY-30056; CGP-14458; Ultravate. C$_{25}$H$_{31}$-ClF$_2$O$_5$; mol wt 484.96. C 61.92%, H 6.44%, Cl 7.31%, F 7.84%, O 16.50%. Trihalogenated corticosteroid structurally related to clobetasol, *q.v.* Prepn: J. Kalvoda, G. Anner, **DE 2743069**; *eidem*, **US 4619921** (1978, 1986 both to Ciba-Geigy). Clinical pharmacology in psoriatic patients: W. A. Watson *et al., Pharmacotherapy* **10**, 107 (1990). Clinical trial in localized vitiligo: L. Xunquan *et al., Int. J. Dermatol.* **29**, 295 (1990). Review of clinical trials in plaque psoriasis: A. M. Rivera, S. Hsu, *Am. J. Clin. Dermatol.* **6**, 311-316 (2005).

White to off-white powder. Crystals from methylene chloride + ether, mp 220-221°. Freely sol in acetone, dichloromethane. Practically insol in water.

THERAP CAT: Anti-inflammatory; antipsoriatic.

4630. Halofantrine. [69756-53-2] 1,3-Dichloro-α-[2-(dibutylamino)ethyl]-6-(trifluoromethyl)-9-phenanthrenemethanol; 1-(1,3-dichloro-6-trifluoromethyl-9-phenanthryl)-3-di(*n*-butyl)amino-propanol; γ-(dibutylamino)-1,3-dichloro-6-(trifluoromethyl)-9-phenanthrenepropanol. C$_{26}$H$_{30}$Cl$_2$F$_3$NO; mol wt 500.43. C 62.40%, H 6.04%, Cl 14.17%, F 11.39%, N 2.80%, O 3.20%. Prepn of the hydrochloride: W. T. Colwell *et al., J. Med. Chem.* **15**, 771 (1972); of the β-glycerophosphate salt: J. F. Rossignol, **EP 138374**; *idem*, **US 4507288** (both 1985 to Smithkline Beckman). Antimalarial activity *in vivo*: L. H. Schmidt *et al., Antimicrob. Agents Chemother.* **14**, 292 (1978). HPLC determn in human blood: J. W. Hines *et al., J. Pharm. Sci.* **74**, 433 (1985). Antimalarial activity in clinically-induced infection in humans: J. Rinehart *et al., Am. J. Trop. Med. Hyg.* **25**, 769 (1976); T. M. Cosgriff *et al., ibid.* **31**, 1075 (1982); in naturally acquired disease: J. P. Coulaud *et al., Trans. R. Soc. Trop. Med. Hyg.* **80**, 615 (1986). Review of pharmacology and therapeutic potential: H. M. Bryson, K. L. Goa, *Drugs* **43**, 236-258 (1992).

Hydrochloride. [36167-63-2] SKF-102886; WR-171669; Halfan. C$_{26}$H$_{30}$Cl$_2$F$_3$NO.HCl; mol wt 536.89. Two crystalline forms have been reported: mp 93-96°, mp 203-204°.

β-Glycerophosphate. C$_{29}$H$_{39}$Cl$_2$F$_3$NO$_7$P. White crystals, mp 60-65°.

THERAP CAT: Antimalarial.

4631. Halofenozide. [112226-61-6] 4-Chlorobenzoic acid 2-benzoyl-2-(1,1-dimethylethyl)hydrazide; RH-0345; Mach 2. C$_{18}$-H$_{19}$ClN$_2$O$_2$; mol wt 330.81. C 65.35%, H 5.79%, Cl 10.72%, N 8.47%, O 9.67%. Ecdysone agonist causing premature molting; insect growth regulator. Prepn: R. W. Addor *et al., EP 228564* (1987 to Am Cyanamid). Susceptibility study in insects: R. S. Cowles, M. G. Villani, *J. Econ. Entomol.* **89**, 1556 (1996). LC determn of residues in grass: S. S. Skorczynski, *J. AOAC Int.* **80**, 418 (1997). Review of biological activity and efficacy: T. S. Dhadialla *et al.*, *Annu. Rev. Entomol.* **43**, 545-569 (1998).

mp 198.0-199.0°.
USE: Insecticide.

4632. Halofuginone. [55837-20-2] *rel*-7-Bromo-6-chloro-3-[3-[(2*R*,3*S*)-3-hydroxy-2-piperidinyl]-2-oxopropyl]-4(3*H*)-quinazolinone; (±)-*trans*-7-bromo-6-chloro-3-[3-(3-hydroxy-2-piperidyl)-acetonyl]-4(3*H*)-quinazolinone; 7-bromo-6-chlorofebrifugine; HAL. $C_{16}H_{17}BrClN_3O_3$; mol wt 414.68. C 46.34%, H 4.13%, Br 19.27%, Cl 8.55%, N 10.13%, O 11.57%. Halogenated deriv of febrifugine, *q.v.* Prepn: E. Waletzky *et al.*, **US 3320124** (1967 to Am. Cyanamid). *In vivo* and *in vitro* activity: J. G. Ryley *et al.*, *Parasitology* **70**, 203 (1975). Toxicity to fresh water organisms: *Bull. Environ. Contam. Toxicol.* **15**, 720 (1976). HPLC determn in chicken feed: A. Anderson *et al.*, *J. Chromatogr.* **168**, 471 (1979); in chicken serum: R. C. Beier *et al.*, *J. Liq. Chromatogr.* **17**, 2961 (1994). Evaluation as coccidiostat: P. E. Waibel *et al.*, *Poult. Sci.* **66**, 1629 (1987). Inhibition of collagen synthesis and effect on skin tearing: I. Granot *et al.*, *ibid.* **70**, 1559 (1991).

Relative stereochemistry

Hydrobromide. [64924-67-0] RU-19110; Stenorol. $C_{16}H_{17}Br$-ClN_3O_3.HBr; mol wt 495.60. Crystals, mp 247° (dec).
THERAP CAT (VET): Antiprotozoal (coccidiostat).

4633. Halometasone. [50629-82-8] (6α,11β,16α)-2-Chloro-6,9-difluoro-11,17,21-trihydroxy-16-methylpregna-1,4-diene-3,20-dione; 2-chloroflumethasone; C-48401-Ba. $C_{22}H_{27}ClF_2O_5$; mol wt 444.90. C 59.39%, H 6.12%, Cl 7.97%, F 8.54%, O 17.98%. Synthetic corticosteroid. Prepn: G. G. Anner *et al.*, **CH 540244**; *idem*, **US 3652554** (1973, 1972 both to Ciba-Geigy). Series of clinical trials in eczema, psoriasis, dermatomycoses, bacterial infections: *J. Int. Med. Res.* **11**, Suppl. 1 (1983).

mp 220-222° (dec). $[\alpha]_D^{20}$ +40° (c = 0.97 in dioxane).
Monohydrate. Sicorten. $C_{22}H_{27}ClF_2O_5.H_2O$; mol wt 462.91.
THERAP CAT: Glucocorticoid; topical anti-inflammatory.

4634. Haloperidol. [52-86-8] 4-[4-(4-Chlorophenyl)-4-hydroxy-1-piperidinyl]-1-(4-fluorophenyl)-1-butanone; 4-[4-(*p*-chlorophenyl)-4-hydroxypiperidino]-4'-fluorobutyrophenone; 4'-fluoro-4-(4-hydroxy-4-*p*-chlorophenylpiperidino)butyrophenone; 1-(3-*p*-fluorobenzoylpropyl)-4-*p*-chlorophenyl-4-hydroxypiperidine; R-1625; Aloperidin; Bioperidolo; Brotopon; Dozic; Einalon S; Euky-

stol; Haldol; Halosten; Keselan; Linton; Peluces; Serenace; Serenase; Sigaperidol. $C_{21}H_{23}ClFNO_2$; mol wt 375.87. C 67.11%, H 6.17%, Cl 9.43%, F 5.05%, N 3.73%, O 8.51%. Prepn: P. A. J. Janssen *et al.*, *J. Med. Pharm. Chem.* **1**, 281 (1959); P. A. J. Janssen, **BE 577977** (1959), *C.A.* **54**, 4630c (1960); *idem*, **GB 895309**; *idem*, **US 3438991** (1962, 1969 both to Janssen). Toxicity: E. I. Goldenthal, *Toxicol. Appl. Pharmacol.* **18**, 185 (1971); A. J. Collins, M. Horlington, *Br. J. Pharmacol.* **37**, 140 (1969). Metabolism in man: A. Forsman *et al.*, *Curr. Ther. Res.* **21**, 606 (1977). Comprehensive description: C. A. Janicki, C. Y. Ko, *Anal. Profiles Drug Subs.* **9**, 341-369 (1980). Review of pharmacology and therapeutic efficacy of decanoate in psychosis: R. Beresford, A. Ward, *Drugs* **33**, 31-49 (1987).

White to faintly yellow amorphous crystals, mp 148.0-149.4°. uv max (9:1 0.1*M* HCl:methanol): 247, 221 nm (ε 13300, 15000). pKa 8.3. Soly in water: 1.4 mg/100 ml. Freely sol in methanol, acetone, benzene, dil acids; sol in chloroform; sparingly sol in alc; slightly sol in ether. LD_{50} orally in rats: 165 mg/kg (Goldenthal); i.p. in mice: 60 mg/kg (Collins, Horlington).
Hydrochloride. [1511-16-6] $C_{21}H_{23}ClFNO_2.HCl$. Crystals, mp 226-227.5°. Soly in water: 300 mg/100 ml.
Decanoate. [74050-97-8] KD-136; Haldol Decanoate; Halomonth; Neoperidole. $C_{31}H_{41}ClFNO_3$; mol wt 530.12.
THERAP CAT: Antidyskinetic (in Gilles de la Tourette's disease); antipsychotic.

4635. Halostachine. [495-42-1] (α*R*)-α-[(Methylamino)-methyl]benzenemethanol; (−)-α-(methylaminomethyl)benzyl alcohol; 1-hydroxy-1-phenyl-2-methylaminoethane; (−)-phenyl(methylaminomethyl)carbinol; 2-methylamino-1-phenylethanol. C_9H_{13}-NO; mol wt 151.21. C 71.49%, H 8.67%, N 9.26%, O 10.58%. Isoln from *Halostachis caspica (Halostachis caspia)*: Menshikov, Rubinshtein, *J. Gen. Chem. USSR* **13**, 801 (1943), *C.A.* **39**, 1172 (1945). Synthesis: Menshikov, Borodina, *J. Gen. Chem. USSR* **17**, 1569 (1947), *C.A.* **42**, 2245 (1948). Configuration: Lukes *et al.*, *Collect. Czech. Chem. Commun.* **26**, 466 (1961). As a possible cause of ryegrass staggers: Aasen *et al.*, *Aust. J. Agric. Res.* **20**, 71 (1969). TLC determn in rye and fescue grasses: L. P. Bush, J. A. D. Jeffreys, *J. Chromatogr.* **111**, 165 (1975). Vasoactive potential: C. B. Davis *et al.*, *Vet. Hum. Toxicol.* **25**, 6 (1983). One pot synthesis: P. Zandbergen *et al.*, *Tetrahedron* **48**, 3977 (1992).

Crystals, mp 43-45°. $[\alpha]_D^{20}$ −47.03°. Soluble in water, alc, ether.
Hydrochloride. Crystals from alcohol, mp 113-114°. $[\alpha]_D^{20}$ −52.46°. Freely sol in water.

4636. Halosulfuron-methyl. [100784-20-1] 3-Chloro-5-[[[[(4,6-dimethoxy-2-pyrimidinyl)amino]carbonyl]amino]sulfonyl]-1-methyl-1*H*-pyrazole-4-carboxylic acid methyl ester; methyl 3-chloro-5-(4,6-dimethoxypyrimidin-2-ylcarbamoylsulfamoyl)-1-methylpyrazole-4-carboxylate; MON-12000; MON-12037; NC-319; Manage; Permit. $C_{13}H_{15}ClN_6O_7S$; mol wt 434.81. C 35.91%, H 3.48%, Cl 8.15%, N 19.33%, O 25.76%, S 7.37%. Sulfonylurea herbicide. Prepn: S. Yamamoto *et al.*, **JP Kokai 85 208977**; *idem*, **US 4668277** (1985, 1987 both to Nissan); S. Yamamoto *et al.*, *ACS Symp. Ser.* **504**, 34 (1992). Physical properties and field trials in corn: K. Suzuki *et al.*, *Brighton Crop Prot. Conf. - Weeds* **1991**, 31. Field trials in landscape plants: R. T. Hurt, W. K. Vencill, *J. Environ. Hort.* **12**, 135 (1994).

White powder, mp 172-173°. Vapor pressure (25°): 2.8×10^{-12} mm Hg. Soly in water (25°): 36 mg/l. LD_{50} orally in rats: 8865 mg/kg; dermally in rabbits: >2000 mg/kg; LC_{50} (4 hr) in rats: >4.3 mg/l (Suzuki).

USE: Herbicide.

4637. Halothane. [151-67-7] 2-Bromo-2-chloro-1,1,1-trifluoroethane; bromochlorotrifluoroethane; 1,1,1-trifluoro-2,2-chlorobromoethane; Fluothane; Rhodialothan. $C_2HBrClF_3$; mol wt 197.38. C 12.17%, H 0.51%, Br 40.48%, Cl 17.96%, F 28.88%. $CF_3CHClBr$. Prepn from a mixt of F_3CCH_2Cl and F_3CCBr_2Cl: Suckling, Raventos, **US 2921098** (1960 to I.C.I.); by rearrangement of $F_2BrCCHFCl$: Scherer, Kühn, **US 2959624** (1960 to Hoechst); from BrClCHCBrCl$_2$: Chapman, McGinty, **GB 805764** (1958 to I.C.I.); from Br$_2$ClCCF$_3$: McGinty, **US 3082263** (1963 to I.C.I.); Madai, Muller, *J. Prakt. Chem.* **19**, 83 (1963). Enantiomeric resolution: J. Meinwald *et al.*, *Science* **251**, 560 (1991). Comprehensive description: R. D. Daley, *Anal. Profiles Drug Subs.* **1**, 119-147 (1972).

Non-flammable, highly volatile liquid. Characteristic, sweetish, not unpleasant odor. d_4^{20} 1.871. bp 50.2°; bp$_{243}$ 20°. n_D^0 1.3697. Sensitive to light, may be stabilized with 0.01% thymol. Miscible with alc, chloroform, ether, petr ether, fixed oils. Soly in water 0.345%.

Caution: Potential symptoms of overexposure are irritation of eyes, skin, respiratory system; confusion, drowsiness, dizziness, nausea, analgesia, anesthesia; cardiac arrhythmia; liver and kidney damage; decreased audio visual performance. *See NIOSH Pocket Guide to Chemical Hazards* (DHHS/NIOSH 97-140, 1997) p 156.

THERAP CAT: Anesthetic (inhalation).

THERAP CAT (VET): Anesthetic (inhalation).

4638. Haloxazolam. [59128-97-1] 10-Bromo-11b-(2-fluorophenyl)-2,3,7,11b-tetrahydrooxazolo[3,2-d][1,4]benzodiazepin-6(5H)-one; CS-430; Somelin. $C_{17}H_{14}BrFN_2O_2$; mol wt 377.21. C 54.13%, H 3.74%, Br 21.18%, F 5.04%, N 7.43%, O 8.48%. Sleep-inducing agent, related structurally to oxazolam and cloxazolam, q.q.v. Prepn of analogous compounds: T. Miyadera *et al.*, *J. Med. Chem.* **14**, 520 (1971). Structure-activity study: M. Yoshimoto *et al.*, *Chem. Pharm. Bull.* **25**, 1378 (1977). Pharmacological studies: T. Kamioka *et al.*, *Arzneim.-Forsch.* **28**, 838 (1978). Metabolism in animals: R. Hayashi *et al.*, *Oyo Yakuri* **17**, 617 (1979), *C.A.* **91**, 204161s (1979).

Colorless cryst, mp 185°. Sparingly sol in water. LD_{50} in mice: 1850 mg/kg orally (Kamioka).

Note: This is a controlled substance (depressant): **21 CFR**, 1308.14.

THERAP CAT: Sedative, hypnotic.

4639. Hamamelis. Witch hazel; winter bloom; snapping hazel; striped alder; spotted alder; tobacco wood. Dried leaves of *Hamamelis virginiana* L., *Hamamelidaceae*, collected in autumn. History of hamamelis extract and distillate: Lloyd, Lloyd, *J. Am. Pharm. Assoc.* **24**, 220 (1935). *Habit.* North America (New Eng-

land to Minnesota, southward to Louisiana). *Constit.* Hamamelitannin, gallic acid, volatile oil, bitter principle.

THERAP CAT: Astringent.

THERAP CAT (VET): Has been used as astringent, hemostatic, sedative.

4640. Hamamelitannin. [469-32-9] 2-C-[[(3,4,5-Trihydroxybenzoyl)oxy]methyl]-D-ribose 5-(3,4,5-trihydroxybenzoate); 2-C-(hydroxymethyl)-D-ribofuranose 2′,5-digallate. $C_{20}H_{20}O_{14}$; mol wt 484.37. C 49.59%, H 4.16%, O 46.24%. From bark of *Hamamelis virginiana* L., *Hamamelidaceae*: Grüttner, *Arch. Pharm.* **236**, 278 (1898); from bark of *Castanea dentata* (Marsh.) Borkh., *Fagaceae*: Mayer, Kunz, *Naturwissenschaften* **46**, 206 (1959). Structure: Mayer *et al.*, *Ann.* **688**, 232 (1965). Synthesis: Ezekiel *et al.*, *Carbohydr. Res.* **11**, 233 (1969).

Prisms from water, mp 145-147°. $[\alpha]_D^{19}$ +32.6° (c = 1.5).

Hexamethyl ether. $C_{26}H_{32}O_{14}$. Needles from dioxane + water, mp 73-76°. $[\alpha]_{578}^{25}$ +10.0° (c = 2.0 in dioxane). Sol in acetone, dioxane, methanol. Practically insol in water.

4641. Hamamelose. [4573-78-8] 2-C-(Hydroxymethyl)-D-ribose. $C_6H_{12}O_6$; mol wt 180.16. C 40.00%, H 6.71%, O 53.28%. From tannin of witch hazel (*Hamamelis virginiana* L., *Hamamelidaceae*): Fischer, Freudenberg, *Ber.* **45**, 2709 (1912); Freudenberg, Peters, *ibid.* **53**, 953 (1920); Anderson, *U.S.A.E.C.* **UCRL-8870**, 114 (1959). Structure: Schmidt, *Ann.* **476**, 250 (1929). Configuration: Schmidt, Heintz, *ibid.* **515**, 77 (1934). Synthesis: Novák, Sorm, *Collect. Czech. Chem. Commun.* **30**, 3303 (1965); Paulsen *et al.*, *Ber.* **105**, 1978 (1972). Synthesis of L-hamamelose: Burton *et al.*, *Proc. Chem. Soc. London* **1962**, 181. Synthesis and ^{1}H, ^{13}C NMR study of aqueous equilibrium: W. A. Szarek *et al.*, *Can. J. Chem.* **61**, 461 (1983).

D-Form. Crystals from abs ethanol, mp 111°. $[\alpha]_D^{21}$ −7.4° (equilib in water).

L-Form. Crystals from ethanol + ethyl acetate, mp 110-111°. $[\alpha]_D^{22}$ +1.3° (3 min) → +7.3° (equilib after 17 min).

4642. Hamycin. [1403-71-0] Primamycin. Polyene antibiotic complex produced by *Streptomyces pimprina*: M. J. Thirumalachar *et al.*, *Hind. Antibiot. Bull.* **3**, 136 (1961), *C.A.* **55**, 27515i (1961); M. J. Thirumalachar, **US 3261751** (1966 to Hindustan Antibiot.). Formerly believed to be identical to trichomycin and candicidin, q.q.v.: H. Burrows, D. Calam, *J. Chromatogr.* **53**, 566 (1970). Subsequent HPLC studies have shown the three polyene antibiotics to be different entities: P. Helboe *et al.*, *ibid.* **189**, 249 (1980). Structural study showing hamycin to be a mixture of components A, B, C, D: R. C. Pandey, K. L. Rinehart, *16th Interscience Conference on Antimicrob. Ag. Chemother.*, Chicago, 1976, *Abstracts of Papers*, no. 41. Absorption, excretion, tissue concentration: M. G. Phatak *et al.*, *Indian J. Exp. Biol.* **12**, 284 (1974), *C.A.* **82**, 118703j (1975). Toxicity studies: Williams *et al.*, *Antimicrob. Agents Chemother.* **1964**, 737; antifungal activity: A. C. Parekh, C. V. Dave, *Life Sci.* **19**, 1737 (1976). HPLC study: W. Mechlinski, C. P. Schaffner, *J. Antibiot.* **33**, 591 (1980). Review of clinical use in mycoses: M. J.

Thirumalachar, *J. Sci. Ind. Res.* **31**, 542-544 (1972), *C.A.* **79**, 13308 (1973).

Yellow amorphous powder, no definite mp, decomp 160°. $[\alpha]_D^{25}$ +216°. uv max (80% methanol): 383 nm ($E_{1cm}^{1\%}$ 916). Amphoteric. Almost insol in water, benzene, chloroform, dry lower aliphatic alcohols, ether. Sol in basic solvents such as pyridine, collidine, and in aq lower alcohols. In concd H_2SO_4 gives stable blue color; no coloration with ferric chloride or with HCl. LD_{50} i.v. in mice (using sodium carboxymethyl cellulose as a vehicle): 6.16 mg/kg (Williams). Also reported as LD_{50} i.v. in Swiss mice (using a colloidal suspension): 1.20 mg/kg (Parekh, Dave).

Hamycin A. mp >300° (dec), $[\alpha]_D^{24}$ +181.1° (c = 0.6 in DMF). uv max: 380 nm ($E_{1cm}^{1\%}$ 989).

THERAP CAT: Antifungal.

4643. Hanatoxins. [253687-10-4] HaTx. Peptide toxins isolated from the venom of the Chilean tarantula, *Phrixotrichus spatulata*, originally named *Grammostola spatulata*; modifies voltage-dependent K^+ channels by shifting channel opening to more depolarized voltages. 35 amino acids long with 3 disulfide linkages, mol wt ~4.1 kDa. Isoln and characterization: K. J. Swartz, R. MacKinnon, *Neuron* **15**, 941 (1995). Mechanism of action and identification of receptor binding sites: *eidem, ibid.* **18**, 665, 675 (1997). Use as probe for determn of channel structure: Y. Li-Smerin, K. J. Swartz, *Proc. Natl. Acad. Sci. USA* **95**, 8585 (1998); V. Ruta *et al., Nature* **422**, 180 (2003). Reagent for measurement of channel current in myocytes: H. M. Himmel *et al., Am. J. Physiol.* **277**, H107 (1999).

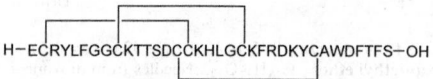

H—ECRYLFGGCKTTSDCCKHLGCKFRDKYCAWDFTFS—OH

Hanatoxin 1. [281212-01-9] (native); [170780-00-4] (reduced). HaTx₁. NMR determn of 3d-structure: H. Takahashi *et al., J. Mol. Biol.* **297**, 771 (2000).

Hanatoxin 2. [170780-01-5] (reduced). HaTx₂.

USE: Inhibition of voltage-gated K^+ channels.

4644. Haplophytine. [16625-20-0] 3,4-Didehydro-19-hydroxy-16,17-dimethoxy-1-methyl-15-[(3aS,7R)-2,3,5,6-tetrahydro-11-hydroxy-4-methyl-1,13-dioxo-1H-3a,7-methanopyrrolo[1,2-a]-[1,3]benzodiazocin-7(4H)-yl]-aspidospermidin-21-oic acid γ-lactone. $C_{37}H_{40}N_4O_7$; mol wt 652.75. C 68.08%, H 6.18%, N 8.58%, O 17.16%. Insecticidal alkaloid from the Mexican plant *Haplophyton cimicidum* A. DC., *Apocynaceae:* Rogers *et al., J. Am. Chem. Soc.* **74**, 1987 (1952); *ibid.* **80**, 3708 (1958). Partial structure: Snyder *et al., ibid.* **80**, 3708 (1958). Structure: Rae *et al., ibid.* **89**, 3061 (1967). Crystal structure and absolute config.: Zacharias, *Acta Crystallogr.* **26B**, 1455 (1970). Synthetic approaches: P. Yates, D. A. Schwartz, *Can. J. Chem.* **61**, 509 (1983). *Review:* Yates *et al., J. Am. Chem. Soc.* **95**, 7842 (1973).

Crystals from ethanol + chloroform, mp 290-293° (rapid heating, starting at 250°). Also reported as mp 300-302° (Yates). $[\alpha]_D^{25}$ +109.0° (chloroform). uv max (ethanol): 220, 265, 305 nm (ε 48500, 14300, 4500). Very sol in chloroform, benzene, dioxane, ethyl acetate. Moderately sol in acetone, methanol, somewhat less in ethanol. Practically insol in water, ether, petr ether. Readily sol in dil acids or alkalies.

Dihydrochloride. $C_{27}H_{31}N_3O_5$·2HCl; mol wt 550.48. mp 208-218° (dec); darkens at 200°.

O-**Methylhaplophytine.** $C_{28}H_{33}N_3O_5$; mol wt 491.59. Crystals from ether + ethanol, mp 288-291° (dec). $[\alpha]_D^{24}$ +12° (c = 4.37 in chloroform).

4645. Haptoglobins. α_2-Globulins which combine with hemoglobin to form a weak peroxidase and which constitute about one-quarter of the α_2-globulin fraction of human plasma: Jayle, Conas, *Bull. Soc. Chim. Biol.* **34**, 65 (1952). Prepn: Herman-Boussier *et al., ibid.* **48**, 817, 837 (1960). For each individual there are three genetic types of haptoglobins, each reacting identically with antibodies to each of the others. Structure of this serum glycoprotein is a tetramer composed of α- and β-polypeptide chains, the α-chain varying in the three phenotypes, the β-chains being identical. Amino acid sequence of α-chains: Black, Dixon, *Nature* **218**, 736 (1968); Malchy, Dixon, *Can. J. Biochem.* **51**, 321 (1973). C- and N-terminal sequences of the β-chain: Barnett *et al., Biochemistry* **11**, 1189 (1972); A. Kurosky *et al., Comp. Biochem. Physiol.* **55B**, 453 (1976); *eidem, Biochemistry* **15**, 5326 (1976). Lowered levels of haptoglobins found in individuals with acute hepatitis and pernicious anemia. *Review:* Laurell, Grönvall, *Adv. Clin. Chem.* **5**, 135-172 (1962). Monograph: Kirk, *The Haptoglobin Groups in Man* (S. Karger, New York, 1968) 77 pp. Review on genetics, biochemistry and physiology of human haptoglobins: J. Javid, *Curr. Top. Hematol.* **1**, 151-192 (1978).

4646. Harmaline. [304-21-2] 4,9-Dihydro-7-methoxy-1-methyl-3H-pyrido[3,4-b]indole; 1-methyl-7-methoxy-3,4-dihydro-β-carboline; 3,4-dihydroharmine; harmidine; harmalol methyl ether; *O*-methylharmalol. $C_{13}H_{14}N_2O$; mol wt 214.27. C 72.87%, H 6.59%, N 13.07%, O 7.47%. CNS stimulant from seeds of *Peganum harmala* L., *Zygophyllaceae:* Goebel, *Ann.* **38**, 363 (1841); from *Banisteria caapi* Spruce, *Malpighiaceae:* Hochstein, Paradies, *J. Am. Chem. Soc.* **79**, 5735 (1957). Structure: Manske *et al., J. Chem. Soc.* **1927**, 1. Synthesis: Späth, Lederer, *Ber.* **63**, 120, 2102 (1930); Spenser, *Can. J. Chem.* **37**, 1851 (1959). Identity with harmidine: Robinson, *Chem. Ind. (London)* **1965**, 605. Pharmacology: Fuentes, Longo, *Neuropharmacology* **10**, 15 (1971). Metabolism: Ho *et al., Biochem. Pharmacol.* **20**, 1313 (1971). Review of structure and synthesis work: Hofmann, *Sven. Farm. Tidskr.* **75**, 933 (1971), *C.A.* **76**, 149609g (1972).

Orthorhombic bipyramidal prisms, tablets from methanol, rhombic octahedra from ethanol, mp 229-231°. Solns fluoresce blue. pK 4.2. uv max (methanol): 218, 260, 376 nm (log ε 4.27, 3.90, 4.02). Slightly sol in water, alcohol, ether; quite sol in hot alcohol, dil acids.

Hydrochloride dihydrate. Slender, yellow needles, moderately sol in water, alcohol.

N-**Acetylharmaline.** Needles, mp 204-205°.

4647. Harmalol. [525-57-5] 4,9-Dihydro-1-methyl-3H-pyrido[3,4-b]indol-7-ol; 3,4-dihydro-1-methyl-9H-pyrido[3,4-b]indol-7-ol. $C_{12}H_{12}N_2O$; mol wt 200.24. C 71.98%, H 6.04%, N 13.99%, O 7.99%. From seeds of *Peganum harmala* L., *Zygophyllaceae:* Göbel, *Ann.* **38**, 363 (1841); Fischer, *Ber.* **18**, 400 (1885); by demethylation of harmaline: Coulthard *et al., Biochem. J.* **27**, 727 (1933). HPLC determn in *Peganum harmala* L. seeds: M. Kartal *et al., J. Pharm. Biomed. Anal.* **31**, 263 (2003). *In vitro* neuroprotective effects: D. H. Kim *et al., Eur. J. Neurosci.* **13**, 1861 (2001).

Trihydrate. [6027-99-2] Red needles from water. Dec 212° (anhydr). Readily sol in hot water, acetone, chloroform, alkali hydroxides, but not carbonates. Aq solns are yellow with green fluorescence.

Lactate. [6028-08-6] $C_{12}H_{12}N_2O.C_3H_6O_3$. mp 174-176°; monohydrate as bright yellow leaflets, mp 116-120°. Sol in water and alcohol.

O-Ethylharmalol. [6010-17-9] $C_{14}H_{16}N_2O$. Brown needles from alc, mp 237-239°.

O-n-Propylharmalol. [6028-01-9] $C_{15}H_{18}N_2O$. Brown needles from alc, mp 195-197°.

O-n-Butylharmalol. [6197-42-8] $C_{16}H_{20}N_2O$. Colorless needles from alc, mp 173°.

O-Methylharmalol see Harmaline.

4648. Harman. [486-84-0] 1-Methyl-9*H*-pyrido[3,4-*b*]indole; 3-methyl-4-carboline; 2-methyl-*β*-carboline; aribine; loturine; passiflorin. $C_{12}H_{10}N_2$; mol wt 182.23. C 79.09%, H 5.53%, N 15.37%. From bark of *Sickingia rubra* (Mart.) K. Schum. *(Arariba rubra* Mart.), *Rubiaceae; Symplocus racemosa* Roxb., *Symplocaceae;* and *Passiflora incarnata* L., *Passifloraceae:* Rieth, Wohler, *Ann.* **120**, 247 (1861); Späth, *Monatsh. Chem.* **40**, 351; *idem, ibid.* **41**, 401 (1920); Neu, *Arzneim.-Forsch.* **4**, 601 (1954). Structure: Neu, *ibid.* **6**, 94 (1956). Synthesis: Harvey, Robson, *J. Chem. Soc.* **1938**, 97; Snyder *et al., J. Am. Chem. Soc.* **70**, 222 (1948); Clemo, Holt, *J. Chem. Soc.* **1953**, 1313; Kametani *et al., J. Chem. Soc. C* **1968**, 1006. Isoln from cigarette smoke: Poindexter, Carpenter, *Chem. Ind. (London)* **1962**, 176. Toxicity study: E. B. Sigg *et al., Arch. Int. Pharmacodyn.* **149**, 164 (1964).

Bitter orthorhombic crystals from heptane + cyclohexane, mp 237-238°. Exhibits bright blue fluorescence in uv light. pKa: 7.37, 14.6. uv max (methanol): 234, 287, 347 nm (log ε 4.57, 4.21, 3.66). Absorption and fluorescence spectra: O. S. Wolfbeis *et al., Monatsh. Chem.* **113**, 509 (1982). Practically insol in water. Sol in dil acids. LD_{50} i.p. in mice: 50 mg/kg (Sigg).

Hydrochloride. $C_{12}H_{10}N_2.HCl$. Rosettes of needles from ethanol + 20% HCl in water, sublimes at 120-130°.

4649. Harmine. [442-51-3] 7-Methoxy-1-methyl-9*H*-pyrido[3,4-*b*]indole; banisterine; yageine; telepathine; leucoharmine. $C_{13}H_{12}N_2O$; mol wt 212.25. C 73.57%, H 5.70%, N 13.20%, O 7.54%. From seeds of *Peganum harmala* L., *Zygophyllaceae:* Göbel, *Ann.* **38**, 363 (1841); Reinhard *et al., Phytochemistry* **7**, 503 (1968); from *Banisteria caapi* Spruce, *Malpighiaceae:* Hochstein, Paradies, *J. Am. Chem. Soc.* **79**, 5735 (1957); from *Banisteriopsis inebrians* Morton, *Malpighiaceae:* O'Connell, Lynn, *J. Am. Pharm. Assoc.* **42**, 753 (1953). Structure: Manske *et al., J. Chem. Soc.* **1927**, 1, 240. Synthesis: Späth, Lederer, *Ber.* **63B**, 120 (1930); Akaboro, Saito, *ibid.* 2245; Hahn *et al., Ber.* **67B**, 2031 (1934); *Ann.* **520**, 107, 123 (1935); *Ber.* **71B**, 2163, 2175 (1938); Harvey, Robson, *J. Chem. Soc.* **1938**, 97. Metabolism: Slotkin, DiStefano, *J. Pharmacol. Exp. Ther.* **174**, 456 (1970). Toxicity study: K. K. Chen *et al., J. Pharmacol. Exp. Ther.* **79**, 127 (1943).

Slender, orthorhombic prisms from methanol, mp 261° (dec). Sublimes. pKa 7.70. uv max (methanol): 241, 301, 336 nm (log ε 4.61, 4.21, 3.69). Absorption and fluorescence spectra: O. S. Wolfbeis, E. Fürlinger, *Z. Phys. Chem.* **129**, 171 (1982). Slightly sol in water, alcohol, chloroform, ether.

Hydrochloride dihydrate. Crystals, mp 262° (dec), mp 321° when anhydrous. Sol in 40 parts water, freely sol in hot water. Aq solns have blue fluorescence. LD_{50} i.v. in mice: 38 mg/kg (Chen).

4650. Harpin. [151438-54-9] Harpin Ea; Messenger. Protein product of the *hrp*N gene of *Erwinia amylovora*, a phytopathogenic

bacterium that causes fire blight disease. Composed of 403 amino acid residues; mol wt ~44 kDa. Member of a class of glycine-rich, acidic, heat stable proteins that elicit the hypersensitive response as part of the plant's defense mechanism vs pathogenic bacteria. Induces systemic acquired resistance to fire blight and many other plant pathogens, enhances plant growth, and promotes insect repellency. Isoln: Z.-M. Wei *et al., Science* **257**, 85 (1992). Prepn: S. V. Beer *et al.,* **WO 9401546**; *eidem,* **US 5849868** (1994, 1998 both to Cornell Research Foundation). Elicitation of hypersensitive response: S. V. Beer *et al.* in *Curr. Plant Sci. Biotechnol. Agric.* vol. **14**, E. W. Nester, D. P. S. Verma, Eds. (Kluwer, Dordrecht, 1993) pp 281-286. Induction of plant resistance: Z.-M. Wei, S. V. Beer, *Acta Hortic.* **411**, 223 (1996); and mode of action of disease resistance: H. Dong *et al., Plant J.* **20**, 207 (1999). Effects on tobacco leaf cell membranes: S. M. Pike *et al., Physiol. Mol. Plant Pathol.* **53**, 39 (1998). Blue mold resistance study in apples: G. de Capdeville *et al., Plant Dis.* **87**, 39 (2003). Review of *hrp* genes and harpin proteins: J. F. Kim, S. V. Beer in *Fire Blight*, J. Vanneste, Ed. (CABI Publishing, Wallingford, 2000) pp. 141-161; of commercial product: *Biopesticide Regulatory Action Document: Harpin Protein* (U.S. EPA, 2002) 32 pp.

pI 4.3. Mean bulk density: 0.452 g/ml. uv max: 204 nm (pH 7). LD_{50} in rats orally; rabbits dermally (g/kg): >2; >6. LC_{50} in rainbow trout (mg/l): >3270 (U.S. EPA).

USE: Biodegradable plant health regulator effective in vegetables, field crops, small grains, trees, vines, specialty crops, turf and ornamentals.

4651. Hassium. [54037-57-9] Element 108; unniloctium. Hs, Uno; at. no. 108. Group VIII (8). No stable nuclides. Prepn of isotope $^{265}108$ ($T_{1/2}$ ~1.8 msec, α-decay, rel. at. mass 265.1306) by ^{208}Pb ($^{58}Fe,n$): G. Münzenberg *et al., Z. Phys.* **A317**, 235 (1984); of isotopes $^{263}108$ ($T_{1/2}$ ~1.9 sec) by ^{209}Bi ($^{55}Mn,n$) and $^{264}108$ by ^{207}Pb ($^{58}Fe,n$) or $^{208}Pb(^{58}Fe,2n)$: Y. T. Oganessian *et al., ibid.* **A319**, 215 (1984); of isotope $^{264}108$ ($T_{1/2}$ ~76 μsec, α-decay) by $^{207}Pb(^{58}Fe,n)$: G. Münzenberg *et al., ibid.* **A324**, 489 (1986). Review of synthesis: *eidem, ibid.* **A328**, 49-59 (1987); G. N. Flerov, G. M. Ter-Akopian, *Prog. Particle Nucl. Phys.* **19**, 197-239 (1987); G. T. Seaborg in *Transuranium Elements: A Half Century*, L. R. Morss, J. Fuger, Eds. (Am. Chem. Soc., Washington DC, 1992) p 40-41.

4652. Hasubanonine. [1805-85-2] 7,8-Didehydro-3,4,7,8-tetramethoxy-17-methylhasubanan-6-one. $C_{21}H_{27}NO_5$; mol wt 373.45. C 67.54%, H 7.29%, N 3.75%, O 21.42%. Alkaloid having a modified morphinan skeleton with a five membered heterocyclic ring, hitherto unknown in natural sources. Alkaloids with this unique skeleton are named *hasubanan* alkaloids. Synthesis of hasubanan skeleton: S. Shiotani, T. Kometani, *Tetrahedron Lett.* **1976**, 767. Isolated from *Stephania japonica* Miers, *Menispermaceae:* Kondo *et al., Annu. Rep. Itsuu Lab.* **2**, 35 (1951); Satomi, *ibid.* **3**, 37 (1953); Kondo, Satomi, *ibid.* **8**, 41 (1957); *C.A.* **48**, 2728c (1954); **51**, 17956i (1957). Structure: Tomita *et al., Tetrahedron Lett.* **1964**, 2937; *eidem, Chem. Pharm. Bull.* **13**, 538 (1965). Total synthesis: Ibuka *et al., Tetrahedron Lett.* **1970**, 4811; *Chem. Pharm. Bull.* **22**, 782 (1974). Biosynthesis: A. R. Battersby *et al., Chem. Commun.* **1974**, 773; *eidem, J. Chem. Soc. Perkin Trans. 1* **1981**, 2010, 2016, 2030. *Review:* Y. Inubushi, T. Ibuka in *The Alkaloids* vol. **XVI**, R. H. F. Manske, Ed. (Academic Press, New York, 1977) pp 393-428.

Prisms from methanol, mp 116°. $[\alpha]_D$ −219° (ethanol).

4653. HATU. [148893-10-1] 1-[Bis(dimethylamino)methylene]-1*H*-1,2,3-triazolo[4,5-*b*]pyridinium 3-oxide hexafluorophosphate(1−) (1:1); N-HATU. $C_{10}H_{15}F_6N_6OP$; mol wt 380.24. C 31.59%, H 3.98%, F 29.98%, N 22.10%, O 4.21%, P 8.15%. Coupling reagent used for both soln and solid-phase peptide synthesis. Originally reported compd, *O-(7-azabenzotriazol-1-yl)-N,N,N'N'-tetramethyluronium hexafluorophosphate*, was structurally represented as the uronium salt (*O*-form); subsequently determined to be the guanidinium salt (*N*-form) in both the crystalline state and in soln. The structure is dependent upon prepn conditions; organic bases promote isomerization of the uronium form to the guanidinium form. The uronium salt also functions as a peptide coupling reagent. Prepn: L. A. Carpino, *J. Am. Chem. Soc.* **115**, 4397 (1993). Peptide coupling racemization studies: *idem et al.*, *Tetrahedron Lett.* **35**, 2279 (1994). X-ray structure determination: I. Abdelmoty *et al.*, *Lett. Pept. Sci.* **1**, 57 (1994). Semi-empirical structural studies: J. M. Bofill, F. Albericio, *J. Chem. Res. Synop.* **6**, 302 (1996). Comparison of prepn, structure, and reactivity of uronium and guanidinium compounds: L. A. Carpino *et al.*, *Angew. Chem. Int. Ed.* **41**, 441 (2002). *Review*: A. Speicher *et al.*, *J. Prakt. Chem. Chem.-Ztg.* **340**, 581-583 (1998).

White to light brown crystalline solid. *Irritant.* Sol in DMF. Violent decompn can occur when dried at elevated temp.

USE: Reagent in synthetic organic chemistry.

4654. HBTU. [94790-37-1] 1-[Bis(dimethylamino)methylene]-1*H*-benzotriazolium 3-oxide hexafluorophosphate(1−) (1:1); N-HBTU. $C_{11}H_{16}F_6N_5OP$; mol wt 379.25. C 34.84%, H 4.25%, F 30.06%, N 18.47%, O 4.22%, P 8.17%. Coupling reagent used for both soln and solid-phase peptide synthesis. Originally reported compd, *O-(benzotriazol-1-yl)-N,N,N',N'-tetramethyluronium hexafluorophosphate*, was structurally represented as the uronium salt (*O*-form); subsequently determined to exist as the guanidinium salt (*N*-form) in both the crystalline state and in soln. The structure is dependent upon prepn conditions; organic bases promote isomerization of the uronium form to the guanidinium form. The uronium salt also functions as a peptide coupling reagent. Prepn and use: V. Dourtoglou *et al.*, *Tetrahedron Lett.* **19**, 1269 (1978); V. Dourtoglou, B. Gross, *Synthesis* **1984**, 572. Solid phase synthesis applications: C. G. Fields *et al.*, *Pept. Res.* **4**, 95 (1991). X-ray structure determination: I. Abdelmoty *et al.*, *Lett. Pept. Sci.* **1**, 57 (1994). Comparison of prepn, structure, and reactivity of uronium and guanidinium compounds: L. A. Carpino *et al.*, *Angew. Chem. Int. Ed.* **41**, 441 (2002). *Review*: A. Speicher *et al.*, *J. Prakt. Chem. Chem.-Ztg.* **340**, 581-583 (1998).

White solid from acetonitrile, mp 254° (dec). *Irritant.* Sol in DMF. Violent decompn can occur when dried at elevated temp.

USE: Reagent in synthetic organic chemistry.

4655. HCS. [11085-36-2] Lactogen (human placental); lactogenic hormone (placental human); human chorionic somatomammotropin; chorionic growth hormone-prolactin; CGP; human placental lactogen; HPL. Growth hormone from human placenta. Ex-

hibits lactogenic as well as growth-promoting activity: Josimovich, MacLearen, *Endocrinology* **71**, 209 (1962); Kaplan, Grumbach, *J. Clin. Endocrinol. Metab.* **24**, 80 (1964). Structure is a single polypeptide chain containing 190 amino acid residues: Li *et al.*, *Science* **173**, 56 (1971); Sherwood *et al.*, *Nature New Biol.* **233**, 59 (1971). Structural revision to 191 amino acid residues and comparison with human growth hormone and human prolactin: Li *et al.*, *Arch. Biochem. Biophys.* **155**, 95 (1973); Niall *et al.*, *Recent Prog. Horm. Res.* **29**, 387 (1973). Purification: Parcells, Dahlgren, US 3687833 (1972 to Upjohn). Terminology: Li *et al.*, *Experientia* **24**, 1288 (1968). *Review*: M. Chatterjee, H. N. Munro in *Vitamins and Hormones* vol. **35**, P. L. Munson *et al.*, Eds. (Academic Press, New York, 1977) pp 149-208.

4656. Hecameg®. [115457-83-5] Methyl-α-D-glucopyranoside 6-(heptylcarbamate); 6-*O*-(*N*-heptylcarbamoyl)-methyl-α-D-glucopyranoside. $C_{15}H_{29}NO_7$; mol wt 335.40. C 53.72%, H 8.72%, N 4.18%, O 33.39%. Glycolipid surfactant. Prepn: D. Plusquellec *et al.*, *Anal. Biochem.* **179**, 145 (1989). Kinetics of surfactant exchange; effect on micelle aggregation: M. Frindi *et al.*, *J. Phys. Chem.* **96**, 8137 (1992). Crystal structure: S. B. Engelsen *et al.*, *Carbohydr. Res.* **264**, 161 (1994). Assessment as detergent in biochemical applications: M. B. Ruiz *et al.*, *Biochim. Biophys. Acta* **1193**, 301 (1994). Use in protein purification: M. C. Sanders *et al.*, *J. Biol. Chem.* **271**, 2651 (1996); P. Bron *et al.*, *J. Mol. Biol.* **287**, 117 (1999).

Orthorhombic white needles, mp 108-110°. d 1.229 g/cm³. $[\alpha]_D^{22}$ +89 ±2° (c = 9.42 x 10^{-3} in water). Neutral, non-ionic.

USE: Detergent, primarily for protein purification.

4657. Hecogenin. [467-55-0] (3β,5α,25R)-3-Hydroxyspirostan-12-one. $C_{27}H_{42}O_4$; mol wt 430.63. C 75.31%, H 9.83%, O 14.86%. A steroidal sapogenin which has been isolated from plants, particularly from numerous *Agave* species. Isoln: Marker *et al.*, *J. Am. Chem. Soc.* **69**, 2167 (1947); Rubin, US 3303186 (1967). Synthesis and configuration: Mazur *et al.*, *J. Am. Chem. Soc.* **82**, 5889 (1960). Separation of an optically inactive product from sapogenin mixtures: Cardenas, US 3013010 (1961 to Searle). *Review*: L. F. Fieser, M. Fieser, *Steroids* (Reinhold, New York, 1959) pp 667-671 *sqq.*

Crystals from acetone, mp 264-266°. $[\alpha]_D$ +8° (CHCl₃) (Mazur *et al.*); mp 245°, 253°, 268° (Marker *et al.*); mp 240-245°, 245-250°. $[\alpha]_D$ ±0° (CHCl₃) (Cardenas).

Pseudohecogenin. [11005-20-2] (3β,5α,25R)-3,26-Dihydroxyfurost-20(22)-en-12-one. $C_{27}H_{42}O_4$; mol wt 430.63. Prepn from hecogenin: Cameron *et al.*, *J. Chem. Soc.* **1955**, 2807; Wall, Serota, US 2870143 (1959 to USDA). Plates from aq acetone, mp 190-191°. $[\alpha]_D^{20}$ +103° (c = 1.5 in chloroform), +96° (c = 1 in dioxane). uv max (ethanol): 213 nm (ε 6400).

USE: In prepn of steroidal hormones.

4658. Hectorite. [12173-47-6] Strese & Hofmann's hectorite. Swelling and gelling clay of the montmorillonite group. Approx formula: $Na_{0.67}(Mg,Li)_6Si_8O_{20}(OH,F)_4$. Some analyses show neither Li nor F, whether by oversight or by variation in the mineral is uncertain: M. H. Hey, *Mineral Species and Varieties* (Brit. Museum, London, 2nd ed., 1962) p 205.

Caution: Dust can be irritating to respiratory tract.

USE: In the chill-proofing of beer: Shaler *et al.*, **US 3100707** (1963 to American Tansul).

4659. Hederagenin. [465-99-6] ($3\beta,4\alpha$)-3,23-Dihydroxy-olean-12-en-28-oic acid; caulosapogenin; melanthigenin. $C_{30}H_{48}O_4$; mol wt 472.71. C 76.23%, H 10.24%, O 13.54%. Occurs as glycoside in many saponins, *see* α-Hederin. Isoln and structure: van der Haar, *Arch. Pharm.* **250**, 424 (1912); Ruzicka *et al.*, *Helv. Chim. Acta* **28**, 380 (1945); Haynes *et al.*, *J. Chem. Soc.* **1963**, 744. Identity with melanthigenin: Mustafa, Soliman, *ibid.* **1943**, 70. Identity with caulosapogenin: McShefferty, Stenlake, *ibid.* **1956**, 2314. Molecular and crystal structure: R. Roques *et al.*, *Acta Crystallogr.* **B34**, 1634 (1978). *See also* α- and β-Amyrin and Oleanolic Acid.

Crystals from alc, mp 332-334°. $[\alpha]_D^{20}$ +81° (c = 0.7 in pyridine). Freely sol in pyridine; sol in chloroform-alcohol mixtures; slowly sol in alcohol; practically insol in water; sol in dil alcoholic NaOH, but not in water solns of alkalies.

Diacetate. mp 172-174°. $[\alpha]_D^{20}$ +64° ($CHCl_3$).
Methyl ester. mp 240°. $[\alpha]_D^{23}$ +76° (c = 0.8 in $CHCl_3$).

4660. α-Hederin. [27013-91-8] ($3\beta,4\alpha$)-3-[2-*O*-(6-Deoxy-α-L-mannopyranosyl)-α-L-arabinopyranosyl]oxy]-23-hydroxy-olean-12-en-28-oic acid; helixin (the saponin). $C_{41}H_{66}O_{12}$; mol wt 750.97. C 65.58%, H 8.86%, O 25.57%. Isoln from ivy leaves *(Hedera helix* L., *Araliaceae):* van der Haar, *Arch. Pharm.* **250**, 424 (1912); **251**, 632, 650; *idem, Biochem. Z.* **76**, 335 (1916); *idem, Ber.* **54**, 3142 (1921). Isoln and structure: Tschesche *et al.*, *Z. Naturforsch.* **20b**, 708 (1965).

Precipitated from ethanol by addition of ether, mp 256-259°. $[\alpha]_D^{20}$ +14.5° (c = 0.92 in methanol).

4661. Helenalin. [6754-13-8] (3a*S*,4*S*,4a*R*,7a*R*,8*R*,9a*R*)-3,-3a,4,4a,7a,8,9,9a-Octahydro-4-hydroxy-4a,8-dimethyl-3-methyl-eneazuleno[6,5-*b*]furan-2,5-dione; 6α,8β-dihydroxy-4-oxoambrosa-2,11(13)-dien-12-oic acid 12,8-lactone. $C_{15}H_{18}O_4$; mol wt 262.31. C 68.68%, H 6.92%, O 24.40%. Pseudoguaianolide sesquiterpenoid lactone from *Helenium autumnale* L., *H. amarum* (Raf.) H. Roch, *H. microcephalum* DC., *Compositae.* Isoln: Clark, *J. Am. Chem. Soc.* **58**, 1982 (1936); and characterization: Adams, Herz, *ibid.* **71**, 2546 (1949). Structure: Büchi, Rosenthal, *ibid.* **78**, 3860 (1956); Barton,

de Mayo, *Q. Rev. Chem. Soc.* **11**, 189 (1957); Herz, *J. Org. Chem.* **27**, 4043 (1962). Abs config studies: Herz, Kagan, *ibid.* **32**, 216 (1967). Synthesis of racemic form: Y. Ohfune *et al.*, *J. Am. Chem. Soc.* **100**, 5946 (1978); M. R. Roberts, R. H. Schlessinger, *ibid.* **101**, 7626 (1979); C. H. Heathcock *et al.*, *ibid.* **104**, 1907 (1982). Toxicity study: D. A. Witzel *et al.*, *Am. J. Vet. Res.* **37**, 859 (1976).

Bitter, sternutative crystals from benzene, mp 167-168°. $[\alpha]_D^{25}$ −102.8° (c = 3.64 in 95% ethanol). uv max: 223 nm (ε 11900). Slightly sol in water; sol in alcohol, chloroform, hot benzene. LD_{50} in mice, rats (mg/kg): 150, 125 orally (Witzel).

Acetylhelenalin. $C_{17}H_{20}O_5$. Crystals from aq methanol, mp 184°. uv max: 221, 316 nm (ε 12600, 61).

4662. Helenynolic Acid. [7309-58-2] (9*S*,10*E*)-9-Hydroxy-10-octadecen-12-ynoic acid. $C_{18}H_{30}O_3$; mol wt 294.44. C 73.43%, H 10.27%, O 16.30%. From seed oil of *Helichrysum bracteatum* Andr., *Compositae:* R. G. Powell *et al.*, *J. Am. Oil Chem. Soc.* **42**, 165 (1965). Structure: *eidem, J. Org. Chem.* **30**, 610 (1965). Absolute configuration: J. C. Craig *et al.*, *ibid.* 4342. Partial synthesis of racemic helenynolic acid: H. R. S. Conacher, Gunstone, *Lipids* **5**, 137 (1970). Synthesis of (±)-form: T. B. Patrick, G. F. Melm, *J. Org. Chem.* **44**, 645 (1979).

Methyl ester. [870-09-7] $C_{19}H_{32}O_3$. uv max (isooctane): 228, 238 nm (ε 17400, 14300). $[\alpha]_{600}^{26}$ −7° (c = 3.6 in ethanol).

4663. Helicin. [618-65-5] 2-(β-D-Glucopyranosyloxy)benz-aldehyde; salicylaldehyde β-D-glucoside. $C_{13}H_{16}O_7$; mol wt 284.26. C 54.93%, H 5.67%, O 39.40%. Prepn by the oxidation of salicin with dil nitric acid: Schiff, *Ann.* **154**, 19 (1870); from salicylaldehyde + *O*-tetraacetyl-α-glucosidyl bromide: Robertson, Waters, *J. Chem. Soc.* **1930**, 2729. ORD and stereochemical studies: Tsuzuki *et al.*, *Bull. Chem. Soc. Jpn.* **44**, 526 (1971).

Needles with 0.75 mol H_2O from H_2O; mp 175-176° when dried at 100°. $[\alpha]_D^{20}$ −60° (c = 1.4). One gram dissolves in 55 ml water; freely sol in hot water, alcohol. Forms compds with urea, thiourea, certain amino acids.

Tetraacetate. $C_{21}H_{24}O_{11}$. Needles from alc, mp 142°. $[\alpha]_D^{20}$ −37° (acetone).

Note: The same formula is ascribed to **spirein**, found in *Spiraea camtschatica* Pall. and in *S. ulmaria* L., *Rosaceae.* Emulsin hydrolyzes helicin and spirein, yielding D-glucose and salicylaldehyde.

4664. Heliosupine. [32728-78-2] 1,5-Dideoxy-4-*C*-methyl-3-*C*-[[[(1*S*,7a*R*)-2,3,5,7a-tetrahydro-1-[[(2*Z*)-2-methyl-1-oxo-2-buten-1-yl]oxy]-1*H*-pyrrolizin-7-yl]methoxy]carbonyl] D-erythro-pentitol; 2-methyl-2-butenoic acid 7-[[2,3-dihydroxy-2-(1-hydroxyethyl)-3-methyl-1-oxobutoxy]methyl]-2,3,5,7a-tetrahydro-1*H*-pyrrolizin-7-yl ester; cynoglossophine. $C_{20}H_{31}NO_7$; mol wt 397.47. C 60.44%, H 7.86%, N 3.52%, O 28.18%. Hepatotoxic pyrrolizidine alkaloid isolated from *Heliotropium supinum* L., *Boraginaceae:* Denisova *et al.*, *Dokl. Akad. Nauk SSSR* **93**, 59 (1953); from *Cynog-*

lossum officinale L., *Boraginaceae:* Man'ko, Borisyuk, *Ukr. Khim. Zh.* **23**, 362 (1957), *C.A.* **52**, 2188a (1958); Man'ko, Marchenko, *Khim. Prir. Soedin.* **7**, 537 (1971), *C.A.* **75**, 126598t (1971). Identity with cynoglossophine: Man'ko, *Ukr. Khim. Zh.* **25**, 627 (1959), *C.A.* **54**, 12494d (1960). Structure: Crowley, Culvenor, *Aust. J. Chem.* **12**, 694 (1959). Biosynthesis studies: Crout, *J. Chem. Soc. C* **1967**, 1233. *Reviews: see* Lasiocarpine.

Colorless gum. $[\alpha]_D^{20}$ $-4.3°$ (c = 5.1 in ethanol).
Methyl ether *see* Lasiocarpine.

4665. Helium. [7440-59-7] He; at. wt 4.002602; at. no. 2. Group VIIIA (18), also known as Group 0. A noble gas characterized by an electronic structure in which the outer p subshell is entirely filled. Natural isotopic mixture (mass numbers): 4 (99.999862%), 3 (1.38 × 10^{-4}%). Known artificial radioactive isotopes: 5-8. Longest-lived isotope: ^{6}He (T$_{1/2}$ 806.7 msec, β-emitter). Abundance in igneous rock of earth's crust: 3 × 10^{-3} ppm by wt; concentration in air: 5.24 ppm by vol. Identified in the spectrum of the sun's chromosphere by Lockyer and Frankland in 1868. Obtained by Hillebrand in 1890 by heating uranium minerals and identified by Ramsay in 1895. Found in natural gas from which it is extracted on a commercial scale. Produced in the decay of radioactive elements: 1 kg of uranium in its conversion into 865 g of lead forms 756 L of helium; also produced in nature by the bombardment of beryllium, lithium, and other light elements with cosmic rays, x-rays and high-speed protons and deuterons. Monograph: G. A. Cook, *Argon, Helium and the Rare Gases* (Interscience, New York, 1961). *Reviews:* Cockett, Smith, "The Monatomic Gases" in *Comprehensive Inorganic Chemistry* vol. 1, J. C. Bailar, Jr. *et al.*, Eds. (Pergamon Press, Oxford, 1973) pp 139-211; E. Cook, *Science* **206**, 1141-1146 (1979); S.-C. Hwang, W. R. Weltmer, Jr. in *Kirk-Othmer Encyclopedia of Chemical Technology* vol. **13** (John Wiley & Sons, 4th ed., 1995) pp 1-38; *Chemistry of the Elements* N. N. Greenwood, A. Earnshaw, Eds. (Pergamon Press, New York, 1984) pp 1042-1059. Review of use in inductively coupled plasma-mass spectrometry: S. F. Durrant, *Fresenius J. Anal. Chem.* **347**, 389-392 (1993).

Colorless, odorless, tasteless, monatomic, inert gas. *Non-flammable.* Will form compds with highly electronegative elements such as O, F, Cl. Cannot be frozen by lowering the temp at ordinary press.; no triple point. Very slightly sol in water (ml/100 ml): 0.97 at 0°; 1.08 at 50°. Trouton's const 4.64.

^{4}**He.** Critical temp 5.2014 K, critical press 227.5 kPa, critical d 69.64 kg/m^3. Gas: d^0 (101.3 kPa) 0.17850 kg/m^3, d (normal bp) 16.89 kg/m^3. Liquid: normal bp $-268.926°$, d (normal bp) 125.0 kg/m^3, heat of vaporization (normal bp) 81.70 J/mol. Two liquid forms exist: He I above ~2.2 K; He II below ~2.2 K. He II is a superconducting liquid; has very low viscosity; superfluid.

^{3}**He.** Critical temp 3.324 K, critical press 116.4 kPa, critical d 41.3 kg/m^3. Gas: d^0 (101.3 kPa) 0.1347 kg/m^3, d (normal bp) 23.64 kg/m^3. Liquid: normal bp $-269.959°$, d (normal bp) 58.9 kg/m^3, heat of vaporization (normal bp) 25.48 J/mol.

Caution: Can act as a simple asphyxiant by displacing air. *See: Matheson Gas Data Book* (Matheson Co., Inc., 4th ed., East Rutherford, NJ, 1966) pp 249-253.

USE: Liquid helium (the most volatile liq known) as cryogen for the production of low temps, in MRI machines. Gas in manuf of semiconducting devices; in detection devices for leaks in vacuum systems; in Ne-He lasers; in gas mixtures as the working fluid in plasma devices; in mixtures with Ne and Ar in Geiger counters. *Q-Gas*, a mixture of 98.7% He and 1.3% butane, has been used as a filling for gas-flow Geiger counters. Gas as a shield in gas tungsten-arc welding, in metal processing; substitute for N$_2$ in synthetic breathing gas for deep sea divers and workers in high pressure conditions; coolant in high temp nuclear reactors; carrier gas in gas-liquid and gas-solid chromatography; inert diluent; to create inert

atmosphere; to fill balloons and airships, lifting power is 0.93 if hydrogen is taken as 1.00.

4666. Helonias. False unicorn; blazing star; starwort. Rhizome and roots of *Chamaelirium luteum* (L.) A. Gray, *Liliaceae. Habit.* Ontario and Eastern U.S.

4667. Helveticoside. [630-64-8] (3β,5β)-3-[(2,6-Dideoxy-β-D-*ribo*-hexopyranosyl)oxy]-5,14-dihydroxy-19-oxocard-20(22)-enolide; erisimin; erysimin; alleoside A. C$_{29}$H$_{42}$O$_9$; mol wt 534.65. C 65.15%, H 7.92%, O 26.93%. A β-glycoside consisting of one mole strophanthidin and one mole D-digitoxose. Isoln from *Erysimum helveticum* (Jacq.) DC. and *Erysimum crepidifolium* Reichenb., *Cruciferae*, and structure: Nagata *et al.*, *Helv. Chim. Acta* **40**, 41 (1957); *eidem*, *Festschr. Arthur Stoll* **1957**, 715; Gmelin, **DE 1221764** (1966). From *Erysimum* spp.: Kowalewski, *Helv. Chim. Acta* **43**, 1314 (1960). Toxicity study: Graebner, Giesel, *Arzneim.-Forsch.* **22**, 1854 (1972).

Dihydrate. Needles from dil methanol, mp 153-157°. $[\alpha]_D^{25}$ $+30.7°$ (c = 1.5 in methanol); $[\alpha]_D^{24}$ $+26.0°$ (c = 1.1 in chloroform). uv max (ethanol): 217, 304 nm (log ε 4.27, 1.55). *Poisonous.* LD$_{100}$ i.v. in cats: 0.104 mg/kg (Graebner, Giesel).

3′,4′-Diacetate. C$_{33}$H$_{46}$O$_{11}$. Prisms from methanol + ether, mp 237-244°. $[\alpha]_D^{25}$ $+41.3°$ (chloroform).

4668. Helvolic Acid. [29400-42-8] (4α,6β,8α,9β,13α,14β,-16β,17Z)-6,16-Bis(acetyloxy)-3,7-dioxo-29-nordammara-1,17-(20),24-trien-21-oic acid; (Z)-6β,16β-dihydroxy-3,7-dioxo-29-nor-8α,9β,13α,14β-dammara-1,17(20),24-trien-21-oic acid diacetate; fumigacin. C$_{33}$H$_{44}$O$_8$; mol wt 568.71. C 69.70%, H 7.80%, O 22.51%. Antibiotic substance of the fusidane class. Produced by *Aspergillus fumigatus:* Waksman *et al.*, *J. Bacteriol.* **45**, 233 (1943); by *A. fumigatus* mut. *helvola:* Chain *et al.*, *Br. J. Exp. Pathol.* **24**, 108 (1943). Structural studies: Okuda *et al.*, *Chem. Pharm. Bull.* **12**, 121 (1964); Oxley, *Chem. Commun.* **1966**, 729; Okuda *et al.*, *Tetrahedron Lett.* **1967**, 2295. Revised structure and stereochemistry: Iwaki *et al.*, *Chem. Commun.* **1970**, 1119. Chemical and biological data: Reshetova, *Antibiotiki* **14**, 554 (1969). Review of literature until 1960: Wilson, "Miscellaneous *Aspergillus* Toxins" in *Microbial Toxins* vol. **VI**, A. Ciegler *et al.*, Eds. (Academic Press, New York, 1971) p 265.

Needles from methanol, mp 215°. $[\alpha]_D^{25}$ $-121°$ (chloroform). uv max (ethanol): 231 nm (log ε 4.24). Very slightly sol in water; slightly sol in petr ether, methanol, ethanol. More sol in hot methanol, ethanol; sol in chloroform, acetone, ethyl acetate, benzene, pyridine, glacial acetic acid, dioxane, ether.

Methyl ester. Methyl helvolate. $C_{34}H_{46}O_8$. Crystals from methanol, mp 257°. $[\alpha]_D^{25} -140°$ (chloroform).

4669. Hematein. [475-25-2] 6a,7-Dihydro-3,4,6a,10-tetrahydroxybenz[b]indeno[1,2-d]pyran-9(6H)-one; hydroxybrasilein; hydroxybrazilein. $C_{16}H_{12}O_6$; mol wt 300.27. C 64.00%, H 4.03%, O 31.97%. Not to be confused with hematin. From hematoxylin or logwood extract and NH_3 by treatment with air: Engels et al., J. Chem. Soc. 93, 1115 (1908); Rolland, Teintex 3, 261, 322, 460 (1938); Justin-Mueller, Melliand Textilber. 30, 26, 63 (1949), C.A. 46, 8375h (1952).

Reddish-brown crystals with yellowish-green metallic luster. mp >200°; also stated as 250° with decompn. Sol in about 1700 parts water; slightly sol in alcohol, ether; insol in benzene, chloroform; freely sol in ammonia with brownish-violet color and in dil NaOH with bright red color. Forms salts with heavy metals.
USE: As an indicator like hematoxylin; for staining animal tissue, particularly cell nuclei.

4670. Hematin. [15489-90-4] (SP-5-13)-[7,12-Diethenyl-3,-8,13,17-tetramethyl-21H,23H-porphine-2,18-dipropanoato(4−)-$\kappa N^{21},\kappa N^{22},\kappa N^{23},\kappa N^{24}$]hydroxyferrate(2−) hydrogen (1:2); [dihydrogen 3,7,12,17-tetramethyl-8,13-divinyl-2,18-porphinedipropionato(2−)]hydroxyiron; hydroxy[dihydrogen protoporphyrin IX-ato-(2−)]iron; ferriheme hydroxide; ferriporphyrin hydroxide; ferriprotoporphyrin basic; hydroxyhemin; phenodin. $C_{34}H_{33}FeN_4O_5$; mol wt 633.51. C 64.46%, H 5.25%, Fe 8.82%, N 8.84%, O 12.63%. Found in the body in pathological conditions, e.g., after phosgene poisoning, in pernicious anemia. Prepn: Fischer-Orth, Die Chemie des Pyrrols II, 1, 386 (Leipzig, 1937). Brief review including structure: J. E. Falk, Porphyrins and Metalloporphyrins (Elsevier, New York, 1964) pp 17, 23, 46. Toxicology: D. L. Lips et al., Toxicol. Lett. 2, 329 (1978). Use in treatment of hepatic porphyria: C. J. Watson et al., Ann. Intern. Med. 79, 80 (1973); G. J. Dhar et al., ibid. 83, 20 (1975); J. M. Lamon et al., Medicine 58, 252 (1979).

Solvated crystals from pyridine, dry at 40° in vacuo. Freely sol in dil solns of alkali hydroxides, slightly sol in hot pyridine. Unstable. Absorption max (10% aq NaOH): 580 nm (E_{mM} 10.5). LD_{50} i.v. in rats: 4.32 mg/100 g (Lips).

4671. Hematoporphyrin. [14459-29-1] 7,12-Bis(1-hydroxyethyl)-3,8,13,17-tetramethyl-21H,23H-porphine-2,18-dipropanoic acid; 7,12-bis(1-hydroxyethyl)-3,8,13,17-tetramethyl-2,18-porphinedipropionic acid; 1,3,5,8-tetramethyl-2,4-bis(α-hydroxyethyl)-porphine-6,7-dipropionic acid; hematoporphyrin IX; Photodyn. $C_{34}H_{38}N_4O_6$; mol wt 598.70. C 68.21%, H 6.40%, N 9.36%, O 16.03%. Prepd from hemin by the action of hydrobromic acid in glacial acetic acid: Fischer-Orth, Die Chemie des Pyrrols II, 1, 421 (Leipzig, 1937). Review and prepn: J. E. Falk, Porphyrins and Metalloporphyrins (Elsevier, New York, 1964) pp 175-177.

Deep red crystals. Absorption max (0.1N KOH): 615.5, 565, 534.4, 499.5 nm. Insol in water. Sol in alc. Sparingly sol in ether, chloroform.
Hydrochloride monohydrate. [6033-08-5] Sensibion. $C_{34}H_{38}$-N_4O_6·HCl·H_2O; mol wt 653.17. Prepn of a stable product: Woods, Steigman, US 2858320 (1958 to Baxter Labs.). Dark red crystals.
THERAP CAT: Antidepressant.

4672. Hematoxylin. [517-28-2] (6aS,11bR)-7,11b-Dihydrobenz[b]indeno[1,2-d]pyran-3,4,6a,9,10(6H)-pentol; hematoxiline; hydroxybrazilin; hydroxybrasilin. $C_{16}H_{14}O_6$; mol wt 302.28. C 63.58%, H 4.67%, O 31.76%. From the heart-wood of logwood (Haematoxylon campechianum Linn., Leguminosae): Chevreul, Ann. Chim. Phys. 82, 54, 126 (1810). Structure: Perkin, Robinson, J. Chem. Soc. 93, 489 (1908). Synthesis: Dann, Hofmann, Angew. Chem. 75, 1125 (1963); Morsingh, Robinson, Tetrahedron 26, 281 (1970); Kirkiacharian, Billet, Bull. Soc. Chim. Fr. 1972, 3292. Stereochemistry: Craig et al., J. Org. Chem. 30, 1573 (1965). Spectroscopic and physico-chemical properties: C. Bettinger, H. W. Zimmerman, Histochemistry 95, 279 (1991). Prepn for staining: R. B. Bosma et al., J. Histotechnol. 16, 371 (1993). Use in determn of ruthenium: A.-A. Y. El-Sayed, Fresenius J. Anal. Chem. 349, 830 (1994). Review: Robinson, Bull. Soc. Chim. Fr. 1958, 125. Review of theory and application as histological stain: B. B. Hrapchak, Am. J. Med. Technol. 42, 371-379 (1976).

Trihydrate. White to yellowish crystals; redden on exposure to light, mp 100-140°; also stated as 140°. Slightly sol in cold water, ether; sol in hot water, hot alc, also in alkali hydroxides, borax, glycerol. Its solns darken on standing.
USE: Chiefly as a stain in microscopy; also in manufacture of ink.

4673. Hematoxylon. Logwood. Heart-wood of Haematoxylon campechianum L., Leguminosae. Habit. Central America; grown in the West Indies. Constit. About 10% hematoxylin; tannin, resin.
USE: Largely as dye.

4674. Heme. [14875-96-8] (SP-4-2)-[7,12-Diethenyl-3,8,-13,17-tetramethyl-21H,23H-porphine-2,18-dipropanoato(4−)-$\kappa N^{21},\kappa N^{22},\kappa N^{23},\kappa N^{24}$]ferrate(2−) hydrogen (1:2); [dihydrogen 3,-7,12,17-tetramethyl-8,13-divinyl-2,18-porphinedipropionato(2−)]-iron; 1,3,5,8-tetramethyl-2,4-divinylporphine-6,7-dipropionic acid ferrous complex; ferroheme; hem; protoheme; protoheme IX; reduced hematin; ferroprotoporphyrin. $C_{34}H_{32}FeN_4O_4$; mol wt 616.50. C 66.24%, H 5.23%, Fe 9.06%, N 9.09%, O 10.38%. Heme occurs free in tissues in the presence of certain pathological conditions, and in normal tissues; it occurs as the prosthetic group of a number of hemoproteins. It has been identified as the prosthetic group of hemoglobins, erythrocruorins (the hemoglobin analog of many invertebrates), myoglobins, some peroxidases, catalases, and cytochromes b. It is the color-furnishing portion of hemoglobin. Obtained when a soln of hematin in alkali is reduced in absence of nitrogenous substances: Bertin-Sans, de Moitessier, *Compt. Rend.* **114**, 923 (1892); Dhéré *et al.*, *ibid.* **165**, 515 (1917). Synthesis: Fischer-Orth, *Die Chemie des Pyrrols* **II**, 1, 384 (Leipzig, 1937). Biosynthesis: Shemin, *Naturwissenschaften* **57**, 185 (1970). *Review:* J. E. Falk, *Porphyrins and Metalloporphyrins* (Elsevier, New-York, 1964) pp 8, 94, 183. Comprehensive monograph: Chance *et al.*, *Hemes and Hemoproteins* (Academic Press, New York, 1966) 624 pp; *Handbook of Experimental Pharmacology* **vol. 44**, entitled "Heme and Hemoproteins", F. DeMatteis, W. N. Aldridge, Eds. (Springer-Verlag, New York, 1978) 449 pp.

Fine brown needles with a dark violet sheen. Absorption max in phosphate buffer at pH 7: ~550, 575 nm (E_{mM}^{572} 5.5). Sparingly sol in glacial acetic acid; freely sol in the presence of oxygen. Very unstable.

4675. Hementin. [87041-58-5] Anticoagulant enzyme from the salivary glands of the South American giant leech *Haementeria ghiliarii.* Extraction: R. T. Sawyer *et al.*, US 4390630 (1983 to Univ. California, Berkeley). Anticoagulant, fibrinolytic properties: A. Z. Budzynski *et al.*, *Proc. Soc. Exp. Biol. Med.* **168**, 266 (1981). Purification and identification as single polypeptide chain with mol wt approx 120000: S. M. Malinconico *et al.*, *J. Lab. Clin. Med.* **103**, 44 (1984). Proteolytic degradation of fibrinogen by hementin: *eidem, ibid.* **104**, 842 (1984). Protease inhibition: E. H. Murer *et al.*, *Thromb. Haemostasis* **51**, 24 (1984).

4676. Hemerythrin. A non-heme, oxygen-carrying protein found in members of four invertebrate phyla: sipunculids, polychaetes, priapulids and brachiopods. Oxygenated form is called *oxyhemerythrin.* Isoln from the sipunculid, *Golfingia goldii* (or *Phascolosoma goldii*): Klotz *et al.*, *Arch. Biochem. Biophys.* **68**, 284 (1957). Consists of eight subunits, each containing two ferrous ions which form a complex with one molecule of oxygen: Klotz, Keresztes-Nagy, *Biochemistry* **2**, 445, 923 (1963). Each subunit is built up from 113 amino acids. Amino acid sequence: Klippenstein *et al.*, *ibid.* **7**, 3868 (1968); Klippenstein, *ibid.* **11**, 372 (1972). Structure of trimeric hemerythrin: J. L. Smith *et al.*, *Nature* **303**, 86 (1983). *Reviews:* Okamura, Klotz, "Hemerythrin" in *Inorganic Biochemistry* **vol. 1**, G. L. Eichhorn, Ed. (Elsevier, New York, 1973) pp 320-343; Klotz *et al.*, *Science* **192**, 335-344 (1976); J. S. Loehr, T. M. Loehr in *Advances in Inorganic Biochemistry* **vol. 1**, G. L. Eichhorn, L. G. Marzilli, Eds. (Elsevier, New York, 1979) pp 235-252.

Hemerythrin is colorless. Oxyhemerythrin forms violet-pink crystals. Spectral data: Garbett *et al.*, *Arch. Biochem. Biophys.* **135**, 419 (1969).

4677. Hemicelluloses. Large group of polysaccharides found, in association with lignin, in the primary and secondary cell walls of all plants and of some seaweeds. Whether lignin and hemicellulose are chemically bonded or lignin mechanically entraps hemicellulose molecules is still unknown. There are variations in the hemicellulosic composition of plants and even between the organs of the same plant. Hemicellulosic composition of plants changes with growth and maturation and is influenced also by environmental factors during growth. Principal sugar residues present in hemicellulose are: D-xylose, D-glucose, D-galactose, D-mannose, L-arabinose, D-glucuronic acid, D-galacturonic acid, 4-*O*-methyl-D-glucuronic acid, L-rhamnose and L-fucose. The most ubiquitous and abundant hemicelluloses are the **xylans** which are composed of linear and/or branched chains of β-(1 → 4) linked D-xylopyranosyl units. Hemi-

celluloses from woods: T. E. Timell, *Adv. Carbohydr. Chem.* **19**, 247 (1964); *idem, ibid.* **20**, 409 (1965). Hemicelluloses from grasses and cereals: K. C. B. Wilke, *ibid.* **36**, 215 (1979).

4678. Hemicholinium. [16478-59-4] 2,2'-[1,1'-Biphenyl]-4,4'-diylbis[2-hydroxy-4,4-dimethylmorpholinium]; 2,2'-(4,4'-biphenylylene)bis[2-hydroxy-4,4-dimethylmorpholinium]. [$C_{24}H_{34}$-N_2O_4]$^{2+}$. Specific, competitive inhibitor of the sodium-dependent high-affinity choline uptake system (HACU) localized on cholinergic presynaptic nerve terminals, resulting in the inhibition of acetylcholine synthesis. Prepn of seco tautomer: J. P. Long, F. W. Schueler, *J. Am. Pharm. Assoc.* **43**, 79 (1954). *In vitro* pharmacology and revised hemiacetal structure: F. W. Schueler, *J. Pharmacol. Exp. Ther.* **115**, 127 (1955). Structure-activity studies: R. P. DiAugustine, V. B. Haarstad, *Biochem. Pharmacol.* **19**, 559 (1970); J. G. Cannon *et al.*, *Pharm. Res.* **5**, 359 (1988). Pharmacology: J. J. Freeman *et al.*, *J. Pharmacol. Exp. Ther.* **210**, 91 (1979); J. J. Hagan *et al.*, *Psychopharmacology* **98**, 347 (1989). Receptor binding studies in human brain: J. Pascual *et al.*, *J. Neurochem.* **54**, 792 (1990); in rat brain: *eidem, Pharmacol. Res.* **24**, 345 (1991). Potential uses of tritium labeled compound: T. W. Vickroy *et al.* in *Synth. Appl. Isotop. Labeled Compd. 1985*, R. R. Muccino, Ed. (Elsevier, Amsterdam, 1986) pp 145-150; H. K. Happe, L. C. Murrin, *J. Neurochem.* **60**, 1191 (1993).

Dibromide. [312-45-8] Hemicholinium-3; HC-3. $C_{24}H_{34}Br_2$-N_2O_4. mp 180°. Moderately soluble in water, soluble in ethanol. Occurs as the monohydrate, white solid from ethanol/ether, mp 226-228° (dec). LD$_{50}$ i.p. in female mice: 0.048-0.082 mg/kg (DiAugustine, Haarstad).

USE: Research tool as cholinergic probe, to deplete acetylcholine stores.

4679. Hemin. [16009-13-5] (*SP*-5-13)-Chloro[7,12-diethenyl-3,8,13,17-tetramethyl-21*H*,23*H*-porphine-2,18-dipropanoato-(4−)-κN^{21},κN^{22},κN^{23},κN^{24}]ferrate(2−) hydrogen (1:2); chloro[dihydrogen 3,7,12,17-tetramethyl-8,13-divinyl-2,18-porphinedipropionato(2−)]iron; chlorohemin; 1,3,5,8-tetramethyl-2,4-divinylporphine-6,7-dipropionic acid ferrichloride; Teichmann's crystals; ferriheme chloride; ferriprotoporphyrin chloride; ferriporphyrin chloride. $C_{34}H_{32}ClFeN_4O_4$; mol wt 651.95. C 62.64%, H 4.95%, Cl 5.44%, Fe 8.57%, N 8.59%, O 9.82%. Prepd from hemoglobin soln by heating with acetic acid and sodium chloride. Practical procedure for its prepn from ox blood: Schalfejeff, *J. Russ. Phys. Chem. Soc.* 1885, 30; *Ber.* **18**, 232 (1885); Gattermann-Wieland, *Praxis des Organischen Chemikers* 23rd ed., p 407; Fischer-Orth, *Die Chemie des Pyrrols* **II**, 1, 377 (Leipzig, 1937); H. Fischer, *Org. Synth.* **coll. vol. III**, 442 (1955); Labbe, Nishida, *Biochim. Biophys. Acta* **26**, 437 (1957). Biosynthesis: Karlzeile, *Angew. Chem.* **66**, 729 (1954). *Review:* Stoll, *Experientia* **4**, 6 (1948); H. H. Inhoffen, *Naturwissenschaften* **55**, 457 (1968); W. S. Caughey in *Inorganic Biochemistry* **vol. 2**, G. L. Eichhorn, Ed. (Elsevier, New York, 1973) pp 797-831.

Long, thin blades from glacial acetic acid or from chloroform-pyridine-acetic acid, appearing brown in transmitted light and steel-blue in reflected light. Sinters at 240° but is not melted at 300°. Absorption spectrum: Fischer-Orth, *loc. cit.* Freely sol in dil ammonia water, also in solns of sodium hydroxide with hematin formation, *i.e.*, the chlorine is displaced by an OH group. Practically insol in carbonate solns, dil acid solns. Sol in strong organic bases such as trimethylamine, *p*-toluidine, dimethylaniline. Sol in concd H_2SO_4 with loss of Fe. Sparingly sol in 70-80% alc. Practically insol but stable in water.

Dimethyl ester. $C_{36}H_{36}ClFeN_4O_4$. Needles from benzene, not melted at 300°. Freely sol in acetic acid, benzene, chloroform, acetone.

Diethyl ester. $C_{38}H_{40}ClFeN_4O_4$. Crystals, freely sol in chloroform.

4680. Hemipyocyanine. [528-71-2] 1-Phenazinol; α-hydroxyphenazine; 1-hydroxy-5,10-diazoanthracene; pyoxanthose. $C_{12}H_8N_2O$; mol wt 196.21. C 73.46%, H 4.11%, N 14.28%, O 8.15%. Found in old cultures of *Pseudomonas pyocyanea.* Isoln: Fordos, *Compt. Rend.* **56**, 1128 (1863). Prepd from pyocyanine by the action of 2% NaOH: Wrede, Strack, *Z. Physiol. Chem.* **140**, 12 (1924); by the reaction of pyrogallol monomethyl ether and *o*-phenylenediamine followed by demethylation of the formed α-methoxyphenazine: Surrey, *Org. Synth.* **coll. vol. III**, 753 (1955). Alternate syntheses: Hegedüs, *Helv. Chim. Acta* **33**, 766 (1950); Vivian, *Nature* **178**, 753 (1956).

Yellow needles from benzene, mp 159-160°. Sublimes easily *in vacuo.* Slightly sol in hot water; freely sol in the usual organic solvents except petr ether. Sol in aq alkaline solns with purplish-red color which turns yellow on neutralization. Forms red salts with mineral acids.

4681. Hemisulfur Mustard. [693-30-1] 2-[(2-Chloroethyl)-thio]ethanol; 2-chloro-2′-hydroxydiethyl sulfide; 2-chloroethyl 2-hydroxyethyl sulfide; β-chloroethyl β-hydroxyethyl thioether; mustard chlorohydrin; semi-mustard; semisulfur mustard. C_4H_9ClOS; mol wt 140.63. C 34.16%, H 6.45%, Cl 25.21%, O 11.38%, S 22.80%. Prepn: Ogston *et al., Trans. Faraday Soc.* **44**, 45 (1948); Grant, Kinsey, *J. Am. Chem. Soc.* **68**, 2075 (1946); Fuson, Ziegler, *J. Org. Chem.* **11**, 510 (1946); Rueggeberg *et al., ibid.* **13**, 110 (1948); Tsou *et al., ibid.* **26**, 4987 (1961).

Oily, somewhat hygroscopic liquid. $bp_{0.6}$ 100° (Fuson); $bp_{0.5-0.75}$ 87°; bp_6 44.5° (Rueggeberg). n_D^{20} 1.5188; $n_D^{24.5}$ 1.5205. Sol in water; miscible with bis(2-chloroethyl) sulfide, thiodiglycol. Unstable, forming sulfonium salts at room temp or in solns with ethanol or chloroform.

4682. Hemocyanins. A non-heme, oxygen carrying copper protein found in arthropods and mollusca of which **keyhole-limpet hemocyanin (KLH)** is an example. Dissolved in the hemolymph; not found in blood cells. Mol wt ranges from 4.5×10^5 to 1.3×10^7. One molecule of oxygen is bound by two copper atoms. The oxygenated form, **oxyhemocyanin** is blue, while **deoxyhemocyanin** is colorless. The copper can be removed reversibly to form **apohemocyanin.** Structure of snail, *Helix pomatia,* hemocyanin: J. E. Mellema, A. Klug, *Nature* **239**, 146 (1972). The role of copper in hemocyanin: R. Lontie, L. Vanquickenborne, *Met. Ions Biol. Syst.* **3**, 183 (1974). *Reviews:* K. E. van Holde, E. F. J. van Bruggen in *Subunits in Biological Systems* **vol. 5**, pt. A, S. N. Timasheff, G. D. Fasman, Eds. (Dekker, New York, 1971) pp 1-53; R. Lontie, R. Witters in *Inorganic Biochemistry* **vol. I**, G. L. Eichhorn, Ed. (Elsevier, New York, 1973) pp 344-358; E. F. J. van Bruggen, *Trends Biochem. Sci.* **5**, 185-8 (1980). Review of respiratory function: J. Bonaventura, C. Bonaventura, *Am. Zool.* **20**, 7-17 (1980); C. P. Mangnum, *ibid.* 19-38.

USE: KLH as an experimental antigen in animals.

4683. Hemoglobin. Hb; ferrohemoglobin. The major component of red blood cells which transports oxygen from the lungs to body tissues and facilitates the return transport of carbon dioxide. Mammalian hemoglobins have mol wts of about 64,500. Composed of four peptide chains called globins, each of which is bound to a heme, *q.q.v.* Normal human hemoglobin is composed of a pair of two identical chains. Iron is coordinated to four pyrrole nitrogens of protoporphyrin IX, and to an imidazole nitrogen of a histidine residue from the globin side of the porphyrin. The sixth coordination position is available for binding with oxygen and other small molecules. Called **oxyhemoglobin,** HbO_2, in the oxygenated form and **carboxyhemoglobin,** HbCO, when oxygen is displaced by carbon monoxide. Binds reversibly with oxygen while the heme iron remains in the ferrous state. Autoxidation is prevented by the cover of hydrophobic groups of the globin. When the iron in hemoglobin is oxidized from the ferrous to the ferric state the compd is called methemoglobin, *q.v.* and is accompanied by a loss of oxygen-binding capacity. Hemoglobin is usually prepd by separating the red blood corpuscles from the lighter plasma by centrifuging; the plasma is siphoned off, and on adding ether to the blood corpuscle paste, the cells burst. After another centrifugation to remove the ruptured cell envelopes, a clear red soln of the protein is obtained [Fieser, Fieser, *Org. Chem.* (New York, 3rd ed., 1956) p 455]. Prepn of cryst HbO_2 from washed horse or dog erythrocytes: Heidelberger, *J. Biol. Chem.* **53**, 31 (1922); *see also* Ferry, Green, *ibid.* **81**, 175 (1929); from human blood: Drabkin, *ibid.* **164**, 703 (1946). Structure studies: Muirhead, Perutz, *Nature* **199**, 633 (1963); Perutz *et al., ibid.* **219**, 131 (1968); Perutz, *Proc. R. Soc. London* **173B**, 113 (1969). Respiratory properties of hemoglobin and its function as a carrier of oxygen and carbon dioxide: Peters, Van Slyke, *Quantitative Clinical Chemistry* **vol. I** (Baltimore, 1932). Mechanism of action: Arnone in *Annu. Rev. Med.* **25**, 123-130 (1974). *Reviews:* Lemberg, Legge, *Hematin Compounds and Bile Pigments* (New York, 1949); F. W. Sunderman, *Hemoglobin: its Precursors and Metabolites* (Lippincott, Philadelphia, 1964); H. Lehmann, R. G. Huntsman, *Man's Hemoglobins* (ibid. 1966); Huisman, Schroeder, *New Aspects of the Structure, Function, and Synthesis of Hemoglobins* (Butterworth, London, 1971); M. F. Perutz, *Annu. Rev. Biochem.* **48**, 327-386 (1979); G. Fermi, M. F. Perutz, *Haemoglobin & Myoglobin* (Oxford Univ. Press, New York, 1982) 104 pp; R. E. Dickerson, I. Geis, *Hemoglobin: Structure, Function, Evolution, and Pathology* (Benjamin-Cummings, Menlo Park, Calif., 1983) 176 pp.

Crystal form, solubility, affinity for oxygen, absorption spectra differ quantitatively in hemoglobins of different species, due to the variation in amino acid sequence of the protein moiety since the same heme group is present in all vertebrate and many invertebrate hemoglobins. Human HbO_2 = tetragonal crystals; horse HbO_2 = orthorhombic crystals from citrated blood, monoclinic crystals from oxalated blood. Absorption spectra, *see* Lemberg, Legge, *loc. cit.,* 228. Oxyhemoglobin is an article of commerce where it is called hemoglobin. Brownish-red powder or scales. Soluble in about 7 parts water, slowly sol in glycerol.

4684. Hemozoin. [39404-00-7] β-Hematin; malaria pigment; haemozoin; HZ. Polymerized heme produced as a detoxification response by malaria parasites and by blood-sucking insects. It is the end product of hemoglobin digestion in intraerythrocytic malaria parasites. Constituents include hematin and protein but the exact composition remains unclear. Identification: T. Carbone, *Giorn. Regia. Acad. Med. Torino* **39**, 901 (1891). Biochemical characterization: P. Goldie *et al., Am. J. Trop. Med. Hyg.* **43**, 584 (1990). Identification in tissue: A D. Sullivan, S. R. Meshnick, *Parasitol. Today* **12**, 161 (1996); of formation in insects: M. F. Oliveira *et al., Nature* **400**, 517 (1999). Depression of cellular immunity: F. Turrini *et al., ibid.* **9**, 297 (1993). Effect of antimalarials on depolymerization: A. V. Pandey, B. L. Tekwani, *FEBS Lett.* **402**, 236 (1997). *Review:* P. Arese, E. Schwarzer, *Ann. Trop. Med. Parasitol.* **91**, 501 (1997).

Insoluble, black-brown, crystalline pigment.

4685. Hendrickson's Reagent. [72450-51-2] μ-Oxohexaphenyldiphosphorus(2+) 1,1,1-trifluoromethanesulfonic acid (1:2); triphenylphosphonium anhydride trifluoromethanesulfonate; POP. $C_{36}H_{30}OP_2.2CF_3O_3S$; mol wt 838.71. C 54.42%, H 3.61%, O 13.35%, P 7.39%, F 13.59%, S 7.65%. Alternative to Mitsunobu reagents for the esterification of primary alcohols. Synthesis method: J. B. Hendrickson, S. M. Schwartzman, *Tetrahedron Lett.*

Consult the Name Index before using this section.

16, 277 (1975). Prepn: A. Aaberb *et al., ibid.* 20, 2263 (1979). Synthetic applications as a dehydration agent: J. B. Hendrickson, Md. S. Hussoin, *J. Org. Chem.* 52, 4137 (1987); *eidem et al., Synlett* 1996, 661; and X-ray crystal structure determn: S.-L. You *et al., Angew. Chem. Int. Ed.* 42, 83 (2003). Mechanism of dehydration reactions vs the Mitsunobu reaction: K. E. Elson *et al., Org. Biomol. Chem.* 1, 2958 (2003).

White crystalline solid, mp 74-75°. Sol in pyridine, THF. *Water and air sensitive.*

USE: Dehydration reagent in organic synthesis.

4686. Henna. Dried powdered leaves of *Lawsonia alba* Lam., *L. inermis* L., and *L. spinosa* L., *Lythraceae*. Obtained from North Africa or India. Contains about 1% of lawsone, *q.v.* Ref: Cox, *Analyst* 63, 397 (1938); Talaat, *Br. Med. J.* II, 944 (1960).

USE: For dyeing hair and nails auburn to red, in the Orient together with "reng," the dried, powdered leaves of the indigo plant, in order to produce darker and even bluish-black shades. For relatively permanent dyeing the pH must be about 5.5; this is achieved by the addition of citric, boric, or adipic acid. Ingredient of many commercial hair rinses.

4687. Heparamine. [53260-52-9] *N*-Desulfoheparin. The $-NHSO_3H$ groups in heparin are replaced by $-NH_2$ groups. Prepn of sodium salt by methanolysis of sodium heparinate and saponification: Velluz *et al., Compt. Rend.* 247, 1521 (1958); from sodium heparinate heated with HCl, neutralized with $NaHCO_3$ and dialyzed against running water: Foster *et al., J. Chem. Soc.* 1961, 1204; from sodium heparinate demineralized with Dowex 50 resin: GB 863235 (1961 to UCLAF).

Sodium salt. Solid, sol in water. $[\alpha]_D^{20}$ +67 ± 2°.

N-**Acyl derivatives.** Heparides. Hepadides. Similar to heparin in clearing alimentary lipemia but are practically devoid of its anticoagulant activity.

4688. Heparin. [9005-49-6] Heparinic acid; Arteven; Leparan. Glycosaminoglycan with anticoagulant activity. Heterogenous mixture of variably sulfated polysaccharide chains composed of repeating units of D-glucosamine and either L-iduronic or D-glucuronic acids. Mol wt ranges from 6000-30000 Da. Biosynthesized and stored in mast cells of various animal tissues, particularly liver, lung or gut. Commercial heparin is isolated from beef lung or pork intestinal mucosa. Isoln from mammalian tissue: Howell, *Am. J. Physiol.* 63, 434 (1922-23); 71, 553 (1924-25); Korn, *J. Biol. Chem.* 234, 1325 (1959); L. B. Jaques, *Can. J. Biochem. Physiol.* 37, 1183 (1959); J. A. Bush *et al.,* US 2884358 (1959 to So. Calif. Gland). Purification: G. Nominé *et al.,* US 2989438 (1961 to UCLAF); Toccaceli, US 3016331 (1962 to Ormonoterapia Richter); L. Roden *et al., Methods Enzymol.* 26, 73 (1972). Structural studies: M. L. Wolfrom, *J. Am. Chem. Soc.* 72, 5796 (1950); Velluz *et al., Compt. Rend.* 247, 1521 (1958); M. L. Wolfrom *et al., J. Org. Chem.* 29, 540 (1964). Configuration of glycosidic linkages: M. L. Wolfrom *et al., ibid.* 31, 1173 (1966); A. S. Perlin *et al., Can. J. Chem.* 48, 2260 (1970); T. Helting, U. Lindahl, *J. Biol. Chem.* 246, 5442 (1971). Identification of L-iduronic acid residues: A. S. Perlin *et al., Carbohydr. Res.* 7, 369 (1968). Antithrombotic activity results from the binding and activation of *antithrombin III*, a plasma protein which inhibits several enzymes in the coagulation cascade: R. D. Rosenberg, *Fed. Proc.* 36, 10 (1977). Anticoagulant activity is related to the mol wt of the polysaccharide fragments; low molecular weight components exhibit decreased hemorrhagic effects while retaining antithrombin binding ability: L.-O. Andersson *et al., Thromb. Res.* 15, 531 (1979); T. W. Barrowcliffe *et al., Br. J. Haematol.* 41, 573 (1979); J. Hirsch *et al., Semin. Thromb. Hemostasis* 11, 13 (1985).

Characterization of the antithrombin binding site: U. Lindahl *et al., Proc. Natl. Acad. Sci. USA* 76, 3198 (1979); J. Choay *et al., Thromb. Res.* 18, 573 (1980). Synthesis of the pentasaccharide corresponding to the binding site sequence: *eidem, Biochem. Biophys. Res. Commun.* 116, 492 (1983). Symposium on structure, activity and clinical applications: *Fed. Proc.* 36, 9-116 (1977). Review of mechanism of action: I. Björk, U. Lindahl, *Mol. Cell. Biochem.* 48, 161-182 (1982); of structure-activity relationships and prepn of low mol wt fractions: B. Casu, *Adv. Carbohydr. Chem. Biochem.* 43, 51-134 (1985); of biosynthesis: U. Lindahl *et al., Trends Biochem. Sci.* 11, 221-225 (1986). Comprehensive description: F. Nachtmann *et al., Anal. Profiles Drug Subs.* 12, 215-276 (1983). Overview of clinical results in pulmonary embolism and venous thrombosis: R. Collins *et al., N. Engl. J. Med.* 318, 1162-1170 (1988); J. Hirsh, *ibid.* 324, 1565-1574 (1991); of clinical studies with low mol wt heparinoids: H. ten Cate *et al., Am. J. Hematol.* 27, 146-153 (1988).

Antithrombin binding site of Heparin

Heparin has a rotation of $[\alpha]_D^{20}$ +55°.

Calcium salt. [37270-89-6] Calciparine; Ecasolv.

Magnesium salt. [54479-70-8] Magnesium heparinate; Cutheparine. Sol in water. Insol in organic solvents.

Potassium salt. [9005-48-5] Clarin (formerly).

Sodium salt. [9041-08-1] Heparin sodium; Hepsal; Lipo-Hepin; Liquémin; Longheparin; Monoparin; Panheprin; Pularin; Liquaemin Sodium; Minihep; Thromboliquine; Thrombophob; Unihep. White to grayish-brown amorphous powder. Odorless, hygroscopic. $[\alpha]_D^{25}$ +47° (c = 1.5 in water). One gram dissolves in 20 ml water. Sol in saline soln. Practically insol in alcohol, acetone, benzene, chloroform, ether. pH of 1% aq soln = 6.0 to 7.5. Absorption spectrum: Burson *et al., J. Am. Chem. Soc.* 78, 5874 (1956). Ampuled solns may be stored at room temp for at least 12 months. Commercially available ampuled, sterile solns contain 0.5% phenol or chlorobutanol as preservative.

THERAP CAT: Anticoagulant.

THERAP CAT (VET): Anticoagulant.

4689. HEPES. [7365-45-9] 4-(2-Hydroxyethyl)-1-piperazineethanesulfonic acid; *N*-(2-hydroxyethyl)piperazine-*N'*-2-ethanesulfonic acid. $C_8H_{18}N_2O_4S$; mol wt 238.30. C 40.32%, H 7.61%, N 11.76%, O 26.86%, S 13.45%. One of several zwitterionic *N*-substituted aminosulfonic acids known as *Good Buffers*, active in the pH range of 6-8.5. Prepn: N. E. Good *et al., Biochemistry* 5, 467 (1966). Buffering characteristics: W. J. Ferguson *et al., Anal. Biochem.* 104, 300 (1980). Temperature effects on pKa and pH: R. N. Roy *et al., Cryo Letters* 6, 285 (1985). Interaction with hydroxyl radicals: M. Hicks, J. M. Gebicki, *FEBS Lett.* 199, 92 (1986). Interference with Lowry protein determination: H. M. Himmel, W. Heller, *J. Clin. Chem. Clin. Biochem.* 25, 909 (1987). Use as biological buffer: H. Eagle, *Science* 174, 500 (1971); K. V. Rao, *Indian J. Exp. Biol.* 15, 552 (1977); E. D. Lalague *et al., J. Microsc.* 127, 307 (1982); G. I. McFadden, M. Melkonian, *Phycologia* 25, 551 (1986).

Crystals from alcohol + water, mp 234°. $pKa_1 \approx 3$, pKa_2 (20°) 7.55. ΔpKa/°C −0.014. Saturated aqueous soln is 2.25M at 0°.

USE: Biological buffer.

4690. Heptabarbital. [509-86-4] 5-(1-Cyclohepten-1-yl)-5-ethyl-2,4,6(1*H*,3*H*,5*H*)-pyrimidinetrione; 5-(1-cyclohepten-1-yl)-5-ethylbarbituric acid; 5-ethyl-5-cycloheptenylbarbituric acid; heptabarb; Heptadorm; Medomin. $C_{13}H_{18}N_2O_3$; mol wt 250.30. C 62.38%, H 7.25%, N 11.19%, O 19.18%. Prepn: **FR 870714** (1942 to Geigy); Taub, **US 2501551** (1950).

Crystals, mp 174°. Slightly bitter taste. uv max (0.2N NaOH): 218.5, 254 nm. Very sparingly sol in water, more sol in alcohol. At 25° 100 ml of soln contains: 4.0 g in alcohol; 5.7 g in acetone; 1.4 g. in chloroform. Soluble in alkaline solns. Forms water-soluble sodium, magnesium, and calcium salts.

Note: This is a controlled substance (depressant): **21 CFR,** 1308.13.

THERAP CAT: Sedative, hypnotic.

4691. Heptachlor. [76-44-8] 1,4,5,6,7,8,8-Heptachloro-3a,-4,7,7a-tetrahydro-4,7-methano-1H-indene; E-3314; Velsicol 104; Drinox; Heptamul. $C_{10}H_5Cl_7$; mol wt 373.30. C 32.18%, H 1.35%, Cl 66.47%. Organochlorine pesticide for soil and wood treatment. Prepn: Bluestone *et al.*, **US 2576666** (1951 to Julius Hyman); Mc-Kenna *et al.*, **US 2661377-8** (1953 to Shell); Kleiman, Tapas, **US 2904599** (1959 to Velsicol). GC-MS determn in sediment: M.-S. Kim *et al.*, *J. Chromatogr. A* **1208**, 25 (2008); in cereals and animal feed: S. Walorczyk, *ibid.* **202**. Toxicity data: T. B. Gaines, *Toxicol. Appl. Pharmacol.* **14**, 515 (1969). Review of toxicology and human exposure: *Toxicological Profile for Heptachlor/Heptachlor Epoxide* (PB2008-100005, 2007) 203 pp.

Crystals, mp 95-96°. Camphor-like odor. Vapor pressure at 25° = 3×10^{-4} mm Hg. Soly in g/100 ml solvent at 27°: acetone 75, benzene 106, carbon tetrachloride 112, cyclohexanone 119, alcohol 4.5, xylene 102. LD_{50} in male, female rats (mg/kg): 100, 162 orally (Gaines).

Caution: Potential symptoms of overexposure are tremors, convulsions, liver damage. Potential occupational carcinogen. *See: NIOSH Pocket Guide to Chemical Hazards* (DHHS/NIOSH 2005-149, 2007) p 157.

Note: Heptachlor is listed as a persistent organic pollutant (POP) in Annex A of the *Stockholm Convention on Persistent Organic Pollutants* (United Nations, Stockholm, 2001) 43 pp; amended (Geneva, 2009) 63 pp.

USE: Formerly as insecticide for control of cotton boll weevil.

4692. Heptaminol. [372-66-7] 6-Amino-2-methyl-2-heptanol; 2-methyl-6-amino-2-heptanol; 6-methyl-2-amino-6-heptanol. $C_8H_{19}NO$; mol wt 145.25. C 66.15%, H 13.19%, N 9.64%, O 11.01%. Prepn: H. T. F. Givens, R. M. Herbst, **US 2457656** (1948 to Bilhuber); of hydrochloride: J. Doeuvre, J. Poizat, *Compt. Rend.* **224**, 286 (1947). Pharmacokinetics and metabolism: F. Chanoine *et al.*, *Arzneim.-Forsch.* **31**, 1430 (1981). HPLC determn in biological fluids: R. R. Brodie *et al.*, *J. Chromatogr.* **274**, 179 (1983). Clinical trial in hypotension: D. Milon *et al.*, *Fundam. Clin. Pharmacol.* **4**, 695 (1990). Review of mode of action: B. Pourrias *Ann. Pharm. Fr.* **49**, 127-138 (1991).

Colorless liquid, bp$_7$ 92-93°.

Hydrochloride. [543-15-7] RP-2831; Cortensor; Eoden; Hept-a-myl; Heptylon. $C_8H_{19}NO.HCl$; mol wt 181.70. Crystals, mp

150°. Freely sol in water. Sol in alcohol. Practically insol in acetone, benzene, ether. pH of 2% aq soln: 4.5-5.5.

5′-Adenylate. [57249-13-5] Ampecyclal.

THERAP CAT: Antihypotensive.

4693. Heptanal. [111-71-7] Heptaldehyde; Aldehyde C-7; heptylaldehyde; oenanthal; enanthal; oenanthol; oenanthaldehyde; enanthaldehyde. $C_7H_{14}O$; mol wt 114.19. C 73.63%, H 12.36%, O 14.01%. Obtained by distilling castor oil under reduced pressure: Rogers, *J. Am. Pharm. Assoc. Sci. Ed.* **12**, 503 (1923); Dominguez *et al.*, *J. Chem. Educ.* **29**, 446 (1952). Catalytic dehydration of ricinoleic acid methyl ester yields heptanal as a cleavage product in almost quantitative yield: Panjutin, *Chem. Zentralbl.* **1928**, II, 747.

Liquid. Penetrating fruity odor. d_4^0 0.83423; d_4^{15} 0.82162; d_4^{30} 0.80902. mp −43.3°. bp$_{760}$ 152.8°; bp$_{30}$ 59.6°; bp$_{10}$ 42.5°. n_D^{20} 1.42571. Viscosity at 15°: 0.977 cP; at 30°: 0.791 cP. Surface tension (γ) at 30° = 25.68. Surface tension against water at 30°: 14.41. Heat of combustion (liq) −1062.4 kcal/mol. Slightly sol in water. Misc with alc, ether. Sol in 3 vols of 60% alc.

USE: Manufacture of 1-heptanol; ethyl oenanthate.

4694. n-Heptane. [142-82-5] Heptane. C_7H_{16}; mol wt 100.21. C 83.90%, H 16.09%. A hydrocarbon from petroleum. Toxicity data: Lazarew, *Arch. Exp. Pathol. Pharmakol.* **143**, 223 (1929).

Volatile, flammable liquid. d_4^{20} 0.684. bp 98.4°. mp −90.7°. Flash pt, open cup: 30°F (−1°C); closed cup: 25°F (−4°C). n_D^{25} 1.3855. Insol in water. Sol in alcohol, chloroform, ether. LC (2 hr in air) in mice: 75 mg/l (Lazarew).

Caution: Potential symptoms of overexposure are lightheadedness, giddiness, stupor, vertigo, incoordination; loss of appetite, nausea; unconsciousness. Direct contact may cause dermatitis; aspiration of liquid may cause chemical pneumonia. *See NIOSH Pocket Guide to Chemical Hazards* (DHHS/NIOSH 97-140, 1997) p 156.

USE: As standard in testing knock of gasoline engines.

4695. Heptanoic Acid. [111-14-8] Enanthic acid; oenanthic acid; oenanthylic acid; n-heptoic acid; n-heptylic acid. $C_7H_{14}O_2$; mol wt 130.19. C 64.58%, H 10.84%, O 24.58%. Found in the various fusel oils in appreciable amounts. Has been observed in rancid oils. Prepd by the oxidation of heptaldehyde with potassium permanganate in dil sulfuric acid: Ruhoff, *Org. Synth.* **coll. vol. II,** 315 (1943). Toxicity study: L. Orö, A. Wretlind, *Acta Pharmacol. Toxicol.* **18**, 141 (1961).

Oily liquid. Disagreeable, rancid odor. Faint, tallow-like odor when spectroscopically pure. d_4^0 0.9345; d_4^{15} 0.9222; d_4^{20} 0.9181; d_4^{30} 0.9099. mp −7.5°. bp$_{760}$ 223.01°, also reported as 221.9°; bp$_{256}$ 187.5°; bp$_{64}$ 150.8°. Specific heat 0.54 cal/g. Heat of combustion −986.1 cal/g (20°). n_D^{20} 1.42162. Viscosity 3.40 cP at 30°; 0.82 cP at 120°. Interfacial tension against water: 7.0 dynes/cm. pka (25°): 4.4. Soly in water (15°): 0.2419 g/100 ml H_2O. Sol in ethanol, ether, DMF, DMSO. LD_{50} i.v. in mice: 1200±56 mg/kg (Orö, Wretlind).

Methyl ester. [106-73-0] $C_8H_{16}O_2$; mol wt 144.21. Liquid, d_4^{20} 0.8815; mp −55.8°; bp 173.8°. n_D^{20} 1.41152.

Ethyl ester see Ethyl Oenanthate.

4696. 1-Heptanol. [111-70-6] n-Heptyl alcohol; enanthic alcohol; 1-hydroxyheptane. $C_7H_{16}O$; mol wt 116.20. C 72.36%, H 13.88%, O 13.77%. Prepd from heptaldehyde by reduction with iron filings in dil acetic acid: Clarke, Dreger, *Org. Synth.* **6**, 52 (1926); *cf.* Noller, Bannerot, *ibid.* **14**, 91 (1934); *ibid.* **coll. vol. I** (2nd ed., 1941) p 304. Other methods include the reaction between pentane and ethylene oxide in the presence of anhydr aluminum bromide: I. G. Farbenind, **FR 716604**, *C.A.* **26**, 2198 (1932); and the action of amyl magnesium bromide on ethylene oxide: Vaughn *et al.*, *J. Am.*

Chem. Soc. **55**, 4207 (1933). Colloidal water solns: Traube, Klein, *Kolloid-Z.* **29**, 236 (1921), *Chem. Zentralbl.* **1922**, I, 233.

Liquid. d_4^{25} 0.8187. mp −34.6°. bp_{760} 175.8°; bp_{400} 155.6°; bp_{200} 136.6°; bp_{100} 119.5°; bp_{40} 9.98°; bp_{20} 85.8°; bp_{10} 74.7°; bp_5 64.3°; $bp_{1.0}$ 42.4°. n_D^{25} 1.4224. One liter of water dissolves 1.0 g at 18°; 2.85 g at 100°; 5.15 g at 130°. Miscible with alcohol, ether.

4697.   2-Heptanol. [543-49-7] Amylmethylcarbinol; (±)-2-heptanol; 2-hydroxyheptane. $C_7H_{16}O$; mol wt 116.20. C 72.36%, H 13.88%, O 13.77%. Occurs in oil of cloves: Masson, *Compt. Rend.* **149**, 630 (1910). Prepd by the action of amylmagnesium bromide on acetaldehyde: Henry, de Wael, *Rec. Trav. Chim.* **28**, 446 (1909); by the reduction of methyl amyl ketone with sodium in alcoholic soln: Thoms, Mannich, *Ber.* **36**, 2544 (1903); Pickard, Kenyon, *J. Chem. Soc.* **99**, 58 (1911); **105**, 849 (1914); Whitmore, Otterbacher, *Org. Synth.* **10**, 60 (1930); by the action of *Penicillium palitans* on coconut oil: Stokoe, *Biochem. J.* **22**, 82, 84 (1928). Toxicity data: Smyth *et al.*, *Arch. Ind. Hyg. Occup. Med.* **10**, 61 (1954).

Liquid. d^0 0.8344; d^{20} 0.8193. bp_{760} 158-160°. n_D 1.42131. Almost insol in water: 3.5 g/l. Sol in alc, ether, benzene. LD_{50} orally in rats: 2.58 g/kg (Smyth).

(2S)-Form. [6033-23-4] Liquid. d_4^{20} 0.8190; d_4^{35} 0.8050; d_4^{51} 0.7920; d_4^{64} 0.7815; d_4^{110} 0.7417. bp_{20} 73.5°. $[\alpha]_D^{20}$ +11.45° (1.039 g in 20 ml abs ethanol); $[\alpha]_4^{20}$ +13.71° (0.992 in 20 ml benzene).

(2R)-Form. [6033-24-5] Liquid. d_4^{20} 0.8184. bp_{23} 74.5°. $[\alpha]_D^{17}$ −10.48°.

4698.   2-Heptanone. [110-43-0] Methyl amyl ketone. C_7H_{14}-O; mol wt 114.19. C 73.63%, H 12.36%, O 14.01%. Found in oil of cloves and in cinnamon-bark oil. Responsible for the "peppery" odor in cheeses of the Roquefort type: Hammer, Bryant, *Iowa State Coll. J. Sci.* **11**, 281 (1937). Prepd by the ketone decompn of ethyl butylacetoacetate: Drake, Riemenschneider, *J. Am. Chem. Soc.* **52**, 5005 (1930); Dehn, Jackson, *ibid.* **55**, 4285 (1933); Johnson, Hager, *Org. Synth.* **coll. vol. I** (2nd ed., 1941) p 351; by hydration of 1-heptyne and 2-heptyne: Thomas *et al.*, *J. Am. Chem. Soc.* **60**, 719 (1938). Toxicity study: Smyth *et al.*, *Am. Ind. Hyg. Assoc. J.* **23**, 95 (1962).

Liquid. Penetrating fruity odor. d_4^0 0.8324; d_4^{15} 0.8197; d_4^{30} 0.8068. bp_{760} 151.5°; bp_{21} 111°. n_D^{15} 1.41156; n_D^{25} 1.40729. Very slightly sol in water. Sol in alc, ether. LD_{50} orally in rats: 1.67 g/kg (Smyth).
Caution: Potential symptoms of overexposure are irritation of eyes, skin and mucous membranes; headache; narcosis, coma; dermatitis. *See NIOSH Pocket Guide to Chemical Hazards* (DHHS/NIOSH 97-140, 1997) p 200.

USE: In perfumery as constituent of artificial carnation oils; as industrial solvent.

4699.   Heptenophos. [23560-59-0] Phosphoric acid 7-chlorobicyclo[3.2.0]hepta-2,6-dien-6-yl dimethyl ester; 7-chlorobicyclo[3.2.0]hepta-2,6-dien-6-yl dimethylphosphate; *O,O*-dimethyl *O*-[7-chlorobicyclo[3.2.0]hepta-2,6-dien-6-yl]phosphate; HOE-2982; Ragadan; Hostaquick. $C_9H_{12}ClO_4P$; mol wt 250.61. C 43.13%, H 4.83%, Cl 14.15%, O 25.54%, P 12.36%. Organophosphate insecticide; cholinesterase inhibitor. Prepn: B. Böhner, K. Rüfenacht, **ZA 6706947**; *eidem,* **US 3600474** (1968, 1971 both to Geigy). Activity in food crops: R. T. Hewson, *Proc. 8th Br. Insectic. Fungic. Conf.* **2**, 697 (1975). Efficacy vs ectoparasites: W. Bonin, *ibid.* 705. Degradation in soil: W. Schwab *et al.*, *J. Agric. Food Chem.* **42**, 1578 (1994).

Pale amber liquid, $bp_{0.001}$ 94-95°. d_4^{20} 1.294. Vapor pressure at 20°: 7.5 × 10^{-4} mm Hg. Sol in xylene, acetone, methanol. LD_{50} orally in rats: 96-117 mg/kg (Hewson).

USE: Insecticide.

THERAP CAT (VET): Ectoparasiticide.

4700.   Heptoxime. [530-97-2] 1,2-Cycloheptanedione 1,2-dioxime. $C_7H_{12}N_2O_2$; mol wt 156.19. C 53.83%, H 7.74%, N 17.94%, O 20.49%. Prepn from cycloheptanone: Vander *et al.*, *J. Org. Chem.* **14**, 836 (1949); Belcher *et al.*, *J. Chem. Soc.* **1958**, 2743. *Review:* Banks, Nicholas, *USAEC* **ISC-737** (1956).

Crystals from benzene, mp 182°. pK_1: 10.65 ± 0.2; pK_2: 12.21 ±0.2.
Monohydrate. Crystals from water, mp 179-180°.
USE: As reagent in quantitative determination of nickel.

4701.   D-*manno*-Heptulose. [3615-44-9] D-*manno*-2-Heptulose; D-*manno*-ketoheptose. $C_7H_{14}O_7$; mol wt 210.18. C 40.00%, H 6.71%, O 53.28%. Isoln from the wet pulp of the fruit of the avocado tree, *Persea gratissima* Gaertn. f. (*P. americana* Mill.), *Lauraceae:* LaForge, *J. Biol. Chem.* **28**, 511 (1917); Montgomery, Hudson, *J. Am. Chem. Soc.* **61**, 1654 (1939); Richtmyer in *Methods in Carbohydrate Chemistry* R. L. Whistler *et al.*, Eds. **vol. 1**, 173 (1962).

Large, transparent prisms from methanol, mp 151-152°. $[\alpha]_D^{20}$ +29° (c = 2 in H_2O). Sol in water. Usually accompanied by perseitol. Although perseitol is much more insol in water and methanol than D-*manno*-heptulose, clean separation of the two substances is not always easy and may require patient fractional crystn.

4702.   Hercynine. [534-30-5] (αS)-α-Carboxy-*N,N,N*-trimethyl-1*H*-imidazole-4-ethanaminium inner salt; L-(1-carboxy-2-imidazol-4-ylethyl)trimethyl ammonium hydroxide inner salt; histidine-betaine; histidine trimethylbetaine. $C_9H_{15}N_3O_2$; mol wt 197.24. C 54.81%, H 7.67%, N 21.30%, O 16.22%. Present in many fungi, especially in *Amanita muscaria* Fr. and *Agaricus campestris* L., *Agaricaceae*. Isoln: Kutscher, *Zentralbl. Physiol.* **24**, 775 (1910); **26**, 569 (1912); Barger, Ewins, *J. Chem. Soc.* **99**, 2340 (1911); Küng, *Z. Physiol. Chem.* **91**, 249 (1914). Occurrence in the *Limulus polyphemus* L. (king crab), and prepn: Ackermann, List, *ibid.* **313**, 30 (1958). Synthesis: Reinhold *et al.*, *J. Med. Chem.* **11**, 258 (1968). Biosynthesis by fungi and *Actinomycetales:* Genghof, *J. Bacteriol.* **103**, 475 (1969).

White crystals from methanol and ether, mp 237-238° (dec). $[\alpha]_D^{22}$ +44.5° (5*N* HCl). Sol in water, alc.

4703. Herqueinone. [26871-30-7] (7aS,9R)-8,9-Dihydro-4,-6,7a-trihydroxy-5-methoxy-1,8,8,9-tetramethyl-3H-phenaleno[1,2-b]furan-3,7(7aH)-dione. $C_{20}H_{20}O_7$; mol wt 372.37. C 64.51%, H 5.41%, O 30.08%. Red fungal pigment isolated from *Penicillium herquei*: Stodola *et al.*, *Nature* **167**, 773 (1951); Galarraga *et al.*, *Biochem. J.* **61**, 456 (1955); Harman *et al.*, *J. Org. Chem.* **20**, 1260 (1955). Structure: Cason *et al.*, *Tetrahedron* **18**, 839 (1962); Paul, Sim, *Proc. Chem. Soc. London* **1962**, 352; Cason *et al.*, *J. Org. Chem.* **35**, 179 (1970); J. S. Brooks, G. A. Morrison, *Tetrahedron Lett.* **1970**, 963; *eidem*, *J. Chem. Soc. Perkin Trans. 1* **1972**, 421; **1974**, 2114. Crystal structure and abs config: A. Quick *et al.*, *Chem. Commun.* **1980**, 1051. ^{13}C-NMR study: T. Suga *et al.*, *Chem. Lett.* **1981** 1063.

Red needles from alc, dec 226°. Sublimes in high vacuum at 175-190°. Absorption max (ethanol): 220, 250, 314, 416 nm (log ε 4.29, 4.09, 4.47, 3.66). Sol in acetone, DMF, 2N NaOH, concd HCl, cold concd H_2SO_4; fairly sol in methanol, ethanol, ether, ethyl acetate, chloroform; slightly sol in benzene, carbon disulfide. Practically insol in petr ether, carbon tetrachloride, water, aq Na_2CO_3.

4704. Herrmann-Beller Catalyst. [172418-32-5] Bis[μ-(acetato-κO:$\kappa O'$)]bis[[2-[bis(2-methylphenyl)phosphino-κP]phenyl]methyl-κC]dipalladium stereoisomer; *trans*-di(μ-acetato)bis[o-(di-o-tolylphosphino)benzyl]dipalladium(II); Herrmann's palladacycle. $C_{46}H_{46}O_4P_2Pd_2$; mol wt 937.66. C 58.92%, H 4.95%, O 6.83%, P 6.61%, Pd 22.70%. Highly efficient palladacycle catalyst; exists in a monomer/dimer equilibrium in soln. Prepn: M. Beller *et al.*, *DE* **4421753** (1995 to Hoechst); *eidem*, *US* **5831107** (1998 to Aventis); and catalytic applications: W. A. Herrmann *et al.*, *Angew. Chem. Int. Ed.* **34**, 1844 (1995); *eidem*, *Chem. Eur. J.* **3**, 1357 (1997); W. A. Herrmann *et al.*, *J. Chem. Educ.* **77**, 92 (2000). EXAFS characterization: S. G. Fiddy *et al.*, *Chem. Commun.* **2003**, 2682. Catalysis mechanism in the Heck reaction: V. P. W. Böhm, W. A. Herrmann, *Chem. Eur. J.* **7**, 4191 (2001). Review of catalytic applications: A. Speicher *et al.*, *J. Prakt. Chem.* **341**, 605-608 (1999).

Yellow crystals from toluene/hexane or CH_2Cl_2/hexane, mp 229-231° (dec). Air stable; thermally stable; becomes catalytically active at ~80°.
USE: Homogeneous catalyst in carbon-carbon bond forming reactions such as the Heck and Suzuki reactions.

4705. Hesperetin. [520-33-2] (2S)-2,3-Dihydro-5,7-dihydroxy-2-(3-hydroxy-4-methoxyphenyl)-4H-1-benzopyran-4-one;

3',5,7-trihydroxy-4'-methoxyflavanone; cyanidanon 4'-methyl ether 1626. $C_{16}H_{14}O_6$; mol wt 302.28. C 63.58%, H 4.67%, O 31.76%. The aglucon of hesperidin. Prepd by hydrolysis of hesperidin or by synthesis: Shinoda, Kawagoye, *C.A.* **23**, 2957 (1929); Seka, Prosche, *Monatsh. Chem.* **69**, 284 (1936). Sepn of isomers: Arthur *et al.*, *J. Chem. Soc.* **1956**, 632. Structure and configuration: Arakawa, Nakazaki, *Ann.* **636**, 111 (1960). *See also* Bioflavonoids.

(±)-Form, prisms from alc, mp 226-228°. Freely sol in alc, moderately in ether. Slightly sol in water, chloroform, benzene; sol in dil alkalies. Precipitated by carbonates.

4706. Hesperidin. [520-26-3] (2S)-7-[[6-O-(6-Deoxy-α-L-mannopyranosyl)-β-D-glucopyranosyl]oxy]-2,3-dihydro-5-hydroxy-2-(3-hydroxy-4-methoxyphenyl)-4H-1-benzopyran-4-one; hesperetin 7-rhamnoglucoside; cirantin; hesperetin-7-rutinoside. $C_{28}H_{34}O_{15}$; mol wt 610.57. C 55.08%, H 5.61%, O 39.31%. Predominant flavonoid in lemons and sweet oranges *(Citrus sinensis)*. Extraction procedures: Higby, *J. Am. Pharm. Assoc. Sci. Ed.* **30**, 629 (1941); *US* **2421061** (1947); Baier, *US* **2442110** (1948 to Calif. Fruit Growers Exchange). Structure: King, Robertson, *J. Chem. Soc.* **1931**, 1704; Arthur *et al.*, *ibid.* **1956**, 632; Horowitz, Gentili, *Tetrahedron* **19**, 773 (1963). Synthesis: Zemplen, Bognar, *Ber.* **75**, 1043 (1943); **76**, 773 (1943). Identity with cirantin: Manwaring *et al.*, *Phytochemistry* **7**, 1881 (1968). *See also* Bioflavonoids, Rutinose.

Fine, dendritic needles by precipitation at pH 6-7, mp 258-262° (softens at 250°). $[\alpha]_D^{20}$ −76° (c = 2 in pyridine). One gram dissolves in 50 l water. Sol in formamide, DMF at 60°. Slightly sol in methanol, hot glacial acetic acid. Almost insol in acetone, benzene, chloroform. Freely sol in dil alkalies, pyridine.

4707. Hetacillin. [3511-16-8] (2S,5R,6R)-6-[(4R)-2,2-Dimethyl-5-oxo-4-phenyl-1-imidazolidinyl]-3,3-dimethyl-7-oxo-4-thia-1-azabicyclo[3.2.0]heptane-2-carboxylic acid; 6-(2,2-dimethyl-5-oxo-4-phenyl-1-imidazolidinyl)penicillanic acid; phenazacillin; BRL-804; Versapen. $C_{19}H_{23}N_3O_4S$; mol wt 389.47. C 58.59%, H 5.95%, N 10.79%, O 16.43%, S 8.23%. Semi-synthetic antibiotic related to penicillin. Prepn and structure: Hardcastle *et al.*, *J. Org. Chem.* **31**, 897 (1966); Johnson, Panetta, *US* **3198804** (1965 to Bristol-Myers). Pharmacology: Kirby, Kind, *Ann. N.Y. Acad. Sci.* **145**, 291 (1967); Ueda *et al.*, *J. Antibiot.* **20B**, 206 (1967), *C.A.* **69**, 95016w (1968). Stability studies: Saccani, Pansera, *Boll. Chim. Farm.* **107**, 640 (1968). Epimerization at C-6 to *epihetacillin*: Johnson *et al.*, *Tetrahedron Lett.* **1968**, 1903.

Rectangular plates from water + methyl isobutyl ketone, dec 182.8-183.9°; also reported as mp 189.2-191.0°. $[\alpha]_D^{25}$ +366° (pyridine). Practically insol in most organic solvents and water. Sol in dil aq NaOH soln (pH 7-8), DMF, DMSO, pyridine, methanol (with dec).

Potassium salt. [5321-32-4] Hetacin-K; Natacillin; Versatrex. $C_{19}H_{22}KN_3O_4S$; mol wt 427.56.

Methyl ester. $C_{20}H_{25}N_3O_4S$. Crystals from carbon tetrachloride, mp 101.5-102°. Sol in most organic solvents.

Epihetacillin. Crystals, mp 164-165°. $[\alpha]_D^{23}$ +232° (pyridine).

THERAP CAT: Antibacterial.

THERAP CAT (VET): Antibacterial.

4708. Hetastarch. [9004-62-0] Cellulose 2-hydroxyethyl ether; hydroxyethyl starch; HES; 6-H.E.S.; Hespan; Hespander; Hestar; Plasmasteril. A starch derivative of undetermined exact composition. Comprised of more than 90% amylopectin and etherified to the extent that an average of 7 to 8 of the OH groups present in every ten D-glucopyranose units of the polymer have been converted to OCH_2CH_2OH groups. Preparation by treating starch with pyridine and ethylene chlorohydrin: Mima, Yokoyama, *JP 70 6556* (1970 to Green Cross), *C.A.* **73**, 36822r (1970). Structure evaluation studies: Banks *et al.*, *Br. J. Pharmacol.* **47**, 172 (1973). Physicochemical and biological properties: Tamada *et al.*, *Chem. Pharm. Bull.* **19**, 286 (1971); Banks *et al.*, *Staerke* **24**, 181 (1972). Pharmacology: Irikura *et al.*, *Oyo Yakuri* **6**, 985 sqq (1972). Metabolism: Bogan *et al.*, *Toxicol. Appl. Pharmacol.* **15**, 206 (1969); Ryan *et al.*, *Xenobiotica* **2**, 141 (1972). Toxicity studies: Irikura *et al.*, *Oyo Yakuri* **6**, 1023 (1972). Introduction as a plasma expander: Wiedersheim, *Arch. Int. Pharmacodyn.* **111**, 353 (1957). Review of production and uses: Hjermstad, *Starch Chem. Technol.* **2**, 423-432 (1967).

R or R' = H or CH$_2$CH$_2$OH

White powder. Very sol in water. Insol in alcohol.

Amylopectin derivative. The sequence is frequently interrupted by a unit in which R' is the residue of an additional O-hydroxyethylated α-D-glucopyranosyl moiety that constitutes the first unit in a branch or sub-branch of the polymer.

USE: Cryoprotective agent for erythrocytes.

THERAP CAT: Plasma volume expander.

THERAP CAT (VET): Plasma volume expander.

4709. 5-HETE. [70608-72-9] (5S,6E,8Z,11Z,14Z)-5-Hydroxy-6,8,11,14-eicosatetraenoic acid; (S)-5-hydroxy-6-*trans*-8,-11,14-*cis*-eicosatetraenoic acid. $C_{20}H_{32}O_3$; mol wt 320.47. C 74.96%, H 10.07%, O 14.98%. Important intermediate in a series of biosynthetic processes leading from arachidonic acid, *q.v.*, to a number of biologically active compounds. First discovered in the transformation of arachidonic acid by rabbit peritoneal polymorphonuclear leukocytes: P. Borgeat *et al.*, *J. Biol. Chem.* **251**, 7816 (1976). Revised description: *eidem*, *ibid.* **252**, 8772 (1977). It has also been found from metabolism of arachidonic acid by human peripheral blood polymorphonuclear leukocytes: P. Borgeat, B. Samuelsson, *Proc. Natl. Acad. Sci. USA* **76**, 2148 (1979); and has been shown to be strongly chemotactic for human eosinophils and neutrophils: E. J. Goetzl *et al.*, *Immunology* **39**, 491 (1980); E. J. Goetzl, *N. Engl. J. Med.* **303**, 822 (1980). 5-HETE is formed via *5-HPETE (5-hydroperoxy-6,8,11,14-eicosatetraenoic acid)*, a precursor of the leukotrienes, *q.v.* Chemical and enzymic syntheses of 5-HETE and 5-HPETE: E. J. Corey *et al.*, *J. Am. Chem. Soc.* **102**, 1435 (1980). Synthesis via phenylselenylation of arachidonic acid: J. E. Baldwin *et al.*, *Tetrahedron* **37**, Suppl. 1, 263 (1981). Large-scale synthesis of 5-HETE: E. J. Corey, S. Hashimoto, *Tetrahedron Lett.* **22**, 299 (1981). Biosynthetic study: R. C. Murphy, *Diss. Abstr. B* **42**, 2839 (1982). Selected synthesis of octadeuterated (±)-5-HETE: W. C. Hubbard *et al.*, *Prostaglandins* **23**, 61 (1982). Stereospecific syntheses of 5-(S)- and 5-(R)-HETE and transformation to (±)-5-HPETE: R. Zamboni, J. Rokach, *Tetrahedron Lett.* **24**, 999 (1983). Review

of syntheses: J. G. Atkinson, J. Rokach, in *Handbook of Prostaglandins and Related Lipids: The Eicosanoids*, A. L. Willis *et al.*, Eds. (CRC Press, Boca Raton, in press).

Methyl ester. $C_{21}H_{34}O_3$. Colorless oil. $[\alpha]_D^{23}$ +14.0°; $[\alpha]_{436}^{23}$ +35.7° (c = 2.02 in benzene).

4710. Hexaaminecobalt Trichloride. [10534-89-1] (*OC*-6-11)-Hexaammine-cobalt(3+) chloride (1:3); luteocobaltic chloride; hexammino-cobalt chloride. $Cl_3CoH_{18}N_6$; mol wt 267.47. Cl 39.76%, Co 22.03%, H 6.78%, N 31.42%. $[Co(NH_3)_6]Cl_3$. Prepd from $CoCl_2$, NH_4Cl, NH_3 and O_2: Bjerrum, McReynolds, *Inorg. Synth.* **2**, 217 (1946).

Wine-red or brownish-orange-red, monoclinic crystals. Sol in water; on long boiling with water, NH_3 is evolved and $Co(OH)_3$ pptd.

USE: As reagent for pyrophosphoric acid, for the estimation of phosphate.

4711. Hexaborane(10). [23777-80-2] Hexaboron decahydride; borohexane. B_6H_{10}; mol wt 74.94. B 86.55%, H 13.45%. Prepd by the reaction of magnesium boride with hydrochloric or phosphoric acid: Stock, Kuss, *Ber.* **56B**, 789 (1923).

Liquid. mp −62.3°; bp 108°; vapor pressure (0°): 7.5 mm: Burg, Kratzer, *Inorg. Chem.* **1**, 725 (1962). d^0 0.69. Slowly dec at room temp. Hydrolyzes in water after long heating.

4712. Hexabromocyclododecane. [3194-55-6] 1,2,5,6,9,10-Hexabromocyclododecane; HBCD; FR-1206; Saytex HP-900. $C_{12}H_{18}Br_6$; mol wt 641.70. C 22.46%, H 2.83%, Br 74.71%. Brominated flame retardant; produced by bromination of 1,5,9-cyclododecatriene. Technical grade consists of a mixture of isomers, predominantly the α-, β-, and γ- enantiomeric pairs. Prepn of 2 crystalline forms: L. I. Zakharkin, V. V. Korneva, *Dokl. Akad. Nauk SSSR* **132**, 1078 (1960); G. Eglinton *et al.*, *J. Chem. Soc. C* **1969**, 474. Manufacturing processes: M. Minsinger *et al.*, *DE 1147574* (1963 to BASF); J. K. Kendall, *US 6506952* (2003 to Albemarle Corp.). Absolute configuration of isomers: N. V. Heeb *et al.*, *Chemosphere* **68**, 940 (2007). Thermally-induced isomerization: R. Köppen *et al.*, *ibid.* **71**, 656 (2008). Biodegradation: J. W. Davis *et al.*, *Environ. Sci. Technol.* **40**, 5395 (2006). Enantiomer specific LC/MS/MS determn in food samples: B. Gómara *et al.*, *Anal. Chim. Acta* **605**, 53 (2007). Review of environmental fate: A. Covaci *et al.*, *Environ. Sci. Technol.* **40**, 3679-3688 (2006).

(+)-γ-Form

White powder or granules, mp 175-185°. Dec 220°. Vapor pressure: 6.3×10^{-5} Pa. Log P (octanol/water): 5.62. Insol in water. Sol in styrene, acetone.

α-HBCD. [134237-50-6] Pair of enantiomers having the relative configuration (1R,2R,5S,6R,9R,10S). Soly in water: 48.8 μg/l. d 2.338.

β-HBCD. [134237-51-7] Pair of enantiomers having the relative configuration (1R,2S,5R,6R,9R,10S). Soly in water: 14.7 μg/l. d 2.412.

γ-HBCD. [134237-52-8] Pair of enantiomers having the relative configuration (1R,2R,5R,6S,9S,10R). Predominant form in technical grade product; isomerizes to α-form at temperatures above 160°. Soly in water: 2.1 μg/l. d 2.390.

USE: Additive flame retardant in extruded polystyrene foams, polystyrene and polypropylene resins, injection-molded polystyrene;

used in thermal insulation building materials, upholstery textiles, electronics.

4713. Hexachlorobenzene. [118-74-1] 1,2,3,4,5,6-Hexachlorobenzene; perchlorobenzene; Anticarie; Bunt-cure; Bunt-nomore. C_6Cl_6; mol wt 284.77. C 25.31%, Cl 74.69%. Organochlorine pesticide for seed treatment. Produced as a by-product during the manufacture of chlorinated solvents and pesticides. Prepn: F. Becke, H. Sperber, US 2792434 (1957 to BASF). Teratogenicity study: K. D. Courtney *et al., Toxicol. Appl. Pharmacol.* **35**, 239 (1976). Carcinogenicity studies: J. R. P. Cabral *et al., Nature* **269**, 510 (1977); D. L. Arnold *et al., Food Chem. Toxicol.* **23**, 779 (1985); (corr. *ibid.* **26**, 169 (1988)). uv spectrum: H. Conrad-Billroth, *Z. Phys. Chem.* **19**, 76 (1932); O. Schnepp, R. Kopelman, *J. Chem. Phys.* **30**, 868 (1959). GC-MS determn in soil: A. Rashid *et al., J. Chromatogr. A* **1217**, 2933 (2010). Toxicology: R. Ockner, R. Schmid, *Nature* **189**, 499 (1961); G. Vettorazzi, *Residue Rev.* **56**, 107 (1975). Review of toxicology and human exposure: *Toxicological Profile for Hexachlorobenzene* (PB2003-100139, 2002) 300 pp.; of environmental fate and global distribution: J. L. Barber *et al., Sci. Total Environ.* **349**, 1-44 (2005); of environmental toxicology and health effects: L. Reed *et al., Rev. Environ. Health* **22**, 213-243 (2007).

Crystalline needles. d^{23} 2.044. d^{20} 1.5691. mp 231°. bp 323-326° (sublimes). Flash point 242°C. Vapor pressure at 20°: 1.09×10^{-5} mm Hg. Log P (octanol/water): 5.5. *Poisonous.* Soly in water (25°): 0.0062 mg/l. Sol in benzene, chloroform, ether, carbon disulfide, boiling alc; sparingly sol in cold alc.

Note: Not to be confused with benzene hexachloride, *see* Lindane.

Caution: Potential symptoms of chronic oral overexposure are dermal lesions, hypertrichosis, anorexia, weight loss, focal alopecia, corneal opacity, atrophic hands, hepatomegaly, porphyria, skeletal muscle wasting. *See Clinical Toxicology of Commercial Products,* R. E. Gosselin *et al.,* Eds. (Williams & Wilkins, Baltimore, 5th ed., 1984) Section II, pp 170-171; *Patty's Industrial Hygiene and Toxicology* vol. **2B**, G. D. Clayton, F. E. Clayton, Eds. (John Wiley & Sons, New York, 4th ed., 1994) pp 1477-1492. This substance is reasonably anticipated to be a human carcinogen: *Report on Carcinogens, Twelfth Edition* (PB2011-111646, 2011) p 224.

Note: Hexachlorobenzene is listed as a persistent organic pollutant (POP) in Annex A and Annex C of the *Stockholm Convention on Persistent Organic Pollutants* (United Nations, Stockholm, 2001) 43 pp; amended (Geneva, 2009) 63 pp.

USE: In organic syntheses. Formerly as agricultural fungicide.

4714. Hexachlorobutadiene. [87-68-3] 1,1,2,3,4,4-Hexachloro-1,3-butadiene; perchlorobutadiene. C_4Cl_6; mol wt 260.74. C 18.43%, Cl 81.58%. Prepn: F. Krafft, *Ber.* **10**, 801 (1877); E. T. McBee, R. E. Hatton, *Ind. Eng. Chem.* **41**, 809 (1949); J. Jelinek, M. Hudlicky, *Collect. Czech. Chem. Commun.* **22**, 651 (1957). GC determn in blood: P. E. Kastl, E. A. Hermann, *J. Chromatogr.* **280**, 390 (1983). Toxicity study: D. Gradiski *et al., Eur. J. Toxicol. Environ. Hyg.* **8**, 180 (1975). Review of toxicology and human exposure: *Toxicological Profile for Hexachlorobutadiene* (PB95-100160, 1994) 135 pp. Review: G. Choudhary, *Environ. Carcinog. Ecotoxicol. Rev.* **C13**, 179-203 (1995).

Colorless liquid with faint turpentine-like odor. mp −21°; bp_{760} 215°. d_4^{20} 1.6820; n_D^{20} 1.5542. Viscosity (cP): 2.446 at 37.8°; 1.131 at 98.9°. *Poisonous.* Soly in water (20°): 0.0005%. Sol in ethanol, ether. LD_{50} in male, female mice, rats (mg/kg): 105, 76, 216, 175 i.p.; 80, 65, 250, 270 orally (Gradiski).

Caution: Potential symptoms of overexposure in exptl animals are irritation of eyes, skin, respiratory system; kidney damage. Potential occupational carcinogen. *See: NIOSH Pocket Guide to Chemical Hazards* (DHHS/NIOSH 97-140, 1997) p. 158; *Patty's Industrial Hygiene and Toxicology* vol. **2E**, G. D. Clayton, F. E. Clayton, Eds. (John Wiley & Sons, New York, 4th ed., 1994) p 4240-4247.

USE: Intermediate in the manuf of rubber compds, chlorofluorocarbons, and lubricants. Hydraulic fluid, fluid for gyroscopes, heat transfer fluid, solvent, laboratory reagent. Soil fumigant for vineyards.

4715. Hexachloroethane. [67-72-1] 1,1,1,2,2,2-Hexachloroethane; carbon hexachloride; perchloroethane. C_2Cl_6; mol wt 236.72. C 10.15%, Cl 89.85%. CCl_3CCl_3. Prepn: *Beilstein* **1**, 87 (1918) and suppls. Toxicity data: Barsoum, Saad, *Q. J. Pharm. Pharmacol.* **7**, 205 (1934). Review of toxicology and human exposure: *Toxicological Profile for Hexachloroethane* (PB98-101041, 1997) 190 pp.

Crystals; camphorous odor. d 2.09. Readily sublimes without melting. bp 186.8° (triple point). Heat of sublimation 12.2 kcal/mol. Sol in alc, benzene, chloroform, ether, oils. Insol in water. MLD i.v. in dogs: 325 mg/kg (Barsoum, Saad).

Caution: Potential symptoms of overexposure are irritation of eyes, skin, mucous membranes. *See NIOSH Pocket Guide to Chemical Hazards* (DHHS/NIOSH 97-140, 1997) p 158. *See also Patty's Industrial Hygiene and Toxicology* vol. **2E**, G. D. Clayton, F. E. Clayton, Eds. (John Wiley & Sons, New York, 4th ed., 1994) p 4144-4151. This substance is reasonably anticipated to be a human carcinogen: *Report on Carcinogens, Twelfth Edition* (PB2011-111646, 2011) p 227.

USE: In metallurgy for refining aluminum alloys, removing impurities from molten metals, recovering metal from ores or smelting products. Degassing agent for magnesium; to inhibit explosiveness of methane and combustion of ammonium perchlorate. Smoke generator in grenades; in pyrotechnics. Ignition suppressant, in fire extinguishing fluids, polymer additive, flame-proofing agent, vulcanizing agent. In prodn of synthetic diamonds.

THERAP CAT (VET): Anthelmintic (flukicide).

4716. Hexachlorophene. [70-30-4] 2,2′-Methylenebis[3,4,6-trichlorophenol]; 2,2′-dihydroxy-3,3′,5,5′,6,6′-hexachlorodiphenylmethane; bis(3,5,6-trichloro-2-hydroxyphenyl)methane; AT-7; G-11; Bilevon; Dermadex; Exofene; Hexosan; pHisohex; Surgi-Cen; Surofene. $C_{13}H_6Cl_6O_2$; mol wt 406.89. C 38.37%, H 1.49%, Cl 52.27%, O 7.86%. Prepd by the condensation of 2 mols of 2,4,5-trichlorophenol with 1 mol formaldehyde in the presence of concd sulfuric acid: Gump, US 2250480 (1941 to Burton T. Bush). Improved procedures: US 2435593 (1948) and US 2812365 (1957 to Givaudan). Acute toxicity: T. B. Gaines, R. E. Linder, *Fundam. Appl. Toxicol.* **7**, 299 (1986).

Crystals from benzene, mp 164-165°. *Poisonous.* Freely sol in acetone, alcohol, ether; sol in chloroform, propylene glycol, polyethylene glycols, olive oil, cottonseed oil, dil aq solns of the alkalies. Insol in water. Forms salts with alkalies and alkaline earths. Phenol coefficient ∼125 (monopotassium salt). Incompatible with Tweens from bacteriological point of view. LD_{50} in adult male, female rats (mg/kg): 66, 57 orally (Gaines, Linder).

Monophosphate. Hepadist.

Caution: Excessive dosage to animals results in symptoms of neurotoxicity. Reversible vacuolar changes mainly affecting the myelin of the brain and spinal cord have been reported. Because of potential neurotoxicity in humans, the FDA has regulated use. *See:* Lockhart, *Pediatrics* **50**, 229 (1972).

USE: Chiefly in the manuf of germicidal soaps.

THERAP CAT: Antiseptic; disinfectant.

THERAP CAT (VET): Anthelmintic (flukicide).

4717. Hexaconazole. [79983-71-4] α-Butyl-α-(2,4-dichloro-phenyl)-1H-1,2,4-triazole-1-ethanol; (RS)-2-(2,4-dichlorophenyl)-1-(1H-1,2,4-triazol-1-yl)hexan-2-ol; PP-523; ICI-A-0523; Anvil; Planete. $C_{14}H_{17}Cl_2N_3O$; mol wt 314.21. C 53.52%, H 5.45%, Cl 22.56%, N 13.37%, O 5.09%. Ergosterol biosynthesis inhibitor. Prepn: H. K. Spencer, **BE 886128** (1981 to Sandoz); and resolution of isomers: P. A. Worthington, *Pestic. Sci.* **31**, 457 (1991). Properties and fungicidal activity in vines and food crops: M. C. Shephard *et al.*, *Proc. Br. Crop Prot. Conf. - Pests Dis.* **1986**, 19. Review of analytical methods: K. J. Harradine *et al.* in *Comprehensive Analytical Profiles of Important Pesticides*, J. Sherma, T. Cairns, Eds. (CRC Press, Boca Raton, 1993) pp 43-57.

White crystalline solid, mp 111°. d^{25} 1.29. Log P (octanol/water): -3.9 at 20°. Vapor pressure (20°): 2×10^{-8} kPa. Soly at 20° (g/l): water 0.017; methanol 246; acetone 164; toluene 59; hexane 0.8. LD_{50} orally in mallard ducks, male rats, female rats: >4000, 2189, 6071 mg/kg; dermally in rats: >2000 mg/kg. LC_{50} (96 hour) in rainbow trout: >6.7 mg/l (Shephard).
USE: Agricultural fungicide.

4718. Hexadimethrine Bromide. [28728-55-4] Poly[(di-methyliminio)-1,3-propanediyl(dimethyliminio)-1,6-hexanediyl bromide (1:2)]; N,N,N',N'-tetramethyl-1,6-hexanediamine polymer with 1,3-dibromopropane; polymer of N,N,N',N'-tetramethylhexamethylenediamine and trimethylene bromide; poly(N,N,N',N'-tetramethyl-N-trimethylenehexamethylenediammonium dibromide); Polybrene. $(C_{13}H_{30}Br_2N_2)_x$. Toxicity study: Kimura *et al.*, *Toxicol. Appl. Pharmacol.* **1**, 185 (1959).

White, hygroscopic, amorphous polymer. Soluble in water up to 10%. pH of 1% saline soln 5-9. Stable in soln and when autoclaved. Polymers with mol wt of 5000-10,000 have LD_{50} i.v. in mice: 25-40 mg/kg (Kimura).
THERAP CAT: Heparin antagonist.

4719. Hexaflumuron. [86479-06-3] N-[[[3,5-Dichloro-4-(1,-1,2,2-tetrafluoroethoxy)phenyl]amino]carbonyl]-2,6-difluorobenz-amide; 1-[3,5-dichloro-4-(1,1,2,2-tetrafluoroethoxy)phenyl]-(2,6-di-fluorobenzoyl)urea; DE-473; XRD-473; Consult; Recruit; Recruit II. $C_{16}H_8Cl_2F_6N_2O_3$; mol wt 461.14. C 41.67%, H 1.75%, Cl 15.37%, F 24.72%, N 6.07%, O 10.41%. Insect growth regulator; inhibits chitin synthesis. Prepn: R. H. Rigterink, R. J. Sbragia, **EP 71279**; *eidem*, **US 4468405** (1983, 1984 both to Dow). Physical properties and activity: R. J. Sbragia *et al.*, *Proc. 10th Int. Congr. Plant Prot.* **1**, 417 (1983). Chromatographic determn in soil: A. Khoshab, R. Teasdale, *J. Chromatogr. A* **660**, 195 (1994). Environmental distribution: D. Yon *et al.*, *Brighton Crop Prot. Conf. - Pests Dis.* **1992**, 907. Field trials in food crops: K. N. Komblas, R. C. Hunter, *Proc. Br. Crop Prot. Conf. - Pests Dis.* **1986**, 907; vs subterranean termites: N.-Y. Su, *J. Econ. Entomol.* **87**, 389 (1994).

White solid, mp 197-199°. Soly in water (23°): 0.7 mg/l. vapor pressure (298 K): 5.87×10^{-9}.
USE: Insecticide.

4720. Hexafluorenium Bromide. [317-52-2] N^1,N^6-Di-9H-fluoren-9-yl-N^1,N^1,N^6,N^6-tetramethyl-1,6-hexanediaminium bromide (1:2); hexamethylenebis[9-fluorenyldimethylammonium bromide]; hexamethylenebis(dimethyl-9-fluorenylammonium bromide); Mylaxen. $C_{36}H_{42}Br_2N_2$; mol wt 662.55. C 65.26%, H 6.39%, Br 24.12%, N 4.23%. Neuromuscular blocking agent with pseudocholinesterase inhibitory activity. Prepn: Cavallito *et al.*, *J. Am. Chem. Soc.* **76**, 1862 (1954); Cavallito, Gray, **US 2783237** (1957 to Irwin, Neisler). Clinical trial for prolongation of succinylcholine muscular block: L. F. Walts *et al.*, *Anesthesiology* **33**, 503 (1970). Review: R. M. Britton, M. Figueroa, *Anesth. Analg.* **52**, 100-105 (1973).

Crystals from *n*-propanol, mp 188-189°.
THERAP CAT: Succinylcholine synergist.

4721. Hexafluoroacetone. [684-16-2] 1,1,1,3,3,3-Hexafluoro-2-propanone. C_3F_6O; mol wt 166.02. C 21.70%, F 68.66%, O 9.64%. Synthon in organic and inorganic reactions with an exceptionally electron deficient carbonyl group. Prepn: N. Fukuhara, L. A. Bigelow, *J. Am. Chem. Soc.* **63**, 788 (1941). Comprehensive review: C. G. Krespan, W. J. Middleton in *Fluorine Chem. Rev.* **vol. 1**, P. Tarrant, Ed. (Marcel Dekker, New York, 1967) pp 145-196. Review of reactivity: H. W. Roesky, *J. Fluorine Chem.* **30**, 123-139 (1985); of use in inorganic chemistry: M. Witt *et al.*, *Adv. Inorg. Chem. Radiochem.* **30**, 223-312 (1986); in polymer chemistry: P. E. Cassidy *et al.*, *J. Macromol. Sci. Rev. Macromol. Chem. Phys.* **C29**, 365-429 (1989); in peptide chemistry: K. Burger *et al.*, *Amino Acids* **8**, 195-199 (1995); of toxicology: G. L. Kennedy, Jr., *Crit. Rev. Toxicol.* **21**, 149-170 (1990).

bp_{760} $-27.4°$. mp $-122°$. $d^{23.3}$ 1.323; $d^{44.4}$ 1.149. uv max: 302 nm (ε 16.9). Critical pressure: 411 psia. Critical temp: 84.1°. Heat capacity (Btu/lb/°F) of liquid at 0°: 0.192; of vapor: 0.23. *Poisonous, corrosive.* Reacts with water to form hydrates. LD_{50} (mg/kg) in rats: 190 orally (trihydrate), 670 dermally (sesquihydrate); in mice: 250 i.p., 180 i.v.; in rabbits: 113 dermally (trihydrate) (Kennedy).
USE: Protecting and activating reagent in peptide chemistry; in synthesis of high performance fluoropolymers, pharmaceutical and agricultural chemicals; in ^{19}F NMR. Solvent for polyamides, polyesters, polyacetals, polyols.

4722. Hexafluorobenzene. [392-56-3] 1,2,3,4,5,6-Hexafluorobenzene; perfluorobenzene. C_6F_6; mol wt 186.06. C 38.73%, F 61.27%. Prepn: E. T. McBee *et al.*, *Ind. Eng. Chem.* **39**, 378 (1947); J. A. Godsell *et al.*, *Nature* **178**, 199 (1956). Toxicology: C. F. B. Nhachi, *Toxicology* **39**, 317 (1986). Mechanistic study of metabolite formation: I. M. C. M. Rietjens, J. Vervoort, *Chem. Res. Toxicol.* **5**, 10 (1992).

mp −13 to −11°. bp$_{743}$ 81.0-82.0° (McBee); also reported as bp 80° (Godsell). n_D^{20} 1.3760; n_D^{18} 1.3746. d$_4^{25}$ 1.612.

USE: Solvent; intermediate in chemical synthesis.

4723. Hexalure. [23192-42-9] (7Z)-7-Hexadecen-1-ol 1-acetate; cis-7-hexadecenyl acetate; cis-1-acetoxy-7-hexadecene; hexalene. C$_{18}$H$_{34}$O$_2$; mol wt 282.47. C 76.54%, H 12.13%, O 11.33%. Synthetic sex pheromone for pink bollworm moths, *Pectinophora gossypiella* (Saunders). Discovery and prepn: N. Green et al., *Experientia* **25**, 682 (1969); N. Green, J. C. Keller, **DE 1960155**; *eidem*, **US 3586712** (1970, 1971 both to U.S. Sec. Agric.). Field trials: J. C. Keller et al., *J. Econ. Entomol.* **62**, 1520 (1969). Acute toxicity study: M. Beroza et al., *Toxicol. Appl. Pharmacol.* **31**, 421 (1975). *See also* Gossyplure, Propylure.

Clear oily liquid, bp$_{0.001}$ 100-104°. n_D^{25} 1.4484. Insol in water. Sol in hexane, ether, acetone, benzene. LD$_{50}$ in rats (mg/kg): >34600 orally; in rabbits (mg/kg): >2025 dermally (Beroza).

USE: Insect attractant.

4724. Hexamethonium. [60-26-4] N^1,N^1,N^1,N^6,N^6,N^6-Hexamethyl-1,6-hexanediaminium; hexamethylenebis(trimethylammonium); α,ω-bis(trimethylammonium)hexane; hexathonide; hexamethone. [C$_{12}$H$_{30}$N$_2$]$^{2+}$. Prepn: H. J. Barber, **US 2641610** (1953 to May & Baker); H. J. Barber, K. Gaimster, *J. Pharm. Pharmacol.* **3**, 663 (1951). Ganglion blocking agent: R. Wien, D. F. J. Mason, *Br. J. Pharmacol.* **6**, 611 (1951).

Bromide. [55-97-0] C-6; Bistrium bromide; Esametina; Gangliostat; Hexameton bromide; Hexanium bromide; Simpatoblock; Vegolysen, Vegolysin. C$_{12}$H$_{30}$Br$_2$N$_2$; mol wt 362.19. Hygroscopic crystals, mp 274-276°. Sol in water, alc. Insol in acetone, chloroform, ether. pH of 1% soln: 6.2-7.0.

Chloride. [60-25-3] Bistrium chloride; Chloor-hexaviet; Hestrium chloride; Hexameton chloride; Hexone chloride; Hiohex chloride; Methium chloride; Meton. C$_{12}$H$_{30}$Cl$_2$N$_2$; mol wt 273.29. Hygroscopic crystals, dec 289-292°. Sol in water, alc. Practically insol in chloroform, ether. pH of 10% soln: 5.5-6.5.

Iodide. [870-62-2] Hexathide. C$_{12}$H$_{30}$I$_2$N$_2$; mol wt 456.19. Hygroscopic cryst powder. Sol in water. Practically insol in alc.

Tartrate. [2079-78-9] Vegolysen-T. C$_{20}$H$_{40}$N$_2$O$_{12}$; mol wt 500.54. Hygroscopic powder, dec 186°. Sol in water. Practically insol in alc. pH of 10% soln: 3.8.

THERAP CAT: Antihypertensive.

4725. Hexamethyldisilazane. [999-97-3] 1,1,1-Trimethyl-N-(trimethylsilyl)silanamine; HMDS. C$_6$H$_{19}$NSi$_2$; mol wt 161.40. C 44.65%, H 11.87%, N 8.68%, Si 34.80%. Silylating agent. Prepn: R. O. Sauer, *J. Am. Chem. Soc.* **66**, 1707 (1944). Surface tension and density as a function of temperature: R. S. Myers, H. L. Clever, *J. Chem. Eng. Data* **14**, 161 (1969). Mass spec. study: J. Tamás, P. Miklos, *Org. Mass Spectrom.* **10**, 859 (1975). Raman, IR and NMR spectra: K. Hamada, H. Morishita, *Spectrosc. Lett.* **16**, 717 (1983). Molecular structure and conformation: T. Fjeldberg, *J. Mol. Struct.* **112**, 159 (1984). Use in deactivation of chromatography columns: T. Welsch et al., *J. Chromatogr.* **241**, 41 (1982); R. C. Kong et al., *Chromatographia* **18**, 362 (1984). Use as adhesion promoter for photoresists: J. N. Helbert, H. G. Hughes in *Proc. Symp. Adhesion Aspects Polymeric Coatings*, K. L. Mittal, Ed. (Plenum Press, New York, 1983) pp 499-508; D. Freeman, W. Kern, *J. Vac. Sci. Technol. A* **7**, 1446 (1989). Physical properties, toxicity and technical data: Petrarch Systems Inc. Product Information Sheets. Brief review of chemistry and applications: R. J. De Pasquale, *Am. Lab.* **5**(6), 35-39 (1973).

Colorless liquid with an ammonia-like odor. bp 126.0°. d$_4^{20}$ 0.7742. n_D^{20} 1.4080. Flash pt, closed cup: 81°F (27°C). pKa 7.55. Vapor pressure at 50°C: 79 mm. Viscosity: 0.90 cSt. Ignition pt: 380°. Dielectric constant (1000 Hz): 2.27. Insol in water, reacts slowly. LD$_{LO}$ i.p. in mice: 650 mg/kg (Petrarch Systems Inc. Product Information Sheets).

Caution: Causes severe burns to eyes and irritation of the skin (Petrarch Systems Inc. Product Information Sheets).

USE: Deactivation of chromatographic support materials. In electronic industry as an adhesion promoter for photoresists on silicon.

4726. Hexamethyldisiloxane. [107-46-0] 1,1,1,3,3,3-Hexamethyldisiloxane; bis(trimethylsilyl) ether; bis(trimethylsilyl) oxide; HMDO; HMDS; HMDSO; Me$_3$SiOSiMe$_3$; TMSOTMS. C$_6$H$_{18}$OSi$_2$; mol wt 162.38. C 44.38%, H 11.17%, O 9.85%, Si 34.59%. Reagent in organic synthesis used to introduce the trimethylsilyl group. Prepn: R. O. Sauer, *J. Am. Chem. Soc.* **66**, 1707 (1944). Molecular structure: M. J. Barrow et al., *Acta Crystallogr. B* **35**, 2093 (1979); K. B. Borisenko et al., *J. Mol. Struct.* **406**, 137 (1997). Use in the trimethylsilyl protection of alcohols: H. W. Pinnick et al., *Tetrahedron Lett.* **19**, 4261 (1978); of carboxylic acids: H. Matsumoto et al., *Chem. Lett.* **1980**, 1475. Synthetic utility in thionation reactions: T. J. Curphey, *J. Org. Chem.* **67**, 6461 (2002); in multicomponent reactions: M. Anary-Abbasinejad et al., *Synth. Commun.* **38**, 3706 (2008). Application as NMR spectroscopy internal standard: B. C. Hamper et al., *J. Org. Chem.* **63**, 708 (1998).

Colorless liquid with mild odor. *Flammable.* mp −59°. bp$_{757}$ 100.4°. d$_4^{20}$ 0.7638. n_D^{20} 1.3772. Flash pt, closed cup: 33.1°F (0.6°C). Ignition temp: 644°F (340°C). Sol in organic solvents. Insol in water. Hygroscopic. Vapors may form explosive mixture with air. Store under inert gas.

USE: Reagent in synthetic organic chemistry. Internal standard in ^{1}H NMR spectroscopy. In silicone fluids and solvents. Component of liquid bandages.

4727. Hexamethylenediamine. [124-09-4] 1,6-Hexanediamine; 1,6-diaminohexane; HMD. C$_6$H$_{16}$N$_2$; mol wt 116.21. C 62.01%, H 13.88%, N 24.11%. Diamine intermediate derived from adiponitrile, *q.v.*; utilized industrially in the production of aliphatic and semi-aromatic polyamides. Prepn: T. Curtius, H. Clemm, *J. Prakt. Chem.* **62**, 189 (1900); K. H. Slotta, R. Tschesche, *Ber.* **62**, 1398 (1929); S. Alini et al., *J. Mol. Catal. A* **206**, 363 (2003). Manuf: *Faith, Keyes & Clark's Industrial Chemicals*, F. A. Lowenheim, M. K. Moran, Eds. (Wiley-Interscience, New York, 4th ed., 1975) pp 442-444. *Review:* R. A. Smiley in *Ullmann's Encyclopedia of Industrial Chemistry*, **vol. 16** (Wiley-VCH, Weinheim, 6th ed., 2003) pp 433-437.

Colorless platelets, leaflets. Sublimes as long needles. *Corrosive.* Odor of piperidine. mp 42°. bp 205°; bp$_{20}$ 100°. d^{25} 0.854. Flash pt, closed cup: 176°F (80°C); open cup: 201°F (94°C). pK$_1$ 11.11; pK$_2$ 10.01. Vapor pressure (kPa): 6.7 (117.7°); 13.3 (135.1°); 26.7 (154.3°); 60.0 (181.2°); 101.3 (200.6°). Heat of combustion (25°): 40208 kJ/kg. Soly in 100 g water (30°): 960 g. Sol in alcohols, benzene, aromatic solvents; poorly sol in aliphatic hydrocarbons. Hygroscopic. Absorbs carbon dioxide from air. Store under inert gas.

Dihydrochloride. [6055-52-3] C$_6$H$_{16}$N$_2$.2HCl; mol wt 189.12. Needles from water or alcohol, mp 248°. Freely sol in water.

USE: Intermediate in the manuf of nylon and other polyamides.

4728. Hexamethylene Glycol. [629-11-8] 1,6-Hexanediol; 1,6-dihydroxyhexane. C$_6$H$_{14}$O$_2$; mol wt 118.18. C 60.98%, H

11.94%, O 27.08%. Prepd by reduction of ethyl adipate with copper chromite: Lazier *et al.*, *Org. Synth.* **coll. vol. II**, 325 (1943); from 2,5-tetrahydrofurandimethanol: Utne *et al.*, **US 3070633** (1962 to Merck & Co.). Toxicity study: C. P. Carpenter *et al.*, *Toxicol. Appl. Pharmacol.* **28**, 313 (1974).

HO~~~~~~OH

Crystals, mp 42.8°. bp$_{760}$ 208°; bp$_{10}$ 134°. n_D^{25} 1.4579. Dipole moment: 2.48 (dioxane). Sol in water, alcohol; sparingly sol in hot ether. LD$_{50}$ orally in rats: 3.73 g/kg (Carpenter).
Bis(3,5-dinitrobenzoate). [99871-46-2] $C_{20}H_{18}N_4O_{12}$. Crystals from dioxane + ethanol, mp 169-171°.
USE: Intermediate in the production of nylon; to make hexamethylenediamine, polyesters, polyurethans; in gasoline refining; as plasticizer.

4729.   Hexamethylolmelamine. [531-18-0] 1,1',1'',1''',-1'''',1'''''-(1,3,5-Triazine-2,4,6-triyltrinitrilo)hexakismethanol; (*s*-triazine-2,4,6-triyltrinitrilo)hexamethanol; hexakis(hydroxymethyl)melamine; Resloom M 75. $C_9H_{18}N_6O_6$; mol wt 306.28. C 35.29%, H 5.92%, N 27.44%, O 31.34%. Prepn: Gams *et al.*, *Helv. Chim. Acta* **24**, 302E (1941); Widmer, Fisch, **US 2387547** (1945 to Ciba).

Amorphous to cryst, white mass, mp 135-139°. Sol in water. Easily forms insol polymerization products. For the prepn of aq solns double distilled water and hard glass vessels are recommended. A 1.4-2% aq soln is thus possible.
USE: In fireproofing and creaseproofing of cottons, rayons.

4730.   Hexamidine. [3811-75-4] 4,4'-[1,6-Hexanediylbis(oxy)]bisbenzenecarboximidamide; 4,4'-(hexamethylenedioxy)dibenzamidine; 4,4'-diamidino-α,ω-diphenoxyhexane. $C_{20}H_{26}N_4O_2$; mol wt 354.45. C 67.77%, H 7.39%, N 15.81%, O 9.03%. Prepn: A. J. Ewins *et al.*, **GB 507565** (1939 to May & Baker); and trypanocidal activity: J. N. Ashley *et al.*, *J. Chem. Soc.* **1942**, 103. Activity in fibrinolytic systems: J. D. Geratz, *Thromb. Diath. Haemorrh.* **29**, 154 (1973). Antibacterial activity: G. Michel *et al.*, *J. Int. Med. Res.* **14**, 205 (1986). Antifungal activity: M. C. Reynaud, C. Chauve, *Bull. Soc. Fr. Mycol. Med.* **15**, 269 (1986). HPLC determn in pharmaceutics: P. Taylor *et al.*, *J. Pharm. Sci.* **72**, 1477 (1983); in cosmetics: B. Wyhowski de Bukanski, M. O. Masse, *Int. J. Cosmet. Sci.* **6**, 283 (1984). Clinical use as a topical antiseptic: M. J. Fénelon, *Bordeaux Med.* **3**, 867 (1970). Use in treatment of acne: P. Taylor, A. A. Levy, **EP 93186** (1983 to Richardson-Vicks). Amebicidal effects: D. Perrine *et al.*, *Antimicrob. Agents Chemother.* **39**, 339 (1995).

Isethionate. [659-40-5] RF-2535; Desomedine; Hexomedin. $C_{24}H_{38}N_4O_{10}S_2$; mol wt 606.71. Prisms from HCl, mp 246-247° (dec).
USE: Preservative in cosmetics.
THERAP CAT: Antiseptic (topical).

4731.   Hexaminolevulinate. [140898-97-1] 5-Amino-4-oxopentanoic acid hexyl ester; 5-aminolevulinic acid hexyl ester; hexyl 5-aminolevulinate. $C_{11}H_{21}NO_3$; mol wt 215.29. C 61.37%, H 9.83%, N 6.51%, O 22.29%. Prodrug of δ-aminolevulinic acid, *q.v.*; induces production of intracellular protoporphyrin IX. Prepn: H. Takeya, **JP Kokai 92 9360** (1992 to Cosmo Sogo Kenkyusho); and use as photosensitizer: K. E. Gierskcky *et al.*, **WO 9628412** (1996 to Norwegian Radium Res. Found.); *eidem*, **US 6034267** (2000 to PhotoCure). Synthesis and pharmacology: C. J. Whitaker *et al.*, *Anti-Cancer Drug Des.* **15**, 161 (2000). Clinical evaluation for photodetection of bladder cancer: N. Lange *et al.*, *Br. J. Cancer* **80**, 185 (1999).

H_3C~~~~~O~~~~~NH$_2$

Hydrochloride. [140898-91-5] P-1206; Hexvix. $C_{11}H_{21}NO_3$.-HCl; mol wt 251.75. Crystals from methanol/ethyl acetate, mp 94°.
THERAP CAT: Diagnostic aid (fluorescence imaging agent).

4732.   *n*-Hexane. [110-54-3] Hexane. C_6H_{14}; mol wt 86.18. C 83.62%, H 16.38%. Chief constituent of petr ether, ligroin, and commercial hexane. Review of toxicity and metabolism: D. Couri, M. Milk, *Annu. Rev. Pharmacol. Toxicol.* **22**, 145-166 (1982). Book: *Technology and Solvents for Extracting Oilseeds and Nonpetroleum Oil*, P. J. Wan, P. J. Wakelyn, Eds. (AOCS Press, Champaign, 1987) 350 pp.

H_3C~~~~~CH_3

Colorless, very volatile liquid; faint, peculiar odor. d_4^{20} 0.6591. bp 69°. mp −100 to −95°. n_D^{20} 1.375. Flash point: −18.0°C. Vapor pressure (mmHg): 186.1 at 30°, 400.6 at 50°. Partition coefficient (1-octanol/water) at 25°: >4100. Spec heat at 22.0°: 0.536 cal/g/°C. Viscosity at 15°: 0.337 cP. Heat of vaporization: 79.4 cal/g. Surface tension at 20°: 18.41 dyne/cm. Insol in water. Miscible with alcohol, chloroform, ether. LC$_{50}$ (4 hr) in mice by inhalation: 48000 ppm; LD$_{50}$ orally in rats: 32.0 g/kg (Couri, Milks).
Caution: Potential symptoms of overexposure are lightheadedness; giddiness; nausea; headache; peripheral neuropathy; numbness of extremities, muscle weakness; irritation of eyes and nose; dermatitis; aspiration of liquid may cause chemical pneumonia. *See NIOSH Pocket Guide to Chemical Hazards* (DHHS/NIOSH 97-140, 1997) p 162.
USE: Determining refractive index of minerals. Filling for thermometers instead of mercury, usually with a blue or red dye. Extraction solvent for oilseed processing. Organic solvent for high performance liquid chromatography.

4733.   Hexanitrohexaazaisowurtzitane. [135285-90-4] Octahydro-1,3,4,7,8,10-hexanitro-5,2,6-(iminomethenimino)-1*H*-imidazo[4,5-*b*]pyrazine; CL-20; HNIW; 2,4,6,8,10,12-hexanitro-2,-4,6,8,10,12-hexaazatetracyclo[5.5.0.0^{5,9}.0^{3,11}]dodecane. $C_6H_6N_{12}$-O_{12}; mol wt 438.19. C 16.45%, H 1.38%, N 38.36%, O 43.81%. Symmetric polyazacyclic nitramine explosive. Polymorphs α, β, γ, ε, ζ have been identified and confirmed by x-ray diffraction or IR techniques. A sixth polymorph δ has been identified, but not confirmed. First synthesized by A. T. Nielsen in 1987; modified by R. Wardle and J. C. Hinshaw. Gas phase dissociation: R. J. Doyle, Jr., *Org. Mass Spectrom.* **26**, 723 (1991). EPR spectra: M. D. Pace, *J. Phys. Chem.* **95**, 5858 (1991). Identification of δ polymorph by high-pressure phase transition and verification of ζ by ir: T. P. Russell *et al.*, *ibid.* **96**, 5509 (1992). Pressure-temperature phase diagrams: *eidem*, *ibid.* **97**, 1993 (1993). Thermal decomposition: J. C. Oxley *et al.*, *ibid.* **98**, 7004 (1994). High-pressure/thermal shock matrix isolation deflagration and detonation study: J. K. Rice, T. P. Russell, *Chem. Phys. Lett.* **234**, 195 (1995). Review: S. Borman, *Chem. Eng. News* **72**, 18-22 (Jan. 17, 1994).

Explosive. Crystalline.

USE: Rocket propellants. Munitions. Construction or demolition explosive.

4734. 1-Hexanol. [111-27-3] *n*-Hexyl alcohol; amylcarbinol; pentylcarbinol; 1-hydroxyhexane. $C_6H_{14}O$; mol wt 102.18. C 70.53%, H 13.81%, O 15.66%. Occurs as the acetate in seeds and fruits of *Heracleum sphondylium* L. and *H. giganteum* Fisch., *Umbelliferae.* Lab prepn by the action of butylmagnesium bromide on ethylene oxide: Dreger, *Org. Synth.* **coll. vol. I,** 306 (2nd ed., 1941). Reduction of 1,3-hexadienal with iron wire in the presence of nickel acetate: Zeisel, Neuwirth, *Ann.* **433,** 127 (1923); *cf.* Baumgarten, Glatzel, *Ber.* **59,** 2659 (1926); Kuhn, Hoffer, *Ber.* **63,** 2165 (1930). Industrial prepn by reducing ethyl caproate with sodium in abs alcohol: Bouveault, Blanc, **DE 164294** (1903). Toxicity study: H. F. Smyth *et al., Arch. Ind. Hyg. Occup. Med.* **4,** 119 (1951).

$$H_3C\diagup\diagdown\diagup\diagdown OH$$

Liquid. d_4^{25} 0.8153; d_4^{35} 0.8082. mp $-51.6°$. bp_{760} 157°; bp_{400} 138°; bp_{200} 119.6°; bp_{100} 102.8°; bp_{60} 92°; bp_{40} 83.7°; bp_{20} 70.3°; bp_{10} 58.2°; bp_5 47.2°; $bp_{1.0}$ 24.4°. n_D^{25} 1.4162. Absorption spectrum: Massol, Faucon, *Bull. Soc. Chim.* [4] **11,** 932. Flash pt, closed cup: 145°F (63°C). Slightly sol in water; miscible with alcohol, ether. LD_{50} orally in rats: 4.59 g/kg (Smyth).

USE: Manuf antiseptics, hypnotics.

4735. Hexazinone. [51235-04-2] 3-Cyclohexyl-6-(dimethylamino)-1-methyl-1,3,5-triazine-2,4(1*H*,3*H*)-dione; DPX-3674; Velpar. $C_{12}H_{20}N_4O_2$; mol wt 252.32. C 57.12%, H 7.99%, N 22.21%, O 12.68%. Broad spectrum, pre- and post-emergence herbicide effective against woody and herbaceous weeds. Prepn: **NL 7307218**; J. J. Fuchs, J. B. Wommack, **US 3850924**; K. Lin, **US 3983116** (1973, 1974, 1976 all to Du Pont). Herbicidal activity: D. A. Allison, T. D. Joyce, *Proc. Brit. Weed Control Conf.* **1974,** 279; J. D. Riggleman, *Proc. South. Weed Sci. Soc.* **31,** 141 (1978). Field trials: S. J. B. Hay, R. G. Jones, *Proc. Symp. Methods Weed Control* **1977,** 139; D. E. Yarborough *et al., Weed Sci.* **34,** 723 (1986). HPLC determn in soil and water: D. C. Bouchard, T. L. Lavy, *J. Chromatogr.* **270,** 396 (1983). Residues in ground water: D. G. Neary, *South. J. Appl. For.* **7,** 217 (1983). Retention and degradation in soil: K. I. N. Jensen, E. R. Kimball, *Bull. Environ. Contam. Toxicol.* **38,** 232 (1987).

White crystalline solid. mp 97-100.5°. Very soluble in water, 330 g/l at 25°. LD_{50} orally in rats: 1690 mg/kg (Neary).

USE: Herbicide.

4736. 3-Hexen-1-ol. [544-12-7] Leaf alcohol; Blätteralkohol. $C_6H_{12}O$; mol wt 100.16. C 71.95%, H 12.08%, O 15.97%. CH_3-$CH_2CH{=}CHCH_2CH_2OH$. Occurs in leaves of odoriferous plants (including shrubs and trees). Isoln from Japanese oil of peppermint: Walbaum *J. Prakt. Chem.* [2] **96,** 245 (1917); Walbaum, Rosenthal, *Ber. Schimmel* 1929 Jubiläums-Ausg., p 205; from raspberry juice: Bohnsack, *Ber.* **75,** 72 (1942). The natural product has the *cis* configuration: Crombie, Harper, *J. Chem. Soc.* **1950,** 873; Harper, Smith, *ibid.* **1955,** 1512.

Liquid. Strong odor resembling that of isoamyl alc, approaching the odor of green leaves when highly dil. d_{15}^{22} 0.846. bp 156-157°; bp_9 55-56°. n_D^{20} 1.4389.

4737. Hexestrol. [84-16-2] *rel*-4,4′-[(1*R*,2*S*)-1,2-Diethyl-1,2-ethanediyl]bisphenol; 4,4′-(1,2-diethylethylene)diphenol; *meso*-3,4-bis(*p*-hydroxyphenyl)-*n*-hexane; 4,4′-dihydroxy-γ,δ-diphenylhexane; 4,4′-dihydroxy-α,β-diethyldiphenylethane; dihydrodiethylstilbestrol; hexoestrol; Synthovo; Cycloestrol; Hexanoestrol; Hormoestrol; Syntrogène. $C_{18}H_{22}O_2$; mol wt 270.37. C 79.96%, H 8.20%,

O 11.83%. From anethole HBr: Campbell *et al., Proc. Roy. Soc.* **B128,** 253 (1940); Bernstein, Wallis, *J. Am. Chem. Soc.* **62,** 2871 (1940); **US 2357985** (1944); Buu-Hoi, Hoán, *J. Org. Chem.* **14,** 1023 (1949). By the reduction of Grignard compds with cobaltous chloride: Kharasch *et al., J. Am. Chem. Soc.* **65,** 491 (1943). Proof of meso configuration: Wessely, Welleba, *Ber.* **74,** 777 (1941). Comprehensive description: H. Y. Aboul-Enein *et al., Anal. Profiles Drug Subs.* **11,** 347-374 (1982).

Relative stereochemistry

Needles from benzene, thin plates from dil alc, mp 185-188°. Freely sol in ether; sol in acetone, alcohol, methanol; slightly sol in benzene, chloroform. Sol in vegetable oils upon slight warming, also in dil solns of alkali hydroxides. Practically insol in water and in dil mineral acids.

Diacetate. Retalon-Lingual. Crystals, mp 137-139°.

Dipropionate. Retalon Oleosum. Crystals from petr ether, mp 127-128°.

Diphosphate. [14188-82-0] Hexestrol 4,4′-diphosphoric ester; Cytostatin. $C_{18}H_{24}O_8P_2$; mol wt 430.33. Prepn: **GB 593480** (1947 to Roche); Atherton *et al.,* **US 2490573** (1949 to Hoffmann-La Roche).

THERAP CAT: Estrogen; antineoplastic (hormonal).

THERAP CAT (VET): Estrogen.

4738. Hexestrol Bis(β-diethylaminoethyl ether). [2691-45-4] 2,2′-[(1,2-Diethyl-1,2-ethanediyl)bis(4,1-phenyleneoxy)]bis[*N*,-*N*-diethylethanamine]; 2,2‴-[(1,2-diethylethylene)bis(*p*-phenyleneoxy)]bis(triethylamine); 3,4-bis[*p*-(β-diethylaminoethoxy)phenyl]hexane; α,α′-diethyl-4,4′-bis(β-diethylaminoethoxy)bibenzyl. $C_{30}H_{48}$-N_2O_2; mol wt 468.73. C 76.87%, H 10.32%, N 5.98%, O 6.83%. Prepn: Lowe *et al., J. Chem. Soc.* **1951,** 3286.

Dihydrochloride. [69-14-7] Coralgil; Coralgina. $C_{30}H_{48}N_2$-O_2.2HCl; mol wt 541.64. Needles from alc + ethyl acetate, mp 226-227°. Freely sol in water, methanol, chloroform, hot alc.

THERAP CAT: Vasodilator (coronary).

4739. Hexetidine. [141-94-6] 1,3-Bis(2-ethylhexyl)hexahydro-5-methyl-5-pyrimidinamine; 5-amino-1,3-bis(2-ethylhexyl)hexahydro-5-methylpyrimidine; 1,3-bis(β-ethylhexyl)-5-methyl-5-aminohexahydropyrimidine; 5-amino-1,3-di(β-ethylhexyl)hexahydro-5-methylpyrimidine; Glypesin; Hexigel; Hexocil; Hexoral; Hextril; Oraldene; Oraseptic; Sterisil; Steri/Sol. $C_{21}H_{45}N_3$; mol wt 339.61. C 74.27%, H 13.36%, N 12.37%. Prepn: Senkus, *J. Am. Chem. Soc.* **68,** 1611 (1946); **US 2415047**; Bell, Necker, **US 3054797** (1947, 1962 both to Comm. Solvents Corp.). Comprehensive description: G. Satzinger *et al., Anal. Profiles Drug Subs.* **7,** 277-295 (1978).

Liquid. d_{20}^{20} 0.8889. $bp_{0.4}$ 160°. n_D^{20} 1.4668. Sol in petr ether, methanol, benzene, acetone, ethanol, *n*-hexane, chloroform. Practically insol in water. pKa 8.3. Has good thermal stability.

THERAP CAT: Antiseptic.

THERAP CAT (VET): Antiseptic.

4740. Hexobarbital. [56-29-1] 5-(1-Cyclohexen-1-yl)-1,5-dimethyl-2,4,6(1*H*,3*H*,5*H*)-pyrimidinetrione; 5-(1-cyclohexen-1-yl)-1,5-dimethylbarbituric acid; 5-cyclohexenyl-3,5-dimethylbarbituric acid; methylhexabital; methexenyl; enhexymal; hexobarbitone; Citodon; Citopan; Cyclonal; Dorico; Evipal; Evipan; Hexanastab Oral; Noctivane; Sombucaps; Sombulex; Somnalert. $C_{12}H_{16}N_2O_3$; mol wt 236.27. C 61.00%, H 6.83%, N 11.86%, O 20.31%. Prepd by condensation of monomethylurea with methyl cyclohexenylmethylcyanoacetate in abs alcohol in the presence of sodium: **US 1947944**; *cf.* **DE 595175**; **FR 753178**. Comparative clinical trial in anesthesia: D. W. Barron *et al.*, *Br. J. Anaesth.* **38**, 802 (1966). GC determn in human plasma: D. D. Breimer, J. M. Van Rossum, *J. Chromatogr.* **88**, 235 (1974). Pharmacokinetics: N. P. E. Vermeulen *et al.*, *Br. J. Clin. Pharmacol.* **15**, 459 (1983); M. Van der Graaff *et al.*, *Biopharm. Drug Dispos.* **7**, 265 (1986).

Prismatic crystals, practically tasteless, mp 145-147°. Practically insol in water. One gram dissolves in about 3 liters of water. Sol in methanol, hot ethanol, ether, chloroform, acetone, benzene, aq solns of alkali hydroxides, but not carbonates.

Sodium salt. [50-09-9] Hexobarbital soluble; Cyclonal Sodium; Dorico Soluble; Evipal Sodium; Evipan Sodium; Hexanastab; Hexenal; Methexenyl Sodium; Noctivane Sodium; Narcosan Soluble; Privenal. $C_{12}H_{15}N_2NaO_3$; mol wt 258.25. White, bitter, very hygroscopic powder. Very sol in water, freely sol in alcohol, methanol, acetone. Practically insol in chloroform, ether, benzene. A soln in water absorbs carbon dioxide from the air, causing precipitation of the insol free acid. Aq and alcoholic solns are alkaline to litmus and phenolphthalein. The pH of a 10% w/v soln of hexobarbital soluble is about 11.5.

Note: This is a controlled substance (depressant): **21 CFR, 1308.13.**

THERAP CAT: Sedative, hypnotic. Anesthetic (intravenous).

THERAP CAT (VET): Sedative, hypnotic.

4741. Hexobendine. [54-03-5] 3,4,5-Trimethoxybenzoic acid 1,1'-[1,2-ethanediylbis[(methylimino)-3,1-propanediyl]] ester; 3,4,5-trimethoxybenzoic acid diester with 3,3'-[ethylenebis(methylimino)]di-1-propanol; *N*,*N*'-dimethyl-*N*,*N*'-bis[3-(3',4',5'-trimethoxybenzoxy)propyl]ethylenediamine; hexabendin. $C_{30}H_{44}N_2O_{10}$; mol wt 592.69. C 60.80%, H 7.48%, N 4.73%, O 26.99%. Prepn: O. Kraupp, K. Schlögl, **AT 231432**; *eidem*, **US 3267103** (1964, 1966 both to OSSW). Pharmacology: Rudolph *et al.*, *Arzneim.-Forsch.* **20**, 637 (1970); Kolassa, Pfleger, *Biochem. Pharmacol.* **20**, 490 (1971); H. Rameis *et al.*, *Arzneim.-Forsch.* **30**, 671 (1980).

mp 75-77°.

Dihydrochloride. [50-62-4] ST-7090; Reoxyl; Ustimon; Andiamine. $C_{30}H_{44}N_2O_{10}.2HCl$; mol wt 665.60. Crystals, mp 170-174°. uv max: 267 nm. Freely sol in water, less sol in alc. Practically insol in ether.

THERAP CAT: Vasodilator (coronary).

4742. Hexocyclium Methyl Sulfate. [115-63-9] 4-(2-Cyclohexyl-2-hydroxy-2-phenylethyl)-1,1-dimethylpiperazinium methyl sulfate (1:1); *N*-(β-cyclohexyl-β-hydroxy-β-phenylethyl)-*N*¹-methylpiperazine methosulfate; *N*-(β-cyclohexyl-β-hydroxy-β-phenylethyl)-*N*¹-methylpiperazine dimethylsulfate; 4-(β-cyclohexyl-β-hydroxy-β-phenethyl)-1,1-dimethylpiperazinium methyl sulfate; Tral; Tralin. $C_{21}H_{36}N_2O_5S$; mol wt 428.59. C 58.85%, H 8.47%, N 6.54%, O 18.66%, S 7.48%. Anticholinergic. Description: Helgren *et al.*, *J. Am. Pharm. Assoc. Sci. Ed.* **46**, 639 (1957). Prepn: Weston, **US 2907765** (1959 to Abbott).

Crystals, mp 200-210°. uv max (0.1*N* H₂SO₄): 252, 257, 263 nm. Soly in water about 50% w/v. Slightly sol in chloroform. Insol in ether.

THERAP CAT: Antispasmodic.

4743. Hexoprenaline. [3215-70-1] 4,4'-[1,6-Hexanediylbis-[imino(1-hydroxy-2,1-ethanediyl)]]bis-1,2-benzenediol; α,α'-[hexamethylenebis(iminomethylene)]bis[3,4-dihydroxybenzyl alcohol]; *N*,*N*'-bis[2-(3,4-dihydroxyphenyl)-2-hydroxyethyl]hexamethylenediamine; BYK 1512. $C_{22}H_{32}N_2O_6$; mol wt 420.51. C 62.84%, H 7.67%, N 6.66%, O 22.83%. β₂-Adrenergic agonist. Prepn: O. Schmid *et al.*, **AT 241436**; *eidem*, **US 3329709** (1965, 1967, both to OSSW); Schmid *et al.*, **DE 1215729** (1966 to Lentia GmbH). Pharmacology: Thiede *et al.*, *Arzneim.-Forsch.* **21**, 416 (1971). Clinical studies: Schindl, *ibid.* **20**, 1755 (1970); several authors in *Wien. Klin. Wochenschr.* **83**, 75, 80, 101, 114, 117, 130 (1971). Clinical evaluation for interruption of labor: J. Lipshitz *et al.*, *J. Reprod. Med.* **31**, 1023 (1986).

Crystals, mp 162-165° (hemihydrate).

Dihydrochloride. [4323-43-7] ST-1512; Ipradol. $C_{22}H_{32}N_2$-$O_6.2HCl$; mol wt 493.42. Crystals from methanol-ether, mp 197.5-198°.

Sulfate. [32266-10-7] Bronalin; Delaprem; Etoscol; Gynipral; Ipradol; Leanol. $C_{22}H_{32}N_2O_6.H_2SO_4$; mol wt 518.58. Crystals from water-alcohol, mp 222-228°.

THERAP CAT: Bronchodilator; tocolytic.

4744. Hexylcaine Hydrochloride. [532-76-3] 1-(Cyclohexylamino)-2-propanol 2-benzoate hydrochloride (1:1); D-109; Cyclaine. $C_{16}H_{24}ClNO_2$; mol wt 297.82. C 64.53%, H 8.12%, Cl 11.90%, N 4.70%, O 10.74%. Prepn: Cope, Hancock, *J. Am. Chem. Soc.* **66**, 1453 (1944); Cope, **US 2486374** (1949 to Sharp & Dohme). Pharmacokinetics: R. M. Rodgers *et al.*, *Res. Commun. Chem. Pathol. Pharmacol.* **29**, 99 (1980).

Crystals from abs alc, mp 177-178.5°. Soluble in water to the extent of about 12% w/w. A 1% soln is stable to boiling and autoclaving for sterilization purposes.

THERAP CAT: Anesthetic (local).

THERAP CAT (VET): Anesthetic (local).

4745. 2-Hexyldecanoic Acid. [25354-97-6] $C_{16}H_{32}O_2$; mol wt 256.43. C 74.94%, H 12.58%, O 12.48%. Prepn: Lederer *et al.*, *Bull. Soc. Chim. Fr.* **1952**, 413.

Viscous oil, bp$_{0.02}$ 140-150°. n_D^{24} 1.4432.

Sodium salt. [536-37-8] Devaricin. White powder. Sol in water.

THERAP CAT: Sclerosing agent.

4746. Hexylene Glycol. [107-41-5] 2-Methyl-2,4-pentanediol; α,α,α'-trimethyltrimethyleneglycol; pinakon. $C_6H_{14}O_2$; mol wt 118.18. C 60.98%, H 11.94%, O 27.08%. Prepn: Franke, *Monatsh. Chem.* **22**, 1067 (1901); Leopold, **DE 486767** (1925 to I. G. Farben), *Frdl.* **16**, 679; Adkins, Cramer, *J. Am. Chem. Soc.* **52**, 4349 (1930); Arundale, Mikeska, **US 2367324** (1945 to Standard Oil). Toxicity study: Smyth, Carpenter, *J. Ind. Hyg. Toxicol.* **30**, 63 (1948).

Liquid. Mild sweetish odor. d$_{15}^{15}$ 0.924. bp$_{760}$ 198°; bp$_{10}$ 97°. Flash pt (open cup): 93°C (200°F). n_D^{20} 1.4276. Dipole moment: 2.8. Viscosity at 20° = 34 cP. Sol in water, alcohol, ether, lower aliphatic hydrocarbons. LD$_{50}$ orally in rats: 4.70 g/kg (Smyth, Carpenter).

Diacetate. $C_{10}H_{18}O_4$. Liquid. bp$_{12}$ 95°.

Caution: Potential symptoms of overexposure are irritation of eyes, skin, respiratory system; headache, dizziness, nausea, incoordination, CNS depression; dermatitis, skin sensitization. *See NIOSH Pocket Guide to Chemical Hazards* (DHHS/NIOSH 97-140, 1997) p 164.

USE: Cosmetics, hydraulic brake fluids (as coupling agent to castor oil).

4747. Hexyl Methyl Ketone. [111-13-7] 2-Octanone; methyl hexyl ketone. $C_8H_{16}O$; mol wt 128.22. C 74.94%, H 12.58%, O 12.48%.

Liquid; apple odor; camphor taste. d$_4^{20}$ 0.820. mp −16°. bp 172-173°. n_D^{20} 1.41512. Insol in water. Miscible with alc, ether.

4748. 4-Hexylresorcinol. [136-77-6] 4-Hexyl-1,3-benzenediol; 4-hexyl-1,3-dihydroxybenzene; ST-37; Ascaryl; Caprokol; Crystoids; Gelovermin; Sucrets; Worm-Agen. $C_{12}H_{18}O_2$; mol wt 194.27. C 74.19%, H 9.34%, O 16.47%. Prepd by reduction of hexanoylresorcinol with zinc amalgam + dil HCl: Dohme *et al.*, *J. Am. Chem. Soc.* **48**, 1688 (1926); Twiss, *ibid.* 2206; **DE 488419** and **DE 489117** (both 1929 to Sharp & Dohme). Toxicity: Lamson *et al.*, *J. Pharmacol. Exp. Ther.* **53**, 198 (1935).

Pale yellow, heavy liquid becoming solid on standing at room temp. Needles from benzene or petr ether, mp 67.5-69°. bp$_{6-7}$ 178-180°; bp$_{13-14}$ 198-200°; bp$_{760}$ 333-335°. Pungent odor; sharp astringent taste. Sol in ether, chloroform, acetone, alcohol, vegetable oils;

slightly sol in petr ether; sol in about 2000 parts water. LD$_{50}$ orally in rats: 550 mg/kg (Lamson).

Caution: Concd solns can cause irritation of skin, mucous membranes: *Clinical Toxicology of Commercial Products*, R. E. Gosselin *et al.*, Eds. (Williams & Wilkins, Baltimore, 5th ed., 1984) Section II, p 190.

THERAP CAT: Anthelmintic (Nematodes). Antiseptic (topical).

THERAP CAT (VET): Anthelmintic.

4749. Hexythiazox. [78587-05-0] *rel*-(4*R*,5*R*)-5-(4-Chlorophenyl)-*N*-cyclohexyl-4-methyl-2-oxo-3-thiazolidinecarboxamide; *trans*-4-methyl-5-(4-chlorophenyl)-3-cyclohexylcarbamoyl-2-thiazolidone; HTZ; NA-73; DPX-Y5893-9; Nissorun; Acariflor; Cesar; Savey; Zeldox. $C_{17}H_{21}ClN_2O_2S$; mol wt 352.88. C 57.86%, H 6.00%, Cl 10.05%, N 7.94%, O 9.07%, S 9.09%. Prepn: I. Iwataki *et al.*, **DE 3037105**; *eidem*, **US 4442116** (1981, 1984 to Nippon Soda). Miticidal activity: M. A. Hoy, Y.- L. Ouyang, *J. Econ. Entomol.* **79**, 1377 (1986); C. Welty *et al.*, *ibid.* **81**, 586 (1988). Residue determn in crops: M. Tokieda *et al.*, *J. Pestic. Sci.* **12**, 711 (1987), *C.A.* **108**, 162770b (1988). Brief description of physical properties, toxicity and efficacy: Nippon Soda Co., Ltd., *Jpn. Pestic. Inf.* **44**, 21 (1984).

Relative stereochemistry

White odorless crystals, mp 105.5°. Vapor pressure at 20°: 2.54 × 10^{-8} mm Hg. Soly at 20° (g/100 ml): acetone 16, chloroform 137.9, methanol 2.06, *n*-hexane 0.39, xylene 36.2, acetonitrile 2.86; water 0.5 ppm. LD$_{50}$ in male, female mice, male, female rats (mg/kg): all >5000 orally; all >5000 dermally (Nippon Soda Co.).

USE: Acaricide.

4750. HFC-134a. [811-97-2] 1,1,1,2-Tetrafluoroethane; 1,2,-2,2-tetrafluoroethane; R134a; HFA-134a. $C_2H_2F_4$; mol wt 102.03. C 23.54%, H 1.98%, F 74.48%. F_3CCH_2F. Environmentally acceptable alternative to chlorofluorocarbons used as a refrigerant, solvent and propellant. Overview of commercial development: L. E. Manzer, *Catal. Today* **13**, 13-22 (1992). General description: D. Mayer *et al.*, *Jt. Assess. Commod. Chem.* **1995**, 1-36. Dielectric data: Y. Tanaka *et al.*, *Fluid Phase Equilib.* **80**, 107 (1992); pressure-volume data: M.-S. Zhu *et al.*, *ibid.* 149. Density measurements of HFC-134a and its mixtures: P. S. Fialho *et al.*, *High Temp. - High Pressures* **27/28**, 569 (1996). Chemistry of atmospheric degradation: J. Sehested, T. J. Wallington, *Environ. Sci. Technol.* **27**, 146 (1993). Catalytic synthesis and kinetics: F. H. Ribeiro *et al.*, *Catal. Lett.* **45**, 149 (1997). Review of catalytic synthesis: L. E. Manzer, V. N. M. Rao, *Adv. Catal.* **39**, 329-350 (1993). Review of toxicology: D. J. Alexander, S. E. Libretto, *Hum. Exp. Toxicol.* **14**, 715-720 (1995). Review of viscosity measurements: C. M. B. Oliveira, W. A. Wakeham, *High Temp. - High Pressures* **27/28**, 91-98 (1996). Review of use as refrigerant: E. Preisegger, R. Henrici, *Int. J. Refrig.* **15**, 326-331 (1992); as blowing agent: R. E. Morgan *et al.*, *UTECH Asia '97, Int. Polyurethanes Conf. Exhib. Asia-Pac.* (Crain Communications, London, UK, 1997) pp 1-8.

Non-flammable, colorless gas with a faint ethereal odor. bp$_{1atm}$ −26.15°. mp −101°. Critical temp: 101.05°. Critical pressure: 40.64 atm. Critical density: 0.508 g/cm^3. d$_{25}$ 1.202. log P (octanol/water): 1.06. Heat of vaporization (0°): 47.52 cal/g. Viscosity (25°): 0.204 cP. Surface tension (25°): 8.466 dynes/cm. LC$_{50}$ (15 min) in rats: 3400000 mg/m^3 (>800000 ppm); LC$_{50}$ (4 hr) in rats: 2215000 mg/m^3 (>500000 ppm) (Mayer).

USE: Refrigerant, propellant for pharmaceuticals; blowing agent for foams.

4751. HHCB. [1222-05-5] 1,3,4,6,7,8-Hexahydro-4,6,6,7,-8,8-hexamethylcyclopenta[*g*]-2-benzopyran; 6-oxa-1,1,2,3,3,3,8-hexamethyl-2,3,5,6,7,8-hexahydro-1*H*-benz[*f*]-indene; Galaxolide. $C_{18}H_{26}O$; mol wt 258.41. C 83.66%, H 10.14%, O 6.19%. Isochroman musk odorant; scent is primarily due to the (−)-(4*S*,7*R*) and (−)-(4*S*,7*S*) isomers. Prepn: L. G. Heeringa, M. G. J. Beets, **GB 991146**; *eidem*, **US 3360530** (1965, 1967 both to International Flavors & Fragrances). Prepn and olfactory characterization of enantiomers: G. Fráter *et al.*, *Helv. Chim. Acta* **82**, 1656 (1999). Synthesis of olfactorally active stereoisomers: A. Ciappa *et al.*, *Tetrahedron: Asymmetry* **13**, 2193 (2002). HPLC determn: W. Schüssler, L. Nitschke, *Fresenius J. Anal. Chem.* **361**, 220 (1998). Subchronic toxicity study: A. M. Api, R. A. Ford, *Toxicol. Lett.* **111**, 143 (1999).

(4*S*,7*R*)-Form

Colorless crystals from methanol, mp 57-58°. bp$_{0.8}$ 129°. Strong musk odor. n_D^{20} 1.5342. d$_4^4$ 1.0054. Crystal density: 1.087 g/cm³.

(−)-(4*S*,7*R*)-Form. [252332-95-9] Colorless crystals from methanol, mp 78-78.5°. $[\alpha]_D^{24}$ −23.5°; $[\alpha]_{546}^{24}$ −27.3° (c = 5.0 in ethanol). Powerful typical musk odor.

(−)-(4*S*,7*S*)-Form. [172339-62-7] Colorless crystals from hexane, mp 57-58°. $[\alpha]_D^{23}$ −25.1°; $[\alpha]_{546}^{23}$ −29.2° (c = 5.0 in ethanol). Musky odor with dry aspects; slightly less powerful than (−)-(4*S*,7*R*)-form.

(+)-(4*R*,7*S*)-Form. [172339-63-8] Colorless crystals from methanol, mp 77-78°. $[\alpha]_D^{24}$ +24.1°; $[\alpha]_{546}^{24}$ +28.0° (c = 5.2 in ethanol). Weak, musky, uncharacteristic odor.

(+)-(4*R*,7*R*)-Form. [252332-96-0] Colorless crystals from hexane, mp 52-53°. $[\alpha]_D^{23}$ +25.2°; $[\alpha]_{546}^{23}$ +29.3° (c = 5.0 in ethanol). Very weak, mainly fruity odor.

USE: Fragrance ingredient in perfumes, soaps, cosmetics, and detergents.

4752. Himbacine. [6879-74-9] (3*S*,3a*R*,4*R*,4a*S*,8a*R*,9a*S*)-4-[(1*E*)-2-[(2*R*,6*S*)-1,6-Dimethyl-2-piperidinyl]ethenyl]decahydro-3-methylnaphtho[2,3-*c*]furan-1(3*H*)-one. C$_{22}$H$_{35}$NO$_2$; mol wt 345.53. C 76.47%, H 10.21%, N 4.05%, O 9.26%. Alkaloid isolated from the bark of the Australian magnolia, *Galbulimima baccata*, *Himantandraceae*, also known as *Himantandra baccata* Bail. Potent muscarinic M$_2$/M$_4$-receptor antagonist. Isoln: R. F. C. Brown *et al.*, *Aust. J. Chem.* **9**, 283 (1956). Crystal structure: J. Fridrichsons, A. M. Mathieson, *Acta Crystallogr.* **15**, 119 (1962). Receptor binding and selectivity: J. H. Miller *et al.*, *J. Pharmacol. Exp. Ther.* **263**, 663 (1992). Structure-activity studies: D. Doller *et al.*, *Bioorg. Med. Chem. Lett.* **9**, 901 (1999). Stereoselective synthesis of enantiomers: M. Takadoi *et al.*, *Tetrahedron* **58**, 9903 (2002). Biomimetic total synthesis: K. Tchabanenko *et al.*, *Org. Lett.* **7**, 585 (2005).

Large, glistening needles from heptane, mp 132°. $[\alpha]_D^{14}$ +63° (c = 1.04 in chloroform). $[\alpha]_D^{24}$ +56° (c = 0.21 in chloroform). Sol in ethanol.

4753. Hippuric Acid. [495-69-2] *N*-Benzoylglycine; benzoylaminoacetic acid; benzamidoacetic acid. C$_9$H$_9$NO$_3$; mol wt 179.18. C 60.33%, H 5.06%, N 7.82%, O 26.79%. Present in the urine of herbivorous animals; also in smaller amounts in human urine. Prepd from benzoyl chloride and glycine in NaOH soln: Ingersoll, Babcock, *Org. Synth.* **coll. vol. II**, 328 (1943).

Crystals. mp 187-188°. One gram dissolves in about 250 ml cold water, 1000 ml chloroform, 400 ml ether, 60 ml amyl alcohol; slightly sol in cold, freely in hot alcohol or hot water; also sol in aq soln of sodium phosphate. Practically insol in benzene, carbon disulfide, petr ether.

Ammonium salt. [532-93-4] C$_9$H$_{12}$N$_2$O$_3$; mol wt 196.21. Crystals. Freely sol in water; sol in alcohol.

Potassium salt monohydrate. [6487-47-4] C$_9$H$_8$KNO$_3$.H$_2$O; mol wt 235.28. Cryst powder. Very sol in water; sol in alcohol.

4754. Hirsutic Acid C. [3650-17-7] (1a*R*,2*R*,3a*R*,3b*R*,5*S*,-6a*R*,7a*S*)-Decahydro-2-hydroxy-3a,5-dimethyl-3-methylenecyclopenta[4,5]pentaleno[1,6a-*b*]oxirene-5-carboxylic acid; 6β,6aβ-epoxy-2,3,3aα,3b,4,5,6,6a,7,7aα-decahydro-5β-hydroxy-2β,3bβ-dimethyl-4-methylene-1*H*-cyclopenta[*a*]pentalene-2-carboxylic acid. C$_{15}$H$_{20}$O$_4$; mol wt 264.32. C 68.16%, H 7.63%, O 24.21%. Antibiotic metabolite isolated from the fungus *Stereum hirsutum*: Heatley *et al.*, *Br. J. Exp. Pathol.* **28**, 35 (1947). Structure and stereochemistry: Comer *et al.*, *Chem. Commun.* **1965**, 310; Comer, Trotter, *J. Chem. Soc. B* **1966**, 11; Comer *et al.*, *Tetrahedron* **23**, 4761 (1967). Synthesis of (±)-form: Hashimoto *et al.*, *Tetrahedron Lett.* **1974**, 3745; M. Yamazaki *et al.*, *Chem. Lett.* **1981**, 1245. Stereosynthesis: B. M. Trost *et al.*, *J. Am. Chem. Soc.* **101**, 1284 (1979); of (+)-form: M. Shibasaki *et al.*, *Tetrahedron Lett.* **23**, 5311 (1982).

Prisms from ethanol, mp 179-182°. $[\alpha]_D^{23}$ +116° (c = 1.05).

Methyl ester. C$_{16}$H$_{22}$O$_4$. Colorless prisms, mp 161-162°. $[\alpha]_D^{20}$ +119° (c = 2.25).

4755. Hirudin. [8001-27-2] Exhirud; Hirudex. Anticoagulant protein extracted from the salivary glands of the medicinal leech, *Hirudo medicinalis* Linn. Hirudin is a single chain polypeptide containing 65 amino acids; mol wt approximately 7000 daltons. Several isoforms have been identified. A specific inhibitor of thrombin, *q.v.*, hirudin does not require the presence of other coagulation factors. Isoln: J. B. Haycraft, *Arch. Exp. Pathol. Pharmakol.* **18**, 209 (1884). Initial characterization: F. Markwardt, *Naturwissenschaften* **42**, 537 (1955). Improved isoln: P. Walsmann, F. Markwardt, *Thromb. Res.* **40**, 563 (1985). Amino acid sequence: T. E. Petersen *et al.* in *Protides of the Biological Fluids* **vol. 23**, H. Peeters, Ed. (Pergamon Press, London, 1976) pp 145-149. Antithrombin activity: F. Markwardt, *Methods Enzymol.* **19**, 924 (1970). Pharmacokinetics in humans: F. Markwardt *et al.*, *Thromb. Haemostasis* **52**, 160 (1984). Chromogenic assay in plasma: U. Griessbach *et al.*, *Thromb. Res.* **37**, 347 (1985). Series of articles on development, pharmacology, and therapeutic potential of natural and recombinant hirudins: *Semin. Thromb. Hemostasis* **15**, 261-333 (1989). Review of pharmacology and therapeutic potential: P. H. Johnson, *Annu. Rev. Med.* **45**, 165-177 (1994); of therapeutic use in myocardial infarction: U. Zeymer, K.-L. Neuhaus, *Drug Saf.* **12**, 234-239 (1995).

Gray or white flakes or powder. pI 3.9. Sol in water, physiol saline soln, in pyridine. Practically insol in alcohol, ether, acetone, benzene. Deteriorates on storage in sealed ampuls, on exposure to heat and when in soln with dil acids.

Desirudin. [120993-53-5] 63-Desulfohirudin (*Hirudo medicinalis* isoform HV1); ds-hirudin; CGP-39393; Revasc. Identical in amino acid sequence to natural hirudin variant 1 (HV1) except that it lacks a sulfate group on Tyr at residue 63. Production by recombinant DNA technology: M. Liersch *et al.*, **EP 168342** (1986 to

Ciba-Geigy); *eidem*, **US 5422249** (1995 to Ciba-Geigy; UCP Gen-Pharma); E. Fortkamp *et al.*, *DNA* **5**, 511 (1986); J. Dodt *et al.*, *FEBS Lett.* **202**, 373 (1986). Physicochemical properties: A. Electricwala *et al.*, *Thromb. Haemostasis* **63**, 499 (1990). Clinical pharmacology: M. Verstraete *et al.*, *J. Am. Coll. Cardiol.* **22**, 1080 (1993). Clinical trial in coronary angioplasty: A. A. van den Bos *et al.*, *Circulation* **88**, 2058 (1993); in acute coronary syndromes: E. J. Topol *et al.*, *N. Engl. J. Med.* **335**, 775 (1996).

Lepirudin. [138068-37-8] 1-L-Leucine-2-L-threonine-63-desul-fohirudin (*Hirudo medicinalis* isoform HV1); HBW-023; Refludan. Peptide of 65 amino acids produced by recombinant DNA technology. Prepn: P. Crause *et al.*, **EP 324712**; *eidem*, **US 5180668** (1989, 1993 both to Hoechst AG). Clinical pharmacology: H. J. Roethig *et al.*, *Int. Congr. Ser.* **944**, 227 (1991). Clinical trial as adjunctive therapy with TPA: U. Zeymer *et al.*, *Am. J. Cardiol.* **76**, 997 (1995); in heparin-induced thrombocytopenia: F. Schiele *et al.*, *Am. J. Hematol.* **50**, 20 (1995).

THERAP CAT: Antithrombotic.

4756. Histamine. [51-45-6] 1*H*-Imidazole-5-ethanamine; 2-(4-imidazolyl)ethylamine; 4-imidazoleethylamine; 5-imidazoleethylamine; β-aminoethylimidazole; β-aminoethylglyoxaline. $C_5H_9N_3$; mol wt 111.15. C 54.03%, H 8.16%, N 37.81%. Biogenic amine widely distributed in nature; decarboxylation product of histidine, *q.v.* Present in most mammalian tissues; primarily stored in mast cells and basophils. Exhibits multiple biological effects through at least 3 specific receptors. Induces bronchoconstriction and vasodilation; stimulates gastric acid secretion; and acts as a neurotransmitter. Synthesis: A. Windaus, W. Vogt, *Ber.* **40**, 3691 (1907); F. L. Pyman, *J. Chem. Soc.* **101**, 530 (1912). Identification as vasodilator: H. H. Dale, P. P. Laidlaw, *J. Physiol.* **41**, 318 (1910). Toxicity data: K. Nagai *et al.*, *Arzneim.-Forsch.* **17**, 1575 (1967). Diagnostic use in pheochromocytoma: T. Nakai, R. Yamada, *J. Clin. Endocrinol. Metab.* **57**, 19 (1983); in airway responsiveness: A. James, G. Ryan, *Respirology* **2**, 97 (1997). Clinical efficacy in combination with interleukin-2, *q.v.*, in treatment of acute myelogenous leukemia: K. Hellstrand *et al.*, *Leuk. Lymphoma* **27**, 429 (1997). Mechanism of antineoplastic activity: *eidem*, *Acta Oncol.* **37**, 347 (1998). Review of assay methods for determn in plasma: E. Oosting *et al.*, *Clin. Exp. Allergy* **20**, 349-357 (1990). Review of role in allergic disease: M. V. White, *J. Allergy Clin. Immunol.* **86**, 599-605 (1990); in gastric acid secretion: K. J. Obrink, *Scand. J. Gastroenterol.* **26**, Suppl. 180, 4-8 (1991); in homeostatic energy metabolism: T. Sakata *et al.*, *Nutrition* **13**, 403-411 (1997). Review of pharmacology and classification of receptors: S. J. Hill *et al.*, *Pharmacol. Rev.* **49**, 253-278 (1997).

Deliquescent needles from chloroform, mp 83-84°. bp$_{18}$ 209-210°. Freely sol in water, alcohol, hot chloroform. Practically insol in ether. LD$_{50}$ i.p. in mice: 2020 mg/kg (Nagai).

Dihydrochloride. [56-92-8] Ceplene. $C_5H_9N_3.2HCl$; mol wt 184.06. Prisms from ethanol, mp 244-246°. Freely sol in water, methanol; sol in ethanol.

Phosphate. [51-74-1] $C_5H_9N_3.2H_3PO_4$. Clear, colorless prisms from water, mp 132-133°. Freely sol in water.

THERAP CAT: Antineoplastic (immunomodulator); diagnostic aid (gastric secretion, pheochromocytoma, bronchial hyperreactivity).

4757. Histatins. Histidine-rich proteins; parotid basic proteins; post-parotid basic proteins. Group of small, cationic, salivary proteins with antimicrobial and wound healing properties. Histatins bind to hydroxyapatite and have a role in the acquired enamel pellicle of human teeth. Produced and secreted by the parotid and submandibular salivary glands; histatins 1, 3, and 5 are the most abundant. Histatins 1 and 3 are the primary gene products; other histatins are proteolytic fragments of the 2 parent proteins. Isoln of histidine-rich proteins from human parotid saliva: D. I. Hay, *Arch. Oral Biol.* **20**, 553 (1975); B. J. Baum *et al.*, *Arch. Biochem. Biophys.* **177**, 427 (1976). Characterization and structures of histatins 1, 3, 5: F. G. Oppenheim *et al.*, *J. Biol. Chem.* **263**, 7472 (1988); of 2, 4, and 6 - 12: R. F. Troxler *et al.*, *J. Dent. Res.* **69**, 2 (1990). Identification of

HIS genes: L. M. Sabatini, E. A. Azen, *Biochem. Biophys. Res. Commun.* **160**, 495 (1989). Improved isoln via zinc precipitation: B. Flora *et al.*, *Protein Expression Purif.* **23**, 198 (2001). Review of structures, activity, and therapeutic potential as antifungal agents: H. Tsai, L. A. Bobek, *Crit. Rev. Oral Biol. Med.* **9**, 480-497 (1998); K. Kavanagh, S. Dowd, *J. Pharm. Pharmacol.* **56**, 285-289 (2004). Wound-closure stimulating properties: M. J. Oudhoff *et al.*, *FASEB J.* **22**, 3805 (2008).

```
1                                                          16
Asp-Ser-His-Ala-Lys-Arg-His-His-Gly-Tyr-Lys-Arg-Lys-Phe-His-Glu
                                                           |
32                                  24                     |
Asn-Asp-Tyr-Leu-Tyr-Asn-Ser-Arg-Tyr-Gly-Arg-His-Ser-His-His-Lys
```

Histatin 3

Histatin 1. [115966-66-0] (unspecified); [101056-53-5] (human parotid saliva). HRP-1; PPb. $C_{217}H_{298}N_{69}O_{64}P$; mol wt 4928.16. Primary translation product of the HIS1 gene; contains 38 amino acid residues, including a phosphorylated serine at position 2. pI 7.04.

Histatin 3. [115966-67-1] (unspecified); [112844-49-2] (human parotid saliva). HRP-3; Pbe. $C_{178}H_{258}N_{64}O_{48}$; mol wt 4062.42. Primary translation product of the HIS2 gene; contains 32 amino acid residues. pI >9.5.

Histatin 5. [115966-68-2] HRP-5; Pbb. $C_{133}H_{195}N_{51}O_{33}$; mol wt 3036.35. Proteolytic cleavage product of histatin 3 corresponding to residues 1-24.

4758. Histidine. [71-00-1] L-Histidine; His; H; (*S*)-α-amino-1*H*-imidazole-4-propanoic acid; α-amino-4(or 5)-imidazolepropionic acid; glyoxaline-5-alanine. $C_6H_9N_3O_2$; mol wt 155.16. C 46.45%, H 5.85%, N 27.08%, O 20.62%. Essential amino acid for human development; precursor to histamine and a component of carnosine, *q.v.* Isoln: A. Kossel and S. G. Hedin, *Zeit. Physiol.* **22**, 176, 191 (1896). Early chemistry and biochemistry: *Amino Acids and Proteins*, D. M. Greenberg, Ed. (Charles C. Thomas, Springfield, IL, 1951) 950 pp., *passim*; J. P. Greenstein, M. Winitz, *Chemistry of the Amino Acids* vols **1-3** (John Wiley and Sons, Inc., New York, 1961) pp. 1971-1995, *passim*. NMR spectra: D. H. Sachs *et al.*, *J. Biol. Chem.* **246**, 6576 (1971). HPLC determn in dipeptides: E. Kasziba *et al.*, *J. Chromatogr.* **432**, 315 (1988). Often found at the active site of proteins: W. T. Morgan *et al.*, *J. Biol. Chem.* **268**, 6256 (1993); J.-W. P. Boots *et al.*, *Biochim. Biophys. Acta* **1248**, 27 (1995). Role in proton translocation – a model: J. E. Morgan *et al.*, *J. Bioenerg. Biomembr.* **26**, 599 (1994). Review of metabolism: F. B. Stifel, R. H. Herman, *Am. J. Clin. Nutr.* **24**, 207-217 (1971). Review of role in nutrition: *eidem*, *ibid.* **25**, 182-185 (1972); L. P. Mercer *et al.*, *Nutrition* **6**, 273-277 (1990).

Sweet needles or plates, dec 287° (softens at 277°). $[\alpha]_D^{20}$ −10.9° (c = 0.77 in 0.5*N* NaOH); $[\alpha]_D^{25}$ −38.95° (c = 0.75 to 3.77); $[\alpha]_D^{25}$ +13.34° (c = 1.00-4.05 in 6.1*N* HCl). pK$_1$ 1.82; pK$_2$ 6.00; pK$_3$ 9.17. Soly in water at 25°: 41.9 g/l. Very slightly sol in alc. Insol in ether and other common neutral solvents.

Monohydrochloride. Ecristidine; Laristine; Larostidin; Plexamine. $C_6H_9N_3O_2.HCl$; mol wt 191.62. Rhombic crystals, dec 251-252° (also forms a monohydrate, mp 80°, anhydrous, mp at 140°). $[\alpha]_D^{26}$ +8.0° (c = 2 in 3 mols HCl). Fairly sol in water. Insol in alcohol, ether.

DL-Form. Quadrilateral plates or tetragonal prisms, dec 285°. Sol in water.

4759. Histones. Small chromosomal proteins (mol wt 12,000-20,000) possessing an open, unfolded structure, attached to DNA of cell nuclei by ionic linkages. First isolated from bird erythrocytes: Kossel, *Z. Physiol. Chem.* **8**, 511 (1884). Prepd also from cell nuclei of calf thymus, spleen, and liver, liver and spleen of leukemic rats; bovine spermatozoa: Hnilica, *Biochem. J.* **82**, 123 (1962); Berry, Mayer, *Exp. Cell Res.* **20**, 116 (1960). Classification is based on relative amounts of lysine and arginine: Histone I is very rich in lysine and has several subtypes; histone II, moderately rich in lysine with two subtypes; histone III, moderately rich in arginine, contains cysteine; histone IV, very rich in arginine and in glycine. Histones of the same type obtained from various plant and animal sources are very similar in amino acid sequence. Amino acid composition of animal histones: Hnilica et al., *Biochim. Biophys. Acta* **124**, 109 (1966); of plant histones and similarity of plant and animal histones: Fambrough, Bonner, *Biochemistry* **5**, 2563 (1966). Complete amino acid sequence of calf thymus histone IV: DeLange et al., *J. Biol. Chem.* **244**, 319 (1969); Ogawa et al., *ibid.* 4387; of histone III: DeLange et al., *Proc. Natl. Acad. Sci. USA* **69**, 882 (1972); eidem, *J. Biol. Chem.* **248**, 3261 (1973); Hooper et al., *ibid.* 3275; of the two subtypes of histone II: Iwai et al., *Nature* **226**, 1056 (1970); Yeoman et al., *J. Biol. Chem.* **247**, 6018 (1972). Toxicity study: Starbuck et al., *Arch. Int. Pharmacodyn. Ther.* **165**, 374 (1967). Review of structural studies: DeLange, *Acc. Chem. Res.* **5**, 368 (1972). Review of histone in the chromosome structure: Bradbury et al., *Ann. N.Y. Acad. Sci.* **222**, 266 (1973). Review of biological functions: Binner, Garrard, *Life Sci.* **14**, 209 (1974). General review: Butler et al., *Prog. Biophys. Mol. Biol.* **18**, 211 (1968); R. J. DeLange, E. L. Smith in *Proteins* vol. **4**, H. Neurath, R. L. Hill, Eds. (Academic, New York, 3rd ed., 1979) pp 134-243.

Histones are susceptible to enzymatic cleavage; sol in an Hg_2SO_4-H_2SO_4 medium. Infrared spectra: de Lozé, *Compt. Rend.* **246**, 417 (1958). LD_{50} i.v. in rats of lysine-rich, slightly lysine-rich, and arginine-rich histones: 90, 60-70, 60 mg/kg (Starbuck).

4760. Histrelin. [76712-82-8] 6-[1-(Phenylmethyl)-D-histidine]-9-(*N*-ethyl-L-prolinamide)-1-9-luteinizing hormone-releasing factor (swine); 6-[1-(phenylmethyl)-D-histidine]-9-(*N*-ethyl-L-prolinamide)-10-deglycinamideluteinizing hormone-releasing factor (pig); L-pyroglutamyl-L-histidyl-L-tryptophyl-L-seryl-L-tyrosyl-D-N^{im}-benzylhistidyl-L-leucyl-L-arginyl-L-proline ethylamide; [(im-Bzl)-D-His^6,Pro^9-NEt]-gonadotropin-releasing hormone; ORF-17070; RWJ-17070. $C_{66}H_{86}N_{18}O_{12}$; mol wt 1323.53. C 59.89%, H 6.55%, N 19.05%, O 14.51%. Synthetic nonapeptide agonist analog of gonadotropin-releasing hormone, *q.v.* Prepn: J. E. F. Rivier, W. W. Vale, Jr., **EP 21620**; eidem, **US 4244946** (both 1981 to Salk Inst.). *See also:* J. E. Rivier, W. W. Vale, *Life Sci.* **23**, 869 (1978). Pharmacology: Y-Q. Cao et al., *Int. J. Androl.* **5**, 158 (1982); J. W. Gunnet et al., *J. Endocrinol.* **131**, 211 (1991). HPLC purification: J. Rivier et al., *J. Chromatogr.* **288**, 303 (1984). Amino acid sequence: C. J. Shaw, M. L. Cotter, *Chromatographia* **21**, 197 (1986). Review of pharmacology and therapeutic use in central precocious puberty (CPP): L. B. Barradell, D. McTavish, *Drugs* **45**, 570-588 (1993). Efficacy and safety of subdermal implant in CPP: E. A. Eugster et al., *J. Clin. Endocrinol. Metab.* **92**, 1697 (2007). Review of pharmacology and pharmacokinetics in prostate cancer: P. Schlegel, *BJU Int.* **103**, Suppl. 2, 7-13 (2009); of clinical experience: E. D. Crawford, *ibid.* 14-22.

$[\alpha]_D^{20}$ −33.9° (c = 1 in acetic acid).

Acetate. [220810-26-4] Supprelin LA; Vantas. $C_{66}H_{86}N_{18}O_{12} \cdot xC_2H_4O_2$

THERAP CAT: Antineoplastic (hormonal). In treatment of central precocious puberty.

4761. Histrionicotoxin. [34272-51-0] (2*S*,6*R*,7*S*,8*S*)-7-(1*Z*)-1-Buten-3-yn-1-yl-2-(2*Z*)-2-penten-4-yn-1-yl-1-azaspiro[5.5]undecan-8-ol; HTX; (−)-HTX 1; (−)-histrionicotoxin. $C_{19}H_{25}NO$; mol wt 283.42. C 80.52%, H 8.89%, N 4.94%, O 5.64%. One of a family of piperidine alkaloids isolated from the skin of the poison arrow frog *Dendrobates histrionicus*. Noncompetitive inhibitor of nicotinic acetylcholine receptors characterized by an unusual spirocyclic *cis*-enyne structure. Virtually nontoxic, the family was originally misnamed because of the source. Isoln: J. W. Daly et al., *Proc.*

Natl. Acad. Sci. USA **68**, 1870 (1971). Isoforms: *idem et al.*, *Helv. Chim. Acta* **60**, 1128 (1977). Acetylcholine antagonism: M. Glavinovic et al., *Can. J. Physiol. Pharmacol.* **52**, 1220 (1974); as allosteric inhibitor: S. M. Sine, P. Taylor, *J. Biol. Chem.* **257**, 8106 (1982). Total synthesis: S. C. Carey et al., *Tetrahedron Lett.* **26**, 5887 (1985); stereospecific synthesis: G. M. Williams et al., *J. Am. Chem. Soc.* **121**, 4900 (1999). *Review:* J. W. Daly, *J. Nat. Prod.* **61**, 162-172 (1997).

Viscous oil; on storage at −15°C crystals slowly form, mp 75-76°. uv max: 224 nm (ε 15500). $[\alpha]_D^{20}$ −112° (c = 0.34 in ethanol).

USE: Biochemical probe for neuromuscular transmission.

4762. HLö-7. [120103-35-7] 1-[[[4-(Aminocarbonyl)pyridinio]methoxy]methyl]-2,4-bis[(hydroxyimino)methyl]pyridinium iodide (1:2); HLoe-7. $C_{15}H_{17}I_2N_5O_4$; mol wt 585.14. C 30.79%, H 2.93%, I 43.38%, N 11.97%, O 10.94%. Bisquaternary Hagedorn oxime; acetylcholinesterase reactivator. *In vitro* reactivation of tabun and soman acetylcholinesterase inhibition: L. P. A. de Jong et al., *Biochem. Pharmacol.* **38**, 633 (1989). Stability and decomposition: P. Eyer et al., *Arch. Toxicol.* **63**, 59 (1989). Synthesis, pharmacology and toxicity: eidem, *ibid.* **66**, 603 (1992). Pharmacokinetics: U. Spöhrer et al., *ibid.* **68**, 480 (1994). Efficacy in nerve agent poisoning: B. P. C. Melchers et al., *Pharmacol. Biochem. Behav.* **49**, 781 (1994). MS determn: P. A. D'Agostino et al., *Rapid Commun. Mass Spectrom.* **10**, 805 (1996). Comparative reactivation studies in human acetylcholinesterase inhibition: F. Worek et al., *Biochem. Pharmacol.* **68**, 2237 (2004).

Crystals from hot water, mp 160° (dec). Soly in water: 20 mg/ml. Aq solns are unstable and liberate iodine even when protected from light. uv max (pH 2): 298, 252, 224 nm (log ε 4.20, 4.20, 4.15); (pH 11) 368 nm (log ε 4.35). pKa_1 7.04, pKa_2 8.52. LD_{50} i.m. in female mice: 744 μmol/kg (Eyer, 1992).

Dimethanesulfonate. [145613-73-6] $C_{15}H_{17}N_5O_4 \cdot 2CH_3O_3S$; mol wt 521.52. Crystals from warm methanol, 173-174 (dec). pH (1% aq soln) ~4. Soly in water: 0.68 g/ml. LD_{50} i.m. in female mice: 799 μmol/kg (Eyer, 1992).

THERAP CAT: Antidote (organophosphate poisoning).

4763. HMPA. [680-31-9] *N,N,N′,N′,N″,N″*-Hexamethylphosphoric triamide; hexamethylphosphoramide; hexametapol; hempa; HMPT; ENT-50882. $C_6H_{18}N_3OP$; mol wt 179.20. C 40.22%, H 10.13%, N 23.45%, O 8.93%, P 17.28%. Prepn: Saul, Godfrey, **US 2752392** (1956 to Monsanto); Godfrey, **US 2852550** (1958 to Monsanto); Miller, Lomonte, **US 3084190** (1963 to Dow); Vetter, Noeth, *Ber.* **96**, 1308 (1963). Solvation effects: J. E. Dubois, A. Bienvenue, *Tetrahedron Lett.* **1966**, 1809. Carcinogenicity studies: J. A. Zapp, *Science* **190**, 422 (1975); *IARC Monographs* **15**, 211 (1977). *Review:* H. Normant, *Angew. Chem. Int. Ed.* **6**, 1047 (1967).

Colorless liquid, completely miscible with water. mp 7.20°. bp$_{760}$ 235°. bp$_{11}$ 105-107°, bp$_6$ 97-99°; bp$_{2.5}$ 78°. n_D^{21} 1.4572. d$_{20}$ 1.03. Dipole moment at 25°: 5.37D. LD$_{50}$ in male, female rats (mg/kg): 2650, 3360 orally. *See:* T. B. Gaines, *Toxicol. Appl. Pharmacol.* **14**, 515 (1969).

Caution: Potential symptoms of overexposure are irritation of eyes, skin, respiratory system; dyspnea; abdominal pain. *See NIOSH Pocket Guide to Chemical Hazards* (DHHS/NIOSH 97-140, 1997) p 162. This substance is reasonably anticipated to be a human carcinogen: *Report on Carcinogens, Twelfth Edition* (PB2011-111646, 2011) p 229.

USE: Aprotic solvent in organic synthesis. Solvent for polymers, gases; polymerization catalyst; thermal stabilizer in polystyrene; protective additive for polyvinyl and polyolefin resins against uv light degradation. De-icing additive for jet fuels. Chemosterilant for a number of insect pests; chemical mutagen.

4764. HNE. [18286-49-2] (*E*-form); [29343-52-0] (unspecified stereo). (2*E*)-4-Hydroxy-2-nonenal; 4-HNE. C$_9$H$_{16}$O$_2$; mol wt 156.23. C 69.19%, H 10.32%, O 20.48%. Toxic aldehyde formed during lipid peroxidation in response to oxidative stress. Reacts primarily with the amino acids His or Lys to form an adduct which may alter protein function. Increased levels of the HNE-adduct are seen in Alzheimer's disease and after ozone exposure. Identification of cytotoxic activity of lipid oxidative products: E. Schauenstein, H. Esterbauer, *Monatsh. Chem.* **94**, 164 (1963). Prepn: H. Esterbauer, W. Weger, *ibid.* **98**, 1994 (1967). Synthesis: H. W. Gardner *et al., Lipids* **27**, 686 (1992). Quantitative analysis by uv, GC/MS, HPLC: M. Kinter in *Free Radicals* N. A. Punchard, F. J. Kelly, Eds. (IRL Press, Oxford, UK, 1996) pp 133-145. Cytotoxicity: A. Benedetti *et al., Biochim. Biophys. Acta* **620**, 281 (1980). Structural definition of adduct sites: D. V. Nadkarni, L. M. Sayre, *Chem. Res. Toxicol.* **8**, 284 (1995); by MS: M. S. Bolgar, S. J. Gaskell, *Anal. Chem.* **68**, 2325 (1996). Ozone-induced formation: A. Kirichenko *et al., Toxicol. Appl. Pharmacol.* **141**, 416 (1996). Role in Alzheimer's disease: R. J. Mark *et al., J. Neurochem.* **68**, 255 (1997); L. M. Sayre *et al., ibid.* 2092.

bp$_{0.4\ torr}$ 84-87°. n_D 1.471. d$_4^{20}$ 0.944. uv max (water): 447 nm; (ethanol) 452.5 nm; (hexane) 465 nm (ε 13750, 13100, 14400). Soly at 20° (g/l): water 6.6. Partition coefficient K$_{20}$ (H$_2$O/CHCl$_3$): 0.04.

4765. Holmium. [7440-60-0] Ho; at. wt 164.93032; at. no. 67; valence 3. A rare earth metal of the yttrium group; member of the lanthanide series. Naturally occurring isotope (mass number): 165; known artificial radioactive isotopes: 146-164; 166-170. Estimated abundance in earth's crust 1.15-1.4 ppm; occurs in rare earth minerals. Discovered by Soret in 1878 and independently by Cleve in 1879: Cleve, *Compt. Rend.* **89**, 478, 708 (1879). Purification of salts: Holmberg, *Z. Anorg. Chem.* **71**, 226 (1911). Separation from erbium: Driggs, Hopkins, *J. Am. Chem. Soc.* **47**, 363 (1925); from other rare earths: Spedding *et al., ibid.* **69**, 2812 (1947); **76**, 2557 (1954). Absorption spectrum: Severin, *Z. Phys.* **125**, 455 (1949). Analysis by means of emission spectra: Smith, Wiggins, *Analyst* **74**, 95 (1949). Toxicity study: Haley *et al., Toxicol. Appl. Pharmacol.* **8**, 37 (1966). Reviews of prepn, properties and compds: *The Rare Earths,* F. H. Spedding, A. H. Daane, Eds. (Krieger, Huntington, N.Y., 1971, reprint of 1961 ed.) 641 pp; Hulet, Bode, "Separation Chemistry of the Lanthanides and Transplutonium Actinides" in *MTP Int. Rev. Sci.: Inorg. Chem., Ser. One* vol. 7, K. W. Bagnall, Ed. (University Park Press, Baltimore, 1972) pp 1-45; Moeller, "The Lanthanides" in *Comprehensive Inorganic Chemistry* vol. 4, J. C. Bailar, Jr. *et al.,* Eds. (Pergamon Press, Oxford, 1973) pp 1-101; F. H. Spedding in *Kirk-Othmer Encyclopedia of Chemical Technology* vol. 19, (John Wiley & Sons, New York, 3rd ed., 1982) pp 833-854; *Chemistry of the Elements,* N. N. Greenwood, A. Earnshaw, Eds. (Pergamon Press, New York, 1984) pp 1423-1449. Brief review of properties: G. T. Seaborg, *Radiochim. Acta* **61**, 115-122 (1993).

Metal; hexagonal close-packed crystals. d 8.7947. mp 1474°. bp 2700°. Heat of fusion: 16.874 kJ/mol. Heat of sublimation (25°): 300.8 kJ/mol. Forms yellow-green salts.

Oxide. [12055-62-8] Holmia. Ho$_2$O$_3$; mol wt 377.86. Yellow solid. Obtained by igniting the hydroxide, nitrate, sulfate, oxalate. Dissolves in acids with formation of a yellow salt.

Chloride. [10138-62-2] Cl$_3$Ho; mol wt 271.28. Bright yellow, crystalline solid, mp 718°. Formed by heating the hydrated salt in a current of hydrogen chloride at 350°. LD$_{50}$ in mice: 560 mg/kg i.p.; 7.2 g/kg orally (Haley).

Bromide. [13825-76-8] Br$_3$Ho; mol wt 404.64. mp 914°.

Iodide. [13813-41-7] HoI$_3$; mol wt 545.64. Obtained by passing hydrogen iodide over the anhyd chloride at 600°: Jantsch *et al., Z. Anorg. Chem.* **207**, 353 (1932). Light-yellow solid. mp 1010 ±10°.

4766. Holomycin. [488-04-0] *N*-(4,5-Dihydro-5-oxo-1,2-dithiolo[4,3-*b*]pyrrol-6-yl)acetamide; *N*-(4,5-dihydro-5-oxo-1,2-dithiolo[4,3-*b*]pyrrol-6-yl)-*N*-methylformamide; 6-acetamido-1,2-dithiolo[4,3-*b*]pyrrol-5(4*H*)-one; *N*-demethylthiolutin. C$_7$H$_6$N$_2$O$_2$S$_2$; mol wt 214.26. C 39.24%, H 2.82%, N 13.07%, O 14.93%, S 29.93%. Antibiotic substance produced by a strain of *Streptomyces griseus* (Krainski) Waksman et Henrici. Isoln and structure: Ettlinger *et al., Helv. Chim. Acta* **42**, 563 (1959). Crystal structure: Jensen, *J. Antibiot.* **22**, 231 (1969). Activity: Von Daehne *et al., ibid.* 233. Manuf process: Gaumann *et al.,* US 3014922 (1961 to Ciba). Total synthesis: Schmidt, Geiger, *Ann.* **664**, 168 (1963); Büchi, Lukas, *J. Am. Chem. Soc.* **86**, 5654 (1964); Hagio, Yoneda, *Bull. Chem. Soc. Jpn.* **47**, 1484 (1974); J. E. Ellis *et al., J. Org. Chem.* **42**, 2891 (1977).

Orange-yellow flakes from methanol + ethyl acetate, dec 268-270. uv max: 245, 302, 290 nm (log ε 3.78, 3.51, 4.05). LD$_{50}$ i.v. in mice: 5-10 mg/kg (Von Daehne).

4767. Homarine. [445-30-7] 2-Carboxy-1-methylpyridinium inner salt; 1-methyl-2-pyridinium carboxylate; picolinic acid *N*-methylbetaine; *N*-methylpicolinic acid. C$_7$H$_7$NO$_2$; mol wt 137.14. C 61.31%, H 5.15%, N 10.21%, O 23.33%. Quaternary ammonium base occurring in tissues of marine animals such as sea urchin, *Arabacia pustulosa,* jellyfish, *Velella spirans,* lugworm, *Arenicola marina.* Isoln: Hoppe-Seyler, *Z. Physiol. Chem.* **222**, 105 (1933). Prepn: Kosower, Patton, *J. Org. Chem.* **26**, 1318 (1961); Quast, Schmitt, *Ann.* **732**, 64 (1970). Review of distribution and function in animals: Brodzicki, *Kosmos (Warsaw)* **16A**, 431- 438 (1967), *C.A.* **68**, 18617v (1968).

Crystals from methanol. Does not have a melting point, slowly carbonizes when heated. Absorption spectrum: Gasteiger *et al., Ann. N.Y. Acad. Sci.* **90**, 624 (1962).

Hydrochloride. C$_7$H$_7$NO$_2$.HCl. Fine needles, dec 170-175°. Freely sol in water, less sol in methanol and ethanol.

4768. Homatropine. [87-00-3] α-Hydroxybenzeneacetic acid (3-*endo*)-8-methyl-8-azabicyclo[3.2.1]oct-3-yl ester; 1α*H*,5α*H*-tropan-3α-ol mandelate; mandelyltropeine; tropine mandelate. C$_{16}$H$_{21}$NO$_3$; mol wt 275.35. C 69.79%, H 7.69%, N 5.09%, O 17.43%. Ophthalmic anticholinergic. Prepd from mandelic acid and tropine: Ladenburg, *Ann.* **217**, 82 (1883); Chemnitius, *J. Prakt. Chem.* **117**, 142 (1927). Prepn of D(−)- and L(+)-forms: Werner, Miltenberger, *Ann.* **631**, 163 (1960). Comprehensive description: F. J. Muhtadi *et al., Anal. Profiles Drug Subs.* **16**, 245-290 (1987). Toxicity: R. L. Cahen, K. Tvede, *J. Pharmacol. Exp. Ther.* **105**, 166 (1952).

Prisms from ether, mp 99-100°. *Poisonous*. Hygroscopic, but only slightly sol in water; sol in alcohol, benzene, chloroform, ether, acetone, dil acids.

Hydrobromide. [51-56-9] Homatrisol; Bufopto Homatrocel. $C_{16}H_{21}NO_3 \cdot HBr$; mol wt 356.26. Orthorhombic, bipyramidal prisms from water, mp ~212° (partial decompn). *Poisonous*. One gram dissolves in 6 ml water, 40 ml alcohol (12 ml at 60°), 420 ml chloroform. Insol in ether. pH (1% aq soln): 5.4.

Hydrochloride. [637-21-8] $C_{16}H_{21}NO_3 \cdot HCl$. Prisms from water, mp 217-220° (dec). *Poisonous*. Freely sol in water, alcohol. pH (1% aq soln): 5.4.

Methylbromide. [80-49-9] Arkitropin; Homapin; Malcotran; Mesopin; Novatrin; Novatropine; Sethyl. $C_{16}H_{21}NO_3 \cdot CH_3Br$; mol wt 370.29. Minute needles from alc + ether, mp 191-192° (slight dec). *Poisonous*. Slowly darkens on exposure to light. Very sol in water; freely sol in dil alc and acetone containing about 20% water; slightly sol in abs alc. Practically insol in ether, acetone. pH (1% aq soln) 5.9; pH (10% aq soln) 4.5. LD_{50} in mice (mg/kg): 1400 orally, 60 i.p. (Cahen, Tvede).

THERAP CAT: Mydriatic.

4769. Homidium. [3546-21-2] 3,8-Diamino-5-ethyl-6-phenylphenanthridinium; 2,7-diamino-9-phenyl-10-ethylphenanthridinium; 2,7-diamino-10-ethyl-9-phenylphenanthridinium; ethidium; RD-1572; Novidium; Babidium. $[C_{21}H_{20}N_3]^+$. Prepn: T. I. Watkins, *J. Chem. Soc.* **1952**, 3059; Short *et al.*, US 2662082 (1953 to Boots Pure Drug). uv spectrum: B. Hudson, R. Jacobs, *Biopolymers* **14**, 1309 (1975). Inhibition of DNA synthesis: B. A. Newton, *J. Gen. Microbiol.* **17**, 718 (1957); of DNA polymerase: W. H. Elliott, *Biochem. J.* **86**, 562 (1963). Intercalation of double-stranded DNA: M. J. Waring, *J. Mol. Biol.* **13**, 269 (1965); J.-B. LePecq, C. Paoletti, *ibid.* **27**, 87 (1967). Mutagenicity studies: J. T. MacGregor, I. J. Johnson, *Mutat. Res.* **48**, 103 (1977); G. S. Probst *et al.*, *Environ. Mutagen.* **3**, 11 (1981). Review of interactions with nucleic acids: J.-B. LePecq in *Methods of Biochemical Analysis* **vol. 20**, D. Glick, Ed. (Wiley-Interscience, New York, 1971) pp 41-86; M. Waring in *Antibiotics* vol. 3, J. W. Corcoran, F. E. Hahn, Eds. (Springer-Verlag, New York, 1975) pp 141-165.

Bromide. [1239-45-8] Dromilac. $C_{21}H_{20}BrN_3$; mol wt 394.32. Bitter tasting dark red crystals from alc, mp 238-240°. uv max in water: 210, 285, 316, 343 nm (ε 200-500, 5000-10000, 50000, 40000). Sol in 20 parts water and 750 parts chloroform at 20°.

Chloride. $C_{21}H_{20}ClN_3$. Dark red cryst powder. Crystallizes with 1 mol of ethanol. Sol in 5 parts of water at room temp.

USE: In sepn and determn of nucleic acids.

THERAP CAT (VET): Antiprotozoal (Trypanosoma).

4770. Homochelidonine. [476-33-5] (4bR,5S,11bS)-4b,5,6,-11b,12,13-Hexahydro-1,2-dimethoxy-12-methyl[1,3]benzodioxolo-[5,6-c]phenanthridin-5-ol; α-homochelidonine. $C_{21}H_{23}NO_5$; mol wt

369.42. C 68.28%, H 6.28%, N 3.79%, O 21.65%. From root of *Chelidonium majus* L., *Papaveraceae:* Schmidt, Selle, *Arch. Pharm.* **228**, 441 (1890). Structure: Späth, Kuffner, *Ber.* **64**, 1123 (1931); H. W. Bersch, *Arch. Pharm.* **291**, 491 (1958). Total synthesis: I. Ninomiya *et al.*, *Heterocycles* **7**, 137 (1977).

Orthorhombic prisms from ethyl acetate, mp 182°. $[\alpha]_D$ +118° (alc). Very sol in chloroform. Sol in alcohol, acetic acid, ethyl acetate, dil mineral acids; sparingly sol in ether.

4771. Homochlorcyclizine. [848-53-3] 1-[(4-Chlorophenyl)-phenylmethyl]hexahydro-4-methyl-1H-1,4-diazepine; N-(p-chlorobenzhydryl)-N'-methylhomopiperazine; 1-(4-chlorodiphenylmethyl)-4-methyl-2,3,4,5,6,7-hexahydro-1,4-diazepine; 1-(p-chloro-α-phenylbenzyl)-4-methylhomopiperazine; SA-97. $C_{19}H_{23}ClN_2$; mol wt 314.86. C 72.48%, H 7.36%, Cl 11.26%, N 8.90%. Prepn: Weston, Sommers, US 2655498 (1953 to Abbott); *eidem, J. Am. Chem. Soc.* **76**, 5805 (1954). HPLC determn of enantiomers in urine: M. Nishikata *et al.*, *J. Chromatogr.* **612**, 239 (1993).

$bp_{0.8}$ 177°. n_D^{25} 1.5804.

Dihydrochloride. [1982-36-1] Curosajin; Homoclomin. $C_{19}H_{23}ClN_2 \cdot 2HCl$; mol wt 387.77. Crystals from ethanol, mp 227-228°.

THERAP CAT: Antihistaminic.

4772. Homocysteine. [6027-13-0] L-Homocysteine; (S)-2-amino-4-mercaptobutanoic acid; Hcy. $C_4H_9NO_2S$; mol wt 135.18. C 35.54%, H 6.71%, N 10.36%, O 23.67%, S 23.72%. Sulfur containing amino acid; intermediate in the biosynthesis of cysteine from methionine. Elevated serum levels have been associated with increased risk of cardiovascular disease. Readily oxidizes to the disulfide form, homocystine, *q.v.*, or cyclizes to form the thiolactone. Produced by the demethylation of methionine: L. W. Butz, V. du Vigneaud, *J. Biol. Chem.* **99**, 135 (1932). Prepn from S-benzylhomocysteine: B. Riegel, V. du Vigneaud, *ibid.* **112**, 149 (1935); from cystathionine: Binkley, *Methods Enzymol.* **2**, 314 (1955). Prepn of D- and L-forms: V. du Vigneaud, B. G. Brown, *Biochem. Prep.* **5**, 93 (1957); H. Miyazaki *et al.*, *Bull. Chem. Soc. Jpn.* **66**, 536 (1993). Asymmetric synthesis of L-form: M. Adamczyk *et al.*, *Tetrahedron: Asymmetry* **10**, 4151 (1999). LC-ESI-MS/MS determn in plasma: S. Li *et al.*, *J. Chromatogr. B* **870**, 63 (2008). Review of metabolism and role in vascular disease: R. Castro *et al.*, *J. Inherit. Metab. Dis.* **29**, 3-20 (2006); of mechanisms of pathogenesis: J. Perla-Kaján *et al.*, *Amino Acids* **32**, 561-572 (2007).

Crystals from ethanol, mp 245-247° (dec). $[\alpha]_D^{20}$ +26.8° (c = 1.00 in HCl); $[\alpha]_D^{24}$ +27.8° (c = 0.0027 in 1N HCl).

DL-Form. [454-29-5] Platelets from dil ethanol, mp 232-233°. pK$_1$ 2.22; pK$_2$ 8.87; pK$_3$ 10.86.

D-Form. [6027-14-1] Crystals from ethanol, mp 246-249° (dec). [α]$_D^{20}$ −26.8° (c = 1.00 in HCl).

DL-Form thiolactone hydrochloride. [6038-19-3] 3-Aminohydro-2(3H)-thiophenone hydrochloride. C$_4$H$_7$NOS.HCl; mol wt 153.62. Crystals from abs ethanol.

D-Form thiolactone hydrochloride. [1120-77-0] [α]$_D^{26}$ −21.5° (c = 1).

L-Form thiolactone hydrochloride. [31828-68-9] [α]$_D^{26}$ +21.5° (c = 1).

4773. L-Homocystine. [626-72-2]; [462-10-2] (unspecified stereo). (2S,2'S)-4,4'-Dithiobis[2-aminobutanoic acid]; L-4,4'-dithiobis[2-aminobutyric acid]. C$_8$H$_{16}$N$_2$O$_4$S$_2$; mol wt 268.35. C 35.81%, H 6.01%, N 10.44%, O 23.85%, S 23.89%. Naturally occurring, dimeric form of homocysteine, q.v. Appearance in urine is indicative of the metabolic disorder, cystathionine β-synthase deficiency. Prepn of DL-form from methionine: L. W. Butz, V. du Vigneaud, J. Biol. Chem. **99**, 135 (1932). Synthesis by the "malonate" method: W. I. Patterson, V. du Vigneaud, ibid. **111**, 393 (1935); from 3,6-bis(β-chloroethyl)-2,5-dioxopiperazine: H. R. Snyder, G. W. Cannon, J. Am. Chem. Soc. **66**, 511 (1944). Prepn of D- and L-forms: V. du Vigneaud, W. I. Patterson, J. Biol. Chem. **109**, 97 (1935); H. Miyazaki et al., Bull. Chem. Soc. Jpn. **66**, 536 (1993). Asymmetric synthesis: M. Adamczyk et al., Tetrahedron: Asymmetry **10**, 4151 (1999). Isotachophoretic determn in urine: N. Mizobuchi et al., J. Chromatogr. B **382**, 321 (1986). LC-MS/MS determn in plasma: M. Tomaiuolo et al., ibid. **842**, 64 (2006).

Crystals, dec 281-284°. [α]$_D^{21}$ −16° (H$_2$O); [α]$_D^{26}$ +77° (1.0N HCl).

DL-Form. [870-93-9] Bis(γ-amino-γ-carboxypropyl)disulfide. Platelets from water, dec 263-265°. pK$_1$ 1.59; pK$_2$ 2.54; pK$_3$ 8.52; pK$_4$ 9.44. Soly in water: 1 part per 5000.

D-Form. [6027-15-2] Crystals, dec 281-284°. [α]$_D^{21}$ +16° (H$_2$O); [α]$_D^{26}$ −77° (1.0N HCl).

USE: Biomarker for inherited cystathionine β-synthase deficiency.

4774. Homoeriodictyol. [446-71-9] (2S)-2,3-Dihydro-5,7-dihydroxy-2-(4-hydroxy-3-methoxyphenyl)-4H-1-benzopyran-4-one; 4',5,7-trihydroxy-3'-methoxyflavanone; eriodictyonone; cyanidanon-3-methyl ether 1625. C$_{16}$H$_{14}$O$_6$; mol wt 302.28. C 63.58%, H 4.67%, O 31.76%. From Eriodictyon californicum (H. & A.) Torr., and E. angustifolium Nutt., Hydrophyllaceae: Geissman, J. Am. Chem. Soc. **62**, 3258 (1940); Hadley, Gisvold, J. Am. Pharm. Assoc. **33**, 275 (1944). Purification: Seka, Prosche, Monatsh. Chem. **69**, 284 (1936). Structure: Shinoda, Sato, J. Pharm. Soc. Jpn. **49**, 64 (1929), C.A. **23**, 4210 (1929). Synthesis: Farooq et al., Naturwissenschaften **46**, 76 (1959).

Crystals from 70% acetic acid. Needles by sublimation in high vacuum (0.003-0.005 mm Hg) at 190-195°. Plates from dil alc, dec 225° (after drying in vacuo at 110°). [α]$_D^{20}$ −28° (alc). uv max (alc): 290, 328 nm (log ε 2.26, 2.33). Moderately sol in alcohol, acetic acid. Nearly insol in water, ethyl acetate; practically insol in benzene, chloroform.

Triacetate. mp 115-116°.

4775. Homogentisic Acid. [451-13-8] 2,5-Dihydroxybenzeneacetic acid; 2,5-dihydroxyphenylacetic acid; 2,5-dihydroxy-α-

toluic acid. C$_8$H$_8$O$_4$; mol wt 168.15. C 57.14%, H 4.80%, O 38.06%. An important intermediate in metabolism of tyrosine and phenylalanine, q.q.v. Occurs in plants and in the urine of alkaptonurics: Garrod, Inborn Errors of Metabolism (Oxford Medical Publications, London, 1923, and later). Isoln from alkaptonuric urine: Mörner, Z. Physiol. Chem. **117**, 85 (1921). Synthesis from 2,5-dihydroxymandelic acid or from 2,5-dihydroxyphenylglyoxylic acid by boiling with fuming HI: Neubauer, Flatow, ibid. **52**, 395 (1907). Alternate syntheses: L. DeForrest Abbott, J. D. Smith, J. Biol. Chem. **179**, 365 (1949); S. B. Bostock, A. H. Renfrew, Synthesis **1978**, 66; J. L. Bloomer, K. M. Damodaran, ibid. **1980**, 111. Biosynthesis from tyrosine: Davies et al., J. Chem. Soc. **1964**, 3126. Metabolic studies: W. E. Knox, M. LeMay-Knox, Biochem. J. **49**, 686 (1951); B. N. LaDu, V. G. Zannoni, J. Biol. Chem. **217**, 777 (1955).

Monohydrate. Prisms from water. Anhydrous leaflets from hot alcohol + chloroform, mp 152°. Freely sol in water, alcohol, ether; insol in chloroform, benzene. Easily dehydrated to the lactone. Aq solns are stable.

Dimethyl ether. C$_{10}$H$_{12}$O$_4$. mp 124.5°.

Methyl ester dimethyl ether. C$_{11}$H$_{14}$O$_4$. mp 45°.

4776. Homoharringtonine. [26833-87-4] Cephalotaxine 4-methyl (2R)-2-hydroxy-2-(4-hydroxy-4-methylpentyl)butanedioate (ester); omacetaxine mepesuccinate; CGX-635; Ceflatonin; Myelostat; Omapro. C$_{29}$H$_{39}$NO$_9$; mol wt 545.63. C 63.84%, H 7.20%, N 2.57%, O 26.39%. Naturally occurring, antitumor alkaloid from the Japanese plum yew, Cephalotaxus harringtonia, Cephalotaxaceae, and related species. Inhibits protein synthesis and induces apoptosis. Identification as ester of cephalotaxine: R. G. Powell et al., Tetrahedron Lett. **11**, 815 (1970). Extraction method and quantification in plant material: idem, Phytochemistry **11**, 1467 (1972). Structure and activity: idem et al., J. Pharm. Sci. **61**, 1227 (1972). Synthesis by esterification of cephalotaxine: S. Hiranuma, T. Hudlicky, Tetrahedron Lett. **23**, 3431 (1982); J.-P. Robin et al, ibid. **40**, 2931 (1999). Manuf process: idem et al., US 7169774 (2007 to Stragen). HPLC determn in plasma: Y.-P. Chan et al., J. Chromatogr. **496**, 155 (1989). Clinical pharmacokinetics in patients with leukemia: V. Lévy et al., Br. J. Cancer **95**, 253 (2006). Review of discovery and development: H. M. Kantarjian et al., Cancer **92**, 1591-1605 (2001); of pharmacology and clinical experience in chronic myelogenous leukemia: A. Quintás-Cardama, J. Cortes, Expert Opin. Pharmacother. **9**, 1029-1037 (2008); and in other myeloid malignancies: eidem, IDrugs **11**, 356-372 (2008).

White crystalline solid, mp 143-145°. [α]$_D^{20}$ −125° (c = 0.24 in CHCl$_3$). uv max (ethanol): 290 nm (log ε 3.62). Sol in chloroform, dichloromethane.

THERAP CAT: Antineoplastic.

4777. Homosalate. [118-56-9] 2-Hydroxybenzoic acid 3,3,5-trimethylcyclohexyl ester; salicylic acid 3,3,5-trimethylcyclohexyl ester; 3,3,5-trimethylcyclohexyl salicylate; homomenthyl salicylate; Eusolex HMS; Neo Heliopan HMS; Parsol HMS. C$_{16}$H$_{22}$O$_3$; mol wt 262.35. C 73.25%, H 8.45%, O 18.30%. UV-B absorber in cosmetics and sunscreens. Prepn: Stockelbach, **US 2369084** (1945 to Fries Bros.).

bp$_4$ 161-165°. n^{20} 1.516 to 1.518. d$_{25}^{25}$ 1.045. Sol in oil.

THERAP CAT: Ultraviolet screen.

4778. Homoserine. [672-15-1] L-Homoserine; (S)-2-amino-4-hydroxybutanoic acid; L-2-amino-4-hydroxybutyric acid; α-amino-γ-hydroxy-n-butyric acid. $C_4H_9NO_3$; mol wt 119.12. C 40.33%, H 7.62%, N 11.76%, O 40.29%. Principal free amino acid occurring in pea plants: A. I. Virtanen, *Acta Chem. Scand.* **7**, 1423 (1953); J.A. Bakhuis, *Nature* **180**, 713 (1957). Prepn: Fischer, Blumenthal, *Ber.* **40**, 106 (1907); Armstrong *J. Am. Chem. Soc.* **70**, 1756 (1948); Birnbaum, Greenstein, *Arch. Biochem. Biophys.* **42**, 212 (1953); M. Frankel, Y. Knobler, *J. Am. Chem. Soc.* **80**, 3147 (1958). Review of homoserine production by fermentation: T. Nara in *Microbial Prod. Amino Acids*, K. Yamada, Ed. (Wiley, New York, 1972) pp 417-434.

L-form

Flat prisms from 90% alc. Dec 203°. [M]$_D$ +21.8° (5N HCl), [M]$_D$ +14.3° (glacial acetic acid). [α]$_D^{26}$ −8.8° (c = 5 in H$_2$O); [α]$_D^{26}$ +18.3° (c = 2 in 2N HCl). On standing for 8 hrs at 26° the [α]$_D$ of the HCl soln decreases to nearly zero as the corresponding levorotatory-γ-butyrolactone is formed.

L-Homoserine γ-lactone monohydrochloride. Prepd by refluxing L-homoserine with 2N HCl for 2 hrs, crystals, [α]$_D^{26}$ −27.0° (c = 5).

D-Homoserine. Crystals, dec 203°. [α]$_D^{26}$ +8.8° (c = 5).

DL-Homoserine. Crystals from dil ethanol, dec 186-187°.

4779. Homovanillic Acid. [306-08-1] 4-Hydroxy-3-methoxybenzeneacetic acid; (4-hydroxy-3-methoxyphenyl)acetic acid; 4-hydroxy-3-methoxy-α-toluic acid. $C_9H_{10}O_4$; mol wt 182.18. C 59.34%, H 5.53%, O 35.13%. Metabolite found in human urine. Prepn: F. Tiemann, N. Nagai, *Ber.* **10**, 201 (1877); F. Mauthner, *Ann.* **370**, 368 (1909); H. E. Fisher, H. Hibbert, *J. Am. Chem. Soc.* **69**, 1208 (1947). End product of dopamine, *q.v.*: K. Shaw *et al.*, *J. Biol. Chem.* **226**, 255 (1957); M. Goodall, H. Alton, *Biochem. Pharmacol.* **25**, 2635 (1968).

White crystals, mp 143°. Sol in water, benzene. Slightly sol in alcohol, ether. Insol in cyclohexane.

4780. HON. [26911-39-7]; [4439-84-3] (unspecified stereo). 5-Hydroxy-4-oxo-L-norvaline; L-2-amino-5-hydroxylevulinic acid; (S)-2-amino-5-oxo-5-hydroxypentanoic acid; δ-hydroxy-γ-oxo-L-norvaline. $C_5H_9NO_4$; mol wt 147.13. C 40.82%, H 6.17%, N 9.52%, O 43.50%. Antitubercular antibiotic substance isolated from *Streptomyces akiyoshiensis nova* sp.: S. Tatsuoka *et al.*, *J. Antibiot.* **14A**, 39 (1961). Synthesis and structure: A. Miyake, *Chem. Pharm. Bull.* **8**, 1071, 1074, 1079 (1960).

Needles from water + acetone, no definite mp. [α]$_D^{20}$ −6° (c = 1 in water); [α]$_D^{17}$ −8.2° (c = 3.4 in water). pKa 2.91. uv max (water): 271 nm (ε 24). Stable in pure, dry state. Less stable in basic than in acidic solns. Colors red when heated in caustic soln, losing antibiotic activity. The DL-form is half as active as L-HON. LD$_{50}$ in mice (mg/kg): 5200 i.v.; 8000 s.c.; 7600 orally (Tatsuoka).

4781. Honokiol. [35354-74-6] 3′,5-Di-2-propenyl-[1,1′-biphenyl]-2,4′-diol; 3′,5-diallyl-2,4′-biphenyldiol; 5,3′-diallyl-2,4′-dihydroxydiphenyl. $C_{18}H_{18}O_2$; mol wt 266.34. C 81.17%, H 6.81%, O 12.01%. Bioactive principle isolated from the bark of *Magnolia obovata*, Thunb., *Magnoliaceae* and other *Magnolia* species used in Japanese and Chinese traditional medicine. Name derived from "Honoki", the Japanese name of *M. obovata*. Isomeric with magnolol, *q.v.* Identification in methanolic extracts of *M. obovata* bark: M. Fujita *et al.*, *Chem. Pharm. Bull.* **20**, 212 (1972). Isoln from seeds of *M. grandiflora* L.: F. S. El-Feraly, W.-S. Li, *Lloydia* **41**, 442 (1978). Synthesis: T. Takeya *et al.*, *Chem. Pharm. Bull.* **34**, 2066 (1986). CNS depressant effects *in vivo*: K. Watanabe *et al.*, *Planta Med.* **49**, 103 (1983). HPLC determn in *M. officinalis* extracts: T.-H. Tsai, C.-F. Chen, *J. Chromatogr.* **598**, 143 (1992).

Colorless needles, mp 86-86.5° (El-Feraly, Li); also reported as mp 87.5° (Fujita). Sol in usual organic solvents and caustic alkali. uv max (ethanol): 294 nm (ε 8200).

4782. Hoodia. Genus of succulent plants in the family *Apocynaceae*, primarily *Hoodia gordonii* (Masson) Sweet ex Decne. or *H. pilifera* (L.f.) Plowes (syn. *Trichocaulan piliferum*); used traditionally by the Bushman of the Kalahari desert as an appetite suppressant, thirst quencher and as a cure for digestive disturbances. *Habit.* Angola, Botswana, Namibia, and South Africa. Brief review: B.-E. van Wyk, M. Wink, *Medicinal Plants of the World* (Timber Press, Portland, 2004) p 171. Description of the genus: D. Court, *Succulent Flora of Southern Africa* (Balkema Press, Rotterdam, 2000) pp 169-172.

P57AS3

P57AS3. [384329-61-7] (3β,12β,14β)-3-[(O-6-Deoxy-3-O-methyl-β-D-glucopyranosyl-(1 → 4)-O-2,6-dideoxy-3-O-methyl-β-D-*ribo*-hexopyranosyl-(1 → 4)-2,6-dideoxy-3-O-methyl-β-D-*ribo*-hexopyranosyl)oxy]-14-hydroxy-12-[[(2E)-2-methyl-1-oxo-2-butenyl]oxy]pregn-5-en-20-one; P57. $C_{47}H_{74}O_{15}$; mol wt 879.09. Steroid glycoside thought to be the anorectic principle. Isoln and activity: F. R. Van Heerden *et al.*, **WO 9846243**; *eidem*, **US 6376657** (1998, 2002 both to CSIR). Mechanism of action study: D. B. Mac Lean, L.-G. Luo, *Brain Res.* **1020**, 1 (2004). mp 147-152°. [α]$_D^{20}$ +12.67° (c = 3 in chloroform).

USE: Appetite suppressant.

4783. Hopantenic Acid. [18679-90-8] 4-[[(2R)-2,4-Dihydroxy-3,3-dimethyl-1-oxobutyl]amino]butanoic acid; D-(+)-4-(2,4-dihydroxy-3,3-dimethylbutyramido)butyric acid; D-homopantothenic acid. $C_{10}H_{19}NO_5$; mol wt 233.26. C 51.49%, H 8.21%, N 6.00%, O 34.29%. Homolog of pantothenic acid, *q.v.* Prepn: R. Fuerst, L. L. Li, *Biochim. Biophys. Acta* **86**, 26 (1964); C. M. DeSha, R. Fuerst, *ibid.* 33. Prepn of the calcium salt: Y. Nishizawa *et al.*, **JP 66 732** (1966 to Tanabe), *C.A.* **64**, 12555c (1965). Series of articles on analysis, physicochemical properties, electrophysiology, pharmacology, toxicity studies, teratology: *Bitamin* **33**, 603-632 (1966), *C.A.* **65**, 5949a-g (1966). Colorimetric determn, properties: T. Kodama *et al.*, *J. Vitaminol.* **13**, 298 (1967), *C.A.* **68**, 62755n (1968). Peri- and postnatal studies in rats: Y. Asano *et al.*, *Oyo Yakuri* **19**, 1011 (1980), *C.A.* **94**, 58160 (1981). Mechanism of action: V. M. Arakumov, M. A. Kovler, *Farmakol. Toksikol.* **44**, 30 (1981), *C.A.* **94**, 114577 (1981). Determn in plasma: Y. Umeno *et al.*, *J. Chromatogr.* **226**, 333 (1981). Toxicology studies: Y. Nishizawa *et al.*, *J. Vitaminol.* **15**, 26 (1969). Pharmacology and acute toxicity study: Y. Nishizawa, *Med. J. Osaka Univ.* **35**, 41 (1984).

Sol in water; stable at pH 5-6. $[\alpha]_D^{20}$ +23.8°. pKa 4.52 at 25°. LD$_{50}$ in male, female mice, male, female rats (mg/kg): 850, 954, 1575, 1458 i.p.; 2063, 2495, 5940, 7348 s.c.; 6297, 7935, 16810, 13350 orally (Nishizawa).

Calcium salt hemihydrate. Calcium homopantothenate; hopantenate calcium; pantogam; Hopate. $C_{20}H_{36}CaNO_5 \cdot \frac{1}{2}H_2O$; mol wt 419.60. White powder, mp 155-165°. $[\alpha]_D^{20}$ +24.19° (c = 2 in water). Bitter taste. Sol in water, slightly sol in methanol. Practically insol in organic solvents. Stable at pH 5-6.

THERAP CAT: Cerebral activator.

4784. Hops. Carefully dried strobiles of *Humulus lupulus* L., *Moraceae*, bearing their glandular trichomes. *Habit.* Europe, Asia, North America, cultivated widely. *Constit.* Volatile oil (0.3-1%), bitter acids incl. lupulone, humulone; flavonoids incl. xanthohumol; ferulic, caffeic and chlorogenic acids; tannins, resins. Review of chemistry and constituents: R. Stevens, *Chem. Rev.* **67**, 19-71 (1967); of botany, chemistry and use in beer brewing: M. Moir, *J. Am. Soc. Brew. Chem.* **58**, 131-146 (2000); of medicinal uses: J. Barnes *et al.*, *Herbal Medicines* (Pharmaceutical Press, London, 2nd Ed., 2002) pp 290-292.

Volatile oil. [8007-04-3] Oil of hops. *Constit.* Complex mixture of more than 100 terpenoid components, primarily humulene, β-caryophyllene, farnescene. Light yellow to green-yellow liquid. Darkens and becomes viscous with aging. d_{25}^{25} 0.825-0.926. n_D^{20} 1.470-1.494. Acid value not more than 11.0. Saponification value between 14 and 69. Sol in most fixed oils. Practically insol in glycerin, propylene glycol. *Keep well closed, cool and protected from light.*

USE: In beer brewing; oil in hair rinses and bath prepns.

THERAP CAT: Sedative.

4785. Hordenine. [539-15-1] 4-[2-(Dimethylamino)ethyl]-phenol; *N,N*-dimethyltyramine; *p*-hydroxy-*N,N*-dimethylphenethylamine; anhaline; eremursine; peyocactine. $C_{10}H_{15}NO$; mol wt 165.24. C 72.69% H 9.15%, N 8.48%, O 9.68%. Isoln from barley germs: Leger, *Compt. Rend.* **142**, 108 (1906); **143**, 234, 916 (1906); **144**, 488 (1907); Erspamer, Falconieri, *Naturwissenschaften* **39**, 431 (1952). Structure: Leger, *Bull. Soc. Chim. Fr.* [3] **35**, 868 (1906); [4] **1**, 148 (1907); Gaebel, *Arch. Pharm.* **244**, 441 (1906). Synthesis from phenethyl alcohol: Barger, *J. Chem. Soc.* **95**, 2193 (1909); from tyrosine: Raoul, *Compt. Rend.* **204**, 74 (1937); from *p*-(β-hydroxyethyl)anisole: Cheng *et al.*, *J. Am. Chem. Soc.* **73**, 4081 (1951).

Orthorhombic prisms from alcohol or from benzene + petr ether, needles from water, mp 117-118°. bp$_{11}$ 173°. Sublimes 140-150°. Very sol in alcohol, chloroform, ether. 7 grams dissolve in 1000 ml water. Sparingly sol in benzene, toluene, xylene. Practically insol in petr ether.

Hydrochloride. [6027-23-2] $C_{10}H_{15}NO \cdot HCl$. Needles from alcohol, mp 177°. Very sol in water.

4786. Horehound. Hoarhound. *Marrubium vulgare* (Tourn.) L., *Labiatae*. *Habit.* Europe, Asia, naturalized in the U.S. *Constit.* Bitter principle marrubiin, volatile oil.

THERAP CAT: Expectorant.

4787. Horse Chestnut. Deciduous, flowering tree, *Aesculus hippocastanum* L., *Hippocastanaceae*, bearing spiny, globular fruit capsules that contain up to 3 reddish brown, inedible seeds. Medicinal portions are the dried leaves and an extract prepared from the seeds. *Habit.* Balkan peninsula; widely cultivated in Northern hemisphere as a shade tree. *Constit.* Mixture of triterpene saponins known as escin, *q.v.* (3-5%); coumarins such as esculetin, fraxin, scopolin; flavonoids incl. astragalin, isoquercetin, rutin, leucocyanidin; tannins, starch, fatty acids; also contains the toxic glycoside, esculin. Comprehensive description: N. Tiffany *et al.*, *J. Herb. Pharmacother.* **2**, 71-85 (2002). Review of constituents and uses: J. Barnes *et al.*, *Herbal Medicines* (Pharmaceutical Press, London, 2nd Ed., 2002) pp 296-299; A. Y. Leung, S. Foster, *Encyclopedia of Common Natural Ingredients*, (Wiley-Interscience, Hoboken, 2nd Ed., 2003) pp 304-306; J. Gruenwald *et al.*, *PDR for Herbal Medicines* (Thomson PDR, Montvale, 3rd Ed., 2004) pp 445-448.

Dried extract. [8053-39-2] Horse chestnut seed extract; Aescorin; Aescusan; Essaven; Venalot; Venoplant; Venopyronum; Venostasin. Principle constituent is escin which is used to calculate potency. Clinical pharmacokinetics and bioavailability of formulations: D. Bässler *et al.*, *Adv. Ther.* **20**, 295 (2003). Review of clinical trials in chronic venous insufficiency: U. Siebert *et al.*, *Int. Angiol.* **21**, 305-315 (2002).

Caution: Potential symptoms of toxicity following ingestion of leaves, bark, flowers or raw seeds are vomiting, diarrhea, headache, stupor, coma, paralysis (Tiffany).

USE: In shampoos, skin care products, body and hand creams, lotions.

THERAP CAT: Seed extract in treatment of chronic venous insufficiency.

4788. Horse-radish. Raphanus rusticanus; Kren; Maliner Kren. The root of *Radicula armoracia* (L.) Robinson [*Cochlearia armoracia* L.; *Armoracia lopathifolia* Gilib.; *Roripa armoracia* Hitch.; *Nasturtium armoracia* Fries], *Cruciferae*. *Habit.* Europe, naturalized in North America. Contains ascorbic acid and sinigrin which yields allyl isothiocyanate on hydrolysis with peroxidase or myrosinase, an enzyme from black mustard: Stoll, Seebeck, *Helv. Chim. Acta* **31**, 1432 (1948).

Active ingredient of *Rasapen*, a urinary antiseptic.

USE: Condiment.

4789. HPA-23. [89899-81-0] Ammonium antimony sodium tungsten oxide $((NH_4)_{17}Sb_9Na_2W_{21}O_{86})$; hexaoctacontaoxononaantimonatehenieicosatungstate(19−) heptadecaammonium disodium; ammonium 5-tungsto-2-antimonate; 5-tungsto-2-antimonate; 21-tungsto-9-antimonate. $H_{68}Na_2N_{17}O_{86}Sb_9W_{21}$; mol wt 6685.04. H 1.03%, Na 0.69%, N 3.56%, O 20.58%, Sb 16.39%, W 57.75%. $(NH_4)_{17}Na(NaSb_9W_{21}O_{86})$. One of the condensed mineral heteropolyanions, also called polyoxotungstates. Synthesis: M. Michelon, G. Hervé, *C. R. Seances Acad. Sci. Ser. C* **274**, 209 (1972); M. Michelon *et al.*, *J. Inorg. Nucl. Chem.* **42**, 1583 (1980). Crystal structure: J. Fischer *et al.*, *J. Am. Chem. Soc.* **98**, 3050 (1976). Laser Raman studies of intracellular localization: C. Cibert, C. Jasmin, *Biochem. Biophys. Res. Commun.* **108**, 1424 (1982). Antiviral activity in mice: C. Jasmin *et al.*, *J. Natl. Cancer Inst.* **53**, 469 (1974); G. H. Werner *et al.*, *J. Gen. Virol.* **31**, 59 (1976). In vitro inhibition of rabies virus: H. Tsiang *et al.*, *J. Gen. Virol.* **40**, 665 (1978). Effects in scrapie model systems: R. H. Kimberlin, C. A. Walker, *Lancet* **2**, 591 (1979); *eidem*, *Arch. Virol.* **78**, 9 (1983). Structure-activity study: M. Hervé *et al.*, *Biochem. Biophys. Res. Commun.* **116**, 222 (1983). Clinical evaluation in AIDS: W. Rozenbaum *et al.*, *Lancet* **1**, 450 (1985); B. L. Moskovitz, *Antimicrob. Agents Chemother.* **32**, 1300 (1988).

Crystals from water. Solution has pH near 6.7, and is stable around neutrality. LD$_{50}$ i.p. in mice: 750 mg/kg (Werner).

Note: Formerly known as *ammonium antimony tungsten oxide.*

4790. HQNO. [341-88-8] 2-Heptyl-4-quinolinol 1-oxide; 2-heptyl-4-hydroxyquinoline *N*-oxide. $C_{16}H_{21}NO_2$; mol wt 259.35. C

74.10%, H 8.16%, N 5.40%, O 12.34%. Inhibitor of electron transport through the cytochrome bc_l segment of the respiratory chain. Isoln of the naturally occurring antagonist to dihydrostreptomycin from *Pseudomonas pyocyanea:* Hays *et al., J. Biol. Chem.* **159**, 725 (1948); J. Lightbown, *J. Gen. Microbiol.* **11**, iv (1954). Properties: J. W. Cornforth, A. T. James, *Biochem. J.* **58**, xlviii (1954). Synthesis: *eidem, ibid.* **63**, 124 (1956); D. E. Ames *et al., J. Chem. Soc.* **1956**, 3079. Inhibition of electron transport: J. W. Lightbown, F. L. Jackson, *Biochem. J.* **63**, 130 (1956); M. Avron, *ibid.* **78**, 735 (1961); N. J. Jacobs, M. J. Wolin, *Biochim. Biophys. Acta* **69**, 29 (1963). Effect on proton permeability of the mitochondrial membrane: K. Krab, M. Wikström, *Biochem. J.* **186**, 637 (1980). Sites of inhibition: G. Izzo *et al., FEBS Lett.* **93**, 320 (1978); M. Droppa *et al., Z. Naturforsch.* **36C**, 109 (1981).

Crystals from methylethyl ketone, mp 156-157°.

4791. Humic Acids. Allomelanins found in soils, coals, and peat, resulting from the decompn of organic matter, particularly dead plants. Consists of a mixture of complex macromolecules having polymeric phenolic structures with the ability to chelate with metals, esp iron. *Review:* R. A. Nicolaus, *Melanins* (Hermann, Paris, 1968) pp 147-153; W. Flaig *et al.* in *Soil Components* **vol. 1**, J. E. Gieseking, Ed. (Springer, New York, 1975) pp 1-211.

Chocolate-brown, dust-like powder. Slightly sol in water, usually with much swelling; sol in alkali hydroxides and carbonates; also sol in hot concd HNO_3 with dark-red color.

USE: In mud baths, drilling muds, pigments for printing inks, fertilizers, growth hormones for plants, transporters of trace minerals in soil: Steelnick, *J. Chem. Educ.* **40**, 379 (1963).

4792. Humulene. [6753-98-6] (1E,4E,8E)-2,6,6,9-Tetramethyl-1,4,8-cyclodecatriene; α-humulene; α-caryophyllene. $C_{15}H_{24}$; mol wt 204.36. C 88.16%, H 11.84%. Sesquiterpenoid isomer of caryophyllene, *q.v.* occurring in many essential oils, especially oil of hops *(Humulus lupulus* L. *Moraceae)* and leaves of *Lindera strychnifolia* (F.) Will *Lauraceae.* Occurs in nature as a mixture with β-humulene. Isolation of mixture: A. C. Chapman, *J. Chem. Soc.* **67**, 54, 780 (1895). Identity with α-caryophyllene: F. Sorm *et al., Collect. Czech. Chem. Commun.* **14**, 693, 699, 716 (1949). Structure: F. Sorm *et al., ibid.* **19**, 570 (1954). Stereochemistry: A. T. McPhail, G. A. Sim, *J. Chem. Soc. B* **1966**, 112. Synthesis: E. J. Corey, E. Hanamaka, *J. Am. Chem. Soc.* **89**, 2758 (1967). Stereoselective synthesis: Y. Kitagawa *et al., ibid.* **99**, 3864 (1977); E. J. Corey *et al., Tetrahedron Lett.* **34**, 3675 (1993). Chromatographic conversion to β-humulene: V. Benesova *et al., Collect. Czech. Chem. Commun.* **26**, 1832 (1961). *Reviews:* F. Sorm in *Fortschr. Chem. Org. Naturst.* **19**, 1-32 (1961); *Rodd's Chemistry of Carbon Compounds* **vol. IIC**, S. Coffey, Ed., (Elsevier, New York, 2nd ed., 1969) pp 282-283.

Liquid. bp_5 106-107°. n_D^{30} 1.5004, *see:* N. P. Damodaran, S. Dev, *Tetrahedron* **24**, 4113 (1968). Also reported as bp_{10} 123°. n_D^{25} 1.5015. d_4^{25} 0.8865, *see:* R. P. Hildebrand *et al., Chem. Ind. (London)* **1959**, 489. NMR spectrum: S. Dev *et al., J. Am. Chem. Soc.* **90**, 1246 (1968).

Silver nitrate complex. $C_{15}H_{24}\cdot2AgNO_3$. Crystals from aq ethanol, mp 175°.

β-Humulene. [116-04-1] (E,E)-1,4,4-Trimethyl-8-methylene-1,5-cyclodecadiene. Liquid. n_D^{20} 1.5014. d_4^{20} 0.8905.

4793. Humulon. [26472-41-3] (6R)-3,5,6-Trihydroxy-4,6-bis(3-methyl-2-buten-1-yl)-2-(3-methyl-1-oxobutyl)-2,4-cyclohexadien-1-one; α-bitter acid; α-lupulic acid; humulone. $C_{21}H_{30}O_5$; mol wt 362.47. C 69.59%, H 8.34%, O 22.07%. Antibiotic constituent of hops *(Humulus lupulus* L., *Moraceae). See also* Lupulon. Isoln from commercial hops: Bungener, *Bull. Soc. Chim.* [2] **45**, 487 (1886); Barth, Lintner, *Ber.* **31**, 2022 (1898); Wollmer, *Ber.* **49**, 780 (1916); Lewis *et al., J. Clin. Invest.* **28**, 916 (1949). Structure: Riedl, *Ber.* **85**, 692 (1952); Carson, *J. Am. Chem. Soc.* **73**, 4652 (1951). Absolute configuration and structure of preferred isomer: DeKeukeleire, Verzele, *Tetrahedron* **26**, 385 (1970).

Crystals from ether, mp 65-66.5°. Bitter taste, esp in alcoholic soln. More stable to air than lupulon. Monobasic acid. $[\alpha]_D^{20} -212°$ (1.0 g in 15.5 g 96% alc). uv max (ethanol): 237, 282 nm (ε 13,760; 8330). Soluble in the usual organic solvents. Slightly sol in boiling water from which it separates as a milky precipitate on cooling. Forms a sodium salt which is readily sol in water. Suffers no loss of bacteriostatic potency against *Staphylococcus aureus* upon autoclaving 40 ppm in phosphate buffer at pH 6.5 or 8.5. The presence of ascorbic acid in low concns extends the duration of bacteriostatic action.

4794. Huperzine A. [102518-79-6] (5R,9R,11E)-5-Amino-11-ethylidene-5,6,9,10-tetrahydro-7-methyl-5,9-methanocycloocta-[b]pyridin-2(1H)-one; selagine; HUP. $C_{15}H_{18}N_2O$; mol wt 242.32. C 74.35%, H 7.49%, N 11.56%, O 6.60%. Reversible alkaloid inhibitor of AChE which crosses the blood-brain barrier. Occurs as the (−)-form in the vegetative part of clubmosses; teas brewed from these mosses have traditionally been used in China to alleviate memory problems. Isolated as selagine from *Lycopodium selago* L., *Lycopodiaceae:* J. Muszynski, *Q. J. Pharm. Pharmacol.* **21**, 34 (1948); from *Huperzia serrata* as huperzine: Chin. Co-op Res. Group, *J. Tradit. Chin. Med.* **2**, 45 (1982). Original structure: Z. Valenta *et al., Tetrahedron Lett.* **1960**, 26; revised structure as huperzine A: J.-S. Liu *et al., Can. J. Chem.* **64**, 837 1986. Identity with selagine: W. A. Ayer *et al., ibid.* **67**, 1538 (1989). Synthesis of (±)-form: A. P. Kozikowski *et al., J. Org. Chem.* **56**, 4636 (1991); G. Campiani *et al., ibid.* **58**, 7660 (1993). NMR spectra: B. N. Zhou *et al., Phytochemistry* **34**, 1425 (1993). HPLC determn in serum: J. Ye *et al., J. Chromatogr. B* **817**, 187 (2005). Anticholinesterase activity: Y.-E. Wang *et al., Acta Pharmacol. Sin.* **7**, 110 (1986); binding profile of enantiomers: M. McKinney *et al., Eur. J. Pharmacol.* **203**, 303 (1991); binding specificity: Y. Ashani *et al., Mol. Pharmacol.* **45**, 555 (1994). Clinical evaluation in senile dementia: R.-W. Zhang *et al., Acta Pharmacol. Sin.* **12**, 259 (1991). Brief review of chemistry and clinical use: D. Bai, *Pure Appl. Chem.* **65**, 1103-1112 (1993).

Monoclinic crystals from acetone, mp 214-215°. $[\alpha]_D -147°$ (c = 0.36 in CH_3OH) (Ayer). Also reported as mp 230°. $[\alpha]_D^{24.5} -150.4°$ (c = 0.498 in MeOH) (Liu). uv max (EtOH): 231, 313 nm (log ε 4.01, 3.89).

THERAP CAT: In treatment of memory disorders.

4795. Hyalobiuronic Acid. [499-15-0] 2-Amino-2-deoxy-3-O-β-D-glucopyranuronosyl-D-glucose; 3-O-(β-D-glucopyranosyluronic acid)-2-amino-2-deoxy-D-glucose. $C_{12}H_{21}NO_{11}$; mol wt 355.30. C 40.57%, H 5.96%, N 3.94%, O 49.53%. Disaccharide unit of hyaluronic acid. Isoln from hyaluronic acid: Rapport *et al.*, *Nature* **168**, 996 (1951). Structure: Weissman, Meyer, *J. Am. Chem. Soc.* **76**, 1753 (1954). Synthesis: Takanashi *et al.*, *ibid.* **84**, 3029 (1962).

Rectangular prisms from water, darken at 190° with no characteristic melting or dec point. $pK_1' = 2.6$, $pK_2' = 7.1$. Shows mutarotation: $[\alpha]_D^{20} +34° \rightarrow +30°$ (c = 1.08 in 0.1N HCl). Sparingly sol in hot water, dilute HCl, dil NaHCO$_3$. Practically insol in water, glacial acetic acid, ethanol, methanol and pyridine.

N-Acetylhyalobiuronic acid. $C_{14}H_{23}NO_{12}$. Amorphous. $pK' = 3.3$. $[\alpha]_D^{24} -32°$ (c = 2.0 in water).

4796. Hyaluronic Acid. [9004-61-9] Unbranched high molecular weight polysaccharide made up of alternating glucuronic acid and *N*-acetyl glucosamine units. Present in the connective tissue of all vertebrates as the hyaluronate; in man high concentrations are found in skin, cartilage, in the umbilical cord, in vitreous body and in synovial fluid. Isoln and characterization: K. Meyer, J. W. Palmer, *J. Biol. Chem.* **107**, 629 (1934); *eidem, ibid.* **114**, 689 (1936). Structure: K. Meyer, *Fed. Proc.* **17**, 1075 (1958). Crystal structure: I. C. M. Dea *et al.*, *Science* **179**, 560 (1973); E. D. T. Atkins, J. K. Sheehan, *ibid.* 562. *Reviews:* Tauber, *Chemistry and Technology of Enzymes* (New York, 1946); Meyer, Rapport in *Adv. Enzymol.* **13**, 199 (1952); R. L. Whistler, E. J. Olson in *Adv. Carbohydr. Chem.* **12**, 299 (1957). Review of role in various developmental processes: B. P. Toole, *Cell Biology of Extracellular Matrix*, E. D. Hay, Ed. (Plenum Press, New York, 1981) pp 259-288.

Sodium salt. [9067-32-7] ARTZ; Connettivina; Equron; Healon; Healonid; Hyacid; Hyalgan; Hyalovet; Hyonate; Ial; Opegan; Provisc; Synacid. $[\alpha]_D^{25} -74°$ (c = 0.25 in water): Rapport *et al.*, *J. Am. Chem. Soc.* **73**, 2416 (1951).

USE: Surgical aid (ophthalmological).

THERAP CAT (VET): Adjunct in treatment of noninfectious synovitis. Osteoarthritis in dogs and horses.

4797. Hyaluronidases. [9001-54-1] Spreading factor; diffusing factor; invasin; Alidase; Apertase; Diffusin; Enzodase; Harodase; Hyalase; Hyalozima; Hyalidase; Hyasmonta; Hyason; Hyazyme; Jalovis; Kinaden; Kinetin; Luronase; Permease; Rondase; Ronidase; Thiomucase; Unidasa; Wydase. Enzymes which have in common the cleavage of glycosidic bonds of hyaluronic acid, *q.v.*, and, to a variable degree, of some other acid mucopolysaccharides of connective tissue. The skin is probably the largest store of hyaluronidase in the body; the enzyme although generally present in an inactive form, may be supposed to regulate the velocity of water and metabolite exchange by decreasing the viscosity of the intercellular matrix. Also has a physiological role in fertilization: The sperm is rich in the enzyme and can thus advance better in the cervical canal and reach the ovum. Found in the type II pneumococci, in group A and C hemolytic streptococci, *Staphylococcus aureus* and *Clostridium welchii:* Linker *et al.*, *J. Biol. Chem.* **219**, 13 (1956); in heads of leeches: Linker *et al.*, *Nature* **180**, 810 (1957); in snake venoms: Favilli, *ibid.* **145**, 866 (1940); in testes: Hahn, *Biochem. Z.* **315**, 83 (1943); Högberg, *Acta Chem. Scand.* **8**, 1098 (1954). Biochemical properties: D. Platt, *Arzneim.-Forsch.* **20**, 1836 (1970). *Review:*

Meyer, Rapport in *Adv. Enzymol.* **13**, 199-236 (1952); Meyer *et al.*, *The Enzymes* vol. 4, P. D. Boyer *et al.*, Eds. (Academic Press, New York, 2nd ed., 1960) pp 447-460; Meyer, *ibid.* **vol. 5** (3rd ed., 1971) pp 307-320. Reviews of clinical trials in myocardial infarction: G. S. May *et al.*, *Prog. Cardiovasc. Dis.* **25**, 335-359 (1983); A. B. Saunders, *Emerg. Med. Clin. North Am.* **6**, 361-372 (1988). Hyaluronidase manufacturers define their product in terms of turbidity-reducing (TR) units or in viscosity units. Prepd solns for injection usually contain 150 turbidity-reducing units or 500 viscosity units dissolved in 1 ml of isotonic NaCl soln.

USE: Pharmaceutic aid (diffusing agent in s.c. injections).

THERAP CAT: Spreading agent.

THERAP CAT (VET): To promote diffusion, absorption, resorption.

4798. Hycanthone. [3105-97-3] 1-[[2-(Diethylamino)ethyl]-amino]-4-(hydroxymethyl)-9*H*-thioxanthen-9-one. $C_{20}H_{24}N_2O_2S$; mol wt 356.48. C 67.39%, H 6.79%, N 7.86%, O 8.98%, S 8.99%. Metabolite of lucanthone, *q.v.*: Rosi *et al.*, *Nature* **208**, 1005 (1965). Prepn by oxidative fermentation of lucanthone and schistosomicidal activity: Rosi *et al.*, *J. Med. Chem.* **10**, 867 (1967); NL **6410359**, and Rosi, Peruzzotti, US **3294803**; US **3312598** (1965, 1966, 1967 all to Sterling Drug). Alternate synthesis: Laidlaw *et al.*, *J. Org. Chem.* **38**, 1743 (1973).

Crystals, mp 100.6-102.8°. Absorption max (ethanol): 233, 258, 329, 438 nm (ε 19400, 37000, 9700, 6600). Extremely sensitive to acid.

Hydrochloride. mp 173-176° (dec).

Mesylate. [23255-93-8] Etrenol.

THERAP CAT: Anthelmintic (Schistosoma).

4799. Hydantoin. [461-72-3] 2,4-Imidazolidinedione; 2,4-(3*H*,5*H*)-imidazoledione; glycolylurea. $C_3H_4N_2O_2$; mol wt 100.08. C 36.00%, H 4.03%, N 27.99%, O 31.97%. Prepn: Baeyer, *Ann.* **130**, 129 (1864). Manuf: Gresham, Schweitzer, US **2402134** (1946 to du Pont); White, Wysong, US **2663713** (1953 to Dow). *Review:* J. H. Bateman in *Kirk-Othmer Encyclopedia of Chemical Technology* vol. 12 (Wiley-Interscience, New York, 3rd ed., 1980) pp 692-711.

Needles from methanol, mp 220°. Slightly sol in water or ether; sol in alcohol, in solns of fixed alkali hydroxides.

4800. Hydnocarpic Acid. [459-67-6] (1*R*)-2-Cyclopentene-1-undecanoic acid; 11-(2-cyclopenten-1-yl)undecanoic acid. $C_{16}H_{28}O_2$; mol wt 252.40. C 76.14%, H 11.18%, O 12.68%. Component of chaulmoogra oil; naturally occurring in *d*-form. Isoln from seeds of *Hydnocarpus wightiana* Blume or *H. anthelmintica* Pierre, *Flacourtiaceae*, or from the seeds of *Taraktogenos kurzii* King, *Bixaceae:* B. Power, M. Barrowcliff, *J. Chem. Soc.* **87**, 884 (1905); *ibid.* **91**, 557 (1907). Structure: R. L. Shriner, R. Adams, *J. Am. Chem. Soc.* **47**, 2727 (1925). Synthesis of *dl*-hydnocarpic acid: D. G. M. Diaper, J. C. Smith, *Biochem. J.* **42**, 581 (1948). Toxicity data: Anderson, *Int. J. Lepr.* **2**, 99 (1934). Antimicrobial spectrum: P. L. Jacobsen, L. Levy, *Proc. West. Pharmacol. Soc.* **15**, 44 (1972). Mechanism of action: *eidem, Antimicrob. Agents Chemother.* **3**, 373 (1973). Chromatographic determn in seed oils: W. W. Christie *et al.*, *Lipids* **24**, 116 (1989).

Colorless, glistening leaflets from petr ether + ethyl acetate, mp 59-60°. $[\alpha]_D$ +68.3° (chloroform). Sparingly sol in usual organic solvents; sol in chloroform.

dl-Form. Pearly plates from alcohol, ethyl actate or petr ether + ethyl acetate, mp 59-59.5°.

Sodium salt. Sodium hydnocarpate; hydnocarpate sodium; sodium gynocardate. Yellowish powder. Sol in water, alc. The aq soln is alkaline. MLD in rats (mg/kg): 100-125 i.v. (Anderson).

THERAP CAT: Antibacterial (leprostatic).

4801. Hydracrylic Acid. [503-66-2] 3-Hydroxypropanoic acid; β-hydroxypropionic acid; ethylene lactic acid. $C_3H_6O_3$; mol wt 90.08. C 40.00%, H 6.71%, O 53.28%. Prepd by alkaline hydrolysis of the nitrile: R. R. Read, *Org. Synth.* **coll. vol. I**, 321 (2nd ed., 1941).

Viscous liq. Strong acid, pK (25°): 4.51. On distn or boiling with 50% H_2SO_4 dec into water and acrylic acid. Very sol in water, sol in alcohol, miscible with ether.

Sodium salt. [6487-38-3] $C_3H_5NaO_3$; mol wt 112.06. Deliquescent crystals, mp 143°.

Calcium salt dihydrate. [6556-13-4] $C_6H_{10}CaO_6.2H_2O$; mol wt 254.25. Prisms, mp 140-145°. Freely sol in cold water.

4802. Hydralazine. [86-54-4] 1-Hydrazinylphthalazine; 1(2H)-phthalazinone hydrazone; 1-hydrazinophthalazine; Ciba 5968; Präparat 5968; C-5968. $C_8H_8N_4$; mol wt 160.18. C 59.99%, H 5.03%, N 34.98%. Prepd by the action of hydrazine hydrate on 1-chloro- or 1-phenoxyphthalazine: Hartmann, Druey, **US 2484029** (1949 to Ciba); Druey, Ringier, *Helv. Chim. Acta* **34**, 204 (1951). Metabolism: Z. H. Israile, P. G. Dayton, *Drug Metab. Rev.* **6**, 283 (1977); K. Schmid *et al.*, *Arzneim.-Forsch.* **31**, 1143 (1981). Pharmacology: J. L. Cangiano *et al.*, *J. Lab. Clin. Med.* **92**, 516 (1978). Clinical paper: R. F. Albrecht *et al.*, *Int. Anesthesiol. Clin.* **16**, 299 (1978). Acute toxicity: L. Dorigotti *et al.*, *Pharmacol. Res. Commun.* **8**, 295 (1976). Comprehensive description: C. E. Orzech *et al.*, *Anal. Profiles Drug Subs.* **8**, 283-314 (1979).

Yellow needles from methanol, mp 172-173° (rapid heating). One gram dissolves in 3 ml 2N acetic acid, in 12 ml warm methanol. Forms a red compd (phthalazinylhydrazone) with acetone at 60° in presence of 2N acetic acid. LD_{50} in mice, rats (mg/kg): 122, 90 orally; 101, 40 i.p. (Dorigotti).

Hydrochloride. [304-20-1] Alphapress; Apresoline. $C_8H_8N_4.$HCl; mol wt 196.64. Yellow crystals, dec 273°. Soly in water (g/100 ml) at 15°: 3.01; at 25°: 4.42; in 95% ethanol: 0.2 g/100 ml. Very slightly sol in ether. pH of a 2% aq soln 3.5 to 4.5. uv max (0.001% aq soln): 211, 240, 260, 304, 315 nm. Aq solns containing 20 mg/ml may be preserved with 0.5% chlorobutanol.

THERAP CAT: Antihypertensive.

4803. Hydrallostane. [516-41-6] (5α,11β)-11,17,21-Trihydroxypregnane-3,20-dione; 11β,17α,21-trihydroxyallopregnane-3,20-dione; allodihydrohydrocortisone; allopregnane-11β,17α,21-triol-3,20-dione; 4,5α-dihydrocortisol; allodihydro F. $C_{21}H_{32}O_5$; mol wt 364.48. C 69.20%, H 8.85%, O 21.95%. Isoln from beef and hog adrenals: Neher, Wettstein, *Helv. Chim. Acta* **39**, 2062 (1956). Prepn: Pataki *et al.*, *J. Biol. Chem.* **195**, 753 (1952); Fukushima, Daum, *J. Org. Chem.* **26**, 520 (1961); Gould, Oliveto, **US 2783254**; **US 2897216** (1957; 1959 to Schering); Gould, Herzog, **US 2783226** (1957 to Schering); **GB 742888** (1956 to Syntex).

Crystals, mp 234-240°. $[\alpha]_D^{25}$ +83° (acetone). Practically insol in water. Sol in methanol, acetone, chloroform.

21-Acetate. $C_{23}H_{34}O_6$. Crystals from hexane + ethyl acetate, mp 211-213°. $[\alpha]_D^{20}$ +69° (chloroform).

21-tert-Butylacetate. $C_{27}H_{42}O_6$. Crystals from ethanol, mp 256-264° (also crystallizes as a hydrate). Sol in chloroform, oils, fats.

4804. Hydramethylnon. [67485-29-4] 1,5-Bis[4-(trifluoromethyl)phenyl]-1,4-pentadien-3-one 2-(1,4,5,6-tetrahydro-5,5-dimethyl-2-pyrimidinyl)hydrazone; 1,5-bis(α,α,α-trifluoro-p-tolyl)-1,4-pentadiene-3-one (1,4,5,6-tetrahydro-5,5-dimethyl-2-pyrimidinyl)hydrazone; tetrahydro-5,5-dimethyl-2(1H)-pyrimidinone[3-[4-(trifluoromethyl)phenyl]-1-[2-[4-(trifluoromethyl)phenyl]ethenyl]-2-propenylidene]hydrazone; AC-217300; Amdro; Combat; Maxforce. $C_{25}H_{24}F_6N_4$; mol wt 494.49. C 60.72%, H 4.89%, F 23.05%, N 11.33%. Slow-activating stomach poison insecticide. Prepn: J. B. Lovell, **US 4087525** (1975 to Am. Cyanamid). Activity: *idem*, *Proc. Br. Crop Prot. Conf. - Pests Dis.* **1979**, 575. Field trials for control of red imported fire ant: D. P. Harlan *et al.*, *Southwest. Entomol.* **6**, 150 (1981); W. A. Banks *et al.*, *ibid.* 158; for control of cockroaches: J. F. Milio *et al.*, *J. Econ. Entomol.* **79**, 1280 (1986).

Crystals from isopropanol, mp 189-191°. Insol in water. Sol in alc, acetone.

USE: Insecticide.

4805. Hydrangea. Seven barks. Dried rhizome and roots of *Hydrangea arborescens* L., Saxifragaceae. Habit. Eastern U.S. Constit. The glucoside hydrangin, saponin, resins, fixed and volatile oils, starch.

4806. Hydrastine. [118-08-1] (3S)-6,7-Dimethoxy-3-[(5R)-5,6,7,8-tetrahydro-6-methyl-1,3-dioxolo[4,5-g]isoquinolin-5-yl]-1(3H)-isobenzofuranone; l-β-hydrastine. $C_{21}H_{21}NO_6$; mol wt 383.40. C 65.79%, H 5.52%, N 3.65%, O 25.04%. Isoln of naturally occurring l-β-form from *Hydrastis canadensis* L., Ranunculaceae together with berberine and canadine. Synthesis of diastereoisomeric mixtures of hydrastines: Hope *et al.*, *J. Chem. Soc.* **1931**, 236; Marshall *et al.*, *ibid.* **1934**, 1315; M. Hanaoka *et al.*, *Chem. Pharm. Bull.* **27**, 1947 (1979); J. R. Falck, S. Manna, *Tetrahedron Lett.* **22**, 619 (1981). Resolution to the l-β-form: Haworth, Pinder, *J. Chem. Soc.* **1950**, 1776; Haworth *et al.*, *Nature* **165**, 549 (1950). Structure: Knabe, *Arch. Pharm.* **293**, 121 (1960). Abs config: Ohta *et al.*, *Tetrahedron Lett.* **1963**, 859; Blaha *et al.*, *Collect. Czech. Chem. Commun.* **29**, 2328 (1964); Snatzke *et al.*, *Tetrahedron* **25**, 5059 (1969). Biosynthesis: Gear, Spenser, *Can. J. Chem.* **41**, 783 (1963).

Orthorhombic prisms from alc, mp 132°. $[\alpha]_D^{20}$ −50° (c = 0.3 in abs alc). uv max (ethanol): 202, 218, 238, 298, 316 nm (log ε 4.79, 4.53, 4.15, 3.86, 3.63). pK 7.8. Freely sol in acetone and benzene; insol in water. The salts hydrolyze easily and do not crystallize well.

Hydrochloride. [5936-28-7] $C_{21}H_{21}NO_6 \cdot HCl$. Prepn: R. Paech, M. V. Tracy, *Modern Methods of Plant Analysis* vol. IV (Springer-Verlag, Berlin, 1955) p 383. Powder, mp 116°. $[\alpha]_D^{17}$ +127° (c = 4 in dil HCl). Very sol in water, alcohol; slightly sol in CHCl₃; very slightly sol in ether. pH (0.5% aq soln): 4.2. Hygroscopic. *Keep well closed.*

THERAP CAT: Hydrochloride formerly as uterine hemostatic, antiseptic.

4807. Hydrastinine. [6592-85-4] 5,6,7,8-Tetrahydro-6-methyl-1,3-dioxolo[4,5-g]isoquinolin-5-ol; 1-hydroxy-6,7-methylenedioxy-2-methyl-1,2,3,4-tetrahydroisoquinoline. $C_{11}H_{13}NO_3$; mol wt 207.23. C 63.76%, H 6.32%, N 6.76%, O 23.16%. By oxidation of hydrastine: Freund, Will, *Ber.* **20**, 88 (1887). From berberine: Freund, **DE 241136** (1910), *C.A.* **6**, 2145 (1912). From cotarnine: Pyman, Remfry, *J. Chem. Soc.* **101**, 1595 (1912); Topchiev, *J. Appl. Chem. USSR* **6**, 529 (1933). From formylhomopiperonylamine: Decker *et al.*, *Ann.* **395**, 299, 321, 328 (1913); Rosenmund, *Ber. Dtsch. Pharm. Ges.* **29**, 200 (1919). From safrole: Kindler, Peschke, *Arch. Pharm.* **270**, 353 (1932). Structure study: Schneider, Müller, *Ann.* **615**, 34 (1958).

Needles from petr ether, mp 117°. Freely sol in alcohol, chloroform, ether, dil acids; practically insol in cold, moderately sol in hot water. Solns with water or alcohol are yellow and fluorescent, those with nonpolar organic solvents are colorless. pK 2.62. Absorption spectra: Dobbie, Tinkler, *J. Chem. Soc.* **85**, 1005 (1904).

THERAP CAT: The hydrochloride as cardiotonic; uterine hemostatic.

4808. Hydrastis. Golden seal; orange root; yellow root; yellow puccoon; Indian turmeric. Dried rhizome and roots of *Hydrastis canadensis* L., *Ranunculaceae*, contg not less than 2.5% ether-soluble alkaloids. *Habit.* North America. *Constit.* 2-4% hydrastine, 2-3% berberine; canadine, volatile oil, resin.

4809. Hydrazine. [302-01-2] Hydrazine anhydrous; diamide; diamidogen; nitrogen hydride. H_4N_2; mol wt 32.05. H 12.58%, N 87.41%. Simplest diamine; versatile reagent chemical with numerous applications. Prepn and isoln of derivatives: T. Curtius, *Ber.* **20**, 1632 (1887); T. Curtius, R. Jay, *J. Prakt. Chem.* **39**, 27 (1889); F. Raschig, *Ber.* **40**, 4580 (1907); P. W. Schenk in *Handbook of Preparative Inorganic Chemistry* vol. **1**, G. Brauer, Ed. (Academic Press, New York, 1963) pp 469. Review of prepn methods: H. Hayashi, *Res. Chem. Intermed.* **24**, 183-196 (1998). Review of synthetic utility: E. A. A. Hafez *et al.*, *Heterocycles* **22**, 1821-1877 (1984). Toxicity data: Witkin, *Arch. Ind. Health* **13**, 34 (1956). Toxicology study: Back, Thomas, *Annu. Rev. Pharmacol.* **10**, 395 (1970). Review of carcinogenic risk: *IARC Monographs* **4**, 127-136 (1974); of toxicology: R. von Burg, T. Stout, *J. Appl. Toxicol.* **11**, 447-450 (1991); and human exposure: *Toxicological Profile for Hydrazines* (PB98-101025, 1997) 224 pp. Books: L. F. Audrieth, B. A. Ogg, *The Chemistry of Hydrazine* (Wiley, New York, 1951); C. C. Clark, *Hydrazine* (Mathieson Chem., Baltimore, 1953). *Reviews:* Troyan, *Ind. Eng. Chem.* **45**, 2608-2612 (1953); Zimmer, *Chem. Ztg.* **79**, 599-605 (1955); Hudson *et al.*, "Hydrazine" in *Mellor's* vol. **VIII**, suppl. II, *Nitrogen* (Part 2), 69-114 (1967); Jones in *Comprehensive Inorganic Chemistry* vol. **2**, J. C. Bailar, Jr. *et al.*, Eds. (Pergamon Press, Oxford, 1973) pp 250-265; E. F. Rothgery in *Kirk-Othmer Encyclopedia of Chemical Technology* vol. **13**, A. Seidel, Ed. (John Wiley & Sons, Hoboken, 5th ed., 2005) pp 562-607.

Colorless oily liq, fuming in air. Penetrating odor resembling that of ammonia. *Corrosive, flammable, poisonous.* mp 2.0°. bp₇₆₀ 113.5°; bp₇₁ 56°; bp₅ ₐₜₘ 170°; bp₁₀ ₐₜₘ 200°; bp₂₀ ₐₜₘ 236°. d_4^{-5} 1.146; d_0^0 1.0253; d_4^2 1.024; d_4^{15} 1.011; d_4^{25} 1.0036; d_4^{35} 0.9955. One gallon of commercial product weighs 8.38 lbs. $n_D^{22.3}$ 1.46979; n_D^{35} 1.46444. Flash pt, closed cup: 126°F (52°C). Dipole moment 1.83-1.90. Dielectric constant (25°): 51.7. Latent heat of fusion (mp): 3.025 kcal/mole; latent heat of vaporization (bp): 9760 kcal/mole (calc). Crit temp 380°; crit pressure 14 atm. Diacidic base. pK_1 (25°): ~6.05. Misc with water, methyl, ethyl, propyl, isobutyl alcohols. Forms salts with inorganic acids. Highly polar solvent. Powerful reducing agent. Dissolves many inorganic substances. Hygroscopic. Forms an azeotropic mixture with water, bp₇₆₀ 120.3°, which contains 55 mole-% (68.5 weight-%) N_2H_4. Contracts on freezing. Burns with violet flame. Explodes during distn if traces of air are present, also affected by uv and metal ion catalysts. Can be stored for years if sealed in glass and kept in a cool, dark place. LD_{50} in mice (mg/kg): 57 i.v.; 59 orally (Witkin).

Monohydrate. [7803-57-8] Hydrazinium hydroxide. $H_4N_2 \cdot H_2O$; mol wt 50.06. Prepn and properties: A. E. Tutton, *Nature* **43**, 205 (1891). X-ray study: M. Zocchi *et al.*, *Acta Crystallogr.* **15**, 803 (1962). Fuming refractive liquid with faint characteristic odor. *Poisonous, corrosive.* mp −51.7° or below −65° (two eutectics). bp₇₄₀ 118-119°; bp₂₆ 47°. d^{21} 1.03. n_D^{20} 1.42842. Flash pt, closed cup: 205°F (96°C). Miscible with water, alc. Insol in chloroform, ether, benzene. Very powerful reducing agent. Strong base. Attacks glass, rubber, cork, but not stainless V₂A steel or Allegheny stainless 304 and 347. Molybdenum steels such as Allengheny stainless 316 should not be used.

Dihydrochloride. [5341-61-7] $H_4N_2 \cdot 2HCl$; mol wt 104.96. White crystalline powder, mp 198°. d 1.42. *Poisonous, corrosive.* Freely sol in water, slightly in alc.

Caution: Potential symptoms of overexposure to hydrazine are irritation of eyes, skin, nose and throat; temporary blindness; dizziness, nausea; dermatitis; burns skin and eyes. See *NIOSH Pocket Guide to Chemical Hazards* (DHHS/NIOSH 2005-149, 2007) p 166. See also *Patty's Industrial Hygiene and Toxicology* vol. **2E**, G. D. Clayton, F. E. Clayton, Eds. (John Wiley & Sons, Inc., New York, 4th ed., 1994) pp 3435-3441. Hydrazine is reasonably anticipated to be a human carcinogen: *Report on Carcinogens, Twelfth Edition* (PB2011-111646, 2011) p 234.

USE: Reagent in synthetic organic chemistry; used in manuf of agricultural and pharmaceutical chemicals, spandex fibers, chemical blowing agents, and antioxidants. In water treatment for corrosion protection; in rocket fuels. Dihydrochloride as chlorine scavenger for HCl gas streams. Monohydrate as solvent for inorganic materials, as rocket engine propellant when mixed with methanol, and in manuf of "Helman" catalyst, consisting of 80% hydrazine hydrate, 19.5% ethanol, 0.5 to 0.05% copper, used to dec hydrogen peroxide in V-2 type rockets.

4810. Hydrazine Sulfate. [10034-93-2] Hydrazine sulfate (1:1); hydrazinium sulfate; hydrazonium sulfate; hydrazine dihydrogen sulfate (1:1); Sehydrin. $H_6N_2O_4S$; mol wt 130.12. H 4.65%, N 21.53%, O 49.18%, S 24.64%. $H_2NNH_2 \cdot H_2SO_4$. Sulfuric acid salt of hydrazine, *q.v.* Initial prepn and isoln: T. Curtius, *Ber.* **20**, 1632 (1887); and physical properties: T. Curtius, R. Jay, *J. Prakt. Chem.* **39**, 27 (1889). Prepn: R. Adams, B. K. Brown, *Org. Synth.* coll. vol. **I**, 309 (1941); Audrieth, Nickles, *Inorg. Synth.* **1**, 90 (1939). Industrial prepn: *BIOS Final Report* **369**; Moncrieff, *Manuf. Chem.* **18**, 177 (1947). Revised lab procedures: Pfeiffer, Simons, *Ber.* **80**, 127 (1947). Crystal structure: Nitta *et al.*, *Acta Crystallogr.* **4**, 289 (1951); Jönsson, Hamilton, *ibid.* **26B**, 536 (1970). Clinical study in advanced cancer patients: V. A. Filov *et al.*, *Invest. New Drugs* **13**, 89 (1995). Review of activity and clinical studies in cancer cachexia: J. Gold, *Nutr. Cancer* **9**, 59-66 (1987).

White orthorhombic crystals, glass-like plates or prisms. mp 254°. d 1.378; d^7 2.016. *Poisonous, corrosive, irritant, skin sensitizer.* Freely sol in hot water; sol in about 33 parts water. Insol in alc. pH of 0.2 molar aq soln 1.3.

Caution: This substance is reasonably anticipated to be a human carcinogen: *Report on Carcinogens, Twelfth Edition* (PB2011-111646, 2011) p 234.

USE: Reagent in synthetic organic chemistry. In the gravimetric estimation of nickel, cobalt and cadmium; in the refining of rare metals; as antioxidant in soldering flux for light metals; as reducing agent in the analysis of minerals and slags; in separating polonium from tellurium; in tests for blood; for destroying fungi and molds.

THERAP CAT: Antineoplastic.

4811. 4-Hydrazinobenzenesulfonic Acid. [98-71-5] 4-Hydrazinylbenzenesulfonic acid; p-sulfophenylhydrazine; phenylhydrazine-p-sulfonic acid. $C_6H_8N_2O_3S$; mol wt 188.20. C 38.29%, H 4.28%, N 14.89%, O 25.50%, S 17.04%. Prepn by sulfonation of phenylhydrazine: L. Claisen, P. Roosen, *Ann.* **278**, 296 (1894); by the reduction of p-diazobenzenesulfonic acid: Th. Zincke, A. Kuchenbecker, *Ann.* **330**, 1 (1903); L. V. Lazeeva *et al.*, SU **1057493** (1983 to Tambov Pigment), *C.A.* **100**, 138755q (1984). Used in resoln of 2-pyrazoline cmpds: M. Mukai *et al.*, *Can. J. Chem.* **57**, 360 (1979); in isoln of volatile ketones: W. Treibs, H. Röhnert, *Ber.* **84**, 433 (1951); in analysis of trace amounts of selenium: T. Kawashima *et al.*, *Anal. Chim. Acta* **49**, 443 (1970); *eidem, ibid.* **89**, 65 (1977).

$$H_2N-\overset{\displaystyle H}{N}-\!\!\!\!\bigcirc\!\!\!\!-SO_3H$$

Needles from water, mp 286°. Slightly sol in water, alcohol.

4812. Hydrazoic Acid. [7782-79-8] Hydrogen azide; hydronitric acid; triazoic acid; azidic acid; stickstoffwasserstoffsäure (German). HN_3; mol wt 43.03. H 2.34%, N 97.66%. Produced by the action of sulfuric acid on sodium azide: L. F. Audrieth, C. F. Gibbs, *Inorg. Synth.* **1**, 77 (1939); using stearic acid: Günther, Meyer, *Z. Elektrochem.* **41**, 541 (1935). Prepn of water and ether solns of hydrazoic acid: W. S. Frost *et al.*, *J. Am. Chem. Soc.* **55**, 3516 (1933); L. F. Audrieth, C. F. Gibbs, *loc. cit.*; P. W. Schenk in *Handbook of Preparative Inorganic Chemistry* vol. **1**, G. Brauer, Ed. (Academic Press, New York, 2nd ed., 1963) pp 472-474. GC determn: J. M. Zehner, R.A. Simonaitis, *J. Chromatogr. Sci.* **14**, 493 (1976). Toxicity study: Graham *et al.*, *J. Ind. Hyg. Toxicol.* **30**, 98 (1948). Review of toxicology: C. F. Reinhardt, M. R. Brittelli, in *Patty's Industrial Hygiene and Toxicology* Vol. **2A**, G. D. Clayton, F. E. Clayton, Eds. (Wiley-Interscience, New York, 1981) pp 2779-2784. *Reviews:* Mason in *Mellor's* vol. **VIII**, suppl. II, *Nitrogen* (part II), 1-15 (1967); Jones in *Comprehensive Inorganic Chemistry* vol. **2**, J. C. Bailar Jr. *et al.*, Eds. (Pergamon Press, Oxford, 1973) pp 276-293.

Mobile liquid. Intolerable pungent odor. *Extremely explosive.* mp −80°. bp 37°. pKa 4.72. LD_{50} in mice (mg/kg): 21.5 i.p. (Graham).

Caution: Acute exposure: eye irritation, cough, headache, fall in blood pressure, weakness, collapse. *Chronic exposure:* hypotension, weakness, palpitation, ataxia (Graham).

USE: Industrially in prepn of heavy metal azides for shell detonators.

4813. Hydrindantin. [5103-42-4] 2,2′-Dihydroxy-[2,2′-bi-1*H*-indene]-1,1′,3,3′ (2*H*,2′*H*)-tetrone; 2,2′-dihydroxy-[2,2′-biindan]-1,1′,3,3′-tetrone; reduced ninhydrin. $C_{18}H_{10}O_6$; mol wt 322.27. C 67.09%, H 3.13%, O 29.79%. Formed by the action of potassium cyanide on ninhydrin: Bruice, Richards, *J. Org. Chem.* **23**, 145 (1958). Convenient prepn by reduction of ninhydrin with ascorbic acid: Moore, Stein, *J. Biol. Chem.* **211**, 907 (1954).

Dihydrate. Prisms from acetone, anhydr at 100°, turns reddish-brown at 200°, dec 249-254°. Very sparingly sol in hot water; sol in Methyl Cellosolve. Sol with decompn in aq Na_2CO_3 solns (deep red color), and in NaOH solns (deep blue color). Can be precipitated from carbonate solns by the addn of acid. Deep purple color with ammonia and blue color with amino acids.

USE: Reagent for the photometric determination of amino acids and similar compds.

4814. Hydriodic Acid. [10034-85-2] HI; mol wt 127.91. A soln of hydrogen iodide in water. Marketed in various concns, *e.g.*, 57% HI, d 1.7; 47%, d 1.5; 10%, d 1.1. Prepd by absorption of hydrogen iodide gas in water or by the action of iodine on hydrogen sulfide according to the eq $H_2S + I_2 \rightarrow 2HI + S$: Heisig, Frykholm, *Inorg. Synth.* **1**, 157 (1939). *See also* Hydrogen Iodide.

Colorless when freshly made, but rapidly turns yellowish or brown on exposure to light and air. This discoloration can be prevented by the addition of about 1.5% hypophosphorous acid (H_3PO_2). Concd solns that have been stored for some time are usually opaque from oxidation, they may be regenerated with hypophosphorous acid: Foster, Nahas, Jr., *Inorg. Synth.* **2**, 210 (1946). *Corrosive. Keep protected from light and air, preferably not above 30°.* Miscible with water or alcohol. Dissolves iodine. The azeotrope (constant-boiling acid) bp_{760} 127°, d 1.70, contains 56.9% HI. Strong, corrosive acid, attacks natural rubber. 0.1 molar soln, pH 1.0.

Caution: Strong irritant.

USE: Reducing agent, manuf of inorganic iodides, pharmaceuticals, disinfectants. The 57% acid is also used for analytical purposes, such as methoxyl determinations.

THERAP CAT: Expectorant.

4815. Hydrobenzoin. [492-70-6] 1,2-Diphenyl-1,2-ethanediol; diphenylethyleneglycol. $C_{14}H_{14}O_2$; mol wt 214.26. C 78.48%, H 6.59%, O 14.93%. Prepn of d-, l-, dl- and meso-forms: Forst, Zincke, *Ann.* **182**, 262, 275 (1876); Irvine, Weir, *J. Chem. Soc.* **91**, 1390 (1907); Buck, Jenkins, *J. Am. Chem. Soc.* **51**, 2163 (1929); L. F. Fieser, *Organic Experiments* (D. C. Heath, Boston, 1964) pp 210, 214, 216-217, 229-231. Improved method for prepn of dl-form from the meso isomer: Collet, *Synthesis* **1973**, 664.

dl-Form. Crystals from ether + petr ether, mp 120°. Resolved into d- and l-forms by slow, repeated crystallizations from ether.
d-Form. $[\alpha]_D^{20}$ +97.6° (chloroform).
l-Form. $[\alpha]_D^{20}$ −97.0° (chloroform).
meso-Form. Monoclinic leaflets from alcohol or water, mp 139°. Soly in water: 0.25% at 20°, 1.25% at 100°. Freely sol in hot alcohol, chloroform. Infrared absorption (chloroform): 2.82, 2.96μ. Dipole moment 2.67.

4816. Hydrobromic Acid. [10035-10-6] HBr; mol wt 80.91. A soln of hydrogen bromide gas in water. Marketed in various concns, e.g., 50% HBr, d 1.517; 40% HBr, d 1.38; 34%, d 1.31; 10%, d 1.08. Lab prepn according to the eq $H_2SO_4 + KBr \rightarrow KHSO_4 + HBr$: Heisig, Amdur, *Inorg. Synth.* **1**, 155 (1939). *See also* Hydrogen Bromide.

Colorless or faintly yellow; slowly darkens on exposure to air and light. Miscible with water, alcohol. *Corrosive. Keep protected from light.* When dil hydrobromic acid is distilled, a weaker acid comes over first and when a very concd acid is boiled, HBr gas chiefly distills over first; in both cases a "constant boiling" acid contg about 47.5% HBr remains which distills unchanged at 126°. The bp and compn of the azeotrope varies with pressure: at 100 mm, bp 74.12°, compn 49.80% HBr; at 400 mm, bp 107.00°, compn 48.47%; at 700 mm, bp 122°, compn 47.74%; at 800 mm, bp 125.79°, compn 47.56%: Bonner *et al.*, *J. Am. Chem. Soc.* **55**, 1406 (1933). Aq solns are strongly acid. The satd aq soln contains 68.85% HBr at 0° and 66% at 25°.

Caution: Strong irritant.

USE: The concd acid is used principally in analytical chemistry and organic prepns. Catalyst.

THERAP CAT: Sedative.

THERAP CAT (VET): Has been used as a sedative.

4817. Hydrocarbostyril. [553-03-7] 3,4-Dihydro-2(1*H*)-quinolinone; 3,4-dihydrocarbostyril; 3,4-dihydro-2-quinolinol; o-aminohydrocinnamic acid lactam; 2-oxo-1,2,3,4-tetrahydroquinoline; dihydro-α-quinolone. C_9H_9NO; mol wt 147.18. C 73.45%, H 6.16%, N 9.52%, O 10.87%. Prepd by the catalytic reduction of onitrocinnamic acid: Blout, Silverman, *J. Am. Chem. Soc.* **66**, 1442 (1944).

Prisms from methanol + water, mp 165-166.5°. Freely sol in alcohol, ether, dimethylformamide. Practically insol in water. Sol in hot aq NaOH solns.

4818. Hydrochloric Acid. [7647-01-0] Muriatic acid. HCl; mol wt 36.46. A soln of hydrogen chloride gas (HCl) in water. Prepn and reviews: *see* Hydrogen Chloride.

Fumes in air. *Corrosive.* May be colored yellow by traces of iron, chlorine, and organic matter. Reagent grade concd hydrochloric acid contains close to 38.0% HCl. 83 ml of concd HCl poured into sufficient water to make 1 liter yields approx 1.0*N* HCl. The pH of 1.0*N* HCl is 0.10; of 0.1*N* = 1.10; of 0.01*N* = 2.02; of 0.001*N* = 3.02; of 0.0001*N* = 4.01. n_D^{18} (1.0*N* soln) 1.34168. d_4^{15} 1.05 (10.17% w/w soln); 1.10 (20%); 1.15 (29.57%); 1.20 (39.11%). Freezing pt: −17.14° (10.81% soln); −62.25° (20.69%); −46.2° (31.24%); −25.4° (39.17%), *Gmelins, Chlorine* (8th ed.) **6**, 136-137 (1927). Constant boiling azeotrope with water bp₇₆₀ 108.58° contg 20.22% HCl, d_4^{15} 1.096. Boiling weaker or stronger aq solns results in loss of either component until the constant boiling acid is obtained.

Caution: Corrosive burns may result from the inhalation of acid fumes and from skin contact with or the ingestion of strong acid. Symptoms after ingestion or skin contact include immediate pain and ulceration of all membranes and tissues which come in contact with the acid. Ingestion may be associated with nausea, vomiting and intense thirst; corrosion of the stomach may lead within a few hours or a few days to gastric perforation and peritonitis. Late esophageal, gastric and pyloric strictures and stenoses should be anticipated. Contact of conc acid with the eye can cause extensive necrosis of the conjunctiva and corneal epithelium, resulting in perforation or opaque scarring. Chemical pneumonitis can be expected after respiratory exposure to acid vapors or after tracheobronchial aspiration of ingested acid. Death may occur due to complications such as circulatory shock, asphyxia due to glottic or laryngeal edema, perforation of the stomach with peritonitis, gastic hemorrhage, infection or anition due to stricture formation. *See: Clinical Toxicology of Commercial Products*, R. E. Gosselin *et al.*, Eds. (Williams & Wilkins, Baltimore, 5th ed., 1984) Section III, pp 8-11.

USE: In the production of chlorides; refining ore in the production of tin and tantalum; for the neutralization of basic systems; as laboratory reagent; hydrolyzing of starch and proteins in the prepn of various food products; pickling and cleaning of metal products; as catalyst and solvent in organic syntheses. Also used for oil- and gas-well treatment and in removing scale from boilers and heat-exchange equipment. Trace metal analysis. Titrant; acidification; digestion. Pharmaceutic aid (acidifier).

THERAP CAT (VET): Has been used as gastric acidifier.

4819. Hydrochlorothiazide. [58-93-5] 6-Chloro-3,4-dihydro-2*H*-1,2,4-benzothiadiazine-7-sulfonamide 1,1-dioxide; 6-chloro-3,4-dihydro-7-sulfamoyl-2*H*-1,2,4-benzothiadiazine 1,1-dioxide; 6-chloro-7-sulfamyl-3,4-dihydro-1,2,4-benzothiadiazine 1,1-dioxide; 3,4-dihydrochlorothiazide; chlorsulfonamidodihydrobenzothiadiazine dioxide; chlorosulthiadil; Dichlotride; Disalunil; Esidrex; Esidrix; HydroDiuril; Hydrosaluric; Oretic. $C_7H_8ClN_3O_4S_2$; mol wt 297.73. C 28.24%, H 2.71%, Cl 11.91%, N 14.11%, O 21.49%, S 21.54%. Prepn: de Stevens *et al.*, *Experientia* **14**, 463 (1958); Werner *et al.*, *J. Am. Chem. Soc.* **82**, 1161 (1960); Jones, Novello, **US 3025292** (1962 to Merck & Co.); de Stevens, Werner, **US 3163645** (1964 to Ciba); Klosa, **DE 1163332** (1964); J. S. Irons, T. M. Cook, **US 3164588** (1965 to Merck & Co.). Purification: Downing, **US 3043840** (1962 to Merck & Co.). LC/MS/MS determn in plasma: F. Liu *et al.*, *J. Pharm. Biomed. Anal.* **44**, 1187 (2007). Toxicity data: J. J. Piala *et al.*, *J. Pharmacol. Exp. Ther.* **134**, 273 (1961). Comprehensive description: H. P. Deppeler, *Anal. Profiles Drug Subs.* **10**, 405-441 (1981). Review of pharmacology and clinical experience of combination with valsartan in hypertension: A. J. Wagstaff, *Drugs* **66**, 1881-1901 (2006).

White, or practically white, crystalline powder, mp 273-275°. uv max (methanol + trace HCl): 317, 271, 226 nm ($A_{1cm}^{1\%}$ 130, 654, 1280). pKa 7.9, 9.2. Sol in sodium hydroxide solution; slightly sol in water. LD₅₀ in mice (mg/kg): 590 i.v.; >8000 orally (Piala).

THERAP CAT: Diuretic.

THERAP CAT (VET): Diuretic.

4820. Hydrocinchonidine. [485-64-3] (8α,9*R*)-10,11-Dihydrocinchonan-9-ol; (−)-dihydrocinchonidine; cinchamidine. $C_{19}H_{24}N_2O$; mol wt 296.41. C 76.99%, H 8.16%, N 9.45%, O 5.40%. Found in cinchona barks: Forst, Böhringer, *Ber.* **14**, 1270 (1881); Hesse, *ibid.* 1683; *idem, Ann.* **214**, 1 (1882); Skita, Nord, *Ber.* **45**, 3312 (1912); Heidelberger, Jacobs, *J. Am. Chem. Soc.* **41**, 817 (1919). Configuration: Ochiai *et al.*, *J. Pharm. Soc. Jpn.* **67**, 211 (1947). Stereospecific synthesis from secologanin: R. T. Brown, D. Curless, *Tetrahedron Lett.* **27**, 6005 (1986). Total synthesis: M. Ihara *et al.*, *J. Chem. Soc. Perkin Trans. 1* **1988**, 1277. Stereoisomer of hydrocinchonine, *q.v.*

Needles or leaflets, mp 230°. $[\alpha]_D^{15}$ −98° (alcohol). Practically insol in water. Sol in alcohol, chloroform; slightly sol in ether.

4821. Hydrocinchonine. [485-65-4] (9*S*)-10,11-Dihydrocinchonan-9-ol; cinchotine; cinconifine; (+)-dihydrocinchonine; pseudocinchonine. $C_{19}H_{24}N_2O$; mol wt 296.41. C 76.99%, H 8.16%, N 9.45%, O 5.40%. From cinchona barks: Caventou, Willm, *Compt. Rend.* **69**, 284 (1869). Prepn from cinchonine: Hesse, *Ann.* **300**, 46 (1898); Pum, *Monatsh. Chem.* **16**, 68 (1895); Arlt, *ibid.* **20**, 426, 439 (1899); Heidelberger, Jacobs, *J. Am. Chem. Soc.* **41**, 817 (1919). Structure: Rabe, *Ber.* **55**, 522 (1922). Conversion of hydroquinidine to hydrocinchonine: King, *J. Chem. Soc.* **1946**, 523. HPLC determn: C.-T. A. Chung, E. J. Staba, *J. Chromatogr.* **295**, 276 (1984). Stereospecific synthesis from secologanin: R. T. Brown, D. Curless, *Tetrahedron Lett.* **27**, 6005 (1986). Total synthesis: M. Ihara *et al.*, *J. Chem. Soc. Perkin Trans. 1* **1988**, 1277.

Prisms, mp 268-269°. $[\alpha]_D^{14}$ +204° (c = 0.6 in alc). Almost insol in water, ether. Sol in alcohol.

Hydrochloride. $C_{19}H_{24}N_2O \cdot HCl$. Crystals, mp 220-221°. $[\alpha]_D^{23}$ +155° (c = 0.8 in water).

4822. Hydrocinnamic Acid. [501-52-0] Benzenepropanoic acid; 3-phenylpropionic acid; β-phenylpropionic acid; benzylacetic acid. $C_9H_{10}O_2$; mol wt 150.18. C 71.98%, H 6.71%, O 21.31%.

Prepn by reduction of cinnamic acid: Ingersoll, *Org. Synth.* **9**, 42 (1929); from propiophenone: Schwenk, Papa, *J. Org. Chem.* **11**, 798 (1946); by Mauer oxidation of 1-phenyl-3-propanol: Langenbeck, Richter, *Ber.* **89**, 202 (1956); by chromate oxidation of propylbenzene: Reitsema, Allphin, *J. Org. Chem.* **27**, 27 (1962).

White, cryst powder, mp 47-48°. bp 280°; bp_{75} 194-197°; bp_{18} 145-147°; bp_6 125-129°. Sol in 170 parts cold, more sol in hot water, in alcohol, benzene, chloroform, ether, glacial acetic acid, petr ether, carbon disulfide.

4823. Hydrocodone. [125-29-1] (5α)-4,5-Epoxy-3-methoxy-17-methylmorphinan-6-one; dihydrocodeinone. $C_{18}H_{21}NO_3$; mol wt 299.37. C 72.22%, H 7.07%, N 4.68%, O 16.03%. Semisynthetic opioid analgesic. Prepn from dihydrothebaine: M. Freund, E. Speyer, *Ber.* **53**, 2250 (1920); by hydrogenation of codeinone: C. Mannich, H. Löwenheim, *Arch. Pharm.* **258**, 295 (1920). Prepn from codeine: H. Rapoport *et al., J. Org. Chem.* **15**, 1103 (1950); A. Stein, *Pharmazie* **10**, 180 (1955). Asymmetric synthesis: C. Y. Hong *et al., J. Am. Chem. Soc.* **115**, 11028 (1993); K. A Parker, D. Fokas, *J. Org. Chem.* **71**, 449 (2006). LC/MS/MS determn in plasma: Y.-L. Chen *et al., J. Chromatogr. B* **769**, 55 (2002). Pharmacology and toxicology: N. B. Eddy, J. G. Reid, *J. Pharmacol. Exp. Ther.* **52**, 468 (1934). Evaluation of abuse liability: J. P. Zacny *et al., Drug Alcohol Depend.* **78**, 243 (2005). Clinical comparison with oxycodone for acute pain: C. A. Marco *et al., Acad. Emerg. Med.* **12**, 282 (2005).

Prisms from alcohol, mp 197-198°. $[\alpha]_D^{25}$ −203° (c = 0.41 in $CHCl_3$). Sol in alcohol, acetone, ethyl acetate, chloroform. Insol in water. uv max: 280 nm (ε 1310). LD_{50} s.c. in mice: 85.7 mg/kg (Eddy, Reid).

Hydrochloride. [25968-91-6] $C_{18}H_{21}NO_3 \cdot HCl$. Monohydrate, crystals, mp 185-186° dec. $[\alpha]_D^{27}$ −130° (c = 2.877). Very sol in water.

Bitartrate hemipentahydrate. [34195-34-1]; [143-71-5] (anhydrous). Dicodid (tabl.); Hycodan. $C_{18}H_{21}NO_3 \cdot C_4H_6O_6 \cdot 2\frac{1}{2}H_2O$; mol wt 494.49. Component of *Lortab, Vicodin, Vicoprofen, Zydone*. Fine white crystals or crystalline powder. mp 118-128°. One gram dissolves in 16 ml water, in 150 g 95% ethanol. Insol in ether, chloroform. pH of a 2% aq soln about 3.6. *Protect from light.*

Note: This is a controlled substance (opiate): **21 CFR**, 1308.12.

THERAP CAT: Analgesic; antitussive.

4824. Hydrocortisone. [50-23-7] (11β)-11,17,21-Trihydroxypregn-4-ene-3,20-dione; cortisol; 4-pregnene-11β,17α,21-triol-3,20-dione; 17-hydroxycorticosterone; anti-inflammatory hormone; Kendall's compound F; Reichstein's substance M; Ala-Cort; Cetacort, Cort-Dome; Cortef; Cortenema; Cortril; Dermacort; Dioderm; Efcortelan; Ficortril; Hydracort; Hydrocort; Hydrocortisyl; Hydrocortone; Hytone; Mildison; Nutracort; Proctocort. $C_{21}H_{30}O_5$; mol wt 362.47. C 69.59%, H 8.34%, O 22.07%. Principal glucocorticoid hormone produced by the adrenal cortex. Biosynthesis stimulated by ACTH, *q.v.* Circulates in plasma primarily bound to *corticosteroid-binding globulin*, also known as *transcortin*, and to albumin. Isoln from adrenal glands: Reichstein, *Helv. Chim. Acta* **20**, 953 (1937); Mason *et al., J. Biol. Chem.* **124**, 459 (1938); from urine: Mason, Sprague, *ibid.* **175**, 451 (1948); from blood: Reich *et al., ibid.* **187**, 411 (1950). Configuration: von Euw, Reichstein, *Helv. Chim. Acta* **25**, 988 (1942); **30**, 205 (1947). Synthesis: N. L. Wen-

dler *et al., J. Am. Chem. Soc.* **72**, 5793 (1950). Biosynthesis by isolated adrenal glands: O. Hechter *et al., Arch. Biochem. Biophys.* **25**, 457 (1950); A. Zaffaroni *et al., J. Am. Chem. Soc.* **73**, 1390 (1951). Prepn by microbial transformation: H. C. Murray, D. H. Peterson, US 2602769 (1952 to Upjohn); D. R. Colingsworth *et al., J. Biol. Chem.* **203**, 807 (1953). Total biosynthesis in yeast: F. M. Szczebara *et al., Nat. Biotechnol.* **21**, 143 (2003). Comprehensive description: K. Florey, *Anal. Profiles Drug Subs.* **12**, 277-324 (1983). Review of clinical use in dermatoses: A. M. Kligman, K. H. Kaidbey, *Cutis* **22**, 232-244 (1978). Review of clinical assays in serum and urine: A. Moore *et al., Ann. Clin. Biochem.* **22**, 435-454 (1985). Physiological role in immunity: W. M. Jefferies, *Med. Hypotheses* **34**, 198-208 (1991); in fetal maturation: G. C. Liggins, *Reprod. Fertil. Dev.* **6**, 141-150 (1994).

White, bitter-tasting crystalline powder, striated blocks from abs ethanol or isopropanol, mp 217-220° with some decompn. $[\alpha]_D^{22}$ +167° (abs ethanol). uv max: 242 nm ($E_{1cm}^{1\%}$ 445). Soly (mg/ml) at 25°: water 0.28; ethanol 15.0; methanol 6.2; acetone 9.3; chloroform 1.6; propylene glycol 12.7; ether about 0.35. Sol in concd sulfuric acid with intense green fluorescence.

21-Acetate. [50-03-3] Colifoam; Colofoam; Cortaid; Cortifoam; Lanacort; Lenirit; Sigmacort; Sintotrat. $C_{23}H_{32}O_6$; mol wt 404.50. Monoclinic, sphenoidal, tabular crystals from dil acetone. Tasteless. d_4^{20} 1.289; dec 223°. $[\alpha]_D^{25}$ +166° (c = 0.4 in dioxane); $[\alpha]_D^{25}$ +150.7° (c = 0.5 in acetone). uv max (methanol): 242 nm ($E_{1cm}^{1\%}$ 390). Somewhat hygroscopic. Soly in water: 1 mg/100 ml; in ethanol: 0.45 g/100 ml; in methanol: 3.9 mg/ml; in acetone: 1.1 mg/g; in ether: 0.15 mg/ml. One gram dissolves in about 200 ml chloroform. Very sol in DMF; sol in dioxane.

17-Butyrate. [13609-67-1] Alfason; Locoid; Plancol. $C_{25}H_{36}O_6$; mol wt 432.56. White, practically odorless, crystalline powder. Freely sol in chloroform; sol in methanol, alc, acetone; slightly sol in ether. Practically insol in water.

21-Phosphate disodium salt. [6000-74-4] Hydrocortisone sodium phosphate; Cleiton; Efcortesol. $C_{21}H_{29}Na_2O_8P$; mol wt 486.41. White to pale yellow powder. uv max (methanol): 242 nm ($A_{1cm}^{1\%}$ 298-341). $[\alpha]_D^{25}$ +120° (H_2O). Exceedingly hygroscopic. Soly in water (25°): >500 mg/ml. pH of a 1% aq soln: 7.5-8.5. Slightly sol in alc. Practically insol in chloroform, dioxane, ether.

21-Sodium succinate. [125-04-2] Hydrocortisone hemisuccinate sodium salt; Corlan; Efcortelan; Saxizon; Solu-Cortef. $C_{25}H_{33}NaO_8$; mol wt 484.52. Amorphous, hygroscopic, white powder, mp 169.0-171.2°. Soly in water: ~500 mg/ml. Similarly sol in methanol, ethanol; sparingly sol in chloroform.

17-Valerate. [57524-89-7] Hydrocortisone valerate; Westcort. $C_{26}H_{38}O_6$; mol wt 446.58.

THERAP CAT: Glucocorticoid.

THERAP CAT (VET): Glucocorticoid.

4825. Hydrocotarnine. [550-10-7] 5,6,7,8-Tetrahydro-4-methoxy-6-methyl-1,3-dioxolo[4,5-g]isoquinoline; 8-methoxy-5,6-methylenedioxy-2-methyl-1,2,3,4-tetrahydroisoquinoline. $C_{12}H_{15}NO_3$; mol wt 221.26. C 65.14%, H 6.83%, N 6.33%, O 21.69%. Found in mother liquors from morphine extraction. It is not certain whether it is formed from narcotine during the extraction or whether it exists in the poppy plant. May also be prepd by reduction of cotarnine: Topchiev, *J. Appl. Chem. USSR* **6**, 529 (1933), *C.A.* **28**, 2718 (1934); Schneider, Müller, *Ann.* **615**, 34 (1958); Knabe, *Arch. Pharm.* **292**, 652 (1959). Reduction of hydrocotarnine with sodium in alcohol leads to replacement of the methoxyl group by hydrogen, with formation of hydrohydrastinine. Review and bibliography: Small, Lutz, "Chemistry of the Opium Alkaloids," Suppl. No. 103, *Public Health Reports*, Washington (1932).

Hemihydrate. Plates from petr ether, mp 56°. Loses water of crystn at 60°. May be distilled with little decompn at 100°: Hesse, *Ber.* **4**, 693 (1871). Absorption spectrum: Hantzsch, *Ber.* **44**, 1816 (1911); Steiner, *Compt. Rend.* **176**, 244, 1379 (1923); Csokán, *Z. Anal. Chem.* **124**, 344 (1942). Almost insol in water, alkaline solns. Sol in alcohol, acetone, chloroform, benzene, ether.

Hydrochloride monohydrate. Prisms; sol in water.
Hydrobromide. Crystals, mp 237°; sparingly sol in water.
Hydriodide. Needles from methanol, mp 196°; sol in hot water.
Methiodide. Needles from water, plates from alc, mp 206°.
Methobromide. Needles from chloroform, dec 221°.

4826. Hydroflumethiazide. [135-09-1] 3,4-Dihydro-6-(trifluoromethyl)-2*H*-1,2,4-benzothiadiazine-7-sulfonamide 1,1-di-oxide; 6-trifluoromethyl-3,4-dihydro-2*H*-1,2,4-benzo-thiadiazine 1,1-dioxide; 3,4-dihydro-7-sulfamyl-6-trifluoromethyl-1,2,4-benzothiadiazine 1,1-dioxide; trifluoromethylhydrothiazide; dihydroflumethiazide; methforylthiazidine; metflorylthiazidine; Diu-cardin; Elodrine; Finuret; Hydol; Hydrenox; Leodrine; NaClex; Ro-diuran; Rontyl; Saluron; Sisuril; Vergonil. $C_8H_8F_3N_3O_4S_2$; mol wt 331.28. C 29.01%, H 2.43%, F 17.20%, N 12.68%, O 19.32%, S 19.36%. Synthesis: Holdrege *et al., J. Am. Chem. Soc.* **81**, 4807 (1959); Close *et al., ibid.* **82**, 1132 (1960); Yale *et al., ibid.* 2042; Novello *et al., J. Org. Chem.* **25**, 970 (1960). Numerous patents, *e.g.*, Lund *et al.*, US 3254076 (1966 to Lövens Kemiske Fabrik). Pharmacology: J. J. Piala *et al., J. Pharmacol. Exp. Ther.* **134**, 273 (1961). Comprehensive description: C. E. Orzech *et al., Anal. Profiles Drug Subs.* **7**, 297-317 (1978).

Crystals, mp 272-273°. uv max (methanol): 272.5 nm (log ε 4.286). Soly in mg/ml at 25°: acetone >100; methanol 58; acetonitrile 43; water 0.3; ether 0.2; benzene <0.1. pK_1 8.9; pK_2 10.7. Forms water-sol salts with bases. LD_{50} in mice (mg/kg): >8000 orally, 750 i.v., 6280 i.p. (Piala).

THERAP CAT: Antihypertensive; diuretic.

4827. Hydrofluoric Acid. [7664-39-3] Fluohydric acid. HF; mol wt 20.01. Soln of hydrogen fluoride gas in water. Obtained by distilling calcium fluoride with H_2SO_4. Vapor pressure data: Brosheer *et al., Ind. Eng. Chem.* **39**, 423 (1947). Compn of liq and vapor: Munter *et al., ibid.* 427. Review of toxicology and human exposure: *Toxicological Profile for Fluorides, Hydrogen Fluoride, and Fluorine* (PB2004-100002, 2003) 404 pp. *See also* Hydrogen Fluoride.

Colorless or almost colorless, fuming liquid. *Corrosive, poisonous.* Miscible with water. Weak acid: pKa 3.19. The 38.2% (w/w HF) soln is a binary azeotrope; bp 112.2°. Attacks glass or stoneware, dissolving the silica. *Keep in plastic, lead, wax, or paraffin paper bottles.* Has been marketed in concns of about 47% and 53%. d 1.15-1.18.

Caution: Potential symptoms of overexposure are pulmonary edema, skin and eye burns, rhinitis, bronchitis, bone changes. Direct contact may cause irritation of eyes, skin, nose and throat. *See NIOSH Pocket Guide to Chemical Hazards* (DHHS/NIOSH 97-140, 2003) p 168. *See also Patty's Industrial Hygiene and Toxicology* **vol. 2B**, G. D. Clayton, F. E. Clayton, Eds. (Wiley-Interscience, New York, 3rd ed., 1981) pp 2945-2948.

USE: Cleaning cast iron, copper, brass; removing efflorescence from brick and stone, or sand particles from metallic castings; working over too heavily weighted silks; frosting, etching glass and enamel; polishing crystal glass; decomposing cellulose; enameling and

galvanizing iron; increasing porosity of ceramics. Its salts are used as insecticides and to arrest undesirable fermentation in brewing. Trace metal analysis. Also used in analytical work to determine SiO_2, etc.

4828. Hydrofuramide. [494-47-3] 1-(2-Furanyl)-*N*,*N'*-bis(2-furanylmethylene)methanediamine; *N*,*N'*-difurfurylidene-2-furanmethanediamine; furfuramide. $C_{15}H_{12}N_2O_3$; mol wt 268.27. C 67.16%, H 4.51%, N 10.44%, O 17.89%. Prepn: Hartley, Dobbie, *J. Chem. Soc.* **73**, 598 (1898); Taniyama, *J. Chem. Soc. Jpn. Ind. Chem. Sect.* **51**, 33 (1948); Kapur *et al., J. Sci. Ind. Res.* **19B**, 509 (1960); Kamal *et al., Tetrahedron* **19**, 869 (1963). Structure: Soundararajan, Anantakrishnan, *Proc. Indian Acad. Sci.* **38A**, 176 (1953).

Brownish crystals from abs alcohol, mp 117°. bp ~250° with decompn. uv max: 259, 215 nm (log ε 4.18, 4.16). Practically insol in water; freely sol in alc, ether; readily dec by acids.

USE: Vulcanization accelerator.

4829. Hydrogen. [1333-74-0] Protium. H; at. wt [1.00784; 1.00811]; conventional at. wt 1.008; at. no. 1; valence 1. Group IA (1). Elemental state: H_2. Exists in two forms, distinguished by the nuclear spins of the atoms: *ortho* has parallel spins, *para* has antiparallel spins. Normal hydrogen is a 3:1 equilibrium ratio of *ortho* to *para* at rm temp. Naturally occurring isotopes: 1 (protium 99.985%); 2 (deuterium 0.015%); 3 (tritium, traces only). The most abundant element in the known universe. Occurrence in the earth's atmosphere 0.00005% H_2. First recognized as an element by Cavendish in 1766; named by Lavoisier. Obtained by passing H_2O vapors over heated iron; by electrolysis of water or by action of HCl or H_2SO_4 on Fe or Zn; by hydrolysis of metal hydrides. Produced industrially by steam reforming, partial oxidation, coal gasification and water electrolysis. *Reviews: Nouveau Traité de Chimie Minérale* **vol. 1**, P. Pascal, Ed. (Masson, Paris, 1956) pp 565-675; Mackay in *Comprehensive Inorganic Chemistry* **vol. 1**, J. C. Bailar, Jr. *et al.*, Eds. (Pergamon Press, Oxford, 1973) pp 1-76; *Chemistry of the Elements* N. N. Greenwood, A. Earnshaw, Eds. (Pergamon Press, New York, 1984) pp 38-74; T. A. Czuppon *et al.* in *Kirk-Othmer Encyclopedia of Chemical Technology* **vol. 13** (Wiley-Interscience, New York, 4th ed., 1995) pp 838-894. *See also* Deuterium and Tritium.

Colorless, odorless, tasteless gas; *flammable or explosive when mixed with air, oxygen, chlorine, etc.* mp −259.2° (13.96 K) at 54 mm (triple point). bp −252.77° (20.39 K). d_{gas} 0.069 (air = 1); d_{liq} 0.0700 (at bp); d_{sol} 0.0763 (13 K). A liter of the gas at 0° weighs 0.08987 g. Crit. temp −239.9°; crit press. 12.8 atm. Sol in about 50 vols of water at 0°. Ionization potential of H atom 13.59 eV.

Caution: Can act as an asphyxiant by displacing air. *See: Matheson Gas Data Book* (Matheson, 6th ed., Lyndhurst, NJ, 1980) pp 366-371.

USE: In oxy-hydrogen blowpipe (welding) and limelight; autogenous welding of steel and other metals; manuf ammonia, synthetic methanol, HCl, NH_3; hydrogenation of oils, fats, naphthalene, phenol; in balloons and airships; in metallurgy to reduce oxides to metals; in petroleum refining; in thermonuclear reactions (ionizes to form protons, deuterons (D) or tritons (T)). Liq hydrogen used in bubble chambers to study subatomic particles; as a coolant.

4830. Hydrogen Bromide. [10035-10-6] Hydrobromic acid; anhydrous hydrobromic acid. BrH; mol wt 80.91. Br 98.76%, H 1.25%. Prepd commercially by direct combination of the elements at 375° preferably over a catalyst such as platinized silica gel or platinized asbestos: Richards, Hönigschmid, *J. Am. Chem. Soc.* **32**, 1581 (1910); Smyth, Hitchcock, *ibid.* **55**, 1830 (1933); Schneider, Johnson, *Inorg. Synth.* **1**, 152 (1939). Lab procedure from tetrahydronaphthalene and bromine: Müller, *Monatsh. Chem.* **49**, 29 (1928); Duncan, *Inorg. Synth.* **1**, 151 (1939); Schmeisser in *Handbook of Preparative Inorganic Chemistry* **vol. 1**, G. Brauer, Ed. (Academic Press, New York, 2nd ed., 1963) pp 282-286. Detailed description of laboratory methods of prepn: Houben-Weyl, *Methoden der organischen Chemie* **vol 5/4** (Thieme, Stuttgart, 4th ed., 1960) p 16-20. Review of prepn and properties of HBr and other hydrogen halides: Woolf in *Mellor's* **vol. II**, Suppl I (originally published as Suppl II, part 1) 724-741 (1956); John in *Bromine and its Compounds*, Z. E. Jolles, Ed. (Ernest Benn, London, 1966) pp 81-105; Downs, Adams in *Comprehensive Inorganic Chemistry* **vol. 2**, J. C. Bailar, Jr. *et al.*, Eds. (Pergamon Press, Oxford, 1973) pp 1280-1329.

Colorless, nonflammable gas. Acrid odor. Fumes in moist air forming clouds which have a sour taste. d 2.71 (air = 1.00). mp $-86.9°$. bp$_{760}$ $-66.8°$; bp$_{11.0\ atm}$ $-4.8°$; bp$_{17.1\ atm}$ $12°$; bp$_{30.0\ atm}$ $36°$; bp$_{59.2\ atm}$ $70°$. Crit temp $89.8°$; crit press. 84.5 atm. Sp heat (cal/g/°C): solid $(-91°)$ 0.152; liquid 0.176; gas $(27°)$ 0.085. Heat of fusion at mp: 7.44 cal/g. Heat of vaporization at bp: 51.3 cal/g. *Poisonous, corrosive.* Freely sol in water: One vol H_2O dissolves 600 vols HBr gas at $0°$. Also sol in alc. Soly in organic solvents: Fernandes, *J. Chem. Eng. Data* **17**, 377 (1972); Gerrard, *Chem. Ind. (London)* **1969**, 295; Ahmed *et al.*, *J. Appl. Chem.* **20**, 109 (1970). Aq solns are strongly acid. The satd aq soln contains 68.85% HBr at $0°$ and 66% at $25°$. The boiling point of a constant-boiling mixture is $122.5°$ at 740 mm and $126°$ at 760 mm. The composition of the constant-boiling mixture is 47.38% HBr at 752 mm. For complete tables: Bonner *et al.*, *J. Am. Chem. Soc.* **55**, 1406 (1943). *See also* Hydrobromic Acid. Anhydr HBr is marketed in steel cylinders in the form of a gas over liquid. LC$_{50}$ in mice, rats: 814, 2858 ppm by inhalation, K. C. Back *et al.*, *Reclassification of Materials Listed as Transportation Health Hazards* (TSA-20-72-3, PB 214-270, 1972).

Caution: Potential symptoms of overexposure are irritation of eyes, skin, nose and throat; direct contact with solutions may cause skin and eye burns; direct contact with liquid may cause frostbite. *See NIOSH Pocket Guide to Chemical Hazards* (DHHS/NIOSH 97-140, 1997) p 166.

USE: Mfg organic and inorganic bromides, hydrobromic acid, as reducing agent and as catalyst in controlled oxidations, in the alkylation of aromatic compds, in the isomerization of conjugated diolefins.

4831. **Hydrogen Chloride.** [7647-01-0] Hydrochloric acid; anhydrous hydrochloric acid. ClH; mol wt 36.46. Cl 97.23%, H 2.76%. HCl. *See also* Hydrochloric Acid. Produced industrially by the interaction of NaCl and H_2SO_4; from NaCl, SO_2, air and water vapor; by controlled combination of the elements; or as a by-product of the synthesis of chlorinated hydrocarbons: A. C. Cumming, *Hydrochloric Acid and Salt Cake* (Gurney and Jackson, London, 1923); N. A. Laury, *Hydrochloric Acid and Sodium Sulfate* (Chem. Catalog Co., New York, 1927); Maude, *Chem. Eng. Prog.* **44**, 179 (1948); *Faith, Keyes & Clark's Industrial Chemicals*, F. A. Lowenheim, M. K. Moran, Eds. (Wiley-Interscience, New York, 4th ed., 1975) pp 454-461. Prepn of pure HCl for research purposes: Hönigschmid *et al.*, *Z. Anorg. Allg. Chem.* **163**, 315 (1927); Kemp, *J. Chem. Educ.* **37**, 142 (1960); Schmeisser in *Handbook of Preparative Inorganic Chemistry* **vol. 1**, G. Brauer, Ed. (Academic Press, New York, 2nd ed., 1963) pp 280-282. Toxicity: K. I. Darmer *et al.*, *Am. Ind. Hyg. Assoc. J.* **35**, 623 (1974). Reviews of prepn and properties: Addison, Lewis in *Mellor's* **vol. II**, suppl. I (originally published as suppl. II, part I) 402-475 (1956); Downs, Adams in *Comprehensive Inorganic Chemistry* **vol. 2**, J. C. Bailar, Jr. *et al.*, Eds. (Pergamon Press, Oxford, 1973) pp 1280-1329; D. S. Rosenberg in *Kirk-Othmer Encyclopedia of Chemical Technology* **vol. 12** (Wiley-Interscience, New York, 3rd ed., 1980) pp 983-1015.

Colorless, nonflammable gas. Characteristic pungent odor. Fumes in air. *Poisonous, corrosive.* d 1.268 (air = 1.000). d 1.639 g/l. mp $-114.22°$. bp$_{760}$ $-85.05°$; bp$_{100}$ $-114.61°$; bp$_{10}$ $-137.77°$; bp$_{1.0}$ $-154.37°$. Critical temp $51.4°$; critical pressure 81.6 atm; critical density 0.42 g/ml. n_D^{20} (liquid under pressure) 1.256. Heat capacity at constant volume ($15°$): 0.1939 cal/g/°C. Heat capacity at constant pressure ($15°$): 0.1375 cal/g/°C. Heat of vaporization at $-85°$: 3860 cal/mole; heat of soln (infinite dilution) -17.88 kcal/mole; heat of formation at gas at $25°$: -22.063 kcal/mole. Dielectric constant (gas at $0°$) 1.0046; dipole moment 1.07. Soly in water (g/100 g H_2O): 82.3 ($0°$); 67.3 ($30°$); 63.3 ($40°$); 59.6 ($50°$); 56.1 ($60°$). Forms a const boiling mixture: 20.22 g/100 g soln; *see* Hydrochloric Acid. Soly in methanol (g/100 g soln): 54.6 ($-10°$); 51.3 ($0°$); 47.0 ($20°$); 43.0 ($30°$); in ethanol: 45.4 ($0°$); 42.7 ($10°$); 41.0

($20°$); 38.1 ($30°$); in ether: 37.52 ($-10°$); 35.6 ($0°$); 24.9 ($20°$); 19.47 ($30°$). LC$_{50}$ (30 min) in mice, rats: 2142, 5666 ppm (Darmer).

Caution: Potential symptoms of overexposure are irritation of nose, throat and larynx; coughing, choking; dermatitis; direct contact with solutions may cause eye and skin burns; direct contact with liquid may cause frostbite. *See NIOSH Pocket Guide to Chemical Hazards* (DHHS/NIOSH 97-140, 1997) p 166. *See also Patty's Industrial Hygiene and Toxicology* **vol. 2B**, G. D. Clayton, F. E. Clayton, Eds. (Wiley-Interscience, New York, 3rd ed., 1981) pp 2959-2961.

USE: In the manuf of pharmaceutical hydrochlorides, vinyl chloride from acetylene, alkyl chlorides from olefins, and arsenious chloride from arsenious oxide. In the chlorination of rubber, as a gaseous flux for babbitting operations. In organic reactions involving isomerization, polymerization, and alkylation. For making chlorine where economical.

4832. **Hydrogen Cyanide.** [74-90-8] Hydrocyanic acid; formonitrile; prussic acid; AC. CHN; mol wt 27.03. C 44.44%, H 3.73%, N 51.82%. HCN. Volatile poison; inhibits cytochrome oxidase and other metalloenzymes by binding to the metal cofactor. Natural sources include cyanogen glycosides in cassava root, lima beans, and seeds of fruits such as apples, peaches, or apricots. Combustion product of nitrogen-containing materials; significant contributor to toxicity from smoke inhalation. Prepd on a large scale by the catalytic oxidation of ammonia-methane mixtures (Andrussow Process): *see* Andrussow, *Angew. Chem.* **48**, 593 (1935); Maffezzoni, *Chim. Ind. (Milan)* **34**, 460 (1952); *Faith, Keyes & Clark's Industrial Chemicals*, F. A. Lowenheim, M. K. Moran, Ed. (Wiley-Interscience, New York, 4th ed., 1975) pp 482-486. Prepd in the lab by acidifying NaCN or K$_4$[Fe(CN)$_6$]: Glemser in *Handbook of Preparative Inorganic Chemistry* **vol. 1**, G. Brauer, Ed. (Academic Press, New York, 2nd ed., 1963) pp 658-660. GS-MS determn in air samples: Y. Juillet *et al.*, *Analyst* **130**, 977 (2005). Oxidation kinetics: P. Dagaut *et al.*, *Prog. Energy Combust. Sci.* **34**, 1 (2008). Review of synthetic applications: G. Romeder in *Handbook of Reagents for Organic Synthesis*, **Vol. 1** (Wiley, New York, 1999) pp 415-416. Review of analytical methods for determn in blood: A. E. Lindsay *et al.*, *Anal. Chim. Acta* **511**, 185-195 (2004); of toxicology and human exposure: *Toxicological Profile for Cyanide* (PB2007-100674, 2006) 341 pp.

Colorless gas or liquid; characteristic odor of bitter almonds. *Intensely poisonous even when mixed with air. Flammable.* Burns in air with a blue flame. Flashpoint (closed cup): $0°F$ ($-17.8°C$). mp $-13.4°$. bp $25.6°$. d(gas) 0.941 (air = 1); d(liq) 0.687. Weak acid, pKa ($25°$): 9.21. Miscible with water, alc; slightly sol in ether. LC$_{50}$ in rats, mice, dogs: 544 ppm (5 min), 169 ppm (30 min), 300 ppm (3 min) by inhalation, K. C. Back *et al.*, *Reclassification of Materials Listed as Transportation Health Hazards* (TSA-20-72-3; PB214-270, 1972).

Caution: Potential symptoms of acute overexposure are abdominal pain, nausea, vomiting; dizziness, headache, confusion; tachypnea, hyperpnea; loss of consciousness, convulsions, cardiorespiratory failure. *See NIOSH Pocket Guide to Chemical Hazards* (PB2005-149, 2005) p 168; F. J. Baud, *Hum. Exp. Toxicol.* **26**, 191-201 (2007).

USE: Reagent in organic synthesis; rodenticide; insect fumigant; chemical warfare agent.

4833. **Hydrogen Fluoride.** [7664-39-3] Hydrofluoric acid; hydrofluoric acid gas; fluohydric acid gas; fluoric acid; anhydr hydrofluoric acid. FH; mol wt 20.01. F 94.94%, H 5.04%. HF. Obtained by the action of sulfuric acid on fluorspar (calcium fluoride): *Faith, Keyes & Clark's Industrial Chemicals*, F. A. Lowenheim, M. K. Moran, Eds. (Wiley-Interscience, New York, 4th ed., 1975) pp 462-467; prepn of pure HF: Simons, *Inorg. Synth.* **1**, 134 (1939); Shamir, Netzer, *J. Sci. Instrum.* (Ser. 2) **1**, 770 (1968). Exists as hydrogen-bonded polymers: Simons, Hildebrand, *J. Am. Chem. Soc.* **46**, 2183 (1924); Jarry, Davis, *J. Phys. Chem.* **57**, 600 (1953); Atoji, Lipscomb, *Acta Crystallogr.* **7**, 173 (1954). pKa determn: N. E. Vanderborgh, *Talanta* **15**, 1009 (1968). Cryoscopic determn: R. J. Gillespie, D. A. Humphreys, *J. Chem. Soc. A* **1970**, 2311. Toxicity study: M. J. Rosenholtz *et al.*, *Am. Ind. Hyg. Assoc. J.* **24**, 253 (1963). Review of prepn, properties and chemistry: Simons in *Fluorine Chemistry* **vol. 1**, J. H. Simons, Ed. (Academic Press, New York, 1950); Hyman, Katz, "Liquid Hydrogen Fluoride" in *Non-*

aqueous Solvent Systems, T. C. Waddington, Ed. (Academic Press, New York, 1965) pp 47-81; O'Donnell in *Comprehensive Inorganic Chemistry* vol. 2, J. C. Bailar, Jr. *et al.*, Eds. (Pergamon Press, Oxford, 1973) pp 1038-1054; J. F. Gall in *Kirk-Othmer Encyclopedia of Chemical Technology* vol. 10 (Wiley-Interscience, New York, 3rd ed., 1980) pp 733-753; of toxicology and human exposure: *Toxicological Profile for Fluorides, Hydrogen Fluoride, and Fluorine* (PB2004-100002, 2003) 404 pp.

Colorless gas. Fumes in air. *Corrosive, poisonous.* d^{34} 1.27 (air = 1); d_4^0 1.002. mp $-83.57°$. bp 19.51°; bp_{400} 2.5°; bp_{200} $-13.2°$; bp_{100} $-28.2°$; bp_{40} $-45.0°$; bp_{20} $-56.0°$; bp_5 $-74.7°$. Very sol in water and alcohol. Slightly sol in ether. Sol in many organic solvents; soly (wt % at 5°): benzene 2.54; toluene 1.80; *m*-xylene 1.28; tetralin 0.27. Many compds are sol in HF. Anhydr HF is one of the most acidic substances known; Hammett acidity function (H_0) -10.98. Weak acid in aqueous soln. pKa 3.189. Forms a constant boiling mixture with water, *see* hydrofluoric acid. Dissolves silica, silicic acid, glass. *Store in steel cylinders.* LC_{50} (15 min.) in rats, guinea pigs: 2689, 4327 ppm (Rosenholtz).

Caution: Potential symptoms of overexposure are irritation of eyes, skin, nose and throat; pulmonary edema; skin and eye burns; rhinitis; bronchitis; bone changes. *See NIOSH Pocket Guide to Chemical Hazards* (DHHS/NIOSH 97-140, 1997) p 168. *See also Patty's Industrial Hygiene and Toxicology* vol. 2B, G. D. Clayton, F. E. Clayton, Eds. (Wiley-Interscience, New York, 3rd ed., 1981) pp 2945-2948.

USE: Catalyst, especially in the petroleum industry (paraffin alkylation); in fluorination processes, especially in the aluminum industry; in the manuf of fluorides; for separating uranium isotopes; in making fluorine contg plastics; in dye chemistry.

4834. Hydrogen Iodide. [10034-85-2] Hydriodic acid; anhydrous hydriodic acid. HI; mol wt 127.91. H 0.79%, I 99.21%. Prepd by catalytic union of the elements: Caley, Burford, *Inorg. Synth.* 1, 159 (1939); Powell, Campbell, *J. Am. Chem. Soc.* 69, 1227 (1947). May also be prepd by treating concd hydriodic acid solns with P_2O_5: Schmeisser in *Handbook of Preparative Inorganic Chemistry* vol. 1, G. Brauer, Ed. (Academic Press, New York, 2nd ed., 1963) pp 286-289. Lab prepn: Hoffman, *Inorg. Synth.* 7, 180 (1963). Purification: A. Klemenc, *Die Behandlung und Reindarstellung von Gasen* (Vienna, 2nd ed., 1948) p 239; Irving, Wilson, *Chem. Ind. (London)* 1964, 653. Reviews of prepn and properties of HI and other hydrogen halides: Hills in *Mellor's* vol. II, suppl. 1 (originally published as suppl. II, part 1) 857-869 (1956); Downs, Adams, in *Comprehensive Inorganic Chemistry* vol. 2, J. C. Bailar, Jr. *et al.*, Eds. (Pergamon Press, Oxford, 1973) pp 1280-1329.

Colorless, acrid, non-flammable gas. Fumes in moist air. Decomposed by light. mp $-50.8°$. bp_{760} $-35.1°$; $bp_{2\ atm}$ $-18.9°$; $bp_{5\ atm}$ 7.3°; $bp_{10\ atm}$ 32.0°; $bp_{60\ atm}$ 127.5°. d^0 5.66; d^{25} 5.23 g/l. Crit temp 151.0°, crit press. 82.0 atm. Sp heat (25°) 0.0545 cal/g/°C. *Poisonous.* Extremely sol in water. Soly (g/100 g H_2O): 234 (10°); 900 (0°). Soly in organic solvents: Gerrard, *Chem. Ind. (London)* 1969, 295; Ahmed *et al.*, *J. Appl. Chem.* 20, 109 (1970). Forms an azeotrope with water, *see* Hydriodic Acid. Reacts with the lower aliphatic alcohols forming the corresponding iodo compds. Forms a colorless liquid at atm pressure when cooled with dry ice and ether or similar cooling mixture. Attacks natural rubber.

Caution: Strong irritant.

USE: Manuf of hydriodic acid, organic iodo compds, to remove iodine from iodo compds.

4835. Hydrogen Peroxide. [7722-84-1] Hydrogen peroxide (H_2O_2); hydrogen dioxide; hydroperoxide; Albone; Hioxyl; Lensan A; Mirasept; Oxysept; Pegasyl. H_2O_2; mol wt 34.01. H 5.93%, O 94.08%. First reported by Thenard in 1818; prepd by treating barium peroxide with acid. Manuf of aqueous solns: *Faith, Keyes & Clark's Industrial Chemicals*, F. A. Lowenheim, M. K. Moran, Eds. (Wiley-Interscience, New York, 4th ed., 1975) pp 487-495; R. Powell, *Hydrogen Peroxide Manufacture* (Noyes Dev. Corp., Park Ridge, N.J., 1968) 221 pp. Production of anhydr hydrogen peroxide by continuous fractional crystn: Crewson, Ryan, *US 2724640* (1955 to Becco). Production and green reactions in liquid CO_2, *q.v.*: D. Hâncu *et al.*, *Acc. Chem. Res.* 35, 757 (2003). Use in green oxidation reactions: R. Noyori *et al.*, *Chem. Commun.* 2003, 1977. Properties of 90% hydrogen peroxide: E. S. Shanley, F. P. Greenspan, *Ind. Eng. Chem.* 39, 1536 (1947). *Reviews: ACS Monograph Series* no.

128, entitled "Hydrogen Peroxide," W. C. Schumb. Ed. (Reinhold, New York, 1955) 759 pp; Ebsworth *et al.*, in *Comprehensive Inorganic Chemistry* vol. 2, J. C. Bailar, Jr. *et al.*, Eds. (Pergamon Press, Oxford, 1973) pp 771-778; J. R. Kirchner in *Kirk-Othmer Encyclopedia of Chemical Technology* vol. 13 (Wiley-Interscience, New York, 3rd ed., 1981) pp 12-38.

Colorless, rather unstable liquid; bitter taste. Distillable in high vacuum. May dec violently if traces of impurities are present. d^0 1.463. mp $-0.43°$. bp 152°. Misc with water; sol in ether. Insol in petr ether. Decomposed into water and oxygen by many organic solvents. Marketed as a soln in water in concns of 3-90% by wt. Solns of hydrogen peroxide gradually deteriorate and are usually stabilized by the addition of acetanilide or similar organic materials. Agitation or contact with rough surfaces, metals or many other substances accelerates decompn. Rapidly dec by alkalies, finely divided metals; the presence of mineral acid renders it more stable. *Oxidizer, corrosive. Keep protected from light and in a cool place.*

3% Solution. Oxydol. Contains 2.5-3.5% by wt of $H_2O_2 = 8$-12 vols oxygen. Colorless, slightly acid liq. d ~1.00.

30% Solution. Contains 30% by wt of $H_2O_2 = 100$ vols of oxygen. Clear, colorless liquid. d ~1.11. Miscible with water. Now replacing the 3% soln for industrial uses; diluted to the required strength immediately before use. It also is used for making the 3% soln.

Caution: Potential symptoms of overexposure are irritation of eyes, nose and throat; corneal ulceration; erythema, vesicles on skin; bleaching of hair. *See NIOSH Pocket Guide to Chemical Hazards* (DHHS/NIOSH 97-140, 1997) p 168.

USE: Environmentally friendly oxidant. Bleaching agent in foods, textiles, and personal care products; oxidant in wastewater treatment. Catalyst. Used in analytical chemistry for trace metal analysis. A 90% soln is used in rocket propulsion.

THERAP CAT: Antiseptic; disinfectant.

THERAP CAT (VET): Topical antiseptic and cleansing agent (as a dilute soln).

4836. Hydrogen Selenide. [7783-07-5] Selenium hydride. H_2Se; mol wt 80.98. H 2.49%, Se 97.51%. Prepd by heating selenium and hydrogen in a sealed tube at 440°: Hautefeuille, *Bull. Soc. Chim.* 7 (2), 198 (1867); by passing a mixture of hydrogen and selenium vapor over pumice stone at 440°: Corenwinder, *Ann. Chim. Phys.* 34 (3), 77 (1852); by warming potassium or ferrous selenide with hydrochloric acid: Berzelius, *Acad. Handl. Stockholm* 39, 13 (1818); by the action of water on aluminum selenide: Fonzes-Diacon, *Traité de Chimie Minérale, Paris* 1, 469 (1904); Waitkins, Shutt, *Inorg. Synth.* 2, 183 (1946). Review of chemistry and toxicity: *Medical and Biologic Effects of Environmental Pollutants: Selenium* (Nat. Acad. Sci., Washington D.C., 1976) 203 p.

Gas. Disagreeable odor. Highly toxic and reactive gas that decomposes rapidly in presence of oxygen to form elemental Se and H_2O. d_4^{-42} 2.12. bp $-41.3°$. Liquefies at 0° under a pressure of 6.6 atm; at 18°, 8.6 atm; at 52°, 21.5 atm; at 100°, 47.1 atm; at the crit temp 137°, 91.0 atm. mp $-65.73°$. v.p. at $-30°$, 1.75 atm; v.p. at 0.2°, 4.5 atm; v.p. at 30.8°, 12 atm. K_1 at 25° = 1.30×10^{-4}; K_2 at 25° = 1×10^{-11}. *Poisonous, flammable.* Soly in water (ml/100 ml): 377 (4°); 270 (22.5°). Sol in carbonyl chloride and carbon disulfide. Unites directly with most metals to form metal selenides.

Caution: Potential symptoms of overexposure are irritation of eyes, nose and throat; nausea, vomiting and diarrhea; metallic taste, garlic breath; dizziness, lassitude and fatigue; direct contact with liquid may cause frostbite. *See NIOSH Pocket Guide to Chemical Hazards* (DHHS/NIOSH 97-140, 1997) p 168.

4837. Hydrogen Sulfide. [7783-06-4] Sulfureted hydrogen; sulfur hydride; "hydrosulfuric acid". H_2S; mol wt 34.08. H 5.92%, S 94.07%. Evolved from numerous environmental natural sources such as bacterial decomposition of vegetable and animal proteinaceous material. Occurs naturally as a component of crude petroleum, natural gas, volcanic gas and sulfur springs. Also a pollutant released into the environment as a by-product of a variety of industrial operations. Lab prepn: Bickford, Wilkinson, *Inorg. Synth.* 1, 111 (1939). Purification: Ward *et al.*, *ibid.* 3, 14 (1950). Toxicity studies: E. H. Vernot *et al.*, *Toxicol. Appl. Pharmacol.* 42, 417 (1977); M. F. Tansy *et al.*, *J. Toxicol. Environ. Health* 8, 71-88 (1981). Review of toxicity and properties: R. O. Beauchamp, Jr., *et al.*, *Crit. Rev. Toxicol.* 13, 25-97 (1984); of toxicology: R. J. Reiffenstein *et* *al., Annu. Rev. Pharmacol. Toxicol.* **32**, 109-134 (1992); and human exposure: *Toxicological Profile for Hydrogen Sulfide* (PB2007-100675, 2006) 253 pp.

Gas with characteristic odor of rotten eggs, perceptible in air at concns of 0.02-0.13 ppm, sweetish taste. *Flammable, poisonous.* Burns in air with pale blue flame. Ignition temp 260°. Explosive limits when mixed with air: lower limit 4.3% by vol, upper limit 46% by vol. mp −85.49°; bp −60.33°: Giauque, Blue, *J. Am. Chem. Soc.* **58**, 831 (1936). Heavier than air; 1.5392 g/l (0°; 760 mm). d^{gas} 1.19 (air = 1.00). Vapor pressure 18.75×10^5 Pa. One gram H_2S dissolves in 187 ml water at 10°, in 242 ml water at 20°, in 314 ml water at 30°; in 94.3 ml abs alcohol at 20°; in 48.5 ml ether at 20°. Sol in glycerol, gasoline, kerosene, carbon disulfide, crude oil. Water solns of H_2S are not stable, absorbed oxygen causes the formation of elemental sulfur, and the solns become turbid rapidly. In a 50:50 v/v mixture of glycerol and water the precipitation of sulfur is retarded considerably. pH of freshly prepd satd water soln 4.5. pKa_1 7.04; pKa_2 11.96. LC_{50} in mice, rats (ppm): 634, 712 (1 hr inhalation) (Vernot). LC_{50} in rats (ppm): 444 (4 hr inhalation) (Tansy).

Caution: Highly toxic irritant and chemical asphyxiant; overexposure can be fatal. Insidious poison, since sense of smell may be fatigued and fail to give warning of high concns. Direct contact with gas may cause irritation of eyes and respiratory tract resulting in keratoconjunctivitis, photophobia, lacrimation, corneal opacity; rhinitis, laryngitis, cough, bronchopneumonia. Direct contact with solution may cause skin irritation, erythema. Potential symptoms of overexposure by inhalation include salivation, GI disturbances; giddiness, headache, vertigo, confusion, unconsciousness; tachypnea, tachycardia, sweating, fatigue. Exposure to very high vapor concentrations may result in systemic intoxication leading to paralysis of respiratory center of brain, apnea and sudden collapse. *See Clinical Toxicology of Commercial Products*, R. E. Gosselin *et al.*, Eds. (Williams & Wilkins, Baltimore, 5th ed., 1984) Section III, pp 198-202; *NIOSH Pocket Guide to Chemical Hazards* (DHHS/NIOSH 97-140, 1997) p 170.

USE: To produce elemental sulfur and sulfuric acid; in manuf of heavy water and other chemicals; in metallurgy; as analytical reagent.

4838. **Hydrogen Telluride.** [7783-09-7] Tellurium hydride. H_2Te; mol wt 129.62. H 1.56%, Te 98.44%. Prepd by the action of H_2O or HCl on aluminum telluride; by electrolysis of a 50% soln of sulfuric or phosphoric acid with a Te cathode: Dennis, Anderson, *J. Am. Chem. Soc.* **36**, 882 (1914); Fehér in *Handbook of Preparative Inorganic Chemistry* vol. **1**, G. Brauer, Ed. (Academic Press, New York, 2nd ed., 1963) pp 438-441.

Colorless gas. Offensive, garlic-like odor. *Highly poisonous!* mp −49°. bp −2°. d_{liq}^{-12} 2.68. Wt of one liter of the gas: 6.234 g. Liquid H_2Te is dec immediately by light. The dry gas is stable to light, but dec in the presence of dust, traces of moisture, rubber, cork, etc. Sol in water with fairly quick decompn. A satd aq soln is about 0.1*N*.

Caution: Imparts offensive odor to breath. Symptoms similar to hydrogen selenide, *q.v.*

4839. **Hydrohydrastinine.** [494-55-3] 5,6,7,8-Tetrahydro-6-methyl-1,3,dioxolo[4,5-*g*]isoquinoline. $C_{11}H_{13}NO_2$; mol wt 191.23. C 69.09%, H 6.85%, N 7.32%, O 16.73%. Prepn from cotarnine, *q.v.*: Topchiev, *J. Appl. Chem. USSR* **6**, 529 (1933), *C.A.* **28**, 2718 (1934); Clayson, *J. Chem. Soc.* **1949**, 2016; Schneider, Müller, *Ann.* **615**, 34 (1958); Knabe, *Arch. Pharm.* **292**, 652 (1959). *See also:* Hydrastinine, Hydrocotarnine.

Crystals from petr ether, mp 66°. bp$_{752}$ 303°. Sol in alc, ether, acetone, benzene, CS_2, ethyl acetate. Absorption spectrum: Dobbie, Tinkler, *J. Chem. Soc.* **85**, 1007 (1904).

Hydrochloride. [5985-04-6] $C_{11}H_{13}NO_2$.HCl. Crystals from water or alcohol, dec 278°. Sol in water.
Hydrobromide. [5985-05-7] $C_{11}H_{13}NO_2$.HBr. Needles from water, dec 272°.

Hydriodide. $C_{11}H_{13}NO_2$.HI. Crystals from water, dec 242°. Sol in water or alc.
Platinichloride. [5985-06-8] $(C_{11}H_{13}NO_2)_2$.H$_2$PtCl$_6$. Yellow tablets, dec 222°. Sparingly sol in alc.

4840. **Hydromorphone.** [466-99-9] (5α)-4,5-Epoxy-3-hydroxy-17-methylmorphinan-6-one; dihydromorphinone. $C_{17}H_{19}$-NO$_3$; mol wt 285.34. C 71.56%, H 6.71%, N 4.91%, O 16.82%. Semisynthetic opioid analgesic. Prepn from morphine: **DE 365683**; **DE 623821** (1922, 1936 both to Knoll); by oxidation of dihydromorphine: H. Rapoport *et al.*, *J. Org. Chem.* **15**, 1103 (1950). Crystal structure: Steinmetz, *Z. Kristallogr.* **67**, 434 (1928). Pharmacology: N. B. Eddy, J. G. Reid, *J. Pharmacol. Exp. Ther.* **52**, 468 (1934). Toxicity data: M. E. Buchwald, G. S. Eadie, *ibid.* **71**, 197 (1941). GC-MS determn in blood: R. Meatherall, *J. Anal. Toxicol.* **29**, 301 (2005). Review of pharmacology and clinical efficacy: N. Sarhill *et al.*, *Support. Care Cancer* **9**, 84-96 (2001); A. Murray, N. A. Hagen, *J. Pain Symptom Manage.* **29**, S57-S66 (2005).

Crystals from ethanol, mp 266-267°. $[\alpha]_D^{25}$ −194° (c = 0.98 in dioxane).

Hydrochloride. [71-68-1] Dilaudid; Jurnista; Opidol; Palladon; Palladone; Sophidone. $C_{17}H_{19}NO_3$.HCl; mol wt 321.80. Fine white, or practically white, odorless, crystalline powder, mp 305-315° with decompn (evacuated tube). $[\alpha]_D^{25}$ −133.0° (c = 1 in water). Sol in 3 parts water; sparingly sol in alcohol. Practically insol in ether. LD_{50} in mice (mg/kg): 61-96 i.v. (Buchwald, Eadie). *Protect from light.*

Note: This is a controlled substance (opiate): **21 CFR**, 1308.12.
THERAP CAT: Analgesic.

4841. **Hydroorotic Acid.** [155-54-4] Hexahydro-2,6-dioxo-4-pyrimidinecarboxylic acid; 4,5-dihydroorotic acid. $C_5H_6N_2O_4$; mol wt 158.11. C 37.98%, H 3.83%, N 17.72%, O 40.48%. Prepn from carbethoxyasparagine: Miller *et al.*, *J. Am. Chem. Soc.* **75**, 6086 (1953); **US 2773872** (1956 to Merck & Co.).

L-Form. Crystals, dec 266°. $[\alpha]_D^{25.3}$ +33.23° (c = 1.992 in 1% NaHCO$_3$).
D-Form. Crystals from water, dec 266°. $[\alpha]_D^{25.3}$ −31.54° (c = 2.01 in 1% NaHCO$_3$).
DL-Form. Crystals from water, dec 259°. Forms a water-sol salt.
USE: The L- and DL-dihydroorotic acids are precursors to the biological pyrimidines, such as thymine, uracil, cytosine and hydroxymethylcytosine. When employed in large amounts D-dihydroorotic acid is an antimetabolite of L-dihydroorotic acid in pyrimidine utilization. It has also been found to inhibit the orotic acid utilization of those organisms which have a requirement for it in order to grow. In general it may be said to inhibit bacterial growth.

4842. **Hydroprene.** [41096-46-2] (2*E*,4*E*)-3,7,11-Trimethyl-2,4-dodecadienoic acid ethyl ester; ethyl (2*E*,4*E*)-3,7,11-trimethyl-dodeca-2,4-dienoate; OMS-1696; SHA-486300; ZR-512; Gencor; Gentrol. $C_{17}H_{30}O_2$; mol wt 266.43. C 76.64%, H 11.35%, O 12.01%. Juvenile hormone mimic. Prepn: C. A. Henrick, **US 4021461** (1977 to Zoecon); *idem et al.*, *J. Agric. Food Chem.* **21**, 354 (1973); and insecticidal activity of enantiomers: *idem et al.*, *ibid.* **26**, 542 (1978). Effect on sweet potato weevil: G. M. Ram, A. C. Sekhar, *Indian J. Comp. Anim. Physiol.* **2**, 87 (1991); cockroach:

B. L. Reid, G. W. Bennett, *J. Econ. Entomol.* **87**, 1537 (1994). Effect of (S)-enantiomer on cockroach: J. P. Edwards, J. E. Short, *ibid.* **86**, 436 (1993). Review and use in food industries: M. Martinez, *Cereal Foods World* **38**, 818-820 (1993).

$bp_{0.03}$ 95°. uv max (hexane): 262 nm (ε 28300). LD_{50} orally in rats: >34000 mg/kg (Edwards).

(S)-form. [65733-18-8] $bp_{0.05}$ 80°. $[\alpha]_D^{25}$ +2.9° (c = 0.020 in methanol).

USE: Insect growth regulator; insecticide.

4843. Hydroquinidine. [1435-55-8] (9S)-10,11-Dihydro-6'-methoxycinchonan-9-ol; dihydroquinidine; hydroconchinine. $C_{20}H_{26}N_2O_2$; mol wt 326.44. C 73.59%, H 8.03%, N 8.58%, O 9.80%. An alkaloid of cinchona, stereoisomeric with hydroquinine. Usually prepd by hydrogenation of quinidine: Heidelberger, Jacobs, *J. Am. Chem. Soc.* **41**, 826 (1919). Conversion to dihydrocinchonine by removal of the methoxy group: King, *J. Chem. Soc.* **1946**, 523. Manuf pat.: Gutzwiller, Uskokovic, **DE 1933599** (1970 to Hoffmann-La Roche), *C.A.* **72**, 90696v (1970). Pharmacology: Cosnier *et al.*, *Therapie* **26**, 97 (1971).

Plates from ether, needles from alcohol, mp 169°. $[\alpha]_D^{20}$ +231° (c = 2.02 in alc); +299° (c = 0.82 in 0.1N H_2SO_4). Readily sol in hot alcohol; slightly sol in water and ether.

Hydrochloride. [1476-98-8] Serecor. $C_{20}H_{26}N_2O_2$.HCl; mol wt 362.90. Rhombic plates, mp 273-274°. $[\alpha]_D^{26}$ +184° (c = 1.3). Freely sol in methanol, chloroform; less readily in water or abs alcohol; difficultly sol in dry acetone.

THERAP CAT: Antiarrhythmic.

4844. Hydroquinine. [522-66-7] (8α,9R)-10,11-Dihydro-6'-methoxycinchonan-9-ol; dihydroquinine. $C_{20}H_{26}N_2O_2$; mol wt 326.44. C 73.59%, H 8.03%, N 8.58%, O 9.80%. An alkaloid of cinchona, found in quinine sulfate mother liquors. Stereoisomeric with hydroquinidine, *q.v.* Usually prepd by careful hydrogenation of quinine: Heidelberger, Jacobs, *J. Am. Chem. Soc.* **41**, 819 (1919). Total synthesis: Rabe *et al.*, *Ber.* **64B**, 2487 (1931). Synthesis of isomers: Rubtsov, *J. Gen. Chem. USSR* **9**, 1493 (1939), *C.A.* **34**, 2850 (1940); *ibid.* **13**, 593, 702 (1943), *C.A.* **39**, 705 (1945). LC determn in quinine beverages: L. P. Valenti, *J. Assoc. Off. Anal. Chem.* **68**, 782 (1985).

Needles from ether or benzene, mp 172°. $[\alpha]_D^{18}$ −142° (alc); $[\alpha]_D^{20}$ −236° (c = 0.82 in 0.1N H_2SO_4). pK_1 = 5.33. Freely sol in acetone, alcohol, chloroform, ether, petr ether; fairly sol in ammonia water. Almost insol in water (290 mg/l).

Hydrochloride hemihydrate. $C_{20}H_{27}ClN_2O_2.\frac{1}{2}H_2O$. Prisms from water, mp 208° (Heidelberger). $[\alpha]_D^{21}$ −124° (c = 1.1). pH of 0.005 molar soln 5.85. Freely sol in water, alcohol, methanol, acetone. Almost insol in ether. Crystallizes also with 2 mols H_2O. Infrared spectrum: J. Suszko, Z. Dega-Szafran, *Bull. Acad. Pol. Sci. Ser. Sci. Chim.* **12**, 607 (1964), *C.A.* **62**, 7819b (1965).

THERAP CAT: Depigmentor.

4845. Hydroquinone. [123-31-9] 1,4-Benzenediol; *p*-dihydroxybenzene; hydroquinol; quinol; Aida; Black and White Bleaching Cream; Eldoquin; Eldopaque; Tecquinol. $C_6H_6O_2$; mol wt 110.11. C 65.45%, H 5.49%, O 29.06%. Prepd by the oxidation of aniline: L. Gattermann, T. Wieland, *Die Praxis des Organischen Chemikers* (de Gruyter, Berlin, 40th ed., 1961) p 266; by reduction of quinone: Kitchen, **US 1322580** (1920); Seyewetz, Miodon, *Bull. Soc. Chim. Fr.* **33**, 449 (1923); by Elbs persulfate oxidation of phenol: Baker, Brown, *J. Chem. Soc.* **1948**, 2303; Forrest, Petrow, *ibid.* **1950**, 2340; from acetylene + CO: Howk, Sauer, **US 3055949** (1962 to du Pont). Toxicity data: Woodard *et al.*, *Fed. Proc.* **8**, 348 (1949). Toxicology and carcinogenicity study: F. W. Kari *et al.*, *Food Chem. Toxicol.* **30**, 737 (1992). *Review:* J. Varagnat in *Kirk-Othmer Encyclopedia of Chemical Technology* **vol. 13** (Wiley-Interscience, New York, 3rd ed., 1981) pp 39-69.

Crystals, mp 170-171°. d_{15} 1.332. bp 285-287°. Sol in 14 parts water. Freely sol in alc, ether; slightly sol in benzene. Its soln becomes brown in the air due to oxidation; the oxidation is very rapid in presence of alkali. *Poisonous. Keep well closed and protected from light. Handle with caution.* LD_{50} orally in rats: 320 mg/kg (Woodard).

Caution: Harmful effects may occur following overexposure to dust and vapors by inhalation or by contact with skin or eyes. Contact with skin may cause dermatitis. Contact with eyes may cause eye irritation, keratitis, discoloration of conjunctiva and corneal changes. Ingestion may cause tinnitus, nausea, dizziness, sense of suffocation, increased respiration rate, vomiting, pallor, muscle twitching, headache, dyspnea, cyanosis, delirium, and collapse. Urine is usually green or brownish green in color and continues to darken on standing. *See Patty's Industrial Hygiene and Toxicology* **vol. 2B**, G. D. Clayton, F. E. Clayton, Eds. (Wiley-Interscience, New York, 4th ed., 1994) pp 1590-1592. *See also NIOSH Pocket Guide to Chemical Hazards* (DHHS/NIOSH 97-140, 1997) p 170.

USE: As photographic reducer and developer; as reagent in the determination of small quantities of phosphate; as antioxidant.

THERAP CAT: Depigmentor.

4846. Hydroxocobalamin. [13422-51-0] *Co*-Hydroxycobinamide *f*-(dihydrogen phosphate) inner salt 3'-ester with (5,6-dimethyl-1-α-D-ribofuranosyl-1*H*-benzimidazole-*κN³*); cobinamide hydroxide dihydrogen phosphate (ester) inner salt 3'-ester with 5,6-dimethyl-1-α-D-ribofuranosyl-1*H*-benzimidazole; α-(5,6-dimethylbenzimidazolyl)hydroxocobamide; vitamin B_{12a}; hydroxocobemine; OHB_{12}; Cobalin-H; Dodécavit; Duradoce; Neo-Cytamen; Redisol H. $C_{62}H_{89}CoN_{13}O_{15}P$; mol wt 1346.38. C 55.31%, H 6.66%, Co 4.38%, N 13.52%, O 17.82%, P 2.30%. Physiological analog of vitamin B_{12} where the CN group is replaced with OH. Exists in aq soln as an equilibrium mixture of the hydroxy isomer and the ionic aqua isomer (aquacobalamin). Precursor of the coenzymes methylcobalamin and cobamamide, *q.q.v.* Prepn: E. A. Kaczka *et al.*, *J. Am. Chem. Soc.* **71**, 1514 (1949); *eidem, ibid.* **73**, 335 (1951); *eidem*, **US 2738301** (1956 to Merck & Co.). Acids replace the hydroxo group with the anion of the acid: *eidem, Science* **112**, 354 (1950). Prepn from cyanocobalamin: R. Bieganowski, G. Klar, *Chem. Ztg.* **106**, 235 (1982). HPLC determn in plasma: A. Astier, F. J. Baud, *J. Chromatogr. B* **667**, 129 (1995). Clinical evaluation as antidote to cyanide in smoke inhalation: S. W. Borron *et al.*, *Ann. Emerg. Med.* **49**, 794 (2007). *Review:* "Vitamin B_{12}" in *Vitamins*, W. Friedrich, Ed. (de Gruyter, Berlin, 1988) pp 837-928. Review of clinical use for acute cyanide poisoning: G. Shepherd, L. I. Velez, *Ann. Pharmacother.* **42**, 661-669 (2008).

Dark red, orthorhombic needles or platelets from water + acetone. Darkens at 200°, but not melted at 300°. Absorption max (H$_2$O): 279, 325, 359, 516, 537 nm (ε 19000, 11400, 20600, 8900, 9500). The anhydrous form is very hygroscopic. Sparingly sol in water, methanol, in lower aliphatic alcohols. Practically insol in acetone, ether, petr ether, chloroform, halogenated hydrocarbons, benzene.

Aquacobalamin. [13422-52-1] Co-Aquacobinamide dihydrogen phosphate (ester) inner salt 3′-ester with (5,6-dimethyl-1-α-D-ribofuranosyl-1H-benzimidazole-κN^3) ion(1+) hydroxide (1:1); aquocobalamin; vitamin B$_{12b}$; vitamin B$_{12d}$. C$_{62}$H$_{90}$CoN$_{13}$O$_{15}$P.-OH; mol wt 1364.39. Absorption max (H$_2$O): 274, 317, 351, 499, 527 nm (ε 20600, 6100, 26500, 8100, 8500).

Acetatocobalamin. [22465-48-1] Co-(Acetato-κO)-cobinamide dihydrogen phosphate (ester) inner salt 3′-ester with (5,6-dimethyl-1-α-D-ribofuranosyl-1H-benzimidazole-κN^3); Depogamma; Docelan; Novidroxin. C$_{64}$H$_{91}$CoN$_{13}$O$_{16}$P; mol wt 1388.41. Prepn of stable solns: Marcus et al., **FR 1336671** (1963 to Merck & Co.), C.A. **60**, 380a (1964).

Nitrocobalamin. [20623-13-6] Co-(Nitrito-κO)-cobinamide dihydrogen phosphate (ester) inner salt 3′-ester with (5,6-dimethyl-1-α-D-ribofuranosyl-1H-benzimidazole-κN^3); nitritocobalamin; vitamin B$_{12c}$. C$_{62}$H$_{88}$CoN$_{14}$O$_{16}$P; mol wt 1375.37. Prepn: E. A. Kaczka et al., J. Am. Chem. Soc. **73**, 3569 (1951). Red crystalline solid. Absorption max in water: 352, 527.5 nm (E$_{1cm}^{1\%}$ 153.2, 59.5); in 0.01N NaOH: 357, 535 nm (E$_{1cm}^{1\%}$ 139, 62.5).

Sulfitocobalamin. [15671-27-9] Co-(Sulfito-κO)-cobinamide dihydrogen phosphate (ester) inner salt 3′-ester with (5,6-dimethyl-1-α-D-ribofuranosyl-1H-benzimidazole-κN^3); cobalaminsulfonic acid. C$_{62}$H$_{89}$CoN$_{13}$O$_{17}$PS; mol wt 1410.43. Prepn: Fricke, **US 2721162** (1955 to Abbott). Absorption max: 275, 365, 418, 516 nm (E$_{1cm}^{1\%}$ 328, 130, 49, 61).

THERAP CAT: Vitamin (hematopoietic). Hydroxocobalamin as antidote to cyanide poisoning.

THERAP CAT (VET): Vitamin (hematopoietic).

4847. Hydroxyamphetamine. [1518-86-1] 4-(2-Aminopropyl)phenol; dl-p-hydroxy-α-methylphenethylamine; dl-1-p-hydroxyphenyl-2-propylamine; p-hydroxyphenylisopropylamine; α-methyltyramine; Paredrine. C$_9$H$_{13}$NO; mol wt 151.21. C 71.49%, H 8.67%, N 9.26%, O 10.58%. Prepn from oxime of p-methoxyphenyl acetone: Mannich, Jacobsohn, Ber. **43**, 189 (1910); **DE 243546**. From p-nitrobenzyl chloride and a salt of nitroethane: Hoover, Hass, J. Org. Chem. **12**, 501 (1947). Diagnostic use in Horner's syndrome: H. L. Van der Wiel, J. Van Gijn, J. Neurol. Sci. **59**, 229 (1983).

Crystals (rosettes) from benzene, mp 125-126°. Sol in water, alcohol, chloroform, ethyl acetate.

Iodide. C$_9$H$_{14}$INO. Stout prisms, mp 155°. Freely sol in water, alcohol, acetone.

Hydrochloride. C$_9$H$_{14}$ClNO. Crystals from HCl, mp 171-172°. Sol in water, alcohol. Practically insol in ether.

Hydrobromide. [306-21-8] C$_9$H$_{14}$BrNO. White crystals. Freely sol in water, alcohol, acetone; slightly sol in chloroform. Practically insol in ether.

THERAP CAT: Mydriatic. Diagnostic aid (Horner syndrome).

4848. 1-Hydroxy-7-azabenzotriazole. [39968-33-7] 3-Hydroxy-3H-1,2,3-triazolo[4,5-b]pyridine; 7-aza-1-hydroxybenzotriazole; HOAt. C$_5$H$_4$N$_4$O; mol wt 136.11. C 44.12%, H 2.96%, N 41.16%, O 11.75%. Peptide coupling additive that suppresses racemization of chiral centers and improves upon reaction rate and yield. Prepn: G. A. Mokrushina et al., Chem. Heterocycl. Compd. **11**, 880 (1975); D. A. Williamson, B. E. Bowler, Tetrahedron **52**, 12357 (1996). Crystal structure studies: F. Hoffmann et al., J. Mol. Struct. **476**, 289 (1999); of second polymorph: H. Nowell et al., Acta Crystallogr. B **62**, 642 (2006). Uses in peptide bond synthesis: L. A. Carpino, J. Am. Chem. Soc. **115**, 4397 (1993); C. Han et al., ibid. **127**, 10039 (2005); H. U. Vora, T. Rovis, ibid. **129**, 13796 (2007). Review: S. A. Kates et al. in Encyclopedia of Reagents for Organic Synthesis **4**, (Wiley, New York, 1995) pp 2784-2786.

Fluffy light yellow solid from water, mp 217° (Williamson); also reported as solid from water, mp 205-206° (Mokrushina). Crystal density 1.616 (polymorph I), 1.636 (polymorph II). uv max: 220, 280, 325 nm (log ε 4.10, 3.78, 3.40). Sol in DMF. Violent decompn may occur when dried at elevated temp.

USE: Reagent and catalyst in synthetic organic chemistry.

4849. p-Hydroxybenzaldehyde. [123-08-0] 4-Hydroxybenzaldehyde; 4-formylphenol. C$_7$H$_6$O$_2$; mol wt 122.12. C 68.85%, H 4.95%, O 26.20%. Widely distributed in plants in very small amounts. Obtained as a byproduct from the Reimer-Tiemann reaction for salicylaldehyde from phenol: L. F. Fieser, M. Fieser, Organic Chemistry (Reinhold, New York, 3rd ed., 1956) p 681; cf. Reimer, Tiemann, Ber. **9**, 824 (1876); Herzfeld, Tiemann, Ber. **10**, 64 (1877).

Needles from water. Slight, agreeable, aromatic odor, mp 116°. Irritant. Sublimes at atmospheric pressure without decomposition. Reacts like a weak monobasic acid: Ka at 25° = 2.2×10^{-8}. Dipole moment: 4.19. Sparingly sol in cold water: 1.38 g/100 ml H$_2$O at 30.5°; more sol in hot water. Freely sol in alc, ether. Slightly sol in benzene: 3.68 g/100 ml C$_6$H$_6$ at 65°.

USE: Reagent in synthetic organic chemistry.

4850. p-Hydroxybenzoic Acid. [99-96-7] 4-Hydroxybenzoic acid; 4-carboxyphenol. C$_7$H$_6$O$_3$; mol wt 138.12. C 60.87%, H 4.38%, O 34.75%. Prepn from p-bromophenol: Gilinan, Arntzen, J. Am. Chem. Soc. **69**, 1537 (1947); from p-hydroxybenzaldehyde: Pearl, J. Org. Chem. **12**, 85 (1947); from phenol + potassium ethyl carbonate: Jones, *Chem. Ind. (London)* **1958**, 228; from potassium phenolate + CO$_2$: **GB 942418** (1963 to Inventa). A metabolic product of *Penicillium patulum:* Tanenbaum, Bassett, *Biochim. Biophys. Acta* **28**, 21 (1958).

Crystals, mp 213-214°. d 1.46. Soluble in about 125 parts water, freely in alcohol, slightly in chloroform; sol in ether, acetone. Practically insol in carbon disulfide. Ferric chloride does not color its aq soln.

USE: In org. syntheses; intermediate for dyes, fungicides.

4851. 1-Hydroxy-1*H*-benzotriazole. [2592-95-2]; [123333-53-9] (hydrate). Benzazimidol; *N*-hydroxybenzotriazole; HOBt. C$_6$H$_5$N$_3$O; mol wt 135.13. C 53.33%, H 3.73%, N 31.10%, O 11.84%. Additive used in peptide coupling reactions to suppress epimerization of chiral centers. Prepn: R. Nietzki, E. Braunschweig, *Ber.* **27**, 3381 (1894); T. Zinke, P. Schwarz, *Ann.* **311**, 329 (1900). Crystal structure of tautomeric forms: R. Bosch *et al., Acta Crystallogr. C* **39**, 1089 (1983). Use in peptide synthesis: W. König, R. Geiger, *Ber.* **103**, 788 (1970). Peptide coupling kinetics: L. C. Chan, B. G. Cox, *J. Org. Chem.* **72**, 8863 (2007). *Review:* B. Lygo in *Encyclopedia of Reagents for Organic Synthesis* **4**, L. A. Paquette, Ed. (Wiley, NY, 1995) pp 2752-2755.

White solid, mp 157° (depends upon degree of hydration). *Flammable.* Heating may cause an explosion. Crystal density 1.452. Sol in hot water, alc, glacial acetic acid; slightly sol in cold water. Practically insol in ether, benzene, petr ether, chloroform.

USE: Reagent in synthetic organic chemistry.

4852. α-Hydroxybenzylphosphinic Acid. [52705-43-8] *P*-(Hydroxyphenylmethyl)phosphinic acid; (α-hydroxybenzyl)phosphonous acid. C$_7$H$_9$O$_3$P; mol wt 172.12. C 48.85%, H 5.27%, O 27.89%, P 18.00%. Prepd from benzaldehyde (3 moles) and H$_3$PO$_2$ (1 mole): Ville, *Ann. Chim. (Paris)* [6] **23**, 305 (1891); *Beilstein* **7**, 232; Viout, *J. Rech. Cent. Nat. Rech. Sci.* no. **28**, 15-31 (1954), *C.A.* **50**, 7078d (1956).

Leaflets from butanol, mp 110°. Freely sol in water, alcohol; also sol in benzene. Practically insol in anhydr ether.

Sodium salt. Phos; Phoselit; Phosilite. Very freely sol in water. The pH is on the acid side, but aq solns are suitable for injection.

THERAP CAT: Sodium salt as nutrient.

4853. 4'-Hydroxybutyranilide. [101-91-7] *N*-(4-Hydroxyphenyl)butanamide; *p*-hydroxybutyranilide; *N*-butyroyl-*p*-aminophenol; Suconox-4. C$_{10}$H$_{13}$NO$_2$; mol wt 179.22. C 67.02%, H 7.31%, N 7.82%, O 17.85%. Prepn: Kuhn *et al., Z. Physiol. Chem.* **247**, 197 (1937); Fierz-David, Kuster, *Helv. Chim. Acta* **22**, 82 (1939); Rohmann, Friedrich, *Arch. Pharm.* **278**, 456 (1940).

Needles from water, mp 139-140°. Slightly sol in cold water, more sol in hot water, sol in alcohol.

USE: As antioxidant: Young, Cottle, **US 2654722** (1953 to Standard Oil).

4854. γ-Hydroxybutyrate. [591-81-1] 4-Hydroxybutanoic acid; γ-hydroxybutyric acid; 4-hydroxybutyrate; gamma-hydroxybutyrate; GHB. C$_4$H$_8$O$_3$; mol wt 104.11. C 46.15%, H 7.75%, O 46.10%. Endogenous constituent of mammalian brain; thought to function as a neurotransmitter or neuromodulator. Biosynthesized from γ-aminobutyric acid, *q.v.*; freely crosses the blood-brain barrier. Prepn of salts: A. Saytzeff, *Ann.* **171**, 258 (1874); C. S. Marvel, E. R. Birkhimer, *J. Am. Chem. Soc.* **51**, 260 (1929). Acute toxicity: B. Bruguerolle *et al., Therapie* **32**, 375 (1977). GC-MS determn in biological fluids: S. D. Ferrara *et al., J. Pharm. Biomed. Anal.* **11**, 483 (1993). Clinical pharmacokinetics: *idem et al., Br. J. Clin. Pharmacol.* **34**, 231 (1992). Clinical studies in narcolepsy: L. Scrima *et al., Sleep* **13**, 479 (1990); in opiate withdrawal: L. Gallimberti *et al., Neuropsychopharmacology* **9**, 77 (1993). Review of efficacy in alcoholism: G. Biggio *et al., Adv. Biochem. Psychopharmacol.* **47**, 281-288 (1992). Review of potential role as neurotransmittor: G. Tunnicliff, *Gen. Pharmacol.* **23**, 1027-1034 (1992); of neuropharmacology and abuse potential: R. Bernasconi *et al., Trends Pharmacol. Sci.* **20**, 135-141 (1999); of clinical experience in narcolepsy: M. J. Thorpy, *Expert Opin. Pharmacother.* **6**, 329-335 (2005).

Sodium salt. [502-85-2] γ-OH; sodium oxybate; sodium γ-oxybutyrate; Wy-3478; NSC-84223; Somsanit; Gamma-OH; Xyrem. C$_4$H$_7$NaO$_3$; mol wt 126.09. Crystals from alcohol. Highly sol in aq solns. LD$_{50}$ in male, female rats (mg/kg): 2,000, 1,650 i.p. (Bruguerolle).

Note: This is a controlled substance (depressant): **21 CFR, 1308.11.**

THERAP CAT: Anesthetic (intravenous). In treatment of narcolepsy; in treatment of alcoholism.

4855. β-Hydroxybutyric Acid. [300-85-6] 3-Hydroxybutanoic acid. C$_4$H$_8$O$_3$; mol wt 104.11. C 46.15%, H 7.75%, O 46.10%. Prepd from acetoacetic ester by the action of sodium amalgam: Wislicenus, *Ann.* **149**, 207 (1869); Marian, *Biochem. Z.* **150**, 283 (1924); by the oxidation of aldol: Wurtz, *Compt. Rend.* **76**, 1167 (1873); from crotonic acid by heating with dil acid: Wacker, **DE 441003**; *Frdl.* **15**, 135; by heating crotonitrile with KOH soln: Bruylants, *Bull. Soc. Chim. Belg.* **31**, 182 (1922).

Hygroscopic syrup. Volatile with steam. Sol in water, alcohol, ether. On distn it dec into crotonic acid and water.

d-Form. [6168-83-8] (3*S*)-3-Hydroxybutanoic acid. Prepd by the action of *Aspergillus griseus* on the *dl*-form: McKenzie, Harden, *J. Chem. Soc.* **83**, 430 (1903). Crystals. [α]$_D^{10}$ +24.3° (c = 2.226). Sol in water, alcohol, ether.

l-Form. [625-72-9] (3*R*)-3-Hydroxybutanoic acid. Found in the urine of diabetics (as much as 30 g per day). Isoln: Fischer, Scheibler, *Ber.* **42**, 1221 (1909); Shaffer, Marriott, *J. Biol. Chem.* **16**, 268 (1913). Hygroscopic, monoclinic crystals, mp 45.5-48°. [α]$_D^{25}$ −24.5° (c = 5). K at 22° = 3.86×10^{-5}. Freely sol in water, alcohol, ether. Sparingly sol in benzene. On distn it dec into crotonic acid and water.

4856. 3-Hydroxycamphor. [10373-81-6] 3-Hydroxy-1,7,7-trimethylbicyclo[2.2.1]heptan-2-one; oxycamphor; Oxaphor. C$_{10}$H$_{16}$O$_2$; mol wt 168.24. C 71.39%, H 9.59%, O 19.02%. Prepn: Lucius, Brüning, **DE 91718** (1897); Manasse, *Ber.* **30**, 659 (1897); Lapworth, Chapman, *J. Chem. Soc.* **79**, 377 (1901).

Needles from benzene + petr ether, mp 205-206°. Sol in 50 parts water, more in hot water; very sol in alcohol, chloroform, ether.

THERAP CAT: Antipruritic (topical).

4857. Hydroxychloroquine. [118-42-3] 2-[[4-[(7-Chloro-4-quinolinyl)amino]pentyl]ethylamino]ethanol; 7-chloro-4-[4-[ethyl(2-hydroxyethyl)amino]-1-methylbutylamino]quinoline; 7-chloro-4-[4-(N-ethyl-N-β-hydroxyethylamino)-1-methylbutylamino]quinoline; 7-chloro-4-[5-(N-ethyl-N-2-hydroxyethylamino)-2-pentyl]aminoquinoline; oxychloroquine; oxichlorochine. $C_{18}H_{26}ClN_3O$; mol wt 335.88. C 64.37%, H 7.80%, Cl 10.55%, N 12.51%, O 4.76%. Prepd by reacting a mixture of 4,7-dichloroquinoline, phenol and N'-ethyl-N'-β-hydroxyethyl-1,4-pentadiamine at 125-130°: Surrey, Hammer, J. Am. Chem. Soc. **72**, 1814 (1950); Surrey, US 2546658 (1951 to Sterling Drug). Use in combination with cyclophosphamide and azathioprine, q.q.v. in the treatment of rheumatoid arthritis: D. J. McCarty, G. F. Carrera, J. Am. Med. Assoc. **248**, 1718 (1982). Reassessment in the treatment of rheumatoid arthritis: Am. J. Med. **75**, no. 1A, 1-56 (1983). Series of articles on clinical use: ibid. **85**, Suppl. 4A, 1-71 (1988).

Crystals from ethylene dichloride and Skellysolve B; mp 89-91°.

Diphosphate. $C_{18}H_{26}ClN_3O.2H_3PO_4$. Recrystallized from ethanol, mp 168-170° (dec).

Sulfate. [747-36-4] Ercoquin; Plaquenil Sulfate; Quensyl. $C_{18}H_{26}ClN_3O.H_2SO_4$; mol wt 433.95. White crystalline powder; odorless but has a bitter taste. pH of aq solns about 4.5. Exists in two forms, the usual form mp ~240°, the other mp ~198°. Freely sol in water. Practically insol in alcohol, chloroform, ether.

THERAP CAT: Antimalarial; antirheumatic; lupus erythematosus suppressant.

4858. 1α-Hydroxycholecalciferol. [41294-56-8] (1R,3R)-5-[(2E)-2-[(1R,3aS,7aR)-1-[(1R)-1,5-Dimethylhexyl]octahydro-7a-methyl-4H-inden-4-ylidene]ethylidene]-4-methylene-1,3-cyclohexanediol; (1α,3β,5Z,7E)-9,10-secocholesta-5,7,10(19)-triene-1,3-diol; 1α-hydroxyvitamin D_3; 1α-OH-CC; alfacalcidol; Alfarol; Alpha D_3; EinsAlpha; Etalpha; One-Alpha; Vetalpha. $C_{27}H_{44}O_2$; mol wt 400.65. C 80.94%, H 11.07%, O 7.99%. Synthetic analog of calcitriol, q.v., the hormonal form of vitamin D_3, which shows identical potency with respect to stimulation of intestinal calcium absorption and bone mineral mobilization. This common activity is thought to be due either to the presence in both of a 1α-hydroxy group or more likely, to the conversion of the 1α-OH-CC in vivo to 1α,25-dihydroxycholecalciferol: Haussler et al., Proc. Natl. Acad. Sci. USA **70**, 2248 (1973). Synthesis from cholesterol: Holick et al., Science **180**, 190 (1973); Barton et al., J. Am. Chem. Soc. **95**, 2748 (1973); Fürst et al., Helv. Chim. Acta **56**, 1708 (1973); M. Morisaki et al., Chem. Pharm. Bull. **23**, 3272 (1975); T. Sato et al., ibid. **26**, 2933 (1979). Total synthesis: R. G. Harrison et al., Tetrahedron Lett. **1973**, 3649; eidem, J. Chem. Soc. Perkin Trans. 1 **1974**, 2654; P. J. Kocienski, B. Lythgoe, ibid. **1980**, 1400. Synthesis of the 1β-epimer from the 1α-form and biological properties: H. E. Paaren et al., Chem. Commun. **1977**, 890.

mp 134-136° (Harrison); also reported as mp 138-139.5° (Fürst). $[\alpha]_D^{25}$ +28° (ether). uv max (ether): 264 nm (ε 18000) (Harrison); also reported as ε 20200 (Barton).

1β-Epimer. 1β-Hydroxycholecalciferol; 1β-hydroxyvitamin D_3.
THERAP CAT: Vitamin D source.
THERAP CAT (VET): Vitamin D source.

4859. 24-Hydroxycholesterol. [474-73-7] (3β,24S)-Cholest-5-ene-3,24-diol. $C_{27}H_{46}O_2$; mol wt 402.66. C 80.54%, H 11.52%, O 7.95%. Two stereoisomers exist: *cholest-5-ene-3β,24β-diol* and *cholest-5-ene-3β,24α-diol*. Isoln of the naturally occurring 24β-epimer, from horse brain: Ercoli et al., Boll. Soc. Ital. Biol. Sper. **29**, 494 (1953), C.A. **49**, 4744i (1955); from beef spinal cord: Fieser et al., J. Org. Chem. **22**, 1380 (1957). Prepn and separation of isomers: Ercoli, Ruggieri, Gazz. Chim. Ital. **83**, 720 (1953); eidem, J. Am. Chem. Soc. **75**, 3284 (1953). Configurations at C-24: Klyne, Stokes, J. Chem. Soc. **1954**, 1979.

24β-Epimer. Cerebrostenediol; cerebrosterol. Needles from acetone, mp 175-176°. $[\alpha]_D^{20}$ −48.2° (c = 1.06 in $CHCl_3$).

24β-Epimer dibenzoate. $C_{41}H_{54}O_4$. Crystals from ether + acetone, mp 182-183°. α_D −19° (c = 1.20 in $CHCl_3$).

24α-Epimer. Crystals, mp 182-183°. $[\alpha]_D^{20}$ −26.8° (c = 0.672 in $CHCl_3$).

24α-Epimer dibenzoate. Crystals, mp 141-142°. $[\alpha]_D^{26}$ −11.8° (c = 1 in $CHCl_3$).

4860. 25-Hydroxycholesterol. [2140-46-7] (3β)-Cholest-5-ene-3,25-diol; Δ^5-cholestene-3β,25-diol. $C_{27}H_{46}O_2$; mol wt 402.66. C 80.54%, H 11.52%, O 7.95%. An important intermediate in the synthesis of 25-hydroxycholecalciferol, q.v. Synthesis: A. Ryer et al., J. Am. Chem. Soc. **72**, 4247 (1950); W. G. Dauben, H. L. Bradlow, ibid. 4248. Isoln from autoxidation products of cholesterol, q.v.: L. F. Fieser et al., J. Org. Chem. **22**, 1380 (1957). Prepn from pregnenolone, q.v.: T. A. Narwid, M. R. Uskokovic, US 3856780 (1974). Alternate syntheses: M. Morisaki et al., Chem. Pharm. Bull. **21**, 457 (1973); A. Rotman, Y. Mazur, Chem. Commun. **1974**, 15; J. J. Partridge et al., Helv. Chim. Acta **57**, 764 (1974); T. A. Narwid et al., ibid. 771; W. G. Salmond et al., Tetrahedron Lett. **1977**, 987, 1237, 1695; K. Ochi et al., Chem. Pharm. Bull. **27**, 252 (1979); M. Riediker, J. Schwartz, Tetrahedron Lett. **22**, 4655 (1981).

Colorless needles from methanol, mp 178-180°. $[\alpha]_D^{25}$ −39.0° (c = 1.05 in chloroform).

20(S)-isomer. Crystals from chloroform, mp 189.5-190.5°. $[\alpha]_D^{25}$ −41.50° (c = 0.9278 in chloroform).

4861. Hydroxycitric Acid. [27750-10-3]; [6205-14-7] (unspecified stereo). 3-C-Carboxy-2-deoxy-D-erythro-pentaric acid; (1S,2S)-1,2-dihydroxy-1,2,3-propanetricarboxylic acid; (−)-2-hydroxycitric acid; garcinia acid. $C_6H_8O_8$; mol wt 208.12. C 34.63%, H 3.87%, O 61.50%. Principal acid found in the fruit of various species of garcinia, q.v., especially the Malabar tamarind, Garcinia cambogia Desr., Guttiferae (alt. Clusiaceae). Nutraceutical used in weight loss preparations. Competitive inhibitor of ATP-citrate lyase, a key enzyme in lipogenesis. Synthesis (unspecified stereo): H. Kiliani, P. Loeffler, Ber. **37**, 3612 (1904); of all four stereoisomers: C. Martius, R. Maué, Z. Phys. Chem. **269**, 33 (1941). Isoln from G. cambogia: Y. S. Lewis, S. Neelakantan, Phytochemistry **4**, 619

(1965); Y. S. Lewis, *Methods Enzymol.* **13**, 613 (1969). Abs config of lactone: J. P. Glusker *et al., Arch. Biochem. Biophys.* **132**, 573 (1969). Enzyme inhibition: J. A. Watson *et al., ibid.* **135**, 209 (1969); and effect on lipogenesis: A. C. Sullivan *et al., ibid.* **150**, 183 (1972). HPLC determn in commercial extracts: G. K. Jayaprakasha, K. K. Sakariah, *J. Liq. Chromatogr. Relat. Technol.* **23**, 915 (2000). GC-MS determn in blood: Y. C. Loe *et al., Anal. Biochem.* **292**, 148 (2001). Review of chemistry and biological effects: B. S. Jena *et al., J. Agric. Food Chem.* **50**, 10-22 (2002).

$[\alpha]_D^{20}$ $-20°$. Rapidly cyclizes to the lactone form.

Calcium potassium salt. [449158-84-3] Super Citrimax. Unspecified, mixed calcium and potassium salts. Description and safety assessment: M. G. Soni *et al., Food Chem. Toxicol.* **42**, 1513 (2004). Review of clinical experience for weight management: H. G. Preuss *et al., J. Med.* **35**, 33-48 (2004). Highly sol in water. LD_{50} orally in rats: >5000 mg/kg; dermally in rabbits: >2000 mg/kg (Preuss).

Hydroxycitric acid lactone. [27750-13-6] Garcinia lactone. $C_6H_6O_7$; mol wt 190.11. Slightly hygroscopic needles, mp 178°. $[\alpha]_D^{20}$ $+100°$ (c = 2 in water). Very sol in alcohol, water. Fairly sol in ether.

allo-**Hydroxycitric acid.** [27750-11-4] 3-*C*-Carboxy-2-deoxy-D-*threo*-pentaric acid; allohydroxycitric acid; hibiscus acid. Naturally occurring isomer found in hibiscus flowers. Isoln from *Hibiscus sabdariffa* L., *Malvaceae*: C. Griebel, *Z. Unters. Lebensm.* **77**, 561 (1939). $[\alpha]_D^{20}$ $+31°$.

allo-**Hydroxycitric acid lactone.** [469-72-7] Hibiscus lactone. $C_6H_6O_7$; mol wt 190.11. Hygroscopic needles, mp 183°. $[\alpha]_D^{20}$ $+122°$ (c = 2 in water). Very sol in water, alcohol; slightly sol in ether.

USE: Dietary supplement; weight loss aid.

4862. Hydroxycodeinone. [508-54-3] (5α)-7,8-Didehydro-4,5-epoxy-14-hydroxy-3-methoxy-17-methylmorphinan-6-one; 14-hydroxycodeinone. $C_{18}H_{19}NO_4$; mol wt 313.35. C 69.00%, H 6.11%, N 4.47%, O 20.42%. Preparation from thebaine: Freund, Speyer, *J. Prakt. Chem.* **94**, 135 (1916). From codeine: Merck, **DE 411530**; *Frdl.* **15**, 1516; K. W. Bentley, *The Chemistry of the Morphine Alkaloids* (Oxford, 1954). Improved synthesis: F. M. Hauser *et al., J. Med. Chem.* **17**, 1117 (1974). Isoln from *Papaver bracteatum* Lindl.: H. G. Theuns *et al., Phytochemistry* **16**, 753 (1977).

Plates from 96% alcohol + few drops chloroform, dec 275°. Freely sol in $CHCl_3$, methyl Cellosolve, petr ether, ethyl acetate; slightly sol in alcohol. Practically insol in water, ether; also insol in aq alkaline solns.

Hydrochloride monohydrate. $C_{18}H_{19}NO_4 \cdot HCl \cdot H_2O$. Rods from water, dec 285-286°. $[\alpha]_D^{20}$ $-150°$ (c = 2.5).

Note: This is a controlled substance (opiate): **21 CFR**, 1308.12.

4863. *p*-Hydroxyephedrine. [365-26-4] *rel*-(αS)-4-Hydroxy-α-[(1R)-1-(methylamino)ethyl]benzenemethanol; *p*-hydroxy-α-[1-(methylamino)ethyl]benzyl alcohol; 1-(4-hydroxyphenyl)-2-methylaminopropanol; α-(1-methylaminoethyl)-*p*-hydroxybenzyl alcohol; oxyephedrine; Carnigen; Edornat; Methylsympatol; Suprifen. $C_{10}H_{15}NO_2$; mol wt 181.24. C 66.27%, H 8.34%, N 7.73%, O 17.66%. Prepn: Stolz, **US 1878021** (1932 to Winthrop). Prepn of hydrochlorides of *d*- and *l*-forms: Takamatsu, Minaki, *J. Pharm. Soc. Jpn.* **76**, 1230 (1956), *C.A.* **51**, 4305e (1957).

Relative stereochemistry

Cryst powder, mp 152-154°. Sparingly sol in water, alcohol, ether; readily sol in NaOH soln and dil acids.

Hydrochloride. $C_{10}H_{15}NO_2 \cdot HCl$. Crystals, mp 209-211°. Sol in 3 parts water, 5 parts glycerol, 10 parts 90% alcohol. Sparingly sol in abs alcohol, acetone.

d-**Form hydrochloride.** Crystals, dec 223-224°. $[\alpha]_D^{22}$ $+33.8°$.

l-**Form hydrochloride.** Crystals, dec 222-223°. $[\alpha]_D^{22}$ $-35.5°$.

THERAP CAT: Adrenergic.

4864. Hydroxyglutamic Acid. [533-62-0] (unspecified stereo). β-Hydroxyglutamic acid; α-amino-β-hydroxyglutaric acid; 2-amino-3-hydroxypentanedioic acid. $C_5H_9NO_5$; mol wt 163.13. C 36.81%, H 5.56%, N 8.59%, O 49.04%. Naturally occuring derivative of glutamic acid; has 4 possible stereoisomers. Synthesis from ethyl α-isonitrosoacetone dicarboxylate: C. R. Harington, S. S. Randall, *Biochem. J.* **25**, 1917 (1931). Configuration and prepn of diastereomeric racemates: W. J. Leanza, K. Pfister, *J. Biol. Chem.* **201**, 377 (1953). Resolution of isomers: T. Kaneko *et al., Nippon Kagaku Zasshi* **80**, 316 (1959); T. Kamiya, *Chem. Pharm. Bull.* **17**, 890 (1969). Prepn of L-form isomers: L. Tamborini *et al., Tetrahedron* **65**, 6083 (2009).

L-*erythro*-β-hydroxyglutamic acid

L-*erythro*-**Isomer.** [6209-00-3] (3S)-3-Hydroxy-L-glutamic acid. Absolute configuration: (2S,3S). mp 187° (dec). $[\alpha]_D^{10}$ $+8.4°$ (c = 2.61 in water); $[\alpha]_D^{20}$ $+30.8°$ (c = 1.98 in 20% aq HCl).

D-*erythro*-**Isomer.** [6208-99-7] (3R)-3-Hydroxy-D-glutamic acid. Absolute configuration: (2R,3R). mp 188° (dec). $[\alpha]_D^{10}$ $-7.7°$ (c = 2.85 in water); $[\alpha]_D^9$ $-27.2°$ (c = 0.77 in 20% aq HCl).

L-*threo*-**Isomer.** [6208-98-6] (3R)-3-Hydroxy-L-glutamic acid. Absolute configuration: (2S,3R). mp 203° (dec). $[\alpha]_D^{29}$ $+6.1°$ (c = 1.86 in water); $[\alpha]_D^{12}$ $+19.6°$ (c = 1.83 in 20% aq HCl).

D-*threo*-**Isomer.** [6208-97-5] (3S)-3-Hydroxy-D-glutamic acid. Absolute configuration: (2R,3S). mp 203° (dec). $[\alpha]_D^{30}$ $-5.5°$ (c = 2.74 in water); $[\alpha]_D^{22}$ $-18.5°$ (c = 1.68 in 20% aq HCl).

DL-*erythro*-**Form.** [50896-27-0] Fine powdery crystals from alcohol, mp 198° (dec). Soly in water: ~2.3%.

DL-*threo*-**Form.** [17093-76-4] *allo*-β-hydroxy-DL-glutamic acid. Crystals from hot water, mp 193-196° (dec). Soly in water: ~2%.

4865. 8-Hydroxy-7-iodo-5-quinolinesulfonic Acid. [547-91-1] 7-Iodo-8-hydroxyquinoline-5-sulfonic acid; *m*-iodo-*o*-hydroxyquinolineanasulfonic acid; Ferron; Loretin. $C_9H_6INO_4S$; mol wt 351.11. C 30.79%, H 1.72%, I 36.14%, N 3.99%, O 18.23%, S 9.13%. Obtained from the potassium salt of 8-hydroxy-5-quinolinesulfonic acid by the action of KI, bleaching powder, and HCl: Claus, *Arch. Pharm.* **231**, 704 (1893); **DE 72942** (1893); by the action of KI and Ca(OCl)$_2$ on 8-hydroxy-5-quinolinesulfonic acid: Botton, *C.A.* **52**, 18407e (1958).

Sulfur yellow, almost odorless and tasteless, crystalline powder. mp 260-270° (dec). One gram dissolves in 500 ml cold, 170 ml boiling water; slightly sol in alcohol. Practically insol in ether or oils.

Compd with chloroquine (2:1). [7270-12-4] Chloquinate; chlochinate; cloquinate; Resotren. Prepn: Koenig, Andersag, US 2650224 (1953 to Schenley Industries).

Mixture with sodium bicarbonate. Chiniofon; quiniofon; Quinoxyl; Sefona; Yatren; Yochinol.

USE: As colorimetric reagent for ferric ion.

THERAP CAT: Antiamebic; antiseptic.

4866. 4-Hydroxyisophthalic Acid. [636-46-4] 4-Hydroxy-1,3-benzenedicarboxylic acid. $C_8H_6O_5$; mol wt 182.13. C 52.76%, H 3.32%, O 43.92%. Prepd by high pressure carbonation of potassium phenate (Kolbe-Schmitt reaction): Baine et al., J. Org. Chem. **19**, 510 (1954); by boiling salicylic acid with carbon tetrachloride in alkaline medium in the presence of Cu powder: **DE 258887** (1913 to Zeltner & Landau); Chem. Zentralbl. **1913**, I, 1641; Frdl. **11**, 210. Major constituent of the brown dust residue from sublimation of salicylic acid: Hunt et al., Chem. Ind. (London) **1955**, 417. Analgesic and antipyretic activity: G. B. Chesher et al., Nature **175**, 206 (1955).

Branched needles from water, platelets from dil alc. Dec 314-315°. One gram dissolves in 3 liters of water at 24°, in 160 ml at 100°. Freely sol in alcohol, ether.

Dimethyl ester. Crystals, mp 97.5°.
Diethyl ester. Crystals, mp 54-55°.

4867. Hydroxylamine. [7803-49-8] H_3NO; mol wt 33.03. H 9.16%, N 42.41%, O 48.44%. Prepn as the hydrochloride: W. L. Semon, Org. Synth. **coll. vol. I**, 318 (1932). Prepn: Hurd, Inorg. Synth. **1**, 87 (1939); Benson et al., J. Am. Chem. Soc. **78**, 4202 (1956). Crystal structure: Meyers, Lipscomb, Acta Crystallogr. **5**, 583 (1955). Toxicity data: Riemann, Acta Pharmacol. Toxicol. **6**, 285 (1950); R. P. Smith, W. R. Layne, J. Pharmacol. Exp. Ther. **165**, 30 (1969). Mutagenic action: Phillips, Brown in Prog. Nucleic Acid Res. Mol. Biol. **7**, 349-368 (1967). Reviews: Mason, "Hydroxylamine" in Mellor's **vol. VIII**, supplement 2, Nitrogen (part 2), 115-157 (1967); Jones in Comprehensive Inorganic Chemistry **vol. 2**, J. C. Bailar, Jr. et al., Eds. (Pergamon Press, Oxford, 1973) pp 265-276.

Unstable, large white flakes or needles, mp 33°, bp_{22} 58°. d_4^0 1.2255; d_4^{40} 1.204. pK (20°) 7.97. Very sol in water, liquid ammonia and methanol. The soly in the higher alcohols decreases with increasing mol wt. Sparingly sol in ether, benzene, carbon disulfide, chloroform. Very hygroscopic. Dec by hot water. Undergoes rapid decompn at room temp esp in the presence of atm moisture and CO_2. Detonates in test tube heated with flame. LD_{50} i.p. in mice: 1.83 mmol/kg (Smith, Layne).

Hydrochloride. [5470-11-1] Oxammonium hydrochloride. H_3-NO.HCl; mol wt 69.49. Monoclinic columnar crystals; slowly dec when moist. d_{17} 1.67. mp about 151°. One gram dissolves in about 1 ml water (83 g in 100 ml water at 17°); 19 ml alcohol; 8 ml methanol. Sol in glycerol, propylene glycol. Insol in cold ether. pH of 0.2 molar aq soln 3.2. Keep well closed. LD_{50} orally in mice: 408 mg/kg (Riemann).

Sulfate. [10039-54-0] Oxammonium sulfate. $(H_3NO)_2.H_2SO_4$; mol wt 164.13. Crystals, mp about 170°. Freely sol in water.

Caution: Skin irritant. May cause methemoglobinemia, sulfhemoglobinemia, cyanosis, convulsions, hypotension and coma. See Clinical Toxicology of Commercial Products, R. E. Gosselin et al., Eds. (Williams & Wilkins, Baltimore, 5th ed., 1984) Section II, p 117.

USE: As reducing agent in photography. Reagent in synthetic and analytical chemistry. Purification of aldehydes and ketones. Detection of mercury and silver in water. As antioxidant for fatty acids and soaps. As dehairing agent for hides.

4868. Hydroxylupanine. [15358-48-2] (2S,7S,7aR,14S,-14aS)-Dodecahydro-2-hydroxy-7,14-methano-2H,11H-dipyrido-[1,2-a:1',2'-e][1,5]diazocin-11-one; octalupine; oxylupanine; 13-

hydroxylupanine. $C_{15}H_{24}N_2O_2$; mol wt 264.37. C 68.15%, H 9.15%, N 10.60%, O 12.10%. From seed of Lupinus perennis L., L. angustifolius L. and L. polyphyllus Lindl., Leguminosae: Rink, Schäfer, Arch. Pharm. **287**, 290 (1954); Winterfeld, Pies, ibid. **290**, 537 (1957); Bohlmann, Winterfeldt, Ber. **93**, 1956 (1960). Structure: Galinovsky et al., Monatsh. Chem. **81**, 77 (1950). Biosynthesis: Schuette et al., Arch. Pharm. **296**, 438 (1963).

Crystals from dry acetone, mp 169-170°. $[\alpha]_D^{20}$ +45.6° (c = 1.49 in ethanol); $[\alpha]_D^{20}$ +65.1° (c = 1.09 in water). Sol in water, alc, chloroform; slightly sol in benzene, ether.

Methiodide. $C_{15}H_{24}N_2O_2.CH_3I$. Needles from methanol + ether, mp 254-255°.

4869. Hydroxymercurichlorophenols. Prepared from o-chlorophenol and mercuric oxide or mercuric salts: F. C. Whitmore, Organic Compounds of Mercury (Chemical Catalog Co., New York, 1921) p 269; **DE 234851** (1910 to Bayer), Frdl. **10**, 1272.

2-Chloro-4-(hydroxymercuriphenol). [538-04-5] (3-Chloro-4-hydroxyphenyl)hydroxymercury; Semesan. $C_6H_5ClHgO_2$; mol wt 345.14. (n = 1).

USE: Fungicide.

4870. 1-(Hydroxymethyl)-5,5-dimethylhydantoin. [116-25-6] 1-(Hydroxymethyl)-5,5-dimethyl-2,4-imidazolidinedione; monomethyloldimethylhydantoin; MDMH; methylol dimethylhydantoin. $C_6H_{10}N_2O_3$; mol wt 158.16. C 45.57%, H 6.37%, N 17.71%, O 30.35%. Prepd by treating 5,5-dimethylhydantoin with formaldehyde: Mackey, **US 2762708** (1956 to GAF).

Crystals, mp 100°. Freely sol in water, methanol, ethanol, acetone; slightly sol in ethyl acetate. Practically insol in ether, trichloroethylene, carbon tetrachloride.

USE: As preservative in cosmetic prepns. The compound liberates formaldehyde steadily at a very slow rate in the presence of water at pH 6. The available formaldehyde is about 19% (w/w), and 0.1-10% is usuallly mixed with the cosmetic product. When heated in the dry state MDMH forms a water-soluble **dimethylhydantoin formaldehyde resin**, compatible with gelatin, polyvinyl acetate, ethyl cellulose. The resin is used in hair lacquers.

4871. 5-(Hydroxymethyl)-2-furaldehyde. [67-47-0] 5-(Hydroxymethyl)-2-furancarboxaldehyde; 5-(hydroxymethyl)-2-furancarbonal; 5-(hydroxymethyl)-2-furfural; HMF; 5-hydroxymethyl-2-formylfuran. $C_6H_6O_3$; mol wt 126.11. C 57.15%, H 4.80%, O 38.06%. Prepn from the fructose portion of the sugar molecule in 57% yield: Haworth, Jones, J. Chem. Soc. **1944**, 667; Haworth, Wiggins, **GB 591858** (1947); **GB 600871** (1948). Improved process: Garber, Jones, **US 2929823** (1960 to Merck & Co.). Purification: Jones, Lange, **US 2994645** (1961 to Merck & Co.). Outline of process using cornstarch, glucose and sucrose as raw materials: Medwick, Chem. Eng. News **75** (Sept. 11, 1961). Prepn from molasses: Jones, Lange, **US 3066150** (1962 to Merck & Co.); Hales et al., **US 3071599** (1963 to Atlas Chem.).

Needles from ether + petr ether, mp 31.5°. Avoid contact with eyes. Produces harmless yellow stains on skin. Odor of chamomile flowers. Very slightly volatile with steam (as compared with furfural). d_4^{25} 1.2062. $bp_{0.02}$ 110°. n_D^{18} 1.5627. uv max: 283 nm. Heat of combustion: 665 kcal/mol. Freely sol in water, methanol, ethanol, acetone, ethyl acetate, dimethylformamide. Sol in ether, benzene, chloroform. Less sol in carbon tetrachloride. Sparingly sol in petr ether. *Keep protected from light and air.*

USE: In the synthesis of dialdehydes, glycols, ethers, aminoalcohols, acetals. Aq acid catalyzes ring opening.

4872. N-(Hydroxymethyl)nicotinamide. [3569-99-1] *N*-(Hydroxymethyl)-3-pyridinecarboxamide; 3-pyridinecarboxylic acid hydroxymethylamide; pyridine-3-carboxylic acid *N*-methylolamide; Bilamid(e); Felosan; Nikoform; Choligen. $C_7H_8N_2O_2$; mol wt 152.15. C 55.26%, H 5.30%, N 18.41%, O 21.03%. Prepn: Graf, *J. Prakt. Chem.* **138**, 292 (1933); Chechelska, Urbanski, *Rocz. Chem.* **27**, 396 (1953), *C.A.* **49**, 1033f (1955).

Crystals, mp 141-142°. Freely sol in hot alcohol and water; sparingly sol in cold alcohol and water.

Hydrochloride. $C_7H_8N_2O_2 \cdot HCl$. Crystals, mp >120° (dec).

THERAP CAT: Cholagogue.

4873. 1-Hydroxy-2-naphthoic Acid. [86-48-6] 1-Hydroxy-2-naphthalenecarboxylic acid; 1-naphthol-2-carboxylic acid; 2-carboxy-1-naphthol. $C_{11}H_8O_3$; mol wt 188.18. C 70.21%, H 4.29%, O 25.51%. Obtained by the action of CO_2 on the sodium salt of α-naphthol under pressure at 120-140°.

White to reddish crystals. mp 191-192°. Almost insol in cold water; freely sol in alcohol, benzene, ether, alkalies.

USE: Reagent in synthetic organic chemistry.

4874. 3-Hydroxy-2-naphthoic Acid. [92-70-6] 3-Hydroxy-2-naphthalenecarboxylic acid; 3-naphthol-2-carboxylic acid. $C_{11}H_8O_3$; mol wt 188.18. C 70.21%, H 4.29%, O 25.51%. Obtained by the action of CO_2 on the sodium salt of β-naphthol under pressure at 280-290°: Schmitt, Burkard, *Ber.* **20**, 2702 (1887); **DE 50341**; *Frdl.* **2**, 133; **DE 436034**; *Chem. Zentralbl.* **1926**, I, 1717; *Frdl.* **15**, 295; **DE 436524**; *Chem. Zentralbl.* **1927**, I, 182; *Frdl.* **15**, 298; Schwenk, *Chem. Ztg.* **53**, 335 (1929).

Very pale yellow crystals, mp 222-223°. Practically insol in cold water. Slightly sol in hot water. Sol in benzene, chloroform. Freely sol in alc, ether. Sol in alkaline solns.

Magnesium salt. Regacholyl.

USE: The water-sol sodium salt has been used to solubilize riboflavin: Arnold *et al.*, *Poult. Sci.* **31**, 350 (1952).

4875. N-(4-Hydroxyphenyl)glycine. [122-87-2] *p*-Hydroxyphenylaminoacetic acid; *p*-hydroxyanilinoacetic acid; 4-(carboxymethylamino)phenol; photoglycine; Glycin; Iconyl; Monazol. C_8-

H_9NO_3; mol wt 167.16. C 57.48%, H 5.43%, N 8.38%, O 28.71%. Prepd from *p*-aminophenol and chloracetic acid: Vater, *J. Prakt. Chem.* **29**, 291 (1884); Meldola *et al.*, *J. Chem. Soc.* **111**, 552 (1917); Galatis, *Helv. Chim. Acta* **4**, 576 (1921).

Shiny leaflets from water, browns at 200°, begins to melt at 220°, completely melted at 245-247° (decompn). Sparingly sol in water, alcohol, acetone, ether, chloroform, ethyl acetate, benzene, glacial acetic acid. Sol in alkalies and mineral acids. Freely sol in warm 20% hydrochloric acid.

USE: Photographic developer. In determination of iron; detection and determination of phosphorus and silicon. Acid indicator in bacteriology.

4876. N-Hydroxyphthalimide. [524-38-9] 2-Hydroxy-1*H*-isoindole-1,3(2*H*)-dione; NHPI. $C_8H_5NO_3$; mol wt 163.13. C 58.90%, H 3.09%, N 8.59%, O 29.42%. Mediates oxidative catalytic transformations of organic compounds with molecular oxygen. Prepn: L. Cohn, *Ann.* **205**, 295 (1880); H. Gross, I. Keitel, *J. Prakt. Chem.* **311**, 692 (1969). Crystal structure: F.-M. Miao *et al.*, *Acta Crystallogr.* **C51**, 712 (1995). Transition metal free oxidation reactions: C. Einhorn *et al.*, *Chem. Commun.* **1997**, 447; epoxidation of alkenes: T. Iwahama *et al.*, *Heterocycles* **52**, 693 (2000). Transition metal cocatalysis in alkane oxidation: Y. Ishii *et al.*, *J. Org. Chem.* **61**, 4520 (1996); in hydroxylation of fatty acids: M. A. Oakley *et al.*, *J. Mol. Catal. A* **150**, 105 (1999). Description of catalytic mechanisms: Y. Ishii, S. Sakaguchi, *Catal. Surv. Jpn.* **3**, 27 (1999).

Yellowish needles from water, mp 227-229°. Also reported as colorless monoclinic crystals from water. d 1.59 mg m^{-3}.

USE: Catalyst for oxidation reactions.

4877. 17α-Hydroxyprogesterone. [68-96-2] 17-Hydroxy-pregn-4-ene-3,20-dione; 4-pregnen-17α-ol-3,20-dione. $C_{21}H_{30}O_3$; mol wt 330.47. C 76.32%, H 9.15%, O 14.52%. Isoln from adrenal glands: Pfiffner, North, *J. Biol. Chem.* **132**, 459 (1940); **139**, 855 (1941); von Euw, Reichstein, *Helv. Chim. Acta* **24**, 879 (1941). Prepn: Julian *et al.*, **US 2648662** (1953 to Glidden); Ringold *et al.*, **US 2802839**; Stork *et al.*, **US 2805203** (both 1957 to Syntex); Chemerda *et al.*, **US 2777843**; Cutler, Chemerda, **US 2786856**; **US 2786857**; Dulaney, McAleer, **US 2813060** (all 1957 to Merck & Co.); Cutler *et al.*, *J. Org. Chem.* **24**, 1629 (1959); Pederson, **US 3000883** (1961 to Upjohn).

Rhombic or hexagonal leaflets from acetone or alcohol, mp 222-223° (rapid heating). With slow heating the substance undergoes molecular rearrangement accompanied by partial resolidification and becomes completely molten only at 276°. $[\alpha]_D^{17}$ +105.6° (c = 1.0417 in chloroform).

Acetate. 17α-Acetoxyprogesterone. $C_{23}H_{32}O_4$. Crystals from chloroform + methanol, mp 239-240°. uv max: 240 nm (log ε 4.33).

Caproate. [630-56-8] 17α-Hydroxyprogesterone hexanoate; Lentogest; Proge; Proluton Depot. $C_{27}H_{40}O_4$; mol wt 428.61.

Prepn: Kaspar *et al.*, **US 2753360** (1956 to Schering AG). Comprehensive description: K. Florey, *Anal. Profiles Drug Subs.* **4** 209-224 (1975). Dense needles from isopropyl ether or methanol, mp 119-121°. $[\alpha]_D^{20}$ +61° (c = 1 in chloroform). Soly (mg/ml): sesame oil 25-29; levulinic acid butyl ester 350-400.

THERAP CAT: Progestogen.

THERAP CAT (VET): Estrus regulator.

4878. Hydroxyproline. [51-35-4] (4*R*)-4-Hydroxy-L-proline; Hyp; L$_s$-hydroxyproline; (−)-(2*S*,4*R*)-4-hydroxyproline; *trans*-4-hydroxyproline; 4(*R*)-hydroxy-2(*S*)-pyrrolidinecarboxylic acid. $C_5H_9NO_3$; mol wt 131.13. C 45.80%, H 6.92%, N 10.68%, O 36.60%. Amino acid analog of proline, *q.v.*; non-essential for human development. Shows *cis-trans* isomerism. Constituent of collagen, *q.v.* Isoln from gelatin: E. Fischer, *Ber.* **35**, 2660 (1902). Synthesis: H. Leuchs, *Ber.* **38**, 1937 (1905); and structure determn: H. Leuchs, J. F. Brewster, *Ber.* **46**, 986 (1913). Stereochemistry: Hudson, Neuberger, *J. Org. Chem.* **15**, 24 (1950). Early chemistry and biochemistry: *Amino Acids and Proteins*, D. M. Greenberg, Ed. (Charles C. Thomas, Springfield, IL, 1951) 950 pp., *passim*; J. P. Greenstein, M. Winitz, *Chemistry of the Amino Acids* vols **1-3** (John Wiley and Sons, Inc., New York, 1961) pp. 2018-2042, *passim*. Flow sheets of four different syntheses: *Chem. Eng. News* **40**, 40 (Nov. 12, 1962). One pot synthesis: S. G. Ramaswamy, E. Adams, *J. Org. Chem.* **42**, 3440 (1977). Proof of occurence in plants: D. Ashford, A. Neuberger, *Trends Biochem. Sci.* **5**, 245 (1980). Effects of *cis-trans* isomerism on protein structure: N. Panasik, Jr. *et al.*, *Int. J. Pept. Protein Res.* **44**, 262 (1994). *Review:* R. Kuttan, A. N. Radhakrishnan, *Adv. Enzymol.* **37**, 273-348 (1973). Review of metabolism: E. Adams, L. Frank, *Annu. Rev. Biochem.* **49**, 1005-1061 (1980).

Rhombs or needles from water, mp 274°. $[\alpha]_D$ −76.5° (c = 2.5 in water). pK_1' 1.82; pK_2' 9.65. Soly in water at 0°: 288.6 g/l; at 25°: 361.1 g/l; at 50°: 451.8 g/l; at 65°: 516.7 g/l. Very slightly sol in alcohol. Insol in ether.

***cis*-Form.** [618-27-9] Allohydroxyproline; (2*S*,4*S*)-4-hydroxyproline. mp 238-241°. $[\alpha]_D^{18}$ −58.1° (c = 5.2 in water).

4879. Hydroxypropyl Cellulose. [9004-64-2] Cellulose 2-hydroxypropyl ether; oxypropylated cellulose; Klucel; Lacrisert. Nonionic water soluble ether of cellulose, *q.v.* that produces solns having a wide range of viscosity (200-2500 cp). Prepn: **NL 6401036**; E. D. Klug, **US 3278520, US 3278521** (1964, 1966, 1966 all to Hercules). Use in the treatment of dry eye syndrome: T. P. Werblin *et al.*, *Ophthalmology* **88**, 78 (1981); P. Huguet *et al.*, *Bull. Soc. Ophthalmol. Fr.* **81**, 1173 (1981). Review of chemistry, physical properties and uses: E. D. Klug in *Encyclopedia of Polymer Science and Technology* vol. **15** (Interscience, New York, 1971) pp 307-314; A. J. Desmarais, *Industrial Gums*, R. L. Whistler, Ed. (Academic Press, New York, 2nd ed., 1973) pp 649-672.

Off-white, odorless, tasteless powder, softens at 130°. Sol in many polar organic solvents. Ppts from water at 40-45°. Thermoplastic.

USE: As emulsifier, stabilizer, whipping aid, protective colloid, film former or thickener in foods; as binder in ceramics and glazes; in hair and cosmetic prepns; in vacuum-formed containers and blow-molded bottles; as suspending agent in PVC polymerization. Pharmaceutic aid (tablet coating agent).

THERAP CAT: Topical protectant.

4880. Hydroxypropyl Methylcellulose. [9004-65-3] Cellulose 2-hydroxypropyl methyl ether; hypromellose; HPMC; Artelac; Benecel MP; GenTeal; Gonak; Goniosol; Methocel (E, F, J, K); Ocucoat. Derivative of methylcellulose, *q.v.*, that also contains 3-12% hydroxypropyl substitution. Prepn: A. B. Savage, **US 2949452** (1960 to Dow). Review of chemistry, physical properties and use: *idem*, *Encyclopedia of Polymer Science and Technology* vol. **3** (Interscience, New York, 1965); pp 496-511; G. K. Greminger, A. B. Savage, *Industrial Gums*, R. L. Whistler, Ed. (Academic Press, New York, 1973) pp 619-647. Clinical efficacy as ocular lubricant in Sjogren's syndrome: I. Toda *et al.*, *Cornea* **15**, 120 (1996); in con-

tact lens solutions: P. A. Simmons *et al.*, *CLAO J.* **27**, 192 (2001). Review of production and uses in foods: P. de Mariscal, D. A. Bell in *Handbook of Fat Replacers*, S. Roller, S. A. Jones, Eds. (CRC Press, Boca Raton FL, 1996) pp 145-159. Review of drug release from HPMC-based pharmaceutical systems: J. Siepmann, N. A. Peppas, *Adv. Drug Delivery Rev.* **48**, 139-157 (2001).

White or off white, granular powder. Sol in most polar organics. Dissolves slowly in cold water. Swells in water and produces a clear to opalescent, viscous, colloidal mixture. Insol in hot water, dehydrated alc, ether, chloroform. Solns have a wide range of viscosity (400-60,000 mPa·s).

USE: As emulsifier, film former, protective colloid, stabilizer, suspending agent, or thickener in foods. Pharmaceutic aid (suspending agent; tablet excipient; demulcent; viscosity increasing agent; coating agent); hydrophilic carrier in drug delivery systems. In adhesives, asphalt emulsions, caulking compounds, tile mortars, plastic mixes, cements, paints.

THERAP CAT: Ocular lubricant.

4881. 8-Hydroxyquinoline. [148-24-3] 8-Quinolinol; oxyquinoline; hydroxybenzopyridine; oxybenzopyridine; phenopyridine; oxychinolin; oxine. C_9H_7NO; mol wt 145.16. C 74.47%, H 4.86%, N 9.65%, O 11.02%. Prepn: Z. H. Skraup, *Monatsh. Chem.* **1**, 316 (1880); **3**, 536 (1882); R. H. F. Manske *et al.*, *Can. J. Res.* **27F**, 359 (1949). Toxicity study: Bernstein *et al.*, *Toxicol. Appl. Pharmacol.* **5**, 599 (1963). *Review:* J. P. Phillips, *Chem. Rev.* **56**, 271-297 (1956). Book: R. G. W. Hollingshead, *Oxine and Its Derivatives* I-IV (Butterworth, London, 1954/56).

White crystals or crystalline powder. mp 76°. bp ~267°. Almost insol in water, ether. Freely sol in alc, acetone, chloroform, benzene, aq mineral acids. LD$_{50}$ i.p. in mice: 48 mg/kg (Bernstein).

Sulfate. [134-31-6] Chinosol. $(C_9H_7NO)_2 \cdot H_2SO_4$; mol wt 388.39. Pale yellow, cryst powder; slight saffron odor; burning taste. mp 175-178°. Very sol in water; freely sol in methanol; sol in about 100 parts glycerol; slightly sol in alc. Practically insol in acetone, ether.

Aluminum sulfate. [153-77-5] Nyxolan; Aloxyn. $C_{27}H_{24}AlN_3O_{15}S_3$; mol wt 753.66.

USE: As fungistat; complexing agent; chelating agent in determn of trace metal ions. Aluminum sulfate as antiperspirant, deodorant.

THERAP CAT: Antiseptic. Aluminum sulfate as anthelmintic (Nematodes).

THERAP CAT (VET): Antiseptic.

4882. 8-Hydroxy-5-quinolinesulfonic Acid. [84-88-8] $C_9H_7NO_4S$; mol wt 225.22. C 48.00%, H 3.13%, N 6.22%, O 28.41%, S 14.23%. Prepn: K. Matsumura, *J. Am. Chem. Soc.* **49**, 810 (1927); N. K. Chawla, M. M. Jones, *Inorg. Chem.* **3**, 1549 (1964).

Pale yellow, needle-like crystals or cryst powder; odorless. mp 322-324°. Freely sol in water, slightly in organic solvents.

USE: In determn of trace metal ions.

4883. Hydroxystilbamidine. [495-99-8] 4-[2-[4-(Aminoiminomethyl)phenyl]ethenyl]-3-hydroxybenzenecarboximidamide; 2-hydroxy-4,4'-stilbenedicarboxamidine; 2-hydroxy-4,4'-diaminidostilbene; 2-hydroxy-4,4'-diguanylstilbene; 2-hydroxystilbamide. $C_{16}H_{16}N_4O$; mol wt 280.33. C 68.55%, H 5.75%, N 19.99%, O 5.71%. Prepn: J. N. Ashley, J. O. Harris, *J. Chem. Soc.* **1946**, 567; A. J. Ewins *et al.*, **GB 574486**; A. J. Ewins, **US 2510047** (1946, 1950 both to May & Baker). Organ and tissue distribution in animals: I. Snapper *et al.*, *Cancer* **4**, 1246 (1951). Pharmacology and antiprotozoal activity: I. Snapper *et al.*, *Trans. N.Y. Acad. Sci.* **14**, 269 (1952). Probe for studying nucleic acid conformation: B. Festy, *C. R. Seances Acad. Sci. Ser. D* **266**, 1433 (1968); B. Festy, M. Daune, *Biochemistry* **12**, 4827 (1973); B. Festy *et al.*, *Biochim. Biophys. Acta* **407**, 24 (1975). Crystal structure: C. Courseille *et al.*, *C. R. Seances Acad. Sci. Ser. C* **274**, 1921 (1972). Use as a fluorochrome for selective staining of nuclei: L. B. Murgatroyd, *Histochemistry* **74**, 107 (1982). *Review:* B. Festy in *Antibiotics* vol. 5, pt. 2, F. E. Hahn, Ed. (Springer-Verlag, New York, 1979) pp 223-235.

Yellow microcrystals from nitrobenzene, mp 235°. LD$_{50}$ in mice (mg/g); 0.027 i.v.; 0.14 s.c. (Ewins, 1950).

Isethionate. [533-22-2] C$_{20}$H$_{28}$N$_4$O$_9$S$_2$. Yellow crystals, discolored by light, mp 286° (dec). Freely sol in water. Soly in alcohol ~1.0 g/100 ml. Practically insol in ether. pH of a 1% aq soln 3.3 to 5.3. Solns show strong yellow fluorescence under uv light. Solns should be freshly prepared. Although the hydroxy compd is more stable in soln, such solns should not be stored and must not be used if cloudy.

THERAP CAT: Antiprotozoal (Leishmania).

4884. N-Hydroxysuccinimide. [6066-82-6] 1-Hydroxy-2,5-pyrrolidinedione; HOSu; NHS; NHSI. C$_4$H$_5$NO$_3$; mol wt 115.09. C 41.74%, H 4.38%, N 12.17%, O 41.70%. Reagent utilized in amide bond synthesis and other coupling reactions. Prepn: G. Errera, *Gazz. Chim. Ital.* **25 II**, 25 (1895). Crystal structure: P. G. Jones, *Acta Crystallogr.* **E59**, 1951 (2003). Early use in peptide synthesis: G. W. Anderson *et al.*, *J. Am. Chem. Soc.* **85**, 3039 (1963); and alternative prepn method: *eidem, ibid.* **86**, 1839 (1964). Synthetic applications: A. Cecchetto *et al.*, *Tetrahedron Lett.* **43**, 3605 (2002); A. Schulze, A. Giannis, *Adv. Synth. Catal.* **346**, 252 (2004); M. A. Mironov *et al.*, *Tetrahedron Lett.* **46**, 3957 (2005); H.-C. Ma, X.-Z. Jiang, *J. Org. Chem.* **72**, 8943 (2007). Reviews: D. W. Knight in *Encyclopedia of Reagents for Organic Synthesis* **4**, L. A. Paquette, Ed. (John Wiley & Sons, New York, 1995) pp 2780-2781; P. A. Wang, *Synlett* **2008**, 2384-2385.

White crystalline solid, mp 99-100°. Crystal density: 1.563 mg/m^3. Absorption max (ethanol): 265.5 (ε 625). pKa 6.0. Sol in water, DMF, alcohols, ethyl acetate. Insol in cold ether.

USE: Reagent in synthetic organic chemistry.

4885. Hydroxytetracaine. [490-98-2] 4-(Butylamino)-2-hydroxybenzoic acid 2-(dimethylamino)ethyl ester; *p*-butylaminosalicylic acid 2-dimethylaminoethyl ester; 2-dimethylaminoethyl *p*-butylaminosalicylate; hydroxamethocaine; Rhenocain; Salicain. C$_{15}$H$_{24}$N$_2$O$_3$; mol wt 280.37. C 64.26%, H 8.63%, N 9.99%, O 17.12%. Prepn: **GB 736960** (1955); **GB 760003** (1956 to Rheinpreussen AG); Grimme, Schmitz, *Ber.* **84**, 734 (1951).

Hydrochloride. C$_{15}$H$_{24}$N$_2$O$_3$.HCl. Crystals from water, mp 157°. Soly in water at 20°: about 4%.

Hemihydrate. Prisms from ligroin, mp 48°.

THERAP CAT: Anesthetic (local).

4886. 5-Hydroxytryptophan. [56-69-9] 5-HTP. C$_{11}$H$_{12}$N$_2$O$_3$; mol wt 220.23. C 59.99%, H 5.49%, N 12.72%, O 21.79%. Precursor of serotonin. Synthesis from 5-benzyloxyindole: Ek, Witkop, *J. Am. Chem. Soc.* **76**, 5579 (1954); Shaw, Morris, *Biochem. Prep.* **9**, 92 (1962); from 5-benzyloxytryptophan: Frangatos, Chubb, *Can. J. Chem.* **37**, 1374 (1959); Frangatos, **CA 619472** (1961 to Frank W. Horner); Ash, **GB 845034** (1960 to May & Baker); from tryptophan: Renson *et al.*, *Biochem. Biophys. Res. Commun.* **6**, 20 (1961). Prepn of L- and D-forms: A. J. Morris, M. D. Armstrong, *J. Org. Chem.* **22**, 306 (1957). Crystal and molecular structure of DL-form: Wakahara *et al.*, *Tetrahedron Lett.* **1970**, 3003. Use of L-5HTP in treatment of myoclonus: M. H. Van Woert, D. Rosenbaum, *Adv. Neurol.* **26**, 107 (1979). Clinical trial in migraine: F. Titus *et al.*, *Eur. Neurol.* **25**, 327 (1986). *Review:* M. H. Van Woert, *Orphan Drugs*, F. E. Karch, Ed. (Marcel Dekker, New York, 1982) pp 13-31. Review of antidepressant efficacy: W. F. Byerley *et al.*, *J. Clin. Psychopharmacol.* **7**, 127-137 (1987); of metabolism and clinical experience: T. C. Birdsall, *Altern. Med. Rev.* **3**, 271-280 (1998).

L-form

Minute rods or needles from ethanol, dec 298-300°. uv max (H$_2$O at pH 6.0): 278 nm. Soly in water at 5°: 1.0 g/100 ml; at 100°: 5.5 g/100 ml. Soly in 50% boiling alc: 2.5 g/100 ml. Aq solns are stable at low pH.

L-Form. [4350-09-8] Oxitriptan; L-5HTP; Cincofarm; Levothym; Lévotonine; Oxyfan; Telesol; Tript-Oh. Crystals, [α]$_D^{20}$ −32.5° (H$_2$O); [α]$_D^{20}$ +16.0° (4*N* HCl).

D-Form. [4350-07-6] Crystals, [α]$_D^{20}$ +32.2° (H$_2$O).

THERAP CAT: Antidepressant; in treatment of myoclonus.

4887. Hydroxytyrosol. [10597-60-1] 4-(2-Hydroxyethyl)-1,2-benzenediol; 3,4-dihydroxy-1-benzeneethanol; 2-(3,4-dihydroxyphenyl)ethanol. C$_8$H$_{10}$O$_3$; mol wt 154.17. C 62.33%, H 6.54%, O 31.13%. Powerful phenolic antioxidant found in olive oil and other foods; exists in olive oil primarily in esterified form with elenolic acid as oleuropein aglycone. Prepn: C. Schöpf *et al.*, *Ann.* **563**, 86 (1949); P. G. Baraldi *et al.*, *ibid.* **1983**, 684. Synthesis and recovery from olive oil mill waste waters: R. Capasso *et al.*, *J. Agric. Food Chem.* **47**, 1745 (1999). Photochemical synthesis from tyrosol, *q.v.*: S. Azabou *et al.*, *ibid.* **55**, 4877 (2007). GC/MS determn in wine: D. Di Tommaso *et al.*, *J. High Resolut. Chromatogr.* **21**, 549 (1998). Metabolism study using capillary GC/MS determn in plasma and urine: E. Miro-Casas *et al.*, *Clin. Chem.* **49**, 945 (2003). Antioxidant activity: M. De Lucia *et al.*, *Tetrahedron* **62**, 1273 (2006); S. J. Rietjens *et al.*, *J. Agric. Food Chem.* **55**, 7609 (2007).

Colorless oil. uv max (methanol): 218, 281 nm (ε 4300, 1970). Log P (octanol/water): 0.03 ±0.01 (pH 5.5); 0.02 ±0.01 (pH 7.4).

Acetate. [69039-02-7] 4-[2-(Acetyloxy)ethyl]-1,2-benzenediol; 3,4-dihydroxyphenylethyl acetate. C$_{10}$H$_{12}$O$_4$; mol wt 196.20. Prepn and antioxidant activity: M. H. Gordon *et al.*, *J. Agric. Food Chem.* **49**, 2480 (2001). White solid from CH$_2$Cl$_2$/CCl$_4$ (1:3), mp 81-83°. Log P (octanol/water): 0.82 ±0.01 (pH 5.5); 0.82 ±0.05 (pH 7.4).

4888. Hydroxyurea. [127-07-1] Hydroxycarbamide; Droxia; Hydrea; Litalir; Siklos. CH$_4$N$_2$O$_2$; mol wt 76.06. C 15.79%, H 5.30%, N 36.83%, O 42.07%. Ribonucleotide reductase inhibitor; blocks DNA synthesis and repair. Induces fetal hemoglobin production in patients with sickle cell anemia. Prepn: W. F. C. Dressler, R. Stein, *Ann.* **150**, 242 (1869). Alternate route: P. J. Graham, **US**

2705727 (1955 to Du Pont). GC-MS determn in plasma: T. Kettani *et al., J. Chromatogr. B* **877**, 446 (2009). Symposium on mechanism of action and clinical experience in cancer: *Semin. Oncol.* **19**, Suppl. 9, 1-116 (1992). Clinical trial in sickle cell anemia: S. Charache *et al., N. Engl. J. Med.* **332**, 1317 (1995). Review of pharmacology and potential therapeutic indications: P. Navarra, P. Preziosi, *Crit. Rev. Oncol. Hematol.* **29**, 249-255 (1999); of clinical experience in pediatric sickle cell disease: J. J. Strouse *et al., Pediatrics* **122**, 1332-1342 (2008).

Needles from alc, mp 133-136°. Somewhat hygroscopic. Freely sol in water, hot alc.

THERAP CAT: Antineoplastic; in treatment of sickle cell anemia.

4889. Hydroxyzine. [68-88-2] 2-[2-[4-[(4-Chlorophenyl)-phenylmethyl]-1-piperazinyl]ethoxy]ethanol; 1-(*p*-chloro-α-phenylbenzyl)-4-(2-hydroxyethoxyethyl)piperazine; 1-(*p*-chlorodiphenylmethyl)-4-[2-(2-hydroxyethoxy)ethyl]piperazine; *N*-(4-chlorobenzhydryl)-*N* '-(hydroxyethyloxyethyl)piperazine; 1-(*p*-chlorobenzhydryl)-4-[2-(2-hydroxyethoxy)ethyl]diethylenediamine; UCB-4492; Tran-Q; Tranquizine. $C_{21}H_{27}ClN_2O_2$; mol wt 374.91. C 67.28%, H 7.26%, Cl 9.46%, N 7.47%, O 8.53%. H_1 receptor antagonist. Outline of commercial prepn: *Chem. Week* **79** (5), 70 (Aug. 4, 1956); Morren, US **2899436** (1959 to UCB). Pharmacology and metabolism: Cannizaro, *Boll. Chim. Farm.* **104**, 39 (1965); Close *et al., Ind. Chim. Belge* **33**, 94 (1968); *eidem, Proc. Eur. Soc. Study Drug Toxic.* **9**, 144 (1968); K. F. Pong, C. L. Huang, *J. Pharm. Sci.* **63**, 1527 (1974). Pharmacokinetics and antihistaminic activity: F. E. R. Simons *et al., J. Allergy Clin. Immunol.* **73**, 69 (1984); S. Ting *et al., ibid.* **75**, 63 (1985). Clinical trials of efficacy in allergic rhinitis: L. Wong *et al., ibid.* **67**, 223 (1981); in urticaria: R. P. Harvey *et al., ibid.* **68**, 262 (1981); as anti-emetic: R. McKenzie *et al., Anesth. Analg.* **60**, 783 (1981); as pre-surgical sedative: G. Wallace, L. J. Mindlin, *ibid.* **63**, 571 (1984). Toxicity data: E. I. Goldenthal, *Toxicol. Appl. Pharmacol.* **18**, 185 (1971). Comprehensive description: J. Tsau, N. DeAngelis, *Anal. Profiles Drug Subs.* **7**, 319-341 (1978).

Dihydrochloride. [2192-20-3] Alamon; Atarax; Aterax; Durrax; Orgatrax; Quiess; Vistaril Parenteral. $C_{21}H_{27}ClN_2O_2 \cdot 2HCl$; mol wt 447.83. White crystals, mp 193°. Bitter taste. Very sol in water; sol in chloroform; slightly sol in acetone. Practically insol in ether. Solns are unstable to intense uv light. LD_{50} in rats (mg/kg): 126 i.p.; 950 orally (Goldenthal).

Pamoate. [10246-75-0] Equipose; Masmoran; Paxistil; Vistaril Pamoate. $C_{21}H_{27}ClN_2O_2 \cdot C_{23}H_{16}O_6$; mol wt 763.28. Pale yellow crystals. Freely sol in DMF. Practically insol in water, methanol.

THERAP CAT: Anxiolytic. Antihistaminic.

THERAP CAT (VET): Has been used as a tranquilizer.

4890. Hygrine. [496-49-1] 1-[(2*R*)-1-Methyl-2-pyrrolidinyl]-2-propanone; (+)-*N*-methyl-2-acetonylpyrrolidine. $C_8H_{15}NO$; mol wt 141.21. C 68.05%, H 10.71%, N 9.92%, O 11.33%. Occurs in leaves of *Erythroxylon coca* Lam., *Erythroxylaceae* of diverse origin: Liebermann, *Ber.* **22**, 677 (1889). Synthesis: Galinovsky *et al., Monatsh. Chem.* **82**, 551 (1951); Lukes *et al., Collect. Czech. Chem. Commun.* **24**, 2433 (1959); Leonard, Cook, *J. Am. Chem. Soc.* **81**, 5627 (1959). Enzymatic synthesis: Tuppy, Faltaous, *Monatsh. Chem.* **91**, 167 (1960). Stereochemistry: Galinovsky *et al., ibid.* **84**, 798 (1953). Absolute configuration: Lukes *et al., Collect. Czech. Chem. Commun.* **25**, 483 (1960).

Liquid. bp_{11} 76.5°; bp_{14} 81°. n_D^{20} 1.4555. Sol in alcohol, chloroform, dil acids; slightly sol in water.

Oxime. $C_8H_{16}N_2O$. Crystals from ether, mp 123-124°.

4891. Hygromycin. [6379-56-2] 5-Deoxy-5-[[(2*E*)-3-[4-[(6-deoxy-β-D-*arabino*-hexofuranos-5-ulos-1-yl)oxy]-3-hydroxyphenyl]-2-methyl-1-oxo-2-propen-1-yl]amino]-1,2-*O*-methylene-D-*neo*-inositol; homomycin; hygromycin A; 1703-18B; ST-4331. $C_{23}H_{29}$-NO_{12}; mol wt 511.48. C 54.01%, H 5.72%, N 2.74%, O 37.54%. Antibiotic substance produced by *Streptomyces hygroscopicus* (Jensen) Waksman & Henrici, from forest soil near Indianapolis, Ind: R. L. Mann *et al., Antibiot. Chemother.* **3**, 1279 (1953). Produced also by *Streptomyces noboritoensis* from soil collected in Kanawaga Prefecture. Identity of homomycin and hygromycin: K. Isono *et al., J. Antibiot.* **10A**, 21 (1957). Structure: *eidem, ibid.* 160; R. L. Mann, D. O. Woolf, *J. Am. Chem. Soc.* **79**, 120 (1957). Abs config: K. Kakinuma, Y. Sakagami, *Agric. Biol. Chem.* **42**, 279 (1978). Synthesis of the sugar component: M. Nakajima, S. Takahashi, *ibid.* **31**, 1082 (1967). Total synthesis: N. Chida *et al., Chem. Commun.* **1989**, 436. Peptide synthesis inhibitor: M. Guerrero, J. Modolell, *Eur. J. Biochem.* **107**, 409 (1980).

Amorphous white powder. mp 105-109°, dec above 160°. Also reported as mp 110-112° (dec) (Chida). Weakly acidic. pKa 8.9. $[\alpha]_D^{25}$ −126° (c = 1 in water). Freely sol in water, alc. Practically insol in the less polar solvents. uv max (dil HCl): 272, 214 nm ($E_{1cm}^{1\%}$ 306, 416).

4892. Hygromycin B. [31282-04-9] *O*-6-Amino-6-deoxy-L-*glycero*-D-*galacto*-heptopyranosylidene-(1 → 2-3)-*O*-β-D-talo-pyranosyl-(1 → 5)-2-deoxy-*N*^3-methyl-D-streptamine; Hygromix. $C_{20}H_{37}N_3O_{13}$; mol wt 527.52. C 45.54%, H 7.07%, N 7.97%, O 39.43%. Antibiotic substance produced by *Streptomyces hygroscopicus* (Jensen) Waksman & Henrici, from soil samples from Nebraska and Indiana: Mann, Bromer, *J. Am. Chem. Soc.* **80**, 2714 (1958). Prodn: McGuire, US **3018220** (1962 to Lilly). Structure: Neuss *et al., Helv. Chim. Acta* **53**, 2314 (1970). Studies on mode of action: A. Gonzalez *et al., Biochim. Biophys. Acta* **521**, 459 (1978).

Amorphous powder, dec 160-180°. Weakly basic. pKa 7.1, 8.8. $[\alpha]_D^{20}$ +20.2° (c = 1 in water). Freely sol in water, methanol and ethanol. Practically insol in less polar solvents. Absorption spectrum: Mann, Bromer, *loc. cit.*

THERAP CAT (VET): Anthelmintic (swine and chicken).

4893. Hygrophylline. [3573-82-8] (1α,14α)-1,2-Dihydro-12,14-dihydroxysenecionan-11,16-dione. $C_{18}H_{27}NO_6$; mol wt 353.42. C 61.17%, H 7.70%, N 3.96%, O 27.16%. Pyrrolizidine alkaloid isolated from *Senecio hygrophyllus* Dyer and Sm., *Compositae:* Richardson, Warren, *J. Chem. Soc.* **1943**, 452. Structure and stereochemistry: Schlosser, Warren, *ibid.* **1965**, 5707. NMR spectrum: S. E. Drewes *et al., J. Chem. Soc. Perkin Trans. 1* **1981**, 287.

X-ray crystallography: M. F. Mackay *et al.*, *Acta Crystallogr.* **C41**, 395 (1985).

Prisms from acetone, mp 173-174°. $[\alpha]_D^{20}$ −67.3° (c = 2.9 in ethanol).

4894. Hymecromone. [90-33-5] 7-Hydroxy-4-methyl-2*H*-1-benzopyran-2-one; 7-hydroxy-4-methylcoumarin; 7-hydroxy-4-methyl-2-oxo-3-chromene; 4-methylumbelliferone; β-methylumbelliferone; imecromone; 4-MU; Cantabilin; Cantabiline; Cholestil; Cholonerton; Cholspasmin. $C_{10}H_8O_3$; mol wt 176.17. C 68.18%, H 4.58%, O 27.24%. Prepd from resorcinol and ethyl acetoacetate: H. Von Pechmann, C. Duisberg, *Ber.* **16**, 2119 (1883); A. Russell, J. R. Frye, *Org. Synth.* **21**, 23 (1941); L. L. Woods, J. Sapp, *J. Org. Chem.* **27**, 3703 (1962). Use in determn of nitric acid: F. J. Welcher, *Organic Analytical Reagents* vol. 1 (Van Nostrand, New York, 1947) pp 214-215. Use in fluorometric assays: J. A. R. Mead *et al.*, *Biochem. J.* **61**, 569 (1955); in detection of chemical warfare agents: L. Malosse *et al.*, *Analyst* **133**, 588 (2008). Clinical pharmacokinetics: E. R. Garrett *et al.*, *Biopharm. Drug Dispos.* **14**, 13 (1993). Clinical trial for treatment of motor disorders of the bile duct: A. Abate *et al.*, *Drugs Exp. Clin. Res.* **27**, 223 (2001).

Crystals from alc, mp 194-195° (Woods, Sapp), also reported as 185-186° (Pechmann, Duisberg). uv max (methanol): 221, 251, 322.5 nm. Exhibits strong blue fluorescence in alcohol or in mildly alkaline solutions with maximum intensity at pH 10. Excitation max: ≃370 nm; emission max ≃450 nm (pH 10). Sol in methanol, glacial acetic acid; slightly sol in ether, chloroform, toluene. Practically insol in cold water.

USE: Analytical reagent; fluorescent indicator.

THERAP CAT: Choleretic; biliary antispasmodic.

4895. Hymexazol. [10004-44-1] 5-Methyl-3(2*H*)-isoxazolone; 3-hydroxy-5-methylisoxazole; F-319; RTY-319; Tachigaren. $C_4H_5NO_2$; mol wt 99.09. C 48.49%, H 5.09%, N 14.14%, O 32.29%. Prepn: **NL 6511925**; I. Iwai *et al.*, **US 3544584** (1966, 1970 both to Sankyo); I. Iwai, N. Nakamura, *Chem. Pharm. Bull.* **14**, 1277 (1966). Metabolism in plants: S. Kamimura *et al.*, *Phytopathology* **64**, 1273 (1974). Properties and analytical methods: T. Nakamura *et al.*, *Anal. Methods Pestic. Plant Growth Regul.* **10**, 215 (1978). GC determn in plants and soils: *idem et al.*, *Annu. Rep. Sankyo Res. Lab.* **28**, 130 (1976). Plant growth promoting activity: Y. Ota, *Jpn. Agric. Res. Q.* **9**, 1 (1975). Mode of fungicidal action: T. Nakanishi, H. D. Sisler, *J. Pestic. Sci.* **8**, 173 (1983). Field studies: P. A. Payne, G. E. Williams, *Crop Prot.* **9**, 371 (1990). Series of articles on chemistry, biological activity and metabolism: *Annu. Rep. Sankyo Res. Lab.* **25**, 1-51 (1973). Brief review: R. Ishitsuka, *Jpn. Pestic. Inf.* **18**, 27-30 (1974).

White needles from *n*-hexane, mp 84-85°. Also reported as mp 86-87° (Nakamura, 1983). Vapor pressure (25°) 1.01 × 10^{-3} Torr. Soly at 25° (g/100 ml): ~8.5 in water. Very sol in methanol, etha-

nol, isopropanol, acetone, methyl isobutyl ketone, tetrahydrofuran, dioxane, DMF, ethylene glycol, chloroform. Fairly sol in diethyl ether, benzene, xylene, trichloroethylene. Slightly sol in *n*-hexane, carbon disulfide. LD_{50} in male, female mice, rats (mg/kg): 2148, 1968, 4678, 3909 orally; 1297, 1167, 1924, 1884 s.c.; 445, 514, >1000, >1000 i.v.; in rats, rabbits (mg/kg): >10000, >2000 dermally (Nakamura, 1978).

USE: Agricultural fungicide and plant growth regulator.

4896. Hyodeoxycholic Acid. [83-49-8] (3α,5β,6α)-3,6-Dihydroxycholan-24-oic acid; 3α,6α-dihydroxy-5β-cholanic acid; α-hyodeoxycholic acid; hyodesoxycholic acid. $C_{24}H_{40}O_4$; mol wt 392.58. C 73.43%, H 10.27%, O 16.30%. Isoln from pig bile: Wieland, Gumlich, *Z. Physiol. Chem.* **215**, 18 (1933); Trickey, **US 2547726** (1951 to U.S. Rubber); Fogle, **US 2745849** (1956 to Armour); Buckley, Ziegler, **US 2758120** (1956 to Canada Packers); Liebig, **US 3006927** (1961 to Riedel-de Haen). Configuration: Moffett, Hoehn, *J. Am. Chem. Soc.* **69**, 1995 (1947).

Crystals from ethyl acetate, mp 196-197°. $[\alpha]_D^{20}$ +8° (alc). Moderately sol in ethanol, glacial acetic acid. Less sol in ether, acetone, ethyl acetate, benzene.

4897. Hyoscyamine. [101-31-5] (α*S*)-α-(Hydroxymethyl)-benzeneacetic acid (3-*endo*)-8-methyl-8-azabicyclo[3.2.1]oct-3-yl ester; [3(*S*)-*endo*]-α-(hydroxymethyl)benzeneacetic acid 8-methyl-8-azabicyclo[3.2.1]oct-3-yl ester; 1α*H*,5α*H*-tropan-3α-ol(−)-tropate (ester); 3α-tropanyl *S*-(−)-tropate; *l*-tropic acid ester with tropine; *l*-tropine tropate; daturine; duboisine; *l*-hyoscyamine; Cystospaz; Levsin. $C_{17}H_{23}NO_3$; mol wt 289.38. C 70.56%, H 8.01%, N 4.84%, O 16.59%. Anticholinergic. From *Hyoscyamus niger* L., *Atropa belladonna* L., *Datura stramonium* L., and other *Solanaceae*: Ladenburg, *Ann.* **206**, 274 (1881). Identity with duboisine and daturine: Beckurts, *Apoth. Ztg.* **27**, 683 (1912). Obtained by resolution of atropine: Werner, Miltenberger, *Ann.* **631**, 161 (1960). Prepd from (−)-acetyltropoyl chloride and atropine hydrochloride: Fodor *et al.*, *Acta Chim. Acad. Sci. Hung.* **28**(4), 409 (1961), *C.A.* **61**, 1903g (1964). Configuration: G. Fodor, G. Csepreghy, *Tetrahedron Lett.* **1** (7), 16 (1959). Comprehensive description: F. J. Muhtadi, *Anal. Profiles Drug Subs. Excip.* **23**, 153-228 (1994).

Silky, tetragonal needles from evaporating alc. *Keep well closed and protect from light and heat;* easily racemized. mp 108.5°. $[\alpha]_D^{20}$ −21.0° (alc). pKa (21°) 9.7. One gram dissolves in 281 ml water (pH 9.5), 69 ml ether, 150 ml benzene, 1 ml chloroform. Freely sol in alcohol, dil acids. uv max (methanol): 247, 252, 258, 264 nm ($A_{1cm}^{1\%}$ 5.18, 5.70, 6.12, 5.10). Absorption spectra: Dobbie, Fox, *J. Chem. Soc.* **103**, 1194, 1195 (1913).

Hydrobromide. [306-03-6] $C_{17}H_{23}NO_3\cdot HBr$. Deliquescent crystals, mp 152°. pH 5.4 (1 in 100). Levorotatory. Freely sol in water. One gram dissolves in 3 ml alcohol, 1.2 ml chloroform, 2260 ml ether.

Hydrochloride. $C_{17}H_{23}NO_3 \cdot HCl$. Crystals. *Poisonous.* mp 149-151°. Freely sol in water, alcohol.

Methyl bromide. *N*-Methylhyoscyaminium bromide. $C_{17}H_{23}$-$NO_3 \cdot CH_3Br$. Crystals, mp 210-212°. Freely sol in water, dil alc; slightly sol in abs alc.

Sulfate dihydrate. [6835-16-1] Egacene; Egazil Duretter; Peptard. $(C_{17}H_{23}NO_3)_2 \cdot H_2SO_4 \cdot 2H_2O$; mol wt 712.85. Needles from alcohol, mp 206° when dry. $[\alpha]_D^{15} -29°$ (c = 2). pH 5.3 (1 in 100). One gram dissolves in 0.5 ml water, about 5.0 ml alcohol. Very slightly sol in chloroform. Practically insol in ether.

THERAP CAT: Antispasmodic.

4898.   **Hyoscyamus.** Henbane; hog's bean; insane root; poison tobacco; black henbane. Dried leaves, with or without the tops, of *Hyoscyamus niger* L., *Solanaceae*, yielding 0.040% hyoscyamus alkaloids (chiefly hyoscyamine and scopolamine). The leaves and seeds have been used as smooth muscle relaxants. The alkaloid content of the seeds is less than that of the leaves. Egyptian henbane *(Hyoscyamus muticus* L., *Solanaceae)* contains about 0.5% alkaloids. *Habit.* Europe, Africa, Asia, naturalized in U.S.; cultivated in England. *Constit.* Leaves: scopolamine, hyoscyamine, hyoscipicrin, choline.

4899.   **Hypalon**®. [9008-08-6] Elastomer made by substituting chlorine and sulfonyl chloride groups into polyethylene. Prepd from polyethylene of average mol wt of about 20,000; contains approx one chlorine atom for each seven carbon atoms and one sulfonyl chloride group for every ninety carbon atoms. Three types of Hypalon are available: 20, 30, and 40 with sp gr of 1.12, 1.28, and 1.18, respectively. Prepn: Reed, Horn, **US 2046090** (1933). *Reviews: Ullmanns Encyklopädie der technischen Chemie* **vol. 9,** 342 (1957); Keeley, "Hypalon Synthetic Rubber" in *Introduction to Rubber Technology*, Morton, Ed. (Reinhold, New York, 1959) p 349.

4900.   **Hypaphorine.** [487-58-1] (α*S*)-α-Carboxy-*N*,*N*,*N*-trimethyl-1*H*-indole-3-ethanaminium inner salt; 1-trimethylammonio-3-(3-indolyl)propionate. $C_{14}H_{18}N_2O_2$; mol wt 246.31. C 68.27%, H 7.37%, N 11.37%, O 12.99%. The trimethylbetaine of tryptophan. Poisonous alkaloid from seeds of *Erythrina americana* Mill., *E. sandwicensis* Degener, *E. crista-galli* L., and *Abrus precatorius* Linn., *Leguminosae:* Folkers, Koniuszy, *J. Am. Chem. Soc.* **61,** 1232 (1939); **62,** 1677 (1940); Deulofeu *et al., J. Chem. Soc.* **1939,** 1841; Tung, Lao, *C.A.* **55,** 17770i (1961). Structure and synthesis: von Romburgh, Barger, *J. Chem. Soc.* **99,** 2068 (1911).

Crystals from dil alc, dec 237°; after purification through the nitrate, dec 255°. $[\alpha]_D^{27} +113°$ (c = 0.52). Sol in 12 parts water, slightly in alc. Almost insol in other usual solvents.

Hydrochloride. $C_{14}H_{18}N_2O_2 \cdot HCl$. Crystals from water, dec 231-232°, $[\alpha]_D^{32} +90°$ (c = 0.5). Moderately sol in water.

Nitrate. $C_{14}H_{18}N_2O_2 \cdot HNO_3$. Dec 224°, $[\alpha]_D^{20} +95°$. Sol in 170 parts water.

4901.   **Hyperforin.** [11079-53-1] (1*R*,5*S*,6*R*,7*S*)-4-Hydroxy-6-methyl-1,3,7-tris(3-methyl-2-buten-1-yl)-5-(2-methyl-1-oxopropyl)-6-(4-methyl-3-penten-1-yl)bicyclo[3.3.1]non-3-ene-2,9-dione. $C_{35}H_{52}O_4$; mol wt 536.80. C 78.31%, H 9.76%, O 11.92%. Principal acylphloroglucinol component in St. John's wort with concentrations of 2-4% in the fresh herb. Inhibits synaptosomal uptake of serotonin, norephephrine and dopamine; thought to be responsible for the antidepressant activity of medicinal preparations of hypericum, *q.v.* Isoln and antibacterial activity: A. I. Gurevich *et al., Antibiotiki* **16,** 510 (1971), *C.A.* **75,** 95625 (1971). Improved isoln procedure: C. A. J. Erdelmeier, *Pharmacopsychiatry* **31,** Suppl 1, 2 (1998). Structure: N. S. Bystrov *et al., Tetrahedron Lett.* **1975,** 2791. Abs config: I. Brondz *et al., Acta Chem. Scand.* **A37,** 263 (1983). HPLC determn in plasma: J. D. Chi, M. Franklin, *J. Chromatogr. B* **735,** 285 (1999). Series of articles on psychopharmacol-

ogy and neurotransmitter reuptake inhibition: *Pharmacopsychiatry* **31,** Suppl 1, 7-29 (1998).

mp 79-80°. $[\alpha]_D^{18} +41°$ (in ethanol). pKa 4.8 (50% aq ethanol). uv max (methanol): 275.5 nm (log ε 3.95). Unstable when exposed to air and light.

4902.   **Hypericin.** [548-04-9] 1,3,4,6,8,13-Hexahydroxy-10,11-dimethylphenanthro[1,10,9,8-*opqra*]perylene-7,14-dione stereoisomer; 4,5,7,4′,5′,7′-hexahydroxy-2,2′-dimethylnaphthodianthrone; hypericum red. $C_{30}H_{16}O_8$; mol wt 504.45. C 71.43%, H 3.20%, O 25.37%. Characteristic principle of hypericum, *q.v.* (St. John's wort). Isoln from *Hypericum perforatum* L., *Hypericaceae:* Brockmann *et al., Ann.* **553,** 1 (1942). Synthesis from bromoemodin trimethylether: Brockmann, Muxfeld, *Naturwissenschaften* **40,** 411 (1953). Structure: Brockmann *et al., ibid.* **37,** 540 (1950); Brockmann, Sanne, *ibid.* **40,** 509 (1953). Total synthesis: Brockmann, Kluge, **US 2707704** (1955 to Schenley); Brockmann *et al., Ber.* **90,** 2302 (1957); Brockmann, Eggers, *ibid.* **91,** 547 (1958). Clinical pharmacokinetics and evaluation of phototoxic potential: J. Brockmöller *et al., Pharmacopsychiatry* **30,** Suppl 2, 94 (1997). HPLC determn in plasma: J. D. Chi, M. Franklin, *J. Chromatogr. B* **724,** 195 (1999).

Solvated blue-black needles from pyridine + methanolic HCl, dec 320°. Freely sol in pyridine and other organic bases yielding cherry-red solns with red fluorescence. Almost insol in most other organic solvents. Sol in alkaline aq solns; below pH 11.5 solns are red, above pH 11.5 they are green with red fluorescence. Absorption and fluorescence spectra: Scheibe, Schöntag, *Ber.* **75,** 2019 (1942); Brockmann, Neef, *loc. cit.*

4903.   **Hypericum.** St. John's wort; hyperici herba. Herbaceous, perennial plant, *Hypericum perforatum* L. (*Hypericaceae*), widely used in traditional medicine. *Habit.* Europe, Asia, North Africa; naturalized in the U.S. *Constit.* Hyperforin, hypericin, *q.q.v.*, pseudohypericin; flavonoids such as hyperin, rutin, quercetin; biapigenins; essential oil; tannins; procyanidins. Medicinal formulations are prepared from the dried flowering tops. Description of botany, constituents and medicinal uses: E. Bombardelli, P. Morazzoni, *Fitoterapia* **66,** 43-68 (1995). HPLC analysis and effect of cultivation on components: B. Büter *et al., Planta Med.* **64,** 431 (1998). HPLC determn of constituents in dietary supplements: W. Li, J. F. Fitzloff, *J. Chromatogr. B* **765,** 99 (2001). Pharmacology of extracts: S. Perovic, W. E. G. Müller, *Arzneim.-Forsch.* **45,** 1145 (1995); J. T. Neary, Y. Bu, *Brain Res.* **816,** 358 (1999). Clinical trial in comparison with imipramine for depression: M. Philipp *et al., Br. Med. J.* **319,** 1534 (1999). Reviews of clinical experience: E. Ernst *et al.,*

Eur. J. Clin. Pharmacol. **54**, 589-594 (1998); B. Gaster, J. Holroyd, *Arch. Intern. Med.* **160**, 152-156 (2000).

THERAP CAT: Antidepressant.

4904. Hypochlorous Acid. [7790-92-3] ClHO; mol wt 52.46. Cl 67.58%, H 1.92%, O 30.50%. HClO. Known in aq soln only. Formed by the action of water on chlorine: Cady, *Inorg. Synth.* **5**, 160 (1957). A 25% soln of hypochlorous acid is obtained when chlorine hydrate and mercuric oxide are distilled at low pressure: Goldschmidt, *Ber.* **52**, 753 (1919). *Review:* J. A. Wojtowicz, "Chlorine Monoxide, Hypochlorous Acid, and Hypochlorites", in *Kirk-Othmer Encyclopedia of Chemical Technology* **vol. 5** (Wiley-Interscience, New York, 3rd ed., 1979) pp 580-611.

Aq soln (25%), greenish yellow liquid. May be stored for a few days if kept at $-20°$. Dec slowly to Cl_2, O_2 and $HClO_4$. Very weak acid. pK ($25°$) 7.49. Strong oxidizing agent. When a 25% soln is evapd in a vacuum, chlorine monoxide, Cl_2O, is given off. Cl_2O in soln with HClO may be removed by shaking with carbon tetrachloride, but will form again. *Protect from light.*

Note: An aqueous soln of the potassium salt is known as *Javelle water* or *Eau de Javelle* (also used for sodium hypochlorite, *q.v.*).

USE: Disinfectant.

4905. Hypoglycine A. [156-56-9] ($\alpha S,1R$)-α-Amino-2-methylenecyclopropanepropanoic acid; 2-methylenecyclopropanealanine; 2-amino-4,5-methylenehex-5-enoic acid; α-amino-β-(2-methylenecyclopropyl)propionic acid; hypoglycin; hypoglycin A. $C_7H_{11}NO_2$; mol wt 141.17. C 59.56%, H 7.85%, N 9.92%, O 22.67%. Hypoglycemic principle from the akee plant, *Blighia sapida* Kon., *Sapindaceae:* Hassall *et al.*, *Nature* **173**, 356 (1954); Hassall, Reyle, *Biochem. J.* **60**, 334 (1955). Structure: de Ropp *et al.*, *J. Am. Chem. Soc.* **80**, 1004 (1958); Renner *et al.*, *Helv. Chim. Acta* **41**, 589 (1958); Ellington *et al.*, *J. Chem. Soc.* **1959**, 80. Synthesis: Carbon *et al.*, *J. Am. Chem. Soc.* **80**, 1002 (1958); Black, Landor, *Tetrahedron Lett.* **1963**, 1065.

Yellow plates from methanol + water, mp 280-284°. $[\alpha]_D^{32}$ +9.2°.

Methyl ester hydrochloride. $C_8H_{13}NO_2 \cdot HCl$. Needles from methanol + ether, mp 151-152°. $[\alpha]_D^{22}$ +36° (c = 2.0).

Hypoglycine B. [502-37-4] *N*-L-γ-Glutamyl-3-(2-methylenecyclopropyl)alanine; γ-L-glutamylhypoglycine. $C_{12}H_{18}N_2O_5$. Structure: Jöhl, Stoll, *Helv. Chim. Acta* **42**, 156 (1959); Hassall, John, *J. Chem. Soc.* **1960**, 4112. Synthesis: Jöhl, Stoll, *Helv. Chim. Acta* **42**, 716 (1959). Yellow needles from acetone + water, mp 194-195°, 200-206°. $[\alpha]_D^{32}$ +9.6° (c = 1.12). Neutralization equivalent 175.

4906. Hypophosphoric Acid. [7803-60-3] $H_4O_6P_2$; mol wt 161.97. H 2.49%, O 59.27%, P 38.25%. Prepn from its salts: Salzer, *Ann.* **187**, 322 (1877); **211**, 1 (1882). *Review:* Ohashi, "Lower Oxo Acids of Phosphorus and Their Salts" in *Topics in Phosphorus Chemistry* **Vol. 1**, M. Grayson, E. J. Griffith, Eds. (Interscience, New York, 1964) pp 113-187.

Orthorhombic plates, mp 70° (easily forms a dihydrate, mp 55°), usually available only in aq soln. The aq acid is colorless and odorless; dec on concn at atm pressure. Readily forms crystallizable normal and acid sodium salts.

USE: Acid sodium salts in baking powders.

4907. Hypophosphorous Acid. [6303-21-5] Phosphinic acid; hydroxyphosphine oxide. H_3O_2P; mol wt 66.00. H 4.58%, O 48.48%, P 46.93%. Conveniently prepd by treating NaH_2PO_2 with an ion-exchange resin: Klement, *Z. Anorg. Allg. Chem.* **260**, 267 (1949). *Review:* Ohashi, "Lower Oxo Acids of Phosphorus and Their Salts" in *Topics in Phosphorus Chemistry* **Vol. 1**, M. Grayson, E. J. Griffith, Eds. (Interscience, New York, 1964) pp 113-187.

Water-free acid forms deliquesc crystals; supercools to a colorless, odorless, oily liquid. d 1.493. mp 26.5°. Dec by heat into H_3PO_4 and spontaneously flammable PH_3. Oxidized by hot H_2SO_4; SO_2 and S formed. Powerful reducing agent. Miscible with water, alcohol, ether. Marketed in aq solns of various concns, *e.g.*, 50%, d 1.274; 30-32%, d 1.13; 10%, d 1.04. pK_1 1.1.

4908. Hypoxanthine. [68-94-0] 1,9-Dihydro-6*H*-purin-6-one; 1,7-dihydro-6*H*-purin-6-one; purin-6(1*H*)-one; sarcine; sarkin. $C_5H_4N_4O$; mol wt 136.11. C 44.12%, H 2.96%, N 41.16%, O 11.75%. Desmotropic forms: purin-6-ol; 1*H*-purin-6-ol; 3*H*-purin-6-ol; 9*H*-purin-6-ol; purin-6(3*H*)-one; 9*H*-purin-6(1*H*)-one. Formed in the animal body during the breakdown of nucleic acids; formation from adenosine continues after death. Also widely distributed in the vegetable kingdom. Numerous syntheses, *e.g.*, from 2,6,8-trichloropurine: Fischer, *Ber.* **30**, 2226 (1897); by oxidation of adenine: Krüger, *Z. Physiol. Chem.* **18**, 445 (1894). By reduction of uric acid: Sundwik, *ibid.* **76**, 486 (1912); by condensing ethyl cyanoacetate and thiourea in the presence of Na-ethoxide: Traube, *Ann.* **331**, 64 (1904); from mercapto-4-hydroxy-6-aminopyrimidine: Taylor, Cheng. *J. Org. Chem.* **25**, 148 (1960). *Reviews:* Levene, Bass, *ACS Monograph Series* **no. 56**, entitled "Nucleic Acids" (New York, 1931); Chargaff, Davidson, *Nucleic Acids* **vols. I & II** (Academic Press, New York, 1955).

Small octahedra, needles from water, dec 150° without melting. Sol in 1400 parts water, 70 parts boiling water; sol in dil acids, alkalies. Combines with one equivalent of acid or with two equivalents of base. pK_b (25°): 8.7. Absorption spectrum: Dhere, *Compt. Rend.* **141**, 720.

4909. Hypusine. [34994-11-1] N^6-[(2*R*)-4-Amino-2-hydroxybutyl]-L-lysine; (2*S*,9*R*)-2,11-diamino-9-hydroxy-7-azaundecanoic acid; N^6-(4-amino-2-hydroxybutyl)-2,6-diaminohexanoic acid. $C_{10}H_{23}N_3O_3$; mol wt 233.31. C 51.48%, H 9.94%, N 18.01%, O 20.57%. Naturally occurring L-amino acid formed by the post-translational modification of lysine. Name coined to indicate the condensation product of hydroxyputrescine with lysine. Isoln: T. Shiba *et al.*, *Biochim. Biophys. Acta* **244**, 523 (1971); from archaebacteria: H. Schümann, F. Klink, *Syst. Appl. Microbiol.* **11**, 103 (1989). In eukaryotes occurs only at a single position in one protein, eukaryotic protein synthesis initiation factor 5A, (eIF-5A, formerly eIF-4D): M. H. Park *et al.*, *Proc. Natl. Acad. Sci. USA* **78**, 2869 (1981); H. L. Cooper *et al.*, *ibid.* **80**, 1854 (1983). Biosynthesis: M. H. Park *et al.*, *J. Biol. Chem.* **257**, 7217 (1982); M. H. Park *et al.*, *ibid.* **259**, 12123 (1984). Synthesis and stereochemistry: T. Shiba *et al.*, *Bull. Chem. Soc. Jpn.* **55**, 899 (1982). Total synthesis: C. M. Tice, B. Ganem, *J. Org. Chem.* **48**, 5043, 5048 (1983); R. J. Bergeron *et al.*, *ibid.* **58**, 6804 (1993). Biological significance in eukaryotes: J. W. B. Hershey *et al.*, *Biochim. Biophys. Acta* **1050**, 160 (1990). HPLC determn in eukaryotic samples: S. Beninati *et al.*, *Anal. Biochem.* **184**, 16 (1990). Chromatographic determn in archaebacteria: D. Bartig, F. Klink, *J. Chromatogr.* **606**, 43 (1992). Review of biological significance: M. H. Park *et al.*, *Trends Biochem. Sci.* **18**, 475-479 (1993).

Dihydrochloride. [82310-93-8] $C_{10}H_{23}N_3O_3 \cdot 2HCl$. White crystals, mp 239-241° (dec). $[\alpha]_D^{23}$ +8.3° (c = 0.96 in 6*M* HCl).

I

4910. IACFT. [180468-34-2] (1*R*,2*S*,3*S*,5*S*)-3-(4-Fluoro-phenyl)-8-[(2*E*)-3-(iodo-2-propen-1-yl)]-8-azabicyclo[3.2.1]octane-2-carboxylic acid methyl ester; 2β-carboxymethoxy-3β-(4-fluoro-phenyl)-*N*-(1-iodoprop-1-en-3-yl)nortropane; Altropane. $C_{18}H_{21}$-FINO$_2$; mol wt 429.27. C 50.36%, H 4.93%, F 4.43%, I 29.56%, N 3.26%, O 7.45%. Cocaine analog with dopamine transport inhibitor activity. Radiolabeled form designed for SPECT brain imaging. Prepn: D. R. Elmaleh *et al.*,*WO 9511901*; *eidem*, **US 5493026** (1995, 1996 both to Gen. Hosp. Corp., Harvard College, Organix and Northeastern Univ.); and receptor binding studies: *eidem*, *J. Nucl. Med.* **37**, 1197 (1996). Series of articles on pharmacology and binding studies: B. K. Madras *et al.*, *Synapse* **29**, 93-127 (1998). Clinical evaluation in Parkinson's disease: A. J. Fischman *et al.*, *ibid.* 128. CE determn in plasma: K. Hettiarachchi *et al.*, *J. Chromatogr. A* **895**, 87 (2000).

White solid, mp 115-116°,
^{123}I-Labeled form. [208517-65-1]
^{125}I-Labeled form. [180268-78-4]
THERAP CAT: ^{123}I-labeled form as diagnostic aid (radioactive imaging agent).

4911. Ibafloxacin. [91618-36-9] 9-Fluoro-6,7-dihydro-5,8-dimethyl-1-oxo-1*H*,5*H*-benzo[*ij*]quinolizine-2-carboxylic acid; R-835; S-25930. $C_{15}H_{14}FNO_3$; mol wt 275.28. C 65.45%, H 5.13%, F 6.90%, N 5.09%, O 17.44%. Tricyclic fluorinated quinolone-type antibacterial; related to flumequine, *q.v.* Prepn: R. M. Stern, **EP 109284**; *idem*, **US 4472405** (both 1984 to Riker Labs.). Antimicrobial spectrum: L. J. V. Piddock *et al.*, *Eur. J. Clin. Microbiol.* **5**, 303 (1986); K. V. I. Rolston *et al.*, *Eur. J. Clin. Microbiol. Infect. Dis.* **7**, 681 (1988). Synthesis, absolute configuration and antibacterial activity of enantiomers: J. F. Gerster *et al.*, *J. Med. Chem.* **30**, 839 (1987).

Off-white solid, mp 269-272°.
***S*-Form.** Colorless crystals from DMF, mp 274-276°. $[\alpha]_D$ −107.4° (c = 0.1109 g/10mL in dichloroethane).
***R*-Form.** Colorless crystals from DMF, mp 274-276°. $[\alpha]_D$ +106.8° (c = 0.1165 g/10mL in dichloroethane).
THERAP CAT (VET): Antibacterial.

4912. Ibandronic Acid. [114084-78-5] *P,P′*-[1-Hydroxy-3-(methylpentylamino)propylidene]bisphosphonic acid. C_9H_{23}-NO$_7$P$_2$; mol wt 319.23. C 33.86%, H 7.26%, N 4.39%, O 35.08%, P 19.41%. Bisphosphonate antiresorptive agent. Prepn: R. Gall, E. Bosies, **DE 3623397**; *eidem*, **US 4927814** (1988, 1990 both to Boehringer Mann.). Inhibition of bone resorption: R. C. Mühlbauer *et al.*, *J. Bone Miner. Res.* **6**, 1003 (1991). Mechanism of action study: C. Vitté *et al.*, *Endocrinology* **137**, 2324 (1996). Clinical trial in cancer-associated hypercalcemia: M. Pecherstorfer *et al.*, *J. Clin. Oncol.* **14**, 268 (1996); in osteoporosis: P. Ravn *et al.*, *Bone* **19**, 527 (1996). Review of pharmacology and clinical efficacy: M. Dooley,

J. A. Balfour, *Drugs* **57**, 101-108 (1999); of clinical experience in osteoporosis: R. D. Chapurlat, P. D. Delmas, *Expert Opin. Pharmacother.* **4**, 391-396 (2003); of development and therapeutic potential: L. Gennari, *IDrugs* **8**, 155-169 (2005).

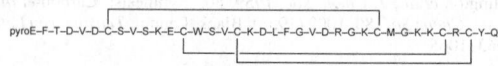

Dec 84°.
Sodium salt. [138926-19-9] (monohydrate); [138844-81-2] (anhydrous). Ibandronate sodium; BM-21.0955; Bondronat; Boniva; Bonviva. $C_9H_{22}NNaO_7P_2$; mol wt 341.21. Occurs as the monohydrate.
THERAP CAT: Bone resorption inhibitor.

4913. Iberiotoxin. [129203-60-7] IbTX. Peptide inhibitor of K$^+$ channels isolated from the venom of the scorpion, *Buthus tamulus*. Single polypeptide chain of 37 amino acids with 3 disulfide bridges; mol wt 4.23 kDa. Isoln, purification and characterization: A. Galvez *et al.*, *J. Biol. Chem.* **265**, 11083 (1990). 3D structure determn by NMR: B. A. Johnson, E. E. Sugg, *Biochemistry* **31**, 8151 (1992). Mode of action: S. Candia *et al.*, *Biophys. J.* **63**, 583 (1992). Synthesis of labeled mutant: A. Koschak *et al.*,*Biochemistry* **36**, 1943 (1997). Use as probe: W. A. A. Kunze *et al.*, *Pfluegers Arch.* **428**, 300 (1994); G. Fauaz *et al.*, *Vascul. Pharmacol.* **40**, 127 (2003); Y. Tanaka *et al.*, *Arch. Pharmacol.* **367**, 35 (2003).

pyroE–F–T–D–V–D–C–S–V–S–K–E–C–W–S–V–C–K–D–L–F–G–V–D–R–G–K–C–M–G–K–K–C–R–C–Y–Q

USE: Selective probe for high-conductance calcium-activated potassium channels.

4914. Ibogaine. [83-74-9] 12-Methoxyibogamine. $C_{20}H_{26}$-N$_2$O; mol wt 310.44. C 77.38%, H 8.44%, N 9.02%, O 5.15%. Indole alkaloid of the *iboga* group. Isoln from root (1.27%), root-bark (2 to 6%), stems (1.95%) and leaves (0.35%) of the shrub *Tabernanthe iboga* Baill., *Apocynaceae*, found in Africa: Dybowski, Landrin, *Compt. Rend.* **133**, 748 (1901); Haller, Heckel, *ibid.* 850, 1236; from other *Apocynaceae*: H. Achenbach, B. Raffelsberger, *Z. Naturforsch.* **35B**, 219, 885 (1980); N. Ghorbel *et al.*, *J. Nat. Prod.* **44**, 717 (1981); T. Mulamba *et al.*, *ibid.* 184; B. Richard *et al.*, *ibid.* **46**, 283 (1983). Purification: Schlittler *et al.*, *Helv. Chim. Acta* **36**, 1341 (1953). Revised extraction procedure: Dickel *et al.*, *J. Am. Chem. Soc.* **80**, 123 (1958). Review of early isolation work: Lebeau, Janot, *Traité de Pharmacie Chimique* **vol. 4** (Masson et Cie., Paris, 1956) pp 2982-2988. Structure: Bartlett *et al.*, *J. Am. Chem. Soc.* **80**, 126 (1958). Mass spectrum: Biemann, Friedmann-Spiteller, *ibid.* **83**, 4805 (1961). Synthesis: Büchi *et al.*, *ibid.* **88**, 3099 (1966); Rosenmund *et al.*, *Ber.* **108**, 1871 (1975). Derivs: Taylor, **US 2877229** (1959 to Ciba). Absolute configuration: K. Blàha *et al.*, *Tetrahedron Lett.* **1972**, 2763. Interatomic distances similar to those of serotonin: J. M. Kelley, R. H. Adamson, *Pharmacology* **10**, 28 (1973). NMR spectrum: E. Wenkert *et al.*, *Helv. Chim. Acta* **59**, 2437 (1976). Determn in biological fluids: E. Bertol *et al.*, *J. Chromatogr.* **117**, 239 (1976). Iboga extracts said to be used by African natives while stalking game, to enable them to remain motionless for as long as 2 days while retaining mental alertness. Neuropharmacological studies: Schneider, Sigg, *Ann. N.Y. Acad. Sci.* **66**, 765 (1957); S. Gershon, W. J. Lang, *Arch. Int. Pharmacodyn. Ther.* **135**, 31 (1962). Cardiovascular effects: J. A. Schneider, R. K. Rinehart, *ibid.* **110**, 92 (1957). Serotonergic properties: R. S. Sloviter *et al.*, *J. Pharmacol. Exp. Ther.* **214**, 231 (1980). Experimental use in treatment of heroin addiction: H. S. Lotsof, **US 4499096** (1985). *Reviews:* W. I. Taylor, "The Iboga and Voacanga Alkaloids" in *The Alkaloids, Chemistry and Physiology* **Vol. 8**, R. H. F. Manske, Ed. (Academic Press, New York, 1965) p 203-235, *idem*, *ibid.* **Vol. 11** (1968), pp 79-98.

Prismatic needles from abs ethanol, mp 152-153°. Sublimes$_{0.01}$ 150°. $[\alpha]_D^{20}$ −53° (in 95% ethanol). pKa 8.1 in 80% methylcellosolve. uv max (methanol): 226, 298 nm (log ε 4.39, 3.93). Sol in ethanol, ether, chloroform, acetone, benzene. Practically insol in water.

Hydrochloride. $C_{20}H_{26}N_2O \cdot HCl$. Crystals. Dec 299-300°. $[\alpha]_D^{25}$ −63° (ethanol); $[\alpha]_D^{25}$ −49° (H_2O). Soluble in water, methanol, ethanol. Slightly sol in acetone, chloroform. Practically insol in ether.

Note: This is a controlled substance (hallucinogen): **21 CFR,** 1308.11.

4915. Ibopamine. [66195-31-1] 2-Methylpropanoic acid 1,1'-[4-[2-(methylamino)ethyl]-1,2-phenylene] ester; 4-[2-(methylamino)ethyl]-*o*-phenylene diisobutyrate; *N*-methyldopamine diisobutyric ester; 3,4-di-*o*-isobutyryl epinine. $C_{17}H_{25}NO_4$; mol wt 307.39. C 66.43%, H 8.20%, N 4.56%, O 20.82%. Inotropic agent with dopaminergic and adrenergic agonist activities; converted to active metabolite, deoxyepinephrine, *q.v.* Prepn: C. Casagrande, G. Ferrari, **DE 2734678;** *eidem,* **US 4218470** (1978, 1980 both to Simes). Pharmacology: G. F. Melloni *et al., Curr. Ther. Res.* **25**, 406 (1979); *eidem, ibid.* **26**, 466 (1979). Series of articles on synthesis, pharmacology, clinical efficacy: *Arzneim.-Forsch.* **36**, 285-408 (1986). Review of pharmacodynamics, pharmacokinetics and therapeutic efficacy: J. M. Henwood, P. A. Todd, *Drugs* **36**, 11-31 (1988). Series of articles on clinical use in congestive heart failure: *Cardiology* **77**, Suppl. 5, 1-95 (1990). Clinical efficacy and safety in heart failure: J. R. Hampton *et al., Lancet* **349**, 971 (1997). Clinical evaluation in diagnosis of glaucoma: G. Marchini *et al., J. Ocul. Pharmacol. Ther.* **17**, 215 (2001); in treatment of ocular hypertony: L. C. Ugahary *et al., Am. J. Ophthalmol.* **141**, 571 (2006).

Hydrochloride. [75011-65-3] SB-7505; Inopamil; Scandine; Trazyl; $C_{17}H_{25}NO_4 \cdot HCl$; mol wt 343.85. Crystals from ethyl acetate, mp 132°.

THERAP CAT: Cardiotonic. Diagnostic aid (glaucoma).

4916. Ibotenic Acid. [2552-55-8] α-Amino-2,3-dihydro-3-oxo-5-isoxazoleacetic acid; α-amino-3-hydroxy-5-isoxazoleacetic acid; amino-(3-hydroxy-5-isoxazolyl)acetic acid. $C_5H_6N_2O_4$; mol wt 158.11. C 37.98%, H 3.83%, N 17.72%, O 40.48%. Fly-killing and narcosis-potentiating amino acid structurally similar to kainic acid, *q.v.*, extracted from poisonous mushroom species. Isoln from *Amanita pantherina* (DC.) Fr., and *A. muscaria* (L.) Fr., *Agaricaceae:* Takemoto *et al., J. Pharm. Soc. Jpn.* **84**, 1233 (1964); Eugster *et al., Tetrahedron Lett.* **1965**, 1813. Structure: Takemoto *et al., J. Pharm. Soc. Jpn.* **84**, 1186, 1232 (1964). Syntheses: Gagneux *et al., Tetrahedron Lett.* **1965**, 2081; Sirakawa *et al., Chem. Pharm. Bull.* **14**, 89 (1966); Kishida *et al., ibid.* **14**, 92 (1966); **15**, 1025 (1967). Improved synthesis: Nakamura, *ibid.* **19**, 46 (1971). Industrial pats: **BE 665249,** *C.A.* **65**, 2266e (1966); Gagneux *et al.,* **US 3459862** (1965, 1969 both to Geigy); Kishida *et al.,* **JP 68 15975;** **JP 69 25780** (1968, 1969 both to Sankyo), *C.A.* **70**, 77944p (1969); **72**, 13054g (1970). Pharmacology: Theobald *et al., Arzneim.-Forsch.* **18**, 311 (1968); Johnston *et al., Biochem. Pharmacol.* **17**, 2488 (1968). Exhibits potent neuroexcitatory activity: *eidem, Na-*

ture **248**, 804 (1974). Chemistry review: Eugster, *Fortschr. Chem. Org. Naturst.* **27**, 261-321 (1969); Catalfomo, Eugster, *Bull. Narc.* **22**, 33-41 (1970). Excitatory and possible sedative actions on spinal neurons: D. R. Curtis *et al., J. Physiol.* **291**, 19 (1979); in cerebral cortex: E. Puil, *Can. J. Physiol. Pharmacol.* **59**, 1025 (1981). Use as experimental neurotoxic agent: A. Contestabile *et al., Experientia* **40**, 524 (1984).

Crystals from water or methanol, mp 151-152° (anhydrous); mp 144-146° (monohydrate). LD_{50} in mice, rats (mg/kg): 15, 42 i.v.; 38, 129 orally (Theobald).

USE: Neurobiological tool.

4917. Ibritumomab Tiuxetan. [206181-63-7] Anti-(human CD20 (antigen)) immunoglobulin G1 (mouse monoclonal IDEC-Y2B8 γ_1-chain) disulfide with mouse monoclonal IDEC-Y2B8 κ-chain, dimer, *N*-[2-[bis(carboxymethyl)amino]-3-(4-isothiocyanatophenyl)propyl]-*N*-[2-[bis(carboxymethyl)amino]propyl]glycine conjugate. Murine monoclonal antibody targeted against CD-20 antigen on mature B lymphocytes, conjugated to linker-chelator tiuxetan. Radiolabeled form designed for antibody-targeted treatment of non-Hodgkin's lymphoma (NHL). Prepn of monoclonal antibody: M. E. Reff *et al., Blood* **83**, 435 (1994). Prepn of radiolabeled conjugate: D. R. Anderson *et al.,* **WO 9411026;** *eidem,* **US 5736137** (1994, 1998 both to IDEC); P. C. Chinn *et al., Int. J. Oncol.* **15**, 1017 (1999). Clinical pharmacology: S. J. Knox *et al., Clin. Cancer Res.* **2**, 457 (1996). Clinical evaluation in NHL: T. E. Witzig *et al., J. Clin. Oncol.* **17**, 3793 (1999). Review of development and therapeutic potential: G. A. Wiseman *et al., Clin. Cancer Res.* **5** Suppl., 3281s-3286s (1999); of clinical experience: A. J. Grillo-López, *Expert Rev. Anticancer Ther.* **2**, 485-493 (2002).

^{90}Y-Chelate. Yttrium Y 90 ibritumomab tiuxetan; Y2B8-MX-DTPA; IDEC-Y2B8; Zevalin.

THERAP CAT: ^{90}Y-chelate as antineoplastic.

4918. Ibudilast. [50847-11-5] 2-Methyl-1-[2-(1-methylethyl)pyrazolo[1,5-*a*]pyridin-3-yl]-1-propanone; 3-isobutyryl-2-isopropylpyrazolo[1,5-*a*]pyridine; KC-404; Ketas. $C_{14}H_{18}N_2O$; mol wt 230.31. C 73.01%, H 7.88%, N 12.16%, O 6.95%. Leukotriene D_4 antagonist. Prepn: T. Irikura *et al.,* **DE 2315801;** *eidem,* **US 3850941** (1973, 1974 both to Kyorin). Pharmacology and antiallergic activity: K. Nishino *et al., Jpn. J. Pharmacol.* **33**, 267 (1983); H. Nagai *et al., ibid.* 1215. *In vitro* cerebral vasodilating activity: M. Ohashi *et al., Arch. Int. Pharmacodyn.* **280**, 216 (1986); *in vivo* activity: W. M. Armstead *et al., J. Pharmacol. Exp. Ther.* **244**, 138 (1988). Bronchodilating activity in animals: S. Mue *et al., Arch. Int. Pharmacodyn.* **283**, 153 (1986). Antiplatelet activity in animals: M. Ohashi *et al., ibid.* 321; M. Ohashi *et al., Gen. Pharmacol.* **17**, 385 (1986).

Crystals from hexane, mp 53.5-54°. Slightly sol in water, freely sol in organic solvents. LD_{50} i.v. in mice: 260 mg/kg (Irikura, 1973).

THERAP CAT: Antiallergic; antiasthmatic; vasodilator (cerebral).

4919. Ibuprofen. [15687-27-1] α-Methyl-4-(2-methylpropyl)benzeneacetic acid; *p*-isobutylhydratropic acid; (±)-2-(4-isobutylphenyl)propionic acid; RD-13621; Advil; Brufen; Brufort; Dolgit; Dolocyl; Epobron; Fenbid; Gynofug; Ibumetin; Ibutop; Ipren; Motrin; Napacetin; Novogent; Nuprin; Nurofen; Opturem; Proflex; Solufen; Tabalon; Urem. $C_{13}H_{18}O_2$; mol wt 206.29. C 75.69%, H 8.80%, O 15.51%. Nonsteroidal anti-inflammatory drug (NSAID); activity resides primarily in the (S)-isomer. Prepn: J. S. Nicholson, S. S. Adams, **GB 971700**; *eidem,* **US 3385886** (1964, 1968 both to Boots Pure Drug); T. Shiori, N. Kawai, *J. Org. Chem.* **43**, 2936 (1978); J. T. Pinhey, B. A. Rowe, *Tetrahedron Lett.* **21**, 965 (1980). Methods for resolution of enantiomers: R. Bhushan, J. Martens, *Biomed. Chromatogr.* **12**, 309 (1998). HPLC determn in plasma: R. Canaparo *et al., ibid.* **14**, 219 (2000). Pharmacology: S. S. Adams *et al., Arch. Int. Pharmacodyn. Ther.* **178**, 115 (1969). Acute toxicity: G. Orzalesi *et al., Arzneim.-Forsch.* **27**, 1006 (1977). Clinical trial for closure of patent ductus arteriosus: B. Van Overmeire *et al., N. Engl. J. Med.* **343**, 674 (2000). Review of pharmacology and clinical experience: M. Busson *et al., J. Int. Med. Res.* **14**, 53-62 (1986); of clinical pharmacokinetics and metabolism: N. M. Davies, *Clin. Pharmacokinet.* **34**, 101-154 (1998).

Colorless, crystalline stable solid, mp 75-77°. Very sol in alc, methanol, acetone, chloroform; readily sol in most organic solvents; slightly sol in ethyl acetate. Practically insol in water. LD_{50} in male mice, rats (mg/kg): 495, 626 i.p.; 1255, 1050 orally (Orzalesi).

Lysine salt. Arfen; Imbun. $C_{13}H_{18}O_2 \cdot C_6H_{14}N_2O_2$; mol wt 352.48.

S-Form L-lysine salt monohydrate. [141505-32-0] Dexibuprofen lysine; MK-223. $C_{13}H_{18}O_2 \cdot C_6H_{14}N_2O_2 \cdot H_2O$; mol wt 370.49.

THERAP CAT: Anti-inflammatory; analgesic; antipyretic.

THERAP CAT (VET): Anti-inflammatory.

4920. Ibuproxam. [53648-05-8] N-Hydroxy-α-methyl-4-(2-methylpropyl)benzeneacetamide; *dl*-2-(4-isobutylphenyl)propionohydroxamic acid; G-277; Ibudros. $C_{13}H_{19}NO_2$; mol wt 221.30. C 70.56%, H 8.65%, N 6.33%, O 14.46%. Hydroxylamine deriv of ibuprofen, *q.v.*, to which it is converted *in vivo.* Prepn: G. Orzalesi, R. Selleri, **DE 2400531**; *eidem,* **US 4082707** (1974, 1978 both to Manetti & Roberts). Metabolism in rats: G. Orzalesi *et al., Arzneim.-Forsch.* **27**, 1012 (1977); in humans: *eidem, ibid.* **30**, 1607 (1980). Pharmacological study: *eidem, ibid.* **27**, 1006 (1977). Thermal decomposition: S. Chimichi *et al., J. Pharm. Sci.* **69**, 521 (1980). Physico-chemical properties: M. Mannelli *et al., Boll. Chim. Farm.* **119**, 203 (1980).

Crystals from acetone/petr ether, mp 119-121°. Sol in methanol, ethanol, acetone, ethyl ether. Practically insol in water and petr ether. LD_{50} in mice, rats (g/kg): >2, >3 orally (Orzalesi, Selleri, 1978).

THERAP CAT: Anti-inflammatory.

4921. Ibutilide. [122647-31-8] N-[4-[4-(Ethylheptylamino)-1-hydroxybutyl]phenyl]methanesulfonamide. $C_{20}H_{36}N_2O_3S$; mol wt 384.58. C 62.46%, H 9.44%, N 7.28%, O 12.48%, S 8.34%. Methanesulfonanilide antiarrhythmic agent; prolongs myocardial action potential duration, predominantly by activation of slow inward sodium current. Prepn: J. B. Hester, **EP 164865**; *idem,* **US 5155268** (1985, 1992 both to Upjohn); *idem et al., J. Med. Chem.* **34**, 308 (1991). HPLC determn of enantiomers: C. L. Hsu, R. R. Walters, *J. Chromatogr. B* **667**, 115 (1995); determn of racemate in plasma: L. Tian *et al., J. Chromatogr. B* **816**, 81 (2005). Review of electrophysiology and pharmacology: G. V. Naccarelli *et al., Am. J. Cardiol.* **78**, Suppl. 8A, 12-16 (1996); of pharmacology and clinical potential: R. H. Foster *et al., Drugs* **54**, 312-330 (1997). Clinical trial in atrial flutter or fibrillation: B. S. Stambler *et al., Circulation* **94**, 1613 (1996); J. T. VanderLugt *et al., ibid.* **100**, 369 (1999); in elderly patients: R. M. Gowda *et al., Am. J. Ther.* **11**, 95 (2004).

Fumarate. [122647-32-9] U-70226E; Corvert. $(C_{20}H_{36}N_2O_3S)_2 \cdot C_4H_4O_4$; mol wt 885.23. Crystals from acetone, mp 117-119°. uv max (95% ethanol): 228, 267 nm (ε 16670, 894). Soly (mg/ml): aqueous >100 (pH ≤ 7).

THERAP CAT: Antiarrhythmic (class III).

4922. Icatibant. [130308-48-4] D-Arginyl-L-arginyl-L-prolyl-(4R)-4-hydroxy-L-prolylglycyl-3-(2-thienyl)-L-alanyl-L-seryl-(3R)-1,2,3,4-tetrahydro-3-isoquinolinecarbonyl-(2S,3aS,7aS)-octahydro-1H-indole-2-carbonyl-L-arginine; D-Arg-[Hyp³,Thi⁵,D-Tic⁷,-Oic⁸]bradykinin; H-D-Arg-Arg-Pro-Hyp-Gly-Thia-Ser-D-Tic-Oic-Arg-OH. $C_{59}H_{89}N_{19}O_{13}S$; mol wt 1304.54. C 54.32%, H 6.88%, N 20.40%, O 15.94%, S 2.46%. Synthetic peptidomimetic; bradykinin B_2 receptor antagonist. Prepn: S. Henke *et al.,* **DE 3938751**; *eidem,* **US 5648333** (1990, 1997 both to Hoechst). *In vitro* and *in vivo* pharmacology studies: F. J. Hock *et al., Br. J. Pharmacol.* **102**, 769 (1991); K. Wirth *et al., ibid.* 774. Pharmacology and receptor binding studies: N.-E. Rhaleb *et al., Eur. J. Pharmacol.* **210**, 115 (1992); M. Félétou *et al., ibid.* **274**, 57 (1995); F. Bellucci *et al., ibid.* **491**, 121 (2004). Metabolism and HPLC determn in synovial fluid: A. P. Bond *et al., Agents Actions Suppl.* **38**, 582 (1992). Chiral amino acid analysis: J. Ermer *et al., Arch. Pharm.* **328**, 635 (1995). Computational conformation analysis: M. Filizola *et al., J. Biomol. Struct. Dyn.* **15**, 639 (1998). Clinical effects on bradykinin-induced vasodilation: J. R. Cockcroft *et al., Br. J. Clin. Pharmacol.* **38**, 317 (1994). Clinical evaluation in hereditary angioedema: K. Bork *et al., J. Allergy Clin. Immunol.* **119**, 1497 (2007).

Acetate. [138614-30-9] HOE-140; JE-049; Firazyr.

THERAP CAT: In treatment of hereditary angioedema.

4923. Iceland Moss. *Cetraria islandica* (L.) Ach., *Parmeliaceae,* a lichen growing in all northern countries. Exported from Iceland, Norway, and Sweden. The gum from the powdered plant appears to be a hemicellulose contg uronic acid, galactose, mannose, and glucose; *cf.* Mantell, *The Water-Soluble Gums,* New York, 1947. About 60% of dried Iceland moss dissolves when boiled with water contg a little sodium bicarbonate. The soln forms a jelly when cold.

USE: Manuf sea biscuits which are somewhat more resistant to weevil infestation than when wheat flour alone is used. In foods for convalescents. Manuf sizing agents for rayon; hair-setting lotions, other cosmetics.

4924. Ichthammol. [8029-68-3] Ammonium bituminosulfonate; ammonium ichthosulfonate; ammonium sulfobituminate; ammonium sulfoichthyolate; bitumol; bituminol; ichthammonium; ammonium bithiolicum; ichthosulfol; Ichthyol; Hirathiol; Ichden; Ichtammon; Ichthadone; Ichthymall; Ichthysalle; Ichthalum; Ichthium; Ichtopur; Ichthosan; Ichthynat; Ichthyopon; Lithol; Petrosulpho; Perichthol; Piscarol; Pisciol; Saurol; Subitol; Sulfogenol; Thilaven; Thiolin; Thiozin; Trasulphane; Tumenol; Leukochthol; Ichthosauran; Amsubit; Bitulan. Obtained by sulfation and ammoniation of a distillate from mineral deposits (bituminous schists) originally found near Seefeld, Tyrol. Contains satd and unsatd hydrocarbons, nitrogenous bases, acids, and several thiophene derivs. Analysis shows at least 2.5% NH_3 and at least 10% S. Also contains traces of some 20 minerals and *"zoomelanoidic" acids*. Method of prepn: Schröter, **DE 35216** (1885); Helmers, **DE 76128** (1892). Similar deposits occur in Asia east of Lake Baikal where the oil is known as *stone oil, barakshin, Asil*; sold in India for remedial purposes as *saladjidi*: Gerbrein, *Photo-Journal* (Montreal, 1969, July 2-9) p 19. *Review:* Wernicke, *Chem. Ztg.* **60**, 85-87 (1936).

Pale yellow or (usually) brownish-black, thick, viscous liquid. Bituminous odor. Miscible with water, glycerol, propylene glycol, fats, oils, carbowaxes, lanolin. Partially sol in alcohol, ether. An injectable form is marketed as *Adnexol*.

THERAP CAT: Anti-infective (topical).

THERAP CAT (VET): Demulcent, emollient, antiseptic.

4925. Icilin. [36945-98-9] 3,6-Dihydro-1-(2-hydroxyphenyl)-4-(3-nitrophenyl)-2(1*H*)-pyrimidinone; 1-(2'-hydroxyphenyl)-4-(3"-nitrophenyl)-1,2,3,6-tetrahydropyrimidine-2-one; AG-3-5. $C_{16}H_{13}N_3O_4$; mol wt 311.30. C 61.73%, H 4.21%, N 13.50%, O 20.56%. Peripheral cold channel agonist. Induces cold sensations when applied topically in humans, and intense shivering behavior in animal studies. Significantly more potent than menthol, *q.v.* Prepn: C. Podesva, J. M. Do Nascimento, **DE 2142385**; *eidem*, **US 3821221** (1972, 1974 both to Delmar Chemicals). Description of cold producing properties: E. T. Wei, D. A. Seid, *J. Pharm. Pharmacol.* **35**, 110 (1983). Mechanism of action studies: H. Chuang *et al.*, *Neuron* **43**, 859 (2004); J. L. Werkheiser *et al.*, *Eur. J. Pharmacol.* **547**, 101 (2006).

Crystals from DMF + ethyl acetate, mp 228-230°. LD_{50} i.p. in rats: >1500 mg/kg (Wei, Seid).

4926. Iclaprim. [192314-93-5] 5-[(2-Cyclopropyl-7,8-dimethoxy-2*H*-1-benzopyran-5-yl)methyl]-2,4-pyrimidinediamine; AR-100. $C_{19}H_{22}N_4O_3$; mol wt 354.41. C 64.39%, H 6.26%, N 15.81%, O 13.54%. Dihydrofolate reductase (DHFR) inhibitor; derivative of 2,4-diaminopyrimidine. Administered as a racemic mixture of two equipotent enantiomers. Prepn: R. Masciadri, **WO 9720839**; *eidem*. **US 5773446** (1997, 1998 both to Hoffmann-La Roche). Enantioselective synthesis of isomers: C. Tahtaoui *et al.*, *J. Org. Chem.* **75**, 3781 (2010). Crystallographic study of binding to DHFR: C. Oefner *et al.*, *Acta Crystallogr. D* **65**, 751 (2009). Comparative antibacterial spectrum *in vitro* against gram-positive pathogens: H. S. Sader *et al.*, *Antimicrob. Agents Chemother.* **53**, 2171 (2009). Clinical evaluation in skin infections: D. Krievins *et al.*, *ibid.* 2834. Review of development and clinical experience: S. A. Kohlhoff, R. Sharma, *Expert Opin. Invest. Drugs* **16**, 1441-1448 (2007); C. A. Sincak, J. M. Schmidt, *Ann. Pharmacother.* **43**, 1107-1114 (2009).

White crystals from ethanol, mp 229°.

Methanesulfonate. [474793-41-4] Iclaprim mesylate; Mersarex. $C_{19}H_{22}N_4O_3 \cdot CH_4O_3S$; mol wt 450.51. White crystalline powder, mp 204°.

THERAP CAT: Antibacterial.

4927. Idarubicin. [58957-92-9] (7*S*,9*S*)-9-Acetyl-7-[(3-amino-2,3,6-trideoxy-α-L-*lyxo*-hexopyranosyl)oxy]-7,8,9,10-tetrahydro-6,9,11-trihydroxy-5,12-naphthacenedione; (1*S*,3*S*)-3-acetyl-1,-2,3,4,6,11-hexahydro-3,5,12-trihydroxy-6,11-dioxo-1-naphthacenyl-3-amino-2,3,6-trideoxy-α-L-*lyxo*-hexopyranoside; 4-demethoxydaunomycin; 4-demethoxydaunorubicin; DMDR; IMI-30; NSC-256439. $C_{26}H_{27}NO_9$; mol wt 497.50. C 62.77%, H 5.47%, N 2.82%, O 28.94%. Orally active anthracycline; analog of daunorubicin, *q.v.* Prepn: B. Patelli *et al.* **DE 2525633**; *eidem*, **US 4046878** (1976, 1977 both to Soc. Farmac. Ital.); and antitumor activity: F. Arcamone *et al.*, *Cancer Treat. Rep.* **60**, 829 (1976). Total synthesis for larger scale preparation: M. J. Broadhurst *et al.*, *Chem. Commun.* **1982**, 158. Synthesis of optically pure isomers: Y. Kimura *et al.*, *Bull. Chem. Soc. Jpn.* **59**, 423 (1986). Metabolism and biodistribution in rats: G. Zini *et al.*, *Cancer Chemother. Pharmacol.* **16**, 107 (1986). HPLC determn in plasma: S. S. N. De Graaf *et al.*, *J. Chromatogr.* **491**, 501 (1989). Clinical pharmacokinetics: H. C. Gillies *et al.*, *Br. J. Clin. Pharmacol.* **23**, 303 (1987). Clinical evaluation of cardiac toxicity: F. Villani *et al.*, *Eur. J. Cancer Clin. Oncol.* **25**, 13 (1989). Reviews of pharmacology and antitumor efficacy: A. M. Casazza, *Cancer Treat. Rep.* **63**, 835-844 (1979); F. Ganzina *et al.*, *Invest. New Drugs* **4**, 85-105 (1986). Symposium on clinical experience in acute leukemias: *Semin. Oncol.* **17**, Suppl. 2, 1-36 (1989).

Hydrochloride. [57852-57-0] Idamycin; Zavedos. $C_{26}H_{27}-NO_9 \cdot HCl$; mol wt 533.96. Orange crystalline powder, mp 183-185° (Arcamone); also reported as mp 172-174° (Broadhurst). $[\alpha]_D^{20}$ +205° (c = 0.1 in methanol) (Arcamone); also reported as $[\alpha]_D^{20}$ +188° (c = 0.10 in methanol) (Kimura). Sol in methanol; slightly sol in water. Insol in acetone, ethyl ether.

THERAP CAT: Antineoplastic.

4928. Idazoxan. [79944-58-4] 2-(2,3-Dihydro-1,4-benzodioxin-2-yl)-4,5-dihydro-1*H*-imidazole; 2-[2-(1,4-benzodioxanyl)]-2-imidazoline. $C_{11}H_{12}N_2O_2$; mol wt 204.23. C 64.69%, H 5.92%, N 13.72%, O 15.67%. α_2-Adrenergic blocker. Prepn: J. Krapcho, W. A. Lott, **US 2979511** (1961 to Olin Mathieson); and confirmation of structure: C. B. Chapleo, P. L. Myers, *Tetrahedron Lett.* **22**, 4839 (1981). Structure-activity study: C. B. Chapleo *et al.*, *J. Med. Chem.* **26**, 823 (1983). Pharmacology: H. Dabiré, *J. Pharmacol.* **17**, 113 (1986). Clinical evaluation in progressive supranuclear palsy: D. G. Cole, J. H. Growdon, *J. Neural Transm. Suppl.* **42**, 283 (1994); as adjunct to fluphenazine in schizophrenia: R. E. Litman *et al.*, *Br. J. Psychiatry* **168**, 571 (1996).

Hydrochloride. [79944-56-2] RX-781094. $C_{11}H_{12}N_2O_2 \cdot HCl$; mol wt 240.69. White, crystalline solid from isopropanol, mp 207-208°.

4929. Idebenone. [58186-27-9] 2-(10-Hydroxydecyl)-5,6-dimethoxy-3-methyl-2,5-cyclohexadiene-1,4-dione; 6-(10-hydroxydecyl)-2,3-dimethoxy-5-methyl-1,4-benzoquinone; 2,3-dimethoxy-

5-methyl-6-(10'-hydroxydecyl)-1,4-benzoquinone; 6-(10-hydroxy-decyl)ubiquinone; CV-2619; Avan; Catena; Daruma; Lucebanol; Mnesis; Prevage; Sovrima. $C_{19}H_{30}O_5$; mol wt 338.44. C 67.43%, H 8.94%, O 23.64%. Antioxidant, short chain analog of ubiquinone, q.v.; exhibits protective effects against cerebral ischemia. Prepn: H. Morimoto et al., DE 2519730; eidem, US 4271083 (1975, 1981 both to Takeda); K. Okamoto et al., Chem. Pharm. Bull. 30, 2797 (1982); C.-A. Yu, L. Yu, Biochemistry 21, 4096 (1982). Effect on ischemia-induced amnesia in rats: N. Yamazaki et al., Jpn. J. Pharmacol. 36, 349 (1984). Review of chemistry, toxicology and pharmacology: I. Zs-Nagy, Arch. Gerontol. Geriatr. 11, 177-186 (1990); of pharmacology and clinical efficacy in age-related cognitive disorders: J. C. Gillis et al., Drugs Aging 5, 133-152 (1994). Clinical trial in Alzheimer's disease: H. Gutzmann et al., Pharmacopsychiatry 35, 12 (2002); in Friedreich's ataxia: N. A. Di Prospero et al., Lancet Neurol. 6, 878 (2007). Clinical effect on photodamaged skin: D. H. McDaniel et al., J. Cosmet. Dermatol. 4, 167 (2005).

Orange needles from ligroin, mp 46-50° (Morimoto); also reported as crystals from hexane + ethyl acetate, mp 52-53° (Okamoto). Sol in organic solvents. Practically insol in water.

USE: Topical antioxidant in cosmetic formulations.

THERAP CAT: Nootropic.

4930. Idose. [2152-76-3] $C_6H_{12}O_6$; mol wt 180.16. C 40.00%, H 6.71%, O 53.28%. Prepn of D-idose by reduction of D-idonolactone: Fischer, Fay, Ber. 28, 1975 (1895); from D-galactose: Sorking, Reichstein, Helv. Chim. Acta 28, 1 (1945); from tri-O-acetyl-1,6-anhydro-β-D-idopyranose: eidem, ibid. 662. Prepn of L-idose by reduction of L-idonolactone: Fischer, Fay, loc. cit.; by hydrolysis of 1,2-O-isopropylidene-L-idofuranose: Meyer, Reichstein, Helv. Chim. Acta 29, 152 (1946); von Vargha, Ber. 87, 1351 (1954). Improved synthesis: M. Blanc-Muesser, J. Defaye, Synthesis 1977, 568. Structure: S. F. Dyke, The Carbohydrates (Interscience, New York, 1960) p 45. Conformation: Reeves, J. Am. Chem. Soc. 72, 1499 (1950). Review: R. L. Whistler, M. L. Wolfrom, Methods in Carbohydrate Chemistry (Academic Press, New York, 1962) pp 140-145.

β-D-Idose

D-Form. Syrup. $[\alpha]_D^{13}$ +15.8° (c = 2.3).

Phenylosazone. $C_{18}H_{22}N_4O_4$. Yellow needles from alc, mp 168-169°.

L-Form. Syrup. $[\alpha]_D^{20}$ -17.4° (c = 3.6).

1,2-O-Isopropylidene-L-idofuranose. Plates from ethyl acetate, mp 113-114°. $[\alpha]_D^{26}$ -20° (c = 2.7 in methanol).

4931. Idoxuridine. [54-42-2] 2'-Deoxy-5-iodouridine; 1-(2-deoxy-β-D-ribofuranosyl)-5-iodouracil; 5-iodo-2'-deoxyuridine; IdU; IdUR; IUdR; IdUrd; Dendrid; Emanil; Herpes-Gel; Herplex; Idexur; Idoxene; Idulea; Iduridin; Kerecid; Ophthalmadine; Stoxil; Virudox. $C_9H_{11}IN_2O_5$; mol wt 354.10. C 30.53%, H 3.13%, I 35.84%, N 7.91%, O 22.59%. Cytotoxic nucleoside with antiviral and antineoplastic activity. Prepn: Prusoff, Biochim. Biophys. Acta 32, 295 (1959); Cheong et al., J. Biol. Chem. 235, 1441 (1960); Chang, Welch, J. Med. Chem. 6, 428 (1963); Amiard, Torelli, FR 1336866 (1963 to Roussel-UCLAF), C.A. 60, 3082g (1964); GB 1024156; Prystas, Sorm, Collect. Czech. Chem. Commun. 29, 121

(1964). Crystal and molecular structure: Camerman, Trotter, Acta Crystallogr. 18, 203 (1965). Review: W. H. Prusoff et al. in Antibiotics vol. 5 (pt. 2), F. E. Hahn, Ed. (Springer-Verlag, New York, 1979) pp 236-261.

Crystals from water, triclinic, dec 160° (Prusoff; Chang, Welch), 190-195° (Cheong et al.), 240° (Amiard, Torelli), over 175° (Prystas, Sorm). uv max (water): 288 nm (log ε 3.87). $[\alpha]_D^{25}$ +7.4° (c = 0.108 in water); $[\alpha]_D^{20}$ +29° (N soda). Physical properties: Ravin, Gulesich, J. Am. Pharm. Assoc. [NS] 4, 122 (1964). pKa 8.25. pH of 0.1% aq soln, about 6. Soly at 25° in mg/ml: 2.0 in water; 2.0 in 0.2N HCl; 74.0 in 0.2N NaOH; 4.4 in methanol; 2.6 in alc; 0.014 in ether; 0.003 in chloroform; 1.6 in acetone; 1.8 in ethyl acetate; 5.7 in dioxane. LD_{50} i.p. in mice: 2.5 g/kg (Prusoff, 1979).

α-Anomer. 1-(2-Deoxy-α-D-erythro-pentofuranosyl)-5-iodouracil; 1-(2-deoxy-α-D-ribofuranosyl)-5-iodouracil; α-2'-deoxy-5-iodouridine. Crystals from water, dec 170°. $[\alpha]_D^{25}$ +21.8° (c = 0.170). uv max (water): 288 nm (log ε 3.88).

THERAP CAT: Antiviral.

4932. Idraparinux. [162610-17-5] Methyl O-2,3,4-tri-O-methyl-6-O-sulfo-α-D-glucopyranosyl-(1 → 4)-O-2,3-di-O-methyl-β-D-glucopyranuronosyl-(1 → 4)-O-2,3,6-tri-O-sulfo-α-D-glucopyranosyl-(1 → 4)-O-2,3-di-O-methyl-α-L-idopyranuronosyl-(1 → 4)-α-D-glucopyranoside 2,3,6-tris(hydrogen sulfate). $C_{38}H_{64}O_{49}S_7$; mol wt 1529.30. C 29.84%, H 4.22%, O 51.26%, S 14.67%. Synthetic pentasaccharide that inhibits factor Xa; structurally analogous to the antithrombin binding site of heparin, q.v. Prepn: M. Petitou, C. A. van Boeckel, EP 0529715; eidem, US 5378829 (1993, 1995 both to Akzo; Sanofi); P. Westerduin et al., Bioorg. Med. Chem. 2, 1267 (1994); C. Chen, B. Yu, Bioorg. Med. Chem. Lett. 19, 3875 (2009). Biochemical and pharmacological properties: J. M. Herbert et al., Blood 91, 4197 (1998). Clinical pharmacology and reversal by factor VIIa: N. R. Bijsterveld et al., Br. J. Haematol. 124, 653 (2004). Clinical pharmacokinetics: C. Veyrat-Follet et al., J. Thromb. Haemost. 7, 559 (2009). Review of clinical development: P. Prandoni et al., Expert Opin. Invest. Drugs 17, 773-777 (2008).

$[\alpha]_D^{23}$ +54.2° (c = 1.0 in water).

Nonasodium salt. [149920-56-9] Idraparinux sodium; Org-34006; SANORG-34006; SR-34006. $C_{38}H_{55}Na_9O_{49}S_7$; mol wt 1727.14. $[\alpha]_D^{20}$ +55° (c = 1 in water).

THERAP CAT: Antithrombotic.

4933. Idrocilamide. [6961-46-2] N-(2-Hydroxyethyl)-3-phenyl-2-propenamide; N-(2-hydroxyethyl)cinnamamide; LCB-29; Brolitène; Srilane. $C_{11}H_{13}NO_2$; mol wt 191.23. C 69.09%, H 6.85%, N 7.32%, O 16.73%. Prepn: M. Bayssat et al., DE 2015447; eidem, US 3659014 (1970, 1972 to LIPHA); Chim. Ther. 8, 202

(1973). Pharmacology: Grand *et al., Eur. J. Med. Chem.* **9**, 205 (1974). Metabolism: Belleville *et al., Therapie* **29**, 829 (1974).

White crystals from ethyl acetate or acetone, mp 100-102°. Soluble in alcohol; slightly sol in water. LD_{50} orally in mice, rats: >2950, >3000 mg/kg (Grand).

THERAP CAT: Muscle relaxant (skeletal).

4934. Iduronate-2-sulfatase. [50936-59-9] L-Iduronosulfatase; L-iduronate-2-sulfate 2-sulfohydrolase; idunorate-2-sulfate sulfatase; sulfoiduronate sulfatase; Hunter corrective factor; EC 3.1.6.13. Lysosomal enzyme that hydrolyzes the 2-sulfate esters of terminal L-iduronate-2-sulfate residues of dermatan sulfate and heparan sulfate. Genetic deficiency of the enzyme results in the lysosomal storage disorder, mucopolysaccharidosis-II (MPS-II), also known as Hunter syndrome. The human enzyme is translated as a 550 amino acid prepro-enzyme containing a 25 amino acid signal sequence which is cleaved in the secreted proenzyme. Further post-translational modification is required for bioactivity. Purification from human urine and role in Hunter syndrome: M. Cantz *et al., J. Biol. Chem.* **247**, 5456 (1972). Characterization of bioactivity: G. Bach *et al., Proc. Natl. Acad. Sci. USA* **70**, 2134 (1973). Purification and characterization of 2 isoforms from human liver: J. Bielicki *et al., Biochem. J.* **271**, 75 (1990). Isoln and sequence of cDNA from human epithelial cells and putative amino acid sequence: P. J. Wilson *et al., Proc. Natl. Acad. Sci. USA* **87**, 8531 (1990). Prepn of recombinant form in Chinese hamster ovary cells: J. Bielicki *et al., Biochem. J.* **289**, 241 (1993). Review of Hunter syndrome pathology and treatment: M. Beck, *Curr. Pharm. Biotechnol.* **12**, 861-866 (2011).

Idursulfase. Elaprase. Highly glycosylated form of human iduronate-2-sulfatase produced by recombinant DNA technology in human cells. Glycoprotein comprised of 525 amino acids with 8 asparagine-linked glycosylation sites; mol wt ~76 kDa. Amino acid sequence is identical to the human proenzyme form. Post-translational modification of cysteine-59 to formylglycine is required for enzymatic activity. Prepn: P. J. Wilson *et al., US 5932211; idem, US 6153188* (1999, 2000 both to Women's and Children's Hospital). Characterization and evaluation in MPS-II patients: J. Muenzer *et al., Genet. Med.* **8**, 465 (2006). Review of pharmacology and therapeutic potential: L. A. Clarke, *Expert Opin. Pharmacother.* **9**, 311-317 (2008). Clinical trials in MPS-II: J. Muenzer *et al., Genet. Med.* **13**, 95 (2011); *idem et al., ibid.,* 102.

THERAP CAT: Enzyme replacement therapy for mucopolysaccharidosis-II.

4935. α-L-Iduronidase. [9073-56-7] α-L-Iduronide iduronohydrolase; IDUA; Hurler corrective factor; EC 3.2.1.76. Lysosomal enzyme that hydrolyzes nonreducing terminal α-L-iduronide glycosidic bonds in heparan sulfate and dermatan sulfate. Genetic deficiency of the enzyme results in the lysosomal storage disorder, mucopolysaccharidosis-I (MPS-I), also known as Hurler and Scheie syndromes. Identification in human fibroblasts and liver: R. Matalon *et al., Biochem. Biophys. Res. Commun.* **42**, 340 (1971); in human urine: R. W. Barton, E. F. Neufeld, *J. Biol. Chem.* **246**, 7773 (1971). Localization in lysosomes: B. Weissmann, R. Santiago, *Biochem. Biophys. Res. Commun.* **46**, 1430 (1972). Identification as enzymatic defect in Hurler's syndrome: R. Matalon, A. Dorfman, *ibid.* **47**, 959 (1972); G. Bach *et al., Proc. Natl. Acad. Sci. USA* **69**, 2048 (1972). Purification and characterization of multiple forms: P. R. Clements *et al., Biochem. J.* **259**, 199 (1989). Cloning and expression in Chinese hamster ovary cells: H. S. Scott *et al., Proc. Natl. Acad. Sci. USA* **88**, 9695 (1991).

Laronidase. [250378-38-2] Alronidase; Aldurazyme. Human α-L-iduronidase produced by recombinant DNA technology in Chinese hamster ovary cells. Prepn and pharmacology: E. D. Kakkis *et al., Protein Expression Purif.* **5**, 225 (1994). *See also:* E. D. Kakkis, B. Tanamachi, *WO 9958691* (1999 to Harbor-UCLA). Clinical evaluation in MPS-I: *idem et al., N. Engl. J. Med.* **344**, 182 (2001).

THERAP CAT: Enzyme replacement therapy for mucopolysaccharidosis-I.

4936. Ifenprodil. [23210-56-2] α-(4-Hydroxyphenyl)-β-methyl-4-(phenylmethyl)-1-piperidineethanol; 4-benzyl-α-(*p*-hydroxyphenyl)-β-methyl-1-piperidineethanol; 2-(4-benzylpiperidino)-1-(4-hydroxyphenyl)-1-propanol; 1-methyl-2-hydroxy-2-(4-hydroxyphenyl)ethyl-1-(4-benzylpiperidine); RC-61-91. $C_{21}H_{27}NO_2$; mol wt 325.45. C 77.50%, H 8.36%, N 4.30%, O 9.83%. Prepn and pharmacology: M. Carron *et al., FR M5733; US 3509164* (1968, 1970 to Robert et Carrière); C. Carron *et al., Arzneim.-Forsch.* **21**, 1992 (1971).

mp 114°.

Neutral tartrate. Cerocral; Dilvax; Vadilex. $(C_{21}H_{27}NO_2)_2 \cdot C_4H_6O_6$; mol wt 800.99. Crystals from methanol, mp 178-180°. Sol in alc, water. Very slightly sol in acetone, chloroform. Practically insol in ether. LD_{50} in male Swiss mice (mg/kg): 17 i.v.; 120 i.p.; 275 orally (Carron, 1970).

THERAP CAT: Vasodilator (cerebral and peripheral).

4937. Ifosfamide. [3778-73-2] *N*,3-Bis(2-chloroethyl)tetrahydro-2*H*-1,3,2-oxazaphosphorin-2-amine 2-oxide; 3-(2-chloroethyl)-2-[(2-chloroethyl)amino]tetrahydro-2*H*-1,3,2-oxazaphosphorin-2-oxide; iphosphamid(e); isoendoxan; isophosphamide; A-4942; Asta Z-4942; MJF-9325; NSC-109724; Z-4942; Holoxan; Ifex; Ifomide; Mitoxana. $C_7H_{15}Cl_2N_2O_2P$; mol wt 261.08. C 32.20%, H 5.79%, Cl 27.16%, N 10.73%, O 12.26%, P 11.86%. Cytostatic agent, related structurally to cyclophosphamide, *q.v.* Prepn: **FR 1530962** (1968 to Asta), *C.A.* **71**, 49998m (1969); H. Arnold *et al., US 3732340* (1973 to Asta). Chemical properties: H. Arnold, *Proc. 5th Int. Congr. Chemother. Vienna* (Verhandlungen, Vienna, 1967) **2**, pp 751-754. Pharmacology: N. Brock, *ibid.* pp 155-161. Molecular structure and conformation: H. A. Brassfield *et al., J. Am. Chem. Soc.* **97**, 4143 (1975). Mass spectrometry: M. Przybylski *et al., Biomed. Mass Spectrom.* **4**, 209 (1977). Mechanism of action: S. Tomita *et al., Chemotherapy (Tokyo)* **25**, 3014 (1977). Metabolism: A. Takamizawa *et al., Chem. Pharm. Bull.* **25**, 2900 (1977); *eidem, J. Med. Chem.* **17**, 1237 (1974). Toxicity studies: R. Marcy *et al., IRCS Med. Sci. Libr. Compend.* **5**, 427, 478 (1977). Mutagenicity studies: D. Wald, *J. Mutat. Res.* **56**, 319 (1978); G. R. Mohn, J. Ellenberger, *ibid.* **32**, 331 (1976). Clinical studies: P. J. Creaven *et al., Cancer Treat. Rep.* **60**, 445, 451 (1976); J. Schnitker *et al., Arzneim.-Forsch.* **26**, 1793 (1976). Symposium on clinical efficacy and comparison with cyclophosphamide: *Cancer Chemother. Pharmacol.* **18**, Suppl. 2, S1-S58 (1986). Reviews of pharmacology, toxic effects and clinical activity: M. Zalupski, L. H. Baker, *J. Natl. Cancer Inst.* **80**, 556-566 (1988); S. E. Schoenike, W. J. Dana, *Clin. Pharm.* **9**, 179-191 (1990).

Crystals from anhyd ether, mp 39-41°. Very sol in alc, ethyl acetate, isopropyl alc, methanol, methylene chloride; freely sol in water; very slightly sol in hexanes. LD_{50} in rats (mg/kg): 160 i.p. (Arnold, 1973); also reported as 150 i.p. (Brock).

THERAP CAT: Antineoplastic.

4938. Ignatia. Ignatius bean; St. Ignatius' bean. Dried, ripe seed of *Strychnos ignatii* Berg., *Loganiaceae. Habit.* Philippine Islands, naturalized in Vietnam, Cambodia. *Constit.* 2-3% strychnine, about 1% brucine, igasuric acid, loganin.

THERAP CAT: Bitter tonic.

4939. Ilimaquinone. [71678-03-0] 3-[[(1*R*,2*S*,4a*S*,8a*S*)-Decahydro-1,2,4a-trimethyl-5-methylene-1-naphthalenyl]methyl]-2-hydroxy-5-methoxy-2,5-cyclohexadiene-1,4-dione; IQ. $C_{22}H_{30}O_4$; mol wt 358.48. C 73.71%, H 8.44%, O 17.85%. Sesquiterpenoid quinone isolated from the marine sponge, *Hippiospongia metachromia*; naturally occurring as (−)-form. Inhibits cellular secretions by selective breakdown of the Golgi apparatus. Isoln and structure determn: R. T. Luibrand *et al.*, *Tetrahedron* **35**, 609 (1979). Revised stereochemistry: R. J. Capon, J. K. MacLeod, *J. Org. Chem.* **52**, 5059 (1987). Total stereosynthesis: S. D. Bruner *et al.*, *ibid.* **60**, 1114 (1995); increased yield: S. Poigny *et al.*, *ibid.* **63**, 5890 (1998). Photoaffinity study on cellular targets: H. S. Radeke, M. N. L. Snapper, *Bioorg. Med. Chem.* **6**, 1227 (1998). Mechanism of action studies: P. A. Takizawa *et al.*, *Cell* **73**, 1079 (1993); H. S. Radeke *et al.*, *Chem. Biol.* **6**, 639 (1999). Use as inhibitor of protein secretion: P. A. Feldman *et al.*, *J. Membr. Biol.* **155**, 275 (1997).

Orange needles from hexane, mp 113-114°. $[\alpha]_D^{23}$ −23.2° (c = 1.12 in CHCl$_3$).

USE: Biological probe for intracellular communications and vesicle-mediated transport.

4940. Illudins. Anti-tumor antibiotic substances produced by the poisonous basidiomycetes, *Clitocybe illudens* (now called *Omphalotus illudens*): Anchel *et al.*, *Proc. Natl. Acad. Sci. USA* **36**, 300 (1950); **38**, 927 (1952); and *Lampteromyces japonicus*: Nakanishi *et al.*, *Nature* **197**, 292 (1963); Endo *et al.*, *Chem. Commun.* **1970**, 309. Structure: McMorris, Anchel, *J. Am. Chem. Soc.* **87**, 1594 (1965); Matsumoto *et al.*, *Tetrahedron* **21**, 2671 (1965); Tada *et al.*, *Chem. Pharm. Bull.* **12**, 853 (1964). Stereochemistry: Nakanishi *et al.*, *ibid.* 856; *Tetrahedron* **21**, 1231 (1965); Matsumoto *et al.*, *Bull. Chem. Soc. Jpn.* **37**, 1716 (1964). Abs config of illudin S: Harada, Nakanishi, *Chem. Commun.* **1970**, 310. Total synthesis of illudin M: Matsumoto *et al.*, *J. Am. Chem. Soc.* **90**, 3280 (1968); *Tetrahedron Lett.* **1970**, 1171; of illudin S: *eidem*, *ibid.* **1971**, 2049. *In vitro* antitumor activity in human cancer cells: M. J. Kelner *et al.*, *J. Natl. Cancer Inst.* **82**, 1562 (1990).

Illudin M   R = H
Illudin S   R = OH

Illudin M. [1146-04-9] (3′*S-trans*)-2′,3′-Dihydro-3,6-dihydroxy-2′,2′,4′,6′-tetramethylspiro[cyclopropane-1,5′-[5*H*]inden]-7′(6′*H*)-one. $C_{15}H_{20}O_3$; mol wt 248.32. Rods, mp 120-122.5° (Matsumoto); from ethanol-water, mp 128-130° (McMorris). uv max (ethanol): 228, 318 nm (ε 13900, 3600).
Monoacetate. $C_{17}H_{22}O_4$. mp 75-76° from petr ether.
Illudin S. [1149-99-1] [2′*S*-(2′α,3′β,6′α)]-2′,3′-Dihydro-3′,6′-dihydroxy-2′-(hydroxymethyl)-2′,4′,6′-trimethylspiro[cyclopropane-1,5′-[5*H*]inden]-7′(6′*H*)-one; lampterol; lunamycin (obsolete). $C_{15}H_{20}O_4$; mol wt 264.32. Needles from acetone, mp 124-126°. uv max (ethanol): 233, 319 nm (ε 13200, 3600). Sol in polar organic solvents.

Diacetate. $C_{19}H_{24}O_5$. Crystals from petr ether, mp 99-100°. uv max (ethanol): 227, 313 nm (ε 12900, 3400).

4941. Iloperidone. [133454-47-4] 1-[4-[3-[4-(6-Fluoro-1,2-benzisoxazol-3-yl)-1-piperidinyl]propoxy]-3-methoxyphenyl]ethanone; HP-873; ILO-522; Fanapt. $C_{24}H_{27}FN_2O_4$; mol wt 426.49. C 67.59%, H 6.38%, F 4.45%, N 6.57%, O 15.01%. Combined dopamine (D$_2$) and serotonin (5HT$_2$) receptor antagonist. Prepn: J. T. Strupczewski *et al.*, **EP 402644**; *eidem*, **US 5364866** (1990, 1994 both to Hoechst-Roussel); *eidem*, *J. Med. Chem.* **38**, 1119 (1995). Pharmacology: M. R. Szewczak *et al.*, *J. Pharmacol. Exp. Ther.* **274**, 1404 (1995). Clinical pharmacokinetics: S. M. Sainati *et al.*, *J. Clin. Pharmacol.* **35**, 713 (1995). HPLC determn in plasma: A. E. Mutlib, J. T. Strupczewski, *J. Chromatogr. B* **669**, 237 (1995). Receptor binding study: S. Kongsamut *et al.*, *Eur. J. Pharmacol.* **317**, 417 (1996). Review of clinical experience in schizophrenia: L. Citrome, *Int. J. Clin. Pract.* **63**, 1237-1248 (2009).

Crystals from ethanol, mp 118-120°. Very slightly sol in 0.1 N HCl; freely insol in chloroform, ethanol, methanol, acetonitrile. Practically insol in water.

THERAP CAT: Antipsychotic.

4942. Iloprost. [78919-13-8] (5*E*)-5-[(3a*S*,4*R*,5*R*,6a*S*)-Hexahydro-5-hydroxy-4-[(1*E*,3*S*)-3-hydroxy-4-methyl-1-octen-6-yn-1-yl]-2(1*H*)-pentalenylidene]pentanoic acid; 5-(*E*)-(1*S*,5*S*,6*R*,7*R*)-7-hydroxy-6-[(*S*-(3*S*,4*RS*)-3-hydroxy-4-methyl-1-octenyl]bicyclo-[3.3.0]octan-3-ylidenepentanoic acid; (*E*)-(3a*S*,4*R*,5*R*,6a*S*)-hexahydro-5-hydroxy-4-[(*E*)-(3*S*,4*RS*)-3-hydroxy-4-methyl-1-octen-6-ynyl]-Δ$^{2(1H),δ}$-pentalenevaleric acid; ciloprost; ZK-36374. $C_{22}H_{32}O_4$; mol wt 360.49. C 73.30%, H 8.95%, O 17.75%. Prostacyclin analog; 1:1 mixture of 16α- and 16β-methyl diastereomers. Prepn: W. Skuballa *et al.*, **DE 2845770**; *eidem*, **US 4692464** (1980, 1987 both to Schering AG); W. Skuballa, H. Vorbrüggen, *Adv. Prostaglandin Thromboxane Leukotriene Res.* **11**, 299 (1983). Antithrombotic and cardiovascular pharmacology: K. Schrör *et al.*, *Arch. Pharmacol.* **316**, 252 (1981). Review of clinical pharmacology and pharmacokinetics: S. M. Grant, K. L. Goa, *Drugs* **43**, 889-924 (1992). Clinical study in limb ischemia: J. Dormandy, *Therapie* **46**, 319 (1991); in Raynaud phenomenon: F. M. Wigley *et al.*, *Ann. Intern. Med.* **120**, 199 (1994); in pulmonary hypertension: H. Olschewski *et al.*, *N. Engl. J. Med.* **347**, 322 (2002).

Colorless oil.
Tromethamine. [697225-02-8] Endoprost; Ilomedin; Ventavis. $C_{22}H_{32}O_4 \cdot C_4H_{11}NO_3$; mol wt 481.63.
THERAP CAT: Vasodilator (peripheral). In treatment of pulmonary hypertension.

4943. Imatinib. [152459-95-5] 4-[(4-Methyl-1-piperazinyl)-methyl]-*N*-[4-methyl-3-[[4-(3-pyridinyl)-2-pyrimidinyl]amino]-phenyl]benzamide; *N*-[5-[4-(4-methylpiperazinomethyl)benzoylamido]-2-methylphenyl]-4-(3-pyridyl)-2-pyrimidineamine. $C_{29}H_{31}N_7O$; mol wt 493.62. C 70.56%, H 6.33%, N 19.86%, O 3.24%. Tyrosine kinase inhibitor; highly specific for BCR-ABL, the enzyme associated with chronic myelogenous leukemia (CML) and certain forms of acute lymphoblastic leukemia (ALL). Also shown to inhibit the transmembrane receptor KIT and platelet-derived growth factor (PDGF) receptors. Prepn: J. Zimmermann, **EP 564409**; *idem*, **US 5521184** (1993, 1996 both to Ciba-Geigy); *idem et al., Bioorg. Med. Chem. Lett.* **7**, 187 (1997). Structural mechanism of ABL specificity: T. Schindler *et al., Science* **289**, 1938 (2000). Activity vs KIT and PDGF receptor kinases: E. Buchdunger *et al., J. Pharmacol. Exp. Ther.* **295**, 139 (2000). Clinical trial in CML: H. Kantarjian *et al., N. Engl. J. Med.* **346**, 645 (2002); in gastrointestinal stromal tumors related to KIT: G. D. Demetri *et al., ibid.* **347**, 472 (2002). Review of clinical experience: D. G. Savage, K. H. Antman, *ibid.* **346**, 683-693 (2002); and pharmacology: V. K. Pindolia *et al., Pharmacotherapy* **22**, 1249-1265 (2002); and development of therapeutic target: B. J. Druker, *Adv. Cancer Res.* **91**, 1-30 (2004).

mp 211-213°. pKa$_1$ 8.07; pKa$_2$ 3.73; pKa$_3$ 2.56; pKa$_4$ 1.52.
Methanesulfonate. [220127-57-1] STI-571; CGP-57148B; Gleevec; Glivec. C$_{29}$H$_{31}$N$_7$O.CH$_3$SO$_3$H; mol wt 589.72. Prepn of crystalline form: J. Zimmermann *et al.*, **WO 9903854** (1999 to Novartis). Occurs in 2 crystalline modifications. α-form, begins to melt at 226°; β-form, mp 217°. Lipophilic at pH 7.4. Soly in water: >100 g/l (pH 4.2); 49 mg/l (pH 7.4).

THERAP CAT: Antineoplastic.

4944. Imazamethabenz. [81405-85-8] 2-[4,5-Dihydro-4-methyl-4-(1-methylethyl)-5-oxo-1*H*-imidazol-2-yl]-4(or 5)-methylbenzoic acid methyl ester; imazamethabenz methyl; imazethabenz; AC-222293; AC-293; CL-222293; Assert; Dagger. C$_{16}$H$_{20}$N$_2$O$_3$; mol wt 288.35. C 66.65%, H 6.99%, N 9.72%, O 16.65%. Selective, post-emergence imidazolinone herbicide; mixture of *methyl 2-(4-isopropyl-4-methyl-5-oxo-2-imidazolin-2-yl)-p-toluate* and *methyl 6-(4-isopropyl-4-methyl-5-oxo-2-imidazolin-2-yl)-m-toluate* (approx 3:2). Prepn of isomeric mixture: M. Los, **US 4188487** (1980 to American Cyanamid). Activity, physical properties and toxicity: D. L. Shaner *et al., Proc. Br. Crop Prot. Conf. - Weeds* **1982**, 25; K. Hedlund, L. Andersson, *Weeds Weed Control* **28**, 1 (1987). Activity of component isomers: D. L. Shaner *et al., Proc. Br. Crop Prot. Conf. - Weeds* **1982**, 333. Mechanism of action: J. B. Pillmoor, J. C. Caseley, *Pestic. Biochem. Physiol.* **27**, 340 (1987). Persistence and mobility in soil: R. Allen, J. C. Caseley, *Proc. Br. Crop Prot. Conf. - Weeds* **1987**, 569. Field studies: K. Kirkland, N. E. Shafer, *ibid.* **1982**, 33; A. A. Hudson, S. C. E. Townsend, *ibid.* **1985**, 923; in food crops: S. D. Miller, H. P. Alley, *Weed Technol.* **1**, 29 (1987).

m - form

p - form

Off-white fine powder with a tendency to form easily friable aggregates; slight musty odor. Softening begins at 108-117°, melting starts at 113-122° and is completed at 144-153°. Soly (g/100 ml) at 25°: acetone 18.2, DMSO 23.8, distilled water 0.13 (*p*-isomer), 0.22 (*m*-isomer), *n*-heptane 0.04, isopropyl alcohol 14.4, methanol 24.4, methylene chloride 30.0, toluene 3.9. Soly (g/100 g) at 25°: xylene <5, DMF 30. Partition coefficient (*n*-octanol/water): 35 (*p*-isomer), 66 (*m*-isomer). LD$_{50}$ in rats, rabbits (mg/kg): >5000, 4500 orally, >2000, >2000 dermally (Hedlund, Anderson).

USE: Herbicide.

4945. Imazamox. [114311-32-9] 2-[4,5-Dihydro-4-methyl-4-(1-methylethyl)-5-oxo-1*H*-imidazol-2-yl]-5-(methoxymethyl)-3-pyridinedicarboxylic acid; 2-(4-isopropyl)-4-methyl-5-oxo-2-imidazolin-2-yl-5-(methoxymethyl)nicotinic acid; AC-299263; CL-299263; Raptor. C$_{15}$H$_{19}$N$_3$O$_4$; mol wt 305.33. C 59.01%, H 6.27%, N 13.76%, O 20.96%. Acetohydroxyacid synthase (AHAS) inhibitor. Prepn: R. F. Doehner, Jr., **EP 254951**; *idem et al.,* **US 5334576** (1988, 1994 both to Am. Cyanamid). Chemical and physical properties: G. R. Glover, *Proc. South. Weed Sci. Soc.* **48**, 269 (1995). Metabolism study: B. Tecle *et al., Brighton Crop Prot. Conf. - Weeds* **1997**, 605. CE determn in beans: K. Ohba *et al., J. Pestic. Sci.* **22**, 277 (1997). Field trials: K. A. Nelson, K. A. Renner, *Weed Technol.* **12**, 137 (1998); R. E. Blackshaw, *ibid.* **64**. Soil persistence: T. Cobucci *et al., Weed Sci.* **46**, 258 (1998). *Review:* T. M. Brady *et al., ACS Symp. Ser.* **686**, 30-37 (1998).

mp 166.0-166.7°. Soly (g/100 ml): hexane 0.0006; methanol 6.68; acetonitrile 1.85; toluene 0.21; acetone 2.93; dichloromethane 14.3; ethyl acetate 1.02. Soly in water: 4160 ppm. Volatility: <1.0 × 10^{-7} torr. Partition coefficient (octanol/water): 0.004 (pH 7). pK$_1$ 2.3; pK$_2$ 3.3. LD$_{50}$ (technical grade) orally in rats: >5000 mg/kg; dermally in rabbits: >4000 mg/kg; LC$_{50}$ by inhalation in rats: >6.3 mg/l (Glover).

USE: Herbicide.

4946. Imazapyr. [81334-34-1] 2-[4,5-Dihydro-4-methyl-4-(1-methylethyl)-5-oxo-1*H*-imidazol-2-yl]-3-pyridinecarboxylic acid; 2-(4-isopropyl-4-methyl-5-oxo-2-imidazolin-2-yl)nicotinic acid. C$_{13}$H$_{15}$N$_3$O$_3$; mol wt 261.28. C 59.76%, H 5.79%, N 16.08%, O 18.37%. Inhibitor of acetolactate synthase, a key enzyme in branched chain amino acid biosynthesis. Prepn: M. Los, **EP 41623**; *idem*, **US 4798619** (1981, 1989 both to Am. Cyanamid). Description: R. Paxman *et al., Proc. 38th N. Z. Weed Pest Control Conf.,* 73 (1985). Mode of action: D. Shaner *et al., Proc. Br. Crop Prot. Conf. - Weeds* **1985**, 147. Determn in soil: O. Nováková, *Chromatographia* **39**, 62 (1994). Persistence in soil: S. Vizantinopoulos, P. Lolos, *Bull. Environ. Contam. Toxicol.* **52**, 404 (1994). Field trials in sod: K. A. Griffin *et al., Crop Sci.* **34**, 202 (1994).

Crystals from acetone + hexane, mp 170-172.5°. pK$_1$ 1.9; pK$_2$ 3.6. Soly in water (pH 7): ~1.0 g/l. LD$_{50}$ orally in rats: >5000 mg/kg; dermally in rabbits: >2000 mg/kg (Paxman).

Isopropylamine salt. [81510-83-0] AC-252925; CL-252925; Arsenal; Chopper. C$_{13}$H$_{15}$N$_3$O$_3$.C$_3$H$_9$N; mol wt 320.39. mp 160-180° (dec).

USE: Herbicide.

4947. Imazaquin. [81335-37-7] 2-[4,5-Dihydro-4-methyl-4-(1-methylethyl)-5-oxo-1*H*-imidazol-2-yl]-3-quinolinecarboxylic acid; 2-(5-isopropyl-5-methyl-4-oxo-2-imidazolin-2-yl)-3-quinolinecarboxylic acid; AC-252214. C$_{17}$H$_{17}$N$_3$O$_3$; mol wt 311.34. C 65.58%, H 5.50%, N 13.50%, O 15.42%. Pre- and post-emergence imidazolinone herbicide especially for use in soybean crops. Prepn: M. Los, **EP 41623** (1981 to Am. Cyanamid), *C.A.* **96**, 199687q (1982). Alternate prepn: D. R. Maulding, R. F. Doehner, Jr., **US 4459408** (1984 to Am. Cyanamid). Inhibition of branched-chain amino acid biosynthesis: D. L. Shaner *et al., Plant Physiol.* **76**, 545 (1984). Metabolism by plants: D. L. Shaner, P. A. Robson, *Weed Sci.* **33**, 469 (1985). Persistence in soil: G. Basham *et al., ibid.* **35**, 576 (1987). Herbicidal activity: M. W. Beale *et al., Proc. Annu. Meet. Northeast. Weed Sci. Soc.* **38**, 36 (1984); W. F. Congleton *et al., Weed Technol.* **1**, 186 (1987).

Crystals from hexane + ethyl acetate, mp 219-222° (dec). Slightly sol in some organic solvents. Soly in water at 25°: 60-120 ppm. LD$_{50}$ orally in rats: 5000 mg/kg; LC$_{50}$ (96 hr) in rainbow trout: 100 mg/l (Congleton).

Ammonium salt. Scepter. C$_{17}$H$_{20}$N$_4$O$_3$; mol wt 328.37. Sol in water.

USE: Herbicide.

4948. Imazethapyr. [81335-77-5] 2-[4,5-Dihydro-4-methyl-4-(1-methylethyl)-5-oxo-1H-imidazol-2-yl]-5-ethyl-3-pyridinecarboxylic acid; (±)-5-ethyl-2-(4-isopropyl-4-methyl-5-oxo-1H-imidazolin-2-yl)nicotinic acid; AC-263499; CL-263499. C$_{15}$H$_{19}$N$_3$O$_3$; mol wt 289.34. C 62.27%, H 6.62%, N 14.52%, O 16.59%. Selective imidazolinone herbicide. Inhibitor of acetolactate synthase, a key enzyme in the biosynthesis of branched chain amino acids. Prepn: M. Los, **EP 41623**; idem, **US 4798619** (1981, 1989 both to Am. Cyanamid). Comprehensive description: T. R. Peoples et al., Proc. Br. Crop Prot. Conf. - Weeds **1985**, 99-106. Determn in soil: O. Nováková, Chromatographia **39** 62 (1994). Metabolism in corn: N. M. Mallipudi et al., J. Agric. Food Chem. **42**, 1213 (1994).

White to off-white crystalline solid, mp 172-175°. pK$_1$ 2.1; pK$_2$ 3.9. Soly in water (25°): 1415 ppm. LD$_{50}$ orally in rats and mice: >5000 mg/kg; dermally in rabbits: >2000 mg/kg (Peoples).

Ammonium salt. [101917-66-2] Pivot; Pursuit. C$_{15}$H$_{22}$N$_4$O$_3$; mol wt 306.37.

USE: Herbicide.

4949. Imazosulfuron. [122548-33-8] 2-Chloro-N-[[(4,6-dimethoxy-2-pyrimidinyl)amino]carbonyl]imidazo[1,2-a]pyridine-3-sulfonamide; 1-(2-chloroimidazo[1,2-a]pyridin-3-ylsulfonyl)-3-(4,6-dimethoxypyrimidin-2-yl)urea; TH-913; Brazzos; Takeoff. C$_{14}$H$_{13}$ClN$_6$O$_5$S; mol wt 412.81. C 40.73%, H 3.17%, Cl 8.59%, N 20.36%, O 19.38%, S 7.77%. Acetolactate-synthase (ALS) inhibitor; sulfonylurea herbicide for weed control in rice crops. Prepn: Y. Ishida et al., **EP 238070**; eidem, **US 5017212** (1987, 1991 both to Takeda); eidem, J. Pestic. Sci. **18**, 175 (1993). Mechanism of action study: Y. Tanaka, H. Yoshikawa, J. Weed Sci. Technol. **43**, 291 (1998). Soil degradation studies: K. Mikata et al., J. Pestic. Sci. **26**, 376 (2001). LC determn in water and soil: M. Ventriglia et al., J. Chromatogr. A **857**, 327 (1999); in rice: Y. Akiyama et al., J. Food Hyg. Soc. Jpn. **43**, 99 (2002). Review of development: H. Nagase, Agrochem. Jpn. **64**, 15-16 (1994); and biological activity: Y. Ishida et al., J. Pestic. Sci. **21**, 247-258 (1996).

Crystalline solid, mp 183-184° (dec). d^{25} 1.574. pK$_a$ 4.0. Log P (octanol/water): 0.05. Vapor pressure (25°): 3.4 × 10^{-10} mmHg. Soly at 25° (mg/l): water 308 (pH 7.0), 67 (pH 6.1), 5 (pH 5.1), acetonitrile 2500, ethyl acetate 2200, acetone 7600, dichloromethane 12800, xylene 400.

USE: Herbicide.

4950. Imibenconazole. [86598-92-7] N-(2,4-Dichlorophenyl)-1H-1,2,4-triazole-1-ethanimidothioic acid (4-chlorophenyl)methyl ester; 1,2,4-triazol-1-yl-isothioacetic acid 2',4'-dichloroanilide S-p-chlorobenzyl ether; 4-chlorobenzyl-N-(2,4-dichlorophenyl)-2-(1H-1,2,4-triazol-1-yl)thioacetimidate; HF-6305; HF-8505; Manage. C$_{17}$H$_{13}$Cl$_3$N$_4$S; mol wt 411.73. C 49.59%, H 3.18%, Cl 25.83%, N 13.61%, S 7.79%. Broad spectrum triazole fungicide. Prepn: H. Ohyama et al., **FR 2514766**; eidem, **US 4512989** (1983, 1985 both to Hokko Chem. Ind.). HPLC determn of metabolite in fruits and vegetables: A. Kaihara et al., J. Health Sci. **46**, 336 (2000). Review of physical properties, biological activity and field trials: H. Ohyama et al., Brighton Crop Prot. Conf. - Pests Dis. **1988**, 519-526; and mode of action: Y. Ogawa, Agrochem. Jpn. **1995**, no. 67, 20-21.

White crystalline solid, mp 89.5-90°. Vapor pressure (25°): 85 nPa. Soly in water (25°): 1.7 mg/l. Soly at 25° (g/l): acetone 1063; benzene 580; xylene 250; methanol 120. LD$_{50}$ in male, female rats, male, female mice (mg/kg): 2800, 3000, >5000, >5000 orally; in male, female rats (mg/kg): >2000, >2000 dermally; in honey bees (μg/bee): >125 orally. LC$_{50}$ (48 hr) in carp: 1.02 ppm. LC$_{50}$ (6 hr) in water fleas: >102 ppm (Ogawa).

USE: Fungicide.

4951. Imidacloprid. [138261-41-3] (2E)-1-[(6-Chloro-3-pyridinyl)methyl]-N-nitro-2-imidazolidinimine; 1-[(6-chloro-3-pyridinyl)methyl]-4,5-dihydro-N-nitro-1H-imidazol-2-amine; 1-[(6-chloro-3-pyridinyl)methyl]-N-nitro-2-imidazolidinimine; 1-(6-chloro-3-pyridylmethyl)-N-nitroimidazolidin-2-ylideneamine; 1-(2-chloro-5-pyridylmethyl)-2-(nitroimino)imidazolidine; BAY NTN 33893; Admire; Advantage; Confidor; Gaucho; Marathon; Merit; Nuprid; Premier; Provado; Trimax. C$_9$H$_{10}$ClN$_5$O$_2$; mol wt 255.66. C 42.28%, H 3.94%, Cl 13.87%, N 27.39%, O 12.52%. Chloronicotinyl insecticide; targets the nicotinic acetylcholine receptor. Prepn: K. Shiokawa et al., **EP 192060**; eidem, **US 4742060** (1986, 1988 both to Nihon Tokushu Noyaku Seizo KK). Properties and bioactivity: A. Elbert et al., Brighton Crop Prot. Conf. - Pests Dis. **1990**, 21. Field trials: H. E. Schmeer et al., ibid. 29. Field study in combination with moxidectin vs. feline heartworm: L. Venco et al., Vet. Parasitol. **154**, 67 (2008). LC-MS determn in fruits and vegetables: D. Zywitz et al., Dtsch. Lebensm. Rundsch. **99**, 188 (2003). Degradation study: V. Kitsiou et al., Appl. Catal. B **86**, 27 (2009). Review of mode of action and agricultural uses: W. Leicht, Pestic. Outlook **4**, 17-21 (1993); of chemistry, toxicology, and veterinary use vs flea infestation: N. Mencke, P. Jeschke, Curr. Top. Med. Chem. **2**, 701-715 (2002).

Crystals, mp 143.8°; second crystalline modification, mp 136.4°. Vapor pressure (20°): 2.0 × 10^{-9} mbar. Soly in water (20°): 0.51 g/l. Log P (octanol/water): 0.57. LD$_{50}$ in rats (mg/kg): ~450 orally; >5000 dermally (Elbert).

USE: Insecticide.

THERAP CAT (VET): Ectoparasiticide.

4952. Imidapril. [89371-37-9] (4S)-3-[(2S)-2-[[(1S)-1-(Ethoxycarbonyl)-3-phenylpropyl]amino]-1-oxopropyl]-1-methyl-2-oxo-4-imidazolidinecarboxylic acid; (4S)-3-[(2S)-2-[N-[(1S)-1-(ethoxycarbonyl)-3-phenylpropyl]amino]propionyl]-1-methyl-2-oxoimidazolidine-4-carboxylic acid. $C_{20}H_{27}N_3O_6$; mol wt 405.45. C 59.25%, H 6.71%, N 10.36%, O 23.68%. Angiotensin converting enzyme (ACE) inhibitor. Prepn: N. Yoneda et al., **EP 95163**; eidem, **US 4508727** (1983, 1985 both to Tanabe Seiyaku); K. Hayashi et al., J. Med. Chem. **32**, 289 (1989). Mechanism of action at MMP-9 active site: D. Yamamoto et al., J. Mol. Cell. Cardiol. **43**, 670 (2007). Series of articles on pharmacology, pharmacokinetics and metabolism in animals: Arzneim.-Forsch. **42**, 446-512 (1992). HPLC determn in plasma and urine: K. Tagawa et al., J. Chromatogr. **617**, 95 (1993). Clinical pharmacology: P. Démolis et al., Fundam. Clin. Pharmacol. **8**, 80 (1994). Clinical trial in hypertension: M. J. Vandenburg et al., Br. J. Clin. Pharmacol. **37**, 265 (1994).

Colorless crystals from ethyl acetate + n-hexane, mp 139-140°. $[\alpha]_D^{20}$ −71.7° (c = 0.5 in ethanol).

Monohydrochloride. [89396-94-1] TA-6366; Tanatril. $C_{20}H_{27}N_3O_6 \cdot HCl$; mol wt 441.91. Colorless crystals, mp 214-216° (dec). $[\alpha]_D^{20}$ −64.1° (c = 0.5 in ethanol).

Diacid. [89371-44-8] Imidaprilat. $C_{18}H_{23}N_3O_6$. Crystals, mp 239-241° (dec). $[\alpha]_D^{19}$ −88.4° (c = 1 in 5% $NaHCO_3$).

THERAP CAT: Antihypertensive.

4953. Imidazole. [288-32-4] 1H-Imidazole; glyoxaline; 1,3-diazole; iminazole; miazole; pyrro[b]monazole; 1,3-diaza-2,4-cyclopentadiene. $C_3H_4N_2$; mol wt 68.08. C 52.93%, H 5.92%, N 41.15%. Prepd by the action of ammonia on glyoxal: Debus, Ann. **107**, 204 (1858); from glyoxal, ammonia, and formaldehyde: Radziszewski, Ber. **15**, 1493 (1882); Behrend, Schmitz, Ann. **277**, 338 (1893); vapor phase synthesis from formamide and ethylenediamine in presence of a dehydrogenation catalyst: Green, **US 3255200** (1966 to Air Products and Chemicals). Crystal structure: B. M. Craven et al., Acta Crystallogr. B **33**, 2585 (1977). Acute toxicity: Nishie et al., Toxicol. Appl. Pharmacol. **14**, 301 (1969). Review: Pyman, J. Soc. Dyers Colour. **36**, 107 (1920). Monograph: K. Hofmann, Imidazole and Its Derivatives (Interscience, New York, 1953). Review of imidazole chemistry: Grimmett, Adv. Heterocycl. Chem. **12**, 103-183 (1970).

Stout prisms from benzene. mp 90-91°. bp_{760} 257°; bp_{20} 165-168°; bp_{12} 138.2°. Weak base. pK (25°): 6.92. Absorption spectrum: Rosanov, J. Russ. Phys. Chem. Soc. **48**, 1241 (1916); Chem. Zentralbl. **1923**, III, 1080. Freely sol in water, alcohol, ether, chloroform, pyridine; slightly sol in benzene; very sparingly sol in petr ether. LD_{50} in mice (mg/kg): 610 i.p.; 1880 orally (Nishie).

USE: Karl Fischer reagent in analytical chemistry. Reagent in synthetic organic chemistry.

4954. Imidazole Salicylate. [36364-49-5] 2-Hydroxybenzoic acid compd with 1H-imidazole (1:1); mono(2-hydroxybenzoate)-1H-imidazole; salizolo; ITF-182; Flogozen; Selezen. $C_{10}H_{10}N_2O_3$; mol wt 206.20. C 58.25%, H 4.89%, N 13.59%, O 23.28%. Prepn from equimolar amounts of salicylic acid with imidazole: M. Brissemoret, Bull. Soc. Chim. Fr. [3] **35**, 316 (1906). Use as anti-inflammatory agent: **BE 889704**; G. Sportoletti, **US 4329340** (1981, 1982

both to Italfarmaco). Pharmacology: P. G. Pagella et al., Arzneim.-Forsch. **33**, 716 (1983). Penetration of inflamed sites: eidem, ibid. **34**, 208 (1984). Pharmacokinetics: H. P. Kuemmerle et al., Int. J. Clin. Pharmacol. Ther. Toxicol. **22**, 521 (1984). Series of clinical studies: Boll. Chim. Farm. **122**, 37S-63S (1983). Review: R. Fantozzi, Drugs Exp. Clin. Res. **10**, 853-856 (1984).

Crystals from methanol-ether, mp 123-124°. uv max: 300 nm ($E_{1cm}^{1\%}$ 182.5). Soly in water >100 mg/cc. LD_{50} in male, female rats, mice (mg/kg): 763, 724, 595, 685 s.c.; 422, 434, 462, 435 i.v.; 1211, 1430, 1034, 1091 orally (Pagella).

THERAP CAT: Anti-inflammatory; antipyretic; analgesic.

4955. 2-Imidazolidinone. [120-93-4] 2-Imidazolidone; ethylene urea. $C_3H_6N_2O$; mol wt 86.09. C 41.86%, H 7.03%, N 32.54%, O 18.58%. Prepd from ethylenediamine and carbon dioxide under the influence of heat and pressure: Mulvaney, Evans, Ind. Eng. Chem. **40**, 393 (1948); from ethylenediamine and urea: Schweitzer, J. Org. Chem. **15**, 471 (1950). Systematic survey and bibliography: Klaus Hofmann, Imidazole, Part I (Interscience, New York, 1953).

Needles from chloroform, mp 131°. Very sol in water and in hot alc; difficultly sol in ether.

USE: Manuf high polymers, finishing agents for textiles and leather. In the formulation of plasticizers, lacquers, and adhesives. Insecticide: Simkover, **US 3242044** (1966 to Shell).

4956. Imidocarb. [27885-92-3] N,N'-Bis[3-(4,5-dihydro-1H-imidazol-2-yl)phenyl]urea; 3,3'-di-2-imidazolin-2-ylcarbanilide. $C_{19}H_{20}N_6O$; mol wt 348.41. C 65.50%, H 5.79%, N 24.12%, O 4.59%. Prepn: **GB 1007334**; R. Fischer, R. Hirt, **US 3338917** (1965, 1967 both to Wander). Babesicidal effect in mice and rats: G. Schmidt et al., Res. Vet. Sci. **10**, 530 (1969); E. Beveridge, ibid. 534. Effect on exptl anaplasmosis in calves: T. O. Roby, ibid. **13**, 519 (1972). Comparison of the dipropionate and tetracycline, q.v., in canine ehrlichiosis: J. E. Price, T. T. Dolan, Vet. Rec. **107**, 275 (1980). Efficacy in Babesia felis infection: F. T. Potgieter, J. S. Afr. Vet. Assoc. **52**, 289 (1981). Effect on B. ovis infection in sheep: S. A. Michael, A. H. El Refaii, Trop. Anim. Health Prod. **14**, 1 (1982), C.A. **96**, 192991 (1982).

Dihydrochloride. [5318-76-3] Imizocarb; 4A65. $C_{19}H_{20}N_6O \cdot 2HCl$; mol wt 421.33. Solid, mp 350° (dec). LD_{50} in mice, rats (mg/kg): 107, 150 s.c. (Beveridge).

Dipropionate. Imizol; Imizad Equine Injection. $C_{25}H_{32}N_6O_7$; mol wt 528.57.

THERAP CAT (VET): Antiprotozoal (Babesia).

4957. Imidurea. [39236-46-9] N,N''-Methylenebis[N'-[3-(hydroxymethyl)-2,5-dioxo-4-imidazolidinyl]urea]; methanebis[N,N'-(5-ureido-2,4-diketotetrahydroimidazole)-N,N-dimethylol]; imidazolidinyl urea; Abiol; Biopure 100; Germall 115; Sepicide CI; Tri-Stat IU; Unicide U-13. $C_{11}H_{16}N_8O_8$; mol wt 388.30. C 34.03%, H 4.15%, N 28.86%, O 32.96%. Condensation product of formaldehyde and allantoin, q.q.v. Member of a family of heterocyclic substituted urea compounds with formaldehyde releasing activity. Prepn: P. A. Berke, **US 3248285** (1966 to Sutton). Description, toxicity and antimicrobial activity: P. A. Berke, W. E. Rosen, *Am. Perfum. Cosmet.* **85**, 55 (1970). Kinetics of formaldehyde release: M. Johansen, H. Bundgaard, *Arch. Pharm. Chemi Sci. Ed.* **9**, 117 (1981). Determn in cosmetic products: G. M. Michalakis, E. F. Barry, *J. Soc. Cosmet. Chem.* **45**, 193 (1994). Review of use and safety assessment: *J. Environ. Pathol. Toxicol.* **4**, 133-146 (1980); of properties and preservative activity: W. E. Rosen, P. A. Berke, *Cosmet. Sci. Technol. Ser.* **1**, 191-205 (1984).

Exists as a monohydrate. White, free-flowing fine powder, dec at temps >160°. Odorless, tasteless. Non-uv absorbing. Polar, hydrophilic. Soly (g/100 g solvent): water 200; ethylene glycol 150; propylene glycol 120; glycerin 100; methanol 0.05; ethanol <0.05; sesame oil <0.05. Insol in most organic solvents.

USE: Antimicrobial preservative in cosmetics and topical pharmaceutical preparations.

4958. Imiglucerase. [154248-97-2] 495-L-Histidineglucosylceramidase (human placental isoenzyme protein moiety reduced); rGCR; Cerezyme. Mannose-terminated human glucocerebrosidase produced by recombinant DNA technology in Chinese hamster ovary (CHO) cells. Glycoprotein containing 497 amino acids with 4 *N*-linked glycosylation sites. Sequence differs from endogenous glucocerebrosidase, *q.v.*, by the substitution of histidine for arginine at position 495. Oligosaccharide chains are modified by sequential deglycosylation to expose mannose residues which enhances uptake by tissue macrophages where glucosylceramide accumulates in Gaucher's disease. Mol wt 60.43 kDa. Prepn of rGCR: J. Rasmussen *et al.*, **WO 9007573**; *eidem*, **US 5236838** (1990, 1993 both to Genzyme); of oligosaccharide-modified rGCR: B. Friedman, M. Hayes, **US 5549892** (1996 to Genzyme). Structural analysis: Y. Kacher *et al.*, *Biol. Chem.* **389**, 1361 (2008). Clinical evaluation in Gaucher's disease: A. Zimran *et al.*, *Lancet* **345**, 1479 (1995). Review of pharmacology and clinical experience: N. J. Weinreb, *Expert Opin. Pharmacother.* **9**, 1987-2000 (2008); D. Elstein, A. Zimran, *Biologics* **3**, 407-417 (2009).

THERAP CAT: Enzyme replacement therapy for Gaucher's disease.

4959. Iminodiacetic Acid. [142-73-4] *N*-(Carboxymethyl)-glycine; iminodiethanoic acid; diglycine; IDA. $C_4H_7NO_4$; mol wt 133.10. C 36.10%, H 5.30%, N 10.52%, O 48.08%. Obtained from nitrilotriacetic acid, $N(CH_2COOH)_3$, by HCl-hydrolysis in a bomb tube: Schwarzenbach *et al.*, *Helv. Chim. Acta* **28**, 1133 (1945); by oxygenation in presence of palladium/carbon catalyst: Tetenbaum, Stone, *Chem. Commun.* **1970**, 1699. Iminodiacetic acid and nitrilotriacetic acid are formed upon boiling chloroacetic acid with concd aq ammonia: Heintz, *Ann.* **149**, 88 (1869). *See also* Martell, Bersworth, *J. Org. Chem.* **15**, 46 (1950).

Orthorhombic crystals, dec 247.5° (commercial grade, mp 220-250°). pKa$_1$ 2.98; pKa$_2$ 9.89. Forms salts with acids and bases. Soly in water at 5°: 2.43 g/100 ml. Practically insol in acetone, methanol, ether, benzene, carbon tetrachloride, heptane.

Hydrochloride. [2802-06-4] $C_4H_7NO_4$·HCl; mol wt 169.56. Crystals, dec 238-239°. Concd aq solns yield the free acid when adjusted to pH 2 with NaOH.

Sodium salt monohydrate. [6011-32-1] $C_4H_6NNaO_4$·H_2O; mol wt 173.10. Freely sol in water. Forms complexes with Mg, Ca, Ba.

USE: Has been suggested as intermediate in the manuf of surface active agents, complex salts, chelating agents.

4960. Iminodisuccinic Acid. [7408-20-0]; [70543-06-5] (unspecified stereo). *N*-[(1*S*)-1,2-Dicarboxyethyl]-L-aspartic acid; 2,2'-

iminodisuccinic acid. $C_8H_{11}NO_8$; mol wt 249.18. C 38.56%, H 4.45%, N 5.62%, O 51.37%. Synthesis of ester from fumaric acid, ammonia and water: G. Stadnikoff, *Ber.* **44**, 44 (1911). Industrial processes: **GB 1306331** (1971 to Pfizer); T. Groth *et al.*, **US 6107518** (2000 to Bayer). Synthesis and complex forming properties: E. D. Malakhaev *et al.*, *J. Gen. Chem. USSR* **48**, 2361 (1978). Thermochemical study: V. P. Vasil'ev *et al.*, *Russ. J. Phys. Chem.* **63**, 1566 (1989). Formation constants with metal cations: C. K. Gangoda, D. R. Williams, *Chem. Speciation Bioavailability* **9**, 101 (1997).

White solid, mp 206-208°. pK$_1$ 2.96 ±0.05; pK$_2$ 3.84 ±0.02; pK$_3$ 4.83 ±0.05; pK$_4$ 10.12 ±0.04 (25°). Heat of combustion: 2974.3 ±0.6 kJ/mol. Sol in water; insol in alcohol, ether, acetone, benzene.

Tetrasodium salt. [144538-83-0] Baypure CX 100. Review of commercial chemical product: *NICNAS New chemical assessment: Aspartic acid, N-(1,2-dicarboxyethyl)-tetrasodium salt*, (Sydney, Nat. Ind. Chem. Notif. Assess. Scheme, 2002) 22 pp. White solid, mp >300°. Bulk density (kg/m^3): ~740. Vapor pressure at 20°: 1.5×10^{-16} kPa. Sol at 20° (g/l): water 564 (pH 13.1). LD$_{50}$ in rats (mg/kg): >2000 orally; >2000 dermally (NICNAS).

USE: Biodegradable metal chelator in detergents, cleaners and personal care products; in photographic applications; in trace nutrient agricultural fertilizers. In mfr of textiles and newspaper. Scale deposit remover, water softener, bleaching agent stabilizer.

4961. Imipenem. [64221-86-9] (5*R*,6*S*)-6-[(1*R*)-1-Hydroxyethyl]-3-[[2-[(iminomethyl)amino]ethyl]thio]-7-oxo-1-azabicyclo-[3.2.0]hept-2-ene-2-carboxylic acid; *N*-formimidoylthienamycin; imipemide. $C_{12}H_{17}N_3O_4S$; mol wt 299.35. C 48.15%, H 5.72%, N 14.04%, O 21.38%, S 10.71%. Broad-spectrum semi-synthetic carbapenem antibiotic; first stable derivative of thienamycin, *q.v.* Prepn: W. J. Leanza *et al.*, *J. Med. Chem.* **22**, 1435 (1979); B. G. Christensen *et al.*, **BE 848545**; *eidem*, **US 4194047** (1977, 1980 both to Merck & Co.); of crystalline form: T. W. Miller, **EP 6639**; *eidem*, **US 4260543** (1980, 1981 both to Merck & Co.). Total synthesis without formation of thienamycin: I. Shinkai *et al.*, *Tetrahedron Lett.* **23**, 4903 (1982). Antimicrobial activity: S. Ishihara *et al.*, *Int. J. Antimicrob. Agents* **19**, 565 (2002). HPLC determn in serum: C. M. Myers, J. L. Blumer, *Antimicrob. Agents Chemother.* **26**, 78 (1984); in pharmaceutical formulations: R. J. Forsyth, D. P. Ip, *J. Pharm. Biomed. Anal.* **12**, 1243 (1994). Pharmacokinetics in dogs: C. W. Barker *et al.*, *Am. J. Vet. Res.* **64**, 694 (2003). Series of articles on *in vitro* activity, pharmacokinetics, clinical efficacy of combination with cilastatin sodium, *q.v.*, a renal dehydropeptidase I inhibitor: *J. Antimicrob. Chemother.* **12**, Suppl. D, 1-155 (1983); *Rev. Infect. Dis.* **7**, Suppl. 3, S389-S536 (1985); *Am. J. Med.* **78**, Suppl. 6A, 1-167 (1985). Comprehensive description: E. R. Oberholtzer, *Anal. Profiles Drug Subs.* **17**, 73-114 (1988); M. M. Buckley *et al.*, *Drugs* **44**, 408-444 (1992).

Monohydrate. [74431-23-5] MK-787. $C_{12}H_{17}N_3O_4S$·H_2O; mol wt 317.36. Off-white, nonhygroscopic crystals from water-ethanol. $[\alpha]_D^{25}$ +86.8° (c = 0.05 in 0.1*M* phosphate, pH 7). uv max (water): 299 nm (ε 9670, 98% NH$_2$OH ext). pKa$_1$ ~3.2, pKa$_2$ ~9.9. Soly (mg/ml): water 10, methanol 5, ethanol 0.2, acetone < 0.1, DMF < 0.1, DMSO 0.3.

Combination with cilastatin sodium. [85960-17-4] Imipem; Primaxin; Tenacid; Tienam; Zienam.

THERAP CAT: Antibacterial.

THERAP CAT (VET): Antibacterial.

4962. Imipramine. [50-49-7] 10,11-Dihydro-*N*,*N*-dimethyl-5*H*-dibenz[*b*,*f*]azepine-5-propanamine; 5-(3-dimethylaminopropyl)-10,11-dihydro-5*H*-dibenz[*b*,*f*]azepine; *N*-(γ-dimethylaminopropyl)iminodibenzyl; imizin; G-22355. $C_{19}H_{24}N_2$; mol wt 280.42. C 81.38%, H 8.63%, N 9.99%. Tricyclic antidepressant. Prepn: Haefliger, Schindler, **US 2554736** (1951 to Geigy); *eidem*, *Helv. Chim. Acta* **37**, 472 (1954). Reviews of pharmacology: Crismon, *Psychopharmacol. Bull.* **4**, 151 pp (Oct. 1967); Glassman, Perel, *Arch. Gen. Psychiatry* **28**, 649 (1973). Comprehensive description: D. N. Kender, R. E. Schiesswohl, *Anal. Profiles Drug Subs.* **14**, 37-75 (1985). Comparative clinical trials in depression: R. S. Lipman *et al.*, *Arch. Gen. Psychiatry* **43**, 68 (1986); in anxiety: R.J. Kahn *et al.*, *ibid.* **79**. Clinical trial for panic disorder: D. H. Barlow *et al.*, *J. Am. Med. Assoc.* **283**, 2529 (2000). Toxicity data: A. Tobe *et al.*, *Arzneim.-Forsch.* **31**, 1278 (1981).

Free base, $bp_{0.1}$ 160°.
Hydrochloride. [113-52-0] Chrytemin; Imidol; Melipramine; Pryleugan; Tofranil. $C_{19}H_{24}N_2 \cdot HCl$; mol wt 316.87. Crystals from acetone, mp 174-175°. Acquires a yellow to reddish discoloration under the influence of light. Freely sol in water, alc; sol in acetone. Insol in ether, benzene. LD_{50} in mice, rats (mg/kg): 400, 490 orally; 110, 90 i.p. (Tobe).
Pamoate. [10075-24-8] Tofranil-PM. $(C_{19}H_{24}N_2)_2 \cdot C_{23}H_{16}O_6$; mol wt 949.21. Fine yellow, tasteles, odorless powder. Sol in ethanol, acetone, ether, chloroform, carbon tetrachloride. Insol in water.
THERAP CAT: Antidepressant.
THERAP CAT (VET): In treatment of cataplexy, urinary incontinence, narcolepsy, ejaculatory dysfunction, and behavioral disorders.

4963. Imipramine *N*-Oxide. [6829-98-7] 10,11-Dihydro-*N*,*N*-dimethyl-5*H*-dibenz[*b*,*f*]azepine-5-propanamine *N*-oxide; 5-[3-(dimethylamino)propyl]-10,11-dihydro-5*H*-dibenz[*b*,*f*]azepine *N*-oxide; *N*-(γ-dimethylaminopropyl)iminodibenzyl *N*-oxide; IPNO. $C_{19}H_{24}N_2O$; mol wt 296.41. C 76.99%, H 8.16%, N 9.45%, O 5.40%. Identification as a metabolite of imipramine, *q.v.*: V. Fishman, H. Goldenberg, *Proc. Soc. Exp. Biol. Med.* **110**, 187 (1962). Prepn: **FR M2508**; H. Dyrsting, J. B. Pedersen, **US 3574852** (1964, 1971, both to Dumex). Pharmacology: W. Theobald *et al.*, *Med. Pharmacol. Exp.* **15**, 187 (1966). Metabolism: M. H. Bickel, M. Baggiolini, *Biochem. Pharmacol.* **15**, 1155 (1966); R. Minder *et al.*, *Arch. Pharmakol.* **268**, 334 (1971). GLC determn in biological samples: H. J. Weder, M. H. Bickel, *J. Chromatogr.* **37**, 181 (1968); HPLC determn: E. Koyama *et al.*, *Ther. Drug Monit.* **15**, 224 (1993). Clinical trial: W. Rapp *et al.*, *Acta Psychiatr. Scand.* **49**, 77 (1973). Clinical pharmacokinetics: A. Nagy, T. Hansen, *Acta Pharmacol. Toxicol.* **42**, 58 (1978).

White needle-shaped crystals, mp 120-123° (dec). Sol in methanol, ether, acetone and benzene. Strongly hygroscopic.
Monohydrate. mp 75-79°.
Hydrochloride. [20438-98-6] Imiprex. $C_{19}H_{24}N_2O \cdot HCl$; mol wt 332.87. Colorless crystalline powder, mp 167-174°; also reported as mp 153-155° (dec). Very bitter taste. An aq. soln is moderately acidic. Sol in 75 parts water, in 12 parts 95% ethanol, 4.5 parts chloroform. Nearly insol in ether. LD_{50} in rats, mice (mg/kg): 90, 150 i.p. (Dyrsting, Pedersen).
THERAP CAT: Antidepressant.

4964. Imiquimod. [99011-02-6] 1-(2-Methylpropyl)-1*H*-imidazo[4,5-*c*]quinolin-4-amine; 4-amino-1-isobutyl-1*H*-imidazo[4,5-*c*]quinoline; R-837; S-26308; Aldara; Zyclara. $C_{14}H_{16}N_4$; mol wt 240.31. C 69.97%, H 6.71%, N 23.31%. Toll-like receptor 7 agonist that activates immune cells; stimulates production of interferon-α, *q.v.* Prepn: J. F. Gerster, **EP 145340**; *idem*, **US 4689338** (1985, 1987 both to Riker). Mechanism of action study: K. Megyeri *et al.*, *Mol. Cell. Biol.* **15**, 2207 (1995). Clinical pharmacology: P. L. Witt *et al.*, *Cancer Res.* **53**, 5176 (1993). Clinical trial for genital warts: L. Edwards *et al.*, *Arch. Dermatol.* **134**, 25 (1998). Series of articles on pharmacology and clinical experience in cutaneous viral infections: *J. Am. Acad. Dermatol.* **43**, S1-S30 (2000). Clinical trial in basal cell carcinoma: T. K. Eigentler *et al.*, *ibid.* **57**, 616 (2007); in actinic keratoses: C. W. Hanke *et al.*, *ibid.* **62**, 573 (2010); in dermal melanoma: K. Turza *et al.*, *J. Cutan Pathol.* **37**, 94 (2010).

Crystals from DMF, mp 292-294°.
THERAP CAT: Antiviral; immunomodulator; antineoplastic.

4965. Imisopasem Manganese. [218791-21-0] (*PB*-7-11-2344′3′)-Dichloro[(4a*R*,13a*R*,17a*R*,21a*R*)-1,2,3,4,4a,5,6,12,13,-13a,14,15,16,17,17a,18,19,20,21,21a-eicosahydro-11,7-nitrilo-7*H*-dibenzo[*b*,*h*][1,4,7,10]tetraazacycloheptadecine-κN^5,κN^{13},κN^{18},-κN^{21},κN^{22}]manganese; manganese(II)dichloro[(4*R*,9*R*,14*R*,19*R*)-3,-10,13,20,26-pentaazatetracyclo[20.3.1.0.^{4,9}^{14,19}]hexacosa-1(26),22(23),24-triene]; M-40403; SC-72325. $C_{21}H_{35}Cl_2MnN_5$; mol wt 483.38. C 52.18%, H 7.30%, Cl 14.67%, Mn 11.37%, N 14.49%. Synthetic catalyst that mimics the bioactivity of superoxide dismutase (SOD), *q.v.*; selectively removes superoxide anions, does not react with nitric oxide, peroxynitrite or hydrogen peroxide. Prepn and SOD activity: K. W. Ashton *et al.*, **WO 9302090** (1993 to Monsanto); D. P. Riley *et al.*, **US 5637578** (1997); D. Salvemini *et al.*, *Science* **286**, 304 (1999). Pharmacological effect on the inflammatory cascade: *idem et al.*, *Br. J. Pharmacol.* **132**, 815 (2001); on inflammatory pain: Z. Wang *et al.*, *J. Pharmacol. Exp. Ther.* **309**, 869 (2004). Review of pharmacology and therapeutic potential: M. Di Napoli, F. Papa, *IDrugs* **8**, 67-76 (2005).

Crystals from water.
THERAP CAT: Anti-inflammatory.

4966. IMPA. [1832-54-8] *P*-Methylphosphonic acid mono(1-methylethyl) ester; *i*PMPA; *O*-isopropyl methyl phosphonic acid; neutralized sarin. $C_4H_{11}O_3P$; mol wt 138.10. C 34.79%, H 8.03%, O 34.76%, P 22.43%. Degradation product of the nerve gas, sarin, *q.v.* Identification as hydrolysis product: F. C. G. Hoskin, *Can. J. Biochem. Physiol.* **34**, 75 (1956). Prepn as silver salt: *idem*, *Can. J. Chem.* **35**, 581 (1957); of ester: J. I. G. Cadogan *et al.*, *J. Chem. Soc. B* **1971**, 1988. Biodisposition in mice: P. J. Little *et al.*, *Toxicol. Appl. Pharmacol.* **83**, 412 (1986). Biological monitoring in human urine after sarin exposure: M. Minami *et al.*, *J. Toxicol. Sci.* **23**, Suppl. II, 250 (1998). Biodegradation: Y. Zhang *et al.*, *Biotechnol. Bioeng.* **64**, 221 (1999). Determn in ground water by GC with flame photometric detection: G. A. Sega *et al.*, *J. Chromatogr. A* **790**, 143 (1997); LC-MS rapid determn: R. W. Read, R. M. Black, *ibid.* **862**, 169 (1999).

bp$_{0.02}$ 88°. n_D^{25} 1.4210.

USE: Marker for the detection of sarin.

4967. Imperatorin. [482-44-0] 9-[(3-Methyl-2-buten-1-yl)-oxy]-7H-furo[3,2-g][1]benzopyran-7-one; 6-hydroxy-7-(3-methyl-2-butenyloxy)-5-benzofuranacrylic acid δ-lactone; 8-isoamylenoxy-psoralen; marmelosin; ammidin; pentosalen. C$_{16}$H$_{14}$O$_4$; mol wt 270.28. C 71.10%, H 5.22%, O 23.68%. From roots of *Imperatoria osthruthium* L., *Umbelliferae*: Späth, Holzen, *Ber.* **66**, 1137 (1933); from seeds of *Angelica archangelica* L.: Späth, Vierhapper, *ibid.* **71**, 1667 (1938); from fruit of *Pastinaca sativa* L., *Umbelliferae*: Soine *et al.*, *J. Am. Pharm. Assoc.* **45**, 426 (1956). Identity with marmelosin: Späth *et al.*, *Ber.* **70**, 1021 (1937). Identity with ammidin: Fahmy, Abu-Shady, *Q. J. Pharm. Pharmacol.* **21**, 499 (1948). Partial synthesis: Späth, Holzen, *Ber.* **68**, 1123 (1935). Synthesis: Späth, Vierhapper, *ibid.* **70**, 248 (1937).

Prisms from ether, long fine needles from hot water, mp 102°. Practically insol in cold water. Very sparingly sol in boiling water; freely sol in chloroform; sol in benzene, alcohol, ether, petr ether, alkali hydroxides. uv max: 302, 265, 250 nm (log ε 3.95, 4.00, 4.24).

4968. Imperialine. [18059-10-4] (3β,5α)-3,20-Dihydroxy-cevan-6-one; raddeamine; sipeimine. C$_{27}$H$_{43}$NO$_3$; mol wt 429.65. C 75.48%, H 10.09%, N 3.26%, O 11.17%. Isolation from bulb of *Fritillaria imperialis* L., *Liliaceae*: Fragner, *Ber.* **21**, 3284 (1888); Boit, *ibid.* **87**, 472 (1954); from *Petilium eduardi*: Shakirov *et al.*, *C.A.* **60**, 13280h (1964). Identity with sipeimine and raddeamine: Chu *et al.*, *C.A.* **53**, 7503e (1959); Nurridinov, Yunusov, *C.A.* **57**, 15165h (1962). Structure: Nuriddinov *et al.*, *Khim. Prir. Soedin.* **3**, 316 (1967); *C.A.* **68**, 69168g (1968). Prepn of the β-D-glucopyranoside: Shakirov *et al.*, *C.A.* **63**, 1858e (1965). Isolation and structure of the β-D-glucopyranoside: *eidem*, *C.A.* **63**, 3007f (1965). Absolute configuration: S. Itô *et al.*, *Tetrahedron Lett.* **17**, 3161 (1976).

Prisms from methanol, mp 267°. [α]$_D^{22}$ −38.5° (c = 1.5 in chloroform).

β-D-Glucopyranoside. Edpetiline. C$_{33}$H$_{53}$NO$_8$. Crystals from methanol, mp 272-276°. [α]$_D$ −57.89° (c = 0.449 in methanol).

4969. Incadronic Acid. [124351-85-5] P,P'-[(Cycloheptyl-amino)methylene]bisphosphonic acid. C$_8$H$_{19}$NO$_6$P$_2$; mol wt 287.19. C 33.46%, H 6.67%, N 4.88%, O 33.43%, P 21.57%. Bisphosphonate antiresorptive agent. Prepn: Y. Isomura *et al.*, EP

325482; *eidem*, US 4970335 (1989, 1990 both to Yamanouchi); M. Takeuchi *et al.*, *Chem. Pharm. Bull.* **41**, 688 (1993). HPLC determn in plasma, urine and bone: T. Usui *et al.*, *J. Chromatogr.* **584**, 213 (1992). Effect on bone mineral density in dogs: H. Motoie *et al.*, *J. Bone Miner. Res.* **10**, 910 (1995). Toxicology: A. Okazaki *et al.*, *J. Toxicol. Sci.* **20**, Suppl. 1, 15 (1995). Clinical pharmacokinetics: T. Usui *et al.*, *Int. J. Clin. Pharmacol. Ther.* **35**, 239 (1997). Clinical evaluation in hypercalcemia of malignancy: S. Fukumoto *et al.*, *J. Clin. Endocrinol. Metab.* **79**, 165 (1994).

Crystals from methanol-water, mp 232-233°.

Disodium salt. [138330-18-4] Incadronate sodium; cimadronate sodium; YM-175; Bisphonal. C$_8$H$_{17}$NNa$_2$O$_6$P$_2$; mol wt 331.15. Occurs as the monohydrate. White, odorless crystalline powder. Easily sol in water. LD$_{50}$ in male, female rats (mg/kg): 23, 21 i.v. (Okazaki).

THERAP CAT: Bone resorption inhibitor.

4970. Indacaterol. [312753-06-3] 5-[(1R)-2-[(5,6-Diethyl-2,3-dihydro-1H-inden-2-yl)amino]-1-hydroxyethyl]-8-hydroxy-2(1H)-quinolinone; (R)-5-[2-(5,6-diethylindan-2-ylamino)-1-hydroxyethyl]-8-hydroxy-1H-quinolin-2-one; QAB-149. C$_{24}$H$_{28}$N$_2$O$_3$; mol wt 392.50. C 73.44%, H 7.19%, N 7.14%, O 12.23%. Long-acting β$_2$-adrenoceptor agonist. Prepn: B. Cuenoud *et al.*, WO 0075114; *eidem*. US 6878721 (2000, 2005 both to Novartis). Pharmacology: C. Battram *et al.*, *J. Pharmacol. Exp. Ther.* **317**, 762 (2006). Effect on mast cell mediator release: A.-M. Scola *et al.*, *Br. J. Pharmacol.* **158**, 267 (2009). Clinical evaluation of bronchodilation in persistent asthma: F. Kanniess *et al.*, *J. Asthma* **45**, 887 (2008); in COPD: G. Feldman *et al.*, *BMC Pulm. Med.* **10**, 11 (2010). Review of pharmacology and clinical experience: D. P. Tashkin, *Expert Opin. Pharmacother.* **11**, 2077-2085 (2010).

Maleate. [753498-25-8] QAB149-AFA; Arcapta Neohaler; Onbrez Breezhaler. C$_{24}$H$_{28}$N$_2$O$_3$.C$_4$H$_4$O$_4$; mol wt 508.57. White to very slightly grayish or yellowish powder. Freely sol in N-methyl-pyrrolidone, DMF; slightly sol in methanol, ethanol, propylene glycol; very slightly sol in water, isopropanol. Practically insol in ethyl acetate, n-octanol.

THERAP CAT: Bronchodilator.

4971. Indan. [496-11-7] 2,3-Dihydro-1H-indene; hydrindene. C$_9$H$_{10}$; mol wt 118.18. C 91.47%, H 8.53%. Occurs in coal tar: Krämer, Spilker, *Ber.* **29**, 552 (1896). Prepn: *eidem*, *ibid.* **33**, 2257 (1900); Grosse *et al.*, *Ind. Eng. Chem.* **38**, 1041 (1946); Carpenter, Easter, *J. Org. Chem.* **19**, 87 (1954); Juday, *ibid.* **22**, 532 (1957); Hunter, Aldridge, US 3082267 (1963 to Esso); Bestmann *et al.*, *Ann.* **718**, 33 (1968).

Liquid. d$_4^{20}$ 0.9639; d$_4^{50}$ 0.9378. mp −51.4°. bp$_{762}$ 176.5°; bp$_{12}$ 61-64°. n_D^{20} 1.5383. uv max (isooctane): 272, 265, 258 nm (log ε 3.18, 3.08, 2.90). Insol in water. Sol in organic solvents.

1-Methylindan. $C_{10}H_{12}$. Liq, d_4^{20} 0.9402. bp_{760} 186-190°; bp_{10} 60°. n_D^{20} 1.5222.

4972. Indanazoline. [40507-78-6] *N*-(2,3-Dihydro-1*H*-inden-4-yl)-4,5-dihydro-1*H*-imidazol-2-amine; *N*-(2-imidazolin-2-yl)-*N*-(4-indanyl)amine. $C_{12}H_{15}N_3$; mol wt 201.27. C 71.61%, H 7.51%, N 20.88%. Prepn: H. J. May, A. Berg, **DE 2136325**; *eidem*, **US 3882229** (1973, 1975 both to Nordmark); H. J. May, *Arzneim.-Forsch.* **30**, 1733 (1980). Toxicity data: W. Worstmann *et al.*, *ibid.* 1760. Series of articles on galenical development, pharmacology, pharmacokinetics and toxicity: *ibid.* 1738-1787.

Crystals from petr ether, mp 109-113°.
Hydrochloride. [40507-80-0] E-VA-16; Farial. $C_{12}H_{15}N_3$.HCl; mol wt 237.73. Crystals from isopropanol, mp 182-184°. LD_{50} in male, female mice (mg/kg): 179, 233 orally; 22.3, 26.9 i.v.; in male, female rats: 481, 542 orally; 16.3, 17.6 i.v. (Worstmann).

THERAP CAT: Decongestant (nasal).

4973. Indanofan. [133220-30-1] 2-[[2-(3-Chlorophenyl)-2-oxiranyl]methyl]-2-ethyl-1*H*-indene-1,3(2*H*)-dione; MK-243; MX-70906; NH-502; Trebiace. $C_{20}H_{17}ClO_3$; mol wt 340.80. C 70.49%, H 5.03%, Cl 10.40%, O 14.08%. Herbicide developed for use in paddy rice and turf. Prepn: T. Jikihara *et al.*, **EP 398258**; *eidem*, **US 5076830** (1990, 1991 both to Mitsubishi); K. Tanaka *et al.*, *Synthesis* **1999**, 249. Review of physical and biological properties: K. Yagi, K. Inoue, *Agrochem. Jpn.* **76**, 12-14 (2000).

Colorless crystals from methanol-water, mp 60.4-61.3°; also reported as off-white crystals. Soly in water at 25°: 17.1 mg/l. Vapor pressure at 25°: 2.8×10^{-6} Pa. LD_{50} male, female rats (mg/kg): 631, 460 orally; >2000, >2000 dermally. LC_{50} in rats (mg/l): 1.57 by inhalation (Yagi).

USE: Herbicide.

4974. Indanthrene. [81-77-6] 6,15-Dihydro-5,9,14,18-anthrazinetetrone; *N,N'*-dihydro-1,2,1',2'-anthraquinonazine. $C_{28}H_{14}N_2O_4$; mol wt 442.43. C 76.01%, H 3.19%, N 6.33%, O 14.46%. Vat dye discovered by René Bohn (1901). Prepn: Fierz-David, Blangey, *Grundlegende Operationen der Farbenchemie* (Vienna, 5th ed., 1943) pp 304-305; Thielert, Bauman, **US 2693469** (1954 to Bayer); Sutter, Fioroni, **US 2831860** (1958 to Ciba); Kastner, **US 3138612** (1964 to Allied Chem.). Structure: Weinstein, Merrit, *J. Am. Chem. Soc.* **81**, 3759 (1959).

Blue powder, dec 470-500°. uv max on cellophane film: 278 nm. Practically insol in organic solvents. Sol in concd H_2SO_4, in dil alkali solns. *Indanthrene Blue R*, the usual commercial grade, is extremely stable to light and heat, but sensitive to chlorine. A purer grade, *Indanthrene Brilliant Blue FF*, is not as sensitive to chlorine.

USE: Mainly to dye cotton.

4975. Indapamide. [26807-65-8] 3-(Aminosulfonyl)-4-chloro-*N*-(2,3-dihydro-2-methyl-1*H*-indol-1-yl)benzamide; 4-chloro-*N*-(2-methyl-1-indolinyl)-3-sulfamoylbenzamide; *N*-(3-sulfamyl-4-chlorobenzamido)-2-methylindoline; S-1520; SE-1520; Bajaten; Fludex; Indaflex; Indamol; Lozol; Noranat; Tandix; Veroxil. $C_{16}H_{16}ClN_3O_3S$; mol wt 365.83. C 52.53%, H 4.41%, Cl 9.69%, N 11.49%, O 13.12%, S 8.76%. Prepn: Beregi *et al.*, **FR 2003311**; *eidem*, **US 3565911** (1969, 1971, both to Sci. Union et Cie, Soc. Franc. Recherche Med.). LC/MS determn in plasma: F. Albu *et al.*, *J. Chromatogr. B* **816**, 35 (2005). Pharmacology: Leary *et al.*, *Curr. Ther. Res.* **15**, 571 (1973); D. B. Campbell, R. A. Moore, *Postgrad. Med. J. Suppl.* **57**, 7 (1981). Acute toxicity data: J. Kyncl *et al.*, *Arzneim.-Forsch.* **25**, 1491 (1975). Symposium on pharmacology and clinical efficacy: *Am. J. Med.* **84**, Suppl. 1B, 1-111 (1988); *Am. J. Cardiol.* **65**, Suppl. H, 1H-80H (1990). Comprehensive description: T. J. DiFeo, J. E. Shuster, *Anal. Profiles Drug Subs. Excip.* **23**, 229-268 (1994).

Crystals from isopropanol/water, mp 160-162°. uv max (methanol): 242, 278, 286 nm ($A_{1cm}^{1\%}$ 630, 98, 100). Sol in methanol, alc, glacial acetic acid, acetonitrile, ethyl acetate; very slightly sol in chloroform, ether. Practically insol in water. pKa (25°) 8.8 ± 0.2. LD_{50} in rats, mice, guinea pigs (mg/kg): 393-421, 410-564, 347-416 i.p.; 394-440, 577-635, 272-358 i.v.; >3000 all species orally (Kyncl).

Hemihydrate. Damide; Ipamix; Natrilix; Pressural. $C_{16}H_{16}ClN_3O_3S.\frac{1}{2}H_2O$; mol wt 374.84.

THERAP CAT: Antihypertensive; diuretic.

4976. 1*H*-Indazole. [271-44-3] Isoindazole; benzopyrazole. $C_7H_6N_2$; mol wt 118.14. C 71.17%, H 5.12%, N 23.71%. Prepn: Stephenson, *Org. Synth.* **coll. vol. III**, 475 (1955); Ainsworth, *Org. Synth.* **39**, 27 (1959); Huisgen, Bast, *ibid.* **42**, 69 (1962).

Needles from hot water, mp 146.5°. bp_{743} 267-270°. Sol in hot water, alcohol, ether.

4977. Indecainide. [74517-78-5] 9-[3-[(1-Methylethyl)amino]propyl]-9*H*-fluorene-9-carboxamide; 9-carbamoyl-9-(3-isopropylaminopropyl)fluorene; 9-[3-(isopropylamino)propyl]-9-(aminocarboyl)fluorene; ricainide. $C_{20}H_{24}N_2O$; mol wt 308.43. C 77.88%, H 7.84%, N 9.08%, O 5.19%. Prepn: W. B. Lacefield, R. L. Simon, **US 4197313** and **US 4452745** (1980, 1984 both to Lilly). Pharmacokinetics in animals: T. L. Lindstrom, G. W. Whitaker, *Drug Metab. Dispos.* **12**, 683 (1984). Metabolism: *eidem*, *ibid.* 691. LC determn in biological fluids: K. Z. Farid *et al.*, *J. Chromatogr.* **337**, 329 (1985). Toxicological studies: G. E. Sandusky, Jr., D. B. Meyers, *Fundam. Appl. Toxicol.* **5**, 175 (1985). Clinical evaluation in cardiac arrhythmias: P. F. Nestico *et al.*, *Am. J. Cardiol.* **59**, 1332 (1987); P. J. Podrid *et al.*, *ibid.* **61**, 764 (1988).

Crystals from Skelly B, mp 94-95°.

Hydrochloride. [73681-12-6] LY-135837; Decabid. $C_{20}H_{24}$-$N_2O.HCl$; mol wt 344.88. Crystals from chloroform, mp 216.5-217°; also reported as crystals from fresh ethanol and diethyl ether, mp 203-204°. LD_{50} orally in male, female mice, rats: 100, 96, 103, 82 mg/kg (Sandusky, Meyers).

THERAP CAT: Antiarrhythmic (cardiac depressant).

4978. Indene. [95-13-6] 1H-Indene; indonaphthene. C_9H_8; mol wt 116.16. C 93.06%, H 6.94%. Found in the tars from coal, lignite, and crude petr. Isoln from coal tar: Weissgerber, *Ber.* **42**, 569 (1909); *Brennst.-Chem.* **5**, 208 (1924); Weissgerber, Seidler, *Ber.* **60**, 2088 (1927). Prepn from acetylene over activated charcoal at 625°: Zelinsky, *Ber.* **57**, 264 (1924). Manuf from tetrahydronaphthalene by passing over SiO_2-Al_2O_3 catalyst at 670°: **GB 578083** (1946). Prepn from o-$BrCH_2C_6H_4CH_2CH_2Br$ and triphenylphosphine: Bestmann *et al., Ann.* **718**, 33 (1968).

Liquid. d_4^4 1.0081; d_4^{20} 0.9968; d_4^{50} 0.9692. mp $-1.8°$. bp_{760} 181.6°; bp_{400} 157.8°; bp_{200} 135.6°; bp_{100} 114.7°; bp_{60} 100.8°; bp_{40} 90.7°; bp_{20} 73.9°; bp_{10} 58.5°; bp_5 44.3°; $bp_{1.0}$ 16.4°. $n_D^{18.5}$ 1.5773. Absorption spectrum: Morton, de Gouveia, *J. Chem. Soc.* **1934**, 911. Insol in water. Miscible with most organic solvents. Polymerizes and oxidizes on standing. Concd H_2SO_4 forms metaindene, $(C_9H_8)_{16-22}$.

Caution: Potential symptoms of overexposure in exptl animals are irritation of eyes, skin, mucous membranes; dermatitis; skin sensitization; liver, kidney and spleen injury; aspiration of liquid may cause chemical pneumonia. *See NIOSH Pocket Guide to Chemical Hazards* (DHHS/NIOSH 97-140, 1997) p 170.

4979. Indenolol. [60607-68-3] 1-[1H-Inden-4(or 7)-yloxy]-3-[(1-methylethyl)amino]-2-propanol; (±)-1-[inden-4(or 7)-yloxy]-3-(isopropylamino)-2-propanol; YB-2; Sch-28316Z. $C_{15}H_{21}NO_2$; mol wt 247.34. C 72.84%, H 8.56%, N 5.66%, O 12.94%. Nonselective β-adrenergic blocker. Prepn: M. Murakami *et al.,* **DE 1955229**; *eidem,* **US 4045482** (1970, 1977 both to Yamanouchi); K. Murase *et al., Yakugaku Zasshi* **92**, 1358 (1972), *C.A.* **78**, 71723 (1973). Indenolol is a tautomeric mixture of the 7- and 4-indenyloxy isomers in a 2:1 ratio, respectively. Unambiguous synthesis of the two isomers: *eidem, Chem. Pharm. Bull.* **24**, 552 (1976). β-Blocking and cardiovascular properties: T. Takenaka, S. Tachikawa, *Arzneim.-Forsch.* **22**, 1864 (1972). General pharmacology: M. Takeda *et al., Oyo Yakuri* **7**, 469 (1973), *C.A.* **80**, 44015 (1974). Comparative pharmacological study with other β-blockers: W. Bartsch *et al., Arzneim.-Forsch.* **27**, 1022 (1977). Dose-response studies: F. E. Okupa *et al., Clin. Pharmacol. Ther.* **29**, 434 (1981). Pharmacokinetics: R. Sega *et al., J. Clin. Pharmacol.* **25**, 337 (1985). Clinical evaluation in hypertension: B. Trimarco *et al., ibid.* 328; L. Poggesi *et al., Clin. Pharmacol. Ther.* **41**, 344 (1987).

Crystals from n-hexane/ether, mp 88-89°.

Hydrochloride. [81789-85-7] Pulsan; Securpres. $C_{15}H_{21}NO_2.$-HCl; mol wt 283.80. Crystals from ethanol/ether, mp 147-148°. LD_{50} in mice: 26 mg/kg i.v. (Bartsch).

THERAP CAT: Antihypertensive, antiarrhythmic, antianginal.

4980. Indican. [487-60-5] 1H-Indol-3-yl-β-D-glucopyranoside; 3-(β-glucosido)indole; indoxyl-β-D-glucoside; plant indican. $C_{14}H_{17}NO_6$; mol wt 295.29. C 56.95%, H 5.80%, N 4.74%, O 32.51%. Naturally occuring precursor of the blue dye, indigo, *q.v.*; isolated from *Polygonum tinctorium* Ait., *Polygonaceae* and from various *Indigofera* spp., *Fabaceae*. Enzymatically hydrolyzed to the aglycone, indoxyl, which readily oxidizes and dimerizes to form the dye. Extraction procedure: Hoogewerff, Meulen, *Rec. Trav. Chim.*

19, 166 (1900); A. G. Perkin, W. P. Bloxam, *J. Chem. Soc. Trans.* **91**, 1715 (1907); A. G. Perkin, F. Thomas, *ibid.* **95**, 793 (1909). Synthesis: A. Robertson, *J. Chem. Soc.* **1927**, 1937; A. Robertson, R. B. Waters, *ibid.* **1933**, 30; K. Freudenberg *et al., Ber.* **85**, 641 (1952). Quantification in *P. tinctorum* leaves by HPLC: L. G. Angelini *et al., Biotechnol. Prog.* **19**, 1792 (2003). Effect of field conditions on indican content and indigo production: *idem et al., J. Agric. Food Chem.* **52**, 7541 (2004).

Colorless prisms from warm ethanol + benzene, mp 176-178°. $[\alpha]_D^{19}$ $-77.64°$ (c = 1.0 in water). Sol in water, methanol, alcohol, acetone; slightly sol in ether, benzene, chloroform, ethyl acetate, carbon disulfide; sparingly sol in anhydr alcohol.

Trihydrate. Colorless, silky needles from water, mp 57-58°. $[\alpha]_{546}^{19}$ $-65.63°$ (c = 1.0 in water). After heating *in vacuo* (oil pump) over H_2SO_4 for 48 hrs the hemihydrate is obtained, mp 101°.

Note: The term indican has also been used to refer to indoxyl sulfate, *q.v.*

USE: Starting material for the production of indigo.

4981. Indigo. [64784-13-0]; [482-89-3] (unspecified stereo). (2E)-2-(1,3-Dihydro-3-oxo-2H-indol-2-ylidene)-1,2-dihydro-3H-indol-3-one; indigotin; indigo blue; D & C Blue No. 6; C.I. Pigment Blue 66; C.I. Vat Blue 1; C.I. 73000. $C_{16}H_{10}N_2O_2$; mol wt 262.27. C 73.27%, H 3.84%, N 10.68%, O 12.20%. Primary constituent of the natural indigo dye known since antiquity for its distinctive, dark-blue color. Originally obtained from *Indigofera tinctoria* L., *Fabaceae* and various other plants that produce the precursor glycoside known as indican, *q.v.* One of the first dyes to be prepared synthetically for commercial use. Isoln from powdered, natural indigo: J. Fritzsche, *J. Prakt. Chem.* **28**, 193 (1843). Synthesis: A. Baeyer, *Ber.* **11**, 1296 (1878); *idem, ibid.* **12**, 456 (1879). Commercial process: K. Heumann, *Ber.* **23**, 3043 (1890); *idem, ibid.,* 3431. Mechanism of formation from indoxyl: G. A. Russell, G. Kaupp, *J. Am. Chem. Soc.* **91**, 3851 (1969). One pot synthesis from indole: Y. Yamamoto *et al., Bull. Chem. Soc. Jpn.* **84**, 82 (2011). Crystal structure: H. von Eller, *Acta Crystallogr.* **5**, 142 (1952); P. Süsse *et al., Z. Kristallogr.* **184**, 269 (1988). Conformation of the indigo chromophore: M. Klessinger, *Dyes Pigm.* **3**, 235 (1982). Photostability studies: J. S. Seixas de Melo *et al., Angew. Chem. Int. Ed.* **46**, 2094 (2007); I. Iwakura *et al., Chem. Lett.* **38**, 1020 (2009). Review of industrial dying processes: R. S. Blackburn *et al., Color. Technol.* **125**, 193-207 (2009).

Dark-blue powder with coppery luster. Sublimes at about 300°; dec 390°. d (cryst) 1.48. Practically insol in water, alcohol, ether, and dil acids. Dissolves in nonpolar solvents with red and in polar solvents with blue color. With fuming H_2SO_4 it forms a sol sulfonic acid.

USE: Dye for sutures and textiles, especially denim.

4982. Indigo Carmine. [860-22-0] 2-(1,3-Dihydro-3-oxo-5-sulfo-2H-indol-2-ylidene)-2,3-dihydro-3-oxo-1H-indole-5-sulfonic acid sodium salt (1:2); 3,3'-dioxo-[$\Delta^{2,2'}$-biindoline]-5,5'-disulfonic acid disodium salt; disodium 5,5'-indigotin disulfonate; sodium indigo disulfonate; soluble indigo blue; indigotine; Acid Blue 74; C.I. Acid Blue 74; C.I. Food Blue 1; FD & C Blue No. 2; C.I. 73015. $C_{16}H_8N_2Na_2O_8S_2$; mol wt 466.35. C 41.21%, H 1.73%, N 6.01%, Na 9.86%, O 27.45%, S 13.75%. Synthesis and structure determination: Vorlander, Schubart, *Ber.* **34**, 1860 (1901). Prepn and properties: Matthews, *Color Trade J.* **6**, 96 (1920). *See also Colour Index* vol. 4 (3rd ed., 1971) p 4597.

Dark-blue powder with coppery luster. Sensitive to light. Its solns have a blue or bluish-purple color. Changes color from blue to yellow between pH 11.5-14. One gram dissolves in about 100 ml water at 25°. Slightly sol in alcohol. Practically insol in most other organic solvents. It is also marketed as a paste with water, the dye contents varying according to specification or requirements of the user. It almost always contains sodium chloride or sulfate used for "salting" it out. Indigo carmine is very sensitive to oxidizing agents. The color is readily discharged by nitric acid, chlorates, etc. The color of the aq soln fades on standing.

USE: Colorant for nylon, surgical sutures, foods and ingested drugs. Histological stain used for collagen, plant cell nuclei, and fluorescent staining of eosinophil granules. In vivo stain for endoscopic diagnosis; detection of sentinel lymph nodes; angiographic tracer. As a reagent for functional kidney tests, for detection of nitrates, chlorates and in testing milk.

4983. Indinavir. [150378-17-9] 2,3,5-Trideoxy-*N*-[(1*S*,2*R*)-2,3-dihydro-2-hydroxy-1*H*-inden-1-yl]-5-[(2*S*)-2-[[(1,1-dimethylethyl)amino]carbonyl]-4-(3-pyridinylmethyl)-1-piperazinyl]-2-(phenylmethyl)-D-*erythro*-pentonamide; (α*R*,γ*S*,2*S*)-α-benzyl-2-(*tert*-butylcarbamoyl)-γ-hydroxy-*N*-[(1*S*,2*R*)-2-hydroxy-1-indanyl]-4-(3-pyridylmethyl)-1-piperazinevaleramide; *N*-(2*R*)-hydroxy-1(*S*)-indanyl]-2(*R*)-(phenylmethyl)-4(*S*)-hydroxy-5-[1-[4-(3-pyridylmethyl)-2(*S*)-(*N*-*tert*-butylcarbamoyl)piperazinyl]]pentanamide. $C_{36}H_{47}N_5O_4$; mol wt 613.80. C 70.45%, H 7.72%, N 11.41%, O 10.43%. Member of the novel hydroxyaminopentane amide class of HIV-1 protease inhibitors. Prepn: J. P. Vacca *et al.*, **EP 541168**; *eidem*, **US 5413999** (1993, 1995 both to Merck & Co.); B. D. Dorsey *et al.*, *J. Med. Chem.* **37**, 3443 (1994). Diastereoselective synthesis: D. Askin *et al.*, *Tetrahedron Lett.* **35**, 673 (1994); P. E. Maligres *et al.*, *ibid.* **36**, 2195 (1995). Crystal structure of HIV-II protease complex: Z. Chen *et al.*, *J. Biol. Chem.* **269**, 26344 (1994). Antiviral activity and pharmacokinetics: J. P. Vacca *et al.*, *Proc. Natl. Acad. Sci. USA* **91**, 4096 (1994). HPLC determn in plasma and urine: E. Woolf *et al.*, *J. Chromatogr. A* **692**, 45 (1995). Metabolism in humans: S. K. Balani *et al.*, *Drug Metab. Dispos.* **23**, 266 (1995). Review of clinical pharmacology and therapeutic efficacy: G. L. Plosker, S. Noble, *Drugs* **58**, 1165-1203 (1999).

Occurs as monohydrate, crystals from wet ethyl or isopropyl acetate, loss of water below 100° followed by recrystallization of anhydrous form with mp 153-154°; second anhydrous form, mp 167.5-168°. Soly in water (mg/ml): 0.015 (unbuffered), >1.5 (pH 4.0). $[\alpha]_D^{22}$ +24.1° (c = 0.0133 in chloroform).

Sulfate. [157810-81-6] MK-639; Crixivan. $C_{36}H_{47}N_5O_4 \cdot H_2 \cdot SO_4$; mol wt 711.88. White to off-white, hygroscopic, crystalline powder. Occurs as the monoethanolate, crystals from absolute ethanol, softens at 135°, mp 150-153° (dec). Converts to hydrate upon loss of ethanol and exposure to moist air. Sol in water and methanol.

THERAP CAT: Antiretroviral.

4984. Indiplon. [325715-02-4] *N*-Methyl-*N*-[3-[3-(2-thienylcarbonyl)pyrazolo[1,5-*a*]pyrimidin-7-yl]phenyl]acetamide; CL-285489; NBI-34060. $C_{20}H_{16}N_4O_2S$; mol wt 376.43. C 63.82%, H

4.28%, N 14.88%, O 8.50%, S 8.52%. Short-acting, nonbenzodiazepine sedative; high-affinity α_1-selective allosteric potentiator of GABA$_A$ receptor function. Prepn: J. P. Dusza *et al.*, **US 4521422** (1985 to Am. Cyanamid). *See also: idem et al.*, **US 6399621** (2002 to Am. Cyanamid). Receptor binding studies: S. K. Sullivan *et al.*, *J. Pharmacol. Exp. Ther.* **311**, 537 (2004). *In vivo* pharmacology: A. C. Foster *et al.*, *ibid.* 547. Review of clinical development: D. N. Neubauer, *Expert Opin. Invest. Drugs* **14**, 1269-1276 (2005).

Yellow solid.
THERAP CAT: Sedative, hypnotic.

4985. Indirubin. [906748-38-7]; [479-41-4] (unspecified stereo). (3*Z*)-3-(1,3-Dihydro-3-oxo-2*H*-indol-2-ylidene)-1,3-dihydro-2*H*-indol-2-one; indigo red; indigopurpurin; C.I. 73200. $C_{16}H_{10}N_2O_2$; mol wt 262.27. C 73.27%, H 3.84%, N 10.68%, O 12.20%. Naturally occuring constituent of Tyrian purple, a dye obtained from the marine mollusk, *Hexaplex trunculus* L., *Muricidae*, and of natural indigo dyes extracted from various plants. Structural isomer of blue indigo, *q.v.* Identified as the antileukemic principle in the traditional Chinese medicine, Danggui Luhui Wan; inhibits cyclin-dependent kinases (CDK) and glycogen synthase kinase-3 (GSK-3). Prepn from indican: E. Schunck, *J. Prakt. Chem.* **66**, 321 (1855). Prepn by condensation of isatin with indoxyl in alkaline soln: A. Baeyer, *Ber.* **14**, 1741 (1881). Isoln from natural Bengal indigo: A. G. Perkin, W. P. Bloxam, *J. Chem. Soc. Trans.* **91**, 279 (1907). Crystal structure: H. Pandraud, *Acta Crystallogr.* **14**, 901 (1961). Isoln with *cis*-isomer from *Isatis tinctoria*: T. Maugard *et al.*, *Phytochemistry* **58**, 897 (2001). Prepn, properties, and determn in natural pigments: M. Puchalska *et al.*, *J. Mass Spectrom.* **39**, 1441 (2004). Determn in *Isatis indigotica* by LC-MS/MS: P. Zou, H. L. Koh, *Rapid Commun. Mass Spectrom.* **21**, 1239 (2007); in Tyrian purple by HPLC: W. Nowik *et al.*, *J. Chromatogr. A* **1218**, 1244 (2011). Effect on cyclin-dependent kinases: R. Hoessel *et al.*, *Nat. Cell Biol.* **1**, 60 (1999); and antiproliferative activity in human tumor cells: D. Marko *et al.*, *Br. J. Cancer* **84**, 283 (2001). Review of use in treatment of leukemia: Z. Xiao *et al.*, *Leuk. Lymphoma* **43**, 1763-1768 (2002); of molecular mechanisms of action: G. Eisenbrand *et al.*, *J. Cancer Res. Clin. Oncol.* **130**, 627-635 (2004).

Fine dark-red needles from nitrobenzene, mp 345-347°. Abs max (DMSO): 540 nm. Insol in water. Sol in boiling acetic acid with violet-red color. Forms violet solns in alcohol, deep crimson in conc H_2SO_4.

3*E*-Form. [397242-72-7] Isoindirubin; *cis*-indirubin. Dark violet crystals. Abs max (DMSO): 552 nm.

4986. Indisulam. [165668-41-7] N^1-(3-Chloro-1*H*-indol-7-yl)-1,4-benzenedisulfonamide; *N*-(3-chloro-7-indolyl)-1,4-benzenedisulfonamide; *N*-(3-chloro-1*H*-indol-7-yl)-4-sulfamoylbenzenesulfonamide; E-7070. $C_{14}H_{12}ClN_3O_4S_2$; mol wt 385.84. C 43.58%, H 3.13%, Cl 9.19%, N 10.89%, O 16.59%, S 16.62%. Synthetic antitumor sulfonamide derivative capable of inducing apoptosis. Prepn: H. Yoshino *et al.*, **WO 9507276**; *eidem*, **US 5721246** (1995, 1998 both to Eisai); T. Owa *et al.*, *J. Med. Chem.* **42**, 3789 (1999). Mechanism of action: K. Fukuoka *et al.*, *Invest. New Drugs* **19**, 219 (2001). LC/MS/MS determn in plasma, urine, and feces: J. Hendrik Beumer *et al.*, *Rapid Commun. Mass Spectrom.* **18** 2839 (2004). Clinical pharmacokinetics and pharmacodynamics: R. I. Haddad *et al.*, *Clin. Cancer Res.* **10**, 4680 (2004); D. C. Talbot *et al.*, *ibid.* **13**, 1816 (2007). Review of development, pharmacology, and clinical experience: C. T. Supuran, *Expert Opin. Invest. Drugs* **12**, 283-287 (2003).

Colorless, solid from EtOH-H$_2$O, mp 242-243° (dec). Sol in methanol.

THERAP CAT: Antineoplastic.

4987. Indium. [7440-74-6] In; at. wt 114.818; at. no. 49; valence 3, 2, 1. Group IIIA(13). Natural isotopes: 115 (95.77%); 113 (4.23%); ^{115}In is a β^- emitter, T$_{\frac{1}{2}}$ 6 × 10^{14} years. Artificial radioactive isotopes: 107-112; 114; 116-124. Occurrence in the earth's crust: 1 × 10^{-5}%. Discovered in sphalerite ore by Reich and Richter in 1863. Generally found in zinc blendes. Monograph: M. T. Ludwick, *Indium* (Indium Corp. of America, Utica, N.Y., 1950). *Review:* Wade, Banister in *Comprehensive Inorganic Chemistry* **vol. 1**, J. C. Bailar, Jr. *et al.*, Eds. (Pergamon Press, Oxford, 1973) pp 997-1000, 1065-1117; E. F. Milner, C. E. T. White in *Kirk-Othmer Encyclopedia of Chemical Technology* **vol. 13** (Wiley-Interscience, New York, 3rd ed., 1981) pp 207-212.

Soft, white metal with bluish tinge. Emits a "tin cry" on bending. Ductile, malleable, softer than lead, leaves a mark on paper. Quite stable in air. Crystallizes and is diamagnetic. d^{20} 7.3. mp 155°. bp 2000°. Sp heat: 0.0568 cal/g/°C. Hardness (Mohs') = 1.2. Unaffected by water; attacked by mineral acids. Very resistant to alkalies.

Caution: Indium salts are relatively nontoxic when administered orally; highly toxic when administered subcutaneously or intravenously. Experimental animal poisoning has produced injury to blood, heart, liver, kidneys: E. Browning, *Toxicity of Industrial Metals* (Appleton-Century-Crofts, New York, 2nd ed., 1969) pp 164-168.

USE: In bearing alloys; as a thin film on moving surfaces made from other metals. In dental alloys. In semiconductor research. In nuclear reactor control rods (in the form of an Ag-In-Cd alloy).

4988. Indium Antimonide. [1312-41-0] Antimony compd with indium (1:1). InSb; mol wt 236.58. In 48.53%, Sb 51.47%. Prepd by melting together stoichiometric amounts of indium and antimony in evacuated ampuls or in zone-refining apparatus: Kleppa, *J. Am. Chem. Soc.* **77**, 897 (1955); Harmon, *J. Electrochem. Soc.* **103**, 128 (1956). *Reviews:* Welker, Weiss in *Solid State Physics* **vol. 3** (Academic Press, New York, 1956); Minden, *Sylvania Technol.* **11**, no. 1, 13-25 (Jan. 1958); Hulme, Mullin, *Solid State Electron.* **5** (Pergamon Press, 1962) 211-247.

Crystals (zinc blende structure). mp 535°. d at mp 5.74 (solid); 6.48 (liq). Dielectric constant = 15.9. Energy gap at 25° = 0.18 ev. Hole mobility 1250 cm^2/volt-sec. Electron mobility approx 80,000 cm^2/volt-sec.

USE: In semiconductor electronics. Grown p-n junctions have been made by doping a melt with an acceptor impurity such as zinc or cadmium, and dipping in an n-type crystal. Rate-grown junctions have also been made. Broad-area surface junctions have been produced by out-diffusing antimony in vacuum from the surface of an n-type crystal, producing a p-n junction just inside the surface. Also has photoconductive, photoelectromagnetic, and magnetoresistive properties. Useful as an infrared detector and filter, and in Hall effect devices.

4989. Indium Arsenide. [1303-11-3] AsIn; mol wt 189.74. As 39.49%, In 60.51%. InAs. Prepd by fusion of the elements in an

evacuated, sealed tube: Gans *et al.*, *Compt. Rend.* **237**, 310 (1953); Talley, Enright, *Phys. Rev.* **94**, 1931 (1954); Harmon *et al.*, *ibid.* **104**, 1562 (1956); Minden, *Sylvania Technol.* **11**, no. 1, 17 (Jan. 1958).

Metallic appearance. Small single crystals have been grown by a modification of the Czochralski technique: Gremmelmaier, *Z. Naturforsch.* **11A**, 463 (1956). mp 943°. Hardly attacked by mineral acids. Energy gap: 0.35 ev. Electron mobility: approx 33,000 cm^2/volt-sec. Hole mobility: 460 cm^2/volt-sec. Dielectric constant 11.7.

USE: In semiconductor electronics.

4990. Indium Gallium Aluminum Phosphide. [108424-49-3]; [108730-13-8]. InGaAlP. Crystal lattice matched to GaAs substrates for production of visible (670-690 nm) light lasers. Conforms to the general formula In$_{0.5}$(Ga$_{(1-x)}$Al$_x$)$_{0.5}$P. Prepn by molecular beam epitaxy: H. Asahi *et al.*, *J. Appl. Phys.* **53**, 4928 (1982); M. J. Hafich *et al.*, *J. Vac. Sci. Technol. B* **10**, 969 (1992). Refractive indices: H. Tanaka *et al.*, *J. Appl. Phys.* **59**, 985 (1986). Photoluminescence studies: S. Naritsuka *et al.*, *J. Electron. Mater.* **20**, 687 (1991). Schottky barrier energy: A. Nanda *et al.*, *Appl. Phys. Lett.* **61**, 81 (1992). Thermal analysis: G. Hatakoshi *et al.*, *Trans. Inst. Electron. Inf. Commun. Eng.* **E71**, 315 (1988); K. Itaya *et al.*, *IEEE J. Quantum Electron.* **29**, 2068 (1993). Use in lasers: G. Hatakoshi *et al.*, *Jpn. J. Appl. Phys.* **31**, 501 (1992); J. Rennie *et al.*, *IEEE J. Quantum Electron.* **29**, 1857 (1993).

USE: Laser diode arrays for high-density optical systems and information processing systems such as barcode readers.

4991. Indium Nitride. [25617-98-5] Indium mononitride. InN; mol wt 128.83. In 89.12%, N 10.87%. Semiconductor material. Prepn and structure determn: R. Juza, H. Hahn, *Z. Anorg. Allg. Chem.* **239**, 282 (1938). Low-temperature organometallic chemical vapor deposition: R. A. Fischer *et al.*, *Chem. Mater.* **8**, 1356 (1996). Prepn and characterization of nanocrystalline forms: J. Xiao *et al.*, *Inorg. Chem.* **42**, 107 (2003). Physical properties of films: G. V. Samsonov, A. F. Andreeva, *Sci. Sintering* **12**, 155 (1980). Thermal properties: S. Krukowski *et al.*, *J. Phys. Chem. Solids* **59**, 289 (1998). Optical and electronic properties: V. V. Sobolev, M. A. Zlobina, *Semiconductors* **33**, 385 (1999).

Usually crystallizes as a hexagonal wurtzite lattice. T$_{dec}$ 427-550°. mp ~1900°. Bandgap 1.89 eV. *n* (4.80 eV) = 2.78. d 6.78±0.05.

USE: In manuf of optoelectronic devices such as light-emitting diodes, laser diodes, and solar cells.

4992. Indium Oxide. [1312-43-2] Indium oxide (In$_2$O$_3$); indium sesquioxide; diindium trioxide; trioxodiindium. In$_2$O$_3$; mol wt 277.63. In 82.71%, O 17.29%.

White to pale-yellow powder. d 7.18. Volatilizes at 850°. Insol in water. Sol in hot mineral acids.

USE: In glass manufacture.

4993. Indium Phosphide. [22398-80-7] Indium monophosphide. InP; mol wt 145.79. In 78.76%, P 21.25%. Prepd from white phosphorus and indium iodide at 400°: Thiel, Koelsch, *Z. Anorg. Chem.* **66**, 319 (1910); from phosphorus vapor and heated indium metal: Jandelli, *Gazz. Chim. Ital.* **71**, 58 (1941). Synthesis in zone melting furnace at 1010° from a non-stoichiometric melt: Minden, *Sylvania Technol.* **11**, no. 1, 18 (Jan. 1958).

Brittle mass with metallic appearance, not easily attacked by mineral acids. mp 1070°. Dielectric constant: 10.8. Energy gap: 1.3 ev at 25°. Electron mobility: approx 4600 cm^2/volt-sec. Hole mobility: approx 150 cm^2/volt-sec. Solid solns of InP can cover the energy gap continuously from 0.3 to 1.3 ev. Rectification has been observed in InP although it is more characteristic of a Schottky type barrier than the minority carrier injection phenomenon observed in germanium.

USE: In electronics for research on semiconductors.

4994. Indium Selenide. [1312-42-1] InSe; mol wt 193.78. In 59.25%, Se 40.75%. Prepd from the elements: Klemm *et al.*, *Z. Anorg. Allg. Chem.* **219**, 45 (1934); Schubert *et al.*, *Naturwissenschaften* **41**, 448 (1954).

Black crystals, rhombohedric structure, mp 660°.

USE: In semiconductor research.

4995. Indium Sulfate. [13464-82-9] Sulfuric acid indium-(3+) salt (3:2); diindium trisulfate; indium sesquisulfate; indium trisulfate. $In_2O_{12}S_3$; mol wt 517.80. In 44.35%, O 37.08%, S 18.57%. $In_2(SO_4)_3$.

White, hygroscopic powder. d 3.44. Sol in water. *Keep well closed.* MLD in rabbits: 1.8 g/kg orally; 0.67 mg/kg i.v.; *see:* McCord *et al., J. Ind. Hyg. Toxicol.* **24**, 243 (1942).

4996. Indium Telluride. [1312-45-4] Indium telluride (In_2-Te_3); diindium tritelluride; indium sesquitelluride. In_2Te_3; mol wt 612.44. In 37.50%, Te 62.50%. Prepn of black brittle crystals by double furnace technique: Inuzuka, Sugaike, *Proc. Jpn. Acad.* **30**, 383 (1954), *C.A.* **49**, 2922e (1955).

Pycnometric d 5.78; x-ray d 5.798. mp 667°: Klemm, Ulrich, *Z. Anorg. Allg. Chem.* **219**, 45 (1934), *C.A.* **29**, 1730 (1935). Activation energies: 0.94 ev at 300-600 K, 2.4 ev at 600-900 K.

USE: In semiconductor research.

4997. Indium Tribromide. [13465-09-3] Indium bromide ($InBr_3$); tribromoindium. Br_3In; mol wt 354.53. Br 67.61%, In 32.39%. $InBr_3$. Water-stable, green Lewis acid catalyst. Prepn: G. P. Baxter, C. M. Alter, *J. Am. Chem. Soc.* **55**, 1943 (1933). Hydrolysis constants: T. Moeller, *ibid.* **64**, 953 (1942). Raman spectroscopy of aqueous solns: M. A. Marques *et al., J. Chem. Soc. Faraday Trans.* **86**, 3883 (1990). Chemical bonding analysis of binary indium bromides: R. Dronskowski, *Inorg. Chem.* **33**, 6201 (1994). Catalytic applications in organic synthesis: J. S. Yadav, B. V. S. Reddy, *Synthesis* **2002**, 511; *eidem et al., Synlett* **2003**, 396; C. Peppe *et al., ibid.* **2004**, 1723; L. Yin *et al., ibid.* 1727; Z.-H. Zhang *et al., Adv. Synth. Catal.* **348**, 184 (2006). Review of synthetic applications: Z.-H. Zhang, *Synlett* **2005**, 711-712.

White crystalline solid.

USE: Regio-, chemo-, and stereoselective Lewis acid catalyst in organic synthesis.

4998. Indium Trichloride. [10025-82-8] Indium chloride ($InCl_3$); trichloroindium. Cl_3In; mol wt 221.17. Cl 48.09%, In 51.91%. $InCl_3$.

Yellowish, deliquesc crystals. d 4.0. mp 586°; sublimes at 500°. Freely sol in water. *Keep tightly closed.* MLD s.c. in rats: 10.2 mg/kg; MLD i.v. in rabbits: 0.64 mg/kg; *see:* McCord *et al., J. Ind. Hyg. Toxicol.* **24**, 243 (1942).

USE: In electroplating using a soln of the salt with dextrose and NaCN. This soln is stable, though it turns dark on standing and deposits a mud which, however, contains no indium.

4999. Indium Trifluoride. [7783-52-0] Indium fluoride (InF_3); indic fluoride. F_3In; mol wt 171.81. F 33.17%, In 66.83%. InF_3. Prepd by heating In_2O_3 or $(NH_4)_3InF_6$ in a current of fluorine: Hannebohn, Klemm, *Z. Anorg. Allg. Chem.* **229**, 342 (1936); Ensslin, Dreyer, *ibid.* **249**, 119 (1942); Kwasnik in *Handbook of Preparative Inorganic Chemistry* vol. **1**, G. Brauer, Ed. (Academic Press, New York, 2nd ed., 1963) pp 228-229; from In and HF: Brewer *et al., J. Inorg. Nucl. Chem.* **9**, 56 (1959); Muetterties, Castle, *ibid.* **18**, 148 (1961).

Colorless substance, d 4.39. mp 1170°. bp >1200°. Soly in water (25°) 0.040 g/100 ml. Freely sol in dil acids. Stable in hot and cold water.

Trihydrate. Crystals. Soly in water (22°) 8.49 g/100 ml, indicating a complex.

5000. Indo-1. [96314-96-4] 2-[4-[Bis(carboxymethyl)amino]-3-[2-[2-[bis(carboxymethyl)amino]-5-methylphenoxy]ethoxy]phenyl]-$1H$-indole-6-carboxylic acid; 1-[2-amino-5-(6-carboxyindol-2-yl)phenoxy]-2-(2'-amino-5-methylphenoxy)ethane-N,N,N',N'-tetraacetic acid. $C_{32}H_{31}N_3O_{12}$; mol wt 649.61. C 59.17%, H 4.81%, N 6.47%, O 29.55%. Fluorescent calcium imaging agent; structurally related to BAPTA, *q.v.* Ca^{2+} binding causes a spectral shift to shorter wavelengths; ratio of the fluorescence at each wavelength is related to changes in Ca^{2+} binding and independent of dye concentration. Synthesis and characterization: G. Grynkiewicz *et al., J. Biol. Chem.* **260**, 3440 (1985). Acid-base and calcium binding properties: F. Bancel *et al., J. Photochem. Photobiol. A* **53**, 397 (1990); spectral properties: H. Szmacinski *et al., Biophys. J.* **70**, 547 (1996). Practical aspects of use in intracellular measurements: K. R. Sipido, G. Callewaert, *Cardiovasc. Res.* **29**, 717 (1995). Detection of other metal cations: G. A. Peeters *et al., Am. J. Physiol.* **256**, C351 (1989); J. R. Jefferson *et al., Anal. Biochem.* **187**, 328 (1990).

Review of use in plant cells: D. S. Bush, R. L. Jones, *Plant Physiol.* **93**, 841-845 (1990).

uv max absorption (100mM KCl): 349 nm free dye; 331 nm Ca^{2+} complex; emission maxima: 485 nm free dye; 410 nm Ca^{2+} complex. pKa_1: 6.2; pKa_2: 4.35.

USE: Fluorescent probe primarily for measuring calcium in biological systems.

5001. Indobufen. [63610-08-2] 4-(1,3-Dihydro-1-oxo-$2H$-isoindol-2-yl)-α-ethylbenzeneacetic acid; ($\pm$)-2-[p-(1-oxo-2-isoindolinyl)phenyl]butyric acid; 1-oxo-2-[p-[(α-ethyl)carboxymethyl]phenyl]isoindoline; 2-[4-(1-carboxypropyl)phenyl]-1-isoindolinone; K-3920; Ibustrin. $C_{18}H_{17}NO_3$; mol wt 295.34. C 73.20%, H 5.80%, N 4.74%, O 16.25%. Platelet aggregation inhibitor. Prepn: R. W. J. Carney, G. de Stevens, **DE 2034240** (1971 to Ciba), *C.A.* **74**, 12547p (1971); P. N. Giraldi *et al.,* **DE 2154525**; *eidem,* **US 4118504** (1972, 1978 both to Carlo Erba); G. Nannini *et al., Arzneim.-Forsch.* **23**, 1090 (1973). Pharmacology: M. Bergameaschi *et al., Pharmacol. Res. Commun.* **16**, 979 (1984). HPLC determn in plasma: E. Wahlin-Boll *et al., Eur. J. Clin. Pharmacol.* **20**, 375 (1981). Review of clinical pharmacokinetics and efficacy in vascular disease: L. R. Wiseman *et al., Drugs* **44**, 445-464 (1992); in atherothrombosis: N. Bhana, K. J. McClellan, *Drugs Aging* **18**, 369-388 (2001).

Crystals from ethanol, mp 182-184°.

THERAP CAT: Antithrombotic.

5002. Indocyanine Green. [3599-32-4] 2-[7-[1,3-Dihydro-1,1-dimethyl-3-(4-sulfobutyl)-$2H$-benz[e]indol-2-ylidene]-1,3,5-heptatrien-1-yl]-1,1-dimethyl-3-(4-sulfobutyl)-$1H$-benz[e]indolium inner salt sodium salt (1:1); anhydro-3,3,3',3'-tetramethyl-1,1'-bis(4-sulfobutyl)-4,5,4',5'-dibenzoindotricarbocyanine hydroxide inner salt sodium salt; Fox green; Cardio-Green. $C_{43}H_{47}N_2NaO_6S_2$; mol wt 774.97. C 66.64%, H 6.11%, N 3.61%, Na 2.97%, O 12.39%, S 8.27%. Prepn: Heseltine, Brooker, **US 2895955** (1959 to Kodak).

A tricarbocyanine type of dye with infrared absorbing properties; peak absorption at about 800 nm. Has little or no absorption in the visible. Commercial product contains moisture and about 5% sodium iodide as contaminant. Sol in water, methanol. Practically insol in most other organic solvents.

USE: In infrared photography; in prepn of Wratten filters.

THERAP CAT: Diagnostic aid (blood volume determination, cardiac output, hepatic function).

5003. Indole. [120-72-9] 1*H*-Indole; 2,3-benzopyrrole; 1-azaindene; 1-benzazole; benzo[*b*]pyrrole. C$_8$H$_7$N; mol wt 117.15. C 82.02%, H 6.02%, N 11.96%. Obtained from the 240-260° fraction from coal tar: Weissgerber, *Ber.* **43**, 3520 (1910); from feces: Bergeim, *J. Biol. Chem.* **32**, 17 (1917). Prepn from *o*-formotoluide: Tyson, *Org. Synth.* **coll. vol. III**, 479 (1955); by dehydrocyclizing ortho alkyl anilines: Erner *et al.*, **US 2953575** (1960 to Houdry Process); from *N*-(2-tolyl)-*N'*-methyl-*N'*-phenylformamidine: Lorenz *et al.*, *J. Org. Chem.* **30**, 2531 (1965). Toxicity data: Smyth *et al.*, *Am. Ind. Hyg. Assoc. J.* **23**, 95 (1962). Comprehensive review: Van Order, Lindwall, *Chem. Rev.* **30**, 69 (1942); D. W. Bannister in *Kirk-Othmer Encyclopedia of Chemical Technology* **vol. 13** (Wiley-Interscience, New York, 3rd ed., 1981) pp 213-222.

Leaflets, mp 52°. bp$_{762}$ 253°; bp$_{28}$ 128-133°. Intense fecal odor. Volatile with steam. Soluble in hot water, hot alcohol, ether, benzene. LD$_{50}$ orally in rats: 1 g/kg (Smyth).

USE: In highly dil solns the odor is pleasant, hence indole has been used in perfumery.

5004. Indoleacetic Acid. [87-51-4] 1*H*-Indole-3-acetic acid; 3-(carboxymethyl)indole; 3-indolylmethylcarboxylic acid; heteroauxin; IAA. C$_{10}$H$_9$NO$_2$; mol wt 175.19. C 68.56%, H 5.18%, N 8.00%, O 18.26%. Plant hormone; recognized as the principal auxin of higher plants. Prepd by the reaction of indole with potassium glycolate at 250°: Johnson, Crosby, *J. Org. Chem.* **28**, 1246 (1963). From indole and chloroacetic acid: Shagalov *et al.*, **US 3320281** (1967). *Reviews:* Leopold in *The Hormones* **vol. IV**, G. Pincus *et al.*, Eds. (Academic Press, New York, 1964) pp 1-66; Thimann in *The Physiology of Plant Growth and Development*, M. B. Wilkins, Ed. (McGraw-Hill, New York, 1969) pp 1-45.

Leaflets or cryst powder from water, mp 168-170°. pK 4.75. Sparingly sol in water or chloroform; freely sol in alc; sol in acetone, ether.

USE: Plant growth regulator.

5005. Indolebutyric Acid. [133-32-4] 1*H*-Indole-3-butanoic acid; indole-3-butyric acid; 4-(3-indolyl)butyric acid; Seradix. C$_{12}$H$_{13}$NO$_2$; mol wt 203.24. C 70.92%, H 6.45%, N 6.89%, O 15.74%. Prepn: Jackson, Manske, *J. Am. Chem. Soc.* **52**, 5029 (1930); by heating indole, γ-butyrolactone, and sodium hydroxide, followed by acidification of the product: Fritz, **US 3051723** (1962 to Union Carbide); by decarboxylation of 2-carboxyindole-3-butyric acid: Bowman, Islip, *Chem. Ind. (London)* **1971**, 154. Toxicity studies: Anderson *et al.*, *Proc. Soc. Exp. Biol. Med.* **34**, 138 (1936); Pesonen *et al.*, *Acta Endocrinol.* **5**, 409 (1950). Hypoglycemic effect in rats: Mirsky *et al.*, *Endocrinology* **59**, 715 (1956).

White or slightly yellow crystals. Slight characteristic odor. mp 123-125°. Practically insol in water, chloroform. Sol in alc, ether, acetone. LD$_{50}$ i.p. in mice: 100 mg/kg (Anderson).

USE: To stimulate root formation of plant clippings.

5006. Indolicidin. [140896-21-5] L-Isoleucyl-L-leucyl-L-prolyl-L-tryptophyl-L-lysyl-L-tryptophyl-L-prolyl-L-tryptophyl-L-tryptophyl-L-prolyl-L-tryptophyl-L-arginyl-L-argininamide. C$_{100}$H$_{132}$N$_{26}$O$_{13}$; mol wt 1906.33. C 63.01%, H 6.98%, N 19.10%, O 10.91%. Cationic antimicrobial peptide of the cathelicidin family. Isolated from bovine neutrophils; has physiological role in host defense. Active vs Gram-negative and -positive bacteria, fungi, and protozoa. Named for its indole-rich structure, composed of 39% tryptophan and 23% proline. Identification: M. E. Selsted *et al.*, *J. Biol. Chem.* **267**, 4292 (1992). Synthesis: R. J. Van Abel *et al.*, *Int. J. Pept. Protein Res.* **45**, 401 (1995). Structural study in detergent micelles: A. Rozek *et al.*, *Biochemistry* **39**, 15765 (2000). Interaction with microbial cell membrane: T. J. Falla *et al.*, *J. Biol. Chem.* **271**, 19298 (1996); J. E. Shaw *et al.*, *J. Struct. Biol.* **154**, 42 (2006); with target cell DNA: C.-H. Hsu *et al.*, *Nucleic Acids Res.* **33**, 1053 (2005).

Ile–Leu–Pro–Trp–Lys–Trp–Pro–Trp–Trp–Pro–Trp–Arg–Arg–NH$_2$

5007. Indolmycin. [21200-24-8] (5*S*)-5-[(1*R*)-1-(1*H*-Indol-3-yl)ethyl]-2-(methylamino)-4(5*H*)-oxazolone; (1*R*,5*S*)-5-(1-indol-3-ylethyl)-2-(methylamino)-2-oxazolin-4-one; 2-methylamino-5α-(β-indolyl)ethyl-2-oxazolin-4-one; PA-155A. C$_{14}$H$_{15}$N$_3$O$_2$; mol wt 257.29. C 65.36%, H 5.88%, N 16.33%, O 12.44%. Antibiotic substance produced by *Streptomyces albus*: Rao, *Antibiot. Chemother.* **10**, 312 (1960); Marsh *et al.*, *ibid.* 316; **GB 862685** (1961 to Pfizer). Structure: M. Schach von Wittenau, H. Els, *J. Am. Chem. Soc.* **83**, 4678 (1961). Total synthesis: *eidem, ibid.* **85**, 3425 (1963); Preobrazhenskaya *et al.*, *Tetrahedron* **24**, 6131 (1968); T. Takeda, T. Mukaiyama, *Chem. Lett.* **1980**, 163. Abs config: T. H. Chan, R. K. Hill, *J. Org. Chem.* **35**, 3519 (1970). Biosynthetic studies: Hornemann *et al.*, *Chem. Commun.* **1969**, 245; *eidem, J. Am. Chem. Soc.* **93**, 3028 (1971).

Long rectangular prisms from methanol or ethyl acetate, mp 209-210°. [α]$_D^{25}$ −214° (c = 2 in methanol). uv max: 218 nm (E$_{1cm}^{1\%}$ 1960). Weakly basic, stable to heat. Slightly sol in water, benzene, ether; moderately in lower alcohols, acetone.

5008. 3-Indolylacetone. [1201-26-9] 1-(1*H*-Indol-3-yl)-2-propanone; 3-acetonylindole; 3-(2-oxopropyl)indole. C$_{11}$H$_{11}$NO; mol wt 173.22. C 76.27%, H 6.40%, N 8.09%, O 9.24%. Prepn: Brown *et al.*, *J. Chem. Soc.* **1952**, 3172; Williamson, *ibid.* **1962**, 2834; Morris *et al.*, **GB 974895** (1964 to Parke, Davis).

Brownish rhombs from benzene or needles from aq methanol. mp 115-117.5°. uv max (ethanol): 221, 280, 289 nm (ε 35100, 6400, 5300).

5009. Indomethacin. [53-86-1] 1-(4-Chlorobenzoyl)-5-methoxy-2-methyl-1*H*-indole-3-acetic acid; 1-(*p*-chlorobenzoyl)-5-methoxy-2-methyl-3-indolylacetic acid; indometacin; Bonidon; Catlep; Chrono-Indocid; Confortid; Dolcidium; Elmetacin; Idomethine; Inacid; Indacin; Indocid; Indocin; Indomed; Indomee; Indomod; Indoptic; Indoptol; Indoxen; Inflazon; Inteban; Mikametan; Rheumacin LA; Serastar; Vonum. C$_{19}$H$_{16}$ClNO$_4$; mol wt 357.79. C 63.78%, H 4.51%, Cl 9.91%, N 3.91%, O 17.89%. Nonsteroidal anti-inflammatory agent (NSAID); inhibits prostaglandin synthesis. Prepn: T. Y. Shen *et al.*, *J. Am. Chem. Soc.* **85**, 488 (1963); T. Y. Shen, **US 3161654** (1964 to Merck & Co.). Alternate process: **BE 679678** (1966 to Sumitomo). HPLC determn in plasma and urine: T. B. Vree *et al.*, *J. Chromatogr.* **616**, 271 (1993). Anti-inflammatory and antipyretic activity: C. A. Winter *et al.*, *J. Pharmacol. Exp. Ther.* **141**, 369 (1963). Metabolic studies: D. W. Yesair *et al.*, *Biochem. Pharmacol.* **19**, 1579 (1970). Toxicity: C. D. Klaassen, *Toxicol. Appl. Pharmacol.* **38**, 127 (1976). Review of discovery, pharmacology and mechanism of action: T. Y. Shen, C. A. Winter in *Advances in Drug Research* **vol. 12**, A. B. Simmons, Ed. (Academic Press, New York, 1977) pp 89-245; of clinical trials in osteoarthritis: W. M. O'Brien, *Clin. Pharmacol. Ther.* **9**, 94-107 (1968); of role in ductus closure: S. M. Douidar *et al.*, *Dev. Pharmacol. Ther.* **11**, 196-212 (1988).

Crystals exhibiting polymorphism, mp for one form ~155°, for the other ~162°. uv max (ethanol): 230, 260, 319 nm (ε 20800, 16200, 6290). pKa 4.5. Sol in acetone, castor oil; sparingly sol in alc, chloroform, ether. Practically insol in water. Stable in neutral or slightly acidic media; dec by strong alkali. LD$_{50}$ i.p. in rats: 13 mg/kg (Klaassen).

Sodium salt trihydrate. [74252-25-8] Indocin I.V.; Indocid PDA. $C_{19}H_{15}ClNNaO_4.3H_2O$. Pale yellow crystalline powder. pH of 1% soln: 8.4. Very sol in methanol; sol in water, ethanol. Very slightly sol in chloroform, acetone.

Meglumine salt. Liometacen.

THERAP CAT: Anti-inflammatory; analgesic. For closure of patent ductus arteriosus in preterm infants.

5010. Indoramin. [26844-12-2] *N*-[1-[2-(1*H*-Indol-3-yl)ethyl]-4-piperidinyl]benzamide; 3-[2-(4-benzamidopiperidino)ethyl]indole; Wy-21901. $C_{22}H_{25}N_3O$; mol wt 347.46. C 76.05%, H 7.25%, N 12.09%, O 4.60%. α$_1$-Adrenergic blocking agent with antihypertensive and bronchodilating activity. Prepn: J. L. Archibald, J. L. Jackson, **ZA 6803204**; *eidem*, **US 3527761** (1969, 1970 both to Wyeth); J. L. Archibald *et al.*, *J. Med. Chem.* **14**, 1054 (1971). Pharmacological study: R. B. Royds *et al.*, *Clin. Pharmacol. Ther.* **13**, 380 (1972). Pharmacokinetics: G. H. Draffan *et al.*, *Br. J. Clin. Pharmacol.* **3**, 489 (1976). Antihypertensive activity: G. S. Stokes *et al.*, *Clin. Pharmacol. Ther.* **25**, 783 (1979). Cardiovascular effects in man: A. J. Coleman *et al.*, *J. Int. Med. Res.* **7**, 511 (1979). Effects on respiration in guinea pigs: C. Hamer, D. M. Temple, *Agents Actions* **10**, 399 (1980). Determn of therapeutic concentrations: A. J. Swaisland, *Analyst* **106**, 717 (1981). Symposium on pharmacology and clinical studies: *Br. J. Clin. Pharmacol.* **12**, Suppl. 1, 1S-140S (1981). Review of pharmacology, therapeutic efficacy: B. Holmes, E. M. Sorkin, *Drugs* **31**, 467-499 (1986).

Crystals from ethanol, mp 208-210°.

Hydrochloride. [38821-52-2] Baratol; Doralese; Vidora; Wydora; Wypresin. $C_{22}H_{25}N_3O.HCl$; mol wt 383.92. Crystals, mp 230-232°. Recryst from isopropanol gives different cryst modification, mp 258-260°.

THERAP CAT: Antihypertensive.

5011. Indospicine. [16377-00-7] (2*S*)-2,7-Diamino-7-iminoheptanoic acid; L-6-amidinonorleucine; L-2-amino-6-amidinohexanoic acid; L-α-amino-ε-amidinocaproic acid. $C_7H_{15}N_3O_2$; mol wt 173.22. C 48.54%, H 8.73%, N 24.26%, O 18.47%. Naturally occurring analog of arginine. Teratogenic and hepatotoxic factor found in extracts of *Indigofera spicata* Forsk (*Indigofera endecaphylla* Jacq.), *Leguminosae*. Isoln: M. P. Hegarty, A. W. Pound, *Nature* **217**, 354 (1968). Isoln, structure and biological studies: *eidem*, *Aust. J. Biol. Sci.* **23**, 831 (1970). Total synthesis: Culvenor *et al.*, *Aust. J. Chem.* **24**, 371 (1971). *Review:* M. P. Hegarty, *Australas. J. Dermatol.* **14**, 35-38 (1973).

Monohydrochloride monohydrate. [76467-71-5] $C_7H_{15}N_3$-$O_2.HCl.H_2O$; mol wt 227.69. Needle-like crystals from aq ethanol, mp 131-134°. [α]$_D^{22}$ +18° (c = 1.1 in 5*N* HCl).

5012. Indoxacarb. [173584-44-6] (4a*S*)-7-Chloro-2,5-dihydro-2-[[(methoxycarbonyl)[4-(trifluoromethoxy)phenyl]amino]-carbonyl]indeno[1,2-*e*][1,3,4]oxadiazine-4a(3*H*)-carboxylic acid methyl ester; DPX-KN128; Advion; Avaunt; Provaunt; Steward. $C_{22}H_{17}ClF_3N_3O_7$; mol wt 527.84. C 50.06%, H 3.25%, Cl 6.72%, F 10.80%, N 7.96%, O 21.22%. Oxadiazine pro-insectide; bioactivated in lepidoptera to the decarbomethoxylated metabolite that inhibits neuronal sodium channels. Prepn: G. D. Annis *et al.*, **WO 9211249** (1992 to DuPont); *eidem*, **US 5462938** (1995). Physical properties and field trials: H. H. Harder *et al.*, *Brighton Crop Prot. Conf. - Pests Dis.* **1996**, 449. Efficacy of *DPX-MP062*, a 3:1 mixture of active (*S*)- to inactive (*R*)-isomer: U. Pluschkell *et al.*, *Pestic. Sci.* **54**, 85 (1998). Metabolism and mode of action: K. D. Wing *et al.*, *Crop Prot.* **19**, 537 (2000); B. Lapied *et al.*, *Br. J. Pharmacol.* **132**, 587 (2001).

(RS)-**Form.** [144171-61-9] DPX-JW062. mp 140-141°. Partition coefficient (octanol/water): ~40000. Vapor pressure (20-25°): <10^{-5} Pa. Soly (mg/l): water <0.5; 1-octanol 480; methanol 390; acetonitrile 76000; acetone 140000. LD$_{50}$ orally in rat: >5000 mg/kg; dermally in rabbit: >2000 mg/kg; LC$_{50}$ in rat by inhalation: >2 mg/l; LC$_{50}$ (96 hr) in bluegill sunfish, rainbow trout: >1.0, >0.5 mg/l (Harder).

USE: Insecticide.

5013. Indoxyl Sulfate. [487-94-5] 1*H*-Indol-3-ol 3-(hydrogen sulfate); 1*H*-indol-3-ol hydrogen sulfate ester; indol-3-yl sulfate; 3-indoxylsulfuric acid; indican; metabolic indican. $C_8H_7NO_4S$; mol wt 213.21. C 45.07%, H 3.31%, N 6.57%, O 30.02%, S 15.04%. Uremic toxin produced as a result of bacterial degradation of dietary tryptophan in the gut. Elevated levels have been detected in serum of patients with chronic kidney disease; elevated urinary levels may be diagnostic for certain malabsorption syndromes. Unstable as the free acid. Prepn of potassium salt from aqueous indoxyl and potassium bisulfate: Baeyer, *Ber.* **14**, 1741 (1881); from chlorosulfonic acid and *N*-acetylindoxyl in pyridine: Schwenk, Jolles, *Biochem. Z.* **69**, 467 (1915); by persulfate oxidation of indole: Boyland *et al.*, *Biochem. J.* **62**, 546 (1956). HPLC determn in serum of patients with renal failure: L. A. Stanfel *et al.*, *Clin. Chem.* **32**, 938 (1986). Use as biomarker for malabsorption syndromes and bacterial contamination of the bowel: B. H. Novis *et al.*, *S. Afr. Med. J.* **45**, 1167 (1971). Review of toxicology and role in chronic kidney disease: T. Niwa, *Nagoya J. Med. Sci.* **72**, 1-11 (2010).

OSO₃H

Potassium salt. [2642-37-7] Indol-3-yl potassium sulfate; urinary indican; potassium indoxyl sulfate. $C_8H_6KNO_4S$; mol wt 251.30. Light brown plates from aq alc, dec 179-180° with sublimation. Very sol in water. Practically insol in cold alcohol.

Note: Do not confuse with plant indican, *q.v.*, which is indoxyl-β-D-glucose.

5014. Infliximab. [170277-31-3] Anti-(human tumor necrosis factor) immunoglobulin G (human-mouse monoclonal cA2 heavy chain); potassium indoxyl sulfate. chain, dimer; chimeric A2 antibody; cA2; Remicade. Chimeric monoclonal antibody that binds and neutralizes soluble and transmembrane tumor necrosis factor α (TNFα). Consists of constant regions of human (Hu)IgG1κ coupled to Fv region of a high-affinity neutralizing murine anti-HuTNFα antibody (A2). Prepn: J. Le *et al.*, **WO 9216553** (1992 to New York Univ.; Centocor); D. M. Knight *et al.*, *Mol. Immunol.* **30**, 1443 (1993). Pharmacology: S. A. Siegel *et al.*, *Cytokine* **7**, 15 (1995). Mechanism of action: B. J. Scallon *et al.*, *ibid.* 251. Clinical study in rheumatoid arthritis: R. Maini *et al.*, *Lancet* **354**, 1932 (1999); in psoriasis: U. Chaudhari *et al.*, *ibid.* **357**, 1842 (2001); in Crohn's disease: B. E. Sands *et al.*, *N. Engl. J. Med.* **350**, 876 (2004). Review of use in early rheumatoid arthritis: R. C. Geletka, E. W. St. Clair, *Expert Opin. Biol. Ther.* **5**, 405-417 (2005); of use in Crohn's disease: M. Bewtra, G. R. Lichtenstein, *ibid.*, 589-599 (2005).

THERAP CAT: Anti-inflammatory; antipsoriatic.

5015. Infusorial Earth. Siliceous earth; diatomaceous earth; fossil flour; kieselguhr; Celite; Super-Cel. Siliceous frustules and fragments of various species of diatoms. *See also* Silicon Dioxide.

White to light gray to pale buff powder. Insol in water, acids, or dil alkalies. Capable of taking up and holding about four times its wt of water.

USE: Clarifying agent. Largely used as an absorbent for liquids and for dispensing fluid extracts in powder form; also in cataplasms and as constituent of and excipient for pill masses. Clarifying oils, varnishes; filtering liquids; manuf heat insulators, fire brick, and fire- and acid-proof packing materials; filler for paper, paints; absorbent dynamite; in metal polishes, dentifrices, nail polishes; in chromatography.

5016. Inhibins. Gonadal polypeptide hormones produced by Sertoli cells in males and granulosa cells in females, which selectively inhibits the secretion of FSH, *q.v.* Dimer, mol wt 32,000 Da, isolated in two isoforms, inhibin A and inhibin B, which have the same α subunit linked to similar but distinct β subunits. Dimers of the β subunits have been isolated which stimulate the secretion of FSH, *see* activins. Name "inhibin" coined upon identification as a water-soluble, testicular factor with pituitary inhibitory activity: D. R. McCullagh, *Science* **76**, 19 (1932). Specific inhibition of FSH: H. F. Klinefelter *et al.*, *J. Clin. Endocrinol.* **2**, 615 (1942); B. P. Setchell, F. Jacks, *J. Endocrinol.* **62**, 675 (1974). Identification in ovarian follicular fluid: F. H. de Jong, R. M. Sharpe, *Nature* **263**, 71 (1976); N. B. Schwartz, C. P. Channing, *Proc. Natl. Acad. Sci. USA* **74**, 5721 (1977). Localization of production to Sertoli cells: A. Steinberger, E. Steinberger, *Endocrinology* **99**, 918 (1976); to granulosa cells: G. F. Erickson, A. J. W. Hsueh, *ibid.* **103**, 1960 (1978). Immunohistochemical confirmation of production sites: P. Cuevas *et al.*, *Biochem. Biophys. Res. Commun.* **142**, 23 (1987). Identification of inhibin-like proteins from human seminal plasma: N. G. Seidah *et al.*, *FEBS Lett.* **175**, 349 (1984); C. H. Li *et al.*, *Proc. Natl. Acad. Sci. USA* **82**, 4041 (1985). Isolation of inhibins and higher mol wt precursors: K. Miyamoto *et al.*, *Biochem. Biophys. Res. Commun.* **129**, 396 (1985). Purification and characterization of inhibins A and B: J. Rivier *et al.*, *ibid.* **133**, 120 (1985); N. Ling *et al.*, *Proc. Natl. Acad. Sci. USA* **82**, 7217 (1985). Each subunit is synthesized as a larger precursor molecule, then processed to the mature form. Complete amino acid sequence of mature form α, βA and βB subunits: A. J. Mason *et al.*, *Nature* **318**, 659 (1985); A. J. Mason *et*

al., *Biochem. Biophys. Res. Commun.* **135**, 957 (1986). Demonstration of negative feedback mechanism between secretion of FSH and inhibin: S.-Y. Ying *et al.*, *Proc. Natl. Acad. Sci. USA* **84**, 4631 (1987). Proposed nomenclature: H. G. Burger, *J. Endocrinol.* **117**, 159 (1988). Approach to nonradiometric assays: R. Schwall *et al.*, *Prog. Clin. Biol. Res.* **285**, 205 (1988). Review of early bioassays: B. Hudson *et al.*, *J. Reprod. Fertil. Suppl.* **26**, 17-29 (1979). Reviews of purification and physiology: C. P. Channing *et al.*, *Proc. Soc. Exp. Biol. Med.* **178**, 339-361 (1985); F. H. de Jong, D. M. Robertson, *Mol. Cell. Endocrinol.* **42**, 95-103 (1985). Review of possible use as contraceptive: A. R. Sheth, S. B. Moodbidri, *Adv. Contracep.* **2**, 131-139 (1986); P. Franchimont, *Male Contraception: Advances and Future Prospects*, G. I. Zatuchni *et al.*, Eds. (Harper & Row, Philadelphia, 1986) pp 408-418. *Reviews:* F. H. de Jong, *Oxford Rev. Reprod. Biol.* **9**, 1-53 (1987); N. Ling *et al.*, *Vitam. Horm.* (New York) **44**, 1-46 (1988).

5017. Iniparib. [160003-66-7] 4-Iodo-3-nitrobenzamide; BSI-201. $C_7H_5IN_2O_3$; mol wt 292.03. C 28.79%, H 1.73%, I 43.46%, N 9.59%, O 16.44%. Poly(ADP-ribose) polymerase-1 (PARP-1) inhibitor. Prodrug that is selectively reduced by tumor cells to the 3-nitroso derivative and induces tumor apoptosis. Prepn: E. Kun *et al.*, **WO 9426730**; *eidem*, **US 5464871** (1994, 1995 both to Octamer); and cytotoxic activity: J. Mendeleyev *et al.*, *Biochem. Pharmacol.* **50**, 705 (1995). Effect on PARP-1: P. I. Bauer *et al.*, *ibid.* **63**, 455 (2002). Mechanism of selective tumoricidal action: E. Kun *et al.*, *Mol. Med. Rep.* **2**, 739 (2009). Review of pharmacology and therapeutic potential: H. Liang, A. R. Tan, *IDrugs* **13**, 646-656 (2010).

Yellow crystals from acetonitrile, mp 152-155°. uv max (ethanol): 308, 242, 208 nm (ε 1590, 13100, 14500).

THERAP CAT: Antineoplastic.

5018. Inosine. [58-63-9] Hypoxanthine riboside; 9-β-D-ribofuranosylhypoxanthine; 1,9-dihydro-9-β-D-ribofuranosyl-6H-purin-6-one; hypoxanthosine; Inosie; Oxiamine; Ribonosine; Trophicardyl. $C_{10}H_{12}N_4O_5$; mol wt 268.23. C 44.78%, H 4.51%, N 20.89%, O 29.82%. In meat and meat extracts, in sugar beets. Prepd from adenosine by incubation with purified adenosine deaminase from intestine: Kalckar, *J. Biol. Chem.* **167**, 445 (1947); also by the action of sodium nitrite and acetic acid on adenosine: Levene, Jacobs, *Ber.* **43**, 3161 (1910); by the use of barium nitrite and H_2SO_4: Reiff *et al.*, **US 3049536** (1962 to Zellstoff-Fabrik Waldhof). Fermentation method: Motozaki *et al.*, **US 3111459** (1963 to Ajinomoto). Structure: Bredereck, *Ber.* **66**, 198 (1933); *Z. Physiol. Chem.* **223**, 61 (1934); Gulland, Holiday, *J. Chem. Soc.* **1936**, 765.

Dihydrate. Long rectangular plates from water, mp 90°. Anhydrous needles from 80% alc, dec 218° (rapid heating). $[\alpha]_D^{18}$ −49.2° (c = 0.9 in H_2O). $[\alpha]_{white}^{20}$ −73° (0.5 g + 2 ml N NaOH + 3 ml H_2O). 100 ml of the satd water soln at 20° contain 1.6 g inosine. Absorption spectrum: Kalckar, *loc. cit.* uv max (pH 6.0): 248.5 nm (ε 12200). Boiling with 0.1N H_2SO_4 yields hypoxanthin and D-ribose.

THERAP CAT: Activates cellular functions.

5019. Inosine Pranobex. [36703-88-5] Inosine compd with 1-(dimethylamino)-2-propanol 4-(acetylamino)benzoate (1:3:3);

inosine acedobene dimepranol; inosiplex; methisoprinol; NP-113; NPT-10381; Aviral; Delimmun; Imunoviral; Isoprinosin; Isoprinosina; Isoprinosine; Isoviral; Modimmunal; Pranosina; Pranosine; Viruxan. $C_{52}H_{78}N_{10}O_{17}$; mol wt 1115.25. C 56.00%, H 7.05%, N 12.56%, O 24.39%. Immunostimulant complex formed from the *p*-acetamidobenzoate salt of dimethylaminoisopropanol and inosine in a 3:1 molar ratio. Prepn: P. Gordon, **DE 1965431**; *idem*, **US 3646007** (1971, 1972 both to Newport Pharm.). Antiviral activity: E. R. Brown, P. Gordon, *Can. J. Microbiol.* **18**, 1463 (1972); R. L. Muldoon *et al.*, *Antimicrob. Agents Chemother.* **2**, 224 (1972). Stimulatory effect on T-cell function: L. Binderup, *Int. J. Immunopharmacol.* **7**, 93 (1985). Pharmacological and therapeutic potential: D. M. Campoli-Richards *et al.*, *Drugs* **32**, 383 (1986). Clinical immunopharmacology: A. J. Glasky, J. F. Gordon, *Cancer Detect. Prev. Suppl.* **1**, 597 (1987). Clinical trial in subacute sclerosing panencephalitis (SSPE): C. E. Jones *et al.*, *Lancet* **1**, 1034 (1982); G. Gascon *et al.*, *Brain Dev.* **15**, 346 (1993). Clinical trial in pre-AIDS patients: C. Pedersen *et al.*, *N. Engl. J. Med.* **322**, 1757 (1990). Review of efficacy in HIV infection: C. De Simone *et al.*, *Int. J. Immunopharmacol.* **13**, Suppl. 1, 19-27 (1991).

Neutral water-soluble solid. LD_{50} in mice and rats (mg/kg): >4000 orally and i.p. (Gordon).

THERAP CAT: Immunomodulator; antiviral.

5020. Inosinic Acid. [131-99-7] 5′-Inosinic acid; 5-inosinic acid; muscle inosinic acid; t-inosinic acid; hypoxanthine riboside-5-phosphoric acid; IMP. $C_{10}H_{13}N_4O_8P$; mol wt 348.21. C 34.49%, H 3.76%, N 16.09%, O 36.76%, P 8.90%. Prepn from meat extract: Levene, Bass, *Nucleic Acids* (New York, 1931) p 229; from dried sardines: Yoshida, Kageyama, **JP 56 732** (1956 to Ajinomoto), *C.A.* **51**, 3870b (1957). Structure: Levene, Bass, *op. cit.*, pp 187-192; Bredereck, *Ber.* **66**, 198 (1933); Levene, Tipson, *J. Biol. Chem.* **111**, 313 (1935). Also prepd from muscle by enzymatic deamination of muscle adenylic acid: Ostern, *Biochem. Z.* **254**, 65 (1932); by hydrolysis of inosine triphosphate: Kleinzeller, *Biochem. J.* **36**, 729 (1942). Studies on the enzymatic synthesis: Greenberg, *J. Biol. Chem.* **190**, 611 (1951); Korn *et al.*, *ibid.* **217**, 875 (1955). Microbial fermentation method using mutant strains of *Micrococcus glutamicus:* Kinoshita *et al.*, **US 3232844** (1966 to Kyowa).

Syrup, solidifies to a glass when dried over H_2SO_4. Agreeable sour taste. $pK_1 = 2.4$; $pK_2 = 6.4$. Absorption spectrum: Kalckar, *J. Biol. Chem.* **167**, 445 (1947). Freely sol in water, in formic acid; very sparingly sol in alcohol, ether. On boiling with acid hydrolyzes to 1 mol H_3PO_4, 1 mol hypoxanthine, 1 mol D-ribose.

Disodium salt dihydrate. $C_{10}H_{11}N_4Na_2O_8P \cdot 2H_2O$. Barely sol in alcohol, ether, acetone; soly in water at 20° about 13 g/100 ml. Kawasaki, *New Food Ind.* **3**, no. 1, 17 (1961).

Barium salt. $C_{10}H_{11}BaN_4O_8P$. Hemipentadecahydrate, lustrous leaflets. Becomes anhydr at 100° *in vacuo*. $[\alpha]_D^{20}$ −18.5° (0.3 g of anhydr Ba salt in 10 ml of 2.5% HCl).

USE: Its salts as flavor intensifiers, like sodium glutamate. Examples of mixtures of sodium inosinate and sodium glutamate or other salts: Toi *et al.*, **US 3109741** (1963 to Ajinomoto).

5021. Inosital. [87-89-8] *myo*-Inositol; *meso*-inositol; *i*-inositol; hexahydroxycyclohexane; cyclohexanehexol; cyclohexitol; meat sugar; inosite; mesoinosite; phaseomannite; dambose; nucite; bios I; rat antispectacled eye factor; mouse antialopecia factor. $C_6H_{12}O_6$; mol wt 180.16. C 40.00%, H 6.71%, O 53.28%. Widely distributed in plants and animals. Growth factor for animals and microorganisms. Isoln from heart muscle: Scherer, *Ann.* **73**, 322 (1850); from liver: Woolley, *J. Biol. Chem.* **139**, 29 (1941). Synthesis: Wieland, Wishart, *Ber.* **47**, 2082 (1914); Anderson, Wallis, *J. Am. Chem. Soc.* **70**, 2931 (1948). Obtained commercially from corn steep liquor, since inositol is present as phytic acid in corn: Bartow, Walker, *Ind. Eng. Chem.* **30**, 300 (1938); **US 2112553** (1938); Hoglan, Bartow, *J. Am. Chem. Soc.* **62**, 2397 (1940); Elkin, Meadows, **US 2414365** (1947); **GB 601273** (1948 to Corn Prod. Refining). Nine possible stereoisomers: Seven are optically inactive or *meso*. Two optically active forms, the racemic form, and several *cis,trans*-isomers occur naturally. The prevalent natural form is *cis*-1,2,3,5-*trans*-4,6-cyclohexanehexol which is described here. *Reviews:* R. Beckmann, *m-Inosit* (Editio Cantor, Aulendorf, 1953); several authors in *The Vitamins* vol. 2, W. H. Sebrell, Jr., R. S. Harris, Eds. (Academic Press, New York, 1954) pp 321-386; *ibid.* **vol. 3** (2nd ed., 1971) pp 340-415.

Anhydr, non-hygroscopic crystals from water or acetic acid above 80°. Sweet taste. d 1.752. mp 225-227°. Optically inactive. Soly in water at 25°: 14 g/100 ml soln; at 60°: 28 g/100 ml soln. Practically insol in alc absolute, ether, and other common organic solvents. Aq solns are neutral to litmus.

Dihydrate. Efflorescent crystals from water below 50°. d 1.524. mp 218°. Becomes anhydr at 100°.

Monophosphate. [573-35-3] $C_6H_{13}O_9P$. Prepn: Posternak, Posternak, *Helv. Chim. Acta* **12**, 1165 (1929); McCormick, Carter, *Biochem. Prep.* **2**, 65 (1952). Crystals from water + alcohol, dec 195-197°. Titrates as a dibasic acid. Freely soluble in water (1 g dissolves in 3 ml H_2O). Practically insol in abs ethanol, ether. Remarkably resistant to hydrolysis by boiling with strong alkali. May be hydrolyzed by boiling with 6N HCl for 14 hrs.

THERAP CAT: Vitamin B complex; lipotropic.

5022. Inositol Niacinate. [6556-11-2] *myo*-Inositol hexa-3-pyridinecarboxylate; hexanicotinoyl inositol; hexanicotinyl *cis*-1,2,-3,5-*trans*-4,6-cyclohexane; inositol hexanicotinate; *meso*-inositol hexanicotinate; Dilcit; Dilexpal; Mesotal; Esantene; Hämovannid; Hexanicit; Hexopal; Linodil; Mesonex; Palohex. $C_{42}H_{30}N_6O_{12}$; mol wt 810.73. C 62.22%, H 3.73%, N 10.37%, O 23.68%. Prepn: Badgett, Woodward, *J. Am. Chem. Soc.* **69**, 2907 (1947).

Crystals, mp 254.3-254.9°. Practically insol in water. Sol in dil acids.

THERAP CAT: Vasodilator (peripheral).

5023. Insulin. [9004-10-8]; [11061-68-0] (human). Polypeptide hormone produced by pancreatic beta cells that regulates carbohydrate homeostasis. Converted by proteolysis from the single chain proinsulin, q.v., to the active dimer composed of 51 amino acid residues; mol wt ~6000. Regulates carbohydrate and lipid metabolism, and influences protein synthesis. Insulin was the first protein for which the chemical structure and mol wt were determined. Also the first commercial health care product produced by recombinant DNA technology. Because of its solubility at physiological pH, insulin is rapidly absorbed after subcutaneous injection. Various complexes with protamine and/or zinc have been prepd to improve drug delivery. In addition to biological source (human, porcine or bovine), insulin formulations for therapeutic use are classified according to onset and duration of action. Isoln: F. G. Banting, C. H. Best, *J. Lab. Clin. Med.* **7**, 251 (1921-22). Crystallization: Abel, *Proc. Natl. Acad. Sci. USA* **12**, 132 (1926). Purification and properties: J. Lens, *Biochim. Biophys. Acta* **2**, 76 (1948). Complete amino acid sequence of bovine insulin: F. Sanger, H. Tuppy, *Biochem. J.* **49**, 463, 481 (1951); F. Sanger, E. O. P. Thompson, ibid. **53**, 353, 366 (1953). Identification of 2 chain structure: A. P. Ryle et al., ibid. **60**, 541 (1955). Review of structure determination: F. Sanger, *Science* **129**, 1340 (1959). Structure of human insulin: D. Nichol, L. F. Smith, *Nature* **187**, 483 (1960). Crystal structure: D. C. Hodgkin, *Verh. Schweiz. Naturforsch. Ges.* **150**, 93 (1970). Synthesis of human insulin: P. G. Katsoyannis et al., *J. Am. Chem. Soc.* **88**, 164, 166 (1966); by the enzymatic modification of porcine insulin: M. A. Ruttenberg, *Science* **177**, 623 (1972). Review of synthetic insulins: P. G. Katsoyannis, *Recent Prog. Horm. Res.* **23**, 505-563 (1967). Synthesis of human insulin gene: H. M. Hsiung et al., *Nucleic Acids Res.* **6**, 1371 (1979); **7**, 2199 (1979); **8**, 5753 (1980); S. A. Narang et al., *Nucleic Acids Symp. Ser.* **7**, 377 (1980). Review of the development and production of human insulin by recombinant DNA technology: I. S. Johnson, *Science* **219**, 632-637 (1983). Molecular basis of insulin action: M. P. Czech, *Annu. Rev. Biochem.* **46**, 359 (1977). History: M. Bliss, *The Discovery of Insulin* (Univ. Chicago Press, Chicago, 1982) 304 pp. Review of biosynthesis: D. F. Steiner et al., *Recent Prog. Horm. Res.* **25**, 207-282 (1969). Review of the structure and function of the insulin receptor: J. Lee, P. F. Pilch, *Am. J. Physiol.* **266**, C319-C334 (1994). Symposium on the physiological regulation of insulin secretion and the pathogenesis of diabetes: *Diabetologia* **37**, Suppl. 2, S1-S187 (1994). Review of bioactivity, pharmacokinetics and therapeutic efficacy of human insulin: R. N. Brogden, R. C. Heel, *Drugs* **34**, 350-371 (1987). Review of insulin formulations and therapy: J. A. Galloway, R. E. Chance, *Horm. Metab. Res.* **26**, 591-598 (1994); of analogs and alternative delivery methods: J. E. Gerich, *Am. J. Med.* **113**, 308-316 (2002); of clinical development of inhaled formulations: L. D. Mastrandrea, T. Quattrin, *Adv. Drug Delivery Rev.* **58**, 1061-1075 (2006).

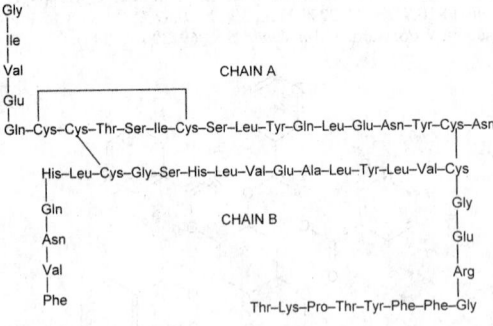

HUMAN INSULIN

Crystals, hexagonal system, usually obtained as flat rhombohedra and contg 0.4% Zn. Sol in dil acids and alkalies. Isoelectric point 5.30 to 5.35.

Bovine insulin. [11070-73-8] Hypurin.

Porcine insulin. [12584-58-6] Iletin II. Differs from human insulin by a single amino acid substitution.

Recombinant human insulin. Biosynthetic human insulin; insulin (prb); Huminsulin; Humulin; Insuman Rapid; Velosulin. Human insulin prepd by recombinant DNA technology. Clinical evaluation of inhaled intrapulmonary delivery in type 1 diabetes: J. S. Skyler et al., *Lancet* **357**, 331 (2001); in type 2 diabetes: W. T. Cefalu et al., *Ann. Intern. Med.* **134**, 203 (2001).

Semi-synthetic human insulin. Insulin (emp); Actrapid; Exubera; Humalog; Novolin. Human insulin prepd by enzymatic modification of porcine insulin.

Zinc insulin. [8049-62-5] Monotard; Ultratard; Vetsulin. Crystalline prepn of insulin containing 0.45-0.9% zinc. Formulated as suspensions in physiological saline; size of the particles determines the duration of action. Formulations are designated as prompt (or semilente), lente and extended (or ultralente).

Protamine zinc insulin. [9004-17-5] PZI insulin; PZI Vet. Suspensions of insulin modified by the addition of zinc chloride and protamine sulfate. White or almost white suspension, pH 7.1-7.4. Onset of action occurs from 4-6 hrs after s.c. injection; duration of action is 36 hrs.

Isophane insulin. NPH insulin; neutral protein Hagedorn insulin; Insulatard. Crystallized prepn of protamine, zinc and insulin. Prepn: H. C. Hagedorn et al., *J. Am. Med. Assoc.* **106**, 177 (1936). *Review:* P. Felig, ibid. **251**, 393-396 (1984). White suspensions of rod-shaped crystals ~30 nm in length, pH 7.1-7.4. Onset of action is 3-4 hrs following s.c. injection; duration of action is 18-28 hrs.

Insulin [131]**I.** Radio-iodinated insulin. Prepn: Burrows et al., *J. Clin. Invest.* **36**, 393 (1957); Grodsky et al., *Arch. Biochem. Biophys.* **81**, 264 (1959).
USE: Insulin ^{131}I used in the study of insulin binding factors from insulin resistant sera.

THERAP CAT: Antidiabetic.

THERAP CAT (VET): Antidiabetic.

5024. Insulinase. [9013-83-6] Insulin protease; insulin-degrading enzyme; insulysin; EC 3.4.24.56. An enzyme that hydrolyzes insulin and is prepd from hog pancreas: Brink, Lewis, US **2957809** (1960 to Merck & Co.). May be obtained from commercial pancreatin or trypsin. Even the purified crystals contain large amounts of elastase. *Review:* Thomas, *Postgrad. Med. J. Suppl.* **49**, 940 (1973).

5025. Insulin Aspart. [116094-23-6] 28B-L-Aspartic acid-insulin (human); AspB28-insulin (human); B28-asp-insulin; INA-X14; Novorapid. Rapid-acting insulin analog produced by recombinant DNA technology. Identical to human insulin except for one amino acid substitution. Prepn: J. Brange et al., *Nature* **333**, 679 (1988). Pharmacology and safety: V. Dall, *Arzneim.-Forsch.* **49**, 463 (1999). Clinical pharmacokinetics and dynamics: S. R. Mudaliar et al., *Diabetes Care* **22**, 1501 (1999). Clinical trial for postprandial glycemic control in type 1 diabetics: P. Raskin et al., ibid. **23**, 583 (2000). Review of pharmacology and clinical experience: K. L. Simpson, C. M. Spencer, *Drugs* **57**, 759-765 (1999).

THERAP CAT: Antidiabetic.

5026. Insulin Detemir. [169148-63-4] 29B-[N6-(1-Oxotetradecyl)-L-lysine](1A-21A),(1B-29B)-insulin (human); 29B-(N6-myristoyl-L-lysine)-30B-de-L-threonineinsulin (human); LysB29-tetradecanoyl, des(B30)-insulin; NN-304; Levemir. $C_{267}H_{402}N_{64}O_{76}S_6$; mol wt 5916.89. C 54.20%, H 6.85%, N 15.15%, O 20.55%, S 3.25%. Soluble long-acting basal insulin analog. Prepn: S. Havelund et al., **WO 9507931** (1995 to Novo Nordisk); and albumin binding affinity: P. Kurtzhals et al., *Biochem. J.* **312**, 725 (1995). Clinical pharmacokinetics: T. Danne et al., *Diabetes Care* **26**, 3087 (2003). Pharmacodynamic profile: J. Plank et al., ibid. **28**, 1107 (2005). Clinical comparison with NPH insulin in type 1 diabetes: P. Home et al., ibid. **27**, 1081 (2004). Review of pharmacology and clinical experience: P. Home, P. Kurtzhals, *Expert Opin. Pharmacother.* **7**, 325-343 (2006).

THERAP CAT: Antidiabetic.

5027. Insulin Glargine. [160337-95-1] 21A-Glycine-30Ba-L-arginine-30Bb-L-arginine-insulin (human); [Gly(A21),Arg(B31),Arg(B32)]insulin (human); HOE-901; Lantus. Long-acting analog of human insulin produced by recombinant DNA technology. Prepn: M. Dörschug, DE **3837825**; idem, US **5656722** (1990, 1997 both to Hoechst). Characterization of receptor interaction: L. Berti *et al., Horm. Metab. Res.* **30**, 123 (1998). Clinical pharmacodynamics: L. Heinemann *et al., Diabetes Care* **23**, 644 (2000); pharmacokinetics: D. R. Owens *et al., ibid.* 813. Clinical trial in type 1 diabetics: R. E. Ratner *et al., ibid.* 639. Review of clinical experience: P. S. Gillies *et al., Drugs* **59**, 253-260 (2000).

pI 6.7. Sol in acid pH. Insol at physiological pH.

THERAP CAT: Antidiabetic.

5028. Insulin Glulisine. [207748-29-6] 3^B-L-Lysine-29^B-L-glutamic acid-insulin (human); [Lys(B3),Glu(B29)]-insulin (human); HMR-1964; Apidra. $C_{258}H_{384}N_{64}O_{78}S_6$; mol wt 5822.64. C 53.22%, H 6.65%, N 15.40%, O 21.43%, S 3.30%. Rapid-acting insulin analog produced in *E. coli* by recombinant DNA technology. Identical to human insulin except for 2 amino acid substitutions. Prepn: J. Ertl *et al.,* **EP 885961** (1998 to Hoechst Marion Roussel); *eidem,* **US 6221633** (2001 to Aventis). Evaluation of mitogenic potential: I. Rakatzi *et al., Diabetes* **52**, 2227 (2003); of β-cell protective effect: *idem et al., Biochem. Biophys. Res. Commun.* **310**, 852 (2003). Clinical trial in type 2 diabetes: G. Dailey *et al., Diabetes Care* **27**, 2363 (2004). Review of clinical pharmacokinetics: R. H. A. Becker, A. D. Frick, *Clin. Pharmacokinet.* **47**, 7-20 (2008); of pharmacology and clinical experience: K. P. Garnock-Jones, G. L. Plosker, *Drugs* **69**, 1035-1057 (2009).

THERAP CAT: Antidiabetic.

5029. Insulin-like Growth Factors. IGFs. Family of conserved peptide hormones structurally homologous with insulin. Two major circulating forms mediate the growth promoting effects of somatotropin, *q.v.* IGF-I regulates both prenatal and postnatal growth; IGF-II is a key factor in fetal development. IGFs are produced primarily in the liver under the regulation of growth hormone; also produced locally by most tissues. Transported in the serum by *IGF binding proteins* (or *IGFBP*) that prolong the half-life and regulate the metabolic effects of IGFs. Discovered and termed *sulphation factors* because of their ability to stimulate the incorporation of sulfate by cartilage: W. D. Salmon, W. H. Daughaday, *J. Lab. Clin. Med.* **49**, 825 (1957). These peptides have also been referred to as *NSILA-S*, or non-suppressible insulin-like acting substance. The designation *somatomedin* was proposed to connote the intermediary relationship to somatotropin: W. H. Daughaday *et al., Nature* **235**, 107 (1972). Discussion of nomenclature: *idem et al., J. Clin. Endocrinol. Metab.* **65**, 1075 (1987). Isoln, chemical characterization, biological properties of IGF-I and IGF-II: E. Rinderknecht, R. E. Humbel, *Proc. Natl. Acad. Sci. USA* **73**, 2365, 4379 (1976). Amino acid sequence of IGF-I: *eidem, J. Biol. Chem.* **253**, 2769 (1978); of IGF-II: *eidem, FEBS Lett.* **89**, 283 (1978). Total synthesis of human IGF-I: C. H. Li *et al., Proc. Natl. Acad. Sci. USA* **80**, 2216 (1983); of human IGF-II: *eidem, Biochem. Biophys. Res. Commun.* **127**, 420 (1985). Review of molecular biology: W. H. Daughaday, P. Rotwein, *Endocr. Rev.* **10**, 68-91 (1989); of mechanism of action: A. Spagnoli, R. C. Rosenfeld, *Endocrinol. Metab. Clin. North Am.* **25**, 615-631 (1996). Review of IGF binding proteins: D. R. Clemmons, *Cytokine Growth Factor Rev.* **8**, 45-62 (1997). *Reviews:* C. E. H. Stewart, P. Rotwein, *Physiol. Rev.* **76**, 1005-1026 (1996); D. Le Roith, *N. Engl. J. Med.* **336**, 633-640 (1997).

Insulin-like Growth Factor I. [67763-96-6] IGF-I; somatomedin 1; somatomedin C; SM-C. Single chain, basic protein containing 70 amino acid residues. Review of physiology and potential therapeutic uses: E. R. Froesch *et al., Diabetes Metab. Rev.* **12**, 195-215 (1996).

Insulin-like Growth Factor II. [67763-97-7] IGF-II; multiplication-stimulating activity III-2; MSA III-2. Single chain, slightly acidic protein containing 66 or 67 amino acid residues depending on the species.

Human insulin-like growth factor I *see* Mecasermin.

5030. Insulin Lispro. [133107-64-9] 28^B-L-Lysine-29^B-L-prolineinsulin (human); [Lys(B28),Pro(B29)]-insulin (human); LY-275585; Humalog. Rapid-acting insulin analog produced in *E. coli* by recombinant DNA technology. Identical to human insulin except for the transposition of proline and lysine at positions 28 and 29 on the B chain. Prepn: R. E. Chance *et al.,* **EP 383472;** *eidem,* **US 5514646** (1990, 1996 both to Lilly). Study of immunogenicity: C. M. Zwickl *et al., Arzneim.-Forsch.* **45**, 524 (1995). General pharmacology: D. R. Helton *et al., ibid.* **46**, 91 (1996). Clinical compar-

ison with regular human insulin: J. H. Anderson, Jr. *et al., Diabetes* **46**, 265 (1997). Review of development and pharmacokinetics: F. Holleman, J. B. L. Hoekstra, *N. Engl. J. Med.* **337**, 176-183 (1997); of clinical trials: V. A. Koivisto, *Ann. Med.* **30**, 260-266 (1998). Series of articles on pharmacology and clinical experience: *Acta Clin. Belg.* **54**, 233-254 (1999).

THERAP CAT: Antidiabetic.

5031. Intedanib. [656247-17-5] (3Z)-2,3-Dihydro-3-[[[4-[methyl[2-(4-methyl-1-piperazinyl)acetyl]amino]phenyl]amino]-phenylmethylene]-2-oxo-1*H*-indole-6-carboxylic acid methyl ester; 3Z-[1-[4-[*N*-[(4-methylpiperazin-1-yl)methylcarbonyl]-*N*-methyl-amino]anilino]-1-phenylmethylene]-6-methoxycarbonyl-2-indolinone; BIBF-1120; Vargatef. $C_{31}H_{33}N_5O_4$; mol wt 539.64. C 69.00%, H 6.16%, N 12.98%, O 11.86%. Tyrosine kinase inhibitor acting on vascular endothelial growth factor receptor (VEGFR), platelet-derived growth foctor receptor (PDGFR), and fibroblast growth factor receptor (FGFR). Prepn: A. Heckel *et al.,* **WO 0127081;** G. J. Roth *et al.,* **US 6762180** (2001, 2004 both to Boehringer Ingelheim); and triple angiokinase inhibition: *eidem, J. Med. Chem.* **52**, 4466 (2009). Preclinical antitumor activity: F. Hilberg *et al., Cancer Res.* **68**, 4774 (2008). Clinical pharmacokinetics: I. Okamoto *et al., Mol. Cancer Ther.* **9**, 2825 (2010). Clinical evaluation in non-small cell lung cancer: M. Reck *et al., Ann. Oncol.* **22**, 1374 (2011); in ovarian cancer: J. A. Ledermann *et al., J. Clin. Oncol.* **29**, 3798 (2011). Review of development and therapeutic potential: M. Reck, *Expert Opin. Invest. Drugs* **19**, 789-794 (2010).

THERAP CAT: Antineoplastic.

5032. Integrins. Family of transmembrane glycoproteins involved in cellular adhesion and signal transduction. Name derived from their ability to "integrate" activities of the extracellular matrix (ECM) and the cytoskeleton. Integrins exhibit widespread evolutionary distribution: identified in mammals and other vertebrates, insects, and yeast; homologues have been identified in plants. At least 20 have been identified in mammals; composed of αβ heterodimers selected from among 16 α and 8 β subunits. The α subunits vary in size from 120-200 kDa with ~1000 amino acid residues and usually consist of a heavy and light chain joined by a disulfide bond; β subunits generally range from 90-120 kDa with ~800 amino acids. Three major subfamilies have been characterized. Most integrins bind to the Arg-Gly-Asp (RGD) amino acid sequence found on components of the ECM, such as fibronectin, *q.v.* Others bind to cell membrane proteins, such as the intercellular cell adhesion molecules (ICAMs), or to soluble ligands, such as fibrinogen, *q.v.* Several recognize more than one ligand. Integrins anchor cells to the ECM or to adjacent cells, regulate cell spreading and motility, and transduce extracellular stimuli into a variety of intracellular signals. Implicated in a variety of physiological processes including embryological development, wound healing, immune functions, thrombosis, and metastasis. Identification of membrane glycoproteins involved in cell adhesion: D. E. Wylie *et al., J. Cell Biol.* **80**, 385 (1979). Identification of the transmembrane link between the ECM and the cytoskeleton: J. W. Tamkun *et al., Cell* **46**, 271 (1986). Description of integrin family: R. O. Hynes, *ibid.* **48**, 549 (1987); C. A. Buck, A. F. Horwitz, *Annu. Rev. Cell Biol.* **3**, 179-205 (1987). Role of RGD sequence in cell adhesion: E. Ruoslahti, M. D. Pierschbacher, *Science* **238**, 491 (1987). Review of integrin structure and ligand binding: A. Sonnenberg, *Curr. Top. Microbiol. Immunol.* **184**, 7-35 (1993); D. S. Tuckwell, M. J. Humphries, *Crit. Rev. Oncol. Hematol.* **15**, 149-171 (1993). Review of pharmacology: D. Cox *et al., Med. Res. Rev.* **14**, 195-228 (1994). Review of role in signal transduction: A. Richardson, J. T. Parsons, *BioEssays* **17**, 229-236 (1995); E. A. Clark, J. S. Brugge, *Science* **268**, 233-239 (1995).

β₁-Integrins. VLA antigens; VLA integrins; very late activation antigens. Widely distributed on various cell types. Share a common β_1 chain complexed with various α chains. Bind to ECM components such as fibronectin, collagen, laminin. Ca^{2+}/Mg^{2+}-dependent. Review: L. G. M. Baldini, L. M. Cro, *Leuk. Lymphoma* **12**, 197-203 (1994).

β₂-Integrins. Leukocyte integrins; Leu-Cam proteins; leukocyte adhesion molecules. Contain the β_2 chain; found on leukocytes. Bind to ICAMs, complement 3, and fibrinogen. Play an important role in regulating the immune system.

β₃-Integrins. Cytoadhesins. Contain the β_3 subunit; found on platelets, megakaryocytes and some melanoma cells. Ligands include fibrinogen, fibronectin, von Willebrand factor, vitronectin, and thrombospondin. Includes $\alpha_{IIb}\beta_3$, also known as **platelet glycoprotein IIb/IIIa (GPIIb-IIIa)**, a fibrinogen receptor involved in platelet aggregation. Review: M. H. Ginsberg *et al.*, *Thromb. Haemostasis* **70**, 87-93 (1993).

5033. Interferon. IFN. A family of species-specific vertebrate proteins that confer non-specific resistance to a broad range of viral infections, affect cell proliferation and modulate immune responses. Discovered by A. Isaacs and J. Lindenmann, *Proc. Roy. Soc.* **B147**, 258 (1957) while studying viral interference. Originally produced by the interaction of inactivated influenza virus with chick chorioallantoic membranes; subsequently found to be inducible by viable virus in a variety of cells: D. C. Burke, A. Isaacs, *Br. J. Exp. Pathol.* **39**, 452 (1958); in human leukocytes: I. Gresser, *Proc. Soc. Exp. Biol. Med.* **108**, 799 (1961). Host cell specificity and physical properties: D. A. J. Tyrrell, *Nature* **184**, 452 (1959); T. C. Merigan, *Science* **145**, 811 (1964). Production of acid-labile interferon by mitogen-stimulated human leukocytes: E. F. Wheelock, *ibid.* **149**, 310 (1966). Three major interferons, alpha, beta, and gamma, *q.q.v.*, have been identified based on antigenic and physicochemical properties, the nature of the inducer, and the cellular source from which they are derived, *cf. Nature* **286**, 110 (1980). Known collectively as type I interferon, IFNs-α and -β are structurally related, are stable at pH 2, and compete for the same cell surface receptor. IFN-γ, also known as type II interferon, is structurally unrelated to type I IFNs, is acid labile, and has a different receptor. Receptor binding study: A. A. Branca, C. Baglioni, *Nature* **294**, 768 (1981). *Reviews:* S. Baron, F. Dianzoni, Eds., *Tex. Rep. Biol. Med.* **35**, 1-573 (1977); *Ann. N.Y. Acad. Sci.* **350**, entitled "Regulatory Function of Interferons", J. Vilcek *et al.*, Eds. (1980) 643 pp; S. Baron *et al.*, Eds., *Tex. Rep. Biol. Med.* **41**, 1-715 (1982). Review of immunobiology and clinical significance: E. R. Stiem *et al.*, *Ann. Intern. Med.* **96**, 80-93 (1982); of pharmacology and toxicology: G. J. Mannering, L. B. Deloria, *Annu. Rev. Pharmacol. Toxicol.* **26**, 455-515 (1986); of IFN-gene family: C. Weissmann, H. Weber, *Prog. Nucleic Acid Res. Mol. Biol.* **33**, 251-300 (1986). Symposium on antiviral activity of natural and recombinant IFNs: *Antiviral Res.* **5**, Suppl. 1, 131-257 (1985). Review of structure, function and nomenclature: K. C. Zoon, *Interferon* **9**, 1-12 (1987). Books: *Interferons and Interferon Inducers*, N. Finter, Ed. (North Holland Publ. Co., Amsterdam, 1973) 598 pp; W. E. Stewart, *The Interferon System* (Springer, New York, 1979) 421 pp; *Interferons*, T. C. Merigan, R. M. Friedman, Eds. (Academic Press, New York, 1982) 481 pp; *Interferon*, K. Munk, H. Kirchner, Eds. (S. Karger, New York, 1982) 233 pp; H. Strander, *Interferon Treatment of Human Neoplasia* (Academic Press, New York, 1986) 265 pp.

5034. Interferon-α. Alfa-interferon; alpha-interferon; IFN-α; LeIF; leukocyte interferon; lymphoblastoid interferon. Cytokine with antiviral, antiproliferative, and immunomodulatory activity. Produced by leukocytes stimulated by virus, bacteria, or protozoa as part of the immune response; artificially induced by double-stranded RNA. Inhibits protein synthesis and viral replication; enhances cytotoxic activity. Multiple subtypes have been identified which contain ~166 amino acids and are variably glycosylated; mol wt 16-27.5 kDa. Identification of interferon produced by human leukocytes exposed to virus: I. Gresser, *Proc. Soc. Exp. Biol. Med.* **108**, 799 (1961). Purification and characterization: M. Rubinstein *et al.*, *Proc. Natl. Acad. Sci. USA* **76**, 640 (1979). Prepn by recombinant DNA technology: S. Nagata *et al.*, *Nature* **284**, 316 (1980); D. V. Goeddel *et al.*, *ibid.* **287**, 411 (1980). Comparison of structures and activities of HuIFN-α subtypes: M. Streuli *et al.*, *Science* **209**, 1343 (1980); D. V. Goeddel *et al.*, *Nature* **290**, 20 (1981). *Review:* A.

Meager in *Cytokines*, A. R. Mire-Sluis, R. Thorpe, Eds. (Academic Press, San Diego, 1998) pp 361-389. Review of pharmacology and clinical efficacy in malignant and viral disease: R. T. Dorr, *Drugs* **45**, 177-211 (1993); of use in metastatic renal cell carcinoma: A. Ravaud, M.-S. Dilhuydy, *Expert Opin. Biol. Ther.* **5**, 749-762 (2005); in chronic hepatitis C: M. Moriyama, Y. Arakawa, *Expert Opin. Pharmacother.* **7**, 1163-1179 (2006).

Interferon alfa-2a. [77907-69-8] Interferon-αA (human leukocyte protein moiety); IFN-αA; Ro-22-8181; Canferon; Roferon-A. Recombinant HuIFN-α produced in *E. coli*. Contains 165 amino acids; mol wt ~19 kDa. Symposium on clinical antineoplastic activity: *Semin. Oncol.* **12**, Suppl. 5, 1-34 (1985). Review of clinical efficacy in hepatitis C: C. M. Perry, M. I. Wilde, *BioDrugs* **10**, 65-89 (1998).

Interferon alfa-2b. [98530-12-2] Interferon α2 (human leukocyte clone pM21 protein moiety reduced); interferon α2 (human leukocyte clone Hif-SN206 protein moiety reduced); IFN-α_2; Sch-30500; YM-14090; Intron A; Viraferon. Recombinant HuIFN-α produced in *E. coli*. Mol wt ~19 kDa. Symposium on clinical antineoplastic activity: *Invest. New Drugs* **5**, Suppl, S1-S77 (1987). Clinical trial in hepatitis C: S. Schenker *et al.*, *J. Interferon Cytokine Res.* **17**, 665 (1997).

Interferon alfa-2c. [142192-09-4] Interferon α2 (human clone pAD19B-IFN protein moiety reduced). Clinical trial in chronic myelogenous leukemia: J. Thaler *et al.*, *Leuk. Res.* **21**, 75 (1997); in lung cancer: C. Prior *et al.*, *Eur. Respir. J.* **10**, 392 (1997).

Interferon alfacon-1. [118390-30-0] *N*-L-Methionyl-22-L-arginine-76-L-alanine-78-L-aspartic acid-79-L-glutamic acid-86-L-tyrosine-90-L-tyrosine-156-L-threonine-157-L-asparagine-158-L-leucine-interferon α1 (human lymphoblast reduced); consesus interferon; CIFN; Inferax; Infergen. Synthetic, recombinant interferon derived by assigning the most commonly observed amino acid in each position of several alpha IFN subtypes to create a consensus sequence. Contains 166 amino acids; mol wt 19.5 kDa. Molecular characterization and bioactivity: L. M. Blatt *et al.*, *J. Interferon Cytokine Res.* **16**, 489 (1996). Clinical trial in hepatitis C: M. J. Tong *et al.*, *Hepatology* **26**, 747 (1997).

Interferon alfa-n1. NSC-339140; Wellferon. Mixture of natural human α-interferons derived from lymphoblastoid cells exposed to Sendai virus. Review of clinical trials in hepatitis C: G. C. Farrell, *Hepatology* **26**, Suppl. 1, 96S-100S (1997); of pharmacology and clinical efficacy: C. M. Perry, A. J. Wagstaff, *BioDrugs* **9**, 125-154 (1998).

Interferon alfa-n3. Alferon. Mixture of natural, human α-interferons produced by human leukocytes exposed to Sendai virus. mol wt 16-27 kDa. Clinical trial in hepatitis C: D. M. Simon *et al.*, *Hepatology* **25**, 445 (1997).

THERAP CAT: Antiviral; antineoplastic.

THERAP CAT (VET): Antiviral.

5035. Interferon-β. Beta-interferon; fibroblast interferon; FIF; IFN-β; IFN-β_1. Cytokine with antiviral, antiproliferative and immunomodulatory activity produced by fibroblasts in response to stimulation by live or inactivated virus or by certain synthetic polynucleotides. One of the type I interferons. Glycoprotein containing 166 amino acids; mol wt ~20 kDa. Production by human fibroblast cell cultures: E. A. Havell, J. Vilcek, *Antimicrob. Agents Chemother.* **2**, 476 (1972). Purification and initial characterization: E. Knight, Jr., *Proc. Natl. Acad. Sci. USA* **73**, 520 (1976). Amino acid analysis, partial sequence: *idem et al.*, *Science* **207**, 525 (1980); S. Stein *et al.*, *Proc. Natl. Acad. Sci. USA* **77**, 5716 (1980). Production by recombinant DNA technology: T. Taniguchi *et al.*, *Proc. Natl. Acad. Sci. USA* **77**, 5230 (1980); R. Derynck *et al.*, *Nature* **285**, 542 (1980); D. V. Goeddel *et al.*, *Nucleic Acids Res.* **8**, 4057 (1980). Review of therapeutic activity in viral disease and in multiple sclerosis (MS): J. J. Alam, *Curr. Opin. Biotechnol.* **6**, 688-691 (1995).

Interferon Beta (human). Feron; Fiblaferon; Frone. Natural interferon produced by cell cultured human fibroblasts. Clinical evaluation in MS: L. Jacobs *et al.*, *Arch. Neurol.* **44**, 589 (1987); in herpes simplex infections: M. Glezerman *et al.*, *Lancet* **1**, 150 (1988). Review of industrial prepn: M. Morandi, A. Valeri, *Adv. Biochem. Eng. Biotechnol.* **37**, 57-72 (1988).

Interferon Beta-1a. [145258-61-3] Avonex; Rebif. Recombinant human IFN-β produced in Chinese Hamster ovary cells. Glycoprotein consisting of 166 amino acid residues; contains ~11%

carbohydrate by weight. Clinical trial in MS: L. D. Jacobs *et al.*, *Ann. Neurol.* **39**, 285 (1996).

Interferon Beta-1b. [145155-23-3] IFN-beta$_{ser}$; Betaferon; Betaseron; Extavia. Nonglycosylated polypeptide produced in *E. coli*. Consists of 165 amino acid residues; mol wt ~18.5 kDa. Synthetic mutein having a serine substituted for the cysteine residue at position 17 of the native molecule. Prepn: D. F. Mark *et al.*, *Proc. Natl. Acad. Sci. USA* **81**, 5662 (1984). Clinical pharmacokinetics: O. A. Khan *et al.*, *Neurology* **46**, 1639 (1996). Clinical trial of long-term effect in MS: L. Kappos *et al.*, *Lancet Neurol.* **8**, 987 (2009). Review of pharmacology and clinical efficacy in MS: D. E. Goodkin, *Lancet* **344**, 1057-1060 (1994); in newly emerging MS: K. McKeage, *CNS Drugs* **9**, 787-792 (2008).

THERAP CAT: Antiviral; immunomodulator.

5036. Interferon-γ. [82115-62-6] Gamma-interferon; IFN-γ; immune IFN; ImIFN. Immunomodulatory cytokine produced by T-cells and natural killer (NK) cells as part of the immune response. Stimulates cellular immunity: activates macrophages, promotes antigen presentation, and enhances NK-cell activity. Regulates production of immunoglobulin G2a by B cells. Natural human IFN-γ is a glycoprotein containing 143 amino acids; mol wt 16-25 kDa. Forms non-covalent homodimers under physiological conditions. A type II interferon, IFN-γ is structurally unrelated to interferons-α and β, *q.q.v.*, is acid-labile, and has a different cell surface receptor. Discovery of interferon produced by non-infected human leukocytes stimulated by mitogen: E. F. Wheelock, *Science* **149**, 310 (1965). Production by antigen-stimulated leukocytes: J. A. Green *et al.*, *ibid.* **164**, 1415 (1969). Large-scale production by cell culture: M. P. Langford *et al.*, *Infect. Immun.* **26**, 36 (1979). Purification and characterization: Y. K. Yip *et al.*, *Proc. Natl. Acad. Sci. USA* **79**, 1820 (1982). Prepn of recombinant human IFN-γ: P. W. Gray *et al.*, *Nature* **295**, 503 (1982). Amino acid sequence: E. Rinderknecht *et al.*, *J. Biol. Chem.* **259**, 6790 (1984). Review of biological activities: E. M. Bonnem, R. K. Oldham, *J. Biol. Response Modif.* **6**, 275-301 (1987); U. Boehm *et al.*, *Annu. Rev. Immunol.* **15**, 749-795 (1997). Review of clinical experience in antimicrobial defense: H. W. Murray, *Intensive Care Med.* **22**, Suppl. 4, S456-S461 (1996). *Review:* E. De Maeyer, J. De Maeyer-Guignard, in *Cytokines* (Academic Press, San Diego, 1998) pp 391-400.

Interferon Gamma-1a. [98059-18-8] N^2-[*N*-(*N*-L-Cysteinyl-L-tyrosyl)-L-cysteinyl]-interferon γ (human lymphocyte protein moiety reduced); Cys-Tyr-Cys-interferon γ; S-6810; Immuneron; Polyferon. Nonglycosylated HuIFN-γ produced in *E. coli* by recombinant DNA technology. Consists of 146 amino acid residues; mol wt ~17 kDa. Clinical trial in rheumatoid arthritis: E. M. Lemmel *et al.*, *Rheumatol. Int.* **8**, 87 (1988).

Interferon Gamma-1b. [98059-61-1] N^2-L-Methionyl-1-139-interferon γ (human lymphocyte protein moiety reduced); interferon gamma-2a; Actimmune; Immukin. Nonglycosylated HuIFN-γ produced in *E. coli* by recombinant DNA technology. Consists of 140 amino acid residues; mol wt ~16.5 kDa. Review of pharmacology and therapeutic potential in chronic granulomatous disease: P. A. Todd, K. L. Goa, *Drugs* **43**, 111-122 (1992).

THERAP CAT: Immunomodulator.

5037. Interleukin-1. IL-1; lymphocyte activating factor; LAF; B-cell activating factor; BAF; T-cell replacing factor; TRF; endogenous pyrogen; leukocytic endogenous mediator; mononuclear cell factor. Immunoenhancing, pyrogenic, polypeptide factor produced in blood and in a variety of tissues by mononuclear phagocytes responding to antigens or inflammatory agents. Responsible for a wide variety of bioactivities. Elicits a non-antigen specific amplification of cellular and humoral immune responses. Induces production of interleukin-2 (IL-2), collagenase and prostaglandins, *q.q.v.* Initial identification of factor inducing T-cell proliferation: I. Gery *et al.*, *J. Exp. Med.* **136**, 128, 143 (1972); of factor enhancing antibody production: D. D. Wood, S. L. Gaul, *J. Immunol.* **113**, 925 (1974). Isoln and preliminary chemical characterization: I. Gery, R. E. Handschumacher, *Cell. Immunol.* **11**, 162 (1974); D. D. Wood, *J. Immunol.* **123**, 2395 (1979). Definition and nomenclature: L. A. Aarden *et al.*, *ibid.* 2928. Identity with endogenous pyrogen: L. J. Rosenwasser, C. A. Dinarello, *Cell. Immunol.* **63**, 134 (1981). Purification of murine IL-1: S. B. Mizel, D. Mizel, *J. Immunol.* **126**, 834 (1981); of human IL-1: J. A. Schmidt, *J. Exp. Med.* **160**, 772 (1984). Exists in several biochemically distinct forms exhibiting charge het-

erogeneity and differing in amino acid sequence. Produced as ~30,-000 mol wt precursor which is subsequently converted to low mol wt form. Cloning and expression of cDNA for murine precursor IL-1: P. T. Lomedico *et al.*, *Nature* **312**, 458 (1984); for human precursor IL-1: P. E. Auron *et al.*, *Proc. Natl. Acad. Sci. USA* **81**, 7907 (1984); for 2 distinct forms of human IL-1 (IL-1α and IL-1β): C. J. March *et al.*, *Nature* **315**, 641 (1985). Amino acid sequence of dominant species corresponding to IL-1β: P. Cameron *et al.*, *J. Exp. Med.* **162**, 790 (1985). Crystallization of recombinant human IL-1β: D. B. Carter *et al.*, *Proteins Struct. Funct. Genet.* **3**, 121 (1988). Characterization of membrane-associated IL-1: E. A. Kurt-Jones *et al.*, *Proc. Natl. Acad. Sci. USA* **82**, 1204 (1985). Induction of IL-2 production by IL-1: K. A. Smith *et al.*, *J. Exp. Med.* **151**, 1551 (1980); S. Gillis, S. B. Mizel, *Proc. Natl. Acad. Sci. USA* **78**, 1122 (1981). Effect on B-cells in antibody production: D. D. Wood, *J. Immunol.* **123**, 2400 (1979); P. Lipsky, *Contemp. Top. Mol. Immunol.* **10**, 195 (1985). Effect in inflammatory response: S. B. Mizel *et al.*, *Proc. Natl. Acad. Sci. USA* **78**, 2474 (1981). Role in the pathogenesis of the acute-phase response: C. A. Dinarello, *N. Engl. J. Med.* **311**, 1413 (1984). Comparison of bioactivity of 4 forms of human IL-1: D. D. Wood *et al.*, *J. Immunol.* **134**, 895 (1985). Review of inter-relationship with lymphokines: N. M. Kouttab *et al.*, *Clin. Chem.* **30**, 1539-1545 (1984). *Reviews:* J. J. Oppenheim *et al.*, *Fed. Proc.* **41**, 257-262 (1982); S. B. Mizel, *Immunol. Rev.* **63**, 51-72 (1982); K. Bendtzen, *Allergy* **38**, 219-226 (1983); C. A. Dinarello, *Rev. Infect. Dis.* **6**, 51-95 (1984).

Isoelectric point (human): 6.8-7.3 (dominant species); 5.3-5.8 (minor species). Isoelectric point (murine): 4.5-5.5. Sensitive to trypsin, chymotrypsin, sodium dodecyl sulfate. Insensitive to 2-mercaptoethanol, neuraminidase, sodium periodate, iodacetamide. Stable at pH 3-10; stable at 56° for 60 min; unstable after 2 min at 100°.

5038. Interleukin-2. [85898-30-2] T-Cell growth factor; TCGF; thymocyte stimulating factor; costimulator; lymphocyte mitogenic factor; IL-2. Pleiotropic cytokine produced by T-lymphocytes following activation by antigens or mitogens in the presence of IL-1, *q.v.* Native, human IL-2 is a glycoprotein containing 133 amino acid residues; mol wt ~15 kDa. Potent immunomodulator with important role in activating and maintaining the immune response. Induces growth and proliferation of CD4$^+$ and CD8$^+$ T cells; enhances natural killer cell activity; stimulates antigen secretion; potentiates the release of γ-interferon and other cytokines. Also serves to down-regulate the immune system to prevent autoimmunity. Identification of soluble mitogenic factor produced by T-cells in response to antigen: R. S. Geha, E. Merler, *Cell. Immunol.* **10**, 86 (1974). Identification of T-cell growth promoting factor produced by mitogen stimulated lymphocytes: D. A. Morgan *et al.*, *Science* **193**, 1007 (1976). Definition and nomenclature: L. A. Aarden *et al.*, *J. Immunol.* **123**, 2928 (1979). Induction of IL-2 production by IL-1: K. A. Smith *et al.*, *J. Exp. Med.* **151**, 1551 (1980). Review of early studies: J. Watson, D. Mochizuki, *Immunol. Rev.* **51**, 257 (1980). Purification and partial amino acid sequence: R. J. Robb *et al.*, *Proc. Natl. Acad. Sci. USA* **80**, 5990 (1983). Expression of cloned cDNA for human IL-2 and predicted amino acid sequence: T. Taniguchi *et al.*, *Nature* **302**, 305 (1983). Isolation of human IL-2 gene: S. Mita *et al.*, *Biochem. Biophys. Res. Commun.* **117**, 114 (1983). Purification and cloning of the T-cell surface receptor for IL-2: W. J. Leonard *et al.*, *Nature* **311**, 626 (1984). Review of recombinant IL-2 muteins: L. T. Vlasveld, E. M. Rankin, *Cancer Treat. Rev.* **20**, 275-311 (1994). Review of pharmacology and therapeutic use in cancer: R. Whittington, D. Faulds, *Drugs* **46**, 446-514 (1993); of function, production and clinical applications: S. L. Gaffen, K. D. Liu, *Cytokine* **28**, 109-123 (2004).

Isoelectric point: 6.5-6.8 (human); 3.9-5.0 (mouse). Sensitive to trypsin, chymotrypsin, sodium dodecyl sulfate. Stable in 2-mercaptoethanol, urea (2-4 mol/liter), neuraminidase. Stable at pH 3-10; at 56°C for 60 min. Unstable after 30 min. at 70°C.

Aldesleukin. [110942-02-4] 125-L-Serine-2-133-interleukin 2 (human reduced); Proleukin. Nonglycosylated interleukin-2 mutein produced by expressing the human IL-2 gene from Jurkat leukemia cells in *E. coli*. Production and bioactivity: S. A. Rosenberg *et al.*, *Science* **223**, 1412 (1984). Clinical trial in metastatic melanoma: M. B. Atkins *et al.*, *J. Clin. Oncol.* **17**, 2105 (1999). Review of clinical trials in HIV-infection: A. K. Pau, J. A. Tavel, *Curr. Opin. Pharmacol.* **2**, 433-439 (2002); in renal cell carcinoma: M. Schmidinger *et al.*, *Expert Rev. Anticancer Ther.* **4**, 957-980 (2004).

Teceleukin. [136279-32-8] *N*-L-Methionylinterleukin 2 (human); BG-8301; Ro-23-6019; Imunace. Human interleukin-2 produced by recombinant DNA technology. Clinical evaluation of combination with interferon-α in renal cell carcinoma: H. Miyake *et al.*, *Int. J. Oncol.* **10**, 338 (2005).

THERAP CAT: Antineoplastic; immunomodulator.

5039. Interleukin-3. Multipotent colony stimulating factor; IL-3; multi-CSF. Hematopoietic growth factor that promotes the survival, differentiation, and proliferation of multipotent hematopoietic stem cells and of committed progenitor cells of the megakaryocyte, granulocyte-macrophage, erythroid, eosinophil, basophil and mast cell lineages. Enhances thrombopoiesis, phagocytosis, antibody-mediated cellular cytotoxicity, and eosinophil metabolism. Species-specific, variably glycosylated peptide; human: 14-28 kDa, 133 amino acids; murine: 22-32 kDa, 140 amino acids. Produced primarily by activated T-lymphocytes; also by natural killer cells and mast cells. Undetectable in normal serum, postulated to be involved in the immune response. Purification from murine myelomonocytic leukemia cells (WEHI-3): J. N. Ihre *et al.*, *J. Immunol.* **129**, 2431 (1982). Cloning of murine IL-3: M. C. Fung *et al.*, *Nature* **307**, 233 (1984); T. Yokota *et al.*, *Proc. Natl. Acad. Sci. USA* **81**, 1070 (1984); of human: Y.-C. Yang *et al.*, *Cell* **47**, 3 (1986). Review of molecular and cellular biology: C. F. Morris *et al.*, *Immunol. Ser.* **49**, 177-214 (1990); of structure and function: A. Lindemann, R. Mertelsmann, *Cancer Invest.* **11**, 609-623 (1993). Review of clinical pharmacology: A. Ganser, *ibid.* 212-218.

Muplestim. [148641-02-5] SDZ-ILE-964; Hemokine. Human IL-3 produced in *E. coli* by recombinant DNA technology. Clinical pharmacokinetics: B. Biesma *et al.*, *Cancer Res.* **53**, 5915 (1993). Clinical evaluation in chemotherapy induced neutropenia: S. Kudoh *et al.*, *Cancer Chemother. Pharmacol.* **38**, Suppl., S89 (1996).

Fusion protein with GM-CSF. [137463-76-4] PIXY321; GM-CSF/IL-3 fusion protein. Hybrid molecule consisting of the active domains of human GM-CSF and IL-3 coupled by a flexible amino acid sequence. Construction and expression in yeast: B. M. Curtis *et al.*, *Proc. Natl. Acad. Sci.* **88**, 5809 (1991). Evaluation in cancer patients: S. Vadhan-Raj *et al.*, *J. Clin. Oncol.* **12**, 715 (1994). Review: *idem, Stem Cells* **12**, 253-261 (1994).

THERAP CAT: Hematopoietic; antineutropenic.

5040. Interleukin-4. B-cell growth factor; BCGF-1; B-cell stimulatory factor-1; BSF-1; IL-4. Pleiotropic cytokine produced primarily by activated T-cells; also produced by mast cells and basophils. Directs the immune response toward humoral immunity and mediates the allergic response. Enhances the development of T$_{H2}$ helper cells; regulates the B-cell immunoglobulin class switch to IgE and IgG$_4$. Also promotes mast cell development and eosinophil adherence. Exhibits anti-inflammatory effects on macrophages and monocytes. Mature human IL-4 is a glycoprotein containing 129 amino acids; mol wt ~20 kDa. Identification as a growth factor for B-cells: M. Howard *et al.*, *J. Exp. Med.* **155**, 914 (1982). Initial characterization: J. J. Farrar *et al.*, *J. Immunol.* **131**, 1838 (1983). Review of molecular characterization and function of IL-4 and its receptor: W. E. Paul, *Blood* **77**, 1859-1870 (1991); R. K. Puri, *Cancer Treat. Res.* **80**, 143-185 (1995). Inhibitory role in hematopoiesis and therapeutic potential in hematologic malignancies: K. Akashi, *Leuk. Lymphoma* **9**, 205 (1993). Mechanism of antitumor activity: R. I. Tepper, *J. Allergy Clin. Immunol.* **94**, 1225 (1994). Clinical evaluation with IL-2 in malignancy: T. Olencki *et al.*, *J. Immunother.* **19**, 69 (1996). Review of role in asthma: M. Wills-Karp *et al.*, *Adv. Exp. Med. Biol.* **409**, 343-347 (1996); in allergy: J. J. Ryan, *J. Allergy Clin. Immunol.* **99**, 1-5 (1997).

5041. Interleukin-5. Eosinophil differentiation factor; T-cell replacing factor; B-cell growth factor II; IL-5; EDF; TRF; BCGFII. Essential cytokine for the maturation, growth, activation, and survival of eosinophils; also acts on basophils. Produced primarily by activated T cells; other sources include natural killer cells, bone marrow endothelial cells, and eosinophils. Exists as a variably glycosylated homodimer; mol wt 45-60 kDa. The mature, human monomer consists of 115 amino acid residues. Originally characterized as a T cell replacing factor: K. Takatsu *et al.*, *J. Immunol.* **125**, 2646 (1980). Identity with BCGFII: N. Harada *et al.*, *ibid.* **134**, 3944 (1985); with EDF: C. J. Sanderson *et al.*, *Proc. Natl. Acad. Sci. USA* **83**, 437 (1986). Review of discovery and characterization of IL-5

and its receptor: K. Takatsu *et al.*, *Adv. Immunol.* **57**, 145-190 (1994). Review of bioactivity and relevance to allergic disease: R. W. Egan *et al.*, *Allergy* **51**, 71-81 (1996); of potential as therapeutic target: F. Cuss, *ibid.* **53**, 89-92 (1998). *Reviews:* S. Karlen *et al.*, *Int. Rev. Immunol.* **16**, 227-247 (1998); G. J. Roboz, S. Rafii, *Curr. Opin. Hematol.* **6**, 164-168 (1999).

5042. Interleukin-6. B-cell stimulatory factor-2; BSF-2; hepatocyte stimulating factor; hybridoma growth factor; interferon-β$_2$; IFN-β$_2$; IL-6. Pleiotropic cytokine produced by T-cells, monocytes, fibroblasts, endothelial cells, and keratinocytes. Glycoprotein occurring in at least 5 isoforms; mol wt 23-30 kDa. Human IL-6 contains 184 amino acid residues. Critical factor in host defense, regulating immune and inflammatory responses. Stimulates B-cell differentiation and antibody production; induces production of hepatic acute-phase protein synthesis; regulates bone metabolism; and synergizes with IL-3, *q.v.*, in megakaryocyte development and platelet production. Identification of factor produced by fibroblasts under conditions for IFN-β production: P. B. Sehgal, A. D. Sagar, *Nature* **288**, 95 (1980). Description of multiple identities and proposed designation as IL-6: J. Van Damme *et al.*, *Eur. J. Biochem.* **168**, 543 (1987); P. Poupart *et al.*, *EMBO J.* **6**, 1219 (1987). Use as predictor of mortality in coronary artery disease: E. Lindmark *et al.*, *J. Am. Med. Assoc.* **286**, 2107 (2001). Review of discovery and biological significance: M. C. J. Wolvekamp, R. L. Marquet, *Immunol. Lett.* **24**, 1-10 (1990). Role in aging processes: W. B. Ershler, *J. Am. Geriatr. Soc.* **41**, 176-181 (1993). Clinical potential in chemotherapy-induced thrombocytopenia: G. J. Veldhuis *et al.*, *Leuk. Lymphoma* **20**, 373-379 (1996). Comprehensive review: M. Lotz, *Cancer Treat. Res.* **80**, 209-233 (1995). Review of role in lymphoproliferative disorders: J. F. Seymour, R. Kurzrock, *ibid.* **84**, 167-206 (1996).

5043. Interleukin-10. Cytokine synthesis inhibitory factor; CSIF; IL-10. Immunomodulatory cytokine produced by a wide variety of mammalian cell types including macrophages, monocytes, T cells, B cells, and keratinocytes. Inhibits the synthesis of proinflammatory cytokines, such as interleukin-1 and tumor necrosis factor α. Upregulates humoral immune responses and attenuates cell-mediated immune reactions. Protein composed of 160 amino acid residues; mol wt 18.5 kDa. Exists as a homodimer; the native human form has little or no glycosylation. A viral analog has been identified that is produced by Epstein Barr virus (EBV). Discovery: D. F. Fiorentino *et al.*, *J. Exp. Med.* **170**, 2081 (1989). Cloning and expression of murine form and homology with product of EBV gene BCRFI: K. W. Moore *et al.*, *Science* **248**, 1230 (1990). Review of discovery, properties and biological function: T. R. Mosmann, *Adv. Immunol.* **56**, 1-26 (1994). Review of potential clinical applications in inflammatory disease: I. Lalani *et al.*, *Ann. Allergy Asthma Immunol.* **79**, 469-483 (1997); in infectious diseases: S. M. Opal *et al.*, *Clin. Infect. Dis.* **27**, 1497-1507 (1998); in inflammatory bowel disease: M. W. Leach *et al.*, *Toxicol. Pathol.* **27**, 123-133 (1999). *Review:* K. Asadullah *et al.*, *Pharmacol. Rev.* **55**, 241-269 (2003).

Ilodecakin. [149824-15-7] Interleukin-10 (human clone pH15C); rhIL-10; Sch-52000; Tenovil. Human form produced in *E. coli* by recombinant DNA technology. Production: T. R. Mosmann *et al.*, *EP 405980*; *eidem, US 5231012* (1991, 1993 both to Schering). Clinical pharmacokinetics: E. Radwanski *et al.*, *Pharm. Res.* **15**, 1895 (1998). Clinical effect on circulating leukocyte populations: R. D. Huhn *et al.*, *Immunopharmacology* **41**, 109 (1999).

5044. Interleukin-11. IL-11; adipogenesis inhibitory factor; AGIF. Multifunctional cytokine produced by stromal cells such as fibroblasts, epithelial cells, and osteoblasts. Expressed in a wide variety of tissues, including CNS, thymus, lung, bone, and connective tissue. Nonglycosylated protein containing 178 amino acid residues; mol wt 19 kDa. Important regulator of hematopoiesis; stimulates growth of myeloid, erythroid, and megakaryocyte progenitor cells. Regulates bone metabolism; inhibits production of proinflammatory mediators; and protects against gastromucosal injury. Identification: S. R. Paul *et al.*, *Proc. Natl. Acad. Sci. USA* **87**, 7512 (1990). Identity with AGIF: I. Kawashima *et al.*, *FEBS Lett.* **283**, 199 (1991). Review of biological activity and clinical studies: X. Du, D. A. Williams, *Blood* **89**, 3897-3908 (1997); A. J. Dorner *et al.*, *BioDrugs* **8**, 418-429 (1997). *Review:* P. F. Schendel, K. J. Turner in *Cytokines*, A. R. Mire-Sluis, R. Thorpe, Eds. (Academic Press, San Diego, 1998) pp 169-182.

Highly basic protein, pI 11.7. Rich in leucine (23%) and proline (12%). Has no disulfide bonds as it does not contain cysteine. Heat stable; melting temp 90°. Highly sol at physiologic pH.

Oprelvekin. [145941-26-0] 2-178-Interleukin 11 (human clone pXM/IL-11); Neumega. Human IL-11 produced in *E. coli* by recombinant DNA technology. Differs from naturally occurring protein only in the absence of the amino-terminal proline residue. Clinical pharmacokinetics: K. Aoyama *et al.*, *Br. J. Clin. Pharmacol.* **43**, 571 (1997). Clinical trial in chemotherapy-induced thrombocytopenia: I. Tepler *et al.*, *Blood* **87**, 3607 (1996); C. Isaacs *et al.*, *J. Clin. Oncol.* **15**, 3368 (1997). *Review:* J. A. Kaye, *Curr. Opin. Hematol.* **3**, 209-215 (1996).

THERAP CAT: Antithrombocytopenic.

5045. Interleukin-1 Receptor Antagonist. IL-1ra; interleukin-1 inhibitor protein; IRAP. Endogenous inhibitor of interleukin-1, *q.v.*; modulates the normal, acute inflammatory response to infection and injury. First known naturally occurring protein that functions as a specific receptor antagonist of a cytokine. Variably glycosylated, 152 amino acid protein; mol wt 17-25 kDa. Secreted (glycosylated) and intracellular (nonglycosylated) forms have been identified. Structurally homologous with IL-1 and produced by the same cells. Isoln of an IL-1 inhibitor produced by human monocytes: W. P. Arend *et al.*, *J. Immunol.* **134**, 3868 (1985). Isoln from human urine: P. Seckinger *et al.*, *ibid.* **139**, 1541 (1987). Purification: C. H. Hannum *et al.*, *Nature* **343**, 336 (1990). Cloning and expression of human IL-1ra: S. P. Eisenberg *et al.*, *ibid.* 341. Study of biological function: E. Hirsch *et al.*, *Proc. Natl. Acad. Sci. USA* **93**, 11008 (1996). Therapeutic potential in IL-1-mediated disease: R. C. Thompson *et al.*, *Int. J. Immunopharmacol.* **14**, 475 (1992); C. A. Dinarello, *Nutrition* **11**, 492 (1995). *Reviews:* W. P. Arend, *Adv. Immunol.* **54**, 167-227 (1993); A. C. Lennard, *Crit. Rev. Immunol.* **15**, 77-105 (1995). Prognostic value in percutaneous coronary intervention: G. Patti *et al.*, *Am. J. Cardiol.* **89**, 372 (2002).

Anakinra. [143090-92-0] N^2-L-Methionyl-interleukin 1 receptor antagonist (human isoform x reduced); Kineret. Nonglycosylated, recombinant human IL-1ra expressed in *E. coli*; mol wt 17 kDa. Prepn: C. H. Hannum *et al.*, **EP 343684**; *eidem*, **US 5075222** (1989, 1991 both to Synergen). Biological properties: W. P. Arend *et al.*, *J. Clin. Invest.* **85**, 1694 (1990). Clinical trial in sepsis syndrome: C. J. Fisher, Jr. *et al.*, *Crit. Care Med.* **22**, 12 (1994); in rheumatoid arthritis: R. M. Fleischmann *et al.*, *Arthritis Rheum.* **48**, 927 (2003). Series of articles on safety and clinical experience in rheumatoid arthritis: *Rheumatology* **42**, Suppl. 2, ii22-ii40 (2003).

THERAP CAT: Anti-inflammatory.

5046. Interleukin-4 Receptor. IL-4R. Endogenous receptor for interleukin-4, *q.v.*, expressed on the surface of B and T lymphocytes, mast cells and macrophages, that mediates the bioactivities of the cytokine. mol wt ~140 kDa. Also secreted as a soluble, truncated form (*sIL-4R*) which lacks the transmembrane and intracellular domains and is capable of binding and sequestering circulating IL-4. Identification of membrane bound receptors: J. Ohara, W. E. Paul, *Nature* **325**, 537 (1987). Characterization of murine: L. S. Park *et al.*, *Proc. Natl. Acad. Sci. USA* **84**, 1669 (1987); of human: *idem et al.*, *J. Exp. Med.* **166**, 476 (1987). Cloning of murine and identification of soluble form: B. Mosley *et al.*, *Cell* **59**, 335 (1989). Detection of sIL-4R in biological fluids: R. Fernandez-Botran, E. S. Vitetta, *Proc. Natl. Acad. Sci. USA* **87**, 4204 (1990). Immunoregulatory effects of recombinant sIL-4R: C. R. Maliszewski *et al.*, *Proc. Soc. Exp. Biol. Med.* **206**, 233 (1994). Review of signaling mechanisms of membrane receptor: A. D. Keegan *et al.*, *Immunol. Today* **15**, 423-432 (1994). Review of bioactivity and therapeutic potential of sIL-4R: K. Enssle *et al.*, *Behring Inst. Mitt.* **96**, 103-117 (1995).

Altrakincept. [239076-74-5] Recombinant human soluble interleukin-4 receptor; Nuvance. Soluble form of the human IL-4 receptor produced by recombinant DNA technology in Chinese hamster ovary cells. Glycoprotein; mol wt 54 kDa. Prepn: D. J. Cosman *et al.*, **EP 367566**; B. Mosley *et al.*, **US 5599905** (1990, 1997 both to Immunex). Clinical evaluation in atopic asthma: L. C. Borish *et al.*, *Am. J. Respir. Crit. Care Med.* **160**, 1816 (1999). *Review:* R. Lange, *Curr. Opin. Cardiovasc. Pulm. Renal Invest. Drugs* **1**, 526-531 (1999).

THERAP CAT: Soluble form as antiasthmatic.

5047. Inula. Elecampane; scabwort; elfwort; horseheal. Dried rhizome and roots of *Inula helenium* L., *Compositae. Habit.* Central

Asia, Europe; naturalized in U.S. *Constit.* Inulin, volatile oil, alantol, helenin, alantic acid, acrid resin.

5048. Inulin. [9005-80-5] Dahlin; alantin; alant starch. Mol wt approx 5000. Polysaccharide of *Compositae* which partially or completely replaces starch as a reserve food. Isoln from dahlia tubers: McDonald, "Polyfructosans and Difructose Anhydrides" in *Advan. Carbohyd. Chem.* **vol. 2**, 254 (1946); from Jerusalem artichoke tubers: Bacon, Edelman, *Biochem. J.* **48**, 114 (1951). Structure: E. G. V. Percival, *Structural Carbohydrate Chemistry* (J. Garnet Miller, London, 2nd ed., 1962) p 274.

Spherical crystals from water. Hygroscopic in moist air. $[\alpha]_D^{20}$ −40° (c = 2) for the anhydr. Sol in hot water; slightly sol in cold water and organic solvents. Yields D-fructose and D-glucose upon acid hydrolysis.

Acetate. Fine powder from methanol. $[\alpha]_D^{20}$ −34° (c = 1.5 in chloroform); $[\alpha]_D^{20}$ −43° (c = 1.8 in acetic acid).

Trimethylinulin. Powder from hot water + acetone, mp 140°. $[\alpha]_D^{20}$ −55° (c = 1.03 in chloroform); $[\alpha]_D^{20}$ −54° (c = 1.09 in benzene).

THERAP CAT: Diagnostic aid (renal function).

5049. Invertase. [9001-57-4] β-Fructofuranosidase; invertin; saccharase; sucrase; EC 3.2.1.26. Enzymes, obtained primarily from yeast as well as other sources, which catalyze the hydrolysis of sucrose into fructose and glucose. Since sucrose is both a β-fructofuranoside and an α-glucoside, it is split by two different types of enzymes, **β-fructofuranosidases** or **β-h-fructosidases** and certain **α-glucosidases** or **α-n-glucosidoinvertases**, which attack the sucrose molecule from the fructose and glucose end, respectively. β-Fructofuranosidase, generally obtained from yeast, is characterized by its ability to hydrolyze raffinose, while α-glucosidase is inactive toward raffinose. *Reviews:* K. Myrbäck, "Invertases" in *The Enzymes* **vol. 4**, P. D. Boyer *et al.*, Eds. (Academic Press, New York, 1960) pp 379-396; Lampen, "Yeast and Neurospora Invertases", *ibid.* **vol. 5** (3rd ed., 1971) pp 291-305; Cochrane, *Soc. Chem. Ind. Monogr.* **11**, 25-31 (1959) (Pub. 1961); Meister, *Wallerstein Lab. Commun.* **28**, 7 (1965).

USE: For preparation of invert sugar from sucrose; as analytical reagent for sucrose.

5050. Invert Sugar. [8013-17-0] Nulomoline; Calorose; Invesol; Insubeta; Travert. A mixture of about 50% glucose (dextrose) and 50% fructose (levulose) obtained by hydrolysis of sucrose. Hydrolysis of the sucrose may be carried out with acids or enzymes. Invert sugar is slightly levorotatory, reduces Fehling's soln and can be fermented. Honey is mostly invert sugar. Due to the levulose, it is somewhat sweeter than sucrose. The commercial product is obtained by inversion of a 96% cane sugar soln. The inversion is carried out at pH 3 to 4 by means of invertase and dil HCl. The acid is usually neutralized with sodium carbonate to pH 6.5. At this point, the dextrose crystallizes and the entire mass is beaten into a creamy, plastic product.

USE: In food products, in confectionery. As a humectant, like glycerol, to hold moisture and to prevent drying out. In brewing.

THERAP CAT: Nutrient (parenteral).

5051. Iobenguane. [80663-95-2] *N*-[(3-Iodophenyl)methyl]-guanidine; *m*-iodobenzylguanidine; MIBG. $C_8H_{10}IN_3$; mol wt 275.09. C 34.93%, H 3.66%, I 46.13%, N 15.28%. Norepinephrine analog with specific affinity for tissues of sympathetic nervous system and related tumors; prepd as ^{123}I and ^{131}I labeled forms. Prepn and imaging studies: D. M. Wieland *et al.*, *J. Nucl. Med.* **21**, 349 (1980); *eidem*, **US 4584187** (1986). Improved synthesis: P. A. P. M. van Doremalen, A. G. M. Janssen, *J. Radioanal. Nucl. Chem. Lett.* **96**, 97 (1985). Metabolism in man: T. J. Mangner *et al.*, *J. Nucl. Med.* **27**, 37 (1986). HPLC determn in serum and urine: D. Schwabe *et al.*, *J. Chromatogr.* **487**, 177 (1989). Radiopharmacokinetics: S. Ertl *et al.*, *Nucl. Med. Commun.* **8**, 643 (1987). Clinical evaluation of myocardial imaging: D. Fagret *et al.*, *Eur. J. Nucl. Med.* **15**, 624 (1989). Diagnostic use in pheochromocytoma: B. Shapiro *et al.*, *J. Nucl. Med.* **26**, 576 (1985); therapeutic use: M. Krempf *et al.*, *J. Clin. Endocrinol. Metab.* **72**, 455 (1991). Symposia on therapeutic and diagnostic use in neuroblastoma: *Advances in Neuroblastoma Research* **2**, A. E. Evans *et al.*, Eds. (Alan R. Liss, Inc., New York, 1988) p 643-726; *Med. Pediatr. Oncol.* **15**, 157-228 (1987). Review of pharmacology: J. C. Sisson, D. M. Weiland, *Am. J. Physiol. Imaging* **1**, 96-103 (1986); of biodistribution and clinical studies: A. R. Wafelman *et al.*, *Eur. J. Nucl. Med.* **21**, 545-559 (1994); of therapeutic use in neural crest tumors: L. Troncone, V. Rufini, *Anticancer Res.* **17**, 1823-1832 (1997).

Sulfate. $(C_8H_{10}IN_3)_2 \cdot H_2SO_4$. Colorless crystals from water + ethanol, mp 166-167°.

THERAP CAT: Radiolabeled forms as antineoplastic; diagnostic aid (radioactive imaging agent).

5052. Iobitridol. [136949-58-1] N^1,N^3-Bis(2,3-dihydroxypropyl)-5-[[3-hydroxy-2-(hydroxymethyl)-1-oxopropyl]amino]-2,4,6-triiodo-N^1,N^3-dimethyl-1,3-benzenedicarboxamide; 5-[3-hydroxy-2-(hydroxymethyl)-propionamido]-*N*,*N'*-dimethyl-*N*,*N'*-bis-(2,3-dihydroxypropyl)-2,4,6-triiodoisophthalamide; Xenetix. C_{20}-$H_{28}I_3N_3O_9$; mol wt 835.17. C 28.76%, H 3.38%, I 45.59%, N 5.03%, O 17.24%. Nonionic, monomeric, low-osmolality x-ray contrast medium for urography and angiography. Prepn: M. Schaefer *et al.*, **EP 437144**; *eidem*, **US 5043152** (both 1991 to Guerbet). Biodistribution studies: P. Bourrinet *et al.*, *Invest. Radiol.* **29**, 1057 (1994). HPLC determn in biological fluids: *idem et al.*, *J. Chromatogr. B* **670**, 369 (1995). Toxicology: A.-M. Donadieu *et al.*, *Acta Radiol.* **37**, Suppl. 400, 17 (1996). Series of articles on pharmacology, pharmacokinetics and clinical studies: *eidem*, *ibid.* 1-92.

Hydrophilic. Log P (octanol/water): −2.63 at 37°. LD_{50} in male, female mice (g I/kg): 16.8, 16.6 i.v. (Donadieu).

THERAP CAT: Diagnostic aid (radiopaque medium).

5053. Iocarmic Acid. [10397-75-8] 3,3′-[(1,6-Dioxo-1,6-hexanediyl)diimino]bis[2,4,6-triiodo-5-[(methylamino)carbonyl]benzoic acid]; 5,5′-(adipoyldiimino)bis[2,4,6-triiodo-*N*-methylisophthalamic acid]; Myelotrast. $C_{24}H_{20}I_6N_4O_8$; mol wt 1253.87. C 22.99%, H 1.61%, I 60.73%, N 4.47%, O 10.21%. Prepn: **GB 1033695**; **US 3290366** (both 1966 to Mallinckrodt); G. B. Hoey *et al.*, *J. Med. Chem.* **9**, 964 (1966). Clinical studies: Kunze, Schiefer, *Dtsch. Med. Wochenschr.* **97**, 245 (1972). Diagnostic use in myelog-

raphy: P. Ahlgren, *Acta Radiol. Diagn.* **13**, 753 (1972); in hysterosalpingography: H. E. Schutte, *Diagn. Imaging* **51**, 277 (1982). Spinal toxicity in rats: I. O. Skalpe, A. Torvik, *Invest. Radiol.* **10**, 154 (1975).

Crystals from dimethylformamide, mp 302° (dec).

Di-*N*-methylglucamine salt. [54605-45-7] Dimeglumine iocarmate; LM-280; Dimeray; Dimer-X; Dirax. $C_{38}H_{54}I_6N_6O_{18}$; mol wt 1644.30. Soly in water (25°): 65 g/100 ml.

THERAP CAT: Diagnostic aid (radiopaque medium).

5054. Iocetamic Acid. [16034-77-8] 3-[Acetyl-(3-amino-2,4,6-triiodophenyl)amino]-2-methylpropanoic acid; *N*-acetyl-*N*-(3-amino-2,4,6-triiodophenyl)-2-methyl-β-alanine; *N*-acetyl-*N*-(2,4,6-triiodo-3-aminophenyl)-β-aminoisobutyric acid; DRC-1201; MP-620; Cholebrine; Cholimil. $C_{12}H_{13}I_3N_2O_3$; mol wt 613.96. C 23.48%, H 2.13%, I 62.01%, N 4.56%, O 7.82%. Prepn: **NL 6515305**; **NL 6607275** (both 1967 to Dagra), *C.A.* **67**, 108422m (1967); **69**, 10263b (1968); Korver, *Rec. Trav. Chim.* **87**, 308 (1968). Pharmacology and toxicology: J. M. Janbroers *et al.*, *Toxicol. Appl. Pharmacol.* **14**, 232, 246 (1969). Metabolic studies: Neleman, *Pharm. Weekbl.* **102**, 1039 (1967). Clinical studies: Hekster, *Radiol. Clin. Biol.* **37**, 338 (1968). Use in cholecystography: B. Goldberg, *Radiol. Clin.* **46**, 42 (1977).

White to light cream-colored powder, mp 224-225° (**NL 6515305**). Also reported as mp range 191-212° (Korver). Practically insol in water. Very slightly sol in ether, ethanol, benzene; slightly sol in acetone, chloroform (Janbroers). LD_{50} in rats (g/kg): 7.1 orally; 0.70 i.v. (Janbroers). Consists of 2 isomers having mps 232° and 200-201° and pKa's of 4.25 and 4.0 respectively. The lower melting compd is approx two times as soluble.

THERAP CAT: Diagnostic aid (radiopaque medium).

5055. Iodamide. [440-58-4] 3-(Acetylamino)-5-[(acetylamino)methyl]-2,4,6-triiodobenzoic acid; α,5-diacetamido-2,4,6-triiodo-*m*-toluic acid; 3-acetamido-5-(acetamidomethyl)-2,4,6-triiodobenzoic acid; ametriodinic acid; SH-926. $C_{12}H_{11}I_3N_2O_4$; mol wt 627.94. C 22.95%, H 1.77%, I 60.63%, N 4.46%, O 10.19%. Prepn: Felder, Pitre, **FR 1382277** and **US 3360436** (1964 and 1967 to Eprova); Felder *et al.*, *Helv. Chim. Acta* **48**, 259 (1965). Pharmacology and toxicology: F. Bonati *et al.*, *Arzneim.-Forsch.* **15**, 222 (1965); Z. B. Zsebök, L. Szlavy, *ibid.* **17**, 1380 (1967); *eidem*, *Int. Z. Klin. Pharmakol. Ther. Toxikol.* **3**, 157 (1970). Clinical trials: Grothuesmann, *Arzneim.-Forsch.* **15**, 233 (1965). Comprehensive description: D. Pitré, *Anal. Profiles Drug Subs.* **15**, 337-365 (1986).

Crystals from acetic acid, mp 255-257°. Soly in water: 0.3 g/100 ml (22°).

N-Methyl-D-glucamine salt. [18656-21-8] Iodamide meglumine; Isteropac E.R.; Jodomiron; Opacist E.R.; Renovue-65; Renovue-DIP; Uromiro. $C_{19}H_{28}I_3N_3O_9$; mol wt 823.16. Sparingly sol in methanol, slightly sol in water, ethanol. Practically insol in ether, chloroform. LD_{50} in mice, rats, rabbits (g/kg): 9.0, 11.4, 13.2 i.v.; in rats, guinea pigs (g/kg): 17.9, 15.0 i.p. (Bonati).

THERAP CAT: Diagnostic aid (radiopaque medium).

5056. Iodic Acid. [7782-68-5] Iodic acid (HIO_3); hydrogen iodate. HIO_3; mol wt 175.91. H 0.57%, I 72.14%, O 27.28%. Prepd by the oxidation of iodine with nitric acid, perchloric acid, or hydrogen peroxide: Baxter, Tilley, *Z. Anorg. Allg. Chem.* **61**, 295 (1909); Lamb *et al.*, *J. Am. Chem. Soc.* **42**, 1643 (1920); Bray, Caulkins, *ibid.* **53**, 44 (1931). Electrolytic method: Willard, Ralston, *Trans. Am. Electrochem. Soc.* **62**, 239 (1932).

Orthorhombic crystals. Darkens upon exposure to light. d_4^0 4.629. mp 110° (dec). Decompn to $HIO_3 \cdot I_2O_5$ starts at 70° and decompn to I_2O_5 is complete at 220°. Not hygroscopic, yet very freely sol in water: 269 g/100 ml H_2O at 20°; 295 g/100 ml H_2O at 40°. Density of aq solns (w/w) at 18°: 1% = 1.0071; 10% = 1.0900; 20% = 1.1969; 40% = 1.4640; 75% = 2.4710. Strong electrolyte, especially when dil. Sol in nitric acid, dil alcohol. Insol in abs alc, ether, chloroform. *Keep well closed and protected from light.*

USE: Titration standard in analytical chemistry.

THERAP CAT: Astringent, disinfectant.

5057. Iodine. [7553-56-2] I; at. wt 126.90447; at. no. 53; valences −1, 1, 3, 5, 7. A halogen; Group VIIA (17). Does not exist as elemental state, I, in nature. Occurs as diatomic molecule, I_2. Abundance in igneous rocks: 3×10^{-5}% by wt; in seawater: 5×10^{-8}% by wt. Naturally occurring stable isotope (mass number): 127 (100%); known artificial radioactive isotopes: 108 to 141; commonly used radioactive tracer elements: 123 ($T_{1/2}$ 13.27 hrs, EC decay), 125 ($T_{1/2}$ 59.408 days, EC decay), 129 ($T_{1/2}$ 1.57×10^7 years, β^- decay), 131 ($T_{1/2}$ 8.02070 days, β^- decay). Discovered in 1811 by Courtois. Classed among the rarer elements. Trace element essential to animal and plant life. Extracted commercially from natural subterranean brines, and Chilean nitrate-bearing earth (caliche); formerly from seaweed. *Reviews: MTP Int. Rev. Sci.: Inorg. Chem., Ser. One* **Vol. 3**, V. Gutmann, Ed. (Butterworths, London, 1972); A. J. Downs, C. J. Adams, "Chlorine, Bromine, Iodine and Astatine" in *Comprehensive Inorganic Chemistry* **vol. 2**, J. C. Bailar, Jr. *et al.*, Eds. (Pergamon Press, Oxford, 1973) p 1107-1573; *Chemistry of the Elements* N. N. Greenwood, A. Earnshaw, Eds. (Pergamin Press, New York, 1984) pp 920-1041; A. Lauterbach, G. Ober in *Kirk-Othmer Encyclopedia of Chemical Technology* **vol. 14** (Wiley-Interscience, New York, 4th ed., 1995) pp 709-737. Review of geochemistry: R. Fuge, C. C. Johnson, *Environ. Geochem. Health* **8**, 31-54 (1986); of human toxicity: *Iodine Toxicity* (PB89-183016, 1989) 107 pp; of metabolism and nutritional deficiency: J. T. Dunn in *Contemporary Endocrinology: Diseases of the Thyroid*, L. E. Braverman, Ed. (Humana Press Inc., Totowa, NJ, 1997) pp 349-360; of toxicology and human exposure: *Toxicological Profile for Iodine* (PB2004-104399, 2004) 580 pp.

Bluish-black scales or plates; diatomic; metallic luster; characteristic odor; sharp acrid taste. Readily sublimes at rm temp, forming violet corrosive vapor. mp 113.60°. bp 185.24°. d (solid, 25°) 4.93, d (liq 120°) 3.960. Vapor pressure (solid): 0.030 mm (0°); 0.305 mm (25°): 2.154 mm (50°); 26.78 mm (90°). Heat capacity at constant pressure (25°) 13.011 cal/mole/°C: Shirley, Giauque, *J. Am. Chem. Soc.* **81**, 4778 (1959). Total soly in water (25°) 0.0013 moles/l with negligible formation of HOI (6.4×10^{-6} moles/l); freely sol in aq solns of HI or iodides. Soly in organic solvents (g I_2/100 g soln, 25°): benzene 14.09; CS_2 16.47; ethanol 21.43; ethyl ether 25.20; cyclohexane 2.719; heptane 1.702; CCl_4 (35°) 2.603: Hildebrand, Jenks, *J. Am. Chem. Soc.* **42**, 2180 (1920); Hildebrand *et al.*, *ibid.* **72**, 1017 (1950). Soly in water is increased by alkali bromides, but decreased by sulfates and nitrates. Freely sol in carbon disulfide, chloroform, carbon tetrachloride; sol in glacial acetic acid, glycerol oils; sparingly sol in glycerin. Solutions of iodine in aq solns of inorganic iodides are brown or deep brown, depending on the concn of the iodine. Solvents contg nitrogen atoms, such as pyridine, amines, or quinoline form brown solns. Aliphatic hydrocarbons, chloroform, carbon tetrachloride, carbon disulfide, phosphorus trichloride and fluorinated amines give violet solns. Aromatic hydrocarbons give pink or reddish-brown solns. Alcohols and ethers give deep brown solns. Less reactive than bromine; E° (aq) ½I$_2$/I$^-$ 0.535 V dissociation energy (25°): 36.115 kcal. Iodine stains may be removed with sodium thiosulfate soln or ammoniated alc.

Tamed Iodine *see* Iodophors.

Caution: Potential symptoms of overexposure are irritation of eyes, skin, and nose; lacrimation; headache; tight chest; skin burns, rash; cutaneous hypersensitivity. *See NIOSH Pocket Guide to Chemical Hazards* (DHHS/NIOSH 97-140, 1997) p 172. Ingestion of large quantities causes abdominal pain, vomiting and diarrhea due to the highly corrosive action of iodine on the GI tract. *See Clinical Toxicology of Commercial Products*, R. E. Gosselin *et al.*, Eds. (Williams & Wilkins, Baltimore, 5th ed., 1984) Sect. III, pp 213-214.

USE: Manuf of organic and inorganic chemicals, pharmaceuticals, radiopaque contrast agents, animal feed supplements, disinfectants, stabilizers, inks, colorants, photographic chemicals. Catalyst in the alkylation and condensation of aromatic amines; in sulfations and sulfonations; for prodn of synthetic rubber. Microbicide for drinking water and swimming pools. Artificial radioisotopes as tracers in biological, biochemical and chemical research; in medical imaging. Iodometric standard in analytical chemistry.

THERAP CAT: Antihyperthyroid; anti-infective (topical).

THERAP CAT (VET): Internally for goiter, hypothyroidism, in iodine deficiency. Topically as antiseptic, disinfectant, counterirritant and to promote absorption.

5058. Iodine Heptafluoride. [16921-96-3] Iodine fluoride (IF_7). F_7I; mol wt 259.89. F 51.17%, I 48.83%. IF_7. Prepd by passing fluorine through liq IF_5 at 90°, then heating the vapors to 270°: Ruff, Keim, *Z. Anorg. Allg. Chem.* **193**, 176 (1930); Kwasnik in *Handbook of Preparative Inorganic Chemistry* **vol. 1**, G. Brauer, Ed. (Academic Press, New York, 2nd ed., 1963) pp 160-161. May also be prepd from fluorine and dried PdI_2 or KI to minimize formation of the impurity, IOF_5: Bartlett, Levchuk, *Proc. Chem. Soc. London* **1963**, 342; Selig *et al.*, *J. Phys. Chem.* **71**, 2739 (1967). *Reviews:* Kemmitt, Sharp, *Adv. Fluorine Chem.* **4**, 246-247 (1965); Stein, "Physical and Chemical Properties of Halogen Fluorides" in *Halogen Chemistry* **Vol. 1**, V. Gutmann, Ed. (Academic Press, New York, 1967) pp 133-224; Meinert, *Z. Chem.* **7**, 41-57 (1967).

Colorless gas. Crystals when solid. Moldy, acrid odor. *Attacks glass and quartz.* d (liq; 6°) 2.8. mp 6.45°. Sublimes at 4.77°. Easily forms a supercooled liquid which boils at +4.5°. Trouton constant 26.4. Water dissolves the gas without violence, some decompn. Readily absorbed by NaOH soln.

Caution: Highly irritating to skin, mucous membranes. *See also* Fluorine.

USE: Fluorinating agent.

5059. Iodine Monobromide. [7789-33-5] Iodine bromide (IBr); bromine monoiodide. BrI; mol wt 206.81. Br 38.64%, I 61.36%. IBr. Prepd from the elements, excess Br is removed by heating to 50° in a current of CO_2: Bornemann, *Ann.* **189**, 202 (1877); Terwogt, *Z. Anorg. Allg. Chem.* **47**, 203 (1905); *Gmelins, Iodine* (8th ed.) **8**, p 631 (1933); Schmeisser in *Handbook of Preparative Inorganic Chemistry* **vol. 1**, G. Brauer, Ed. (Academic Press, New York, 2nd ed., 1963) p 291-292. Crystal structure: Swink, Carpenter, *Acta Crystallogr.* **24B**, 429 (1968).

Brownish-black crystals or very hard solid. d 4.416. mp 40°. bp 116° (decompn). Sol in water, alcohol, ether, carbon disulfide, glacial acetic acid. May be stored in brown glass-stoppered bottles. *Keep in a cool place.*

Caution: Vapors are corrosive to the eyes and mucous membranes.

5060. Iodine Monochloride. [7790-99-0] Iodine chloride (ICl); Wijs' chloride. ClI; mol wt 162.35. Cl 21.84%, I 78.17%. ICl. Prepd from the elements: Cornog, Karges, *Inorg. Synth.* **1**, 165 (1939); Buckles, Bader, *ibid.* **9**, 130 (1967).

Black crystals or reddish-brown liquid. The crystals occur in two modifications: α-form (stable), black needles (ruby red by transmitted light), mp 27.2°; β-form (labile), black platelets (brownish red by transmitted light), mp 13.9°. bp 97° (decompn). d_4^{29} 3.10. *Corrosive.* Sol in water, alc, ether, CS_2, acetic acid. May be stored in brown glass-stoppered bottles. Not hygroscopic, but forms I_2O_5 in the presence of air.

Caution: Attacks the skin, forming dark, painful patches (effective countermeasure: immediate washing with 20% HCl).

USE: In Wijs' soln (iodine monochloride in glacial acetic acid), used to determine iodine values of fats and oils.

THERAP CAT: Anti-infective (topical).

5061. Iodine Pentafluoride. [7783-66-6] Iodine fluoride (IF$_5$); pentafluoroiodine. F$_5$I; mol wt 221.90. F 42.81%, I 57.19%. IF$_5$. Prepd by passing fluorine over iodine with cooling: Gore, *Philos. Mag.* [4] **41**, 309 (1871); Moissan, *Compt. Rend.* **135**, 563 (1902); *Bull. Soc. Chim. Fr.* [3] **29**, 6 (1903); Ruff, Keim, *Z. Anorg. Allg. Chem.* **201**, 245 (1931); Ruff, Braida, *ibid.* **220**, 43 (1934); Kwasnik in *Handbook of Preparative Inorganic Chemistry* **vol. 1**, G. Brauer, Ed. (Academic Press, New York, 2nd ed., 1963) pp 159-160. *Reviews:* Kemmitt, Sharp, *Adv. Fluorine Chem.* **4**, 247-248 (1965); Stein, "Physical and Chemical Properties of Halogen Fluorides" in *Halogen Chemistry* **Vol. 1**, V. Gutmann, Ed. (Academic Press, New York, 1967) pp 133-224; Meinert, *Z. Chem.* **7**, 41-57 (1967).

Liquid. mp 9.43°. bp 100.5°. d^{25} 3.19. *Oxidizer, poisonous, corrosive.* Thermodynamic data: Osborne *et al., J. Chem. Phys.* **54**, 3790 (1971). *Fumes in air!* Attacks glass, especially when hot. Violent reaction with water. Organic compds carbonize on contact, sometimes with conflagration. Instant reaction with S, P, Si, Bi, W, As, usually with incandescence.

USE: Mild fluorinating agent.

5062. Iodine Pentoxide. [12029-98-0] Iodine oxide (I$_2$O$_5$); diiodine pentoxide; iodic anhydride. I$_2$O$_5$; mol wt 333.80. I 76.04%, O 23.96%. Prepd by thermal dehydration of iodic acid at 195°: Lamb *et al., J. Am. Chem. Soc.* **42**, 1644 (1920). Crystal structure: Selte, Kjekshus, *Acta Chem. Scand.* **24**, 1912 (1970).

Needle shaped, hygroscopic crystals (an impure product may have a pink cast). d^{25} 5.08. Decompn into I$_2$ and O$_2$ begins at 275° and proceeds rapidly at 350°. Very sol in water (187.4 g/100 ml at 13°) forming HIO$_3$. Insol in abs alc, ether, chloroform, carbon disulfide. Dissolves in nitric acid from which it crystallizes as I$_2$O$_5$ if the HNO$_3$ concn is >50%. Strong oxidizing agent, may detonate when triturated with oxidizable matter. *Keep well closed and protected from light.*

USE: Oxidizes carbon monoxide to carbon dioxide with the formn of iodine (Ditte's reaction, which proceeds rapidly at 65° or above, but slowly at room temp). This reaction is used in gas analysis and for removing carbon monoxide from the air (in respirators preferably in presence of some H$_2$SO$_4$).

5063. Iodine Trichloride. [865-44-1] Iodine chloride (ICl$_3$); trichloroiodine. Cl$_3$I; mol wt 233.25. Cl 45.59%, I 54.41%. ICl$_3$. Prepd by adding finely powdered iodine to an excess of liq chlorine, which is then boiled away: Thomas, Depuis, *Compt. Rend.* **143**, 282 (1906); Booth, Morris, *Inorg. Synth.* **1**, 167 (1939). Alternate process: Birk, *Angew. Chem.* **41**, 751 (1928); *Z. Anorg. Allg. Chem.* **172**, 399 (1928); Wilke-Dörfurt, Wolff, *ibid.* **185**, 333 (1930).

Long yellow or brownish needles or fluffy powder. Pungent, irritating odor. *Corrosive to human skin. Must be stored in well closed amber glass bottles, preferably in refrigerator.* d^{-4} 3.203. mp ~33°. Volatile at room temp. mp 101° at 16 atm (pressure of its satd vapor). A pressure of 1 atm is reached at 64°. Complete decompn into ICl and Cl$_2$ at 77°. Generally used as a 20 to 35% soln in concd HCl.

Caution: Concd solns are strongly irritating.

USE: Chlorinating and oxidizing agent.

THERAP CAT: Anti-infective (topical).

5064. Iodinin. [68-81-5] 1,6-Phenazinediol 5,10-dioxide; 1,5-dihydroxyphenazine *N,N'*-dioxide. C$_{12}$H$_8$N$_2$O$_4$; mol wt 244.21. C 59.02%, H 3.30%, N 11.47%, O 26.21%. Antibiotic pigment from *Chromobacterium iodinum:* Clemo, McIlwain, *J. Chem. Soc.* **1938**, 479; Hegedüs in *Emil Barell Jubilee Volume* (1946), p 388; from *Waksmania aerata* and *Pseudomonas iodinum:* Gerber, Lechevalier, *Biochemistry* **3**, 598 (1964). Structure: Clemo, Daglish, *J. Chem. Soc.* **1950**, 1481. Synthesis: Matsumura, Takeda, *Nippon Kagaku Zasshi* **81**, 515 (1960), *C.A.* **56**, 470a (1962). Biosynthetic studies on incorporation of shikimic acid, *q.v.,* into iodinin: U. Hollstein *et al., Tetrahedron Lett.* **1978**, 2987; R. B. Herbert *et al., J. Chem. Soc. Perkin Trans. 1* **1979**, 2411; T. Etherington *et al., ibid.* 2416.

Purple crystals with coppery sheen from chloroform, dec 236°. pK 12.5. Stable in acid, unstable in alkali. Sol in benzene, toluene, xylene, carbon disulfide, chloroform, ethyl acetate. Slightly sol in hot alc. Practically insol in cold alc, ether, acetic acid, petr ether, amyl alcohol. Insol in water. Sol in concd sulfuric acid and in glacial acetic acid with red color. Sol in NaOH solns giving brilliantly blue solns which deposit green crystals of the unstable sodium deriv.

5065. Iodipamide. [606-17-7] 3,3'-[(1,6-Dioxo-1,6-hexanediyl)diimino]bis[2,4,6-triiodobenzoic acid]; 3,3'-(adipoyldiimino)bis[2,4,6-triiodobenzoic acid]; *N,N'*-adipylbis(3-amino-2,-4,6-triiodobenzoic acid); adipic acid di(3-carboxy-2,4,6-triiodoanilide); adipiodone; Cholografin; Cholospect. C$_{20}$H$_{14}$I$_6$N$_2$O$_6$; mol wt 1139.77. C 21.08%, H 1.24%, I 66.81%, N 2.46%, O 8.42%. Prepn: Priewe, Rutkowski, *US 2776241* (1957 to Schering AG); Kotler-Brajtburg *et al., Rocz. Chem.* **36**, 763 (1962), *C.A.* **58**, 5568a (1963). Purification: Cassebaum, Drux, *DD 33738* (1964), *C.A.* **63**, 11441b (1965). Pharmacology: E. Fischer, F. Varga, *Acta Physiol. Acad. Sci. Hung.* **38**, 135 (1970). Metabolic studies: Kiyono *et al., Radioisotopes* **20**, 78 (1971). Pharmacology and toxicity study: F. J. Rosenberg *et al., Invest. Radiol.* **15**, S142 (1980). Comprehensive description: H. H. Lerner, *Anal. Profiles Drug Subs.* **3**, 333-363 (1974).

White crystalline powder, dec 306-308°. $n_D^{21.5}$ 1.3294 (c = 0.445 in methanol). Soly at 20°: methanol 0.8%; ethanol 0.3%; acetone 0.2%; ether 0.1%. Very slightly sol in water, chloroform. Practically insol in benzene. LD$_{50}$ in mice, rats (mg/kg): 2380 ±290, 4430 ±310 i.v. (Rosenberg).

Disodium salt. [2618-26-0] C$_{20}$H$_{12}$I$_6$N$_2$Na$_2$O$_6$. Prepd by dissolving the free acid in a dil aq soln of NaOH and buffering to pH 6.5- 7.7. Used as 20% soln.

Bis[*N*-methylglucamine] salt. [3521-84-4] Iodipamide meglumine; Biligrafin; Endocistobil; Endografin; Intrabilix; Transbilix. C$_{34}$H$_{48}$I$_6$N$_4$O$_{16}$; mol wt 1530.20. More sol in water than the disodium salt. Used as 40% soln.

THERAP CAT: Diagnostic aid (radiopaque medium—cholecystographic).

THERAP CAT (VET): Diagnostic aid (radiopaque medium).

5066. Iodixanol. [92339-11-2] 5,5'-[(2-Hydroxy-1,3-propanediyl)bis(acetylimino)]bis[*N,N'*-bis(2,3-dihydroxypropyl)-2,-4,6-triiodo-1,3-benzenedicarboxamide]; 5,5'-[(2-hydroxytrimethylene)bis(acetylimino)]bis[*N,N'*-bis(2,3-dihydroxypropyl)-2,4,6-triiodoisophthalamide]; 1,3-bis(acetylamino)-*N,N'*-bis[3,5-bis(2,3-dihydroxypropylaminocarbonyl)-2,4,6-triiodophenyl]-2-hydroxypropane; 2-5410-3A; Visipaque. C$_{35}$H$_{44}$I$_6$N$_6$O$_{15}$; mol wt 1550.19. C 27.12%, H 2.86%, I 49.12%, N 5.42%, O 15.48%. Nonionic, dimeric x-ray contrast medium. Prepn: P. E. Hansen *et al., EP 108638* (1984 to Nyegaard), *C.A.* **101**, 151599g (1984). HPLC determn in plasma: H. Nomura *et al., J. Chromatogr.* **572**, 333 (1991). Clinical pharmacokinetics: M. G. Svaland *et al., Invest. Radiol.* **27**, 130 (1992). Clinical studies in cerebral arteriography: Y. Palmers *et al., Eur. J. Radiol.* **17**, 203 (1993); in cardiac angiography: J. A. Hill *et* *al.*, *Am. J. Cardiol.* **74**, 57 (1994); in urography: R. M. Conroy *et al.*, *Clin. Radiol.* **49**, 337 (1994). Use as density gradient medium: T. Ford *et al.*, *Anal. Biochem.* **220**, 360 (1994); J. Graham *et al.*, *ibid.* 367.

White to off-white amorphous powder. Hygroscopic. mp 240-250°. Freely sol in water. Viscosity (37°) 11.1 mPa·s; osmolality 290 mOsm/kg H_2O; pH 7.2-7.6 (c = 320 mgI/ml). 50% aqueous soln (w/v): viscosity 8.3 mPa·s; n 1.4128 ±0.0003; d 1.266 ±0.003; osmolality 200 ±5 mOsm. LD_{50} i.v. in mice: 21 gI/kg (Palmers).
THERAP CAT: Diagnostic aid (radiopaque medium).

5067. Iodized Oil. [8001-40-9] Iodinated vegetable fats and glyceridic oils; Lipiodol. An iodine addition product of vegetable oils, contg 38-42% organically combined iodine.
Thick, viscous, oily liq; alliaceous odor; oleaginous taste; dec on exposure to air and light, becoming brown. *Keep in small, nearly full containers, protected from light.*
THERAP CAT: Diagnostic aid (radiopaque medium).
THERAP CAT (VET): X-ray contrast medium.

5068. Iodoacetic Acid. [64-69-7] 2-Iodoacetic acid. $C_2H_3IO_2$; mol wt 185.95. C 12.92%, H 1.63%, I 68.25%, O 17.21%. Prepd by treating chloroacetic acid in acetone soln with NaI.

Colorless or white crystals. mp 82-83°. *Corrosive, poisonous, skin sensitizer.* Sol in water, alc; very slightly sol in ether. Light sensitive. Store under inert gas at 2-8°C.
USE: Reagent.

5069. Iodoalphionic Acid. [577-91-3] 4-Hydroxy-3,5-diiodo-α-phenylbenzenepropanoic acid; β-(4-hydroxy-3,5-diiodo-phenyl)-α-phenylpropionic acid; 3,5-diiodo-α-phenylphloretic acid; Coletrast; Priodax; Pheniodol; Dikol; Iodobil; Jodobil; Biliognost; Tenicid; Biliselectan. $C_{15}H_{12}I_2O_3$; mol wt 494.07. C 36.47%, H 2.45%, I 51.37%, O 9.71%. Prepd by iodinating β-(4-hydroxyphen-yl)-α-phenylpropionic acid with iodine in an aq soln of alkali iodide in the presence of NH_3 or with iodine in acetic anhydride as the iodinating agent, *cf.* **GB 559024** (1944), *C.A.* **40**, 1883 (1946). Prepn of optically active forms: Tullar, Hoppe, **US 2552696** (1951 to Sterling Drug). Toxicity data: J. O. Hoppe, S. Archer, *Am. J. Roentgenol. Radium Ther.* **69**, 630 (1953).

*dl-*Form. Faintly yellowish powder. dec 157-162°. Crystals from acetic acid, mp 163.5-163.8°. Insol in water. Sol in alcohol, ether, alkali carbonate and in alkali hydroxide solns; slightly sol in benzene and chloroform. Forms a water-sol disodium salt. LD_{50} in mice (g/kg): 3.8 orally (Hoppe, Archer).
*l-*Form. Hydrated crystals, mp 80-85°, $[\alpha]_D^{25}$ −59° (c = 1.5 in 95% alcohol).
*l-*Form disodium salt. mp >230°, $[\alpha]_D^{26}$ −56.6° (c = 1.5 in water).
*d-*Form. Hydrated crystals, mp 80-85°, $[\alpha]_D^{26}$ +59° (c = 1.5 in 95% alcohol).

*d-*Form disodium salt. mp >230°, $[\alpha]_D^{26}$ +56.6° (c = 1.5 in water).
THERAP CAT: Diagnostic aid (radiopaque medium—cholecysto-graphic).

5070. *p*-Iodoaniline. [540-37-4] 4-Iodobenzeneamine; 4-io-dophenylamine; *p*-aminoiodobenzene. C_6H_6NI; mol wt 219.03. C 32.90%, H 2.76%, N 6.40%, I 57.94%.

White crystals. mp 67-68°. Slightly sol in water, freely in alcohol, chloroform, ether.

5071. *o*-Iodoanisole. [529-28-2] 1-Iodo-2-methoxybenzene; 2-methoxyiodobenzene; 2-methoxyphenyl iodide. C_7H_7IO; mol wt 234.04. C 35.92%, H 3.01%, I 54.22%, O 6.84%. Prepd by diazo-tizing *o*-anisidine and decomposing the diazonium salt with KI.

Yellow liquid; darkens on exposure to air. d^{20} 1.80. bp_{730} 238-240°. Insol in water; miscible with alcohol, chloroform, ether. *Protect from light.*

5072. Iodobenzene. [591-50-4] Benzene iodide; phenyl io-dide. C_6H_5I; mol wt 204.01. C 35.32%, H 2.47%, I 62.21%. Ob-tained by diazotizing aniline and then treating with an aq soln of KI, or by the action of HNO_3 on a mixture of C_6H_6 and iodine.

Colorless liquid; rapidly becomes yellow; characteristic odor. d_4^{15} 1.8384. bp 188-189°. mp −30°. n_D^{18} 1.621. Insol in water; miscible with alcohol, chloroform, ether. *Protect from light.*

5073. Iodobenzene Dichloride. [932-72-9] (Dichloroiodo)-benzene; benzene iododichloride. $C_6H_5Cl_2I$; mol wt 274.91. C 26.21%, H 1.83%, Cl 25.79%, I 46.16%. First reported hypervalent iodine compd; utilized primarily as chlorinating reagent. Prepn from iodobenzene and chlorine gas: C. Willgerodt, *Chem. Zentralbl.* **56**, 833 (1885); *idem*, *J. Prakt. Chem.* **33**, 154 (1886); H. L. Lucas, E. R. Kennedy, *Org. Synth.* coll. vol. **III**, 482 (1955). Large scale prepn: A. Zanka *et al.*, *Org. Process Res. Dev.* **2**, 270 (1998). Syn-thetic applications: D. H. R. Barton *et al.*, *Tetrahedron* **48**, 8881 (1992); and new prepn methodology: X.-F. Zhao, C. Zhang, *Synthe-sis* **2007**, 551. Crystal structure: E. M. Archer, T. G. D. Van Schalk-wyk, *Acta Crystallogr.* **6**, 88 (1953); J. V. Carey *et al.*, *J. Chem. Res. Synop.* **1996**, 358. Review: Z. Wu, *Synlett* **2008**, 782-783.

Yellow monoclinic needles, mp 112-113° (Zhou). Also reported as mp 113-117° (dec) (Zanka). Crystal density: 2.2 g/cm³. Decom-poses fairly rapidly when exposed to air and uv light.
USE: In synthetic organic chemistry, primarily as a chlorinating and oxidizing reagent. Also used in other synthetic transformations.

5074. *o*-Iodobenzoic Acid. [88-67-5] 2-Iodobenzoic acid. $C_7H_5IO_2$; mol wt 248.02. C 33.90%, H 2.03%, I 51.17%, O 12.90%. Prepd by diazotizing *o*-aminobenzoic acid and treating the resulting diazonium compound with aq KI.

White needles. d 2.25. mp 162°. Sparingly sol in water; sol in alcohol, ether.

5075. Iodochlorhydroxyquin. [130-26-7] 5-Chloro-7-iodo-8-quinolinol; 5-chloro-8-hydroxy-7-iodoquinoline; chloroiodoquin; iodochlorohydroxyquinoline; iodochloroxyquinoline; clioquinol; Amebil; Alchloquin; Amoenol; Bactol; Barquinol; Budoform; Chinoform; Cliquinol; Eczecidin; Enteroquinol; Entero-Septol; Enterozol; Entrokin; Hi-Enterol; Iodoenterol; Nioform; Quinambicide; Rometin; Vioform. C_9H_5ClINO; mol wt 305.50. C 35.38%, H 1.65%, Cl 11.60%, I 41.54%, N 4.58%, O 5.24%. Prepn: **DE 117767** (1899 to Ciba); **US 641491** (1900); A. Das, S. L. Mukherji, *J. Org. Chem.* **22**, 1111 (1957). Toxicity study: Davis *et al.*, *Am. J. Trop. Med.* **24**, 29 (1944). Comprehensive description: G. Padmanabhan *et al.*, *Anal. Profiles Drug Subs.* **18**, 57-90 (1989).

Brownish-yellow, bulky powder. Practically odorless. Dec ~178-179°. uv max (water/conc. HCl): 266 nm ($A_{1cm}^{1\%}$ 990); (0.1N methanolic NaOH): 269 nm ($A_{1cm}^{1\%}$ 1120); (ethanol): 255 nm ($A_{1cm}^{1\%}$ 1570). One part dissolves in 43 parts of boiling alc, 128 parts chloroform, 17 parts of boiling ethyl acetate, 170 parts of cold acetic acid, 13 parts of boiling acetic acid. Almost insol in water, cold alc, ether. Soly at room temp (mg/ml): water <0.01, methanol 1.9, ethanol 1.3, ether 5.4, chloroform 14.9, 0.1N NaOH 17.3, 0.1N HCl 0.02, acetonitrile between 1.7-2.5, tetrahydrofuran between 20-100, ethyl acetate between 6.7-10, carbon disulfide between 10-20, dimethyl sulfoxide between 20-100, dimethyl acetamide between 20-100. Darkens upon exposure to light. LD₅₀ orally in cats: 400 mg/ kg (Davis).

Caution: Has been linked with the occurrence of subacute myelo-optic neuropathy (S.M.O.N. syndrome) in Japan: T. Tsubaki *et al.*, *Lancet* **1**, 696 (1971); *see also ibid.* **1**, 534 (1977); J. Tateishi in *Drug-Induced Sufferings* T. Soda, Ed. (Excerpta Medica, New York, 1980) pp 464-472.

THERAP CAT: Anti-infective (topical); antiamebic.

THERAP CAT (VET): Has been used as an intestinal anti-infective.

5076. Iodoform. [75-47-8] Triiodomethane. CHI_3; mol wt 393.73. C 3.05%, H 0.26%, I 96.69%. Prepn from acetone, sodium hypochlorite, potassium iodide, and sodium hydroxide: Glass, *Q. J. Pharm. Pharmacol.* **8**, 351 (1935); from chloroform + methyl iodide: Soroos, Hinkamp, *J. Am. Chem. Soc.* **67**, 1642 (1945); by electrolysis: Glasstone, *Ind. Chem.* **7**, 315 (1931). Description of the iodoform reaction: Seelye, Turney, *J. Chem. Educ.* **36**, 572 (1959). Toxicity study: Kutob, Plaa, *Toxicol. Appl. Pharmacol.* **4**, 354 (1962).

Yellow powder or crystals; mp ~120°; dec at high temp with evolution of iodine. Unctuous touch; characteristic, disagreeable odor. Volatile with steam. d 4.1. One gram dissolves in 80 ml glycerol, 60 ml cold alc, 34 ml olive oil, 16 ml boiling alc, 10 ml chloroform, 7.5 ml ether, 3 ml carbon disulfide. Freely sol in benzene, acetone; sparingly sol in glycerin; slightly sol in petr ether. Practically insol in water. LD₅₀ s.c. in mice: 1.6 mmoles/kg (Kutob, Plaa).

Caution: Potential symptoms of overexposure are irritation of eyes and skin; lightheadedness, dizziness, nausea, incoordination, CNS depression; dyspnea; liver, kidney and heart damage; visual disturbances. *See NIOSH Pocket Guide to Chemical Hazards* (DHHS/NIOSH 97-140, 1997) p 172.

THERAP CAT: Anti-infective (topical).

THERAP CAT (VET): Antiseptic, disinfectant for superficial lesions and in the female reproductive tract.

5077. _o_-Iodohippurate Sodium. [133-17-5] *N*-(2-Iodobenzoyl)glycine sodium salt (1:1); sodium *o*-iodohippurate; Hippodin; Jodairol. $C_9H_7INNaO_3$; mol wt 327.05. C 33.05%, H 2.16%, I 38.80%, N 4.28%, Na 7.03%, O 14.68%. Prepd by the condensation of *o*-iodobenzoic acid with glycine, and treatment with NaOH soln: A. P. Sachs, **US 2135474** (1938 to Zonite Prods.). Diagnostic use in post-transplant renography: H. Huland, K. Bischoff, *J. Urol.* **129**, 925 (1983); J.-I. D. Jorgensen, J. Ladefoged, *Dan. Med. Bull.* **34**, 50 (1987).

Dihydrate. [5990-94-3] Crystals. Freely sol in water; sol in alcohol and in dil solns of alkalies. An aq soln is neutral or faintly alkaline to litmus.

¹³¹I-Labeled form. [881-17-4] Iodohippurate sodium I 131; Hippuran I 131; Hipputope.

¹²³I-Labeled form. [56254-07-0] Iodohippurate sodium I 123; Nephroflow.

THERAP CAT: Diagnostic aid (radiopaque medium); radiolabeled forms as diagnostic aid (radioactive imaging agent).

5078. _o_-Iodophenol. [533-58-4] 2-Iodophenol. C_6H_5IO; mol wt 220.01. C 32.76%, H 2.29%, I 57.68%, O 7.27%. Prepd by the action of iodine on *o*-chloromercuriphenol in chloroform: Whitmore, Hanson, *Org. Synth.* **4**, 37 (1925).

Plates from ligroin, mp 43°. d⁸⁰ 1.8757. bp₁₆₀ 186-187°. Sol in alcohol, ether, chloroform, carbon disulfide, benzene; fairly sol in hot water.

5079. _p_-Iodophenol. [540-38-5] 4-Iodophenol; 4-hydroxy-iodobenzene; 4-hydroxyphenyl iodide; 1-iodo-4-hydroxybenzene. C_6H_5IO; mol wt 220.01. C 32.76%, H 2.29%, I 57.68%, O 7.27%. Obtained by action of iodine in alkaline KI upon phenol; or by diazotization of *p*-aminophenol and subsequent replacement of the diazonium group by iodine.

White or reddish crystals; characteristic odor. mp 93-94°. Slightly sol in water, freely sol in alc, ether. *Protect from light.*

Note: The designation "phenol iodide", or more properly "iodized phenol", is applied to a soln of 1 part iodine in 4 parts phenol.

5080. Iodophors. Tamed Iodine. The term iodophor may be applied to any product in which surface active agents (such as nonoxynol, *q.v.*) or polymers (such as polyvinylpyrrolidone, *cf.* povidone-iodine) act as carriers and solubilizing agents for iodine. Other examples: *Biopal CVL-10*; *Dermevan*; *Idonyx*; *Iobac*; *Ioprep*; *Iosan*; *Kleenodyne*; *Rhudane*; *Showersan*; *Wescodyne*; *Westamine X.* An iodophor usually enhances the bactericidal activity of iodine, reduces vapor pressure and odor. Staining is almost nonexistent and wide dilution with water is possible. Examples of prepn: M. V. Shelanski, M. W. Winicov, **US 2710277** (1955 to West Labs); B. J. Scheib *et al.*, **US 2977315** (1961 to Lazarus Labs). Germicidal activity: C. A. Lawrence *et al.*, *J. Am. Pharm. Assoc.* **46**, 500 (1957). Reviews of manufacture, properties and use: D. R. Batey, *Aust. J. Dairy Technol.* **31**, 5-7 (1976); A. Leitmanova *et al.*, *Pharmazie* **42**, 809-813 (1987).

USE: Germicide.

THERAP CAT: Antiseptic; disinfectant.

THERAP CAT (VET): Antiseptic; disinfectant.

Consult the Name Index before using this section.

5081. Iodophthalein Sodium. [2217-44-9] 3,3-Bis(4-hydroxy-3,5-diiodophenyl)-1(3H)-isobenzofuranone sodium salt (1:2); soluble iodophthalein; tetraiodophenolphthalein sodium; T.I.P.P.S.; tetraiodophthalein sodium; tetiothalein sodium; Iodeikon; Cholepulvis; Keraphen; Shadocol; Bilitrast; Iodognost; Stipolac; Tetraiode; Foriod; Iodtetragnost; Antinosin; Cholumbrin; Iodorayoral; Opacin; Photobiline; Piliophen; Videophel; Radiotetrane; Nosophene sodium; Iodophene sodium. $C_{20}H_8I_4Na_2O_4$; mol wt 865.88. C 27.74%, H 0.93%, I 58.62%, Na 5.31%, O 7.39%. Isomeric with phentetiothalein sodium, q.v. Prepd by the action of iodine (in KI soln) on phenolphthalein in aq alk soln: Classen, Löb, *Ber.* **28**, 1603 (1895); **DE 85930** (1894); **DE 88390** (1895). Also obtained by treating an alk soln of phenolphthalein with iodine monochloride: Kalle, **DE 143596** (1900). Prepn of free acid: Orndorff, Mahood, *J. Am. Chem. Soc.* **40**, 937 (1918). Toxicity data: J. O. Hoppe *et al.*, *J. Med. Chem.* **13**, 997 (1970). Metabolism in rabbits: D. Pitre, L. Fumagalli, *Farmaco Ed. Sci.* **32**, 76 (1977).

Trihydrate. Pale-blue to violet crystals. Saline and astringent taste. Somewhat hygroscopic. Gradually decomp on exposure to air due to absorption of CO_2 and becomes incompletely sol. One g dissolves in about 7 ml water, forming a clear, deep-blue dichroic soln. Slightly sol in alc. *Keep well closed.* LD_{50} in mice (mg/kg): 360 i.v., 3800 orally (Hoppe).

Free acid. [386-17-4] Iodophthalein; tetraiodophthalein; Iodophene; Nosophen. $C_{20}H_{10}I_4O_4$; mol wt 821.91. Light-yellow, odorless, tasteless powder; dec at about 200°. Practically insol in water. Slightly sol in alcohol; sol in alkalies, chloroform, ether.

THERAP CAT: Diagnostic aid (radiopaque medium—cholecystographic).

5082. Iodopsin. [1415-94-7] One of 3 visual pigments found in retinal cone cells. Absorption maximum approx 560 nm. Biological activity is similar to that of rhodopsin, q.v. Composed of the chromophore, 11-*cis* retinal, q.v., bound to photopsin, the specific protein component of cone pigments. (*See* Opsins). Isoln from chicken retinas: G. Wald, *Nature* **140**, 545 (1937). Prepn from 11-*cis* retinal and opsin: G. Wald *et al.*, *J. Gen. Physiol.* **38**, 623 (1955). Methods for purification, prepn, and assay: R. Hubbard *et al.*, *Methods Enzymol.* **18**, 615-653 (1971). Exposure to light initiates the conversion of iodopsin through a series of distinct intermediates to yield opsin and *trans*-retinal. Photochemistry: R. Hubbard, A. Kropf, *Nature* **183**, 448 (1959); T. Yoshizawa, G. Wald, *ibid.* **214**, 566 (1967). Studies on the iodopsin binding site: H. Matsumoto *et al.*, *Biochim. Biophys. Acta* **404**, 300 (1975). Three distinct pigments are responsible for color vision: W. B. Marks *et al.*, *Science* **143**, 1181 (1964); P. K. Brown, G. Wald, *ibid.* **144**, 45 (1964). The variation in the absorption maximum of the 3 pigments is regulated by differences in the apoprotein portion of the molecule: R. Hubbard, L. Sperling, *Exp. Eye Res.* **17**, 581 (1973); B. Honig *et al.*, *J. Am. Chem. Soc.* **101**, 7084 (1979). *Reviews:* G. Wald, *Science* **162**, 230-239 (1968); H. Matsumoto, T. Yoshizawa, *Methods Enzymol.* **81**, 154-160 (1982). General review of the visual process: P. S. Zurer, *Chem. Eng. News* **61**, 24-35 (Nov. 28, 1983).

5083. Iodopyracet. [300-37-8] 3,5-Diiodo-4-oxo-1(4H)-pyridineacetic acid compound with 2,2'-iminobis[ethanol] (1:1); bis(hydroxyethyl)ammonium 3,5-diiodo-4-pyridone-N-acetate; diethanolamine 3,5-diiodo-4-pyridone-N-acetate; 3,5-diiodo-4-pyridone-N-acetic acid diethanolamine salt; diodone; RP-3203; Diatrast; Diodrast; Iopyracil; Nosylan; Neo-Methiodal; Neo-Skiodan; Neo-Tenebryl; Nosydrast; Oparenol; Per-Abrodil; Per-Radiographol; Pyelosil; Pylumbrin; Savac; Umbradil; Uriodone; Vasiodone; Xumbradil. $C_{11}H_{16}I_2N_2O_5$; mol wt 510.07. C 25.90%, H 3.16%, I 49.76%, N

5.49%, O 15.68%. Prepn: **FR 728634** (1931 to I. G. Farben); J. Reitmann, **US 1993039** (1935 to Winthrop). Pharmacology, toxicology and radiographic studies: R. St. A. Heathcote, R. A. Gardner, *Br. J. Radiol.* **6**, 304 (1933). HPLC determn in biological fluids: P. Hekman, C. A. M. Van Ginneken, *J. Chromatogr.* **182**, 492 (1980).

Odorless powder, dec 155-157°. Soly: about 36% in water; about 12% in methanol, much less sol in ice-cold methanol. Practically insol in acetone, ether, chloroform. pH of aq solns: 5 to 8, depending on concn. 3,5-Diiodo-4-pyridone-N-acetic acid separated by pptn with dil HCl and dried at 100°, mp 245-249°, and contains not less than 61.5% and not more than 63.5% iodine: *N.F.* **XI.**

Iodopyracet Injection. 35% (w/v) soln of the diethanolamine salt of 3,5-diiodo-4-pyridone-N-acetic acid. Contains about 22% I (w/v). d_4^{25} 1.185. Neutral to litmus.

Iodopyracet Concentrated Solution. Diodrast Concentrated Solution. 70% soln (w/v) of the diethanolamine salt of 3,5-diiodo-4-pyridone-N-acetic acid. Contains about 44% I (w/v). Neutral to litmus.

Iodopyracet Compound Solution. Diodrast Compound Solution. Soln contg approx 40.5% (w/v) of the diethanolamine salt and approx 9.5% (w/v) of the diethylamine salt of 3,5-diiodo-4-pyridone-N-acetic acid. Contains about 25% I. Neutral to litmus. d_4^{25} 1.270.

THERAP CAT: Diagnostic aid (radiopaque medium—urographic).
THERAP CAT (VET): Diagnostic aid (radiopaque medium).

5084. Iodopyrrole. [87-58-1] 2,3,4,5-Tetraiodo-1H-pyrrole; Iodol. C_4HI_4N; mol wt 570.68. C 8.42%, H 0.18%, I 88.95%, N 2.45%. Prepd by iodination of pyrrole: Potts, *J. Chem. Soc.* **1953**, 3711; Treibs, Kolm, *Ann.* **614**, 176 (1958). Isoln from shale-oil naphta: Janssen *et al.*, *J. Am. Chem. Soc.* **73**, 4040 (1951).

Needles from ethanol, dec 162-164°, also reported as dec 150-160° and 140-150°. One gram dissolves in 4900 ml water, 9 ml alcohol, 1.5 ml ether, 105 ml chloroform, 155 ml glycerol; sol in fixed oils.

THERAP CAT: Antiseptic (local).
THERAP CAT (VET): Antiseptic, disinfectant for superficial lesions and in the female reproductive tract.

5085. Iodoquinol. [83-73-8] 5,7-Diiodo-8-quinolinol; diiodohydroxyquin; diiodo-oxyquinoline; 5,7-diiodo-8-hydroxyquinoline; SS-578; Diodoquin; Disoquin; Floraquin; Dyodin; Dinoleine; Searlequin; Diodoxylin; Rafamebin; Ioquin; Direxiode; Stanquinate; Yodoxin; Zoaquin; Enterosept; Embequin. $C_9H_5I_2NO$; mol wt 396.95. C 27.23%, H 1.27%, I 63.94%, N 3.53%, O 4.03%. Prepd by the action of iodine monochloride on 8-hydroxyquinoline: Papesch, Burtner, *J. Am. Chem. Soc.* **58**, 1314 (1936); by the action of KIO_3 on 8-hydroxyquinoline: Zeifman, *C.A.* **34**, 3745. Electrolytic prepn: Brown, Berkowitz, *Trans. Electrochem. Soc.* **75**, 385 (1939). *See also:* Claus, **DE 78880**; Passek, **DE 411050**; Matsumura, *C.A.* **21**, 1461 (1927); Pirrone, Cherubino, *C.A.* **28**, 3073 (1934).

Crystals from xylene. The medicinal grade is a yellowish-brown powder. mp 200-215° (extensive decompn). Almost insol in water. Sparingly sol in alcohol, ether, and acetone; sol in hot pyridine and in hot dioxane.

THERAP CAT: Antiamebic.

5086.   N-Iodosaccharin. [86340-94-5] 2-Iodo-1,2-benziso-thiazol-3(2*H*)-one 1,1-dioxide; NISac. $C_7H_4INO_3S$; mol wt 309.08. C 27.20%, H 1.30%, I 41.06%, N 4.53%, O 15.53%, S 10.37%. Iodination reagent. Prepn: M. Papadopoulou, A. Varvoglis, *J. Chem. Res. Synop.* **1983**, 66. Synthesis and use for iodination of aromatics and alkenes: D. Dolenc, *Synlett* **2000**, 544. As an activator of thioglycosides: M. Aloui, A. J. Fairbanks, *ibid.* **2001**, 797; *eidem, Chem. Commun.* **2001**, 1406. Iodination of enol acetates and diones: D. Dolenc, *Synth. Commun.* **33**, 2917 (2003); of isatins: S. P. L. de Souza *et al., Heterocycl. Commun.* **9**, 31 (2003).

Crystallizes in small pale yellow crystals as the monohydrate, mp 206-208°. Reprecipitation from THF, then drying produces a white powder which can be stored indefinitely in a dark freezer. pKa: 1.30. Sol in common organic solvents (better in more polar ones such as acetone, acetonitrile). Insol in water.

USE: Iodination reagent; activator of thioglycosides.

5087.   Iodosobenzene. [536-80-1] Iodosylbenzene; PhIO. C_6H_5IO; mol wt 220.01. C 32.76%, H 2.29%, I 57.68%, O 7.27%. Versatile oxidizing agent. Prepn: C. Willgerodt, *Ber.* **26**, 357 (1893). Synthesis: H. Saltzman, J. G. Sharefkin, *Org. Synth.* **5**, 658 (1973). Mechanistic study: R. M. Moriarty *et al., J. Am. Chem. Soc.* **103**, 686 (1981). Use as oxidant: S. Mukerjee *et al., ibid.* **119**, 8097 (1997); G. Mielniczak, A. Lopusinski, *Synlett* **2001**, 505; as mediator: P. Dauban *et al., J. Am. Chem. Soc.* **123**, 7707 (2001); C.-C. Guo *et al., Appl. Catal. A* **230**, 53 (2002). Brief review: A. Minatti, *Synlett* **2003**, 140-141.

Yellowish amorphous powder. *Explosive* if heated to 210° due to disproportion into phenyliodide and iodylbenzene.

USE: Oxygen transfer reagent for stiochiometric or catalytic cross-functionalization of alkenes, alcohols, sulfides, and organometallo compds.

5088.   N-Iodosuccinimide. [516-12-1] 1-Iodo-2,5-pyrrolidinedione; succiniodimide; NIS. $C_4H_4INO_2$; mol wt 224.99. C 21.35%, H 1.79%, I 56.40%, N 6.23%, O 14.22%. Prepd by iodination of silver succinimide: Djerassi, Lenk, *J. Am. Chem. Soc.* **75**, 3493 (1953). Structural studies: K. Padmanabhan *et al., Acta Crystallogr.* **C46**, 88 (1990). Use as synthetic reagent for cysteine peptides: H. Shih, *J. Org. Chem.* **58**, 3003 (1993); in oxidation and halogenation reactions: A. Speicher, T. Eicher, *J. Prakt. Chem.* **340**, 278 (1998).

Needles from dioxane + carbon tetrachloride. mp 200-201°. Sol in acetone, methanol; moderately sol in dioxane. Practically insol in CCl_4, ether. Dec in water. *Protect from light.*

Caution: Avoid contact with skin and mucous membranes.

USE: Iodination of ketones and aldehydes.

5089.   Iodosulfuron-methyl-sodium. [144550-36-7] 4-Iodo-2-[[[[(4-methoxy-6-methyl-1,3,5-triazin-2-yl)amino]carbonyl]ami-

no]sulfonyl]benzoic acid methyl ester sodium salt (1:1); methyl 4-iodo-2-[3-(4-methoxy-6-methyl-1,3,5-triazin-2-yl)ureidosulfonyl]-benzoate sodium salt; AE-F115008. $C_{14}H_{13}IN_5NaO_6S$; mol wt 529.24. C 31.77%, H 2.48%, I 23.98%, N 13.23%, Na 4.34%, O 18.14%, S 6.06%. Sulfonylurea herbicide for post-emergence use in cereals; acetolactate synthase inhibitor. Prepn: O. Ort *et al.,* **WO 9213845**; *eidem,* **US 5688745** (1992, 1997 both to Hoechst). Comprehensive description and use with the safener, mefenpyr-diethyl: E. Hacker *et al., Brighton Crop Prot. Conf. - Weeds* **1999**, 15-22. Field trials in wheat: H. L. Crooks *et al., Weed Technol.* **18**, 93 (2004).

Almost colorless to light beige crystalline powder, mp 152°. Vapor pressure (25°): 6.7×10^{-9} Pa. Soly in water at 20° (g/l): 0.16 (pH 5); 25 (pH 7); 65 (pH 9). Log P (octanol/water): 1.07 (pH 5); −0.70 (pH 7); −1.22 (pH 9). LD_{50} in rats (mg/kg): 2678 orally; >2000 dermally (Hacker).

Combination with mefenpyr-diethyl. Husar; Hussar.

USE: Herbicide.

5090.   3-Iodotyrosine. [70-78-0] 3-Iodo-L-tyrosine; 3-iodo-4-hydroxyphenylalanine; monoiodotyrosine. $C_9H_{10}INO_3$; mol wt 307.09. C 35.20%, H 3.28%, I 41.32%, N 4.56%, O 15.63%. Prepn of L-form: Hillman, Hillman-Elies, *Z. Physiol. Chem.* **305**, 177 (1956); Pitt-Rivers, *Chem. Ind. (London)* **1956**, 21. Prepn of L- and DL-form: Harington, Rivers, *Biochem. J.* **38**, 320 (1944).

Crystals from water, dec 204-206°. $[\alpha]_D^{20}$ −4.4° (c = 5 in 1*N* HCl). Dissolves in 15 parts boiling water.

DL-Form monohydrate. Crystals from water, dec 200-201°.

5091.   o-Iodoxybenzoic Acid. [64297-64-9]; [61717-82-6] (cyclic tautomer). 2-Iodylbenzoic acid; 1-hydroxy-1,2-benziodoxol-3(1*H*)-one 1-oxide; IBX; 2-iodoxybenzoic acid. $C_7H_5IO_4$; mol wt 280.02. C 30.03%, H 1.80%, I 45.32%, O 22.85%. Hypervalent 10-I-4 iodine reagent; cyclic tautomer is the predominant molecular structure. Prepn: C. Hartmann, V. Meyer, *Ber.* **26**, 1727 (1893); F. R. Greenbaum, *Am. J. Pharm.* **108**, 17 (1936); and conversion to stable 12-I-5 periodinanes: D. B. Dess, J. C. Martin, *J. Am. Chem. Soc.* **113**, 7277 (1991). Improved prepn: M. Frigerio *et al., J. Org. Chem.* **64**, 4537 (1999). Prepn of a stabilized, nonexplosive formulation: D. Depernet, B. Francois, **US 6462227** (2002 to Simafex). IR analysis: R. Bell, K. J. Morgan, *J. Chem. Soc.* **1960**, 1209. Crystal structure: J. Z. Gougoutas, *Acta Crystallogr.* **10**, 489 (1981). ^{13}C NMR analysis: A. R. Katritzky *et al., Magn. Reson. Chem.* **27**, 1007 (1989). Oxidizing properties and reaction conditions: M. Frigerio *et al., J. Org. Chem.* **60**, 7272 (1995). Kinetics and mechanism of oxidation reactions: S. De Munari *et al., ibid.* **61**, 9272 (1996). Characterization of crystallographic forms: P. J. Stevenson *et al., J. Chem. Soc. Perkin Trans. 2* **1997**, 589. Brief review of chemistry: T. Wirth, *Angew. Chem. Int. Ed.* **40**, 2812-2814 (2001).

White crystalline solid, mp 233° (dec). Virtually insol in common organic solvents; insol in water. Soly in DMSO: 1.5 M. Sol in aq pyridine, aq sodium hydroxide (dec).

Caution: Explosive under impact or heating >200°: J. B. Plumb, D. J. Harper, *Chem. Eng. News* **68**, 3 (July 16, 1990).

USE: Mild, chemoselective, environmentally friendly oxidizing agent. In prepn of Dess-Martin periodinane, *q.v.*

5092. Iofetamine I 123. [75917-92-9] 4-(Iodo-^{123}I)-α-methyl-N-(1-methylethyl)benzeneethanamine; (±)-*p*-iodo-[123]I-*N*-isopropyl-α-methylphenethylamine; [^{123}I] (±)-*N*-isopropyl-*p*-iodoamphetamine. $C_{12}H_{18}$[123]IN. Lipid soluble radioactive brain imaging agent. Prepn: R. M. Baldwin *et al.*, **EP 11858** (1980 to Hoffmann-La Roche); *eidem*, **US 4360511** (1982 to Medi-Physics). Detailed synthesis: L. Carlsen, K. Andresen, *Eur. J. Nucl. Med.* **7**, 280 (1982). Modified synthesis for improved yields: A. Najafi, *J. Labelled Compd. Radiopharm.* **24**, 1167 (1987). Preparative HPLC purification: J. Mertens *et al.*, *Nucl. Med. Commun.* **5**, 705 (1984). Brain uptake and localization: H. S. Winchell *et al.*, *J. Nucl. Med.* **21**, 940 (1980); H. S. Winchell *et al.*, *ibid.* 947. Distribution kinetics in animals: P. Som *et al.*, *Int. J. Nucl. Med. Biol.* **12**, 185 (1985); in humans: B. L. Holman *et al.*, *J. Nucl. Med.* **25**, 25 (1984). Diagnostic use for cerebral function in Alzheimer's disease: K. A. Johnson *et al.*, *Arch. Neurol.* **45**, 392 (1988). Reviews of radiopharmaceutical production and diagnostic use with single photon emission computed tomography (SPECT) imaging: M. B. Cohen *et al.*, *Appl. Radiat. Isot.* **37**, 749-763 (1986); of diagnostic applications in stroke: C. H. Park *et al.*, *RadioGraphics* **8**, 305-326 (1988); of metabolism and kinetics: R. M. Baldwin, J.-L. Wu, *J. Nucl. Med.* **29**, 122-124 (1988).

$bp_{0.05}$ 91-92°. Also reported as $bp_{0.6}$ 98-99° (Najafi).

Hydrochloride. [85068-76-4] IMP; ^{123}I labeled IMP; ^{123}I-M123; Perfusamine; Spectamine. $C_{12}H_{18}$[123]IN.HCl. Crystals from ether, mp 156-158°.

THERAP CAT: Diagnostic aid (radioactive imaging agent).

5093. Ioflupane I 123. [155798-07-5] (1*R*,2*S*,3*S*,5*S*)-8-(3-Fluoropropyl)-3-[4-(iodo[123]I)phenyl]-8-azabicyclo[3.2.1]octane-2-carboxylic acid methyl ester; [1*R*-(*exo*,*exo*)]-8-(3-fluoropropyl)-3-[4-(iodo[123]I)phenyl]-8-azabicyclo[3.2.1]octane-2-carboxylic acid methyl ester; *N*-(3-fluoropropyl)-2β-carbomethoxy-3β-(4-^{123}I-iodophenyl)nortropane; methyl 8-(3-fluoropropyl)-3β-(*p*-iodo-^{123}I-phenyl)-1α*H*,5α*H*-nortropane-2β-carboxylate; ^{123}I-FP-CIT; DaTSCAN. $C_{18}H_{23}FINO_2$; mol wt 431.29. C 50.13%, H 5.38%, F 4.41%, I 29.42%, N 3.25%, O 7.42%. Analog of cocaine, *q.v.*, that binds specifically to the dopamine transporters (DAT) in the membrane of the presynaptic dopaminergic neuron; designed as SPECT imaging agent to determine nigrostriatal neuronal integrity. Prepn: J. L. Neumeyer *et al.*, **US 5310912** (1994 to Research Biochemicals); *idem et al.*, *J. Med. Chem.* **37**, 1558 (1994). HPLC determn in biological fluids: K. A. Bergstroem *et al.*, *Hum. Psychopharmacol.* **11**, 483 (1996). Binding study: J. L. Neumeyer *et al.*, *J. Med. Chem.* **39**, 543 (1996). Clinical biodistribution: J. Booij *et al.*, *Eur. J. Nucl. Med.* **25**, 24 (1998). Clinical imaging for tremor differentiation: H. T. S. Benamer *et al.*, *Mov. Disord.* **15**, 503 (2000); for dementia differentiation: D. C. Costa *et al.*, *ibid.* **18**, Suppl. 7, S34 (2003). Functional imaging for dopaminergic deficit: H. T. S. Benamer *et al.*, *ibid.* **18**, 977 (2003). Clinical use in diagnosis of parkinsonism: A. M. Catafau *et al.*, *ibid.* **19**, 1175 (2004). Review of role in dopaminergic imaging: P. M. Kemp, *Nucl. Med. Commun.* **26**, 87-96 (2005).

mp 82-83°. $[α]_D^{24}$ −12.8°.

THERAP CAT: Diagnostic aid (radioactive imaging agent).

5094. Ioglycamic Acid. [2618-25-9] 3,3'-[Oxybis[(1-oxo-2,1-ethanediyl)imino]]bis[2,4,6-triiodobenzoic acid]; 3,3'-[oxybis(methylenecarbonylimino)]bis[2,4,6-triiodobenzoic acid]; 3,3'-(diglycoloyldiimino)bis(2,4,6-triiodobenzoic acid); diglycolic acid bis-(2,4,6-triiodo-3-carboxanilide); *N,N*-(oxydiacetyl)bis[3-amino-2,-4,6-triiodobenzoic acid]. $C_{18}H_{10}I_6N_2O_7$; mol wt 1127.71. C 19.17%, H 0.89%, I 67.52%, N 2.48%, O 9.93%. Prepn: H. Priewe *et al.*, *Ber.* **87**, 651 (1954); H. Priewe, R. Rutkowski, **US 2776241**; **US 2853424** (1957, 1958 both to Schering AG).

Occurs in three cryst modifications: Neudert, Röpke, *Helv. Chim. Acta* **41**, 855 (1958). mp with baking at 222°, sintering at 227°, splitting of iodine at 245°, dec 281°.

Meglumine salt. [14317-18-1] Dimeglumine ioglycamide; Biligram. $C_{18}H_{10}I_6N_2O_7.2(C_7H_{17}NO_5)$; mol wt 1518.14. Water soluble.

THERAP CAT: Diagnostic aid (radiopaque medium—cholecystographic).

5095. Iohexol. [66108-95-0] 5-[Acetyl(2,3-dihydroxypropyl)amino]-N^1,N^3-bis(2,3-dihydroxypropyl)-2,4,6-benzenedicarboxamide; *N,N'*-bis(2,3-dihydroxypropyl)-5-[*N*-(2,3-dihydroxypropyl)acetamido]-2,4,6-triiodoisophthalamide; Win-39424; Compd 545; Omnipaque. $C_{19}H_{26}I_3N_3O_9$; mol wt 821.14. C 27.79%, H 3.19%, I 46.36%, N 5.12%, O 17.54%. Nonionic radio-contrast medium. Prepn: V. Nordal, H. Holtermann, **DE 2726196**; *eidem*, **US 4250113** (1977, 1981 both to Nyegaard). HPLC-UV determn in plasma: R. S. Soman *et al.*, *J. Chromatogr. B* **816**, 339 (2005). Pharmacology and toxicology: *Acta Radiol.* **Suppl. 362**, 1-134 (1980). Acute toxicity: S. Salvesen, *ibid.* **73**. Fibrillatory potential in dogs: G. L. Wolf *et al.*, *Invest. Radiol.* **16**, 320 (1981). Comparative clinical studies in coronary angiography: G. B. J. Mancini *et al.*, *Am. J. Cardiol.* **51**, 1218 (1983); I. D. Sullivan *et al.*, *Br. Heart J.* **51**, 643 (1984); M. A. Bettmann *et al.*, *Radiology* **153**, 583 (1984). *Review:* T. Almén, *Acta Radiol.* **Suppl. 366**, 9-19 (1983).

Crystals from butanol, mp 174-180°. Hygroscopic. Very sol in water, methanol. Practically insol or insol in ether, chloroform. Stable in aqueous solutions. Viscosity (cP): 6.2 at 37°; 12.6 at 20° (c = 200 mg Iodine/ml). LD_{50} in male, female rats, mice (g Iodine/kg): 15.0, 12.3, 24.3, 25.1 i.v. (Salvesen).

THERAP CAT: Diagnostic aid (radiopaque medium).

5096. Iomeglamic Acid. [25827-76-3] 5-[(3-Amino-2,4,6-triiodophenyl)methylamino]-5-oxopentanoic acid; 3'-amino-2',-4',6'-triiodo-*N*-methylglutaranilic acid; *N*-methyl-*N*-(3-amino-2,4,6-triiodophenyl)glutaramic acid; *N*-methyl-*N*-(2,4,6-triiodo-3-aminophenyl)glutaramidic acid; RG-270; Falignost. $C_{12}H_{13}I_3N_2O_3$; mol wt 613.96. C 23.48%, H 2.13%, I 62.01%, N 4.56%, O 7.82%. Prepn: Cassebaum, Dierbach, **DD 67209** (1969), *C.A.* **72**, 66654j (1970). Biotransformation: Pfeifer *et al.*, *Pharmazie* **27**, 403 (1972). Pharmacology: H. Bekker *et al.*, *ibid.* 411. Clinical evaluation: Barke, *Zentralbl. Pharm. Pharmakother. Laboratoriums-diagn.* **110**, 1117 (1971).

Yellow to brown crystals from acetic acid, mp 169°. Freely sol in DMF; slightly sol in ethanol, chloroform. Practically insol in water. LD_{50} i.v. in mice: 500 mg/kg (Bekker).

THERAP CAT: Diagnostic aid (radiopaque medium—cholecysto-graphic).

5097. Iomeprol. [78649-41-9] N^1,N^3-Bis(2,3-dihydroxypro-pyl)-5-[(2-hydroxyacetyl)methylamino]-2,4,6-triiodo-1,3-benzene-dicarboxamide; N,N'-bis(2,3-dihydroxypropyl)-2,4,6-triiodo-5-(N-methylglycolamido)isophthalamide; B-16880; Imeron; Iomeron. $C_{17}H_{22}I_3N_3O_8$; mol wt 777.09. C 26.28%, H 2.85%, I 48.99%, N 5.41%, O 16.47%. Nonionic x-ray contrast media. Prepn: E. Felder, D. Pitrè, **EP 026281**; *eidem,* **US 4352788** (1981, 1982 both to Bracco). HPLC determn in plasma and urine: V. Lorussu *et al., J. Chromatogr.* **525**, 401 (1990). Series of articles on properties, pharmacology, pharmacokinetics and clinical experience in angiography and urography: *Eur. J. Radiol.* **18**, Suppl. 1, S1-S124 (1994). Toxicology study: A. Morisetti *et al., ibid.* S21.

Crystalline powder, mp 285-291° (dec). Extremely sol in water, very sol in methanol, poorly sol in ethanol. Practically insol in chloroform. Aqueous soln of 150 mg I/ml: d_4^{20} 1.166, d_4^{37} 1.161; osmolality (37°) 0.27 osmol/kg H_2O; viscosity at 20° (mPa·s) 1.9, at 37° 1.3; n_D^{20} 1.3828. Aqueous soln of 300 mg I/ml: d_4^{20} 1.334, d_4^{37} 1.329; osmolality (37°) 0.52 osmol/kg H_2O; viscosity at 20° (mPa·s) 8.4, at 37° 4.8; n_D^{20} 1.4327. Aqueous soln of 400 mg I/ml: d_4^{20} 1.446, d_4^{37} 1.441; osmolality (37°) 0.72 osmol/kg H_2O; viscosity at 20° (mPa·s) 28.9, at 37° 13.9; n_D^{20} 1.4660. LD_{50} (400 mgI/ml soln) i.v. in mice, rats (gI/kg): 19.9, 14.5; LD_{50} (370 mgI/ml soln) i.v. in dogs (gI/kg): >12.5 (Morisetti).

THERAP CAT: Diagnostic aid (radiopaque medium).

5098. Ionic Liquids. Liquids composed entirely of ions; molten salts with melting points below 100°. Commonly consist of organic cations and inorganic anions. The cations most frequently used to generate ionic liquids are alkyl derivatives of imidazolium, pyridinium, ammonium and phosphonium. Early development: S. Sugden, H. Wilkins, *J. Chem. Soc.* **1929**, 1291. Environmental risk assessment: B. Jastorff *et al., Green Chem.* **5**, 136 (2003). Comprehensive description: R. D. Rogers, K. R. Seddon, Eds., *ACS Symp. Ser.* **818**, 1-474 (2002). Review of prepn, properties and uses in synthesis and catalysis: T. Welton, *Chem. Rev.* **99**, 2071-2083 (1999); P. Wasserscheid, W. Keim, *Angew. Chem. Int. Ed.* **39**, 3772-3789 (2000); of applications in organic synthesis: H. Zhao, S. V. Malhotra, *Aldrichim. Acta* **35**, 75-83 (2002); of use as reaction media for catalytic reactions: R. Sheldon, *Chem. Commun.* **2001**, 2399-2407; for biocatalytic transformations: F. van Rantwijk *et al., Trends Biotechnol.* **21**, 131-138 (2003); of commercial uses: J. H. Davis, Jr., P. A. Fox, *Chem. Commun.* **2003**, 1209-1212.

Many are colorless, some are pale yellow to orange in color. No measurable vapor pressure. Thermally stable; many have liquid ranges of more than 300°. Dissolve a wide range of organic and inorganic compounds; soly of gases is generally good. Immisc with many organic solvents, forming biphasic systems.

USE: Highly polar, noncoordinating solvents with a broad range of green chemistry applications in organic synthesis and catalysis.

5099. Ionone. [8013-90-9] Irisone. Family of naturally occurring, cyclic terpenoids; fragrance components of many volatile oils. The term, ionone, commonly refers to a mixture of isomers in which the major constituents are α-ionone and β-ionone. The latter is a key intermediate in the synthesis of vitamin A. Isoln from the volatile oil of *Boronia megastigma* Nees., *Rutaceae:* Y. R. Naves, G. R. Parry, *Helv. Chim. Acta* **30**, 419 (1946). Prepn by condensing citral with acetone: F. Tiemann, *Ber.* **33**, 3703 (1900); H. Hibbert, L. T. Cannon, *J. Am. Chem. Soc.* **46**, 119 (1924); H. J. V. Krishna, B. N. Joshi, *J. Org. Chem.* **22**, 224 (1957). Prepn from isobutylene and vinyl methyl ketone: Pasedach, Seefelder, **DE 1000374** (1957 to BASF), *C.A.* **54**, 1595g (1960). Prepn by cyclization of pseudo-ionone with H_2SO_4: W. Kimel *et al., J. Org. Chem.* **22**, 1611 (1957); Kaiser, Kimel, **US 2877271** (1959 to Hoffmann-La Roche). *Reviews:* Y. R. Naves, *J. Soc. Cosmet. Chem.* **22**, 439 (1971); H. Surburg, J. Panten, *Common Fragrance and Flavor Materials* (Wiley, Weinheim, 5th Ed., 2006) pp 66-70; J. Lalko *et al., Food Chem. Toxicol.* **45**, S251-S257 (2007).

α-Ionone   β-Ionone

Mixed isomers, pale yellow to yellow liquid. d_{25}^{25} 0.933-0.937. n_D^{20} 1.503-1.508. bp_{12} 126-128°. Miscible with abs alc. Soluble in 2 to 3 parts of 70% alcohol, ether, chloroform, benzene. Very slightly sol in water.

α-**Ionone.** [127-41-3] (3E)-4-(2,6,6-Trimethyl-2-cyclohexen-1-yl)-3-buten-2-one; α-cyclocitrylideneacetone. $C_{13}H_{20}O$; mol wt 192.30. Review of toxicology and uses: J. Lalko *et al., Food Chem. Toxicol.* **45**, S235-S240 (2007). Colorless to slightly yellow liquid with sweet-floral odor, reminiscent of violets. bp_{11} 123-124°. d_4^{20} 0.9319. n_D^{20} 1.4982. uv max (alc): 228.5 nm (ε 14300). Sol in alc, most fixed oils, propylene glycol. Insol in glycerin, water.

β-**Ionone.** [79-77-6] (3E)-4-(2,6,6-Trimethyl-1-cyclohexen-1-yl)-3-buten-2-one. $C_{13}H_{20}O$; mol wt 192.30. Review of toxicology and uses: J. Lalko *et al., Food Chem. Toxicol.* **45**, S241-S247 (2007). Colorless to pale, straw-colored liquid. Odor reminiscent of cedar wood; violet-like upon dilution. bp_{10} 127-128.5°. d_4^{20} 0.9461. n_D^{20} 1.5202. uv max (alc): 293.5 nm (ε 8700). Sol in alc, most fixed oils, propylene glycol. Insol in glycerin, water.

USE: Fragrance ingredient in perfumery, in cosmetics and personal care products, in household cleaners and detergents. Flavoring agent in beverages, baked goods, and candies.

5100. Iopamidol. [60166-93-0] N^1,N^3-Bis[2-hydroxy-1-(hydroxymethyl)ethyl]-5-[[(2S)-2-hydroxy-1-oxopropyl]amino]-2,4,6-triiodo-1,3-benzenedicarboxamide; (S)-N,N'-bis[2-hydroxy-1-(hydroxymethyl)ethyl]-2,4,6-triiodo-5-lactamidoisophthalamide; 5-(α-hydroxypropionylamino)-2,4,6-triiodoisophthalic acid di(1,3-dihydroxyisopropylamide); iomapidol; B-15000; SQ-13396; Iopamiro; Iopamiron; Isovue; Jopamiro; Niopam; Solutrast. $C_{17}H_{22}I_3N_3O_8$; mol wt 777.09. C 26.28%, H 2.85%, I 48.99%, N 5.41%, O 16.47%. Nonionic radiocontrast medium. Prepn: E. Felder, D. Pitre, **DE 2547789**; *eidem,* **US 4001323** (1976, 1977 both to Savac). Physicochemical properties, preclinical studies: E. Felder *et al., Farmaco Ed. Sci.* **32**, 835 (1977). Determn of optical purity: *eidem, Farmaco Ed. Prat.* **32**, 3 (1982). *In vitro* effects: B. Schulze, H. K. Beyer, *Arzneim.-Forsch.* **31**, 1067 (1981). Development, initial testing: M. Sovak *et al., Radiology* **142**, 115 (1982). Clinical studies in angiography: A. J. Molyneux, P. W. Sheldon, *Br. J. Radiol.* **55**, 117 (1982); G. H. Whitehouse, S. L. Snowdon, *Clin. Radiol.* **33**, 231 (1982); in myelography: B. Drayer *et al., Am. J. Neuroradiol.* **3**, 59 (1982). Comprehensive description: E. Felder *et al., Anal. Profiles Drug Subs.* **17**, 115-154 (1988).

White, odorless powder. Dec at about 300° without melting. $[\alpha]_D^{20}$ −2.01° (c = 10 in water). pKa (25°) 10.70. Miscible with boiling ethanol. Very sol in water; sparingly sol in methanol. Practically insol in alc, chloroform. LD_{50} in mice, rats, rabbits, dogs (g/kg): 44.5, 28.2, 19.6, 34.7 i.v. (Felder). Also reported as: LD_{50} in mice (mg iodine/kg body wt): 21,800 i.v.; 20,000 i.p.; 1500 intracerebral (Felder, Pitre).

THERAP CAT: Diagnostic aid (radiopaque medium).

5101. Iopanoic Acid. [96-83-3] 3-Amino-α-ethyl-2,4,6-triiodobenzenepropanoic acid; 3-amino-α-ethyl-2,4,6-triiodohydrocinnamic acid; iodopanoic acid; 3-(3-amino-2,4,6-triiodophenyl)-2-ethylpropanoic acid; 3-(3-amino-2,4,6-triiodophenyl)-2-ethylpropionic acid; Cistobil; Colepax; Telepaque; Teletrast. $C_{11}H_{12}I_3NO_2$; mol wt 570.94. C 23.14%, H 2.12%, I 66.68%, N 2.45%, O 5.60%. Oral radiocontrast medium. Prepn: S. Archer, **US 2705726** (1955 to Sterling Drug). Optical resolution: Pitré, Boveri, *J. Med. Chem.* **11**, 406 (1968). Toxicity data: P. Tirone, G. Rosati, *Farmaco Ed. Prat.* **31**, 397 (1976). Review of pharmacology and clinical efficacy in oral cholecystography: R. N. Berk *et al.*, *N. Engl. J. Med.* **290**, 204 (1974). Comprehensive description: D. Pitre, *Anal. Profiles Drug Subs.* **14**, 181-206 (1985).

Cream-colored powder. Sol in alc, chloroform, ether, and solutions of alkali hydroxides and carbonates. Insol in water.

dl-Form. Cream-colored solid, mp 155.2-157°. pKa 4.8. Insol in water. Sol in dil alkali, in 95% alc, in other organic solvents. LD_{50} in mice, rats (mg/kg): 285, 320 i.v.; 1540, 2870 orally (Tirone).

l-Form. [17879-97-9] Crystals, mp 162-163°. $[\alpha]_D^{20}$ −5.2 ± 0.1° (c = 2 in ethanol).

d-Form. [17879-96-8] Crystals, mp 162°. $[\alpha]_D^{20}$ +5.1 ± 0.1° (c = 2 in ethanol).

Sodium salt. [2497-78-1] Bilijodon-Natrium.

THERAP CAT: Diagnostic aid (radiopaque medium—cholecystographic).

5102. Iopentol. [89797-00-2] 5-[Acetyl(2-hydroxy-3-methoxypropyl)amino]-N^1,N^3-bis(2,3-dihydroxypropyl)-2,4,6-triiodo-1,3-benzenedicarboxamide; N,N'-bis(2,3-dihydroxypropyl)-5-[N-(2-hydroxy-3-methoxypropyl)acetamido]-2,4,6-triiodoisophthalamide; Imagopaque. $C_{20}H_{28}I_3N_3O_9$; mol wt 835.17. C 28.76%, H 3.38%, I 45.59%, N 5.03%, O 17.24%. Nonionic, water soluble vascular contrast medium. Prepn: K. Wille, **EP 105752** (1984 to Nyegaard A/S). Synthesis and preparative LC purification: W. Skjold, A. Berg, *J. Chromatogr.* **366**, 299 (1986). Effects on blood-brain barrier in animals: A. A. Michelet, *Acta Radiol.* **28**, 329 (1987). Series of articles on clinical experience: *Eur. Radiol.* **7**, Suppl. 4, S104-S161 (1997).

Osmolality (300 mg I/ml): 0.64 mol/kg H_2O. Viscosity at 20° (300 mg I/ml): 13.2 mPa·s. Partition coefficient (1-octanol/water): 0.007.

THERAP CAT: Diagnostic aid (radiopaque medium).

5103. Iophendylate. [99-79-6] 4-Iodo-ι-methylbenzenedecanoic acid ethyl ester; ethyl 10-(p-iodophenyl)undecylate; ethyl 10-(p-iodophenyl)hendecanoate. $C_{19}H_{29}IO_2$; mol wt 416.34. C 54.81%, H 7.02%, I 30.48%, O 7.69%. Chief component of *Ethiodan, Mulsopaque, Myodil, Neurotrast, Pantopaque*; may also contain other isomers including 11-(p-iodophenyl)undecylate. Prepn: W. H. Strain *et al.*, *J. Am. Chem. Soc.* **64**, 1436 (1942); *eidem*, **US 2348231** (1944 to Noned Corp. and Kodak); W. Baker *et al.*, *J. Soc. Chem. Ind.* **63**, 223 (1944). Pharmacology and toxicity: T. B. Steinhausen *et al.*, *Radiology* **43**, 230 (1944). Clinical evaluation of diagnostic use in myelography: G. H. Ramsey *et al.*, *ibid.* 236. Clinical comparison with metrizamide, *q.v.*, in diagnostic radiology: S. A. Keiffer *et al.*, *ibid.* **129**, 695 (1978). Evaluation of use in MR imaging of the spine: I. F. Braun *et al.*, *Am. J. Neuroradiol.* **7**, 997 (1986); A. K. Anand *et al.*, *Comput. Radiol.* **11**, 165 (1987).

Colorless to pale yellow, viscous liquid. Darkens slowly in air. d_{20}^{20} 1.240-1.263. n_D^{25} 1.5230-1.5280. Saponif equiv 395-420. Freely sol in alcohol, benzene, chloroform, ether; very slightly sol in water. LD_{50} in mice, rats (g/kg): 4.6, 19 i.p. (Steinhausen).

THERAP CAT: Diagnostic aid (radiopaque medium).

THERAP CAT (VET): Diagnostic aid (radiopaque medium).

5104. Iophenoxic Acid. [96-84-4] α-Ethyl-3-hydroxy-2,4,6-triiodobenzenepropanoic acid; α-ethyl-3-hydroxy-2,4,6-triiodohydrocinnamic acid; α-(2,4,6-triiodo-3-hydroxybenzyl)butyric acid; α-ethyl-β-(3-hydroxy-2,4,6-triiodophenyl)propionic acid; α-ethyl-β-(2,4,6-triiodo-3-hydroxyphenyl)propionic acid; triiodoethionic acid; Teridax. $C_{11}H_{11}I_3O_3$; mol wt 571.92. C 23.10%, H 1.94%, I 66.57%, O 8.39%. Prepn: Papa *et al.*, *J. Am. Chem. Soc.* **75**, 1107 (1953); **GB 726987** (1955 to Sterling Drug). Toxicity data: J. O. Hoppe *et al.*, *J. Med. Chem.* **13**, 997 (1970).

Crystals from benzene + petr ether, mp 143-144°. LD_{50} i.v. in mice: 374 mg/kg (Hoppe).

THERAP CAT: Diagnostic aid (radiopaque medium—cholecystographic).

5105. **Iopromide.** [73334-07-3] N^1,N^3-Bis(2,3-dihydroxy-propyl)-2,4,6-triiodo-5-[(2-methoxyacetyl)amino]-N^1-methyl-1,3-benzenedicarboxamide; N,N'-bis(2,3-dihydroxypropyl)-2,4,6-tri-iodo-5-(2-methoxyacetamido)-N-methylisophthalamide; 5-meth-oxyacetylamino-2,4,6-triiodoisophthalic acid [(2,3-dihydroxy-N-methylpropyl)-(2,3-dihydroxypropyl)]diamide; Ultravist. $C_{18}H_{24}I_3$-N_3O_8; mol wt 791.12. C 27.33%, H 3.06%, I 48.12%, N 5.31%, O 16.18%. Nonionic, injectable radio-contrast medium. Prepn: U. Speck *et al.*, **DE 2909439**; *eidem*, **US 4364921** (1980, 1982 both to Schering AG). Physicochemical properties and pharmacological profile: W. Muetzel, U. Speck, *Am. J. Neuroradiol.* **4**, 350 (1983); P. Dawson *et al.*, *Acta Radiol. Diagn.* **25**, 253 (1984). Clinical experience: K.-J. Wolf *et al.*, *ibid.* **24**, 55 (1983).

Colorless solid. Stable in aqueous solutions. Viscosity (cP): 4.8 at 37°; 10.2 at 20° (c = 300 mg iodine/ml). LD_{50} in mice, rats (g iodine/kg body weight): 16.5, 11.4 i.v. (Muetzel).

THERAP CAT: Diagnostic aid (radiopaque medium).

5106. **Iopronic Acid.** [37723-78-7] 2-[[2-[3-(Acetylamino)-2,4,6-triiodophenoxy]ethoxy]methyl]butanoic acid; (±)-2-[[2-(3-acetamido-2,4,6-triiodophenoxy)ethoxy]methyl]butyric acid; 3-[2-(3-acetylamino-2,4,6-triiodophenoxy)ethoxy]-2-ethylpropionic acid; B-11420; SQ-21983; Bilimiro; Bilimiron; Oravue; Videobil. C_{15}-$H_{18}I_3NO_5$; mol wt 673.02. C 26.77%, H 2.70%, I 56.57%, N 2.08%, O 11.89%. Oral cholecystographic agent. Prepn: E. Felder, D. Pitre, **DE 2128902**; *eidem*, **US 3842124** (1972, 1974 both to Bracco); E. Felder *et al.*, *Farmaco Ed. Sci.* **31**, 349 (1976). Pharmacology, toxicology: P. Tirone, G. Rosati, *Farmaco Ed. Prat.* **31**, 397, 437 (1976). Series of articles on metabolism: *ibid.* 516-546, 755. Intestinal absorption: J. R. Amberg *et al.*, *Invest. Radiol.* **15**, S136 (1980). Effect on bile flow and composition: J. L. Barnhart *et al.*, *ibid.* S124. Toxicity: P. Tirone, G. Rosati, *Farmaco Ed. Prat.* **31**, 397 (1976).

Crystals from 50% ethanol, mp 130°. LD_{50} in mice, rats, dogs (mg/kg): 1950, 5650, >3000 orally; 1090, 1000, 835 i.v. (Tirone).

THERAP CAT: Diagnostic aid (radiopaque medium—cholecystographic).

5107. **Iothalamic Acid.** [2276-90-6] 3-(Acetylamino)-2,4,6-triiodo-5-[(methylamino)carbonyl]benzoic acid; 5-acetamido-2,4,6-triiodo-N-methylisophthalamic acid. $C_{11}H_9I_3N_2O_4$; mol wt 613.92. C 21.52%, H 1.48%, I 62.01%, N 4.56%, O 10.42%. Synthesis: Hoey *et al.*, *J. Med. Chem.* **6**, 24 (1963).

Crystals, dec about 285°.

N-**Methylglucamine salt.** [13087-53-1] Meglumine iothalamate; Conray; Contrix "28"; Cysto-Conray. $C_{18}H_{26}I_3N_3O_9$; mol wt 809.13.

Sodium salt. [1225-20-3] Sodium iothalamate; Angio-Conray; Angio-Contrix "48"; Conray-400; Medio-Contrix "38". $C_{11}H_8I_3$-N_2NaO_4; mol wt 635.90.

^{131}**I-Labeled sodium salt.** [15845-98-4] Glofil-131.

THERAP CAT: Diagnostic aid (radiopaque medium).

5108. **Iotrolan.** [79770-24-4] 5,5'-[(1,3-Dioxo-1,3-pro-panediyl)bis(methylimino)]bis[N,N'-bis[2,3-dihydroxy-1-(hydroxy-methyl)propyl]-2,4,6-triiodo-1,3-benzenedicarboxamide]; 5,5'-[ma-lonylbis(methylimino)]bis[N,N'-bis[2,3-dihydroxy-1-(hydroxy-methyl)propyl]-2,4,6-triiodoisophthalamide]; iotrol; DL-3117; SH-437; ZK-39482; Isovist. $C_{37}H_{48}I_6N_6O_{18}$; mol wt 1626.24. C 27.33%, H 2.98%, I 46.82%, N 5.17%, O 17.71%. Nonionic, iso-tonic contrast medium. Prepn: M. Sovak, R. Ranganathan, **EP 33426**; *eidem*, **US 4341756** (1981, 1982 both to Univ. Cal., Berkeley). Neurotoxicity study: M. Sovak *et al.*, *Radiology* **142**, 115 (1982). Neuropharmacology: M. Sovak *et al.*, *Am. J. Neuroradiol.* **4**, 319 (1983); and cerebral distribution: J.-C. Castel *et al.*, *Neuro-radiology* **29**, 206 (1987). Arthrographic evaluation in animals: J. Guerra *et al.*, *Invest. Radiol.* **19**, 228 (1983). Cardiovascular effects in animals: P. Lanzer *et al.*, *ibid.* **20**, 746 (1985). Clinical study of patient tolerance and diagnostic adequacy for intrathecal use: B. Hammer, E. Deisenhammer, *Neuroradiology* **27**, 337 (1985); B. Hoffmann *et al.*, *ibid.* **29**, 380 (1987); M. D. Malnor *et al.*, *Radiology* **158**, 845 (1986).

Colorless glassy solid. Soly at 20° (mgI/ml): water >400. LD_{50} in mice, rats (gI/kg): >26, 12.7 i.v. (Sovak, 1982).

THERAP CAT: Diagnostic aid (radiopaque medium).

5109. **Ioversol.** [87771-40-2] N^1,N^3-Bis(2,3-dihydroxypro-pyl)-5-[(2-hydroxyacetyl)(2-hydroxyethyl)amino]-2,4,6-triiodo-1,3-benzenedicarboxamide; N,N'-bis(2,3-dihydroxypropyl)-5-[N-(2-hy-droxyethyl)glycolamido]-2,4,6-triiodoisophthalamide; MP-328; Op-tiray. $C_{18}H_{24}I_3N_3O_9$; mol wt 807.12. C 26.79%, H 3.00%, I 47.17%, N 5.21%, O 17.84%. Nonionic, low osmolality, radio-graphic contrast agent. Prepn: Y. Lin, **EP 83964**; *idem*, **US 4396598** (both 1983 to Mallinckrodt). Clinical pharmacokinetics: R. A. Wilkins *et al.*, *Invest. Radiol.* **24**, 781 (1989). Comparative clinical study in intravenous urography: A. J. Kaufman *et al.*, *Urol. Radiol.* **12**, 56 (1990). Toxicity studies: W. H. Ralston *et al.*, *Invest. Radiol.* **24**, Suppl. 1, S2 (1989). Symposium on pharmacology, tox-icology and diagnostic use in contrast enhanced computed tomo-graphic scanning and angiography: *ibid.* S1-S76.

mp 186-198°. Soly in water >125% wt/vol. Partition coefficient (octanol/water): 0.0004 (c = 0.01 mgI/ml). Osmolality: 702 mOsm/kg H_2O (c = 320 mgI/ml). Viscosity (cP): 11.6 (20°); 9.9 (25°); 5.8 (37°) (c = 320 mgI/kg). d^{37} 1.371 (c = 320 mgI/ml). LD_{50} in mice, rats, rabbits, dogs (gI/kg): 17, 15, >25, >12 i.v.; in male, female rats (mgI/kg): 1000, >1200 intracisternally (Ralston).

THERAP CAT: Diagnostic aid (radiopaque medium).

5110. Ioxaglic Acid. [59017-64-0] 3-[[2-[[3-(Acetylmethyl-amino)-2,4,6-triiodo-5-[(methylamino)carbonyl]benzoyl]amino]-acetyl]amino]-5-[[(2-hydroxyethyl)amino]carbonyl]-2,4,6-triiodo-benzoic acid; *N*-(2-hydroxyethyl)-2,4,6-triiodo-5-[2-[2,4,6-triiodo-3-(*N*-methylacetamido)-5-(methylcarbamoyl)benzamido]acet-amido]isophthalamic acid; P-286. $C_{24}H_{21}I_6N_5O_8$; mol wt 1268.89. C 22.72%, H 1.67%, I 60.01%, N 5.52%, O 10.09%. Ionic contrast medium of low osmolality. Prepn: G. Tilly *et al.*, **DE 2523567**; *eidem*, **US 4014986** (1975, 1977 both to Guerbet). Effect on periph-eral arterial blood flow in dogs: R. M. Steiner *et al.*, *Clin. Radiol.* **31**, 621 (1980). Effect on thyroid function: R. G. Grainger, G. W. Pennington, *Br. J. Radiol.* **54**, 768 (1981). Clinical study in femoral angiography: S. Suzuki *et al.*, *Acta Radiol.* **23**, 87 (1982); in cardiac radiology: R. L. Feldman, *Am. J. Cardiol.* **61**, 1334 (1988). Toxic-ity study: M. Sovak *et al.*, *Invest. Radiol.* **16**, 438 (1981). Mutation study: W. Hadnagy *et al.*, *Mutat. Res.* **104**, 249 (1982).

Mixture of meglumine and sodium salts. [76820-74-1] MP-302; Hexabrix.

THERAP CAT: Diagnostic aid (radiopaque medium).

5111. Ioxilan. [107793-72-6] 5-[Acetyl(2,3-dihydroxypro-pyl)amino]-N^1-(2,3-dihydroxypropyl)-N^3-(2-hydroxyethyl)-2,4,6-triiodo-1,3-benzenedicarboxamide; *N*-(2,3-dihydroxypropyl)-5-[*N*-(2,3-dihydroxypropyl)acetamido]-*N'*-(2-hydroxyethyl)-2,4,6-tri-iodoisophthalamide; ioxitol; Oxilan. $C_{18}H_{24}I_3N_3O_8$; mol wt 791.12. C 27.33%, H 3.06%, I 48.12%, N 5.31%, O 16.18%. Nonionic io-dinated contrast medium. Prepd not claimed: M. Sovak, R. Rangan-athan, **WO 8700757**; *eidem*, **US 5035877** (1987, 1991 both to Cook Imaging). HPLC analysis of isomers: S. J. Foster, M. Sovak, *Invest. Radiol.* **23**, Suppl. 1, S106 (1988). Pharmacology: R. W. Katzberg *et al.*, *ibid.* **25**, 46 (1990). Clinical evaluation: M. R. Callantine *et al.*, *ibid.* Suppl. 1, S107; C. L. McIntosh, R. Reed, *ibid.* **29**, Suppl. 2, S40 (1994). Brief review of physical properties: M. Sovak, *ibid.* **23**, Suppl. 1, S79-S83 (1988).

White to off-white powder. Sol in water, methanol. Osmolality (mOsm/kg H_2O): 690 (c = 350 mgI/ml); 570 (c = 300 mgI/ml). Viscosity (37°): 4.6-4.7 cP (c = 300 mgI/ml).

THERAP CAT: Diagnostic aid (radiopaque medium).

5112. IPBC. [55406-53-6] *N*-Butylcarbamic acid 3-iodo-2-propy-1-nyl ester; 3-iodo-2-propynyl butylcarbamate; Glycacil; Po-lyphase P-100. $C_8H_{12}INO_2$; mol wt 281.09. C 34.18%, H 4.30%, I 45.15%, N 4.98%, O 11.38%. Prepn: **BE 817530**; W. Singer, **US 3923870** (both 1975 to Troy Chemical). Aquatic toxicities: A. P. Farrell *et al.*, *Arch. Environ. Contam. Toxicol.* **35**, 472 (1998); fate and behavior: L. Juergensen *et al.*, *Environ. Toxicol.* **15**, 201 (2000). HPLC-MS determn in cosmetic formulations: M. Frauen *et al.*, *J. Pharm. Biomed. Anal.* **25**, 965 (2001). GC-MS determn of degra-dation products: K. G. Karaisz, N. H. Snow, *J. Microcol. Separat.* **13**, 1 (2001). Review of use in wood protection: J. Hansen, *Mod. Paint Coat.* **74**, 50-56, 90 (1984); of microbial activities and com-merical uses: R. Gruening, *Cosmet. Toiletries* **112**, 59-65 (1997).

White crystalline powder with a specific odor. mp 66°. Specific gravity: 1.575 g mL^{-1}. Vapor pressure (Pa): <0.002 (20°); 0.007 (30°). pH 7.01. Good soly in ionic and non-ionic surfactants, emul-sifying agents and polar solvents. LD$_{50}$ orally in rats: 1580 mg/kg; dermally in rabbit: >2000 mg/kg. LC$_{50}$ (96 hr) in bluegill sunfish, rainbow trout, bobwhite quail: 1.12, 0.31, >7683 ppm. LC$_{50}$ (8 day) in mallard duck: >7182 ppm (Hansen).

USE: Fungicide; mildewcide; preservative in cosmetics, paints and coatings, metal working fluids; wood protection.

5113. Ipecac. Ipecacuanha. Dried rhizome and roots of *Ce-phaelis ipecacuanha* (Brot.) A. Rich., *Rubiaceae*, known as Rio or Brazilian Ipecac; or of *Cephaelis acuminata* Karsten, known as Car-tagena, Nicaragua or Panama Ipecac. Habit. Brazil to Bolivia; cul-tivated in India and Malaysia. *Constit.* Emetine, cephaeline, eme-tamine, psychotrine, methyl psychotrine, protoemetine and resin. Cartagena ipecac contains ~2.2-2.5% total alkaloids, with twice as much cephaeline as emetine. Determn of major alkaloids in ipecac-uanha root: A. G. Davidson, S. M. Hassan, *J. Pharm. Biomed. Anal.* **2**, 441 (1984); in pharmaceutical preparations by HPLC: D. A. El-vidge *et al.*, *J. Chromatogr.* **463**, 107 (1989). Review of distribution and chemistry of ipecac alkaloids: M.-M. Janot in *The Alkaloids* **vol. 3**, R. H. F. Manske, H. L. Holmes, Eds. (Academic Press, New York, 1953) pp 363-394; A. Brossi, S. Teitel, *ibid.* **vol. 13**, 189-212 (1971). Review of use in acute poisoning: W. D. King, *Clin. Toxicol.* **17**, 353-358 (1980).

THERAP CAT: Emetic; expectorant.

THERAP CAT (VET): Emetic; expectorant.

5114. Ipilimumab. [477202-00-9] Anti-(human CTLA-4 (antigen)) immunoglobulin G1 (human γ_1-chain) disulfide with hu-man κ-chain, dimer; MDX-010; Yervoy. Recombinant, human monoclonal antibody produced in Chinese hamster ovary (CHO) cells. Binds to the cytotoxic T-lymphocyte-associated antigen 4 (CTLA-4) and blocks the interaction of CTLA-4 with its ligands, CD80/CD86. Mol wt ~148 kDa. Prepn: A. J. Korman *et al.*, **WO 0114424**; *eidem*, **US 6984720** (2001, 2006 both to Medarex). Clin-ical pharmacology and pharmacokinetics: J. S. Weber *et al.*, *J. Clin. Oncol.* **26**, 5950 (2008). Clinical trial in metastic melanoma: F. S. Hodi *et al.*, *N. Engl. J. Med.* **363**, 711 (2010). Review of develop-ment and clinical experience in melanoma: A. Hoos *et al.*, *Semin. Oncol.* **37**, 533-546 (2010); M. E. Culver *et al.*, *Ann. Pharmacother.* **45**, 510-519 (2011).

THERAP CAT: Antineoplastic

5115. Ipodate. [5587-89-3] 3-[[(Dimethylamino)methylene]-amino]-2,4,6-triiodobenzenepropanoic acid; 3-[(dimethylamino-methylene)amino]-2,4,6-triiodohydrocinnamic acid; 2,4,6-triiodo-3-[(dimethylaminomethylene)amino]hydrocinnamic acid; β-(3-di-methylaminomethyleneamino-2,4,6-triiodophenyl)propionic acid. $C_{12}H_{13}I_3N_2O_2$; mol wt 597.96. C 24.10%, H 2.19%, I 63.67%, N 4.68%, O 5.35%. Prepn: Priewe, Poljak, *Ber.* **93**, 2347 (1960). Toxicity data: J. O. Hoppe *et al.*, *J. Med. Chem.* **13**, 997 (1970).

Crystals, mp 168-169° (dec 225°). Practically insol in water. Very sol in methanol, ethanol, chloroform, acetone, dil sulfuric acid. Es-timated pK 5-5.5.

Sodium salt. [1221-56-3] Sodium ipodate; Biloptin; Oragrafin Sodium. $C_{12}H_{12}I_3N_2NaO_2$; mol wt 619.94. Bitter leaflets from wa-ter + acetone, mp 303-304° (decompn with evolution of iodine). Soly in DMF and DMSO about 33 g/100 ml; in dimethylacetamide about 66 g/100 ml. Freely sol in water, methanol, ethanol; very slightly sol in chloroform. Practically insol in acetone, ether. LD_{50} in mice (mg/kg): 290 i.v.; 2570 orally (Hoppe).

Calcium salt. [1151-11-7] Calcium ipodate; Solu-Biloptin; Orografin Calcium. $C_{24}H_{24}CaI_6N_4O_4$; mol wt 1233.98. Crystals, mp 298-302°. Soly in water at 20°: 0.1%. Sol in chloroform, dimethylformamide, hot propylene glycol.

Ethyl ester. SH-617L. $C_{14}H_{17}I_3N_2O_2$; mol wt 626.02.

THERAP CAT: Diagnostic aid (radiopaque medium—cholecystographic).

5116. Ipomea. Mexican scammony (root); Orizaba jalap root. Dried root of *Ipomoea orizabensis* Ledenois, *Convolvulaceae.* Active constituent is the resin. Yields not less than 15% total ipomea resins. Different from *Ipomoea violacea* var. *Pearly Gates* Hort., *Convolvulaceae* and *Ipomoea rubrocoerulea* var. *praecox*, **morning-glory, ololiuqui**, which contain ergot alkaloids. Occurrence of lysergic acid derivatives and of ergolines in Ipomea: A. Hofmann, H. Tscherter, *Experientia* **16**, 414 (1960); D. Stauffacher *et al., Helv. Chim. Acta* **48**, 1379 (1965).

THERAP CAT: Cathartic.

5117. Ipratropium Bromide. [22254-24-6] (3-*endo*,8-*syn*)-3-(3-Hydroxy-1-oxo-2-phenylpropoxy)-8-methyl-8-(1-methylethyl)-8-azoniabicyclo[3.2.1]octane bromide (1:1); 3α-hydroxy-8-isopropyl-1αH,5αH-tropanium bromide (±)-tropate; 8-isopropylnoratropine methobromide; *N*-isopropylnoratropinium bromomethylate; Sch-1000; Atem; Atrovent; Bitrop; Itrop; Narilet; Respontin; Rinatec. $C_{20}H_{30}BrNO_3$; mol wt 412.37. C 58.25%, H 7.33%, Br 19.38%, N 3.40%, O 11.64%. Anticholinergic. Prepn: K. Zeile *et al.*, ZA **6707766**; *eidem*, US **3505337** (1968, 1970, both to Boehringer, Ing.); W. Schulz *et al., Arzneim.-Forsch.* **26**, 960 (1976). Chemistry and pharmacokinetics: W. Deckers, *Postgrad. Med. J.* **51**, Suppl. 7, 76 (1975). Pharmacology and toxicology: A. Engelhardt, H. Klupp, *ibid.* 82. Series of articles on pharmacology, toxicology, pharmacokinetics and clinical studies: *Arzneim.-Forsch.* **26**, 974-985, 989-1020 (1976). Toxicity data: L. Sarafana *et al., ibid.* 985. *Review:* R. Bauer *et al.*, in *Pharmacological and Biochemical Properties of Drug Substances* vol. 2, M. E. Goldberg, Ed. (Am. Pharm. Assoc., Washington, DC, 1979) pp 489-515. Symposium on pharmacology, toxicology and clinical efficacy: *Am. J. Med.* **81**(5A), 1-102 (1986). Review of clinical toxicology: J. D. Truwit, *Crit. Care Clin.* **7**, 639-657 (1991).

White crystals from *n*-propanol, mp 230-232°. Freely sol in methanol and other lower alcohols; sol in water; slightly sol in alc. Insol in ether, chloroform and fluorohydrocarbons. Fairly stable in neutral and acid solns; rapidly hydrolyzed in alkaline soln. LD_{50} in male, female mice (mg/kg): 1001, 1083 orally; 12.29, 14.97 i.v.; 300, 340 s.c. LD_{50} in male, female rats (mg/kg): 1663, 1779 orally; 15.89, 15.70 i.v. (Sarafana).

THERAP CAT: Bronchodilator; antiarrhythmic.

5118. Ipriflavone. [35212-22-7] 7-(1-Methylethoxy)-3-phenyl-4H-1-benzopyran-4-one; 7-isopropoxy-3-phenyl-4H-1-benzopyran-4-one; 7-isopropoxy-3-phenylchromone; 7-isopropoxyisoflavone; FL-113; TC-80; Iprosten; Osten; Osteofix; Yambolap. $C_{18}H_{16}O_3$; mol wt 280.32. C 77.13%, H 5.75%, O 17.12%. Isoflavone derivative with anti-anginal and anti-osteopenic activity. Prepn: L. Feuer *et al.*, DE **2125245** (1971 to Chinoin), *C.A.* **76**, 72407e (1972). Use as anabolic in animals: *eidem*, US **3833730**; in humans; *eidem*, US **3949085** (1974, 1976 both to Chinoin). Cardiovascular proper-

ties in stable angina: V. Grubich *et al., Lancet* **1**, 211 (1979). Cardiological effects in animals: L. Feuer *et al., Arzneim.-Forsch.* **31**, 953 (1981). Metabolism and disposition in rats: K. Yoshida *et al., Radioisotopes* **34**, 612, 618 (1985). Effect in rats on estrogen-stimulated calcitonin secretion: I. Yamazaki, *Life Sci.* **38**, 757 (1986); I. Yamazaki, M. Kinoshita, *ibid.* 1535; on glucocorticoid-induced osteoporosis: I. Yamazaki *et al., ibid.* 951.

Crystals from acetone, mp 115-117°.

THERAP CAT: Calcium regulator.

5119. Iprindole. [5560-72-5] 6,7,8,9,10,11-Hexahydro-*N*,*N*-dimethyl-5H-cyclooct[b]indole-5-propanamine; 5-[3-(dimethylamino)propyl]-6,7,8,9,10,11-hexahydro-5H-cyclooct[b]indole; 1-(3-dimethylaminopropyl)-2,3-hexamethyleneindole; pramindole (obsolete); Tertran. $C_{19}H_{28}N_2$; mol wt 284.45. C 80.23%, H 9.92%, N 9.85%. Prepn: L. M. Rice, M. E. Freed, BE **623933**; *eidem*, US **3282942** (1963, 1966, both to Am. Home Prods.); Rice *et al., J. Med. Chem.* **7**, 313 (1964). Pharmacology: Gluckman, Baum, *Psychopharmacologia* **15**, 169 (1969); Baum *et al., Eur. J. Pharmacol.* **13**, 287 (1971). Mechanism of action: Miller *et al., Experientia* **26**, 863 (1970); Lemberger *et al., Biochem. Pharmacol.* **19**, 3021 (1970); Freeman, Sulser, *J. Pharmacol. Exp. Ther.* **183**, 307 (1972).

Hydrochloride. [20432-64-8] Wy-3263; Prondol; Galatur. $C_{19}H_{28}N_2$·HCl; mol wt 320.91. Crystals from methanol + acetone, mp 146-147°.

THERAP CAT: Antidepressant.

5120. Iproclozide. [3544-35-2] 2-(4-Chlorophenoxy)acetic acid 2-(1-methylethyl)hydrazide. $C_{11}H_{15}ClN_2O_2$; mol wt 242.70. C 54.44%, H 6.23%, Cl 14.61%, N 11.54%, O 13.18%. Monoamine oxidase inhibitor. Prepn: Libermann, Denis, *Bull. Soc. Chim. Fr.* **1961**, 1952. GC determn in urine: R. M. DeSagher *et al., Anal. Chem.* **47**, 1144 (1976).

Crystals, mp 93-94°.

THERAP CAT: Antidepressant.

5121. Iprodione. [36734-19-7] 3-(3,5-Dichlorophenyl)-*N*-(1-methylethyl)-2,4-dioxo-1-imidazolidinecarboxamide; glycophene; promidione; RP-26019; ROP-500F; NRC-910; LFA-2043; FA-2071; Rovral; CHIPCO-26019. $C_{13}H_{13}Cl_2N_3O_3$; mol wt 330.17. C 47.29%, H 3.97%, Cl 21.47%, N 12.73%, O 14.54%. Prepn: M. Sauli, DE **2149923**; *idem*, US **3755350** (1972, 1973 to Rhône-Poulenc). Biological properties: L. Lacroix *et al., Phytiatr.-Phytopharm.* **23**, 165 (1974). Structural rearrangement: B. K. Cooke *et al., Pestic. Sci.* **10**, 393 (1979). Thermal degradn: J. Gomez *et al., J. Agric. Food Chem.* **30**, 180 (1982). HPLC determn in surface water: C. E. Goewie, E. A. Hogendoorn, *Sci. Total Environ.* **47**, 349 (1985). GC-MS determn in chicory and leek: F. Rouberty, J. Fournier, *Chromatographia* **41**, 693 (1995). Field study: T. B. Brenneman *et al.*, *Plant Dis.* **71**, 546 (1987). Degradation in soil: A. Walker, *Pestic. Sci.* **21**, 219 (1987). Review: L. Lacroix *et al.*, *Anal. Methods Pestic. Plant Growth Regul.* **11**, 247-261 (1980).

Colorless, odorless, nonhygroscopic crystals, mp ~136°. Vapor pressure at 20°: $<10^{-6}$ mm Hg. Soly in water at 20°: 13 mg/l. Soly at 20° (g/l): ethanol 25; methanol 25; acetone 300; dichloromethane 500; DMF 500. LD_{50} in mice, rats (g/kg): 4, 3.5 orally (Lacroix, 1980).

USE: Fungicide.

5122. Iproniazid. [54-92-2] 4-Pyridinecarboxylic acid 2-(1-methylethyl)hydrazide; isonicotinic acid 2-isopropylhydrazide; 1-isonicotinoyl-2-isopropylhydrazine; 1-isonicotinyl-2-isopropylhydrazine. $C_9H_{13}N_3O$; mol wt 179.22. C 60.32%, H 7.31%, N 23.45%, O 8.93%. Monoamine oxidase inhibitor. Prepd by treating isoniazid with isopropyl bromide in abs alcohol in the presence of sodium: McMillan *et al.*, *J. Am. Pharm. Assoc. Sci. Ed.* **42**, 457 (1953). Alternate synthesis: Fox, Gibas, *J. Org. Chem.* **18**, 994 (1953). Comprehensive description: F. Belal, H. Abdel-Aliem, *Anal. Profiles Drug Subs.* **20**, 337-368 (1991).

Needles from benzene + ligroin, mp 112.5-113.5°. Freely sol in water, alcohol; pH of aq solns about 6.7.

Dihydrochloride. $C_9H_{13}N_3O.2HCl$. Rhomboid crystals from isopropanol, dec at 227-228°. Freely sol in water; aq solns are very acidic.

Phosphate. [305-33-9] Marsilid. $C_9H_{13}N_3O.H_3PO_4$; mol wt 277.22. White to slightly yellowish powder, melting range 175-184°. pH of a 5% soln: 2.7-3.6. uv spectrum: J. Kracmar, J. Kracmarova, *Pharmazie* **34**, 27 (1979). uv max (methanol): 265 nm ($A_{1cm}^{1\%}$ 166); (water): 264 nm ($A_{1cm}^{1\%}$ 176); (0.1N HCl): 267 nm ($A_{1cm}^{1\%}$ 179); (0.1N KOH): 244, 272, 308 nm ($A_{1cm}^{1\%}$ 113, 121, 135). Soly at 25° (mg/ml): water ~188; methanol ~21; ethanol (96%) ~9; chloroform ~0.6; diethyl ether <0.1; hexane <0.1.

THERAP CAT: Antidepressant.

5123. Ipronidazole. [14885-29-1] 1-Methyl-2-(1-methylethyl)-5-nitro-1*H*-imidazole; 2-isopropyl-1-methyl-5-nitroimidazole; Ro-7-1554; Ipropran. $C_7H_{11}N_3O_2$; mol wt 169.18. C 49.70%, H 6.55%, N 24.84%, O 18.91%. Prepn: **GB 1119636**; M. Hoffer, M. Mitrovic, **US 3634446** (1968, 1972 both to Hoffmann-La Roche); K. Butler, **ZA 6607466** (1968 to Pfizer). Structure-activity relationship: K. Butler *et al.*, *J. Med. Chem.* **10**, 891 (1967). Activity studies as an antihistomonal agent: M. Mitrovic *et al.*, *Antimicrob. Agents Chemother.* **1968**, 445; M. Mitrovic, E. G. Schildknecht, *Poult. Sci.* **49**, 86 (1970); as a growth promotant: W. L. Marusich *et al.*, *ibid.* **98**. Toxicity studies: *eidem, ibid.* **92**.

White plates, mp 60°. LD_{50} orally in poults: 640 ± 25 mg/kg (Marusich).

Hydrochloride. mp 177-182°. Water-soluble.

THERAP CAT (VET): Antiprotozoal (Histomonas).

5124. Iprovalicarb. [140923-17-7] *N*-[(1*S*)-2-Methyl-1-[[[1-(4-methylphenyl)ethyl]amino]carbonyl]propyl]carbamic acid 1-methylethyl ester; isopropyl-2-methyl-1-[(1-*p*-tolyethyl)carbamoyl]-(*S*)-propylcarbamate; *N*-(isopropyloxycarbonyl)-L-valine-4-methylphenylethylamide; SZX-722; Melody. $C_{18}H_{28}N_2O_3$; mol wt 320.43. C 67.47%, H 8.81%, N 8.74%, O 14.98%. Fungicide for use in food crops; commercial product is a mixture of two diastereomers (*SS, SR*). Prepn: T. Seitz *et al.*, **DE 4026966**; *eidem*, **US 5453531** (1992, 1995 both to Bayer). Comprehensive description: K. Stenzel *et al.*, *Brighton Crop Prot. Conf. - Pests Dis.* **2**, 367 (1998). Series of articles on prepn, structure, field studies, ecological profile and metabolism: *Pflanzenschutz-Nachr. Bayer* **52**, 5-114 (1999). LC-MS determn: R. Brennecke *ibid.* **53**, 5 (2000). ELISA method for residue detection: J. K. Lee *et al.*, *J. Agric. Food Chem.* **52**, 6680 (2004).

(*S,R*)-form

White powder, mp 163-165° (mixture), 183° (*SR*), 199° (*SS*). d^{20} 1.11. Log P (octanol/water): 3.2. Vapor pressure: 7.7×10^{-10} hPa (20°), 2.1×10^{-9} hPa (25°). LD_{50} in rats (mg/kg): >5000 orally; >5000 dermally; in bobwhite quail (mg/kg): >2000 orally. LC_{50} in rats (mg/m³): >4977 by inhalation (Stenzel).

USE: Fungicide.

5125. Ipsapirone. [95847-70-4] 2-[4-[4-(2-Pyrimidinyl)-1-piperazinyl]butyl]-1,2-benzisothiazolin-3(2*H*)-one 1,1-dioxide; isapirone. $C_{19}H_{23}N_5O_3S$; mol wt 401.49. C 56.84%, H 5.77%, N 17.44%, O 11.95%, S 7.99%. Nonbenzodiazepine anxiolytic; 5-hydroxytryptamine (5-HT$_1$) receptor agonist. Prepn: W. Dompert *et al.*, **DE 3321969**; P.-R. Seidel *et al.*, **US 4818756** (1984, 1989 both to Troponwerke). Pharmacology and serotonin (5-HT$_1$) receptor binding study: J. Traber *et al.*, *Brain Res. Bull.* **12**, 741 (1984). 5-HT$_1$ specific binding activity: W. U. Dompert *et al.*, *Arch. Pharmacol.* **328**, 467 (1985); T. Glaser, J. Traber, *ibid.* **329**, 211 (1985). Behavioral effects in mice: R. J. DeSouza *et al.*, *Br. J. Pharmacol.* **89**, 377 (1986). Neurophysiological effects in comparison with buspirone: M. J. Rowan, R. Anwyl, *Eur. J. Pharmacol.* **132**, 93 (1987). HPLC determn in biological samples: G. Bianchi, S. Caccia, *J. Chromatogr.* **431**, 477 (1988).

Crystals from isopropanol, mp 137-138°.

Hydrochloride. [92589-98-5] Bay q 7821; TVX Q 7821. $C_{19}H_{23}N_5O_3S.HCl$; mol wt 437.94. Crystals from isopropanol, mp 221-222°.

THERAP CAT: Anxiolytic.

5126. IPTG. [367-93-1] 1-Methylethyl 1-thio-β-D-galactopyranoside; isopropyl β-D-thiogalactopyranoside. $C_9H_{18}O_5S$; mol wt 238.30. C 45.36%, H 7.61%, O 33.57%, S 13.45%. Artificial, slow-hydrolyzing inducer of the lactose (*lac*) operon. Prepn: B. Helferich, D. Türk, *Ber.* **89**, 2215 (1956); U. Carlsson *et al.*, *Protein Eng.* **4**, 1019 (1991). Induction of β-galactosidase: V. Paces *et al.*, *Collect. Czech. Chem. Commun.* **38**, 2983 (1973). Binding to β-D-galactosidase: C. K. De Bruyne, M. Yde, *Carbohydr. Res.* **56**, 153 (1977). Concentration effects on *lac* operon induction in *E. coli*: S.

Cho *et al.*, *Biochem. Biophys. Res. Commun.* **128**, 1268 (1985). Effect of lactose permease on induction by IPTG: L. H. Hansen *et al.*, *Curr. Microbiol.* **36**, 341 (1998).

mp 109.5-110.5°. $[\alpha]_D^{22}$ −31.4° (c = 5 in water). Also reported as crystals from dioxane, mp 118-119°. $[\alpha]_D$ −26.4° (c = 1 in water) (Carlsson). Log P (octanol/water): −1.26.

USE: Induces protein synthesis in *E. coli* when transcription is controlled by the *lac*-promoter. In conjunction with X-gal, *q.v.*, in detection of *lac* gene activity during cloning experiments.

5127. Irbesartan. [138402-11-6] 2-Butyl-3-[[2'-(2*H*-tetrazol-5-yl)[1,1'-biphenyl]-4-yl]methyl]-1,3-diazaspiro[4.4]non-1-en-4-one; 2-butyl-3-[*p*-(*o*-1*H*-tetrazol-5-ylphenyl)benzyl]-1,3-diazaspiro[4.4]non-1-en-4-one; 2-*n*-butyl-4-spirocyclopentane-1-[(2'-(tetrazol-5-yl)biphenyl-4-yl)methyl]-2-imidazolin-5-one; BMS-186295; SR-47436; Aprovel; Avapro. $C_{25}H_{28}N_6O$; mol wt 428.54. C 70.07%, H 6.59%, N 19.61%, O 3.73%. Angiotensin II type 1 (AII$_1$)-receptor antagonist. Prepn: C. Bernhart *et al.*, **WO 9114679**; *eidem*, **US 5270317** (1991, 1993 both to Sanofi); *eidem et al.*, *J. Med. Chem.* **36**, 3371 (1993). HPLC determn in plasma and urine: S.-Y. Chang *et al.*, *J. Chromatogr. B* **702**, 149 (1997). Metabolism: T. J. Chando *et al.*, *Drug Metab. Dispos.* **26**, 408 (1998). Review of pharmacology and clinical efficacy: J. C. Gillis, A. Markham, *Drugs* **54**, 885-902 (1997). Renoprotective effect in diabetic nephropathy: E. J. Lewis *et al.*, *N. Engl. J. Med.* **345**, 851 (2001); H.-H. Parving *et al.*, *ibid.* 870.

Crystals from 96% ethanol, mp 180-181°. Slightly sol in alc, methylene chloride. Practically insol in water.

THERAP CAT: Antihypertensive.

5128. Irgarol®. [28159-98-0] N^2-Cyclopropyl-N^4-(1,1-dimethylethyl)-1,3,5-triazine-2,4-diamine; 2-(*tert*-butylamino)-4-(cyclopropylamino)-6-(methylthio)-*s*-triazine. $C_{11}H_{19}N_5S$; mol wt 253.37. C 52.15%, H 7.56%, N 27.64%, S 12.65%. Photosystem-II (PSII) herbicide, inhibits photosynthetic electron transport in chloroplasts. Prepn: D. Berrer, C. Vogel, **DE 1914014** (1969 to Agripat); *eidem*, **US 3629256** (1971 to Geigy). Photodegradation and persistence in water: H. Okamura *et al.*, *J. Environ. Sci. Health* **B34**, 225 (1999). Review of analytical methods: N. Voulvoulis *et al.*, *Chemosphere* **38**, 3503-3516 (1999). Ecological risk assessment: L. W. Hall, Jr. *et al.*, *Crit. Rev. Toxicol.* **29**, 367-437 (1999).

Crystals from water, mp 128-133°. Soly (ppm): 9.0 in water; 5.9 in 0.3 mol/l salinity; 1.8 in 0.6 mol/l salinity. Vapor pressure 6.6×10^{-7} mm Hg. Log P (*n*-octanol/water): 3.95; also reported as 2.8 (Hall). LC$_{50}$ (96 hr in salt water) in mysid shrimp, inland silverside, sheepshead minnow (ng/l): 400000, 1580000, 3500000; LC$_{50}$ (96 hr in fresh water) in rainbow trout, bluegill sunfish (ng/l): 790000, 2600000 (Hall).

USE: Booster algicide in antifouling paint.

5129. Iridium. [7439-88-5] Ir; at. wt 192.217; at. no. 77; valences 1, 3, 4; also 2, 5, 6. Group VIII (9). Two naturally occurring isotopes: 191 (38.5%); 193 (61.5%); artificial, radioactive isotopes: 182-190; 191; 192; 194-198. Occurrence in earth's crust about 0.001 ppm. Discovered in 1804 by Tennant. Occurs in nature in the metallic state, usually as a natural alloy with osmium (osmiridium); found in small quantities alloyed with native platinum (platinum mineral) or with native gold. Recovery and purification from osmiridium: Deville, Debray, *Ann. Chim. Phys.* **61**, 84 (1861); from the platinum mineral: Wichers, *J. Res. Natl. Bur. Stand.* **10**, 819 (1933). Reviews of prepn, properties and chemistry of iridium and other platinum metals: Gilchrist, *Chem. Rev.* **32**, 277-372 (1943); W. P. Griffith, *The Chemistry of the Rarer Platinum Metals* (John Wiley, New York, 1967) pp 1-41, 227-312; Livingstone in *Comprehensive Inorganic Chemistry* vol. 3, J. C. Bailar Jr. *et al.*, Eds. (Pergamon Press, Oxford, 1973) pp 1163-1189, 1254-1274.

Silver-white, very hard metal; face-centered cubic lattice. mp 2450°; bp ~4500°. d_4^{20} 22.65; highest sp gr of all elements. Specific heat 0.0307 cal/g/°C. Mohs' hardness 6.5. Pure iridium is not attacked by any acids including aqua regia; only slightly by fused (non-oxidizing) alkalies. It is superficially oxidized on heating in the air. Is attacked by fluorine and chlorine at a red heat; attacked by potassium sulfate or by a mixture of potassium hydroxide and nitrate on fusion; attacked by lead, zinc or tin. The powdered metal is oxidized by air or oxygen at a red heat to the dioxide, IrO$_2$, but on further heating the dioxide dissociates into its constituents.

USE: In manufacturing crucibles; in hardening platinum; in making nibs for fountain-pen points.

5130. Iridium Hexafluoride. [7783-75-7] (*OC*-6-11)-Iridium fluoride (IrF$_6$). F$_6$Ir; mol wt 306.21. F 37.23%, Ir 62.77%. IrF$_6$. Prepd by direct fluorination of iridium: Ruff, Fischer, *Z. Anorg. Allg. Chem.* **179**, 161 (1929); Robinson, Westland, *J. Chem. Soc.* **1956**, 4481. Crystal structure: Siegel, Northrop, *Inorg. Chem.* **5**, 2187 (1966).

Golden yellow, cubic crystals. d (calc) 4.82. Very hygroscopic. mp 44.4°. bp 53°. Volatilizes on slow heating. Reduced to IrF$_4$ by halogens. Decomposes in aqueous soln. Does not react with dry glass below 150°. May be stored in evacuated quartz ampuls.

5131. Iridium Sesquioxide. [1312-46-5] Iridium oxide (Ir$_2$O$_3$); diiridium trioxide. Ir$_2$O$_3$; mol wt 432.43. Ir 88.90%, O 11.10%.

Blue-black powder. Decomposes at about 1000° into metal and oxygen. Insol in water; slowly dissolved by boiling HCl. Oxidized to IrO$_2$ by HNO$_3$.

5132. Iridium Trichloride. [10025-83-9] Iridium chloride (IrCl$_3$); trichloroiridium. Cl$_3$Ir; mol wt 298.57. Cl 35.62%, Ir 64.38%. IrCl$_3$. Prepn: Dillamore, Edwards, *J. Inorg. Nucl. Chem.* **31**, 2427 (1969). Several known modifications. Crystal structure: Brodersen *et al.*, *Naturwissenschaften* **52**, 205 (1966); Babel, Deigner, *Z. Anorg. Allg. Chem.* **339**, 57 (1965).

α-**Form.** Brown, monoclinic crystals. β-Form; red, orthorhombic crystals. Olive-green modification also prepd. Decomposes at 763°. Insol in water, acids, alkalies.

USE: Catalyst for chlorination and polymerization reactions.

5133. Iridomyrmecin. [485-43-8] (4*S*,4*aS*,7*S*,7*aR*)-Hexahydro-4,7-dimethylcyclopenta[*c*]pyran-3(1*H*)-one; 2-(hydroxymethyl)-α,3-dimethylcyclopentaneacetic acid δ-lactone; iridomyrmexin. $C_{10}H_{16}O_2$; mol wt 168.24. C 71.39%, H 9.59%, O 19.02%. Iridolactone isolated from the Argentine ant *Iridomyrmex humilis* Mayr., *Dolichoderinae*. Exhibits antibacterial and insect antifeedant activity. Isoln: Pavan, *Ric. Sci.* **19**, 1011 (1949); *Chim. Ind. (Milan)* **37**, 625 (1955); Fusco *et al.*, *ibid.* **251**, 958; Cavill *et al.*, *Aust. J. Chem.* **9**, 288 (1956); **10**, 352 (1957). Synthesis: Korte *et al.*, *Tetrahedron*

6, 201 (1959); Sisido *et al., J. Org. Chem.* **29**, 3361 (1964); R. S. Matthews, J. K. Whitesell, *ibid.* **40**, 3312 (1975); Y. Yamada *et al., Chem. Lett.* **1978**, 1405; P. Callant *et al., Tetrahedron* **37**, 2085 (1981). Stereochemistry: Korte *et al., Ber.* **94**, 1952 (1961); McConnell *et al., Tetrahedron Lett.* **1962**, 445; Büchi, Manning, *Tetrahedron* **18**, 1049 (1962). Activity as bactericide and insecticide: M. Pavan, *Z. Hyg. Infektionskrankh.* **134**, 136 (1952), *C.A.* **46**, 9796f (1952).

Prisms from petr ether. Aromatic odor, resembling catnip. Salty taste. mp 60-61°. Sublimes$_{0.01}$ 50-55°. bp$_{1.5}$ 104-108°. $[\alpha]_D^{20}$ +210° (c = 4 in ethanol); $[\alpha]_D^{17}$ +205° (c = 0.223 in carbon tetrachloride). n_D^{65} 1.4607. Sparingly sol in water: 0.2 g/100 ml H$_2$O (25°). Sol in fats and fat solvents. Freely sol in ether.

5134. Irigenin. [548-76-5] 5,7-Dihydroxy-3-(3-hydroxy-4,5-dimethoxyphenyl)-6-methoxy-4H-1-benzopyran-4-one; 3',5,7-trihydroxy-4',5',6-trimethoxyisoflavone. C$_{18}$H$_{16}$O$_8$; mol wt 360.32. C 60.00%, H 4.48%, O 35.52%. From rhizome of *Iris florentina* L., *Iridaceae:* de Laire, Tiemann, *Ber.* **26**, 2010 (1893). Structure: Baker, *J. Chem. Soc.* **1928**, 1022. Synthesis: Baker *et al., Tetrahedron Lett.* no. **5**, 6 (1960); Farkas, Várady, *Ber.* **93**, 2685 (1960); Baker *et al., J. Chem. Soc. C* **1970**, 1219.

Yellow plates or needles from dil alc, mp 185°. uv max (abs ethanol): 267 nm. Sol in warm alcohol, benzene, chloroform; practically insol in water, ether, petr ether.

Triacetate. C$_{24}$H$_{22}$O$_{11}$. Prisms from dil acetic acid, mp 127-128°.

7-Glucoside. Iridin; 7-(glucosyloxy)-3',5-dihydroxy-4',5',6-trimethoxyisoflavone. C$_{24}$H$_{26}$O$_{13}$. Monohydrate, needles from dil alc, mp 208°. When anhydrous, mp 217°. One gram of anhydrous iridin dissolves in 500 ml water, 43 ml acetone. Sol in hot alcohol; practically insol in ether, ethyl acetate, benzene, chloroform.

Note: There also is an oleoresin called *iridin* (Extractum Iridis) from *Iris versicolor* (blue flag) by extraction with 60% alc. Another substance named *iridine* is a protamine from the sperm of rainbow trout: Felix, Mager, *Z. Physiol. Chem.* **249**, 124 (1937).

5135. Irinotecan. [97682-44-5] [1,4'-Bipiperidine]-1'-carboxylic acid (4S)-4,11-diethyl-3,4,12,14-tetrahydro-4-hydroxy-3,14-dioxo-1H-pyrano[3',4':6,7]indolizino[1,2-b]quinolin-9-yl ester; 7-ethyl-10-[4-(1-piperidino)-1-piperidino]carbonyloxycamptothecin; (+)-7-ethyl-10-hydroxycamptothecine 10-[1,4'-bipiperidine]-1'-carboxylate; (+)-(4S)-4,11-diethyl-4-hydroxy-9-[(4-piperidinopiperidino)carbonyloxy]-1H-pyrano[3',4':6,7]indolizino[1,2-b]quinoline-3,14-(4H,12H)-dione. C$_{33}$H$_{38}$N$_4$O$_6$; mol wt 586.69. C 67.56%, H 6.53%, N 9.55%, O 16.36%. DNA topoisomerase I inhibitor. Semisynthetic derivative of camptothecin, *q.v.*; de-esterified *in vivo* to the active metabolite, **7-ethyl-10-hydroxycamptothecin** (**SN-38**). Prepn: **JP Kokai 85 19790;** T. Miyasaka *et al.,* **US 4604463** (1985, 1986 both to Yakult Honsha); S. Sawada *et al., Chem. Pharm. Bull.* **39**, 1446 (1991). Antitumor activity and toxicity data: T. Kunimoto *et al., Cancer Res.* **47**, 5944 (1987). Mechanism of action study: Y. Kawato *et al., ibid.* **51**, 4187 (1991). Clinical trial in colorectal cancer: J. Y. Douillard *et al., Lancet* **355**, 1041 (2000). Review of pharmacology and clinical experience: N. Ma-

suda *et al., Crit. Rev. Oncol. Hematol.* **24**, 3-26 (1996); S. O'Reilly, E. K. Rowinsky, *ibid.* 47-70; of clinical toxicity: K. Seiter, *Expert Opin. Drug Saf.* **4**, 45-53 (2005); of clinial trials in colorectal cancer: C. Fuchs *et al., Cancer Treat. Rev.* **32**, 491-503 (2006).

Pale yellow powder, mp 222-223°.

Hydrochloride trihydrate. [136572-09-3]; [100286-90-6] (hydrochloride). CPT-11; DQ-2805; Campto; Camptosar. C$_{33}$H$_{38}$N$_4$O$_6$.HCl.3H$_2$O; mol wt 677.19. Slightly pale yellow needles or crystalline powder from water, mp 256.5°. $[\alpha]_D^{20}$ +67.7° (c = 1 in water). uv max (ethanol): 221, 254, 359, 372 nm (ε 53800, 36600, 26200, 25300). Slightly sol in water and organic solvents. pH of 2% aqueous soln = 4. LD$_{50}$ in mice (mg/kg): 177.5 i.p.; 765.3 orally (Kunimoto).

THERAP CAT: Antineoplastic.

5136. Irofulven. [158440-71-2] (6'R)-6'-Hydroxy-3'-(hydroxymethyl)-2',4',6'-trimethylspiro[cyclopropane-1,5'-[5H]inden]-7'(6'H)-one; (hydroxymethyl)acylfulvene; HMAF; MGI-114; NSC-683863. C$_{15}$H$_{18}$O$_3$; mol wt 246.31. C 73.15%, H 7.37%, O 19.49%. Semi-synthetic antitumor agent derived from illudin S, *q.v.* Inhibits DNA synthesis and induces apoptosis in tumor cells. Prepn: M. J. Kelner *et al.,* **US 5523490** (1996 to Univ. California); T. C. McMorris *et al., J. Nat. Prod.* **59**, 896 (1996). Enantioselective synthesis: *idem et al., J. Org. Chem.* **69**, 619 (2004). Effect on DNA integrity and apoptosis: J. M. Woynarowski *et al., Biochem. Pharmacol.* **54**, 1181 (1997). Clinical pharmacokinetics and toxicology: J. Alexandre *et al., Clin. Cancer Res.* **10**, 3377 (2004). Clinical evaluation in hormone-refractory prostate cancer: N. Senzer *et al., Am. J. Clin. Oncol.* **28**, 36 (2005). *Review:* M. Baekelandt, *Curr. Opin. Investig. Drugs* **3**, 1517-1526 (2002).

Crystals from ether, mp 127-129°. d 1.262 g/cm³. $[\alpha]_D^{25}$ −639° (c = 0.096 in ethanol). uv max (ethanol): 235, 330 nm (ε 15100, 9200) with tailing to 480 nm.

THERAP CAT: Antineoplastic.

5137. Iron. [7439-89-6] Fe; at. no. 26; at. wt 55.845; valences 2, 3; seldom 1, 4, 6. Group VIII (8). Four naturally occurring isotopes: 54 (5.82%); 56 (91.66%); 57 (2.19%); 58 (0.33%); artificial, radioactive isotopes: 52; 53; 55; 59-61. Second most abundant metal in earth's crust after aluminum: about 5%. The earth's core is believed to consist mainly of iron. Important ores include hematite (Fe$_2$O$_3$), magnetite (Fe$_3$O$_4$), limonite [FeO(OH).nH$_2$O] and *siderite* (FeCO$_3$). Essential dietary nutrient. Study of iron and its compds by Mössbauer spectroscopy: Danon, "^{57}Fe: Metal, Alloys and Inorganic Compounds" in *Chemical Applications of Mössbauer Spectroscopy,* V. I. Goldanskii, R. H. Herber, Eds. (Academic Press, New York, 1968) p 159-313. Ions involved in oxygen transport, electron transport, nitrogen fixation and a number of other biological processes: Nielands, "Evolution of Biological Iron Binding Centers" in *Struct. Bonding* **11**, 145-170 (1972). Review of biology, pharmacology and toxicity of iron compounds: several authors, *Clin. Toxicol.* **4**, 525-642 (1971); of metabolism and homeostasis: G. Papanikolaou, K. Pantopoulos, *Toxicol. Appl. Pharmacol.* **202**, 199-211 (2005). Comprehensive reviews: Feldmann, Schenck in *Ullmanns Encyklopädie der technischen Chemie* vol. **6** (München-Berlin, 1955) pp 261-407; Nicholls in *Comprehensive Inorganic Chemistry* vol. **3**, J. C. Bailar, Jr. *et al.*, Eds. (Pergamon Press, Oxford, 1973) pp 979-1051; W. A. Knepper in *Kirk-Othmer Encyclopedia of Chemical Technology* vol. **13** (Wiley-Interscience, New York, 3rd ed., 1981) pp 735-753. Review of use in cereal flour fortification: R. Hurrell *et al.*, *Nutr. Rev.* **60**, 391-406 (2002).

Silvery-white or gray, soft, ductile, malleable, somewhat magnetic metal. Holds magnetism only after hardening (as alloy steel, *e.g.*, Alnico). Supplied as ingots, powder, wire, sheets, etc. Takes a bright polish; can be rolled, hammered, bent, particularly when red hot. Stable in dry air but readily oxidizes in moist air, forming "rust" (chiefly oxide, hydrated). In powder form it is black to gray. Commercial iron usually contains some C, P, Si, S and Mn. d pure 7.86; cast 7.76; wrought 7.25-7.78; steel 7.6-7.78. mp pure 1535°; cast 1000-1300°; wrought 1500°; steel 1300°. bp 3000°. Electrical resistivity (20°): 9.71 microhm-cm. Readily attacked by dil mineral acids and attacked or dissolved by organic acids; not appreciably attacked by cold concd H_2SO_4 or HNO_3, but is attacked by the hot acids.

USE: Alloyed with C, Mn, Cr, Ni, and other elements to form steels. Nutritional supplement in wheat flours, corn meal, grits and other cereal products. [55]Fe and ^{59}Fe used in tracer studies; the former in biological studies. Standard for elemental analysis.

5138. Iron Dextran. [9004-66-4] Dextran iron complex; DexFerrum; Imferon; INFeD. Complex of ferric hydroxide with dextran, *q.q.v.* Prepn: E. London, G. D. Twigg, US 2820740 (1958 to Benger); US RE 24642 (1959); J. R. Herb, US 2885393 (1959 to Laros). Use in treatment of iron-deficiency anemia: I. M. Baird, D. A. Podmore, *Lancet* **264**, 942 (1954). Acute toxicity: R. P. Beliles, *Toxicol. Appl. Pharmacol.* **23**, 537 (1972). Comparison of different formulations: R. Lawrence, *PDA J. Pharm. Sci. Technol.* **52**, 190 (1998). Mössbauer spectroscopy and x-ray diffraction study: B. Knight *et al.*, *J. Inorg. Biochem.* **73**, 227 (1999). Use as an iron supplement for anemic piglets: A. K. Egeli, T. Framstad, *Res. Vet. Sci.* **66**, 179 (1999). Clinical evaluation in iron deficiency: J. C. Barton *et al.*, *Am. J. Med.* **109**, 27 (2000). Review of carcinogenicity studies: *IARC Monographs* **2**, 161-178 (1973); of clinical experience: D. L. Burns *et al.*, *Nutrition* **11**, 163-168 (1995); of toxicity profile: D. L. Burns, J. J. Pomposelli, *Kidney Int.* **55**, S119-S124 (1999).

Commercial product is a dark brown, slightly viscous soln. Very sol in water, DMSO. Dec in 95% ethanol and acetone. LD_{50} i.v. in mice: 2240 mg Fe/kg (Beliles).

Caution: This substance is reasonably anticipated to be a human carcinogen: *Report on Carcinogens, Twelfth Edition* (PB2011-111646, 2011) p 246.

THERAP CAT: Hematinic.

THERAP CAT (VET): Nutritional factor (parenteral). Used in iron deficiency anemia, chiefly in pigs.

5139. Irone. [1335-94-0] 6-Methylionone. The fragrant principle of violets, best isolated from the rhizomes of iris or from orris oil: F. Tiemann, P. Kruger, *Ber.* **26**, 2675 (1893); Ruzicka *et al.*, *Helv. Chim. Acta* **16**, 1143 (1933); Naves, Mazuyer, *Les Parfums Naturels* (Paris, 1939). Irone, as isolated, is an isomeric mixture of α-, β-, and γ-irones, *q.q.v.*, with each having the same molecular formula, $C_{14}H_{22}O$. Synthesis of isomeric mixture: Eschinazi, US 3019265 (1962 to Givaudan).

Light yellow, slightly viscous liquid. Very stable, does not discolor. In dil alcoholic soln the characteristic odor of violets is best appreciated.

USE: In perfumery for orris and violet compositions.

5140. α-Irone. [79-69-6] 4-(2,5,6,6-Tetramethyl-2-cyclohexen-1-yl)-3-butene-2-one. $C_{14}H_{22}O$; mol wt 206.33. C 81.50%, H 10.75%, O 7.75%. Main perfume ingredient of violets. Both the *d*-(1,5)-*cis*- and *d*-(1,5)-*trans*-forms occur in nature. Isoln from irone mixture and synthesis: Naves, *Helv. Chim. Acta* **31**, 893, 1103 (1948); Ruzicka *et al.*, *ibid.* **31**, 257 (1948). Synthesis: Barton, Mousseron-Canet, *J. Chem. Soc.* **1960**, 271. Industrial preparation:

Kaiser, Kimel, US 2877271 (1959 to Hoffmann-La Roche); Eschenmoser *et al.*, US 3413351; US 3470241 (1968, 1969 to Firmenich & Cie). Nomenclature in early papers: Naves, *Helv. Chim. Acta* **32**, 969 (1949). Stereochemistry: Tribolet, Schinz, *ibid.* **37**, 2184 (1954); Naves, Ardizio, *Bull. Soc. Chim. Fr.* **1955**, 1479; V. Rautenstrauch, G. Ohloff, *Helv. Chim. Acta* **54**, 1776 (1971); V. Rautenstrauch *et al.*, *ibid.* **67**, 325 (1984). Stereoselective synthesis of *cis*- and *trans*-α-irones: S. Torii *et al.*, *J. Org. Chem.* **45**, 16 (1980); of (±)-*cis*-α-irone: C. Nussbaumer, G. Fráter, *ibid.* **52**, 2096 (1987).

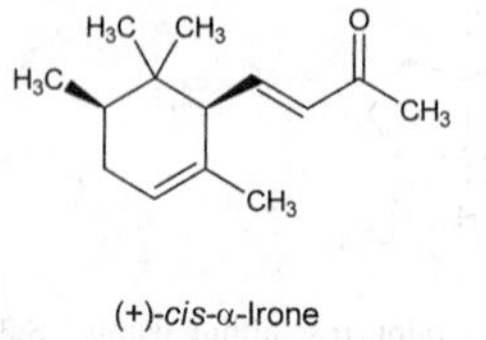

(+)-*cis*-α-Irone 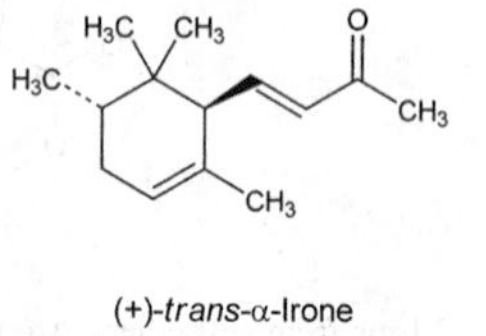   (+)-*trans*-α-Irone

All forms are viscous liquids with characteristic odors.
*dl-cis-α-*Irone. d_4^{20} 0.9360; n_D^{20} 1.50098. uv max (ethanol): 227 nm (ε 15400). *See:* Naves, Bachmann, *Helv. Chim. Acta* **32**, 394 (1949).
*d-cis-α-*Irone. [35124-13-1] $[α]_D^{20}$ +109° (CH_2Cl_2).
*dl-trans-α-*Irone. d_4^{20} 0.9347; n_D^{20} 1.50119. uv max (ethanol): 229 (ε 15450).
*d-trans-α-*Irone. [90242-81-2] $[α]_D^{20}$ +420° (CH_2Cl_2).

5141. β-Irone. [79-70-9] 4-(2,5,6,6-Tetramethyl-1-cyclohexen-1-yl)-3-buten-2-one. $C_{14}H_{22}O$; mol wt 206.33. C 81.50%, H 10.75%, O 7.75%. Frequent byproduct in synthesis of α-irone. Stereoselective synthesis of (±)-form: S. Torii *et al.*, *J. Org. Chem.* **45**, 16 (1980). For general refs *see* α-irone.

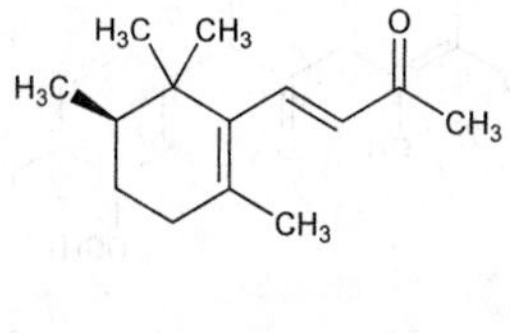

(+)-β-Irone

Oil with odor similar to that of β-ionone. $bp_{0.1}$ 85-90°; $bp_{0.7}$ 99-104°; bp_{11} 125°. $[α]_D^{20}$ +59° (CH_2Cl_2). d_4^{21} 0.9434; n_D^{21} 1.5178; n_D^{25} 1.5162. uv max: 295 nm (log ε 4.05).

5142. γ-Irone. [79-68-5] 4-(2,2,3-Trimethyl-6-methylenecyclohexan-1-yl)-3-buten-2-one. $C_{14}H_{22}O$; mol wt 206.33. C 81.50%, H 10.75%, O 7.75%. Principal fragrant constituent of natural iris oil. Isoln from irone mixture: Ruzicka *et al.*, *Helv. Chim. Acta* **30**, 1807 (1947); Ruzicka *et al.*, *ibid.* **31**, 257 (1948). Synthesis of *dl*-γ-irone: Favre, Schinz, *ibid.* **41**, 1368 (1958). Synthesis of (±)-*cis*- and *trans*-γ-irone: F. Leyendecker, M.-T. Comte, *Tetrahedron* **43**, 85 (1987); T. Kawanobe *et al.*, *Agric. Biol. Chem.* **51**, 791 (1987). Stereoselective synthesis of (±)-*cis*-γ-irone: C. Nussbaumer, G. Fráter, *Helv. Chim. Acta* **71**, 619 (1988). Stereochemistry: *See* α-irone.

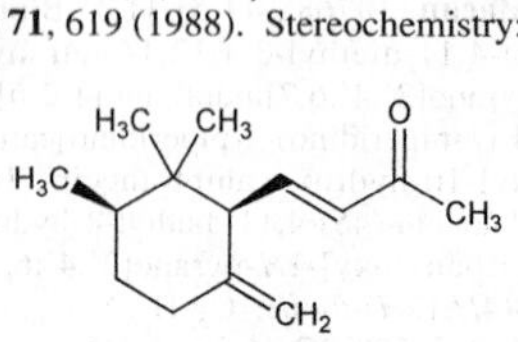

(+)-*cis*-γ-Irone

Oil. $bp_{0.06}$ 85-88°. bp_2 114-116°. d_4^{15} 0.939; n_D^{15} 1.505. $[α]_D^{20}$ +2° (methylene chloride). uv max: 230 nm (log ε 4.2).
USE: In perfumery.

5143. Iron Pentacarbonyl. [13463-40-6] (*TB*-5-11)-Iron carbonyl (Fe(CO)$_5$); iron carbonyl; pentacarbonyliron. C_5FeO_5; mol wt 195.90. C 30.66%, Fe 28.51%, O 40.83%. Fe(CO)$_5$. Prepn from

CO and Fe ore: Wallis, Townshend, **US 2378053** (1945 to International Nickel); from CO and Fe or FeSO$_4$.7H$_2$O: Reppe *et al.*, *Ann.* **582**, 116 (1953); from CO and Fe amalgams: Ettmayer, Jangg, *Monatsh. Chem.* **92**, 834 (1961); from CO and steel turnings: Shipman, **GB 897204** (1962 to ICI). Convenient lab prepn of small quantities of Fe(CO)$_5$ from Fe(CO)$_4$I$_2$: Hieber, Lagally, *Z. Anorg. Allg. Chem.* **245**, 295 (1940). Thermodynamic data: Cotton *et al.*, *J. Am. Chem. Soc.* **81**, 800 (1959); Leadbetter, Spice, *Can. J. Chem.* **37**, 1923 (1959). Toxicity: Sunderman *et al.*, *Arch. Ind. Health* **19**, 11 (1959); Gage, *Br. J. Ind. Med.* **27**, 1 (1970). *Reviews:* Cable, Sheline, *Chem. Rev.* **56**, 1 (1956); Wender *et al.*, *The Chemistry and Catalytic Properties of Cobalt and Iron Carbonyls* (U.S. Govt. Printing Office, Washington, 1962) 83 pp; H. Alper, "Organic Syntheses with Iron Pentacarbonyl" in *Organic Syntheses via Metal Carbonyls* **vol. 2**, I. Wender, P. Pino, Eds. (John Wiley, New York, 1977) pp 545-593. Review of *iron tetracarbonyl*, the photochemically produced intermediate: M. Poliakoff, *Chem. Soc. Rev.* **7**, 527-540 (1978).

Colorless to yellow, oily liquid. *Pyrophoric in air;* burns to Fe$_2$O$_3$. Dec by light to Fe$_2$(CO)$_9$ and CO. mp −20°. bp 103°. d$_4^{20}$ 1.46-1.52; n$_D^{22}$ 1.453. *Poisonous, flammable.* Critical temp 285-288°; critical pressure 29.6 atm. Flash pt −15°C. Heat capacity at constant pressure (14°) 56.9 cal/mole/°C. Latent heat of fusion 3161 cal/mol; latent heat of vaporization 9.6 kcal/mole. Heat of combustion −386.9 kcal/mole; heat of formation [Fe(CO)$_5$(liq)] −182.6 kcal/mole. Practically insol in water, liquid ammonia. Readily sol in most organic solvents including ether, benzene, petr ether, acetone, ethyl acetate, carbon tetrachloride, carbon disulfide; slightly sol in alcohol. *Protect from light and air.* LD$_{50}$ in mice, rats (mg/l): 2.19, 0.91 inhalation for 30 min. (Sunderman).

Caution: Potential symptoms of overexposure are irritation of eyes, mucous membranes, respiratory system; headache, dizziness, nausea, vomiting; fever, cyanosis, cough, dyspnea; liver, kidney, lung injury; degenerative CNS changes. *See NIOSH Pocket Guide to Chemical Hazards* (DHHS/NIOSH 97-140, 1997) p 174.

USE: To make finely divided iron, so-called carbonyl iron, which is used in the manuf of powdered iron cores for high frequency coils used in the radio and television industry; as antiknock agent in motor fuels; as catalyst and reagent in organic reactions.

5144. Iron Sorbitex. [62765-90-6] Glucitol iron complex, compd with citric acid; iron sorbitol; iron sorbitol citrate; Jectofer. Sterile soln (pH 7.2-7.9) of a complex of iron, sorbitol, and citric acid, stabilized with dextrin in a dil soln of sorbitol. Contains 50 ±2 mg/ml of elemental iron. Average mol wt <5000. Prepn: Lindvall, Andersson, *Br. J. Pharmacol. Chemother.* **17**, 358 (1961).

The prepn is hypertonic, stable in serum and rapidly absorbed from muscle, does not produce hemolysis, affects coagulation only at very high concns.

THERAP CAT: Hematinic.

5145. Iron Sucrose. [8047-67-4] Iron saccharate; iron(III) hydroxide sucrose complex; saccharated ferric oxide; Ferrivenin; Venofer. Composed of a polynuclear ferric hydroxide inner sphere surrounded by sucrose molecules; mol wt 34 to 60 kDa. Prepn: C. Mannich, C. A. Rojahn, *Ber. Dtsch. Pharm. Ges.* **32**, 158 (1922); and use in i.v. treatment of anemia: H. G. B. Slack, J. F. Wilkinson, *Lancet* **256**, 11, 163 (1949). Histological properties: P. Geisser *et al.*, *Arzneim.-Forsch.* **42**, 1439 (1992). Pharmacokinetics: B. G. Danielson *et al.*, *ibid.* **46**, 615 (1996). Clinical evaluation in dialysis-associated anemia: C. Charytan *et al.*, *Am. J. Kidney Dis.* **37**, 300 (2001). Review of clinical experience: J. Yee, A. Besarab, *ibid.* **40**, 1111-1121 (2002).

Brown powder. Sol in water (pH 10.5 to 11.1; 1430 mOsm/l). LD$_{50}$ in mice (mg Fe/kg): >2500 orally; >200 i.v. (Geisser).

THERAP CAT: Hematinic.

5146. Irsogladine. [57381-26-7] 6-(2,5-Dichlorophenyl)-1,-3,5-triazine-2,4-diamine; 2,4-diamino-6-(2,5-dichlorophenyl)-*s*-triazine; dicloguamine. C$_9$H$_7$Cl$_2$N$_5$; mol wt 256.09. C 42.21%, H 2.76%, Cl 27.69%, N 27.35%. Gastric cytoprotectant with benzoguanamine skeleton. Prepn: H. Murai *et al.*, **DE 2506814**; *eidem*, **US 3966728** (1975, 1976 both to Nippon Shinyaku). Series of articles on pharmacology and toxicology: F. Ueda *et al.*, *Arzneim.-Forsch.* **34**, 474-491 (1984). Tissue distribution studies: T. Ando *et al.*, *ibid.* **36**, 1221 (1984). Metabolism: M. Sugiyama *et al.*, *ibid.*

1229. Phase I study: M. Nakashima *et al.*, *ibid.* **34**, 492 (1984). Mechanism of antiulcer action study: F. Ueda *et al.*, *Oyo Yakuri* **33**, 157 (1987), *C.A.* **107**, 17604a (1987).

Colorless crystals from dioxane, mp 268-269°.
Maleate. [84504-69-8] MN-1695; Gaslon. C$_9$H$_7$Cl$_2$N$_5$.C$_4$H$_4$O$_4$; mol wt 372.16. Crystals from dioxane, mp 205° (dec). LD$_{50}$ in male, female mice, rats (mg/kg): 6035, 5697, 3898, 2917 orally; 2841, 3216, 1600, 1524 s.c.; 775, 1006, 558, 545 i.p. (Ueda).

THERAP CAT: Antiulcerative.

5147. Isanic Acid. [506-25-2] 17-Octadecene-9,11-diynoic acid; bolecic acid; erythrogenic acid. C$_{18}$H$_{26}$O$_2$; mol wt 274.40. C 78.79%, H 9.55%, O 11.66%. Found in vegetable oils obtained from tropical and equatorial regions. Isoln from boleko oil: De Vries, *Oleagineux* **12**, 427 (1957); **US 2789993** (1957 to UCB). Structure: Steger, Van Loon, *Rec. Trav. Chim.* **59**, 1156 (1940); Gunstone, Sealy, *J. Chem. Soc.* **1963**, 5772; Morris, *ibid.* 5779. Synthesis: Black, Weedon, *ibid.* **1953**, 1785.

Crysts from petr ether, mp 39.5°. d$_4^{45}$ 0.9309; d$_4^{60}$ 0.91966; d$_4^{78}$ 0.9095; n$_D^{50}$ 1.49148; n$_D^{78}$ 1.4860. Turns pink on exposure to air and light. Tendency to polymerize and explode, especially when heated above 250°. Sol in acetone, ethanol, isopropanol. Moderately sol in petr ether.

5148. Isatin. [91-56-5] 1*H*-Indole-2,3-dione; 2,3-indolinedione; 2,3-diketoindoline; 2,3-dioxoindoline. C$_8$H$_5$NO$_2$; mol wt 147.13. C 65.31%, H 3.43%, N 9.52%, O 21.75%. May be obtained by oxidation of indigo or of oxygenated indoles such as indoxyl, oxindole, or dioxindole: Erdmann, *J. Prakt. Chem.* [1] **24**, 1 (1841); Laurent, *ibid.* **25**, 430 (1842); **DE 229815**; *Frdl.* **10**, 353 (1910); **JP 152932** (1942 to ICI); *C.A.* **44**, 1544d (1950). Synthesis: Sandmeyer, *Helv. Chim. Acta* **2**, 234 (1919); **GB 128122**; Marvel, Hiers, *Org. Synth.* **coll. vol. I** (2nd ed., 1941) p 327; Wibaut, Gerling, *Rec. Trav. Chim.* **50**, 41 (1931); Neunhoeffer, Lehmann, *Ber.* **94**, 2960 (1961); Ziegler *et al.*, *Monatsh. Chem.* **94**, 453 (1963). May be isolated from the urine of rabbits that are fed *o*-nitrophenylglyoxylic acid: Bohm, *Z. Physiol. Chem.* **265**, 210 (1940). Pharmacology: Singh, *Indian Vet. J.* **48**, 672 (1971). *Reviews:* Heller, *Ueber Isatin, Isatyd, Dioxindol und Indophenin* (F. Enke, Stuttgart, 1931); Sumpter, *Chem. Rev.* **34**, 393 (1944). Discussion of chemistry of isatin: Morton, *Chemistry of Heterocyclic Compounds* (New York, 1946) pp 126-132.

Orange-colored monoclinic prisms. mp 203.5° (partial sublimation). Absorption spectrum: Hartley, Dobbie, *J. Chem. Soc.* **75**, 647, 656. Freely sol in boiling alcohol; sol in ether and in boiling water with reddish-brown color; sol in alkali hydroxide solns with a violet color becoming yellow on standing. The alc soln imparts a persistent, disagreeable odor to the human skin. Extremely weak base, forms a crystalline perchlorate, C$_8$H$_5$NO$_2$.HClO$_4$.2H$_2$O.

USE: Manuf vat dyes. In analytical chemistry as a reagent for cuprous ions, mercaptans, thiophene, and indican.

5149. Isavuconazole. [241479-67-4] 4-[2-[(1*R*,2*R*)-2-(2,5-Difluorophenyl)-2-hydroxy-1-methyl-3-(1*H*-1,2,4-triazol-1-yl)propyl]-4-thiazolyl]benzonitrile; (2*R*,3*R*)-3-[4-(4-cyanophenyl)thiazol-2-yl]-2-(2,5-difluorophenyl)-1-(1*H*-1,2,4-triazol-1-yl)butan-2-ol; BAL-4815; RO-0094815. $C_{22}H_{17}F_2N_5OS$; mol wt 437.47. C 60.40%, H 3.92%, F 8.69%, N 16.01%, O 3.66%, S 7.33%. Triazole antifungal; ergosterol synthesis inhibitor. Prepn: T. Hayase *et al.*, **WO 9945008** (1999 to Hoffmann-LaRoche); *eidem*, **US 6300353** (2001 to Basilea Pharm.); of triazolium chloride prodrug: H. Fukuda *et al.*, **WO 0132652** (2001 to Hoffmann-LaRoche); **US 6812238** (2004 to Basilea Pharm.); J. Ohwada *et al.*, *Bioorg. Med. Chem. Lett.* **13**, 191 (2003). LC-MS/MS determn in blood: F. Farowski *et al.*, *Antimicrob. Agents Chemother.* **54**, 1815 (2010). Pharmacology: J. Majithiya *et al.*, *J. Antimicrob. Chemother.* **63**, 161 (2009). Clinical pharmacokinetics: A. Schmitt-Hoffmann *et al.*, *Antimicrob. Agents Chemother.* **50**, 279, 286 (2006). Toxicology study: K. Kobayashi, I. Horii, *J. Toxicol. Sci.* **27**, 107 (2002). Review of development and therapeutic potential: J. Guinea, E. Bouza, *Future Microbiol.* **3**, 603-615 (2008); of *in vitro* antifungal spectrum: G. R. Thompson, N. P. Wiederhold, *Mycopathologia* **170**, 291-313 (2010).

Isavuconazonium chloride. [338990-84-4] *N*-Methylglycine [2-[[[1-[1-[(2*R*,3*R*)-3-[4-(4-cyanophenyl)-2-thiazolyl]-2-(2,5-difluorophenyl)-2-hydroxybutyl]-4*H*-1,2,4-triazolium-4-yl]ethoxy]carbonyl]methylamino]-3-pyridinyl]methyl ester chloride; 1-[[*N*-methyl-*N*-3-[(methylamino)acetoxymethyl]pyridin-2-yl]carbamoyloxy]-ethyl-1-[(2*R*,3*R*)-2-(2,5-difluorophenyl)-2-hydroxy-3-[4-(4-cyanophenyl)thiazol-2-yl]butyl]-1*H*-[1,2,4]triazol-4-ium chloride. $C_{35}H_{35}ClF_2N_8O_5S$; mol wt 753.22.
Isavuconazonium chloride hydrochloride. [497235-79-7] BAL-8557; RO-0098557. $C_{35}H_{35}ClF_2N_8O_5S \cdot HCl$; mol wt 789.68. Water soluble prodrug cleaved by plasma esterases to the active moiety. White powder. Soly in water (mg/ml): >1000.
THERAP CAT: Antifungal.

5150. Iseganan. [257277-05-7] Cyclic (5 → 14),(7 → 12)-L-arginylglycylglycyl-L-leucyl-L-cysteinyl-L-tyrosyl-L-cysteinyl-L-arginylglycyl-L-arginyl-L-phenylalanyl-L-cysteinyl-L-valyl-L-cysteinyl-L-valylglycyl-L-argininamide; IB-367. $C_{78}H_{126}N_{30}O_{18}S_4$; mol wt 1900.30. C 49.30%, H 6.68%, N 22.11%, O 15.15%, S 6.75%. Cationic antibacterial cyclic peptide composed of 17 amino acid residues with 2 disulfide linkages; synthetic protegrin analog based on the sequence of Protegrin-1 as isolated from porcine neutrophils. Prepn: C. Chang *et al.*, **WO 9718826**; *eidem*, **US 5994306** (1997, 1999 both to IntraBiotics). CE characterization and purification: J. Chen *et al.*, *J. Chromatogr. A* **853**, 197 (1999). Structure activity study: *eidem et al.*, *Biopolymers* **55**, 88 (2000). *In vitro* and *in vivo* antibacterial activity: D. A. Mosca *et al.*, *Antimicrob. Agents Chemother.* **44**, 1803 (2000). Clinical study in oral mucositis: F. J. Giles *et al.*, *Leuk. Lymphoma* **44**, 1165 (2003). Review of pharmacology and clinical experience: *idem et al.*, *Expert Opin. Invest. Drugs* **11**, 1161-1170 (2002).

Arg–Gly–Gly–Leu–Cys–Tyr–Cys–Arg–Gly–Arg–Phe–Cys–Val–Cys–Val–Gly–ArgNH₂

Hydrochloride. [256475-21-5]; [244015-05-2] (hydrate). IB-367-03
THERAP CAT: Antibacterial.

5151. Isepamicin. [58152-03-7] *O*-6-Amino-6-deoxy-α-D-glucopyranosyl-(1 → 4)-*O*-[3-deoxy-4-*C*-methyl-3-(methylamino)-β-L-arabinopyranosyl-(1 → 6)]-N^1-[(2*S*)-3-amino-2-hydroxy-1-oxopropyl]-2-deoxy-D-streptamine; 1-*N*-[(*S*)-3-amino-2-hydroxypropionyl]gentamicin B; HAPA-B; Sch-21420. $C_{22}H_{43}N_5O_{12}$; mol wt 569.61. C 46.39%, H 7.61%, N 12.30%, O 33.71%. Aminoglycoside antibiotic; semisynthetic derivative of gentamicin B. Prepn: M. J. Weinstein *et al.*, **BE 818431**; J. J. Wright, P. J. L. Daniels, **BE 824657** (both 1975 to Sherico); J. J. Wright *et al.*, **US 4002742** (1977 to Schering Corp.); T. L. Nagabhushan *et al.*, *J. Antibiot.* **31**, 681 (1978). Antibacterial spectrum *in vitro* and *in vivo*: G. H. Miller *et al.*, *ibid.* 688. Comparison with amikacin of nephrotoxic potential: L. I. Rankin *et al.*, *Antimicrob. Agents Chemother.* **16**, 491 (1979). Series of articles on antibacterial activity, toxicology and clinical efficacy: *Chemotherapy (Tokyo)* **35**, Suppl 5, 1-610 (1985); on toxicology: *Jpn. J. Antibiot.* **39**, 3164-3328 (1986), *C.A.* **107**, 32666r-32673r (1987). Acute toxicity: T. Morino *et al.*, *ibid.* 3164, *C.A.* **107**, 32667s (1987).

Sulfate. [67814-76-0] Exacin; Isepacin; Isépalline. $[\alpha]_D^{26}$ +110.9° (c = 1 in water as the disulfate hydrate). LD_{50} in male, female mice, rats (mg/kg): 234, 236, 489, 476 i.v.; 3312, 3320, 3451, 3392 s.c.; >5000 orally both species (Morino).
THERAP CAT: Antibacterial.

5152. Isethionic Acid. [107-36-8] 2-Hydroxyethanesulfonic acid; hydroxyethylsulfonic acid. $C_2H_6O_4S$; mol wt 126.13. C 19.05%, H 4.80%, O 50.74%, S 25.42%. Obtained by heating ethylene, chlorosulfonic acid, and water: Klason, *J. Prakt. Chem.* [2] **19**, 234 (1879); from ether and SO_3: Hubner, *Ann.* **223**, 198 (1884); from ethylenesulfite + $NaHCO_3$: Smith, **US 2899461** (1959 to Dow).

Colorless, syrupy, strongly acid liquid. Kept over H_2SO_4, it solidifies to a very hygroscopic, cryst mass. Miscible with water, alc. Forms readily cryst salts with organic bases.
Caution: Irritating to skin, mucous membranes.

5153. Ishikawa Reagent. [309-88-6] *N*,*N*-Diethyl-1,1,2,3,-3,3-hexafluoro-1-propanamine; PPDA. $C_7H_{11}F_6N$; mol wt 223.16. C 37.68%, H 4.97%, F 51.08%, N 6.28%. Reagent for the fluorination of alcohols and esters; prepd from perfluoropropene and diethylamine. Prepn: I. L. Knunyants *et al.*, *Bull. Acad. Sci. USSR Div. Chem. Sci.* **1953**, 255; D. C. England *et al.*, *J. Am. Chem. Soc.* **82**, 5116 (1960); A. Y. Yakubovich *et al.*, *Dokl. Akad. Nauk SSSR (Engl. Transl.)* **161**, 399 (1965); and use as a fluorinating agent: A. Takaoka *et al.*, *Bull. Chem. Soc. Jpn.* **52**, 3377 (1979). Synthetic applications: S. Watanabe *et al.*, *J. Fluorine Chem.* **31**, 135, 247 (1986); **36**, 361 (1987); **38**, 243 (1988); **39**, 17 (1988); **47**, 187 (1990); K. Ogu *et al.*, *Tetrahedron Lett.* **39**, 305 (1998); G. Koch *et al.*, *Synlett* **2004**, 693.

bp$_{50}$ 54° (England). Also reported as bp$_{20}$ 38-39° (Yakubovich). n_D^{24} 1.2160. d$_4^{24}$ 1.2127.

USE: Fluorinating agent.

5154. Isinglass. Ichthyocolla; fish glue. The inner membrane of the swimming bladder of *Acipenser huso* and other species of sturgeon and of hake. *Constit.* Chiefly glutin.

Thin, white or yellowish, semi-transparent, pearly iridescent, horny sheets. Sol in hot water, hot dil alcohol. In cold water, softens and becomes adhesive.

USE: As nutrient and food instead of gelatin; as adhesive, in "court plaster"; for clarifying (fining) wine and beer; as protective colloid in manuf various chemicals; in glass and porcelain cements; for lustering and stiffening silks and other fabrics; with pyroxylin for waterproofing fabrics.

5155. Isoaminile. [77-51-0] α-[2-(Dimethylamino)propyl]-α-(1-methylethyl)benzeneacetonitrile; 4-(dimethylamino)-2-isopropyl-2-phenylvaleronitrile; 3-cyano-5-dimethylamino-3-phenyl-2-methylhexane; α-(β-dimethylaminopropyl)-α-isopropylphenylacetonitrile. C$_{16}$H$_{24}$N$_2$; mol wt 244.38. C 78.64%, H 9.90%, N 11.46%. Prepn: **GB 822695**; Stühmer, Funke, **US 2934557** (1959, 1960, both to Kali-Chemie).

Liquid, bp$_3$ 138-146°.

Citrate. [28416-66-2] Peracon. C$_{16}$H$_{24}$N$_2$.C$_6$H$_8$O$_7$; mol wt 436.51. Crystals from alcohol + ether, mp 63-64°.

Cyclamate. [10075-36-2] C$_{22}$H$_{37}$N$_3$O$_3$S; mol wt 423.62. Prepn: Dickinson, **US 3074996** (1963 to Abbott).

Tartrate. C$_{16}$H$_{24}$N$_2$.C$_4$H$_6$O$_6$. Crystals, mp 64°.

THERAP CAT: Antitussive.

5156. Isoamyl Acetate. [123-92-2] 3-Methyl-1-butanol 1-acetate; amylacetic ester. C$_7$H$_{14}$O$_2$; mol wt 130.19. C 64.58%, H 10.84%, O 24.58%. The technical product is also known as *pear oil* or *banana oil*.

Colorless, neutral liq; pear-like odor and taste. d$_4^{15}$ 0.876. Pure isoamyl acetate bp 142°. n_D^{21} 1.400; the ordinary grade of commerce boils between 120-145°. Flash pt, closed cup: 92°F (33°C); open cup: 100°F (38°C). Sol in 400 parts water; miscible with alcohol, ether, ethyl acetate, amyl alcohol. Soly of water in isoamyl acetate (25°) 1.6% by volume.

Caution: Potential symptoms of overexposure are irritation of eyes, skin, nose and throat; dermatitis. *See NIOSH Pocket Guide to Chemical Hazards* (DHHS/NIOSH 97-140, 1997) p 174.

USE: In alcohol solution as a pear flavor in mineral waters and syrups; as solvent for old oil colors, for tannins, nitrocellulose, lacquers, celluloid, and camphor; swelling bath sponges; covering unpleasant odors, perfuming shoe polish; manuf artificial silk, leather or pearls, photographic films, celluloid cements, waterproof varnishes, bronzing liquids, and metallic paints; dyeing and finishing textiles. A special grade of the amyl acetate has been used for burning in the Hefner lamp serving as a photometric standard.

5157. Isoamylamine. [107-85-7] 3-Methyl-1-butanamine; isopentylamine; isobutylcarbylamine; 3-methylbutylamine. C$_5$H$_{13}$-N; mol wt 87.17. C 68.89%, H 15.03%, N 16.07%. Prepn from acetamide + isoamylbromide: Erickson, *Ber.* **59B**, 2665 (1926); from α-methylhydroxylamine + isoamylmagnesium chloride: Sheverdina, Kocheshkov, *J. Gen. Chem. USSR* **8**, 1825 (1938). Isoln from *Clostridium sordelli:* Prevot, Thouvenot, *Ann. Inst. Pasteur* **103**, 925 (1962).

Colorless liquid; strong ammonia odor. d^{18} 0.751. bp 95°. n_D^{18} 1.4096. Miscible with water, alcohol, chloroform, ether.

Hydrochloride. [541-23-1] C$_5$H$_{13}$N.HCl; mol wt 123.62. Crystals from acetone, mp 215°. One gram dissolves in 0.5 ml water, 20 ml chloroform; sol in alcohol.

Caution: Irritating to skin, mucous membranes.

5158. Isoamyl Benzoate. [94-46-2] 3-Methyl-1-butanol 1-benzoate. C$_{12}$H$_{16}$O$_2$; mol wt 192.26. C 74.97%, H 8.39%, O 16.64%.

Colorless liquid, d$_4^{15}$ 0.993. bp 260-262°. Insol in water; miscible with alcohol.

USE: In perfumery and cosmetics.

5159. Isoamyl Bromide. [107-82-4] 1-Bromo-3-methylbutane. C$_5$H$_{11}$Br; mol wt 151.05. C 39.76%, H 7.34%, Br 52.90%. Prepd from isoamyl alcohol, bromine and phosphorus or by refluxing isoamyl alcohol with HBr in presence of H$_2$SO$_4$: Kamm, Marvel, *Org. Synth.* **vol. 1**, p 4 (1921).

Colorless liquid. d$_4^{15}$ 1.210. bp 120-121°. mp −112°. n_D^{15} 1.4433. Slightly sol in water (0.02 g/100 ml at 16.5°); miscible with alcohol, ether.

USE: In organic synthesis.

5160. Isoamyl Butyrate. [106-27-4] Butanoic acid 3-methylbutyl ester. C$_9$H$_{18}$O$_2$; mol wt 158.24. C 68.31%, H 11.47%, O 20.22%.

Colorless liq; aromatic pear-like odor. d$_{15}^{19}$ 0.866. bp 179°. Slightly sol in water; miscible with alcohol, ether.

USE: Manuf artficial rum and fruit essences.

5161. Isoamyl Chloride. [107-84-6] 1-Chloro-3-methylbutane. C$_5$H$_{11}$Cl; mol wt 106.59. C 56.34%, H 10.40%, Cl 33.26%. Prepd by refluxing isoamyl alcohol and HCl in presence of ZnCl$_2$; also by direct chlorination of isopentane.

Liquid. d^0 0.893. bp ~100°. n_D^{20} 1.4103. Slightly sol in water; miscible with alcohol, ether.

5162. Isoamyl Cyanide. [542-54-1] 4-Methylpentanenitrile; isocapronitrile. C$_6$H$_{11}$N; mol wt 97.16. C 74.17%, H 11.41%, N 14.42%. Prepd from KCN and isoamyl chloride in boiling alcohol.

Liquid, very disagreeable odor. d$_4^{20}$ 0.806. bp 155-156°. mp −51°. n_D^{20} 1.406. Insol in water; miscible with alc, ether.

5163.   **Isoamyl Ether.** [544-01-4] 1,1′-Oxybis[3-methylbutane]; diisoamyl ether; diisopentyl ether; di-3-methylbutyl ether; isoamyl oxide. $C_{10}H_{22}O$; mol wt 158.29. C 75.88%, H 14.01%, O 10.11%.

Colorless liquid; pleasant, fruity odor. d_4^{12} 0.783. bp 172°. n_D^{20} 1.408. Insol in water; miscible with alc, chloroform, ether.

USE: Solvent in Grignard reaction; also as solvent of odorous principles; manuf lacquers; regenerating rubber.

5164.   **Isoamyl Formate.** [110-45-2] 3-Methyl-1-butanol 1-formate; isopentyl formate; 3-methylbutyl formate. $C_6H_{12}O_2$; mol wt 116.16. C 62.04%, H 10.41%, O 27.55%.

Colorless liquid; fruity odor. d^{20} 0.877. bp 123-124°. n_D^{20} 1.391. Sol in 300 parts water; miscible with alc, ether. LD_{50} orally in rats: 9840 mg/kg; *see:* P. M. Jenner *et al., Food Cosmet. Toxicol.* **2**, 327 (1964).

USE: Artificial fruit syrups.

5165.   **Isoamyl Iodide.** [541-28-6] 1-Iodo-3-methylbutane; isopentyl iodide; 3-methylbutyliodide. $C_5H_{11}I$; mol wt 198.05. C 30.32%, H 5.60%, I 64.08%. Obtained from isoamyl alcohol, iodine, and phosphorus, or by distillation of the alcohol with HI.

Liquid; quickly becomes brown on exposure to air and light. d_{15}^{18} 1.515. bp 147°. Slightly sol in water; miscible with alc, ether. *Keep well closed and protected from light.*

5166.   **Isoamyl Isovalerate.** [659-70-1] 3-Methylbutanoic acid 3-methylbutyl ester; amyl isovalerate; amyl valerate; apple oil; isoamyl valerianate; isopentyl 3-methylbutanoate; 3-methylbutyl 3-methylbutanoate. $C_{10}H_{20}O_2$; mol wt 172.27. C 69.72%, H 11.70%, O 18.57%.

Colorless liquid; apple-like odor. d_4^{19} 0.858. bp 191-194°. n_D^{19} 1.413. Very slightly sol in water; miscible with alc, ether.

USE: As apple essence for flavoring liqueurs and candy.

5167.   **Isoamyl Nitrate.** [543-87-3] 3-Methyl-1-butanol 1-nitrate; isopentyl nitrate; 3-methylbutyl nitrate. $C_5H_{11}NO_3$; mol wt 133.15. C 45.10%, H 8.33%, N 10.52%, O 36.05%.

Colorless liquid. d_4^{22} 0.996. bp 147-148°. n_D^{22} 1.4122. Slightly sol in water; miscible with alcohol, ether.

Caution: Do not confuse with isoamyl nitrite, *q.v.*

5168.   **Isoamyl Nitrite.** [110-46-3] Nitrous acid 3-methylbutyl ester; Isopentyl nitrite. $C_5H_{11}NO_2$; mol wt 117.15. C 51.26%, H 9.46%, N 11.96%, O 27.31%. Prepd by nitrosation of isopentyl alcohol: Bevillard, Choucroun, *Bull. Soc. Chim. Fr.* **1957**, 337.

Yellowish, transparent, flammable liquid; penetrating fragrant, somewhat fruity odor. Unstable and dec on exposure to air and light. The pure nitrite has d_{25}^{25} 0.875. bp 97-99°, but volatilizes readily at much lower temps. n_D^{21} 1.3871. Very slightly sol in water; miscible with alc, chloroform, ether. *Forms an explosive mixture with air or oxygen. Keep in tightly closed containers protected from light and in cool place. Incompat.* Alcohol, antipyrine, caustic alkalies, alkaline carbonates, potassium iodide, bromides, ferrous salts.

5169.   **Isoamyl Phthalate.** [605-50-5] 1,2-Benzenedicarboxylic acid 1,2-bis(3-methylbutyl) ester; diisoamyl phthalate; amyl phthalate. $C_{18}H_{26}O_4$; mol wt 306.40. C 70.56%, H 8.55%, O 20.89%.

Colorless, practically odorless liquid. d 1.028. bp_{40} 225°. Insol in water; sol in organic solvents.

USE: Plasticizer for nitrocellulose and resin lacquers; preventing foam in manuf of glue; in rubber cements.

5170.   **Isoamyl Salicylate.** [87-20-7] 2-Hydroxybenzoic acid 3-methylbutyl ester; isoamyl 2-hydroxybenzoate; 3-methylbutyl 2-hydroxybenzoate. $C_{12}H_{16}O_3$; mol wt 208.26. C 69.21%, H 7.74%, O 23.05%. Review of toxicological profile as a fragrance ingredient: A. Lapczynski *et al., Food Chem. Toxicol.* **45**, Suppl. 1, S418-S423 (2007).

Colorless liquid; sweet, clover-like odor. d_{15}^{19} 1.048. bp 274-278°. n_D^{20} 1.506. Almost insol in water, miscible with alcohol, chloroform, ether. LD_{50} orally in rats: >5.0 g/kg (Lapczynski).

USE: In perfumery and soaps.

5171.   **Isoascorbic Acid.** [89-65-6] D-*erythro*-Hex-2-enonic acid γ-lactone; D-araboascorbic acid; erythorbic acid; isovitamin C; saccharosonic acid; glucosaccharonic acid; D-*erythro*-3-ketohexonic acid lactone; D-*erythro*-3-oxohexonic acid lactone; erycorbin; Mercate "5"; Neo-Cebicure. $C_6H_8O_6$; mol wt 176.12. C 40.92%, H 4.58%, O 54.50%. Epimer of L-ascorbic acid. Has one-twentieth of the vitamin C activity of L-ascorbic acid. Prepd by treating methyl 2-keto-D-gluconate with sodium methoxide: Maurer, Schiedt, *Ber.* **66**, 1054 (1933); *eidem, ibid.* **67**, 1239 (1934); Ohle *et al., ibid.* **67**, 324 (1934); Baird *et al., J. Chem. Soc.* **1934**, 63; Reichstein *et al., Helv. Chim. Acta* **17**, 516 (1934). Synthesis from sucrose: Heimann, Reiff, *Pharm. Zentralhalle* **93**, 97 (1954). Production by *Penicillium* spp: Takahashi *et al., Nature* **188**, 411 (1960); **US 3052609** (1962 to Sankyo); *Bull. Agric. Chem. Soc. Jpn.* **24**, 533 (1960), *C.A.* **55**, 1788 (1961). Conformation: Matsui *et al., Agric. Biol. Chem.* **27**, 185 (1963).

Shiny granular crystals from water or dioxane. mp 164-171°; dec at 174°. $[\alpha]_D^{16.5}$ $-17°$ (c = 1.8 in 0.01N HCl); $[\alpha]_D^{20}$ $-16.6°$ (H_2O). One g sol in $\sim$ 2.5 ml water, in $\sim$ 20 ml alc. Soluble in water, alc, pyridine. Moderately sol in acetone. Slightly sol in glycerol.

Sodium salt. Sodium erythorbate; Mercate "20"; Neo-Cebitate. Crystals, sol in water. pH of aq solns of the sodium salt between 5 and 6. A 10% soln, made from commercial grade, may have a pH of 7.2 to 7.9. The free acid is more sol in water (40 g/100 ml H_2O) than the sodium salt (16 g/100 ml H_2O).

USE: Antioxidant and antimicrobial preservative for foods: Kadin, Osadca, *J. Agric. Food Chem.* **7**, 358 (1959).

5172. Isobenzan. [297-78-9] 1,3,4,5,6,7,8,8-Octachloro-1,-3,3a,4,7,7a-hexahydro-4,7-methanoisobenzofuran; 1,3,4,5,6,7,-10,10-octachloro-4,7-endomethylene-4,7,8,9-tetrahydrophthalan; telodrin; SD-4402; R-6700. $C_9H_4Cl_8O$; mol wt 411.73. C 26.25%, H 0.98%, Cl 68.88%, O 3.89%. Chlorinated hydrocarbon pesticide. Prepn: **GB 772212** (1957 to Ruhrchemie); Korte, Stiasni, *Ann.* **656**, 140 (1962); Feichtinger, Linden, *Chem. Ind. (London)* **1965**, 1938. Toxicology study: A. N. Worden, *Toxicol. Appl. Pharmacol.* **14**, 556 (1969).

Crystals from heptane, mp 120-122°. Sol in acetone, benzene, toluene, ether, xylene, heavy aromatic naphtha. LD_{50} in male, female rats, mice (mg/kg): 11.1, 8.9, 10, 10 orally (Worden).

USE: Insecticide.

5173. Isoborneol. [124-76-5] rel-(1R,2R,4R)-1,7,7-Trimethylbicyclo[2.2.1]heptan-2-ol; exo-1,7,7-trimethylbicyclo[2.2.1]heptan-2-ol; exo-2-bornanol; exo-2-camphanol; (±)-isoborneol; dl-isoborneol. $C_{10}H_{18}O$; mol wt 154.25. C 77.87%, H 11.76%, O 10.37%. Prepn: Pickard, Littlebury, *J. Chem. Soc.* **91**, 1973 (1907); Truett, Moulton, *J. Am. Chem. Soc.* **73**, 5913 (1951); Ziegler, **GB 803178** (1958). Resolution: Pickard, Littlebury, *loc. cit.*; Kenyon, Priston, *ibid.* **127**, 1472 (1925). Prepn of the (1R,2R,4R)-form by reduction of d-camphor with lithium aluminum hydride: Trevoy, Brown, *J. Am. Chem. Soc.* **71**, 1675 (1949). Configuration (isoborneol = exo-form; borneol = endo-form): Toivonen *et al.*, *Acta Chem. Scand.* **3**, 991 (1949). Separation of isoborneol from borneol via the p-nitrobenzoate deriv: Truett, Moulton, *loc. cit. Review:* J. L. Simonsen, *The Terpenes* vol. II (University Press, Cambridge, 2nd ed., 1949) pp 365-367; A. R. Pinder, *The Chemistry of the Terpenes* (Chapman & Hall, London, 1960) pp 22-24, 101, 103, 105-107, 111.

(1R,2R,4R)-form

Crystals from petr ether. Sublimes on heating, mp 212° (in a sealed tube). Practically insol in water. Readily sol in alcohol, ether, chloroform.

(1S,2S,4S)-Form. [16725-71-6] (+)-isoborneol; d-isoborneol. Crystals from petr ether, mp 214°. Approx $[\alpha]_D$ +34.3° in alc soln: Picard, Littlebury, *loc. cit.*

(1R,2R,4R)-Form. [10334-13-1] (−)-isoborneol; l-isoborneol. Crystals from petr ether, mp 214°. Approx $[\alpha]_D$ −34.3° in alc soln.

5174. Isobornyl Thiocyanoacetate. [115-31-1] 2-Thiocyanatoacetic acid rel-(1R,2R,4R)-1,7,7-trimethylbicyclo[2.2.1]hept-2-yl ester; terpinyl thiocyanoacetate; Thanite. $C_{13}H_{19}NO_2S$; mol wt 253.36. C 61.63%, H 7.56%, N 5.53%, O 12.63%, S 12.65%. *See*

also Borneol. May be prepared by treating isoborneol with chloroacetyl chloride and KCNS: J. N. Borglin, **US 2217611** (1940 to Hercules).

Relative stereochemistry

Technical grade, yellow oily liquid with terpene-like odor. Contains 82% or more of isobornyl thiocyanoacetate with other terpenes. $bp_{0.06}$ 95°. Flash pt 82°C (180°F). d_4^{25} 1.1465. Acid no. 1.19. n_D^{25} 1.512. Very sol in alcohol, benzene, chloroform, ether. Practically insol in water.

USE: Insecticide, especially in cattle sprays.

5175. Isobutyl Acetate. [110-19-0] Acetic acid 2-methylpropyl ester; 2-methylpropyl acetate. $C_6H_{12}O_2$; mol wt 116.16. C 62.04%, H 10.41%, O 27.55%. Prepn from isobutyl alcohol + acetic acid: **BE 505023** (1951 to Soc. Belge de l'Azote et des Prod. Chim. du Marly); from methyl isobutyl ketone: White, Emmons, *Tetrahedron* **17**, 31 (1962). Isoln from wood-rotting fungus, *Endoconidiophora coerulescens* Münch.: Birkinshaw, Morgan, *Biochem. J.* **47**, 55 (1950).

Colorless liquid. d_4^{20} 0.871. bp 118°. mp −99°. Flash pt, closed cup: 64°F (18°C). n_D^{19} 1.3907. *Flammable.* Sol in 180 parts water, freely in alc.

Caution: Potential symptoms of overexposure are headache; drowsiness; irritation of eyes, upper respiratory system and skin; anesthesia. *See NIOSH Pocket Guide to Chemical Hazards* (DHHS/NIOSH 97-140, 1997) p 176.

USE: Flavoring, solvent.

5176. Isobutyl Alcohol. [78-83-1] 2-Methyl-1-propanol; isopropylcarbinol; 1-hydroxymethylpropane; isobutanol; fermentation butyl alcohol. $C_4H_{10}O$; mol wt 74.12. C 64.82%, H 13.60%, O 21.59%. Present in fusel oil; also produced by fermentation of carbohydrates: Baraud, Genevois, *Compt. Rend.* **247**, 2479 (1958); Sukhodol, Chatskii, *Spirt. Promst.* **28**, 35 (1962), *C.A.* **57**, 5124e (1962). Prepn: Wender *et al.*, *J. Am. Chem. Soc.* **71**, 4160 (1949); Schreyer, **US 2564130** (1951 to du Pont); Pistor, **US 2753366**; Harrer, Rühl, **DE 1011865**; Himmler, Schiller, **US 2787628** (1956, 1957, 1957 all to BASF). Toxicity study: Smyth *et al.*, *Arch. Ind. Hyg. Occup. Med.* **10**, 61 (1954).

Colorless, refractive liq; odor like that of amyl alcohol, but weaker. d^{15} 0.806. d^{25} 0.79761. bp 108°. mp −108°. Flash pt, closed cup: 82°F (28°C). n_D^{15} 1.3976. n_D^{25} 1.39370. *Flammable.* Sol in about 20 parts water; misc with alcohol, ether. LD_{50} orally in rats: 2.46 g/kg (Smyth).

Caution: Potential symptoms of overexposure are irritation of eyes and throat; headache, drowsiness; skin cracking. *See NIOSH Pocket Guide to Chemical Hazards* (DHHS/NIOSH 97-140, 1997) p 176.

USE: Manuf esters for fruit flavoring essences; organic solvent in paint, varnish removers.

5177. Isobutylamine. [78-81-9] 2-Methyl-1-propanamine; 2-methylpropylamine; 1-amino-2-methylpropane. $C_4H_{11}N$; mol wt

73.14. C 65.69%, H 15.16%, N 19.15%. Prepn from isobutyl alcohol + NH₃: **GB 847799** (1960 to Cellulose-Polymères et Dérivés "Cepede"); Shirley, Speranza, **US 3128311** (1964 to Jefferson Chem.). Isoln from fungi: von Kamienski, *Planta* **50**, 331 (1958). Acute toxicity: K. L. Cheever *et al.*, *Toxicol. Appl. Pharmacol.* **62**, 150 (1982).

Liquid. d_4^{25} 0.724. mp −85°. bp 68-69°. n_D^{17} 1.3988. *Flammable, corrosive.* Miscible with water, alcohol, ether. LD₅₀ (14 day) in male, female rats (mg/kg): 224.4, 231.8 orally (Cheever).

Caution: Skin contact can result in erythema, blistering. Inhalation causes headache, dryness of nose and throat.

5178.   Isobutyl *p*-Aminobenzoate. [94-14-4] 4-Aminobenzoic acid 2-methylpropyl ester; Cycloform; Isobutyl Kelo-form; Isocaine. C₁₁H₁₅NO₂; mol wt 193.25. C 68.37%, H 7.82%, N 7.25%, O 16.56%. Prepn: Adams *et al.*, *J. Am. Chem. Soc.* **48**, 1758 (1926).

Crystals from benzene, mp 65°. Slightly sol in water; sol in alcohol, benzene, ether, acetone, olive oil.

USE: Sunscreen.

THERAP CAT: Anesthetic (topical).

5179.   Isobutylbenzene. [538-93-2] (2-Methylpropyl)benzene; 2-methyl-1-phenylpropane. C₁₀H₁₄; mol wt 134.22. C 89.49%, H 10.51%. Prepd from bromobenzene, isobutyl iodide, and sodium preferably in the presence of benzene: Wreden, Znatovicz, *Ber.* **9**, 1606 (1876); Schramm, *Monatsh. Chem.* **9**, 616 (1888); by passing vapors of dimethylbenzyl carbinol and hydrogen over activated charcoal: Zelinsky, Gawerdowskaya, *Ber.* **61**, 1052 (1928).

Liquid. d_4^{20} 0.8673. bp₇₆₀ 170.5°; bp₄₀₀ 145.2°; bp₂₀₀ 120.7°; bp₁₀₀ 99.0°; bp₆₀ 84.1°; bp₄₀ 73.2°; bp₂₀ 54.7°; bp₁₀ 37.3°; bp₅ 21.1°; bp₁.₀ −9.8°. n_D^{20} 1.4928. Critical temp 377.1°; crit pressure 23,636 mm Hg.

5180.   Isobutyl Bromide. [78-77-3] 1-Bromo-2-methylpropane; 2-methylpropyl bromide. C₄H₉Br; mol wt 137.02. C 35.06%, H 6.62%, Br 58.32%. Obtained from isobutyl alc and PBr₃: Noller, Dinsmore, *Org. Synth.* **coll. vol. II**, 358 (1943).

Colorless liquid. d^{15} 1.272. bp 91.5°. mp −119°. n_D^{15} 1.4391. Slightly sol in water (0.6 g/l); miscible with alcohol, ether.

5181.   Isobutyl *n*-Butyrate. [539-90-2] Butanoic acid 2-methylpropyl ester; isobutyl butanoate; 2-methylpropyl butanoate. C₈H₁₆O₂; mol wt 144.21. C 66.63%, H 11.18%, O 22.19%.

Liquid. d 0.866. bp 157°. n_D^{20} 1.4035. Slightly sol in water; miscible with alcohol, ether.

5182.   Isobutyl Carbamate. [543-28-2] Carbamic acid 2-methylpropyl ester. C₅H₁₁NO₂; mol wt 117.15. C 51.26%, H 9.46%, N 11.96%, O 27.31%. Prepd from isobutyl chloroformate and NH₃ in benzene soln.

Crystals, d 0.956. mp 67°; also stated at 61°. bp 206-207°. Insol in water. Sol in alcohol, ether.

5183.   Isobutyl Chloride. [513-36-0] 1-Chloro-2-methylpropane. C₄H₉Cl; mol wt 92.57. C 51.90%, H 9.80%, Cl 38.30%. Formed by the action of HCl or PCl₅ on isobutyl alcohol.

Liquid. d^{15} 0.883. bp 68-69°. mp −131°. n_D^{15} 1.40096. Insoluble in water. Miscible with alcohol, ether.

5184.   Isobutyl Chlorocarbonate. [543-27-1] Carbonochloridic acid 2-methylpropyl ester; chloroformic acid isobutyl ester; isobutyl chloroformate; 2-methylpropyl chloroformate. C₅H₉ClO₂; mol wt 136.58. C 43.97%, H 6.64%, Cl 25.96%, O 23.43%. Prepd from isobutyl alcohol and phosgene.

Clear liquid; vapors irritate eyes and mucous membranes. d 1.040. bp 130°. Gradually dec by water and alcohol. Miscible with benzene, chloroform, ether.

5185.   Isobutyl Cyanoacrylate. [1069-55-2] 2-Cyano-2-propenoic acid 2-methylpropyl ester; 2-cyanoacrylic acid isobutyl ester; bucrilate; bucrylate; IBC; IBCA. C₈H₁₁NO₂; mol wt 153.18. C 62.73%, H 7.24%, N 9.14%, O 20.89%. Monomer rapidly polymerizes upon contact with fluid or tissue. Preparative method: A. E. Ardis, **US 2467926** (1949 to B. F. Goodrich). Development of polymer nanoparticles as a magnetic drug carrier: A. Ibrahim *et al.*, *J. Pharm. Pharmacol.* **35**, 59 (1983). Clinical experience in treatment of varicocele-associated infertility by transcatheter embolization: F. H. Comhaire, M. Kunnen, *Fertil. Steril.* **43**, 781 (1985); in treatment of arteriovenous malformations: H. V. Vinters *et al.*, *N. Engl. J. Med.* **314**, 477 (1986). Review of 2-cyanoacrylic ester polymers: J. T. O'Connor in *Kirk-Othmer Encyclopedia of Chemical Technology* **vol. 1** (Wiley-Interscience, New York, 4th ed., 1991) pp 344-352; of use in treatment of bleeding gastric varices: R. Kind *et al.*, *Endoscopy* **32**, 512-519 (2000).

Liquid, bp 71-73°. d 0.99. n_D^{20} 1.4352. Viscosity: 2.0 cP. Flash pt: 199°F (93°C). Heat of polymerization: 66.9 kJ/mol.

Polymer. [26809-38-1] Poly(isobutyl 2-cyanoacrylate); PIBCA. LD₅₀ i.v. in mice (mg/kg): 242 (Ibrahim).

USE: Adhesive; polymer nanoparticles as pharmaceutic aid for controlled release drug delivery.

THERAP CAT: Tissue adhesive.

5186.   Isobutylene. [115-11-7] 2-Methyl-1-propene; isobutene. C₄H₈; mol wt 56.11. C 85.62%, H 14.37%. Obtained from refinery streams by absorption on 65% H₂SO₄ at about 15°: Packie, Rupp, **US 2424186**, **US 2509885**; Draeger, **US 2456260**; Steele,

Epps Jr., **US 2497191** (1947, 1950, 1948, 1950, all to Standard Oil); Peters, Gothman; Edwards, Wesselhoft, **US 2962537**; **US 3129265** (1960, 1964, both to Esso). Separation from a mixed C_4 stream using 50% H_2SO_4: **GB 824573**, **GB 858645** and **FR 1337232** (1959, 1961 and 1963 to Compagnie Francaise de Raffinage); Valet *et al.*, *Hydrocarbon Process. Petr. Refin.* **41**, No. 5, 119 (1962); Martel, *Chem. Eng.* **72**, No. 7, 66 (1965). Prepn: Verdol, **US 3170000** (1965 to Sinclair). *Review:* Kennedy, Kirshenbaum, "Isobutylene" in *Vinyl and Diene Monomers* (part 2), E. C. Leonard, Ed. (Wiley-Interscience, New York, 1971) pp 691-756.

Gas. bp_{760} −6.900°; bp_{100} −49.309°; bp_{30} −67.90°; bp_{10} −81.95°; bp_1 −105.06°. d_4^{20} 0.5942; d_4^{25} 0.5879; d_4^{30} 0.5815. Practically insol in water. Very sol in alc, ether, sulfuric acid.

Caution: Simple asphyxiant.

USE: Primarily used to produce diisobutylene, trimers, butyl rubber, and other polymers; also to produce antioxidants for foods, packaging, food supplements, and for plastics: Hatch, *Pet. Refin.* **39**, No. 6, 207 (1960).

5187. Isobutyl Ether. [628-55-7] 1,1′-Oxybis[2-methylpropane]; diisobutyl ether. $C_8H_{18}O$; mol wt 130.23. C 73.78%, H 13.93%, O 12.29%.

Colorless liquid, characteristic odor. d^{15} 0.761. bp 122-124°. Insol in water. Miscible with alcohol, ether.

5188. Isobutyl Formate. [542-55-2] Formic acid 2-methylpropyl ester; tetryl formate. $C_5H_{10}O_2$; mol wt 102.13. C 58.80%, H 9.87%, O 31.33%.

Liquid. d_4^{20} 0.885. bp 98°. mp −95°. n_D^{20} 1.3858. *Flammable.* Sol in 100 parts water; miscible with alc, ether.

5189. Isobutyl Iodide. [513-38-2] 1-Iodo-2-methylpropane. C_4H_9I; mol wt 184.02. C 26.11%, H 4.93%, I 68.96%. Obtained by distilling isobutyl alcohol and HI, or from isobutyl alcohol, iodine and phosphorus.

Liquid; becomes brown on exposure. d^{20} 1.605. bp 120°. mp −93°. n_D^{20} 1.4960. Insol in water. Miscible with alcohol, ether. *Keep well closed and protected from light.*

5190. Isobutyl Isobutyrate. [97-85-8] 2-Methylpropanoic acid 2-methylpropyl ester; isobutyl isobutanoate; 2-methylpropyl 2-methylpropanoate. $C_8H_{16}O_2$; mol wt 144.21. C 66.63%, H 11.18%, O 22.19%. Occurs naturally in fruits such as apples, bananas, mangoes, melons and strawberries, and in honey, rum and wine.

Liquid, odor and taste reminiscent of pineapple. d_4^0 0.875. bp 147°. mp −81°. n_D^{20} 1.3999. Flash pt, closed cup: 37°C. *Flammable.* Sol in organic solvents, alcohol, ethers. Insol in water.

Caution: Potential symptoms of overexposure are drowsiness, dermatitis; inhalation or direct contact may cause irritation of eyes, respiratory tract, skin.

USE: Flavoring agent in foods and beverages. Pharmaceutic aid (flavor).

5191. Isobutyl Isovalerate. [589-59-3] 3-Methylbutanoic acid 2-methylpropyl ester; isobutyl isopentanoate; isobutyl valerate; 2-methylpropyl 3-methylbutanoate. $C_9H_{18}O_2$; mol wt 158.24. C 68.31%, H 11.47%, O 20.22%.

Liquid; ethereal odor. d^{20} 0.853. bp 170-172°. n_D^{20} 1.4064. Insol in water. Miscible with alcohol, ether.

USE: Flavoring and manuf fruit essences.

5192. Isobutyl Mercaptan. [513-44-0] 2-Methyl-1-propanethiol; isobutyl thiol; 2-methylpropyl mercaptan. $C_4H_{10}S$; mol wt 90.18. C 53.28%, H 11.18%, S 35.55%. May be prepd from isobutyl bromide and KHS in alcohol, *see ref under* other butyl mercaptans.

Mobile liq. Heavy skunk odor. mp −79°. bp 88°. *Flammable.* d_4^{20} 0.8357. n_D^{20} 1.43859. Slightly sol in water; very sol in alcohol, ether, liquid hydrogen sulfide.

5193. Isobutyl Nitrate. [543-29-3] Nitric acid 2-methylpropyl ester. $C_4H_9NO_3$; mol wt 119.12. C 40.33%, H 7.62%, N 11.76%, O 40.29%.

Liquid. d_4^{20} 1.015. bp 123-125°. n_D^{20} 1.4028. Insol in water. Miscible with alcohol, ether.

5194. Isobutyl Nitrite. [542-56-3] Nitrous acid 2-methylpropyl ester. $C_4H_9NO_2$; mol wt 103.12. C 46.59%, H 8.80%, N 13.58%, O 31.03%. Prepd from isobutyl alcohol, $NaNO_2$, and dil H_2SO_4.

Colorless liq. d_4^{22} 0.870. bp 67°. n_D^{22} 1.3715. Slightly sol and gradually dec by water. Miscible with alcohol.

5195. Isobutyl Propionate. [540-42-1] Propanoic acid 2-methylpropyl ester. $C_7H_{14}O_2$; mol wt 130.19. C 64.58%, H 10.84%, O 24.58%.

Liquid; agreeable, ethereal odor. d_4^0 0.888. bp 137°. mp −71°. n_D^{20} 1.3975. *Flammable.* Insol in water. Miscible with alc.

USE: Manuf fruit essences.

5196. Isobutyl Stearate. [646-13-9] Octadecanoic acid 2-methylpropyl ester; isobutyl octadecanoate; stearic acid isobutyl ester. $C_{22}H_{44}O_2$; mol wt 340.59. C 77.58%, H 13.02%, O 9.39%.

Paraffin-like, crystal substance at low temp, mp about 20°.

USE: Waterproof coatings, polishes, face creams, rouges, ointments, soaps, rubber manuf, dye solns, inks, lubricants.

5197. Isobutyl Sulfide. [592-65-4] 1,1'-Thiobis[2-methylpropane]; diisobutyl sulfide. $C_8H_{18}S$; mol wt 146.29. C 65.68%, H 12.40%, S 21.92%. Obtained by heating isobutyl chloride or potassium isobutyl sulfate with K_2S.

Liquid. d_4^{10} 0.836. bp 171-173°. Insol in water. Miscible with alcohol, ether.

5198. Isobutyl Urethane. [539-89-9] N-(2-Methylpropyl)-carbamic acid ethyl ester; ethyl isobutylcarbamate; isobutylcarbamic acid ethyl ester. $C_7H_{15}NO_2$; mol wt 145.20. C 57.90%, H 10.41%, N 9.65%, O 22.04%. Obtained by shaking isobutylamine with ethyl chloroformate and aq KOH in the cold.

Liquid; apple-like odor. d_4^{20} 0.943. Solidif below −65°. bp_{17} 96°. n_D^{20} 1.4288. Insol in water. Sol in alcohol.

5199. Isobutyraldehyde. [78-84-2] 2-Methylpropanal; isobutylaldehyde; isobutyric aldehyde; 2-methylpropionaldehyde; 2-propanecarboxaldehyde. C_4H_8O; mol wt 72.11. C 66.63%, H 11.18%, O 22.19%. Manuf by the oxo process from propylene, carbon monoxide, and hydrogen at 130-160° and 1500-3000 psi in the presence of a cobalt catalyst: Roelen, US 2327066 (1943); also prepared by air oxidation of isobutyl alcohol: Fossek, Monatsh. Chem. 2, 614 (1881); Lipp, Ann. 205, 2 (1880). Toxicity study: Smyth et al., Arch. Ind. Hyg. Occup. Med. 10, 61 (1954). Review: P. D. Sherman in Kirk-Othmer Encyclopedia of Chemical Technology vol. 4 (Wiley-Interscience, New York, 3rd ed., 1978) pp 376-386.

Pungent odor. d_4^{20} 0.7938. mp −65.9°. bp_{760} 64°. n_D^{20} 1.3730. Heat of combustion 599.9 kcal. Flash pt (open cup) <20°F. Soly in water at 20°: 11 g/100 ml H_2O. Flammable. Miscible with ethanol, ether, carbon disulfide, acetone, benzene, toluene, chloroform. Forms an azeotrope with water contg 94% isobutyraldehyde. The azeotrope bp_{760} 59°. Weight per gallon of isobutyraldehyde: 6.55 lbs. Oxidizes slowly on exposure to air, forming isobutyric acid. LD_{50} orally in rats: 3.7 g/kg (Smyth).

USE: In the synthesis of pantothenic acid, valine, leucine, cellulose esters, perfumes, flavors, plasticizers, resins, gasoline additives.

5200. Isobutyric Acid. [79-31-2] 2-Methylpropanoic acid; dimethylacetic acid; isobutanoic acid; isopropylformic acid. $C_4H_8O_2$; mol wt 88.11. C 54.53%, H 9.15%, O 36.32%. Prepn from 1-nitroisobutane: Lippincott, Hass, Ind. Eng. Chem. 31, 118 (1939); from methallyl chloride: Towle, Hall, US 2667508 (1954 to Standard Oil Co. of Indiana); from propylene: Alderson, US 3020314 (1962 to du Pont); from 2-methylpropane + CO with HF catalyst: Friedman, Cotton, J. Org. Chem. 27, 481 (1962).

Liquid; pungent odor like that of butyric acid, but not as unpleasant. bp_{760} 152-155°. mp −47°. d_4^{20} 0.950. n_D^{20} 1.3930. Flash pt, open cup; 170°F (77°C). Flammable, corrosive. Sol in 6 parts of water; misc with alcohol, chloroform, ether.

5201. Isobutyronitrile. [78-82-0] 2-Methylpropanenitrile; 2-cyanopropane; dimethylacetonitrile; isopropyl cyanide. C_4H_7N; mol wt 69.11. C 69.52%, H 10.21%, N 20.27%. Prepn: E. A. Letts, Ber. 5, 669 (1872); A. E. Arbusow, ibid. 43, 2296 (1910); R. E. Kent et al., Org. Synth. coll. vol. III, 493 (1955); A. Ahmad, Synthesis 1976, 418. Acute toxicity and human industrial experience: H. Zeller et al., Zentralbl. Arbeitsmed. Arbeitsschutz 19, 225 (1969), C.A. 71, 128340 (1969). Acute toxicity and metabolism: H. Tanii, K. Hashimoto, Arch. Toxicol. 55, 47 (1984).

Colorless liquid. bp_{740} 99-102°. bp 101-103° (Kent); also reported as bp 103-103.5° (Arbusow), bp 104° (Ahmad), bp 107-108° (Letts). $d_0^{16.25}$ 0.7731. n_D^{25} 1.3713. Flammable, poisonous. Log P (n-octanol/water): 0.46. LD_{50} (mg/kg): 25 i.p. in mice; 200 orally in rats (Zeller). LD_{50} orally in male mice: 0.3652 mmol/kg (Tanii).

Caution: Potential symptoms of overexposure are irritation of eyes, skin, nose, throat; headache, dizziness, weakness, giddiness, confusion, convulsions; dyspnea; abdominal pain, nausea, vomiting. See NIOSH Pocket Guide to Chemical Hazards (DHHS/NIOSH 97-140, 1997) p 178.

USE: Solvent.

5202. Isocarboxazid. [59-63-2] 5-Methyl-3-isoxazolecarboxylic acid 2-(phenylmethyl)hydrazide; 3-(N-benzylhydrazinocarbonyl)-5-methylisoxazole; 1-benzyl-2-(5-methyl-3-isoxazolylcarbonyl)hydrazine; Ro-5-0831; Marplan. $C_{12}H_{13}N_3O_2$; mol wt 231.26. C 62.32%, H 5.67%, N 18.17%, O 13.84%. Monoamine oxidase inhibitor. Prepn: Gardner, Wenis, US 2908688 (1959 to Hoffmann-La Roche); J. Med. Pharm. Chem. 2, 133 (1960). Comprehensive description: B. C. Rudy, B. Z. Senkowski, Anal. Profiles Drug Subs. 2, 295-314 (1973).

Crystals from methanol, practically tasteless. mp 105-106°. Very sparingly sol in hot water (0.05%), somewhat more (1 to 2%) in 95% alc, in glycerol, in propylene glycol.

THERAP CAT: Antidepressant.

5203. Isochondrodendrine. [477-62-3] (12aR,24aR)-2,3,-12a,13,14,15,24,24a-Octahydro-5,17-dimethoxy-1,13-dimethyl-8,-11:20,23-dietheno-1H,12H-[1,10]dioxacyclooctadecino[2,3,4-ij:11,12,13-i'j']diisoquinoline-6,18-diol; $O^7,O^{7'}$-didemethylcycleanine; isobebeerine. $C_{36}H_{38}N_2O_6$; mol wt 594.71. C 72.71%, H 6.44%, N 4.71%, O 16.14%. From Chondodendron tomentosum Ruiz and Pav., Menispermaceae: Faltis, Monatsh. Chem. 33, 873 (1912); Faltis, Neumann, ibid. 42, 311 (1921); Dutcher, J. Am. Chem. Soc. 68, 419 (1946); from the drug Radix pareirae bravae: King, J. Chem. Soc. 1940, 737. Structure: Faltis, Dietreich, Ber. 67, 231 (1934); Jeffreys, J. Chem. Soc. 1956, 4451.

Needles from methanol, dec 288°. $[\alpha]_D^{20}$ −29° (c = 1.3 in chloroform). Sol in alcohol, benzene, chloroform.

5204. Isocinchomeronic Acid. [100-26-5] 2,5-Pyridinedicarboxylic acid. $C_7H_5NO_4$; mol wt 167.12. C 50.31%, H 3.02%, N 8.38%, O 38.29%. Prepd by treating 5-ethyl-2-picoline (aldehyde-collidine) with suitable oxidizing agents such as HNO_3, $KMnO_4$, or H_2SeO_3: Meyer, Staffen, *Monatsh. Chem.* **34**, 517 (1913); Jordan, *Ind. Eng. Chem.* **44**, 332 (1952); Kato, *Bull. Chem. Soc. Jpn.* **34**, 636 (1961); by oxidation of 5-acetyl-2-methylpyridine: Binns, Swan, *J. Chem. Soc.* **1962**, 2831; from the lutidine fraction of coal tar lutidine: Lukes *et al.*, *Collect. Czech. Chem. Commun.* **26**, 3044 (1961).

Triclinic leaflets or prisms from dilute HCl, mp 254°. Sublimes as nicotinic acid when heated above mp. Practically insol in cold water, alcohol, ether, benzene. Slightly sol in boiling water, boiling alcohol. Appreciably sol in hot, dil aq solns of mineral acids.
Monohydrate. Crystals from water or alcohol, dec 238°.
Dihydrazide. Crystals, mp 268-269°.
USE: Intermediate in the manuf of nicotinic acid.

5205. Isoconazole. [27523-40-6] 1-[2-(2,4-Dichlorophenyl)-2-[(2,6-dichlorophenyl)methoxy]ethyl]-1H-imidazole; 1-[2,4-dichloro-β-[(2,6-dichlorobenzyl)oxy]phenethyl]imidazole. $C_{18}H_{14}Cl_4N_2O$; mol wt 416.12. C 51.96%, H 3.39%, Cl 34.08%, N 6.73%, O 3.84%. Prepn: E. F. Godefroi *et al.*, *J. Med. Chem.* **12**, 784 (1969); *eidem*, *DE 1940388*; *eidem*, *US 3717655* and *US 3839574* (1970, 1973, 1974 all to Janssen). *In vitro* activity: H. J. Kessler, *Arzneim.-Forsch.* **29**, 1344 (1979). Animal study: H. J. Kessler *et al.*, *ibid.* 1352. Antimicrobial activity in humans: H. Wendt, H. J. Kessler, *ibid.* 846. Bioavailability: U. Täuber, M. Rzadkiewicz, *Mykosen* **22**, 201 (1979).

Nitrate. [24168-96-5] R-15454; Fazol; Gyno-Travogen; Travogen; Travogyn. $C_{18}H_{14}Cl_4N_2O.HNO_3$; mol wt 479.14. Solid, mp 182-183°.
THERAP CAT: Antibacterial; antifungal.

5206. Isocorybulbine. [22672-74-8] (13S,13aR)-5,8,13,13a-Tetrahydro-3,9,10-trimethoxy-13-methyl-6H-dibenzo[a,g]quinolizin-2-ol; (13S-trans)-5,8,13,13a-tetrahydro-3,9,10-trimethoxy-13-methyl-6H-dibenzo[a,g]quinolizin-2-ol; 3,9,10-trimethoxy-13α-methyl-13aβ-berbin-2-ol; 2-hydroxy-13-methyl-3,9,10-trimethoxyberbine. $C_{21}H_{25}NO_4$; mol wt 355.43. C 70.97%, H 7.09%, N 3.94%, O 18.01%. From tubers of *Corydalis cava* (L.) Schweigg. & Korte *(C. tuberosa* DC., *Fumariaceae):* Gadamer, *Arch. Pharm.* **240**, 19 (1902). Structure: Späth, Dobrowsky, *Ber.* **58**, 1274 (1925). Synthesis: Späth, Holter, *ibid.* **59**, 2800 (1926). Abs configuration: P. W. Jeffs, *Experientia* **21**, 690 (1965); K. Iwasa *et al.*, *J. Org. Chem.* **46**, 4744 (1981).

Leaflets, mp 179-180°; also given as 187-188°. $[\alpha]_D^{15}$ +301° in chloroform. Sol in alcohol, dil acids.
Methiodide. $C_{21}H_{25}NO_4.CH_3I$. Crystals, mp 218-221°.

5207. Isocorydine. [475-67-2] (6aS)-5,6,6a,7-Tetrahydro-1,2,10-trimethoxy-6-methyl-4H-dibenzo[de,g]quinolin-11-ol; 1,2,10-trimethoxy-6aα-aporphin-11-ol; 11-hydroxy-1,2,10-trimethoxyaporphine; artabotrine; luteanine. $C_{20}H_{23}NO_4$; mol wt 341.41. C 70.36%, H 6.79%, N 4.10%, O 18.74%. From tubers of *Corydalis cava* (L.) Schweigg. & Korte *(C. tuberosa* DC., *Fumariaceae); Artabotrys suaveolens* Blume, *Anonaceae; Papaver oreophilum* Rupr.; *Phoebe clemensii* Allen *(Lauraceae)* and others. Isoln: Gadamer, *Arch. Pharm.* **249**, 669 (1911); Johns, Lamberton, *Aust. J. Chem.* **20**, 1277 (1967); Pfeifer, Mann, *Pharmazie* **23**, 82 (1968). Structure: Späth, Berger, *Ber.* **64**, 2038 (1931); Gulland *et al.*, *J. Chem. Soc.* **1931**, 2885; Barger, Sargent, *ibid.* **1939**, 991. Identity with artabotrine: Schlittler, Huber, *Helv. Chim. Acta* **35**, 111 (1952); with luteanine: Manske, *Can. J. Res.* **21B**, 13 (1943). Total synthesis: Kametani *et al.*, *Tetrahedron* **27**, 5367 (1971). Pharmacology: Berezhinskaya *et al.*, *Farmakol. Toksikol. (Moscow)* **31**, 44 (1968), *C.A.* **68**, 94521z (1968).

Plates from ethanol or acetone, mp 185°. $[\alpha]_D^{20}$ +195° (chloroform). Sol in ether, chloroform, alcohol, acetone, alkali hydroxide. Practically insol in water, alkali carbonates.

5208. Isocorypalmine. [53447-14-6] (R)-5,8,13,13a-Tetrahydro-3,9,10-trimethoxy-6H-dibenzo[a,g]quinolizin-2-ol; 3,9,10-trimethoxy-13aα-berbin-2-ol; d-tetrahydrocolumbamine. $C_{20}H_{23}NO_4$; mol wt 341.41. C 70.36%, H 6.79%, N 4.10%, O 18.74%. From tubers of *Corydalis cava* (L.) Schweigg. & Korte *(C. tuberosa* DC.) and *C. nobilis* Pers., *Fumariaceae:* Gadamer *et al.*, *Arch. Pharm.* **265**, 675 (1927); Manske, *Can. J. Res.* **18B**, 288 (1940). Structure: Späth, Burger, *Ber.* **59**, 1486 (1926). Synthesis of *dl*-form: Govindachari *et al.*, *ibid.* **92**, 1654 (1959). Configuration: Corrodi, Hardegger, *Helv. Chim. Acta* **39**, 889 (1956).

Crystals, mp 239-241°. $[\alpha]_D^{20}$ +30.3° (c = 0.4 in chloroform). Sol in alcohol, ether.
dl-Form. [6487-33-8] Crystals from methanol, mp 216°.

5209. Isocrotonic Acid. [503-64-0] (2Z)-2-Butenoic acid; cis-2-butenoic acid; cis-crotonic acid. $C_4H_6O_2$; mol wt 86.09. C 55.81%, H 7.03%, O 37.17%. Prepd by reacting acetaldehyde with malonic acid in the presence of pyridine: v. Auwers *et al.*, *Ann.* **432**, 46 (1923); from ethyl acetoacetate: Hatch, Nesbitt, *J. Am. Chem. Soc.* **72**, 727 (1950); from but-2-ynoic acid: Allan *et al.*, *J. Chem. Soc.* **1955**, 1862; from $CH_3CH=CHMgBr + CO_2$: Normant, Maitte, *Bull. Soc. Chim. Fr.* **1956**, 1439. Structure: v. Auwers, Wissebach, *Ber.* **56**, 715 (1923); Plisov, Bogatskii, *Zh. Obshch. Khim.* **27**, 360 (1957), *C.A.* **51**, 15401c (1957). *Review:* W. Blau *et al.*, in *Ullmann's Encyclopedia of Industrial Chemistry* **vol. A8** (VCH, Weinheim, 5th ed., 1987) pp 83-89.

Liquid. bp_5 55-56°; bp_{10} 62-64°; bp_{760} 168-169°. mp 15°; crystallized from pentane. d_4^{20} 1.0267. n_D^{14} 1.4483. n_D^{20} 1.4450. uv max (95% ethanol): 205.5 nm (ε 13500).

Methyl ester. [4358-59-2] $C_5H_8O_2$; mol wt 100.12. Liquid. bp 118°. n_D^{20} 1.4175. uv max (95% ethanol): 205.5 nm (ε 14000).

Benzyl ester. [92758-75-3] $C_{11}H_{12}O_2$; mol wt 176.22. Liquid. bp_{10} 121-122°. n_D^{20} 1.5110.

Caution: Can cause severe irritation of skin, mucous membranes.

5210. Isocyanic Acid. [75-13-8] Carbimide. CHNO; mol wt 43.03. C 27.91%, H 2.34%, N 32.55%, O 37.18%. HN=C=O. Prepn: J. Liebig, F. Wöhler, *Ann.* **39**, 29 (1846); Steyermark, *J. Org. Chem.* **28**, 586 (1963); R. Vorhoeve, L. E. Trimble, *Science* **202**, 525 (1978). Crystal structure: Von Dohlen, Carpenter, *Acta Crystallogr.* **8**, 646 (1955). Review of prepn and properties: D. J. Belson, A. N. Strachan, *Chem. Soc. Rev.* **11**, 41-56 (1982). *See also* Cyanic Acid.

Store in dil solns of carbon tetrachloride or ether at −30° to slow polymerization. The free acid in the vapor phase or in ether soln gives no indication of being a mixture, and all the evidence supports the iso structure, HNCO, whereas in aq soln cyanic acid HOCN is present: N. V. Sidgwick, *The Chemical Elements and Their Compounds* **vol. I** (Oxford, 1950) p 673.

Caution: Strongly acidic, will blister skin.

5211. Isodurene. [527-53-7] 1,2,3,5-Tetramethylbenzene. $C_{10}H_{14}$; mol wt 134.22. C 89.49%, H 10.51%. Occurs in coal tar. Prepd by the action of dimethyl sulfate on 2,4,6-trimethylphenyl-magnesium bromide: Smith, *Org. Synth.* **11**, 66 (1931).

Liquid. d_4^0 0.8961; d_4^{20} 0.8906. mp −24°. bp_{760} 197.9°; bp_{400} 173.7°; bp_{200} 149.9°; bp_{100} 128.3°; bp_{60} 115.4°; bp_{40} 105.8°; bp_{20} 91.0°; bp_{10} 77.8°; bp_5 65.8°; $bp_{1.0}$ 40.6°. n_D^{20} 1.5134; n_{He}^{20} 1.51126. Insol in water. Sol in alcohol; very sol in ether.

5212. Isoestradiol. [517-04-4] (8α,17β)-Estra-1,3,5(10)-triene-3,17-diol; $\Delta^{1,3,5}$-8α-epiestratriene-3,17β-diol; 8α-estradiol; 8-epiestradiol; 8-isoestradiol-17β. $C_{18}H_{24}O_2$; mol wt 272.39. C 79.37%, H 8.88%, O 11.75%. Prepn by high pressure hydrogenation of dihydroequilin: Serini, Logemann, *Ber.* **71**, 186 (1938); by hydrogenation of equilenin with Raney nickel at 2800 psi and 85°: Dauben, Ahramjian, *J. Am. Chem. Soc.* **78**, 633 (1956). Prepn of *dl*-form from *dl*-equilenin and stereochemistry: Johnson *et al.*, *ibid.* **80**, 661 (1958).

d-**Form.** Crystals from dil methanol + chloroform, mp 181°. $[\alpha]_D^{20}$ +18° (16 mg in 2 ml dioxane).

3-Benzoate. $C_{25}H_{28}O_3$. Crystals from ethyl acetate, mp 190°. $[\alpha]_D^{20}$ +9.5° (17 mg in 2 ml dioxane).

dl-**Form.** Needles from dil methanol, mp 213.5-214°.

5213. 8-Isoestrone. [517-06-6] (8α)-3-Hydroxyestra-1,3,5-(10)-trien-17-one; $\Delta^{1,3,5}$-8-epiestratrien-3-ol-17-one; 8α-estrone; 8-epiestrone. $C_{18}H_{22}O_2$; mol wt 270.37. C 79.96%, H 8.20%, O 11.83%. Prepn: Serini, Logemann, *Ber.* **71**, 186 (1938); Johnson *et al.*, *J. Am. Chem. Soc.* **80**, 661 (1958); Ananchenko, Torgov, *Tetrahedron Lett.* **1963**, 1553.

dl-**Form.** Prisms from methanol, mp 254-255°.

dl-**Methyl ether.** $C_{19}H_{24}O_2$. Blades from methanol, mp 153-155°.

dl-**Benzoate.** $C_{25}H_{26}O_3$. Prisms from methanol, mp 197-198°.

5214. Iso E Super®. [54464-57-2] (unspecified stereo); [59056-94-9] (*trans*-form). 1-(1,2,3,4,5,6,7,8-Octahydro-2,3,8,8-tetramethyl-2-naphthalenyl)ethanone. $C_{16}H_{26}O$; mol wt 234.38. C 81.99%, H 11.18%, O 6.83%. Synthetic woody odorant. Prepn: J. B. Hall, J. M. Sanders, DE 2408689; *eidem*, US 3911018 (1974, 1975 both to Int. Flavor Fragrance). Patch testing for contact allergies: P. J. Frosch *et al.*, *Contact Dermatitis* **33**, 333 (1995). Adherence and duration to fabric: S. Widder, *Dragoco Rep.* **1999**, 69. Isoln of major odor producing "impurity": C. Nussbaumer *et al.*, *Helv. Chim. Acta* **82**, 1016 (1999). Brief description: M. Gras, *Perfum. Flavor.* **17**, 2-12 (1992). Review of synthesis and use: G. Fráter *et al.*, *Tetrahedron* **54**, 7633-7703 (1998).

trans-Form

Rich, warm-woody odor with a shade of amber. $bp_{2.8\ mm}$ 134-135°.

USE: Scent in perfumes, laundry products and cosmetics.

5215. Isoetharine. [530-08-5] 4-[1-Hydroxy-2-[(1-methylethyl)amino]butyl]-1,2-benzenediol; 3,4-dihydroxy-α-[1-(isopropylamino)propyl]benzyl alcohol; α-(1-isopropylaminopropyl)protocatechuyl alcohol; *N*-isopropylethylnorepinephrine; 1-(3,4-dihydroxyphenyl)-2-isopropylamino-1-butanol; etyprenaline; isoetarine; Win-3046; Dilabron; Neoisuprel. $C_{13}H_{21}NO_3$; mol wt 239.32. C 65.24%, H 8.85%, N 5.85%, O 20.06%. Prepn: Bockmühl *et al.*, DE 638650 (1936 to I. G. Farben.). Pharmacology: Lands *et al.*, *J. Pharmacol. Exp. Ther.* **99**, 45 (1950); *eidem*, *J. Am. Pharm. Assoc.* **47**, 744 (1958).

Hydrochloride. [2576-92-3] Numotac. $C_{13}H_{21}NO_3 \cdot HCl$; mol wt 275.77. Crystals from methanol + ether. mp 212-213° (dec). Sol in water; sparingly sol in alc. Practically insol in ether.

THERAP CAT: Bronchodilator.

5216. Isoeugenol. [97-54-1] 2-Methoxy-4-(1-propen-1-yl)-phenol; 4-hydroxy-3-methoxy-1-propenylbenzene; 4-propenyl-guaiacol. $C_{10}H_{12}O_2$; mol wt 164.20. C 73.15%, H 7.37%, O 19.49%. Occurs in ylang-ylang and other essential oils. Prepd from eugenol: West, *J. Soc. Chem. Ind.* **59**, 275 (1940); Pal'gi, *Zh. Obshch. Khim.* **28**, 2239 (1958). Stereochemistry: von Auwers, *Ber.* **68**, 1346 (1935); Puxeddu, Rattu, *Gazz. Chim. Ital.* **67**, 647 (1937). Toxicity study: P. M. Jenner *et al.*, *Food Cosmet. Toxicol.* **2**, 327 (1964).

trans-form

Oily liquid; easily becomes somewhat yellow. d_4^{25} 1.080. bp 266°; bp_8 128-130°. mp -10°. n_D^{19} 1.5739. Slightly sol in water; misc with alcohol, ether. LD_{50} orally in rats: 1560 mg/kg (Jenner).

trans-**Form.** [5932-68-3] Crystals, mp 33°; bp_{12} 140°. d_4^{20} 1.087. n_D^{20} 1.5778.

cis-**Form.** [5912-86-7] Liquid, bp_{11} 133°. d_4^{20} 1.088. n_{He}^{20} 1.5724.
USE: Manuf vanillin.

5217. Isofagomine. [169105-89-9] (3R,4R,5R)-5-(Hydroxymethyl)-3,4-piperidinediol; (3R,4R,5R)-3,4-dihydroxy-5-hydroxymethylpiperidine. $C_6H_{13}NO_3$; mol wt 147.17. C 48.97%, H 8.90%, N 9.52%, O 32.61%. Glycosidase inhibitior; iminosugar analog of glucose. Binds to the active site of glucocerebrosidase, the lysosomal enzyme deficient in Gaucher's disease; acts as a pharmacological chaperone to improve stability and delivery of mutant forms of the enzyme. Prepn: T. M. Jespersen *et al.*, *Angew. Chem. Int. Ed. Engl.* **33**, 1778 (1994); *eidem*, *Tetrahedron* **50**, 13449 (1994). Improved synthesis: H. Ouchi *et al.*, *Tetrahedron Lett.* **45**, 7053 (2004). Glycosidase inhibition profile: W. Dong *et al.*, *Biochemistry* **35**, 2788 (1996). Effect on glucocerebrosidase trafficking to the lysosome: R. A. Steet *et al.*, *Proc. Natl. Acad. Sci. USA* **103**, 13813 (2006); J.-S. Shen *et al.*, *Biochem. Biophys. Res. Commun.* **369**, 1071 (2008).

$[\alpha]_D^{20}$ +19.6° (c = 0.85 in ethanol).
Tartrate. [919364-56-0] AT-2101; Plicera. $C_6H_{13}NO_3 \cdot C_4H_6O_6$; mol wt 297.26.
THERAP CAT: In treatment of Gaucher's disease.

5218. Isofenphos. [25311-71-1] 2-[[Ethoxy[(1-methylethyl)amino]phosphinothioyl]oxy]benzoic acid 1-methylethyl ester; salicylic acid isopropyl ester O-ester with O-ethyl isopropylphosphoramidothioate; 1-methylethyl 2-[[ethoxy[(1-methylethyl)amino]phosphinothioyl]oxy]benzoate; O-ethyl O-2-isopropoxycarbonylphenyl isopropylphosphoramidothioate; isophenphos; Amaze; Bay 92114; SRA-12869; Lighter; Oftanol. $C_{15}H_{24}NO_4PS$; mol wt 345.39. C 52.16%, H 7.00%, N 4.06%, O 18.53%, P 8.97%, S 9.28%. Selective soil insecticide. Prepn: G. Schrader *et al.*, **FR 1600932**; *eidem*, **US 3621082** (1970, 1971 both to Bayer). Physical and biological properties: B. Homeyer, *Meded. Fac. Landbouwwet. Rijksuniv. Gent* **39**, 789 (1974). Determn of residues: M. J. Brown, I. H. Williams, *Pestic. Sci.* **7**, 545 (1976). MS determn: T. Cairns *et al.*, *Anal. Chem.* **56**, 2547 (1984). Persistence in soil: A. Felsot, *J. Environ. Sci. Health* **B19**, 13 (1984).

Oil, $bp_{0.01}$ 120°. d_4^{20} 1.339. Vapor pressure at 20°: 3×10^{-6} Torr; at 40°: 3.8×10^{-5} Torr. Soly in water at 20°: 23.8 ppm. Sol in dichloromethane, cyclohexanone, acetone, alc, ether, benzene. LD_{50} orally in rats, mice (mg/kg): 30-40, 90-130 (Homeyer).
USE: Insecticide.

5219. Isoflavone. [574-12-9] 3-Phenyl-4H-1-benzopyran-4-one; 3-phenylchromone. $C_{15}H_{10}O_2$; mol wt 222.24. C 81.07%, H 4.54%, O 14.40%. Prepn from o-hydroxyphenyl benzyl ketone, sodium dust and ethyl formate: Joshi, Venkataraman, *J. Chem. Soc.* **1934**, 513. Structure: Warburton, *Q. Rev. Chem. Soc.* **8**, 67 (1954).

Leaflets and needles from petr ether, mp 148°. uv max: 245, 307 nm (log ε 4.41, 3.82).

5220. Isoflupredone. [338-95-4] (11β)-9-Fluoro-11,17,21-trihydroxypregna-1,4-diene-3,20-dione; 1-dehydro-9α-fluorohydrocortisone; 9-fluoroprednisolone. $C_{21}H_{27}FO_5$; mol wt 378.44. C 66.65%, H 7.19%, F 5.02%, O 21.14%. Prepn: J. Fried *et al.*, *J. Am. Chem. Soc.* **77**, 4181 (1955). Microbiological prepn: A. Nobile *et al.*, *J. Am. Chem. Soc.* **77**, 4184 (1955); E. Vischer *et al.*, *Helv. Chim. Acta* **38**, 1502 (1955); T. H. Stoudt *et al.*, *Arch. Biochem. Biophys.* **59**, 304 (1955). Prepn of the 21-acetate: R. F. Hirschmann *et al.*, *ibid.* 3166; J. A. Hogg *et al.*, *ibid.* 4438; Ch. Meystre *et al.*, *Helv. Chim. Acta* **39**, 734 (1956); A. Wellstein *et al.*, **DE 1020329** (1957 to Ciba); **GB 843214** (1960 to Schering). Absorption spectra: L. L. Smith, W. H. Muller, *J. Org. Chem.* **23**, 960 (1958). Mass spectrometry: P. Toft *et al.*, *Can. J. Pharm. Sci.* **7**, 53 (1972). Urinary excretion in horses: D. I. Chapman *et al.*, *Vet. Rec.* **100**, 447 (1977).

Crystals from acetone, mp 263-266° (dec). $[\alpha]_D^{23}$ +108° (c = 0.611 in ethanol). uv max (ethanol): 240 nm (ε 15800) (Vischer), also reported as mp 274-275° (dec); $[\alpha]_D^{23}$ +94° (alcohol)(Fried).
21-Acetate. [338-98-7] U-6013; Predef. $C_{23}H_{29}FO_6$; mol wt 420.48. Crystals from acetone/isopropyl ether or methanol, mp 244-246° (dec). $[\alpha]_D^{23}$ +108° (c = 0.735 in dioxane). uv max (ethanol): 240 (ε 16250).
THERAP CAT (VET): Anti-inflammatory.

5221. Isoflurane. [26675-46-7] 2-Chloro-2-(difluoromethoxy)-1,1,1-trifluoroethane; 1-chloro-2,2,2-trifluoroethyl difluoromethyl ether; compd 469; Aerrane; Forane; Forene. $C_3H_2ClF_5O$; mol wt 184.49. C 19.53%, H 1.09%, Cl 19.22%, F 51.49%, O 8.67%. Prepn: Croix, Terrell, **DE 1814962** (1969); Terrell, **US 3535388**; **US 3535425** (both 1970 to Air Reduction); Terrell *et al.*, *J. Med. Chem.* **14**, 517 (1971). Series of articles on pharmacology: *Anesthesiology* **35**, 8-53 (1971); Byles *et al.*, *Can. Anaesth. Soc. J.* **18**, 376-407 (1971). Enantiomeric resolution: J. Meinwald *et al.*, *Science* **251**, 560 (1991).

Clear, colorless liquid having a slight odor. bp 48.5°. Vapor pressure at 25°: 330 mm. sp gr 1.45. n_D^{20} 1.3002. Nonflammable; soda lime stable. Easily miscible with organic liquids including fats and oils. Insol in water.
USE: Solvent and dispersant for fluorinated materials.
THERAP CAT: Anesthetic (inhalation).
THERAP CAT (VET): Anesthetic (inhalation).

5222. Isoflurophate. [55-91-4] Phosphorofluoridic acid bis(1-methylethyl) ester; phosphorofluoridic acid diisopropyl ester; isopropyl fluophosphate; diisopropyl fluorophosphonate; diisopropyl fluorophosphate; diisopropylphosphorofluoridate; fluostigmine; isofluorphate; DFP; Diflupyl; Dyflos; Floropryl; Fluropryl. $C_6H_{14}FO_3P$; mol wt 184.15. C 39.13%, H 7.66%, F 10.32%, O 26.06%, P 16.82%. Prepd by the action of PCl_3 on isopropanol, chlorinating the resulting intermediate, and converting the diisopropyl chlorophosphate by means of sodium fluoride: Hardy, Kosolapoff, **US 2409039** (1946); *see also* Edgewood Arsenal, *Chemical Warfare Service TDMR* **832** (April, 1944); **GB 601210** (1948); Lange, von Krueger, *Ber.* **65**, 1598 (1932); Saunders, Stacey, *J. Chem. Soc.* **1948**, 695; B. C. Saunders, *Some Aspects of the Chemistry and Toxic Action of Organic Compounds Containing Phosphorus and Fluorine*

(Cambridge, 1957) p 46. Toxicity data: R. G. Horton *et al.*, *J. Pharmacol. Exp. Ther.* **87**, 414 (1946).

Clear to pale yellow liquid. Forms HF in presence of moisture. d 1.055. mp −82°. bp$_5$ 46°; bp$_9$ 62°; bp$_{760}$ 183° (by extrapolation). Vapor pressure at 20°: 0.579 mm. n_D^{25} 1.3830. Soly in water at 25°: 1.54% w/w (dec; pH about 2.5). Sol in alc, vegetable oils. Not very sol in mineral oils. The anhydr compd or oil solns are stable in glass containers at room temp. LD$_{50}$ in mice (mg/kg): 3.71 s.c.; 36.8 orally (Horton).

Caution: Highly toxic; cholinesterase inactivator. Do not inhale vapor. Vapor is extremely irritating to eyes and mucous membranes. Avoid contact with skin.

THERAP CAT: Cholinergic (ophthalmic).

THERAP CAT (VET): Has been used as a miotic.

5223. L-Isoglutamine. [636-65-7] (4*S*)-4,5-Diamino-5-oxopentanoic acid; 4-amino-L-glutaramic acid; glutamic acid α-amide. $C_5H_{10}N_2O_3$; mol wt 146.15. C 41.09%, H 6.90%, N 19.17%, O 32.84%. Prepn by the action of NH_3 on *N*-carbobenzyloxy-L-glutamic acid anhydride: Kozo Narita, *J. Chem. Soc. Jpn. Pure Chem. Sect.* **74**, 832 (1953); from 1-tosylpyroglutamide by reduction with Na in NH_3: Swan, du Vigneaud, *J. Am. Chem. Soc.* **76**, 3110 (1954); from the γ-methyl ester of *N*-tritylglutamic acid: Amiard, Heymes, **US 2927118** (1960 to UCLAF).

Crystals from acetone, dec 181°. [α]$_D^{21}$ +20.5° (c = 6.1 in H_2O). pK$_1'$ 3.81; pK$_2'$ 7.88. Sol in water. Sparingly sol in organic solvents.

5224. Isolan. [119-38-0] *N,N*-Dimethylcarbamic acid 3-methyl-1-(1-methylethyl)-1*H*-pyrazol-5-yl ester; 1-isopropyl-3-methyl-5-pyrazolyl dimethylcarbamate; G-23611. $C_{10}H_{17}N_3O_2$; mol wt 211.27. C 56.85%, H 8.11%, N 19.89%, O 15.15%. Carbamate insecticide: Gerguson, Alexander, *J. Agric. Food Chem.* **1**, 288 (1953); Muller, Spindler, *Experientia* **10**, 91 (1954). Prepd by treating the K salt of the enol form of 1-isopropyl-3-methyl-5-pyrazolone with dimethylcarbamoyl chloride: **CH 279553** and **CH 282655; GB 681376** (all 1952 to Geigy); Kost, Sagitullin, *Zh. Obshch. Khim.* **33**, 867 (1963), *C.A.* **59**, 8724 (1963). Toxicity studies: T. B. Gaines *et al.*, *Nature* **209**, 88 (1966); T. B. Gaines, *Toxicol. Appl. Pharmacol.* **14**, 515 (1969).

Liquid, bp$_{0.7}$ 103°, bp$_{2.5}$ 117.5-118°. d 1.07. Vapor pressure at 20° = 0.001 mm Hg. Unlike most insecticides, Isolan is more toxic to rats dermally than orally (Gaines, 1966). LD$_{50}$ in male, female rats (mg/kg): 23, 13 orally; 5.6, 6.2 dermally (Gaines, 1969).

USE: Insecticide.

5225. Isoleucine. [73-32-5] L-Isoleucine; Ile; I; (2*S*,3*S*)-α-amino-β-methylvaleric acid; (2*S*,3*S*)-2-amino-3-methylpentanoic acid; *erythro*-3-methyl-L-norvaline. $C_6H_{13}NO_2$; mol wt 131.18. C 54.94%, H 9.99%, N 10.68%, O 24.39%. An essential amino acid for human development. Isoln from beet-sugar mother liquors: F. Ehrlich, *Ber.* **37**, 1809 (1904). Structure and identification of isomers: *idem, Ber.* **40**, 2538 (1907); J. P. Greenstein *et al., J. Biol. Chem.* **204**, 307 (1953). Early chemistry and biochemistry: *Amino*

Acids and Proteins, D. M. Greenberg, Ed. (Charles C. Thomas, Springfield, IL, 1951) 950 pp., *passim;* J. P. Greenstein, M. Winitz, *Chemistry of the Amino Acids* **vol 1-3** (John Wiley and Sons, Inc., New York, 1961) pp. 2043-2074, *passim.* TLC determ in serum from patients with maple sugar urine disease (MSDU): R. J. Allen *et al., Clin. Chem.* **18**, 413 (1972). Metabolism of Ile and isomers in MSDU: U. Wendel *et al., Pediatr. Res.* **25**, 11 (1989); and clinical consequences: T. Mogos *et al., Rev. Roum. Med. Interne* **32**, 57 (1994). Review of biosynthesis: M. Iaccarino *et al., Curr. Top. Cell. Regul.* **14**, 29-73 (1978).

Waxy, shiny, rhombic leaflets from alc. Bitter taste. Sublimes 168-170°. Dec 284°. [α]$_D^{20}$ +40.61° (c = 4.6 in 6.1*N* HCl); +11.09° (c = 3.3 in 0.33*N* NaOH). [M]$_D$ +53.5° (5*N* HCl); +64.2° (glacial acetic acid). pK$_1$ 2.36; pK$_2$ 9.68. Soly at 23.7°: 33.85 g/kg water. Sparingly sol in hot alc (0.13% w/w at 80°), hot acetic acid. Insol in ether.

DL-Form. [443-79-8] (±)-*erythro*-2-Amino-3-methylpentanoic acid. Glistening rhombic or monoclinic plates from dil alc. Dec 292°. pK$_1$ 2.32; pK$_2$ 9.76. Soly in water (g/l): 18.3 at 0°; 22.3 at 25°; 30.3 at 50°; 46.1 at 75°; 78.0 at 100°.

L-allo-Form. [1509-34-8] α-Ile; L-(+)-alloisoleucine; (+)-*threo*-2-amino-3-methylpentanoic acid; *threo*-3-methyl-L-norvaline. Waxy leaflets. Sweet taste. Dec 280°. [α]$_D^{20}$ +14.0° (c = 2); +38.1° (c = 2 in 6*N* HCl). [M]$_D$ +53.1° (5*N* HCl); +55.7° (glacial acetic acid). pK$_1$ 2.27; pK$_2$ 9.62. One part dissolves in 34.2 parts water at 20°; 0.82 part dissolves in 100 parts 80% alcohol at 20°; 1.97 parts dissolve in 100 parts 80% alcohol; 0.19 part dissolves in 100 parts abs alcohol at 90°.

5226. Isolysergic Acid. [478-95-5] (8α)-9,10-Didehydro-6-methylergoline-8-carboxylic acid. $C_{16}H_{16}N_2O_2$; mol wt 268.32. C 71.62%, H 6.01%, N 10.44%, O 11.93%. Parent compd of the ergotinine group of alkaloids from ergot. Prepn: Smith, Timmis, *J. Chem. Soc.* **1936**, 1440; Craig *et al., J. Biol. Chem.* **125**, 289 (1938). Differs from lysergic acid by the α-configuration at C-8. Stereochemical studies: Stenlake, *J. Chem. Soc.* **1955**, 1626; Leemann, Fabbri, *Helv. Chim. Acta* **42**, 2696 (1959).

Dihydrate from water, mp 218° (dec). [α]$_D^{20}$ +281° (pyridine). pKa 3.44, pKb 8.61. More sol in water and pyridine than lysergic acid.

Methyl ester. Rods from benzene, mp 170-174°. [α]$_D^{20}$ +179° (c = 0.5 in chloroform).

5227. Isomaltol. [3420-59-5] 1-(3-Hydroxy-2-furanyl)ethanone; 3-hydroxy-2-furyl methyl ketone. $C_6H_6O_3$; mol wt 126.11. C 57.15%, H 4.80%, O 38.06%. Trace constituent of bread. Obtained by the action of enzymes on starch: Backe, *Compt. Rend.* **151**, 78 (1910). Synthesis from α-lactose and a secondary amine: Hodge, Nelson, **US 3054805** (1962 to USDA). Structure: Hodge, Nelson, *Cereal Chem.* **38**, 207 (1961); Fischer, Hodge, *J. Org. Chem.* **29**, 776 (1964).

Crystals from water or ether, mp 98-103°. uv max (abs methanol): 280 nm (E$_{1cm}^{1\%}$ 1270). Sol in alcohols, acetone, chloroform, benzene, ethyl acetate, hot water. Practically insol in cold water and petr ether.

Benzoate. C$_{13}$H$_{10}$O$_4$. Crystals from benzene-toluene, mp 100-101°.

p-Nitrophenylhydrazone. C$_{12}$H$_{11}$N$_3$O$_4$. Deep red crystals from nitrobenzene, mp 100-101°.

5228. Isomaltulose. [13718-94-0]; [58024-13-8] (monohydrate). 6-*O*-α-D-Glucopyranosyl-D-fructose; 6-*O*-α-D-glucopyranosyl-D-fructofuranose; Palatinose. C$_{12}$H$_{22}$O$_{11}$; mol wt 342.30. C 42.11%, H 6.48%, O 51.41%. Reducing disaccharide occurring naturally in honey and sugar cane juice; commercially available in crystalline (42% of the sweetness of sucrose) and molasses (70% of the sweetness of sucrose) forms. Prepn: D. Weidenhagen, S. Lorenz, **DE 1049800** (1959 to Süddeutsche Zucker AG); *eidem*, *Z. Zuckerind.* **7**, 533 (1957); E. S. Sharpe *et al.*, *J. Org. Chem.* **25**, 1062 (1960). Clinical effects on glucose levels in diabetic patients: K. Kawai *et al.*, *Horm. Metab. Res.* **21**, 338 (1989). Review of cariological studies: D. Birkhed *et al.*, *Dtsch. Zahnaerztl. Z.* **42**, S124-S127 (1987). Review of development and properties: W. E. Irwin, P. J. Sträter in *Alternative Sweeteners*, L. O'Brien Nabors, R. C. Gelardi, Eds. (Marcel Dekker, Inc., New York, 1991) pp 299-307; and production and clinical activity: I. Takazoe in *Progress in Sweeteners*, T. H. Grenby, Ed. (Elsevier, Barking, UK, 1989) pp 143-167. Review of toxicology and metabolism: B. A. R. Lina *et al.*, *Food Chem. Toxicol.* **40**, 1375-1381 (2002).

Crystals as monohydrate, mp 123-124°. [α]$_D^{20}$ +97.2° (Weidenhagen, 1957); also reported as [α]$_D^{20}$ +103° (c = 1.9 in water) (Sharpe). Nonhygroscopic, acid stable, low glycemic index.

USE: Non-cariogenic sweetener; sugar substitute in foods and beverages.

5229. Isometamidium Chloride. [34301-55-8] 3-Amino-8-[3-[3-(aminoiminomethyl)phenyl]-2-triazen-1-yl]-5-ethyl-6-phenylphenanthridinium chloride (1:1); 8-[3-(*m*-amidinophenyl)-2-triazeno]-3-amino-5-ethyl-6-phenylphenanthridinium chloride; 7-*m*-amidinophenyldiazoamino-2-amino-10-ethyl-9-phenylphenanthridinium chloride; M & B 4180 A; Samorin. C$_{28}$H$_{26}$ClN$_7$; mol wt 496.02. C 67.80%, H 5.28%, Cl 7.15%, N 19.77%. Prepn: Wragg *et al.*, *Nature* **182**, 1005 (1958). Structure: Berg, *ibid.* **188**, 1106 (1960).

Red crystals from aq methanol, dec 244-245°.

THERAP CAT (VET): Antitrypanosomal agent.

5230. Isomethadone. [466-40-0] 6-(Dimethylamino)-5-methyl-4,4-diphenyl-3-hexanone; 6-dimethylamino-4,4-diphenyl-5-methyl-3-hexanone; 1-dimethylamino-2-methyl-3,3-diphenyl-4-hexanone; isoamidone. C$_{21}$H$_{27}$NO; mol wt 309.45. C 81.51%, H 8.79%, N 4.53%, O 5.17%. Synthetic opioid analgesic. Prepn of *dl*-form: Easton *et al.*, *J. Am. Chem. Soc.* **70**, 76 (1948); Walton *et al.*, *J. Chem. Soc.* **1949**, 648; Larsen, Tullar, **US 2773901** (1956 to

Sterling Drug). Prepn of *d*- and *l*-forms and resolution of *dl*-form: Larsen *et al.*, *J. Am. Chem. Soc.* **70**, 4194 (1948); Howe, Sletzinger, *ibid.* **71**, 2935 (1949); Larsen, Tullar, *loc. cit.* Configuration of *l*-form, the isomer with greater analgesic activity: Beckett *et al.*, *Chem. Ind. (London)* **1960**, 1418. Pharmacology: Eddy *et al.*, *J. Pharmacol. Exp. Ther.* **98**, 121 (1950); Winter, Flataker, *ibid.* 305.

dl-**Form.** [116836-09-0] Slightly yellow, very viscous liq, bp$_{12}$ 215°.

dl-**Form hydrobromide.** C$_{21}$H$_{27}$NO.HBr. Crystals from water, mp 149-150°.

d-**Form.** [26594-41-2] Oil, [α]$_D^{25}$ +20.8°.

d-**Form hydrochloride.** [63814-06-2] C$_{21}$H$_{27}$NO.HCl. mp 231-232°. [α]$_D^{25}$ +70°.

d-**Form hydrochloride monohydrate.** C$_{21}$H$_{27}$NO.HCl.H$_2$O, mp 176-177°. [α]$_D^{25}$ +66°.

l-**Form.** [561-10-4] Oily liq, bp$_{0.6}$ 162-165°. [α]$_D^{25}$ −20° (c = 1.5 in 95% ethanol).

l-**Form hydrobromide.** Crystals, mp 217-218°. [α]$_D^{25}$ −59° (c = 1.5).

l-**Form hydrochloride.** [7487-81-2] mp 231-233°. [α]$_D^{25}$ −70° (c = 1.5 in water), −90° (methanol). Sol in water, alcohol. pH of 1% aq soln: 5 to 6.5. Solutions and tablets are stable. LD$_{50}$ s.c. in mice: 21 mg/kg (Winter, Flataker).

l-**Form hydrochloride monohydrate.** mp 173-174°.

Note: This is a controlled substance (opiate): **21 CFR**, 1308.12.

5231. Isometheptene. [503-01-5] *N*,6-Dimethyl-5-hepten-2-amine; *N*,1,5-trimethyl-4-hexenylamine; 6-methylamino-2-methylheptene; 2-methyl-6-methylamino-2-heptene; methylisooctenylamine; methyloctenylamine; Octin; Octon; Octanil. C$_9$H$_{19}$N; mol wt 141.26. C 76.52%, H 13.56%, N 9.92%. Prepn: Klavehn, Wolf, **US 2230753**; **US 2230754** (both 1941 to E. Bilhuber). Toxicity data: Walton *et al.*, *J. Pharmacol. Exp. Ther.* **92**, 214 (1948).

Colorless, oily liq; characteristic amine odor. Strong base. Volatile with steam. d 0.795. bp 176-178°; bp$_7$ 58-59°. n$_D^{15}$ 1.4472. Practically insol in water. Freely sol in alc, ether, acetone, chloroform.

Hydrochloride. [6168-86-1] C$_9$H$_{19}$N.HCl; mol wt 177.72. Crystals, very hygroscopic, mp 68-69°. Sol in water, alc. LD$_{50}$ in mice (mg/kg): 17.5 i.v.; 171 s.c. (Walton).

Bitartrate (acid tartrate). [5984-50-9] C$_{13}$H$_{25}$NO$_6$; mol wt 291.34. Crystals, mp 78-80° (preliminary sintering). Freely sol in water, alcohol.

Mucate. [7492-31-1] C$_{24}$H$_{48}$N$_2$O$_8$; mol wt 492.65. Bitter crystals, mp 152° (rapid heating). Freely sol in water; soly in alc about 5 g/100 ml. Practically insol in ether, chloroform. LD$_{50}$ orally in dogs: 148 mg/kg (Walton).

THERAP CAT: Adrenergic.

THERAP CAT (VET): Sympathomimetic. Antispasmodic for gut and urinary tract.

5232. Isoniazid. [54-85-3] 4-Pyridinecarboxylic acid hydrazide; isonicotinic acid hydrazide; isonicotinoylhydrazine; isonicotinylhydrazine; INH; rimitsid; tubazid; RP-5015; FSR-3; Isonex; Isotamine; Isozid; Neoteben; Nicizina; Nicozid; Nydrazid; Rimifon; Tebesium; Tibinide; Valifol. C$_6$H$_7$N$_3$O; mol wt 137.14. C 52.55%, H 5.15%, N 30.64%, O 11.67%. Prepn: Meyer, Mally, *Monatsh. Chem.* **33**, 400 (1912); Lock, *Pharm. Ind.* **14**, 366 (1952); Urbanski *et al.*, *Rocz. Chem.* **27**, 161 (1953); Gasson, **US 2830994** (1958 to Distillers). Compositions for combating tuberculosis: H. H. Fox,

US 2596069 (1952 to Hoffmann-La Roche). Pharmacokinetics: W. W. Weber, D. W. Hein, *Clin. Pharmacokinet.* **4**, 401 (1979). Mechanism of action study: K. Johnsson *et al.*, *J. Am. Chem. Soc.* **117**, 5009 (1995). Toxicity data: E. H. Jenney, C. C. Pfeiffer, *J. Pharmacol. Exp. Ther.* **122**, 110 (1958). Evaluation of carcinogenic risk: *IARC Monographs* **4**, 159 (1974). Comprehensive description: G. A. Brewer, *Anal. Profiles Drug Subs.* **6**, 183-258 (1977). Clinical trial in HIV-positive patients: J. W. Pape *et al.*, *Lancet* **342**, 268 (1993). Review of use in prevention and treatment of TB: American Thoracic Society, *Am. Rev. Respir. Dis.* **134**, 355-363 (1986); T. J. Jordan *et al.*, *ibid.* **144**, 1357-1360 (1991).

Crystals from alc, mp 171.4°. uv max (water): 266 nm (E$_{1cm}^{1\%}$ 378); (0.01*N* HCl): 265 nm (E$_{1cm}^{1\%}$ ~420). Soly in water at 25°: about 14%; at 40°: about 26%; in alc at 25°: about 2%, in boiling alc: about 10%; in chloroform: about 0.1%. Practically insol in ether, benzene. pH of a 1% aq soln 5.5 to 6.5. Aqueous solns may be sterilized at 120° for 30 min. LD$_{50}$ in mice (mg/kg): 151 i.p., 149 i.v. (Jenney, Pfeiffer).

4-Aminosalicylate. [2066-89-9] Pasiniazide; GEWO-399. C$_{13}$H$_{14}$N$_4$O$_4$; mol wt 290.28. An equimolecular salt produced by dissolving the components in hot dil alc, followed by cooling and evapn: Charonnat, Boime, *Compt. Rend.* **236**, 2140 (1953). Prepn: **CH 303085** (1955 to Hoffmann-La Roche). Yellow crystals from methanol or ethanol, mp 142-144°. Sparingly sol in water. uv max 272, 303 nm (E$_{1cm}^{1\%}$ 550, 445).

Methanesulfonate. [13447-95-5] 4-Pyridinecarboxylic acid 2-(sulfomethyl)hydrazide; methaniazide. C$_7$H$_9$N$_3$O$_4$S; mol wt 231.23. Prepn: Logemann, **US 2759944** (1956 to Carlo Erba). Crystals, dec 187-189°.

Methanesulfonate sodium. [3804-89-5] Sodium isonicotinylhydrazide methanesulfonate. C$_7$H$_8$N$_3$NaO$_4$S; mol wt 253.21. Yellow crystals from water, dec 164-167°.

THERAP CAT: Antibacterial (tuberculostatic).

THERAP CAT (VET): Antibacterial (tuberculostatic); anti-actinomycotic agent.

5233. Isonicotinic Acid. [55-22-1] 4-Pyridinecarboxylic acid; γ-pyridinecarboxylic acid; γ-picolinic acid. C$_6$H$_5$NO$_2$; mol wt 123.11. C 58.54%, H 4.09%, N 11.38%, O 25.99%. Prepd by permanganate oxidation of γ-picoline: Gilman, Broadbent, *J. Am. Chem. Soc.* **70**, 2757 (1948); Fields, **US 2946801** (1960 to Standard Oil); from 4-ethylpyridine: Wibaut, Arens, *Rec. Trav. Chim.* **60**, 137 (1941); from citric acid to citrazinic acid to 2,6-dichloroisonicotinic acid: Behrmann, Hofmann, *Ber.* **17**, 2681 (1884); Wibaut, *Rec. Trav. Chim.* **63**, 141 (1944). Prepn of ethyl ester: Burrus, Powell, *J. Am. Chem. Soc.* **67**, 1469 (1945).

Platelets, mp 319°. Sublimes at 260° and 15 mm pressure. pK (25°) 4.96. pH of satd aq soln 3.6. Sparingly sol in cold water (0.52 g/100 ml at 20°), more sol in hot water. Practically insol in benzene, ether, boiling alcohol.

Methyl ester. C$_7$H$_7$NO$_2$. Liq, slight odor, similar to mint or oil of wintergreen. mp 8.5°. bp$_{21}$ 104°; bp$_{760}$ 209° (slight decompn).

Ethyl ester. Liq, ester-like odor. Needles, when cooled by salt-ice mixture, mp 23°. d$_4^{15}$ 1.0091 (liq). bp$_5$ 78.5°; bp$_{15}$ 110°; bp$_{760}$ 220°. Sol in alcohol, ether, chloroform, benzene. Practically insol in water.

5234. Isonicotinic Acid Diethylamide. [530-40-5] *N*,*N*-Diethyl-4-pyridinecarboxamide; *N*,*N*-diethylisonicotinamide; pyridine-4-carboxylic acid diethylamide. C$_{10}$H$_{14}$N$_2$O; mol wt 178.24.

C 67.39%, H 7.92%, N 15.72%, O 8.98%. Prepd by the action of phosphorus oxychloride upon a mixture of isonicotinic acid and diethylamine: Carrara, **US 2858317** (1958 to Lepetit).

Slightly viscous liq. bp$_{1.0}$ 119-120°. n_D^{20} 1.525. Miscible with water, ether, chloroform, acetone, alcohol.

5235. Isonipecotic Acid. [498-94-2] 4-Piperidinecarboxylic acid; hexahydroisonicotinic acid. C$_6$H$_{11}$NO$_2$; mol wt 129.16. C 55.80%, H 8.58%, N 10.84%, O 24.77%. Prepd by reduction of isonicotinic acid in glacial acetic acid in the presence of platinum oxide: Wibaut, *Rec. Trav. Chim.* **63**, 141 (1944); by reduction of isonicotinic acid in H$_2$O and concd HCl with PtO$_2$: Sperber *et al.*, **US 2739968** (1956 to Schering); Freifelder, **US 3159639** (1964 to Abbott).

Needles. Darkens at about 300°; mp 336°.

Hydrochloride. C$_6$H$_{11}$NO$_2$.HCl. mp 300° (Wibaut); mp 293° with decompn (Sperber). Sol in ethanol and methanol; freely sol in water.

5236. Isonitrosoacetone. [306-44-5] 2-Oxopropanal 1-oxime; methylglyoxal aldoxime; pyruvaldoxime; propanone 1-oxime. C$_3$H$_5$NO$_2$; mol wt 87.08. C 41.38%, H 5.79%, N 16.09%, O 36.75%. Prepd from acetone by treatment with sodium nitrite in acetic acid at 0°: Küster, *Z. Physiol. Chem.* **155**, 174 (1926); by the action of methyl nitrite on acetone in ether in the presence of HCl: Slater, *J. Chem. Soc.* **117**, 589 (1920).

Leaflets from ether + petr ether or from carbon tetrachloride, mp 69°. Sublimes easily forming shiny needles. Volatile with steam. Freely sol in water, ether. Moderately sol in warm chloroform, benzene, carbon tetrachloride. Practically insol in petr ether. Sol in alkalies forming yellow solns. Faintly acid to litmus. pK (25°) 8.39.

5237. Isonitrosoacetophenone. [532-54-7] α-Oxobenzeneacetaldehyde aldoxime; phenylglyoxal 2-oxime; benzoylformaldoxime. C$_8$H$_7$NO$_2$; mol wt 149.15. C 64.42%, H 4.73%, N 9.39%, O 21.45%.

Plates or prisms, mp 126-128°. Slightly sol in cold water; more sol in hot water; sol in alkalies and alkali carbonates.

USE: As a reagent for the detection of ferrous ions with which it gives a blue color sol in chloroform.

5238. Isonixin. [57021-61-1] *N*-(2,6-Dimethylphenyl)-1,2-dihydro-2-oxo-3-pyridinecarboxamide; 2-pyridone-3-carboxylic acid 2,6-xylidide; 2-hydroxy-2',6'-nicotinoxylidide; Nixyn. C_{14}-$H_{14}N_2O_2$; mol wt 242.28. C 69.40%, H 5.82%, N 11.56%, O 13.21%. Non-steroidal anti-inflammatory agent. Prepn: J.-A. Canicio Chimeno, **BE 820578**; *idem*, **US 4031105** (1975, 1977 both to Hermes). Prepn, pharmacology and toxicity: R. Cadena *et al.*, *Arzneim.-Forsch.* **27**, 1457 (1977). Toxicological study: M. T. Mitjavila *et al.*, *ibid.* 1460. Absorption, excretion studies: G. Carrera *et al.*, *ibid.* **29**, 1401 (1979). Tolerance in humans: T. M. Serra *et al.*, *Therapie* **35**, 173 (1980). HPLC determn in plasma: F. González Lopez *et al.*, *Farmaco Ed. Prat.* **38**, 273 (1983).

White crystalline powder from ethanol, mp 266-267°. Sol in chloroform and strong alkali. Practically insol in acids and water. LD_{50} in male, female mice, male, female rats (mg/kg): 7000, 8000, >6000, >6000 orally; all >2000 i.p. (Cadena).

THERAP CAT: Analgesic, anti-inflammatory.

5239. Isooctane. [540-84-1] 2,2,4-Trimethylpentane; isobutyltrimethylmethane. C_8H_{18}; mol wt 114.23. C 84.12%, H 15.88%. Produced by the refining of petroleum.

Mobile liquid, odor of gasoline. *Highly flammable.* Antiknock octane no. 100. mp −107.45°. bp 99.3°. d_4^{20} 0.69194. n_D^{20} 1.39157. Dipole moment: 0. Flash pt, closed cup: 10°F (−12°C). Practically insol in water; somewhat sol in abs alcohols; sol in benzene, toluene, xylene, chloroform, ether, carbon disulfide, carbon tetrachloride, DMF and oils, except castor oil.

Note: The name isooctane has also been applied to 2-methylheptane.

USE: In determining octane numbers of fuels; in spectrophotometric analysis; as solvent and thinner.

5240. Isooctyl Alcohol. [26952-21-6] Isooctanol. $C_8H_{18}O$; mol wt 130.23. C 73.78%, H 13.93%, O 12.29%. A mixture of closely related isomeric branched-chain primary alcohols: RCH_2OH where R represents a branched heptyl radical. The branching consists mostly of methyl groups located in the 3-, 4-, or 5-positions.

Caution: Potential symptoms of overexposure are irritation of eyes, skin, nose, throat; eye and skin burns. *See NIOSH Pocket Guide to Chemical Hazards* (DHHS/NIOSH 97-140, 1997) p 178.

5241. Isopentyl Alcohol. [123-51-3] 3-Methyl-1-butanol; isoamyl alcohol; isobutyl carbinol; primary isoamyl alcohol; fermentation amyl alcohol. $C_5H_{12}O$; mol wt 88.15. C 68.13%, H 13.72%, O 18.15%. A major component of commercial amyl alcohol, fusel oil, *q.v.*, and ***potato-spirit oil.*** Prepn from butadiene, CO, H_2O, plus catalyst: Alderson, **US 3020314** (1962 to du Pont); from methylbutenes: Brown, Zweifel, *J. Am. Chem. Soc.* **82**, 1504 (1960). Early review of metabolism and toxicity: H. W. Haggard *et al.*, *J. Ind. Hyg. Toxicol.* **27**, 1 (1945). Review of manuf by fractionation of fusel oil and via chlorination of pentanes, and properties: W. L. Faith *et al.*, Eds., *Industrial Chemicals* (John Wiley, New York, 2nd ed., 1957) pp 107-114. Acute toxicity data: Smyth *et al.*, *Am. Ind. Hyg. Assoc. J.* **30**, 470 (1969).

Liquid; characteristic, disagreeable odor; pungent, repulsive taste. bp_{760} 130.5°. mp −117.2°. d_4^{15} 0.813. d^{25} 0.80631. n_D^{20} 1.4075. n_D^{25} 1.40519. Flash pt, closed cup: 114°F (45°C); open cup: 132°F (55°C). Slightly sol in water (2 g/100 ml at 14°); misc with alcohol,

ether, benzene, chloroform, petr ether, glacial acetic acid, oils. LD_{50} orally in rats: 7.07 ml/kg (Smyth).

Caution: Potential symptoms of overexposure are irritation of eyes, skin, nose and throat; headache, dizziness; coughing, dyspnea, nausea, vomiting and diarrhea; skin cracking. *See NIOSH Pocket Guide to Chemical Hazards* (DHHS/NIOSH 97-140, 1997) p 174. *See also Patty's Industrial Hygiene and Toxicology* **vol. 2C**, G. D. Clayton, F. E. Clayton, Eds. (Wiley-Interscience, New York, 3rd ed., 1982) pp 4588-4599.

USE: Organic solvent for fats, resins, alkaloids, etc.; manuf isoamyl (amyl) compds, isovaleric acid, mercury fulminate, pyroxylin, artificial silk, lacquers, smokeless powders; in microscopy; for dehydrating celloidin solns; for determining fat in milk.

5242. Isophorone. [78-59-1] 3,5,5-Trimethyl-2-cyclohexen-1-one; α-isophorone; 1,5,5-trimethyl-3-oxocyclohexene; isoforon; isoacetophorone. $C_9H_{14}O$; mol wt 138.21. C 78.21%, H 10.21%, O 11.58%. Prepn: E. Knoevenagel, *Ann.* **297**, 113 (1897). Synthesis: H. Ueda *et al.*, *Agric. Biol. Chem.* **30**, 1004 (1966); D. S. Torok, W. J. Scott, *Tetrahedron Lett.* **34**, 3067 (1993). Colorimetric field test for determn in air: P. Andrew, R. Wood, *Analyst* **95**, 691 (1970). Solid-phase microextraction/GC determn in aq. samples: J.-Y. Horng, S.-D. Huang, *J. Chromatogr. A* **678**, 313 (1994). Brief description: P. D. Sherman, Jr., A. J. Papa in *Kirk-Othmer Encyclopedia of Chemical Technology* **vol. 13** (Wiley-Interscience, New York, 3rd ed., 1981) pp 918-922. Review of prepn and purification: G. S. Salvapati, M. Janardanarao, *J. Sci. Ind. Res.* **42**, 261-267 (1983). Review of toxicology and human exposure: *Toxicological Profile for Isophorone* (PB90-180225, 1989) pp 101.

Clear liquid with peppermint-like odor. bp 215.3°. Freezing pt −8.1°. Flash pt (open cup): 184°F (84°C). Autoignition pt: 864°F (462°C). d_{20}^{20} 0.9229. d_4^{20} 0.9613. n_D^{20} 1.4778. uv max (MeOH): 235.5 nm (ε 14300). Vapor pressure at 20°: 0.3 mm Hg. Soly in water (mg/l): 12000 (20°); 14500 (25°). Sol in ether, acetone alcohol. LD_{50} in male, female rats and male mice (mg/kg): 2700 ±200, 2100 ±200, 2200 ±200 orally (PB90-180225).

Caution: Potential symptoms of overexposure are irritation of eyes, nose, throat; headache, nausea, dizziness, fatigue, malaise, narcosis; dermatitis. *See NIOSH Pocket Guide to Chemical Hazards* (DHHS/NIOSH 97-140, 1997) p 178.

USE: Solvent in some printing inks, paints, lacquers and adhesives.

5243. Isophthalic Acid. [121-91-5] 1,3-Benzenedicarboxylic acid; *m*-phthalic acid. $C_8H_6O_4$; mol wt 166.13. C 57.84%, H 3.64%, O 38.52%. Obtained by the oxidation of *m*-xylene: Smith, *J. Am. Chem. Soc.* **43**, 1920 (1921); Weisemann, Fragen, *Proc. 5th World Petrol. Congr.* **4**, 197 (1960); Brill, *Ind. Eng. Chem.* **52**, 837 (1960); Bhattacharyya, Ganguly, *J. Indian Chem. Soc.* **38**, 463 (1961); Saffer, Barker, **US 3089906** (1963 to Mid-Century Corp.); Hay, **GB 951192** (1964 to General Electric).

Cryst powder, mp 345-348°; sublimes without decompn. Sol in 8000 parts cold water, 460 boiling water; freely sol in alcohol, glacial acetic acid; practically insol in benzene, petr ether.

5244. Isophytol. [505-32-8] 3,7,11,15-Tetramethyl-1-hexadecen-3-ol; 2,6,10,14-tetramethylhexadec-15-en-14-ol; 2,6,10-trimethyl-14-vinylpentadecan-14-ol. $C_{20}H_{40}O$; mol wt 296.54. C 81.01%, H 13.60%, O 5.40%. Decompn product of chlorophyll.

Synthesis from linalool or citral: Fischer, Löwenberg, *Ann.* **475**, 183 (1929); Karrer *et al.*, *Helv. Chim. Acta* **26**, 1741 (1943); Sarycheva *et al.*, *Zh. Obshch. Khim.* **28**, 647 (1958); Maurit *et al.*, *ibid.* **32**, 2483 (1962); from acetylene: Nazarov *et al.*, *ibid.* **28**, 1444 (1958); Blaha, Weichet, *CS* **88887** (1959), *C.A.* **54**, 2167c (1960); from pseudoionone and propargyl alcohol: Sato *et al.*, *J. Org. Chem.* **28**, 45 (1963).

Oily liquid. d_4^{20} 0.8519. n_D^{20} 1.4571. $bp_{0.01}$ 107-110°; $bp_{0.06}$ 125-128°. Practically insol in water; sol in the usual organic solvents.
USE: Prepn of vitamins E and K_1.

5245. Isopilosine. [491-88-3] (3*S*,4*R*)-Dihydro-3-[(*R*)-hydroxyphenylmethyl]-4-[(1-methyl-1*H*-imidazol-5-yl)methyl]-2(3*H*)-furanone; carpiline; carpidine. $C_{16}H_{18}N_2O_3$; mol wt 286.33. C 67.12%, H 6.34%, N 9.78%, O 16.76%. This compound was originally called pilosine (the *cis*-isomer of isopilosine): H. W. Voigtländer, W. Rosenberg, *Arch. Pharm.* **292**, 579 (1959). Isoln from leaves of *Pilocarpus microphyllus* Stapf, *Rutaceae:* F. L. Pyman, *J. Chem. Soc.* **101**, 2260 (1912). Synthesis and abs config: Link, Bernauer, *Helv. Chim. Acta* **55**, 1053 (1972). Crystal structure: W. E. Oberhaensli, *Cryst. Struct. Commun.* **1**, 203 (1972).

Needles from alcohol, mp 182-182.5°. $[\alpha]_D^{20}$ +37.6° (alcohol).
Pilosine. [13640-28-3] Needles from alcohol, mp 179°. $[\alpha]_D^{20}$ +83.9° (alcohol). uv max (alcohol): 210 nm (log ε 4.10).

5246. Isopimaric Acid. [5835-26-7] (1*R*,4a*R*,4b*S*,7*S*,10a*R*)-7-Ethenyl-1,2,3,4,4a,4b,5,6,7,8,10,10a-dodecahydro-1,4a,7-trimethyl-1-phenanthrenecarboxylic acid; 13β-methyl-13-vinylpodocarp-7-ene-15-oic acid; miropinic acid. $C_{20}H_{30}O_2$; mol wt 302.46. C 79.42%, H 10.00%, O 10.58%. Isoln from *Dacrydium biforme* Pilg., *Podocarpaceae:* Hosking, Brandt, *Ber.* **68**, 1311 (1935); from the bled resin of *Podocarpus ferrugineus*, *Podocarpaceae:* Brandt, Neubauer, *J. Chem. Soc.* **1940**, 683; from *Pimus palustris* Mill., *Pinaceae:* Harris, Sanderson, *J. Am. Chem. Soc.* **70**, 2079, 2081 (1948). Identity of isopimaric and miropinic acids: Brossi, Jeger, *Helv. Chim. Acta* **33**, 722 (1950). Structure: Antkowiak *et al.*, *J. Org. Chem.* **27**, 1930 (1962). Stereochemistry: Ireland, Newbould, *ibid.* 1931; Antkowiak *et al.*, *Can. J. Chem.* **43**, 1257 (1965); Bose *et al.*, *Indian J. Chem.* **5**, 228 (1967).

Needles from methanol or ethanol, mp 160°. $[\alpha]_D^{16}$ −3.6° (10.4% soln in 1:1 alcohol:chloroform). Freely sol in chloroform, benzene; moderately sol in ether, alcohol; slightly sol in light petroleum. Absorption spectrum: Brossi, Jeger.

5247. Isoprene. [78-79-5] 2-Methyl-1,3-butadiene. C_5H_8; mol wt 68.12. C 88.16%, H 11.84%. Isoln from the products of the pyrolysis of natural rubber: Williams, *Trans. R. Soc. London* **150**, 241 (1860); Boonstra, van Amerongen, *Ind. Eng. Chem.* **41**, 161 (1949). Structure: Euler, *Ber.* **30**, 1989 (1897). Prepn from turpentine: Tilden, *J. Chem. Soc.* **45**, 410 (1884); Bibb, *US* **2386537** (1945

to Newport Ind.); from terpenes: Davis *et al.*, *Ind. Eng. Chem.* **38**, 53 (1946); Bourbon, *US* **2547684** (1951 to Manufacture de Caoutchouc Michelin); by condensation of isobutylene and formaldehyde in acetic acid: Blomquist, Verdol, *J. Am. Chem. Soc.* **77**, 78 (1955); Chaffe *et al.*, *FR* **1294716** (1962 to Inst. Francais du Petrole); by dehydrogenation of isoamylenes from cracked gasolines: **GB 875346** (Apr. 21, 1960 to Shell Int. Res. Maatschappij); Voge *et al.*, **US 3110746** (1963 to Shell); by dimerization of propylene followed by pyrolysis of the resulting 2-methyl-2-pentene: Gorin, Oblad, **US 2404056** (1946 to Socony-Vacuum Oil); **GB 913852** (1962 to Goodyear); by dehydrogenation of isopentane: Dempsey, **US 2914588** (1959 to Sun Oil); Owen, **US 2982795** (1961 to Phillips Petroleum); Stevenson, **US 3088986** (1963 to Air Products & Chem.); by reacting isoamylenes with methanol and pyrolizing and dehydrogenating the resulting tertiary ethers: Verdol, Walker, **US 2972645** (1961 to Sinclair Refining). Toxicity: Gostinskii, *C.A.* **62**, 15338a (1965). Reviews of isoprene and its polymers: Bean *et al.*, "Isoprene Polymers" in *Encyclopedia of Polymer Science and Technology* vol. 7 (Interscience, New York, 1967) pp 782-855; Bailey, "Isoprene" in *Vinyl and Diene Monomers* (part 2), E. C. Leonard, Ed. (Wiley-Interscience, New York, 1971) pp 997-1148; W. M. Saltman in *Kirk-Othmer Encyclopedia of Chemical Technology* vol. 13 (Wiley-Interscience, New York, 3rd ed., 1981) pp 818-837; of carcinogenic risk: *IARC Monographs* **60**, 215-232 (1994).

Colorless, volatile liquid. Unstable, oxidizable. *Flammable.* bp_{760} 34.067°. mp −145.95°. d_4^{20} 0.681; d_{20}^{20} 0.6805. n_D^{20} 1.42160. Practically insol in water. Miscible with alcohol or ether. LD_{50} for mice: 144 mg isoprene vapors/l air (Gostinskii).
Caution: Respiratory irritation is a potential symptom of overexposure. May act as a CNS depressant and asphyxiant at high concns. See *Patty's Industrial Hygiene and Toxicology* vol. 2B, G. D. Clayton, F. E. Clayton, Eds. (John Wiley & Sons, New York, 4th ed., 1994) p 1252-1253. This substance is reasonably anticipated to be a human carcinogen: *Report on Carcinogens, Twelfth Edition* (PB2011-111646, 2011) p 247.
USE: Manuf "synthetic" natural rubber, butyl rubber. Copolymer in the production of synthetic elastomers.

5248. Isopropamide Iodide. [71-81-8] γ-(Aminocarbonyl)-*N*-methyl-*N*,*N*-bis(1-methylethyl)-γ-phenylbenzenepropanaminium iodide (1:1); (3-carbamoyl-3,3-diphenylpropyl)diisopropylmethylammonium iodide; 2,2-diphenyl-4-diisopropylaminobutyramide methiodide; R-79; Darbid; Priamide; Tyrimide. $C_{23}H_{33}IN_2O$; mol wt 480.43. C 57.50%, H 6.92%, I 26.41%, N 5.83%, O 3.33%. Anticholinergic. Prepn: P. Janssen *et al.*, *Arch. Int. Pharmacodyn.* **103**, 82 (1955); M. E. Speeter, **US 2823233** (1958 to Bristol). Crystal and molecular structure: N. Datta *et al.*, *J. Chem. Soc. Perkin Trans. 2* **1977**, 781. Comprehensive description: R. S. Santoro *et al.*, *Anal. Profiles Drug Subs.* **2**, 315-338 (1973); A. Post, R. S. Santoro, *ibid.* **12**, 721-732 (1983).

Crystals or amorphous powder, mp 198-201° (dec) (Janssen); also reported as mp 189.0-191.5° (Santoro). Sensitive to light. Freely sol in boiling water, in methanol, ethanol, chloroform; sparingly sol in water; very slightly sol in benzene, ether.
Free base. $C_{22}H_{30}N_2O$. mp 84-86°.
Note: Ingredient of **Stelabid** (also contg trifluoperazine, *q.v.*).
THERAP CAT: Antispasmodic.
THERAP CAT (VET): Antiemetic; antidiarrheal.

5249. **Isopropenyl Acetate.** [108-22-5] 1-Propen-2-ol 2-acetate; 1-propen-2-yl acetate. $C_5H_8O_2$; mol wt 100.12. C 59.98%, H 8.05%, O 31.96%. Prepn from ketene + acetone: Hull, Agett; Hagemeyer, US 2481669; US 2476860 (both 1949 to Eastman Kodak); Young, US 2461016; US 2511423 (1949, 1950 both to Carbide & Carbon Chem.); Mawer, US 2684980 (1954 to I.C.I.); Buttner, Enk, US 2867653 (1959 to Wacker-Chemie). Use as reagent for acylation of potential enols: Hagemeyer, Hull, *Ind. Eng. Chem.* **41**, 2920 (1949); Hull, Agett, US 2482066 (1949 to Eastman Kodak); Jeffery, Satchell, *J. Chem. Soc.* **1962**, 1876, 1906. Toxicity study: Smyth *et al., J. Ind. Hyg. Toxicol.* **31**, 60 (1949).

Liquid. bp 97°; bp_{200} 58-60°. n_D^{20} 1.4001. LD_{50} orally in rats: 3.0 g/kg (Smyth).

Caution: Mild irritant. Narcotic in high concns.

USE: Reagent for acylation of potential enols.

5250. **4-(5-Isopropenyl-2-methyl-1-cyclopenten-1-yl)-2-butanone.** [87-45-6] 4-[2-Methyl-5-(1-methylethenyl)-1-cyclopenten-1-yl]-2-butanone; 4-(2-methyl-5-isopropenyl-1-cyclopenten-1-yl)-2-butanone; Pentione. $C_{13}H_{20}O$; mol wt 192.30. C 81.20%, H 10.48%, O 8.32%. Prepn: Kimel, Sax, US 2799706 (1957 to Hoffmann-La Roche).

Pale yellow liquid. Citrus-like odor with woody background. d_{25}^{25} 0.9218. $bp_{0.5}$ 72-75°. n_D^{25} 1.4800. Freely sol in diethyl phthalate. At least 10% soly in 70% ethanol, in mineral oil and in corn oil.

USE: In perfumery, i.e., in milled soaps, in modified eau de Cologne, in neroli and orange-blossom compositions. Blends well with ionones.

5251. **Isopropyl Acetate.** [108-21-4] Acetic acid 1-methylethyl ester; 2-acetoxypropane; 1-methylethyl acetate. $C_5H_{10}O_2$; mol wt 102.13. C 58.80%, H 9.87%, O 31.33%. Toxicity study: Smyth *et al., Arch. Ind. Hyg. Occup. Med.* **10**, 61 (1954).

Colorless liquid. d_4^{20} 0.870. bp 89°. Flash pt, open cup: 40°F (4°C); closed cup: 36°F (2°C). *Flammable.* Sol in 23 parts water at 27°; miscible with alc, ether. LD_{50} orally in rats: 6.75 g/kg (Smyth).

Caution: Potential symptoms of overexposure are irritation of eyes, nose and skin; dermatitis. *See NIOSH Pocket Guide to Chemical Hazards* (DHHS/NIOSH 97-140, 1997) p 180.

USE: Solvent for cellulose derivatives, plastics, oils and fats; in perfumery.

5252. **Isopropyl Acetoacetate.** [542-08-5] 3-Oxobutanoic acid 1-methylethyl ester; acetoacetic acid isopropyl ester; isopropyl 3-oxobutanoate. $C_7H_{12}O_3$; mol wt 144.17. C 58.32%, H 8.39%, O 33.29%.

Liquid. d_4^{25} 0.957. bp ~205° with decompn. Slightly sol in water, freely in alcohol, ether.

5253. **Isopropylacetone.** [108-10-1] 4-Methyl-2-pentanone; methyl isobutyl ketone; hexone. $C_6H_{12}O$; mol wt 100.16. C 71.95%, H 12.08%, O 15.97%. Manuf: *Faith, Keyes & Clark's Industrial Chemicals*, F. A. Lowenheim, M. K. Moran, Eds. (Wiley-Interscience, New York, 4th ed., 1975) pp 543-546. Toxicity data: Smyth *et al., Arch. Ind. Hyg. Occup. Med.* **4**, 119 (1951).

Colorless liquid; faint, ketonic and camphor odor. mp −84.7°. d_4^{20} 0.801. bp 117-118°. n_D^{20} 1.396. Flash pt, closed cup: 57.2°F (14°C). *Flammable.* Autoignition temp 459°. Log P (octanol/water): 1.31. Vapor pressure (20°): 15 mmHg. Slightly sol in water (17 g/l at 20°); misc with alcohol, benzene, ether, acetone. LD_{50} orally in rats: 2.08 g/kg (Smyth).

Caution: Potential symptoms of overexposure are irritation of eyes, skin and mucous membranes; headache; narcosis, coma; dermatitis. *See NIOSH Pocket Guide to Chemical Hazards* (DHHS/ NIOSH 97-140, 1997) p 164.

USE: Solvent for gums, resins, nitrocellulose, paints, varnishes, and lacquers. Denaturant for rubbing alcohol. In manuf of methyl amyl ketone. In dry-cleaning preparations.

5254. **Isopropyl Alcohol.** [67-63-0] 2-Propanol; isopropanol; secondary propyl alcohol; dimethyl carbinol; petrohol; IPA. C_3H_8-O; mol wt 60.10. C 59.96%, H 13.42%, O 26.62%. Manuf from propylene: *Faith, Keyes & Clark's Industrial Chemicals*, F. A. Lowenheim, M. K. Moran, Eds. (Wiley-Interscience, New York, 4th ed., 1975) pp 496-501. Monograph: L. F. Hatch, *Isopropyl Alcohol* (McGraw-Hill, New York, 1961) 184 pp. Toxicity: Smyth, Carpenter, *J. Ind. Hyg. Toxicol.* **30**, 63 (1948).

Liquid with slight odor resembling that of a mixture of ethanol and acetone. Slightly bitter taste (*not potable!*). mp −88.5°; fp −89.5°. bp_{760} 82.5°; bp_{400} 67.8°; bp_{200} 53.0°; bp_{100} 39.5°; bp_{60} 30.5°; bp_{40} 23.8°, bp_{20} 12.7°; bp_{10} 2.4°; bp_5 −7.0°; $bp_{1.0}$ −26.1°. d_4^{20} 0.78505; d_4^{25} 0.78084; d_4^{83} 0.728. Flash pt, closed cup: 11.7°C (53°F). Autoignition temp 455.6°C (852°F). Lower explosive limit in air: 2.5% (v/v). n_D^8 1.3852; n_D^{15} 1.3802; n_D^{20} 1.37723; n_D^{25} 1.3749. Absorption spectrum: Brode, *J. Phys. Chem.* **30**, 61 (1926). *Flammable.* Miscible with water, alcohol, ether, chloroform. Insol in salt solns. May be recovered from aq mixtures by salting out with sodium chloride, sodium sulfate, sodium hydroxide, etc. Forms an azeotrope with water (bp_{760} 80.37°, d_4^0 0.83361) isopropanol 87.7% (w/w). Freezing points of mixtures with water (v/v, per cent by volume): Isopropanol 15% = −3.3°; 25% = −6.7°; 30% = −8.3°; 35% = −11.1°; 40% = −13°; 45% = −17.8°; 50% = −23°; 60% = −32°. LD_{50} orally in rats: 5.8 g/kg (Smyth, Carpenter).

Caution: Potential symptoms of overexposure are irritation of eyes, nose and throat; drowsiness, dizziness and headache; dry cracking skin. *See NIOSH Pocket Guide to Chemical Hazards* (DHHS/ NIOSH 97-140, 1997) p 180.

USE: In antifreeze compositions; as solvent for gums, shellac, essential oils; in the extraction of alkaloids; in quick-drying oils; in quick-drying inks; in denaturing ethyl alcohol; in body rubs; hand lotions, after-shave lotions and similar cosmetics. Organic solvent for creosote, resins, gums; in manuf of acetone, glycerol, isopropyl acetate. Pharmaceutic aid (solvent).

THERAP CAT: Antiseptic.

THERAP CAT (VET): Antiseptic, rubefacient.

5255. **Isopropylamine.** [75-31-0] 2-Propanamine; 2-amino-propane; 1-methylethylamine. C_3H_9N; mol wt 59.11. C 60.96%, H 15.35%, N 23.70%. Prepn from acetone + NH_3: Skita, Keil, *Ber.* **61**, 1682 (1928); Norton *et al., J. Org. Chem.* **19**, 1054 (1954); from acetone oxime: C. F. Winans, H. Adkins, *J. Am. Chem. Soc.* **55**, 2051 (1933). Toxicity study: H. F. Smyth *et al., Arch. Ind. Hyg. Occup. Med.* **4**, 119 (1951).

$$H_3C \overset{\displaystyle CH_3}{\underset{\displaystyle}{\diagup}} NH_2$$

Colorless liquid; ammonia odor; strong base. *Flammable.* d_4^{15} 0.694. mp $-101°$. bp 33-34°. n_D^{15} 1.3770. Flash pt, open cup: $-15°F\ (-26°C)$. Miscible with water, alcohol, ether. LD_{50} orally in rats: 820 mg/kg (Smyth).

Caution: Potential symptoms of overexposure are irritation of eyes, nose, throat and skin; pulmonary edema; visual disturbance; skin and eye burns; dermatitis. *See NIOSH Pocket Guide to Chemical Hazards* (DHHS/NIOSH 97-140, 1997) p 180.

5256. Isopropyl Bromide. [75-26-3] 2-Bromopropane. C_3-H_7Br; mol wt 122.99. C 29.30%, H 5.74%, Br 64.97%. Obtained by heating isopropyl alcohol with HBr.

$$H_3C \overset{\displaystyle CH_3}{\underset{\displaystyle}{\diagup}} Br$$

Colorless liquid. d_4^{20} 1.31. mp $-89°$. bp 59-60°. n_D^{20} 1.4251. Slightly sol in water; miscible with alcohol, benzene, chloroform, ether.

5257. Isopropyl Chloride. [75-29-6] 2-Chloropropane. C_3-H_7Cl; mol wt 78.54. C 45.88%, H 8.98%, Cl 45.14%. Prepd by refluxing isopropyl alcohol with concd HCl and $ZnCl_2$.

$$H_3C \overset{\displaystyle CH_3}{\underset{\displaystyle}{\diagup}} Cl$$

Liquid. d^{15} 0.868. mp $-117°$. bp 35-36°. Flash pt, closed cup: $-26°F\ (-32°C)$. *Flammable.* Slightly sol in water; miscible with alcohol, ether.

5258. Isopropyl Ether. [108-20-3] 2,2′-Oxybispropane; 2-isopropoxypropane; diisopropyl ether; DIPE. $C_6H_{14}O$; mol wt 102.18. C 70.53%, H 13.81%, O 15.66%. Prepn: E. Erlenmeyer, *Ann.* **126**, 305 (1863). Physical properties and solubility: H. R. Fife, E. W. Reid, *Ind. Eng. Chem.* **22**, 513 (1930). Manuf process: M. N. Harandi, H. Owen, *US 5208387* (1993 to Mobil). GC determn in gasolines: N. G. Johansen, *J. High Resolut. Chromatogr. Chromatogr. Commun.* **7**, 487 (1984). Acute toxicity: E. T. Kimura *et al.,* *Toxicol. Appl. Pharmacol.* **19**, 699 (1971). Comparison with MTBE, ETBE and TAME, *q.q.v.,* as gasoline oxygenate: M. J. McNally *et al., Oil Gas J.* **90**, 39 (May 25, 1992).

$$H_3C \overset{\displaystyle CH_3 \quad CH_3}{\underset{\displaystyle}{\diagdown O \diagup}} CH_3$$

Liquid. bp_{760} 68.27°. fp $-85.89°$. d_4^{20} 0.72813. n_D^{20} 1.36888. Surface tension in air (23°): 32 dynes/cm. Viscosity (25°): 0.379 cP. Vapor pressure (20°): 158 mm Hg. Specific heat (22-27°): 0.526 cal/g. Flash pt, closed cup: $-6.7°F$. Latent heat of vaporization (67.5°): 68.2 cal/g. Soly of commercial grade in water (19°): 1.71 wt%. Misc with most organic solvents. Ethanolamines, glycols, glycerol and their monoesters have limited soly; diesters are sol. Readily forms peroxides when exposed to oxygen unless stabilized with inhibitors such as hydroquinone or BHT. LD_{50} in 14 day old, young adult, adult rats (ml/kg): 6.4, 16.5, 16.0 orally (Kimura).

Caution: Potential symptoms of overexposure are irritation of eyes, skin and nose; respiratory discomfort; dermatitis. *See NIOSH Pocket Guide to Chemical Hazards* (DHHS/NIOSH 97-140, 1997) p 182.

USE: As organic solvent; fuel additive.

5259. Isopropylidene Glycerol. [100-79-8] 2,2-Dimethyl-1,3-dioxolane-4-methanol; 2,2-dimethyl-4-hydroxymethyl-1,3-dioxolane; glycerol dimethylketal; acetone glycerol; Solketal. C_6-$H_{12}O_3$; mol wt 132.16. C 54.53%, H 9.15%, O 36.32%. Prepn: Fisher, *Ber.* **28**, 1167 (1895); Fisher, Pfähler, *Ber.* **53**, 1606 (1920);

Hibbert, Morazain, *Can. J. Res.* **2**, 38 (1930); Smith, Lindberg, *Ber.* **64**, 505 (1931); M. M. Maglio, C. A. Burger, *J. Am. Chem. Soc.* **68**, 529 (1946); M. Renoll, M. S. Newman, *Org. Synth.* **coll. vol. III**, 502 (1955); Mikschik, **AT 180926** (1955 to Chemomedica Chemikalien), *C.A.* **49**, 15951 (1955); Williams, **GB 802022** (1958 to Peter Spence & Sons); **GB 819835** (1959 to Bayer); Perez *et al.,* **ES 499129** (1982 to Calipe), *C.A.* **97**, 55794v (1982). Synthesis of D(+)-form: E. Baer, H. Fischer, *J. Biol. Chem.* **128**, 463 (1939); of L(−)-form: *eidem, J. Am. Chem. Soc.* **61**, 761 (1939). Conversion of D(+)- to L(−)-form: *eidem, ibid.* **67**, 944 (1945).

Practically odorless liquid of medium mobility. d_4^{20} 1.064. bp_{10} 82°. n_D^{20} 1.4383. Viscosity (20°): 11 cP. Practically nonflammable at ordinary storage temps: Flash pt 90°C (194°F). Evaporation no. about 600 (ether = 1). Miscible with water, alcohols, acetals, esters, ethers, aromatic hydrocarbons, chlorinated hydrocarbons, gasolines, petr ether, turpentine, oils. LD_{50} in rats (g/kg): 7 orally; 3 i.p. (Perez).

D(+)-Form. $[\alpha]_D$ +13.6°; $[\alpha]_D$ +10.8° (c = 15.19 in benzene).
L(−)-Form. $[\alpha]_D$ −13.6°; $[\alpha]_D$ −10.8° (c = 15.19 in benzene).
USE: Versatile solvent and plasticizer. Pharmaceutic aid (solubilizing and suspending agent).

5260. Isopropyl Iodide. [75-30-9] 2-Iodopropane. C_3H_7I; mol wt 169.99. C 21.20%, H 4.15%, I 74.65%. Prepared by distilling isopropyl alcohol with HI; or from glycerol, iodine, water, and phosphorus.

$$H_3C \overset{\displaystyle CH_3}{\underset{\displaystyle}{\diagup}} I$$

Colorless liquid but readily discolors in air and light. d_4^{20} 1.703. mp $-90°$. bp 89-90°. n_D^{20} 1.5026. Sol in 720 parts water; miscible with alcohol, benzene, chloroform, ether.

5261. Isopropyl Myristate. [110-27-0] Tetradecanoic acid 1-methylethyl ester; isopropyl tetradecanoate; 1-methylethyl tetradecanoate; myristic acid isopropyl ester. $C_{17}H_{34}O_2$; mol wt 270.46. C 75.50%, H 12.67%, O 11.83%. The commercial product may contain small amounts of esters of palmitic and other satd fatty acids. Physical properties: Bonhorst *et al., Ind. Eng. Chem.* **40**, 2379 (1948).

Clear, oily liquid of low viscosity, practically odorless. mp ~3°. bp_2 140.2°; bp_{20} 192.6°; dec 208°. d^{20} 0.8532; d^{99} 0.7942. n_D^{25} 1.432-1.434. Withstands oxidation and does not readily become rancid. Freely sol in 90% alc. Insol in water, glycerin, propylene glycol. Dissolves many waxes, cholesterol, lanolin. Miscible with castor oil, cottonseed oil, mineral oil, acetone, chloroform, ethyl acetate, toluene and most other organic solvents and fixed oils.

USE: In cosmetic and topical medicinal prepns where good absorption through the skin is desired.

5262. Isopropyl Nitrite. [541-42-4] Nitrous acid 1-methylethyl ester; 2-propanol nitrite; nitrous acid isopropyl ester. C_3H_7-NO_2; mol wt 89.09. C 40.45%, H 7.92%, N 15.72%, O 35.92%. Prepd by treating isopropyl alcohol with nitrosyl chloride: Kornblum *et al., J. Am. Chem. Soc.* **77**, 5531 (1955); by adding concd sulfuric acid and isopropyl alcohol to sodium nitrite: Levin, Hartung, *Org. Synth.* **24**, 26 (1944).

Pale yellow oil. d_4^0 0.856; d_4^{25} 0.844. bp$_{752}$ 39-39.5°; bp$_{745}$ 39-40°. n_D^{20} 1.3520.

Caution: Can cause vasodilation with fall in blood pressure, tachycardia, headache. Large doses can cause methemoglobinuria with cyanosis. Severe poisoning results in shock which can end fatally.

USE: Jet propellant.

5263. Isoproterenol. [7683-59-2] 4-[1-Hydroxy-2-[(1-methylethyl)amino]ethyl]-1,2-benzenediol; 3,4-dihydroxy-α-[(isopropylamino)methyl]benzyl alcohol; α-(isopropylaminomethyl)protocatechuyl alcohol; isoprenaline; isopropylarterenol; 1-(3,4-dihydroxyphenyl)-2-isopropylaminoethanol; isopropylaminomethyl-(3,4-dihydroxyphenyl)carbinol; *N*-isopropyl-β-dihydroxyphenyl-β-hydroxyethylamine; dihydroxyphenylethanolisopropylamine; *N*-isopropylnoradrenaline; epinephrine isopropyl homolog; A-21; Aludrine; Aleudrin; Isuprel; Norisodrine; Asiprenol; Asmalar; Neo-Epinine; Novodrin; Isupren; Neodrenal; Isopropydrin; Assiprenol; Respifral; Bellasthman; Saventrine; Proternol; Isorenin; Vapo-N-Iso; Isonorin. $C_{11}H_{17}NO_3$; mol wt 211.26. C 62.54%, H 8.11%, N 6.63%, O 22.72%. β-Adrenergic agonist. Catecholamine derivative prepd by the reaction of 3,4-dihydroxy-α-haloacetophenone with an excess of isopropylamine: Scheuing, Thomä, **DE 723278** (1942 to Boehringer, Ing.); **US 2308232** (1943); from guaiacol and chloral hydrate: Beke *et al.*, *Pharm. Zentralhalle* **92**, 237 (1953). Resolution: Kerschbaum, Benedikt, *Monatsh. Chem.* **83**, 1090 (1952); Beccari *et al.*, *Science* **118**, 249 (1953); Delmar *et al.*, **US 2715141** (1955 to Delmar Chem.). Configuration: Pratesi *et al.*, *Farmaco Ed. Sci.* **15**, 3 (1960). Toxicity: E. I. Goldenthal, *Toxicol. Appl. Pharmacol.* **18**, 185 (1971). Effect on gastric acid secretion: M. J. Daly, *Scand. J. Gastroenterol.* **89**, 3 (1984). Use in treatment of primary pulmonary hypertension: D. A. Pietro *et al.*, *N. Engl. J. Med.* **310**, 1032 (1984). Review of pharmacology and comparison with other β-adrenoceptor agonists: V. T. Popa, *J. Asthma* **21**, 183-207 (1984). Comprehensive description: M. Tariq, A. A. Al-Badr, *Anal. Profiles Drug Subs.* **14**, 391-422 (1985).

dl-**Form.** Crystals from alc, mp 155.5°. pKa 8.64. LD$_{50}$ in male, female rats (mg/kg): 3675, 4282 orally (Goldenthal).

dl-**Form hydrochloride.** Aerolone; Aerotrol; Euspiran; Isomenyl; Isovon; Mistarel; Suscardia. $C_{11}H_{17}NO_3$.HCl; mol wt 247.72. Crystals from alc, mp 170-171°. Soly: one gram dissolves in 3 ml water, in 50 ml ethanol (95%). Sparingly sol in alc, less sol in dehydrated alc. Practically insol in benzene. Insol in chloroform, ether. pH of 1% aq soln about 5. Aq solns turn brownish-pink upon prolonged exposure to air or upon addition of alkali. LD$_{50}$ orally in rats: 2221 ±93 mg/kg (Goldenthal).

dl-**Form sulfate dihydrate.** [6700-39-6] Aludrin; Isomist; Propal. $(C_{11}H_{17}NO_3)_2.H_2SO_4.2H_2O$; mol wt 556.62. Crystals from acetone + methanol, mp 128° (some decomp). One gram dissolves in about 4 ml water. Slightly sol in alc. Practically insol in chloroform, ether, benzene. pH of a 1% aq soln about 5.

l-**Form.** Crystals, mp 164-165°. $[\alpha]_D^{19}$ −45.0° (c = 2 in 2*N* HCl).

l-**Form hydrochloride.** Crystals, dec 162-164°. $[\alpha]_D^{20}$ −50°.

l-**Form *d*-bitartrate dihydrate.** Isolevin. Crystals, mp 80-83° (sinters at 78°). $[\alpha]_D^{19}$ −14.9° (c = 2.31).

THERAP CAT: Bronchodilator.

THERAP CAT (VET): Sympathomimetic. Chiefly as bronchodilator.

5264. Isoproturon. [34123-59-6] *N,N*-Dimethyl-*N'*-[4-(1-methylethyl)phenyl]urea; 3-*p*-cumenyl-1,1-dimethylurea; *N*-(4-isopropylphenyl)-*N',N*-dimethylurea; I.P.U.; HOE-16410; CGA-18731; Arelon; Protugan. $C_{12}H_{18}N_2O$; mol wt 206.29. C 69.87%, H 8.80%, N 13.58%, O 7.76%. Pre- and post-emergence herbicide for control of annual grasses and broad-leaved weeds. Prepn and herbicidal properties: C. W. Todd, **US 2655447** (1953 to du Pont); **BE 770928** (1971 to Pepro); A. Thizy *et al.*, **US 4295877** (1981 to Philagro). Physical properties, herbicidal activity: *eidem*, *C. R. Seances Acad. Sci. Ser. D* **274**, 2053 (1972); J. Rognon *et al.*, *Meded. Fac. Landbouwwet. Rijksuniv. Gent* **37**, 663 (1972). Microbial deg-

radation of labelled isoproturon: J.-C. Fournier *et al.*, *Chemosphere* **4**, 207 (1975). Efficacy in winter cereals: R. T. Hewson, *Proc. 12th Br. Weed Control Conf.*, 75 (1974); P. Gonzales *et al.*, *Weed Res.* **23**, 39 (1983). HPLC determn in soil: G. Kulshrestha, R. Khazanchi, *J. Chromatogr.* **318**, 144 (1985).

White crystals from benzene, mp 158°. Soly in water at 22°: 70 mg/l. LD$_{50}$ in mice, rats (mg/kg): 3350, 3600 orally (Thizy, 1972).

USE: Herbicide.

5265. Isopyrocalciferol. [474-70-4] (3β,9β,22*E*)-Ergosta-5,-7,22-trien-3-ol; 9β-ergosterol. $C_{28}H_{44}O$; mol wt 396.66. C 84.78%, H 11.18%, O 4.03%. Differs sterically from pyrocalciferol at C-10, from ergosterol at C-9. *See ref under* Pyrocalciferol. Stereochemistry: Castells *et al.*, *J. Chem. Soc.* **1959**, 1159. Selenium dehydrogenation gives Diel's hydrocarbon (3'-methyl-1,2-cylopentenophenanthrene).

Prisms from ether-methanol. mp 112-115°. uv max: 262, 280 nm. $[\alpha]_D^{20}$ +332°, $[\alpha]_{546}^{20}$ +415° (c = 1.5 in chloroform). Precipitated by digitonin.

Acetate. $C_{30}H_{46}O_2$. mp 113-115°, $[\alpha]_D^{20}$ +333°, $[\alpha]_{546}^{20}$ +423° (c = 2 in chloroform).

5266. Isoquassin. [21293-20-9] (14α)-2,12-Dimethoxypicrasa-2,12-diene-1,11,16-trione; picrasmin. $C_{22}H_{28}O_6$; mol wt 388.46. C 68.02%, H 7.27%, O 24.71%. Bitter principle from Jamaica quassia, *Picrasma excelsa* (Sw.) Planch., *Simaroubaceae*. Isoln: Clark, *J. Am. Chem. Soc.* **60**, 1146 (1938). Identity of isoquassin and picrasmin: Adams, Whaley, *ibid.* **72**, 375 (1950). Isomer of quassin, *q.v.:* Valenta *et al.*, *Tetrahedron* **18**, 1433 (1962).

Crystals from methanol, mp 222-225°. $[\alpha]_D^{20}$ +45.4° (chloroform). uv max: 258 nm (ε 12500).

5267. Isoquercitrin. [482-35-9] 2-(3,4-Dihydroxyphenyl)-3-(β-D-glucofuranosyloxy)-5,7-dihydroxy-4*H*-1-benzopyran-4-one; 3,3',4',5,7-pentahydroxyflavone-3-glucoside; quercetin-3-glucoside; isotrifoliin; trifoliin. $C_{21}H_{20}O_{12}$; mol wt 464.38. C 54.32%, H 4.34%, O 41.34%. From flowers of *Gossypium herbaceum* L., *Malvaceae*: Perkin, *J. Chem. Soc.* **95**, 2181 (1909); from flowers of *Aesculus hippocastanum* L., *Hippocastanaceae*: Hörhammer, Wagner, *Arch. Pharm.* **290**, 224 (1957); from *Tropaeolum majus* L., *Tropaeolaceae*: Delaveau, *Compt. Rend.* **252**, 1510 (1961); from *Arnica montana* L., *Compositae*: Friedrich, *Naturwissenschaften*

49, 541 (1962). Structure: Attree, Perkin, *J. Chem. Soc.* **1927**, 234. Identity with trifoliin and isotrifoliin: Hattori *et al.*, *J. Chem. Soc. Jpn.* **58**, 844 (1937). Synthesis: Ice, Wender, *J. Am. Chem. Soc.* **74**, 4606 (1952); *eidem*, **US 2727890** (1955 to the USAEC).

Yellow needles from water, mp 225-227°. uv max: 257, 369 nm. Practically insol in cold but sparingly sol in boiling water. Sol in alkaline solns with a deep yellow tint.

5268. Isoquinoline. [119-65-3] 2-Benzazine; benzo[c]pyridine; 2-azanaphthalene. C_9H_7N; mol wt 129.16. C 83.69%, H 5.46%, N 10.84%. Occurs in coal tar: Hoogewerff, van Dorp, *Rec. Trav. Chim.* **4**, 125, 285 (1885); Weissgerber, *Ber.* **47**, 3175 (1914); Harris, Pope, *J. Chem. Soc.* **121**, 1029 (1922). Isoln by a freeze-out method: Kjellman, **US 2483420** (1946 to Koppers). Purification: Freiser, Glowacki, *J. Am. Chem. Soc.* **71**, 514 (1949). Synthesis: D. L. Boger *et al.*, *Tetrahedron* **37**, 3977 (1981). Toxicity data: Smyth *et al.*, *Arch. Ind. Hyg. Occup. Med.* **4**, 119 (1951). Comprehensive survey of isoquinoline alkaloids: Gensler, *Heterocyclic Compounds* **vol. 4**, R. C. Elderfield, Ed. (Wiley, New York, 1952) pp 344-490; T. Kametani, *Total Synthesis of Natural Products* **vol. 3**, J. ApSimon, Ed. (Wiley, New York, 1977) pp 1-272.

Liquid. Pungent odor resembling that of a mixture of anise oil and benzaldehyde. Hygroscopic platelets when solid. d_4^{30} 1.09101; d_4^{80} 1.05143. mp 26.48°. n_D^{30} 1.62078. Viscosity (cP): 3.2528 (30°); 1.0230 (100°); 0.4223 (200°). bp_{743} 242.2°; bp_{760} 243.25°. Dipole moment: 2.49. More basic than quinoline, pKa (25°): 8.60. Almost insol in water. Miscible with many organic solvents. Sol in dil acids. LD_{50} orally in rats: 360 mg/kg (Smyth).

Sulfate. $C_9H_7N \cdot H_2SO_4$. Crystals, mp 209-209.5°. Sol in water.

USE: Synthesis of dyes, insecticides, antimalarials, rubber accelerators.

5269. Isosafrole. [120-58-1] 5-(1-Propen-1-yl)-1,3-benzodioxole; 1,2-(methylenedioxy)-4-propenylbenzene. $C_{10}H_{10}O_2$; mol wt 162.19. C 74.06%, H 6.21%, O 19.73%. Purification and separation from safrole: Balbiano, *Ber.* **42**, 1502 (1911); Hoering, Baum, *ibid.* 3082. Prepn: Bert, *Compt. Rend.* **213**, 873 (1941); Naves, Ardizio, *Bull. Soc. Chim. Fr.* **1957**, 1053; Fengeas, *ibid.* **1964**, 1892; Cabiddu *et al.*, *Ann. Chim. (Rome)* **52**, 1261 (1962). Review and evaluation of studies of carcinogenic action in laboratory animals: *IARC Monographs* **10**, 231-241 (1976).

trans-Form.

***trans*-Form.** [4043-71-4] β-Isosafrole. Liquid, odor of anise. bp_{760} 253°; bp_{100} 179.5°; bp_{20} 135.6°; $bp_{3.4}$ 85-86°. mp 8.2°. d_4^{20} 1.1206. n_D^{20} 1.5782. uv max (96% alc): 305, 267, 259.5 nm (ε 5340; 11600; 12160). Miscible with alc, ether, benzene. Sol in 8 parts of 90% alcohol.

***cis*-Form.** [17627-76-8] α-Isosafrole. Liquid. $bp_{3.5}$ 77-79°. mp −21.5°. d_4^{20} 1.1182. n_D^{20} 1.5691. uv max (96% alc): 296.5, 259 nm (ε 4450; 10000).

USE: Manuf heliotropin; in perfumes.

5270. Isosorbide. [652-67-5] 1,4:3,6-Dianhydro-D-glucitol; 1,4:3,6-dianhydrosorbitol; AT-101; NSC-40725; Hydronol; Ismotic; Isobide. $C_6H_{10}O_4$; mol wt 146.14. C 49.31%, H 6.90%, O 43.79%.

Prepd by acid dehydration of D-glucitol: Haworth, Wiggins, **GB 600870** (1948). Purification: Hartmann, **US 3160641** (1964 to Atlas Chemical Industries). Review of prepn, structure, properties: Wiggins, *Adv. Carbohydr. Chem.* **5**, 191-228 (1950). Stereochemistry: Cope, Shen, *J. Am. Chem. Soc.* **78**, 3177 (1956). Conformation studies: Hopton, Thomas, *Can. J. Chem.* **47**, 2395 (1969). Pharmacodynamics in man: Nodine *et al.*, *Clin. Pharmacol. Ther.* **14**, 196 (1973).

Crystals, mp 61-64°. $[\alpha]_D$ +44°.

THERAP CAT: Diuretic.

5271. Isosorbide Dinitrate. [87-33-2] 1,4:3,6-Dianhydro-D-glucitol 2,5-dinitrate; dinitrosorbide; 1,4:3,6-dianhydrosorbitol 2,5-dinitrate; Carvasin; Cedocard; Corovliss; Dignonitrat; Dilatrate; Diniket; Frandol; Imtack; Isocard; Isoket; IsoMack; Isorbid; Isordil; Isostenase; Isotrate; Langoran; Maycor; Myorexon; Nitorol; Nitrol; Nitrosorbon; Rifloc; Risordan; Sorbangil; Sorbichew; Sorbidilat; Sorbid SA; Sorbitrate; Vascardin; Vasorbate. $C_6H_8N_2O_8$; mol wt 236.14. C 30.52%, H 3.41%, N 11.86%, O 54.20%. Coronary vasodilator. Prepn from sorbitol: Goldberg, *Acta Physiol. Scand.* **15**, 173 (1948); Kochergin, Titkova, *C.A.* **54**, 8647h (1960). Absorption and disposition in man: H. Laufen *et al.*, *Arzneim.-Forsch.* **33**, 980 (1983). Comprehensive description: L. A. Silvieri, N. J. DeAngelis, *Anal. Profiles Drug Subs.* **4**, 225-244 (1975).

Hard colorless crystals, mp 70°. $[\alpha]_D^{20}$ +135° (alc). Sparingly sol in water (1.089 mg/ml, also given as 1.0 g dissolves in 900 ml H_2O). Freely sol in organic solvents, such as acetone, alc, ether.

Isosorbide-5-mononitrate. [16051-77-7] Corangin; Elan; Elantan; Imdur; Ismo; Isomonat; Monicor; Monit; Mono-Cedocard; Monoclair; Monoket; Mono Mack; Monosorb; Olicard; Pentacard. $C_6H_9NO_6$; mol wt 191.14. Metabolite of isosorbide dinitrate.

THERAP CAT: Antianginal.

5272. Isothebaine. [568-21-8] (6aS)-5,6,6a,7-Tetrahydro-2,11-dimethoxy-6-methyl-4H-dibenzo[de,g]quinolin-1-ol; 2,11-dimethoxy-6aα-aporphin-1-ol; 1-hydroxy-2,11-dimethoxyaporphine. $C_{19}H_{21}NO_3$; mol wt 311.38. C 73.29%, H 6.80%, N 4.50%, O 15.41%. Occurs as d-form in the root and whole plant of *Papaver orientale* L., *Papaveraceae*, collected in late autumn. Isoln: Gadamer, *Arch. Pharm.* **249**, 39 (1911). Structure: Bentley, Blues, *J. Chem. Soc.* **1956**, 1732; Bentley, Dyke, *J. Org. Chem.* **22**, 429 (1957). Synthesis of dl-form: Battersby, Brown, *Proc. Chem. Soc. London* **1964**, 85; H. Hara *et al.*, *Chem. Pharm. Bull.* **29**, 1083 (1981). Synthesis of natural and racemic forms: Battersby *et al.*, *J. Chem. Soc.* **1965**, 4550. Biosynthesis: *eidem*, *Chem. Commun.* **1965**, 230.

Light-sensitive, rhombic crystals. mp 203-204°. $[\alpha]_D^{18}$ +285° in alc. Sol in alcohol, chloroform, slightly in ether.

5273. Isothipendyl. [482-15-5] N,N,α-Trimethyl-10H-pyrido[3,2-b][1,4]benzothiazine-10-ethanamine; 10-(2-dimethylamino-2-methylethyl)-10H-pyrido[3,2-b][1,4]benzothiazine; 10-(2-dimethylaminopropyl)-1-azaphenothiazine; N-dimethylaminoisopropyl-thiophenylpyridylamine; D-201. $C_{16}H_{19}N_3S$; mol wt 285.41. C 67.33%, H 6.71%, N 14.72%, S 11.23%. Prepn: H. L. Yale, F. Sowinski, *J. Am. Chem. Soc.* **80**, 1651 (1958). Pharmacology: A. von Schlichtegroll, *Arzneim.-Forsch.* **8**, 489 (1958). Clinical evaluation: N. W. Klehr *et al.*, *Therapiewoche* **34**, 6050 (1984); E. Milsmann, P. Rohdewald, *Dermatologica* **170**, 230 (1985). Spectroscopic determn: M. H. Abdel-Hay *et al.*, *J. Pharm. Belg.* **45**, 259 (1990). HPLC determn: A. Turcant *et al.*, *Clin. Chem.* **37**, 1210 (1991).

bp$_{0.4mm}$ 171-174°. LD$_{50}$ in mice (mg/kg): 65 ±2 i.p.; 222 ±18 orally (Schlichtegroll).
Hydrochloride. [1225-60-1] Andantol; Andanton; Nilergex. $C_{16}H_{19}N_3S$·HCl; mol wt 321.87. Sol in water.
THERAP CAT: Antihistaminic.

5274. Isotretinoin. [4759-48-2] 13-*cis*-Retinoic acid; 2-*cis*-vitamin A acid; neovitamin A acid; Ro-4-3780; Accutane; Contracne; Curacne; Isotrex; Oratane; Procuta; Roaccutane. $C_{20}H_{28}O_2$; mol wt 300.44. C 79.96%, H 9.39%, O 10.65%. Naturally occurring metabolite of vitamin A, *q.v.*; inhibits sebum production. Prepn: C. D. Robeson *et al.*, *J. Am. Chem. Soc.* **77**, 4111 (1955). Stereoselective process: R. Lucci, **EP 111325**; *idem*, **US 4556518** (1984, 1985 both to Hoffmann-La Roche). Toxicology and teratogenicity study: J. J. Kamm, *J. Am. Acad. Dermatol.* **6**, 652 (1982). Identification as endogenous metabolite of all-*trans*-retinoic acid: M. E. Cullum, M. H. Zile, *J. Biol. Chem.* **260**, 10590 (1985). HPLC determn in serum: G. Tang, R. M. Russell, *J. Lipid Res.* **31**, 175 (1990). GC determn in pharmaceutical formulations: E. M. Lima *et al.*, *J. Pharm. Biomed. Anal.* **38**, 678 (2005). Review of pharmacology and clinical efficacy in acne: A. R. Shalita *et al.*, *Cutis* **42**, Suppl. 6A, 1-19 (1988). Symposium on clinical experience: *Dermatology* **195**, Suppl. 1, 1-37 (1997). Review of safety profile: A. Charakida *et al.*, *Expert Opin. Drug Saf.* **3**, 119-129 (2004).

Yellow crystals or reddish-orange plates from isopropyl alcohol, mp 174-175°. uv max: 354 nm (ε 39800). Sol in chloroform; sparingly sol in alcohol, isopropyl alcohol, polyethylene glycol 400. Practically insol in water. LD$_{50}$ (20 day) in mice, rats (mg/kg): 904, 901 i.p.; 3389, >4000 orally (Kamm).
THERAP CAT: Antiacne.
THERAP CAT (VET): In treatment of dermatologic conditions.

5275. Isovaleraldehyde. [590-86-3] 3-Methylbutanal; isopentanal; isovaleral; isovaleric aldehyde. $C_5H_{10}O$; mol wt 86.13. C 69.73%, H 11.70%, O 18.58%. Occurs in orange, lemon, peppermint, eucalyptus and other oils. Made by oxidation of isoamyl alcohol with $Na_2Cr_2O_7$ and H_2SO_4.

Colorless liquid; pungent apple-like odor. d$_{20}^{20}$ 0.785. mp −51°. bp 92-93°. n_D^{20} 1.3902. Sparingly sol in water; miscible with alcohol, ether.
USE: In artificial flavors and perfumes.

5276. Isovaleramide. [541-46-8] 3-Methylbutanamide; isopentanamide; isovaleric acid amide. $C_5H_{11}NO$; mol wt 101.15. C 59.37%, H 10.96%, N 13.85%, O 15.82%.

Crystals. d 0.965. mp 135-137°. bp 232°. Sol in water, alcohol.

5277. Isovaleric Acid. [503-74-2] 3-Methylbutanoic acid; delphinic acid; isopentanoic acid; isopropylacetic acid; isovalerianic acid. $C_5H_{10}O_2$; mol wt 102.13. C 58.80%, H 9.87%, O 31.33%. Occurs in hop oil, tobacco and several other plants: Schmeltz *et al.*, *J. Assoc. Off. Agric. Chem.* **46**, 779 (1963). Prepn: Caldwell, **US 2484486** (1949 to Kodak); Gustak, Sajko, *Ark. Kemi* **24**, 11 (1952), *C.A.* **49**, 163e (1955); L'vov *et al.*, *Zh. Prikl. Khim.* **35**, 700 (1962). "Valeric" acid of commerce is isovaleric acid. Toxicity data: L. Orö, A. Wretlind, *Acta Pharmacol. Toxicol.* **18**, 141 (1961).

Liquid. Acid taste; disagreeable, rancid-cheese odor. d$_4^{20}$ 0.931. mp −37°. bp 175-177°. n_D^{20} 1.4043. Sol in 24 parts water; sol in alcohol, chloroform, ether. *Keep tightly closed.* LD$_{50}$ i.v. in mice: 1120±30 mg/kg (Orö, Wretlind).
USE: In flavors, perfumes, manuf sedatives.

5278. Isovaleryl Chloride. [108-12-3] 3-Methylbutanoyl chloride; isopentanoyl chloride; isovaleric acid chloride. C_5H_9ClO; mol wt 120.58. C 49.81%, H 7.52%, Cl 29.40%, O 13.27%.

Liquid. d$_4^{20}$ 0.985. bp 114-116°. n_D^{20} 1.4156. Dec by water, alcohol.

5279. 2-Isovalerylindane-1,3-dione. [83-28-3] 2-(3-Methyl-1-oxobutyl)-1H-indene-1,3(2H)-dione; 2-isovaleryl-1,3-indandione; PMP; Valone. $C_{14}H_{14}O_3$; mol wt 230.26. C 73.03%, H 6.13%, O 20.84%. Prepn: Shapiro *et al.*, *J. Org. Chem.* **25**, 1860 (1960).

Yellow solid from methanol, mp 68-69°. Sol in most organic solvents. Practically insol in water.
USE: Insecticide, rodenticide.

5280. Isovaline. [595-39-1] DL-2-Amino-2-methylbutyric acid; α-amino-α-methylbutyric acid. $C_5H_{11}NO_2$; mol wt 117.15. C 51.26%, H 9.46%, N 11.96%, O 27.31%. Prepn: Kurono, *Biochem. Z.* **134**, 427 (1922); Levene, Steiger, *J. Biol. Chem.* **76**, 299 (1928); Bucherer, Steiner, *J. Prakt. Chem.* **141**, 5 (1934); Upham, Dermer, *J. Org. Chem.* **22**, 799 (1957). Optical enantiomorphs: Ehrlich, Wendel, *Biochem. Z.* **8**, 438 (1908); Fischer, von Gravenitz, *Ann.* **406**, 5 (1914); Baker *et al.*, *J. Am. Chem. Soc.* **74**, 4701 (1952); Greenstein *et al.*, *J. Biol. Chem.* **204**, 307 (1953).

Monoclinic prisms, mp 307-308° (closed tube). Sublimes at around 300°. Soly in cold water about 39 g/100 ml; in alc about 6.6 g/100 g alc at 75°. Slightly sol in ether.
L-Form. [595-40-4] Crystals. $[\alpha]_D^{25}$ +11.13° (c = 5 in H_2O). [M]$_D$ +9.7° (5N HCl); [M]$_D$ +26.3° (glacial acetic acid).

D-Form. [3059-97-0] Long needles from water + acetone. $[\alpha]_D^{25}$ $-11.28°$ (c = 5 in H_2O).

5281. Isoxaben. [82558-50-7] N-[3-(1-Ethyl-1-methylpropyl)-5-isoxazolyl]-2,6-dimethoxybenzamide; benzamizole; EL-107; NA-8318; Flexidor; Gallery. $C_{18}H_{24}N_2O_4$; mol wt 332.40. C 65.04%, H 7.28%, N 8.43%, O 19.25%. Selective, pre-emergent herbicide for broadleaf weeds and annual grasses. Prepn: K. W. Burow, **GB 2084140**; *see also: idem*, **US 4636243** (1982, 1987 both to Lilly). Physical properties and herbicidal activity: F. Huggenberger *et al.*, *Proc. Br. Crop Prot. Conf. - Weeds* **1982**, 47. Mobility and degradation in soil: F. Huggenberger, P. J. Ryan, *ibid.* **1985**, 947. Effect on cellular growth and metabolism: A. Lefebvre *et al.*, *Weed Res.* **27**, 125 (1987). Field trials: F. Huggenberger, F. Gueguen, *Crop Prot.* **6**, 75 (1987); F. O. Colbert, D. H. Ford, *Proc. West. Soc. Weed Sci.* **40**, 155 (1987).

White crystalline solid, mp 176-179°. Soly at 25° (mg/ml): methanol 50-100; ethyl acetate 50-100; acetonitrile 30-50; toluene 4-5; hexane 0.07-0.08; water 0.001-0.002. Vapor pressure at 30°: <3.9 $\times 10^{-7}$ mm Hg. LD_{50} in mice, rats (mg/kg): >10000 orally both species; LC_{50} in rats by inhalation: >1.99 mg/l (Huggenberger, 1982).

USE: Herbicide.

5282. Isoxadifen-ethyl. [163520-33-0] 4,5-Dihydro-5,5-diphenyl 3-isoxazolecarboxylic acid ethyl ester; ethyl 4,5-dihydro-5,5-diphenyl-1,2-oxazole-3-carboxylate; AE-F122006. $C_{18}H_{17}NO_3$; mol wt 295.34. C 73.20%, H 5.80%, N 4.74%, O 16.25%. Safener used to prevent the phytotoxic effects of herbicides on crops. Prepn: **DE 4331448**; L. Willms *et al.*, **US 5516750** (1995, 1996 both to Hoechst Schering AgrEvo); G. Cremonesi *et al.*, *Tetrahedron: Asymmetry* **19**, 2850 (2008). Effect on herbicide metabolism in corn: W. Schulte, H. Köcher, *Bayer CropSci. J.* **62**, 35 (2009). LC-MS/MS screening method for determn in drinking water: K. Greulich, L. Alder, *Anal. Bioanal. Chem.* **391**, 183 (2008). GC-MS determn in rice: L. Lucini, G. P. Molinari, *Pest Manag. Sci.* **66**, 621 (2010). Review of use in combination with tembotrione, *q.v.*: M. Wegener, H. Roos, *J. Plant Dis. Prot.* **Sp. Iss. 21**, 629-634 (2008).

Colorless solid from diethyl ether, mp 87-88°.
USE: Herbicide safener.

5283. Isoxaflutole. [141112-29-0] (5-Cyclopropyl-4-isoxazolyl)[2-(methylsulfonyl)-4-(trifluoromethyl)phenyl]methanone; 5-cyclopropyl-1,2-oxazol-4-yl α,α,α-trifluoro-2-mesyl-p-tolyl ketone; 5-cyclopropylisoxazol-4-yl 2-mesyl-4-trifluoromethylphenyl ketone; 5-cyclopropyl-4-(2-methanesulfonyl-4-trifluoromethylbenzoyl)isoxazole; 264-EUP-99; RPA-201772; Balance; Merlin. $C_{15}H_{12}F_3NO_4S$; mol wt 359.32. C 50.14%, H 3.37%, F 15.86%, N 3.90%, O 17.81%, S 8.92%. Pre-emergent herbicide designed for use in maize. Disrupts pigment biosynthesis via an inhibition of 4-hydroxyphenylpyruvate dioxygenase. Prepn: P. A. Cain *et al.*, **EP 527036** (1993 to Rhone-Poulenc). Properties and biological activity: B. M. Luscombe *et al.*, *Brighton Crop Prot. Conf. - Weeds* **1995**, 35. Weed control in maize: B. M. Luscombe, K. E. Pallett, *Pestic. Outlook* **7**, 29 (1996). Mechanism of action: K. E. Pallett *et al.*, *Pestic. Sci.* **50**, 83 (1997). Soil persistence: J. Rouchaud *et al.*, *Bull. Environ. Contam. Toxicol.* **60**, 577 (1998).

Off-white or pale yellow solid, mp 138-138.5° (Cain); also reported as mp 140° (Luscombe). Vapor pressure at 25°: 1×10^{-6} Pa. Soly in water: 6.2 mg/l. LD_{50} in rats, quails, mallard ducks (mg/kg): >5000, >2150, >2150 orally; in rabbits (mg/kg): >2000 dermally (Luscombe).

USE: Herbicide.

5284. Isoxepac. [55453-87-7] 6,11-Dihydro-11-oxodibenz[b,e]oxepin-2-acetic acid; oxepinac; HP-549; P-720549; Artil. $C_{16}H_{12}O_4$; mol wt 268.27. C 71.64%, H 4.51%, O 23.86%. Nonsteroidal anti-inflammatory with analgesic and antipyretic activity. Prepn: K. Ueno *et al.*, **JP Kokai 74 124086** (1974 to Daiichi), *C.A.* **82**, 170735d (1975); G. C. Helsley *et al.*, **DE 2442060**; *eidem*, **US 4585788** (1975, 1986 both to Hoechst); K. Ueno *et al.*, *J. Med. Chem.* **19**, 941 (1976); D. E. Aultz *et al.*, *ibid.* **20**, 66 (1977). Pharmacology: H. B. Lassman *et al.*, *Arch. Int. Pharmacodyn. Ther.* **227**, 142 (1977). Disposition and metabolism in animals and humans: H. P. A. Illing, J. M. Fromson, *Drug Metab. Dispos.* **6**, 510 (1978). HPLC determn in plasma: J. A. Slack, *J. Chromatogr.* **221**, 431 (1980); H. K. L. Hundt, L. W. Brown, *ibid.* **225**, 482 (1981). Clinical studies: S. Sasaki, *Arzneim.-Forsch.* **28**, 462 (1978); L. B. Svendsen *et al.*, *Scand. J. Rheumatol.* **10**, 186 (1981); J. Scott, E. C. Huskisson, *Rheumatol. Rehabil.* **21**, 48 (1982).

Crystals from ethyl acetate, mp 131-132.5°. LD_{50} orally in rats: 199 mg/kg (Ueno).

THERAP CAT: Anti-inflammatory.

5285. Isoxsuprine. [395-28-8] 4-Hydroxy-α-[1-[(1-methyl-2-phenoxyethyl)amino]ethyl]benzenemethanol; p-hydroxy-α-[1-[(1-methyl-2-phenoxyethyl)amino]ethyl]benzyl alcohol; p-hydroxy-N-(1-methyl-2-phenoxyethyl)norephedrine; 1-(p-hydroxyphenyl)-2-(1-methyl-2-phenoxyethylamino)-1-propanol; 2-(3-phenoxy-2-propylamino)-1-(p-hydroxyphenyl)-1-propanol. $C_{18}H_{23}NO_3$; mol wt 301.39. C 71.73%, H 7.69%, N 4.65%, O 15.93%. Prepn: Moed, van Dijk, *Rec. Trav. Chim.* **75**, 1215 (1956); **GB 832286** and **GB 832287** (both 1960 to N. V. Philip Gloeilampenfabrieken); Moed, **US 3056836** (1962 to N. Am. Philips). Configuration: Dirkx, *Rec. Trav. Chim.* **83**, 535 (1964). Toxicity data: Goldenthal, *Toxicol. Appl. Pharmacol.* **18**, 185 (1971). Clinical evaluation in Raynaud's phenomenon: H. Wesseling *et al.*, *Eur. J. Clin. Pharmacol.* **20**, 329 (1981); in intermittent claudication: S. H. Skotnicki *et al.*, *Angiology* **35**, 685 (1984). Veterinary trial in navicular disease: A. S. Turner, C. M. Tucker, *Equine Vet. J.* **21**, 338 (1989).

Crystals, mp 102.5-103.5°.
Hydrochloride. [579-56-6] Dilavase; Duvadilan; Duviculine; Navilox; Suprilent; Vadosilan; Vasodilan; Vasotran. $C_{18}H_{23}NO_3$·HCl; mol wt 337.84. Bitter crystals from water, mp 203-204°. Sparingly sol in alc. Soly in water at 25° about 2.0% w/v. LD_{50} in rats (mg/kg): 1750 orally; 164 i.p. (Goldenthal).

THERAP CAT: Vasodilator (peripheral).

THERAP CAT (VET): In treatment of navicular disease.

5286. Ispronicline. [252870-53-4] (2*S*,4*E*)-*N*-Methyl-5-[5-(1-methylethoxy)-3-pyridinyl]-4-penten-2-amine; RJR-1734; TC-1734. $C_{14}H_{22}N_2O$; mol wt 234.34. C 71.76%, H 9.46%, N 11.95%, O 6.83%. Specific $\alpha_4\beta_2$ nicotinic receptor agonist. Prepn: W. S. Caldwell *et al.*, **WO 9965876** (1999 to R. J. Reynolds); *eidem*, **US 6958399** (2005 to Targacept). Effect on cortical acetylcholine release in rats: M. C. Obinu *et al.*, *Prog. Neuro-Psychopharmacol. Biol. Psychiatry* **26**, 913 (2002). Review of pharmacology: G. J. Gatto *et al.*, *CNS Drug Rev.* **10**, 147-166 (2004); and clinical experience: H. Geerts, *Curr. Opin. Investig. Drugs* **7**, 60-69 (2006).

Colorless oil, $bp_{0.5}$ 90-100°.

Hemigalactarate. [252870-54-5] $C_{28}H_{44}N_4O_2 \cdot C_6H_{10}O_8$; mol wt 678.82. White crystalline powder, mp 140-143°.

THERAP CAT: Nootropic.

5287. Isradipine. [75695-93-1] 4-(2,1,3-Benzoxadiazol-4-yl)-1,4-dihydro-2,6-dimethyl-3,5-pyridinedicarboxylic acid 3-methyl 5-(1-methylethyl) ester; 4-(4-benzofurazanyl)-1,4-dihydro-2,6-dimethyl-3,5-pyridinedicarboxylic acid methyl 1-methylethyl ester; isopropyl 4-(2,1,3-benzoxadiazol-4-yl)-1,4-dihydro-5-methoxycarbonyl-2,6-dimethyl-3-pyridinecarboxylate; isrodipine; PN-200-110; Clivoten; DynaCirc; Esradin; Lomir; Prescal. $C_{19}H_{21}N_3O_5$; mol wt 371.39. C 61.45%, H 5.70%, N 11.31%, O 21.54%. Dihydropyridine calcium channel blocker. Prepn: P. Neumann, **DE 2949491**; *idem*, **US 4466972** (1980, 1984 both to Sandoz). Prepn of enantiomers: A. Vogel, **DE 3320616** (1983 to Sandoz). Comparative study of *in vitro* effects on human and canine cerebral arteries: E. Müller-Schweinitzer, P. Neumann, *J. Cereb. Blood Flow Metab.* **3**, 354 (1983). Effect on α-adrenoceptor mediated vasoconstriction in rats: K. Jie *et al.*, *Arch. Int. Pharmacodyn.* **278**, 72 (1985). Pharmacokinetics: F. L. S. Tee, J. M. Jaffe, *Eur. J. Clin. Pharmacol.* **32**, 361 (1987). Clinical evaluation in angina and coronary artery disease: C. E. Handler, E. Sowton, *ibid.* **27**, 415 (1984); in hypertension: E. B. Nelson *et al.*, *Clin. Pharmacol. Ther.* **40**, 694 (1986). Comparison of hemodynamic effects of enantiomers: R. P. Hof *et al.*, *J. Cardiovasc. Pharmacol.* **8**, 221 (1986). Series of articles on pharmacology and clinical use: *Am. J. Med.* **86**, 1-146 (1989).

mp 168-170°.

S(+)-**Form.** PN-205-033. Crystals from ether + hexane, mp 142°. $[\alpha]_D^{20}$ +6.7° (c = 1.5 in ethanol).

R(−)-**Form.** PN-205-034. Crystals from ether + hexane, mp 140°. $[\alpha]_D^{20}$ −6.7° (c = 1.67 in ethanol).

THERAP CAT: Antihypertensive; antianginal.

5288. Israpafant. [117279-73-9] 4-(2-Chlorophenyl)-6,9-dimethyl-2-[2-[4-(2-methylpropyl)phenyl]ethyl]-6*H*-thieno[3,2-*f*][1,2,4]triazolo[4,3-*a*][1,4]diazepine; (±)-4-(*o*-chlorophenyl)-2-(*p*-isobutylphenethyl)-6,9-dimethyl-6*H*-thieno[3,2-*f*]-*s*-triazolo[4,3-*a*][1,4]diazepine; Y-24180; Pafnol. $C_{28}H_{29}ClN_4S$; mol wt 489.08. C 68.76%, H 5.98%, Cl 7.25%, N 11.46%, S 6.56%. Platelet activating factor (PAF) antagonist. Prepn: T. Tahara *et al.*, **EP 268242**; *eidem*, **US 4820703** (1988, 1989 both to Yoshitomi). Pharmacology: M. Terasawa *et al.*, *Prostaglandins* **40**, 553 (1990). Receptor binding study: S. Takehara *et al.*, *ibid.* 571. Clinical evaluation in asthma: S. Hozawa *et al.*, *Am. J. Respir. Crit. Care Med.* **152**, 1198 (1995).

Colorless crystals from isopropyl ether, mp 129.5-131.5°. Sol in propylene glycol.

THERAP CAT: Antiasthmatic.

5289. Istradefylline. [155270-99-8] 8-[(1*E*)-2-(3,4-Dimethoxyphenyl)ethenyl]-1,3-diethyl-3,7-dihydro-7-methyl-1*H*-purine-2,6-dione; (*E*)-8-(3,4-dimethoxystyryl)-1,3-diethyl-7-methylxanthine; KW-6002. $C_{20}H_{24}N_4O_4$; mol wt 384.44. C 62.49%, H 6.29%, N 14.57%, O 16.65%. Selective adenosine A_{2A} receptor antagonist. Prepn: F. Suzuki *et al.*, **EP 590919**; *eidem*, **US 5484920** (1994, 1996 both to Kyowa). Improved prepn: J. Hockemeyer *et al.*, *J. Org. Chem.* **69**, 3308 (2004). Receptor binding study: L. K. Harper *et al.*, *Pharmacol. Biochem. Behav.* **83**, 114 (2006). Neuroprotective effects in MPTP-treated mice: M. Pierri *et al.*, *Neuropharmacology* **48**, 517 (2005). Clinical experience as dual therapy with levadopa: W. Bara-Jimenez *et al.*, *Neurology* **61**, 293 (2003); as monotherapy in advanced Parkinson's disease: R. A. Hauser *et al.*, *ibid.* 297. Review of development and clinical experience: P. Jenner, *Expert Opin. Invest. Drugs* **14**, 729-738 (2005).

Pale yellow needles from isopropanol, mp 190.4-191.3°.

THERAP CAT: Antiparkinsonian.

5290. Itaconic Acid. [97-65-4] 2-Methylenebutanedioic acid; methylenesuccinic acid; propylenedicarboxylic acid. $C_5H_6O_4$; mol wt 130.10. C 46.16%, H 4.65%, O 49.19%. Obtained by dry distillation of citric acid and subsequent treatment of the anhydride with water. Produced on a large scale by submerged aerobic fermentation using *Aspergillus terreus* and low cost carbohydrates from beet or cane: Kane *et al.*, **US 2385283** (1945 to Pfizer). Synthesis from propargyl chloride, carbon monoxide, nickel carbonyl and water: Chiusoli, **US 3025320** (1962 to Montecatini).

Hygroscopic crystals; characteristic odor. d 1.63. mp 162-164° with decompn. Also reported as mp 172°. *See:* Kinoshita, *Acta Phytochim.* **5**, 273 (1931). One gram dissolves in 12 ml water, 5 ml alcohol; very slightly sol in benzene, chloroform, ether, carbon disulfide, petr ether. *Keep well closed.*

5291. Itopride. [122898-67-3] *N*-[[4-[2-(Dimethylamino)-ethoxy]phenyl]methyl]-3,4-dimethoxy benzamide; *N*-[*p*-[2-(dimethylamino)ethoxy]benzyl]veratramide. $C_{20}H_{26}N_2O_4$; mol wt 358.44. C 67.02%, H 7.31%, N 7.82%, O 17.85%. Dopamine D_2-receptor antagonist with anticholinesterase activity. Prepn: Y. Itoh *et al.*, **EP 306827**; *eidem*, **US 4983633** (1989, 1991 both to Hokuriku Pharm.); J. Sakaguchi *et al.*, *Chem. Pharm. Bull.* **40**, 202 (1992). Pharmacology: Y. Iwanaga *et al.*, *Jpn. J. Pharmacol.* **56**, 261 (1991). Toxicity studies: M. H. Barker *et al.*, *Yakuri to Chiryo* **21**, 1685, 1669 (1993). Metabolism study: T. Mushiroda *et al.*, *Drug Metab. Dispos.* **28**, 1231 (2000). HPLC determn in serum and clinical pharmacokinetics: S. S. Singh *et al.*, *J. Chromatogr. B* **818**, 213 (2005). Clinical study in postoperative ileus: R. Gürlich *et al.*, *Chir. Gastroenterol. Interdiszip.* **20**, 61 (2004); in dyspepsia: D. N. Amarapurkar, P. Rane, *J. Indian Med. Assoc.* **102**, 735 (2004).

Crystals from ethanol + isopropyl ether, mp 111-112°.

Hydrochloride. [122892-31-3] HSR-803; HC-803; Itax; Ganaton. $C_{20}H_{26}N_2O_4$.HCl; mol wt 394.90. Crystals from ethanol, mp 194-195°. Also reported as prisms from ethanol, mp 190.5-191.5° (Sakaguchi). Readily sol in water.

THERAP CAT: Gastroprokinetic.

5292. Itraconazole. [84625-61-6] 4-[4-[4-[4-[[2-(2,4-Dichlorophenyl)-2-(1*H*-1,2,4-triazol-1-ylmethyl)-1,3-dioxolan-4-yl]methoxy]phenyl]-1-piperazinyl]phenyl]-2,4-dihydro-2-(1-methylpropyl)-3*H*-1,2,4-triazol-3-one; (±)-1-*sec*-butyl-4-[*p*-[4-[*p*-[[(2*R**,4*S**)-2-(2,4-dichlorophenyl)-2-(1*H*-1,2,4-triazol-1-ylmethyl)-1,3-dioxolan-4-yl]methoxy]phenyl]-1-piperazinyl]phenyl]-Δ²-1,2,4-triazolin-5-one; oriconazole; R-51211; Itrizole; Sempera; Siros; Sporanox; Triasporin. $C_{35}H_{38}Cl_2N_8O_4$; mol wt 705.64. C 59.57%, H 5.43%, Cl 10.05%, N 15.88%, O 9.07%. Orally active antimycotic structurally related to ketoconazole, *q.v.* Has 3 chiral centers; mixture of 4 *cis*-isomers. Prepn: J. Heeres, L. J. J. Backx, **EP 6711**; *eidem*, **US 4267179** (1980, 1981 both to Janssen); J. Heeres *et al.*, *J. Med. Chem.* **27**, 894 (1984). *In vitro* activity: A. Espinel-Ingroff *et al.*, *Antimicrob. Agents Chemother.* **26**, 5 (1984). HPLC determn in biological samples: R. Woestenborghs *et al.*, *J. Chromatogr.* **413**, 332 (1987). Symposium on pharmacology and clinical efficacy: *Rev. Infect. Dis.* **9**, Suppl 1, S1-S152 (1987). Toxicity data: H. Van Cauteren *et al.*, *ibid.* S43. Review of clinical pharmacokinetics: J. Heykants *et al.*, *Mycoses* **32**, Suppl 1, 67-87 (1989); A. G. Prentice, A. Glasmacher, *J. Antimicrob. Chemother.* **56**, i17-i22 (2005); of clinical efficacy in dermatophytosis: P. De Doncker, G. Cauwenbergh, *Br. J. Clin. Pract.* **Suppl. 71**, 118-122 (1990). Review: A. M. Sugar, *Curr. Clin. Top. Infect. Dis.* **13**, 74-98 (1993); R. Caputo, *Expert Rev. Anti Infect. Ther.* **1**, 531-542 (2003).

Relative stereochemistry

Crystals from toluene, mp 166.2°. pKa 3.7. Lipophilic; partition coefficient (*n*-octanol/aq buffer of pH 8.1): 5.66. Practically insol in water and dil acidic solns. LD_{50} (14 day) in mice, rats, dogs (mg/kg): >320, >320, >200 orally (Van Cauteren).

THERAP CAT: Antifungal.

THERAP CAT (VET): Antifungal.

5293. Itramin Tosylate. [13445-63-1] 2-Aminoethanol 1-nitrate 4-methylbenzenesulfonate (1:1); 2-aminoethanol nitrate mono-*p*-toluenesulfonate; 2-nitroethylaminotoluene-*p*-sulfonate; Cardisan; Tostram; Nilatil. $C_9H_{14}N_2O_6S$; mol wt 278.28. C 38.85%, H 5.07%, N 10.07%, O 34.50%, S 11.52%. Prepn: **SE 168308** (1959 to Aktiebolaget Pharmacia), *C.A.* **54**, 24405d (1960).

Crystals from ethanol, mp 132-133°.

THERAP CAT: Vasodilator.

5294. Ivabradine. [155974-00-8] 3-[3-[[[(7*S*)-3,4-Dimethoxybicyclo[4.2.0]octa-1,3,5-trien-7-yl]methyl]methylamino]propyl]-1,3,4,5-tetrahydro-7,8-dimethoxy-2*H*-3-benzazepin-2-one; 7,8-dimethoxy-3-[3-[[(1*S*)-(4,5-dimethoxybenzocyclobutan-1-yl)methyl]methylamino]propyl]-1,3,4,5-tetrahydro-2*H*-benzazepin-2-one; S-16257. $C_{27}H_{36}N_2O_5$; mol wt 468.59. C 69.21%, H 7.74%, N 5.98%, O 17.07%. Selective bradycardic agent with direct effect on the pacemaker I_f current of the sinoatrial node. Prepn: J.-L. Peglion *et al.*, **EP 534859**; *eidem*, **US 5296482** (1993, 1994 both to ADIR). *In vitro* electrophysiological effects: C. Thollon *et al.*, *Br. J. Pharmacol.* **112**, 37 (1994). LC-MS determn in plasma: M. François-Bouchard *et al.*, *J. Chromatogr. B* **745**, 261 (2000). Clinical pharmacokinetics: I. Ragueneau *et al.*, *Clin. Pharmacol. Ther.* **64**, 192 (1998). Clinical trial in stable angina: J. S. Borer *et al.*, *Circulation* **107**, 817 (2003); in chronic heart failure: K. Swedberg *et al.*, *Lancet* **376**, 875 (2010). Review of mechanism of action: D. DiFrancesco, J. A. Camm, *Drugs* **64**, 1757-1765 (2004); of clinical experience and therapeutic potential: G. Riccioni, *Expert Opin. Pharmacother.* **12**, 443-450 (2011).

Hydrochloride. [148849-67-6] S-16257-2; Coralan; Coraxam; Corlentor; Procoralan. $C_{27}H_{36}N_2O_5$.HCl; mol wt 505.05. Crystals from acetonitrile, mp 135-140°. $[\alpha]_{589}^{21}$ +7.8°; $[\alpha]_{365}^{21}$ +27.8° (c = 1% in DMSO).

(±)-**Form.** [149470-60-0] $C_{27}H_{36}N_2O_5$. Crystals from ethyl acetate, mp 101-103°.

THERAP CAT: Antianginal.

5295. Ivacaftor. [873054-44-5] *N*-[2,4-Bis(1,1-dimethylethyl)-5-hydroxyphenyl]-1,4-dihydro-4-oxo-3-quinolinecarboxamide; *N*-(5-hydroxy-2,4-di-*tert*-butylphenyl)-4-oxo-1*H*-quinoline-3-carboxamide; VX-770; Kalydeco. $C_{24}H_{28}N_2O_3$; mol wt 392.50. C 73.44%, H 7.19%, N 7.14%, O 12.23%. Potentiator of the cystic fibrosis transmembrane conductance regulator (CFTR) protein. Prepn: S. Hadida-Ruah *et al.*, **WO 06002421**; *eidem*, **US 7495103** (2006, 2009 both to Vertex). *In vitro* pharmacology in epithelial cells: F. Van Goor *et al.*, *Proc. Natl. Acad. Sci. USA* **106**, 18825 (2009). Clinical pharmacology and safety: F. J. Accurso *et al.*, *N. Engl. J. Med.* **363**, 1991 (2010). Clinical trial in cystic fibrosis patients with the G551D mutation: B. W. Ramsey *et al.*, *ibid.* **365**, 1663 (2011).

THERAP CAT: In treatment of cystic fibrosis.

5296. Ivermectin. [70288-86-7] 22,23-Dihydroabamectin; 22,23-dihydroavermectin B₁; 22,23-dihydro C-076B₁; MK-933; Acarexx; Eqvalan; Heartgard; Ivomec; Mectizan; Noromectin; Stromectol. Semi-synthetic deriv of abamectin, *q.v.*; consists of a mixture of not less than 80% component B_{1a} and not more than 20% component B_{1b}. Prepn: **JP Kokai 79 61198**; J. C. Chabala, M. H. Fisher, **US 4199569** (1979, 1980 both to Merck & Co.); J. C. Chabala *et al.*, *J. Med. Chem.* **23**, 1134 (1980). Irreversible effects on adult parasites in onchocerciasis: A. P. Plaisier *et al.*, *J. Infect. Dis.* **172**, 204 (1995). Metabolism in animals: S.-H. L. Chiu *et al.*, *Drug*

     Consult the Name Index before using this section.

Metab. Dispos. **14**, 590 (1986). Pharmacokinetics in horses and sheep: S. E. Marriner *et al., J. Vet. Pharmacol. Ther.* **10**, 175 (1987). HPLC determn in cattle and sheep tissues: P. C. Tway *et al., J. Agric. Food Chem.* **29**, 1059 (1981); in human plasma and milk: R. Chiou *et al., J. Chromatogr.* **416**, 196 (1987). Review of early chemistry and biology: W. C. Campbell *et al., Science* **221**, 823-828 (1983). Review of clinical use and pharmacology: W. C. Campbell, G. W. Benz, *J. Vet. Pharmacol. Ther.* **7**, 1-16 (1984); T. B. Barragry, *Can. Vet. J.* **28**, 512-517 (1987); of pharmacology: J. L. Bennett *et al., Parasitol. Today* **4**, 226-228 (1988). Book: *Ivermectin and Abamectin,* W. C. Campbell, Ed. (Springer-Verlag, New York, 1989) 363 pp. Review in filariasis: D. Richard-Lenoble *et al., Fundam. Clin. Pharmacol.* **17**, 199-203 (2003); in dermatologic applications: A. L. Dourmishev *et al., Int. J. Dermatol.* **44**, 981-988 (2005). *See also:* Avermectins.

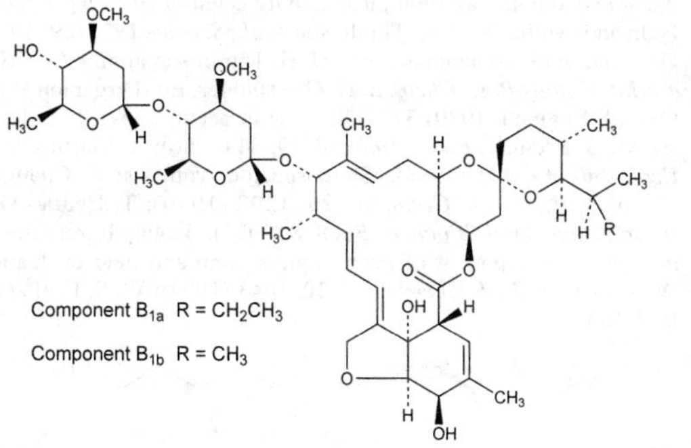

Component B₁ₐ R = CH₂CH₃

Component B₁ᵦ R = CH₃

White to yellowish-white, slightly hygroscopic, crystalline powder, mp ~ 155°. [α]$_D$ +71.5 ± 3° (c = 0.755 in chloroform). uv max (methanol): 238, 245 nm (ε 27100, 30100). Soly in water: ~4 μg/ml. Very sol in methyl ethyl ketone, propylene glycol, polyethylene glycol. Freely sol in methanol, methylene chloride; sol in acetone, acetonitrile and in 95% ethanol. Practically insol in satd hydrocarbons such as cyclohexane; insol in hexane. LD$_{50}$ in dogs, rhesus monkeys (mg/kg): ~ 80, > 24 orally. (MSDS Merck & Co. Inc., 2003).

Component B₁ₐ. [71827-03-7] 5-*O*-Demethyl-22,23-dihydroavermectin A₁ₐ; 22,23-dihydroavermectin B₁ₐ; 22,23-dihydro C-076B₁ₐ. $C_{48}H_{74}O_{14}$; mol wt 875.11. Crystals from ethanol/water, mp 155-157°.

Component B₁ᵦ. [70209-81-3] 5-*O*-Demethyl-25-de(1-methylpropyl)-22,23-dihydro-25-(1-methylethyl)avermectin A₁ₐ; 22,23-dihydroavermectin B₁ᵦ. $C_{47}H_{72}O_{14}$; mol wt 861.08.

THERAP CAT: Anthelmintic (Onchocerca).

THERAP CAT (VET): Anthelmintic, insecticide, acaricide.

5297. Ixabepilone. [219989-84-1] (1*S*,3*S*,7*S*,10*R*,11*S*,12*S*,-16*R*)-7,11-Dihydroxy-8,8,10,12,16-pentamethyl-3-[(1*E*)-1-methyl-2-(2-methyl-4-thiazolyl)ethenyl]-17-oxa-4-azabicyclo[14.1.0]hepta-decane-5,9-dione; azaepothilone B; BMS-247550; NSC-710428; Ixempra. $C_{27}H_{42}N_2O_5S$; mol wt 506.70. C 64.00%, H 8.36%, N 5.53%, O 15.79%, S 6.33%. Microtubule stabilizing semisynthetic lactam analog of epothilone B, *q.v.* Prepn: G. D. Vite *et al.,* **WO 9902514**; *eidem,* **US 6605599** (1999, 2003 both to Bristol-Myers Squibb); R. M. Borzilleri *et al., J. Am. Chem. Soc.* **122**, 8890 (2000). Antineoplastic activity *in vitro* and *in vivo*: F. Y. F. Lee *et al., Clin. Cancer Res.* **7**, 1429 (2001). Clinical evaluation with estramustine in prostate cancer: J. E. Rosenberg *et al., Cancer* **106**, 58 (2006). Clinical pharmacokinetics and evaluation in solid tumors: S. Mani *et al., Clin. Cancer Res.* **10**, 1289 (2004). Clinical evaluation in breast cancer: J. A. Low *et al., J. Clin. Oncol.* **23**, 2726 (2005); in prostate cancer: M. Hussain *et al., ibid.* 8724. Review of pharmacology and clinical development: N. Lin *et al., Curr. Opin. Investig. Drugs* **4**, 746-756 (2003).

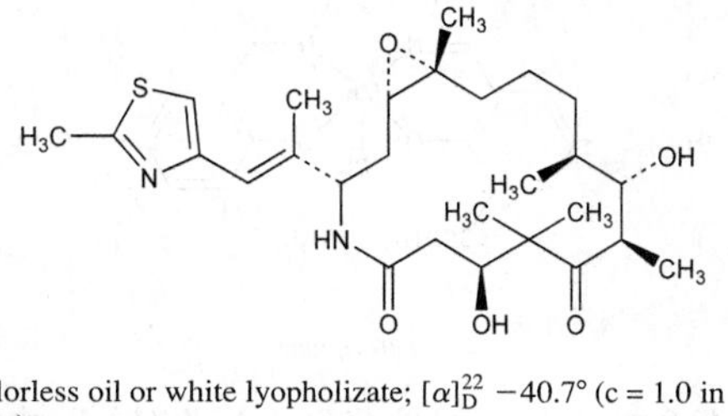

Colorless oil or white lyopholizate; [α]$_D^{22}$ −40.7° (c = 1.0 in chloroform).

THERAP CAT: Antineoplastic.

5298. Ixbut. A native name for *Euphorbia lancifolia* Schlecht, *Euphorbiaceae.* Used as a galactagogue: Serrano, *Time,* August 1st, **1949**, p 38; F. Rosengarten, Jr., *Bot. Mus. Leafl. Harv. Univ.* **26** (9-10), 277-309 (Nov.-Dec. 1978).

5299. Ixodin. Hirudin-like anticoagulant from *Ixodes ricinus* L., *Acari* and other blood-sucking ticks: Sabbatani, *Arch. Ital. Biol.* **31**, 37 (1899). Thrombokinase inhibitor. Extraction procedure: F. Markwardt, *Blutgerinnungshemmende Wirkstoffe aus blutsaugenden Tieren* (Jena, 1963) pp 88-92.

White powder. Soluble in 80% ethanol, 90% methanol and 75% acetone. Practically insol in ether. Stable at pH 1.0-12.0. Adsorbed on Amberlite IRC 50 (H-form). Acts like a high-molecular weight polypeptide. Attacked by papain.

J

5300. Jacobsen's Catalyst. [149656-63-3] (*SP*-5-13)-Chloro[[*rel*-2,2′-[(1*R*,2*R*)-1,2-cyclohexanediylbis[(nitrilo-κ*N*)methylidyne]]bis[4,6-bis(1,1′-dimethylethyl)phenolato-κ*O*]](2−)]manganese. C$_{36}$H$_{52}$ClMnN$_2$O$_2$; mol wt 635.21. C 68.07%, H 8.25%, Cl 5.58%, Mn 8.65%, N 4.41%, O 5.04%. Manganese-salen Schiff base complex used as a catalyst in asymmetric syntheses. Prepn: E. N. Jacobsen *et al.*, *J. Am. Chem. Soc.* **113**, 7063 (1991); large-scale: J. F. Larrow *et al.*, *J. Org. Chem.* **59**, 1939 (1994). Synthesis and enantioselectivity of dimer: K. B. M. Janssen *et al.*, *Tetrahedron: Asymmetry* **8**, 3481 (1997). HPLC determn of enantiomeric purity in commercial samples: J. Zukowski, *Chirality* **10**, 362 (1998). Mechanism of asymmetric epoxidation of indenes: D. L. Hughes *et al.*, *J. Org. Chem.* **62**, 2222 (1997). Use in epoxidation of isoflavones: W. Adam *et al.*, *Tetrahedron: Asymmetry* **9**, 1121 (1998).

(R,R)-form

(*R,R*)-form. [138124-32-0] (*R,R*)-[*N,N′*-Bis(3,5-di-*tert*-butylsalicylidene)-1,2-cyclohexanediamine]manganese(III) chloride. mp 324-326°. [α]$_D^{23}$ +580° (c = 0.01 in ethanol).

USE: Chiral catalyst for epoxidation of olefins.

5301. Jalap. Dried tuberous root of *Exogonium purga* (Hayne) Lindl. (*E. jalapa* Baill., *Ipomoea purga* Hayne), Convolvulaceae. *Habit.* Mexico; culivated in India. *Constit.* 7-12% resin, gum, sugar.

THERAP CAT: Cathartic.

5302. Jambul. Jamboo; Java plum; jumbul. Bark, fruit and seeds of *Syzygium jambolanum* (Lam.) DC. (*Eugenia jambolana* Lam.), *Myrtaceae. Habit.* East Indies. *Constit.* Bark: resin, tannin. Fruit: volatile and fixed oils, resin, tannin. Seeds: resin, fat, gallic acid, albumin.

THERAP CAT: Antidiarrheal.

5303. Janus Green B. [2869-83-2] 3-(Diethylamino)-7-[2-[4-(dimethylamino)phenyl]diazenyl]-5-phenylphenazinium chloride (1:1); C.I. 11050. C$_{30}$H$_{31}$ClN$_6$; mol wt 511.07. C 70.51%, H 6.11%, Cl 6.94%, N 16.44%. Prepd by diazotizing 3-amino-7-(diethylamino)-5-phenylphenazinium chloride (*N,N*-diethylphenosafranine) and coupling resulting diazo compd with *N,N*-dimethylaniline: *Colour Index* vol. **4** (3rd ed., 1971) p 4015.

USE: As dye for cotton, wool; as biological stain; in electrodeposition of copper, Brown, Fellows, US **2882209** (1959 to Udylite Res. Corp.).

5304. Japan Wax. [8001-39-6] Vegetable wax; sumach wax; Japan tallow. A fat expressed from mesocarp of the fruit of *Rhus succedanea* L., Anacardiaceae. *Habit.* Japan and China. *Constit.* 10-15% palmitin; stearin, olein; 1% japanic acid and homologs. Brief review: C. S. Letcher in *Kirk-Othmer Encyclopedia of Chem-*

ical Technology vol. **24** (Wiley-Interscience, New York, 3rd ed., 1984) p 470.

Pale yellow, flat cakes, disks or squares with a greasy feel; somewhat tallow-like, rancid odor and taste. d 0.97-0.98. mp 53.5-55°. Acid number 22-23. Saponification number 217-237. Iodine number 10-15. Insol in water or cold alcohol; sol in benzene, carbon disulfide, petr ether, ether, hot alc, alkalies.

USE: As a substitute for beeswax in wax varnishes or candles, ingredients in plasters, ointments; floor waxes, furniture polish. As a plasticizer in dental impression compounds.

5305. Japonilure. [64726-91-6] (5*R*)-5-(1*Z*)-1-Decen-1-yldihydro-2(3*H*)-furanone; [(4*R*,5*Z*)]-tetradecen-4-olide; (*R,Z*)-5-(dec-1-enyl)oxacyclopentan-2-one; IN-60. C$_{14}$H$_{24}$O$_2$; mol wt 224.34. C 74.95%, H 10.78%, O 14.26%. Sex pheromone produced by the female Japanese beetle, *Popillia japonica* Newman, *Scarabaeidae* (Coleoptera); biological activity is inhibited by its antipode. Isoln and synthesis: J. H. Tumlinson *et al.*, *Science* **197**, 789 (1977). Description of attractant activity: J. H. Tumlinson in *Adv. Pest. Sci.*, *4th Int. Congr. Pest. Chem.* **2**, H. Geissbühler, Ed (Pergamon Press, Oxford, England, 1979) 315-322. Use in beetle traps: T. L. Ladd, Jr., M. G. Klein, *J. Econ. Entomol.* **79**, 84 (1986); A. Martins *et al.*, *Ecol. Bull.* **39**, 101 (1988). Stereospecific synthesis: S. Chattopadhyay *et al.*, *Synth. Commun.* **20**, 1299 (1990); T. Ebata *et al.*, *Biosci. Biotechnol. Biochem.* **56**, 818 (1992). Isolation from *Anomala spp.* as component of pheromone system and field evaluation: W. S. Leal *et al.*, *J. Chem. Ecol.* **20**, 1643 (1994); W. S. Leal *et al.*, *ibid.* 1667.

bp$_{0.1mm}$ 130-134°. [α]$_D^{21}$ −69.0° (c = 1.42 in CHCl$_3$). [α]$_D^{25}$ −70.2° (c = 0.5 in CHCl$_3$).

5306. Jasmolins. Active insecticidal constituents of pyrethrum flowers. Isoln and structure: Godin *et al.*, *J. Econ. Entomol.* **58**, 548 (1965); Godin *et al.*, *J. Chem. Soc. C* **1966**, 322. Stereochemistry: Begley *et al.*, *Chem. Commun.* **1972**, 1276. Review of toxicology and human exposure: *Toxicological Profile for Pyrethrins and Pyrethroids* (PB2004-100004, 2003) 332 pp.

Jasmolin I   R = CH$_3$
Jasmolin II   R = COOCH$_3$

Jasmolin I. [4466-14-2] (1*R*,3*R*)-2,2-Dimethyl-3-(2-methyl-1-propenyl)cyclopropanecarboxylic acid (1*S*)-2-methyl-4-oxo-3-(2*Z*)-2-pentenyl-2-cyclopenten-1-yl ester; 4′,5′-dihydropyrethrin I. C$_{21}$H$_{30}$O$_3$; mol wt 330.47. Liquid. uv max (Spectrosol hexane): 219 nm (ε 21500).

Jasmolin II. [1172-63-0] (1*R*,3*R*)-3-[(1*E*)-3-Methoxy-2-methyl-3-oxo-1-propenyl]-2,2-dimethylcyclopropanecarboxylic acid (1*S*)-2-methyl-4-oxo-3-(2*Z*)-2-pentenyl-2-cyclopenten-1-yl ester; 4′,5′-dihydropyrethrin II. C$_{22}$H$_{30}$O$_5$; mol wt 374.48. Liquid. uv max (Spectrosol hexane): 229 nm (ε 22900).

5307. Jasmone. [488-10-8] 3-Methyl-2-(2*Z*)-2-penten-1-yl-2-cyclopenten-1-one. C$_{11}$H$_{16}$O; mol wt 164.25. C 80.44%, H 9.82%, O 9.74%. Found in the volatile portion of oil from jasmine flowers. Natural jasmone is the *cis*-ketone. Isoln and structure: Ruzicka, Pfeiffer, *Helv. Chim. Acta* **16**, 1208 (1933). Stereochemistry: Crombie, Harper, *J. Chem. Soc.* **1952**, 869. Synthesis of *cis*-jasmone: H. Hunsdiecker, *Ber.* **75**, 447 (1942); Stork, Borch, *J. Am. Chem. Soc.* **86**, 936 (1964); Büchi, Wuest, *J. Org. Chem.* **31**, 977 (1966); Sakan *et al.*, *Chem. Lett.* **1973**, 713; P. Bakuzis, M. L. F. Bakuzis, *J. Org. Chem.* **42**, 2362 (1977). Synthesis of dihydrojasmone and *cis*-jasmone: C. S. Subramaniam *et al.*, *J. Chem. Soc. Perkin Trans. 1* **1979**, 2346; T. Kato *et al.*, *Chem. Pharm. Bull.* **28**,

349 (1980). Synthesis of *trans*-jasmone: Sisido *et al.*, *J. Org. Chem.* **29**, 2290 (1964). Comprehensive synthetic reviews: R. A. Ellison, *Synthesis* **1973**, 397; T. L. Ho, *Synth. Commun.* **4**, 265 (1974).

cis - Form

Oil, odor of jasmine, bp$_{27}$ 146°. n_D^{20} 1.4978. uv max: 235 nm (ε 12,000).

trans-Form. [6261-18-3] Oil, odor of jasmine, bp$_{23}$ 142°. n_D^{20} 1.4974. uv max: 234 nm (ε 12300).

USE: In perfumery.

5308. Jasmonic Acid. [6894-38-8] (1*R*,2*R*)-3-Oxo-2-(2*Z*)-2-penten-1-ylcyclopentaneacetic acid; (1*S*,2*S*)-3-oxo-2-(2'-*cis*-pentenyl)cyclopentan-1-acetate; JA. C$_{12}$H$_{18}$O$_3$; mol wt 210.27. C 68.55%, H 8.63%, O 22.83%. Naturally occurring plant growth inhibitor. Orginally isolated from jasmin flowers as the methyl ester, the major aroma component: E. Demole *et al.*, *Helv. Chim. Acta* **45**, 675 (1962). Isoln of the free acid from *Lasiodiplodia theobromae*: D. C. Aldridge *et al.*, *J. Chem. Soc.* **1971**, 1623; by fermentation: D. Broadbent *et al.*, *GB 1286266* (1972 to ICI). Occurrence in plants: A. Meyer *et al.*, *J. Plant Growth Regul.* **3**, 1 (1984). NMR characterization: A. Husain *et al.*, *J. Nat. Prod.* **56**, 2008 (1993). Quantification by GC/MS: M. J. Mueller, W. Brodschelm, *Anal. Biochem.* **218**, 425 (1994). Resoln of racemate: R. Kramell *et al.*, *Phytochem. Anal.* **7**, 209 (1996). Review of role in plant development: Y. Koda, *Int. Rev. Cytol.* **135**, 155-199 (1992); of biochemisty and physiology: G. Sembdner, B. Parthier, *Annu. Rev. Plant Physiol. Plant Mol. Biol.* **44**, 569-589 (1993); of signaling in plants: P. Staswick, *Curr. Top. Plant Physiol.* **11**, 14-23 (1993).

Colorless oil. n_D^{23} 1.486. [α]$_D^{25}$ −73° (c = 1 in CH$_3$OH).

Methyl ester. [1211-29-6] (−)-Methyl *cis*-2-pent-2'-enyl-3-oxocyclopentylacetate; JA-ME; MeJA; methyl jasmonate. C$_{13}$H$_{20}$O$_3$; mol wt 224.30. Colorless liquid with a sharp persistent odor. $n_D^{21.8}$ 1.4730. [α]$_D$ −76.5° (c = 3.4 in CH$_3$OH). d$_4^{22.6}$ 1.021.

USE: Methyl ester in perfumes.

5309. Jatrorrhizine. [3621-38-3] 5,6-Dihydro-3-hydroxy-2,9,10-trimethoxydibenzo[*a,g*]quinolizinium; 7,8,13,13a-tetradehydro-3-hydroxy-2,9,10-trimethoxyberbinium; jateorrhizine; neprotin. [C$_{20}$H$_{20}$NO$_4$]$^+$. From root of *Jateorhiza palmata* (DC.) Miers *(J. columba* Miers), *Menispermaceae*: Feist, *Arch. Pharm.* **245**, 586 (1907); from *Berberis asiatica* Roxb. ex DC. and *B. thunbergii* DC., *Berberidaceae*: Chatterjee *et al.*, *J. Indian Chem. Soc.* **31**, 83 (1954); Tomita, Kikuchi, *J. Pharm. Soc. Jpn.* **76**, 597 (1956); from *Coptis teeta* Wall., *Ranunculaceae*: Chatterjee *et al.*, *J. Indian Chem. Soc.* **29**, 97 (1952); from *Mahonia acanthifolia* Wall., *M. borealis* Takeda and *M. simonsii* Takeda, *Berberidaceae*: Chatterjee, Guha, *J. Am. Pharm. Assoc.* **39**, 577 (1950); Chatterjee *et al.*, *ibid.* **40**, 36 (1951). Structure: Späth, Duschinsky, *Ber.* **58**, 1939 (1925).

Iodide. C$_{20}$H$_{20}$INO$_4$. Reddish-yellow needles, mp 208-210°. Sol in water and alcohol.

5310. Javanicin. [476-45-9] 5,8-Dihydroxy-6-methoxy-2-methyl-3-(2-oxopropyl)-1,4-naphthalenedione; 3-acetonyl-5,8-dihydroxy-6-methoxy-2-methyl-1,4-naphthoquinone. C$_{15}$H$_{14}$O$_6$; mol wt 290.27. C 62.07%, H 4.86%, O 33.07%. Antibiotic substance produced by *Fusarium javanicum*: Arnstein, Cook, *J. Chem. Soc.* **1947**, 1021. Prepd by reduction of fusarubin: Ruelius, Gauhe, *Ann.* **569**, 38 (1950). Structure: Birch, Donovan, *Chem. Ind. (London)* **1954**, 1047; Whalley, *ibid.* **1958**, 131; Hardegger *et al.*, *Helv. Chim. Acta* **47**, 2027 (1964).

Red crystals with a coppery luster from ethanol, decomp 207.5-208°. Absorption max (alc): 303, 305 nm (log ε 3.97, 3.90); in chloroform: 307, 510 nm (log ε 3.99, 3.86).

Diacetyljavanicin. C$_{19}$H$_{18}$O$_8$. Needles from dil acetone, dec 207-208°. Absorption max (alc): 221, 290, 426 nm (log ε 4.57, 4.17, 3.79).

5311. Jerusalem Artichoke. Topinambur. The subterranean stem tuber of *Helianthus tuberosus* L., *Compositae*, a kind of sunflower native to North America. Used as food by American Indians. Contains up to 20% (wet basis) of polysaccharides usable as sweetening agents: Rubin, *US 2782123* (1957).

5312. Jervine. [469-59-0] (2'*R*,3*S*,3'*R*,3'a*S*,6'*S*,6a*S*,6b*S*,-7'a*R*,11a*S*,11b*R*)-2,3,3'a,4,4',5',6,6',6a,6b,7,7',7'a,8,11a,11b-Hexadecahydro-3-hydroxy-3',6',10,11b-tetramethylspiro[9*H*-benzo[*a*]fluorene-9,2'(3'*H*)-furo[3,2-*b*]pyridin]-11(1*H*)-one; (3β,-23β)-17,23-epoxy-3-hydroxyveratraman-11-one. C$_{27}$H$_{39}$NO$_3$; mol wt 425.61. C 76.20%, H 9.24%, N 3.29%, O 11.28%. Steroidal alkaloid found in *Veratrum grandiflorum* (Maxim.) Loes F., *V. album* L., *V. viride* Sol., *Liliaceae*: Saito, *Bull. Chem. Soc. Jpn.* **9**, 15 (1934), *C.A.* **28**, 2463 (1934); Poethke, *Arch. Pharm.* **275**, 357, 571 (1937); *ibid.* **276**, 170 (1938); *ibid.* **282**, 56 (1944); Seiferle *et al.*, *J. Econ. Entomol.* **35**, 35 (1942); Jacobs, Craig, *J. Biol. Chem.* **160**, 555 (1945). Structure: Fried *et al.*, *J. Am. Chem. Soc.* **73**, 2970 (1951). Reviews: Wintersteiner in Graff, *Essays in Biochemistry* (Wiley, New York, 1956) pp 308-321; *idem, Festschrift Arthur Stoll* (Basel, 1957) pp 166-176; L. F. Fieser, M. Fieser, *Steroids* (Reinhold, New York, 1959) pp 870-877. Stereochemistry: Sicher, Tichy, *Tetrahedron Lett.* no. 12, 6 (1959); Augustine, *Chem. Ind. (London)* **1961**, 1448; Mitsuhashi, Shimizu, *Tetrahedron* **19**, 1027 (1963); Bailey *et al.*, *Tetrahedron Lett.* **1963**, 555; Masamune *et al.*, *ibid.* **1965**, 489. Revised stereochemistry: Scott *et al.*, *ibid.* **1967**, 2381; Kupchan, Suffness, *J. Am. Chem. Soc.* **90**, 2730 (1968); Sprague *et al.*, *Tetrahedron* **27**, 4857 (1971). Total synthesis: Kutney *et al.*, *Can. J. Chem.* **53**, 1796 (1975). Comparative toxicity: O. Krayer *et al.*, *J. Pharmacol. Exp. Ther.* **82**, 167 (1944); K. Tanaka, *ibid.* **113**, 89 (1955).

Needles from methanol + water, mp 243.5-244.5° (Saito). [α]$_D^{20}$ −150° (ethanol) (Saito); [α]$_D^{20}$ −167.6° (chloroform) (Poethke). uv max: 250, 360 nm (ε 15000, 60). LD$_{50}$ i.v. in mice: 9.3 mg/kg (Krayer). LD$_{50}$ s.c. in male mice: 29 mg/kg (Tanaka).

Diacetyljervine. [7622-06-2] C$_{31}$H$_{43}$NO$_5$. mp 173-175°. [α]$_D$ −112°. uv max (ethanol): 250, 360 nm (ε 16400, 80).

Hydrochloride. [903508-18-9] $C_{27}H_{39}NO_3 \cdot HCl$; mol wt 462.07. mp 300-302°.

5313. Jesaconitine. [16298-90-1] (1α,3α,6α,14α,15α,16β)-20-Ethyl-1,6,16-trimethoxy-4-(methoxymethyl)aconitane-3,8,13,-14,15-pentol 8-acetate 14-(4-methoxybenzoate). $C_{35}H_{49}NO_{12}$; mol wt 675.77. C 62.21%, H 7.31%, N 2.07%, O 28.41%. From *Aconitum fischeri* Reich. and *A. sachalinense* F. Schmidt, *Ranunculaceae:* Makoshi, *Arch. Pharm.* **247**, 243 (1909); Majima, Morio, *Ber.* **57**, 1472 (1924). Structure: Ochiai *et al., J. Pharm. Soc. Jpn.* **75**, 545 (1955); Keith, Pelletier, *Chem. Commun.* **1967**, 993; corrected in: *ibid.* **1968**, 1739; *eidem, J. Org. Chem.* **33**, 2497 (1968).

Amorphous powder, dec 128-131°. Sol in ether, dil acids.
Perchlorate. $C_{35}H_{49}NO_{12} \cdot HClO_4$. Prisms from alcohol, dec 230-232°. $[\alpha]_D$ −16.7° (methanol).

5314. Jojoba Oil. Oil of jojoba. A liquid wax ester mixture extracted from ground or crushed seeds from *Simmondsia chinensis* and *S. californica* Nutt. *Buxaceae,* desert shrubs native to Arizona, California, and northern Mexico: Greene, Foster, *Bot. Gaz.* **94**, 826 (1933); Green *et al., J. Chem. Soc.* **1936**, 1750; McKinney, Jamieson, *Oil Soap (Chicago)* **13**, 289 (1936). Similar to sperm whale oil, it is composed essentially of C_{20} and C_{22} straight chain monoethylene acids and alcohols in the form of esters: Molaison *et al., J. Am. Oil Chem. Soc.* **36**, 379 (1959); Miwa, *ibid.* **48**, 259 (1971). Solvent effects in extraction: Knoepfler *et al., ibid.* **36**, 644 (1959). Comparison of sulfurized jojoba and sperm whale oils as high pressure lubricants: T. K. Miwa *et al., ibid.* **56**, 765 (1979). Potential chemical utilization studies: Fore *et al., ibid.* **37**, 387 (1960); J. D. Johnson, C. W. Hinman, *Science* **208**, 460 (1980). Possible uses: J. H. Brown, *Manuf. Chem.* **50**(6), 47 (1979). *Reviews:* Knoepfler, Vix, *J. Agric. Food Chem.* **6**, 118 (1958); *Products from Jojoba: A Promising New Crop for Arid Lands,* Committee on Jojoba Utilization, Natl. Res. Council, 1975; *Jojoba: New Crop for Arid Lands, New Raw Material for Industry* Natl. Res. Council, 1985.

Liquid wax. fp 10.6 to 7°; mp 6.8 to 7°; bp$_{757}$ (under N_2) 398°. Fire point 338°. Pour point 10°. d^{25} 0.8642. n^{25} 1.4648. Iodine no. 81.7. Saponification value 92.2. Acid value 0.32. Avg mol wt of wax esters: 606. Highly stable, and resistant to bacterial degradation; can be stored for years without becoming rancid.

USE: Potentially as lubricant, fuel, chemical feedstock, substitute for sperm whale oil. For other potential uses *see* Fore *et al., loc. cit.,* Knoepfler, Vix, *loc. cit,* Johnson, Hinman, *loc. cit.*

5315. Josamycin. [16846-24-5] Leucomycin V 3-acetate 4^B-(3-methylbutanoate); leucomycin A₃; EN-141; Iosalide; Jomybel; Josamina. $C_{42}H_{69}NO_{15}$; mol wt 828.01. C 60.92%, H 8.40%, N 1.69%, O 28.98%. Macrolide antibiotic produced by *Streptomyces narbonensis* var. *josamyceticus* nov. var. Isoln and characterization: T. Osono *et al., J. Antibiot.* **20A**, 174 (1967). Manuf process: H. Umezawa, T. Osono, **JP 66 21759** (1966 to Microbiochemical Res. Found.), *C.A.* **66**, 54258w (1967). Prepn: Y. Oka *et al.,* **JP Kokai 76 41497** (1976 to Yamanouchi); *C.A.* **85**, 121788b (1976). Identification with leucomycin A₃: S. Omura *et al., J. Antibiot.* **23**, 511 (1970). Structure: *eidem, ibid.* **27**, 366 (1974). Absolute configuration: A. Ducruix *et al., Chem. Commun.* **1976**, 947. Stereospecific total synthesis: K. Tatsuta *et al., Tetrahedron Lett.* **1980**, 2837. Retrosynthetic studies: K. C. Nicolaou *et al., J. Am. Chem. Soc.* **103**, 1222 (1981). A 17-membered aglycone was proposed at one time, *see* T. Osono, H. Umezawa in *Drug Action and Drug Resistance in Bacteria* **1**, S. Mitsuhashi, Ed. (University Park Press, Baltimore, 1971) pp 41-120; T. Osono *et al., J. Antibiot.* **27**, 366 (1974). Pharmacology: K. Kuriaki *et al., Jpn. J. Antibiot.* **22**, 232 (1969).

$$R = $$

Colorless needles from benzene, mp 130-133° (after drying under reduced pressure at 100° for 5 hrs). $[\alpha]_D^{25}$ −70° (c = 1 in ethanol). uv max (0.001*N* HCl): 232 nm ($E_{1cm}^{1\%}$ 325). pKa 7.1 (40% aq methanol). Very sol in methanol, ethanol, acetone, chloroform, ethyl acetate, dioxane, acidic water. Sol in butanol, ether, CCl_4, benzene, toluene. Practically insol in water, petr ether, ligroin, *n*-hexane.
Propionate. [51016-68-3] Josacine; Josamy; Josaxin; Wilprafen. $C_{45}H_{73}NO_{16}$; mol wt 884.07.
THERAP CAT: Antibacterial.
THERAP CAT (VET): Antibacterial.

5316. Juglans. Butternut; white walnut; lemon walnut; oil nut. Dried inner root bark of *Juglans cinerea* L., *Juglandaceae,* collected in the autumn. *Habit.* North America. *Constit.* Resinoid juglandin, juglone, juglandic acid, fixed and volatile oils, tannin.

5317. Juglone. [481-39-0] 5-Hydroxy-1,4-naphthalenedione; 5-hydroxy-1,4-naphthoquinone; 8-hydroxy-1,4-naphthoquinone; C.I. 75500; C.I. Natural Brown 7; nucin; regianin. $C_{10}H_6O_3$; mol wt 174.16. C 68.97%, H 3.47%, O 27.56%. Coloring matter occurring in various *Juglandaceae* spp: Brissemoret, Combes, *Compt. Rend.* **141**, 838 (1905). Isoln from walnut shells: Combes, *Bull. Soc. Chim.* **1**(4), 800 (1907). Isoln and description of sedative properties in fish, mammals: B. A. Westfall *et al., Science* **134**, 1617 (1961). Isoln from pecans and identification as inhibitory agent of mycelial growth of *Fusicladium effusum:* P. A. Hedin *et al., J. Agric. Food Chem.* **28**, 340 (1980). Synthesis: Bernthsen, Semper, *Ber.* **20**, 938 (1887); Willstätter, Wheeler, *Ber.* **47**, 2798 (1914); Teuber, Götz, *Ber.* **87**, 1236 (1954); C. Grundmann, *Synthesis* **1977**, 644. Tumor-promoting activity study: B. L. Van Duuren *et al., J. Med. Chem.* **21**, 26 (1978). Use as pH indicator: S. S. Sawhney, B. M. L. Bhatia, *J. Indian Chem. Soc.* **57**, 438 (1980).

Yellow needles from benzene + petr ether, mp 155°. Sublimes. Absorption max (methanol): 420 nm (log ε 3.56). Volatile with steam. Slightly sol in hot water; freely sol in chloroform, benzene; sol in alcohol, ether; sol in aq solns of alkalies giving a purplish-red soln.

5318. Juniper. Dioecious, evergreen shrub, *Juniperus communis* L., *Cupressaceae,* bearing tangy, bitter, blue berries. Medicinal portions inlude the ripe berries, berry cones and essential oil. *Habit.* Europe, northern Africa, northern Asia, North America. *Constit.* Volatile oil (1-2%), diterpenes, catechin tannins, flavonoids, inverted sugar (20-30%), oligomeric proanthocyanidins. Analysis by GLC of the oil of six western North American species: F. C. Vasec, R. W. Scora, *Am. J. Bot.* **54**, 781 (1967). Botanical description and medicinal uses: J. Gruenwald *et al., PDR for Herbal Medicines* (Medical Economics, Montvale, 2nd Ed., 2000) pp 440-

441. Review of pharmacology: J. Barnes *et al.*, *Herbal Medicines* (Pharmaceutical Press, London, 2nd Ed., 2002) pp 317-319.

Volatile oil. [8002-68-4] Juniperberry oil; oil of juniper. Obtained by steam distillation of the dried ripe fruit (berries). *Constit.* Primarily monoterpenes incl. α-pinene, β-myrcene, sabinene, camphene, terpineol. Colorless, faintly green or yellow liquid with characteristic odor and aromatic, bitter taste. Odor and taste as of the berries. d_{25}^{25} 0.854-0.879. $[\alpha]_D^{25}$ 0 to −15°. n_D^{20} 1.474-1.484. Sol in most fixed oils, mineral oil. Insol in glycerin, propylene glycol. *Keep cool in well-closed and well-filled bottles, protected from light.*

USE: Volatile oil as flavor component of gin; in perfumery.

THERAP CAT: Diuretic; carminative.

5319. Juniper Tar. Oil of cade; empyreumatic oil of juniper; oil of juniper tar; Haarlem oil; Harlem oil; Tilly drops; Holland balsam; silver drops; silver balsam; Kaparlem; Caparlem. Volatile oil from wood of *Juniperus oxycedrus* L., *Cupressaceae*. *Constit.* Chiefly cadinene. Toxicity data: P. M. Jenner *et al.*, *Food Cosmet. Toxicol.* **2**, 327 (1964). *Review:* D. L. J. Opdyke, *ibid.* **13**, 733-734 (1975).

Dark brown, more or less viscid liq; smoky odor; acrid, slightly aromatic taste. d_{25}^{25} 0.950-1.055. n_D^{20} 1.510-1.530. Very slightly sol in water; sol in 3 vols ether, in chloroform, amyl alcohol, glacial acetic acid, oil turpentine; partly sol in alcohol or petr ether. LD_{50} orally in rats: 8014 mg/kg (Jenner).

USE: As a gin-like flavor; in perfumery.

THERAP CAT: Anti-eczematic (topical).

5320. Justicidins. Lignans from various species of *Justicia*, *Acanthaceae*. Isoln and structures for justicidins A, B, C, D, E and F have been elucidated. Isoln of A and B: Munakata *et al.*, *Tetrahedron Lett.* **1965**, 4167; Ohta *et al.*, *Agric. Biol. Chem.* **33**, 610 (1969); Okigawa *et al.*, *Tetrahedron* **26**, 4301 (1970). Structure of A: Govindachari *et al.*, *Tetrahedron Lett.* **1967**, 3517; Horii *et al.*, *Chem. Pharm. Bull.* **19**, 535 (1971). Synthesis of B: Munakata *et al.*, *Tetrahedron Lett.* **1967**, 2831; Munakata, Katsura, **GB 1178341** (1970); T. Momose *et al.*, *Chem. Pharm. Bull.* **26**, 3195 (1978). Isoln and structures of C and D: Ohta, Munakata, *Tetrahedron Lett.* **1970**, 923. Synthesis of C: Horii *et al.*, *Chem. Pharm. Bull.* **17**, 1878 (1969). Structure of E: Wada, Munakata, *Tetrahedron Lett.* **1970**, 2017. Synthesis of E: Holmes, Stevenson, *J. Org. Chem.* **36**, 3450 (1971). Synthesis of D, E, and F: Z. I. Horii *et al.*, *Chem. Pharm. Bull.* **25**, 1803 (1977).

Justicidin A R = OCH₃
Justicidin B R = H

Justicidin A. [25001-57-4] 9-(1,3-Benzodioxol-5-yl)-4,6,7-trimethoxynaphtho[2,3-*c*]furan-1(3*H*)-one; diphyllin methyl ether. $C_{22}H_{18}O_7$; mol wt 394.38. Crystals, mp 263°. uv max ($CHCl_3$): 265, 295, 315, 335 nm (log ε 4.35, 4.13, 4.13, 3.33).

Justicidin B. [17951-19-8] 9-(1,3-Benzodioxol-5-yl)-6,7-dimethoxynaphtho[2,3-*c*]furan-1(3*H*)-one; dehydrocollinusin. $C_{21}H_{16}O_6$; mol wt 364.35. Crystals, mp 240°. uv max ($CHCl_3$): 260, 295, 310, 350 nm (log ε 4.52, 4.13, 4.13, 3.41).

USE: Piscicidal agent.

5321. Juvenile Hormones. JH. Family of hormones, secreted by the corpora allata, controlling the larval metamorphosis of insects; so named since they induce the retention of insects' juvenile characteristics and prevent maturation. Isoln from the abdomen of the male wild silk moth, *Hyalophora cecropia* L.: C. M. Williams, *Nature* **178**, 212 (1956); H. Röller, J. S. Bjerke, *Life Sci.* **4**, 1617 (1965). Structure elucidation of JH I: H. Röller *et al.*, *Angew. Chem. Int. Ed.* **6**, 179 (1967). Racemic synthesis of JH I: K. H. Dahm *et al.*, *J. Am. Chem. Soc.* **89**, 5292 (1967). Toxicity of racemic JH I: J. B. Siddall, M. Slade, *Nature New Biol.* **229**, 158 (1971). Abs config of JH I: K. Nakanishi *et al.*, *Chem. Commun.* **1971**, 1235; A. S. Meyer *et al.*, *Proc. Natl. Acad. Sci. USA* **68**, 2312 (1971). Stereoselective syntheses of JH I: E. J. Corey *et al.*, *ibid.* **90**, 5618 (1968); W. S. Johnson *et al.*, *ibid.* 6225. Enantioselective syntheses of JH I: P. Loew, W. S. Johnson, *ibid.* **93**, 3765 (1971); D. J. Faulkner, M. R. Petersen, *ibid.* 3766. Isoln, structure, and abs config of JH II and III: K. J. Judy *et al.*, *Proc. Natl. Acad. Sci. USA* **70**, 1509 (1973). Isoln and structure of *JH 0*: B. J. Bergot *et al.*, *Science* **210**, 336 (1980); and *iso-JH 0* (MeJH I): *idem et al.* in *Juvenile Hormone Biochemistry*, G. E. Pratt, G. T. Brooks, Eds. (Elsevier-North Holland Biomedical Press, Amsterdam, 1981) pp 33-45. Biosynthesis, isoln, and structure elucidation of *JH III bisepoxide* (JHB₃): D. S. Richard *et al.*, *Proc. Natl. Acad. Sci. USA* **86**, 1421 (1989). Abs config of JHB₃: A. J. Herlt *et al.*, *Chem. Commun.* **1993**, 1497. Enantioselective synthesis of JH III: K. Mori, H. Mori, *Tetrahedron* **43**, 4097 (1987); of JH I and II: K. Mori, M. Fujiwhara, *ibid.* **44**, 343 (1988); of JHB₃: R. W. Rickards, R. D. Thomas, *Tetrahedron Lett.* **34**, 8369 (1993). Study of regulatory role: M. Cusson *et al.*, *Arch. Insect Biochem. Physiol.* **25**, 329 (1994). Book: *The Juvenile Hormones*, L. I. Gilbert, Ed. (Plenum Press, New York, 1976) 572 pp. Review of biosynthesis: D. A. Schooley, F. C. Baker in *Comprehensive Insect Physiology, Biochemistry and Pharmacology* vol. 8, G. A. Kerkut, L. I. Gilbert, Eds. (Pergamon Press, New York, 1985) pp 363-389. Review of techniques for identification and quantification: F. C. Baker in *Morphogenetic Hormones of Arthropods* Part 1, A. P. Gupta, Ed. (Rutgers Univ. Press, New Brunswick, 1990) pp 389-453. Review of discovery and identification of JHB₃: C.-M. Yin, *Zool. Stud.* **33**, 237-245 (1994). Review of molecular mechanisms of action: G. Jones, *Annu. Rev. Entomol.* **40**, 147-169 (1995).

	R	R'
JH I	CH₂CH₃	CH₂CH₃
JH II	CH₂CH₃	CH₃
JH III	CH₃	CH₃

Juvenile hormone I. [13804-51-8] [2*R*-[2α(2*E*,6*E*),3α]]-7-Ethyl-9-(3-ethyl-3-methyloxiranyl)-3-methyl-2,6-nonadienoic acid methyl ester; methyl (2*E*,6*E*,10*R*,11*S*)-10,11-epoxy-7-ethyl-3,11-dimethyl-2,6-tridecadienoate; C-18 JH. $C_{18}H_{30}O_3$; mol wt 294.44. n_D^{24} 1.4732. $[\alpha]_D^{23}$ +14.9° (c = 0.935 in chloroform). $[\alpha]_D^{22.5}$ +14.5° (c = 0.78 in methanol).

Juvenile hormone II. [34218-61-6] [2*R*-[2α(2*E*,6*E*),3α]]-9-(3-Ethyl-3-methyloxiranyl)-3,7-dimethyl-2,6-nonadienoic acid methyl ester; methyl (2*E*,6*E*,10*R*,11*S*)-10,11-epoxy-3,7,11-trimethyl-2,6-tridecadienoate; C-17 JH. $C_{17}H_{28}O_3$; mol wt 280.41. $n_D^{23.5}$ 1.4774. $[\alpha]_D^{24.5}$ +17.6° (c = 0.590 in methanol).

Juvenile hormone III. [22963-93-5] [*R*-(*E*,*E*)]-9-(3,3-Dimethyloxiranyl)-3,7-dimethyl-2,6-nonadienoic acid methyl ester; methyl (2*E*,6*E*,10*R*)-10,11-epoxy-3,7,11-trimethyl-2,6-dodecadienoate; C-16 JH. $C_{16}H_{26}O_3$; mol wt 266.38. Colorless oil. n_D^{24} 1.4736. $[\alpha]_D^{24}$ +6.71° (c = 0.57 in methanol).

K

5322. Kaempferol. [520-18-3] 3,5,7-Trihydroxy-2-(4-hydroxyphenyl)-4*H*-1-benzopyran-4-one; 3,4',5,7-tetrahydroxyflavone; nimbecetin; pelargidenolon 1497; populnetin; rhamnolutein; robigenin; swartziol; trifolitin. $C_{15}H_{10}O_6$; mol wt 286.24. C 62.94%, H 3.52%, O 33.54%. Plant flavonoid isolated from *Delphinium consolida* L., *Ranunculaceae:* A. G. Perkin, E. J. Wilkinson, *J. Chem. Soc.* **81**, 585 (1902); from grapefruit *(Citrus paradisi* Macf., *Rutaceae):* Dunlap, Wender, *Anal. Biochem.* **4**, 110 (1962); from stems and seeds of *Cuscuta reflexa* Roxb., *Convolvulaceae:* K. W. Gopinath *et al., J. Sci. Ind. Res.* **21B**, 601 (1962). Identity with swartziol: M. R.-R. Paris, L. Bézanger-Beauquesne, *Compt. Rend.* **242**, 1761 (1956). Structure: S. von Kostanecki, A. Rozycki, *Ber.* **34**, 3721 (1901). Total synthesis: Y.-H. Lu *et al., Yao Hsueh Hsueh Pao* **15**, 477 (1980); *C.A.* **94**, 174808a (1981). Convenient synthesis: M. Ichikawa *et al., Org. Prep. Proced. Int.* **14**, 183 (1982). Mutagenicity studies: J. T. MacGregor, L. Jurd, *Mutat. Res.* **54**, 297 (1978); A. A. Hardigree, J. L. Epler, *ibid.* **58**, 231 (1978). HPLC determination in soybean: J. D. Gaynor *et al., Chromatographia* **25**, 1049 (1988).

Yellow needles, mp 276-278°; also reported as light yellow powder from ethanol-water, mp 278-280° (dec) (Ichikawa). uv max: 265, 365 nm. Slightly sol in water; sol in hot alcohol, ether or alkalies.

3,7-Dirhamnoside. Kaemferitrin; lespedin. $C_{27}H_{30}O_{14}$. Isoln from the leaves of *Indigo arrecta* Benth., *Leguminosae:* Perkin, *J. Chem. Soc.* **91**, 435 (1907); from leaves of *Trichosanthus cucumeroides* Maxim., *Cucurbitaceae:* Nakaoki, Morita, *J. Pharm. Soc. Jpn.* **77**, 108 (1957). Identity with lespedin: Hatlori, *Nature* **168**, 788 (1952). Crystals from dil alc, mp 201-203°. Slightly sol in boiling water and cold alcohol.

3-Glucoside. Astragalin. $C_{21}H_{20}O_{11}$. From *Podophyllum peltatum* L. and *P. emodi* Wall., *Berberidaceae.* Isoln and structure: von Wartburg, Kuhn, *Experientia* **21**, 67 (1965). Crystals from methanol, mp 175-178°. $[\alpha]_D^{20} -16.9°$ (c = 0.45 in methanol).

5323. Kahalalide F. [149204-42-2] *N*-(5-Methyl-1-oxohexyl)-D-valyl-L-threonyl-L-valyl-D-valyl-D-prolyl-L-ornithyl-D-alloisoleucyl-D-allothreonyl-D-alloisoleucyl-D-valyl-L-phenylalanyl-(2Z)-2-amino-2-butenoyl-L-valine (13 → 8)-lactone; PM-92102. $C_{75}H_{124}N_{14}O_{16}$; mol wt 1477.90. C 60.95%, H 8.46%, N 13.27%, O 17.32%. One of a family of cyclic depsipeptides isolated from the Hawaiian marine mollusk, *Elysia rufescens.* First reported diet-derived chemical defense peptide; production of which depends on consumption of the green alga, *Bryopsis pennata.* Isoln: P. J. Schauer *et al.,* **EP 610078** (1994 to Pharma Mar); M. T. Hamann, P. J. Scheuer, *J. Am. Chem. Soc.* **115**, 5825 (1993); of the family: M. T. Hamann *et al., J. Org. Chem.* **61**, 6594 (1996). Synthesis: A. López-Macià *et al., J. Am. Chem. Soc.* **123**, 11398 (2001). Stereochemistry: G. Goetz *et al., Tetrahedron* **55**, 7739 (1999). Updated stereochemistry: I. Bonnard *et al., J. Nat. Prod.* **66**, 1466 (2003). Identification as defensive peptide: M. A. Becerro *et al., J. Chem. Ecol.* **27**, 2287 (2001). LC/MS/MS determn in plasma: E. Stokvis *et al., J. Mass Spectrom.* **37**, 992 (2002). Toxicology: A. P. Brown *et al., Cancer Chemother. Pharmacol.* **50**, 333 (2002). *In vitro* cytotoxic activity: Y. Suárez *et al., Mol. Cancer Ther.* **2**, 863 (2003). Mechanism of cytotoxicity: J. M. Sewell *et al., Eur. J. Cancer* **41**, 1637 (2005). Clinical pharmacology in prostate cancer: J. M. Rademaker-Lakhai *et al., Clin. Cancer Res.* **11**, 1854 (2005). Review of development and therapeutic potential: M. T. Hamann, *Curr. Opin. Mol. Ther.* **6**, 657-665 (2004).

White amorphous powder. $[\alpha]_D -8°$ (c = 4.32 in CH_3OH). LD_{50} in male, female rats (μg/kg): 375, 600 i.v. (Brown).

THERAP CAT: Antineoplastic.

5324. Kainic Acid. [487-79-6] (2*S*,3*S*,4*S*)-2-Carboxy-4-(1-methylethenyl)-3-pyrrolidineacetic acid; 2-carboxy-3-carboxymethyl-4-isopropenylpyrrolidine; digenic acid; α-kainic acid; L_S-*xylo*-kainic acid; Digenin; Helminal. $C_{10}H_{15}NO_4$; mol wt 213.23. C 56.33%, H 7.09%, N 6.57%, O 30.01%. Excitotoxic amino acid used to identify a specific subset of EAA receptors. Consequently the receptors are known as kainate receptors. Isolated as an anthelmintic principle from the dried red alga *Digenea simplex* (Wulf.) Ag., *Rhodomelaceae:* Murakami *et al., J. Pharm. Soc. Jpn.* **73**, 1026 (1953); **JP 54 4947** (1954), *C.A.* **49**, 13604i (1955); Katsuya *et al.,* **JP 64 1942** (1964 to New-Japan Pharmaceutical Co). Structure: Watase, Nitta, *Bull. Chem. Soc. Jpn.* **30**, 889 (1957); Watase *et al., ibid.* **31**, 714 (1958). Eight theoretical stereoisomers: Nitta *et al., Nature* **181**, 761 (1958); **GB 795750** (1958 to Takeda). Synthesis: Ueno, **US 2902492**; Tatsuoka *et al.,* **US 2954384** (1959, 1960 both to Takeda); W. Oppolzer, H. Andres, *Helv. Chim. Acta* **62**, 2282 (1979). Neurotoxic activity: J. V. Nadler, *Life Sci.* **24**, 289 (1979). Mechanism of neurotoxicity: E. G. McGeer *et al., Adv. Neurol.* **23**, 577 (1979); J. T. Coyle *et al., ibid.* 593; J. W. Ferkany *et al., Nature* **298**, 757 (1982); J. Garthwaite, G. Garthwaite, *ibid.* **305**, 138 (1983). Induces epileptogenic lesions: J. V. Nadler, *Neurosci. Res. Program Bull.* **19**, 369 (1981). Autoradiographic characterization of binding sites: J. T. Greenamyre *et al., J. Pharmacol. Exp. Ther.* **233**, 254 (1985). *Book: Kainic Acid as a Tool in Neurobiology,* E. G. McGeer *et al.,* Eds. (Raven Press, New York, 1978). *Review:* J. T. Coyle, *Ciba Found. Symp.* **126**, 186-203 (1987).

Needles, dec 251°. $[\alpha]_D^{24} -14.8°$ (c = 1.01). Intense absorption at 6.05 and 11.2 μ. Sol in water. Insol in ethanol. Stable in boiling aq solns.

USE: Neurobiological tool.

THERAP CAT: Anthelmintic (Nematodes).

5325. Kalkitoxin. [247184-89-0] (2*R*)-*N*-[(3*S*,5*S*,6*R*)-7-[(4*R*)-4-Ethenyl-4,5-dihydro-2-thiazolyl]-3,5,6-trimethylheptyl]-*N*,2-dimethylbutanamide. $C_{21}H_{38}N_2OS$; mol wt 366.61. C 68.80%, H 10.45%, N 7.64%, O 4.36%, S 8.74%. Neurotoxic lipopeptide isolated from the Caribbean marine algae, *Lyngbya majuscula.* Isoln, structure determn and synthesis: M. Wu *et al., J. Am. Chem. Soc.* **122**, 12041 (2000). NMDA-receptor based neurotoxicity: F. W. Berman *et al., Toxicon* **37**, 1645 (1999). Total synthesis: T. Asano *et al., Tennen Yuki Kagobutsu Toronkai Koen Yoshishu* **2000**, 691. Brief review: J. Fricker, *Drug Discov. Today* **6**, 223-224 (2001).

$[\alpha]_D^{25}$ +16° (c = 0.07 in CHCl$_3$). uv max (methanol): 250 nm (ε 2600). LC$_{50}$ in common goldfish, brine shrimp: 700, 150 nM (Wu).

5326. Kallidin. [342-10-9] N^2-L-Lysylbradykinin; kallidin-10; kallidin II. C$_{56}$H$_{85}$N$_{17}$O$_{12}$; mol wt 1188.40. C 56.60%, H 7.21%, N 20.04%, O 16.16%. A hypotensive and smooth muscle-stimulating principle formed by the proteolysis of kininogen by glandular and other kallikreins, q.v. The decapeptide structure is a homologue of bradykinin, q.v.: Werle et al., Z. Physiol. Chem. **326**, 174 (1961). Synthesis: Nicolaides et al., Biochem. Biophys. Res. Commun. **6**, 210 (1961); Pless et al., Helv. Chim. Acta **45**, 394 (1962).

Lys–Arg–Pro–Pro–Gly–Phe–Ser–Pro–Phe–Arg

Amorphous precipitate. $[\alpha]_D^{21}$ -57° (c = 1 in N acetic acid). R$_f$ value in a butanol/acetic acid/water system (70:10:20) 0.15.

THERAP CAT: Vasodilator.

5327. Kallikrein. [9001-01-8] Kallidinogenase; Callicrein; Padreatin; Padukrein; Glumorin; Depot-Glumorin; Circuletin; Kalirechin; Onokrein P; Padutin; Prokrein; Promotin. Hypotensive enzyme which releases kinins from plasma proteins. Major sources in the body are blood plasma, glandular tissues, and urine, occurring abundantly in the pancreas, parotid and submaxillary glands, in intestinal wall, in feces, in duodenal juice, and to a lesser degree in kidney. Isoln from mammalian pancreas or urine: Abelous, Bardier, C. R. Seances Soc. Biol. Ses Fil. **64**, 848 (1908); Webster et al., Proc. Soc. Exp. Biol. Med. **93**, 181 (1956). Separation into two components, kallikreins A and B: E. Habermann, Z. Physiol. Chem. **328**, 15 (1962); F. Fiedler, E. Werle, ibid. **348**, 1087 (1967); C. Kutzbach, G. Schmidt-Kastner, ibid. **353**, 1099 (1972). Prepn of high purity material: Werle, Trautschold, **DE 1102973** (1960 to Bayer). Review on pig pancreatic kallikrein: F. Fiedler, Methods Enzymol. **45B**, 289-303 (1976); on human kallikrein and prekallikrein: R. W. Colman, A. Bagdasarian, ibid. 303-322. Plasma kallikrein differs from glandular or urinary kallikrein. The latter two liberate kallidin, q.v.; the former releases bradykinin, q.v., both from the common precursor, kininogen. Pharmacology: Franz, Marquardt, Arzneim.-Forsch. **10**, 779 (1960). Reviews: Schachter, Physiol. Rev. **49**, 509 (1969); Suzuki et al., Adv. Exp. Med. Biol. **8**, 15 (1970).

THERAP CAT: Vasodilator.

5328. Kamala. Kamila; kameela; spoonwood. Glands and hairs covering the fruits of Mallotus philippinensis (Lam.) Muell. Arg., (also known as Rottlera tinctoria Roxb.) Euphorbiaceae. Habit. Philippine Islands, India, China, Australia. Constit. Rottlerin, isoallorottlerin, flavanoids, resins, wax. Determn of constituents: Khorana, Motiwala, Indian J. Pharm. **11**, 37 (1949); M. Lounasmaa et al., Planta Med. **28**, 16 (1975); C. J. Widén, H. S. Puri, ibid. **40**, 284 (1980); V. K. Ahluwalia et al., Indian J. Chem. **27B**, 238 (1988). Anthelmintic effects: S. S. Gupta et al., Indian J. Physiol. Pharmacol. **28**, 63 (1984).

THERAP CAT: Anthelmintic (Cestodes).

THERAP CAT (VET): Has been used as anthelmintic.

5329. Kanamycin. [8063-07-8] Antibiotic complex produced by Streptomyces kanamyceticus Okami & Umezawa from Japanese soil: Umezawa et al., J. Antibiot. **10A**, 181 (1957); **US 2931798** (1960). Comprised of three components, kanamycin A, the major component (usually designated as kanamycin) and kanamycins B and C, two minor congeners. Isoln and purification of kanamycins A and B and their salts: Johnson et al., and Johnson, Hardcastle, **US 2936307** and **US 2967177** (1960, 1961 both to Bristol-Myers). Separation process: Rothrock, Putter, **US 3032547** (1962 to Merck & Co.). Prepn of kanamycin C: Murase et al., J. Antibiot. **14A**, 156 (1961). Studies on kanamycin B: Wakazawa et al., ibid. 180, 187. Structure of kanamycin A: Ogawa et al., ibid. **11A**, 169 (1958); Cron et al., J. Am. Chem. Soc. **80**, 4741 (1958). Structure of kanamycin B: Ito et al., J. Antibiot. **17A**, 189 (1964). Structure of kanamycin C: Murase, ibid. **14A**, 367 (1961). Abs config of kanamycin

A: Hichens, Rinehart, J. Am. Chem. Soc. **85**, 1547 (1963); Umezawa et al., Bull. Chem. Soc. Jpn. **39**, 1244 (1966). Crystal structure of kanamycin A: Koyama et al., Tetrahedron Lett. **1968**, 1875. Monograph: Ann. N.Y. Acad. Sci. **76**, 17-408 (1958). Synthesis of kanamycin A: Umezawa et al., J. Antibiot. **21**, 367 (1968); Nakajima et al., Tetrahedron Lett. **1968**, 623; Umezawa et al., Bull. Chem. Soc. Jpn. **42**, 533 (1969). Synthesis of kanamycin B: eidem, J. Antibiot. **21**, 424 (1968); Bull. Chem. Soc. Jpn. **42**, 537 (1969). Chemical conversion of kanamycin B to kanamycin C: S. Toda et al., J. Antibiot. **30**, 1002 (1977). Synthesis of kanamycin C: Umezawa et al., Bull. Chem. Soc. Jpn. **41**, 533 (1968); J. Antibiot. **21**, 162 (1968). Effects on protein synthesis: Suzuki et al., ibid. **23**, 99 (1970). Toxicity data (kanamycin A sulfate): Zel'tser et al., Antibiotiki **19**, 552 (1974). Comprehensive description: P. J. Claes et al., Anal. Profiles Drug Subs. **6**, 259-296 (1977).

	R	R'
Kanamycin A	NH$_2$	OH
Kanamycin B	NH$_2$	NH$_2$
Kanamycin C	OH	NH$_2$

Kanamycin A. [59-01-8] O-3-Amino-3-deoxy-α-D-glucopyranosyl-(1 → 6)-O-[6-amino-6-deoxy-α-D-glucopyranosyl-(1 → 4)]-2-deoxy-D-streptamine. C$_{18}$H$_{36}$N$_4$O$_{11}$; mol wt 484.50. Crystals from methanol + ethanol. $[\alpha]_D^{24}$ +146° (0.1N H$_2$SO$_4$). LD$_{50}$ i.v. in mice: 583 mg/kg (Wakazawa).

Kanamycin A sulfate. [25389-94-0] Cantrex; Cristalomicina; Enterokanacin; Kamycine; Kamynex; Kanacedin; Kanamytrex; Kanasig; Kanatrol; Kanicin; Kannasyn; Kantrex; Klebcil; Otokalixin; Resistomycin; Kanescin; Kanaqua. (U.S.P. requires that kanamycin sulfate contains not less than 75% kanamycin A on an anhydr basis). Irregular prisms, dec over a wide range above 250°C. Freely sol in water. Practically insol in the common alcohols and nonpolar solvents. LD$_{50}$ in mice: 20.7 g/kg orally; 1450 mg/kg i.p. (Zel'tser).

Kanamycin B. [4696-76-8] Bekanamycin; aminodeoxykanamycin; NK-1006. C$_{18}$H$_{37}$N$_5$O$_{10}$; mol wt 483.52. Crystals, mp 178-182° (dec). $[\alpha]_D^{18}$ +130° (c = 0.5 in water). $[\alpha]_D^{21}$ +114° (c = 0.98 in water). Sol in water, formamide; slightly sol in chloroform, isopropyl alcohol. Practically insol in the common alcohols and nonpolar solvents. LD$_{50}$ i.v. in mice: 136 mg/kg (Wakazawa).

Kanamycin B sulfate. [29701-07-3] Coltericin; Kanendomycin; Kanendos. Pharmacokinetics: F. Di Nola et al., Minerva Med. **70**, 1803 (1979).

Kanamycin C. [2280-32-2] C$_{18}$H$_{36}$N$_4$O$_{11}$; mol wt 484.50. Crystals from methanol + ethanol, dec above 270°. $[\alpha]_D^{20}$ +126° (H$_2$O). Sol in water; slightly sol in formamide. Practically insol in the common alcohols and nonpolar solvents.

THERAP CAT: Antibacterial.

THERAP CAT (VET): Antibacterial.

5330. Kaolin. Bolus alba; China clay; porcelain clay; white bole; argilla. Essentially a hydrated aluminum silicate, approximately H$_2$Al$_2$Si$_2$O$_8$.H$_2$O. Prepared for pharmaceutical and medicinal purposes by levigating with water to remove sand, etc.

White or yellowish-white, earthy mass or white powder; unctuous when moist. Insol in water, cold acids or in alkali hydroxides.

Caution: Potential symptoms of overexposure are chronic pulmonary fibrosis, stomach granuloma. See NIOSH Pocket Guide to Chemical Hazards (DHHS/NIOSH 97-140, 1997) p 182.

USE: Manuf porcelain, pottery, bricks, Portland cement; ultramarine, color lakes; refractory mortar; plaster material; filler for paper; electric and heat insulators; clarifying liquids; drying and emollient agent.

THERAP CAT: Adsorbent.

THERAP CAT (VET): Topical and G.I. adsorbent. Poultice.

5331. Karanjin. [521-88-0] 3-Methoxy-2-phenyl-4*H*-furo-[2,3-*h*]-1-benzopyran-4-one. $C_{18}H_{12}O_4$; mol wt 292.29. C 73.97%, H 4.14%, O 21.89%. From *Pongamia glabra* Vent., *Leguminosae:* Beal, Katti, *J. Am. Pharm. Assoc.* **14**, 1086 (1925); Rao, Rao, *J. Indian Chem. Soc.* **17**, 526 (1940); Bhat *et al.*, *J. Am. Oil Chem. Soc.* **33**, 197 (1956). Structure: Limaye, *Rasayanam* **1**, 1 (1936), *C.A.* **31**, 2206[9] (1937); Manjunath *et al.*, *Ber.* **72B**, 39 (1939). Synthesis: Seshadri, Venkateswarlu, *Proc. Indian Acad. Sci.* **13A**, 404 (1941); *ibid.* **17A**, 16 (1943); Kawase *et al.*, *Bull. Chem. Soc. Jpn.* **28**, 273 (1955); Rao, Seshadri, *Proc. Indian Acad. Sci.* **33A**, 168 (1951); Aneja *et al.*, *Tetrahedron* **2**, 203 (1958); Raizada *et al.*, *J. Sci. Ind. Res.* **19B**, 76 (1960).

Needles from methanol, mp 157-158°. Sol in methanol, ethanol, chloroform, benzene, ether, concd H_2SO_4, HNO_3, HOAc, HCl. Practically insol in petr ether, dil mineral acids.

5332. Karaya Gum. [9000-36-6] Gum karaya; kadaya; katilo; kullo; kuteera; sterculia; Indian tragacanth; mucara. The dried exudate of the tree *Sterculia urens* Roxb., *Sterculiaceae*, found in India, especially in the Gujerat region and in the central provinces: Toothaker, *The Soluble Gums* (Philadelphia, 1921); Mantell, *The Water-Soluble Gums* (New York, 1947). Constituents and structure: Hirst, Dunstan, *J. Chem. Soc.* **1953**, 2332. Structure is a partially acetylated polysaccharide containing about 8% acetyl groups and about 37% uronic acid residues. *Reviews:* F. Smith, R. Montgomery, *The Chemistry of Plant Gums and Mucilages* (Reinhold, New York, 1959); Goldstein, Alter, in *Industrial Gums*, R. L. Whistler, Ed. (Academic Press, New York, 2nd ed., 1973) pp 273-287.

Finely ground white powder, faint odor of acetic acid. Acid to litmus. Absorbs water rapidly to form viscous mucilages at low concs. Viscosity decreases on addn of acid or alkali. Color of the soln lightens in acidic media and darkens in alkaline soln due to the presence of tannins. Gum karaya loses viscosity forming ability when stored in the dry state, the loss being greater for a powdered material than for the crude gum. Cold storage inhibits this degradation.

Note: Karaya gum occurring in broken irregular pieces having a somewhat crystalline appearance has been referred to commercially as 'crystal' gum.

USE: As denture adhesive; as binder in paper manuf; as stabilizer, thickener, texturizer, emulsifier in foods; as thickening agent for dyes in textile industry. A substitute for gum tragacanth.

THERAP CAT: Cathartic.

5333. Karsil. [2533-89-3] *N*-(3,4-Dichlorophenyl)-2-methylpentanamide; 3',4'-dichloro-2-methylvaleranilide; Niagara 4562. $C_{12}H_{15}Cl_2NO$; mol wt 260.16. C 55.40%, H 5.81%, Cl 27.25%, N 5.38%, O 6.15%. Prepd from 3,4-dichloroaniline and 2-methylvaleryl chloride: Dorschner *et al.*, **GB 869169** (1961 to FMC).

Crystals, mp 106-107°.

USE: Herbicide.

5334. Kava. Kava-kava; ava-ava; kawa. Dried rhizome and roots of *Piper methysticum* Forst., *Piperaceae*. *Habit.* Polynesia. Most important constituents are: kawain, dihydrokawain, methysticin, dihydromethysticin, and yangonin: Borsche, Lewinsohn,

Ber. **66**, 1792 (1933). Chemical and pharmacological investigation of the kava constituents: Klohs *et al.*, *J. Med. Pharm. Chem.* **1**, 95 (1959); Meyer, Kretzschmar, *Klin. Wochenschr.* **44**, 902 (1966). Review of chemistry, pharmacology and historical sketch: *U.S. Public Health Service Publ. No. 1645*, D. H. Efron, Ed., pp 103-181 (1967).

Note: Kava is also the popular name for the intoxicating drink prepared from the plant's roots.

5335. Kawain. [500-64-1] (6*R*)-5,6-Dihydro-4-methoxy-6-[(1*E*)-2-phenylethenyl]-2*H*-pyran-2-one; 5-hydroxy-3-methoxy-7-phenyl-2,6-heptadienoic acid δ-lactone; 4-methoxy-6-(β-phenylvinyl)-5,6-dihydro-α-pyrone; 4-methoxy-6-styryl-5,6-dihydro-α-pyrone; kavain; gonosan. $C_{14}H_{14}O_3$; mol wt 230.26. C 73.03%, H 6.13%, O 20.84%. From the rhizome and roots of *Piper methysticum* Forst., *Piperaceae* (kava): Borsche, Peitzsch, *Ber.* **63**, 2414 (1930); Hänsel, Beiersdorff, *Naturwissenschaften* **45**, 573 (1958); *eidem*, *Arzneim.-Forsch.* **9**, 581 (1959). Synthesis: Fowler, Henbest, *J. Chem. Soc.* **1950**, 3642; Kostermans, *Nature* **166**, 788 (1950); *idem*, *Rec. Trav. Chim.* **70**, 79 (1951); Z. H. Israili, E. E. Smissman, *J. Org. Chem.* **41**, 4070 (1976). Abs config: Snatzke, Hänsel, *Tetrahedron Lett.* **1968**, 1797. Crystal and molecular structure: A. Yoshino, W. Nowacki, *Z. Kristallogr. Kristallgeom. Kristallphys. Kristallchem.* **136**, 66 (1972), *C.A.* **78**, 63591z (1973).

Rods from methanol + ether, mp 105-106°. bp$_{0.1}$ 195-197°. $[\alpha]_D^{20}$ +105° (abs alc). uv max (methanol): 210, 245, 282 nm (log ε 4.38, 4.44, 2.81). Practically insol in water. Sol in acetone, ether, methanol; slightly sol in hexane.

(±)-Form. [3155-48-4] Needles from methanol, mp 146-147°.

Dihydrokawain. [587-63-3] Marindinin. $C_{14}H_{16}O_3$; mol wt 232.28. Crystals from ether, mp 58-60°. $[\alpha]_D^{24}$ +31° (methanol). uv max (methanol): 236 nm (log ε 4.14). Sol in alc, chloroform; moderately sol in ether. Practically insol in petr ether, water.

5336. Kebuzone. [853-34-9] 4-(3-Oxobutyl)-1,2-diphenyl-3,5-pyrazolidinedione; 1,2-diphenyl-4-(γ-ketobutyl)-3,5-pyrazolidinedione; 1,2-diphenyl-4-(3'-oxobutyl)-3,5-dioxopyrazolidine; ketophenylbutazone; KPB; Ketazon. $C_{19}H_{18}N_2O_3$; mol wt 322.36. C 70.79%, H 5.63%, N 8.69%, O 14.89%. Nonsteroidal anti-inflammatory agent related to phenylbutazone, *q.v.* Prepn: Deuss *et al.*, **US 2910481** (1959 to Geigy). Review of pharmacology: Horakova *et al.*, *Pharmacotherapeutica* **1950-1959**, 335-350 (1963); *C.A.* **60**, 6072g (1964). Metabolism: Nemecek *et al.*, *Arzneim.-Forsch.* **16**, 1339 (1966); Queisnerova, Nemecek, *Cesk. Farm.* **20**, 55 (1971); *C.A.* **75**, 47077u (1971).

Crystals, mp 115.5-116.5° or 127.5-128.5° depending on cryst form.

THERAP CAT: Anti-inflammatory.

5337. Kefir Fungi. Kefir grains; kefir seeds. A conglomeration of various fungi, including *Dispora caucasica*, *Schizomycetes* and a species of *Saccharomyces*.

Grayish-yellow lumps, irregular in size; firm, toughly gelatinous consistency, becoming cartilaginous and brittle when dry.

Consult the Name Index before using this section.

USE: Preparing a nutritious beverage of fermented milk known as kefir.

5338. Kenaf. Both the annual herbaceous plant and the resulting fiber are known as Kenaf. Believed to have orginated in Western Sudan around 4000 BC, the fiber, primarily produced from *Hibiscus cannibinus* L., was traditionally used for ropes, canvas, sacks and carpets. Structural characterization of fibers: S. M. A. Shah *et al.*, *Pak. J. Sci. Ind. Res.* **23**, 213 (1980); of bark and lignins: A. M. L. Seca *et al.*, *J. Agric. Food Chem.* **46**, 3100 (1998). Genetic diversity: Z. Cheng *et al.*, *Hereditas* **136**, 231 (2002). Evaluation in laminated products: G. N. Ramaswamy *et al.*, *Indust. Crops Products* **17**, 1 (2003). Brief review: C. L. Webber, III, R. E. Bledsoe, "Kenaf: Production, Harvesting, Processing, and Products" in *New Crops*, J. Janick, J. E. Simon, Eds. (John Wiley and Sons, New York, 1993) pp 416-421. *Review:* H. P. Stout, "Jute and Kenaf" in *Handbook of Fiber Science and Technology* **4**, M. Lewin, E. M. Pearce, Eds. (Marcel Dekker, New York, 1985) pp 701-726; R. K. Maiti, N. Samajpati, *Indian Agric.* **44**, 105-146 (2000).

USE: Traditionally in fiber production; more recently in pulping and papermaking, potting and filtration media, animal feed and as a jute substitute.

5339. Keratin. A protein obtained from hair, wool, horn, nails, claws, beaks, scales, membranes of egg shells and nerve tissue. There are two types of keratins—hard keratin of hair, horn, nails, etc., and soft keratin (pseudokeratin) of the epidermis and whalebone: Rudall, *Adv. Protein Chem.* **7**, 253-290 (1952). The keratins contain all of the common amino acids and differ from other fibrous structural proteins chiefly by their high cystine content. Sequence studies revealed no preferable grouping or periodicity. Amino acid compn of a few keratins: Tristram in H. Neurath, K. Bailey, *The Proteins* vol. **1A** (Academic Press, New York, 1953) p 220. Stability of the protein is due to frequent primary valence cross-links (disulfide bonds) and secondary valence cross-links (hydrogen bonds) between neighboring polypeptide chains. Prepn of different forms for different purposes: Grassmann, **DE 673203** and **DE 682257** (both 1939). Molecular structure of α-keratin: Fraser *et al.*, *Nature* **203**, 1231 (1964); of β-keratin: M. L. Huggins, *Macromolecules* **13**, 465 (1980). *Reviews:* Ward, Lundgren, *Adv. Protein Chem.* **9**, 243-297 (1954); Crewther *et al.*, *ibid.* **20**, 191 (1965); Bradbury, *ibid.* **27**, 111 (1973).

Characteristic properties ascribed to keratins: *(a)* insolubility in water, including aq solns of salts, hydrotropic substances, and dil acids and bases at temps not much above room temp; *(b)* resistance to proteolytic enzymes; *(c)* resistance to hydrolysis; *(d)* solvolysis by mixtures of substances which break the —S—S— bonds and the hydrogen bonds.

USE: Coating "enteric" pills which are unaffected in the stomach, but dissolved by the alkaline intestinal secretions. Detailed coating instructions: Dale, *Pharm. J.* **129**, 494 (1932). Other uses are in formulations of foam type extinguishers and in the production of protein hydrolyzates.

5340. Keratinase. [37341-53-0] Family of proteolytic enzymes that degrade the insoluble protein, keratin, *q.v.*; also degrade soluble proteins, such as casein and gelatin. Produced by a wide range of microorganisms, especially gram-positive bacteria, actinomycetes, and pathogenic fungi. Mol wts range from 18 to 200 kDa, but are generally <50 kDa. Production by a strain of *Streptomyces fradiae*: W. J. Nickerson, J. J. Noval, **US 2988487** (1961 to Rutgers Res. and Educ. Found.); W. J. Nickerson *et al.*, *Biochim. Biophys. Acta* **77**, 73 (1963). Use in the dehairing of hides and skins: R. S. Robison, W. J. Nickerson, **US 2988488** (1961 to Mearl). Isoln from the dermatophytic fungus, *Trichophyton mentagrophytes*: R. J. Yu *et al.*, *J. Bacteriol.* **96**, 1435 (1968). Review of microbial sources, properties, and industrial applications: R. Gupta, P. Ramnani, *Appl. Microbiol. Biotechnol.* **70**, 21-33 (2006); A. Brandelli *et al.*, *ibid.* **85**, 1735-1750 (2010). Review of use in the management of keratinous waste: T. Kornillowicz-Kowalska, J. Bohacz, *Waste Management* **31**, 1689-1701 (2011).

Soluble in water. Stable over wide pH range, generally showing optimum efficiency at pH 7.5-9.0.

USE: For degradation of keratin in the production of animal feeds and fertilizers, in waste management, in detergents, in the leather industry to dehair animal skins.

5341. Kermesic Acid. [18499-92-8] 9,10-Dihydro-3,5,6,8-tetrahydroxy-1-methyl-9,10-dioxo-2-anthracenecarboxylic acid; C.I. Natural Red 3; C.I. 75460. $C_{16}H_{10}O_8$; mol wt 330.25. C 58.19%, H 3.05%, O 38.76%. Principle constituent of kermes, one of the oldest known insect dyes. Isoln and characterization: O. Dimroth, *Ber.* **43**, 1387 (1910); O. Dimroth, W. Scheurer, *Ann.* **399**, 43 (1913). Structure: O. Dimroth, R. Fick, *Ann.* **411**, 315 (1916). Revised structure: D. D. Gadgil *et al.*, *Tetrahedron Lett.* **1968**, 2223. Synthesis: D. W. Cameron *et al.*, *Chem. Commun.* **1978**, 688; *eidem*, *Aust. J. Chem.* **34**, 2401 (1981).

Dark red rosettes from acetic acid, mp >320° (dec). Absorption max: 276, 312, 498 nm (log ε 4.52, 4.12, 3.96). Slightly sol in cold water; sol in hot water (yellowish-red soln). Violet-red in conc H_2SO_4, turning blue on addn of boric acid. Violet in aq NaOH.

USE: As brilliant scarlet dye.

5342. Kerosene. [8008-20-6] Kerosine (petroleum); kerosine. A mixture of petroleum hydrocarbons, chiefly of the methane series having from 10 to 16 carbon atoms per molecule. It constitutes the fifth fraction in the distillation of petroleum (after the petr ethers and before the oils). A typical analysis of the kerosene fraction from Midcontinent crude includes *n*-dodecane, three alkyl derivatives of benzene, naphthalene, 1- and 2-methyl-5,6,7,8-tetrahydronaphthalene. Toxicity study: W. B. Deichmann *et al.*, *Ann. Intern. Med.* **21**, 803 (1944). Toxicological studies and recommended treatment of kerosene poisoning: H. W. Gerarde, *Toxicol. Appl. Pharmacol.* **1**, 462 (1959).

Pale yellow or water-white, mobile, oily liquid. Characteristic, not altogether disagreeable odor. d ~0.80. bp 175-325°. Flash pt 150-185°F. (65-85°C). *Flammable.* Insol in water. Misc with other petroleum solvents. Kerosene may be deodorized and decolorized by washing with (fuming) sulfuric acid, followed by sodium plumbite soln and sulfur (Doctor sweetening). LD_{50} orally in rabbits: 28 ml/kg (Deichmann).

Caution: Potential symptoms of overexposure are irritation of eyes, skin, nose, throat; burning sensation in chest; headache, nausea, weakness, restlessness, incoordination, confusion, drowsiness; vomiting, diarrhea; dermatitis; aspiration of liquid may cause chemical pneumonia. *See NIOSH Pocket Guide to Chemical Hazards* (DHHS/NIOSH 97-140, 1997) p 184.

USE: In kerosene lamps, flares, and stoves; as degreaser and cleaner; in jet fuels.

5343. Ketamine. [6740-88-1] 2-(2-Chlorophenyl)-2-(methylamino)cyclohexanone. $C_{13}H_{16}ClNO$; mol wt 237.73. C 65.68%, H 6.78%, Cl 14.91%, N 5.89%, O 6.73%. Prepn: C. L. Stevens, **BE 634208**; *idem*, **US 3254124** (1963, 1966 both to Parke, Davis). Isoln of optical isomers: T. W. Hudyma *et al.*, **DE 2062620** (1971 to Bristol-Myers). Clinical pharmacology of racemate and enantiomers: P. F. White *et al.*, *Anesthesiology* **52**, 231 (1980). Toxicity: E. J. Goldenthal, *Toxicol. Appl. Pharmacol.* **18**, 185 (1971). Enantioselective HPLC determn in plasma: G. Geisslinger *et al.*, *J. Chromatogr.* **568**, 165 (1991). Comprehensive description: W. C. Sass, S. A. Fusari, *Anal. Profiles Drug Subs.* **6**, 297-322 (1977). Review of pharmacology and use in veterinary medicine: M. Wright, *J. Am. Vet. Med. Assoc.* **180**, 1462-1471 (1982). Review of pharmacology and clinical experience: D. L. Reich, G. Silvay, *Can. J. Anaesth.* **36**, 186-197 (1989); in pediatric procedures: S. M. Green, N. E. Johnson, *Ann. Emerg. Med.* **19**, 1033-1046 (1990).

Crystals from pentane-ether, mp 92-93°. uv max (0.01N NaOH in 95% methanol): 301, 276, 268, 261 nm ($A_{1cm}^{1\%}$ 5.0, 7.0, 9.8, 10.5). pKa 7.5. pH of 10% aq soln 3.5.

Hydrochloride. [1867-66-9] CI-581; Ketalar; Ketanest; Ketaset; Ketavet; Vetalar. $C_{13}H_{16}ClNO·HCl$; mol wt 274.19. White crystals, mp 262-263°. Soly in water: 20 g/100 ml. Freely sol in methanol; sol in alc; sparingly sol in chloroform. LD_{50} in adult mice, rats (mg/kg): 224 ±4, 229 ±5 i.p. (Goldenthal).

Note: This is a controlled substance (depressant): **21 CFR,** 1308.13.

THERAP CAT: Anesthetic (intravenous).

THERAP CAT (VET): Anesthetic (intravenous).

5344. Ketanserin. [74050-98-9] 3-[2-[4-(4-Fluorobenzoyl)-1-piperidinyl]ethyl]-2,4[1*H*,3*H*]-quinazolinedione; R-41468; Ketensin; Serefrex; Taseron. $C_{22}H_{22}FN_3O_3$; mol wt 395.43. C 66.82%, H 5.61%, F 4.80%, N 10.63%, O 12.14%. Specific serotonin (5HT$_2$)-receptor antagonist. Prepn: J. Vandenbeck *et al.*, **EP 13612;** *eidem,* **US 4335127** (1980, 1982 both to Janssen). X-ray structure: O. M. Peeters *et al.*, *Cryst. Struct. Commun.* **11**, 375 (1982). Receptor binding profile: J. E. Leysen *et al.*, *Life Sci.* **28**, 1015 (1981). Pharmacology: J. M. Van Neuten *et al.*, *J. Pharmacol. Exp. Ther.* **218**, 217 (1981). HPLC determn in plasma: A. T. Kacprowicz *et al.*, *J. Chromatogr.* **272**, 417 (1983). Clinical efficacy in intermittent claudication: J. De Cree *et al.*, *Lancet* **2**, 775 (1984); in Raynaud's phenomenon: J. R. Seibold, A. H. M. Jageneau, *Arthritis Rheum.* **27**, 139 (1984); in hypertension: A. Amery *et al.*, *J. Cardiovasc. Pharmacol.* **6**, 182 (1984). Series of articles on pharmacology and clinical studies: *ibid.* **7**, Suppl. 7, S1-S182 (1985); on pharmacokinetics and metabolism: *Arzneim.-Forsch.* **38**, 775-800 (1988). Review of pharmacology and clinical efficacy in hypertension and vascular disease: R. N. Brogden, E. M. Sorkin, *Drugs* **40**, 903-949 (1990).

Crystals from 4-methyl-2-pentanone, mp 227-235°. Soly (g/100 ml): 0.001 in water; 0.038 in ethanol; 2.34 in DMF. pKa 7.5.

Tartrate. [83846-83-7] R-49945; Ket; Perketan; Serepress; Sufrexal. $C_{22}H_{22}FN_3O_3·C_4H_6O_6$; mol wt 545.52.

THERAP CAT: Antihypertensive.

5345. Ketazolam. [27223-35-4] 11-Chloro-8,12b-dihydro-2,8-dimethyl-12b-phenyl-4*H*-[1,3]oxazino[3,2-*d*][1,4]benzodiazepine-4,7(6*H*)-dione; U-28774; Anseren; Unakalm. $C_{20}H_{17}ClN_2O_3$; mol wt 368.82. C 65.13%, H 4.65%, Cl 9.61%, N 7.60%, O 13.01%. Prepn: J. Szmuszkovicz, **DE 1947226;** *idem,* **US 3573282** (1970, 1971 both to Upjohn). Synthesis and structure: J. Szmuszkovicz *et al.*, *Tetrahedron Lett.* **1971**, 3665. Analysis by HPLC: D. J. Weber, *J. Pharm. Sci.* **61**, 1797 (1972). Pharmacology: V. H. Sethy, *Arch. Exp. Pathol. Pharmakol.* **301**, 157 (1978). Clinical studies: L. F. Fabre, *J. Int. Med. Res.* **4**, 50 (1976); L. A. Gottschalk, J. B. Cohn, *Psychopharmacol. Bull.* **14**, 39 (1978).

Colorless prisms from chloroform/ether, mp 182-183.5° (sinters at 170°). uv max (ethanol): 202, 241 nm (ε 40600; 18400).

Note: This is a controlled substance (depressant): **21 CFR,** 1308.14.

THERAP CAT: Anxiolytic.

5346. Ketene. [463-51-4] Ethenone; carbomethene. C_2H_2O; mol wt 42.04. C 57.14%, H 4.80%, O 38.06%. $CH_2\!=\!C\!=\!O$. Prepd by the thermal decompn of acetone, diketene or acetic anhydride: Hurd, *Org. Synth.* **coll. vol. I**, 330 (2nd ed., 1941); S. Andreades, H. D. Carlson, *ibid.* **coll. vol. V**, 679 (1973). Structure of ketene dimer as 3-buteno-β-lactone: Blomquist, Baldwin, *J. Am. Chem. Soc.* **70**, 29 (1948); Hurd, Blanchard, *ibid.* **72**, 1461 (1950); Katz, Lipscomb, *J. Org. Chem.* **17**, 515 (1953). Toxicology: J. F. Treon *et al.*, *J. Ind. Hyg. Toxicol.* **31**, 209 (1949). *Review* on the prepn of stable ketenes: R. S. Ward in *The Chemistry of Ketenes, Allenes and Related Compounds* Part 1, S. Patai, Ed. (Wiley, New York, 1980) pp 223-277; on synthetic uses of ketenes: W. T. Brady, *ibid.* pp 279-308.

Gas. Penetrating odor. mp −150°. bp −56°. Electron diffraction pictures: Beach, Stevenson, *J. Chem. Phys.* **6**, 75 (1938). Fairly sol in acetone. All operations with ketene should be carried out in an efficient hood (Andreades, Carlson).

Caution: Potential symptoms of overexposure are irritation of skin, eyes, nose, throat and respiratory system; pulmonary edema. *See NIOSH Pocket Guide to Chemical Hazards* (DHHS/NIOSH 97-140, 1997) p 184.

USE: For the conversion of higher acids into their anhydrides; for acetylation in the manuf of cellulose acetate and aspirin.

5347. Kethoxal. [27762-78-3] 3-Ethoxy-1,1-dihydroxy-2-butanone; β-ethoxy-α-ketobutyraldehyde monohydrate; 3-ethoxy-2-oxobutyraldehyde hydrate; U-2032. $C_6H_{12}O_4$; mol wt 148.16. C 48.64%, H 8.16%, O 43.19%. Antiviral glyoxal derivative. Specifically reacts with unpaired guanine residues of nucleic acids. Prepn: L. Rappen, *J. Prakt. Chem.* **157**, 177 (1941); and antiviral activity: B. D. Tiffany *et al.*, *J. Am. Chem. Soc.* **79**, 1682 (1957). Reaction with guanine residues: M. Staehelin, *Biochim. Biophys. Acta* **31**, 448 (1959); R. Shapiro, J. Hachmann, *Biochemistry* **5**, 2799 (1966). Use as RNA probe: M. Litt, V. Hancock, *ibid.* **6**, 1848 (1967); M. Balzer, R. Wagner, *Anal. Biochem.* **256**, 240 (1998). Use in peptide mapping to identifiy arginine residues: O. T. Akinsiku *et al.*, *J. Mass Spectrom.* **40**, 1372 (2005).

Yellow oil. bp$_{11}$ 54-58°. bp$_{760}$ 145°. Initial $n_D^{23.7}$ 1.4348. Miscible with alc; sol in benzene with yellow color; one part dissolves in ten parts of water.

Bis(thiosemicarbazone). [2507-91-7] 2,2′-[1-(1-Ethoxyethyl)-1,2-ethanediylidene]bishydrazinecarbothioamide; gloxazone; BW-356-C-61; NSC-82116. $C_8H_{16}N_6OS_2$; mol wt 276.38.

USE: RNA footprinting probe.

5348. Ketobemidone. [469-79-4] 1-[4-(3-Hydroxyphenyl)-1-methyl-4-piperidinyl]-1-propanone; 4-(*m*-hydroxyphenyl)-1-methyl-4-piperidyl ethyl ketone. $C_{15}H_{21}NO_2$; mol wt 247.34. C 72.84%, H 8.56%, N 5.66%, O 12.94%. Opioid analgesic. Prepn: **GB 591992** (1947 to Ciba); O. Eisleb, **DE 752755** (1952 to I. G. Farben); H. Kägi, K. Miescher, *Helv. Chim. Acta* **32**, 2489 (1949). LC/MS/MS determn in plasma: M. Lampinen *et al.*, *J. Chromatogr. B* **789**, 347 (2003). Pharmacokinetics in critically ill patients: A. Al-Shurbaji, L. Tokics, *Br. J. Clin. Pharmacol.* **54**, 583 (2002). Pediatric trial for postoperative pain: L. Jylli *et al.*, *Acta Anaesthesiol. Scand.* **48**, 1256 (2004).

Fine needles from ethyl acetate, mp 156-157°.

Hydrochloride. [5965-49-1] Hoechst 10720; Win-1539; Cliradon; Ketodur; Ketogan Novum; Ketorax. $C_{15}H_{21}NO_2.HCl$; mol wt 283.80. Crystals, mp 201-202°. Sol in water; slightly sol in alcohol. Aq solns may be sterilized by boiling for short periods.

Note: This is a controlled substance (opiate): **21 CFR**, 1308.11.

THERAP CAT: Analgesic.

5349. Ketoconazole. [65277-42-1] *rel*-1-[4-[4-[[(2R,4S)-2-(2,4-Dichlorophenyl)-2-(1H-imidazol-1-ylmethyl)-1,3-dioxolan-4-yl]methoxy]phenyl]-1-piperazinyl]ethanone; *cis*-1-acetyl-4-[4-[[2-(2,4-dichlorophenyl)-2-(1H-imidazol-1-ylmethyl)-1,3-dioxolan-4-yl]methoxy]phenyl]piperazine; R-41400; Fungoral; Ketoderm; Ketoisdin; Nizoral; Panfungol; Terzolin; Triatop. $C_{26}H_{28}Cl_2N_4O_4$; mol wt 531.43. C 58.76%, H 5.31%, Cl 13.34%, N 10.54%, O 12.04%. Orally active, broad-spectrum antimycotic. Prepn: J. Heeres *et al.*, **DE 2804096**, *eidem*, **US 4144346** and **US 4223036** (1978, 1979, 1980, all to Janssen); *eidem*, *J. Med. Chem.* **22**, 1003 (1979). Pharmacokinetics: E. W. Gascoigne *et al.*, *Clin. Res. Rev.* **1**, 177 (1981); C. Brass *et al.*, *Antimicrob. Agents Chemother.* **21**, 151 (1982). HPLC determn in human serum: V. L. Pascucci *et al.*, *J. Pharm. Sci.* **72**, 1467 (1983). Series of articles on animal and human studies: *Rev. Infect. Dis.* **2**, 519-692 (1980). Effect on hepatic enzymes *in vitro* and *in vivo*: K. N. Buchi *et al.*, *Biochem. Pharmacol.* **35**, 2845 (1986); J. K. Ritter, M. R. Franklin, *Toxicol. Lett.* **36**, 51 (1987). Case reports of hepatic toxicity: J. K. Heiberg, E. Svejgaard, *Br. Med. J.* **283**, 825 (1981); R. Rollman, L. Loof, *Br. J. Dermatol.* **108**, 376 (1983). Controlled clinical trials: E. A. Petersen *et al.*, *Ann. Intern. Med.* **93**, 791 (1980); W. T. Hughes *et al.*, *J. Infect. Dis.* **147**, 1060 (1983); H. W. Jolly *et al.*, *Cutis* **31**, 208 (1983). Clinical evaluation as inhibitor of steroid synthesis: N. Sonino, *N. Engl. J. Med.* **317**, 812 (1987). Review of pharmacology and therapeutic efficacy: C. A. Sohn, *Clin. Pharm.* **1**, 217 (1982); R. C. Heel *et al.*, *Drugs* **23**, 1-36 (1982). Series of articles on clinical efficacy and therapeutic experience: *Drugs Exp. Clin. Res.* **12**, 397-427 (1986).

Relative stereochemistry

Crystals from 4-methyl-2-pentanone, mp 146°. LD_{50} in mice, rats, guinea pigs, dogs (mg/kg): 44, 86, 28, 49 i.v.; 702, 227, 202, 780 orally (Heel).

THERAP CAT: Antifungal.

THERAP CAT (VET): Antifungal.

5350. α-Ketoglutaric Acid. [328-50-7] 2-Oxopentanedioic acid; 2-oxoglutaric acid; 2-oxo-1,5-pentanedioic acid. $C_5H_6O_5$; mol wt 146.10. C 41.11%, H 4.14%, O 54.75%. Plays an important role in amino acid metabolism (transamination): Severo Ochoa, "Enzymic Mechanisms in the Citric Acid Cycle" in *Adv. Enzymol.* **15**, 183-270 (1954). Prepn: Friedman, Kosower, *Org. Synth.* **coll. vol. III**, 510 (1955); Bottorff, Moore, *ibid.* **coll. vol. V**, 687 (1973). Microbial synthesis using a strain of *Pseudomonas*: Lockwood *et al.*, **US 2443919** (1948); Berger, Witt, **US 2841616** (1958).

Crystals from acetone-benzene, mp 113.5°. Freely sol in water, alcohol. Very sparingly sol in ether.

Diethyl ester. [5965-53-7] $C_9H_{14}O_5$; mol wt 202.21. Liq, bp_{23} 160°, bp_{13} 144°.

Compound with L(+)-ornithine. L(+)-Ornithine α-ketoglutarate; Ornicetil. $C_{10}H_{18}N_2O_7$; mol wt 278.26.

USE: Can be converted to L-glutamic acid by *Aeromonas* spp. *see* Good, **US 2933434** (1960 to International Minerals & Chem. Corp.).

5351. 2-Keto-L-gulonic Acid. [526-98-7] L-*xylo*-2-Hexulosonic acid; 2-oxo-L-gulonic acid. $C_6H_{10}O_7$; mol wt 194.14. C 37.12%, H 5.19%, O 57.69%. Important intermediate in vitamin C manufacture. Prepn from sorbitol: Reichstein, Grüssner, *Helv. Chim. Acta* **17**, 311 (1934); Reichstein, **US 2301811** (1942); from L-sorbose: Haworth *et al.*, **GB 443901** (1936); by microbial conversion: Huang, **US 3043749** (1962 to Pfizer); Motizuki *et al.*, **US 3234105** (1966 to Takeda); Z. Y. Yuan *et al.*, *Ann. N.Y. Acad. Sci.* **672**, 628 (1992). Biosynthesis and metabolism in bacteria: S. Makover *et al.*, *Biotechnol. Bioeng.* **17**, 1485 (1975). NMR structure determn: T. C. Crawford *et al.*, *J. Am. Chem. Soc.* **102**, 2220 (1980). HPLC separation in fermentation broth: R. A. Lazarus, J. L. Seymour, *Anal. Biochem.* **157**, 360 (1986). Review of early syntheses: T. C. Crawford, S. A. Crawford, *Adv. Carbohydr. Chem. Biochem.* **37**, 79-159 (1980).

Crystals from water (may be washed with acetone). mp 171° (slight dec). $[\alpha]_D^{18}$ −48° (c = 1). Moderately sol in water. Strong acid. Reduces boiling Fehling's soln rapidly.

Methyl ester. [3031-98-9] $C_7H_{12}O_7$; mol wt 208.17. Crystals, mp 155-157°. $[\alpha]_D^{18}$ −25° (c = 1 in methanol).

Ethyl ester. [5965-50-6] $C_8H_{14}O_7$; mol wt 222.19. Crystals, $[\alpha]_D^{20}$ −14.5° (c = 0.63 in abs alc).

5352. Ketoprofen. [22071-15-4] 3-Benzoyl-α-methylbenzeneacetic acid; *m*-benzoylhydratropic acid; 2-(3-benzoylphenyl)-propionic acid; RP-19583; Alreumat; Alrheumun; Capisten; Epatec; Fastum; Iso-K; Ketofen; Ketopron; Menamin; Meprofen; Orudis; Oruvail; Profenid; Toprec; Toprek. $C_{16}H_{14}O_3$; mol wt 254.29. C 75.57%, H 5.55%, O 18.87%. Prepn: D. Farge *et al.*, **ZA 6800524**; *eidem*, **US 3641127** (1968, 1972 both to Rhône-Poulenc); G. A. Pinna *et al.*, *Farmaco Ed. Sci.* **35**, 684 (1980). Resolution of isomers: S. Rendic *et al.*, *Chimia* **29**, 170 (1975). Enantioselective synthesis and absolute configuration of (+)-form: G. Comisso *et al.*, *Gazz. Chim. Ital.* **110**, 123 (1980). Pharmacology of enantiomers: P. J. Hayball *et al.*, *Chirality* **4**, 484 (1992). HPLC determn in plasma: R. Lovlin *et al.*, *J. Chromatogr. B* **679**, 196 (1996). Toxicity data: K. Ueno *et al.*, *J. Med. Chem.* **19**, 941 (1976). Comprehensive description: G. G. Liversidge, *Anal. Profiles Drug Subs.* **10**, 443-471 (1981). Review of pharmacokinetics: F. Jamali, D. R. Brocks, *Clin. Pharmacokinet.* **19**, 197-217 (1990); of clinical experience: E. M. Veys, *Scand. J. Rheumatol.* **Suppl. 90**, 3-44 (1991).

Crystals from 6:20 benzene-petr ether, mp 94°. uv max (methanol): 255 nm (log ε 4.33). Sol in ether, alc, acetone, chloroform, DMF, ethyl acetate. Slightly sol in water. LD_{50} orally in rats: 101 mg/kg (Ueno).

Lysine salt. [57469-78-0] Artrosilene. $C_{16}H_{14}O_3.C_6H_{14}N_2O_2$; mol wt 400.48.

(S)-(+)-Form tromethamine salt. [156604-79-4] Dexketoprofen trometamol; Enantyum; Keral. $C_{16}H_{14}O_3.C_4H_{11}NO_3$; mol wt 375.42. Prepn: G. Carganico *et al.*, **WO 9411332** (1994 to Menarini). Clinical trial: C. Gay *et al.*, *Clin. Drug Invest.* **11**, 320 (1996). White crystalline solid from ethanol-ethyl acetate, mp 104.8-105.1°. $[\alpha]_D^{20}$ −5.2° (c = 1.47 in methanol).

THERAP CAT: Anti-inflammatory; analgesic.

THERAP CAT (VET): Anti-inflammatory; analgesic.

5353. Ketorolac. [74103-06-3] 5-Benzoyl-2,3-dihydro-1H-pyrrolizine-1-carboxylic acid; 5-benzoyl-1,2-dihydro-3H-pyrrolo-[1,2-*a*]pyrrole-1-carboxylic acid; RS-37619. $C_{15}H_{13}NO_3$; mol wt 255.27. C 70.58%, H 5.13%, N 5.49%, O 18.80%. Non-selective cyclooxygenase (COX) inhibitor. Prepn and separation of isomers:

BE 856681; J. M. Muchowski, A. F. Kluge, **US 4089969** (both 1978 to Syntex). Alternate processes: J. M. Muchowski, R. Greenhouse, **US 4347186** (1982 to Syntex); F. Franco *et al.*, *J. Org. Chem.* **47**, 1682 (1982); J. B. Doherty, **US 4496741** (1985 to Merck & Co.). Absolute configuration of isomers: A. Guzman *et al.*, *J. Med. Chem.* **29**, 589 (1986). Structure-activity relationships: J. M. Muchowski *et al.*, *ibid.* **28**, 1037 (1985). Pharmacology: W. H. Rooks *et al.*, *Agents Actions* **12**, 684 (1982). Clinical comparison with acetaminophen in post-operative pain: H. J. McQuay *et al.*, *Clin. Pharmacol. Ther.* **39**, 89 (1986). Review of clinical efficacy in ocular inflammation: H. D. Perry, E. D. Donnenfeld, *Expert Opin. Pharmacother.* **7**, 99-107 (2006). Review of pharmacokinetics and efficacy of delivery formulations: V. R. Sinha *et al.*, *Expert Opin. Drug Deliv.* **6**, 961-975 (2009).

Crystals from ethyl acetate + ether, mp 160-161°. uv max in methanol: 245, 312 nm (ε 7080, 17400). pKa 3.49 ±0.02. LD_{50} orally in mice: ~200 mg/kg (Rooks).

(±)-Form tromethamine salt. [74103-07-8] Acular; Acuvail; Dolac; Lixidol; Sprix; Tarasyn; Toradol. $C_{19}H_{24}N_2O_6$; mol wt 376.41. White or almost white, crystalline powder. mp 165-170° (dec). Freely sol in water, methanol; slightly sol in alc, dehydrated alc, tetrahydrofuran. Practically insol in acetone, methylene chloride, dichloromethane, toluene, ethyl acetate, dioxane, hexane, butyl alc, acetonitrile.

(+)-Form. Crystals from hexane + ethyl acetate, mp 174° (Guzman); also reported as mp 154-156° (Muchowski, Kluge). $[\alpha]_D$ +173° (c = 1 in methanol).

(−)-Form. Crystals from hexane + ethyl acetate, mp 169-170° (Guzman); also reported as mp 153-155° (Muchowski, Kluge). $[\alpha]_D$ −176° (c = 1 in methanol).

THERAP CAT: Analgesic; anti-inflammatory.

5354. Ketotifen. [34580-13-7] 4,9-Dihydro-4-(1-methyl-4-piperidinylidene)-10*H*-benzo[4,5]cyclohepta[1,2-*b*]thiophen-10-one. $C_{19}H_{19}NOS$; mol wt 309.43. C 73.75%, H 6.19%, N 4.53%, O 5.17%, S 10.36%. Prepn: J. P. Bourquin *et al.*, **DE 2111071**; *eidem*, **US 3682930** (1971, 1972 both to Sandoz); E. Waldvogel *et al.*, *Helv. Chim. Acta* **59**, 866 (1976). Pharmacology: U. Martin, D. Roemer, *Arzneim.-Forsch.* **28**, 770 (1978). Clinical studies: B. Wüthrich, P. Radielovic, *Dtsch. Med. Wochenschr.* **103**, 1865 (1978); H. Gmür, M. Scherrer, *Schweiz. Med. Wochenschr.* **109**, 881 (1979). *Review:* U. Martin *et al.*, in *Pharmacological and Biochemical Properties of Drug Substances* **vol. 3**, M. E. Goldberg, Ed. (Am. Pharm. Assoc., Washington, DC, 1981) pp 424-460. Review of clinical experience in asthma and allergy: S. M. Grant *et al.*, *Drugs* **40**, 412-448 (1990).

Crystals from ethyl acetate, mp 152-153°.
Fumarate. [34580-14-8] HC-20511; Totifen; Zaditen; Zaditor; Zasten. $C_{19}H_{19}NOS \cdot C_4H_4O_4$; mol wt 425.50. Crystals, mp 192° (dec).

THERAP CAT: Antiasthmatic; antiallergic.

5355. Khat. Chat; quat. Leaves of *Catha edulis* Forsk., Celastraceae. *Habit.* East Africa, Arabia. *Constit.* Cathinone, norpseu-

doephedrine, *q.q.v.*, (−)-norephedrine, cathidine, cathedulin. Traditionally used in E. Africa and Yemen as an amphetamine-like stimulant. Isoln and characterization of constituents: O. Wolfes, *Arch. Pharm.* **268**, 81 (1930); H. Friebel, R. Brilla, *Naturwissenschaften* **50**, 354 (1963); M. Cais *et al.*, *Tetrahedron* **31**, 2727 (1975); R. L. Baxter *et al.*, *Chem. Commun.* **1976**, 463. Isoln of cathinone, the major psychoactive alkaloid: X. Schorno, E. Steinegger, *Experientia* **35**, 572 (1979). Review of botany, cultivation and use: A. Getahun, A. D. Krikorian, *Econ. Bot.* **27**, 353-377 (1973); of chemistry: *eidem, ibid.* 378-389; of pharmacology and abuse potential: P. Nencini, A. M. Ahmed, *Drug Alcohol Depend.* **23**, 19-29 (1989); P. Kalix, *Pharm. World Sci.* **18**, 69-73 (1996).

5356. Khellin. [82-02-0] 4,9-Dimethoxy-7-methyl-5*H*-furo[3,2-*g*][1]benzopyran-5-one; 5,8-dimethoxy-2-methyl-4′,5′-furo-6,7-chromone; 5,8-dimethoxy-2-methyl-6,7-furanochromone; 4,9-dimethoxy-7-methyl-5-oxofuro[3,2-*g*]-1,2-chromene; 4,9-dimethoxy-7-methyl-5-oxo-1,8-dioxabenz[*f*]indene; visammin; Ammivin; Lynamine; Vasokellina. $C_{14}H_{12}O_5$; mol wt 260.25. C 64.61%, H 4.65%, O 30.74%. Found in seeds of *Ammi visnaga* Lam., *Umbelliferae*; one of the active principles of the Egyptian traditional medicine known as "khella." *See also:* visnadine, visnagin. Isoln and structure: Späth, Gruber, *Ber.* **71**, 106 (1938). Synthesis: Baxter *et al.*, *J. Chem. Soc.* **1949**, S 30; Gardner *et al.*, *J. Org. Chem.* **15**, 841 (1950); Schönberg, Sina, *J. Am. Chem. Soc.* **72**, 1611, 3396 (1950); M. W. Reed, H. W. Moore, *J. Org. Chem.* **53**, 4166 (1988). Crystal structure: J. P. Beale, *Cryst. Struct. Commun.* **2**, 125 (1973). Toxicity data: E. Busch *et al.*, *Arzneim.-Forsch.* **11**, 915 (1961). GC determn in serum: A. S. Carlin *et al.*, *J. Chromatogr.* **614**, 324 (1993). Clinical evaluation in vitiligo: B. Ortel *et al.*, *J. Am. Acad. Dermatol.* **18**, 693 (1988). Comprehensive description: M. A. Hassan, M. U. Zubai, *Anal. Profiles Drug Subs.* **9**, 371-396 (1980).

Crystals from methanol. Bitter taste. mp 154-155°. $bp_{0.05}$ 180-200°. uv max (alc): 250, 338 nm ($E_{1cm}^{1\%}$ 1600, 200). Soly in g/100 ml at 25°: water 0.025; acetone 3.0; methanol 2.6; isopropanol 1.25; ether 0.5. Much more sol in hot water and hot methanol. LD_{50} in mice, rats (mg/kg): 30.6, 34.4 i.v.; 50.8, 68.8 orally (Busch).

THERAP CAT: Vasodilator (coronary). Photosensitizer in treatment of vitiligo.

5357. Kifunensine. [109944-15-2] (5*R*,6*R*,7*S*,8*R*,8a*S*)-Hexahydro-6,7,8-trihydroxy-5-(hydroxymethyl)imidazo[1,2-*a*]pyridine-2,3-dione; FR-900494. $C_8H_{12}N_2O_6$; mol wt 232.19. C 41.38%, H 5.21%, N 12.07%, O 41.34%. Specific α-mannosidase I inhibitor produced by the actinomycete, *Kitasatosporia kifunense*. Isoln: M. Iwami *et al.*, *J. Antibiot.* **40**, 612 (1987). Structure: H. Kayakiri *et al.*, *J. Org. Chem.* **54**, 4015 (1989). Synthesis: *idem et al.*, *Tetrahedron Lett.* **31**, 225 (1990). Mannosidase I inhibition and effect on glycoprotein processing: A. D. Elbein *et al.*, *J. Biol. Chem.* **265**, 15599 (1990). Characterization of high mannose type oligosaccharides produced in the presence of kifunensine: *idem et al.*, *Arch. Biochem. Biophys.* **288**, 177 (1991). Review of use in the study of glycoprotein processing: *idem*, *FASEB J.* **5**, 3055-3063 (1991). Review of use to control glycosylation in human monoclonal antibody production: Q. Zhou *et al.*, *Biotechnol. Bioeng.* **99**, 652-665 (2008); in the production of glycoproteins in human cell culture: J. E. Nettleship *et al.*, *Methods Mol. Biol.* **498**, 245-263 (2009).

Colorless prisms, mp >280°. $[\alpha]_D$ +58° (c = 0.1 in water). Readily sol in water; slightly sol in methanol, ethanol. Insol in acetone, ethyl acetate, chloroform.

USE: Biological reagent to control glycoprotein processing.

5358. Kiku Oil. Obtained by distillation from leaves and flowers of *Chrysanthemum indicum* L., *Compositae*. Produced in Japan: Perrier, *Bull. Soc. Chim. Fr.* [3] **23**, 216 (1900). *Constit. l*-Camphene, camphor carvone, xanthophyll, coumarin, angelic acid esters. Used as a folk remedy in Japan in a manner comparable with the use of chamomile and mint in Europe, but also against intestinal worms.

Colorless or greenish oil. Odor reminiscent of oil of Eucalyptus. d_4^{15} 0.932. n_D^{18} 1.4931. Acid to moist litmus paper. Ten grams dissolves in 100 grams abs alc at 95°. Almost insol in alc at 70°.

5359. Kinetin. [525-79-1] *N*-(2-Furanylmethyl)-9*H*-purin-6-amine; N^6-furfuryladenine; 6-furfurylaminopurine. $C_{10}H_9N_5O$; mol wt 215.22. C 55.81%, H 4.22%, N 32.54%, O 7.43%. A cell division factor found in various plant parts and in yeast. Isoln from autoclaved water slurries of deoxyribonucleic acid: C. O. Miller *et al., J. Am. Chem. Soc.* **77**, 1392 (1955). Structure and synthesis from furfurylamine and 6-methylmercaptopurine: *eidem, ibid.* 2662; *eidem*, **78**, 1375 (1956); US 2903455 (1959 to Wisc. Alumni Res. Found.). Physiologically active at very great dilutions, but only in presence of auxin, *see* Indoleacetic Acid. Crystal structure: M. Soriano-Garcia, R. Parthasarathy, *Acta Crystallogr.* **33B**, 2674 (1977). *Review:* Miller, *Annu. Rev. Plant Physiol.* **12**, 395 (1961).

Platelets from abs ethanol, mp 266-267° (sealed tube). Sublimes at 220°. pKa_1 2.7; pKa_2 9.9. uv max (ethanol): 268 nm (ε 18,650). Slightly soluble in cold water, methanol, ethanol. Freely soluble in dil aq HCl or NaOH. Can be extracted from neutral aq solns by shaking with ether.

USE: Plant growth regulator.

5360. Kininogens. [12244-26-7] Precursor molecules of the vasoactive kinins. Also serve as endogenous cysteine proteinase inhibitors known as type 3 cystatins and are essential cofactors for the contact activation of blood clotting. Single chain glycoproteins synthesized by hepatocytes and secreted into plasma. Three forms have been identified in mammals: H-kininogen (HK) having a molecular mass of 88-115 kDa, L-kininogen (LK) with molecular mass of 50-68 kDa, and T-kininogen, mol wt 68 kDa, which has been found only in rats. Structures consist of an amino-terminal heavy chain portion linked to a carboxy terminal light chain by the kinin segment which is released by kallikrein, *q.v.* The heavy chain contains 3 tandemly repeated domains that are structurally homologous to the type-2 cystatins, *q.v.* The HK light chain contains 2 additional domains that are responsible for the procoagulant activity. Purification from bovine serum: E. Habermann *et al., Z. Phys. Chem.* **332**, 121 (1963); *eidem, Biochem. Z.* **337**, 440 (1963). Isoln of 2 forms from human plasma: F. M. Habal *et al., Biochem. Pharmacol.* **23**, 2291 (1974). Review of purification methods and structure: H. Kato *et al., Methods Enzymol.* **80**, Part C, 172-199 (1981). Identity with α-cysteine proteinase inhibitors: W. Müller-Esterl *et al., FEBS Lett.* **182**, 310 (1985). Review of structure and function: R. A. DeLa Cadena, R. W. Colman, *Trends Pharmacol. Sci.* **12**, 272-275 (1991); of biological role: K. D. Bhoola *et al., Pharmacol. Rev.* **44**, 1-80 (1992); C. Blais, Jr. *et al., Peptides* **21**, 1903-1940 (2000).

H-Kininogen. [308068-04-4] α_1-CPI; α_1-cysteine proteinase inhibitor; high molecular weight kininogen. Preferred substrate for plasma kallikrein to release bradykinin, *q.v.* α-globulin; isoelectric point 4.3. Plasma conc in humans: 90 μg/ml.

L-Kininogen. α_2-CPI; α_2-cysteine proteinase inhibitor; low molecular weight kininogen. Preferred substrate for tissue kallikrein to release kallidin, *q.v.* β-globulin; isoelectric point 4.7. Plasma conc in humans: 170 μg/ml.

5361. Kino. [8052-27-5] Resin kino; gum kino. Dried juice from trunk of *Pterocarpus marsupium* Roxb., *Leguminosae. Habit.*

Western Africa, East India, Ceylon. *Constit.* 70-80% kinotannic acid; kino-red, pyrocatechol, kinoin, gum. Contains not less than 45% alcohol-soluble and not less than 80% water-soluble material.

THERAP CAT: Astringent.

THERAP CAT (VET): Has been used as astringent.

5362. Kistrin. [127829-86-1] Kistrin (*Agkistrodon rhodostoma* reduced). Member of a homologous family of low molecular weight, cysteine-rich proteins known as *disintegrins*. Isolated from various viper venoms; characterized by an Arg-Gly-Asp (RGD) binding site and by inhibition of platelet aggregation by antagonism of glycoprotein IIb-IIIa integrin receptors. A single chain polypeptide, mol wt ~7318, consists of 68 amino acids with six intramolecular disulfide bonds; no regular secondary structure. Isolated from the venom of the Malayan pit viper *Agkistrodon rhodostoma*. Purification, chemical and biological characterization: M. S. Dennis *et al., Proc. Natl. Acad. Sci. USA* **87**, 2471 (1989); R. A. Lazarus, M. S. Dennis, WO 9015072 (1990 to Genentech). Preliminary structural study: M. Adler *et al., Science* **253**, 445 (1991). NMR structural study: M. Adler, G. Wagner, *Biochemistry* **31**, 1031 (1992). Enhancement of coronary arterial thrombolysis in dogs: T. Yasuda *et al., Circulation* **83**, 1038 (1991). Binding interaction with glycoprotein IIb-IIIa: M. S. Dennis *et al., Proteins* **15**, 312 (1993). Brief review: R. J. Gould *et al., Proc. Soc. Exp. Biol. Med.* **195**, 168-171 (1990).

5363. Kitol. [4626-00-0] $[1\alpha,2\alpha,5\alpha(1E,3E),6\beta(1E,3E,5E)]$-3,6-Dimethyl-5-[2-methyl-4-(2,6,6-trimethyl-1-cyclohexen-1-yl)-1,3-butadienyl]-6-[4-methyl-6-(2,6,6-trimethyl-1-cyclohexen-1-yl)-1,3,5-hexatrienyl]-3-cyclohexene-1,2-dimethanol. $C_{40}H_{60}O_2$; mol wt 572.92. C 83.86%, H 10.56%, O 5.59%. One of the provitamins A. Obtained from mammalian liver oil: Embree, Shantz, *J. Am. Chem. Soc.* **65**, 910 (1943); Clough *et al., Science* **105**, 436 (1947); Barua, Morton, *Biochem. J.* **45**, 309 (1949); Chatan, Fridenson, *Compt. Rend.* **234**, 1094 (1952). Purification: Tawara, Fukazawa, *C.A.* **47**, 12483e (1953). Conversion to vitamin A: Libermann, Grundland, *Compt. Rend.* **224**, 1033 (1947). Structure: Burger *et al., Chem. Commun.* **1965**, 588; Giannotti *et al., ibid.* **1966**, 28; Giannotti *et al., Bull. Soc. Chim. Fr.* **1966**, 3299; Burger, Garbers, *J. Chem. Soc. Perkin Trans. 1* **1973**, 590.

Crystals from methanol, mp 88-90° (Embree); 98-100° (Chatan); 72° (Tawara). $[\alpha]_D$ −2.6° (c = 1.1 in chloroform). Labile in light, air, petr ether. Kitol has no vitamin A activity. uv max: 290 nm ($E_{1cm}^{1\%}$ 586).

Diphenylazobenzoate. Exists in two forms, probably geometric isomers, one mp 125-126°, the other 153-155°.

5364. Kodel®. Polyester staple and filament fiber. Prepn from dimethyl terephthalate and 1,4-cyclohexanedimethanol: Kibler *et al.*, US 2901466 (1959 to Eastman-Kodak). Review and structure: R. W. Moncrieff, *Man-Made Fibres* (John Wiley, New York, 4th ed., 1963) pp 389-393.

Solid, mp 290-295°. Burns slowly. Sp gr 1.22. Fiber has good resistance to acids and alkalies. Treatment of fabric with trichloroethylene and methylene chloride causes it to shrink; boiling water and perchloroethylene induce slight shrinkage. Fabric has good crease resistance.

5365. Kojic Acid. [501-30-4] 5-Hydroxy-2-(hydroxymethyl)-4H-pyran-4-one; 5-hydroxy-2-(hydroxymethyl)-4-pyrone; 2-hydroxymethyl-5-hydroxy-γ-pyrone. $C_6H_6O_4$; mol wt 142.11. C 50.71%, H 4.26%, O 45.03%. Antibiotic substance produced in an aerobic process by a variety of microorganisms from a wide range of carbon sources. Isoln from *Aspergillus oryzae*: Saito, *Bot. Mag. Tokyo* **21**, 249 (1907). Structure: Yabuta, *J. Chem. Soc.* **125**, 575 (1924); Heyns, Vogelsang, *Ber.* **87**, 13 (1954). Synthesis: Stacey, Turton, *J. Chem. Soc.* **1946**, 661; Lichtenthaler, Heidel, *Angew. Chem. Int. Ed.* **8**, 978 (1968). Industrial prepn: **GB 826244** (1959 to Pfizer). Mutagenicity study: C. I. Wei *et al.*, *Toxicol. Lett.* **59**, 213 (1991). *Review:* Beélik, *Adv. Carbohydr. Chem.* **11**, 145-183 (1956); Wilson, "Miscellaneous *Aspergillus* Toxins" in *Microbial Toxins* vol. **VI**, A. Ciegler *et al.*, Eds. (Academic Press, New York, 1971) pp 235-250.

Prismatic needles from acetone, ethanol + ether or methanol+ ethyl acetate, mp 153-154°. pKa 7.90, 8.03. Freely sol in water, ethanol, acetone; sparingly sol in ether, ethyl acetate, chloroform, pyridine. Absorption spectrum: Stacey, Turton.

USE: Converted to maltol and ethyl maltol, flavor-enhancing additives. Food additive to inhibit tyrosinase.

5366. Kola. Cola; Soudan coffee; Bissy nuts; gooroo nuts; guru nuts. Dried cotyledons of *Cola nitida* Schott and Endl. or of other species of *Cola, Sterculiaceae*. Habit. West Africa; naturalized in West Indies, India, Ceylon. *Constit.* About 1.5% caffeine; theobromine, kola-red, kolatin; glucoside kolanin; kolazyme; kolatannin, glucose, gum.

THERAP CAT: Analeptic.

5367. Konjac. Konnyaku. Flowering perennial plant with tuberous roots, *Amorphophallus konjac* K. Koch, *Araceae* or closely related species such as *A. bulbifer* B., *A. oncophylls* Prain ex Hook. f., or *A. variabilis* Blume. *Habit.* Southeast Asia. The tubers are ground to produce a flour used in traditional Asian cooking. *Constit.* Konjac mannan (8-10%), starch, lipid, minerals. Review of plant cultivation, production of flour and properties of konjac mannan: S. Takigami in *Handbook of Hydrocolloids*, G. O. Phillips, P. A. Williams, Eds. (CRC Press, Boca Raton, 2000) pp 413-424.

Konjac mannan. [37220-17-0] Konjac glucomannan. Nonionic, slightly branched polysaccharide. Consists of β-1,4 linked D-glucose and D-mannose in a ratio of approx 1 to 1.6, some branching at C3 of mannose, and an acetyl group attached to one per 19 sugar residues. Average mol wt 200-2000 kDa. Review of properties and applications: R. J. Tye, *Food Technol.* **45**, 82-92 (1991); K. Nishinari in *Novel Macromolecules in Food Systems*, G. Doxastakis, V. Kiosseoglou, Eds. (Elsevier Science, Amsterdam, 2000) pp 309-330. Light tan powder. Dispersible in hot or cold water; forms highly viscous solutions. Viscosity (cP): 31,600 (1% in water); 341,000 (2% in water). A heat-stable gel is formed in the presence of mild alkali.

USE: Source of dietary fiber; as thickening and gelling agent in foods.

5368. Kopsine. [559-48-8] 3-Hydroxy-22-oxokopsan-1-carboxylic acid methyl ester. $C_{22}H_{24}N_2O_4$; mol wt 380.44. C 69.46%, H 6.36%, N 7.36%, O 16.82%. From *Kopsia fructicosa* A.D., *Apocynaceae*: Bhattacharya *et al.*, *J. Am. Chem. Soc.* **71**, 3370 (1949). Structure: Spiteller, *Monatsh. Chem.* **93**, 1220 (1962); Govindachari *et al.*, *Helv. Chim. Acta* **45**, 1146 (1962); **46**, 572 (1963); Guggisberg *et al.*, *ibid.* **52**, 76 (1969).

Crystals from alc, dec 217-218°. $[\alpha]_D^{27}$ −14.3 ± 1° (c = 2 in chloroform). uv max (ethanol): 240, 278, 285-286 nm (log ε 4.08, 3.37, 3.35). Sol in chloroform; sparingly sol in methanol, ethanol, ethyl acetate, benzene, ether. Practically insol in petr ether, water.

Methiodide. $C_{23}H_{27}IN_2O_4$. Crystals from methanol, dec 194-196°.

Oxalate. $C_{24}H_{26}N_2O_8$. Prisms from alcohol + acetone, dec 154°. Sol in water.

5369. Koser's Reagent. [27126-76-7] Hydroxy(4-methylbenzenesulfonato-κO)phenyliodine; [hydroxy(tosyloxy)iodo]benzene; HTIB; phenyliodoso hydroxide tosylate; phenylhydroxytosyloxyiodine. $C_{13}H_{13}IO_4S$; mol wt 392.21. C 39.81%, H 3.34%, I 32.36%, O 16.32%, S 8.17%. Unsymmetrical aryl λ^3-iodane. Prepn: O. Y. Neiland, B. Y. Karele, *J. Org. Chem. USSR* **6**, 889 (1970). Solid state structure determn by single-crystal x-ray analysis: G. F. Koser *et al.*, *J. Org. Chem.* **41**, 3609 (1976). Review of use in organic synthesis: R. M. Moriarty *et al.*, *Synlett* **1990**, 365-383; of chemistry: G. F. Koser, *Aldrichim. Acta* **34**, 89-102 (2001).

Stable, nonhygroscopic crystalline solid. Colorless fine needles from dichloroethane, mp 140-142° (Neiland), also reported as colorless prisms from diethyl ether + methanol, mp 136-138.5° (Koser *et al.*). Sparingly sol in acetonitrile and dichloromethane at room temp. Readily dissolves in acetonitrile near the reflux temp. to give a yellow soln. Sol in dimethylformamide, alcohols and water with the formation of a yellow color. Sol in chloroform and acetic acid to give a colorless soln. Soly in water at 22°: ca. 1 g/42 mL.

USE: Versatile synthetic reagent in phenyliodination and/or tosylation of a range of organic substrates; in oxidative transformations including oxidative rearrangements.

5370. Kosins. Phloroglucinols from flowers of *Hagenia abyssinica* J. J. Gmel. (*Brayera anthelmintica* Kunth.), *Rosaceae*. Isoln and early characterization: M. Leichsenring, *Arch. Pharm.* **232**, 50 (1894); A. Lobeck, *ibid.* **239**, 672 (1901); B. A. Hems, A. R. Todd, *J. Chem. Soc.* **1937**, 562. Structure of α-kosin: W. Riedl, *Ber.* **89**, 2600 (1956). Orginally identified as α- and β-kosins; ultimately shown to be a mixture of compds differing only in their acyl side chains. Structural elucidation of kosins and protokosin: M. Lounasmaa *et al.*, *Acta Chem. Scand.* **B28**, 1200 (1974). Mass spectrometry: *idem*, P. Varenne, *Planta Med.* **34**, 153 (1978). Identification and proportions in male and female flowers: T. Z. Woldemariam *et al.*, *Anal. Proc.* **27**, 178 (1990). Antitumor activity: *idem, et al.*, *J. Pharm. Biomed. Anal.* **10**, 555 (1992).

α-Kosin

α-Kosin. 5,5′-Methylenebis[4,6-dihydroxy-2-methoxy-3-methylisobutyrophenone]. $C_{25}H_{32}O_8$. Yellow needles from ethanol, mp 160-160.5°. uv max: 227, 290 nm (ε 30800, 24400). Soluble in alcohol, benzene, chloroform, ether, glacial acetic acid, alkalies.

Protokosin. [1392-97-8] Structurally related phloroglucinol. Exists in several tautomeric forms and is a mixt of isobutyryl, isovaleryl, and 2-methylbutyryl side chain homologues. Colorless needles from acetone, mp 181-183°. $[\alpha]_D^{25}$ +13.9° (c = 0.610 in chloroform). uv max (cyclohexane): 224, 285 nm (ε 28200, 36400). Practically insol in water. Slightly sol in alcohol, light petroleum; freely sol in ether, acetone, ethyl acetate, chloroform.

THERAP CAT: Has been used as anthelmintic (Cestodes).

5371. Krebiozen. [9008-19-9] A white powder "chemically separated from horses' serum after stimulation of their cell network by the injection of *Actinomyces bovis*." Claimed to be a lipopolysaccharide. Developed by Stevan Durovic. Used experimentally in the treatment of cancer: Szujewski, *J. Am. Med. Assoc.* **148**, 929 (1952). Quackery Congress Symposium: James F. Holland, "The Krebiozen Story", *J. Am. Med. Assoc.* **200**, 213-218 (1967). *See also*: W. F. Janssen, *Anal. Chem.* **50**, 197A (1978); G. A. Curt, "Unsound Methods of Cancer Treatment" in *Cancer: Principles and Practice of Oncology*, V. T. DeVita, Jr. *et al.*, Eds. (J. B. Lippincott, Philadelphia, 4th ed., 1993) pp 2734-2747.

5372. Kresoxim-methyl. [143390-89-0] (αE)-α-(Methoxyimino)-2-[(2-methylphenoxy)methyl]benzeneacetic acid methyl ester; methyl-(E)-methoxyimino[α-(o-tolyloxy)-o-tolyl]acetate; BAS-490 F; BAS-490-02F; Candit; Sovran; Stroby. $C_{18}H_{19}NO_4$; mol wt 313.35. C 69.00%, H 6.11%, N 4.47%, O 20.42%. Synthetic strobilurin fungicide which inhibits mitochondrial respiration by blocking electron transfer at bc_1 complex. Prepn: B. Wenderoth *et al.*, **DE 3623921**; *eidem*, **US 4829085** (1988, 1989 both to BASF). Description of chemical properties and biological activities: E. Ammermann *et al.*, *Brighton Crop Prot. Conf. - Pests Dis.* **1992**, 403. Fungicidal effects on spore germination: R. E. Gold, G. M. Leinhos, *Pestic. Sci.* **43**, 250 (1995). Field trials against scab and powdery mildew on fruit: P. Creemers *et al.*, *Meded. Fac. Landbouwwet. Univ. Gent* **61**, 431 (1996); A. Brunelli *et al.*, *Brighton Crop Prot. Conf. - Pests Dis.* **1996**, 137. *Review*: R. E. Gold *et al.*, *Mod. Fungic. Antifungal Compd.*, *11th Int. Symp.* **1996**, 79-92.

Colorless, odorless crystals, mp 97.2-101.7°. Vapor pressure at 20°: 2.3×10^{-8} mbar. Soly at 20° (mg/l): 2 water. log P at 25° (n-octanol/water, pH 7): 3.4. LD_{50} in rats (mg/kg): >5000 orally; >2000 dermally (Ammerman).

USE: Agricultural fungicide.

5373. Krypton. [7439-90-9] Kr; at. wt 83.798; at. no. 36. Group VIIIA (18), also known as Group 0. A noble gas characterized by an electronic structure in which the outer *p* subshell is entirely filled. Naturally occurring stable isotopes (mass numbers): 78 (0.35%); 80 (2.25%); 82 (11.6%); 83 (11.5%); 84 (57.0%); 86 (17.3%); known artificial radioactive isotopes: 71-77; 79; 81; 85; 87-95; 97. Isoln from the final residues obtained after evaporation of liq air: Ramsay, Travers, *Proc. Roy. Soc.* **63** [A], 405 (1898). Concentration in air: 1.14 ppm by vol. Obtained commercially from the atmosphere by distillation-liquefaction process. Prepn: Lepape, *Compt. Rend.* **187**, 231 (1928); Claude, *ibid.* 581. Monograph: *Argon, Helium and the Rare Gases* vols. **1** 2,, G. A. Cook, Ed. (Interscience, New York, 1963) 818 pp. Diagnostic use of ^{81m}Kr in lung ventilation imaging: A. Zwijnenburg *et al.*, *Clin. Phys. Physiol. Meas.* **9**, 147 (1988). *Reviews*: A. H. Cockett, K. C. Smith, "The Monatomic Gases" in *Comprehensive Inorganic Chemistry* vol. **1**, J. C. Bailar, Jr. *et al.*, Eds. (Pergamon Press, Oxford, 1973) pp 139-211; N. Bartlett, F. O. Sladky, "The Chemistry of Krypton, Xenon

and Radon" *ibid.* pp 213-249; S.-C. Hwang, W. R. Weltmer, Jr. in *Kirk-Othmer Encyclopedia of Chemical Technology* vol. **13** (John Wiley & Sons, 4th ed., 1995) pp 1-38; G. J. Schrobilgen, J. M. Whalen, *ibid.* pp 38-53; *Chemistry of the Elements*, N. N. Greenwood, A. Earnshaw, Eds. (Pergamon Press, New York, 1984) pp 1042-1059.

Colorless, odorless, inert, monatomic gas; non-flammable. Will form compds with highly electronegative elements such as O, F, Cl. Condenses to a colorless liquid. Triple pt temp 115.95 K, press 73.15 kPa. Critical temp 209.4 K, critical press 5502 kPa, critical d 908 kg/m³. Gas: d^0 (101.3 kPa) 3.7493 kg/m³, d (normal bp) 8.6 kg/m³. Liquid: normal bp $-153.35°$, d (normal bp) 2415 kg/m³, d (triple pt) 2451 kg/m³, heat of vaporization (normal bp) 9.050 kJ/mol. Solid: d (triple pt) 2826 kg/m³, heat of vaporization (triple pt) 10.77 kJ/mol, heat of fusion (triple pt) 1.64 kJ/mol; solid form exists as face-centered cubic crystals at normal pressure. Excitation potentials: 9.91; 10.03; 10.56; 10.64 ev; ionization potentials: 14.00; 14.66 ev: Brocklehurst, *Q. Rev. Chem. Soc.* **22**, 147 (1968). Soly of gas in water (20°): 59.4 cm³/kg water. Slightly soluble in water with formation of a hydrate; ideal formula Kr.5.75H₂O. A hexadeuterate has been prepd. Soluble in liquid oxygen. Emission spectra: T. Jacksier, R. M. Barnes, *Appl. Spectrosc.* **48**, 65 (1994).

Krypton difluoride. [13773-81-4] F_2Kr; mol wt 121.79. Prepn: Turner, Pimentel, *Science* **140**, 974 (1963). Colorless solid; decomposes rapidly at room temperature. d (calc) 3.24. Vapor pressure (mm Hg): 10 ± 1 at $-15.5°$; 29 ± 2 at 0°; 73 ± 3 at 15°.

Caution: Can act as a simple asphyxiant by displacing air. *See: Matheson Gas Data Book* (Matheson Co., Inc., 4th ed., East Rutherford, NJ, 1966) pp 313-314.

USE: As gas filler in special purpose electric bulbs. To initiate and sustain the arc in metal-vapor discharge tubes, such as mercury and sodium vapor lamps. To determine surface areas of fine solids by adsorption techniques.

THERAP CAT: ^{81m}Kr as diagnostic aid (radioactive imaging agent).

5374. Kyanmethin. [461-98-3] 2,6-Dimethyl-4-pyrimidinamine; 4-amino-2,6-dimethylpyrimidine; 6-amino-2,4-dimethylpyrimidine. $C_6H_9N_3$; mol wt 123.16. C 58.51%, H 7.37%, N 34.12%. Prepn from acetonitrile and potassium methoxide: Ronzio, Cook, *Org. Synth.* **coll. vol. III**, 71 (1955).

Needles from alcohol or scales from benzene, mp 182-183°. One gram dissolves in 0.64 ml water at 18° or in 5.25 parts alcohol at 18°.

5375. Kynurenic Acid. [492-27-3] 4-Hydroxy-2-quinolinecarboxylic acid; 4-hydroxyquinaldic acid. $C_{10}H_7NO_3$; mol wt 189.17. C 63.49%, H 3.73%, N 7.40%, O 25.37%. Found in the urine of some animals as a metabolic product of tryptophan; its excretion is stepped up in avitaminoses B_1, B_2 and B_6. Isoln and syntheses: Späth, *Monatsh. Chem.* **42**, 89 (1921); Besthorn, *Ber.* **54**, 1330 (1921). Alternate syntheses: Benassi, *Gazz. Chim. Ital.* **91**, 1097 (1961); Wald, Joullie, *J. Org. Chem.* **31**, 3369 (1966); Jordanides, *Ann.* **729**, 244 (1969).

Yellow needles, mp 282-283°. Soly in water: about 0.9% at 100°. Sol in hot alc. Insol in ether.

Methyl ester. $C_{11}H_9NO_3$. Yellow crystals, mp 224°. *Cf.* xanthurenic acid.

USE: In nutrition studies, specifically in vitamin B deficiency diseases.

5376. Kynurenine. [343-65-7] α,2-Diamino-γ-oxobenzene-butanoic acid; 3-anthraniloylalanine. $C_{10}H_{12}N_2O_3$; mol wt 208.22. C 57.68%, H 5.81%, N 13.45%, O 23.05%. An amino acid produced in the body from tryptophan. Isoln from urine of rabbits that had been fed tryptophan: Matsuoka, Yoshimatsu, *Z. Physiol. Chem.* **143**, 206 (1925); Butenandt *et al., ibid.* **279**, 27 (1943); Heidelberger *et al., J. Biol. Chem.* **179**, 143 (1949). Structure and synthesis: Butenandt *et al., loc. cit.* Laboratory prepn by oxidation of L-tryptophan with a *Pseudomonas* sp.: Hayaishi, Meister, *Biochem. Prep.* **3**, 108 (1953). From acetyltryptophan: Warnell, Berg, *J. Am. Chem. Soc.* **76**, 1708 (1954); Auerbach, Knox, *Methods Enzymol.* **3**, 620 (1957).

L-Kynurenine hydrate. $3C_{10}H_{12}N_2O_3.H_2O$. Leaflets from water, dec 180-190°. $[\alpha]_D^{20}$ −29° (c = 0.4). Slightly sol in water (more sol than the DL-form). Forms a molecular compound with sucrose, $C_{22}H_{34}N_2O_{14}.H_2O$, rosettes; dec 145-153°, $[\alpha]_D^{21}$ +14.5° (c = 0.7). Soluble in water.

L-Kynurenine sulfate monohydrate. $C_{10}H_{12}N_2O_3.H_2SO_4.$-$H_2O$. Needles from water + alcohol. Darkens at 165°, dec 195°. $[\alpha]_D^{20}$ +7.3° (c = 1). uv max (pH 7.0): 230, 257, 360 nm (ε 18,900, 7500, 4500). Soluble in water, slightly in alcohol. Kynurenine sulfate requires 3 molecules alkali for neutralization in alcoholic soln.

L-Kynurenine diacetate. $C_{14}H_{16}N_2O_5$. Obtained by acetylating L-kynurenine with ketene. Needles, mp 198°.

Anhydro-L-kynurenine monoacetate. $C_{12}H_{12}N_2O_3$. Obtained by acetylating L-kynurenine with acetic anhydride in pyridine. Needles from alc, darkens at 215°, dec 237°.

DL-Kynurenine sulfate. $C_{10}H_{12}N_2O_3.H_2SO_4$. Crystals from water + alcohol. Darkens at 166°, dec 194°. Soluble in water. Slightly sol in alcohol.

USE: In biochemical investigations.

L

5377. Labetalol. [36894-69-6] 2-Hydroxy-5-[1-hydroxy-2-[(1-methyl-3-phenylpropyl)amino]ethyl]benzamide; 5-[1-hydroxy-2-[(1-methyl-3-phenylpropyl)amino]ethyl]salicylamide; ibidomide. $C_{19}H_{24}N_2O_3$; mol wt 328.41. C 69.49%, H 7.37%, N 8.53%, O 14.61%. Specific competitive antagonist at both α- and β-adrenergic receptor sites. Prepn: L. H. Lunts, D. T. Collin, **DE 2032642**; *eidem*, **US 4012444** (1971, 1977 both to Allen & Hanburys). Synthesis of labetalol and enantiomers: J. E. Clifton *et al.*, *J. Med. Chem.* **25**, 670 (1982); and comparison of cardiovascular properties: E. H. Gold *et al.*, *ibid.* 1363. Abs config of dilevalol: P. Murray-Rust *et al.*, *Acta Crystallogr. C* **40**, 825 (1984). Adrenoceptor blocking properties: E. J. Sybertz *et al.*, *J. Pharmacol. Exp. Ther.* **218**, 435 (1981). HPLC determn in serum or plasma: T. F. Woodman, B. Johnson, *Ther. Drug Monit.* **3**, 371 (1981). Metabolism in animals and man: R. Hopkins *et al.*, *Biochem. Soc. Trans.* **4**, 726 (1976). Toxicity: K. Shimpo *et al.*, *Hokkaido Igaku Zasshi* **53**, 15 (1978), *C.A.* **90**, 66465v (1974). Review of pharmacology: R. Donnelly, G. J. A. Macphee, *Clin. Pharmacokinet.* **21**, 95-109 (1991); of therapeutic applications in hypertension and ischemic heart disease: K. L. Goa *et al.*, *Drugs* **37**, 583-627 (1989).

Hydrochloride. [32780-64-6] AH-5158A; Sch-15719W; Amipress; Ipolab; Labelol; Labrocol; Normodyne; Presdate; Pressalolo; Trandate. $C_{19}H_{24}N_2O_3 \cdot HCl$; mol wt 364.87. White crystalline solid from ethanol-ethyl acetate, mp 187-189°. Sol in water, ethanol. Insol in ether, chloroform. LD_{50} in male, female mice, male, female rats (mg/kg): 114, 120, 113, 107 i.p.; 47, 54, 60, 53 i.v.; 1450, 1800, 4550, 4000 orally (Shimpo).

(R,R)-Form hydrochloride. [75659-08-4]; [75659-07-3] (free base). Dilevalol hydrochloride; Sch-19927. Polymorphic crystals from ethanol, mp 133-134° (dec); mp 192-193.5° (dec). $[\alpha]_D^{26}$ $-30.6°$ (c = 1.0 in ethanol).

THERAP CAT: Antihypertensive.

5378. Laccaic Acid. C.I. Natural Red 25; lac dye; C.I. 75450. Pigment found in the lac resin produced by the insect *Coccus laccae (Laccifer lacca Kerr)* on certain trees in India. Isoln: R. E. Schmidt, *Ber.* **20**, 1285 (1887). Chemistry: O. Dimroth, S. Goldschmidt, *Ann.* **399**, 62 (1913). Originally thought to be one compound, laccaic acid has been separated into four components: A (major), B, C and D. Isoln of A: R. Burwood *et al.*, *J. Chem. Soc. C* **1965**, 6067. Structure of A: *eidem*, *Tetrahedron Lett.* **1966**, 3059; *eidem*, *J. Chem. Soc. C* **1967**, 842; E. D. Pandhare *et al.*, *Indian J. Chem.* **7**, 977 (1969). Isoln and structure of B: *eidem*, *Tetrahedron Lett.* **1967**, 2437; N. S. Bhide *et al.*, *Indian J. Chem.* **7**, 987 (1969); of C: A. V. R. Rao *et al.*, *ibid.* 188; of D: A. R. Mehandale *et al.*, *Tetrahedron Lett.* **1968**, 2231. Synthesis of D: D. W. Cameron *et al.*, *Chem. Commun.* **1978**, 688; *eidem*, *Aust. J. Chem.* **34**, 2401 (1981).

Laccaic Acid A R = CH₂CH₂NHCOCH₃
Laccaic Acid B R = CH₂CH₂OH
Laccaic Acid C R = CH₂CH(NH₂)COOH
Laccaic Acid D

Laccaic acid A. Laccaic acid A_1. $C_{26}H_{19}NO_{12}$; mol wt 537.43. Red platelets from methanol, chars at 230°. Absorption max (conc H_2SO_4): 302, 361, 518, 558 nm (log ε 4.33, 4.11, 4.32, 4.37).

Laccaic acid B. $C_{24}H_{16}O_{12}$; mol wt 496.38. Red needles from methanol.

Laccaic acid C. $C_{25}H_{17}NO_{13}$; mol wt 539.41. Dark red needles from methanol, dec >360°.

Laccaic acid D. Xanthokermesic acid. $C_{16}H_{10}O_7$; mol wt 314.25. Yellow needles from water, dec >300°.

USE: As crimson dye.

5379. Lacidipine. [103890-78-4] 4-[2-[(1E)-3-(1,1-Dimethylethoxy)-3-oxo-1-propen-1-yl]phenyl]-1,4-dihydro-2,6-dimethyl-3,5-pyridinedicarboxylic acid 3,5-diethyl ester; diethyl (E)-4-[2-[2-(tert-butoxycarbonyl)vinyl]phenyl]-2,6-dimethyl-1,4-dihydropyridine-3,5-dicarboxylate; 4-[o-[(E)-2-carboxyvinyl]phenyl]-1,4-dihydro-2,6-dimethyl-3,5-pyridinedicarboxylic acid 4-tert-butyl diethyl ester; GR-43659X; GX-1048; Caldine; Lacimen; Lacipil; Lacirex; Midotens; Motens. $C_{26}H_{33}NO_6$; mol wt 455.55. C 68.55%, H 7.30%, N 3.07%, O 21.07%. Dihydropyridine calcium channel blocker. Prepn: C. Semeraro *et al.*, **DE 3529997**; *eidem*, **US 4801599** (1986, 1989 both to Glaxo). Clinical pharmacology: M. Safar *et al.*, *Clin. Pharmacol. Ther.* **46**, 94 (1989). HPTLC determn in urine: V. R. Kharat *et al.*, *J. Pharm. Biomed. Anal.* **28**, 789 (2002). Clinical trial in atherosclerosis: A. Zanchetti *et al.*, *Circulation* **106**, 2422 (2002). Review of use in hypertension: P. L. McCormack, A. J. Wagstaff, *Drugs* **63**, 2327-2356 (2003).

Crystalline solid from ethyl acetate, mp 174-175°.

THERAP CAT: Antihypertensive.

5380. Lacmoid. Resorcin blue; Iris Blue B; Fluorescent Blue; C.I. 51400. Prepn: Traub, Hock, *Ber.* **17**, 2615 (1884); Musso *et al.*, *Angew. Chem.* **73**, 434 (1961). Discussion of structure controversy: *H. J. Conn's Biological Stains*, R. D. Lillie, Ed. (Williams & Wilkins, Baltimore, 8th ed., 1969) pp 283-284. *See also: Colour Index* vol. 4 (3rd ed., 1971) p 4468.

Dark-violet, lustrous scales or granules. Freely sol in methanol, ethanol, amyl alcohol, glacial acetic acid, acetone, phenol. Sparingly sol in water, ether. Practically insol in chloroform, benzene, petr ether.

USE: As acid-base indicator in 0.2% soln in alcohol. pH: 4.4 red; 6.4 blue. Satisfactory for titrating mineral acids, strong bases, many alkaloids; determining alkalinity and temporary hardness in water analysis. Biological stain; dye for wool, silk. Not adapted for carbonates, weak inorganic and organic acids, weak bases. Lacmoid is more sensitive than litmus, particularly in form of test paper.

5381. Lacosamide. [175481-36-4] (2R)-2-(Acetylamino)-3-methoxy-N-(phenylmethyl)propanamide; (R)-N-benzyl-2-acetamido-3-methoxypropionamide; erlosamide; harkoseride; ADD-234037; SPM-927; Vimpat. $C_{13}H_{18}N_2O_3$; mol wt 250.30. C 62.38%, H 7.25%, N 11.19%, O 19.18%. Functionalized amino acid with anti-epileptic and analgesic activity. Enhances slow activation of voltage-gated sodium channels and binds to collapsin response mediator protein-2 (CRMP-2). Prepn: D. Choi *et al.*, *J. Med. Chem.* **39**, 1907 (1996); H. Kohn, **WO 9733861**; *idem*, **US 5773475** (1997, 1998 both to Research Corp. Technol.). Improved prepn and anticonvulsant activity: S. V. Andurkar *et al.*, *Tetrahedron: Asymmetry* **9**, 3841 (1998). Review of pharmacology, mechanisms of action, and clinical efficacy in treatment of diabetic neuropathic pain: V. Biton, *Expert Rev. Neurother.* **8**, 1649-1660 (2008); in partial-onset seizures: A. Beydoun *et al.*, *ibid.* **9**, 33-42 (2009).

White solid, mp 143-144°. $[\alpha]_D^{23}$ +16.0° (c = 1 in CH_3OH). Sparingly sol in water; slightly sol in acetonitrile, ethanol.

Note: This is a controlled substance (depressant): **21 CFR, 1308.15**

THERAP CAT: Anticonvulsant; analgesic in neuropathic pain.

5382. Lactate Dehydrogenase. [9001-60-9] Lactic dehydrogenase; serum lactic dehydrogenase; EC 1.1.1.27. Enzyme found in almost all animal tissues, in microorganisms, and in plants. Catalyzes the equilibrium reaction of pyruvic acid to lactic acid with the concomitant interconversion of NADH and NAD^+. Isoln from heart muscle, rat skeletal muscle, and Jensen sarcoma: Straub, *Biochem. J.* **34**, 483 (1940); Kubowitz, Ott, *Biochem. Z.* **314**, 94 (1943); Meister, *Biochem. Prep.* **2**, 18 (1952). Structure is a tetramer of mol wt about 140,000. Consists of units of mol wt about 35,000. Two types of subunits are distinguishable: M (muscle) type and H (heart) type. Lactate dehydrogenases of heart and muscle are mainly H_4 and M_4; all other possible hybrids have been found in various tissues. Elevations of lactate dehydrogenase activity have been found in myocardial infarction, hepatocellular necrosis, metastatic carcinoma, diabetic ketosis, sickle cell anemia, malignant lymphoma, infectious mononucleosis, and cerebral infarction: Standjord *et al., J. Am. Med. Assoc.* **182**, 1099 (1962). Comprehensive reviews: Everse, Kaplan, *Adv. Enzymol. Relat. Areas Mol. Biol.* **37**, 61 (1973); Holbrook *et al.,* in *The Enzymes* vol. XI (part A), P. D. Boyer, Ed. (Academic Press, New York, 3rd ed., 1975) pp 191-292.

USE: In the determination of pyruvate (used in conjunction with reduced coenzyme). In the diagnosis of myocardial infarction and leukemia.

5383. D-Lactic Acid. [10326-41-7] (2R)-2-Hydroxypropanoic acid; D(−)-lactic acid; (R)-(−)-lactic acid; levorotatory lactic acid; l-lactic acid; D-Milchsäure. $C_3H_6O_3$; mol wt 90.08. C 40.00%, H 6.71%, O 53.28%. Obtained by resolution of DL-lactic acid: Purdie, Walker, *J. Chem. Soc.* **61**, 754 (1892); Borsook *et al., J. Biol. Chem.* **102**, 449 (1933). Convenient laboratory prepn from glucose using *Lactobacillus leichmannii:* Brin, *Biochem. Prep.* **3**, 61 (1953).

Crystals from ether + isopropyl ether, mp 52.8°. $[\alpha]_{546}^{21.5}$ −2.6° (c = 8). pK = 3.83. Sol in water, alcohol, acetone, ether, glycerol. Practically insol in chloroform. Forms salts with many metals. Most of these salts are dextrorotatory.

Zinc D(+)-lactate dihydrate. $Zn(C_3H_5O_3)_2 \cdot 2H_2O$; mol wt 279.58. Crystals, $[\alpha]_D^{14}$ +8.18° (c = 2.5).

5384. DL-Lactic Acid. [50-21-5] 2-Hydroxypropanoic acid; 2-hydroxy-2-methylacetic acid; racemic lactic acid; ordinary lactic acid; α-hydroxypropionic acid; Milchsäure; Lactovagan; Tonsillosan (Lösung). $C_3H_6O_3$; mol wt 90.08. C 40.00%, H 6.71%, O 53.28%. Occurs in sour milk as a result of lactic acid bacteria; also found in molasses due to partial conversion of sugars, in apples and other fruits, tomato juice, beer, wines, opium, ergot, foxglove, and several higher plants, especially during germination. Lactic acid is prepd technically by "lactic acid fermentation" of carbohydrates such as glucose, sucrose, lactose with *Bacillus acidi lacti* or related organisms such as *Lactobacillus delbrueckii, L. bulgaricus* etc. The fermentation is carried out at relatively high temps. Produced commercially by fermentation of whey, cornstarch, potatoes, molasses. Review on the production of lactic acid by fermentation: S. C. Prescott, C. G. Dunn, *Industrial Microbiology* (McGraw-Hill, New York, 3rd ed., 1959) pp 304-331. Chem prepns from acetaldehyde and CO in dil H_2SO_4 at 130-200° and 900 atm: Loder, **US 2265945** (1938 to

du Pont); by hydrolysis of hexoses with NaOH: Lock, **US 2382889** (1943). Prepn of crystalline lactic acid: Borsook *et al., J. Biol. Chem.* **102**, 449 (1933). Toxicity data: Smyth *et al., J. Ind. Hyg. Toxicol.* **23**, 259 (1941). Comprehensive description: F. J. Al-Shammary *et al., Anal. Profiles Drug Subs. Excip.* **22**, 263-316 (1993).

Crystals, mp 16.8°. bp_{14-15} 122°; $bp_{0.5-1}$ 82-85°. pKa at 25° 3.86. n_D^{20} 1.4262. Heat of combustion at constant pressure 3615 cal/kg. Volatile with superheated steam. Sol in water, alc, furfurol; less sol in ether. Practically insol in chloroform, petr ether, carbon disulfide. Caustic in concd solns. LD_{50} orally in rats: 3.73 g/kg (Smyth).

Barium salt. [533-91-5] Barium lactate. $C_6H_{10}BaO_6$; mol wt 315.47. Powder. *Poisonous.* Sol in water, dil alcohol.

Copper salt dihydrate. [5893-64-1] Cupric lactate. C_6H_{10}-$CuO_6 \cdot 2H_2O$; mol wt 277.72. Green to blue crystals. Readily sol in water. Practically insol in alcohol.

USE: In dyeing baths, as mordant in printing woolen goods, solvent for water-insoluble dyes (alcohol-soluble induline, nigrosine, spirit-blue). Reducing chromates in mordanting wool. Food preservative. Manuf cheese, confectionery. Component of babies' milk formulas; acidulant in beverages; for acidulating worts in brewing. In prepn of sodium lactate injections. Ingredient of cosmetics. Component of spermatocidal jellies. For removing *Clostridium butyricum* in manuf of yeast; dehairing, plumping, and decalcifying hides. Solvent for cellulose formate. Flux for soft solder. Manuf lactates which are used in food products, in medicine, and as solvents. Plasticizer, catalyst in the casting of phenolaldehyde resins.

THERAP CAT: Acidulant.

THERAP CAT (VET): Has been used as a caustic, and in dilute solutions to irrigate tissues; as an intestinal antiseptic and antiferment.

5385. L-Lactic Acid. [79-33-4] (2S)-2-Hydroxypropanoic acid; L(+)-lactic acid; (S)-(+)-lactic acid; dextrorotatory lactic acid; d-lactic acid; sarcolactic acid; paralactic acid; Fleishmilchsäure; L-Milchsäure. $C_3H_6O_3$; mol wt 90.08. C 40.00%, H 6.71%, O 53.28%. Occurs in small quantities in the blood and muscle fluid of man and animals. The lactic acid concn increases in muscle and blood after vigorous activity. L(+)-Lactic acid is also present in liver, kidney, thymus gland, human amniotic fluid, and other organs and body fluids. Obtained by resolution of DL-lactic acid: Purdie, Walker, *J. Chem. Soc.* **61**, 754 (1892); Borsook *et al., J. Biol. Chem.* **102**, 449 (1933). Convenient laboratory prepn from glucose by fermentation by *Lactobacillus delbrueckii:* Brin, *Biochem. Prep.* **3**, 61 (1953). Prepn from hexoses using *B. dextrolacticus:* Andersen, Greaves, *Ind. Eng. Chem.* **34**, 1522 (1942). Monograph: M. Brin, R. H. Dunlop, "Chemistry and Metabolism of L- and D-Lactic Acids" in *Ann. N.Y. Acad. Sci.* **119**, 851-1165 (1965).

Crystals from acetic acid or chloroform, mp 53°. $[\alpha]_{546.1}^{21-22}$ +2.6° (c = 2.5). pK at 25°, 3.79. Forms salts with many metals. The salts are more sol in water than the salts of the racemic acid. Most of the salts are levorotary.

Zinc L(−)-lactate dihydrate. $Zn(C_3H_5O_3)_2 \cdot 2H_2O$; mol wt 279.58. Prisms. $[\alpha]_D^{25}$ −8.2° (c = 2.5 in water).

5386. Lactic Acid Homopolymer. [26023-30-3] Poly[oxy(1-methyl-2-oxo-1,2-ethanediyl)]; poly(lactic acid); PLA; EcoPLA; Lacea; Lactron; Lacty. Biodegradable thermoplastic polymer which can be derived from entirely renewable resources; degrades *in vivo* to lactic acid. Properties are dependent on enantiomeric content, polymer length, and method of preparation. Prepn: W. H. Carothers et al., *J. Am. Chem. Soc.* **54**, 761 (1932); J. Kleine, H.-H. Kleine, *Makromol. Chem.* **30**, 23 (1959). Synthesis of L-form by lactide ring-opening polymerization: H. R. Kricheldorf, S.-R. Lee, *Polymer* **36**, 2995 (1995); by lactic acid melt/solid polycondensation: R. Miyoshi *et al.*, *Int. Polym. Process.* **11**, 320 (1996); S.-I. Moon *et al.*, *High Perform. Polym.* **13**, S189 (2001). Kinetics and mechanism of ring-opening polymerization: W. Dittrich, R. C. Schulz, *Angew. Makromol. Chem.* **15**, 109 (1971). Comparison of properties from different synthetic methods: M. Ajioka *et al.*, *J. Environ. Polym. Degrad.* **3**, 225 (1995). Comparison of melt-spun and solution-spun L-form fibers: B. Eling *et al.*, *Polymer* **23**, 1587 (1982). Thermal properties: K. Jamshidi *et al.*, *ibid.* **29**, 2229 (1988). ^{13}C solid state NMR study of crystallinity and morphology: K. A. M. Thakur *et al.*, *Macromolecules* **29**, 8844 (1996). ^{1}H NMR determn of lactide composition: *idem et al.*, *Anal. Chem.* **69**, 4303 (1997). Review of properties and uses: M. Vert *et al.*, *J. Macromol. Sci. Pure Appl. Chem.* **A32**, 787-796 (1995); of material design systems: M. Spinu *et al.*, *ibid.* **A33**, 1497-1530 (1996); of mfr, properties and commercial applications: J. Lunt, *Polym. Degrad. Stab.* **59**, 145-152 (1998); A. Södergard, M. Stolt, *Prog. Polym. Sci.* **27**, 1123-1163 (2002); K. M. Nampoothiri *et al.*, *Bioresource Technol.* **101**, 8493-8501 (2010).

Glassy material. mp 175°. Glass transition temperature: 55-60°. d 1.25 g/cm^3. n 1.45. Contact angle: 0.254.

D-Form. [26917-25-9] Poly-D-lactic acid; poly[(+)-lactic acid]; D-lactide homopolymer; PLDA.

L-Form. [26161-42-2] Poly-L-lactic acid; poly-L-lactide; PLLA; Sculptra. Clinical trial in HIV-associated facial lipoatrophy: C. M. Burgess, R. M. Quiroga, *J. Am. Acad. Dermatol.* **52**, 233 (2005); in esthetic rejuvenation: K. R. Beer, M. I. Rendon, *Semin. Cutan. Med. Surg.* **25**, 127 (2006). Semicrystalline polymer. mp 180°. Glass transition temperature: ~55°. Sol in chloroform, furan, 1,4-dioxane, 1,3-dioxolane, pyridine. Insol in water, alcohols.

USE: In packaging (primarily food), biomedical, film and fiber applications.

THERAP CAT: In treatment of facial lipoatrophy. Surgical aid (bioabsorbable sutures).

5387. Lactic Acid Lactate. [617-57-2] 2-Hydroxypropanoic acid 1-carboxyethyl ester; 2-(lactoyloxy)propanoic acid; 2-(2-hydroxypropanoyloxy)propanoic acid. $C_6H_{10}O_5$; mol wt 162.14. C 44.45%, H 6.22%, O 49.34%. Prepd by heating lactic acid at 120° for 10 hours: Dietzel, Krug, *Ber.* **58**, 1307 (1925).

Pale yellow, clear, odorless oil. Sol in water and in the usual organic solvents.

Methyl ester. $C_7H_{12}O_5$. Prepn: Claborn, US 2371281 (1945 to the people of the U.S.). bp$_{7.8}$ 107°; n_D^{20} 1.4313.

USE: The methyl ester as a solvent or plasticizer.

5388. Lactisole. [13794-15-5] 2-(4-Methoxyphenoxy)propanoic acid; 2-(*p*-methoxyphenoxy)propionic acid; PMP; HPMP. $C_{10}H_{12}O_4$; mol wt 196.20. C 61.22%, H 6.17%, O 32.62%. Inhibitor of human sweet taste perception; binds to the transmembrane domains of human taste receptor T1R3. Bioactivity resides in the *S*-isomer which is a naturally occurring constituent of coffee beans. Prepn: H. Sobotka, J. Austin, *J. Am. Chem. Soc.* **74**, 3813 (1952). Use in foods as flavor modifier: M. G. Lindley, E. B. Rathbone, **EP 0159864** (1985 to Tate & Lyle); *eidem*, **US 5045336** (1991 to Amstar Sugar). Isoln from roasted coffee beans: E. B. Rathbone *et al.*, *J. Agric. Food Chem.* **37**, 54 (1989); and activity of isomers: *eidem*, *ibid.*, **58**. Crystal structure: M. Mathlouthi *et al.*, *J. Mol. Struct.* **326**, 25 (1994). Effect on taste of natural and artifical sweeteners: S. S. Schiffman *et al.*, *Chem. Senses* **24**, 439 (1999). T1R3 binding study: P. Jiang *et al.*, *J. Biol. Chem.* **280**, 15238 (2005).

Monoclinic crystals, d 1.3031. mp 90°. bp$_1$ 173-175°.

Sodium salt. [150436-68-3] Sodium 2-(4-methoxyphenoxy)propanoate; NaPMP; ORP-178. $C_{10}H_{11}NaO_4$; mol wt 218.18. White to pale cream, crystalline solid, mp 190°. Sol in water, propylene glycol.

S-Isomer. [4276-74-8] mp 65-66°. $[\alpha]_D^{20}$ $-42.7°$ (c = 0.95 in ethanol).

USE: Biological tool for taste perception studies. Sweetness reducing agent for use in foods.

5389. Lactitol. [585-86-4] 4-*O*-β-D-Galactopyranosyl-D-glucitol; β-galactoside sorbitol; lactit; lactit M; lactite; lactobiosit; lactosit; lactositol. $C_{12}H_{24}O_{11}$; mol wt 344.31. C 41.86%, H 7.03%, O 51.11%. Polyol sweetener; relative sweetness compared to sucrose is 36%. Prepd by hydrogenation of lactose, *q.v.*: M. J. B. Senderens, *Compt. Rend.* **170**, 47 (1920); M. L. Wolfrom *et al.*, *J. Am. Chem. Soc.* **60**, 571 (1938). Pharmacology: D. H. Patil *et al.*, *Br. J. Nutr.* **57**, 195 (1987). Crystal structure: J. A. Kanters *et al.*, *Acta Crystallogr.* **C46**, 2408 (1990); J. Kivikoski *et al.*, *Carbohydr. Res.* **223**, 45 (1992). Toxicology: E. J. Sinkeldam *et al.*, *J. Am. Coll. Toxicol.* **11**, 165 (1992). Clinical trial in chronic hepatic encephalopathy: O. Riggio *et al.*, *Hepatogastroenterology* **37**, 524 (1990); as a laxative: L. Goovaerts, G. P. Ravelli, *Acta Ther.* **19**, 61 (1993). Review of properties and applications: J. A. van Velthuijsen, *J. Agric. Food Chem.* **27**, 680-686 (1979); of chemistry and use in foods: C. H. den Uyl, *Dev. Sweeteners* **3**, 65-81 (1987).

Crystals from absolute ethanol, mp 146°. $[\alpha]_D^{23}$ +14° (c = 4 in water). Sol in water, dimethyl sulfoxide, *N,N*-dimethylformamide; slightly sol in ethanol, ether. Strongly hygroscopic.

Monohydrate. [81025-04-9] Importal; Portolac. White, sweet, odorless, crystalline solid. Non-hygroscopic. mp 94-97° (van Velthuijsen), water of crystallization evaporates 145°-185°; also reported as mp 120° (den Uyl). $[\alpha]_D^{22}$ +12.3°. Soly at 25° (g/100 g solvent): water 206; ethanol 0.75; ether 0.4; DMSO 233; DMF 39; at 50°: water 512; ethanol 0.88; at 75°: water 917.

Dihydrate. [81025-03-8] Lacty. White, sweet, odorless, crystalline powder. Data for food grade, mp 75°. $[\alpha]_D^{25}$ +13.5-15.0°. pH of 10% solution 4.5 - 8.5. 140 g will dissolve in 100 ml water at 25°.

USE: Sweetener in food.

THERAP CAT: Laxative. In treatment of hepatic encephalopathy.

5390. Lactobacillic Acid. [19625-10-6] (1*R*,2*S*)-2-Hexylcyclopropanedecanoic acid; 11,12-methyleneoctadecanoic acid; phytomonic acid. $C_{19}H_{36}O_2$; mol wt 296.50. C 76.97%, H 12.24%, O 10.79%. A lipid constituent of various microorganisms. Isoln from *Lactobacillus arabinosus*: K. Hofmann, R. A. Lucas, *J. Am. Chem. Soc.* **72**, 4328 (1950); from *Agrobacterium tumefaciens* and identity of phytomonic acid with lactobacillic acid: K. Hofmann, F. Tausig, *J. Biol. Chem.* **213**, 425 (1955). Structure: K. Hofmann *et al.*, *J. Am. Chem. Soc.* **80**, 5717 (1958). Abs config: J. F. Tocanne, *Tetrahedron* **28**, 363 (1972).

Crystals from acetone, mp 27.8-28.8°. Soluble in acetone, petr ether.

Methyl ester. $C_{20}H_{38}O_2$. Liq, bp$_3$ 187-187.5°. Soluble in many fat solvents.

Amide. Lactobacillamide. $C_{19}H_{37}NO$. Crystals, mp 79.4-81.5°. Soluble in dimethylformamide.

5391. Lactobionic Acid. [96-82-2] 4-*O*-β-D-Galactopyranosyl-D-gluconic acid; 4-(β-D-galactosido)-D-gluconic acid. $C_{12}H_{22}O_{12}$; mol wt 358.30. C 40.23%, H 6.19%, O 53.58%. Obtained by oxidation of lactose: Fischer, Meyer, *Ber.* **22**, 362 (1889); Ruff, Ollendorff, *ibid.* **33**, 1806 (1900); Isbell, *J. Res. Natl. Bur. Stand.* **11**, 713 (1933); Margariello, *US 2746916* (1956 to Nat. Dairy Res. Labs.); Eddy, *Nature* **181**, 904 (1958); Nishizuka *et al.*, *J. Biol. Chem.* **235**, PC13 (1960). Manuf from lactose: Y. Sato *et al.*, *DE 2038230* (1971 to Hayashibara Co.). Crystal structure of calcium salt: W. J. Cook, C. E. Bugg, *Acta Crystallogr. B* **29**, 215 (1973). NMR studies: T. Taga *et al.*, *Bull. Chem. Soc. Jpn.* **51**, 2278 (1978). For therapeutic use *see* Erythromycin Lactobionate.

COOH
H—C—OH
HO—C—H
H—C—O
HO—C—OH (with ring O, OH, CH$_2$OH, OH structure)

Syrup. Freely sol in water, slightly sol in methanol, ethanol, glacial acetic acid. Dehydration by distillation with dioxane yields *lactobionic δ-lactone*, $C_{12}H_{20}O_{11}$, non-deliquescent crystals, dec 195-196°. Shows mutarotation. $[\alpha]_D^{20}$ +53.0° initial (c = 8.8) $\rightarrow$ $[\alpha]_D^{20}$ +22.6° final (240 minutes).

Calcium salt. Calcium lactobionate. $C_{24}H_{42}CaO_{24}$. Pentahydrate, hairlike needles in brushlike groups. When anhyd, slender needles from small amts of anhydr ethanol. $[\alpha]_D^{20}$ +23.7° (c = 6.28). n_D^{20} 1.4583 (concd syrup just before crystallization). Freely sol in water.

5392. Lactose. [63-42-3] 4-*O*-β-D-Galactopyranosyl-D-glucose; 4-(β-D-galactosido)-D-glucose; milk sugar. $C_{12}H_{22}O_{11}$; mol wt 342.30. C 42.11%, H 6.48%, O 51.41%. Present in milk of mammals: human 6.7%; cow's 4.5%. Milk at body temp contains lactose as an equilibrium mixture of 2 parts of α-lactose and 3 parts of β-lactose. By-product of the cheese industry, produced from whey: Davis, *Can. Dairy and Ice Cream J.* **19**, 52 (1940); *Milk Trade Gaz.* **12**, 4 (1941); F. Ullmann, *Encyklopädie der Technischen Chemie*, **VII**, 579 (2nd ed., 1931). Structure and configuration: Zemplén, *Ber.* **59**, 2402 (1926); Levene, Sobotka, *J. Biol. Chem.* **71**, 471 (1926); Levene, Wintersteiner, *ibid.* **75**, 315 (1927); Haworth, Long, *J. Chem. Soc.* **1927**, 544; Hudson, *J. Am. Chem. Soc.* **52**, 1712 (1930); Hassid, Ballou in *The Carbohydrates*, W. Pigman, Ed. (Academic Press, New York, 1957) p 495. Synthesis: Haskins *et al.*, *J. Am. Chem. Soc.* **64**, 1852 (1942). *Reviews:* Whittier, *Chem. Rev.* **2**, 85-125 (1926); *J. Dairy Sci.* **27**, 505-537 (1944); Weisberg, *ibid.* **37**, 1106-1115 (1954); L. A. W. Thelwall, *Dev. Food Carbohydr.* **2**, 275-326 (1980). Comprehensive description: H. G. Brittain *et al.*, *Anal. Profiles Drug Subs.* **20**, 369-398 (1991).

HO CH$_2$OH / HO OH / CH$_2$OH O OH (disaccharide structure)

On hydrolysis with 2% H_2SO_4 or with emulsin lactose yields 1 mol D-glucose and 1 mol D-galactose. Reduces Fehling's soln.

α-Lactose monohydrate. Is the usual milk sugar and the lactose of pharmacy. Monoclinic sphenoidal crystals from water. Faintly sweet taste. Stable in air, but readily absorbs odors. d^{20} 1.53. Becomes anhydrous at 120°. mp 201-202° (rapid heating). Shows mutarotation. $[\alpha]_D^{20}$ +92.6° $\rightarrow$ +83.5° (10 min.) $\rightarrow$ +69° (50 min)

$\rightarrow$ +52.3° (22 hrs, c = 4.5). The final value is obtained instantly in the presence of a trace of NH_3. U.S.P. requires +54.4° to +55.9° (c = 10). One gram dissolves in 5 ml water, in 2.6 ml boiling water. Practically insol in alc. Insol in chloroform, ether. Ka at 16.5° = 6.0×10^{-13}. d$_4^{20}$ of aq solns calcd for the monohydrate: 5.2% = 1.018; 10.2% = 1.038; 20.0% = 1.078; 30.2% = 1.123; 50.9% = 1.226; 60.8% = 1.281; 69.1% = 1.330.

β-Lactose. $C_{12}H_{22}O_{11}$. Obtained by crystallizing concd solns of α-lactose above 93.5°. Somewhat sweeter than the α-form. $[\alpha]_D^{25}$ +34° (3 min) $\rightarrow$ +39° (6 min) $\rightarrow$ +46° (1 hr) $\rightarrow$ +52.3° (22 hrs). One gram dissolves in 2.2 ml water at 15°, in 1.1 ml boiling water. After a few days crystals of the less sol α-monohydrate appear from satd solns.

USE: Both forms of lactose are employed, with the α-form predominating: as a nutrient in preparing modified milk and food for infants and convalescents (Whittier). In baking mixtures. Pharmaceutic aid (tablet and capsule excipient and diluent). To produce lactic acid fermentation in ensilage and food products. As chromatographic adsorbent in analytical chemistry. In culture media. For many additional uses *see* Weisberg.

THERAP CAT (VET): Added to cow's milk for feeding orphan foals.

5393. Lactucarium. Lettuce opium. Dried milky juice or latex of several species of lettuce, primarily *Lactuca virosa* L., Asteraceae (wild lettuce); garden lettuce, *L. sativa* L., and related species, *L. serriola* and *L. sagittata* have also been used. Thought to possess sleep-inducing and pyschotropic properties. *Constit.* Lactucin, lactucopicrin, lactucerol, lactucic acid. Brief description: V. E. Tyler in *The Honest Herbal* (Pharmaceutical Products Press, New York, 3rd ed., 1993) pp 193-195.

Similar to opium in odor, taste and appearance.

5394. Lactucin. [1891-29-8] (3a*R*,4*S*,9a*S*,9b*R*)-3,3a,4,5,9a,-9b-Hexahydro-4-hydroxy-9-(hydroxymethyl)-6-methyl-3-methyleneazuleno[4,5-*b*]furan-2,7-dione. $C_{15}H_{16}O_5$; mol wt 276.29. C 65.21%, H 5.84%, O 28.95%. From various *Lactuca* spp and *Cichorium intybus* L., *Compositae*. Isoln: Schenck, Graf, *Arch. Pharm.* **274**, 537 (1936); **275**, 36 (1937); Schenck *et al.*, *ibid.* **294**, 17 (1961). Purification: Späth *et al.*, *Monatsh. Chem.* **82**, 114 (1951). Structure: Dolejs *et al.*, *Collect. Czech. Chem. Commun.* **23**, 2195 (1958); Barton, Narayanan, *J. Chem. Soc.* **1958**, 963; Michl, Högenauer, *Monatsh. Chem.* **89**, 317 (1958). Revised stereochemistry: Bachelor, Itô, *Can. J. Chem.* **51**, 3626 (1973).

(chemical structure: OH, O, CH$_2$, OH, H$_3$C, O)

Crystals from methanol, sinters at 218°, mp 228-233°. $[\alpha]_D$ +49° (c = 0.90 in methanol), +77.9° (c = 3.44 in pyridine). uv max: 257 nm (ε 14,000). Soluble in water, ethanol, methanol, ethyl acetate, dioxane anisol.

p-Hydroxyphenylacetate hydrate. Intybin; lactucopicrin. $C_{23}H_{22}O_7$. Identity with intybin: Schmitt, *Bot. Arch.* **40**, 516 (1940); *C.A.* **36**, 5616 (1942). Structure: Michl, Högenauer, *Monatsh. Chem.* **91**, 500 (1960). Crystals from water, dec 148-151°. $[\alpha]_D^{17.5}$ +67.3° (pyridine).

5395. Lactulose. [4618-18-2] 4-*O*-β-D-Galactopyranosyl-D-fructose; 4-D-galactopyranosyl-4-D-fructofuranose; 4-*O*-β-D-galactosyl-D-fructose; 4-β-D-galactosido-D-fructose; Cholac; Constilac; Duphalac; Generlac; Lactuflor; Laevilac; Normase. $C_{12}H_{22}O_{11}$; mol wt 342.30. C 42.11%, H 6.48%, O 51.41%. Prebiotic, synthetic disaccharide composed of galactose and fructose. Prepn: E. M. Montgomery, C. S. Hudson, *J. Am. Chem. Soc.* **52**, 2101 (1930); Oosten, *Rec. Trav. Chim.* **86**, 673 (1967); K. B. Hicks, F. W. Parrish, *Carbohydr. Res.* **82**, 393 (1980). Clinical evaluation of effects on fecal bifidobacteria and selected metabolic indices: Y. Bouhnik *et al.*, *Eur. J. Clin. Nutr.* **58**, 462 (2004).

Hexagonal clustered plates from methanol, mp 168-171°. Sweeter than lactose, but not as sweet as sucrose. Shows mutarotation; constant value after 24 hrs: $[\alpha]_D^{22}$ −51.5°. Soly in water (w/w) at 30°: 76.4%; at 60°: 81%; at 90°: >86%. Acid hydrolysis yields galactose and fructose.

THERAP CAT: Laxative.

THERAP CAT (VET): Reducer of blood ammonia levels; in treatment of hepatic encephalopathy; laxative.

5396. Lafutidine. [118288-08-7] (+)-2-[(2-Furanylmethyl)-sulfinyl]-*N*-[(2*Z*)-4-[[4-(1-piperidinylmethyl)-2-pyridinyl]oxy]-2-buten-1-yl]acetamide; 2-(furfurylsulfinyl)-*N*-[(*Z*)-4-[[4-(piperidino-methyl)-2-pyridyl]oxy]-2-butenyl]acetamide; FRG-8813; Protecadin; Stogar. $C_{22}H_{29}N_3O_4S$; mol wt 431.55. C 61.23%, H 6.77%, N 9.74%, O 14.83%, S 7.43%. Second generation histamine H_2-receptor antagonist. Prepn of racemate: N. Hirakawa *et al.*, **EP 282077**; *eidem*, **US 4912101** (1988, 1990 both to Fujirebio); and pharmacology: *eidem*, *Chem. Pharm. Bull.* **46**, 616 (1998). Pharmacology: S. Onodera *et al.*, *Jpn. J. Pharmacol.* **68**, 161 (1995). Mode of action study: M. Umeda *et al.*, *J. Gastroenterol. Hepatol.* **14**, 859 (1999). Gastroprotective effects in rats: H. Ajioka *et al.*, *Pharmacology* **61**, 83 (2000). Clinical pharmacokinetics: S. Haruki *et al.*, *Yakuri to Chiryo* **23**, 3049 (1995). Toxicology study: A. Broadmeadow *et al.*, *Oyo Yakuri* **50**, 167 (1995).

Prepd as the (±) mixture, crystals from benzene-hexane, mp 92.7-94.9°. Slightly bitter taste. Freely sol in DMF, glacial acetic acid; sol in methanol; sparingly sol in dehydrated ethanol; very slightly sol in ether. Practically insol in water.

THERAP CAT: Antiulcerative.

5397. Laidlomycin. [56283-74-0] 16-Deethyl-3-*O*-demethyl-16-methyl-3-*O*-(1-oxopropyl)monensin. $C_{37}H_{62}O_{12}$; mol wt 698.89. C 63.59%, H 8.94%, O 27.47%. Polyether ionophore antibiotic; structurally related to monensin, *q.v.* Isoln from *Streptomyces spp.* and antimicrobial activity: F. Kitame *et al.*, *J. Antibiot.* **27**, 884 (1974). Structure determn: F. Kitame, N. Ishida, *ibid.* **29**, 759 (1976). Prepn of esters: A. F. Kluge, R. D. Clark, **EP 24189** (1981 to Syntex). ^{13}C-NMR study and enhancement of biological activities: R. D. Clark *et al.*, *J. Antibiot.* **35**, 1527 (1982). Mechanism of action study: U. Gräfe *et al.*, *J. Basic Microbiol.* **29**, 391 (1989). HPLC determn in fermentation broth: M. Beran *et al.*, *Chromatographia* **31**, 603 (1991). Effect on cattle growth: M. T. Van Koevering *et al.*, *1991 Animal Science Research Report* (Agricultural Experiment Station, Oklahoma State University, Stillwater, OK, 1991) 241; M. L. Galyean *et al.*, *J. Anim. Sci.* **70**, 2950 (1992).

Colorless prisms from chloroform-ethyl acetate. mp 151-153°. $[\alpha]_D^{22}$ +51.3° (c=0.2 in CHCl$_3$). Sol in CHCl$_3$, acetone, alcohols. Insol in water, *n*-hexane. LD$_{50}$ in mice (mg/kg): 5 i.p.; 1 i.v.; 2.5 s.c. (Kitame).

Propionate potassium. [84799-02-0] RS-11988; Cattlyst. $C_{40}H_{65}KO_{13}$; mol wt 793.05. mp 190-192°.

THERAP CAT (VET): Growth promotant.

5398. Lambda-cyhalothrin. [91465-08-6] *rel*-(1*S*,3*S*)-3-[(1*Z*)-2-Chloro-3,3,3-trifluoro-1-propen-1-yl]-2,2-dimethylcyclo-propanecarboxylic acid (*R*)-cyano(3-phenoxyphenyl)methyl ester; λ-cyhalothrin; PP-321; Icon; Karate; Oxyfly; Saber; Warrior. $C_{23}H_{19}ClF_3NO_3$; mol wt 449.85. C 61.41%, H 4.26%, Cl 7.88%, F 12.67%, N 3.11%, O 10.67%. Synthetic pyrethroid insecticide comprised of the bioactive, *Z*-(1*R*,3*R*) *S*-isomer of cyhalothrin, *q.v.*, in a 1:1 mixture with its enantiomer, the *Z*-(1*S*,3*S*) *R*-isomer. Prepn: M. J. Robson, J. Crosby, **EP 106469**; J. Crosby, **US 4510098** (1984, 1985 both to ICI); M. J. Robson *et al.*, *Br. Crop Prot. Conf. - Pests Dis.* **1984**, 853. Description of chemical properties and biological activities: A. R. Jutsum *et al.*, *ibid.* 421. Voltammetric determn in soil and water: H. C. Oudou *et al.*, *Anal. Chim. Acta* **523**, 69 (2004). Separation and determn of isomers: R. N. Rao *et al*, *Anal. Sci.* **20**, 1745 (2004). Field trials for use in horticultural crops: D. Wilson, J. Trevenna, *Br. Crop Prot. Conf. - Pests Dis.* **1986**, 323; for mosquito control: A. A. Weathersbee *et al.*, *J. Am. Mosq. Control Assoc.* **7**, 238 (1991). Veterinary trial as ectoparasiticide: R. G. Endris *et al.*, *Vet. Ther.* **3**, 387 (2002). Review of chemistry and environmental fate: L. M. He *et al.*, *Rev. Environ. Contam. Toxicol.* **195**, 71-91 (2008). Evaluation of aquatic toxicity: J. M. Giddings *et al.*, *Ecotoxicology* **18**, 239 (2009).

Z-(1*R*,3*R*) *S*-isomer

White odorless solid, mp 49.2°. bp$_{0.2}$ 187-190°. d^{25} 1.33. Log P (octanol/water): 7.00. Vapor pressure at 20°: 0.0002 mPa. Soly in water (20°): 0.005 mg/l. Sol in common range of solvents at 21°C. LD$_{50}$ (technical grade) in male, female rats (mg/kg): 79, 56 orally; 632, 696 dermally (Jutsum). LC$_{50}$ (96 hr) in rainbow trout, zebra fish (ng/L): 214, 640 (Giddings).

Z-(1R,3R) S-isomer. [76703-62-3] Gamma-cyhalothrin; (*S*)-α-cyano-3-phenoxybenzyl (1*R*,3*R*)-3-[(*Z*)-2-chloro-3,3,3-trifluoropropenyl]-2,2-dimethylcyclopropanecarboxylate; Fentrol; Nexide; Proaxis; Trojan; Vantex. Bioactive isomer of cyhalothrin. Manuf process: S. M. Brown, B. D. Gott, **US 7468453** (2008 to Syngenta). Properties of formulations: H. K. Frederiksen *et al.*, *J. Controlled Release* **86**, 243 (2003). White crystals from hexane, mp 55-58°. LC$_{50}$ (96 hr) in rainbow trout, zebra fish (ng/L): 111, 270 (Giddings).

USE: Insecticide.

THERAP CAT (VET): Ectoparasiticide.

5399. Lamifiban. [144412-49-7] 2-[[1-[(2*S*)-2-[[4-(Amino-iminomethyl)benzoyl]amino]-3-(4-hydroxyphenyl)-1-oxopropyl]-4-piperidinyl]oxy]acetic acid; [[1-[*N*-(*p*-amidinobenzoyl)-L-tyrosyl]-4-piperidinyl]oxy]acetic acid; Ro-44-9883. $C_{24}H_{28}N_4O_6$; mol wt 468.51. C 61.53%, H 6.02%, N 11.96%, O 20.49%. Specific nonpeptide platelet fibrinogen receptor (GPIIb/IIIa) antagonist. Prepn: L. Alig *et al.*, **EP 505868**; *eidem*, **US 5378712** (1992, 1995 both to Hoffmann-La Roche); L. Alig *et al.*, *J. Med. Chem.* **35**, 4393 (1992). Pharmacology: J.-P. Carteaux *et al.*, *Thromb. Haemostasis* **70**, 817 (1993); Y. Takiguchi *et al.*, *ibid.* **73**, 683 (1995).

Crystals (zwitterionic form) from water, mp above 200° (dec). $[\alpha]_D^{20}$ +29.8° (c = 0.86 in 1N HCl).

Trifluoroacetate salt. [144412-50-0] $C_{24}H_{28}N_4O_6 \cdot C_2HF_3O_2$. mp 125-130° (dec). LD_{50} i.v. in mice: 250 mg/kg (Alig, 1995).

THERAP CAT: Antithrombotic.

5400. Laminaran. [9008-22-4] Laminarin. A polysaccharide found in brown seaweed and occurring principally in the *Laminaria* spp. Linear polymer composed of β-(1 → 3)-linked glucose residues; may contain small amounts of β-(1 → 6) linkages as interresidue linkages or as branch points and 2-3% D-mannitol as end groups. Two forms of laminaran are recognized; they are referred to as soluble and insoluble laminaran: Percival, Ross, *J. Chem. Soc.* **1951**, 720. Structure: Peat *et al.*, *ibid.* **1958**, 724, 729; **1960**, 175; Goldstein *et al.*, *Chem. Ind. (London)* **1959**, 124; Annan *et al.*, *ibid.* **1962**, 984; Annan *et al.*, *J. Chem. Soc.* **1965**, 885; Maeda, Nisizawa, *Carbohydr. Res.* **7**, 97 (1968). Structure of soluble laminaran from *Eisenia bicyclis*: T. Usui *et al.*, *Agric. Biol. Chem.* **43**, 603 (1979). NMR studies of laminaran: D. Gagnaire, *Org. Magn. Reson.* **11**, 344 (1978); H. Friebolin *et al.*, *ibid.* **12**, 216 (1979). *Review:* W. A. P. Black, E. T. Dewar in *Industrial Gums*, R. L. Whistler, Ed. (Academic Press, New York, ed., 1973) pp 137-145.

Water-insoluble laminaran. Isolated from *L. cloustoni* Edmondst., *Laminariaceae*, is pptd spontaneously from the aq acid extract of the plant. Has lower degree of branching than the sol form. Typical analysis of the dry material: 92.5% polyglucose, 0.4% nonvolatile matter; $[\alpha]_D^{16}$ −13.4° (c = 0.9). Amorphous triacetate, $(C_{12}H_{16}O_8)_n$, $[\alpha]_D^{16}$ −63.5° (c = 0.4 in chloroform).

Soluble form. Isolated from *L. digitata*, is separated from the acidified extract only after addition of a precipitant such as ethanol. Typical analysis (on dry basis): 91.2% polyglucose, 1.0% non-volatile matter; $[\alpha]_D^{18}$ −11.9° (c = 2.1).

Sulfate. Laminaran hydrogen sulfate. Laminaran can be sulfated to varying degrees. Highly sulfated products have anticoagulant properties comparable to heparin, while laminarans with few sulfate groups are antilipemic only: Besterman, Evans, *Br. Med. J.* **1**, 310 (1957).

5401. Laminin. Abundant structural component of the basal lamina; critical to the stability of the extracellular matrix and to the adhesion of cells to the basement membrane. Family of heterotrimeric glycoproteins composed of a heavy chain, designated α (also known as A) and 2 light chains, designated β (B1) and γ (B2), which are linked by disulfide bonds to form an asymmetrical cross-shaped structure. Eight genetically distinct laminin subunits have been identified: α1, α2, α3, β1, β2, β3, γ1, and γ2. Seven different assembly forms (laminins-1 to -7) are known and appear to be tissue specific and developmentally regulated. Exhibits diverse biological activities. Influences adhesion, growth, morphology and differentiation of a variety of cells via specific receptors including several of the integrin type. Isoln from murine Engelbreth-Holm-Swarm (EHS) tumor: R. Timpl *et al.*, *J. Biol. Chem.* **254**, 9933 (1979). Review of biological activities: H. K. Kleinman *et al.*, *J. Cell. Biochem.* **27**, 317-325 (1985); of role in neural development: V. Nurcombe, *Pharmacol. Ther.* **56**, 247 (1992). Tissue distribution: E. Engvall *et al.*, *Cell Regul.* **1**, 731 (1990). Review of laminin binding proteins and receptors: R. P. Mecham, *Annu. Rev. Cell Biol.* **7**, 71-91 (1991); and role in metastasis: V. Castronovo, *Invasion Metastasis* **13**, 1-30 (1993). Structure and function of laminin isoforms: E. Engvall, *Kidney Int.* **43**, 2-6 (1993); K. Tryggvason, *Curr. Opin. Cell Biol.* **5**, 877-882 (1993). Nomenclature: R. E. Burgeson *et al.*, *Matrix Biol.* **14**, 209 (1994). Review of structure: R. Timpl, J. C. Brown, *ibid.* 275-281.

Laminin-1. EHS-laminin. Prototype laminin produced by murine EHS tumor. Contains α1 (also known as A or Ae), β1 (B1, B1e) and γ1 (B2, B2e) subunits.

Laminin-2. Merosin; laminin M. Variant found in striated muscle, placental trophoblast and Schwann cell basement membranes. Contains α2 (also known as M or Am), β1 and γ1 subunits. Identification: I. Leivo, E. Engvall, *Proc. Natl. Acad. Sci. USA* **85**, 1544 (1988); K. Ehrig *et al.*, *ibid.* **87**, 3264 (1990).

Laminin-3. S-laminin; synaptic laminin. Contains α1, β2 (also known as s or B1s), and γ1 chains. Identification in neuromuscular junction: D. D. Hunter *et al.*, *Nature* **338**, 229 (1989).

Laminin-5. Kallinin; nicein. Contains α3, β3, and γ2 subunits. Isoln from human keratinocytes: P. Rousselle *et al.*, *J. Cell Biol.*

114, 567 (1991). Review of role in progression of cancer: J. Lohi, *Int. J. Cancer* **94**, 763-767 (2001).

5402. Lamivudine. [134678-17-4] 4-Amino-1-[(2R,5S)-2-(hydroxymethyl)-1,3-oxathiolan-5-yl]-2(1H)-pyrimidinone; (−)-2′-deoxy-3′-thiacytidine; (−)-1-[(2R,5S)-2-(hydroxymethyl)-1,3-oxathiolan-5-yl]cystosine; 3′-thia-2′,3′-dideoxycytidine; 3TC; (−)-BCH-189; GR-109714X; Epivir; Zeffix. $C_8H_{11}N_3O_3S$; mol wt 229.25. C 41.91%, H 4.84%, N 18.33%, O 20.94%, S 13.98%. Reverse transcriptase inhibitor. Prepn: J. A. V. Coates *et al.*, **WO 9117159**. Synthesis of enantiomers: J. W. Beach *et al.*, *J. Org. Chem.* **57**, 2217 (1992); of (−)-enantiomer: D. C. Humber *et al.*, *Tetrahedron Lett.* **33**, 4625 (1992). HPLC determn in urine: D. M. Morris, K. Selinger, *J. Pharm. Biomed. Anal.* **12**, 255 (1994). Clinical trial in hepatitis B: F. Nevens *et al.*, *Gastroenterology* **113**, 1258 (1997). Review of pharmacology and clinical efficacy in HIV infection: C. M. Perry, D. Faulds, *Drugs* **53**, 657-680 (1997); of clinical experience in chronic hepatitis B and cirrhosis: C. Hanché, J-P Villeneuve, *Expert Opin. Pharmacother.* **7**, 1835-1843 (2006).

Crystals from boiling ethanol, mp 160-162°. $[\alpha]_D^{21}$ −135° (c = 0.38 in methanol). Soly in water (20°): ~70 mg/ml.

THERAP CAT: Antiretroviral.

5403. Lamotrigine. [84057-84-1] 6-(2,3-Dichlorophenyl)-1,-2,4-triazine-3,5-diamine; 3,5-diamino-6-(2,3-dichlorophenyl)-1,2,4-triazine; LTG; BW-430C; Lamictal. $C_9H_7Cl_2N_5$; mol wt 256.09. C 42.21%, H 2.76%, Cl 27.69%, N 27.35%. Prepn: M. G. Baxter *et al.*, **EP 21121** (1981 to Wellcome Foundation); D. A. Sawyer *et al.*, **US 4602017** (1986). HPLC determn in plasma: C.-L. Cheng *et al.*, *J. Chromatogr. B* **817**, 199 (2005). Anticonvulsant activity: A. A. Miller *et al.*, *Epilepsia* **27**, 483 (1986). Mechanism of action studies: M. J. Leach *et al.*, *ibid.* 490; X. Xie, R. M. Hagan, *Neuropsychobiology* **38**, 119 (1998). Series of articles on clinical pharmacology, antiepileptic efficacy and safety: *Epilepsia* **32**, Suppl. 2, S1-S21 (1991). Clinical trial in bipolar depression: J. R. Calabrese *et al.*, *J. Clin. Psychiatry* **60**, 79 (1999). Review of clinical experience in epilepsy: H. Choi, M. J. Morrell, *Expert Opin. Pharmacother.* **4**, 243-251 (2003); in bipolar disorder: Z. Bhagwagar, G. M. Goodwin, *Expert Opin. Pharmacother.* **6**, 1401-1408 (2005).

White to pale cream-colored powder. Crystals from isopropanol, mp 216-218° (uncorr). pKa 5.7. Soly at 25° (mg/ml): water 0.17; 0.1M HCl 4.1. LD_{50} in mice, rats (mg/kg): 250, >640 orally (Sawyer).

THERAP CAT: Anticonvulsant. In treatment of bipolar depression.

5404. Lanatosides. Family of four glycosides, A, B, C, D, isolated from various species of *Digitalis* including *D. lanata* Ehrh., *Scrophulariaceae*: Stoll, Kreis, *Helv. Chim. Acta* **16**, 1049 (1933); Ligeti, *Pharmazie* **12**, 433 (1957); **14**, 162 (1959); Angliker *et al.*, *Ann.* **607**, 131 (1957); *D. Lutea* L., *Scrophulariaceae*: Cole, Gisvold, *J. Am. Pharm. Assoc. Sci. Ed.* **47**, 654 (1958). Structure: Tschesche *et al.*, *Ber.* **92**, 2258 (1959); Uskert, *Ann.* **638**, 199 (1960); Kuhn *et al.*, *Helv. Chim. Acta* **45**, 881 (1962). Absorption spectrum: Bell, *J. Am. Pharm. Assoc. Sci. Ed.* **49**, 277 (1960). Cardioactivity of lanatoside C: K.-O. Haustein, *Pharmacology* **11**, 117 (1974). Mitogenicity: L. L. G. Hammarström, C. I. E. Smith, *J. Immunol.* **120**, 694 (1978). HPLC determn: V. Y. Davydov *et al.*, *J. Chromatogr.* **248**, 49 (1982); of lanatoside C: F. Orosz *et al.*, *Anal. Biochem.* **156**, 171 (1986).

Lanatoside A	R = digitoxigenin
Lanatoside B	R = gitoxigenin
Lanatoside C	R = digoxigenin
Lanatoside D	R = diginatigenin

Lanatoside A. [17575-20-1] (3β,5β)-3-[(*O*-β-D-Glucopyranosyl-(1 → 4)-*O*-3-*O*-acetyl-2,6-dideoxy-β-D-*ribo*-hexopyranosyl-(1 → 4)-*O*-2,6-dideoxy-β-D-*ribo*-hexopyranosyl-(1 → 4)-2,6-dideoxy-β-D-*ribo*-hexopyranosyl)oxy]-14-hydroxycard-20(22)-enolide; digilanide A; Adigal. $C_{49}H_{76}O_{19}$; mol wt 969.13. Long, flat prisms from methanol, dec 245-248°. $[\alpha]_D^{20}$ +31.6° (0.48 g in 25 ml 95% alc); $[\alpha]_D^{20}$ +23.2° (0.95 g in 25 ml dioxane). Soluble in 20 parts methanol, 40 parts alcohol, 225 parts chloroform, 16,000 parts water.

Lanatoside B. [17575-21-2] (3β,5β,16β)-3-[(*O*-β-D-Glucopyranosyl-(1 → 4)-*O*-3-*O*-acetyl-2,6-dideoxy-β-D-*ribo*-hexopyranosyl-(1 → 4)-*O*-2,6-dideoxy-β-D-*ribo*-hexopyranosyl-(1 → 4)-2,6-dideoxy-β-D-*ribo*-hexopyranosyl)oxy]-14,16-dihydroxycard-20-(22)-enolide; digilanide B. $C_{49}H_{76}O_{20}$; mol wt 985.13. Long, flat prisms from alcohol, dec 245-248° after drying in high vacuum at 150°. $[\alpha]_D^{20}$ +36.7° (0.47 g in 25 ml 95% alc); $[\alpha]_D^{20}$ +31.8° (0.25 g in 14.1 ml dioxane). Soluble in 20 parts methanol, 40 parts alcohol, 550 parts chloroform. Nearly insol in water.

Lanatoside C. [17575-22-3] (3β,5β,12β)-3-[(*O*-β-D-Glucopyranosyl-(1 → 4)-*O*-3-*O*-acetyl-2,6-dideoxy-β-D-*ribo*-hexopyranosyl-(1 → 4)-*O*-2,6-dideoxy-β-D-*ribo*-hexopyranosyl-(1 → 4)-2,6-dideoxy-β-D-*ribo*-hexopyranosyl)oxy]-12,14-dihydroxycard-20-(22)-enolide; digilanide C; Allocor; Cedilanid; Ceglunat; Celadigal; Cetosanol; Lanimerck. $C_{49}H_{76}O_{20}$; mol wt 985.13. Long, flat prisms from alcohol, dec 248-250°, after drying in vacuum at 150°. $[\alpha]_D^{20}$ +33.4 to +33.7° (200 mg dry weight in 10 ml alcohol). One gram dissolves in 20,000 ml methanol, in 2000 ml chloroform. Freely sol in pyridine, dioxane. Practically insol in ether, petr ether.

Lanatoside D. [11030-31-2] (3β,5β,12β,16β)-3-[(*O*-β-D-Glucopyranosyl-(1 → 4)-*O*-3-*O*-acetyl-2,6-dideoxy-β-D-*ribo*-hexopyranosyl-(1 → 4)-*O*-2,6-dideoxy-β-D-*ribo*-hexopyranosyl-(1 → 4)-2,6-dideoxy-β-D-*ribo*-hexopyranosyl)oxy]-12,14,16-trihydroxycard-20(22)-enolide; digilanide D. $C_{49}H_{76}O_{21}$; mol wt 1001.13. Needles from methanol and water, dec 242-250°. $[\alpha]_D^{20}$ +40.5° (c = 5.95 in methanol). uv max: 220 nm (log ε 4.16).

THERAP CAT: Cardiotonic.

5405. Landiolol. [133242-30-5] 4-[(2*S*)-2-Hydroxy-3-[[2-[(4-morpholinylcarbonyl)amino]ethyl]amino]propoxy]benzenepropanoic acid [(4*S*)-2,2-dimethyl-1,3-dioxolan-4-yl]methyl ester; (−)-2,2-dimethyl-1,3-dioxolan-4*S*-ylmethyl 3-[4-[3-[2-(morpholinocarbonylamino)ethylamino]-2*S*-hydroxypropoxy]phenyl]propionate. $C_{25}H_{39}N_3O_8$; mol wt 509.60. C 58.92%, H 7.71%, N 8.25%, O 25.12%. Cardioselective β-adrenergic blocker. Prepn: S. Iguchi *et al.*, **EP** 397031; **US** 5013734 (1990, 1991 both to Ono Pharm.); *idem et al.*, *Chem. Pharm. Bull.* **40**, 462 (1992). HPLC determn in blood: Y. Hashimoto *et al.*, *Biol. Pharm. Bull.* **32**, 121 (2009). Electrophysiology: A. Sugiyama *et al.*, *J. Cardiovasc. Pharmacol.* **34**, 70 (1999). Clinical pharmacology: A. Kitamura *et al.*, *Eur. J. Clin. Pharmacol.* **51**, 467 (1997). Clinical pharmacokinetics in patients with cardiac tachyarrhythmias: H. Atarashi *et al.*, *Clin. Pharmacol. Ther.* **68**, 143 (2000). Clinical trial in postoperative supraventricular arrythmia: S. Wariishi *et al.*, *Interact. Cardiovasc. Thorac. Surg.* **9**, 811 (2009).

Hydrochloride. [144481-98-1] ONO-1101; Onoact. $C_{25}H_{39}N_3O_8$·HCl; mol wt 546.06. mp 125.4°. Sol in ethanol. LD_{50} in mice (mg/kg): 290 i.v. (Iguchi).

THERAP CAT: Antiarrhythmic.

5406. Laninamivir. [203120-17-6] 5-(Acetylamino)-4-[(aminoiminomethyl)amino]-2,6-anhydro-3,4,5-trideoxy-7-*O*-methyl-D-*glycero*-D-*galacto*-non-2-enonic acid; (2*R*,3*R*,4*S*)-3-acetamido-2-[(1*R*,2*R*)-2,3-dihydroxy-1-methoxypropyl]-4-guanidino-3,4-dihydro-2*H*-pyran-6-carboxylic acid; 5-acetamido-4-guanidino-2,3,4,-5,7-pentadeoxy-7-methoxy-D-*glycero*-D-*galacto*-non-2-enopyranosoic acid; 7-*O*-methyl-4-guanidino-Neu5Ac2en; R-125489. $C_{13}H_{22}N_4O_7$; mol wt 346.34. C 45.08%, H 6.40%, N 16.18%, O 32.34%. Neuraminidase inhibitor; analog of the sialic acids. Prepn: T. Honda *et al.*, **EP** 823428; *eidem*, **US** 6340702 (1998, 2002 both to Sankyo); and anti-influenza activity: T. Masuda *et al.*, *Chem. Pharm. Bull.* **51**, 1386 (2003); of the octanoyl ester prodrug: T. Honda *et al.*, *Bioorg. Med. Chem. Lett.* **19**, 2938 (2009). Comparative antiviral spectrum: M. Yamashita *et al.*, *Antimicrob. Agents Chemother.* **53**, 186 (2009). Efficacy in mouse-model of influenza: S. Kubo *et al.*, *ibid.* **54**, 1256 (2010). Clinical evaluation in treatment of pediatric influenza: N. Sugaya, Y. Ohashi, *ibid.*, 2575. Review of pharmacology and clinical efficacy in influenza: H. Ikematsu, N. Kawai, *Expert Rev. Anti Infect. Ther.* **9**, 851-857 (2011).

Trifluoroacetate salt. [203120-18-7] $C_{13}H_{22}N_4O_7$·$C_2HF_3O_2$; mol wt 460.36. Colorless solid. $[\alpha]_D^{25}$ −9.3° (c = 0.1 in methanol).

Octanoic acid ester. [203120-46-1]; [371755-92-9] (hydrate). (4*S*,5*R*,6*R*)-5-Acetamido-4-guanidino-6-[(1*R*,2*R*)-2-hydroxy-1-methoxy-3-(octanoyloxy)propyl]-5,6-dihydro-4*H*-pyran-2-carboxylic acid; laninamivir octanoate; CS-8958; Inavir. $C_{21}H_{36}N_4O_8$; mol wt 472.54.

THERAP CAT: Antiviral.

5407. Lankamycin. [30042-37-6] (3*R*,4*S*,5*R*,6*S*,7*S*,9*S*,11*R*,-12*S*,13*S*,14*R*)-4-[(4-*O*-Acetyl-2,6-dideoxy-3-*C*-methyl-3-*O*-methyl-α-L-*xylo*-hexopyranosyl)oxy]-12-(acetyloxy)-6-[(4,6-dideoxy-3-*O*-methyl-β-D-*xylo*-hexopyranosyl)oxy]-9-hydroxy-14-[(1*S*,2*S*)-2-hydroxy-1-methylpropyl]-3,5,7,9,11,13-hexamethyloxacyclotetradecane-2,10-dione; Kujimycin B. $C_{42}H_{72}O_{16}$; mol wt 833.02. C 60.56%, H 8.71%, O 30.73%. Macrolide antibiotic produced by *Streptomyces violaceoniger* from soil of Ceylon. Isolation: Gaumann *et al.*, *Helv. Chim. Acta* **45**, 138 (1962). Structure: *eidem, ibid.* **47**, 78 (1964); Egan, Martin, *J. Am. Chem. Soc.* **92**, 4129 (1970). Stereochemistry: Muntwyler, Keller-Schierlein, *Helv. Chim. Acta* **55**,

On hydrolysis an aglucone, monoacetyllankolid, $C_{26}H_{48}O_{10}$, and two sugar-like substances, lankavose (D-chalcose, *q.v.*) and acetylarcanose: Keller-Schierlein, Roncari, *ibid.* **45**, 138 (1962). Structure: *eidem, ibid.* **47**, 78 (1964); Egan, Martin, *J. Am. Chem. Soc.* **92**, 4129 (1970). Stereochemistry: Muntwyler, Keller-Schierlein, *Helv. Chim. Acta* **55**,

460 (1972). Identity with kujimycin B: Omura *et al.*, *J. Antibiot.* **22**, 629 (1969).

Crystals from ether + petr ether. Double mp, 147-150° and 181-182°. $[\alpha]_D^{20}$ −94° (c = 1.23 in alc.). uv max: 289 nm (log ε 1.50).

5408. Lanoconazole. [101530-10-3] (αE)-α-[4-(2-Chlorophenyl)-1,3-dithiolan-2-ylidene]-1*H*-imidazole-1-acetonitrile; latoconazole; TJN-318; NND-318; Astat. $C_{14}H_{10}ClN_3S_2$; mol wt 319.83. C 52.58%, H 3.15%, Cl 11.08%, N 13.14%, S 20.05%. Prepn: A. Soe *et al.*, **JP Kokai 85 218387**; *idem et al.*, **US 4636519** (1985, 1987 both to Nihon Nohyaku). *In vivo* antifungal activity: H. Oka *et al.*, *Arzneim.-Forsch.* **42**, 345 (1992); Y. Niwano *et al.*, *Antimicrob. Agents Chemother.* **38**, 2204 (1994). Quantitation of *Z*-isomer impurity in bulk drug: A. P. Kumar *et al.*, *J. Pharm. Biomed. Anal.* **50**, 535 (2009). Toxicity study: P. L. Munt *et al.*, *Oyo Yakuri* **43**, 195 (1992). *In vivo* efficacy vs *Cryptococcus neoformans* encephalitis: K. Furukawa *et al.*, *J. Antimicrob. Chemother.* **46**, 443 (2000).

Light yellow crystals, mp 141.5°. LD_{50} in male, female mice, rats (mg/kg): 3224, 2715, 993, 652 orally; 2158, 1743, 1655, 2596 i.p.; >5000 both species s.c.; LD_{50} dermally in rats: >5000 mg/kg (Munt).

THERAP CAT: Antifungal.

5409. Lanolin. [8006-54-0] Wool fat; anydrous lanolin; Emery 1600. Refined form of wool wax, the unctuous secretion of the sebaceous glands of sheep which is deposited onto the wool fibers. Known since antiquity for its emollient properties. Chemically a wax consisting of a complex mixture of esters and polyesters of lanolin alcohols, *q.v.*, and higher fatty acids, including α- and ω-hydroxy acids. Lanolin can be separated by solvent fractionation into lanolin oil and lanolin wax. Analytical methods: C. W. Spilker, T. B. Richey, *Cosmet. Perfum.* **88**, 43 (Sept. 1973). Discussion of lanolin allergy: R. Wolf, *Dermatology* **192**, 198 (1996). Review of prepn, derivatives and uses: G. Barnett, *Cosmet. Toiletries* **101**, 23-44 (Mar. 1986); J. Thewlis, *Agro-Food-Ind. Hi-Tech* **8**, no. 3, 14-20 (1997).

Translucent, pale yellow, tenacious, unctuous mass; slight odor or practically odorless. mp 38-42°. Sp gr at 25° 0.935. Iodine no. 27. Hydroxyl no. 32. Freely sol in chloroform, ether; sparingly sol in cold, more in hot alcohol. Insol in water but mixes without separation with about twice its weight of water.

Hydrous lanolin. Ointment base prepared by the addition of about 25-30% water. Yellowish-white unctuous mass; slight odor. Practically insol in water. Sol in chloroform or ether with the separation of the water.

Acetylated lanolin. [61788-48-5] Acylan; Fancol ACEL; Modulan; Ritacetyl. Prepn: L. I. Conrad, K. Motiuk, *J. Soc. Cosmet. Chem.* **6**, 344 (1955). Almost odorless, pale yellow, soft, unctuous solid. mp 36°. Sp gr at 25° 0.935. Iodine no. 27. Hydroxyl no. 2. Sol in mineral oils, in some vegetable oils. Readily dispersed in oil-in-water emulsions.

Lanolin oil. Liquid lanolin; dewaxed lanolin; Lanogene; Lantrol 1673; Ritalan. Amber viscous liquid. Odorless, tasteless. Sol in mineral oil. Rich in lower molecular weight, branched aliphatic acids and alcohols.

Lanolin wax. Lanfrax. Light yellow waxy solid. Insol in water; emulsifies in ~3 parts water.

USE: Pharmaceutic aid (ointment base). Emulsifier, emollient, conditioner and lubricant in cosmetics and toiletries.

5410. Lanolin Alcohols. [8027-33-6] Wool alcohols; Amerchol 400; Ceralan; Emery 1780; Eucerin; Fancol LA; Hartolan; Ritawax. Complex combination of organic alcohols obtained by the saponification of lanolin. Composed of monohydric and dihydric aliphatic alcohols and sterols, primarily cholesterol and lanosterol, *q.q.v.* Prepn: L. I. Conrad, K. Motiuk, *J. Soc. Cosmet. Chem.* **6**, 344 (1955). Composition: M. L. Schlossman, J. P. McCarthy, *Contact Dermatitis* **5**, 65 (1979). Clinical evaluation of moisturizing capacity: J. Sindhvananda *et al.*, *J. Soc. Cosmet. Chem.* **44**, 279 (1993).

Golden-brown solid, plastic when warm, but brittle when cold; faint characteristic odor. mp >56°. Practically insol in water. Very slightly sol in alcohol; sol in ether, chloroform and light petroleum.

Acetylated Lanolin Alcohols. [61788-49-6] Acetulan; Fancol ALA. Pale yellow, practically odorless liquid. Sp gr at 25°: 0.867. Neutral to litmus. Acid no. 0.35. Hydroxyl no. 2.0. Saponification no. 190.0. Hydrophobic, practically insol in water with no emulsification. Miscible with mineral oil, castor oil, vegetable oils, isopropanol, 95% ethanol, isopropyl myristate, isopropyl palmitate, butyl stearate.

USE: Emulsifier, stabilizer, emollient and lubricant in cosmetics, toiletries and pharmaceuticals.

5411. Lanosterol. [79-63-0] (3β)-Lanosta-8,24-dien-3-ol; kryptosterol. $C_{30}H_{50}O$; mol wt 426.73. C 84.44%, H 11.81%, O 3.75%. The core steroid from which all others are derived by biological modification. From wool fat of sheep: Windaus, Tschesche, *Z. Physiol. Chem.* **190**, 51 (1930). Identity with kryptosterol: Ruzicka *et al.*, *Helv. Chim. Acta* **28**, 759 (1945). Structure: Voser *et al.*, *ibid.* **35**, 2414 (1952); Barnes *et al.*, *J. Chem. Soc.* **1953**, 571. Stereochemistry: *eidem, ibid.* **1953**, 576. Prepn from isocholesterol: Bloch, Urech, *Biochem. Prep.* **6**, 32 (1958). Prepn by cyclization of squalene: Cornforth *et al.*, *Ciba Found. Symp. Terpenes Sterols* **1958**, 119; van Tamelen *et al.*, *J. Am. Chem. Soc.* **88**, 4752 (1966). Mechanism of the squalene to lanosterol conversion: *eidem, ibid.* **104**, 6479, 6480 (1982).

Crystals, mp 138-140°. $[\alpha]_D^{20}$ +62.0° (chloroform). .

Lanosteryl acetate. Crystals, mp 131.5-133°, $[\alpha]_D^{20}$ +62.5° (c = 1.12 in chloroform).

5412. Lanreotide. [108736-35-2] 3-(2-Naphthalenyl)-D-alanyl-L-cysteinyl-L-tyrosyl-D-tryptophyl-L-lysyl-L-valyl-L-cysteinyl-L-threoninamide cyclic (2 → 7)-disulfide; angiopeptin; DC 13-116; BIM-23014. $C_{54}H_{69}N_{11}O_{10}S_2$; mol wt 1096.33. C 59.16%, H 6.34%, N 14.05%, O 14.59%, S 5.85%. Octapeptide somatostatin analog. Prepn: D. H. Coy *et al.*, **EP 215171**; *eidem*, **US 4853371** (1987, 1989 both to Tulane Educational Fund). Clinical pharmacokinetics and hormonal effects: J. M. Kuhn *et al.*, *Br. J. Clin. Phar-*

macol. **38**, 213 (1994). Clinical evaluation in acromegaly: P. Caron *et al., Eur. J. Endocrinol.* **132**, 320 (1995).

H₂N ... Cys–Tyr–D-Trp–Lys–Val–Cys–Thr–NH₂

Acetate. [127984-74-1] BIM-23014C; Somatuline LP. $C_{54}H_{69}$-$N_{11}O_{10}S_2.xC_2H_4O_2$
THERAP CAT: Antineoplastic.

5413. Lansoprazole. [103577-45-3] 2-[[[3-Methyl-4-(2,2,2-trifluoroethoxy)-2-pyridinyl]methyl]sulfinyl]-1*H*-benzimidazole; 2-(2-benzimidazolylsulfinylmethyl)-3-methyl-4-(2,2,2-trifluoroethoxy)pyridine; A-65006; AG-1749; Agopton; Lansox; Lanzor; Limpidex; Ogast; Ogastoro; Prevacid; Takepron; Zoton. $C_{16}H_{14}F_3N_3$-O_2S; mol wt 369.36. C 52.03%, H 3.82%, F 15.43%, N 11.38%, O 8.66%, S 8.68%. Gastric proton-pump inhibitor. Prepn: A. Nohara, Y. Maki, **EP 174726**; *eidem*, **US 4628098** (both 1986 to Takeda). HPLC determn in plasma: T. Uno *et al., J. Chromatogr. B* **816**, 309 (2005). Pharmacology: H. Satoh *et al., J. Pharmacol. Exp. Ther.* **248**, 806 (1989). Mechanism of action study: H. Nagaya *et al., ibid.* **252**, 1289 (1990). Clinical pharmacology and effect on human gastric acid secretion: P. Müller *et al., Aliment. Pharmacol. Ther.* **3**, 193 (1989). Review of pharmacology and clinical experience: H. D. Langtry, M. I. Wilde, *Drugs* **54**, 473-500 (1997). Comparative clinical trial with esomeprazole in erosive esophagitis: C. W. Howden *et al., Clin. Drug Invest.* **22**, 99 (2002).

White to brownish-white powder. mp 178-182° (dec). Freely sol in DMF. Practically insol in water.
R-**Form.** [138530-94-6] Dexlansoprazole; Kapidex. $C_{16}H_{14}F_3$-N_3O_2S; mol wt 369.36. Clinical trial in non-erosive reflux disease: R. Fass *et al., Aliment. Pharmacol. Ther.* **29**, 1261 (2009). White to nearly white crystalline powder, mp 140° (dec). Freely sol in DMF, methanol, dichloromethane, ethanol, ethyl acetate; sol in acetonitrile; slightly sol in ether; very slightly sol in water. Practically insol in hexane.
THERAP CAT: Antiulcerative.

5414. Lanthanum. [7439-91-0] La; at. wt 138.90547; at. no. 57; valence 3. Group IIIB (3). A rare earth metal; member of the lanthanide series. Naturally occurring isotopes (mass numbers): 139 (99.91%); 138 (0.09%), radioactive, $T_{1/2}$ 1.06×10^{11} years; known artificial radioactive isotopes: 123-137; 140-149. Estimated abundance in earth's crust: 18-35 ppm. Found in association with cerium and other light lanthanons. Sources of commercial importance are the rare earth minerals monazite and bastnaesite; also found in cerite. Discovery and isoln: Mosander, *Pogg. Ann.* **47**, 207 (1839). Sepn: James, *J. Am. Chem. Soc.* **34**, 757 (1912). Prepn of metal: Mazzi, *Atti X Congr. Int. Chim.* **3**, 604 (1938); Spedding *et al., Ind. Eng. Chem.* **44**, 553 (1952). Toxicity study: Cochran *et al., Arch. Ind. Hyg. Occup. Med.* **1**, 637 (1950). Reviews of prepn, properties and compds: *The Rare Earths* F. H. Spedding, A. H. Daane, Eds. (Krieger, Huntington, N.Y., 1971, reprint of 1961 ed.) 641 pp; Hulet, Bode, "Separation Chemistry of the Lanthanides and Transplutonium Actinides" in *MTP Int. Rev. Sci.: Inorg. Chem., Ser. One* Vol. **7**, K. W. Bagnall, Ed. (University Park Press, Baltimore, 1972) pp 1-45; Vickery, "Scandium, Yttrium, Lanthanum" in *Comprehensive Inorganic Chemistry* vol. **3**, J. C. Bailar Jr. *et al.*, Eds. (Pergamon Press, Oxford, 1973) pp 329-353; Moeller, "The Lanthanides" *ibid.* vol. **4**, 1-101; F. H. Spedding in *Kirk-Othmer Encyclopedia of*

Chemical Technology vol. **19** (John Wiley & Sons, New York, 3rd ed., 1982) pp 833-854; *Chemistry of the Elements* N. N. Greenwood, A. Earnshaw, Eds. (Pergamon Press, New York, 1984) pp 1102-1110, 1423-1449. Brief review of properties: G. T. Seaborg, *Radiochim. Acta* **61**, 115-122 (1993).

White, malleable metal; tarnishes in air. Three crystalline forms: hexagonal α-form, d 6.162, transforms to β-form at 310°; face-centered cubic β-form, d 6.19, transforms to γ-form at 864°; body-centered cubic γ-form, d 5.97, exists at >864°. mp 920°. bp 3464°. Heat of fusion: 6.201 kJ/mol. Heat of sublimation (25°): 431.0 kJ/mol. E°(aq) La^{3+}/La −2.52 V (calc). Very active; dec water slowly in the cold, more readily on heating. Readily attacked by mineral acids; not attacked by cold concd H_2SO_4. Burns in air at about 450° producing a mixture of oxide and nitride; forms the hydride on heating in hydrogen. Forms alloys with several metals.
Oxide. [1312-81-8] Lanthanum oxide (La_2O_3); lanthana; lanthanum sesquioxide; lanthanum trioxide. La_2O_3; mol wt 325.81. Almost white, amorphous powder. d 6.51. mp >2000°. Insol in water. Sol in dil mineral acids with formation of salts. Absorbs CO_2 from the air.
Hydroxide. [14507-19-8] Lanthanum hydroxide (La(OH)₃). H_3LaO_3; mol wt 189.93. White, amorphous precip, prepd by adding excess of caustic alkali to a lanthanum salt soln. Strongly basic, displaces ammonia from ammonium salts, absorbs CO_2 from air. On dehydration yields $La_2O_3.H_2O$.
Chloride. [10099-58-8]; [10025-84-0] (heptahydrate). Lanthanum chloride (LaCl₃). Cl₃La; mol wt 245.26. Heptahydrate, triclinic crystals. Sol in water or alc. On heating in HCl the anhydr salt (mp 852°) is formed. LD_{50} in rats: 4.2 g/kg orally; 350 mg/kg i.p. (Cochran).
Sulfate. [10099-60-2]; [10294-62-9] (nonahydrate). Sulfuric acid lanthanum(3+) salt (3:2). $La_2O_{12}S_3$; mol wt 565.98. Nonahydrate, hexagonal prisms. Prepd by treating a lanthanum salt with a slight excess of sulfuric acid. Dec at white heat. Is the least sol of the rare earth sulfates; soly in water decreases with increase in temp. Insol in alc. Forms double salts with alkali or ammonium hydroxide. Anhydr salt prepd by heating hydrate. LD_{50} in rats: >5.0 g/kg orally; 275 mg/kg i.p. (Cochran).
Nitrate. [10099-59-9]; [10277-43-7] (hexahydrate). Nitric acid lanthanum(3+) salt (3:1). LaN_3O_9; mol wt 324.92. Hexahydrate, white deliquesc crystals, mp ~40°, at higher temp forms a basic salt. bp 126°. Very sol in water, alc. Forms double salts with bivalent ion nitrates and ammonium nitrates. *Keep well closed.* LD_{50} in rats: 4.5 g/kg orally; 450 mg/kg i.p. (Cochran).
USE: Oxide in glass to improve optical properties. Chloride heptahydrate as matrix modifier in analytical chemistry. La^{3+} used in experimental biology as a specific antagonist of calcium: Weiss, *Annu. Rev. Pharmacol.* **14**, 343 (1974).

5415. Lanthanum Carbonate. [587-26-8] Carbonic acid lanthanum(3+) salt (3:2); lanthanum sesquicarbonate; Fosrenol; Foznol. $C_3La_2O_9$; mol wt 457.83. C 7.87%, La 60.68%, O 31.45%. La₂(CO₃)₃. Binds dietary phosphate to reduce hyperphosphatemia in patients with chronic renal failure. Occurs in various hydrated forms. Prepn of hydrated forms: M. L. Salutsky, L. L. Quill, *J. Am. Chem. Soc.* **72**, 3306 (1950); and use of hydrated pharmaceutical compositions: B. A. Murrer, N. A. Powell, **US 5968976** (1999 to AnorMed Inc.). Crystal structure of octahydrate: D. B. Shinn, H. A. Eick, *Inorg. Chem.* **7**, 1340 (1968). Soly equilibrium of octahydrate: D. Ferri, F. Salvatore, *Acta Chem. Scand.* **37**, 531 (1983). Clinical trial in hyperphosphatemia: M. S. Joy, W. F. Finn, *Am. J. Kidney Dis.* **42**, 96 (2003). Review of clinical development: T. S. Harrison, L. J. Scott, *Drugs* **64**, 985-996 (2004); and clinical efficacy: F. Albaaj, A. J. Hutchison, *Expert Opin. Pharmacother.* **6**, 319-328 (2005).
Octahydrate. [6487-39-4] Artificial lanthanite. $C_3La_2O_9.8H_2$-O; mol wt 601.95. White, crystalline powder. Practically insol in water. Freely sol in dil mineral acids.
THERAP CAT: In treatment of hyperphosphatemia.

5416. Lanthionine. [922-55-4] S-[(2R)-2-Amino-2-carboxyethyl]-L-cysteine; 3,3′-thiodi-L-alanine; L-lanthionine. $C_6H_{12}N_2$-O_4S; mol wt 208.23. C 34.61%, H 5.81%, N 13.45%, O 30.73%, S 15.40%. A rare amino acid found in proteins; sulfide analog of cystine. First isolated as artifacts of wool hydrolysates. Isoln: M. J.

Horn *et al.*, *J. Biol. Chem.* **138**, 141 (1941); from chick embryo: N. H. Sloane, K. G. Untch, *Biochemistry* **5**, 2658 (1966); from silkworm and Japanese oak moth: D. R. Rao *et al.*, *ibid.* **6**, 1208. Synthesis: V. du Vigneaud, G. B. Brown, *J. Biol. Chem.* **138**, 151 (1941). Structure studies: I. W. Stapleton, O. A. Weber, *Int. J. Pept. Protein Res.* **3**, 243 (1971). Crystal structure: G. R. Desiraju, D. R. Rao, *Acta Crystallogr.* **C46**, 627 (1990).

Elongated hexagonal plates. mp 295-296° (dec). $[\alpha]_D^{25}$ +9.4° (c = 1.4 in 2.4N NaOH); $[\alpha]_D^{22}$ +8.6° (c = 5 in 2.4N NaOH); $[\alpha]_D^{22}$ +6.0° (c = 1 in 1N NaOH).

D-Form. [5965-92-4] *S*-[(2S)-2-Amino-2-carboxyethyl]-D-cysteine. Elongated hexagonal plates. Darkens at 245°, dec 293-295°. $[\alpha]_D^{21}$ −8.0° (c = 5 in 2.4N NaOH).

DL-Form. [3183-08-2] Elongated hexagonal plates. Chars at 240°, dec 286-292°.

***meso*-Form.** [922-56-5] *S*-[(2R)-2-Amino-2-carboxyethyl]-D-cysteine. Six-sided plates having a triangular appearance from dil NH_3. Softens at 270°, dec 304°. Stable to alkalies. Sol in dil acids and alkalies. Sparingly sol in water. Insol in alcohol, ether, chloroform, acetone.

5417. Lapachol. [84-79-7] 2-Hydroxy-3-(3-methyl-2-buten-1-yl)-1,4-naphthalenedione; lapachic acid; taiguic acid; tecomin; greenhartin; NSC-11905. $C_{15}H_{14}O_3$; mol wt 242.27. C 74.37%, H 5.82%, O 19.81%. Yellow crystalline material derived from the heartwood of Asian and South American bignoniaceous plants, esp Surinam greenheart, Taigu wood, Lapacho heartwood and Bethabarra wood: Arnoudon, *Compt. Rend.* **41**, 1152 (1857); Stein, *J. Prakt. Chem.* **99**, 1 (1866); Paterno, *Gazz. Chim. Ital.* **9**, 506 (1879); Greene, Hooker, *Am. Chem. J.* **11**, 267 (1889). Structure: Paterno, *Gazz. Chim. Ital.* **12**, 337 (1882); Hooker *J. Chem. Soc.* **61**, 611 (1892); **69**, 1355 (1896); *J. Am. Chem. Soc.* **58**, 1168 (1936). Synthesis: Fieser, *ibid.* **49**, 857 (1927); Hooker, *ibid.* **58**, 1181 (1936); G. R. Pettit, L. E. Houghton, *J. Chem. Soc. C* **1971**, 509. Although it is related structurally to vitamin K, *q.v.*, it does not possess antihemorrhagic activity: H. J. Almquist, A. A. Klose, *J. Am. Chem. Soc.* **61**, 1923 (1939); L. F. Fieser *et al.*, *J. Biol. Chem.* **137**, 659 (1941). It is reported to be an inhibitor of respiratory processes: E. G. Ball *et al.*, *ibid.* **168**, 257 (1947). Lapachol has also exhibited antitumor activity vs Walker 256 carcinoma: K. V. Rao *et al.*, *Proc. Am. Assoc. Cancer Res.* **8**, 55 (1967). Mass spectrometry: T. A. Elwood *et al.*, *Org. Mass Spectrom.* **3**, 841 (1970). Chromatographic detection: M. H. Simatupang *et al.*, *J. Chromatogr.* **52**, 180 (1970). Pharmacology: S. M. Sieber *et al.*, *Cancer Treat. Rep.* **60**, 1127 (1976). Toxicity study: R. K. Morrison *et al.*, *Toxicol. Appl. Pharmacol.* **17**, 1 (1970). Review of antitumor activity: K. V. Rao, *Cancer Chemother. Rep. Part 2* **4** (4), 11-17 (1974).

Yellow prisms from alcohol or ether, mp 140°. uv max: 251.5, 278, 331 nm (log ε 4.38, 4.28, 3.43). Soluble in alcohol, chloroform, benzene, acetic acid, slightly sol in ether, hot water. Sol in aq NaOH solns forming a bright red sodium salt. LD_{50} in male, female BALB/c mice (g/kg): 0.487; 0.792 orally (Morrison).

5418. Lapatinib. [231277-92-2] *N*-[3-Chloro-4-[(3-fluorophenyl)methoxy]phenyl]-6-[5-[[[2-(methylsulfonyl)ethyl]amino]methyl]-2-furanyl]-4-quinazolinamine; GW-572016. $C_{29}H_{26}$-$ClFN_4O_4S$; mol wt 581.06. C 59.95%, H 4.51%, Cl 6.10%, F 3.27%, N 9.64%, O 11.01%, S 5.52%. Reversible dual inhibitor of ErbB1 (EGFR) and ErbB2 (HER-2) tyrosine kinases. Prepn: M. C. Carter *et al.*, **WO 9935146** (1999 to Glaxo); *eidem*, **US 6727256**

(2004 to SmithKline Beecham). Mechanism of action study: W. Xia *et al.*, *Oncogene* **21**, 6255 (2002); and crystal structure of complex with epidermal growth factor receptor (EGFR, ErbB1): E. R. Wood *et al.*, *Cancer Res.* **64**, 6652 (2004). LC-MS/MS determn in plasma: F. Bai *et al.*, *J. Chromatogr. B* **831**, 169 (2006). Pharmacokinetics and clinical activity in metastatic carcinomas: H. A. Burris III *et al.*, *J. Clin. Oncol.* **23**, 5305 (2005). Clinical trial in combination with capecitabine in HER2-positive breast cancer: C. E. Geyer *et al.*, *N. Engl. J. Med.* **355**, 2733 (2006). Review of clinical experience: F. Montemurro *et al.*, *Expert Opin. Biol. Ther.* **7**, 257-268 (2007).

Ditoluenesulfonate monohydrate. [388082-78-8]; [388082-77-7] (anhydrous). Lapatinib ditosylate; GW-572016F; Tykerb; Tycerb; Tyverb. $C_{29}H_{26}ClFN_4O_4S.2C_7H_8O_3S.H_2O$; mol wt 943.47. Yellow solid. Soly at 25° (mg/ml): 0.007 in water; 0.001 in 0.1N HCl.

THERAP CAT: Antineoplastic.

5419. Lappa. Burdock; clotbur; bardana. Dried first-year root of *Arctium lappa* L., or of *Arctium minus* Bernh., *Compositae.* Habit. Europe, Northern Asia; naturalized in N. America. Constit. Volatile oil, bitter principle, inulin, tannin.

THERAP CAT: Dermatologic.

5420. Lappaconitine. [32854-75-4] (1α,14α,16β)-20-Ethyl-1,14,16-trimethoxyaconitane-4,8,9-triol 4-[2-(acetylamino)benzoate]. $C_{32}H_{44}N_2O_8$; mol wt 584.71. C 65.73%, H 7.59%, N 4.79%, O 21.89%. From tubers and herb of *Aconitum septentrionale* Kölle, *A. orientale* Mill., *A. excelsum* Reichb., *A. ranunculaefolium*, *Ranunculaceae*: Schulze, Ulfert, *Arch. Pharm.* **260**, 230 (1922); **28**, 258 (1958), *C.A.* **50**, 1852e (1956); Kuzovkov, Massagetov; Platonova *et al.*, *Zh. Obshch. Khim.* **25**, 178 (1955), *C.A.* **52**, 12883i (1958); Mollov *et al.*, *C. R. Acad. Bulg. Sci.* **17**, 251 (1964), *C.A.* **61**, 12324g (1964). Structure: Khaimova *et al.*, *Tetrahedron Lett.* **1964**, 2711. Revised structure: V. A. Tel'nov *et al.*, *Khim. Prir. Soedin.* **6**, 583 (1970), *C.A.* **74**, 42527k (1971). Analgesic activity studies in mice and rats: N. Ono, J. Satoh, *Arzneim.-Forsch.* **38**, 892 (1988). Toxicology: F. Dybing *et al.*, *Acta Pharmacol. Toxicol.* **7**, 337 (1951).

Bitter crystals, mp 217-218°. $[\alpha]_D^{18}$ +27° (chloroform). Sol in benzene; slightly sol in alcohol, ether. Practically insol in water. LD_{50} in mice (mg/kg): 6.9 i.v.; 9.1 i.p.; approx 20 orally (Dybing).

5421. Lapyrium Chloride. [6272-74-8] 1-[2-Oxo-2-[[2-[(1-oxododecyl)oxy]ethyl]amino]ethyl]pyridinium chloride (1:1); 1-[(2-dodecanoyloxyethylcarbamoyl)methyl]pyridinium chloride; 1-[[(2-hydroxyethyl)carbamoyl]methyl]pyridinium chloride laurate (ester); *N*-(lauroylcolaminoformylmethyl)pyridinium chloride; *N*-(acylcolaminoformylmethyl)pyridinium chloride; emulsept (obsolete); E-607; DG-6; Emcol E-607. $C_{21}H_{35}ClN_2O_3$; mol wt 398.97. C 63.22%, H 8.84%, Cl 8.89%, N 7.02%, O 12.03%. Cationic surfactant. Prepn: A. K. Epstein, B. R. Harris, **US 2290173** (1942). Bactericidal potency: A. K. Epstein *et al.*, *Proc. Soc. Exp. Biol. Med.*

53, 238 (1943). Toxicity study: A. E. Vivino, T. Koppanyi, *J. Am. Pharm. Assoc. Sci. Ed.* **35**, 169 (1946). Chemoenzymatic synthesis: E. M. Rustoy, A. Baldessari, *Eur. J. Org. Chem.* **2005**, 4628.

Powder, mp 141-144°.

Stearoyl analog. [14492-68-3] Quaternium-7; steapyrium chloride; Emcol E-607S. $C_{27}H_{47}ClN_2O_3$; mol wt 483.13.

USE: Surfactant ingredient in personal care products; in wastewater treatment; in corrosion inhibition formulations. Pharmaceutic aid (surfactant).

THERAP CAT: Antiseptic, disinfectant.

5422. Laquinimod. [248281-84-7] 5-Chloro-*N*-ethyl-1,2-dihydro-4-hydroxy-1-methyl-2-oxo-*N*-phenyl-3-quinolinecarboxamide; *N*-ethyl-*N*-phenyl-1,2-dihydro-4-hydroxy-5-chloro-1-methyl-2-oxo-quinoline-3-carboxamide; ABR-215062. $C_{19}H_{17}ClN_2O_3$; mol wt 356.81. C 63.96%, H 4.80%, Cl 9.94%, N 7.85%, O 13.45%. Immunomodulator. Prepn: A. Björk *et al.*, WO 9955678; *eidem*, US 6077851 (1999, 2000 both to Active Biotech AB). Large scale synthesis: K. Jansson *et al.*, *J. Org. Chem.* **71**, 1658 (2006); J. Wennerberg *et al.*, *Org. Process Res. Dev.* **11**, 674 (2007). Structure activity studies: S. Jönsson *et al.*, *J. Med. Chem.* **47**, 2075 (2004). Determn in plasma by HPLC: K. Edman *et al.*, *J. Chromatogr. B* **785**, 311 (2003); by LC/MS/MS: C. J. Sennbro *et al.*, *Rapid Commun. Mass Spectrom.* **20**, 3313 (2006). Role of cytochrome P450 3A4 in metabolism: H. Tuvesson *et al.*, *Drug Metab. Dispos.* **33**, 866 (2005). Clinical evaluation in relapsing multiple sclerosis (MS): C. Polman *et al.*, *Neurology* **64**, 987 (2005). Review of pharmacology and clinical experience in MS: J. Thöne, R. Gold, *Expert Opin. Drug Metab. Toxicol.* **7**, 365-370 (2011).

White to off-white crystals from *n*-heptane, mp 201° (dec). pKa 4.2.

Sodium salt. [248282-07-7] $C_{19}H_{17}ClN_2NaO_3$; mol wt 379.80. White, non-hygroscopic crystals from ethanol.

THERAP CAT: Immunomodulator; in treatment of multiple sclerosis.

5423. Lard. Adeps; axungia porci. Purified internal fat from abdomen of the hog.

Soft, white unctuous mass; slight characteristic odor, bland taste. d 0.917. mp 36-42°. Insoluble in water. Very slightly sol in alcohol, freely in benzene, chloroform, ether, carbon disulfide, petr ether. Sapon. no. 195-203. Iodine no. 46-70. *Keep cool and in tight containers.*

Benzoinated lard. Prepd by heating lard with 1% benzoin powder at temps not above 60°. Less likely to become rancid than pure lard.

Lard oil. The oil expressed from lard at a low temp. *Constit.* Olein, stearin. Colorless or pale yellow liq. d 0.905-0.915. Solidif −2° to +4°. n_D^{20} 1.470-1.472. Sapon no. 195-197. Iodine no. 56-74. Acid no. 1.5-2.5. Insol in water or cold alc. Freely sol in benzene.

USE: As vehicle for medicinal agents and in manufacture of ointments. Oil as lubricant, manuf soap, oiling wool, illuminant.

5424. Larkspur. Delphinium; knight's spur; lark's-heel; lark's-claw; staggerweed. Dried ripe seeds of *Delphinium ajacis L.*, Ranunculaceae. *Habit.* Central Europe, cultivated in U.S. *Constit.* Calcatripine, volatile oil, gum, resin, fixed oil, gallic and aconitic acids. Identification and toxicity of alkaloids: G. D. Manners *et al.*, *J. Range Manage.* **45**, 63 (1992).

Caution: Toxic. Poisoning from percutaneous absorption may occur.

THERAP CAT: Pediculicide.

5425. Laromustine. [173424-77-6] 1-(2-Chloroethyl)-2-[(methylamino)carbonyl]-2-(methylsulfonyl)hydrazide methanesulfonic acid; 1,2-bis(methylsulfonyl)-1-(2-chloroethyl)-2-[(methylamino)carbonyl]hydrazine; VNP-40101M; 101M; Cloretazine. $C_6H_{14}ClN_3O_5S_2$; mol wt 307.76. C 23.42%, H 4.59%, Cl 11.52%, N 13.65%, O 25.99%, S 20.83%. DNA alkylating and crosslinking agent. Bifunctional sulfonylhydrazine prodrug that generates a chloroethylating species, VPN-4090CE, and methyl isocyanate, *q.v.*, a carbamoylating agent. Prepn: A. Sartorelli *et al.*, WO 9702029; *eidem*, US 5637619 (both 1997 to Yale); K. Shyam *et al.*, *J. Med. Chem.* **39**, 796 (1996). Analytical methods and formulation: G. Krishna *et al.*, *AAPS PharmSciTech* **2** (3), 39-47 (2001). LC-ESI-MS/MS determn in plasma: F. Bai *et al.*, *J. Chromatogr. B* **853**, 97 (2007). Mechanism of action: K. Ishiguro *et al.*, *Mol. Cancer Ther.* **5**, 969 (2006); P. G. Penketh *et al.*, *Leuk. Res.* **32**, 1546 (2008). Clinical evaluation in recurrent glioblastoma multiforme: M. A. Badruddoja *et al.*, *Neuro Oncol.* **9**, 70 (2007); in acute myeloid leukemia: F. Giles *et al.*, *J. Clin. Oncol.* **25**, 25 (2007).

Crystals from ethanol, mp 146-147.5°. Soly (mg/ml): water 0.66, propylene glycol 1.4, polyethylene glycol 300 16.8, ethyl alcohol 1.4.

VPN-4090CE. [127792-84-1] 2-(2-Chloroethyl)-2-(methylsulfonyl)hydrazide methanesulfonic acid; 1,2-bis(methylsulfonyl)-1-(2-chloroethyl)hydrazine; 90CE. $C_4H_{11}ClN_2O_4S_2$; mol wt 250.71. Prepn: K. Shyam *et al.*, *J. Med. Chem.* **33**, 2259 (1990). Light yellow solid from ethanol, mp 138-139°.

THERAP CAT: Antineoplastic.

5426. Laropiprant. [571170-77-9] (3*R*)-4-[(4-Chlorophenyl)methyl]-7-fluoro-1,2,3,4-tetrahydro-5-(methylsulfonyl)cyclopent[*b*]indole-3-acetic acid; [(3*R*)-4-(4-chlorobenzyl)-7-fluoro-5-(methylsulfonyl)-1,2,3,4-tetrahydrocyclopenta[*b*]indol-3-yl]-acetic acid; MK-0524. $C_{21}H_{19}ClFNO_4S$; mol wt 435.89. C 57.87%, H 4.39%, Cl 8.13%, F 4.36%, N 3.21%, O 14.68%, S 7.36%. Selective prostaglandin D$_2$ (PGD$_2$) receptor subtype 1 (DP$_1$) antagonist. Prepn: C. Berthelette *et al.*, WO 03062200; *eidem*, US 7317036 (2003, 2008 both to Merck Frosst). Synthesis and pharmacokinetics: C. F. Sturino *et al.*, *J. Med. Chem.* **50**, 794 (2007). Asymmetric synthesis: K. R. Campos *et al.*, *J. Org. Chem.* **70**, 268 (2005). HPLC/MS/MS determn in plasma: M. S. Schwartz *et al.*, *J. Chromatogr. B* **837**, 116 (2006). Clinical pharmacokinetics: Y.-H. Wang *et al.*, *J. Clin. Pharmacol.* **51**, 406 (2011). Clinical effect on niacin-induced vasodilation: E. Lai *et al.*, *Clin. Pharmacol. Ther.* **81**, 849 (2007); D. Maccubbin *et al.*, *Int. J. Clin. Pract.* **62**, 1959 (2008). Review of pharmacology and clinical experience: A. G. Olsson *Expert Opin. Pharmacother.* **11**, 1715-1726 (2010); of the combination with niacin and simvastatin for treatment of dyslipidemia: K.-H. Yiu *et al.*, *Expert Opin. Invest. Drugs* **19**, 437-449 (2010).

White solid, mp 175°. $[\alpha]_D^{21} = -29.3°$ (c = 1.0 in MeOH).

Combination with nicotinic acid. [1046050-73-0] MK-0524A; Cardaptive; Tradaptive.

THERAP CAT: Reduction of niacin-induced flushing in the management of hyperlipidemia.

5427. Lasalocid A. [25999-31-9] 6-[(3*R*,4*S*,5*S*,7*R*)-7-[(2*S*,-3*S*,5*S*)-5-Ethyl-5-[(2*R*,5*R*,6*S*)-5-ethyltetrahydro-5-hydroxy-6-methyl-2*H*-pyran-2-yl]tetrahydro-3-methyl-2-furanyl]-4-hydroxy-3,5-dimethyl-6-oxononyl]-2-hydroxy-3-methylbenzoic acid; 3-methyl-6-[7-ethyl-4-hydroxy-3,5-dimethyl-6-oxo-7-[5-ethyl-3-methyl-5-(5-ethyl-5-hydroxy-6-methyl-2-tetrahydropyranyl)-2-tetrahydrofuryl]-heptyl]salicylic acid; antibiotic X-537A; ionophore X-4537A; Ro-2-2985; X-537A; Bovatec. C$_{34}$H$_{54}$O$_8$; mol wt 590.80. C 69.12%, H 9.21%, O 21.66%. Ionophorous (transport-inducing) antibiotic isolated from an unidentified *Streptomyces* from soil samples of Hyde Park, Mass.: Berger *et al.*, *J. Am. Chem. Soc.* **73**, 5295 (1951). Prepn and activity of title compd and derivs: A. Stempel, J. Westley, **DE 2040998**; *eidem*, **US 3715372** (1971, 1973 to Hoffmann-La Roche); Westley *et al.*, *J. Med. Chem.* **16**, 397 (1973). Structure: *eidem*, *Chem. Commun.* **1970**, 71. Crystal and molecular structure studies: Johnson *et al.*, *ibid.* **72**; *J. Am. Chem. Soc.* **92**, 4428 (1970). Total synthesis: T. Nakata *et al.*, *ibid.* **100**, 2933 (1978); R. E. Ireland *et al.*, *ibid.* **102**, 1155 (1980). Complete assignment of ^{13}C-NMR spectra of lasalocid A and its sodium salt: H. Seto *et al.*, *J. Antibiot.* **31**, 289 (1978). Effect in coccidiosis: G. M. J. Horton, P. H. G. Stockdale, *Am. J. Vet. Res.* **42**, 433 (1981). Biosynthesis: Westley *et al.*, *Chem. Commun.* **1970**, 1467; C. R. Hutchinson *et al.*, *J. Am. Chem. Soc.* **103**, 5953, 5956 (1981). Mode of action studies: Lin, Kun, *Biochem. Biophys. Res. Commun.* **50**, 820 (1973). Isoln and structure of four homologs, lasalocids B, C, D, E from *Streptomyces lasaliensis*: J. W. Westley *et al.*, *J. Antibiot.* **27**, 744 (1974).

Crystals, mp 110-114°; also reported as mp 100-109° (unsharp). [α]$_D^{25}$ −7.55° (methanol). uv max (50% aq isopropanol): 248, 318 nm (ε 6750, 4200). Sol in organic solvents. Insol in water. LD$_{50}$ in mice (mg/kg): 40 i.p. (Berger); also reported as LD$_{50}$ in mice (mg/kg): 146 orally, 64 i.p., J. W. Westley, "Antibiotics (Polyether)" in *Kirk-Othmer Encyclopedia of Chemical Technology* Vol. 3 (Interscience, New York, 3rd ed., 1978) p 61.

Sodium salt. Avatec. C$_{34}$H$_{53}$NaO$_8$; mol wt 612.78. Crystals from benzene-ligroin, mp (open capillary) 191-192° (dec), also reported as mp 168-171°. [α]$_D^{25}$ −30° (c = 1 in methanol). uv max (50% aq isopropanol): 308 nm (ε 4100).

THERAP CAT (VET): Coccidiostat (for poultry).

5428. Laserpitin. [7067-12-1] (2*Z*,2′*Z*)-2-Methyl-2-butenoic acid 1,1′-[(1*R*,3a*S*,4*S*,6*R*,8*S*,8a*S*)-decahydro-1,6-dihydroxy-3a,6-dimethyl-1-(1-methylethyl)-5-oxo-4,8-azulenediyl] ester. C$_{25}$H$_{38}$O$_7$; mol wt 450.57. C 66.64%, H 8.50%, O 24.86%. From root of *Laserpitium latifolium* L., *Umbelliferae*: Külz, *Arch. Pharm.* **221**, 161 (1883); Morgenstern, *Monatsh. Chem.* **33**, 709 (1912). Structure: M. Holub *et al.*, *Tetrahedron Lett.* **1965**, 1441, 2855. Absolute configuration: M. Holub *et al.*, *Collect. Czech. Chem. Commun.* **35**, 3597 (1970).

Crystals, mp 118°. Sol in alcohol, chloroform, ether, petr ether, fatty oils. Practically insol in water.

USE: Used as seasoning and flavoring agent in antiquity (Greece and Rome).

5429. Lasiocarpine. [303-34-4] (2*Z*)-2-Methyl-2-butenoic acid (1*S*,7a*R*)-7-[[(2*R*)-2,3-dihydroxy-2-[(1*S*)-1-methoxyethyl]-3-methyl-1-oxobutoxy]methyl]-2,3,5,7a-tetrahydro-1*H*-pyrrolizin-1-yl ester. C$_{21}$H$_{33}$NO$_7$; mol wt 411.50. C 61.30%, H 8.08%, N 3.40%, O 27.22%. Hepatotoxic pyrrolizidine alkaloid isolated from *Heliotropium lasiocarpum* Fish. et C. Mey. *Boraginaceae*: G. Menschikoff, *Ber.* **65**, 974 (1932); G. Menschikoff, J. Schdanowitsch, *Ber.* **69**, 1110 (1936). Structure: L. J. Drummond, *Nature* **167**, 41 (1951); R. Adams, B. L. VanDuvren, *J. Am. Chem. Soc.* **76**, 6379 (1954). Review and evaluation of studies of carcinogenicity and toxicity in laboratory animals: *IARC Monographs* **10**, 281-290, 333-342 (1976). Comprehensive reviews: L. Bull *et al.*, *The Pyrrolizidine Alkaloids* (North-Holland, Amsterdam, 1965) 293 pp; F. L. Warren in *The Alkaloids* vol. **12**, R. H. F. Manske, Ed. (Academic Press, New York, 1970) pp 245-331; D. J. Robins, *Fortschr. Chem. Org. Naturst.* **41**, 115-203 (1982).

Colorless leaflets from petr ether. mp 94-95.5°. [α]$_D$ −4° (10% alc). Sol on ether, alc, benzene; difficultly sol in water.

5430. Lasofoxifene. [180916-16-9] (5*R*,6*S*)-5,6,7,8-Tetrahydro-6-phenyl-5-[4-[2-(1-pyrrolidinyl)ethoxy]phenyl]-2-naphthalenol; (−)-*cis*-6-phenyl-5-[4-(2-pyrrolidin-1-ylethoxy)phenyl]-5,6,-7,8-tetrahydronaphthalen-2-ol. C$_{28}$H$_{31}$NO$_2$; mol wt 413.56. C 81.32%, H 7.56%, N 3.39%, O 7.74%. Selective estrogen receptor modulator (SERM). Partial estrogen agonist. Prepn: K. O. Cameron, P. A. Jardine, **WO 9621656**; K. O. Cameron *et al.*, **US 5552412** (1996, 1996 both to Pfizer). Structure-activity relationship: R. L. Rosati *et al.*, *J. Med. Chem.* **41**, 2928 (1998). Improved synthesis: X. Yang *et al.*, *Org. Lett.* **2**, 4025 (2000); Y. Sano *et al.*, *Chem. Lett.* **36**, 40 (2007). Crystal structure of complex with estrogen receptor α domain: F. F. Vajdos *et al.*, *Protein Sci.* **16**, 897 (2007). Clinical comparison with raloxifene, *q.v.*, in postmenopausal osteoporosis: M. R. McClung *et al.*, *Menopause* **13**, 377 (2006). Review of preclinical studies: H. Z. Ke *et al.*, *J. Am. Aging Assoc.* **25**, 87-100 (2002); of pharmacology and clinical experience: L. Gennari *et al.*, *Expert Opin. Invest. Drugs* **15**, 1091-1103 (2006).

Hydrochloride. [180915-85-9] C$_{28}$H$_{31}$NO$_2$.HCl; mol wt 450.02. Crystals from acetonitrile/methylene chloride, mp 260-263°. [α]$_D$ −330.6° (c = 0.05 in CH$_2$Cl$_2$).

Tartrate. [190791-29-8] CP-336156; Fablyn; Oporia. C$_{28}$H$_{31}$-NO$_2$.C$_4$H$_6$O$_6$; mol wt 563.65. Prepn: C. K. Chiu, M. Meltz, **US 5948809** (1999 to Pfizer). HPLC analysis of dissolution properties: J. S. Space *et al.*, *J. Pharm. Biomed. Anal.* **44**, 1064 (2007). White solid from 95% ethanol.

THERAP CAT: Antiosteoporotic.

5431. Latanoprost. [130209-82-4] (5Z)-7-[(1R,2R,3R,5S)-3,5-Dihydroxy-2-[(3R)-3-hydroxy-5-phenylpentyl]cyclopentyl]-5-heptenoic acid 1-methylethyl ester; isopropyl (Z)-7-[(1R,2R,3R,5S)-3,5-dihydroxy-2-[(3R)-3-hydroxy-5-phenylpentyl]cyclopentyl]-5-heptenoate; 13,14-dihydro-17-phenyl-18,19,20-trinor-PGF$_{2\alpha}$-isopropyl ester; PhXA-41; Xalatan. C$_{26}$H$_{40}$O$_5$; mol wt 432.60. C 72.19%, H 9.32%, O 18.49%. Prostaglandin analog; prodrug of active acid metabolite. Prepd (not claimed): J. Ivanics *et al.*, **WO 9300329** (1993 to Kabi Pharmacia; Chinoin). Prepn: B. Resul *et al.*, *J. Med. Chem.* **36**, 243, 2242 (corr.) (1993). Clinical pharmacology: C. B. Toris *et al.*, *Ophthalmology* **100**, 1297 (1993). Clinical evaluation in glaucoma: P. Rácz *et al.*, *Arch. Ophthalmol.* **111**, 657 (1993); as single therapy and in combination with timolol, *q.v.*: A. H. Rulo *et al.*, *Br. J. Ophthalmol.* **78**, 899 (1994). Review of preclinical pharmacology: J. Stjernschantz *et al.*, *Adv. Prostaglandin Thromboxane Leukotriene Res.* **23**, 513-518 (1995).

Colorless oil. [α]$_D^{20}$ +31.57° (c = 0.91 in acetonitrile). Very sol in acetonitrile; freely sol in acetone, ethanol, ethyl acetate, isopropanol, methanol, octanol. Practically insol in water.

THERAP CAT: Antiglaucoma.

THERAP CAT (VET): Antiglaucoma.

5432. α-Latrotoxin. [65988-34-3] Neurotoxic protein isolated from the venom of black widow spiders, *Latrodectus sp.* Mol wt of monomer ~130 KDa. Active tetrameric form induces a massive release of neurotransmitters from neurosecretory cells; its mode of action involves receptor-mediated calcium dependent binding and calcium independent binding as well as pore formation. Identification of toxin as a peptide: S. Bettini, N. Toschi-Frontali, *Proc. 11th Int. Congr. Entomol.* **1960**, 115-121. Purification: N. Frontali *et al.*, *J. Cell Biol.* **68**, 462 (1976). 3D-structure: E. V. Orlova *et al.*, *Nat. Struct. Biol.* **7**, 48 (2000). Possible mechanism of transmitter release: M. Khvotchev, T. C. Südhof, *EMBO J.* **19**, 3250 (2000). Brief review: A. Malgaroli, J. Meldolesi, "α-Latrotoxin (black widow spider)" in *Guide to Protein Toxins and Their Use in Cell Biology*, R. Rappuoli, C. Montecucco, Eds. (Oxford University Press, Oxford, 1997) pp 233-235. Review: A. Grasso *et al.*, *Cell Mol. Mech. Toxin Action* **2**, 333-355 (1998); of mechanisms of action: A. W. Henkel, S. Sankaranarayanan, *Cell Tissue Res.* **296**, 229-233 (1999).

USE: Neurobiological tool for studying exocytosis, toxin-binding sites and pre-synaptic organization.

5433. Latrunculins. Highly potent toxins which disrupt microfilament organization by binding the actin monomer. Isolated from the Red Sea sponge *Latruncula magnifica* Keller; also found in the Pacific nudibranch *Chromodoris elisabethina* and the Fijian sponge *Spongia mycofijiensis*. Isoln and biological activities: I. Néeman *et al.*, *Mar. Biol.* **30**, 293 (1975). First family of natural products possessing the 2-thiazolidinone moiety; latrunculins A-D are 14- or 16- membered macrolides, infrastructures previously unknown in marine isolates. Structure elucidation of A and B: A. Groweiss *et al.*, *J. Org. Chem.* **48**, 3512 (1983); of A through D: Y. Kashman *et al.*, *Tetrahedron* **41**, 1905 (1985). Chemistry of A and B: D. Blasberger *et al.*, *Ann.* **1989**, 1171. Total synthesis of B: R. Zibuck *et al.*, *J. Am. Chem. Soc.* **108**, 2451 (1986); of A: J. D. White, M. Kawasaki, *J. Am. Chem. Soc.* **112**, 4991 (1990); of A, B, C and M: A. B. Smith, III *et al.*, *ibid.* **114**, 2995 (1992). Comparative study with cytochalasin D, *q.v.*, on the effects actin organization and cell processes: I. Spector *et al.*, *Cell Motil. Cytoskeleton* **13**, 127 (1989). Inhibition of the actin monomer: W. M. Morton *et al.*, *Nat. Cell Biol.* **2**, 376 (2000). Review of use as research tool: K. Ayscough, *Methods Enzymol.* **298**, 18-25 (1998).

Latrunculin A   Latrunculin B

Latrunculin A. [76343-93-6] (4R)-4-[(1R,4Z,8E,10Z,12S,15R,17R)-17-Hydroxy-5,12-dimethyl-3-oxo-2,16-dioxabicyclo[13.3.1]-nonadeca-4,8,10-trien-17-yl]-2-thiazolidinone; LAT-A. C$_{22}$H$_{31}$-NO$_5$S; mol wt 421.55. Foam. [α]$_D^{24}$ +152° (c = 1.2 in chloroform). uv max (methanol): 218 nm (ε 23500).

Latrunculin B. [76343-94-7] (4R)-4-[(1R,4Z,8Z,10S,13R,15R)-15-Hydroxy-5,10-dimethyl-3-oxo-2,16-dioxabicyclo[11.3.1]heptadeca-4,8-dien-15-yl]-2-thiazolidinone; LAT-B. C$_{20}$H$_{29}$NO$_5$S; mol wt 395.51. Major toxin. Localization within the Red Sea sponge: O. Gillor *et al.*, *Mar. Biotechnol.* **2.**, 213 (2000). Crystals. [α]$_D^{24}$ +112° (c = 0.48 in chloroform). uv max (methanol): 212 nm (ε 17200).

USE: In elucidation of molecular mechanisms of motile processes.

5434. Laudanidine. [301-21-3] 2-Methoxy-5-[[(1R)-1,2,3,4-tetrahydro-6,7-dimethoxy-2-methyl-1-isoquinolinyl]methyl]phenol; *l*-laudanine; tritopine. C$_{20}$H$_{25}$NO$_4$; mol wt 343.42. C 69.95%, H 7.34%, N 4.08%, O 18.63%. The *l*-form of laudanine, *q.v.* Traces in opium: Hesse, *Ann.* **282**, 208 (1894); Kauder, *Arch. Pharm.* **228**, 419 (1890). Structure and synthesis: Späth, Bernhauer, *Ber.* **58**, 200 (1925); Späth, Seka, *ibid.* 1272; Späth, Burger, *Monatsh. Chem.* **47**, 733 (1926); Frydman *et al.*, *Tetrahedron* **4**, 342 (1958). Configuration: Corrodi, Hardegger, *Helv. Chim. Acta* **39**, 889 (1956).

Prisms from alcohol, mp 185°. [α]$_D^{15}$ −88° (p = 5 in chloroform). [α]$_D^{22}$ −94.7 ± 1.3°. The hydrochloride is more sol than laudanine hydrochloride and makes possible a separation of the two alkaloids.

5435. Laudanine. [85-64-3] 2-Methoxy-5-[(1,2,3,4-tetrahydro-6,7-dimethoxy-2-methyl-1-isoquinolinyl)methyl]phenol; *dl*-laudanidine. C$_{20}$H$_{25}$NO$_4$; mol wt 343.42. C 69.95%, H 7.34%, N 4.08%, O 18.63%. In the alkaline mother liquors from morphine extraction, in amounts corresp to about 0.005% of the opium employed: L. F. Small, R. E. Lutz, *Chemistry of the Opium Alkaloids*, Supplement No. 103, Public Health Reports, Washington (1932) p 34. Structure: Späth, *Monatsh. Chem.* **41**, 297 (1920). Synthesis: Späth, Lang, *ibid.* **42**, 273 (1921); Frydman *et al.*, *Tetrahedron* **4**, 342 (1958).

Orthorhombic prisms from alcohol or chloroform, mp 167°. d_4^{20} 1.26. uv max: 284 nm (log ε 3.78). Sol in benzene, chloroform, hot alcohol; sparingly sol in ether. Practically insol in water.

5436. Laudanosine. [2688-77-9] (1*S*)-1-[(3,4-Dimethoxyphenyl)methyl]-1,2,3,4-tetrahydro-6,7-dimethoxy-2-methylisoquinoline; 1,2,3,4-tetrahydro-6,7-dimethoxy-2-methyl-1-veratrylisoquinoline; *N*-methyltetrahydropapaverine. $C_{21}H_{27}NO_4$; mol wt 357.45. C 70.56%, H 7.61%, N 3.92%, O 17.90%. Occurs in opium (0.0008%). It is the last alkaloid to be separated from morphine extraction mother liquors; occurs as (+)-form. Synthesis: Pictet, Finkelstein, *Ber.* **42**, 1979 (1909); Frydman *et al.*, *Tetrahedron* **4**, 342 (1958); Elliott, *J. Heterocycl. Chem.* **9**, 853 (1972). Asymmetric synthesis: M. Konda *et al.*, *Chem. Pharm. Bull.* **23**, 1025 (1975); of (*R*)-(−)-laudanosine: M. Konda *et al.*, *ibid.* **25**, 69 (1977); R. E. Gawley, G. A. Smith, *Tetrahedron Lett.* **29**, 301 (1988). Configuration: Leithe, *Ber.* **64**, 2827 (1931); Faltis, Adler, *Arch. Pharm.* **284**, 281 (1951); Corrodi, Hardegger, *Helv. Chim. Acta* **39**, 889 (1956).

Crystals from light petr (30-60°), mp 89°. $[\alpha]_D^{16}$ +106° (c = 1.6 in 97% alc); $[\alpha]_D^{16}$ +130° (chloroform); $[\alpha]_D^{22}$ +52.2 ±1.3° (chloroform). $[\alpha]_D^{20}$ +82.5° (ethanol). Absorption spectrum: Dobbie, Lauder, *J. Chem. Soc.* **83**, 626 (1903). Practically insol in water. Freely sol in alcohol, chloroform, ether, hot petr ether.

(−)-**Form.** Colorless needles from ethanol, mp 83-85°. $[\alpha]_D^{20}$ −84.8° (c = 0.466 in ethanol).

(±)-**Form.** Crystals from dil alc, mp 114-115.5°.

5437. Laurel. Bay laurel; Roman laurel; Turkish laurel; sweet bay. Evergreen shrub, *Laurus nobilis* L., Lauraceae, bearing pale yellow flowers, shiny black berries and glossy aromatic foliage. Known since ancient times as a medicinal plant and spice; often used as a symbol of victory or achievement. *Habit.* Mediterrean region. *Constit.* Leaves: volatile oil (1-3%), costunolide, laurenobiolide, catechins, proanthocyanidins, alkaloids, plant acids. Fruit: fatty oil (25-55%), sesquiterpene lactones, volatile oil (1-4%). Comprehensive description: S. Kumar *et al.* in *Handbook of Herbs and Spices*, **Vol. 1**, K. V. Peter, Ed., (Woodhead Publishing, Cambridge, 2001) pp 52-61. Composition of essential oil: M. Ozcan, J.-C. Chalchat, *J. Med. Food*, **8**, 408 (2005). Reviews of composition and uses: A. Y. Leung, S. Foster, *Encyclopedia of Common Natural Ingredients*, (Wiley-Interscience, Hoboken, 2nd Ed., 2003) pp 69-72; J. Gruenwald *et al.*, *PDR for Herbal Medicines* (Thomson PDR, Montvale, 3rd Ed., 2004) p 500.

Laurel berry oil. [8002-41-3] Fixed oil from fresh fruit. *Constit.* Chiefly lauric, palmitic, oleic acids, volatile oil, sesquiterpene lactones. Greenish, fatty solid. d ~0.88. mp ~40°. n_D^{25} 1.4783. Sapon no. 198-199. Iodine no. 68-80. Insol in water. Sparingly sol in alcohol; sol in benzene, ether, carbon disulfide.

Laurel leaf oil. Oil of sweet bay; bay leaf oil; volatile oil of laurel. Obtained by steam distillation from leaves of *Laurus nobilis* L., Lauraceae. *Constit.* Eucalyptol (30-60%), α- and β-pinene, α-terpineol, sabinene, linalool. Light yellow to yellow liquid with aromatic spicy odor. d_{25}^{25} 0.905-0.929. n_D^{20} 1.465-1.470. Rotation −10° to −19°. Acid no. not >3.0. Sapon no. 15-45; 36-85 after acetylation. Sol in most fixed oils, in 1 vol 80% alcohol, with cloudiness in mineral oil, propylene glycol. Insol in glycerin. *Keep well closed, cool and protected from light.*

Note: Should not be confused with West Indian bay, *Pimenta racemosa*, or California laurel, *Umbellularia californica*.

USE: Culinary herb (bay leaf), component of bouquet garni; flavoring agent. Fragrance component in soaps, creams, lotions, detergents, perfumes.

THERAP CAT: Rubefacient; in treatment of rheumatism and gout; carminative.

5438. Laureline. [81-38-9] (7a*R*)-6,7,7a,8-Tetrahydro-11-methoxy-7-methyl-5*H*-benzo[*g*]-1,3-benzodioxolo[6,5,4-*de*]quino-

line; 10-methoxy-1,2-(methylenedioxy)aporphine. $C_{19}H_{19}NO_3$; mol wt 309.37. C 73.77%, H 6.19%, N 4.53%, O 15.51%. From bark of *Laurelia novae-zelandiae* A. Cunn., Lauraceae: Aston, *J. Chem. Soc.* **97**, 1381 (1910). Synthesis: Schlittler, *Helv. Chim. Acta* **15**, 394 (1932); Faltis *et al.*, *Ber.* **77B**, 686 (1945); Gibson *et al.*, *J. Chem. Soc. C* **1970**, 2234; Govindachari *et al.*, *Indian J. Chem.* **8**, 475 (1970).

Cubes from petr ether. $[\alpha]_D^{20}$ −99.2° (c = 0.736 in 50% alc). Absorption spectrum: Girardet, *J. Chem. Soc.* **1931**, 2636. Feebly basic. Readily oxidizes in air. Sol in alcohol, ether. Practically insol in water.

Hydrochloride. $C_{19}H_{19}NO_3 \cdot HCl$. Crystals, mp 280°. $[\alpha]_D^{20}$ −57°. Sparingly sol in water.

(*S*)-**Form.** [65981-49-9] Cubes from petr ether, mp 114-115°. $[\alpha]_D^{20}$ +97.9° (c = 1.12 in abs alc); $[\alpha]_D^{20}$ +100° (c = 0.560 in CS_2).

(±)-**Form.** [3749-97-1] Rough needles from petr ether, mp 115-116°.

Nitrate. $C_{19}H_{19}NO_3 \cdot HNO_3$. Crystals, mp 238-240°. Sparingly sol in water.

Hydriodide. $C_{19}H_{19}NO_3 \cdot HI$. Crystals. Practically insol in water.

Oxalate. $(C_{19}H_{19}NO_3)_2 \cdot (COOH)_2$. Moderately sol in water.

5439. Lauric Acid. [143-07-7] Dodecanoic acid; dodecoic acid; laurostearic acid; 1-undecanecarboxylic acid. $C_{12}H_{24}O_2$; mol wt 200.32. C 71.95%, H 12.08%, O 15.97%. Isoln from coconut oil: Dale, Meara, *J. Sci. Food Agric.* **6**, 162 (1955); Naudet *et al.*, *Bull. Soc. Chim. Fr.* **1959**, 718; from arecanut fat: Pathak, Mathur, *J. Sci. Food Agric.* **5**, 461 (1954); from *Holoptelea integrifolia* seed oil: Badami, *ibid.* **13**, 297 (1962). Synthesis: Ballard *et al.*, US **2572238** (1951 to Shell); Langenbeck, Richter, *Ber.* **89**, 202 (1956); Sprowls, US **2782214** (1957 to Baker Castor Oil); Zapesochnoya *et al.*, *Zh. Obshch. Khim.* **33**, 2552 (1963). Toxicity study: L. Orö, A. Wretlind, *Acta Pharmacol. Toxicol.* **18**, 141 (1961).

White, crystalline powder; slight odor of bay oil, mp 44°; also reported as 48°. d_4^4 0.869. bp$_{100}$ 225°; bp$_{20}$ 160-165°. n_D^{82} 1.4183. Insol in water. 1 g dissolves in 1 ml alcohol, 2.5 ml propyl alcohol; freely sol in benzene, ether. LD$_{50}$ i.v. in mice: 131 ±5.7 mg/kg (Orö, Wretlind).

5440. Laurocapram. [59227-89-3] 1-Dodecylhexahydro-2*H*-azepin-2-one; 1-dodecylazacycloheptan-2-one; *N*-dodecyl-ε-caprolactam; N-0252; Azone. $C_{18}H_{35}NO$; mol wt 281.48. C 76.81%, H 12.53%, N 4.98%, O 5.68%. Caprolactam derivative used to enhance percutaneous absorption of physiologically active agents. Also exhibits intrinsic anti-inflammatory activity. Prepn: A. P. Swain *et al.*, *J. Org. Chem.* **18**, 1087 (1953). Improved synthesis: V. J. Rajadhyaksha *et al.*, EP **95096**; *eidem*, US **4422970** (both 1983 to Nelson). Use as skin penetrant: V. J. Rajadhyaksha, US **3989816**; US **4405616** (1976, 1983 to Nelson). Anti-inflammatory activity: E. L. Nelson, US **4310525** (1982 to Nelson). Pharmacology: R. B. Stoughton, *Arch. Dermatol.* **118**, 474 (1982). Comprehensive description: *idem*, *Drug Dev. Ind. Pharm.* **9**, 725 (1983). Topical antiviral activity *in vivo*: M. F. Leonard *et al.*, *Chemotherapy (Basel)* **33**, 151 (1987).

Clear, colorless liquid, mp $-7°$. $bp_{50\mu}$ 160°. d 0.91; n 1.4701. Insol in water. Freely sol in most organic solvents. LD_{50} rats, mice (g/kg): 8 i.v., i.p. (Stoughton).

USE: Pharmaceutic aid (excipient).

5441. Lauroguadine. [135-43-3] N,N'''-[4-(Dodecyloxy)-1,3-phenylene]bisguanidine; 1,1'-[4-(dodecyloxy)-m-phenylene]diguanidine; 2,4-diguanidinophenyl lauryl ether; 2,4-diguanidino-1-dodecyloxybenzene; 2,4-diguanidino-1-lauryloxybenzene; 2,4-diguanidinophenyl dodecyl ether; dodecyl 2,4-diguanidinophenyl ether. $C_{20}H_{36}N_6O$; mol wt 376.55. C 63.79%, H 9.64%, N 22.32%, O 4.25%. Prepn of the dihydrochloride: Pasini, *Farmaco Ed. Sci.* **8**, 646 (1953); idem, *Rend. Sci. Farmitalia* **1**, 405 (1954); **GB 730394** (1955 to Farmitalia).

Dihydrochloride monohydrate. P-7; Farmidril. $C_{20}H_{36}N_6O\cdot2HCl\cdot H_2O$; mol wt 467.48. Crystals from water + ethanol. Sinters from 135-210°, dec 250°. Soly (g/100 ml): 0.2 in cold water; ~10 in boiling water; 84.2 in methanol (25°). Practically insol in acetone, benzene. Aq solns are neutral to litmus.

THERAP CAT: Antiprotozoal (Trichomonas).

5442. Laurolinium Acetate. [146-37-2] 4-Amino-1-dodecyl-2-methylquinolinium acetate (1:1); 4-amino-1-dodecylquinaldinium acetate; 1-dodecyl-4-aminoquinaldinium acetate; Laurodin. $C_{24}H_{38}N_2O_2$; mol wt 386.58. C 74.57%, H 9.91%, N 7.25%, O 8.28%. Prepn: D. Caldwell *et al., J. Pharm. Pharmacol.* **13**, 554 (1961); D. Caldwell, L. R. Rowe, **US 2997476** (1961 to Allen & Hanburys). Toxicity data: W. A. Cox, P. F. D'Arcy, *J. Pharm. Pharmacol.* **15**, 129 (1963).

Crystals from acetone, mp 170-171°. Freely sol in water. About 1 g dissolves in 2 ml of water at 20°. LD_{50} in mice (mg/kg): 131 ±36.2 orally; 30.2 ±5.6 s.c.; 6.0 ±04 i.v.; 2.3 ±0.2 i.p. (Cox, D'Arcy).

THERAP CAT: Antiseptic.

5443. Laurotetanine. [128-76-7] (6aS)-5,6,6a,7-Tetrahydro-1,2,10-trimethoxy-4H-dibenzo[de,g]quinolin-9-ol; 1,2,10-trimethoxy-6aα-noraporphin-9-ol; Litsoeine. $C_{19}H_{21}NO_4$; mol wt 327.38. C 69.71%, H 6.47%, N 4.28%, O 19.55%. From the bark of *Litsea citrata* Blume (*Tetranthera citrata* (Blume) Nees), *Lauraceae* and allied plants. Isoln: Greshoff, *Ber.* **23**, 3537 (1890); Filippo, *Arch. Pharm.* **236**, 601 (1898). Structure: Barger *et al., Ber.* **66**, 450 (1933). Synthesis: Kikkawa, *C.A.* **53**, 17163i (1959).

Monohydrate. Needles, mp 125°. $[\alpha]_D^{25}$ +98.5°. Practically insol in water. Freely sol in alcohol, chloroform, ethyl acetate, slightly in ether.

5444. Lauryl Bromide. [143-15-7] 1-Bromododecane; dodecyl bromide. $C_{12}H_{25}Br$; mol wt 249.24. C 57.83%, H 10.11%, Br 32.06%. Prepd by the action of hydrobromic acid on primary n-lauryl alcohol in the presence of sulfuric acid: Kamm, Marvel, *Org. Synth.* **1**, 7 (1921).

Liquid. bp_{45} 175-180°. Insol in water. Sol in alc, ether.

5445. Lavender. Garden lavender; true lavender. Aromatic, woody perennial, *Lavandula angustifolia* Mill. (syn. *L. officinalis* Chaix.), *Labiatae*. Parts used are the fresh or dried flowering tops and the essential oil. Also used for oil production are *spike lavender* (*L. latifolia*, syn. *L. spica*) and *lavandin* referring to hybrid varieties including *L. abrialis* and *L. hybridia*. *Habit.* Mediterranean region; cultivated widely. *Constit.* Volatile oil (1-3%), tannins, coumarin, umbelliferone, herniarin, flavonoids. Comprehensive description: M. T. Lis-Balchin in *Handbook of Herbs and Spices*, **Vol. 2**, K. V. Peter, Ed., (Woodhead Publishing, Cambridge, 2004) pp 179-195. Characterization of essential oil: R. Shellie *et al., J. Chromatogr. A* **970**, 225 (2002); A. R. Fakhari *et al., ibid.* **1098**, 14 (2005). Review of clinical studies of psychological effects: M. Kirk-Smith, *Int. J. Aromather.* **13**, 82-89 (2003); of medicinal uses: J. Gruenwald *et al., PDR for Herbal Medicines* (Thomson PDR, Montvale, 3rd Ed., 2004) pp 285-288.

Lavender oil. [8000-28-0] Oil of lavender. Volatile oil obtained by steam distillation of fresh flowering tops of *L. angustifolia*. *Constit.* Complex mixture of components, chiefly linalool (20-50%), linalyl acetate (30-40%), terpinen-4-ol, lavandulyl acetate, lavandulol. Colorless or yellow liquid. d_{25}^{25} 0.875-0.888. n_D^{20} 1.459-1.470. Rotation: $-3°$ to $-10°$. Slightly sol in water; sol in 4 vols 70% alcohol. Insol in propylene glycol. *Keep well closed, cool and protected from light.*

Lavandin oil. [8022-15-9] Abrial lavandin oil. Volatile oil obtained from flowering tops of *L. abrialis* or other hybrid varieties. *Constit.* Linalool, linalyl acetate, eucalyptol, camphor. Pale yellow to yellow liquid with a slight, camphoraceous odor. d_{25}^{25} 0.885-0.893. n_D^{20} 1.460-1.464. Rotation: $-2°$ to $-5°$. Sol in most fixed oils, propylene glycol, with opalescence in mineral oil, in 2 vols 70% alcohol. Relatively insol in glycerin.

Spike lavender oil. [8016-78-2] Oil of spike. Volatile oil from dried flowers of *L. latifolia*. *Constit.* 40-60% eucalyptol and camphor, linalool, linalyl acetate (1%). Pale yellow to yellow liquid; camphoraceous lavender odor. d_{25}^{25} 0.893-0.909. Rotation: $-5°$ to +5°. n_D^{20} 1.463-1.468. Sol in most fixed oils, propylene glycol, in 3 vols 70% alcohol; slightly sol in glycerin, mineral oil. *Keep well closed, cool and protected from light.*

USE: Fragrance component in cosmetics, soaps, lotions, perfumes. Flavor in beverages, desserts, aromatic vinegars. Oil in aromatherapy for a soothing, calming effect.

THERAP CAT: Carminative, astringent, antiseptic.

5446. Lawesson's Reagent. [19172-47-5] 2,4-Bis(4-methoxyphenyl)-1,3,2,4-dithiadiphosphetane 2,4-disulfide; anisyldithiophosphinic anhydride. $C_{14}H_{14}O_2P_2S_4$; mol wt 404.45. C 41.58%, H 3.49%, O 7.91%, P 15.32%, S 31.71%. Thionating agent for conversion of carbonyls to thiocarbonyls. Prepn: H. Z. Lecher *et al., J. Am. Chem. Soc.* **78**, 5018 (1956); and delineation of use: B. S. Pedersen *et al., Bull. Soc. Chim. Belg.* **87**, 223 (1978); S. Scheibye *et al., ibid.* 229. Mechanistic study: T. B. Rauchfuss, G. A. Zank, *Tetrahedron Lett.* **27**, 3445 (1986). Use in thionations: A. Z.-Q. Khan, J. Sandström, *J. Chem. Soc. Perkin Trans. 1* **1988**, 2085; S. Araki *et al., Bull. Chem. Soc. Jpn.* **61**, 2977 (1988); as reducing agent for sulfoxides: H. Bartsch, T. Erker, *Tetrahedron Lett.* **33**, 199 (1992). Review of thionation reactions: M. P. Cava, M. I. Levinson, *Tetrahedron* **41**, 5061-5087 (1985); of applications in organic and organometallic syntheses: M. Jesberger *et al., Synthesis* **2003**, 1929-1958.

mp 229°, should be stored under anhydrous conditions.
USE: Thiation reagent.

5447. Lawrencium. [22537-19-5] Lr; formerly Lw; at. no. 103; valence 3. Man-made radioactive element. No stable nuclides; known isotopes (mass numbers): 253-262. Longest-lived known isotope: 262 ($T_{1/2}$ 216 minutes, rel. at. mass 262.110). Prepn of first isotope ^{257}Lr ($T_{1/2}$ 0.65 sec, α-emitter): A. Ghiorso *et al.*, *Phys. Rev. Lett.* **6**, 473 (1961). Prepn of isotope originally assigned mass number 257, later changed to 258 ($T_{1/2}$ 4.3 seconds, α-emitter) by bombardment of Cf with boron ions: Eskola *et al.*, *Phys. Rev. C* **4**, 632 (1971). Prepn of isotopes including ^{256}Lr ($T_{1/2}$ 28 seconds) by irradiating ^{243}Am with ^{18}O ions: Donets *et al.*, *At. Energ.* **19**, 109 (1965), *C.A.* **64**, 1542c (1966). *Reviews:* C. Keller, *The Chemistry of the Transuranium Elements* (Verlag Chemie, Weinheim, English Ed., 1971) pp 609-612; Silva, "Trans-Curium Elements" in *MTP Int. Rev. Sci.: Inorg. Chem., Ser. One* vol. 8, A. G. Maddock, Ed. (University Park Press, Baltimore, 1972) pp 71-105; Ghiorso, *Handb. Exp. Pharmakol.* **36**, 691-715 (1973); Taylor, *ibid.*, *Phys. Rev.* 717-738; R. J. Silva in *The Chemistry of the Actinide Elements* vol. 2, J. J. Katz *et al.*, Eds. (Chapman and Hall, New York, 1986) pp 1099-1103; *The Elements Beyond Uranium*, G. T. Seaborg, W. D. Loveland, Eds. (John Wiley & Sons, Inc., New York, 1990) p 49-51.
Caution: Radiation hazard; handling requires special equipment and shielding facilities (Katz *et al.*, *loc. cit.* **vol. 2**, p. 1128).

5448. Lawsone. [83-72-7] 2-Hydroxy-1,4-naphthalenedione; 2-hydroxy-1,4-naphthoquinone; C.I. 75480; C.I. Natural Orange 6. $C_{10}H_6O_3$; mol wt 174.16. C 68.97%, H 3.47%, O 27.56%. From leaves of *Lawsonia inermis* L. and *L. alba* Lam., *Lythraceae:* Latif, *Indian J. Agric. Sci.* **29**, No. 2-3, 147 (1959), *C.A.* **55**, 14828g (1961). Synthesis: Fieser, *J. Am. Chem. Soc.* **70**, 3165 (1948); Jain, Seshadri, *Proc. Indian Acad. Sci.* **35A**, 233 (1952); Eistert, Müller, *Ber.* **92**, 2071 (1959).

Yellow prisms from acetic acid, dec 195-196°.
THERAP CAT: Ultraviolet screen.

5449. Lazabemide. [103878-84-8] *N*-(2-Aminoethyl)-5-chloro-2-pyridinecarboxamide; *N*-(2-aminoethyl)-5-chloropicolinamide. $C_8H_{10}ClN_3O$; mol wt 199.64. C 48.13%, H 5.05%, Cl 17.76%, N 21.05%, O 8.01%. Member of a novel class of potent, reversible, and extremely selective monoamine oxidase type B (MAO-B) inhibitors. Prepn: R. Imhof, E. Kyburz, **GB 2163746**; *eidem*, **US 4764522** (1986, 1988 both to Hoffmann-LaRoche). HPLC determn in plasma: R. Wyss, W. Philipp, *J. Chromatogr.* **507**, 187 (1990). Neurochemical profile: W. E. Haefely *et al.*, *Adv. Neurol.* **53**, 505 (1990). Enzyme binding studies: J. Saura *et al.*, *J. Neurosci.* **12**, 1977 (1992). Clinical trial: Parkinson Study Group, *Ann. Neurol.* **33**, 350 (1993).

Monohydrochloride. [103878-83-7] Ro-19-6327. $C_8H_{10}ClN_3$-O.HCl; mol wt 236.10. Crystals from methanol/ether, mp 193-195°. Partition coefficient (*n*-octanol/water) ~0.1. pKa 8.9. LD_{50} orally in mice: 1000-2000 mg/kg (Imhof, Kyburz, 1986).
THERAP CAT: Antiparkinsonian.

5450. Lazurite. [1302-83-6] Lapis lazuli; lasurite. Composition: $(Na,Ca)_4(AlSiO_4)_3(SO_4,S,Cl)$, E. S. Dana, *A System of Mineralogy* (John Wiley, New York, 6th ed., 1901) pp 432-433; C. S. Hurlbut, Jr., *Dana's Manual of Mineralogy* (John Wiley, New York, 17th ed., 1959) p 503.

Blue, blue-violet or greenish-blue, translucent, cubic or dodecahedral crystals. d 2.4. Dec by HCl with pptn of SiO_2 and evolution of H_2S.
USE: In manuf of vases, ornamental furniture, mosaics; in paints, jewelry.

5451. Lead. [7439-92-1] Pb; at. wt 207.2; at. no. 82; valence 2, 4. Group IVA (14). Four naturally occurring isotopes: 204 (1.40%); 206 (25.2%); 207 (21.7%); 208 (51.7%); artificial, radioactive isotopes: 195-203; 205; 209-214. One of the metals known to the ancient world. Extent of occurrence in earth's crust about 15 g/ton, also expressed as 0.002% (depth of crust: 16 km). Occurs chiefly as sulfide in *galena*, other minerals include *anglesite* $(PbSO_4)$, *cerussite* $(PbCO_3)$, *mimetite* $[PbCl_2.3Pb_3(AsO_4)_2]$ and *pyromorphite* $[PbCl_2.3Pb_3(PO_4)_2]$. Recovery from ore and purification: Heuser, *Metall.* **9**, 675 (1955); Ziegfeld, *Eng. Min. J.* **153**, 82 (1952). Prepn of high purity lead: Piontelli, Fagnani, *Chim. Ind. (Milan)* **34**, 629 (1952), *C.A.* **47**, 12062 (1953); Giesen, *Technik* **2**, 393 (1947), *C.A.* **42**, 852 (1948); Hughes *J. Electrochem. Soc.* **101**, 267 (1954); Baralis, Marone, *Met. Ital.* **59**, 494 (1967), *C.A.* **67**, 119613a (1967). Review of uses, corrosion metallurgy: Mullarkey, *Ind. Eng. Chem.* **49**, 1607 (1957). Reviews of lead, its alloys and compds: W. Hofmann, *Lead and Lead Alloys, Properties and Technology* (Springer, New York, Eng. Ed., 1970) 551 pp; Abel in *Comprehensive Inorganic Chemistry* vol. 2, J. C. Bailar, Jr. *et al.*, Eds. (Pergamon Press, Oxford, 1973) pp 105-146; H. E. Howe in *Kirk-Othmer Encyclopedia of Chemical Technology* vol. 14 (Wiley-Interscience, New York, 3rd ed., 1981) pp 98-139. Review of carcinogenic risk: *IARC Monographs* **23**, 325-415 (1980); of toxicology and human exposure: *Toxicological Profile for Lead* (PB2008-100007, 2008) 582 pp.
Bluish-white, silvery, gray metal. Highly lustrous when freshly cut, tarnishes upon exposure to air. Very soft and malleable, easily melted, cast, rolled, and extruded. Cubic crystal structure. mp 327.4°; bp 1740°. d_4^{20} 11.34; d (at mp) 10.65: Schneider *et al.*, *Naturwissenschaften* **41**, 326 (1954). Heat of vaporization (1740°) 206 cal/g. Heat capacity (20°): 0.031 cal/g/°C. Resistivity (μ-ohm-cm) at 20°: 20.65; at 100°: 27.02; at 320°: 54.76; at 330°: 96.74. Vapor pressure at 1000°: 1.77 mm Hg. E° (aq) Pb/Pb^{2+} +0.126 v. Coefficient of linear expansion (0-100°) 29×10^{-6}, (20-300°) 31.3×10^{-6}, ($-183°$ to $+14°$) 27×10^{-6}; thermal conductivity varies from 0.083 at 50° to 0.077 at 225°: Francl, Kingery, *J. Am. Ceram. Soc.* **37**, 80 (1954). Viscosity of molten lead (cP): 3.2 (327.4°); 2.32 (400°); 1.54 (600°); 1.23 (800°). Heat capacity and heat of fusion study: Douglas, Dever, *J. Am. Chem. Soc.* **76**, 4824 (1954); hardness 1 on Mohs' scale; Brinell hardness (high purity Pb) 4.0: McLellan, *Am. Mineral.* **30**, 635 (1945). Reacts with hot concd nitric acid, with boiling concd hydrochloric or sulfuric acid. Attacked by pure water, weak organic acids in the presence of oxygen. Resistant to tap water, hydrofluoric acid, brine, solvents.
Caution: Potential symptoms of overexposure are lassitude, insomnia; facial pallor; anorexia, weight loss, malnutrition; constipation, abdominal pain, colic; anemia; gingival lead line; tremor; paralysis of wrists or ankles; kidney disease; hypotension. Symptoms in children include encephalopathy, behavioral and learning deficits; in adults, peripheral motor neuropathy. *See NIOSH Pocket Guide to Chemical Hazards* (PB2003-100121, 2003) p 184; *Patty's Industrial Hygiene and Toxicology* vol 2C, G. D. Clayton, F. E. Clayton, Eds. (Wiley-Interscience, New York, 4th ed., 1994) pp 2065-2087. Lead and lead compounds are reasonably anticipated to be human carcinogens: *Report on Carcinogens, Twelfth Edition* (PB2011-111646, 2011) p 251.
USE: Construction material for tank linings, piping, and other equipment handling corrosive gases and liqs used in the manuf of sulfuric acid, petr refining, halogenation, sulfonation, extraction, condensation; for x-ray and atomic radiation protection; manuf of tetraethyllead, pigments for paints, and other organic and inorganic lead compds; bearing metal and alloys; storage batteries; in ceramics, plastics, and electronic devices; in building construction; in solder and other lead alloys; in the metallurgy of steel and other metals.

5452. Lead Acetate. [301-04-2] Acetic acid lead(2+) salt (2:1); neutral lead acetate; lead(II) acetate; plumbous acetate; normal lead acetate; sugar of lead; salt of Saturn. $C_4H_6O_4Pb$; mol wt 325.29. C 14.77%, H 1.86%, O 19.67%, Pb 63.70%. Pb(CH$_3$-COO)$_2$. Toxicity data: W. R. Bradley, W. G. Fredrick, *Ind. Med.*

10, Ind. Hyg. Sect. **2**, 15 (1941). Review of safety assessment in hair coloring: A. J. Cohen, F. J. C. Roe, *Food Chem. Toxicol.* **29**, 485-507 (1991); of toxicology and human exposure: *Toxicological Profile for Lead* (PB2008-100007, 2008) 582 pp.

White or colorless crystals or flakes. d 3.12.

Trihydrate. [6080-56-4] $C_4H_6O_4Pb.3H_2O$; mol wt 379.33. Colorless crystals or white granules or powder; slight acetic odor; slowly effloresces. *Poisonous*. Takes up CO_2 from air and becomes incompletely sol. d 2.55. mp 75° when rapidly heated; at a little above 100° it begins to lose acetic acid; dec completely above 200°. One gram dissolves in 1.6 ml water, 0.5 ml boiling water, 30 ml alcohol; freely sol in glycerol. Aq solns of lead acetate dissolve lead monoxide. pH of 5% aq soln at 25° = 5.5-6.5. *Keep well closed.* LD_{50} i.p. in rats: 15 mg Pb/100g (Bradley, Fredrick).

Caution: This substance is reasonably anticipated to be a human carcinogen: *Report on Carcinogens, Twelfth Edition* (PB2011-111646, 2011) p 251.

USE: Mordant in cotton dyes; lead coating for metals; drier in paints, varnishes and pigment inks; colorant in hair dyes. Weighting silks; manuf lead salts, chrome-yellow; as analytical reagent for detection of sulfide, determination of CrO_3, MoO_3.

THERAP CAT: Astringent.

THERAP CAT (VET): Astringent and sedative (usually in lotions) for bruises and superficial inflammation. Has been used internally in diarrheas.

5453. Lead Antimonate(V). [13510-89-9] Antimony lead oxide ($Sb_2Pb_3O_8$); antimonic acid (H_3SbO_4) lead(2+) salt (2:3); Naples yellow.

Orange-yellow powder. Insol in water, dil acids.

Caution: Lead and all lead compounds are listed as reasonably anticipated to be human carcinogens: *Report on Carcinogens, Twelfth Edition* (PB2011-111646, 2011) p 251.

USE: As pigment in oil painting, staining glass, crockery and porcelain.

5454. Lead Arsenate. [7784-40-9] Arsenic acid (H_3AsO_4) lead(2+) salt (1:1); acid lead arsenate; lead hydroarsenate; lead hydrogen arsenate. $HAsO_4Pb$; mol wt 347.13. H 0.29%, As 21.58%, O 18.44%, Pb 59.69%. $PbHAsO_4$. Common form of lead arsenate; occurs in nature as the mineral, schultenite. Prepd by adding excess lead nitrate to a soln of sodium arsenate: H. V. Tartar, R. H. Robinson, *J. Am. Chem. Soc.* **36**, 1843 (1914); *Mellor's* **Vol. IX**, 193 (1929). Insecticidal properties: L. B. Norton, *Ind. Eng. Chem.* **40**, 691 (1948). Industrial prepn: R. J. Thrift, US 2549945 (1951 to DuPont). Ferroelectric properties: B. B. Lavrencic, J. Petzelt, *J. Chem. Phys.* **67**, 3890 (1977); N. Kida *et al.*, *J. Electron Spectrosc. Relat. Phenom.* **101-103**, 603 (1999). Synthesis and characterization of nanocrystals: Z. Xiu *et al.*, *Mater. Res. Bull.* **39**, 2019 (2004). Analysis and distribution in contaminated soils: K. Newton *et al.*, *Environ. Pollut.* **143**, 197 (2006).

White, heavy powder. *Poisonous*. d_4^{20} 5.786. At ~280° loses H_2O and is converted into pyroarsenate. Insol in water. Sol in HNO_3, caustic alkalies. LD_{50} in rats, rabbits (mg/kg): 825, 125 orally; *see* J. L. Voight *et al.*, *J. Am. Pharm. Assoc. Sci. Ed.* **37**, 122 (1948).

Schultenite. [14758-11-3] Description: L. J. Spencer, *Nature* **118**, 411 (1926); R. Falls *et al.*, *Mineral. Mag.* **49**, 65 (1985). Colorless through white to yellowish, transparent, euhedral crystals with white streak. Hardness: 2.5. Refractive index: 1.9255. d 6.06.

Caution: Lead and all lead compounds are listed as reasonably anticipated to be human carcinogens: *Report on Carcinogens, Twelfth Edition* (PB2011-111646, 2011) p 251.

USE: Formerly in insecticidal sprays for fruit orchards.

5455. Lead Arsenite. [10031-13-7] Arsenous acid lead(2+) salt. $As_2H_2O_4Pb$; mol wt 423.06. As 35.42%, H 0.48%, O 15.13%, Pb 48.98%. $Pb(AsO_2)_2$.

White powder. *Poisonous*. d 5.85. Sol in dil HNO_3. Insol in water.

Caution: Lead and all lead compounds are listed as reasonably anticipated to be human carcinogens: *Report on Carcinogens, Twelfth Edition* (PB2011-111646, 2011) p 251.

USE: As insecticide like the arsenate.

5456. Lead Azide. [13424-46-9] Lead azide ($Pb(N_3)_2$). N_6Pb; mol wt 291.24. N 28.86%, Pb 71.14%. $Pb(N_3)_2$. Prepd from sodium azide and lead nitrate: Schenk in *Handbook of Preparative Inorganic Chemistry* vol. 1, G. Brauer, Ed. (Academic Press, New York, 2nd ed., 1963) p 763. Review: B. T. Fedoroff *et al.*, *Encyclopedia of Explosives and Related Items* **vol. 1** (Picatinny Arsenal, Dover, N.J., 1960) pp A545-A587. Review of toxicology and human exposure: *Toxicological Profile for Lead* (PB2008-100007, 2008) 582 pp.

Needles or white powder. Explodes at 350° or on percussion. Heat of formation (25°): +110.5 kcal/mol. Freely sol in acetic acid. Soly in water: 0.023% at 18°; 0.09% at 70°. Insol in NH_4OH.

Caution: Lead and all lead compounds are listed as reasonably anticipated to be human carcinogens: *Report on Carcinogens, Twelfth Edition* (PB2011-111646, 2011) p 251.

USE: As primer in explosives. In the form of dextrinated lead azide.

5457. Lead Borate. [14720-53-7]; [10214-39-8] (monohydrate). Boric acid (HBO_2) lead(2+) salt (2:1). B_2O_4Pb; mol wt 292.82. B 7.38%, O 21.86%, Pb 70.76%. $Pb(BO_2)_2$.

White powder. *Poisonous*. Sol in dil HNO_3. Insol in water.

Caution: Lead and all lead compounds are listed as reasonably anticipated to be human carcinogens: *Report on Carcinogens, Twelfth Edition* (PB2011-111646, 2011) p 251.

USE: Drier for varnishes and paints; with other metals (e.g., Ag) in galvanoplasty for production of conducting coatings on glass, pottery, porcelain, and chinaware.

5458. Lead Bromide. [10031-22-8] Lead bromide ($PbBr_2$); lead(II) bromide; plumbous bromide. Br_2Pb; mol wt 367.01. Br 43.54%, Pb 56.46%. $PbBr_2$. Review of toxicology and human exposure: *Toxicological Profile for Lead* (PB2008-100007, 2008) 582 pp.

White, cryst powder. *Poisonous*. d 6.66. mp 373°. On solidifying forms a horn-like mass. Sol in about 200 parts cold water, 20 parts boiling water. Insol in alc.

Caution: Lead and all lead compounds are listed as reasonably anticipated to be human carcinogens: *Report on Carcinogens, Twelfth Edition* (PB2011-111646, 2011) p 251.

5459. Lead Chloride. [7758-95-4] Lead chloride ($PbCl_2$); lead(2+) chloride; lead(II) chloride; plumbous chloride. Cl_2Pb; mol wt 278.10. Cl 25.49%, Pb 74.51%. $PbCl_2$. Occurs in nature as the mineral, *cotunnite*. Toxicity data: Tartler, *Arch. Hyg.* **125**, 273 (1941). Review of toxicology and human exposure: *Toxicological Profile for Lead* (PB2008-100007, 2008) 582 pp.

White, cryst powder. *Poisonous*. d 5.85. mp 501°. bp 950°. Sol in 93 parts cold water, 30 parts boiling water; readily sol in soln of NH_4Cl, NH_4NO_3, alkali hydroxides; slowly in glycerol. MLD in guinea pigs (mg/kg): 1500-2000 orally (Tartler).

Caution: Lead and all lead compounds are listed as reasonably anticipated to be human carcinogens: *Report on Carcinogens, Twelfth Edition* (PB2011-111646, 2011) p 251.

USE: Manuf Pattison's white lead, Verona Yellow, Turner's Patent Yellow, lead oxychloride; as solder and flux.

5460. Lead Chromate(VI). [7758-97-6] Chromic acid (H_2CrO_4) lead(2+) salt (1:1); Chrome yellow; Cologne yellow; King's yellow; Leipzig yellow; Paris yellow; C.I. Pigment Yellow 34; lead chromium oxide ($PbCrO_4$); plumbous chromate; C.I. 77600. CrO_4Pb; mol wt 323.19. Cr 16.09%, O 19.80%, Pb 64.11%. $PbCrO_4$. Occurs in nature as the minerals, *crocoite* and *phoenicochroite*. Ref: *Colour Index* **vol. 4** (3rd ed., 1971) p 4677. Evaluation of carcinogenic risk: *IARC Monographs* **23**, 205-324 (1980); *ibid.* **49**, 49-256 (1990). Review of environmental exposure and bioavailability: D. Waldron, *Am. Ink Maker* **72**, 50-60 (1994); of toxicology and human exposure: *Toxicological Profile for Lead* (PB2008-100007, 2008) 582 pp.

Yellow or orange-yellow powder. d 6.3. mp 844°. It is one of the most insol salts (0.2 mg/l H_2O). Sol in solns of fixed alkali hydroxides, in dil HNO_3. Insol in acetic acid.

Caution: Chromium hexavalent (VI) compounds are listed as known human carcinogens: *Report on Carcinogens, Twelfth Edition* (PB2011-111646, 2011) p 106.

USE: Pigment for paints and inks; in oil and water colors; printing fabrics, decorating china and porcelain; in chemical analysis of organic substances; in traffic paints.

5461. Lead Chromate(VI) Oxide. [18454-12-1] Lead chromate oxide ($Pb_2(CrO_4)O$); chromic acid lead(2+) salt (1:2); basic lead chromate; red lead chromate; chrome red; chromium lead oxide; Persian red; Austrian cinnabar. $CrPb_2O_5$; mol wt 546.39. Cr 9.52%, Pb 75.84%, O 14.64%. $PBCrO_4.PbO$. *See: Colour Index* **vol. 4,** (3rd ed., 1971) p 4677.

Red powder. Insol in water.

Caution: Lead and all lead compounds are listed as reasonably anticipated to be human carcinogens: *Report on Carcinogens, Twelfth Edition* (PB2011-111646, 2011) p 251.

USE: As pigment.

5462. Lead Dioxide. [1309-60-0] Lead oxide (PbO_2); lead-(IV) oxide; lead oxide brown; lead peroxide; lead superoxide; plumbic oxide. O_2Pb; mol wt 239.20. O 13.38%, Pb 86.62%. Occurs in nature as the mineral, *plattnerite*. Lab prepn from lead acetate and calcium hypochlorite: Newell, Maxson, *Inorg. Synth.* **1**, 45 (1939); by hydrolysis of lead acetate: Kuhn, Hammer, *Ber.* **83**, 413 (1950). Review of toxicology: B. Venugopal, T. D. Luckey, *Environ. Qual. Saf.* Suppl. 1, 4-73 (1975).

Dark-brown powder; evolves oxygen when heated, first forming Pb_3O_4, at high temp PbO. d 9.38. *Oxidizer.* Sol in HCl with evolution of Cl; in dil HNO_3 in presence of H_2O_2, oxalic acid, or other reducers; sol in alkali iodide solns with liberation of iodine; soluble in hot caustic alkali solns. Insol in water. LD_{50} i.p. in guinea pigs: 220 mg/kg (Venugopal, Luckey).

Caution: Lead and all lead compounds are listed as reasonably anticipated to be human carcinogens: *Report on Carcinogens, Twelfth Edition* (PB2011-111646, 2011) p 251.

USE: Electrodes in batteries; oxidizing agent in manuf dyes; as discharge in dyeing with indigo; manuf rubber substitutes; with amorphous phosphorus as ignition surface for matches; pyrotechny; manuf pigments; in analytical chemistry.

5463. Lead Fluoride. [7783-46-2] Lead fluoride (PbF_2); lead difluoride; plumbous fluoride. F_2Pb; mol wt 245.20. F 15.50%, Pb 84.50%. PbF_2. Prepd by treating lead carbonate or hydroxide with hydrogen fluoride and evaporating the soln; by mixing solns of potassium fluoride and lead acetate; by precipitation from the soln of a lead salt by HF; by the action of fluorine on lead: Ruff, *Die Chemie des Fluors* (Springer, Berlin, 1920) p 33; Eméléus in *Fluorine Chemistry* **vol. 1,** J. H. Simons, Ed. (Academic Press, New York, 1950) p 51; Kwasnik in *Handbook of Preparative Inorganic Chemistry* **vol. 1,** G. Brauer, Ed. (Academic Press, New York, 1963) pp 218-219. *Review:* Kemmitt, Sharp, *Adv. Fluorine Chem.* **4**, 188 (1965).

White to colorless crystals. *Poisonous.* Dimorphous: orthorhombic, converted to cubic above 316°. d (orthorhombic) 8.445; d (cubic) 7.750. mp 824°. bp 1293°. Soly in water (g/100 ml): 0.057 (0°); 0.065 (20°). The soly increases in the presence of HNO_3 or nitrates.

Caution: Lead and all lead compounds are listed as reasonably anticipated to be human carcinogens: *Report on Carcinogens, Twelfth Edition* (PB2011-111646, 2011) p 251.

5464. Lead Formate. [811-54-1] Formic acid lead(2+) salt (2:1); lead diformate; lead(II) formate. $C_2H_2O_4Pb$; mol wt 297.23. C 8.08%, H 0.68%, O 21.53%, Pb 69.71%. $Pb(CHO_2)_2$.

White, lustrous prisms or needles. *Poisonous.* d 4.63. Dec at 190°. Sol in 65 parts cold water, 6 parts boiling water. Insol in alc.

Caution: Lead and all lead compounds are listed as reasonably anticipated to be human carcinogens: *Report on Carcinogens, Twelfth Edition* (PB2011-111646, 2011) p 251.

5465. Lead Hexafluorosilicate. [25808-74-6] Hexafluorosilicate(2−) lead(2+) (1:1); lead fluosilicate; lead silicofluoride. F_6Pb-Si; mol wt 349.28. F 32.64%, Pb 59.32%, Si 8.04%. $PbSiF_6$.

Dihydrate. Colorless crystals. Sol in water. *Poisonous.*

Caution: Lead and all lead compounds are listed as reasonably anticipated to be human carcinogens: *Report on Carcinogens, Twelfth Edition* (PB2011-111646, 2011) p 251.

USE: In refining lead by electrolytic methods.

5466. Lead Iodide. [10101-63-0] Lead iodide (PbI_2); lead(II) iodide; plumbous iodide. I_2Pb; mol wt 461.01. I 55.05%, Pb 44.94%. PbI_2. Review of toxicology and human exposure: *Toxicological Profile for Lead* (PB2008-100007, 2008) 582 pp.

Bright yellow, heavy odorless powder. *Poisonous.* d 6.16. mp 402°. One gram dissolves in 1350 ml cold, 230 ml boiling water; freely sol in soln of sodium thiosulfate; sol in 200 parts cold 90 parts hot aniline, in concd solns of alkali iodides. Insol in alc or cold HCl. *Protect from light.*

Caution: Lead and all lead compounds are listed as reasonably anticipated to be human carcinogens: *Report on Carcinogens, Twelfth Edition* (PB2011-111646, 2011) p 251.

USE: Bronzing, gold pencils, mosaic gold, printing, photography.

5467. Lead Molybdate(VI). [10190-55-3] Lead molybdenum oxide ($PbMoO_4$); plumbous molybdate. MoO_4Pb; mol wt 367.15. Mo 26.13%, O 17.43%, Pb 56.43%. $PbMoO_4$. Occurs as the mineral *wulfenite*.

Yellow powder. Sol in HNO_3 or NaOH when freshly pptd. Insol in water.

Caution: Lead and all lead compounds are listed as reasonably anticipated to be human carcinogens: *Report on Carcinogens, Twelfth Edition* (PB2011-111646, 2011) p 251.

USE: In pigments.

5468. Lead Monoxide. [1317-36-8] Lead oxide (PbO); lead-(II) oxide; lead oxide yellow; plumbous oxide; litharge; massicot; lead protoxide. OPb; mol wt 223.20. O 7.17%, Pb 92.83%. PbO. Prepn of high purity PbO: Kwestroo, Huizing, *J. Inorg. Nucl. Chem.* **27**, 1591 (1965); Kwestroo *et al., ibid.* **29**, 39 (1967). Mfg processes: *Faith, Keyes & Clark's Industrial Chemicals,* F. A. Lowenheim, M. K. Moran, Eds. (Wiley-Interscience, New York, 4th ed., 1975) pp 509-513. Toxicity study: W. R. Bradley, W. G. Fredrick, *Ind. Med.* **10**, Ind. Hyg. Sect. **2**, 15 (1941). Review of toxicology and human exposure: *Toxicological Profile for Lead* (PB2008-100007, 2008) 582 pp.

Exists in two forms: red to reddish-yellow, tetragonal crystals, stable at ordinary temp; yellow, orthorhombic crystals, stable above 489°: Petersen, *J. Am. Chem. Soc.* **63**, 2617 (1941). *Poisonous.* At 300-450° in the air, it converts slowly into Pb_3O_4, but at higher temp reverts to PbO. d 9.53. mp 888°. Sol in acetic acid, dil HNO_3, in warm solns of fixed alkali hydroxides. Insol in water, alc. LD_{50} i.p. in rats: 40 mg Pb/100g (Bradley, Fredrick).

Caution: Lead and all lead compounds are listed as reasonably anticipated to be human carcinogens: *Report on Carcinogens, Twelfth Edition* (PB2011-111646, 2011) p 251.

USE: In ointments, plasters; preparing soln of lead subacetate. Glazing pottery; glass flux for painting on porcelain and glass; lead glass; varnishes; with glycerol as metal cement; producing iridescent colors on brass and bronze; coloring sulfur-containing substances, *e.g.,* hair, nails, wool, horn; manuf artificial tortoise shell and horn; pigment for rubber; manuf boiled linseed oil; in assay of gold and silver ores.

5469. Lead Nitrate. [10099-74-8] Nitric acid lead(2+) salt (2:1); plumbous nitrate. N_2O_6Pb; mol wt 331.21. N 8.46%, O 28.98%, Pb 62.56%. $Pb(NO_3)_2$. Review of toxicology and human exposure: *Toxicological Profile for Lead* (PB2008-100007, 2008) 582 pp.

White or colorless translucent crystals. *Oxidizer; poisonous.* d 4.53. One g dissolves in 2 ml cold, 0.75 ml boiling water, in 2500 ml abs alcohol, 75 ml abs methanol. Insol in concd HNO_3. The aq soln is slightly acid. pH of 20% aq soln at 25° = 3.0-4.0.

Caution: Lead and all lead compounds are listed as reasonably anticipated to be human carcinogens: *Report on Carcinogens, Twelfth Edition* (PB2011-111646, 2011) p 251.

USE: Manuf matches and special explosives; as mordant in dyeing and printing on textiles; mordant for staining horn, mother-of-pearl; oxidizer in dye industry; sensitizer in photography; process engraving. Titrant for complexometric titration.

THERAP CAT (VET): Has been used as a caustic in equine canker.

5470. Lead Oxalate. [814-93-7] Ethanedioic acid lead(2+) salt (1:1); lead(II) oxalate. C_2O_4Pb; mol wt 295.22. C 8.14%, O 21.68%, Pb 70.18%. $Pb(COO)_2$.

White, heavy powder. Dec at 300°. d 5.28. *Poisonous.* Sol in dil HNO_3, fixed alkali hydroxides; sparingly sol in acetic acid. Insol in water.

Caution: Lead and all lead compounds are listed as reasonably anticipated to be human carcinogens: *Report on Carcinogens, Twelfth Edition* (PB2011-111646, 2011) p 251.

5471. Lead Phosphate. [7446-27-7] Phosphoric acid lead-(2+) salt (2:3); lead(II) phosphate. $O_8P_2Pb_3$; mol wt 811.54. O 15.77%, P 7.63%, Pb 76.60%. $Pb_3(PO_4)_2$. Review of toxicology and human exposure: *Toxicological Profile for Lead* (PB2008-100007, 2008) 582 pp.
White powder. d 6.9. mp 1014°. *Poisonous.* Sol in HNO_3, fixed alkali hydroxides. Insol in water, alc.
Caution: This substance is reasonably anticipated to be a human carcinogen: *Report on Carcinogens, Twelfth Edition* (PB2011-111646, 2011) p 251.
USE: Stabilizer for styrene and casein plastics; in special glasses.

5472. Lead Sesquioxide. [1314-27-8] Lead oxide (Pb_2O_3); lead trioxide; plumbous plumbate. O_3Pb_2; mol wt 462.40. O 10.38%, Pb 89.62%. Pb_2O_3.
Reddish-yellow powder; converted at 370° in air to Pb_3O_4, dec at ~530° to PbO. Dec by concd HCl or H_2SO_4 with the liberation of Cl or oxygen, respectively. Insol in water.
Caution: Lead and all lead compounds are listed as reasonably anticipated to be human carcinogens: *Report on Carcinogens, Twelfth Edition* (PB2011-111646, 2011) p 251.

5473. Lead Stearate. [1072-35-1] Octadecanoic acid lead-(2+) salt (2:1); lead distearate; stearic acid lead salt. $C_{36}H_{70}O_4Pb$; mol wt 774.15. C 55.85%, H 9.11%, O 8.27%, Pb 26.76%. $Pb(C_{18}H_{35}O_2)_2$.
White powder, mp ~125°. *Poisonous.* Sol in hot alc. Insol in water.
Caution: Lead and all lead compounds are listed as reasonably anticipated to be human carcinogens: *Report on Carcinogens, Twelfth Edition* (PB2011-111646, 2011) p 251.
USE: In extreme pressure lubricants; as drier in varnishes.

5474. Lead Subacetate. [1335-32-6] Bis(acetato-κO)tetra-hydroxytrilead; lead monosubacetate; monobasic lead acetate. $C_4H_{10}O_8Pb_3$; mol wt 807.72. C 5.95%, H 1.25%, O 15.85%, Pb 76.96%. $Pb(C_2H_3O_2)_2.2Pb(OH)_2$.
White, heavy powder. *Poisonous.* Sol in 16 parts cold, 4 parts boiling water with alkaline reaction. On exposure to air absorbs CO_2 and becomes incompletely sol. *Keep well closed.*
Caution: Lead and all lead compounds are listed as reasonably anticipated to be human carcinogens: *Report on Carcinogens, Twelfth Edition* (PB2011-111646, 2011) p 251.
USE: In sugar analysis to remove coloring matters, etc., from solns before polarizing; for clarifying and decolorizing other solns of organic substances.

5475. Lead Sulfate. [7446-14-2] Sulfuric acid lead(2+) salt (1:1); lead(II) sulfate. O_4PbS; mol wt 303.26. O 21.10%, Pb 68.32%, S 10.57%. $PbSO_4$. Occurs as the minerals, *anglesite* and *lanarkite*. Review of toxicology and human exposure: *Toxicological Profile for Lead* (PB2008-100007, 2008) 582 pp.
White, heavy, cryst powder. *Poisonous.* d 6.2. mp 1170°. Sol in about 2225 parts water; more soluble in dil HCl or HNO_3, less in dil H_2SO_4; sol in NaOH, ammonium acetate or tartrate soln; sol in concd hydriodic acid. Insol in alc.
Caution: Lead and all lead compounds are listed as reasonably anticipated to be human carcinogens: *Report on Carcinogens, Twelfth Edition* (PB2011-111646, 2011) p 251.
USE: As pigment instead of white lead; manuf of galvanic and lead-acid batteries; manuf minium, in lithography; preparing rapidly drying oil varnishes; weighting fabrics.

5476. Lead Sulfide. [1314-87-0] Lead sulfide (PbS); lead(II) sulfide; plumbous sulfide. PbS; mol wt 239.26. Pb 86.60%, S 13.40%. Occurs as the mineral, *galena*. Toxicity study: W. R. Bradley, W. G. Fredrick, *Ind. Med.* **10**, Ind. Hyg. Sect. 2, 15 (1941). Review of toxicology and human exposure: *Toxicological Profile for Lead* (PB2008-100007, 2008) 582 pp.
Black powder. Sol in HNO_3, hot, dil HCl. Insol in water. LD_{50} i.p. in rats: 160 mg Pb/100 g (Bradley, Fredrick).
Caution: Lead and all lead compounds are listed as reasonably anticipated to be human carcinogens: *Report on Carcinogens, Twelfth Edition* (PB2011-111646, 2011) p 251.
USE: Glazing earthenware.

5477. Lead Telluride. [1314-91-6] Lead telluride (PbTe); lead(2+) telluride. PbTe; mol wt 334.80. Pb 61.89%, Te 38.11%.

Found in nature as the mineral, *altaite*. Prepd from lead nitrate, sodium carbonate and powdered tellurium: Montignie, *Bull. Soc. Chim. Fr.* **1947**, 750. Prepn of single crystals by heating stoichiometric quantities of the elements in a graphite cup or fused quartz tube: Brady, *J. Electrochem. Soc.* **101**, 466 (1954).
Silver-gray cubic crystals. d_4^{20} 8.16. mp 905°. Most of the crystal is *p*-type, the *n*-type material being present in the surface layer. Energy gap 0.27 ev. Electron mobility 2240 cm^2/volt-sec. Hole mobility 860 cm^2/volt-sec. Resistivity 0.005 ohm-cm (*p*-type), 0.00090 ohm-cm (*n*-type). Not attacked by hydrochloric, hydrofluoric, perchloric and acetic acids or their mixtures; not attacked by solns of 30% potassium hydroxide or of alkali metal sulfides. Dil nitric acid turns the surface black, while concd nitric acid produces lighter gray surface and turns the black surface to gray. Hot concd sulfuric acid produces a reddish-violet surface.
Caution: Lead and all lead compounds are listed as reasonably anticipated to be human carcinogens: *Report on Carcinogens, Twelfth Edition* (PB2011-111646, 2011) p 251.
USE: In photoconductor cells; in semiconductor research.

5478. Lead Tetraacetate. [546-67-8] Acetic acid lead(4+) salt (4:1); lead(IV) acetate; plumbic acetate. $C_8H_{12}O_8Pb$; mol wt 443.38. C 21.67%, H 2.73%, O 28.87%, Pb 46.73%. $Pb(CH_3COO)_4$. Prepd from Pb_3O_4 and glacial acetic acid preferably in the presence of some acetic anhydride: Dimroth, Schweizer, *Ber.* **56**, 1375 (1923); Bailar, *Inorg. Synth.* **1**, 47 (1939); Baudler in *Handbook of Preparative Inorganic Chemistry* vol. 1, G. Brauer, Ed. (Academic Press, New York, 2nd ed., 1963) p 767. Prepn by electrolysis: Fioshin, Gus'kov, *Dokl. Akad. Nauk SSSR* **112**, 303 (1957), *C.A.* **51**, 16146 (1957); Sataev *et al.*, *Khim. Prom. (Moscow)* **46**, 892 (1970), *C.A.* **74**, 49005x (1971). Reviews of prepn and use as oxidizing agent: Criegee "Oxidations with Lead Tetraacetate" in *Oxidation in Organic Chemistry*, Part A, K. B. Wiberg, Ed. (Academic Press, New York, 1965) pp 277-366; Zyka, *Pure Appl. Chem.* **13**, 569-581 (1966).
Colorless monoclinic prisms from glacial acetic acid. Turns pink easily. Unstable in air. Hydrolyzed by water with the formation of brown lead dioxide and acetic acid. *Avoid contact with skin.* d_4^{17} 2.228. mp 175-180°. Sol in hot glacial acetic acid, benzene, chloroform, tetrachloroethane, nitrobenzene. Dissolves in concd halogen acids with the formation of haloplumbic acids, H_2PbX_6. The dry material can be stored in sealed, evacuated ampuls.
Caution: Lead and lead compounds are listed as reasonably anticipated to be human carcinogens: *Report on Carcinogens, Twelfth Edition* (PB2011-111646, 2011) p 251.
USE: Selective oxidizing agent in organic syntheses: Criegee, *Angew. Chem.* **53**, 321 (1940); *Newer Methods of Preparative Organic Chemistry* (Interscience, N. Y., 1948) pp 1-17.

5479. Lead Tetrafluoride. [7783-59-7] Tetrafluoroplumbane; plumbic fluoride. F_4Pb; mol wt 283.19. F 26.83%, Pb 73.17%. PbF_4. Prepd by passing fluorine diluted with CO_2 or N_2 over PbF_2 at 300°: v. Wartenberg, *Z. Anorg. Allg. Chem.* **244**, 339 (1940). *Review:* Kemmitt, Sharp, *Adv. Fluorine Chem.* **4**, 187 (1965).
White, tetragonal crystals, d 6.7. mp ~600°. Readily hydrolyzes and turns brown (forms PbO_2) in the presence of moisture.
Caution: Lead and lead compounds are listed as reasonably anticipated to be human carcinogens: *Report on Carcinogens, Twelfth Edition* (PB2011-111646, 2011) p 251.
USE: Has been proposed as a fluorinating agent for hydrocarbons.

5480. Lead Tetroxide. [1314-41-6] Lead oxide (Pb_3O_4); Lead oxide red; red lead; minium; lead orthoplumbate; mineral orange; mineral red; Paris red; Saturn red; C.I. Pigment Red 105; C.I. 77578. O_4Pb_3; mol wt 685.60. O 9.33%, Pb 90.67%. Pb_3O_4. The article of commerce contains about 90% Pb_3O_4; the remainder being chiefly lead monoxide. Prepn: M. Baudler in *Handbook of Preparative Inorganic Chemistry* vol. 1, G. Brauer, Ed. (Academic Press, New York, 1963) pp 755-757. Structure: S. T. Gross, *J. Am. Chem. Soc.* **65**, 1107 (1943). *Review: Mellor's* vol. 7 (1930) pp 672-680. Book: *Biologic Effects of Atmospheric Pollutants: Lead* (Nat. Acad. Sci., Washington, D.C., 1972) 330 p.
Bright-red, heavy powder. *Poisonous.* Dec at ~500° with evolution of oxygen. d 9.1. Sol in excess glacial acetic acid, in hot HCl with evolution of Cl, in dil HNO_3 in presence of H_2O_2. Insol in water or alc. LD_{50} i.p. in rats: 45 mg Pb/100 g (*Lead*, 1972).

Caution: Lead and lead compounds are listed as reasonably anticipated to be human carcinogens: *Report on Carcinogens, Twelfth Edition* (PB2011-111646, 2011) p 251.

USE: Plasters and ointments; manuf colorless glass; glaze for faience; flux for porcelain painting, protective paint for iron and steel; oil-color for ship paints, varnishes; coloring rubber; cement for glass, gas and steam pipes; storage batteries; pencils for writing on glass; manuf lead peroxide; matches.

5481. Lead Thiocyanate. [592-87-0] Thiocyanic acid lead-(2+) salt (2:1); lead(II) thiocyanate; lead sulfocyanate. $C_2N_2PbS_2$; mol wt 323.36. C 7.43%, N 8.66%, Pb 64.08%, S 19.83%. Pb-(SCN)$_2$. Prepn: Gardner, Weinberger, *Inorg. Synth.* **1**, 85 (1939).

White, odorless powder. d 3.82. *Poisonous.* Sol in ~200 parts cold, 50 parts boiling water; also sol in alkali hydroxide and thiocyanate solns.

Caution: Lead and lead compounds are listed as reasonably anticipated to be human carcinogens: *Report on Carcinogens, Twelfth Edition* (PB2011-111646, 2011) p 251.

USE: Reverse dyeing with aniline black; manufacture of safety matches and cartridges.

5482. Lead Vanadate(V). [10099-79-3] Lead vanadium oxide (PbV$_2$O$_6$); lead metavanadate. O$_6$PbV$_2$; mol wt 405.08. O 23.70%, Pb 51.15%, V 25.15%. Pb(VO$_3$)$_2$.

Yellow powder. Insol in water. Dec by HNO$_3$.

Caution: Lead and lead compounds are listed as reasonably anticipated to be human carcinogens: *Report on Carcinogens, Twelfth Edition* (PB2011-111646, 2011) p 251.

USE: Manuf other vanadium compds; as pigment.

5483. Lecithins. [8002-43-5] Phosphatidylcholine; Lecithol; Vitellin; Kelecin; Granulestin. Phosphatides found in all living organisms (plants and animals). Significant constituent of nervous tissue and brain substance. A mixture of the diglycerides of stearic, palmitic, and oleic acids, linked to the choline ester of phosphoric acid. Commercial grades contain 2.2% P. Isoln from eggs: Sinclair, *Can. J. Res.* **26B**, 777 (1948). Product of commerce is predominantly soybean lecithin obtained as a by-product in the manuf of soybean oil: Stanley in K. S. Markley, *Soybeans* vol. **II** (Interscience, New York, 1951) pp 593-647. Soybean lecithin contains palmitic acid 11.7%, stearic 4.0%, palmitoleic 8.6%, oleic 9.8%, linoleic 55.0%, linolenic 4.0%, C$_{20}$ to C$_{22}$ acids (includes arachidonic) 5.5%. Synthesis of a mixed acid α-lecithin: de Haas, van Deenen, *Tetrahedron Lett.* **1960** (no. 9), 1. Synthetic L-α-(distearoyl)lecithin is identical with hydrogenated egg yolk lecithin and L-α-(dipalmitoyl)lecithin is identical with colfosceril palmitate, *q.v.*, a natural phosphatide of brain, lung, and spleen. Commercial grades of natural lecithin are reported to contain a potent vasodepressor substance: McQuarrie, Andersen, **US 2931818** (1960 to Cutter Labs.). Comprehensive monograph: G. B. Ansell, J. N. Hawthorne, *Phospholipids* (Elsevier, New York, 1964) 439 pp; A. Wendel, "Lecithin" in *Kirk-Othmer Encyclopedia of Chemical Technology* vol. **15** (Wiley-Interscience, New York, 4th ed., 1995) pp 192-210.

Waxy mass when the acid value is about 20. Pourable, thick fluid when the acid value is around 30. Color is nearly white when freshly made, but rapidly becomes yellow to brown in air. d$_4^{24}$ 1.0305. Iodine value 95; saponification value 196. Insoluble but swells up in water and in NaCl soln forming a colloidal suspension. Soluble in about 12 parts cold, abs alcohol; sol in chloroform, ether, petr ether, in mineral oils and fatty acids; sparingly sol in benzene. Insol in acetone; practically insol in cold vegetable and animal oils.

USE: Edible and digestible surfactant and emulsifier of natural origin. Used in margarine, chocolate and in the food industry in general. In pharmaceuticals and cosmetics. Many other industrial uses, e.g. treating leather and textiles.

THERAP CAT: Lipotropic.

5484. Lectins. Agglutinins; affinitins; phytoagglutinins; phasins; protectins. A group of proteins, widely distributed in nature, that have the ability to agglutinate erythrocytes and many other types of cells. Their existence has been known since 1899, when Stillmark isolated a hemagglutinin from castor beans. The term "lectin" (from the Latin *legere*, to choose) was first introduced by W. C. Boyd and E. Slapleigh in *Science* **119**, 419 (1954). It is now used to designate "a sugar-binding protein or glycoprotein of non-immune origin which agglutinates cells and/or precipitates glycoconjugates": I. J. Goldstein *et al.*, *Nature* **285**, 66 (1980). Lectins are found primarily in seeds of plants, but also occur in roots, leaves and bark. In addition, they are present in invertebrates, such as clams, snails, and horseshoe crabs, and in several vertebrate species. The term *phytohemagglutinin* is used to refer to plant lectins. Important members of the lectin family include concanavalin A, abrin, ricin, *q.q.v.*, as well as *soybean agglutinin* or *SBA* and *wheat germ agglutinin* or *WGA*. Lectins vary considerably in chemical and physical properties; only a limited number have been purified. Mol wts of 17,000 to 400,000 have been reported and most lectins have been found to contain Mn^{2+} and Ca^{2+}. Nearly all lectins can be inhibited by free oligo- or monosaccharides of appropriate specificity. Although their physiological functions in plants or in other organisms are unknown, lectins exhibit a variety of unusual biological properties. Some are specific in their reactions with human blood groups; some induce mitosis in lymphocytes. WGA from wheat germ lipase has been shown to agglutinate mouse tumor cells more readily than cells from normal tissue: J. C. Aub *et al.*, *Proc. Natl. Acad. Sci. USA* **50**, 613 (1963); M. M. Burger, A. R. Goldberg, *ibid.* **57**, 359 (1967). Soybean agglutinin and concanavalin A have been shown to agglutinate cell lines transformed by viral or chemical carcinogens: M. Inbar, L. Sachs, *Nature* **223**, 710 (1969); *eidem, Proc. Natl. Acad. Sci. USA* **63**, 1418 (1969); B. A. Sela *et al.*, *J. Membr. Biol.* **3**, 267 (1970). Some plant lectins mimic the direct effects of insulin on nuclear envelope phosphorylation: F. Purrello *et al.*, *Science* **221**, 462 (1983). Soybean agglutinin has also been used in bone marrow transplants in patients with severe combined immunodeficiency: Y. Reisner *et al.*, *Blood* **61**, 341 (1983). *Reviews:* N. Sharon, H. Lis, *Science* **177**, 949-955 (1972); *eidem, Annu. Rev. Biochem.* **42**, 541-574 (1973); L. Sequeira, *Annu. Rev. Phytopathol.* **16**, 453-481 (1978). Book: *Lectins: Biology, Biochemistry, Clinical Biochemistry* vol. **1**, T. C. Bog-Hansen, Ed. (de Gruyter, New York, 1981) 414 pp.

USE: As tools for studying cell surface properties; in cancer research.

5485. Ledol. [577-27-5] (1aS,4S,4aR,7S,7aR,7bR)-Decahydro-1,1,4,7-tetramethyl-1H-cyclopropp[*e*]azulen-4-ol; "Ledum camphor". C$_{15}$H$_{26}$O; mol wt 222.37. C 81.02%, H 11.79%, O 7.19%. Occurs in the essential oil from leaves of *Ledum palustre* L.: Grassmann, *Repert. Pharm.* **38**, 53 (1931); Hjelt, *Ber.* **28**, 3087 (1895); from *L. groenlandicum* Veder; *L. columbianum* Piper, *Ericaceae:* Cain, Lynn, *J. Am. Pharm. Assoc.* **23**, 666 (1934); Penfold, *J. Proc. R. Soc. N.S.W.* **59**, 206 (1925). Structure: G. Büchi *et al.*, *Tetrahedron Lett.* **1959** (no. 6), 14. Stereochemistry: L. Dolejs, F. Sorm, *Tetrahedron Lett.* **1959** (no. 17), 1; revised stereochemistry: G. Büchi *et al.*, *J. Am. Chem. Soc.* **91**, 6473 (1969).

Needles from alc, mp 104-105°. Sublimes easily, even below the mp. bp$_{760}$ 292°. n$_D^{110}$ 1.4667. [α]$_D^{20}$ +28° (c = 10 in chloroform). Practically insol in water. Sol in alc (about 10% w/v). Soluble in other organic solvents.

Chromate. C$_{30}$H$_{50}$O$_4$Cr. Ruby-red prisms, mp 92°. [α]$_{671}^{20}$ +30° (c = 2 in chloroform).

5486. Leflunomide. [75706-12-6] 5-Methyl-*N*-[4-(trifluoromethyl)phenyl]-4-isoxazolecarboxamide; α,α,α-trifluoro-5-methyl-

4-isoxazolecarboxy-*p*-toluidide; 5-methylisoxazole-4-carboxylic acid trifluoromethylanilide; HWA-486; Arava. $C_{12}H_9F_3N_2O_2$; mol wt 270.21. C 53.34%, H 3.36%, F 21.09%, N 10.37%, O 11.84%. Disease modifying antirheumatic drug (DMARD) with immunosuppressant activity. Converted *in vivo* to its active open ring metabolite, teriflunomide, *q.v.* Inhibits *de novo* pyrimidine biosynthesis in immune cells. Prepn: F. J. Kaemmerer, R. Schleyerbach, **DE 2854439**; *eidem*, **US 4284786** (1980, 1981 both to Hoechst); and pharmacology: P. Fossa *et al.*, *Farmaco* **46**, 789 (1991). HPLC determn of active metabolite in plasma: V. C. Dias *et al.*, *Ther. Drug Monit.* **17**, 84 (1995). Review of mechanisms of action: R. I. Fox, *J. Rheumatol.* **25**, Suppl. 53, 20-26 (1998); of clinical pharmacology: B. Rozman, *ibid.* 27-32. Clinical trial in rheumatoid arthritis: V. Strand *et al.*, *Arch. Intern. Med.* **159**, 2542 (1999). *Reviews:* C. Miceli-Richard, M. Dougados, *Expert Opin. Pharmacother.* **4**, 987-997 (2003); J. P. Kaltwasser, F. Behrens, *ibid.* **6**, 787-801 (2005).

Crystals from toluene, mp 166.5°. pKa 10.8. Soly in water (25°): 25 - 27 mg/l. Freely sol in methanol, ethanol, acetone, 2-propanol, ethyl acetate, acetonitrile, chloroform.

THERAP CAT: Antirheumatic.

5487. Leghemoglobin. Legoglobin. Hemoglobin-like red pigment present in the root nodules of leguminous plants. Isolation from soya beans: Keilin, Wang, *Nature* **155**, 227 (1945); Appleby, *Biochim. Biophys. Acta* **60**, 226 (1962). Mol wt is approx one-fourth that of hemoglobin: Ehrenberg, Ellfolk, *Acta Chem. Scand.* **17**, S343 (1963). Resolved into four components on DEAE-cellulose column: Ellfolk, *ibid.* **14**, 609 (1960). Suggested to act as an oxido-reduction catalyst in the symbiotic nitrogen fixation: *idem, ibid.* **15**, 975 (1961). Primary structure of soybean leghemoglobin: Ellfolk, Sievers, *ibid.* **25**, 3532 (1971).

5488. Lemon Peel. Outer rind of fresh ripe fruit of *Citrus limonum* (L.) Risso *(C. medica* var. *limon* L.), *Rutaceae. Habit.* Northern India; cultivated in California, West Indies, Italy, Spain. *Constit.* Volatile oil, hesperidin, bitter extractive.

USE: As a flavor in medicines; also in beverages, confectionery, and cooking.

5489. Lenacil. [2164-08-1] 3-Cyclohexyl-6,7-dihydro-1*H*-cyclopentapyrimidine-2,4-(3*H*,5*H*)-dione; 3-cyclohexyl-5,6-tri-methyleneuracil; 3-cyclohexyl-1,5,6,7-tetrahydro-2*H*-cyclopenta-pyrimidine-2,4(3*H*)-dione; Du Pont 634; Venzar. $C_{13}H_{18}N_2O_2$; mol wt 234.30. C 66.64%, H 7.74%, N 11.96%, O 13.66%. Prepn: S. Senda, H. Fujimura, **JP 62 4892** (1962), *C.A.* **59**, 642g (1963); and herbicidal activity: E. J. Soboczenski, **US 3235360** (1966 to Du-Pont). Synthesis: S. Senda *et al.*, *Chem. Pharm. Bull.* **21**, 1894 (1973). Mode of action: C. E. Hoffmann, *Pestic. Chem., Proc. 2nd Int. Congr. Pestic.* **5**, A. S. Tahori, Ed., (Gordon and Breach Science Publishers, New York, 1972) pp 65-85. Field evaluation: J. G. Hilton, W. E. Bray, Proc. *12th Brit. Weed Contr. Conf.* 485 (1974). GC determn in soils: D. J. Caverly, R. C. Denney, *Analyst* **102**, 576 (1977). Soil persistence: M. Tena *et al.*, *Weed Res.* **23**, 245 (1982). Field tolerance: D. V. Clay, *Tests Agrochem. Cultiv.* **11**, 80 (1990).

Crystals from methanol or dioxane, mp >290° (Senda, Fujimura). Also reported as light grey crystals from DMF, mp 310-313° (Sobo-czenski). Technical product properties: odorless, white to tan crystalline solid. d^{20} 1.32 kg/l. Vapor pressure (25°): 0.2 μPa. Soly in water 6 mg/l; in xylene: 2 g/l. LD$_{50}$ (80% wettable powder) orally in rats: >11,000 mg/kg. LC$_{50}$ 96 hr in bluegill sunfish; rainbow trout: 100-1000; 135 mg/l. LC$_{50}$ 8 day in bobwhite quail, Peking duck (mg/kg): 2300, >5620. (DuPont technical data sheet).

USE: Herbicide.

5490. Lenalidomide. [191732-72-6] 3-(4-Amino-1,3-dihy-dro-1-oxo-2*H*-isoindol-2-yl)-2,6-piperidinedione; 1-oxo-2-(2,6-dioxopiperidin-3-yl)-4-aminoisoindoline; CC-5013; Revlimid. $C_{13}H_{13}N_3O_3$; mol wt 259.27. C 60.22%, H 5.05%, N 16.21%, O 18.51%. Immunomodulatory drug; analog of thalidomide, *q.v.* Prepn: G. W. Muller *et al.*, **US 5635517** (1997 to Celgene); and *in vitro* TNF-α inhibition: *eidem, Bioorg. Med. Chem. Lett.* **9**, 1625 (1999). LC-MS determn in plasma: T. M. Tohnya *et al.*, *J. Chromatogr. B* **811**, 135 (2004). Clinical evaluation in multiple myeloma: P. G. Richardson *et al.*, *Blood* **100**, 3063 (2002); in myelo-dysplastic syndromes: A. List *et al.*, *N. Engl. J. Med.* **352**, 549 (2005). Review of development, pharmacology and therapeutic potential: C. S. Mitsiades, N. Mitsiades, *Curr. Opin. Investig. Drugs* **5**, 635-647 (2004); of clinical experience in multiple myeloma: J. B. Zeldis *et al. Expert Opin. Pharmacother.* **11**, 829-842 (2010).

Off-white to pale-yellow solid powder. Sol in organic solvent/water mixtures and buffered aqueous solvents. More sol in organic solvents and low pH solutions. Soly significantly lower in less acidic buffers (~0.4-0.5 mg/ml).

THERAP CAT: Immunomodulator.

5491. Lenampicillin. [86273-18-9] (2*S*,5*R*,6*R*)-6-[[(2*R*)-2-Amino-2-phenylacetyl]amino]-3,3-dimethyl-7-oxo-4-thia-1-azabi-cyclo[3.2.0]heptane-2-carboxylic acid (5-methyl-2-oxo-1,3-dioxol-4-yl)methyl ester; ampicillin (5-methyl-2-oxo-1,3-dioxolen-4-yl)-methyl ester; 6-[D(−)-α-aminophenylacetamido]penicillanic acid (5-methyl-2-oxo-1,3-dioxol-4-yl)methyl ester. $C_{21}H_{23}N_3O_7S$; mol wt 461.49. C 54.66%, H 5.02%, N 9.11%, O 24.27%, S 6.95%. Orally active ampicillin prodrug. Prepn: F. Sakamoto *et al.*, **EP 39086**; *eidem*, **US 4342693** (1981, 1982 both to Kanebo); F. Saka-moto *et al.*, *Chem. Pharm. Bull.* **32**, 2241 (1984); S. Ikeda *et al.*, *ibid.* 4316. Metabolism in man, dogs, rats: N. Awata *et al.*, *Jpn. J. Antibiot.* **38**, 1776, 1785, *C.A.* **104**, 161479u, 122550r (1985). Human pharmacokinetics: A. Saito, M. Nakashima, *Antimicrob. Agents Chemother.* **29**, 948 (1986). Series of articles on antibacterial activity, mutagenicity, pharmacology, clinical trials: *Chemotherapy (Tokyo)* **32**, Suppl. 8, pp 1-772 (1984). Acute toxicity data: F. Ogino *et al., ibid.* 31.

Hydrochloride. [80734-02-7] KB-1585; KBT-1585; Takacillin; Varacillin. $C_{21}H_{23}N_3O_7S$·HCl; mol wt 497.95. Crystals from iso-propanol-ethyl acetate, mp 145° (dec). LD$_{50}$ in male, female rats, male, female mice (mg/kg): approx 10000, approx 10000, 8294, 8492 orally; 4362, 4471, 3576, 4284 s.c.; 876, 838, 711, 775 i.v.; in dogs (mg/kg): >300 orally (Ogino).

THERAP CAT: Antibacterial.

5492. Lenthionine. [292-46-6] 1,2,3,5,6-Pentathiepane. C_2-H_4S_5; mol wt 188.35. C 12.75%, H 2.14%, S 85.11%. Odorous principle from the edible mushroom *Shiitake Lentinus edodes* (Berk.) Sing. Isoln and synthesis: Morita, Kobayashi, *Tetrahedron Lett.*

1966, 573. Simple, efficient synthesis: I. Still, G. W. Kutney, *ibid.* **22**, 1939 (1981).

Crystals from methylene chloride, mp 60-61°.

5493. **Lentinan.** [37339-90-5] Biomoduline; LC-33. $(C_6H_{10}O_5)_n$; mol wt 400,000-800,000. Neutral polysaccharide isolated from the edible mushroom, *Lentinus edodes* (Berk.) Sing. Primary structure is a β-1,3-D-glucan having 2 β-1,6-glucopyranoside branchings for every 5 β-1,3 linear linkages. Isoln and antitumor activity: G. Chihara *et al., Nature* **222**, 637 (1969); *eidem, Cancer Res.* **30**, 2776 (1970). Immunostimulant activity: Y. Y. Maeda, G. Chihara, *Nature* **229**, 634 (1971). Prepn of stable aqueous solution: M. Fujii *et al.*, **NL 7601114**; *eidem*, **US 4207312** (1976, 1980 both to Ajinomoto; Morishita). Structural study: T. Sasaki *et al., Carbohydr. Res.* **47**, 99 (1976). Preclinical evaluation and toxicity studies: G. Chihara, *Adv. Exp. Med. Biol.* **166**, 189 (1983). Mechanism of action studies: H. Miyakoshi, T. Aoki, *Int. J. Immunopharmacol.* **6**, 365, 373 (1984). Clinical trial in gastric and colorectal cancer: T. Taguchi *et al., Adv. Exp. Med. Biol.* **166**, 181 (1983); K. Okuyama *et al., Cancer* **55**, 2498 (1985). *Review:* G. Chihara, *Int. J. Tissue Res.* **4**, 207-225 (1982).

White powder, dec 250°. Sol in aqueous alkali, formic acid. Slightly sol in hot water, DMSO. Insol in cold water, alcohol, ether, chloroform, pyridine, hexamethylphosphoramide. Stable against sulfuric and hydrochloric acids. $[\alpha]_D^{20}$ +13.5-14.5° (in 2% NaOH); +19.5-21.5° (in 10% NaOH).

THERAP CAT: Antineoplastic; immunomodulator.

5494. **Lenvatinib.** [417716-92-8] 4-[3-Chloro-4-[[(cyclopropylamino)carbonyl]amino]phenoxy]-7-methoxy-6-quinolinecarboxamide; ER-203492-00. $C_{21}H_{19}ClN_4O_4$; mol wt 426.86. C 59.09%, H 4.49%, Cl 8.30%, N 13.13%, O 14.99%. Multiple receptor tyrosine kinase inhibitor. Inhibits angiogenesis through suppression of VEGF receptors. Prepn: Y. Funahashi *et al.*, **WO 0232872**; *eidem*, **US 7253286** (2002, 2007 both to Eisai). *In vivo* antitumor activity in human sarcoma xenografts: S. Bruheim *et al., Int. J. Cancer* **129**, 742 (2011). Anti-angiogenic activity in exptl lung cancer metastases: H. Ogino *et al., Mol. Cancer Ther.* **10**, 1218 (2011). Clinical pharmacokinetics and evaluation in solid tumors: K. Yamada *et al., Clin. Cancer Res.* **17**, 2528 (2011).

White crystals from ethyl-acetate + hexane.

Methanesulfonate. [857890-39-2] Lenvatinib mesylate; E-7080. $C_{21}H_{19}ClN_4O_4.CH_4O_3S$; mol wt 522.96.

THERAP CAT: Antineoplastic.

5495. **Leonurine.** [24697-74-3] 4-Hydroxy-3,5-dimethoxybenzoic acid 4-[(aminoiminomethyl)amino]butyl ester; 4-hydroxy-3,5-dimethoxybenzoic acid δ-guanidinobutyl ester; syringic acid δ-guanidinobutyl ester; [4-(4-hydroxy-3,5-dimethoxybenzoyloxy)butyl]guanidine; 4-guanidino-1-butanol syringate. $C_{14}H_{21}N_3O_5$; mol wt 311.34. C 54.01%, H 6.80%, N 13.50%, O 25.69%. Isoln from leaves of *Leonurus sibiricus* L., *Labiatae:* Kubota, Nakajima, *C.A.* **25**, 771 (1931). Structure: Goto *et al., Tetrahedron Lett.* **1962**, 545. Revised structure: Kishi *et al., ibid.* **1968**, 637. Structure and synthesis: Sugiura *et al., Tetrahedron* **25**, 5155 (1969). Alternate synthesis: K. F. Cheng *et al., Experientia* **35**, 571 (1979). Biological effect: H. W. Yeung *et al., Planta Med.* **31**, 51 (1977).

Hydrochloride monohydrate. $C_{14}H_{21}N_3O_5.HCl.H_2O$. mp 193-194°. pKa 7.9 in water.

5496. **Lepidine.** [491-35-0] 4-Methylquinoline; cincholepidine. $C_{10}H_9N$; mol wt 143.19. C 83.88%, H 6.34%, N 9.78%. Prepn from 2-chlorolepidine: Neumann *et al., Org. Synth.* **26**, 45 (1946); from aniline hydrochloride: Campbell, **US 2451611** (1948 to du Pont); Bach, Rast, *J. Prakt. Chem.* **17**, 63 (1962); from ethylacetanilide: Ardashev, Minkin, *Zh. Obshch. Khim.* **28**, 1578 (1958).

Colorless, oily liq; quinoline odor; turns reddish-brown in light. bp 261-263°; bp$_{14-15}$ 126-127°; bp$_{6-7}$ 115-120°; bp$_{1.5-2}$ 90-95°. d_4^{20} 1.0826. n_D^{20} 1.6190. mp 0°. Slightly sol in water; miscible with alc, benzene, ether. *Protect from light.*

5497. **Leptandra.** Culver's root; black root. Dried rhizome and roots of *Veronicastrum virginicum* (L.) Farw. (*Leptandra virginica* (L.) Nutt.), Scrophulariaceae. Habit. North America. Constit. A sterol; esters of cinnamic, methoxycinnamic and fatty acids; resin saponin, tannin, sugars.

THERAP CAT: Cathartic.

5498. **Leptin.** OB protein. Peptide hormone produced by adipose tissue that is thought to act on receptors in the brain to regulate body weight and fat deposition. Name derived from the Greek word "leptos" meaning "thin." Protein product of the *obese (ob)* gene, mutation of which causes marked obesity and diabetes in mice. Secreted by adipocytes; mol wt 16 kDa. Identification of weight-regulating substance in blood of normal mice: D. L. Coleman, *Diabetologia* **9**, 294 (1973). Cloning of mouse and human *ob* genes: Y. Zhang *et al., Nature* **372**, 425 (1994). Biological activity as regulator of body weight in mice: M. A. Pelleymounter *et al., Science* **269**, 540 (1995); J. L. Halaas *et al., ibid.* 543. Identification of brain as target tissue: L. A. Campfield *et al., ibid.* 546. Increased *ob* gene expression in obese humans: R. V. Considine *et al., J. Clin. Invest.* **95**, 2986 (1995). *Review of discovery and role in energy homeostasis:* M. Rosenbaum, R. L. Leibel, *Trends Endocrinol. Metab.* **9**, 117-124 (1998).

5499. **Leptomycin B.** [87081-35-4] (2*E*,5*S*,6*R*,7*S*,9*R*,10*E*,-12*E*,15*R*,16*Z*,18*E*)-19-[(2*S*,3*S*)-3,6-Dihydro-3-methyl-6-oxo-2*H*-pyran-2-yl]-17-ethyl-6-hydroxy-3,5,7,9,11,15-hexamethyl-8-oxo-2,10,12,16,18-nonadecapentaenoic acid; elactocin; CI-940; CL-1957A; NSC-364372; PD-114720. $C_{33}H_{48}O_6$; mol wt 540.74. C 73.30%, H 8.95%, O 17.75%. Antibiotic compound with antifungal and antitumor activity. Inhibits the CRM1/exportin1 pathway of nuclear transport in eukaryotic cells, facilitating its use in the study of nuclear export. Isoln from *Streptomyces* sp. ATS1287: T. Hamamoto *et al., J. Antibiot.* **36**, 639 (1983); from actinomycete ATCC

39366: G. C. Hokanson *et al.*, **EP 139458**; *eidem*, **US 4771070** (1985, 1988 both to Warner Lambert). Antitumor activity: B. J. Roberts *et al.*, *Cancer Chemother. Pharmacol.* **16**, 95 (1986); and antimicrobial activity: J. B. Tunac *et al.*, *J. Antibiot.* **38**, 460 (1985). Synthesis and abs config: M. Kobayashi *et al.*, *Tetrahedron Lett.* **39**, 8291 (1998). HPLC determn in fermentation extracts: M. Stadler *et al.*, *J. Chromatogr. A* **818**, 187 (1998). Review of use as inhibitor of nuclear export: M. Yoshida *et al.*, *Actinomycetologica* **12**, 120-128 (1998).

Isolated as yellow sticky oil (Hamamoto). Sol in methanol, ethanol, ethyl acetate and ethyl ether. Insol in *n*-hexane, water. $[\alpha]_D$ −24.5° (c = 0.70 in MeOH). uv max (ethanol): 225 nm (ε 20000); shoulder 240 nm (ε 15000). Purified as a pale yellow solid foam, mp 41-44°, with prior softening (Hokanson). $[\alpha]_D^{23}$ −157 ° (c = 0.7% in chloroform).

USE: Biological tool for studying nuclear localization and protein trafficking in eukaryotic cells.

5500. Lercanidipine. [100427-26-7] 1,4-Dihydro-2,6-dimethyl-4-(3-nitrophenyl)-3,5-pyridinedicarboxylic acid 3-[2-[(3,3-diphenylpropyl)methylamino]-1,1-dimethylethyl] 5-methyl ester; methyl 1,1,*N*-trimethyl-*N*-(3,3-diphenylpropyl)-2-aminoethyl 1,4-dihydro-2,6-dimethyl-4-(3-nitrophenyl)pyridine-3,5-dicarboxylate; methyl 1,1-dimethyl-2-[*N*-(3,3-diphenylpropyl)-*N*-methylamino]-ethyl 2,6-dimethyl-4-(3-nitrophenyl)-1,4-dihydropyridine-3,5-dicarboxylate; masnidipine. $C_{36}H_{41}N_3O_6$; mol wt 611.74. C 70.68%, H 6.76%, N 6.87%, O 15.69%. Dihydropyridine calcium channel blocker. Prepn: D. Nardi *et al.*, **EP 153016**; *eidem*, **US 4705797** (1985, 1987 both to Recordati). Pharmacology: G. Bianchi *et al.*, *Pharmacol. Res.* **21**, 193 (1989). Clinical evaluation in hypertension: E. Rimoldi *et al.*, *Acta Ther.* **20**, 23 (1994).

Hydrochloride. [132866-11-6] Rec-15-2375; R-75; Lerdip; Zanidip. $C_{36}H_{41}N_3O_6$.HCl; mol wt 648.20. Prepd as the hemihydrate, mp 119-123°. LD_{50} in mice (mg/kg): 83 i.p.; 657 orally (Nardi).

THERAP CAT: Antihypertensive.

5501. Lesopitron. [132449-46-8] 2-[4-[4-(4-Chloro-1*H*-pyrazol-1-yl)butyl]-1-piperazinyl]pyrimidine. $C_{15}H_{21}ClN_6$; mol wt 320.83. C 56.16%, H 6.60%, Cl 11.05%, N 26.20%. Selective serotonin $5HT_{1A}$-receptor agonist. Prepn: A. Colombo Pinol *et al.*, **EP 382637**; *eidem*, **US 5128343** (1990, 1992 both to Esteve). Receptor binding studies and pharmacology: B. Costall *et al.*, *J. Pharmacol. Exp. Ther.* **262**, 90 (1992); M. Ballarin *et al.*, *Br. J. Pharmacol.* **113**, 425 (1994). Gastroprotective effects: A. J. Farré *et al.*, *J. Pharmacol. Exp. Ther.* **272**, 832 (1995).

Dihydrochloride. [132449-89-9] E-4424. $C_{15}H_{21}ClN_6$.2HCl; mol wt 393.74. Crystals, mp 194-197.5°.

THERAP CAT: Anxiolytic.

5502. Letrozole. [112809-51-5] 4,4′-(1*H*-1,2,4-Triazol-1-ylmethylene)bisbenzonitrile; 1-[bis(4-cyanophenyl)methyl]-1,2,4-triazole; 4-[1-(4-cyanophenyl)-1-(1,2,4-triazol-1-yl)methyl]benzonitrile; CGS-20267; Femara. $C_{17}H_{11}N_5$; mol wt 285.31. C 71.57%, H 3.89%, N 24.55%. Nonsteroidal aromatase inhibitor; structurally related to fadrozole, *q.v.* Prepn: R. M. Bowman *et al.*, **EP 236940**; *eidem*, **US 4978672** (1987, 1990 both to Ciba-Geigy). Pharmacology: A. S. Bhatnagar *et al.*, *J. Steroid Biochem. Mol. Biol.* **37**, 1021 (1990). Clinical suppression of estrogen biosynthesis: L. M. Demers *et al.*, *ibid.* **44**, 687 (1993). EIA and HPLC determn in biological fluids: C. U. Pfister *et al.*, *J. Pharm. Sci.* **83**, 520 (1994). Clinical evaluation in breast cancer: A. Lipton *et al.*, *Cancer* **75**, 2132 (1995). Review of clinical efficacy: P. E. Goss, R. E. Smith, *Expert Rev. Anticancer Ther.* **2**, 249-260 (2002).

White to yellowish, crystalline powder. mp 181-183°. Freely sol in dichloromethane; slightly sol in alc. Practically insol in water.

THERAP CAT: Antineoplastic.

5503. Leucine. [61-90-5] L-Leucine; Leu; L; 2-amino-4-methylvaleric acid; α-aminoisocaproic acid; (*S*)-2-amino-4-methylpentanoic acid. $C_6H_{13}NO_2$; mol wt 131.18. C 54.94%, H 9.99%, N 10.68%, O 24.39%. An essential amino acid for human development. Discovery of leucine is attributed to Proust in 1819 who reported its separation from fermented milk curds. Early chemistry and biochemistry: *Amino Acids and Proteins*, D. M. Greenberg, Ed. (Charles C. Thomas, Springfield, IL, 1951) 950 pp., *passim*; J. P. Greenstein, M. Winitz, *Chemistry of the Amino Acids* **vols 1-3** (John Wiley and Sons, Inc., New York, 1961) pp. 2075-2096, *passim*. Chromatographic determn of specific activity in biological fluids: E. P. Donahue *et al.*, *J. Chromatogr.* **571**, 29 (1991); separation of isotopes: P. Q. Baumann *et al.*, *ibid.* **573**, 11 (1992). Metabolism in humans: K. J. Motil *et al.*, *Metabolism* **30**, 783 (1981); in children receiving parenteral nutrition: O. Goulet *et al.*, *Am. J. Physiol.* **265**, E540 (1993). Kinetics modelling in humans: C. Cobelli *et al.*, *ibid.* **261**, E539 (1991).

White glistening hexagonal plates from aq alc. d_{18} 1.293. Sublimes at 145-148°. Dec 293-295° (rapid heating, sealed tube). $[M]_D$ +21.0° (5*N* HCl); $[M]_D$ +29.5° (glacial acetic acid). $[\alpha]_D^{25}$ −10.8° (c = 2.2); $[\alpha]_D^{26}$ +15.1° in 6*N* HCl (38 mols HCl per mol leucine); $[\alpha]_D^{20}$ +7.6° in 3*N* NaOH (30 mols NaOH per mol leucine). R_f value 0.79. Soly in water (g/l): 22.7 (0°); 24.26 (25°); 28.87 (50°); 38.23 (75°); 56.38 (100°); in 99% alcohol: 0.72; in acetic acid: 10.9. Insol in ether.

DL-Form. Leaflets fom water. Sweet taste. Dec 332° (also reported as 290°). Sublimes. pK_1 2.36; pK_2 9.60. Soly in water (g/l): 7.97 (0°); 9.91 (25°); 14.06 (50°); 22.76 (75°); 42.06 (100°); in 90% alcohol: 1.3. Insol in ether.

Hydrochloride. $C_6H_{13}NO_2$.HCl. Crystals. Freely sol in water.

USE: Nutrient.

5504. Leucocyanidin. [480-17-1] (2*R*,3*S*)-2-(3,4-Dihydroxyphenyl)-3,4-dihydro-2*H*-1-benzopyran-3,4,5,7-tetrol; 3,3′,4,4′,5,7-flavanhexol; 3,3′,4,4′,5,7-hexahydroxyflavane; leucocyanidol; Flavan; Hamaméliode P; Résivit. $C_{15}H_{14}O_7$; mol wt 306.27. C 58.83%, H 4.61%, O 36.57%. From petals of Asiatic cotton flowers (*Gossypium* spp): Stephens, *Arch. Biochem. Biophys.* **18**, 449

(1948); from *Butea frondosa* Koen. ex Roxb., *Leguminosae:* Ganguly, Seshadri, *Tetrahedron* **6**, 21 (1959). Prepn from quercetin: Bauer *et al.*, *Chem. Ind. (London)* **1954**, 433; from taxifolin: Freudenberg, Weinges, *Ann.* **613**, 61 (1958). Metabolism: Claveau, Masquelier, *Can. J. Pharm. Sci.* **1**, 74 (1966). *See also* Bioflavonoids.

Monohydrate. Crystals from ethyl acetate + petr ether, mp above 355°. uv max (ethanol): 285 nm. Soluble in water, alcohol, acetone. Practically insol in ether, chloroform, petr ether.

THERAP CAT: Capillary protectant.

5505. Leucomycin. [1392-21-8] Kitasamycin; C-637; Ayermicina; Sineptina; Stereomycine; Syneptine. Macrolide antibiotic complex similar to carbomycin, *q.v.* and erythromycin, *q.v.*, produced by *Streptomyces kitasatoensis* Hata: T. Hata *et al.*, *J. Antibiot.* **6A**, 87 (1953); Sano, *ibid.* **7A**, 93 (1954). Early work performed on a mixture of at least six components, A_1, A_2, B_1-B_4, of which A_1 was the most biologically active: J. Abe *et al.*, *J. Chem. Soc. Jpn. Pure Chem. Sect.* **81**, 969 (1960); T. Watanabe *et al.*, *Bull. Chem. Soc. Jpn.* **33**, 1100, 1104 (1960). Leucomycin isolated from an improved strain of *S. kitasatoensis* is a mixture of at least eight biologically active components, A_1 and new substances A_3-A_9: Hata *et al.*, *US 3535309* (1970 to Kitasato Institute and Toyo Jozo). Leucomycins A_1, A_3-A_9, U and V are substituted variations of the structure represented below. Structure of A_1: T. Watanabe *et al.*, *Angew. Chem.* **76**, 792 (1964); T. Hata *et al.*, *Chem. Pharm. Bull.* **15**, 358 (1967); of A_3: S. Omura *et al.*, *Tetrahedron Lett.* **1967**, 609, 1267; of A_4 through A_9: *eidem*, *J. Antibiot.* **20A**, 234 (1967); **21**, 272 (1968). Conformation: *eidem*, *Tetrahedron* **28**, 2839 (1972). Revised configuration at C-9: L. A. Freiberg *et al.*, *J. Org. Chem.* **39**, 2474 (1974). Liquid chromatography of A_1, A_3-A_9, U, V: S. Omura *et al.*, *J. Antibiot.* **26**, 795 (1973). Antibacterial and antimycoplasmal activity: Iwata, Akiba, *ibid.* **15A**, 258 (1962); Omura *et al.*, *ibid.* **21**, 532 (1968); **25**, 105 (1972). *In vitro* and preliminary clinical studies: B. C. Stratford, S. Dixson, *Med. J. Aust.* **1**, 1029 (1974). Antimicrobial transformation of A_5 to V: K. Singh, S. Rakhit, *ibid.* **32**, 78 (1979). Biosynthesis: C. Kitao *et al.*, *ibid.* **32**, 1055 (1979); S. Omura *et al.*, *ibid.* **36**, 611 (1983). *Reviews:* Desvignes, *Ann. Pharm. Fr.* **21**, 569 (1963); Toju, Omura, "Chemical and Biological Studies on Leucomycins (Kitasamycins)" in *Drug Action and Drug Resistance in Bacteria* vol. I, S. Mitsuhashi, Ed. (University Park Press, Baltimore, 1971) pp 267-291; Keller-Schierlein in *Fortschr. Chem. Org. Naturst.* **30**, 313-460 (1973); S. Omura, A. Nakagawa, *J. Antibiot.* **28**, 401-433 (1975).

Leucomycin A₁

Leucomycin A_1, A_2, B_1-B_4 complex. (Sano, *loc. cit.*). Powder from acetone; stable at room temp; dec 125°. $[\alpha]_D^{20}$ −67.1° (ethanol). uv max (ethanol): 232, 285 nm ($E_{1cm}^{1\%}$ 228, 8.6). Freely sol in most organic solvents. Insol in petr ether. Slightly sol in water.

Leucomycin A_3-A_9 complex. White powder; mp 128-145°. uv max (methanol): 231 nm ($E_{1cm}^{1\%}$ 353). $[\alpha]_D^{25}$ −53° (c = 1 in chloroform). Sol in methanol, ethanol, ethyl acetate, butyl acetate, acetone, benzene, and chloroform; difficultly sol in water. Insol in petr ether. LD_{50} in mice (mg/kg): >650 i.v., >1000 orally (*US 3535309*).

Leucomycin A_1. [16846-34-7] $C_{40}H_{67}NO_{14}$; mol wt 785.97. $[\alpha]_D^{25}$ −66.0° ($CHCl_3$). uv max (methanol): 232 nm ($E_{1cm}^{1\%}$ 400). pKa′ (50% ethanol): 6.69.

Triacetylleucomycin A_1. $C_{46}H_{73}NO_{17}$. Crystals, mp 125-126°. $[\alpha]_D^{25}$ −82.5° (c = 1.3 in $CHCl_3$).

Leucomycin A_3 see Josamycin.

THERAP CAT: Antibacterial.

5506. Leucopterin. [492-11-5] 2-Amino-5,8-dihydro-4,6,7-(3H)-pteridinetrione; 2-amino-4,6,7-pteridinetriol; 2-amino-4,6,7-trihydroxypteridine; 2-amino-4,6,7-trihydroxypyrimido[4,5-*b*]pyrazine. $C_6H_5N_5O_3$; mol wt 195.14. C 36.93%, H 2.58%, N 35.89%, O 24.60%. A colorless substance found in the wings of butterflies (especially of white-winged butterflies). May be obtained from xanthopterin by dehydrogenation: Wieland, Purrmann, *Ann.* **544**, 172 (1940). Synthesis by fusing 2,4,5-triamino-6-hydroxypyrimidine with an excess of oxalic acid: Purrmann, *ibid.* 188. *See also* Xanthopterin. ¹³C NMR studies: G. Müller, W. von Philipsborn, *Helv. Chim. Acta* **56** 2680 (1973). Review on pterins: Purrmann, *Fortschr. Chem. Org. Naturst.* **4**, 64 (1945).

Fine colorless crystals forming yellow Na and Ag salts, and a sparingly sol NH_4 salt. Soluble in alkaline solns with blue fluorescence.

5507. Leukotrienes. LTs. A family of endogenous metabolites of arachidonic acid via the lipoxygenase pathway, chemically related to the prostaglandins and thromboxanes, *q.q.v.* The name leukotrienes was applied because of their origin in leukocytes and their conjugated triene structures. [For a description of the nomenclature of individual leukotrienes see B. Samuelsson, S. Hammarström, *Prostaglandins* **19**, 645 (1980)]. Members of the group are potent bronchoconstrictors that play an important pathophysiological role in immediate hypersensitivity reactions; they have been proposed as mediators of the inflammatory process and some are potent chemotactic agents. Initial studies of novel arachidonate metabolites in rabbit polymorphonuclear leukocytes: P. Borgeat *et al.*, *J. Biol. Chem.* **251**, 7816 (1976); **252**, 8772 (1977). Subsequent studies and structure of a dihydroxyeicosatetraenoic acid (leukotriene B, LTB_4): P. Borgeat, B. Samuelsson, *ibid.* **254**, 2643 (1979). LTB_4 is formed from an unstable intermediate oxido-eicosatetraenoic acid, leukotriene A or LTA_4: *eidem*, *Proc. Natl. Acad. Sci. USA* **76**, 3213 (1979). Stereochemistry and enzymatic conversion of LTA_4 to LTB_4: O. Radmark *et al.*, *Biochem. Biophys. Res. Commun.* **92**, 954 (1980). Synthesis of the four optical isomers of LTA_4: J. Rokach *et al.*, *Tetrahedron Lett.* **22**, 2759, 2763 (1981). Total synthesis of LTB_4: E. J. Corey *et al.*, *J. Am. Chem. Soc.* **102**, 7984 (1980); Y. Guindon *et al.*, *Tetrahedron Lett.* **23**, 739 (1982). Formation of the 20-hydroxy and 20-carboxy metabolites of LTB_4: G. Hansson *et al.*, *FEBS Lett.* **130**, 107 (1981); total synthesis: R. Zamboni, J. Rokach, *Tetrahedron Lett.* **23**, 4751 (1982). Earlier studies had described a "slow-reacting substance of anaphylaxis" (*SRS-A* or *SRS*) released from guinea pig and cat lung by cobra venom and in guinea pig lung after anaphylactic shock, *cf.* W. S. Feldberg, C. H. Kellaway, *J. Physiol.* **94**, 187 (1938); C. H. Kellaway, E. R. Trethewie, *Q. J. Exp. Physiol.* **30**, 121 (1940). The relationship between SRS and members of the leukotriene family was established following publication of the uv spectrum of purified SRS-A, which showed the presence of the conjugated triene, *cf.* H. R. Morris *et al.*, *FEBS Lett.* **87**, 203 (1978). Purification and structure of LTC_4, an SRS from mouse mastocytoma cells: R. C. Murphy *et al.*, *Proc. Natl. Acad. Sci. USA* **76**, 4275 (1979); S. Hammarström *et al.*, *Biochem. Biophys. Res. Commun.* **91**, 1266 (1979). Total synthesis of LTC_4: E. J. Corey *et al.*, *J. Am. Chem. Soc.* **102**, 1436, 3663 (1980); J. Rokach *et al.*, *Tetrahedron Lett.* **21**, 1485 (1980). Identity of synthetic LTC_4 with SRS from mouse mastocytoma cells: S. Hammarström *et al.*, *Biochem. Biophys. Res. Commun.* **92**, 946 (1980). Structure of the SRS from rat basophil leukemia cells (RBL-1) and identification as a leukotriene (LTD$_4$): H. R. Morris *et al.*, *Prostaglandins* **19**, 185 (1980). It was subsequently proposed that LTC$_4$ was an intermediate in the biosynthesis of LTD$_4$. Identity of SRS-A released in sensitized guinea pig lung perfusates and LTD$_4$: *eidem*, *Nature* **285**, 104 (1980). Assignment of stereochemistry: *eidem*, *Prostaglandins* **20**, 601 (1980). Detection of LTA$_4$ as an intermediate in the biosynthesis of LTC$_4$ and LTD$_4$: S. Hammarström, B. Samuelsson, *FEBS Lett.* **122**, 83 (1980). The sulfone of LTC$_4$, which has also been proposed as a natural product, *cf.* H. Ohnishi *et al.*, *Prostaglandins* **20**, 655 (1980), has been found to be as potent as LTC$_4$: T. Jones *et al.*, *ibid.* **24**, 279 (1982). Synthesis: Y. Girard *et al.*, *Tetrahedron Lett.* **23**, 1023 (1982). Discovery of LTF$_4$: M. E. Anderson *et al.*, *Proc. Natl. Acad. Sci. USA* **79**, 1088 (1982). Synthesis: F. Ellis *et al.*, *Tetrahedron Lett.* **23**, 3735 (1982). It is now known that SRS-A is made up of varying amounts of cysteine-containing members of the leukotrienes, *i.e.* leukotrienes C$_4$, D$_4$, and E$_4$. These three LTs are generally found to be 100 to 1000 times more potent, on a molar basis, than histamine or prostaglandins in their effects on pulmonary airways. Review of chemistry and structure elucidation: D. A. Clark, A. Marfat, *Annu. Rep. Med. Chem.* **17**, 291-300 (1982). Comprehensive review of synthesis of leukotrienes and other lipoxygenase-derived products: J. G. Atkinson, J. Rokach, in *Handbook of Eicosanoids: Prostaglandins and Related Lipids* IB, A. L. Willis, Ed. (CRC Press, Boca Raton, 1987) pp 175-263. General reviews: P. Sirois, P. Borgeat, *Int. J. Immunopharmacol.* **2**, 281-293 (1980); P. Borgeat, P. Sirois, *J. Med. Chem.* **24**, 121-126 (1981); B. Samuelsson, *Int. Arch. Allergy Appl. Immunol.* **66**, Suppl. 1, 98-106 (1981); L. S. Wolfe, *J. Neurochem.* **38**, 1-14 (1982); J. L. Marx, *Science* **215**, 1380-1384 (1982); B. Samuelsson, S. Hammarström, *Vitam. Horm.* **39**, 1-30 (1982). Books: *SRS-A and Leukotrienes*, P. J. Piper, Ed. (Research Studies Press, London, 1981) 279 pp; *Advances in Prostaglandin, Thromboxane, and Leukotriene Research* vol. 9, B. Samuelsson, R. Paoletti, Eds. (Raven Press, New York, 1982) 341 pp.

LTA$_4$

LTB$_4$

LTC$_4$ R =

LTD$_4$ R =

LTE$_4$ R =

Leukotriene A$_4$. [72059-45-1] (2S,3S)-3-(1E,3E,5Z,8Z)-1,3,5,8-Tetradecatetraenyloxiranebutanoic acid; leukotriene A; LTA$_4$. C$_{20}$-

H$_{30}$O$_3$; mol wt 318.46. Unstable. Characterized as its methyl ester: mp 28-32°. [α]$_D^{25}$ -27° (hexane). uv max (methanol): 270, 278, 290 nm (ε 43900, 56700, 43100), *cf.* J. Rokach *et al.*, *Tetrahedron Lett.* **22**, 2759 (1981); I. Ernest *et al.*, *ibid.* **23**, 167 (1982).

Leukotriene B$_4$. [71160-24-2] (5S,6Z,8E,10E,12R,14Z)-5,12-Dihydroxy-6,8,10,14-eicosatetraenoic acid; leukotriene B; LTB$_4$. C$_{20}$H$_{32}$O$_4$; mol wt 336.47. uv max (methanol): 260, 270.5, 281 nm (ε 38000, 50000, 39000).

Leukotriene C$_4$. [72025-60-6] L-γ-Glutamyl-S-[(1R,2E,4E,6Z,-9Z)-1-[(1S)-4-carboxy-1-hydroxybutyl]-2,4,6,9-pentadecatetraenyl]-L-cysteinylglycine; leukotriene C; leukotriene C$_1$; LTC$_4$. C$_{30}$-H$_{47}$N$_3$O$_9$S; mol wt 625.78. uv max (methanol): 270, 280, 290 nm (ε 32000, 40000, 31000). Can be stored for several days without appreciable decomposition in frozen (-20°) pH 6.8 phosphate buffer under argon or as the tripotassium salt frozen in water. Biological activity destroyed after incubation with soybean lipoxygenase and uv max shifts to 308 nm.

Leukotriene D$_4$. [73836-78-9] S-[(1R,2E,4E,6Z,9Z)-1-[(1S)-4-Carboxy-1-hydroxybutyl]-2,4,6,9-pentadecatetraenyl]-L-cysteinylglycine; leukotriene D; LTD$_4$. C$_{25}$H$_{40}$N$_2$O$_6$S; mol wt 496.66. uv max (methanol): 270, 280, 290 nm (ε 31000, 40000, 31000). Storage, destruction of biological activity, uv shift are the same as for LTC$_4$.

Leukotriene E$_4$. [75715-89-8] (5S,6R,7E,9E,11Z,14Z)-6-[[(2R)-2-Amino-2-carboxyethyl]thio]-5-hydroxy-7,9,11,14-eicosatetraenoic acid; leukotriene E; LTE$_4$. C$_{23}$H$_{37}$NO$_5$S; mol wt 439.61. uv max (ethanol) of methyl ester: 269, 280, 291 nm (ε 28200, 35200, 28900).

5508. Leupeptins. Class of modified tripeptide protease inhibitors produced by various species of *Actinomycetes*. Two major components, ***leupeptin Ac-LL*** and ***leupeptin Pr-LL***, consisting of L-leucyl-L-leucyl-DL-argininal modified at the amino terminal by acetyl- or propionyl-, respectively, have been isolated. Various minor analogs, in which valine or isoleucine replaces either or both leucines, have also been found. Isolation from *Actinomycetes*: T. Aoyagi *et al.*, *J. Antibiot.* **22**, 283 (1969); S.-I. Kondo *et al.*, *Chem. Pharm. Bull.* **17**, 1896 (1969). Structure and synthesis of leupeptins Ac-LL and Pr-LL: K. Kawamura *et al.*, *ibid.* 1902; K. Maeda *et al.*, *J. Antibiot.* **24**, 402 (1971). Improved purification of leupeptin Ac-LL: M. C. Y. Ning, R. J. Beynon, *Int. J. Biochem.* **18**, 813 (1986). Inhibition of proteases: T. Aoyagi *et al.*, *J. Antibiot.* **22**, 558 (1969); and synthesis of analogs: G. Borin *et al.*, *Z. Physiol. Chem.* **362**, 1435 (1981). Effects on protein degradation in normal and diseased muscle: P. Libby, A. L. Goldberg, *Science* **199**, 534 (1978); I. Nonaka *et al.*, *Acta Neuropathol.* **58**, 279 (1982); R. P. Hummel, III *et al.*, *J. Surg. Res.* **45**, 140 (1988). HPLC determn of leupeptin Ac-LL in serum and muscle: M. Kai *et al.*, *J. Chromatogr.* **345**, 259 (1985).

Note: In some sources leupeptin refers only to leupeptin Ac-LL.

USE: Enzyme inhibitor in biological preparations.

5509. Leuprolide. [53714-56-0] 6-D-Leucine-9-(*N*-ethyl-L-prolinamide)-1-9-luteinizing hormone-releasing factor (swine); 6-D-leucine-9-(*N*-ethyl-L-prolinamide)-10-deglycinamideluteinizing hormone-releasing factor (pig); leuprorelin; (D-Leu^6)-des-Gly^{10}-LH-RH-ethylamide. C$_{59}$H$_{84}$N$_{16}$O$_{12}$; mol wt 1209.42. C 58.59%, H 7.00%, N 18.53%, O 15.87%. Synthetic nonapeptide analog of gonadotropin-releasing hormone, *q.v.* Prepn: M. Fujino *et al.*, DE **2446005** (1975 to Takeda), *C.A.* **83**, 10895y (1975); R. L. Gendrich *et al.*, US **4005063** (1977 to Abbott). Synthesis: J. A. Vilchez-Martinez *et al.*, *Biochem. Biophys. Res. Commun.* **59**, 1226 (1974); M. Fujino *et al.*, *ibid.* **60**, 406 (1974); E. Nicolás *et al.*, *Tetrahedron* **53**, 3179 (1997). Comparison of biological activity with natural GnRH: D. H. Coy *et al.*, *Biochem. Biophys. Res. Commun.* **67**, 576 (1975). Pharmacokinetics: L. T. Sennello *et al.*, *J. Pharm. Sci.* **75**, 158 (1986). Clinical efficacy in prostatic carcinoma: M. B. Garnick *et al.*, *N. Engl. J. Med.* **311**, 1281 (1984); in benign prostatic hypertrophy: L. M. Eri, K. J. Tveter, *J. Urol.* **150**, 359 (1993). Clinical trial in endometriosis: J. M. Wheeler *et al.*, *Am. J. Obstet. Gynecol.* **167**, 1367 (1992). Review of extended release formulation adjuvant therapy in prostate cancer: R. Perez-Marrero, R. C. Tyler, *Expert Opin. Pharmacother.* **5**, 447-457 (2004).

5-oxoPro–His–Trp–Ser–Tyr–D-Leu–Leu–Arg–ProNHC$_2$H$_5$

Fluffy solid. [α]$_D^{25}$ -31.7° (c = 1 in 1% acetic acid).

Monoacetate. [74381-53-6] Leuprolide acetate; Abbott 43818; A-43818; TAP-144; Carcinil; Eligard; Enantone; Leuplin; Lucrin; Lupron; Prostap; Viadur. $C_{59}H_{84}N_{16}O_{12}.C_2H_4O_2$; mol wt 1269.47. Fine or fluffy, white to off-white powder. pKa 9.6. Very sol in water, ethanol, propylene glycol.

THERAP CAT: Antineoplastic (hormonal).

THERAP CAT (VET): Antineoplastic.

5510. Levallorphan. [152-02-3] 17-(2-Propen-1-yl)morphinan-3-ol; *l*-*N*-allyl-3-hydroxymorphinan; *l*-3-hydroxy-*N*-allylmorphinan. $C_{19}H_{25}NO$; mol wt 283.42. C 80.52%, H 8.89%, N 4.94%, O 5.64%. Prepn from 3-hydroxy-*N*-methylmorphinan: Schnider, Grüssner, *Helv. Chim. Acta* **34**, 2211 (1951). Alternate route: Hellerbach *et al.*, *ibid.* **39**, 429 (1956). Comprehensive description: B. C. Rudy, B. Z. Senkowski, *Anal. Profiles Drug Subs.* **2**, 339-361 (1973).

Crystals from dilute ethanol, mp 180-182°. $[\alpha]_D^{20}$ −88.9° (c = 3 in methanol).

Tartrate. [71-82-9] Lorfan. $C_{19}H_{25}NO.C_4H_6O_6$; mol wt 433.50. Crystals from ethanol, mp 176-177°. $[\alpha]_D^{16}$ −39°. Soluble in water.

THERAP CAT: Narcotic antagonist.

THERAP CAT (VET): Narcotic antagonist.

5511. Levamisole. [14769-73-4] (6*S*)-2,3,5,6-Tetrahydro-6-phenylimidazo[2,1-*b*]thiazole; (−)-6-phenyl-2,3,5,6-tetrahydroimidazo[2,1-*b*]thiazole; Levovermax; Totalon. $C_{11}H_{12}N_2S$; mol wt 204.29. C 64.67%, H 5.92%, N 13.71%, S 15.69%. Biological response modifier with anthelmintic activity. Prepn of racemate: A. H. M. Raeymaekers *et al.*, **US 3274209** (1966 to Janssen); *eidem, J. Med. Chem.* **9**, 545 (1966). Pharmacology: D. Thienpoint *et al.*, *Nature* **209**, 1084 (1966). Prepn and abs config of isomers: A. H. M. Raeymaekers *et al.*, *Tetrahedron Lett.* **1967**, 1467. *See also:* M. W. Bullock *et al.*, *J. Med. Chem.* **11**, 169 (1968); **US 3565907** (1971 to Am. Cyanamid); Dewar *et al.*, **US 3579530** (1971 to ICI). Stereospecific inhibition of succinic dehydrogenase: H. Van den Bossche, P. A. J. Janssen, *Biochem. Pharmacol.* **18**, 35 (1969). Exptl effect on IL-1 production: E. S. Kimball, *Ann. N.Y. Acad. Sci.* **685**, 259 (1993). Review of pharmacology: H. Schneiden, *Int. J. Immunopharmacol.* **3**, 9 (1981); of clinical use in parasitic infections: M. J. Miller, *Drugs* **20**, 122-130 (1980). Determn in animal tissues by ELISA: J. J. Silverlight, R. Jackman: *Analyst* **119**, 2705 (1994); by LC-MS: A. Cannavan *et al.*, *ibid.* **120**, 331 (1995). Clinical trial with fluorouracil in colon cancer: C. G. Moertel *et al.*, *Ann. Intern. Med.* **122**, 321 (1995). Review of immunopharmacology and clinical trials in cancer treatment: H. C. Stevenson *et al.*, *J. Clin. Oncol.* **9**, 2052-2066 (1991); M. De Brabander *et al.*, *Anticancer Res.* **12**, 177-188 (1992); W. K. Amery, J. P. Bruynseels, *Int. J. Immunopharmacol.* **14**, 481-492 (1992).

mp 60-61.5°. $[\alpha]_D^{25}$ −85.1° (c = 10 in chloroform).

Hydrochloride. [16595-80-5] R-12564; Ascaridil; Decaris; Ergamisol; Levacide; Levadin; Levasole; Meglum; Nemicide; Nilverm; Ripercol; Solaskil; Spartakon; Tramisol. $C_{11}H_{12}N_2S.HCl$; mol wt 240.75. White or almost white, crystalline powder. mp 227-229°. $[\alpha]_D^{20}$ −124 ± 2° (c = 0.9, water). Freely sol in water; sol in alc; slightly sol in methylene chloride. Practically insol in ether. Stable in aq acid media.

DL-Form. [5036-02-2] Tetramisole; tetramizole. Crystals, mp 87-89°.

DL-Form hydrochloride. [5086-74-8] Bayer 9051; McN-JR-8299; R-8299. Crystals, mp 264-265°. Sol in water (21 g/100 ml at 20°), methanol, propylene glycol; sparingly sol in ethanol. Slightly sol in chloroform, hexane, acetone. LD_{50} in mice, rats (mg/kg): 22, 24 i.v.; 84, 130 s.c.; 210, 480 orally (Thienpoint).

D-(+)-Form. Dexamisole. mp 60-61.5°. $[\alpha]_D^{25}$ +85.1° (c = 10 in chloroform).

D-(+)-Form hydrochloride. R-12563. mp 227-227.5°. $[\alpha]_D^{20}$ +125° (c = 0.7 in water).

THERAP CAT: Anthelmintic (nematodes); immunomodulator.

THERAP CAT (VET): Anthelmintic (Nematodes).

5512. Levcromakalim. [94535-50-9] (3*S*,4*R*)-3,4-Dihydro-3-hydroxy-2,2-dimethyl-4-(2-oxo-1-pyrrolidinyl)-2*H*-1-benzopyran-6-carbonitrile; (3*S*,4*R*)-3-hydroxy-2,2-dimethyl-4-(2-oxo-1-pyrrolidinyl)-6-chromancarbonitrile; (−)-6-cyano-3,4-dihydro-2,2-dimethyl-*trans*-4-(2-oxo-1-pyrrolidinyl)-2*H*-benzo[*b*]pyran-3-ol; lemakalim; (−)-cromakalim; BRL-38227. $C_{16}H_{18}N_2O_3$; mol wt 286.33. C 67.12%, H 6.34%, N 9.78%, O 16.76%. Potassium channel opener. Prepn of racemate: J. M. Evans *et al.*, **EP 76075**; *eidem,* **US 4446113** (1983, 1984 both to Beecham); of isomers: E. Faruk, **EP 120428** (1984 to Beecham); V. A. Ashwood *et al.*, *J. Med. Chem.* **29**, 2194 (1986). Mechanism of action and hemodynamic effects: R. P. Hof *et al.*, *Circ. Res.* **62**, 679 (1988). Pharmacology: J. C. Clapham *et al.*, *Arzneim.-Forsch.* **41**, 385 (1991). Clinical evaluation in hypertension: S. Suzuki *et al.*, *ibid.* **45**, 859 (1995).

Crystals from ethyl acetate, mp 242-244°. $[\alpha]_D^{26}$ −52.2° (c = 1 in chloroform).

(+)-Form. Crystals from ethyl acetate, mp 243-245°. $[\alpha]_D^{26}$ +53.5° (c = 1 in chloroform).

(±)-Form. [94470-67-4] Cromakalim; BRL-34915. Crystals from ethyl acetate, mp 230-231°.

THERAP CAT: Antihypertensive.

5513. Levetiracetam. [102767-28-2] (α*S*)-α-Ethyl-2-oxo-1-pyrrolidineacetamide; 2(*S*)-(2-oxopyrrolidin-1-yl)butyramide; UCB-L059; SIB-S1; Keppra. $C_8H_{14}N_2O_2$; mol wt 170.21. C 56.45%, H 8.29%, N 16.46%, O 18.80%. (*S*)-Enantiomer of the ethyl analog of piracetam, *q.v.* Prepn: J. Gobert *et al.*, **EP 162036**; *eidem,* **US 4943639** (1985, 1990 both to UCB); F. Boschi *et al.*, *Tetrahedron: Asymmetry* **16**, 3739 (2005). HPLC-UV determn in plasma: J. Martens-Lobenhoffer, S. M. Bode-Böger, *J. Chromatogr. B* **819**, 197 (2005). Clinical evaluation in refractory partial seizures: S. D. Shorvon *et al.*, *Epilepsia* **41**, 1179 (2000); E. Ben-Menachem, U. Falter, *ibid.* 1276. Review of pharmacokinetics: P. N. Patsalos, *Pharmacol. Ther.* **85**, 77-85 (2000); of clinical efficacy: E. Ben-Menachem, *Expert Opin. Pharmacother.* **4**, 2079-2088 (2003); of safety and tolerability: D. E. Briggs, J. A. French, *Expert Opin. Drug Saf.* **3**, 415-424 (2004).

Crystals from ethyl acetate, mp 117°. $[\alpha]_D^{25}$ −90.0° (c = 1 in acetone). Soly (g/100 ml): water 104.0; chloroform 65.3; methanol 53.6; ethanol 16.5; acetonitrile 5.7. Practically insol in *n*-hexane. LD_{50} in male mice, male rats (mg/kg): 1081, 1038 i.v. (Gobert, 1990).

THERAP CAT: Anticonvulsant.

THERAP CAT (VET): Anticonvulsant.

Consult the Name Index before using this section.

5514. Levobunolol. [47141-42-4] 5-[(2S)-3-[(1,1-Dimethylethyl)amino]-2-hydroxypropoxy]-3,4-dihydro-1(2H)-naphthalenone; (−)-5-[3-(tert-butylamino)-2-hydroxypropoxy]-3,4-dihydro-1(2H)-naphthalenone; l-bunolol; W-6421A. C$_{17}$H$_{25}$NO$_3$; mol wt 291.39. C 70.07%, H 8.65%, N 4.81%, O 16.47%. Nonselective β-adrenoceptor antagonist. Prepn of racemate: C. F. Schwender et al., J. Med. Chem. **13**, 684 (1970); J. Shavel, Jr., S. Farber, DE 1948144; eidem, US 3641152 (1970, 1972 both to Warner-Lambert). Resolution of enantiomers: J. Shavel, Jr., C. F. Schwender, DE 2046043; eidem, US 3649691 (1971, 1972 both to Warner-Lambert); R. D. Dennis et al., US 4463176 (1984 to Bristol Myers, Mead Johnson). Cardiovascular pharmacology, anti-arrhythmic activity of racemate and isomers: R. D. Robson, H. R. Kaplan, J. Pharmacol. Exp. Ther. **175**, 157, 168 (1970). Adrenoceptor blocking activity: H. R. Kaplan et al., Eur. J. Pharmacol. **16**, 237 (1971). Metabolism in humans: F. J. DiCarlo et al., Clin. Pharmacol. Ther. **22**, 858 (1977); F.-J. Leinweber et al., Pharmacology **16**, 70 (1978). Corneal permeability: R. D. Schoenwald, H.-S. Huang, J. Pharm. Sci. **72**, 1266, 1272 (1983). HPLC determn in biological fluids: H. Hengy, E. U. Kolle, J. Chromatogr. **338**, 444 (1985); D. D. S. Tang-Liu et al., J. Liq. Chromatogr. **9**, 2237 (1986). Clinical evaluation in hypertension: E. Arce-Gomez et al., Curr. Ther. Res. **19**, 386 (1976); comparison with timolol, q.v., in glaucoma and ocular hypertension: A. Cinotti et al., Am. J. Ophthalmol. **99**, 11 (1985); D. Long et al., ibid. 18. Review: H. R. Kaplan, "Levobunolol" in Pharmacology of Antihypertensive Drugs, A. Scriabine, Ed. (Raven Press, New York, 1980) pp 317-323. Brief review of long-term treatment of glaucoma: G. D. Novack, Gen. Pharmacol. **17**, 373-377 (1986).

LD$_{50}$ in male, female rats, mice (mg/kg): 700, 800, 1530, 1220 orally; 25, 28, 78, 84 i.v.; LD$_{50}$ in male, female hamsters, dogs (mg/kg): 435, 500, 100, 100 orally (Kaplan, 1980).

Hydrochloride. [27912-14-7] W-7000A; Betagan; Vistagan. C$_{17}$H$_{25}$NO$_3$.HCl; mol wt 327.85. Crystals from methanol-ether, mp 209-211°. [α]$_{589}^{24}$ −19.6±0.7° (c = 2.90 in methanol). uv max (NaOH): 221, 253, 310 nm (ε 24700, 9000, 2400). Sol in water, methanol; slightly sol in alc, chloroform.

THERAP CAT: Antiglaucoma.

5515. Levocabastine. [79516-68-0] (3S,4R)-1-[cis-4-Cyano-4-(4-fluorophenyl)cyclohexyl]-3-methyl-4-phenyl-4-piperidinecarboxylic acid; (−)-trans-1-[cis-4-cyano-4-(p-fluorophenyl)cyclohexyl]-3-methyl-4-phenylisonipecotic acid; (−)-cabastine. C$_{26}$H$_{29}$FN$_2$O$_2$; mol wt 420.53. C 74.26%, H 6.95%, F 4.52%, N 6.66%, O 7.61%. Histamine H$_1$-receptor antagonist. Prepn: R. Stokbroekx et al., EP 34415; eidem, US 4369184 (1981, 1983 both to Janssen); and pharmacology: R. A. Stokbroekx et al., Drug Dev. Res. **8**, 87 (1986). Pharmacology: K. Tasaka et al., Arzneim.-Forsch. **40**, 1295 (1990). Clinical pharmacology: N. Rombaut et al., Ann. Allergy **67**, 75 (1991). Clinical trial in children with allergic conjunctivitis: M. Rimas et al., Allergy **45**, 18 (1990); in allergic rhinitis: M. Schata et al., J. Allergy Clin. Immunol. **87**, 873 (1991). Review of pharmacology and therapeutic use: K. L. Dechant, K. L. Goa, Drugs **41**, 202-224 (1991).

Hydrochloride. [79547-78-7] R-50547; Levophta; Livostin. C$_{26}$H$_{29}$FN$_2$O$_2$.HCl; mol wt 456.99.
THERAP CAT: Antihistaminic.

5516. Levodopa. [59-92-7] 3-Hydroxy-L-tyrosine; (−)-3-(3,4-dihydroxyphenyl)-L-alanine; L-dopa; β-(3,4-dihydroxyphenyl)-L-alanine; (−)-2-amino-3-(3,4-dihydroxyphenyl)propanoic acid; Bendopa; Deadopa; Dopaflex; Dopal; Dopaidan; Dopalina; Dopar; Doparkine; Doparl; Dopasol; Dopaston; Dopastral; Cidandopa; Doprin; Eldopal; Eldopar; Eldopatec; Eurodopa; Laradopa; Maipedopa; Larodopa; Ledopa; Parda; Levopa; Veldopa (formerly Weldopa). C$_9$H$_{11}$NO$_4$; mol wt 197.19. C 54.82%, H 5.62%, N 7.10%, O 32.45%. Naturally occurring form of dopa, q.v., the biological precursor of the catecholamines. Prepn from l-3-nitrotyrosin: Wasser, Lewandowski, Helv. Chim. Acta **4**, 657 (1921); from 3-(3,4-methylenedioxyphenyl)-L-alanine: Yamada et al., Chem. Pharm. Bull. **10**, 693 (1962); from L-tyrosine: Vorbrüggen, Krolikiewicz, Ber. **105**, 1168 (1972); Bretschneider et al., Helv. Chim. Acta **56**, 2857 (1973); from Vicia faba beans: Wysong, US 3253023 (1966 to Dow Chem.); by fermentation of L-tyrosine: Sih et al., J. Am. Chem. Soc. **91**, 6204 (1969); Florent, Renaut, DE 2102793 (1971 to Rhône-Poulenc), C.A. **75**, 108505f (1971). Sepn from racemate: Vogler, Baumgartner, Helv. Chim. Acta **35**, 1776 (1952); NL 6514950; US 3405159 (1966, 1968 both to Merck & Co.). Molecular conformation: Becker et al., Biochem. Biophys. Res. Commun. **41**, 444 (1970). Metabolism studies: Shaw et al., J. Biol. Chem. **226**, 255 (1957); Calne et al., Br. J. Pharmacol. **37**, 57 (1969). Hemodynamic effects in congestive heart failure: S. I. Rajfer et al., N. Engl. J. Med. **310**, 1357 (1984). Series of articles on clinical efficacy in Parkinson's disease: Adv. Neurol. **45**, 457-510 (1986). Reviews on L-dopa and parkinsonism: Barbeau, Can. Med. Assoc. J. **101**, 791 (1969); Pletscher et al., Schweiz. Med. Wochenschr. **100**, 797 (1970); Calne, Sandler, Nature **226**, 21 (1970); L-Dopa and Parkinsonism, A. Barbeau, Ed. (F. A. Davis, Philadelphia, 1970). Review of acute toxicity data: W. G. Clark et al., Toxicol. Appl. Pharmacol. **28**, 1-7 (1974). Comprehensive description: R. Gomez et al., Anal. Profiles Drug Subs. **5**, 189-223 (1976).

Colorless to white, odorless and tasteless crystals or crystalline powder. Needles from water, mp 276-278° (dec) (Yamada); also reported as mp 284-286° (Wysong). [α]$_D^{13}$ −13.1° (c = 5.12 in 1N HCl). uv max (0.001N HCl): 220.5, 280 nm (log ε 3.79, 3.42). Readily sol in dil HCl and formic acid. Soly in water: 66 mg/40 ml. Practically insol in ethanol, benzene, chloroform and ethyl acetate. In the presence of moisture, L-dopa is rapidly oxidized by atmospheric oxygen and darkens. LD$_{50}$ in mice (mg/kg): 3650 ±327 orally, 1140 ±66 i.p., 450 ±42 i.v., >400 s.c.; in male, female rats (mg/kg): >3000, >3000 orally; 624, 663 i.p.; >1500, >1500 s.c. (Clark).

THERAP CAT: Antiparkinsonian.

5517. Levoglucosenone. [37112-31-5] (1S,5R)-6,8-Dioxabicyclo[3.2.1]oct-2-en-4-one; 1,6-anhydro-3,4-dideoxy-β-D-glycerohex-3-enopyranos-2-ulose; 1,6-anhydro-3,4-dideoxy-Δ^3-β-D-pyranosen-2-one. C$_6$H$_6$O$_3$; mol wt 126.11. C 57.15%, H 4.80%, O 38.06%. (−)-Form is a pyrolysis product of cellulose and cellulose-containing materials including pulp and paper waste products. Prepn from cellulosic materials: Y. Halpern et al., J. Org. Chem. **38**, 204 (1973); and decompn reactions: F. Shafizadeh, P. P. S. Chin, Carbohydr. Res. **46**, 149 (1976). Prepn from Kraft paper: eidem, ibid. **58**, 79 (1977). Synthesis of (−)-form: M. Shibagaki et al., Chem. Lett. **1990**, 307; of (+) and (−) enantiomers: T. Taniguchi et al., Synlett **1996**, 971. MS analysis: Y. Halpern, J. P. Hoppesch, J. Org. Chem. **50**, 1556 (1985). Electrochemistry: C. Z. Smith et al., J. Chem. Res. Synop. **1987**, 88. Stereoselective reactivity: F. Shafizadeh et al., Carbohydr. Res. **71**, 169 (1979). Cycloaddition reactions: idem et al., ibid. **114**, 71 (1983); A. J. Blake et al., Tetrahedron **48**, 8053 (1992). Michael addition reactions: M. G. Essig, Carbohydr. Res. **156**, 225 (1986); A. V. Samet et al., ibid. **91**, 8786 (1996). Applications in chiral carbohydrate synthesis: Y. Gelas-Mialhe et al., Heterocycles **24**, 931 (1986); Z. J. Witczak, Pure Appl. Chem. **66**, 2189 (1994); R. Blattner, D. M. Page, J. Carbohydr. Chem. **13**, 27 (1994).

Faintly greenish-yellow liquid. n_D^{25} 1.5084. uv max (*n*-hexane): 211, 275 nm (log $E_{1cm}^{1\%}$ 2.82, 1.5). uv max (95% ethanol): 218, 275 nm (log $E_{1cm}^{1\%}$ 2.78, 1.5). $[\alpha]_D^{25}$ $-460°$ (c = 1.0 in CHCl$_3$) (Halpern, 1973); also reported as $[\alpha]_D^{31}$ $-514.8°$ (c = 0.3 in CHCl$_3$) (Taniguchi).

USE: Chiral building block in organic synthesis.

5518. Levomepromazine. [60-99-1] (*βR*)-2-Methoxy-*N*,*N*,-*β*-trimethyl-10*H*-phenothiazine-10-propanamine; (−)-10-(3-dimethylamino-2-methylpropyl)-2-methoxyphenothiazine; methotrimeprazine; 2-methoxytrimeprazine; levomeprazine; RP-7044. C$_{19}$H$_{24}$N$_2$OS; mol wt 328.47. C 69.48%, H 7.37%, N 8.53%, O 4.87%, S 9.76%. Phenothiazine derivative with antipsychotic and analgesic properties. Prepn: Courvoisier *et al.*, *C. R. Seances Soc. Biol. Ses Fil.* **151**, 1378 (1957); R. M. Jacob, J. G. Robert, US 2837518 (1958 to Rhône-Poulenc). HPLC determn in plasma: T. Loennechen, S. G. Dahl, *J. Chromatogr.* **503**, 205 (1990). Determn in pharmaceutical preparations: E. R. M. Kedor-Hackmann *et al.*, *Drug Dev. Ind. Pharm.* **26**, 261 (2000). Clinical pharmacokinetics: S. G. Dahl, *Clin. Pharmacol. Ther.* **19**, 435 (1976). Evaluation in treatment-resistant schizophrenia: S. Lal *et al.*, *J. Psychiatry Neurosci.* **31**, 271 (2006). Review of pharmacology and clinical efficacy in psychiatry: B. Green *et al.*, *Curr. Med. Res. Opin.* **20**, 1877-1881 (2004).

Crystals from heptane or acetone, mp 116-118°. pKa 8.8. $[\alpha]_D^{20}$ $-12°$ (c = 5 in chloroform). Store in airtight containers; protect from light.

Maleate. [7104-38-5] Milezin; Neurocil (tabl.); Nozinan (tabl.); Sinogan (tabl.); Tisercin. C$_{19}$H$_{24}$N$_2$OS.C$_4$H$_4$O$_4$; mol wt 444.55. Crystals, darkened by light. Dec about 190°. Sparingly sol in water (0.3% at 20°) and in ethanol (0.4%). pH of a 0.3% aq soln is 4.3.

Hydrochloride. [1236-99-3] Neurocil (inj.); Nozinan (inj.); Sinogan (inj.). C$_{19}$H$_{24}$N$_2$OS.HCl; mol wt 364.93. White or slightly yellow, slightly hygroscopic crystalline powder. Freely sol in water, alcohol.

THERAP CAT: Antipsychotic.

5519. Levomethadyl Acetate. [1477-40-3] (*αS*)-*β*-[(2*S*)-2-(Dimethylamino)propyl]-*α*-ethyl-*β*-phenylbenzeneethanol 1-acetate; (−)-6-(dimethylamino)-4,4-diphenyl-3-heptanol acetate (ester); *α*-*l*-acetylmethadol; *levo*-*α*-acetylmethadol; LAAM. C$_{23}$H$_{31}$NO$_2$; mol wt 353.51. C 78.15%, H 8.84%, N 3.96%, O 9.05%. Longest-acting enantiomer of methadyl acetate, *q.v.* Duration of action due to active metabolites. Prepn: A. Pohland *et al.*, *J. Am. Chem. Soc.* **71**, 460 (1949). Synthesis of metabolites: F. I. Carroll *et al.*, *J. Org. Chem.* **41**, 3521 (1976). Metabolism in rats: G. L. Henderson *et al.*, *Drug Metab. Dispos.* **5**, 321 (1977); M. Man *et al.*, *ibid.* **8**, 55 (1980); in man: B. S. Finkle *et al.*, *J. Anal. Toxicol.* **6**, 100 (1982). Determn of LAAM and metabolites by HPLC: C.-H. Kiang *et al.*, *J. Chromatogr.* **222**, 81 (1981); by GLC: K. Verebey *et al.*, *ibid.* **343**, 339 (1985). Analgesic activity in mice: N. B. Eddy *et al.*, *J. Org. Chem.* **17**, 321 (1952); in man: A. S. Keats, H. K. Beecher, *J. Pharmacol. Exp. Ther.* **105**, 210 (1952). Pharmacology in monkeys: S. J. Mule, A. L. Misra, *Ann. N.Y. Acad. Sci.* **311**, 199 (1978). Comparison with methadone in the treatment of heroin addiction: T. J. Crowley *et al.*, *Psychopharmacology* **86**, 458 (1985). Review of use in treatment of narcotic addiction: J. D. Blaine *et al.*, *Ann. N.Y. Acad. Sci.* **362**, 101-115 (1981).

Hydrochloride. [43033-72-3] ORLAAM. C$_{23}$H$_{31}$NO$_2$.HCl; mol wt 389.96. Crystals from ethanol-ether, mp 215°. $[\alpha]_D^{25}$ $-60°$ (c = 0.2). Sol in water. LD$_{50}$ in mice (mg/kg): 110.0 s.c., 172.8 orally (Eddy).

Note: This is a controlled substance (opiate): **21 CFR**, 1308.12.

THERAP CAT: In treatment of narcotic addiction.

5520. Levomethorphan. [125-70-2] 3-Methoxy-17-methyl-morphinan; *l*-methorphan. C$_{18}$H$_{25}$NO; mol wt 271.40. C 79.66%, H 9.29%, N 5.16%, O 5.89%. Opioid analgesic. Prepn of racemate: O. Schnider, A. Grüssner, *Helv. Chim. Acta* **32**, 821 (1949); *eidem*, US 2524856 (1950 to Hoffmann-La Roche); of optically active forms: *eidem*, *Helv. Chim. Acta* **34**, 2211 (1951); *eidem*, US 2676177 (1954 to Hoffmann-La Roche). Configuration of *l*-form: H. Corrodi *et al.*, *Helv. Chim. Acta* **42**, 212 (1959). Analgesic activity and metabolism: M. J. Cooper, M. W. Anders, *Life Sci.* **15**, 1665 (1974). LC-MS/MS-determn in biological samples: R. Kikura-Hanajiri *et al.*, *Anal. Bioanal. Chem.* **400**, 165 (2011).

Crystals from alcohol + water, mp 109-111°. $[\alpha]_D^{20}$ $-49.3°$ (c = 1.5 in alcohol).

Hydrobromide. [125-68-8] Ro-1-5470/6; Ro-1-7788. C$_{18}$H$_{25}$-NO.HBr; mol wt 352.32. Crystals as the monohydrate, mp 124-126°. $[\alpha]_D^{20}$ $-26.3°$ (c = 1.5 in water).

dl-Form. [510-53-2] Racemethorphan; *dl-cis*-1,3,4,9,10,10a-hexahydro-6-methoxy-11-methyl-2*H*-10,4a-iminoethanophenan-threne; deoxydihydrothebacodine. Crystals, mp 81-83°.

dl-Form Hydrobromide. [6031-86-3] Racemethorphan hydrobromide; Ro-1-5470. Crystals, mp 239-240°; as the monohydrate, mp 92-94°.

d-Form *see* Dextromethorphan.

Note: Levomethorphan and racemethorphan are controlled substances (opiates): **21 CFR**, 1308.12.

5521. Levophacetoperane. [24558-01-8] *rel*-(*αR*,2*R*)-(−)-*α*-Phenyl-2-piperidinemethanol 2-acetate; *threo*-1-acetoxy-1-phenyl-1-(2-piperidyl)methane; acetic acid *α*-phenyl-2-piperidylmethyl ester; 1-phenyl-1-(2′-piperidyl)-1-acetoxymethane; phacetoperane; RP-8228. C$_{14}$H$_{19}$NO$_2$; mol wt 233.31. C 72.07%, H 8.21%, N 6.00%, O 13.71%. Prepn: Jacob, Joseph, US 2928835 (1960 to Rhône-Poulenc).

Relative stereochemistry

Hydrochloride. [23257-56-9] Lidepran. $C_{14}H_{19}NO_2$·HCl; mol wt 269.77. Crystals from acetone + ether, mp 229-230°. Levorotatory.

THERAP CAT: Antidepressant; anorexic.

5522. Levopimaric Acid. [79-54-9] (1R,4aR,4bS,10aR)-1,2,-3,4,4a,4b,5,9,10,10a-Decahydro-1,4a-dimethyl-7-(1-methylethyl)-1-phenanthrenecarboxylic acid; 13-isopropylpodocarpa-8(14),12-dien-15-oic acid; $\Delta^{6,8(14)}$-abietadienoic acid; l-pimaric acid; β-pimaric acid; l-sapietic acid. $C_{20}H_{30}O_2$; mol wt 302.46. C 79.42%, H 10.00%, O 10.58%. Isolation from American pine oleoresin: Palkin, Harris, J. Am. Chem. Soc. **55**, 3677 (1933); from French galipot, from Pinus maritima: Ruzicka, Bacon, Helv. Chim. Acta **20**, 1542 (1937); from Pinus palustris: Harris, Sanderson, J. Am. Chem. Soc. **70**, 334, 3671 (1948). Structure: Ruzicka, Kaufmann, Helv. Chim. Acta **23**, 1346 (1940); cf. Arbuzov, Chem. Zentralbl. **1942**, II, 893. Stereochemistry: Schuller, Lawrence, J. Am. Chem. Soc. **83**, 2563 (1961); Burgstahler et al., ibid. **83**, 4660 (1961); Weiss et al., Chem. Ind. (London) **1962**, 1286; Dauben, Coates, J. Org. Chem. **28**, 1698 (1963). Conformation: Burgstahler et al., Chem. Commun. **1971**, 121; Weiss et al., ibid. **1972**, 17.

Orthorhombic crystals, mp 150°; $[\alpha]_D^{20}$ −280.4° (c = 0.7 in alcohol); $[\alpha]_D^{14}$ −266.6° (c = 0.4 in chloroform). Absorption max 273 nm: Kraft, Ann. **520**, 133 (1935). Sol in most organic solvents. Practically insol in water. Forms a crystalline ammonium salt.

Methyl ester. $C_{21}H_{32}O_2$. Crystals from methanol, mp 64°. $[\alpha]_D^{20}$ −268° (c = 1 in ether).

Molecular compound with quinone. $C_{26}H_{34}O_4$. Yellow prisms from methanol, mp 214°. $[\alpha]_D^{20}$ −148° (c = 0.7 in chloroform).

5523. Levopropoxyphene. [2338-37-6] (αR)-α-[(1S)-2-(Dimethylamino)-1-methylethyl]-α-phenylbenzeneethanol 1-propanoate; α-l-4-dimethylamino-3-methyl-1,2-diphenyl-2-butanol propionate; α-l-4-dimethylamino-1,2-diphenyl-3-methyl-2-butanol propionate; l-propoxyphene. $C_{22}H_{29}NO_2$; mol wt 339.48. C 77.84%, H 8.61%, N 4.13%, O 9.43%. Prepn: Pohland, Sullivan, J. Am. Chem. Soc. **77**, 3400 (1955). Stereoselective synthesis: Pohland et al., J. Org. Chem. **28**, 2483 (1963). Toxicity data: E. I. Goldenthal, Toxicol. Appl. Pharmacol. **18**, 185 (1971). See also Propoxyphene.

Crystals from petr ether, mp 75-76°. $[\alpha]_D^{25}$ −68.2° (c = 0.6 in chloroform).

Hydrochloride. [1596-70-9] $C_{22}H_{29}NO_2$·HCl; mol wt 375.94. Crystals from methanol + ethyl acetate, mp 163-164°. $[\alpha]_D^{20}$ −60.1° (c = 0.7).

2-Naphthalenesulfonate. [5714-90-9] Levopropoxyphene napsylate; Novrad; Letusin; Contratuss. $C_{22}H_{29}NO_2$·$C_{10}H_8SO_3$; mol wt 547.71. LD_{50} orally in female rats: 1455 ±77 mg/kg (Goldenthal).

THERAP CAT: Antitussive.

5524. Levorphanol. [77-07-6] 17-Methylmorphinan-3-ol; (−)-3-hydroxy-N-methylmorphinan; levorphan; lemoran; Ro-1-

5431. $C_{17}H_{23}NO$; mol wt 257.38. C 79.33%, H 9.01%, N 5.44%, O 6.22%. Orally active synthetic morphine analog. Prepn of racemate from 2-methyl-1-benzyl-1,2,3,4,5,6,7,8-octahydroisoquinoline: Grewe, Naturwissenschaften **33**, 333 (1946); Angew. Chem. **A59**, 198 (1947); Grewe, Mondon, Ber. **81**, 279 (1948); **CH 280674** (1952 to Hoffmann-La Roche). Prepn of isomers: Schnider, Grüssner, Helv. Chim. Acta **34**, 2211 (1951); Vogler, **US 2744112** (1956 to Hoffmann-La Roche). Absolute configuration: Corrodi et al., Helv. Chim. Acta **42**, 212 (1959). Analgesic activity and toxicity data: L. O. Randall, G. Lehmann, J. Pharmacol. Exp. Ther. **99**, 163 (1950). HPLC determn in plasma: R. Lucek, R. Dixon, J. Chromatogr. **341**, 239 (1985). Clinical pharmacokinetics: R. Dixon et al., Res. Commun. Chem. Pathol. Pharmacol. **41**, 3 (1983).

Crystals, mp 198-199°. $[\alpha]_D^{20}$ −56° (c = 3 in absolute alcohol).

Tartrate dihydrate. [5985-38-6] Ro-1-5431/7; Dromoran; Levo-Dromoran. $C_{17}H_{23}NO$·$C_4H_6O_6$·2H_2O; mol wt 443.49. Crystals, mp 113-115° (when anhydrous, mp 206-208°). $[\alpha]_D^{20}$ −14° (c = 3 in water). pH of a 0.2% aq soln 3.4 to 4.0. One gram dissolves in 45 ml water, in 110 g alcohol, in 50 g ether. Insol in chloroform.

dl-Form. [297-90-5] Racemorphan; methorphinan. Crystals from anisole and dil alcohol, mp 251-253°.

dl-Form hydrobromide. [5985-35-3] Nu-2206. $C_{17}H_{23}NO$·HBr; mol wt 338.29. Crystals, mp 193-195°. Sol in water; sparingly sol in alcohol. Practically insol in ether. LD_{50} i.v. in mice: 41 mg/kg (Randall, Lehmann).

d-Form. [125-73-5] Dextrorphan; Ro-1-6794. HPLC-MS/MS determn in urine: M. L. Constanzer et al., J. Chromatogr. B **816**, 297 (2005). Crystals, mp 198-199°. $[\alpha]_D^{20}$ +56.3° (c = 3 in abs alcohol).

d-Form tartrate monohydrate. $C_{17}H_{23}NO$·$C_4H_6O_6$·H_2O. Crystals, mp 183-185°. $[\alpha]_D^{20}$ +34.6° (c = 3 in water). Sol in water.

Note: Levorphanol and racemorphan are controlled substances (opiates): **21 CFR,** 1308.12.

THERAP CAT: Analgesic.

5525. Levosimendan. [141505-33-1] 2-[2-[4-[(4R)-1,4,5,6-Tetrahydro-4-methyl-6-oxo-3-pyridazinyl]phenyl]hydrazinylidene]-propanedinitrile; [[4-[(4R)-1,4,5,6-tetrahydro-4-methyl-6-oxo-3-pyridazinyl]phenyl]hydrazono]propanedinitrile; mesoxalonitrile (−)-[p-[(R)-1,4,5,6-tetrahydro-4-methyl-6-oxo-3-pyridazinyl]phenyl]-hydrazone; (R)-simendan; OR-1259; Simdax. $C_{14}H_{12}N_6O$; mol wt 280.29. C 59.99%, H 4.32%, N 29.98%, O 5.71%. Bioactive enantiomer of racemate, simendan. Positive inotropic agent with vasodilating activity. Binds to and sensitizes the myocardial contractile protein, troponin C, to calcium, thereby stabilizing the troponin conformation needed to trigger muscle contraction. Prepn: R. J. Backstrom et al., **GB 2251615**; P. Nore et al., **US 5569657** (1992, 1996 both to Orion). Pharmacology: A. F. E. Rump et al., Pharmacol. Toxicol. **74**, 244 (1994). Binding studies: P. Pollesello et al., J. Biol. Chem. **269**, 28584 (1994). HPLC determn in plasma: M. Karlsson et al., Biomed. Chromatogr. **11**, 54 (1997). Clinical pharmacokinetics: E.-P. Sandell et al., J. Cardiovasc. Pharmacol. **26**, Suppl. 1, S57 (1995). Review of mechanism of action: H. Haikala, I.-B. Lindén, ibid. S10-S19 (1995); of pharmacology, toxicology and clinical evaluation: P. S. Pagel et al., Cardiovasc. Drug Rev. **14**, 286-316 (1996). Comparison with dobutamine, q.v., in heart failure: F. Follath et al., Lancet **360**, 196 (2002).

Yellow crystalline powder, mp 210-214°. $[\alpha]_D^{25}$ −566° (tetrahydrofurane/methanol). Soluble in water. pKa 6.3. LD_{50} in male, female mice, male rats (mg/kg): 156, 152, 103 orally; 32, 50, 57 i.v. (Pagel).

THERAP CAT: Cardiotonic.

5526. Levulinic Acid. [123-76-2] 4-Oxopentanoic acid; β-acetylpropionic acid; laevulinic acid; 4-oxovaleric acid. $C_5H_8O_3$; mol wt 116.12. C 51.72%, H 6.94%, O 41.33%. Laboratory procedure from starch or cane sugar by boiling with HCl: McKenzie, *Org. Synth.* **coll. vol. I**, 335 (1941). Produced commercially from low grade cellulose. By-product of furfural manuf. Extensive review: Leonard, *Ind. Eng. Chem.* **48**, 1331 (1956).

Plates or leaflets (commercial product is yellow), mp 33-35°. bp 245-246°. d 1.1447. n_D^{16} 1.442. Freely sol in water, alcohol, ether; essentially insol in aliphatic hydrocarbons. *Protect from light.*

Phenylhydrazone. [588-60-3] Antithermin. $C_{11}H_{14}N_2O_2$; mol wt 206.25. Leaflets, mp 108°. Slightly sol in cold water, more sol in hot water; freely sol in alcohol, chloroform, ether, dil acids.

USE: In organic syntheses; in the manuf of nylon, synthetic rubbers, plastics, and medicinals.

5527. Lexipafant. [139133-26-9] *N*-Methyl-*N*-[[4-[(2-methyl-1*H*-imidazo[4,5-*c*]pyridin-1-yl)methyl]phenyl]sulfonyl]-L-leucine ethyl ester; *N*-methyl-*N*-[[α-(2-methyl-1*H*-imidazo[4,5-*c*]-pyridin-1-yl)-*p*-tolyl]sulfonyl]-L-leucine ethyl ester; BB-882; Zacutex. $C_{23}H_{30}N_4O_4S$; mol wt 458.58. C 60.24%, H 6.59%, N 12.22%, O 13.96%, S 6.99%. Platelet activating factor (PAF) antagonist. Prepn: M. Whittaker, A. Miller, **WO 9203422**; *eidem*, **US 5200412** (1992, 1993 both to British Bio-Technology). Structure-activity report: M. Whittaker *et al.*, *J. Lipid Mediators Cell Signalling* **10**, 151 (1994). Pharmacology: F. M. Abu-Zidan *et al.*, *Pharmacol. Toxicol.* **78**, 23 (1996). Clinical evaluation in acute pancreatitis: A. N. Kingsnorth *et al.*, *Br. J. Surg.* **82**, 1414 (1995).

White crystalline solid from ethyl acetate, mp 105°. $[\alpha]_D^{20}$ −6.7° (c = 2.0 in CDCl₃).

THERAP CAT: Anti-inflammatory.

5528. Liarozole. [115575-11-6] 6-[(3-Chlorophenyl)-1*H*-imidazol-1-ylmethyl]-1*H*-benzimidazole; (±)-5-(*m*-chloro-α-imidazol-1-ylbenzyl)benzimidazole. $C_{17}H_{13}ClN_4$; mol wt 308.77. C 66.13%, H 4.24%, Cl 11.48%, N 18.15%. Retinoic acid metabolism blocking agent. Inhibits cytochrome P450-dependent enzymes involved in steroid biosynthesis and retinoic acid catabolism. Prepn: A. H. M. Raeymaekers *et al.*, **EP 260744**; *eidem*, **US 4859684** (1988, 1989 both to Janssen). *In vivo* antitumor activity: R. Van Ginckel *et al.*, *Prostate* **16**, 313 (1990). Pharmacology and effect on steroid synthesis: J. Bruynseels *et al.*, *ibid.* 345; and effect on retinoic acid: R. De Coster *et al.*, *J. Steroid Biochem. Mol. Biol.* **43**, 197 (1992). Clinical evaluation in prostate cancer: C. Mahler *et al.*, *Cancer* **71**, 1068 (1993); in psoriasis: P. Dockx *et al.*, *Br. J. Dermatol.* **133**, 426 (1995); in treatment of ichthyosis: C. J. Verfaille *et al.*, *Br. J. Dermatol.* **156**, 965 (2007).

mp 108.2°.

Fumarate. [145858-52-2] R-85246; Liazal. $2C_{17}H_{13}ClN_4 \cdot 3C_4H_4O_4$; mol wt 965.75.

Hydrochloride. [145858-50-0] R-75251. $C_{17}H_{13}ClN_4 \cdot HCl$; mol wt 345.23.

5529. Liatris. Deer's tongue; vanilla plant. Leaves of *Trilisa odoratissima* (Walt.) Cass. (*Liatris odoratissima* (Walt.) Willd.), *Compositae. Habit.* U.S., Virginia to Florida and Louisiana. *Constit.* Volatile oil, coumarin.

USE: In perfumery and for perfuming smoking, chewing and snuff tobacco.

5530. Licheniformins. Antibiotic substances produced by *Bacillus licheniformis.* Isoln: Callow *et al.*, *Br. J. Exp. Pathol.* **28**, 418 (1947). Separation of licheniformins A, B, and C: Callow, Work, *Biochem. J.* **51**, 558 (1952). Mol wt as measured by sedimentation: licheniformin A, 4400; licheniformin B, 3800; and licheniformin C, 4800: Ogston, *ibid.* **51**, 569 (1952).

All three constituents in the form of hydrochlorides are white, amorphous, slightly hygroscopic powders, melting with decompn at indefinite temperatures.

5531. Lichenin. [1402-10-4] Lichenan; moss starch. $C_6H_{10}O_5$; mol wt 162.14. C 44.45%, H 6.22%, O 49.34%. Polyglucan from *Cetraria islandica* (L.) Ach., *Parmeliaceae* (Iceland Moss). Isoln: Peat *et al.*, *J. Chem. Soc.* **1957**, 3916. Linear polysaccharide structure composed of regular sequences of two (1 → 4)- and one (1 → 3)-β-D-glucopyranosyl residues: Perlin, Suzuki, *Can. J. Chem.* **40**, 50 (1962); Fleming, Manners, *Biochem. J.* **100**, 4, 24 (1966). Exhibits antineoplastic activity: Shibata, **JP 71 17147** (1971), *C.A.* **75**, 67474z (1971); Tadahiro *et al.*, *Chem. Pharm. Bull.* **20**, 2445 (1972). NMR spectrum: D. Gagnaire, M. Vincedon, *Bull. Soc. Chim. Fr.* **1977**, 479.

White powder. Sol in boiling water, in HCl. $[\alpha]_D$ +18.4°.

5532. Licochalcones. Oxygenated chalcones isolated from licorice root, a traditional medicine prepared from various species of *Glycyrrhiza* (Leguminosae). Exhibit antiparasitic and antitumor activity and appear to interfere with mitochondrial respiration. Four have been identified, A-D, as well as a closely related compound, **echinatin.** Isoln of A and B from *G. glabra* L. and structure: T. Saitoh, S. Shibata, *Tetrahedron Lett.* **50**, 4461 (1975); of C and D from *G. inflata*: K. Kajiyama *et al.*, *Phytochemistry* **31**, 3229 (1992). Mechanism of action study: H. Haraguchi *et al.*, *ibid.* **48**, 125 (1998); L Zhai *et al.*, *J. Antimicrob. Chemother.* **43**, 793 (1999).

Licochalcone A

Licochalcone A. [58749-22-7] (2*E*)-3-[5-(1,1-Dimethyl-2-propenyl)-4-hydroxy-2-methoxyphenyl]-1-(4-hydroxyphenyl)-2-propen-1-one. $C_{21}H_{22}O_4$; mol wt 338.40. Synthesis: A. Islam, M. A. Hossain, *Indian J. Chem.* **32B**, 713 (1993). HPLC determn in biological fluids: L. Nadelmann *et al.*, *J. Chromatogr. B* **695**, 389 (1997). Antileishmanial activity: M. Chen *et al.*, *Antimicrob. Agents Chemother.* **38**, 1339 (1994). Antimalarial activity: *eidem*, *ibid.* 1470. Effect on human cancer cell lines: E. J. Park *et al.*, *Planta Med.* **64**, 464 (1998). Yellow needles from methanol-water, mp 100°. uv max in methanol: 264, 308, 378 nm (log ε 3.96, 3.97, 4.31).

Licochalcone B. [58749-23-8] (2*E*)-3-(3,4-Dihydroxy-2-methoxyphenyl)-1-(4-hydroxyphenyl)-2-propen-1-one. $C_{16}H_{14}O_5$; mol wt 286.28. Yellow needles from methanol-water, mp 197°. uv max in methanol: 262, 366 nm (log ε 3.71, 4.27).

Licochalcone C. [144506-14-9] (2*E*)-3-[4-Hydroxy-2-methoxy-3-(3-methyl-2-butenyl)phenyl]-1-(4-hydroxyphenyl)-2-propen-1-one. $C_{21}H_{22}O_4$; mol wt 338.40. Amorphous. uv max in methanol: 250, 308sh, 358 nm (log ε 3.88, 3.95, 4.24).

Licochalcone D. [144506-15-0] (2*E*)-3-(3,4-Dihydroxy-2-methoxyphenyl)-1-[4-hydroxy-3-(3-methyl-2-butenyl)phenyl]-2-propen-1-one. $C_{21}H_{22}O_5$; mol wt 354.40. Pale yellow needles from methanol-water, mp 113°. uv max in methanol: 254, 359 nm (log ε 3.86, 4.30).

5533. Licofelone. [156897-06-2] 6-(4-Chlorophenyl)-2,3-dihydro-2,2-dimethyl-7-phenyl-1*H*-pyrrolizine-5-acetic acid; ML-3000. $C_{23}H_{22}ClNO_2$; mol wt 379.88. C 72.72%, H 5.84%, Cl 9.33%, N 3.69%, O 8.42%. Dual inhibitor of cyclooxygenase and 5-lipoxygenase. Prepn: G. Dannhardt *et al.*, **EP 397175**; *eidem*, **US 5260451** (1990, 1993 both to Merckle); S. A. Laufer *et al.*, *J. Med. Chem.* **37**, 1894 (1994); J. Cossy, D. Belotti, *Tetrahedron* **55**, 5145 (1999). Mechanism of action: S. Tries *et al.*, *Inflammation Res.* **51**, 135 (2002). Experience in experimental dog osteoarthritis: D. Lajeunesse *et al.*, *Ann. Rheum. Dis.* **63**, 78 (2004). Gastrointestinal safety profile in dogs: M. Moreau *et al.*, *J. Vet. Pharmacol. Ther.* **28**, 81 (2005). Review of clinical development: S. Laufer, *Inflammopharmacology* **9**, 101-112 (2001); J. M. Alvaro-Gracia, *Rheumatology* **43**, Suppl. 1, i21-i25 (2004).

Slightly yellowish solid, mp 162-163°.
THERAP CAT: Anti-inflammatory.
THERAP CAT (VET): Anti-inflammatory.

5534. Lidamidine. [66871-56-5] *N*-(2,6-Dimethylphenyl)-*N'*-[imino(methylamino)methyl]urea; 1-(2,6-dimethylphenyl)-3-methylamidinourea. $C_{11}H_{16}N_4O$; mol wt 220.28. C 59.98%, H 7.32%, N 25.43%, O 7.26%. Amidinourea with antisecretory, antimotility properties. Prepn as hydrochloride: **BE 844832** (1977 to Rorer). *See also:* J. Diamond, G. H. Douglas, **US 4147804** (1979 to Rorer). Prepn, structure activity relationship: G. H. Douglas *et al.*, *Arzneim.-Forsch.* **28**, 1435 (1978). Physical-chemical properties: J. J. Zalipsky *et al.*, *ibid.* 1441. Effect on α_2-adrenergic receptors and electrolyte absorption: T. Durbin *et al.*, *Gastroenterology* **82**, 1352 (1982). Series of articles on pharmacology, metabolism, pharmacokinetics: *Arzneim.-Forsch.* **28**, 1448-1480 (1978). Toxicity: B. J. Chou *et al.*, *ibid.* 1471. Clinical comparison with loperamide, *q.v.*, in acute diarrhea: G. Gasbarrini *et al.*, *ibid.* **36**, 1843 (1986). Brief review: G. Friedman, *Am. J. Gastroenterol.* **80**, 143 (1985).

Hydrochloride. [65009-35-0] WHR-1142A; Lidarral; Smodin. $C_{11}H_{16}N_4O \cdot HCl$; mol wt 256.73. White powder, mp 194-197°. uv max (H_2O): 262, 271 nm (ε 626, 524). Soly at 25° (mg/ml): water

153.55, methanol 297.94, ethanol 88.55, chloroform 4.62, hexane 0.01. LD_{50} in male mice, male, female rats (mg/kg): 260, 267, 160 orally; in mice (mg/kg): 56 i.v. (Chou).
THERAP CAT: Antiperistaltic; antidiarrheal.

5535. Lidocaine. [137-58-6] 2-(Diethylamino)-*N*-(2,6-dimethylphenyl)acetamide; 2-diethylamino-2',6'-acetoxylidide; ω-diethylamino-2,6-dimethylacetanilide; lignocaine; Cuivasil; Lidoderm; LidoPosterine; Vagisil; Versatis. $C_{14}H_{22}N_2O$; mol wt 234.34. C 71.76%, H 9.46%, N 11.95%, O 6.83%. Long-acting, membrane stabilizing agent against ventricular arrhythmia. Originally developed as a local anesthetic. Prepn: N. M. Löfgren, B. J. Lundqvist, **US 2441498** (1948 to Astra); A. D. H. Self, A. P. T. Easson, **GB 706409** (1954 to May & Baker); I. P. S. Hardie, E. S. Stern, **GB 758224** (1956 to J. F. Macfarlane & Co.); Zhuravlev, Nikolaev, *Zh. Obshch. Khim.* **30**, 1155 (1960). Toxicity studies: E. R. Smith, B. R. Duce, *J. Pharmacol. Exp. Ther.* **179**, 580 (1971); G. H. Kronberg *et al.*, *J. Med. Chem.* **16**, 739 (1973). Review of pharmacokinetics: N. L. Benowitz, W. Meister, *Clin. Pharmacokinet.* **3**, 177 (1978). Review of action as local anesthetic: Löfgren, *Studies on Local Anesthetics: Xylocaine, A New Synthetic Drug* (Hoeggstroms, Stockholm, 1948); Cooper, *Pharm. J.* **171**, 68 (1953). Reviews of antiarrhythmic agents: J. L. Anderson *et al.*, *Drugs* **15**, 271 (1978); L. H. Opie, *Lancet* **1**, 861 (1980); E. Carmeliet, *Ann. N.Y. Acad. Sci.* **427**, 1 (1984). Comprehensive description: K. Groningsson *et al.*, *Anal. Profiles Drug Subs.* **14**, 207-243 (1985); M. F. Powell, *ibid.* **15**, 761-779 (1986). Review of use in treatment of postherpetic neuralgia: P. S. Davies, B. S. Galer, *Drugs* **64**, 937-947 (2004).

Needles from benzene or alcohol, mp 68-69°. bp_4 180-182°; bp_2 159-160°. Very sol in alcohol, chloroform; freely sol in ether, benzene. Practically insol in water. Dissolves in oils. Partition coefficient (octanol/water, pH 7.4): 43.
Hydrochloride. [73-78-9]; [6108-05-0] (monohydrate). Basicaina; Batixim; Dynexan; Heweneural; Licain; Lidesthesin; Lidofast; Lidoject; Lidrian; Odontalg; Sedagul; Xylocaine; Xylocard; Xylocitin; Xyloneural. $C_{14}H_{22}N_2O \cdot HCl$; mol wt 270.80. White crystals, mp 127-129°; monohydrate, mp 77-78°. Very sol in water, alcohol; sol in chloroform. Insol in ether. pH of 0.5% aq soln: 4.0-5.5. LD_{50} in mice (mg/kg): 292 orally (Smith, Duce); 105 i.p.; 19.5 i.v. (Kronberg).
THERAP CAT: Anesthetic (local); antiarrhythmic (class IB).
THERAP CAT (VET): Anesthetic (local).

5536. Lidofenin. [59160-29-1] *N*-(Carboxymethyl)-*N*-[2-[(2,6-dimethylphenyl)amino]-2-oxoethyl]glycine; *N*-(2,6-dimethylacetanilide)iminodiacetic acid; *N*-(2,6-dimethylphenylcarbamoylmethyl)iminodiacetic acid; [[(2,6-xylylcarbamoyl)methyl]imino]diacetic acid; HIDA. $C_{14}H_{18}N_2O_5$; mol wt 294.31. C 57.13%, H 6.16%, N 9.52%, O 27.18%. Iminodiacetic acid analog. Prepn of compd and ^{99m}Tc-complex: M. D. Loberg *et al.*, *eidem*, **US 4017596** (1976, 1977 both to Research Corp.); and imaging studies: M. D. Loberg *et al.*, *J. Nucl. Med.* **17**, 633 (1976). Structural studies: M. D. Loberg, A. T. Fields, *Int. J. Appl. Radiat. Isot.* **29**, 167 (1978). Chromatographic determn of radiochemical purity: A. M. Zimmer *et al.*, *Eur. J. Nucl. Med.* **7**, 88 (1982). Biodistribution and clinical kinetics: J. Ryan *et al.*, *J. Nucl. Med.* **18**, 995 (1977). Clinical studies of diagnostic use as hepatobiliary imaging agent: H. S. Weissmann *et al.*, *Am. J. Roentgenol.* **132**, 523 (1979); R. W. Nicholson *et al.*, *Br. J. Radiol.* **53**, 878 (1980).

Crystals from water, mp 201-203°. LD$_{50}$ i.v. in male mice: 168 mg/kg (Ryan).

Complex with ^{99m}Tc. Technetium Tc 99m lidofenin; ^{99m}Tc-HIDA; TechneScan HIDA.

THERAP CAT: ^{99m}Tc complex as diagnostic aid (radioactive imaging agent).

5537. Lidoflazine. [3416-26-0] 4-[4,4-Bis(4-fluorophenyl)-butyl]-N-(2,6-dimethylphenyl)-1-piperazineacetamide; 4-[4,4-bis(p-fluorophenyl)butyl]-1-piperazineaceto-2',6'-xylidide; 1-[4,4-di(4-fluorophenyl)butyl]-4-[(2,6-dimethylanilinocarbonyl)methyl]piperazine; McN-JR-7094; R-7904; Angex; Clinium; Klinium; Ordiflazine; Corflazine. C$_{30}$H$_{35}$F$_2$N$_3$O; mol wt 491.63. C 73.29%, H 7.18%, F 7.73%, N 8.55%, O 3.25%. Calcium blocking agent. Prepn: NL 6507312; H. K. F. Hermans, W. K. Schaper, US 3267104 (1965, 1966, both to Janssen). Crystal structure: G. Germain *et al.*, *Acta Crystallogr.* **33B**, 1971 (1977). Tissue specificity of calcium-blocking properties: J. M. Van Neuten, D. Wellens, *Arch. Int. Pharmacodyn. Ther.* **242**, 329 (1979). Cardioprotective effects: W. Daenen, W. Flameng, *Angiology* **32**, 543 (1981). Effects in angina: F. L. Gobel *et al.*, *Circulation* **65**, 1 Pt. 2, 127 (1982); W. Shapiro *et al.*, *ibid.* 143.

Crystals, mp 159-161°. Almost insol in water (<0.01%), very sol in chloroform (>50%), but much less sol in other common organic solvents: W. K. Schaper *et al.*, *J. Pharmacol. Exp. Ther.* **152**, 265 (1966).

THERAP CAT: Vasodilator (coronary).

5538. Light Green SF Yellowish. [5141-20-8] N-Ethyl-N-[4-[[4-[ethyl[(3-sulfophenyl)methyl]amino]phenyl](4-sulfophenyl)-methylene]-2,5-cyclohexadien-1-ylidene]-3-sulfobenzenemethan-aminium inner salt sodium salt (1:2); C.I. Acid Green 5; ethyl[4-[p-[ethyl(m-sulfobenzyl)amino]-α-(p-sulfophenyl)benzylidene]-2,5-cyclohexadien-1-ylidene](m-sulfobenzyl)ammonium hydroxide inner salt disodium salt; C.I. 42095; FD & C Green No. 2; Lissamine Green SF. C$_{37}$H$_{34}$N$_2$Na$_2$O$_9$S$_3$; mol wt 792.84. C 56.05%, H 4.32%, N 3.53%, Na 5.80%, O 18.16%, S 12.13%. Prepn and refs: *Colour Index* vol. 4 (3rd ed., 1971) p 4385. Biological use: *H. J. Conn's Biological Stains*, R. D. Lillie, Ed. (Williams & Wilkins, Baltimore, 9th ed., 1977) pp 257-258, 582. Toxicity data: F. C. Lu, A. Lavalle, *Can. Pharm. J.* **97**, 30 (1964). Chronic toxicity study: W. H. Hansen *et al.*, *Food Cosmet. Toxicol.* **4**, 389 (1966).

A reddish-brown powder. Sol in water to a green soln which turns yellowish-brown with HCl and then gradually fades. Addition of NaOH almost completely decolorizes the soln yielding a dull violet ppt. LD$_{50}$ orally in rats: >2 g/kg (Lu, Lavalle).

USE: As dye, biological stain.

5539. Lignans. Plant products of low molecular weight formed primarily by the oxidative coupling of p-hydroxyphenylpro-

pene units in which the two units may be linked by an oxygen bridge. The monomeric precursor units are cinnamic acid, cinnamyl alcohol, propenylbenzene and allylbenzene. The term lignan or *Haworth lignan* is applied to compounds derived by coupling acid and/or alcohol while the compounds derived by coupling propenyl and/or allyl derivatives are called *neolignans*: O. R. Gottlieb, *Fortschr. Chem. Org. Naturst.* **35**, 1-72 (1978). Lignans occur widely and have been obtained from roots, heartwood, foliage, fruit and resinous exudates of plants. Lignans are optically active compounds. They represent the dimer stage intermediate between monomeric propylphenol units and lignin. Naturally occurring trimers and tetramers have not been reported. Occurrence of lignans, enterolactone, *q.v.*, and enterodiol, in man and animal species: S. R. Stitch *et al.*, *Nature* **287**, 238 (1980); K. D. R. Setchell, *ibid.* 740. Synthesis of first lignans found in man and animals: G. Cooley *et al.*, *Tetrahedron Lett.* **22**, 349 (1981).

5540. Lignin. [9005-53-2] The most abundant natural aromatic organic polymer found in all vascular plants. Lignin together with cellulose and hemicellulose, *q.q.v.*, are the major cell wall components of the fibers of all wood and grass species. Lignin is composed of coniferyl, p-coumaryl and sinapyl alcohols in varying ratios in different plant species. Monographs: F. E. Brauns, D. A. Brauns, *The Chemistry of Lignin*, Supplement Volume covering the literature 1949-1958 (Academic Press, New York, 1960) 804 pp; I. A. Pearl, *The Chemistry of Lignin* (Marcel Dekker, New York, 1967) 360 pp. Structural aspects and applications: H. Veeramani, G. A. Wani, *Chem. Ind. Dev.* **11**, 13-25 (1977). Chemistry and structure: C. A. Reddy, L. Forney, *Dev. Ind. Microbiol.* **19**, 27-34 (1978). *Reviews*: Nord, Shubert, *Sci. Am.* **199**, no. 4, 104-113 (Oct. 1958); I. A. Pearl, "Lignin as a Raw Material for the Production of Pure Chemicals," *J. Chem. Educ.* **35**, 502 (1958); D. W. Goheen, C. H. Hoyt in *Kirk-Othmer Encyclopedia of Chemical Technology* **vol. 14**, (Wiley-Interscience, New York, 3rd ed., 1981) pp 294-312.

USE: Source of vanillin, syringic aldehyde, dimethyl sulfoxide. Extender for phenolic plastics, to strengthen rubber (esp for shoe soles), as oil mud additive, to stabilize asphalt emulsions, to precipitate proteins.

5541. Lignoceric Acid. [557-59-5] Tetracosanoic acid. C$_{24}$H$_{48}$O$_2$; mol wt 368.65. C 78.19%, H 13.12%, O 8.68%. Obtained from beechwood tar or by the distillation of rotten oak wood: Sullivan, *Ind. Eng. Chem.* **8**, 1027 (1916). Most natural fats contain small amounts (0.2-1%). The seed fat of the Indian tree *Adenanthera pavonina* is said to contain 25%. Synthesis: Fieser, Szmuszkovicz, *J. Am. Chem. Soc.* **70**, 3352 (1948).

Crystals, mp 84.15°. n_D^{100} 1.4287. Soly in 91.53% ethanol: 0.182 g/100 ml. Neutralization value 152.2.

Methyl ester. C$_{25}$H$_{50}$O$_2$; mol wt 382.67. Platelets, mp 58-59.8°.

5542. Ligroin. [8032-32-4] Ligroine; V.M.&P. naphtha; varnish makers' and painters' naphtha; refined solvent naphtha; solvent naphtha; Benzoline; Canadol. Term that has been applied to petroleum fractions of the same nature as described for petroleum benzin, *q.v.*, but of higher density, higher boiling range, and higher flash pt. Defined by ASTM prior to 1950 as synonymous with petroleum benzin and petroleum ether: *ASTM Standard Specifications* D 288-49, 865-867 (1949).

Refined solvent, mobile, flammable liquid. d$_{15.6}^{15.6}$ 0.850 to 0.870. Distillation range at 760 mm: percentage recovered at 130° = not more than 5; percentage recovered at 145° = not less than 90. End point (dry point) = not above 155°. Technical benzin (high boiling petr ether) usually has a d$_{15.6}^{15.6}$ 0.730-0.750 and bp$_{760}$ 90-120°.

Caution: Potential symptoms of overexposure are irritation of eyes, upper respiratory system; dermatitis, CNS depression; aspiration of liquid may cause chemical pneumonia. *See NIOSH Pocket Guide to Chemical Hazards* (DHHS/NIOSH 97-140, 1997) p 332.

5543. Limaprost. [74397-12-9] (2E)-7-[(1R,2R,3R)-3-Hydroxy-2-[(1E,3S,5S)-3-hydroxy-5-methyl-1-nonen-1-yl]-5-oxocyclopentyl]-2-heptenoic acid; (2*E*,11α,13*E*,15*S*,17*S*)-11,15-dihy-droxy-17,20-dimethyl-9-oxoprosta-2,13-dien-1-oic acid; 17*S*,20-di-methyl-*trans*-2,3-didehydro-PGE₁; 9-oxo-11α,15α-dihydroxy-17*S*,20-dimethylprosta-*trans*-2,*trans*-13-dienoic acid; 17*S*-methyl-ω-homo-*trans*-Δ²-PGE₁; ONO-1206; OP-1206; Opalmon; Prorenal. $C_{22}H_{36}O_5$; mol wt 380.53. C 69.44%, H 9.54%, O 21.02%. Derivative of prostaglandin E₁, *q.v.* Prepn: M. Hayashi *et al.*, **DE 3002677** (1980 to Ono); *eidem*, **US 4294849** (1981 to Warner-Lambert). Cardiovascular pharmacology in animals: T. Tsuboi *et al.*, *Arch. Int. Pharmacodyn.* **247**, 89 (1980). Effect on smooth muscle *in vitro*: P. G. Adaikan, S. M. M. Karim, *Prostaglandins Med.* **6**, 449 (1981). Coronary vasodilating effects in primates: S. R. Kottegoda *et al.*, *Prostaglandins Leukotrienes Med.* **8**, 343 (1982). Clinical hemodynamics in chronic lung disease: T. Ishizaki *et al.*, *Chest* **85**, 382 (1984).

White crystals, mp 97-100°.
THERAP CAT: Antianginal.

5544. Limestone. Natural calcium carbonate; agricultural limestone; Agstone; lithographic stone; Solnhofen stone. A term originally applied only to minerals consisting largely of $CaCO_3$, such as Portland stone, dolomite, marble, and chalk, now used indiscriminately to designate technical and agricultural grades of calcium carbonate. *Review:* R. S. Boynton in *Kirk-Othmer Encyclopedia of Chemical Technology* **vol. 14** (Wiley-Interscience, New York, 3rd ed., 1981) pp 343-382.
Caution: Potential symptoms of overexposure are irritation of eyes, skin, mucous membranes; cough, sneezing, rhinorrhea; lacrimation. *See NIOSH Pocket Guide to Chemical Hazards* (DHHS/NIOSH 97-140, 1997) p 186.

5545. Limettin. [487-06-9] 5,7-Dimethoxy-2*H*-1-benzopyran-2-one; 5,7-dimethoxycoumarin; citropten. $C_{11}H_{10}O_4$; mol wt 206.20. C 64.07%, H 4.89%, O 31.04%. From rind of fruit of *Citrus lima* Lunan (*C. limetta* Auth.), *Rutaceae:* Tilden, Beck, *J. Chem. Soc.* **57**, 323 (1890); from W. Indian lime oil: Caldwell, Jones, *ibid.* **1945**, 570; from citrus oils: Stanley, Vannier, **US 2889337** (1959 to U.S.D.A.). Synthesis: Schmidt, *Arch. Pharm.* **242**, 288 (1904); Heyes, Robertson, *J. Chem. Soc.* **1936**, 1831.

Needles from methanol, mp 147-148°. uv max (alcohol): 222, 247, 250.5, 324 nm (log ε 4.03, 3.84, 3.84, 4.18). Almost insol in boiling water, ether, petr ether. Freely sol in alcohol, chloroform, acetone.

5546. Limonene. [138-86-3] 1-Methyl-4-(1-methylethenyl)-cyclohexene; *p*-mentha-1,8-diene; cinene; cajeputene; kautschin. $C_{10}H_{16}$; mol wt 136.24. C 88.16%, H 11.84%. Occurs in various ethereal oils, particularly in oils of lemon, orange, caraway, dill and bergamot. Isoln of *d*-limonene from mandarin peel oil (*Citrus reticulata* Blanco, *Rutaceae*): Kugler, Kováts, *Helv. Chim. Acta* **46**, 1480 (1963). *Review:* J. L. Simonsen, *The Terpenes* **vol. I** (University Press, Cambridge, 2nd ed., 1947) pp 143-165.

dl-**Form.** Inactive limonene; dipentene. Liquid. Pleasant lemon-like odor. bp_{763} 175.5-176.5°. $d_4^{20.85}$ 0.8402. n_D 1.4744. Practically insol in water. Miscible with alcohol. With dry HCl or HBr it forms monohalides, and with aq HCl or HBr, the dihalide.
d-**Form.** Liquid. bp_{763} 175.5-176°. d_4^{21} 0.8402. n_D^{21} 1.4743. $[α]_D^{19.5}$ +123.8°.
l-**Form.** Liquid. bp_{763} 175.5-176.5°. $d_4^{20.5}$ 0.8407. n_D^{21} 1.474. $[α]_D^{19.5}$ −101.3°.
Caution: Skin irritant, sensitizer.
USE: Solvent, manuf resins; wetting and dispersing agent.

5547. Limonin. [1180-71-8] (2a*R*,4a*R*,4b*R*,5a*S*,8*S*,8a*S*,-10a*R*,10b*R*,14a*S*)-8-(3-Furanyl)decahydro-2,2,4a,8a-tetramethyl-11*H*,13*H*-oxireno[*d*]pyrano[4',3':3,3a]isobenzofuro[5,4-*f*][2]ben-zopyran-4,6,13(2*H*,5a*H*)-trione; limonoic acid di-δ-lactone; limonoic acid 3,19:16,17-dilactone. $C_{26}H_{30}O_8$; mol wt 470.52. C 66.37%, H 6.43%, O 27.20%. Tetracyclic triterpenoid; bitter principle of citrus fruits. Isoln: Bernays, *Ann.* **40**, 317 (1841). Improved isoln and purification: P. G. Pifferi *et al.*, *Ital. J. Food Sci.* **5**, 269 (1993). Structure: A. Melera *et al.*, *Helv. Chim. Acta* **40**, 1420 (1957). Stereochemistry: D. Arigoni *et al.*, *Experientia* **16**, 41 (1960); S. Arnott *et al.*, *ibid.* 49. Partial synthesis: C. Lüthy *et al.*, *Helv. Chim. Acta* **57**, 1060 (1974). HPLC determn in citrus juice: W. W. Widmer, *J. Agric. Food Chem.* **39**, 1472 (1991). Review of chemistry: V. P. Maier *et al.*, *ACS Symp. Ser.* **143**, 63-82 (1980); and removal of juice bitterness: R. F. H. Dekker, *Aust. J. Biotechnol.* **2**, 65 (1988).

Bitter crystals from methylene chloride + isopropanol or acetic acid, mp 298°. $[α]_D$ −128° (c = 1.21 in acetone). uv max: 207, 285 nm (ε 7000, 38). Slightly sol in water, ether; sol in alcohol, glacial acetic acid.

5548. Linaclotide. [851199-59-2] L-Cysteinyl-L-cysteinyl-L-α-glutamyl-L-tyrosyl-L-cysteinyl-L-cysteinyl-L-asparaginyl-L-pro-lyl-L-alanyl-L-cysteinyl-L-threonylglycyl-L-cysteinyl-L-tyrosine cyclic (1 → 6),(2 → 10),(5 → 13)-tris(disulfide); [9-L-tyrosine]heat-stable enterotoxin (*Escherichia coli*)-(6-19)-peptide. $C_{59}H_{79}N_{15}$-$O_{21}S_6$; mol wt 1526.73. C 46.42%, H 5.22%, N 13.76%, O 22.01%, S 12.60%. Synthetic 14 amino acid peptide containing 3 disulfide bridges; structurally analogous to the diarrhea-causing, heat-stable enterotoxins produced by *E. coli*. Guanylate cyclase-C (GC-C) receptor agonist that acts locally on intestinal epithelial cells; increases cGMP production which triggers a signal transduction cascade leading to increased fluid secretion and accelerated colonic transport. Prepn: M. G. Currie, S. Mahajan-Miklos, **WO 04069165**; *idem et al.*, **US 7304036** (2004, 2007 both to Microbia). Optimized solid-phase synthesis: M. Góngora-Benítez *et al.*, *Pept. Sci.* **96**, 69 (2011). Structural and receptor binding studies: R. W. Busby *et al.*, *Eur. J. Pharmacol.* **649**, 328 (2010). Pharmacology: A. P. Bryant *et al.*, *Life Sci.* **86**, 760 (2010). Clinical evaluation in irritable bowel syndrome: J. M. Johnston *et al.*, *Gastroenterology* **139**, 1877 (2010). Clinical trial in chronic constipation: A. J. Lembo *et al.*, *N. Engl. J. Med.* **365**, 527 (2011). Review of development and clinical experience: N. Lee, A. Wald, *Expert Opin. Drug Metab. Toxicol.* **7**, 651-659 (2011).

Cys–Cys–Glu–Tyr–Cys–Cys–Asn–Pro–Ala–Cys–Thr–Gly–Cys–Tyr

Acetate. [851199-60-5] MM-416775; MD-1100. $C_{59}H_{79}N_{15}$-$O_{21}S_6 \cdot C_2H_4O_2$; mol wt 1586.78.

THERAP CAT: In treatment of irritable bowel syndrome and chronic constipation.

5549. Linagliptin. [668270-12-0] 8-[(3*R*)-3-Amino-1-piperidinyl]-7-(2-butyn-1-yl)-3,7-dihydro-3-methyl-1-[(4-methyl-2-quinazolinyl)methyl]-1*H*-purine-2,6-dione; 8-[3(*R*)-aminopiperidin-1-yl]-7-(2-butynyl)-3-methyl-1-(4-methylquinazolin-2-ylmethyl)xanthine; BI-1356; Ondero; Tradjenta. $C_{25}H_{28}N_8O_2$; mol wt 472.55. C 63.54%, H 5.97%, N 23.71%, O 6.77%. Dipeptidyl peptidase IV inhibitor. Prepn: F. Himmelsbach *et al.*, **DE 10238243**; *eidem*, **US 7407955** (2004, 2008 both to Boehringer Ing.); M. Eckhardt *et al.*, *J. Med. Chem.* **50**, 6450 (2007). Pharmacology: L. Thomas *et al.*, *J. Pharmacol. Exp. Ther.* **325**, 175 (2008). Clinical pharmacokinetics: T. Heise *et al.*, *Diabetes Obes. Metab.* **11**, 786 (2009). Clinical trial in type 2 diabetes: S. Del Prato *et al.*, *ibid.* **13**, 258 (2011).

White to yellow, slightly hygroscopic solid. Also reported as crystalline solid, mp 202°. pKa 1.9; 8.6. Sol in methanol; sparingly sol in ethanol; very slightly sol in isopropanol, acetone.

THERAP CAT: Antidiabetic.

5550. Linalool. [78-70-6] 3,7-Dimethyl-1,6-octadien-3-ol; 2,6-dimethyl-2,7-octadien-6-ol; linalol; linalyl alcohol. $C_{10}H_{18}O$; mol wt 154.25. C 77.87%, H 11.76%, O 10.37%. Fragrant monoterpene alcohol found in the essential oils of numerous aromatic plants. Both isomers are naturally occurring but have different olfactory properties. The *l*-form is a chief constituent of lavender, bergomot, and rosewood oils; the *d*-form of coriander oil. Also produced synthetically as a mixture of isomers: F. Tiemann, *Ber.* **31**, 808 (1898); Y.-R. Naves, *Helv. Chim. Acta* **42**, 1692 (1959). Absolute configuration of isomers: G. Ohloff, E. Klein, *Tetrahedron* **18**, 37 (1962). Synthesis from methylheptenone: L. Ruzicka, V. Fornasir, *Helv. Chim. Acta* **2**, 182 (1919); from citral: G. V. Nair, G. D. Pandit, *Tetrahedron Lett.* **7**, 5097 (1966); from α-pinene: V. A. Semikolenov *et al.*, *Appl. Catal. A* **211**, 91 (2001). GC/MS determn of enantiomers in lavender oils: S. Bilke, A. Mosandl, *Eur. Food Res. Technol.* **214**, 532 (2002). Review of properties, uses, and toxicology: C. S. Letizia *et al.*, *Food Chem. Toxicol.* **41**, 943-964 (2003); G. Kamatou, A. Viljoen, *Nat. Prod. Commun.* **3**, 1183-1192 (2008).

R-form

Colorless to very pale yellow liquid with flowery, fresh scent. *Irritant.* bp_{720} 194-197°; bp_{14} 89-91°. d^{15} 0.865. Flash pt, closed cup: 172.4°F (78°C).

R-(−)-**Form.** [126-91-0] Licareol; *l*-linalool. Colorless liquid with odor of lavender. bp_{760} 198°; bp_{25} 98-98.3°; bp_{14} 86-87°. d^{20} 0.8622. n_D^{22} 1.4604. $[α]_D^{20}$ −20.1°. Practically insol in water. Miscible with alcohol, ether.

S-(+)-**Form.** [126-90-9] Coriandrol; *d*-linalool. Liquid with odor of petitgrain. bp_{760} 198-200°; bp_{26} 114-114.5°; $bp_{15.5}$ 93-94°; bp_{12} 86°. d_4^{20} 0.8733. n_D^{20} 1.4673. $[α]_D^{20}$ +19.3°. Soluble in 10 vol 50% alc, 4 vol 60% alc.

USE: Fragrance component in perfumes, cosmetics, soaps, and detergents; flavoring agent in foods. Synthetic intermediate.

5551. Linalyl Acetate. [115-95-7] 3,7-Dimethyl-1,6-octadien-3-ol 3-acetate; linalool acetate; bergamol. $C_{12}H_{20}O_2$; mol wt 196.29. C 73.43%, H 10.27%, O 16.30%. Fragrant ester of linalool. The *l*-form is a chief constituent of lavender and bergamot oils; also found in essential oils of many other plants. Isoln from lavender and bergamot oils: F. W. Semmler, F. Tiemann, *Ber.* **25**, 1180 (1892). Prepn from linalool: M. E. Aeschbach, **US 2423545** (1947 to Norda Essential Oil & Chem. Co.); and thermodynamic study: A. Martin *et al.*, *Chem. Eng. Technol.* **30**, 726 (2007). Industrial prepn of (±)-form: J. A. Birbiglia *et al.*, **US 2797235** (1957 to Hoffmann-La Roche). Synthesis of *R*-form: G. Vidari *et al.*, *Tetrahedron: Asymmetry* **10**, 3547 (1999). Enantiomeric distribution in various oils: B. Weinreich, S. Nitz, *Chem. Mikrobiol. Technol. Lebensm.* **14**, 117 (1992); H. Casabianca *et al.*, *J. High Resolut. Chromatogr.* **21**, 107 (1998). Review of properties, uses and toxicology: C. S. Letizia *et al.*, *Food Chem. Toxicol.* **41**, 965-979 (2003).

R-form

Colorless liquid with characteristic bergamot-lavender odor and sweet, acrid taste. d_4^{20} 0.895. bp 220°. $bp_{0.2mm}$ 44°. n_D^{25} 1.4479-1.4480. Flash point, closed cup: 185°F (85°C). Insol in water, glycerol. Miscible with alcohol, ether. LD_{50} orally in rats, mice: 14550, 13360 mg/kg (Letizia).

R-form. [16509-46-9] *l*-Linalyl acetate. Colorless oil. d^{20} 0.8972. $[α]_D^{20}$ −3.0° (c = 1.0 in methylene chloride).

USE: Fragrance component in perfumes, cosmetics, and soaps; flavoring agent in foods.

5552. Linamarin. [554-35-8] 2-(β-D-Glucopyranosyloxy)-2-methylpropanenitrile; phaseolunatin. $C_{10}H_{17}NO_6$; mol wt 247.25. C 48.58%, H 6.93%, N 5.67%, O 38.82%. From the seed skins or embryos of flax: Jorissen, Hairs, *Bull. Acad. Roy. Sci. Belg.* [3] **21**, 529 (1891); André *et al.*, *Compt. Rend.* **231**, 590 (1950); Lüdtke, *Biochem. Z.* **323**, 428 (1953). Synthesis: Fischer, Anger, *Ber.* **52**, 854 (1919). Biosynthesis in white clover: Butler, Butler, *Nature* **187**, 780 (1960).

Bitter needles, mp 142-143°. $[α]_D^{18}$ −29°. Freely sol in water, cold alcohol, hot acetone; slightly in hot ethyl acetate, ether, benzene, chloroform. Practically insol in petr ether. Evolves HCN with linseed meal but not with emulsin.

Tetraacetate. $C_{18}H_{25}NO_{10}$. Needles from alcohol, mp 140-141°. $[α]_D^{14}$ −10.8° (acetone). Sol in acetone, ethyl acetate, chloroform, glacial acetic acid, benzene, warm methanol and ethanol. Practically insol in petr ether.

5553. Linarin. [480-36-4] 7-[[6-*O*-(6-Deoxy-α-L-mannopyranosyl)-β-D-glucopyranosyl]oxy]-5-hydroxy-2-(4-methoxyphenyl)-4*H*-1-benzopyran-4-one; acacetin-β-rutinoside; linarigeninglucoside; 5,7-dihydroxy-4′-methoxyflavone-D-glucosido-L-rhamnoside; buddleoflavonoloside. $C_{28}H_{32}O_{14}$; mol wt 592.55. C 56.76%, H 5.44%, O 37.80%. From the flowers of *Linaria vulgaris* Mill., *Scrophulariaceae*: Merz, Wu, *Arch. Pharm.* **274**, 126 (1936); from *Cirsium oleraceum* Scop., *Compositae*: Wagner *et al.*, *ibid.* **293**, 1053 (1960). Structure: Baker *et al.*, *J. Chem. Soc.* **1951**, 691. Synthesis: Zemplén, Bognàr, *Ber.* **74**, 1818 (1941).

Monohydrate. Needles from methanol, mp 268-270°. $[\alpha]_D^{26}$ −100° (0.07 g in 10 ml glacial acetic acid); $[\alpha]_D^{24}$ −87° (0.05 g in pyridine). Practically insol in water and the usual organic solvents. Sol in nitrobenzene, phenol, aniline, pyridine, concd acids and alkalies. The water of crystn cannot be removed at 100° *in vacuo* over P_2O_5 (Merz); may be removed at 138° in high vacuum (Zemplén). Hydrolysis gives 5,7-dihydroxy-4′-methoxyflavone, D-glucose, and L-rhamnose.

5554. Linatine. [10139-06-7] 1-[[(4S)-Amino-4-carboxy-1-oxobutyl]amino]-D-proline; *N*-(D-2-carboxy-1-pyrrolidinyl)-L-glutamine; 1-[*N*-(γ-L-glutamyl)amino]-D-proline. $C_{10}H_{17}N_3O_5$; mol wt 259.26. C 46.33%, H 6.61%, N 16.21%, O 30.86%. A vitamin B_6 antagonist. The first reported naturally occurring hydrazino acid having antibacterial properties. Isoln from flaxseed (*Linum usitatissimum*), characterization, and synthesis: Klosterman *et al.*, *Biochemistry* **6**, 170 (1967); Lamoureux, *Diss. Abstr. B* **28**, 4908 (1968). Bacterial inhibition: Parsons *et al.*, *Antimicrob. Agents Chemother.* **1967**, 415.

Amorphous powder. $[\alpha]_D^{25}$ +46.4° (c = 2.8 in water). Attempts to obtain well-defined crystals have failed; melts with dec over a wide temperature range. Very sol in water. Practically insol in anhydr organic solvents.

(−)-**Isomer.** [10139-07-8] (S)-1-[4-(Amino-4-carboxy-1-oxobutyl)amino]-L-proline. Amorphous solid. $[\alpha]_D^{24}$ −34.6° (c = 2 in water).

5555. Lincomycin. [154-21-2] Methyl 6,8-dideoxy-6-[[[(2S,4R)-1-methyl-4-propyl-2-pyrrolidinyl]carbonyl]amino]-1-thio-D-erythro-α-D-galacto-octopyranoside; lincolnensin; U-10149; NSC-70731. $C_{18}H_{34}N_2O_6S$; mol wt 406.54. C 53.18%, H 8.43%, N 6.89%, O 23.61%, S 7.89%. Antibiotic produced by *Streptomyces lincolnensis* var. *lincolnensis*. Isoln: D. J. Mason *et al.*, *Antimicrob. Agents Chemother.* **1962**, 555; R. R. Herr, M. E. Bergy, *ibid.* 560; M. E. Bergy *et al.*, **US 3086912** (1963 to Upjohn). Structure: H. Hoeksema *et al.*, *J. Am. Chem. Soc.* **86**, 4223 (1964). Synthesis: B. J. Magerlein, *Tetrahedron Lett.* **11**, 33 (1970). Stereoselective synthesis: S. Knapp, P. J. Kukkola, *J. Org. Chem.* **55**, 1632 (1990). Mechanism of action: F. N. Chang in *Antibiotics* vol. 5(pt. 1), F. E. Hahn, Ed. (Springer-Verlag, New York, 1979) pp 127-134. Clinical study: F. Puleo *et al.*, *Gazz. Med. Ital.* **138**, 401 (1979). Toxicology: J. E. Gray *et al.*, *Toxicol. Appl. Pharmacol.* **6**, 476 (1964). Comprehensive description: H. Y. Muti, F. H. Al-Hajjar, *Anal. Profiles Drug Subs. Excip.* **23**, 269-319 (1994).

Sol in methanol, lower alcohols, ethyl acetate, acetone, chloroform. Slightly sol in water.

Hydrochloride monohydrate. [7179-49-9]; [859-18-7] (anhydrous). Albiotic; Lincocin; Lincomix. $C_{18}H_{34}N_2O_6S.HCl.H_2O$; mol wt 461.01. White or almost white, crystalline powder. Odorless or with a slight mercaptan-like odor and a bitter taste. mp 156-158°. $[\alpha]_D^{25}$ +143° (water). pKa 7.60. pH of a 1% aq. soln: 4.8. Sol in 2 parts water, in 40 parts ethanol, in 20 parts DMF. Very slightly sol in acetone. Practically insol in chloroform, ether. LD_{50} orally in rats: >4 g/kg; i.p. in mice: 1 g/kg (Gray).

THERAP CAT: Antibacterial.

THERAP CAT (VET): Antibacterial.

5556. Lindane. [58-89-9] (1α,2α,3β,4α,5α,6β)-1,2,3,4,5,6-Hexachlorocyclohexane; γ-HCH; γ-benzene hexachloride; gamma benzene hexachloride; gamma hexachlor; γ-BHC; ENT-7796; Hexit; Jacutin; Lorexane; Quellada; Scabecid. $C_6H_6Cl_6$; mol wt 290.81. C 24.78%, H 2.08%, Cl 73.14%. Organochlorine pesticide. Active isomer among the eight well-described stereoisomers of hexachlorocyclohexane. Prepn: T. Hardie, **US 2218148** (1940 to ICI); R. E. Slade, *Chem. Ind. (London)* **1945**, 314; F. A. Gunther, *ibid.* **1946**, 399. Metabolism: R. Engst *et al.*, *Residue Rev.* **68**, 59 (1977); **72**, 71 (1979). Acute toxicity: T. B. Gaines, *Toxicol. Appl. Pharmacol.* **14**, 515 (1969). GC-MS determn in plasma: M. Moreno Frías *et al.*, *J. Chromatogr. B* **760**, 1 (2001); in river sediments: P. N. Carvalho *et al.*, *Talanta* **76**, 1124 (2008). Review of isomers: I. Hornstein, *Science* **121**, 206 (1955); S. H. Safe in *Kirk-Othmer Encyclopedia of Chemical Technology* vol. 6 (John Wiley & Sons, New York, 4th ed., 1993) pp 135-139. Reviews of carcinogenic risk: M. D. Reuber, *Environ. Res.* **19**, 460-481 (1979); S. D. Vesselinovitch, F. W. Carlborg, *Toxicol. Pathol.* **11**, 12 (1983); of toxicology and human exposure: *Toxicological Profile for Alpha, Beta, Gamma, and Delta-Hexachlorocyclohexane* (PB2006-100003, 2005) 377 pp.

Crystals, mp 112.5°. Slight musty odor. Vapor pressure at 20°: 9.4 × 10⁻⁶ mm Hg. n_D^{20} 1.644. Soly in g/100 g at 20°: acetone 43.5, benzene 28.9, CHCl$_3$ 24.0, ether 20.8, ethanol 6.4. Insol in water. LD_{50} in male, female rats (mg/kg): 88, 91 orally (Gaines).

Caution: Poisoning may occur by ingestion, inhalation, or percutaneous absorption. Potential symptoms of overexposure are irritation of eyes, skin, nose, throat; dizziness, headache, nausea, vomiting, diarrhea, tremors, weakness, convulsions, dyspnea, cyanosis; aplastic anemia; muscle spasms. *See NIOSH Pocket Guide to Chemical Hazards* (DHHS/NIOSH 97-140, 1997) p 186; *Clinical Toxicology of Commercial Products*, R. E. Gosselin *et al.*, Eds. (Williams & Wilkins, Baltimore, 5th ed., 1984) Section III, pp 239-241. Lindane and other hexachlorocyclohexane isomers are reasonably anticipated to be human carcinogens: *Report on Carcinogens, Twelfth Edition* (PB2011-111646, 2011) p 256.

Note: Lindane is listed as a persistent organic pollutant (POP) in Annex A of the *Stockholm Convention on Persistent Organic Pollutants* (United Nations, Stockholm, 2001) 43 pp; amended (Geneva, 2009) 63 pp.

USE: Insecticide.

THERAP CAT: Pediculicide; scabicide.

THERAP CAT (VET): Ectoparasiticide.

5557. Lindlar Catalyst. [53092-86-7] Pd-Pb-CaCO$_3$. Prepn: Lindlar, *Helv. Chim. Acta* **35**, 446 (1952); Lindlar, Dubuis cited by Fieser, Fieser, *Reagents for Organic Synthesis* (New York, 1967) p 566.

USE: In selective hydrogenation of triple bonds to *cis*-double bonds.

5558. Lineatin. [65035-34-9] (1R,2S,5R,7R)-1,3,3-Trimethyl-4,6-dioxatricyclo[3.3.1.0²·⁷]nonane; [1R-(1α,2β,5α,7β)]-3,3,7-trimethyl-4,9-dioxatricyclo[3.3.1.0²·⁷]nonane; 3,3,7-trimethyl-2,9-dioxatricyclo[3.3.1.0⁴·⁷]nonane; 4,6,6-lineatin. $C_{10}H_{16}O_2$; mol wt

168.24. C 71.39%, H 9.59%, O 19.02%. Isoln of the unique tricyclic aggregation pheromone from ambrosia beetles, *Trypodendron lineatum* (Olivier): J. G. MacConnell *et al.*, *J. Chem. Ecol.* **3**, 549 (1977). Synthesis of (±)-form: K. Mori, M. Sasaki, *Tetrahedron Lett.* **1979**, 1329; K. N. Slessor *et al.*, *J. Org. Chem.* **45**, 2290 (1980); K. Mori *et al.*, *Tetrahedron Lett.* **23**, 1921 (1982); L. Skattebol, Y. Stenstrom, *ibid.* **24**, 3021 (1983); B. D. Johnston *et al.*, *J. Org. Chem.* **50**, 114 (1985). Synthesis of racemate and optical isomers: K. Mori, M. Sasaki, *Tetrahedron* **36**, 2197 (1980). Short stereoselective synthesis: I. Aljancic-Solaja *et al.*, *Helv. Chim. Acta* **70**, 1302 (1987). Comparative activity of the isomers: J. H. Borden *et al.*, *Can. Entomol.* **112**, 107 (1980).

Oil, bp$_{10}$ 70°. $[\alpha]_D^{24}$ +66.3° (c = 3.1 in CHCl$_3$).

USE: Insect sex attractant.

5559. Linezolid. [165800-03-3] *N*-[[(5*S*)-3-[3-Fluoro-4-(4-morpholinyl)phenyl]-2-oxo-5-oxazolidinyl]methyl]acetamide; PNU-100766; U-100766; Zyvox; Zyvoxid. C$_{16}$H$_{20}$FN$_3$O$_4$; mol wt 337.35. C 56.97%, H 5.98%, F 5.63%, N 12.46%, O 18.97%. Prototype of the oxazolidinone antimicrobials; inhibits bacterial mRNA translation. Prepn: M. R. Barbachyn *et al.*, **WO 9507271** (1995 to Upjohn); *eidem*, **US 5688792** (1997 to Pharmacia & Upjohn); S. J. Brickner *et al.*, *J. Med. Chem.* **39**, 673 (1996). Antibacterial spectrum: C. W. Ford *et al.*, *Antimicrob. Agents Chemother.* **40**, 1508 (1996). Mechanism of action study: D. L. Shinabarger *et al.*, *ibid.* **41**, 2132 (1997). HPLC determn in plasma: C. Buerger *et al.*, *J. Chromatogr. B* **796**, 155 (2003). Clinical comparison with vancomycin, *q.v.*, for MRSA infections: D. L. Stevens *et al.*, *Clin. Infect. Dis.* **34**, 1481 (2002). Review of pharmacology: L. D. Dresser, M. J. Rybak, *Pharmacotherapy* **18**, 456-462 (1998); and clinical experience: R. Norrby, *Expert Opin. Pharmacother.* **2**, 293-302 (2001).

White crystals from ethyl acetate and hexanes, mp 181.5-182.5°. $[\alpha]_D^{20}$ −9° (c = 0.919 in chloroform).

THERAP CAT: Antibacterial.

5560. Linoleic Acid. [60-33-3] (9*Z*,12*Z*)-9,12-Octadecadienoic acid; 9,12-linoleic acid; linolic acid. C$_{18}$H$_{32}$O$_2$; mol wt 280.45. C 77.09%, H 11.50%, O 11.41%. An essential fatty acid. Major constituent of many vegetable oils, *e.g.*, cottonseed, soybean, peanut, corn, sunflower seed, safflower, poppy seed, linseed, and perilla oils, where it occurs as a glyceride. Characteristic ingredient of semidrying oils. Isoln: Swern, Parker, *J. Am. Oil Chem. Soc.* **30**, 5 (1953); Parker *et al.*, *Biochem. Prep.* **4**, 86 (1955); McCutcheon, *Org. Synth.* **coll. vol. III**, 526 (1955). Summary of work on structure: T. P. Hilditch, *The Chemical Constitution of Natural Fats* (Chapman & Hall, London, 2nd ed. 1956). Synthesis: Raphael, Sondheimer, *J. Chem. Soc.* **1950**, 2102; Gensler, Thomas, *J. Am. Chem. Soc.* **73**, 4601 (1951); Walborsky *et al.*, *ibid.* 2590; Nigam, Weedon, *J. Chem. Soc.* **1956**, 4052; Osbond, Wickens, *Chem. Ind. (London)* **1959**, 1288. Review of physiological role in mammals: H. S. Hansen, *Trends Biochem. Sci.* **11**, 263 (1986).

Colorless oil. Easily oxidized by air, cannot be distilled without decompn. Storage in ester form is recommended. d$_4^{18}$ 0.9038; d$_4^{22}$ 0.9007. mp −12°. bp$_{1.4}$ 202°; bp$_{16}$ 230°. $n_D^{11.5}$ 1.4715; n_D^{20} 1.4699; $n_D^{21.5}$ 1.4683; n_D^{50} 1.4588. Iodine value: 181.1. Thiocyanogen value 96.7. Freely sol in ether. Sol in abs alc. One ml dissolves in 10 ml petr ether. Miscible with dimethylformamide, fat solvents, oils.

Aluminum salt. [645-17-0] Al(C$_{18}$H$_{31}$O$_2$)$_3$. Yellow lumps or powder; linseed oil odor. Practically insol in water. Sol in oils, fixed alkali hydroxides.

Methyl ester *see* Methyl Linoleate.

Ethyl ester *see* Ethyl Linoleate.

Cyclohexylamide. [3207-50-9] Linolexamide; *N*-cyclohexyl-linoleamide; Clinolamide. C$_{24}$H$_{43}$NO; mol wt 361.61.

USE: Manuf paints, coatings, emulsifiers, vitamins. Aluminum salt used to manuf lacquers.

THERAP CAT: Nutrient (essential fatty acid).

THERAP CAT (VET): Dietary supplement.

5561. Linolenic Acid. [463-40-1] (9*Z*,12*Z*,15*Z*)-9,12,15-Octadecatrienoic acid; α-linolenic acid. C$_{18}$H$_{30}$O$_2$; mol wt 278.44. C 77.65%, H 10.86%, O 11.49%. An essential fatty acid. Occurs as the glyceride in most drying oils. Synthesis: Nigam, Weedon, *J. Chem. Soc.* **1956**, 4049; Osbond, Wickens, *Chem. Ind. (London)* **1959**, 1288. Biosynthetic studies: C. G. Kannangara *et al.*, *Biochem. Biophys. Res. Commun.* **52**, 648 (1973); B. S. Jacobson *et al.*, *ibid.* 1190; C. J. Bedord *et al.*, *Arch. Biochem. Biophys.* **185**, 15 (1978). Effects on lipid metabolism in rat tissue: M. L. Garg *et al.*, *Lipids* **23**, 847 (1988). Review of dietary linolenic acid in mammals: J. Tinoco *et al.*, *ibid.* **14**, 166-171 (1979); in man: N. Zöllner, *Prog. Lipid Res.* **25**, 177-180 (1986).

Colorless liquid. d$_4^{18}$ 0.914. bp$_1$ 230-232°. Insol in water. Sol in organic solvents.

THERAP CAT: Nutrient (essential fatty acid).

5562. γ-Linolenic Acid. [506-26-3] (6*Z*,9*Z*,12*Z*)-6,9,12-Octadecatrienoic acid; *cis*-6,*cis*-9,*cis*-12-octadecatrienoic acid; gamolenic acid; GLA; Viacutan. C$_{18}$H$_{30}$O$_2$; mol wt 278.44. C 77.65%, H 10.86%, O 11.49%. Polyunsaturated fatty acid produced in the body as the Δ^6-desaturase metabolite of linoleic acid, *q.v.* Converted to dihomo-γ-linolenic acid, a biosynthetic precursor of monoenoic prostaglandins such as PGE$_1$. Present to varying extents in the fatty acid fraction of evening primrose oil (7-10%), in borage oil (18-26%), in black currant oil (15-20%) and in oils from different fungal sources (6-24%). Isoln from evening primrose oil, *q.v.*: A. Heiduschka, K. Luft, *Arch. Pharm.* **257**, 33 (1919). Proposed structure: Eibner *et al.*, *Chem. Umschau.* **34**, 312 (1927). Confirmation of structure: J. P. Riley, *J. Chem. Soc.* **1949**, 2728. Discussion of occurrence, esp. in fungi: R. Shaw, *Biochim. Biophys. Acta* **98**, 230 (1965). Synthesis: J. M. Osbond *et al.*, *J. Chem. Soc.* **1961**, 2779; J. M. Osbond, *ibid.* 5270. Metabolism studies: J. F. Mead, D. R. Howton, *J. Biol. Chem.* **229**, 575 (1957); K. J. Stone *et al.*, *Lipids* **14**, 174 (1979). Effect of source on essential fatty acid and prostanoid metabolite formation: D. K. Jenkins *et al.*, *Med. Sci. Res.* **16**, 525 (1988).

Hexabromide derivative. C$_{18}$H$_{30}$Br$_6$O$_2$. Crystals from ethyl methyl ketone, mp 201-202°.

USE: Nutrient.

THERAP CAT: In treatment of atopic eczema.

THERAP CAT (VET): Dermatological.

5563. Linseed. Flaxseed; linum. Dried ripe seeds of *Linum usitatissimum* L., Linaceae. *Constit.* 30-40% oil, about 6% mucilage, about 25% proteins and linamarin. Brief review of medicinal uses: M. Wichtl, N. G. Bisset, *Herbal Drugs and Phytopharmaceuticals*, English Ed. (CRC Press, Boca Raton, 1994) pp 298-300

Linseed oil. A drying oil obtained by expression of linseed. *Constit.* Glycerides of linolenic, linoleic, oleic, stearic, palmitic and myristic acids. *Ref:* T. P. Hilditch, *The Chemical Constitution of Natural Fats* (London, 3rd ed., 1956) p 175 sqq; E. W. Eckey, *Veg-*

etable Fats and Oils (New York, 1954) pp 535-547. Yellowish liquid, peculiar odor, bland taste. Exposed to air it gradually thickens, becomes darker, and acquires a more pronounced odor and taste. d 0.925-0.935. n_D^{40} 1.4725-1.4750. Does not congeal above −20°. Sapon no.: 187-195. Iodine no. not below 170. Unsaponifiable matter not over 1.5%. Slightly sol in alcohol, miscible with chloroform, ether, petr ether, carbon disulfide, oil turpentine.

USE: Emollient. Oil in varnishes, paints, putty, oilcloths, linoleum, printing inks, artificial rubber, tracing cloth, tanning and enameling leather; applied to paper and fabrics to render them waterproof and tough.

THERAP CAT: Laxative; externally as poultice.

THERAP CAT (VET): Laxative.

5564. Linuron. [330-55-2] *N'*-(3,4-Dichlorophenyl)-*N*-methoxy-*N*-methylurea; methoxydiuron; Du Pont Herbicide 326; HOE-2810; Afalon; Linurex; Lorox. $C_9H_{10}Cl_2N_2O_2$; mol wt 249.09. C 43.40%, H 4.05%, Cl 28.46%, N 11.25%, O 12.85%. Selective pre- and post-emergence herbicide. Preparation: Sherer, Heller, US 2960534 (1960 to Hoechst). Degradn in soil: G. F. Kempson-Jones, R. J. Hance, *Pestic. Sci.* **10**, 449 (1979). Toxicity study: G. W. Bailey, J. L. White, *Residue Rev.* **10**, 97 (1965).

mp 93-94°. Vapor pressure at 24°: 1.5×10^{-5} mm Hg. Soly in water: 75 ppm. Partially sol in acetone, alcohol, benzene, toluene, xylene. LD_{50} orally in rats: 1500 mg/kg (Bailey, White).

USE: Herbicide.

5565. Liothyronine. [6893-02-3] *O*-(4-Hydroxy-3-iodophenyl)-3,5-diiodo-L-tyrosine; L-3-[4-(4-hydroxy-3-iodophenoxy)-3,5-diiodophenyl]alanine; 4-(3-iodo-4-hydroxyphenoxy)-3,5-diiodophenylalanine; 3,5,3'-triiodothyronine; T_3. $C_{15}H_{12}I_3NO_4$; mol wt 650.98. C 27.68%, H 1.86%, I 58.48%, N 2.15%, O 9.83%. One of the hormones produced by the thyroid gland that is involved in the maintenance of metabolic homeostasis. Also produced in peripheral tissues as the active metabolite of thyroxine, *q.v.* Identification in serum: J. Gross, R. Pitt-Rivers, *Lancet* **I**, 439 (1952). Prepn from diiodothyronine: J. Roche *et al.*, *Biochim. Biophys. Acta* **11**, 215 (1953). Isoln from thyroid gland and synthesis: J. Gross, R. Pitt-Rivers, *Biochem. J.* **53**, 645 (1953). Bioactivity: *eidem, ibid.* 652. Elevated levels of T_3 have been noted in victims of sudden infant death syndrome: G. Kocsard-Varo, *Med. J. Aust.* **2**, 789 (1973); M. A. Chacon, J. T. Tildon, *J. Pediatr.* **99**, 758 (1981). Direct determn in serum by RIA: I. J. Chopra *et al.*, *Thyroid* **6**, 255 (1996). Clinical trial in combination with thyroxine: R. Bunevicius *et al.*, *N. Engl. J. Med.* **340**, 424 (1999). Review of pharmacology and clinical uses: E. Sypniewski, *Ann. Thorac. Surg.* **56**, S2-S8 (1993); of clinical trials in depression: R. Aronson *et al.*, *Arch. Gen. Psychiatry* **53**, 842-848 (1996); of use in heart transplantation: V. Jeevanandam, *Thyroid* **7**, 139-145 (1997). Review of T_3-receptors and molecular mechanism of action: M. A. Lazar, *Endocr. Rev.* **14**, 184-193 (1993); G. A. Brent, *N. Engl. J. Med.* **331**, 847-853 (1994). Review of biological effects: H. C. Freake, J. H. Oppenheimer, *Annu. Rev. Nutr.* **15**, 263-291 (1995).

Crystals, dec 236-237°. $[\alpha]_D^{29.5}$ +21.5° (c = 4.75 in a mixture of 1 part *N* HCl + 2 parts ethanol). Insol in water, alc, propylene glycol. Sol in dil alkalies with the formation of a brownish, water-soluble, sodium salt.

Sodium salt. [55-06-1] Liothyronine sodium; sodium L-triiodothyronine; Cytobin; Cytomel; Cynomel; Tertroxin. $C_{15}H_{11}I_3$-

$NNaO_4$; mol wt 672.96. Light tan, odorless, crystalline powder. Slightly sol in alc; very slightly sol in water. Practically insol in most organic solvents.

Hydrochloride. [6138-47-2] Thybon. $C_{15}H_{12}I_3NO_4 \cdot HCl$; mol wt 687.44. Long birefringent needles, dec 202-203°. $[\alpha]_D^{29.5}$ +21.5° (c = 4.75 in a mixture of 1 vol *N* HCl and 2 vols ethanol).

THERAP CAT: Thyroid hormone.

THERAP CAT (VET): Thyroid hormone.

5566. Lipase. [9001-62-1] Triacylglycerol lipase. An enzyme (or more exactly a group of enzymes) belonging to the esterases. Hydrolyzes fat (present in ester form, such as glycerides) yielding fatty acids and glycerol. Catalyzes digestion. Widely distributed in the plant world, also in molds, bacteria, milk and milk products, and in animal tissues, especially in the pancreas. Isoln from castor beans: H. Gibian in *Ullmanns Encyklopädie der technischen Chemie*, 3rd ed., **vol. 7**, 406-407 (1956). Purification of pancreatic lipase: Marchis-Mouren *et al.*, *Arch. Biochem. Biophys.* **83**, 309 (1959). Review of milk lipases: Chandan, Shahani, *J. Dairy Sci.* **47**, 471 (1964). Comprehensive reviews: Wills, *Adv. Lipid Res.* **3**, 197-240 (1965); Desnuelle in *The Enzymes* **vol. 7**, P. D. Boyer, Ed. (Academic Press, New York, 3rd ed., 1972) pp 575-616.

The optimum temp for enzyme action is between 35° and 37° at pH 5-6. Lipase contains sulfhydryl groups and is inactivated by substances that inhibit such compds. It is activated by substances that keep SH groups in the reduced state, such as glutathione, cysteine, and ascorbic acid. The addition of acid activates lipase preparations. Castor-oil lipase is activated by sulfuric, oxalic, formic, acetic and butyric acids. Acetic, salicylic and hydrochloric acids increase the action of lipase derived from various organs of the pig. Caprylic and caproic acids increase the action of lipase derived from certain mold fungi. Almost all organic solvents decrease lipase activity, petr ether being an exception.

USE: To split fats without damaging sensitive constituents, such as vitamins or unsaturated fatty acids. In food processing for flavor improvement; in detergents for the improvement of cleaning action. For review of industrial applications of microbial lipases, *see:* Seitz, *J. Am. Oil Chem. Soc.* **51**, 12 (1974).

THERAP CAT: Digestive enzyme.

5567. Lipoprotein Lipase. [9004-02-8] Diacylglycerol lipase; clearing factor; EC 3.1.1.34. A specific lipase which preferentially hydrolyzes triglycerides in the presence of a lipoprotein complex, forming glycerol and unesterified fatty acids: Korn, *Methods Enzymol.* **5**, 542 (1962). The normal substrate is turbid lipemic or chylomicron-containing plasma. This substance is clarified by the enzyme, hence the term clearing factor. First detected in the plasma of animals injected intravenously with heparin: Hahn, *Science* **98**, 19 (1943); Robinson, French, *Pharmacol. Rev.* **12**, 241 (1960). Prepn from post heparin human plasma: Baskys, *Diss. Abstr.* **20**, 1146 (1959). It has since been prepd from several tissues of normal, untreated animals. Prepn from chicken adipose tissue: Korn, Quigley, *J. Biol. Chem.* **226**, 833 (1957); from human heart: Schnatz *et al.*, *Am. J. Physiol.* **205**, 401 (1963). Role in the metabolism of triglycerides by adipose tissues: Rodbell, Scow in K. Rodahl, *Fat as a Tissue* (McGraw-Hill, New York, 1964) pp 110-126. *Reviews:* Jensen, *Prog. Chem. Fats Other Lipids* **11**, 347 (1971); Desnuelle in *The Enzymes* **vol. 7**, P. D. Boyer, Ed. (Academic Press, New York, 3rd ed., 1972) pp 606-609.

5568. Lipotropic Hormone. [9035-55-6] Lipotropin; pituitary lipotropic hormone; lipid-mobilizing hormone; adipokinetic hormone; lipolytic hormone; lipotrophin; LPH. Hypophyseal hormone which stimulates the release of fatty acids from adipose tissue. *β-LPH* is a single chain polypeptide containing 89-93 amino acid residues and has been proposed as the biosynthetic precursor for *β*-MSH and *β*-endorphin, *q.q.v.*, *see:* M. Chretien *et al.*, *Can. J. Biochem.* **57**, 1111 (1979). *γ-LPH* consists of 58 amino acids and is identical to the sequence of the first 58 residues of *β*-LPH. Both contain sequences common to ACTH and *β*-melanotropin, *q.q.v.* Sequences differ slightly among mammalian species. Isoln and proposed structure of ovine *β*-LPH: C. H. Li, *Nature* **201**, 924 (1964); Y. Birk, C. H. Li, *J. Biol. Chem.* **239**, 1048 (1964); C. H. Li *et al.*, *Nature* **208**, 1093 (1965); C. H. Li, *Arch. Biol. Med. Exp.* **5**, 55 (1968); revised structure: Chretien *et al.*, *Int. J. Pept. Protein Res.* **4**, 263 (1972). Isoln and characterization of porcine *β*-LPH: Gilardeau, Chretien, *Can. J. Biochem.* **48**, 1017 (1970). Amino acid sequence of porcine γ-LPH: Graf *et al.*, *Acta Biochim. Biophys.* **5**, 305 (1970). Isoln, characterization of bovine β-LPH: P. Lohmar, C. H. Li, *Biochem. Biophys. Res. Commun.* **77**, 1088 (1977); of rat β-LPH: M. Rubinstein *et al.*, *Proc. Natl. Acad. Sci. USA* **74**, 3052 (1977); of whale β-LPH: H. Kawauchi *et al.*, *Int. J. Pept. Protein Res.* **15**, 171 (1980); of turkey β-LPH: W. C. Chang *et al.*, *ibid.* 261. Initial isoln, partial structure of human β-LPH: G. Cseh *et al.*, *FEBS Lett.* **2**, 42 (1968); *eidem, ibid.* **21**, 344(1972). Isoln, characterization, amino acid sequence of human β-LPH: C. H. Li, D. Chung, *Int. J. Pept. Protein Res.* **17**, 131 (1981). *Review:* D. T. Krieger *et al.*, *Recent Prog. Horm. Res.* **36**, 277-344 (1980).

5569. Liraglutide. [204656-20-2] L-Histidyl-L-alanyl-L-α-glutamylglycyl-L-threonyl-L-phenylalanyl-L-threonyl-L-seryl-L-α-aspartyl-L-valyl-L-seryl-L-seryl-L-tyrosyl-L-leucyl-L-α-glutamylglycyl-L-glutaminyl-L-alanyl-L-alanyl-N^6-[N-(1-oxohexadecyl)-L-γ-glutamyl]-L-lysyl-L-α-glutamyl-L-phenylalanyl-L-isoleucyl-L-alanyl-L-tryptophyl-L-leucyl-L-valyl-L-arginylglycyl-L-arginylglycine; N^{26}-(hexadecanoyl-γ-glutamyl)-(34-arginine)GLP-1-(7-37)-peptide; Lys^{26}(N^ε-(γ-glutamyl-(N^α-hexadecanoyl))), Arg^{34}-GLP-1(7-37); NN-2211; Victoza. $C_{172}H_{265}N_{43}O_{51}$; mol wt 3751.26. C 55.07%, H 7.12%, N 16.06%, O 21.75%. Acylated derivative of glucagon-like peptide 1, *q.v.* Prepn: L. B. Knudsen *et al.*, **WO 9808871**; *idem et al.*, **US 6268343** (1998, 2001 both to Novo Nordisk); *eidem, J. Med. Chem.* **43**, 1664 (2000). Clinical effect on glucose and insulin homeostasis: H. Agerso, P. Vicini, *Eur. J. Pharm. Sci.* **19**, 141 (2003). Clinical pharmacokinetics: B. Elbrond *et al.*, *Diabetes Care* **25**, 1398 (2002). Clinical comparison with exenatide: J. B. Buse *et al.*, *Lancet* **374**, 39 (2009). Review of clinical effect on glycemic control: P. Raskin, P. F. Mora, *Int. J. Clin. Pract.* **64**, 21-27 (2010); of clinical experience in type 2 diabetes: R. Kela *et al.*, *Expert Opin. Biol. Ther.* **11**, 951-959 (2011).

Glu-Gly-Thr-Phe-Thr-Ser-Asp-Val-Ser-Ser-Tyr-Leu-Glu-Gly-Gln-Ala-Ala-Lys-Glu
|
Ala   Phe
|
His   Gly-Arg-Gly-Arg-Val-Leu-Trp-Ala-Ile

THERAP CAT: Antidiabetic.

5570. Liranaftate. [88678-31-3] *N*-(6-Methoxy-2-pyridinyl)-*N*-methylcarbamothioic acid *O*-(5,6,7,8-tetrahydro-2-naphthalenyl) ester; *O*-5,6,7,8-tetrahydro-2-naphthyl *N*-(6-methoxy-2-pyridyl)-*N*-methylthiocarbamate; piritetrate; M-732; Zefnart. $C_{18}H_{20}N_2O_2S$; mol wt 328.43. C 65.83%, H 6.14%, N 8.53%, O 9.74%, S 9.76%. Squalene epoxidase inhibitor. Prepn: **BE 897021**; T. Takematsu *et al.*, **US 4554012** (1983, 1985 both to Toyo Soda Manufacturing Co., Ltd.). Mode of action study: T. Morita *et al.*, *J. Med. Vet. Mycol.* **27**, 17 (1989). Antifungal spectrum: K. Iwata *et al.*, *Antimicrob. Agents Chemother.* **33**, 2118 (1989). Physicochemical properties and stability: H. Awano *et al.*, *Iyakuhin Kenkyu* **23**, 558 (1992).

White crystalline powder, mp 98.5-99.5°. Soly (ml/g): acetone 10; chloroform 2; dehydrated ethanol 152; dichloromethane 2; ether 21; hexane 233; methanol 175; water >10000. Soly (mg/l): 0.10 (distilled water); 0.07 (buffer pH 1); 0.08 (buffer pH 4); 0.07 (buffer pH 7); 0.08 (buffer pH 11). Sol in benzene, dimethyl sulfoxide, *N,N*-dimethylformamide.

THERAP CAT: Antifungal.

5571. Lisdexamfetamine. [608137-32-2] (2*S*)-2,6-Diamino-*N*-[(1*S*)-1-methyl-2-phenylethyl]hexanamide; L-lysine-*d*-amphetamine. $C_{15}H_{25}N_3O$; mol wt 263.39. C 68.40%, H 9.57%, N 15.95%, O 6.07%. Inactive prodrug in which dextroamphetamine is covalently bound to L-lysine; the amide bond is hydrolyzed during metabolism to release active *d*-amphetamine. Prepn: T. Mickle *et al.*, **WO 05032474**; *eidem*, **US 05054561** (both 2005 to New River Pharm.). Evaluation of cytochrome P450 interations: S. Krishnan, S. Moncrief, *Drug Metab. Dispos.* **35**, 180 (2007). Pharmacokinetics in adults: S. Krishnan, Y. Zhang, *J. Clin. Pharmacol.* **48**, 293 (2008). Clinical efficacy in pediatric ADHD: J. Biederman *et al.*, *Clin. Ther.* **29**, 450 (2007). Review of pediatric pharmacology and clinical experience: S. K. A. Blick, G. M. Keating, *Pediatr. Drugs* **9**, 129-135 (2007); of development and clinical experience: S. V. Faraone, *Expert Opin. Pharmacother.* **9**, 1565-1574 (2008).

Golden colored solid from MeOH, mp 120-122°.

Dimethanesulfonate. [608137-33-3] Lisdexamfetamine dimesylate; NRP-104; Vyvanse. $C_{15}H_{25}N_3O.2CH_4O_3S$; mol wt 455.59. White to off-white powder, mp 120-122°. Soly in water: 792 mg/ml.

Note: This is a controlled substance (opiate): **21 CFR**, 1308.12.

THERAP CAT: In treatment of attention deficit hyperactivity disorder (ADHD).

5572. Lisinopril. [76547-98-3] N^2-[(1*S*)-1-Carboxy-3-phenylpropyl]-L-lysyl-L-proline. $C_{21}H_{31}N_3O_5$; mol wt 405.50. C 62.20%, H 7.71%, N 10.36%, O 19.73%. Orally active angiotensin-converting enzyme (ACE) inhibitor. Prepn: A. A. Patchett *et al.*, **EP 12401**; E. E. Harris *et al.*, **US 4374829** (1980, 1983 both to Merck & Co.); M. T. Wu *et al.*, *J. Pharm. Sci.* **74**, 352 (1985). Comprehensive description: D. P. Ip *et al.*, *Anal. Profiles Drug Subs. Excip.* **21**, 233-276 (1992). HPLC determn in urine: Y. C. Wong, B. G. Charles, *J. Chromatogr. B* **673**, 306 (1995); in pharmaceutical formulations: C. A. Beasley *et al.*, *J. Pharm. Biomed. Anal.* **37**, 559 (2005). Series of articles in hypertension and congestive heart failure: *Am. J. Med.* **85**, Suppl. 3B, 1-59 (1988). Review of clinical efficacy in myocardial infarction: K. L. Goa *et al.*, *Drugs* **52**, 564-588 (1996); in diabetic complications: *idem et al.*, *ibid.* **53**, 1081-1105 (1997); in congestive heart failure: K. Simpson, B. Jarvis, *ibid.* **59**, 1149-1167 (2000).

Dihydrate. [83915-83-7] MK-521; Acerbon; Alapril; Carace; Coric; Corprilin; Lisitril; Novatec; Prinil; Prinivil; Vivatec; Zestril. $C_{21}H_{31}N_3O_5.2H_2O$; mol wt 441.53. White to off-white, odorless, crystalline powder. Crystals from methanol:ethyl acetate, mp 159-160°. pKa$_1$ (25°) 2.5; pKa$_2$ 4.0; pKa$_3$ 6.7; pKa$_4$ 10.1. Partition coefficient at room temp (*n*-octanol/0.1*M* pH 7 phosphate buffer): 10.2 ±0.5. Approx uv max (0.1*N* NaOH): 246, 254, 258, 261, 267 nm (A$^{1\%}_{1cm}$ 4.0, 4.5, 5.1, 5.1, 3.7); (0.1*N* HCl): 246, 253, 258, 264, 267 nm (A$^{1\%}_{1cm}$ 3.2, 3.9, 4.5, 5.0, 2.8). [α]$^{25}_D$ −23.3° (c=1.0 in CH$_3$-OH). [α]$^{25}_{405}$ −120° (c = 1 in 0.25*M* pH 6.4 zinc acetate); [α]$^{25}_{436}$ −96° (c = 1 in 0.25*M* pH 6.4 zinc acetate). Sol in water; sparingly sol in methanol. Practically insol in ethanol, acetone, acetonitrile, chloroform, DMF.

THERAP CAT: Antihypertensive.

THERAP CAT (VET): Antihypertensive.

5573. Lisofylline. [100324-81-0] 3,7-Dihydro-1-[(5R)-5-hydroxyhexyl]-3,7-dimethyl-1H-purine-2,6-dione; 1-[(R)-5-hydroxyhexyl]theobromine; 1-(5R-hydroxyhexyl)-3,7-dimethylxanthine; CT-1501R; ProTec. $C_{13}H_{20}N_4O_3$; mol wt 280.33. C 55.70%, H 7.19%, N 19.99%, O 17.12%. Methylxanthine that inhibits production of phosphatidic acid during the inflammatory response. Identification as metabolite of pentoxifylline: H.-J. Hinze *et al.*, *Arzneim.-Forsch.* **22**, 1144 (1972). Enantioselective process: W. Aretz *et al.*, **DE 3942872**; *eidem*, **US 5310666** (1991, 1994 both to Hoechst). Asymmetric synthesis: J. P. Klein *et al.*, **WO 9531450** (1995 to Cell Therapeutics). Study of mechanism of action in septic shock: G. C. Rice *et al.*, *Proc. Natl. Acad. Sci. USA* **91**, 3857 (1994). Enhancement of hematopoietic recovery following 5-fluorouracil treatment in mice: E. Clarke *et al.*, *Cancer Res.* **56**, 105 (1996).

mp 110°. $[\alpha]_D^{20}$ −5.6° (c = 6.7 in ethanol).
THERAP CAT: Immunomodulator.

5574. Lisuride. [18016-80-3] N′-[(8α)-9,10-didehydro-6-methylergolin-8-yl]-N,N-diethylurea; 9-(3,3-diethylureido)-4,6,6a,-7,8,9-hexahydro-7-methylindolo[4,3-f,g]quinoline; 1,1-diethyl-3-(D-6-methylisoergolen-8-yl)urea; N-(D-6-methyl-8-isoergolenyl)-N,N-diethylurea; methylergol carbamide; lysuride. $C_{20}H_{26}N_4O$; mol wt 338.46. C 70.97%, H 7.74%, N 16.55%, O 4.73%. Dopamine D_2-receptor agonist. Prepn: V. Zikan, M. Semonsky, *Collect. Czech. Chem. Commun.* **25**, 1922 (1960); *eidem*, *Pharmazie* **23**, 147 (1968). Pharmacology and toxicity: Z. Votava, I. Lamplova, *Physiol. Bohemoslov.* **12**, 37 (1963), *C.A.* **59**, 9221d (1963).

Crystals from benzene, mp 186°. $[\alpha]_D^{20}$ +313° (c = 0.60 in pyridine).
Maleate. [19875-60-6] Apodel; Cuvalit; Dopergin; Eunal; Lysenyl; Revanil. $C_{20}H_{26}N_4O.C_4H_4O_4$; mol wt 454.53. Prisms from ethanol, mp 200° (dec). $[\alpha]_D^{20}$ +288° (c = 0.5 in methanol). uv max (methanol): 313 nm. LD_{50} i.v. in mice: 14.4 mg/kg (Votava, Lamplova).

THERAP CAT: Antimigraine; prolactin inhibitor; antiparkinsonian.

5575. Lita®. Protein-based fat mimetic consisting of water-dispersible microparticles of zein, *q.v.*, a hydrophobic prolamine derived from corn. Prepn: L. E. Stark, A. T. Gross, **US 5021248** (1991 to Enzytech); *eidem*, **US 5145702** (1992 to Opta Food Ingred.). Brief description: R. Iyengar, A. Gross in *Biotechnology and Food Ingredients*, I. Goldberg, R. Williams, Eds. (Van Nostrand Reinhold, New York, 1991) pp 287-313; M. S. Miller in *Protein Functionality in Food Systems*, N. S. Hettiarachchy, G. R. Zeigler, Eds. (Marcel Dekker, New York, 1994) pp 435-465.

Hydrophobic, smooth spheroidal particles. Size distribution: 0.3-3.0 μm. Insol in water. Sol in ethanol. Colloidal suspensions stable to 95°.
USE: Fat substitute for food products.

5576. Lithium. [7439-93-2] Li; at. wt [6.938; 6.997]; conventional at. wt 6.94; at. no. 3; valence 1. Group IA (1). Alkali metal.

Occurrence in earth's crust: 20 ppm. Naturally occurring isotopes: 7 (92.5%); 6 (7.5%); artificial radioactive isotopes: 5, 8-11; all are unstable ($T_{\frac{1}{2}}$ < 1 sec). Discovered as salt in 1817 by Arfvedson: *Ann. Chim. Phys.* [2] **10**, 82 (1819); metal prepared independently by Davy and Brandé in 1818. Occurs in a number of minerals: *spodumene* ($LiAlS_2O_6$), *lepidolite* [K(Li,Al)$_3$(Si,Al)$_4O_{10}$(F,OH)$_2$], *petalite* ($LiAlSi_4O_{10}$), *amblygonite* (AlPO$_4$,LiF), and *triphylite* (Li-FePO$_4$). Also recovered from natural brines. Prepn of the metal by electrochemical processes: Guntz, *Compt. Rend.* **117**, 732 (1893); Ruff, Johannsen, *Z. Elektrochem.* **12**, 186 (1906); by reduction of the oxide with magnesium or aluminum: Warren, *Chem. News* **74**, 6 (1896); Hanson, **US 2028390** (1936). Reviews of biology, pharmacology and toxicity of lithium ion: Schou, "Lithium in Psychiatry: A Review" in *Psychopharmacology, A Review of Progress 1957-1967*, D. H. Efron, Ed. (Public Health Service Publication No. 1836, 1968) pp 701-718; Doig *et al.*, *J. Chem. Educ.* **50**, 343-345 (1973); Samuel, Gottesfeld, *Endeavour* **32**, 122-128 (1973); Saran, Gaind, *Clin. Toxicol.* **6**, 257-269 (1973); Schou, *Annu. Rev. Pharmacol. Toxicol.* **16**, 231-243 (1976). *Reviews:* Hart, Beumel, "Lithium and its Compounds" in *Comprehensive Inorganic Chemistry* vol. **1**, J. C. Bailar, Jr. *et al.*, Eds. (Pergamon Press, Oxford, 1973) pp 331-367; C. W. Kamienski *et al.* in *Kirk-Othmer Encyclopedia of Chemical Technology* vol. **15** (Wiley-Interscience, New York, 4th ed., 1995) pp 434-463; *Chemistry of the Elements* N. N. Greenwood, A. Earnshaw, Eds. (Pergamon Press, New York, 1984) pp 75-116; *Patty's Industrial Hygiene and Toxicology* vol. **2C**, G. D. Clayton, F. E. Clayton, Eds. (John Wiley & Sons, Inc., New York, 4th ed., 1994) pp 2087-2097. Review of pharmacology: N. J. Birch, J. D. Phillips, *Adv. Inorg. Chem.* **36**, 49-75 (1991); and clinical use: L. H. Price, G. R. Heninger, *N. Engl. J. Med.* **331**, 591-598 (1994).

Silvery-white metal; body-centered cubic structure; becomes yellowish on exposure to moist air. mp 180.54°. bp 1347°. Also reported as bp 1336 ± 5°. *See:* Hartman, Schneider, *Z. Anorg. Chem.* **180**, 275 (1929). d^{20} 0.534. Hardest of the alkali metals; Mohs' hardness 0.6. Heat capacity at constant pressure (25°): 5.892 cal/mole deg. *See:* Douglass *et al.*, *J. Am. Chem. Soc.* **77**, 2144 (1955). E° (aqueous) Li/Li⁺ 3.045 V. Reacts with H_2O forming the hydroxide and H_2. Reacts violently with inorganic acids; reacts slowly with cold H_2SO_4. Sol in liquid ammonia forming a blue soln. Does not react with oxygen at room temp; forms Li_2O when heated to 100° or higher. Emits characteristic crimson color (670.8 nm) in flame. *Flammable; dangerous when wet. Keep under mineral oil or other liquid free from oxygen or water.* Li metal ignites in air near its mp; presents fire and explosion risk when exposed to water, N, acids or oxidizing reagents.

Caution: Potential symptoms of mild overexposure to the Li ion by ingestion of lithium salts are impaired concentration, lethargy, irritability, muscle weakness; moderate overexposure may cause disorientation, confusion, drowsiness; severe overexposure may cause impaired consciousness with progression to coma, convulsions, and impaired renal function (Price, Heninger). Direct contact with metal may be corrosive and cause skin and eye burns. *See: Fire Protection Guide to Hazardous Materials* (National Fire Protection Assoc., Quincy, MA, 12th ed., 1997) Section 49, p 82. *See also: Clinical Toxicology of Commercial Products*, R. E. Gosselin *et al.*, Eds. (Williams & Wilkins, Baltimore, 5th ed., 1984) Section III, pp 241-245.

USE: In prodn of organometallic alkyl and aryl lithium compounds; in prodn of high-strength, low-density aluminum alloys for the aircraft industry; extremely tough, low-density alloys with aluminum and magnesium used for armour plate and aerospace components. In polymerization catalysts for the polyolefin plastics industry; manuf of high-strength glass and glass-ceramics. As anode in electrochemical cells and batteries; as chemical intermediate in organic syntheses. Lithium stearate as thickener and gelling agent to transform oils into lubricating greases.

5577. Lithium Acetate. [546-89-4] Acetic acid lithium salt (1:1); Quilonum. $C_2H_3LiO_2$; mol wt 65.98. C 36.41%, H 4.58%, Li 10.52%, O 48.50%. Prepn: *Gmelins, Lithium* (8th Ed.) **20**, pp 230-232 (1927). Prepn and solubilities: N. V. Sidgwick, J. A. H. R. Gentle, *J. Chem. Soc.* **121**, 1837 (1922). Crystal structure of anhydrous form: C. Saunderson, R. B. Ferguson, *Acta Crystallogr.* **14**, 321 (1961). Pharmacokinetics in psychiatric patients: J.-L. Evrard *et al.*, *Acta Psychiatr. Scand.* **58**, 67 (1978).

mp 286° (slight dec). Changes to dihydrate at 56.5°.

Dihydrate. [6108-17-4] Quilonorm. $C_2H_3LiO_2.2H_2O$; mol wt 102.01. Rhombic crystals. Begins to melt at 49° (Gmelin). Congruent mp 57.8° (Sidgwick). Freely sol in water or alcohol. Its aq soln is practically neutral.

THERAP CAT: Antimanic.

5578. Lithium Amide. [7782-89-0] Lithamide. H_2LiN; mol wt 22.96. H 8.78%, Li 30.23%, N 61.01%. $LiNH_2$. Prepd from Li + NH_3: Titherley, *J. Chem. Soc.* **65**, 517 (1894); Ruff, Geisel *Ber.* **39**, 840 (1906); **44**, 505 (1911); Campbell *et al.*, *Proc. Indiana Acad. Sci.* **50**, 123 (1940); Schenk in *Handbook of Preparative Inorganic Chemistry* vol. 1, G. Brauer, Ed. (Academic Press, New York, 2nd ed., 1963) p 463. *See also: Inorg. Synth.* **2**, 135. Crystal structure: Juza, Opp, *Z. Anorg. Allg. Chem.* **266**, 313 (1951).

Tetragonal crystal structure, d 1.18. mp 380-400°. Sublimes in NH_3 current. When heated *in vacuo* to 450°, NH_3 is given off and Li_2NH forms, which dec ~750-800°. Heat of formation: 42 kcal/mole at 18° and 760 mm Hg. Stable in air-tight containers at room temp. Dec by water to form $LiOH + NH_3$, slowly when in lumps, faster as the particle size decreases. Insol in anhyd ether, benzene, toluene.

USE: In Claisen condensations, alkylation of nitriles and ketones, synthesis of ethynyl compds, acetylenic carbinols.

5579. Lithium Benzoate. [553-54-8] Benzoic acid lithium salt (1:1). $C_7H_5LiO_2$; mol wt 128.06. C 65.65%, H 3.94%, Li 5.42%, O 24.99%.

White, cryst powder. One gram dissolves in 3 ml water, 13 ml alcohol, 10 ml boiling alcohol. Soly in water is increased by sodium benzoate. The aq soln is slightly alkaline to litmus; pH ~8.

USE: Has been used as a lubricant for compressing tablets.

5580. Lithium Borate. [12007-60-2] Boron lithium oxide $(B_4Li_2O_7)$; lithium biborate; lithium tetraborate. $B_4Li_2O_7$; mol wt 169.11. B 25.57%, Li 8.21%, O 66.22%. $Li_2B_4O_7$. Optical properties: T. Sugawara *et al.*, *Solid State Commun.* **107**, 233 (1998).

Pentahydrate. [1303-94-2] $B_4Li_2O_7.5H_2O$; mol wt 259.19. White, cryst powder. Slightly sol in water. Practically insol in alcohol.

USE: In enamels, X-ray fluorescence spectrometry, and medical dosimetry. Light acidic flux used mainly with alkaline samples such as cements and steel; flux component of eutectic mixtures.

5581. Lithium Borohydride. [16949-15-8] Lithium tetrahydroborate(1−) (1:1). BH_4Li; mol wt 21.78. B 49.63%, H 18.51%, Li 31.86%. $LiBH_4$. Prepd by the action of diborane on ethyllithium; by the action of aluminum borohydride on ethyllithium: Schlesinger, Brown, *J. Am. Chem. Soc.* **62**, 3429-35 (1940). Review of lithium and other metal tetrahydroborates: James, Wallbridge, *Prog. Inorg. Chem.* **11**, 99-231 (1970).

Orthorhombic crystals, mp 268°; dec 380°. d 0.66. Stable under ordinary conditions, but dec in moist air. *Dangerous when wet.* Reacts with hydrogen chloride to form hydrogen, diborane, and lithium chloride; forms lithium boromethoxide and hydrogen with methyl alcohol. Sol in water at pH >7. Aq solns dec slowly if the pH is >7, and rather vigorously when acidified. Sol in ether, tetrahydrofuran, aliphatic amines.

USE: Strong reducing agent. Used in the reduction of compds contg ketonic, aldehydic, or ester carbonyls and a nitrile group, where reduction of the carbonyl, but not of the nitrile group, is desired. In the determn of free carboxyl groups in peptides and proteins; after esterification and acetylation, only the ester groups, and none of the peptide bonds are reduced.

5582. Lithium Bromide. [7550-35-8] Lithium monobromide. BrLi; mol wt 86.84. Br 92.01%, Li 7.99%. LiBr. Usually contains water and 85-90% LiBr.

White, very deliquesc, slightly bitter, granular powder. mp 547° when anhyd. Sol in 0.6 part water, 0.4 part boiling water; freely sol in alcohol, glycol; sol in ether, amyl alcohol. The aq soln is neutral or only slightly alkaline to litmus. *Keep well closed.*

Caution: Large doses may cause CNS depression. Chronic absorption may cause skin eruptions and CNS disturbances due to bromide. May also cause disturbed blood electrolyte balance.

USE: As humectant, in air-conditioning systems.

THERAP CAT: Sedative, hypnotic.

5583. Lithium Carbonate. [554-13-2] Carbonic acid lithium salt (1:2); Camcolit; Carbolith; Eskalith; Hypnorex; Limas; Liskonum; Lithane; Lithicarb; Lithobid; Plenur; Priadel; Quilonum-retard; Téralithe. CLi_2O_3; mol wt 73.89. C 16.26%, Li 18.78%, O 64.96%. Li_2CO_3. Purification: Caley, Elving, *Inorg. Synth.* **1**, 1 (1939). Initial report of effect in mental illness: J. F. J. Cade, *Med. J. Aust.* **2**, 349 (1949). Early review of biology and pharmacology of lithium: M. Schou, *Pharmacol. Rev.* **9**, 17-58 (1957); of toxicity: *idem, Acta Pharmacol. Toxicol.* **15**, 70-84 (1958). Thermal behavior: A. Reisman, *J. Am. Chem. Soc.* **80**, 3558 (1958). Acute toxicity data: H. F. Smyth *et al.*, *Am. Ind. Hyg. Assoc. J.* **30**, 470 (1969). Toxicological studies in mammals: Gralla, McIlhenny, *Toxicol. Appl. Pharmacol.* **21**, 428 (1972). Pharmacokinetics: J. S. Ku *et al.*, *Int. J. Clin. Pharmacol. Ther. Toxicol.* **25**, 648 (1987). Clinical studies: J. Rybakowski *et al.*, *Int. Pharmacopsychiatry* **15**, 86 (1980); B. F. Kjellman *et al.*, *Acta Psychiatr. Scand.* **62**, 32 (1980). Review of development of use: G. Chouinard, *Union Med. Can.* **109**, 221-224, 226, 304 (1980). Comprehensive description: H. C. Stober, *Anal. Profiles Drug Subs.* **15**, 367-391 (1986).

White, light, alkaline powder. d 2.11. mp 720 ±1°. One gram dissolves in 78 ml cold, 140 ml boiling water; dissolved by dil acids. Practically insol in alcohol. LD_{50} orally in rats: 0.71 g/kg (Smyth).

USE: In the production of glazes on ceramic and electrical porcelain. Reagent in synthetic organic chemistry. Matrix modifier.

THERAP CAT: Antimanic.

5584. Lithium Chloride. [7447-41-8] Lithium chloride (LiCl). ClLi; mol wt 42.39. Cl 83.63%, Li 16.37%. LiCl. Prepn: Pray, *Inorg. Synth.* **5**, 153 (1957). Review of prepn and properties: Oliver in *Mellor's* vol. II, supplement II, *The Alkali Metals* (part 1) 179-215 (1961). Pharmacology: M. Wielosz, *Pol. J. Pharmacol. Pharm.* **26**, 399 (1974).

Deliquesc, cubic crystals, granules or cryst powder; sharp saline taste. mp 613° when anhyd; bp 1360°; d 2.07. One gram dissolves in 1.3 ml cold, 0.8 ml boiling water; sol in alcohol, acetone, amyl alcohol, pyridine. The aq soln is neutral or slightly alkaline. *Keep well closed.* LD_{50} in mice (mg/kg): 990 i.p.; in rats (mg/kg): 600 i.p., 4.8 i.v. (Wielosz).

Caution: Prolonged absorption may cause disturbed electrolyte balance, impaired renal function, CNS disturbances.

USE: Manuf mineral waters; in pyrotechnics; soldering aluminum; in refrigerating machines. Reagent in synthetic organic chemistry. Matrix modifier in analytical chemistry.

THERAP CAT: Antidepressant.

5585. Lithium Chromate(VI). [14307-35-8] Chromic acid (H_2CrO_4) lithium salt (1:2). $CrLi_2O_4$; mol wt 129.87. Cr 40.04%, Li 10.69%, O 49.28%. Li_2CrO_4.

The anhyd salt is appreciably sol in methanol and ethanol. Eutectic with water below −60°. *Keep well closed.*

Dihydrate. [7789-01-7] $CrLi_2O_4.2H_2O$; mol wt 165.90. Yellow, deliquesc, cryst powder. Becomes anhyd at 74.6°. Very sol in water.

Caution: Chromium hexavalent (VI) compounds are listed as known human carcinogens: *Report on Carcinogens, Twelfth Edition* (PB2011-111646, 2011) p 106.

USE: Corrosion inhibitor for water-cooled atomic reactors. Soln as low temp heat transfer medium.

5586. Lithium Citrate. [919-16-4] 2-Hydroxy-1,2,3-propanetricarboxylic acid lithium salt (1:3); citric acid trilithium salt; Litarex; Lithonate S. $C_6H_5Li_3O_7$; mol wt 209.92. C 34.33%, H 2.40%, Li 9.92%, O 53.35%.

Tetrahydrate. [6080-58-6] $C_6H_5Li_3O_7.4H_2O$; mol wt 281.98. White granules or cryst powder; feebly alkaline taste; deliquesces on exposure to moist air; loses all of its water at 105°. Sol in 1.5 parts water, slightly in alc. pH ~8. *Keep well closed.*

USE: Formerly in soft drinks.

THERAP CAT: Antidepressant.

5587. Lithium Fluoride. [7789-24-4] Lithium fluoride (LiF); lithium monofluoride. FLi; mol wt 25.94. F 73.24%, Li 26.75%. LiF. Prepd from LiOH and HF or by dissolving Li_2CO_3 in excess HF, evaporating to dryness, and heating to red heat: v. Wartenberg, Schultz, *Z. Elektrochem.* **27**, 568 (1921); Kwasnik in *Handbook of Preparative Inorganic Chemistry* vol. 1, G. Brauer, Ed. (Academic

Press, New York, 2nd ed., 1963) p 235. Review of toxicology of fluoride compounds: G. L. Waldbott, *Acta Med. Scand. Suppl.* **400**, 1-44 (1963).

Cubic crystals (NaCl lattice) or white fluffy powder. d^{20} 2.640. mp 848°; bp 1681°. Volatilizes at 1100-1200°. Soly in water (25°): 0.13 g/100 ml. Soluble in acids. With hydrofluoric acid it forms lithium bifluoride, $LiHF_2$. With lithium hydroxide it forms a double salt LiF.LiOH, mp 462°. LD in guinea pigs (mg/kg): 200 orally, 2000 s.c. (Waldbott).

USE: As flux for soldering and welding aluminum, in the manuf of vitreous enamels and glazes. Lithium fluoride prisms are used in infrared spectrophotometers.

5588. Lithium Formate. [556-63-8] Formic acid lithium salt (1:1). $CHLiO_2$; mol wt 51.96. C 23.12%, H 1.94%, Li 13.36%, O 61.58%. HCOOLi.

Monohydrate. [6108-23-2] $CHLiO_2.H_2O$; mol wt 69.97. Colorless or white crystals. d 1.46. Sol in 3 parts water. The aq soln is practically neutral.

5589. Lithium Hydride. [7580-67-8] Lithium hydride (LiH); lithium monohydride. HLi; mol wt 7.95. H 12.68%, Li 87.30%. LiH. Prepd by the direct combination of hydrogen and lithium: Gibb, *Trans. Electrochem. Soc.* **93**, 198-211 (1948). Review of prepn and properties: Truter in *Mellor's* **vol. II**, supplement II, *The Alkali Metals* (part 1) 131-145 (1961).

Cubic crystals, darkens rapidly on exposure to light, the commercial product is usually gray. mp 680°. d 0.76-0.77. *Dangerous when wet.* No solvent known. Rapidly dec in water to form lithium hydroxide and hydrogen. Reacts with the lower alcohols, carboxylic acids, chlorine, and ammonia at 400° to liberate hydrogen.

Caution: Potential symptoms of overexposure are irritation and burns of eyes and skin; burns of mouth and esophagus (if ingested); nausea; muscle twitches; mental confusion; blurred vision. *See NIOSH Pocket Guide to Chemical Hazards* (DHHS/NIOSH 97-140, 1997) p 186.

USE: Reducing agent; condensing agent with ketones and acid esters; desiccant; in hydrogen generators; 1 g in water liberates ~2.8 liters of hydrogen at STP.

5590. Lithium Hydroxide. [1310-65-2] Lithium hydroxide (Li(OH)); lithium hydrate. HLiO; mol wt 23.95. H 4.21%, Li 28.98%, O 66.80%. LiOH. Prepn: Barnes, *J. Chem. Soc.* **1931**, 2605; Cohen, *Inorg. Synth.* **5**, 3 (1957); Bravo, *ibid.* **7**, 1 (1963). Review of prepn, properties and uses: Oliver in *Mellor's* **vol. II**, supplement II, *The Alkali Metals* (part 1) 159-171 (1961). Review of synthetic applications: F. de Carvalho da Silva, *Synlett* **2006**, 1451-1452.

Granular, free-flowing powder; acrid, strongly alkaline. Readily absorbs CO_2 and water from air. d 2.54. mp 471°. Sol in water, slightly sol in alcohol. *Corrosive. Keep tightly closed.*

Monohydrate. $LiOH.H_2O$; mol wt 41.96. Small monoclinic crystals. d_4^{20} 1.51. Heat of formation -188.9 kcal/mol at 25°. Heat of soln -0.87 kcal/mol at 25°. Soly in water (w/w) at 0°: 10.7%; at 20°: 10.9%; at 100°: 14.8%. Slightly sol in alcohol. pH of a 1.0N soln about 14.

Caution: Strongly alkaline and hence caustic. Very irritating to skin. Systemic toxicity similar to other lithium compds. *See* Lithium.

USE: In photographic developers; in alkaline storage batteries; in prepn of other lithium salts where use of carbonate is not practical; as catalyst in the production of alkyd resins, in esterifications. In the production of lithium soaps, sulfonates, greases. Reagent in synthetic organic chemistry.

5591. Lithium Iodide. [10377-51-2] Lithium iodide (LiI); lithium monoiodide. ILi; mol wt 133.84. I 94.82%, Li 5.19%. LiI.

Trihydrate. [7790-22-9] $ILi.3H_2O$; mol wt 187.89. White, deliquescent granules or fused masses; becomes yellow on exposure to air, due to liberation of iodine. mp 73°; mp 446° when anhydr. Sol in about 0.5 part water or alcohol, freely in amyl alcohol or acetone. The aq soln is neutral or slightly alkaline. *Keep tightly closed and protected from light.*

USE: In photography.

5592. Lithium Nitrate. [7790-69-4] Nitric acid lithium salt (1:1). $LiNO_3$; mol wt 68.94. Li 10.07%, N 20.32%, O 69.62%.

Colorless, deliquesc granules. d 2.38. mp ~255°. Sol in ~2 parts water; sol in alcohol. The aq soln is neutral. *Oxidizer. Keep well closed.*

5593. Lithium Oxalate. [553-91-3] Ethanedioic acid lithium salt (1:2); dilithium oxalate. $C_2Li_2O_4$; mol wt 101.90. C 23.57%, Li 13.62%, O 62.80%. $Li_2C_2O_4$.

White crystals. d 2.12. Sol in 15 parts water.

5594. Lithium Oxide. [12057-24-8] Lithium oxide (Li_2O); dilithium oxide. Li_2O; mol wt 29.88. Li 46.45%, O 53.54%. Prepd from lithium peroxide: Cohen, *Inorg. Synth.* **5**, 5 (1957); from lithium hydroxide: Bravo, *ibid.* **7**, 3 (1963). Review of prepn and properties: Oliver in *Mellor's* **vol. II**, supplement II, *The Alkali Metals* (part 1) 146-158 (1961). Physical properties: van Arkel *et al.*, *Can. J. Chem.* **31**, 1009 (1953); Brewer, Margrave, *J. Phys. Chem.* **59**, 421 (1955).

Finely divided powder or crusty material, mp 1570° (van Arkel); also reported as mp 1427° (Brewer). d_4^{25} 2.013. Readily absorbs carbon dioxide and water from the atm. At elevated temp attacks glass, silica, many metals.

5595. Lithium Perchlorate. [7791-03-9] Perchloric acid lithium salt (1:1). $ClLiO_4$; mol wt 106.39. Cl 33.32%, Li 6.52%, O 60.15%. $LiClO_4$.

Small crystals. d_4^{25} 2.43. mp 236°. Decompn starts at ~400° and becomes rapid at 430° yielding lithium chloride and oxygen. Heat of formation: -99.94 kcal/mol at 25°. Soly in water (w/w) at 0°: 29.9%; at 25°: 37.5%; at 100°: 71.5%. Appreciably sol in alcohol, acetone, ether, ethyl acetate.

Trihydrate. [13453-78-6] $ClLiO_4.3H_2O$; mol wt 160.43.

Caution: May be irritating on contact with skin, mucous membranes.

USE: Oxidizing agent.

5596. Lithium Silicate. [10102-24-6] Silicic acid (H_2SiO_3) lithium salt (1:2); lithium metasilicate. Li_2O_3Si; mol wt 89.96. Li 15.43%, O 53.35%, Si 31.22%. Li_2SiO_3. Prepd by fusing Li_2CO_3 with SiO_2: Schwarz, Sturm, *Ber.* **47**, 1737 (1914).

Orthorhombic needles. d_4^{25} 2.52. mp 1201°. Heat of formation (solid): -434.9 kcal/mol. Heat of formation (liq) -374.6 kcal/mol. Latent heat of fusion (1177°) = 7.24 kcal/mol, also reported as -80.2 cal/g. Insol in cold water, dec by boiling water, dilute hydrochloric acid.

USE: To calibrate thermoelements.

5597. Lithium Sulfate. [10377-48-7] Sulfuric acid lithium salt (1:2); Lithiofor. Li_2O_4S; mol wt 109.94. Li 12.63%, O 58.21%, S 29.16%. Li_2SO_4.

Monohydrate. [10102-25-7] $Li_2O_4S.H_2O$; mol wt 127.95. Colorless crystals; loses the water at 130°. d 2.06. Sol in 2.6 parts water. Almost insol in alcohol. The aq soln is neutral.

THERAP CAT: Antidepressant.

5598. Lithium Tetracyanoplatinate(II). [14402-73-4] Platinous lithium cyanide; lithium platinocyanide. $C_4Li_2N_4Pt$; mol wt 313.04. C 15.35%, Li 4.43%, N 17.90%, Pt 62.32%. $Li_2Pt(CN)_4$.

Pentahydrate. Greenish-yellow crystals. Slightly sol in water.

USE: In x-ray photography.

5599. Lithium Tetrafluoroborate. [14283-07-9] Lithium tetrafluoroborate(1−) (1:1); lithium fluoroboride; lithium fluoborate. BF_4Li; mol wt 93.74. B 11.53%, F 81.07%, Li 7.40%. $LiBF_4$. Mild Lewis acid. Prepn: P. Baumgarten, W. Bruns, *Ber.* **72**, 1753 (1939); I. Shapiro, H. G. Weiss, *J. Am. Chem. Soc.* **75**, 1753 (1953). NMR structural study: E. C. Reynhardt, J. A. J. Lourens, *J. Chem. Phys.* **80**, 6240 (1984); ^{19}F NMR study of soln behavior: V. N. Plakhotnik *et al.*, *J. Fluorine Chem.* **98**, 133 (1999). Catalytic use in Diels-Alder reactions: D. A. Smith, K. N. Houk, *Tetrahedron Lett.* **32**, 1549 (1991); J. S. Yadav *et al.*, *Synthesis* **2001**, 1065. Review of applications in organic synthesis: B. S. Babu, K. K. Balasubramanian, *Acros Org. Acta* **7**, 1-3 (2000).

White, crystalline, hygroscopic solid. Sol in water, methanol, acetonitrile. Soly in ether at 25°: 1.3 g/100 ml.

USE: Catalyst in organic synthesis; electrolyte for lithium ion batteries.

5600. Lithium Triethylborohydride. [22560-16-3] (*T*-4)-Lithium triethylhydroborate(1−) (1:1); Super-Hydride. $C_6H_{16}BLi$;

mol wt 105.94. C 68.03%, H 15.22%, B 10.20%, Li 6.55%. LiBH-$(C_2H_5)_3$. Commercial product is available in THF solution. Prepn: H. C. Brown *et al.*, *J. Am. Chem. Soc.* **75**, 192 (1953); P. Binger *et al.*, *Ann.* **717**, 21 (1968); H. C. Brown *et al.*, *Inorg. Chem.* **16**, 2229 (1977). Mechanism of reduction of organic halides: E. C. Ashby *et al.*, *J. Org. Chem.* **49**, 4505 (1984). Use as a reducing agent in organic and inorganic synthesis: H. C. Brown *et al.*, *J. Org. Chem.* **45**, 1 (1980); S. Krishnamurthy, H. C. Brown, *ibid.* **48**, 3085 (1983); H. C. Brown, S.-C. Kim, *ibid.* **49**, 1064 (1984); B. E. Blough, F. I. Carroll, *Tetrahedron Lett.* **34**, 7239 (1993); C. K. Yee *et al.*, *Langmuir* **15**, 3486 (1999); H. Tanaka, K. Ogasawara, *Tetrahedron Lett.* **43**, 4417 (2002).

Colorless crystals, mp 66-67° (Binger). Also reported as white needles from benzene, mp 78-83° (dec) (Brown, 1977). Moderately sol in benzene, toluene, *n*-hexane. *Unstable in air. Reacts violently with water.*

USE: Powerful and selective reducing agent.

5601. Lithocholic Acid. [434-13-9] $(3\alpha,5\beta)$-3-Hydroxycholan-24-oic acid; 3α-hydroxycholanic acid; 17β-(1-methyl-3-carboxypropyl)etiocholan-3α-ol. $C_{24}H_{40}O_3$; mol wt 376.58. C 76.55%, H 10.71%, O 12.75%. Found in ox bile, human bile, rabbit bile, and in ox and pig gallstones. Isoln: Fischer, *Z. Physiol. Chem.* **73**, 234 (1911). Characterization: Wieland, Weyland, *ibid.* **110**, 123 (1920). Prepn from cholic or from desoxycholic acid: Hoehn, Mason, *J. Am. Chem. Soc.* **62**, 569 (1940); Sarel, Yanuka, *J. Org. Chem.* **24**, 2018 (1959). Crystal and molecular structure by x-ray diffraction: S. K. Arora *et al.*, *Acta Crystallogr.* **32B**, 415 (1976).

Hexagonal leaflets from alcohol, prisms from acetic acid, mp 184-186°. $[\alpha]_D^{20}$ +33.7° (c = 1.5 in abs ethanol); $[\alpha]_D^{19}$ +23.3° (Wieland); $[\alpha]_D^{20}$ +32.1° (Fischer). Freely sol in hot alc. Sol in ether than cholic or desoxycholic acid. Sol in about 10 times its weight of ethyl acetate. Slightly sol in glacial acetic acid (about 0.2 g in 3 ml). More sol in benzene than desoxycholic acid. Insol in petr ether, gasoline, ligroin, water.

Methyl ester. $C_{25}H_{42}O_3$. Crystallizes with ½ mol methanol, mp 125-127°.

Ethyl ester. $C_{26}H_{44}O_3$. Crystals, mp 92-93°.

Benzyl ester. $C_{31}H_{46}O_3$. Crystals, mp 145-148°.

Acetyllithocholic acid. $C_{26}H_{42}O_4$. Crystals, mp 169°.

Acetyllithocholic acid methyl ester. $C_{27}H_{44}O_4$. Flat needles from pentane, mp 123-130°.

Acetyllithocholic acid ethyl ester. $C_{28}H_{46}O_4$. Crystals, mp 90-91°.

5602. Lithopone. [1345-05-7] C.I. Pigment White 5; Griffith's zinc white. A white pigment consisting of a mixture of zinc sulfide, barium sulfate and some zinc oxide. Several grades of lithopone are commercially available. The zinc sulfide content varies from 26 to 60%. The commercial importance of lithopone has decreased since the introduction of titanium pigments.

USE: In water and oil paints to provide thixotropy, improve gloss and flow.

5603. Litmus. [1393-92-6] Lacmus; tournesol; turnsole; lacca musica; lacca coerulea. Blue coloring matter from various species of lichens, particularly *Variolaria*, *Lecanora*, and *Roccella*. *Habit.* Scandinavia, shores of Mediterranean, Azores, California, East India, Madagascar. *Constit.* Chiefly azolitmin and erythrolitmin combined with alkalies: lecanoric acid, orcein, erythrolein; small amounts α,β,γ-amino and hydroxyorcein. Structure studies: Beecken *et al.*, *Angew. Chem.* **73**, 665 (1961). *Review:* M. A. P. Segurado *et al.*, *Chem. Educ.* **16**, 99-108 (2011).

Blue powder, lumps or cubes. Partly soluble in water or alcohol.

USE: As acid-base indicator; pH: 4.5 red, 8.3 blue. For preparing litmus papers; in microscopy to color culture media for diagnostic purposes. Has been used for coloring beverages.

5604. Livetins. Major water-soluble proteins found in egg yolk: Shepard, Hottle, *J. Biol. Chem.* **179**, 349 (1949). Prepn and mol wts of α- and β-livetin: Martin *et al.*, *Can. J. Biochem. Physiol.* **35**, 241 (1957). Prepn and mol wt of γ-livetin: Martin, Cook, *ibid.* **36**, 153 (1958). Separation of four fractions, α_1-, α_2-, β-, and γ-livetins, and preparation of β- and γ-livetins: Oberdorfer, *Z. Physiol. Chem.* **331**, 280 (1963). Protein constituents of the livetin fraction of egg yolk were identified as follows: α-livetin identified as serum albumin; β-livetin as an α_2-glycoprotein with a sedimentation coeff of 2.95; and γ-livetin as serum γ-globulin: Williams, *Biochem. J.* **83**, 346 (1962). Heterogeneity of livetin proteins in egg yolk: S. F. Hui, R. H. Common, *Can. J. Biochem.* **44**, 1357 (1966); W. M. McIndoe, J. Culbert, *Int. J. Biochem.* **10**, 659 (1979).

5605. Lixisenatide. [320367-13-3] L-Histidylglycyl-L-α-glutamylglycyl-L-threonyl-L-phenylalanyl-L-threonyl-L-seryl-L-α-aspartyl-L-leucyl-L-seryl-L-lysyl-L-glutaminyl-L-methionyl-L-α-glutamyl-L-α-glutamyl-L-α-glutamyl-L-alanyl-L-valyl-L-arginyl-L-leucyl-L-phenylalanyl-L-isoleucyl-L-α-glutamyl-L-tryptophyl-L-leucyl-L-lysyl-L-asparaginylglycylglycyl-L-prolyl-L-seryl-L-serylglycyl-L-alanyl-L-prolyl-L-prolyl-L-prolyl-L-seryl-L-lysyl-L-lysyl-L-lysyl-L-lysyl-L-lysinamide; des-38-proline-exendine-4 (*Heloderma suspectum*)-(1-39)-peptidylpenta-L-lysyl-L-lysinamide; AVE-0010; ZP-10A. $C_{215}H_{347}N_{61}O_{65}S$; mol wt 4858.56. C 53.15%, H 7.20%, N 17.59%, O 21.40%, S 0.66%. Glucagon-like peptide 1 (GLP-1) receptor agonist; exendin-4 analog with 44 amino acids, modified by deleting one proline residue and adding six lysine residues to the C-terminus. Prepn: B. D. Larsen *et al.*, WO 0104156 (2001 to Zealand). Receptor binding study and effect on insulin mRNA expression: C. Thorkildsen *et al.*, *J. Pharmacol. Exp. Ther.* **307**, 490 (2003). Pharmacology: U. Werner *et al.*, *Regul. Pept.* **164**, 58 (2010). Clinical trial in type 2 diabetes: R. E. Ratner *et al.*, *Diabet. Med.* **27** 1024 (2010). Review of pharmacology and clinical experience: M. Christensen *et al.*, *Expert Opin. Invest. Drugs* **20**, 549-557 (2011); M. Christensen *et al.*, *Expert Rev. Endocrinol. Metab.* **6**, 513-525 (2011).

THERAP CAT: Antidiabetic.

5606. Lixivaptan. [168079-32-1] *N*-[3-Chloro-4-(5*H*-pyrrolo[2,1-*c*][1,4]benzodiazepine-10(11*H*)-ylcarbonyl)phenyl]-5-fluoro-2-methylbenzamide; CL-347985; VPA-985. $C_{27}H_{21}ClFN_3O_2$; mol wt 473.93. C 68.43%, H 4.47%, Cl 7.48%, F 4.01%, N 8.87%, O 6.75%. Arginine vasopressin V_2-receptor antagonist; increases water excretion and normalizes serum sodium levels. Prepn: J. D. Albright *et al.*, EP 636625; *eidem*, US 5516774 (1995, 1996 both to Am. Cyan.); J. D. Albright *et al.*, *J. Med. Chem.* **41**, 2442 (1998). Metabolism: A. J. Molinari *et al.*, *Bioorg. Med. Chem. Lett.* **17**, 5796 (2007). Clinical effect on serum sodium concentration in hyponatremia: F. Wong *et al.*, *Hepatology* **37**, 182 (2003). Aquaretic effect in chronic heart failure: W. T. Abraham *et al.*, *J. Am. Coll. Cardiol.* **47**, 1615 (2006). Review of pharmacology and clinical experience: H. D. Zmily *et al.*, *Curr. Opin. Investig. Drugs* **20**, 831-848 (2011).

Crystals from ethyl acetate, mp 191-195°.
THERAP CAT: In treatment of congestive heart failure.

5607. Lobaplatin. [135558-11-1] (*SP*-4-3)-[*rel*-(1*R*,2*R*)-1,2-Cyclobutanedimethanamine-$\kappa N,\kappa N$'][(2*S*)-2-(hydroxy-κO)propanoato(2−)-κO]platinum; cis-[trans-1,2-cyclobutanebis(methyl-amine)][(S)-lactato-O¹,O²]platinum; D-19466. $C_9H_{18}N_2O_3Pt$; mol wt 397.34. C 27.21%, H 4.57%, N 7.05%, O 12.08%, Pt 49.10%. Water-soluble platinum compound; mixture of two trans diastereo-isomers. Prepn: W. Schumacher et al., EP 324154; eidem, US 5023335 (1989, 1991 both to ASTA Pharma); and pharmacology: R. Voegeli et al., J. Cancer Res. Clin. Oncol. 116, 439 (1990). HPLC determn of diastereoisomers in plasma: J. Welink et al., J. Chromatogr. B 675, 107 (1996). Clinical evaluation: M. Degardin et al., Invest. New Drugs 13, 253 (1995); and pharmacokinetics: J. A. Gietema et al., Br. J. Cancer 71, 1302 (1995). Evaluation in canine osteosarcoma: J. Kirpensteijn et al., Anticancer Res. 22, 2765 (2002). Review of clinical development: M. J. McKeage, Expert Opin. Invest. Drugs 10, 119-128 (2001).

Relative stereochemistry

Crystals from diethyl ether, mp 220° (dec). Sol in water. LD_{50} i.p. in mice: 46 mg/kg (Voegeli).

THERAP CAT: Antineoplastic.

THERAP CAT (VET): Antineoplastic.

5608. Lobelia. Indian tobacco; wild tobacco; emetic herb; asthma weed; bladder pod; vomit wort. Dried leaves and tops of Lobelia inflata L., Lobeliaceae. (The seeds are also used.) Habit. Canada, U.S. Constit. Leaves and tops: chiefly lobeline, also lobel-idine, lobelanine, lobelanidine, and other alkaloids. Seeds: lobeline, fixed oil.

THERAP CAT: Expectorant.

THERAP CAT (VET): Has been used as an expectorant.

5609. Lobeline. [90-69-7] 2-[(2R,6S)-6-[(2S)-2-Hydroxy-2-phenylethyl]-1-methyl-2-piperidinyl]-1-phenylethanone; 2-[6-(β-hydroxyphenethyl)-1-methyl-2-piperidyl]acetophenone; α-lobeline; inflatine. $C_{22}H_{27}NO_2$; mol wt 337.46. C 78.30%, H 8.06%, N 4.15%, O 9.48%. High affinity nicotinic ligand. From herb and seeds of Lobelia inflata L., Lobeliaceae (Indian tobacco): H. Wie-land, Ber. 54, 1784 (1921). Structure: H. Wieland, O. Dragendorff, Ann. 473, 83 (1929). Synthesis: H. Wieland, I. Drishaus, ibid. 102; G. Schering, L. Winterhalder, ibid. 126. Absolute configuration: Schöpf, Müller, Ann. 687, 241 (1965). Pharmacology: M. I. Damaj et al., J. Pharmacol. Exp. Ther. 282, 410 (1997). Receptor binding study: D. Flammia et al., J. Med. Chem. 42, 3726 (1999). Compre-hensive description: F. J. Muhtadi, Anal. Profiles Drug Subs. 19, 261-313 (1990). Review of mechanism of action and therapeutic potential in psychostimulant abuse: L. P. Dwoskin, P. A. Crooks, Biochem. Pharmacol. 63, 89-98 (2002).

Needles from alc, ether, benzene, mp 130-131°. $[α]_D^{15}$ −43° (alc). Very slightly sol in water or in petr ether; sol in hot alcohol, or in chloroform, benzene, ether.

Hydrochloride. [134-63-4] $C_{22}H_{27}NO_2$·HCl; mol wt 373.92. Rosettes of slender needles from alcohol, mp 178-180°. $[α]_D^{20}$ −43° (c = 2). uv max (methanol) 245, 280 nm (log ε 4.08, 3.05). One gram dissolves in 40 ml water, 12 ml alcohol; very sol in chloroform; very slightly sol in ether. Water solns are slightly acid to litmus. A 1% soln in water has a pH of 4.0-6.0.

Sulfate. [134-64-5] $(C_{22}H_{27}NO_2)_2$·H_2SO_4; mol wt 773.00. Hy-groscopic. Crystals from alc. $[α]_D^{20}$ −25° (c = 2). Soluble in about 30 parts water, slightly in alcohol.

(±)-Form. [134-65-6] Lobelidine. Prisms, mp 110°.

THERAP CAT: CNS stimulant.

THERAP CAT (VET): Has been used as a respiratory stimulant, ruminatoric.

5610. Lobenzarit. [63329-53-3] 2-[(2-Carboxyphenyl)ami-no]-4-chlorobenzoic acid; N-(2-carboxyphenyl)-4-chloroanthranilic acid; 4-chloro-2,2′-iminodibenzoic acid; CCA. $C_{14}H_{10}ClNO_4$; mol wt 291.69. C 57.65%, H 3.46%, Cl 12.15%, N 4.80%, O 21.94%. Disease modifying antirheumatic drug (DMARD) with immunomo-dulating activity. Prepn: M. Tanemura et al., DE 2526092; eidem, US 4092426 (1976, 1978 both to Chugai). HPLC determn in plasma: F. J. Schwende, S. W. Turner, Pharm. Res. 8, 523 (1991). Immunosuppression of B cell activity: S. Hirohata et al., Arthritis Rheum. 35, 168 (1992). Clinical trial in rheumatoid arthritis: Y. Shiokawa et al., J. Rheumatol. 11, 615 (1984); in systemic lupus erythematosus: S. Hirohata et al., Clin. Exp. Rheumatol. 12, 261 (1994). Review of pharmacology: R. González, D. Remírez, Invest. Allergol. Clin. Immunol. 7, 77-82 (1997).

Crystals, mp >306°. LD_{50} in male, female rats (mg/kg): 2100, 2600 orally (Tanemura).

Disodium salt. [64808-48-6] Carfenil. $C_{14}H_8ClNNa_2O_4$; mol wt 335.65. Solid, mp 388° (dec).

THERAP CAT: Antirheumatic.

5611. Lochnericine. [72058-36-7] (5α,6α,7α,12β,19α)-2,3-Didehydro-6,7-epoxyaspidospermidine-3-carboxylic acid methyl es-ter. $C_{21}H_{24}N_2O_3$; mol wt 352.43. C 71.57%, H 6.86%, N 7.95%, O 13.62%. Indole alkaloid. From Vinca rosea Linn. (Catharanthus roseus G. Don.), Apocynaceae: Gorman et al., J. Am. Pharm. Assoc. 48, 256 (1959); Svoboda et al., ibid. 48, 659 (1959); Moza, Trojánek, Collect. Czech. Chem. Commun. 28, 1419 (1963). Structure: Moza et al., Tetrahedron Lett. 1964, 2561.

Crystals from methanol, dec 190-193°. $[α]_D^{27}$ −432° (chloroform). pKa in 66% DMF: 4.2. uv max (ethanol): 227, 299, 328 nm (log ε 4.10, 4.15, 4.32).

5612. Locust Bean Gum. Carob flour; Johannisbrotmehl; Arobon. The ground kernel endosperms of tree pods of Ceratonia siliqua L., Leguminosae (St. John's bread). Consists of proteins such as albumins, globulins, prolamins, gluteline; carbohydrates such as reducing sugars, sucrose, dextrins, pentosans; ash; fat; crude fiber; moisture. Refs: Plaut et al., Bull. Res. Counc. Isr. 3, No. 11, 129 (1953), C.A. 48, 5397 (1954); Griffiths, Manuf. Chem. 20, 321 (1949); Coit, Econ. Bot. 5, 82 (1951). Use as coffee, chocolate, cocoa substitute, extender: W. A. Meer, Manuf. Confect. 59, 41 (1979). Caffeine, theobromine content: W. J. Craig, T. T. Nguyen, J. Food Sci. 49, 302 (1984). Review: F. Rol in Industrial Gums, R. L. Whistler, Ed. (Academic Press, New York, 2nd ed., 1973) pp 323-337.

Yellow-green color, is odorless and tasteless, but acquires a legu-minous taste when boiled in water.

USE: Stabilizer, thickener, and binder in foods and cosmetics. Coffee, chocolate, cocoa substitute. Sizing and finishing agent in textiles. As fiber bonding in paper manuf. Drilling mud additive.

THERAP CAT: Adsorbent-demulcent.

5613. Lodoxamide. [53882-12-5] 2,2′-[(2-Chloro-5-cyano-1,3-phenylene)diimino]bis(2-oxoacetic acid); N,N′-(2-chloro-5-cyano-*m*-phenylene)dioxamic acid. $C_{11}H_6ClN_3O_6$; mol wt 311.63. C 42.40%, H 1.94%, Cl 11.38%, N 13.48%, O 30.80%. Mast cell stabilizer. Prepn: C. M. Hall, J. B. Wright, **DE 2362409**; *eidem*, **US 3993679** (1974, 1976 both to Upjohn); J. B. Wright *et al.*, *J. Med. Chem.* **21**, 930 (1978). HPLC determn in human plasma: I. L. Honigberg *et al.*, *J. Chromatogr.* **276**, 213 (1983). Mechanism of action study: M. K. Church, J. Hiroi, *Br. J. Pharmacol.* **90**, 421 (1987). Clinical evaluation in asthma: R. L. Case *et al.*, *J. Am. Med. Assoc.* **247**, 661 (1982); in allergic eye disease: G. T. Fahy *et al.*, *Eur. J. Ophthalmol.* **2**, 144 (1992).

mp 212° (dec). uv max (0.1*N* NaOH) 239.5 (ε 23800).

Diethyl ester. [53882-13-6] Diethyl *N,N'*-(2-chloro-5-cyano-*m*-phenylene)dioxamate; lodoxamide ethyl; U-42718. $C_{15}H_{14}ClN_3O_6$; mol wt 367.74. Crystals from ethanol, mp 177-179°.

Tromethamine salt. [63610-09-3] Lodoxamide tromethamine; lodoxamide trometamol; U-42585E; Alomide. $C_{11}H_6ClN_3O_6 \cdot 2C_4H_{11}NO_3$; mol wt 553.91.

THERAP CAT: Antiallergic; topically in allergic conjunctivitis.

5614. Lofepramine. [23047-25-8] 1-(4-Chlorophenyl)-2-[[3-(10,11-dihydro-5*H*-dibenz[*b,f*]azepin-5-yl)propyl]methylamino]-ethanone; 4'-chloro-2-[[3-(10,11-dihydro-5*H*-dibenz[*b,f*]azepin-5-yl)propyl]methylamino]acetophenone; *N*-methyl-*N*-(4-chloroben-zoylmethyl)-3-(10,11-dihydro-5*H*-dibenzo[*b,f*]azepin-5-yl)propyl-amine; lopramine. $C_{26}H_{27}ClN_2O$; mol wt 418.97. C 74.54%, H 6.50%, Cl 8.46%, N 6.69%, O 3.82%. Psychotropic drug related to imipramine, *q.v.* Prepn: E. Eriksoo *et al.*, **GB 1177525**; *eidem*, **US 3637640** (1970, 1972 both to AB Leo). Chemistry and pharmacology: E. Eriksoo, O. Rohte, *Arzneim.-Forsch.* **20**, 1561 (1970). Absorption and metabolism: J. R. Tulic *et al.*, *Acta Pharmacol. Toxicol.* **32**, 304 (1973). Distribution and excretion: G. Plym Forshell, *Xenobiotica* **5**, 73 (1975). Pharmacokinetics: G. Plym Forshell *et al.*, *Eur. J. Clin. Pharmacol.* **9**, 291 (1976). Clinical study: S. Wright, L. Herrmann, *Arzneim.-Forsch.* **26**, 1167 (1976). Review of clinical experience in depression: S. G. Lancaster, J. P. Gonzalez, *Drugs* **37**, 123-140 (1989).

Crystals from methanol or acetone, mp 104-106°. Easily oxidized by air and other oxidizing agents to desipramine and *p*-chlorobenzoic acid.

Hydrochloride. [26786-32-3] Leo 640; Amplit; Gamanil; Gamonil; Lomont; Tymelyt. $C_{26}H_{27}ClN_2O \cdot HCl$; mol wt 455.42. Crystals from butanone, mp 152-154°. Sol in methanol, ethanol, chloroform. Practically insol in water. LD_{50} in mice, rats (mg/kg): >2500, >1000 orally; 920, >1000 i.p.; >1000, >1000 s.c. (Eriksoo, Rohte).

THERAP CAT: Antidepressant.

5615. Lofexidine. [31036-80-3] 2-[1-(2,6-Dichlorophen-oxy)ethyl]-4,5-dihydro-1*H*-imidazole; 2-[1-(2,6-dichlorophenoxy)-ethyl]-2-imidazoline. $C_{11}H_{12}Cl_2N_2O$; mol wt 259.13. C 50.99%, H 4.67%, Cl 27.36%, N 10.81%, O 6.17%. α_2-Adrenoceptor agonist related structurally to clonidine, *q.v.* Prepn of the HCl salt: H. Baganz, H. J. May, **ZA 6800850**; *eidem*, **US 3966757** (1968, 1976 both to Nordmark); of the free base: *eidem*, **DE 1935479** (1971 to Nordmark). Pharmacological studies: J. Velly, *J. Pharmacol.* **8**, 351 (1977); B. Jarrot *et al.*, *Biochem. Pharmacol.* **28**, 141 (1979). NMR

data and cardiovascular effects: P. B. M. Timmermans, P. A. Van Zwieten, *Eur. J. Med. Chem.* **15**, 323 (1980). Hypotensive and sedative properties: P. Birch *et al.*, *Br. J. Pharmacol.* **68**, 107 (1980). Effects in hypertension: N. D. Vlachakis *et al.*, *Fed. Proc.* **39**, 4844 (1980). Series of articles on pharmacology, toxicology, clinical studies: *Arzneim.-Forsch.* **32**, 915-993 (1982). Toxicity studies: T. H. Tsai *et al.*, *ibid.* 955. Review of clinical trials in treatment of opiate withdrawal: J. Strang *et al.*, *Am. J. Addict.* **8**, 337-348 (1999).

Crystals, mp 126-128°.

Hydrochloride. [21498-08-8] MDL-14042A; Ba-168; Brit-Lofex; Lofetensin. $C_{11}H_{12}Cl_2N_2O \cdot HCl$; mol wt 295.59. Crystals from ethanol/ether or 2-propanol, mp 221-223° (**U.S.** patent); also reported as mp 230-232° (**Ger.** patent). Very sol in water, ethanol. Slightly sol in 2-propanol. Practically insol in ether. LD_{50} in mice, rats, dogs (mg/kg): between 74-147 orally (all species); between 8-18 i.v. (all species) (Tsai).

THERAP CAT: In treatment of opioid withdrawal symptoms; anti-hypertensive.

5616. Loflucarban. [790-69-2] *N*-(3,5-Dichlorophenyl)-*N'*-(4-fluorophenyl)thiourea; 3,5-dichloro-4'-fluorothiocarbanilide; Fluonilid. $C_{13}H_9Cl_2FN_2S$; mol wt 315.19. C 49.54%, H 2.88%, Cl 22.49%, F 6.03%, N 8.89%, S 10.17%. Prepd from *p*-fluorophenyl isothiocyanate and 3,5-dichloroaniline or from 3,5-dichlorophenyl isothiocyanate and *p*-fluoroaniline: **BE 613154** (1962 to Madan).

Crystals from ethanol, mp 148°. Soluble in ethyl oleate, isopropyl myristate.

THERAP CAT: Antifungal.

5617. Loganin. [18524-94-2] (1*S*,4a*S*,6*S*,7*R*,7a*S*)-1-(β-D-Glucopyranosyloxy)-1,4a,5,6,7,7a-hexahydro-6-hydroxy-7-methyl-cyclopenta[*c*]pyran-4-carboxylic acid methyl ester; 7-hydroxy-6-desoxyverbenalin. $C_{17}H_{26}O_{10}$; mol wt 390.39. C 52.30%, H 6.71%, O 40.98%. Key intermediate in the biosynthesis of indole alkaloids. First isolated from the seeds but chiefly from the pulp of the fruit of *Strychnos nux-vomica* L., *Loganiaceae:* Dunstan, Short, *Pharm. J.* **14**, 1025 (1883); Merz, Krebs, *Arch. Pharm.* **275**, 217 (1937); Merz, Lehmann, *ibid.* **290**, 543 (1957). Structure: Sheth *et al.*, *Tetrahedron Lett.* **1961**, 394; Büchi, Manning, *Tetrahedron* **18**, 1049 (1962). Crystal structure: Lentz, Rossmann, *Chem. Commun.* **1969**, 1269; P. G. Jones *et al.*, *Acta Crystallogr.* **B36**, 481 (1980). Abs config: Inouye *et al.*, *Tetrahedron* **26**, 3905 (1970). Total synthesis: Büchi *et al.*, *J. Am. Chem. Soc.* **92**, 2165 (1970); Partridge *et al.*, *ibid.* **95**, 532 (1973); Büchi *et al.*, *ibid.* 540; B.-W. Au-Yeung, I. Fleming, *Chem. Commun.* **1977**, 81; I. Fleming, B.-W. Au-Yeung, *Tetrahedron* **37**, Suppl. 9, 13 (1981); K. Hiroi *et al.*, *Chem. Lett.* **1981**, 559. Biosynthetic studies: Battersby, "Biosynthesis II: Terpenoid Indole Alkaloids", in *The Alkaloids* **vol. 1**, The Chemical Society (Burlington House, London, 1971) pp 31-47.

Crystals, mp 222-223°. $[\alpha]_D^{20}$ −82.1° (water). Freely sol in water; less sol in 96% alcohol; sparingly in abs alcohol. Practically insol in ether, petr ether, ligroin, ethyl acetate, acetone, chloroform.

5618. Loline. [25161-91-5] (2R,3R,3aS,4S,6aS)-Hexahydro-N-methyl-2,4-methano-4H-furo[3,2-b]pyrrol-3-amine; festucine; methyl-N-depropionyldecorticasine. $C_8H_{14}N_2O$; mol wt 154.21. C 62.31%, H 9.15%, N 18.17%, O 10.37%. One of a group of pyrrolizidine alkaloids whose common feature is a unique oxygen bridge. Isoln from seeds of *Lolium cuneatum* Nevski, *Gramineae:* S. Y. Yunusov, S. T. Akramov, *J. Gen. Chem. USSR* **25**, 1765 (1955); from tall fescue hay, *Festuca arundinaceae* Schreb., *Gramineae:* Yates, Tookey, *Aust. J. Chem.* **18**, 53 (1965). Identity with festucine: Aasen, Culvenor, *ibid.* **22**, 2021 (1969). Identity with methyl-N-depropionyldecorticasine: Ribas *et al.*, *An. Quim.* **64**, 516 (1968), *C.A.* **69**, 109764c (1968). Abs config: R. B. Bates, S. R. Morehead, *Tetrahedron Lett.* **1972**, 1629. Synthesis of bisquaternary salts: N. P. Abdullaev *et al.*, *Khim. Prir. Soedin.* **3**, 371 (1977), *C.A.* **88**, 23216 (1978). Synthetic and structural studies: S. R. Wilson *et al.*, *J. Org. Chem.* **46**, 3887 (1981).

Dihydrochloride. [25161-92-6] $C_8H_{14}N_2O.2HCl$. Needles from abs ethanol, dec 237-242°. $[\alpha]_D^{25}$ +4.6° (c = 4.37 in water). pKa 8.25 and 2.5-3.0.

5619. Lomefloxacin. [98079-51-7] 1-Ethyl-6,8-difluoro-1,4-dihydro-7-(3-methyl-1-piperazinyl)-4-oxo-3-quinolinecarboxylic acid. $C_{17}H_{19}F_2N_3O_3$; mol wt 351.35. C 58.11%, H 5.45%, F 10.81%, N 11.96%, O 13.66%. Fluorinated quinolone antibacterial. DNA gyrase antagonist. Prepn: Y. Itoh *et al.*, **DE 3433924**; *eidem*, **US 4528287** (both 1985 to Hokuriku Pharm.). *In vitro* and *in vivo* activity: T. Hirose *et al.*, *Antimicrob. Agents Chemother.* **31**, 854 (1987). Comparative antibacterial spectrum *in vitro:* R. Wise *et al.*, *ibid.* **32**, 617 (1988). HPLC determn in urine and bile and preliminary pharmacokinetics in rats: A. Saito *et al.*, *ibid.* 156. Photochemistry study: E. Fasani *et al.*, *Eur. J. Org. Chem.* **2004**, 5075. Supplement on antibacterial spectrum, pharmacokinetics and clinical efficacy: *Chemotherapy (Tokyo)* **36**, Suppl. 2, 1-1418 (1988). Comprehensive description: Y. D. Sanzgiri *et al.*, *Anal. Profiles Drug Subs. Excip.* **23**, 321-369 (1994).

Colorless needles from ethanol, mp 239-240.5°. Soly at 25° (mg/ml): 1.03 at ionic strength 0.15M NaCl. uv max (0.15M acetate buffer, pH 5): 226, 288, 320 nm ($a_M \times 10^{-3}$ 15.5, 36.7, 13.5); uv max (0.05M phosphate buffer, pH 7): 282, 326 nm ($a_M \times 10^{-3}$ 31.5, 13.3); uv max 0.15M borate buffer, μ = 0.15M NaCl, pH 9): 282, 328 nm ($a_M \times 10^{-3}$ 29.3, 13.2). LD_{50} in mice (mg/kg): 245.6 i.v.; >4000 orally (Itoh).

Monohydrochloride. [98079-52-8] NY-198; SC-47111; Bareon; Chimono; Lomebact; Maxaquin; Okacin; Okacyn. $C_{17}H_{19}F_2$-$N_3O_3.HCl$; mol wt 387.81. Colorless needles from water, mp 290-300° (dec).

THERAP CAT: Antibacterial.

5620. Lomerizine. [101477-55-8] 1-[Bis(4-fluorophenyl)-methyl]-4-[(2,3,4-trimethoxyphenyl)methyl]piperazine; 1-[bis(p-fluorophenyl)methyl]-4-(2,3,4-trimethoxybenzyl)piperazine; 1-(2,-3,4-trimethoxybenzyl)-4-[bis(4-fluorophenyl)methyl]piperazine. $C_{27}H_{30}F_2N_2O_3$; mol wt 468.54. C 69.21%, H 6.45%, F 8.11%, N 5.98%, O 10.24%. Diphenylpiperazine calcium channel blocker; selective cerebral vasodilator. Prepn: H. Ohtaka *et al.*, **EP 159566**;

eidem, **US 4663325** (1985, 1987 both to Kanebo); *eidem et al.*, *Chem. Pharm. Bull.* **35**, 3270 (1987). Structure-activity study: H. Ohtaka, G. Tsukamoto, *ibid.* 4117. HPLC determn in plasma and brain tissue: H. Waki, S. Ando, *J. Chromatogr.* **494**, 408 (1989). Pharmacology: T. Kanazawa *et al.*, *J. Cardiovasc. Pharmacol.* **16**, 430 (1990). Protective effect in cerebral ischemia: H. Hara *et al.*, *Arch. Int. Pharmacodyn.* **304**, 206 (1990); V. Danielisova, M. Chavko, *Neurochem. Res.* **19**, 1503 (1994).

Dihydrochloride. [101477-54-7] KB-2796; Terranas. $C_{27}H_{30}$-$F_2N_2O_3.2HCl$; mol wt 541.46. Colorless crystals from acetonitrile, mp 214-218° (dec). Also reported as mp 204-207° (dec). LD_{50} orally in mice: 300 mg/kg (Ohtaka).

THERAP CAT: Antimigraine.

5621. Lomustine. [13010-47-4] N-(2-Chloroethyl)-N'-cyclohexyl-N-nitrosourea; 1-(2-chloroethyl)-3-cyclohexyl-1-nitrosourea; CCNU; NSC-79037; RB-1509; Cecenu; CeeNU; CiNU. C_9H_{16}-ClN_3O_2; mol wt 233.70. C 46.26%, H 6.90%, Cl 15.17%, N 17.98%, O 13.69%. Chloroethylnitrosourea derivative with antitumor activity. Synthesis: T. P. Johnston *et al.*, *J. Med. Chem.* **9**, 892 (1966). Clinical review: S. K. Carter, J. W. Newman, *Cancer Chemother. Rep. Part 3* **1**, 136 (1968). Physiological disposition: V. T. Oliverio *et al.*, *Cancer Res.* **30**, 1330 (1970). Toxicology and pharmacology: G. R. Thompson, R. E. Larson, *Toxicol. Appl. Pharmacol.* **21**, 405 (1972); V. T. Oliverio, *Cancer Chemother. Rep. Part 3* **4**, 13 (1973). Use in canine epitheliotropic lymphoma: R. E. Risbon *et al.*, *J. Vet. Intern. Med.* **20**, 1389 (2006). Comprehensive description: F. J. Al-Shammary, *Anal. Profiles Drug Subs.* **19**, 315-340 (1990).

Yellow powder, mp 90°. Freely sol in chloroform; sol in ethanol, acetone. Soly in water, 0.1N NaOH, 0.1N HCl or 10% ethanol: <0.05 mg/ml; in absolute ethanol: 70 mg/ml. The bulk drug should be stored in a deep freeze and protected from moisture. LD_{50} in male mice (mg/kg): 51 orally; 56 i.p.; 61 s.c. (Thompson, Larson).

Caution: This substance is reasonably anticipated to be a human carcinogen: *Report on Carcinogens, Twelfth Edition* (PB2011-111646, 2011) p 326.

THERAP CAT: Antineoplastic.

THERAP CAT (VET): Antineoplastic.

5622. Lonafarnib. [193275-84-2] 4-[2-[4-[(11R)-3,10-Dibromo-8-chloro-6,11-dihydro-5H-benzo[5,6]cyclohepta[1,2-b]pyridin-11-yl]-1-piperidinyl]-2-oxoethyl]-1-piperidinecarboxamide; Sch-66336; Sarasar. $C_{27}H_{31}Br_2ClN_4O_2$; mol wt 638.83. C 50.76%, H 4.89%, Br 25.02%, Cl 5.55%, N 8.77%, O 5.01%. Tricyclic farnesyl protein transferase inhibitor. Prepn: R. J. Doll *et al.*, **WO 9723478**; *eidem*, **US 5874442** (1997, 1999 both to Schering-Plough); F. G. Njoroge *et al.*, *J. Med. Chem.* **41**, 4890 (1998). Improved prepn: S.-C. Kuo *et al.*, *J. Org. Chem.* **68**, 4984 (2003). Crystal structure of complex with farnesyl protein transferase: C. L. Strickland *et al.*, *J. Med. Chem.* **42**, 2125 (1999). LC/MS/MS determn in plasma: N. M. G. M. Appels *et al.*, *Rapid Commun. Mass Spectrom.* **19**, 2187 (2005). Antitumor activity in human xenograft and transgenic mouse cancer models: M. Liu *et al.*, *Cancer Res.* **58**, 4947 (1998). Clinical evaluation in cancer: A. A. Adjei *et al.*, *ibid.* **60**, 1871 (2000). Inhibition of P-glycoprotein mediated drug efflux: E.

5623   **Lonazolac**

Wang *et al.*, *ibid.* **61**, 7525 (2001). Clinical evaluation with pacli-taxel in non-small cell lung cancer: E. S. Kim *et al.*, *Cancer* **104**, 561 (2005). Review of development and clinical experience: F. Morgillo, H.-Y. Lee, *Expert Opin. Invest. Drugs* **15**, 709-719 (2006).

White solid. Crystals from acetone, mp 214.5-215.9° (monohydrate); anhydrous form mp 222-223°. $[\alpha]_D^{25}$ = +49.1° (c = 0.21 in methanol).

THERAP CAT: Antineoplastic.

5623. Lonazolac. [53808-88-1] 3-(4-Chlorophenyl)-1-phenyl-1H-pyrazole-4-acetic acid. $C_{17}H_{13}ClN_2O_2$; mol wt 312.75. C 65.29%, H 4.19%, Cl 11.33%, N 8.96%, O 10.23%. Prepn: R. A. Newberry, **GB 1373212** (1974 to Wyeth); G. Rainer, **US 4146721** (1979 to Byk Gulden); G. Rainer *et al.*, *Arzneim.-Forsch.* **31**, 649 (1981). Pharmacology: R. Riedel, *ibid.* 655. Clinical studies: G. Lonauer *et al.*, *Z. Rheumatol.* **40**, 161 (1981); W. Siegmeth, P. Placheta, *Wien. Klin. Wochenschr.* **94**, 145 (1982).

Crystals from ethanol/water, mp 150-151°. uv max (methanol): 281 nm (ε 24800); (0.1N NaOH): 281 nm (ε 23700). pKa 4.3. LD$_{50}$ in male mice, rats (mg/kg): 195, 165 i.v. (Riedel).

Calcium salt. [75821-71-5] Argun L; Irritren. $C_{34}H_{24}CaCl_2$-N_4O_4; mol wt 663.57. Solid, melts between 270-290° (dec). uv max (0.1N NaOH): 280 nm (ε 46400). LD$_{50}$ in male, female mice, male, female rats (mg/kg): 670, 845, 730, 1000 orally (Riedel).

THERAP CAT: Anti-inflammatory.

5624. Longifolene. [475-20-7] (1S,3aR,4S,8aS)-Decahydro-4,8,8-trimethyl-9-methylene-1,4-methanoazulene; junipene; kuromatsuene. $C_{15}H_{24}$; mol wt 204.36. C 88.16%, H 11.84%. Tricyclic sesquiterpene present to the extent of 5-10% in Indian turpentine oil which is produced commercially from the Himalayan pine, *Pinus longifolia* Roxb., *Pinaceae*: J. L. Simonsen, *J. Chem. Soc.* **117**, 570 (1920). Structure: R. H. Moffett, D. Rogers, *Chem. Ind. (London)* **1953**, 916; P. Naffa, G. Ourisson, *ibid.* 917. Abs config: G. Ourisson, *Bull. Soc. Chim. Fr.* **1955**, 895. Identity of longifolene with junipene and kuromatsuene: S. Akiyoshi *et al.*, *Tetrahedron* **9**, 237 (1960). Total synthesis of naturally occurring (+)-form: E. J. Corey *et al.*, *J. Am. Chem. Soc.* **86**, 478 (1964); of the (±)-form: *eidem*, *ibid.* **83**, 1251 (1961); J. E. McMurray, S. J. Isser, *ibid.* **94**, 7132 (1972); R. A. Volkmann *et al.*, *ibid.* **97**, 4777 (1975). Review of chemistry of longifolene and its derivs: S. Dev, *Acc. Chem. Res.* **14**, 82-88 (1981); *idem*, *Fortschr. Chem. Org. Naturst.* **40**, 50-104 (1981).

Viscous oil, bp$_{706}$ 254-256°, bp$_{15}$ 126-127°. $[\alpha]_D^{18}$ +42.73°. d_4^{18} 0.9319; n_D^{20} 1.5040. Sol in benzene. Insol in water.

Borane derivative. Dilongifolylborane. $C_{30}H_{51}B$. Heavy, snow-white, shiny plates, mp 160-161° (sealed, evacuated capillary). Strongly dimeric. Sparingly sol in common organic solvents. Prepn and use in hydroboration: P. K. Jadhav, H. C. Brown, *J. Org. Chem.* **46**, 2988 (1981).

USE: Borane deriv as a chiral hydroborating agent.

5625. Lonidamine. [50264-69-2] 1-[(2,4-Dichlorophenyl)-methyl]-1H-indazole-3-carboxylic acid; 1-(2,4-dichlorobenzyl)-1H-indazole-3-carboxylic acid; diclondazolic acid; DICA; AF-1890; Doridamina. $C_{15}H_{10}Cl_2N_2O_2$; mol wt 321.16. C 56.10%, H 3.14%, Cl 22.08%, N 8.72%, O 9.96%. Antitumor agent with antispermatogenic activity; inhibits cellular respiration in neoplastic cells. Prepn: G. Palazzo, B. Silvestrini, **DE 2310031**; *eidem*, **US 3895026** (1973, 1975 both to Angelini Francesco); G. Corsi, G. Palazzo, *J. Med. Chem.* **19**, 778 (1976). Antispermatogenic activity: T. J. Lobl, *Arch. Androl.* **2**, 353 (1979). *In vitro* and *in vivo* antitumor activity: B. Silvestrini *et al.*, *Br. J. Cancer* **47**, 221 (1983). Mechanism of action in neoplastic cells: A. Floridi *et al.*, *Exp. Mol. Pathol.* **42**, 293 (1985). Pharmacology: V. Cioli *et al.*, *Arzneim.-Forsch.* **34**, 455 (1984). HPLC determn in plasma and urine: R. Leclaire *et al.*, *J. Chromatogr.* **277**, 427 (1983). Toxicological study: R. Heywood *et al.*, *Chemotherapy (Basel)* **27**, Suppl. 2, 91 (1981). Series of articles on mechanism of action, pharmacokinetics and clinical applications in spermatogenesis and tumors: *ibid.* 5-120; on pharmacology and clinical evaluation in cancers: *Oncology* **41**, Suppl. 1, 1-121 (1984). Symposium on clinical evaluations in cancers: *Semin. Oncol.* **18**, Suppl 4, 1-72 (1991).

Crystals from ethanol, mp 207°. Sol in methanol, acetic acid. LD$_{50}$ in mice, rats (mg/kg): 900, 1700 orally; 435, 525 i.p. (Heywood).

THERAP CAT: Antineoplastic.

5626. Lonomycins. Polyether antibiotics active against grampositive bacteria, esp. coccidia. The major component is lonomycin A (initially named lonomycin). Lonomycins B and C are minor congeners. Isoln from *Streptomyces ribosidificus* strain TM-481 as sodium salt: M. Shibata *et al.*, **JP Kokai 75 49495** (1975 to Taisho Pharm.); S. Omura *et al.*, *J. Antibiot.* **29**, 15 (1976). Identity of A with antibiotic DE-3936; fermentation, isolation, physicochemical properties and biological activity: M. Ohshima *et al.*, *ibid.* 354. Structure of A: N. Otake, M. Koenuma, *Tetrahedron Lett.* **1975**, 4147; C. Riche, C. Pascard-Billy, *Chem. Commun.* **1975**, 951; of B, C: H. Seto *et al.*, *J. Antibiot.* **31**, 929 (1978); T. Mizutani *et al.*, *ibid.* **33**, 1224 (1980). ^{13}C-NMR spectrum of A: H. Seto *et al.*, *ibid.* **33**. Ionophorous activity of A: M. Mitani, N. Otake, *ibid.* **31**, 750 (1978). Cardiovascular effects of A: K. Kaneko *et al.*, *Res. Commun. Chem. Pathol. Pharmacol.* **67**, 17 (1990); K. Tsuchida *et al.*, *Arch. Int. Pharmacodyn.* **314**, 25 (1991). Total synthesis of A: D. A. Evans *et al.*, *J. Am. Chem. Soc.* **117**, 3448 (1995).

Lonomycin A

Sodium salts are sol in lower alcohols, acetone, ethyl acetate, benzene, chloroform, ether. Slightly sol in n-hexane, petr ether. Insol in water.

Lonomycin A. [58785-63-0] Antibiotic DE-3936; emericid. $C_{44}H_{76}O_{14}$. Colorless prisms from benzene-petr ether, mp 109-114°. $[\alpha]_D^{20}$ +66.6° (c = 1 in CHCl$_3$); $[\alpha]_D^{23}$ +57.5° (c = 0.40 in

CH$_2$Cl$_2$). pKa′ (66% acetone) 5.9. Sol in most organic solvents. Insol in water.

Lonomycin A sodium salt. [58845-80-0] C$_{44}$H$_{75}$NaO$_{14}$. Colorless prisms from benzene-petr ether, mp 173-176° (dec) (Ohshima); also reported as mp 188-189° (Omura). $[\alpha]_D^{25}$ +49.8° (c = 1 in CH$_3$OH), $[\alpha]_D^{25}$ +67.0° (c = 1 in CHCl$_3$) (Ohshima); also reported as $[\alpha]_D^{25}$ +47° (c = 1 in CH$_3$OH) (Omura). LD$_{50}$ in mice (mg/kg): 45.8 orally; 13.0 i.p.; 37.5 s.c. (Ohshima).

Lonomycin B sodium salt. C$_{44}$H$_{75}$NaO$_{14}$. White prisms from *n*-hexane-benzene, mp 181-182°. $[\alpha]_D$ +48.2° (c = 0.5 in CH$_3$OH).

Lonomycin C sodium salt. C$_{43}$H$_{73}$NaO$_{14}$. White prisms from *n*-hexane-benzene, mp 186-187°. $[\alpha]_D$ +49.2° (c = 0.5 in CH$_3$OH).

5627. Looplure. [14959-86-5] (7Z)-7-Dodecen-1-ol acetate; *cis*-7-dodecenyl acetate; *cis*-1-acetoxy-7-dodecene; ENT-33266. C$_{14}$H$_{26}$O$_2$; mol wt 226.36. C 74.29%, H 11.58%, O 14.14%. Sex pheromone of the female cabbage looper moth, *Trichoplusia ni* (Hübner). Isoln, identification and synthesis: R. S. Berger, *Ann. Entomol. Soc. Am.* **59**, 767 (1966). Synthesis and activity of isomers: N. Green *et al., J. Med. Chem.* **10**, 533 (1967); H. H. Toba *et al., J. Econ. Entomol.* **63**, 1048 (1970). Alternate syntheses: W. Seidel *et al., Ber.* **110**, 3544 (1977); H. J. Bestmann *et al., Ber.* **112**, 1923 (1979); M. Horiike, C. Hirano, *Agric. Biol. Chem.* **44**, 2229 (1980); D. Basavaiah *et al., J. Org. Chem.* **47**, 1792 (1982). Acute toxicity study: M. Beroza *et al., Toxicol. Appl. Pharmacol.* **31**, 421 (1975).

Oil, bp$_{0.05}$ 98-100°. n_D^{25} 1.4420. LD$_{50}$ in rats (mg/kg): >13430 orally; in rabbits (mg/kg): >2025 dermally; LC$_{50}$ (24 hr) in rainbow trout, bluegill sunfish (ppm): >100, >100 (Beroza).

USE: Insect attractant.

5628. Loperamide. [53179-11-6] 4-(4-Chlorophenyl)-4-hydroxy-*N,N*-dimethyl-α,α-diphenyl-1-piperidinebutanamide; 4-(*p*-chlorophenyl)-4-hydroxy-*N,N*-dimethyl-α,α-diphenyl-1-piperidinebutyramide. C$_{29}$H$_{33}$ClN$_2$O$_2$; mol wt 477.05. C 73.02%, H 6.97%, Cl 7.43%, N 5.87%, O 6.71%. Prepn: Janssen *et al.*, *FR 2100711*; *eidem*, *US 3714159* (1972, 1973 to Janssen); Stokbroekx *et al., J. Med. Chem.* **16**, 782 (1973). LC-MS/MS determn in plasma: B. Streel *et al., J. Chromatogr. B* **814**, 263 (2005). Series of articles on pharmacology, toxicology, metabolism, and clinical studies: *Arzneim.-Forsch.* **24**, 1640-1665 (1974). Toxicity studies: C. J. E. Niemegeers *et al., ibid.* 1633, 1636. *Review:* D. A. Shriver *et al.*, in *Pharmacological and Biochemical Properties of Drug Substances* **vol. 3**, M. E. Goldberg, Ed. (Am. Pharm. Assoc., Washington, DC, 1981) pp 461-476. Comprehensive description: J. Van Rompay, J. E. Carter, *Anal. Profiles Drug Subs.* **19**, 341-365 (1990).

Hydrochloride. [34552-83-5] PJ-185; R-18553; Arret; Blox; Brek; Dissenten; Fortasec; Imodium; Imosec; Imossel; Lopemid; Lopemin; Loperyl; Suprasec; Tebloc. C$_{29}$H$_{33}$ClN$_2$O$_2$.HCl; mol wt 513.50. Crystals from isopropanol, mp 222-223°. uv max (0.1*N* HCl/2-propanol, 10/90 v/v): 253, 259, 265, 273 nm (ε 532, 648, 581, 233). Freely sol in chloroform; slightly sol in dilute acids; very slightly sol in isopropyl alc. Soly (g/100ml): water (pH 1.7) 0.14; citrate-phosphate (pH 6.1) 0.008; citrate-phosphate (pH 7.9) <0.001; methanol 28.6; ethanol 5.37; 2-propanol 1.11; dichloromethane 35.1; acetone 0.20; ethyl acetate 0.035; diethyl ether <0.001; hexane <0.001; toluene 0.001; *N,N*-dimethylformamide 10.3; tetrahydrofuran 0.32; 4-methyl-2-pentanone 0.020; propylene glycol 5.64; polyethylene glycol 400 1.40; dimethylsulfoxide 20.5; 2-butanone 0.18. pKa 8.66. Practically insol in H$_2$O at physiological pH

(0.002%). Stable, can be stored for several years under normal conditions; not hygroscopic; not affected by light. LD$_{50}$ in mice (mg/kg): 75 s.c.; 28 i.p.; 105 orally; in rats (mg/kg): 185 orally (Niemegeers).

THERAP CAT: Antidiarrheal.

5629. Lophotoxin. [78697-56-0] (1*R*,2*S*,4*S*,10*R*,12*R*,14*R*,-15*R*)-2-(Acetyloxy)-12-methyl-4-(1-methylethenyl)-17-oxo-11,16,-18,19-tetraoxapentacyclo[12.2.2.1^{6,9}.0^{1,15}.0^{10,12}]nonadeca-6,8-diene-7-carboxaldehyde; LTX. C$_{22}$H$_{24}$O$_8$; mol wt 416.43. C 63.45%, H 5.81%, O 30.74%. Neuromuscular toxin isolated from several spp of Pacific gorgonians (sea fans and whips) of the genus *Lophogorgia*. Although it belongs to the cembrene class of diterpenoids, the ability of LTX to cause irreversible postsynaptic blockade at the neuromuscular junction resembles the mode of action of the snake venom protein, α-bungarotoxin (*cf.* bungarotoxins). Isoln and structure: W. Fenical *et al., Science* **212**, 1512 (1981). Molecular mode of action: J. R. Smythies, *Med. Hypotheses* **7**, 1457 (1981). Toxicity study: P. Culver, R. S. Jacobs, *Toxicon* **19**, 825 (1981). Description of the cytotoxic, ichthyotoxic, and antibacterial activity of alcoholic extracts of gorgonians: R. S. Jacobs *et al., Fed. Proc.* **40**, 26 (1981); V. J. Paul, W. Fenical, *J. Org. Chem.* **45**, 3401 (1981).

White needles, mp 164-166°. $[\alpha]_D^{27}$ +14.2° (c = 1.7 in chloroform). LD$_{50}$ in mice (mg/kg): 8.9 s.c. (Culver, Jacobs).

5630. Lopinavir. [192725-17-0] (α*S*)-*N*-[(1*S*,3*S*,4*S*)-4-[[(2,6-Dimethylphenoxy)acetyl]amino]-3-hydroxy-5-phenyl-1-(phenylmethyl)pentyl]tetrahydro-α-(1-methylethyl)-2-oxo-1(2*H*)-pyrimidineacetamide; (α*S*)-tetrahydro-*N*-[(α*S*)-α-[(2*S*,3*S*)-2-hydroxy-4-phenyl-3-[2-(2,6-xylyloxy)acetamido]butyl]phenethyl]-α-isopropyl-2-oxo-1(2*H*)-pyrimidineacetamide; (2*S*,3*S*,5*S*)-2-(2,6-dimethylphenoxyacetyl)amino-3-hydroxy-5-[2*S*-(1-tetrahydropyrimid-2-onyl)-3-methylbutanoyl]amino-1,6-diphenylhexane; ABT-378; A-157378. C$_{37}$H$_{48}$N$_4$O$_5$; mol wt 628.81. C 70.67%, H 7.69%, N 8.91%, O 12.72%. Peptidomimetic HIV protease inhibitor. Prepn: H. L. Sham *et al., WO 9721685; idem et al., US 5914332* (1997, 1999 both to Abbott); E. J. Stoner *et al., Org. Process Res. Dev.* **3**, 145 (1999). Protease inhibition and pharmacokinetics: H. L. Sham *et al., Antimicrob. Agents Chemother.* **42**, 3218 (1998). *In vitro* metabolism study: G. N. Kumar *et al., Drug Metab. Dispos.* **27**, 86 (1999). Pharmacokinetic enhancement by ritonavir: G. N. Kumar *et al., ibid.* 902. Clinical trial with ritonavir in HIV infection: S. Walmsley *et al., N. Engl. J. Med.* **346**, 2039 (2002).

Colorless solid from ethyl acetate, mp 124-127°.
Mixture with ritonavir. Kaletra.
THERAP CAT: Antiretroviral.

5631. Loprazolam. [61197-73-7] (2*Z*)-6-(2-Chlorophenyl)-2,4-dihydro-2-[(4-methyl-1-piperazinyl)methylene]-8-nitro-1*H*-

imidazo[1,2-*a*][1,4]benzodiazepin-1-one. $C_{23}H_{21}ClN_6O_3$; mol wt 464.91. C 59.42%, H 4.55%, Cl 7.63%, N 18.08%, O 10.32%. Annelated 1,4-benzodiazepine deriv. Prepn: J. B. Taylor, D. R. Harrison, **DE 2605652**; *eidem*, **US 4044142** (1976, 1977 both to Roussel-UCLAF). CNS activity: J. R. Agar *et al.*, *J. Med. Chem.* **20**, 1035 (1977). Pharmacological profile: T. G. Johns *et al.*, *Arch. Int. Pharmacodyn. Ther.* **240**, 53 (1979). Effects on sleep in normal volunteers: I. Hindmarch, C. A. Clyde, *Drugs Exp. Clin. Res.* **2**, 61 (1980). Review of pharmacology and efficacy in insomnia: B. G. Clark *et al.*, *Drugs* **31**, 500-516 (1986).

Crystals from chloroform/ether, mp 214-215°. LD_{50} in mice: >1000 mg/kg orally (Agar).

Methanesulfonate. [70111-54-5] Loprazolam mesylate; HR-158; RU-31158; Dormonoct; Havlane; Somnovit; Sonin. $C_{23}H_{21}ClN_6O_3 \cdot CH_3SO_3H$; mol wt 561.01. Crystals from methylene chloride, mp 205-210°.

Note: This is a controlled substance (depressant): **21 CFR**, 1308.14.

THERAP CAT: Sedative, hypnotic.

5632. Loprinone. [106730-54-5] 1,2-Dihydro-5-imidazo-[1,2-*a*]pyridin-6-yl-6-methyl-2-oxo-3-pyridinecarbonitrile; 1,2-dihydro-5-imidazo[1,2-*a*]pyridin-6-yl-6-methyl-2-oxonicotinonitrile; olprinone. $C_{14}H_{10}N_4O$; mol wt 250.26. C 67.19%, H 4.03%, N 22.39%, O 6.39%. Inotropic agent which inhibits cAMP phosphodiesterase. Prepn: M. Yamanaka *et al.*, **JP Kokai 86 218589**; *eidem*, **US 4751227** (1986, 1988 both to Eisai). Synthesis: M. Yamanaka *et al.*, *J. Labelled Compd. Radiopharm.* **31**, 125 (1992). Pharmacology: M. Tajimi *et al.*, *Arch. Pharmacol.* **344**, 602 (1991). Pharmacokinetics in dogs: Y. Uemura *et al.*, *J. Pharm. Pharmacol.* **45**, 1077 (1993). Review of pharmacology, toxicology and clinical evaluation: M. Endoh, *Cardiovasc. Drug Rev.* **11**, 432-450 (1993).

Crystals from DMF, mp >300°.

Hydrochloride monohydrate. E-1020. $C_{14}H_{10}N_4O \cdot HCl \cdot H_2O$; mol wt 304.73. White to yellowish-white odorless crystalline powder, mp >300°. Sol in water (pH 3-5) up to 3 mg/ml. LD_{50} in male, female rats, mice (mg/kg): 7,804, >10,000, >10,000, >10,000 orally; 176, 240, 242, 269 i.v.; 2,133, 2,890, 3,898, 4,479 s.c. (Endoh).

THERAP CAT: Cardiotonic.

5633. Loracarbef. [76470-66-1] (6*R*,7*S*)-7-[[(2*R*)-2-Amino-2-phenylacetyl]amino]-3-chloro-8-oxo-1-azabicyclo[4.2.0]oct-2-ene-2-carboxylic acid; (6*R*,7*S*)-3-chloro-7-[(*R*)-phenylglycinamido]-8-oxo-1-azabicyclo[4.2.0]oct-2-ene-2-carboxylic acid; carbacefaclor; LY-163892. $C_{16}H_{16}ClN_3O_4$; mol wt 349.77. C 54.94%, H 4.61%, Cl 10.14%, N 12.01%, O 18.30%. Carbacephem analog of cefaclor, *q.v.* Prepn: T. Hirata *et al.*, **EP 14476**; *eidem*, **US 4708956** (1980, 1987 both to Kyowa Hakko); I. Matsukura *et al.*, *Chem. Pharm. Bull.* **37**, 1239 (1989). Prepn of the crystalline monohydrate: C. E. Pasini, **EP 311366** (1989 to Lilly). Enantioselective synthesis: C. C. Bodurow *et al.*, *Tetrahedron Lett.* **30**, 2321 (1989). Antibacterial spectrum: K. Sato *et al.*, *J. Antibiot.* **42**, 1844 (1989). β-

lactamase stability: R. N. Jones, A. L. Barry, *Eur. J. Clin. Microbiol.* **6**, 570 (1987). Pharmacology in animals: T. Shetler *et al.*, *Arzneim.-Forsch.* **43**, 60 (1993). Pharmacokinetics in children: J. D. Nelson *et al.*, *Antimicrob. Agents Chemother.* **32**, 1738 (1988). Review of pharmacology and therapeutic efficacy: R. N. Brogden, D. McTavish, *Drugs* **45**, 716-736 (1993); R. W. Force, M. C. Nahata, *Ann. Pharmacother.* **27**, 321-329 (1993).

mp >300° (with browning).

Monohydrate. [121961-22-6] KT-3777; Lorabid; Lorax. $C_{16}H_{16}ClN_3O_4 \cdot H_2O$; mol wt 367.79. White crystalline solid, mp 205-215° (dec). $[\alpha]_D^{21}$ +34.0° (c = 0.35 in water).

THERAP CAT: Antibacterial.

5634. Lorajmine. [47562-08-3] (17*R*,21α)-Ajmalan-17,21-diol 17-(2-chloroacetate); 17-monochloroacetylajmaline; 17-chloroacetylajmaline; MCAA. $C_{22}H_{27}ClN_2O_3$; mol wt 402.92. C 65.58%, H 6.75%, Cl 8.80%, N 6.95%, O 11.91%. Semi-synthetic ajmaline derivative. Prepn from ajmaline: **BE 622395** (1962 to Inverni Della Beffa); A. Bonati, A. Bocchia, *Farmaco Ed. Sci.* **18**, 84 (1963). Improved prepn: A. Bonati, **DE 2023949**; *eidem*, **US 3741972** (1970, 1973 both to Inverni Della Beffa). *In vitro* and *in vivo* anti-arrhythmic activity: C. Capra, *Farmaco Ed. Sci.* **19**, 865 (1964); cardiac action: Y. Katano *et al.*, *Arzneim.-Forsch.* **23**, 483 (1973); cardiovascular effects and toxicity: C. Capra *et al.*, *Farmaco Ed. Prat.* **35**, 49 (1980). Determination in plasma by TLC fluorescence: L. J. Dombrowski *et al.*, *J. Pharm. Sci.* **64**, 643 (1975). GC-MS determination in blood: St. D. Clemans *et al.*, *Arzneim.-Forsch.* **27**, 1128 (1977). Comparison with other ajmaline esters: M. Salmona *et al.*, *J. Pharm. Sci.* **64**, 1561 (1975); P. Jaillon *et al.*, *Arch. Int. Pharmacodyn.* **224**, 310 (1976). Clinical studies: M. Cafiero *et al.*, *Minerva Cardioangiol.* **32**, 59 (1984). Book: G. G. Gensini, Ed., *Concepts on the Mechanism and Treatment of Arrhythmias* (Futura Publishing Co., Mount Kisco, N.Y., 1974) 281 pp.

Crystals, mp 232-238°. $[\alpha]_D^{20}$ +27.5° (c = 1 in chloroform).

Hydrochloride. [40819-93-0] Win-11831; Nevergor; Ritmos Elle. $C_{22}H_{27}ClN_2O_3 \cdot HCl$; mol wt 439.38. Crystals from methyl ethyl ketone, mp 243-246°. $[\alpha]_D^{20}$ +29° (c = 1 in ethanol). LD_{50} in mice, rats (mg/kg): 176, 139 i.p.; 370, 480 orally (Capra).

THERAP CAT: Antiarrhythmic (cardiac depressant).

5635. Loratadine. [79794-75-5] 4-(8-Chloro-5,6-dihydro-11*H*-benzo[5,6]cyclohepta[1,2-*b*]pyridin-11-ylidene)-1-piperidinecarboxylic acid ethyl ester; 11-[*N*-(ethoxycarbonyl)-4-piperidylidene]-8-chloro-6,11-dihydro-5*H*-benzo[5,6]cyclohepta[1,2-*b*]pyridine; Sch-29851; Alavert; Claritin; Clarityn; Clarytin; Lisino. $C_{22}H_{23}ClN_2O_2$; mol wt 382.89. C 69.01%, H 6.06%, Cl 9.26%, N 7.32%, O 8.36%. Nonsedating-type histamine H_1-receptor antagonist. Prepn: F. J. Villani, **US 4282233** (1981 to Schering); F. J. Villani *et al.*, *Arzneim.-Forsch.* **36**, 1311 (1986). Pharmacology: A. Barnett *et al.*, *Agents Actions* **14**, 590 (1984). Evaluation of CNS effects in humans: C. M. Bradley, A. N. Nicholson, *Eur. J. Clin. Pharmacol.* **32**, 419 (1987). Clinical pharmacokinetics: E. Radwanski *et al.*, *J. Clin. Pharmacol.* **27**, 530 (1987). Clinical trial in allergic rhinitis: R. J. Dockhorn *et al., Ann. Allergy* **58**, 407 (1987); in comparison with terfenadine, *q.v.*: F. Horak *et al., Arzneim.-Forsch.* **38**, 124 (1988); G. Bruttmann *et al., J. Allergy Clin. Immunol.* **83**, 411 (1989). HPLC determn in plasma: O. Q. P. Yin *et al., J. Chromatogr. B* **796**, 165 (2003). Review of pharmacology and therapeutic efficacy: M. Haria *et al., Drugs* **48**, 617-637 (1994).

Crystals from acetonitrile, mp 134-136°. Freely sol in acetone, chloroform, methanol, toluene. Insol in water.

THERAP CAT: Antihistaminic.

5636. Lorazepam. [846-49-1] 7-Chloro-5-(2-chlorophenyl)-1,3-dihydro-3-hydroxy-2*H*-1,4-benzodiazepin-2-one; Wy-4036; Ativan; Emotival; Lorax; Lorsilan; Pro Dorm; Psicopax; Punktyl; Quait; Sedatival; Sedazin; Somagerol; Tavor; Temesta; Wypax. $C_{15}H_{10}Cl_2N_2O_2$; mol wt 321.16. C 56.10%, H 3.14%, Cl 22.08%, N 8.72%, O 9.96%. Prepn: S. C. Bell, **BE 621819**; *idem*, **US 3296249** (1963, 1967 both to Am. Home Prod.); and structure-activity study: *idem et al., J. Med. Chem.* **11**, 457 (1968). Determn by HPLC: I. Jane, A. McKinnon, *J. Chromatogr.* **323**, 191 (1985). Series of articles on pharmacology, metabolism, clinical studies: *Arzneim.-Forsch.* **21**, 1047-1102 (1971). Toxicity study: G. Owen *et al., ibid.* 1065. Comprehensive description: J. G. Rutgers, C. M. Shearer, *Anal. Profiles Drug Subs.* **9**, 397-426 (1980). Review of clinical pharmacology and therapeutic use: B. Ameer, D. J. Greenblatt, *Drugs* **21**, 161-200 (1981). Clinical trial to prevent recurrent alcohol-related seizures: G. D'Onofrio *et al., N. Engl. J. Med.* **340**, 915 (1999).

Crystals, mp 166-168°. uv max (methanol): 229 nm, (1*N* NaOH): 233 nm, (1*N* HCl): 237 nm. Soly (mg/ml): water 0.08, chloroform 3, alcohol 14, propylene glycol 16, ethyl acetate 30. pK_1 1.3; pK_2 11.5. LD_{50} in mice, rats (mg/kg): 3178, >5000 orally (Owen).

Note: This is a controlled substance (depressant): **21 CFR,** 1308.14.

THERAP CAT: Anxiolytic; anticonvulsant.

5637. Lorcainide. [59729-31-6] *N*-(4-Chlorophenyl)-*N*-[1-(1-methylethyl)-4-piperidinyl]benzeneacetamide; 4′-chloro-*N*-(1-isopropyl-4-piperidyl)-2-phenylacetanilide. $C_{22}H_{27}ClN_2O$; mol wt 370.92. C 71.24%, H 7.34%, Cl 9.56%, N 7.55%, O 4.31%. Prepn: S. Sanczuk, H. K. F. Hermans, **DE 2642856** (1977 to Janssen); *idem*, **US 4126689** (1978 to Janssen). Determn of lorcainide and its metabolites by GLC: R. Woestenborghs *et al., J. Chromatogr.* **164**, 169 (1979). Disposition and anti-arrhythmic effects: U. Klotz *et al., Int. J. Clin. Pharmacol. Biopharm.* **17**, 152 (1979). Pharmacokinetics and tissue distribution: U. Klotz, E. Golbs, *Arzneim.-Forsch.* **30**, 619 (1980). Clinical study: P. Somani, S. di Giorgi, *Chest* **78**, 658 (1980). Toxicology: G. J. Barton, R. Marsboom, "A Brief Summary of Animal Toxicology Studies on Lorcainide", in *Prognosis and Pharmacotherapy of Life-Threatening Arrhythmias*, Jähnchen *et al.*, Eds. (Royal Soc. Med. London, 1981) pp 23-25. Review of pharmacology and efficacy: C. E. Eriksson, R. N. Brogden, *Drugs*

27, 279-300 (1984). Symposium: *Am. J. Cardiol.* **54**, 1B-54B (1984).

LD_{50} in male mice, male and female rats (mg/kg): 18.8, 19.3, 18.6 i.v.; 483, 395, 435 orally (Barton, Marsboom).

Hydrochloride. [58934-46-6] R-15889; Ro-13-1042; Lopantrol; Lorivox; Remivox. $C_{22}H_{27}ClN_2O \cdot HCl$; mol wt 407.38. Crystals from 2-propanone and 2-propanol, mp 263°.

THERAP CAT: Antiarrhythmic (class IC).

5638. Lorcaserin. [616202-92-7] (1*R*)-8-Chloro-2,3,4,5-tetrahydro-1-methyl-1*H*-3-benzazepine. $C_{11}H_{14}ClN$; mol wt 195.69. C 67.52%, H 7.21%, Cl 18.12%, N 7.16%. Serotonin 5-HT$_{2c}$ receptor agonist. Prepn: J. Smith, B. Smith, **WO 03086306**; *eidem*, **US 6953787** (2003, 2005 both to Arena); B. M. Smith *et al. J. Med. Chem.* **51**, 305 (2008). Pharmacology and receptor binding: W. J. Thomsen *et al., J. Pharmacol. Exp. Ther.* **325**, 577 (2008). Clinical trial for weight management in obese patients: S. R. Smith *et al., N. Engl. J. Med.* **363**, 245 (2010).

Hydrochloride. [846589-98-8] APD-356; Lorqess. $C_{11}H_{14}$-ClN·HCl; mol wt 232.15. Prepn: R. K. Agarwal *et al.*, **WO 06069363** (2006 to Arena). Hygroscopic crystals, mp 201°. Readily converts to a crystalline hemihydrate.

THERAP CAT: Antiobesity agent.

5639. Lormetazepam. [848-75-9] 7-Chloro-5-(2-chlorophenyl)-1,3-dihydro-3-hydroxy-1-methyl-2*H*-1,4-benzodiazepin-2-one; *N*-methyllorazepam; Wy-4082; Ergocalm; Loramet; Noctamid. $C_{16}H_{12}Cl_2N_2O_2$; mol wt 335.18. C 57.34%, H 3.61%, Cl 21.15%, N 8.36%, O 9.55%. Analog of lorazepam, *q.v.* Prepn: S. C. Bell, **BE 621819**; *idem*, **US 3296249** (1963, 1967 both to Am. Home Products); *idem, J. Med. Chem.* **11**, 457 (1968); A. Nudelman *et al., J. Pharm. Sci.* **63**, 1886 (1974). Pharmacokinetics and biotransformation in humans: M. Hümpel *et al., Eur. J. Drug Metab. Pharmacokinet.* **4**, 237 (1979). Hypnotic effect: A. Doenicke *et al., Anaesthesist* **28**, 578 (1979). Comparative study: H. Ott *et al., ibid.* 29. Absorption, distribution, excretion of ^{14}C-lormetazepam: R. Girkin *et al., Xenobiotica* **10**, 401 (1980). Radioimmunologic study: M. Hümpel *et al., Clin. Pharmacol. Ther.* **28**, 673 (1980).

Crystals from ethanol/THF, mp 205-207°.

Note: This is a controlled substance (depressant): **21 CFR,** 1308.14.

THERAP CAT: Sedative, hypnotic.

5640. Lornoxicam. [70374-39-9] 6-Chloro-4-hydroxy-2-methyl-*N*-2-pyridinyl-2*H*-thieno[2,3-*e*]-1,2-thiazine-3-carboxamide

1,1-dioxide; 6-chloro-4-hydroxy-2-methyl-3-(2-pyridylcarbamoyl)-2H-thieno[2,3-*e*]-1,2-thiazine-1,1-dioxide; chlortenoxicam; Ro-13-9297; TS-110; Xefo. $C_{13}H_{10}ClN_3O_4S_2$; mol wt 371.81. C 42.00%, H 2.71%, Cl 9.53%, N 11.30%, O 17.21%, S 17.25%. Cyclooxygenase inhibitor; structurally similar to tenoxicam, *q.v.* Prepn: R. Pfister *et al.*, **DE 2838851**; *idem*, **US 4180662** (both 1979 to Hoffmann-La Roche). Clinical pharmacokinetics: S. I. Ankier *et al.*, *Postgrad. Med. J.* **64**, 752 (1988). Symposium on pharmacology and clinical experience: *ibid.* **66**, Suppl. 4, S1-S50 (1990). Overview of pharmacology and safety assessment: T. P. Pruss *et al.*, *ibid.* S18.

Orange to yellow crystals, mp 225-230° (dec). pKa₂ 4.7. uv max: 371 nm. Partition coefficient (*n*-octanol/pH 7.4 buffer): 1.8. LD₅₀ orally in mice, rats, rabbits, dogs, monkeys: >10 mg/kg (Pruss).

THERAP CAT: Anti-inflammatory; analgesic.

5641. Losartan. [114798-26-4] 2-Butyl-4-chloro-1-[[2'-(2H-tetrazol-5-yl)[1,1'-biphenyl]-4-yl]methyl]-1H-imidazole-5-methanol; 2-*n*-butyl-4-chloro-5-hydroxymethyl-1-[[2'-(1H-tetrazol-5-yl)biphenyl-4-yl]methyl]imidazole; 2-butyl-4-chloro-1-[*p*-(*o*-1H-tetrazol-5-ylphenyl)benzyl]imidazole-5-methanol. $C_{22}H_{23}ClN_6O$; mol wt 422.92. C 62.48%, H 5.48%, Cl 8.38%, N 19.87%, O 3.78%. Non-peptide angiotensin II receptor antagonist. Prepn: D. J. Carini, J. J. V. Duncia, **EP 253310**; D. J. Carini *et al*, **US 5138069** (1988, 1992 both to Du Pont); D. J. Carini *et al.*, *J. Med. Chem.* **34**, 2525 (1991). HPLC determn in plasma: A. Zarghi *et al.*, *Arzneim.-Forsch.* **55**, 569 (2005). Angiotensin receptor binding study: A. T. Chiu *et al.*, *J. Pharmacol. Exp. Ther.* **252**, 711 (1990); receptor subtype specificity: A. T. Chiu *et al.*, *Biochem. Biophys. Res. Commun.* **172**, 1195 (1990). Pharmacology of active metabolite, *EXP-3174*: P. C. Wong *et al.*, *J. Pharmacol. Exp. Ther.* **255**, 211 (1990). Symposium on pharmacology: *Am. J. Hypertens.* **4**, 271S-353S (1991). Clinical effect on mortality in heart failure: B. Pitt *et al.*, *Lancet* **355**, 1582 (2000). Renoprotective effect in diabetic nephropathy: B. M. Brenner *et al.*, *N. Engl. J. Med.* **345**, 861 (2001). Review of clinical experience in hypertension: M. McIntyre *et al.*, *Pharmacol. Ther.* **74**, 181-194 (1997); of clinical pharmacokinetics: D. A. Sica *et al.*, *Clin. Pharmacokinet.* **44**, 797-814 (2005).

Light yellow solid, mp 183.5-184.5°. pKa: 5-6.

Monopotassium salt. [124750-99-8] Du Pont 753; DUP-753; MK-954; Cozaar; Lortaan; Lorzaar; Losaprex; Neo-Lotan. $C_{22}H_{22}ClKN_6O$; mol wt 461.01. White to off-white free-flowing crystalline powder. Freely sol in water; sol in alcohols; slightly sol in common organic solvents, such as acetonitrile, methy ethyl ketone.

Monopotassium salt mixture with hydrochlorothiazide. Fortzaar; Hyzaar.

THERAP CAT: Antihypertensive.

5642. Loteprednol Etabonate. [82034-46-6] (11β,17α)-17-[(Ethoxycarbonyl)oxy]-11-hydroxy-3-oxoandrosta-1,4-diene-17-carboxylic acid chloromethyl ester; chloromethyl 17α-ethoxycarbonyloxy-11β-hydroxyandrosta-1,4-diene-3-one-17β-carboxylate; 17α-ethoxycarbonyloxy-Δ'-cortienic acid chloromethyl ester; CDDD-5604; HGP-1; P-5604; Alrex; Lotemax. $C_{24}H_{31}ClO_7$; mol wt 466.96. C 61.73%, H 6.69%, Cl 7.59%, O 23.98%. Ophthalmic

corticosteroid. Prepn: N. S. Bodor, **BE 889563** (1981 to Otsuka); *idem*, **US 4996335** (1991). Physicochemical properties: M. Alberth *et al.*, *J. Biopharm. Sci.* **2**, 115 (1991). HPLC determn in plasma and urine: G. Hochhaus *et al.*, *J. Pharm. Sci.* **81**, 1210 (1992). NMR structural studies: S. Rachwal *et al.*, *Steroids* **61**, 524 (1996); *idem et al.*, *ibid.* **63**, 193 (1998). Metabolism and transdermal permeability: N. Bodor *et al.*, *Pharm. Res.* **9**, 1275 (1992). Evaluation of effect on intraocular pressure: J. D. Bartlett *et al.*, *J. Ocul. Pharmacol.* **9**, 157 (1993). Clinical trial in keratoconjunctivitis sicca: S. C. Pflugfelder *et al.*, *Am. J. Ophthalmol.* **138**, 444 (2004). Review of ophthalmic clinical studies: J. F. Howes, *Pharmazie* **55**, 178-183 (2000).

Crystals from THF + hexane, mp 220.5-223.5°. Soly at 25° (mg/ml): 0.0005 in water; 0.037 in 50% propylene glycol + water. Lipophilicity (log K): 3.04.

THERAP CAT: Anti-inflammatory (topical).

5643. Lotrifen. [66535-86-2] 2-(4-Chlorophenyl)-[1,2,4]triazolo[5,1-*a*]isoquinoline; 2-(*p*-chlorophenyl)-*s*-triazolo[5,1-*a*]isoquinoline; L-12717; DL-717-IT; Canocenta; Privaprol. $C_{16}H_{10}$-ClN₃; mol wt 279.73. C 68.70%, H 3.60%, Cl 12.67%, N 15.02%. Non-hormonal antifertility agent. Prepn: **BE 815498**; A. Omodei-Salé *et al.*, **US 4075341** (1974, 1978 both to Lepetit). Pharmacokinetics: G. GalliZani *et al.*, *J. Pharmacobio-Dyn.* **5**, 55 (1981). Pregnancy-terminating effect in dogs: G. Galliani, A. Omodei-Salé, *J. Small Anim. Pract.* **23**, 295 (1982). Effect on subsequent fertility: G. Galliani *et al.*, *IRCS Med. Sci.* **12**, 433, 435 (1984). *Review:* A. Assandri *et al.*, *Rev. Drug Metab. Drug Interact.* **4**, 237 (1982).

Crystals mp 238-240°.

THERAP CAT (VET): Abortifacient.

5644. Lovastatin. [75330-75-5] (2S)-2-Methylbutanoic acid (1S,3R,7S,8S,8aR)-1,2,3,7,8,8a-hexahydro-3,7-dimethyl-8-[2-[(2R,-4R)-tetrahydro-4-hydroxy-6-oxo-2H-pyran-2-yl]ethyl]-1-naphthalenyl ester; (1S,3R,7S,8S,8aR)-1,2,3,7,8,8a-hexahydro-3,7-dimethyl-8-[2-[(2R,4R)-tetrahydro-4-hydroxy-6-oxo-2H-pyran-2-yl]ethyl]-1-naphthalenyl (S)-2-methylbutyrate; 1,2,6,7,8,8a-hexahydro-β,δ-dihydroxy-2,6-dimethyl-8-(2-methyl-1-oxobutoxy)-1-naphthaleneheptanoic acid δ-lactone; 2β,6α-dimethyl-8α-(2-methyl-1-oxobutoxy)mevinic acid lactone; mevinolin; 6α-methylcompactin; monacolin K; MK-803; Mevacor; Mevinacor; Mevlor; Sivlor. $C_{24}H_{36}O_5$; mol wt 404.55. C 71.26%, H 8.97%, O 19.77%. Fungal metabolite; potent inhibitor of HMG-CoA reductase, the rate controlling enzyme in cholesterol biosynthesis. Isoln from *Monascus ruber:* A. Endo, *J. Antibiot.* **32**, 852 (1979); from *Aspergillus terreus:* R. L. Monaghan *et al.*, **US 4231938** (1980 to Merck & Co.). Structure and biochemical properties: A. W. Alberts *et al.*, *Proc. Natl. Acad. Sci. USA* **77**, 3957 (1980). Total synthesis: M. Hirama, M. Iwashita, *Tetrahedron Lett.* **24**, 1811 (1983). Review of syntheses: T. Rosen, C. H. Heathcock, *Tetrahedron* **42**, 4909-4951 (1986). Biosynthesis: M. D. Greenspan, J. B. Yudkovitz, *J. Bacteriol.* **162**, 704 (1985); R. N. Moore *et al.*, *J. Am. Chem. Soc.* **107**, 3694 (1985). HPLC determn in plasma and bile: R. J. Stubbs *et al.*, *J. Chromatogr.* **383**, 438 (1986). Clinical pharmacology: S. M. Grundy, G. L. Vega, *J. Lipid Res.* **26**, 1464 (1985). Clinical comparison with gemfibrozil, *q.v.:* M. J. Tikkanen *et al., Am. J. Cardiol.* **62**, 35J (1988). Review of clinical experience: J. A. Tobert, *Am. J. Cardiol.* **62**, 28J-34J (1988). Comprehensive description: G. S. Brenner *et al., Anal. Profiles Drug Subs. Excip.* **21**, 277-305 (1992). Prevention of acute coronary events in men and women with average cholesterol levels: J. R. Downs *et al., J. Am. Med. Assoc.* **279**, 1615 (1998). Review of development and clinical experience with extended-release niacin, *q.v.:* H. E. Bays, *Expert Rev. Cardiovasc. Ther.* **2**, 485-501 (2004). Review of therapeutic efficacy in high risk populations: A. Frisinghelli, A. Mafrici, *Clin. Drug Invest.* **27**, 591-604 (2007).

White crystals, mp (under N$_2$): 174.5°. [α]$_D^{25}$ +323° (c = 0.5 g in 100 ml acetonitrile). uv max: 231, 238, 247 nm (A$^{1\%}$ 532, 621, 418). Soly at room temp (mg/ml): acetone 47, acetonitrile 28, *n*-butanol 7, *i*-butanol 14, chloroform 350, *N,N*-dimethylformamide 90, ethanol 16, methanol 28, *n*-octanol 2, *n*-propanol 11, *i*-propanol 20, water 0.4 × 10^{-3}. Practically insol in hexane. LD$_{50}$ orally in mice: >1000 mg/kg (Endo).

THERAP CAT: Antilipemic.

5645. Loxapine. [1977-10-2] 2-Chloro-11-(4-methyl-1-piperazinyl)dibenz[*b*, *f*][1,4]oxazepine; oxilapine; CL-62362; S-805; SUM-3170. C$_{18}$H$_{18}$ClN$_3$O; mol wt 327.81. C 65.95%, H 5.53%, Cl 10.81%, N 12.82%, O 4.88%. Prepn: **NL 6406089**; Schmutz *et al.*, **US 3546226** (1964, 1970 both to Wander); *eidem Helv. Chim. Acta* **50**, 245 (1967); Coppola, **US 3412193** (1968 to Am. Cyanamid). Crystal structure: D. B. Cosulich, F. M. Lovell, *Acta Crystallogr.* **33B**, 1147 (1977). Pharmacology: Schmutz *et al., Chim. Ther.* **2**, 424 (1967); Latimer, *J. Pharmacol. Exp. Ther.* **166**, 151 (1969). Toxicity data: Stille *et al., Arzneim.-Forsch.* **15**, 841 (1965). Toxicity studies: Mineshita *et al., Oyo Yakuri* **4**, 293 (1970), *C.A.* **76**, 81145v (1972). Review of pharmacology and therapeutic efficacy: R. C. Heel *et al., Drugs* **15**, 198-217 (1978).

Pale yellowish crystals from petr ether, mp 109-110°. LD$_{50}$ orally in mice: 65 mg/kg (Stille).

Succinate. [27833-64-3] CL-71563; Loxapac; Loxitane. C$_{18}$H$_{18}$ClN$_3$O.C$_4$H$_6$O$_4$; mol wt 445.90.

THERAP CAT: Anxiolytic.

5646. Loxiglumide. [107097-80-3] 4-[(3,4-Dichlorobenzoyl)amino]-5-[(3-methoxypropyl)pentylamino]-5-oxopentanoic acid; (±)-4-(3,4-dichlorobenzamido)-*N*-(3-methoxypropyl)-*N*-pentylglutaramic acid; CR-1505. C$_{21}$H$_{30}$Cl$_2$N$_2$O$_5$; mol wt 461.38. C 54.67%, H 6.55%, Cl 15.37%, N 6.07%, O 17.34%. Cholecystokinin type-1 (CCK-1) antagonist. Prepn: F. Makovec *et al.*, **WO 8703869**; *eidem*, **US 4769389** (1987, 1988 both to Rotta). Pharmacology and receptor binding: I. Setnikar *et al., Arzneim.-Forsch.* **37**, 703 (1987). Pharmacokinetics: *idem et al., ibid.* **38**, 716 (1988). Effect on bilio-pancreatic secretion: W. E. Schmidt *et al., Digestion* **46**, Suppl. 2, 232 (1990). Clinical evaluation in irritable bowel syndrome: P. A. Cann *et al., Ann. N.Y. Acad. Sci.* **713**, 449 (1994); in nonulcer dyspepsia: A. S. B. Chua *et al. ibid.* 451; in pancreatitis: K. Shiratori *et al., Pancreas* **25**, e1 (2002).

(R)-Form

Crystals from acetone, mp 113-115°. pKa ~5. Soly in water: 0.01%.

(R)-Form. [119817-90-2] Dexloxiglumide; CR-2017. HPLC determn in plasma: R. Brodie *et al., J. Chromatogr. B* **784**, 91 (2003). *In vitro* biopharmaceutical properties: S. Tolle-Sander *et al., J. Pharm. Sci.* **92**, 1968 (2003). Clinical pharmacokinetics: C. Webber *et al., Xenobiotica* **33**, 625 (2003). Clinical evaluation in irritable bowel syndrome: F. Cremonini *et al., Am. J. Gastroenterol.* **100**, 652 (2005). Soly (μg/ml): 33 (pH 3.4), 533 (pH 7.5). pKa 4.48.

THERAP CAT: Gastroprokinetic.

5647. Loxoprofen. [68767-14-6] α-Methyl-4-[(2-oxocyclopentyl)methyl]benzeneacetic acid; (±)-*p*-[(2-oxocyclopentyl)methyl]hydratropic acid. C$_{15}$H$_{18}$O$_3$; mol wt 246.31. C 73.15%, H 7.37%, O 19.49%. Non-steroidal anti-inflammatory pro-drug; the active metabolite is the *trans*-cyclohydroxypentane. Prepn of racemate: A. Terada *et al.*, **DE 2814556**; *eidem*, **US 4161538** (1978, 1979 both to Sankyo); of enantiomers: A. Terada, E. Misaka, **EP 55588**; *eidem*, **US 4400534** (1982, 1983 both to Sankyo). Use as topical anti-inflammatory: A. Terada *et al.*, **BE 889149** (1981 to Sankyo). Metabolism: Y. Tanaka *et al., Chem. Pharm. Bull.* **31**, 3656 (1983); *eidem, ibid.* **32**, 1040 (1984). Structure of metabolites: S. Naruto *et al., ibid.* 258. Optical inversion of (2*R*)- to (2*S*)-isomer after oral administration to rats: H. Nagashima *et al., ibid.* 251. Inhibition of *in vivo* and *in vitro* prostaglandin synthesis: K. Matsuda *et al., Biochem. Pharmacol.* **33**, 2473 (1984).

Colorless oil. bp$_{0.3}$ 190-195°. mp 108.5-111°.

Sodium salt. [80382-23-6] CS-600; Loxonin. C$_{15}$H$_{17}$NaO$_3$; mol wt 268.29.

THERAP CAT: Anti-inflammatory; analgesic.

5648. Lubeluzole. [144665-07-6] (α*S*)-4-(2-Benzothiazolylmethylamino)-α-[(3,4-difluorophenoxy)methyl]-1-piperidineethanol; R-87926; Prosynap. C$_{22}$H$_{25}$F$_2$N$_3$O$_2$S; mol wt 433.52. C 60.95%, H 5.81%, F 8.76%, N 9.69%, O 7.38%, S 7.40%. Modulates the glutamate-activated nitric oxide synthase (NOS) pathway. Prepn: R. A. Stokbroekx, G. A. J. Grauwels, **EP 501552**; **US 5434168** (1992, 1995 both to Janssen). Mechanism of action study: M. De Ryck *et al., J. Pharmacol. Exp. Ther.* **279**, 748 (1996); A. S. Lesage *et al., ibid.* 759. Clinical pharmacokinetics: J. De Keyser *et al., Clin. Ther.* **19**, 1340 (1997). Clinical trial in acute ischemic stroke: J. Grotta *et al., Stroke* **28**, 2338 (1997).

Crystals from 2,2'-oxybispropane, mp 65.8°. [α]$_D^{20}$ +4.38° (c = 1 in methanol).

THERAP CAT: Neuroprotective.

5649. Lubiprostone. [333963-40-9] (11α,15R)-11,15-Epoxy-16,16-difluoro-15-hydroxy-9-oxoprostan-1-oic acid; 7-[(1R,-3R,6R,7R)-3-(1,1-difluoropentyl)-3-hydroxy-2-oxabicyclo[4.3.0]-nonane-8-one-7-yl]heptanoic acid; RU-0211; SPI-0211; Amitiza. C$_{20}$H$_{32}$F$_2$O$_5$; mol wt 390.47. C 61.52%, H 8.26%, F 9.73%, O 20.49%. Prostaglandin E$_1$ derivative that enhances intestinal fluid secretion. Locally activates type 2 chloride channels (ClC-2) on apical surface of intestinal epithelial cells in a protein kinase A-independent manner. Exists in a keto-hemiacetal equilibrium; favors tautomeric bicyclic conformation in the absence of water. Prepn: R. Ueno *et al.*, **EP 0284180**; *eidem*, **US 5284858** (1988, 1994 to K. K. Ueno Seiyaku Oyo Kenkyujo). Mechanism of action: J. Cuppoletti *et al.*, *Am. J. Physiol. Cell Physiol.* **287**, C1173 (2004). Clinical trial in irritable bowel syndrome with constipation: D. A. Drossman *et al.*, *Aliment. Pharmacol. Ther.* **29**, 329 (2009). Review of development and clinical experience in chronic constipation: A. Rivkin, L. Chagan, *Clin. Ther.* **28**, 2008-2021 (2006); B. E. Lacy, L. Campbell Levy, *J. Clin. Gastroenterol.* **41**, 345-351 (2007).

White, odorless crystals or crystalline powder. Very sol in ether and ethanol. Practically insol in hexane and water.

Monocyclic tautomer. [136790-76-6] (11α)-16,16-Difluoro-11-hydroxy-9,15-dioxoprostan-1-oic acid; 13,14-dihydro-15-keto-16,16-difluoroprostaglandin E$_1$.

THERAP CAT: Laxative.

5650. Lucanthone. [479-50-5] 1-[[2-(Diethylamino)ethyl]-amino]-4-methyl-9H-thioxanthen-9-one; 1-(2-diethylaminoethyl-amino)-4-methylthioxanthen-9-one. C$_{20}$H$_{24}$N$_2$OS; mol wt 340.49. C 70.55%, H 7.11%, N 8.23%, O 4.70%, S 9.42%. Prepd by the reaction of 1-chloro-4-methylthioxanthone with *asym*-diethylethylenediamine: Mauss, *Naturwissenschaften* **33**, 253 (1946); *idem*, *Ber.* **81**, 19 (1948); Sharp, *J. Chem. Soc.* **1951**, 2961; Archer, Suter, *J. Am. Chem. Soc.* **74**, 4296 (1952). Review of mode of action: Weinstein, Hirschberg, *Prog. Mol. Subcell. Biol.* **2**, 232 (1971). *Review:* Hirschberg in *Antibiotics* vol. 3, J. W. Corcoran, F. E. Hahn, Eds. (Springer-Verlag, New York, 1975) pp 274-303.

mp 64-65°. Sol in the usual organic solvents.

Hydrochloride. [548-57-2] MS-752; RP-3735; Miracil D; Miracol; Nilodin; Tixantone. C$_{20}$H$_{24}$N$_2$OS.HCl; mol wt 376.94. Yellow crystals from alcohol, mp 195-196°. Freely sol in water. Aq soln (orange) is neutral. Slightly sol in alcohol.

THERAP CAT: Anthelmintic (Schistosoma).

5651. Lucensomycin. [13058-67-8] 22-[(3-Amino-3,6-dideoxy-β-D-mannopyranosyl)oxy]-12-butyl-1,3,26-trihydroxy-10-oxo-6,11,28-trioxatricyclo[22.3.1.0^{5,7}]octacosa-8,14,16,18,20-pentaene-25-carboxylic acid; Antibiotic FI 1163; FI-1163; Etruscomicina; Etruscomycin. C$_{36}$H$_{53}$NO$_{13}$; mol wt 707.81. C 61.09%, H 7.55%, N 1.98%, O 29.38%. Polyene antifungal antibiotic isolated from cultures of *Streptomyces lucensis:* Arcamone *et al.*, *G. Microbiol.* **4**,

119 (1957); Arcamone, Perego, *Ann. Chim. (Rome)* **49**, 345 (1959); Marini, Pennella, *Proc. Symp. Antibiotics Prague* (May 1959) p 148; Arcamone *et al.*, **US 3170837** (1965 to Farmitalia). Structure: Guadiano *et al.*, *Tetrahedron Lett.* **1966**, 3559, 3567; *Gazz. Chim. Ital.* **96**, 1470 (1966); *Chim. Ind. (Milan)* **48**, 1327 (1966). Revised structure: R. C. Pandey, K. L. Rinehart, *J. Antibiot.* **29**, 1035 (1976).

Crystalline powder. [α]$_D^{20}$ +296° (pyridine), +50° (methanolic 0.1N HCl). uv max: 218, 278, 290, 303, 318 nm (E$_{1cm}^{1\%}$ 300, 370, 780, 1170, 1098). Practically insol in water, anhydr alcohol, nonpolar solvents. Sol in pyridine, dimethylformamide. Unstable beyond pH 6-8, and to heat, light, or air.

THERAP CAT: Antifungal.

5652. Luciferin. A generic term referring to a substrate which, upon oxidation by the enzyme luciferase, produces bioluminescence. Luciferins isolated from different species may vary greatly in structure, although in many cases identical structures have been found in widely diverse animals. The most widely studied luciferins are those isolated from the sea pansey, *Renilla reniformis*, the ostracod, *Cypridina hilgendorfii*, the limpet, *Latia neritoides*, and the firefly, *Photinus pyralis (see* separate entry, Firefly Luciferin). Structure of *Renilla reniformis* luciferin: K. Hori *et al.*, *Proc. Natl. Acad. Sci. USA* **74**, 4285 (1977). *Review:* M. J. Cormier *et al.*, *Fortschr. Chem. Org. Naturst.* **30**, 1-60 (1973).

5653. Lucifer Yellows. Water soluble fluorescent dyes. Description of properties and potential uses: W. W. Stewart, *Cell* **14**, 741 (1978). Prepn: *idem*, *J. Am. Chem. Soc.* **103**, 7615 (1981). Use in flux kinetics: C. L. Bowman, H. Tedeschi, *Biochim. Biophys. Acta* **731**, 261 (1983); in intercellular communications: M. H. El-Fouly *et al.*, *Exp. Cell Res.* **168**, 422 (1987); as covalent marker: K. M. Lowe, R. E. McCarty, *Biochemistry* **37**, 2507 (1998).

Lucifer Yellow CH   R =

Lucifer Yellow VS   R =

Lucifer Yellow CH. [67769-47-5] 6-Amino-2-[(hydrazinocarbonyl)amino]-2,3-dihydro-1,3-dioxo-1H-benz[*de*]isoquinoline-5,8-disulfonic acid dilithium salt; lucifer yellow carbohydrazide; 4-amino-N-[(hydrazinocarbonyl)amino]-2,3-naphthalimide-3,6-disulfonate. C$_{13}$H$_9$Li$_2$N$_5$O$_9$S$_2$; mol wt 457.24. Fluffy orange hygroscopic powder. uv max (water, excitation max): 280, 428 nm (ε 24200, 11900); emission max: 540 nm. Quantum yield: 0.21.

Lucifer Yellow VS. [71231-14-6] 6-Amino-2-[3-(ethenylsulfonyl)phenyl]-2,3-dihydro-1,3-dioxo-1H-benz[*de*]isoquinoline-5,8-disulfonic acid dilithium salt; lucifer yellow vinylsulfone. $C_{20}H_{12}Li_2$-$N_2O_{10}S_3$; mol wt 550.38. Crystals. uv max (water, excitation max): 280, 428 nm (ε 24600, 12200); emission max: 540 nm. Quantum yield: 0.24.

USE: Fluoroprobe for biological tracing.

5654. Lufenuron. [103055-07-8] *N*-[[[2,5-Dichloro-4-(1,1,-2,3,3,3-hexafluoropropoxy)phenyl]amino]carbonyl]-2,6-difluoro-benzamide; 1-[2,5-dichloro-4-(1,1,2,3,3,3-hexafluoropropoxy)phenyl]-3-(2,6-difluorobenzoyl)urea; fluphenacur; CGA-184699; Program. $C_{17}H_8Cl_2F_8N_2O_3$; mol wt 511.15. C 39.95%, H 1.58%, Cl 13.87%, F 29.73%, N 5.48%, O 9.39%. Chitin synthesis inhibitor. Prepn: J. Drabek, M. Böger, **EP 179022**; *eidem*, **US 4798837** (1986, 1989 both to Ciba-Geigy). Effect on fleas following oral administration in cats: W. F. Hink *et al.*, *J. Med. Entomol.* **28**, 424 (1991). HPLC determn: M. Zakson *et al.*, *Pestic. Sci.* **35**, 117 (1992). Review of properties and field trials in crops: F. Buholzer *et al.*, *Meded. Fac. Landbouwwet. Univ. Gent* **57**, 781-790 (1992).

Colorless, odorless, crystalline solid, mp 174.1°. Vapor pressure (25°): $<4 \times 10^{-6}$ Pa. Soly (20°): water <0.1 ppm; acetone 40%; cyclohexanone 35%; xylene 5%; methanol 4.5%. Log P (*n*-octanol/water): 5.12. LD_{50} in rats (mg/kg): >2000 orally; LC_{50} in rats (mg/m³): >2350 (Buholzer).

THERAP CAT (VET): Ectoparasiticide.

5655. Luliconazole. [187164-19-8] (αE)-α-[(4*R*)-4-(2,4-Di-chlorophenyl)-1,3-dithiolan-2-ylidene]-1*H*-imidazole-1-acetonitrile; (*R*)-(−)-(*E*)-[4-(2,4-dichlorophenyl)-1,3-dithiolan-2-ylidene]-1-imidazolylacetonitrile; NND-502; PR-2699; Lulicon. $C_{14}H_9Cl_2$-N_3S_2; mol wt 354.27. C 47.46%, H 2.56%, Cl 20.01%, N 11.86%, S 18.10%. Antimycotic imidazole that blocks ergosterol biosynthesis by inhibiting sterol 14α-demethylase. Structural analog of lanoconazole, *q.v.* Prepn of (±)-form: N. Hasagawa *et al.*, **JP 60218387**; A. Seo *et al.*, **US 4636519** (1985, 1987 both to Nihon Nohyaku); of *R*-(−)-form: H. Kodama *et al.*, **WO 9702821**; *eidem*, **US 5900488** (1997, 1999 both to Nihon Nohyaku). *In vitro* antifungal activity: Y. Niwano *et al.*, *Antimicrob. Agents Chemother.* **42**, 967 (1998); K. Uchida *et al.*, *J. Infect. Chemother.* **10**, 216 (2004). Mechanism of action, pharmacology, and comparison with lanoconazole: Y. Niwano *et al.*, *Curr. Med. Chem. Anti-infect. Agents* **2**, 147-160 (2003). Comparative clinical trial for topical treatment of tinea pedis: S. Watanabe *et al.*, *Mycoses* **49**, 236 (2006).

Crystals from ethyl acetate + *n*-hexane.
(±)-(*E*)-**Form.** [101530-21-6] Yellow viscous substance, mp 100.4°.

THERAP CAT: Topical antifungal.

5656. Lumazine. [487-21-8] 2,4(1*H*,3*H*)-Pteridinedione. C_6-$H_4N_4O_2$; mol wt 164.12. C 43.91%, H 2.46%, N 34.14%, O 19.50%. Prepn from 4,5-diamino-2,6-dihydoxypyrimidine and glyoxal sodium bisulfate in the presence of ammonia: Weijlard *et al.*, *J. Am. Chem. Soc.* **67**, 804 (1945); **US 2479442** (1950 to Merck & Co.); from 5,6-diaminouracil hydrosulfite + ethylene glycol disulfonic acid disodium salt + HCl: Dallacker, Steiner, *Ann.* **660**, 98 (1962).

Yellow to orange needles from water, mp 348-349°. Soly in water at 25° = 0.125%. Sol in acetic acid. Aq solns have a bluish-green fluorescence when neutral, a green fluorescence when alkaline, and a blue fluorescence when acid.

5657. Lumefantrine. [82186-77-4] (9*Z*)-2,7-Dichloro-9-[(4-chlorophenyl)methylene]-α-[(dibutylamino)methyl]-9*H*-fluorene-4-methanol; 2-dibutylamino-1-[2,7-dichloro-9-(4-chlorobenzylidene)-9,11-fluoren-4-yl]ethanol; *dl*-benflumelol; benflumetol; BFL; CPG-56695. $C_{30}H_{32}Cl_3NO$; mol wt 528.94. C 68.12%, H 6.10%, Cl 20.11%, N 2.65%, O 3.02%. Racemic aryl alcohol originally synthesized in the 1970's by the Academy of Military Medical Sciences in Beijing, China. Inhibits hemozoin formation. Prepn: R. Deng *et al.*, **CN 1042535** (1990 to Acad. Military Med. Sci., Microbiol. & Epidemic Dis. Instit.). LC determn in plasma: A. Annerberg *et al.*, *J. Chromatogr. B* **822**, 330 (2005). In vitro activity against *Plasmodium falciparum*: B. Pradines *et al.*, *Antimicrob. Agents Chemother.* **43**, 418 (1999).

Odorless, yellow powder. Poorly sol in water, oil, and most organic solvents. Sol in unsaturated fatty acids.

Co-artemether. [141204-94-6] CPG-56697; Coartem; Riamet. Fixed 6:1 mixture with artemether, *q.v.* Clinical pharmacokinetics and bioavailability: F. Ezzet *et al.*, *Br. J. Clin. Pharmacol.* **46**, 553 (1998). Clinical trial in children against *P. falciparum* malaria: C. Hatz *et al.*, *Trop. Med. Int. Health* **3**, 498 (1998); in adults: S. Looareesuwan *et al.*, *Am. J. Trop. Med. Hyg.* **60**, 238 (1999). Review of comparative clinical trials in malaria: A. A. Omari *et al.*, *Trop. Med. Int. Health* **9**, 192-199 (2004).

THERAP CAT: Antimalarial.

5658. Lumichrome. [1086-80-2] 7,8-Dimethylbenzo[*g*]-pteridine-2,4-(1*H*,3*H*)-dione; 7,8-dimethylalloxazine; 6,7-dimethylalloxazine. $C_{12}H_{10}N_4O_2$; mol wt 242.24. C 59.50%, H 4.16%, N 23.13%, O 13.21%. Irradiation product of riboflavine. Isoln: Karrer *et al.*, *Helv. Chim. Acta* **17**, 1010 (1934). Prepn: Tishler, Wellman, **US 2417143** (1947 to Merck & Co.); Cresswell, Wood, *J. Chem. Soc.* **1960**, 4768; Bardos *et al.*, **US 3057865** (1962 to Armour Pharm. Co.); Seng, Ley, *Angew. Chem.* **84**, 1061 (1972).

Pale yellow crystals from pyridine or pyridine + alcohol, mp $>300°$. Sparingly sol in methanol and 90% hot ethanol, water and chloroform. The aqueous, alcoholic and chloroformic solns fluoresce blue. Absorption spectrum: Karrer *et al.*, *loc. cit.*

5659. Lumiflavine. [1088-56-8] 7,8,10-Trimethylbenzo[g]-pteridine-2,4(3H,10H)-dione; 7,8,10-trimethylisoalloxazine. $C_{13}H_{12}N_4O_2$; mol wt 256.27. C 60.93%, H 4.72%, N 21.86%, O 12.49%. Irradiation product of riboflavine. Isoln: Warburg, Christian, *Biochem. Z.* **266**, 377 (1933). Structure: Kuhn, Rudy, *Ber.* **67**, 1298 (1934). Prepn: Tishler, Wellman, US 2417143 (1947 to Merck & Co.); Birch, Moye, *J. Chem. Soc.* **1958**, 2622; Yoneda *et al., Chem. Pharm. Bull.* **20**, 1832 (1972).

Orange crystals from 12% acetic acid, mp 320°. Purified by sublimation in vacuum, mp 330°. Freely sol in chloroform; very sparingly sol in water and most organic solvents. Aq and chloroformic solns fluoresce green. uv max: 269, 355, 445 nm (ε 38,800, 11,700, 11,800) in 0.1N NaOH and 264, 373, 440 nm (ε 34,700, 11,400, 10,400) in 0.1N HCl.

5660. Luminol. [521-31-3] 5-Amino-2,3-dihydro-1,4-phthalazinedione; o-aminophthalylhydrazide; 3-aminophthalic hydrazide; o-aminophthaloyl hydrazide. $C_8H_7N_3O_2$; mol wt 177.16. C 54.24%, H 3.98%, N 23.72%, O 18.06%. Prepn from 3-nitrophthalic acid: Huntress *et al., J. Am. Chem. Soc.* **56**, 241 (1934); Redemann, Redemann, *Org. Synth.* **29**, 78, 8 (1949); L. Fieser, *Organic Experiments* (Boston, 1964) pp 240-242. Chemiluminescent: Weber, *Ber.* **75**, 565 (1942); Weber *et al., ibid.* **76**, 366 (1943).

Crystals, mp 319-320°. Oxidation of luminol is accompanied by a striking emission of light.
Note: Not to be confused with Luminal.
USE: Detection of copper, iron, peroxides, cyanides.

5661. Lumiracoxib. [220991-20-8] 2-[(2-Chloro-6-fluorophenyl)amino]-5-methylbenzeneacetic acid; 5-methyl-2-(2'-chloro-6'-fluoroanilino)phenylacetic acid; COX-189; Prexige. $C_{15}H_{13}ClFNO_2$; mol wt 293.72. C 61.34%, H 4.46%, Cl 12.07%, F 6.47%, N 4.77%, O 10.89%. Selective cyclooxygenase-2 (COX-2) inhibitor. Prepn: R. A. Fujimoto *et al.,* WO 9911605; *eidem,* US 6310099 (1999, 2001 both to Novartis). Clinical evaluation of gastroduodenal tolerability: C. Rordorf *et al., Aliment. Pharmacol. Ther.* **18**, 533 (2003). Clinical experience in dysmenorrhea: M. Bitner *et al., Int. J. Clin. Pract.* **58**, 340 (2004); in rheumatoid arthritis: P. Geusens *et al. ibid.* 1033; in tension headache: E. Packman *et al., Headache* **45**, 1163 (2005). Review of clinical pharmacology: C. M. Rordorf *et al., Clin. Pharmacokinet.* **44**, 1247-1266 (2005); of clinical experience in osteoarthritis: F. Berenbaum *et al., J. Int. Med. Res.* **33**, 21-41 (2005); in osteoarthritis, rheumatoid arthritis, and dental pain: T. J. Schnitzer *et al., Curr. Med. Res. Opin.* **21**, 151-161 (2005).

mp 158-159°.
THERAP CAT: Anti-inflammatory.

5662. Lumisterol. [474-69-1] (3β,9β,10α,22E)-Ergosta-5,7,22-trien-3-ol. $C_{28}H_{44}O$; mol wt 396.66. C 84.78%, H 11.18%, O 4.03%. Differs from ergosterol by spatial arrangement involving the methyl group at C-10. Prepd by ultraviolet irradiation of a benzene-alcohol soln of ergosterol: Askew *et al., Proc. R. Soc. London* **B109**, 488 (1932). Stereochemistry: Castells *et al., J. Chem. Soc.* **1959**, 1159. Dehydrogenation with selenium gives Diel's hydrocarbon, 3'-methyl-1,2-cyclopentenophenanthrene.

Needles from acetone-methanol, mp 118°. $[\alpha]_D^{19}$ +191.5°, $[\alpha]_{546}^{19}$ +235.4° (c = 2 in acetone). uv max: 265, 280 nm: Windaus *et al., Ann.* **493**, 259 (1932); Heilbron *et al., J. Chem. Soc.* **1937**, 411. Practically insol in water. Sol in organic solvents. Forms a monomolecular compd with calciferol, mp 122°.
Acetate. $C_{30}H_{46}O_2$. mp 100°. $[\alpha]_D^{19}$ +130.5°, $[\alpha]_{546}^{19}$ +163° (c = 1.8 in acetone).

5663. Lunacridine. [83-58-9] 3-[(2R)-2-Hydroxy-3-methyl-butyl]-4,8-dimethoxy-1-methyl-2(1H)-quinolinone. $C_{17}H_{23}NO_4$; mol wt 305.37. C 66.87%, H 7.59%, N 4.59%, O 20.96%. From bark of *Lunasia costulata* Miq., and *L. quercifolia* K. Schum. et Lauterb., *Rutaceae.* (+)-Form is naturally occurring. Isoln: Boorsma, *Bull. Inst. Bot. Buitenzorg* **21**, 8 (1904). Structure: Goodwin, Horning, *J. Am. Chem. Soc.* **81**, 1908 (1959). Synthesis: Clarke, Grundon, *J. Chem. Soc.* **1964**, 438. Synthesis and stereochemistry: R. M. Bowman *et al., J. Chem. Soc. Perkin Trans. 1* **1973**, 1051. Short synthesis of racemate: M. Ramesh, P. Shanmugam, *Indian J. Chem.* **24B**, 767 (1985).

Crystals from methanol/water, mp 86-87°. $[\alpha]_{589}^{25}$ +28.1°; $[\alpha]_{436}^{25}$ +76.5° (c = 0.935 in ethanol). uv spectra: Goodwin, Horning, *loc. cit.* Very slightly sol in water; freely sol in alcohol, benzene, chloroform, ether, ethyl acetate, carbon disulfide, petr ether.
Hydroperchlorate. $C_{17}H_{24}ClNO_8$. Shiny flakes from methanol/ether, mp 146-148°; remelts at 193-195°. $[\alpha]_{589}^{25}$ +22.3°; $[\alpha]_{436}^{25}$ +60.7° (c = 0.750 in ethanol).
(±)-**Form.** [52306-35-1] Prisms from petr ether, mp 72-74°. uv max (methanol): 239, 258, 285, 292, 333 nm (ε 24000, 26300, 8300, 7800, 3100).

5664. Lunacrine. [82-40-6] (2R)-3,9-Dihydro-8-methoxy-9-methyl-2-(1-methylethyl)furo[2,3-b]quinolin-4(2H)-one. $C_{16}H_{19}NO_3$; mol wt 273.33. C 70.31%, H 7.01%, N 5.12%, O 17.56%. From bark of *Lunasia costulata* Miq., and *L. amara* Blanco, *Rutaceae.* Isoln: Boorsma, *Bull. Inst. Bot. Buitenzorg* **6**, 15 (1900). Structure: Goodwin, Horning, *J. Am. Chem. Soc.* **81**, 1908 (1959). Synthesis: Clarke, Grundon, *J. Chem. Soc.* **1964**, 438. Stereochemistry: Bowman *et al., J. Chem. Soc. Perkin Trans. 1* **1973**, 1051.

Fine needles from ethyl acetate, mp 117-119°. $[\alpha]_{589}^{25}$ −50.4°; $[\alpha]_{436}^{25}$ −116° (c = 0.806 in ethanol). Practically insol in water. Freely sol in alcohol, benzene, chloroform, ether, carbon disulfide, ethyl acetate; slightly sol in petr ether.

dl-Form. [95723-59-4] Prisms from ethyl acetate + pentane, mp 146-148°. $[\alpha]_{589}^{25}$ +2.3°; $[\alpha]_{436}^{25}$ +4.2° (c = 0.864 in ethanol).

5665. Lunasine. [6901-22-0] (2R)-2,3-Dihydro-4,8-dimethoxy-9-methyl-2-(1-methylethyl)furo[2,3-*b*]quinolinium. [$C_{17}H_{22}$-NO_3]$^+$. From bark of *Lunasia costulata* Miq., and *L. quercifolia* K. Schum. et Lauterb., *Rutaceae*. Isoln: Boorsma, *Bull. Inst. Bot. Buitenzorg* **6**, 14 (1900). Structure: Price, *Aust. J. Chem.* **12**, 458 (1959). Stereochemistry: Bowman *et al.*, *J. Chem. Soc. Perkin Trans. 1* **1973**, 1051.

(−)-Form perchlorate. [121543-68-8] $C_{17}H_{22}NO_3.ClO_4$. Microcrystalline powder from methanol + ether, mp 195-196°. $[\alpha]_D^{20}$ −29.3° (c = 0.46 in methanol).

5666. Lunularic Acid. [23255-59-6] 2-Hydroxy-6-[2-(4-hydroxyphenyl)ethyl]benzoic acid; 6-(*p*-hydroxyphenethyl)salicylic acid. $C_{15}H_{14}O_4$; mol wt 258.27. C 69.76%, H 5.46%, O 24.78%. Dihydrostilbene growth inhibitor isolated from liverworts and algae. Isoln from *Lunularia cruciata* (L.) Dum., *Lunulariaceae*, and characterization: I. F. M. Valio *et al.*, *Nature* **223**, 1178 (1969); I. F. M. Valio, W. W. Schwab, *J. Exp. Bot.* **21**, 138 (1970); from more than 70 species of liverwort: J. Gorham, *Phytochemistry* **16**, 249 (1977); from cultures of *Marchantia polymorpha*: S. Abe, Y. Ohta, *ibid.* **22**, 1917 (1983). Synthesis: Y. Arai *et al.*, *Tetrahedron Lett.* **1972**, 1615; S. Haneck, K. Schreiber, **DD 126866** (1977 to Akad. Wissenschaft. D.D.R.). Effect on liverwort growth: J. Gorham, *Phytochemistry* **17**, 99 (1978). Metabolism by *L. cruciata*: R. J. Pryce, *ibid.* **11**, 1355 (1972).

Pale yellow crystals from methanol/water, mp 192°. uv max in neutral ethanol: 280, 287, 308 nm (ε 3300, 3600, 4200); in weakly alkaline ethanol: 300 nm (ε 6600).

USE: Growth inhibitor for lower plants.

5667. Lupanine. (7α,7aα,14α,14aβ)-Dodecahydro-7,14-methano-2H,11H-dipyrido[1,2-*a*:1′,2′-*e*][1,5]diazocin-11-one; dodecahydro-7,14-methano-4H,6H-dipyrido[1,2-*a*:1′,2′-*e*][1,5]-diazocin-4-one. $C_{15}H_{24}N_2O$; mol wt 248.37. C 72.54%, H 9.74%, N 11.28%, O 6.44%. Racemic lupanine is found in white lupins, *d*-lupanine is found in blue lupins, *l*-lupanine has been prepd from the natural racemic form. Structure: Davis, *Arch. Pharm.* **235**, 199, 218, 229 (1897); Clemo, Leitch, *J. Chem. Soc.* **1928**, 1811; Clemo *et al.*, *ibid.* **1931**, 429; *ibid.* **1933**, 644; Ing, *ibid.* **1933**, 504; Karrer *et al.*, *Helv. Chim. Acta* **11**, 1062 (1928); *ibid.* **13**, 1292 (1930); Winterfeld, Holschneider, *Ber.* **64**, 137, 2415 (1931), *ibid.* **66**, 1751

(1933); *Ann.* **499**, 109 (1932); Winterfeld, Kneuer, *Ber.* **64**, 150 (1931); Winterfeld, Hoffman, *Arch. Pharm.* **275**, 5, 65 (1937). Abs config of *l*-form: Okuda *et al.*, *Chem. Ind. (London)* **1961**, 1116. Crystal structure of *dl*-form: H. Doucerain *et al.*, *Acta Crystallogr.* **B32**, 3213 (1976). Biosynthetic study of *d*-form: W. M. Golebewski, I. D. Spenser, *J. Am. Chem. Soc.* **106**, 7925 (1984). *See also* sparteine, *q.v.*

d-form

dl-Form. [4356-43-8] Orthorhombic prisms from acetone, mp 98-99°. bp$_{1.0}$ 185-195°. Sol in water, alcohol, ether, chloroform. Insol in petr ether. *See*: Soldaini, *Arch. Pharm.* **231**, 321 (1893); Couch, *J. Am. Chem. Soc.* **56**, 1423 (1934).

Dihydrochloride. $C_{15}H_{24}N_2O.2HCl$. Deliquescent prisms, mp 185° (dec).

Hydrochloride dihydrate. $C_{15}H_{24}N_2O.HCl.2H_2O$. Crystals, mp 177-178°. mp 250-252° (dry).

d-Form. [550-90-3] 2-Oxosparteine. Syrup crystallizing difficultly in hygroscopic needles, mp 40-44°. bp$_3$ 190-193°. n_D^{24} 1.5444. $[\alpha]_D^{25}$ +84° (c = 4.8 in alc). Freely sol in water, alcohol, chloroform, ether. Sol in petr ether.

Hydrochloride dihydrate. $C_{15}H_{24}N_2O.HCl.2H_2O$. Rhombic crystals from water, mp 127° (dry).

l-Form. [486-88-4] Hydrorhombinine. Viscous oil, bp$_{1.0}$ 186-188°. $[\alpha]_D$ about −61° in acetone.

α-Isolupanine. [486-87-3] *cis-cis*-Isomer of lupanine: Marion, Leonard, *Can. J. Chem.* **29**, 355 (1951). Isoln of *d*-form from *Lupinus sericeus* Pursh., *L. perennis* L. and *L. augustifolius*, *Leguminosae*: Marion *et al.*, *ibid.* **31**, 181 (1953); Rink, Schäfer, *Arch. Pharm.* **287**, 290 (1954); Winterfeld, Pies, *ibid.* **290**, 537 (1957). Needles from petr ether, mp 75-76°. $[\alpha]_D^{26}$ +39° (c = 0.77). $[\alpha]_D^{26}$ +65.9° (c = 3.4 in abs ethanol).

5668. Lupeol. [545-47-1] (3β)-Lup-20(29)-en-3-ol; monogynol B; β-viscol; fagarasterol. $C_{30}H_{50}O$; mol wt 426.73. C 84.44%, H 11.81%, O 3.75%. Abundant plant triterpene. Occurs in the skin of lupin seeds, in chicle, in the latex of fig trees and of rubber plants. Was detected in cocoons of *Bombyx mori*. Isoln: Cohen, *Rec. Trav. Chim.* **28**, 368 (1909); Ruzicka *et al.*, *Helv. Chim. Acta* **20**, 1567 (1937). Identity with β-viscol: Meyer, Jeger, *ibid.* **31**, 1868 (1948). Identity with monogynol B: Chatterji *et al.*, *J. Sci. Ind. Res.* **18B**, 262 (1959). Structure: Ruzicka *et al.*, *Helv. Chim. Acta* **28**, 942 (1945); Ames *et al.*, *J. Chem. Soc.* **1951**, 450. Structure of hydrochloride: Halsall *et al.*, *ibid.* **1952**, 2862. Configuration: Barton, Holmes, *ibid.* **1952**, 78; Djerassi *et al.*, *J. Am. Chem. Soc.* **77**, 5330 (1955). NMR studies: Buckley *et al.*, *Chem. Ind. (London)* **1971**, 298. Total synthesis: Stork *et al.*, *J. Am. Chem. Soc.* **93**, 4945 (1971).

Needles from alcohol or acetone, mp 215°. $[\alpha]_D^{20}$ +27.2° (c = 4.8 in chloroform). Freely sol in ether, benzene, petr ether, warm alcohol. Practically insol in water, dil acid and alkalies.

Acetate. $C_{32}H_{52}O_2$. Needles from acetone, mp 218°. $[\alpha]_D^{20}$ +47.3° (c = 2 in chloroform).

Benzoate. $C_{37}H_{54}O_2$. Prisms from acetone, mp 273-274°. $[\alpha]_D^{20}$ +61° (0.78 g in 25 ml chloroform).

Hydrochloride. $C_{30}H_{50}O \cdot HCl$. Needles from ethanol, mp 211-212°. $[\alpha]$ −31° (c = 1.1 in chloroform).

5669. Lupinine. [486-70-4] (1R,9aR)-Octahydro-2H-quinolizine-1-methanol; l-lupinine; (−)-lupinine. $C_{10}H_{19}NO$; mol wt 169.27. C 70.96%, H 11.31%, N 8.27%, O 9.45%. Naturally occurring l-form isolated from seeds and herb of *Lupinus luteus* L. and other *L.* species, *Leguminosae* also found in *Anabasis aphylla* L., *Chenopodiaceae*. Extraction procedure: J. F. Couch, *J. Am. Chem. Soc.* **56**, 2434 (1934). Structure and synthesis: R. W. Willstätter, E. Fourneau, *Ber.* **35**, 1910 (1902); P. Karrer *et al.*, *Helv. Chim. Acta* **11**, 1062 (1928); K. Winterfeld, F. W. Holschneider, *Ber.* **64B**, 137, 692 (1931); K. Winterfeld, *ibid.* 692. Synthesis of racemic lupinine: F. W. Holschneider, K. Winterfeld, *Arch. Pharm.* **277**, 192 (1939); G. C. Gerrans *et al.*, *Tetrahedron Lett.* **1975**, 4171; T. Iwashita *et al.*, *J. Org. Chem.* **47**, 230 (1982). Synthesis of racemic lupinine and *epi-lupinine*: G. R. Clemo *et al.*, *J. Chem. Soc.* **1937**, 965; H. Takahata *et al.*, *Chem. Pharm. Bull.* **34**, 4523 (1986); of *l-epi-lupinine*: M. L. Bremmer, S. M. Weinreb, *Tetrahedron Lett.* **24**, 261 (1983). Absolute configuration of (−)-form: R. C. Cookson, *Chem. Ind. (London)* **1953**, 339. Crystal structure: A. Koziol *et al.*, *Acta Crystallogr.* **B34**, 3491 (1978). Biosynthesis: Soucek, Schutte, *Angew. Chem.* **74**, 901 (1962); W. M. Golebiewski, I. D. Spenser, *J. Am. Chem. Soc.* **106**, 1441 (1984).

Stout, orthorhombic prisms from acetone, mp 68.5-69.2°. bp$_4$ 160-164°; bp$_{755}$ 269-270°. $[\alpha]_D^{26}$ −25.9° (c = 3 in water); $[\alpha]_D^{28}$ −21° (c = 9.5 in alcohol). Sol in water, alcohol, chloroform, ether. It is a strong base.

l-Form hydrochloride. [6113-09-3] $C_{10}H_{19}NO \cdot HCl$; mol wt 205.73. Orthorhombic prisms, mp 208-213°. $[\alpha]_D$ −14°.

dl-Form. Crystals from acetone, mp 58.5-59.5°.

5670. Luprostiol. [73523-00-9] (5Z)-7-[(1S,2R,3R,5S)-2-[[(2S)-3-(3-Chlorophenoxy)-2-hydroxypropyl]thio]-3,5-dihydroxycyclopentyl]-5-heptenoic acid; 9α,11α,15-trihydroxy-16-m-chlorophenoxy-13-thia-17,18,19,20-tetranor-5-prostenoic acid; prostianol; EMD-34946; Prosolvin; Reprodin. $C_{21}H_{29}ClO_6S$; mol wt 444.97. C 56.68%, H 6.57%, Cl 7.97%, O 21.57%, S 7.20%. Analog of prostaglandin $F_{2\alpha}$, q.v. Prepn: J. Kraemer *et al.*, **DE 2513371** (1976 to E. Merck); *eidem*, **US 4309441** (1982 to E. Merck). Luteolytic study in cattle: H. De Vries, H. Feenstra, *J. Vet. Pharmacol. Ther.* **2**, 223 (1979). Estrus cycle regulation: D. Schams, H. Karg, *Theriogenology* **17**, 499 (1982).

THERAP CAT (VET): Luteolytic.

5671. Lupulin. Glandular trichomes separated from strobiles of *Humulus lupulus* L., *Moraceae* (hops). *Habit.* Europe, Asia, North America; widely cultivated. *Constit.* Lupamaric acid, humulol, about 3% volatile oil; resin, choline, wax (myricin), tannin, asparagin. Not less than 60% of lupulin is sol in ether.

THERAP CAT: Aromatic bitter, sedative.

THERAP CAT (VET): Formerly in nymphomania.

5672. Lupulon. [468-28-0] 3,5-Dihydroxy-2,6,6-tris(3-methyl-2-buten-1-yl)-4-(3-methyl-1-oxobutyl)-2,4-cyclohexadien-1-one; β-bitter acid; β-lupulic acid. $C_{26}H_{38}O_4$; mol wt 414.59. C 75.32%, H 9.24%, O 15.44%. Antimicrobial constituent of hops. Isoln from commercial hops: Bungener, *Bull. Soc. Chim. Fr.* [2] **45**, 487 (1886); Barth, Lintner, *Ber.* **31**, 2022 (1898); Wöllmer, *Ber.* **49**, 780 (1916); Lewis, *et al.*, *J. Clin. Invest.* **28**, 916 (1949). Structure: Wöllmer, *Ber.* **58**, 672 (1925); Wieland, *ibid.* **102**, 2012; Govaert, Verzele, *Bull. Soc. Chim. Belg.* **58**, 432 (1949); Riedl, *Ber.* **85**, 692 (1952). Toxicity study: R. Hänsel, H. H. Wagener, *Arzneim.-Forsch.* **17**, 79 (1967).

Prisms from 90% methanol, mp 92-94°. Bitter taste esp in alc soln. Turns yellow and amorphous within a few days with development of an odor. Perfectly stable *in vacuo* even at 60°. Slightly acid reaction. Monobasic acid. Optically inactive. Soluble in methanol, ethanol, petr ether, hexane, isooctane. Slightly sol in neutral or acidic aq solns. Forms a sodium salt which is readily sol in water. The addition of 0.1% ascorbic acid exerts a marked protective action on the bacteriostatic activity of lupulon steamed or autoclaved at a concentration of 4 ppm in phosphate buffers at pH 6.5 and 8.5. LD_{50} in mice, rats (mg/kg): 525, 100 orally (Hänsel, Wagener).

5673. Lurasidone. [367514-87-2] (3aR,4S,7R,7aS)-2-[[(1R,2R)-2-[[4-(1,2-Benzisothiazol-3-yl)-1-piperazinyl]methyl]cyclohexyl]methyl]hexahydro-4,7-methano-1H-isoindole-1,3(2H)-dione. $C_{28}H_{36}N_4O_2S$; mol wt 492.68. C 68.26%, H 7.37%, N 11.37%, O 6.49%, S 6.51%. Combined serotonin (5HT$_2$) and dopamine (D$_2$) receptor antagonist. Prepn: I. Saji *et al.*, **EP 464846**; *eidem*, **US 5532372** (1992, 1996 both to Sumitomo). Receptor binding and pharmacology: T. Ishibashi *et al.*, *J. Pharmacol. Exp. Ther.* **334**, 171 (2010). Clinical evaluation in acute schizophrenia: M. Nakamura *et al.*, *J. Clin. Psychiatry* **70**, 829 (2009). Review of pharmacology and clinical experience: J. M. Meyer *et al.*, *Expert Opin. Invest. Drugs* **18**, 1715-1726 (2009).

Hydrochloride. [367514-88-3] SM-13496; Latuda. $C_{28}H_{36}N_4O_2S \cdot HCl$; mol wt 529.14. White to off-white powder. mp 215-217°. Sparingly sol in methanol; slightly sol in ethanol; very slightly sol in water, acetone. Practically insol or insol in 0.1N HCl, toluene.

THERAP CAT: Antipsychotic.

5674. Luteinizing Hormone. [9002-67-9] LH; ICSH; interstitial cell stimulating hormone. Gonadotropic hormone secreted by the anterior pituitary under the regulation of LH-RH, q.v. Together with follicle-stimulating hormone, q.v., is critical to the growth and maturity of ovarian follicles. Stimulates androgen production by ovarian theca cells that is converted to estrogen by granulosa cells. Induces ovulation and the development of the corpus luteum. In males, stimulates production of testosterone by Leydig cells of the testes. Structure is a heterodimer of non-covalently linked α and β subunits; analogous with FSH, TSH and chorionic gonadotropin, q.q.v. Isoln from sheep pituitaries: C. H. Li *et al.*, *Endocrinology* **27**, 803 (1940); *eidem*, *Science* **92**, 355 (1940); *eidem*, *J. Am. Chem.*

Soc. **64**, 367 (1942). Isoln from pig pituitaries: T. Shedlovsky *et al.*, *Science* **92**, 178 (1940); B. F. Chow *et al.*, *Endocrinology* **30**, 650 (1942). Complete amino acid sequence of ovine LH subunits: H. Papkoff *et al.*, *J. Am. Chem. Soc.* **93**, 1531 (1971); W.-K. Liu *et al.*, *J. Biol. Chem.* **247**, 4351, 4365 (1972); of human LH subunits: H. T. Keutmann *et al.*, *Endocr. Res. Commun.* **5**, 57 (1978); *eidem*, *Biochem. Biophys. Res. Commun.* **90**, 842 (1979). Reviews of structural studies: D. N. Ward *et al.*, *Recent Prog. Horm. Res.* **29**, 533-561 (1973); H. Papkoff *et al.*, *ibid.* 563-590. Review of physiological roles: M. Filicori, *Fertil. Steril.* **71**, 405-414 (1999); C. V. Rao, *ibid.* **76**, 1097-1100 (2001); of clinical experience in ovulation induction: M. Filicori, G. E. Cognigni, *J. Clin. Endocrinol. Metab.* **86**, 1437-1441 (2001); Z. Shoham, *Fertil. Steril.* **77**, 1170-1177 (2002).

White powder. Isoelectric point 4.6 (sheep), 7.45 (pig). Soluble in water. Chymotrypsin, trypsin and pepsin destroy the gonadotropic action of LH. Picrolonic, flavianic, picric, and trichloroacetic acids precipitate LH with retention of its physiological activity.

Lutropin alfa. [152923-57-4] Luteinizing hormone (human α-subunit reduced) complex with luteinizing hormone (human β-subunit reduced) glycoform α; Luveris. Recombinant human LH produced in genetically engineered Chinese hamster ovary cells. Review of prepn: S. Chappel, *Int. Congr. Symp. Sem. Ser.* **12**, 147-152 (1997). Series of articles on pharmacokinetics and bioavailability: *Fertil. Steril.* **69**, 189-209 (1998). Clinical trial in assisted reproductive procedures: E. Loumaye *et al.*, *J. Clin. Endocrinol. Metab.* **86**, 2607 (2001).

THERAP CAT: Gonad-stimulating principle.

5675. Luteolin. [491-70-3] 2-(3,4-Dihydroxyphenyl)-5,7-dihydroxy-4*H*-1-benzopyran-4-one; 3′,4′,5,7-tetrahydroxyflavone; digitoflavone; cyanidenon 1470. C$_{15}$H$_{10}$O$_6$; mol wt 286.24. C 62.94%, H 3.52%, O 33.54%. Flavonoid found in many fruits, vegetables, and medicinal plants. Exhibits anti-oxidant, anti-inflammatory, and anticancer activities. Structural studies: Perkin, *J. Chem. Soc.* **69**, 800 (1896); Fleischer, *Ber.* **32**, 1186 (1899); Perkin, Horsfall, *J. Chem. Soc.* **77**, 1315 (1900); Hayashi, Inoue, *Acta Phytochim.* **15**, 53 (1949). Identity with digitoflavone: Kiliani, Mayer, *Ber.* **34**, 3577 (1901). Synthesis: Hutchins, Wheeler, *J. Chem. Soc.* **1939**, 91. Review of potential for cancer treatment and prevention: Y. Lin *et al.*, *Curr. Cancer Drug Targets* **8**, 634-646 (2008); of bioactivities and distribution in plants: M. López-Lázaro, *Mini-Rev. Med. Chem.* **9**, 31-59 (2009).

Monohydrate. Yellow needles from alc, dec 328-330°. Sublimes in high vacuum. Sparingly sol in water. Sol in alkalies forming yellow solns.
5-Glucoside. Galuteolin. C$_{21}$H$_{20}$O$_{11}$. From seeds of *Galega officinalis* L., *Leguminosae:* Barger, White, *Biochem. J.* **17**, 836 (1923). Structure: Nakamura, Hukuti, *J. Pharm. Soc. Jpn.* **60**, 449 (1940); *C.A.* **34**, 7910³ (1940); Nordström, Swain, *J. Chem. Soc.* **1953**, 2764. Yellow needles from hot dil alc, dec 280°. Practically insol in water. Slightly sol in abs alcohol; sol in hot dil alcohol.
7-Glucoside. Cynaroside. C$_{21}$H$_{20}$O$_{11}$. From *Achillea millefolium* L., *Compositae:* Hörhammer *et al.*, *Acta Chim. Acad. Sci. Hung.* **40**, 463 (1964). Yellow needles from alc, mp 254-256°. uv max (CH$_3$OH): 350, 255 nm (log ε 4.30, 4.27).

5676. Lutetium. [7439-94-3] Lutecium. Lu; at. wt 174.9668; at. no. 71; valence 3. A lanthanide; belongs to the yttrium group of rare earth metals. Naturally occurring isotopes (mass numbers): 175 (97.41%); 176 (2.59%), radioactive, T½ 3.59 × 10¹⁰ years, β⁻ emitter. Known artificial radioactive isotopes: 151; 153-174; 177-181. Abundance in earth's crust: 0.75-0.8 ppm. Occurs in xenotime, gadolinite and other rare earth minerals. Discovered independently: Urbain, *Compt. Rend.* **145**, 759 (1907); **146**, 406 (1908); **152**, 141 (1911); and called *cassiopeium*: von Welsbach, *Sitzungsber. Akad.*

Wiss. Wien **1907**, 468; *idem*, *Monatsh. Chem.* **29**, 181 (1908); **30**, 695 (1909). Isoln in form of salt: Prandtl, *Z. Anorg. Chem.* **238**, 321 (1938); Marsh, *J. Chem. Soc.* **1952**, 4804. Sepn by ion exchange: Spedding *et al.*, *J. Am. Chem. Soc.* **76**, 2557 (1954). Toxicity study: Haley, *J. Pharm. Sci.* **54**, 663 (1965). Reviews of prepn, properties and compds: *The Rare Earths*, F. H. Spedding, A. H. Daane, Eds. (Krieger, Huntington, N.Y., 1971, reprint of 1961 ed.) 641 pp; Hulet, Bode, "Separation Chemistry of the Lanthanides and Transplutonium Actinides" in *MTP Int. Rev. Sci.: Inorg. Chem., Ser. One* vol. 7, K. W. Bagnall, Ed. (University Park Press, Baltimore, 1972) pp 1-45; Moeller, "The Lanthanides" in *Comprehensive Inorganic Chemistry* vol. 4, J. C. Bailar, Jr. *et al.*, Eds. (Pergamon Press, Oxford, 1973) pp 1-101; F. H. Spedding in *Kirk-Othmer Encyclopedia of Chemical Technology* vol. 19 (John Wiley & Sons, New York, 3rd ed., 1982) pp 833-854; *Chemistry of the Elements*, N. N. Greenwood, A. Earnshaw, Eds. (Pergamon Press, New York, 1984) pp 1423-1449. Brief review of properties: G. T. Seaborg, *Radiochim. Acta* **61**, 115-122 (1993).

Silvery-white metal. Hexagonal close-packed crystals, d 9.8404. mp 1663°. bp 3402°. Heat of fusion: 18.65 kJ/mol. Heat of sublimation (25°): 427.6 kJ/mol.
Oxide. Lu$_2$O$_3$. Cubic crystals.
Chloride. LuCl$_3$. Colorless crystals. Sublimes above 750°. mp 892 ±2°; sol in water. LD$_{50}$ in mice: 315 mg/kg i.p.; 7.1 g/kg orally (Haley).
Sulfate. Lu$_2$(SO$_4$)$_3$. Octahydrate, soly in water (g/100 g): 42.27 (20°); 16.93 (40°).

5677. 2,6-Lutidine. [108-48-5] 2,6-Dimethylpyridine; α,α′-lutidin. C$_7$H$_9$N; mol wt 107.16. C 78.46%, H 8.47%, N 13.07%. Isolated from the basic fraction of coal tar: Heap *et al.*, *J. Am. Chem. Soc.* **43**, 1936 (1921); from a bone oil fraction: Ladenburg, Roth, *Ber.* **18**, 51 (1885). Synthesis from ethyl acetoacetate, formaldehyde, and ammonia: Singer, McElvain, *Org. Synth.* **coll. vol. II**, 214 (1943).

Oily liquid. Odor of pyridine + peppermint. d$_4^{20}$ 0.9252. mp −5.8°. bp$_{760}$ 144°; bp$_{87}$ 79°. n$_D^{20}$ 1.49797. Solubility in water at 45.3° = 27.2% (w/w); at 48.1° = 18.1%; at 57.5° = 12.1%; at 74.5° = 9.5%. Also sol in alcohol, ether. Miscible with dimethylformamide and tetrahydrofuran.

5678. Lyapolate Sodium. [25053-27-4] Ethenesulfonic acid homopolymer sodium salt; sodium lyapolate; sodium apolate; sodium polyethylene sulfonate; polyethylene sodium sulfonate; PES; Peson. [CH$_2$CH(SO$_3$Na)]$_n$
THERAP CAT: Anticoagulant.

5679. Lycoctonine. [26000-17-9] (1α,6β,14α,16β)-20-Ethyl-4-(hydroxymethyl)-1,6,14,16-tetramethoxyaconitane-7,8-diol; royline. C$_{25}$H$_{41}$NO$_7$; mol wt 467.60. C 64.22%, H 8.84%, N 3.00%, O 23.95%. Isomeric with delsoline. Originally isolated from *Aconitum lycoctonum* L., *Ranunculaceae:* Hubschmann, *Schweiz. Wochschr. Pharm.* **3**, 269 (1865). Widely distributed in *Aconitum* and *Delphinium* spp., *Ranunculaceae*. Structure: Edwards *et al.*, *Can. J. Chem.* **34**, 1315 (1956); Anet *et al.*, *ibid.* **35**, 400 (1957). Identity with royline: Edwards, Rodger, *ibid.* **37**, 1187 (1959). Stereochemistry: Przybylska, Marion, *ibid.* 1843. Revised configuration: S. W. Pelletier *et al.*, *J. Am. Chem. Soc.* **103**, 6536 (1981).

Crystals. Bitter taste. mp 143°. $[\alpha]_D^{20}$ +53° (ethanol). Slightly sol in water, ether, benzene; freely sol in alcohol, chloroform.

Ajacine. [509-17-1] *N*-Acetylanthranilic acid ester. $C_{34}H_{48}N_2O_9$; mol wt 628.76. Needles from 70% alcohol, mp 154°. $[\alpha]_D^{22}$ +49.5° (c = 2 in abs alcohol); $[\alpha]_D^{16}$ +53° (c = 0.66 in chloroform). uv max: 223, 252, 310 nm (ε 28400, 16600, 5400). Slightly sol in water; sol in ether, alcohol with blue fluorescence.

Lycaconitine. [25867-19-0] *N*-Succinylanthranilic acid ester. $C_{36}H_{48}N_2O_{10}$. Isoln: Schulze, Bierling, *Arch. Pharm.* **251**, 8 (1913). White, amorphous powder, mp 111-114°. $[\alpha]_D$ +42°; also reported as +31.5°. Insol in water. Sol in alcohol, chloroform, ether.

5680. Lycodine. [20316-18-1] (4a*R*,5*S*,10b*R*,12*R*)-2,3,4,4a,-5,6-Hexahydro-12-methyl-1*H*-5,10b-propano-1,7-phenanthroline. $C_{16}H_{22}N_2$; mol wt 242.37. C 79.29%, H 9.15%, N 11.56%. Isoln from *Lycopodium annotinum* L., *Lycopodiaceae*: F. A. L. Anet, C. R. Eves, *Can. J. Chem.* **36**, 902 (1958). Structure: W. A. Ayer, G. G. Iverach, *ibid.* **38**, 1823 (1960); F. A. L. Anet, M. V. Rao, *Tetrahedron Lett.* **1960**, 9. Synthesis of (±)-form: E. Kleinman, C. H. Heathcock, *ibid.* **1979**, 4125; C. H. Heathcock *et al.*, *J. Am. Chem. Soc.* **104**, 1054 (1982).

Cryst powder, mp 118-119°. $[\alpha]_D$ −10° (c = 1.01 in ethanol). pKa$_1$ 3.97; pKa$_2$ 8.08.

5681. Lycopene. [502-65-8] ψ,ψ-Carotene; (*all-trans*)-lycopene. $C_{40}H_{56}$; mol wt 536.89. C 89.49%, H 10.51%. Carotenoid antioxidant occurring in ripe fruit, especially in tomatoes. One kg of fresh ripe tomatoes yields 0.02 g lycopene. Isoln procedure: R. Willstätter, H. H. Escher, *Z. Physiol. Chem.* **64**, 47 (1910). Chromatographic sepn from other carotenoids: A. Winterstein, *ibid.* **51** (1933); A. Winterstein, G. Stein, *ibid.* **220**, 247 (1933). HPLC determn in plasma; M. V. Vertzoni *et al.*, *J. Chromatogr. B* **819**, 149 (2005). Structure: Willstätter, Escher, *loc. cit.*; Karrer and collaborators: *Helv. Chim. Acta* **11**, 751, 1201 (1928); **12**, 285 (1929); **13**, 1084 (1930); **14**, 435 (1931); Kuhn, Grundmann, *Ber.* **65**, 898, 1880 (1932). Synthesis: P. Karrer *et al.*, *Helv. Chim. Acta* **33**, 1349 (1950); O. Isler *et al.*, *ibid.* **39**, 463 (1956). Commercial prepn: Kabbe *et al.*, **DE 1168890** (1964 to Bayer). Oral toxicity study: W. Mellert *et al.*, *Food Chem. Toxicol.* **40**, 1581 (2002). Review of antioxidant activity and role in human health: A. V. Rao, S. Agarwal, *Nutr. Res.* **19**, 305-323 (1999).

Long, deep red needles from carbon disulfide + ethanol, from methylene chloride + methanol, mp 172-173°. Absorption max, *trans*-form (petroleum ether): 446, 472, 505 nm (E$_{1cm}^{1\%}$ 2250, 3450, 3150). One gram dissolves in 50 ml carbon disulfide, in 3 l boiling ether, in 12 l boiling petr ether, in 14 l hexane at 0°. Sol in chloroform, benzene. Almost insol in methanol, ethanol, water.

15,15′-*cis*-Form. [59092-07-8] mp ~105° then solidifies again. Absorption max (petroleum ether): 361, 444, 470, 502 nm (E$_{1cm}^{1\%}$ 1110, 1280, 1660, 1280).

5682. Lycophyll. [19891-75-9] ψ,ψ-Carotene-16,16′-diol; (*all-trans*)-lycopene-16,16′-diol. $C_{40}H_{56}O_2$; mol wt 568.89. C 84.45%, H 9.92%, O 5.62%. Carotenoid pigment, isoln from *So-*

lanum dulcamara L. and *Lycopersicum esculentum* Mill., *Solanaceae* and structure: Zechmeister, Cholnoky, *Ber.* **69**, 422 (1936). Revised structure: Cholnoky, Szabolcs, *Tetrahedron Lett.* **1968**, 1931. Stereochemistry: Kelly *et al.*, *Acta Chem. Scand.* **25**, 1607 (1971). Synthesis: Kjosen, Liaaen-jensen, *ibid.* 1500.

Purple leaflets from benzene + methanol, needles from benzene + petr ether, mp 179°. Absorption max (benzene): 521, 487, 456 nm. Freely sol in carbon disulfide, less in benzene, ethanol, very slightly sol in petr ether.

Dipalmitate. Purple needles from benzene + methanol, mp 76°.

5683. Lycopodine. [466-61-5] (1*S*,8a*R*,9*R*,11*R*,12a*R*)-Dodecahydro-11-methyl-1,9-ethanobenzo[*i*]quinolizin-14-one; (15*R*)-15-methyllycopodan-5-one. $C_{16}H_{25}NO$; mol wt 247.38. C 77.68%, H 10.19%, N 5.66%, O 6.47%. From herb of *Lycopodium complanatum* L., *Lycopodiaceae*: Bödeker, *Ann.* **208**, 363 (1881). Structure: Harrison *et al.*, *Can. J. Chem.* **39**, 2086 (1961). Stereochemistry: Anet, *Tetrahedron Lett.* **1960**, no. 20, 13. Synthesis of (±)-form: G. Stork *et al.*, *J. Am. Chem. Soc.* **90**, 1647 (1968); W. A. Ayer *et al.*, *ibid.* 1648; C. H. Heathcock *et al.*, *ibid.* **100**, 8036 (1978); **104**, 1054 (1982); D. Schumann *et al.*, *Ann.* **1982**, 1700. Review of synthetic studies: Stork, *Pure Appl. Chem.* **17**, 383 (1968).

Bitter prisms, mp 114-115°; mp (racemate) 130-131°. $[\alpha]_D$ −24° (alc). Sol in water, alc, benzene, chloroform, ether.

5684. Lycopodium. Club-moss spores; Lycopodium seed (spores); vegetable sulfur. Spores of *Lycopodium clavatum* L., *Lycopodiaceae*. The spores of *Lycopodium annotinum* L., and of *L. anceps* Wall., *Lycopodiaceae* can also be used. The spores contain a substance called **selagnine**. *Habit.* North America, Europe, Asia; cultivated in Russia.

The spores are a fine yellowish powder which is highly flammable. Odorless and tasteless. Unctuous to the touch and easily sticking to the fingers. Lycopodium powder is very mobile and when poured on a flat surface should form an even layer, without visible lumps or dimples. When observed in chloral hydrate soln it is seen that the powder consists of unicellular lycopod spores, about 30 μ in diameter, in the shape of triangular pyramids with a convex base and rounded angles; a three-radical suture runs from the top of the pyramid along its facets. After warming and crushing the spores between glass slides, they burst along the suture and yield drops of oil, assuming a red color with alkalies. Adulteration usually consists of the admixture of flour (detected microscopically with iodine soln which stains the starch grains of the flour violet). Other admixtures may be pine pollen and sawdust. When mounted with chloral hydrate, pine pollen is larger than lycopodium, it is oval and has two lateral flying sacs, filled with air, and appearing black at the beginning of the observation. Sawdust is easily detected by the phloroglucinol test.

USE: As covering for pills or suppositories; in explosives, pyrotechnics, flashlight powders; as "dry parting compound" in foundry work for ornamental and nameplate castings. Review: Appel, *Sci. Mon.* **78**, 268 (1954).

THERAP CAT: Adsorbent.

5685. Lycopus. Bugleweed; sweet bugle; water bugle. Whole plant of *Lycopus virginicus* L., *Labiatae*. *Habit*. N. America. *Constit*. Volatile oil, resin, tannin, glucoside.

5686. Lycoramine. [21133-52-8] (4a*S*,6*S*,8a*R*)-4a,5,7,8,9,-10,11,12-Octahydro-3-methoxy-11-methyl-6*H*-benzofuro[3a,3,2-*ef*][2]benzazepin-6-ol; 1,2-dihydrogalanthamine; 3,4-dihydrogalanthamine. $C_{17}H_{23}NO_3$; mol wt 289.38. C 70.56%, H 8.01%, N 4.84%, O 16.59%. One of the minor alkaloids of *Lycoris radiata* Herb., *Amaryllidaceae*: Kondo, Ishiwata, *Ber.* **70B**, 2427 (1937). From *Narcissus* spp.: Boit *et al.*, *ibid.* **90**, 725 2197 (1957). Tentative structure: Kobayashi, Uyeo, *Chem. Ind. (London)* **1956**, 177. Structure elucidation: William, Rogers, *Proc. Chem. Soc. London* **1964**, 357. Synthesis of the naturally occurring (−)-form: Barton, Kirby, *J. Chem. Soc.* **1962**, 806; of the (±)-form: N. Hazama *et al.*, *J. Chem. Soc. C* **1968**, 2947; Y. Misaka *et al.*, *ibid.* **2954**; A. G. Schultz *et al.*, *J. Am. Chem. Soc.* **99**, 8065 (1977); S. F. Martin, P. J. Garrison, *J. Org. Chem.* **46**, 3567 (1981); **47**, 1513 (1982).

Plates from acetone, mp 121°. $[\alpha]_D^{27}$ −98° (alcohol). Freely sol in water; sol in alc, acetone. Its salts hydrolyze easily.

5687. Lycorine. [476-28-8] (1*S*,2*S*,12b*S*,12c*S*)-2,4,5,7,12b,-12c-Hexahydro-1*H*-[1,3]dioxolo[4,5-*j*]pyrrolo[3,2,1-*de*]phenanthridine-1,2-diol; (1α,2β)-3,12-didehydro-9,10-[methylenebis(oxy)]-galanthan-1,2-diol; 3,3a-didehydrolycoran-1α,2β-diol; amarylline; belamarine; narcissine; galanthidine. $C_{16}H_{17}NO_4$; mol wt 287.32. C 66.89%, H 5.96%, N 4.88%, O 22.27%. Alkaloid isolated from the bulbs of *Lycoris radiata* L., *Narcissus pseudonarcissus* L., *N. tazetta* L., from *Buphane disticha* Herb., in *Crinum* spp., *Amaryllis belladonna* L., *Clivia miniata* Regel and other *Amaryllidaceae*. Extraction procedure: Cook, Loudon in Manske-Holmes, *Alkaloids* **vol. II** (Academic Press, 1952) p 336. Identity of lycorine and galanthidine: Proskurnina, *Dokl. Akad. Nauk SSSR* **90**, 565 (1953), *C.A.* **49**, 12500c (1955). Structure: Takeda *et al.*, *J. Am. Chem. Soc.* **80**, 2562 (1958). Stereochemistry: Nakagawa, Uyeo, *J. Chem. Soc.* **1959**, 3736; K. Kotera *et al.*, *Tetrahedron Lett.* **1966**, 2009. Crystal structure: R. Roques, *Acta Crystallogr.* **30B**, 296 (1974). Attempted synthesis: Dyke *et al.*, *Tetrahedron* **29**, 213 (1973). Synthesis: Y. Tsuda *et al.*, *Chem. Commun.* **1975**, 933; *eidem*, *J. Chem. Soc. Perkin Trans. 1* **1979**, 1358; T. Sano *et al.*, *Heterocycles* **14**, 1097 (1980); S. F. Martin, C. Tu, *J. Org. Chem.* **46**, 3763 (1981). Biosynthesis: Archer *et al.*, *Proc. Chem. Soc. London* **1963**, 168; Fugati, Mazza, *Chem. Commun.* **1972**, 936. *Review:* W. C. Wildman in *The Alkaloids* **vol. XI**, R. H. F. Manske, Ed. (Academic Press, New York, 1968) pp 307-400.

Stout prisms from alcohol, mp 275-280° (dec). $[\alpha]_D^{16}$ −129° (c = 0.16 in 98% alc). Alkaline reaction to litmus; salts hydrolyze easily. Sparingly sol in alcohol, chloroform, petr ether. Sol in dilute acids. Practically insol in water, alkalies.

Hydrochloride. [2188-68-3] $C_{16}H_{17}NO_4 \cdot HCl$; mol wt 323.77. Long needles from water, mp 217° (dec with slight preliminary sintering). $[\alpha]_D^{20}$ +43°.

Hydrochloride monohydrate. [6150-58-9] $C_{16}H_{17}NO_4 \cdot HCl \cdot H_2O$; mol wt 341.79. Elongated prisms from water, mp 206°. Soluble in 20 parts water.

Methiodide. [15069-99-5] $C_{16}H_{17}NO_4 \cdot CH_3I$; mol wt 429.25. Two forms exist which may be stereoisomeric about the nitrogen atom. α- Form: polyhedra from alc, dec 247°. $[\alpha]_D^{20}$ −46° (c = 1.52). Freely sol in water; sparingly sol in hot alcohol. β-Form: prisms contg $1H_2O$ from water, dec 198°, after recryst from alcohol, dec 281°. $[\alpha]_D^{20}$ +123° (c = 2.44). Freely sol in water and hot alcohol.

5688. Lycoxanthin. [19891-74-8] ψ,ψ-Carotene-16-ol; (*all-trans*)-lycopen-16-ol. $C_{40}H_{56}O$; mol wt 552.89. C 86.90%, H 10.21%, O 2.89%. Carotenoid isolated from *Solanum dulcamara* L., *Lycopersicum esculentum* Mill., *Solanaceae; Tamus communis* L., *Dioscoreaceae*. Isoln by chromatography and structure: Zechmeister, Cholnoky, *Ber.* **69**, 422 (1936). Structure: Karrer *et al.*, *Helv. Chim. Acta* **13**, 268, 1084 (1930); **14**, 614, 843 (1931); Winterstein, *Angew. Chem.* **72**, 902 (1960). Revised structure: Cholnoky, Szabolcs, *Tetrahedron Lett.* **1968**, 1931. Stereochemistry: Kelly *et al.*, *Acta Chem. Scand.* **25**, 1607 (1971). Total synthesis: Kjosen, Liaaen-jensen, *ibid.* 1500.

Purple needles from carbon disulfide. Reddish-brown round or acicular cryst aggregates from benzene + petr ether, mp 168°. Absorption max (acetone): 448, 474 ($E_{1cm}^{1\%}$ 3080), 505 nm. Sol in carbon disulfide and benzene. Moderately sol in petr ether. Sparingly sol in alc.

Monoacetate. Deep purple needles from benzene+ methanol, mp 137°. Freely sol in carbon disulfide; sparingly sol in alcohol, petr ether.

5689. Lymecycline. [992-21-2] N^6-[[[[(4*S*,4a*S*,5a*S*,6*S*,12a*S*)-4-(Dimethylamino)-1,4,4a,5,5a,6,11,12a-octahydro-3,6,10,12,12a-pentahydroxy-6-methyl-1,11-dioxo-2-naphthacenyl]carbonyl]amino]methyl]-L-lysine; tetracyclinemethylene lysine; *N*-lysinomethyltetracycline; Tetralisal; Tetralysal. $C_{29}H_{38}N_4O_{10}$; mol wt 602.64. C 57.80%, H 6.36%, N 9.30%, O 26.55%. Semi-synthetic antibiotic related to tetracycline, *q.v.* Prepn: R. K. Blackwood, K. J. Brunings, US 3042716 (1962 to Pfizer); F. Lauria, W. Logemann, DE 1134071 (1962 to Carlo Erba); E. Tubaro, E. Raffaldoni, *Boll. Chim. Farm.* **100**, 9 (1961). Clinical pharmacokinetics: A. Schreiner, A. Digranes, *Chemotherapy (Basel)* **31**, 261 (1985). Comparison with doxycycline of phototoxic potential: M. Bjellerup, B. Ljunggren, *Br. J. Dermatol.* **130**, 356 (1994). Clinical trial in acne: L. Bossuyt *et al.*, *Eur. J. Dermatol.* **13**, 130 (2003); of combination with adapalene: W. J. Cunliffe *et al.*, *J. Am. Acad. Dermatol.* **49**, S218 (2003).

Sodium salt. $C_{29}H_{37}N_4NaO_{10}$; mol wt 624.62. uv max (CH_3-OH): 376 nm.

THERAP CAT: Antibacterial.

5690. Lynestrenol. [52-76-6] (17α)-19-Norpregn-4-en-20-yn-17-ol; 17α-ethinyl-17β-hydroxyestr-4-ene; 17α-ethynylestr-4-en-17β-ol; 3-desoxynorlutin; ethinylestrenol; Exluton(a); Exlutena; Orgametril; Orgametil. $C_{20}H_{28}O$; mol wt 284.44. C 84.45%, H

9.92%, O 5.62%. Prepn: de Winter *et al., Chem. Ind. (London)* **1959**, 905. Metabolism: Kamyab *et al., Biochem. J.* **103**, 14P (1967); *J. Endocrinol.* **42**, 337 (1968). Pharmacokinetics and metabolism: H. Kuhl *et al., Contraception* **26**, 303 (1982). Clinical trial of use in normophasic oral contraceptives: N. Dombrowicz *et al., ibid.* **22**, 537 (1980). Review of carcinogenicity studies: *IARC Monographs* **21**, 407-415 (1979).

Solid, mp 158-160°. $[\alpha]_D$ −13° (chloroform).
Mixture with ethinyl estradiol. [8064-76-4] Fysionorm; Minilyn; Ovanon; Ovoresta; Yermonil.
Mixture with mestranol. [8015-14-3] Lyndiol.
THERAP CAT: Progestogen. In combination with estrogen as oral contraceptive.

5691. Lypressin. [50-57-7] 8-L-Lysinevasopressin; Phe³-Lys⁸-oxytocin; Diapid; Postacton; Syntopressin. $C_{46}H_{65}N_{13}O_{12}S_2$; mol wt 1056.23. C 52.31%, H 6.20%, N 17.24%, O 18.18%, S 6.07%. Purification after isolation from hog pituitary glands: D. N. Ward, V. du Vigneaud, *J. Biol. Chem.* **222**, 951 (1956). Structure and synthesis: V. du Vigneaud *et al., J. Am. Chem. Soc.* **75**, 4880 (1953); M. F. Bartlett *et al., ibid.* **78**, 2905 (1956). Synthesis: Bodanszky *et al., ibid.* **82**, 3195 (1960); Boissonnas, Huguenin, *Helv. Chim. Acta* **43**, 182 (1960); Zaoral, *Collect. Czech. Chem. Commun.* **30**, 1853 (1965); Meienhofer, Sano, *J. Am. Chem. Soc.* **90**, 2996 (1968); Meienhofer, Trzeciak, *Proc. Natl. Acad. Sci. USA* **68**, 1006 (1971). Configuration: Schally, Barret, *J. Am. Chem. Soc.* **87**, 2497 (1965). *Review:* E. Schröder, K. Lübke, *The Peptides* **vol. II** (Academic Press, New York, 1966) pp 336-350.

Cys–Tyr–Phe–Gln–Asn–Cys–Pro–Lys–GlyNH₂

THERAP CAT: Antidiuretic; vasopressor.

5692. Lysalbinic Acid. [9006-58-0] Albumins, lysalbinic acids. Colloidal product of the action of caustic alkalies on albumin. Mixture of peptides; mol wt 600-11000, depending on hydrolysis conditions. Prepn: Paal, *Ber.* **35**, 2195 (1902); **DE 129031** (1902); **DE 132322** (1902). Prepn, properties, and surfactact characteristics: P. L. Starokadomskyy, I. Y. Dubey, *Int. J. Pharm.* **308**, 149 (2006). Usually isolated as the Na salt. Readily sol in water, forming a colloidal soln which foams. It is precipitated by alc, and it may be purified by dialysis.
Colloidal gold. Prepd by adding a soln of gold chloride to a soln of lysalbinic acid. The intensely red hydrosol is precipitated with alcohol and dried. The precipitate forms hard, bronze-colored granules which may be stored for a long time, yielding a red, colloidal gold soln upon addn of water. Analysis of dried granules shows about 30% Au. *See:* C. Paal, *Ber.* **35**, 2236-2244 (1902); *eidem*, **US 701605** (1902 to Kalle).
USE: Non-ionic detergent. As a protective colloid for metal sols, especially gold, silver, and mercury sols.

5693. Lysergamide. [478-94-4] (8β)-9,10-Didehydro-6-methylergoline-8-carboxamide; lysergic acid amide; ergine. $C_{16}H_{17}N_3O$; mol wt 267.33. C 71.89%, H 6.41%, N 15.72%, O 5.98%. Isoln from *Rivea corymbosa* (L.) and from *Ipomoea tricolor* Cav., *Convolvulaceae:* Hofmann, Tscherter, *Experientia* **16**, 414 (1964). Prepn from lysergic acid hydrazide: Ainsworth, **US 2756235** (1956 to Lilly); from lysergic acid and phosgene-dimethylformamide complex: Patelli, Bernardi, **US 3141887** (1964 to Farmitalia). Microbiological production: Rutschmann, Kobel, **US 3219545** (1965 to Sandoz).

Prisms from methanol, dec 242°. $[\alpha]_{5461}^{20}$ +15° (c = 0.5 in pyridine).
Methanesulfonate. $C_{16}H_{17}N_3O.CH_3SO_3H$; mol wt 363.43. Prisms from methanol + acetone, dec 232°.
Note: This is a controlled substance (depressant): **21 CFR, 1308.13.**

5694. Lysergic Acid. [82-58-6] (8β)-9,10-Didehydro-6-methylergoline-8-carboxylic acid. $C_{16}H_{16}N_2O_2$; mol wt 268.32. C 71.62%, H 6.01%, N 10.44%, O 11.93%. Lysergic acid and isolysergic acid are the main cleavage products formed on alkaline hydrolysis of the alkaloids which are characteristic of ergot.: Jacobs, Craig *et al., J. Biol. Chem.* **104**, 547 (1934); **125**, 289 (1938); **130**, 399 (1939); **145**, 487 (1942); *J. Org. Chem.* **10**, 76 (1945). High-yield production by *Claviceps paspali:* Arcamone *et al., Proc. R. Soc. London Ser. B* **155**, 26 (1961). Total synthesis: Kornfeld *et al., J. Am. Chem. Soc.* **76**, 5256 (1954); **78**, 3087 (1956); M. Julia *et al., Tetrahedron Lett.* **1969**, 1569; V. W. Armstrong *et al., ibid.* **1976**, 4311; W. Oppolzer *et al., Helv. Chim. Acta* **64**, 478 (1981); R. Ramage *et al., Tetrahedron* **37**, Suppl. 9, 157 (1981); J. Rebek, D. F. Tai, *Tetrahedron Lett.* **24**, 859 (1983). Stereochemistry: Stoll *et al., Helv. Chim. Acta* **37**, 2039 (1954); Stenlake, *J. Chem. Soc.* **1955**, 1626; Leeman, Fabbri, *Helv. Chim. Acta* **42**, 2696 (1959). Absolute configuration: Stadler, Hofmann, *ibid.* **45**, 2005 (1962).

Hexagonal scales, plates with one or two moles H₂O from water, mp 240° (dec). $[\alpha]_D^{20}$ +40° (c = 0.5 in pyridine). Behaves as an acid and base, pKa 3.44, pKb 7.68. Moderately sol in pyridine. Sparingly sol in water and in neutral organic solvents; sol in NaOH, NH₄OH, Na₂CO₃, and HCl solns. Slightly sol in dil H₂SO₄.
Methyl ester. [4579-64-0] $C_{17}H_{18}N_2O_2$; mol wt 282.34. Thin leaflets from benzene, mp 168°.
Note: This is a controlled substance (depressant): **21 CFR, 1308.13.**

5695. Lysergide. [50-37-3] (8β)-9,10-Didehydro-N,N-diethyl-6-methylergoline-8-carboxamide; N,N-diethyl-D-lysergamide; D-lysergic acid diethylamide; LSD; LSD-25; lysergsäure diethylamid. $C_{20}H_{25}N_3O$; mol wt 323.44. C 74.27%, H 7.79%, N 12.99%, O 4.95%. Microbial formation by *Claviceps paspali* over the hydroxyethylamide: Arcamone *et al., Proc. R. Soc. London* **155B**, 26 (1961). Partial synthesis: Stoll, Hofmann, *Helv. Chim. Acta* **26**, 944 (1943); **38**, 421 (1955). Industrial prepn: Pioch, **US 2736728**; Garbrecht, **US 2774763** (both 1956 to Lilly); Patelli, Bernardi, **US 3141887** (1964 to Farmitalia). Isotope-labeled LSD: Stoll *et al., Helv. Chim. Acta* **37**, 820 (1954). LC-MS determn in biological fluids: J. Canezin *et al., J. Chromatogr. B* **765**, 15 (2001). Toxicity data: E. Rothlin, *Ann. N.Y. Acad. Sci.* **66**, 668 (1957). *Review:* Hoffer, *Clin. Pharmacol. Ther.* **6**, 183 (1965). Book: *The Use of LSD in Psychotherapy and Alcoholism*, H. A. Abramson, Ed. (Bobbs-Merrill, Indianapolis, 1967) 697 pp.

Pointed prisms from benzene, mp 80-85°. $[\alpha]_D^{20}$ +17° (c = 0.5 in pyridine). uv max (ethanol): 311 nm ($E_{1cm}^{1\%}$ 257). LD_{50} in mice, rats, rabbits (mg/kg): 46, 16.5, 0.3 i.v. (Rothlin).

D-Tartrate. $C_{46}H_{64}N_6O_{10}$; mol wt 861.05. Solvated, elongated prisms from methanol, mp 198-200°. $[\alpha]_D^{20}$ +30°. Soluble in water.

Note: This is a controlled substance (hallucinogen): **21 CFR,** 1308.11.

USE: In biochemical research as antagonist to serotonin. Has been used experimentally as adjunct in study and treatment of mental disorders.

5696. Lysidine. [534-26-9] 4,5-Dihydro-2-methyl-1*H*-imidazole; 2-methyl-2-imidazoline; methylglyoxalidine. $C_4H_8N_2$; mol wt 84.12. C 57.11%, H 9.59%, N 33.30%. Prepn: Chitwood, Reid, *J. Am. Chem. Soc.* **57**, 2424 (1935); King, McMillan, *ibid.* **68**, 1774 (1946); Kyrides, **US 2392326** and **US 2404299** (both 1946 to Monsanto); Ahrens, **US 2813862** (1957 to Organon).

Needles from benzene, mp 105°. bp 198-200°. Soluble in water, alc, chloroform; less sol in benzene, carbon tetrachloride, petr ether; practically insol in ether. *Keep well closed. Incompat:* Acids, metallic salts, alkaloids.

5697. Lysine. [56-87-1] L-Lysine; Lys; K; (*S*)-2,6-diaminohexanoic acid; α,ε-diaminocaproic acid. $C_6H_{14}N_2O_2$; mol wt 146.19. C 49.30%, H 9.65%, N 19.16%, O 21.89%. An essential amino acid for human development; probably the most limited in the food chain. Identified in a hydrolysate of casein by Drechsel in 1889. Named "lysatine" from the Greek for "loosing", as it produced urea on treatment with barium hydroxide; changed to "lysine" in 1891. Verification of structure: E. Fischer, F. Weigert, *Ber.* **35**, 3772 (1902). Early chemistry and biochemistry: *Amino Acids and Proteins,* D. M. Greenberg, Ed. (Charles C. Thomas, Springfield, IL, 1951) 950 pp, *passim;* J. P. Greenstein, M. Winitz, *Chemistry of the Amino Acids* vols. **1-3** (John Wiley & Sons, Inc., New York, 1961) pp 2097-2124, *passim.* Review of supplementation of cereals: C. Feldberg, C. P. Hetzel, *Food Technol.* **12**, 496 (1958); of wheat proteins: S. B. Vaghefi *et al., Am. J. Clin. Nutr.* **27**, 1231-1246 (1974). Review of metabolism in mammals: F. C. Fellows, M. H. R. Lewis, *Biochem. J.* **136**, 329-334 (1973); of metabolic errors: N. A. J. Carson, *Clin. Endocrinol. Metab.* **3**, 71-86 (1974). Brief review of production by fermentation: *Nutr. Rev.* **43**, 88-90 (1985). Review of synthesis: G. Galili, *Plant Cell* **7**, 899-906 (1995).

Needles from water, hexagonal plates from dil alcohol. Darkens at 210°; dec 224.5°. $[\alpha]_D^{20}$ +14.6° (c = 6.5); $[\alpha]_D^{23}$ +25.9° (c = 2 in 6.0*N* HCl). pK_1 2.18; pK_2 8.95; pK_3 10.53. Very freely sol in water. Insol in common neutral solvents.

Dihydrochloride. $C_6H_{14}N_2O_2 \cdot 2HCl$. Crystals from ethanol + ether, mp 193°. $[\alpha]_D^{20}$ +15.3° (c = 2).

Monohydrochloride. [657-27-2] Darvyl; Enisyl. $C_6H_{14}N_2O_2 \cdot$HCl; mol wt 182.65. Crystals from dil ethanol. mp 263-264° when anhydrous. $[\alpha]_D^{25}$ +14.6° (c = 2 in 0.6*N* HCl).

Monoorotate. Lysortine. $C_{11}H_{18}N_4O_6$; mol wt 302.29. Prepn: **GB 922361** (1962 to A.E.C. Soc. Chim. Org. Biol.). Crystals, dec 315°.

USE: Enrichment of cereals and feeds.

5698. Lysine Acetylsalicylate. [62952-06-1] Lysine 2-(acetyloxy)benzoate (1:1); DL-lysine mono[2-(acetyloxy)benzoate]; lysine monosalicylate acetate; aspirin lysine salt; LAS; Aspidol; Aspisol; Delgesic; Flectadol; Lysal; Quinvet; Venopirin; Vetalgina. $C_{15}H_{22}N_2O_6$; mol wt 326.35. C 55.21%, H 6.80%, N 8.58%, O 29.41%. Water soluble, injectable aspirin derivative. Prepn: **FR 1295304** (1962 to Equilibre Biologique). Prepn of D,L isomers and racemate: **FR 2115060** (1972 to Metabio). Comparison with aspirin: S. Rampon *et al., Rhumatologie* **24**, 141 (1972). Pharmacokinetics: H. von Ross *et al., Klin. Wochenschr.* **56**, 1119 (1978); F. Gentit, *Gaz. Med. Fr.* **86**, 4539 (1979). Veterinary use as analgesic: P. Richez *et al., J. Vet. Pharmacol. Ther.* **2**, 231 (1979); *eidem, ibid.* **3**, 121 (1980). Analgesic activity in post-operative pain: K. Korttila *et al., Br. J. Anaesth.* **52**, 613 (1980); in articular pain: C. Diaz *et al., Clin. Ther.* **4**, 121 (1981). Use in treatment of premature labor: F. Wolff *et al., Arch. Gynecol.* **233**, 15 (1982). Comparison with aspirin in production of gastromucosal damage: J.-F. Bretagne *et al., Gastroenterol. Clin. Biol.* **8**, 28 (1984).

Crystals from ethanol, mp 154-156°. Sol in water; slightly sol in ethanol. Insol in methanol, acetone, ether.

THERAP CAT: Analgesic; antipyretic; anti-inflammatory.

THERAP CAT (VET): Analgesic; antipyretic; anti-inflammatory.

5699. L-Lysine L-Glutamate. [5408-52-6] L-Glutamic acid compd with L-lysine (1:1). $C_{11}H_{23}N_3O_6$; mol wt 293.32. C 45.04%, H 7.90%, N 14.33%, O 32.73%. Prepd from DL-lysine and L-glutamic acid as an intermediate in the resolution of DL-lysine: Emmick, Rogers, **US 2556907; US 2657230** (1951, 1953 both to du Pont).

Monohydrate. $C_{11}H_{23}N_3O_6 \cdot H_2O$; mol wt 311.34. Crystals, $[\alpha]^{22}$ +3.73° (12% in water).

USE: As a flavor and nutritive additive to food: **GB 882163** (1961 to Stamicarbon N.V.).

5700. Lysostaphin. [9011-93-2] Antibiotic protein complex produced by *Staphylococcus staphyloliticus* with highly specific lytic activity against other *Staphylococcus* species. Contains three enzymes: a hexosaminidase, an amidase, and the major component, an endopeptidase which cleaves the polyglycine cross-linkages in the staphylococcal cell wall. Isoln and antibacterial spectrum: Schindler, Schuhardt, *Proc. Natl. Acad. Sci. USA* **51**, 414 (1964). Prepn by fermentation: *eidem,* **US 3278378** (1966); Zygmunt, Browder, **US 3398056** (1968 to Mead Johnson). The endopeptidase is a single polypeptide chain of mol wt 25,000: H. R. Trayer, C. E. Buckley, *J. Biol. Chem.* **245**, 4842 (1970). Identification of active principle: H. P. Browder *et al., Biochem. Biophys. Res. Commun.* **19**, 383 (1965). Characterization of three enzyme components: T. Wadstrom, O. Vesterberg, *Acta Pathol. Microbiol. Scand.* **79**, 248 (1971); O.-J. Iversen, A. Grov, *Eur. J. Biochem.* **38**, 293 (1973). Clinical studies to eradicate *Staph. aureus* from nasal carriers: R. L. Harris *et al., Antimicrob. Agents Chemother.* **1967**, 110; K. E. Quickel *et al., Appl. Microbiol.* **1971**, 446. Use as identification method for *Staphylococcus* sp.: K. H. Schleifer, W. E. Kloos, *J. Clin. Microbiol.* **1**, 337

(1975); P. J. Severance *et al.*, *ibid.* **11**, 724 (1980); disk modification: B. Poutrel, J.-P. Caffin, *ibid.* **13**, 1023 (1982). *Review:* Zygmunt, Tavormina, *Progress in Drug Research* **Vol. 16**, E. Jucker, Ed. (Birkhauser Verlag, Basel, 1972) pp 309-333.

Isoelectric pt: pH 10.4-11.4. uv max: 278 nm. Sedimentation coeff 2.35 S. Destroyed by pepsin or trypsin. Inhibited by Hg^{2+}, Cu^{2+}, Zn^{2+}. LD_{50} (7-day) in mice, rats (mg/kg): 820, 530 i.v. (Zygmunt).

5701. Lysozyme. [9001-63-2] Muramidase; *N*-acetylmuramide glycanohydrolase; *N*-acetylmuramyl hydrolase; globulin G_1. Mol wt about 14,400 ± 100. Mucolytic enzyme with antibiotic properties, first discovered by A. Fleming: *Proc. R. Soc. London* **93B**, 306 (1922). Found in tears, nasal mucus, milk, saliva, blood serum, in a great number of tissues and secretions of different animals, vertebrates and invertebrates, in egg white, in some molds, and in the latex of different plants. Isoln from egg white: Alderton *et al.*, *J. Biol. Chem.* **157**, 43 (1945); Alderton, Fevold, *ibid.* **164**, 1 (1946); *Biochem. Prep.* **1**, 67 (1949); Sophianopoulos *et al.*, *J. Biol. Chem.* **237**, 1107 (1962); from *Ficus* latex: Meyer *et al.*, *ibid.* **163**, 733 (1946). Structure consists of a single polypeptide chain of 129 amino acid subunits of 20 different kinds cross-linked by four disulfide bridges. Chromatographic studies: Goncalves *et al.*, *Arch. Biochem. Biophys.* **60**, 171 (1956); King, Craig, *J. Am. Chem. Soc.* **80**, 3366 (1958). Complete amino acid sequence of egg white lysozyme: Jolles *et al.*, *Biochim. Biophys. Acta* **78**, 668 (1963); Canfield, *J. Biol. Chem.* **238**, 2698 (1963). Synthetic studies of egg white lysozyme: L. E. Barston *et al.* in *Chemistry and Biology of Peptides*, *Proc. 3rd Am. Peptide Symp.*, J. Meienhofer, Ed. (Ann Arbor Science Publishers, Inc., Michigan, 1972) pp 231-233; J. J. Sharp *et al.*, *J. Am. Chem. Soc.* **95**, 6097 (1973). Three dimensional structure: Blake *et al.*, *Nature* **206**, 757 (1965); North, *Sci. J.* **2**(11), 55 (1966); Phillips, *Sci. Am.* **215**, 78 (Nov. 1966); Blake *et al.*, *Proc. R. Soc. London* **167B**, 365 (1967). Primary chemical structure and tentative complete sequence of human milk lysozyme: Jolles, Jolles, *Helv. Chim. Acta* **52**, 2671 (1969). Complete primary structure of human milk lysozyme and comparison with lysozymes of various origins:

eidem, ibid. **54**, 2668 (1971). Dissolves bacterial cell wall mucopolysaccharides by hydrolyzing the β-(1 → 4) linkages between *N*-acetyl-D-muramic acid and 2-acetylamino-2-deoxy-D-glucose residues. Also acts on chitin. *Reviews:* Salton, *Bacteriol. Rev.* **21**, 82 (1957); Acker, Hartsell, *Sci. Am.* **202**, 132 (June 1960); Jolles, "Lysozyme" in P. D. Boyer, H. Lardy, K. Myrback, *The Enzymes* **vol. 4**, (Academic Press, New York, 2nd ed., 1960) pp 431-445; *idem, Angew. Chem.* **76**, 20 (1964); Raftery, Dahlquist, *Fortschr. Chem. Org. Naturst.* **27**, 340 (1969); Hamaguchi, Hayashi, "Lysozyme" in *Proteins, Structure and Function* **vol. 1**, M. Funatsu *et al.*, Eds. (Kodansha, Tokyo, Wiley, New York, 1972) pp 85-222. Book: *Lysozyme*, E. F. Osserman *et al.*, Eds. (Academic Press, New York, 1974).

Crystals. Isoelectric pt: pH 10.5-11.0. Fairly stable in acid soln. Not affected by heat up to 55°: Cotterill, Winter, *Poult. Sci.* **33**, 1185 (1954).

Hydrochloride. [9066-59-5] Acdeam; Antalzyme; Immunozima; Lanzyme; Leftose; Likinozym; Lisozima; Murazyme; Neutase; Neuzyme; Toyolysom-DS.

THERAP CAT: Mucolytic enzyme. Antiviral.

5702. D-Lyxose. [1114-34-7] $C_5H_{10}O_5$; mol wt 150.13. C 40.00%, H 6.71%, O 53.28%. Prepd by the oxidation of calcium D-galactonate: Clark, *J. Biol. Chem.* **31**, 605 (1917); Fletcher *et al.*, *J. Am. Chem. Soc.* **72**, 4546 (1950); by γ-irradiation of lactose: Adachi, *J. Dairy Sci.* **45**, 1427 (1962). Structure of β-D-lyxose: Hordvik, *Acta Chem. Scand.* **15**, 1781 (1961).

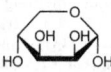

α-D-Lyxose

Hygroscopic monoclinic prisms from ethanol + ether, mp 106-107°. Sweet taste. d^{20} 1.545. Shows mutarotation: $[\alpha]_D^{20}$ +5.5° → −14.0° (c = 0.82 in water). Freely sol in water. One part dissolves in 38 parts abs alc at 17°; 100 ml of 90% alc satd at 20° contain 7.9 g.

M

5703. Mabuterol. [56341-08-3] 4-Amino-3-chloro-α-[[(1,1-dimethylethyl)amino]methyl]-5-(trifluoromethyl)benzenemethanol; 4-amino-α-[(*tert*-butylamino)methyl]-3-chloro-5-(trifluoromethyl)-benzyl alcohol; 1-(4'-amino-3'-chloro-5'-trifluoromethylphenyl)-2-*tert*-butylaminoethanol; ambuterol. $C_{13}H_{18}ClF_3N_2O$; mol wt 310.75. C 50.25%, H 5.84%, Cl 11.41%, F 18.34%, N 9.01%, O 5.15%. Orally active β₂-adrenergic agonist related to clenbuterol, *q.v.* Prepn and resolution of isomers: **BE 808743** (1974 to Thomae); G. Engelhardt *et al.*, **US 4119710** (1978 to Boehringer Ing.); *eidem, Arzneim.-Forsch.* **34**, 1612 (1984). Series of articles on pharmacology, pharmacokinetics, toxicology and clinical studies: *ibid.* 1625-1700. Toxicity data: K. Amemiya *et al., ibid.* 1680. Teratological study: A. M. Hoberman *et al., J. Am. Coll. Toxicol.* **4**, 91 (1985). Determn in human urine and plasma by enzyme immunoassay: I. Yamamoto *et al., J. Immunoassay* **6**, 261 (1985).

dl-**Form hydrochloride.** [95656-48-7] KF-868; PB-868Cl; Broncholin. $C_{13}H_{18}ClF_3N_2O \cdot HCl$; mol wt 347.20. Crystals from ethyl acetate + ether, mp 205-206°. Fairly sol in water. LD_{50} in male, female mice, male, female rats (mg/kg): 41.5, 51.1, 26.4, 28.1 i.v.; 60.3, 60.0, 76.3, 78.3 i.p.; 113.0, 125.7, 117.2, 123.1 s.c.; 220.8, 199.9, 319.3, 305.6 orally (Amemiya).
d-**Form hydrochloride.** [95656-54-5] mp >194° (slow dec). $[\alpha]_{364}^{20}$ +154.9° (c = 1 in methanol).
l-**Form hydrochloride.** [95656-55-6] mp >194° (slow dec). $[\alpha]_{364}^{20}$ −154.8° (c = 1 in methanol).
THERAP CAT: Bronchodilator; antiasthmatic.

5704. Macassar Oil. Kusum oil; Kon oil; Paka oil. From the nut kernels of *Schleichera trijuga* Willd., *Sapindaceae* (Ceylon oak). *Habit.* India, Burma, Ceylon, Java. *Constit.* Glycerides of oleic acid 60%, arachidic 20-25%, palmitic 5-8% stearic 2-6%; free acetic acid 1-2%. Unsaponifiable matter 1.5-3% (this may include cyanide-containing glycosides which are removed by careful refining). *Reviews:* Sen-Gupta, *J. Soc. Chem. Ind. London* **39**, 88T (1920); Dhingra *et al., ibid.* **48**, 281T (1929).
    Yellowish-white oil. Pleasant odor. Melting range: 21-31° (initial to complete transparency). d_{15}^{99} 0.860. n_D^{21} 1.46757; n_D^{27} 1.46655; $n_D^{31.5}$ 1.4646; n_D^{45} 1.4636. Acid value 10-70; saponification value 220-230; iodine value 48-58; Reichert-Meissl value 16. Miscible with ether, chloroform, other vegetable oils, petrolatum, lanolin.
    USE: In hair oil formulations.

5705. Macitentan. [441798-33-0] N-[5-(4-Bromophenyl)-6-[2-[(5-bromo-2-pyrimidinyl)oxy]ethoxy]-4-pyrimidinyl]-N'-propylsulfamide; propylsulfamic acid [5-(4-bromophenyl)-6-[2-(5-bromopyrimidin-2-yloxy)ethoxy]pyrimidin-4-yl]amide; ACT-064992. $C_{19}H_{20}Br_2N_6O_4S$; mol wt 588.28. C 38.79%, H 3.43%, Br 27.17%, N 14.29%, O 10.88%, S 5.45%. Dual endothelin (ET_A and ET_B) receptor antagonist. Prepn: M. Bolli *et al.,* **WO 02053557**; *eidem,* **US 7094781** (2002, 2006 both to Actelion). Pharmacology and receptor binding study: M. Iglarz *et al., J. Pharmacol. Exp. Ther.* **327**, 736 (2008). Clinical pharmacokinetics and dissolution profile of formulations: O. Kummer *et al., Eur. J. Pharm. Sci.* **38**, 384 (2009). Antitumor efficacy in exptl models of ovarian cancer: S.-J. Kim *et al., Neoplasia* **13**, 167 (2011). Review of pharmacology and clinical experience: S. G. Raja, *Curr. Opin. Investig. Drugs* **11**, 1066-1073 (2010).

Powder. Log P (*n*-octanol/aq buffer, pH 7.4): 2.9. pKa 6.2.
THERAP CAT: In treatment of pulmonary arterial hypertension.

5706. Maclurin. [519-34-6] (3,4-Dihydroxyphenyl)(2,4,6-trihydroxyphenyl)methanone; 2,3',4,4',6-pentahydroxybenzophenone; morintannic acid; moritannic acid; laguncurin; kino-yellow; C.I. Natural Yellow 11; C.I. 75240. $C_{13}H_{10}O_6$; mol wt 262.22. C 59.55%, H 3.84%, O 36.61%. From wood of *Chlorophora tinctoria* (L.) Gaud. (*Morus tinctoria* L., *Maclura tinctoria* (L.) D. Don), *Moraceae* (old fustic): Haley, Bassin, *J. Am. Pharm. Assoc.* **40**, 111 (1951); Laidlow, Smith, *Chem. Ind. (London)* **1959**, 1604; from *Morus alba* Linn., *Moraceae:* Spada *et al., Gazz. Chim. Ital.* **86**, 46 (1956). Identity with laguncurin and kino-yellow: Nierenstein, *Q. J. Pharm. Pharmacol.* **16**, 11 (1943). *See also Colour Index* **vol. 4** (3rd ed., 1971) p 4627.

Yellow needles from ethanol, mp 222-222.5°. Sol in 190 parts water; freely sol in alcohol or ether.
USE: Dyeing fabrics.

5707. MacMillan's Enamine Catalyst. [156592-72-2] (2R,5S)-2-(1,1-Dimethylethyl)-3,5-dimethyl-4-imidazolidinone; (2R,5S)-2-*t*-butyl-3,5-dimethylimidazolidin-4-one; MacMillan's photoredox organocatalyst. $C_9H_{18}N_2O$; mol wt 170.26. C 63.49%, H 10.66%, N 16.45%, O 9.40%. Enantioselective organocatalyst that mediates a variety of asymmetric transformations; structurally related to MacMillan's Imidazolidinone Catalysts, *q.v.* Can also be utilized in conjunction with a photoredox catalyst and a photon source in a mode of activation termed photoredox organocatalysis. Prepn: R. Naef, D. Seebach, *Helv. Chim. Acta* **68**, 135 (1985); L Samulis, N. C. O. Tomkinson, *Tetrahedron* **67**, 4263 (2011). Large scale prepn of the hydrochloride: T. H. Graham *et al., Org. Synth.* **88**, 42 (2011). Utility in asymmetric aldehyde chlorination and terminal epoxide formation via singly occupied molecular orbital (SOMO) catalysis: M. Amatore *et al., Angew. Chem. Int. Ed.* **48**, 5121 (2009). Use in enantioselective photoredox organocatalysis for aldehyde alkylations: D. A. Nicewicz, D. W. C. MacMillan, *Science* **322**, 77 (2008); M. Neumann *et al., Angew. Chem. Int. Ed.* **50**, 951 (2011); for aldehyde α-trifluoromethylations: D. A. Nagib *et al., J. Am. Chem. Soc.* **131**, 10875 (2009); for aldehyde α-benzylations: H.-W. Shih *et al., ibid.* **132**, 13600 (2010). Achievement of SOMO activation with the (2S,5S) isomer for styrene carbo-oxidations: T. H. Graham *et al., J. Am. Chem. Soc.* **130**, 16494 (2008); for cycloadditions: N. T. Jui *et al., ibid.* **132**, 10015 (2010).

Colorless oil. $[\alpha]_D^{25}$ −1.8° (c = 0.1 in methanol).

Hydrochloride. [1092799-01-3] $C_9H_{18}N_2O.HCl$; mol wt 206.71. White crystalline solid from ethanol, mp 211-216° (dec). $[\alpha]_D^{23}$ −43.4° (c = 1.0 in methanol). Non-hygroscopic.

USE: Catalyst for asymmetric synthesis.

5708. MacMillan's Imidazolidinone Catalysts. MacMillan's organocatalysts. Broadly general, enantioselective organocatalysts applied to a wide array of chemical transformations, including cycloadditions, conjugate additions, hydrogenations, epoxidations, and cascade reactions. Catalytic mechanisms of action include both lowest unoccupied molecular orbital (LUMO)-lowering iminium activation and singly occupied molecular orbital (SOMO) activation. Prepn of the 2,2-dimethyl compd: I. Solodin *et al.*, *J. Chem. Soc. Chem. Commun.* **1990**, 1321; and its evaluation as a first generation LUMO lowering catalyst: K. A. Ahrendt *et al.*, *J. Am. Chem. Soc.* **122**, 4243 (2000). Design of the 2-*tert*-butyl compd as a second generation catalyst and its use in enantioselective indole alkylations: J. F. Austin, D. W. C. MacMillan, *ibid.* **124**, 1172 (2002). Prepn and use in enantioselective benzene alkylations: N. A. Paras, D. W. C. MacMillan, *ibid.* 7894. Synthetic utility in cycloadditions: R. M. Wilson *et al.*, *ibid.* **127**, 11616 (2005). Applications as a generic mode of SOMO activation: H.-Y. Jang *et al.*, *ibid.* **129**, 7004 (2007); T. D. Beeson *et al.*, *Science* **316**, 582 (2007). Review of iminium activation reactions: G. Lelais, D. W. C. MacMillan, *Aldrichim. Acta* **39**, 79-87 (2006).

First generation catalyst: $R^1 = CH_3$, $R^2 = CH_3$

Second generation catalyst: $R^1 = C(CH_3)_3$, $R^2 = H$

MacMillan's imidazolidinone first generation catalyst. [132278-63-8] (5*S*)-2,2,3-Trimethyl-5-(phenylmethyl)-4-imidazolidinone; (5*S*)-5-benzyl-2,2,3-trimethylimidazolidin-4-one. $C_{13}H_{18}N_2O$; mol wt 218.30. Viscous oil. $[\alpha]_D^{25}$ −48.7° (c = 2.92 in ethanol).

First generation catalyst hydrochloride. [278173-23-2] $C_{13}H_{18}N_2O.HCl$; mol wt 254.76. Colorless crystals from isopropanol.

MacMillan's imidazolidinone second generation catalyst. [346440-54-8] (2*S*,5*S*)-2-(1,1-Dimethylethyl)-3-methyl-5-(phenylmethyl)-4-imidazolidinone; (2*S*,5*S*)-5-benzyl-2-*tert*-butyl-3-methyl-4-imidazolidinone; (2*S*,5*S*)-(−)-2-*tert*-butyl-3-methyl-5-benzyl-4-imidazolidinone. $C_{15}H_{22}N_2O$; mol wt 246.35. White crystalline solid. Air and moisture stable.

Second generation catalyst hydrochloride. [447461-78-1] $C_{15}H_{22}N_2O.HCl$; mol wt 282.81. White crystalline solid from pentane/dichloromethane. $[\alpha]_D$ −39.6° (c = 1.0 in chloroform). Air and moisture stable.

USE: Catalysts for asymmetric synthesis.

5709. Macromerine. [2970-95-8] α-[(Dimethylamino)methyl]-3,4-dimethoxybenzenemethanol; α-[(dimethylamino)methyl]-veratryl alcohol; 3,4-dimethoxy-α-[(dimethylamino)methyl]benzyl alcohol; dimethylaminomethyl 3,4-dimethoxyphenyl carbinol. $C_{12}H_{19}NO_3$; mol wt 225.29. C 63.98%, H 8.50%, N 6.22%, O 21.30%. Prepn of *dl*-form: La Manna *et al.*, *Farmaco Ed. Sci.* **15**, 9 (1960); Chapman *et al.*, *Proc. R. Soc. London Ser. B* **163**, (990), 116 (1965); Hodgkins *et al.*, *Tetrahedron Lett.* **1967**, 1321; of *d*- and *l*-forms: La Manna *et al.*, *loc. cit.* Isoln of *l*-form from cactus, *Coryphantha macromeris* (Engelm.) Lem., Britton and Rose, *Cactaceae:* Hodgkins *et al.*, *loc. cit.*

dl-**Form.** Crystals, mp 46-47°.

l-**Form.** Crystals, mp 66-67.5°. $[\alpha]_D^{25}$ −147.01° (c = 0.0390 g/ml chloroform), −42.61° (c = 0.0200 g/ml abs alcohol).

d-**Form.** Crystals, mp 60-62°.

dl-**Form hydrochloride.** $C_{12}H_{19}NO_3.HCl$. Crystals from ethanol, mp 163-164°.

l-**Form hydrochloride.** Crystals, mp 178-179°. $[\alpha]_D^{18}$ −41.3° (c = 2.04 in 50% ethanol).

d-**Form hydrochloride.** Crystals from ethanol, mp 178- 179°. $[\alpha]_D^{18}$ +51.5° (c = 2.2 in 50% ethanol).

Note: Both the *dl*- and *l*-forms are physiologically active, and caused hallucinogenic reactions when tested in animals (Hodgkins *et al., loc. cit.*).

5710. Macrophage Colony-Stimulating Factor. [81627-83-0] Colony-stimulating factor 1; M-CSF; CSF-1. Species specific hematopoietic growth factor which stimulates the development of committed progenitor cells to monocytes/macrophages. Also potentiates phagocytic activity and monocyte-mediated tumor cell cytotoxicity. Heavily glycosylated homodimer; produced in response to specific activating agents by a variety of cell types including monocytes, fibroblasts and endothelial cells. In humans, several molecular species have been identified which are produced via alternative splicing mechanisms from a single-copy gene. The primary transcripts include a 90 kDa glycoprotein, which is the principle form detected in serum and urine, and a fully bioactive, membrane-associated 40-50 kDa dimer. Murine M-CSF is similar to the human long form. Isoln from human urine: E. R. Stanley *et al.*, *Fed. Proc.* **34**, 2272 (1975); from cultured human pancreatic carcinoma cells: M. Wu *et al.*, *J. Biol. Chem.* **254**, 6226 (1979). Purification of murine M-CSF: E. R. Stanley, P. M. Heard, *ibid.* **252**, 4305 (1977). Characterization of human M-CSF: S. K. Das *et al.*, *Blood* **58**, 630 (1981); K. Motoyoshi *et al.*, *ibid.* **60**, 1378 (1982). Cloning of short form (human clone pcCSF-17): E. S. Kawasaki *et al.*, *Science* **230**, 291 (1985); of long form (human clone p3ACSF-69): G. G. Wong *et al.*, *ibid.* **235**, 1504 (1987). Review of molecular biology: E. S. Kawasaki, M. B. Ladner, *Immunol. Ser.* **49**, 155-176 (1990); of bioactivity and clinical studies: D. H. Munn, N.-K. V. Cheung, *Semin. Oncol.* **19**, 395-407 (1992).

Mirimostim. [121547-04-4] 1-214-Colony-stimulating factor 1 (human clone p3ACSF-69 protein moiety reduced) homodimer; Costilate; Leukoprol.

Lanimostim. [117276-75-2] 4-221-Colony-stimulating factor 1 (human clone p3ACSF-69 protein moiety reduced). Nonglycosylated, 218 amino acid polypeptide produced in *E. coli* by recombinant DNA technology. Prepn and renaturation into bioactive homodimer: R. Halenbeck *et al.*, *Biotechnology* **7**, 710 (1989). Clinical evaluation in invasive fungal infections: J. Nemunaitis *et al.*, *Blood* **82**, 1422 (1993). Immunomodulatory effects and pharmacokinetics: R. M. Bukowski *et al.*, *J. Clin. Oncol.* **12**, 97 (1994).

THERAP CAT: Immunomodulator.

5711. Maduramicin. [84878-61-5] (2*R*,3*S*,4*S*,5*R*,6*S*)-6-[(1*R*)-1-[(2*S*,5*R*,7*S*,8*R*,9*S*)-2-[(2*S*,2′*R*,3′*S*,5*R*,5′*R*)-3′-[(2,6-Dideoxy-3,4-di-*O*-methyl-β-*L*-*arabino*-hexopyranosyl)oxy]octahydro-2-methyl-5′-[(2*S*,3*S*,5*R*,6*S*)-tetrahydro-6-hydroxy-3,5,6-trimethyl-2*H*-pyran-2-yl][2,2′-bifuran]-5-yl]-9-hydroxy-2,8-dimethyl-1,6-dioxaspiro[4.5]dec-7-yl]ethyl]tetrahydro-2-hydroxy-4,5-dimethoxy-3-methyl-2*H*-pyran-2-acetic acid, monoammonium salt; antibiotic X-14868A ammonium salt; CL-273703; Cygro; $C_{47}H_{83}NO_{17}$; mol wt 934.17. C 60.43%, H 8.96%, N 1.50%, O 29.11%. Polyether antibiotic chemically related to the Ionomycins, *q.v.* Isoln as sodium salt from *Nocardia* sp X-14868 and biological activity: C.-M. Liu *et al.*, **US 4278663** (1981 to Hoffmann-La Roche); as free acid hydrate from *Actinomadura yumaense* sp nov.: D. P. Labeda *et al.*, **US 4407946** (1983 to Am. Cyanamid). Fermentation and properties: C.-M. Liu *et al.*, *J. Antibiot.* **36**, 343 (1983). ^{13}C-NMR spectrum: S. Rajan, *ibid.* **37**, 1495 (1984). Biosynthetic studies: H.-R. Tsou *et al.*, *ibid.* 1651; H.-R. Tsou *et al.*, *ibid.* **40**, 94 (1987). Antimalarial activity: L. Oronsky, **US 4496549** (1985 to Am. Cyanamid). Nematocidal activity: I. B. Wood, **US 4510134** (1985 to Am. Cyanamid).

Sodium salt. $C_{47}H_{79}NaO_{17}$. Crystals from ethyl acetate + *n*-hexane, mp 193-195°. $[\alpha]_D$ +40.6° (chloroform). $[\alpha]_D$ +23.8° (methanol).

Note: Not to be confused with **maduramycin** isolated from *Actinomadura rubra*, $C_{28}H_{22}O_{10}$: W. F. Fleck *et al.*, *16th Interscience Conference on Antimicrob. Ag. Chemother.*, Chicago, 1976, *Abstracts of Papers*, no. 51.

THERAP CAT (VET): Coccidiostat.

5712. Mafenide. [138-39-6] 4-(Aminomethyl)benzenesulfonamide; α-amino-*p*-toluenesulfonamide; *p*-(aminomethyl)benzenesulfonamide; 4-homosulfanilamide; maphenide; Marfanil; Mesudrin; Mesudin; Sulfamylon; Homosulfamine; Ambamide; Neofamid; Septicid; Emilene; Homonal; Paramenyl. $C_7H_{10}N_2O_2S$; mol wt 186.23. C 45.15%, H 5.41%, N 15.04%, O 17.18%, S 17.22%. Prepn: Miller *et al.*, *J. Am. Chem. Soc.* **62**, 2099 (1940); Klarer, **US 2288531** (1942 to Winthrop); Bergeim, Braker, *J. Am. Chem. Soc.* **66**, 1459 (1944); Kusami *et al.*, *J. Pharm. Soc. Jpn.* **64**, no. 9a, 51 (1944); Komokina *et al.*, *J. Appl. Chem. USSR* **21**, 681 (1948); Angyal, Jenkin, *Aust. J. Sci. Res.* **3A**, 461 (1950). Toxicity study: T. W. Skulan, J. O. Hoppe, *Life Sci.* **5**, 2279 (1966). Comprehensive description: A. K. Dash, S. Saha, *Anal. Profiles Drug Subs. Excip.* **24**, 277-305 (1996).

Crystals from alc, mp 151-152°. Sol in dil alkali, acid.

Hydrochloride. [138-37-4] $C_7H_{10}N_2O_2S \cdot HCl$. Crystals from 95% alc, mp 256°. Neutral in soln. LD_{50} in rats, mice (mg/kg): 1170, 900 i.v. (Skulan, Hoppe).

Acetate. [13009-99-9] Mafatate; Mefamide. $C_7H_{10}N_2O_2S \cdot C_2H_4O_2$; mol wt 246.28. White to pale yellow, crystalline powder. mp 151-152°. Freely sol in water, methanol. LD_{50} in rats, mice (mg/kg): 2040, 1580 i.v. (Skulan, Hoppe).

Propionate. [12001-72-8] Sulfomyl. $C_7H_{10}N_2O_2S \cdot C_3H_6O_2$; mol wt 260.31. Crystals, mp 158°. Readily sol in water.

THERAP CAT: Antibacterial.

5713. Magainins. A class of potent antimicrobial peptides isolated from the granular gland of clawed toad, *Xenopus laevis*. Name coined by Zasloff from the Hebrew word for "shield". Consists of magainins I and II, both containing 23 amino acid residues and differing only at the 10 (I: Gly, II: Lys) and 22 (I: Lys, II: Asn) positions. Isolation and characterization: M. Zasloff, *Proc. Natl. Acad. Sci. USA* **84**, 5449 (1987). Identification of post-secretory degradation product: T. W. Schwartz *et al.*, *Nature* **329**, 494 (1987). Brief overview: J. Alper, *Chem. Week* 66 (November 4, 1987); M. Cannon, *Nature* **328**, 478 (1987).

Gly–Ile–Gly–Lys–Phe–Leu–His–Ser–Ala–Gly–Lys–Phe
|
Ser–Lys–Met–Ile–Glu–Gly–Val–Phe–Ala–Lys–Gly

Magainin I

5714. Magaldrate. [74978-16-8] Aluminum magnesium hydroxide sulfate ($Al_5Mg_{10}(OH)_{31}(SO_4)_2$) hydrate; magnesium aluminate hydrate; monalium hydrate; AY-5710; Dynese; Riopan; Ripon. Formerly treated as $[Mg(OH)]_4[(HO)_4Al(OH)(HO)Al(OH)_4] \cdot 2H_2O$. Prepd by mixing together and allowing to react at 0° to 50° an alkali aluminate soln (contg 3-5 mols of Na_2O or K_2O per mol Al_2O_3) with a magnesium salt soln in such proportions that 1 g of Al combines with 0.9 to 3.0 g of Mg, and washing the pptd product: Hallmann, **US 2923660** (1960 to Byk-Gulden Lomberg). Characterization: C. L. Peterson *et al.*, *Pharm. Res.* **10**, 998 (1993).

White, odorless, crystalline powder which contains the equivs of not <28.0% and not >39.0% MgO and not <17.0% and not >25.0% Al_2O_3. Sol in dilute solns of mineral acids. Insol in water, alc.

THERAP CAT: Antacid.

5715. Magenta I. [632-99-5] 4-[(4-Aminophenyl)(4-imino-2,5-cyclohexadien-1-ylidene)methyl]-2-methylbenzenamine hydrochloride (1:1); C.I. Basic Violet 14; α^4-(*p*-aminophenyl)-α^4-(4-imino-2,5-cyclohexadien-1-ylidene)-2,4-xylidine hydrochloride; 2-methyl-4,4'-[(4-imino-2,5-cyclohexadien-1-ylidene)methylene]dianiline hydrochloride; fuchsine; magenta; rosaniline hydrochloride; C.I. 42510. $C_{20}H_{20}ClN_3$; mol wt 337.85. C 71.10%, H 5.97%, Cl 10.49%, N 12.44%. Prepn: Fischer, Fischer, *Ann.* **194**, 276, 290 (1878); *Ber.* **13**, 2204 (1880); J. T. Scalan, *J. Am. Chem. Soc.* **57**, 887 (1935). Evaluation of carcinogenicity studies: *IARC Monographs* **4**, 57 (1974). *See also: Colour Index* **vol. 4** (3rd ed., 1971) p 4389.

Metallic green, lustrous crystals, dec above 200°. Absorption max (ethanol): 543 nm (ε 93000). 2.65 parts dissolve in 1000 parts water. Practically insol in ether. Sol in alcohol with a carmine red color.

USE: As a dye or in manuf of other dyes.

THERAP CAT: Antifungal.

5716. Magnalium. [147413-41-0] (50:50); [51542-71-3] (70:30). Aluminum alloy, base, Al,Mg (Magnalium); magnesium-aluminum alloy, aluminum base. Alloy of aluminum with 2-50% magnesium. Description: *Eng. News* **1908**, 13. Glitter effects in pyrotechnics: T. Shimizu, *Pyrotechnica* **14**, 21 (1992); as blue strobe producer: C. Jennings-White, *ibid.* 33. Potential use in fuel-air explosives: F. Ryan, *J. Pyrotech.* **4**, 25 (1996); as solid rocket propellant: H. Habu *et al.*, *32nd Int. Ann. Conf. ICT - Energetic Materials* (Fraunhofer Inst. Chem. Technol., 2001) p 7/1-7/11. Mechanism of noise-producing reactions with lead oxides: T. Shimizu, *Pyrotechnica* **13**, 10 (1990). Safety of fireworks compositions: *idem*, *Proc. 26th Int. Pyrotech. Semin.* **26**, 481 (1999). Pyrophoric at particle sizes <1 μm.

USE: Alloys with high Mg content (~50%) are used in fireworks as a color intensifier and for production of noise and sparks; as solid rocket propellant. Alloys with low Mg content (2-10%) are used in aircraft, automotive, and instrument parts.

5717. Magnesium. [7439-95-4] Mg; at. wt 24.3050; at. no. 12; valence 2. Group IIA (2). Alkaline earth metal. One of the most common elements in the earth's crust: 27,640 ppm. Naturally occurring isotopes: 24 (78.99%); 25 (10.00%); 26 (11.01%). Known radioactive isotopes: 20-23, 27-34. Found naturally only in the form of its compounds in *magnesite*, *carnallite*, *dolomite* [CaMg(CO_3)$_2$], *epsomite*, *kieserite*, and many other minerals; found in seawater, underground natural brines and salt deposits. Essential nutrient for most plant and animal life. Commercial production by electrolytic reduction of magnesium chloride or thermal reduction of magnesium oxide. First obtained in metallic form by Davy in 1808 by electrolyte reduction of magnesium oxide with mercury cathode. Prepn: Deville *et al.* in *Gmelins, Magnesium* (8th ed.) **27A**, 121 (1937). Review of magnesium and its compounds: Goodenough, Stenger, "Magnesium, Calcium, Strontium, Barium and Radium" in *Comprehensive Inorganic Chemistry* **vol. 1**, J. C. Bailar, Jr. *et al.*, Eds. (Pergamon Press, Oxford, 1973) pp 591-664; *Chemistry of the Elements* N. N. Greenwood, A. Earnshaw, Eds. (Pergamon Press, New York, 1984) pp 117-154; C. B. Wilson *et al.* in *Kirk-Othmer Encyclopedia of Chemical Technology* **vol. 15** (Wiley-Interscience, New York, 4th ed., 1995) pp 622-674.

Silvery-white metal; hexagonal close-packed structure. Slowly oxidizes in moist air. Available as bars, ribbons, wire, castings, sheets and powder. mp 651°. bp 1100°. d^{20} 1.738. Specific heat (20°) 0.245 cal/g. Heat of fusion 88 cal/g. Electrical resistivity 4.46 μohm-cm. E° (aq) Mg^{2+}/Mg −2.37 V. *Dangerous when wet; spontaneously combustible.* Reacts very slowly with water at ordinary temp, less slowly at 100°. Reacts readily with dil acids with liberation of hydrogen; reacts with aq solns of ammonium salts, forming a double salt. Reduces carbon monoxide, carbon dioxide, sulfur dioxide, nitric oxide, and nitrous oxide at a red heat. Fine powder, thin sheets, chips and turnings are easily ignited and burn with intense heat and brilliant white flame. Combines directly with nitrogen, sulfur, the halogens, phosphorus, and arsenic. Reacts vigorously with anhydrous methyl alcohol.

Caution: Direct contact may cause irritation of skin, eyes, respiratory system. *See: Fire Protection Guide to Hazardous Materials* (National Fire Protection Assoc., Quincy, MA, 12th ed., 1997) Section 49, p 83.

USE: In alloys to produce light weight structural metals. In aluminum alloys to improve mechanical properties; in Grignard reagents; in dry cell batteries; in pyrotechnics. For hot metal desulfurization, esp. molten iron; prodn of ductile iron; metal reduction to produce elemental boron, titanium, zirconium; corrosion protection of steel structures; sacrificial anodes for corrosion protection.

5718. Magnesium Acetate. [142-72-3] Acetic acid magnesium salt (2:1); cromosan. $C_4H_6MgO_4$; mol wt 142.39. C 33.74%, H 4.25%, Mg 17.07%, O 44.94%. $Mg(C_2H_3O_2)_2$.

Tetrahydrate. [16674-78-5] $C_4H_6MgO_4.4H_2O$; mol wt 214.45. Colorless or white, deliquesc crystals. d 1.45. mp about 80°. Very sol in water or alcohol. The aq soln is neutral or slightly acid. *Keep well closed.*

USE: Buffer. Detection of sodium.

5719. Magnesium Acetylsalicylate. [132-49-0] 2-(Acetyloxy)benzoic acid magnesium salt (2:1); magnesium aspirin; Apyron; Fyracyl; Magisal; Magnespirin; Novacetyl. $C_{18}H_{14}MgO_8$; mol wt 382.61. C 56.51%, H 3.69%, Mg 6.35%, O 33.45%. $(CH_3COOC_6H_4COO)_2Mg$.

White, nonhygroscopic, almost tasteless and odorless powder. Freely soluble in water; less sol in alcohol.

THERAP CAT: Has been used as analgesic, antipyretic.

5720. Magnesium Amide. [7803-54-5] Magnesium diamide. H_4MgN_2; mol wt 56.35. H 7.16%, Mg 43.13%, N 49.71%. $Mg(NH_2)_2$. Prepd by the action of ammonia on an ether soln of magnesium diethyl or on magnesium activated with iodine (at 400°): Terentiew, *Z. Anorg. Chem.* **162**, 351 (1927); Schlenk, Jr., *Ber.* **64**, 738 (1931). Prepd from metal and liquid ammonia at high pressure: Juza, Jacobs, *Angew. Chem. Int. Ed.* **5**, 247 (1966); *eidem, Z. Anorg. Allg. Chem.* **370**, 245 (1969).

White powder or crystals. d_4^{25} 1.39. *Catches fire in air.* Dec on heating. *Reacts violently with water, evolving ammonia.*

USE: Polymerization catalyst.

5721. Magnesium Benzoate. [553-70-8] Benzoic acid magnesium salt (2:1). $C_{14}H_{10}MgO_4$; mol wt 266.54. C 63.09%, H 3.78%, Mg 9.12%, O 24.01%. $Mg(C_7H_5O_2)_2$.

Trihydrate. White, odorless powder. mp about 200°. Sol in 20 parts water, freely in hot water, alcohol. The aq soln is neutral or slightly acid.

5722. Magnesium Borate. [13703-82-7] Boric acid (HBO_2) magnesium salt (2:1); "antifungin"; magnesium metaborate. B_2MgO_4; mol wt 109.92. B 19.67%, Mg 22.11%, O 58.22%. $Mg(BO_2)_2$. Various magnesium borate minerals occur in nature. These

include: *ascharite, camsellite, inderite, kotoite, kurnakovite, paternoite, pinnoite, szaibelyite.*

Octahydrate. White powder. Slightly sol in water.

USE: Antiseptic; fungicide.

5723. Magnesium Bromide. [7789-48-2] Magnesium dibromide. Br_2Mg; mol wt 184.11. Br 86.80%, Mg 13.20%. $MgBr_2$.

Hexahydrate. Colorless, very deliquesc crystals or white granules; bitter taste. mp about 165° with decompn. Sol in 0.3 part water; sol in alcohol. The aq soln is neutral. *Keep well closed.*

USE: In organic syntheses.

THERAP CAT: Sedative, anticonvulsant.

5724. Magnesium Carbonate Hydroxide. [39409-82-0] Carbonic acid magnesium salt (1:1) mixt. with magnesium hydroxide ($Mg(OH)_2$) hydrate; Marinco C. Usually a basic carbonate, approx $(MgCO_3)_4.Mg(OH)_2.5H_2O$; approx mol wt 485; contains 40-42% MgO. Prepn of various hydrates: Pond, Heneghan, **US 3169826** (1965 to Merck & Co.). Magnesium carbonate minerals include *magnesite* and *lansfordite.*

White, odorless, bulky powder or light friable masses; at about 700° is converted into MgO. Sol in about 3300 parts CO_2-free water; more sol in water contg CO_2; sol in dil acids with effervescence. Insol in alcohol. Imparts slight alkaline reaction to water.

Hydrotalcite. [12304-65-3] Altacite; Hi-Ti; Nacid; Talcid. An aluminum magnesium hydroxide carbonate hydrate of the general formula $Al_2O_3.6MgO.CO_2.12H_2O$. Prepn: Kumuru *et al.*, **US 3539306** (1970 to Kyowa).

Caution: Potential symptoms of overexposure to magnesite are irritation of eyes, skin, respiratory system; cough. *See NIOSH Pocket Guide to Chemical Hazards* (DHHS/NIOSH 97-140, 1997) p 188.

USE: Fireproofing, heat insulating; preparing effervescent magnesium citrate; clarifying liqs by filtration; in tooth and face powders, in polishing compds; manuf mineral waters, pigments, paper; filler for rubber.

THERAP CAT: Magnesium carbonate hydroxide as antacid, cathartic; hydrotalcite as antacid.

THERAP CAT (VET): Laxative.

5725. Magnesium Chlorate. [10326-21-3] Chloric acid magnesium salt (2:1). Cl_2MgO_6; mol wt 191.20. Cl 37.08%, Mg 12.71%, O 50.21%. $Mg(ClO_3)_2$.

Hexahydrate. White, very deliquesc crystals or cryst powder; bitter taste; d 1.80. mp ~35°. Sol in 0.9 part water; slightly sol in alcohol. *Oxidizer. Keep well closed.* LD_{50} orally in rats: 5.25 g/kg: Ulrich, *J. Pharmacol. Exp. Ther.* **35**, 1 (1929).

5726. Magnesium Chloride. [7786-30-3] Magnesium chloride ($MgCl_2$); Magnogene. Cl_2Mg; mol wt 95.21. Cl 74.47%, Mg 25.53%. $MgCl_2$. Prepd from magnesium ammonium chloride hexahydrate in the presence of HCl: Bryce-Smith, *Inorg. Synth.* **6**, 9 (1960). Toxicity study: H. F. Smyth *et al.*, *Am. Ind. Hyg. Assoc. J.* **30**, 470 (1969).

Soft, highly deliquescent leaflets. mp 712° (rapid heating). d 2.41, also reported as d 2.325. Slow heating releases chlorine at 300°. Can be distilled in a stream of hydrogen. Attacks fused silica when melted. Very sol in water, with the evolution of much heat, giving a clear soln. Freely sol in alc. Easily forms alcoholates and etherates.

Hexahydrate. [7791-18-6] $Cl_2Mg.6H_2O$; mol wt 203.30. Deliquescent crystals. d 1.56. At 100° loses $2H_2O$ (17.7%); at 110° begins to lose some HCl. By strong ignition is converted into oxychloride. mp, when rapidly heated, ~118° with decompn. One gram dissolves in 0.6 ml water, 0.3 ml boiling water, 2 ml alcohol. Its aq soln is neutral. *Keep well closed.* LD_{50} orally in rats: 8.1 g/kg (Smyth).

USE: Fireproofing wood; in disinfectants; various magnesia cements; fire extinguishers; dressing cotton fabrics; in floor-sweeping compds; carbonizing wool; manuf parchment paper or artificial leather; as addition to casein glue. As a reagent in analytical chemistry for adjustment of ion strength.

THERAP CAT: Cathartic.

THERAP CAT (VET): Has been used in bovine hypomagnesemia.

5727. Magnesium Citrate. [7779-25-1] 2-Hydroxy-1,2,3-propanetricarboxylic acid magnesium salt (1:?); Citramag. $Mg_x(C_6$-

$H_5O_7)_y$. Prepn of stable, soluble form: J. L. Davenport, C. F. De Costa, US 2260004 (1941 to Pfizer). Clinical evaluation of cathartic efficacy in pediatric toxic ingestions: Y.-J. Sue *et al.*, *Ann. Emerg. Med.* **24**, 709 (1994); of colonoscopic prepn with magnesium citrate + bisacodyl vs castor oil, *q.q.v.*: C.-C. Chen *et al.*, *J. Gastroenterol. Hepatol.* **14**, 1219 (1999).

Magnesium citrate dibasic. [144-23-0] Magnesium hydrogen citrate. $C_6H_6MgO_7$; mol wt 214.41.

Magnesium citrate tribasic. [3344-18-1]; [153531-96-5] (nonahydrate); [6150-79-4] (tetradecahydrate). Trimagnesium dicitrate. $C_{12}H_{10}Mg_3O_{14}$; mol wt 451.11. Soly of hydrated forms: A. Apelblat, *J. Chem. Thermodyn.* **25**, 1443 (1993). Thermal decomposition of tetradecahydrate: S. A. A. Mansour, *Thermochim. Acta* **233**, 231 (1994). Soly in water at 298 K (*m*): nonahydrate 0.0482; tetradecahydrate 0.0446.

THERAP CAT: Cathartic.

5728. Magnesium Diboride. [12007-25-9] Magnesium boride (MgB_2). B_2Mg; mol wt 45.93. B 47.07%, Mg 52.92%. MgB_2. Prepn and crystal structure: V. Russell *et al.*, *Acta Crystallogr.* **6**, 870 (1953); M. E. Jones, R. E. Marsh, *J. Am. Chem. Soc.* **76**, 1434 (1954). Use as reagent in metathesis reactions: E. G. Gillan, R. B. Kaner, *Chem. Mater.* **8**, 333 (1996). Superconductive properties: J. Nagamatsu *et al.*, *Nature* **410**, 63 (2001). Brief review: R. J. Cava, *ibid.* 23-24.

Soluble in dilute hydrochloric acid. Becomes superconductive at 39 K.

USE: Chemical reagent.

5729. Magnesium Fluoride. [7783-40-6] Magnesium difluoride; Sellaite. F_2Mg; mol wt 62.30. F 60.99%, Mg 39.01%. MgF_2. Prepd from $MgCO_3$ + HF: Klemm *et al.*, *Z. Anorg. Allg. Chem.* **176**, 13 (1928). Review of toxicology of fluoride compounds: G. L. Waldbott, *Acta Med. Scand. Suppl.* **400**, 1-44 (1963).

Colorless substance (rutile-type lattice). Slight violet fluorescence. d 3.148. Mohs' hardness: 6. mp 1248°. bp 2260°. Very sparingly sol in water 87 mg/l (18°). Slightly sol in dil acids, especially nitric acid. May be stored in glass bottles. LD in guinea pigs (mg/kg): 1000 orally, 3000 s.c. (Waldbott).

USE: In the ceramics and glass industry.

5730. Magnesium Formate. [557-39-1] Formic acid magnesium salt (2:1). $C_2H_2MgO_4$; mol wt 114.34. C 21.01%, H 1.76%, Mg 21.26%, O 55.97%. $Mg(HCOO)_2$.

Dihydrate. Crystals or granules. Sol in water; insol in alcohol. Its aq soln is practically neutral.

5731. Magnesium Germanide. [1310-52-7] Germanium compd. with magnesium (1:2). $GeMg_2$; mol wt 121.24. Ge 59.91%, Mg 40.09%. Mg_2Ge. Prepd by fusion of the components: Winkler, *Helv. Phys. Acta* **28**, 633 (1955).

Crystals, mp 1115°.

USE: In semiconductor research.

5732. Magnesium Hexafluorosilicate. [16949-65-8] Hexafluorosilicate(2−) magnesium (1:1); magnesium silicofluoride; magnesium fluosilicate. F_6MgSi; mol wt 166.38. F 68.51%, Mg 14.61%, Si 16.88%. $MgSiF_6$. Review of toxicology of fluorine compounds: G. L. Waldbott, *Acta Med. Scand. Suppl.* **400**, 1-44 (1963).

LD in guinea pigs (mg/kg): 200 orally, 400 s.c. (Waldbott).

Hexahydrate. [18972-56-0] $F_6MgSi.6H_2O$. White, efflorescent, odorless crystals. d 1.788. At ~120° loses SiF_4. Soluble in water. Insol in alcohol. pH of 1% aq soln: 3.1.

USE: For mothproofing of textile fabrics.

5733. Magnesium Hydride. [7693-27-8] Magnesium dihydride. H_2Mg; mol wt 26.32. H 7.66%, Mg 92.34%. Prepd by thermal decompn of diethylmagnesium at 200° in high vacuum: Wiberg, Bauer, *Ber.* **85**, 593 (1952); from the elements: Ellinger *et al.*, *J. Am. Chem. Soc.* **77**, 2647 (1955).

White, nonvolatile mass or tetragonal crystals. d 1.45. Dec at 280° in high vacuum. Strong reducing agent. Ignites spontaneously on contact with air, forming MgO and H_2O. *Dangerous when wet.* On contact with water a violent evolution of hydrogen takes place. On contact with methanol, magnesium alcoholate and hydrogen are formed. Forms double hydrides with boron hydride and aluminum hydride.

5734. Magnesium Hydroxide. [1309-42-8] Magnesium hydrate; Carmilax; Emgesan; Marinco H. H_2MgO_2; mol wt 58.32. H 3.46%, Mg 41.68%, O 54.87%. $Mg(OH)_2$. Occurs in nature as the mineral *brucite*. Prepn from magnesium chloride or sulfate and NaOH: Perlard, Waldron, US 3127241 (1964 to Dow). Industrial method for producing magnesium hydroxide having variable particle sizes: Chisholm, US 3232708 (1966 to FMC).

Bulky white, amorphous powder. Sol in dil acids. Practically insol in water (1:80,000), alc. Imparts slight alkaline reaction to water. pH of aq slurry: 9.5-10.5. Absorbs CO_2 in the presence of water. *Keep well closed.*

THERAP CAT: Antacid; cathartic.

THERAP CAT (VET): Laxative.

5735. Magnesium Iodide. [10377-58-9] Magnesium diiodide. I_2Mg; mol wt 278.11. I 91.26%, Mg 8.74%. MgI_2.

Octahydrate. White, deliquesc powder; readily discolors in air and light. Very sol in water; sol in alcohol. The aq soln is neutral or slightly alkaline. *Keep well closed and protected from light.*

5736. Magnesium Lactate. [18917-93-6] (*T*-4)-Bis[2-(hydroxy-κO)propanoato-κO]magnesium; 2-hydroxypropanoic acid magnesium salt. $C_6H_{10}MgO_6$; mol wt 202.45. C 35.60%, H 4.98%, Mg 12.01%, O 47.42%. $(CH_3CHOHCOO)_2Mg$.

Trihydrate. White cryst, very bitter powder. One gram dissolves in 25 ml cold water, 3.5 ml boiling water; slightly sol in alcohol. The aq soln is slightly acid.

THERAP CAT: Cathartic.

5737. Magnesium Monoperoxyphthalate. [78948-87-5] (*T*-4)-Bis[2-carboxybenzenecarboperoxoato(2−)-$\kappa OO,\kappa O'$]magnesate(2−) hydrogen (1:2); magnesium bis(monoperoxyphthalate); H48; MMPP. $C_{16}H_{10}MgO_{10}$; mol wt 386.55. C 49.72%, H 2.61%, Mg 6.29%, O 41.39%. Versatile, mild peroxygen product; stable alternative to 3-chloroperoxybenzoic acid, *q.v.*, in oxidation reactions. Prepn: G. J. Hignett, EP 27693; *idem*, US 4385008 (1981, 1983 both to Interox). Thermal decompn studies: B. M. Kariuki, W. Jones, *Mol. Cryst. Liq. Cryst.* **186**, 45 (1990); *eidem*, *ibid.* **248**, 21 (1994). Decontamination reactions with warfare agents and insecticides: C. Lion *et al.*, *Phosphorus Sulfur Silicon Relat. Elem.* **56**, 213 (1991). Antimicrobial properties: M. G. C. Baldry, *J. Appl. Bacteriol.* **57**, 499 (1984). Clinical evaluation as an anti-plaque mouthrinse and dentifrice: C. Scully *et al.*, *J. Clin. Periodontol.* **26**, 234 (1999). Review of synthetic applications in oxidation reactions: P. Brougham *et al.*, *Synthesis* **1987**, 1015-1017; H. Heaney, *Aldrichim. Acta* **26**, 35-45 (1993).

Hexahydrate. [84665-66-7] $C_{16}H_{10}MgO_{10}.6H_2O$; mol wt 494.64. White granular solid, mp 96°. Soly in water (15°): 160 g/l. Sol in low mol wt alcohols; very low soly in common organic solvents.

USE: Oxidizing agent in organic synthesis. Bleaching agent. Decontaminant reagent for nerve gases and insecticides.

5738. Magnesium Nitrate. [10377-60-3] Nitric acid magnesium salt (2:1). MgN_2O_6; mol wt 148.31. Mg 16.39%, N 18.89%, O 64.73%. $Mg(NO_3)_2$. Hydrated form occurs in nature as the mineral *nitromagnesite*. Thermal decompn studies of hexahydrate: F. Paulik *et al.*, *J. Therm. Anal.* **34**, 627 (1988).

Hexahydrate. [13446-18-9] $MgN_2O_6.6H_2O$; mol wt 256.40. Colorless, clear, deliquesc crystals. d 1.464. mp 89°. Begins to boil at 147°. Sol in 0.8 part water; freely sol in alcohol. The aq soln is neutral. *Oxidizer. Keep well closed.*

USE: In pyrotechnics; in the concentration of nitric acid.

5739. Magnesium Oleate. [1555-53-9] (9*Z*)-9-Octadecenoic acid magnesium salt (2:1). $C_{36}H_{66}MgO_4$; mol wt 587.23. C 73.63%, H 11.33%, Mg 4.14%, O 10.90%. $Mg(C_{18}H_{33}O_2)_2$. Commercial form usually contains some stearate and palmitate.

Yellowish powder or mass. Insol in water; partly sol in alcohol, ether, petr ether.

USE: As an addition to naphtha to prevent spontaneous ignition in cleaning establishments.

5740. Magnesium Oxalate. [547-66-0] [Ethanedioato(2−)-$\kappa O^1,\kappa O^2$]-magnesium; ethanedioic acid magnesium salt. C_2MgO_4; mol wt 112.32. C 21.39%, Mg 21.64%, O 56.98%. MgC_2O_4.

Dihydrate. White powder. Sol in about 1500 parts water, in dil mineral acid; insol in alcohol.

5741. Magnesium Oxide. [1309-48-4] Magnesium oxide (MgO); magnesia; calcined magnesia; magnesia usta; Magcal; Maglite. MgO; mol wt 40.30. Mg 60.31%, O 39.70%. Occurs in nature as the mineral *periclase*. Commercial prepn from magnesite ores: Adams, **US 3320029** (1967 to Northwest Magnesite).

White, very fine, odorless powder. Available in a very bulky form termed "Light" or in a dense form termed "Heavy". Takes up CO_2 and H_2O from the air, the light form more readily than the heavy. mp 2800°. Highly reflective in visible and near uv region. Combines with water to form magnesium hydroxide. Sol in dil acids. Practically insol in pure water, soly increased by CO_2; insol in alcohol. Imparts a slight alkaline reaction to water. pH of satd aq soln 10.3. *Keep well closed.*

Caution: Potential symptoms of overexposure to fumes are irritation of eyes and nose; metal fume fever (coughing, chest pain and flu-like fever). *See NIOSH Pocket Guide to Chemical Hazards* (DHHS/NIOSH 97-140, 1997) p 188.

USE: Manuf refractory crucibles, fire bricks, magnesia cements and boiler scale compounds, "powdered" oils, casein glue. Reflector in optical instruments; white color standard. Insulator at low temp. Absorbant for colorants prior to determn. Prepn of Eschka's reagent.

THERAP CAT: Antacid.

THERAP CAT (VET): Antacid, laxative, in hypomagnesemia.

5742. Magnesium Perchlorate. [10034-81-8] Perchloric acid magnesium salt (2:1); Anhydrone; Dehydrite. Cl_2MgO_8; mol wt 223.20. Cl 31.77%, Mg 10.89%, O 57.34%. $Mg(ClO_4)_2$.

White, very hygroscopic, granular or flaky powder; dec above 250°. Dissolves in water with evolution of a considerable amount of heat. Crystallizes from water with $6H_2O$. *Oxidizer. Keep tightly closed.*

Caution: Can cause irritation of skin, mucous membranes.

USE: As a drying agent for gases. The article of commerce may contain an amount of water equivalent to a dihydrate, but even the trihydrate is said to be effective for drying gases.

5743. Magnesium Permanganate. [10377-62-5] $MgMn_2O_8$; mol wt 262.17. Mg 9.27%, Mn 41.91%, O 48.82%. $Mg(MnO_4)_2$. As polymerization catalyst: Pengilly *et al.,* **US 3576790** (1971 to Goodyear). In purification of benzene: Ingwalson *et al.,* **US 3478092** (1969 to Velsicol).

Bluish-black, deliquesc crystals. Freely sol in water. *Keep well closed.*

USE: Catalyst.

5744. Magnesium Peroxide. [14452-57-4] Magnesium dioxide; Magnesium Perhydrol; Magnesium Superoxol. MgO_2; mol wt 56.30. Mg 43.17%, O 56.83%. The article of commerce contains 15-25% MgO_2, the balance being magnesium hydroxide. For medicinal use the 25% article is preferred.

White, tasteless, odorless powder. Insol in water and gradually dec by it with liberation of oxygen; sol in dil acids, forming hydrogen peroxide; when strongly heated loses all the peroxide oxygen. *Oxidizer. Keep well closed.*

THERAP CAT: Antacid, anti-infective.

5745. Magnesium Phosphate, Dibasic. [7757-86-0] Phosphoric acid magnesium salt (1:1); magnesium hydrogen phosphate; secondary magnesium phosphate. $HMgO_4P$; mol wt 120.28. H 0.84%, Mg 20.21%, O 53.21%, P 25.75%. $MgHPO_4$. Trihydrate occurs in nature as the minerals *newberyite, phosphor-roesslerite*.

Trihydrate. White, cryst powder. d 2.13. Slightly sol in water; sol in dil acids.

THERAP CAT: Cathartic.

5746. Magnesium Phosphate, Monobasic. [13092-66-5] Phosphoric acid magnesium salt (2:1); magnesium biphosphate; acid

magnesium phosphate; primary magnesium phosphate. $H_4MgO_8P_2$; mol wt 218.28. H 1.85%, Mg 11.13%, O 58.64%, P 28.38%. Mg-$(H_2PO_4)_2$.

Trihydrate. White powder. Soluble in water.

USE: In fireproofing wood.

5747. Magnesium Phosphate, Tribasic. [7757-87-1] Phosphoric acid magnesium salt (2:3); "neutral" magnesium phosphate; tertiary magnesium phosphate; trimagnesium phosphate. $Mg_3O_8P_2$; mol wt 262.85. Mg 27.74%, O 48.69%, P 23.57%. $Mg_3(PO_4)_2$. Octahydrate occurs in nature as the mineral *bobierrite*.

Pentahydrate. White, cryst powder. Loses last mol of water at ~400°. Readily sol in diluted mineral acids. Almost insol in water.

THERAP CAT: Antacid.

5748. Magnesium Pyrophosphate. [13446-24-7] Diphosphoric acid magnesium salt (1:2); dimagnesium diphosphate. Mg_2-O_7P_2; mol wt 222.55. Mg 21.84%, O 50.32%, P 27.84%. $Mg_2P_2O_7$.

Trihydrate. White powder; loses its water at 100°. d 2.56. Insol in water. Sol in dil mineral acids.

5749. Magnesium Salicylate. [18917-89-0] (*T*-4)-Bis[2-(hydroxy-κO)benzoato-κO]magnesium; 2-hydroxybenzoic acid magnesium salt; Analate; Lorisal; Magan; Mobidin; Triact. $C_{14}H_{10}$-MgO_6; mol wt 298.53. C 56.33%, H 3.38%, Mg 8.14%, O 32.16%. $Mg[C_6H_4(OH)COO]_2$.

Tetrahydrate. White, odorless, efflorescent, crystalline powder. Sol in 13 parts water. The aq soln is slightly acid. Freely sol in methanol; sol in alc; slightly sol in ether.

THERAP CAT: Anti-infective (intestinal).

5750. Magnesium Selenide. [1313-04-8] Magnesium monoselenide. MgSe; mol wt 103.27. Mg 23.54%, Se 76.46%. Prepd by the action of hydrogen selenide on anhydr magnesium chloride at a red heat and by the action of selenium vapor carried in a current of nitrogen on powdered magnesium: Fonzes-Diacon, *Contribution a l'Etude des Séléniures Métalliques* (Montpellier, 1901); by dropping selenium into molten magnesium: Liddell, *Chem. Metall. Eng.* **25**, 102, 263, 453 (1921).

Light brown powder. d 4.21. Unstable in air. Dec in water.

5751. Magnesium Silicates. Several varieties of magnesium silicate are known. *See also* asbestos, talc.

Magnesium metasilicate. [13776-74-4] $MgSiO_3$; mol wt 100.39. Occurs in nature as the minerals *clinoenstatite, enstatite, protoenstatite*. White monoclinic crystals, dec at 1557°. d_4^{25} 3.192. Practically insol in water; very slightly sol in HF.

Magnesium orthosilicate. [10034-94-3] Mg_2SiO_4; mol wt 140.69. Occurs in nature as the mineral *forsterite*. White orthorhombic crystals. mp 1910°; d 3.21. Practically insol in water.

Magnesium trisilicate. [14987-04-3] Magnesium mesotrisilicate; Magnosil; Petimin; Trisomin. $Mg_2Si_3O_8$; mol wt 260.86. Occurs in nature as the minerals *meerschaum, parasepiolite, sepiolite*. Prepd from sodium silicate and magnesium sulfate: Glass, *Q. J. Pharm. Pharmacol.* **9**, 445 (1936); Uyeda, **US 3272594** (1966 to Merck & Co.). White, odorless, tasteless, slightly hygroscopic powder. Usually contains some water of hydration. Insol in water, alc. Readily decomposes with mineral acids.

Magnesium trisilicate pentahydrate. Sellagen.

Serpentine (mineral). $Mg_3O_7Si_2$; mol wt 241.08. Occurs in nature as the dihydrate, $(OH)_4MgSi_2O_5$; exists in two forms: *antigorite*, a platy variety, and *chrysotile*, a fibrous variety. The latter is the most common form of asbestos.

USE: An activated magnesium silicate, *Florisil*, a hard, porous, granular substance, is used in vitamin analysis, chromatography, antibiotic processing: Simons, **US 2393625** (1946 to Floridin). Pharmaceutic aid (glidant or anticaking agent).

THERAP CAT: Trisilicate as antacid.

THERAP CAT (VET): Trisilicate as antacid; gastric sedative.

5752. Magnesium Silicide. [22831-39-6] Dimagnesium silicide. Mg_2Si; mol wt 76.70. Mg 63.38%, Si 36.62%. Prepd by heating finely powdered Mg and Si in a proportion of 20 to 6: Gire, *Compt. Rend.* **196**, 1404 (1933), *C.A.* **27**, 3678 (1933); Winkler, *Helv. Phys. Acta* **28**, 633 (1955), *C.A.* **50**, 6908c (1956). Prepn and conversion to silanes: Emeleus, Maddock, *J. Chem. Soc.* **1946**, 1131.

Slate-blue, cubic crystals. d_4^{20} ~2.0. mp 1085°. Dec at 550° *in vacuo* yielding Mg_3Si_2. Heat of formation: 18.5 cal/mol. Dec by water, HCl. *Dangerous when wet.*

USE: In semiconductor research. Has been used to build Mg-Si rectifiers.

5753. Magnesium Stannide. [1313-08-2] Magnesium compd with tin (2:1) . Mg_2Sn; mol wt 167.32. Mg 29.05%, Sn 70.95%. Prepn by fusion of a stoichiometric mixture of the elements: Winkler, *Helv. Phys. Acta* **28**, 633 (1955); Korenblit, Kolesnikov, *Zh. Tekh. Fiz.* **26**, 941 (1956), *C.A.* **50**, 11475b (1956). Magnetic susceptibility and electrical conductivity measurements: Boltaks, *Dokl. Akad. Nauk SSSR* **64**, 487, 653, (1949), *C.A.* **43**, 4528f, 4533f (1949). Electrical conductivity, magnetoresistance: Frederikse *et al.*, *Phys. Rev.* **103**, 67 (1956).

Bluish-white, metallic substance. Calcium fluoride crystal structure. mp 778°. Resistivity (25°): 42,000 microhm-cm; thermoelectric power 90 microvolts per degree for 15-100°. Energy of activation: 22,100 cal/mole. Soluble in water, dilute HCl.

USE: In semiconductor research.

5754. Magnesium Stearate. [557-04-0] Octadecanoic acid magnesium salt (2:1). $C_{36}H_{70}MgO_4$; mol wt 591.26. C 73.13%, H 11.93%, Mg 4.11%, O 10.82%. $Mg(C_{18}H_{35}O_2)_2$. The commercial preparation also contains palmitate.

White powder. Insoluble in water, alc, ether. Dec by dil acids.

USE: In baby dusting powders; as tablet lubricant.

5755. Magnesium Sulfate. [7487-88-9] Sulfuric acid magnesium salt (1:1). MgO_4S; mol wt 120.36. Mg 20.19%, O 53.17%, S 26.64%. $MgSO_4$. Monohydrate occurs in nature as the mineral *kieserite. Review: Gmelins, Magnesium* (8th ed) **27**, 210-211, 223-226 (1939). Review of clinical experience: R. Swain, B. Kaplan-Machlis, *South. Med. J.* **92**, 1040-1047 (1999); of toxicology: J. W. Van Hook, *Crit. Care Clin.* **7**, 215-223 (1991); R. B. Birrer *et al.*, *J. Emerg. Med.* **22**, 185-188 (2002). Clinical trial in prevention of eclamptic convulsions: L. Duley *et al.*, *Lancet* **359**, 1877 (2002).

Trihydrate. [15320-30-6] $MgO_4S.3H_2O$; mol wt 174.41. Prepn: Bennett, *US 3297413* (1967 to Dow). Odorless crystals.

Heptahydrate. [10034-99-8] Bitter salts; epsom salts; Mg 5-Sulfat. $MgO_4S.7H_2O$; mol wt 246.47. Occurs in nature as the mineral *epsomite.* Efflorescent crystals or powder; bitter, saline, cooling taste. d 1.67. Soly in water (g/100 ml): at 20° = 71; at 40° = 91. Freely (and slowly) sol in glycerin; sparingly sol in alc. Its aq soln is neutral. pH 6-7. On exposure to dry air at ordinary temp it loses ~1H_2O; at 70-80° loses 4H_2O; at 100° loses 5H_2O; at 120° loses 6H_2O, rapidly reabsorbing water when exposed to moist air; loses the last mol of water at ~250°.

Dried magnesium sulfate. Prepd by heating the heptahydrate until ~25% of its weight is lost. Used in *Morison's paste. Keep well closed.*

Caution: Parenteral use or use in presence of renal insufficiency may lead to magnesium intoxication. *See:* M. J. Ellenhorn *et al.*, *Ellenhorn's Medical Toxicology: Diagnosis and Treatment of Human Poisoning* (Williams & Wilkins, Baltimore, 2nd ed., 1997) pp 1013-1014. Potential symptoms of overexposure are pseudocoma, hyporeflexia, muscle weakness, ataxia, confusion, nausea, vomiting, flushing, bradycardia, hypotension, widened QRS, AV block and respiratory arrest (Swain, Kaplan-Machlis).

USE: Weighting cotton and silk; increasing the bleaching action of chlorinated lime; manuf mother-of-pearl and frosted papers; fireproofing fabrics; dyeing and printing calicos; in fertilizers; explosives, matches; mineral water; tanning leather.

THERAP CAT: Anticonvulsant; cathartic; magnesium replenisher.

THERAP CAT (VET): Anticonvulsant; cathartic; magnesium replenisher.

5756. Magnesium Sulfite. [7757-88-2] Sulfurous acid magnesium salt (1:1). MgO_3S; mol wt 104.36. Mg 23.29%, O 45.99%, S 30.72%. $MgSO_3$. Occurs as tri- and hexahydrates.

Hexahydrate. Colorless crystals or white, cryst powder; gradually oxidizes to sulfate on exposure to air; loses all its water at 200°; dec at higher temp. Sol in ~150 parts water, slightly more in boiling water.

USE: A soln of magnesium sulfite in sulfurous acid (magnesium bisulfite) is used in the manuf of paper pulp.

5757. Magnesium Thiocyanate. [306-61-6] Magnesium sulfocyanate. $C_2MgN_2S_2$; mol wt 140.46. C 17.10%, Mg 17.30%, N 19.94%, S 45.65%. $Mg(SCN)_2$.

Tetrahydrate. Colorless or white deliquesc crystals. Freely sol in water or alcohol. *Keep well closed.*

5758. Magnesium Thiosulfate. [10124-53-5] Thiosulfuric acid ($H_2S_2O_3$) magnesium salt (1:1); magnesium hyposulfite; Magnosulf; Antichoc Hipmag. MgO_3S_2; mol wt 136.42. Mg 17.82%, O 35.18%, S 47.00%. MgS_2O_3.

Hexahydrate. Colorless or white crystals; loses 3H_2O at 170°. d 1.82. Sol in 2 parts water; insol in alcohol. Its aq soln is neutral.

5759. Magneson. [74-39-5] 4-[2-(4-Nitrophenyl)diazenyl]-1,3-benzenediol; 4-[(4-nitrophenyl)azo]-1,3-benzenediol; 4-(*p*-nitrophenylazo)resorcinol; 2,4-dihydroxy-4'-nitroazobenzene. $C_{12}H_9N_3O_4$; mol wt 259.22. C 55.60%, H 3.50%, N 16.21%, O 24.69%. *Ref:* Suitzu, Okuma, *J. Soc. Chem. Ind. Jpn.* **29**, 132 (1926); Ruigh, *J. Am. Chem. Soc.* **51**, 1456 (1929); Engel, *ibid.* **52**, 1812 (1930).

Brownish-red powder. Practically insol in water; sol in dil aq NaOH.

USE: For the detection of magnesium with which it yields a bright blue color in alkaline soln. Also used to determine molybdenum with which it forms a red-violet complex: Nikitina, Andrianova, *C.A.* **75**, 136789v (1971).

5760. Magnoflorine. [2141-09-5] (6a*S*)-5,6,6a,7-Tetrahydro-1,11-dihydroxy-2,10-dimethoxy-6,6-dimethyl-4*H*-dibenzo[*de,g*]-quinolinium; thalictrine. $[C_{20}H_{24}NO_4]^+$. From *Magnolia grandiflora* L., *Magnoliaceae.* Isoln and structure: Nakano, *Chem. Pharm. Bull.* **2**, 329 (1954). From *Cocculus trilobus* D.C., *Menispermaceae:* Nakano, *ibid.* **4**, 69 (1956); from *Thalictrum thunbergii* D.C., *Ranunculaceae:* Fujita, Tomimatsu, *ibid.* **6**, 107 (1958); from *Aristolochia clematitis* L., *Aristolochiaceae:* Pailer, Pruckmayr, *Monatsh. Chem.* **90**, 145 (1959). Identity with thalictrine: Gopinath *et al.*, *J. Sci. Ind. Res.* **18B** 444 (1959). Synthesis: Tomita, Kikkawa, *J. Pharm. Soc. Jpn.* **77**, 195 (1957).

Iodide. $C_{20}H_{24}INO_4$. Crystals from methanol + acetone, dec 248-249°. $[\alpha]_D^{15}$ +220.1° (methanol). uv max: 270, 310 nm (log ε 3.75, 3.59).

5761. Magnolol. [528-43-8] 5,5'-Di-2-propen-1-yl-[1,1'-biphenyl]-2,2'-diol; 5,5'-diallyl-2,2'-biphenyldiol; 2,2'-bichavicol; dehydrodichavicol; 2,2'-dihydroxy-5,5'-diallylbiphenyl. $C_{18}H_{18}O_2$; mol wt 266.34. C 81.17%, H 6.81%, O 12.01%. Bioactive constituent of *Magnoliae Cortex*, which is the bark of *Magnolia officinalis*, Rehd. et Wils., *Magnoliaceae*, known in Chinese traditional medicine as *houpo*, or the bark of *M. obovata*, Thunb., called *wakoboku* in Japanese. Isoln: Y. Sugii, *Yakugaku Zasshi* **50**, 183 (1930), *C.A.* **24**, 3505 (1930). Synthesis: H. Erdtman, J. Runeberg, *Acta Chem. Scand.* **11**, 1060 (1957); J. Runeberg, *ibid.* **12**, 188 (1958). Isoln from seeds of *M. grandiflora* L.: F. S. El-Feraly, W.-S. Li, *Lloydia* **41**, 442 (1978). CNS depressant effects: K. Watanabe *et al.*, *Jpn. J. Pharmacol.* **25**, 605 (1975); *eidem, Planta Med.* **49**, 103 (1983). Anti-inflammatory and analgesic effects: J.-P. Wang *et al.*, *Arch. Pharmacol.* **346**, 707 (1992). HPLC determn: T.-H. Tsai, C.-F. Chen, *J. Chromatogr.* **598**, 143 (1992).

mp 101.5-102°. uv max: 293 nm (log ε 3.90). Sol in ethanol.

5762. Maitotoxin. [59392-53-9] MTX. $C_{164}H_{256}Na_2O_{68}S_2$; mol wt 3425.88. C 57.50%, H 7.53%, Na 1.34%, O 31.76%, S 1.87%. Polycyclic ether toxin involved in ciguatera fish poisoning of humans. From the Tahitian for the surgeon fish (maito) from which it was orginally isolated; later found to be a product of the dinoflagellate, *Gamberdiscus toxicus*. One of the most toxic and the largest (mol wt 3422 Da) natural products known to date besides the biopolymers. Isoln from surgeon fish: T. Yasumoto *et al.*, *Bull. Jpn. Soc. Sci. Fish.* **42**, 359 (1976); from dinoflagellate: *idem et al.*, "A New Toxic Dinoflagellate Found in Association with Ciguatera" in *Toxic Dinoflagellate Blooms*, D. L. Taylor, H. Seliger, Eds. (Elsevier/North Holland, Inc., NY, 1979) pp 65-70. Toxicity and channel activation: M. Takahashi *et al.*, *J. Biol. Chem.* **257**, 7287 (1982). Structure: M. Murata *et al.*, *J. Am. Chem. Soc.* **115**, 2060 (1993); and stereochemistry: L. R. Cook *et al., ibid.* **119**, 7928 (1997). Use as activator of calcium influx: J. W. Daly *et al., Biochem. Pharmacol.* **50**, 1187 (1995); H. M. Brereton *et al., Biochim. Biophys. Acta* **1540**, 107 (2001); as receptor differentiator: P. M. Lundy *et al., Eur. J. Pharmacol.* **487**, 17 (2004). Effect of cations on MTX response: V. Morales-Tlalpan, L. Vaca, *Toxicon* **40**, 493 (2002). Determn by capillary zone electrophoresis: N. Bouaïcha *et al., ibid.* **35**, 955 (1997). Review of use as tool for calcium flux: F. Gusovsky, J. W. Daly, *Biochem. Pharmacol.* **39**, 1633-1639 (1990); of mechanism of action and pharmacology: M. Estacion, *Food Sci. Technol.* **103**, 473-503 (2000).

Sol in water, methanol, ethanol, 1-butanol saturated with water. Insol in diethyl ether, acetone and chloroform.

USE: Calcium channel activator; research tool for ion flux.

5763. Malachite Green. [569-64-2]; [510-13-4] (carbinol base). *N*-[4-[[4-(Dimethylamino)phenyl]phenylmethylene]-2,5-cyclohexadien-1-ylidene]-*N*-methylmethanaminium chloride (1:1); bis[*p*-(dimethylamino)phenyl]phenylmethylium chloride; C.I. Basic Green 4; C.I. 42000; aniline green; benzal green; benzaldehyde green; china green; diamond green B; diamond green Bx; diamond green P extra; fast green; light green N; new victoria green extra O; new victoria green extra I; new victoria green extra II; solid green O; victoria green W; victoria green B. $C_{23}H_{25}ClN_2$; mol wt 364.92. C 75.70%, H 6.91%, Cl 9.71%, N 7.68%. Triphenylmethane dye with

fungicidal and limited antiseptic activity. Prepn: *Colour Index* **Vol. 4** (3rd ed., 1971) p 4380. Conformational changes in excited state: S. Saikan, J. Sei, *J. Chem. Phys.* **79**, 4154 (1983). Use in electron microscopy: R. G. Pourcho *et al., Stain Technol.* **53**, 29 (1978); as stain in determn of prostaglandin E and F compounds: E. J. Singh *et al., J. Chromatogr.* **105**, 195 (1975); in determn of inorganic phosphate: W. Hohenwallner, E. Wimmer, *Clin. Chim. Acta* **45**, 169 (1973); S. G. Carter, D. W. Karl, *J. Biochem. Biophys. Methods* **7**, 7 (1982). Antiseptic activity: J. E. Madden *et al., Surg. Forum* **22**, 63 (1971). Toxicology: N. Brock, A. Erhardt, *Arzneim.-Forsch.* **1**, 5 (1951); T. D. Bills, L. L. Marking, *Invest. Fish Control* **75**, 6 (1977); S. Clemmensen *et al., Arch. Toxicol.* **56**, 43 (1984). Reviews of use as fungicide in fish culture: N. C. Nelson, *U.S. NTIS Report* (PB-235450, 1974) 79 pp, *C.A.* **82**, 150006p (1975), (PB-235451, 1974) 33 pp, *C.A.* **83**, 1547j (1975); and as biological stain: *Conn's Biological Stains*, R. W. Horobin, J. A. Kiernan, Eds. (BIOS Scientific Publishers Ltd, Oxford, UK, 10th ed., 2002) 189-190. Review of toxicological effects: S. Srivastava *et al., Aquat. Toxicol.* **66**, 310-329 (2004).

Green crystals with metallic luster. Very sol in water; sol in alcohol, methanol, amyl alcohol. Water solns are blue-green, absorption max 616.9 nm; yellow below pH 2. pK 6.90. LD$_{50}$ in mice (mg/kg): 80 orally; 4.2 i.p. (Brock, Erhardt). LC$_{50}$ (6 hr, 12°C) in bluegill sunfish, channel catfish, rainbow trout, coho salmon (mg/l): 2.19 (pH 7.5), 0.960 (pH 7.5), 6.8 (pH 8.0), 3.0 (pH 7.5) (Bills, Marking).

Oxalate salt. [2437-29-8] $C_{46}H_{50}N_4 \cdot 2HC_2O_4 \cdot C_2H_2O_4$; mol wt 927.02. Sol to 4% in water, 5% in ethanol. Insol in xylene.

Leucomalachite green. [129-73-7] Bis(*p*-dimethylaminophenyl)phenylmethane. $C_{23}H_{26}N_2$; mol wt 330.48. Colorless, reduced form of malachite green. Practically insol in water. Slightly sol in ethanol.

Note: The term malachite green applies to the oxalate as well as the chloride.

USE: For directly dyeing silk, wool, jute and leather; dyeing cotton after mordanting. Biological stain. Clinical reagent (inorganic phosphate assay). As spot test reagent for detecting sulfurous acid and cerium. As acid-base indicator: pH 0.0 yellow, 2.0 green; 11.6 green, 14 colorless.

THERAP CAT (VET): Fungicide and parasiticide in fish.

5764. Malathion. [121-75-5] 2-[(Dimethoxyphosphinothioyl)thio]butanedioic acid 1,4-diethyl ester; diethyl (dimethoxyphosphinothioylthio)succinate; diethyl mercaptosuccinate *S*-ester with *O,O*-dimethyl phosphorodithioate; *S*-(1,2-dicarbethoxyethyl) *O,O*-dimethyl dithiophosphate; insecticide no. 4049; carbofos; malathon; maldison; mercaptothion; phosphothion; ENT-17034; Cythion; Derbac-M; Fyfanon; Ovide; Prioderm. $C_{10}H_{19}O_6PS_2$; mol wt 330.35. C 36.36%, H 5.80%, O 29.06%, P 9.38%, S 19.41%. Organophosphate insecticide; cholinesterase inhibitor. Prepn: Johnson *et al., J. Econ. Entomol.* **45**, 279 (1952); Cassaday, **US 2578652** (1951 to Am. Cyanamid). Purification: Usui, **US 2962521** (1960 to Sumitomo). GC-MS/MS determn in urine: M. C. Márquez *et al., J. Chromatogr. A* **939**, 79 (2001). Clinical trial for head lice infestation: T. L. Meinking *et al., Pediatr. Dermatol.* **24**, 405 (2007). Toxicity data: T. B. Gaines, *Toxicol. Appl. Pharmacol.* **14**, 515 (1969). Review of distribution, transport and fate in the environment: M. S. Mulla *et al., Residue Rev.* **81**, 1-159 (1981); of carcinogenic risk: *IARC Monographs* **30**, 103-129 (1983); of toxicology and human exposure: *Toxicological Profile for Malathion* (PB2004-100003, 2003) 327 pp.

Colorless or slightly yellow liq, mp 2.9°. bp$_{0.7}$ 156-157°. Characteristic odor. d$_4^{25}$ 1.23. n$_D^{25}$ 1.4985. Vapor pressure at 30°: 4 × 10^{-5} mm Hg. Slightly sol in water (145 ppm). Misc with many organic solvents including alcohols, esters, ketones, ethers, aromatic and alkylated aromatic hydrocarbons and vegetable oils. Limited soly in certain paraffin hydrocarbons. Petroleum ether is sol to about 35% in malathion. Hydrolyzed at pH >7.0 or <5.0. Stable in an aq soln buffered to pH 5.26. LD$_{50}$ in female, male rats (mg/kg): 1000, 1375 orally (Gaines).

Caution: Potential symptoms of overexposure are miosis, aching eyes, blurred vision and lacrimation; eye and skin irritation; salivation; anorexia, nausea, vomiting, abdominal cramps, diarrhea, giddiness, confusion and ataxia; rhinorrhea, headache; tightening of chest, wheezing and laryngeal spasms. *See NIOSH Pocket Guide to Chemical Hazards* (DHHS/NIOSH 97-140, 1997) p 188.

USE: Insecticide.

THERAP CAT: Pediculicide.

THERAP CAT (VET): Ectoparasiticide.

5765. Maleamic Acid. [557-24-4] (2Z)-4-Amino-4-oxo-2-butenoic acid; maleic acid monoamide. C$_4$H$_5$NO$_3$; mol wt 115.09. C 41.74%, H 4.38%, N 12.17%, O 41.70%. Prepd by passing ammonia into a mixture of maleic anhydride and an inert solvent such as xylene, dioxane, etc.: Robinson, Humburger, **US 2459964** (1949 to Beck, Koller).

Crystals from alcohol, mp 172-173°. Also reported as mp 178-180° (dec). Very sol in water, hot alcohol. Practically insol in ether, chloroform, benzene.

5766. Maleanilic Acid. [555-59-9] (2Z)-4-Oxo-4-(phenyl-amino)-2-butenoic acid; *N*-phenylmaleamic acid. C$_{10}$H$_9$NO$_3$; mol wt 191.19. C 62.82%, H 4.75%, N 7.33%, O 25.10%. Use of esters as fungicides: Ligett *et al.*, **US 2885319** (1959 to Pittsburgh Coke & Chem.).

Monoclinic yellow crystals, dec 192°. Apparent density 1.418 g/ml at 30°. Soly in g/100 ml soln at 27°: acetonitrile 0.2; acetone 0.3; benzene <0.1; butyl Cellosolve 1.1; carbon tetrachloride <0.1; chloroform <0.1; dimethylformamide 12.9; dioxane 0.9; ethanol 0.2; ether <0.1; methanol 0.3; toluene <0.1; water <0.1. Shows hydrolysis in aq soln. Dissociates to maleic anhydride and aniline when heated above 100° *in vacuo.*

USE: Esters as fungicides.

5767. Maleic Acid. [110-16-7] (2Z)-2-Butenedioic acid; toxilic acid; *cis*-1,2-ethylenedicarboxylic acid. C$_4$H$_4$O$_4$; mol wt 116.07. C 41.39%, H 3.47%, O 55.14%. Prepd by the catalytic oxidation of benzene over vanadium pentoxide: Bhattacharyya, Venkataraman, *J. Appl. Chem.* **8**, 728 (1958); Saffer, Olenberg, **FR 1321416** (1963 to Scientific Design), *C.A.* **59**, 11265d (1963). Crystal structure: Shahat, *Acta Crystallogr.* **5**, 763 (1952). *Review:* W.

D. Robinson, R. A. Mount in *Kirk-Othmer Encyclopedia of Chemical Technology* **vol. 14** (Wiley-Interscience, New York, 3rd ed., 1981) pp 770-793.

White crystals from water, mp 138-139°; from alcohol and benzene, mp 130-131°. Faint, acidulous odor; characteristic repulsive, astringent taste. d 1.59. Is converted in part into the much higher-melting fumaric acid (mp 287°) when heated to a temp slightly above the melting point. Freely sol in water, alcohol; sol in acetone, glacial acetic acid; sparingly sol in ether. Practically insol in benzene.

Caution: Strong irritant.

USE: Manuf artificial resins; to retard rancidity of fats and oils in 1:10,000 (these are said to keep 3 times longer than those without the acid); dyeing and finishing wool, cotton, and silk; preparing the maleate salts of antihistamines and similar drugs.

5768. Maleic Anhydride. [108-31-6] 2,5-Furandione; *cis*-butenedioic anhydride; toxilic anhydride. C$_4$H$_2$O$_3$; mol wt 98.06. C 48.99%, H 2.06%, O 48.95%. May be prepd by sublimation of maleic acid and P$_2$O$_5$ under reduced pressure: Kempf, *J. Prakt. Chem.* (2) **78**, 239 (1908). Commercial production by catalytic vapor-phase oxidation of benzene or other suitable hydrocarbons: Weiss, Downs, *Ind. Eng. Chem.* **12**, 228 (1920); **US 1318633** (1920 to Barrett Co.). Many other syntheses. Review of commercial methods of manufacture: Ashcroft, Clifford, *Chem. Prod.* **24**, 11 (1961), *C.A.* **55**, 9724d (1961); *Faith, Keyes & Clark's Industrial Chemicals*, F. A. Lowenheim, M. K. Moran, Eds. (Wiley-Interscience, New York, 4th ed., 1975) pp 514-518. *Review:* W. D. Robinson, R. A. Mount in *Kirk-Othmer Encyclopedia of Chemical Technology* **vol. 14** (Wiley-Interscience, New York, 3rd ed., 1981) pp 770-793. Book: B. C. Trivedi, B. M. Culbertson, *Maleic Anhydride* (Plenum, New York, 1982) 872 pp.

Orthorhombic needles from chloroform; also readily by sublimation. Commercial grades are furnished in fused form, as briquettes. d 1.48. mp 52.8°. bp$_{760}$ 202.0°; bp$_{400}$ 179.5°; bp$_{200}$ 155.9°; bp$_{100}$ 135.8°; bp$_{60}$ 122.0°; bp$_{40}$ 111.8°; bp$_{20}$ 95.0°; bp$_{10}$ 78.7°; bp$_5$ 63.4°. Specific heat: 0.285 (solid); 0.396 (liq). *Corrosive.* Sol in water, forming maleic acid. Soly at 25° (g/100 g): acetone 227; ethyl acetate 112; chloroform 52.5; benzene 50; toluene 23.4; *o*-xylene 19.4; carbon tetrachloride 0.60; ligroin 0.25. Sol in dioxane. Sol in alc with ester formation.

Caution: Potential symptoms of overexposure are conjunctivitis; photophobia, double vision; nasal and upper respiratory irritation; bronchial asthma; dermatitis. *See NIOSH Pocket Guide to Chemical Hazards* (DHHS/NIOSH 97-140, 1997) p 188.

USE: In Diels-Alder syntheses (as a dienophile), manuf alkyd-type of resins, dye intermediates, pharmaceuticals, agricultural chemicals (maleic hydrazide, malathion), in copolymerization reactions.

5769. Maleic Hydrazide. [123-33-1] 1,2-Dihydro-3,6-pyridazinedione; maleic acid hydrazide; MH; Fazor; Malazide; Regulox. C$_4$H$_4$N$_2$O$_2$; mol wt 112.09. C 42.86%, H 3.60%, N 24.99%, O 28.55%. Prepd by treating maleic anhydride with hydrazine hydrate in alcohol: Arndt *et al.*, *C.A.* **43**, 579 (1949); Curtius, Foerstinger, *J. Prakt. Chem.* (2) **51**, 391 (1895). From maleic acid and a hydrazine salt of a strong inorganic acid: Harris, Schoene, **US 2575954** (1951 to U.S. Rubber). Alternate prepn from hydrazine sulfate and maleic anhydride in aqueous NaOH: Amatsu, Karasawa, *C.A.* **51**, 18014c (1957); from hydrazine hydrate and maleic anhydride in glacial acetic acid: Feuer *et al.*, *J. Am. Chem. Soc.* **80**, 3790 (1958). Has the ability to inhibit growth of plants without killing them: Schoene, Hoffmann, *Science* **109**, 588 (1949). Toxicity data: R. Ben-Dyke *et al.*, *World Rev. Pest Control* **9**, 119 (1970). *Review:* Massey, *Manuf. Chem.* **26**, 197-200 (1955). Review and chromatographic studies: Fishbein, *Chromatography of Environmental Hazards* (Elsevier, New York, 1972) pp 161-166. Review of toxicology: R. Ponnampalam *et al.*, *Regul. Toxicol. Pharmacol.* **3**, 38-47 (1983).

Crystals from water, dec 260°. Also reported as mp >300° (Feuery). Slightly sol in hot alcohol, more sol in hot water. LD₅₀ orally in rats: 3800-6800 mg/kg; dermally in rabbits: >4000 mg/kg (Ben-Dyke).

USE: Experimentally in horticulture and agriculture. To control suckering of tobacco. In the synthesis of pyridazine.

5770. Maleuric Acid. [105-61-3] (2Z)-4-[(Aminocarbonyl)-amino]-4-oxo-2-butenoic acid; N-carbamoylmaleamic acid; maleyl-urea. $C_5H_6N_2O_4$; mol wt 158.11. C 37.98%, H 3.83%, N 17.72%, O 40.48%. Prepd from maleic anhydride and urea: Dunlap, Phelps, *Am. Chem. J.* **19**, 492 (1897); Batt *et al.*, *J. Am. Chem. Soc.* **76**, 3663 (1954).

Crystals from hot water, dec 158.5°. uv max: 235 nm (ε 8720). Sol in hot acetic acid. Practically insol in cold water, cold acetic acid, acetone, ligroin, chloroform, alc, ether.
Methyl ester. Crystals, mp 113-114°. Sol in hot water, methanol, ethanol, acetone, dioxane.
Butyl ester. Crystals, mp 95-98°. Sol in ethanol, acetone, benzene, chloroform.

5771. Malic Acid. [6915-15-7] 2-Hydroxybutanedioic acid; hydroxysuccinic acid; 2-hydroxyethane-1,2-dicarboxylic acid. $C_4H_6O_5$; mol wt 134.09. C 35.83%, H 4.51%, O 59.66%. The naturally occurring isomer is the L-form which has been found in apples and many other fruits and plants. Prepn of D- and DL-forms, and resolution of racemic mixture: McKenzie *et al.*, *J. Chem. Soc.* **123**, 2875 (1923). Solubilities: Descamps, *Bull. Soc. Chim. Belg.* **49**, 91 (1940). Microbial production of L-form: Kitahara; Abe *et al.*, US **2972566**; US **3063910** (1961, 1962, both to Kyowa). Configuration: J. A. Mills, W. Klyne in *Progress in Stereochemistry* **vol. 1**, W. Klyne, Ed. (Academic Press, New York, 1954) pp 182-183; E. L. Eliel, *Stereochemistry of Carbon Compounds* (McGraw-Hill, New York, 1962) pp 97-98; Cymerman-Craig, Roy, *Tetrahedron* **21**, 1847 (1965). Review of uses: *Manuf. Chem. Aerosol News* **35**, 56 (December, 1964). *Review:* S. E. Berger, "Hydroxy Dicarboxylic Acids" in *Kirk-Othmer Encyclopedia of Chemical Technology* **vol 13** (Wiley-Interscience, New York, 3rd ed., 1981) pp 103-110.

L-Malic Acid

Crystals, mp 131-132°. Strongly acid taste. Soly in g/100 g solvent at 20°: methanol 82.70, diethyl ether 0.84, ethanol 45.53, acetone 17.75, dioxane 22.70, water 55.8. Practically insol in benzene.
D-(+)-Form. [636-61-3] Crystals, mp 101°.
L-(−)-Form. [97-67-6] Apple acid. Crystals from acetone, or acetone + CHCl₃, mp 100°. Dec ~140°. [α]_D −2.3° (c = 8.5). Soly in g/100 g solvent at 20°: methanol 197.22, diethyl ether 2.70, ethanol 86.60, acetone 60.66, dioxane 74.35, water 36.35. Practically insol in benzene.

USE: Intermediate in chemical synthesis. Chelating and buffering agent. Flavoring agent, flavor enhancer and acidulant in foods.

5772. Mallow. Common mallow; high mallow; cheeseflower. Leaves of *Malva sylvestris* L., and *M. rotundifolia* L., *Malvaceae*. Habit. Europe, Asia, naturalized in U.S. *Constit.* Pectin, tannin, coloring matter.

5773. Malondialdehyde. [542-78-9] Propanedial; malonaldehyde. $C_3H_4O_2$; mol wt 72.06. C 50.00%, H 5.60%, O 44.40%. Endogenous product of lipid peroxidation and prostaglandin biosynthesis in mammals; also formed by oxidation of polyunsaturated fatty acids in foods. Reacts with functional groups of a variety of cellular compounds including DNA and amino acids, forming genotoxic adducts and crosslinks. Associated with detrimental protein modifications and decreased nutritional value in foods. Prepn: L. Claisen, *Ber.* **36**, 3664 (1903); R. Hüttel, *ibid.* **74**, 1825 (1941). Chemistry and reactions with amino acids: V. Nair *et al.*, *J. Am. Chem. Soc.* **103**, 3030 (1981). NMR spectroscopy: S. H. Bertz, G. Dabbagh, *J. Org. Chem.* **55**, 5161 (1990). Determn methods using thiobarbituric acid: H. H. Draper *et al.*, *Free Radical Biol. Med.* **15**, 353 (1993). HPLC determn in foods: P. Bergamo *et al.*, *J. Agric. Food Chem.* **46**, 2171 (1998); in biological samples: R. Mateos *et al.*, *J. Chromatogr. B* **827**, 76 (2005). Acute toxicity study: D. L. Crawford *et al.*, *Toxicol. Appl. Pharmacol.* **7**, 826 (1965). Clinical use as biomarker in neurodegenerative disease: M. Dib *et al.*, *J. Neurol.* **249**, 367 (2002); in coronary artery disease: H. H. Jung *et al.*, *Am. J. Nephrol.* **24**, 537 (2004). Review of metabolism: H. H. Draper, M. Hadley, *Xenobiotica* **20**, 901-907 (1990); of biochemistry: H. Esterbauer *et al.*, *Free Radical Biol. Med.* **11**, 81-128 (1991). Review of carcinogenic risk: *IARC Monographs* **71**, Part 3, 1037-1047 (1999); of toxicity of DNA adducts: L. J. Marnett, *IARC Sci. Publ.* **150**, 17-27 (1999); and pathological significance: D. Del Rio *et al.*, *Nutr. Metab. Cardiovasc. Dis.* **15**, 316-328 (2005).

Needles, mp 72-74°. pKa (enolate) 4.46. Stable in neutral conditions but not in acidic. LD₅₀ orally in rats: 632 mg/kg (Crawford).
Sodium salt. [24382-04-5] $C_3H_4NaO_2$; mol wt 95.05. White needles from ethanol/ether, mp 246° (dec). uv max (0.01M HCl): 245 nm (ε 12800).

Direct contact may cause irritation of eyes, skin and respiratory system; potential symptom of overexposure is CNS depression. Potential occupational carcinogen. See NIOSH Pocket Guide to Chemical Hazards (2003-100121, 2003) p 190.

USE: Biomarker for oxidative stress. Index of oxidative rancidity in foods.

5774. Malonic Acid. [141-82-2] Propanedioic acid; carboxyacetic acid; methanedicarboxylic acid. $C_3H_4O_4$; mol wt 104.06. C 34.63%, H 3.87%, O 61.50%. Prepd from malic acid: Dessaignes, *Ann.* **107**, 251 (1858). Made by the interaction of monochloroacetic acid and NaCN followed by hydrolysis of the resulting cyanoacetic acid: Weiner, *Org. Synth.* **coll. vol. II**, 376 (1943). Prepn from diethyl malonate: Britton, Monroe, US **2373011** (1945 to Dow); from ligneous wastes: Grangaard, US **2928868** (1960 to Kimberly-Clark); from sodium acetate: Normant, Angelo, *Bull. Soc. Chim. Fr.* **1962**, 810. *Review:* D. W. Hughes in *Kirk-Othmer Encyclopedia of Chemical Technology* **vol. 14** (Wiley-Interscience, New York, 3rd ed., 1981) pp 794-810.

Small crystals, mp ~135° with decompn; sublimes *in vacuo*. *Strong irritant.* d 1.63. One gram dissolves in 0.65 ml water, about 2 ml alcohol, 1.1 ml methanol, 3 ml propyl alcohol, 13 ml ether, 7 ml pyridine.
Diethyl Ester see Ethyl Malonate.
USE: In manuf of barbiturates.

5775. Malononitrile. [109-77-3] Propanedinitrile; methylene cyanide; dicyanomethane; cyanoacetonitrile. $C_3H_2N_2$; mol wt

66.06. C 54.55%, H 3.05%, N 42.41%. Prepn: Henri, *Compt. Rend.* **102**, 1394 (1886); B. B. Corson *et al.*, *Org. Synth.* **coll. vol. 2**, 379 (1943); A. R. Surrey, *J. Am. Chem. Soc.* **65**, 2471 (1943); *idem*, US **2389217** (1945 to Winthrop); *idem*, *Org. Synth.* **coll. vol. 3**, 535 (1955). Toxicity study: G. R. N. Jones, M. S. Israel, *Nature* **228**, 1315 (1970). *Reviews:* F. Freeman, *Chem. Rev.* **69**, 591-624 (1969); A. J. Fatiadi, *Synthesis* **1978**, 165-204, 241-282.

$$NC \diagup \diagdown CN$$

Colorless solid, mp 32°. bp$_{760}$ 218-219°, bp$_{20}$ 109°. d$_4^{20}$ 1.1910; n$_D^{34}$ 1.4146. *Poisonous.* Very sol in alc, ether; sol in water, acetone, benzene. LD$_{50}$ i.p. in mice: 12.9 mg/kg (Jones, Israel).

Caution: Potential symptoms of overexposure are irritation of eyes, skin, nose, throat; headache, dizziness, weakness, giddiness, confusion, convulsions; dyspnea; abdominal pain, nausea, vomiting. *See NIOSH Pocket Guide to Chemical Hazards* (DHHS/NIOSH 97-140, 1997) p 190.

USE: In organic synthesis.

5776. Malotilate. [59937-28-9] 2-(1,3-Dithiol-2-ylidene)propanedioic acid 1,3-bis(1-methylethyl) ester; diisopropyl 1,3-dithiol-2-ylidenemalonate; NKK-105; Hepation; Kantec. C$_{12}$H$_{16}$O$_4$S$_2$; mol wt 288.38. C 49.98%, H 5.59%, O 22.19%, S 22.23%. Prepn from diisopropyl malonate: K. Taninaka *et al.*, **DE 2545569**; *eidem*, US **4035387** (1976, 1977 both to Nihon Nohyaku); from the corresponding ketene mercaptide: H. Matsui *et al.*, US **4327223** (1982 to Nihon Nohyaku). Effect on CCl$_4$-induced liver injury in rats: Y. Imaizumi *et al.*, *Jpn. J. Pharmacol.* **31**, 15 (1981). Enhancement of rat liver protein synthesis: *eidem*, *ibid.* **32**, 369 (1982). Pharmacokinetics and pharmacodynamics: M. Buhrer *et al.*, *Eur. J. Clin. Pharmacol.* **30**, 407 (1986). Clinical evaluation in liver cirrhosis: S. Takase *et al.*, *Gastroenterol. Jpn.* **23**, 639 (1988).

Pale yellow crystals, mp 60.5°. Sol in benzene, cyclohexane, *n*-hexane, ether.

THERAP CAT: Hepatoprotectant.

5777. Maltol. [118-71-8] 3-Hydroxy-2-methyl-4*H*-pyran-4-one; 3-hydroxy-2-methyl-4-pyrone; 3-hydroxy-2-methyl-γ-pyrone; larixinic acid; palatone; veltol. C$_6$H$_6$O$_3$; mol wt 126.11. C 57.15%, H 4.80%, O 38.06%. Found in the bark of young larch trees (*Larix decidua* Mill.), in pine needles (*Abies alba* Mill., *Pinaceae*), in chicory, in wood tars and oils, in roasted malt. Isoln from these sources and structure: Kiliani, Bazlen, *Ber.* **27**, 3115 (1894); Feuerstein, *Ber.* **34**, 1804 (1901); Erdmann, Schaefer, *Ber.* **43**, 2398 (1910); Reichstein, Beitter, *Ber.* **63**, 824 (1930); Peratoner, Tamburello, *Chem. Zentralbl.* **76**, II, 680 (1905). Also obtained by alkaline hydrolysis of streptomycin salts: Schenck, Spielman, *J. Am. Chem. Soc.* **67**, 2276 (1945). Synthesis: Spielman, Freifelder, *ibid.* **69**, 2908 (1947); Chawla, McGonigal, *J. Org. Chem.* **39**, 3281 (1974). Novel high-yield synthesis: T. M. Brennan *et al.*, *Tetrahedron Lett.* **1978**, 331; P. D. Weeks *et al.*, *J. Org. Chem.* **45**, 1109 (1980). History and comparison with isomaltol: Hodge, Nelson, *Cereal Chem.* **38**, 207 (1961).

Monoclinic prisms from chloroform, orthorhombic bipyramidal crystals + monoclinic prisms from 50% alcohol, mp 161-162°. Fragrant, caramel-like odor. Begins to sublime at 93°. Volatile with steam. uv max (0.1*N* HCl): 274 nm (E$_m$ 8400); (0.1*N* NaOH): 317 nm (E$_m$ 7300). pH of 0.5% aq soln 5.3. One gram dissolves in 82 ml water, 80 ml glycerin, 21 ml alc, 28 ml propylene glycol. Freely sol in hot water, chloroform; sparingly sol in benzene, ether, petr ether; sol in aq alkali hydroxides giving yellow solns.

USE: Flavoring agent, to impart "freshly baked" odor and flavor to bread and cakes.

5778. Maltose. [69-79-4] 4-*O*-α-D-Glucopyranosyl-D-glucose; malt sugar; maltobiose; 4-(α-D-glucosido)-D-glucose; Maltos; Martos-10. C$_{12}$H$_{22}$O$_{11}$; mol wt 342.30. C 42.11%, H 6.48%, O 51.41%. Monohydrate obtained in about 80% yield by enzymatic (diastase) degradation of starch: Gore, US **1657079** (1928). Structure: Haworth, Peat, *J. Chem. Soc.* **1926**, 3094; Hassid, Ballou in W. Pigman, *The Carbohydrates* (Academic Press, New York, 1957) p 498. Crystal and molecular structure: F. Takusagawa, R. A. Jacobson, *Acta Crystallogr.* **B34**, 213 (1978). *Review:* E. Tarelli, *Dev. Food Carbohydr.* **2**, 187-227 (1980); R. Khan, *Carbohydr. Chem. Biochem.* **39**, 213-278 (1981).

Monohydrate. Crystals from water or dil alc, mp 102-103°. Does not lose its water of crystn by drying at room temp *in vacuo* over H$_2$SO$_4$ or P$_2$O$_5$. About one-third as sweet as sucrose. *See:* Isbell, Pigman, *J. Res. Natl. Bur. Stand.* **18**, 141 (1937). Shows mutarotation. [α]$_D^{20}$ +111.7° → +130.4° (c = 4). pKa (21°): 12.05. Very sol in ethanol; freely sol in water; slightly sol in methanol. Practically insol in ether.

USE: Nutrient, sweetener, in culture media, in prepd bee food. Parenteral supplement of sugar for diabetics. Fermentable intermediate in brewing. Stabilizer for polysulfides. In pharmaceutical dispensing.

5779. Malvidin Chloride. [643-84-5] 3,5,7-Trihydroxy-2-(4-hydroxy-3,5-dimethoxyphenyl)-1-benzopyrylium chloride (1:1); 3,-4′,5,7-tetrahydroxy-3′,5′-dimethoxyflavylium chloride; enidin; primulidin; syringidin; 3,4′,5,7-tetrahydroxy-3′,5′-dimethoxy-2-phenylbenzopyrylium chloride. C$_{17}$H$_{15}$ClO$_7$; mol wt 366.75. C 55.67%, H 4.12%, Cl 9.67%, O 30.54%. Found as the diglucoside (malvin) in wild malve (*Primula viscosa* All., *Primulaceae*) and as the monoglucoside in blue grapes: Willstätter, Mieg, *Ann.* **408**, 122 (1915); Karrer, Widmer, *Helv. Chim. Acta* **10**, 5 (1927). Structure: Anderson, Nabenhauer, *J. Am. Chem. Soc.* **48**, 2997 (1926). Synthesis: Bradley, Robinson, *J. Chem. Soc.* **1928**, 1541.

Prisms or rhombic tablets, appearing red by transmitted light with a green luster when dry or with a steel-blue luster when in contact with solvent. Usually obtained as the mono- or dihydrate. The anhydr salt is very hygroscopic. Not melted at 300°. Sol in abs alc, giving a violet-red soln. Sparingly sol in water. Also sol in amyl alc. In methanol the substance is first sol with purple color, then begins to separate as red crystals which are violet by transmitted light.

3-β-Glucoside tetrahydrate. 3-(Glucosyloxy)-4′,5,7-trihydroxy-3′,5′-dimethoxyflavylium chloride; enin; cyclamin. C$_{23}$H$_{25}$-ClO$_{12}$.4H$_2$O. Synthesis: Levy *et al.*, *J. Chem. Soc.* **1931**, 2701.

Dark prisms with a green metallic shine, from alcohol + HCl. Sparingly sol in water, alcohol, glycerol. Practically insol in benzene, chloroform, ether.

3,5-Diglucoside. 3,5-Bis(glucosyloxy)-4′,7-dihydroxy-3′,5′-dimethoxyflavylium chloride; malvin; malvoside. $C_{29}H_{35}ClO_{17}$. Synthesis: Robinson, Todd, *ibid.* **1932**, 2299. Reddish-brown prisms or needles with a green shine from dil HCl, dec 165°.

3-Galactoside hemihendecahydrate. Primulin. $C_{23}H_{25}ClO_{12}\cdot5\frac{1}{2}H_2O$. Bronze metallic needles or prisms from methanol + HCl. Bluish-violet by transmitted light. Very sol in methanol, ethanol, cold water, forming deep red solns; slightly sol in acetone. Practically insol in ethyl acetate.

5780. Mancozeb. [8018-01-7] [N-[2-[(Dithiocarboxy)amino]ethyl]carbamodithioato(2−)-κS,κS′]manganese mixt. with [N-[2-[(dithiocarboxy)amino]ethyl]carbamodithioato(2−)-κS,κS′]zinc; [[1,2-ethanediylbis(carbamodithioato)](2−)]manganese mixt. with [[1,2-ethanediylbis(carbamodithioato)](2−)]zinc; ethylenebis(dithiocarbamic acid) manganese zinc complex; manzeb; manganese ethylenebis(dithiocarbamate) (polymeric) complex with zinc salt; zinc manganese ethylenebisdithiocarbamate; Dithane M-45; Fore; Manzate; Manzin; Nemispor; Penncozeb; Vondozeb. Polymeric salt of ethylenebisdithiocarbamic acid containing approx 20% manganese and 2.5% zinc; related to maneb and zineb, *q.q.v.* Displays activity against a wide range of foliage fungal diseases. Prepn: **NZ 131543**; C. B. Lyon *et al.*, **US 3379610** (1965, 1968 both to Rohm & Haas). Metabolic fate in plants, animals: W. R. Lyman, "Metabolic fate of Dithane M-45," in *Pestic. Terminal Residues, Invited Pap. Int. Symp.*, A. S. Tahori, Ed. (Butterworth, New York, 1971) pp 243-256. Residue studies: *WHO Pestic. Residues Ser.* **vol. 4**, (WHO, FAO, 1975) 451; W. H. Newsome, *J. Agric. Food Chem.* **27**, 1188 (1979). HPLC determn: E. Gustafsson *et al.*, *ibid.* **29**, 729 (1981). Demonstrates low phytotoxicity: L. H. Cornford, *Proc. 36th N. Z. Weed Pest Control Conf.* 258 (1983). In potato blight: H. W. Platt, *Can. J. Plant Pathol.* **5**, 38 (1983). *In vitro* activity against *Alternaria alternata:* V. A. Bourbos *et al.*, *Br. Crop Prot. Counc. Monogr.* **2**, No. 31, 461 (1985). Aquatic toxicology: C. J. Van Leeuwen *et al.*, *Aquat. Toxicol.* **7**, 145 (1985).

x : y   10:1

Insol in water.
USE: Fungicide.

5781. Mandelic Acid. [90-64-2] α-Hydroxybenzeneacetic acid; *dl*-mandelic acid; racemic mandelic acid; α-hydroxy-α-toluic acid; α-hydroxyphenylacetic acid; phenylhydroxyacetic acid; phenylglycolic acid; amygdalic acid; amygdalinic acid; paramandelic acid; Uromaline. $C_8H_8O_3$; mol wt 152.15. C 63.15%, H 5.30%, O 31.55%. Prepd by the action of warm, dil alkali upon dichloroacetophenone: *Org. Synth.* **23**, 48 (1943); by hydrolysis of mandelonitrile (prepd from benzaldehyde and hydrogen cyanide or from benzaldehyde, sodium bisulfite, and sodium cyanide): *Org. Synth.* **coll. vol. I**, 336 (1941); *ibid.* **coll. vol. III**, 538 (1955); L. F. Fieser, *Organic Experiments* (Boston, 1964) p 109. May be prepd by boiling amygdalin with HCl.

Orthorhombic plates from water, mp 119°. Darkens and dec on prolonged exposure to light. d 1.30. Can be distilled rapidly *in vacuo* at 2 mm without much decomp. Acid to litmus. pK (25°) 3.37. One gram dissolves in 6.3 ml water, 1 ml alc; freely sol in ether, isopropyl alcohol.

Calcium salt. [134-95-2] Calcium mandelate; Camdelate. $C_{16}H_{14}CaO_6$; mol wt 342.36. Contains 88.88% mandelic acid. White

powder. One gram dissolves in 80 ml boiling water. Slightly sol in cold water. Insol in alc. The aq soln is slightly acid.

Sodium salt. [114-21-6] Sodium mandelate. $C_8H_7NaO_3$; mol wt 174.13. Cryst powder; slight aromatic odor. Very sol in water; sol in alcohol; the aq soln is neutral or slightly alkaline to litmus.

THERAP CAT: Antiseptic (urinary).

5782. Mandelic Acid Isoamyl Ester. [5421-04-5] α-Hydroxybenzeneacetic acid 3-methylbutyl ester; mandelic acid isopentyl ester; isoamyl mandelate; Atractyl; Spasmol; Vermiparin; Spasmostenyl. $C_{13}H_{18}O_3$; mol wt 222.28. C 70.25%, H 8.16%, O 21.59%. Prepd by esterification of DL-mandelic acid with an excess of isoamyl alcohol in the presence of concd H_2SO_4: Rona *et al.*, *Biochem. Z.* **181**, 50 (1927). Improved procedure and spasmolytic activity: N. Brock *et al.*, *Arzneim.-Forsch.* **2**, 165 (1952).

Oily liquid. bp_{11} 172° (Rona); bp_{12} 155° (Brock). Practically insol in water. Miscible with fat solvents. Easily hydrolyzed by esterases.
Caution: The pure, undiluted liquid is a skin irritant.

THERAP CAT (VET): Has been used as an anthelmintic.

5783. Mandelonitrile. [532-28-5] α-Hydroxybenzeneacetonitrile; mandelic acid nitrile; benzaldehyde cyanohydrin. C_8H_7NO; mol wt 133.15. C 72.17%, H 5.30%, N 10.52%, O 12.02%. Isoln of *d*-form from peach flower buds: Jones, Enzie, *Science* **134**, 284 (1961). Prepn of *d*-form by hydrolysis of amygdalin: Auld, *J. Chem. Soc.* **95**, 927 (1909); Smith, *Ber.* **64**, 427 (1931). Resolution of *dl*-form: Feist, *Arch. Pharm.* **247**, 226 (1909). Synthesis of *l*-form using cotton fibers: Bredig, Gerstner, *Biochem. Z.* **250**, 414 (1932). Asymmetric synthesis: Krieble, Wieland, *J. Am. Chem. Soc.* **43**, 164 (1921); Prelog, Wilhelm, *Helv. Chim. Acta* **37**, 1634 (1954); Tsuboyama, *Bull. Chem. Soc. Jpn.* **35**, 1004 (1962). It has been suggested that mandelonitrile may be responsible for the alleged anticancer activity of Laetrile, *q.v.: see* Culliton, *Science* **182**, 1000 (1973). Metabolism study: P. D. A. Singh *et al.*, *Biochem. Pharmacol.* **34**, 2207 (1985).

dl-**Form.** Yellow, oily liquid, mp −10°. Decomp at 170°. d 1.115-1.120. Almost insol in water. Freely sol in alcohol, chloroform, ether.

d-**Form.** $[\alpha]_D^{25}$ +43.75° (c = 5.006 in benzene).
USE: Preparing bitter almond water, by mixing 11 g with 500 g alc and 1489 g water; the mixture contains 0.1% HCN.

5784. Mandelonitrile Glucoside. [138-53-4] α-(β-D-Glucopyranosyloxy)benzeneacetonitrile. $C_{14}H_{17}NO_6$; mol wt 295.29. C 56.95%, H 5.80%, N 4.74%, O 32.51%. Prepn of *d*-form by action of yeast on amygdalin: Fischer, *Ber.* **28**, 1508 (1895); Auld, *J. Chem. Soc.* **93**, 1276 (1908). Isoln of *d*-form from *Prunus serotina*, Ehrk, *P. macrophylla* Sieb et Zucc., *Rosaceae:* Power, Moore, *ibid.* **95**, 243 (1909); **97**, 1099 (1910); Kariyone, Matsushima, *J. Pharm. Soc. Jpn.* **No. 514**, 1061 (1924); from *Eucalyptus corynocalyx* F.v.M., *Myrtaceae:* Finnemore *et al.*, *J. Proc. R. Soc. N.S.W.* **69**, 209 (1936); from *Pteridium aquilinum* (L.) Kühn, *Polypodiaceae:* Kofod, Eyjolfssen, *Tetrahedron Lett.* **1966**, 1289. Isoln of *dl*-form from *Prunus laurocerasus* L. and *Cotoneaster microphylla* Wall., *Rosaceae:* Winkler, Simon, *Ann.* **31**, 263 (1839); Hérissey, *J. Pharm. Chim.* **24**, 537 (1906). Isoln of *l*-form from *Sambucus nigra* L., *Caprifoliaceae:* Guignard, *Compt. Rend.* **141**, 16 (1905); Bourquelot, Danjot, *ibid.* 598 (1905). Prulaurasin is *dl*-mandelonitrile glucoside and sambunigrin is *l*-mandelonitrile glucoside: Caldwell, Courtauld, *J. Chem. Soc.* **91**, 671 (1907). Biosynthesis of *d*-mandelonitrile glucoside: Mentzer, Favre-Bonvin, *Compt. Rend.* **253**,

1072 (1961). Abs config of isomers: U. Schwarzmaier, *Ber.* **109**, 3250 (1976).

d-Form. Prunasin. Needles from chloroform, mp 147-148°. $[\alpha]_D^{14}$ −29.94°. Soluble in water, alcohol, acetone. Converted by alkalies to prulaurasin.

dl-Form. Prulaurasin. Slightly bitter needles from ethyl acetate + ether, mp 123-125°. $[\alpha]_D$ −54° (water). Soluble in water, alcohol. Practically insol in ether. On hydrolysis yields *dl*-mandelic acid.

l-Form. Sambunigrin. Bitter needles from hot ethyl acetate, mp 151-152°. $[\alpha]_D^{18}$ −75.1°. Soluble in water, alcohol, ethyl acetate. On hydrolysis yields *l*-mandelic acid.

5785. Mandipropamid. [374726-62-2] 4-Chloro-*N*-[2-[3-methoxy-4-(2-propyn-1-yloxy)phenyl]ethyl]-α-(2-propyn-1-yloxy)-benzeneacetamide; 2-(4-chlorophenyl)-*N*-[2-[3-methoxy-4-(prop-2-ynyloxy)phenyl]ethyl]-2-(prop-2-ynyloxy)acetamide; (*RS*)-2-(4-chlorophenyl)-*N*-[3-methoxy-4-(prop-2-ynyloxy)phenethyl]-2-(prop-2-ynyloxy)acetamide; NOA-446510; Revus. $C_{23}H_{22}ClNO_4$; mol wt 411.88. C 67.07%, H 5.38%, Cl 8.61%, N 3.40%, O 15.54%. Mandelamide fungicide used to control foliar diseases caused by oomycetes in food crops. Prepn: C. Lamberth *et al.*, **WO 0187822**; *eidem*, **US 6683211** (2001, 2004 both to Syngenta); *eidem*, *Bioorg. Med. Chem.* **16**, 1531 (2008). Properties and field studies: F. Huggenberger *et al.*, *BCPC Int. Conf. - Crop Sci. Technol.* **2005**, 87. Analysis of resistance potential: Y. Cohen *et al.*, *Plant Pathol.* **56**, 836 (2007). Efficacy vs. downy mildew in lettuce: *eidem*, *Eur. J. Plant Pathol.* **122**, 169 (2008). LC-MS/MS determn in fruits: K. Banerjee *et al.*, *J. Agric. Food Chem.* **57**, 4068 (2009). Mechanism of action study: M. Blum *et al.*, *Mol. Plant Pathol.* **11**, 227 (2010).

Light beige powder, mp 96.4-97.3°. dec >200°. d^{22} 1.24. Vapor pressure (25°): <9.4 × 10^{-7} Pa. Log P (octanol/water): 3.2 (25°). Soly at 25° (mg/l): water 4.2, *n*-hexane 42; (g/l): *n*-octanol 4.8, toluene 29, methanol 66, ethyl acetate 120, acetone 300, dichloromethane 400. LD$_{50}$ in rats (mg/kg): >5000 orally, >2000 dermally; LC$_{50}$ (inhalation) in rats: >5000 mg/m^3 (Huggenberger).
USE: Agricultural fungicide.

5786. Maneb. [12427-38-2] [*N*-[2-[(Dithiocarboxy)amino]-ethyl]carbamodithioato(2−)-κS,κS']manganese; [[1,2-ethanediyl-bis[carbamodithioato]](2−)]manganese; [ethylenebis(dithiocarbamato)]manganese; ethylenebis[dithiocarbamic acid] manganous salt; manganous ethylenebis[dithiocarbamate]; Dithane M-22; Manex; Trimangol. $(C_4H_6MnN_2S_4)_n$. Polymeric salt of ethylenebisdithiocarbamic acid; related to zineb and mancozeb, *q.q.v.* Prepd from aq nabam, *q.v.*, by neutralizing with acetic acid and adding MnCl$_2$ soln: A. L. Flenner, **US 2504404** (1950 to Du Pont), *see also* W. F. Hester, **US 2317765** (1943 to Rohm & Haas). Fungicidal activity is due to degradation products, principally *ethylenethiuram monosulfide*: R. A. Ludwig *et al.*, *Can. J. Bot.* **33**, 42 (1955). *In vitro* activity against *Alternaria alternata*: V. A. Bourbos *et al.*, *Br. Crop Prot. Counc. Monogr.* **2**, No. 31, 461 (1985). Decomposition of maneb to ethylenebisdithiocarbamate and ethylene thiourea, *q.v.*: W. J. Trotter, J. Pardue, *J. Food Saf.* **4**, 59 (1982). HPLC determn: K. Gustafsson *et al.*, *J. Agric. Food Chem.* **29**, 729 (1981). Toxicological studies

in rats: D. W. R. Bleyl, H. Seidler, *Nahrung* **29**, 421 (1985); in aquatic organisms: C. J. Van Leeuwen *et al.*, *Aquat. Toxicol.* **7**, 145 (1985).

Yellow powder. Crystals from alcohol. *Spontaneously combustible; dangerous when wet.* Moderately sol in water. Sol in chloroform, pyridine.
USE: Agricultural fungicide.

5787. Mangafodipir. [118248-94-5] (free acid); [155319-91-8] (hexahydrogen). (*OC*-6-13)-[[*N,N'*-1,2-Ethanediylbis[*N*-[[3-(hydroxy-κO)-2-methyl-5-[(phosphonooxy)methyl]-4-pyridinyl]methyl]glycinato-κN,κO]](8−)]manganate(6−); manganese(II)-*N,N'*-dipyridoxylethylenediamine-*N,N'*-diacetate-5,5-bis(phosphonate); manganese dipyridoxal diphosphate; MnDPDP; S-095. $C_{22}H_{24}$-MnN$_4$O$_{14}$P$_2$; mol wt 685.33. C 38.56%, H 3.53%, Mn 8.02%, N 8.18%, O 32.68%, P 9.04%. Paramagnetic manganese (II) chelate designed as a tissue specific imaging agent taken up by normal liver parenchyma. Prepn: S. M. Rocklage, S. C. Quay, **EP 290047**; *eidem*, **US 4933456** (1988, 1990 both to Salutar); *idem et al.*, *Inorg. Chem.* **28**, 477 (1989). Pharmacology, toxicity and image enhancement studies: G. Elizondo *et al.*, *Radiology* **178**, 73 (1991). HPLC determn in plasma: K. G. Toft *et al.*, *J. Pharm. Biomed. Anal.* **15**, 973 (1997). Series of articles on clinical studies, toxicology and physicochemical properties: *Acta Radiol.* **38**, 626-789 (1997). Review of use as contrast agent for liver lesion detection: N. M. Rofsky, J. P. Earls, *MRI Clin. North Am.* **4**, 73-85 (1996).

LD$_{50}$ i.v. in mice: 5.4 mmol/kg (Elizondo).
Trisodium salt. [140678-14-4] Magnafodipir trisodium; Win-59010; Teslascan. $C_{22}H_{27}MnN_4Na_3O_{14}P_2$; mol wt 757.33. Pale yellow, triclinic hygroscopic crystals. d 1.537. uv max (water): 220, 257, 319 nm (ε 37600, 10300, 13400). Soly (g/ml): 0.4596 water, 0.0230 methanol, 0.0008 ethanol, 0.0006 acetone, 0.0011 chloroform. Log P (1-octanol:water) −5.62; (1-butanol: water) −3.68. Prepd as 0.01 mmol/ml aqueous infusion: bright yellow, clear soln, pH 7.5. Viscosity (mPa·s): 1.0 at 20°, 0.7 at 37°. Osmolality (37°): 290 mosmol/kg. d^{20} 1.01 g/ml.
THERAP CAT: Diagnostic aid (MRI contrast agent).

5788. Manganese. [7439-96-5] Mn; at. wt 54.938045; at. no. 25; valences 2, 4, 7; 1, 3, 5, 6 rare. Group VIIB (7). One stable isotope: 55; artificial radioactive isotopes 49-54; 56-58. Widely-distributed, abundant element; constitutes 0.085% of earth's crust. Occurs in the minerals pyrolusite, hausmannite, manganite, *braunite* (3Mn$_2$O$_3$.MnSiO$_3$), *manganosite* (MnO), and in several others; occurs in minute quantities in water, plants and animals. Essential dietary nutrient. First isolated by Gahn in 1774. Prepn of the metal: John *et al.*, cited by Mellor, *A Comprehensive Treatise on Inorganic and Theoretical Chemistry* **12**, 163 (1932); A. H. Sully, *Manganese* (Academic Press, New York, 1955) 305 pp. Review of physical properties: Meaden, *Met. Rev.* **13**, 97-114 (1968). Review of manganese and its compds: Kemmitt in *Comprehensive Inorganic Chemistry* **vol. 3**, J. C. Bailar Jr. *et al.*, Eds. (Pergamon Press, Oxford, 1973) pp 771-876; L. R. Matricardi, J. H. Downing in *Kirk-Othmer Encyclopedia of Chemical Technology* **vol. 14**, (Wiley-Interscience, New York, 3rd ed., 1981) pp 824-843. Review of toxicology and human exposure: *Toxicological Profile for Manganese* (PB2000-108025, 2000) 504 pp.

Steel gray, lustrous, hard, brittle metal. Exists in four allotropic forms: α-form, body-centered cubic, stable below 710° (approx), d^{20} 7.47; β-form, cubic, stable in the range 710-1079°, d^{20} 7.26; γ-form or electrolytic Mn, face-centered cubic, stable in the range 1079-1143°C, d^{1100} 6.37; δ-form, body-centered cubic, stable from 1143° to mp, d^{1143} 6.28. γ-Mn when stabilized at room temp is face-centered tetragonal, d^{20} 7.21. mp 1244°. bp 2095°. Sp heat 0.115 cal/g/°C; latent heat of fusion: 3.5 kcal/g-atom. Mohs' hardness 5.0. Superficially oxidized on exposure to air. Burns with an intense white light when heated in air. Dec water slowly in the cold, rapidly on heating; pure electrolytic manganese is not attacked by water at ordinary temp; slightly attacked by steam. Reacts with dil mineral acids with evolution of hydrogen and formation of divalent manganous salts. Reacts with aq solns of sodium or potassium bicarbonate. When heated in nitrogen above 2000° burns to form a nitride. Converted by fluorine into the di- and trifluoride, by chlorine into the dichloride. In form of powder reduces most metallic oxides on heating. On heating, reacts directly with carbon, phosphorus, antimony or arsenic.

Caution: Potential symptoms of overexposure are parkinsonism; asthenia, insomnia and mental confusion; metal fume fever (dry throat, coughing, tight chest, dyspnea, rales and flu-like fever); lower back pain; vomiting; malaise; fatigue; kidney damage. *See NIOSH Pocket Guide to Chemical Hazards* (DHHS/NIOSH 97-140, 1997) p 190.

USE: In manuf of steel; for rock crushers, railway points and crossings, wagon buffers; as a constituent of several alloys, *e.g.*, ferromanganese, copper manganese, Manganin.

5789. Manganese Acetate. [638-38-0] Acetic acid manganese(2+) salt (2:1); diacetylmanganese. $C_4H_6MnO_4$; mol wt 173.03. C 27.77%, H 3.50%, Mn 31.75%, O 36.99%. $Mn(CH_3COO)_2$.

Tetrahydrate. Pale red, transparent monoclinic crystals. d 1.59. Sol in water or alcohol. LD_{50} orally in rats: 3.73 g/kg: Smyth *et al.*, *Am. Ind. Hyg. Assoc. J.* **30**, 470 (1969).

USE: As mordant in dyeing; manuf bister; drier for paints and varnishes (1-1.5:1000).

5790. Manganese Bromide. [13446-03-2] Manganese dibromide. Br_2Mn; mol wt 214.75. Br 74.42%, Mn 25.58%. $MnBr_2$.

Tetrahydrate. Rose-red, slightly deliquesc crystals. mp 64° with some decompn. Sol in 0.5 parts water; sol in alcohol. The aq soln is slightly acid. *Keep well closed.*

5791. Manganese Carbonate. [598-62-9] Carbonic acid manganese(2+) salt (1:1). $CMnO_3$; mol wt 114.95. C 10.45%, Mn 47.79%, O 41.75%. $MnCO_3$. (Usually contains some H_2O). Occurs in nature as the mineral *rhodochrosite*.

Pink to almost white powder when freshly pptd, but gradually becomes light brown in the air. Rhombohedral, calcite structure. d 3.1. Insol in water or alcohol. Sol in dil acid. *Keep well closed.*

USE: As pigment—"manganese white"; drier for varnishes; in feeds.

5792. Manganese Carbonyl. [10170-69-1] Decacarbonyldimanganese (*Mn–Mn*); manganese pentacarbonyl dimer. $C_{10}Mn_2O_{10}$; mol wt 389.98. C 30.80%, Mn 28.17%, O 41.03%. $Mn_2(CO)_{10}$. First prepd by reduction of MnI_2 with a Grignard reagent under CO pressure: Hurd *et al.*, *J. Am. Chem. Soc.* **71**, 1899 (1949). Improved prepns: Brimm *et al.*, *ibid.* **76**, 3831 (1954); Closson, *ibid.* **80**, 6167 (1958); Podall *et al.*, *ibid.* **82**, 1325 (1960); Calderazzo, *Inorg. Chem.* **4**, 293 (1965). Crystal structure: Dahl, Rundle, *Acta Crystallogr.* **16**, 419 (1963); molecular structure: Almenningen *et al.*, *Acta Chem. Scand.* **23**, 685 (1969). *Reviews: Organic Syntheses via Metal Carbonyls* **vol. 1**, I. Wender, P. Pino, Eds. (Interscience, New York, 1968) *passim;* Anisimov *et al.*, *Usp. Khim.* **37**, 380-408 (1968); *Russ. Chem. Rev.* **37**, 184-197 (1968) (English translation).

Golden yellow, monoclinic crystals. d^{25} 1.75. mp 154-155°. Stable under an atm of CO. In absence of CO, begins to dec at 110°. Sol in organic solvents. Insol in water. Less stable to air, heat and light in soln.

USE: Catalyst; antiknock additive.

5793. Manganese Chloride. [7773-01-5] Manganese chloride (MnCl₂); manganous chloride; manganese dichloride. Cl_2Mn;

mol wt 125.84. Cl 56.34%, Mn 43.66%. $MnCl_2$. Acute toxicity studies: D. J. Holbrook, Jr. *et al.*, *Environ. Health Perspect.* **10**, 95 (1975); P. P. Singh, A. Y. Junnarkar, *Indian J. Pharmacol.* **23**, 153 (1991).

bp 1231°. LD_{50} in mice, rats (mg/kg): 1330.0, 1470.0 orally; 126.0, 147.0 i.p.; 171.0, 92.6 i.v. (Singh, Junnarkar).

Tetrahydrate. [13446-34-9] $Cl_2Mn.4H_2O$; mol wt 197.90. Reddish, slightly deliquesc, monoclinic crystals. d 2.01. mp 58°. Sol in 0.7 part water; sol in alcohol. Insol in ether. pH of 0.2 molar aq soln 5.5. *Keep well closed.* LD_{50} in rats (mmole/kg): 7.5 orally; 0.70 i.p. (Holbrook).

USE: In dyeing (manganese bister); disinfecting; purifying natural gas; linseed oil drier; in electric batteries.

5794. Manganese Difluoride. [7782-64-1] Manganese fluoride (MnF₂); manganous fluoride; manganese fluoride. F_2Mn; mol wt 92.93. F 40.89%, Mn 59.12%. MnF_2. Prepd from manganese carbonate and hydrogen fluoride: Moissan, Venturi, *Compt. Rend.* **130**, 1158 (1900); Kwasnik in *Handbook of Preparative Inorganic Chemistry* **Vol. 1**, G. Brauer, Ed. (Academic Press, New York, 2nd ed., 1963) pp 262-263. Review of toxicology of fluoride compounds: G. L Waldbott, *Acta Med. Scand. Suppl.* **400**, 1-44 (1963).

Pink, quadratic prisms (tetragonal structure, rutile type) or reddish powder. *Poisonous.* d 3.98. mp 856°. Soly in water (g/100 ml): 0.66 (40°); 0.44 (60°); 0.48 (100°). Insol in alc. Sol in dil hydrofluoric acid, concd hydrochloric or nitric acid. LD in guinea pigs (mg/kg): 200 orally, 700 s.c. (Waldbott).

Tetrahydrate. Obtained by dissolving manganese carbonate in hydrofluoric acid, evaporating, and drying *in vacuo.*

5795. Manganese Dioxide. [1313-13-9] Manganese oxide (MnO₂); manganese binoxide; manganese peroxide; manganese superoxide; black manganese oxide. MnO_2; mol wt 86.94. Mn 63.-19%, O 36.80%. Occurs in nature as the mineral *pyrolusite*, or made artificially (pptd). The native product is heavy, steel-gray when in lumps, black when powdered; the pptd product is a brownish-black, fine powder. Both usually contain some Mn_3O_4 and some water. When ignited evolves oxygen, leaving Mn_3O_4. Lab prepn: Moore *et al.*, *J. Am. Chem. Soc.* **72**, 856 (1950); Covington *et al.*, *Trans. Faraday Soc.* **58**, 1975 (1962). Toxicity study: D. J. Holbrook, Jr. *et al.*, *Environ. Health Perspect.* **10**, 95 (1975). Review of use as reagent: J. S. Pizey, *Synthetic Reagents* **vol. 2** (John Wiley, New York, 1974) pp 143-174.

Tetragonal crystals (rutile structure). Insol in water, nitric or cold sulfuric acid. Slowly dissolves in cold HCl with evolution of Cl_2; in presence of hydrogen peroxide or oxalic acid it dissolves in dil H_2-SO_4 or HNO_3. *Strong oxidizer; should not be heated or rubbed with organic matter or other oxidizable substances, e.g., sulfur, sulfides, phosphides, hypophosphites, etc.* LD_{50} orally in rats: >40 mmole/kg (Holbrook).

USE: The mineral is the source of manganese and all its compds; largely used in manuf manganese steel; oxidizer; in alkaline batteries (dry cells); for making amethyst glass, decolorizing glass; painting on porcelain, faience and majolica. The ppt is used in electrotechnics, pigments, browning gun barrels, drier for paints and varnishes, printing and dyeing textiles.

5796. Manganese Hypophosphite. [10043-84-2] Phosphinic acid manganese(2+) salt (2:1). $H_4MnO_4P_2$; mol wt 184.91. H 2.18%, Mn 29.71%, O 34.61%, P 33.50%. $Mn(H_2PO_2)_2$.

Monohydrate. Pink, odorless, almost tasteless crystals or powder. When heated evolves spontaneously flammable phosphine. One gram dissolves in 6.5 ml water, 6 ml boiling water. Insol in alcohol.

5797. Manganese Iodide. [7790-33-2] Manganese diiodide. I_2Mn; mol wt 308.75. I 82.21%, Mn 17.79%. MnI_2.

Tetrahydrate. Rose-red crystals; rapidly becomes brown on exposure to air and light, due to liberation of iodine. Very sol in water with gradual decompn; sol in alcohol. The aq soln is slightly acid. *Keep tightly closed and protected from light.*

5798. Manganese Nitrate. [10377-66-9] Nitric acid manganese(2+) salt (2:1); manganese dinitrate. MnN_2O_6; mol wt 178.95. Mn 30.70%, N 15.65%, O 53.64%. $Mn(NO_3)_2$. Prepn: Dehnicke, Strähle, *Ber.* **97**, 1502 (1964). *Review: Gmelins, Manganese* (8th ed.) **56**, part C3, 267-281 (1975).

Consult the Name Index before using this section.

Colorless; hygroscopic. *Oxidizer.* Sol in water, dioxane, tetrahydrofuran, acetonitrile.

Tetrahydrate. Pink, deliquesc cryst masses at temp below 20°. d 2.129. mp 37.1°. Very sol in water, sol in alcohol. The aq soln is slightly acid.

Hexahydrate. Rose-colored, deliquesc, monoclinic needles (colorless after several crystns). d 1.8. mp 25.0°. Freely sol in water, alcohol.

USE: Intermediate in manuf of reagent grade MnO_2; in prepn of porcelain colorants.

5799. Manganese Oxalate. [640-67-5] [Ethanedioato(2−)-$\kappa O^1, \kappa O^2$]-manganese. C_2MnO_4; mol wt 142.96. C 16.80%, Mn 38.43%, O 44.76%. MnC_2O_4.

Dihydrate. White cryst powder; dec at 150°. Slightly sol in water; sol in acids.

5800. Manganese Oxide. [1317-35-7] Manganomanganic oxide; manganese tetroxide; trimanganese tetroxide. Mn_3O_4; mol wt 228.81. Mn 72.03%, O 27.97%. Occurs in nature as the mineral *hausmannite*. Prepn: Moore *et al., J. Am. Chem. Soc.* **72**, 856 (1950).

Brownish-black powder. d 4.7. Insol in water. Sol in HCl with evolution of chlorine.

Caution: Potential symptoms of overexposure are asthenia, insomnia, mental confusion; low back pain, vomiting, fatigue; kidney damage; pneumonitis. *See NIOSH Pocket Guide to Chemical Hazards* (DHHS/NIOSH 97-140, 1997) p 192.

5801. Manganese Selenide. [1313-22-0] MnSe; mol wt 133.90. Mn 41.03%, Se 58.97%. Prepd by the action of manganese salts on alkali selenides: Berzelius *et al.,* cited by Mellor, *A Comprehensive Treatise on Inorganic and Theoretical Chemistry* **10**, 798 (1930); by the action of a manganese salt on a soln of hydrogen selenide out of contact with air: Moser, Atynsky, *Monatsh. Chem.* **45**, 235 (1925).

Gray-black cubic crystals. d^{15} 5.59. Completely oxidized when heated to redness in oxygen. Insol in cold water. Dec in hot water and dil acids.

5802. Manganese Sesquioxide. [1317-34-6] Manganese oxide (Mn_2O_3); dimanganese trioxide. Mn_2O_3; mol wt 157.87. Mn 69.60%, O 30.40%. Occurs in nature as the mineral *manganite*, $Mn_2O_3.H_2O$. Prepn: Moore *et al., J. Am. Chem. Soc.* **72**, 856 (1950).

Black, fine powder. d 4.50. Insol in water. Sol in HCl with evolution of chlorine.

5803. Manganese Silicate. [7759-00-4] Approx $MnSiO_3$. Occurs in nature as the minerals *rhodonite, manganjustite, tephroite*.

Red crystals or yellowish-red powder. The pptd article is a yellowish-red powder. d of the mineral 3.48. Insol in water.

USE: As color for special glass; producing red glazes on pottery.

5804. Manganese Sulfate. [7785-87-7] Sulfuric acid manganese(2+) salt (1:1). MnO_4S; mol wt 150.99. Mn 36.39%, O 42.38%, S 21.23%. $MnSO_4$. Forms several hydrates. The article of commerce is usually a mixture of the tetra- and pentahydrates.

Monohydrate. [10034-96-5] $MnO_4S.H_2O$; mol wt 169.01. Pale red, slightly efflorescent crystals; loses all water at 400-450°. Sol in about 1 part cold, 0.6 part boiling water. Insol in alc.

USE: In dyeing; for red glazes on porcelain; boiling oils for varnishes; in fertilizers for vines, tobacco; in feeds.

THERAP CAT (VET): Nutritional factor (essential trace element in all animals); prevention of perosis in poultry.

5805. Manganese Sulfide. [18820-29-6] Manganese monosulfide. MnS; mol wt 87.00. Mn 63.15%, S 36.85%. Occurs in nature as the mineral *alabandite* or *manganblende*. Prepn: Mehmed, Haraldsen, *Z. Anorg. Chem.* **235**, 194 (1938).

Pink, green or brown-green powder. Three cryst modifications: α-form, green cubic crystals; β-form, red cubic crystals; γ-form, red hexagonal crystals. Practically insol in water; sol in dil acids. In moist condition it readily oxidizes in air to the sulfate.

5806. Manganese Trifluoride. [7783-53-1] Manganese fluoride (MnF_3); manganic fluoride. F_3Mn; mol wt 111.93. F 50.92%,

Mn 49.08%. MnF_3. Prepd by fluorination of manganese iodide according to the equation $2MnI_2 + 13F_2 \rightarrow 2MnF_3 + 4IF_5$: Moissan, *Compt. Rend.* **130**, 622 (1900); v. Wartenberg, *Z. Anorg. Allg. Chem.* **244**, 346 (1940); Kwasnik in *Handbook of Preparative Inorganic Chemistry* vol. 1, G. Brauer, Ed. (Academic Press, New York, 2nd ed., 1963) pp 263-264.

Red mass; monoclinic crystals. d 3.54. Stable to 600°. Hydrolyzed by water. May be stored in glass ampuls.

USE: Fluorinating agent in organic chemistry.

5807. Mangostin. [6147-11-1] 1,3,6-Trihydroxy-7-methoxy-2,8-bis-(3-methyl-2-buten-1-yl)-9*H*-xanthen-9-one; 1,3,6-trihydroxy-7-methoxy-2,8-di(3-methyl-2-butenyl)xanthone. $C_{24}H_{26}O_6$; mol wt 410.47. C 70.23%, H 6.38%, O 23.39%. From various parts of the mangosteen tree *(Garcinia mangostana* L., Guttiferae): Schmid, *Ann.* **93**, 83 (1855); Dragendorff, *ibid.* **482**, 280 (1930). Structure: Yates, Stout, *J. Am. Chem. Soc.* **80**, 1691 (1958); Scheinmann, *Chem. Commun.* **1967**, 1015; Stout *et al., ibid.* **1968**, 211.

Yellow crystals from benzene, mp 181.6-182.6°. uv max (ethanol): 243, 259, 318, 351 nm (log ε 4.54, 4.44, 4.38, 3.86). Practically insol in water; sol in alcohol, ether, acetone, chloroform, ethyl acetate.

3,6-Dimethylmangostin. $C_{26}H_{30}O_6$. Pale yellow needles from ethanol, mp 123.3-123.8°. uv max (ethanol): 245, 262, 314, 350 nm (log ε 4.50, 4.53, 4.36, 3.81).

5808. Manidipine. [89226-50-6] 1,4-Dihydro-2,6-dimethyl-4-(3-nitrophenyl)-3,5-pyridinedicarboxylic acid 3-[2-[4-(diphenylmethyl)-1-piperazinyl]ethyl] 5-methyl ester; 2-[4-(diphenylmethyl)-1-piperazinyl]ethyl methyl (±)-1,4-dihydro-2,6-dimethyl-4-(*m*-nitrophenyl)-3,5-pyridinecarboxylate; franidipine. $C_{35}H_{38}N_4O_6$; mol wt 610.71. C 68.84%, H 6.27%, N 9.17%, O 15.72%. Dihydropyridine calcium channel blocker. Prepn: K. Meguro, A. Nagaoka, **EP 94159**; *eidem,* **US 4892875** (1983, 1990, both to Takeda); K. Meguro *et al., Chem. Pharm. Bull.* **33**, 3787 (1985). Synthesis and bioactivity of optical isomers: M. Kajino *et al., ibid.* **37**, 2225 (1989). Acute toxicity study: S. Chiba *et al., Yakuri to Chiryo* **17**, Suppl. 4, 961 (1989), *C.A.* **111**, 146499b (1989). Determn of enantiomers in human serum: M. Yamaguchi *et al., J. Chromatogr.* **575**, 123 (1992). Clinical evaluation in hypertension: K. Mizuno *et al., Curr. Ther. Res.* **52**, 248 (1992).

Light yellow crystals from isopropyl ether + hexane, mp 125-128°.

Dihydrochloride. [89226-75-5] CV-4093; Calslot. $C_{35}H_{38}N_4$-$O_6.2HCl$; mol wt 683.63. Two crystalline forms. α-form: yellow crystals, mp 157-163°. β-form: light yellow, fine crystals, mp 174-180°. Also prepd as the monohydrate, mp 167-170°. LD_{50} in male, female mice, male, female rats (mg/kg): 387, 340, 222, 199 s.c.; 62.2, 68.0, 66.5, 48.8 i.p.; 190, 171, 247, 156 orally (Chiba).

THERAP CAT: Antihypertensive.

5809. Manna. Dried exudation of *Fraxinus ornus* L., Oleaceae. *Habit.* Mediterranean Basin, Asia Minor, Spain. *Constit.*

40-60% mannitol; 10-16% mannotetrose; 6-16% mannotriose; glucose, mucilage, fraxin.

One gram dissolves in 5 ml water, 150 ml 90% alcohol.

THERAP CAT: Cathartic.

5810. Mannitol. [69-65-8] D-Mannitol; mannite; manna sugar; cordycepic acid; Manicol; Mannidex; Osmitrol; Osmosal; Resectisol. $C_6H_{14}O_6$; mol wt 182.17. C 39.56%, H 7.75%, O 52.69%. Widespread in plants and plant exudates; obtained from manna and seaweeds: *The Carbohydrates*, W. Pigman, Ed. (Academic Press, New York, 1957) pp 249-250. Forms a stable, equimolar compound with H_2O_2: S. Tanatar, *J. Russ. Phys. Chem. Soc.* **40**, 376, *C.A.* **3**, 883 (1909). Prepd by electrolytic reduction of glucose: Creighton, *Can. Chem. Process Ind.* **26**, 690 (1942), *C.A.* **37**, 1088⁵ (1943); Wolfrom *et al.*, *J. Am. Chem. Soc.* **68**, 578 (1946). Prepn from seaweed: Sorensen, Kristensen, US 2516350 (1950); by hydrogenation of invert sugar, monosaccharides, and sucrose: Kasehagen, and Kasehagen, Luskin, US 2642462, US 2749371, and US 2759024 (1953, 1956, and 1956, all to Atlas Powder). Review of prepn: Pigman, *loc. cit.* Novel synthesis: M. Makkee *et al.*, *Chem. Commun.* **1980**, 930.

$$\begin{array}{c} CH_2OH \\ | \\ HO-CH \\ | \\ HO-CH \\ | \\ HC-OH \\ | \\ HC-OH \\ | \\ CH_2OH \end{array}$$

Orthorhombic needles from alc, mp 166-168°. Sweetish taste. d^{20} 1.52. bp$_{3.5}$ 290-295°. Is inactive or very slightly levorotatory in distilled water. Forms sodium mannitoborate on addition of borax giving a greater rotation: $[\alpha]_D^{20}$ +23° → +24° after 1 hr in a soln of 10 g mannitol +12.8 g borax + sufficient H_2O to make 100 ml. One gram dissolves in ~5.5 ml water, 83 ml alcohol; more sol in hot water. Insol in ether. Sol in pyridine, aniline, aq solns of alkalies. One gram dissolves in 18 ml glycerol (d 1.24). Soly tables: Creighton, Klauder, *J. Franklin Inst.* **195**, 687 (1923). pKa (18°): 13.50.

USE: Used with boric acid in the manuf of dry electrolytic condensers for radio applications; in making artificial resins and plasticizers; in pharmacy as excipient and diluent for solids and liqs. In analytical chemistry for boron determn; titrimetric determn of boric acid. In the manuf of mannitol hexanitrate. Used in the food industry as anticaking and free-flow agent, flavoring agent, lubricant and release agent, stabilizer and thickener and nutritive sweetener.

THERAP CAT: Diuretic. Diagnostic aid (renal function).

5811. D-Mannitol Hexanitrate. [15825-70-4] Mannitol nitrate; nitromannite; nitromannitol; Medemanol; Dilangil; Moloid; Mannitrin; Manexin. $C_6H_8N_6O_{18}$; mol wt 452.15. C 15.94%, H 1.78%, N 18.59%, O 63.69%. Made by nitration of mannitol: Fleury *et al.*, *Mem. Poudres* **31**, 107 (1949).

Long needles in regular clusters from alc. mp 106-108°. Soluble in alc, in ether; insol in water. Explodes on percussion. Its stability at ordinary temps is such that it may be used commercially, but it is distinctly less stable than nitroglycerol at 75°. Its employ for pharmaceutical prepns is only in admixture with carbohydrate substances in dilutions corresponding to 1 part of mannitol hexanitrate to 9 or more parts of carbohydrate. In such dilutions mannitol hexanitrate is considered nonexplosive.

THERAP CAT: Vasodilator.

5812. D-Mannose. [3458-28-4] Seminose; carubinose. $C_6H_{12}O_6$; mol wt 180.16. C 40.00%, H 6.71%, O 53.28%. Prepn of α-form by treating ivory nut shavings with H_2SO_4: Isbell, *J. Res. Natl. Bur. Stand.* **26**, 47 (1941); Isbell, Frush in *Methods in Carbohydrate Chemistry*, R. L. Whistler, M. L. Wolfrom, Eds. (Academic Press, New York, 1962) pp 145-147. Prepn and stability of α- and β-forms: Reeves, *J. Am. Chem. Soc.* **72**, 1499 (1950); J. Sowden in *The Carbohydrates*, W. Pigman, Ed. (Academic Press, New York, 1957) pp 94-95.

α-Form. Crystals from methanol, mp 133°. $[\alpha]_D$ +29.3° → +14.2° (water).

β-Form. Orthorhombic, bisphenoidal needles from alcohol or acetic acid, dec 132°. Sweet taste with bitter aftertaste. d^{20} 1.54. Shows mutarotation. $[\alpha]_D^{20}$ -17.0° → +14.2° (c = 4). One gram dissolves in 0.4 ml water, 120 ml methanol, 250 ml abs ethanol, 3.5 ml pyridine. pKa (18°): 11.98. Reduces Fehling's soln; is fermented by yeast.

Phenylhydrazone. $C_{12}H_{18}N_2O_5$. Crystals from dil ethanol, mp 199-200°. $[\alpha]_D^{20}$ +26.3° → +33.8° (pyridine).

CaCl₂-addition compound tetrahydrate. $C_6H_{12}O_6 \cdot CaCl_2 \cdot 4H_2O$. mp 101-102°. $[\alpha]_D^{20}$ -31.3° → +6.0° (c = 9).

5813. Maprotiline. [10262-69-8] N-Methyl-9,10-ethanoanthracene-9(10H)-propanamine; 9-(γ-methylaminopropyl)-9,10-dihydro-9,10-ethanoanthracene; 1-(3-methylaminopropyl)dibenzo[b,e]bicyclo[2.2.2]octadiene. $C_{20}H_{23}N$; mol wt 277.41. C 86.59%, H 8.36%, N 5.05%. Prepn: P. Schmidt *et al.*, US 3399201 (1968 to Ciba). Synthesis, NMR, mass spectra: M. Wilhelm, P. Schmidt, *Helv. Chim. Acta* **52**, 1385 (1969). Toxicity: R. Hess *et al.*, *Boll. Chim. Farm.* **112**, 782 (1973), *C.A.* **81**, 33479p (1974). Review of pharmacology and therapeutic efficacy: R. M. Pinder *et al.*, *Drugs* **13**, 321 (1977). Comprehensive description: S. K. Suh, J. B. Smith, *Anal. Profiles Drug Subs.* **15**, 393-426 (1986).

mp 92-94°.

Hydrochloride. [10347-81-6] Ba-34276; Deprilept; Ludiomil; Psymion. $C_{20}H_{23}N \cdot HCl$; mol wt 313.87. Crystals from isopropanol, mp 230-232°. Freely sol in methanol and chloroform, slightly sol in water. Practically insol in isooctane. LD$_{50}$ in mice, rats (mg/kg): ~750, ~900 orally (Hess).

THERAP CAT: Antidepressant.

5814. Maraviroc. [376348-65-1] 4,4-Difluoro-N-[(1S)-3-[(3-exo)-3-[3-methyl-5-(1-methylethyl)-4H-1,2,4-triazol-4-yl]-8-azabicyclo[3.2.1]oct-8-yl]-1-phenylpropyl]cyclohexanecarboxamide; N-[(1S)-3-[3-(3-isopropyl-5-methyl-4H-1,2,4-triazol-4-yl)-exo-8-azabicyclo[3.2.1]oct-8-yl]-1-phenylpropyl]-4,4-difluorocyclohexanecarboxamide; UK-427857; Celsentri; Selzentry. $C_{29}H_{41}F_2N_5O$; mol wt 513.68. C 67.81%, H 8.05%, F 7.40%, N 13.63%, O 3.11%. CCR5 chemokine receptor antagonist; inhibits HIV entry by blocking interaction of viral coat protein gp120 with the receptor. Prepn: M. Perros *et al.*, WO 0190106; *eidem*, US 6667314 (2001, 2003 both to Pfizer); D. A. Price *et al.*, *Tetrahedron Lett.* **46**, 5005 (2005). Overview of structure identification and development: A. Wood, D. Armour, *Prog. Med. Chem.* **43**, 239-271 (2005). Pharmacokinetics and metabolism in humans and animals: D. K. Walker *et al.*, *Drug Metab. Dispos.* **33**, 587 (2005). Clinical evaluation in HIV-1 infected patients: G. Fätkenheuer *et al.*, *Nat. Med.* **11**, 1170 (2005). *Review:* N. A. Meanwell, J. F. Kadow, *Curr. Opin. Investig. Drugs* **8**, 669-681 (2007).

White solid from toluene/hexane (2:1), mp 197-198°. pKa 7.3. Moderate lipophilicity, logD 2.1.

THERAP CAT: Antiretroviral.

5815. Marbofloxacin. [115550-35-1] 9-Fluoro-2,3-dihydro-3-methyl-10-(4-methyl-1-piperazinyl)-7-oxo-7H-pyrido[3,2,1-ij][4,-1,2]benzoxadiazine-6-carboxylic acid; Marbocyl; Zeniquin. $C_{17}H_{19}FN_4O_4$; mol wt 362.36. C 56.35%, H 5.29%, F 5.24%, N 15.46%, O 17.66%. Fluorinated quinolone antibacterial. Prepn: M. Aoki et al., **EP 259804**; K. Yokose et al., **US 4801584** (1987, 1989 both to Hoffmann-LaRoche). HPLC determn in plasma: M. A. Garcia et al., J. Chromatogr. B **729**, 157 (1999). Evaluation of experimental surgical infection in dogs: P. Gruet et al., Vet. Rec. **140**, 199 (1997). Pharmacokinetics and efficacy in cattle: C. Eyett-Burton et al., Cattle Practice **5**, 289 (1997). Pharmacokinetics in renally impaired dogs: H. P. Lefebvre et al., J. Vet. Pharmacol. Ther. **21**, 453 (1998).

Crystals from methanol, mp 268-269° (dec).
THERAP CAT (VET): Antibacterial.

5816. Margaric Acid. [506-12-7] Heptadecanoic acid. $C_{17}H_{34}O_2$; mol wt 270.46. C 75.50%, H 12.67%, O 11.83%. Prepn: Kaufmann, Stamm, Ber. **91**, 2121 (1958); Bhattacharyya et al., Chem. Ind. (London) **1959**, 1352; Hünig, Ledle, Ber. **93**, 913 (1960). Metabolism: Boyer, Scheig, Lipids **4**, 615 (1969). Toxicity study: L. Orö, A. Wretlind, Acta Pharmacol. Toxicol. **18**, 141 (1961).

Crystals from alcohol, mp 61°. d 0.853. bp$_{100}$ 227°. n_D^{60} 1.4342. Insol in water. Freely sol in ether; slightly sol in alcohol. LD$_{50}$ i.v. in mice: 360.3 mg/kg (Orö, Wretlind).

5817. Margatoxin. [145808-47-5] MgTX. $C_{178}H_{286}N_{52}O_{50}S_7$; mol wt 4178.98. C 51.16%, H 6.90%, N 17.43%, O 19.14%, S 5.37%. Peptide toxin isolated from the new world scorpion, Centruroides margaritatus, consists of 39 amino acids with 3 disulfide bridges. Selectively inhibits voltage-gated potassium channels. Isoln and characterization: M. Garcia-Calvo et al., J. Biol. Chem. **268**, 18866 (1993). Synthesis: M. A. Bednarek et al., Biochem. Biophys. Res. Commun. **198**, 619 (1994). NMR determn of three dimensional structure: B. A. Johnson et al., Biochemistry **33**, 15061 (1994). Use as an affinity label for potassium channels: H.-G. Knaus et al., ibid. **34**, 13627 (1995); H. S. Fischer, A. Saria, Neurosci. Lett. **263**, 208 (1999).

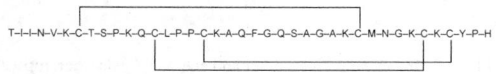

USE: Biochemical probe.

5818. Marimastat. [154039-60-8] (2S,3R)-N^4-[(1S)-2,2-Dimethyl-1-[(methylamino)carbonyl]propyl]-N^1,2-dihydroxy-3-(2-methylpropyl)butanediamide; N^2-[3S-hydroxy-4-(N-hydroxyamino)-2R-isobutylsuccinyl]-L-tert-leucine-N^1-methylamide; BB-2516. $C_{15}H_{29}N_3O_5$; mol wt 331.41. C 54.36%, H 8.82%, N 12.68%, O 24.14%. Synthetic matrix metalloproteinase inhibitor (MMPI). Prepn: J. P. Dickens et al., **WO 9402447**; eidem, **US 5700838** (1994, 1997 both to British Biotech) and structure-activity study: D. E. Levy et al., J. Med. Chem. **41**, 199 (1998). Improved synthesis: R. J. Davenport, R. J. Watson, Tetrahedron Lett. **41**, 7983 (2000). Clinical study in pancreatic cancer: S. R. Bramhall et al., J. Clin. Oncol. **19**, 3447 (2001); in gastric cancer: idem et al., Br. J. Cancer **86**, 1864 (2002). Review of pharmacology: H. S. Rasmussen, P. P.

McCann, Pharmacol. Ther. **75**, 69-75 (1997); and clinical development: A. L. Thomas, W. P. Steward, Expert Opin. Invest. Drugs **9**, 2913-2922 (2000). Review of structure-activity and mechanism of action: M. Whittaker et al., Chem. Rev. **99**, 2735-2776 (1999).

Soly in water: 1.4 mg/ml.
THERAP CAT: Antineoplastic.

5819. Maropitant. [147116-67-4] (2S,3S)-N-[[5-(1,1-Dimethylethyl)-2-methoxyphenyl]methyl]-2-(diphenylmethyl)-1-azabicyclo[2.2.2]octan-3-amine; (2S,3S)-2-benzhydryl-N-(5-tert-butyl-2-methoxybenzyl)quinuclidin-3-amine. $C_{32}H_{40}N_2O$; mol wt 468.69. C 82.01%, H 8.60%, N 5.98%, O 3.41%. Selective neurokinin-1 (NK$_1$) receptor antagonist; blocks the action of substance P on the central nervous system. Prepn: F. Ito et al., **WO 9221677** (1992 to Pfizer); eidem, **US 6222038** (2001). Pharmacokinetics in dogs: H. A. Benchaoui et al., J. Vet. Pharmacol. Ther. **30**, 336 (2007). Efficacy in dogs in treatment and prevention of emesis: V. A. de la Puente-Redondo et al., J. Small Anim. Pract. **48**, 93 (2007); of cisplatin-induced emesis: idem et al., Am. J. Vet. Res. **68**, 48 (2007); D. M. Vail et al., Vet. Comp. Oncol. **5**, 38 (2007).

Citrate monohydrate. [359875-09-5] CJ-11972; Cerenia. $C_{32}H_{40}N_2O.C_6H_8O_7.H_2O$; mol wt 678.82.
THERAP CAT (VET): Antiemetic.

5820. Marrubiin. [465-92-9] (2aS,5aS,6R,7R,8aR,8bR)-6-[2-(3-Furanyl)ethyl]decahydro-6-hydroxy-2a,5a,7-trimethyl-2H-naphtho[1,8-bc]furan-2-one; 15,16-epoxy-6β,9-dihydroxy-8βH-labda-13(16),14-dien-19-oic acid γ-lactone; 5-[2-(3-furyl)ethyl]decahydro-5,8-dihydroxy-1,4a,6-trimethyl-1-naphthoic acid γ-lactone. $C_{20}H_{28}O_4$; mol wt 332.44. C 72.26%, H 8.49%, O 19.25%. Diterpene lactone principle isolated from white horehound, Marrubium vulgare (Tourn.) L., Labiatae: Harms, Arch. Pharm. **83**, 144 (1842); Ludwig, Kromayer, ibid. **158**, 257 (1861); Nicholas, J. Pharm. Sci. **53**, 895 (1964). Alternate view that marrubiin is an artefact generated from **premarrubiin** during the isoln: Henderson, McCrindle, J. Chem. Soc. C **1969**, 2014. Structure and stereochemistry: Cocker et al., J. Chem. Soc. **1953**, 2540; Chem. Ind. (London) **1954**, 1561; **1955**, 1484; Fulke, McCrindle, ibid. **1965**, 647. Total stereochemistry: Wheeler et al., Tetrahedron **23**, 3909 (1967); Appleton et al., J. Chem. Soc. C **1967**, 1943. Synthesis: Mangoni et al., Tetrahedron **28**, 611 (1972).

Crystals from alc, mp 160°. $[\alpha]_D^{20}$ +35.8° (c = 3.1 in chloroform); $[\alpha]_D^{24}$ +45° (acetone). uv max: 208, 212, 216 nm (log ε 3.75, 3.75, 3.70). One gram dissolves in 60 ml alc. Freely sol in chloroform, acetone, hot alc, pyridine; sparingly sol in ether, benzene. Practically insol in water.

5821. Martin Sulfurane. [32133-82-7] (*T*-4)-Bis[α,α-bis(trifluoromethyl)benzenemethanolato-κO]diphenylsulfur; diphenylbis(1,1,1,3,3,3-hexafluoro-2-phenyl-2-propoxy)sulfurane. $C_{30}H_{20}$-$F_{12}O_2S$; mol wt 672.53. C 53.58%, H 3.00%, F 33.90%, O 4.76%, S 4.77%. Mild dehydration reagent in organic synthesis; used in conversion of alcohols to olefins, and diols to epoxides. Prepn: J. C. Martin, R. J. Arhart, *J. Am. Chem. Soc.* **93**, 2341 (1971); *idem et al.*, *Org. Synth.* **coll. vol. VI**, 163 (1988); and chemical properties: R. J. Arhart, J. C. Martin, *J. Am. Chem. Soc.* **94**, 4997 (1972). X-ray crystal structure: I. C. Paul *et al.*, *ibid.* **93**, 6674 (1971); **94**, 5010 (1972). Synthetic applications in dehydration of alcohols: J. C. Martin, R. J. Arhart, *ibid.* **93**, 4327 (1971); R. J. Arhart, J. C. Martin, *ibid.* **94**, 5003 (1972); J. C. Martin *et al.*, *ibid.* **96**, 4604 (1974); F. Yokokawa, T. Shioiri, *Tetrahedron Lett.* **43**, 8679 (2002); in oxidation of secondary amines to imines: J. A. Franz, J. C. Martin, *J. Am. Chem. Soc.* **97**, 583 (1975); in cleavage of secondary amides: J. A. Franz, J. C. Martin, *J. Am. Chem. Soc.* **97**, 6137 (1975).

Crystals from pentane, mp 107-109°. *Corrosive*. Crystal density: 1.54 g/cm^3. uv max (isooctane): 226 nm ($\varepsilon \approx 10^4$).

USE: Dehydrating and oxidizing agent.

5822. Masitinib. [790299-79-5] 4-[(4-Methyl-1-piperazinyl)methyl]-*N*-[4-methyl-3-[[4-(3-pyridinyl)-2-thiazolyl]amino]-phenyl]benzamide; *N*-(3-(4-(pyridin-3-yl)thiazol-2-ylamino)-4-methylphenyl)-4-((4-methylpiperazin-1-yl)methyl)benzamide. C_{28}-$H_{30}N_6OS$; mol wt 498.65. C 67.44%, H 6.06%, N 16.85%, O 3.21%, S 6.43%. Tyrosine kinase inhibitor that selectively targets c-KIT, platelet-derived growth factor (PDGF) and Lyn receptors. Inhibits survival, degranulation, and migration of mast cells. Prepn: M. Ciufolini *et al.*, **WO 04014903**; *eidem*, **US 7423055** (2004, 2008 both to AB Science). Characterization of receptor inhibition: P. Dubreuil *et al.*, *PLoS ONE* **4**, e7258 (2009). Clinical pharmacokinetics and antitumor activity: J. C. Soria *et al.*, *Eur. J. Cancer* **45**, 2333 (2009). Clinical evaluation in asthma: M. Humbert *et al.*, *Allergy* **64**, 1194 (2009); in rheumatoid arthritis: J. Tebib *et al.*, *Arthritis Res. Therapy* **11**, R95 (2009); in mastocytosis: C. Paul *et al.*, *Am. J. Hematol.* **85**, 921 (2010); in GI stromal tumors: A. LeCesne *et al.*, *Eur. J. Cancer* **46**, 1344 (2010); in pancreatic cancer: E. Mitry *et al.*, *Cancer Chemother. Pharmacol.* **66**, 395 (2010); in Alzheimer's disease: F. Piette *et al.*, *Alzheimer's Res. Therapy* **3**, 16 (2011). Veterinary trial in canine mast cell tumors: K. A. Hahn *et al.*, *Am. J. Vet. Res.* **71**, 1354 (2010).

Methanesulfonate. [1048007-93-7] Masitinib mesylate; AB-1010; Kinavet CA1; Masivet. $C_{28}H_{30}N_6OS \cdot CH_4O_3S$; mol wt 594.75. White powder. Freely sol in 0.1*M* HCl, DMSO; sol in water; slightly sol in ethanol, propylene glycol. Practically insol in 0.1*M* NaOH, *n*-hexane.

THERAP CAT: Antineoplastic.

THERAP CAT (VET): Antineoplastic.

5823. Mastic. Balsam tree; pistachia galls; mastiche; mastix; lentisk; Mastisol. Concrete resinous exudation from *Pistacia lentiscus* L., *Anacardiaceae*. *Habit*. Mediterranean Islands, especially Chios. *Constit*. Volatile oil (about 2%); masticinic, masticonic acids; masticoresene.

Pale yellow or greenish-yellow, globular, elongated or pear-shaped tears; slightly balsamic odor and terebene taste. Almost insol in water. Nearly completely sol in alcohol; 1 g dissolves in 0.5 ml chloroform, 0.5 ml ether; partially sol in oil turpentine.

USE: In tooth cements, plasters, lacquers, chewing gums, and incense; also for retouching negatives.

5824. Maté. Paraguay tea; yerba maté; St. Bartholomew's tea; Jesuit's tea. Subtropical, flowering, evergreen tree, *Ilex paraguariensis* St. Hil., *Aquifoliaceae*. Dried leaves and stems are used to prepare the traditional South American beverage. Also used medicinally for its CNS stimulant, diuretic, antioxidant, and weight reducing properties. *Habit*. Brazil, Uruguay, Argentina, and Paraguay. *Constit*. Polyphenols, chlorogenic acid and derivatives; xanthine alkaloids, caffeine, theobromine, theophylline; flavonoids, quercetin, kaempferol, and their glycosides; ursolic acid saponins known as matesaponins; minerals, calcium, potassium, magnesium, manganese, sulfur. Phenolic content and antioxidant activity: L. Bravo *et al.*, *Food Res. Int.* **40**, 393 (2007). Analysis of mineral composition: R. Giulian *et al.*, *J. Agric. Food Chem.* **55**, 741 (2007). Determn of aroma constituents: H. C. Araujo *et al.*, *Phytochem. Anal.* **18**, 469 (2007). Clinical evaluation for weight loss: T. Andersen, J. Fogh, *J. Hum. Nutr. Dietet.* **14**, 243 (2001). Clinical effect on serum lipid and lipoprotein levels: E. C. De Morais *et al.*, *J. Agric. Food Chem.* **57**, 8316 (2009). Review of botany, constituents, and bioactivities: C. I. Heck, E. G. De Mejia, *J. Food Sci.* **72**, R138- 151 (2007); N. Bracesco *et al.*, *J. Ethnopharmacol.* **136**, 378-384 (2011).

USE: Dietary supplement; stimulant beverage.

5825. Matico. Dried leaves of *Piper elongatum* Vahl. (*P. angustifolium* R. & P.), *Piperaceae*. *Habit*. Peru, Bolivia, Brazil, Mexico, Cuba. *Constit*. 1-3.5% volatile oil, maticin (a bitter principle); artanthic acid, tannin, mucilage, resin.

THERAP CAT: Astringent.

5826. Matricarin. [5989-43-5] (3*S*,3a*R*,4*S*,9a*S*,9b*R*)-4-(Acetyloxy)-3,3a,4,5,9a,9b-hexahydro-3,6,9-trimethylazuleno[4,5-*b*]furan-2,7-dione. $C_{17}H_{20}O_5$; mol wt 304.34. C 67.09%, H 6.62%, O 26.28%. Isoln from *Matricaria chamomilla* L. and *Artemisia tilesii* Ledeb, *Compositae*: Z. Cekan *et al.*, *Collect. Czech. Chem. Commun.* **24**, 1554 (1959). Structure: W. Herz, K. Ueda, *J. Am. Chem. Soc.* **83**, 1139 (1961). Molecular conformation: M. Martinez V. *et al.*, *J. Nat. Prod.* **51**, 221 (1988).

Plates from benzene + petr ether and acetone + petr ether, mp 190-191°. $[\alpha]_D^{23}$ +23.5° (c = 0.65 in chloroform). uv max: 255 nm (ε 15,100).

Desacetoxymatricarin. [17946-87-1] $C_{15}H_{18}O_3$. Isoln from *Artemisia* spp.: M. Holub, V. Herout, *Collect. Czech. Chem. Commun.* **27**, 2980 (1962); H. H. A. Linde, M. S. Ragab, *Helv. Chim. Acta* **50**, 1961 (1967). Synthesis and stereochemistry: E. H. White *et al.*, *Tetrahedron* **25**, 2099 (1969). Crystals from ether, mp 205-205.5°, also reported as mp 202-203° (Holub). $[\alpha]_D^{24}$ +52.5° (c = 1 in chloroform).

Desacetylmatricarin monohydrate. $C_{15}H_{18}O_4 \cdot H_2O$. Crystals from benzene + acetone, mp 123-125°, resolidifies and mp 143-146°.

Tetrahydromatricarin. $C_{17}H_{24}O_5$. Needles from benzene + petr ether, mp 175-178°. $[\alpha]_D$ +32.3° (c = 1.92 in chloroform).

5827. Matrine. [519-02-8] (7a*S*,13a*R*,13b*R*,13c*S*)-Dodeca-hydro-1*H*,5*H*,10*H*-dipyrido[2,1-*f*:3′,2′,1′-*ij*][1,6]naphthyridin-10-

one; matridin-15-one; sophocarpidine. $C_{15}H_{24}N_2O$; mol wt 248.37. C 72.54%, H 9.74%, N 11.28%, O 6.44%. Occurs naturally as the (+)-form in the Chinese drug *Kuh Seng* or the Japanese *Shinkyogan*, the dried roots of *Sophora angustifolia* Sieb. & Zucc., *S. flavescens* Ait. and other *Sophora* spp., *Leguminosae*. Isomeric with lupanine, *q.v.* Isoln: Kondo, *Arch. Pharm.* **266**, 1 (1928); Winterfeld, Knener, *Ber.* **64**, 150 (1931); Orechov, Proskurnina, *Ber.* **67**, 77 (1934); Briggs, Ricketts, *J. Chem. Soc.* **1937**, 1795. Identity with sophocarpidine: Orechov *et al.*, *Ber.* **68**, 429 (1935). Alternate method of isoln: F. Bohlmann *et al.*, *Ber.* **91**, 2189 (1958). Structure: Tsuda, Murakami, *Ber.* **69**, 429 (1936). Configuration: Bohlmann *et al.*, *Ber.* **91**, 2176 (1958). Absolute configuration: Okuda *et al.*, *Chem. Pharm. Bull.* **14**, 314 (1966); Cervinka, *Z. Chem.* **7**, 190 (1967). Synthetic studies: Tsuda *et al.*, *J. Org. Chem.* **21**, 1481 (1956); **23**, 1179 (1958). Synthesis of *matridine* (*deoxymatrine*): Mandell, Singh, *J. Am. Chem. Soc.* **83**, 1766 (1961). Total synthesis of (±)-matrine: Mandell *et al.*, *ibid.* **87**, 5234 (1965); J. Chen *et al.*, *Chem. Commun.* **12**, 905 (1986). Synthesis of (+)-matrine: Okuda *et al.*, *Chem. Pharm. Bull.* **14**, 275 (1966). Biosynthesis: Schütte *et al.*, *Arch. Pharm.* **295**, 34 (1962).

Has been obtained in four forms: α-Form: needles or flat prisms, mp 76°. β-Form: orthorhombic prisms, mp 87°, $[\alpha]_D^{20}$ +38° (alc). γ-Form: liquid, bp_6 223°, d_4^{20} 1.088, n_D^{85} 1.5287. δ-Form: prisms, mp 84°. Sol in water, benzene, chloroform, ether, carbon disulfide; slightly sol in petr ether.

5828. Maxacalcitol. [103909-75-7] (1R,3S,5Z)-4-Methylene-5-[(2E)-2-[(1S,3aS,7aS)-octahydro-1-[(1S)-1-(3-hydroxy-3-methyl-butoxy)ethyl]-7a-methyl-4H-inden-4-ylidene]ethylidene]-1,3-cyclohexanediol; (+)-(5Z,7E,20S)-20α-(3-hydroxy-3-methylbutoxy)-9,10-secopregna-5,7,10(19)-trien-1α,3β-diol; 1α,25-dihydroxy-22-oxavitamin D_3; 22-oxa-1α,25-dihydroxyvitamin D_3; 22-oxacalcitriol; 22-oxa-1,25(OH)$_2$D$_3$; MC-1275; Sch-209579; Oxarol; Prezios. $C_{26}H_{42}O_4$; mol wt 418.62. C 74.60%, H 10.11%, O 15.29%. Noncalcemic analog of vitamin D. Prepn: N. Kubodera *et al.*, **EP 184112**; *eidem*, **US 4891364** (1986, 1990 both to Chugai); E. Murayama *et al.*, *Chem. Pharm. Bull.* **34**, 4410 (1986). Industrial synthesis: H. Shimizu *et al.*, *Org. Process Res. Dev.* **9**, 278 (2005). Pharmacology: A. J. Brown *et al.*, *J. Clin. Invest.* **84**, 728 (1989). Immunomodulatory effects: M. Komine *et al.*, *Arch. Dermatol. Res.* **291**, 500 (1999). Metabolism study: M. Ishigai *et al.*, *J. Steroid Biochem. Mol. Biol.* **66**, 281 (1998). LC-MS determn in serum: *idem et al.*, *J. Chromatogr. B* **706**, 261 (1998). Clinical evaluation in psoriasis: J. Barker *et al.*, *Br. J. Dermatol.* **141**, 274 (1999); in secondary hyperparathyroidism: Y. Tsukamoto *et al.*, *Am. J. Kidney Dis.* **35**, 458 (2000).

Colorless crystals from hexane + ethyl acetate, mp 122°. $[\alpha]_D^{20}$ +49.4° (c = 1.00 in ethanol). uv max (ethanol): 263 nm.

THERAP CAT: Antihyperparathyroid; antipsoriatic.

5829. Maytansine. [35846-53-8] Maitansine; NSC-153858. $C_{34}H_{46}ClN_3O_{10}$; mol wt 692.20. C 59.00%, H 6.70%, Cl 5.12%, N

6.07%, O 23.11%. Anti-leukemic ansa macrolide, member of a class of compounds that includes the rifamycins, streptovaricins etc. First ansa compound isolated from a plant rather than produced by a microorganism. Isoln from *Maytenus ovatus* Loes., *Celastraceae* and structure elucidation: S. M. Kupchan *et al.*, *J. Am. Chem. Soc.* **94**, 1354 (1972); S. M. Kupchan, **DE 2241418** (1974); S. M. Kupchan *et al.*, **US 3896111** (1975 to Research Corp.). Crystal structure and absolute configuration of the (3-bromopropyl) ether: R. F. Bryan *et al.*, *J. Chem. Soc. Perkin Trans. 2* **1973**, 897. Synthetic studies: A. I. Meyers *et al.*, *Tetrahedron Lett.* **1974**, 717; **1975**, 1745, 1749; E. J. Corey, M. G. Bock, *ibid.* **1975**, 2643; E. J. Corey *et al.*, *ibid.* **1978**, 1051; A. I. Meyers, J. P. Hudspeth, *ibid.* **22**, 3925 (1981). Total synthesis of (±)-form: A. I. Meyers *et al.*, *J. Am. Chem. Soc.* **102**, 6597 (1980); of naturally occurring (−)-form: E. J. Corey *et al.*, *ibid.* 6613. Cytotoxic action: M. K. Wolpert-Defilippes, *Biochem. Pharmacol.* **24**, 751 (1975). Clinical studies in carcinoma, melanoma: J. A. Neidhart *et al.*, *Cancer Treat. Rep.* **64**, 675 (1980); D. L. Ahmann *et al.*, *ibid.* 721. Toxicity: G. M. Mugera, J. M. Ward, *ibid.* **61**, 1333 (1977).

mp 171-172°. $[\alpha]_D^{26}$ −145° (c = 0.055 in chloroform). uv max (ethanol): 233, 254, 282, 290 nm (ε 29800, 27200, 5690, 5520). LD_{50} in rats (mg/kg): 0.48 s.c. (Mugera, Ward).

5830. Maytansinoid DM1. [139504-50-0] $N^{2'}$-Deactyl-$N^{2'}$-(3-mercapto-1-oxopropyl)maytansine; DM1. $C_{35}H_{48}ClN_3O_{10}S$; mol wt 738.29. C 56.94%, H 6.55%, Cl 4.80%, N 5.69%, O 21.67%, S 4.34%. Cytotoxic, semisynthetic derivative of the ansa macrolide, maytansine, *q.v.* Tubulin polymerization inhibitor used in immunoconjugates designed to improve drug delivery via covalent linkage to a monoclonal antibody (mAb) targeting a specific tumor cell. The term, *mertansine*, refers to an immunoconjugate of DM1 formed via a linker derived from *N*-succinimidyl-4-(2-pyridyldithio)pentanoate (SPP). *Emtansine* refers to a DM1 conjugate using the thioether linker, succinimidyl-4-(*N*-maleimidomethyl)cyclohexane-1-carboxylate (MCC). Prepn, combination with antibodies, and cytotoxic activity: R. J. Chari *et al.*, **EP 425235**; *eidem*, **US 5208020** (1991, 1993 both to ImmunoGen); R. V. J. Chari *et al.*, *Cancer Res.* **52**, 127 (1992); W. C. Widdison *et al.*, *J. Med. Chem.* **49**, 4392 (2006). HPLC determn of free DM1 in immunoconjugate preparations: M. S. Fleming *et al.*, *Anal. Biochem.* **340**, 272 (2005). Structure-activity and optimization of linkers: A. G. Polson *et al.*, *Cancer Res.* **69**, 2358 (2009); B. A. Kellogg *et al.*, *Bioconjugate Chem.* **22**, 717 (2011). Mechanism of cytotoxic action: E. Oroudjev *et al.*, *Mol. Cancer Ther.* **9**, 2700 (2010).

White solid, mp 190-192° (dec). $[\alpha]_D^{25}$ −113.2° (c = 0.306 in chloroform).

THERAP CAT: Antineoplastic.

5831. Mazindol. [22232-71-9] 5-(4-Chlorophenyl)-2,5-dihydro-3*H*-imidazo[2,1-*a*]isoindol-5-ol; 5-(4-chlorophenyl)-2,3-dihydro-5-hydroxy-5*H*-imidazo[2,1-*a*]isoindole; SaH-42548; Magrilon; Mazildene; Sanorex; Teronac. $C_{16}H_{13}ClN_2O$; mol wt 284.74. C 67.49%, H 4.60%, Cl 12.45%, N 9.84%, O 5.62%. Prepn: W. J. Houlihan, DE 1814540 *C.A.* **71**, 81368s (1969); W. J. Houlihan, M. K. Eberle, DE 1930488; *eidem*, US 3597445 (1969, 1970, 1970 all to Sandoz); T. S. Sulkowski, US 3763178 (1973 to American Home Prod.). Clinical trials as appetite depressant: C. Sirtori *et al.*, *Am. J. Med. Sci.* **261**, 341 (1971); A. J. Hadler, *J. Clin. Pharmacol.* **12**, 453 (1972). Effect on human prolactin and growth hormone responses: D. A. Thompson *et al.*, *Metabolism* **30**, 1015 (1981). Clinical evaluation in Duchenne muscular dystrophy: P. J. Collipp *et al.*, *J. Med. Genet.* **21**, 254 (1984).

White crystalline solid from ethanol, mp 215-217° (Sulkowski); also reported as crystals from acetone-hexane, mp 198-199° (Houlihan, Eberle). uv max (95% ethanol): 223, 268.5, 272 nm (ε 19000, 4400, 4400). Sol in ethanol; slightly sol in methanol, chloroform. Insol in water.

Note: This is a controlled substance (stimulant): **21 CFR**, 1308.14.

THERAP CAT: Anorexic; CNS stimulant.

5832. Mazipredone. [13085-08-0] (11β)-11,17-Dihydroxy-21-(4-methyl-1-piperazinyl)pregna-1,4-diene-3,20-dione; 11β,17α-dihydroxy-3,20-dioxo-21-(4-methyl-1-piperazinyl)pregna-1,4-diene; 11β,17-dihydroxy-21-(4-methyl-1-piperazinyl)-Δ^1-progesterone. $C_{26}H_{38}N_2O_4$; mol wt 442.60. C 70.56%, H 8.65%, N 6.33%, O 14.46%. Prepn: Tuba *et al.*, HU 150350 (1963 to Gedeon Richter), *C.A.* **60**, 3057h (1964).

Crystals from tetrahydrofuran + ligroin, dec 199°.
Hydrochloride. [60-39-9] Depersolone. $C_{26}H_{38}N_2O_4 \cdot HCl$; mol wt 479.06. Crystals, dec 246°. Sol in water.

THERAP CAT: Anti-inflammatory.

5833. MCPA. [94-74-6] 2-(4-Chloro-2-methylphenoxy)acetic acid; (4-chloro-*o*-toloxy)acetic acid; 2-methyl-4-chlorophenoxyacetic acid; MCP; Agritox; Agroxone; Cornox; Methoxone. $C_9H_9ClO_3$; mol wt 200.62. C 53.88%, H 4.52%, Cl 17.67%, O 23.92%. Prepn: Synerholm, Zimmerman, *Contrib. Boyce Thompson Inst.* **14**, 91 (1945); Templeman, Sexton, *Proc. Roy. Soc.* **133B**, 300 (1946); Foster, GB 573479 and GB 573510 (both 1945 to ICI); Skeeters, US 2740810 (1956 to Diamond Alkali). Toxicity study: Rowe, Hymas, *Am. J. Vet. Res.* **15**, 622 (1954). Review of carcinogenic risk: *IARC Monographs* **30**, 255-269 (1983); of genotoxicity studies: B. Elliott, *Mutagenesis* **20**, 3-13 (2005).

Plates from benzene, mp 120°. Practically insol in water. LD_{50} orally in rats: 700 mg/kg (Rowe, Hymas).
Sodium salt. [3653-48-3] Chiptox. $C_9H_8ClNaO_3$; mol wt 222.60. Very sol in water.

USE: Herbicide.

5834. MDA. [4764-17-4] α-Methyl-1,3-benzodioxole-5-ethanamine; 3,4-methylenedioxyamphetamine; 3,4-methylenedioxy-α-methyl-β-phenylethylamine; 3,4-methylenedioxyphenylisopropylamine; "Love"; EA-1299. $C_{10}H_{13}NO_2$; mol wt 179.22. C 67.02%, H 7.31%, N 7.82%, O 17.85%. Entactogen; synthetic amphetamine derivative with stimulant and hallucinogenic properties. Metabolite of MDMA, *q.v.* Prepn: C. Mannich, W. Jacobsohn, *Ber.* **43**, 189 (1910); U. Braun *et al.*, *J. Pharm. Sci.* **69**, 192 (1980). Pharmacology: L. E. Markert, D. C. S. Roberts, *Pharmacol. Biochem. Behav.* **39**, 569 (1991); K. M. Hegadoren *et al.*, *Psychopharmacology* **118**, 295 (1995). HPLC determn in whole blood: R. E. Michel *et al.*, *J. Neurosci. Methods* **50**, 61 (1993). GC/MS determn in hair: P. Kintz *et al.*, *J. Chromatogr. B* **670**, 162 (1995). Acute toxicity studies: P. N. Thiessen, D. A. Cook, *Clin. Toxicol.* **6**, 193 (1973); H. F. Hardman *et al.*, *Toxicol. Appl. Pharmacol.* **25**, 299 (1973). Review of toxicity, mechanism of action, and psychoactive effects: R. P. Climko *et al.*, *Int. J. Psychiatry Med.* **16**, 359-372 (1986-1987).

Almost colorless oil. bp_{22} 157°. $bp_{0.2}$ 80-90°.
Hydrochloride. [6292-91-7] $C_{10}H_{13}NO_2 \cdot HCl$. Crystals from isopropanol/ether, mp 187-188° (Braun); also reported as mp 180-181° (Mannich, Jacobsohn). Easily sol in water, alcohol. LD_{50} in mice, rats, guinea pigs (mg/kg): 68, 27, 28 i.p. (Hardman).
Note: This is a controlled substance (hallucinogen): **21 CFR**, 1308.11.

5835. MDE. [82801-81-8] *N*-Ethyl-α-methyl-1,3-benzodioxole-5-ethanamine; (±)-*N*-ethyl-3,4-methylenedioxyamphetamine; *N*-ethyl-3,4-methylenedioxyphenylisopropylamine; 3,4-methylenedioxyethamphetamine; MDEA; "Eve". $C_{12}H_{17}NO_2$; mol wt 207.27. C 69.54%, H 8.27%, N 6.76%, O 15.44%. Entactogen; *N*-ethyl derivative of MDA, *q.v.* Prepn: U. Braun *et al.*, *J. Pharm. Sci.* **69**, 192 (1980); F. T. Noggle, Jr. *et al*, *J. Assoc. Off. Anal. Chem.* **70**, 981 (1987). Pharmacology: M. Johnson *et al.*, *Biochem. Pharmacol.* **38**, 4333 (1989); K. M. Hegadoren *et al.*, *Psychopharmacology* **118**, 295 (1995). Determn in human urine using immunoassays and GC/MS: G. W. Kunsman *et al.*, *J. Anal. Toxicol.* **14**, 149 (1990). HPLC determn in tablets: M. Longo *et al.*, *J. Liq. Chromatogr.* **17**, 649 (1994). Impact on public health and abuse potential: G. P. Dowling *et al.*, *J. Am. Med. Assoc.* **257**, 1615 (1987); R. O. Bost, *J. Forensic Sci.* **33**, 576 (1988); B. Tehan *et al.*, *Anaesthesia* **48**, 507 (1993). Psychological effects in humans: L. Hermle *et al.*, *Neuropsychopharmacology* **8**, 171 (1993). Neuroendocrine and cardiovascular effects in humans: E. Gouzoulis *et al.*, *ibid.* 187.

Viscous, colorless oil. $bp_{0.2}$ 85-95°.
Hydrochloride. [74341-78-9] Fine white granular solid from ethanol-ether, mp 197-198° (Noggle). Also reported as white solid from isopropanol-ether, mp 201-202° (Braun). LD_{50} i.p. in mice: 102 mg/kg (Noggle).
Note: This is a controlled substance (hallucinogen): **21 CFR**, 1308.11.

5836. MDMA. [42542-10-9] *N*,α-Dimethyl-1,3-benzodioxole-5-ethanamine; (±)-3,4-methylenedioxymethamphetamine; *N*-methyl-3,4-methylenedioxyphenylisopropylamine; "Ecstasy"; "E"; "XTC"; "Adam". $C_{11}H_{15}NO_2$; mol wt 193.25. C 68.37%, H 7.82%, N 7.25%, O 16.56%. Entactogen; *N*-methyl derivative of MDA, *q.v.* (*S*)-Form possesses more potent CNS activity. Prepn: DE 274350 (1914 to E. Merck), *C.A.* **8**, 3350 (1914); U. Braun *et al.*, *J. Pharm.*

Sci. **69**, 192 (1980). Spectral and chromatographic analyses: K. Bailey *et al.*, *J. Assoc. Off. Anal. Chem.* **58**, 62 (1975). GC/MS determn in urine: B. K. Gan *et al.*, *J. Forensic Sci.* **36**, 1331 (1991). HPLC determn in whole blood: R. E. Michel *et al.*, *J. Neurosci. Methods* **50**, 61 (1993). Acute toxicity study: H. F. Hardman *et al.*, *Toxicol. Appl. Pharmacol.* **25**, 299 (1973). Series of articles on history, pharmacology, and neurotoxicity: *Ann. N.Y. Acad. Sci.* **600**, 601-715 (1990). Review of biochemical actions: M. Rattray, *Essays Biochem.* **26**, 77-87 (1991); of pharmacology, toxicology, and human experience: T. D. Steele *et al.*, *Addiction* **89**, 539-551 (1994); A. R. Green *et al.*, *Psychopharmacology* **119**, 247-260 (1995); H. Kalant, *Can. Med. Assoc. J.* **165**, 917-928 (2001).

Oil, bp$_{0.4}$ 100-110°.
Hydrochloride. [64057-70-1] C$_{11}$H$_{15}$NO$_2$.HCl. Crystals from isopropanol/*n*-hexane, mp 147-148° (Bailey). Crystals from isopropanol/ether, mp 152-153° (Braun). uv max (ethanol): 286 nm (ε 3843). LD$_{50}$ in mice, rats, guinea pigs (mg/kg): 97, 49, 98 i.p. (Hardman).
Note: This is a controlled substance (hallucinogen): **21 CFR,** 1308.11.

5837. Mebendazole. [31431-39-7] *N*-(6-Benzoyl-1*H*-benzimidazol-2-yl)carbamic acid methyl ester; 5-benzoyl-2-benzimidazolecarbamic acid methyl ester; methyl 5-benzoyl-2-benzimidazolecarbamate; R-17635; Bantenol; Equivurm Plus; Lomper; Mebenvet; Noverme; Ovitelmin; Pantelmin; Telmin; Vermicidin; Vermirax; Vermox. C$_{16}$H$_{13}$N$_3$O$_3$; mol wt 295.30. C 65.08%, H 4.44%, N 14.23%, O 16.25%. Prepn: J. L. H. Van Gelder *et al.*, **DE 2029637**; *eidem*, **US 3657267** (1971, 1972 to Janssen). Activity studies: Walker, Knight, *Vet. Rec.* **90**, 58 (1972). Clinical studies: Brugmans *et al.*, *J. Am. Med. Assoc.* **217**, 313 (1971); Callear, Neave, *Br. Vet. J.* **127**, xli (1971); *eidem, ibid.* **129**, 79 (1973). Comprehensive description: A. A. Al-Badr, M. Tariq, *Anal. Profiles Drug Subs.* **16**, 291-326 (1987).

Crystals from acetic acid and methanol, mp 288.5°. Freely sol in formic acid. Practically insol in water, ethanol, ether, chloroform and dilute solns of mineral acids. LD$_{50}$ orally: >80 mg/kg in sheep; >40 mg/kg in mice, rats and chickens (Van Gelder).
THERAP CAT: Anthelmintic (Nematodes).
THERAP CAT (VET): Anthelmintic.

5838. Mebeverine. [3625-06-7] 3,4-Dimethoxybenzoic acid 4-[ethyl[2-(4-methoxyphenyl)-1-methylethyl]amino]butyl ester; veratric acid 4-[ethyl(*p*-methoxy-α-methylphenethyl)amino]butyl ester; 3,4-dimethoxybenzoic acid 4-[ethyl(*p*-methoxy-α-methylphenethyl)amino]butyl ester; 4-[ethyl(*p*-methoxy-α-methylphenethyl)-amino]butyl 3,4-dimethoxybenzoate; 4-[*N*-[2-(*p*-methoxyphenyl)-1-methylethyl]-*N*-ethylamino]butyl 3,4-dimethoxybenzoate. C$_{25}$H$_{35}$NO$_5$; mol wt 429.56. C 69.90%, H 8.21%, N 3.26%, O 18.62%. Smooth muscle relaxant. Prepn: **BE 609490** *C.A.* **59**, 517b (1963) and T. Kralt *et al.*, **DE 1126889**; *eidem*, **US 3265577** (1962, 1962, 1966 to N. V. Philips). Pharmacology: G. Bertaccini *et al.*, *Farmaco Ed. Sci.* **30**, 823 (1975).

Hydrochloride. [2753-45-9] Colofac; Duspatalin; Duspatal. C$_{25}$H$_{35}$NO$_5$.HCl; mol wt 466.02. Crystals from ethyl methyl ketone, mp 105-107° (Ger. patent); also reported as mp 129-131° (Belg. patent).
THERAP CAT: Antispasmodic.

5839. Mebhydroline. [524-81-2] 2,3,4,5-Tetrahydro-2-methyl-5-(phenylmethyl)-1*H*-pyrido[4,3-*b*]indole; 5-benzyl-2,3,4,5-tetrahydro-2-methyl-1*H*-pyrido[4,3-*b*]indole; 3-methyl-9-benzyl-1,2,3,4-tetrahydro-γ-carboline; 5-benzyl-1,2,3,4-tetrahydro-2-methyl-γ-carboline; *N*-methyl-9-benzyltetrahydro-γ-carboline; Incidal. C$_{19}$H$_{20}$N$_2$; mol wt 276.38. C 82.57%, H 7.29%, N 10.14%. Prepn: Hörlein, *Ber.* **87**, 463 (1954); **GB 721171** (1954 to Bayer); Hörlein, **US 2786059** (1957 to Schenley). Pharmacology of Diazoline: Kharkevich, *Farmakol. Toksikol.* **20**(no. 6), 46 (1957).

Minute crystals, mp 95°. bp$_1$ 207-215°. Practically insol in water. Freely sol in methanol, ethanol, acetone, chloroform. Slightly sol in ether.
1,5-Naphthalenedisulfonate salt. [6153-33-9] Diazoline; Fabahistin; Omeril. (C$_{19}$H$_{20}$N$_2$)$_2$.C$_{10}$H$_8$O$_6$S$_2$; mol wt 841.05. White powder, dec 280°. Practically insol in water. Sparingly sol in hot glacial acetic acid. Sol in hot formamide, giving a yellow soln.
THERAP CAT: Antihistaminic.

5840. MeBmt. [59865-23-5] 4-(2*E*)-2-Buten-1-yl-2,4,5-trideoxy-2-(methylamino)-L-xylonic acid; (2*S*,3*R*,4*R*,6*E*)-3-hydroxy-4-methyl-2-(methylamino)-6-octenoic acid; (4*R*)-4-[(*E*)-2-butenyl]-4,*N*-dimethyl-L-threonine. C$_{10}$H$_{19}$NO$_3$; mol wt 201.27. C 59.68%, H 9.52%, N 6.96%, O 23.85%. Characteristic amino acid of cyclosporins, *q.v.* Identification in cyclosporin A: A. Ruegger *et al.*, *Helv. Chim. Acta* **59**, 1075 (1976). Syntheses: R. M. Wenger, *ibid.* **66**, 2308 (1983); D. A. Evans, A. E. Weber, *J. Am. Chem. Soc.* **108**, 6757 (1986); U. Schmidt, W. Siegel, *Tetrahedron Lett.* **28**, 2849 (1987). Appears to be necessary for full biological activity of cyclosporin: R. M. Wenger, *Angew. Chem. Int. Ed.* **24**, 77 (1985); D. H. Rich *et al.*, *J. Med. Chem.* **29**, 978 (1986).

Crystals from ethanol, mp 240-241° (Wenger). [α]$_D^{20}$ +13.5° (c = 0.50 in water, pH 7). Also reported as mp 242-243° (Evans, Weber). [α]$_D$ +17° (c = 0.51 in 0.4*N* aq HCl).

5841. Mebrofenin. [78266-06-5] *N*-[2-[(3-Bromo-2,4,6-trimethylphenyl)amino]-2-oxoethyl]-*N*-(carboxymethyl)glycine; [[[(3-bromomesityl)carbamoyl]methyl]imino]diacetic acid; *N*-(3-bromo-2,4,6-trimethylacetanilide)iminodiacetic acid; trimethylbromo-IDA; Choletec. C$_{15}$H$_{19}$BrN$_2$O$_5$; mol wt 387.23. C 46.53%, H 4.95%, Br 20.63%, N 7.23%, O 20.66%. Prepn: A. D. Nunn, M. D. Loberg, **BE 891534**; *eidem*, **US 4418208** (1982, 1983 both to Squibb). Physicochemical properties of ^{99m}technetium complex used for hepatobiliary tract imaging: A. D. Nunn *et al.*, *J. Nucl. Med.* **24**, 423 (1983). Pharmacokinetics in rats: A. R. Fritzberg *et al.*, *J. Pharm. Sci.* **73**, 1861 (1984). Evaluation as cholescintigraphic imaging agent in animals: A. van Aswegen *et al.*, *Nucl. Med. Biol.* **13**, 509 (1986); J. Kapuscinski *et al.*, *Nucl. Med.* **25**, 188 (1986); in humans: W. C. Klingensmith *et al.*, *Radiology* **146**, 181 (1983). Toxicity study: M. Jiang *et al.*, *Zhonghua Heyixue Zazhi* **4**, 214 (1984), *C.A.* **102**, 42241j (1985).

Crystals from ethanol, mp 198-200°. LD_{50} in mice, rats (mg/kg): 213.8, 226.4 i.v. (Jiang).

THERAP CAT: ^{99m}Tc complex as diagnostic aid (radioactive imaging agent).

5842. Mebutamate. [64-55-1] 2-Methyl-2-(1-methylpropyl)-1,3-propanediol 1,3-dicarbamate; 2-sec-butyl-2-methyl-1,3-propanediol dicarbamate; carbamic acid 2-sec-butyl-2-methyltrimethylene ester; 2-sec-butyl-2-methyltrimethylenecarbamate; 2-methyl-2-sec-butyl-1,3-propanediol dicarbamate; 2,2-dicarbamoyloxymethyl-3-methylpentane; dicamoylmethane; W-583; Capla; Butatensin; Carbuten; Dormate; Mebutina; Prean; Sigmafon; Vallene; Mega; No-Press; Axiten; Ipotensivo. $C_{10}H_{20}N_2O_4$; mol wt 232.28. C 51.71%, H 8.68%, N 12.06%, O 27.55%. Prepn: Berger, Ludwig, **US 2878280** (1959 to Carter Prod.). Mechanism of action: Kletzkin, *Arch. Int. Pharmacodyn. Ther.* **164**, 71 (1966). Metabolism: Edelson, Douglas, *ibid.* **173**, 182 (1968).

Crystals, mp 77-79°. Soly in water: ~0.1%. Sol in most organic solvents. On heating or boiling with acid or alkali, hydrolyzes to the alcohol, ammonia and carbon dioxide.

Note: This is a controlled substance (depressant): **21 CFR,** 1308.14.

THERAP CAT: Antihypertensive.

5843. Mecamylamine. [60-40-2] N,2,3,3-Tetramethylbicyclo[2.2.1]heptan-2-amine; N,2,3,3-tetramethyl-2-norbornanamine; N,2,3,3-tetramethyl-2-norcamphanamine; 3-methylaminoisocamphane; 2-methylaminoisocamphane; 2-methylamino-2,3,3-trimethylnorbornane; N-methyl-2-isocamphanamine; 3β-methylamino-2,-2,3-trimethylbicyclo[2.2.1]heptane; mecamine. $C_{11}H_{21}N$; mol wt 167.30. C 78.97%, H 12.65%, N 8.37%. Nicotinic receptor antagonist; orally active ganglionic blocker. Prepn: G. A. Stein *et al.*, *J. Am. Chem. Soc.* **78**, 1514 (1956); K. Pfister, III, G. A. Stein, **US 2831027** (1958 to Merck & Co.). Resolution of isomers and structure activity study: C. A. Stone *et al.*, *J. Med. Pharm. Chem.* **5**, 665 (1962). Abs config of isomers: B. Schönenberger *et al.*, *Helv. Chim. Acta* **69**, 12 (1986). Stereoselective receptor binding profile: R. L. Papke *et al.*, *J. Pharmacol. Exp. Ther.* **297**, 646 (2001); N. B. Fedorov *et al.*, *ibid.* **328**, 525 (2009). GC-MS determn in plasma: P. Jacob, III *et al.*, *J. Pharm. Biomed. Anal.* **23**, 653 (2000). Review of pharmacology and clinical experience in hypertension: J. M. Young *et al.*, *Clin. Ther.* **23**, 532-565 (2001). Review of therapeutic potential in neuropsychiatric disorders: I. Bacher *et al.*, *Expert Opin. Pharmacother.* **10**, 2709-2721 (2009).

Oily liquid. $bp_{4.0}$ 72°. n_D^{25} 1.4881. Slightly sol in water.

Hydrochloride. [826-39-1] Inversine. $C_{11}H_{21}N.HCl$; mol wt 203.75. Crystals, dec 245.5-246.5°. Bittersweet. Soly (g/100 ml): water 21.2; ethanol 8.2; glycerol 10.4; isopropanol 2.1. pH of 1% aq soln 6.0-7.5.

S-(+)-Form hydrochloride. [107596-30-5] TC-5214. Review of pharmacology and antidepressant effects in animal models: P. M.

Lippiello *et al.*, *CNS Neurosci. Ther.* **14**, 266-277 (2008). Crystals from 2-propanol, mp 262-264° (dec). $[\alpha]_D^{25}$ +20.6° (c = 1.5 in chloroform). Freely sol in water; sol in alcohol, glycerol; sparingly sol in isopropanol.

R-(−)-Form hydrochloride. [107596-31-6] TC-5213. Crystals from 2-propanol, mp 258° (dec). $[\alpha]_D^{25}$ −20.6° (c = 1.5 in chloroform).

THERAP CAT: Antihypertensive.

5844. Mecasermin. [68562-41-4] Insulin-like growth factor I (human); rhIGF-1; Increlex. $C_{331}H_{512}N_{94}O_{101}S_7$; mol wt 7648.71. C 51.98%, H 6.75%, N 17.21%, O 21.13%, S 2.93%. Recombinant human form of insulin-like growth factor I (IGF-I) produced in *E. coli*; has the same amino acid sequence as the endogenous form. Prepn: J. M. Lee, A. Ulrich, **EP 128733** (1984 to Genentech). Pharmaceutical formulation: R. G. Clark *et al.*, **US 5681814** (1997 to Genentech). Review of clinical experience: P. Norman, *Curr. Opin. Investig. Drugs* **7**, 371-380 (2006); A. L. Rosenbloom, *Adv. Ther.* **26**, 40-54 (2009).

THERAP CAT: Treatment of growth failure in children.

5845. Mecasermin Rinfabate. [478166-15-3] Insulin-like growth factor I (human) complex with insulin-like growth factor-binding protein IGFBP-3 (human); rhIGF-I/rhIGFB-3 complex; Iplex; SomatoKine. Recombinant human insulin-like growth factor I (IGF-I) complexed with its principle binding protein, IGFB-3, in equimolar amounts. The two proteins are produced separately by 2 strains of *E. coli*, purified, and combined into a 1:1 complex. Preparative method and use in promoting bone formation: C. M. Bagi *et al.*, **US 6017885** (2000 to Celtrix). Musculoskeletal effects in osteoporotic patients: S. Boonen *et al.*, *J. Clin. Endocrinol. Metab.* **87**, 1593 (2002). Review of pharmacology and clinical experience: P. Norman, *Curr. Opin. Investig. Drugs* **4**, 466-471 (2003); R. M. Williams *et al.*, *Expert Opin. Drug Metab. Toxicol.* **4**, 311-324 (2008).

THERAP CAT: In treatment of amyotrophic lateral sclerosis.

5846. Mechlorethamine. [51-75-2] 2-Chloro-N-(2-chloroethyl)-N-methylethanamine; 2,2′-dichloro-N-methyldiethylamine; N-methyl-2,2′-dichlorodiethylamine; di(chloroethyl)methylamine; methylbis(β-chloroethyl)amine; methyldi(2-chloroethyl)amine; chlormethine; nitrogen mustard; stickstofflost; MBA; HN2. C_5H_{11}-Cl_2N; mol wt 156.05. C 38.48%, H 7.11%, Cl 45.43%, N 8.98%. Nitrogen mustard alkylating agent; component of the MOPP chemotherapy protocol. Originally developed as a chemical warfare blister agent. Prepd by action of thionyl chloride on 2,2′-(methylimino)diethanol in trichloroethylene: Prelog, Stepan, *Collect. Czech. Chem. Commun.* **7**, 93 (1935); Hanby, Rydon, *J. Chem. Soc.* **1947**, 513; Abrams *et al.*, *J. Soc. Chem. Ind. London* **68**, 280 (1949); T. W. Doyle, D. M. Vyas in *Kirk-Othmer Encyclopedia of Chemical Technology* vol. 5 (John Wiley & Sons, New York, 4th ed., 1993) pp 876-879, 900-901. HPLC determn in pharmaceutical preparations: J. C. Reepmeyer, *J. Chromatogr. A* **1085**, 262 (2005). Chemical weapons screening method and degradation profile in water: H.-C. Chua *et al.*, *ibid.* **1102**, 214 (2006). Environmental fate simulation: S. L. Bartelt-Hunt *et al.*, *Environ. Sci. Technol.* **40**, 4219 (2006). Effects of mustard alkylation on protein synthesis: A. Masta *et al.*, *Nucleic Acids Res.* **23**, 3508 (1995). Review of clinical experience in mycosis fungoides: E. C. Vonderheid, *Int. J. Dermatol.* **23**, 180-186 (1984); Y. H. Kim *et al.*, *Arch. Dermatol.* **139**, 165-173 (2003); in Hodgkin's disease: C. Fermé *et al.*, *N. Engl. J. Med.* **357**, 1916-1927 (2007).

Mobile liquid. Faint odor of herring. *Vesicant, necrotizing irritant. Never use without appropriate gas mask.* Volatility at 25° = 3.581 mg/l. d_4^{25} 1.118. mp −60°. bp_{18} 87°; bp_{10} 75°; bp_5 64°; bp_2 59°. Log P (octanol/water): 0.9. pKa: 6.8. Vapor pressure (25°): 0.427 mm Hg. Very slightly sol in water. Miscible with DMF, CS_2, CCl_4, many other organic solvents and oils. The undiluted liq dec on standing and forms polymeric quaternary ammonium salts which are insol in the free base.

Hydrochloride. [55-86-7] Caryolysine; Mustargen. $C_5H_{11}Cl_2$-$N.HCl$; mol wt 192.51. Light yellow brown, crystalline, hygroscopic leaflets from acetone or chloroform, mp 109-111°. Very sol in water; sol in alcohol. Initial pH of a 2% aq soln: 3.0-4.0. Dry crystals are stable at temps up to 40°. LD_{50} in rats (mg/kg): 1.1 i.v.; 1.9 s.c. *See:* W. P. Anslow *et al., J. Pharmacol. Exp. Ther.* **91**, 224 (1947).

N-Oxide see Mechlorethamine Oxide.

Caution: Direct contact may cause severe skin and eye irritation; injury to deeper ocular structures, particularly the iris and lens. *See Patty's Industrial Hygiene and Toxicology* **vol. 2E,** G. D. Clayton, F. E. Clayton, Eds. (John Wiley & Sons, Inc., New York, 4th ed., 1994) pp 3297-3299. Mechlorethamine hydrochloride is reasonably anticipated to be a human carcinogen: *Report on Carcinogens, Twelfth Edition* (PB2011-111646, 2011) p 297.

USE: Base as a chemical warfare agent.

THERAP CAT: Hydrochloride as an antineoplastic.

THERAP CAT (VET): Hydrochloride has been used as an antineoplastic.

5847. Mechlorethamine Oxide. [126-85-2] 2-Chloro-*N*-(2-chloroethyl)-*N*-methylethanamine *N*-oxide; 2,2'-dichloro-*N*-methyl-diethylamine *N*-oxide; *N*-methyl-2,2'-dichlorodiethylamine *N*-oxide; methylbis(β-chloroethyl)amine *N*-oxide; methyldi(2-chloroethyl)amine *N*-oxide; nitrogen mustard *N*-oxide. $C_5H_{11}Cl_2NO$; mol wt 172.05. C 34.91%, H 6.44%, Cl 41.21%, N 8.14%, O 9.30%. Prepn of hydrochloride by treating 2,2'-dichloro-*N*-methyldiethylamine with hydrogen peroxide and acetic anhydride in ether, followed by shaking with 10% hydrochloric acid: Aiko *et al., J. Pharm. Soc. Jpn.* **72**, 1297 (1952, English Text), *C.A.* **47**, 1289 (1953). *Cf.* Stahmann, Bergmann, *J. Org. Chem.* **11**, 586 (1946).

Hydrochloride. [302-70-5] Nitromin; Mitomen; Mustron. $C_5H_{12}Cl_3NO$; mol wt 208.51. Prisms from acetone, mp 109-110°. Sol in water.

THERAP CAT: Antineoplastic.

5848. Meclizine. [569-65-3] 1-[(4-Chlorophenyl)phenyl-methyl]-4-[(3-methylphenyl)methyl]piperazine; 1-(*p*-chloro-α-phenylbenzyl)-4-(*m*-methylbenzyl)piperazine; 1-(*p*-chlorobenzhydryl)-4-(*m*-methylbenzyl)piperazine; meclozine; parachloramine. $C_{25}H_{27}ClN_2$; mol wt 390.96. C 76.80%, H 6.96%, Cl 9.07%, N 7.17%. Antihistaminic antinauseant. Synthesis: H. Morren, **BE 502889** (1951); *idem,* **US 2709169** (1955 to U.C.B.). Outline of reactions: Grivsky, *Ind. Chim. Belge* **17**, 735 (1952). HPLC determn in tablet formulations: N. H. Foda *et al., Anal. Lett.* **21**, 1177 (1988). Clinical evaluation in vertigo: B. Cohen, J. M. B. Vianney de Jong, *Arch. Neurol.* **27**, 129 (1972); in treatment of postoperative nausea and vomiting: C. M. Forrester *et al., AANA J.* **75**, 27 (2007).

bp_2 230°.

Dihydrochloride monohydrate. [31884-77-2] UCB-5062; Agyrax; Ancolan; Antivert; Bonamine; Bonine; Postafen. $C_{25}H_{27}ClN_2$·2HCl·H_2O; mol wt 481.89. White or slightly yellowish, crystalline powder. mp 215°. Freely sol in chloroform, pyridine, and in acid-alcohol-water mixtures; slightly sol in dil acids, alc. Practically insol in water (0.1 gm/100 ml), ether.

THERAP CAT: Antiemetic; antivertigo; in treatment of motion sickness.

THERAP CAT (VET): Antiemetic.

5849. Meclocycline. [2013-58-3] (4*S*,4a*R*,5*S*,5a*R*,12a*S*)-7-Chloro-4-(dimethylamino)-1,4,4a,5,5a,6,11,12a-octahydro-3,5,10,-

12,12a-pentahydroxy-6-methylene-1,11-dioxo-2-naphthacenecarboxamide; 7-chloro-6-methylene-5-hydroxytetracycline; GS-2989; NSC-78502. $C_{22}H_{21}ClN_2O_8$; mol wt 476.87. C 55.41%, H 4.44%, Cl 7.43%, N 5.87%, O 26.84%. Semi-synthetic antibiotic derived from tetracycline, *q.v.* Prepn: R. K. Blackwood *et al.,* **US 2984686** (1961 to Pfizer); *eidem, J. Am. Chem. Soc.* **83**, 2773 (1961); of the free base and 5-sulfosalicylate salt: *eidem, ibid.* **85**, 3943 (1963). *In vitro* study: L. Lucca, G. Vittadini, *G. Ital. Chemioter.* **26**, 203 (1979). ^{13}C-NMR study: E. Mazzola *et al., J. Pharm. Sci.* **69**, 229 (1980). Toxicity study: F. Bernardi *et al., G. Ital. Chemioter.* **17**, 276 (1970).

uv max (methanol, 0.01*N* HCl): 245, 347 nm (log ε 4.34, 410). LD_{50} in mice (mg/kg): >5000 orally, 425 i.p. (Bernardi).

5-Sulfosalicylate. [73816-42-9] Meclan; Mecloderm; Meclosorb; Meclutin; Traumatociclina. $C_{29}H_{27}ClN_2O_{14}S$; mol wt 695.05. uv max (methanol, 0.01*N* HCl): 239, 268, 346 nm (log ε 4.46, 4.07, 4.11).

THERAP CAT: Antibacterial.

5850. Meclofenamic Acid. [644-62-2] 2-[(2,6-Dichloro-3-methylphenyl)amino]benzoic acid; *N*-(2,6-dichloro-*m*-tolyl)anthranilic acid; meclphenamic acid; CI-583; INF-4668; Arquel. $C_{14}H_{11}Cl_2NO_2$; mol wt 296.15. C 56.78%, H 3.74%, Cl 23.94%, N 4.73%, O 10.80%. Prepn: Scherrer, Short, **DE 1149015** *C.A.* **61**, 1801d (1964) and **US 3313848** (1963, 1967 to Parke, Davis); Juby *et al., J. Med. Chem.* **11**, 111 (1968). Pharmacology and toxicology: Winder *et al., J. Pharmacol. Exp. Ther.* **148**, 422 (1965); Winder, *Ann. Phys. Med.* **1966**, 7; Kaump, *ibid.* 16. Series of articles on structure-activity relationships, mechanism of action, clinical studies: *Arzneim.-Forsch.* **33**, 619-680 (1983). Effect on prostaglandin E receptor binding: M. C. P. Rees *et al., Lancet* **2**, 541 (1988).

White crystals from acetone-water, mp 257-259°; also reported as mp 248-250°. Soly (mg/ml): water 0.03; 0.1*N* NaOH 28. pH of satd aq soln: ~6.9.

Sodium salt monohydrate. Lenidolor; Meclodol; Meclomen; Movens. $C_{14}H_{10}Cl_2NNaO_2·H_2O$; mol wt 336.14. mp 289-291°. Soly in water: 15 mg/ml (slightly turbid). pH 8.7.

THERAP CAT: Anti-inflammatory; antipyretic.

THERAP CAT (VET): Anti-inflammatory.

5851. Meclofenoxate. [51-68-3] 2-(4-Chlorophenoxy)acetic acid 2-(dimethylamino)ethyl ester; dimethylaminoethyl *p*-chlorophenoxyacetate; centrophenoxine; meclofenoxane; acephen; ANP-235; Analux; Cetrexin; Proseryl. $C_{12}H_{16}ClNO_3$; mol wt 257.71. C 55.93%, H 6.26%, Cl 13.76%, N 5.44%, O 18.62%. Prepn: Rumpf, Thuillier, **FR M398** (1962 to Centre Nat'l. Recherche Sci.), *C.A.* **57**, 16768e (1962); Thuillier *et al., Compt. Rend.* **249**, 2081 (1959); Thuillier, Rumpf, *Bull. Soc. Chim. Fr.* **1960**, 1786. Pharmacology: Petkov, *C.A.* **65**, 20717g (1966); Liberman, *Farmakol. Toksikol.* **30**, 409 (1967). Auxin activity: Conti, *Boll. Chim. Farm.* **107**, 325 (1968).

Hydrochloride. [3685-84-5] Cellative; Clocete; Lucidril; Methoxynal; Proserout; Brenal; Marucotol; Helfergin. $C_{12}H_{16}ClNO_3$.-HCl; mol wt 294.17. Crystals from isopropyl alcohol or acetone, mp 135-139°. Sol in cold water. Sparingly sol in cold isopropyl alcohol, acetone. Practically insol in benzene, ether, chloroform. A 5% soln in water has pH 6. LD_{50} in mice (mg/kg): 330 i.v.; 1750 orally; 845 i.p. (Rumpf, Thuillier).

USE: Plant growth regulator.
THERAP CAT: Cerebral stimulant.

5852. Mecloqualone. [340-57-8] 3-(2-Chlorophenyl)-2-methyl-4(3H)-quinazolinone; 2-methyl-3-(2-chlorophenyl)-4-quinazolone; Nubarene. $C_{15}H_{11}ClN_2O$; mol wt 270.72. C 66.55%, H 4.10%, Cl 13.09%, N 10.35%, O 5.91%. Prepd from acetylanthranilic acid and o-chloroaniline in the presence of $POCl_3$: Jackman et al., J. Pharm. Pharmacol. **12**, 529 (1960); Closa, J. Prakt. Chem. [4] **14**, 84 (1961). Metabolism: Dubnick, Towne, Toxicol. Appl. Pharmacol. **22**, 82 (1972); Daenens, Van Bovan, Arzneim.-Forsch. **24**, 195 (1974).

Crystals mp 126-128°.
Hydrochloride. $C_{15}H_{11}ClN_2O.HCl$. Crystals, mp 239-241°.
Note: This is a controlled substance (depressant): **21 CFR, 1308.11.**
THERAP CAT: Sedative; hypnotic.

5853. Meconic Acid. [497-59-6] 3-Hydroxy-4-oxo-4H-pyran-2,6-dicarboxylic acid; oxychelidonic acid. $C_7H_4O_7$; mol wt 200.10. C 42.02%, H 2.01%, O 55.97%. From opium which contains 4 to 6%. Prepn: Thoms, Pietrulla, Ber. Pharm. Ges. **31**, 4 (1921); Wibaut, Kleinpool, Rec. Trav. Chim. **66**, 24 (1947).

Monohydrate. Prisms from concd aq solns. One gram dissolves in 4 ml hot water, 50 ml methanol, 50 ml ethyl acetate, 100 ml acetone; freely sol in ethanol, benzene.
Trihydrate. Orthorhombic pyramidal prisms from dil aq solns. Becomes anhyd when heated to 100-102° for 20 min. Dec with evolution of CO_2 when heated to 120° or when boiled with water. The slight brown tint often characteristic of meconic acid prepns is due to traces of iron.

5854. Meconin. [569-31-3] 6,7-Dimethoxy-1(3H)-isobenzofuranone; 6,7-dimethoxyphthalide; opianyl; meconinic acid lactone. $C_{10}H_{10}O_4$; mol wt 194.19. C 61.85%, H 5.19%, O 32.96%. Isolated from opium by Dublanc in 1832. Occurs also in the root of Hydrastis canadensis L., Ranunculaceae: Freund, Ber. **22**, 456, 459 (1889). Is formed in oxidation of noscapine with HNO_3. Synthesis from o-veratric acid: Edwards et al., J. Chem. Soc. **1925**, 195; Wilson et al., J. Org. Chem. **16**, 792 (1951). From opianic acid: Brown, Newbold, J. Chem. Soc. **1952**, 4878.

White, optically inactive needles; sharp bitter taste. mp 102-103°. uv max: 213, 308 nm (ε 25000, 3800). Sol in 700 parts cold, 22 parts boiling water; sol in alcohol, benzene, chloroform, ether, glacial acetic acid; slowly sol in alkalies with formation of alkali salt of

meconinic acid, $(CH_2OH)(CH_3O)_2C_6H_2COOH$. The acid itself is unstable, rapidly changing to lactone.

5855. Mecoprop. [93-65-2] 2-(4-Chloro-2-methylphenoxy)-propanoic acid; (±)-2-[(4-chloro-o-tolyl)oxy]propionic acid; mechlorprop; MCPP; CMPP; RD-4593; Astix CMPP; Iso-Cornox; Compitox; Compitox Plus; Proponex-Plus. $C_{10}H_{11}ClO_3$; mol wt 214.65. C 55.96%, H 5.17%, Cl 16.52%, O 22.36%. Prepn: M. E. Synerholm, P. W. Zimmerman, Contrib. Boyce Thompson Inst. **14**, 91 (1945). Studies on plant growth regulation: C. H. Fawcett et al., Ann. Appl. Biol. **40**, 231 (1953); and comparison of enantiomers: M. Matell, Kungl. Lantbruks-Hogsk. Ann. **20**, 207 (1953); B. Aberg, ibid. 241. GLC determn: H. G. Higson, D. Butler, Analyst **85**, 657 (1960). Crystal structure: G. Smith et al., Acta Crystallogr. **B36**, 992 (1980). Herbicidal activity: G. B. Lush, Proc. 3rd Br. Weed Control Conf. 625 (1956); E. L. Leafe, ibid. 633; B. Wallgren, Weeds Weed Contr. **24th** Swedish Weed Conf. 30 (1983); of (+)-enantiomer: J. Toll, Weeds Weed Contr. **28th** Swedish Weed Conf. 100 (1987). Degradation in soils: L. Lindholm et al., Acta Agric. Scand. **32**, 429 (1982); A. E. Smith, Bull. Environ. Contam. Toxicol. **34**, 656 (1985). Toxicological studies: M. R. Gurd et al., Food Cosmet. Toxicol. **3**, 883 (1965); H. G. Verschuuren et al., Toxicology **3**, 349 (1975); R. Roll, G. Matthiaschk, Arzneim.-Forsch. **33**, 1479 (1983). EC-GLC determn in tissues and biological fluids: J. De Beer et al., Vet. Hum. Toxicol. **21**, Suppl., 172 (1979). HPLC resolution of enantiomers: B. Blessington et al., J. Chromatogr. **396**, 177 (1987).

Solid, mp 93-94°. LD_{50} in rats (mg/kg): 1210 orally, 402 i.p. (Verschuuren).
(+)-Form. Mecoprop-P; Duplosan KV. Solid, mp 95-96°. $[\alpha]_D^{25}$ +19° (alcohol).
Sodium salt. $C_{10}H_{11}ClNaO_3$. LD_{50} i.p. in rats, mice: 500, 600 mg/kg; orally in mice: 650 mg/kg (Gurd).
Diethylamine salt. Mecopar. $C_{14}H_{22}ClNO_3$; mol wt 287.78. LD_{50} in rats, mice (mg/kg): 1060 ±120, 600 ±35 orally; 350, 400 i.p. (Gurd).
Potassium salt. Mecomec; Hedonal MCPP. $C_{10}H_{11}ClKO_3$; mol wt 253.74.
USE: Herbicide.

5856. Mecysteine. [2485-62-3] L-Cysteine methyl ester; methyl cysteine; methyl β-mercaptoalanine; methyl 2-amino-3-mercaptopropionate. $C_4H_9NO_2S$; mol wt 135.18. C 35.54%, H 6.71%, N 10.36%, O 23.67%, S 23.72%. Prepn of the hydrochloride: M. Bergmann, G. Michalis, Ber. **63**, 987 (1930); L. Zervas, D. M. Theodoropoulos, J. Am. Chem. Soc. **78**, 1359 (1956).

Hydrochloride. [18598-63-5] LJ-48; Visclair. $C_4H_9NO_2S$.-HCl; mol wt 171.64. Crystals from methanol, mp 140-141°. $[\alpha]_D^{20}$ −2.9° (methanol).
THERAP CAT: Mucolytic.

5857. Medazepam. [2898-12-6] 7-Chloro-2,3-dihydro-1-methyl-5-phenyl-1H-1,4-benzodiazepine; Ansilan; Diepin; Medazepol; Megasedan; Narsis; Nobrium; Psiquium; Resmit; Rudotel; Tranquilax. $C_{16}H_{15}ClN_2$; mol wt 270.76. C 70.98%, H 5.58%, Cl 13.09%, N 10.35%. Prepn: L. H. Sternbach et al., J. Org. Chem. **28**, 2456 (1963); G. A. Archer et al., BE 620773 C.A. **59**, 10095b (1963); E. Reeder, L. H. Sternbach, US 3243427 (1963, 1966 both to Hoffmann-La Roche); S. Inaba et al., Chem. Pharm. Bull. **20**, 1628 (1972); M. Mihalic et al., J. Heterocycl. Chem. **14**, 941 (1977). Pharmacology: L. O. Randall et al., Arch. Int. Pharmacodyn. Ther. **185**, 135 (1970). Crystal structure: G. Gilli et al., Acta Crystallogr. **B34**, 3793 (1978).

Colorless prismatic crystals from ether + petr ether, mp 95-97°. LD$_{50}$ in mice (mg/kg): 360 i.p., 1070 orally (Randall).

Hydrochloride. [2898-11-5] C$_{16}$H$_{15}$ClN$_2$.HCl. Orange-red crystalline powder. Freely sol in water, alcohol.

Note: This is a controlled substance (depressant): **21 CFR, 1308.14.**

THERAP CAT: Anxiolytic.

5858. Medetomidine. [86347-14-0] 5-[1-(2,3-Dimethyl-phenyl)ethyl]-1*H*-imidazole; (±)-4-(α,2,3-trimethylbenzyl)imid-azole; 4-[(α-methyl)-2,3-dimethylbenzyl]imidazole. C$_{13}$H$_{16}$N$_2$; mol wt 200.29. C 77.96%, H 8.05%, N 13.99%. α$_2$-Adrenergic agonist. Prepn: A. J. Karjalainen *et al.*, **GB 2101114**; A. J. Karjalainen, K. O. A. Kurkela, **US 4544664** (1983, 1985 both to Farmos). Receptor binding study: R. Virtanen *et al.*, *Eur. J. Pharmacol.* **150**, 9 (1988). LC/MS determn in plasma: H. Kanazawa *et al.*, *J. Chromatogr.* **631**, 215 (1993). Sedative and cardiovascular effects in humans: M. Scheinin *et al.*, *Br. J. Clin. Pharmacol.* **24**, 443 (1987). Veterinary evaluation in cats: D. Stenberg *et al.*, *J. Vet. Pharmacol. Ther.* **10**, 319 (1987); in horses and sheep: C. E. Bryant *et al.*, *Res. Vet. Sci.* **65**, 149 (1998); in elephants: B. Sarma *et al.*, *ibid.* **73**, 315 (2002). Review of pharmacology and use in dogs and cats: L. K. Cullen, *Br. Vet. J.* **152**, 519-535 (1996).

Hydrochloride. [86347-15-1] MPV-785; Domitor; Sedator. C$_{13}$H$_{16}$N$_2$.HCl; mol wt 236.74. White or almost white crystalline substance. Sol in water.

d-**Form** *see* Dexmedetomidine.

THERAP CAT (VET): Sedative; analgesic.

5859. Medicagol. [1983-72-8] 3-Hydroxy-6*H*-[1,3]di-oxolo[4′,5′:5,6]benzofuro[3,2-*c*][1]benzopyran-6-one; 7-hydroxy-11,12-(methylenedioxy)coumestan; 7-hydroxy-5′,6′-methylene-dioxybenzofurano(3′,2′:3,4)coumarin. C$_{16}$H$_8$O$_6$; mol wt 296.23. C 64.87%, H 2.72%, O 32.41%. Occurs in alfalfa having viral leafspot infections. Synthesis: Livingston *et al.*, *J. Org. Chem.* **30**, 2353 (1965); Jurd, *J. Pharm. Sci.* **54**, 1221 (1965); Fukui *et al.*, *Experientia* **24**, 536 (1968).

mp 326-327°. uv max (ethanol): 245, 270, 297, 310, 348 nm (log ε 4.29, 3.91, 3.85, 4.03, 4.46).

5860. Medicarpin. [32383-76-9] (6a*R*,11a*R*)-6a,11a-Dihy-dro-9-methoxy-6*H*-benzofuro[3,2-*c*][1]benzopyran-3-ol; (−)-3-hy-droxy-9-methoxypterocarpan; demethylhomopterocarpin. C$_{16}$-H$_{14}$O$_4$; mol wt 270.28. C 71.10%, H 5.22%, O 23.68%. Antifungal phytoalexin produced by leguminous species. Isoln from the heart-wood of *Swartzia madagascariensis* Desv., *Caesalpinioideae*: S. H. Harper *et al.*, *Chem. Ind. (London)* **1965**, 562; from alfalfa, *Medi-*

cago sativa L., *Leguminosae:* D. G. Smith *et al.*, *Physiol. Plant Pathol.* **1**, 41, (1971); from red clover, *Trifolium pratense* L., *Leguminosae:* V. J. Higgins, D. G. Smith, *Phytopathology* **62**, 235 (1972). ^{13}C-NMR: A. A. Chalmers *et al.*, *Tetrahedron* **33**, 1735 (1977). HPLC: J. Koster *et al.*, *J. Chromatogr.* **270**, 392 (1983). Synthesis of racemic form: W. Cocker *et al.*, *J. Chem. Soc. C* **1965**, 1034. Biosynthetic studies: P. M. Dewick, *Chem. Commun.* **1975**, 656; S. W. Banks *et al.*, *ibid.* **1982**, 157; H. A. M. Al-Ani, P. M. Dewick, *J. Chem. Soc. Perkin Trans. 1* **1984**, 2831. Antifungal properties: L. J. Duczek, V. J. Higgins, *Can. J. Bot.* **54**, 2620 (1976); H. D. Van Etten, *Phytochemistry* **15**, 655 (1976); A. O. Latunde-Dada, J. A. Lucas, *Physiol. Plant Pathol.* **26**, 31 (1985).

Prisms from benzene, mp 127.5-128.5°. [α]$_D^{22}$ −226° (chloroform). uv max: 207, 282, 287, 310 nm (log ε 4.86, 3.97, 4.01, 3.38).

(±)-**Form.** Prisms from ethyl acetate/light petroleum, mp 194-195°.

5861. Medifoxamine. [32359-34-5] *N,N*-Dimethyl-2,2-di-phenoxyethanamine; (dimethylamino)acetaldehyde diphenyl acetal; *N,N′*-dimethyl-2,2-diphenoxyethylamine. C$_{16}$H$_{19}$NO$_2$; mol wt 257.33. C 74.68%, H 7.44%, N 5.44%, O 12.43%. Prepn: **FR M5498** (1967 to Gerda), *C.A.* **72**, 12358x (1970); M. A. Brunet *et al.*, *Bull. Soc. Chim. Fr.* **6**, 2000 (1967). Pharmacology and toxicity: A. Vagne *et al.*, *Therapie* **26**, 553 (1971). Clinical pharmacology: M. A. Randhawa *et al.*, *Hum. Psychopharmacol.* **3**, 195 (1988). Comparative clinical trial with clomipramine, *q.v.*, in depression: H. Scharbach *et al.*, *Psychol. Med.* **18**, 1485 (1986).

Fumarate. [16604-45-8] LG-152; Clédial; Gerdaxyl. C$_{16}$H$_{19}$-NO$_2$.C$_4$H$_4$O$_4$; mol wt 373.41. Crystals from 95% ethanol, mp 128.5°. LD$_{50}$ orally in rats: 750 mg/kg (Vagne).

THERAP CAT: Antidepressant.

5862. Medrogestone. [977-79-7] 6,17-Dimethylpregna-4,6-diene-3,20-dione; 6,17α-dimethyl-6-dehydroprogesterone; AY-62022; Colpro; Colprone; Prothil. C$_{23}$H$_{32}$O$_2$; mol wt 340.51. C 81.13%, H 9.47%, O 9.40%. Prepn: Deghenghi, Gaudry, *J. Am. Chem. Soc.* **83**, 4668 (1961); Deghenghi *et al.*, *Tetrahedron* **19**, 289 (1963); *J. Med. Chem.* **6**, 301 (1963); Deghenghi, **US 3133913, US 3210387**; Morand, Deghenghi, **US 3170936** (1964, 1965, and 1965, all to Am. Home Prod.). HPLC determn in serum: W. T. Robinson, L. Cosyns, *Arzneim.-Forsch.* **29**, 882 (1979). Comparative clinical evaluation in benign prostatic hyperplasia: D. F. Paulson, R. D. Kane, *J. Urol.* **113**, 811 (1975); in endometrial carcinoma: K. C. Podratz *et al.*, *Obstet. Gynecol.* **66**, 106 (1985).

Crystals from ether, mp 144-146°. $[\alpha]_D^{23}$ +79° (c = 1 in chloroform). uv max: 288 nm (ε 25000).

THERAP CAT: Progestogen.

5863. Medronic Acid. [1984-15-2] P,P'-Methylenebisphosphonic acid; methylenediphosphonic acid; methanebisphosphonic acid; methanediphosphonic acid; MDP. $CH_6O_6P_2$; mol wt 176.00. C 6.82%, H 3.44%, O 54.54%, P 35.20%. Prepn: G. Schwarzenbach, J. Zurc, *Monatsh. Chem.* **81**, 202 (1950), *C.A.* **44**, 8205f (1950); V. A. Ginsburg, A. Y. Yakubovich, *J. Gen. Chem. USSR* **28**, 710 (1958), *C.A.* **52**, 17091g (1958); J. A. Cade, *J. Chem. Soc.* **1959**, 2266. Physical properties: K. Moedritzer, R. R. Irani, *J. Inorg. Nucl. Chem.* **22**, 297 (1961). Prepn of ^{99m}Tc-complexes: G. Subramanian, J. G. McAfee, **US 4032625** (1977 to Research Corp.); J. Kroesbergen *et al.*, *Int. J. Nucl. Med. Biol.* **12**, 419 (1986). HPLC separation of ^{99m}Tc-complexes: D. J. Hoch, T. C. Pinkerton, *Appl. Radiat. Isot.* **37**, 593 (1986). Biodistribution, toxicity and clinical evaluation for skeletal imaging: G. Subramanian *et al.*, *J. Nucl. Med.* **16**, 744 (1975). Comparative biodistribution studies: G. Subramanian *et al.*, *Radiology* **149**, 823 (1983). Clinical pharmacokinetics: L. Rosenthall *et al.*, *Clin. Nucl. Med.* **2**, 232 (1977). Clinical studies for bone imaging: U. Büll *et al.*, *Br. J. Radiol.* **50**, 629 (1977); for neuroblastoma imaging: I. Garty *et al.*, *Clin. Nucl. Med.* **14**, 515 (1989).

Crystals from acetic acid, mp 199-200°. LD_{50} in mice, rabbits (mg/kg): 45-50 i.v. (Subramanian, 1975).

Complex with ^{99m}Tc. Technetium Tc 99m medronate; ^{99m}Tc-MDP; Amerscan MDP; Osteolite; TechneScan MDP.

THERAP CAT: ^{99m}Tc complex as diagnostic aid (radioactive imaging agent).

5864. Medroxyprogesterone. [520-85-4] (6α)-17-Hydroxy-6-methylpregn-4-ene-3,20-dione; 17α-hydroxy-6α-methylprogesterone; 6α-methyl-17α-hydroxyprogesterone; 6α-methyl-4-pregnen-17α-ol-3,20-dione. $C_{22}H_{32}O_3$; mol wt 344.50. C 76.70%, H 9.36%, O 13.93%. Orally active progestogen; formerly used in combinations as oral contraceptive. Prepn: J. C. Babcock *et al.*, *J. Am. Chem. Soc.* **80**, 2904 (1958); **GB 866381**; G. B. Spero, **US 3377364** (1961, 1968 both to Upjohn); P. Ruggieri, C. Ferrari, **US 3043832** (1962 to Ormonoterapia Richter); B. Camerino *et al.*, **US 3061616** (1962 to Farmitalia). HPLC determn in plasma: J. Read, G. Mould, *J. Chromatogr.* **341**, 437 (1985). Comparative clinical trial with norethisterone as injectable contraceptive: H. K. Toppozada *et al.*, *Contraception* **28**, 1 (1983). Clinical trial in advanced breast cancer: M. Izuo *et al.*, *Cancer* **56**, 2576 (1985). Review of pharmacology and clinical uses: I. S. Fraser, E. Weisberg, *Med. J. Aust.* **1**, Suppl. 1, 3-19 (1981).

Crystals from chloroform, mp 220-223.5°. $[\alpha]_D^{25}$ +75° (in chloroform). uv max (ethanol): 241 nm (ε 16000).

17-Acetate. [71-58-9] 17α-Acetoxy-6α-methylprogesterone; 6-α-methyl-17α-acetoxyprogesterone; MAP; Amen; Clinovir; Cycrin; Depo-Clinovir; Depo-Prodasone; Depo-Provera; Farlutal; Gestoral; Hysron; Oragest; Perlutex; Provera; Veramix. $C_{24}H_{34}O_4$; mol wt 386.53. Crystals from methanol, mp 207-209°. $[\alpha]_D$ +61° (in chloroform). uv max (ethanol): 240 nm (ε 15900). Freely sol in chlo-

roform; sol in acetone, dioxane; sparingly sol in ethanol, methanol; slightly sol in ether. Insol in water.

THERAP CAT: Progestogen; contraceptive (injectable); antineoplastic (hormonal).

THERAP CAT (VET): Progestogen; estrus regulator.

5865. Medrysone. [2668-66-8] $(6\alpha,11\beta)$-11-Hydroxy-6-methylpregn-4-ene-3,20-dione; 11β-hydroxy-6α-methylprogesterone; 6α-methyl-11β-hydroxyprogesterone; hydroxymesterone; U-8471; HMS. $C_{22}H_{32}O_3$; mol wt 344.50. C 76.70%, H 9.36%, O 13.93%. Prepn and use as an intermediate: Sebek *et al.*, Spero, Thompson, **US 2864837** and **US 2968655** (1958 and 1961, both to Upjohn). In treatment of ocular inflammation: Bedrossian, *Arch. Ophthalmol.* **81**, 184 (1969).

Crystals, mp 155-158°. $[\alpha]_D$ +189° (in chloroform).

THERAP CAT: Glucocorticoid.

5866. Medullipin. [97621-73-3] Medullipin I; antihypertensive neutral renomedullary lipid; ANRL. Antihypertensive lipid prohormone produced by the renomedullary interstitial cells of the renal papilla under the control of the renal artery perfusion pressure. Converted to the active form, ***medullipin II***, by the cytochrome P-450 dependent enzyme system of the liver. Medullipin II acts as as vasodilator that suppresses sympathetic activity, causes diuresis-natriuresis, and has a suppressive action on the CNS. Constitutes a feedback control of the renin-angiotensin system. Extraction from renomedullary interstitial cells: E. E. Muirhead *et al.*, *Lab. Invest.* **35**, 162 (1977); from renal venous effluent: E. E. Muirhead *et al.*, *J. Lab. Clin. Med.* **99**, 64 (1982). Biological activity: E. E. Muirhead *et al.*, *Hypertension* **5**, Suppl I, I-112 (1983); G. Karlström *et al.*, *Acta Physiol. Scand.* **137**, 521 (1989). Activation by the liver: E. E. Muirhead *et al.*, *Trans. Assoc. Am. Physicians* **101**, 226 (1988). Effect in spontaneous hypertensive rats: E. E. Muirhead *et al.*, *Hypertension* **17**, 1092 (1991). Review of medullipin system of blood pressure control: E. E. Muirhead, *Am. J. Hypertens.* **4**, 556S-568S (1991).

5867. Meerwein's Reagent. [368-39-8] Triethyloxonium tetrafluoroborate(1−) (1:1); triethyloxonium fluoborate. $C_6H_{15}BF_4$-O; mol wt 189.99. C 37.93%, H 7.96%, B 5.69%, F 40.00%, O 8.42%. Powerful ethylating agent; converts alcohols to ethyl ethers at neutral pH. Prepn: H. Meerwein *et al.*, *J. Prakt. Chem.* **147**, 257 (1937); *idem et al.*, *ibid.* **154**, 83 (1939); H. Meerwein, *Org. Synth.* **46**, 113 (1966). Synthetic applications: N. Kornblum, G. P. Coffey, *J. Org. Chem.* **31**, 3449 (1966); R. Kreher, *Angew. Chem. Int. Ed.* **12**, 1022 (1973); D. G. McMinn, *Synthesis* **1976**, 824; D. Crich, H. Dyker, *Tetrahedron Lett.* **30**, 475 (1989); Y. Yamamoto *et al.*, *Synthesis* **1995**, 571; A. J. Kiessling, C. K. McClure, *Synth. Commun.* **27**, 923 (1997). *Brief review:* S. Pichlmair, *Synlett* **2004**, 195-196.

Colorless solid, mp 91-92° (dec). Hygroscopic. *Store at 0-5° in dichloromethane or diethyl ether.*

USE: Alkylating agent for nucleophilic functional groups in organic synthesis.

5868. Mefenacet. [73250-68-7] 2-(2-Benzothiazolyloxy)-*N*-methyl-*N*-phenylacetamide; 2-(1,3-benzothiazol-2-yloxy)-*N*-methylacetanilide; FOE-1976; NTN-801; Hinochloa; Rancho. $C_{16}H_{14}$-N_2O_2S; mol wt 298.36. C 64.41%, H 4.73%, N 9.39%, O 10.72%,

S 10.75%. Oxyacetamide herbicide; inhibits cell division and growth in meristematic plant tissue. Prepn: H. Förster *et al.*, **EP 5501**; *eidem*, **US 4509971** (1979, 1985 both to Bayer). Properties and field trials in rice: R. R. Schmidt *et al.*, *Meded. Fac. Landbouwwet. Rijksuniv. Gent* **49**, 1075 (1984). Properties and toxicity studies: *J. Pestic. Sci.* **13**, 633 (1988). Distribution and metabolism: B. Krauskopf *et al.*, *Proc. 4th Symposium on Weed Problems in Mediterranean Climates* **2**, 386 (1989). GC determn in agricultural products: Y. Nakamura *et al.*, *J. Agric. Food Chem.* **42**, 2508 (1994).

Crystals, mp 134.8°. Vapor pressure at 20°: 4.8×10^{-11} mm Hg. Partition coefficient (octanol/water): log P 3.23. Soly at 20° (g/l): 0.004 water; 0.1-1 *n*-hexane; 20-50 toluene; >200 dichloromethane; 5-10 2-propanol. LD_{50} in rats, mice (mg/kg): >5000, >5000 orally; >1000, >1000 s.c.; >5000, >5000 dermally; LC_{50} (4 hr) in rats: 134 mg/m^3 (Schmidt).

USE: Herbicide.

5869. Mefenamic Acid. [61-68-7] 2-[(2,3-Dimethylphenyl)-amino]benzoic acid; *N*-(2,3-xylyl)anthranilic acid; CI-473; INF-3355; Lysalgo; Mefenacid; Parkemed; Ponalar; Ponstan; Ponstel; Ponstyl; Pontal. $C_{15}H_{15}NO_2$; mol wt 241.29. C 74.67%, H 6.27%, N 5.81%, O 13.26%. Prepn: **BE 605302**; R. A. Scherrer, **US 3138636** (1961, 1964 both to Parke-Davis). Pharmacology: C. V. Winder *et al.*, *J. Pharmacol. Exp. Ther.* **138**, 405 (1962); C. V. Winder *et al.*, *Arthritis Rheum.* **12**, 472 (1969); and toxicology: Mokhort, Korkhova, *Farmakol. Toksikol. (Kiev)* **1968**, 85, *C.A.* **71**, 29080c (1969); U. Jahn, R. W. Adrian, *Arzneim.-Forsch.* **19**, 36 (1969). Crystal structure: J. F. McConnell, F. Z. Company, *Cryst. Struct. Commun.* **5**, 861 (1976). Clinical trial in primary dysmenorrhea: P. W. Budoff, *J. Am. Med. Assoc.* **241**, 2713 (1979). HPLC determn in human plasma: I. Niopas, K. Mamzoridi, *J. Chromatogr. B* **656**, 447 (1994).

White to off-white crystals, mp 230-231° (effervescence). pKa 4.2. uv max (0.1*N* NaOH): 285, 340 nm. Soly in H$_2$O, pH 7.1 (g/100 ml): 0.0041 (25°); 0.008 (37°). Soluble in solns of alkali hydroxides; sparingly sol in ether, chloroform; slightly sol in ethanol, methanol. LD_{50} orally in mice, rats: 630, 790 mg/kg (Jahn, Adrian).

Sodium salt. $C_{15}H_{14}NNaO_2$. White powder. Soly in H$_2$O: >5 g/100 ml. LD_{50} in mice (mg/kg): 600 orally; 150 i.p. (Mokhort, Korkhova).

THERAP CAT: Anti-inflammatory; analgesic.

5870. Mefenorex. [17243-57-1] *N*-(3-Chloropropyl)-α-methylbenzeneethanamine; *N*-(3-chloropropyl)-α-methylphenethylamine; 1-phenyl-2-(3-chloropropylamino)propane. $C_{12}H_{18}ClN$; mol wt 211.73. C 68.07%, H 8.57%, Cl 16.74%, N 6.62%. Prepn: Beschke *et al.*, **DE 1210873** (1966 to Hoffmann-La Roche), *C.A.* **64**, 19486c (1966). Clinical pharmacology: P. Netter *et al.*, *Arzneim.-Forsch.* **28**, 1310 (1978). GC-MS determn in urine: A. Franceschini *et al.*, *J. Chromatogr.* **541**, 109 (1991). Pharmacokinetics and metabolism: S. Rendic *et al.*, *Eur. J. Drug Metab. Pharmacokinet.* **19**, 107 (1994). *Review:* J. Engel *et al.*, *Drug Alcohol Depend.* **17**, 229-234 (1986).

Hydrochloride. [5586-87-8] Ro-4-5282; Incital; Pondinil; Pondinol. $C_{12}H_{18}ClN.HCl$; mol wt 248.19. Solid, mp 128-130°.

Note: This is a controlled substance (stimulant): **21 CFR**, 1308.14.

THERAP CAT: Anorexic.

5871. Mefenpyr-diethyl. [135590-91-9] 1-(2,4-Dichlorophenyl)-4,5-dihydro-5-methyl-1*H*-pyrazole-3,5-dicarboxylic acid 3,5-diethyl ester; diethyl 1-(2,4-dichlorophenyl)-5-methyl-2-pyrazoline-3,5-dicarboxylate; AE-F107892; HOE-107892. $C_{16}H_{18}Cl_2N_2O_4$; mol wt 373.23. C 51.49%, H 4.86%, Cl 19.00%, N 7.51%, O 17.15%. Herbicide safener that enhances metabolism of herbicides and exhibits selective control of grass weeds in wheat and barley. Prepn: W. Rösch *et al.*, **WO 9107874**; *eidem*, **US 5700758** (1991, 1997 both to Hoechst). Comprehensive description: E. Hacker *et al.*, *Brighton Crop Prot. Conf. - Weeds* **1999**, 15-22. Effects on glutathione *S*-transferase activity on dark-grown barley: R. Scalla, A. Roulet, *Physiol. Plant.* **116**, 336 (2002). Field study of mixture with iodosulfuron-methyl-sodium in barley: S. R. King, E. S. Hagood, Jr., *Weed Technol.* **19**, 372 (2005).

mp 50-52°. Vapor pressure at 25°: 1.4×10^{-5} Pa. Soly in water (20°): 20 mg/ml (pH 6.2). Log P (octanol/water): 3.83 (pH 6.3, 21°). LD_{50} in rats (mg/kg): >5000 orally, >4000 dermally; LC_{50} in carp (mg/l): 2.4 (Hacker).

USE: Herbicide safener.

5872. Mefloquine. [53230-10-7] *rel*-(α*S*)-α-(2*R*)-2-Piperidinyl-2,8-bis(trifluoromethyl)-4-quinolinemethanol; DL-*erythro*-α-2-piperidyl-2,8-bis(trifluoromethyl)-4-quinolinemethanol; Ro-21-5998. $C_{17}H_{16}F_6N_2O$; mol wt 378.32. C 53.97%, H 4.26%, F 30.13%, N 7.40%, O 4.23%. Quinoline methanol antimalarial agent. Prepn (stereo unspec): C. J. Ohnmacht *et al.*, *J. Med. Chem.* **14**, 926 (1971). Prepn, activity, and abs config of isomers: F. I. Carroll, J. T. Blackwell, *J. Med. Chem.* **17**, 210 (1974). Prepn of *erythro*-form: G. Grethe, T. Mitt, **DE 2806909**; H. Bömches, B. Hardegger, **US 4507482** (1978, 1985 both to Hoffmann-La Roche). X-ray crystal structure: A. Skórska *et al.*, *Bioorg. Med. Chem. Lett.* **16**, 850 (2006). Mechanism of action studies: M. W. Davidson *et al.*, *J. Med. Chem.* **20**, 1117 (1977); R. E. Brown *et al.*, *Life Sci.* **25**, 1857 (1979). Photochemistry: G. A. Epling, U. C. Yoon, *Chem. Lett.* **1982** (2), 211. Pharmacokinetics: D. E. Schwartz *et al.*, *Chemotherapy (Basel)* **28**, 70 (1982). HPLC determn in blood and plasma: I. M. Kapetanovic *et al.*, *J. Chromatogr.* **277**, 209 (1983). Comprehensive description: P. Lim, *Anal. Profiles Drug Subs.* **14**, 157-180 (1985). Review of development: *Bull. WHO* **61**, 169-178 (1983); of clinical trials in malaria prophylaxis: A. Croft, P. Garner, *Br. Med. J.* **315**, 1412-1416 (1997).

Relative stereochemistry

mp 174-176°. Crystal density: 1.432 g/cm^3.

Hydrochloride. [51773-92-3] WR-142490; Lariam. $C_{17}H_{16}F_6N_2O.HCl$; mol wt 414.78. Odorless, white or slightly yellow, crystalline powder; mp 259-260° (dec). Exhibits polymorphism. Freely sol in methanol; sol in ethanol, ethyl acetate; very slightly sol in water. uv max (methanol): 222, 283, 304, 318 nm (ε 46700, 6600, 4000, 3100).

THERAP CAT: Antimalarial.

5873. Mefluidide. [53780-34-0] N-[2,4-Dimethyl-5-[[(trifluoromethyl)sulfonyl]amino]phenyl]acetamide; 5-acetamido-2,4-dimethyltrifluoromethanesulfonanilide; methafluoridamid; MBR-12325; VEL-3973; Vistar; Embark. $C_{11}H_{13}F_3N_2O_3S$; mol wt 310.29. C 42.58%, H 4.22%, F 18.37%, N 9.03%, O 15.47%, S 10.33%. Prepn: T. L. Fridinger, **DE 2406475**; *idem*, **US 3894073** (1974, 1975 to 3M). Metabolism: G. W. Ivie, *J. Agric. Food Chem.* **28**, 1286 (1980).

Cryst solid, mp 183-185°. Vapor press at 25°: $<10^{-4}$ mm Hg. pKa 4.6. Soly at 25° (g/l): water 0.18; benzene 0.31; dichloromethane 2.1; 1-octanol 17; methanol 310; acetone 350. Aq solns decomp in uv light. Mildly corrosive to metals.

USE: Plant growth regulator; herbicide.

5874. Mefruside. [7195-27-9] 4-Chloro-N^1-methyl-N^1-[(tetrahydro-2-methyl-2-furanyl)methyl]-1,3-benzenedisulfonamide; 4-chloro-N^1-methyl-N^1-(tetrahydro-2-methylfurfuryl)-m-benzenedisulfonamide; 2-[(4-chloro-N^1-methyl-3-sulfamoylbenzenesulfonamido)methyl]-2-methyltetrahydrofuran; N-(4-chloro-3-sulfamoylbenzenesulfonyl)-N-methyl-2-furfurylamine; B-1500; Baycaron. $C_{13}H_{19}ClN_2O_5S_2$; mol wt 382.87. C 40.78%, H 5.00%, Cl 9.26%, N 7.32%, O 20.89%, S 16.75%. Prepn: H. Horstmann *et al.*, **GB 1031916**; *idem*, **US 3356692** (1966, 1967 to Bayer AG); of *dl*-, *d*-, and *l*-forms: *idem*, *Arzneim.-Forsch.* **17**, 653 (1967).

dl-**Form.** Crystals, mp 149-150°.
d-**Form.** Crystals, mp 146°. $[\alpha]_{578}^{20}$ +5.4° (c = 2.026 in methanol).
l-**Form.** Crystals, mp 146°. $[\alpha]_{578}^{20}$ −5.5° (c = 2.100 in methanol). More active as a diuretic than the *d*-form.

THERAP CAT: Diuretic.

5875. Megacins. Antibiotic proteins produced by strains of *Bacillus megaterium*, which are highly specific in their antibacterial activity against strains of homologous species. Isoln of megacin A from *B. megaterium*: Ivanovics, Alföldi, *Nature* **174**, 465 (1954); Holland, *Biochem. J.* **78**, 641 (1961). Destroys the cytoplasmic membrane of sensitive bacteria: Nagy, *Acta Microbiol. Acad. Sci. Hung.* **6**, 337 (1959). Isoln of megacin C from *B. megaterium*: Holland, Roberts, *J. Gen. Microbiol.* **35**, 271 (1964). Has a mode of action quite different from megacin A and is more reminiscent of colicins, *q.v.* Review: Holland in *Antibiotics* vol. 1, D. Gottlieb, P. Shaw, Eds. (Springer-Verlag, New York, 1967) pp 688-695.

5876. Megestrol Acetate. [595-33-5] 17-(Acetyloxy)-6-methylpregna-4,6-diene-3,20-dione; 17α-acetoxy-6-methylpregna-4,6-diene-3,20-dione; 6-dehydro-6-methyl-17α-acetoxyprogesterone; 6-methyl-$\Delta^{4,6}$-pregnadien-17α-ol-3,20-dione acetate; Maygace; Megace; Megestat; Megestil; Nia; Niagestin; Ovaban. $C_{24}H_{32}O_4$; mol wt 384.52. C 74.97%, H 8.39%, O 16.64%. Orally active progestogen; formerly used in combinations as oral contraceptive. Prepn: Ringold *et al.*, *J. Am. Chem. Soc.* **81**, 3712 (1959); Dodson, Sollman, **US 2891079** (1959 to Searle); Kirk *et al.*, **GB 870286** *C.A.* **56**, 10248i (1962), and **US 3356573** (1967 to Brit. Drug Houses); Cross, **US 3400137** (1968 to Syntex). Biological effectiveness studies: Chang, Kincl, *Steroids* **12**, 689 (1968). Metabolic studies: Cooper, Kellie, *ibid.* **11**, 133 (1968). Soly and diffusion studies: Sundaram, Kincl, *ibid.* **12**, 517 (1968). Comparative clinical trial with tamoxifen in metastatic breast cancer: J. C. Allergra *et al.*, *Semin. Oncol.* **12**, Suppl. 6, 61 (1985). Evaluation in prevention

of hot flashes: C. L. Loprinzi *et al.*, *N. Engl. J. Med.* **331**, 347 (1994). Review of carcinogenicity studies: *IARC Monographs* **21**, 431-439 (1979).

Crystals from methanol, mp 214-216°. Soly at 37° (μg/ml): water 2; plasma 24. Very sol in chloroform; sol in acetone; sparingly sol in alc; slightly sol in ether and in fixed oils. $[\alpha]_D^{24}$ +5° (chloroform). uv max (ethanol): 287 nm (log ε 4.40).

Mixture with ethinyl estradiol. [8064-66-2] Co-Ervonum; Kombiquens; Noval; Nuvacon; Planovin; Tri-Ervonum; Volidan; Weradys.

Mixture with mestranol. [8064-51-5] Delpregnin.

THERAP CAT: Progestogen. Palliative treatment of breast and endometrial carcinoma.

THERAP CAT (VET): Progestogen. Estrus regulator.

5877. Meglutol. [503-49-1] 3-Hydroxy-3-methylpentanedioic acid; 3-hydroxy-3-methylglutaric acid; dicrotalic acid; medroglutaric acid; HMG; HMGA; CB-337; Lipoglutaren; Mevalon. $C_6H_{10}O_5$; mol wt 162.14. C 44.45%, H 6.22%, O 49.34%. Hypolipidemic agent that decreases the rate of cholesterol synthesis. Prepn via Reformatsky reaction between ethyl acetoacetate and ethyl bromoacetate: R. Adams, B. L. Van Duuren, *J. Am. Chem. Soc.* **75**, 2377 (1953). Synthesis via oxidation of diallylmethylcarbinol with ozone and hydrogen peroxide: H. J. Klosterman, F. Smith, *ibid.* **76**, 1229 (1954); J. L. Rabinowitz, *Biochem. Prep.* **6**, 25 (1958). *Caution:* attempts to increase batch size of the ozonolysis resulted in an explosion, *cf. Chem. Eng. News* **51**(6), 29 (1973). Improved synthesis: A. Yavrouian *et al.*, *Synthesis* **1981**, 791. Hypocholesteremic properties: Z. H. Beg, M. Siddiqi, *Experientia* **23**, 380 (1967); M. Siddiqi, Z. H. Beg, **US 3629449** (1971); Z. H. Beg, P. J. Lupien, *Biochim. Biophys. Acta* **260**, 439 (1972); A. N. K. Yusufi, M. Siddiqi, *Atherosclerosis* **20**, 517 (1974); C. D. Padova *et al.*, *Life Sci.* **30**, 1907 (1982). Biosynthetic studies: L. W. White, H. Rudney, *Biochemistry* **9**, 2713 (1970); L. Hagenfeldt, K. Hellstrom, *Life Sci.* **9**, Pt. 2, 991 (1970). Organ distribution: L. L. Savoie, P. J. Lupien, *Can. J. Physiol. Pharmacol.* **53**, 638 (1975). Effectiveness in familial hypercholesteremia: P. J. Lupien *et al.*, *J. Clin. Pharmacol.* **19**, 120 (1979). GC study in patients with organic acidurias: K. Tanaka, D. J. Hine, *J. Chromatogr.* **239**, 301 (1982). Toxicological study: L. L. Savoie, P. J. Lupien, *Arzneim.-Forsch.* **25**, 1284 (1975).

Cryst from ether/petr ether, mp 108-109°. Sol in water. Stable when stored dry. LD_{50} in mice (g/kg): 7.33 orally; 3.23 i.p. (Savoie, Lupien).

THERAP CAT: Antilipemic.

5878. Meitnerium. [54038-01-6] Element 109; unnilennium. Mt, Une; at. no. 109. Group VIII (9). No stable nuclides. Prepn of isotope ^{266}Mt ($T_{1/2}$ ~3.5 msec, α-decay) by ^{209}Bi(^{58}Fe,n): G. Münzenberg *et al.*, *Z. Phys.* **A309**, 89 (1982); *eidem*, *ibid.* **A315**, 145 (1984). Review of synthesis: G. N. Flerov, G. M. Ter-Akopian, *Prog. Particle Nucl. Phys.* **19**, 197-239 (1987); G. T. Seaborg in *Transuranium Elements: A Half Century*, L. R. Morss, J. Fuger, Eds. (Am. Chem. Soc., Washington DC, 1992) p 41-42.

5879. Melagatran. [159776-70-2] N-[(1R)-2-[(2S)-2-[[[[4-(Aminoiminomethyl)phenyl]methyl]amino]carbonyl]-1-azetidinyl]-1-cyclohexyl-2-oxoethyl]glycine; N-[(R)-[[(2S)-2-[[p-amidinobenzyl]carbamoyl]-1-azetidinyl]carbonyl]cyclohexylmethyl]glycine; H-

319/68. $C_{22}H_{31}N_5O_4$; mol wt 429.52. C 61.52%, H 7.28%, N 16.31%, O 14.90%. Synthetic thrombin inhibitor; active metabolite of ximelagatran, *q.v.* Prepn: K. T. Antonsson *et al.*, **WO 9429336**; *eidem*, **US 5939392** (1994, 1999 both to Astra AB). Pharmacology: D. Gustafsson *et al.*, *Thromb. Haemostasis* **79**, 110 (1998). Clinical evaluation in patients with deep vein thrombosis: H. Eriksson *et al.*, *ibid.* **81**, 358 (1999). LC-MS/MS determn in plasma: K. Dunér *et al.*, *J. Chromatogr. B* **852**, 317 (2007). Review of clinical experience: L. Testa *et al.*, *Expert Opin. Drug Saf.* **6**, 397-406 (2007).

THERAP CAT: Antithrombotic.

5880. Melamine. [108-78-1] 1,3,5-Triazine-2,4,6-triamine; 2,4,6-triamino-*s*-triazine; cyanurotriamide. $C_3H_6N_6$; mol wt 126.12. C 28.57%, H 4.80%, N 66.64%. Usually prepd by heating dicyandiamide, $H_2NC(=NH)NHC\equiv N$, under pressure: Mackay, **US 2737513** (1956 to Am. Cyanamid). Alternate methods starting with urea: Mackey, **US 2760961** (1956 to Am. Cyanamid); Pomot *et al.*, **US 3111519** (1963 to Office Natl. Ind. de l'Azote). X-ray and neutron crystal structure: J. N. Varghese *et al.*, *Acta Crystallogr.* **33B**, 2102 (1977). Review of mfg processes: *Faith, Keyes & Clark's Industrial Chemicals*, F. A. Lowenheim, M. K. Moran, Eds. (Wiley-Interscience, New York, 4th ed., 1975) pp 519-523.

Monoclinic prisms. mp <250°. Sublimes. d^{250} 1.573. Slightly sol in water; very slightly sol in hot alc; insol in ether.
USE: Forms synthetic resins with formaldehyde.

5881. Melanins. Pigments responsible for the dark color of skin, hair, feathers, fur, insect cuticle, soil; found also in fungi, bacteria, and pathological human urine where it is an indication of melanotic tumors. Structures are highly irregular polymers produced in the form of granules which may be bound to protein material. *Allomelanins* are found in the plant kingdom and are produced from nitrogen-free precursors. *See:* aspergillin, humic acid. *Eumelanins* and *phaeomelanins* are found in the animal kingdom. Eumelanins are black or brown, insoluble, nitrogenous pigments produced by the oxidative polymerization of 5,6-dihydroxyindoles derived enzymatically from tyrosine via dopa. One of the best characterized is sepiomelanin, *q.v.* Phaeomelanins are sulfur-containing, alkali-soluble, yellow to reddish-brown pigments produced by oxidative polymerization of cysteinyldopas via 1,4-benzothiazine intermediates. Biosynthesis from tyrosine: H. S. Raper, *Physiol. Rev.* **8**, 245 (1928); H. S. Mason, *J. Biol. Chem.* **172**, 83 (1948). *In vitro* prepn from dopa: L. E. Arnow, *Science* **87**, 308 (1938). Series of articles on structure and biosynthesis: M. Piattelli, R. A. Nicolaus, *Tetrahedron* **15**, 66 (1961); M. Piattelli *et al.*, *ibid.* **18**, 941 (1962); *eidem, ibid.* **19**, 2061 (1963); R. A. Nicolaus *et al.*, *ibid.* **20**, 1163 (1964). NMR-study: G. A. Duff *et al.*, *Biochemistry* **27**, 7112 (1988). Determn of eumelanin metabolites in human urine: S. Pavel, W. van der Slik, *J. Chromatogr.* **375**, 392 (1986). Review of biosynthesis: J. M. Pawelek, A. M. Körner, *Am. Sci.* **70**, 136-145 (1982); of photochemistry and photobiology: M. R. Chedekel, *Photochem. Photobiol.* **35**, 881-885 (1982); M. R. Chedekel, L. Zeise, *Lipids* **23**, 587-591 (1988). *Reviews:* R. A. Nicolaus, *Melanins* (Hermann, Paris, 1968); G. Prota, *Med. Res. Rev.* **8**, 525-556 (1988).

5882. Melanostatin. [9083-38-9] Melanocyte-stimulating hormone release-inhibiting factor; MIF; MRIH; MSH release-inhibiting hormone; melanotropin inhibiting factor. An inhibiting factor which mediates hypothalamic control of MSH, *q.v.*, a pituitary hormone: A. J. Kastin, A. V. Schally, *Gen. Comp. Endocrinol.* **7**, 452 (1966); A. V. Schally, A. J. Kastin, *Endocrinology* **79**, 768 (1966). Although several substances have MIF activity, the exact nature of MSH release inhibition is not known: A. J. Kastin *et al.*, *Yale J. Biol. Med.* **46**, 617 (1973); *eidem, Fed. Proc.* **39**, 2931 (1980). The first factor with melanocyte-release inhibiting activity to be isolated was the side-chain tripeptide of oxytocin, *q.v.*, now referred to as MIF-I: M. E. Celis *et al.*, *Proc. Natl. Acad. Sci. USA* **68**, 1428 (1971). Isoln of bovine MIF-I and identity with the tripeptide (Pro-Leu-GlyNH$_2$): R. M. G. Nair *et al.*, *Biochem. Biophys. Res. Commun.* **43**, 1376 (1971). Synthesis: H. Irie *et al.*, *Chem. Lett.* **1980**, 705. Extrapituitary effects of MIF-I are known, particularly potentiation of DOPA-induced behavioral changes: N. P. Plotnikoff *et al.*, *Life Sci.* **10**, 1279 (1971); *eidem, Neuroendocrinology* **14**, 271 (1974). Clinical studies in Parkinson's disease: A. J. Kastin, A. Barbeau, *Can. Med. Assoc. J.* **107**, 1079 (1972); A. Barbeau, *Lancet* **2**, 683 (1975); V. E. Schneider *et al.*, *Arzneim.-Forsch.* **28**, 1296 (1978). *Reviews:* A. V. Schally *et al.*, *Recent Prog. Horm. Res.* **24**, 497-581 (1968); A. J. Kastin *et al.*, *Life Sci.* **25**, 401-414 (1979); *eidem* in *Polypeptide Hormones*, R. F. Beers, E. G. Bassett, Eds. (Raven Press, New York, 1980) pp 223-224; *eidem, Fed. Proc.* **39**, 2931-2936 (1980).

Pro–Leu–Gly–NH₂

MIF-I

MIF-I. [2002-44-0] L-Prolyl-L-leucylglycinamide; melanostatin I (ox); MSH release-inhibiting factor I. $C_{13}H_{24}N_4O_3$; mol wt 284.36.

5883. Melarsoprol. [494-79-1] 2-[4-[(4,6-Diamino-1,3,5-triazin-2-yl)amino]phenyl]-1,3,2-dithiarsolane-4-methanol; *p*-[(4,6-diamino-*s*-triazin-2-yl)amino]dithiobenzenearsonous acid 3-hydroxypropylene ester; 2-*p*-(4,6-diamino-*s*-triazin-2-ylamino)phenyl-4-hydroxymethyl-1,3,2-dithiarsoline; 2-(4-melamin-2-ylphenyl)-4-hydroxymethyl-1,3-dithia-2-arsolane; Mel B; Arsobal. $C_{12}H_{15}$-AsN_6OS_2; mol wt 398.33. C 36.18%, H 3.80%, As 18.81%, N 21.10%, O 4.02%, S 16.10%. Prepn: Friedheim, **US 2659723** (1953); **US 2772303** (1956).

Practically insol in water, cold ethanol, methanol. Sol in propylene glycol.
THERAP CAT: Antiprotozoal (Trypanosoma).

5884. Melatonin. [73-31-4] *N*-[2-(5-Methoxy-1*H*-indol-3-yl)ethyl]acetamide; *N*-acetyl-5-methoxytryptamine; Regulin. C_{13}-$H_{16}N_2O_2$; mol wt 232.28. C 67.22%, H 6.94%, N 12.06%, O 13.78%. A hormone of the pineal gland, also produced by extrapineal tissues, that lightens skin color in amphibians by reversing the darkening effect of MSH, *q.v.* Melatonin has been postulated as the mediator of photic-induced antigonadotrophic activity in photoperiodic mammals and has also been shown to be involved in thermoregulation in some ectotherms and in affecting locomotor activity rhythms in sparrows. Isoln from the pineal glands of beef cattle: Lerner *et al.*, *J. Am. Chem. Soc.* **80**, 2587 (1958); Wurtman *et al.*, *Science* **141**, 277 (1963). Structure: Lerner *et al.*, *J. Am. Chem. Soc.* **81**, 6084 (1959). Crystal and molecular structure: A. Wakahara, *Chem. Lett.* **1972**, 1139. Synthesis from 5-methoxyindole as starting material by two different routes: Szmuszkovicz *et al.*, *J. Org. Chem.* **25**, 857 (1960). Biochemical role of melatonin: *Chem. Eng. News* **45**, 40 (May 1, 1967). Pharmacological studies: Barchas *et al.*, *Nature* **214**, 919 (1967). Identification of antigonadal action sites in mouse brain: J. D. Glass, G. R. Lynch, *Science* **214**, 821 (1981). Binding studies in human hypothalamus: S. M. Reppert *et al.*, *Science* **242**, 78 (1988). Efficacy in control of estrus in red deer: G. W. Asher, *Anim. Reprod. Sci.* **22**, 145 (1990). *Reviews:* M. K. Vaughn,

Int. J. Rev. Physiol. **24**, 41-95 (1981); D. C.Klein *et al., Life Sci.* **28**, 1975-1986 (1981). Book: *Advan. Biosci.* **vol. 29**, N. Birau, W. Schlott, Eds. (Pergamon Press, New York, 1981) 420 pp. Review of etiological role in clinical disease: A. Miles, D. Philbrick, *Crit. Rev. Clin. Lab. Sci.* **25**, 231-253 (1987); in psychiatric disorders: *eidem, Biol. Psychiatry* **23**, 405-425 (1988).

Pale yellow leaflets from benzene, mp 116-118°. uv max: 223, 278 nm (ε 27550, 6300).

THERAP CAT (VET): Control of estrus.

5885. Meldrum's Acid. [2033-24-1] 2,2-Dimethyl-1,3-dioxane-4,6-dione; malonic acid cyclic isopropylidene ester; isopropylidene malonate; 2,2-dimethyl-4,6-diketo-1,3-dioxane. $C_6H_8O_4$; mol wt 144.13. C 50.00%, H 5.59%, O 44.40%. Prepd from malonic acid and acetone in the presence of acetic anhydride: Meldrum, *J. Chem. Soc.* **93**, 598 (1908). Prepn and structure: Davidson, Bernhard, *J. Am. Chem. Soc.* **70**, 3426 (1948); P. Schuster *et al., Monatsh. Chem.* **95**, 53 (1964); G. H. Bihlmayer *et al., ibid.* **98**, 564 (1967). *Review:* H. McNab, *Chem. Soc. Rev.* **7**, 345-358 (1978).

Crystals from acetone + water, dec 94-95°. pKa 5.1.

USE: In organic synthesis as a substitute for acyclic malonic esters and generally as a C_3O_2 synthon.

5886. Melengestrol Acetate. [2919-66-6] 17-(Acetyloxy)-6-methyl-16-methylenepregna-4,6-diene-3,20-dione; 6α-methyl-6-dehydro-16-methylene-17-acetoxyprogesterone; MGA. $C_{25}H_{32}O_4$; mol wt 396.53. C 75.73%, H 8.13%, O 16.14%. Orally active progestational steroid. Prepn: D. N. Kirk *et al.,* **GB 886619**; *eidem,* **US 3332940** (1962, 1967 both to British Drug Houses). Environmental degradation study: B. Schiffer *et al., Environ. Health Perspect.* **109**, 1145 (2001). Receptor binding study: G. A. Perry *et al., Domest. Anim. Endocrinol.* **28**, 147 (2005). Review of pharmacology in animals: R. G. Zimbelman *et al., J. Am. Vet. Med. Assoc.* **157**, 1528-1536 (1970); of use in control of bovine estrus: D. J. Patterson *et al., J. Anim. Sci.* **67**, 1895-1906 (1989).

Needles from acetone/hexane, mp 224-226°. Freely sol in chloroform, ethyl acetate; slightly sol in alc. Insol in water. $[\alpha]_D^{23}$ −127° (c = 0.31 in chloroform). uv max (ethanol): 287.5 nm (log ε 4.35). LD$_{50}$ (mg/kg): >8000 orally in rats; >2500 i.p. in mice (Zimbelman).

THERAP CAT (VET): Progestogen; growth promotant.

5887. Melezitose. [597-12-6] *O*-α-D-Glucopyranosyl-(1 → 3)-β-D-fructofuranosyl-α-D-glucopyranoside. $C_{18}H_{32}O_{16}$; mol wt 504.44. C 42.86%, H 6.39%, O 50.75%. Trisaccharide built from 2 mols glucose and 1 mol fructose. Acid hydrolysis yields at first glucose and turanose. Occurs in a manna that forms upon the

Douglas fir, jack pine *(Pinus virginiana* Mill., *Pinaceae)* and other trees. During periods of drought, bees collect it from the manna in sufficient quantity to change the character of the honey, making it unsuitable for maintaining the bees, but providing a source of melezitose: Hudson, Sherwood, *J. Am. Chem. Soc.* **42**, 116 (1920). Procedure: Bates, *Natl. Bur. Stand. Circ.* **C440**, p 472 (1942). Structure: Hehre, Carlson, *Arch. Biochem. Biophys.* **36**, 158 (1952); Hehre, *Adv. Carbohydr. Chem.* **8**, 277 (1953). *Review:* E. B. Rathbone, *Dev. Food Carbohydr.* **2**, 145-185 (1980).

Dihydrate. Crystals from water. The water of crystn is given up at 110°. When anhydr, mp 153-154°. $[\alpha]_D^{20}$ +88° (c = 4). Sol in water, very sparingly sol in alc. Not fermented by baker's yeast; does not reduce Fehling's soln.

5888. Melibiose. [585-99-9] 6-*O*-α-D-Galactopyranosyl-D-glucose; 6-(α-D-galactosido)-D-glucose. $C_{12}H_{22}O_{11}$; mol wt 342.30. C 42.11%, H 6.48%, O 51.41%. Prepd from raffinose by fermentation with top yeast which removes the fructose: Hudson, Harding, *J. Am. Chem. Soc.* **37**, 2734 (1915); Fletcher, Diehl, *ibid.* **74**, 5774 (1952). Structure: Haworth, Leitch, *J. Chem. Soc.* **113**, 188 (1918); Charlton *et al., ibid.* **1926**, 99; Charlton *et al., ibid.* **1927**, 1527; Haworth *et al., ibid.* 3146; Levene, Jorpes, *J. Biol. Chem.* **86**, 403 (1930). Synthesis: Helferich, Bredereck, *Ann.* **465**, 166 (1928).

Dihydrate. Monoclinic crystals from water or dil alcohol. mp 84-85°. Shows mutarotation. $[\alpha]_D^{20}$ +111.7° → +129.5° (c = 4). One gram dissolves in 0.4 ml water, 8.5 ml methanol, 220 ml abs alcohol. Dilute acids hydrolyze melibiose to D-glucose and D-galactose. Also split by emulsin and by bottom yeast. Reduces Fehling's soln. 3.5 g melibiose dihydrate are about as sweet as 1.0 g sucrose.

5889. Melilot. Sweet clover; yellow Melilot; yellow sweet clover. Dried leaves and flowering tops of *Melilotus officinalis* (L.) Lam., *Leguminosae. Habit.* Europe, Asia, natural to some extent in U.S. *Constit.* Coumarin, resin, volatile oil.

An extract is marketed as *Esberiven.*

5890. Melinamide. [14417-88-0] (9Z,12Z)-*N*-(1-Phenylethyl)-9,12-octadecadienamide; *N*-(α-methylbenzyl)linoleamide; MBLA; AC-223; Artes. $C_{26}H_{41}NO$; mol wt 383.62. C 81.41%, H 10.77%, N 3.65%, O 4.17%. Linoleic acid derivative. Prepn and use in lowering blood cholesterol levels: **FR 1476596**; T. Seki *et al.,* **US 3621043** (1967, 1971 both to Sumitomo). Physical properties and effect on cholesterol levels in rabbits: K. Toki *et al., J. Atheroscler. Res.* **7**, 708 (1967). Comparative activity of melinamide and its optical isomers: H. Fukushima, K. Nakatani, *ibid.* **9**, 65 (1969); H. Fukushima *et al., ibid.* **10**, 403 (1969). Mechanism of action studies: D. Kritchevsky, S. A. Tepper, *Arzneim.-Forsch.* **21**, 1024 (1971); D. Kritchevsky *et al., Lipids* **12**, 16 (1977); K. Natori *et al., Jpn. J. Pharmacol.* **42**, 517 (1986). Metabolism: A. Hirohashi *et al., Xenobiotica* **6**, 329 (1976).

Oil. mp <4°. bp$_{0.03}$ 200-215°; bp$_{0.07}$ 200-204°. n_D^{23} 1.5050; n_D^{30} 1.4863. Saponification value 0.9. Iodine value 127.0.

THERAP CAT: Antilipemic.

5891. Melitracen. [5118-29-6] 3-(10,10-Dimethyl-9(10*H*)-anthracenylidene)-*N*,*N*-dimethyl-1-propanamine; *N*,*N*,10,10-tetramethyl-$\Delta^{9(10H),\gamma}$-anthracenepropylamine; 9,10-dihydro-10,10-dimethyl-9-(3-dimethylaminopropylidene)anthracene; 9-[3-(dimethylamino)propylidene]-10,10-dimethyl-9,10-dihydroanthracene; *N*,*N*-dimethyl-3-(10,10-dimethyl-9(10*H*)-anthrylidene)propylamine. C$_{21}$H$_{25}$N; mol wt 291.44. C 86.55%, H 8.65%, N 4.81%. Prepn of the hydrochloride: Holm, *Acta Chem. Scand.* **17**, 2437 (1963); *idem*, **GB 939856**; *idem*, **US 3177209** (1963, 1965 both to Kefalas A/S). Crystal structure: J. Lopez de Lerma *et al.*, *Acta Crystallogr.* **B35**, 1739 (1979). Toxicity data: P. V. Petersen *et al.*, *Acta Pharmacol. Toxicol.* **24**, 121 (1966).

Hydrochloride. [10563-70-9] U-24973A; Melixeran; Trausabun; Dixeran. C$_{21}$H$_{25}$N.HCl; mol wt 327.90. Crystals from acetone, mp 245-248°. LD$_{50}$ i.v. in mice: 52 mg/kg (Petersen).

THERAP CAT: Antidepressant.

5892. Melittin. [20449-79-0] Melittin (honeybee); melittin I; mellitin; Forapin. C$_{131}$H$_{229}$N$_{39}$O$_{31}$; mol wt 2846.52. C 55.28%, H 8.11%, N 19.19%, O 17.42%. Strongly basic polypeptide, the principal component of the venom of the honey bee, *Apis mellifica (mellifera)*, comprising 40-50% of the dried venom. A "direct" hemolysin, one of the hemolytic principles present in the venom, the other being "indirect" acting **phospholipase A.** Sepn and isoln: Neumann *et al.*, *Naturwissenschaften* **39**, 286 (1952); Neumann, Haberman, *Arch. Exp. Pathol. Pharmakol.* **222**, 367 (1954); Habermann, Reiz, *Biochem. Z.* **341**, 451 (1965). Melittin is the first polypeptide whose biological effects can be understood on the basis of its primary structure. Elucidation of structure and correlation with activity: E. Habermann, J. Jentsch, *Z. Physiol. Chem.* **348**, 37 (1967). Conformation studies: R. Bazzo *et al.*, *Eur. J. Biochem.* **173**, 139 (1988). About 10% of the melittin is thought to be formylated at the *N*-terminus: Kreil, Kreil-Kiss, *Biochem. Biophys. Res. Commun.* **27**, 275 (1967). Isoln and structure of *N$^\alpha$-formyl melittin*: Lübke *et al.*, *Experientia* **27**, 765 (1971). Synthesis of melittin and related peptides: Lübke, Schröder, *Peptides*, H. C. Beyerman, A. van der Linde, W. M. van den Brink, Eds. (North-Holland Publishing Company, Amsterdam, 1967) pp 271-279; Dorman, Markley, *J. Med. Chem.* **14**, 5 (1970); Schröder *et al.*, *Experientia* **27**, 764 (1971). Solid phase synthesis and purification: M. T. Tosteson *et al.*, *Biochemistry* **26**, 6627 (1987). Review of biochemistry and pharmacology: Habermann, *Science* **177**, 314 (1972).

Gly–Ile–Gly–Ala–Val–Leu–Lys–Val–Leu–Thr–Thr–Gly–Leu–Pro
NH$_2$Gln–Gln–Arg–Lys–Arg–Lys–Ile–Trp–Ser–Ile–Leu–Ala

Cream white, water soluble powder. $[\alpha]_D^{21}$ −89.52° (c = 0.409).

THERAP CAT: Antirheumatic.

5893. Mellitic Acid. [517-60-2] 1,2,3,4,5,6-Benzenehexacarboxylic acid; hexacarboxybenzene; mellic acid. C$_{12}$H$_6$O$_{12}$; mol

wt 342.17. C 42.12%, H 1.77%, O 56.11%. Preparation from carbonaceous material: M. Kiebler, **US 2461740** (1949 to Carnegie Inst. of Tech.); Germain *et al.*, *Bull. Soc. Chim. Fr.* **1962**, 779; from tetrahalophthalic acid: Brusset, Uny, *ibid.* **1951**, 565; Juettner, **US 3067246** (1962).

Crystals. mp 286-288° in sealed tube with decompn. Freely sol in water or alcohol; sol in boiling concd H$_2$SO$_4$ without decompn.

5894. Meloxicam. [71125-38-7] 4-Hydroxy-2-methyl-*N*-(5-methyl-2-thiazolyl)-2*H*-1,2-benzothiazine-3-carboxamide 1,1-dioxide; Metacam; Mobec; Mobic; Mobicox; Movalis; Movatec; Parocin. C$_{14}$H$_{13}$N$_3$O$_4$S$_2$; mol wt 351.40. C 47.85%, H 3.73%, N 11.96%, O 18.21%, S 18.25%. Preferential cyclooxygenase (COX-2) inhibitor. Prepn: G. Trummlitz *et al.*, **DE 2756113** (1979 to Thomae); *eidem*, **US 4233299** (1980 to Boehringer Ingelheim). Pharmacology in horses: P. Lees *et al.*, *Br. Vet. J.* **147**, 97 (1991). Physicochemical properties: R.-S. Tsai *et al.*, *Helv. Chim. Acta* **76**, 842 (1993). Veterinary trial in dogs: A. J. Henderson *et al.*, *Prakt. Tierarzt* **75**, 179 (1994). Series of articles on pharmacology, mechanism of action and clinical efficacy: *Br. J. Rheumatol.* **35**, Suppl. 1, 1-77 (1996). Clinical trials of GI tolerability in arthritis: C. Hawkey *et al.*, *ibid.* **37**, 937 (1998); J. Dequeker *et al.*, *ibid.* 946. Review of clinical experience: R. Fleischmann *et al.*, *Expert Opin. Pharmacother.* **3**, 1501-1512 (2002).

Crystals from ethylene chloride, mp 254° (dec). pKa: 4.08 in water; 4.24 ± 0.01 in water/ethanol (1:1); 4.63 ± 0.03 in water/ethanol (1:4). Log P (octanol/water): 3.02. Sol in DMF; slightly sol in acetone; very slightly sol in methanol, alc. Practically insol in water. LD$_{50}$ orally in mice: 470 mg/kg (Trummlitz, 1980).

THERAP CAT: Anti-inflammatory.

THERAP CAT (VET): Anti-inflammatory.

5895. Melperone. [3575-80-2] 1-(4-Fluorophenyl)-4-(4-methyl-1-piperidinyl)-1-butanone; 4′-fluoro-4-(4-methylpiperidino)butyrophenone; γ-(4-methylpiperidino)-*p*-fluorobutyrophenone; methylperone; flubuperone. C$_{16}$H$_{22}$FNO; mol wt 263.36. C 72.97%, H 8.42%, F 7.21%, N 5.32%, O 6.07%. Neuroleptic agent related structurally to haloperidol, *q.v.* Prepn: **BE 651144**; S. E. H. Hernestam *et al.*, **US 3816433** (1964, 1974 both to Ferrosan). Distribution of ^{14}C melperone: N. Einer-Jensen, E. Hansson, *Acta Pharmacol. Toxicol.* **23**, 65 (1965). Pharmacological and toxicological studies: J. A. Christensen *et al.*, *ibid.* 109; R. Heywood, A. K. Palmer, *Farmaco Ed. Prat.* **29**, 586 (1974). Dopamine-receptor binding in relation to clinical effect: I. Creese *et al.*, *Science* **192**, 481 (1976). Sedative and sleep-inducing properties: R. Kretzschmer *et al.*, *Arzneim.-Forsch.* **26**, 1073 (1976). Clinical studies in anxiety: W. J. Poeldinger, *Therapiewoche* **30**, 4862 (1980); L. F. Fabre, M. J. Napoliello, *Curr. Ther. Res.* **30**, 427 (1981).

Liquid, bp$_{0.1}$ 120-125°.

Hydrochloride. [1622-79-3] FG-5111; Buronil; Eunerpan. C$_{16}$H$_{22}$FNO.HCl; mol wt 299.81. Crystals, mp 209-211°. LD$_{50}$ in rats, mice (mg/kg): 330, 230 orally; 40, 35 i.v. (Christensen).

THERAP CAT: Antipsychotic.

5896. Melphalan. [148-82-3] 4-[Bis(2-chloroethyl)amino]-L-phenylalanine; *p*-di(2-chloroethyl)amino-L-phenylalanine; L-phenylalanine mustard; alanine nitrogen mustard; L-PAM; melfalan; L-sarcolysine; NSC-8806; CB-3025; Alkeran. C$_{13}$H$_{18}$Cl$_2$N$_2$O$_2$; mol wt 305.20. C 51.16%, H 5.94%, Cl 23.23%, N 9.18%, O 10.48%. Bifunctional, non-cell-cycle specific alkylating agent. Syntheses: Bergel, Stock, *J. Chem. Soc.* **1954**, 2409; **1955**, 1223; *eidem,* **US 3032584**; **US 3032585** (both 1962 to NRDC); Larionov, *Lancet* **2**, 169 (1955). Neurotoxicity study: M. G. Donelli *et al., J. Pharm. Pharmacol.* **18**, 760 (1966). HPLC determn in biological fluids: I. Muckenschnabel *et al., Eur. J. Pharm. Sci.* **5**, 129 (1997). Toxicity study: W. C. J. Ross, *Biochem. Pharmacol.* **13**, 969 (1964). Review of carcinogenicity studies: *IARC Monographs* **9**, 167-180 (1975). *Review:* R. L. Furner, R. K. Brown, *Cancer Treat. Rep.* **64**, 559-574 (1980). Clinical trial in multiple myeloma: A. Palumbo *et al., Blood* **104**, 3052 (2004). Review of pharmacology and clinical efficacy: B. L. Samuels, J. D. Bitran, *J. Clin. Oncol.* **13**, 1786-1799 (1995).

Needles from methanol (monosolvate), mp 182-183° (dec). [α]$_D^{25}$ +7.5° (c = 1.33 in 1.0*N* HCl); [α]$_D^{22}$ −31.5° (c = 0.67 in methanol). Soluble in propylene glycol and dilute mineral acids; slightly sol in ethanol, methanol. Practically insol in water, chloroform, ether. LD$_{50}$ i.p. in rats: 14.7 μmol/kg (Ross).

D-Form. [13045-94-8] D-Sarcolysine; medphalan; CB-3026; NSC-35051. Needles from methanol (monosolvate) mp 181.5-182° (dec). [α]$_D^{21}$ −7.5° (c = 1.26 in 1.0*N* HCl).

DL-Form. [531-76-0] Merphalan; sarcolysine. Tiny needles from methanol, mp 180-181°.

Caution: This substance is listed as a known human carcinogen: *Report on Carcinogens, Twelfth Edition* (PB2011-111646, 2011) p 258.

THERAP CAT: Antineoplastic.

THERAP CAT (VET): Antineoplastic.

5897. Memantine. [19982-08-2] 3,5-Dimethyltricyclo-[3.3.1.1^{3,7}]decan-1-amine; 3,5-dimethyl-1-adamantanamine; 1-amino-3,5-dimethyladamantane; DMAA; D-145. C$_{12}$H$_{21}$N; mol wt 179.31. C 80.38%, H 11.81%, N 7.81%. NMDA-receptor antagonist; deriv of adamantane, *q.v.* Prepn of the hydrochloride: K. Gerzon *et al., J. Med. Chem.* **6**, 760 (1963); of the free base and hydrochloride: J. Mills, E. Krumkalns, **US 3391142** (1968 to Lilly). Series of articles on pharmacology and metabolism: *Arzneim.-Forsch.* **27**, 1471-1489 (1977); on neuropharmacology and clinical efficacy: *ibid.* **32**, 1236-1276 (1982). Pharmacodynamics and pharmacokinetics: W. Wesemann *et al., ibid.* **33**, 1122 (1983). NMDA-receptor binding study: J. Kornhuber *et al., Eur. J. Pharmacol.* **166**, 589 (1989); and neuroprotective properties: *idem et al., J. Neural Transm.* **43**, Suppl., 91 (1994). HPLC determn in plasma: S. E. Toker *et al., J. Sep. Sci.* **34**, 2645 (2011). Clinical studies as antispasmodic: H. Rohde, *Fortschr. Med.* **100**, 2023 (1982); in senile dementia: R. Görtelmeyer, H. Erbler, *Arzneim.-Forsch.* **42**, 904 (1992). Review of clinical experience in neurodegenerative disorders: K. K. Jain, *Expert Opin. Invest. Drugs* **9**, 1397-1406 (2000); in treatment of dementia: D. Lo, G. T. Grossberg, *Expert Rev. Neurother.* **11**, 1359-1370 (2011).

Oil, n_D^{25} 1.4941.

Hydrochloride. [41100-52-1] Akatinol; Auxura; Ebixa; Memary; Namenda. C$_{12}$H$_{21}$N.HCl; mol wt 215.77. Fine white to off-white powder. Crystals from alcohol/ether, mp 258° (Mills, Krumkalns); also reported as mp 290-295° (Gerzon). Sol in water.

THERAP CAT: Neuroprotective; antispasmodic; antiparkinsonian.

5898. Menadiol. [481-85-6] 2-Methyl-1,4-naphthalenediol; 2-methyl-1,4-naphthohydroquinone; 2-methyl-1,4-naphthoquinol; dihydrovitamin K$_3$. C$_{11}$H$_{10}$O$_2$; mol wt 174.20. C 75.84%, H 5.79%, O 18.37%. Synthetic naphthoquinol derivative having physiological properties of Vitamin K, *q.v.*; starting material for prepn of menaquinones, *q.v.* Prepn: K. Fries, W. Lohmann, *Ber.* **54**, 2912 (1921); D. W. MacCorquodale *et al., J. Biol. Chem.* **131**, 357 (1939); C. D. Snyder, H. Rapoport, *J. Am. Chem. Soc.* **96**, 8046 (1974). Crystal structure: J. Gaultier, C. Hauw, *Acta Crystallogr.* **25B**, 51 (1969). Bioactivity and tissue distribution: M. J. Thierry, J. W. Suttie, *J. Nutr.* **97**, 512 (1969).

White needles from dil alc, mp 168-170° (MacCorquodale); also reported as mp 181° (Gaultier, Hauw). Slightly sol in benzene, chloroform; easily sol in acetone, alcohol.

Diacetate. [573-20-6] 1,4-Diacetoxy-2-methylnaphthalene; acetomenaphthone; vitamin K$_4$; Kapilin; Prokayvit Oral; Vitavel K. C$_{15}$H$_{14}$O$_4$; mol wt 258.27. Prepn: Sah *et al., Ber.* **73**, 762 (1940); *Rec. Trav. Chim.* **59**, 461 (1940). Crystals, mp 112-114°. Practically insol in water. Slightly sol in cold alc; sol in 3.3 parts boiling alc, in acetic acid.

Sodium diphosphate. [131-13-5] 2-Methyl-1,4-naphthalenediol diphosphoric acid ester tetrasodium salt; Kappadione; Synkavit; Synkayvite. C$_{11}$H$_8$Na$_4$O$_8$P$_2$; mol wt 422.08. Prepn: Fieser, Fry, *J. Am. Chem. Soc.* **62**, 228 (1940). Prepd as the hexahydrate, white to pinkish hygroscopic powder. Salty taste. Very sol in water. Practically insol in methanol, ether, acetone; insol in ethanol.

THERAP CAT: Vitamin (prothrombogenic).

5899. Menadione. [58-27-5] 2-Methyl-1,4-naphthalenedione; 2-methyl-1,4-naphthoquinone; menaphthone; vitamin K$_{2(0)}$; vitamin K$_3$; Kappaxin; Kayquinone; Thyloquinone. C$_{11}$H$_8$O$_2$; mol wt 172.18. C 76.73%, H 4.68%, O 18.58%. Synthetic naphthoquinone derivative having physiologic properties of vitamin K, *q.v.* Prepn: Fieser, *J. Biol. Chem.* **133**, 391 (1940). Toxicity study: Molitor, Robinson, *Proc. Soc. Exp. Biol. Med.* **43**, 725 (1940). Alkylated *in vivo* to the bioactive metabolite, menaquinone-4, *q.v.*: C. Martius, H. O. Esser, *Biochem. Z.* **331**, 1 (1958). Metabolism and tissue distribution: W. V. Taggart, J. T. Matschiner, *Biochemistry* **8**, 1141 (1969). Improved synthesis: W. Adam *et al., Angew. Chem. Int. Ed.* **33**, 2475 (1994).

Bright yellow crystals. Very faint acrid odor. mp 105-107°. Stable in air; dec by sunlight. One gram dissolves in about 60 ml alcohol, in 10 ml benzene, in 50 ml vegetable oils; sparingly sol in chloroform, carbon tetrachloride. Practically insol in water. The alcoholic soln is neutral to litmus. Solutions may be heated to 120° without dec. Destroyed by alkalies and reducing agents. *Keep protected from light.* LD$_{50}$ orally in mice: ~0.5 g/kg (Molitor, Robinson).

Sodium bisulfite. [130-37-0]; [6147-37-1] (trihydrate). 1,2,3,4-Tetrahydro-2-methyl-1,4-dioxo-2-naphthalenesulfonic acid sodium salt; Hykinone; Klotogen; K-Thrombin. $C_{11}H_9NaO_5S$; mol wt 276.24. Prepn: Moore, *J. Am. Chem. Soc.* **63**, 2049 (1941). Structure: Carmack *et al.*, *ibid.* **72**, 844 (1950). Occurs as the trihydrate. White, hygroscopic crystals. Discolors and may turn purple under the influence of light. One gram dissolves in ~2 ml water. Slightly sol in alcohol. Almost insol in ether, benzene.

Dimethylpyrimidinol bisulfite. [14451-99-1] Hetrazeen. $C_{17}H_{18}N_2O_6S$; mol wt 378.40. Prepn: J. B. Nanninga *et al.*, **US 3328169** (1967 to Heterochemical). Efficacy in swine diets: R. W. Seerley *et al.*, *J. Anim. Sci.* **42**, 599 (1976). Cryst powder, mp 215-217°. Soly in water ~1 g/100 ml. Slightly sol in alcohol. Insol in ether, benzene.

THERAP CAT: Vitamin (prothrombogenic).

THERAP CAT (VET): Vitamin (prothrombogenic); antidote to bishydroxycoumarin poisoning, including sweet clover poisoning.

5900. Menadoxime. [573-01-3] 2-[[(3-Methyl-4-oxo-1(4*H*)-naphthalenylidene)amino]oxy]acetic acid ammonium salt (1:1); menadione carboxymethoxime ammonium salt; menaphthone carboxymethoxime ammonium salt; carboxymethylmenadione monoxime ammonium salt; Kapilon injectable. $C_{13}H_{14}N_2O_4$; mol wt 262.27. C 59.54%, H 5.38%, N 10.68%, O 24.40%. Prepn: Holland, **GB 621934** (1949 to Glaxo).

Crystals from alc. Sol in water. Aq solns are neutral. May be sterilized by autoclaving and stored in ampuls if protected from light. Shows high antihemorrhagic activity.

Free acid. $C_{13}H_{11}NO_4$. Yellow platelets from alc, mp 162-163°. Forms water-sol salts.

THERAP CAT: Vitamin (prothrombogenic).

5901. Menaquinones. Vitamin K_2; 2-methyl-3-*all-trans*-polyprenyl-1,4-naphthoquinones. Group of prenylated naphthoquinone derivatives having the physiological activity of vitamin K, *q.v.* Nomenclature is based on the number of isoprene residues comprising the side chain. Compds of varying chain length are produced by bacteria, particularly by normal intestinal flora, and may serve as a source of vitamin K for humans. Birds and animals are capable of alkylating menadione, *q.v.*, to produce menaquinone 4. Originally isolated from putrefied fish meal: R. W. McKee *et al.*, *J. Biol. Chem.* **131**, 327 (1939). Identification of menaquinone 6: S. B. Binkley *et al.*, *ibid.* **133**, 721 (1940). Isoln of menaquinone 7 from cultures of *Bacillus brevis*: M. Tishler, W. L. Sampson, *Proc. Soc. Exp. Biol. Med.* **68**, 136 (1948). Structures and syntheses of series of menaquinones: O. Isler *et al.*, *Helv. Chim. Acta* **41**, 786 (1958). Review of isoln, props, synthesis: O. Isler, O. Wiss, *Vitam. Horm.* **17**, 53-90 (1959); H. Mayer, O. Isler, *Methods Enzymol.* **18C**, 469-547 (1971); of biosynthesis: D. R. Threlfall, *Vitam. Horm.* **29**, 153-200 (1971); R. Bentley, R. Meganathan, *Microbiol. Rev.* **46**, 241-280 (1982). Menaquinones with side chains of up to 15 repeating units have been described: T. Sakano *et al.*, *Chem. Pharm. Bull.* **34**, 4322 (1986). HPLC determn in bone: S. J. Hodges *et al.*, *J. Bone Miner. Res.* **8**, 1005 (1993). Metabolism and distribution: H. H. W. Thijssen, M. J. Drittij-Reijnders, *Br. J. Nutr.* **72**, 415 (1994). Absorption and bioavailability in humans: J. M. Conly *et al.*, *Am. J. Gastroenterol.* **89**, 915 (1994).

Menaquinone 4

Menaquinone 4. [863-61-6] (*E,E,E*)-2-Methyl-3-(3,7,11,15-tetramethyl-2,6,10,14-hexadecatetraenyl)-1,4-naphthalenedione; menatetrenone; MK 4; Vitamin $K_{2(20)}$; Glakay; Kaytwo. $C_{31}H_{40}O_2$; mol wt 444.66. Review of clinical experence in osteoporosis: J.

Iwamoto *et al.*, *Nutr. Rev.* **64**, 509-517 (2006). Yellow crystals from alcohol, mp 35°. uv max: 248 nm ($E_{1cm}^{1\%}$ 439).

Menaquinone 6. [84-81-1] (*all-E*)-2-(3,7,11,15,19,23-Hexamethyl-2,6,10,14,18,22-tetracosahexaenyl)-3-methyl-1,4-naphthalenedione; 2-difarnesyl-3-methyl-1,4-naphthoquinone; farnoquinone; MK 6; Vitamin $K_{2(30)}$. $C_{41}H_{56}O_2$; mol wt 580.90. Yellow crystals from acetone + alcohol or petr ether, mp 50°. uv max (petr ether): 243, 248, 261, 270, 325-328 nm ($E_{1cm}^{1\%}$ 304, 320, 290, 292, 53).

Menaquinone 7. [2124-57-4] (*all-E*)-2-(3,7,11,15,19,23,27-Heptamethyl-2,6,10,14,18,22,26-octacosaheptaenyl)-3-methyl-1,4-naphthalenedione; Vitamin $K_{2(35)}$. $C_{46}H_{64}O_2$; mol wt 649.02. Light yellow microcrystalline plates from petr ether, mp 54°. bp$_{0.0002}$ 200° (some dec). uv max (petr ether): 243, 248, 261, 270, 325-328 nm ($E_{1cm}^{1\%}$ 278, 195, 266, 267, 48). Slightly less sol than vitamin K_1 in the same organic solvents.

THERAP CAT: Vitamin (prothrombogenic).

5902. Menbutone. [3562-99-0] 4-Methoxy-γ-oxo-1-naphthalenebutanoic acid; 3-(4-methoxy-1-naphthoyl)propionic acid; β-(1-methoxy-4-naphthoyl)propionic acid; γ-oxo-4-methoxy-1-naphthalenebutyric acid; Ictéryl. $C_{15}H_{14}O_4$; mol wt 258.27. C 69.76%, H 5.46%, O 24.78%. Prepn: Ruzicka, Waldman, *Helv. Chim. Acta* **15**, 907 (1932); Fieser, Hershberg, *J. Am. Chem. Soc.* **58**, 2314 (1936); Burtner, **US 2623065** (1952 to Searle).

Crystals, mp 172-173°.

Magnesium salt. [16643-66-6] Hepalande. $C_{30}H_{26}MgO_8$; mol wt 538.84.

THERAP CAT: Choleretic.

5903. Mendelevium. [7440-11-1] Md; formerly Mv; at. no. 101; valence 3, 2. Man-made radioactive element. No stable nuclides; known isotopes (mass numbers): 247-252, 254-259. Longest-lived known isotope 258 (T$_{\frac{1}{2}}$ 55 days, α-emitter, rel. at. mass 258.0984). Prepn of isotope ^{256}Md (T$_{\frac{1}{2}}$ 1.27 hrs, rel. at. mass 256.0941; decays by electron capture to ^{256}Fm) by bombarding ^{253}Es with helium ions: A. Ghiorso *et al.*, *Phys. Rev.* **98**, 1518 (1955). Prepn of isotope ^{258}Md by bombardment of ^{255}Es with ^{4}He ions: Fields *et al.*, *Nucl. Phys. A* **154**, 407 (1970). *Reviews:* C. Keller, *The Chemistry of the Transuranium Elements* (Verlag Chemie, Weinheim, English Ed., 1971) pp 595-600; Silva, "Trans-Curium Elements" in *MTP Int. Rev. Sci.: Inorg. Chem., Ser. One* **vol. 8**, A. G. Maddock, Ed. (University Park Press, Baltimore, 1972) pp 71-105; *Comprehensive Inorganic Chemistry* **vol. 5**, J. C. Bailar, Jr. *et al.*, Eds. (Pergamon Press, Oxford, 1973) *passim; Handb. Exp. Pharmakol.* **36**, 689-738 (1973); R. J. Silva in *The Chemistry of the Actinide Elements* **vol. 2**, J. J. Katz *et al.*, Eds. (Chapman and Hall, New York, 1986) pp 1092-1095; *The Elements Beyond Uranium*, G. T. Seaborg, W. D. Loveland, Eds. (John Wiley & Sons, Inc., New York, 1990) p 38-46.

Caution: Radiation hazard; handling requires special equipment and shielding facilities (Katz *et al.*, *loc. cit.* **vol. 2**, p. 1128).

5904. Menhaden Oil. Pogy oil; mossbunker oil. Obtained along the East Coast of North America from the menhaden fish, *Brevoortia tyrannis*, somewhat larger than a herring. The oil contains (in the form of glycerides) about 6% myristic acid, 16% palmitic acid, 30% linoleic acid, 19% C_{20}- and 11% C_{22}-acids (highly unsaturated).

Reddish oil. Characteristic, distasteful fishy odor and taste. d 0.925-0.933. mp 38.5-47.2°. n_D^{20} 1.480. Sapon no. 191-200. Iodine no. 139-180. Acid no. 3.0-11.6. Sol in ether, benzene, petr ether, naphtha, kerosene, CS_2.

USE: Substitute for linseed oil. In leather dressing formulations. Hydrogenated menhaden oil can be used as a substitute for tallow in soap-making.

5905. Menthol. [2216-51-5]; [89-78-1] (*dl*-form). (1*R*,2*S*,-5*R*)-5-Methyl-2-(1-methylethyl)cyclohexanol; 3-*p*-menthanol; *l*-menthol; hexahydrothymol; peppermint camphor. $C_{10}H_{20}O$; mol wt 156.27. C 76.86%, H 12.90%, O 10.24%. Obtained from peppermint oil or other mint oils, or prepd synthetically by hydrogenation of thymol. Chromatographic sepn from mint oils: Chang, **US 2760993** (1956 to Iowa State College). Toxicity data: P. M. Jenner *et al.*, *Food Cosmet. Toxicol.* **2**, 327 (1964).

Crystals or granules; peppermint taste and odor. d 0.890. mp 41-43°. bp 212°. n_D^{25} 1.458. $[\alpha]_D^{18}$ −50° (10% alc soln). Very sol in alcohol, chloroform, ether, petr ether, solvent hexane; freely sol in glacial acetic acid, liquid petrolatum and in mineral, fixed and volatile oils; slightly sol in water. LD_{50} orally in rats: 3180 mg/kg (Jenner). *Incompat:* Butylchloral hydrate, camphor, phenol, chloral hydrate, Exalgine, betanaphthol, resorcinol or thymol in triturations; potassium permanganate, chromium trioxide, pyrogallol.

USE: Flavor in liqueurs, confectionery; perfumery, cigarettes, cough drops, and nasal inhalers.

THERAP CAT: Antipruritic (topical).

THERAP CAT (VET): Has been used as a mild local anesthetic, antiseptic; internally as a carminative and gastric sedative.

5906. *l*-Menthone. [14073-97-3] (2*S*,5*R*)-5-Methyl-2-(1-methylethyl)cyclohexanone; (1*R*,4*S*)-(−)-*p*-menthan-3-one; 1-methyl-4-isopropylcyclohexan-3-one. $C_{10}H_{18}O$; mol wt 154.25. C 77.87%, H 11.76%, O 10.37%. Of the four optically active isomers of menthone, the one occurring most frequently in nature. Found in various volatile oils, such as pennyroyal, peppermint, geranium: Simonsen, *The Terpenes* vol. I (University Press, Cambridge, 2nd ed., 1947) pp 314-327. Prepd by chromic acid oxidation of *l*-menthol: Hussey, Baker, *J. Org. Chem.* **25**, 1434 (1960); Brown, Garg, *J. Am. Chem. Soc.* **83**, 2952 (1961).

Bitter liq; slight peppermint odor. bp 207°. bp_{41} 116-119°. mp −6°. d_4^{20} 0.895. n_D^{20} 1.4505; n_D^{23} 1.4490. $[\alpha]_D^{20}$ −24.8°; $[\alpha]_D^{27}$ −28.9°. Slightly sol in water; sol in organic solvents.

Apinol. Obtained by dry distln of wood of *Pinus palustris* Mill. (*P. australis* Michx.), *Pinaceae* is chiefly *l*-menthone: *J. Pharm. Chim.* **18**, 139, 177, 208 (1918), *C.A.* **13**, 56⁹ (1919). Amber-colored oil, bp ~182.2°, d 0.946.

USE: In perfume and flavor compositions.

5907. Menthyl Acetate. [2623-23-6]; [89-48-5] (*dl*-form). (1*R*,2*S*,5*R*)-5-Methyl-2-(1-methylethyl)cyclohexanol 1-acetate; *l*-menthol acetate. $C_{12}H_{22}O_2$; mol wt 198.31. C 72.68%, H 11.18%, O 16.14%. Present in peppermint oil.

Colorless liquid, characteristic odor. d_4^{20} 0.919. bp 227°. n_D^{20} 1.4468. $[\alpha]_D^{20}$ −79.42°. Slightly sol in water; miscible with alcohol, ether.

USE: In perfumery; emphasizes floral notes, especially that of rose, used in toilet waters having a lavender odor. Has been suggested for flavoring extracts having caraway or mint flavors.

5908. Menthyl Anthranilate. [134-09-8] 5-Methyl-2-(1-methylethyl)cyclohexanol 1-(2-aminobenzoate); anthranilic acid *p*-menth-3-yl ester; menthyl *o*-aminobenzoate; meradimate; Neo Heliopan MA. $C_{17}H_{25}NO_2$; mol wt 275.39. C 74.14%, H 9.15%, N 5.09%, O 11.62%. UV-A absorber in cosmetics and sunscreens. Prepn: H. G. Rule, W. E. MacGillivray, *J. Chem. Soc.* **1929**, 401; M. S. Carpenter, **US 2170185** (1939 to Givaudan-Delawanna). Photophysical properties and use as sunscreen: A. Beeby, A. E. Jones, *Photochem. Photobiol.* **72**, 10 (2000). Photostability study: O. Ozer *et al.*, *Cosmet. Toiletries* **116**, 67 (2001). Review of sunscreen formulations: K. Klein, H. A. Finkelmeier, *ibid.* **105**, 75-77 (1990).

Colorless or pale yellow oil with distinct bluish fluorescence, bp_3 177-179° (Carpenter). Also reported as mp 62.5-63.5°; $bp_{0.33}$ 156° (Rule, MacGillivray). d^{35} 1.037. Sol in oil. Dissolves readily in alcohol, dilute aq acids. uv max (ethanol): 220, 249, 340 nm. Fluorescence emission max (ethanol): 405 nm.

THERAP CAT: Ultraviolet screen.

5909. Menthyl Isovalerate. [28221-20-7]; [89-47-4] (*dl*-form); [16409-46-4] (unspecified stereo). 3-Methylbutanoic acid (1*R*,2*S*,5*R*)-5-methyl-2-(1-methylethyl)cyclohexyl ester; isovaleric acid *l*-menthyl ester; *l*-menthyl valerate. $C_{15}H_{28}O_2$; mol wt 240.39. C 74.95%, H 11.74%, O 13.31%. Naturally occurring component of peppermint oil. Isoln from American peppermint oil: F. B. Power, C. Kleber, *Arch. Pharm.* **232**, 636 (1894). Prepn from *l*-menthol and valeryl chloride: H. Rupe, *Ann.* **369**, 311 (1910). Biosynthesis in *Mentha piperita* L.: H. Rothbächer, *Pharmazie* **23**, 389 (1968). Prepn and NMR spectra: A. Ahmad *et al.*, *J. Essent. Oil Res.* **12**, 775 (2000).

Colorless liquid; menthol and valerian odor; cooling, faintly bitter taste. bp 241°; bp_9 129°. d^{25} 0.910; d^{15} 0.906-0.908. n_D^{20} 1.4485-1.4486. $[\alpha]_D^{26}$ −58.39°. Insol in water. Freely sol in alcohol, chloroform, ether, oils. Dec by alkalies.

Mixture with menthol. Validol. GLC analysis in tablets: G. B. Golubitskii *et al.*, *J. Anal. Chem.* **63**, 65 (2008).

USE: Pharmaceutic aid (flavor). Flavor in beverages, candies, baked goods.

THERAP CAT: Mixture with menthol in treatment of angina.

5910. Menyanthes. Buck bean; bog bean; marsh trefoil; water shamrock. Dried leaves or roots of *Menyanthes trifoliata* L., *Gentianaceae*. *Habit.* Europe, Asia, N. America. *Constit.* Menyanthin.

5911. Meobentine. [46464-11-3] *N*-[(4-Methoxyphenyl)-methyl]-*N′,N″*-dimethylguanidine; 1-(*p*-methoxybenzyl)-2,3-dimethylguanidine. $C_{11}H_{17}N_3O$; mol wt 207.28. C 63.74%, H 8.27%, N 20.27%, O 7.72%. Antidysrhythmic, antifibrillatory deriv of guanidine, *q.v.* Prepn of the sulfate: R. A. Maxwell, E. Walton, **DE 2030693** corresp to **US 3949089** (1971, 1976 both to Burroughs Wellcome). Antidysrhythmic effects and tissue concentration: K. B. Touw *et al.*, *Pharmacologist* **19**, 268 (1977); K. B. Touw, *Diss. Abstr. B* **39**, 5340 (1979). Pharmacokinetics by radioimmunoassay:

J. W. A. Findlay *et al.*, *Pharmacologist* **21**, 337 (1979). Pharmacological study: W. B. Wastila *et al.*, *J. Pharm. Pharmacol.* **33**, 594 (1981).

Sulfate. [58503-79-0] Rythmatine. $(C_{11}H_{17}N_3O)_2.H_2SO_4$; mol wt 512.63. Cryst, mp 273-274°.

THERAP CAT: Antiarrhythmic (class III).

5912. Mepanipyrim. [110235-47-7] 4-Methyl-*N*-phenyl-6-(1-propyn-1-yl)-2-pyrimidinamine; 2-anilino-4-methyl-6-(1-propynyl)pyrimidine; *N*-(4-methyl-6-prop-1-ynylpyrimidin-2-yl)aniline; KIF-3535; KUF-6201; Frupica. $C_{14}H_{13}N_3$; mol wt 223.28. C 75.31%, H 5.87%, N 18.82%. Antifungal for use in food crops; inhibits secretion of host-cell wall degrading enzymes. Prepn: S. Ito *et al.*, **EP 224339**; *eidem*, **US 4814338** (1987, 1989 both to Kumiai; Ihara); and biological activity: S. Hayashi *et al.*, *J. Pestic. Sci.* **22**, 165 (1997). Fungicidal activity and field trials: S. Maeno *et al.*, *Brighton Crop Prot. Conf. - Pests Dis.* **1990**, 415. Mode of action study: I. Miura *et al.*, *Pestic. Biochem. Physiol.* **48**, 222 (1994). GC-MS determn in grapes, must and wine: P. Cabras *et al.*, *J. AOAC Int.* **81**, 1185 (1998). Metabolism in tomato seedlings: M. Ikeda *et al.*, *J. Pestic. Sci.* **23**, 9 (1998).

White solid with two crystalline modifications. mp 125-126° (Maeno); also reported as 132.8° (Hayashi). Specific gravity 1.2025. Soly in water at 20°: 5.58 mg/l. Sol in most organic solvents. Vapor pressure at 20°: 1.03×10^{-5} torr. Log P (octanol/water): 3.42. LD_{50} in mice, rats, bobwhite, mallard (mg/kg): >5000, >5000, >2250, >2250 orally; in rats (mg/kg): >2000 dermally. LC_{50} in bluegill, rainbow trout (mg/l): 3.8, 3.1 (Maeno).

USE: Agricultural fungicide.

5913. Meparfynol. [77-75-8] 3-Methyl-1-pentyn-3-ol; ethyl ethynyl methyl carbinol; 2-ethynyl-2-butanol; methylparafynol; methylpentynol; Allotropal; Dorison; Dormalest; Dormidin; Dormigen; Dormiphen; Dormison; Dormosan; Formison; Hesofen; Hexofen; Imnudorm; Oblivon; Pentadorm; Perlopal; Riposon; Seral; Somnesin. $C_6H_{10}O$; mol wt 98.15. C 73.42%, H 10.27%, O 16.30%. Prepd from methyl ethyl ketone and sodium acetylide in liquid ammonia or by a Grignard reaction: **DE 285770** (1913 to Bayer); **DE 289800**; **DE 291185**; Sung Wouseng, *Ann. Chim.* [10] **1**, 343 (1924); Rupe, Vonaesch, *Ann.* **442**, 80 (1925); Carothers, Coffman, *J. Am. Chem. Soc.* **54**, 4071 (1932); Campbell *et al.*, *ibid.* **60**, 2882 (1938); Campbell, Campbell, *Proc. Indiana Acad. Sci.* **50**, 123-127 (1940); Smith, **US 2385547** (1945 to Commercial Solvents); A. W. Johnson, *The Chemistry of the Acetylenic Compounds* vol. I (Edward Arnold & Co., London, 1946) p 278; Hurd, McPhee, *J. Am. Chem. Soc.* **69**, 239 (1947); P. Piganiol, *Acetylene Homologs and Derivatives* (Brooklyn, 1950); Papa *et al.*, *Arch. Biochem. Biophys.* **33**, 482 (1951); Thiele, Martinez, *Ciencia (Mexico City)* **15**, 70 (1955). An optically active form has been described: Hickmann, Kenyon, *J. Chem. Soc.* **1955**, 2051. Toxicity data: Margolin *et al.*, *Science* **114**, 384 (1951).

Mobile liquid. Acrid odor, burning taste. d_4^{20} 0.8688. d_{20}^{20} 0.8721; 7.28 lbs/U.S. gal. bp_{760} 121-122°; bp_{37} 50°; $bp_{6.5}$ 20°. mp -30.6°. n_D^{20} 1.4318. Flash pt 101.3°F. Surface tension at 25°: 23.8 dynes/cm; 5% aq soln: 34.1 dynes/cm. Soly in water at 25°: 12.8 g/100 ml. Sol in ether. Miscible with acetone, benzene, carbon tetrachloride, Cellosolve, cyclohexanone, diethylene glycol, ethyl acetate, kerosene, methyl ethyl ketone, mineral spirits, ethanolamine, neatsfoot oil, petr ether, soybean oil, Stoddard solvent. LD_{50} orally in mice, rats, guinea pigs: 600-900 mg/kg (Margolin).

Carbamate. [302-66-9] N-Oblivon; Oblivon C; Trusono. $C_7H_{11}NO_2$; mol wt 141.17. Prepn: McLamore *et al.*, *J. Org. Chem.* **20**, 1379 (1955); from 3-methyl-1-pentyn-3-ol, trichloroacetic acid, KOCN: McCrea *et al.*, **GB 761817** (1956 to British Schering). Acute toxicity: Soehring *et al.*, *Arzneim.-Forsch.* **5**, 161 (1955). Crystals from ether + petr ether. mp 55.8-57° (McLamore); from cyclohexane, mp 53.5-55° (McCrea). bp_{16} 120-121°; $bp_{0.01}$ 95°. Solubility in water: 1.6 g/100 ml. LD_{50} (4 hr) s.c. in mice: 0.56 g/kg (Soehring).

THERAP CAT: Hypnotic; sedative.

5914. Mepartricin. [11121-32-7] Partricin methyl ester; methylpartricin; SN-654; SPA-S-160; Ipertrofan; Orofungin; Tricandil; Tricangine. Methyl ester of the heptaene macrolide antibiotic complex, partricin, *q.v.* Prepn: T. Bruzzese *et al.*, *Experientia* **28**, 1515 (1972); T. Bruzzese, R. Ferrari, **DE 2154436** (1972 to SPA); *eidem*, **US 3780173** (1973). Alternate prepn: R. C. Pandey, *J. Antibiot.* **30**, 158 (1977); R. C. Tweit *et al.*, *ibid.* **35**, 997 (1982). Structure of mepartricins A and B: J. Golik *et al.*, *ibid.* **33**, 904 (1980). Antimycotic activity: W. Ritzerfield, *Farmaco Ed. Sci.* **27**, 235 (1972); activity vs *Trichomonas vaginalis*: G. Pucci, S. Ripa, *Farmaco Ed. Prat.* **28**, 293 (1973). Hemolytic activity studies: B. Cybulska *et al.*, *Biochem. Pharmacol.* **33**, 41 (1984). Stability data: S. Mizuba, K. Lee, *Dev. Ind. Microbiol.* **16**, 380 (1975). Use of mepartricin and its complexes in treatment of benign prostatic hypertrophy: T. Bruzzese, L. Ferrari, **US 4237117** (1980 to SPA). Clinical studies: M. De Bernardi *et al.*, *Curr. Ther. Res.* **43**, 1159 (1988); T. Lotti *et al.*, *ibid.* **44**, 402 (1988).

Mepartricin A R' = CH_3
Mepartricin B R' = H

Deep yellow crystalline material. uv max (ethanol): 401, 378, 359, 340 nm. Slightly sol in water, aq alkali; sol in petr ether, benzene. Sol in acetone, alcohols, pyridine, DMF, DMSO. LD_{50} in mice: >2 g/kg orally; 200 mg/kg i.p. (Bruzzese, 1972).

Mepartricin A. [62534-68-3] 40-Demethyl-3,7-dideoxo-3,7-dihydroxy-N^{47}-methyl-5-oxocandicidin D methyl ester cyclic 15,19-hemiacetal; gedamycin methyl ester. $C_{60}H_{88}N_2O_{19}$; mol wt 1141.36. Lemon yellow powder from ether, mp 145-149° (dec). uv max (methanol): 400, 377, 357, 339, 287, 240, 234, 204 nm (ε 79326, 92454, 68094, 51685, 14199, 24612, 26505, 16092).

Mepartricin B. [62534-69-4] $C_{59}H_{86}N_2O_{19}$; mol wt 1127.33. Lemon yellow powder from ether, mp 154-158° (dec). uv max (methanol): 402, 379, 359, 340, 285, 233, 204 nm (ε 81101, 94729, 64171, 41558, 16196, 23835, 21696).

Complex with sodium lauryl sulfate. SPA-S-222; Montricin.

THERAP CAT: Antifungal; antiprotozoal (Trichomonas). In treatment of benign prostatic hypertrophy.

5915. Mepenzolate Bromide. [76-90-4] 3-[(2-Hydroxy-2,2-diphenylacetyl)oxy]-1,1-dimethylpiperidinium bromide (1:1); *N*-methyl-3-piperidyl benzilate methyl bromide; *N*-methyl-3-piperidyl diphenylglycolate methobromide; Cantil; Trancolon. $C_{21}H_{26}$-

BrNO$_3$; mol wt 420.35. C 60.00%, H 6.23%, Br 19.01%, N 3.33%, O 11.42%. Anticholinergic. Prepn: Biel, US 2918408 (1959 to Lakeside Labs.). Prepn of base: Biel *et al.*, *J. Am. Chem. Soc.* 77, 2250 (1955). Crystal structure: J. M. Leger *et al.*, *Acta Crystallogr.* B35, 886 (1979). Toxicity data: E. I. Goldenthal, *Toxicol. Appl. Pharmacol.* 18, 185 (1971).

Crystals, mp 228-229° (dec). LD$_{50}$ orally in rats: 742±47 mg/kg (Goldenthal).

Free base. [25990-43-6] bp$_{0.03}$ 175-176°.

Note: N-Methyl-3-piperidyl benzilate is a controlled substance (hallucinogen): 21 CFR, 1308.11.

THERAP CAT: Antispasmodic.

5916. Meperidine. [57-42-1] 1-Methyl-4-phenyl-4-piperidinecarboxylic acid ethyl ester; 1-methyl-4-phenylisonipecotic acid ethyl ester; N-methyl-4-phenyl-4-carbethoxypiperidine; ethyl 1-methyl-4-phenylpiperidine-4-carboxylate; isonipecaine; pethidine. C$_{15}$H$_{21}$NO$_2$; mol wt 247.34. C 72.84%, H 8.56%, N 5.66%, O 12.94%. Opioid analgesic. Prepn: Eisleb, US 2167351 (1939 to Winthrop); Smissman, Hite, *J. Am. Chem. Soc.* 81, 1201 (1959). Pharmacokinetics: L. E. Mather, P. J. Meffin, *Clin. Pharmacokinet.* 3, 352 (1978). Metabolism: S. Y. Yeh *et al.*, *J. Pharm. Sci.* 70, 867 (1981). GC-MS/MS determn in biological fluids: A. Ishii *et al.*, *J. Chromatogr. B* 792, 117 (2003). Toxicity data: O. W. Barlow, J. R. Lewis, *J. Pharmacol. Exp. Ther.* 103, 147 (1951). Comprehensive description: N. P. Fish, N. J. DeAngelis in *Anal. Profiles Drug Subs.* 1, 175-205 (1972).

Hydrochloride. [50-13-5] Algil; Alodan; Centralgin; Demerol hydrochloride; Dispadol; Dolantin; Dolestine; Dolosal; Mefedina. C$_{15}$H$_{21}$NO$_2$.HCl; mol wt 283.80. White, minute crystals, mp 186-189°. Slightly bitter taste. Stable to air. Just acid to litmus (1:10 soln). Very sol in water; sol in alcohol, acetone, ethyl acetate; sparingly sol in ether; slightly sol in isopropanol. Insol in benzene. Aq solns may be sterilized by boiling for short periods without dec. LD$_{50}$ orally in rats: 170 mg/kg (Barlow, Lewis).

Note: This is a controlled substance (opiate): 21 CFR, 1308.12.

THERAP CAT: Analgesic.

THERAP CAT (VET): Analgesic; sedative; anesthetic.

5917. Mephenesin. [59-47-2] 3-(2-Methylphenoxy)-1,2-propanediol; 3-(o-tolyloxy)-1,2-propanediol; 1,2-dihydroxy-3-(2-methylphenoxy)propane; α-(o-tolyl)glyceryl ether; glyceryl o-tolyl ether; o-cresyl glycerol ether; cresoxypropanediol; cresoxydiol; α,β-dihydroxy-γ-(2-methylphenoxy)propane; BDH-312; Atensin; Avosyl; Avoxyl; Curythan; Daserol; Decontractyl; Dioloxol; Glyotol; Glykresin; Lissephen; Memphenesin; Mepherol; Mephesin; Mephson; Mervaldin; Myanesin; Myanol; Myodetensine; Myolysin; Myopan; Myoserol; Myoten; Oranixon; Prolax; Relaxar; Relaxil; Renarcol; Rhex; Sansdolor; Sinan; Spasmolyn; Stilalgin; Thoxidil; Tolansin; Tolcil; Tolhart; Tolosate; Toloxyn; Tolserol; Tolulexin; Tolulox; Tolyspaz; Walconesin. C$_{10}$H$_{14}$O$_3$; mol wt 182.22. C 65.91%, H 7.74%, O 26.34%. Prepn: P. Morch, *Arch. Pharm. Chemi* 54, 327

(1947), *C.A.* 42, 2058 (1948); GB 589821 (1947 to Carroll and Boake Roberts); of the carbamate: W. A. Lott, E. Pribyl, US 2609386 (1952 to Squibb). Pharmacology: F. M. Berger, W. Bradley, *Br. J. Pharmacol.* 1, 265 (1946); and toxicity data: P. E. Dresel, I. H. Slater, *Proc. Soc. Exp. Biol. Med.* 79, 286 (1952); A. P. Roszkowski, *J. Pharmacol. Exp. Ther.* 129, 75 (1960).

Crystals. Bitter taste. Produces numbness of the tongue. mp 70-71°. uv max (0.005% aq soln): 270 nm (E 0.395). Freely sol in alcohol, propylene glycol, chloroform. At 20° one part dissolves in 85 parts water, in 11 parts ether. Urea and its derivatives, particularly urethan, increase the water soly. One part dissolves in 60 parts of 5% urethan soln, in 40 parts of 10%, in 4.5 parts of 25%. Aq solns are stable, can be sterilized by heating, and are compatible and freely miscible with solns of sodium chloride, glucose, and derivatives of barbituric and thiobarbituric acids. pH of satd aq soln ~6. LD$_{50}$ in mice, rats, hamsters (mg/kg): 471, 283, 322 i.p.; 990, 945, 821 orally (Roszkowski). LD$_{50}$ in mice (mM/kg): 2.83 i.p.; 10.53 orally (Dresel, Slater).

Carbamate. [533-06-2] Tolseram. C$_{11}$H$_{15}$NO$_4$; mol wt 225.24. Crystals from water, mp about 93°; also a hemihydrate, mp 80-84°. uv max (ethanol): 271, 277 nm (A$_{1cm}^{1\%}$ 72.7, 64.1). Has a lower water-solubility and a higher oil-solubility than mephenesin. Soly in water about 0.3%, in chloroform about 2.0%. Freely sol in alcohol. LD$_{50}$ in mice (mM/kg): 2.77 i.p.; 7.67 orally (Dresel, Slater).

THERAP CAT: Muscle relaxant (skeletal).

THERAP CAT (VET): Muscle relaxant.

5918. Mephenoxalone. [70-07-5] 5-[(2-Methoxyphenoxy)-methyl]-2-oxazolidinone; 5-(o-methoxyphenoxymethyl)-2-oxazolidinone; metoxadone; methoxydon(e); methoxadone; AHR-233; OM-518; Dorsiflex; Control-Om; Xerene. C$_{11}$H$_{13}$NO$_4$; mol wt 223.23. C 59.19%, H 5.87%, N 6.27%, O 28.67%. Prepn: Lunsford *et al.*, *J. Am. Chem. Soc.* 82, 1166 (1960); Lunsford, US 2895960 (1959 to A. H. Robins). Toxicity data: E. I. Goldenthal, *Toxicol. Appl. Pharmacol.* 18, 185 (1971). Determn in pharmaceutical formulations: N. Erk, *J. Pharm. Biomed. Anal.* 21, 429 (1999).

Crystals from 95% ethanol, mp 143-145°. Water insol. LD$_{50}$ orally in rats: 3820 ±17 mg/kg (Goldenthal).

THERAP CAT: Muscle relaxant (skeletal).

5919. Mephentermine. [100-92-5] N,α,α-Trimethylbenzeneethanamine; 2-methylamino-2-methyl-1-phenylpropane; N-methyl-ω-phenyl-*tert*-butylamine; N,α,α-trimethylphenethylamine. C$_{11}$H$_{17}$N; mol wt 163.26. C 80.93%, H 10.50%, N 8.58%. α-Adrenergic agonist. Prepd by hydrogenating 1-chloro-2-methylamino-2-methyl-1-phenylpropane hydrochloride in methanol in the presence of palladium barium carbonate: Bruce, Szabo, US 2597445 (1952 to Wyeth). Alternate process: Abel *et al.*, US 2590079 (1952 to Wyeth). Pharmacology: D. K. Eckfeld *et al.*, *J. Am. Pharm. Assoc.* 43, 705 (1954).

Liquid. Fishy amine odor. bp$_{7mm}$ 82-83°. n_D^{20} 1.5110. d$_{20}^{20}$ 0.9231. pKa (30°) 10.35. Alkaline reaction. Freely sol in alc. Sol in ether. Practically insol in water.

Sulfate dihydrate. [6190-60-9] Wyamine; Mephine. $(C_{11}H_{17}-N)_2.H_2SO_4.2H_2O$; mol wt 460.63. Crystals, mp 215-217° (dec). One gram dissolves in 20 ml water, about 150 ml ethanol (95%). Practically insol in chloroform. pH of aq soln about 6. LD_{50} (calc as base) intra-abdominally in mice: 89.4 mg/kg (Eckfeld).

THERAP CAT: Antihypotensive.

5920. Mephenytoin. [50-12-4] 5-Ethyl-3-methyl-5-phenyl-2,4-imidazolidinedione; 5-ethyl-3-methyl-5-phenylhydantoin; 3-methyl-5,5-phenylethylhydantoin; "methyl hydantoin"; phenylethylmethylhydantoin; 3-ethylnirvanol; methoin; Insulton; Mesontoin; Mesantoin; Phenantoin; Sedantoinal; Gerot-Epilan; Sacerno. $C_{12}-H_{14}N_2O_2$; mol wt 218.26. C 66.04%, H 6.47%, N 12.84%, O 14.66%. Prepn: **FR 769667**; **CH 166004** (both 1934 to Sandoz); **CH 179692** (1935 to Sandoz). Pharmacology: I. A. Dzhagatspanyan *et al.*, *Pharm. Chem. J.* **25**, 181 (1991).

Crystals, mp 136-137°. Insol in water. Forms a water-soluble sodium salt which has an alkaline reaction. LD_{50} i.p. in mice: 300 mg/kg (Dzhagatspanyan).

THERAP CAT: Anticonvulsant.

5921. Mephobarbital. [115-38-8] 5-Ethyl-1-methyl-5-phenyl-2,4,6($1H,3H,5H$)-pyrimidinetrione; 5-ethyl-1-methyl-5-phenylbarbituric acid; 5-phenyl-5-ethyl-3-methylbarbituric acid; *N*-methylethylphenylbarbituric acid; methylphenobarbital; Phemiton; Prominal; Mebaral. $C_{13}H_{14}N_2O_3$; mol wt 246.27. C 63.40%, H 5.73%, N 11.38%, O 19.49%. Prepn: Taub, Kropp, **DE 537366** (1929 to I. G. Farben.).

White, tasteless crystals. mp 176°. Freely sol in hot water; sol in chloroform and in sols of fixed alkali hydroxides and carbonates; slightly sol in cold water, alc, ether.

Note: This is a controlled substance (depressant): **21 CFR, 1308.14.**

THERAP CAT: Anticonvulsant; sedative; hypnotic.
THERAP CAT (VET): Anticonvulsant; sedative; hypnotic.

5922. Mephosfolan. [950-10-7] *N*-(4-Methyl-1,3-dithiolan-2-ylidene)phosphoramidic acid diethyl ester; phosphonodithioimidocarbonic acid cyclic propylene *P*,*P*-diethyl ester; 2-diethoxyphosphinylimino-4-methyl-1,3-dithiolane; cyclic propylene (diethoxyphosphinyl)dithioimidocarbonate; EI-47470; ENT-25991; Cytrolane. $C_8H_{16}NO_3PS_2$; mol wt 269.31. C 35.68%, H 5.99%, N 5.20%, O 17.82%, P 11.50%, S 23.81%. Insecticidal contact and stomach poison with systemic activity: D. L. Bull *et al.*, *J. Econ. Entomol.* **57**, 112 (1964). Prepn: J. B. Lovell, **US 3197365** (1965 to Am. Cyanamid). Metabolism: J. Zulalian, R. C. Blinn, *J. Agric. Food Chem.* **25**, 1033 (1977). Photodegradation: C. C. Ku *et al.*, *ibid.* **27**, 1046 (1979).

Yellow to amber liquid, $bp_{0.001}$ 120°. n_D^{26} 1.5354. Soly in water at 25°: 57 g/kg. Sol in acetone, ethanol, benzene, 1,2-dichloroethane. Stable at neutral pH; hydrolyzed by acid or alkali.

USE: Insecticide, acaricide.

5923. Mepindolol. [23694-81-7] 1-[(1-Methylethyl)amino]-3-[(2-methyl-1*H*-indol-4-yl)oxy]-2-propanol; 1-(isopropylamino)-3-[(2-methylindol-4-yl)oxy]-2-propanol. $C_{15}H_{22}N_2O_2$; mol wt 262.35. C 68.67%, H 8.45%, N 10.68%, O 12.20%. β-Adrenergic blocker; the 2-methyl analog of pindolol, *q.v.* Prepn: F. Troxler, **CH 469002** and **CH 472404** (both 1969 to Sandoz); F. Seeman *et al.*, *Helv. Chim. Acta* **54**, 2411 (1971). Pharmacokinetics: R. Gugler *et al.*, *Arzneim.-Forsch.* **25**, 1067 (1975). Pharmacodynamics: J. Bonelli *et al.*, *Eur. J. Clin. Pharmacol.* **15**, 1 (1979). Clinical studies: H. M. Beumer *et al.*, *Int. J. Clin. Pharmacol. Biopharm.* **16**, 249 (1978); M. Sukerman *et al.*, *Curr. Ther. Res.* **25**, 384 (1979).

Crystals from ethyl acetate, mp 100-102° (Seeman). Also reported as 95-97° (Troxler).

Sulfate salt. [56396-94-2] SH-E-222; Betagon; Corindolan; Mepicor. $(C_{15}H_{22}N_2O_2)_2.H_2SO_4$; mol wt 622.78.

THERAP CAT: Antihypertensive; antianginal.

5924. Mepiquat Chloride. [24307-26-4] 1,1-Dimethylpiperidinium chloride (1:1); BAS-083; BAS-85559X; Pix. $C_7H_{16}ClN$; mol wt 149.66. C 56.18%, H 10.78%, Cl 23.69%, N 9.36%. Prepn: B. Zeeh *et al.*, **DE 2207575**; *eidem*, **US 3905798** (1973, 1975 both to BASF). IC determn: A. Fegert *et al.*, *Fresenius J. Anal. Chem.* **339**, 441 (1991). Toxicity study: BASF Agric. Chem. Div., *J. Pestic. Sci.* **17**, S269 (1992). Field study on cotton: J. S. McConnell *et al.*, *J. Plant Nutr.* **15**, 457 (1992); R. K. Boman, R. L. Westerman, *J. Prod. Agric.* **7**, 70 (1994). Review: P. E. Schott, F. R. Rittig, in *Chemical Manipulation of Crop Growth and Development* J. S. McLaren, Ed. (Butterworth Scientific, London, 1982) pp 415-424.

White crystals, mp 285°. Soly at 20° g/100g: water >50; methyl alcohol 25.0; ethyl alcohol 16.2; chloroform 1.1; acetone <0.1; benzene <0.1; cyclohexanone <0.1; diethyl ether <0.1; olive oil <0.1. Log P (*n*-octanol:water): −2.83. LD_{50} in mice, rats (mg/kg): 780, 464 orally; in rats (mg/kg): >2000 dermally (BASF Agric. Chem. Div.).

USE: Plant growth regulator.

5925. Mepitiostane. [21362-69-6] $(2\alpha,3\alpha,5\alpha,17\beta)$-2,3-Epithio-17-[(1-methoxycyclopentyl)oxy]androstane; cyclopentanone $2\alpha,3\alpha$-epithio-5α-androstan-17β-yl methyl acetal; 10364-S; Thiodelone; Thioderon. $C_{25}H_{40}O_2S$; mol wt 404.65. C 74.21%, H 9.96%, O 7.91%, S 7.92%. Anabolic steroid. Orally active deriv of epitiostanol, *q.v.* Prepn: T. Komeno, **ZA 6800565** corresp to **US 3567713** (1968, 1971 both to Shionogi); T. Komeno *et al.*, *Shionogi Kenkyusho Nempo* **19**, 3 (1969), *C.A.* **72**, 28554u (1970). Inhibitory effect on mammary tumors in rats: O. Takatani, S. Kumaoka, *Gann* **68**, 337 (1977), *C.A.* **87**, 112187u (1977). Clinical study in advanced breast cancer: K. Inoue *et al.*, *Cancer Treat. Rep.* **62**, 743 (1978). Toxicity studies: *Oyo Yakuri* **16**, 739-812 (1978), *C.A.* **90**, 146152f, 146153g (1979).

Cryst, mp 98-101°. [α]$_D^{20}$ +22.5 ±0.5° (c = 1 in chloroform).
THERAP CAT: Antineoplastic.

5926. Mepivacaine. [96-88-8] *N*-(2,6-Dimethylphenyl)-1-methyl-2-piperidinecarboxamide; 1-methyl-2′,6′-pipecoloxylidide; *dl-N*-methylpipecolic acid 2,6-dimethylanilide; *dl-N*-methylhexahydropicolinic acid 2,6-dimethylanilide. $C_{15}H_{22}N_2O$; mol wt 246.35. C 73.13%, H 9.00%, N 11.37%, O 6.49%. Prepn: Ekenstam *et al.*, *Acta Chem. Scand.* **11**, 1183 (1957); US 2799679 (1957 to A. B. Bofors). Prepn of other salts: Rinderknecht, *Helv. Chim. Acta* **42**, 1324 (1959); GB 826668 (1960 to Crookes Labs.). Resolution of isomers: Tullar, *J. Med. Chem.* **14**, 891 (1971); Friberger, Aberg, *Acta Pharm. Suec.* **8**, 361 (1971). Pharmacology: Helmy *et al.*, *J. Egypt. Med. Assoc.* **50**, 688 (1967). Metabolism: Reynolds, *Br. J. Anaesth.* **43**, 33 (1971). Toxicity data: G. Aberg, *Acta Pharmacol. Toxicol.* **31**, 273 (1972).

Crystals from ether, mp 150-151°.
Hydrochloride. [1722-62-9] Carbocaina; Carbocaine hydrochloride; Chlorocain; Meaverin; Mepicaton; Mepident; Mepivastesin; Optocain; Scandicain. $C_{15}H_{22}N_2O.HCl$; mol wt 282.81. mp 262-264°. Sol in water. LD$_{50}$ in mice, rats (mg/kg): 280, 500 s.c. (Aberg).
(+)-Form. Dexivacaine.
THERAP CAT: Anesthetic (local).
THERAP CAT (VET): Anesthetic (local).

5927. Mepixanox. [17854-59-0] 3-Methoxy-4-(1-piperidinylmethyl)-9*H*-xanthen-9-one; 3-methoxy-4-piperidinylmethyl-9-oxo-10-oxa-9,10-dihydroanthracene; mepixanthone; Pimexone. $C_{20}H_{21}NO_3$; mol wt 323.39. C 74.28%, H 6.55%, N 4.33%, O 14.84%. Prepn: M. Sparaci, ZA 6902150 (1969 to Mondi), *C.A.* **72**, 111,300d (1970). Alternate process: D. Milani, US 3646030 (1972 to Mondi). CNS activity: P. Da Re *et al.*, *J. Med. Chem.* **13**, 527 (1970). Pharmacology: R. Guira *et al.*, *Minerva Pneumolog.* **21**, 317 (1982). Pharmacokinetics in man: G. Grossi *et al.*, *Curr. Ther. Res.* **38**, 141 (1985). HPLC determn in serum: G. Grossi *et al.*, *J. Chromatogr.* **309**, 214 (1984). Clinical comparison with doxapram, *q.v.*, in chronic bronchopulmonary disease: M. Parziale *et al.*, *G. Ital. Mal. Torace* **38**, 323 (1984). Clinical evaluations in chronic obstructive lung disease: D. Olivieri *et al.*, *Arch. Monaldi Tisiol. Mal. Appar. Respir.* **35**, 117 (1980); L. Bertoli *et al.*, *Minerva Pneumolog.* **23**, 323 (1984).

White crystalline powder from ethyl acetate, mp 159-160°. LD$_{50}$ i.p. in mice: 70.73 mg/kg (Sparaci).
Hydrochloride. $C_{20}H_{22}ClNO_3$. White crystalline solid from ethanol, mp >200° (dec).
THERAP CAT: Respiratory stimulant.

5928. Meprednisone. [1247-42-3] (16β)-17,21-Dihydroxy-16-methylpregna-1,4-diene-3,11,20-trione; 16β-methylprednisone; Betapred; Betapar; Betalone; Deltisona B. $C_{22}H_{28}O_5$; mol wt 372.46. C 70.95%, H 7.58%, O 21.48%. Prepn: E. P. Oliveto *et al.*, *J. Am. Chem. Soc.* **80**, 4428 (1958); D. Taub *et al.*, *ibid.* 4435; D. Taub *et al.*, *ibid.* **82**, 4012 (1960); G. G. Nathansohn *et al.*, *Experientia* **17**, 448 (1961); GB 901092 (1962 to Scherico); R. Rausser, E. P. Oliveto, US 3164618 (1965 to Schering).

Crystals, mp 200-205°. [α]$_D$ +200° (dioxane). uv max (methanol): 239 nm (E$_{1cm}^{1\%}$ 416).
21-Acetate. $C_{24}H_{30}O_6$. mp 232-235°. [α]$_D$ +210° (dioxane). uv max (methanol): 238 nm (E$_{1cm}^{1\%}$ 358).
THERAP CAT: Glucocorticoid.

5929. Meprobamate. [57-53-4] 2-Methyl-2-propyl-1,3-propanediol 1,3-dicarbamate; carbamic acid 2-methyl-2-propyltrimethylene ester; 2,2-di(carbamoyloxymethyl)pentane; 2-methyl-2-propyltrimethylene carbamate; procalmadiol; procalmidol; Andaxin; Artolon; Atraxin; Cyrpon; Ecuanil; Equanil; Mepavlon; Meprodil; Meprospan; Meprotabs; Miltaun; Miltown; Nervonus; Oasil; Perequil; Pertranquil; Probamyl; Quaname; Quanil; Restenil; Trancot; Tranquilan; Urbilat; Visano. $C_9H_{18}N_2O_4$; mol wt 218.25. C 49.53%, H 8.31%, N 12.84%, O 29.32%. Prepn: Ludwig, Piech, *J. Am. Chem. Soc.* **73**, 5779 (1951); Berger, Ludwig, US 2724720 (1955 to Carter Prod.). Pharmacology: F. M. Berger, *J. Pharmacol. Exp. Ther.* **112**, 413 (1954). Comprehensive description: C. Shearer, P. Rulon, *Anal. Profiles Drug Subs.* **1**, 207-232 (1972); C. Shearer, *ibid.* **11**, 587-591 (1982).

Crystals from hot water, mp 104-106°. Characteristic bitter taste. Freely sol in acetone, alc, and most organic solvents. Soly in water at 20°: 0.34% (w/w), at 37°: 0.79% (w/w). Practically insol in ether. Easily forms supersatd solns with hot water. Aq solns are neutral. Stable in dilute acid and alkali. LD$_{50}$ i.p. in mice: 800 mg/kg (Berger).
Note: This is a controlled substance (depressant): 21 CFR, 1308.14.
THERAP CAT: Anxiolytic.

5930. Meptazinol. [54340-58-8] 3-(3-Ethylhexahydro-1-methyl-1*H*-azepin-3-yl)phenol; 1-methyl-3-ethyl-3-(*m*-hydroxyphenyl)hexahydro-1*H*-azepine. $C_{15}H_{23}NO$; mol wt 233.36. C 77.20%, H 9.93%, N 6.00%, O 6.86%. Mixed opioid agonist-antagonist. Prepn of the hydrobromide: J. F. Cavalla, A. C. White, DE 1941534; *eidem*, US 3729465 (1970, 1973 both to Wyeth); of the free base: *eidem*, US 4197241 (1980 to Wyeth). Properties: P. G. Goode, A. C. White, *Br. J. Pharmacol.* **43**, 462 (1971). HPLC determn in plasma: T. Frost, *Analyst* **106**, 999 (1981). Clinical pharmacokinetics: G. Davies, *Eur. J. Clin. Pharmacol.* **23**, 535 (1982). Pharmacology and abuse potential: R. E. Johnson, D. R. Jasinski, *Clin. Pharmacol. Ther.* **41**, 426 (1987). Symposium on clinical studies: *Postgrad. Med. J.* **59**, Suppl. 1, 1-94 (1983). Review of pharmacology and clinical efficacy: B. Holmes, A. Ward, *Drugs* **30**, 285-312 (1985).

Cryst from acetonitrile, mp 127.5-133°.

Hydrochloride. [59263-76-2] Wy-22811; Meptid. $C_{15}H_{23}NO\cdot$HCl; mol wt 269.81.

THERAP CAT: Analgesic.

5931. Mequitazine. [29216-28-2] 10-(1-Azabicyclo[2.2.2]oct-3-ylmethyl)-10*H*-phenothiazine; 10-(3-quinuclidinylmethyl)-phenothiazine; LM-209; Butix; Metaplexan; Mircol; Primalan; Zesulan. $C_{20}H_{22}N_2S$; mol wt 322.47. C 74.49%, H 6.88%, N 8.69%, S 9.94%. Prepn: G. Gueremy *et al.*, **DE 2009555** corresp to **US 3987042** (1970, 1976 to Sogeras). Absorption, distribution, and excretion: A. Uzan *et al.*, *Xenobiotica* **6**, 633 (1976). Biotransformation: *eidem, ibid.* 649. Pharmacologic study: *eidem, J. Pharm. Pharmacol.* **31**, 701 (1979). Clinical studies: J. Blamoutier, *Curr. Med. Res. Opin.* **5**, 366 (1978); P. Laugier, M. Orusco, *ibid.* 371.

Crystals from acetonitrile, mp 130-131°.

THERAP CAT: Antihistaminic.

5932. Meralein Sodium. [4386-35-0] (3′,6′-Dihydroxy-2′,7′-diiodo-1,1-dioxidospiro[3*H*-2,1-benzoxanthiole-3,9′-[9*H*]xanthen]-4′-yl)hydroxymercury sodium salt (1:1); *o*-[6-hydroxy-5-(hydroxymercuri)-2,7-diiodo-3-oxo-3*H*-xanthen-9-yl]benzenesulfonic acid sodium salt; 2,7-diiodo-4-hydroxymercuriresorcinsulfonphthalein monosodium salt; monohydroxymercuridiiodoresorcinsulfonphthalein sodium salt; sodium meralein; Merodicein. $C_{19}H_9HgI_2$-NaO_7S; mol wt 858.72. C 26.58%, H 1.06%, Hg 23.36%, I 29.56%, Na 2.68%, O 13.04%, S 3.73%. Prepn: Dunning, Farinholt, *J. Am. Chem. Soc.* **51**, 804 (1929). Pharmacology and toxicology: Macht, Cook, *J. Pharmacol. Exp. Ther.* **43**, 571 (1931).

Green scales, turn dark red on pulverizing. Sol in water; aq soln slightly fluorescent.

THERAP CAT: Antiseptic, disinfectant.

5933. Meralluride. [8069-64-5] [3-[[[(3-Carboxylato-1-oxopropyl)amino]carbonyl-*κO*]amino]-2-methoxypropyl-*κC*]hydroxymercurate(1−) sodium (1:1) mixture with 3,7-dihydro-1,3-dimethyl-1*H*-purine-2,6-dione; *N*-[2-methoxy-3-[(1,2,3,6-tetrahydro-1,3-dimethyl-2,6-dioxopurin-7-yl)mercuri]propyl]carbamoyl]succinamic acid; Mercuhydrin; Mercuretin. $C_{16}H_{23}HgN_6NaO_8$; mol wt 650.97. C 29.52%, H 3.56%, Hg 30.81%, N 12.91%, Na 3.53%, O 19.66%.

Structure: Pearson, Sigal, *J. Org. Chem.* **15**, 1055 (1950). Toxicity data: E. I. Goldenthal, *Toxicol. Appl. Pharmacol.* **18**, 185 (1971).

White to slightly yellow powder, slowly dec on exposure to light. Saturated soln is acid to litmus. Slightly sol in water; sol in hot water, glacial acetic acid, and solns of alkali hydroxides. Almost insol in alc, chloroform, ether. LD_{50} s.c. in rats: 28±7 mg/kg (Goldenthal).

THERAP CAT: Diuretic.

THERAP CAT (VET): Diuretic.

5934. Merbromin. [129-16-8] (2′,7′-Dibromo-3′,6′-dihydroxy-3-oxospiro[isobenzofuran-1(3*H*),9′-[9*H*]xanthen]-4′-yl)hydroxymercury sodium salt (1:2); [2,7-dibromo-9-(*o*-carboxyphenyl)-6-hydroxy-3-oxo-3*H*-xanthen-4-yl]hydroxymercury disodium salt; mercurochrome; dibromohydroxymercurifluorescein disodium salt; no. 220 sol; Chromargyre; Planochrome; Flavurol; D.O.M.F.; Mercurophage; Mercurocol; Gallochrome; Gynochrome; Mercurome; Asceptichrome; Mercuranine. $C_{20}H_8Br_2HgNa_2O_6$; mol wt 750.66. C 32.00%, H 1.07%, Br 21.29%, Hg 26.72%, Na 6.13%, O 12.79%. Prepd by treating dibromofluorescein with mercuric acetate and sodium hydroxide: White, *J. Am. Chem. Soc.* **42**, 2355 (1920); also prepd by the action of mercuric acetate on dibromofluorescein sodium: Rymill, Corran, *Q. J. Pharm. Pharmacol.* **7**, 543 (1934); *see also* **US 1535003** (1925).

Trihydrate, iridescent green scales or granules. Freely sol in water, giving a carmine-red soln. Very dil solns (1:2000) possess a yellow-green fluorescence. pH of 0.5% soln: 8.8. One gram dissolves in 50 grams of 94% alcohol, in 8.1 grams methanol: Denoel, *J. Pharm. Belg.* **22**, 423 (1940); *ibid.* **23**, 75 (1941). Practically insol in alcohol, acetone, chloroform, ether. *Incompat.* Acids, most alkaloidal salts and most local anesthetics. Colors the skin carmine-red. The stains may be removed by washing first with permanganate soln and then with oxalic acid soln.

THERAP CAT: Antibacterial.

THERAP CAT (VET): Antiseptic.

5935. 2-Mercaptobenzothiazole. [149-30-4] 2(3*H*)-Benzothiazolethione; 2-benzothiazolethiol; MBT; Captax; Dermacid; Mertax; Thiotax. $C_7H_5NS_2$; mol wt 167.24. C 50.27%, H 3.01%, N 8.38%, S 38.34%. Prepd industrially by reacting aniline, carbon disulfide, and sulfur at elevated press. and temps: Kelly, **US 1631871**. Purification by treatment with a per-acid salt in alkaline medium: Weyker, Ebel, **US 2730528** (1956 to Am. Cyanamid). Improved process: Szlatinay, **US 3031073** (1962 to Monsanto).

Pale yellow, monoclinic needles or leaflets. Disagreeable odor. d 1.42. mp 180.2-181.7° (the technical product mp 170-175°). Practically insol in water. Soly at 25° (g/100 ml) in alcohol: 2.0; ether 1.0; acetone 10.0; benzene 1.0; carbon tetrachloride <0.2; naphtha

5936

<0.5. Moderately sol in glacial acetic acid. Sol in alkalies and alkali carbonate solns.

Zinc salt. Bantex. Light yellow powder, d_4^{25} 1.70.

Sodium salt. Nuodex 84.

USE: Rubber vulcanization accelerator. Salts used as fungicide.

5936. **2-Mercaptoethanol.** [60-24-2] β-Mercaptoethanol; 2-hydroxy-1-ethanethiol; 2-hydroxyethyl mercaptan; monothioethyleneglycol; thioglycol. C_2H_6OS; mol wt 78.13. C 30.75%, H 7.74%, O 20.48%, S 41.03%. Prepn: Woodward, *J. Chem. Soc.* **1948**, 1892; Peppel, Signaigo; Jones, US 2402665; US 3394192 (1946, 1968 both to du Pont). Properties: Bennett, *J. Chem. Soc.* **121**, 2139 (1922). IR: Thompson, *J. Am. Chem. Soc.* **61**, 1398 (1939). Toxicology: K. White *et al.*, *J. Pharm. Sci.* **62**, 237 (1973).

Liquid with very strong, disagreeable odor. bp$_{742}$ 157-158° (dec). d_4^{20} 1.1143. n_D^{20} 1.4996. The pure liquid is miscible with water, alc, ether and benzene. LD$_{50}$ in mice (mg/kg): 322.0 i.p.; 344.8 orally (White).

Caution: Irritating to eyes, nose and skin.

USE: In organic synthesis; as biochemical research tool.

5937. **Mercaptomerin Sodium.** [21259-76-7] (T-4)-[3-[[(3-Carboxylato-2,2,3-trimethylcyclopentyl)carbonyl-κO]amino]-2-methoxypropyl-κC][2-(mercapto-κS)acetato(2−)-κO]mercurate(2−) sodium (1:2); N-(γ-carboxymethylmercaptomercuri-β-methoxy)propylcamphoramic acid disodium salt; Diucardyn sodium; Thiomerin sodium. $C_{16}H_{25}HgNNa_2O_6S$; mol wt 606.01. C 31.71%, H 4.16%, Hg 33.10%, N 2.31%, Na 7.59%, O 15.84%, S 5.29%. Prepn: Lehman, US 2576349 (1951 to Wyeth); Wendt, Bruce, *J. Org. Chem.* **23**, 1448 (1958); Wendt, US 2834795 (1958 to Am. Home Prod.).

Hygroscopic white powder. Dec 150-155°. Freely soluble in water. Soluble in alcohol. Practically insoluble in ether, benzene, chloroform.

THERAP CAT: Diuretic.
THERAP CAT (VET): Diuretic.

5938. **6-Mercaptopurine.** [50-44-2] 1,9-Dihydro-6H-purine-6-thione; 1,7-dihydro-6H-purine-6-thione; purine-6-thiol; 6MP; Leukerin; Mercaleukin; Puri-Nethol; Purinethol. $C_5H_4N_4S$; mol wt 152.18. C 39.46%, H 2.65%, N 36.82%, S 21.07%. Prepd from hypoxanthine and phosphorus pentasulfide: G. B. Elion *et al.*, *J. Am. Chem. Soc.* **74**, 411 (1952); G. H. Hitchings, G. B. Elion, US 2697709 (1954 to Burroughs Wellcome). Improved procedure: A. G. Beaman, R. K. Robins, *J. Am. Chem. Soc.* **83**, 4042 (1961). Prepn from 7-aminothiazolo[5,4-d]pyrimidine: G. H. Hitchings, G. B. Elion, US 2721866 (1955 to Burroughs Wellcome). Metabolized in the body to 6-thiouric acid (6-mercapto-2,8-purinediol): G. B. Elion *et al.*, *J. Am. Chem. Soc.* **81**, 3042 (1959). Pharmacology: H. Froberg, M. S. Schencking, *Arch. Toxicol.* **32**, 1 (1974). HPLC determn in body fluids: J. L. Rudy *et al.*, *Ann. Clin. Biochem.* **25**, 504 (1988). Clinical trial with methotrexate in acute lymphoblastic leukemia: S. Koizumi *et al.*, *Cancer* **61**, 1292 (1988). Comprehensive description: S. A. Benezra, P. R. B. Foss, *Anal. Profiles Drug Subs.* **7**, 343-357 (1978). Review in treatment of Crohn's disease and ulcerative colitis: D. H. Present, *Gastroenterol. Clin. North Am.* **18**, 57-71 (1989); in treatment of leukemia: B. A. Kamen, *Semin. Hematol.* **28**, 12-14 (1991).

Monohydrate. Yellow prisms from water. Becomes anhydrous at 140°, dec 313-314°. uv max (0.1N NaOH): 230, 312 nm (ε 14000, 19600); (0.1N HCl): 222, 327 nm (ε 9240, 21300); (methanol): 216, 329 nm (ε 8940, 19300). pKa$_1$ 7.77, pKa$_2$ 11.17. Insol in water, acetone, ether. Sol in hot ethanol, in alkaline solns with slow decompn. LD$_{50}$ in mice, hamsters (mg/kg): 157, 364 i.p. (Froberg, Schencking).

THERAP CAT: Antineoplastic; immunosuppressant.
THERAP CAT (VET): Antineoplastic.

5939. **Mercumallylic Acid.** [86-36-2] [3-(3-Carboxy-2-oxo-2H-1-benzopyran-8-yl)-2-methoxypropyl]hydroxymercurate(1−) hydrogen (1:1); [3-[3-(hydroxymercuri)-2-methoxypropyl]salicylidene]malonic acid δ-lactone; 8-[3-(hydroxymercuri)-2-methoxypropyl]-2-oxo-2H-1-benzopyran-3-carboxylic acid; 8-(2-methoxy-3-hydroxymercuripropyl)coumarin-3-carboxylic acid; 8-[3-(hydroxymercuri)-2-methoxypropyl]-3-carboxycoumarin; 8-[3-(hydroxymercuri)-2-methoxypropyl]coumarin-3-carboxylic acid. $C_{14}H_{14}HgO_6$; mol wt 478.85. C 35.12%, H 2.95%, Hg 41.89%, O 20.05%. Prepn: A. Schlesinger *et al.*, US 2667442 (1954 to Endo); L. H. Werner, C. R. Scholz, *J. Am. Chem. Soc.* **76**, 2453 (1954). Toxicology of mercumatalin sodium: H. Blumberg *et al.*, *J. Pharmacol. Exp. Ther.* **105**, 336 (1952).

Powder, bitter taste. mp 155-160° (Schlesinger); mp 197° (Werner, Scholz). One gram dissolves in about 4.2 parts of 1N NaOH. Slightly sol in acetic acid. Very slightly sol in water, alcohol, chloroform. Practically insol in ether.

Sodium salt compound with theophylline. [8018-15-3] Mercumatilin sodium; Cumertilin Sodium. $C_{21}H_{21}HgN_4NaO_8$; mol wt 681.00. Dry, light yellow powder. Very sol in water. LD$_{50}$ in rats (mg Hg/kg): 9.8 i.v.; 238 orally (Blumberg).

THERAP CAT: Diuretic.

5940. **Mercuric Acetate.** [1600-27-7] Acetic acid mercury-(2+) salt (2:1); bis(acetyloxy)mercury; diacetoxymercury. $C_4H_6HgO_4$; mol wt 318.68. C 15.08%, H 1.90%, Hg 62.94%, O 20.08%. Hg(CH$_3$COO)$_2$. Prepd from HgO and acetic acid: Wagenknecht, Juza in *Handbook of Preparative Inorganic Chemistry* vol. 2, G. Brauer, Ed. (Academic Press, New York, 2nd ed., 1965) pp 1120-1121. Review of toxicity and human exposure: *Toxicological Profile for Mercury* (PB99-142416, 1999) 676 pp.

Crystals or cryst powder; slight acetic odor; sensitive to light. *Poisonous.* d 3.28. mp 178-180° (overheating results in decompn). One g dissolves in 2.5 ml cold, 1 ml boiling water; sol in alc. *Keep well closed and protected from light.* Aq solns decomp on standing, yielding a yellow ppt.

USE: Chiefly for mercuration of organic compounds; for the absorption of ethylene. Determn of nitrate in chromium compounds.

5941. **Mercuric Benzoate.** [583-15-3] Benzoic acid mercury-(2+) salt (2:1). $C_{14}H_{10}HgO_4$; mol wt 442.82. C 37.97%, H 2.28%, Hg 45.30%, O 14.45%. Hg(C$_7$H$_5$O$_2$)$_2$.

Monohydrate. Odorless, cryst powder; sensitive to light. *Poisonous.* Sol in 90 parts cold, 40 parts boiling water; freely sol in NaCl soln; slightly sol in alcohol. When boiled with water or alc it hydrolyzes to a basic salt and free benzoic acid. *Protect from light.*

THERAP CAT: Formerly as antisyphilitic.

5942. **Mercuric Bromide.** [7789-47-1] Mercury bromide (HgBr$_2$). Br$_2$Hg; mol wt 360.40. Br 44.34%, Hg 55.66%. HgBr$_2$. Usually prepd from the elements: Jander, Brodersen, *Z. Anorg. Chem.* **261**, 261 (1950).

White crystals or cryst powder; sensitive to light. *Poisonous.* d 6.05. mp 237°; sublimes at higher temp. Sol in about 200 parts cold, 25 parts boiling water; freely sol in hot alcohol, in methanol, HCl,

HBr, alkali bromide solns; slightly sol in chloroform. *Protect from light.*

USE: Colorimetric detection of arsenic.

5943. Mercuric Chloride. [7487-94-7] Mercury chloride ($HgCl_2$); mercury bichloride; corrosive sublimate; mercury perchloride; corrosive mercury chloride. Cl_2Hg; mol wt 271.49. Cl 26.12%, Hg 73.88%. $HgCl_2$. Review of clinical toxicology of mercury and its compounds: H. B. Gerstner, J. E. Huff, *J. Toxicol. Environ. Health* **2**, 491-526 (1977); of toxicity and human exposure: *Toxicological Profile for Mercury* (PB99-142416, 1999) 676 pp.

Crystals or white granules or powder. *Poisonous. May be fatal if swallowed.* d 5.4. mp 277°; volatilizes unchanged at about 300°; also slightly volatile at ordinary temp; volatilizes appreciably so at 100°. One gram dissolves in 13.5 ml water (soly is increased by HCl or alkali chlorides), in 2.1 ml boiling water, 3.8 ml alc, 1.6 ml boiling alc, 200 ml benzene, 22 ml ether, 12 ml glycerol, 40 ml acetic acid; also sol in methanol, acetone, ethyl acetate; slightly sol in carbon disulfide, pyridine. pH about 4.7. Also reported as 3.2 for 0.2 molar aq soln. Coagulates albumin; produces with NaOH a yellow ppt (difference from calomel, which turns black). *Incompat:* Formates, sulfites, hypophosphites, phosphates, sulfides, albumin, gelatin, alkalies, alkaloid salts, ammonia, lime water, antimony and arsenic, bromides, borax, carbonates; reduced iron; copper, iron, lead, silver salts; infusions of cinchona, columbo, oak bark or senna; tannic acid; vegetable astringents.

Caution: Highly toxic. Corrosive to mucous membranes. Ingestion may cause severe nausea, vomiting, hematemesis, abdominal pain, diarrhea, melena, renal damage, prostration. 1 or 2 g is frequently fatal. Poisoning and death also have occurred from intrauterine douches and application of alcoholic soln to large areas of skin. Do not breathe dust. Keep away from feed or food products. Wash hands before eating or smoking. *Antidote:* Dimercaprol (BAL).

USE: Preserving (kyanizing) wood and anatomical specimens; also embalming; disinfecting; browning and etching steel and iron; intensifier in photography; white reserve in fabric printing; tanning leather; electroplating aluminum; depolarizer for dry batteries; freeing gold from lead; magic photograms; mordant for rabbit and beaver furs; staining wood and vegetable ivory pink; manuf of ink for mercurography; treating seed potatoes; manuf other mercury compds. As an important reagent in analytical chemistry. Amalgamation of zinc for Jones reductor.

THERAP CAT: Topical antiseptic, disinfectant.

THERAP CAT (VET): Caustic, antiseptic, disinfectant.

5944. Mercuric Cyanide. [592-04-1] Mercury cyanide (Hg-$(CN)_2$); cianurina; dicyanomercury; mercury dicyanide. C_2HgN_2; mol wt 252.63. C 9.51%, Hg 79.40%, N 11.09%. $Hg(CN)_2$. Obtained by evaporating a soln of HgO in aq HCN: Biltz, *Z. Anorg. Allg. Chem.* **170**, 161 (1928).

Colorless, odorless, tetragonal crystals or white powder; darkens on exposure to light. *Poisonous.* d 3.996. Dec at 320°. One gram dissolves in 13 ml water, 3 ml boiling water, 13 ml alcohol, 4 ml methanol; slowly sol in glycerol; slightly sol in ether. *Protect from light.*

THERAP CAT: Topical antiseptic.

THERAP CAT (VET): Has been used as a topical antiseptic.

5945. Mercuric Fluoride. [7783-39-3] Mercury fluoride (HgF_2); mercury difluoride. F_2Hg; mol wt 238.59. F 15.93%, Hg 84.07%. HgF_2. Prepd from $HgCl_2$ and F_2: Henne, Midgley, *J. Am. Chem. Soc.* **58**, 886 (1936).

White powder or cubic crystals. Very sensitive to moisture. d^{15} 8.95. mp 645°. bp >650°. Turns yellow and hydrolyzes in the prolonged presence of water.

Dihydrate. Obtained when mercuric oxide is dissolved in excess of 50% HF.

USE: In the fluorination of organic compds.

5946. Mercuric Iodide, Red. [7774-29-0] Mercury iodide (HgI_2); mercury biniodide; diiodomercury. HgI_2; mol wt 454.40. Hg 44.14%, I 55.86%.

Scarlet-red, heavy, odorless, almost tasteless powder; sensitive to light; at 130° becomes yellow, and then red on cooling. *Poisonous.* d 6.28. mp 259°. bp ~350° and sublimes. Soly in water (25°) 0.006

g/100 g. 1 g dissolves in 115 ml alcohol, 20 ml boiling alcohol, about 120 ml ether, about 60 ml acetone, 910 ml chloroform, 75 ml ethyl acetate, 260 ml carbon disulfide, readily in alkali iodides, $HgCl_2$, $Na_2S_2O_3$; sol in 230 ml olive oil, 50 ml castor oil. *Protect from light.*

USE: In analytical chemistry for preparation of Nessler's Reagent (alkaline mercuric potassium iodide solution).

THERAP CAT: Antiseptic (topical).

THERAP CAT (VET): Has been used as a topical antiseptic, counterirritant, vesicant.

5947. Mercuric Nitrate. [10045-94-0] Nitric acid mercury-(2+) salt (2:1); mercury pernitrate. HgN_2O_6; mol wt 324.60. Hg 61.80%, N 8.63%, O 29.57%. $Hg(NO_3)_2$.

Hydrate. White or slightly yellow, deliquesc, cryst powder; odor of nitric acid. *Poisonous.* d 4.3. Sol in a small amount of water; with much water or on boiling with water, an insol basic salt is formed; sol in dil acids. *Keep well closed and protect from light.*

USE: Manufacture of felt; mercury fulminate; destroying phylloxera.

5948. Mercuric Oleate. [1191-80-6] (9Z)-9-Octadecenoic acid mercury(2+) salt (2:1); oleate of mercury. $C_{36}H_{66}HgO_4$; mol wt 763.51. C 56.63%, H 8.71%, Hg 26.27%, O 8.38%. $Hg(C_{17}H_{33}$-$COO)_2$. Commercial product prepd by dissolving yellow mercuric oxide in oleic acid; contains 24-26% HgO and about 10% excess oleic acid.

Yellowish-brown, somewhat transparent, ointment-like mass; odor of oleic acid. *Poisonous.* On keeping, the mercury is partly reduced to mercurous state or metal. Practically insol in water. Slightly sol in alcohol or ether, freely in fixed oils. *Protect from light.*

THERAP CAT: Has been used as ectoparasiticide.

5949. Mercuric Oxide, Red. [21908-53-2] Mercury oxide (HgO); red precipitate. HgO; mol wt 216.59. Hg 92.61%, O 7.39%. Contains 99-99.5% HgO.

Bright red or orange-red, heavy, odorless, crystalline powder or scales; orthorhombic structure; yellow when finely powdered. *Poisonous.* Dec on exposure to light into mercury and oxygen; at 400° becomes almost black, but red again on cooling; at 500° dec into mercury and oxygen. d 11.14. Practically insol in water. Sol in dil HCl or HNO_3 or in solns of alkali cyanides or iodides, slowly in solns of alkali bromides; insol in alcohol. *Protect from light.*

USE: In marine bottom paints, diluting pigments for painting on porcelain, with graphite as depolarizer in dry batteries. In Kjeldahl nitrogen determination; and as reagent for citric acid, thiophene, glucose, aldehyde, urea, acetone. As reagent and catalyst in organic reactions: J. S. Pizey, *Synthetic Reagents* **vol. 1** (John Wiley, New York, 1974) pp 295-319.

THERAP CAT: Antiseptic (topical).

THERAP CAT (VET): Has been used as a topical antiseptic.

5950. Mercuric Oxide, Yellow. [21908-53-2] Mercury oxide (HgO); yellow precipitate. HgO; mol wt 216.59. Hg 92.61%, O 7.39%. Contains 99-99.5% HgO.

Yellow or orange-yellow, heavy, odorless powder, orthorhombic structure. *Poisonous.* Becomes red on heating and yellow again on cooling; more finely divided and more reactive than red mercuric oxide. Other physical properties and solubilities: *see* Mercuric Oxide, Red. *Protect from light. Incompat:* Reducing agents.

USE: Similar to that of the red oxide; in the manuf of organic mercurials. In analytical chemistry for determining Zn or HCN; detecting acetic acid in formic acid, CO in gas mixtures. Catalyst.

THERAP CAT: Topical antiseptic (ophthalmic).

THERAP CAT (VET): Has been used as a topical antiseptic, fungicide, in chronic skin conditions, conjunctivitis, corneal ulcers.

5951. Mercuric Oxycyanide. [1335-31-5] Mercury cyanide oxide ($Hg_2(CN)_2O$). $C_2Hg_2N_2O$; mol wt 469.22. C 5.12%, Hg 85.50%, N 5.97%, O 3.41%. $HgO.Hg(CN)_2$. For reasons explained below, the article of commerce contains about 33% mercuric oxycyanide and about 67% mercuric cyanide.

White, orthorhombic crystals or cryst powder d 4.44. One gram dissolves in 80 ml cold water, more sol in hot water. *Poisonous. It explodes when touched with a flame or by percussion;* hence for commerce it is made with an excess of mercuric cyanide which eliminates the danger of explosion.

THERAP CAT: Antiseptic (topical).

5952. Mercuric Potassium Cyanide. [591-89-9] Tetrakis-(cyano-κC)mercurate(2−) potassium (1:2); dipotassium tetracyanomercurate; mercury potassium cyanide (HgK$_2$(CN)$_4$); potassium tetracyanomercurate(II). C$_4$HgK$_2$N$_4$; mol wt 382.86. C 12.55%, Hg 52.39%, K 20.42%, N 14.63%. K$_2$[Hg(CN)$_4$].

Colorless or white crystals. Sol in water. *Poisonous.* Rapidly decomposed by acids to evolve hydrogen cyanide.

USE: In manuf of mirrors to prevent the silver coating from becoming yellow; as reagent in testing for free acids.

5953. Mercuric Salicylate. [5970-32-1] [2-(Hydroxy-κO)-benzoato(2−)-κO]mercury; mercury subsalicylate. C$_7$H$_4$HgO$_3$; mol wt 336.70. C 24.97%, H 1.20%, Hg 59.58%, O 14.26%. Article of commerce contains 54-58% Hg. Prepn: C. H. Rogers *et al.*, *Inorganic Pharmaceutical Chemistry* (Lea & Febriger, Philadelphia, 4th ed., 1948) pp 409-411.

White or slightly yellowish or pinkish, odorless powder. *Poisonous.* Insol in water or alcohol. Sol in warm solns of alkali halides, fixed alkali hydroxides and carbonates. *Incompat:* Alkali iodides.

THERAP CAT: Topical antiseptic.

5954. Mercuric Sodium *p*-Phenolsulfonate. [535-55-7] 4-Hydroxybenzenesulfonic acid mercury(2+) sodium salt (2:1:2); mercury and sodium phenolsulfonate; mercuriphenoldisulfonate sodium; Hermophényl. C$_{12}$H$_8$HgNa$_2$O$_8$S$_2$; mol wt 590.88. C 24.39%, H 1.36%, Hg 33.95%, Na 7.78%, O 21.66%, S 10.85%. Prepd by the action of the sodium salt of *p*-phenolsulfonic acid on yellow mercuric oxide: Lumière, Chevrottier, *Compt. Rend.* **132**, 145 (1901); Lumière, Perrin, *Pharmacie* **1**, 102 (1943). The commercial product is a mixture of the monosulfonate (80%) and the disulfonate.

White powder. One gram dissolves in 5 ml water. The aq soln does not coagulate albuminous matter.

Caution: Ingestion or percutaneous absorption may cause mercury poisoning.

USE: As germicide in soaps and lotions. Usual concn in soap about 1:100.

THERAP CAT: Antiseptic (local).

5955. Mercuric Sulfate. [7783-35-9] Sulfuric acid mercury-(2+) salt (1:1); mercury bisulfate. HgO$_4$S; mol wt 296.65. Hg 67.62%, O 21.57%, S 10.81%. HgSO$_4$.

White, odorless granules or cryst powder. *Poisonous.* d 6.47. Dec by water into a yellow insol, basic sulfate and free H$_2$SO$_4$; sol in HCl, hot dil H$_2$SO$_4$, concd soln of NaCl. *Protect from light.*

USE: Electrolyte for primary batteries; with NaCl for extracting gold and silver from roasted pyrites. Determn of oxygen demand. Catalyst for Kjeldahl method. Reagent for wine coloring, barbital, and cystine.

5956. Mercuric Sulfide, Black. [1344-48-5] Mercury sulfide (HgS); ethiops mineral. HgS; mol wt 232.65. Hg 86.22%, S 13.78%. Occurs as a mineral in California. Prepd by passing hydrogen sulfide through a soln of mercuric chloride in hydrochloric acid.

Black or grayish-black, heavy, odorless, tasteless, amorphous powder. Also occurs as black, cubic crystals (β-form). Transition temp (red to black) 386°. Black form can exist indefinitely in metastable state at room temp. Insol in water, alcohol, dil mineral acids.

USE: As pigment for horn, rubber, etc.

5957. Mercuric Sulfide, Red. [1344-48-5] Mercury sulfide (HgS); vermilion; chinese red; C.I. Pigment Red 106; C.I. 77766. HgS; mol wt 232.65. Hg 86.22%, S 13.78%. Occurs in nature as the mineral **cinnabar**. Prepd from mercuric acetate, ammonium thiocyanate, glacial acetic acid and hydrogen sulfide: Newell *et al.*, *Inorg. Synth.* **1**, 19 (1939). *See also Colour Index* vol. **4** (3rd ed., 1971) p 4682.

Bright scarlet-red powder, lumps, hexagonal crystals (α-form); blackens on exposure to light, particularly in presence of H$_2$O or

alkali hydroxides. At about 250° becomes brownish, at higher temp black, but red again on cooling. When ignited in air it dec into metal and sulfur, the latter burning to SO$_2$. Practically insol in water; not attacked by HNO$_3$ or cold HCl; dec by hot concd H$_2$SO$_4$; sol in aqua regia with separation of S, in warm hydriodic acid with evolution of H$_2$S. *Protect from light.*

USE: For coloring plastics, sealing wax, and with FeSO$_4$ for marking linen; manuf fancy colored papers; as pigment.

THERAP CAT: Antibacterial.

5958. Mercuric Thiocyanate. [592-85-8] Thiocyanic acid mercury(2+) salt (2:1); mercuric sulfocyanate; mercuric sulfocyanide. C$_2$HgN$_2$S$_2$; mol wt 316.75. C 7.58%, Hg 63.33%, N 8.84%, S 20.24%. Hg(SCN)$_2$. Prepd from Hg(NO$_3$)$_2$ + KSCN: Peters, *Z. Anorg. Allg. Chem.* **77**, 157 (1912).

Odorless powder. When crystalline, usually in radially arranged needles. *Poisonous.* When heated it swells up to many times its original vol, decomposing finally into mercury, nitrogen, etc., at about 165°. Slightly sol in cold water (0.069 g/100 ml H$_2$O at 25°), more sol in boiling water with decompn; sol in dil HCl, in solns of alkali cyanides, chlorides. *Protect from light.*

USE: For Pharaoh's serpents (fireworks); intensifier in photography.

5959. Mercurophen. [52486-78-9] Hydroxy(4-hydroxy-3-nitrophenyl)-mercury sodium salt (1:1); sodium 4-(hydroxymercuri)-2-nitrophenolate; sodium hydroxymercuri-*o*-nitrophenolate; 4-(hydroxymercuri)-2-nitrophenol sodium salt. C$_6$H$_4$HgNNaO$_4$; mol wt 377.68. C 19.08%, H 1.07%, Hg 53.11%, N 3.71%, Na 6.09%, O 16.94%. Prepn from *o*-nitrophenol and Hg(OCOCH$_3$)$_2$: Schamberg *et al.*, **US 1390972** (1922). Structure: Malcolm, *J. Bacteriol.* **22**, 403 (1931).

Brick-red, odorless powder. *Poisonous.* Sol in hot water. Solns are deep amber in color.

THERAP CAT: Antiseptic, disinfectant.

THERAP CAT (VET): Antiseptic, disinfectant.

5960. Mercurous Acetate. [631-60-7] C$_4$H$_6$Hg$_2$O$_4$; mol wt 519.27. C 9.25%, H 1.16%, Hg 77.26%, O 12.32%. Hg$_2$(CH$_3$-COO)$_2$. Prepd from Hg$_2$(NO$_3$)$_2$ + CH$_3$COONa: Wagenknecht, Juza, in *Handbook of Preparative Inorganic Chemistry* vol. **2**, G. Brauer, Ed. (Academic Press, New York, 2nd ed., 1965) p 1120.

Lustrous leaflets or cryst powder. Darkens on exposure to light. Sol in about 100 parts water; sol in dil acetic acid. Practically insol in alcohol, ether. Aq solns dec quickly under the influence of light and heat. *Keep well closed and protected from light.*

THERAP CAT: Antibacterial.

5961. Mercurous Bromide. [10031-18-2] Br$_2$Hg$_2$; mol wt 560.99. Br 28.49%, Hg 71.51%. Hg$_2$Br$_2$.

White, odorless powder; darkens on exposure to light. d 7.3. Sublimes at approx 390° (dec). Insol in water, alcohol, ether; dec by hot HCl or alkali bromides. *Protect from light.*

5962. Mercurous Chloride. [10112-91-1] Mercury chloride (Hg$_2$Cl$_2$); calomel; mild mercury chloride; mercury monochloride; mercury protochloride; mercury subchloride; precipité blanc; Calogreen. Cl$_2$Hg$_2$; mol wt 472.08. Cl 15.02%, Hg 84.98%. Hg$_2$Cl$_2$. Review of toxicity and human exposure: *Toxicological Profile for Mercury* (PB99-142416, 1999) 676 pp.

White, odorless, tasteless, heavy powder; slowly dec by sunlight into mercuric chloride and metallic mercury; sublimes at 400-500° without melting. d 7.15. Practically insol in water (0.00020 g/100 ml H$_2$O at 25°). HCl or alkali and alkaline earth chlorides increase soly in water; insol in alc, ether. Dec by solns of alkali iodides, bromides or cyanides into the mercuric salt and metallic mecury; solns of alkali chlorides act similarly but slowly. Blackened by ammonia, caustic alkali and alkaline earth solns. *Protect from light.*

USE: Dark green Bengal lights; calomel paper, mixed with gold in painting on porcelain; for calomel electrodes; as fungicide; in agriculture to control root maggots on cabbage and onions.

THERAP CAT: Cathartic, diuretic, antiseptic, antisyphilitic.

THERAP CAT (VET): Has been used as cathartic, and locally as an antiseptic and desiccant.

5963. Mercurous Iodide. [15385-57-6] Yellow mercury iodide; mercury protoiodide. Hg_2I_2; mol wt 654.99. Hg 61.25%, I 38.75%.

Bright-yellow, amorphous, heavy, odorless powder. Darkens or becomes greenish on exposure to light, HgI_2 and metallic mercury being formed. d 7.70. mp 290° when rapidly heated with partial decompn into Hg and HgI_2. Insol in water, alcohol or ether. Sol in solns of mercurous or mercuric nitrates; cold ammonia, its solns or alkali iodide dec it into mercury and mercuric iodide. *Protect from light.* "Green" mercury iodide is made from metallic mercury and iodine, the green color being due to presence of some uncombined mercury.

5964. Mercurous Nitrate. [10415-75-5] Nitric acid mercury-(1+) salt (1:1); mercury protonitrate; dimercury dinitrate. $Hg_2N_2O_6$; mol wt 525.19. Hg 76.39%, N 5.33%, O 18.28%. $Hg_2(NO_3)_2$. Normally exists as dihydrate. Prepn of anhyd salt: Potts, Allred, *Inorg. Chem.* **5**, 1066 (1966).

Dihydrate. Colorless crystals, usually with slight odor of HNO_3. mp ~70° with decompn. *Poisonous.* d 4.78. Sol in 13 parts water contg 1% HNO_3; with water alone a basic salt is formed. Blackened by ammonia, caustic alkali, and alkaline earth solns. *Keep well closed and protected from light.*

USE: Fire gilding, blackening brass.

5965. Mercurous Sulfate. [7783-36-0] Hg_2O_4S; mol wt 497.24. Hg 80.68%, O 12.87%, S 6.45%. Hg_2SO_4.

White to slightly yellow cryst powder. Becomes gray on exposure to light with production of mercury and mercuric sulfate. d 7.56. Slightly sol in water (0.06 g/100g at 25°); sol in dil HNO_3.

USE: For making electric batteries; with zinc sulfate in the standard Clark cell and with cadmium sulfate in the standard Weston cell.

5966. Mercury. [7439-97-6] Hydrargyrum; liquid silver; quicksilver. Hg; at. wt 200.59; at. no. 80; valences 1, 2. Group IIB (12). Abundance in earth's crust 0.5 ppm. Natural isotopes: 202 (29.80%); 200 (23.13%); 199 (16.84%); 201 (13.22%); 198 (10.02%); 204 (6.85%); 196 (0.146%); known isotopes range in mass number from 189 to 206. Obtained by roasting cinnabar (mercuric sulfide). General reviews: Roberts, *Adv. Inorg. Chem. Radiochem.* **11** (Academic Press, New York, 1968) pp 309-339; Aylett, "Group IIB" in *Comprehensive Inorganic Chemistry* **vol. 3** (Pergamon Press, Oxford, 1973) pp 187-328; H. J. Drake in *Kirk-Othmer Encyclopedia of Chemical Technology* **vol. 15** (Wiley-Interscience, New York, 3rd ed., 1981) pp 143-156. Review of clinical toxicology: H. B. Gerstner, J. E. Huff, *J. Toxicol. Environ. Health* **2**, 491-526 (1977); and human exposure: *Toxicological Profile for Mercury* (PB99-142416, 1999) 676 pp.

Silver-white, heavy, mobile, liquid metal; slightly volatile at ordinary temp; solid mercury is a tin-white, ductile, malleable mass which may be cut with a knife. mp −38.87°; bp 356.72°; d^{25} 13.534. Heat capacity at constant pressure (25°) 6.687 cal/mole deg. Vapor pressure (25°): 2×10^{-3} mm; heat of vaporization (25°): 14.652 kcal/mole: Busey, Giauque, *J. Am. Chem. Soc.* **75**, 806 (1953). Surface tension (25°): 484 dynes/cm; electrical resistivity (20°): 95.76 μohm cm. *Corrosive; poisonous.* When pure does not tarnish on exposure to air at ordinary temp, but when heated to near the boiling point slowly oxidizes to HgO. Forms alloys with most metals except iron and combines with sulfur at ordinary temperatures. E^0 (aq) Hg/Hg^{2+} −0.854 V; E^0 (aq) 2 Hg/Hg_2^{2+} −0.789 V. Soly in water (25°): 0.28 μmoles/l; data on soly in organic solvents: Spencer, Voigt, *J. Phys. Chem.* **72**, 464 (1968). Reacts with HNO_3 and hot, concd H_2SO_4; does not react with dil HCl, cold H_2SO_4, or alkalies. Reacts with ammonia solns in air to form Hg_2NOH, **Millon's base.** Mercury salts when heated with Na_2CO_3 yield metallic Hg and are reduced to metal by H_2O_2 in the presence of alkali hydroxide. Cu, Fe, Zn and many other metals ppt metallic Hg from neutral or slightly acid solns of mercury salts. Soluble ionized *mercuric* salts give a yellow ppt of HgO with NaOH and a red ppt of HgI_2 with alkali iodide. *Mercurous* salts give a black ppt with alkali hydroxides and a white ppt of calomel with HCl or sol chlorides. They are slowly dec by sunlight.

Caution: Readily absorbed via respiratory tract (elemental mercury vapor, mercury compd dusts), intact skin, and G.I. tract, although occasional incidental swallowing of metallic mercury is without harm. Spilled and heated elemental mercury is particularly hazardous. *Acute:* sol salts have violent corrosive effects on skin and mucous membranes; severe nausea, vomiting, abdominal pain, bloody diarrhea; kidney damage; death usually within 10 days. *Chronic:* inflammation of mouth and gums, excessive salivation, loosening of teeth; kidney damage; muscle tremors, jerky gait, spasms of extremities; personality changes, depression, irritability, nervousness. Phenyl and alkyl mercurials can cause skin burns and be absorbed by the skin. Burning sensation is delayed several hours and thus gives no warning. Alkyls have affinity for brain tissue and may cause permanent damage. Phenyls are no more toxic than inorganic Hg. *Antidote:* Dimercaprol (BAL). *See* E. Browning, *Toxicity of Industrial Metals* (Appleton-Century Crofts, New York, 2nd ed., 1969) pp 226-242.

USE: In barometers, thermometers, hydrometers, pyrometers; in mercury arc lamps producing ultraviolet rays; in switches, fluorescent lamps; in mercury boilers; manuf all mercury salts, mirrors; as catalyst in oxidation of organic compds; extracting gold and silver from ores; making amalgams, electric rectifiers, mercury fulminate; also in dentistry; in determining N by Kjeldahl method, for Millon's reagent; as cathode in electrolysis, electroanalysis, polarography, and many other uses. Also in pharmaceuticals, agricultural chemicals, anti-fouling paints.

5967. Mercury Fulminate. [628-86-4] Bis(fulminato-κC)-mercury; fulminic acid mercury(2+) salt; Knallquecksilber. $C_2HgN_2O_2$; mol wt 284.62. C 8.44%, Hg 70.48%, N 9.84%, O 11.24%. Primary explosive prepd by addition of ethanol to mercury in nitric acid soln: E. Howard, *Phil. Trans. R. Soc.* **90**, 204 (1800). Coulometric determn in explosive compositions: A. J. van der Hulst, *Anal. Chem.* **41**, 207 (1969). Thermal decomposition analysis: M. E. Brown, G. M. Swallowe, *Thermochim. Acta* **49**, 333 (1981). NMR structural studies: F. De Sarlo *et al., Org. Magn. Reson.* **22**, 372 (1984). Crystal structure determn: W. Beck *et al., Z. Anorg. Allg. Chem.* **633**, 1417 (2007).

$$O{\equiv}N{\equiv}C-Hg-C{\equiv}N{\equiv}O$$

Grey crystalline solid. Crystal d 4.467. *Explosive. Poisonous. Highly sensitive to friction and shock. Store under water and protect from light.*

USE: Primary explosive; component of percussion cap and blasting cap detonators.

5968. Merimepodib. [198821-22-6] N-[[3-[[[[3-Methoxy-4-(5-oxazolyl)phenyl]amino]carbonyl]amino]phenyl]methyl] carbamic acid (3S)-tetrahydro-3-furanyl ester; VX-497. $C_{23}H_{24}N_4O_6$; mol wt 452.47. C 61.05%, H 5.35%, N 12.38%, O 21.22%. Inosine 5′-monophosphate dehydrogenase (IMPDH) inhibitor. Prepn (stereochemistry unspecified): D. M. Armistead *et al.,* **WO 9740028**; *eidem,* **US 5807876** (1997, 1998 both to Vertex). Manufacturing process: A. R. Looker *et al., Org. Process Res. Dev.* **12**, 666 (2008). *In vitro* antiviral spectrum: W. Markland *et al., Antimicrob. Agents Chemother.* **44**, 859 (2000); immunosuppressive activity: J. Jain *et al., J. Pharm. Sci.* **90**, 625 (2001). Clinical evaluation in hepatitis C: J. G. McHutchison *et al., Antivir. Ther.* **10**, 635 (2005). Review of structure-activity and mechanism of action: M. D. Sintchak, E. Nimmesgern, *Immunopharmacology* **47**, 163-184 (2000); of development and therapeutic potential: G. Tossing, *IDrugs* **6**, 372-376 (2003).

THERAP CAT: Antiviral.

5969. Merocyanine 540. [62796-23-0] 2-[4-(1,3-Dibutyl-tetrahydro-4,6-dioxo-2-thioxo-5(2*H*)-pyrimidinylidene)-2-buten-1-ylidene]-3(2*H*)-benzoxazolepropanesulfonic acid sodium salt (1:1); (5-[(3-sodium-sulfopropyl-2(3*H*)-benzoxazolylidene)-2-butenyl-idene]-1,3-dibutyl-2-thiobarbituric acid); MC 540. $C_{26}H_{32}N_3$-NaO_6S_2; mol wt 569.67. C 54.82%, H 5.66%, N 7.38%, Na 4.04%, O 16.85%, S 11.26%. Prototype fluorescent dye used to study plasma membrane potential; originally developed for use in photography. Adopts an all-*trans* arrangement in the ground state and isomerizes upon photoexcitation. Preferentially binds to viable electrically excitable cells, leukemic cells, and apoptotic cells. Exhibits phototoxicity towards neoplastic cells and viruses. Preparative method and membrane potential studies in living cells: L. B. Cohen *et al., J. Membr. Biol.* **19**, 1 (1974). Staining of electrically excitable cells: T. G. Easton *et al., Cell* **13**, 475 (1978); of leukemic cells: J. E. Valinsky *et al., ibid.* 487. Photochemical and photophysical properties: J. Davila *et al., Photochem. Photobiol.* **53**, 1 (1991); A. C. Benniston *et al., J. Phys. Chem. A* **107**, 4347 (2003). Spectroscopy studies: B. Cunderlíková, L. Sikurová, *Chem. Phys.* **263**, 415 (2001). Biodistribution and toxicity of photoproducts: S. Pervaiz *et al., Cancer Chemother. Pharmacol.* **31**, 467 (1993). Review of biomedical applications: F. Sieber, *Photochem. Photobiol.* **46**, 1035-1042 (1987). Use in flow cytometry to detect apoptosis: T. Laakko *et al., J. Immunol. Methods* **261**, 129 (2002). Clinical evaluation for purging of autologous stem cell grafts: G. S. Anderson *et al., J. Photochem. Photobiol. B* **69**, 87 (2003).

Solid, mp 285° (dec). *Irritant.* pKa 1.60±0.03. Abs max: 530 nm (water); 554 nm (methanol); 558 nm (ethanol); 559 nm (acetone); 568 nm (chloroform); 565 nm (dioxane). Emission max: 573 nm (water); 577 nm (methanol); 579 nm (ethanol); 580 nm (acetone); 589 nm (chloroform); 584 nm (dioxane). Sol in water, methanol, ethanol, acetone, dioxane, DMSO. LD_{50} (of light activated compd) in mice (mg/kg): 320 i.p.; 160 i.v. (Pervaiz).

USE: Dye to stain living cells. Fluorescent probe to study plasma membrane properties and to detect apoptosis.

5970. Meropenem. [96036-03-2] (4*R*,5*S*,6*S*)-3-[[(3*S*,5*S*)-5-[(Dimethylamino)carbonyl]-3-pyrrolidinyl]thio]-6-[(1*R*)-1-hydroxyethyl]-4-methyl-7-oxo-1-azabicyclo[3.2.0]hept-2-ene-2-carboxylic acid; (1*R*,5*S*,6*S*)-2-[(3*S*,5*S*)-5-(dimethylaminocarbonyl)pyrrolidin-3-ylthio]-6-[(*R*)-1-hydroxyethyl]-1-methylcarbapen-2-em-3-carboxylic acid. $C_{17}H_{25}N_3O_5S$; mol wt 383.46. C 53.25%, H 6.57%, N 10.96%, O 20.86%, S 8.36%. Carbapenem antibiotic. Prepn: M. Sunagawa *et al.,* **EP 126587**; M. Sunagawa, **US 4943569** (1984, 1990 both to Sumitomo). Structure-activity study: M. Sunagawa *et al., J. Antibiot.* **43**, 519 (1990). Crystal structure: K. Yanagi *et al., Acta Crystallogr.* **C48**, 1737 (1992). HPLC determn in serum and bronchial secretions: M. Ehrlich *et al., J. Chromatogr. B* **751**, 357 (2001). Pharmacokinetics: R. Wise *et al., Antimicrob. Agents Chemother.* **34**, 1515 (1990). Series of articles on antimicrobial activity, metabolism: *J. Antimicrob. Chemother.* **24**, Suppl. A, 1-320 (1989); and clinical performance: *ibid.* **36**, Suppl. A, 1-223 (1995). Review of clinical experience in intensive care: M. Hurst, H. M. Lamb, *Drugs* **59**, 653-680 (2000).

Trihydrate. [119478-56-7] ICI-194660; SM-7338; Meronem; Meropen; Merrem. White to pale yellow crystalline powder. Sol in DMF, 5% dibasic potassium phosphate soln; sparingly sol in water, 5% monobasic potassium phosphate soln; very slightly sol in alc. Practically insol in acetone, ether.

THERAP CAT: Antibacterial.

5971. Mersalyl. [492-18-2] [3-[[2-(Carboxylatomethoxy)-benzoyl-κ*O*]amino]-2-methoxypropyl-κ*C*]hydroxymercurate(1−) sodium (1:1); *o*-[[3-(hydroxymercuri)-2-methoxypropyl]carbamoyl]phenoxyacetic acid sodium salt; sodium *o*-[(3-hydroxymercuri-2-methoxypropyl)carbamoyl]phenoxyacetate; *N*-(γ-hydroxymercuri-β-methoxypropyl)salicylamide-*O*-acetic acid sodium salt; mercuramide; Salyrgan. $C_{13}H_{16}HgNNaO_6$; mol wt 505.85. C 30.87%, H 3.19%, Hg 39.65%, N 2.77%, Na 4.54%, O 18.98%. Prepn: Diels, Beccard, *Ber.* **39**, 4125 (1906); Bockmühl, Schwarz, **DE 423031** (1925 to Hoechst); *Frdl.* **15**, 1609. Use as diuretic: F. Brunn, *Wien. Klin. Wochenschr.* **37**, 901 (1924). Toxicity data: E. B. Robbins, K. K. Chen, *J. Am. Pharm. Assoc.* **40**, 249 (1951). Use of acid as physiological sulfhydryl inhibitor: P. Mavier, J. Hanoune, *Eur. J. Biochem.* **59**, 593 (1975).

Bitter crystals. Somewhat deliquescent. Gradually dec by light. One gram dissolves in about 1 ml water, in about 3 ml ethanol (95%), in 2 ml abs methanol. Practically insol in ether, chloroform. Aq solns are alkaline to litmus. LD_{50} i.v. in mice, rats: 72.6, 17.7 mg/kg (Robbins, Chen).

Mersalyl Acid. [486-67-9] $C_{13}H_{17}HgNO_6$; mol wt 483.87. Odorless, white, slightly hygroscopic powder. Slightly sol in water, dil mineral acids; sol in solutions of alkali hydroxides.

USE: Sulfhydryl specific biochemical probe.

THERAP CAT: Diuretic.

THERAP CAT (VET): Diuretic.

5972. MES. [4432-31-9] 4-Morpholineethanesulfonic acid; 2-(4-morpholino)ethyl sulfonate; 2-(*N*-morpholino)ethanesulfonic acid. $C_6H_{13}NO_4S$; mol wt 195.23. C 36.91%, H 6.71%, N 7.17%, O 32.78%, S 16.42%. One of the zwitterionic *N*-substituted aminosulfonic acids known as *Good Buffers*, active in the pH range 6-8. Prepn as sodium salt: S. Malkiel, J. P. Mason, *J. Org. Chem.* **8**, 199 (1943); as acid: N. E. Good *et al., Biochemistry* **5**, 467 (1966). Thermodynamic parameters: C. D. McGlothlin, J. Jordan, *Anal. Lett.* **9**, 245 (1976). Acid-base properties: A. Balikungeri, *Chimia* **43**, 13 (1989). Interference with protein determn: P. L. Lleu, G. Rebel, *Anal. Biochem.* **192**, 215 (1991). Use as buffer: S. C. Lee *et al., J. AOAC Int.* **75**, 395 (1992); P. J. Henney *et al., Anim. Genet.* **25**, 363 (1994).

Colorless crystals from alcohol/water, dec >300°. Apparent pKa at 0.1*M* ionic: 0°, 6.38; 20°, 6.15; 37°, 5.98. At 0.1*M* ionic, 25.0° ±0.1°: pKa_1 1.99; pKa_2 6.21. Δ pKa/°C −0.011. Saturation at 0°: 0.65*M*.

USE: Biological buffer.

5973. Mesaconic Acid. [498-24-8] (2*E*)-2-Methyl-2-butene-dioic acid; methylfumaric acid; *trans*-1-propene-1,2-dicarboxylic acid. $C_5H_6O_4$; mol wt 130.10. C 46.16%, H 4.65%, O 49.19%. Prepd by heating citraconic anhydride with nitric acid: Shriner *et al., Org. Synth.* **11**, 74 (1931); from ethyl itaconate: Jones *et al., J. Chem. Soc.* **1954**, 1865; from *N*,*N*-diethyl-5-diethylamino-2,3-dihydro-3-furamide: Sauer *et al., J. Am. Chem. Soc.* **81**, 693 (1959). Obtained almost instantly by the action of sunlight on citraconic acid in ether + chloroform in the presence of a little bromine: Fittig, Langworthy, *Ber.* **26**, 46 (1893); *Ann.* **304**, 119, 149 (1899); by

Consult the Name Index before using this section.

action of β-radiation on citraconic acid + bromine: Lavigne, Levine, **US 2979445** (1961 to California Res. Corp.).

Orthorhombic needles from alcohol, monoclinic tablets from ethyl acetate, mp 204-205°. d 1.466. Sublimes. bp 250° (decompn). K_1 at 25° = 8.5 × 10⁻⁴; K_2 = 15 × 10⁻⁶. 100 g water dissolves 2.7 g at 18° and 117.9 g at the boiling point; 100 g 90% alc dissolves 30.6 g at 17° and 95.7 g at the boiling point. Sol in ether; sparingly sol in chloroform, carbon disulfide and ligroin.

Dimethyl ester. $C_7H_{10}O_4$. Liq, bp 206°. d_4^{21} vac 1.12011. n_D^{20} 1.45575. Sol in 122 parts water at 15°.

Diethyl ester. $C_9H_{14}O_4$. Liq, bp 229°. d_4^{20} vac 1.04675. n_4^{20} 1.44936.

5974. Mesalamine. [89-57-6] 5-Amino-2-hydroxybenzoic acid; 5-aminosalicylic acid; 5-amino-2-hydroxybenzene-1-carboxylic acid; m-aminosalicylic acid; fisalamine; mesalazine; 5-ASA; Asacol; Asacolitin; Claversal; Lialda; Lixacol; Mesasal; Pentasa; Rowasa; Salofalk. $C_7H_7NO_3$; mol wt 153.14. C 54.90%, H 4.61%, N 9.15%, O 31.34%. Prepn by reduction of m-nitrobenzoic acid with Zn dust and HCl: H. Weil et al., Ber. **55B**, 2664 (1922); by electrolytic reduction: Le Guyader, Peltier, Compt. Rend. **253**, 2544 (1961). Use in manufacture of dyes: **GB 751386** (1956 to J. R. Geigy). Identification as active metabolite of sulfasalazine, q.v.: A. K. Azad Khan et al., Lancet **2**, 892 (1977); P. A. M. Van Hees et al., Gut **21**, 632 (1980). HPLC determn in serum: E. Brendel et al., J. Chromatogr. **385**, 299 (1987). Bioavailability, plasma level and excretion: S. N. Rasmussen et al., Gastroenterology **83**, 1062 (1982). Clinical evaluation in Crohn's disease: eidem, ibid. **85**, 1350 (1983). Clinical trials in ulcerative colitis: M. Campieri et al., Lancet **2**, 270 (1981); L. S. Friedman et al., Am. J. Gastroenterol. **81**, 412 (1986); in comparison with olsalazine, q.v.: S. S. Rao et al., Scand. J. Gastroenterol. **22**, 332 (1987). Brief review: G. Friedman, Am. J. Gastroenterol. **81**, 141-144 (1986). Review of clinical efficacy in colonic diverticulosis: A. Tursi, Expert Opin. Pharmacother. **6**, 69-74 (2005).

White to pinkish crystals, dec ~280°. Sol in dilute hydrochloric acid, dilute alkali hydroxides; slightly sol in water; very slightly sol in methanol, dehydrated alc, acetone. Practically insol in n-butyl alc, chloroform, ether, ethyl acetate, n-hexane, methylene chloride, n-propyl alc.

USE: In manuf of light-sensitive paper, azo and sulfur dyes.

THERAP CAT: Anti-inflammatory (gastrointestinal).

5975. Mescaline. [54-04-6] 3,4,5-Trimethoxybenzeneethanamine; 3,4,5-trimethoxyphenethylamine; mezcaline. $C_{11}H_{17}NO_3$; mol wt 211.26. C 62.54%, H 8.11%, N 6.63%, O 22.72%. Psychotomimetic alkaloid isolated from **peyote** (mescal buttons), the flowering heads of Lophophora williamsii (Lemaire) Coult., Cactaceae. Isoln: A. Heffter, Ber. **29**, 221 (1896). Structure and synthesis: E. Späth, Monatsh. Chem. **40**, 129 (1919); K. H. Slotta, H. Heller, Ber. **63**, 3029 (1930); E. Späth, F. Becke, Monatsh. Chem. **66**, 327 (1935); M. U. Tsao, J. Am. Chem. Soc. **73**, 5495 (1951); K. Banholzer et al., Helv. Chim. Acta **35**, 1577 (1952). Novel synthesis: M. N. Aboul-Enein, A. I. Eid, Acta Pharm. Suec. **16**, 267 (1979). MS determn: S. P. Jindal, T. Lutz, Eur. J. Mass Spectrom. Biochem. Med. Environ. Res. **2**, 117 (1982). Pharmacokinetics in rabbits: C. Van Peteghem et al., Eur. J. Drug Metab. Pharmacokinet. **7**, 1 (1982). Mode of action study: M. E. Trulson et al., Eur. J. Pharmacol. **96**, 151 (1983). Use in evaluating serotonin S_2 antagonists: C. J. E. Niemegeers et al., Drug Dev. Res. **3**, 123 (1983). Evaluation of use with chlorpromazine, q.v., in various psychoses: H. C. B. Denber, S. Merlis: J. Nerv. Ment. Dis. **122**, 463 (1955). Toxicity

data: L. B. Speck, J. Pharmacol. Exp. Ther. **119**, 78 (1957); H. F. Hardman et al., Toxicol. Appl. Pharmacol. **25**, 299 (1973). Reviews: A. R. Patel, Progress in Drug Research vol. **11**, E. Jucker, Ed. (Birkhaüser Verlag, Basel, 1968) pp 11-47; G. J. Kapadia, M. B. E. Fayez, J. Pharm. Sci. **59**, 1699-1727 (1970).

Crystals, mp 35-36°. bp_{12} 180°. Moderately sol in water; sol in alcohol, chloroform, benzene. Practically insol in ether, petr ether. Takes up CO_2 from the air and forms a crystalline carbonate. LD_{50} i.p. in rats: 370 mg/kg (Speck).

Hydrochloride. $C_{11}H_{17}NO_3.HCl$. Needles, mp 181°. Sol in water, alcohol. LD_{50} in mice, rats, guinea pigs (mg/kg): 212, 132, 328 i.p. (Hardman).

Sulfate dihydrate. $(C_{11}H_{17}NO_3)_2.H_2SO_4.2H_2O$. Prisms, mp 183-186°. Sol in hot water, methanol; sparingly sol in cold water, ethanol.

Acid sulfate. $C_{11}H_{17}NO_3.H_2SO_4$. Crystals, mp 158°.

N-Benzoylmescaline. Needles from aq alc, mp 121°. Very sol in alcohol, ether.

N-Methylmescaline. Occurs naturally, bp 130-140°.

N-Acetylmescaline. Occurs naturally, mp 94°.

Note: This is a controlled substance (hallucinogen): **21 CFR**, 1308.11.

5976. Mesembrine. [24880-43-1] (3aS,7aS)-3a-(3,4-Dimethoxyphenyl)octahydro-1-methyl-6H-indol-6-one; 3a-(3,4-dimethoxyphenyl)tetrahydro-1-methyl-6(3aH)-indolinone. $C_{17}H_{23}NO_3$; mol wt 289.38. C 70.56%, H 8.01%, N 4.84%, O 16.59%. Alkaloid used in preparing Channa, a drug of Southwest Africa. Occurs naturally as the (−)-form. From Sceletium expansum L., S. tortuosum L, Aizoaceae: Hartwick, Zwicky, Apoth. Ztg. **29**, 925 (1914); Rimington et al., J. Vet. Sci. Animal Ind. **9**, 187 (1938), C.A. **32**, 4279⁹ (1938). Structure: Popelak et al., Naturwissenschaften **47**, 156 (1960). Configuration: P. W. Jeffs et al., J. Am. Chem. Soc. **91**, 3831 (1969). Synthesis of (±) form: Shamma, Rodriguez, Tetrahedron Lett. **1965**, 4847; O. Hoshino et al., Heterocycles **10**, 61 (1978); of (±)-form and trans isomer: Oh-Ishi, Kugita, Chem. Pharm. Bull. **18**, 299 (1970). Synthesis of (+)-form: Yamada, Otani, Tetrahedron Lett. **1971**, 1133; eidem, Chem. Pharm. Bull. **21**, 2130 (1973). Stereoselective synthesis of (±)-form: Wijnberg, Speckamp, Tetrahedron Lett. **1975**, 3963; eidem, Tetrahedron **34**, 2579 (1978); S. F. Martin et al., J. Org. Chem. **44**, 3391 (1979); S. Takano et al., Chem. Lett. **1981**, 1385. Enantioselective synthesis of natural mesembrine: eidem, Tetrahedron Lett. **22**, 4479 (1981). Biosynthesis: Jeffs et al., J. Am. Chem. Soc. **93**, 3752 (1971); eidem, Chem. Commun. **1977**, 60. Review of mesembrine alkaloids: A. Popelak, G. Lettenbauer in The Alkaloids vol. **IX**, R. H. F. Manske, Ed. (Academic Press, New York, 1967) pp 467-481; R. V. Stevens in The Total Synthesis of Natural Products vol. **3**, J. ApSimon, Ed. (Wiley, New York, 1977) pp 443-453.

Pale yellow oil. $bp_{0.3}$ 186-190°. $[\alpha]_D^{20}$ −55.4° (CH_3OH). Freely sol in alcohol, chloroform, acetone; slightly sol in ether. Practically insol in benzene, petr ether, alkalies.

Hydrochloride. $C_{17}H_{23}NO_3.HCl$. mp 205-206°. $[\alpha]_D^{20}$ −8.4° (CH_3OH).

(+)-**Form.** Pale yellow oil. Partially optically active. $[\alpha]_D^{20}$ +16.1° (c = 1.32 in CH_3OH).

(+)-**Form hydrochloride.** $C_{17}H_{23}NO_3 \cdot HCl$. Crystals from 2-propanol, mp 206.5-207.5°. $[\alpha]_D^{20}$ +7.3° (c = 0.465 in CH_3OH).

(±)-**Form.** Colorless oil. $bp_{0.07}$ 178°.

(±)-**Form hydrochloride.** $C_{17}H_{23}NO_3 \cdot HCl$. mp 179-181°.

5977. Mesitylene. [108-67-8] 1,3,5-Trimethylbenzene; *sym*-trimethylbenzene. C_9H_{12}; mol wt 120.20. C 89.93%, H 10.06%. Occurs in coal tar and in petroleum crudes; prepd by dehydrating acetone with H_2SO_4: Adams, Hufford, *Org. Synth.* **2**, 41 (1922).

Liquid; peculiar odor. d_4^{20} 0.8637. mp −44.8°. n_D^{18} 1.49541. bp_{760} 164.7°; bp_{100} 98.9°; bp_{20} 61°; bp_{10} 47.4°; $bp_{1.0}$ 9.6°. Practically insol in water (100 g H_2O dissolve 0.002 g). Miscible with alcohol, ether, benzene.

Caution: Potential symptoms of overexposure are irritation of eyes, skin, nose, throat, respiratory system; bronchitis; hypochromic anemia, headache, drowsiness, fatigue, dizziness, nausea, incoordination; vomiting, confusion; aspiration of liquid may cause chemical pneumonia. *See NIOSH Pocket Guide to Chemical Hazards* (DHHS/NIOSH 97-140, 1997) p 320.

5978. Mesityl Oxide. [141-79-7] 4-Methyl-3-penten-2-one; 2,2-dimethylvinyl methyl ketone; isopropylideneacetone; 2-methyl-4-oxo-2-pentene. $C_6H_{10}O$; mol wt 98.15. C 73.42%, H 10.27%, O 16.30%. Made by distilling diacetone alcohol with a small amount of iodine: Conant, Tuttle, *Org. Synth.* **1**, 53 (1921). Condensation of acetone to mesityl oxide using sulfonated polystyrene-divinylbenzene resin as ion exchange catalyst: Klein, Banchero, *Ind. Eng. Chem.* **48**, 1278 (1956). Believed to be a mixture of two isomers. Toxicity data: N. F. Izmerov *et al.*, *Toxicometric Parameters of Industrial Toxic Chemicals Under Single Exposure* (Centre Internat. Projects, Moscow, 1982) p 81. Brief review of toxicity: *Dangerous Prop. Ind. Mater. Rep.* **9**, 58-65 (1989).

Colorless, oily liq; honey-like odor. d_4^{15} 0.8592. bp_{760} 130°; bp_{100} 72.1°; bp_{20} 26°; $bp_{1.0}$ −8.7°. mp −41.5°; also reported as mp −59°. Can be made to crystallize at low temp in petr ether. n_D^{22} 1.4425. Absorption spectrum: Morton, *J. Chem. Soc.* **1926**, 719. Sol in about 30 parts water; miscible with most organic liqs. Flash pt: 87°F (30.6°C). LD_{50} in mice (mg/kg): 710 ±85 intragastric; LC_{50} in mice (2 hr), rats (4 hr) (mg/m³): 10000 ±270, 9000 ±600 (Ismerov).

Caution: Potential symptoms of overexposure are irritation of eyes, skin and mucous membranes; narcosis, coma. *See NIOSH Pocket Guide to Chemical Hazards* (DHHS/NIOSH 97-140, 1997) p 194.

USE: Solvent for nitrocellulose, many gums and resins, particularly vinyl resins. In lacquers, varnishes and enamels. In making methyl isobutyl ketone.

5979. Mesna. [19767-45-4] 2-Mercaptoethanesulfonic acid sodium salt (1:1); sodium mercaptoethanesulfonate; D-7093; UCB-3983; Mesnex; Mistabron; Mistabronco; Mucofluid; Uromitexan. $C_2H_5NaO_3S_2$; mol wt 164.17. C 14.63%, H 3.07%, Na 14.00%, O 29.24%, S 39.06%. $[HSCH_2CH_2SO_3]^- Na^+$. Sulfhydryl donor used to reduce the urotoxic effects of antineoplastic alkylating agents; also has mucolytic activity. Prepn: I. M. Lipovich, *J. Appl. Chem. USSR* **18**, 718 (1945); C. H. Schramm *et al.*, *J. Am. Chem. Soc.* **77**, 6231 (1955). Synthesis and properties: V. E. Petrun'kin, *C.A.* **51**, 5693a (1957); *ibid.* **54**, 24379c (1960). Use as mucolytic: **NL 6605816**; H. Morren, **US 3567835** (1966, 1971 both to U.C.B.); as uroprotective agent: N. Brock, **DE 2756018**; *idem*, **US 4220660** (1979, 1980 both to Asta-Werke AG). Pharmacology, toxicity, and uroprotective effects in animals: N. Brock *et al.*, *Eur. J. Cancer Clin. Oncol.* **18**,

1377 (1982). HPLC determn in plasma and urine: C. A. James, H. J. Rogers, *J. Chromatogr.* **382**, 394 (1986). Clinical pharmacokinetics: C. A. James *et al.*, *Br. J. Clin. Pharmacol.* **23**, 561 (1987). Clinical study of mucolytic effects: M. Tekeres *et al.*, *Clin. Ther.* **4**, 56 (1981). Symposium on pharmacology, toxicity and clinical uroprotective efficacy: *Cancer Treat. Rev.* **10**, Suppl. A, 1-192 (1983). *Reviews:* I. C. Shaw, M. I. Graham, *ibid.* **14** 67-86 (1987); S. E. Schoenike, W. J. Dana, *Clin. Pharm.* **9**, 179-191 (1990).

Freely sol in water, sparingly sol in organic solvents. LD_{50} in male, female mice, male, female rats (mg/kg): 1887, 2048, 2098, 1683 i.v.; 2005, 2098, 1529, 1251 i.p.; 6102, >7200, 4440, 4679 orally (Brock).

THERAP CAT: Mucolytic; antineoplastic adjunct (uroprotective).

5980. Mesoridazine. [5588-33-0] 10-[2-(1-Methyl-2-piperidinyl)ethyl]-2-(methylsulfinyl)-10*H*-phenothiazine; thioridazine-2-sulfoxide; TPS-23. $C_{21}H_{26}N_2OS_2$; mol wt 386.57. C 65.25%, H 6.78%, N 7.25%, O 4.14%, S 16.59%. Dopamine receptor blocking agent; analog of thioridazine. Prepn: Renz *et al.*, **US 3084161** (1963 to Sandoz). Pharmacology and toxicology: Loew *et al.*, *Boll. Chim. Farm.* **106**, 332-371 (1967). Effects on dopaminergic function in comparison with sulforidazine and thioridazine, *q.q.v.*: D. M. Niedzwiecki *et al.*, *J. Pharmacol. Exp. Ther.* **228**, 636 (1984); C. D. Kilts *et al.*, *ibid.* **231**, 334 (1984). GLC determn in plasma: E. C. Dinovo *et al.*, *J. Pharm. Sci.* **65**, 667 (1976). Clinical evaluation in sleep disorders: K. Adam *et al.*, *Br. J. Clin. Pharmacol.* **3**, 157 (1976); as antipsychotic: R. Axelsson, *Curr. Ther. Res.* **21**, 587 (1977). Toxicity studies: S. Maruyama *et al.*, *Niigata Igakkai Zasshi* **81**, 611 (1967), *C.A.* **68**, 76856h (1968). Brief description: *J. Am. Med. Assoc.* **216**, 313 (1971).

Oily product.

Benzenesulfonate. [32672-69-8] Mesoridazine besylate; NC-123; Lidanar; Lidanil; Serentil. $C_{27}H_{32}N_2O_4S_3$; mol wt 544.74. LD_{50} in mice (mg/kg): 33 i.v.; 611 s.c.; 346 orally (Maruyama).

Tartrate. $C_{25}H_{32}N_2O_7S_2$. Crystals from ethyl acetate, mp 115-120°.

THERAP CAT: Antipsychotic.

5981. Mesosulfuron-methyl. [208465-21-8] 2-[[[[(4,6-Dimethoxy-2-pyrimidinyl)amino]carbonyl]amino]sulfonyl]-4-[[(methylsulfonyl)amino]methyl]benzoic acid methyl ester; methyl 2-[3-(4,6-dimethoxypyrimidin-2-yl)ureidosulfonyl]-4-methanesulfonamidomethylbenzoate; AE-F130060; Osprey; Silverado. $C_{17}H_{21}N_5O_9S_2$; mol wt 503.50. C 40.55%, H 4.20%, N 13.91%, O 28.60%, S 12.73%. Sulfonylurea herbicide; acetolactate synthase inhibitor. Prepn: K. Lorenz *et al.*, **WO 9510507**; *eidem* **US 5648315** (1995, 1997 both to Hoechst Schering AgrEvo). Comprehensive description: E. Hacker *et al.*, *BCPC Conf. - Weeds* **2001**, 43-48. Greenhouse trial of combination with iodosulfuron-methyl-sodium in winter wheat: W. A. Bailey *et al.*, *Weed Sci.* **51**, 515 (2003). Field trial: L. Kong *et al.*, *Crop Prot.* **28**, 387 (2009).

Cream-colored solid, mp 195.4°. Vapor pressure (25°): 1.1 × 10^{-11} Pa. Log P (octanol/water): 1.39 (pH 5); −0.48 (pH 7); −2.06 (pH 9). Soly in water at 20° (g/l): 7.24 × 10^{-3} (pH 5); 0.483 (pH 7); 15.39 (pH 9). LD_{50} in rats (mg/kg): >5000 orally; >5000 dermally. LC_{50} in rats: >1.33 mg/l air (Hacker).

USE: Herbicide.

5982. Mesotrione. [104206-82-8] 2-[4-(Methylsulfonyl)-2-nitrobenzoyl]-1,3-cyclohexanedione; ZA-1296; Callisto. $C_{14}H_{13}$-NO_7S; mol wt 339.32. C 49.56%, H 3.86%, N 4.13%, O 33.01%, S 9.45%. Triketone herbicide for use in maize; inhibits p-hydroxyphenylpyruvate dioxygenase (HPPD). Synthetic mimic of *leptospermone*, a natural herbicide produced by the bottlebrush plant, *Callistemon citrinus* Stapf. Prepn: C. G. Carter, **EP 186118** (1986 to Stauffer); idem, **US 5006158** (1991 to ICI). Comprehensive description: R. A. Wichert et al., *Brighton Crop Prot. Conf. - Weeds* **1999**, 105-110. HPLC determn in crops, soil, water: P. Alferness, L. Wiebe, *J. Agric. Food Chem.* **50**, 3926 (2002). Adsorption and degradation in soil: J. S. Dyson et al., *J. Environ. Qual.* **31**, 613 (2002). Review: G. Mitchell et al., *Pest Manage. Sci.* **57**, 120-128 (2001).

Opaque solid, mp 165°. pKa (20°): 3.12. Vapor pressure (20°): 4.27 × 10^{-8} mm Hg. Soly in water at 20° (g/l): 2.2 (pH 4.8); 15 (pH 6.9); 22 (pH 9). LD_{50} in rats (mg/kg): >5000 orally; >2000 dermally. LC_{50} in rats (mg/l): >5 by inhalation. LC_{50} (96 hr) in bluegill sunfish, rainbow trout (mg/l): >120, >120 (Wichert).

USE: Herbicide.

5983. Mesoxalic Acid. [473-90-5] 2-Oxopropanedioic acid; ketomalonic acid; oxomalonic acid. $C_3H_2O_5$; mol wt 118.04. C 30.53%, H 1.71%, O 67.77%. Occurs in *Medicago sativa* L., *Leguminosae;* has been found in beet molasses. Prepd by boiling a soln of alloxan and lead acetate: Deichsel, *J. Prakt. Chem.* [1] **93**, 194 (1864). By electrolysis of *d*-tartaric acid in alkaline soln: Chem. *Zentralbl.* **1922**, III, 871. Laboratory prepn from dibromomalonic acid: Conrad, Reinbach, *Ber.* **35**, 1819 (1902); from malonic ester and N_2O_3: Curtiss, *Am. Chem. J.* **35**, 477 (1906) Prepn of diethyl ester: A. W. Dox, *Org. Synth.* **coll. vol. I**, 266 (2nd ed., 1941).

Solid, mp 121°.
Hydrate. [560-27-0] Dihydroxymalonic acid. $C_3H_4O_6$; mol wt 136.06. Begins to melt at 113-114° and is clear at 121°. Very sol in water; sol in alc, ether.
Diethyl ester. [609-09-6] Ethyl mesoxalate; ethyl oxomalonate. $C_7H_{10}O_5$; mol wt 174.15. Golden-yellow liquid. bp_{19} 105-107°; bp_{15} 103-108°. $d_4^{15.6}$ 1.1419. $n_D^{15.6}$ 1.419.

5984. Mesquite Gum. [9000-47-9] Sonora; Prosopis gum. Collected from small thorny trees abundant in the arid regions of the Western United States and as far south as Chile: *Prosopis juliflora* (Swartz) DC., *P. dulcis* Kunth., *P. horrida* Kunth., *P. inermis* H.B.K., *P. glandulosa* Torr., *P. pubescens* Benth., *P. spicigera* L., and other species of *Prosopis, Leguminosae*. Mesquite gum resembles acacia (gum arabic) in its physical and chemical characteristics. Review of structure work: F. Smith, R. Montgomery, *The Chemistry of Plant Gums and Mucilages* (Reinhold, New York, 1959) pp 175, 288-291.

USE: Substitute for acacia. Potential source of L-arabinose and D-glucuronic acid, cf. C. L. Mantell, *The Water-Soluble Gums* (Reinhold, New York, 1947) pp 72-73.

5985. Mestanolone. [521-11-9] (5α,17β)-17-Hydroxy-17-methylandrostan-3-one; 17β-hydroxy-17α-methyl-3-androstanone; 17α-methylandrostan-17β-ol-3-one; 17α-methylandrostan-3-on-17β-ol; Anabo; Antalone; Duramin; Mechiaron; Prohormo; Protenolon; Tantarone. $C_{20}H_{32}O_2$; mol wt 304.47. C 78.90%, H 10.59%, O 10.51%. Prepd by the oxidation of 17-methyl-3,17-androstanediol: Ruzicka et al., *Helv. Chim. Acta* **18**, 1487 (1935); **CH 208080** (1940 to Ciba).

Crystals from ethyl acetate, mp 192-193°. Insol in water. Sol in acetone, alcohol, ether, ethyl acetate.
Note: This is a controlled substance (anabolic steroid): **21 CFR,** 1308.13, as defined in 1300.01.
THERAP CAT: Anabolic.

5986. Mesterolone. [1424-00-6] (1α,5α,17β)-17-Hydroxy-1-methylandrostan-3-one; 1α-methyl-5α-androstan-17β-ol-3-one; 1α-methyl-5α-dihydrotestosterone; Androviron; Proviron; Mestoranum. $C_{20}H_{32}O_2$; mol wt 304.47. C 78.90%, H 10.59%, O 10.51%. Prepn of acetate: R. Wiechert, **DE 1122944** corresp to **US 3361773** (1962, 1968 to Schering, AG).

Crystals from ethyl acetate, mp 203.5-205.0°. $[\alpha]_D^{20}$ +17.6° (c = 0.875 in $CHCl_3$).
Acetate. 17β-Acetoxy-1α-methyl-5α-androstan-3-one. $C_{22}H_{34}$-O_3. Crystals, mp 169-170°. $[\alpha]_D^{25}$ +16.5° (c = 0.88 in $CHCl_3$).
Note: This is a controlled substance (anabolic steroid): **21 CFR,** 1308.13, as defined in 1300.01.
THERAP CAT: Androgen.

5987. Mestranol. [72-33-3] (17α)-3-Methoxy-19-norpregna-1,3,5(10)-trien-20-yn-17-ol; 17α-ethynyl-3-methoxy-1,3,5(10)-estratrien-17β-ol; 17α-ethynylestradiol 3-methyl ether; Menophase; Norquen; Ovastol. $C_{21}H_{26}O_2$; mol wt 310.44. C 81.25%, H 8.44%, O 10.31%. Orally active estrogenic steroid. Prepn: F. B. Colton, **US 2666769** (1954 to Searle); *J. Am. Chem. Soc.* **79**, 1123 (1957). Comprehensive description: H. A. El-Obeid, A. A. Al-Badr, *Anal. Profiles Drug Subs.* **11**, 375-406 (1982). Clinical pharmacokinetics: J. W. Goldzieher et al., *Contraception* **21**, 17 (1980). Effect on carbohydrate metabolism: W. N. Spellacy et al., *Metabolism* **31**, 106 (1982). Randomized, double-blind clinical trials: S. Koetsawang et al., *Contraception* **25**, 231 (1982); A. Sheth et al., ibid. 243. Evaluation of carcinogenic risk: *IARC Monographs* **21**, 257 (1979).

Crystals from methanol or acetone, mp 150-151°. uv max (methanol): 279, 287.5 nm ($E_{1cm}^{1\%}$ 82, 14.4). Freely sol in chloroform; sol in dioxane, ethanol, ether, acetone; sparingly sol in dehydrated alc; slightly sol in methanol. Insol in water.

Note: Also used in combination with chlormadinone acetate, ethynodiol, lynestrenol, norethindrone or norethynodrel, *q.q.v.* Has been used in combination with megestrol acetate, *q.v.*

Caution: This substance is listed as a known human carcinogen: *Report on Carcinogens, Twelfth Edition* (PB2011-111646, 2011) p 184.

THERAP CAT: Estrogen; in combination with progestogen as oral contraceptive.

5988. Mesulfen. [135-58-0] 2,7-Dimethylthianthrene; 2,6-dimethylthianthrene; 2,6-dimethyldiphenylene disulfide; mesulphen; Mitigal; Odylen; Sudermo; Peligal; Neosulfine. $C_{14}H_{12}S_2$; mol wt 244.37. C 68.81%, H 4.95%, S 26.24%. Prepn: Cohen, Skirrow, *J. Chem. Soc.* **75**, 890 (1899); Barber, Smiles, *ibid.* **1928**, 1149; Rumpf, *Bull. Soc. Chim. Fr.* **7**, 632 (1940).

Needles from acetic acid, ethyl acetate, or alcohol, mp 123°. bp_3 184°; bp_{14} 228-231°. Freely sol in acetone, chloroform, ether, petr ether; moderately sol in abs alcohol, ethyl acetate; practically insol in water.

THERAP CAT: Scabicide, antipruritic.

THERAP CAT (VET): Scabicide, antipruritic.

5989. Metachrome Yellow. [584-42-9] 2-Hydroxy-5-[2-(3-nitrophenyl)diazenyl]benzoic acid sodium salt (1:1); C.I. Mordant Yellow 1; C.I. 14025; alizarine yellow GG; salicyl yellow; sodium *m*-nitrobenzeneazosalicylate. $C_{13}H_8N_3NaO_5$; mol wt 309.21. C 50.50%, H 2.61%, N 13.59%, Na 7.44%, O 25.87%. Prepd from diazotized *m*-nitroaniline and salicylic acid in alkaline medium: R. Nietzki, **US 424019**; *Colour Index* **vol. 4** (3rd ed., 1971) p 4058.

Yellow powder. Slightly sol in cold water; more sol in hot water.

USE: Biological stain and acid-base indicator. pH: 10.2 colorless, 12.0 yellow.

5990. Metaclazepam. [84031-17-4] 7-Bromo-5-(2-chlorophenyl)-2,3-dihydro-2-(methoxymethyl)-1-methyl-1*H*-1,4-benzodiazepine; brometazepam; metuclazepam; Ka-2547; KC-2547. $C_{18}H_{18}BrClN_2O$; mol wt 393.71. C 54.91%, H 4.61%, Br 20.30%, Cl 9.00%, N 7.12%, O 4.06%. Benzodiazepine deriv with alkoxymethyl group replacing carbonyl in parent compd. Prepn: W. Milkowski *et al.*, **DE 2221558**; *eidem*, **US 4098786** (1973, 1978 both to Kali-Chemie). Effect on psychomotor performance and memory in volunteers: Z. Subhan, I. Hindmarch, *Drugs Exp. Clin. Res.* **9**, 567 (1983). Interaction with ethanol: H. R. Musch, M. Ruhland, *Arzneim.-Forsch.* **32**, 567 (1982); H. J. Mallach *et al.*, *Blutalkohol* **20**, 196 (1983); V. Schmidt *et al.*, *Med. Welt* **35**, 32 (1984). Clinical studies: G. Laakman *et al.*, *Arzneim.-Forsch.* **30**, 1233 (1980).

Hydrochloride. [61802-93-5] Talis. $C_{18}H_{18}BrClN_2O.HCl$; mol wt 430.17. Crystals from ethanol, mp 193-196°. LD_{50} orally in NMRI white mice: 1578 mg/kg (Milkowski).

THERAP CAT: Anxiolytic.

5991. [2.2]Metacyclophane. [2319-97-3] Tricyclo-[9.3.1.1^{4,8}]hexadeca-1(15),4,6,8(16),11,13-hexaene; di-*m*-xylylene; *m*-dixylylene. $C_{16}H_{16}$; mol wt 208.30. C 92.26%, H 7.74%. Prepd by the action of phenyllithium on *m*-xylene dibromide: Allinger *et al.*, *J. Am. Chem. Soc.* **83**, 1974 (1961); *eidem*, *J. Org. Chem.* **32**, 2272 (1967).

Orthorhombic prisms from ether, mp 132.5°. bp_{760} 290°; bp_{12} 170°. Sparingly sol in alc. More sol in ether, benzene.

5992. Metaflumizone. [139968-49-3] 2-[2-(4-Cyanophenyl)-1-[3-(trifluoromethyl)phenyl]ethylidene]-*N*-[4-(trifluoromethoxy)-phenyl]hydrazinecarboxamide; (*EZ*)-2′-[2-(4-cyanophenyl)-1-(α,α,-α-trifluoro-*m*-tolyl)ethylidene]-4-(trifluoromethoxy)carbanilohydrazide; BAS-320I; Alverde; ProMeris; Siesta. $C_{24}H_{16}F_6N_4O_2$; mol wt 506.41. C 56.92%, H 3.18%, F 22.51%, N 11.06%, O 6.32%. Semicarbazone insecticide dervied from pyrazoline that blocks voltage-dependent sodium channels. Prepn: K. Takagi *et al.*, **EP 462456**; *eidem*, **US 5543573** (1991, 1996 both to Nihon Nohyaku). Laboratory trial against Colorado potato beetle: G. C. Cutler *et al.*, *Resistant Pest Manag. Newsl.* **16**, 33 (2006). Field trial for control of red imported fire ants: X. P. Hu, D. Song, *Sociobiology* **50**, 1107 (2007); against beet armyworm in tomato: J. E. Taylor, D. G. Riley, *J. Entomol Sci.* **42**, 430 (2007). Series of articles on discovery, development, mode of action, efficacy, and pharmacokinetics for veterinary use: *Vet. Parasitol.* **150**, 175-281 (2007). Toxicological profile: K. Hempel *et al.*, *ibid.*, 190. Efficacy of spot-on formulation against KS1 flea strain in cats: M. Dryden *et al.*, *ibid.* **151**, 74 (2008).

E-isomer

Crystals from ether-*n*-hexane, mp 191°. Rapidly isomerizes to a mixture of 90% *E*-isomer and 10% Z-isomer. LD_{50} in rats (mg/kg): >5000, >5000 orally, dermally; LC_{50} (4 hr.) in rats: >5.2 mg/l by inhalation (Hempel).

USE: Insecticide.

THERAP CAT (VET): Ectoparasiticide.

5993. Metalaxyl. [57837-19-1] *N*-(2,6-Dimethylphenyl)-*N*-(2-methoxyacetyl)-alanine methyl ester; methyl *N*-(2-methoxyacetyl)-*N*-(2,6-xylyl)-DL-alaninate; metaxanin; CGA-48988; Ridomil. $C_{15}H_{21}NO_4$; mol wt 279.34. C 64.50%, H 7.58%, N 5.01%, O 22.91%. Phenylamide fungicide for use in food crops, shrubs and turf. Prepn: A. Hubele, **DE 2515091**; *idem*, **US 4151299** (1975, 1979 to Ciba-Geigy). Comprehensive description: P. A. Urech, F. J. Schwinn, *Phytiatr.-Phytopharm.* **27**, 239-247 (1978). Antifungal spectrum: A. Kerkenaar, A. K. Sijpesteijn, *Pestic. Biochem. Physiol.* **15**, 71 (1981). Enzyme-linked immunosorbent assay of residues in foods: W. H. Newsome, *J. Agric. Food Chem.* **33**, 528 (1985).

Whitish crystals, mp 71-72°. Vapor pressure (20°): 2.2×10^{-6} mm Hg. Soly in water (20°): 7.1 g/l. Readily sol in most organic solvents. LD_{50} in rats (mg/kg): 669 orally (Urech).

R-form. [70630-17-0] N-(2,6-Dimethylphenyl)-N-(methoxyacetyl)-D-alanine methyl ester; (R)-[(2,6-dimethylphenyl)methoxyacetylamino]propionic acid methyl ester; metalaxyl-M; mefenoxam; CGA-329351; Apron; Ridomil Gold; Subdue. Enantioselective prepn: H.-U. Blaser, F. Spindler, *Top. Catal.* **4**, 275 (1997). Comprehensive description: C. Nuninger *et al.*, *Brighton Crop Prot. Conf. - Pests Dis.* **1996**, 41-46. Pale yellow to light brown viscous liquid. dec ~270°. Log P (n-octanol/water): 1.71. Soly in water (25°): 26 g/l.

USE: Agricultural fungicide.

5994. Metaldehyde. [9002-91-9] Acetaldehyde homopolymer; metacetaldehyde. Polymer of acetaldehyde. $(C_2H_4O)_n$. Prepd by polymerization of acetaldehyde in the presence of HCl, H_2SO_4 at low temp: Kekule, Zincke, *Ann.* **162**, 125 (1872); in the presence of pyridine and HBr: R. S. Wilder, **US 2426961** (1947 to Publicker Ind.). Use as molluscicide: R. M. Coloso *et al.*, *Crop Prot.* **17**, 669 (1998).

Prisms, mp 246° in sealed tube. Sublimes at about 112°, but dec with partial regeneration of aldehyde above 80°. *Flammable.* Practically insol in water. Sol in benzene, chloroform; sparingly sol in alcohol, ether.

Caution: Potential symptoms of overexposure by ingestion are severe abdominal pain, nausea, vomiting, diarrhea, convulsions, coma. *See Clinical Toxicology of Commercial Products*, R. E. Gosselin *et al.*, Eds. (Williams & Wilkins, Baltimore, 5th ed., 1984) p II-187.

USE: In compressed form as a fuel instead of alcohol; molluscicide.

5995. Metamitron. [41394-05-2] 4-Amino-3-methyl-6-phenyl-1,2,4-triazin-5(4H)-one; BAY DRW 1139; Goltix. $C_{10}H_{10}N_4O$; mol wt 202.22. C 59.40%, H 4.98%, N 27.71%, O 7.91%. Prepn: K. Dickore *et al.*, **DE 2107757**; *eidem*, **US 3847914** (1972, 1974 both to Bayer); W. Draber *et al.*, *Ann.* **1976**, 2206. Chemistry and mode of action: *idem et al.*, *Proc. 8th Int. Plant Prot. Congress* **1**, 203 (1975). Comprehensive description: H. Lembrich, *Pflanzenschutz-Nachr.* **31**, 197 (1978). Degradation in soil: R. Allen, A. Walker, *Pestic. Sci.* **18**, 95 (1987). HPLC determn in surface water: R. B. Geerdink, *J. Chromatogr.* **543**, 244 (1991).

Crystals from isopropanol/water, mp 169°. uv max (methanol): 312 nm (log ε 4.06). Vapor pressure (20-70°): $<1 \times 10^{-4}$ mbar. Soly (g/100g): water 0.18; isopropanol 0-1; methylene chloride 1-5; toluene 0-1; ligroin (80-110°) 0-1. Stable in acid media; rapidly decomp at pH >10. LD_{50} orally in male, female rats, male, female mice: 3343, 1832, 1450, 1463 mg/kg; dermally in rats: >500 mg/kg. LC_{50} (4 hr) in rats, mice, hamsters (mg/m^3): >331, >206, >206 by inhalation (Lembrich).

USE: Herbicide.

5996. Metampicillin. [6489-97-0] (2S,5R,6R)-3,3-Dimethyl-6-[[(2R)-(methyleneamino)phenylacetyl]amino]-7-oxo-4-thia-1-azabicyclo[3.2.0]heptane-2-carboxylic acid; D-6-[α-(methyleneamino)-phenylacetamido]penicillanic acid; methampicillin; Bonopen; Fedacilina; Micinovo; Pravacilin; Ruticina; Suvipen; Viderpen. $C_{17}H_{19}N_3O_4S$; mol wt 361.42. C 56.50%, H 5.30%, N 11.63%, O 17.71%, S 8.87%. Semi-synthetic antibiotic related to penicillin. Prepn: **BE 661232**; Gradnik, **GB 1081093** (1965, 1967 both to E.R.A.S.M.E.), *C.A.* **65**, 3884e (1966), *C.A.* **68**, 114595g (1968). Synthesis: Gradnik *et al.*, *Farmaco Ed. Sci.* **26**, 20 (1971). Antibacterial activity studies: Sutherland *et al.*, *Chemotherapy* **17**, 145 (1972). Pharmacokinetics: Fleischmann *et al.*, *Farmaco Ed. Prat.* **26**, 106

(1971). Clinical studies: Farina, *Minerva Med.* **60**, 1999 (1969); Ginocchi, *ibid.* 2003; Cardinale, Arrotta, *ibid.* 2011.

Sodium salt. [6489-61-8] Ocelina; Magnipen; Venzoquimpe. $C_{17}H_{18}N_3NaO_4S$; mol wt 383.40.

THERAP CAT: Antibacterial.

5997. Metanephrine. [5001-33-2] 4-Hydroxy-3-methoxy-α-(methylaminomethyl)benzenemethanol; α-(methylaminomethyl)-vanillyl alcohol; 3-O-methylepinephrine; 3-O-methyladrenaline; 1-(4-hydroxy-3-methoxyphenyl)-2-methylaminoethanol. $C_{10}H_{15}NO_3$; mol wt 197.23. C 60.90%, H 7.67%, N 7.10%, O 24.34%. A naturally occurring derivative of epinephrine, found in the urine and in certain tissues. Various methods of prepn: Külz, Hornung, **DE 682394** (1939); *Chem. Zentralbl.* **1940**, I, 1078, *C.A.* **36**, 3011 (1942); Axelrod *et al.*, *J. Biol. Chem.* **233**, 697 (1958); Heacock, Hutzinger, *Chem. Ind. (London)* **1961**, 595.

dl-Form hydrochloride. $C_{10}H_{15}NO_3$·HCl. Prisms from ethanol + ether (also reported as crystals from dil acetone). Dec 175°. uv max (ethanol): 231, 280 nm (ε 7600, 3100).

5998. Metanilic Acid. [121-47-1] 3-Aminobenzenesulfonic acid; m-sulfanilic acid; aniline-m-sulfonic acid. $C_6H_7NO_3S$; mol wt 173.19. C 41.61%, H 4.07%, N 8.09%, O 27.71%, S 18.51%. Usually obtained by reduction of 3-nitrobenzenesulfonic acid: Fierz-David, Blangey, *Fundamental Processes of Dye Chemistry* (Interscience, New York, 1949) pp 120-123; A. I. Vogel, *Practical Organic Chemistry* (Longmans, London, 3rd ed., 1959) p 589. Large-scale process: *FIAT Final Rept.* **1313** (I), 187-191 (1948). The industrial reduction with iron filings and dil acid gives up to 90% yields, in the lab it seldom exceeds 55%. Better yields with small amounts are claimed for a hydrazine-Raney nickel reduction (about 75%): Gialdi *et al.*, *Farmaco Ed. Sci.* **14**, 765 (1959); or by using WS_2 (about 94%): Ehrmann, **FR 1336648** (1963 to BASF).

Anhydrous, orthorhombic needles from water. d 1.69. Very slow crystn yields triclinic prisms of the sesquihydrate. Crystallographic data for both forms: Hall, Maslen, *Acta Crystallogr.* **18**, 301-306 (1965). Photomicrograph of the sesquihydrate: *Helv. Chim. Acta* **12** (1929), facing page 666. Both forms dec on heating without melting. pK (25°) 3.70. Soly of the sesquihydrate in water (16.8°): 2.37% (w/w). The soly of the anhydr form in water is given as 0.79% (w/w) at 0° and as 6.50% (w/w) at 85°. Sparingly sol in methanol, ethanol.

Sodium salt. $H_2NC_6H_4SO_3Na$. Minute crystals from water, dec 302-304°.

USE: The sodium salt in the manuf of azo dyes. In the synthesis of certain sulfa drugs.

5999. Metanil Yellow. [587-98-4] 3-[2-[4-(Phenylamino)-phenyl]diazenyl]benzenesulfonic acid sodium salt (1:1); C.I. Acid Yellow 36; m-[(p-anilinophenyl)azo]benzenesulfonic acid sodium salt; sodium salt of metanilylazodiphenylamine; Ext. D & C Yellow

Metaphanine

No. 1; C.I. 13065; Tropaeolin G. $C_{18}H_{14}N_3NaO_3S$; mol wt 375.38. C 57.59%, H 3.76%, N 11.19%, Na 6.12%, O 12.79%, S 8.54%. Prepn: Welcher, *Organic Analytical Reagents* **vol. 4** (Van Nostrand, 1948) p 516; *Colour Index* **vol. 4** (3rd ed., 1971) p 4045.

Brownish-yellow powder. Sol in water, alc; moderately sol in benzene, ether; slightly sol in acetone.

USE: As indicator in 0.1% soln, of which 2 drops are required for 10 ml liquid. pH: 1.2 red to 2.3 yellow.

6000. Metaphanine. [1805-86-3] $(8\beta,10\beta)$-8,10-Epoxy-8-hydroxy-3,4-dimethoxy-17-methylhasubanan-7-one. $C_{19}H_{23}NO_5$; mol wt 345.40. C 66.07%, H 6.71%, N 4.06%, O 23.16%. A member of the hasubanan alkaloids; also possesses an intramolecular hemiketal ring. Isolated from stems of *Stephania japonica* Miers, *Menispermaceae* from which the alkaloids ***stephanine*** and protostephanine, *q.v.*, are also obtained: Kondo, Sanada, *J. Pharm. Soc. Jpn.* **514**, 5 (1924); *Yakugaku Zasshi* **44**, 5, 1034 (1924); **48**, 177, 930 (1927); Kondo, Watanabe, *ibid.* **58**, 268 (1938). Structure and stereochemistry: Tomita *et al.*, *Tetrahedron Lett.* **1964**, 3605; *Chem. Pharm. Bull.* **13**, 695 (1965). Synthesis of *dl*-form: Ibuka *et al.*, *Tetrahedron Lett.* **1972**, 1393; *eidem, Chem. Pharm. Bull.* **22**, 907 (1974).

Needles, mp 232°. Also reported as colorless prisms, mp 205-206° for the *dl*-form (Ibuka). pKa 6.03. Sol in water, alcohol.

Dihydrometaphanine. $C_{19}H_{25}NO_5$. Cryst, mp 211°. $[\alpha]_D^{20}$ +72° (in chloroform). pKa 6.76.

6001. Metapramine. [21730-16-5] 10,11-Dihydro-*N*,5-dimethyl-5*H*-dibenz[*b*,*f*]azepin-10-amine; 10,11-dihydro-5-methyl-10-(methylamino)-5*H*-dibenz[*b*,*f*]azepine; RP-19560; Timaxel. $C_{16}H_{18}N_2$; mol wt 238.33. C 80.63%, H 7.61%, N 11.75%. Psychotropic agent, related structurally to imipramine, *q.v.* Prepn: J. C. Fouche, C. G. A. Gueremy, **ZA 6800345** corresp to **US 3622565** (1968, 1971 both to Rhone-Poulenc). Determn in plasma by GC: A. R. Viala *et al.*, *Anal. Chem.* **49**, 2354 (1977). HPLC method: J. P. Sommadossi *et al.*, *J. Chromatogr.* **228**, 205 (1982). Pharmacological and clinical effects: P. Dick, *Encephale* **4**, 41 (1978). Clinical studies: L. F. Gayral *et al.*, *ibid.* 365; E. J. Caille, J.-P. Brun, *Psychol. Med.* **13**, 1879 (1981). Biotransformation in animals and man: B. Decouvelaere *et al.*, *Therapie* **37**, 249 (1982).

Hydrochloride. $C_{16}H_{19}ClN_2$. Crystals from isopropanol + ether, mp 238-240°. Injectable formulation. The fumarate is used for tablet formulations.

THERAP CAT: Antidepressant.

6002. Metaproterenol. [586-06-1] 5-[1-Hydroxy-2-[(1-methylethyl)amino]ethyl]-1,3-benzenediol; 3,5-dihydroxy-α-[(iso-propylamino)methyl]benzyl alcohol; 1-(3,5-dihydroxyphenyl)-2-isopropylaminoethanol; 1-(3,5-dihydroxyphenyl)-1-hydroxy-2-isopropylaminoethane; orciprenaline. $C_{11}H_{17}NO_3$; mol wt 211.26. C 62.54%, H 8.11%, N 6.63%, O 22.72%. Prepn and resolution of *d*- and *l*-forms: **BE 611502**; O. Thoma, K. Zeile, **US 3341594** (1961, 1967 both to Boehringer, Ing.). Pharmacology: H. H. Pelz, *Am. J. Med. Sci.* **253**, 321 (1967). Metabolism: K. Tatsumi *et al.*, *Yakugaku Zasshi* **90**, 639 (1970); *ibid.* **91**, 680 (1971). Toxicity: E. I. Goldenthal, *Toxicol. Appl. Pharmacol.* **18**, 185 (1971). Review of clinical toxicology: J. D. Truwit, *Crit. Care Clin.* **7**, 639-657 (1991).

Crystals, mp 100°.

Sulfate. [5874-97-5] TH-152; Alotec; Alupent; Metaprel; Novasmasol. $(C_{11}H_{17}NO_3)_2.H_2SO_4$; mol wt 520.59. Crystals from 90% ethanol, mp 202-203°. Freely sol in water. LD_{50} in rats (mg/kg): 42 orally (Goldenthal).

d-Form hydrochloride. Crystals, mp 212-213°. $[\alpha]_D$ +45.2° (c = 5 in methanol).

l-Form hydrochloride. Crystals, mp 212°. $[\alpha]_D$ −45°.

THERAP CAT: Bronchodilator.

6003. Metaraminol. [54-49-9] (αR)-α-[(1*S*)-1-Aminoethyl]-3-hydroxybenzenemethanol; (−)-α-(1-aminoethyl)-*m*-hydroxybenzyl alcohol; *m*-hydroxynorephedrine; *m*-hydroxypropadrine; *m*-hydroxyphenylpropanolamine; *m*-hydroxy-α-(1-aminoethyl)benzyl alcohol; 2-amino-1-(*m*-hydroxyphenyl)-1-propanol; 1-(*m*-hydroxyphenyl)-2-amino-1-propanol; α-(*m*-hydroxyphenyl)-β-aminopropanol; metaradrine. $C_9H_{13}NO_2$; mol wt 167.21. C 64.65%, H 7.84%, N 8.38%, O 19.14%. Sympathomimetic amine. Prepn from *m*-hydroxyisonitrosopropiophenone: **GB 353361** (1930 to I. G. Farben); *l*-form: **GB 396951** (1932 to I. G. Farben); from *l-m*-hydroxyphenylacetylcarbinol: **CH 162367** (1931 to I. G. Farben); *see also* **US 1948162** and **US 1951302**; Hartung, **US 1995709** (1935 to Sharp & Dohme). HPLC determn in commercial formulations: C. J. Martin, S. J. Saxena, *J. Pharm. Sci.* **69**, 1459 (1980). Pharmacology: A. Cession-Fossion, *Arch. Int. Pharmacodyn. Ther.* **172**, 341 (1968). Toxicity data: O. H. Siegmund *et al.*, *J. Pharmacol. Exp. Ther.* **92**, 207 (1948). Mechanism of action: D. C. Harrison *et al.*, *Ann. Intern. Med.* **59**, 297 (1963).

Bitartrate. [33402-03-8] Aramine; Araminon. $C_{13}H_{19}NO_8$; mol wt 317.29. White, crystalline powder, mp 176-177°. Freely sol in water; slightly sol in alcohol. Practically insol in chloroform and ether. pH of 1% aq soln about 3.5.

Hydrochloride. [5967-52-2] $C_9H_{13}NO_2.HCl$. Hygroscopic crystals, $[\alpha]_D^{20}$ −19.75°. Freely sol in water. LD_{50} i.p. in mice: 440 mg/kg (Siegmund).

Oxalate dihydrate. [5967-53-3] $C_9H_{13}NO_2.C_2H_2O_4.2H_2O$. Crystals, mp 190°. $[\alpha]_D^{20}$ −21.66°. Sol in water.

THERAP CAT: Adrenergic.

6004. Metazachlor. [67129-08-2] 2-Chloro-*N*-(2,6-dimethylphenyl)-*N*-(1*H*-pyrazol-1-ylmethyl)acetamide; 2-chloro-*N*-(pyrazol-1-ylmethyl)acet-2′,6′-xylidide; 2-chloro-2′,6′-dimethyl-*N*-(pyrazol-1-ylmethyl)acetanilide; Butisan S; Sultan. $C_{14}H_{16}ClN_3O$; mol wt 277.75. C 60.54%, H 5.81%, Cl 12.76%, N 15.13%, O 5.76%. Chloroacetamide herbicide for use in food crops and ornamental plantings. Prepn: K. Eicken *et al.*, **DE 2648008** (1978 to BASF); R. Thomas *et al.*, **US 4517011** (1985 to Bayer). Characterization of crystal polymorphs: U. J. Griesser *et al.*, *J. Therm. Anal. Calorim.* **77**, 511 (2004). GC determn in waste water: A. Wenner, M. Wort-

Consult the Name Index before using this section.

berg, *J. High Resolut. Chromatogr.* **21**, 661 (1998). Properties and field trials: H. Birkler, *Weeds Weed Control* **24**, 184 (1983). Persistence in soil: J. Rouchaud *et al.*, *Weed Sci.* **40**, 149 (1992).

Polymorphic. Three stable crystalline forms: pentagonal or hexagonal, tabular crystals from ethanol, mp 83°; prisms from *n*-hexane, acetone, methanol or ethanol, mp 80°; short, prismatic crystals, mp 76°. d 1.16. Soly in water: 1000 mg/l. LD_{50} orally in rats: 1000 mg/kg; LC_{50} (4 hr) in rats: 6.2 mg/l; LC_{50} (96 hr) in rainbow trout: 6.8 - 10 mg/l (Birkler).
USE: Herbicide.

6005. Metazocine. [3734-52-9] 1,2,3,4,5,6-Hexahydro-3,-6,11-trimethyl-2,6-methano-3-benzazocin-8-ol; 2′-hydroxy-2,5,9-trimethyl-6,7-benzomorphan; methobenzmorphan. $C_{15}H_{21}NO$; mol wt 231.34. C 77.88%, H 9.15%, N 6.05%, O 6.92%. Mixed opioid agonist-antagonist; prototype of the benzomorphan analgesics. Prepn of (±)-*cis*-form by Grewe cyclization of 3,4-lutidine methiodide: E. L. May, E. M. Fry, *J. Org. Chem.* **22**, 1366 (1957). Prepn of α- (*cis*) and β- (*trans*) diastereoisomers: E. L. May, J. H. Ager, *ibid.* **24**, 1432 (1959). Optical resolution of α-forms: E. L. May, N. B. Eddy, *ibid.* 1435; of β-forms: S. E. Fullerton *et al.*, *ibid.* **27**, 2144 (1962). Abs config: A. F. Casy, A. P. Parulkar, *J. Med. Chem.* **12**, 178 (1969). Enantioselective synthesis of *cis*-(−)-form: B. M. Trost, W. Tang, *J. Am. Chem. Soc.* **125**, 8744 (2003). Opioid receptor profile: I. Berzetei-Gurske, G. H. Loew, *Prog. Clin. Biol. Res.* **328**, 33 (1990).

cis-(−)-Form

(±)-*cis*-**Form.** [25144-79-0] Plates from dil methanol, mp 232-235°.

(±)-*cis*-**Form hydrochloride monohydrate.** [70222-87-6] C_{15}-$H_{21}NO.HCl.H_2O$; mol wt 285.81. Rods from abs alcohol + ether, mp 194-196°. LD_{50} in mice (mg/kg): 175 s.c. (May, Eddy).

(±)-*trans*-**Form.** [25145-09-9] Prisms from alcohol, mp 215-217.5°.

(**2R,6R,11R**)-**Form.** [21286-60-2] α-(−)-Metazocine; *cis*-(−)-metazocine. mp 183-184.5°. $[\alpha]_D^{20}$ −84.8° (c = 0.09 in abs ethanol).

(**2S,6S,11S**)-**Form.** [67009-58-9] α-(+)-Metazocine; *cis*-(+)-metazocine. Crystals from acetone + water, mp 183-184.5°. $[\alpha]_D^{20}$ +84.3° (c = 0.83 in abs ethanol).

(**2R,6R,11S**)-**Form.** [21286-57-7] β-(−)-Metazocine; *trans*-(−)-metazocine. Oblong prisms from methanol, mp 179-181°. $[\alpha]_D^{20}$ −88.5° (c = 0.52 in abs alc).

(**2S,6S,11R**)-**Form.** [64023-94-5] β-(+)-Metazocine; *trans*-(+)-metazocine. Rods or needles from aq methanol, mp 180-181.5°. $[\alpha]_D^{20}$ +88.6° (c = 0.35 in abs alc).

Note: This is a controlled substance (opiate): **21 CFR**, 1308.12.

6006. Metconazole. [125116-23-6] 5-[(4-Chlorophenyl)-methyl]-2,2-dimethyl-1-(1*H*-1,2,4-triazol-1-ylmethyl)cyclopentanol; (1*RS*,5*RS*,1*RS*,5*SR*)-5-(4-chlorobenzyl)-2,2-dimethyl-1-(1*H*-1,-2,4-triazol-1-ylmethyl)cyclopentanol; Caramba. $C_{17}H_{22}ClN_3O$; mol wt 319.83. C 63.84%, H 6.93%, Cl 11.08%, N 13.14%, O 5.00%. Broad spectrum triazole antifungal; mixture of two racemic diastereomers. Prepn: S. Kumazawa *et al.*, **EP 267778**; *eidem*, **US**

4938792 (1987, 1990 both to Kureha); and structure activity study: A. Ito *et al.*, *J. Pestic. Sci.* **24**, 262 (1999). Efficacy in wheat and barley: B. T. Grayson *et al.*, *Pestic. Sci.* **45**, 153 (1995); A. J. Sampson *et al.*, *ibid.* 161. Review: *idem et al.*, *Brighton Crop Prot. Conf. - Pests Dis.* **1992**, 419-426.

cis-**Form.** [115850-27-6] WL-136184. Mixture of (1*R*,5*S*) and (1*S*,5*R*) isomers and is more bioactive than the *trans* form. Colorless crystals from hexane-ethyl acetate, mp 113-114°. Soly in water: 15 mg/kg.

trans-**Form.** [115850-28-7] mp 90-92°.
USE: Agricultural fungicide.

6007. Meteloidine. [526-13-6] *rel*-(2*E*)-2-Methyl-2-butenoic acid (1*R*,3-*endo*,5*S*,6*S*,7*R*)-6,7-dihydroxy-8-methyl-8-azabicyclo-[3.2.1]oct-3-yl ester; (*E*)-1α*H*,5α*H*-tropane-3α,6β,7β-triol 3-(2-methylcrotonate); 3-(3,6,7-tropanetriol) tiglate; 6,7-dihydroxytropinetiglic acid ester; 6,7-dihydroxy-3-tiglyloxytropane. $C_{13}H_{21}NO_4$; mol wt 255.31. C 61.16%, H 8.29%, N 5.49%, O 25.07%. From leaves of *Datura meteloides* DC., *Solanaceae*: Pyman, Reynolds, *J. Chem. Soc.* **93**, 2077 (1908); King, *ibid.* **115**, 487 (1919); from *D. ferox* L.: Evans, Wellendorf, *ibid.* **1959**, 1406. Stereochemistry: Heusner, *Z. Naturforsch.* **9b**, 683 (1954). Synthesis: Zeile, Heusner, *Arch. Pharm.* **292**, 238 (1959).

Relative stereochemistry

Flat needles from benzene, mp 141-142°. uv max: 217 nm (ε 12,200). Freely sol in alcohol, chloroform, acetone; sparingly sol in water, ether, benzene.

Hydrobromide dihydrate. $C_{13}H_{22}BrNO_4.2H_2O$. Chisel-shaped needles, when anhydr, mp 250°.

6008. Metepa. [57-39-6] 1,1′,1″-Phosphinylidynetris[2-methylaziridine]; tris[2-methyl-1-aziridinyl]phosphine oxide; tris[1-methylethylene]phosphoric triamide; methyl aphoxide; methapoxide; MAPO. $C_9H_{18}N_3OP$; mol wt 215.24. C 50.22%, H 8.43%, N 19.52%, O 7.43%, P 14.39%. Prepd by treatment of 2-methylethyleneimine with $POCl_3$ in alk medium: Parke *et al.*, **US 2606902** (1952 to Am. Cyanamid). Toxicity study: Gaines, *Toxicol. Appl. Pharmacol.* **14**, 515 (1969).

Liquid, bp 90-92° (0.15-0.3 mm). LD_{50} in male, female rats (mg/kg): 136, 213 orally (Gaines).
USE: Chemosterilant; in creaseproofing and flameproofing textiles.

6009. Metergoline. [17692-51-2] N-[[(8β)-1,6-Dimethyler-golin-8-yl]methyl]carbamic acid phenylmethyl ester; D-8β-[(carbo-benzoxyamino)methyl]-1,6-dimethyl-10α-ergoline; D-N-carbobenz-oxydihydro-1-methyllysergamine I; D-8β-[(carboxyamino)methyl]-1,6-dimethylergoline I benzyl ester; D-N-carboxydihydro-1-methyl-lysergamine I benzyl ester; D-[(4,6,6a,7,8,9,10,10a-octahydro-4,7-dimethyl-10aα-indolo[4,3-fg]quinolin-9β-yl)methyl]carbamic acid benzyl ester; methergoline; Liserdol; Contralac. $C_{25}H_{29}N_3O_2$; mol wt 403.53. C 74.41%, H 7.24%, N 10.41%, O 7.93%. Serotonin 5HT-receptor antagonist. Prepn: Bernardi et al., Gazz. Chim. Ital. **94**, 936 (1964); Camerino et al., US 3238211 (1966 to Farmitalia). Pharmacology: C. Beretta et al., Nature **207**, 421 (1965). Metabolic studies: Arcamone et al., Boll. Chim. Farm. **110**, 704 (1971). Mode of action study: L. Krulich et al., Endocrinology **108**, 1115 (1981). Clinical antiprolactin activity: F. Scapin et al., Eur. J. Clin. Pharmacol. **22**, 181 (1982); A. Caballero et al., J. Reprod. Med. **32**, 115 (1987).

Crystals from benzene + ether, mp 146-149°. $[\alpha]_D^{28}$ −7 ±2°. uv max: 291 nm ($E_{1cm}^{1\%}$ 165). Very sol in pyridine; sol in alc, acetone, chloroform. Practically insol in benzene, ether, water. LD_{50} in mice (mg/kg): 85 i.p., 430 orally; in rats (mg/kg): >800 orally (Beretta).

THERAP CAT: Prolactin inhibitor.

THERAP CAT (VET): Prolactin inhibitor.

6010. Metformin. [657-24-9] N,N-Dimethylimidodicarbon-imidic diamide; 1,1-dimethylbiguanide; N,N-dimethyldiguanide; N'-dimethylguanylguanidine; DMGG; LA-6023. $C_4H_{11}N_5$; mol wt 129.17. C 37.19%, H 8.58%, N 54.22%. Oral hypoglycemic agent. Prepn: Werner, Bell, J. Chem. Soc. **121**, 1790 (1922); Shapiro et al., J. Am. Chem. Soc. **81**, 3728 (1959). Use as antidiabetic: J. J. Sterne, US 3174901 (1965 to Jan Marcel Didier Aron-Samuel). Toxicity: Rx Bulletin **3**, 25 (1972). Determn in plasma: S. AbuRuz et al., J. Chromatogr. B **798**, 203 (2003). Clinical pharmacokinetics: G. T. Tucker et al., Br. J. Clin. Pharmacol. **12**, 235 (1981). Review of pharmacology: L. S. Hermann, Diabete Metab. **5**, 233-245 (1979). Metabolic effects and mechanism of action study: M. Stumvoll et al., N. Engl. J. Med. **333**, 550 (1995). Review of efficacy in polycystic ovary syndrome: J. M. Lord et al., Br. Med. J. **327**, 951-955 (2003); in type 2 diabetes mellitus: S. M. Setter et al., Clin. Ther. **25**, 2991-3026 (2003).

Hydrochloride. [1115-70-4] Debeone; Diabex; Glucophage; Metiguanide. $C_4H_{11}N_5$·HCl; mol wt 165.63. Prisms from water, mp 232° (Werner, Bell); crystals from propanol, mp 218-220° (un-corr) (Shapiro). Freely sol in water; slightly sol in alc. Practically insol in ether, chloroform, acetone, methylene chloride. LD_{50} in rats (mg/kg): 1000 orally, 300 s.c. (Rx Bulletin).

p-Chlorophenoxyacetate (salt). [25672-33-7] Glucinan. $C_4H_{11}N_5$·$C_8H_7ClO_3$; mol wt 315.76.

Embonate. [34461-22-8] Metformin pamoate; Stagid. $(C_4H_{11}N_5)_2$·$C_{23}H_{16}O_6$; mol wt 646.71.

THERAP CAT: Antidiabetic. In treatment of polycystic ovary syndrome.

6011. Methabenzthiazuron. [18691-97-9] N-2-Benzothia-zolyl-N,N'-dimethylurea; 1-(2-benzothiazolyl)-1,3-dimethylurea; metabenzthiazuron; MBU; Bayer 5633; Bayer 74283; Tribunil. $C_{10}H_{11}N_3OS$; mol wt 221.28. C 54.28%, H 5.01%, N 18.99%, O 7.23%, S 14.49%. Derivative of urea. Prepn and use as pre-emer-gence herbicide: N. E. Searle, US 2756135 (1956 to du Pont). Use as pre- and post-emergence herbicide in wheat and barley: H. Hack et al., GB 1085430 (1967 to Bayer). Herbicidal properties: H. Hack, Pflanzenschutz-Nachr. **22**, 331 (1969). Toxicity studies: G. Kimmerle, E. Löser, ibid. 351. Use in winter cereals: D. C. Clark et al., Proc. 12th Brit. Weed Control Conf. 163 (1974). Mode of action: G. F. Collet, Weed Res. **9**, 340 (1969). Long-term effect on soil: P. L. Huge, Pflanzenschutz-Nachr. **34**, 97 (1981). Brief review: P. Lours, Def. Veg. **24**, 91 (1970).

White crystals from benzene, mp 119-120.5°. Soly in water at 20°: 59 ppm. Sol in organic solvents. Vapor pressure at 20°: <10^{-6} mm Hg. LD_{50} in mice (mg/kg): >1000 orally; in male, female rats (mg/kg): >2500, >2500 orally; 540, 315 i.p. (Kimmerle, Löser).

USE: Selective herbicide.

6012. Methacholine Chloride. [62-51-1] 2-(Acetyloxy)-$N,-N,N$-trimethyl-1-propanaminium chloride (1:1); acetyl-β-methyl-choline chloride; O-acetyl-β-methylcholine chloride; (2-hydroxy-propyl)trimethylammonium chloride acetate; (2-acetoxypropyl)tri-methylammonium chloride; trimethyl-β-acetoxypropylammonium chloride; Amechol; Provocholine. $C_8H_{18}ClNO_2$; mol wt 195.69. C 49.10%, H 9.27%, Cl 18.12%, N 7.16%, O 16.35%. Parasympa-thomimetic bronchoconstrictor. Prepn: R. T. Major, J. K. Cline, US 2040146 (1936 to Merck & Co.). Mechanism of ganglionic block-ade in cats: R. L. Volle, J. Pharmacol. Exp. Ther. **158**, 66 (1967). Clinical diagnostic efficacy in bronchial asthma: S. L. Spector, R. S. Farr, J. Allergy Clin. Immunol. **56**, 308 (1975); J. G. Easton, I. Hir-ata, Ann. Allergy **50**, 171 (1983).

White, hygroscopic needles from ether, mp 172-173°. Slight odor of dead fish. Very sol in water; freely sol in alc, chloroform. Insol in ether. Aq solns are neutral to litmus. Should not be handled in very moist atmosphere. Bromide is less hygroscopic.

Antidote: Atropine.

THERAP CAT: Cholinergic. Diagnostic aid (bronchial asthma).

6013. Methacrylic Acid. [79-41-4] 2-Methyl-2-propenoic acid; α-methylacrylic acid. $C_4H_6O_2$; mol wt 86.09. C 55.81%, H 7.03%, O 37.17%. Occurs in oil from Roman chamomile. Prepd by dehydration of α-hydroxyisobutyric acid: Crawford, US 2143941 (1939 to I.C.I.); by hypochlorite oxidation of methyl α-alkylvinyl ketone: Meitzner, US 2192142 (1940 to Rohm & Haas); by hydro-lysis of acetone cyanohydrin: Crawford, GB 405699 (1932 to I.C.I.); by oxidation of methacrolein: Bauer, US 2153406 (1939 to Rohm & Haas). Toxicity study: W. Deichmann, J. Ind. Hyg. Toxi-col. **23**, 343 (1941). Review: J. W. Nemec, L. S. Kirch in Kirk-Othmer Encyclopedia of Chemical Technology vol. 15 (Wiley-Inter-science, New York, 3rd ed., 1981) pp 346-376.

Long prisms, mp 16°. Acrid, repulsive odor. d_4^{20} 1.0153. bp_{760} 163°; bp_{30} 81°; bp_{12} 63°. n_D^{20} 1.43143. Flash pt, open cup: 170°F (76°C). Corrosive. Sol in warm water; miscible with alc, ether. Polymerizes easily, especially on heating or in the presence of traces of HCl. The polymer forms a ceramic-looking mass, sol in abs alc, from which it is precipitated by ether.

Methyl ester. [80-62-6] Methyl methacrylate. Review of carcinogenic risk: *IARC Monographs* **60**, 445-474 (1994). Polymerizes easily, forming a clear plastic known as *Lucite*, *Plexiglas*, *Perspex*. Sol in methyl ethyl ketone, tetrahydrofuran, esters, aromatic and chlorinated hydrocarbons. LD_{50} orally in rats: 8.4 g/kg (Deichmann).

Caution: Potential symptoms of overexposure to methacrylic acid or methyl methacrylate are irritation of eyes, skin, nose, throat, mucous membranes; dermatitis; eye and skin burns. *See NIOSH Pocket Guide to Chemical Hazards* (DHHS/NIOSH 97-140, 1997) pp 194, 214.

USE: In the manuf of methacrylate resins and plastics.

6014. Methacrylonitrile. [126-98-7] 2-Methyl-2-propenenitrile; 2-cyano-1-propene; isobutenenitrile; isopropene cyanide; isopropenylnitrile; α-methylacrylonitrile. C_4H_5N; mol wt 67.09. C 71.61%, H 7.51%, N 20.88%. Prepd by vapor-phase catalytic oxidation of methallylamine: Peters *et al.*, *Ind. Eng. Chem.* **40**, 2046 (1948); US 2375016 (1945 to Shell); by the dehydration of methacrylamide: Kung, US 2373190 (1945 to Goodrich); from isopropylene oxide and ammonia: Spillane, Kayser, US 2557703 (1951 to Allied Chem.). Toxicity study: Smyth *et al.*, *Am. Ind. Hyg. Assoc. J.* **23**, 95 (1962).

Liquid. d_4^{20} 0.8001; d_4^{30} 0.7896. mp $-35.8°$. bp_{760} 90.3°. n_D^{20} 1.4007; n_D^{30} 1.3954. Flash pt, open cup: 55°F (13°C). Viscosity (20°): 0.392 cP. Surface tension at 20°, 24.4 dynes/cm. *Flammable; poisonous.* Soly in water at 20° = 2.57 wt-%, at 50° = 2.69 wt-%. Soly of water in methacrylonitrile at 20° = 1.62 wt-%, at 50° = 2.83 wt-%. Misc with acetone, octane, toluene at 20-25°. LD_{50} orally in rats: 0.25 ml/kg (Smyth).

Caution: Potential symptoms of overexposure are irritation of eyes, skin; lacrimation. *See NIOSH Pocket Guide to Chemical Hazards* (DHHS/NIOSH 97-140, 1997) p 198.

USE: In prepn of homopolymers and copolymers; as an intermediate in the prepn of acids, amides, amines, esters, nitriles.

6015. Methacycline. [914-00-1] (4S,4aR,5S,5aR,12aS)-4-(Dimethylamino)-1,4,4a,5,5a,6,11,12a-octahydro-3,5,10,12,12a-pentahydroxy-6-methylene-1,11-dioxo-2-naphthacenecarboxamide; 6-methyleneoxytetracycline; 6-methylene-5-hydroxytetracycline; metacycline. $C_{22}H_{22}N_2O_8$; mol wt 442.42. C 59.73%, H 5.01%, N 6.33%, O 28.93%. Broad spectrum, semi-synthetic antibiotic related to tetracycline, *q.v.* Prepn from oxytetracycline, *q.v.*: R. K. Blackwood *et al.*, US 2984686 (1961 to Pfizer); *eidem, J. Am. Chem. Soc.* **83**, 2773 (1961); **85**, 3943 (1963). Solubility data: J. R. Marsh, P. J. Weiss, *J. Assoc. Off. Anal. Chem.* **50**, 257 (1967). Toxicity data: E. I. Goldenthal, *Toxicol. Appl. Pharmacol.* **18**, 185 (1971). Comparative clinical study with ampicillin, *q.v.*, in chronic bronchitis: S. Chodosh *et al.*, *Chest* **69**, 587 (1976).

Hydrochloride. [3963-95-9] Adriamicina; Ciclobiotic; Germiciclin; Metadomus; Metilenbiotic; Londomycin; Optimycin; Physiomycine; Rindex; Rondomycin. $C_{22}H_{22}N_2O_8 \cdot HCl$; mol wt 478.88. Obtained as crystals containing 0.5 mol water and 0.5 mol methanol from methanol + acetone + concd HCl + ether. Yellow, crystalline powder, dec ~205°. Bitter taste. Soluble in water, sparingly sol in alcohol. Practically insol in ether, chloroform. uv max (methanol + 0.01N HCl): 253, 345 nm (log ε 4.37, 4.19). LD_{50} in rats, mice (mg/kg): 252, 288 i.p. (Goldenthal).

THERAP CAT: Antibacterial.

6016. Methadone. [76-99-3] 6-Dimethylamino-4,4-diphenyl-3-heptanone; 1,1-diphenyl-1-(2-dimethylaminopropyl)-2-buta-

none; 4,4-diphenyl-6-dimethylamino-3-heptanone. $C_{21}H_{27}NO$; mol wt 309.45. C 81.51%, H 8.79%, N 4.53%, O 5.17%. Opioid analgesic. Prepn: Eisleb, *Office of Publication Board, Department of Commerce* **Report no. PB-981**, 96A; Schultz *et al.*, *J. Am. Chem. Soc.* **69**, 2454 (1947); Easton *et al.*, *ibid.* 2941 (1947). Resolution: Larsen *et al.*, *ibid.* **70**, 4194 (1948); Howe, Sletzinger, *ibid.* **71**, 2935 (1949); Brode, Hill, *J. Org. Chem.* **13**, 191 (1948); Howe, Tishler, US 2644010 (1953 to Merck & Co.); Zaugg, US 2983757 (1961 to Abbott). Pharmacokinetics in patients with chronic pain: C. E. Inturrisi *et al.*, *Clin. Pharmacol. Ther.* **41**, 392 (1987). Toxicity data: Finnegan *et al.*, *J. Pharmacol. Exp. Ther.* **92**, 269 (1948); Winter, Flataker, *ibid.* **98**, 305 (1950). *Review:* Eddy, *J. Am. Pharm. Assoc. Pract. Pharm. Ed.* **8**, 536 (1947). Comprehensive description: R. H. Bishara, *Anal. Profiles Drug Subs.* **3**, 365-439 (1974). Review of use in opioid dependence: E. C. Senay, *Int. J. Addict.* **20**, 803-821 (1985). Clinical trial of methadone maintenance vs long-term detoxification: K. L. Sees *et al.*, *J. Am. Med. Assoc.* **283**, 1303 (2000).

mp 78°; pptd from solns of pH >6.

Hydrochloride. [1095-90-5] AN-148; Hoechst 10820; Depridol; Dolophine; Fenadone; Heptadon; Heptanon; Ketalgin; Mephenon; Methadose; Physeptone. Platelets from alcohol + ether, mp 235°. Bitter taste. uv max: 292 nm. Soly (g/100 ml): water 12; alcohol 8; isopropanol 2.4. Freely sol in chloroform. Practically insol in ether, glycerol. The pH of a 1% aq soln: 4.5-5.6. LD_{50} orally in rats: 95 mg/kg (Finnegan).

l-**Form Hydrochloride.** [5967-73-7] Levomethadone hydrochloride; L-Polamidon. Crystals, mp 241°. $[\alpha]_D^{20}$ $-145°$ (c = 2.5); $[\alpha]_D^{20}$ $-169°$ (c = 2.1 in alc). LD_{50} s.c. in rats: 44 mg/kg (Winter, Flataker).

Note: This is a controlled substance (opiate): **21 CFR**, 1308.12.

THERAP CAT: Analgesic; in treatment of opioid dependence.

6017. Methadyl Acetate. [509-74-0] β-[2-(Dimethylamino)-propyl]-α-ethyl-β-phenylbenzeneethanol 1-acetate; 6-(dimethylamino)-4,4-diphenyl-3-heptanol acetate (ester); O-acetyl-6-dimethylamino-4,4-diphenyl-3-heptanol; 3-acetoxy-6-dimethylamino-4,4-diphenylheptane; 5-acetoxy-2-dimethylamino-4,4-diphenylheptane; acetylmethadol; acemethadone; amidolacetate; race-acetylmethadol. $C_{23}H_{31}NO_2$; mol wt 353.51. C 78.15%, H 8.84%, N 3.96%, O 9.05%. Congener of methadone, *q.v.* Prepn of β-*dl*-form: M. Bockmühl, G. Ehrhart, *Ann.* **561**, 52 (1948); of α-*dl*-form: M. E. Speeter *et al.*, *J. Am. Chem. Soc.* **71**, 57 (1949); R. L. Clark, US 2565592; US 2668814 (1951, 1954 to Merck & Co.); M. E. Speeter, US 2649445 (1953 to Bristol); of α-*d* and α-*l*-forms: A. Pohland *et al.*, *J. Am. Chem. Soc.* **71**, 460 (1949). The α-*dl*-form is more active and less toxic than the β-*dl*-form. The α-*l*-form, levomethadyl acetate, *q.v.*, is less active than the α-*d*-form but is longer acting. Metabolism of the α-*dl*-form: R. E. McMahon *et al.*, *J. Pharmacol. Exp. Ther.* **149**, 436 (1965). Analgesic, depressant activity of α-*dl*-form in mice, rats: N. B. Eddy *et al.*, *ibid.* **98**, 121 (1950); of enantiomers: N. B. Eddy *et al.*, *J. Org. Chem.* **17**, 321 (1952). Comparison of methadone with α-*dl*-form in the treatment of heroin addiction: J. H. Jaffe *et al.*, *J. Am. Med. Assoc.* **211**, 1834 (1970).

α-dl-Form hydrochloride. $C_{23}H_{32}ClNO_2$. Crystals from ethyl acetate, mp 213-214°. Sol in water. pH 4-5. LD_{50} in mice (mg/kg): 61.0 s.c., 118.3 orally (Eddy, 1952).

α-dl-Form hydrobromide. $C_{23}H_{32}BrNO_2$. Crystals, mp 193-194.5°. Sol in water.

α-d-Form. Alphacetylmethadol.

α-d-Form hydrochloride. Crystals from ethanol-ether, mp 215°. $[\alpha]_D^{25}$ +61.2° (c = 0.2). Sol in water. LD_{50} in mice (mg/kg): 72.2 s.c., 130.4 orally (Eddy, 1952).

α-l-Form see Levomethadyl Acetate.

β-l-Form. Betacetylmethadol.

β-dl-Form hydrochloride. Crystals, mp 215-217°. Sol in water. LD_{50} in mice (mg/kg): 42.0 s.c., 80.2 orally (Eddy, 1952).

Note: This is a controlled substance (opiate): **21 CFR,** 1308.11.

6018. Methallibure. [926-93-2] N^1-Methyl-N^2-(1-methyl-2-propen-1-yl)-1,2-hydrazinedicarbothioamide; 1-methyl-6-(1-methylallyl)-2,5-dithiobiurea; N-methylthiocarbamoyl-N'-[(1-methylallyl)thiocarbamoyl]hydrazine; metallibure; ICI-33828; Aimax. $C_7H_{14}N_4S_2$; mol wt 218.34. C 38.51%, H 6.46%, N 25.66%, S 29.37%. Prepn: Paget *et al.*, **GB 878177** (1961 to I.C.I.). Inhibition of pituitary gonadotropic function in animals: Paget *et al.*, *Nature* **192**, 1191 (1961). Prevention of fetus nidation when given orally to rats: Harper, *J. Reprod. Fertil.* **7**, 211 (1964).

Crystals, dec 198-200°.

THERAP CAT (VET): Anterior pituitary activator (for swine).

6019. Methamidophos. [10265-92-6] Phosphoramidothioic acid O,S-dimethyl ester; O,S-dimethyl phosphoramidothioate; Bayer 71628; ENT-27396; Ortho 9006; SRA-5172; Monitor; Tamaron. $C_2H_8NO_2PS$; mol wt 141.12. C 17.02%, H 5.71%, N 9.93%, O 22.67%, P 21.95%, S 22.72%. Prepn: W. Lorenz *et al.*, **BE 666143** (1965 to Bayer), *C.A.* **65**, 16864f (1966); **NL 6602588**; P. S. Magee, **US 3309266** (both 1967 to Chevron). Activity: I. Hammann, *Pflanzenschutz-Nachr.* **23**, 140 (1970). Acute toxicity: T. B. Gaines, R. E. Linder, *Fundam. Appl. Toxicol.* **7**, 299 (1986). Review of properties and metabolism: A. M. A. Khasawinah *et al.*, *Pestic. Biochem. Physiol.* **9**, 211-221 (1978).

Crystals from ether, mp 54°. n_D^{40} 1.5092; $d^{44.5}$ 1.31. Vapor pressure at 30°: 3×10^{-4} mm Hg. Readily sol in water, ethanol. LD_{50} in adult male, female rats (mg/kg): 25, 27 orally (Gaines, Linder).

USE: Insecticide; acaricide.

6020. Methamphetamine. [537-46-2] (αS)-N,α-Dimethyl-benzeneethanamine; (S)-(+)-N,α-dimethylphenethylamine; d-N-methylamphetamine; d-deoxyephedrine; d-desoxyephedrine; 1-phenyl-2-methylaminopropane; d-phenylisopropylmethylamine; methyl-β-phenylisopropylamine; Norodin. $C_{10}H_{15}N$; mol wt 149.24. C 80.48%, H 10.13%, N 9.39%. Central stimulant. Can be prepd by reducing ephedrine or pseudoephedrine: Emde, *Helv. Chim. Acta* **12**, 365 (1929). Prepn by reducing the condensation product of benzyl methyl ketone and methylamine: A. Ogata, *J. Pharm. Soc. Jpn.* **451**, 751 (1919), *C.A.* **14**, 745 (1920). Synthesis from D-phenylalanine: D. B. Repke *et al.*, *J. Pharm. Sci.* **67**, 1167 (1978). Stereochemistry-pharmacology aspects: Patil *et al.*, *J. Pharmacol. Exp. Ther.* **155**, 1, 13 (1967). Toxicity data: A. M. Lands *et al.*, *J. Pharmacol. Exp. Ther.* **89**, 382 (1947). Review of clinical trials in bulimia: H. G. Pope, Jr., J. I. Hudson, *J. Clin. Psychiatry* **47**, 339 (1986). Review of pharmacology and abuse potential: A. K. Cho, *Science* **249**, 631-634 (1990).

Hydrochloride. [51-57-0] "Speed"; "meth"; "ice"; Amphedroxyn; Desfedrin; Desoxyfed; Desoxyn; Destim; Doxephrin; Drinalfa; Gerobit; Hiropon; Isophen; Madrine; Methampex; Methedrine; Methylisomyn; Pervitin; Soxysympamine; Syndrox; Tonedron. $C_{10}H_{15}N.HCl$; mol wt 185.70. White crystals, mp 170-175°. Bitter taste. $[\alpha]_D^{25}$ +14 to +20°. Freely sol in water, alc, chloroform. Very slightly sol in absolute ether. A 1% aq soln is neutral or slightly acid to litmus. LD_{50} i.p. in mice: 70 mg/kg (Lands).

Note: This is a controlled substance (stimulant): **21 CFR,** 1308.12.

THERAP CAT: Anorexic. In attention deficit disorder with hyperactivity.

THERAP CAT (VET): Sympathomimetic, CNS stimulant.

6021. Metham Sodium. [137-42-8] Methylcarbamodithioic acid sodium salt; methyldithiocarbamic acid sodium salt; N-methylaminodithioformic acid sodium salt; sodium N-methylaminomethanethionothiolate; metam sodium; SMDC; carbathione; trimaton; VPM. $C_2H_4NNaS_2$; mol wt 129.17. C 18.60%, H 3.12%, N 10.84%, Na 17.80%, S 49.64%. $CH_3NHCSSNa$. Prepn: Compin, *Bull. Soc. Chim. Fr.* [4] **27**, 464 (1920). Use as soil fumigant: S. C. Dorman, A. B. Lindquist, **US 2766554** (1956 to Stauffer). Conversion to methyl isothiocyanate, *q.v.*, in soil: N. J. Turner, M. E. Corden, *Phytopathology* **53**, 1388 (1963); J. H. Smelt, M. Leistra, *Pestic. Sci.* **5**, 401 (1974). Field study: J. A. Bunt, *Meded. Fac. Landbouwwet. Rijksuniv. Gent* **42**, 1529 (1977); L. P. Moreno *et al.*, *Ann. Appl. Biol.* No. 17, **128**, (Supplement) 6-13 (1996). Determn of efficacy: J. Krikun, *Can. J. Plant Pathol.* **8**, 345 (1986).

Dihydrate. N-869; Maposol; Sistan; Vapam. Crystals, anhydr at 130°. Unpleasant odor, similar to that of carbon disulfide. Nonflammable. Soly in water at 20°: 72.2 g/100 ml H_2O. Moderately sol in alc. Sparingly sol in other solvents. Concd aq solns are stable.

Caution: Irritating to skin, mucous membranes: *Clinical Toxicology of Commercial Products*, R.E. Gosselin *et al.*, Eds. (Williams & Wilkins, Baltimore, 5th ed., 1984) Section II, pp 312.

USE: Soil fumigant to control weeds and weed seeds, nematodes, fungi, and soil insects.

6022. Methandriol. [521-10-8] $(3\beta,17\beta)$-17-Methylandrost-5-ene-3,17-diol; methylandrostenediol; MAD; mestenediol; Masdiol; Metocryst; Metildiolo; Androdiol; Metidione; Nabadial; Neosteron; Diolandrone; Stenediol; Neostene; Crestabolic; Diolostene; Metendiol; Metandiol; Methandiol; Methanabol; Methostan; Neutromone; Neutrosteron; Androteston-M; Megabion (Japanese); Notandron. $C_{20}H_{32}O_2$; mol wt 304.47. C 78.90%, H 10.59%, O 10.51%. Prepd by the action of methylmagnesium iodide on 3β-hydroxy-5-androsten-17-one, also called 5,6-dehydroandrosterone: Ruzicka *et al.*, *Helv. Chim. Acta* **18**, 1487 (1935); Miescher, Klarer, *ibid.* **22**, 962 (1939). Absorption spectrum in H_2SO_4: Bernstein, Lenhard, *J. Org. Chem.* **18**, 1153 (1953). NMR: Hampel, Kraemer, *Tetrahedron* **22**, 1601 (1966).

Crystals from ethyl acetate, mp 205.5-206.5°. $[\alpha]_D^{20}$ −73° (alc). Insol in water. Slightly sol in some organic solvents.

Diacetate. [2061-86-1] $C_{24}H_{36}O_4$. Crystals from hexane, mp 145-146°. $[\alpha]_D^{21}$ −59° (c = 0.984 in alc).

Dipropionate. [3593-85-9] Probolin. $C_{26}H_{40}O_4$; mol wt 416.60.

Note: This is a controlled substance (anabolic steroid): **21 CFR,** 1308.13, as defined in 1300.01.

THERAP CAT: Anabolic.

6023. Methandrostenolone. [72-63-9] (17β)-17-Hydroxy-17-methylandrosta-1,4-dien-3-one; 17α-methyl-17β-hydroxyandrosta-1,4-dien-3-one; 1-dehydro-17α-methyltestosterone; methandienone; Danabol; Nerobol; Nabolin; Stenolon; Dianabol. $C_{20}H_{28}O_2$; mol wt 300.44. C 79.96%, H 9.39%, O 10.65%. Anabolic steroid. Prepn by microbial dehydrogenation of 17α-methyltestosterone: Vischer *et al.*, *Helv. Chim. Acta* **38**, 1502 (1955); by reduction of 17α-methyltestosterone with selenium dioxide: Meystre *et al.*, *ibid.* **39**, 734 (1956); Wettstein *et al.*, **US 2900398** (1959 to Ciba).

Crystals from acetone + ether. mp 163-164°. uv max: 245 nm (ε 15600).
Note: This is a controlled substance (anabolic steroid): **21 CFR**, 1308.13, as defined in 1300.01.
THERAP CAT: Androgen.
THERAP CAT (VET): Anabolic.

6024. Methane. [74-82-8] Marsh gas; methyl hydride. CH_4; mol wt 16.04. C 74.88%, H 25.14%. Widely distributed in nature. American natural gas is about 85% methane. The earth's atm contains 0.00022% by vol. Major constituent of the atm of the outer planets (Jupiter, Saturn, Uranus, Neptune), exact figures in *Landolt-Börnstein* **vol. III** (Springer, 6th ed., 1952) p 59; G. P. Kuiper, *The Atmospheres of the Earth and the Planets* (University of Chicago Press, 1949). Pure carbon combines directly with pure hydrogen at temperatures above 1100° forming methane. Above 1500° amount of methane formed increases with temperature: Pring, *J. Chem. Soc.* **97**, 498 (1910). Can be prepd from sodium acetate and sodium hydroxide, or from aluminum carbide and water: Matthews, *J. Am. Chem. Soc.* **21**, 647 (1899); Carroll, *J. Phys. Chem.* **22**, 148 (1918). Prepd commercially from natural gas or by fermentation of cellulose and sewage sludge: Cost, **US 2583090** (1952 to Elliott Co.); Le Paige, de Dommartin, **FR 994032** (1951), *C.A.* **51**, 10836i (1957); Oswald, Golueke, *Mech. Eng.* **86**, 40 (1964).
Colorless, odorless, non-poisonous gas. *Flammable*. Burns with a pale, faintly luminous flame. d_4^0 0.554 (air = 1) or 0.7168 g/liter. mp −182.6°. bp −161.4°. Crit temp −82.25°; crit pressure 45.8 atm. Heat of combustion 978 Btu/cu ft at 25° (a kilogram of CH_4 yields 13,300 kcal). Forms exposive mixtures with air, the loudest explosions occur when one vol of methane is mixed with 10 vols of air (or 2 vols of oxygen). Air contg less than 5.53% methane no longer explodes. Air contg more than 14% methane burns without noise. Autoignition temp 650°. Soly in water at 17°: 3.5 ml/100 ml H_2O. Sol in alc, ether, other organic solvents.
Caution: Simple asphyxiant.
USE: Constituent of illuminating and cooking gas, in the manuf of hydrogen, hydrogen cyanide, ammonia, acetylene, formaldehyde, in organic syntheses.

6025. Methanearsonic Acid. [124-58-3] Methylarsonic acid; methylarsinic acid; monomethylarsinic acid. CH_5AsO_3; mol wt 139.97. C 8.58%, H 3.60%, As 53.53%, O 34.29%. Prepd from sodium arsenite and methyl iodide: A. J. Quick, R. Adams, *J. Am. Chem. Soc.* **44**, 805 (1922). The disodium salt is easily prepd by treating sodium arsenite with dimethyl sulfate at 85°: R. H. Uhlinger, R. V. Cook, *Ind. Eng. Chem.* **11**, 105 (1919). Other routes are by the reaction of methyl chloride with sodium arsonate under pressure: Miller *et al.*, **US 2442372** (1948); by the reaction of dimethyl sulfate with a solution of arsenic trioxide in sodium hydroxide: Schwerdle, **US 2889347** (1959). Acute toxicity: T. B. Gaines, R. E. Linder, *Fundam. Appl. Toxicol.* **7**, 299 (1986). Review of toxicology and human exposure: *Toxicological Profile for Arsenic* (PB2008-100002, 2007) 559 pp.

Monoclinic, spear-shaped plates from abs alcohol. Pleasant acid taste. mp 161°. Strong dibasic acid. Freely sol in water; sol in alcohol.

Monosodium salt. [2163-80-6] Monosodium methylarsonate; sodium methanearsonate; MSMA; Ansar 6.6; Bueno; Daconate; Weed Hoe. CH_4AsNaO_3; mol wt 161.95. White solid, mp 130-140°. Soly at 20° (g/l): water 580. Insol in most organic solvents. $pKa_1 = 4.1$, $pKa_2 = 9.02$. Vapor pressure (25°): 7.8×10^{-8} mmHg. LD_{50} in adult male, female rats (mg/kg): 1105, 1059 orally (Gaines, Linder).
Disodium salt. [144-21-8] Disodium methylarsonate; DSMA; Ansar 8100; Arrhenal; Arsinyl; Clout; Crab-E-Rad; Dal-E-Rad; Sodar. $CH_3AsNa_2O_3$; mol wt 183.93. Hydrated crystals contg $5H_2O$ or $6H_2O$. One gram dissolves in about one ml water; slightly sol in alc. LD_{50} in adult male, female rats (mg/kg): 928, 821 orally (Gaines, Linder).
USE: Herbicide.

6026. Methanesulfonic Acid. [75-75-2] Methylsulfonic acid. CH_4O_3S; mol wt 96.10. C 12.50%, H 4.20%, O 49.94%, S 33.36%. Prepd from sulfur trioxide and methane: Snyder, Grosse, **US 2493038** (1950 to Houdry Process); by oxidation of dimethyl disulfide: Johnson, Wolff, **US 2697722** (1954 to Standard Oil of Indiana); Proell *et al.*, *Ind. Eng. Chem.* **40**, 1129 (1948). Other prepns and chemistry: Suter, *The Organic Chemistry of Sulfur* (Wiley, New York, 1944).

Solid. d_4^{18} 1.4812. mp 20°. bp_{10} 167°; bp_1 122°. Soly at 26-28° in wt %: hexane, 0; methylcyclopentane, 0; benzene, 1.50; toluene, 0.38; o-chlorotoluene, 0.23; ethyl disulfide, 0.47. Thermally stable at moderately elevated temps. Not hydrolyzed by boiling water or hot aq alkali. Corrosive to iron, steel, brass, copper, lead.
Ethyl ester *see* Ethyl Methanesulfonate.
Methyl ester *see* Methyl Methanesulfonate.
Caution: Strong irritant.
USE: As catalyst in polymerization, alkylation and esterification reactions; as a solvent.

6027. Methanesulfonyl Chloride. [124-63-0] Mesyl chloride. CH_3ClO_2S; mol wt 114.54. C 10.49%, H 2.64%, Cl 30.95%, O 27.94%, S 27.99%. Reagent used in solution with pyridine to introduce mesyl groups in organic synthesis. Prepn: L. Carius, *Ann.* **114**, 140 (1860); T. B. Johnson, J. M. Sprague, *J. Am. Chem. Soc.* **58**, 1348 (1936); C. R. Noller, P. J. Hearst, *ibid.* **70**, 3955 (1948). Environmentally-friendly synthesis: S. Mukhopadhyay *et al.*, *Chem. Commun.* **2004**, 472. Thermodynamics: J. P. Guthrie *et al.*, *Can. J. Chem.* **76**, 929 (1998). Synthetic applications: J. H. Looker, *J. Org. Chem.* **17**, 510 (1952); W. E. Truce, C. W. Vriesen, *J. Am. Chem. Soc.* **75**, 5032 (1953); Y. Morimoto *et al.*, *Chem. Lett.* **1998**, 829; P. A. Jacobi, E. H. Sessions, Jr., *Synth. Commun.* **33**, 2575 (2003).

Liquid. d_4^{18} 1.4805. bp_{730} 161°; bp_{21} 60°. n_D^{23} 1.451. Practically insol in water. Sol in alcohol, ether.
USE: In the synthesis of photographic and agricultural chemicals, pharmaceutical intermediates. As a stabilizer; catalyst; curing and chlorinating agent; precursor to methanesulfonic acid.

6028. Methanethiol. [74-93-1] Methyl mercaptan; mercaptomethane; thiomethyl alcohol; methyl sulfhydrate. CH_4S; mol wt 48.10. C 24.97%, H 8.38%, S 66.65%. CH_3SH. Occurs in "sour" gas of W. Texas, in coal tar, and in petroleum distillates. Isolated from roots of *Raphanus sativus*. Produced in the intestinal tract by the action of anaerobic bacteria on albumin. Evolved from *Penicillium brevicaule* bread cultures containing disulfides. Prepn from sodium methyl sulfate and KHS: Klason, *Ber.* **20**, 3409 (1887); Arndt, *ibid.* **54**, 2236 (1921); catalytically from methanol and hydrogen sulfide: Kramer, Reid, *J. Am. Chem. Soc.* **43**, 880 (1921); from methyl chloride and sodium hydrosulfide: Scott *et al.*, *Ind. Eng. Chem.* **47**, 876 (1955). Review of occurrence, preparation, properties and reactions: E. E. Reid, *Organic Chemistry of Bivalent Sulfur* vol. I (Chemical Publishing Co., New York, 1958) pp 15-261; of toxicology and human exposure: *Toxicological Profile for Methyl Mercaptan* (PB93-110799, 1992) 92 pp.

Gas; odor of rotten cabbage. *Flammable.* mp $-123°$. bp_{760} 5.95°; d_4^{20} 0.8665; d_4^{25} 0.9600. Critical temp 196.8°. Critical pressure 71.4 atm. Heat capacity (solid at 14.97-146.57 K): 0.773-17.47 cal/deg/mole; (liquid at 154.16-271.06 K): 21.27-21.13 cal/deg/mole, Russell *et al., J. Am. Chem. Soc.* **64**, 165 (1942). Azeotrope with isobutane (14.9% methanethiol) bp $-13.00°$. Soly in water at 20°: 23.30 g/l. Forms a cryst hydrate.

Sodium salt heminonahydrate. $CH_3SNa.4\frac{1}{2}H_2O$. Needles. Freely sol in water, methanol. Practically insol in ether.

Copper salt. CH_3SCu. Pale yellow crystals. Practically insol in water, ethanol, ether, benzene.

Caution: Potential symptoms of overexposure to methanethiol are irritation of eyes, skin and respiratory system; narcosis; cyanosis; convulsions; direct contact with liquid may cause frostbite. *See NIOSH Pocket Guide to Chemical Hazards* (DHHS/NIOSH 97-140, 1997) p 214.

USE: Intermediate in manuf of jet fuels, pesticides, fungicides, plastics; synthesis of methionine.

6029. Methanol. [67-56-1] Methyl alcohol; carbinol; wood spirit; wood alcohol. CH_4O; mol wt 32.04. C 37.49%, H 12.58%, O 49.93%. CH_3OH. Originally obtained by the destructive distillation of wood, now usually manuf from hydrogen and carbon monoxide or carbon dioxide, also by oxidation of hydrocarbons. *Review: Faith, Keyes & Clark's Industrial Chemicals*, F. A. Lowenheim, M. K. Moran, Eds. (Wiley-Interscience, New York, 4th ed., 1975) pp 524-529; L. E. Wade *et al.*, in *Kirk-Othmer Encyclopedia of Chemical Technology* vol. **15** (Wiley-Interscience, New York, 3rd ed., 1981) pp 398-415. Review of metabolism and toxicology: J. Liesivuori, H. Savolainen, *Pharmacol. Toxicol.* **69**, 157-163 (1991).

Mobile liq. *Flammable, poisonous.* Slight alcoholic odor when pure; crude material may have a repulsive, pungent odor. Burns with a non-luminous, bluish flame. d_4^0 0.8100; d_4^{15} 0.7960; d_4^{20} 0.7915; d_4^{25} 0.7866. mp $-97.8°$. bp_{760} 64.7°; bp_{400} 49.9°; bp_{200} 34.8°; bp_{100} 21.2°; bp_{60} 12.1°; bp_{40} 5.0°; bp_{20} $-6.0°$; bp_{10} $-16.2°$; bp_5 $-25.3°$; $bp_{1.0}$ $-44.0°$; n_D^{15} 1.33066; n_D^{20} 1.3292. Vapor density: 1.11 (air = 1). Flash pt, closed cup: 54°F (12°C). Ignition temp 470°C (878°F). Explosive limits (%-vol in air): 6.0 to 36.5. Crit temp 240.0°; crit pressure 78.5 atm. Specific heat at 20-25° = 0.595 to 0.605. Dipole moment 1.69. Miscible with water, ethanol, ether, benzene, ketones, and most other organic solvents. Forms azeotropes with many compds.

Caution: Potential symptoms of overexposure are irritation of eyes, skin, upper respiratory system; dermatitis; headache, drowsiness, dizziness, vertigo, light headedness, nausea, vomiting, anorexia; weakness, fatigue; abdominal, back and leg pain; visual disturbances, dimness of vision, dilated pupils; optic nerve damage, bilateral blindness. *See NIOSH Pocket Guide to Chemical Hazards* (DHHS/NIOSH 97-140, 1997) p 200; *Clinical Toxicology of Commercial Products*, R. E. Gosselin *et al.*, Eds. (Williams & Wilkins, Baltimore, 5th ed., 1984) Section III, pp 275-279.

USE: Industrial organic solvent. Solvent and solvent adjuvant for polymers. Solvent in the manuf of vitamins, hormones, and other pharmaceuticals. Raw material for making formaldehyde and methyl esters of organic and inorganic acids. Antifreeze for automotive radiators and air brakes; ingredient of gasoline and diesel oil antifreezes. Octane booster in gasoline. As fuel for picnic stoves and soldering torches. Extractant for animal and vegetable oils. To denature ethanol. Softening agent for pyroxylin plastics.

6030. Methantheline Bromide. [53-46-3] *N,N*-Diethyl-*N*-methyl-2-[(9*H*-xanthen-9-ylcarbonyl)oxy]ethanaminium bromide (1:1); diethyl(2-hydroxyethyl)methylammonium bromide xanthene-9-carboxylate; β-diethylaminoethyl 9-xanthenecarboxylate methobromide; MTB-51; SC-2910; Banthine Bromide; Avagal; Uldumont; Vagantin; Metaxan; Methanide; Xanteline; Gastron; Gastrosedan; Methanthine Bromide; Vagamin; Metanyl; Doladene; Asabaine. $C_{21}H_{26}BrNO_3$; mol wt 420.35. C 60.00%, H 6.23%, Br 19.01%, N 3.33%, O 11.42%. Anticholinergic. Prepn from 9-xanthenecarboxylic acid and 2-diethylaminoethanol: J. W. Cusic, R. A. Robinson, *J. Org. Chem.* **16**, 1921 (1951); *eidem*, US 2659732 (1952 to Searle). Pharmacology: W. E. Hambourger *et al., J. Pharmacol. Exp. Ther.* **99**, 245 (1950).

Crystals from isopropanol, mp 175-176°. Bitter taste. Very slightly hygroscopic. Freely sol in water, alcohol. Practically insol in ether. pH (2% aq soln): 5.0-5.5. Aq solns tend to hydrolyze after a few days. The corresponding chloride is very hygroscopic. uv max (alc): 246, 282 nm ($E_{1cm}^{1\%}$ 135, 69). LD_{50} i.p. in mice: 76 mg/kg (Hambourger).

THERAP CAT: Antispasmodic. In treatment of urinary incontinence.

6031. Methapyrilene. [91-80-5] N^1,N^2-Dimethyl-N^2-2-pyridinyl-N'-(2-thienylmethyl)-1,2-ethanediamine; 2-[(2-dimethylaminoethyl)-2-thenylamino]pyridine; *N,N*-dimethyl-*N'*-(2-pyridyl)-*N'*-(2-thenyl)ethylenediamine; *N,N*-dimethyl-*N'*-(α-pyridyl)-*N'*-(2-methylthienyl)ethylenediamine; thenylpyramine; AH-42; Thenylene; Pyrathyn; Thionylan; Histadyl (formerly); Restryl; Rest-On; Sleepwell; Paradormalene; Pyrinistab; Pyrinistol; Lullamin. $C_{14}H_{19}N_3S$; mol wt 261.39. C 64.33%, H 7.33%, N 16.08%, S 12.27%. Prepd by heating a 2-thenyl halide with an alkali metal salt of *N,N*-dimethyl-*N'*-(2-pyridyl)ethylenediamine: Kyrides, **US 2581868** (1952 to Monsanto). Alternate syntheses: Weston, *J. Am. Chem. Soc.* **69**, 980 (1947); Clapp *et al., ibid.* 1549. Toxicity data: H. M. Lee *et al., Proc. Soc. Exp. Biol. Med.* **80**, 458 (1952). Carcinogenicity study: W. Lijinsky *et al., Science* **209**, 817 (1980). Study of mechanism of hepatocarcinogenicity: K. L. Steinmetz *et al., Carcinogenesis* **9**, 959 (1988). Clinical pharmacokinetics: E. P. Calandre *et al., Clin. Pharmacol. Ther.* **29**, 527 (1981). HPLC determn in feed and sleep aid tablets: B. Shaikh, M. R. Hallmark, *J. Assoc. Off. Anal. Chem.* **64**, 889 (1981); in feed, urine and wastewater: H. C. Thompson, Jr., C. L. Holder, *J. Chromatogr.* **283**, 251 (1984).

Liquid. $bp_{0.45}$ 125-135°; bp_3 173-175°. n_D^{25} 1.5842 (also reported as 1.5835). LD_{50} in mice, guinea pigs (mg/kg): 182.2 ±12.8, 374.9 ±34.5 orally; in mice (mg/kg): 19.85 ±0.69 i.v. (Lee).

Hydrochloride. [135-23-9] $C_{14}H_{19}N_3S.HCl$. Bitter crystals, mp 162°. uv max: 238 nm ($E_{1cm}^{1\%}$ 623); min: 272 nm. One gram dissolves in about 0.5 ml water, in 5 ml alcohol, in 3 ml chloroform. Practically insol in ether, benzene.

Fumarate. [33032-12-1] $(C_{14}H_{19}N_3S)_2.3C_4H_4O_4$. Prepn: Meyer, **GB 694805** (1953 to Monsanto). Crystals, mp 135-136°.

THERAP CAT: Antihistaminic.

THERAP CAT (VET): Antihistaminic.

6032. Methaqualone. [72-44-6] 2-Methyl-3-(2-methylphenyl)-4(3*H*)-quinazolinone; 2-methyl-3-*o*-tolyl-4(3*H*)-quinazolinone; 3,4-dihydro-2-methyl-4-oxo-3-*o*-tolylquinazoline; metolquizolone; QZ-2; CI-705; R-148; TR-495; Parest; Quaalude; Somnafac. $C_{16}H_{14}N_2O$; mol wt 250.30. C 76.78%, H 5.64%, N 11.19%, O 6.39%. Quinazoline sedative-hypnotic. Prepn: Kacker, Zaheer, *J. Indian Chem. Soc.* **28**, 344 (1951); Klosa, *J. Prakt. Chem.* **14**, 84 (1961); **20**, 283 (1963). Review of syntheses: E. F. van Zyl, *Forensic Sci. Int.* **122**, 142-149 (2001). GC determn in serum: M. A. Evenson, G. L. Lensmeyer, *Clin. Chem.* **20**, 249 (1974). Metabolism: W. G. Stillwell *et al., Drug Metab. Dispos.* **3**, 287 (1975). Toxicity data: E. I. Goldenthal, *Toxicol. Appl. Pharmacol.* **18**, 185 (1971). Reports of abuse potential: D. S. Inaba *et al., J. Am. Med. Assoc.* **224**, 1505 (1973); E. M. Pascarelli, *ibid.* 1512. Review of pharmacology: S. S. Brown, S. Goenechea, *Clin. Pharmacol. Ther.* **14**, 314-324 (1973). Comprehensive description: D. M. Patel *et al., Anal. Profiles Drug Subs.* **4**, 245-267 (1975). Comparison of abuse potential with benzodiazepines: M. H. Orzack *et al., Int. J. Addict.* **23**, 449-467 (1988).

Crystals, mp 120°; also given as mp 114-116°. uv max (ethanol): 225, 263, 304, 316 nm; (0.01N HCl): about 234, 269 nm. Sol in ethanol, ether, chloroform. Practically insol in water. LD$_{50}$ orally in rats: 255 mg/kg (Goldenthal).

Hydrochloride. [340-56-7] C$_{16}$H$_{14}$N$_2$O.HCl; mol wt 286.76. Crystals, mp 255-265°. Sol in ether, ethanol. Practically insol in water.

Note: This is a controlled substance (depressant): **21 CFR,** 1308.11.

6033. Metharbital. [50-11-3] 5,5-Diethyl-1-methyl-2,4,6-(1H,3H,5H)-pyrimidinetrione; 5,5-diethyl-1-methylbarbituric acid; Gemonil. C$_9$H$_{14}$N$_2$O$_3$; mol wt 198.22. C 54.53%, H 7.12%, N 14.13%, O 24.21%. Prepn: Halpern, Jones, *J. Am. Pharm. Assoc.* **38**, 352 (1949); Snyder, Link, *J. Am. Chem. Soc.* **75**, 1881 (1953).

Crystals from benzene + petr ether, mp 155°. Solubility (g/100 ml soln) in water 0.12; in alcohol 4.3; in ether 2.6. pH of satd aq solns: 5.6-5.7. pK 8.45: Fox, Shugar, *Bull. Soc. Chim. Belg.* **61**, 44 (1952).

Note: This is a controlled substance (depressant): **21 CFR,** 1308.13.

THERAP CAT: Anticonvulsant.

6034. Methazolamide. [554-57-4] N-[5-(Aminosulfonyl)-3-methyl-1,3,4-thiadiazol-2(3H)-ylidene]acetamide; N-(4-methyl-2-sulfamoyl-Δ^2-1,3,4-thiadiazolin-5-ylidene)acetamide; 5-acetylamino-4-methyl-Δ^2-1,3,4-thiadiazoline-2-sulfonamide. C$_5$H$_8$N$_4$O$_3$S$_2$; mol wt 236.26. C 25.42%, H 3.41%, N 23.71%, O 20.32%, S 27.14%. Carbonic anhydrase inhibitor. Prepn: Young *et al., J. Am. Chem. Soc.* **78**, 4649 (1956); US 2783241 (1957 to Am. Cyanamid); Pala, *Farmaco Ed. Sci.* **13**, 650 (1958). HPLC determn in biological fluids: G. R. Iyer, D. R. Taft, *J. Pharm. Biomed. Anal.* **16**, 1021 (1998). Effects on intraocular pressure and aqueous humor flow rate in dogs: B. J. Skorobohach *et al., Am. J. Vet. Res.* **64**, 183 (2003).

Crystals from water, mp 213-214°. pKa 7.30. uv max (95% ethanol): 254 nm (log ε 3.66); (0.1N NaOH): 247 nm (log ε 3.61). Sol in DMF; slightly sol in acetone; very slightly sol in water, alcohol.

THERAP CAT: Diuretic; antiglaucoma.

THERAP CAT (VET): Antiglaucoma.

6035. Methcathinone. [5650-44-2] 2-(Methylamino)-1-phenyl-1-propanone; α-methylaminopropiophenone; monomethyl-propion; ephedrone; Jeff; mulka; cat; cosmos; Jee cocktail. C$_{10}$H$_{13}$NO; mol wt 163.22. C 73.59%, H 8.03%, N 8.58%, O 9.80%. Oxidation product of ephedrine possessing amphetamine-like central stimulant properties. Prepn: J. F. Hyde *et al., J. Am. Chem. Soc.* **50**, 2287 (1928); K. Y. Zhingel *et al., J. Forensic Sci.* **36**, 915 (1991).

Prepn of enantiomers and chromatographic analysis: J. DeRuiter *et al., J. Chromatogr. Sci.* **32**, 552 (1994). Comparative pharmacology: R. Young, R. A. Glennon, *Pharmacol. Biochem. Behav.* **45**, 229 (1993). Pharmacology of enantiomers: R. A. Glennon *et al., ibid.* **50**, 601 (1995). Absorption and metabolism: S. L. Markantonis *et al., J. Pharm. Pharmacol.* **38**, 515 (1986). GLC/MS determn in urine: *eidem, Chim. Chron.* **14**, 211 (1985). Impact on public health: T. S. Emerson, J. E. Cisek, *Ann. Emerg. Med.* **22**, 1897 (1993).

Hydrochloride. [49656-78-2] C$_{10}$H$_{13}$NO.HCl. Crystals from alcohol-acetone, mp 176-177°.

l-**Form hydrochloride.** C$_{10}$H$_{13}$NO.HCl. Prepn: Y. J. L'Italien, M. C. Rebstock, US 2802865 (1957 to Parke, Davis). Crystals from ethanol-ether or chloroform-petr ether, mp 182-183°. Pungent taste. [α]$_D^{25}$ -53° (c = 1 in water). Sol in water. LD$_{50}$ in mice, rats (mg/kg): 106, 90 iv; 233, 86 sc; 342, 99 orally (Emerson).

Note: This is a controlled substance (stimulant): **21 CFR,** 1308.11.

6036. Methdilazine. [1982-37-2] 10-[(1-Methyl-3-pyrrolidinyl)methyl]-10H-phenothiazine. C$_{18}$H$_{20}$N$_2$S; mol wt 296.43. C 72.93%, H 6.80%, N 9.45%, S 10.82%. Prepn: Feldkamp, Wu, US 2945855 (1960 to Mead Johnson). Toxicity data: E. I. Goldenthal, *Toxicol. Appl. Pharmacol.* **18**, 185 (1971).

Crystals, mp 87-88°.

Hydrochloride. [1229-35-2] Dilosyn; Disyncran; Tacaryl. C$_{18}$H$_{20}$N$_2$S.HCl; mol wt 332.89. Crystals from isopropyl alc, mp 187.5-189°. Freely sol in water, alc, chloroform. LD$_{50}$ orally in rats: 320 mg/kg (Goldenthal).

THERAP CAT: Antipruritic.

6037. Methemoglobin. Hemiglobin; ferrihemoglobin; met Hb. Oxidation product of the normal blood pigment, hemoglobin, in which the iron is present in the ferric state. Methemoglobin may be formed through the direct action of oxidants, through the coaction of hydrogen donors and atmospheric oxygen, or through the autoxidation of hemoglobin which occurs to a small extent in normal blood. Methemoglobin does not have the capacity to combine with molecular oxygen. *Reviews:* Bodansky, *Pharmacol. Rev.* **3**, 144 (1951); Jaffé in *The Red Blood Cell,* Bishop, Surgenor, Eds. (Academic Press, 1964) pp 397-422.

Cleaved by acids and bases to yield globin and ferriprotoporphyrin.

6038. Methenamine. [100-97-0] 1,3,5,7-Tetraazatricyclo-[3.3.1.1^{3,7}]decane; hexamethylenetetramine; HMT; HMTA; hexamine; 1,3,5,7-tetraazaadamantane; hexamethylenamine. C$_6$H$_{12}$N$_4$; mol wt 140.19. C 51.41%, H 8.63%, N 39.97%. Prepn: A. Butlerow, *Ann.* **115**, 322 (1860); from formaldehyde and ammonia: F. Meissner, E. Schwiedessen, US 2762799; US 2762800 (both 1956); *Faith, Keyes & Clark's Industrial Chemicals,* F. A. Lowenheim, M. K. Moran, Eds. (Wiley-Interscience, New York, 4th ed., 1975) pp 445-448. Crystal property studies: G. H. Heilmeier, *Appl. Opt.* **3**, 1281 (1964). Comprehensive reviews: J. F. Walker, *Formaldehyde* (Reinhold, New York, 3rd ed., 1964) pp 511-551; J. M. Dreyfors *et al., Am. Ind. Hyg. Assoc. J.* **50**, 579-585 (1989). Review of therapeutic use: R. Gleckman *et al., Am. J. Hosp. Pharm.* **36**, 1509-1512 (1979); of toxicology: H. J. Trochimowicz *et al.* in *Patty's Industrial*

Hygiene and Toxicology vol. 2E, G. D. Clayton, F. E. Clayton, Eds. (John Wiley & Sons, New York, 1994) pp 3324-3327.

Colorless, odorless crystalline solid with sweet, metallic taste. Sublimes at about 260° without melting and with partial decompn; somewhat volatile at lower temp. n_D^{25} 1.5911. *Flammable.* Specific heat: 36.4 cal/°C. d 1.331. Freely sol in water. Soly at 20° (g/100g): chloroform 13.4; methanol 7.25; abs ethanol 2.89; acetone 0.65; benzene 0.23; xylene 0.14; ether 0.06. Practically insol in petr ether. pH of 0.2 molar aq sol 8.4.

Anhydromethylenecitrate. [6190-43-8] Helmitol. $C_6H_{12}N_4.$-$C_7H_8O_7$; mol wt 344.32. White, cryst powder. mp 175° with decompn. Sol in 10 parts water; very slightly sol in alcohol, ether.

Hippurate. [5714-73-8] Hiprex; Urex; Urotractan. $C_6H_{12}N_4.$-$C_9H_9NO_3$; mol wt 319.37. Crystals, mp 105-110°. Freely sol in water, alcohol.

Mandelate. [587-23-5] Mandelamine; Uronamin. $C_6H_{12}N_4.$-$C_8H_8O_3$; mol wt 292.34. Crystals or white, crystalline powder. mp 128-130°. Very sol in water; sol in alc, chloroform; slightly sol in acetone, ether.

Sulfosalicylate. [20480-93-7] $C_6H_{12}N_4.C_7H_6O_6S$; mol wt 358.37. Occurs as the monohydrate. Odorless, cryst powder, mp ~190° with decompn. Sol in 8 parts water; slightly sol in alcohol.

USE: In adhesives, coatings, and sealing compounds; as crosslinking agent for hardening phenol-formaldehyde resin and vulcanizing rubber; as corrosion inhibitor for steel; as fuel tablets for camping stoves; as stabilizer for lubricating and insulating oils; for the manufacture of explosive compounds (*see* Cyclonite); antimicrobial food additive; detection of metals.

THERAP CAT: Antibacterial (urinary).

THERAP CAT (VET): Antibacterial (urinary).

6039. Methenolone. [153-00-4] (5α,17β)-17-Hydroxy-1-methylandrost-1-en-3-one; 1-methyl-Δ¹-androsten-17β-ol-3-one; méténolone. $C_{20}H_{30}O_2$; mol wt 302.46. C 79.42%, H 10.00%, O 10.58%. Prepn: Wiechert, Kaspar, *Ber.* **93**, 1710 (1960); Popper, **DE 1023764** (1958 to Schering, AG).

Crystals from isopropyl ether, mp 149.5-152°. mp 160-161° (Popper). $[\alpha]_D$ +58.9°.

17-Acetate. [434-05-9] Primobolan Tablets; Primonabol. $C_{22}H_{32}O_3$; mol wt 344.50. Crystals from isopropyl ether, mp 138-139°. uv max (methanol): 240 nm (ε 13300). Sol in methanol, ether, chloroform.

17-Enanthate. [303-42-4] Methenolone enanthate; 17β-heptanoyloxy-1-methyl-5α-androst-1-en-3-one; Primobolan-Depot; Primonabol Depot. $C_{27}H_{42}O_3$; mol wt 414.63. Used as repository form.

Note: This is a controlled substance (anabolic steroid): **21 CFR**, 1308.13, as defined in 1300.01.

THERAP CAT: Anabolic.

6040. Methestrol. [130-73-4] 4,4'-(1,2-Diethyl-1,2-ethanediyl)bis[2-methylphenol]; 4,4'-(1,2-diethylethylene)di-*o*-cresol; 3,4-bis(3-methyl-4-hydroxyphenyl)hexane; dimethylhexestrol; promethestrol; γ-promethestrol. $C_{20}H_{26}O_2$; mol wt 298.43. C 80.49%, H 8.78%, O 10.72%. Prepd by hydrolysis of the dipropionate: Niederl *et al., J. Am. Chem. Soc.* **70**, 508 (1948). Chromatographic identifi-

cation and determn: R. W. Roos, *J. Pharm. Sci.* **63**, 594 (1974). Pharmacologic study: P. H. Jellinck, A. M. Newcombe, *Biochem. Pharmacol.* **29**, 3031 (1980). Pharmacotherapeutic study: C. B. Hammond, W. S. Maxson, *Fertil. Steril.* **37**, 5 (1982).

Crystals from dil acetic acid, mp 145°.

Dipropionate. Meprane Dipropionate. $C_{26}H_{34}O_4$; mol wt 410.55. Cryst, mp 115°. Freely sol in ether, ethyl acetate, benzene; slightly sol in alc. Practically insol in water, dil acids.

THERAP CAT: Estrogen.

6041. Methicillin Sodium. [132-92-3] (2S,5R,6R)-6-[(2,6-Dimethoxybenzoyl)amino]-3,3-dimethyl-7-oxo-4-thia-1-azabicyclo[3.2.0]heptane-2-carboxylic acid sodium salt (1:1); 6-(2,6-dimethoxybenzamido)penicillanic acid sodium salt; sodium 6-(2,6-dimethoxyphenylpenicillin; 2,6-dimethoxyphenylpenicillin sodium salt; sodium 6-(2,6-dimethoxybenzamido)penicillinate; 2,6-dimethoxybenzoylpenin sodium salt; dimethoxyphenecillin sodium; sodium methicillin; BRL-1241; X-1497; Azapen; Belfacillin; Celpillina; Celbenin; Cinopenil; Flabelline; Penistaph; Staphcillin. $C_{17}H_{19}N_2NaO_6S$; mol wt 402.40. C 50.74%, H 4.76%, N 6.96%, Na 5.71%, O 23.86%, S 7.97%. Semi-synthetic antibiotic related to penicillin. Prepn starting with 6-aminopenicillanic acid: Doyle *et al., J. Chem. Soc.* **1962**, 1457; Doyle *et al.,* **US 2951839** (1960); Glombitza, *Ann.* **673**, 166 (1964). Physical-chemical properties: Cotta-Ramusino, Intonti, *Farmaco Ed. Prat.* **16**, 227 (1961).

Monohydrate. Crystals from acetone, mp 196-197° (dec). $[\alpha]_D^{20}$ +230° (c = 5); +225° (c = 1). uv max: 281 nm ($E_{1cm}^{1\%}$ 55); min 264 nm. Solubilities (mg/ml) at 20°: water >300; ethanol 40; ether <0.03; acetone 0.35; chloroform 0.06; isooctane <0.03.

THERAP CAT: Antibacterial.

THERAP CAT (VET): Antimicrobial.

6042. Methidathion. [950-37-8] Phosphorodithioic acid S-[(5-methoxy-2-oxo-1,3,4-thiadiazol-3(2H)-yl)methyl] O,O-dimethyl ester; phosphorodithioic acid O,O-dimethyl ester S-ester with 4-(mercaptomethyl)-2-methoxy-Δ²-1,3,4-thiadiazolin-5-one; dithiophosphoric acid O,O'-dimethyl-S-[(5-methoxy-1,3,4-thiadiazol-2(3H)-one-3-yl)methyl] ester; O,O'-dimethyl-S-[(2-methoxy-1,3,4-thiadiazol-5(4H)-one-4-yl)methyl] dithiophosphate; GS-13005; Supracide. $C_6H_{11}N_2O_4PS_3$; mol wt 302.32. C 23.84%, H 3.67%, N 9.27%, O 21.17%, P 10.25%, S 31.81%. Organophosphate insecticide; cholinesterase inhibitor. Synthesis and degradation products: Rüfenacht, *Helv. Chim. Acta* **51**, 518 (1968); **FR 1335755** (1963 to Geigy), *C.A.* **60**, 1764g (1964). Metabolism in rats: Esser *et al., Helv. Chim. Acta* **51**, 513 (1968). Acute toxicity: T. B. Gaines, R. E. Linder, *Fundam. Appl. Toxicol.* **7**, 299 (1986). Voltammetric determn in water samples: G. Erdogdu, *J. Anal. Chem.* **61**, 673 (2006). Photodegradation studies in soil: L. Sánchez *et al., Chemosphere* **59**, 969 (2005). Carcinogenicity studies: J. A. Quest *et al., Regul. Toxicol. Pharmacol.* **12**, 117 (1990).

Crystals from methanol, mp 39-40°. Soly in water <1%. Readily sol in benzene, acetone, methanol, xylene and other org solvents. *Poisonous*. LD$_{50}$ in adult male, female rats (mg/kg): 31, 32 orally (Gaines, Linder).

USE: Insecticide, acaricide.

6043. Methimazole. [60-56-0] 1,3-Dihydro-1-methyl-2*H*-imidazole-2-thione; 1-methylimidazole-2-thiol; 1-methyl-2-mercaptoimidazole; mercazolyl; thiamazole; Danantizol; Favistan; Felimazole; Mercazole; Strumazol; Tapazol; Tapazole; Thacapzol; Thyrozol; Tirodril; Tizorol. C$_4$H$_6$N$_2$S; mol wt 114.17. C 42.08%, H 5.30%, N 24.54%, S 28.08%. Antihormone; inhibits iodide integration into tyrosyl residues, inhibiting production of thyroid hormones. Prepd by treating aminoacetaldehyde diethyl acetal with methyl isothiocyanate: Wohl, Marckwald, *Ber.* **22**, 1354 (1889); from thiocyanic acid and *N*-substituted amino acetals: R. G. Jones *et al.*, *J. Am. Chem. Soc.* **71**, 4000 (1949). NMR structural study: M. Garner *et al.*, *Bioorg. Med. Chem. Lett.* **4**, 1357 (1994). HPLC determn in fish homogenates: L. Hollosi *et al.*, *J. Pharm. Biomed. Anal.* **36**, 921 (2004). Metabolism: D. S. Sitar, D. P. Thornhill, *J. Pharmacol. Exp. Ther.* **184**, 432 (1973). Comprehensive description: H. Y. Aboul-Enein, A. A. Al-Badr, *Anal. Profiles Drug Subs.* **8**, 351-370 (1979). Review of pharmacology and clinical experience: D. S. Cooper, *N. Engl. J. Med.* **311**, 1353-1362 (1984). Use in feline hyperthyroidism: G. Hoffmann *et al.*, *J. Feline Med. Surg.* **5**, 77 (2003). Clinical trial in hyperthyroidism: F. Azizi *et al.*, *Eur. J. Endocrinol.* **152**, 695 (2005).

White to pale buff, crystalline powder; faint characteristic odor. Leaflets from alc, mp 146-148°. bp 280° (some decompn). uv max (0.1*N* H$_2$SO$_4$): 211, 251.5 nm (E$_{1cm}^{1\%}$ 593, 1528). Freely sol in water, alc, chloroform; sparingly sol in petr ether, benzene; slightly sol in ether.

USE: In cyanide-free silver electroplating. Model compound for endocrine disruption in physiological studies.

THERAP CAT: Antihyperthyroid.

THERAP CAT (VET): In treatment of feline hyperthyroidism.

6044. Methiocarb. [2032-65-7] 3,5-Dimethyl-4-(methylthio)phenol 1-(*N*-methylcarbamate); methylcarbamic acid 4-(methylthio)-3,5-xylyl ester; 4-(methylthio)-3,5-xylyl methylcarbamate; 4-methylthio-3,5-dimethylphenyl *N*-methylcarbamate; mercaptodimethur; metmercapturon; Bayer 37344; H-321; Draza; Mesurol. C$_{11}$H$_{15}$NO$_2$S; mol wt 225.31. C 58.64%, H 6.71%, N 6.22%, O 14.20%, S 14.23%. Prepn: E. Schegk *et al.*, **GB 912895**; *eidem*, **US 3313684** (1962, 1967 both to Bayer); E. E. Gilbert, J. A. Otto, **US 3358012** (1967 to Allied). Molluscicidal activity: H. H. Crowell, *J. Econ. Entomol.* **60**, 1048 (1967). Bird repellent properties: E. W. Schafer, R. B. Brunton, *J. Wildl. Manage.* **35**, 569 (1971). Toxicity study: T. B. Gaines, *Toxicol. Appl. Pharmacol.* **14**, 515 (1969).

White crystalline powder, mp 121.5°. Insol in water. Sol in organic solvents. Unstable in alk media. LD$_{50}$ in male, female rats (mg/kg): 70, 60 orally (Gaines).

USE: Insecticide; molluscicide; bird repellent.

6045. Methiodal Sodium. [126-31-8] 1-Iodomethanesulfonic acid sodium salt (1:1); sodium iodomethanesulfonate; Skiodan; Abrodil; Radiographol; Segosin; Diagnorenol. CH$_2$INaO$_3$S; mol wt 243.98. C 4.92%, H 0.83%, I 52.01%, Na 9.42%, O 19.67%, S 13.14%. CH$_2$ISO$_3$Na. Prepd by the action of sodium sulfite on

methylene iodide at 70° in water-alcohol soln: Ossenbeck *et al.*, **US 1842626** (1932 to Winthrop); from iodoform + sodium sulfite: Allardt, **US 1867793** (1932 to Schering-Kahlbaum AG).

Crystals. Slightly saline taste followed by sweetish aftertaste. Freely sol in water (70 g/100 ml); slightly sol in alcohol (2.5 g/100 ml), benzene, ether, acetone.

THERAP CAT: Diagnostic aid (radiopaque medium—urographic).

6046. Methionic Acid. [503-40-2] Methanedisulfonic acid. CH$_4$O$_6$S$_2$; mol wt 176.16. C 6.82%, H 2.29%, O 54.49%, S 36.40%. Prepn from methane + sulfur trioxide: Snyder, Grosse, **US 2493038** (1950 to Houdry Process); by H$_2$SO$_4$ oxidation of acetic acid: Schwab, Neuwirth, *Ber.* **90**, 567 (1957); from MeSO$_3$H + SO$_3$: Crowder, Gilbert, **US 2842589** (1958 to Allied Chem.). Prepn of aluminum salt: Christian, Jenkins, *J. Am. Pharm. Assoc.* **39**, 633 (1950); **US 2504107** (1950 to Purdue Res. Found.).

Crystals, mp 96-100°.

Aluminum salt. [52667-15-9] Tris[μ-[methanedisulfonato(2−)]]dialuminum; aluminum methanedisulfonate. C$_3$H$_6$Al$_2$-O$_{18}$S$_6$; mol wt 576.39. Crystals from water + alcohol. Hygroscopic. Soly in water at 27°: 69 w/v. pH of 5% aq soln = 3.5.

USE: Antiperspirant.

THERAP CAT: Aluminum salt as topical astringent.

6047. Methionine. [63-68-3] L-Methionine; Met; M; 2-amino-4-(methylthio)butyric acid; α-amino-γ-methylmercaptobutyric acid; (*S*)-2-amino-4-(methylthio)butanoic acid; γ-methylthio-α-aminobutyric acid; Acimethin. C$_5$H$_{11}$NO$_2$S; mol wt 149.21. C 40.25%, H 7.43%, N 9.39%, O 21.44%, S 21.49%. Essential amino acid for human development. Universal translation start signal although usually missing from mature protein. Isoln from casein: J. H. Mueller, *Proc. Soc. Exp. Biol. Med.* **19**, 161 (1922). Early chemistry and biochemistry: *Amino Acids and Proteins*, D. M. Greenberg, Ed. (Charles C. Thomas, Springfield, IL, 1951) 950 pp., *passim*; J. P. Greenstein, M. Winitz, *Chemistry of the Amino Acids* vols. **1-3** (John Wiley and Sons, Inc., New York, 1961) pp. 2125-2155, *passim*. Determn and distribution in non-protein fractions: J. Giovanelli, S. H. Mudd, *J. Biochem. Biophys. Methods* **11**, 1 (1985). GC-MS determn in biological fluids: S. P. Stabler *et al.*, *Anal. Biochem.* **162**, 185 (1987). Evaluation as tracer in cancer imaging in mice: R. Kubota, *J. Nucl. Med.* **36**, 484 (1995). Clinical evaluation in acetaminophen overdose: A. N. Hamlyn *et al.*, *J. Int. Med. Res.* **9**, 226 (1981). Clinical use as radiolabel in hyperparathyroidism: P. Hellman *et al.*, *Surgery* **116**, 974 (1994). Review of metabolism and clinical significance in man: L. D. Fleisher, G. E. Gaull, *Clin. Endocrinol. Metab.* **3**, 37-55 (1974); and in carcinogenesis: T. L. Gatton-Umphress, *Hosp. Pract.* **28**, 83-90 (1993). Review of toxicity: N. J. Benevenga, *J. Agric. Food Chem.* **22**, 2-9 (1974). Review of biosynthesis: I. G. Old *et al.*, *Prog. Biophys. Mol. Biol.* **56**, 145-185 (1991). Review as translation start signal: T. Meinnel *et al.*, *Biochimie* **75**, 1061-1075 (1993).

Minute hexagonal plates from dil alc, mp 280-282° (dec, sealed capillary). [α]$_D^{25}$ −8.11° (c = 0.8). [α]$_D^{20}$ +23.40° (c = 5.0 in 3*N* HCl). Sol in water, but the crystals are somewhat water-repellent at first; also sol in warm, dil alcohol and dilute mineral acids. Insol in abs alcohol, ether, petr ether, benzene, acetone.

D-Form. [348-67-4] Converted by deamination, followed by transamination with resultant inversion to the L-form. Comparative study with L-form of metabolism in plants: M. Pokorny *et al.*, *Phytochemistry* **9**, 2175 (1970). Evaluation in parenteral nutrition: K J. Printen *et al.*, *Am. J. Clin. Nutr.* **32**, 1200 (1979). *Review*: L. D. Stegink, D-Amino Acids in *Clin. Nutr. Update: Amino Acids* H. L.

Greene *et al.*, Eds. (American Medical Association, Chicago, IL, 1977) pp 198-206. $[\alpha]_D^{25}$ +8.12° (c = 0.8). $[\alpha]_D^{25}$ −21.18° (c = 0.8 in 0.2*N* HCl).

DL-Form. [59-51-8] Racemethionine; Banthionine; Dyprin; Lobamine; Metione; Pedameth; Urimeth. Platelets from alc, mp 281° (decompn). d 1.340. pK₁ 2.28; pK₂ 9.21. pH of 1% aq soln 5.6-6.1. R_f value 0.77. Soly in water (g/l) at 0°: 18.18; at 25°: 33.81; at 50°: 60.70; at 75°: 105.2; at 100°: 176.0. Sol in dil acids, alkalies. Very slightly sol in 95% alcohol. Insol in ether.

THERAP CAT: Hepatoprotectant; antidote (acetaminophen poisoning); urinary acidifier.

THERAP CAT (VET): Nutritional supplement; urinary acidifier.

6048. Methionine Hydroxy Analog. [583-91-5] 2-Hydroxy-4-(methylthio)butanoic acid; 2-hydroxy-4-(methylthio)butyric acid; 2-hydroxy-4-(methylmercapto)butyric acid; Alimet. $C_5H_{10}O_3S$; mol wt 150.19. C 39.99%, H 6.71%, O 31.96%, S 21.35%. Prepn (not claimed) and use as poultry feed additive: E. S. Blake, R. J. Wineman, **US 2745745** (1956 to Monsanto). HPLC determn in feeds: A. Baudichau *et al.*, *J. Sci. Food Agric.* **38**, 1 (1987). Use as a feed additive for livestock: A. Papas *et al.*, *J. Nutr.* **104**, 653 (1974); A. K. Clark, A. H. Rakes, *J. Dairy Sci.* **65**, 1493 (1982); D. H. Reifsnyder *et al.*, *J. Nutr.* **114**, 1705 (1984). Efficacy of calcium salt vs free acid: K. P. Boebel, D. H. Baker, *Poult. Sci.* **61**, 1167 (1982).

Calcium salt. [4857-44-7] MHA. $C_{10}H_{18}CaO_6S_2$; mol wt 338.45.

USE: Dietary supplement in livestock.

6049. Methisazone. [1910-68-5] 2-(1,2-Dihydro-1-methyl-2-oxo-3*H*-indol-3-ylidene)hydrazinecarbothioamide; 1-methylindole-2,3-dione 3-thiosemicarbazone; *N*-methylisatin 3-thiosemicarbazone; BW-33-T-57; Marboran; Viruzona. $C_{10}H_{10}N_4OS$; mol wt 234.28. C 51.27%, H 4.30%, N 23.91%, O 6.83%, S 13.68%. Prepn: Bauer, Sadler, *Br. J. Pharmacol.* **15**, 101 (1960); **GB 975357** (1964 to Wellcome Found.).

Crystals from butanol, mp 245°.
THERAP CAT: Antiviral.

6050. Methitural. [467-43-6] Dihydro-5-(1-methylbutyl)-5-[2-(methylthio)ethyl]-2-thioxo-4,6(1*H*,5*H*)-pyrimidinedione; 5-(1-methylbutyl)-5-[2-(methylthio)ethyl]-2-thiobarbituric acid; 5-(2-methylthioethyl)-5-(2-pentyl)-2-thiobarbituric acid. $C_{12}H_{20}N_2O_2S_2$; mol wt 288.42. C 49.97%, H 6.99%, N 9.71%, O 11.09%, S 22.23%. Prepn: Zima, Von Werder, **US 2802827** (1957 to E. Merck).

Sodium salt. [730-68-7] Methioturiate; AM-109; Sch-3132; Neraval; Thiogenal. $C_{12}H_{19}N_2NaO_2S_2$; mol wt 310.41. Very hygroscopic, yellow crystals. Slight odor of mercaptans. Freely sol in water. pH of a 10% aq soln ~9.5. Water solns are unstable as

evidenced by a deepening of color and formation of a cloudy precipitate upon autoclaving and exposure to light. The addition of sodium carbonate has a stabilizing effect and prevents precipitation for about 24 hrs: Irwin *et al.*, *J. Pharmacol. Exp. Ther.* **116**, 317 (1956).

Note: This is a controlled substance (depressant): **21 CFR,** 1308.13.

THERAP CAT: Sedative, hypnotic.

6051. Methixene. [4969-02-2] 1-Methyl-3-(9*H*-thioxanthen-9-ylmethyl)piperidine; 9-(*N*-methyl-3-piperidylmethyl)thioxanthene. $C_{20}H_{23}NS$; mol wt 309.47. C 77.62%, H 7.49%, N 4.53%, S 10.36%. Anticholinergic. Prepn: Caviezel *et al.*, *Pharm. Acta Helv.* **33**, 447 (1958); J. Schmutz, **US 2905590** (1959 to Wander). Metabolism and toxicity study: H. Lehner *et al.*, *Arzneim.-Forsch.* **14**, 89 (1964). Crystal structure: S. S. C. Chu, *Acta Crystallogr.* **B28**, 3625 (1972). Spectrofluorimetric determn: F. Belal *et al.*, *Anal. Chim. Acta* **255**, 103 (1991). Comprehensive description: E. M. Abdel-Moety *et al.*, *Anal. Profiles Drug Subs. Excip.* **22**, 317-358 (1993).

Slightly yellow viscous liquid, bp$_{0.07}$ 171-175°. Insol in water.

Hydrochloride monohydrate. [7081-40-5] Tremoquil; Methixart; Trest; Tremonil; Tremaril; Tremarit; Cholinfall; Methyloxan. $C_{20}H_{23}NS.HCl.H_2O$; mol wt 363.94. Flakes from ether, mp 215-217°. uv max (dil HCl): 268 nm (ε 10250). Sol in water, alcohol, chloroform. Insol in ether.

THERAP CAT: Antiparkinsonian.

6052. Methocarbamol. [532-03-6] 3-(2-Methoxyphenoxy)-1,2-propanediol 1-carbamate; 3-(*o*-methoxyphenoxy)-2-hydroxypropyl 1-carbamate; 2-hydroxy-3-(*o*-methoxyphenoxy)propyl 1-carbamate; guaiacol glyceryl ether carbamate; AHR-85; Neuraxin; Miolaxene; Lumirelax; Etroflex; Delaxin; Robamol; Traumacut; Tresortil; Relestrid; Robaxin. $C_{11}H_{15}NO_5$; mol wt 241.24. C 54.77%, H 6.27%, N 5.81%, O 33.16%. Prepn from 3-(*o*-methoxyphenoxy)-2-hydroxypropyl chlorocarbonate: Murphey, **US 2770649** (1956 to A. H. Robins). Comprehensive description: S. Alessi-Severini *et al.*, *Anal. Profiles Drug Subs. Excip.* **23**, 371-399 (1994).

Crystals from benzene, mp 92-94°. uv max (water): 222, 274 nm ($E_{1cm}^{1\%}$ 298, 94). 1og P −0.06. Soly in water at 20°: 2.5 g/100 ml. Sol in alc with heating, propylene glycol; sparingly sol in chloroform. Insol in *n*-hexane, benzene.

THERAP CAT: Muscle relaxant (skeletal).

THERAP CAT (VET): Muscle relaxant (skeletal).

6053. Methohexital. [151-83-7] 1-Methyl-5-(1-methyl-2-pentyn-1-yl)-5-(2-propen-1-yl)-2,4,6(1*H*,3*H*,5*H*)-pyrimidinetrione; α-*dl*-5-allyl-1-methyl-5-(1-methyl-2-pentynyl)barbituric acid; α-*dl*-1-methyl-5-(1-methyl-2-pentynyl)-5-allylbarbituric acid; methohexitone. $C_{14}H_{18}N_2O_3$; mol wt 262.31. C 64.11%, H 6.92%, N 10.68%, O 18.30%. Barbituric acid hypnotic. Prepn of the α-racemate: W. J. Doran, **US 2872448** (1959 to Lilly). Resolution of enantiomers: *idem*, *J. Org. Chem.* **25**, 1737 (1960). Enantioselective synthesis and activity of isomers: H. Brunner, *Chirality* **13**, 420 (2001); *idem et al.*, *Eur. J. Org. Chem.* **2003**, 855. HPLC determn in plasma: M. J. Avram, T. C. Krejcie, *J. Chromatogr.* **414**, 484

(1986). Clinical pharmacokinetics: Y. Le Normand *et al.*, *Br. J. Clin. Pharmacol.* **26**, 589 (1988). Review of use in dental anesthesia: C. H. Martone *et al.*, *Anesth. Prog.* **38**, 195-199 (1991); in orthopedic procedural sedation: T. Austin *et al.*, *J. Emerg. Med.* **24**, 315-318 (2002).

Crystals from dil ethanol as the α-*dl*-racemate, mp 96°. Slightly sol in ethanol, chloroform, dil alkalies; very slightly sol in water.

Sodium salt. [309-36-4] Brevital; Brevimytal; Brietal. $C_{14}H_{17}N_2NaO_3$; mol wt 284.29. White crystalline powder. Hygroscopic. Freely soluble in water.

Note: This is a controlled substance (depressant): **21 CFR**, 1308.14.

THERAP CAT: Anesthetic (intravenous).

THERAP CAT (VET): Anesthetic (intravenous).

6054. Methomyl. [16752-77-5] *N*-[[(Methylamino)carbonyl]oxy]ethanimidothioic acid methyl ester; *N*-[(methylcarbamoyl)oxy]thioacetimidic acid methyl ester; *S*-methyl *N*-[(methylcarbamoyl)oxy]thioacetimidate; methyl *O*-(methylcarbamoyl)thiolacetohydroxamate; Insecticide 1179; Lannate; Nudrin. $C_5H_{10}N_2O_2S$; mol wt 162.21. C 37.02%, H 6.21%, N 17.27%, O 19.73%, S 19.76%. Prepn: A. G. Jelinek, **ZA 6800093**; *idem*, **US 3506698** (1968, 1970 both to du Pont). Metabolism studies: J. Harvey *et al.*, *J. Agric. Food Chem.* **21**, 769, 775, 781 (1973). Crystal structure: Sim, Waite, *J. Chem. Soc. B* **1971**, 752. Acetylcholinesterase activity: Pickering, Pickering, *Arch. Toxicol.* **27**, 292 (1971). Toxicity studies: A. M. Kaplan, H. Sherman, *Toxicol. Appl. Pharmacol.* **40**, 1 (1977).

Crystals, mp 78-79°. d_4^{24} 1.2946. Soly at 25° (w/w): water 5.8; methanol 100; ethanol 42; isopropanol 22; acetone 73. LD_{50} orally in male rats: 17 mg/kg (Kaplan, Sherman).

Caution: Potential symptoms of overexposure are eye irritation; blurred vision, miosis; salivation; abdominal cramps, nausea, vomiting; dyspnea; weakness, muscle twitching; liver and kidney damage. *See NIOSH Pocket Guide to Chemical Hazards* (DHHS/NIOSH 97-140, 1997) p 194.

USE: Insecticide.

6055. Methoprene. [40596-69-8] (2*E*,4*E*)-11-Methoxy-3,7,11-trimethyl-2,4-dodecadienoic acid 1-methylethyl ester; isopropyl (2*E*,4*E*,7*RS*)-11-methoxy-3,7,11-trimethyl-2,4-dodecadienoate; ZR-515. $C_{19}H_{34}O_3$; mol wt 310.48. C 73.50%, H 11.04%, O 15.46%. Juvenile hormone mimic. Prepn: **BE 778242**; *see also* C. A. Henrick, **US 3818047** (1972, 1974 both to Zoecon); *idem et al.*, *J. Org. Chem.* **40**, 1 (1975); L. Novák *et al.*, *Ann.* **1982**, 1173. Effect on mustard aphid: P. J. Rup, R. K. Gill, *Insect Sci. Applic.* **14**, 173 (1993). Field trial in a saltwater marsh against mosquitoes: V. L. Kramer *et al.*, *J. Am. Mosq. Control Assoc.* **9**, 127 (1993). Mechanism of action study: E. M. Berger *et al.*, *Dev. Biol.* **151**, 410 (1992). HPLC determn in tobacco: S. S. Yang, *Chromatographia* **33**, 309 (1992). GC determn in aquatic microcosms: D. H. Ross *et al.*, *J. Am. Mosq. Control Assoc.* **10**, 202 (1994). Impact on the fathead minnow: *idem et al.*, *ibid.* 211. Review of properties, metabolism, toxicology, and use: J. B. Siddall, *Environ. Health Perspect.* **14**, 119-126 (1976). Review of analytical methods: L. L. Dunham, W. W. Miller in *Analytical Methods for Pesticides and Plant Growth Regulators* **vol. X**, G. Zweig, Ed. (Academic Press, New York, 1978) pp 95-109.

Amber liquid, $bp_{0.06}$ 135-136°. d^{20} 0.9261 g/ml. Vapor pressure at 25°: 2.37×10^{-5} mm Hg. Soly in water at 25°: 1.39 ppm. Sol in org solvents. Temperature stable up to at least 42° for 24 months. Stable in water, org solvents, and in the presence of dil alkali and acid. LD_{50} orally in rats: >34500 mg/kg (Siddall).

(7S)-Form. [65733-16-6] *S*-Methoprene; ZR-2458; Altosid; Precor; Extinguish; Pharorid.

USE: Insecticide; insect growth regulator.

THERAP CAT (VET): Ectoparasiticide.

6056. Methopterin. [2410-93-7] *N*-[4-[[(2-Amino-3,4-dihydro-4-oxo-6-pteridinyl)methyl]methylamino]benzoyl]-L-glutamic acid; *N*-[4-[[(2-amino-4-hydroxy-6-pteridinyl)methyl]methylamino]benzoyl]glutamic acid; 10-methylpteroylglutamic acid; N^{10}-methylfolic acid. $C_{20}H_{21}N_7O_6$; mol wt 455.43. C 52.75%, H 4.65%, N 21.53%, O 21.08%. Prepn: Cosulich, Smith, *J. Am. Chem. Soc.* **70**, 1922 (1948); Cosulich, **US 2563707** (1951 to Am. Cyanamid).

Monohydrate. Yellow spherulites. uv max (0.1*N* NaOH): 255, 302, 368 nm ($\varepsilon \times 10^{-3}$ 26, 27, 9); (0.1*N* HCl): 307 nm ($\varepsilon \times 10^{-3}$ 25).

6057. Methotrexate. [59-05-2] *N*-[4-[[(2,4-Diamino-6-pteridinyl)methyl]methylamino]benzoyl]-L-glutamic acid; 4-amino-N^{10}-methylpteroylglutamic acid; 4-amino-10-methylfolic acid; methylaminopterin; amethopterin; MTX; CL-14377; Ledertrexate; Maxtrex. $C_{20}H_{22}N_8O_5$; mol wt 454.45. C 52.86%, H 4.88%, N 24.66%, O 17.60%. Folic acid antagonist. Prepn: D. R. Seeger *et al.*, *J. Am. Chem. Soc.* **71**, 1753 (1949); J. M. Smith, D. B. Cosulich, **US 2512572** (1950 to American Cyanamid). Metabolism: M. V. Freeman, *J. Pharmacol. Exp. Ther.* **122**, 154 (1958); E. S. Henderson *et al.*, *Cancer Res.* **25**, 1008, 1018 (1965). Toxicity data: H. R. Scherf *et al.*, *Arzneim.-Forsch.* **20**, 1467 (1970). Comprehensive description: A. R. Chamberlin *et al.*, *Anal. Profiles Drug Subs.* **5**, 283-306 (1976). Review of metabolism and pharmacokinetics: W. E. Evans, *Appl. Pharmacokinet.* **1980**, 518-548; of clinical pharmacology and toxicity: J. R. Bertino, *Cancer Chemother.* **3**, 359-375 (1981); J. Jolivet *et al.*, *N. Engl. J. Med.* **309**, 1094-1104 (1983). Symposium on clinical experience in rheumatoid arthritis: *J. Rheumatol.* **12**, Suppl. 12, 1-44 (1985). Review of use in leukemia: O. G. Jonsson, B. A. Kamen, *Cancer Invest.* **9**, 53-60 (1991). Clinical trials in Crohn's disease: B. G. Feagan *et al.*, *N. Engl. J. Med.* **332**, 292 (1995); *idem et al.*, *ibid.* **342**, 1627 (2000). CE determn in blood: C.-Y. Kuo *et al.*, *J. Chromatogr. A* **1014**, 93 (2003). HPLC determn in blood: H. Li *et al.*, *J. Chromatogr. B* **845**, 164 (2007). Review of pulmonary toxicity: O. Lateef *et al.*, *Expert Opin. Drug Saf.* **4**, 723-730 (2005).

Orange-brown or yellow crystalline powder. Freely sol in dilute solutions of alkali hydroxides and carbonates; slightly sol in 6*N* hydrochloric acid. Practically insol in water, alc, chloroform, ether.

Monohydrate. [6745-93-3] Yellow crystals from dil HCl, dec 185-204° (bath preheated to 160°). uv max (0.1N HCl): 244, 307 nm; (0.1N NaOH): 257, 302, 370 nm. Soluble in alkaline solns with decompn. LD$_{50}$ i.v. in rats: 14 mg/kg (Scherf).

Disodium salt. [7413-34-5] Folex; Mexate; Rheumatrex. C$_{20}$-H$_{20}$N$_8$Na$_2$O$_5$; mol wt 498.41.

THERAP CAT: Antineoplastic; antirheumatic.

THERAP CAT (VET): Antineoplastic.

6058. Methoxamine. [390-28-3] α-(1-Aminoethyl)-2,5-dimethoxybenzenemethanol; α-(1-aminoethyl)-2,5-dimethoxybenzyl alcohol; 2-amino-1-(2,5-dimethoxyphenyl)-1-propanol; β-hydroxy-β-(2,5-dimethoxyphenyl)isopropylamine; β-(2,5-dimethoxyphenyl)-β-hydroxyisopropylamine; 2,5-dimethoxynorephedrine. C$_{11}$-H$_{17}$NO$_3$; mol wt 211.26. C 62.54%, H 8.11%, N 6.63%, O 22.72%. α$_1$-Adrenergic agonist. Prepn: Baltzly *et al.*, US 2359707 (1944 to Burroughs Wellcome). Metabolism: A. Klutch, M. Bordun, *J. Med. Chem.* **10**, 860 (1967). Clinical pharmacology: N. T. Smith, C. Whitcher, *Anesthesiology* **28**, 735 (1967); P. D. Snashall *et al.*, *Clin. Sci. Mol. Med.* **54**, 283 (1978). HPLC determn in plasma: I. A. Al-Meshal *et al.*, *J. Liq. Chromatogr.* **12**, 1589 (1989). Therapeutic use: P. M. C. Wright *et al.*, *Anesth. Analg.* **75**, 56 (1992); L. Cabanes *et al.*, *N. Engl. J. Med.* **326**, 1661 (1992). Comprehensive description: A. M. Al-Obaid, M. M. El-Domiaty, *Anal. Profiles Drug Subs.* **20**, 399-431 (1991).

Hydrochloride. [61-16-5] Vasoxine; Vasoxyl; Vasylox. C$_{11}$-H$_{17}$NO$_3$.HCl; mol wt 247.72. Crystals, mp 212-216°. pKa (25°C) 9.2. Very sol in water: One gram dissolves in 2.5 ml water, in 12 ml ethanol. Practically insol in ether, benzene, chloroform. pH of a 2% aq soln between 4.5 and 5.5.

THERAP CAT: Antihypotensive.

6059. Methoxsalen. [298-81-7] 9-Methoxy-7H-furo[3,2-g]-[1]benzopyran-7-one; 6-hydroxy-7-methoxy-5-benzofuranacrylic acid δ-lactone; 9-methoxypsoralen; 8-methoxy-4′,5′:6,7-furocoumarin; 8-methoxy[furano-3′,2′:6,7-coumarin]; ammoidin; xanthotoxin; 8-methoxypsoralen; 8-MOP; Meladinine; 8-MOP; Oxsoralen. C$_{12}$H$_8$O$_4$; mol wt 216.19. C 66.67%, H 3.73%, O 29.60%. Naturally occurring analog of psoralen, *q.v.*, found in spp. of *Leguminosae*, *Umbelliferae*, and *Rutaceae*. Isolation: Priess, *Ber. Dtsch. Pharm. Ges.* **21**, 227 (1911); Thoms, *Ber.* **44**, 3325 (1911); **45**, 3705 (1912); Jois *et al.*, *J. Indian Chem. Soc.* **10**, 41 (1933); Späth *et al.*, *Ber.* **73**, 1361 (1933); Schonberg, Sina, *Nature* **160**, 468 (1947); **161**, 481 (1948); *J. Am. Chem. Soc.* **72**, 4826 (1950). Synthesis: Späth, Pailer, *Ber.* **69**, 767 (1936); Lagercrantz, *Acta Chem. Scand.* **10**, 647 (1956); Stanley, Vannier, US 2889337 (1959 to USDA); P. Nore, E. Honkanen, *J. Heterocycl. Chem.* **17**, 985 (1980). Use in treatment of psoriasis and mycosis fungoides: J. A. Parrish *et al.*, *Int. J. Dermatol.* **19**, 379 (1980). Acute toxicity data: Hakim *et al.*, *J. Pharmacol. Exp. Ther.* **131**, 394 (1961). Phototoxicity study: A. Kornhauser *et al.*, *Science* **217**, 733 (1982). Comprehensive description: M. A. Loutfy, M. A. Hassan, *Anal. Profiles Drug Subs.* **9**, 427-454 (1980). Review of use in photochemotherapy: T. F. Anderson, J. J. Voorhees, *Annu. Rev. Pharmacol. Toxicol.* **20**, 235-258 (1980); *Acta Derm. Venereol.* **Suppl. 106**, 9-42 (1982).

Silky needles from hot water or benzene + petr ether, long rhombic prisms from alcohol + ether. mp 148°. Odorless. Bitter taste followed by tingling sensation. pH 5.5. uv max: 219, 249, 300 nm (log ε 4.32, 4.35, 4.06). Freely sol in chloroform; sol in boiling alc, acetone, acetic acid, vegetable fixed oils, propylene glycol, benzene; also sol in aq alkalies with ring cleavage, but is reconstituted upon

neutralization; sparingly sol in boiling water, liquid petrolatum, ether. Practically insol in water. LD$_{50}$ i.p. in rats: 470 ±30 mg/kg (Hakim).

Caution: Methoxsalen with ultraviolet A (long wave light) therapy (PUVA) is listed as a known human carcinogen: *Report on Carcinogens, Twelfth Edition* (PB2011-111646, 2011) p 259.

THERAP CAT: Pigmentation agent.

6060. Methoxyamine. [67-62-9] O-Methylhydroxylamine; methoxylamine; α-methylhydroxylamine; hydroxylamine methyl ether. CH$_5$NO; mol wt 47.06. C 25.52%, H 10.71%, N 29.76%, O 34.00%. CH$_3$ONH$_2$. Prepd by treating hydroxylamine disulfonic acid with methyl sulfate: Goldfarb, *J. Am. Chem. Soc.* **67**, 1852 (1945). May also be prepd from hydroxyurethan. Use as reagent for carbonyl compounds: A. H. Blatt, *J. Am. Chem. Soc.* **61**, 3494 (1939).

Mobile liquid. Fishy, amine odor. *Highly poisonous!* bp$_{760}$ 49-50°. Miscible with water, alcohol, ether.

Hydrochloride. [593-56-6] Methoxyammonium chloride. CH$_5$-ON.HCl; mol wt 83.52. Nacreous scales from alcohol + ether, mp 149-151°. Sol in water, alcohol.

Caution: Strong irritant.

USE: Analytical reagent for aldehydes and ketones.

6061. Methoxychlor. [72-43-5] 1,1′-(2,2,2-Trichloroethylidene)bis[4-methoxybenzene]; 1,1,1-trichloro-2,2-bis(*p*-methoxyphenyl)ethane; 2,2-di-*p*-anisyl-1,1,1-trichloroethane; DMDT; methoxy-DDT; Marlate. C$_{16}$H$_{15}$Cl$_3$O$_2$; mol wt 345.64. C 55.60%, H 4.37%, Cl 30.77%, O 9.26%. Commercial prepn by the condensation of anisole with chloral in the presence of sulfuric acid: Schneller, Smith, *Ind. Eng. Chem.* **41**, 1027 (1949). Activity: P. Läuger *et al.*, *Helv. Chim. Acta* **27**, 892 (1944). Toxicity study: H. C. Hodge *et al.*, *J. Pharmacol. Exp. Ther.* **99**, 140 (1950). Review of toxicology and human exposure: *Toxicological Profile for Methoxychlor* (PB2003-100140, 2002) 290 pp.

Dimorphic crystals, mp 78-78.2° or 86-88°. Practically insol in water. Soluble in alc. The solubilities are approx those of DDT, *q.v.* LD$_{50}$ orally in rats: 5.0 g/kg (Hodge).

Caution: Potential symptoms of overexposure in exptl animals are fasciculations; trembling, convulsions; kidney and liver damage. Potential occupational carcinogen. *See NIOSH Pocket Guide to Chemical Hazards* (DHHS/NIOSH 97-140, 1997) p 194. *See also Clinical Toxicology of Commercial Products*, R. E. Gosselin *et al.*, Eds. (Williams & Wilkins, Baltimore, 5th ed., 1984) Section II, p 284.

USE: Insecticide.

THERAP CAT (VET): Ectoparasiticide.

6062. 5-Methoxy-N,N-diisopropyltryptamine. [4021-34-5] 5-Methoxy-N,N-bis(1-methylethyl)-1H-indole-3-ethanamine; 3-[2-(diisopropylamino)ethyl]-5-methoxyindole; 5-MeO-DIPT; "Methoxy Foxy"; "Foxy". C$_{17}$H$_{26}$N$_2$O; mol wt 274.41. C 74.41%, H 9.55%, N 10.21%, O 5.83%. Synthetic hallucinogen. Prepn: D. Desaty, D. Keglevic, *Croat. Chem. Acta* **36**, 103 (1964); M. Julia, P. Manoury, *Bull. Soc. Chim. Fr.* **1965**, 1411; and clinical psychopharmacology: A. T. Shulgin, M. F. Carter, *Commun. Psychopharmacol.* **4**, 363 (1980). GC-MS and HPLC-MS determn in blood and urine: S. P. Vorce, J. H. Sklerov, *J. Anal. Toxicol.* **28**, 407 (2004). Determn in forensic samples: C. Huhn *et al.*, *Electrophoresis* **26**, 2391 (2005). Case reports of intoxication: A. A. Muller, *J. Emerg. Nurs.* **30**, 507 (2004); J. M. Wilson *et al.*, *Forensic Sci. Int.* **148**, 31 (2005).

Hydrochloride. [2426-63-3] $C_{17}H_{26}N_2O \cdot HCl$; mol wt 310.87. Crystals from ethanol + ether, mp 180-181°.

Note: This is a controlled substance (hallucinogen): **21 CFR,** 1308.11.

6063. 2-Methoxyethanol. [109-86-4] Ethylene glycol monomethyl ether; Methyl Cellosolve. $C_3H_8O_2$; mol wt 76.10. C 47.35%, H 10.60%, O 42.05%. Prepn from ethylene oxide + methanol: Finch, Hagemeyer, **US 2748171** (1956 to Kodak); from ethylene glycol + diazomethane: Hesse, Majumdar, *Ber.* **93**, 1129 (1960). Toxicity data: Smyth, *J. Ind. Hyg. Toxicol.* **23**, 259 (1941); Werner, *ibid.* **25**, 157 (1943). Series of articles on toxicology: *Environ. Health Perspect.* **57**, 1-275 (1984).

HO⟍⟍OCH₃

Liquid. *Poisonous; flammable.* bp_{760} 124.43°; bp_{20} 34-41°. d_4^{20} 0.9663. Flash pt 115°F. n_D^{20} 1.4028. Miscible with water, alcohol ether, glycerol, acetone, dimethylformamide. LD_{50} in rats, guinea pigs (mg/kg): 2460, 950 orally (Smyth). LC_{50} (7 hr in air) in mice: 4.6 mg/l (Werner).

Caution: Potential symptoms of overexposure are irritation of eyes, nose, throat; headache, drowsiness, lassitude; ataxia, tremor; anemic pallor. *See NIOSH Pocket Guide to Chemical Hazards* (DHHS/NIOSH 2005-149, 2005) p 202.

USE: Solvent for low-viscosity cellulose acetate, natural resins, some synthetic resins and some alcohol-soluble dyes; in dyeing leather, sealing moistureproof cellophane; in nail polishes, quick-drying varnishes and enamels, wood stains. In modified Karl Fischer reagent: Peters, Jungnickel, *Anal. Chem.* **27**, 450 (1955).

6064. 2-Methoxyethyl Acetate. [110-49-6] 2-Methoxyethanol 1-acetate; 1-acetoxy-2-methoxyethane; ethylene glycol monomethyl ether acetate; Methyl Cellosolve Acetate. $C_5H_{10}O_3$; mol wt 118.13. C 50.84%, H 8.53%, O 40.63%. Toxicity data: H. F. Smyth *et al., J. Ind. Hyg. Toxicol.* **23**, 259 (1941). Series of articles on toxicology: *Environ. Health Perspect.* **57**, 1-275 (1984).

H₃C⟍C(=O)O⟍OCH₃

Colorless liquid. *Flammable.* d_{20}^{20} 1.0067. bp 145°. mp −65.1°. n_D^{20} 1.4019. Flash pt, closed cup: 114.8°F (46°C). Miscible with water, most organic solvents, oils; dissolves gums, resins. LD_{50} orally in rats: 3.4 g/kg (Smyth).

Caution: Potential symptoms of overexposure are kidney damage; brain damage. Direct contact may cause eye, nose and throat irritation. *See NIOSH Pocket Guide to Chemical Hazards* (DHHS/NIOSH 97-140, 1997) p 202. *See also Patty's Industrial Hygiene and Toxicology* vol. **2D**, G. D. Clayton, F. E. Clayton, Eds. (Wiley-Interscience, New York, 4th ed., 1994) pp 2918-2921.

USE: Industrial solvent.

6065. Methoxyethylbenzeneboronic Acid. [159752-39-3] B-[2-[(1R)-1-Methoxyethyl]phenyl]boronic acid. $C_9H_{13}BO_3$; mol wt 180.01. C 60.05%, H 7.28%, B 6.01%, O 26.66%. Synthesis of (+)-form and use: K. Burgess, A. M. Porte, *Angew. Chem. Int. Ed.* **33**, 1182 (1994). Chemoenzymatic stereosynthesis of both enantiomers and use: S. M. Resnick *et al., J. Org. Chem.* **60**, 3546 (1995).

White solid from *n*-hexane, mp 72-74°. $[Ga]_D^{25}$ +27 (c = 0.5 in CH_2Cl_2) (Burgess); $[\alpha]_D$ +27.2° (c = 4.6 in CH_2Cl_2) (Resnick).

(−)-S-Form. [166191-23-7] $[\alpha]_D$ −28.7° (c = 2.3 in CH_2Cl_2).

USE: NMR shift reagent for enantiomeric assay of diols.

6066. Methoxyfenozide. [161050-58-4] 3-Methoxy-2-methylbenzoic acid 2-(3,5-dimethylbenzoyl)-2-(1,1-dimethylethyl)hydrazide; N'-*tert*-butyl-N'-(3,5-dimethylbenzoyl)-3-methoxy-2-methyl-

benzohydrazide; N-*tert*-butyl-N'-(3-methoxy-o-toluoyl)-3,5-xylohydrazide; RH-112485; RH-2485; Intrepid; Runner; Prodigy. $C_{22}H_{28}N_2O_3$; mol wt 368.48. C 71.71%, H 7.66%, N 7.60%, O 13.03%. Synthetic nonsteroidal ecdysone agonist. Prepn: Z. Lidert *et al.*, **EP 602794;** *eidem*, **US 5530028** (1994, 1996 both to Rohm & Haas). Comparative insecticidal activity: A. Trisyono, G. M. Chippendale, *J. Econ. Entomol.* **90**, 1486 (1997). Mechanism of action and metabolism: G. Smagghe *et al., Pestic. Sci.* **55**, 386 (1999). Review of physical properties, biological activity and field studies: D. P. Le *et al., Brighton Crop Prot. Conf. - Pests Dis.* **1996**, 481-486; of physical properties and insecticidal activity: G. R. Carlson *et al., Pest Manage. Sci.* **57**, 115-119 (2001).

White powder, mp 204-205°. Vapor pressure (25°): $<4.0 \times 10^{-8}$ torr. Soly in water: 3.3 mg/l; in acetone: 9%. Soly (g/l): crop oil <10, cyclohexanone 90, DMSO 110, N-methylpyrrolidone 380, xylene <10. Soly in acetone 9%. LD_{50} in rats, mice (mg/kg): >5000, >5000 orally; in rats (mg/kg): >2000 dermally. LC_{50} in rats (mg/l): >4.3 by inhalation. LC_{50} (8 day dietary) in mallard duck, bobwhite quail (mg/kg diet): >5620, >5620. LC_{50} in bluegill sunfish, *Daphnia magna* (mg/l): >4.3 (96 hr); 3.7 (48 hr) (Le).

USE: Insecticide.

6067. Methoxyflurane. [76-38-0] 2,2-Dichloro-1,1-difluoro-1-methoxyethane; 2,2-dichloro-1,1-difluoroethyl methyl ether; 1,1-difluoro-2,2-dichloroethyl methyl ether; DA-759; Metofane; Penthrane; Pentrane. $C_3H_4Cl_2F_2O$; mol wt 164.96. C 21.84%, H 2.44%, Cl 42.98%, F 23.03%, O 9.70%. Prepn and properties: Miller *et al., J. Am. Chem. Soc.* **70**, 431 (1948); Park *et al., ibid.* **73**, 861 (1951); **GB 754976** (1956 to Standard Tele. & Cable).

Clear, practically colorless, mobile liquid, bp 105°; bp_{100} 51°. mp −35°. n_D^{20} 1.3861; n_D^{25} 1.3839. d_4^{20} 1.4262 (Miller); d_4^{20} 1.4226 (Park). Miscible with alc, acetone, chloroform, ether, and with fixed oils.

Caution: Potential symptoms of overexposure are eye irritation; CNS depression, analgesia; anesthesia, seizures, respiratory depression; liver and kidney injury. *See NIOSH Pocket Guide to Chemical Hazards* (DHHS/NIOSH 97-140, 1997) p 196.

THERAP CAT: Anesthetic (inhalation).

THERAP CAT (VET): Anesthetic.

6068. 10-Methoxyharmalan. [3589-73-9] 4,9-Dihydro-6-methoxy-1-methyl-3H-pyrido[3,4-b]indole; 1-methyl-6-methoxy-3,4-dihydro-2-carboline; 3,4-dihydromethoxyharman. $C_{13}H_{14}N_2O$; mol wt 214.27. C 72.87%, H 6.59%, N 13.07%, O 7.47%. Prepn from 5-methoxytryptamine: Späth, Lederer, *Ber.* **63**, 2102 (1930); Petrova *et al., Zh. Obshch. Khim.* **33**, 1333 (1963), *C.A.* **59**, 10149b (1963). Weak serotonin inhibitor: M. M. Airaksinen *et al., Acta Pharmacol. Toxicol.* **46**, 308 (1980).

Crystals, mp 208-209°. R_f 0.74 in 4:1:5 butanol, acetic acid, water.

6069. 3-Methoxy-4-hydroxyphenylglycol. [534-82-7] 1-(4-Hydroxy-3-methoxyphenyl)-1,2-ethanediol; (4-hydroxy-3-methoxyphenyl)ethylene glycol; vanylglycol; MHPG. $C_9H_{12}O_4$; mol wt 184.19. C 58.69%, H 6.57%, O 34.74%. Major metabolite of norepinephrine, *q.v.* Isoln from rat urine: J. Axelrod *et al.*, *Biochim. Biophys. Acta* **36**, 576 (1959); from cat liver and kidney: M. Goldstein, S. B. Gertner, *Nature* **187**, 147 (1960). Metabolism studies on urinary MHPG levels in patients with affective disorders: K. Greenspan *et al.*, *J. Psychiatr. Res.* **7**, 171 (1970); J. W. Maas *et al.*, *Biochem. Pharmacol.* **1974**, Suppl., pt. 2, 907; J. J. Schildkraut *et al.*, *Biol. Markers Psychiatry Neurol.*, *Proc. Conf.* **1981**, E. Usdin, I. Hanin, Eds. (Pergamon, Oxford, 1982) pp 23-33. Studies on MHPG levels in cerebrospinal fluid (CSF): E. Garelis *et al.*, *Brain Res.* **79**, 1 (1974); P. Frattini *et al.*, *Clin. Chim. Acta* **125**, 97 (1982); in CSF of patients with CNS disorders: T. N. Chase *et al.*, *J. Neurochem.* **21**, 581 (1973); M. G. Ziegler *et al.*, *Am. J. Psychiatry* **134**, 565 (1977). Direct method to estimate rate of prodn of MHPG by human brain *in vivo*: J. W. Maas *et al.*, *Science* **205**, 1025 (1979). Review of origin and distribution in body fluids: E. M. DeMet, A. E. Halaris, *Biochem. Pharmacol.* **28**, 3043-3050 (1979). Use as biochemical marker to identify mental disorders: T. H. Maugh, *Science* **214**, 39 (1981).

USE: As biochemical tool.

6070.   **2-Methoxynaphthalene.** [93-04-9] Methyl β-naphthyl ether; nerolin "old"; Yara yara. $C_{11}H_{10}O$; mol wt 158.20. C 83.52%, H 6.37%, O 10.11%. Prepn: Hiers, Hager, *Org. Synth.* **coll. vol. I,** 59 (1941).

Leaflets from ether, mp 72°. bp 272°. Practically insol in water. Sparingly sol in alc; sol in ether, carbon disulfide, benzene.

6071.   **Methoxyphenamine.** [93-30-1] 2-Methoxy-*N*,α-dimethylbenzeneethanamine; *o*-methoxy-*N*,α-dimethylphenethylamine; β-(*o*-methoxyphenyl)isopropylmethylamine; α-(2-methoxyphenyl)-β-methylaminopropane. $C_{11}H_{17}NO$; mol wt 179.26. C 73.70%, H 9.56%, N 7.81%, O 8.93%. β-Adrenergic agonist. Prepn: Woodruff *et al.*, *J. Am. Chem. Soc.* **62**, 922 (1940); Heinzelman, *ibid.* **75**, 921 (1953); Morishita, **JP 61 2921** (1961 to Nippon Shinyaku), *C.A.* **55**, 24677f (1961).

Oil, bp_2 97-99°.

Hydrochloride. [5588-10-3] Orthoxine. $C_{11}H_{17}NO.HCl$; mol wt 215.72. Bitter crystals from ether + alcohol, mp 129-131°. Freely sol in water, alcohol, chloroform. Slightly sol in ether, benzene. pH 5.3-5.7 (5% aq soln).

THERAP CAT: Bronchodilator.

6072.   ***N*-(*p*-Methoxyphenyl)-*p*-phenylenediamine.** [101-64-4] N^1-(4-Methoxyphenyl)-1,4-benzenediamine; 4-amino-4'-methoxydiphenylamine; 4-methoxy-4'-aminodiphenylamine; Variamine Blue base. $C_{13}H_{14}N_2O$; mol wt 214.27. C 72.87%, H 6.59%, N 13.07%, O 7.47%. Prepd from the corresp nitroso compd: Willstätter, Kubli, *Ber.* **42**, 4139 (1909).

Needles from ligroin, mp 102°. bp_{12} 238°. Freely sol in alcohol, ether, benzene. Slightly sol in water, petr ether.

Hydrochloride. [3566-44-7] $C_{13}H_{14}N_2O.HCl$; mol wt 250.73. Gray-blue solid.

Sulfate. [3169-21-9] $(C_{13}H_{14}N_2O)_2.H_2SO_4$; mol wt 526.61. Crystals, sol in water. Addition of ferric chloride produces intensely blue solns.

USE: Hydrochloride as indicator in analytical chemistry.

6073.   **2-Methoxypropene.** [116-11-0] 2-Methoxy-1-propene; isopropenyl methyl ether; methyl 1-methylvinyl ether; methyl isopropenyl ether. C_4H_8O; mol wt 72.11. C 66.63%, H 11.18%, O 22.19%. Reagent for the protection of the hydroxyl group; often utilized in carbohydrate chemistry to protect diols as acetonides. Prepn: L. Claisen, *Ber.* **31**, 1019 (1898). Improved prepn: M. S. Newman, M. C. Vander Zwan, *J. Org. Chem.* **38**, 2910 (1973); S. Krill *et al.*, **WO 0196269** (2001 to Degussa). Purification by distillation: G. Köhler *et al.*, **US 7208640** (2007 to Degussa). Utility in sugar chemistry for the protection of diols: J. Gelas, D. Horton, *Carbohydr. Res.* **45**, 181 (1975); C. Copeland, R. V. Stick, *Aust. J. Chem.* **31**, 1371 (1978); E. Fanton *et al.*, *J. Org. Chem.* **46**, 4057 (1981). Additional diol acetonation procedure: T. Carofiglio *et al.*, *Main Group Met. Chem.* **15**, 247 (1992). Toxicity studies: H. F. Smyth, Jr. *et al.*, *Am. Ind. Hyg. Assoc. J.* **30**, 470 (1969).

Colorless liquid. *Flammable.* bp 38°. d_4^{20} 0.737. n_D^{25} 1.3768. Flash pt, closed cup: −20°F (−29°C). Protect from light. Store under inert gas at 4°C. LD_{50} in rats (ml/kg): 1.87 orally (Smyth).

USE: Reagent in synthetic organic chemistry.

6074.   **6-Methoxy-α-tetralone.** [1078-19-9] 3,4-Dihydro-6-methoxy-1(2*H*)-naphthalenone; 6-methoxy-3,4-dihydro-1(2*H*)-naphthalenone; 1-keto-6-methoxy-1,2,3,4-tetrahydronaphthalene; 5-keto-2-methoxy-5,6,7,8-tetrahydronaphthalene. $C_{11}H_{12}O_2$; mol wt 176.22. C 74.98%, H 6.86%, O 18.16%. Prepn: Papa, *J. Am. Chem. Soc.* **71**, 3246 (1949); Ananchenko *et al.*, *Tetrahedron* **18**, 1355 (1962).

Crystals from methanol or ligroin, mp 80°. bp_1 135-139°, bp_{11} 171°.

USE: In the synthesis of derivatives of estrane and 19-norsteroids (Ananchenko *et al.*, *loc. cit.*).

6075.   **1-Methoxy-3-(trimethylsilyloxy)-1,3-butadiene.** [54125-02-9] (*E*-form); [59414-23-2] (unspecified stereo). [[(2*E*)-3-Methoxy-1-methylene-2-propen-1-yl]oxy]trimethylsilane; Danishefsky's diene. $C_8H_{16}O_2Si$; mol wt 172.30. C 55.77%, H 9.36%, O 18.57%, Si 16.30%. A silyl enol ether used in Diels-Alder reactions; highly reactive and nucleophilic diene. Prepn and ^{1}H-NMR: S. Danishefsky, T. Kitahana, *J. Am. Chem. Soc.* **96**, 7807 (1974); S. Danishefsky *et al.*, *Org. Synth.* **61**, 147 (1983). In synthesis of vernolepin, *q.v.*: *eidem*, *J. Am. Chem. Soc.* **99**, 6066 (1977). Of pentalenolactone: *eidem*, *ibid.* **100**, 6536 (1978). Regiospecific addition to juglone, *q.v.*: R. K. Boeckman *et al.*, *ibid.* 7098. *Review:* S. Danishefsky, *Acc. Chem. Res.* **14**, 400 (1981).

Oil. bp_{23} 78-81°.

USE: In organic syntheses.

6076.   **5-Methoxytryptamine.** [608-07-1] 5-Methoxy-1*H*-indole-3-ethanamine; 3-(2-aminoethyl)-5-methoxyindole; meksamin;

mexamine. $C_{11}H_{14}N_2O$; mol wt 190.25. C 69.45%, H 7.42%, N 14.72%, O 8.41%. Prepn: Supniewski *et al.*, *C.A.* **55**, 15458 (1961). Proposed as potentiator for hypnotics, sedatives. Claimed to be more active than serotonin: Mashkovsky, Arutyunyan, *Farmakol. Toksikol.* **26** (no. 1), 10 (1963).

Crystals from ethanol, mp 121-122°.
Hydrochloride. $C_{11}H_{14}N_2O.HCl$. Crystals, dec 248°.

6077. Methscopolamine Bromide. [155-41-9] (1α,2β,4β,-5α,7β)-7-[(2S)-3-Hydroxy-1-oxo-2-phenylpropoxy]-9,9-dimethyl-3-oxa-9-azoniatricyclo[3.3.1.0²,⁴]nonane bromide (1:1); 6β,7β-epoxy-3α-hydroxy-8-methyl-1αH,5αH-tropanium bromide tropate (ester); *N*-methylscopolammonium bromide; hyoscine methyl bromide; scopolamine methobromide; scopolamine methyl bromide; epoxymethamine bromide; Holopon; Pamine. $C_{18}H_{24}BrNO_4$; mol wt 398.30. C 54.28%, H 6.07%, Br 20.06%, N 3.52%, O 16.07%. Anticholinergic. Prepd by the action of methyl bromide on scopolamine base: Visscher, US 2753288 (1956 to Upjohn). Treatment of duodenal ulcer by transdermally administered methscopolamine bromide: R. P. Walt *et al.*, *Br. Med. J.* **284**, 1736 (1982).

Crystals from ethanol, dec 214-217°. Freely sol in water, in dil ethanol. Slightly sol in abs ethanol.
THERAP CAT: Antispasmodic.

6078. Methsuximide. [77-41-8] 1,3-Dimethyl-3-phenyl-2,5-pyrrolidinedione; *N*,2-dimethyl-2-phenylsuccinimide; 1,3-dimethyl-3-phenyl-2,5-dioxopyrrolidine; mesuximide; Celontin; Petinutin. $C_{12}H_{13}NO_2$; mol wt 203.24. C 70.92%, H 6.45%, N 6.89%, O 15.74%. Prepd by the action of methylamine on α-methyl-α-phenylsuccinic acid: Miller, Long, *J. Am. Chem. Soc.* **73**, 4895 (1951); *eidem*, US 2643257 (1953 to Parke, Davis). Pharmacology: P. G. Marshall, D. K. Vallance, *J. Pharm. Pharmacol.* **6**, 740 (1954).

Crystals from dil alc, mp 52-53°. bp$_{0.1}$ 121-122°. Very sol in chloroform; freely sol in methanol, alc, ether; slightly sol in hot water. LD$_{50}$ orally in mice: 0.9 g/kg (Marshall, Vallance).
THERAP CAT: Anticonvulsant.

6079. Methyclothiazide. [135-07-9] 6-Chloro-3-(chloromethyl)-3,4-dihydro-2-methyl-2*H*-1,2,4-benzothiadiazine-7-sulfonamide 1,1-dioxide; 6-chloro-3-chloromethyl-3,4-dihydro-2-methyl-7-sulfamoyl-1,2,4-benzothiadiazine 1,1-dioxide; 6-chloro-3-chloromethyl-2-methyl-7-sulfamyl-3,4-dihydro-1,2,4-benzothiadiazine 1,1-dioxide; Aquatensen; Duretic; Enduron; Enduronum; Naturon.

$C_9H_{11}Cl_2N_3O_4S_2$; mol wt 360.22. C 30.01%, H 3.08%, Cl 19.68%, N 11.67%, O 17.77%, S 17.80%. Prepn: Close *et al.*, *J. Am. Chem. Soc.* **82**, 1132 (1960). Description: Ford, *Curr. Ther. Res.* **2**, 422 (1960). Comprehensive description: J. A. Raihle, *Anal. Profiles Drug Subs.* **5**, 307-326 (1976).

Crystals from alcohol + water, mp 225°. uv max (methanol): 226, 267, 311 nm (ε 39300, 21250, 3300). pKa 9.4. Freely sol in acetone, pyridine; sparingly sol in methanol; slightly sol in ethanol; very slightly sol in water, chloroform, benzene.
THERAP CAT: Diuretic; antihypertensive.

6080. Methyl Abietate. [127-25-3] (1*R*,4a*R*,4b*R*,10a*R*)-1,2,-3,4,4a,4b,5,6,10,10a-Decahydro-1,4a-dimethyl-7-(1-methylethyl)-1-phenanthrenecarboxylic acid methyl ester; Abalyn. $C_{21}H_{32}O_2$; mol wt 316.49. C 79.70%, H 10.19%, O 10.11%. The article of commerce is a mixture of the methyl esters of the rosin acids.

Colorless to yellow, almost odorless, thick liquid. d_{20}^{20} 1.040. bp 360-365° with decompn. n_D^{20} 1.530. Flash pt 180-218°C. Insol in water. Miscible with usual organic solvents, also with aliphatic hydrocarbons. Dissolves ester gums, rosin, many synthetic resins as well as ethyl cellulose, rubber, etc.
USE: As a solvent for ester gums, rosin, many synthetic resins, ethyl cellulose, rubber, etc.; in the manuf of varnish resins; as ingredient in adhesives.

6081. *N*-Methylacetanilide. [579-10-2] *N*-Methyl-*N*-phenylacetamide; acetomethylanilide; Exalgin. $C_9H_{11}NO$; mol wt 149.19. C 72.46%, H 7.43%, N 9.39%, O 10.72%. Made by the action of acetic anhydride on methylaniline.

Orthorhombic rods from alcohol, plates from ether or petr ether. mp 102-104°. bp$_{712}$ 253°. One gram dissolves in 60 ml water, 2 ml boiling water, 2 ml alcohol, 1.5 ml chloroform, 10 ml ether.

6082. Methyl Acetate. [79-20-9] Acetic acid methyl ester; methyl ethanoate. $C_3H_6O_2$; mol wt 74.08. C 48.64%, H 8.16%, O 43.19%.

Colorless liquid; pleasant odor. mp −98°. bp 56.9°. d_4^{20} 0.9342; d_4^{25} 0.9279. n_D^{20} 1.3614: Mumford, Phillips, *J. Chem. Soc.* **1950**, 75. Flash pt, closed cup: 14°F (−10°C). Sol in water; miscible with alcohol, ether.
Caution: Potential symptoms of overexposure are irritation of eyes, skin, nose and throat; headache, drowsiness; optic nerve atro-

phy; chest tightness. *See NIOSH Pocket Guide to Chemical Hazards* (DHHS/NIOSH 97-140, 1997) p 196.

USE: Solvent for nitrocellulose, acetylcellulose, and many resins and oils; manuf artificial leather.

6083. Methyl Acetoacetate. [105-45-3] 3-Oxobutanoic acid methyl ester; acetoacetic acid methyl ester. $C_5H_8O_3$; mol wt 116.12. C 51.72%, H 6.94%, O 41.33%. Toxicity data: H. F. Smyth, C. P. Carpenter, *J. Ind. Hyg. Toxicol.* **30**, 63 (1948).

Liquid. d 1.078-1.080. bp 169-171° (slight dec). mp −80°. n_D^{20} 1.418. Flash pt 82°C. Sol in 2 parts water; miscible with alc, ether; gives a deep red color with $FeCl_3$. LD_{50} orally in rats: 3.0 g/kg (Smyth, Carpenter).

6084. Methyl Acetylsalicylate. [580-02-9] 2-(Acetyloxy)-benzoic acid methyl ester; acetylsalicylic acid methyl ester; methyl-aspirin; Methylrodin; Methylrhodine. $C_{10}H_{10}O_4$; mol wt 194.19. C 61.85%, H 5.19%, O 32.96%. Prepd from acetylsalicylic acid and diazomethane: Herzig, Tichatschek, *Ber.* **39**, 1559 (1906); from methyl salicylate and acetic anhydride: Erdmann, *J. Prakt. Chem.* [2] **56**, 154 (1897); Thorp, **US 1255950** (1918).

Plates from petr ether. mp 51-52°. bp$_9$ 134-136°. Very sparingly sol in water; sol in alc, ether, chloroform, glycerol, propylene glycol, oils.

USE: Fixative for perfumes.

6085. Methyl Acrylate. [96-33-3] 2-Propenoic acid methyl ester; acrylic acid methyl ester; methyl propenoate. $C_4H_6O_2$; mol wt 86.09. C 55.81%, H 7.03%, O 37.17%. Convenient prepn from ethylene chlorohydrin: **DE 571123** (1928 to Rohm & Haas). For direct syntheses from acetylene and carbon monoxide *see* W. Reppe, *Chemie & Technik der Acetylen-Druck-Reaktionen* (Weinheim, 2nd ed., 1952). Toxicity study: H. F. Smyth, C. P. Carpenter, *J. Ind. Hyg. Toxicol.* **30**, 63 (1948).

Liquid. Acrid odor. *Lacrimator.* d_4^{20} 0.9561; d_{20}^{20} 0.9574; d_4^{0} 0.9702; d_4^{-5} 0.9868; d_4^{-10} 0.9929. Weighs 8.0 lbs/gal. mp −76.5°. bp$_{608}$ 70°; bp$_{428}$ 60°; bp$_{298}$ 50°; bp$_{200}$ 40°; bp$_{88}$ 20°; bp$_{54}$ 10°; bp$_{41.5}$ 5°; bp$_{32}$ 0°; bp$_{24.5}$ −5°; bp$_{18.5}$ −10°. n_D^{20} 1.401. Sp heat at −60° = 0.444 cal/g/°C; heat of vaporization 8.25 kcal/mol; heat of combustion 502.88 kcal/mol. Soly in water at 20° = 6 g/100 ml, at 40° = 5 g/100 ml. Soly of water in methyl acrylate at 20° = 1.8 ml/100 g. Sol in alc, ether. Azeotropes: 9.5% water = bp 73°; 49.0% methanol = bp 61°. Easily polymerizes on standing. The polymerization process can be speeded up by heat, light, and peroxides. If pure, the monomer can be stored below +10° without incurring polymerization. LD_{50} orally in rats: 0.3 g/kg (Smyth, Carpenter).
Polymer. Transparent, elastic substance. Practically no odor. Little adhesive power. Resists the usual solvents.
Caution: Potential symptoms of overexposure to the monomer are irritation of eyes, upper respiratory system and skin. *See NIOSH Pocket Guide to Chemical Hazards* (DHHS/NIOSH 97-140, 1997) p 198. *See also Clinical Toxicology of Commercial Products*, R. E. Gosselin *et al.*, Eds. (Williams & Wilkins, Baltimore, 5th ed., 1984) Section II, p 409.

USE: The monomer in manuf leather finish resins, textile and paper coatings, and plastic films. Produces the hardest resin of the acrylate ester series.

6086. Methylal. [109-87-5] Dimethoxymethane; formal; formaldehyde dimethyl acetal; methylene dimethyl ether. $C_3H_8O_2$; mol wt 76.10. C 47.35%, H 10.60%, O 42.05%. Prepn by catalytic vapor-phase oxidation of methanol in the presence of small amts of HCl; from methanol and formaldehyde: Frevel, Hedelund, **US 2663742**; **US 2691684** (1953, 1954, both to Dow); from paraformaldehyde + methanol in the presence of $CaCl_2$ + HCl: R. Rambaud, D. Besserre, *Bull. Soc. Chim. Fr.* **1955**, 45.

Colorless, clear, volatile, liq; chloroform odor; pungent taste. *Flammable.* bp$_{760}$ 41.6°; bp$_{754}$ 41.5°. mp −105°. d_4^{14} 0.8669; d_4^{20} 0.8593. n_D^{18} 1.3589. Flash pt, closed cup: 0°F (−18°C). Sol in 3 parts water; miscible with alcohol, ether, oils.
Caution: Potential symptoms of overexposure are irritation of eyes, skin and upper respiratory system; anesthesia. *See NIOSH Pocket Guide to Chemical Hazards* (DHHS/NIOSH 97-140, 1997) p 198.

USE: In perfumery; manuf artificial resins; reaction medium for Grignard and Reppe reactions.

6087. Methyl Allyl Trisulfide. [34135-85-8] Methyl-2-propenyl trisulfide; allyl methyl trisulfide; methyl 2-propenyl trisulfane; MATS. $C_4H_8S_3$; mol wt 152.29. C 31.55%, H 5.30%, S 63.16%. Component of garlic oil which inhibits platelet aggregation. Synthesis, identification by GC: D. M. Oaks *et al.*, *Anal. Chem.* **36**, 1560 (1964). Isoln from garlic oil, effect on platelet-rich plasma: T. Ariga *et al.*, *Lancet* **1**, 150 (1981). Improved synthesis: A. W. Mott, G. Barany, *Synthesis* **1984**, 657. Use as antithrombotic agent: **JP Kokai 82 209218** (1982 to Alcon), *C.A.* **98**, 95692n (1983).

Oil, bp$_{0.05}$ 28-30°.

6088. Methylamine. [74-89-5] Methanamine; monomethylamine; aminomethane. CH_5N; mol wt 31.06. C 38.67%, H 16.23%, N 45.10%. CH_3NH_2. Occurs in herring brine, in urine of dogs after eating meat, in certain plants such as *Mentha aquatica*, in crude methanol together with di- and trimethylamine. Made by heating methyl alcohol, ammonium chloride, and zinc chloride to ~300°; by heating ammonium chloride and formaldehyde: Marvel, Jenkins, *Org. Synth.* **3**, 67 (1923); from methanol + ammonia: Smith, **US 2456599** (1948 to Comm. Solvents). Inhalation toxicity studies: L. A. Kinney *et al.*, *Inhalation Toxicol.* **2**, 29 (1990); S. N. Sarkar, M. S. Sastry, *J. Environ. Biol.* **13**, 273 (1992).
Flammable gas at ordinary temp and pressure. Fuming liquid when cooled in ice and salt mixture. $d_4^{-10.8}$ 0.699. mp −93.5°. bp$_{760}$ −6.3°; bp$_{400}$ −19.7°; bp$_{200}$ −32.4°; bp$_{100}$ −43.7°; bp$_{10}$ −73.8°. Flash pt 32.5°F (0°C). Stronger base than ammonia: pK_b (25°): 3.35. Absorption spectrum: Bielecki, Henri, *Compt. Rend.* **156**, 1861 (1913). One vol of water at 12.5° dissolves 1154 vols of the gas, and 959 vols at 25°. 100 ml of a benzene soln satd at 25° contain 10.5 g methylamine. Sol in alc; miscible with ether. LD_{50} orally in rats: 100-200 mg/kg (Kinney). LC_{50} in rats: 0.448 ml/l (Sarkar, Sastry).
Hydrochloride. [593-51-1] $CH_5N.HCl$; mol wt 67.52. Deliquescent tetragonal tablets from alcohol, mp 227-228° with sublimation. bp$_{15}$ 225-230°. Sol in water, in abs alcohol. 100 g of boiling abs alcohol dissolve 23.01 g. Insol in chloroform, acetone, ether, ethyl acetate. *Keep well closed.* LD_{50} orally in rats: 1600-3200 mg/kg (Kinney).
Caution: Potential symptoms of overexposure are irritation of eyes, skin and respiratory system; coughing; skin and mucous membrane burns; dermatitis; conjunctivitis; direct contact with liquid may cause frostbite. *See NIOSH Pocket Guide to Chemical Hazards* (DHHS/NIOSH 97-140, 1997) p 200.

USE: Methylamine is used in tanning. Methylamine and methylamine hydrochloride are used in organic synthesis for introducing the methylamino group.

6089. 2-(Methylamino)ethanol. [109-83-1] Methyl(β-hydroxyethyl)amine; *N*-methylethanolamine; methylethylolamine. C_3H_9NO; mol wt 75.11. C 47.97%, H 12.08%, N 18.65%, O 21.30%. *In vivo* precursor of choline; isoln from mutant of *Neurospora crassa* which has lost its ability to synthesize choline: Horowitz, *J. Biol. Chem.* **162**, 413 (1946). Prepn by mixing ethylene oxide with concd methylamine soln with external cooling: Knorr, Matthes, *Ber.* **31**, 1069 (1898); Lowe *et al.*, **GB 763434** (1956 to Oxirane Ltd.); Nikolaev *et al.*, *Zh. Obshch. Khim.* **33**, 391 (1963). Toxicity studies: R. Hartung, H. H. Cornish, *Toxicol. Appl. Pharmacol.* **12**, 486 (1968); B. Ballantyne, H.-W. Leung, *Vet. Hum. Toxicol.* **38**, 422 (1996).

Viscous liquid. Fishy odor. d^{20} 0.937. bp_{760} 155-156°; bp_{12} 64-65°. n_D^{20} 1.4385. Miscible with water, alcohol, ether. Strong base. Corrosive to cork, metals. LD_{50} in rats (g/kg): 3.36 orally, 1.33 i.p. (Hartung). LD_{50} in male, female rats (mg/kg): 1908, 1391 orally; male, female rabbits (mg/kg): 1880, 1006 dermally (Ballantyne, Leung).
Caution: Direct contact may cause skin and eye irritation, skin corrosioin. *See* Ballantyne, Leung.

6090. Methyl Aminolevulinate. [33320-16-0] 5-Amino-4-oxopentanoic acid methyl ester; P-1202. $C_6H_{11}NO_3$; mol wt 145.16. C 49.65%, H 7.64%, N 9.65%, O 33.06%. Prodrug of δ-aminolevulinic acid, *q.v.*; induces production of intracellular protoporphyrin IX. Prepn: S. I. Zav'yalov, A. G. Zavozin, *Bull. Acad. Sci. USSR Div. Chem. Sci.* **36**, 1663 (1987). Porphyrin enrichment in solar keratoses: C. Fritsch *et al.*, *Photochem. Photobiol.* **68**, 218 (1998). Clinical trial in photodynamic therapy of basal cell carcinomas: A. M. Soler *et al.*, *Br. J. Dermatol.* **145**, 467 (2001); C. Vinciullo *et al.*, *Br. J. Dermatol.* **152**, 765 (2005); in actinic keratosis: R. M. Szeimies *et al.*, *J. Am. Acad. Dermatol.* **47**, 258 (2002). Review of pharmacology and clinical experience: M. A. A. Siddiqui *et al.*, *Am. J. Clin. Dermatol.* **5**, 127-137 (2004).

Hydrochloride. [79416-27-6] Metvix. $C_6H_{11}NO_3 \cdot HCl$; mol wt 181.62. mp 119-121°.
THERAP CAT: Antineoplastic (photosensitizer).

6091. *p*-Methylaminophenol Sulfate. [55-55-0] 4-(Methylamino)phenol sulfate (2:1); monomethyl-*p*-aminophenol sulfate; *p*-hydroxymethylaniline sulfate; Photol; Verol; Rhodol; Armol; Elon; Genol; Graphol; Photo-Rex; Pictol; Planetol; Metol. $C_{14}H_{20}N_2O_6S$; mol wt 344.38. C 48.83%, H 5.85%, N 8.13%, O 27.87%, S 9.31%.

Crystals. Discolors in air. mp ~260° with decompn. Sol in 20 parts cold, 6 parts boiling water; slightly sol in alc; insol in ether. *Keep well closed and protected from light.*
USE: Photographic developer; dyeing furs. Determn of phosphate.

6092. 4-Methylaminorex. [3568-94-3] 4,5-Dihydro-4-methyl-5-phenyl-2-oxazolamine; 2-amino-4-methyl-5-phenyl-2-oxazoline; 4-MAX. $C_{10}H_{12}N_2O$; mol wt 176.22. C 68.16%, H 6.86%, N 15.90%, O 9.08%. Indirect acting sympathomimetic agent related to amphetamine, *q.v.* Has four stereoisomers; the ($\pm$)-*cis* racemate being the drug of abuse. Prepn: G. I. Poos *et al.*, *J. Med. Chem.* **6**, 266 (1963); G. I. Poos, **US 3161650** (1964 to McNeil Labs). Synthesis and chemical properties of isomers: R. F. X. Klein *et al.*, *J. Forensic Sci.* **34**, 962 (1989). Structure-activity comparison of isomers: R. A. Glennon, B. Misenheimer, *Pharmacol. Biochem. Behav.* **35**, 517 (1990). Mechanism of action study with behavioural effects: K. Batsche *et al.*, *J. Pharmacol. Exp. Ther.* **269**, 1029 (1994). Metabolism in rats: G. L. Henderson *et al.*, *J. Anal. Toxicol.* **19**, 563 (1995). Neurochemical effects: Y. Zheng *et al.*, *J. Pharm. Pharmacol.* **49**, 89 (1997).

(+)-*cis*

($\pm$)-*cis* **Form.** [29493-77-4] EU4EA; U4Euh; Ice; McN-822. Crystals from benzene, mp 154.5-156° (Poos, 1963). Also reported as mp 139-142° (Glennon). $[\alpha]_D^{25}$ −244.7° for (−)-form; $[\alpha]_D^{25}$ +240.9° for (+)-form.
Note: The ($\pm$)-*cis* isomers are controlled substances (stimulant): **21 CFR**, 1308.11.

6093. Methylaniline. [100-61-8] *N*-Methylbenzenamine; *N*-monomethylaniline; *N*-phenylmethylamine. C_7H_9N; mol wt 107.16. C 78.46%, H 8.47%, N 13.07%. Made by heating aniline chloride and methyl alcohol under pressure. Toxicology: J. F. Treon *et al.*, *J. Ind. Hyg. Toxicol.* **31**, 1 (1949).

Colorless or slightly yellow liquid; becomes brown on exposure to air. d_4^{20} 0.989. mp −57°. bp 194-196°. $n_D^{21.2}$ 1.5702. Slightly sol in water; sol in alc, ether. LD in rabbits (g/kg): 0.28 orally; in rabbits, cats (mg/kg): 24, 24 i.v. (Treon).
Caution: Potential symptoms of overexposure are weakness, dizziness and headache; dyspnea, cyanosis; methemoglobinemia; pulmonary edema; liver and kidney damage. *See NIOSH Pocket Guide to Chemical Hazards* (DHHS/NIOSH 97-140, 1997) p 218.

6094. Methyl Anthranilate. [134-20-3] 2-Aminobenzoic acid methyl ester; methyl 2-aminobenzoate; neroli oil (artificial). $C_8H_9NO_2$; mol wt 151.17. C 63.56%, H 6.00%, N 9.27%, O 21.17%. Occurs in neroli, ylang-ylang, bergamot, jasmine, other essential oils and in grape juice; also obtained synthetically by esterifying anthranilic acid with CH_3OH in presence of HCl.

Crystals. d 1.168. mp 24-25°. bp_{15} 135.5°. Slightly sol in water; freely sol in alcohol or ether. LD_{50} orally in rats, mice: 2910, 3900 mg/kg: P. M. Jenner *et al.*, *Food Cosmet. Toxicol.* **2**, 327 (1964).
USE: As perfume for ointments; manuf synthetic perfumes.

6095. 2-Methylanthraquinone. [84-54-8] 2-Methyl-9,10-anthracenedione; β-methylanthraquinone. $C_{15}H_{10}O_2$; mol wt 222.24. C 81.07%, H 4.54%, O 14.40%. Occurs in teakwood. Prepd by oxidation of 2-methylanthracene; by formation from phthalic anhydride and toluene: Fieser, *Org. Synth.* **4**, 43 (1925); from 6- and 7-methylanthraquinone-1-carboxylic acids with powdered copper in quinoline: Fieser, Martin, *J. Am. Chem. Soc.* **58**, 1443 (1936); from 1,4-naphthoquinone and isoprene: Carothers, Berchet, *ibid.* **55**, 2813 (1933).

Needles from alcohol, mp 177°. Sublimes. Very sol in benzene, toluene, xylene; sol in alcohol, ether, glacial acetic acid, concd H_2SO_4. Insol in water.

6096. N^G**-Methylarginine.** [17035-90-4] N^5-[Imino(methylamino)methyl]-L-ornithine; tilarginine; targinine; N^G-monomethyl-L-arginine; L-NMA. $C_7H_{16}N_4O_2$; mol wt 188.23. C 44.67%, H 8.57%, N 29.77%, O 17.00%. Nitric oxide synthase (NOS) inhibitor; naturally occurring analog of arginine. Isoln from bovine brain and prepn: T. Nakajima *et al.*, *Biochim. Biophys. Acta* **230**, 212 (1971). Improved prepn: J. L. Corbin, M. Reporter, *Anal. Biochem.* **57**, 310 (1974); S. Pundak, M. Wilchek, *J. Org. Chem.* **46**, 808 (1981); of acetate salt: W. Zhang *et al.*, *Synth. Commun.* **24**, 2789 (1994). Inhibition of NO release by endothelial cells: R. M. J. Palmer *et al.*, *Biochem. Biophys. Res. Commun.* **153**, 1251 (1988). Review of biological significance: J. Leiper, P. Vallance, *Cardiovasc. Res.* **43**, 542-548 (1999). Clinical evaluation in cardiogenic shock: G. Cotter *et al.*, *Circulation* **101**, 1358 (2000).

Crystals from water-ethanol. $[\alpha]_D^{25}$ +26° (c = 1 in 1N HCl).
Acetate. [53308-83-1] $C_7H_{16}N_4O_2 \cdot C_2H_4O_2$; mol wt 248.28. White granular solid. $[\alpha]_D^{22}$ +9.5° (c = 0.10 in water). Sol in water.
USE: Reagent to inhibit nitric oxide production in biological systems.
THERAP CAT: In treatment of cardiogenic shock.

6097. Methylbenzethonium Chloride. [25155-18-4] N,N-Dimethyl-N-[2-[2-[methyl-4-(1,1,3,3-tetramethylbutyl)phenoxy]ethoxy]ethyl]benzenemethanaminium chloride (1:1); benzyldimethyl[2-[2-(p-1,1,3,3-tetramethylbutylcresoxy)ethoxy]ethyl]ammonium chloride; [2-[2-(p-octylcresoxy)ethoxy]ethyl]dimethylbenzylammonium chloride; benzyldimethyl[2-[2-[4-(1,1,3,3-tetramethylbutyl)tolyloxy]ethoxy]ethyl]ammonium chloride. $C_{28}H_{44}ClNO_2$; mol wt 462.12. C 72.78%, H 9.60%, Cl 7.67%, N 3.03%, O 6.92%. Quaternary ammonium salt with antimicrobial action. Germicidal activity: F. Hart, C. L. Huyck, *J. Am. Pharm. Assoc. Sci. Ed.* **37**, 272 (1948). *In vitro* activity vs *Leishmania major:* J. El-On, G. Messer, *Am. J. Trop. Med. Hyg.* **35**, 1110 (1986). Clinical trial in combination with paramomycin, *q.v.*, in cutaneous leishmaniasis: G. Krause, A. Kroeger, *Trans. R. Soc. Trop. Med. Hyg.* **88**, 92 (1994). Review of properties, commercial uses and toxicology: *J. Am. Coll. Toxicol.* **4**, 65-106 (1985).

Monohydrate. [1320-44-1] Diaparene; Hyamine 10X. White, bitter tasting crystals, mp 161-163°. Very sol in water, alc, cello-

solve, ether, hot benzene. Practically insol in chloroform. LD_{50} in rats (mg/kg): 100 s.c. (Hart, Huyck).
USE: In cosmetics as preservative; cationic surfactant.
THERAP CAT: Antiseptic, disinfectant.

6098. Methyl Benzoate. [93-58-3] Benzoic acid methyl ester; methyl benzenecarboxylate; essence of Niobe; oil of Niobe. $C_8H_8O_2$; mol wt 136.15. C 70.58%, H 5.92%, O 23.50%. Toxicity data: Smyth *et al.*, *Arch. Ind. Hyg. Occup. Med.* **10**, 61 (1954).

Colorless, transparent liquid; pleasant odor. d_4^{15} 1.094. mp ~ $-15°$. bp 198-200°. n_D^{15} 1.5205. Flash pt, closed cup: 181°F (82°C). Insol in water; misc with alcohol, ether, methanol. LD_{50} orally in rats: 3.43 g/kg (Smyth).
USE: In perfumes (Peau d'Espagne).

6099. Methyl Benzoylsalicylate. [610-60-6] 2-(Benzoyloxy)benzoic acid methyl ester; benzosalin; salicylic acid methyl ester benzoate. $C_{15}H_{12}O_4$; mol wt 256.26. C 70.31%, H 4.72%, O 24.97%.

Crystals. mp 85°. bp 385°. Insol in water. One gram dissolves in 35 ml alcohol; sol in benzene, chloroform, ether.

6100. α-Methylbenzylamine. [98-84-0] α-Methylbenzenemethanamine; α-phenylethylamine; 1-phenylethylamine; α-aminoethylbenzene. $C_8H_{11}N$; mol wt 121.18. C 79.29%, H 9.15%, N 11.56%. Prepn from acetophenone and ammonium formate: A. W. Ingersoll *et al.*, *J. Am. Chem. Soc.* **58**, 1808 (1936); by reduction of acetophenone in liquid ammonia: J. C. Robinson, H. R. Snyder, *Org. Synth.* **coll. vol. III**, 717 (1955). Stereospecific synthesis of R-(+)-form: H. Takahashi *et al.*, *Chem. Pharm. Bull.* **29**, 3387 (1981). Resolution of *(dl)*-methylbenzylamine: W. Theikecker, H.-G. Winkler, *Ber.* **87**, 690 (1954). Use of (R)-(+)-methylbenzylamine as resolving agent: E. J. Corey, J. Mann, *J. Am. Chem. Soc.* **95**, 6832 (1973). Use as chiral intermediate in synthesis of α-methyl-α-amino nitriles: K. Weinges *et al.*, *Ber.* **110**, 2098 (1977); of β-amino acids: M. Furukawa *et al.*, *Chem. Pharm. Bull.* **26**, 260 (1978). Toxicity: H. F. Smyth *et al.*, *Arch. Ind. Hyg. Occup. Med.* **4**, 119 (1951).

dl-**Form.** Liquid. Aromatic odor. Absorbs CO_2 from air. d_4^{15} 0.9395. bp_{18} 80-81°. Strong base. Soly in water at 20° about 4.2%. Misc with alcohol, ether. LD_{50} orally in rats: 0.94 g/kg (Smyth).
(+)-**Form.** Liquid, bp 184-186°. d_4^{22} 0.950. $[\alpha]_D^{22}$ +40.3° (neat).
($-$)-**Form.** Liquid, bp_{12} 73°. d_4^{22} 0.950. $[\alpha]_D^{22}$ $-40.3°$ (neat).
USE: As resolving agent; chiral intermediate.

6101. 4-Methylbenzylidene Camphor. [36861-47-9] 1,7,7-Trimethyl-3-[(4-methylphenyl)methylene]bicyclo[2.2.1]heptan-2-one; 3-(4'-methylbenzylidene)camphor; 4-MBC; Eusolex 6300. $C_{18}H_{22}O$; mol wt 254.37. C 84.99%, H 8.72%, O 6.29%. Ultraviolet filter in sunscreens. Prepn: J.-C. Richer, A. Rossi, *Can. J. Chem.* **50**, 1376 (1972). TLC identification in cosmetics: D. H.

Liem, L. T. H. Hilderink, *Int. J. Cosmet. Sci.* **1**, 341 (1979); electrokinetic chromatography determn of enantiomers in cosmetic formulations: B. Gómara *et al.*, *Electrophoresis* **26**, 3952 (2005). Photostability: A. Deflandre, G. Lang, *Int. J. Cosmet. Sci.* **10**, 53 (1988); *eidem, Cosmet. Toiletries* **103**, 69 (1988). *In vivo* assessment of photostability: M. Cambon *et al.*, *J. Cosmet. Sci.* **52**, 1 (2001). Uterotrophic activity: H. Tinwell *et al.*, *Environ. Health Perspect.* **110**, 533 (2002). Evaluation of androgenic activity: R. Ma *et al.*, *Toxicol. Sci.* **74**, 43 (2003).

Crystals from absolute ethanol, mp 98.5-99.5°. $[\alpha]_D^{24}$ +429° (c = 1.615 in benzene). uv max (95% ethanol): 226, 301 nm (ε 7300, 20500).

THERAP CAT: Ultraviolet screen.

6102. Methyl Blue. [28983-56-4] [[4-[Bis[4-[(sulfophenyl)-amino]phenyl]methylene]-2,5-cyclohexadien-1-ylidene]amino]benzenesulfonic acid sodium salt (1:2); sodium triphenyl-*p*-rosanilinetrisulfate; brilliant cotton blue; Helvetia blue; C.I. Acid Blue 93; C.I. 42780. $C_{37}H_{27}N_3Na_2O_9S_3$; mol wt 799.79. C 55.57%, H 3.40%, N 5.25%, Na 5.75%, O 18.00%, S 12.03%. Prepn: *Colour Index* **vol. 4** (3rd ed., 1971) p 4403.

Dark blue powder. Sol in water. Absorption max about 607 nm. *Note:* Do not confuse with methylene blue.
USE: As a coloring; dye for cotton and silk; biological stain.
THERAP CAT: Antiseptic.

6103. Methyl Bromide. [74-83-9] Bromomethane; monobromomethane; Embafume. CH_3Br; mol wt 94.94. C 12.65%, H 3.19%, Br 84.16%. Prepd industrially by the action of hydrobromic acid on methanol. Several modifications of the process, *e.g.*, sulfuric acid is added to sodium bromide and methanol, methyl bromide being removed by distillation. Acute toxicity: D. D. Irish *et al.*, *J. Ind. Hyg. Toxicol.* **22**, 218 (1940). Atmospheric emission and potential ozone depletion from biomass burning: S. Manö, M. O. Andreae, *Science* **263**, 1255 (1994). Review of toxicology and human exposure: *Toxicological Profile for Bromomethane* (PB93-110682, 1992) 129 pp.
Colorless gas. Usually odorless; sweetish, chloroform-like odor at high concns. Burning taste. Non-flammable in air, but burns in oxygen. mp −93.66°; bp 3.56°: Egan, Kemp, *J. Am. Chem. Soc.* **60**, 2097 (1938). Vapor press. (20°): 1420 mm Hg. d_4^0 1.730. d_{gas}^{20} 3.974 g/l; n_D^{-20} 1.4432. Viscosity (0°): 0.397 cP. Spec. heat at −96.6°: 0.165 cal/g/°C; at −13.0°: 1.97 cal/g/°C; at 25°: 0.107 cal/g/°C. Crit temp 194°. Soly in water (20°, 748 mm): 1.75 g/100 g of soln. Forms a cryst hydrate, $CH_3Br.20H_2O$, below 4°. Freely sol in alcohol, chloroform, ether, carbon disulfide, carbon tetrachloride, benzene. LC for rats in air (6 hrs): 514 ppm (Irish).
Caution: Potential symptoms of overexposure by inhalation are irritation of respiratory system; headache; visual disturbance; vertigo; nausea, vomiting, anorexia, abdominal pain; malaise, muscle weakness, incoordination, slurring of speech, staggering gait; hand

tremor; convulsions; mental confusion; dyspnea; pulmonary edema; coma; death from respiratory or circulatory collapse. Onset of symptoms may be delayed. Direct contact may cause eye irritation, skin irritation or vesiculation; direct contact with liquid may cause frostbite. Potential occupational carcinogen. *See NIOSH Pocket Guide to Chemical Hazards* (DHHS/NIOSH 97-140, 1997) p 200; *Clinical Toxicology of Commercial Products*, R. E. Gosselin *et al.*, Eds. (Williams & Wilkins, Baltimore, 5th ed., 1984) Section III, pp 280-284.
USE: In ionization chambers. For degreasing wool. Extracting oils from nuts, seeds, flowers. Soil or space fumigant for insects, fungi, rodents. Methylating agent. Has been used as fire extinguishing agent.

6104. 2-Methyl-1-butanol. [137-32-6] Active amyl alcohol; *dl-sec*-butyl carbinol. $C_5H_{12}O$; mol wt 88.15. C 68.13%, H 13.72%, O 18.15%. One of the major components of fusel oil, *q.v.* Prepn: Hawthorne, *J. Org. Chem.* **23**, 1788 (1958); **GB 883375** (1961 to Continental Oil). Isoln of (−)-form by fractional distillation of fusel oil: Milburn, Truter, *J. Chem. Soc.* **1954**, 3344. Prepn of (+)-form: Carnmalm, *Chem. Ind. (London)* **1956**, 1093. Toxicity study: H. F. Smyth *et al.*, *Am. Ind. Hyg. Assoc. J.* **23**, 95 (1962). Review of manuf by fractionation of fusel oil and *via* chlorination of pentanes, and properties: *Industrial Chemicals*, W. L. Faith *et al.*, Eds. (John Wiley, New York, 2nd ed., 1957) pp 107-114.

Liquid, bp 128°. d_4^{20} 0.816. n_D^{25} 1.4104. Flash pt, open cup: 122°F (50°C). Slightly sol in water (3.6 g/100 g at 30°); misc with alcohol, ether. LD_{50} orally in rats: 4.92 ml/kg (Smyth).
(−)-Form. [1565-80-6] (2S)-2-Methyl-1-butanol. $[\alpha]_D^{20}$ −4.75°.
(+)-Form. [616-16-0] (2R)-2-Methyl-1-butanol. d_{20}^{20} 0.826. n_D^{20} 1.411.

6105. 3-Methyl-2-butanol. [598-75-4] 1,2-Dimethyl-1-propanol; *dl-sec*-isoamyl alcohol; *sec*-isopentyl alcohol; isopropyl methyl carbinol. $C_5H_{12}O$; mol wt 88.15. C 68.13%, H 13.72%, O 18.15%. Prepn: Brown, Subba Rao, *J. Am. Chem. Soc.* **81**, 6423 (1959); Cook, *J. Org. Chem.* **27**, 3873 (1962). Review of manuf by fractionation of fusel oil and *via* chlorination of pentanes, and properties: *Industrial Chemicals*, W. L. Faith *et al.*, Eds. (John Wiley, New York, 2nd ed., 1957) pp 107-114.

Liquid, bp 113-114°. bp_{742} 109.5-110.5°. d^{19} 0.819. n_D^{20} 1.4091. Flash pt, closed cup: 103°F (39°C); open cup: 95°F (35°C). Slightly sol in water (2.8 g/100 g at 30°); miscible with alcohol, ether.
Caution: Potential symptoms of overexposure are irritation of eyes, skin, nose and throat; headache, dizziness; cough, dyspnea, nausea, vomiting and diarrhea; skin cracking. *See NIOSH Pocket Guide to Chemical Hazards* (DHHS/NIOSH 97-140, 1997) p 176.

6106. Methyl *tert*-Butyl Ether. [1634-04-4] 2-Methoxy-2-methylpropane; *tert*-butyl methyl ether; MTBE. $C_5H_{12}O$; mol wt 88.15. C 68.13%, H 13.72%, O 18.15%. Prepn: L. Henry, *Rec. Trav. Chim.* **23**, 324 (1904); from methanol and *t*-butyl alcohol: J. F. Norris, G. W. Rigby, *J. Am. Chem. Soc.* **54**, 2088 (1932); from methanol and isobutylene: K. R. Edlund, T. W. Evans, **US 1968601** (1934 to Shell); *eidem, Ind. Eng. Chem.* **28**, 1186 (1936); R. D. Morin, A. E. Bearse, *ibid.* **43**, 1596 (1951); from *t*-butyl alcohol and diazomethane: M. Neeman *et al.*, *Tetrahedron* **6**, 36 (1959). Use as chromatographic eluent: C. J. Little *et al.*, *J. Chromatogr.* **169**, 381 (1979). Experimental use to dissolve cholesterol gallstones *in vivo:* M. J. Allen *et al.*, *Gastroenterology* **88**, 122 (1985); *eidem, N. Engl. J. Med.* **312**, 217 (1985). Method for therapeutic use in dissolving cholesterol calculi: J. L. Thistle, M. J. Allen, **US 4758596** (1988 to Research Corp.). Acute toxicity: D. F. Marsh, C. D. Leake, *Anesthesiology* **11**, 455 (1950). Reviews of toxicity: R. von Burg, *J.*

Appl. Toxicol. **12**, 73-74 (1992); E. Reese, R. D. Kimbrough, *Environ. Health Perspect.* **101**, Suppl. 6, 115-131 (1993); M. G. Costantini, *ibid.* 151-160; and human exposure: *Toxicological Profile for Methyl tert-butyl ether* (PB97-121016, 1996) 268 pp.

Liquid, bp 55.2°. mp −109°. d_4^{20} 0.7404, n_D^{20} 1.3689. Vapor pressure at 25°: 245 mm Hg. Flash pt: −28°C. Ignition temp: 224°C. Soly in water: 4.8 g/100 g. Soly of water in methyl *t*-butyl ether: 1.5 g/100 g. Unstable in acid soln. Volatile; lipophilic. *Flammable.* LC$_{50}$ in mice (15 min): 1.6 mmol/liter of atmosphere (Marsh).
Caution: Potential symptoms of overexposure by inhalation are coughing, burning sensation in nose and throat, headache, nausea, vomiting, dizziness, feeling of spaciness and disorientation (PB97-121016).
USE: Octane booster in gasoline. Organic solvent. Chromatographic eluent esp in HPLC.
THERAP CAT: Cholelitholytic agent.

6107. Methyl Butyl Ketone. [591-78-6] 2-Hexanone; butyl methyl ketone; MBK. $C_6H_{12}O$; mol wt 100.16. C 71.95%, H 12.08%, O 15.97%. Prepn: H. C. Colman, W. H. Perkin, Jr., *J. Chem. Soc., Trans.* **55**, 352 (1889); R. E. Schaad, V. N. Ipatieff, *J. Am. Chem. Soc.* **62**, 178 (1940). Toxicity study: H. F. Smyth *et al.*, *Arch. Ind. Hyg. Occup. Med.* **10**, 61 (1954). Review of toxicology: P. M. J. Bos *et al.*, *Am. J. Ind. Med.* **20**, 175-194 (1991); and human exposure: *Toxicological Profile for 2-Hexanone* (PB93-110773, 1992) 116 pp.

Colorless liquid. d 0.830. bp 127°. *Flammable. Irritant.* Flash pt, closed cup: 73.4°F (23°C). Slightly soluble in water; sol in alcohol, ether. Incompatible with oxidzing agents, bases. LD$_{50}$ orally in rats: 2.59 g/kg (Smyth).
Caution: Potential symptoms of overexposure are irritation of eyes and nose; peripheral neuropathy; weakness, paresthesia; dermatitis; headache; drowsiness. *See NIOSH Pocket Guide to Chemical Hazards* (DHHS/NIOSH 97-140, 1997) p 164.
USE: Reagent and solvent in organic chemistry. In adhesives, lacquers, paint removers, and acrylic coatings.

6108. 2-Methyl-3-butyn-2-ol. [115-19-5] 2-Hydroxy-2-methyl-3-butyne. C_5H_8O; mol wt 84.12. C 71.39%, H 9.59%, O 19.02%. Toxicity data: W. Keil *et al.*, *Arzneim.-Forsch.* **4**, 477 (1954); K. Soehring *et al.*, *ibid.* **5**, 161 (1955).

Liquid. d_{20}^{20} 0.8672. 7.24 lbs/gal. mp 2.6°. bp$_{760}$ 104-105°; bp$_{80}$ 52°; bp$_{12}$ 20°. Flash pt 77°F (25°C). n_D^{20} 1.4211. Surface tension at 25° = 23.8 dynes/cm; 5% aq soln = 41.7 dynes/cm. Miscible with water, acetone, benzene, carbon tetrachloride, Cellosolve, cyclohexanone, diethylene glycol, ethyl acetate, kerosine, methyl ethyl ketone, mineral spirits, monoethanolamine, neatsfoot oil, petr ether, soybean oil, Stoddard solvent. Azeotrope with water, bp 90.7°, contains 28.4% H$_2$O. LD$_{50}$ in mice (mg/kg): 1800 orally (Keil); 2340 s.c. (Soehring).

6109. Methyl Butyrate. [623-42-7] Butanoic acid methyl ester; methyl butanoate. $C_5H_{10}O_2$; mol wt 102.13. C 58.80%, H 9.87%, O 31.33%.

Colorless liquid. d_4^{20} 0.898. mp about −95°. bp 102°. Flash pt 14°C. n_D^{20} 1.3879. *Flammable.* Sol in about 60 parts water; miscible with alcohol, ether.
USE: Manuf artificial rum and fruit essences.

6110. Methyl Carbamate. [598-55-0] Carbamic acid methyl ester; urethylane; methylurethane. $C_2H_5NO_2$; mol wt 75.07. C 32.00%, H 6.71%, N 18.66%, O 42.62%. Prepn from silver or mercuricyanate with H$_2$S and methanol: Birkenbach, Kolb, *Ber.* **68**, 901 (1935). From urea and methanol: **DE 753127** (1940 to I.G. Farbenind.).

White crystals, mp 52-54°. bp 177°. Freely sublimes even at room temp. Freely sol in water, alcohol.

6111. Methylcellulose. [9004-67-5] Cellulose methyl ether; Methocel A; Benecel M; Celevac; Cellucon; Citrucel; Cologel; Tearisol; Tylose M. Cellulose obtained from fibrous plant material and partially etherified with methyl groups. Prepd from cellulose fibers heated with caustic solution and treated with methyl chloride. Commercial methylcellulose has a methoxyl content of 25-33% (degree of substitution 1.5 to 2.0). Review of prepns and properties: Ott, *Cellulose and Cellulose Derivatives* (Wiley-Interscience, New York, 2nd ed., 1954/55); G. K. Greminger, A. B. Savage, in *Industrial Gums*, R. L. Whistler, Ed. (Academic Press, New York, 2nd ed., 1973) pp 619-647. Clinical efficacy as laxative: J. W. Hamilton *et al.*, *Dig. Dis. Sci.* **33**, 993 (1988). Evaluation in lens implantation surgery: J. R. Rojas *et al.*, *Ann. Ophthalmol.* **21**, 389 (1989). Review of production and uses in the food industry: P. de Mariscal, D. A. Bell in *Handbook of Fat Replacers*, S. Roller, S. A. Jones, Eds. (CRC Press, Boca Raton FL, 1996) pp 145-159.
White granules. Odorless, tasteless. Sol in cold water, glacial acetic acid, and in a mixture of equal volumes of alc and chloroform. Insol in hot water, alc, ether, chloroform. An aq soln is best prepd by dispersing the granules in hot (but not boiling) water with stirring and chilling to +5°. The soln is then stable at room temp. Presence of inorganic salts increases the viscosity. The soly is dependent upon degree of substitution. Clear films may be cast from aq soln.
USE: As a substitute for water-soluble gums; to render paper greaseproof, in adhesives, as thickening agent in cosmetics, as protective colloid in emulsions, as binder and stabilizer in foods. As fat replacer in the formulation of dietetic foods. Pharmaceutic aid (coating agent, suspending agent, tablet binder).
THERAP CAT: Laxative; ocular lubricant.
THERAP CAT (VET): Laxative.

6112. Methyl Chloride. [74-87-3] Chloromethane; Freon 40. CH$_3$Cl; mol wt 50.49. C 23.79%, H 5.99%, Cl 70.21%. Known as early as 1835, large scale production started in 1920's. Review of mfg processes: Faith, Keyes & Clark's *Industrial Chemicals*, F. A. Lowenheim, M. K. Moran, Eds. (Wiley-Interscience, New York, 4th ed., 1975) pp 530-538. GC-MS determn in air: D. R. Cronn, D. E. Harsch, *Anal. Lett.* **9**, 1015 (1976). X-ray emission study: D. W. Lindle *et al.*, *Phys. Rev. A* **43**, 2353 (1991). *Review:* M. T. Holbrook in *Kirk-Othmer Encyclopedia of Chemical Technology* vol. 5 (Wiley-Interscience, New York, 4th ed., 1994) pp 1028-1040. Review of toxicology: J. D. Repko. S. M. Lasley, *Crit. Rev. Toxicol.* **6**, 283-302 (1979); and human exposure: *Toxicological Profile for Chloromethane* (PB99-121964, 1998) 288 pp.
Colorless gas of mild odor and sweet taste. mp −97.7°; bp −23.7°; n_D (liq at −23.7°) 1.3712. *Flammable.* Slightly sol in water; misc with chloroform, ether, glacial acetic acid; sol in alcohol. Soly at 20° (ml/100 ml): benzene 4723; carbon tetrachloride 3756; glacial acetic acid 3679; ethanol 3740; at 25° (g/100g): water 0.48.
Caution: Potential symptoms of overexposure are dizziness, nausea and vomiting; visual disturbance; staggering; slurred speech; convulsions, coma; liver and kidney damage; reproductive and teratogenic effects; direct contact with liquid may cause frostbite. Potential occupational carcinogen. *See NIOSH Pocket Guide to Chemical Hazards* (DHHS/NIOSH 97-140, 1997) p 202. *See also Patty's Industrial Hygiene and Toxicology* vol. **2B**, G. D. Clayton, F. E. Clay-

ton, Eds. (Wiley-Interscience, New York, 3rd ed., 1981) pp 3436-3442.

USE: Manuf of silicones, tetramethyleads. Solvent catalyst for butyl rubber. Has been used as a refrigerant.

6113. Methyl Chloroacetate. [96-34-4] 2-Chloroacetic acid methyl ester. $C_3H_5ClO_2$; mol wt 108.52. C 33.20%, H 4.64%, Cl 32.67%, O 29.49%.

Colorless liquid. d_{20}^{20} 1.238. mp −33°. bp 130-132°. *Poisonous; flammable.* Insol in water; miscible with alcohol, ether.

USE: As solvent.

6114. Methyl Chlorocarbonate. [79-22-1] Carbonochloridic acid methyl ester; methyl chloroformate. $C_2H_3ClO_2$; mol wt 94.49. C 25.42%, H 3.20%, Cl 37.52%, O 33.86%. Made from phosgene and methyl alcohol.

Clear liquid. *Flammable, poisonous, corrosive.* bp 71°. d_4^{20} 1.223. Slightly sol in water and gradually dec by it; miscible with alcohol, benzene, chloroform, ether. Store at 2-8°C.

Caution: Vapors strongly irritating to eyes.

USE: Reagent in synthetic organic chemistry.

6115. 3-Methylcholanthrene. [56-49-5] 1,2-Dihydro-3-methylbenz[*j*]aceanthrylene; 20-methylcholanthrene; 3-MECA; 3-MC. $C_{21}H_{16}$; mol wt 268.36. C 93.99%, H 6.01%. From desoxy-cholic acid: Wieland, Schlichting, *Z. Physiol. Chem.* **150**, 267 (1925); Wieland, Wiedersheim, *ibid.* **186**, 229 (1930); Wieland, Dane, *ibid.* **219**, 240 (1933). From cholic acid: Cook, Haslewood, *J. Chem. Soc.* **1934**, 428; Fieser, Seligman; *J. Am. Chem. Soc.* **58**, 2482 (1936); *cf.* Bachmann, *J. Org. Chem.* **1**, 347 (1936). Methyl-cholanthrene has been produced also by an unusual pyrolytic degradation of cholesterol derivatives. Total synthesis: Buchta, Güllich, *Angew. Chem.* **70**, 190 (1958); P. W. Tang, C. A. Maggiulli, *J. Org. Chem.* **46**, 3429 (1981); S. A. Jacobs, R. G. Harvey, *Tetrahedron Lett.* **22**, 1093 (1981). *Review:* E. Clar, *Polycyclic Hydrocarbons* **Vol. 1 & 2** (Academic Press, 1964).

Pale yellow, slender prisms from benzene + ether, mp 179-180°, bp$_{80}$ 280°; d^{20} 1.28. Absorption spectrum: Fieser, Hershberg, *J. Am. Chem. Soc.* **60**, 940 (1938). Sol in benzene, xylene, toluene; slightly sol in amyl alcohol. Insol in water.

USE: Exptlly in cancer research.

6116. Methylcobalamin. [13422-55-4] *Co*-Methylcobin-amide dihydrogen phosphate (ester) inner salt 3′-ester with 5,6-di-methyl-1-α-D-ribofuranosyl-1*H*-benzimidazole-κ*N*³; methyl-5,6-di-methylbenzimidazolylcobalamin; mecobalamin; methyl vitamin B$_{12}$; MeCbl; Algobaz; Hitocobamin-M; Methycobal; Methylcobaz; Xob-aline. $C_{63}H_{91}CoN_{13}O_{14}P$; mol wt 1344.40. C 56.28%, H 6.82%, Co 4.38%, N 13.54%, O 16.66%, P 2.30%. One of the biologically active forms of vitamin B$_{12}$, *q.v.*; differing only by the substitution of a methyl for the cyano group. Coenzyme required in the biosynthesis of methionine from homocysteine. Identification as requirement for methionine biosynthesis by *E. coli*: R. L. Kisliuk, D. D. Woods, *Biochem. J.* **75**, 467 (1960); J. R. Guest *et al., Nature* **195**, 340 (1962). Prepn: O. Müller, G. Müller, *Biochem. Z.* **336**, 299 (1962); D. Dolphin, *Methods Enzymol.* **18**, Pt. C, 34 (1971); D. Au-tissier *et al., Bull. Soc. Chim. Fr.* **1980**, part 2, 192. Interrelationship

with folate metabolism: B. Shane, E. L. R. Stokstad, *Annu. Rev. Nutr.* **5**, 115-141 (1985). General review: "Vitamin B$_{12}$" in *Vita-mins*, W. Friedrich, Ed. (de Gruyter, Berlin, 1988) pp 837-928. Review of role in methyltransferase reactions: R. G. Matthews, *Acc. Chem. Res.* **34**, 681-689 (2001). Review of clinical experience in treatment of hyperhomocysteinemia and peripheral neuropathy: Y.-F. Zhang, G. Ning, *Expert Opin. Invest. Drugs* **17**, 953-964 (2008).

Bright red crystals from water/acetone. uv max (pH 7): 522, 342, 266 nm (ε 9357, 14416, 19897); (0.1*N* HCl): 462, 304, 264 nm (ε 9599, 22855, 24737).

THERAP CAT: Vitamin; enzyme cofactor. In treatment of hyper-homocysteinemia and peripheral neuropathy.

6117. Methyl Cyanoacrylate. [137-05-3] 2-Cyano-2-pro-penoic acid methyl ester; 2-cyanoacrylic acid methyl ester; MCA; mecrilate; mecrylate; methyl 2-cyanoacrylate; methyl α-cyanoacry-late. $C_5H_5NO_2$; mol wt 111.10. C 54.05%, H 4.54%, N 12.61%, O 28.80%. Reacts rapidly with water to form solid polymers. Prepar-ative methods: A. E. Ardis, **US 2467927** (1949 to B. F. Goodrich); C. H. McKeever, **US 2912454** (1959 to Rohm & Haas). Prepn, polymerization and degradation: F. Leonard *et al., J. Appl. Polym. Sci.* **10**, 259 (1966). Clinical evaluation for tubal occlusion in female sterilization: *Contraception* **31**, 243 (1985). Review of chemistry and toxicology: R. Cary, *Concise Int. Chem. Assess. Doc. No. 36* (WHO, Geneva, 2001) 33 pp.

Clear, colorless liquid; strong, acrid odor; bp 48-49° (0.33-0.36 kPa). n_D^{20} 1.4459. n_D^{25} 1.443. d^{20} 1.1044. Surface tension: 37.41 dynes/cm. Vapor pressure at 25°: <0.27 kPa. Sol or partially sol in methyl ethyl ketone, toluene, DMF, acetone, nitromethane.

Polymer. [25067-29-2] Poly(methyl 2-cyanoacrylate).

Caution: Potential symptoms of overexposure are irritation of eyes, skin, nose; blurred vision, lacrimation; rhinitis. *See NIOSH Pocket Guide to Chemical Hazards* (DHHS/NIOSH 97-140, 1997) p 204.

USE: In adhesives; in mfr of plastics, electronics, scientific instru-ments, jewelry, sports equip; in cable joining, manicuring, dentistry, mortuaries, fingerprint development.

THERAP CAT: Tissue adhesive.

6118. Methylcyclohexane. [108-87-2] Hexahydrotoluene; cyclohexylmethane. C_7H_{14}; mol wt 98.19. C 85.63%, H 14.37%.

Prepn: F. Wreden, B. Znatowicz, *Ann.* **187**, 161 (1877); E. Knoevenagel, *ibid.* **297**, 113 (1897); and physical properties: J. P. Wibaut *et al.*, *Rec. Trav. Chim.* **58**, 12 (1939). Toxicology: J. F. Treon *et al.*, *J. Ind. Hyg. Toxicol.* **25**, 199 (1943). Dehydrogenation to produce motor fuel: N. Gruenenfelder *et al.*, *Proc. 20th Intersoc. Energy Convers. Eng. Conf.* **2**, 2.781 (1985). Kinetics of industrial dehydrogenation to toluene: A. Touzani *et al.*, *Stud. Surf. Sci. Catal.* **19**, 357 (1984); use of a membrane reactor to improve process: J. K. Ali, A. Baiker, *Appl. Catal. A* **155**, 41 (1997).

Clear, fragrant liquid, mp −126.3°. bp_{760} 101.2 ± 0.3°; also reported as bp_{760} 103°. $n_D^{18.5}$ 1.41705; n_D^{20} 1.4230; n_D^{25} 1.42063. $d_4^{18.5}$ 0.7662; d_4^{20} 0.76944; d_4^{25} 0.76512. Surface tension: 23.73 dynes/cm. *Flammable.* Insol in water at 20°.

Caution: Potential symptoms of overexposure are irritation of eyes, skin, nose and throat; lightheadedness, drowsiness. *See NIOSH Pocket Guide to Chemical Hazards* (DHHS/NIOSH 97-140, 1997) p 204.

USE: Solvent; starting material in the synthesis of toluene.

6119. 3′-Methyl-1,2-cyclopentenophenanthrene. [549-38-2] 17-Methyl-gona-1,3,5,7,9,11,13-heptaene; 16,17-dihydro-17-methyl-15*H*-cyclopenta[*a*]phenanthrene; Diels' hydrocarbon. $C_{18}H_{16}$; mol wt 232.33. C 93.06%, H 6.94%. Prepd from cholesterol, cholic acid, or suitable sapogenins: Diels, Rickert, *Ber.* **68**, 267 (1935); starting with 2-acetylphenanthrene: Gamble *et al.*, *J. Chem. Soc.* **1935**, 443, 644. Absorption spectrum: Hillemann, *Ber.* **69**, 2610 (1936).

Crystals from acetic acid, mp 126-127°.

6120. 1-Methylcyclopropene. [3100-04-7] 1-MCP; Ethyl-Bloc; Invinsa; SmartFresh. C_4H_6; mol wt 54.09. C 88.82%, H 11.18%. Ethylene inhibiting plant growth regulator. Prepn from 2-methylallyl chloride and $NaNH_2$: F. Fisher, D. E. Applequist, *J. Org. Chem.* **30**, 2089 (1965); from 2-methylallyl chloride and phenyllithium: R. M. Magid *et al.*, *ibid.* **36**, 1320 (1971); and proposed mode of action: E. C. Sisler, M. Serek, *Physiol. Plant.* **100**, 577 (1997). Microwave structural studies: M. K. Kemp, W. H. Flygare, *J. Am. Chem. Soc.* **89**, 3925 (1967); *eidem, ibid.* **91**, 3163 (1969). Kinetics of encapsulation with α-cyclodextrin: T. L. Neoh *et al.*, *J. Agric. Food Chem.* **55**, 11020 (2007). Review of mechanism of action: B. M. Binder, A. B. Bleecker, *Acta Hortic.* **628**, 177-187 (2003); of agricultural applications: S. M. Blankenship, J. M. Dole, *Postharvest Biol. Technol.* **28**, 1-25 (2003); of use on fruits and vegetables: C. B. Watkins, *Biotechnol. Adv.* **24**, 389-409 (2006).

Gas at standard temperature and pressure, bp ≤10°. In some commercial powder formulations, 1-MCP is enclosed in a cyclodextrin cage that releases 1-MCP when mixed with water. LC_{50} in rats by inhalation: >2.5 mg/l (Blankenship, Dole).

USE: Blocks the ripening and aging effects of ethylene in harvested fruits and vegetables, cut and potted flowers, and foliage plants.

6121. Methylcytisine. [486-86-2] (1*R*,5*S*)-1,2,3,4,5,6-Hexahydro-3-methyl-1,5-methano-8*H*-pyrido[1,2-*a*][1,5]diazocin-8-one; caulophylline; 12-methylcytisine; *N*-methylcytisine. $C_{12}H_{16}N_2O$; mol wt 204.27. C 70.56%, H 7.90%, N 13.71%, O 7.83%. Lupine

alkaloid of the quinolizidine type; occurring most abundantly in *Leguminosae.* Bioactive principle of blue cohosh, *q.v.* Isoln from *Caulophyllum thalictroides* Michx., *Berberidaceae:* F. B. Power, A. H. Salway, *J. Chem. Soc.* **103**, 191 (1913); M. S. Flom *et al.*, *J. Pharm. Sci.* **56**, 1515 (1967); from *Baptisia australis* (L.) R. Br., *Leguminosae:* L. Marion, J. Ouellet, *J. Am. Chem. Soc.* **70**, 691 (1948). Absolute configuration: S. Okuda *et al.*, *Chem. Pharm. Bull.* **13**, 491 (1965). Crystal structure: A. A. Freer *et al.*, *Acta Crystallogr.* **C43**, 1119 (1987). Pharmacology: H. C. Ferguson, L. D. Edwards, *J. Am. Pharm. Assoc. Sci. Ed.* **43**, 16 (1954). Toxicity study: R. B. Barlow, L. J. McLeod, *Br. J. Pharmacol.* **35**, 161 (1969). Binding to nicotinic and muscarinic receptors: T. Schmeller *et al.*, *J. Nat. Prod.* **57**, 1316 (1994).

Colorless, prismatic needles from benzene + light petroleum, mp 137°. $[α]_D^{30}$ −224° (c = 1.05 in water). uv max: 234, 309 nm (ε 6860, 7560). Readily sol in water, alcohol, chloroform, benzene. LD_{50} in mice (mg/kg): 21 i.v.; 51 i.p. (Barlow, McLeod).

6122. 5-Methylcytosine. [554-01-8] 6-Amino-5-methyl-2(1*H*)-pyrimidinone; 4-amino-5-methyl-2(1*H*)-pyrimidinone. $C_5H_7N_3O$; mol wt 125.13. C 47.99%, H 5.64%, N 33.58%, O 12.79%. Occurs in a nucleic acid obtained from tubercle bacillus. Isoln: Johnson, Coghill, *J. Am. Chem. Soc.* **47**, 2838 (1925). Synthesis from ethyl bromide addition product of thiourea and from ethyl formyl propionate: Wheeler, Johnson, *Am. Chem. J.* **29**, 492 (1903).

Prisms from water, may contain ½ mol H_2O. mp 270° (effervescence). One gram dissolves in 29 ml water.

6123. Methyl Demeton. [8022-00-2] Phosphorothioic acid *O*-[2-(ethylthio)ethyl] *O*,*O*-dimethyl ester mixt with *S*-[2-(ethylthio)-ethyl] *O*,*O*-dimethyl phosphorothioate; demeton-methyl; methylmercaptophos; methyl systox; Bayer 21/116; Meta-Systox. $C_6H_{15}O_3PS_2$; mol wt 230.28. C 31.29%, H 6.57%, O 20.84%, P 13.45%, S 27.84%. Cholinesterase inhibitor. Isomeric mixture consisting of demeton-*O*-methyl and demeton-*S*-methyl (*O*(and *S*)-[2-(ethylthio)-ethyl] *O*,*O*-dimethyl phosphorothioate). *Ref:* Henglein, Schrader, *Z. Naturforsch.* **10b**, 12 (1955). Prepn: **GB 814332** (1959 to Bayer). Toxicology studies: M. Vandekar, *Br. J. Ind. Med.* **15**, 158 (1958); E. F. Edson, *Pharm. J.* **185**, 361 (1960).

Commercial product is a light yellow liquid. Hydrolyzed by alkali. LD_{50} in rats (mg/kg): 50-75 orally, 300-400 dermally (Edson).

Demeton-*O*-methyl. [867-27-6] Colorless oil, $bp_{0.15}$ 74°. d_4^{20} 1.190. n_D^{20} 1.5063. Soly in water at room temp: 330 ppm. Soluble in organic solvents.

Demeton-*S*-methyl. [919-86-8] Pale yellow oil, $bp_{0.15}$ 89°. d_4^{20} 1.207. n_D^{20} 1.5065. Soly in water at room temp: 3,300 ppm. Sol in organic solvents. LD_{50} in rats (mg/kg): 40 orally, 85 dermally (Edson).

Caution: Potential symptoms of overexposure are irritation of eyes, skin; aching eyes, rhinorrhea; nausea, headache, dizziness, vomiting. *See NIOSH Pocket Guide to Chemical Hazards* (DHHS/NIOSH 97-140, 1997) p 206.

USE: Insecticide.

6124. **Methyl Dihydrojasmonate.** [24851-98-7] 3-Oxo-2-pentylcyclopentaneacetic acid methyl ester; methyl 2-pentyl-3-oxo-cyclopentyl-1-acetate; Hedione. $C_{13}H_{22}O_3$; mol wt 226.32. C 68.99%, H 9.80%, O 21.21%. Fragrance compound identified as a natural product in flowers, tea, herbs, fruits and vegetables. Synthetic commercial formulations contain a near equilibrium mixture of ~10% (±)-*cis* and ~90% (±)-*trans* forms. The (+)-(1R,2S)-form is the most organoleptically active enantiomer. Prepn: E. Demole *et al.*, *Helv. Chim. Acta* **45**, 675, 685, 692 (1962). Fragrance monograph: *Food Chem. Toxicol.* **30**, Suppl. 1, 85S (1992). Enantioselective synthesis of (+)- and (−)-*trans* forms: T. Perrard *et al.*, *Org. Lett.* **2**, 2959 (2000). Review of sensory properties of stereoisomers and enantioselective distribution in plant sources: P. Werkhoff *et al.*, *Food Rev. Int.* **18**, 103-122 (2002).

(+)-(1R,2S)-Form

Pale yellowish or almost colorless oil, bp$_{0.2}$ 109-112°; sweet, floral, jasmine-like odor. Flash point: >200°F (closed cup). Vapor pressure (20°): <0.001 mmHg. n^{20} 1.457-1.462. d^{20}_{20} 0.998-1.006. LD_{50} (g/kg): >5 orally in rats; >5 dermally in rabbits (*Food Chem. Toxicol.*).

(+)-(1R,2S)-Form. [39647-11-5] (+)-Methyl dihydroepijasmonate; (+)-*cis*-methyl dihydrojasmonate; Paradisone. Enantioselective synthesis: D. A. Dobbs *et al.*, *Angew. Chem. Int. Ed.* **39**, 1992 (2000); A. Porta *et al.*, *J. Org. Chem.* **70**, 4876 (2005). Colorless oil; intense floral, jasmine-like, bright, fatty odor. $[\alpha]_D^{20}$ +78° (c = 1.1 in CH_2Cl_2).

(−)-(1R,2R)-Form. [2630-39-9] (−)-*trans*-Methyl dihydrojasmonate. Faint, slightly floral, sweet, jasmine-like odor, bp$_{0.03}$ 100°. n_D^{20} 1.4583. d_4^{21} 0.9968. $[\alpha]_D$ −33.8° (c = 2.9 in $CHCl_3$).

(+)-(1S,2S)-Form. [151716-36-8] (+)-*trans*-Methyl dihydrojasmonate. Colorless oil, bp$_{0.3}$ 108°. Faint, slightly floral, fatty, lemon peel-like odor. $[\alpha]_D^{20}$ +33.8° (c = 0.9 in $CHCl_3$).

USE: Fragrance ingredient in perfumes, toilette and laundry products. Flavor ingredient producing fruit, tea, tobacco, and floral flavors.

6125. **p-Methyldiphenhydramine.** [19804-27-4] N,N-Dimethyl-2-[(4-methylphenyl)phenylmethoxy]ethanamine; N,N-dimethyl-2-[(p-methyl-α-phenylbenzyl)oxy]ethylamine; β-dimethylaminoethyl 4-methylbenzhydryl ether; p-methylbenzhydryl 2-dimethylaminoethyl ether; Neo-Benodine; Toladryl. $C_{18}H_{23}NO$; mol wt 269.39. C 80.25%, H 8.61%, N 5.20%, O 5.94%. Prepn: G. Rieveschl, US 2527962 (1950 to Parke, Davis); GB 683483 (1952 to Fabriken Brocades-Stheeman & Pharmacia).

Hydrochloride. $C_{18}H_{23}NO \cdot HCl$. Crystals from acetone, mp 150-152° (free base bp$_{0.1}$ 143°).

THERAP CAT: Antihistaminic.

6126. **Methyldiphenylamine.** [552-82-9] N-Methyl-N-phenylbenzenamine; diphenylaminomethane; diphenylmethylamine. $C_{13}H_{13}N$; mol wt 183.25. C 85.21%, H 7.15%, N 7.64%. Prepd by heating diphenylamine, methyl alcohol and HCl.

Liquid. d_4^{20} 1.0476. mp −7.6°. bp 296-297°; also given as 282-290°, 293°. n_D^{20} 1.6193. Insol in water; sol in alc, ether.

USE: Manuf dyes; as reagent similar to diphenylamine.

6127. **Methyldopa.** [555-30-6]; [41372-08-1] (sesquihydrate). 3-Hydroxy-α-methyl-L-tyrosine; L-3-(3,4-dihydroxyphenyl)-2-methylalanine; L-α-methyl-3,4-dihydroxyphenylalanine; L-2-amino-2-methyl-3-(3,4-dihydroxyphenyl)propionic acid; L-α-methyldopa; MK-351; Aldomet; Dopamet; Dopegyt; Medopren; Presinol. $C_{10}H_{13}NO_4$; mol wt 211.22. C 56.86%, H 6.20%, N 6.63%, O 30.30%. Prepn of racemate: K. Pfister, G. A. Stein, US 2868818 (1959 to Merck & Co.). Resolution: R. T. Jones *et al.*, US 3158648 (1964 to Merck & Co.); and configuration: E. W. Tristram *et al.*, *J. Org. Chem.* **29**, 2053 (1964). Synthesis from asymmetric intermediates: D. F. Reinhold *et al.*, *ibid.* **33**, 1209 (1968); J. L. León-Romo *et al.*, *Chirality* **14**, 144 (2002). HPLC determn in plasma: G. Bahrami *et al.*, *J. Chromatogr. B* **832**, 197 (2006). Clinical pharmacokinetics: E. Myhre *et al.*, *Clin. Pharmacokinet.* **7**, 221 (1982). Review of mechanism of action and pharmacology: E. D. Frohlich, *Arch. Intern. Med.* **140**, 954-959 (1980); and clinical experience: A. Scriabine, Ed. in *Pharmacology of Antihypertensive Drugs* (Raven, New York, 1980) pp 43-54.

White to yellowish-white, odorless, fine powder. Considerably hygroscopic. Crystals from water as the sesquihydrate, mp 302-304° (dec). $[\alpha]_D^{25}$ −3° (c = 2 in 0.1N HCl). uv max: 281 nm (ε 2780). Very sol in 3N HCl; slightly sol in alcohol. Practically insol in ether. Soly in water at 25°: ~10 mg/ml. pH of satd aq soln about 5.0.

DL-Form. [55-40-3] Crystals from acetone, mp ~300° (dec). Soly in water at 25°: ~20 mg/ml.

L-Form ethyl ester hydrochloride. [2508-79-4] Methyldopate hydrochloride. $C_{12}H_{18}ClNO_4$; mol wt 275.73. Prepn: FR M2153 (1963 to Merck & Co.); of pharmaceutical dosage forms: A. D. Marcus, US 3230143 (1966 to Merck & Co.). White, odorless crystalline powder. Freely sol in water, ethanol, methanol; slightly sol in chloroform. Practically insol in ether.

THERAP CAT: Antihypertensive.

6128. **3-O-Methyldopa.** [300-48-1] 3-Methoxy-L-tyrosine; 3-(4-hydroxy-3-methoxyphenyl)-L-alanine; L-3-methoxy-4-hydroxyphenylalanine; OM-dopa; L-3-MTO; OMD. $C_{10}H_{13}NO_4$; mol wt 211.22. C 56.86%, H 6.20%, N 6.63%, O 30.30%. A major metabolite of L-dopa in man and animals, which has a considerably longer biological half-life than L-dopa: Pletscher *et al.*, *Brain Res.* **4**, 106 (1967); Bartholini, Pletscher, *J. Pharmacol. Exp. Ther.* **161**, 14 (1968); Kuruma *et al.*, *Eur. J. Pharmacol.* **10**, 189 (1970). Proposed as a precursor of dopamine *via* its partial demethylation in the organism: Bartholini *et al.*, *Nature* **230**, 533 (1971); Chalmers *et al.*, *Br. J. Pharmacol.* **43**, 455P (1971); Bartholini *et al.*, *Life Sci.* **14**, 323 (1974). *See also:* Carlsson, Waldeck, *Arch. Pharmacol.* **272**, 441 (1972); Bartholini, Pletscher, *ibid.* **274**, 404 (1972). Metabolic pathways: Bartholini *et al.*, *J. Pharmacol. Exp. Ther.* **183**, 65 (1972). Parkinsonism treatment studies: Calne *et al.*, *Clin. Pharmacol. Ther.* **14**, 386 (1973).

6129. **2-(Methyleneamino)acetonitrile.** [109-82-0] α-Hydroformamine cyanide; N-methyleneglycinonitrile. $C_3H_4N_2$; mol wt 68.08. C 52.93%, H 5.92%, N 41.15%. The actual molecular formula is $C_9H_{12}N_6$: Johnson, Rinehart, *J. Am. Chem. Soc.* **46**, 768, 1653 (1924). Prepd by the action of formaldehyde on a mixture of ammonium chloride, potassium cyanide and acetic acid: Adams, Langley, *Org. Synth.* **coll. vol. I** (2nd ed., 1941) p 355.

Trimeric form. [6865-92-5] Orthorhombic crystals from alc or acetone, mp 129°. Sol in hot water, alc; slightly sol in benzene.

6130. Methylene Azure. Azure I. A somewhat variable mixture obtained by oxidation of methylene blue, contg primarily azure A and azure B, *q.q.v.* Prepn: Bernthsen, *Ann.* **230**, 169 (1885); Kehrmann, *Ber.* **39**, 1804 (1906). *Review:* H. J. Conn's *Biological Stains*, R. D. Lillie, Ed. (Williams & Wilkins, Baltimore, 9th ed., 1977) pp 493-502.

Green glistening crystals. Forms blue soln in water; very sparingly sol in alcohol with a reddish brown fluorescence.

Mixture with methylene blue. Azure II; azure blue II; methylene azure II. Deep-green powder. Sol in water with blue color, less sol in alcohol, slightly sol in chloroform. Insol in ether.

Mixture with methylene blue and eosin. Azure II eosin. Chief ingredient of Giemsa stain. Green powder. Slightly sol in water; sol in alcohol, methanol, glycerol.

USE: Biological stain.

6131. 4,4′-Methylenebis[2-chloroaniline]. [101-14-4] 4,4′-Methylenebis[2-chlorobenzenamine]; 4,4′-diamino-3,3′-dichlorodiphenylmethane; di-(4-amino-3-chlorophenyl)methane; methylenebis(*o*-chloroaniline); MOCA; MBOCA; DACPM. $C_{13}H_{12}Cl_2N_2$; mol wt 267.15. C 58.45%, H 4.53%, Cl 26.54%, N 10.49%. Prepn: Finger, *J. Prakt. Chem.* [2] **79**, 493 (1909); Mayer, *Ber.* **47**, 1161 (1914). Acute toxicity: G. Ya. Kel'man *et al.*, *Kauch. Rezina* **26**(9), 28 (1967), *C.A.* **67**, 120013m (1967). Review of carcinogenicity studies: *IARC Monographs* **vol. 4**, 65-71 (1974); of toxicology and human exposure: *Toxicological Profile for 4,4′-Methylenebis(2-chloroaniline) MBOCA* (PB95-100186, 1994) 146 p.

Flakes from alcohol, mp 110°. Sol in dil acids, ether, alc; slightly sol in water. LD_{50} orally in mice: 880 mg/kg (Kel'man).

Caution: Potential symptoms of overexposure are hematuria, cyanosis, nausea, methemoglobinemia, kidney irritation. *See NIOSH Pocket Guide to Chemical Hazards* (DHHS/NIOSH 97-140, 1997) p 206. This substance is reasonably anticipated to be a human carcinogen: *Report on Carcinogens, Twelfth Edition* (PB2011-111646, 2011) p 262.

USE: Curing agent for polyurethane and epoxy resins. Research tool for studying carcinogens.

6132. Methylene Blue. [61-73-4] 3,7-Bis(dimethylamino)-phenothiazin-5-ium chloride (1:1); C.I. Basic Blue 9; methylthioninium chloride; tetramethylthionine chloride; 3,7-bis(dimethylamino)-phenazathionium chloride; Swiss blue; C.I. 52015; C.I. Solvent Blue 8; Urolene Blue. $C_{16}H_{18}ClN_3S$; mol wt 319.85. C 60.08%, H 5.67%, Cl 11.08%, N 13.14%, S 10.02%. First prepd by Caro in 1876. Provided technically as the zinc double chloride; used medically in a zinc free form. Prepn: H. E. Fierz-David, L. Blangey, *Fundamental Processes of Dye Chemistry* (Interscience, New York, 1949) p 311. *See also Colour Index* **vol. 4** (3rd ed., 1971) p 4470. TLC determn: G. Balansard *et al.*, *Ann. Pharm. Fr.* **46**, 129 (1988). Use as marker for reticulocytes: B. Rudensky, *Scand. J. Clin. Lab. Invest.* **57**, 291 (1997); in melanoma targeted radiotherapy: E. M. Link *et al.*, *Pigment Cell Res.* **7**, 358 (1994); P. J. Blower *et al.*, *Nucl. Med. Biol.* **24**, 305 (1997). Review of use as vital dye: P. Barbosa, T. M. Peters, *Histochem. J.* **3**, 71-93 (1971). History and properties: R. D. Lillie, P. T. Donaldson, *Stain Technol.* **54**, 33-39 (1979). Review of photochemistry: E. M. Tuite, J. M. Kelly, *J. Photochem. Photobiol. B* **21**, 103-124 (1993). Review of use as cationic dye in ceramic quality control: J. E. Funk, D. R. Dinger in *Science of Whitewares*, V. E. Henkes *et al.*, Eds. (American Ceramic Society,

OH, 1996) pp 21-31. Review of use in methemoglobinaemia: J. W. Harvey, A. S. Keitt, *Br. J. Haematol.* **54**, 29-41 (1983).

Trihydrate. [7220-79-3] Dark green crystals or crystalline powder with bronze luster. Odorless; stable in air. Absorption max: 668, 609 nm. Sol in water, chloroform; sparingly sol in alc. Deep blue soln in water or alc. Forms double salts.

USE: Stain in bacteriology; reagent for several chemicals; as mixed indicator; as redox colorimetric agent. Targeting agent for melanoma.

THERAP CAT: Antimethemoglobinemic; antidote (cyanide).

THERAP CAT (VET): Antiseptic; disinfectant; antidote (cyanide and nitrate).

6133. Methylene Bromide. [74-95-3] Dibromomethane. CH_2Br_2; mol wt 173.84. C 6.91%, H 1.16%, Br 91.93%. Prepn: Hartman, Dreger, *Org. Synth.* **coll. vol. I**, 357 (1941). Manuf along with bromochloromethane, from dichloromethane: Lake, Asadorian, *US 2553518* (1951 to Dow).

Liquid, bp 97°. mp −52.7°. d_4^{20} 2.4956. n_D^{20} 1.5419. Soly (g/1000 g water): 11.70 (15°); 11.93 (30°). Miscible with alcohol, ether, acetone.

6134. α-Methylene Butyrolactone. [547-65-9] Dihydro-3-methylene-2(*3H*)-furanone. $C_5H_6O_2$; mol wt 98.10. C 61.22%, H 6.17%, O 32.62%. Isoln from aq extracts of *Erythronium americanum* Ker., *Liliaceae* (dogtooth violet): Cavallito, Haskell, *J. Am. Chem. Soc.* **68**, 2332 (1946).

Liquid. Polymerizes easily when heated above 70°. bp_2 57-60°. Sol in water. pH 4 (150 mg/ml). The lactone is present in all parts of the plant and can be extracted easily with water, but not with ethanol, ether, ethyl acetate or chloroform.

Caution: Irritating to the skin.

6135. Methylene Chloride. [75-09-2] Dichloromethane; methylene dichloride; methylene bichloride. CH_2Cl_2; mol wt 84.93. C 14.14%, H 2.37%, Cl 83.48%. Prepn by chlorination of methane: Lukes *et al.*, *US 2792435* (1957 to Diamond Alkali); Pitt, Bender, *US 2979541* (1961 to Stauffer); Burks, Obrecht, *US 3126419* (1964 to Stauffer). Review of mfg processes: *Faith, Keyes & Clark's Industrial Chemicals*, F. A. Lowenheim, M. K. Moran, Eds. (Wiley-Interscience, New York, 4th ed., 1975) pp 530-538. Toxicity data: E. T. Kimura *et al.*, *Toxicol. Appl. Pharmacol.* **19**, 699 (1971). Review of toxicology and human exposure: *Toxicological Profile for Methylene Chloride* (PB2000-108026, 2000) 313 pp.

Colorless liquid; vapor is not flammable and when mixed with air is not explosive. Soluble in ~50 parts water; miscible with alc, ether, DMF, and with fixed and volatile oils. bp_{760} 39.75°. mp −95°. d_4^0 1.36174; d_4^{15} 1.33479; d_4^{20} 1.3255; d_4^{30} 1.30777. n_D^{20} 1.4244. LD_{50} orally in young adult rats: 1.6 ml/kg (Kimura).

Caution: Potential symptoms of overexposure are fatigue, weakness, sleepiness, lightheadedness; numbness or tingle of limbs; nausea; irritation of eyes and skin. *See NIOSH Pocket Guide to Chemical Hazards* (DHHS/NIOSH 97-140, 1997) p 208. This substance is reasonably anticipated to be a human carcinogen: *Report on Carcinogens, Twelfth Edition* (PB2011-111646, 2011) p 148.

USE: Solvent in paint removers, for cellulose acetate; degreasing and cleaning fluids; as solvent in food processing. Extraction solvent for spice oleoresins, hops, for removal of caffeine from coffee. Pharmaceutic aid (solvent). Aerosol propellant; insecticide.

6136. 5,5′-Methylenedisalicylic Acid. [122-25-8] 3,3′-Methylenebis[6-hydroxybenzoic acid]; 4,4′-dihydroxydiphenylmethane-3,3′-dicarboxylic acid. $C_{15}H_{12}O_6$; mol wt 288.26. C 62.50%, H 4.20%, O 33.30%. Prepd from salicylic acid and formaldehyde in the presence of sulfuric acid: Clemmensen, Heitman, *J. Am. Chem. Soc.* **33**, 737 (1911).

Wedge-like cryst from acetone + benzene. Bitter taste. Dec 238° (higher-melting material may be impure). Turns red at 180° and starts giving off CO_2. Freely sol in methanol, ethanol, ether, acetone, glacial acetic acid. Very slightly sol in hot water. Practically insol in benzene, chloroform, carbon disulfide, petr ether.

Bacitracin salt *see* Bacitracin Methylenedisalicylate.

Diacetyl derivative. $C_{15}H_{10}O_6(CH_3CO)_2$. White powder, mp 142°, practically insol in water, sol in acetone, ethanol.

6137. Methylene Iodide. [75-11-6] Diiodomethane. CH_2I_2; mol wt 267.84. C 4.48%, H 0.75%, I 94.76%. Prepd by the reduction of iodoform with sodium arsenite: Adams, Marvel, *Org. Synth.* **vol. 1**, p 57 (1921). Also by heating iodoform with sodium acetate in alcohol: Bagnara, *Eng. Min. J. Press* **116**, 51 (1923). Absorption spectrum: Lowry, Sass, *J. Chem. Soc.* **1926**, 624; Stepanov, *Acta Physicochim. URSS* **20**, 174 (1945).

Very heavy, highly refractive liq. Darkens on exposure to light, air, and moisture. d_4^{20} 3.32537. Solidifies in leaflets at 5.2° or in thin needles at 5.7°. Usually cooling to 0° is necessary to start crystn. mp 6.0°. bp$_{760}$ 181°; bp$_{70}$ 107°; bp$_{11}$ 68°. $n_D^{10.5}$ 1.7559; n_D^{15} 1.7425. Viscosity (cP): 3.35 (10°); 2.80 (20°); 2.39 (30°). Sol in about 70 parts water; miscible with alcohol, propanol, isopropanol, hexane, cyclohexane, ether, chloroform, benzene. CH_2I_2 dissolves sulfur and phosphorus (more than 1:1 at 25°).

USE: In separating mixtures of minerals. In determining the specific gravity of minerals and other substances. In the manufacture of x-ray contrast media.

6138. Methylenetriphenylphosphorane. [3487-44-3] Triphenylmethylenephosphorane; triphenylphosphine methylene; triphenylphosphine methylide; $C_{19}H_{17}P$; mol wt 276.32. C 82.59%, H 6.20%, P 11.21%. Phosphonium ylide for the methylenation of aldehydes and ketones; generated in situ immediately before use in Wittig-type reactions. Proposed existence in soln: D. D. Coffman, C. S. Marvel, *J. Am. Chem. Soc.* **51**, 3496 (1929). Prepn and use: G. Wittig, G. Geissler, *Ann.* **580**, 44 (1953); G. Wittig, U. Schoellkopf, *Org. Synth.* **coll. vol. V**, 751 (1973). Use as olefin forming reagent: *eidem, Ber.* **87**, 1318 (1954); G. Wittig, W. Haag, *ibid.* **88**, 1654 (1955). Additional synthetic applications: A. P. Uijttewaal *et al., Tetrahedron Lett.* **16**, 1439 (1975); E. J. Corey *et al., ibid.* **26**, 555 (1985); J. A. Murphy *et al., Org. Lett.* **7**, 1427 (2005). X-ray structure: J. C. J. Bart, *J. Chem. Soc. B* **1969**, 350. *Review:* K. C. Lee in *Encyclopedia of Reagents for Organic Synthesis* **5**, L. A. Paquette, Ed. (Wiley, New York, 1995) pp 3492-3496.

Yellow monoclinic crystals, mp 96°. Solid turns white upon exposure to air. Crystal d 1.19. Sol in diethyl ether, THF, DME, DMSO, benzene, toluene. Reacts with water and protic solvents. Solns of ylide vary in color from yellow to orange or red.

USE: Reagent in synthetic organic chemistry.

6139. Methylenomycins. Members of a family of cyclopentenoid antibiotics related structurally to sarkomycins, *q.v.*, and having *in vitro* activity vs gram-positive and gram-negative organisms. Isoln from *Streptomyces violaceoruber*, physical, chemical, biologi-

cal properties: M. Arai *et al.*, **JP Kokai 73 19796** (1973 to Sankyo), *C.A.* **78**, 157861 (1973); T. Haneishi *et al., J. Antibiot.* **27**, 386 (1974). Structures of methylenomycins A and B: *eidem, ibid.* 393. Crystal and molecular structure of (±)-A: B. H. Toder, A. B. Smith, *J. Cryst. Mol. Struct.* **8**, 1 (1979). Stereospecific total synthesis of (±)-A: R. M. Scarborough *et al., J. Am. Chem. Soc.* **99**, 7085 (1977); *eidem, ibid.* **102**, 3904 (1980); and absolute configuration: K. Sakai *et al., Tetrahedron Lett.* **1979**, 2365. Stereospecific total synthesis and absolute configuration of (+)-A: J. Jernow *et al., J. Org. Chem.* **44**, 4210 (1979). Revised structure and total synthesis of B: *eidem, ibid.* 4212. Concise synthesis of B: M. Mikolajczyk, R. Zurawinski, *Synlett* **8**, 575 (1991). Prepn of analogs of A and structure-activity correlations: T. Haneishi *et al., J. Antibiot.* **27**, 400 (1974). Toxicity: *eidem, ibid.* 386. Methylenomycin A is the first example of an antibiotic in which all information required for synthesis is carried by a plasmid, SCP1: L. F. Wright, D. A. Hopwood, *J. Gen. Microbiol.* **95**, 96 (1976). Review of biosynthesis: U. Hornemann, D. A. Hopwood, *Antibiotics* **vol. IV**, J. W. Corcoran, Ed. (Springer-Verlag, New York, 1981) pp 123-131. General review: A. Terehara *et al., Heterocycles* **13**, 353-371 (1979).

Methylenomycin A. [52775-76-5] [1*S*-(1α,2α,5α)]-1,5-Dimethyl-3-methylene-4-oxo-6-oxabicyclo[3.1.0]hexane-2-carboxylic acid. $C_9H_{10}O_4$; mol wt 182.18. Colorless crystals from chloroform/carbon tetrachloride, mp 115° (dec). mp of the (±)-form: 88.5-89°; after subl (70-75°, 0.025 mm Hg), 107.5-108°. $[\alpha]_D^{20}$ +42.3° (c = 1 in chloroform). uv max (methanol): 224 nm (ε 6300). Sol in benzene, chloroform, ethyl acetate, acetone, methanol, water. Slightly sol in *n*-hexane, CCl_4. pKa′ 3.65. LD$_{50}$ in mice (mg/kg): 1500 orally, 75 i.p. (Haneishi).

Methylenomycin B. [52775-77-6] 2,3-Dimethyl-5-methylene-2-cyclopenten-1-one. $C_8H_{10}O$; mol wt 122.17. Neutral colorless oil. uv max (methanol): 240 nm (ε 7650). Sol in ether, benzene, chloroform, ethyl acetate, acetone, alcohols. Slightly sol in *n*-hexane, petr ether. LD$_{50}$ in mice (mg/kg): 260 orally, 245 i.p. (Haneishi).

6140. *N*-Methylephedrine. [552-79-4] (α*R*)-α-[(1*S*)-1-(Dimethylamino)ethyl]benzenemethanol; (1*R*,2*S*)-2-dimethylamino-1-phenylpropanol; L-*N*,*N*-dimethylnorephedrine. $C_{11}H_{17}NO$; mol wt 179.26. C 73.70%, H 9.56%, N 7.81%, O 8.93%. Isoln from *Ephedra distachya* L. (*E. vulgaris* Rich.), and allied *Gnetaceae*: Smith, *J. Chem. Soc.* **1927**, 2056; Wolfes, *Arch. Pharm.* **268**, 327 (1930). Prepn of *dl*-form and resolution of isomers: Nagai, Kanao, *Ann.* **470**, 157 (1929). Prepn of *dl*-form: Pfanz, Müller, *Arch. Pharm.* **288**, 11 (1955).

Crystals from petr ether, mp 87-88°. $[\alpha]_D$ −29.5° (c = 4.5 in methanol).

Hydrochloride. [38455-90-2] Crystals from ethyl acetate or alcohol, mp 192°. $[\alpha]_D^{20}$ −29.8° (c = 4.6). Readily sol in water; less sol in alcohol; sparingly sol in acetone.

***dl*-Form.** [1201-56-5] *erythro*-α-[1-(Dimethylamino)ethyl]benzyl alcohol. Crystals from petr ether or methanol, mp 63.5-64.5°. Readily sol in the usual solvents.

***dl*-Form hydrochloride.** [18760-80-0] Crystals from acetone, mp 207-208°.

***d*-Form.** [42151-56-4] Crystals, mp 87-87.5°. $[\alpha]_D^{20}$ +29.2° (c = 4 in methanol).

d-Form hydrochloride. [54114-10-2] Crystals from ethyl acetate, mp 192°. $[\alpha]_D^{20}$ +30.1°.

6141. *N*-Methylepinephrine. [554-99-4] 4-[2-(Dimethylamino)-1-hydroxyethyl]-1,2-benzenediol; α-[(dimethylamino)methyl]-3,4-dihydroxybenzyl alcohol; α-(3,4-dihydroxyphenyl)-2-dimethylaminoethanol; α-(3,4-dihydroxyphenyl)-α-hydroxy-β-dimethylaminoethane; dimethylaminomethyl-(3,4-dihydroxyphenyl) carbinol; α-(dimethylaminomethyl)protocatechuyl alcohol; *N*-methyladrenaline. $C_{10}H_{15}NO_3$; mol wt 197.23. C 60.90%, H 7.67%, N 7.10%, O 24.34%. Prepn and resolution of racemic mixture: Manna, Campiglio, *Farmaco Ed. Sci.* **14**, 317 (1959). Configuration: Manna, Ghislandi, *ibid.* **19**, 377 (1964).

DL-Form. [6032-14-0] Methadren(e). Crystals from alcohol + ethyl acetate, mp 142-143°.

D(−)-Form. Crystals from ethyl acetate, mp 149-150°. $[\alpha]_D^{18}$ −65.1° (c = 1.41 in 0.5*N* HCl).

L(+)-Form. Crystals, mp 149-150°. $[\alpha]_D^{18}$ +62.3° (c = 1.4).

THERAP CAT: Adrenergic.

6142. Methylergonovine. [113-42-8] (8β)-9,10-Didehydro-*N*-[(1*S*)-1-(hydroxymethyl)propyl]-6-methylergoline-8-carboxamide; *N*-[α-(hydroxymethyl)propyl]-D-lysergamide; D-lysergic acid (+)-butanolamide-(2); *d*-lysergic acid-*d*-1-hydroxybutylamide-2; methylergometrine; methylergobasine. $C_{20}H_{25}N_3O_2$; mol wt 339.44. C 70.77%, H 7.42%, N 12.38%, O 9.43%. Semisynthetic ergot alkaloid; metabolite of methysergide, *q.v.* Prepn: A. Stoll, A. Hofmann, **US 2265207** (1941 to Sandoz); *eidem, Helv. Chim. Acta* **26**, 944 (1943). HPLC determn in plasma: H. T. Smith, N. C. Molinaro, *J. Chromatogr.* **424**, 416 (1988). Clinical trial in prevention of cluster headache: L. Mueller *et al., Headache* **37**, 437 (1997). Review of pharmacology and clinical use in obstetrics and gynecology: A. N. de Groot *et al., Drugs* **56**, 523-535 (1998).

Shiny crystals from benzene, mp 172° (some decompn). $[\alpha]_D^{20}$ −45° (c = 0.4 in pyridine). Sparingly sol in water. Freely sol in alcohol, acetone.

Maleate. [57432-61-8] Basofortina; Methergin; Methergine; Metenarin; Methylergobrevin; Ryegonovin; Spametrin-M. $C_{20}H_{25}$-$N_3O_2 \cdot C_4H_4O_4$; mol wt 455.51. White to pinkish-tan microcryst powder; odorless; bitter taste. Slightly sol in water, alcohol; very slightly sol in chloroform, ether.

THERAP CAT: Oxytocic.

6143. Methyl Ethyl Ketone. [78-93-3] 2-Butanone; ethyl methyl ketone; MEK; 2-oxobutane. C_4H_8O; mol wt 72.11. C 66.63%, H 11.18%, O 22.19%. Prepn from ethyl 2-methylacetoacetate: J. Schramm, *Ann.* **398**, 242 (1913). Manuf by dehydration of 2-butanol and by catalytic oxidation of *n*-butenes: A. J. Papa, P. D. Sherman, Jr. in *Kirk-Othmer Encyclopedia of Chemical Technology* vol. **13** (Wiley-Interscience, New York, 3rd ed., 1981) pp 903-907. Toxicity: H. F. Smyth *et al., Am. Ind. Hyg. Assoc. J.* **23**, 95 (1962). Review of toxicology and human exposure: *Toxicological Profile for 2-Butanone* (PB93-110708, 1992) 144 pp.

Liquid; acetone-like odor. *Flammable.* d_4^{20} 0.805. mp −86°. bp 79.6°. Flash pt, closed cup: 21°F (−6°C). n_D^{15} 1.3814. n_D^{25} 1.3782. Sol in ~4 parts water (27.5%); less sol at higher temp; miscible with alcohol, ether, benzene. Constant boiling mixture with water, bp 73.4°, contains 88.7% methyl ethyl ketone. Soly of water in methyl ethyl ketone: 12.5% at 25°. LD_{50} orally in rats: 6.86 ml/kg (Smyth).

Caution: Potential symptoms of overexposure are irritation of skin, eyes and nose; dermatitis; headache; dizziness; vomiting. *See NIOSH Pocket Guide to Chemical Hazards* (DHHS/NIOSH 97-140, 1997) p 36.

USE: As organic solvent; in the surface coating industry; manuf smokeless powder; colorless synthetic resins.

6144. Methyleugenol. [93-15-2] 1,2-Dimethoxy-4-(2-propen-1-yl)benzene; 4-allyl-1,2-dimethoxybenzene; 4-allylveratrole; 3,4-dimethoxyallylbenzene; eugenol methyl ether. $C_{11}H_{14}O_2$; mol wt 178.23. C 74.13%, H 7.92%, O 17.95%. Natural constituent of the volatile oils of various plants, including cinnamon, clove, nutmeg, basil, and anise. Prepn: K. U. Matsmoto, *Ber.* **11**, 122 (1878). Isoln from the bark of *Cinnamomum oliveri*: G. W. Hargreaves, *J. Chem. Soc.* **109**, 751 (1916). Raman spectroscopy: B. Susz *et al., Helv. Chim. Acta* **19**, 548 (1936). Toxicity: P. M. Jenner *et al., Food Cosmet. Toxicol.* **2**, 327 (1964). HPLC determn in plasma: S. W. Graves, S. Runyon, *J. Chromatogr. B* **663**, 255 (1995). GC/MS determn in basil cultivars: M. Miele *et al., J. Agric. Food Chem.* **49**, 517 (2001). Biosynthesis in sweet basil: E. Lewinsohn *et al., Plant Sci.* **160**, 27 (2000). Safety assessment: R. L. Smith *et al., Food Chem. Toxicol.* **40**, 851 (2002).

Colorless to pale yellow liquid. bp_{30} 146-147°. bp_{760} 244°. n_D^{20} 1.53432. n_D^{27} 1.5305. d_4^{20} 1.0396. LD_{50} orally in rats: 1560 mg/kg (Jenner).

Caution: This substance is reasonably anticipated to be a human carcinogen: *Report on Carcinogens, Twelfth Edition* (PB2011-111646, 2011) p 267.

USE: Fragrance ingredient in perfumes, toiletries and detergents; flavor ingredient in baked goods.

6145. α-Methylfentanyl. [79704-88-4] *N*-[1-(1-Methyl-2-phenylethyl)-4-piperidinyl]-*N*-phenylpropanamide; *N*-[1-(α-methylphenethyl)-4-piperidyl]propionanilide; 1-(1-methyl-2-phenylethyl)-4-(*N*-propanilido)piperidine. $C_{23}H_{30}N_2O$; mol wt 350.51. C 78.81%, H 8.63%, N 7.99%, O 4.56%. Potent derivative of fentanyl, *q.v.* Prepn: P. A. J. Janssen, **FR M2430**; *idem,* **US 3164600** (1964, 1965 both to Janssen). This substance has erroneously been referred to as *"China White"*, the street term for very pure Southeast Asian heroin. Initial identification of "China White" as α-methylfentanyl: T. C. Kram *et al., Anal. Chem.* **53**, 1379 A (1981). Molecular structure determn using tandem mass spectrometry: M. T. Cheng *et al., ibid.* **54**, 2204 (1982). Identification and quantification in tissue: T. J. Gillespie *et al., J. Anal. Toxicol.* **6**, 139 (1982). Confirmation of identity of "China White": S. Suzuki *et al., Chem. Pharm. Bull.* **34**, 1340 (1986). Immunoassay for detection of use in racehorses: J. McDonald *et al., Res. Commun. Chem. Pathol. Pharmacol.* **57**, 389 (1987).

Hydrochloride. [1443-44-3] $C_{23}H_{30}N_2O.HCl$. Crystals from isopropanol, mp 272.8-273.6°.

Note: This is a controlled substance (opiate): **21 CFR**, 1308.11.

6146. Methyl Fluorosulfonate. [421-20-5] Sulfuryl fluoride methyl ester; fluorosulfuric acid methyl ester; methyl fluorosulfate; methyl fluosulfonate; Magic Methyl. CH_3FO_3S; mol wt 114.09. C 10.53%, H 2.65%, F 16.65%, O 42.07%, S 28.10%. Prepn from dimethyl ether and fluosulfonic acid: J. Meyer, G. Schramm, *Z. Anorg. Allg. Chem.* **206**, 24 (1932). Electrochemical prepn: J. P. Coleman, D. Pletcher, *Tetrahedron Lett.* **15**, 147 (1974). Synthetic application as methylating agent: M. G. Ahmed *et al.*, *Chem. Commun.* **1968**, 1533. Toxicity study: M. Hite *et al.*, *Am. Ind. Hyg. Assoc. J.* **40**, 600 (1979). Review: R. W. Alder, *Chem. Ind. (London)* **1973**, 983-986.

Volatile liq, bp 92-94°, mp −95°. *Poisonous.* d 1.412. n_D^{20} 1.3326. Good solvent for most organic compounds. Proton NMR absorption at tau 5.88. LD_{50} orally in mice: <112 mg/kg; LC_{50} 1 hr for rats: 5-6 ppm (Hite).

Caution: Extremely toxic to humans. Exposure can cause fatal pulmonary edema. *See:* D. M. W. vanden Ham, D. van der Meer, *Chem. Eng. News* **54**, 5 (Aug. 30, 1976); *eidem, Chem. Br.* **1976**, 362.

USE: In organic synthesis as methylating agent.

6147. N-Methylformamide. [123-39-7] N-Monomethylformamide; NMF; MMF; NSC-3051. C_2H_5NO; mol wt 59.07. C 40.67%, H 8.53%, N 23.71%, O 27.08%. Prepn: A. Gautier, *Ann.* **151**, 239 (1869); G. F. D'Alelio, E. E. Reid, *J. Am. Chem. Soc.* **59**, 109 (1937); J. A. Marsella, G. P. Pez, *J. Mol. Catal.* **35**, 65 (1986); J. J. Cappon *et al.*, *Recl. Trav. Chim. Pays-Bas* **113**, 318 (1994). Dielectric spectrum: J. Barthel *et al.*, *Chem. Phys. Lett.* **167**, 62 (1990). Toxicology: S. P. Langdon *et al.*, *Toxicology* **34**, 173 (1985). Review of metabolism, toxicity, and pharmacology: G. L. Kennedy, Jr., *Crit. Rev. Toxicol.* **17**, 129-182 (1986). Review of antitumor activity and clinical evaluation: K. Clagett-Carr *et al.*, *J. Clin. Oncol.* **6**, 906-918 (1988).

mp −5.4°. bp 180-185°. bp_{90} 131°. d_4^{25} 0.9961. Sol in acetone, alcohol, water. LD_{50} in mice (mg/kg): 2300 i.p.; 2600 orally; 1580 i.v.; 2700 i.m. (Langdon).

USE: Solvent.

6148. Methyl Formate. [107-31-3] Formic acid methyl ester. $C_2H_4O_2$; mol wt 60.05. C 40.00%, H 6.71%, O 53.29%. $HCOOCH_3$.

Colorless liquid, agreeable odor. d_{15}^{15} 0.987. bp 31.5°. n_D^{20} 1.3440. Flash pt, closed cup: −2°F (−19°C). mp ∼−100°. *Flammable.* Sol in about 3.3 parts water; miscible with alcohol.

Caution: Potential symptoms of overexposure are eye and nose irritation; chest tightness, dyspnea; visual disturbance; CNS depression. *See NIOSH Pocket Guide to Chemical Hazards* (DHHS/NIOSH 97-140, 1997) p 210; *Patty's Industrial Hygiene and Toxicology* vol. **2A**, G. D. Clayton, F. E. Clayton, Eds. (Wiley-Interscience, New York, 3rd ed., 1981) p 2263.

USE: Fumigant and larvicide for tobacco and food crops. Fire hazard is avoided by use with CO_2.

6149. Methyl 2-Furoate. [611-13-2] 2-Furancarboxylic acid methyl ester; methyl furan-2-carboxylate; methyl pyromucate. $C_6H_6O_3$; mol wt 126.11. C 57.15%, H 4.80%, O 38.06%. Flavor ingredient; found naturally in cocoa, coffee, filberts, almonds, and peanuts. Prepn: G. Gennari, *Gazz. Chim. Ital.* **24**, 246 (1894); C. C. Price *et al.*, *J. Am. Chem. Soc.* **63**, 1857 (1941); and heat of vaporization: J. H. Mathews, P. R. Fehlandt, *ibid.* **53**, 3212 (1931). Toxicological profile: D. L. J. Opdyke, *Food Cosmet. Toxicol.* **17**, Suppl., 869 (1979).

Pale yellow liquid; fruity, mushroom-like odor. bp_{20} 76°. bp_{750} 180.5°. bp_{760} 181.8-182.1°. n_D^{20} 1.4875. d_4^{20} 1.1792. Sol in alcohol, ether; slightly sol in water. LD_{50} dermally in rabbits: >1.25 g/kg (Opdyke).

USE: In flavor and fragrance compositions.

6150. N-Methylglucamine. [6284-40-8] 1-Deoxy-1-(methylamino)-D-glucitol; N-methyl-D-glucamine; meglumine. $C_7H_{17}NO_5$; mol wt 195.22. C 43.07%, H 8.78%, N 7.17%, O 40.98%. Prepd from D-glucose and methylamine: Karrer, Herkenrath, *Helv. Chim. Acta* **20**, 83 (1937). Efficacy and toxicity in treatment of leishmaniasis: R. G. Muller *et al.*, *Arch. Inst. Cardiol. Mex.* **52**, 155 (1982); P. Bouree *et al.*, *Pathol. Biol.* **33**, 607 (1985). Antileishmanial activity in dogs: W. L. Chapman *et al.*, *Am. J. Vet. Res.* **45**, 1028 (1984).

Crystals from methanol, mp 128-129°. Does not polymerize or dehydrate unless heated above 150° for prolonged periods. $[α]_D^{18}$ −18.5° (Karrer); $[α]_D^{20}$ −23° (Rhône-Poulenc data sheet). Soly (g/100 ml): water at 25°: ∼100; alcohol at 25°: 1.2; alcohol at 70°: 21. pH of 1% aq soln: 10.5. Forms salts with acids and complexes with metals. Salts with alkyl aryl sulfonic acids act as detergents.

Antimonate. [133-51-7] RP-2168; Glucantim; Glucantime; Protostib. $C_7H_{17}NO_5.HSbO_3$; mol wt 365.98. Powder. Soly in water about 35% ww. Practically insol in alcohol, ether, chloroform. pH of aq solns 6-7.

USE: In the synthesis of surface active agents, pharmaceuticals, dyes.

THERAP CAT: Antimonate as antiprotozoal (Leishmania).

6151. N-Methyl-α-L-glucosamine. [42852-95-9] 2-Deoxy-2-(methylamino)-α-L-glucopyranose. $C_7H_{15}NO_5$; mol wt 193.20. C 43.52%, H 7.83%, N 7.25%, O 41.41%. Together with streptose forms the streptobiosamine moiety of streptomycin: Kuehl *et al.*, *J. Am. Chem. Soc.* **69**, 3032 (1947). Prepn from D-glucose by *Streptomyces griseus:* Silverman, Rieder, *J. Biol. Chem.* **235**, 1251 (1960); from the antibiotic, bluensomycin: Bannister, Argoudelis, *J. Am. Chem. Soc.* **85**, 34 (1963). *Review:* Lemieux, Wolfrom, *Adv. Carbohydr. Chem.* **3**, 337 (1948).

Glass. $[α]_D^{25}$ −62° (c = 1 in methanol).

Hydrochloride. $C_7H_{15}NO_5.HCl$. Needles from ethanol, mp 160-163°. Freely sol in water. Shows mutarotation. $[α]_D^{25}$ −103° → −88° (c = 0.6).

N-Acetyl derivative. mp 165-166°. $[α]_D^{25}$ −51° (c = 0.4).

Pentaacetyl derivative. $C_{17}H_{25}NO_{10}$. mp 160.5-161.5°. $[α]_D^{25}$ −100° (c = 0.7 in chloroform).

6152. α-Methylglucoside. [97-30-3] Methyl-α-D-glucopyranoside. $C_7H_{14}O_6$; mol wt 194.18. C 43.30%, H 7.27%, O 49.44%. Prepd by refluxing finely powdered glucose with methanol-HCl: Fischer, *Ber.* **26**, 2405; **27**, 2987; **28**, 1151 (1895); Helferich, Schäfer, *Org. Synth.* **coll. vol. I**, (2nd ed., 1941) 364. Enzymatic synthesis by means of α-glucosidase from yeast: Bourquelot *et al.*, *Compt. Rend.* **156**, 491 (1913). Prepd industrially by reacting glucose with methanol in the presence of a cation exchange material: *Chem. Eng. News* **33**, 4592 (1955). Monograph: G. N. Bollenback, *Methyl Glucoside, Preparation, Physical Constants, Derivatives* (Academic Press, New York, 1958).

Orthorhombic bisphenoidal crystals, d_4^{30} 1.46. mp 168°. $bp_{0.2}$ 200°. $[\alpha]_D^{20}$ +158.9° (p = 10). pKa (25°): 13.71. Soly at 17° in water 63% (w/w); in 80% alcohol 7.3%; in 90% alcohol 1.6%. Practically insol in ether. Soly also reported at 20° as 108 g/100 g H_2O and as 5.2 g/100 g methanol.

USE: Manuf reconstituted and upgraded drying oils; tall oil esters and varnishes; fatty acid esters; plasticizers; nonionic surface active agents; fast and hard drying alkyd resins, and so-called plasticizing alkyds. In making glucoside hydroxypropyl ethers for polyurethane foam production.

6153. Methylglyoxal. [78-98-8] 2-Oxopropanal; pyruvaldehyde; 2-ketopropionaldehyde; acetylformaldehyde. $C_3H_4O_2$; mol wt 72.06. C 50.00%, H 5.60%, O 44.40%. Physiological metabolite formed by the fragmentation of triosephosphates and by the metabolism of acetone; synthesized by microorganisms. Also occurs in a variety of foods and beverages such as coffee. Reacts with nucleic acids and proteins under physiological conditions serving as a signal for their degradation. Prepn: H. v. Pechmann, *Ber.* **20**, 3213 (1887); H. O. L. Fischer, C. Taube, *ibid.* **57**, 1502 (1924); *eidem, ibid.* **59**, 857 (1926); M. W. Kellum *et al., Anal. Biochem.* **85**, 586 (1978). GC determn in food and biological samples: S. Ohmori *et al., J. Chromatogr.* **415**, 221 (1987). Review of natural occurrance and metabolism in microorganisms: Y. Inoue, A. Kimura, *Adv. Microb. Physiol.* **37**, 177-227 (1995). Review of physiological significance and role in pathogenesis: P. J. Thornalley, *Gen. Pharmacol.* **27**, 565-573 (1996).

Clear, yellow liquid, pungent odor. bp_{760} 72°. d^{24} 1.0455. $n_D^{17.5}$ 1.4002. Hygroscopic. Polymerizes very readily, forming a brittle, resinous mass. Sol in water and alcohol, giving colorless solns.

Oxime see Isonitrosoacetone.

6154. Methyl Green. [14855-76-6] 4-[[4-(Dimethylamino)-phenyl][4-(dimethylimino)-2,5-cyclohexadien-1-ylidene]methyl]-*N*-ethyl-*N,N*-dimethylbenzenaminium bromide chloride (1:1:1); C.I. 42590; ethyl green. $C_{27}H_{35}BrClN_3$; mol wt 516.95. C 62.73%, H 6.82%, Br 15.46%, Cl 6.86%, N 8.13%. Usually sold as the double salt with $ZnCl_2$. Prepn: *Colour Index* **vol. 4** (3rd ed., 1971) p 4394.

Green powder. Sol in water. Yellow soln in conc H_2SO_4, turning green on dilution.
Note: The term methyl green also applies to the trimethyl analog.
USE: Dyeing and printing textiles; as biological stain.

6155. Methylguanidine. [471-29-4] *N*-Methylguanidine. $C_2H_7N_3$; mol wt 73.10. C 32.86%, H 9.65%, N 57.48%. Isoln by Brieger in 1888 from decomposing horsemeat; described as a product of putrefaction: I. Greenwald, *J. Am. Chem. Soc.* **41**, 1109 (1919). Prepn: E. Erlenmeyer, *Ber.* **3**, 896 (1870); E. A. Werner, J. Bell, *J. Chem. Soc.* **121**, 1790 (1922); R. Phillips, H. T. Clarke, *J. Am. Chem. Soc.* **45**, 1755 (1923); E. Philippi, K. Morsch, *Ber.* **60**, 2120 (1927); T. L. Davis, E. N. Rosenquist, *J. Am. Chem. Soc.* **59**, 2112 (1937). Toxicity data: Alles, *J. Pharmacol. Exp. Ther.* **28**, 251 (1926); Fühner, *Arch. Exp. Pathol. Pharmakol.* **166**, 437 (1932).

Colorless, deliquesc, strongly alkaline mass. Very sol in water; sol in alcohol. Reduces permanganate. MLD in rats (mg/kg): 250 s.c. (Alles). LD in frogs, mice (mg/kg): 170-190, 550-600 s.c. (Fühner).

6156. 5-Methyl-3-heptanone. [541-85-5] Amyl ethyl ketone; ethyl amyl ketone; EAK. $C_8H_{16}O$; mol wt 128.22. C 74.94%, H 12.58%, O 12.48%. *Review:* R. F. Buller, *Ind. Eng. Chem.* **48**, 1323-1324 (1956).

Liquid. Mild fruity odor. d_{20}^{20} 0.820-0.824. One gallon weighs 6.83 lbs at 20°. bp_{760} 157-162°. Flash pt 59°C (138°F). Evaporation rate 0.3 (*n*-butyl acetate = 1.0). n_D^{15} 1.4195. Slightly miscible with water. Compatible with alcohols, ketones, ethers, many other organic solvents.
Caution: Potential symptoms of overexposure are irritation of eyes, skin, mucous membranes; headache; narcosis, coma; dermatitis. *See NIOSH Pocket Guide to Chemical Hazards* (DHHS/NIOSH 97-140, 1997) p 210.
USE: Solvent for nitrocellulose-alkyd, nitrocellulose-maleic, and vinyl resins.

6157. Methylhexaneamine. [105-41-9] 4-Methyl-2-hexanamine; 2-amino-4-methylhexane; 1,3-dimethylamylamine; Forthane. $C_7H_{17}N$; mol wt 115.22. C 72.97%, H 14.87%, N 12.16%. Prepn: US 2350318; US 2386273 (1944, 1945 both to Lilly).

Liquid, amine odor. d 0.7620-0.7655. n_D^{25} 1.4150-1.4175. bp_{760} 130-135°. Very slightly sol in water. Freely sol in alc, chloroform, ether, dil acids.
THERAP CAT: Adrenergic.

6158. Methylhydrazine. [60-34-4] Monomethylhydrazine; MMH. CH_6N_2; mol wt 46.07. C 26.07%, H 13.13%, N 60.81%. Early prepns: *Beilstein* **4**, 546; 1st suppl., 560. Prepn of sulfate: Hatt, *Org. Synth.* **coll. vol. I**, 395 (1943); Audrieth, Diamond, *J. Am. Chem. Soc.* **76**, 4869 (1954). Manuf and properties: Knight, *Hydrocarbon Process. Petr. Refin.* **41**, 179 (1962). Toxicity and metabolism: Witkin, *Arch. Ind. Health* **13**, 34 (1956); Dost *et al., Biochem. Pharmacol.* **15**, 1325 (1966); Gregory *et al., Clin. Toxicol.* **4**, 435 (1971); Magee *et al.* in *Proc. Fifth Int. Congress Pharmacology, San Francisco, 1972* **vol. 2**, T. A. Loomis, Ed. (Karger, New York, 1973) pp 140-149.

Clear liquid, odor characteristic of short chain, organic amines. d^{25} 0.874. mp −52.4°. bp 87.5°. *Poisonous; flammable; corrosive.* Heat capacity (25°): 32.25 cal/mole/°C. Flash pt, closed cup: 68°F (20°C). Autoignition temp: 196°. Flammability limits in air (%-

vol): 2.5 to 97 ± 2%. Miscible with water, hydrazine, low mol wt monohydric alcohols. Sol in hydrocarbons. Mildly alkaline base. Strong reducing agent. Ignites spontaneously on contact with strong oxidizing agents such as fluorine, chlorine trifluoride, nitrogen tetroxide, fuming nitric acid. LD_{50} orally in mice, rats: 33.0, 32.5 mg/kg (Witkin); orally in rats: 70.7 mg/kg (Gregory).

Sulfate. [302-15-8] $CH_3NHNH_2 \cdot H_2SO_4$; mol wt 144.15. White plates from 80% ethanol. mp 141-142°.

Caution: Potential symptoms of overexposure to methylhydrazine are irritation of eyes, skin and respiratory system; vomiting; diarrhea; tremors and ataxia; anoxia, cyanosis; convulsions. Potential occupational carcinogen. *See NIOSH Pocket Guide to Chemical Hazards* (DHHS/NIOSH 97-140, 1997) p 210.

USE: In rocket fuel; intermediate in chemical syntheses.

6159. Methyl Iodide. [74-88-4] Iodomethane. CH_3I; mol wt 141.94. C 8.46%, H 2.13%, I 89.41%. Prepd from methyl alcohol, iodine and red phosphorus: King, *Org. Synth.* **coll. vol. II**, 399 (1943); from potassium iodide and methyl sulfate: Hartman, *ibid.* 404. Review of toxicity and carcinogenic risk: *IARC Monographs* **41**, 213-227 (1986); *ibid.* **Suppl. 7**, p 47, 52 (1987); *Evaluation of the Potential Carcinogenicity of Methyl Iodide* (PB93-196343, 1988) 17 pp.

Colorless, transparent liquid; turns brown on exposure to light. *Poisonous.* d_4^{20} 2.28. mp −66.5°. bp 42.5°; dec at 270°. n_D^{21} 1.5293. Sol in about 50 parts water; miscible with alcohol, ether. *Protect from light.* LD_{50} orally in rats: 76 mg/kg; LD_{50} s.c. in mice: 0.78 mmoles/kg (IARC, 1986).

Caution: Potential symptoms of overexposure are nausea, vomiting; vertigo, ataxia; slurred speech, drowsiness; dermatitis; direct contact may cause irritation of eyes, skin and respiratory system. *See NIOSH Pocket Guide to Chemical Hazards* (DHHS/NIOSH 97-140, 1997) p 210. *See also Patty's Industrial Hygiene and Toxicology* **vol. 2E**, G. D. Clayton, F. E. Clayton, Eds. (John Wiley & Sons, Inc., New York, 4th ed., 1994) pp 4030-4034; *Clinical Toxicology of Commercial Products*, R. E. Gosselin *et al.*, Eds. (Williams & Wilkins, Baltimore, 5th ed., 1984) Section II, pp 158-159. This substance has been listed as reasonably anticipated to be a carcinogen: *Fourth Annual Report on Carcinogens* (NTP 85-002, 1985) p 132; delisted due to IARC reevaluation as unclassifiable as to its human carcinogenicity: *Fifth Annual Report on Carcinogens* (NTP 89-239, 1989) p 340.

USE: Methylating agent; in microscopy because of its high refractive index; as imbedding material for examining diatoms; in testing for pyridine. Light sensitive etching agent for electronic circuits; component in fire extinguishers.

6160. Methyl Isobutyrate. [547-63-7] 2-Methylpropanoic acid methyl ester; methyl 2,2-dimethylacetate; methyl 2-methylpropanoate. $C_5H_{10}O_2$; mol wt 102.13. C 58.80%, H 9.87%, O 31.33%.

Colorless, mobile liquid. d_D^{20} 0.891. mp −85 to −84°. bp 93°. n_D^{20} 1.3840. Slightly sol in water; miscible with alcohol, ether.

6161. Methyl Isocyanate. [624-83-9] Isocyanatomethane; isocyanic acid methyl ester; MIC. C_2H_3NO; mol wt 57.05. C 42.11%, H 5.30%, N 24.55%, O 28.04%. Intermediate in the manufacture of insecticides and herbicides including carbaryl, *q.v.* Prepn via Curtius rearrangement: J. W. Boehmer, *Rec. Trav. Chim.* **55**, 379 (1933); by heating *N,N*-diphenyl-*N'*-methylurea: W. Siefken, *Ann.* **562**, 75 (1949); by phosgenation of bis(trimethylsilyl)methylamine: V. F. Mironov *et al.*, *Zh. Obshch. Khim.* **39**, 2598 (1969). Used in prepn of α-aryl-β-methylureas: J. W. Boehmer, *loc. cit.*; of semicarbazides: Ch. C. P. Pacilly, *Rec. Trav. Chim.* **55**, 101 (1936); in conversion of aldoximes to nitriles: J. A. Albright, M. L. Alexander, *Org. Prep. Proced. Int.* **4**, 215 (1972). Toxicity of vapor to rats, mice, humans; physical properties: G. Kimmerle, A. Eben, *Arch. Toxicol.* **20**, 235 (1960). Uptake and distribution studies in animals: J. S. Ferguson *et al.*, *Toxicol. Appl. Pharmacol.* **94**, 104 (1988). Acute oral toxicity: E. H. Vernot *et al.*, *ibid.* **42**, 417 (1977).

Liquid, bp 39-40°. d^{20} 0.96. *Poisonous; flammable.* Vapor press. at 4.2°: 200 torr; at 13.5°: 300 torr; at 20.6°: 400 torr; at 31.2°: 600 torr. LD_{50} in male rats (mg/kg): 140 single oral dose (Vernot). LC_{50} in rats (4 hours exposure to vapor): 5 ppm (Kimmerle, Eben).

Note: An industrial accident during the manufacture of carbaryl in Bhopal, India on December 3, 1984 resulted in the leakage of an unknown amount of methyl isocyanate into the air. Over 2,000 people died and an estimated 200,000 were exposed to the vapor: *Chem. Eng. News* **62**, 6 (Dec. 10, 1984); *ibid.* **63**, 14 (Feb. 11, 1985). Series of articles on follow-up studies on survivors: *Indian J. Exp. Biol.* **26**, 149-176, 201-204 (1988). Review of human toxicity: P. S. Mehta *et al.*, *J. Am. Med. Assoc.* **264**, 2781-2787 (1990).

Caution: Potential symptoms of overexposure are irritation of eyes, skin, nose and throat; respiratory sensitization; coughing, pulmonary secretions, chest pain and dyspnea; asthma; eye and skin injury. *See NIOSH Pocket Guide to Chemical Hazards* (DHHS/NIOSH 97-140, 1997) p 212. Highly volatile. Exposure to 2 ppm for 1-5 min produced tears and irritation of the nose and throat (Kimmerle, Eben).

USE: In organic synthesis; in manufacture of carbamate pesticides.

6162. Methyl Isopropyl Ketone. [563-80-4] 3-Methyl-2-butanone; 2-acetylpropane; isopropyl methyl ketone. $C_5H_{10}O$; mol wt 86.13. C 69.73%, H 11.70%, O 18.58%. Prepn: E. Frankland, B. F. Duppa, *Ann.* **138**, 328 (1866). Large scale prepn: T. W. Mears *et al.*, *J. Res. Natl. Bur. Stand.* **44**, 299 (1950). Miscibility profile: W. M. Jackson, J. S. Drury, *Ind. Eng. Chem.* **51**, 1491 (1959). Thermodynamic properties: J. L. Hales *et al.*, *Trans. Faraday Soc.* **63**, 1876 (1967). Acute toxicity: H. Tanii *et al.*, *Toxicol. Lett.* **30**, 13 (1986).

Colorless, liquid with an acetone-like odor. *Flammable.* bp_{760} 94.2°. fp −94.4°. d^{20} 0.8100; d^{25} 0.8061. n_D^{20} 1.3887; n_D^{25} 1.3861. Log P (*n*-octanol/water): 0.56. LD_{50} orally in male mice: 29.86 mmol/kg (Tanii).

Caution: Potential symptoms of overexposure are irritation of eyes, skin, mucous membranes, respiratory system; cough. *See NIOSH Pocket Guide to Chemical Hazards* (DHHS/NIOSH 97-140, 1997) p 212; *Patty's Industrial Hygiene and Toxicology* **vol 2C**, G. D. Clayton, F. E. Clayton, Eds. (Wiley-Interscience, New York, 4th ed., 1994) pp 1776-1779.

USE: Solvent.

6163. Methylisothiazolinone. [2682-20-4] 2-Methyl-3(2*H*)-isothiazolone; MI; MIT; Neolone 950. C_4H_5NOS; mol wt 115.15. C 41.72%, H 4.38%, N 12.16%, O 13.89%, S 27.84%. Marketed in combination with methylchloroisothiazolinone as a broad spectrum biocide. Prepn: W. D. Crow, N. J. Leonard, *Tetrahedron Lett.* **5**, 1477 (1964); *eidem, J. Org. Chem.* **30**, 2660 (1965); S. N. Lewis *et al.*, *J. Heterocycl. Chem.* **8**, 571 (1971). HPLC determn in cosmetic products: R. Matissek, *Chromatographia* **28**, 34 (1989). *In vitro* neurotoxicity study: S. Du *et al.*, *J. Neurosci.* **22**, 7408 (2002). Safety assessment: *J. Am. Coll. Toxicol.* **11**, 75-128 (1992). Review of epidemiology of allergic reactions: A. C. de Groot, J. W. Weyland, *J. Am. Acad. Dermatol.* **18**, 350-358 (1988); J. Fewings, T. Menné, *Contact Dermatitis* **41**, 1-13 (1999).

Colorless prisms, mp 50-51°. $bp_{0.03}$ 93°. uv max (diethyl ether): 281 nm (ε 6550). uv max (95% ethanol): 275 nm (ε 7250).

Methylchloroisothiazolinone. [26172-55-4] 5-Chloro-2-methyl-3(2*H*)-isothiazolone; MCI. C_4H_4ClNOS; mol wt 149.59. Crystals from ligroin (60-90°), mp 54-55°. uv max (methanol): 277 nm (log ε 3.82).

Mixture with methylchloroisothiazolinone. [55965-84-9] Kathon CG. Commercial product is supplied as a clear, light amber liquid, fp −18° to −21.5°. d^{20} 1.19. Viscosity (23°): 5.0 ±0.2 cP. Readily misc in water, lower alcohols, glycols, other hydrophilic organic solvents. Insol in petrolatum. LD_{50} orally in rabbits (mg/kg): 30; i.p. in male, female rats: 4.6, 4.3. LC_{50} (6 day) in trout, sunfish (mg/l): 0.14, 0.54. LC_{50} (8 day dietary) in Bobwhite quail, Peking duck (mg/kg/day): >60, >100 (*J. Am. Coll. Toxicol.*).

USE: Antimicrobial preservative in cosmetics, hygeine products, paints, emulsions, cutting oils, paper coatings, and water storage and cooling units.

6164. Methyl Isothiocyanate. [556-61-6] Isothiocyanatomethane; methyl mustard oil; Trapex. C_2H_3NS; mol wt 73.11. C 32.86%, H 4.14%, N 19.16%, S 43.85%. Obtained by the action of CS_2 on methylamine. Antibacterial activity: P. Klesse, P. Lukoschek, *Arzneim.-Forsch.* **5**, 505 (1955). Toxicity data: E. H. Vernot *et al., Toxicol. Appl. Pharmacol.* **42**, 417 (1977). Headspace determn in wine: N. Gandini, R. Riguzzi, *J. Agric. Food Chem.* **45**, 3092 (1997). Soil degradation study: A. Frick *et al., J. Environ. Qual.* **27**, 1158 (1998).

H_3C—N=C=S

Crystals, mp 35-36°. bp 119°. n_D^{37} 1.5258. *Poisonous; flammable.* Slightly sol in water; freely sol in alcohol, ether. LD_{50} in rats, mice (mg/kg): 220, 110 orally; in rabbits (mg/kg): 33 dermally (Vernot). LD_{50} s.c. in mice: 0.05 g/kg (Klesse, Lukoschek).

Caution: Highly irritating.
USE: Pesticide; soil fumigant.

6165. Methyl Isovalerate. [556-24-1] 3-Methylbutanoic acid methyl ester; isopentanoic acid methyl ester; methyl 3-methylbutanoate. $C_6H_{12}O_2$; mol wt 116.16. C 62.04%, H 10.41%, O 27.55%.

H_3CO—C(=O)—CH_2—CH(CH_3)—CH_3

Liquid; valerian odor. d_4^{20} 0.881. bp 116-117°. *Flammable.* Slightly sol in water; miscible with alcohol, ether.

6166. Methyl Lactate. [547-64-8] 2-Hydroxypropanoic acid methyl ester; lactic acid methyl ester; methyl 2-hydroxypropanoate. $C_4H_8O_3$; mol wt 104.11. C 46.15%, H 7.75%, O 46.10%. Prepd by heating 1 mole lactic acid condensation polymer with 2.5-5 moles of methanol and a small quantity of H_2SO_4 at 100° for 1-4 hours in a heavy-walled bottle: Rehberg, *Org. Synth.* **coll. vol. III**, 47-48 (1955).

H_3CO—C(=O)—CH(OH)—CH_3

Colorless, transparent liquid. d^{19} 1.09. bp 144-145°. n_D^{16} 1.4156. Soluble in alcohol, ether; dec by water.

Caution: Mild irritant.
USE: Cellulose acetate solvent.

6167. Methyl Linoleate. [112-63-0] (9Z,12Z)-9,12-Octadecadienoic acid methyl ester; linoleic acid methyl ester. $C_{19}H_{34}O_2$; mol wt 294.48. C 77.50%, H 11.64%, O 10.87%. Prepn from safflower-seed oil: Parker *et al., Biochem. Prep.* **4**, 88 (1955).

H_3C—(chain)—C(=O)—OCH_3

Colorless oil. d_4^{18} 0.8886. mp −35°. bp_{16} 212°; bp_4 192°. n_D^{25} 1.4593. Iodine value 172.4. More stable to air oxidation than linoleic acid. Miscible with dimethylformamide, fat solvents, oils.

USE: In the vitamin industry.

6168. Methyl Malonate. [108-59-8] Propanedioic acid 1,3-dimethyl ester; dimethyl malonate; dimethyl 1,3-propanedioate; malonic acid dimethyl ester. $C_5H_8O_4$; mol wt 132.12. C 45.45%, H 6.10%, O 48.44%.

H_3CO—C(=O)—CH_2—C(=O)—OCH_3

Liquid. d_4^{20} 1.154. mp −62°. bp 180-181°. n_D^{17} 1.4149. Slightly sol in water; miscible with alcohol, ether, oils.

6169. Methyl Methanesulfonate. [66-27-3] Methanesulfonic acid methyl ester; methyl methanesulfonic acid; *as*-dimethyl sulfite; methyl mesylate; MMS. $C_2H_6O_3S$; mol wt 110.13. C 21.81%, H 5.49%, O 43.58%, S 29.11%. Prepn: A. Arbusow, P. Pischtschimuka, *Chem. Zentralbl.* **II**, 685 (1909); W. E. Bissenger *et al., J. Am. Chem. Soc.* **70**, 3940 (1948). Metabolism: E. A. Barnsley, *Biochem. J.* **106**, 18p (1968). Mutagenicity studies: U. H. Ehling *et al., Mutat. Res.* **5**, 417 (1968); J. Moutschen, *ibid.* **8**, 581 (1969). Review of carcinogenicity studies: *IARC Monographs* **7**, 253-260 (1974). Review of comparative mutagenicity of MMS and ethyl methanesulfonate, *q.v.*: S. Kondo, *Environ. Sci. Res.* **24**, 743-785 (1981).

H_3C—S(=O)(=O)—OCH_3

Liquid, bp_{753} 203°, $bp_{0.6}$ 59°. d_4^{20} 1.2943; n_D^{20} 1.4140. Soly in water about 1:5, in DMF about 1:1. Slightly sol in nonpolar solvents.

Caution: This substance is reasonably anticipated to be a human carcinogen: *Report on Carcinogens, Twelfth Edition* (PB2011-111646, 2011) p 268.

USE: Exptlly as mutagen, teratogen, brain carcinogen.

6170. *S*-Methylmethionine. [4727-40-6] [(3*S*)-3-Amino-3-carboxypropyl]dimethyl sulfonium inner salt; *S*-methyl-L-methionine; MeMet; vitamin U. $C_6H_{13}NO_2S$; mol wt 163.24. C 44.15%, H 8.03%, N 8.58%, O 19.60%, S 19.64%. Naturally occurring sulfonium compound found in a large number of plants. Identification of antiulcer vitamin in cabbage leaves: G. Cheney, *Calif. Med.* **77**, 248 (1952). Isoln from cabbage juice: R. A. McRorie *et al., J. Am. Chem. Soc.* **76**, 115 (1954). Activity: V. Z. Szabo, G. Vargha, *Arzneim.-Forsch.* **10**, 23 (1960). Determn in plant tissue: P. K. Macnicol, *Anal. Biochem.* **158**, 93 (1986). Review of biosynthesis and physiological role in plants: J. Giovanelli *et al.*, "Sulfur Amino Acids in Plants" in *The Biochemistry of Plants* **5**, B. J. Miflin, Ed. (Academic Press, New York, 1980) pp 453-505.

H_3C—S$^+$(CH_3)—(chain)—CH(NH_2)—COO^-

Bromide. [33515-32-1] Methylmethioninesulfonium bromide. $C_6H_{14}BrNO_2S$. White platelets from ethanol, mp 139° (dec).

DL-Form chloride. [3493-12-7]; [1115-84-0] (L-form chloride). Methylmethioninesulfonium chloride; MMSC; Cabagin-U; Vitas-U. $C_6H_{14}ClNO_2S$; mol wt 199.69. Prepn: R. O. Atkinson, F. Poppelsdorf, *J. Chem. Soc.* **1951**, 1378. HPLC determn in pharmaceutical prepns: C. P. Leung, W. K. H. Leung, *J. Chromatogr.* **479**, 361 (1989). Clinical evaluation in ulcerative colitis: A. S. Salim, *Pharmacology* **45**, 307 (1992). Crystals from hot ethanol, mp 134° (dec). Very hygroscopic; hydrolysed rapidly in alkaline media.

THERAP CAT: Antiulcerative.

6171. *N*-Methylmorpholine *N*-Oxide. [7529-22-8] 4-Methylmorpholine 4-oxide; NMO; NMMO; MMNO. $C_5H_{11}NO_2$; mol wt 117.15. C 51.26%, H 9.46%, N 11.96%, O 27.31%. Morpholine derivative that serves as a mild oxidant and co-oxidant in synthesis reactions. Prepn and use in hydroxylation reactions: W. P. Schneider, A.V. McIntosh, Jr., **US 2769824** (1956 to Upjohn); V. Van-Rheenen *et al., Org. Synth.* **coll. vol. VI**, 342 (1988). Additional oxidation reaction applications: W. P. Griffith *et al., J. Chem. Soc. Chem. Commun.* **1987**, 1625; E. N. Jacobsen *et al., J. Am. Chem. Soc.* **110**, 1968 (1988); W. P. Griffith *et al., Synth. Commun.* **22**, 1967 (1992). Use in cellulose dissolution: K. E. Perepelkin, *Fibre Chem.* **39**, 163 (2007); in cellulose swelling: P. Noé, H. Chanzy, *Can. J. Chem.* **86**, 520 (2008). Crystal structure: E. Maia *et al., Acta Crystallogr. B* **37**, 1858 (1981). *Review:* M. R. Sivik in *Encyclopedia of Reagents for Organic Synthesis* **5**, L. A. Paquette, Ed. (Wiley, New York, 1995) pp 3545-3547.

White crystalline powder, mp 184.2°. *Irritant.* d 1.25. Sol in water, acetone, acetonitrile, ethanol, diethyl ether. Hygroscopic. Store at 2-8°C.
Monohydrate. [70187-32-5] $C_5H_{11}NO_2.H_2O$; mol wt 135.16. Colorless crystals, mp 75-76°. *Irritant.* d 1.28.
USE: Reagent in synthetic organic chemistry; in commercial processes for swelling and dissolution of cellulose.

6172. Methylnaltrexone Bromide. [73232-52-7]; [83387-25-1] (methylnaltrexonium). (5α)-17-(Cyclopropylmethyl)-4,5-epoxy-3,14-dihydroxy-17-methyl-6-oxomorphinanium bromide; *N*-cyclopropylmethyl-noroxymorphone methobromide; naltrexone methobromide; Relistor. $C_{21}H_{26}BrNO_4$; mol wt 436.35. C 57.80%, H 6.01%, Br 18.31%, N 3.21%, O 14.67%. Peripheral opioid receptor antagonist; quaternary derivative of naltrexone, *q.v.*, that does not readily cross the blood brain barrier. Prepn: L. I. Goldberg *et al.*, **US 4176186** (1979 to Boehringer Ing.). Determn by HPLC and solid phase extraction in plasma and urine: J. Osinski *et al., J. Chromatogr. B* **780**, 251 (2002). Clinical evaluation in methadone-induced constipation: C.-S. Yuan *et al., J. Am. Med. Assoc.* **283**, 367 (2000). Clinical pharmacokinetics: *idem et al., J. Clin. Pharmacol.* **45**, 538 (2005). Review of therapeutic potential in opioid-induced bowel dysfunction: J. F. Foss, *Am. J. Surg.* **182**, Suppl. 5A, 19S-26S (2001); of preclinical and clinical experience: C.-S. Yuan, *J. Support. Oncol.* **2**, 111-117 (2004).

Crystals from methanol/ether, mp 253°.
THERAP CAT: Narcotic antagonist. In treatment of opioid bowel dysfunction.

6173. Methyl Nicotinate. [93-60-7] 3-Pyridinecarboxylic acid methyl ester; nicotinic acid methyl ester; Midalgan. $C_7H_7NO_2$; mol wt 137.14. C 61.31%, H 5.15%, N 10.21%, O 23.33%. Obtained by passing HCl gas into a hot methanol soln of nicotinic acid: Engler, *Ber.* **27**, 1787 (1894); by treatment with tetramethylammonium hydroxide in methanol: Prelog, Piantanida, *Z. Physiol. Chem.* **244**, 56 (1936); by heating a mixture of sulfuric acid, selenium and quinoline, followed by refluxing with methanol: Kaufman, *J. Am. Chem. Soc.* **67**, 497 (1945); from nicotinyl chloride hydrochloride and methanol in pyridine: Charonnat *et al., Bull. Soc. Chim. Fr.* **1948**, 1014.

Crystals, mp 39°. bp$_{760}$ 209°; bp$_3$ 70-72°. Soluble in water, alcohol, benzene.
THERAP CAT: Rubefacient.

6174. Methyl Nitrate. [598-58-3] Nitric acid methyl ester. CH_3NO_3; mol wt 77.04. C 15.59%, H 3.93%, N 18.18%, O 62.30%. CH_3ONO_2. Prepn from methanol and nitric acid in the presence of sulfuric acid: Black, Babers, *Org. Synth.* **coll. vol. II**, 412 (1943).
Liquid. Solid at −83°. bp$_{760}$ 64.6° (explodes). d_4^{20} 1.2075. n_D^{20} 1.3748. Slightly sol in water. Sol in alcohol, ether.
USE: Has been used as rocket propellant. Does not need external oxygen for combustion.

6175. Methyl *p*-Nitrobenzenesulfonate. [6214-20-6] 4-Nitrobenzenesulfonic acid methyl ester; *p*-nitrobenzenesulfonic acid methyl ester. $C_7H_7NO_5S$; mol wt 217.20. C 38.71%, H 3.25%, N 6.45%, O 36.83%, S 14.76%. Reagent for selective methylation of cysteine residues in chemical modification of proteins: Nakagawa, Bender, *J. Am. Chem. Soc.* **91**, 1566 (1969). Procedure: Heinrikson, *Biochem. Biophys. Res. Commun.* **41**, 967 (1970). Prepd by the general method of reacting *p*-nitrobenzenesulfonyl chloride with alcohols. *See:* Morgan, Cretcher, *J. Am. Chem. Soc.* **70**, 375 (1948).

White to cream crystals, mp 93-95°.
USE: Methylating agent.

6176. *N*-Methyl-*N'*-nitro-*N*-nitrosoguanidine. [70-25-7] *N*-Methyl-*N*-nitroso-*N'*-nitroguanidine; MNNG. $C_2H_5N_5O_3$; mol wt 147.09. C 16.33%, H 3.43%, N 47.61%, O 32.63%. Prepn: A. F. McKay, G. F. Wright, *J. Am. Chem. Soc.* **69**, 3028 (1947). Reviews of carcinogenicity and mutagenicity studies: *IARC Monographs* **4**, 183-195 (1974); U. Sinha, B. B. Chattoo, *J. Sci. Ind. Res.* **3**, 499-505 (1975).

Yellow crystals from methanol, mp 118° (dec). Reacts with aq KOH to form diazomethane: A. F. McKay, *J. Am. Chem. Soc.* **70**, 1974 (1948). Reacts at acid pH to give methylnitroguanidine.
Caution: This substance is reasonably anticipated to be a human carcinogen: *Report on Carcinogens, Twelfth Edition* (PB2011-111646, 2011) p 302.
USE: Exptlly as carcinogen and mutagen. Formerly in prepn of diazomethane.

6177. Methyl Nonyl Ketone. [112-12-9] 2-Undecanone; 2-hendecanone; nonyl methyl ketone. $C_{11}H_{22}O$; mol wt 170.30. C 77.58%, H 13.02%, O 9.39%. Primary constituent of rue oils distilled mainly from *Ruta montana* L. and *R. graveolens.* Prepn: E. v. Gorup-Besanez, F. Grimm, *Ann.* **157**, 275 (1871); V. Fiandanese *et al., Tetrahedron Lett.* **25**, 4805 (1984); T. Aoyama, T. Shioiri, *Synthesis* **1988**, 228. LC determn in formulations: R. J. Bushway, *J. Assoc. Off. Anal. Chem.* **73**, 743 (1990). Physical properties: E. Guenther, D. Althausen, *The Essential Oils* **vol. 2** (D. Van Nostrand, New York, 1949) p 377. Brief review: D. L. J. Opdyke, *Food Cosmet. Toxicol.* **13**, 869-870 (1975).

Oily liquid. Strong odor. mp 12.1°. bp$_{761}$ 231.5-232.5°. bp$_7$ 99°. d$_4^{20}$ 0.8260-0.8263. n$_D^{30}$ 1.42527. LD$_{50}$ dermally in rabbits: >5 g/kg; LD$_{50}$ orally in rats, mice: >5, 3.88 g/kg (Opdyke).

USE: In the compounding of some synthetic essential oils. As fragrance additive in soaps, detergents, creams, lotions, and perfume. As dog and cat repellent.

6178. Methyl Orange. [547-58-0] 4-[2-[4-(Dimethylamino)-phenyl]diazenyl]benzenesulfonic acid sodium salt (1:1); 4-[[(4-di-methylamino)phenyl]azo]benzenesulfonic acid sodium salt; sodium *p*-dimethylaminoazobenzenesulfonate; helianthine B; C.I. Acid Orange 52; C.I. 13025; Orange III; Gold Orange; Tropaeolin D. C$_{14}$-H$_{14}$N$_3$NaO$_3$S; mol wt 327.33. C 51.37%, H 4.31%, N 12.84%, Na 7.02%, O 14.66%, S 9.79%. Monoazo dye. Prepn from sulfanilic acid sodium nitrite + dimethylaniline: L. Gattermann, *Die Praxis des Organischen Chemikers* (de Gruyter, Berlin, 40th ed., 1961) pp 260-261. *See also Colour Index* **vol. 4** (3rd ed., 1971) p 4043. Chromoisomerism studies: W. M. Dehn, L. McBride, *J. Am. Chem. Soc.* **39**, 1348 (1917).

Orange-yellow powder or cryst scales. Sol in 500 parts water; more sol in hot water. Practically insol in alcohol.

USE: As indicator in 0.1% aq soln. pH: 3.1 red, 4.4 yellow. Employed for titrating most mineral acids, strong bases, estimating alkalinity of waters; useless for organic acids. In assays for bromide ions, oxalic acid, and malonic acid. In dyeing and printing of textiles. In film dosimeters.

6179. Methyl Oxalate. [553-90-2] Ethanedioic acid 1,2-di-methyl ester; dimethyl oxalate; oxalic acid dimethyl ester. C$_4$H$_6$O$_4$; mol wt 118.09. C 40.68%, H 5.12%, O 54.19%.

Colorless crystals. d^{54} 1.148. mp 54°. bp 163-164°. n$_D^{82.1}$ 1.379. Sol in 17 parts water, in alcohol, ether.

6180. Methylparaben. [99-76-3] 4-Hydroxybenzoic acid methyl ester; methyl *p*-hydroxybenzoate; Nipagin M; Tegosept M; Methyl Chemosept; Methyl Parasept. C$_8$H$_8$O$_3$; mol wt 152.15. C 63.15%, H 5.30%, O 31.55%. Prepn: Ladenburg, Fitz, *Ann.* **141**, 247 (1867), Zbarskii, *C.A.* **33**, 9312^3 (1939). Identification in the vaginal secretions of female dogs in estrus: M. Goodwin *et al.*, *Science* **203**, 559 (1979).

White needles, mp 131°. bp 270-280° (dec). Freely sol in alc, methanol, acetone, ether. One gram dissolves in 400 ml water, 40 ml warm oil, about 70 ml warm glycerol. Soly in water (w/v): 0.25% at 20°, 0.30% at 25°.

USE: As preservative in foods, beverages and cosmetics.

6181. Methyl Parathion. [298-00-0] Phosphorothioic acid *O*,*O*-dimethyl *O*-(4-nitrophenyl) ester; *O*,*O*-dimethyl *O*-*p*-nitro-phenyl phosphorothioate; *O*,*O*-dimethyl *O*-*p*-nitrophenyl thiophos-phate; dimethyl parathion; parathion-methyl; metaphos; E-601;

ENT-17292; Folidol-M; Metacide; Penncap M. C$_8$H$_{10}$NO$_5$PS; mol wt 263.20. C 36.51%, H 3.83%, N 5.32%, O 30.39%, P 11.77%, S 12.18%. Organophosphate insecticide; cholinesterase inhibitor. Prepn: Fletcher *et al.*, *J. Am. Chem. Soc.* **72**, 2461 (1950). Manuf: *Faith, Keyes & Clark's Industrial Chemicals*, F. A. Lowenheim, M. K. Moran, Eds. (Wiley-Interscience, New York, 4th ed., 1975) pp 552-555. Toxicity data: T. B. Gaines, *Toxicol. Appl. Pharmacol.* **14**, 515 (1969). Environmental fate: M. T. Moore *et al.*, *Environ. Pollut.* **142**, 288 (2006). Review of carcinogenic risk: *IARC Monographs* **30**, 131-152 (1983); of toxicology and human exposure: *Toxicological Profile for Methyl Parathion* (PB2001-109105, 2001) 267 pp.

Crystals from cold methanol, mp 37-38°. n$_D^{25}$ 1.5367. d$_4^{20}$ 1.358. Log P (octanol/water): 3.5. Soly in water: 55 mg/l; sol in most organic solvents. *Poisonous.* LD$_{50}$ in male, female rats (mg/kg): 14, 24 orally; 67, 67 dermally (Gaines).

Caution: Potential symptoms of overexposure are irritation of eyes and skin; nausea, vomiting, abdominal cramps, diarrhea, salivation; headache, giddiness, vertigo, weakness; rhinorrhea, chest tightness; blurred vision, miosis; cardiac irregularities; muscle fasciculations; dyspnea. *See NIOSH Pocket Guide to Chemical Hazards* (DHHS/NIOSH 97-140, 1997) p 214.

USE: Insecticide.

6182. N-Methylphenazonium Methosulfate. [299-11-6] 5-Methylphenazinium methyl sulfate (1:1); phenazine methosulfate. C$_{14}$H$_{14}$N$_2$O$_4$S; mol wt 306.34. C 54.89%, H 4.61%, N 9.14%, O 20.89%, S 10.47%. Prepn: Kehrmann, *Ber.* **46**, 341 (1913); Hillemann, *ibid.* **71**, 34 (1938); Dickens, McIlwain, *Biochem. J.* **32**, 1615 (1938). Mutagenicity study: S. Venitt, C. Crofton-Sleigh, *Mutat. Res.* **68**, 107 (1979).

Flat yellow to brown parallelepipeds from alc, mp 155-157° (Hillemann); mp 167° (Dickens). Oxidation-reduction potential at 30° and pH 7: E$_h$ = +0.080 v.

USE: As an electron carrier in place of the flavine enzyme of Warburg in the hexosemonophosphate system: Dickens, *loc. cit.* In the prepn of succinic dehydrogenase: Green *et al.*, *J. Biol. Chem.* **217**, 551 (1955).

6183. Methylphenidate. [113-45-1] α-Phenyl-2-piperidine-acetic acid methyl ester; methyl phenidylacetate; methyl α-phenyl-α-(2-piperidyl)acetate; methylphenidan. C$_{14}$H$_{19}$NO$_2$; mol wt 233.31. C 72.07%, H 8.21%, N 6.00%, O 13.71%. Piperidine derivative related structurally to dextroamphetamine, *q.v.* Prepn: L. Panizzon, *Helv. Chim. Acta* **27**, 1748 (1944); M. Hartmann, L. Panizzon, *US 2507631* (1950 to Ciba). Sepn of isomers: R. Rometsch, *US 2957880* (1960 to Ciba). LC/MS/MS determn in dog plasma: R. Bakhtiar *et al.*, *Anal. Chim. Acta* **469**, 261 (2002). Comprehensive description: G. R. Padmanabhan, *Anal. Profiles Drug Subs.* **10**, 473-497 (1981). Pharmacokinetics: N. R. Srinivas *et al.*, *Pharm. Res.* **10**, 14 (1993). Toxicity data: E. N. Greenblatt, A. C. Osterberg *et al.*, *J. Pharmacol. Exp. Ther.* **131**, 115 (1961). Clinical efficacy in attention deficit-hyperactivity disorder (ADHD): W. E. Pelham, Jr. *et al.*, *J. Consult. Clin. Psychol.* **61**, 506 (1993); R. G. Klein, *Encephale* **19**, 89 (1993). Review of pharmacokinetics in ADHD: M. L. Wolraich, M. A. Doffing, *CNS Drugs* **18**, 243-250 (2004); of clinical experience in ADHD and pharmacology of isomers: D. J. Heal, D. M. Pierce, *ibid.* **20**, 713-738 (2006).

bp$_{0.6mm}$ 135-137°. Sol in alcohol, ethyl acetate, ether. Practically insol in water, petr ether.

Hydrochloride. [298-59-9] Ciba 4311b; Concerta; Equasym; Metadate; Ritalin. $C_{14}H_{19}NO_2 \cdot HCl$; mol wt 269.77. Crystals, mp 224-226°. pKa 8.9. Freely sol in water, methanol; sol in alcohol; slightly sol in chloroform, acetone. Solns are acid to litmus. LD$_{50}$ orally in mice: 190 mg/kg (Greenblatt, Osterberg)..

***d-threo*-Form.** [40431-64-9] ($\alpha R,2R$)-α-Phenyl-2-piperidine-acetic acid methyl ester; dexmethylphenidate. Enantioselective synthesis: J. M. Axten *et al.*, *J. Am. Chem. Soc.* **121**, 6511 (1999).

***d-threo*-Form hydrochloride.** [19262-68-1] Dexmethylphenidate hydrochloride; Focalin. Clinical trials in ADHD: L. E. Arnold *et al.*, *J. Child Adolesc. Psychopharmacol.* **14**, 542 (2004); R. Silva *et al.*, *ibid.* 555. White to off-white powder. Freely sol in water, methanol; sol in alcohol; slightly sol in chloroform, acetone.

Note: This is a controlled substance (stimulant): **21 CFR,** 1308.12.

THERAP CAT: CNS stimulant.

THERAP CAT (VET): In treatment of canine behavioral problems.

6184. Methylprednisolone. [83-43-2] ($6\alpha,11\beta$)-11,17,21-Trihydroxy-6-methylpregna-1,4-diene-3,20-dione; 1-dehydro-6α-methylhydrocortisone; Δ^1-6α-methylhydrocortisone; 6α-methyl-11β,17α,21-triol-1,4-pregnadiene-3,20-dione; Medrate; Medrol; Medrone. $C_{22}H_{30}O_5$; mol wt 374.48. C 70.56%, H 8.08%, O 21.36%. Prepn: Spero *et al.*, *J. Am. Chem. Soc.* **78**, 6213 (1956); Fried, *ibid.* **81**, 1235 (1959); Sebek, Spero, **US 2897218** (1959 to Upjohn); Gould, **US 3053832** (1962 to Schering). Review of clinical toxicology: J. D. Truwit, *Crit. Care Clin.* **7**, 639-657 (1991); of neuroprotective pharmacology: E. D. Hall, *J. Neurosurg.* **76**, 13-22 (1992); of pharmacokinetics and clinical efficacy in multiple sclerosis: O. R. Hommes *et al.*, *Mult. Scler.* **1**, 327-328 (1996). Clinical trial in acute spinal cord injury: M. B. Bracken *et al.*, *J. Neurosurg.* **89**, 699 (1998); in carpel tunnel syndrome: J. W. H. H. Dammers *et al.*, *Br. Med. J.* **319**, 884 (1999).

Crystals, mp 228-237°. $[\alpha]_D^{20}$ +83° (dioxane). uv max (95% ethanol): 243 nm (α_M 14875). Sparingly sol in alcohol, dioxane, methanol; slightly sol in acetone, chloroform; very slightly sol in ether. Practically insol in water.

21-Acetate. [53-36-1] Depo-Medrol; Depo-Medrone; Vetacortyl. $C_{24}H_{32}O_6$; mol wt 416.51. Crystals, mp 205-208°. $[\alpha]_D^{20}$ +101° (dioxane). uv max (95% ethanol): 243 nm (α_M 14825). Sol in dioxane; sparingly sol in acetone, alcohol, chloroform, methanol; slightly sol in ether. Practically insol in water.

21-Succinate sodium salt. [2375-03-3] Solu-Medrol; Solu-Medrone. $C_{26}H_{33}NaO_8$; mol wt 496.53. White, or nearly white, odorless, hygrosopic, amorphous solid. Very sol in water, alcohol; very slightly sol in acetone. Insol in chloroform.

Aceponate. [86401-95-8] Advantan. $C_{27}H_{36}O_7$; mol wt 472.58.

THERAP CAT: Glucocorticoid.

THERAP CAT (VET): Glucocorticoid.

6185. Methyl Propanoate. [554-12-1] Propanoic acid methyl ester; methyl propionate. $C_4H_8O_2$; mol wt 88.11. C 54.53%, H 9.15%, O 36.32%.

Colorless liquid. d$_4^{20}$ 0.915. mp −87°. bp 79.7°. Flash pt −2°C. n_D^{19} 1.3769. *Flammable.* Sol in 16 parts water; miscible with alcohol, ether.

USE: In organic synthesis.

6186. Methyl Propyl Ether. [557-17-5] 1-Methoxypropane; propyl methyl ether; Neothyl. $C_4H_{10}O$; mol wt 74.12. C 64.82%, H 13.60%, O 21.59%. Prepd by passing a mixture of methyl and propyl alcohol through a hot layer of β-naphthalenesulfonic or benzenesulfonic acids: Krafft, *Ber.* **26**, 2832 (1893); **DE 69115**; *Frdl.* **III**, 11. Laboratory prepn from sodium propylate and methyl iodide: Chancel, *Ann.* **151**, 305 (1869); Michael, Wilson, *Ber.* **39**, 2573 (1906); and water solubility: G. M. Bennett, W. G. Phillip, *J. Chem. Soc.* **1928**, 1930. Clinical trial as anesthetic: E. A. Parmenter, *Can. Anaesth. Soc. J.* **9**, 424 (1962).

Mobile liquid. *Flammable.* d$_4^0$ 0.7494; d$_4^{13}$ 0.7356. bp$_{761}$ 38.8°. $n_D^{14.3}$ 1.36019. Soly in water (25°): 5 ml/100 ml H$_2$O. Miscible with alc, ether. Absorbs some water. Oil/water coefficient 10 ± 1. Anesthetic index 2.5.

THERAP CAT: Anesthetic (inhalation).

6187. Methyl Propyl Ketone. [107-87-9] 2-Pentanone; ethyl acetone; 4-methyl-2-butanone; propyl methyl ketone; MPK. C_5H_{10}O; mol wt 86.13. C 69.73%, H 11.70%, O 18.58%. Prepn: R. E. Schaad, V. N. Ipatieff, *J. Am. Chem. Soc.* **62**, 178 (1940); C. L. Wilson, *ibid.* **70**, 1313 (1948). Toxicity study: H. F. Smyth *et al.*, *Am. Ind. Hyg. Assoc. J.* **23**, 95 (1962).

Colorless liquid. d$_4^{20}$ 0.809. mp −78°. bp 102°. n_D^{20} 1.3895. *Flammable Irritant.* Flash pt, closed cup: 45°F (7°C). Slightly soluble in water. Misc with alcohol, ether. Incompatible with strong bases, oxidizing agent, reducing agents. LD$_{50}$ orally in rats: 3.73 g/kg (Smyth).

Caution: Potential symptoms of overexposure are irritation of eyes, skin and mucous membranes; headache; dermatitis; narcosis, coma. *See NIOSH Pocket Guide to Chemical Hazards* (DHHS/NIOSH 97-140, 1997) p 244.

USE: Reagent and solvent in organic chemistry. As cleaner for flux and ink processes. In coatings for plastics.

6188. 5-Methylpyrazole-3-carboxylic Acid. [402-61-9] 5-Methyl-1*H*-pyrazole-3-carboxylic acid. $C_5H_6N_2O_2$; mol wt 126.12. C 47.62%, H 4.80%, N 22.21%, O 25.37%. Prepn: Knorr, MacDonald, *Ann.* **279**, 219 (1894); Lehninger, *J. Am. Chem. Soc.* **64**, 2507 (1942). Hypoglycemic metabolite of 3,5-dimethylpyrazole: Smith *et al.*, *J. Med. Chem.* **8**, 350 (1965).

Crystals from water, mp 236-237°.

Methyl ester. $C_6H_8N_2O_2$. Crystals, mp 75-78°.

Ethyl ester. $C_7H_{10}N_2O_2$. Crystals, mp 83°.

Caution: A powerful hypoglycemic agent in rats (characteristic of pyrazoles).

6189. Methyl Pyridyl Ketone. [350-03-8] 1-(3-Pyridinyl)-ethanone; 3-acetylpyridine; β-acetylpyridine; methyl 3-pyridyl ketone; methyl β-pyridyl ketone. C_7H_7NO; mol wt 121.14. C 69.40%, H 5.82%, N 11.56%, O 13.21%. A nicotinic acid antagonist. Prepd in dry distillation of Ca-nicotinate with Ca-acetate. The ketone is isolated through the phenylhydrazone: Engler, Kiby, *Ber.* **22**, 597 (1889).

Liquid, bp 220°. Freely sol in acids.

Ketoxime. $NC_5H_4C(CH_3)=NOH$. Crystals from alcohol or benzene, mp 112°.

Ketoxime hydrochloride. Crystals from alcohol, mp 204°.

Phenylhydrazone. $NC_5H_4C(CH_3)=NNHC_6H_5$. Yellow needles from alcohol, mp 137°.

6190. 1-Methylpyrrolidone. [872-50-4] 1-Methyl-2-pyrrolidinone; N-methyl-α-pyrrolidinone; N-methyl-γ-butyrolactone; NMP; 1-methylazacyclopentan-2-one; MP; M-Pyrol. C_5H_9NO; mol wt 99.13. C 60.58%, H 9.15%, N 14.13%, O 16.14%. Dipolar aprotic solvent. Commercially prepd by condensation of butyrolactone with methylamine. Prepn: J. Tafel, O. Wassmuth, *Ber.* **40**, 2831 (1907); E. Späth, J. Lintner, *ibid.* **69**, 2727 (1936). Use in extraction of unsaturated/aromatic compounds: E. Mueller, G. Hoehfeld, *Proc. 8th World Pet. Congr.* **4** (Applied Sci. Publishers, London, 1971) pp 213-219; H. Klein, H. M. Weitz, *Hydrocarbon Process.* **47**, 135 (November, 1968); A. Sequeira *et al.*, *ibid.* **58**, 155 (September, 1979). Physical properties and solubility: *Industrial Solvents Handbook* (3rd Ed.) 580-582 (1985). Reference guide to properties and use: *M-Pyrol®, N-Methyl-2-Pyrrolidone Handbook*, ISP (subsidiary of GAF Corp.), 1988. Acute toxicity: W. Bartsch *et al.*, *Arzneim.-Forsch.* **26**, 1581 (1976). *Review:* E. V. Hort, L. R. Anderson in *Kirk-Othmer Encyclopedia of Chemical Technology* **vol. 19** (Wiley-Interscience, New York, 3rd ed., 1982) pp 514-520.

Clear liquid. mp −24.4°. bp$_{760}$ 202°. n_D^{25} 1.4690. d_4^{25} 1.027. Flash point, closed cup: 199°F; open cup: 204°F. Viscosity (25°): 1.65 cP. Surface tension at 25°: 40.7 dynes/cm. Dipole moment at 25°: 4.09 D. Dielectric constant at 25°: 32.2. Vapor pressure at 23.2°, 65°: 0.33, 5 mm Hg. Heat of combustion: 719 kcal/mol; of vaporization: 127.3 kcal/kg. Misc with water, alcohol, ether, acetone, ethyl acetate, chloroform, benzene. LD$_{50}$ in mice, rats (ml/kg): 3.5, 2.2 i.v.; 4.3. 2.4 i.p.; 7.5, 3.8 orally (Bartsch).

USE: Industrial solvent primarily used in the extraction of aromatics from lube oils. Other applications include recovery and purification of acetylenes, olefins and diolefins, gas purification, and aromatics extraction from feedstocks. Also, used as polymer solvent.

6191. N-Methylpyrroline. [554-15-4] 2,5-Dihydro-1-methyl-1H-pyrrole; N-methyl-2,5-dihydropyrrole. C_5H_9N; mol wt 83.13. C 72.24%, H 10.91%, N 16.85%. In leaves of *Nicotiana tabacum* L. and *Atropa belladonna* L. By reducing N-methylpyrrole with zinc granules in dil acetic or hydrochloric acid.

Almost colorless liquid. Unpleasant, ammonia-like odor. Fumes in air. bp 79-80°. Strong base. Miscible with water; sol in alcohol, ether, chloroform.

6192. Methyl Red. [493-52-7] 2-[2-[4-(Dimethylamino)-phenyl]diazenyl]benzoic acid; 2-[[4-(dimethylamino)phenyl]azo]-benzoic acid; C.I. Acid Red 2; C.I. 13020. $C_{15}H_{15}N_3O_2$; mol wt 269.30. C 66.90%, H 5.61%, N 15.60%, O 11.88%. Prepn by diazotization of anthranilic acid and coupling with dimethylaniline: Clarke, Kirner, *Org. Synth.* **2**, 47 (1922). *See also Colour Index* **vol. 4** (3rd ed., 1971) p 4043.

Glistening, violet crystals from toluene. mp 181-182°. d 1.31. pKa$_1$ 2.5; pKa$_2$ 9.5; pKb 4.8. Almost insol in water; sol in alcohol and in acetic acid.

Sodium salt. [845-10-3] $C_{15}H_{14}N_3NaO_2$; mol wt 291.29.

Hydrochloride. [63451-28-5] $C_{15}H_{15}N_3O_2$·HCl; mol wt 305.76.

USE: As indicator in 0.1% alcoholic soln; pH: 4.4 red, 6.2 yellow. Used for titrating NH$_3$, weak organic bases, *e.g.*, alkaloids; not suitable for organic acids, except oxalic and picric acid. Methyl red is easily reduced, thereby losing its color, and readings should be made promptly. It is gradually being replaced by sulfonphthalein indicators, such as bromcresol green, which are more stable and exhibit a sharper change in color. Prepn of agar plates in microbiology. Food science applications include assessment of food penetration into packaging and pork quality during cold storage.

6193. Methyl Salicylate. [119-36-8] 2-Hydroxybenzoic acid methyl ester; wintergreen oil; betula oil; sweet birch oil; teaberry oil. $C_8H_8O_3$; mol wt 152.15. C 63.15%, H 5.30%, O 31.55%. Present in leaves of *Gaultheria procumbens* L., *Ericaceae*, in the bark of *Betula lenta* L., *Betulaceae;* prepd by esterification of salicylic acid with methanol. Toxicity data: P. M. Jenner *et al.*, *Food Cosmet. Toxicol.* **2**, 327 (1964).

Colorless, yellowish or reddish, oily liq; odor and taste of gaultheria. mp −8.6°. bp 220-224°. d$_4^{25}$ 1.184. d of the natural ester is ~1.180. n_D^{20} 1.535-1.538. Flash pt, closed cup: 210°F (99°C). Sol in chloroform, ether, alc, glacial acetic acid; slightly sol in water (one gram in about 1500 ml). LD$_{50}$ orally in rats: 887 mg/kg (Jenner).

Caution: Potential symptoms of overexposure are hyperpnea, apathy, lassitude; anorexia, nausea, vomiting, thirst; headache, dizziness, tinitis, difficulty with hearing and vision; confusion, irritability, mania, generalized convulsions; deep coma, death due to respiratory failure or cardiovascular collapse. Direct contact may cause irritation of skin and mucous membranes. *See Clinical Toxicology of Commercial Products*, R. E. Gosselin *et al.*, Eds. (Williams & Wilkins, Baltimore, 5th ed., 1984) Section II, p 205; Section III, pp 368-375.

USE: In perfumery; for flavoring candies, etc.

THERAP CAT: Counterirritant.

THERAP CAT (VET): Counterirritant.

6194. Methyl Silicone Resins. Tympanol. Polymeric methyl silicon oxides, probably of a cross-linked siloxane structure: Rochow, Gilliam, *J. Am. Chem. Soc.* **63**, 798 (1941). Prepd by hydrolyzing dimethyldichlorosilane or its esters followed by oxidation with air and a catalyst to the desired CH$_3$/Si ratio: Hyde, DeLong, *ibid.* 1194; by hydrolysis of dimethyldichlorosilane mixed with methyltrichlorosilane or silicon tetrachloride followed by cocondensation of the products: Rochow, Gilliam, *loc. cit.;* by partially methylating silicon tetrachloride to the desired CH$_3$/Si ratio and hydrolyzing the reaction mixture: US 2258218 (1941).

Characterized by excellent thermal stability and good resistance to oxidation. Oxidize slowly in air at 300° to silica. Can be heated to 550° *in vacuo* or to 500° in hydrogen without dec or melting. CH$_3$/Si = 1.2. Solid. d 1.20. n_D^{20} 1.425. Hardens in 2 hours at 100°. CH$_3$/Si = 1.3. Solid. d 1.15. n_D^{20} 1.422. Hardens in 1.5 hours at 120°. CH$_3$/Si = 1.4. Solid. d 1.08. n_D^{20} 1.421. Hardens in 4 hours at 141°. CH$_3$/Si = 1.5. Solid. d 1.06. n_D^{20} 1.418. Hardens in 24 hours at 100°. CH$_3$/Si <1.2. Sticky syrups. Harden at room temp. CH$_3$/Si >1.5. Oily liquids. Set to a soft gel after several days or weeks at 200°.

USE: In electrical insulation; in heat-resistant paints and varnishes; in protective and decorative finishes.

6195. Methyl Sulfate. [75-93-4] Sulfuric acid monomethyl ester; methylsulfuric acid; methyl hydrogen sulfate; monomethyl sulfate; methyl bisulfate. CH$_4$O$_4$S; mol wt 112.10. C 10.71%, H 3.60%, O 57.09%, S 28.60%. CH$_3$OSO$_2$OH. Prepn from methanol and methyl chlorosulfonate: Levaillant, Simon, *Compt. Rend.* **169**, 855 (1919); from methanol and sulfuric acid: Dumas, Peligot, *Ann.* **15**, 40 (1835); from sulfur trioxide and methanol below 0°: Chamberlain *et al.*, *BIOS Final Report* No. 1482, pp 6 and 8 (1946). Review: C. M. Suter, *The Organic Chemistry of Sulfur* (John Wiley, 1944) pp 18-23.

Oily liquid; does not solidify at −30°. d$_4^{15}$ 1.45-1.47. Freely sol in water, less sol in alcohol. Miscible with ether.
Barium salt. [513-17-7] Barium methyl sulfate. C$_2$H$_6$BaO$_8$S$_2$; mol wt 359.51. Dihydrate, efflorescent crystals. Sol in water, alc.
Calcium salt. [563-33-7] Calcium methyl sulfate; calcium sulfomethylate. C$_2$H$_6$CaO$_8$S$_2$; mol wt 262.26. Crystals. Soluble in water.
See also Dimethyl Sulfate.
USE: In sulfonation; as solvent in bromination of indigo dyes.

6196. N-Methyltaurine. [107-68-6] 2-(Methylamino)ethanesulfonic acid; β-methylaminoethane-α-sulfonic acid. C$_3$H$_9$NO$_3$S; mol wt 139.17. C 25.89%, H 6.52%, N 10.06%, O 34.49%, S 23.04%. Prepn: E. Dittrich, *J. Prakt. Chem.* **18**, 63 (1878); J. W. Schick, E. F. Degering, *Ind. Eng. Chem.* **39**, 906 (1947).

Prisms, mp 241-242°. Freely sol in water. Insol in alcohol, ether.
Sodium salt. [4316-74-9] N-Methylaminoethane sodium sulfonate; sodium N-methyltaurate. C$_3$H$_8$NNaO$_3$S. Prepd as a 35% aq soln. d$_4^{25}$ 1.21. At its freezing point (−28°) the soln becomes a suspension of crystals.
USE: Intermediate in the manuf of surface-active agents.

6197. 17-Methyltestosterone. [58-18-4] (17β)-17-Hydroxy-17-methylandrost-4-en-3-one; 17α-methyl-Δ⁴-androsten-17β-ol-3-one; Android; Metandren; Neohombreol M; Oreton Methyl; Testred. C$_{20}$H$_{30}$O$_2$; mol wt 302.46. C 79.42%, H 10.00%, O 10.58%. Prepn from 17-methyl-Δ⁵,⁶-androstene-3,17-diol: Oppenauer, *Rec. Trav. Chim.* **56**, 137 (1937); US 2384335 (1945 to Alien Property Custodian); from 17-methyl-Δ⁴-androsten-17-ol with CrO$_3$: Miescher, Wettstein, US 2374370; US 2374369 (both 1945 to Ciba); from androstenedione 3-enol ethyl ether: Miescher, US 2386331 (1945 to Ciba); from dehydroandrosterone: Julian *et al.*, US 2435013 (1945 to Glidden).

Crystals from hexane. mp 161-166°. Stable in air. [α]$_D^{25}$ +69 to +75° (dioxane). Soluble in alcohol, methanol, ether and in other organic solvents. Sparingly sol in vegetable oils. Practically insol in water.
Note: This is a controlled substance (anabolic steroid): **21 CFR,** 1308.13, as defined in 1300.01.
THERAP CAT: Androgen.
THERAP CAT (VET): Has been used as an androgenic agent.

6198. 2-Methyltetrahydrofuran. [96-47-9] Tetrahydro-2-methylfuran; tetrahydrosylvan; 2-MeTHF; MeTHF. C$_5$H$_{10}$O; mol wt 86.13. C 69.73%, H 11.70%, O 18.58%. Environmentally friendly organic solvent derived from renewable biomass. Alternative to tetrahydrofuran for organometallic reactions and dichloromethane for biphasic reactions. Prepn: A. Lipp, *Ber.* **22**, 2567 (1889); and physical properties: E. V. Whitehead *et al.*, *J. Am. Chem. Soc.* **73**, 3632 (1951). Prepn by catalytic hydrogenation of biomass containing furfural: I. Ahmed, WO 0234697; US 7064222 (2002, 2006 both to Pure Energy Corp.). Review of green chemistry properties: B. Comanita, *Spec. Chem.* **26**, 44-45 (2006); of solvent applications in organic synthesis: D. F. Aycock, *Org. Process Res. Dev.* **11**, 156-159 (2007).

Colorless liquid. *Flammable.* bp 79.9-80.1°; bp$_{716}$ 77-78°. fp −137.2°. d^{20} 0.8604. n_D^{20} 1.4063. Dielectric constant at 25°: 7. Flash pt, closed cup: 10.4°F (−12°C). Ignition pt: 270°C. Vaporization energy: 89.7 kcal/kg. Forms azeotrope with water (10.6%), bp 71°. Limited miscibility with water. Very stable to basic conditions; stable to acid concentrations of most synthetic processes.
USE: Solvent in synthetic organic chemistry.

6199. 4-Methyl-5-thiazoleethanol. [137-00-8] 5-Hydroxyethyl-4-methylthiazole; 5-(2-hydroxyethyl)-4-methylthiazole; 4-methyl-5-(β-hydroxyethyl)thiazole. C$_6$H$_9$NOS; mol wt 143.20. C 50.33%, H 6.34%, N 9.78%, O 11.17%, S 22.39%. The thiazole moiety of thiamine (vitamin B$_1$). Several syntheses, *e.g.*, by condensing thioformamide with bromoacetopropanol: Buchman, *J. Am. Chem. Soc.* **58**, 1803 (1936); or with γ,γ-dichloro-γ,γ-diacetodipropyl ether: Stein, Stevens, FR 945198 and GB 641426 (1947 and 1950 to Merck & Co.); Londergan, Schmitz, US 2654760 (1953 to du Pont).

Oily, viscous liquid. d$_4^{24}$ 1.196. Characteristic disagreeable odor of thiazole compds, becoming somewhat pleasant at extreme dilutions and imparting a nut-like flavor. bp$_7$ 135°; bp$_3$ 123-124°; bp$_1$ 103°. Sol in alcohol, ether, benzene, chloroform. Very sol in water.
Hydrochloride. C$_6$H$_9$NOS.HCl. Hygroscopic crystals, sol in water, alcohol.
USE: Intermediate in the synthesis of vitamin B$_1$.

6200. Methyl Thiocyanate. [556-64-9] Thiocyanic acid methyl ester; methyl sulfocyanate. C$_2$H$_3$NS; mol wt 73.11. C 32.86%, H 4.14%, N 19.16%, S 43.85%. Prepd from dimethyl sulfate and aq barium thiocyanate. Toxicology: W. F. von Oettingen *et al.*, *J. Ind. Hyg. Toxicol.* **18**, 310 (1936). Toxicity data: H. Ohkawa *et al.*, *Pestic. Biochem. Physiol.* **2**, 85 (1972).

H₃C—S—CN

Colorless liquid; onion odor. d^{20} 1.068. mp $-51°$. bp 130-133°. n_D^{20} 1.4697. Very slightly sol in water; miscible with alcohol, ether. LD_{50} i.p. in mice: 315 nmoles/g (Ohkawa).

6201. Methylthiouracil. [56-04-2] 2,3-Dihydro-6-methyl-2-thioxo-4(1H)-pyrimidinone; 6-methyl-2-thiouracil; 4-methyl-2-thiouracil; MTU; Alkiron; Antibason; Basecil; Basethyrin; Methiacil; Methicil; Methiocil; Muracil; Prostrumyl; Strumacil; Thimecil; Thyreostat I. $C_5H_6N_2OS$; mol wt 142.18. C 42.24%, H 4.25%, N 19.70%, O 11.25%, S 22.55%. Prepn: List, *Ann.* **236**, 1 (1886). Toxicity data: Simon, *Boll. Soc. Ital. Biol. Sper.* **24**, 803 (1948).

H₃C—[pyrimidine ring with NH, S, NH, O]

Crystals, bitter taste, dec 326-331°. Sublimes readily. Very slightly sol in cold water, ether. One part dissolves in about 150 parts boiling water. Slightly sol in alcohol, acetone. Practically insol in benzene, chloroform. Freely sol in aq solns of ammonia and alkali hydroxides. A satd aq soln is neutral or slightly acid to litmus. MLD in rabbits (mg/kg): 2486 orally (Simon).

THERAP CAT: Thyroid inhibitor.

THERAP CAT (VET): Antithyroid substance to promote growth and finishing.

6202. Methylthymol Blue. [3778-22-1] N,N'-[(1,1-Dioxido-3H-2,1-benzoxathiol-3-ylidene)bis[[6-hydroxy-2-methyl-5-(1-methylethyl)-3,1-phenylene]methylene]]bis[N-(carboxymethyl)glycine]; 3,3'-bis[N,N-di(carboxymethyl)aminomethyl]thymolsulfonephthalein; thymolsulfonphthalein-3',3''-bis(methyliminodiacetic acid). $C_{37}H_{44}N_2O_{13}S$; mol wt 756.82. C 58.72%, H 5.86%, N 3.70%, O 27.48%, S 4.24%. Multidentate cation chelant ligand; indicator that forms soluble, colored complexes with metal ions. Prepn from thymol blue, *q.v.*: J. Körbl, R. Pribil, *Chem. Ind. (London)* **1957**, 233; J. Körbl, *Collect. Czech. Chem. Commun.* **22**, 1789 (1957). Chromatographic purification: T. Yoshino *et al., Talanta* **16**, 151 (1969); S. Nakada *et al., Bull. Chem. Soc. Jpn.* **53**, 3365 (1980). Spectroscopy studies: S. Kiciak, M. A. Mehdi, *Talanta* **39**, 265 (1992); *idem et al., ibid.* **42**, 1245 (1995). Acid equilibrium and complex formation constants: T. Yoshino *et al., ibid.* **21**, 211 (1974). Metal chelate studies: B. Karadakov *et al., ibid.* **15**, 525 (1968); C. Bremer, E. Grell, *Inorg. Chim. Acta* **241**, 13 (1996). Chelation ion chromatography applications: B. Paull *et al., Anal. Commun.* **35**, 17 (1998); *eidem, Anal. Chim. Acta* **375**, 117 (1998). Reviews of analytical uses: B. Budesinsky in *Chelates in Analytical Chemistry,* H. A. Flaschka, A. J. Barnard, Jr., Eds. (Marcel Dekker, Inc., 1967) pp 15-47; S. Rani *et al., Chem. Era* **13**, 372-388 (1978).

[chemical structure of Methylthymol Blue]

Solid, mp 252-256° (dec). Absorption max: 435 nm. pKa_1 -1.76; pKa_2 -1.11; pKa_3 0.78; pKa_4 1.13; pKa_5 2.60; pKa_6 3.24; pKa_7 7.2; pKa_8 11.1; pKa_9 13.4. Color of aq soln transitions between yellow, blue, and violet; color is dependent upon protonation state.

Tetrasodium salt. [1945-77-3] $C_{37}H_{40}N_2Na_4O_{13}S$; mol wt 844.75. Crystalline black powder of dull metallic luster. Dec ~300° with swelling. Sol in 2-3 parts cold water. Insol in organic solvents.

Absorption (0.5M sulfuric acid): 455 nm (ε 18900); 485 nm (ε 11800); 545 nm (ε 4450).

Note: Not to be confused with methyl thymol blue, a mixed acid-base indicator composed of methyl red and thymol blue.

USE: Chromogenic indicator for complexometric and spectrophotometric determn of numerous metal ions, including iron, zinc, and magnesium. Also to detect sulfate, fluoride, oxalate, and polyphosphate ions. In chelation ion chromatography. As pH indicator.

6203. Methyltrienolone. [965-93-7] (17β)-17-Hydroxy-17-methylestra-4,9,11-trien-3-one; 17β-hydroxy-17-methyl-19-norandrosta-4,9,11-trien-3-one; 17α-methyl-4,9,11-estratrien-17β-ol-3-one; 17α-methyl-17β-hydroxyestra-4,9,11-trien-3-one; metribolone; R-1881. $C_{19}H_{24}O_2$; mol wt 284.40. C 80.24%, H 8.51%, O 11.25%. Prepn: Velluz *et al., Compt. Rend.* **257**, 569 (1963); **NL 6401555** (1964 to Roussel-UCLAF), *C.A.* **62**, 10498d (1965).

[chemical structure of Methyltrienolone]

Crystals from diisopropyl ether, mp 170°. $[α]_D^{20}$ $-58.7°$ (c = 0.5 in ethanol).

Note: This is a controlled substance (anabolic steroid): **21 CFR,** 1308.13, as defined in 1300.01.

6204. Methyl Trifluoromethanesulfonate. [333-27-7] 1,1,1-Trifluoromethanesulfonic acid methyl ester; methyl triflate; MeOTf. $C_2H_3F_3O_3S$; mol wt 164.10. C 14.64%, H 1.84%, F 34.73%, O 29.25%, S 19.54%. Methylating reagent in organic synthesis. Prepn: T. J. Brice, P. W. Trott, **US 2732398** (1956 to Minnesota Mining & Manuf). Improved prepn: C. D. Beard *et al., J. Org. Chem.* **38**, 3673 (1973). Molecular structure: F. Trautner *et al., Inorg. Chem.* **38**, 3051 (1999). Synthetic applications in methylation reactions: J. Arnarp *et al., Carbohydr. Res.* **44**, C5 (1975); M. Ravenscroft *et al., J. Chem. Soc. Perkin Trans.* 2 **1982**, 1569; M. J. O'Donnell *et al., Tetrahedron Lett.* **25**, 3651 (1984); K. Maruoka *et al., J. Am. Chem. Soc.* **114**, 4422 (1992). *Review:* P. J. Stang *et al., Synthesis* **1982**, 85-126.

[chemical structure of methyl triflate, F₃C—S(=O)(=O)—OCH₃]

Clear liquid. *Flammable. Corrosive.* bp 98-99°; bp_{736} 97-97.5°. n_D^{25} 1.3238; n_D^{20} 1.3260. Flash pt, closed cup: 100°F (38°C). Air sensitive. Turns dark and rapidly hydrolyzes upon exposure to moisture. Store at 2-8°C.

USE: Reagent in synthetic organic chemistry.

6205. Methyl(trifluoromethyl)dioxirane. [115464-59-0] TFDO; TFD. $C_3H_3F_3O_2$; mol wt 128.05. C 28.14%, H 2.36%, F 44.51%, O 24.99%. Cyclic peroxide used for the epoxidation of unreactive alkenes and the oxidation of deactivated hydrocarbons and alcohols. Prepn from the oxidation of 1,1,1-trifluoroacetone with potassium peroxymonosulfate and use in oxidation and epoxidation reactions: R. Mello *et al., J. Org. Chem.* **53**, 3890 (1988). Improved prepn and use in oxidation and epoxidation reactions: *idem et al., Synlett* **2007**, 47. Synthetic utility as oxidizing agent: *idem et al., J. Am. Chem. Soc.* **111**, 6749 (1989); *idem et al., Tetrahedron Lett.* **31**, 6097 (1990); *idem et al., J. Am. Chem. Soc.* **113**, 2205 (1991); L. D'Accolti *et al., J. Org. Chem.* **69**, 8510 (2004).

[chemical structure of Methyl(trifluoromethyl)dioxirane]

Obtained as a dilute yellow soln in 1,1,1-trifluoroacetone with a concn range of 0.6 to 0.8M; can be stored in soln at $-20°C$ for up to one week. Volatile. uv max: 347 nm (ε ~9, 0°C). Sol in most organic solvents, but slowly reacts with them. Keep away from light and traces of heavy metals.

USE: Reagent in synthetic organic chemistry.

6206. Methyltrioxorhenium. [70197-13-6] (*T*-4)-Methyl-trioxorhenium; MTO. CH_3O_3Re; mol wt 249.24. C 4.82%, H 1.21%, O 19.26%, Re 74.71%. CH_3ReO_3. Acts as an oxidation catalyst by transferring an oxygen atom from one species to another; hydrogen peroxide is often the oxygen source. Prepn: I. R. Beattie, P. J. Jones, *Inorg. Chem.* **18**, 2318 (1979); W. A. Herrmann *et al.*, *Angew. Chem. Int. Ed.* **27**, 394 (1988). Catalytic uses: *idem et al.*, *ibid.* **30**, 1636, 1638, 1641 (1991). Photochemistry studies: H. Kunkely *et al.*, *Organometallics* **10**, 2090 (1991); L. J. Morris *et al.*, *ibid.* **20**, 2344 (2001). Review of catalytic chemistry: J. H. Espenson, *Chem. Commun.* **1999**, 479-488; F. E. Kühn, W. A. Herrmann, *Chemtracts Org. Chem.* **14**, 59-83 (2001).

Colorless solid, mp 110°. d^{23} 4.21. Sol in acetonitrile, benzene, chloroform, ethanol, ether. Sparingly sol in carbon disulfide, hexane. Soly at 25° (g/l): water ~50; forms a golden ppt, poly-MTO. uv max (hexane): 205, 231, 260 nm (ε 1600, 1500, 1020). uv max (aqueous): 239, 270 nm (ε 1900, 1300). Air and moisture stable, T_{dec} >300°.

Poly-MTO. First known "polymeric" organometallic oxide, $[H_{0.5}[(CH_3)_{0.92}ReO_3]]_n$. Formed by MTO in aqueous soln. Prepn: W. A. Herrmann, R. W. Fischer, *J. Am. Chem. Soc.* **117**, 3223 (1995). Structural study: *idem et al.*, *ibid.* 3231. Golden solid with graphite-like consistence and reflectance. d^{23} 4.38. Insol in practically all inorganic and organic solvents. Electric resistivity (25°): 6×10^{-3} Ω cm.

USE: Versatile organometallic catalyst active in oxidation reactions, olefin isomerization, epoxidation and metathesis.

6207. α-Methyltryptamine. [299-26-3] α-Methyl-1*H*-indole-3-ethanamine; 3-(2-aminopropyl)indole; AMT; IT-290. C_{11}-$H_{14}N_2$; mol wt 174.25. C 75.82%, H 8.10%, N 16.08%. Synthetic hallucinogen; the (*S*)-(+)-form possesses more potent bioactivity. Prepn: H. R. Synder, L. Katz, *J. Am. Chem. Soc.* **69**, 3140 (1947); A. S. F. Ash, W. R. Wragg, *J. Chem. Soc.* **1958**, 3887; A. G. Terzian *et al.*, *Experientia* **17**, 493 (1961); D. H. Lloyd, D. E. Nichols, *J. Org. Chem.* **51**, 4294 (1986). CNS effects in animals: J. R. Vane *et al.*, *Nature* **191**, 1068 (1961). Clinical pharmacology: H. B. Murphree *et al.*, *Clin. Pharmacol. Ther.* **2**, 722 (1961). Synthesis and activity of enantiomers: D. B. Repke, W. J. Ferguson, *J. Heterocycl. Chem.* **13**, 775 (1976); R. A. Glennon *et al.*, *Biochem. Pharmacol.* **32**, 1267 (1983). Case report of intoxication: H. Long *et al.*, *Vet. Hum. Toxicol.* **45**, 149 (2003). GC-MS determn in blood and urine: T. Ishida *et al.*, *J. Chromatogr. B* **823**, 47 (2005).

(*S*)-Form

White powder. Crystals from diethyl ether, mp 101-102° (Lloyd, Nichols). Also reported as mp 98-99° (Terzian).

(*S*)-**Form.** [7795-51-9] Crystals from ethyl acetate + hexane, mp 125-126°. $[\alpha]_D^{25}$ +34.9° (c = 0.9 in methanol).

(*R*)-**Form.** [7795-52-0] mp 126-127°. $[\alpha]_D^{25}$ −32.1° (c = 1.0 in methanol).

Note: This is a controlled substance (hallucinogen): 21 CFR, 1308.11.

6208. α-Methyl-*m*-tyrosine. [305-96-4] 3-Hydroxy-α-methylphenylalanine; α-methyl-3-(*m*-hydroxyphenyl)alanine; α-MMT. $C_{10}H_{13}NO_3$; mol wt 195.22. C 61.53%, H 6.71%, N 7.17%, O 24.59%. Inhibitor of catecholamine synthesis. Prepn of DL-form: Stein *et al.*, *J. Am. Chem. Soc.* **77**, 700 (1955); Pfister, Stein, **US 2868818** (1959 to Merck & Co.).

Crystals from methanol, dec 296-297°.

6209. 6-Methyluracil. [626-48-2] 6-Methyl-2,4(1*H*,3*H*)-pyrimidinedione; 4-methyluracil. $C_5H_6N_2O_2$; mol wt 126.12. C 47.62%, H 4.80%, N 22.21%, O 25.37%. Prepn: Boese, **US 2138756** (1938 to Carbide and Carbon Chem.); Gleason, **US 2174239** (1939 to Standard Oil). Synthesis from urea and ethyl acetoacetate: J. J. Donleavy, M. A. Kise, *Org. Synth.* **coll. vol. II**, 422 (1943); by isomerization of 2-hydroxy-4-methoxy-6-methylpyrimidine: McOmie *et al.*, *J. Chem. Soc.* **1957**, 1830; from urea and diketene: Khromov-Borisov, Karlinskaya, *J. Gen. Chem. USSR* **26**, 1728 (1956); **SU 101690** (1955), *C.A.* **51**, 12989f (1957).

Crystals from glacial acetic acid, dec >300°. uv max: 277 nm (log ε 3.83). pH of aq soln 13.

6210. Methyl Vinyl Ketone. [78-94-4] 3-Buten-2-one; Δ^3-2-butenone; methylene acetone; acetyl ethylene; δ-oxo-α-butylene; vinyl methyl ketone. C_4H_6O; mol wt 70.09. C 68.55%, H 8.63%, O 22.83%. Prepn by condensation of acetone and formaldehyde to 3-ketobutanol and dehydration to methyl vinyl ketone: Merling, Köhler, *J. Soc. Chem. Ind.* **29**, 1037 (1910); White, Haward, *J. Chem. Soc.* **1943**, 25. Prepn from vinylacetylene: Conaway, **US 1967225** (1934 to DuPont). For review and polymerization characteristics *see* "Vinyl Ketone Polymers" in *Encyclopedia of Polymer Science and Technology* **vol. 14** (Interscience, New York, 1971) pp 617-636.

Liquid with pungent odor. bp$_{760}$ 81.4°. n_D^{20} 1.4086. d_4^{20} 0.8636; d_4^{25} 0.8407. *Poisonous; flammable; corrosive*. Easily soluble in water, methanol, ethanol, ether, acetone, glacial acetic acid. Slightly sol in hydrocarbons. Forms a binary azeotrope with water, bp$_{760}$ 75° (12% water). uv spectrum and electric moments: Rogers, *J. Am. Chem. Soc.* **69**, 2544 (1947). Polymerizes on standing.

Caution: Readily absorbed through skin causing general poisoning of the organism. Irritating to mucous membranes and respiratory tract.

USE: Alkylating agent; commercial starting material for plastics; as intermediate in the synthesis of steroids and vitamin A.

6211. Methymycin. [497-72-3] (3*R*,4*S*,5*S*,7*R*,9*E*,11*S*,12*R*)-12-Ethyl-11-hydroxy-3,5,7,11-tetramethyl-4-[[3,4,6-trideoxy-3-(dimethylamino)-β-D-*xylo*-hexopyranosyl]oxy]oxacyclododec-9-ene-2,8-dione. $C_{25}H_{43}NO_7$; mol wt 469.62. C 63.94%, H 9.23%, N 2.98%, O 23.85%. Antibiotic substance produced by a streptomycete from soil near Oswego, N.Y. Has a macrolide structure (12-membered lactone ring, *compare* Picromycin). Isoln and antibacterial activity: Donin *et al.*, *Antibiot. Annu.* **1**, 179 (1953-4). Production using *Streptomyces venezuelae* cultures: Dutcher *et al.*, **US 2916483** (1959 to Olin Mathieson). Structure: C. Djerassi, J. A. Zderic, *J. Am. Chem. Soc.* **78**, 6390 (1956). Absolute configuration: Rickards, Smith, *Tetrahedron Lett.* **1970**, 1025; Manwaring *et al.*, *ibid.* 1029. Biosynthesis: Birch *et al.*, *J. Chem. Soc.* **1964**, 5274. Synthesis: Masamune *et al.*, *J. Am. Chem. Soc.* **97**, 3512 (1975). Possible mechanism of action: Wilhelm *et al.*, *Antimicrob. Agents Chemother.* **1967**, 236.

Prisms from abs ethanol, mp 195.5-197°. $[\alpha]_D^{22}$ +61° (c = 0.7 in methanol), +74° (c = 1.1 in chloroform). Basic substance, pKb' 5.7. uv max (methanol): 223, 322 nm (ε 10500, 47). Very slightly sol in water, hexane. Sol in methanol, acetone, chloroform, dil acids. Moderately sol in ethanol, ether.

Sulfate. $(C_{25}H_{43}NO_7)_2.H_2SO_4$. Crystals from methanol + acetone. Sol in water, methanol.

Acid sulfate. $C_{25}H_{43}NO_7.H_2SO_4$. Plates from acetone. Sol in water, acetone.

6212. Methyprylon. [125-64-8] 3,3-Diethyl-5-methyl-2,4-piperidinedione; 2,4-dioxo-3,3-diethyl-5-methylpiperidine; 3,3-diethyl-2,4-dioxo-5-methylpiperidine; Noctan; Dimerin; Noludar. $C_{10}H_{17}NO_2$; mol wt 183.25. C 65.54%, H 9.35%, N 7.64%, O 17.46%. Prepn: Frick, Lutz, **US 2680116** (1954 to Hoffmann-La Roche). cf. **DE 930206** (1953). Comprehensive description: B. C. Rudy, B. Z. Senkowski, *Anal. Profiles Drug Subs.* **2**, 363-382 (1973).

Crystals. Bitter taste. mp 74-77°. uv max (alcohol): 295 nm ($A_{1cm}^{1\%}$ 2.0). Sol in water, alcohol, benzene, chloroform.

Caution: This is a controlled substance (depressant): **21 CFR**, 1308.13.

THERAP CAT: Sedative, hypnotic.

6213. Methysergide. [361-37-5] (8β)-9,10-Didehydro-N-[(1S)-1-(hydroxymethyl)propyl]-1,6-dimethylergoline-8-carboxamide; N-[1-(hydroxymethyl)propyl]-1-methyl-D-lysergamide; 1-methylmethylergonovine; 1-methyl-d-lysergic acid butanolamide; 1-methyl-d-lysergic acid (+)-1-hydroxy-2-butylamide; UML-491. $C_{21}H_{27}N_3O_2$; mol wt 353.47. C 71.36%, H 7.70%, N 11.89%, O 9.05%. Serotonin receptor antagonist. Prepn: **GB 854569** (1960); A. Hofmann, F. Troxler, **US 3113133** (1963 to Sandoz). Comparative pharmacology: Z. Votava et al., *Arzneim.-Forsch.* **16**, 220 (1966); P. N. Chambers, P. B. Marshall, *J. Pharm. Pharmacol.* **19**, 65 (1967). Mechanism of action: D. A. Curran et al., *Res. Clin. Stud. Headache* **1**, 74 (1967); J. E. Hardebo et al., *Neurology* **28**, 64 (1978); S. W. J. Lamberts, R. M. Mac Leod, *Endocrinology* **103**, 287 (1978).

Crystals, mp 194-196°. $[\alpha]_D^{20}$ −45° (c = 0.5 in pyridine).

Tartrate. $C_{46}H_{60}N_6O_{10}$. Crystals, sparingly sol in water.

Dimaleate. Dec ~165°. Sol in methanol, less sol in water (1:250). Practically insol in abs ethanol.

Hydrogen maleate. [129-49-7] Methysergide maleate; Deseril; Désernil; Sansert. $C_{21}H_{27}N_3O_2.C_4H_4O_4$; mol wt 469.54.

THERAP CAT: Antimigraine.

6214. Methysticin. [495-85-2] (6R)-6-[(1E)-2-(1,3-Benzodioxol-5-yl)ethenyl]-5,6-dihydro-4-methoxy-2H-pyran-2-one; 5-hydroxy-3-methoxy-7-[3,4-(methylenedioxy)phenyl]-2,6-heptadienoic acid δ-lactone; 4-methoxy-6-[β-(3',4'-methylenedioxyphenyl)vinyl]-5,6-dihydro-α-pyrone; 6-(3',4'-methylenedioxystyryl)-4-methoxy-5,6-dihydro-2H-pyran-2-one; kavahin; kavatin. $C_{15}H_{14}O_5$; mol wt 274.27. C 65.69%, H 5.15%, O 29.17%. One of the active com-

ponents of the intoxicating beverage, Kava, from the Pacific Islands. From root of *Piper methysticum* Forst., *Piperaceae* (kava): Pomeranz, *Monatsh. Chem.* **10**, 783 (1889); Hänsel, Beiersdorff, *Arzneim.-Forsch.* **9**, 581 (1955). Structure: Borsche et al., *Ber.* **60**, 2113 (1927). Synthesis of racemate: Klohs et al., *J. Org. Chem.* **24**, 1829 (1959). Absolute config: Snatzke, Hänsel, *Tetrahedron Lett.* **1968**, 1797. Crystal structure: P. Engel, W. Nowacki, *Z. Kristallogr.* **136**, 437 (1972). ^{13}C NMR: A. Banerji et al., *Org. Magn. Reson.* **13**, 345 (1980). HPLC determn: R. M. Smith et al., *J. Chromatogr.* **283**, 303 (1984). Anticonvulsant effects: Kretzschmar, Meyer, *Arch. Int. Pharmacodyn. Ther.* **177**, 261 (1969). Neuroprotective activity: C. Backhauss, J. Krieglstein, *Eur. J. Pharmacol.* **215**, 265 (1992).

Crystals from methanol, mp 132-134°. uv max (alcohol): 226, 267, 306 nm (log ε 4.40, 4.14, 3.93). Practically insol in water. Sol in alcohol, ether, acetone.

(±)-Dihydromethysticin. [3155-57-5] $C_{15}H_{16}O_5$. Prisms from methanol, mp 118°. uv max (methanol): 232, 288 nm (log ε 4.18, 3.56).

6215. Metiazinic Acid. [13993-65-2] 10-Methyl-10H-phenothiazine-2-acetic acid; (10-methyl-2-phenothiazinyl)acetic acid; N-methyl-3-phenothiazinylacetic acid; methiazic acid; methiazinic acid; metiazic acid; RP-16091; Soridermal; Soripal. $C_{15}H_{13}NO_2S$; mol wt 271.33. C 66.40%, H 4.83%, N 5.16%, O 11.79%, S 11.82%. Prepn: **FR M4163**; Farge et al., **NL 6614516**; **US 3424748** (1966, 1967, 1969 all to Rhône-Poulenc); Messer et al., *C. R. Seances Acad. Sci. Ser. C* **265** (14), 758 (1967). Series of articles on prepn, pharmacology, toxicology and metabolism: *Arzneim.-Forsch.* **19**, 1193-1221 (1969). Toxicity: L. Joulou et al., *ibid.* 1198.

Crystals from benzene, mp 146°. Sol in acetone, ether, chloroform. Forms a water-sol sodium salt. uv max (0.1N NaOH): 253, 305 nm; min 280 nm. LD_{50} in mice, rats (mg/kg): 800, ~500 orally (Joulou).

THERAP CAT: Anti-inflammatory.

6216. Metipranolol. [22664-55-7] 4-[2-Hydroxy-3-[(1-methylethyl)amino]propoxy]-2,3,6-trimethylphenol 1-acetate; 1-(4-hydroxy-2,3,5-trimethylphenoxy)-3-(isopropylamino)-2-propanol 4-acetate; methypranol; trimepranol; VUFB-6453; Betamet; Betanol; Disorat; Glauline; Glausyn; Turoptin. $C_{17}H_{27}NO_4$; mol wt 309.41. C 65.99%, H 8.80%, N 4.53%, O 20.68%. β-Adrenergic blocker. Prepn: L. Blaha et al., **CS 128471** (1968), *C.A.* **71**, 3129a (1969). Pharmacology: W. Bartsch et al., *Arzneim.-Forsch.* **27**, 1022 (1977). Plasma levels and pharmacodynamics in man: O. Mayer et al., *Int. J. Clin. Pharmacol. Ther. Toxicol.* **18**, 113 (1980). Metabolic effects: R. P. Faupel, R. Gotzen, *Med. Klin.* **74**, 929 (1979). Comparative effect in hypertension: K. Hayduk et al., *Therapiewoche* **29**, 7528, 7530 (1979). Review of pharmacology and clinical trials in glaucoma and ocular hypertension: P. E. Battershill, E. M. Sorkin, *Drugs* **36**, 601-615 (1988). Comprehensive description: J. Dohnal et al., *Anal. Profiles Drug Subs.* **19**, 367-396 (1990).

Crystals from cyclohexane, mp 105-107° (Blaha), also reported as 108.5-110.5° (Florey). uv max (methanol): 278, 274 nm ($A_{1cm}^{1\%}$ 51.3, 50.5). Freely sol in 95% ethanol, chloroform, benzene; slightly sol in ether. Practically insol in water. LD_{50} in mice (mg/kg): 31 i.v. (Bartsch).

Hydrochloride. [36592-77-5] Betamann; Optipranolol. C_{17}-$H_{27}NO_4$.HCl; mol wt 345.86. Sol in water.

THERAP CAT: Antihypertensive; antiglaucoma agent.

6217. Metitepine. [20229-30-5] 1-[10,11-Dihydro-8-(methylthio)dibenzo[b,f]thiepin-10-yl]-4-methylpiperazine; 8-methylthio-10-(4-methylpiperazino)-10,11-dihydrodibenzo[b,f]thiepine; methiothepine; methiothepin; Ro-8-6837. $C_{20}H_{24}N_2S_2$; mol wt 356.55. C 67.37%, H 6.79%, N 7.86%, S 17.98%. Serotonin (5-HT_2) receptor antagonist; also exhibits affinity for 5-HT_1-receptors. Prepn: **NL 6608618**; M. Protiva *et al.*, **US 3379729** (1966, 1968 both to SPOFA); and pharmacology: K. Pelz *et al.*, *Collect. Czech. Chem. Commun.* **33**, 1895 (1968); J. O. Jilek *et al.*, *ibid.* **39**, 3338 (1974). Receptor-blocking study: M.-A. Monachon *et al.*, *Arch. Pharmacol.* **274**, 192 (1972). Use in classification of 5-HT receptors: P. B. Bradley *et al.*, *Neuropharmacology* **25**, 563 (1986); E. J. Mylecharane, *Clin. Exp. Pharmacol. Physiol.* **16**, 517 (1989).

Crystals from ethanol, mp 88-89°.

Maleate. [19728-88-2] $C_{20}H_{24}N_2S_2$.$C_4H_4O_4$. Crystals from ethanol, mp 171-173°. LD_{50} in mice (mg/kg): 51 i.v.; 94 orally (Jilek).

USE: Biochemical tool in serotonin receptor binding studies.

6218. Metobromuron. [3060-89-7] N'-(4-Bromophenyl)-N-methoxy-N-methylurea; 3-(p-bromophenyl)-1-methoxy-1-methylurea; Ciba 3126; Pattonex; Patoran. $C_9H_{11}BrN_2O_2$; mol wt 259.10. C 41.72%, H 4.28%, Br 30.84%, N 10.81%, O 12.35%. Prepd by bromination of N-phenyl-N'-methoxy-N'-methylurea: H. Martin *et al.*, **CH 405821**; *eidem*, **US 3288851** (both 1966 to Ciba). Toxicity study: S. Novakova, S. Dinoeva, *Egeszsegtudomany* **17**, 233 (1973), *C.A.* **80**, 56258j (1974).

Crystals from cyclohexane, mp 95-96°. Vapor pressure at 20°: 3×10^{-6} mm Hg. Soly in water at 20° = 320 ppm. Sol in methanol, ethanol, acetone, chloroform. LD_{50} orally in rats: 3875 mg/kg (Novakova, Dinoeva).

USE: Herbicide.

6219. Metoclopramide. [364-62-5] 4-Amino-5-chloro-N-[(2-diethylamino)ethyl]-2-methoxybenzamide; 4-amino-5-chloro-N-[2-(diethylamino)ethyl]-o-anisamide; 4-amino-5-chloro-2-methoxy-N-(β-diethylaminoethyl)benzamide; DEL-1267; Elieten; Eucil; Imperan; Reliveran. $C_{14}H_{22}ClN_3O_2$; mol wt 299.80. C 56.09%, H 7.40%, Cl 11.82%, N 14.02%, O 10.67%. Substituted benzamide with neuroleptic activity. Prepn: **BE 620543**; Thominet, **US 3177252** (1962, 1965 to Soc. d'Etudes Sci. Ind. de l'Ile-de-France); R. Pakula *et al.*, *Arch. Pharm.* **313**, 297 (1980). Colorimetric determn in plasma and solubility: G. Pitel, T. Luce, *Ann. Pharm. Fr.* **23**, 673 (1965). Dopamine D_2 receptor antagonist (adenylate cyclase independent receptors): J. W. Kebabian, D. B. Calne, *Nature* **277**, 93 (1979). Pharmacology: M. A. Smith, F. J. Salter, *Drug Intell. Clin. Pharm.* **14**, 169 (1980). Pharmacokinetics: H. Vergin *et al.*,

Arzneim.-Forsch. **33**, 458 (1983). Review of efficacy in psychiatric disorders: E. D. Peselow, M. Stanley, *Adv. Biochem. Psychopharmacol.* **35**, 163-194 (1982); of pharmacology and clinical use: R. A. Harrington *et al.*, *Drugs* **25**, 451-494 (1983); *ibid.* Suppl. 1, 1-88. Comprehensive description: D. Pitré, R. Stradi, *Anal. Profiles Drug Subs.* **16**, 327-361 (1987).

mp 146.5-148°. Soly at 25° (g/100 ml): water 0.02; 95% ethanol 2.30; abs ethanol 1.90; benzene 0.10; chloroform 6.60.

Dihydrochloride monohydrate. Emetid; Gastronerton; Primperan. $C_{14}H_{22}ClN_3O_2$.2HCl.H_2O; mol wt 390.73. Crystals, dec 145°. Soly at 25° (g/100 ml): water 48; ethanol (95%) 6; benzene 0.10; chloroform 0.10. Stable in acids solns. Unstable in strongly alkaline solns.

Monohydrochloride monohydrate. [54143-57-6] AHR-3070-C; Cerucal; Clopromate; Duraclamid; Emperal; Gastrese; Gastrobid; Gastromax; Gastrosil; Gastro-Tablinen; Maxeran; Maxolon; Meclopran; Metamide; Metoclol; Metocobil; Metramid; Moriperan; Mygdalon; Parmid; Paspertin; Peraprin; Plasil; Pramiel; Reglan. $C_{14}H_{22}$-ClN_3O_2.HCl.H_2O; mol wt 354.27. White crystalline powder, mp 182.5-184°. Sol in water.

THERAP CAT: Antiemetic.

6220. Metocurine Iodide. [7601-55-0] (13aR,25aS)-2,3,-13a,14,15,16,25,25a-Octahydro-9,18,19,29-tetramethoxy-1,1,14,14-tetramethyl-13H-4,6:21,24-dietheno-8,12-metheno-1H-pyrido[3',-2':14,15][1,11]dioxacycloeicosino[2,3,4-ij]isoquinolinium iodide (1:2); 6,6',7',12'-tetramethoxy-2,2,2',2'-tetramethyltubocurararanium diiodide; (+)-O,O'-dimethylchondrocurarine diiodide; d-tubocurarine iodide dimethyl ether; dimethyl tubocurarine iodide; tubocurarine dimethyl ether iodide; Metubine Iodide. $C_{40}H_{48}I_2N_2O_6$; mol wt 906.64. C 52.99%, H 5.34%, I 27.99%, N 3.09%, O 10.59%. Prepd by methylation of d-tubocurarine with methyl iodide.

Crystals, dec 257-267°. $[\alpha]_D^{22}$ +148 to +158° (c = 0.25). uv max: 280 nm. ($E_{1cm}^{1\%}$ 74). Slightly soluble in water (about 300 mg/100 ml), in dil HCl, in dil NaOH. Very slightly sol in alcohol; practically insol in benzene, chloroform, ether.

Trihydrate. Large prisms from water, appreciably sol in methanol.

THERAP CAT: Neuromuscular blocking agent.

6221. Metofenazate. [388-51-2] 3,4,5-Trimethoxybenzoic acid 2-[4-[3-(2-chloro-10H-phenothiazin-10-yl)propyl]-1-piperazinyl]ethyl ester; 2-[4-[3-(2-chlorophenothiazin-10-yl)propyl]-1-piperazinyl]ethyl 3,4,5-trimethoxybenzoate; N-[β-(3,4,5-trimethoxybenzoyloxy)ethyl]-N'-[γ-(3-chloro-10-phenothiazinyl)propyl]piperazine; methophenazine; perphenazine 3,4,5-trimethoxybenzoate. $C_{31}H_{36}ClN_3O_5S$; mol wt 598.16. C 62.25%, H 6.07%, Cl 5.93%, N 7.03%, O 13.37%, S 5.36%. Prepn: *Magy. Kem. Lapja* **17**, 169 (1962), *C.A.* **57**, 11314d (1962); Toldy *et al.*, *Acta Chim. Acad. Sci. Hung.* **42**, 351 (1964).

Crystals from ethanol, mp 102-107°.

Difumarate. [522-23-6] Frenolon. $C_{31}H_{36}ClN_3O_5S.2C_4H_4O_4$; mol wt 830.30.

THERAP CAT: Antipsychotic.

6222. Metofluthrin. [240494-70-6] 2,2-Dimethyl-3-(1-propen-1-yl)cyclopropanecarboxylic acid [2,3,5,6-tetrafluoro-4-(methoxymethyl)phenyl]methyl ester; Eminence; SumiOne. $C_{18}H_{20}F_4O_3$; mol wt 360.35. C 60.00%, H 5.59%, F 21.09%, O 13.32%. Synthetic pyrethroid insecticide; mixture of isomers, primarily (Z)-(1R,3R)-form. Prepn: K. Ujihara, T. Iwasaki, **EP 939073**; *eidem*, **US 6225495** (1999, 2001 both to Sumitomo); and activity vs mosquitoes: K. Ujihara *et al.*, *Biosci. Biotechnol. Biochem.* **68**, 170 (2004); T. Mori *et al.*, *J. Fluorine Chem.* **128**, 1174 (2007). GC-MS determn in indoor air samples: T. Yoshida, *J. Chromatogr. A* **1216**, 5069 (2009). Field trial vs mosquitoes: H. Kawada *et al.*, *Am. J. Trop. Med. Hyg.* **75**, 1153 (2006); J. R. Lucas *et al.*, *J. Am. Mosq. Control Assoc.* **23**, 47 (2007).

(Z)-(1R,3R)-form

Pale yellow, clear oily liquid; slight characteristic odor. d_4^{20} 1.21. Vapor pressure (25°): 1.96×10^{-3}. Flash point (open cup): 352°F (178°C). Viscosity (20°): 19.3 mm^2/s. Soly in water (20°): 0.73 mg/L. Sol in acetonitrile, DMSO, methanol, ethanol, acetone, hexane.

USE: Insecticide.

6223. Metolachlor. [51218-45-2] 2-Chloro-N-(2-ethyl-6-methylphenyl)-N-(2-methoxy-1-methylethyl)acetamide; 2-chloro-6'-ethyl-N-(2-methoxy-1-methylethyl)acet-o-toluidide; α-chloro-2'-ethyl-6'-methyl-N-(1-methyl-2-methoxyethyl)acetanilide; metelilachlor; CGA-24705; Dual. $C_{15}H_{22}ClNO_2$; mol wt 283.80. C 63.48%, H 7.81%, Cl 12.49%, N 4.94%, O 11.27%. Selective pre-emergence herbicide. Prepn: C. Vogel, R. Aebi, **DE 2328340**; *eidem*, **US 3937730** (1973, 1976 both to Ciba-Geigy). Synthesis of stereoisomers: B. T. Cho, Y. S. Chun, *Tetrahedron: Asymmetry* **3**, 337 (1992). Metabolism: L. L. McGahen, J. M. Tiedje, *J. Agric. Food Chem.* **26**, 414 (1978). ELISA determn: T. S. Lawruk *et al.*, *J. Agric. Food Chem.* **41**, 1426 (1993). Review: G. Chesters *et al.*, *Rev. Environ. Contam. Toxicol.* **110**, 1-74 (1989); of biological activity and field trials: H. R. Gerber *et al.*, *Proc. 12th Br. Weed Control Conf.* 787-794 (1974).

Colorless liquid. bp$_{0.001}$ 100°. n_D^{20} 1.5301. Vapor pressure at 20°: 1.3×10^{-5} mm Hg. Soly in water at 20°: 530 ppm. Sol in most organic solvents. Partition coefficient (octanol/water): 2,800. LD$_{50}$ in rats (mg/kg): 2780 orally; >3170 dermally (Gerber).

S-Form. [87392-12-9] Dual Gold.

USE: Herbicide.

6224. Metolazone. [17560-51-9] 7-Chloro-1,2,3,4-tetrahydro-2-methyl-3-(2-methylphenyl)-4-oxo-6-quinazolinesulfonamide; 7-chloro-1,2,3,4-tetrahydro-2-methyl-4-oxo-3-o-tolyl-6-quinazolinesulfonamide; 2-methyl-3-o-tolyl-6-sulfamyl-7-chloro-1,2,3,4-tetrahydro-4-quinazolinone; SR-720-22; Metenix; Mykrox; Xuret; Zaroxolyn. $C_{16}H_{16}ClN_3O_3S$; mol wt 365.83. C 52.53%, H 4.41%, Cl 9.69%, N 11.49%, O 13.12%, S 8.76%. Prepn: B. V. Shetty, **US 3360518** (1967 to Wallace & Tiernan); B. V. Shetty *et al.*, *J. Med. Chem.* **13**, 886 (1970). Pharmacology: E. Belair *et al.*, *Arch. Int. Pharmacodyn. Ther.* **177**, 71 (1969); E. J. Belair, *Res. Commun. Chem. Pathol. Pharmacol.* **2**, 98 (1971). HPLC determn in urine: D. Farthing *et al.*, *J. Chromatogr.* **534**, 228 (1990). Clinical evaluations as diuretic in chronic renal disease: R. Schoonees *et al.*, *N. Y. State J. Med.* **71**, 566 (1971); in hypertension: C. L. Curry *et al.*, *Clin. Ther.* **9**, 47 (1986); in congestive heart failure: A. Kiyingi *et al.*, *Lancet* **335**, 29 (1990).

Crystals from ethanol or butanol, mp 252-254°. LD$_{50}$ in mice (mg/kg): >5000 orally; >1500 i.p. (Shetty, 1967).

THERAP CAT: Diuretic; antihypertensive.

6225. Metomidate. [5377-20-8] 1-(1-Phenylethyl)-1H-imidazole-5-carboxylic acid methyl ester; 1-(α-methylbenzyl)imidazole-5-carboxylic acid methyl ester; methyl 1-(α-methylbenzyl)imidazole-5-carboxylate; methoxymol; methomidate. $C_{13}H_{14}N_2O_2$; mol wt 230.27. C 67.81%, H 6.13%, N 12.17%, O 13.90%. Injectable hypnotic related to etomidate, *q.v.* Prepn: **BE 662474**; E. F. Godefroi, C. A. M. van der Eijcken, **US 3354173** (1965, 1967 both to Janssen); E. F. Godefroi *et al.*, *J. Med. Chem.* **8**, 220 (1965). Anaesthetic effect of combination with azaperone, *q.v.*: J. F. F. Callear, J. F. E. Van Gestel, *Vet. Rec.* **92**, 284 (1973); of combination with fentanyl, *q.v.*: W. Erhardt, *J. Small Anim. Pract.* **19**, 401 (1978); C. J. Green, *Lab. Anim.* **15**, 171 (1981).

Hydrochloride. R-7315; Hypnodil. $C_{13}H_{14}N_2O_2 \cdot HCl$; mol wt 266.73. Crystals from methanol-ether, mp 173-174°. LD$_{50}$ i.v. in rats: 50 mg/kg (Godefroi).

THERAP CAT (VET): Hypnotic.

6226. Metopimazine. [14008-44-7] 1-[3-[2-(Methylsulfonyl)-10H-phenothiazin-10-yl]propyl]-4-piperidinecarboxamide; 1-[3-[2-(methylsulfonyl)phenothiazin-10-yl]propyl]isonipecotamide; 10-[3-(4-carbamoylpiperidino)propyl]-2-(methanesulfonyl)phenothiazine; 1-[3-[2-(methylsulfonyl)phenothiazin-10-yl]propyl]-4-piperidinecarboxamide; EXP-999; RP-9965; Vogalene. $C_{22}H_{27}N_3O_3S_2$; mol wt 445.60. C 59.30%, H 6.11%, N 9.43%, O 10.77%, S 14.39%. Prepn: Jacob, Robert, **DE 1092476** (to Rhône-Poulenc), *C.A.* **56**, 8723h (1962). Toxicity study: E. I. Goldenthal, *Toxicol. Appl. Pharmacol.* **18**, 185 (1971).

Solid, mp 170-171°. LD$_{50}$ in male rats (mg/kg): 976 orally, 1080 s.c. (Goldenthal).

THERAP CAT: Antiemetic.

6227. Metopon. [143-52-2] (5α)-4,5-Epoxy-3-hydroxy-5,17-dimethylmorphinan-6-one; methyldihydromorphinone. C$_{18}$H$_{21}$NO$_3$; mol wt 299.37. C 72.22%, H 7.07%, N 4.68%, O 16.03%. Semisynthetic opioid analgesic. Prepn from thebaine: L. Small *et al., J. Am. Chem. Soc.* **58**, 1457 (1936). From dihydrocodeinone enol acetate: L. Small *et al., J. Org. Chem.* **3**, 204 (1938). Structure: G. Stork, L. Bauer, *J. Am. Chem. Soc.* **75**, 4373 (1953). Opioid receptor binding study: J. P. McLaughlin *et al., Eur. J. Pharmacol.* **294**, 201 (1995).

Needles from alc; sinters 235°, mp 243-245° (evac tube). [α]$_D^{24}$ −140.7° (c = 1 in alc). Slightly sol in organic solvents.

Hydrochloride. [124-92-5] C$_{18}$H$_{21}$NO$_3$.HCl. Crystals from alc, mp 315-318° with decompn (evac tube). [α]$_D^{24}$ −104.8° (c = 1.002 in water). Freely sol in water. Sparingly sol in alc. Slightly sol in chloroform. Very slightly sol in ether. Insol in benzene. A 1% aq soln has a pH of ~5.0.

Note: This is a controlled substance (opiate): **21 CFR,** 1308.12.

6228. Metoprolol. [37350-58-6] 1-[4-(2-Methoxyethyl)-phenoxy]-3-[(1-methylethyl)amino]-2-propanol; (±)-1-(isopropyl-amino)-3-[*p*-(β-methoxyethyl)phenoxy]-2-propanol; CGP-2175; H-93/26. C$_{15}$H$_{25}$NO$_3$; mol wt 267.37. C 67.38%, H 9.43%, N 5.24%, O 17.95%. β$_1$-Adrenergic blocker lacking intrinsic sympathomimetic activity. Prepn: A. E. Brandstrom *et al.,* **DE 2106209**; *eidem,* **US 3998790** (1971, 1976 both to AB Hässle). Pharmacology: B. Ablad *et al., Life Sci.* **12**, 107 (1973); Johansson, *Eur. J. Pharmacol.* **24**, 194 (1973). Toxicology: N.-O. Bodin *et al., Acta Pharmacol. Toxicol.* **36**, Suppl. V, 96 (1975). HPLC determn in dog plasma: J. Fang *et al., J. Chromatogr. B* **809**, 9 (2004). Use in dogs with acquired cardiac disease: J. E. Rush *et al., J. Vet. Cardiol.* **4**, 23 (2002). Clinical trials in acute myocardial infarction: L. Rydén *et al., N. Engl. J. Med.* **308**, 614 (1983); *Drugs* **29**, Suppl. 1, 2-8 (1985); in chronic heart failure: A. Hjalmarson *et al., J. Am. Med. Assoc.* **283**, 1295 (2000). Comprehensive description: J. R. Luch, *Anal. Profiles Drug Subs.* **12**, 325-356 (1983). Review of pharmacology, therapeutic efficacy: P. Benfield *et al., Drugs* **31**, 376-429 (1986); of use in chronic heart failure: A. Prakash, A. Markham, *ibid.* **60**, 647-678 (2000).

Succinate. [98418-47-4] Selozok; Toprol-XL. (C$_{15}$H$_{25}$NO$_3$)$_2$.C$_4$H$_6$O$_4$; mol wt 652.83. White to off-white powder. Freely sol in

water; sol in methanol; sparingly sol in alcohol; slightly sol in isopropyl alcohol, dichloromethane. Practically insol in ethyl acetate, acetone, ether, heptane.

Tartrate. [56392-17-7] Beloc; Betaloc; Lopressor; Lopresor; Metropress; Prelis; Seloken; Selopral. (C$_{15}$H$_{25}$NO$_3$)$_2$.C$_4$H$_6$O$_6$; mol wt 684.82. White, crystalline powder. Soly (mg/ml) at 25°: water >1000; methanol >500; chloroform 496; acetone 1.1; acetonitrile 0.89; hexane 0.001. Freely sol in methylene chloride, alc. Insol in ether. uv max (water): 223 nm (ε 23400). LD$_{50}$ in female mice, male rats (mg/kg): 118, ~90 i.v.; 2090, 3090 orally (Bodin).

THERAP CAT: Antihypertensive; antianginal; antiarrhythmic (class II).

THERAP CAT (VET): Antihypertensive; antianginal; antiarrhythmic (class II).

6229. Metoquinone. [622-91-3] 1,4-Benzenediol compd with 4-(methylamino)phenol (1:2); hydroquinone compd with *p*-(methylamino)phenol (1:2). C$_{20}$H$_{24}$N$_2$O$_4$; mol wt 356.42. C 67.40%, H 6.79%, N 7.86%, O 17.96%. A salt-like compound of 1 mol hydroquinone and 2 mols *p*-monomethylaminophenol.

White crystals. mp ~135° with decompn and blackening. Sol in 100 parts water; less sol in alcohol, acetone; very slightly sol in benzene, chloroform, ether.

USE: As a photographic developer.

6230. Metrafenone. [220899-03-6] (3-Bromo-6-methoxy-2-methylphenyl)(2,3,4-trimethoxy-6-methylphenyl)methanone; 3-bromo-2′,3′,4′,6-tetramethoxy-2,6′-dimethylbenzophenone; 5-bromo-6,6′-dimethyl-2,2′,3′,4′-tetramethoxybenzophenone; AC-375839; BAS-560F; BAS-560-02F; Attenzo; Flexity; Vivando. C$_{19}$H$_{21}$BrO$_5$; mol wt 409.28. C 55.76%, H 5.17%, Br 19.52%, O 19.55%. Benzophenone fungicide against powdery mildew in cereals and grapes. Prepn: J. Curtze *et al.,* **EP 897904**; *eidem,* **US 5945567** (both 1999 to Am. Cyanamid). Mode of action study: K. S. Opalski *et al., Pest Manag. Sci.* **62**, 393 (2006); and biological activity: M. R. Schmitt *et al., ibid.* 383. Activity on powdery mildew: F. Felsenstein *et al., Gesunde Pflanzen* **62**, 29 (2010).

White crystals from diisopropyl ether, mp 99.2-100.8°. d^{20} 1.45. Log P (octanol/water) at 25°: 4.3 (pH 4). Vapor pressure at 20°: 1.15 × 10^{-6} mm Hg. Soly in water (mg/l): 0.474 (deionized); 0.552 (pH >5); 0.492 (pH 7); 0.457 (pH 9). Soly (g/l): acetone 403; acetonitrile 165; dichloromethane 1950; ethyl acetate 261; *n*-hexane 4.8; methanol 26.1; toluene 363.

USE: Agricultural fungicide.

6231. Metralindole. [54188-38-4] 2,4,5,6-Tetrahydro-9-methoxy-4-methyl-1*H*-3,4,6a-triazafluoranthene; 3-methyl-8-methoxy-3*H*-1,2,5,6-tetrahydropyrazino[1,2,3-*ab*]-β-carboline. C$_{15}$H$_{17}$N$_3$O; mol wt 255.32. C 70.56%, H 6.71%, N 16.46%, O 6.27%. Heterocyclic β-carboline derivative used as an antidepressant. Prepn: R. G. Glushkov *et al.,* **DE 2357320**; *eidem,* **US 4088647** (1974, 1978 both to All Union Khim.-Farm Instit. USSR); *eidem, Khim. Farm. Zh.* **16**, 1054 (1982), *C.A.* **98**, 53831b (1983). Psychotropic activity, toxicity: M. D. Mashkovsky *et al.,* **DE 2356091**; *eidem,* **US 3959470** (1974, 1976 both to All Union Khim.-Farm. Instit. USSR). Pharmacology: N. I. Andreeva, M. D. Mashkovsky, *Farmakol. Toksikol. (Moscow)* **43**, 133 (1980). Effect on the EEG of the brain cortex: L. F. Roshchina, *ibid.* 349. Effect on adrenergic neurotransmission: L. V. Panasiuk *et al.*, *ibid.* **45**, 13 (1982). Clinical studies: M. D. Mashkovsky, N. I. Andreeva, *Zh. Nevropatol. Psikhiatr.* **84**, 410 (1984).

Crystals, mp 164-165°.

Hydrochloride. [53734-79-5] Incazan. $C_{15}H_{17}N_3O.HCl$; mol wt 291.78. Off-white odorless powder with bitter taste. mp 305-308°. Sol in water. LD_{50} orally in mice: 445 mg/kg (Mashkovsky, 1976).

THERAP CAT: Antidepressant.

6232. Metreleptin. [186018-45-1] *N*-Methionylleptin (human); r-metHuLeptin. Recombinant human form of the adipocyte hormone, leptin, *q.v.* Prepn: J. M. Friedman *et al.*, **GB 2292382**; *eidem*, **US 6001968** (1996, 1999 both to Rockefeller Univ.). Clinical pharmacokinetics: J. L. Chan *et al.*, *J. Clin. Endocrinol. Metab.* **92**, 2307 (2007). Clinical evaluation in patients with lipodystrophy: E. D. Javor *et al.*, *Diabetes* **54**, 1994 (2005); of combination with pramlintide in obesity: E. Ravussin *et al.*, *Obesity* **17**, 1736 (2009).

THERAP CAT: Leptin replacement therapy in treatment of lipodystrophy.

6233. Metribuzin. [21087-64-9] 4-Amino-6-(1,1-dimethylethyl)-3-(methylthio)-1,2,4-triazin-5(4*H*)-one; 4-amino-6-*tert*-butyl-3-(methylthio)-*as*-triazin-5(4*H*)-one; Bay 94337; Lexone; Sencor; Sencoral. $C_8H_{14}N_4OS$; mol wt 214.29. C 44.84%, H 6.59%, N 26.15%, O 7.47%, S 14.96%. Pre- and post-emergence triazone herbicide. Prepn: **BE 697083**; K. Westphal *et al.*, **US 3671523** (1967, 1972 both to Bayer AG). Physicochemical properties, toxicology and herbicidal activity: L. Eue, *Pflanzenschutz-Nachr.* **25**, 175 (1972). Toxicity data: E. Loeser, G. Kimmerle, *ibid.* 186. HPLC determn in plant tissues: C. E. Parker, G. H. Degen, *J. Liq. Chromatogr.* **6**, 725 (1983).

White crystalline solid, mp 125-126.5°. d_4^{20} 1.28. Vapor pressure at 20°: $<10^{-5}$ mm Hg. Sol in methanol, ethanol. Soly in water: 1200 ppm. LD_{50} orally in rats: 2200 mg/kg; LC_{50} in rainbow trout: >10 ppm (Loeser, Kimmerle).

Caution: Potential symptoms of overexposure in exptl animals are CNS depression; thyroid, liver and enzyme changes. *See NIOSH Pocket Guide to Chemical Hazards* (DHHS/NIOSH 97-140, 1997) p 216.

USE: Herbicide.

6234. Metrizamide. [31112-62-6] 2-[[3-(Acetylamino)-5-(acetylmethylamino)-2,4,6-triiodobenzoyl]amino]-2-deoxy-D-glucose; 2-[3-acetamido-2,4,6-triiodo-5-(*N*-methylacetamido)benzamido]-2-deoxy-D-glucose; 2-[3-acetamido-2,4,6-triiodo-5-(*N*-methylacetamido)benzamido]-2-deoxy-D-glucopyranose; Win-39103; Amipaque. $C_{18}H_{22}I_3N_3O_8$; mol wt 789.10. C 27.40%, H 2.81%, I 48.25%, N 5.33%, O 16.22%. X-ray contrast agent related to metrizoic acid, *q.v.* Prepn: H. O. Torsten *et al.*, **DE 2031724**; *eidem*, **US 3701771** (1971, 1972 both to Nyegaard). Metabolism: J. G. Johansen, S. Kolmannskog, *Invest. Radiol.* **13**, 93 (1978). Pharmacology: T. W. Morris *et al.*, *ibid.* 74; G. F. Dibona, *Proc. Soc. Exp. Biol. Med.* **157**, 453 (1978). Series of articles on clinical use: *Ann. Radiol.* **22**, 195-206 (1979). Toxicity studies: S. Salvesen, *Radiology* **123**, 241 (1977); M. Sovak *et al.*, *ibid.* 242; P. Aspelin, *Invest. Radiol.* **11**, 309 (1976). *Review:* L. Hol *et al.*, in *Pharmacological and Biochemical Properties of Drug Substances* **vol. 1**, M. E. Goldberg, Ed. (Am. Pharm. Assoc., Washington, DC, 1977) pp 387-412.

White crystals from isopropyl alcohol, mp 230° (dec), after drying in vacuo at room temp and 70°. $[\alpha]_D^{20}$ +18.0° (c = 10% in 0.1*N* HCl; equilibrated with respect to mutarotation). Very sol in water (50% w/v) at room temp. LD_{50} i.v. in mice: 15 g/kg (Torsten); 18.6 g/kg (Salveson); 11.5 g/kg (Sovak); 17.3 g/kg (Aspelin).

THERAP CAT: Diagnostic aid (radiopaque medium).

6235. Metrizoic Acid. [1949-45-7] 3-(Acetylamino)-5-(acetylmethylamino)-2,4,6-triiodobenzoic acid; 3-acetamido-2,4,6-triiodo-5-(*N*-methylacetamido)benzoic acid; *N*-methyl-3,5-diacetamido-2,4,6-triiodobenzoic acid. $C_{12}H_{11}I_3N_2O_4$; mol wt 627.94. C 22.95%, H 1.77%, I 60.63%, N 4.46%, O 10.19%. Prepn: Pitré, Fumigalli, *Farmaco Ed. Sci.* **17**, 340 (1962); **GB 973881** (1964 to Nyegaard), *C.A.* **62**, 16139g (1965). Series of articles on use in separation of leukocytes from blood and bone marrow: A. Boyum, *Scand. J. Clin. Lab. Invest.* **21**, Suppl. 97, 7-106 (1968).

Crystals, mp 281-282°.

Sodium salt. [7225-61-8] Metrizoate sodium; Isopaque; Triosil. THERAP CAT: Diagnostic aid (radiopaque medium).

6236. Metronidazole. [443-48-1] 2-Methyl-5-nitro-1*H*-imidazole-1-ethanol; 1-(2-hydroxyethyl)-2-methyl-5-nitroimidazole; 1-(β-ethylol)-2-methyl-5-nitro-3-azapyrrole; Bayer 5630; RP-8823; Acea Gel; Anabact; Arilin; Clont; Deflamon; Elyzol; Flagyl; Fossyol; Klion; MetroGel; Metrolag; Metrolyl; Metrosa; Metrotop; Orvagil; Rathimed; Rosaced; Rozacrème; Rozagel; Rozex; Trichocide; Vagilen; Vagimid; Zadstat; Zidoval. $C_6H_9N_3O_3$; mol wt 171.16. C 42.10%, H 5.30%, N 24.55%, O 28.04%. Prepn: Jacob *et al.*, **US 2944061** (1960 to Rhône-Poulenc); Cossar *et al.*, *Arzneim.-Forsch.* **16**, 23 (1966). Activity studies: Bock, *ibid.* **11**, 587 (1961). Metabolism: Ings *et al.*, *Biochem. Pharmacol.* **15**, 515 (1966). Comprehensive description: L. L. Wearley, G. D. Anthony, *Anal. Profiles Drug Subs.* **5**, 327-344 (1976). Review of activity, pharmacokinetics and use in anaerobic infection: R. N. Brogden *et al.*, *Drugs* **16**, 387-417 (1978); in rosacea: L. K. Schmadel, G. K. McEvoy, *Clin. Pharm.* **9**, 94-101 (1990). *Review:* W. D. Hager, R. P. Rapp, *Obstet. Gynecol. Clin. North Am.* **19**, 497-510 (1992).

Cream-colored crystals, mp 158-160°. Soly at 20° (g/100 ml): water 1.0; ethanol 0.5; ether <0.05; chloroform <0.05. Sol in dil acids; sparingly sol in DMF. pH of satd aq soln: 5.8.

Hydrochloride. [69198-10-3] SC-326421. $C_6H_9N_3O_3.HCl$; mol wt 207.61.

Caution: Metronidazole is reasonably anticipated to be a human carcinogen: *Report on Carcinogens, Twelfth Edition* (PB2011-111646, 2011) p 269.

THERAP CAT: Antiprotozoal (Trichomonas); antibacterial; in treatment of rosacea.

THERAP CAT (VET): Antiprotozoal (Trichomonas); antiamebic; antibacterial.

6237. Metsulfuron-methyl. [74223-64-6] 2-[[[[(4-Methoxy-6-methyl-1,3,5-triazin-2-yl)amino]carbonyl]amino]sulfonyl]benzoic acid methyl ester; *N*-[(4-methoxy-6-methyl-1,3,5-triazin-2-yl)aminocarbonyl]-2-methoxycarbonylbenzenesulfonamide; DPX-T6376; Ally; Allie. $C_{14}H_{15}N_5O_6S$; mol wt 381.36. C 44.09%, H 3.96%, N 18.36%, O 25.17%, S 8.41%. Sulfonyl urea used as pre- and post-emergence herbicide. Prepn and herbicidal activity: G. Levitt, **EP 7687** (1980 to Du Pont); *idem*, **US 4383113** (1983 to Du Pont). Comparison with other herbicides in cereals: A. Aamisepp, *Weeds Weed Control* **26**, 26 (1985). Efficacy in control of gorse: A. I. Popay *et al.*, *Proc. 38th N. Z. Weed Pest Control Conf.* **1985**, 94. Review of properties, herbicidal activity: P. Lefebvre, *Def. Veg.* **38**, 331 (1984).

White crystals, mp 163-166°. Vapor pressure at 25°: 5.8×10^{-5} mm Hg. Soly in water: 0.27 g/l at pH 4.59; 9.5 g/l at pH 6.11. LD_{50} in male and female rats, in male and female rabbits (mg/kg): >5000, >2000 orally. LC_{50} (96 hour) in American perch, rainbow trout: 150 ppm. LD_{50} in wild ducks: 2510 mg/kg (Lefebvre).

USE: Herbicide.

6238. Metyrapone. [54-36-4] 2-Methyl-1,2-di-3-pyridyl-1-propanone; 2-methyl-1,2-bis(3-pyridyl)-1-propanone; methopyrapone; mepyrapone; metopyrone; methbipyranone; Su-4885; Metopiron(e). $C_{14}H_{14}N_2O$; mol wt 226.28. C 74.31%, H 6.24%, N 12.38%, O 7.07%. Prepn: Chart *et al.*, *Experientia* **14**, 151 (1958); W. L. Bencze, M. J. Allen, *J. Am. Chem. Soc.* **81**, 4015 (1959); *eidem*, **US 2966493** (1960 to Ciba).

Crystals from ether and pentane; darkens on exposure to light. mp 50-51°. Sol in methanol, chloroform; sparingly sol in water. Forms water-sol salts with acids.

THERAP CAT: Diagnostic aid (pituitary function).

6239. Metyridine. [114-91-0] 2-(2-Methoxyethyl)pyridine; methyridine; Dekelmin; Promintic. $C_8H_{11}NO$; mol wt 137.18. C 70.05%, H 8.08%, N 10.21%, O 11.66%. Prepn: Arnall, Greenhalgh, **GB 889748** (1962 to ICI and Midland Tar Distillers); **US 3223710** (1965).

Sweet smelling liquid, bp_{17} 94-96°. d^{20} 0.988. n_D^{20} 1.4975. pKa = 5.5. Very sol in water and common solvents.

Sulfate. Mintic.

Hydrochloride. mp 104-105°.

THERAP CAT (VET): Anthelmintic.

6240. Metyrosine. [672-87-7] α-Methyl-L-tyrosine; α-methyl-*p*-tyrosine; α-methyltyrosine; 4-hydroxy-α-methylphenylalanine; α-methyl-3-(*p*-hydroxyphenyl)alanine; metirosine; L-α-MT; α-MPT; MK-781; Demser. $C_{10}H_{13}NO_3$; mol wt 195.22. C 61.53%, H 6.71%, N 7.17%, O 24.59%. Inhibitor of tyrosine hydroxylase, the enzyme catalyzing the first reaction in catecholamine biosynthesis (the hydroxylation of tyrosine to dopa). Prepn: **NL 6607757** (1966 to Merck & Co.), *C.A.* **67**, 91108p (1967). Prepn of DL-form: Stein *et al.*, *J. Am. Chem. Soc.* **77**, 700 (1955); Potts, *J. Chem. Soc.* **1955**, 1632; Pfister, Stein, **US 2868818** (1959 to Merck & Co.); Saari, *J. Org. Chem.* **32**, 4074 (1967). HPLC determn in serum: M. M. Hefnawy, J. T. Stewart, *J. Liq. Chromatogr. Relat. Technol.* **20**, 3009 (1997). Chiral resolution by CE and micellar electrokinetic capillary chromatography: *idem*, *ibid.* **28**, 439 (2005). Clinical pharmacology and metabolism: Engelman *et al.*, *J. Clin. Invest.* **47**, 568, 577 (1968). Review of pharmacology and clinical use: R. N. Brogden *et al.*, *Drugs* **21**, 81-89 (1981); of use prior to surgical resection of pheochromocytoma: R. R. Perry *et al.*, *Ann. Surg.* **212**, 621-628 (1990).

White crystals, mp 310-315°. Sol in acidic aq solns and alkaline aq soln, although subject to oxidative degradation in the latter; very slightly sol in water, acetone, and methanol. Insol in chloroform and benzene.

DL-Form. [672-87-7] Crystals from water, dec 320° (Stein); also reported as dec 330-332° (Potts). Soly in water at room temp: 0.57 mg/ml.

THERAP CAT: Antihypertensive in pheochromocytoma.

6241. Mevaldic Acid. [541-07-1] 3-Hydroxy-3-methyl-5-oxopentanoic acid; 3-hydroxy-3-methylglutaraldehydic acid. $C_6H_{10}O_4$; mol wt 146.14. C 49.31%, H 6.90%, O 43.79%. Prepn by acid hydrolysis of 3-hydroxy-3-methyl-5,5-dimethoxypentanoic acid: Shunk *et al.*, *J. Am. Chem. Soc.* **79**, 3294 (1957). Exists in equilibrium with *β,δ-dihydroxy-β-methyl-δ-valerolactone*.

Unstable and reactive compd. Isolated in soln only.

6242. Mevalonic Acid. [150-97-0] 3,5-Dihydroxy-3-methylpentanoic acid; 3,5-dihydroxy-3-methylvaleric acid; β,δ-dihydroxy-β-methylvaleric acid; hiochic acid. $C_6H_{12}O_4$; mol wt 148.16. C 48.64%, H 8.16%, O 43.19%. Precursor in the biosynthesis of cholesterol. Occurs in equilibrium with the δ-lactone. Isoln from distillers' solns: Wright *et al.*, *J. Am. Chem. Soc.* **78**, 5273 (1956). Synthesis: Wolf *et al.*, *ibid.* **79**, 1486 (1957); Hoffmann *et al.*, *ibid.* 2316; Eggerer *et al.*, *Ann.* **608**, 71 (1957); **US 2915398**; **US 2915531**; **US 2915532**; **US 2915533**; **US 2915551** (all 1959 to Merck & Co.); Shunk *et al.*, **US 3014963** (1961 to Merck & Co.); Hulcher, Hosick, **US 3119842** (1964); F. C. Huang *et al.*, *J. Am. Chem. Soc.* **97**, 4144 (1975). Resolution: Shunk *et al.*, *ibid.* **79**, 3294 (1957); **US 2945059** (1960 to Merck & Co.). Synthesis of the δ-lactone: R. H. Cornforth *et al.*, *Tetrahedron* **18**, 1351 (1962); E. L. Eliel, K. Soai, *Tetrahedron Lett.* **22**, 2859 (1981); A. Banerji, G. P. Kalena, *Synth. Commun.* **12**, 225 (1982). Review of organic and biochemistry: Wagner, Folkers, *Endeavour* **20**, 177-187 (Oct. 1961). Prepn of 5-phosphate: Robinson, Wittreich, **US 3014057**

(1961 to Merck & Co.). Review of role of mevalonic acid in sterol biosynthesis: G. J. Schroepfer, Jr., *Annu. Rev. Biochem.* **50**, 585-621 (1981); E. Caspi, *Tetrahedron* **42**, 3-50 (1986).

Oily liquid. Very sol in water, but also sol in organic solvents, especially polar organic solvents.

δ-Lactone. Mevalolactone; mevalonic lactone; β-hydroxy-β-methyl-δ-valerolactone. $C_6H_{10}O_3$. Hygroscopic crystals, mp 28°.

Benzhydrylamine. $C_{19}H_{23}NO_3$. Crystals from benzene + petr ether, mp 96-97°. $[\alpha]_D^{20}$ −2.0° (c = 2 in ethanol).

6243. Mevastatin. [73573-88-3] (2*S*)-2-Methylbutanoic acid (1*S*,7*S*,8*S*,8a*R*)-1,2,3,7,8,8a-hexahydro-7-methyl-8-[2-[(2*R*,4*R*)-tetrahydro-4-hydroxy-6-oxo-2*H*-pyran-2-yl]ethyl]-1-naphthalenyl ester; 7-[1,2,6,7,8,8a-hexahydro-2-methyl-8-(methylbutyryloxy)-naphthyl]-3-hydroxyheptan-5-olide; 2β-methyl-8α-(2-methyl-1-oxobutoxy)mevinic acid lactone; compactin; 6-demethylmevinolin; CS-500; ML-236 B. $C_{23}H_{34}O_5$; mol wt 390.52. C 70.74%, H 8.78%, O 20.48%. Fungal metabolite which is a potent inhibitor of HMG-CoA reductase, the rate controlling enzyme in cholesterol biosynthesis. Isoln from *Penicillium citrinum*: A. Endo *et al.*, **DE 2524355** corresp to **US 3983140** (1975, 1976 to Sankyo). Isoln from *P. brevicompactum*, crystal and molecular structure: A. G. Brown *et al.*, *J. Chem. Soc. Perkin Trans. 1* **1976**, 1165. Inhibition of HMG-CoA reductase activity: A. Endo *et al.*, *FEBS Lett.* **72**, 323 (1976); M. S. Brown *et al.*, *J. Biol. Chem.* **253**, 1121 (1978). Therapeutic effects in primary hypercholesterolemia: A. Yamamoto *et al.*, *Atherosclerosis* **35**, 259 (1980). Total synthesis: N. Y. Wang *et al.*, *J. Am. Chem. Soc.* **103**, 6538 (1981); M. Hirama, M. Uei, *ibid.* **104**, 4251 (1982); N. N. Girotra, N. L. Wendler, *Tetrahedron Lett.* **23**, 5501 (1982); C.-T. Hsu *et al.*, *J. Am. Chem. Soc.* **105**, 593 (1983); P. A. Grieco *et al.*, *ibid.* 1403; D. L. J. Clive *et al.*, *J. Am. Chem. Soc.* **110**, 6914 (1988). Review of syntheses: T. Rosen, C. H. Heathcock, *Tetrahedron* **42**, 4909-4951 (1986). Review of mevastatin and related compounds: A. Endo, *J. Med. Chem.* **28**, 401-405 (1985).

Crystals from aq ethanol, mp 152°. $[\alpha]_D^{22}$ +283° (c = 0.48 in acetone). uv max: 230, 237, 246 nm (log ε 4.28, 4.30, 4.11).

6244. Mevinphos. [7786-34-7] 3-[(Dimethoxyphosphinyl)-oxy]-2-butenoic acid methyl ester; 3-hydroxycrotonic acid methyl ester dimethyl phosphate; 1-methoxycarbonyl-1-propen-2-yl dimethyl phosphate; methyl 3-(dimethoxyphosphinyloxy)crotonate; *O*,*O*-dimethyl 1-carbomethoxy-1-propen-2-yl phosphate; 2-carbomethoxy-1-methylvinyl dimethyl phosphate; ENT-22374; OS-2046; Phosdrin. $C_7H_{13}O_6P$; mol wt 224.15. C 37.51%, H 5.85%, O 42.83%, P 13.82%. Organophosphate insecticide; cholinesterase inhibitor. Commercial product is a mixture of isomers, containing ~60% of the more active (*E*)- (also referred to as *cis*-) isomer. Prepn: A. R. Stiles, **US 2685552** (1954 to Shell). Activity: R. A. Corey *et al.*, *J. Econ. Entomol.* **46**, 386 (1953); of isomers: J. E. Casida *et al.*, *J. Agric. Food Chem.* **4**, 236 (1956). Configuration of isomers: T. R. Fukuto *et al.*, *J. Org. Chem.* **26**, 4620 (1961). Metabolism and degradn: K. I. Beynon *et al.*, *Residue Rev.* **47**, 55 (1973). Toxicity study: T. B. Gaines, *Toxicol. Appl. Pharmacol.* **14**, 515 (1969). GC-MS determn in animal tissue: W. J. Allender, J. Keegan, *J. Chromatogr.* **609**, 315 (1992).

(E)-isomer

Yellow liquid. d_4^{20} 1.25. n_D^{20} 1.4494. bp$_1$ 106-107.5°. *Poisonous.* Misc with water, acetone, benzene, carbon tetrachloride, chloroform, ethyl and isopropyl alcohols, toluene and xylene. One gram dissolves in 20 ml carbon disulfide and 20 ml kerosene. Practically insol in hexane. LD_{50} in female, male rats (mg/kg): 3.7, 6.1 orally; 4.2, 4.7 dermally (Gaines).

Caution: Potential symptoms of overexposure are miosis; rhinorrhea; headache; tight chest, wheezing, laryngeal spasm, salivation and cyanosis; anorexia, nausea, vomiting, abdominal cramps and diarrhea; paralysis; ataxia, convulsions; low blood pressure, cardiac irregularities; irritation of skin, eyes and respiratory system. *See NIOSH Pocket Guide to Chemical Hazards* (DHHS/NIOSH 97-140, 1997) p 252.

USE: Insecticide.

6245. Mexazolam. [31868-18-5] 10-Chloro-11b-(2-chloro-phenyl)-2,3,7,11b-tetrahydro-3-methyloxazolo[3,2-*d*][1,4]benzo-diazepin-6(5*H*)-one; 10-chloro-11b-(*o*-chlorophenyl)-2,3,7,11b-tetrahydro-3-methyloxazolo[3,2-*d*][1,4]benzodiazepin-6(5*H*)-one; CS-386; Melex; Melium; Sedoxil. $C_{18}H_{16}Cl_2N_2O_2$; mol wt 363.24. C 59.52%, H 4.44%, Cl 19.52%, N 7.71%, O 8.81%. Prepn from 2-amino-2′,5-dichlorobenzophenone: R. Tachikawa *et al.*, **DE 1812252**; *eidem*, **US 3772371** (1969, 1973 to Sankyo). Synthesis via Schiff base, physical properties: T. Miyadera *et al.*, *J. Med. Chem.* **14**, 520 (1971). Acid-base equilibrium: M. Ikeda, T. Nagai, *Chem. Pharm. Bull.* **30**, 3810 (1982). Effect on socially induced suppression in monkeys: T. Kamioka *et al.*, *Psychopharmacology* **52**, 17 (1977). Toxicity studies: H. Masuda *et al.*, *Sankyo Kenkyu-sho Nempo* **30**, 175 (1978), *C.A.* **90**, 180217d. Clinical effects on anxiety and pyschomotor performance in generalised anxiety disorder: L. Ferreira *et al.*, *Clin. Drug Invest.* **23**, 235 (2003).

Crystals, mp 172-175°. pKa 6.69. LD_{50} in male, female mice; male, female rats (mg/kg): 4687, 4571, 810, 4500 orally; >6000, >6000, >4000, >4000 i.p. or s.c. (Masuda).

THERAP CAT: Anxiolytic.

6246. Mexenone. [1641-17-4] (2-Hydroxy-4-methoxyphen-yl)(4-methylphenyl)methanone; 2-hydroxy-4-methoxy-4′-methyl-benzophenone; benzophenone-10; Uvistat. $C_{15}H_{14}O_3$; mol wt 242.27. C 74.37%, H 5.82%, O 19.81%. Prepn: Hardy, Forster, **US 2773903** (1956 to Am. Cyanamid). Reversed phase HPLC determn: L.-H. Wang, *Chromatographia* **50**, 565 (1999). Clinical evaluation of skin protection efficacy: T. M. MacLeod, W. Frain-Bell, *Br. J. Dermatol.* **84**, 266 (1971).

THERAP CAT: Ultraviolet screen.

6247. Mexicain. [52500-59-1] A cysteine proteinase obtained from the fruit of *Pileus mexicanus* probably identical with *Leucopremna mexicana* (A.DC.) Standl., *Caricaceae:* Castaneda-Agullo et al., *Science* **96**, 365 (1942); *J. Biol. Chem.* **159**, 751 (1945). Immunological studies: Estrada-Parra *et al.*, *Rev. Latinoam. Microbiol. Parasitol.* **1969**, 145, *C.A.* **72**, 1744u (1970). Mol wt about 31,100; lysine and alanine believed to be the *N*- and *C*-terminal amino acids resp: Soriano *et al.*, *Rev. Latinoam. Quim.* **1975**, 143, *C.A.* **84**, 56489t (1976).

Similar to but not identical with papain. Mexicain presents some advantage over papain, since it is more stable in solns and does not require cysteine. Isoelectric pt 9.12.

6248. Mexiletine. [31828-71-4] 1-(2,6-Dimethylphenoxy)-2-propanamine; 1-(2,6-xylyloxy)-2-propylamine; 1-(2′,6′-dimethylphenoxy)-2-aminopropane; 1-methyl-2-(2,6-xylyloxy)ethylamine. $C_{11}H_{17}NO$; mol wt 179.26. C 73.70%, H 9.56%, N 7.81%, O 8.93%. Orally active class Ib antiarrhythmic agent; blocks the voltage-dependent fast sodium channel. Prepn: **FR 1551055**; H. Köppe *et al.*, **US 3954872** (1968, 1976 both to Boehringer, Ing.). Stereospecific synthesis: A. Carocci *et al.*, *Tetrahedron: Asymmetry* **11**, 3619 (2000). Pharmacology: B. N. Singh, E. M. Vaughan Williams, *Br. J. Pharmacol.* **44**, 1 (1972); J. D. Allen *et al.*, *ibid.* **45**, 561 (1972). Characterization of metabolites by mass spec and ^{13}C-NMR: K. N. Scott *et al.*, *Drug Metab. Dispos.* **1**, 506 (1973). Toxicological study: A. Kast *et al.*, *Iyakuhin Kenkyu* **13**, 922 (1982), *C.A.* **97**, 192878q (1982). GC-MS determn in serum: A. Dasgupta, O. Yousef, *J. Chromatogr. B* **705**, 283 (1998). Comprehensive description: M. A. Abounassif *et al.*, *Anal. Profiles Drug Subs.* **20**, 433-474 (1991). Review of pharmacology and clinical efficacy in arrhythmia: C. Y. C. Chew *et al.*, *Drugs* **17**, 161-181 (1979); in diabetic neuropathy: B. Jarvis, A. J. Coukell, *ibid.* **56**, 691-707 (1998). Review of clinical pharmacokinetics: L. Labbé, J. Turgeon, *Clin. Pharmacokinet.* **37**, 361-384 (1999).

Hydrochloride. [5370-01-4] Ko-1173; Mexitil; Ritalmex. $C_{11}H_{17}NO.HCl$; mol wt 215.72. White to off-white crystals from ethanol-ether, mp 203-205°. pKa: 9.0. Slightly bitter taste. Freely sol in water, dehydrated alcohol; slightly sol in acetonitrile. Practically insol in ether. LD_{50} in male, female rats, mice, rabbits (mg/kg): 350, 400, 310, 400, 180, 160 orally; in male, female rats, mice (mg/kg): 27, 30, 43, 50 i.v. (Kast).

THERAP CAT: Antiarrhythmic (class IB).

THERAP CAT (VET): Antiarrhythmic (class IB).

6249. Mezereum. Mezereon; olive spurge; dwarf bay; magell; paradise plant; spurge flax; wild pepper. Dried bark (also seed, though not official) from aerial portions of *Daphne mezereum* L., *D. gnidium* L., and *D. laureola* L., Thymelaeaceae. *Habit.* New England, Canada, mountainous Europe, Siberia. *Constit.* Bark: Mezerein and acrid resin, daphnin, umbelliferone, fixed oil. Seed: Fixed oil, acrid resin.

THERAP CAT: Vesicant.

6250. Mezlocillin. [51481-65-3] (2S,5R,6R)-3,3-Dimethyl-6-[[(2R)-2-[[[3-(methylsulfonyl)-2-oxo-1-imidazolidinyl]carbonyl]amino]-2-phenylacetyl]amino]-7-oxo-4-thia-1-azabicyclo[3.2.0]-heptane-2-carboxylic acid; 6R-[2-[3-(methylsulfonyl)-2-oxo-1-imidazolidine carboxamido]-2-phenylacetamido]penicillanic acid; Bay f 1353. $C_{21}H_{25}N_5O_8S_2$; mol wt 539.58. C 46.75%, H 4.67%, N 12.98%, O 23.72%, S 11.88%. Semisynthetic, broad-spectrum antibiotic related to penicillins and azlocillin, *q.v.* Prepn: H. B. König *et al.*, **DE 2152967**; *eidem*, **US 3974142** (1973, 1976 both to Bayer). *In vitro* study: G. P. Bodey, T. Pan, *Antimicrob. Agents Chemother.* **11**, 74 (1977). Pharmacokinetics: H. Lode *et al.*, *Infection* **5**, 163 (1977); S. J. Pancoast, H. C. Neu, *Clin. Pharmacol. Ther.* **24**, 108 (1978). Clinical pharmacology: B. F. Issell *et al.*, *Antimicrob. Agents Chemother.* **13**, 180 (1978). Tissue concentration and efficacy: E. Helwing *et al.*, *Med. Klin.* **74**, 112 (1979). Series of articles on antibacterial activity, pharmacology, and clinical trials: *Arzneim.-Forsch.* **29**, 1915-2032 (1979); *Infection* **10**, Suppl. 3, S121-S266 (1982); *J. Antimicrob. Chemother.* **11**, Suppl. C, 1-108 (1983).

Sodium salt monohydrate. [59798-30-0] Baycipen; Baypen; Mezlin. $C_{21}H_{24}N_5NaO_5S_2.H_2O$; mol wt 531.58. Pale yellow cryst; sol in water, methanol, DMF. Insol in acetone and ethanol.

THERAP CAT: Antibacterial.

6251. Mianserin. [24219-97-4] 1,2,3,4,10,14b-Hexahydro-2-methyldibenzo[c,f]pyrazino[1,2-a]azepine; 2-methyl-1,2,3,4,10,-14b-hexahydro-2H-pyrazino[1,2-f]morphanthridine. $C_{18}H_{20}N_2$; mol wt 264.37. C 81.78%, H 7.63%, N 10.60%. Serotonin receptor antagonist. Prepn: **NL 6603256**; van der Burg, Delobelle, **US 3534041** (1967, 1970, both to Organon); van der Burg *et al.*, *J. Med. Chem.* **13**, 35 (1970). Pharmacology: Saxena *et al.*, *Eur. J. Pharmacol.* **13**, 295 (1971); Vargaftig *et al.*, *ibid.* **16**, 336 (1971); Van Riezen, *Arch. Int. Pharmacodyn. Ther.* **198**, 256 (1972). Molecular and crystal structure: van Rij, Feil, *Tetrahedron* **29**, 1891 (1973). Review of pharmacology and therapeutic efficacy in depressive illness: R. N. Brogden *et al.*, *Drugs* **16**, 273-301 (1978). 5HT-receptor binding study: B. S. Alexander, M. D. Wood, *J. Pharm. Pharmacol.* **39**, 664 (1987). *Review:* H. van Riezen *et al.*, in *Pharmacological and Biochemical Properties of Drug Substances* vol. 3, M. E. Goldberg, Ed. (Am. Pharm. Assoc., Washington, DC, 1981) pp 56-93; *Br. J. Clin. Pharmacol.* **15**, Suppl. 2, S141-S342 (1983).

Hydrochloride. [21535-47-7] GB-94; Org-GB-94; Athymil; Bolvidon; Lantanon; Norval; Tetramide; Tolvin; Tolvon. $C_{18}H_{20}N_2.HCl$; mol wt 300.83. mp 282-284°. LD_{50} in male, female mice (mg/kg): 365, 390 orally; 32.5, 31.0 i.v. (van Riezen).

THERAP CAT: Antidepressant.

6252. Mibefradil. [116644-53-2] Methoxyacetic acid (1S,-2S)-2-[2-[[3-(1H-benzimidazol-2-yl)propyl]methylamino]ethyl]-6-fluoro-1,2,3,4-tetrahydro-1-(1-methylethyl)-2-naphthalenyl ester; (1S,2S)-2-[2-[[3-(2-benzimidazolyl)propyl]methylamino]ethyl]-6-fluoro-1,2,3,4-tetrahydro-1-isopropyl-2-naphthyl methoxyacetate. $C_{29}H_{38}FN_3O_3$; mol wt 495.64. C 70.28%, H 7.73%, F 3.83%, N 8.48%, O 9.68%. Selective T-type calcium channel blocker. Prepn: Q. Branca *et al.*, **EP 268148**; *eidem*, **US 4808605** (1988, 1989 both to Hoffmann-La Roche). Calcium channel selectivity: S. K. Mishra, K. Hermsmeyer, *Circ. Res.* **75**, 144 (1994). Metabolism in rats: H. R. Wiltshire *et al.*, *Xenobiotica* **22**, 837 (1992). Review of pharmacology, toxicology and clinical evaluation: J.-P. Clozel *et al.*, *Cardiovasc. Drug Rev.* **9**, 4-17 (1991); of pharmacology and pharmacokinetics: D. R. Abernethy, *Am. J. Cardiol.* **80**, 4C-11C (1997); of clinical trials in hypertension: S. Oparil, *Am. J. Hypertens.* **11**, 88S-94S (1998); in chronic stable angina pectoris: B. M. Massie, *ibid.* 95S-102S. Post-marketing safety assessment: J. Riley *et al.*, *Int. J. Clin. Pharmacol. Ther.* **40**, 241 (2002).

Dihydrochloride. [116666-63-8] Ro-40-5967; Posicor. C_{29}-$H_{38}FN_3O_3.2HCl$; mol wt 568.56. Bitter tasting, odorless, white crystalline powder, mp 128°. Chemically stable, light insensitive and water soluble. pKa: 4.8; 5.5. LD_{50} in mice, rats (mg/kg): 35, 23 i.v.; >800, >800 orally (Clozel).

THERAP CAT: Antihypertensive.

6253. Mibolerone. [3704-09-4] $(7\alpha,17\beta)$-17-Hydroxy-7,17-dimethylestr-4-en-3-one; $7\alpha,17\alpha$-dimethyl-19-nortestosterone; U-10997; Cheque. $C_{20}H_{30}O_2$; mol wt 302.46. C 79.42%, H 10.00%, O 10.58%. Synthetic anabolic steroid related to testosterone, *q.v.* Prepn: **BE 610385**; J. C. Babcock, J. A. Campbell, **US 3341557** (1962, 1967 both to Upjohn). Biological properties: J. A. Campbell *et al., Steroids* **1**, 317 (1963). Dissolution and solubility rates: W. E. Hamlin *et al., J. Pharm. Sci.* **54**, 1651 (1965). Pharmacodynamics: J. H. Sokolowski, C. W. Kasson, *Am. J. Vet. Res.* **39**, 837 (1978). Efficacy evaluation in dogs: J. H. Sokolowski, *J. Am. Vet. Med. Assoc.* **173**, 983 (1978).

Cryst solid. Soly in deionized water: 0.0454 mg/ml at 37°.

Note: This is a controlled substance (anabolic steroid): **21 CFR,** 1308.13, as defined in 1300.01.

THERAP CAT (VET): Antigonadotropic used to suppress estrus.

6254. Micafungin. [235114-32-6] 1-[(4R,5R)-4,5-Dihydroxy-N^2-[4-[5-[4-(pentyloxy)phenyl]-3-isoxazolyl]benzoyl]-L-ornithine]-4-[(4S)-4-hydroxy-4-[4-hydroxy-3-(sulfooxy)phenyl]-L-threonine]pneumocandin A0. $C_{56}H_{71}N_9O_{23}S$; mol wt 1270.28. C 52.95%, H 5.63%, N 9.92%, O 28.97%, S 2.52%. Semisynthetic echinocandin antifungal; inhibits 1,3-β-glucan synthase. Prepn (stereochem unspec): H. Ohki *et al.*, **WO 9611210**; *eidem*, **US 6107458** (1996, 2000 both to Fujisawa). Synthesis and antifungal activity: M. Tomishima *et al., J. Antibiot.* **52**, 674 (1999). Antifungal spectrum *in vitro*: S. Tawara *et al., Antimicrob. Agents Chemother.* **44**, 57 (2000). Pharmacokinetics and antifungal efficacy: V. Petraitis *et al., ibid.* **46**, 1857 (2002). Review of pharmacology and therapeutic potential: A. H. Groll, T. J. Walsh, *Curr. Opin. Anti-Infect. Invest. Drugs* **2**, 405-412 (2000); A. H. Groll *et al., Expert Opin. Invest. Drugs* **14**, 489-509 (2005).

Sodium salt. [208538-73-2] FK-463; Mycamine. $C_{56}H_{70}N_9$-$NaO_{23}S$; mol wt 1292.27. Water soluble lipopeptide.

THERAP CAT: Antifungal.

6255. Michler's Base. [101-61-1] 4,4'-Methylenebis[N,N-dimethylbenzenamine]; 4,4'-methylenebis(N,N-dimethylaniline);

bis(p-dimethylaminophenyl)methane; 4,4'-tetramethyldiaminodiphenylmethane; tetra-base; Arnold's base. $C_{17}H_{22}N_2$; mol wt 254.38. C 80.27%, H 8.72%, N 11.01%. Prepd by heating dimethylaniline with 40% formaldehyde and concd HCl: Mekel, **DE 1026322** (1958 to BASF); Hey, Sanderson, *J. Chem. Soc.* **1960**, 3203; from dimethylaniline + diacetylperoxide: Horner *et al., Ann.* **626**, 1 (1959); from dimethylaniline + *tert*-butylperbenzoate: Sosnovsky, Yang, *J. Org. Chem.* **25**, 899 (1960).

Lustrous leaflets, mp 90-91°. Sublimes without decompn. bp 390°; $bp_{0.1}$ 155-157°. Sol in benzene, ether, carbon disulfide, acids; slightly sol in cold alc; more sol in hot alc. Insol in water.

Caution: This substance is reasonably anticipated to be a human carcinogen: *Report on Carcinogens, Twelfth Edition* (PB2011-111646, 2011) p 264.

USE: Has been used in manufacture of dyes. As reagent for lead: *Dtsch. Med. Wochenschr.* **52**, 1855 (1926).

6256. Michler's Ketone. [90-94-8] Bis[4-(dimethylamino)phenyl]methanone; 4,4'-bis(dimethylamino)benzophenone; tetramethyldiaminobenzophenone. $C_{17}H_{20}N_2O$; mol wt 268.36. C 76.09%, H 7.51%, N 10.44%, O 5.96%. Prepn from dimethylaniline + phosgene: Michler, *Ber.* **9**, 716 (1876); Michler, Dupertius, *ibid.* **1900**; from dimethylaniline, $AlCl_3$ and CCl_4: Fierz, Koechlin, *Helv. Chim. Acta* **1**, 218 (1918).

White to greenish leaflets, mp 172°. bp >360° with decompn. Sol in alc, warm benzene; very slightly sol in ether. Practically insol in water.

Caution: This substance is reasonably anticipated to be a human carcinogen: *Report on Carcinogens, Twelfth Edition* (PB2011-111646, 2011) p 270.

USE: In manuf of dyes and pigments.

6257. Miconazole. [22916-47-8] 1-[2-(2,4-Dichlorophenyl)-2-[(2,4-dichlorophenyl)methoxy]ethyl]-1H-imidazole; 1-[2,4-dichloro-β-[(2,4-dichlorobenzyl)oxy]phenethyl]imidazole; Daktarin; Florid; Loramyc. $C_{18}H_{14}Cl_4N_2O$; mol wt 416.12. C 51.96%, H 3.39%, Cl 34.08%, N 6.73%, O 3.84%. Ergosterol biosynthesis inhibitor. Prepn: E. F. Godefroi *et al., J. Med. Chem.* **12**, 784 (1969); E. F. Godefroi, J. Heeres, **DE 1940388**; *eidem*, **US 3717655** (1970, 1973 to Janssen). HPLC determn in plasma: L. A. Sternson *et al., J. Chromatogr.* **227**, 223 (1982). Clinical evaluation: Brugmans *et al., Arch. Dermatol.* **102**, 428 (1970); Godts *et al., Arzneim.-Forsch.* **21**, 256 (1971). Review of pharmacology and clinical experience: A. Barasch, A. V. Griffin, *Future Microbiol.* **3**, 265-269 (2008).

White to pale cream powder. May exhibit polymorphism. Freely sol in alc, methanol, isopropyl alc, acetone, propylene glycol, chloroform, DMF; sol in ether. Insol in water.

Nitrate. [22832-87-7] R-14889; Albistat; Conofite; Daktar; Deralbine; Dermonistat; Epi-Monistat; Fungisdin; Gyno-Daktarin; Micatin; Miconal Ecobi; Micotef; Monistat; Prilagin. $C_{18}H_{14}Cl_4N_2$-O.HNO_3; mol wt 479.14. White crystals, mp 170.5° (Godefroi, Heeres, 1970); 184-185° (Godefroi). Freely sol in dimethyl sulfoxide; sol in DMF; sparingly sol in methanol; slightly sol in alc, chloroform, propylene glycol; very slightly sol in water, isopropyl alc. Insol in ether.

(+)-**Form nitrate.** mp 135.3°. $[\alpha]_D^{20}$ +59° (methanol).
(−)-**Form nitrate.** mp 135°. $[\alpha]_D^{20}$ −58° (methanol).

THERAP CAT: Antifungal (topical).
THERAP CAT (VET): Antifungal (topical).

6258. Micranthine. [36104-64-0] 6',7-Epoxy-6-methoxy-2-methyl-oxyacanthan-12'-ol. $C_{34}H_{32}N_2O_5$; mol wt 548.64. C 74.43%, H 5.88%, N 5.11%, O 14.58%. Minor alkaloid from bark of *Daphnandra micrantha* (Tul.) Benth., *Monimiaceae.* Isoln: Pyman, *J. Chem. Soc.* **105**, 1679 (1914). Structure: Bick, Todd, *ibid.* **1950**, 1606. Revised structure and stereochemistry: Bick *et al.*, *Tetrahedron Lett.* **1972**, 33. Biogenesis: I. Ralph, C. Bick, *Heterocycles* **16**, 2105 (1981).

Needles from ethyl acetate or methanol, mp 194-196°. $[\alpha]_D^{22}$ −231° (c = 0.7 in chloroform). Sol in chloroform, ethanol; sparingly sol in methanol, ethyl acetate, benzene; practically insol in ether.

6259. Micrococcin P. [1392-46-7] Micrococcin. Antibiotic substance produced by a species of *Micrococcus:* cf. *Br. J. Exp. Pathol.* **29**, 473 (1948). Isoln of micrococcin P from *B. pumilus:* Heatley, Doery, *Biochem. J.* **50**, 247 (1951). Probable identity of micrococcin and micrococcin P: Abraham *et al.*, *Nature* **178**, 44 (1956). Structure studies: Brookes *et al.*, *J. Chem. Soc.* **1957**, 689; Brookes *et al.*, *ibid.* **1960**, 916; Dean *et al.*, *ibid.* **1961**, 3394; James, Watson, *J. Chem. Soc. C* **1966**, 1361. The antibiotic is a mixture of two components, P_1 (major) and P_2. Proposed total structure of P_1: J. Walker *et al.*, *Chem. Commun.* **1977**, 706. Revised structure of P_1 and structure of P_2: B. W. Bycroft, M. S. Gowland, *ibid.* **1978**, 256. Mechanism of action: E. Cundliffe, J. Thompson, *Eur. J. Biochem.* **118**, 47 (1981). *Review:* Pestka in *Antibiotics* vol. 3, J. W. Corcoran, F. E. Hahn, Eds. (Springer-Verlag, New York, 1975) pp 480-486.

Micrococcin P_1

Needles from ethanol, dec 222-228°. $[\alpha]_D^{21}$ +116° (c = 0.10 in 90% ethanol). uv max (ethanol): 345 nm. Sol in ethanol, chloroform, acetone, glacial acetic acid, pyridine. Insol in ether, benzene, amyl acetate, glycerol. Very slightly sol in water.

6260. Micronomicin. [52093-21-7] O-2-Amino-2,3,4,6-tetradeoxy-6-(methylamino)-α-D-*erythro*-hexopyranosyl-(1 → 4)-O-[3-deoxy-4-C-methyl-3-(methylamino)-β-L-arabinopyranosyl-(1 → 6)]-2-deoxy-D-streptamine; 6'-N-methylgentamicin C_{1a}; gentamicin C_{2b}; sagamicin (formerly); antibiotic KW-1062; KW-1062; XK-62-2. $C_{20}H_{41}N_5O_7$; mol wt 463.58. C 51.82%, H 8.91%, N 15.11%, O 24.16%. Gentamicin C complex antibiotic produced by *Micromonospora sagamiensis* var. *nonreducans:* T. Nara *et al.*, DE 2326781 (1973 to Kyowa), *C.A.* **80**, 58389b (1974); *eidem*, US 4045298 (1977 to Abbott). Isoln, physicochemical and antibacterial properties: R. Okachi *et al.*, *J. Antibiot.* **27**, 793 (1974). Structure: R. S. Egan *et al.*, *ibid.* **28**, 29 (1975). Identity with gentamicin C_{2b}: P. J. L. Daniels *et al.*, *ibid.* 35. Pharmacology: T. Hashimoto *et al.*, *Jpn. J. Antibiot.* **30**, 362 (1977), *C.A.* **89**, 17113z (1978). Toxicity studies: T. Hara *et al.*, *ibid.* 386-449, *C.A.* **89**, 209205c-8f (1978). Biosynthesis: H. Kase *et al.*, *J. Antibiot.* **35**, 1 (1982); *eidem*, *Agric. Biol. Chem.* **46**, 515 (1982). Series of articles on pharmacology, metabolism and clinical studies: *Chemotherapy (Tokyo)* **25**, 1801-2287 (1977). *See also* Gentamicin.

White amorphous powder, mp 260° (dec). $[\alpha]_D^{20}$ +116° (c = 1 in water). Sol in water, methanol. Insol in chloroform, ethyl acetate, benzene, petr ether. LD_{50} i.v. in mice: 93 mg/kg (Okachi).

Sulfate. 6'-N-Methylgentamicin C_{1a} hemipentasulfate; Sagamicin; Santemycin. $(C_{20}H_{41}N_5O_7)_2.5H_2SO_4$; mol wt 1417.51.

THERAP CAT: Antibacterial.

6261. Midazolam. [59467-70-8] 8-Chloro-6-(2-fluorophenyl)-1-methyl-4H-imidazo[1,5-a][1,4]benzodiazepine. $C_{18}H_{13}$-$ClFN_3$; mol wt 325.77. C 66.37%, H 4.02%, Cl 10.88%, F 5.83%, N 12.90%. Short-acting deriv of diazepam, q.v. Prepn: R. I. Fryer, A. Walser, DE 2540522; *eidem*, US 4280957 (1976, 1981 both to Hoffmann-La Roche); A. Walser *et al.*, *J. Org. Chem.* **43**, 936 (1978). LC/MS determn in plasma: E. R. Lepper *et al.*, *J. Chromatogr. B* **806**, 305 (2004); as a probe for cytochrome P450 3A4 activity: H. Kanazawa *et al.*, *J. Chromatogr. A* **1031**, 213 (2004). Series of articles on pharmacology, metabolism, pharmacokinetics, clinical experience: *ibid.* 2177-2288; *Br. J. Clin. Pharmacol.* **16**, Suppl. 1, 1S-199S (1983). Review of pharmacology and therapeutic use: J. W. Dundee *et al.*, *Drugs* **28**, 519-543 (1984); in treatment of status epilepticus: D. F. Hanley, J. F. Kross, *Clin. Ther.* **20**, 1093-1105 (1998). Clinical evaluation for intranasal treatment of febrile seizures in children: E. Lahat *et al.*, *Br. Med. J.* **321**, 83 (2000).

Colorless crystals from ether/methylene chloride/hexane, mp 158-160°. uv max (2-propanol): 220 nm (ε 30000).

Maleate. [59467-94-6] Ro-21-3981/001; Dormicum. $C_{18}H_{13}$-$ClFN_3.C_4H_4O_4$; mol wt 441.84. Crystals from ethanol/ether, mp 114-117° (solvated). LD_{50} in male mice (mg/kg): 760 orally; 86 i.v. *See* L. Pieri *et al., Arzneim.-Forsch.* **31**, 2180 (1981).

Hydrochloride. [59467-96-8] Ro-21-3981/003; Hypnovel; Ipnovel; Versed. $C_{18}H_{13}ClFN_3.HCl$; mol wt 362.23. Sol in aqueous solns.

Note: This is a controlled substance (depressant): **21 CFR**, 1308.14.

USE: Probe to measure cytochrome P450 3A4 activity.

THERAP CAT: Anesthetic (intravenous); anticonvulsant; sedative; hypnotic.

THERAP CAT (VET): Anesthetic (intravenous); anticonvulsant; sedative; hypnotic.

6262. Midecamycins. Macrolide antibiotic complex produced by *Streptomyces mycarofaciens* nov. sp. Isoln and characterization of the major component, midecamycin A_1: T. Tsuruoka *et al.*, **JP 71 28834**; *eidem*, **US 3761588** (1971, 1973 both to Meiji); *eidem, J. Antibiot.* **24**, 319, 452 (1971). Structure: I. Shigeharu *et al., ibid.* 460. Isoln, properties, structures of the minor components, midecamycins A_2, A_3, A_4: T. Tsuruoka *et al., ibid.* 476, 526. Enzymatic interconversion of A_1 and A_3: Y. Matsuhashi *et al., ibid.* **32**, 777 (1979). Metabolism: T. Shomura *et al., Chem. Pharm. Bull.* **22**, 2427 (1974). Pharmacokinetics: K. Fukaya *et al., Chemotherapy (Tokyo)* **21**, 692 (1973). Pharmacology: K. Mashimo *et al., ibid.* 729. Antibacterial activity: T. Watanabe *et al., ibid.* **25**, 1624 (1977). Toxicity and teratogenicity study in mice: M. Moriguchi *et al., Jpn. J. Antibiot.* **15**, 187, 193 (1972), *C.A.* **78**, 11652a, 11653b (1973).

Midecamycin A_1

Midecamycin A_1. [35457-80-8] Leucomycin V 3,4^B-dipropanoate; espinomycin A; mydecamycin; turimycin P3; SF-837; YL-704B1; Aboren; Medemycin; Midecin; Momicine; Myoxam; Normicina; Rubimycin. $C_{41}H_{67}NO_{15}$; mol wt 813.98. Colorless needles from benzene, mp 155-156° (after drying at 80° for 8 hrs in vacuo). $[\alpha]_D^{23}$ −67° (c = 1 in ethanol). uv max (ethanol): 232 nm ($E_{1cm}^{1\%}$ 325). Weakly basic; pKa' 6.9 in 50% aq ethanol. Sol in methanol, ethanol, acetone, chloroform, ethyl acetate, benzene, ethyl ether, acidic water. Practically insol in *n*-hexane, petr ether, water.

Midecamycin A_3. [36025-69-1] 9-Deoxy-9-oxoleucomycin V 3,4^B-dipropanoate; antibiotic SF-837A3. $C_{41}H_{65}NO_{15}$; mol wt 811.96. White powder, mp 122-125°. $[\alpha]_D^{22}$ −44° (c = 1 in ethanol). uv max (ethanol): 280 nm ($E_{1cm}^{1\%}$ 295). pKa' 7.0 in 50% aq ethanol. Solubilities similar to midecamycin A_1.

THERAP CAT: Antibacterial.

6263. Midodrine. [42794-76-3] 2-Amino-*N*-[2-(2,5-dimethoxyphenyl)-2-hydroxyethyl]acetamide; 1-(2′,5′-dimethoxyphenyl)-2-glycinamidoethanol. $C_{12}H_{18}N_2O_4$; mol wt 254.29. C 56.68%, H 7.14%, N 11.02%, O 25.17%. α-Adrenergic agonist. Prepn: K. Wismayr *et al.*, **AT 241435** (1965, 1967 both to OSSW). Hemodynamic effects: G. Hitzenberger *et al., Int. J. Clin. Pharmacol. Ther. Toxicol.* **7**, 323 (1973). Pharmacodynamic actions: H. Pittner *et al., Arzneim.-Forsch.* **26**, 2145 (1976). Plasma levels: N. Kolassa *et al., Arch. Int. Pharmacodyn. Ther.* **238**, 96 (1979). Use in ejaculation disorders: D. Jonas *et al., Eur. Urol.* **5**, 184 (1979); in circulatory disorders: H. Sazovsky, H. Pittner, *Fortschr. Med.* **97**, 733 (1979). Synthesis of potential metabolites: T. Kappe, W. Witoszynaky, *Arch. Pharm.* **308**, 339 (1975).

Hydrochloride. [3092-17-9] ST-1085; Amatine; Gutron; Hipertan; Metligine; ProAmatine. $C_{12}H_{18}N_2O_4.HCl$; mol wt 290.74. Crystals, mp 192-193°.

THERAP CAT: Antihypotensive.

6264. Midostaurin. [120685-11-2] *N*-[(9S,10R,11R,13R)-2,-3,10,11,12,13-Hexahydro-10-methoxy-9-methyl-1-oxo-9,13-epoxy-1*H*,9*H*-diindolo[1,2,3-*gh*:3′,2′,1′-*lm*]pyrrolo[3,4-*j*][1,7]benzodiazonin-11-yl]-*N*-methylbenzamide; *N*-benzoylstaurosporine; CGP-41251; PKC-412. $C_{35}H_{30}N_4O_4$; mol wt 570.65. C 73.67%, H 5.30%, N 9.82%, O 11.21%. Protein kinase C inhibitor; derivative of the naturally occurring alkaloid, staurosporine, *q.v.* Prepn: G. Caravatti, A. Fredenhagen, **EP 296110**; *eidem*, **US 5093330** (1988, 1992 both to Ciba-Geigy); G. Caravatti *et al., Bioorg. Med. Chem. Lett.* **4**, 399 (1994). HPLC determn in plasma: R. van Gijn *et al., J. Chromatogr. B* **667**, 269 (1995). Review of pharmacology: D. Fabbro *et al., Pharmacol. Ther.* **82**, 293-301 (1999); of mechanism of action and therapeutic potential: A. Gescher, *Crit. Rev. Oncol. Hematol.* **34**, 127-135 (2000). Clinical pharmacokinetics and tolerability: D. J. Propper *et al., J. Clin. Oncol.* **19**, 1485 (2001). Clinical evaluation in diabetic macular edema: P. A. Campochiaro *et al., Invest. Ophthalmol. Visual Sci.* **45**, 922 (2004); in acute myeloid leukemia: R. M. Stone *et al., Blood* **105**, 54 (2005).

mp 235-247° with darkening.

THERAP CAT: Antineoplastic.

6265. Mifamurtide. [83461-56-7] *N*-(*N*-Acetylmuramoyl)-L-alanyl-D-α-glutaminyl-*N*-[(7R)-4-hydroxy-4-oxido-10-oxo-7-[(1-oxohexadecyl)oxy]-3,5,9-trioxa-4-phosphapentacos-1-yl]-L-alaninamide; *N*-acetylmuramyl-L-alanyl-D-isoglutaminyl-L-alanine-2-(1′,2′-*O*-dipalmitoyl-*sn*-glycero-3′-hydroxyphosphoryloxy)ethylamide; muramyl tripeptide phosphatidylethanolamine; MTP-cephalin; MTP-PE; CGP-19835. $C_{59}H_{109}N_6O_{19}P$; mol wt 1237.52. C 57.26%, H 8.88%, N 6.79%, O 24.56%, P 2.50%. Synthetic, lipophilic analog of muramyl dipeptide, *q.v.*, a naturally occuring component of bacterial cell walls. Nonspecific immunomodulator that activates monocytes and macrophages and increases tumoricidal activity. Prepn: L. Tarcsay *et al.*, **EP 25495**; *eidem*, **US 4406890** (1980, 1983 both to Ciba-Geigy). Synthesis of radiolabeled form: D. E. Brundish, R. Wade, *J. Labelled Compd. Radiopharm.* **22**, 29 (1985). Liposomal encapsulation: R. Natar *et al., Methods Enzymol.* **132**, 594 (1986). Effect on monocyte-mediated cytotoxicity *in vitro*: E. S. Kleinerman *et al., Cancer Res.* **43**, 2010 (1983). Immunoassay in plasma: B. Gay *et al., J. Biolumin. Chemilumin.* **6**, 73 (1991). Comparative pharmacokinetics of free and liposomal forms: B. Gay *et al., J. Pharm. Sci.* **82**, 997 (1993). Clinical trial as adjuvant therapy in osteosarcoma: P. A. Meyers *et al., J. Clin. Oncol.* **26**, 633 (2008). Review of pharmacology and mechanism of action: A. Nardin *et al., Curr. Cancer Drug Targets* **6**, 123-133 (2006). Review of clinical experience as immunoadjuvant in treatment of osteosarcoma: J. E. Frampton, *Pediatr. Drugs* **12**, 141-153 (2010); K. Ando *et al.*, *Expert Opin. Pharmacother.* **12**, 285-292 (2011).

Sodium salt. [90825-43-7]; [838853-48-8] (hydrate). $C_{59}H_{108^-}N_6NaO_{19}P$; mol wt 1259.50.

Liposomal formulation. L-MTP-PE; CGP 19835A; Junovan; Mepact. Sodium salt encapsulated into liposomes composed of the naturally occuring phospholipids, 1-palmityl-2-oleophosphocholine (POPC) and 1,2-dioleoylphosphatidylserine (OOPS). Ratio of active ingredient to phospholipids is 1:250.

THERAP CAT: Antineoplastic (immunomodulator).

6266. Mifepristone. [84371-65-3] (11β,17β)-11-[4-(Dimethylamino)phenyl]-17-hydroxy-17-(1-propyn-1-yl)estra-4,9-dien-3-one; 11β-[4-(*N,N*-dimethylamino)phenyl]-17α-(prop-1-ynyl)-$\Delta^{4,9}$-estradiene-17β-ol-3-one; RU-486; RU-38486; Corlux; Mifegyne; Mifeprex. $C_{29}H_{35}NO_2$; mol wt 429.60. C 81.08%, H 8.21%, N 3.26%, O 7.45%. Combined glucocorticoid and progesterone receptor antagonist. Prepn: J. G. Teutsch *et al.*, **EP 57115**; *eidem*, **US 4386085** (1982, 1983 both to Roussel-UCLAF). Pharmacology: W. Herrmann *et al.*, *C. R. Seances Acad. Sci. Ser. 3* **294**, 933 (1982). Pituitary and adrenal responses in primates: D. L. Healy *et al.*, *J. Clin. Endocrinol. Metab.* **57**, 863 (1983). LC-MS determn in plasma: N. Z. M. Homer *et al.*, *J. Chromatogr. B* **877**, 497 (2009). Clinical study as abortifacient: B. Couzinet *et al.*, *N. Engl. J. Med.* **315**, 1565 (1986); as postcoital contraceptive: A. Glasier *et al.*, *ibid.* **327**, 1041 (1992). Review of mechanism of action and gynecological applications: E. E. Baulieu, *Science* **245**, 1351-1357 (1989). *Reviews*: I. M. Spitz, C. W. Bardin, *N. Engl. J. Med.* **329**, 404-412 (1993); E. A. Schaff, *Contraception* **81**, 1-7 (2010). Review of antiglucocorticoid activity and clinical experience in Cushing's syndrome: F. Castinetti *et al.*, *Neuroendocrinology* **92**, Suppl. 1, 125-130 (2010).

mp 150°. $[\alpha]_D^{20}$ +138.5° (c = 0.5 in chloroform). Very sol in methanol, chloroform and acetone.

THERAP CAT: Abortifacient.

6267. Miglitol. [72432-03-2] (2*R*,3*R*,4*R*,5*S*)-1-(2-Hydroxyethyl)-2-(hydroxymethyl)-3,4,5-piperidinetriol; *N*-(2-hydroxyethyl)-moranoline; 1,5-dideoxy-1,5-[(2-hydroxyethyl)imino]-D-glucitol; *N*-(β-hydroxyethyl)-1-deoxynojirimycin; Bay m 1099; Diastabol; Glyset. $C_8H_{17}NO_5$; mol wt 207.23. C 46.37%, H 8.27%, N 6.76%, O

38.60%. α-Glucosidase inhibitor; hydroxyethyl derivative of 1-deoxynojirimycin, *q.v.* Prepn: B. Junge *et al.*, **DE 2758025**; *eidem*, **US 4639436** (1979, 1987 both to Bayer AG). Inhibition of α-glucosidase: B. Lembcke *et al.*, *Digestion* **31**, 120 (1985); and hypoglycemic activity: Y. Yoshikuni *et al.*, *J. Pharmacobio-Dyn.* **11**, 356 (1988). Clinical evaluations in insulin dependent diabetes: J. Gerard *et al.*, *Int. J. Clin. Pharmacol. Ther.* **25**, 483 (1987); F. P. Kennedy, J. E. Gerich, *Clin. Pharmacol. Ther.* **42**, 455 (1987). Review of pharmacology, clinical efficacy, and therapeutic potential in non-insulin dependent diabetes: L. J. Scott, C. M. Spencer, *Drugs* **59**, 521-549 (2000).

Crystals from ethanol, mp 114°. Sol in water. pKa 5.9.

THERAP CAT: Antidiabetic.

6268. Miglustat. [72599-27-0] (2*R*,3*R*,4*R*,5*S*)-1-Butyl-2-(hydroxymethyl)-3,4,5-piperidinetriol; *N*-butyldeoxynojirimycin; *N*-butylmoranoline; OGT-918; SC-48334; Zavesca. $C_{10}H_{21}NO_4$; mol wt 219.28. C 54.77%, H 9.65%, N 6.39%, O 29.18%. Amino sugar; inhibitor of glucosyltransferase, an enzyme involved in the biosynthesis of glycosphingolipids. Prepn: B. Junge *et al.*, **DE 2758025**; *see also, eidem*, **US 4639436** (1979, 1987 both to Bayer). Improved synthesis: E. W. Baxter, A. B. Reitz, *J. Org. Chem.* **59**, 3175 (1994); C. R. R. Matos *et al.*, *Synthesis* **1999**, 571. *In vitro* efficacy vs HIV: A. Karpas *et al.*, *Proc. Natl. Acad. Sci. USA* **85**, 9229 (1988). Inhibition of glycolipid biosynthesis: F. M. Platt *et al.*, *J. Biol. Chem.* **269**, 8362 (1994). Clinical evaluation in Gaucher's disease: T. Cox *et al.*, *Lancet* **355**, 1481 (2000). Review of pharmacology and clinical development in Gaucher's disease: P. L. McCormack, K. L. Goa, *Drugs* **63**, 2427-2434 (2003).

White to off-white crystalline solid, mp 125-126°. $[\alpha]_D^{25}$ −15.9° (c = 0.77 in water). Highly sol in water (>1000 mg/ml).

THERAP CAT: In treatment of inherited glycosphingolipid lysosomal storage disorders.

6269. Milbemectin. E-187; Koromite; Milbeknock. Mixture of milbemycins, *q.v.*, in a 3:7 ratio of milbemycin A$_3$ and milbemycin A$_4$. Isolated from fermentation broth of *Streptomyces hygroscopicus* subsp. *aureolacrimosus* SANK60576; structurally related to the avermectins, *q.v.*: A. Aoki *et al.*, **DE 2329486**; *eidem*, **US 3950360** (1973, 1976 both to Sankyo); Y. Takiguchi *et al.*, *J. Antibiot.* **33**, 1120 (1980). Synthesis of A$_3$: M. Hirama *et al.*, *J. Am. Chem. Soc.* **113**, 1830 (1991); S. V. Ley *et al.*, *Tetrahedron Lett.* **34**, 7479 (1993). Metabolism in rats: S. Sadakane *et al.*, *J. Pestic. Sci.* **17**, 147 (1992). Review: A. Aoki *et al.*, *ibid.* **19**, 245-247 (1994).

Milbemycin A$_3$ R = CH$_3$
Milbemycin A$_4$ R = CH$_2$CH$_3$

Stable to heat; relatively unstable under direct sunlight. Relatively unstable to acid; unstable to alkali. LD_{50} in male, female rats, mice (mg/kg): 762, 456, 324, 313 orally; in rats (mg/kg): >5000 dermally (Aoki, 1994). LC_{50} by inhalation in male, female rats (mg/l): 1.90, 2.80 (Aoki, 1994).

Milbemycin A₃. [51596-10-2] (6R,25R)-5-O-Demethyl-28-deoxy-6,28-epoxy-25-methylmilbemycin; milbemycin α_1; antibiotic B-41A3. $C_{31}H_{44}O_7$; mol wt 528.69. White crystalline powder, mp 212-215°. $[\alpha]_D^{20}$ +106° (c = 0.25 in acetone) (Aoki). Also reported as mp 192-194°. $[\alpha]_D^{20}$ +106° (acetone) (Hirama). uv max (ethanol): 238, 244 nm (ε 27800, 30500). Soly in water (20°): 7.2 ppm; methanol 64.8 g/l; ethanol 41.9 g/l; acetone 66.1 g/l; n-hexane 1.4 g/l; benzene 143.1 g/l; ethyl acetate 69.5 g/l. Vapor pressure (20°): <1 × 10^10 mm Hg.

Milbemycin A₄. [51596-11-3] Milbemycin α_3; antibiotic B-41A4. $C_{32}H_{46}O_7$; mol wt 542.71. White crystalline powder, mp 212-215°. $[\alpha]_D^{20}$ +106° (c = 0.25 in acetone). uv max (ethanol): 238, 244 nm (ε 27800, 30500). Soly in water (20°): 0.88 ppm; methanol 458.8 g/l; ethanol 234.0 g/l; acetone 365.3 g/l; n-hexane 6.5 g/l; benzene 524.2 g/l; ethyl acetate 320.4 g/l. Vapor pressure (20°): <1 × 10^10 mm Hg.

USE: Acaricide.

6270. Milbemycin Oxime. [129496-10-2] CGA-179246; Interceptor. Mixture of milbemycin oximes containing ~80% **milbemycin A₄ 5-oxime** ($C_{32}H_{45}NO_7$) and 20% **milbemycin A₃ 5-oxime** ($C_{31}H_{43}NO_7$). Prepn: J. Ide et al., EP 110667; eidem, US 4547520 (1984, 1985 both to Sankyo). Anthelmintic activity in experimentally induced infections in dogs: D. D. Bowman et al., Am. J. Vet. Res. **49**, 1986 (1988); D. D. Bowman et al., ibid. **52**, 64 (1991); in cats: B. L. Blagburn et al., The OVMA Newsletter (February, 1990), 1; in naturally acquired infections in dogs: D. D. Bowman et al., Am. J. Vet. Res. **51**, 487 (1990). Physical data and toxicity: Ciba-Geigy Material Data Safety Sheet.

Component A₃   R = CH₃
Component A₄   R = CH₂CH₃

Practically odorless white to pale yellow powder, mp 169.6-177.4°. pH of aqueous solution: 6.3. LD_{50} orally in mice: 1000-1500 mg/kg (Ciba-Geigy).

THERAP CAT (VET): Anthelmintic.

6271. Milbemycins. [51570-36-6] Antibiotic B-41. Family of macrolide antibiotics with insecticidal and acaricidal activity; structurally related to the avermectins, q.v. Isoln from Streptomyces hygroscopicus subsp. aureolacrimosus and properties: A. Aoki et al., DE 2329486; eidem, US 3950360 (1973, 1976 both to Sankyo); of α_1-α_{10}, β_1-β_3, D-H: Y. Takiguchi et al., J. Antibiot. **33**, 1120 (1980). Isoln and properties of D-H: Y. Takiguchi et al., J. Antibiot. **36**, 502 (1983); of J and K: M. Ono et al., ibid. 509. Structures of milbemycins α_1-α_{10}, β_1: M. Kurabayashi et al., 18th Symp. Chem. Natural Products (Kyoto, 1974) nos. 309-316; of β_1-β_3: H. Mishima et al., Tetrahedron Lett. **1975**, 711; of D-H, J, K: H. Mishima et al., J. Antibiot. **36**, 980 (1983). Total synthesis of (±)-milbemycin β_3: A. B. Smith et al., J. Am. Chem. Soc. **104**, 4015 (1982); of the (+)-form: D. R. Williams et al., ibid. 4708. Prepn of milbemycins from avermectins: H. Mrozik et al., Tetrahedron Lett. **1983**, 5333. Total synthesis studies of D: M. T. Crimmins et al., ibid. **28**, 3651

(1987). Prophylactic use of milbemycin D against Dirofilaria immitis infection in dogs: M. Tagawa et al., Jpn. J. Vet. Sci. **47**, 787 (1985). Toxicity data: N. Matsunuma et al., Sankyo Kenkyusho Nempo **35**, 71 (1983), C.A. **101**, 32870d (1984). Review of syntheses: T. Blizzard et al., in Recent Prog. Chem. Synth. Antibiotics, G. Lukacs, M. Ohno, Eds (Springer-Verlag, Berlin, 1990) pp 66-102. Review of structure-activity relationships: W. L. Shoop et al., Vet. Parasitol. **59**, 139-156 (1995).

Milbemycin D

Sol in n-hexane, benzene, acetone, ethanol, methanol, chloroform; very slightly sol in water.

Milbemycin β₁. [51596-16-8] 25α-Methylmilbemycin B; antibiotic B-41A1. $C_{32}H_{48}O_7$; mol wt 544.73. Amorphous solid. $[\alpha]_D^{20}$ +160° (c = 0.25 in acetone). uv max (ethanol): 245 nm (ε 26500).

Milbemycin D. [77855-81-3] (6R,25R)-5-O-Demethyl-28-deoxy-6,28-epoxy-25-(1-methylethyl)milbemycin B; antibiotic B-41D. $C_{33}H_{48}O_7$; mol wt 556.74. Colorless needles from hexane-ethyl acetate (20:1), mp 186-188°. $[\alpha]_D^{27}$ +107° (c = 0.25 in acetone). uv max (ethanol): 244 nm (31000).

Mixture of milbemycins A₃ and A₄ see Milbemectin.

THERAP CAT (VET): Antiparasitic.

6272. Mildiomycin. [67527-71-3] (S)-4-Amino-1-[4-[[(2S)-amino-3-hydroxy-1-oxopropyl]amino]-9-[(aminoiminomethyl)amino]-6-C-carboxy-2,3,4,7,9-pentadeoxy-α-L-talo-non-2-enopyranosyl]-5-(hydroxymethyl)-2(1H)-pyrimidinone. $C_{19}H_{30}N_8O_9$; mol wt 514.50. C 44.36%, H 5.88%, N 21.78%, O 27.99%. A nucleoside antibiotic with anti-mildew activity and low toxicity in mammals and fish. Isoln from Streptoverticillium rimofaciens B-98891 and characterization: S. Harada, T. Kishi, J. Antibiot. **31**, 519 (1978). Taxonomy and fermentation study: T. Iwasa et al., ibid. 511. Structure: S. Harada et al., J. Am. Chem. Soc. **100**, 4895 (1978); Tetrahedron **37**, 1317 (1981). Anti-mildew spectrum: K. Suetomi, T. Kusaka, Nippon Noyaku Gakkaishi **4**, 349 (1979), C.A. **92**, 17012c (1980).

Hygroscopic solid, mp >300° (monohydrate). $[\alpha]_D^{20}$ +100° (c = 0.5 in water); +78.5° (c = 0.5 in 0.1N HCl). pKa': 2.8, 4.3, 7.2, >12. uv max (pH 7 and 0.1N NaOH): 271 nm ($E_{1cm}^{1\%}$ 164); (0.1N HCl): 280 nm ($E_{1cm}^{1\%}$ 247). LD_{50} in rats, mice: ~500-1000 mg/kg i.v.; ~2.5-5.0 g/kg orally (Harada, Kishi).

6273. Milk. Cow's milk. Compn and physical properties (a study of samples representing 8 million quarts of milk from 8 cities):

Dahlberg, *Proc. Annu. Conv. Milk Ind. Found. Lab. Sect.* **46**, 44 (1953). *Review:* Patton, *Sci. Am.* **221** (1), 58 (July 1969).

White liquid. d 1.032. mp −0.54°. pH 6.62. Fat 3.82%; total solids 12.43%; protein 3.25%; lactose 4.64%; ash 0.73%. Vitamins and minerals in 100 g: Vitamin A 152 I.U.; riboflavin 0.156 mg; thiamine 0.043 mg; ascorbic acid 1.14 mg; vitamin B_{12} 0.56 γ; vitamin D 2.4 I.U.; vitamin E 0.1 mg; nicotinic acid 0.085 mg; pantothenic acid 0.35 mg; pyridoxine 0.048 mg; biotin 0.0035 mg; folic acid 0.2 γ; inositol 13 mg; choline 13 mg; essential fatty acids 99 mg; Ca 113 mg; P 90 mg; Fe 0.032 mg. Caloric value: 65 cal/100 g.

6274. Milk Thistle. Mary thistle; holy thistle. Annual or biennial herb, *Silybum marianum* (L.) Gaertn., *Asteraceae*. Ripe seeds (fruit) used in traditional medicine for hepatoprotective properties. *Habit.* Mediterranean Europe. *Constit.* Fruit contains silymarin flavolignans, flavonoids, and fatty oil (20-30%). Leaves contain flavonoids including apigenin, luteolin, kaempferol, and their glycosides, sterols, fumaric acid. Isoln and characterization of phenolic compounds from *Silybum marianum*: B. Janiak, R. Hänsel, *Planta Med.* **8**, 71 (1960). Comprehensive description and medicinal uses: J. Barnes *et al.*, *Herbal Medicines* (Pharmaceutical Press, London, 3rd Ed., 2007) pp 429-435. Review of therapeutic use in liver diseases: K. Ball, K. V. Kowdley, *J. Clin. Gastroenterol.* **39**, 520-528 (2005); A. Rambaldi *et al.*, *Am. J. Gastroenterol.* **100**, 2583-2591 (2005). Review of therapeutic potential: J. Post-White *et al.*, *Integr. Cancer Ther.* **6**, 104-109 (2007); of milk thistle nomenclature and clinical applications: D. J. Kroll *et al.*, *ibid.* 110-119; of clinical development: C. Tamayo, S. Diamond, *ibid.* 146-157.

Silymarin. [65666-07-1] Apihepar; Laragon; Legalon; Pluropon; Silarine; Silirex; Silliver; Silmar. Antihepatotoxic principle isolated from seeds. Commercial product consists of 70-80% silymarin flavonolignans: silybin and isosilybin diastereomers A and B, *q.q.v.*, **silydianin**, and **silycristin**. LC/MS determn of active constituents in commercial product: J. I. Lee *et al.*, *J. Chromatogr. B* **845**, 95 (2007). Large-scale purification of flavonolignans: T. N. Graf *et al.*, *Planta Med.* **73**, 1495 (2007). Review of use in treatment of liver diseases: R. Saller *et al.*, *Drugs* **61**, 2035-2063 (2001); of anticancer properties: R. Agarwal *et al.*, *Anticancer Res.* **26**, 4457-4498 (2006).

THERAP CAT: Hepatoprotectant.

THERAP CAT (VET): Hepatoprotectant.

6275. Milnacipran. [92623-85-3] *rel*-(1*R*,2*S*)-2-(Aminomethyl)-*N*,*N*-diethyl-1-phenylcyclopropanecarboxamide; 1-phenyl-1-(diethylaminocarbonyl)-2-(aminomethyl)cyclopropane; midalcipran. $C_{15}H_{22}N_2O$; mol wt 246.35. C 73.13%, H 9.00%, N 11.37%, O 6.49%. Serotonin and norepinephrine reuptake inhibitor (SNRI). Prepn: G. Mouzin *et al.*, **EP 68999**; *eidem*, **US 4478836** (1983, 1984 both to Pierre Fabre); B. Bonnaud *et al.*, *J. Med. Chem.* **30**, 318 (1987). LC determn with spectrofluorimetric detection in plasma: C. Puozzo *et al.*, *J. Chromatogr. B* **806**, 221 (2004). Mechanism of action study: C. Moret *et al.*, *Neuropharmacology* **24**, 1211 (1985). Clinical inhibition of monoamine uptake: C. Palmier *et al.*, *Eur. J. Clin. Pharmacol.* **37**, 235 (1989). Clinical trial in fibromyalgia: D. J. Clauw *et al.*, *Clin. Ther.* **30**, 1988 (2008). Review of pharmacology and clinical efficacy in depression: C. M. Spencer, M. I. Wilde, *Drugs* **56**, 405-427 (1998).

Relative stereochemistry

Hydrochloride. [101152-94-7] F-2207; Ixel; Savella; Toledomin. $C_{15}H_{22}N_2O$·HCl; mol wt 282.81. Crystals from ethanol-ethyl ether, mp 179-181°. LD_{50} orally in mice: 237 mg/kg (Bonnaud).

THERAP CAT: Antidepressant.

6276. Milorganite®. [8049-99-8] Activated sewage sludge prepd by microbial treatment of sewage, useful as fertilizer and as a source of vitamin B_{12}: Schendel, Johnson, *J. Agric. Food Chem.* **2**, 23 (1954). Isoln of B_{12}-active product: Miner, Wolnak, **US 2646386** (1953 to Sewerage Commission of Milwaukee); **US 2941933** (1960 to UCLAF).

Caution: The dust can cause irritation of eyes, respiratory tract.

6277. Miloxacin. [37065-29-5] 5,8-Dihydro-5-methoxy-8-oxo-1,3-dioxolo[4,5-*g*]quinoline-7-carboxylic acid; 6,7-methylene-dioxy-1-methoxy-4-oxo-1,4-dihydroquinoline-3-carboxylic acid; antibiotic AB 206; AB-206; Fuldazin. $C_{12}H_9NO_6$; mol wt 263.21. C 54.76%, H 3.45%, N 5.32%, O 36.47%. Quinolone antibacterial; analog of oxolinic acid, *q.v.* Prepn: T. Nakagome *et al.*, **DE 2134451**; *eidem*, **US 3799930** (1972, 1974 both to Sumitomo). Mode of action: T. Nagate *et al.*, *Antimicrob. Agents Chemother.* **17**, 763 (1980). Series of articles on antibacterial activity, absorption and excretion in animals, determn of metabolites by HPLC: *ibid.* **18**, 37-49 (1980). Reproduction studies in rats: T. Yamada *et al.*, *Oyo Yakuri* **19**, 651, 815, 833 (1980), *C.A.* **93**, 197897x, 215570s, 215571t (1980).

Colorless prisms from DMF, mp 264° (dec).

THERAP CAT: Antibacterial.

6278. Milrinone. [78415-72-2] 1,6-Dihydro-2-methyl-6-oxo-[3,4′-bipyridine]-5-carbonitrile; 1,2-dihydro-6-methyl-2-oxo-5-(4-pyridinyl)nicotinonitrile; Win-47203; Corotrope; Milrila. $C_{12}H_9N_3O$; mol wt 211.22. C 68.24%, H 4.30%, N 19.89%, O 7.57%. Selective phosphodiesterase inhibitor with vasodilating and positive inotropic activity. Prepn: G. Y. Lesher *et al.*, **BE 886336**; G. Y. Lesher, R. E. Philion, **US 4313951** (1981, 1982 both to Sterling). Improved process: B. Singh, **US 4413127** (1983 to Sterling); *idem*, *Heterocycles* **23**, 1479 (1985). Pharmacological profile: A. A. Alousi *et al.*, *J. Cardiovasc. Pharmacol.* **5**, 792, 804 (1983). HPLC determn in body fluids: J. Edelson *et al.*, *J. Chromatogr.* **276**, 456 (1983). Clinical studies in congestive heart failure: C. S. Maskin *et al.*, *Circulation* **67**, 1065 (1983); L. S. Sinoway *et al.*, *J. Am. Coll. Cardiol.* **2**, 327 (1983); D. S. Baim *et al.*, *N. Engl. J. Med.* **309**, 748 (1983).

Crystals from DMF + water or from ethanol; hygroscopic. mp >300°. Freely sol in dimethyl sulfoxide; very slightly sol in methanol. Practically insol in water, chloroform.

Lactate. [100286-97-3] Primacor. $C_{12}H_9N_3O$·x$C_3H_6O_3$

THERAP CAT: Cardiotonic.

6279. Miltefosine. [58066-85-6] 2-[[(Hexadecyloxy)hydroxyphosphinyl]oxy]-*N*,*N*,*N*-trimethylethanaminium inner salt; choline phosphate hexadecyl ester, hydroxide, inner salt; hexadecyl 2-(*N*,*N*,*N*-trimethylamino)ethyl phosphate; *n*-hexadecylphosphorylcholine; hexadecylphosphocholine; HPC; D-18506; Impavido; Miltex. $C_{21}H_{46}NO_4P$; mol wt 407.58. C 61.89%, H 11.38%, N 3.44%, O 15.70%, P 7.60%. Synthetic phospholipid; affects cell-signaling pathways and membrane synthesis. Synthesis: U. Kaatze *et al.*, *Chem. Phys. Lipids* **27**, 263 (1980); W. J. Hansen *et al.*, *Lipids* **17**, 453 (1982). *See also:* H. Eibl, **EP 225608**; *idem*, **US 4837023** (1987, 1989 both to Max-Planck Ges. Wissensch.). Toxicity and anticancer properties: C. Muschiol *et al.*, *Lipids* **22**, 930 (1987). Structure-activity study: M. R. Berger *et al.*, *Cancer Treat. Rev.* **17**, 143 (1990). HPTLC determn in plasma: I. Rustenbeck, S. Lenzen, *J. Chromatogr.* **525**, 85 (1990). Clinical evaluation in topical treatment of cutaneous breast cancer metastases: C. Unger *et al.*, *Cancer Treat. Rev.* **17**, 243 (1990). Clinical trial in visceral leishmaniasis: S. Sundar *et al.*, *N. Engl. J. Med.* **347**, 1739 (2002). *Review:* H. Eibl, C. Unger, *Cancer Treat. Rev.* **17**, 233-242 (1990).

mp 232-234° (dec); discoloration at 225°. LD$_{50}$ in rats (mg/kg): 246 orally (Muschiol).

THERAP CAT: Antineoplastic; antiprotozoal (Leishmania).

6280.　Mimosine. [500-44-7] (α*S*)-α-Amino-3-hydroxy-4-oxo-1(4*H*)-pyridinepropanoic acid; 3-hydroxy-4-oxo-1(4*H*)-pyridinealanine; β-[*N*-(3-hydroxy-4-pyridone)]-α-aminopropionic acid; leucenol; leucenine; leucaenine; leucaenol. C$_8$H$_{10}$N$_2$O$_4$; mol wt 198.18. C 48.49%, H 5.09%, N 14.14%, O 32.29%. A naturally occurring amino acid found in large quantities in the seeds and foliage of the legume genera *Mimosa* and *Leucena*. Isoln from *Leucena glauca* (Willd.) Benth., *Leguminosae:* Renz, *Z. Physiol. Chem.* **244**, 153 (1936); Mascré, *Compt. Rend.* **204**, 890 (1937); improved large-scale isoln: N. K. Hart *et al.*, *Heterocycles* **7**, 265 (1977). Structure: R. Adams, V. V. Jones, *J. Am. Chem. Soc.* **69**, 1803 (1947). Synthesis: R. Adams, J. L. Johnson, *ibid.* **71**, 705 (1949). Identity of mimosine and leucenol: Kleipool, Wibaut, *Rec. Trav. Chim.* **69**, 37 (1950); Wibaut, Schuhmacher, *ibid.* **71**, 1017 (1952). Crystal structure: A. Mostad *et al.*, *Acta Chem. Scand.* **27**, 164 (1973). Shown to cause inhibition of hair growth and loss of hair in mice: Crounse *et al.*, *Nature* **194**, 694 (1962). Use as depilatory agent: Hegarty *et al.*, *Aust. J. Agric. Res.* **15**, 153 (1964). Explanation of the toxicity in animals: J. F. Thompson *et al.*, *Annu. Rev. Biochem.* **38**, 137 (1969).

White needles from water, mp 228-229° (dec). [α]$_D^{22}$ −20°.

dl-**Form.** [2116-55-4] White needles from water, mp 235-236° with decompn. Slightly sol in water, much less sol in methanol and ethanol. Practically insol in the higher alcohols, dioxane, ethyl acetate, ether, benzene, chloroform, glacial acetic acid, pyridine, Cellosolve. Sol in dil acids or bases and may be recovered from these solns by adjusting the pH so that it is just acid to bromcresol green. uv max: 282 nm (log ε 4.23).

dl-**Hemihydrate.** C$_8$H$_{10}$N$_2$O$_4$.½H$_2$O; mol wt 207.19. White needles, darkens at 215-226°, mp 227-228° (dec).

Hydrochloride. [62766-26-1] C$_8$H$_{10}$N$_2$O$_4$.HCl; mol wt 234.64. Dec 175°. Sol in water.

Hydrobromide. C$_8$H$_{10}$N$_2$O$_4$.HBr; mol wt 279.09. Dec 179.5°. Sol in water.

USE: Depilatory agent.

6281.　Minaprine. [25905-77-5] *N*-(4-Methyl-6-phenyl-3-pyridazinyl)-4-morpholineethanamine; 4-[2-[(4-methyl-6-phenyl-3-pyridazinyl)amino]ethyl]morpholine. C$_{17}$H$_{22}$N$_4$O; mol wt 298.39. C 68.43%, H 7.43%, N 18.78%, O 5.36%. Prepn: H. Laborit, **DE 2229215**; *idem*, **US 4169158** (1973, 1979 to Centre Etudes Exper. Clin. Physio-Biologie); C.-G. Wermuth, A. Exinger, *Agressologie* **13**, 285 (1972). Pharmacology: H. Laborit *et al.*, *ibid.* 291. Metabolism of ^{14}C-minaprine: A. G. Rico *et al.*, *J. Pharmacol.* **9**, 170 (1978). Pharmacokinetics: J. P. Jeanniot *et al.*, *ibid.* 169. Clinical evaluation: A. Garcia-Maffla, M. H. de Garcia, *Pharmatherapeutica* **2**, 265 (1979). Pharmacological evaluation in depression: K. Bizière *et al.*, *Arzneim.-Forsch.* **32**, 824 (1982). Toxicologic study: A. G. Mazure *et al.*, *Agressologie* **13**, 319 (1972).

Fine buff-colored needles from isopropanol, mp 122°. Insol in water. Slightly sol in cold ethanol, sol in hot ethanol, chloroform.

Dihydrochloride. [25953-17-7] Agr-1240; CB-30038; Brantur; Cantor. C$_{17}$H$_{22}$N$_4$O.2HCl; mol wt 371.31. Crystals from abs ethanol, mp 182°.

THERAP CAT: Antidepressant.

6282.　Mineral Spirits. [64475-85-0] Petroleum spirits; white spirits; turpentine substitutes. Name applied to various types of hydrocarbon solvents, primarily petroleum distillates, which have flash points above 100°F (38°C) and distillation ranges between 300°F (149°C) and 415°F (213°C). *See: ASTM Standard Specifications* **D 235-83,** 71-73 (1983). Review of toxicology and human exposure: *Toxicological Profile for Stoddard Solvent* (PB95-264263, 1995) 169 pp.

Type I. [8052-41-3] Stoddard solvent; Texsolve S; Varsol 1. Regular mineral spirits. Clear liquid. Flash pt (min): 38°C (100°F). d$_{15.6}^{15.6}$ 0.754-0.820. Distillation range: initial bp (min), 149°C (300°F); 50% recovered (max), 182°C (360°F); dry pt (max), 208°C (407°F).

Type II. High flash point mineral spirits. Flash pt (min): 60°C (140°F). d$_{15.6}^{15.6}$ 0.768-0.820. Distillation range: initial bp (min), 177°C (350°F); 50% recovered (max), 196°C (385°F); dry pt (max), 211°C (412°F).

Type III. Odorless mineral spirits. Flash pt (min): 38°C (100°F). d$_{15.6}^{15.6}$ 0.775 (max). Distillation range: initial bp (min), 149°C (300°F); 50% recovered (max), 196°C (385°F); dry pt (max), 213°C (415°F).

Type IV. Texsolve S-2; Varsol 3. Low dry point mineral spirits. Flash pt (min): 38°C (100°F). d$_{15.6}^{15.6}$ 0.754-0.800. Distillation range: initial bp (min), 149°C (300°F); 50% recovered (max), 174°C (345°F); dry pt (max), 185°C (365°F).

Caution: Potential symptoms of overexposure to stoddard solvent are irritation of eyes, nose, throat; dizziness; dermatitis; aspiration of liquid may cause chemical pneumonia. *See NIOSH Pocket Guide to Chemical Hazards* (DHHS/NIOSH 97-140, 1997) p 286.

USE: Solvent; paint thinner. In the coatings and dry cleaning industries.

6283.　Minocycline. [10118-90-8] (4*S*,4a*S*,5a*R*,12a*S*)-4,7-Bis(dimethylamino)-1,4,4a,5,5a,6,11,12a-octahydro-3,10,12,12a-tetrahydroxy-1,11-dioxo-2-naphthacenecarboxamide; 7-dimethylamino-6-demethyl-6-deoxytetracycline. C$_{23}$H$_{27}$N$_3$O$_7$; mol wt 457.48. C 60.39%, H 5.95%, N 9.19%, O 24.48%. Second generation tetracycline antibiotic. Prepn: J. Petisi, J. H. Boothe, **US 3226436** (1965 to Am. Cyanamid); M. J. Martell, J. H. Boothe, *J. Med. Chem.* **10**, 44 (1967); R. F. R. Church *et al.*, *J. Org. Chem.* **36**, 723 (1971). HPLC determn in tissue and serum: W. R. Wrightson *et al.*, *J. Chromatogr. B* **706**, 358 (1998). Comprehensive description: V. Zbinovsky, G. P. Chrekian, *Anal. Profiles Drug Subs.* **6**, 323-339 (1977). Series of articles on toxicology and metabolism: *Toxicol. Appl. Pharmacol.* **11**, 128-183 (1967). Review of antibacterial activity and pharmacology: M. Jonas, B. A. Cunha, *Ther. Drug Monit.* **4**, 137-145 (1982); of clinical safety and efficacy in acne: K. Freeman, *Br. J. Clin. Pract.* **43**, 112-115 (1989). Clinical evaluation in periodontal disease: R. J. Oringer *et al.*, *J. Periodontol.* **73**, 835 (2002); in Huntington disease: Huntington Study Group, *Neurology* **63**, 547 (2004). Review of clinical potential in neurodegenerative disorders: D. Blum *et al.*, *Neurobiol. Dis.* **17**, 359-366 (2004); in acute neurologic injury: H. F. Elewa *et al.*, *Pharmacotherapy* **26**, 515-521 (2006).

Bright yellow-orange amorphous solid. [α]$_D^{25}$ −166° (c = 0.524). uv max (0.1*N* HCl): 352, 263 nm (log ε 4.16, 4.23); (0.1*N* NaOH): 380, 243 nm (log ε 4.30, 4.38).

Hydrochloride. [13614-98-7] Arestin; Dynacin; Klinomycin; Minocin; Vectrin. C$_{23}$H$_{27}$N$_3$O$_7$.HCl; mol wt 493.94. Yellow crystalline powder. Slightly hygroscopic; sensitive to light and to surface oxidation. Sol in solns of alkali hydroxides and carbonates; sparingly sol in water; slightly sol in alc. Practically insol in chloroform, ether.

THERAP CAT: Antibacterial.

THERAP CAT (VET): Antibacterial.

6284. Minodronic Acid. [180064-38-4] *P,P'*-(1-Hydroxy-2-imidazo[1,2-*a*]pyridin-3-ylethylidene)bisphosphonic acid; minodronate. $C_9H_{12}N_2O_7P_2$; mol wt 322.15. C 33.56%, H 3.75%, N 8.70%, O 34.76%, P 19.23%. Third generation bisphosphonate antiresorptive agent. Prepn: Y. Isomura *et al.*, **EP 354806**; *eidem*, **US 4990503** (1990, 1991 both to Yamaouchi). Structure-activity and X-ray crystallography: M. Takeuchi *et al.*, *Chem. Pharm. Bull.* **46**, 1703 (1998). HPLC determn in biological fluids and bone: T. Usui *et al.*, *J. Chromatogr. B* **652**, 67 (1994). *In vivo* antiresorptive activity: Y. Yoshida *et al.*, *J. Bone Miner. Res.* **13**, 1011 (1998). Clinical trial in postmenopausal osteoporosis: H. Hagino *et al.*, *Bone* **44**, 1078 (2009). Mechanism of action study: H. Tokuda *et al.*, *Mol. Med. Rep.* **3**, 167 (2010).

Monohydrate. [155648-60-5] ONO-5920; YM-529; Bonoteo; Recalbon. $C_9H_{12}N_2O_7P_2 \cdot H_2O$; mol wt 340.16. Triclinic crystals from 1*N* HCl, d 1.823. Hemihydrate as needle shaped crystals from water-methanol, mp 222-224° (dec).

THERAP CAT: Bone resorption inhibitor.

6285. Minoxidil. [38304-91-5] 6-(1-Piperidinyl)-2,4-pyrimidinediamine 3-oxide; 6-amino-1,2-dihydro-1-hydroxy-2-imino-4-piperidinopyrimidine; 2,3-dihydro-3-hydroxy-2-imino-6-(1-piperidinyl)-4-pyrimidinamine; 2,4-diamino-6-piperidinopyrimidine 3-oxide; 6-piperidino-2,4-diaminopyrimidine 3-oxide; PDP; U-10858; Alopexil; Alostil; Loniten; Lonolox; Minoximen; Normoxidil; Prexidil; Regaine; Rogaine; Tricoxidil. $C_9H_{15}N_5O$; mol wt 209.25. C 51.66%, H 7.23%, N 33.47%, O 7.65%. Prepn: **NL 6615385**; W. C. Anthony *et al.*, **US 3382247** (1967, 1968 both to Upjohn); J. M. McCall *et al.*, *J. Org. Chem.* **40**, 3304 (1975). *See also* W. C. Anthony, **US 3644364** (1972 to Upjohn). Metabolism: R. C. Thomas *et al.*, *J. Pharm. Sci.* **64**, 1360 (1975). Pharmacology and pharmacokinetics: D. T. Lowenthal *et al.*, *J. Clin. Pharmacol.* **18**, 500 (1978). Percutaneous absorption and excretion: T. J. Franz, *Arch. Dermatol.* **121**, 203 (1985). Clinical studies: O. Andersson, R. Sivertsson, *Acta Med. Scand.* **205**, 213 (1979); M. Moser, *Adv. Cardiol.* **26**, 38 (1979). Clinical trial in early male pattern baldness: E. A. Olsen *et al.*, *J. Am. Acad. Dermatol.* **13**, 185 (1985). Toxicology: R. G. Carlson, E. S. Feenstra, *Toxicol. Appl. Pharmacol.* **39**, 1 (1977). Review of pharmacology and therapeutic use: V. M. Campese, *Drugs* **22**, 257-278 (1981). Review of topical application in baldness: E. Novak *et al.*, *Int. J. Dermatol.* **24**, 82 (1985). Comprehensive description: D. K. J. Gorecki, *Anal. Profiles Drug Subs.* **17**, 185-219 (1988).

Crystals from methanol-acetonitrile, mp 248°, dec 259-261° (Anthony, 1972). pKa 4.61. uv max (ethanol): 230, 261, 285 nm (ε 35210, 11210, 11790); (0.01*N* H_2SO_4): 232, 280 nm (ε 26350, 23850); (0.01*N* KOH): 231, 261.5, 285 nm (ε 36100, 11400,

12040). Soly (mg/ml): propylene glycol 75, methanol 44, ethanol 29, 2-propanol 6.7, dimethylsulfoxide 6.5, water 2.2, chloroform 0.5, acetone <0.5, ethylacetate <0.5, diethyl ether <0.5, benzene <0.5, acetonitrile <0.5. Practically insol in hexane. LD_{50} in rats, mice (mg/kg): 49, 51 i.v. (Carlson, Feenstra).

THERAP CAT: Antihypertensive. Antialopecia agent.

6286. Miokamycin. [55881-07-7] Leucomycin V 3^B,9-diacetate 3,4^B-dipropanoate; 9,3″-diacetylmidecamycin; MOM; ponsinomycin; Miocamycin. $C_{45}H_{71}NO_{17}$; mol wt 898.05. C 60.19%, H 7.97%, N 1.56%, O 30.29%. Macrolide antibiotic; deriv of midecamycin A_1. Prepn: S. Inoue *et al.*, **JP Kokai 74 124087**; *eidem*, **US 4017607** (1974, 1977 both to Meiji Seika); S. Omoto *et al.*, *J. Antibiot.* **29**, 536 (1976); T. Nakamura *et al.*, *Chem. Lett.* **1978**, 1293. Prepn of noncrystalline form and comparison of physico-pharmaceutical properties: T. Sato *et al.*, *Chem. Pharm. Bull.* **29**, 2675 (1981). Antibacterial spectrum: K. Kawaharajo *et al.*, *J. Antibiot.* **34**, 436 (1981). Metabolism: T. Shomura *et al.*, *Chem. Pharm. Bull.* **29**, 2413 (1981). Laboratory and clinical pediatric studies: Y. Toyonaga *et al.*, *Jpn. J. Antibiot.* **35**, 1475 (1982), *C.A.* **98**, 216y (1983). Clinical bioavailability: F. Fraschini *et al.*, *Int. J. Clin. Pharmacol. Res.* **10**, 293 (1989).

Tasteless crystals from isopropanol, mp ~220° (with coloration). $[\alpha]_D^{25}$ −53° (c = 1.0 in chloroform). $[\alpha]_D^{20}$ −74° (c = 1 in methanol). uv max (methanol): 231 nm ($E_{1cm}^{1\%}$ 342). Sol in methanol, acetone, chloroform. Very slightly sol in water.

THERAP CAT: Antibacterial.

6287. Mipafox. [371-86-8] *N,N'*-Bis(1-methylethyl)phosphorodiamidic fluoride; *N,N'*-diisopropylphosphorodiamidic fluoride; bis(isopropylamino)fluorophosphine oxide. $C_6H_{16}FN_2OP$; mol wt 182.18. C 39.56%, H 8.85%, F 10.43%, N 15.38%, O 8.78%, P 17.00%. Prepn: Pound *et al.*, **GB 688787** (1953 to Fisons); Heath, *J. Chem. Soc.* **1956**, 3796.

Crystals, mp 65°.

Caution: Cholinesterase inhibitor.

USE: Insecticide.

6288. Mipomersen. [1000120-98-8] d(*P*-Thio)([2'-*O*-(2-methoxyethyl)]rG-[2'-*O*-(2-methoxyethyl)]m5rC-[2'-*O*-(2-methoxyethyl)]m5rC-[2'-*O*-(2-methoxyethyl)]m5rU-[2'-*O*-(2-methoxyethyl)]m5rC-A-G-T-m5C-T-G-m5C-T-T-m5C-[2'-*O*-(2-methoxyethyl)]rG-[2'-*O*-(2-methoxyethyl)]m5rC-[2'-*O*-(2-methoxyethyl)]-rA-[2'-*O*-(2-methoxyethyl)]m5rC-[2'-*O*-(2-methoxyethyl)]m5rC)-DNA. Second-generation, 2'-*O*-methoxyethyl-modified antisense oligonucleotide (ASO); designed to inhibit human apolipoprotein B-100 (apoB-100), the principal structural component of atherogenic lipoproteins including very low density lipoprotein (VLDL), intermediate-density lipoprotein, and low-density lipoprotein cholesterol (LDL-C). The 20-mer oligonucleotide is complementary to the coding region for human apoB-100 mRNA; hybridization to apoB-100 mRNA inhibits translation of the apoB gene. Prepn: R. M Crooke *et al.*, **WO 04044181** (2004 to ISIS); *eidem*, **US 7511131** (2009 to Genzyme). Effects in lipoprotein(a) transgenic mice: E. Merki *et al.*, *Circulation* **118**, 743 (2008). Pharmacokinetics in rodents, monkeys, and humans: R. Z. Yu *et al.*, *Drug Metab. Dispos.* **35**, 460

(2007). Clinical trial in familial hypercholesterolemia: F. J. Raal *et al.*, *Lancet* **375**, 998 (2010). Review of clinical experience: D. A. Bell *et al.*, *Expert Opin. Invest. Drugs* **20**, 265-272 (2011).

Sodium salt. [629167-92-6] ISIS-301012. $C_{230}H_{305}N_{67}Na_{19}$-$O_{122}P_{19}S_{19}$; mol wt 7594.76.

THERAP CAT: Antilipemic.

6289. Miraculin. Taste-modifying protein; Miralin. A basic glycoprotein of probable mol wt 44,000. Responsible for the taste-changing properties of **Miracle fruit** (*agbayun*), the red berries of *Synsepalum dulcificum* (Schum.) Daniell, *Sapotaceae* [alternate name: *Richardella dulcifica* (Schum. and Thonn) Baehni], a shrub indigenous to tropical W. Africa. Sour materials taste sweet after the tasteless mucilaginous pulp of the fruit has been applied to the tongue. First description: Daniell, *Pharm. J.* **11**, 445 (1852). Isolation studies: Inglett *et al.*, *J. Agric. Food Chem.* **13**, 284 (1965); Kurihara, Beidler, *Science* **161**, 1241 (1968). Isoln and purif of miraculin: Brouwer *et al.*, *Nature* **220**, 373 (1968); **NL 6911954** corresp to Brouwer *et al.*, **US 3682880** (1970, 1972 to Lever Bros.). Max sweetening effect occurs when $4 \times 10^{-7} M$ (in 0.02 M citric acid) is held in the mouth for 3 mins. The protein probably binds to receptors of the taste buds and modifies their function. Mechanism of action studies: Kurihara, Beidler, *Nature* **222**, 1176 (1969). *Reviews:* Henning, *Pharm. Weekbl.* **106**, 271 (1971); Cagan, *Science* **181**, 32 (1973).

USE: Sweetening agent.

6290. Mirex. [2385-85-5] 1,1a,2,2,3,3a,4,5,5,5a,5b,6-Dodecachlorooctahydro-1,3,4-metheno-1*H*-cyclobuta[*cd*]pentalene; perchloropentacyclo[5.2.1.0^{2,6}.0^{3,9}.0^{5,8}]decane; hexachloropentadiene dimer; GC-1283; ENT-025719; Dechlorane. $C_{10}Cl_{12}$; mol wt 545.51. C 22.02%, Cl 77.98%. Prepn: Prins, *Rec. Trav. Chim.* **65**, 455 (1946); Newcomer, McBee, *J. Am. Chem. Soc.* **71**, 952 (1949); Gilbert, **US 2702305** (1955 to Allied Chem.); Johnson, **US 2724710** (1955 to Hooker Electrochem.). Structure: McBee *et al.*, *J. Am. Chem. Soc.* **78**, 1511 (1956). Degradation in the environment to Kepone and related compds: D. A. Carlson *et al.*, *Science* **194**, 939 (1976). GC-MS determn in sediment: M.-S. Kim *et al.*, *J. Chromatogr. A* **1208**, 25 (2008). Carcinogenicity study: B. M. Ullard *et al.*, *J. Natl. Cancer Inst.* **58**, 133 (1977). Toxicity data: T. B. Gaines, *Toxicol. Appl. Pharmacol.* **14**, 515 (1969). *Reviews:* Ungnade, McBee, *Chem. Rev.* **58**, 249-320 (1958); Alley, *J. Environ. Qual.* **2**, 52-61 (1973); E. M. Waters *et al.*, *Environ. Res.* **14**, 212-222 (1977); K. Kaiser, *Environ. Sci. Technol.* **12**, 520-528 (1978). Review of toxicology and human exposure: *Toxicological Profile for Mirex and Chlordecone* (PB95-264354, 1995) 362 pp.

Snow-white odorless crystals from benzene, dec 485°. Supports combustion. Practically insol in water. Soly at room temp: dioxane 15.3%; xylene 14.3%; benzene 12.2%; carbon tetrachloride 7.2%; methyl ethyl ketone 5.6%. Practically noncorrosive to metals. LD$_{50}$ orally in female rats (corn oil suspension): 600 mg/kg (Gaines).

Caution: This substance is reasonably anticipated to be a human carcinogen: *Report on Carcinogens, Twelfth Edition* (PB2011-111646, 2011) p 273.

Note: Mirex is listed as a persistent organic pollutant (POP) in Annex A of the *Stockholm Convention on Persistent Organic Pollutants* (United Nations, Stockholm, 2001) 43 pp; amended (Geneva, 2009) 63 pp.

USE: Formerly as fire retardant for plastics, rubber, paint, paper, electrical goods; insecticide for fire ants, termintes, mealybugs.

6291. Mirtazapine. [61337-67-5] 1,2,3,4,10,14b-Hexahydro-2-methylpyrazino[2,1-*a*]pyrido[2,3-*c*][2]benzazepine; 2-methyl-1,2,3,4,10,14b-hexahydrobenzo[*c*]pyrazino[1,2-*a*]pyrido[3,2-*f*]-

azepine; 6-azamianserin; mepirzepine; Org-3770; Remeron; Zispin. $C_{17}H_{19}N_3$; mol wt 265.36. C 76.95%, H 7.22%, N 15.84%. α_2-Adrenergic blocker; analog of mianserin, *q.v.* Prepn: W. J. Van der Burg, **DE 2614406**; *idem*, **US 4062848** (1976, 1977 both to AKZO). GC determn in plasma: J. E. Paanakker, H. J. M. Van Hal, *J. Chromatogr.* **417**, 203 (1987). Pharmacology and mechanism of action study: T. De Boer *et al.*, *Neuropharmacology* **27**, 399 (1988). Clinical comparison with amitriptyline, *q.v.*: W. T. Smith *et al.*, *Psychopharmacol. Bull.* **26**, 191 (1990). Review of clinical experience in mood and anxiety disorders: M. J. Ostacher *et al.*, *Expert Rev. Neurother.* **3**, 425-433 (2003); and of clinical pharmacokinetics, safety and efficacy: A. Szegedi, N. Schwertfeger, *Expert Opin. Pharmacother.* **6**, 631-641 (2005).

White crystals from petr ether, mp 114-116°. Freely sol in methanol, toluene; sol in chloroform, ethyl ether; sparingly sol in *n*-hexane. Practically insol in water.

THERAP CAT: Antidepressant.

6292. Misch Metal. [8049-20-5] Alloy of mixed rare earth metals. Prepd by electrolysis of fused rare earth chlorides. Obtained from the lanthanide ore, **bastnaesite**. A typical composition is 48-50% cerium, 32-34% lanthanum, 13-14% neodymium, 4-5% praseodymium and 1.5% other rare earth metals. Average mol wt is 140. Uses in organic synthesis: M.-I. Lannou *et al.*, *Tetrahedron* **59**, 10551 (2003).

Pyrophoric powder. Store under argon.

USE: In the manuf of flints for gas and cigarette lighters. Coreductant or promoter in organic synthesis. In metallurgy: decreases microporosity and improves stress resistance, thermomalleability and heat exchange in refined steel.

6293. Misoprostol. [59122-46-2] (11α,13E)-($\pm$)-11,16-Dihydroxy-16-methyl-9-oxoprost-13-en-1-oic acid methyl ester; ($\pm$)-methyl-(1R,2R,3R)-3-hydroxy-2-[(E)-(4RS)-4-hydroxy-4-methyl-1-octenyl]-5-oxocyclopentaneheptanoate; ($\pm$)-15-deoxy-(16RS)-16-hydroxy-16-methyl-PGE$_1$ methyl ester; SC-29333; Cytotec; Gymiso; Misodex. $C_{22}H_{38}O_5$; mol wt 382.54. C 69.08%, H 10.01%, O 20.91%. Cytoprotective prostaglandin PGE$_1$ analog; also exhibits uterotonic and cervical-ripening actions. Double racemate comprised of the (+)- and (−)-enantiomers of the 16R- and 16S-forms. The pharmacologically active form is the (11R,16S)-enantiomer. Prepn: P. W. Collins, R. Pappo, **BE 827127**; *eidem*, **US 3965143** (1975, 1976 both to Searle); P. W. Collins *et al.*, *Tetrahedron Lett.* **48**, 4217 (1975). Prepn, activity, NMR data: P. Collins *et al.*, *J. Med. Chem.* **20**, 1152 (1977). HPLC resolution of enantiomers: D. A. Roston, R. Wijayaratne, *Anal. Chem.* **60**, 948 (1988). Mechanism of gastric secretory inhibition: D. G. Colton *et al.*, *Arch. Int. Pharmacodyn. Ther.* **236**, 86 (1978). Symposium on pharmacology and clinical efficacy: *Dig. Dis. Sci.* **30**, Suppl. 11, 114S-205S (1985). Toxicology profile: F. N. Kotsonis *et al.*, *ibid.* 142S. Clinical trial in prevention of NSAID-induced ulcer: D. Y. Graham *et al.*, *Ann. Intern. Med.* **119**, 257 (1993); to induce labor: H. Fletcher *et al.*, *Obstet. Gynecol.* **83**, 244 (1994). Review of clinical experience in pregnancy: A. B. Goldberg *et al.*, *N. Engl. J. Med.* **344**, 38-47 (2001).

(11*R*, 16*S*)-Form

Light yellow oil. Sol in water. LD$_{50}$ in rats, mice (mg/kg): 40-62, 70-160 i.p.; 81-100, 27-138 orally (Kotsonis).

THERAP CAT: Antiulcerative.

THERAP CAT (VET): Antiulcerative.

6294. Mistletoe. Mistel. Common name for a diverse group of hemiparasitic, evergreen plants that grow within the branches of various trees such as apple, hawthorn, pine, and oak; capable of photosynthesis but obtaining water and nutrients from the host. Bears white, waxy berries with sticky pulp. European mistletoe, *Viscum album* L., *Viscaceae*, has been used in traditional medicine to treat epilepsy, hypertension, rheumatism, and cancer. Medicinal formulations are prepd from the leaves and young stems, and occasionally the berries. *Habit.* Temperate regions of Europe, Great Britain. *Constit.* Toxic mistletoe lectins; polypeptide viscotoxins; flavonoids such as quercetin, isorhamnetin, sakuranetin; terpenoids such as α- and β-amyrin, betulinic acid, oleanolic acid; phenolpropanoids and lignins, such as syringin; phytosterols. Constituents thought to be dependent upon the host plant. Prominant North American species is *Phoradendron leucarpum*, *Viscaceae*. Other species occur in subtropical and tropical climates. Botanical description: D. M. Watson, *Annu. Rev. Ecol. Syst.* **32**, 219 (2001); of *V. album*: D. Zuber, *Flora* **199**, 181 (2004). Pharmacology of standardized medicinal preparations: U. Mengs *et al.*, *Anticancer Res.* **22**, 1399 (2002). Review of constituents: U. Pfüller, in *Mistletoe, The Genus Viscum* (Harwood Acad. Publ., Amsterdam, 2000) pp 101-122; of toxicology: G. M. Stein, *ibid.* pp 183-194. Comprehensive description and medicinal uses: J. Barnes *et al.*, *Herbal Medicines* (Pharmaceutical Press, London, 3rd Ed., 2007) pp 436-446.

Mistletoe extract. [8031-76-3] Iscador; Isorel; Lektinol; Plenosol. Aqueous extract containing lectins and viscotoxins; exhibits cytotoxic and immunostimulant effects. Analysis of constituents of fermented extract: K. Urech *et al.*, *Arzneim.-Forsch.* **56**, 428 (2006). Clinical trial in cancer therapy: R. Grossarth-Maticek *et al.*, *Altern. Ther. Health Med.* **7**, 57-76 (2001).

THERAP CAT: Anti-inflammatory; immunostimulant adjuvant in cancer treatment.

6295. Mistletoe Lectins. Toxic glycoproteins isolated from European mistletoe, *Viscum album* L., *Viscaceae*, and related species. Type II ribosome inactivating proteins, similar to ricin, *q.v.*; inhibit protein synthesis and induce apoptosis. Comprised of 2 polypeptide chains, the cytotoxic A chain is linked to the carbohydrate-binding B chain via a disulfide bridge. Three isoforms have been identified that differ in carbohydrate binding specificity. Isoln of ML-I: P. Ziska *et al.*, *Experientia* **34**, 123 (1978); P. Luther *et al.*, *Int. J. Biochem.* **11**, 429 (1980); S. Olsnes *et al.*, *J. Biol. Chem.* **257**, 13263 (1982); of I, II and III: H. Franz *et al.*, *Biochem. J.*, **195**, 481 (1981). Characterization and bioactivities: *eidem*, *Oncology* **43**, Suppl. 1, 22 (1986). Amino acid sequence of ML-I A chain: M. H. Soler *et al.*, *FEBS Lett.* **399**, 153 (1996); of B chain: *idem et al.*, *Biochem. Biophys. Res. Commun.* **246**, 596 (1998). Crystal structure of ML-I: R. Krauspenhaar *et al.*, *ibid.* **257**, 418 (1999). Structure of ML-III A chain: R. Wacker *et al.*, *J. Pept. Sci.* **10**, 138 (2004); of B chain: *idem et al.*, *ibid.* **11**, 289 (2005). Review of cytotoxic and immunostimulant activities: T. Hajtó *et al.*, *Evid. Based Complement. Alternat. Med.* **2**, 59-67 (2005).

Mistletoe Lectin I. ML-I; *Viscum album* agglutinin I; VAA-I; viscumin. Mol wt 66 kDa; also exists as a dimer in soln, mol wt 115 kDa. Mol wt of A chain, 29 kDa; of B chain, 34 kDa. Carbohydrate specificity: sialic acid and D-galactose. Isoelectric point: pH 6.1. Sol in phosphate buffered saline. LD$_{50}$ i.p. in mice: 28 μg/kg (Franz, 1986).

Mistletoe Lectin II. ML-II. Mol wt 60 kDa. Mol wt of A chain, 27 kDa; of B chain, 32 kDa. Carbohydrate specificity: D-galactose and *N*-acetyl-D-galactosamine. LD$_{50}$ i.p. in mice: 1.5 μg/kg (Franz, 1986).

Mistletoe Lectin III. ML-III. Mol wt 50 kDa. Mol wt of A chain, 25 kDa; of B chain, 30 kDa. Carbohydrate specificity: *N*-acetyl-D-galactosamine. LD$_{50}$ i.p. in mice: 55 μg/kg (Franz, 1986).

Aviscumine. [223577-45-5] Toxin ML-I (mistletoe lectin I) (*Viscum album*); rML-I. Nonglycosylated ML-I produced by recombinant technology in *E. coli*. mol wt 57 kDa. Cloning and characterization: J. Eck *et al.*, *Eur. J. Biochem.* **264**, 775 (1999); *idem et al.*, *ibid.* **265**, 788 (1999). Determn in plasma by immuno-polymerase chain reaction (IPCR) assay: M. Adler *et al.*, *J. Pharm. Biomed.*

Anal. **39**, 972 (2005). Clinical pharmacology in patients with solid tumors: P. Schöffski *et al.*, *Eur. J. Cancer* **41**, 1431 (2005).

6296. Mitiglinide. [145375-43-5] (α*S*,3a*R*,7a*S*)-Octahydro-γ-oxo-α-(phenylmethyl)-2*H*-isoindole-2-butanoic acid; (2*S*)-2-benzyl-3-(*cis*-hexahydroisoindolin-2-ylcarbonyl)propionic acid. C$_{19}$H$_{25}$NO$_3$; mol wt 315.41. C 72.35%, H 7.99%, N 4.44%, O 15.22%. Insulin secretogogue for treatment of type 2 diabetes; inhibits ATP-sensitive potassium (K$_{ATP}$) channels. Prepn: F. Sato *et al.*, **EP 507534**; *eidem*, **US 5202335** (1992, 1993 both to Kissei); T. Yamaguchi *et al.*, *Chem. Pharm. Bull.* **45**, 1518 (1997); *eidem*, *ibid.* **46**, 337 (1998). Improved synthesis: J. Liu *et al.*, *Helv. Chim. Acta* **87**, 1935 (2004). 2D-NMR conformation study: L. Lins *et al.*, *Biochem. Pharmacol.* **52**, 1155 (1996). Effect on K$_{ATP}$ channels: F. Reimann *et al.*, *Br. J. Pharmacol.* **132**, 1542 (2001); on insulin secretion in normal and diabetic β-cells: N. Kaiser *et al.*, *ibid.* **146**, 872 (2005). HPLC/ESI-MS determn in plasma: J. Liang *et al.*, *J. Mass Spectrom.* **42**, 171 (2007). Clinical evaluation in type 2 diabetes: R. Assaloni *et al.*, *Diabetologia* **48**, 1919 (2005). Review of pharmacology and clinical experience: W. J. Malaisse, *Expert Opin. Pharmacother.* **9**, 2691-2698 (2008).

Viscous oil. [α]$_D^{24}$ −3.2° (c = 1.04 in methanol); [α]$_D^{18}$ −3.5° (c = 1.00 in methanol).

Calcium salt dihydrate. [207844-01-7]; [145525-41-3] (anhydrous). KAD-1229; S-21403; Glufast. C$_{38}$H$_{48}$CaN$_2$O$_6$.2H$_2$O; mol wt 704.92. Colorless crystals from 5% aqueous ethanol, mp 179-185° C. [α]$_D^{18}$ +5.7° (c = 1.0 in methanol).

THERAP CAT: Antidiabetic.

6297. Mitobronitol. [488-41-5] 1,6-Dibromo-1,6-dideoxy-D-mannitol; 1,6-dideoxy-1,6-dibromo-D-mannitol; dibromomannitol; DBM; Myebrol; Myelobromol. C$_6$H$_{12}$Br$_2$O$_4$; mol wt 307.97. C 23.40%, H 3.93%, Br 51.89%, O 20.78%. Cytostatic agent; the first dibromohexitol introduced into clinical use. Manuf: **GB 959407** (1964 to Chinoin). Pharmaco-biochemical studies: J. Szabo *et al.*, *Neoplasma* **20**, 13 (1973). Clinical studies: D. F. Chiuten *et al.*, *Cancer* **47**, 442 (1981). Comparative study with busulfan, *q.v.*: R. T. Silver *et al.*, *Cancer* **60**, 1442 (1987).

Crystals from methanol + dichloroethane, mp 176-178°.

THERAP CAT: Antineoplastic.

6298. Mitoguazone. [459-86-9] 2,2′-(1-Methyl-1,2-ethanediylidene)bis[hydrazinecarboximidamide]; 1,1′-[(methylethanediylidene)dinitrilo]diguanidine; methylglyoxal bis(guanylhydrazone); pyruvaldehyde bis(amidinohydrazone). C$_5$H$_{12}$N$_8$; mol wt 184.21. C 32.60%, H 6.57%, N 60.83%. Polyamine biosynthesis inhibitor. Prepn: F. Baiocchi *et al.*, *J. Med. Chem.* **6**, 431 (1963); V. T. Oliverio, C. Denham, *J. Pharm. Sci.* **52**, 202 (1963). Experimentally effective against myelogenous leukemia: Freireich *et al.*, *Cancer Chemother. Rep.* **16**, 183 (1962). HPLC determn in plasma: C.-L. Cheng *et al.*, *J. Chromatogr. B* **793**, 281 (2003). Review of clinical experience: D. D. Von Hoff, *Ann. Oncol.* **5**, 487-493 (1994).

Prepd as the hemihydrate; off-white crystals from isopropyl alcohol, dec 225°. uv max at pH 1: 283 nm (ε 38400); at pH 11: 325 nm (ε 33500). pKa: 7.63, 9.05.

Dihydrochloride. [7059-23-6] NSC-32946; Methyl-GAG. $C_5H_{12}N_8 \cdot 2HCl$; mol wt 257.12. Crystals from acetone, dec 256-257°.

THERAP CAT: Antineoplastic.

6299. Mitolactol. [10318-26-0] 1,6-Dibromo-1,6-dideoxy-galactitol; 1,6-dibromo-1,6-dideoxydulcitol; dibromodulcitol; dibromodulcit; DBD; NSC-104800; Elobromol. $C_6H_{12}Br_2O_4$; mol wt 307.97. C 23.40%, H 3.93%, Br 51.89%, O 20.78%. Diastereoisomer of mitobronitol, q.v., with myelosuppressive effects. Prepn and anticancer activity: **NL 6600395**; P. Horváth-Lengyel et al., **US 3993781** (1966, 1976 both to Chinoin); B. Kellner et al., Nature **213**, 402 (1967). Structure-activity correlations: L. Institoris et al., Arzneim.-Forsch. **17**, 145 (1967). Pharmacological study: T. H. Corbett et al., Cancer **40**, Suppl. 5, 2660 (1977). Metabolism, pharmacokinetics: I. P. Horváth et al., Eur. J. Cancer **15**, 337 (1979). Clinical studies: W. Medina, J. M. Kirkwood, Cancer Treat. Rep. **66**, 195 (1982); D. C. Tormey et al., Am. J. Clin. Oncol. **5**, 33 (1982).

$$CH_2Br$$
$$HC-OH$$
$$HO-CH$$
$$HO-CH$$
$$HC-OH$$
$$CH_2Br$$

Crystals from methanol, aq alc, or aq acetone, mp 187-188° (dec). LD_{50} in rats (mg/kg): 1400 orally; 470 i.p. (Kellner).

THERAP CAT: Antineoplastic.

6300. Mitomycins. [1404-00-8] A group of antitumor antibiotics produced by Streptomyces caespitosus (griseovinaceseus). Isoln of mitomycins A and B: T. Hata et al., J. Antibiot. **9A**, 141 (1956). Isoln of C: S. Wakaki et al., Antibiot. Chemother. **8**, 228 (1958); **GB 830874**; T. Hata et al., **US 3660578** (1960, 1972 both to Kyowa and Kitasato Kenkyusho). Production of the complex: T. Yamamoto, H. Umezawa, **JP 56 2898** (1956 to Nippon Antibiot. Subs. Sci. Assoc.), C.A. **51**, 9100 (1957). Improved method: A. Gourevitch et al., **US 3042582** (1962 to Bristol-Myers). Isoln from S. verticillatus, constituents: D. V. Lefemine et al., J. Am. Chem. Soc. **84**, 3184 (1962). Structures: J. S. Webb et al., ibid. 3185, 3187. Total synthesis of mitomycins A and C: T. Fukuyama et al., Tetrahedron Lett. **1977**, 4295. Toxicity: S. Kinoshita et al., J. Med. Chem. **14**, 13 (1971). Review and evaluation of studies of carcinogenic action of mitomycin C in laboratory animals: IARC Monographs **10**, 171-179 (1976). Review of syntheses: Y. Kishi, J. Nat. Prod. **42**, 549-568 (1979). General review: R. W. Franck, Fortschr. Chem. Org. Naturst. **38**, 1-41 (1979).

Mitomycin C

Mitomycin A. [4055-39-4] $C_{16}H_{19}N_3O_6$; mol wt 349.34. The 6-methoxy analog of mitomycin C. X-ray crystallographic study: A. Tulinsky, J. Am. Chem. Soc. **84**, 3188 (1962); A. Tulinsky, J. H. van den Hende, ibid. **89**, 2905 (1967). Red-violet crystals from acetone + carbon tetrachloride, dec 159-161°. Absorption max (water): 215, 318, 530 nm ($E_{1cm}^{1\%}$ 234, 122, 118.8). Sol in water, benzene, toluene, trichloroethylene, nitrobenzene and many organic solvents. Practically insol in xylene, carbon tetrachloride, carbon disulfide,

petr ether, ligroin, cyclohexane. LD_{50} i.v. in mice: 2 mg/kg (Hata, 1956; Kinoshita).

Mitomycin B. [4055-40-7] $C_{16}H_{19}N_3O_6$; mol wt 349.34. Molecular structure: R. Yahashi, I. Matsubara, J. Antibiot. **29**, 104 (1976); ibid. **31**(6), correction (1978). Violet crystals from acetone + carbon tetrachloride, dec 182-184°. Absorption max (water): 220, 320, 550 nm ($E_{1cm}^{1\%}$ 117.5, 55.0, 9.9). Sol in water and many organic solvents. Practically insol in xylene, carbon tetrachloride, carbon disulfide, petr ether, ligroin, cyclohexane, benzene, toluene, trichloroethylene, nitrobenzene. LD_{50} i.v. in mice: 10 mg/kg (Hata, 1956); also reported as 3 mg/kg (Kinoshita).

Mitomycin C. [50-07-7] [1aS-(1aα,8β,8aα,8bα)]-6-Amino-8-[[(aminocarbonyl)oxy]methyl]-1,1a,2,8,8a,8b-hexahydro-8a-methoxy-5-methylazirino[2',3':3,4]pyrrolo[1,2-a]indole-4,7-dione; MMC; Ametycine; Mitocin-C; Mutamycin. $C_{15}H_{18}N_4O_5$; mol wt 334.33. Chemistry and structure: C. L. Stevens et al., J. Med. Chem. **8**, 1 (1965). Crystal and molecular structure: K. Ogawa et al., Bull. Chem. Soc. Jpn. **52**, 2334 (1979). Revised absolute configuration: K. Shirahata, N. Hirayama, J. Am. Chem. Soc. **105**, 7199 (1983). Mechanism of action: C. Rodigheriero, Farmaco Ed. Sci. **33**, 651 (1978). Comprehensive description: J. H. Beijnen et al., Anal. Profiles Drug Subs. **16**, 361-401 (1986). Blue-violet crystals, does not melt below 360°. Absorption max (methanol): 216, 360, 560 nm ($E_{1cm}^{1\%}$ 742, 742, 0.06). Sol in water, methanol, acetone, butyl acetate and cyclohexanone; slightly sol in benzene, carbon tetrachloride, ether. Practically insol in petr ether. LD_{50} i.v. in mice: 5 mg/kg (Wakaki); also reported as 9 mg/kg (Kinoshita).

N-Methylmitomycin C see Porfiromycin.

THERAP CAT: Mitomycin C as antineoplastic.

6301. Mitotane. [53-19-0] 1-Chloro-2-[2,2-dichloro-1-(4-chlorophenyl)ethyl]benzene; 1,1-dichloro-2-(o-chlorophenyl)-2-(p-chlorophenyl)ethane; 2-(2-chlorophenyl)-2-(4-chlorophenyl)-1,1-dichloroethane; 2,4'-dichlorodiphenyldichloroethane; 2,2-bis(2-chlorophenyl-4-chlorophenyl)-1,1-dichloroethane; o,p'-DDD; CB-313; Lysodren. $C_{14}H_{10}Cl_4$; mol wt 320.03. C 52.54%, H 3.15%, Cl 44.31%. Constituent of commercial DDD, q.v., which contains about 10% of this o,p'-isomer. Prepn from 2,2-dichloro-1-(o-chlorophenyl)ethanol with chlorobenzene in presence of H_2SO_4: H. L. Haller et al., J. Am. Chem. Soc. **67**, 1591 (1945). Isoln from technical grade DDD: Cueto, Brown, Endocrinology **62**, 326 (1958). HPLC determn in plasma: A. Andersen et al., Ther. Drug Monit. **21**, 355 (1999). Use in treatment of hyperadrenocorticism in dogs: P. P. Kintzer, M. E. Peterson, J. Am. Vet. Med. Assoc. **205**, 54 (1994). Review of pharmacology and clinical efficacy in adrenocortical carcinoma: M. L. Gutierrez, S. T. Crooke, Cancer Treat. Rev. **7**, 49-55 (1980); S. Hahner, M. Fassnacht, Curr. Opin. Investig. Drugs **6**, 386-394 (2005).

Crystals from pentane or methanol, mp 76-78°. Slight aromatic odor. Sol in ethanol, isooctane, carbon tetrachloride, ether, hexane, fixed oils, fats. Practically insol in water.

THERAP CAT: Antineoplastic.

THERAP CAT (VET): In treatment of pituitary-dependent hyperadrenocorticism in dogs.

6302. Mitoxantrone. [65271-80-9] 1,4-Dihydroxy-5,8-bis-[[2-[(2-hydroxyethyl)amino]ethyl]amino]-9,10-anthracenedione; DHAQ; NSC-279836. $C_{22}H_{28}N_4O_6$; mol wt 444.49. C 59.45%, H 6.35%, N 12.61%, O 21.60%. Immunosuppressive and cytostatic anthracenedione. Prepn: R. K. Y. Zee-Cheng, C. C. Cheng, J. Med. Chem. **21**, 291 (1978); K. C. Murdock et al., ibid. **22**, 1024 (1979); eidem, **DE 2835661**; K. C. Murdock, F. E. Durr, **US 4197249** (1979, 1980 both to Am. Cyanamid). Antitumor activity: R. E. Wallace et al., Cancer Res. **39**, 1570 (1979). HPLC determn in plasma: K. T. Lin et al., J. Chromatogr. **465**, 75 (1989). Clinical evaluation in indolent lymphomas: F. Hagemeister et al., Oncologist **10**, 150

(2005); in neuromyelitis optica: B. Weinstock-Guttman *et al.*, *Arch. Neurol.* **63**, 957 (2006). Comprehensive description: J. H. Beijnen *et al.*, *Anal. Profiles Drug Subs.* **17**, 221-258 (1988). Review of toxicology: M. McDonald *et al.*, *Drugs Exp. Clin. Res.* **10**, 745-752 (1984); of pharmacology and clinical efficacy in cancer: J. Koeller, M. Eble, *Clin. Pharm.* **7**, 574-581 (1988); of pharmacokinetics and metabolism: G. Ehninger *et al.*, *Clin. Pharmacokinet.* **18**, 365-380 (1990); of pharmacology and clinical efficacy in multiple sclerosis: E. J. Fox, *Clin. Ther.* **28**, 461-474 (2006); O. Neuhaus *et al.*, *Pharmacol. Ther.* **109**, 198-209 (2006).

Crystals from ethanol/hexane, mp 160-162°. Abs max (ethanol): 244, 279, 525, 620, 660 nm (log ε 4.64, 4.31, 3.70, 4.37, 4.38). Sparingly sol in water, slightly sol in methanol. Practically insol in acetonitrile, chloroform, acetone.

Dihydrochloride. [70476-82-3] DHAD; CL-232315; NSC-301739; Elsep; Novantron; Novantrone; Onkotrone; Ralenova. $C_{22}H_{28}N_4O_6 \cdot 2HCl$; mol wt 517.40. Hygroscopic blue-black solid from water/ethanol, mp 203-205°. Abs max (water): 241, 273, 608, 658 nm (ε 41000, 12000, 19200, 20900). Sparingly sol in water, slightly sol in methanol. Practically insol in acetonitrile, chloroform, acetone.

THERAP CAT: Antineoplastic; immunomodulator.

THERAP CAT (VET): Antineoplastic.

6303. Mitragynine. [4098-40-2] (αE,2S,3S,12βS)-3-Ethyl-1,2,3,4,6,7,12,12b-octahydro-8-methoxy-α-(methoxymethylene)indolo[2,3-a]quinolizine-2-acetic acid methyl ester; (16E,20β)-16,17-didehydro-9,17-dimethoxycorynan-16-carboxylic acid methyl ester; (E)-16,17-didehydro-9,17-dimethoxy-17,18-seco-20α-yohimban-16-carboxylic acid methyl ester; 9-methoxycorynantheidine. $C_{23}H_{30}N_2O_4$; mol wt 398.50. C 69.32%, H 7.59%, N 7.03%, O 16.06%. Major alkaloid of *Mitragyna speciosa* Korth., *Rubiaceae*: Field, *J. Chem. Soc.* **119**, 887 (1921); Ing, Raison, *ibid.* **1939**, 986. Structural studies: Hendrickson, *Chem. Ind. (London)* **1961**, 713; Joshi *et al.*, *ibid.* **1963**, 573. Revised structure: Zacharias *et al.*, *Acta Crystallogr.* **18**, 1039 (1965). Pharmacology: Macko *et al.*, *Arch. Int. Pharmacodyn. Ther.* **198**, 145 (1972).

White, amorphous powder, mp 102-106°. bp$_5$ 230-240°. [α]$_D$ +39° (chloroform). uv max: 226, 292 nm (ε 41,150, 6600). Soluble in alcohol, chloroform, acetic acid.

6304. Mitratapide. [179602-65-4] 4-[4-[4-[4-[[(2S,4R)-2-(4-Chlorophenyl)-2-[[(4-methyl-4H-1,2,4-triazol-3-yl)thio]methyl]-1,3-dioxolan-4-yl]methoxy]phenyl]-1-piperazinyl]phenyl]-2,4-dihydro-2-[(1R)-1-methylpropyl]-3H-1,2,4-triazol-3-one; R-103757; Yarvitan. $C_{36}H_{41}ClN_8O_4S$; mol wt 717.29. C 60.28%, H 5.76%, Cl 4.94%, N 15.62%, O 8.92%, S 4.47%. Microsomal triglyceride transfer protein inhibitor; blocks the assembly and release of lipoprotein particles into the bloodstream. Prepn: J. Heeres *et al.*, WO 9613499; *eidem*, US 5521186 (both 1996 to Janssen). Development of a bioavailable formulation: G. Verreck *et al.*, *J. Pharm. Sci.* **93**, 1217 (2004).

mp 157°. pKa 4.1. Soly in water (mg/ml): <0.0005 (neutral pH); in 0.1N HCl: 0.42. Log P (octanol/water): >5.

THERAP CAT (VET): Canine antiobesity agent.

6305. Mivacurium Chloride. [106861-44-3] (1R,1′R)-2,2′-[[(4E)-1,8-Dioxo-4-octene-1,8-diyl]bis(oxy-3,1-propanediyl)]-bis[1,2,3,4-tetrahydro-6,7-dimethoxy-2-methyl-1-[(3,4,5-trimethoxyphenyl)methyl]]isoquinolinium chloride (1:2); (R)-1,2,3,4-tetrahydro-2-(3-hydroxypropyl)-6,7-dimethoxy-2-methyl-1-(3,4,5-trimethoxybenzyl)isoquinolinium chloride, (E)-4-octenedioate (2:1); BW-B1090U; Mivacron. $C_{58}H_{80}Cl_2N_2O_{14}$; mol wt 1100.18. C 63.32%, H 7.33%, Cl 6.44%, N 2.55%, O 20.36%. Non-depolarizing neuromuscular blocking agent. Prepn: R. A. Swaringen, Jr. *et al.*, DD 235638; *eidem*, US 4761418 (1986, 1988 both to Burroughs Wellcome). Clinical pharmacology: H. H. Ali *et al.*, *Br. J. Anaesth.* **61**, 541 (1988); J. J. Savarese *et al.*, *Anesthesiology* **68**, 723 (1988). *In vitro* metabolism: D. R. Cook *et al.*, *Anesth. Analg.* **68**, 452 (1989). Clinical evaluation in adults: S. Weber *et al.*, *ibid.* **67**, 495 (1988); R. P. From *et al.*, *Br. J. Anaesth.* **64**, 193 (1990); in children: J. K. Alifimoff, N. G. Goudsouzian, *ibid.* **63**, 520 (1989).

Amorphous solid. [α]$_D^{20}$ −62.7° (c = 1.9 in water).

THERAP CAT: Muscle relaxant (skeletal).

6306. Mivazerol. [125472-02-8] 2-Hydroxy-3-(1H-imidazol-5-ylmethyl)benzamide; α-imidazol-4-yl-2,3-cresotamide. $C_{11}H_{11}N_3O_2$; mol wt 217.23. C 60.82%, H 5.10%, N 19.34%, O 14.73%. α$_2$-Adrenergic agonist. Prepn: E. Cossement *et al.*, EP 341231; *eidem*, US 4923865 (1989, 1990 both to UCB); *idem et al.*, *J. Pharm. Belg.* **49**, 206 (1994). Specificity for α$_2$ adrenoceptors and binding studies: M. Noyer *et al.*, *Neurochem. Int.* **24**, 221 (1994). Clinical trial to improve perioperative hemodynamic stability: D. T. Mangano *et al.*, *Anesthesiology* **86**, 346 (1997). Mode of action study: M. Guyaux *et al.*, *Acta Anaesthesiol. Scand.* **42**, 238 (1998).

mp 197.6°.

Hydrochloride. [127170-73-4] UCB-22073. $C_{11}H_{11}N_3O_2 \cdot$ HCl; mol wt 253.69. mp 287.8°. LD$_{50}$ i.p. in mice: 760 mg/kg (Cossement).

THERAP CAT: To reduce cardiac complications of surgery.

6307. Mizolastine. [108612-45-9] 2-[[1-[1-[(4-Fluorophenyl)methyl]-1H-benzimidazol-2-yl]-4-piperidinyl]methylamino]-4(3H)-pyrimidinone; 2-[[1-[1-(p-fluorobenzyl)-2-benzimidazolyl]-4-piperidinyl]methylamino]-4(3H)-pyrimidinone; SL-85.0324; Mizollen; Zolim. $C_{24}H_{25}FN_6O$; mol wt 432.50. C 66.65%, H 5.83%, F 4.39%, N 19.43%, O 3.70%. Nonsedating-type histamine H$_1$-receptor antagonist. Prepn: JP Kokai 87 61979; P. Manoury *et al.*, US 4912219 (1987, 1990 both to Synthelabo). Receptor binding study: J. Benavides *et al.*, *Arzneim.-Forsch.* **45**, 551 (1995). HPLC determn in plasma: V. Ascalone *et al.*, *J. Chromatogr.* **619**, 275 (1993). Clinical pharmacokinetics: P. Rosenzweig *et al.*, *Ann. Allergy* **69**, 135 (1992). Effect on psychomotor performance: J. S. Kerr *et al.*, *Eur. J. Clin. Pharmacol.* **47**, 331 (1994). Clinical trial in allergic rhinitis: F. Leynadier *et al.*, *Ann. Allergy Asthma Immunol.* **76**, 163 (1996).

Crystals from ethanol, mp 217°. Sol in methanol.
THERAP CAT: Antihistaminic.

6308. Mizoribine. [50924-49-7] 5-Hydroxy-1-β-D-ribofuranosyl-1*H*-imidazole-4-carboxamide; 4-carbamoyl-1-β-D-ribofuranosylimidazolium-5-olate; HE-69; Bredinin. $C_9H_{13}N_3O_6$; mol wt 259.22. C 41.70%, H 5.06%, N 16.21%, O 37.03%. Nucleoside antibiotic produced by *Eupenicillium brefedianum* with cytotoxic and immunosuppressive activity. Isoln: K. Mizuno *et al.*, **BE 799805**; *eidem*, **US 3888843** (1973, 1975 to Toyo Jozo); *eidem*, *J. Antibiot.* **27**, 775 (1974). Synthesis: M. Hayashi *et al.*, *Chem. Pharm. Bull.* **23**, 245 (1975); K. Fukukuwa *et al.*, *ibid.* **32**, 1644 (1984). Prepn of the aglycone: T. Atsumi *et al.*, **JP Kokai 76 88965** (1976 to Sumitomo), *C.A.* **86**, 106582g (1977). HPLC determn in human serum: K. Takada *et al.*, *J. Chromatogr.* **222**, 156 (1981). Anti-arthritic activity: H. Iwata *et al.*, *Experientia* **33**, 502 (1977). Cytotoxic effect and comparison with aglycone: K. Sakaguchi *et al.*, *J. Antibiot.* **28**, 798 (1975). Antitumor spectrum of aglycone: N. Yoshida *et al.*, *Cancer Res.* **43**, 5851 (1983). Pharmacokinetics: K. Takada, *Eur. J. Pharmacol.* **24**, 457 (1983). Clinical trials in renal transplantation: T. Inou *et al.*, *Transplant. Proc.* **12**, 526 (1980); *eidem*, *ibid.* **13**, 315 (1981); A. Tajima *et al.*, *Transplantation* **38**, 116 (1984).

Crystals from methanol, mp >200° (dec). $[\alpha]_D^{27}$ −35° (c = 0.8 in H_2O). pKa 6.75. uv max (H_2O): 245, 279 nm (E 250, 580). Sol in water. Slightly sol in methanol, ethanol. Insol in most organic solvents. LD_{50} in mice (g/kg): >1.5 i.v., >2.4 i.p. (Mizuno, 1975).

Aglycone. 5-Hydroxy-1*H*-imidazole-4-carboxamide; 4-carbamoyl-5-hydroxyimidazole; 4-carbamoylimidazolium-5-olate; SM-108. $C_4H_5N_3O_2$; mol wt 127.10.
THERAP CAT: Immunosuppressant.

6309. MMT. [12108-13-3] Tricarbonyl[(1,2,3,4,5-η)-1-methyl-2,4-cyclopentadien-1-yl]manganese; methylcyclopentadienylmanganese tricarbonyl; MCMT; tricarbonyl(η⁵-methylcyclopentadienyl)manganese(I); TCMn; AK-33X. $C_9H_7MnO_3$; mol wt 218.09. C 49.57%, H 3.24%, Mn 25.19%, O 22.01%. Organometallic antiknock compound used to increase the octane quality of fuel. Prepn: J. E. Brown *et al.*, **US 2818417** (1957 to Ethyl Corp.). Evaluation of antiknock properties: R. J. Riggs *et al.*, *Proc. 37th Natural Gas Assoc. Am.* **1958**, 51. Effect on auto emissions: W. R. Pierson *et al.*, *J. Air Pollut. Control. Assoc.* **28**, 692 (1978). Use as dopant in Mn deposition: K. Hirabayashi, H. Kozawaguchi, *Jpn. J. Appl. Phys.* **25**, 711 (1986). Pyrolysis mechanism in deposition: S. Wen-

bin *et al.*, *J. Cryst. Growth* **113**, 1 (1991); I. M. T. Davidson *et al.*, *J. Mater. Chem.* **4**, 13 (1994). GC determn in gasoline: W. A. Aue *et al.*, *Anal. Chem.* **62**, 2453 (1990); and in air: V. S. Gaind *et al.*, *Analyst* **117**, 161 (1992). Review as octane enhancer: J. D. Bailie, *Oil Gas J.* **74**, 69-72 (1976); D. P. Hollrah, A. M. Burns, *ibid.* **89**, 86-90 (1991). Review of toxicology: J. Stara *et al.*, *Proc. 4th Ann. Conf Environment. Toxicol.* **AD-781 031**, 251-270 (1973); P. J. Abbott, *Sci. Total Environ.* **67**, 247-255 (1987); and environmental assessment: D. R. Lynam *et al.*, *ibid.* **93**, 107-114 (1990).

Yellow orange liquid, fp −0.75°. n_D^{20} 1.5873. d_{20}^4 1.3942. Vapor pressure ranges from 8 mm at 100° to 360.6 mm at 200°. Readily sol in hydrocarbons and the usual organic solvents including hexane, alchohols, ether, acetone, ethylene glycol, lubricating oils, and hydrocarbon fuels, such as gasoline and diesel fuel. LD_{50} orally in mice, guinea pig, rabbit: 352; 905; 95 mg/kg; orally in male, female rats: 175 ±33; 89 ±14; LC_{50} inhalation in rat: 0.22 mg/l (Stara).
USE: Antiknock fuel additive; dopant for Mn.

6310. αMNP. [63628-25-1] α-Methoxy-α-methyl-1-naphthaleneacetic acid; 2-methoxy-2-(1-naphthyl)propanoic acid. $C_{14}H_{14}O_3$; mol wt 230.26. C 73.03%, H 6.13%, O 20.84%. Chiral derivatization reagent. Prepn and application to amino acids: J. Goto *et al.*, *Chem. Pharm. Bull.* **25**, 847 (1977); *eidem*, *J. Chromatogr.* **152**, 413 (1978). Resolution of enantiomers and abs config: N. Harada *et al.*, *Tetrahedron: Asymmetry* **11**, 1249 (2000). Use in alcohol resolution: A. Ichikawa, *Chirality* **11**, 70 (1999).

(R)-(−)-form

Colorless plates from acetone-hexane, mp 161-162°. uv max (methanol): 263, 271, 281, 287, 292 nm (log ε 3.36, 3.53, 3.57, 3.46, 3.45).
(−)-(R)-Form. [63628-26-2] Colorless plates from ether-hexane, mp 111-112°. $[\alpha]_D^{13}$ −106.3° (c = 0.16 in $CHCl_3$); $[\alpha]_D^{13}$ −128.8° (c = 0.10 in methanol).
(+)-(S)-Form. [102691-93-0] $[\alpha]_D^{26}$ +67.4° (c = 1.39 in $CHCl_3$).
Hydroxy acid. [6341-58-8] (unspecified stereo); [255881-88-0] (S-(+)-form). α-Hydroxy-α-methyl-1-naphthaleneacetic acid; αHNP. $C_{13}H_{12}O_3$. Prepn and abs config: A. Ichikawa *et al.*, *Tetrahedron: Asymmetry* **10**, 4075 (1999). Colorless plates from acetone-hexane, mp 110-112°. $[\alpha]_D^{29}$ +40° (c = 0.21 in $CHCl_3$).
USE: For enantioresolution in organic syntheses.

6311. Mocimycin. [50935-71-2] (aS,2R,3R,4R,6S)-N-[(2E,-4E,6S,7R)-7-[(2S,3S,4R,5R)-5-[(1E,3E,5E)-7-(1,2-Dihydro-4-hydroxy-2-oxo-3-pyridinyl)-6-methyl-7-oxo-1,3,5-heptatrien-1-yl]-tetrahydro-3,4-dihydroxy-2-furanyl]-6-methoxy-5-methyl-2,4-octadien-1-yl]-α-ethyltetrahydro-2,3,4-trihydroxy-5,5-dimethyl-6-(1E,-3Z)-1,3-pentadien-1-yl-2H-pyran-2-acetamide; antibiotic MYC 8003; delvomycin; kirromycin; MYC-8003. $C_{43}H_{60}N_2O_{12}$; mol wt 796.96. C 64.81%, H 7.59%, N 3.52%, O 24.09%. Antibiotic produced by *Streptomyces ramocissimus*. Isoln and properties: C. Vos, J. Den Admirant, **DE 2140674**; *eidem*, **US 3927211**; **US 3923981** (1972, 1975, 1975 all to Gist-Brocades). Structural studies: C. Vos, P. E. J. Verwiel, *Tetrahedron Lett.* **1973**, 2823, 5173. Total structure and identity with kirromycin: H. Maehr *et al.*, *J. Am. Chem. Soc.* **95**, 8449 (1973). Absolute stereochemistry: *eidem*, *ibid.* **96**, 4034 (1974). Conversion to aurodox: H. Maehr *et al.*, *J. Antibiot.* **32**, 361 (1979). Biological studies: H. Wolf *et al.*, *Proc. Natl. Acad. Sci. USA* **71**, 4910 (1974); J. A. M. Van De Klundeit *et al.*, *FEBS Lett.* **81**, 303 (1977). Review of chemistry: H. Maehr *et al.*, *Can. J.*

Chem. **58**, 501-526 (1980). Review of properties and actions: A. Parmeggiani, G. Sander, *Topics in Antibiotic Chemistry* **vol. 5**, P. G. Sammes, Ed. (Halsted Press, New York, 1980) pp 159-221.

Yellow solid. Weakly acidic. $[\alpha]_D^{22}$ $-60°$ (c = 1 in methanol). uv max (methanol/water): 233, 276, 286, 327 nm. Sol in chloroform, methyl isobutyl ketone, butyl acetate, ethyl acetate, acetone, methanol, alkaline soln. Slightly sol in CCl_4, benzene. Insol in diethyl ether, petr ether, water, acid soln. Stable for 4 hrs in 50% methanol at pH 3-12. The solid antibiotic stored at 25° and 37° and low relative humidity shows no activity loss for 5 mos. Stable for 3 mos at 25°, 100% relative humidity; for 2 mos at 37°, 100% relative humidity. LD_{50} i.p. in mice: >1000 mg/kg (DE 2140674).

1-Methyl derivative *see* Aurodox.

5,6-Dihydromocimycin. $C_{43}H_{62}N_2O_{12}$; mol wt 798.97. Isoln from *S. ramocissimus* and properties: H. Jongsma *et al.*, **DE 2658977** (1977 to Gist-Brocades), *C.A.* **87**, 135078y (1977); *see also:* C. Vos *et al.*, **US 4062948** (1977 to Gist-Brocades). Pale yellow solid, dec begins at 123°. $[\alpha]_D^{20}$ $-85°$ (1% methanolic soln). uv max (methanol/water): 233.5, 267, 291, 333 nm (ε 63000, 23000, 19000, 18000). Solubility similar to mocimycin in most solvents.

USE: Animal growth promotant.

6312. Moclobemide. [71320-77-9] 4-Chloro-*N*-[2-(4-morpholinyl)ethyl]benzamide; *p*-chloro-*N*-(2-morpholinoethyl)benzamide; Ro-11-1163; Aurorix; Manerix; Moclamine. $C_{13}H_{17}$-ClN_2O_2; mol wt 268.74. C 58.10%, H 6.38%, Cl 13.19%, N 10.42%, O 11.91%. Reversible monoamine oxidase A (MAO-A) inhibitor. Prepn: W. Burkard, P.-C. Wyss, **DE 2706179**; **US 4210754** (1977, 1980 both to Hoffmann-La Roche). Gas chromatography: K. P. Maguire *et al.*, *J. Chromatogr.* **278**, 429 (1983). Human pharmacokinetics: F. A. Wiesel *et al.*, *Eur. J. Clin. Pharmacol.* **28**, 89 (1985). Clinical study: J. K. Larsen *et al.*, *Acta Psychiatr. Scand.* **70**, 254 (1984). Review of pharmacology: W. P. Burkard *et al.*, *J. Pharmacol. Exp. Ther.* **248**, 391-399 (1989); of neurochemical profile: M. DaPrada *et al.*, *ibid.* 400-414.

Crystals from isopropanol, mp 137°. LD_{50} in rats (mg/kg): 707 orally (Burkard, Wyss).

Hydrochloride. $C_{13}H_{18}Cl_2N_2O_2$. Crystals from isopropanol, mp 208°.

THERAP CAT: Antidepressant.

6313. Modacrylic Fibers. A generic term for acrylic fibers composed of a long chain synthetic polymer containing 35-85% of acrylonitrile units such as **Dynel** and **Verel**. *See:* R. W. Moncrieff, *Man-Made Fibres* (Wiley, New York, 4th ed., 1963) p 9. *Review:* Kennedy in *Encyclopedia of Polymer Science and Technology* **vol. 8**, N. M. Bikales, Ed. (Interscience, New York, 1968) pp 812-839.

USE: In pile fabrics, carpeting, paint rollers.

6314. Modafinil. [68693-11-8] 2-[(Diphenylmethyl)sulfinyl]acetamide; 2-(benzhydrylsulfinyl)acetamide; CRL-40476; Attenace; Provigil. $C_{15}H_{15}NO_2S$; mol wt 273.35. C 65.91%, H 5.53%, N 5.12%, O 11.71%, S 11.73%. α_1-Adrenergic agonist. Prepn: L. Lafon, **DE 2809625**; *eidem*, **US 4177290** (1978, 1979 both to Lab. Lafon). Effects on vigilance: F. Goldenberg, J. S. Weil, *Sleep (Stuttgart)* **8**, 343 (1988); on normal sleep: B. Saletu *et al.*,

Int. J. Clin. Pharmacol. Res. **9**, 183 (1989). Clinical evaluation in hypersomnia and narcolepsy: H. Bastuji, M. Jouvet, *Prog. Neuro-Psychopharmacol. Biol. Psychiatry* **12**, 695 (1988); in adult attention-deficit/hyperactivity disorder: D. C. Turner *et al.*, *Biol. Psychiatry* **55**, 1031 (2004).

White crystals from methanol, mp 164-166°. Sparingly sol in methanol; sparingly to slightly sol in acetone; slightly sol in absolute alc; very slightly sol in water. Practically insol in cyclohexane.

Note: This is a controlled substance (stimulant): **21 CFR,** 1308.14.

THERAP CAT: CNS Stimulant.

6315. Moexipril. [103775-10-6] (3*S*)-2-[(2*S*)-2-[[(1*S*)-1-(Ethoxycarbonyl)-3-phenylpropyl]amino]-1-oxopropyl]-1,2,3,4-tetrahydro-6,7-dimethoxy-3-isoquinolinecarboxylic acid; RS-10085. $C_{27}H_{34}N_2O_7$; mol wt 498.58. C 65.04%, H 6.87%, N 5.62%, O 22.46%. Angiotensin converting enzyme (ACE) inhibitor; dimethoxy analog of quinapril, *q.v.* Prepn: M. L. Hoefle, S. Klutchko, **EP 49605**; *eidem*, **US 4344949** (both 1982 to Warner-Lambert); S. Klutchko *et al.*, *J. Med. Chem.* **29**, 1953 (1986). Pharmacology: O. Edling *et al.*, *J. Pharmacol. Exp. Ther.* **275**, 854 (1995). GC-MS determn in plasma: W. Hammes *et al.*, *J. Chromatogr. B* **670**, 81 (1995). Clinical trials in hypertension: W. B. White *et al.*, *J. Hum. Hypertens.* **8**, 917 (1994); M. Stimpel *et al.*, *Cardiology* **87**, 313 (1996).

Hydrochloride. [82586-52-5] CI-925; RS-10085-197; SPM-925; Fempress; Perdix; Univasc. $C_{27}H_{34}N_2O_7$·HCl; mol wt 535.03. Crystals from ethanol + ethyl ether, mp 141-161°. $[\alpha]_D^{23}$ +34.2° (c = 1.1 in ethanol).

Diacid hydrochloride. [82586-57-0] Moexiprilat hydrochloride. $C_{25}H_{30}N_2O_7$·HCl. Prepd as the monohydrate; crystals from THF + ethanol, mp 145-170°. $[\alpha]_D^{23}$ +37.8° (c = 1.1 in methanol).

THERAP CAT: Antihypertensive.

6316. Mofarotene. [125533-88-2] 4-[2-[4-[(1*E*)-2-(5,6,7,8-Tetrahydro-5,5,8,8-tetramethyl-2-naphthalenyl)-1-propen-1-yl]-phenoxy]ethyl]morpholine; Ro 40-8757. $C_{29}H_{39}NO_2$; mol wt 433.64. C 80.32%, H 9.07%, N 3.23%, O 7.38%. Anticancer aromatic analog of retinoic acid (arotinoid). Prepn: M. Klaus *et al.*, **EP 331983**; *eidem*, **US 4940707** (1989, 1990 both to Hoffmann-La Roche). Mode of action study: J. F. Eliason *et al.*, *Blood* **86**, 4516 (1995). Pharmacokinetics: B. Vallès *et al.*, *Eur. J. Drug Metab. Pharmacokinet.* **20**, 49 (1995). HPLC determn in plasma: R. Wyss *et al.*, *J. Chromatogr. A* **729**, 315 (1996). Mechanism of growth inhibition on pancreatic cancer cells: S. Kawa *et al.*, *Int. J. Cancer* **72**, 906 (1997). Clinical pharmacology in combination with cisplatin and etoposide: L. van Zuylen *et al.*, *Anti-Cancer Drugs* **10**, 361 (1999).

White crystals from ethyl acetate/hexane, mp 107-109°.
THERAP CAT: Antineoplastic.

6317. Mofebutazone. [2210-63-1] 4-Butyl-1-phenyl-3,5-pyrazolidinedione; 4-butyl-1-phenyl-3,5-dioxopyrazolidine; 2-phenyl-3,5-dihydroxy-4-butylpyrazolidine; monophenylbutazone; Arcomonol Tablets; Mobutazon; Mobuzon; Mofesal; Monazan; Monobutyl; Monorheumetten; Reumatox. $C_{13}H_{16}N_2O_2$; mol wt 232.28. C 67.22%, H 6.94%, N 12.06%, O 13.78%. Prepn: Büchi *et al.*, *Helv. Chim. Acta* **36**, 75 (1953); **GB 839057** (1960 to Comm. Farm. Milanese). Toxicity data: Schoetensack, *Arch. Exp. Pathol. Pharmakol.* **233**, 365 (1958).

Crystals from ethanol + water, mp 102-103°. uv max (ethanol): 240, 275 nm ($E^{1\%}_{1cm}$ 443, 245). LD_{50} i.v. in mice: 600 mg/kg (Schoetensack).
THERAP CAT: Anti-inflammatory.

6318. Mofezolac. [78967-07-4] 3,4-Bis(4-methoxyphenyl)-5-isoxazoleacetic acid; 3,4-di(*p*-methoxyphenyl)-5-isoxazoleacetic acid; N-22; Disopain. $C_{19}H_{17}NO_5$; mol wt 339.35. C 67.25%, H 5.05%, N 4.13%, O 23.57%. Prostaglandin biosynthetase inhibitor. Prepn: R. G. Micetich *et al.*, **EP 26928**; *eidem*, **US 4327222** (1981, 1982 both to CDC Life Sci.). Physico-chemical properties: T. Ushio *et al.*, *Iyakuhin Kenkyu* **22**, 892 (1991). HPLC determn in plasma and urine: T. Marunaka, M. Maniwa, *J. Chromatogr.* **422**, 227 (1987). Pharmacology: K. Tanaka *et al.*, *Yakuri to Chiryo* **18**, 3347 (1990), *C.A.* **114**, 135911 (1991). Acute toxicity: K. Satoh *et al.*, *J. Toxicol. Sci.* **15**, Suppl 2, 1 (1990).

White crystalline powder; slight characteristic odor. mp 147.5°. pKa' ~3.3. uv max (methanol): 236 nm ($\varepsilon = 1.83 \times 10^4$). Soly at 20° (w/v%): DMF 68.7; chloroform 19.5; ethyl acetate 16.5; acetone 7.32; methanol 3.93; anhydr ethanol 3.72; ether 2.90; hexane 2.90 × 10^{-5}; water 4.85 × 10^{-3}. Partition coefficient (chloroform/water): 16.6 (pH 6); 1.87 (pH 7); 0.178 (pH 8). LD_{50} in male, female mice, male, female rats (mg/kg): 1528, 1740, 920, 887 orally; 275, 321, 378, 342 i.p.; 612, 545, 572, 510 s.c. (Satoh).
THERAP CAT: Anti-inflammatory; analgesic.

6319. Molindone. [7416-34-4] 3-Ethyl-1,5,6,7-tetrahydro-2-methyl-5-(4-morpholinylmethyl)-4H-indol-4-one; 3-ethyl-6,7-dihydro-2-methyl-5-(morpholinomethyl)indol-4(5H)-one. $C_{16}H_{24}$-N_2O_2; mol wt 276.38. C 69.53%, H 8.75%, N 10.14%, O 11.58%. Prepn: **BE 670798**; I. J. Pachter, K. Schoen, **US 3491093** (1966, 1970 both to Endo). Pharmacology: Sugerman, Herrmann, *Clin. Pharmacol. Ther.* **8**, 261 (1967); Claghorn, *Curr. Ther. Res.* **11**, 524 (1969); Guerrero-Figueroa *et al.*, *ibid.* **15**, 508 (1973). Toxicity study: E. I. Goldenthal, *Toxicol. Appl. Pharmacol.* **18**, 185 (1971).

Crystals, mp 180-181°.
Hydrochloride. [15622-65-8] EN-1733A; Lidone; Moban. LD_{50} orally in rats: 261 mg/kg (Goldenthal).
THERAP CAT: Antipsychotic.

6320. Molsidomine. [25717-80-0] 5-[(Ethoxycarbonyl)amino]-3-(4-morpholinyl)-1,2,3-oxadiazolium inner salt; N-(ethoxycarbonyl)-3-(4-morpholinyl)sydnone imine; N-carboxy-3-morpholinosydnonimine ethyl ester; morsydomine; SIN-10; Corvaton; Corvasal; Molsidolat; Morial; Motazomin. $C_9H_{14}N_4O_4$; mol wt 242.24. C 44.62%, H 5.83%, N 23.13%, O 26.42%. Coronary vasodilator; member of a class of non-benzene aromatic, heterocyclic and meso-ionic type of compounds known as sydnone imines. Developmental work: Brookes, Walker, *J. Chem. Soc.* **1957**, 4409. Prepn: K. Masuda *et al.*, **JP 70 6265**; *eidem*, **US 3769283** (1970, 1973 both to Takeda); *eidem*, *Chem. Pharm. Bull.* **19**, 72 (1971). Stability studies: Asahi *et al.*, *ibid.* **19**, 1079 (1971). Pharmacological studies: Kikuchi *et al.*, *Jpn. J. Pharmacol.* **20**, 102, 187, 253 (1970); Hashimoto *et al.*, *Arzneim.-Forsch.* **21**, 1329 (1971); T. Nakaguchi *et al.*, *Toho Igakkai Zasshi* **17**, 26 (1970). Metabolism: S. Tanayama *et al.*, *Xenobiotica* **4**, 175 (1974). Pharmacokinetics: R. Bergstrand *et al.*, *Eur. J. Clin. Pharmacol.* **27**, 203 (1984). Clinical trial in angina pectoris: P. A. Majid *et al.*, *N. Engl. J. Med.* **302**, 1 (1980); in hypertension: J. Melei *et al.*, *Eur. J. Clin. Pharmacol.* **18**, 231 (1980). Synergism with penbutolol in angina pectoris: A. E. Balestrini *et al.*, *ibid.* **27**, 1 (1984). Review of pharmacology: R.-E. Nitz, V. B. Fiedler, *Pharmacotherapy* **7**, 28 (1987). *Review: Jpn. Med. Gaz.* **8**(9), 10 (1971); E. Bassenge, W. R. Kukovetz in *New Drugs Annual: Cardiovascular Drugs* Vol. 2, A. Scriabine, Ed. (Raven Press, New York, 1984) pp 177-191.

Colorless crystals or white cryst powder, practically tasteless and odorless, mp 140-141° (toluene). Freely sol in $CHCl_3$. Sol in dil HCl, ethanol, ethyl acetate, methanol; sparingly sol in water, acetone, benzene. Very slightly sol in ether, petr ether. pK (100°) 3.0 ± 0.1. Most stable in aq solns pH 5-7; least stable in very alkaline solns. uv max ($CHCl_3$): 326 nm. Sensitive to light of $\lambda <320$ mμ. LD_{50} in male, female mice, male, female rats (mg/kg): 780, 750, 1380, 1350 s.c.; 860, 800, 830, 760 i.v.; 700, 760, 1250, 1250 i.p.; 830, 840, 1050, 1200 orally (Nakaguchi).
THERAP CAT: Antianginal.

6321. Molybdenum. [7439-98-7] Mo; at. wt 95.96; at. no. 42; valences 2,3,4,5,6. Group VIB (6). Naturally occurring isotopes: 98 (23.75%); 96 (16.5%); 95 (15.7%); 92 (15.86%); 94 (9.12%); 100 (9.62%); 97 (9.45%); artificial radioactive isotopes: 88-91; 93; 99; 101-105. Its most important ores are molybdenite, MoS_2, and wulfenite, $PbMoO_4$. Occurrence in the earth's crust: 1-1.5 ppm. Discovered in 1778 by Scheele; isolated in 1782 by Hjelm. Methods of preparation: L. Northcott, *Molybdenum* (Academic Press, New York, 1956) 222 pp; Hein, Herzog, in *Handbook of Preparative Inorganic Chemistry* vol. 2, G. Brauer, Ed. (Academic Press, New York, 2nd ed., 1965) pp 1401-1402. Important trace element; participates in biochemical redox reactions such as N_2-fixation: Spence, *Coord. Chem. Rev.* **4**, 475 (1969). Physical properties: Worthing, *Phys. Rev.* [2] **25**, 846 (1925); D. R. Stoll, G. C. Sinke, "Thermodynamic Properties of the Elements", *Advances in Chemistry Series* **18**, (American Chemical Society, Washington, 1956) pp 23, 130-131. Review of molybdenum and its compds: Rollinson, "Chromium, Molybdenum and Tungsten" in *Comprehensive Inorganic Chemistry* vol. 3, J. C. Bailar Jr. *et al.*, Eds. (Pergamon Press, Oxford, 1973) pp 622-623, 700-742; R. Q. Barr in *Kirk-Othmer Encyclopedia of Chemical Technology* vol. 15 (Wiley-Interscience, New York, 3rd ed., 1981) pp 670-682. Biochemical review: *Bioinorg. Chem.* **II**, K. N. Raymond, Ed. (A.C.S., Washington, 1977) pp 353-430. Symposium on the chemistry and uses of molybdenum and its cmpds: *Polyhedron* **5**, 1-606 (1986).

Dark-gray or black powder with metallic luster or coherent mass of silver-white color; body-centered cubic structure. mp 2622°

(Worthing). bp ~4825°. d 10.28. Spec heat 5.68 cal/g-atom/deg; heat of fusion: 6.6 kcal/g-atom; heat of vaporization: 142 kcal/g-atom (Stoll, Sinke). Fairly stable at ordinary temp; oxidized to the trioxide at a red heat; slowly oxidized by steam. Not attacked by water, by dil acids or by concd hydrochloric acid. Practically insol in alkali hydroxides or fused alkalies. Reacts with nitric acid, hot concd sulfuric acid, fused potassium chlorate or nitrate. Attacked by fluorine at ordinary temp, by chlorine or bromine at a red heat.

Caution: Potential symptoms of overexposure to Mo metal in exptl animals are irritation of eyes, nose, throat; anorexia, diarrhea, weight loss; listlessness; liver and kidney damage. Potential symptoms of overexposure to soluble Mo compds in exptl animals are irritation of eyes, nose, throat; anorexia; incoordination; dyspnea, anemia. *See NIOSH Pocket Guide to Chemical Hazards* (DHHS/NIOSH 97-140, 1997) p 218.

USE: In the form of ferromolybdenum for manufg special steels for tools, boiler plate, rifle barrels, propeller shafts; electrical contacts, spark plugs, x-ray tubes, filaments, screens and grids for radio tubes; in the production of tungsten; glass-to-metal seals; nonferrous alloys; in colloidal form as lubricant additive.

6322. Molybdenum Disulfide. [1317-33-5] Molybdenum sulfide (MoS$_2$). MoS$_2$; mol wt 160.07. Mo 59.94%, S 40.06%. Occurs as the mineral *molybdenite*, which is the principal source of molybdenum. Lab prepn: Bell, Herfert, *J. Am. Chem. Soc.* **79**, 3351 (1957).

Lead-gray, lustrous powder; the artificially prepd sulfide is black and lustrous. d$_{15}^{15}$ 5.06; mp 2375°. Begins to sublime at 450°. Insol in water or dil acids.

USE: Dry lubricant and lubricant additive. Hydrogenation catalyst.

6323. Molybdenum Hexafluoride. [7783-77-9] (*OC*-6-11)-Molybdenum fluoride (MoF$_6$). F$_6$Mo; mol wt 209.94. F 54.30%, Mo 45.70%. MoF$_6$. Prepd by direct fluorination of powdered molybdenum: Ruff, Ascher, *Z. Anorg. Allg. Chem.* **196**, 418 (1931); from MoO$_3$ and SF$_4$: Oppengard *et al.*, *J. Am. Chem. Soc.* **82**, 3825 (1960).

Volatile, white, cubic crystals. Very hygroscopic. d$_{liq}^{19}$ 2.543. mp 17.5°. bp 35.0°. Hydrolyzed by water. Forms blue-white clouds in moist air. Soly in anhydr HF: 1.5 moles/1000 g HF, Frlec, Hyman, *Inorg. Chem.* **6**, 1596 (1967). Should be stored in quartz ampuls.

6324. Molybdenum Sesquioxide. [1313-29-7] Molybdenum oxide (Mo$_2$O$_3$); dimolybdenum trioxide. Mo$_2$O$_3$; mol wt 239.90. Mo 79.99%, O 20.01%.

Grayish-black powder. Very slightly sol in acids.
Combination with ferrous sulfate. Mol-Iron (obsolete).

THERAP CAT: Combination with ferrous sulfate as hematinic.

6325. Molybdenum Trioxide. [1313-27-5] Molybdenum oxide (MoO$_3$); molybdenum(VI) oxide; molybdic anhydride. MoO$_3$; mol wt 143.95. Mo 66.66%, O 33.34%. Prepn from ammonium molybdate: Schumb, Hartford, *J. Am. Chem. Soc.* **56**, 2613 (1934). Toxicity studies: L. T. Fairhall *et al.*, "The Toxicity of Molybdenum," *U.S. Public Health Service Bulletin No. 293*, Washington DC (1945) 36 pp.

White or slightly yellow to slightly bluish powder or granules. d$_4^{26}$ 4.696. mp 795°. Melts to dark-yellow liquid which solidifies to a yellowish-white cryst mass; sublimes at higher temp. bp 1155°. Sol in water (28°) 0.490 g/liter. Sol in concd mineral acids, in solns of alkali hydroxides, ammonia or potassium bitartrate; after strong ignition it is very slightly sol in acids.

USE: Chiefly as a reagent for chemical and trace metal analysis.

6326. Molybdic(VI) Acid. [7782-91-4] (*T*-4)-Hydrogen molybdate (MoO$_4^{2-}$) (2:1). H$_2$MoO$_4$; mol wt 161.96. H 1.24%, Mo 59.24%, O 39.51%. Prepd as the monohydrate from ammonium molybdate and nitric acid: Hein, Herzog in *Handbook of Preparative Inorganic Chemistry* vol. 2, G. Brauer, Ed. (Academic Press, New York, 2nd ed., 1965) pp 1412-1413.
Monohydrate. [25942-34-1] White powder. d 3.1. Sparingly soluble in cold, more in hot water; soluble in NH$_4$OH, H$_2$SO$_4$, fixed alkalies.

USE: Determn of phosphate.

6327. Mometasone Furoate. [83919-23-7] (11β,16α)-9,21-Dichloro-17-[(2-furanylcarbonyl)oxy]-11-hydroxy-16-methylpreg-

na-1,4-diene-3,20-dione; 9,21-dichloro-11β,17-dihydroxy-16α-methylpregna-1,4-diene-3,20-dione 17-(2-furoate); Sch-32088; Asmanex; Elocon. C$_{27}$H$_{30}$Cl$_2$O$_6$; mol wt 521.43. C 62.19%, H 5.80%, Cl 13.60%, O 18.41%. Topical corticosteroid. Prepn: E. L. Shapiro, **EP 57401**; *idem*, **US 4472393** (1982, 1984 both to Schering Corp.); E. L. Shapiro *et al.*, *J. Med. Chem.* **30**, 1581 (1987). LC-MS-MS determn in plasma: S. Sahasranaman *et al.*, *J. Chromatogr. B* **819**, 175 (2005). Clinical trial in psoriasis: R. S. Medansky *et al.*, *Semin. Dermatol.* **6**, 94 (1987); in seasonal allergic rhinitis: D. Graft *et al.*, *J. Allergy Clin. Immunol.* **98**, 724 (1996).

White crystals from aq methanol, mp 218-220°. [α]$_D^{26}$ +58.3° (dioxane). uv max (methanol): 247 nm (ε 26300). Sol in acetone, methylene chloride; moderately sol in ethanol; slightly sol in octanol. Practically insol in water.
Monohydrate. [141646-00-6] Nasonex. C$_{27}$H$_{30}$Cl$_2$O$_6$.H$_2$O; mol wt 539.45. White powder. Freely sol in THF; sol in acetone, chloroform; slightly sol in methanol, ethanol, isopropanol. Practically insol in water.

THERAP CAT: Anti-inflammatory.

6328. Monacetin. [26446-35-5] 1,2,3-Propanetriol monoacetate; acetin; monoacetin; glyceryl monoacetate. C$_5$H$_{10}$O$_4$; mol wt 134.13. C 44.77%, H 7.52%, O 47.71%. C$_3$H$_7$O$_2$.OOCCH$_3$. Obtained along with the di- and triacetates by heating glycerol with glacial acetic acid.

Colorless, very hygroscopic liquid; characteristic odor. The commercial product is pale yellow. d$_4^{20}$ 1.206. bp$_{17}$ ~58°; bp$_3$ 129-131°. Soluble in water, alcohol, slightly in ether. Insol in benzene.

USE: In manuf of smokeless powder and dynamite; as a solvent for basic dyes; in tanning leather.

6329. Monarda. American horsemint; wild bergamot. Leaves of *Monarda punctata* L., *Labiatae*. *Habit.* New York to Florida, west to Texas and Wisconsin. *Constit.* Volatile oils: thymol, carvacrol, neryl formate, geranyl formate, cineole, γ-terpinene, α- and β-pinene. Gas chromatographic analysis of the *oil of monarda*: R. W. Scora, *J. Chromatogr.* **19**, 601 (1965).

USE: Source for thymol.
THERAP CAT: Aromatic stimulant, carminative.

6330. Monastrol. [254753-54-3] (4*S*)-1,2,3,4-Tetrahydro-4-(3-hydroxyphenyl)-6-methyl-2-thioxo-5-pyrimidinecarboxylic acid ethyl ester. C$_{14}$H$_{16}$N$_2$O$_3$S; mol wt 292.35. C 57.52%, H 5.52%, N 9.58%, O 16.42%, S 10.97%. Small molecule which inhibits mitotic kinesin motility protein, *Eg5*, thus preventing spindle assembly. Identification of activity: T. U. Mayer *et al.*, *Science* **286**, 971 (1999). Description of preparative technique: K. Lewandowski *et al.*, *J. Comb. Chem.* **1**, 105 (1999). X-ray structure determn and abs config: C. O. Kappe *et al.*, *Tetrahedron* **56**, 1859 (2000). Use as probe for mitotic mechanisms: T. M. Kapoor *et al.*, *J. Cell Biol.* **150**, 975 (2000).

Colorless crystals from ethanol-hexane, mp 184-186°. Sol in DMSO.

USE: Inhibitor of mammalian cell mitosis.

6331. Monellins. Intensely sweet principle from the fruit of the tropical plant *Dioscoreophyllum cumminsii* Diels, *Menispermaceae* ("Serendipity Berry"). First believed to consist of a single polypeptide chain of approx 91 amino acids, mol wt about 10,700. Preliminary data and extraction: Inglett, May, *J. Food Sci.* **34**, 408 (1969); Van der Wel, *FEBS Lett.* **21**, 88 (1971); Van der Wel, Loeve, *ibid.* **29**, 181 (1973). Purification: Morris, Cagan, *Biochim. Biophys. Acta* **261**, 114 (1972). Characterized by scientists at the Univ. of Pennsylvania's Monell Chemical Senses Center as the first protein to elicit a sweet taste in man, a "chemostimulatory protein". *See:* Morris *et al., J. Biol. Chem.* **248**, 534 (1973); Cagan, *Science* **181**, 32 (1973). Monellin is composed of two nonidentical subunits: Z. Bohak, S. Li, *Biochim. Biophys. Acta* **427**, 153 (1976); the presence of both subunits is necessary for the sweet taste to occur: B. Jirgenson, *ibid.* **446**, 255 (1976). Structure of subunit I: G. Hudson, K. Biemann, *Biochem. Biophys. Res. Commun.* **71**, 212 (1976). Complete amino acid sequence of subunits A and B consisting of 44 and 50 amino acids, resp.: G. Frank, H. Zuber, *Z. Physiol. Chem.* **357**, 585 (1976). Crystallographic studies: A. Wlodawer, K. Hodgson, *Proc. Natl. Acad. Sci. USA* **72**, 398 (1975); G. E. Tomlinson, S. H. Kim, *J. Biol. Chem.* **256**, 12476 (1981).

Water soluble. Approximately 3000 times sweeter than sucrose on a weight basis. Essentially free from carbohydrate: <5 μg/mg protein. uv max (0.01M sodium phosphate buffer): 277 nm.

USE: Potential low-calorie sweetener.

6332. Monensin. [17090-79-8] 2-[5-Ethyltetrahydro-5-[tetrahydro-3-methyl-5-[tetrahydro-6-hydroxy-6-(hydroxymethyl)-3,5-dimethyl-2H-pyran-2-yl]-2-furyl]-2-furyl]-9-hydroxy-β-methoxy-α,-γ,2,8-tetramethyl-1,6-dioxaspiro[4.5]decane-7-butyric acid; monensic acid (obsolete); A-3823A. $C_{36}H_{62}O_{11}$; mol wt 670.88. C 64.45%, H 9.32%, O 26.23%. Polyether antibiotic. Major factor in antibiotic complex isolated from *Streptomyces cinnamonensis*. Discovery and isolation: Haney, Hoehn, *Antimicrob. Agents Chemother.* **1967**, 349. Production: Haney, Hoehn, **US 3501568** (1970 to Lilly). Structure: Agtarap *et al., J. Am. Chem. Soc.* **89**, 5737 (1967). Crystal structure studies: Lutz *et al., Helv. Chim. Acta* **53**, 1732 (1970); *ibid.* **54**, 1103 (1971). Fermentation studies: Stark *et al., Antimicrob. Agents Chemother.* **1967**, 353. Chemistry: Agtarap, Chamberlin, *ibid.* 359. Stereocontrolled total synthesis: T. Fukuyama *et al., J. Am. Chem. Soc.* **101**, 262 (1979); D. B. Collum *et al., ibid.* **102**, 2117, 2118, 2120 (1980). ^{13}C-NMR study: J. A. Robinson, D. L. Turner, *Chem. Commun.* **1982**, 148. Biosynthesis: Day *et al., Antimicrob. Agents Chemother.* **4**, 410 (1973). *Review:* Stark, "Monensin, A New Biologically Active Compound Produced by a Fermentation Process", in *Fermentation Advances*, Pap. Int. Ferment. Symp., 3rd, **1968**, D. Perlman, Ed. (Academic Press, New York, 1969) pp 517-540.

Crystals, mp 103-105° (monohydrate). $[\alpha]_D$ +47.7°. pKa 6.6 (in 66% DMF). Very stable under alkaline conditions. Very sol in organic solvents; slightly sol in water. LD_{50} of monensin complex in mice, chicks (mg/kg): 43.8 ± 5.2, 284 ± 47 orally (Haney, Hoehn).

Sodium salt. Coban; Romensin; Rumensin. $C_{36}H_{61}NaO_{11}$; mol wt 692.86. Off-white to tan, crystalline powder. mp 267-269°. $[\alpha]_D$ +57.3° (methanol). Sol in chloroform, methanol, and other organic solvents; slightly sol in water. Practically insol in solvent hexane.

THERAP CAT (VET): Coccidiostat. Feed additive to improve feed efficiency in ruminants.

6333. Monoamine Oxidase. [9001-66-5] Amine oxidase; adrenaline oxidase; tyraminase; MAO. Since its first description by Hare in 1928 as an enzyme which catalyzed the oxidative deamination of tyramine, it has been found to be widely distributed in animals. This enzyme, which catalyzes the oxidative deamination of a variety of biogenic amines, such as serotonin, norepinephrine, epinephrine, tyramine and dopamine, deaminates compounds in which the amine group is attached to the terminal carbon group. Enzymes other than the classical amine oxidase which catalyze the oxidative deamination of amines, have been found in animals, plants and bacteria. In contrast to the classical amine oxidase, they are inhibited by carbonyl reagents and do not act on N-substituted amines. Enzymes of this group include: oxidases in microorganisms, *spermine oxidase*, *mescaline oxidase* (of rabbit liver), *plant amine oxidase*, and histaminase, *q.v.* Structural studies: Achee, *Biochemistry* **7**, 4329 (1968). Multiple forms of rat brain monoamine oxidase: M. B. H. Youdim *et al., Nature* **223**, 626 (1969). *Review:* H. Blaschko "Amine Oxidase" in *The Enzymes* **vol. 8**, P. D. Boyer *et al.*, Eds. (Academic Press, New York, 2nd ed., 1963) pp 337-351. Book: *Monoamine Oxidase: Structure, Function and Altered Functions*, T. P. Singer *et al.*, Eds. (Academic Press, New York, 1980) 557 pp.

6334. Monobenzone. [103-16-2] 4-(Phenylmethoxy)phenol; p-(benzyloxy)phenol; hydroquinone monobenzyl ether; hydroquinone benzyl ether; p-hydroxyphenyl benzyl ether; benzyl hydroquinone; monobenzyl hydroquinone; Benoquin; Depigman; Pigmex; Benzoquin; Agerite. $C_{13}H_{12}O_2$; mol wt 200.24. C 77.98%, H 6.04%, O 15.98%. Prepd from hydroquinone and benzyl bromide in alcoholic KOH: Schiff, Pellizzari, *Ann.* **221**, 370 (1883). Clinical evaluation: S. W. Becker, Jr., M. C. Spencer, *J. Am. Med. Assoc.* **180**, 279 (1962). Toxicity study: A. Takahashi, E. A. Inglis, *Int. J. Polymer Sci. Technol.* **3**, 93 (1976).

Lustrous leaflets from water, mp 122.5°. Practically insol in cold water. Soly in boiling water ~1.0 g/100 ml. Sol in alcohol, ether, benzene. LD_{50} i.p. in mice: >600 mg/kg (Takahashi).

THERAP CAT: Depigmentor.

6335. Monobutyrin. [557-25-5] Butanoic acid 2,3-dihydroxypropyl ester; 1-butyrylglycerol; α-monobutyrin; glycerol α-monobutyrate; glyceryl 3-monobutyrate. $C_7H_{14}O_4$; mol wt 162.19. C 51.84%, H 8.70%, O 39.46%. Naturally occurring lipid that stimulates angiogenesis. Secreted by differentiating adipocytes; stimulates blood vessel formation and causes vasodilation. The (R)-isomer is bioactive. Prepn of optical isomers: E. Abderhalden, E. Eichwald, *Ber.* **48**, 1847 (1915); of L-form: E. Baer, H. O. L. Fischer, *J. Am. Chem. Soc.* **67**, 2031 (1945); S. Gronowitz *et al., Chem. Phys. Lipids* **14**, 174 (1975). Purification from adipocytes and identification of angiogenic activity: D. E. Dobson *et al., Cell* **61**, 223 (1990). Biosynthesis: W. O. Wilkison *et al., J. Biol. Chem.* **266**, 16886 (1991); W. O. Wilkison, B. M. Spiegelman, *ibid.* **268**, 2844 (1993). Effect on retinal circulation and probable role in pathology of diabetes: Y.-D. C. Halvorsen *et al., J. Clin. Invest.* **92**, 2872 (1993).

(*R*)-**Form.** [5309-42-2] 3-Butyryl-*sn*-glycerol; L-α-butanoyl glycerol. Cryst from pure ether at −70°. n_D^{20} 1.4493. sp wt 0.8600. $[\alpha]_D^{18}$ −0.63°; $[\alpha]_D^{25}$ −7.40° (c = 2.04 in pyridine).

6336. Monocrotaline. [315-22-0] (3R,4R,5R,13aR,13bR)-4,-5,8,10,12,13,13a,13b-Octahydro-4,5-dihydroxy-3,4,5-trimethyl-2H-[1,6]dioxacycloundecino[2,3,4-gh]pyrrolizine-2,6(3H)-dione; (13α,14α)-14,19-dihydro-12,13-dihydroxy-20-norcrotolanan-11,15-dione; crotaline; MCT; NSC-28693; NCI-C56462. $C_{16}H_{23}NO_6$; mol wt 325.36. C 59.07%, H 7.13%, N 4.31%, O 29.50%. Toxic pyrrolizidine alkaloid isolated from *Crotalaria spp.* Isoln: R. Adams, E. F. Rogers, *J. Am. Chem. Soc.* **61**, 2815 (1939); R. B. Tinker, W. M. Lauter, *Econ. Bot.* **10**, 254 (1956); C. C. J. Culvenor, L. W.

Smith, *Aust. J. Chem.* **16**, 239 (1963). Structure: R. Adams *et al.*, *J. Am. Chem. Soc.* **74**, 5612 (1952). Stereochemistry: D. J. Robins, D. H. G. Crout, *J. Chem. Soc. C* **1969**, 1386. Crystal structure: H. Stoeckli-Evans, *Acta Crystallogr. B* **35**, 231 (1979). Stereoselective synthesis: H. Niwa *et al.*, *Tetrahedron* **48**, 10531 (1992). ELISA measurment: M. A. Bober *et al.*, *Toxicon* **27**, 1059 (1989). Toxicology: P. M. Newberne *et al.*, *Toxicol. Appl. Pharmacol.* **18**, 387 (1971); R. A. Roth *et al.*, *ibid.* **60**, 193 (1981). Review of carcinogenicity studies: *IARC Monographs* **10**, 291-302, 333-342 (1976). Comprehensive reviews: L. Bull *et al.*, *The Pyrrolizidine Alkaloids* (North-Holland, Amsterdam, 1968) 293 pp; D. J. Robins, *Fortschr. Chem. Org. Naturst.* **41**, 115-203 (1982). Review of pulmonary toxicity: D. W. Wilson *et al.*, *Crit. Rev. Toxicol.* **22**, 307-325 (1992).

White prisms from abs ethanol, mp 197-198° (dec). $[\alpha]_D^{26}$ −54.7° (c = 5.054 in chloroform). Also reported as colorless crystals from ethanol, mp 187-190° (déc). $[\alpha]_D^{12}$ −55.0° (c = 0.16 in CHCl$_3$). uv max (96% ethanol): 217 nm (log ε 3.32); *see:* Simánek *et al.*, *Collect. Czech. Chem. Commun.* **34**, 1832 (1969). LD$_{50}$ orally in rats: 71 mg/kg (Newberne).
Hydrochloride. C$_{16}$H$_{23}$NO$_6$.HCl. mp 184° (dec). $[\alpha]_D^{28}$ −38.4° (c = 5.2 in water).
Methiodide. C$_{16}$H$_{23}$NO$_6$.CH$_3$I. mp 205° (dec). $[\alpha]_D^{28}$ +23.4° (c = 3.1 in methanol).
USE: For inducing pulmonary diseases in rats.

6337. Monocrotophos. [6923-22-4] Phosphoric acid dimethyl (1*E*)-1-methyl-3-(methylamino)-3-oxo-1-propenyl ester; phosphoric acid dimethyl ester, ester with (*E*)-3-hydroxy-*N*-methylcrotonamide; 3-(dimethoxyphosphinyloxy)-*N*-methyl-*cis*-crotonamide; dimethyl (*E*)-1-methyl-2-(methylcarbamoyl)vinyl phosphate; 3-hydroxy-*N*-methyl-*cis*-crotonamide dimethyl phosphate; azodrin; C-1414; ENT-27129; SD-9129; Phoskill. C$_7$H$_{14}$NO$_5$P; mol wt 223.16. C 37.68%, H 6.32%, N 6.28%, O 35.85%, P 13.88%. Organophosphate insecticide; cholinesterase inhibitor. General prepn: Whetstone, Stiles, US 2802855; *see also* Ward, Morales, US 3400177 (1957, 1968 both to Shell). Metabolite of dicrotophos, *q.v.*: Menzer, Casida, *J. Agric. Food Chem.* **13**, 102 (1965). Toxicity study: T. B. Gaines, *Toxicol. Appl. Pharmacol.* **14**, 515 (1969). Determn of isomers in technical formulations: N. Ismail *et al.*, *J. Chromatogr. A* **903**, 255 (2000).

Crystals, mp 54-55°. *Poisonous.* Commercial product is a reddish-brown solid, mp 25-30°. Vapor pressure at 20°: 7 × 10^6 mm Hg. Misc with water; sol in acetone, ethanol. Practically insol in diesel oils, kerosine. LD$_{50}$ in male, female rats (mg/kg): 17, 20 orally; 126, 112 dermally (Gaines).
Caution: Potential symptoms of overexposure are eye irritation, miosis, blurred vision; dizziness, convulsions; dyspnea; salivation, abdominal cramps, nausea, diarrhea, vomiting. *See NIOSH Pocket Guide to Chemical Hazards* (DHHS/NIOSH 97-140, 1997) p 218.
USE: Insecticide.

6338. Monoctanoin. Monooctanoin; Capmul 8210; Moctanin. Cholesterol solvent used to dissolve gallstones by direct biliary infusion. Semisynthetic mixture of mono- and di-glycerides of octanoic and decanoic acids of which glycerol 1-octanoate is the major component. Description of prepn from coconut oil and use in treatment of gallstones: V. K. Babayan, US 4205086 (1980 to

Stokeley-Van Camp). Physical properties and clinical application: J. L. Thistle *et al.*, *Gastroenterology* **78**, 1016 (1980). Factors affecting cholesterol dissolution rate: J. B. Bogardus, *J. Pharm. Sci.* **73**, 906 (1984). Clinical trial: K. R. Palmer, A. F. Hofmann, *Gut* **27**, 196 (1986). Reviews of cholelitholytic agents: G. D. Bell, *Pharmacol. Ther.* **23**, 79-108 (1983); R. J. Fitzgibbons *et al.*, *Surg. Annu.* **21**, 237-262 (1989).

Glycerol 1-octanoate. [502-54-5] Octanoic acid 2,3-dihydroxypropyl ester; caprylic acid α-monoglyceride; α-monocaprylin. C$_{11}$H$_{22}$O$_4$; mol wt 218.29. Prepn: L. Hartman, *J. Chem. Soc.* **1959**, 4134. Crystals from light petroleum, mp 39.5-40.5°.
THERAP CAT: Cholelitholytic agent.

6339. Monoethyl Tartrate. [608-89-9] (2*R*,3*R*)-2,3-Dihydroxybutanedioic acid 1-ethyl ester; ethyl acid tartrate. C$_6$H$_{10}$O$_6$; mol wt 178.14. C 40.45%, H 5.66%, O 53.89%. Prepn from tartaric acid: Guérin-Varry, *Ann.* **22**, 248 (1837); P. Frankland, J. McCrae, *J. Chem. Soc., Trans.* **73**, 307 (1898). Preparative method from diethyl tartrate: D. S. Kalonia, A. P. Simonelli, *J. Chromatogr.* **455**, 355 (1988). Occurrence in wine: A. D. Webb *et al.*, *J. Agric. Food Chem.* **15**, 334 (1967).

Colorless, very hygroscopic crystals. Sweet taste. mp 90°. Sol in water, alcohol; insol in ether. *Keep well closed.*
USE: Printing with Indol Blue and Crystal Fast Blue on fustian, etc.

6340. Monomethylauristatin E. [474645-27-7] *N*-Methyl-L-valyl-*N*-[(1*S*,2*R*)-4-[(2*S*)-2-[(1*R*,2*R*)-3-[[(1*R*,2*S*)-2-hydroxy-1-methyl-2-phenylethyl]amino]-1-methoxy-2-methyl-3-oxopropyl]-1-pyrrolidinyl]-2-methoxy-1-[(1*S*)-1-methylpropyl]-4-oxobutyl]-*N*-methyl-L-valinamide; MMAE. C$_{39}$H$_{67}$N$_5$O$_7$; mol wt 717.99. C 65.24%, H 9.41%, N 9.75%, O 15.60%. Synthetic analog of dolastatin 10. Tubulin polymerization inhibitor, conjugated to tumor targeting antibodies to form antineoplastic agents. Prepn: S. Doronina *et al.*, WO 02088172; P. D. Senter *et al.*, US 6884869 (2002, 2005 both to Seattle Genetics). Synthesis of antibody drug conjugates (ADC): S. O. Doronina *et al.*, *Nat. Biotechnol.* **21**, 778 (2003); J. A. Francisco *et al.*, *Blood* **102**, 1458 (2003). *In vitro* cytotoxic activity: N. M. Okeley *et al.*, *Clin. Cancer Res.* **16**, 888 (2010).

Off-white solid or white powder. uv max: 215 nm.

6341. Monorden. [12772-57-5] (1a*R*,2*Z*,4*E*,14*R*,15a*R*)-8-Chloro-1a,14,15,15a-tetrahydro-9,11-dihydroxy-14-methyl-6*H*-oxireno[*e*][2]benzoxacyclotetradecin-6,12(7*H*)-dione; (+)-5-chloro-6-(7,8-epoxy-10-hydroxy-2-oxo-3,5-undecadienyl)-β-resorcylic acid *μ*-lactone; radicicol. C$_{18}$H$_{17}$ClO$_6$; mol wt 364.78. C 59.27%, H 4.70%, Cl 9.72%, O 26.32%. Antifungal antibiotic isolated from *Monosporium bonorden:* P. Delmotte, J. Delmotte-Plaquée, *Nature* **171**, 344 (1953). Identity with radicicol: R. N. Mirrington *et al.*, *Tetrahedron Lett.* **5**, 365 (1964). Structure: F. McCapra, A. I. Scott, *ibid.* 869; R. N. Mirrington *et al.*, *Aust. J. Chem.* **19**, 1265 (1966). Absolute configuration: H. G. Cutler *et al.*, *Agric. Biol. Chem.* **51**, 3331 (1987). Total synthesis: M. Lampilas, R. Lett, *Tetrahedron Lett.* **33**, 777 (1992).

Crystals from chloroform, alcohol or benzene, mp 193.5°. $[\alpha]_D^{20}$ +203° (chloroform). uv max (ethanol): 264, 272 nm (ε 13200, 13100).

Diacetate. $C_{22}H_{21}ClO_8$. Crystals, mp 189°. uv max: 275 nm (ε 13000).

6342. Monosodium Glutamate. [142-47-2] L-Glutamic acid sodium salt (1:1); sodium glutamate; MSG; RL-50; Accent; Aji-no-moto. $C_5H_8NNaO_4$; mol wt 169.11. C 35.51%, H 4.77%, N 8.28%, Na 13.59%, O 37.84%. Prepn: K. Ikeda, S. Suzuki, **GB 9440** (1910), *C.A.* **5**, 836 (1910). Commercial prepn: P. R. Shildneck, **US 2306646** (1942 to Staley Manuf.). Determn in food by HPLC: A. T. Rhys Williams, S. A. Winfield, *Analyst* **107**, 1092 (1982); by LC: O. W. Lau, C. S. Mok, *Anal. Chim. Acta* **302**, 45 (1995). Review of use and biological effects: *Food Technol.* **41**, 143-154 (1987). *Review:* T. Kawakita, "L-Monosodium Glutamate (MSG)" in *Kirk-Othmer Encyclopedia of Chemical Technology* **Vol. 2** (John Wiley & Sons, New York, 4th ed., 1992) pp 571-579. Review of toxicology: O. U. Eka *et al.*, *Biokemistri* **4**, 57-73 (1994). Book: *Glutamic Acid: Advances in Biochemistry and Physiology* L. J. Filer *et al.*, Eds. (Raven Press, New York, 1979) 416 pp.

White, practically odorless, free flowing crystals often as the monohydrate, dec 225°. Below −0.8°, crystallizes as a pentahydrate which upon filtration and exposure to air becomes the monohydrate. $[\alpha]^{20}$ −3.5° (10% soln in 5° Bé HCl). $[\alpha]_D^{20}$ +25.16° (10g MSG/100ml 2*N* HCl). d^{20} (saturated water soln): 1.620. pH of 5% water soln: 6.8; of 3% soln (25°): 7.0. Soly at 20° (g/100g): 41.7 water. Sparingly sol in alc. LD_{50} i.g. in mice: 19.9 g/kg (Eka).

USE: Flavor enhancer.

6343. Monotropein. [5945-50-6] (1*S*,4a*S*,7*R*,7a*S*)-1-(β-D-Glucopyranosyloxy)-1,4a,7,7a-tetrahydro-7-hydroxy-7-(hydroxymethyl)cyclopenta[*c*]pyran-4-carboxylic acid. $C_{16}H_{22}O_{11}$; mol wt 390.34. C 49.23%, H 5.68%, O 45.09%. From herb of *Monotropa hypopitys* L., *Ericaceae:* Bridel, *Compt. Rend.* **176**, 1742 (1923); *Bull. Soc. Chim. Biol.* **5**, 722 (1923). Structure: Inouye *et al.*, *Tetrahedron Lett.* **1963**, 1031; Bobbitt *et al.*, *Chem. Ind. (London)* **1964**, 931. Stereochemistry: Norio *et al.*, *Tetrahedron Lett.* **1967**, 2367. GC-mass spec studies: H. Inouye *et al.*, *J. Chromatogr.* **118**, 201 (1976). Purgative activity: *eidem*, *Planta Med.* **25**, 285 (1976).

Monohydrate. Needles from water, dec 170-173°. $[\alpha]_D^{18}$ −127.7° (water). uv max: 235 nm (log ε 3.98). Soluble in water, alcohol. Practically insol in ethyl acetate.

Pentaacetate. $C_{26}H_{32}O_{16}$. Crystals, mp 173-174.5°. $[\alpha]_D^{18}$ −82.5° (ethanol).

6344. Monsel's Solution. [1310-45-8] Ferric subsulfate solution. First described in 1856 by L. Monsel. Prepd by oxidizing

ferrous sulfate with nitric acid in the presence of sulfuric acid. Prepn: W. Procter, Jr., *Am. J. Pharm.* **31**, 403 (1859). Review of chemistry and mechanism of action: E. Epstein, H. I. Maibach, *Arch. Dermatol.* **90**, 226-228 (1964); of hemostatic applications: A. B. Jetmore *et al.*, *Dis. Colon Rectum* **36**, 866-867 (1993); of history: A. P. Garrett *et al.*, *J. Low. Genit. Tract Dis.* **6**, 225-227 (2002).

Reddish-brown liquid. Almost odorless; strongly astringent taste. May be dried at >95°F to produce brilliant red scales. Sol in water, alcohol. Acid to litmus. d^{25} 1.548.

THERAP CAT: Hemostatic.

THERAP CAT (VET): Hemostatic, astringent.

6345. Montan Wax. [8002-53-7] Lignite wax. Obtained by extraction from lignite. Asphalt and resin content and physical properties vary with source of lignite. Brief review: C. S. Letcher in *Kirk-Othmer Encyclopedia of Chemical Technology* **vol. 24** (Wiley-Interscience, New York, 3rd ed., 1984) pp 471-472.

Dark-brown lumps; white or nearly white when bleached. mp 80-86°. Saponification number 88-112. Insol in water. Sol in benzene, chloroform, carbon tetrachloride, hot petr ether; incompletely sol in hot ether or boiling alcohol.

USE: Electric cable insulators; in polishes instead of carnauba wax; manuf candles, waterproof paints and varnishes.

6346. Montelukast. [158966-92-8] 1-[[[(1*R*)-1-[3-[(1*E*)-2-(7-Chloro-2-quinolinyl)ethenyl]phenyl]-3-[2-(1-hydroxy-1-methylethyl)phenyl]propyl]thio]methyl]cyclopropaneacetic acid. $C_{35}H_{36}$-$ClNO_3S$; mol wt 586.19. C 71.71%, H 6.19%, Cl 6.05%, N 2.39%, O 8.19%, S 5.47%. Selective cysteinyl leukotriene type 1 (Cys-LT_1) receptor antagonist. Prepn: M. L. Belley *et al.*, **EP 480717**; *eidem*, **US 5565473** (1992, 1996 both to Merck Frosst); M. Labelle *et al.*, *Bioorg. Med. Chem. Lett.* **5**, 283 (1995). Pharmacological profile: T. R. Jones *et al.*, *Can. J. Physiol. Pharmacol.* **73**, 191 (1995). LC determn in human plasma: R. D. Amin *et al.*, *J. Pharm. Biomed. Anal.* **13**, 155 (1995). Clinical trial in pediatric asthma: B. Knorr *et al.*, *J. Am. Med. Assoc.* **279**, 1181 (1998). Review of pharmacology and clinical applications in asthma and allergy: Z. Diamant, A. P. Sampson, *J. Drug Eval. Respir. Med.* **1**, 53-88 (2002). Review of clinical experience in allergic rhinitis: J. A. Lagos, G. D. Marshall, *Ther. Clin. Risk Manage.* **3**, 327-332 (2007); and asthma: A. Nayak, R. B. Langdon, *Drugs* **67**, 887-901 (2007).

Monosodium salt. [151767-02-1] MK-476; Singulair. $C_{35}H_{35}$-$ClNNaO_3S$; mol wt 608.17. Hygroscopic, white to off-white powder. Freely sol in ethanol, methanol, water. Practically insol in acetonitrile.

THERAP CAT: Antiasthmatic; antiallergic.

6347. Monteplase. [122007-85-6] 84-L-Serine-plasminogen activator (human tissue-type protein moiety reduced); E-6010; Cleactor. Genetically engineered variant of human tissue plasminogen activator (t-PA), *q.v.*, constructed with a single amino acid substitution in the epidermal growth factor domain. Prepn: E. R. Mulvihill *et al.*, **EP 293936** (1988 to ZymoGenetics; Novo; Eisai); *idem et al.*, **US 5486471** (1996 to ZymoGenetics). Thrombolytic properties: S. Suzuki *et al.*, *J. Cardiovasc. Pharmacol.* **17**, 738 (1991). Pharmacology and pharmacokinetics: *eidem*, *ibid.* **22**, 834 (1993). Clinical trial in treatment of myocardial infarction: C. Kawai *et al.*, *J. Am. Coll. Cardiol.* **29**, 1447 (1997); prior to angioplasty: T. Inoue *et al.*, *Am. J. Cardiol.* **95**, 506 (2005).

THERAP CAT: Thrombolytic.

6348. Montmorillonite. [1318-93-0] A clay forming the principal constituent of bentonite and fuller's earth. Approximate formula: $R_{0.33}^+(Al,Mg)_2Si_4O_{10}(OH)_2 \cdot nH_2O$ where R^+, in natural material, includes one or more of the cations Na^+, K^+, Mg^{2+}, and Ca^{2+}, and possibly others.

Caution: Inhalation of the dust can cause respiratory irritation.

USE: In industrial chromatographic technique, *e.g.*, **US 2626888** and **GB 697060** (1953 to Merck & Co.), describing the elution of vitamin B$_{12}$-active substances adsorbed on montmorillonite adsorbents. Widely used in the petroleum industry. Also as catalyst carrier.

6349. Monuron. [150-68-5] N'-(4-Chlorophenyl)-N,N-dimethylurea; 1,1-dimethyl-3-(p-chlorophenyl)urea; CMU; Karmex Monuron Herbicide; Telvar. C$_9$H$_{11}$ClN$_2$O; mol wt 198.65. C 54.42%, H 5.58%, Cl 17.85%, N 14.10%, O 8.05%. Prepd by reacting p-chlorophenyl isocyanate with dimethylamine: Bucha, Todd, *Science* **114**, 493 (1951); *see also* **US 2655444**; **US 2655445**; **US 2655446**; **US 2655447**. Toxicity study: G. W. Bailey, J. L. White, *Residue Rev.* **10**, 97 (1965). *Review:* McCall, *Agric. Chem.* **7**, 40 (1952).

Thin rectangular prisms from methanol, mp 170.5-171.5° (the commercial product melts at 176-177°). Vapor pressure at 25°: 5 × 10^{-7} mm; at 100°: 178 × 10^{-5} mm. Slight odor. Stable toward oxygen and moisture under ordinary conditions at neutral pH; elevated temps and more acid or alkaline conditions appreciably raise rate of hydrolysis. Very slightly sol in water and in no. 3 Diesel oil: About 230 ppm at 25°. pH of satd aq soln 6.26. Moderately sol in methanol, ethanol, acetone. Practically insol in hydrocarbon solvents. LD$_{50}$ orally in rats: 3700 mg/kg (Bailey, White).
Trichloroacetate. Monuron TCA; Urox. C$_9$H$_{11}$ClN$_2$O.C$_2$-HCl$_3$O$_2$; mol wt 362.03. Prepn: Gilbert *et al.*, **US 2782112** (1957 to Allied Chem.). Effective in control of both weeds and grasses. mp 78-81°.
Caution: Anemia and methemoglobinemia have been produced in experimental animals.
USE: Herbicide.

6350. Moore's Ketene. [29342-22-1] 2-Carbonyl-3,3-dimethylbutanenitrile; *tert*-butylcyanoketene. C$_7$H$_9$NO; mol wt 123.16. C 68.27%, H 7.37%, N 11.37%, O 12.99%. Reacts via ene insertion reactions into σ bonds. Prepn: H. W. Moore, W. Weyler, Jr., *J. Am. Chem. Soc.* **92**, 4132 (1970). Reaction with cyclopropanes: M. D. Gheorghiu *et al.*, *J. Org. Chem.* **55**, 3713 (1990). Cycloaddition reactions with vinylarenes: S. V. Sereda *et al.*, *Struct. Chem.* **4**, 333 (1993); M. D. Gheorghiu *et al.*, *Rev. Roum. Chim.* **40**, 653 (1995).

USE: Reagent in cycloaddition reactions.

6351. Moperone. [1050-79-9] 1-(4-Fluorophenyl)-4-[4-hydroxy-4-(4-methylphenyl)-1-piperidinyl]-1-butanone; p-fluoro-4-(4′-hydroxy-4′-p-tolylpiperidino)butyrophenone; p-fluoro-4-(4′-hydroxy-4′-p-methylphenylpiperidino)butyrophenone; 1-(3′-p-fluorobenzoylpropyl)-4-hydroxy-4-p-tolylpiperidine; ω-(4-hydroxy-4-p-tolylpiperidino)-p-fluorobutyrophenone; methylperidol; R-1658. C$_{22}$H$_{26}$FNO$_2$; mol wt 355.45. C 74.34%, H 7.37%, F 5.34%, N 3.94%, O 9.00%. Prepn: Janssen *et al.*, *J. Med. Pharm. Chem.* **1**, 281 (1959); Janssen, **GB 881893** (1961).

Crystals, mp 118-119.5°. uv max: 246.5 nm (ε 12200).

Hydrochloride. [3871-82-7] Luvatren. C$_{22}$H$_{26}$FNO$_2$.HCl; mol wt 391.91. Crystals from isopropyl ether, mp 216-218°.
THERAP CAT: Antipsychotic.

6352. MOPS. [1132-61-2] 4-Morpholinepropanesulfonic acid; 3-(N-morpholino)propanesulfonic acid; N-(3-sulfopropyl)morpholine. C$_7$H$_{15}$NO$_4$S; mol wt 209.26. C 40.18%, H 7.23%, N 6.69%, O 30.58%, S 15.32%. One of the zwitterionic amino acids known as "Good" buffers; active in the pH range of 6.5-8.0. Prepn: C. F. H. Allen *et al.*, *Anal. Chem.* **37**, 156 (1965); N. E. Good, S. Izawa, "Hydrogen Ion Buffers," in *Methods Enzymol.* **24**, 53-65 (1972). Temperature effects on pK: M. Sankar, R. G. Bates, *Anal. Chem.* **50**, 1922 (1978); and pH: R. N. Roy *et al.*, *Cryo Letters* **6**, 139 (1985). Effects on "Lowry" protein determn: H. M. Himmel, W. Heller, *J. Clin. Chem. Clin. Biochem.* **25**, 909 (1987). Uses as a biological buffer: E. P. Paques *et al.*, *Eur. J. Biochem.* **107**, 447 (1980); L. A. Sonna *et al.*, *J. Biol. Chem.* **263**, 6625 (1988).

White powder, mp 283.5-284.5°. *Irritant.* pKa 7.15. pKa$_2$ (25°): 7.184; (37°): 7.041. ΔpKa/°C −0.013. Very sol in water. Is not completely stable when autoclaved in the presence of glucose.
USE: Biological buffer.

6353. Morantel. [20574-50-9] 1,4,5,6-Tetrahydro-1-methyl-2-[(1E)-2-(3-methyl-2-thienyl)ethenyl]pyrimidine. C$_{12}$H$_{16}$N$_2$S; mol wt 220.33. C 65.42%, H 7.32%, N 12.71%, S 14.55%. Prepn: Austin *et al.*, **GB 1120587** (1968 to Pfizer).

Crystals from chloroform + benzene, mp 239-241°.
Tartrate. [26155-31-7] CP-12009-18; Paratect; Rumatel; Suiminth. C$_{12}$H$_{16}$N$_2$S.C$_4$H$_6$O$_6$; mol wt 370.42. White or pale yellow, crystalline powder. Very sol in water, alc. Practically insol in ethyl acetate.
THERAP CAT (VET): Anthelmintic.

6354. Moricizine. [31883-05-3] N-[10-[3-(4-Morpholinyl)-1-oxopropyl]-10H-phenothiazin-2-yl]carbamic acid ethyl ester; 10-(3-morpholinopropionyl)phenothiazine-2-carbamic acid ethyl ester; ethmosine; ethmozin; moracizine; EN-313. C$_{22}$H$_{25}$N$_3$O$_4$S; mol wt 427.52. C 61.81%, H 5.89%, N 9.83%, O 14.97%, S 7.50%. Sodium channel blocker. Prepn: A. Gritsenko *et al.*, **DE 2014201** (1971 to Acad. Med. Sci., USSR); *eidem*, **US 3740395** (1973). Pharmacology and toxicity: N. V. Kaverina *et al.*, *Russ. Pharmacol. Toxicol.* **35**, 74 (1972). Mechanism of action study: T. Yamane *et al.*, *Br. J. Pharmacol.* **108**, 812 (1993). HPLC determn in plasma: C. C. Whitney *et al.*, *J. Pharm. Sci.* **70**, 462 (1981). Symposium on pharmacology, pharmacokinetics, clinical safety and efficacy: *Am. J. Cardiol.* **65**, 1D-71D (1990). *Review:* C. A. Clyne *et al.*, *N. Engl. J. Med.* **327**, 255 (1992).

Crystals, mp 156-157°. pKa 6.4.
Hydrochloride. [29560-58-5] Ethmozine. C$_{22}$H$_{25}$N$_3$O$_4$S.HCl; mol wt 463.98. White crystals from dichloroethane, mp 189° (dec). Sol in water, alc. LD$_{50}$ in mice, rats (mg/kg): 36, 12 i.v.; in mice (mg/kg): 131 i.p. (Kaverina).
THERAP CAT: Antiarrhythmic (class I).

6355. Morin. [480-16-0] 2-(2,4-Dihydroxyphenyl)-3,5,7-trihydroxy-4*H*-1-benzopyran-4-one; 2′,3,4′,5,7-pentahydroxyflavone; 2′-hydroxypelargidenolon 1522; C.I. Natural Yellow 8; C.I. Natural Yellow 11; C.I. 75660. $C_{15}H_{10}O_7$; mol wt 302.24. C 59.61%, H 3.34%, O 37.05%. In wood of old fustic *(Chlorophora tinctoria* (L.) Gaud., *Moraceae)*, also called Cuba wood, or yellow Brazil wood. The wood of the Osage orange tree also contains morin and maclurin. Isoln: Perkin, Pate, *J. Chem. Soc.* **67**, 649 (1895); Rolland, *Teintex* **3**, 460 (1938). Synthesis by condensing phloracetophenone dimethyl ether with 2,4-dimethoxybenzaldehyde: von Kostanecki *et al., Ber.* **39**, 625, 4014, 4022 (1906). Comprehensive list of refs in *Colour Index* **vol. 4** (3rd ed., 1971) p 4636.

Crystallizes with 1 or 2 mols water. Anhydr needles from abs alcohol, dec 285-290°. One gram dissolves in 4 liters water at 20°, in 1060 ml boiling water. Freely sol in alcohol; slightly sol in ether, acetic acid. Sol in aq alkaline solns with intense yellow color which turns brown on exposure to air. Absorpion spectrum: Grinbaumowna, Marchlewski, *Biochem. Z.* **290**, 261 (1937).

Calico yellow. Obtained by the action of bisulfite on fustic extract consists mainly of the bisulfite compound of morin, $C_{15}H_{10}O_7$ + $NaHSO_3$, and is used in calico printing. The calico yellow of commerce is a good source of morin for the laboratory. The bisulfite is removed with HCl.

USE: As a spot test reagent for Al, Be, Zn, Ga, In, and Sc salts: Feigl, *Spot Tests* (New York, 1954); Beck, *Mikrochim. Acta* **2**, 287 (1937). As luminescence indicator: Kocsis, Zádor, *Z. Anal. Chem.* **124**, 42-45 (1942). As a textile dye, morin dyes wool mordanted with chromium: olive yellow; mordanted with aluminum: yellow; mordanted with tin: lemon yellow; mordanted with iron: deep olive brown.

6356. Morindin. [60450-21-7] 1,5-Dihydroxy-2-methyl-6-[(6-*O*-β-D-xylopyranosyl-β-D-glucopyranosyl)oxy]-9,10-anthracenedione; 2-[[6-*O*-(6-deoxy-α-L-mannopyranosyl)-β-D-glucopyranosyl]oxy]-1,5-dihydroxy-6-methyl-9,10-anthracenedione. $C_{27}H_{30}O_{14}$; mol wt 578.52. C 56.06%, H 5.23%, O 38.72%. From bark of *Coprosma australis* Forst. (*C. grandifolia* Hook.), *Rubiaceae:* Briggs, Dacre, *J. Chem. Soc.* **1948**, 564. Structure: Briggs, Le Quesne, *ibid.* **1963**, 3471. On acid hydrolysis yields the aglucone *morindone.* Synthesis and structure proof: B. Vermes *et al., Phytochemistry* **19**, 119 (1980).

Yellow needles from glacial acetic acid, mp 169-171°. $[\alpha]_D^{20}$ −90.0° (c = 0.054 in dioxane). Absorption max (ethanol): 230, 261, 448 nm (log ε 4.15, 3.93, 3.57). Sol in cold water but on boiling forms a water insol form, *β-morindin.* Sol in dioxane, pyridine, acetone, methanol; slightly sol in ethanol, glacial acetic acid. Practically insol in ether, chloroform, benzene, petr ether.

6357. Moroctocog Alfa. (1-742)-(1637-1648)-Blood coagulation factor VIII (human reduced) complex with 1649-2332-blood coagulation factor VIII (human reduced); r-VIII SQ; Refacto. B-domain deleted form of factor VIII produced in Chinese hamster ovary cells by recombinant DNA technology. Prepn: A. Almstedt *et al.,* **WO 9109122** (1991 to KabiVitrum); P. Lind *et al., Eur. J. Biochem.* **232**, 19 (1995). Stability of albumin-free preparation: T.

Osterberg *et al., Pharm. Res.* **14**, 892 (1997). Clinical pharmacokinetics: K. Fijnvandraat *et al., Thromb. Haemostasis* **77**, 298 (1997). Review of development and clinical experience: E. Berntorp, *ibid.* **78**, 256-260 (1997).

THERAP CAT: Antihemophilic factor.

6358. Moroxydine. [3731-59-7] *N*-(Aminoiminomethyl)-4-morpholinecarboximidamide; 4-morpholinecarboximidoylguanidine; N^1,N^1-anhydrobis(β-hydroxyethyl)biguanide; abitilguanide; abitylguanide; ABOB; Virusmin. $C_6H_{13}N_5O$; mol wt 171.20. C 42.09%, H 7.65%, N 40.91%, O 9.35%. Prepn: **GB 776176** (1957 to AB Kabi). Proposed as influenza suppressant: Melander, *Antibiot. Chemother.* **10**, 39 (1960).

Hydrochloride. [3160-91-6] Flumidin; Virustat. $C_6H_{13}N_5O$.-HCl; mol wt 207.66. Crystals.

THERAP CAT: Antiviral.

6359. Morphazinamide. [952-54-5] *N*-(4-Morpholinylmethyl)-2-pyrazinecarboxamide; *N*-morpholinomethylpyrazinamide; morfazinamide; morinamide; B-2310. $C_{10}H_{14}N_4O_2$; mol wt 222.25. C 54.04%, H 6.35%, N 25.21%, O 14.40%. Prepn from pyrazinamide, morpholine, and formalin: Felder, Tiepolo, **DE 1129492** (1962 to Bracco).

Crystals, mp 118.5-119.5°. uv max (ethanol): 269, 317 nm (log ε 3.95, 2.77). One gram dissolves in 3 ml water, 30 ml ethanol, 30 ml benzene, 2.5 ml chloroform.

Hydrochloride. [1473-73-0] B-2311; Piazofolina; Piazolin. $C_{10}H_{14}N_4O_2$.HCl; mol wt 258.71. Crystals, mp 196°. One gram dissolves in 2 ml water, 350 ml ethanol, 2000 ml chloroform.

THERAP CAT: Antibacterial (tuberculostatic).

6360. Morphinan. [468-10-0] (4a*R*,10*R*,10a*R*)-1,3,4,9,10,-10a-Hexahydro-2*H*-10,4a-(iminoethano)phenanthrene. $C_{16}H_{21}N$; mol wt 227.35. C 84.53%, H 9.31%, N 6.16%. Parent substance of the morphine alkaloids, codeine, thebaine, etc. Prepd from 1-benzyl-1,2,3,4,5,6,7,8-octahydro-2-methylisoquinoline: Grewe, Mondon, *Ber.* **81**, 279 (1948); from β-cyclohexenyl-*N*-phenacetylethylamine: Grewe *et al., Ber.* **84**, 527 (1951).

Oily liq, $bp_{0.05}$ 115°.
Hydrochloride. [1071557-77-1] Crystals from acetone + ether, dec 229°.
Sulfate. [5985-36-4] Crystals from alcohol + ether, dec 195°.
***N*-Methylmorphinan.** [3882-38-0] Crystals, mp 61°.

6361. Morphine. [57-27-2]; [6009-81-0] (monohydrate). (5α,6α)-7,8-Didehydro-4,5-epoxy-17-methylmorphinan-3,6-diol; morphium; morphia. $C_{17}H_{19}NO_3$; mol wt 285.34. C 71.56%, H 6.71%, N 4.91%, O 16.82%. Principal alkaloid of opium, *q.v.*; also found endogenously in various animal tissues including brain, blood and liver. Isoln: F. W. A. Sertürner, *Trommsdorff′s J. Pharm.* **13**,

234 (1805); W. Gregory, *Ann.* **7**, 261 (1833). Extraction procedure: L. B. Achor, E. M. K. Geiling, *Anal. Chem.* **26**, 1061 (1954). Structure: J. M. Gulland, R. Robinson, *J. Chem. Soc.* **123**, 980 (1923). Total synthesis: M. Gates, G. Tschudi, *J. Am. Chem. Soc.* **74**, 1109 (1952); **78**, 1380 (1956). Biogenesis: E. Leete, *J. Am. Chem. Soc.* **81**, 3948 (1959). Cystallographic study: M. Mackay, D. C. Hodgkin, *J. Chem. Soc.* **1955**, 3261. Configuration: J. Kalvoda *et al.*, *Helv. Chim. Acta* **38**, 1847 (1955). Toxicity data: M. E. Buchwald, G. S. Eadie, *J. Pharmacol. Exp. Ther.* **71**, 197 (1941). GC-MS determn in biological fluids: R. Wasels, F. Belleville, *J. Chromatogr. A* **674**, 225 (1994). Comprehensive description: F. J. Muhtadi, *Anal. Profiles Drug Subs.* **17**, 259-366 (1988). Review of syntheses: T. Hudlicky *et al.* in *Studies in Natural Products Chemistry*, **Vol. 18**, Atta-ur-Rahman, Ed. (Elsevier, Amsterdam, 1996) pp 43-154; J. Zezula, T. Hudlicky, *Synlett* **2005**, 388-405. Review of physiological role of endogenous morphine: G. B. Stefano *et al.*, *Trends Neurosci.* **23**, 436-442 (2000). Review of clinical experience in pain management: S. Donnelly *et al.*, *Support. Care Cancer* **10**, 13-35 (2002); of pharmacokinetics and metabolism in long term treatment: G. Andersen *et al.*, *J. Pain Symptom Manage.* **25**, 74-91 (2003).

HO-structure

Small rhombic prisms or needles from dil alc as the monohydrate. Becomes anhydrous at 100°, mp 253-254° with decompn. d_4^{20} 1.32. $[\alpha]_D^{25}$ −132° (methanol). pKb at 20° = 6.13; pKa 9.85. pH of satd soln, 8.5. uv max in acid: 285 nm; in alkali: 298 nm. One gram dissolves in about 5000 ml water, 1100 ml boiling water, 210 ml alcohol, 98 ml boiling alc, 1220 ml chloroform, 6250 ml ether, 114 ml amyl alc, 10 ml boiling methanol, 525 ml ethyl acetate. Freely sol in solns of fixed alkali and alkaline earth hydroxides, in phenol, cresols; moderately sol in mixtures of chloroform with alcohols; slightly sol in ammonia, benzene.

Hydrochloride. [52-26-6] $C_{17}H_{19}NO_3 \cdot HCl$; mol wt 321.80. Occurs as the trihydrate. White flakes or crystalline powder; bitter taste. Loses its water of crystn at about 100° and usually becomes yellowish. mp about 200° (dec). $[\alpha]_D^{25}$ −113.5° (c = 2.2 in H_2O, anhydr basis). One gram dissolves in 17.5 ml water, 0.5 ml boiling water, 52 ml alc, 6 ml alc at 60°; slowly sol in glycerol. Insol in chloroform, ether. pH about 5. *Protect from light.* LD_{50} in mice (mg/kg): 226-318 i.v. (Buchwald, Eadie).

Sulfate Pentahydrate. [6211-15-0]; [64-31-3] (anhydrous). Morphine hemisulfate; Avinza; Dolcontin; Kadian; Kapanol; Morcap; Moscontin; MS Contin; MSIR; MST Continus; Oramorph; Sevredol. $2C_{17}H_{19}NO_3 \cdot H_2SO_4 \cdot 5H_2O$; mol wt 758.83. White, fine odorless crystals or powder, or cubical masses. Loses some H_2O at ordinary temp; about $3H_2O$ (7.1%) at 100°, the remainder at 130°. Discolors on exposure to light. mp ~250° with decompn when anhydr. $[\alpha]_D^{25}$ −108.7° (c = 4 in H_2O, anhydr basis). pKa 7.9. Partition coefficient (*n*-octanol/water): 1.42 (at physiological pH). One gram dissolves in 15.5 ml water at 25°, 0.7 ml water at 80°, 565 ml alcohol, 240 ml alcohol at 60°. Insol in chloroform, ether. pH about 4.8. *Keep well closed and protected from light.*

Note: This is a controlled substance (opiate): **21 CFR**, 1308.12.

THERAP CAT: Analgesic.

THERAP CAT (VET): Analgesic, preanesthetic, antitussive, antiperistaltic.

6362. Morpholine. [110-91-8] Tetrahydro-2*H*-1,4-oxazine; tetrahydro-1,4-oxazine; diethylene oximide; diethylene imidoxide. C_4H_9NO; mol wt 87.12. C 55.15%, H 10.41%, N 16.08%, O 18.36%. Prepd by dehydrating diethanolamine: Knorr, *Ann.* **301**, 1 (1898); Jones, Burns, *J. Am. Chem. Soc.* **47**, 2966 (1925); Hampton, Pollard, *ibid.* **58**, 2338 (1936). Toxicity study: H. F. Smyth *et al.*, *Arch. Ind. Hyg. Occup. Med.* **10**, 61 (1954). Review of morpholine and derivatives: A. L. Wilson, *Ind. Eng. Chem.* **27**, 867-871 (1935).

Monograph: Morpholine Technical Bulletin, Jefferson Chemical Co. (New York, 1953).

Mobile, hygroscopic liquid. Characteristic amine odor. mp −4.9°. bp_{760} 128.9°; bp_6 20.0°. d_4^{20} 1.007. n_D^{20} 1.4540. *Corrosive; flammable.* Volatile with steam. Does not form an azeotrope with water. Flash pt, open cup: 100°F (38°C). Surface tension (20°): 37.5 dynes/cm. Viscosity (20°): 2.23 cP. Dipole moment 1.58. Strong base, pKb 5.6. Immiscible with concd NaOH solns. Miscible with water with evolution of some heat, with acetone, benzene, ether, castor oil, methanol, ethanol, ethylene glycol, 2-hexanone, linseed oil, turpentine, pine oil. Will dissolve 109% dimethylamine; 34% trimethylamine; 33% methylamine; >5% naphtha; <1% paraffin oil; <5% sulfur. LD_{50} orally in female rats: 1.05 g/kg (Smyth).

Hydrochloride. [10024-89-2] $C_4H_9NO \cdot HCl$; mol wt 123.58. Crystals, dec 175-176°. Sol in water.

Salicylate. [147-90-0] Retarcyl; Deposal. $C_{11}H_{15}NO_4$; mol wt 225.24. Crystals from ethanol, mp 110-111°. Soluble in water, methanol, ethanol, ethyl acetate, acetone, benzene, chloroform. Practically insol in toluene, xylene, petr ether, ether, carbon tetrachloride.

N-Methylmorpholine. [109-02-4] $C_5H_{11}NO$; mol wt 101.15. Mobile liquid, characteristic ammonia odor. bp_{764} 116-117°; bp_{760} 115.4°. d_4^{20} 0.9168. n_D^{20} 1.4332. Miscible with water, alcohol, ether.

Caution: Potential symptoms of overexposure to morpholine are irritation of eyes, skin, nose and respiratory system; visual disturbance; coughing. *See NIOSH Pocket Guide to Chemical Hazards* (DHHS/NIOSH 97-140, 1997) p 220.

USE: Cheap solvent for resins, waxes, casein, dyes. Morpholine fatty acid salts are used as surface-active agents and emulsifiers. Other morpholine compds are used as corrosion inhibitors, antioxidants, plasticizers, viscosity improvers, insecticides, fungicides, herbicides, local anesthetics and antiseptics.

THERAP CAT: Salicylate as analgesic; antipyretic; anti-inflammatory.

6363. Mosapramine. [89419-40-9] 1'-[3-(3-Chloro-10,11-dihydro-5*H*-dibenz[*b*,*f*]azepin-5-yl)propyl]hexahydrospiro[imidazo[1,2-*a*]pyridine-3(2*H*),4'-piperidin]-2-one; 1'-[3-(3-chloro-10,11-dihydro-5*H*-dibenz[*b*,*f*]azepin-5-yl)propyl]-1,2,3,5,6,7,8,8a-octahydro-2-oxoimidazo[1,2-*a*]-pyridine-3-spiro-4'-piperidine; 3-chloro-5-[3-(2-oxo-1,2,3,5,6,7,8,8a-octahydroimidazo[1,2-*a*]-pyridine-3-spiro-4'-piperidino)propyl]-10,11-dihydro-5*H*-dibenz[*b*,*f*]-azepine; clospipramine. $C_{28}H_{35}ClN_4O$; mol wt 479.07. C 70.20%, H 7.36%, Cl 7.40%, N 11.70%, O 3.34%. Dopamine receptor antagonist with adrenoceptor blocking activity; metabolite of clocapramine, *q.v.* Prepn: C. Tashiro, I. Horii, **US 4337260** (1982 to Yoshitomi); C. Tashiro *et al.*, *Yakugaku Zasshi* **109**, 93 (1989), *C.A.* **112**, 35749j (1990). X-ray crystallographic analysis: I. Ueda, C. Tashiro, *Acta Crystallogr.* **C40**, 422 (1984). Receptor binding studies: M. Setoguchi *et al.*, *Eur. J. Pharmacol.* **112**, 313 (1985). Pharmacokinetics and metabolism: J. Ishigooka *et al.*, *Psychopharmacology* **97**, 303 (1989). Toxicity data: H. Horizoe *et al.*, *Oyo Yakuri* **37**, 529 (1989); *C.A.* **111**, 167174g (1989). HPLC determn in plasma: K. Hikida *et al.*, *Anal. Sci.* **6**, 367 (1990).

structure

Dihydrochloride. [98043-60-8] Y-516; Cremin. $C_{28}H_{35}ClN_4$·O.2HCl; mol wt 551.98. White crystals, mp 271°. LD_{50} in male and female mice (mg/kg): 1008, 1293 orally; 74, 116 i.p.; 1147, 1264 s.c. (Horizoe).

THERAP CAT: Antipsychotic.

6364. Mosapride. [112885-41-3] 4-Amino-5-chloro-2-ethoxy-*N*-[[4-[(4-fluorophenyl)methyl]-2-morpholinyl]methyl]benzamide; (±)-4-amino-5-chloro-2-ethoxy-*N*-[[4-(4-fluorobenzyl)-2-morpholinyl]methyl]benzamide. $C_{21}H_{25}ClFN_3O_3$; mol wt 421.90. C 59.78%, H 5.97%, Cl 8.40%, F 4.50%, N 9.96%, O 11.38%. Selective 5-HT$_4$ receptor agonist. Prepn: T. Kon *et al.*, **EP 243959**; *eidem*, **US 4870074** (1987, 1989 both to Dainippon); S. Kato *et al.*, *J. Med. Chem.* **34**, 616 (1991). Pharmacology: N. Yoshida *et al.*, *Arch. Int. Pharmacodyn. Ther.* **300**, 51 (1989). Series of articles on pharmacokinetics: *Arzneim.-Forsch.* **43**, 859-872 (1993); on pharmacology and metabolism: *ibid.* 1083-1108. HPLC determn in plasma: I. Yokoyama *et al.*, *J. Pharm. Biomed. Anal.* **15**, 1527 (1997). Clinical evaluation in gastro-esphageal reflux disease: M. Ruth *et al.*, *Aliment. Pharmacol. Ther.* **12**, 35 (1998). Toxicity study: S. Yatera *et al.*, *Jpn. Pharmacol. Ther.* **21**, 21 (1993).

Crystals from ethanol, mp 151-153°.

Citrate. [112885-42-4] AS-4370; Gasmotin. $C_{21}H_{25}ClFN_3O_3$·$C_6H_8O_7$; mol wt 614.02. Crystals from ethanol, mp 143-145°. LD_{50} in male, female mice, rats (mg/kg): >3000, >3000, >3000, 1905 orally; >1000, 914, >1000, >1000 i.p.; all species >1000 s.c.; in male, female dogs (mg/kg): both >400 orally (Yatera).

THERAP CAT: Gastroprokinetic.

6365. Mosher's Reagent. [81655-41-6] α-Methoxy-α-(trifluoromethyl)benzeneacetic acid; (±)-Mosher's acid; MTPA; 2-methoxy-2-(trifluoromethyl)phenylacetic acid. $C_{10}H_9F_3O_3$; mol wt 234.17. C 51.29%, H 3.87%, F 24.34%, O 20.50%. Chiral derivatizing reagent. Prepn: D. L. Dull, H. S. Mosher, *J. Am. Chem. Soc.* **89**, 4230 (1967); J. A. Dale *et al.*, *J. Org. Chem.* **34**, 2543 (1969). Microscale prepn: D. E. Ward, C. K. Rhee, *Tetrahedron Lett.* **32**, 7165 (1991). Absolute configuration: S. S. Oh *et al.*, *J. Org. Chem.* **54**, 4499 (1989); of the acid chloride: B. S. Joshi, S. W. Pelletier, *Heterocycles* **51**, 183 (1999). Enantiomeric purity of commerical samples: W. A. König *et al.*, *Tetrahedron Lett.* **31**, 6867 (1990). Use in determn of abs config: T. R. Hoye, M. K. Renner, *J. Org. Chem.* **61**, 2056 (1996); A. Ichikawa, *Enantiomer* **3**, 255 (1998); L. F. Tietze *et al.*, *Eur. J. Org. Chem.* **2000**, 2247.

(*R*)-(+)-form

(*R*)-(+)-**Form.** [20445-31-2] bp$_{1.5}$ 116-118°. $[\alpha]_D^{25}$ +68.5 ±1.3° (c = 1.49 in CH$_3$OH).

(*S*)-(−)-**Form.** [17257-71-5] bp$_{1.5}$ 115-117°. $[\alpha]_D^{24}$ −71.8 ±0.6° (c = 3.28 in CH$_3$OH).

(*R*)-(−)-**Acid chloride.** [39637-99-5] (*R*)-MTPA-chloride. bp$_1$ 54-56°. $[\alpha]_D^{25}$ −10.0 ±0.1° (neat, l = 1).

(*S*)-(+)-**Acid chloride.** [20445-33-4] bp$_1$ 54-56°. $[\alpha]_D^{24}$ +129.0 ±0.2° (c = 5.17 in CCl$_4$).

USE: Resolution of enantiomers primarily of alcohols and amines.

6366. Motesanib. [453562-69-1] *N*-(2,3-Dihydro-3,3-dimethyl-1*H*-indol-6-yl)-2-[(4-pyridinylmethyl)amino]-3-pyridinecarboxamide. $C_{22}H_{23}N_5O$; mol wt 373.46. C 70.76%, H 6.21%, N 18.75%, O 4.28%. Multikinase inhibitor that targets VEGF, PDGF, and Kit receptor tyrosine kinases; inhibits angiogenesis in solid tu-

mors. Prepn: G. Chen *et al.*, **WO 02066470**; B. Askew *et al.*, **US 6878714** (2002, 2005 both to Amgen); of pharmaceutical salt forms: B. B. Liu *et al.*, **WO 07087026**; G. Alva *et al.*, **WO 07092178** (both 2007 to Amgen). Kinase inhibition and bioactivity: A. Polverino *et al.*, *Cancer Res.* **66**, 8715 (2006). Clinical pharmacokinetics: L. S. Rosen *et al.*, *J. Clin. Oncol.* **25**, 2369 (2007). Clinical trial in thyroid cancer: S. I. Sherman *et al.*, *N. Engl. J. Med.* **359**, 31 (2008). Review of clinical trials in non-small cell lung cancer: K. P. S. Raghav, G. R. Blumenschein, *Expert Opin. Invest. Drugs* **20**, 859-869 (2011).

Diphosphate. [857876-30-3] AMG-706. $C_{22}H_{23}N_5O$·$2H_3PO_4$; mol wt 569.45. Crystalline solid, mp 215° (dec). Forms stable monophosphate dihydrate in boiling water.

THERAP CAT: Antineoplastic.

6367. Motexafin Gadolinium. [246252-06-2]; [156436-89-4] (hydrate). (*PB*-7-11-233′2′4)-Bis(acetato-*κO*)[9,10-diethyl-20,21-bis[2-[2-(2-methoxyethoxy)ethoxy]ethoxy]-4,15-dimethyl-8,11-imino-3,6:16,13-dinitrilo-1,18-benzodiazacycloeicosine-5,14-dipropanolato-*κN*^{1},*κN*^{18},*κN*^{23},*κN*^{24},*κN*^{25}]gadolinium; 4,5-diethyl-10,23-dimethyl-9,24-bis(3-hydroxypropyl)-16,17-bis[2-[2-(2-methoxyethoxy)ethoxy]ethoxy]pentaazapentacyclo[20.2.1.1^{3,6}.1$^{8.11}$.0^{14,19}]heptacosa-1,3,5,7,9,11(27),12,14,16,18,20,22(25),23-tridecaene gadolinium(III) complex; gadolinium texaphyrin; Gd-Tex; Gd-T2BET; NSC-695238; PCI-0120; Xcytrin. $C_{52}H_{72}GdN_5O_{14}$; mol wt 1148.42. C 54.39%, H 6.32%, Gd 13.69%, N 6.10%, O 19.50%. Synthetic metal-coordinating expanded porphyrin. Paramagnetic redox mediator and radiation enhancer. Prepn: J. L. Sessler *et al.*, **US 5801229** (1998 to Board of Regents Univ. of Texas); and redox chemistry: J. L. Sessler *et al.*, *J. Phys. Chem. A* **103**, 787 (1999). Tumor selective radiosensitizing activity and MRI characteristics: S. W. Young *et al.*, *Proc. Natl. Acad. Sci. USA* **93**, 6610 (1996). HPLC determn in plasma: R. A. Parise *et al.*, *J. Chromatogr. B* **749**, 145 (2000). Mechanism of action study: D. Magda *et al.*, *Int. J. Radiat. Oncol. Biol. Phys.* **51**, 1025 (2001); S. I. Hashemy *et al.*, *J. Biol. Chem.* **281**, 10691 (2006). Clinical pharmacokinetics: D. I. Rosenthal *et al.*, *Clin. Cancer Res.* **5**, 739 (1999). Clinical evaluation with radiation in brain metastases: M. P. Mehta *et al.*, *J. Clin. Oncol.* **20**, 3445 (2002); C. A. Meyers *et al.*, *J. Clin. Oncol.* **22**, 157 (2004). Review of chemistry: T. D. Mody, J. L. Sessler, *J. Porphyrins Phthalocyanines* **5**, 134-142 (2001); of therapeutic potential in brain tumors: D. Khuntia, M. Mehta, *Expert Rev. Anticancer Ther.* **4**, 981-989 (2004).

Prepd as hemihydrate, dark green microcrystalline solid. uv max (methanol): 350, 414, 473, 739 nm (log ε 4.33, 4.67, 5.06, 4.59).

THERAP CAT: Antineoplastic adjunct (radiosensitizer).

6368. Motexafin Lutetium. [246252-04-0]; [156436-90-7] (hydrate). (*PB*-7-11-233′2′4)-Bis(acetato-κ*O*)[9,10-diethyl-20,21-bis[2-[2-(2-methoxyethoxy)ethoxy]ethoxy]-4,15-dimethyl-8,11-imino-3,6:16,13-dinitrilo-1,18-benzodiazacycloeicosine-5,14-dipropanolato-κ*N*¹,κ*N*¹⁸,κ*N*²³,κ*N*²⁴,κ*N*²⁵]lutetium; 4,5-diethyl-10,23-dimethyl-9,24-bis(3-hydroxypropyl)-16,17-bis[2-[2-(2-methoxyethoxy)ethoxy]ethoxy]pentaazapentacyclo[20.2.1.1³,⁶.1⁸,¹¹.0¹⁴,¹⁹]heptacosa-1,3,5,7,9,11(27),12,14,16,18,20,22(25),23-tridecaene lutetium(III) complex; lutetium texaphyrin; Lu-Tex; PCI-0123; Antrin; Lutrin; Optrin. $C_{52}H_{72}LuN_5O_{14}$; mol wt 1166.14. C 53.56%, H 6.22%, Lu 15.00%, N 6.01%, O 19.21%. Synthetic, metal-coordinating, expanded porphyrin. Diamagnetic photosensitizer that is activated using far-red light to generate cytotoxic singlet oxygen. Prepn: J. L. Sessler *et al.*, US 5801229 (1998 to Board of Regents Univ. of Texas); S. W. Young *et al.*, *Photochem. Photobiol.* **63**, 892 (1996). Electrochemical characteristics: J. L. Sessler *et al.*, *J. Alloys Compd.* **249**, 146 (1997); and singlet oxygen generation: L. I. Grossweiner *et al.*, *Photochem. Photobiol.* **70**, 138 (1999). LC-MS/MS and ICP-AES determn in plasma: D. Miles *et al.*, *AAPS PharmSci.* **5**, 1 (2003). Preclinical studies in cancer and atheromatous plaque: K. W. Woodburn *et al.*, *J. Clin. Laser Med. Surg.* **14**, 343 (1996); in ocular angiography: M. S. Blumenkranz *et al.*, *Am. J. Ophthalmol.* **129**, 353 (2000). Clinical evaluation in coronary arterial photodynamic therapy: D. J. Kereiakes *et al.*, *Circulation* **108**, 1310 (2003). Review of chemistry and therapeutic potential: T. D. Mody, J. L. Sessler, *J. Porphyrins Phthalocyanines* **5**, 134-142 (2001).

Prepd as hydrate, shiny green solid. Sol in water, acetonitrile, methanol, *N*,*N*-dimethylformamide. Aqueous solns (100 μM) stable at 2<pH<11 for several months. uv max (methanol): 354, 414, 474, 732 nm (log ε 4.33, 4.67, 5.10, 4.62).

THERAP CAT: Antiatherosclerotic, antineoplastic (photosensitizer). In treatment of age-related macular degeneration.

6369. Motilin. [52906-92-0] Peptide hormone produced by endocrine cells of the duodeno-jujenal mucosa; increases gastric motility, stimulates pepsin output, and regulates the interdigestive myoelectric complex. Structure contains 22 amino acid residues and is highly conserved across species. Human motilin is identical to porcine. Postulation of existence biological chemical stimulation of canine duodenum: J. C. Brown *et al.*, *Gastroenterology* **50**, 333 (1966). Study of motor effects: U. Strunz *et al.*, *ibid.* **68**, 1485 (1975). Isoln of porcine motilin from mucosa of hog small intestine: J. C. Brown *et al.*, *Can. J. Physiol. Pharmacol.* **49**, 399 (1971). Purification, amino acid composition, *C*-terminal residue: *eidem*, *Gastroenterology* **62**, 401 (1972). Amino acid sequence: *eidem*, *Can. J. Biochem.* **51**, 533 (1973); revised sequence: H. Schubert, J. C. Brown, *ibid.* **52**, 7 (1974). First synthetic work: E. Wünsch *et al.*, *Z. Naturforsch.* **28**C, 235 (1973). Synthesis of the complete docosapeptide corresponding to porcine motilin: H. Yajima *et al.*, *Chem. Commun.* **1975**, 159. Alternate syntheses: Y. Kai *et al.*, *Chem. Pharm. Bull.* **23**, 2346 (1975); S. Yamada *et al.*, *J. Am. Chem. Soc.* **97**, 7174 (1975); E. Izeboud, H. C. Beyerman, *Rec. Trav. Chim.* **99**, 124 (1980). Solid phase synthesis: N. Ikota *et al.*, *Chem. Pharm. Bull.* **28**, 3347 (1980); D. H. Coy *et al.*, *Peptides* **3**, 137 (1982). Radioimmunoassay: J. R. Dryburgh, J. C. Brown, *Gastro-*

enterology **68**, 1169 (1975). Identification of motilin endocrine cells: J. M. Polak *et al.*, *Gut* **16**, 225 (1975). Purification and amino acid sequence of human motilin: P. De Clercq *et al.*, *Regul. Pept.* **55**, 79 (1995). Identification of the human motilin receptor: S. D. Feighner *et al.*, *Science* **284**, 2184 (1999). Review of discovery, synthesis, and bioactivities: W. Domschke, *Dig. Dis.* **22**, 454-461 (1977); and potential clinical applications: Z. Itoh, *Peptides* **18**, 593-608 (1997); B. De Smet *et al.*, *Pharmacol. Ther.* **123**, 207-223 (2009).

Human Motilin

Human motilin. [9072-41-7] Motilin (swine); L-phenylalanyl-L-valyl-L-prolyl-L-isoleucyl-L-phenylalanyl-L-threonyl-L-tyrosyl-glycyl-L-α-glutamyl-L-leucyl-L-glutaminyl-L-arginyl-L-methionyl-L-glutaminyl-L-α-glutamyl-L-lysyl-L-α-glutamyl-L-arginyl-L-asparaginyl-L-lysylglycyl-L-glutamine. $C_{120}H_{188}N_{34}O_{35}S$; mol wt 2699.09.

6370. Motretinide. [56281-36-8] (2*E*,4*E*,6*E*,8*E*)-*N*-Ethyl-9-(4-methoxy-2,3,6-trimethylphenyl)-3,7-dimethyl-2,4,6,8-nonatetraenamide; (all-*E*)-9-(4-methoxy-2,3,6-trimethylphenyl)-3,7-dimethylnona-2,4,6,8-tetraen-1-oic acid ethyl amide; Ro-11-1430; Tasmaderm. $C_{23}H_{31}NO_2$; mol wt 353.51. C 78.15%, H 8.84%, N 3.96%, O 9.05%. Aromatic analog of retinoic acid, *q.v.* Prepn: W. Bollag *et al.*, DE 2414619; *eidem*, US 4215215 (1974, 1980 both to Hoffmann-La Roche). Description of properties: *idem*, *Chemotherapy* **21**, 236 (1975). ¹³C-NMR study: G. Englert, *Helv. Chim. Acta* **58**, 2367 (1975). Pharmacology: L. J. Wilkoff *et al.*, *Cancer Res.* **36**, 964 (1976); S. S. Shapiro *et al.*, *ibid.* 3702. Inhibitory effect on radiation-induced neoplasms: L. Harisiades *et al.*, *Nature* **274**, 486 (1978). Use in treatment of acne: K. Nordin *et al.*, *Dermatologica* **162**, 104 (1981).

Crystals from ethanol, mp 179-180°.
THERAP CAT: Antiacne.

6371. Moveltipril. [85856-54-8] *N*-(Cyclohexylcarbonyl)-D-alanyl-(2*S*)-3-mercapto-2-methylpropanoyl-L-proline; *N*-[3-(*N*-cyclohexanecarbonyl-D-alanylthio)-2-methylpropanoyl]-L-proline; (−)-*N*-[(*S*)-[3-(*N*-cyclohexylcarbonyl-D-alanyl)thio]-2-methylpropionyl]-L-proline; altiopril. $C_{19}H_{30}N_2O_5S$; mol wt 398.52. C 57.26%, H 7.59%, N 7.03%, O 20.07%, S 8.04%. Angiotensin-converting enzyme (ACE) inhibitor; structural analog of captopril. Prepn: S. Tanaka *et al.*, BE 893553 (1982 to Chugai), *C.A.* **98**, 215995n (1983). Pharmacology: K. Sakai *et al.*, *Tohoku J. Exp. Med.* **152**, 363 (1987). Hemodynamic effects in dogs: K. Noguchi *et al.*, *Jpn. J. Pharmacol.* **40**, 373 (1986). Antihypertensive activity: J. Aono *et al.*, *Arch. Int. Pharmacodyn.* **292**, 203 (1988). Tissue specific ACE inhibitory activity: K. Sakai *et al.*, *ibid.* **294**, 228 (1988). Determn by radioimmunoassay: Y. Hinohara *et al.*, *J. Pharmacobio-Dyn.* **11**, 411 (1988).

Solid, mp 113-116°. [α]$_D$ +14.2° (c = 1.05 in methanol).

Calcium salt. [85921-53-5] MC-838; Lowpres. $C_{38}H_{58}CaN_4$-$O_{10}S_2$; mol wt 835.10. Practically white powder, mp ~190°. Bitter taste. $[\alpha]_D^{20}$ −48 to −52° (c = 1 in methanol). Freely sol in water, methanol; sol in ethanol, chloroform. Practically insol in acetone, ethyl acetate. 10% aq soln has pH 5.5-6.5. Stable as aq soln and powder at room temp. LD_{50} in male, female mice, male, female rats (g/kg): all >10.0 orally; 2.1, 2.3, 1.3, 1.3 i.p.; 3.0, 3.8, 3.4, 3.9 s.c.; and in male, female dogs (g/kg): >6.0, >6.0 orally (Sakai).

THERAP CAT: Antihypertensive.

6372. Moxalactam. [64952-97-2] (6*R*,7*R*)-7-[[(2*R*)-2-Car-boxy-2-(4-hydroxyphenyl)acetyl]amino]-7-methoxy-3-[[(1-methyl-1*H*-tetrazol-5-yl)thio]methyl]-8-oxo-5-oxa-1-azabicyclo[4.2.0]oct-2-ene-2-carboxylic acid; *N*-[(6*R*,7*R*)-2-carboxy-7-methoxy-3-[[(1-methyl-1*H*-tetrazol-5-yl)thio]methyl]-8-oxo-5-oxa-1-azabicyclo-[4.2.0]oct-2-en-7-yl]-2-(*p*-hydroxyphenyl)malonamic acid; 7β-[2-carboxy-2-(4-hydroxyphenyl)acetamido]-7α-methoxy-3-[[(1-meth-yl-1*H*-tetrazol-5-yl)thio]methyl]-1-oxa-1-dethia-3-cephem-4-car-boxylic acid; lamoxactam; latamoxef. $C_{20}H_{20}N_6O_9S$; mol wt 520.47. C 46.15%, H 3.87%, N 16.15%, O 27.67%, S 6.16%. Oxa-substituted third generation cephalosporin antibiotic (oxacephalo-sporin). Prepn: M. Narisada, W. Nagata, **DE 2713370**; *eidem*, **US 4138486** (1977, 1979 both to Shionogi); M. Narisada *et al.*, *J. Med. Chem.* **22**, 757 (1979). Laboratory evaluation: T. Yoshida *et al.*, *Antimicrob. Agents Chemother.* **17**, 302 (1980). Mechanism of action study: Y. Komatsu, T. Nishikawa, *ibid.* 316. Pharmacokinetics: R. Wise *et al.*, *ibid.* **18**, 369 (1980). Comparative *in vitro* activity: T. O. Kurtz *et al.*, *ibid.* 645. Series of articles on pharmacology, teratology, toxicity studies: *Chemotherapy (Tokyo)* **28**, Suppl. 7, 1002-1235 (1980). Disc diffusion method to detect methicillin resistance in staphylococci: O. F. Join-Lambert *et al.*, *J. Antimicrob. Chemother.* **59**, 763 (2007). Clinical evaluation in neonates and infants: U. B. Schaad *et al.*, *J. Pediatr.* **98**, 129 (1981). Review of pharmacology and therapeutic use: A. A. Carmine *et al.*, *Drugs* **26**, 279-333 (1983).

Colorless powder, mp 117-122° (dec). $[\alpha]_D^{25}$ −15.3 ±2.6° (c = 0.216 in methanol). uv max (methanol): 276 nm (ε 10200).

Disodium salt. [64953-12-4] LY-12735; S-6059; Shiomarin. $C_{20}H_{18}N_6Na_2O_9S$; mol wt 564.44. $[\alpha]_D^{22}$ −45° (water). uv max (water): 270 nm (ε 12000).

THERAP CAT: Antibacterial.

6373. Moxastine. [3572-74-5] 2-(1,1-Diphenylethoxy)-*N*,*N*-dimethylethanamine; α-methylbenzhydryl 2-dimethylaminoethyl ether; α-methyldiphenhydramine; mephenhydramine; Spofa 325; Alfadryl. $C_{18}H_{23}NO$; mol wt 269.39. C 80.25%, H 8.61%, N 5.20%, O 5.94%. Prepn: Protiva *et al.*, *Chem. Listy* **43**, 257 (1949); **CS 86516** (1957), *C.A.* **53**, 1261 (1959); Z. Vejdelek *et al.*, *Collect. Czech. Chem. Commun.* **49**, 2649 (1984).

bp$_{0.15}$ 129-136°.

Hydrochloride. $C_{18}H_{23}NO.HCl.$ mp 168°.

THERAP CAT: Antihistaminic.

6374. Moxaverine. [10539-19-2] 3-Ethyl-6,7-dimethoxy-1-(phenylmethyl)isoquinoline; 1-benzyl-3-ethyl-6,7-dimethoxyiso-quinoline. $C_{20}H_{21}NO_2$; mol wt 307.39. C 78.15%, H 6.89%, N 4.56%, O 10.41%. Prepn of hydrochloride: **FR 1362765** (1964 to Orgamol, SA), *C.A.* **62**, 536f (1965). Physical properties and solution stability: E. Pawelczyk, M. Zajac, *Diss. Pharm. Pharmacol.* **24**, 5 (1972). HPLC determn in pharmaceutical formulations: E. H. Girgis, *J. Pharm. Sci.* **82**, 503 (1993).

Crystals, mp 71-72.5°. uv max (0.1 *M* H_2SO_4): 253 nm ($A_{1cm}^{1\%}$ 2164).

Hydrochloride. [1163-37-7] Eupaverin; Eupaverina; Kollateral. $C_{20}H_{21}NO_2.HCl$; mol wt 343.85. Crystals from ethanol, mp 208-210° (dec). Sol in hot water, hot alc, and many other organic solvents; very sparingly sol in cold water.

THERAP CAT: Antispasmodic.

6375. Moxestrol. [34816-55-2] (11β,17α)-11-Methoxy-19-norpregna-1,3,5(10)-trien-20-yne-3,17-diol; 3,17-dihydroxy-11β-methoxy-19-nor-17α-pregna-1,3,5-trien-20-yne; 17α-ethynyl-11β-methoxyestra-1,3,5(10)-triene-3,17β-diol; 11β-methoxy-17α-eth-ynyl-Δ$^{1,3,5(10)}$-estratriene-3,17β-diol; 11β-methoxy-17α-ethynyles-tradiol; R-2858; Surestryl. $C_{21}H_{26}O_3$; mol wt 326.44. C 77.27%, H 8.03%, O 14.70%. Prepn: **BE 699394**; D. Bertin, A. Pierdet, **US 3579545** (1967, 1971 both to Roussel-UCLAF); Azadian-Boulanger, Bertin, *Chim. Ther.* **8**, 451 (1973). Activity studies: Raynaud, *Steroids* **21**, 249 (1973); Raynaud *et al.*, *Mol. Pharmacol.* **9**, 520 (1973).

Crystals, mp 280°. $[\alpha]_D^{20}$ +29° (c = 0.6 in ethanol). uv max (ethanol): 280 nm ($E_{1cm}^{1\%}$ 58.4).

THERAP CAT: Estrogen.

6376. Moxidectin. [113507-06-5] (6*R*,23*E*,25*S*)-5-*O*-De-methyl-28-deoxy-25-(1,3-dimethyl-1-buten-1-yl)-6,28-epoxy-23-(methoxyimino)milbemycin B; 23-methyloxime LL-F282249α; (2a*E*,4*E*,5'*R*,6*R*,6'*S*,8*E*,11*R*,13*S*,15*S*,17a*R*,20*R*,20a*R*,20b*S*)-6'-[(*E*)-1,3-dimethyl-1-butenyl]-5',6,6',7,10,11,14,15,17a,20,20a,20b-do-decahydro-20,20b-dihydroxy-5',6,8,19-tetramethylspiro[11,15-methano-2*H*,13*H*,17*H*-furo[4,3,2-*pq*]benzodioxacyclooctadecin-13,2'-[2*H*]pyran]-4',17(3'*H*)-dione 4'-(*E*)-(*O*-methyloxime); CL-301423; Cydectin; Equest; Proheart 6; Quest; Vetdectin. $C_{37}H_{53}$-NO_8; mol wt 639.83. C 69.46%, H 8.35%, N 2.19%, O 20.00%. Semisynthetic macrolide antibiotic; 23-methyloxime analog of ne-madectin, *q.v.* Prepn: B. M. Bain *et al.*, **EP 237339** (1987 to Glaxo); G. Asato, D. J. France, **EP 259779** (1988 to Am. Cyanamid). Efficacy in cattle against lice: J. D. Webb *et al.*, *J. Econ. Entomol.* **84**, 1266 (1991); against gastrointestinal nematodes: G. L. Zimmerman *et al.*, *Am. J. Vet. Res.* **53**, 1409 (1992); J. C. Williams *et al.*, *Vet. Rec.* **131**, 345 (1992); against psoroptic mange: J. F. Lonneux, B. Losson, *Vet. Parasitol.* **45**, 147 (1992). Environmental effects on dung-burying beetles: G. T. Fincher, G. T. Wang, *Southwest. Entomol.* **17**, 303 (1992).

Consult the Name Index before using this section.

THERAP CAT (VET): Anthelmintic; antiparasitic.

6377. Moxifloxacin. [151096-09-2] 1-Cyclopropyl-6-fluoro-1,4-dihydro-8-methoxy-7-[(4aS,7aS)-octahydro-6H-pyrrolo[3,4-b]pyridin-6-yl]-4-oxo-3-quinolinecarboxylic acid. $C_{21}H_{24}FN_3O_4$; mol wt 401.44. C 62.83%, H 6.03%, F 4.73%, N 10.47%, O 15.94%. Fluorinated quinolone antibacterial. Prepn: U. Petersen et al., **EP 550903** (1993 to Bayer). Antibacterial spectrum in vitro: J. M. Woodcock et al., Antimicrob. Agents Chemother. **41**, 101 (1997). Activity vs Mycobacterium tuberculosis: B. Ji et al., ibid. **42**, 2066 (1998). HPLC determn in serum: C. M. Tobin et al., J. Antimicrob. Chemother. **42**, 278 (1998). Clinical pharmacokinetics: H. Stass et al., Antimicrob. Agents Chemother. **42**, 2060 (1998). Review of pharmacology and clinical experience in respiratory infections: M. Miravitlles, A. Anzueto, Expert Opin. Pharmacother. **9**, 1755-1772 (2008).

mp 203-208° (dec). $[\alpha]_D^{23}$ −193°.
Hydrochloride. [186826-86-8] Bay-12-8039; Actira; Avalox; Avelox; Izilox; Moxeza; Moxivig; Octegra; Proflox; Vigamox. $C_{21}H_{24}FN_3O_4 \cdot HCl$; mol wt 437.90. Slightly yellow to yellow crystalline powder, mp 324-325° (dec). $[\alpha]_D^{25}$ −256° (c = 0.5 in water).
THERAP CAT: Antibacterial.
THERAP CAT (VET): Antibacterial.

6378. Moxisylyte. [54-32-0] 4-[2-(Dimethylamino)ethoxy]-2-methyl-5-(1-methylethyl)phenol 1-acetate; 5-(2-dimethylaminoethoxy)carvacrol acetate; 6-acetoxythymol 2-(dimethylamino)ethyl ether; (6-acetoxythymoxy)ethyldimethylamine; 4-(2-dimethylaminoethoxy)-5-isopropyl-2-methylphenyl acetate; thymoxamine. $C_{16}H_{25}NO_3$; mol wt 279.38. C 68.79%, H 9.02%, N 5.01%, O 17.18%. α_1-Adrenergic blocker. Prepn: H. Pahlicke, **DE 905738** (1954 to Diwag) C.A. **52**, 16294i (1958); A. Buzas et al., Bull. Soc. Chim. Fr. **1959**, 839. Pharmacology: K. Greeff, H. J. Schümann, Arzneim.-Forsch. **3**, 341 (1953); J. Mercier et al., Therapie **26**, 785 (1971). Toxicology: J. Roquebert, J. Canellas, ibid. 775. α-Adrenergic antagonist activity: A. T. Birmingham, J. Szolcsanyi, J. Pharm. Pharmacol. **17**, 449 (1965). α_1-receptor selectivity: J. Roquebert et al., Arch. Int. Pharmacodyn. **266**, 282 (1983). Series of articles on metabolism: K.-O. Vollmer et al., Eur. J. Drug Metab. Pharmacokinet. **10**, 61, 71, 139 (1985). Clinical pharmacokinetics: P. Costa et al., J. Pharm. Sci. **82**, 729, 968 (1993). HPLC determn of metabolites in plasma and urine: C. Marquer, F. Bressolle, J. Chromatogr. B **691**, 389 (1997). Clinical trial in Raynaud's disease and chilblains: G. V. Jaffe, J. J. Grimshaw, Br. J. Clin. Pract. **34**, 343 (1980). Review of clinical studies in ophthalmology: M. Wand, W. M. Grant,

Surv. Ophthalmol. **25**, 75-84 (1980). Clinical trial in impotence: P. Costa et al., J. Urol. **149**, 301 (1993).

Hydrochloride. [964-52-3] Arlitene; Carlytene; Erecnos; Icavex; Opilon; Uroalpha. $C_{16}H_{25}NO_3 \cdot HCl$; mol wt 315.84. Shiny beige needles from ethyl acetate/methanol, mp 208-210°. Sol in water, alcohol, chloroform. Insol in ether. LD_{50} in mice, rats (mg/kg): 265 ±19, 740 ±51 orally; 200 ±15, 190 ±19 s.c. (Roquebert, Canellas).
THERAP CAT: Vasodilator (peripheral). In treatment of male erectile dysfunction.

6379. Moxonidine. [75438-57-2] 4-Chloro-N-(4,5-dihydro-1H-imidazol-2-yl)-6-methoxy-2-methyl-5-pyrimidinamine; 4-chloro-6-methoxy-2-methyl-5-(2-imidazolin-2-yl)aminopyrimidine; BDF-5895; Cynt; Physiotens. $C_9H_{12}ClN_5O$; mol wt 241.68. C 44.73%, H 5.00%, Cl 14.67%, N 28.98%, O 6.62%. α_2-Adrenoceptor agonist. Prepn: **NL 7908192**; W. Stenzel et al., **US 4323570** (1980, 1982 both to Beiersdorf AG). Series of articles on pharmacology and binding studies: Arzneim.-Forsch. **38**, 1426-1445 (1988). Pharmacokinetics: D. Trenk et al., J. Clin. Pharmacol. **27**, 988 (1987); W. Kirch et al., ibid. **30**, 1088 (1990). Clinical comparison with clonidine, q.v.: V. Plänitz, ibid. **27**, 46 (1987).

mp 217-219° (dec).
Hydrochloride. [75438-58-3] Crystals from isopropanol/ether, mp 189°.
THERAP CAT: Antihypertensive.

6380. MPTP. [28289-54-5] 1,2,3,6-Tetrahydro-1-methyl-4-phenylpyridine; 1-methyl-4-phenyl-1,2,3,6-tetrahydropyridine. $C_{12}H_{15}N$; mol wt 173.26. C 83.19%, H 8.73%, N 8.08%. Piperidine derivative which causes irreversible symptoms of parkinsonism in humans, monkeys. Prepn as hydrochloride by Grignard reaction: A. Ziering et al., J. Org. Chem. **12**, 894 (1947). Alternative prepn: C. J. Schmidle, R. C. Mansfield, J. Am. Chem. Soc. **78**, 425 (1956). Identification as impurity in "synthetic heroin" and effect on drug users: J. W. Langston et al., Science **219**, 979 (1983). Selective destruction of dopaminergic neurons in primates: R. S. Burns et al., Proc. Natl. Acad. Sci. USA **80**, 4546 (1983). In vitro metabolism by rat brain monoamine oxidase to 1-methyl-4-phenylpyridinium ion (MPP^+): K. Chiba et al., Biochem. Biophys. Res. Commun. **120**, 574 (1984). Studies on mechanism of neurotoxicity: J. W. Langston et al., Science **225**, 1480 (1984); S. P. Markey et al., Nature **311**, 464 (1984). Binding studies in rat brain: C. M. Wieczorek et al., Eur. J. Pharmacol. **98**, 453 (1984); B. Parsons, T. C. Rainbow, ibid. **102**, 375 (1984); in rat, human brain: J. A. Javitch et al., Proc. Natl. Acad. Sci. USA **81**, 4591 (1984). Comparison of idiopathic and MPTP-induced parkinsonism in humans: R. S. Burns et al., N. Engl. J. Med. **312**, 1418 (1985). Review: T. P. Singer et al., Trends Biochem. Sci. **12**, 266-270 (1987); L. M. Sayre, Toxicol. Lett. **48**, 121-149 (1989).

Crystals from heptane, mp 40-42°. $bp_{0.8}$ 85-90°.

6381. MSH. Melanotropin; melanophore-affecting hormone; melanocyte-stimulating hormone; melanophore hormone; melanophore dilating hormone; melanophore expanding hormone; melanophore-stimulating hormone; melanotropic hormone; chromatophorotropic hormone; melanosome-dispersing hormone; pigmentation hormone. First known as *Intermedin(e)* or *B hormone*, a pituitary factor causing color changes in fish and amphibia: B. Zondek, H. Kron, *Klin. Wochenschr.* **11**, 849 (1932). Subsequently, the hormone was identified as two linear peptides designated as α- and β-MSH. γ-MSH, a third peptide has also been identified but has no significant melanocyte stimulating activity. The peptides are secreted by the pars intermedia of the pituitary gland. The brain may be a secondary site of synthesis. α-MSH derives from ACTH and β-MSH from β-lipotrophic hormone, *q.q.v.* These MSH precursors derive from two sections of pro-opiomelanocortin, *q.v.* A third portion of pro-opiomelanocortin may serve as the precursor of γ-MSH. Biosynthesis: B. G. Jenks *et al.*, *J. Endocrinol.* **98**, 19 (1983); G. J. M. Martens *et al.*, *Gen. Comp. Endocrinol.* **49**, 73 (1983). Distribution and biosynthesis in the brain: D. F. Swaab *et al.*, "The Distribution of MSH and ACTH in the Rat and Human Brain and its Relation to Pituitary Stores," in *Endogenous Peptides and Learning and Memory Processes*, J. L. Martinez, Jr. *et al.*, Eds. (Academic Press, New York, 1981) pp 7-36. The mechanism of control of MSH release from the pituitary is still being investigated. There is evidence for inhibitory control by the brain via dopaminergic neurons, inhibitory control by the peptide melanostatin (MIF-1), *q.v.* and stimulatory control by β-adrenergic agents and the peptide CRF, *q.v.*: G. Schmitt *et al.*, *Neuroendocrinology* **33**, 306 (1981); P. G. Smelik *et al.*, "The Role of Catecholamines in the Control of the Secretion of Pro-Opiocortin-Derived Peptides from the Anterior and Intermediate Lobes and its Implications in the Response to Stress," in *The Anterior Pituitary Gland*, A. S. Bhatnagar, Ed. (Raven Press, New York, 1983) pp 113-125; L. Proulx-Ferland *et al.*, *J. Steroid Biochem.* **19**, 439 (1983). Although the role of MSH in adaptive color change in lower vertebrates in well known, its physiological significance in mammals appears to be extrapigmentary. Effects on reproduction: M. E. Celis, M. Volosin, *Prog. Clin. Biol. Res.* **87**, 113 (1982). Reviews of behavioral and neurochemical effects: P. C. Datta, M. G. King, *Neurosci. Biobehav. Rev.* **6**, 297-310 (1982); B. E. Beckwith, C. A. Sandman, *Peptides* **3**, 411-420 (1982). General reviews: Li, *Adv. Protein Chem.* **12**, 270-295 (1957); Novales, *Neuroendocrinology* **2**, 241 (1967); R. Schwyzer, *Proc. R. Soc. London Ser. B* **210**, 5-20 (1980); F. L. Strand, C. M. Smith, *Pharmacol. Ther.* **11**, 509-533 (1980). Book: A. J. Thody, *The MSH Peptides* (Academic Press, New York, 1980) 162 pp.

α-Melanotropin. Contains 13 amino acids. Mammalian α-MSH is the most potent melanocyte-stimulating peptide known. Mechanism of action: T. K. Sawyer *et al.*, *Am. Zool.* **23**, 1983. Isoln from hog pituitary gland: T. H. Lee, A. B. Lerner, *J. Biol. Chem.* **221**, 943 (1956); from pig pituitary gland: Harris, Lerner, *Nature* **179**, 1346 (1957); Harris, *Biochem. J.* **71**, 451 (1959); from horse pituitary gland: Dixon, Li, *J. Am. Chem. Soc.* **82**, 4568 (1960). Structure: Lee *et al.*, *J. Biol. Chem.* **236**, 1390 (1961). Synthesis: Schwyzer *et al.*, *Helv. Chim. Acta* **46**, 870 (1963). Synthesis and biological activity of dogfish α-MSH: A. Eberle *et al.*, *Helv. Chim. Acta* **61**, 2360 (1978). Activity as centrally administered antipyretic agent: M. T. Murphy *et al.*, *Science* **221**, 192 (1983). Cardiovascular effects: D. J. Diz, D. M. Jacobowitz, *Brain Res.* **270**, 265 (1983). Review of the α-melanotropinergic system and its role in the CNS: T. L. O'Donohue, D. M. Jacobowitz in *Polypeptide Hormones*, R. F. Beers, E. G. Bassett, Eds. (Raven Press, New York, 1980) pp 203-222. $[\alpha]_D^{25}$ -58.5 ± 2.5° (c = 0.38 in 10% acetic acid).

β-Melanotropin. Contains 18-22 amino acids differing slightly from one species to another: Lee *et al.*, *J. Biol. Chem.* **236**, 1390 (1961). Isoln from hog pituitary gland: T. H. Lee, A. B. Lerner, *loc. cit.*; from pig pituitary gland: Geschwind *et al.*, *J. Am. Chem. Soc.* **78**, 4494 (1956); **79**, 620 (1957); Harris, Roos, *Biochem. J.* **71**, 434 (1959); from bovine pituitary gland: Geschwind *et al.*, *J. Am. Chem. Soc.* **79**, 6394 (1957); from human pituitary gland: Dixon, *Biochim. Biophys. Acta* **37**, 38 (1960); Harris, *Nature* **184**, 167 (1959). Synthesis of derivatives: Hofmann, *Ann. N.Y. Acad. Sci.* **88**, art. 3, p 689 (1960). Synthesis of bovine β-melanotropin: Schwyzer *et al.*, *Helv. Chim. Acta* **46**, 1975 (1963); S. Lemaire *et al.*, *J. Med. Chem.* **19**, 373 (1976). Solid-phase synthesis of human and monkey β-melanotropins: Wang *et al.*, *Int. J. Pept. Protein Res.* **5**, 33 (1973);

of camel β_{c1}-MSH: Li *et al.*, *Biochemistry* **14**, 953 (1975); of camel β_{c2}-MSH: S. Lemaire *et al.*, *loc. cit.*; of dogfish β-MSH: H. Yajima *et al.*, *Chem. Pharm. Bull.* **26**, 571 (1978); of equine β-MSH: J. Izdebski *et al.*, *Int. J. Pept. Protein Res.* **19**, 327 (1982). It is probable that the human pituitary does not produce β-MSH. The substance reported as human β-MSH in early studies seems to have been an artifact formed by enzymatic degradation of β-LPH during the extraction procedure: K. Tanaka *et al.*, *J. Clin. Invest.* **62**, 94 (1978); A. J. Thody, *loc. cit.*

γ-Melanotropin. Contains 12 amino acids. Discovered during nucleotide sequencing of cloned DNA for bovine pro-opiomelanocortin: S. Nakanishi *et al.*, *Nature* **278**, 423 (1979). May be an antagonist of β-endorphin or a partial agonist/antagonist of ACTH. Synthesis and biological activity: N. Ling *et al.*, *Life Sci.* **25**, 1773 (1979); K. Okamoto *et al.*, *Chem. Pharm. Bull.* **28**, 2839 (1980); W. A. Bijl *et al.*, *Rec. Trav. Chim.* **100**, 123 (1981). Radioimmunoassay: T. Shibasaki *et al.*, *Life Sci.* **26**, 1781 (1980). Behavioral profile in rats: J. M. Van Ree *et al.*, *ibid.* **28**, 2875 (1981). Comparison with α-MSH: T. L. O'Donohue *et al.*, *Peptides* **2**, 101 (1981). Review: J. M. Van Ree *et al.*, *Ciba Found. Symp.* **81**, 263-276 (1981). $[\alpha]_D^{20}$ -38.0° (c 0.5 in 10% acetic acid). $[\alpha]_D^{23}$ -33.4° (c = 1.0 in 1% acetic acid).

6382. Mucins. High molecular weight glycoproteins, major constituents of saliva, gastric juice, intestinal juice, and other secretions: F. Haurowitz, *Chemistry and Biology of Proteins* (Academic Press, New York, 1950) p 199. Mucins are capable of forming viscous solutions and thereby act as lubricants or protectants in cavities of the body or on body surfaces. For prepn (*e.g.* from snails) and their biochemical role see P. A. Levene, *The Hexosamines and Mucoproteins* (Longmans, Green, London, 1925). Ovine submaxillary mucin consists of a single polypeptide chain with about 800 disaccharide units attached, each of which is *N*-acetylneuraminyl(2 → 6) *N*-acetylgalactosamine. Carbohydrate accounts for about 45% of the molecular weight. Structure studies: Pigman, Tanaka, *148th Am. Chem. Soc. Meet.* (Chicago, Aug.-Sept., 1964), Abstracts of Papers, p 11C; Ozeki, Yosizawa, *Arch. Biochem. Biophys.* **142**, 177 (1971); Huser *et al.*, *Z. Physiol. Chem.* **354**, 749 (1973). Glycoproteins in bovine cervical mucus: F. A. Meyer *et al.*, *Adv. Exp. Med. Biol.* **89**, 239 (1977). Review on physical and chemical properties, biosynthesis and function of mucin: several authors in *Mucus in Health and Disease*, M. Elstein, D. V. Parke, Eds. (Plenum Press, New York, 1977) pp 171-311.

Mucins are obtained as greenish-gray or yellow powders forming very viscous solns in water; generally sol in dil alkalies. Insol in acetic acid. Acidic substances with an isoelectric point between pH 3 and 5.

USE: Demulcent; adsorbent.

6383. Mucochloric Acid. [87-56-9] (2Z)-2,3-Dichloro-4-oxo-2-butenoic acid; dichloromalealdehydic acid; α,β-dichloro-β-formylacrylic acid; 2,3-dichloromaleic aldehyde acid. $C_4H_2Cl_2O_3$; mol wt 168.96. C 28.44%, H 1.19%, Cl 41.96%, O 28.41%. Prepn from β,γ-dichloropyromucic acid and bromine water or dil nitric acid: Hill, Jackson, *Am. Chem. J.* **12**, 43 (1890); by heating furfural with manganese dioxide and hydrochloric acid: Simonis, *Ber.* **32**, 2085 (1899); Beattie *et al.*, *J. Chem. Soc.* **1932**, 264; Mowry, *J. Am. Chem. Soc.* **72**, 2535 (1950); by chlorination of butyne-1,4-diol: Dury, *Angew. Chem.* **72**, 864 (1960).

Monoclinic prisms from ether and ligroin, mp 127°. Slightly sol in cold water; sol in hot water; hot benzene, alc. LD_{50} in rats: 0.5-1.0 g/kg orally; 10-25 mg/kg i.p., Fassett in *Industrial Hygiene and Toxicology* vol. **II**, F. A. Patty, Ed. (Interscience, New York, 2nd ed., 1962) p 1977.

Caution: Strong irritant to skin and eyes; a potent skin sensitizer.

6384. Mucochloric Anhydride. [4412-09-3] 5,5'-Oxybis-[3,4-dichloro-2(5H)-furanone]; bis[3,4-dichloro-2(5)-furanonyl] ether; GC-2466. $C_8H_2Cl_4O_5$; mol wt 319.90. C 30.04%, H 0.63%,

Cl 44.33%, O 25.01%. Prepd by refluxing mucochloric acid with benzenesulfonic acid in a mixture of benzene and dioxane: Mowry, *J. Am. Chem. Soc.* **72**, 2535 (1950).

Crystals from benzene + dioxane, α-isomer (racemate): mp 141-143°, β-isomer (*meso* form): mp 180°. Substantially insol in water. Sol in many organic solvents, such as acetone, xylene, cyclohexanone, methylnaphthalenes.
USE: Fungicide. *Ref:* Gilbert, **US 2861919** (1958 to Allied Chem.).

6385. Muconic Acid. [505-70-4] 2,4-Hexadienedioic acid; 1,3-butadiene-1,4-dicarboxylic acid. $C_6H_6O_4$; mol wt 142.11. C 50.71%, H 4.26%, O 45.03%. Prepd by oxidation of phenol and peracetic acid: Boeseken, Engelberts, *C.A.* **26**, 2970 (1932); by treatment of ethyl 1,4-dibromoadipate with alcoholic potassium hydroxide: Guha, Sankaran, *Org. Synth.* **26**, 57 (1946); by isomerization of 3-hydroxy-4-carbomethoxybut-1-ene-1-carboxylic acid lactone: Elvidge *et al., J. Chem. Soc.* **1950**, 2235; by carbonylation of acetylene: Tsuji *et al., J. Am. Chem. Soc.* **86**, 2095 (1964). Configuration and separation of isomers: Boeseken, Kerkhoven, *Rec. Trav. Chim.* **51**, 964 (1932); Elvidge *et al., J. Chem. Soc.* **1953**, 708.

(E,E)-form

(E,E)-Form. [3588-17-8] Prisms from water, mp 301°. uv max (0.1N NaOH): 251, 259, 264 nm (ε 25600, 29100, 25600). One gram dissolves in about 5 liters water at 15°. Quite sol in hot alcohol and glacial acetic acid.
(E,E)-Form dimethyl ester. [1119-43-3] $C_8H_{10}O_4$; mol wt 170.16. Crystals from alcohol, mp 159°. Insol in ether.
(E,E)-Form diethyl ester. [6032-74-2] $C_{10}H_{14}O_4$; mol wt 198.22. Needles from alcohol, mp 64°; d_4^{100} 0.9829. n_{He}^{99} 1.4675.
(Z,Z)-Form. [1119-72-8] Prisms from ethanol, mp 194-195°. uv max (0.1N NaOH): 251, 258, 264 nm (ε 15600, 17000, 15300). Freely sol in boiling water; sparingly sol in ether.
(Z,Z)-Form dimethyl ester. [692-91-1] $C_8H_{10}O_4$; mol wt 170.16. Needles from petr ether, mp 75°. Soluble in ether.
(Z,Z)-Form diethyl ester. [6032-77-5] $C_{10}H_{14}O_4$; mol wt 198.22. Minute crystals from alcohol at −25°; mp 13°. Changes to the *trans-trans*-form on storage at room temp.
(2E,4Z)-Form. [1119-73-9] Needles from hot water, mp 190-191°. uv max (0.1N NaOH): 251, 259, 265 nm (ε 23400, 25600, 23400).
(2E,4Z)-Form dimethyl ester. [692-92-2] $C_8H_{10}O_4$; mol wt 170.16. Needles from aqueous methanol, mp 75°.

6386. Muira Puama. Potency wood. Wood of *Liriosma ovata* Miers, *Oleaceae,* or according to Rebourgeon, *Acanthea virilis* Wehmer, *Acanthaceae. Habit.* Brazil. *Constit.* Aromatic resin, muirapuamine, fat. Studies of components: Auterhoff, Pankow, *Arch. Pharm.* **301**, 481 (1968); **302**, 209 (1969). *Reviews:* Gaebler, *Dtsch. Apoth.* **22**, 94 (1970); Steinmetz, *Q. J. Crude Drug Res.* **11**, 1787 (1971).

6387. Mukaiyama Reagent. [14338-32-0] 2-Chloro-1-methylpyridinium iodide (1:1); 2-chloropyridine methiodide; 1-methyl-2-chloropyridinium iodide. C_6H_7ClIN; mol wt 255.48. C 28.21%, H 2.76%, Cl 13.88%, I 49.67%, N 5.48%. Acid-activating agent for prepn of carboxylic esters. Prepn: M. Liveris, J. Miller, *Aust. J. Chem.* **2**, 297 (1958); and reactions with hydroxide ions: G. B. Barlin, J. A. Benbow, *J. Chem. Soc. Perkin Trans. 2* **1974**, 790. Use in synthesis of carboxylic esters: T. Mukaiyama *et al., Chem. Lett.* **1975**, 1045; T. Mukaiyama, *Angew. Chem. Int. Ed.* **18**, 707 (1979). Kinetic resolution of carboxylic acids and alcohols: A. Mazón *et al., Tetrahedron: Asymmetry* **3**, 1455 (1992). Kinetics of nucleophilic substitution reactions with primary and secondary amines: A. Awwal *et al., Indian J. Chem.* **40B**, 32 (2001); with phenols: M.

Kabir *et al., ibid.* **43B**, 1779 (2004). Use in synthesis of ketenes: R. L. Funk *et al., Synlett* **1989**, 36; of β-lactams: C. M. L. Delpiccolo *et al., J. Comb. Chem.* **5**, 208 (2003); of esters: D. Donati *et al., Tetrahedron Lett.* **46**, 2817 (2005).

Almost colorless needles, mp 207° (Liveris, Miller). Also reported as pale yellow crystals, mp 205-206° (dec) (Barlin, Benbow). uv max (pH 7.0): 213, 274 nm (log ε 3.66, 3.86).
USE: Condensation reagent for the synthesis of esters and ketenes, and for the kinetic resolution of carboxylic acids and alcohols.

6388. Mupirocin. [12650-69-0] (2E)-5,9-Anhydro-2,3,4,8-tetradeoxy-8-[[(2S,3S)-3-[(1S,2S)-2-hydroxy-1-methylpropyl]oxiranyl]methyl]-3-methyl-L-*talo*-non-2-enonic acid, 8-carboxyoctyl ester; pseudomonic acid A; *trans*-pseudomonic acid; BRL-4910A; Bactoderm; Bactroban; Turixin. $C_{26}H_{44}O_9$; mol wt 500.63. C 62.38%, H 8.86%, O 28.76%. Major component of the pseudomonic acids, *q.v.,* an antibiotic complex produced by *Pseudomonas fluorescens* NCIB 10586. Isoln and characterization: A. T. Fuller *et al., Nature* **234**, 416 (1971); K. D. Barrow, G. Mellows, **DE 2227739**; *eidem,* **US 3977943**; *eidem,* **US 4071536** (1973, 1976, 1978 all to Beecham). Purification: P. J. O'Hanlon *et al.,* **DE 2842358**; *eidem,* **US 4222942** (1979, 1980 both to Beecham). Structure: E. B. Chain, G. Mellows, *Chem. Commun.* **1974**, 847; *eidem, J. Chem. Soc. Perkin Trans. 1* **1977**, 294. Absolute configuration: R. G. Alexander *et al., ibid.* **1978**, 561. Prepn from methyl pseudomonate C: *eidem, Tetrahedron Lett.* **22**, 2059 (1981). Total syntheses of (±)-form: B. B. Snider, G. B. Phillips, *J. Am. Chem. Soc.* **104**, 113 (1982); B. B. Snider *et al., J. Org. Chem.* **48**, 3003 (1983). Biosynthesis: T. C. Feline *et al., J. Chem. Soc. Perkin Trans. 1* **1977**, 309. Effect of pH on chemical stability and antibacterial activity: J. P. Clayton *et al., ibid.* **1979**, 838. Inhibition of bacterial protein synthesis: J. Hughes, G. Mellows, *J. Antibiot.* **31**, 330 (1978); of isoleucyl-tRNA synthetase: *eidem, Biochem. J.* **176**, 305 (1978); *eidem, ibid.* **191**, 209 (1980). *In vitro* antibacterial spectrum: R. Sutherland *et al., Antimicrob. Agents Chemother.* **27**, 495 (1985); M. W. Casewell, R. L. R. Hill, *J. Antimicrob. Chemother.* **15**, 523 (1985). Antimycoplasmal activity *in vitro:* R. M. Banks *et al., J. Antibiot.* **41**, 609 (1988). Clinical evaluations: G. D. Reilly, R. C. Spencer, *ibid.* **13**, 295 (1984); M. W. Casewell, R. L. R. Hill, *ibid.* **17**, 365 (1986). *Reviews:* A. Ward, D. M. Campoli-Richards, *Drugs* **32**, 425-444 (1986); M. W. Casewell, R. L. R. Hill, *J. Antimicrob. Chemother.* **19**, 1-5 (1987).

Crystals from ether, mp 77-78°. $[\alpha]_D^{20}$ −19.3° (c = 1 in methanol). uv max (ethanol): 222 nm (ε 14500). Freely sol in acetone, chloroform, dehydrated alc, methanol; slightly sol in ether; very slightly sol in water.
THERAP CAT: Topical antibacterial.
THERAP CAT (VET): Topical antibacterial.

6389. Muramic Acid. [1114-41-6] (R)-2-Amino-3-O-(1-carboxyethyl)-2-deoxy-D-glucose; 3-O-α-carboxyethyl-D-glucosamine. $C_9H_{17}NO_7$; mol wt 251.24. C 43.03%, H 6.82%, N 5.58%, O 44.58%. Amino sugar found (as the *N*-acetyl derivative) in *peptidoglycan,* the main skeletal component of the bacterial cell wall. Discovery: J. T. Park, *J. Biol. Chem.* **194**, 885 (1952). Isoln from spores of *Bacillus megatherium*: R. E. Strange, F. A. Dark, *Nature* **177**, 186 (1956). Identification and synthesis: R. E. Strange, L. H. Kent, *Biochem. J.* **71**, 333 (1959). Stereospecific synthesis: Y. Matsushima, J. T. Park, *J. Org. Chem.* **27**, 3581 (1962); *eidem, Biochem.*

Prep. **10**, 109 (1963); T. Osawa, R. W. Jeanloz, *J. Org. Chem.* **30**, 448 (1965). Review of peptidoglycan structure: H. J. Rogers, *Ann. N.Y. Acad. Sci.* **235**, 29-51 (1974). Use to determine bacterial levels in mammalian tissues: J. Gilbart *et al.*, *J. Microbiol. Methods* **5**, 271 (1986); in airborne dust: A. Fox *et al.*, *Appl. Environ. Microbiol.* **59**, 4354 (1993).

Crystals from water, mp 152-154° (dec); $[\alpha]_D^{25}$ +103° (c = 0.26 in water) (Matsushima, Park). Also reported as crystals from 90% ethanol, mp 160-162° (dec); $[\alpha]_D^{22}$ +146° (6 minutes) → +116° (31 hrs) (c = 0.57 in water) (Osawa, Jeanloz).

***N*-Acetylmuramic acid.** [10597-89-4] (*R*)-2-(Acetylamino)-3-*O*-(1-carboxyethyl)-2-deoxy-D-glucose. $C_{11}H_{19}NO_8$; mol wt 293.27. Crystals from ethyl acetate + methanol, mp 119-121°. $[\alpha]_D^{20}$ +56° (10 minutes) → +40° (24 hrs) (c = 0.68 in water).

USE: As chemical marker for the detection of bacterial contamination.

6390. Muramyl Dipeptide. [53678-77-6] *N*-(*N*-Acetylmuramoyl)-L-alanyl-D-α-glutamine; *N*-acetylmuramyl-L-alanyl-D-isoglutamine; 2-acetamido-3-deoxy-3-*O*-(D-2-propionyl-L-alanyl-D-isoglutamine)-D-glucopyranose; MDP. $C_{19}H_{32}N_4O_{11}$; mol wt 492.48. C 46.34%, H 6.55%, N 11.38%, O 35.74%. Synthetic immunoadjuvant corresponding to the smallest immunologically active glycopeptide subunit of the bacterial cell wall. Prepn: A. Adam *et al.*, **DE 2450355**; *eidem*, **US 4235771** (1975, 1980 both to Agen. Nat. Valor. Recher.); C. Merser *et al.*, *Biochem. Biophys. Res. Commun.* **66**, 1316 (1975); P. Lefrancier *et al.*, *Int. J. Pept. Protein Res.* **9**, 249 (1977). Structure-activity relationships: F. Ellouz *et al.*, *Biochem. Biophys. Res. Commun.* **59**, 1317 (1974); A. Adam *et al.*, *ibid.* **72**, 339 (1976). Review of immunoregulating activity: L. Chedid *et al.*, *Prog. Allergy* **25**, 63-105 (1978). Tissue distribution in mice: M. Parant *et al.*, *Int. J. Immunopharmacol.* **1**, 35 (1979). Potentiation of antitumor immunity: S. Sone *et al.*, *J. Biol. Response Modif.* **3**, 185 (1984); A. E. Eggers, *ibid.* **7**, 229 (1988).

Crystals from methanol-acetone-ether. $[\alpha]_D^{25}$ +44° (acetic acid).
USE: Immunological adjuvant.

6391. Murexide. [3051-09-0] 5-[(Hexahydro-2,4,6-trioxo-5-pyrimidinyl)imino]-2,4,6(1*H*,3*H*,5*H*)-pyrimidinetrione ammonium salt (1:1); 5,5′-nitrilodibarbituric acid monoammonium salt; acid ammonium purpurate; ammonium purpurate. $C_8H_8N_6O_6$; mol wt 284.19. C 33.81%, H 2.84%, N 29.57%, O 33.78%. Prepn from alloxan + NH_3: Hartley, *J. Chem. Soc.* **87**, 1791 (1905); Schwartz, Handritschk, **DD 17589** (1959), *C.A.* **55**, 3630a (1961); from alloxantin + ammonium acetate: Davidson, *J. Am. Chem. Soc.* **58**, 1821 (1936). Structure: Hartley, *J. Chem. Soc.* **87**, 1796 (1905); Winslow, *J. Am. Chem. Soc.* **61**, 2089 (1939); Schreiber, *Mitt. Dtsch. Pharm. Ges.* **28**, 20 (1958).

Purple-red crystals with green metallic luster. Absorption max in water: 520 nm. Sparingly sol in cold water, more in hot water; practically insol in alcohol, ether. The H_2O soln is deep purple and the aq NaOH soln is deep blue.
Monohydrate. [6032-80-0] $C_8H_8N_6O_6 \cdot H_2O$; mol wt 302.20. Crystal structure: R. L. Martin *et al.*, *J. Chem. Soc. Dalton Trans.* **1977**, 1336. d 1.72.
USE: Indicator for complexometric titrations.

6392. Murexine. [20284-40-6] 2-[[3-(1*H*-Imidazol-4-yl)-1-oxo-2-propenyl]oxy]-*N,N,N*-trimethylethanaminium; choline imidazole-4-acrylate (ester); β-(4-imidazolyl)acrylcholine; urocanylcholine. $[C_{11}H_{18}N_3O_2]^+$. Neuromuscular blocker found, often in very large quantities, in the median zone of the hypobranchial body of *Murex trunculus* and of other related species of mollusks: Erspamer, Dordoni, *Ric. Sci. Ricostr.* **16**, 1114 (1946); *Arch. Int. Pharmacodyn.* **74**, 263 (1947); Erspamer, *ibid.* **76**, 308 (1948). Isoln: *idem, Experientia* **4**, 226 (1948); M. Roseghini *et al.*, *Eur. J. Biochem.* **12**, 468 (1970); J. E. Blankenship *et al.*, *Comp. Biochem. Physiol.* **51C**, 129 (1975). Synthesis: Pasini *et al.*, *Ann.* **578**, 6 (1952); Pasini, Coda, **US 2956061** (1960 to Società Farmaceutici). Structure: V. Erspamer, O. Benati, *Science* **117**, 161 (1953). Effect on vertebrate and invertebrate muscles: J. C. de Freitas, *Comp. Biochem. Physiol.* **56C**, 57 (1977).

The base is instantly hydrolyzed by water; unstable both in acid or alkaline media.
Chloride. [6209-43-4] Choline chloride β-(4-imidazolyl)acrylate. $C_{11}H_{18}ClN_3O_2$; mol wt 259.73. Extremely hygroscopic crystals. uv max (pH 4.5): 280-282 nm.
Chloride hydrochloride. [6032-82-2] $C_{11}H_{18}ClN_3O_2 \cdot HCl$; mol wt 296.19. Hygroscopic microcrystalline powder, mp 219-221° (dec); shows the same uv max as the chloride.

6393. Muromonab CD3. [140608-64-6] OKT3; Orthoclone OKT3. Murine monoclonal IgG_{2a} antibody directed against the CD3 surface antigen on mature T-cells; designed as a specific T-cell inhibitor for treatment of transplant rejection. Prepn: P. C. Kung *et al.*, *Science* **206**, 347 (1979); P. C.-S. Kung, G. Goldstein, **EP 18795**; *eidem*, **US 4361549** (1980, 1982 both to Ortho Pharm.). Clinical study in renal transplant: G. Goldstein *et al.*, *N. Engl. J. Med.* **313**, 337 (1985); in heart transplant: M. P. Macris *et al.*, *J. Heart Transplant.* **8**, 281 (1989). Review of development and mechanism of action: G. Goldstein, *Transplant. Proc.* **19**, Suppl. 1, 1-6 (1987); of pharmacology and clinical use in transplantation: M. A. Hooks *et al.*, *Pharmacotherapy* **11**, 26-37 (1991); M. I. Wilde, K. L. Goa, *Drugs* **51**, 865-894 (1996).

THERAP CAT: Immunosuppressant.

6394. Muscalure. [27519-02-4] (9*Z*)-9-Tricosene; *cis*-tricos-9-ene. $C_{23}H_{46}$; mol wt 322.62. C 85.63%, H 14.37%. Sex pheromone of the female common house fly, *Musca domestica* L. Isoln and synthesis: Carlson *et al.*, *Science* **174**, 76 (1971). Alternate syntheses: Eiter, *Naturwissenschaften* **59**, 468 (1972); Cargill, Rosenblum, *J. Org. Chem.* **37**, 3971 (1972); Gribble *et al.*, *Chem. Commun.* **1973**, 735; Ho, Wong, *Can. J. Chem.* **52**, 1923 (1974); K. Abe *et al.*, *Bull. Chem. Soc. Jpn.* **50**, 2792 (1977); V. N. Odinokov *et al.*, *Tetrahedron Lett.* **23**, 1371 (1982). Acute toxicity study: M. Beroza *et al.*, *Toxicol. Appl. Pharmacol.* **31**, 421 (1975).

bp$_{0.1}$ 157-158°. n_D^{26} 1.4517. LD$_{50}$ in rabbits (mg/kg): >2025 dermally; in rats (mg/kg): >23070 orally (Beroza).

USE: Insect attractant.

6395. Muscarine. [300-54-9] 2,5-Anhydro-1,4,6-trideoxy-6-(trimethylammonio)-D-*ribo*-hexitol; [2S-(2α,4β,5α)]-tetrahydro-4-hydroxy-*N,N,N*,5-tetramethyl-2-furanmethanaminium. [C$_9$H$_{20}$-NO$_2$]$^+$. Alkaloid from the red variety of *Amanita muscaria* (L.) Pers., *Agaricaceae*, the fly fungus, a poisonous mushroom. Also found in some other fungi: *Inocybe patouillardi; I. fastigiata, I. umbrina; I. rimosa*. Isoln procedure for the naturally occurring L-(+)-form: Kuehl *et al., J. Am. Chem. Soc.* **77**, 6663 (1955); Eugster, *Helv. Chim. Acta* **39**, 1002 (1956). Structure and synthesis of racemate: Kögl *et al., Rec. Trav. Chim.* **76**, 109 (1957); Kögl *et al., Experientia* **13**, 137, 138 (1957); Cox *et al., Helv. Chim. Acta* **41**, 229 (1958). Alternate syntheses: **GB 828395** (1960 to Hoffmann-La Roche); Matsumoto *et al., Tetrahedron* **25**, 5889 (1969); W. C. Still, J. A. Schneider, *J. Org. Chem.* **45**, 3375 (1980). Synthesis of muscarine: Whiting *et al., Can. J. Chem.* **50**, 3322 (1972); A. M. Mubarak, D. M. Brown, *Tetrahedron Lett.* **21**, 2453 (1980); *eidem, J. Chem. Soc. Perkin Trans. I* **1982**, 809; S. Pochet, T. Huynhdinh, *J. Org. Chem.* **47**, 193 (1982). Chemistry and pharmacology: Waser, *Pharmacol. Rev.* **13**, 465-515 (1961). Toxicity study: P. J. Fraser, *Br. J. Pharmacol.* **12**, 47 (1957). Reviews: C. H. Eugster, *Adv. Org. Chem.* **2**, 427-455 (1960); S. Wilkinson, *Q. Rev. Chem. Soc.* **15**, 153-171 (1961). Configurational relationship in the muscarine series: Bollinger, Eugster, *Helv. Chim. Acta* **54**, 2704 (1971).

Chloride. C$_9$H$_{20}$ClNO$_2$. Stout prisms from ethanol + acetone, mp 180-181°. Extremely hygroscopic. [α]$_D^{25}$ +8.1° (c = 3.5 in ethanol). Very sol in water, ethanol. Slightly sol in chloroform, ether, acetone. Aq solns are stable. LD$_{50}$ i.v. in mice: 0.23 mg/kg (Fraser).

Caution: Potential symptoms of toxicity following ingestion are profuse sweating, increased salivation, visual disturbances, nausea, vomiting, abdominal colic, diarrhea, headache, and bronchospasm. Very high doses may produce lacrimation, incontinence, bradycardia, hypotension and shock. *See Clinical Toxicology of Commercial Products*, R. E. Gosselin *et al.*, Eds. (Williams & Wilkins, Baltimore, 5th ed., 1984) Section II, p 247.

THERAP CAT: Cholinergic.

6396. Muscazone. [2255-39-2] α-Amino-2,3-dihydro-2-oxo-5-oxazoleacetic acid; α-amino-2-oxo-4-oxazoline-5-acetic acid. C$_5$H$_6$N$_2$O$_4$; mol wt 158.11. C 37.98%, H 3.83%, N 17.72%, O 40.48%. From *Amanita muscaria* (L.) Fr., *Agaricaceae*: Eugster *et al., Tetrahedron Lett.* **1965**, 1813. Structure: Fritz *et al., ibid.* **1965**, 2075; Reiner, Eugster, *Helv. Chim. Acta* **50**, 128 (1967). Synthesis: Göth *et al., ibid.* 137.

Crystals, decomp above 190°. uv max at pH 2-7: 212 nm (ε 8700); at pH 12: 220 nm (ε 7500).

6397. Muscimol. [2763-96-4] 5-(Aminomethyl)-3(2*H*)-isoxazolone; 5-(aminomethyl)-3-isoxazolol; 5-aminomethyl-3-hydroxyisoxazole; 3-hydroxy-5-aminomethylisoxazole; agarin; pantherine. C$_4$H$_6$N$_2$O$_2$; mol wt 114.10. C 42.11%, H 5.30%, N 24.55%, O 28.04%. Potent CNS depressant and GABA agonist isolated from

Amanita muscaria (L.) Fr., *Agaricaceae*. Structural identity with the insecticidal substances, pantherine and agarin: Onda *et al., Chem. Pharm. Bull.* **12**, 751 (1964); Eugster *et al., Tetrahedron Lett.* **1965**, 1813; Bowden, Drysdale, *ibid.* 727. Synthesis: Gagneux *et al., ibid.* 2077; Göth *et al., Helv. Chim. Acta* **50**, 137 (1967); Bowden *et al., J. Chem. Soc. C* **1968**, 172. Improved syntheses: Nakamura, *Chem. Pharm. Bull.* **19**, 46 (1971); P. Krogsgaard-Larsen, S. B. Christensen, *Acta Chem. Scand. B* **B30**, 281 (1976); A. Barco *et al., J. Chem. Res. Synop.* **1979**, 176; B. E. McCarry, M. Savard, *Tetrahedron Lett.* **22**, 5153 (1981); V. Jager, M. Frey, *Ann.* **1982**, 817. Conformational structure: L. Brehm *et al., ibid.* **26**, 1972). Industrial patents: Hafliger, Gagneux, **US 3242190**, **US 3397209** (1966, 1968 both to Geigy). Pharmacology: Theobald *et al., Arzneim.-Forsch.* **18**, 311 (1968). Review of pharmacology: F. V. DeFeudis, *Neurochem. Res.* **5**, 1047-1068 (1980).

Crystals, mp 175° (dec). LD$_{50}$ in mice (mg/kg): 3.8 s.c., 2.5 i.p.; in rats (mg/kg): 4.5 i.v., 45 orally (Theobald).

USE: As a molecular probe to study GABA receptors.

6398. Muscone. [10403-00-6] (3*R*)-3-Methylcyclopentadecanone; muskone; (−)-methylexaltone. C$_{16}$H$_{30}$O; mol wt 238.42. C 80.60%, H 12.68%, O 6.71%. The odorous principle of musk. Characterization of naturally occurring (−)-form: Ruzicka, *Helv. Chim. Acta* **9**, 715 (1926). Structure: *ibid.* 1008. Synthesis of (±)-form: Ziegler, Weber, *Ann.* **512**, 164 (1934); Ruzicka, Stoll, *Helv. Chim. Acta* **17**, 1308 (1934); Hunsdiecker, *Ber.* **75B**, 1197 (1942). Review of syntheses: S. Abe *et al., Cosmet. Perfum.* **88**, 67 (1973). Recent syntheses: Stork, Macdonald, *J. Am. Chem. Soc.* **97**, 1264 (1975); H. G. Fliri *et al., Monatsh. Chem.* **110**, 245 (1979); C. Fehr *et al., Helv. Chim. Acta* **62**, 2655 (1979); K. H. Schulte-Elte *et al., ibid.* 2673; G. Cantoni *et al., J. Org. Chem.* **45**, 1906 (1980). Chiral synthesis of (*R*)-(−)- and (*S*)-(+)-forms: Q. Branca, A. Fischli, *Helv. Chim. Acta* **60**, 925 (1977).

Oily liquid. Musk odor. bp 328°; bp$_{0.5}$ 130°. d$_4^{17}$ 0.9221. n_D^{17} 1.4802. [α]$_D^{17}$ −13°. Very slightly sol in water; miscible with alc.

(±)-Form. [541-91-3] Oily liquid. bp$_{0.01}$ 90°. n_D^{26} 1.4767. uv max (ethanol): 285 nm (log ε 1.54).

6399. Musks. Fixatives for perfumes having a characteristic persistent musky aroma. Obtained from plant and animal sources and by chemical synthesis. Natural sources of musk include the musk glands of the male musk deer, *Moschus moschiferus*, the civet cat, *Viverra civetta*, the Louisiana muskrat, ambrette seeds, *Abelmoschus moschatus*, and angelica roots, *Angelica archangelica*. Musk odor is exhibited by several categories of chemical compounds. Natural musks are macrocyclic ketones or lactones having approximately fifteen carbons in their ring structures. Synthetic musks are of greater industrial importance and include nitro and non-nitro benzenes, indans, and tetralins; derivatives of hydrindacene, isochroman, naphthindan, and coumarin. Purification of natural and synthetic musks: Thomas, Stephens, **US 3415813** (1968 to Pfizer). Review of macrocyclic musks: Berends, *Am. Perfum. Cosmet.* **80**, 35 (1965). Review of chemical studies of synthetic musks: T. F. Wood, "Chemistry of the Aromatic Musks" (Givaudan Corp.), company literature.

USE: In perfumery.

6400. Mustard. Annual plant with bright yellow flowers; varieties of *Brassica* spp. (also known as *Sinapis* spp) Cruciferae. *Habit.* Europe, Asia, naturalized in U.S. *Constit.* Sinigrin (potassium myronate), myrosin, sinapine sulfocyanate, fixed oil; erucic, behenic, and sinapolic acids. Parts used are the seeds and the oils produced from them. Only black mustard yields a volatile oil which consists almost entirely of allyl isothiocyanate, *q.v.* Brief review: S. Arctander in *Perfume and Flavor Materials of Natural Origin* (Elizabeth, NJ, 1960) pp 424-425.

Black mustard. Brown mustard; red mustard. Dried ripe seeds of *B. nigra* (L.) Koch.

White mustard. Yellow mustard. Dried ripe seeds of *B. alba* (L.) Boiss.

Fixed oil. Vegetable oil expressed from the seeds of both black and white mustard. *Constit.* Chiefly the glycerides of oleic acid and other fatty acids, including arachidic. Straw-colored or brownish-yellow, or greenish-brown liquid. d_{15}^{15} 0.914-0.916. Solidif $-8°$ to $-16°$. n_D^{40} 1.4655-1.4670. Sapon no. 170-174. Iodine no. 92-97. Insol in water; slightly sol in alcohol; miscible with chloroform, ether, petr ether.

USE: Flavoring agent for foods. Fixed oil as cooking oil.

THERAP CAT: Counterirritant; has also been used as an emetic.

6401. Mustard Gas. [505-60-2] 1,1'-Thiobis[2-chloroethane]; bis(2-chloroethyl)sulfide; β,β'-dichloroethyl sulfide; 2,2'-dichlorodiethyl sulfide; bis(β-chloroethyl)sulfide; 1-chloro-2-(β-chloroethylthio)ethane; sulfur mustard; yellow cross liquid; Kampfstoff "Lost"; Yperite. $C_4H_8Cl_2S$; mol wt 159.07. C 30.20%, H 5.07%, Cl 44.57%, S 20.15%. War gas prepd by treating ethylene with sulfur chloride (Levinstein process): Mann, Pope, *J. Chem. Soc.* **121**, 594 (1922); by treating β,β'-dihydroxyethyl sulfide with HCl gas (German process): Meyer, *Ber.* **19**, 3260 (1886); *Ann.* **240**, 310 (1887); Gomberg, *J. Am. Chem. Soc.* **41**, 1427 (1919). Reactions and derivatives: Helfrich, Reid, *ibid.* **42**, 1208 (1920). Toxicity: Anslow *et al.*, *J. Pharmacol. Exp. Ther.* **93**, 1 (1948). Review of carcinogenic risk: *IARC Monographs* **9**, 181-192 (1975); of toxicology and human exposure: *Toxicological Profile for Sulfur Mustard (Mustard Gas)* (PB2004-100006, 2003) 287 pp.

Oily liquid. *Deadly vesicant.* Weak, sweet, agreeable odor. On cooling it forms prisms, mp 13-14°. d^{13} 1.338 (solid); d_4^{20} 1.2741 (liq). bp_{760} 215-217°; bp_{10} 98°. Volatile with steam. n_D^{20} 1.53125. Very sparingly sol in water; sol in fat solvents, other common organic solvents. High lipid soly. Vapor pressure: 0.025 mm (0°), 0.90 mm (30°). Hydrolyzed by alkalies. Recommended neutralizing agent and inactivator: Bleaching powder, sodium hypochlorite. LD_{50} in rats, mice (mg/kg): 3.3, 8.6 i.v. (Anslow).

Caution: Overexposure may cause severe irritation and tissue damage to eyes, skin, and respiratory and GI tracts; symptoms may be delayed in onset. This substance is listed as a known human carcinogen: *Report on Carcinogens, Twelfth Edition* (PB2011-111646, 2011) p 275.

USE: Formerly in chemical warfare. In biological studies of alkylating agents.

6402. Mycaminose. [519-21-1] 3,6-Dideoxy-3-(dimethylamino)-D-glucose. $C_8H_{17}NO_4$; mol wt 191.23. C 50.25%, H 8.96%, N 7.32%, O 33.47%. Part of the carbomycin molecule. Synthesis and stereochemistry: Richardson, *Proc. Chem. Soc. London* **1961**, 430; Foster *et al.*, *Chem. Ind. (London)* **1962**, 142. Stereochemistry: Grisebach, Hofheinz, *Angew. Chem.* **74**, 499 (1962).

Hydrochloride monohydrate. Crystals from moist isopropanol, mp 116-118°. $[\alpha]_D$ +31° (equilib value, c = 0.96).

6403. Mycarose. [6032-92-4] 2,6-Dideoxy-3-C-methyl-L-*ribo*-hexose; 2,6-didesoxy-3-C-methyl-L-*ribo*-hexose; 2-(4,5-*cis*)-trihydroxy-4,6-dimethyltetrahydropyran. $C_7H_{14}O_4$; mol wt 162.19. C 51.84%, H 8.70%, O 39.46%. Constituent of the macrolides carbomycin A and B, spiramycin A, B and C and tylosin: Woodward,

Angew. Chem. **69**, 50 (1957); Paul, Tchelitcheff, *Bull. Soc. Chim. Fr.* **1957**, 443; *eidem, ibid.* **1960**, 150; Hamill *et al., Antibiot. Chemother.* **11**, 328 (1961). Part of carbomycin molecule: Regna *et al., J. Am. Chem. Soc.* **75**, 4625 (1953). Stereochemistry: Grisebach, Hofheinz, *Angew. Chem.* **74**, 499 (1962); Foster *et al., Proc. Chem. Soc. London* **1962**, 254; Hofheinz *et al., Tetrahedron* **18**, 1265 (1962). Synthesis of DL-mycarose: Korte *et al., ibid.* 1257; Grisebach *et al., Ber.* **96**, 1823 (1963). Synthesis of L-mycarose: Lemal *et al., Tetrahedron* **18**, 1275 (1962); Koto *et al., Bull. Chem. Soc. Jpn.* **45**, 532 (1972). Synthesis of D-mycarose: Flaherty *et al., J. Chem. Soc. C* **1966**, 398. Isoln from aureolic acid: Berlin *et al., Khim. Prir. Soedin.* **8**, 535 (1972), *C.A.* **78**, 4432a (1973).

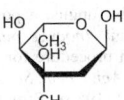

α-L-Pyranose form

Needles from boiling acetone + chloroform, mp 128.5-130.5°, $[\alpha]_D^{25}$ $-31.1°$ (c = 4). Sol in water. Exhibits only end absorption in the ultraviolet.

α-L-Pyranose Form. [18423-82-0] 2,6-Dideoxy-3C-methyl-α-L-*ribo*-hexopyranose.

DL-Form. [92345-04-5] Crystals from acetone + petr ether, mp 110-111°.

D-Form. [6752-46-1] Small needles, mp 129.5-130.5°. $[\alpha]_D$ +32° (c = 1.7 in water).

3-O-Methylmycarose. [3758-45-0] 2,6-Dideoxy-3-C-methyl-3-O-methyl-L-*ribo*-hexose; cladinose. $C_8H_{16}O_4$; mol wt 176.21. Constituent of erythromycin A and B. Liquid, $bp_{0.25}$ at bath temp 120-132°. $[\alpha]_D^{25}$ $-23.1°$ (c = 2.6 in water). Sol in water, alcohol, acetone, ether, benzene, chloroform, carbon tetrachloride; slightly sol in petr ether. Dec by strong acids.

6404. Mycelianamide. [22775-52-6] (6*S*)-3-[[4-[[(2*E*)-3,7-Dimethyl-2,6-octadien-1-yl]oxy]phenyl]methylene]-1,4-dihydroxy-6-methyl-2,5-piperazinedione. $C_{22}H_{28}N_2O_5$; mol wt 400.48. C 65.98%, H 7.05%, N 7.00%, O 19.97%. Antibiotic substance found in the mycelium of *Penicillium griseofulvum:* Anslow, Raistrick, *Biochem. J.* **25**, 39 (1931); Oxford *et al., ibid.* **29**, 1102 (1935); **33**, 240 (1939); Oxford, Raistrick, *ibid.* **42**, 323 (1948). Structure: Birch *et al., J. Chem. Soc.* **1956**, 3717; Bates, Schauble, *Tetrahedron Lett.* **1963**, 1683. Configuration: Gallina *et al., Gazz. Chim. Ital.* **94**, 1301 (1964). Partial synthesis: Brown, Meehan, *Aust. J. Chem.* **21**, 1581 (1968). Total synthesis: N. Shinmon *et al., Chem. Commun.* **1980**, 1020.

Crystals from ethyl acetate, dec 170-172°. $[\alpha]_{546}^{19}$ $-217°$; $[\alpha]_{579}^{19}$ $-182°$. Weak acid. Freely soluble in acetone, dioxane; sparingly sol in other organic solvents; sol in aq sodium carbonate solution, but not in sodium bicarbonate soln. Quickly destroyed by acids and alkalies.

6405. Mycetins. Antibiotic substances produced by *Actinomyces violaceus:* Fainshmidt, Koreniako, *Biokhimiya* **9**, 147 (1944); Krassilnikov, Koreniako, *Mikrobiologiya* **14**, 80 (1945). Chromatographic and structure studies: Yakubov *et al., ibid.* **31**, 526 (1962); Blinov *et al., J. Chromatogr.* **8**, 522 (1962); Yakubov *et al., Antibiotiki* **10**, 771 (1965).

6406. Myclobutanil. [88671-89-0] α-Butyl-α-(4-chlorophenyl)-1*H*-1,2,4-triazole-1-propanenitrile; 2-(4-chlorophenyl)-2-(1*H*-1,2,4-triazol-1-ylmethyl)hexanenitrile; RH-3866; Eagle; Laredo; Nova; Rally; Systhane. $C_{15}H_{17}ClN_4$; mol wt 288.78. C 62.39%, H 5.93%, Cl 12.28%, N 19.40%. Broad spectrum systemic triazole fungicide; ergosterol-biosynthesis inhibitor. Prepn: T. T. Fujimoto, **EP 145294** (1985 to Rohm & Haas), *C.A.* **103**, 160519z (1985). Physical properties and field trials: C. Orpin *et al.*, *Proc. Br. Crop Prot. Conf. - Pests Dis.* **1986**, 55; A. Perrot, *Def. Veg.* **40**, 9 (1986). Antifungal activity: J. A. Quinn *et al.*, *Pestic. Sci.* **17**, 357 (1986).

Light yellow crystals, mp 63-68°. bp$_{1.0}$ 202-208°. Vapor pressure at 25°: 1.6×10^{-6} Torr. Soly at 25°: water 142 ppm. Sol in common organic solvents such as ketones, esters, alcohols, aromatic hydrocarbons. Insol in aliphatic hydrocarbons. LD$_{50}$ in male, female rats (mg/kg): 1600, 2229 orally; LD$_{50}$ in rabbits (mg/kg): 7500 dermally (Orpin).

USE: Agricultural fungicide.

6407. Mycobacidin. [539-35-5] 4-Oxo-2-thiazolidinehexanoic acid; ε-[2-(4-thiazolidone)]hexanoic acid; 4-thiazolidone-2-caproic acid; 2-(5-carboxypentyl)-4-thiazolidone; cinnamonin; actithiazic acid; acidomycin. C$_9$H$_{15}$NO$_3$S; mol wt 217.28. C 49.75%, H 6.96%, N 6.45%, O 22.09%, S 14.76%. Antibiotic substance produced by *Streptomyces lavendulae*, *Streptomyces virginiae* and several other *Streptomyces* spp. Isoln: Tejera *et al.*, *Antibiot. Chemother.* **2**, 333 (1952). Production by fermentation: Grundy *et al.*, **US 2678929** (1954 to Abbott). Synthesis: McLamore *et al.*, *J. Am. Chem. Soc.* **75**, 105 (1953); Nienburg, Friedrichsen, **DE 931651** (1955 to BASF). Toxicity studies: Miyake *et al.*, *Pharm. Bull.* **1**, 84, 89 (1953). *Review:* Caltrider, *Antibiotics* **vol. I**, D. Gottlieb, P. D. Shaw, Eds. (Springer-Verlag, New York, 1967) pp 666-668.

l-Form. Needles from water, ethanol or ethyl acetate, mp 139-140°. $[\alpha]_D^{25}$ −54° (c = 1 in methanol); $[\alpha]_D^{25}$ −60° (c = 1 in ethanol). pK 5.1. Soluble in water, acetone, alcohol, ethylene dichloride, glacial acetic acid. Aq solns are stable over a wide pH range at room temp. Rapidly loses optical activity in dil alkali to give the racemate. Has *in vitro* activity against *Mycobacterium tuberculosis*, *M. ranae* and *M. phlei*. LD$_{50}$ in mice (g/kg): 3.5 i.v., 2 s.c. (Miyake).
Methyl ester. C$_{10}$H$_{17}$NO$_3$S. Needles from ether + hexane, mp 53-54°. $[\alpha]_D^{25}$ −50.9° (methanol).
d-Form. Crystals, mp 138-139°. $[\alpha]_D^{25}$ +57° (methanol).
dl-Form. Crystals from water, mp 123°.

6408. Mycobacillin. [18524-67-9] C$_{65}$H$_{85}$N$_{13}$O$_{30}$; mol wt 1528.46. C 51.08%, H 5.61%, N 11.91%, O 31.40%. Antifungal polypeptide antibiotic isolated from culture filtrates of *Bacillus subtilis*: Majumdar, Bose, *Nature* **181**, 134 (1958). Structural studies: eidem, *Biochem. J.* **74**, 596 (1960); *Arch. Biochem. Biophys.* **90**, 154 (1960). Amino acid configuration: Banerjee, Bose, *Nature* **200**, 471 (1963). Nature of the peptide linkages: Sengupta *et al.*, *Biochem. J.* **121**, 839 (1971). *Review:* Banerjee *et al.*, *Antibiotics* **vol. II**, D. Gottlieb, P. D. Shaw, Eds. (Springer-Verlag, New York, 1967) pp 271-275, 445-446.

Ala–D-α-Asp–Pro–D-α-Asp–D-γ-Glu–Tyr
| |
| α-Asp
| |
D-α-Asp–D-γ-Glu–Leu–D-α-Asp–Ser–Tyr

Needles, mp 235-240°. uv max 277 nm in a solution of *n*-butanol-water-acetic acid-95% ethanol (10:10:2:2).

6409. Mycobactins. Growth factors for *Mycobacterium paratuberculosis (M. johnei)*, the organism responsible for Johne's disease in cattle. Mycobactins are structurally complex *siderophores* (microbial iron chelators). At least nine mycobactins, designated with the letters A, F, H, M, N, P, R, S, and T, have been isolated from various non-pathogenic species of *Mycobacterium*. Earlier studies were actually on *mycobactin P*; *see* isoln from *M. phlei*: Francis *et al.*, *Nature* **163**, 365 (1949); eidem, *Biochem. J.* **55**, 596 (1953); structure: Snow, *ibid.* **94**, 160 (1965). Separation and identification of mycobactins: White, Snow, *ibid.* **108**, 593 (1968); **111**, 785 (1969). Chemical and biological properties: eidem, *ibid.* **115**, 1031 (1969). On saponification, mycobactin splits into *mycobactic acid* and *cobactin*. Synthetic studies: Black *et al.*, *Aust. J. Chem.* **25**, 2155 (1972). Synthesis of mycobactin S2: P. J. Maurer, M. J. Miller, *J. Am. Chem. Soc.* **105**, 240 (1983). Review of early studies: Rose, Snow, "Mycobactin: A Growth Factor for Acid-Fast Bacilli" in G. E. W. Wolstenholme, M. P. Cameron, C. M. O'Connor, *Ciba Found. Symp. Exp. Tuberc. Bacillus Host* **1955**, 41-54. Comprehensive review: Snow, *Bacteriol. Rev.* **34**, 99 (1970).

R^2, R^3, R^5 = H or CH$_3$
R^1, R^4 = CH$_3$, CH$_2$CH$_3$ or long alkyl chain

Able to chelate with metals, showing strong selectivity toward iron. Isolated as octahedral ferric complexes in the form of red-brown glasses; purified to metal-free mycobactins as white microcrystalline powders. Characterizations, esp mass spectral studies, carried out on the aluminum complexes. Mycobactin solids have definite melting points, are stable to air and heat up to about 100°, and show apple-green fluorescence in uv light. Characteristic uv max (methanol): 250, 311 nm for mycobactins having a methyl substituent on the benzene ring; 243, 249, 304 nm for mycobactins without the methyl substituent. Very sol in chloroform; fairly sol in ethanol; less sol in other alcoholic solvents. Slightly sol in benzene, ether, aliphatic hydrocarbons. Stable to acids; easily broken down under alk. conditions.

USE: In the taxonomy of mycobacteria; in development of iron chelators for clinical use.

6410. Mycolic Acids. High mol wt, α-branched, β-hydroxy fatty acids; components of the cell envelopes of all *Mycobacteria*. All known mycolic acids have the basic structure R^{2}CH(OH)-CHR^{1}COOH where R^1 is a C$_{20}$ to C$_{24}$ linear alkane and R^2 is a more complex structure of 30 to 60 carbon atoms that may contain various numbers of carbon-carbon double bonds and/or cyclopropane rings, methyl branches or oxygen functions such as C=O, CH$_3$OCH=, COOH. The structure of mycolic acids varies by families and species. Three principal categories are known: (1) *corynomycolic acids* ranging from C$_{28}$ to C$_{40}$ found mostly in *Corynebacteria;* (2) *nocardic acids*, also called *nocardomycolic acids*, ranging from C$_{40}$ to C$_{60}$, produced by strains of *Nocardia;* and (3) mycobacterial mycolic acids ranging from C$_{60}$ to C$_{90}$. First obtained from a human strain of *Mycobacterium tuberculosis* and studied in an impure form called "unsaponifiable wax": R. J. Anderson, *J. Biol. Chem.* **85**, 351 (1929). Isoln of the first representative of this group of acids from the human tubercle bacillus and derivation of the name "mycolic acids": F. H. Stodola *et al.*, *ibid.* **126**, 505 (1938). Historical reviews: R. J. Anderson, *Fortschr. Chem. Org. Naturst.* **3**, 145-202

(1939); J. Asselineau, E. Lederer, *ibid*. **10**, 170-273 (1953); J. Asselineau, *The Bacterial Lipids* (Hermann-Holden Day, Paris-New York, 1962) pp 82-148; E. Lederer, *Pure Appl. Chem*. **25**, 135 (1971). Major mycolic acids of *Mycobacterium smegmatis:* M. Y. H. Wong *et al*., *J. Biol. Chem*. **254**, 5734 (1979); M. Y. H. Wong, G. R. Gray, *ibid*. 5741. Total synthesis of mycolic acids from *M. smegmatis:* H. C. Huang *et al*., *J. Org. Chem*. **47**, 4018 (1982). Evaluation as immunotherapeutic agents in the treatment of cancer in animal models: M. V. Pimm *et al*., *Int. J. Cancer* **24**, 780 (1979). Clinical evaluation in humans: G. J. Vosika, *Cancer* **44**, 495 (1979). Mycolic acids are acid fast. They are always isolated in the form of more or less complex mixtures. They occur in nature esterified with carbohydrates: with arabinose in the cell wall, with trehalose in cord factor, *q.q.v*.

6411. Mycomycin. [544-51-4] (3*E*,5*Z*)-3,5,7,8-Tridecatetraene-10,12-diynoic acid. $C_{13}H_{10}O_2$; mol wt 198.22. C 78.77%, H 5.09%, O 16.14%. Acetylenic antibiotic produced by *Nocardia acidophilus*. Prepn: E. A. Johnson, K. L. Burdon, *J. Bacteriol*. **54**, 281 (1947). Characterization and structure: W. D. Celmer, I. A. Solomons, *J. Am. Chem. Soc*. **74**, 1870, 2245 (1952); **75**, 1372 (1953). Synthesis of *trans,trans*-mycomycins: F. Bohlmann, W. Sucrow, *Angew. Chem*. **76**, 611 (1964).

Needles from methylene chloride at −40°. Very thermolabile. Complete retention of activity is possible only by storage at −40° or lower. mp 75°. *Explosive at melting point.* $[\alpha]_D^{25}$ −130° (c = 0.4 in ethanol). uv max (ether): 267, 281 nm (ε 61000, 67000). Sol in alcohol, ether, amyl acetate, methylene chloride. Sparingly sol in water. Relatively stable in extremely dil aq solns (approx 0.005 mg/ml), half-life inversely related to concn. Undergoes an unusual rearrangement in normal aq KOH at 27°, involving an allene to acetylene isomerization accompanied by migration of the existing acetylenic bonds to form isomycomycin.

Isomycomycin. [505-94-2] 3,5-Tridecadien-7,9,11-triynoic acid. $C_{13}H_{10}O_2$; mol wt 198.22. CH_3—C≡C—C≡C—C≡C—CH=CHCH=CHCH$_2$COOH. Structure and properties: W. D. Celmer, I. A. Solomons, *J. Am. Chem. Soc*. **74**, 3838 (1952). Colorless needles from ether-hexane. Decomposes slowly above 140°. Optically inactive. uv max (ether): 257.5, 267 (ε 58000, 110000).

6412. Mycophenolic Acid. [24280-93-1] (4*E*)-6-(1,3-Dihydro-4-hydroxy-6-methoxy-7-methyl-3-oxo-5-isobenzofuranyl)-4-methyl-4-hexenoic acid; 6-(4-hydroxy-6-methoxy-7-methyl-3-oxo-5-phthalanyl)-4-methyl-4-hexenoic acid. $C_{17}H_{20}O_6$; mol wt 320.34. C 63.74%, H 6.29%, O 29.97%. Antibiotic produced by *Penicillium brevi-compactum, P. stoloniferum* and related spp. Selectively inhibits lymphocyte proliferation by blocking inosine monophosphate dehydrogenase (IMPDH), an enzyme involved in the *de novo* synthesis of purine nucleotides. Isoln: C. L. Alsberg, O. F. Black, *USDA Bur. Plant Ind. Bull*. **270**, 7 (1912); *C.A*. **7**, 3992 (1913); P. W. Clutterbuck *et al*., *Biochem. J*. **26**, 1441 (1932); and antimicrobial activity: H. W. Florey *et al*., *Lancet* **1**, 46 (1946). Structure: J. H. Birkinshaw *et al*., *Biochem. J*. **50**, 630 (1952); W. R. Logan, G. T. Newbold, *J. Chem. Soc*. **1957**, 1946. Total synthesis: A. J. Birch, J. J. Wright, *Aust. J. Chem*. **22**, 2635 (1969). Biosynthesis: L. Canonica *et al*., *J. Chem. Soc. Perkin Trans. 1* **1972**, 2639. Fermentation, isoln and biological properties: R. H. Williams *et al*., *Antimicrob. Agents Chemother*. **1968**, 229. Immunosuppressive effect: A. Mitsui, S. Suzuki, *J. Antibiot*. **22**, 358 (1969). Mechanism of action: A. C. Allison, E. M. Eugui, *Transplant. Proc*. **26**, 3205 (1994); J. T. Ransom, *Ther. Drug Monit*. **17**, 681 (1995). HPLC determn in plasma: J. Shen *et al*., *J. Chromatogr. B* **817**, 207 (2005); D. Indjova *et al*., *ibid*. 327. Evaluation of risk in pregnancy: N. M. Sifontis *et al*., *Transplantation* **82**, 1698 (2006).

Needles from hot water, mp 141°. pKa 4.5. Partition coefficient (*n*-octanol/water): 570 (pH 2); 1.6 (pH 7.4). Sol in alc. Almost insol in cold water. LD_{50} in mice (mg/kg): >1250 orally; 972.9±77 i.p. (Williams).

2-(4-Morpholinyl)ethyl ester. [128794-94-5] Mycophenolate mofetil; RS-61443; CellCept. $C_{23}H_{31}NO_7$; mol wt 433.50. Prepn: P. H. Nelson *et al*., *US 4753935* (1988 to Syntex); and bioavailability: W. A. Lee *et al*., *Pharm. Res*. **7**, 161 (1990). Pharmacokinetics: R. E. S. Bullingham *et al*., *Transplant. Proc*. **28**, 925 (1996). Clinical trial in renal transplantation: P. A. Keown *et al*., *Transplantation* **61**, 1029 (1996). *Reviews:* D. O. Taylor *et al*., *J. Heart Lung Transplant*. **13**, 571-582 (1994); J. J. Lipsky, *Lancet* **348**, 1357-1359 (1996). Review of clinical experience in inflammatory bowel disease: A. C. Ford *et al*., *Aliment. Pharmacol. Ther*. **17**, 1365-1369 (2003). White to off-white crystalline powder, mp 93-94°. pKa 5.6, 8.5. Partition coefficient (*n*-octanol/water): 0.0085 (pH 2); 238 (pH 7.4). Freely sol in acetone; sol in methanol; sparingly sol in ethanol. Soly in water (mg/kg): 0.043 (pH 7.4), 4.27 (pH 3.6).

Sodium salt. [37415-62-6] Mycophenolate sodium; Myfortic. $C_{17}H_{19}NaO_6$; mol wt 342.32. Clinical trial in renal transplantation: M. Salvadori *et al*., *Am. J. Transplant*. **4**, 231 (2003); K. Budde *et al*., *ibid*. 237. Review of pharmacology and clinical development of enteric-coated formulation: *idem et al*., *Expert Opin. Pharmacother*. **5**, 1333-1345 (2004). White to off-white, crystalline powder. Highly sol in aqueous media at physiological pH. Practically insol in 0.1*N* HCl. LD_{50} in mice (mg/kg): 1176±151 orally; 568±53 i.p. (Williams).

THERAP CAT: Immunosuppressant.

6413. Mycosamine. [527-38-8] 3-Amino-3,6-dideoxymannose; 3-amino-3-deoxyrhamnose; 3,6-dideoxy-3-amino-D-mannopyranose; 3-amino-3,6-dideoxy-D-aldohexose. $C_6H_{13}NO_4$; mol wt 163.17. C 44.17%, H 8.03%, N 8.58%, O 39.22%. An amino sugar which represents the nitrogen-contg moiety of the antifungal antibiotics, nystatin and amphotericin B, *q.q.v*. Isoln and structure: Dutcher *et al*., *Antibiot. Annu*. **1956-1957**, 866; Walters *et al*., *J. Am. Chem. Soc*. **79**, 5076 (1957); Dutcher *et al*., *J. Org. Chem*. **28**, 995 (1963). Stereochemistry: von Saltza *et al*., *J. Am. Chem. Soc*. **83**, 2785 (1961); *eidem, J. Org. Chem*. **28**, 999 (1963).

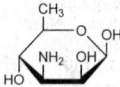

N-Acetylmycosamine. $C_8H_{15}NO_5$. Needles from methanol + acetone, mp 195-197°. $[\alpha]_D^{22}$ −46° (ethanol).

Triacetylmycosamine. $C_{12}H_{19}NO_7$. Prisms from acetone or ethanol, mp 185-187°. $[\alpha]_D^{22}$ +85° (ethanol).

Tetraacetylmycosamine. $C_{14}H_{21}NO_8$. Needles from benzene, mp 159-161°. $[\alpha]_D^{23}$ +39.3° (ethanol).

Hydrochloride. $C_6H_{13}NO_4$.HCl. Prismatic rods from ethanol+ ether, mp 162°. $[\alpha]_D^{24}$ −11.5°.

6414. Mylabris. Chinese cantharides; Chinese blistering flies. Dried insect, *Mylabris sidae (Phalerata)* Pallas, Coleoptera. *Habit*. China and East India. *Constit*. Cantharidin, said to be present in larger proportions than in cantharides (1-1.2%).

Black, cylindrical body, rounded above, flattish below; smaller than cantharides; black wing cases marked with a spot at point of insertion and has three tawny bands.

Caution: Highly toxic! *See* Cantharides.

USE: Source for cantharidin.

6415. Myosin. Mol wt approx 500,000. The main protein in muscle, which, by hydrolyzing ATP, provides the energy necessary for muscular contraction. Individual molecules are rod-shaped and consist of two globular heads attached flexibly to a tail of approx 1500Å length. All muscle myosins have a similar subunit structure composed of two heavy chains and two distinct pairs of light chains. One of the light chains is involved in a Ca^{2+}-dependent regulatory process in the vertebrate skeletal muscle system. Biological prepn and review: A. G. Szent-Györgyi, *Chemistry of Muscular Contraction* (Academic Press, New York, 2nd ed., 1951) pp 38-57, 146-148. Other reviews: *idem, Adv. Enzymol. Relat. Subj. Biochem.* **16**, 313-360 (1955); A. G. Szent-Györgyi *et al.,* in *Sulfur in Proteins,* R. Benesch *et al.,* Eds. (Academic Press, New York, 1959) pp 291-295; J. Gergely, *Biochemistry of Muscle Contraction* (Little, Brown, Boston, 1964) pp 3-115; several authors in *Contractile Proteins and Muscle,* K. Laki, Ed. (Dekker, New York, 1971); several authors in *Methods Enzymol.* **85**, Part B, 55-130 (1982). Review on structure and function of myosin: W. F. Harrington, "Contractile Proteins of Muscle" in *The Proteins* vol. **4**, H. Neurath, R. L. Hill, Eds. (Academic Press, New York, 1979) p 245-409. Polymorphism of mammalian myosins and distribution of isoforms in different muscle fibers reviewed by A. G. Weeds in *Plasticity of Muscle,* D. Pette, Ed. (DeGruyter, Berlin, 1980) pp 55-68. Role of myosin light chains in calcium regulation: Kendrick-Jones, *Nature* **249**, 631 (1974). Tertiary crystal structure of globular heads: I. Rayment *et al., Science* **261**, 50 (1993). *See also* Tropomyosin.

6416. β-Myrcene. [123-35-3] 7-Methyl-3-methylene-1,6-octadiene; 2-methyl-6-methylene-2,7-octadiene. $C_{10}H_{16}$; mol wt 136.24. C 88.16%, H 11.84%. Found in oil of bay, verbena, hop, and others. Isoln: Power, Kleber, *Pharm. Rundsch.* **13**, 60 (1895); *see also* Ruzicka, Stoll, *Helv. Chim. Acta* **7**, 272 (1924); Goulding, Roberts, *J. Chem. Soc.* **105**, 2614 (1914); Booker *et al., ibid.* **1940**, 1453; Kugler, Kováts, *Helv. Chim. Acta* **46**, 1480 (1963). Obtained by pyrolysis of β-pinene: Goldblatt, Palkin, *J. Am. Chem. Soc.* **63**, 3517 (1941); Rummelsburg, **US 2444790** (1948 to Hercules Powder); Houlihan *et al., J. Am. Chem. Soc.* **81**, 4692 (1959). Separation of isomers: Ohloff *et al., Ann.* **675**, 83 (1964). Synthesis: O. P. Vig *et al., Indian J. Chem.* **7**, 450 (1969); **11**, 104 (1973); **13**, 1244 (1975); T. Mandai *et al., Tetrahedron Lett.* **22**, 763 (1981). Formation from isoprene: K. Takabe *et al., Synthesis* **1977**, 307.

Oil. Pleasant odor. d_4^{20} 0.794. n_D^{20} 1.4709. uv max (ethanol): 226 nm (ε 16,100). Practically insol in water. Sol in alcohol, chloroform, ether, glacial acetic acid.

α-Myrcene. 2-Methyl-6-methylene-1,7-octadiene. bp_{10} 44°. n_D^{25} 1.4661. d_{25}^{25} 0.7959. uv max (isooctane): 224.5 nm (ε 18,600). Not found in nature. Prepn: Mitzner *et al., J. Org. Chem.* **30**, 646 (1965); Vig *et al., J. Indian Chem. Soc.* **50**, 329 (1973).

USE: β-Myrcene as an intermediate in the manuf of perfume chemicals.

6417. Myricetin. [529-44-2] 3,5,7-Trihydroxy-2-(3,4,5-trihydroxyphenyl)-4*H*-1-benzopyran-4-one; 3,3′,4′,5,5′,7-hexahydroxyflavone; cannabiscetin; delphidenolon 1575. $C_{15}H_{10}O_8$; mol wt 318.24. C 56.61%, H 3.17%, O 40.22%. From the bark of *Myrica nagi* Thunb., *Myriaceae:* Perkin, Hummel, *J. Chem. Soc.* **69**, 1287 (1896). Structure: Perkin, *ibid.* **81**, 203 (1902). Identity with cannabiscetin: Seshadri, Venkateswarlu, *Proc. Indian Acad. Sci.* **23A**, 296 (1946); *C.A.* **40**, 6447 (1946). Synthesis: Kalff, Robinson, *J. Chem. Soc.* **127**, 181 (1925); Rao *et al., J. Sci. Ind. Res.* **8B**, No. 6, 113 (1949). Occurrence in *Hamamelidaceae* and *Anacardiaceae:* Reznek, Egger, *Z. Naturforsch.* **15b**, 247 (1960). Metabolism: Smith, Griffiths, *Biochem. J.* **118**, 53p (1970).

Yellow needles from dil alc, mp 357°. uv max (ethanol): 375, 255 nm. Sparingly sol in boiling water; sol in alcohol. Practically insol in chloroform, acetic acid.

Hexaacetate. $C_{27}H_{22}O_{14}$. Crystals, mp 213°.

Hexaethyl ether. $C_{27}H_{34}O_8$. Needles from alcohol, mp 149-151°.

3-Rhamnoside. Myricitrin. $C_{21}H_{20}O_{12}$. Structure: Hattori, Hayashi, *Acta Phytochim.* **5**, 213 (1931), *C.A.* **26**, 990[8] (1932). Pale yellow leaflets from water, mp 199-200°. uv max (alc): 262, 352 nm. Sparingly sol in water, abs alcohol.

6418. Myriocin. [35891-70-4] (2S,3R,4R,6E)-2-Amino-3,4-dihydroxy-2-(hydroxymethyl)-14-oxo-6-eicosenoic acid; thermozymocidin; ISP-1. $C_{21}H_{39}NO_6$; mol wt 401.54. C 62.82%, H 9.79%, N 3.49%, O 23.91%. Immunosuppressant and antifungal antibiotic produced by certain thermophilic fungi. Also found in cordyceps, *q.v.,* a traditional Chinese medicine derived from parasitic fungi of the genus *Cordyceps.* Structurally related to sphingosine, *q.v.;* inhibits serine palmitoyltransferase, a key enzyme involved in sphingolipid biosynthesis. Isoln from *Myriococcum albomyces:* D. Kluepfel *et al., J. Antibiot.* **25**, 109 (1972); from *Mycelia sterilia:* R. Craveri *et al., Experientia* **28**, 867 (1972). Immunosuppressant activity and isoln from *Isaria sinclairii,* the anamorph of *Cordyceps sinclairii:* T. Fujita *et al., J. Antibiot.* **47**, 208 (1994). Effect on sphingolipid biosynthesis: Y. Miyake *et al., Biochem. Biophys. Res. Commun.* **211**, 396 (1995). Structure: F. Aragozzini *et al., Experientia* **28**, 881 (1972); and stereochemistry: J. F. Bagli *et al., J. Org. Chem.* **38**, 1253 (1973). Crystal structure: R. Destro, A. Colombo, *J. Chem. Soc. Perkin Trans. 2* **1979**, 896. Absolute configuration: D. R. Payette, G. Just, *Can. J. Chem.* **59**, 269 (1981). Total synthesis from D-fructose: L. Banfi *et al., J. Chem. Soc. Chem. Commun.* **1982**, 488; *eidem, J. Chem. Soc. Perkin Trans. 1* **1983**, 1613. Review of syntheses: H.-S. Byum *et al., Synthesis* **2006**, 2447-2474. HPLC determn in medicinal cordyceps preparations: J. Yu *et al., Anal. Sci.* **25**, 855 (2009).

White to off-white crystalline solid from methanol, mp 180-181°. *Poisonous.* $[\alpha]_D^{25}$ +4° (c = 1 in DMSO); $[\alpha]_D^{24}$ +10.3° (c = 0.386 in methanol). Soly in methanol: 2 mg/ml. Sol in NaOH, hydrochloric acid, conc formic acid, acetic acid. Slightly sol in DMSO, lower alcohols, chloroform, pyridine. Practically insol in water and nonpolar solvents. LD_{50} i.p. in mice, rats: 5-10, 2-5 mg/kg; orally in mice: 300-400 mg/kg (Kluepfel).

USE: Biochemical tool to deplete cells of sphingolipids.

6419. Myristic Acid. [544-63-8] Tetradecanoic acid; 1-tridecanecarboxylic acid. $C_{14}H_{28}O_2$; mol wt 228.38. C 73.63%, H 12.36%, O 14.01%. Occurs in nutmeg butter *(Myristica fragrans* Houtt.) to the extent of 70-80%; predominates in the fats of the *Myristicaceae;* in palm seed fats it may comprise 20% of the total fatty acids; in milk fats between 8-12% of the total acids. Occurs in most animal and vegetable fats; has been found in considerable amounts (up to 15%) in sperm whale oil, *see* Markley, *Fatty Acids* (New York, 1947). Prepn: G. D. Beal, *Org. Synth.* **coll. vol. I**, 379 (2nd ed., 1941). Prepn from tall-oil fatty acids: Segessmann, Molnar, **US 2481356** (1949); from 9-ketotetradecanoic acid: Ames *et al., J. Chem. Soc.* **1950**, 174; by electrolysis of methyl hydrogen adipate + decanoic acid: Greaves *et al., ibid.* **1950**, 3326; by Maurer oxidation of myristyl alc: Langenbeck, Richter, *Ber.* **89**, 202 (1956); from cetanol: Selwitz, **US 2969380** (1961 to Gulf). Wide-line NMR spectrum: A. V. Bailey, R. A. Pittman, *J. Am. Oil Chem. Soc.* **48**, 775 (1971). Toxicity study: L. Orö, A. Wretlind, *Acta Pharmacol. Toxicol.* **18**, 141 (1961).

Crystals or solid from methanol, mp 58.5°. bp_{100} 250.5°; bp_{16} 199°; bp_4 184°. d_4^{54} 0.8622. d_4^{70} 0.8528. d_4^{90} 0.8394. n_D^{60} 1.4305; n_D^{70} 1.4273. Soluble in alc, methanol, ether, petr ether, benzene, chloroform. Practically insol in water. Neutralization value: 245.68. Absorption spectrum: Markley, *op. cit.* LD_{50} i.v. in mice: 432.6 mg/kg (Orö, Wretlind).

Ethyl ester. [124-06-1] Ethyl myristate; ethyl tetradecanoate. $C_{16}H_{32}O_2$; mol wt 256.43. Liquid, mp 12°; bp 295°; bp_{30} 195°. d^{20} 0.856. Sol in alcohol; slightly sol in ether. Insol in water.

USE: As ingredient in soaps and shaving creams; in lubricants; in coatings for anodized aluminum. Pharmaceutic aid (antifoaming agent).

6420. Myristicin. [607-91-0] 4-Methoxy-6-(2-propen-1-yl)-1,3-benzodioxole; 5-allyl-1-methoxy-2,3-(methylenedioxy)benzene. $C_{11}H_{12}O_3$; mol wt 192.21. C 68.74%, H 6.29%, O 24.97%. Aromatic ether extracted from nutmeg, parsley, carrots. Isoln from nutmeg, *Myristica fragrans* Houtt., *Myristicaceae* and characterization: F. B. Power, A. H. Salway, *J. Chem. Soc.* **91**, 2037 (1907); A. T. Shulgin, *Nature* **197**, 379 (1963); from carrots, *Daucus carota* L., *Umbelliferae:* S. G. Yates, R. E. England, *J. Agric. Food Chem.* **30**, 317 (1982); from parsley, *Petroselinum hortense* Hoffman, *Umbelliferae:* M. Balogh *et al., Herba Hung.* **17**, 39 (1978), *C.A.* **91**, 2543e (1979). Synthesis: V. M. Trikojus, D. E. White, *Nature* **144**, 1016 (1939); F. Dallacker, R. Sluysmans, *Monatsh. Chem.* **100**, 560 (1969). HPLC sepn: L. W. Wulf *et al., J. Chromatogr.* **161**, 271 (1978); A. W. Archer, *ibid.* **438**, 117 (1988). GC-MS study: G. M. Sammy, W. W. Nawar, *Chem. Ind.* **1968**, 1279. MS-MS study of nutmeg extract: D. V. Davis, R. G. Cooks, *J. Agric. Food Chem.* **30**, 495 (1982). Possible psychotropic properties: A. T. Shulgin, *Nature* **210**, 380 (1966).

Colorless oil, bp_{40} 173°. n_D^{20} 1.54032. d_{20}^{20} 1.1437.

6421. Myristyl Alcohol. [112-72-1] 1-Tetradecanol; 1-hydroxytetradecane; tetradecyl alcohol. $C_{14}H_{30}O$; mol wt 214.39. C 78.43%, H 14.11%, O 7.46%. Prepn by sodium reduction of fatty acid esters: Hansley, *Ind. Eng. Chem.* **39**, 55 (1947); by LiAlH$_4$ reduction of fatty acids: Watanabe, *Bull. Chem. Soc. Jpn.* **34**, 398 (1961); from acetaldehyde + dimethylamine: Langenbeck *et al., J. Prakt. Chem.* **8**, 112 (1959).

White crystals, mp 38°. d 0.824. bp_{15} 167°. Practically insol in water. Sol in ether, slightly in alcohol.

USE: As emollient for cold creams, etc., also for making the sulfated alcohol whose sodium salt is applicable as a "wetter" in textiles.

6422. Myristyltrimethylammonium Bromide. [1119-97-7] *N,N,N*-Trimethyl-1-tetradecanaminium bromide (1:1); trimethyltetradecylammonium bromide; tetradonium bromide; tetradecyltrimethylammonium bromide; myrtrimonium bromide; Morpan T;

Mytab. $C_{17}H_{38}BrN$; mol wt 336.40. C 60.70%, H 11.39%, Br 23.75%, N 4.16%. Cationic germicidal detergent. Prepn and antibacterial activity: R. S. Shelton *et al., J. Am. Chem. Soc.* **68**, 753 (1946). Toxicity and pharmacology: B. Isomaa, K. Bjondahl, *Acta Pharmacol. Toxicol.* **47**, 17 (1980).

White powder, mp 245-250°. Sol in 5 parts water. *Corrosive.* LD_{50} in mice, rats (mg/kg): 12.0, 15.0 i.v. (Issomaa, Bjondahl).
USE: Disinfectant; deodorant; laboratory reagent.

6423. Myrrh. [9000-45-7] Gum-resin myrrh. From *Commiphora abyssinica* (Berg) Eng. or from other species of *Commiphora, Burseraceae.* Yields not less than 30% alcohol-soluble extract. *Habit.* Nubia, Somaliland, Arabia. *Constit.* 20-25% resin, 57-61% gum, 7-17% volatile oil and a bitter principle.

THERAP CAT: Carminative; astringent.

6424. Myrtol. [8002-55-9] Gelomyrtol. The fraction of the volatile oil from *Myrtus communis* L., *Myrtaceae* distilling between 166-180° and consisting chiefly of eucalyptol and *dextro*-pinene with a small quantity of an undefined camphor.
Colorless or slightly yellow, strongly dextrorotatory liquid (due to *dextro*-pinene); pleasant odor resembling that of turpentine and eucalyptol. d ~0.895. $[\alpha]_D^{20}$ +10°. n_D^{20} ~1.465. Freely sol in alcohol, ether.

6425. Myxin. [13925-12-7] 6-Methoxy-1-phenazinol 5,10-dioxide; 1-hydroxy-6-methoxyphenazine 5,10-dioxide; 3C antibiotic. $C_{13}H_{10}N_2O_4$; mol wt 258.23. C 60.47%, H 3.90%, N 10.85%, O 24.78%. Potent broad-spectrum antibiotic capable of inhibiting a wide range of gram-positive and gram-negative bacteria, various fungi and yeasts. Isoln from a *Sorangium* species (strain 3C), a soilborne myxobacter: Peterson *et al., Can. J. Microbiol.* **12**, 221 (1966). Proposed structure: Edwards, Gillespie, *Tetrahedron Lett.* **1966**, 4867. Revised structure and synthesis: Weigele, Leimgruber, *ibid.* **1967**, 715; Sigg, Toth, *Helv. Chim. Acta* **50**, 716 (1967). Activity: E. Grunberg *et al., Chemotherapia* **12**, 272 (1967). Crystal structure: Hanson, *Acta Crystallogr.* **B24**, 1084 (1968). *Review:* Edwards, Gillespie, *Proc. Int. Symp. Drug Res.* **1967**, 236; Leimgruber, Grunberg, *ibid.* 240; Baker, Vezina, *ibid.* 242.

Red needles from acetone, mp 120-130° (dec) (Sigg, Toth). Strongly exothermic at 149°. May explode on drying with heat: Rachlin, *Chem. Eng. News* **45**, 32 (Sept. 4, 1967). uv max (0.1 *N* HCl): 283, 340 nm (ε 97000, 5400); absorption max (0.1*N* HCl): 505 nm (ε 6500). Stable in water after 6 hours at 37° in 0.05*N* acetate buffer at pH 5.0, in 0.05*N* phosphate at pH 6.0 and 7.0, or in 0.005*N* tris-HCl at pH 8.0 and 9.0. Stable at 70° except in phosphate buffer. Acetate, methanol, or aq tris buffer solns remain stable indefinitely at 4°. LD_{50} in mice (mg/kg): >2000 orally, >2000 s.c., 133 i.p. (Grunberg).

Copper(II) complex. [39238-36-3] Cuprimyxin; Ro-7-4488/1; UNITOP. $C_{26}H_{18}CuN_4O_8$; mol wt 578.00. Prepn: W. Leimgruber *et al., DE 1931466; eidem, US 3586674* (1970, 1971 both to Hoffmann-La Roche). Activity studies: Maestrone *et al., Am. J. Vet. Res.* **33**, 185 (1972). Dark green, fine crystals. uv max (DMSO): 287, 300, 356 nm (ε 68500, 63200, 10000); absorption max (DMSO): 408, 610 nm (ε 10400, 9500).

THERAP CAT (VET): Antibacterial, antifungal.

N

6426. Nabam. [142-59-6] N,N'-1,2-Ethanediylbiscarbamodithioic acid sodium salt (1:2); ethylenebis[dithiocarbamic acid] disodium salt; disodium ethylenebis[dithiocarbamate]; Dithane D-14; Parzate. $C_4H_6N_2Na_2S_4$; mol wt 256.33. C 18.74%, H 2.36%, N 10.93%, Na 17.94%, S 50.03%. Prepd by treating ethylenediamine with CS_2 in the presence of NaOH: Hester, **US 2317765** (1943 to Rohm & Haas); Flenner, **US 2504404** (1950 to du Pont). Fluorimetric determn in water and cereals: T. Pérez-Ruiz et al., Talanta **43**, 193 (1996). Degradation study: E. Milanova et al., Pulp Paper Can. **100**, 67 (1999). Toxicology study: R. B. Smith, Jr. et al., Fed. Proc. **11**, 391 (1952).

LD$_{50}$ orally in rats: 395 ±12 mg/kg (Smith).
Hexahydrate. Crystals from alc. Moderately sol in water. Forms a continuous film on plant surfaces, which is said to become insol in water.
Caution: Direct contact may cause skin irritation. *See Clinical Toxicology of Commercial Products*, R. E. Gosselin et al., Eds. (Williams & Wilkins, Baltimore, 5th ed., 1984) Section II, p 312.
USE: Agricultural fungicide; biocide in pulp and paper industry.

6427. Nabilone. [51022-71-0] rel-(6aR,10aR)-3-(1,1-Dimethylheptyl)-6,6a,7,8,10,10a-hexahydro-1-hydroxy-6,6-dimethyl-9H-dibenzo[b,d]pyran-9-one; Cesamet; LY-109514; Cesamet. $C_{24}H_{36}O_3$; mol wt 372.55. C 77.38%, H 9.74%, O 12.88%. Synthetic cannabinoid with antiemetic, antiglaucoma, and CNS activity. Prepn: R. A. Archer, **DE 2451934**, idem, **US 3968125** (1975, 1976 both to Lilly); R. A. Archer et al., J. Org. Chem. **42**, 2277 (1977). Pharmacology: P. Stark, R. A. Archer, Pharmacologist **17**, 210 (1975); R. M. Orzelek-O'Neil et al., Toxicol. Appl. Pharmacol. **54**, 493 (1980). Physiological disposition: A. Rubin et al., Clin. Pharmacol. Ther. **22**, 85 (1977). Pharmacokinetics: H. R. Sullivan et al., Biomed. Mass Spectrom. **5**, 296 (1978). Behavioral effects: P. Stark, P. B. Dews, J. Pharmacol. Exp. Ther. **214**, 124 (1980). Clinical studies as antiemetic in cancer patients: N. Steele et al., Cancer Treat. Rep. **64**, 219 (1980); C. J. Williams et al., Cancer Clin. Trials **3**, 363 (1980). Clinical evaluation of anxiolytic effects: L. F. Fabre et al., J. Clin. Pharmacol. **21**, Suppl 8-9, 377S (1981); R. M. Glass et al., ibid. 383S. Review of pharmacology and efficacy as antiemetic: A. Ward, B. Holmes, Drugs **30**, 127-144 (1985). Toxicity study in neonatal rats: C. L. Moss et al., Toxicol. Appl. Pharmacol. **48**, A120 (1979). Comprehensive description: R. W. Souter, Anal. Profiles Drug Subs. **10**, 499-512 (1981).

Relative stereochemistry

White crystals from ethyl acetate/hexane, mp 159-160°. uv max (ethanol): 207, 280 nm (ε 47000, 250). pKa in 66% DMF: 13.5.
Note: This is a controlled substance (hallucinogen): **21 CFR**, 1308.12.
THERAP CAT: Antiemetic.

6428. Nabumetone. [42924-53-8] 4-(6-Methoxy-2-naphthalenyl)-2-butanone; 4-(6-methoxy-2-naphthyl)-butan-2-one; BRL-14777; Arthaxan; Balmox; Consolan; Nabuser; Relafen; Relifen; Relifex. $C_{15}H_{16}O_2$; mol wt 228.29. C 78.92%, H 7.06%, O 14.02%. Nonacidic, lipophillic prodrug; metabolized *in vivo* to 6-methoxy-2-

naphthylacetic acid which inhibits prostaglandin synthesis. Prepn: A. W. Lake, C. J. Rose, **DE 2442305**; eidem, **US 4061779** (1975, 1977 both to Beecham); A. C. Goudie et al., J. Med. Chem. **21**, 1260 (1978). Pharmacology: E. A. Boyle et al., J. Pharm. Pharmacol. **34**, 562 (1982); B. Nunn, P. D. Chamberlain, ibid. 576. Metabolism: R. E. Haddock et al., Xenobiotica **14**, 327 (1984). HPLC determn in plasma: J. E. Ray, R. O. Day, J. Chromatogr. **336**, 234 (1984). Symposia on pharmacology, pharmacokinetics and clinical efficacy: Int. Congr. Symp. Ser. - R. Soc. Med. **69**, 1-215 (1985); J. Rheumatol. **19**, Suppl. 36, 1-91 (1992). Review: S. L. Dahl, Ann. Pharmacother. **27**, 456-463 (1993).

Crystals from ethanol, mp 80°. Freely sol in acetone; sparingly sol in alc, methanol. Practically insol in water.
THERAP CAT: Anti-inflammatory; analgesic.

6429. NAD. [53-84-9] Adenosine 5'-(trihydrogen diphosphate) $P' \rightarrow 5'$-ester with 3-(aminocarbonyl)-1-β-D-ribofuranosylpyridinium inner salt; 3-carbamoyl-1-β-D-ribofuranosylpyridinium hydroxide 5' $\rightarrow$ 5'-ester with adenosine 5'-(trihydrogen pyrophosphate) inner salt; nicotinamide-adenine dinucleotide; diphosphopyridine nucleotide; DPN; cozymase; coenzyme I; Co I; codehydrogenase; β-NAD; NAD$^+$; nadide. $C_{21}H_{27}N_7O_{14}P_2$; mol wt 663.43. C 38.02%, H 4.10%, N 14.78%, O 33.76%, P 9.34%. One of the biologically active forms of nicotinic acid, q.v. Occurs in living cells primarily in the oxidized state. Serves as a coenzyme of the dehydrogenases, especially in the dehydrogenation of primary and secondary alcohols. NAD usually acts as a hydrogen acceptor, forming NADH which then serves as a hydrogen donor in the respiratory chain. Isoln from erythrocytes: O. Warburg, W. Christian, Biochem. Z. **287**, 291 (1936); from rabbit muscle: S. Ochoa, ibid. **292**, 68 (1937); from yeast: H. v. Euler, F. Schlenk, Z. Physiol. Chem. **246**, 64 (1937); G. A. LePage, J. Biol. Chem. **168**, 623 (1947); Biochem. Prep. **1**, 28 (1949). Manuf by fermentation: **GB 1190079** (1970 to Kyowa), C.A. **73**, 54631g (1970). Synthetic approach: N. A. Hughes et al., J. Chem. Soc. **1957**, 3733. Nomenclature: M. Dixon, Science **132**, 1548 (1960). Occurs in 2 forms, α-NAD and β-NAD, distinguished by the configuration of the ribosyl nicotinamide linkage. Only the β-anomer is bioactive. Stereochemistry: R. U. Lumieux, J. W. Lown, Can. J. Chem. **41**, 889 (1963). NMR studies: N. J. Oppenheimer et al., Proc. Natl. Acad. Sci. USA **68**, 3200 (1971); R. H. Sarma, R. J. Mynott, Chem. Commun. **1972**, 977; M. Blumenstein, M. A. Raftery, Biochemistry **11**, 1643 (1972). Biosynthesis: Nutr. Rev. **30**, 139 (1972); J. W. Foster, A. G. Moat, Microbiol. Rev. **44**, 83 (1980). Combined bioluminescent assay of NAD and NADH: M. T. Karp, P. I. Vuorinen, Methods Enzymol. **122**, 147 (1986). Review: "Niacin: Nicotinic Acid, Nicotinamide, NAD(P)" in Vitamins, W. Friedrich, Ed. (Walter de Gruyter, Berlin, 1988) pp 473-542.

Very hygroscopic white powder. Freely sol in water without residue. A 1% soln has a pH of ~2. Aq solns are stable for ~1 week. $[\alpha]_D^{20}$ −31.5° (c = 1.2 in water). uv max: 260 nm (ε 18100).
Reduced form. [58-68-4] NADH; DPNH; 1,4-dihydronicotinamide adenine dinucleotide. $C_{21}H_{29}N_7O_{14}P_2$. uv max: 260, 340 nm (ε 15000, 6290).

6430. Nadifloxacin. [124858-35-1] 9-Fluoro-6,7-dihydro-8-(4-hydroxy-1-piperidinyl)-5-methyl-1-oxo-1H,5H-benzo[ij]quinolizine-2-carboxylic acid; jinofloxacin; OPC-7251; Acuatim. $C_{19}H_{21}$-FN_2O_4; mol wt 360.39. C 63.32%, H 5.87%, F 5.27%, N 7.77%, O 17.76%. Fluorinated quinolone antibacterial. Prepn: H. Ishikawa *et al.*, **BE 891046**; *eidem*, **US 4399134** (1982, 1983 both to Otsuka); *eidem*, *Chem. Pharm. Bull.* **37**, 2103 (1989). Toxicity data: K. Hashimoto *et al.*, *Iyakuhin Kenkyu* **21**, 671 (1990), *C.A.* **114**, 156625r (1991). *In vitro* antibacterial activity: K. Vogt *et al.*, *Eur. J. Clin. Microbiol. Infect. Dis.* **11**, 943 (1992). HPLC determn: M. Koike *et al.*, *J. Chromatogr.* **526**, 235 (1990). Clinical trial in treatment of acne: I. Kurokawa *et al.*, *J. Am. Acad. Dermatol.* **25**, 674 (1991).

Colorless prisms from EtOH-H_2O, mp 245-247° (dec). LD_{50} male, female mice and rats (mg/kg): 376.5, 420.6, 225.7, 240.5 i.v. (Hashimoto).

THERAP CAT: Antibacterial (topical).

6431. Nadolol. [42200-33-9] 5-[3-[(1,1-Dimethylethyl)amino]-2-hydroxypropoxy]-1,2,3,4-tetrahydro-2,3-naphthalenediol; 1-(*tert*-butylamino)-3-[(5,6,7,8-tetrahydro-*cis*-6,7-dihydroxy-1-naphthyl)oxy]-2-propanol; (2*R*,3*S*)-5-[3-(*tert*-butylamino)-2-hydroxypropoxy]-1,2,3,4-tetrahydronaphthalene-2,3-diol; 2,3-*cis*-1,2,3,4-tetrahydro-5-[2-hydroxy-3-(*tert*-butylamino)propoxy]-2,3-naphthalenediol; SQ-11725; Corgard; Solgol. $C_{17}H_{27}NO_4$; mol wt 309.41. C 65.99%, H 8.80%, N 4.53%, O 20.68%. β-Adrenergic blocker. Prepn: F. P. Hauck *et al.*, **DE 2258995**; *eidem*, **US 3935267** (1973, 1976, both to Squibb). Resolution of isomers: F. P. Hauck, J. E. Sundeen, **DE 2421549** (1974 to Squibb), *C.A.* **82**, 57481e (1975). Metabolism: J. Dreyfuss *et al.*, *J. Clin. Pharmacol.* **17**, 300 (1977). Toxicology: P. L. Sibley *et al.*, *Toxicol. Appl. Pharmacol.* **44**, 379 (1978). Review of pharmacology: R. C. Heel *et al.*, *Drugs* **1**, 1-23 (1980); M. J. Antonaccio, D. B. Evans, in *Pharmacology of Antihypertensive Drugs*, A. Scriabine, Ed. (Raven Press, New York, 1980) pp 295-301. Comprehensive description: L. Slusarek, K. Florey, *Anal. Profiles Drug Subs.* **9**, 455-485 (1980). Book: *International Experience with Nadolol, a Long-Acting β-Blocking Agent*, F. Gross, Ed. (Grune & Stratton, New York, 1981) 229 pp.

Crystalline powder, mp 124-136°. uv max (methanol): 270, 278 nm ($E_{1cm}^{1\%}$ 37.5, 39.1). pKa 9.67. Freely sol in ethanol, methanol, propylene glycol; sol in water (at pH 2), hydrochloric acid; slightly sol in chloroform, methylene chloride, isopropyl alc, water (between pH 7 and pH 10). Insol in acetone, benzene, ether, hexane, trichloroethane. LD_{50} in mice, rats (mg/kg): 4500, 5300 orally (Antonaccio, Evans).

THERAP CAT: Antihypertensive, antianginal.

6432. NADP. [53-59-8] Adenosine 5′-(trihydrogen diphosphate) 2′-(dihydrogen phosphate) P' → 5′-ester with 3-(aminocarbonyl)-1-β-D-ribofuranosylpyridinium inner salt; 3-carbamoyl-1-β-D-ribofuranosylpyridinium hydroxide 5′ → 5′-ester with adenosine 2′-(dihydrogen phosphate) 5′-(trihydrogen pyrophosphate) inner salt; nicotinamide-adenine dinucleotide phosphate; triphosphopyridine nucleotide; TPN; coenzyme II; Co II; phosphocozymase; codehydrase II; codehydrogenase II. $C_{21}H_{28}N_7O_{17}P_3$; mol wt 743.41. C 33.93%, H 3.80%, N 13.19%, O 36.59%, P 12.50%. One of the biologically active forms of nicotinic acid, *q.v.* Differs from NAD, *q.v.*, by an additional phosphate group at the 2′-position of the adenosine moiety. Serves as a coenzyme of hydrogenases and dehydrogenases. Present in living cells primarily in the reduced form

(NADPH) and is involved in synthetic reactions. Isoln from horse blood: O. Warburg *et al.*, *Biochem. Z.* **282**, 157 (1935); from hog liver by adsorption on charcoal: G. A. LePage, G. C. Mueller, *J. Biol. Chem.* **180**, 975 (1949); from sheep liver by extraction with hot water: A. Kornberg, B. L. Horecker, *Biochem. Prep.* **3**, 24 (1953); from human red blood cells by anion-exchange chromatography: G. R. Bartlett, *J. Biol. Chem.* **234**, 449, 459 (1959). Chemical and enzymatic synthesis: D. R. Walt *et al.*, *J. Am. Chem. Soc.* **106**, 234 (1984). Nomenclature: M. Dixon, *Science* **132**, 1548 (1960). Review of synthesis and metabolism: Y. Nishizuka in *Method. Chim.* **vol. 11**, F. Korte, Ed. (Academic Press, New York, 1977) pp 84-87. *Review:* "Niacin: Nicotinic Acid, Nicotinamide, NAD(P)" in *Vitamins*, W. Friedrich, Ed. (Walter de Gruyter, Berlin, 1988) pp 473-542.

Grayish-white powder. pKa_1 3.9; pKa_2 6.1. uv max: 260 nm (ε 18000). Sol in water, methanol. Much less sol in ethanol. Practically insol in ether, ethyl acetate.

Reduced form. [53-57-6] NADPH; $NADPH_2$; TPNH; dihydrocodehydrogenase II. $C_{21}H_{30}N_7O_{17}P_3$. uv max: 260, 340 nm (ε 18000, 6220).

6433. Nadroparin. Tedegliparin. Low molecular weight fraction of heparin, *q.v.*, prepd from porcine mucosal heparin by selective ethanol precipitation. Mean mol wt 4500 daltons. Prepn: J.-C. Lormeau *et al.*, **DE 2944792**; *eidem*, **US 4486420**; **US 4692435** (1980, 1984, 1987 all to Choay); J. Choay *et al.*, *Thromb. Res.* **18**, 573 (1980). Comparison with unfractionated heparin: G. F. Gensini *et al.*, *Haemostasis* **14**, 466 (1984); M. D. Vandenbroek, B. Bordes, *J. Pharm. Clin.* **5**, 35 (1986); C. Doutremepuich *et al.*, *Thromb. Res.* **43**, 691 (1986). Pharmacokinetics and bioavailability in humans: J. Harenberg *et al.*, *ibid.* **44**, 549 (1986). Clinical trial in prophylaxis of postoperative thromboembolism: V. V. Kakkar, *Nouv. Rev. Fr. Hematol.* **26**, 277 (1984); V. V. Kakkar, W. J. G. Murray, *Br. J. Surg.* **72**, 786 (1985); in acute ischemic stroke: R. Kay *et al.*, *N. Engl. J. Med.* **333**, 1588 (1995).

Calcium salt. CY-216; Fraxiparine.

THERAP CAT: Antithrombotic.

6434. Nafamostat. [81525-10-2] 4-[(Aminoiminomethyl)-amino]benzoic acid 6-(aminoiminomethyl)-2-naphthalenyl ester; 6-amidino-2-naphthyl-4-guanidinobenzoate; nafamstat. $C_{19}H_{17}N_5O_2$; mol wt 347.38. C 65.69%, H 4.93%, N 20.16%, O 9.21%. Nonpeptide protease inhibitor. Prepn: S. Fujii *et al.*, **EP 48433**; *eidem*, **US 4454338** (1982, 1984 both to Torii & Co.); T. Aoyama *et al.*, *Chem. Pharm. Bull.* **33**, 1458 (1985). Inhibitory effects on trypsin, thrombin, kallikrein, plasmin and complement-mediated hemolysis: S. Fujii, Y. Hitomi, *Biochim. Biophys. Acta* **661**, 342 (1981); T. Aoyama *et al.*, *Jpn. J. Pharmacol.* **35**, 203 (1984). Effect in experimental acute pancreatitis in rats: M. Iwaki *et al.*, *ibid.* **41**, 155 (1986). Spectrofluorometric determn in biological material: T. Aoyama *et al.*, *Chem. Pharm. Bull.* **33**, 2142 (1985). Clinical anticomplement activity: Y. Miyamoto *et al.*, *Trans. Am. Soc. Artif. Intern. Organs* **31**, 508 (1985).

Dimethanesulfonate. [82956-11-4] Nafamostat mesylate; FUT-175; Futhan. $C_{19}H_{17}N_5O_2.2CH_3SO_3H$; mol wt 539.58. Crystals from ethyl ether, mp 217-220° (Fujii). Also reported as colorless powder from water, mp 260° (dec) (Aoyama).

THERAP CAT: Enzyme inhibitor (protease).

6435. Nafarelin. [76932-56-4] 6-[3-(2-Naphthalenyl)-D-ala-nine]luteinizing hormone-releasing factor (swine); 5-oxo-L-prolyl-L-histidyl-L-tryptophyl-L-seryl-L-tyrosyl-3-(2-naphthyl)-D-alanyl-L-leucyl-L-arginyl-L-prolylglycinamide; [6-[3-(2-naphthyl)-D-ala-nine]]gonadorelin; D-nal(2)^6-LHRH; NAG. $C_{66}H_{83}N_{17}O_{13}$; mol wt 1322.50. C 59.94%, H 6.33%, N 18.01%, O 15.73%. Synthetic peptide agonist analog of gonadotropin-releasing hormone, *q.v.* Prepn: J. J. Nestor *et al.*, US 4234571 (1980 to Syntex); *eidem, J. Med. Chem.* **25**, 795 (1982). Suppression of luteal and placental function in baboons: B. H. Vickery *et al.*, *Fertil. Steril.* **36**, 664 (1981). Inhibition of ovulation in humans: J. A. Gudmundsson *et al.*, *Contraception* **30**, 107 (1984). Radioimmunoassay in plasma and serum: C. Nerenberg *et al.*, *Anal. Biochem.* **141**, 10 (1984). Kinetics and tissue disposition in animals: N. I. Chu *et al.*, *Drug Metab. Dispos.* **13**, 560 (1985). Clinical evaluation in benign prostatic hyperplasia: C. A. Peters, P. C. Walsh, *N. Engl. J. Med.* **317**, 599 (1987); in endometriosis: M. R. Henzl *et al.*, *ibid.* **318**, 485 (1988). Review of clinical studies: P. G. Hoffman *et al.*, *J. Androl.* **8**, Suppl., S17-S22 (1987).

5-oxoPro–His–Trp–Ser–Tyr–N---C---C–Leu–Arg–Pro–GlyNH₂

Acetate hydrate. [86220-42-0]; [76932-60-0] (anhydrous). RS-94991-298; Nasanyl; Synarel. $C_{66}H_{83}N_{17}O_{13}.xC_2H_4O_2.yH_2O$, where x=1-2 and y=2-8.

mp 188-190°. $[\alpha]_D^{25}$ −27.4° (c = 0.9 in acetic acid).

THERAP CAT: In treatment of endometriosis and central precocious puberty.

6436. Nafcillin. [147-52-4] (2*S*,5*R*,6*R*)-6-[[(2-Ethoxy-1-naphthalenyl)carbonyl]amino]-3,3-dimethyl-7-oxo-4-thia-1-azabi-cyclo[3.2.0]heptane-2-carboxylic acid; 6-(2-ethoxy-1-naphthami-do)penicillin. $C_{21}H_{22}N_2O_5S$; mol wt 414.48. C 60.85%, H 5.35%, N 6.76%, O 19.30%, S 7.73%. Semi-synthetic antibiotic related to penicillin. Prepn: F. P. Doyle *et al.*, GB 880400; F. P. Doyle, J. H. C. Nayler, US 3157639 (1961, 1964 both to Beecham). Prepn of crystalline monohydrate: D. Flitter *et al.*, US 3506645 (1970 to American Home Prod.). HPLC determn: P. Fan-Havard, M. C. Nahata, *Ther. Drug Monit.* **11**, 105 (1989). Antibacterial activity: J. M. Steckelberg *et al.*, *Antimicrob. Agents Chemother.* **37**, 554 (1993). Clinical efficacy: D. L. Palmer *et al.*, *Ann. Thorac. Surg.* **59**, 626 (1995).

Sodium salt monohydrate. [7177-50-6]; [985-16-0] (anhydrous). Sodium 6-(2-ethoxy-1-naphthamido)penicillanate; Wy-3277; Nafcil; Nallpen; Unipen. $C_{21}H_{21}N_2NaO_5S.H_2O$; mol wt 454.46. Freely sol in water, chloroform; sol in alc. Insol in acetone.

THERAP CAT: Antibacterial.

THERAP CAT (VET): Antibacterial.

6437. Nafronyl. [31329-57-4] Tetrahydro-α-(1-naphthalen-ylmethyl)-2-furanpropanoic acid 2-(diethylamino)ethyl ester; tetra-hydro-α-(1-naphthylmethyl)-2-furanpropionic acid 2-(diethylami-no)ethyl ester; 3-(1-naphthyl)-2-tetrahydrofurfurylpropionic acid 2-(diethylamino)ethyl ester; α-tetrahydrofurfuryl-1-naphthalenepro-pionic acid 2-(diethylamino)ethyl ester; *N*-diethylaminoethyl β-(1-naphthyl)-β-tetrahydrofuryl isobutyrate; naftidrofuryl; Dubimax; Gevatran. $C_{24}H_{33}NO_3$; mol wt 383.53. C 75.16%, H 8.67%, N 3.65%, O 12.51%. Prepn: Szarvasi, Bayssat, FR 1363948 corresp to US 3334096 (1964, 1967 both to LIPHA); Szarvasi *et al.*, *Compt. Rend.* **260**, 3095 (1965); *eidem, Bull. Soc. Chim. Fr.* **1966**, 1838. Activity studies: Fontaine *et al.*, *C. R. Seances Acad. Sci. Ser. D* **262**, 719 (1966). Metabolism and toxicology: Fontaine *et al.*, *Chim. Ther.* **4**, 44 (1969); Fontaine *et al.*, *J. Eur. Toxicol.* **11**, 40 (1969). Acute toxicity: P. Bessin *et al.*, *Eur. J. Med. Chem.* **10**, 291 (1975). Clinical studies: C. A. Clyne *et al.*, *Br. J. Surg.* **67**, 347 (1980); R. M. Greenhalgh, *ibid.* **68**, 265 (1981).

bp$_{0.5}$ 190°. d$_4^{31}$ 1.0465. n$_D^{29}$ 1.5513. LD$_{50}$ in mice (mg/kg): 365 orally (Bessin).

Acid oxalate. [3200-06-4] EU-1806; LS-121; Citoxid; Di-Ac-tane; Dusodril; Praxilene. $C_{24}H_{33}NO_3.C_2H_2O_4$; mol wt 473.57. Crystals from ethyl acetate, mp 110-111°. Slightly hygroscopic, sol in water.

THERAP CAT: Vasodilator.

6438. Naftalofos. [1491-41-4] 2-[(Diethoxyphosphinyl)oxy]-1*H*-benz[*de*]isoquinoline-1,3(2*H*)-dione; *N*-hydroxynaphthalimide diethyl phosphate; *O,O*-diethyl *O*-naphthaloximide phosphate; naph-thalophos; phthalophos; Bay 9002; Bayer 25820; ENT-25567; S-940; Maretin; Rametin. $C_{16}H_{16}NO_6P$; mol wt 349.28. C 55.02%, H 4.62%, N 4.01%, O 27.48%, P 8.87%. Prepn: Lorenz, Wegler, DE 962608 and AT 193894 (both 1957 to Bayer).

Minute crystals. The commercial product has a brown to tan color. mp 174-179°. Practically insol in water. Sparingly sol in the usual organic solvents. Fairly sol in methylene chloride.

THERAP CAT (VET): Anthelmintic.

6439. Naftifine. [65472-88-0] *N*-Methyl-*N*-[(2*E*)-3-phenyl-2-propen-1-yl]-1-naphthalenemethanamine; (*E*)-*N*-cinnamyl-*N*-meth-yl-1-naphthalenemethylamine; *N*-methyl-*N*-(1-naphthylmethyl)-3-phenylpropen-1-amine; naftifungin. $C_{21}H_{21}N$; mol wt 287.41. C 87.76%, H 7.37%, N 4.87%. Antimycotic allylamine. Prepn: D. Berney, BE 853976; *idem,* US 4282151 (1977, 1981 both to Sandoz); A. Blade Fond, E. Mendoza-Villela, ES 504432 (1982 to Lab. Frumtost-Prem S.A.), *C.A.* **99**, 38173r (1983); H. Loibner *et al.*, *Tetrahedron Lett.* **25**, 2535 (1984). In vitro activity against dermatophytes and *Candida* spp: A. Georgopoulos *et al.*, *Antimicrob. Agents Chemother.* **19**, 386 (1981). Morphological changes induced in *Trichophyton mentagrophytes:* J. C. Meingassner *et al.*, *J. Invest. Dermatol.* **77**, 444 (1981). Specific inhibitor of squalene epoxidase, a key enzyme in fungal ergosterol biosynthesis: F. Paltauf *et al.*, *Biochim. Biophys. Acta* **712**, 268 (1982); N. S. Ryder, *Antimicrob. Agents Chemother.* **25**, 483 (1984). HPLC determn in human plasma and urine: F. Schatz, H. Haberl, *Arzneim.-Forsch.* **36**, 1850 (1986). Clinical trials in dermatophytosis: U. Ganzinger *et al.*, *Clin. Trials J.* **19**, 342 (1982).

Colorless, viscous oil, $bp_{0.015 \text{ torr}}$ 162-167°.

Hydrochloride. [65473-14-5] AW-105-843; SN-105-843; Exoderil; Naftin; Suadian. $C_{21}H_{21}N.HCl$; mol wt 323.86. mp 177° (from propanol).

THERAP CAT: Antifungal (topical).

6440. Naftopidil. [57149-07-2] 4-(2-Methoxyphenyl)-α-[(1-naphthalenyloxy)methyl]-1-piperazineethanol; RS-1-[4-(2-methoxyphenyl)-1-piperazinyl]-3-(1-naphthoxy)-2-propanol; 1-(2-methoxyphenyl)-4-[3-(naphth-1-yloxy)-2-hydroxypropyl]-piperazine; KT-611; Avishot; Flivas. $C_{24}H_{28}N_2O_3$; mol wt 392.50. C 73.44%, H 7.19%, N 7.14%, O 12.23%. α_1-Adrenergic blocker and serotonin ($5HT_{1A}$) receptor agonist. Prepn: E. C. Witte *et al.*, **DE 2408804**; *eidem*, **US 3997666** (1975, 1976 both to Boehringer Mann.). Clinical pharmacodynamics: R. Kirsten *et al.*, *Eur. J. Clin. Pharmacol.* **46**, 271 (1994). Clinical pharmacokinetics: M. J. G. Farthing *et al.*, *Postgrad. Med. J.* **70**, 363 (1994). HPLC determn in human plasma: G. Niebch *et al.*, *J. Chromatogr.* **534**, 247 (1990). Clinical evaluation in BPH: K. Yasuda *et al.*, *Prostate* **25**, 46 (1994). Review of pharmacology and clinical experience: H. M. Himmel, *Cardiovasc. Drug Rev.* **12**, 32-47 (1994).

Crystals from isopropanol, mp 125-126°; also reported as colorless crystals, mp 125-129°. Insol in water. Partition coefficient (octanol/water): 75. LD_{50} in mice, rats (g/kg): 1.3, 6.4 orally (Himmel).

Dihydrochloride. [57149-08-3] $C_{24}H_{28}N_2O_3.2HCl$. Crystals from methanol/ethanol (1:2), mp 212-213°.

THERAP CAT: Antihypertensive; α-blocker in treatment of symptomatic benign prostate hypertrophy.

6441. Nalbuphine. [20594-83-6] (5α,6α)-17-(Cyclobutylmethyl)-4,5-epoxymorphinan-3,6,14-triol; N-cyclobutylmethyl-14-hydroxydihydronormorphine. $C_{21}H_{27}NO_4$; mol wt 357.45. C 70.56%, H 7.61%, N 3.92%, O 17.90%. Mixed opioid agonist-antagonist. Prepn: **GB 1119270**; I. J. Pachter, Z. Matossian, **US 3393197** (both 1968 to Endo). HPLC determn in plasma: F. de Cazanove *et al.*, *J. Chromatogr. B* **690**, 203 (1997). Opioid receptor binding profile: J.-C. Chen *et al.*, *Life Sci.* **52**, 389 (1993). Review of pharmacology and abuse potential: W. K. Schmidt *et al.*, *Drug Alcohol Depend.* **14**, 339-362 (1985); of analgesic efficacy and clinical experience: M. W. Gunion *et al.*, *Acute Pain* **6**, 29-39 (2004).

Cryst from methanol-acetone, mp 230.5°.

Hydrochloride. [23277-43-2] EN-2234A; Nubain. $C_{21}H_{27}NO_4.HCl$; mol wt 393.91. pKa_1 8.71; pKa_2 9.96. Soly in water at 25°: 35.5 mg/ml. Sol in ethanol (0.8%). Insol in chloroform, ether.

THERAP CAT: Analgesic.

6442. Naled. [300-76-5] Phosphoric acid 1,2-dibromo-2,2-dichloroethyl dimethyl ester; 1,2-dibromo-2,2-dichloroethyl dimethyl phosphate; bromchlophos; ENT-24988; OMS-75; RE-4355; Bromex; Dibrom; Lucanal. $C_4H_7Br_2Cl_2O_4P$; mol wt 380.78. C 12.62%, H 1.85%, Br 41.97%, Cl 18.62%, O 16.81%, P 8.13%. Organophosphate insecticide; cholinesterase inhibitor. Prepn: **GB 855157**; Ospenson, Kohn, **US 2971882** (1960, 1961 both to California Res. Corp.). Toxicity study: T. B. Gaines, *Toxicol. Appl. Pharmacol.* **14**, 515 (1969). GC determn of environmental residue: H. Zhong, M. Latham, *J. Am. Mosq. Control Assoc.* **17**, 225 (2001).

Usually obtained as a liquid, slightly pungent odor, $bp_{0.5}$ 110°. *Poisonous. Corrosive.* d_4^{25} 1.96. Has been crystallized, mp 26.5-27.5°. Vapor pressure at 20° about 2×10^{-3} mm Hg. Practically insol in water. Completely hydrolyzed by water within 48 hrs. Freely sol in aromatic and chlorinated hydrocarbons, ketones, alcohols. Sparingly sol in petroleum solvents and mineral oils. LD_{50} in male rats (mg/kg): 250 orally; 800 dermally (Gaines).

Caution: Potential symptoms of overexposure are miosis, lacrimation; headache; chest tightness, wheezing and laryngeal spasm; salivation; cyanosis; anorexia, nausea, vomiting, abdominal cramps and diarrhea; weakness, twitching and paralysis; giddiness, ataxia and convulsions; low blood pressure; cardiac irregularities; irritation of skin and eyes. *See NIOSH Pocket Guide to Chemical Hazards* (DHHS/NIOSH 97-140, 1997) p 114.

USE: Insecticide; acaricide.

6443. Nalfurafine. [152657-84-6] (2E)-N-[(5α,6β)-17-(Cyclopropylmethyl)-4,5-epoxy-3,14-dihydroxymorphinan-6-yl]-3-(3-furanyl)-N-methyl-2-propenamide; (E)-N-[17-(cyclopropylmethyl)-4,5α-epoxy-3,14-dihydroxymorphinan-6β-yl]-3-(furan-3-yl)-N-methylprop-2-enamide; (−)-17-(cyclopropylmethyl)-3,14β-dihydroxy-4,5α-epoxy-6β-[N-methyl-*trans*-3-(3-furyl)acrylamido]morphinan. $C_{28}H_{32}N_2O_5$; mol wt 476.57. C 70.57%, H 6.77%, N 5.88%, O 16.79%. Selective κ-Opioid receptor agonist. Prepn: H. Nagase *et al.*, **WO 9315081**; *eidem*, **US 6277859** (1993, 2001 both to Toray); and receptor agonist activity: *eidem*, *Chem. Pharm. Bull.* **46**, 366 (1988). Synthesis and structure-activity study: K. Kawai *et al.*, *Bioorg. Med. Chem.* **16**, 9188 (2008). Receptor binding study: T. Seki *et al.*, *Eur. J. Pharmacol.* **376**, 159 (1999). Mechanism of antipruritic activity: Y. Togashi *et al.*, *ibid.* **435**, 259 (2002). Clinical pharmacokinetics: H. Kumagai *et al.*, in *Itch: Basic Mechanisms and Therapy* (Marcel Dekker, New York, 2004) pp 279-286. Clinical study in uremic pruritis: H. Kumagai *et al.*, *Nephrol. Dial. Transplant.* **25**, 1251 (2010).

Crystals from diethyl ether, mp 243.0-254.0° (dec).

Hydrochloride. [152658-17-8] AC-820; TRK-820; Remitch. $C_{28}H_{32}N_2O_5.HCl$; mol wt 616.50. mp 207-217°.

THERAP CAT: Antipruritic; in treatment of hemodialysis related uremic pruritis.

6444. Nalidixic Acid. [389-08-2] 1-Ethyl-1,4-dihydro-7-methyl-4-oxo-1,8-naphthyridine-3-carboxylic acid; 3-carboxy-1-ethyl-7-methyl-1,8-naphthyridin-4-one; 1-ethyl-7-methyl-1,8-naphthyridin-4-one-3-carboxylic acid; Win-18320; Betaxina; Eucistin; Innoxalon; Nalidicron; Nalitucsan; Narigix; NegGram; Negram;

Nevigramon; Nicelate; Nogram; Poleon; Specifin; Uriben; Uriclar; Uralgin; Urodixin; Uroman; Uroneg; Uropan; Wintomylon. $C_{12}H_{12}N_2O_3$; mol wt 232.24. C 62.06%, H 5.21%, N 12.06%, O 20.67%. Prepn: G. Y. Lesher et al., J. Med. Pharm. Chem. **5**, 1063 (1962); G. Y. Lesher, M. D. Gruett, **BE 612258**; eidem, **US 3590036** (1962, 1971 both to Sterling Drug). Mechanism of action studies: G. J. Bourguignon et al., Antimicrob. Agents Chemother. **4**, 479 (1973); W. A. Goss, T. M. Cook, Antibiotics vol. 3, J. W. Corcoran, F. E. Hahn, Eds. (Springer-Verlag, New York, 1975) pp 174-196; A. M. Pedrini, ibid. vol. **5** (pt. 1), F. E. Hahn, Ed. (1979) pp 154-175; H. T. Wright et al., Science **213**, 455 (1981). Comprehensive description: P. E. Grubb, Anal. Profiles Drug Subs. **8**, 371-397 (1979).

Pale buff, crystalline powder, mp 229-230°. Soly at 23° (mg/ml): chloroform 35; toluene 1.6; methanol 1.3; ethanol 0.9; water 0.1; ether 0.1. Sol in methylene chloride, in solns of fixed alkali hydroxides and carbonates; slightly sol in acetone. LD_{50} in mice (mg/kg): 3300 orally; 500 s.c.; 176 i.v. (Lesher, 1962).

THERAP CAT: Antibacterial.

THERAP CAT (VET): Antibacterial.

6445. Nalmefene. [55096-26-9] (5α)-17-(Cyclopropylmethyl)-4,5-epoxy-6-methylenemorphinan-3,14-diol; 6-desoxy-6-methylenenaltrexone; nalmetrene; ORF-11676. $C_{21}H_{25}NO_3$; mol wt 339.44. C 74.31%, H 7.42%, N 4.13%, O 14.14%. Structural analog of naltrexone, q.v., with opiate antagonist activity. Prepn: J. Fishman, **US 3814768** (1974 to M. J. Lewenstein); E. F. Hahn et al., J. Med. Chem. **18**, 259 (1975). Improved synthesis: P. C. Meltzer, J. W. Coe, **EP 140367**; eidem, **US 4535157** (both 1985 to Key Pharm.). Comparative evaluation of efficacy in mice: R. D. Heilman et al., Res. Commun. Chem. Pathol. Pharmacol. **13**, 635 (1976). HPLC determn in human plasma: J. Hsiao, R. Dixon, ibid. **42**, 449 (1983). Receptor binding activity in rat brain: M. E. Michel et al., Methods Find. Exp. Clin. Pharmacol. **7**, 175 (1985). Clinical evaluation of narcotic antagonism: T. J. Gal, C. A. DiFazio, Anesthesiology **64**, 175 (1986). Pharmacokinetics in humans: R. Dixon et al., Clin. Pharmacol. Ther. **39**, 49 (1986); eidem, J. Clin. Pharmacol. **27**, 233 (1987). Use as antagonist to opioid-induced immobilization in large animals: T. J. Kreeger et al., J. Wildl. Dis. **23**, 619 (1987). Comprehensive description: H. G. Brittain, Anal. Profiles Drug Subs. Excip. **24**, 351-395 (1996).

Crystals from ethyl acetate, mp 188-190°.

Hydrochloride. [58895-64-0] Revex. $C_{21}H_{25}NO_3 \cdot HCl$; mol wt 375.89. White to off-white crystalline powder, mp 180-185°. Soly (mg/ml): water (pH 2.25, 5.71, 6.15, 6.25, 7.85, 8.5, 9.15, 10.4): 128; 131; 133; 124; 1.09; 0.18; 0.09; 0.23; methanol 319; ethanol 86.2; acetonitrile 1.07; acetone 0.23; chloroform 0.13. log P (octanol/water): −1.125. pKa 7.63. uv max: 211, 230 nm (a_m 1977, 778 litre-mol/cm).

THERAP CAT: Narcotic antagonist.

THERAP CAT (VET): Narcotic antagonist.

6446. Nalorphine. [62-67-9] (5α,6α)-7,8-Didehydro-4,5-epoxy-17-(2-propen-1-yl)morphinan-3,6-diol; N-allylnormorphine; al-

lorphine; antorphine; NANM. $C_{19}H_{21}NO_3$; mol wt 311.38. C 73.29%, H 6.80%, N 4.50%, O 15.41%. Prepd from normorphine: McCawley et al., J. Am. Chem. Soc. **63**, 314 (1941); amended procedure: Hart, McCawley, J. Pharmacol. **82**, 339 (1944); Weijlard, Erickson, J. Am. Chem. Soc. **64**, 869 (1942); **US 2364833** (1944); Weijlard, **US 2891954** (1959 to Merck & Co.). Toxicity data: E. I. Goldenthal, Toxicol. Appl. Pharmacol. **18**, 185 (1971). Comprehensive description: M. U. Zubair et al., Anal. Profiles Drug Subs. **18**, 195-219 (1989).

Crystals from ether, mp 208-209° (not 92-93° as first given by McCawley). $[\alpha]_D^{25}$ −155.3° (c = 3 in methanol). Sparingly sol in water, ether. Sol in alc, acetone, chloroform, dil alkalies. uv max (in acid): 285 nm; (in alkali): 298 nm.

Hydrobromide. [1041-90-3] Lethidrone; Norfin. $C_{19}H_{21}NO_3 \cdot HBr$; mol wt 392.29. Crystals from alc, dec 258-259°. Sol in water.

Hydrochloride. [57-29-4] Nalline. $C_{19}H_{21}NO_3 \cdot HCl$; mol wt 347.84. Crystals from alc, mp 260-263°. Sol in water. Moderately sol in alc. uv max (water): 285 nm. IR and polarographic data: Seagers et al., J. Am. Pharm. Assoc. Sci. Ed. **41**, 640 (1952). pH of 0.5% aq soln: 5.0. LD_{50} s.c. in rats: 1460 mg/kg (Goldenthal).

Note: This is a controlled substance: 21 CFR, 1308.13.

THERAP CAT: Narcotic antagonist.

THERAP CAT (VET): Narcotic antagonist.

6447. Naloxone. [465-65-6] (5α)-4,5-Epoxy-3,14-dihydroxy-17-(2-propen-1-yl)morphinan-6-one; 17-allyl-4,5α-epoxy-3,14-dihydroxymorphinan-6-one; (−)-N-allyl-14-hydroxynordihydromorphinone; N-allylnoroxymorphone. $C_{19}H_{21}NO_4$; mol wt 327.38. C 69.71%, H 6.47%, N 4.28%, O 19.55%. Specific opioid antagonist. Prepn: **GB 939287** (1963 to Sankyo); M. J. Lewenstein, J. Fishman, **US 3254088** (1966); R. A. Olofson et al., Tetrahedron Lett. **1977**, 1567. Clinical pharmacology and abuse potential: D. R. Jasinski et al., J. Pharmacol. Exp. Ther. **157**, 420 (1967). Metabolism: J. M. Fujimoto, Proc. Soc. Exp. Biol. Med. **133**, 317 (1970). Synthesis of (+)-form and stereospecific activity: I. Iijima et al., J. Med. Chem. **21**, 398 (1978). Clinical trial in Alzheimer's disease: B. Reisberg et al., N. Engl. J. Med. **308**, 721 (1983). Comprehensive description: M. M. A. Hasson et al., Anal. Profiles Drug Subs. **14**, 453-489 (1985). Review of pharmacology and therapeutic uses: L. F. McNicholas, W. R. Martin, Drugs **27**, 81-93 (1984); of clinical experience in Alzheimer's disease: C. Stowe, M. L. Gora, Ann. Pharmacother. **27**, 447-448 (1993); in drug overdose: J. M. Chamberlain, B. L. Klein, Am. J. Emerg. Med. **12**, 650-660 (1994).

Crystals from ethyl acetate, mp 184° (Lewenstein), 177-178° (Sankyo Co.). $[\alpha]_D^{20}$ −194.5° (c = 0.93 in CHCl₃). Sol in chloroform. Practically insol in petr ether.

Hydrochloride. [357-08-4]; [51481-60-8] (dihydrate). EN-15304; Nalone; Narcan; Narcanti. $C_{19}H_{21}NO_4 \cdot HCl$; mol wt 363.84. Crystals from ethanol + ether, mp 200-205°. Sol in water, dilute acids, strong alkali; slightly sol in alc. Practically insol in ether, chloroform.

THERAP CAT: Narcotic antagonist.

THERAP CAT (VET): Narcotic antagonist.

6448.   **Naltrexone.** [16590-41-3] (5α)-17-(Cyclopropylmethyl)-4,5-epoxy-3,14-dihydroxymorphinan-6-one; *N*-cyclopropylmethyl-14-hydroxydihydromorphinone; UM-792. $C_{20}H_{23}NO_4$; mol wt 341.41. C 70.36%, H 6.79%, N 4.10%, O 18.74%. Nonselective opioid receptor antagonist; congener of naloxone, *q.v.* Prepn: H. Blumberg *et al.*, US 3332950 (1967 to Endo). Metabolism study: E. J. Cone, *Tetrahedron Lett.* **1973**, 2607. Pharmacology and toxicity: R. P. Maickel *et al.*, *Ann. N.Y. Acad. Sci.* **281**, 321 (1976). Clinical pharmacology: E. R. Gritz *et al.*, *Clin. Pharmacol. Ther.* **19**, 773 (1976); K. Verebey *et al.*, *ibid.* **20**, 315 (1976). GC determn in urine: H. E. Dayton, C. E. Inturrisi, *Drug Metab. Dispos.* **4**, 474 (1976). LC-MS/MS determn in plasma: H. Yun *et al.*, *Talanta* **71**, 1553 (2007). Review of pharmacokinetics and therapeutic use in opioid dependence: J. P. Gonzales, R. N. Brogden, *Drugs* **35**, 192-213 (1988). Clinical trials in alcoholism: B. J. Berg *et al.*, *Drug Saf.* **15**, 274 (1996); R. S. Croop *et al.*, *Arch. Gen. Psychiatry* **54**, 1130 (1997). Review of clinical experience in alcoholism: H. M. Pettinati *et al.*, *J. Clin. Psychopharmacol.* **26**, 610-625 (2006); of depot preparations: B. A. Johnson, *Expert Opin. Pharmacother.* **7**, 1065-1073 (2006).

Crystals from acetone, mp 168-170°. Sol in ethanol. Insol in water. LD_{50} in mice (mg/kg): 586 s.c. (Maickel).

Hydrochloride. [16676-29-2] EN-1639A; Antaxone; Nalorex; ReVia; Vivitrol. $C_{20}H_{23}NO_4 \cdot HCl$; mol wt 377.87. Crystals from methanol, mp 274-276°.

THERAP CAT: Narcotic antagonist. In treatment of alcoholism.

THERAP CAT (VET): Adjunct in treatment of behavioral disorders.

6449.   **Nandinine.** [572-76-9] (13a*S*)-5,8,13,13a-Tetrahydro-10-methoxy-6*H*-benzo[*g*]-1,3-benzodioxolo[5,6-*a*]quinolizin-9-ol; 10-methoxy-2,3-(methylenedioxy)berbin-9-ol; 5,6,13,13a-tetrahydro-9-hydroxy-10-methoxy-2,3-(methylenedioxy)-8*H*-dibenzo[*a,g*]quinolizine; (+)-tetrahydroberberrubine. $C_{19}H_{19}NO_4$; mol wt 325.36. C 70.14%, H 5.89%, N 4.31%, O 19.67%. From root bark of *Nandina domestica* Thunb., *Berberidaceae:* Eijkman, *Ber.* **17**, 441 (1884); Kitasato, *J. Pharm. Soc. Jpn.* **522**, 1 (1925). Structure: *idem, Acta Phytochim.* **3**, 175 (1927), *C.A.* **22**, 1779 (1928); Späth, Leithe, *Ber.* **63**, 3007 (1930). Configuration: Corrodi, Hardegger, *Helv. Chim. Acta* **39**, 889 (1956). Synthesis: Kametani *et al.*, *J. Chem. Soc. C* **1969**, 2036; *eidem, J. Chem. Soc. Perkin Trans. 1* **1977**, 1151.

Crystals from chloroform + ether, mp 195-196°.

dl-**Form.** [17388-74-9] Colorless needles from ethanol, mp 183-185°.

l-**Form.** Crystals from ether. mp 195-196°. $[\alpha]_D^{15}$ −304° (c = 0.513 in chloroform).

6450.   **Nandrolone.** [434-22-0] (17β)-17-Hydroxyestr-4-en-3-one; 17β-hydroxy-4-estren-3-one; 4-estren-17β-ol-3-one; 17β-hydroxy-19-nor-4-androsten-3-one; 19-nortestosterone. $C_{18}H_{26}O_2$; mol wt 274.40. C 78.79%, H 9.55%, O 11.66%. Prepd from alkyl ethers of estradiol: Birch, *Q. Rev. Chem. Soc.* **4**, 69 (1950); Wilds,

Nelson, *J. Am. Chem. Soc.* **75**, 5366 (1953). Alternate syntheses: H. Ueberwasser *et al.*, *Helv. Chim. Acta* **46**, 344 (1963); I. Shimizu *et al.*, *Tetrahedron Lett.* **21**, 487 (1980).

Dimorphic crystals, mp 112° and 124°. $[\alpha]_D^{22}$ +55° (c = 0.93 in chloroform). uv max (ethanol): 241 nm (ε 17000). Sol in alc, ether, chloroform.

Benzoate. $C_{25}H_{30}O_3$. Needles from 75% alc, mp 174-175°. $[\alpha]_D^{20}$ +104.5° (alc).

Cyclohexanecarboxylate. [18470-94-5] 19-Nortestosterone hexahydrobenzoate. $C_{25}H_{36}O_3$; mol wt 384.56. Small, elongated prisms from petr ether, mp 88-89°, $[\alpha]_D^{20}$ +50° (c = 0.5 in chloroform).

Decanoate. [360-70-3] 19-Nortestosterone decanoate; Deca-Durabolin; Retabolil; Retabolil. $C_{28}H_{44}O_3$; mol wt 428.66. White to yellow crystals, mp 32-35°. Freely sol in ether; sol in ethanol, acetone, chloroform, vegetable oils. Practically insol in water.

Dodecanoate. [26490-31-3] Nandrolone laurate; Laurabolin. $C_{30}H_{48}O_3$; mol wt 456.71.

p-**Hexyloxyphenylpropionate.** [52279-57-9] 19-Nortestosterone-3-(*p*-hexyloxyphenyl)propionate. $C_{33}H_{46}O_4$; mol wt 506.73. Crystals, mp 53-55°. $[\alpha]_D$ +45° (c = 1.0 in dioxane).

Phenpropionate. [62-90-8] 19-Nortestosterone β-phenylpropionate; Durabolin; Superanabolon; Nandrolin. $C_{27}H_{34}O_3$; mol wt 406.57. Crystals, mp 95-96°. $[\alpha]_D$ +58° (chloroform).

Propionate. [7207-92-3] 19-Nortestosterone propionate. $C_{21}H_{30}O_3$; mol wt 330.47. Crystals from aqueous methanol or isopropyl ether, mp 55-60°. $[\alpha]_D^{23.5}$ +58° in chloroform (Rao). uv max: 240 nm (ε 16650).

Note: This is a controlled substance (anabolic steroid): **21 CFR,** 1308.13, as defined in 1300.01.

THERAP CAT: Anabolic.

THERAP CAT (VET): Phenpropionate as anabolic steroid.

6451.   **Napalm.** [8031-21-8] Gelled gasoline. A coprecipitated aluminum soap from naphthenic acids and the fatty acids of coconut oil developed early in 1942 (Fieser, Harris, Hershberg, Morgana, Novello, Putnam) for prepn of gasoline gels for incendiary munitions: US 2606107 (1952); Herron, US 2684339 (1954 to Safety Fuel & Chem. Corp.). The name was derived from the *na*phthenic and *palm*itic acids which are its major constituents. Structure and mfg problems: *Chem. Eng. News* **32**, 2690 (1954). Historical account: L. F. Fieser, *The Scientific Method* (Reinhold, New York, 1964); Bruce, "Chemical Warfare: Flame" in *Kirk-Othmer Encyclopedia of Chemical Technology* (Interscience, New York, 1964) p 888.

USE: Gasoline thickener. In chemical warfare (fire bombs, flame throwers, flame land mines).

6452.   **Napelline.** [5008-52-6] (3*R*,6*S*,6a*R*,6b*R*,8*S*9*R*,11*R*,-11a*R*,12*R*,12a*R*,14*R*)-1-Ethyltetradecahydro-3-methyl-10-methylene-12,3,6a-ethanylylidene-9,11a-methanoazulene[2,1-*b*]azocine-6,8,11-triol; (1α,12α,15β)-21-ethyl-4-methyl-16-methylene-7,20-cycloveatchane-1,12-15-triol; luciculine. $C_{22}H_{33}NO_3$; mol wt 359.51. C 73.50%, H 9.25%, N 3.90%, O 13.35%. Isoln from *Aconitum napellus* L., *Ranunculaceae:* Freudenberg, Rogers, *J. Am. Chem. Soc.* **59**, 2572 (1937); Jacobs, Craig, *J. Biol. Chem.* **143**, 611 (1942). The product obtained by these isoln procedures is a mixture of at least three compounds. Structure: Wiesner, Valenta, *Fortschr. Chem. Org. Naturst.* **16**, 53 (1958). Related stereochemical studies: Okamoto *et al.*, *Chem. Pharm. Bull.* **13**, 1270 (1963). Total synthesis of racemic napelline: Wiesner *et al.*, *Can. J. Chem.* **52**, 2353, 2355 (1974); S. P. Sethi *et al.*, *ibid.* **58**, 1889 (1980).

Rectangular plates from ether + petr ether.
Hydrochloride. $C_{22}H_{34}ClNO_3$. Solvated crystals, dec 200-222°. $[\alpha]_D^{22}$ −93.9° (c = 5). Sol in water.

6453. Naphazoline. [835-31-4] 4,5-Dihydro-2-(1-naphthalenylmethyl)-1H-imidazole; 2-(1-naphthylmethyl)imidazoline. C_{14}-$H_{14}N_2$; mol wt 210.28. C 79.97%, H 6.71%, N 13.32%. α-Adrenergic agonist. Prepd by reacting the acetic acid anhydride of naphthoimidic acid with ethylenediamine: A. Sonn, **US 2161938** (1939 to Soc. Chem. Ind. Basle); by reacting naphthylthioacetamide with ethylenediamine: **DK 62889** (1944 to Lovens Kemiske), *C.A.* **40**, 4398 (1946). Toxicity data: J. Gylfe *et al.*, *Fed. Proc.* **9**, 280 (1950). Comprehensive description: G. M. Wall, *Anal. Profiles Drug Subs. Excip.* **21**, 307-344 (1992).

Hydrochloride. [550-99-2] Ak-Con; Albalon; Clera; Coldan; Iridina Due; Naphcon; Niazol; Opcon; Privine; Rhinantin; Rhinoperd; Sanorin; Sanorin-Spofa; Strictylon. $C_{14}H_{14}N_2 \cdot HCl$; mol wt 246.74. Bitter crystals, mp 255-260°. uv max (ethanol): 223, 270, 280, 287, 291 nm ($E_{1cm}^{1\%}$ 3622, 239, 286, 196, 198). pKa (25°C) 10.35 ±0.02, (35°C) 10.13 ±0.02, (45°C) 9.92 ±0.03. pKa (25°C) 10.35 ±0.02, (35°C) 10.13 ±0.02, (45°C) 9.92 ±0.03. Freely sol in water (40 g dissolve in 100 ml), alc. Very slightly sol in chloroform. Practically insol in ether; insol in benzene. A 1% aq soln has a pH of ~6.2. LD_{50} s.c. in rats: 385 mg/kg (Gylfe).
THERAP CAT: Adrenergic (vasoconstrictor); decongestant.

6454. Naphthacene. [92-24-0] Tetracene; 2,3-benzanthracene; rubene; chrysogen. $C_{18}H_{12}$; mol wt 228.29. C 94.70%, H 5.30%. Occurs in coal tar. Contaminates commercial anthracene to which it imparts a yellow color. Isoln: Cook *et al.*, *Proc. R. Soc. London* **B111**, 455 (1932). Isoln from crude anthracene by chromatography: Winterstein *et al.*, *Z. Physiol. Chem.* **230**, 159 (1934). Synthesis by condensing succinic acid and phthalic anhydride in the presence of sodium acetate: Gabriel, Michael, *Ber.* **10**, 1559, 2207 (1877); **11**, 1682 (1878); Roser, *Ber.* **17**, 2744 (1884); Nathanson, *Ber.* **26**, 2582 (1893); Gabriel, Leupold, *Ber.* **31**, 1159, 1272 (1898); Wanag, *Ber.* **70**, 274 (1937); from 1-naphthol and phthalic anhydride: Deichler, Weizmann, *Ber.* **36**, 547, 719 (1903); **DE 298345** (1916); from 1,5-dihydroxynaphthalene and phthalic anhydride: Bentley *et al.*, *J. Chem. Soc.* **91**, 411, 1588 (1907); from tetralin and phthalic anhydride: Schroeter, *Ber.* **54**, 2242 (1921); **DE 346673** (1918); *cf.* Fieser, *J. Am. Chem. Soc.* **53**, 2329 (1931). Other syntheses: Coulson, *J. Chem. Soc.* **1935**, 77; Weizmann *et al.*, *ibid.* **1939**, 398. Absorption spectrum: Clar, *Ber.* **69**, 607 (1936). Fluorescence maxima: Krishman, Seshan, *Z. Kristallogr.* **89**, 538 (1934).

Orange leaflets from xylene. d 1.35. Sublimes *in vacuo*. mp 341° (open capillary tube), mp 357° (copper block). Difficultly sol in most solvents. Solns show slight green fluorescence in daylight. Does not form a picrate.

6455. Naphthalene. [91-20-3] Naphthalin; naphthene; tar camphor. $C_{10}H_8$; mol wt 128.17. C 93.71%, H 6.29%. Major constituent of coal tar. Dry coal tar contains about 11%. Crystallizes from the middle or "carbolic oil" fraction of the distilled tar. Purified by hot pressing, which may be followed by washing with H_2SO_4, NaOH, and water, then by fractional distillation or by sublimation. Manuf: *Faith, Keyes & Clark's Industrial Chemicals*, F. A. Lowenheim, M. K. Moran, Eds. (Wiley-Interscience, New York, 4th ed., 1975) pp 556-562. *Review:* R. M. Gaydos in *Kirk-Othmer Encyclopedia of Chemical Technology* vol. **15** (Wiley-Interscience, New York, 3rd ed., 1981) pp 698-719. Review of toxicology and human exposure: *Toxicological Profile for Naphthalene, 1-Methylnaphthalene, and 2-Methylnaphthalene* (PB2006-100004, 2005) 347 pp.

Monoclinic prismatic plates from ether or by sublimation; also sold as white scales, powder, balls, or cakes, mp 80.2°. Odor of moth balls. Volatilizes appreciably at room temp. d_4^{20} 1.162. d_4^{100} 0.9628. Sublimes appreciably at temps above mp; volatile with steam. bp_{760} 217.9°; bp_{400} 193.2°; bp_{200} 167.7°; bp_{100} 145.5°; bp_{60} 130.2°; bp_{40} 119.3°; bp_{20} 101.7°; bp_{10} 85.8°. Flash pt, open cup 174°F (79°C); closed cup 190°F (88°C). Autoignition temp 1053°F (567°C). n_D^{100} 1.58212. Purple fluorescence in Hg light (petr ether soln). Ultraviolet absorption: Several characteristic bands between 217.5 and 320 nm in hexane. One gram dissolves in 13 ml methanol or ethanol, in 3.5 ml benzene or toluene, in 8 ml olive oil or turpentine, in 2 ml chloroform or carbon tetrachloride, in 1.2 ml carbon disulfide. Very sol in ether, hydronaphthalenes, in fixed and volatile oils. Insol in water.
Caution: Potential symptoms of overexposure are eye irritation; corneal damage, optical neuritis; dermatitis; headache, confusion, excitement, malaise; nausea, vomiting, abdominal pain; bladder irritation; profuse sweating; acute intravascular hemolysis, anemia, leukocytosis, fever; jaundice; hematuria, hemoglobinuria, renal shutdown. *See NIOSH Pocket Guide to Chemical Hazards* (DHHS/NIOSH 97-140, 1997) p 220; *Clinical Toxicology of Commercial Products*, R. E. Gosselin *et al.*, Eds. (Williams & Wilkins, Baltimore, 4th ed., 1984) Section III, pp 307-311. This substance is reasonably anticipated to be a human carcinogen: *Report on Carcinogens, Twelfth Edition* (PB2011-111646, 2011) p 276.
USE: Manuf phthalic and anthranilic acids which are used in making indigo, indanthrene, and triphenylmethane dyes. Manuf of hydroxyl (naphthols), amino (naphthylamines), sulfonic acid and similar compds used in the dye industries. Manuf of synthetic resins, celluloid, lampblack, smokeless powder. Manuf of hydronaphthalenes (Tetralin, Decalin) which are used as solvents, in lubricants, and in motor fuels. Moth repellent and insecticide.
THERAP CAT: Has been used as antiseptic (topical and intestinal); anthelmintic (Cestodes).
THERAP CAT (VET): Has been used in dusting powders, as an insecticide and internally as an intestinal antiseptic and vermicide.

6456. 1-Naphthaleneacetic Acid. [86-87-3] α-Naphthaleneacetic acid; naphthylacetic acid; NAA; Fruitone-N; Planofix; Tre-Hold. $C_{12}H_{10}O_2$; mol wt 186.21. C 77.40%, H 5.41%, O 17.18%. Prepn from naphthalene + chloroacetic acid: Ogata, Ishiguro, *J. Am. Chem. Soc.* **72**, 4302 (1950); Southwick *et al.*, *ibid.* **76**, 754 (1954); **US 2655531** (1953 to FMC); from naphthylacetonitrile: Wenner, **US 2489348** (1949 to Hoffmann-La Roche); *J. Org. Chem.* **15**, 548 (1950). Activity: F. E. Gardiner *et al.*, *Science* **90**, 208 (1939). Crystal structure: S. S. Rajan, *Acta Crystallogr.* **B34**, 998 (1978). Toxicity study: G. W. Bailey, J. L. White, *Residue Rev.* **10**, 97 (1965).

Needles from water, mp 134.5-135.5°. Sol in about 30 parts alcohol; freely sol in acetone, ether, chloroform. Soly in water at 17°: 0.38 g/l. LD$_{50}$ orally in rats: 1000 mg/kg (Bailey, White).
USE: Plant growth regulator.

6457. 1,8-Naphthalenediamine. [479-27-6] 1,8-Diaminonaphthalene. C$_{10}$H$_{10}$N$_2$; mol wt 158.20. C 75.92%, H 6.37%, N 17.71%. Prepd by reducing 1,8-dinitronaphthalene with phosphorus triiodide: Meyer, Müller, *Ber.* **30**, 775 (1897).

Crystals from dil alc, mp 66.5°. bp$_{12}$ 205°; $n_D^{99.4}$ 1.6828; d$_4^{99.4}$ 1.1265. Sublimable. Turns brown on standing. Soluble in alcohol or ether; slightly sol in water or chloroform.
Dihydrochloride. C$_{10}$H$_{12}$Cl$_2$N$_2$. Leaflets, mp 280°.
USE: Antioxidant for lubricating oils. Detection of selenium and nitrites.

6458. 1,6-Naphthalenedisulfonic Acid. [525-37-1] Ewer-Pick acid; 1,6-disulfonaphthalene. C$_{10}$H$_8$O$_6$S$_2$; mol wt 288.29. C 41.66%, H 2.80%, O 33.30%, S 22.24%. Prepn: Fierz-David, Hasler, *Helv. Chim. Acta* **6**, 1134 (1923).

Crystals. Very sol in water; sol in alcohol; practically insol in ether.

6459. 2,6-Naphthalenedisulfonic Acid. [581-75-9] Ebert-Merz β-acid. C$_{10}$H$_8$O$_6$S$_2$; mol wt 288.29. C 41.66%, H 2.80%, O 33.30%, S 22.24%. Prepn: Fierz-David, Richter, *Helv. Chim. Acta* **28**, 257 (1945).

Deliquesc crystals. Very sol in water, alcohol; practically insol in ether.

6460. 2,7-Naphthalenedisulfonic Acid. [92-41-1] Ebert-Merz α-acid; 2,7-disulfonaphthalene. C$_{10}$H$_8$O$_6$S$_2$; mol wt 288.29. C 41.66%, H 2.80%, O 33.30%, S 22.24%. Prepn from naphthalene and concd H$_2$SO$_4$: Fierz-David, Blangey, *Fundamental Processes of Dye Chemistry* (Interscience, New York, 1949) p 209.

Very deliquesc crystals. Very sol in water, alcohol; practically insol in ether.
USE: In dye chemistry.

6461. 1-Naphthalenesulfonic Acid. [85-47-2] α-Naphthalenesulfonic acid; 1-sulfonaphthalene. C$_{10}$H$_8$O$_3$S; mol wt 208.23. C 57.68%, H 3.87%, O 23.05%, S 15.40%. Made by sulfonating naphthalene with H$_2$SO$_4$ at 0°.

Crystals. mp 90° (dihydrate). Freely sol in water or alcohol, slightly in ether.
USE: Manuf α-naphthol. The sodium salt is used for rendering phenols sol in water.

6462. 2-Naphthalenesulfonic Acid. [120-18-3] β-Naphthalenesulfonic acid; 2-sulfonaphthalene. C$_{10}$H$_8$O$_3$S; mol wt 208.23. C 57.68%, H 3.87%, O 23.05%, S 15.40%. Made by sulfonating naphthalene with H$_2$SO$_4$ at 160°.

Monohydrate. White to slightly brownish, cryst leaflets; very hygroscopic. mp 124-125°. mp 91° when anhydr. Freely sol in water.
USE: Manuf β-naphthol and intermediates.

6463. 1-Naphthalenethiol. [529-36-2] α-Thionaphthol; 1-thionaphthol; 1-mercaptonaphthalene; 1-naphthyl mercaptan. C$_{10}$H$_8$S; mol wt 160.23. C 74.96%, H 5.03%, S 20.01%. Prepd by catalytic sulfurization of naphthalene with S, S$_2$Cl$_2$, etc., followed by hydrogenation: W. A. Lazier *et al.* **US 2402645**; F. K.Signaigo, **US 2402686** (both 1946 to du Pont).

Liquid, strong mercaptan odor. Solidif on cooling. d$_4^0$ 1.1729; d$_4^{20}$ 1.607; d$_4^{23}$ 1.1549. bp$_{760}$ 285°; bp$_{10.3}$ 144.8°; bp$_2$ 138-140°. n_D^{20} 1.6802. Sol in ethanol, ether; sparingly sol in aq alkalies. Volatile with steam.

6464. 2-Naphthalenethiol. [91-60-1] β-Thionaphthol; 2-thionaphthol; 2-mercaptonapthalene; 2-naphthyl mercaptan. C$_{10}$H$_8$S; mol wt 160.23. C 74.96%, H 5.03%, S 20.01%. Prepd by catalytic hydrogenation of a sulfonic acid derivative of naphthalene: Lazier, Signaigo, **US 2402641** (1946 to du Pont); by reduction of naphthalenesulfonyl chloride with zinc: Holt, **US 2216840** (1940 to du Pont).

Crystals from ethanol, disagreeable odor. mp 81°. bp$_{760}$ 286°; bp$_{10.3}$ 146.3°. Very sol in ethanol, ether, petr ether; sparingly sol in water. Slightly volatile with steam.

6465. Naphthalic Acid. [518-05-8] 1,8-Naphthalenedicarboxylic acid. C$_{12}$H$_8$O$_4$; mol wt 216.19. C 66.67%, H 3.73%, O 29.60%. Made by the oxidation of acenaphthene with chromic acid mixture, etc.: Graebe, Gfeller, *Ber.* **25**, 652 (1895); Ogilvie, Wilder, **US 2379032** (1945 to Allied Chem.); Karishin, Fedorenko, *Zh. Prikl. Khim.* **29**, 955 (1956), *C.A.* **50**, 14677b (1956).

Crystals, mp 270°. Practically insol in water; freely sol in warm alcohol, slightly in ether.

6466. 1-Naphthoic Acid. [86-55-5] 1-Naphthalenecarboxylic acid; α-naphthoic acid. C$_{11}$H$_8$O$_2$; mol wt 172.18. C 76.73%, H 4.68%, O 18.58%. Prepn from α-bromonaphthalene by Grignard reaction: Gilman *et al.*, *Org. Synth.* **coll. vol. II**, 425 (1943); from 4-aminonaphthalenesulfonic acid by Sandmeyer reaction: Bassilios,

Bull. Soc. Chim. Fr. **1950**, 757; from β-nitronaphthalene by von Richter reaction: Bunnett, Rauhut, *J. Org. Chem.* **21**, 934, 944 (1956); by oxidation of 1-methylnaphthalene: Aries, **US 2930802** (1960); Barker, **US 2963508** (1960 to Mid-Century).

COOH

Crystals from hot toluene, mp 160.5-162°. bp 300°. uv max (ethanol): 293 nm (log ε_m 3.9). Slightly sol in hot water; freely sol in hot alcohol, ether.

Caution: Moderately irritating to skin, mucous membranes.

6467. 2-Naphthoic Acid. [93-09-4] 2-Naphthalenecarboxylic acid; β-naphthoic acid; isonaphthoic acid. $C_{11}H_8O_2$; mol wt 172.18. C 76.73%, H 4.68%, O 18.58%. Prepn from β-acetylnaphthalene + NaOCl: Newman, Holmes, *Org. Synth.* **coll. vol. II**, 428 (1943); from naphthalene + octylsodium: Morton *et al.*, *J. Am. Chem. Soc.* **64**, 2250 (1942); from 2-aminonaphthalenesulfonic acid by Sandmeyer reaction: Wahl, Bassilios, *Bull. Soc. Chim. Fr.* **1947**, 482; by carboxylation of naphthalene: Prichard, **US 2729673** (1956 to du Pont).

COOH

Crystals from 95% alcohol, mp 184-185°. bp >300°. uv max (ethanol): 235, 280, 335 nm (log ε_m 4.7, 3.8, 3.1). Slightly sol in hot water; sol in alcohol, ether.

Caution: Moderately irritating to skin, mucous membranes.

6468. 1-Naphthol. [90-15-3] 1-Naphthalenol; α-naphthol; alpha-naphthol; α-hydroxynaphthalene. $C_{10}H_8O$; mol wt 144.17. C 83.31%, H 5.59%, O 11.10%. Prepd by fusing the sodium salt of α-naphthalenesulfonic acid with NaOH: Tyrer, **US 2407044; US 2407055; US 2451996** (1946, 1946, 1948); by oxidation of naphthalene: Loeb, **US 3033903** (1962 to Union Carbide). Toxicity data: H. F. Smyth *et al.*, *Am. Ind. Hyg. Assoc. J.* **23**, 95 (1962).

OH

Prisms, mp 96°. Phenolic odor; disagreeable, burning taste. bp 288°; bp$_{40}$ 184°. d$_4^{98.7}$ 1.0954. uv max: 297, 310, 324 nm. Sublimable; volatile with steam. Darkens in light; reduces ammoniacal silver nitrate. Slightly sol in water, freely in alcohol, benzene, chloroform, ether, alkali hydroxide solns. *Protect from light.* LD$_{50}$ orally in rats: 2.59 g/kg (Smyth).

Caution: See 2-Naphthol.

USE: Manuf dyes, intermediates, synthetic perfumes; also in microscopy.

6469. 2-Naphthol. [135-19-3] 2-Naphthalenol; β-naphthol; beta-naphthol; β-hydroxynaphthalene; isonaphthol; C.I. Azoic Coupling Component 1; C.I. Developer 5; C.I. 37500. $C_{10}H_8O$; mol wt 144.17. C 83.31%, H 5.59%, O 11.10%. Prepn from sodium naphthalene-2-sulfonate: Schoeffel, Barton, **US 2760992** (1956 to Sterling); Stevens, Harris, **US 2831895** (1958 to Dow); **FR 1326175** (1963 to Ciba); by oxidation of naphthalene: Simons, **US 2530369** (1950 to Phillips Petroleum); from 2-bromonaphthalene + *tert*-butyl hydroperoxide: Lawesson, Yang, *J. Am. Chem. Soc.* **81**, 4230 (1959).

OH

Crystals, mp 121-123°. bp 285-286°. d 1.22. Flash pt 161°C. Slight phenolic odor. Darkens with age on exposure to light. Sublimes when heated, distillable *in vacuo;* volatile with vapors of alcohol or water; reduces ammoniacal silver nitrate. uv max (95% ethanol): 226, 265, 275, 286, 320, 331 nm (ε 91,194, 3911, 4559, 3301, 1861, 2163). One gram dissolves in 1000 ml water, 80 ml boiling water, 0.8 ml alcohol, 17 ml chloroform, 1.3 ml ether; sol in glycerol, olive oil, solns of alkali hydroxides. *Protect from light. Incompat.:* Antipyrine, camphor, phenol, ferric salts, menthol, potassium permanganate and other oxidizing agents, urethane.

Sodium salt. [875-83-2] Sodium β-naphtholate; sodium naphthol; Microcidin. $C_{10}H_7NaO$; mol wt 166.15. Grayish-white powder; becomes reddish or brownish on exposure to light and air. Soluble in 3 parts water. *Keep well closed and protected from light.*

Caution: Potential symptoms of overexposure to 2-naphthol include crampy abdominal pain, nausea, vomiting and sometimes convulsions. Intestinal or percutaneous absorption may lead to severe nephritis, liver injury and acute hemolytic anemia. Lens opacities and retinal changes may occur. *See Clinical Toxicology of Commercial Products,* R. E. Gosselin *et al.*, Eds. (Williams & Wilkins, Baltimore, 5th ed., 1984) Section II, p 188.

USE: Manuf medicinal organics, dyes, perfumes; the largest single use is probably in making antioxidants for the synthetic rubber industry.

THERAP CAT: Formerly as anthelmintic (Nematodes).

THERAP CAT (VET): Has been used as antiseptic, anthelmintic and counter-irritant in alopecia.

6470. 1-Naphthol-8-amino-3,6-disulfonic Acid. [90-20-0] 4-Amino-5-hydroxy-2,7-naphthalenedisulfonic acid; 8-amino-1-naphthol-3,6-disulfonic acid; H acid. $C_{10}H_9NO_7S_2$; mol wt 319.30. C 37.62%, H 2.84%, N 4.39%, O 35.07%, S 20.08%. Prepn: Willard, *Color Trade J.* **15**, 40 (1924); Mow, **US 2272272** (1942 to Allied Chem.); Hayashi, *Yamaguchi J. Sci.* **2**, 67 (1951), *C.A.* **49**, 2390f (1955); Roos *et al.*, **US 2875243** (1959 to Bayer).

HO$_3$S, SO$_3$H, OH, NH$_2$

White crystals or gray powder. Slightly sol in water, alcohol, ether; sol in alkalies.

USE: Prepn of azo dyes.

6471. 1-Naphthol-4,8-disulfonic Acid. [117-56-6] 4-Hydroxy-1,5-naphthalenedisulfonic acid; α-naphtholdisulfonic acid S; Schöllkopf's acid. $C_{10}H_8O_7S_2$; mol wt 304.29. C 39.47%, H 2.65%, O 36.80%, S 21.07%. Prepd by sulfonation of 1-naphthol-8-sulfonic acid sultone: I. G. Farbenindustrie, **PB 74197**, frames 828-829.

SO$_3$H, SO$_3$H, OH

Crystals. Soluble in water. The sodium salt is very soluble in water.

6472. 2-Naphthol-3,6-disulfonic Acid. [148-75-4] 3-Hydroxy-2,7-naphthalenedisulfonic acid; R acid. $C_{10}H_8O_7S_2$; mol wt 304.29. C 39.47%, H 2.65%, O 36.80%, S 21.07%. Prepn from β-naphthol and concd H_2SO_4: Fierz-David, Blangey, *Fundamental Processes of Dye Chemistry* (Interscience, New York, 1949) p 197.

HO$_3$S, SO$_3$H, OH

Deliquesc needles. Very sol in water, alcohol. Practically insol in ether. The sodium salt is known as *R salt*.

USE: In dye chemistry.

6473.   2-Naphthol-6,8-disulfonic Acid. [118-32-1] 7-Hydroxy-1,3-naphthalenedisulfonic acid; 2-hydroxynaphthalene-6,8-disulfonic acid; G acid. $C_{10}H_8O_7S_2$; mol wt 304.29. C 39.47%, H 2.65%, O 36.80%, S 21.07%. Prepd by heating β-naphthol with concd H_2SO_4, 2-naphthol-3,6-disulfonic acid (R acid) being obtained as a byproduct: Fierz-David, Blangey, *Fundamental Processes of Dye Chemistry* (Interscience, New York, 1949) p 197; Forster, Keyworth, *J. Soc. Chem. Ind. London* **46**, 27T (1927).

Sodium salt. G salt. Platelets or prisms. Freely sol in water, sol in dil alcohol.

Barium salt octahydrate. $C_{10}H_6BaO_7S_2.8H_2O$. Minute prisms. Freely sol in water. Practically insol in alcohol, even when much dil with water.

USE: In the manuf of azo dyes, *see:* Crossley, Resenvelt, *Ind. Eng. Chem.* **16**, 271 (1924).

6474. α-Naphtholphthalein. [596-01-0] 3,3-Bis(4-hydroxynaphthalenyl)-1(3H)-isobenzofuranone; 3,3-bis(4-hydroxy-1-naphthyl)phthalide; di-*p*-α-naphtholphthalide; *p*-α-naphtholphthalein. $C_{28}H_{18}O_4$; mol wt 418.45. C 80.37%, H 4.34%, O 15.29%. Prepn: Werner, *J. Chem. Soc.* **113**, 20 (1918). Commercial prepn contains a small amount of the isomeric 3,3-bis(2-hydroxy-1-naphthyl)phthalide.

Colorless powder when pure; usually grayish-red. mp 253-255°. Practically insol in water. Sol in alcohol.

USE: As indicator in 0.1% or 0.04% soln in alc. pH 7.3 colorless to reddish; 8.7 greenish to blue. Particularly adapted for weak acids in strong alcoholic soln.

6475.   1-Naphthol-2-sulfonic Acid. [567-18-0] 1-Hydroxy-2-naphthalenesulfonic acid; Baum's acid; Schaeffer's α-acid. $C_{10}H_8O_4S$; mol wt 224.23. C 53.57%, H 3.60%, O 28.54%, S 14.30%. Prepd by sulfonation of α-naphthol: Hodgson, Hathaway, *J. Soc. Dyers Colour.* **63**, 109 (1947).

Deliquesc crystals. Slightly sol in cold water; sol in boiling water. Practically insol in ether.

6476.   1-Naphthol-4-sulfonic Acid. [84-87-7] 4-Hydroxy-1-naphthalenesulfonic acid; Nevile and Winther's acid. $C_{10}H_8O_4S$; mol wt 224.23. C 53.57%, H 3.60%, O 28.54%, S 14.30%. Prepn from 1,4-diazonaphthalenesulfonic acid + sulfuric acid: Erdmann, *Ann.* **247**, 341 (1888); by sulfonation of α-naphthylcarbonate: Reverdin, *Ber.* **27**, 3458 (1894); from chlorosulfonic acid + α-naph-

thol: Gebauer-Fülnegg, *Monatsh. Chem.* **49**, 195 (1928); Baddiley *et al.*, US **1452481** (1923); from sodium naphthionate + SO_2: Binns, Lurie, US **1880701** (1933 to Virginia Smelting).

Crystals, dec 170°. K_{OH} 3×10^{-9}. Very sol in water. Salt solns give a blue color with ferric chloride.

USE: Preparation of azo dyes.

6477.   2-Naphthol-6-sulfonic Acid. [93-01-6] 6-Hydroxy-2-naphthalenesulfonic acid; Schaeffer's β-acid. $C_{10}H_8O_4S$; mol wt 224.23. C 53.57%, H 3.60%, O 28.54%, S 14.30%. Prepn from β-naphthol and concd H_2SO_4: Fierz-David, Blangey, *Fundamental Processes of Dye Chemistry* (Interscience, New York, 1949) p 194.

Leaflets. mp 125°. Very sol in water, alcohol. Practically insol in ether.

Sodium salt. Schaeffer's salt. $C_{10}H_7NaO_4S$. Light yellow to pink powder. Freely sol in water; slightly sol in alcohol.

USE: In dye chemistry.

6478.   Naphthol Yellow S. [846-70-8] 8-Hydroxy-5,7-dinitro-2-naphthalenesulfonic acid sodium salt (1:2); Citronin A; Sulfur Yellow S; Acid Yellow S; FD & C Yellow no. 1; Ext. D & C Yellow no. 7; C.I. Acid Yellow 1; C.I. 10316. $C_{10}H_4N_2Na_2O_8S$; mol wt 358.19. C 33.53%, H 1.13%, N 7.82%, Na 12.84%, O 35.73%, S 8.95%. Prepn: *Colour Index* **vol. 4** (3rd ed., 1971) p 4004.

Greenish yellow powder, sol in water to a yellow soln. The free acid is known as *flavianic acid*.

USE: Dye for wool, silk.

6479.   1,2-Naphthoquinone. [524-42-5] 1,2-Naphthalenedione; β-naphthoquinone. $C_{10}H_6O_2$; mol wt 158.16. C 75.94%, H 3.82%, O 20.23%. Prepd by oxidation of 1-amino-2-naphthol with ferric chloride: Fieser, *Org. Synth.* **17**, 68 (1937); by oxidation of naphthalene: Milas, US **2395638** (1946 to Research Corp.); by bacterial dissimilation of naphthalene: Murphy, Stone, *Can. J. Microbiol.* **1**, 579 (1955); from tetrachloro-*o*-benzoquinone: Horner, Dürckheimer, *Z. Naturforsch.* **14b**, 741 (1959). Structure: Hodgson, Hathaway, *Trans. Faraday Soc.* **41**, 115 (1945). Spectroscopic properties: Oliver *et al.*, *Tetrahedron* **24**, 4067 (1968).

Golden yellow needles, dec 145-147°. µv max (abs alc): 250, 340, 405 nm (log ε 4.35, 3.40, 3.40). Sol in alc, benzene, ether, 5% NaOH, 5% $NaHCO_3$, concd H_2SO_4 with green color. Practically insol in water.

2-Semicarbazone. [31853-38-0] Naftazone; Haemostop Injection; Mediaven; Karbinon. $C_{11}H_9N_3O_2$; mol wt 215.21.

USE: As reagent for resorcinol and thalline.

THERAP CAT: 2-Semicarbazone as hemostatic.

6480. 1,4-Naphthoquinone. [130-15-4] 1,4-Naphthalenedione; α-naphthoquinone; 1,4-dihydro-1,4-diketonaphthalene. $C_{10}H_6O_2$; mol wt 158.16. C 75.94%, H 3.82%, O 20.23%. Prepn: Fieser, *Org. Synth.* **coll. vol. I,** 383 (2nd ed., 1941); Braude, Fawcett, *ibid.* **coll. vol. IV,** 698 (1963). Substituted 1,4-naphthoquinones occur in nature, *e.g.,* phthiocol, *q.v.,* various pigments, and the vitamins K.

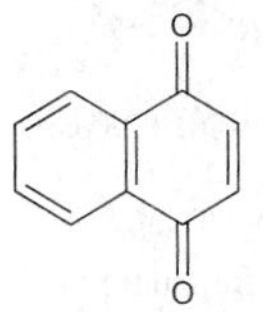

Yellow triclinic needles from alcohol or petr ether. Odor like that of benzoquinone. d 1.422. mp 126°. Begins to sublime below 100°. Sublimation rates: Kempf, *J. Prakt. Chem.* [2] **78,** 236, 257. Easily volatile with steam. uv spectrum: Purvis, *J. Chem. Soc.* **101,** 1318 (1912). Sparingly sol in cold water, slightly in petr ether, freely in hot alcohol, ether, benzene, chloroform, carbon bisulfide, acetic acid; sol in alkali hydroxide solns giving a reddish-brown soln.

6481. Naphthoresorcinol. [132-86-5] 1,3-Naphthalenediol; 1,3-dihydroxynaphthalene. $C_{10}H_8O_2$; mol wt 160.17. C 74.99%, H 5.03%, O 19.98%. Prepn by cyclization of ethyl phenylacetoacetate: Soliman, West, *J. Chem. Soc.* **1944,** 53; by heating 1-amino-3-hydroxy-4-naphthalenesulfonic acid in acidic soln: Meyer, Bloch, *Org. Synth.* **25,** 73 (1945); from 1,3-naphtholsulfonic acid: Kozlov, Odintsov, *J. Appl. Chem. USSR* **17,** 219 (1944); Kozlov *et al., Zh. Prikl. Khim.* **35,** 880 (1962).

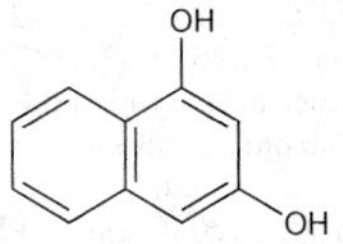

Leaflets, mp 124-125°. Freely sol in water, alcohol, ether.

USE: Reagent for sugars, oils, and for glucuronic acid in urine: Forsyth, *Nature* **161,** 239 (1948); Heyns, Kelch, *Z. Anal. Chem.* **139,** 339 (1953).

6482. 2-Naphthoxyacetic Acid. [120-23-0] 2-(2-Naphthalenyloxy)acetic acid; β-naphthoxyacetic acid; O-(2-naphthyl)glycolic acid. $C_{12}H_{10}O_3$; mol wt 202.21. C 71.28%, H 4.98%, O 23.74%. Prepd by treating β-naphthol with chloroacetic acid in an alkaline medium: Spitzer, *Ber.* **34,** 3192 (1901); Shirley, *Organic Intermediates* (New York, 1951) p 209.

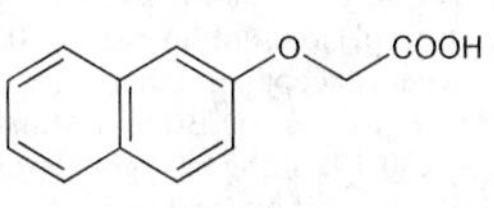

Prisms from hot water or benzene, mp 156°. Moderately sol in hot water. Soluble in alcohol, ether, acetic acid.

Ethyl ester. $C_{14}H_{14}O_3$. Leaflets from alcohol, mp 49°. Sol in alcohol, ether.

USE: As plant hormone, to promote growth of roots on clippings, to prevent fruit from falling prematurely; causes stunted growth when used in excess.

6483. 1-Naphthylamine. [134-32-7] 1-Naphthalenamine; 1-aminonaphthalene; α-naphthylamine; naphthalidine. $C_{10}H_9N$; mol wt 143.19. C 83.88%, H 6.34%, N 9.78%. Prepd by reducing α-nitronaphthalene with Fe and HCl: West, *J. Chem. Soc.* **127,** 494 (1925); from 1-naphthalenecarboxylic acid and hydroxylamine:

Snyder *et al., J. Am. Chem. Soc.* **75,** 2014 (1953). Review of carcinogenic risk: *IARC Monographs* **4,** 87-96 (1974).

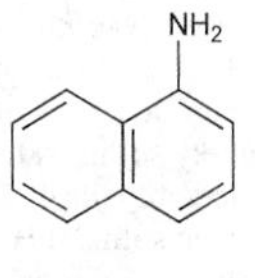

Needles, becoming red on exposure to air, or a reddish, cryst mass; unpleasant odor. Sublimes; volatile with steam. d 1.13. mp 50°. bp 301°. Flash pt 157°C. Sol in 590 parts water; freely sol in alc, ether. Reduces warm ammoniacal silver nitrate. *Keep well closed and protected from light.*

Hydrochloride. $C_{10}H_9N.HCl$. Cryst powder; becomes bluish on exposure to air and light. Sol in about 27 parts water; sol in alcohol, ether. *Protect from light.*

Sulfate dihydrate. $(C_{10}H_9N)_2.H_2SO_4.2H_2O$. White to yellowish cryst powder. Slightly sol in water or alcohol.

Caution: Potential symptoms of overexposure to 1-naphthylamine are dermatitis; hemorrhagic cystitis; dyspnea, ataxia and methemoglobinemia; hematuria; dysuria. Potential occupational carcinogen. *See NIOSH Pocket Guide to Chemical Hazards* (DHHS/NIOSH 97-140, 1997) p 222.

USE: Manufacturing dyes; toning prints made with cerium salts; the hydrochloride with sulfanilic acid is a reagent for nitrate.

6484. 2-Naphthylamine. [91-59-8] 2-Naphthalenamine; 2-aminonaphthalene; β-naphthylamine. $C_{10}H_9N$; mol wt 143.19. C 83.88%, H 6.34%, N 9.78%. Prepd by heating β-naphthol with ammonium sulfite and NH_4OH at 150°: **DE 117471** (1900); *Frdl.* **6,** 190; from 2-naphthalenecarboxylic acid and hydroxylamine: Snyder *et al., J. Am. Chem. Soc.* **75,** 2014 (1953). Review of carcinogenic risk: *IARC Monographs* **4,** 97-111 (1974).

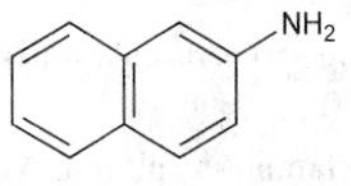

White to reddish crystals; volatile with steam. d_4^{98} 1.061. mp 111-113°. bp 306°; also stated as 294°. Sol in hot water, alc, ether. Reduces warm ammoniacal silver nitrate.

Acetate. $C_{10}H_9N.CH_3COOH$. White to yellowish scales or flakes; slight odor of acetic acid. Very sol in water or alc; sol in ether. *Keep well closed and protected from light.*

Caution: Potential symptoms of overexposure to 2-naphthylamine are dermatitis; hemorrhagic cystitis; dyspnea; ataxia; methemoglobinemia, hematuria; dysuria. *See NIOSH Pocket Guide to Chemical Hazards* (DHHS/NIOSH 97-140, 1997) p 222. 2-Naphthylamine is listed as a known human carcinogen: *Report on Carcinogens, Twelfth Edition* (PB2011-111646, 2011) p 278.

USE: Formerly in manuf of dyes, as antioxidant in rubber.

6485. 1-Naphthylamine-4-sulfonic Acid. [84-86-6] 4-Amino-1-naphthalenesulfonic acid; naphthionic acid; Piria's acid. $C_{10}H_9NO_3S$; mol wt 223.25. C 53.80%, H 4.06%, N 6.27%, O 21.50%, S 14.36%. Prepd on a large-scale by heating α-naphthylamine with equimolar amounts of concd H_2SO_4 at 180-200° or by heating α-naphthylamine acid sulfate: Fierz-David, Blangey, *Fundamental Processes of Dye Chemistry* (Interscience, New York, 1949).

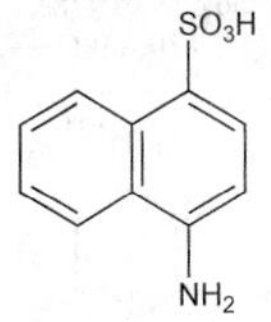

Sesquihydrate. Shiny needles from water. d_4^{25} 1.673. Dec on heating without melting (the amide mp 206°). K at 25° = 2.1×10^{-3}. One gram dissolves in 3.45 l of water at 10°, in 3.22 l at 20°, in 1.69 l at 50°, in 438.5 ml at 100°. Very sparingly sol in alcohol, ether.

Practically insol in acetic acid and acetic anhydride, but becomes sol when some pyridine is added. Sol in dil solns of alkali hydroxides or carbonates with a blue fluorescence.

Sodium salt tetrahydrate. Naphthionine; 101-E. $C_{10}H_8NNa-O_3S.4H_2O$; mol wt 317.29. *Ref:* Piria, *Ann.* **78**, 41 (1851). Large monoclinic prisms from water, loses 3½ mols water at 80°, anhydr at 130°. Sweet aftertaste. Freely sol in water with a blue fluorescence. Also sol in 95% alc. Practically insol in ether. Sparingly sol in concd aq and alcoholic caustic solns. pH of a 1% aq soln 6.8. Even ampuled aq solns discolor under the influence of light.

Sodium salt glucoside. Naphthionine *N*-glucoside; 101-G. $C_{16}-H_{20}NNaO_8S$; mol wt 409.38. *Ref:* Estève *et al.*, *Ann. Pharm. Fr.* **10**, 680 (1952). White powder, bitter taste. Dec 255°. $[\alpha]_D^{23.5} -113°$ (c = 10). Sol in water; slightly sol in alcohol. Practically insol in ether.

USE: The sodium salt is an important dye intermediate in the manuf of Congo Red, Fast Red A, Azo Rubine, and similar azo dyes.

THERAP CAT: Sodium salt as hemostatic.

6486. 1-Naphthylamine-5-sulfonic Acid. [84-89-9] 5-Amino-1-naphthalenesulfonic acid; Laurent's acid. $C_{10}H_9NO_3S$; mol wt 223.25. C 53.80%, H 4.06%, N 6.27%, O 21.50%, S 14.36%. Prepn from 5-chloro-1-naphthalenesulfonic acid + ammonia: Oehler, **DE 72336** (1893); *Frdl.* **3**, 435; from α-naphthylamine + concd sulfuric acid: Erdmann, *Ann.* **275**, 193, 200 (1893); Sixma, *Rec. Trav. Chim.* **73**, 235 (1954); from 2-naphthylamine-8-sulfonic acid: Blangey, *Helv. Chim. Acta* **39**, 977 (1956).

White crystals. Sol in 950 parts cold water; more sol in hot water.

USE: In dye chemistry.

6487. 1-Naphthylamine-8-sulfonic Acid. [82-75-7] 8-Amino-1-naphthalenesulfonic acid; Peri acid. $C_{10}H_9NO_3S$; mol wt 223.25. C 53.80%, H 4.06%, N 6.27%, O 21.50%, S 14.36%. Prepn: Fierz, Weissenbach, *Helv. Chim. Acta* **3**, 310 (1920); Martin, Z. *Farben-Ind.* **20**, 9, 76 (1928); *C.A.* **22**, 2665 (1928); Shebuev *et al.*, **SU 165746** (1964 to State Scientific Res. Inst. of Org. Intermediates & Dyes), *C.A.* **62**, 10392c (1965).

Needles. Sol in 4800 parts cold, 240 parts boiling water; freely sol in glacial acetic acid.

Caution: Irritating to skin.

6488. 2-Naphthylamine-1-sulfonic Acid. [81-16-3] 2-Amino-1-naphthalenesulfonic acid; Tobias acid. $C_{10}H_9NO_3S$; mol wt 223.25. C 53.80%, H 4.06%, N 6.27%, O 21.50%, S 14.36%. Prepd by sulfonation of 2-naphthylamine: Tinker, Hansen, **US 1969189** (1934 to du Pont). Utilization in viral biology study: N. Sakota *et al.*, *J. Ferment. Technol.* **56**, 53 (1978). TLC study: J. Franc, V. Koudelkova, *J. Chromatogr.* **170**, 89 (1979).

Anhydr scales from hot water; hydrated needles from cold water. Slightly sol in cold water; more sol in hot water; very slightly sol in alcohol, ether.

Diethylammonium salt. $C_{10}H_9NO_3S.(C_2H_5)_2NH$. mp 180°.

USE: Dyestuff intermediate.

6489. 2-Naphthylamine-5-sulfonic Acid. [81-05-0] 6-Amino-1-naphthalenesulfonic acid; Dahl's acid. $C_{10}H_9NO_3S$; mol wt 223.25. C 53.80%, H 4.06%, N 6.27%, O 21.50%, S 14.36%. Prepn by reduction of 6-nitro-1-naphthalenesulfonic acid: Kappeler, *Ber.* **45**, 633 (1912); by sulfonation of β-naphthylamine and separation from mixt of sulfonic acids obtained: Green, Vakil, *J. Chem. Soc.* **113**, 35 (1918).

Needles from water. Sol in 3025 parts water at 20°. Almost insol in alcohol.

USE: Dyestuff intermediate.

6490. 2-Naphthyl Benzoate. [93-44-7] 2-Naphthalenol 2-benzoate; betanaphthol benzoate; benzonaphthol; benzoylnaphthol; 2-naphthol benzoate; Lintrin; Haertolan. $C_{17}H_{12}O_2$; mol wt 248.28. C 82.24%, H 4.87%, O 12.89%.

White, crystalline powder; darkens with age. mp 107-110°. Almost insol in water. Freely sol in hot alcohol, chloroform, glycerol, oils; slightly sol in ether.

USE: Hardening agent for paraffin.

THERAP CAT: Antiseptic (intestinal).

6491. *N*-(1-Naphthyl)ethylenediamine. [551-09-7] N^1-1-Naphthalenyl-1,2-ethanediamine; 1-amino-2-(α-naphthylamino)ethane. $C_{12}H_{14}N_2$; mol wt 186.26. C 77.38%, H 7.58%, N 15.04%. Chromophore in colorimetric assay methods; dihydrochloride forms a dye complex with diazonium salts of primary aromatic amines. Prepn from α-naphthylamine and bromoethylphthalimide: A. C. Bratton, E. K. Marshall, Jr., *J. Biol. Chem.* **128**, 537 (1939). Review of pharmaceutical assay development: A. J. Glazko, *Ther. Drug Monit.* **9**, 53-60 (1987).

Straw-yellow, viscous liquid; bp_9 204°; bp_{760} ~320° (dec); n_D^{25} 1.6648; d_4^{25} 1.114. The soly in water is about 0.2 g in 100 ml at 25°; more sol in hot water. pH of satd aq soln: 10.5. Readily sol in common organic solvents except petr ether.

Dihydrochloride. [1465-25-4] Bratton-Marshall reagent. $C_{12}H_{14}N_2.2HCl$; mol wt 259.17. Long hexagonal prisms; mp 188-190°. Readily sol in 95% alcohol, dil hydrochloric acid, hot water; slightly sol in cold water, acetone, abs alc.

USE: Dihydrochloride in colorimetric determn of sulfanilamide in body fluids; also in determn of potassium, nitrites, and sulfates.

6492. 1-Naphthylisocyanate. [86-84-0] 1-Isocyanatonaphthalene; isocyanic acid 1-naphthyl ester. $C_{11}H_7NO$; mol wt 169.18. C 78.09%, H 4.17%, N 8.28%, O 9.46%.

Colorless liquid; almost odorless at ordinary temp; vapors have a pungent odor characteristic of isocyanates (*e.g.*, phenylisocyanate). d 1.181. bp 270°. Sol in chloroform, ether, petr ether, alcohol. *Keep well closed and protected from light.*

6493. 1-Naphthylisothiocyanate. [551-06-4] 1-Isothiocyanatonaphthalene; isothiocyanic acid 1-naphthyl ester; ANIT. C_{11}-H_7NS; mol wt 185.24. C 71.32%, H 3.81%, N 7.56%, S 17.31%. Prepn: Cymerman-Craig *et al.*, *Org. Synth.* **coll. vol. IV**, 700 (1963); Jochims, *Ber.* **101**, 1746 (1968). Has been used with pyrethrum as insecticide. Toxicity studies: Schwartz, Warren, *Public Health Rep.* **54**, 1426 (1939); Ambrose, Miller, *Fed. Proc.* **2**, 74 (1943); Becker, Plaa, *Toxicol. Appl. Pharmacol.* **7**, 804 (1965); Capizzo, Roberts, *ibid.* **17**, 262 (1970).

White needles; practically odorless and tasteless, mp 58°. Insol in water. Freely sol in ether, benzene, hot alcohol, acetone, carbon tetrachloride, olive oil, petr ether. LD_{50} orally in mice: 245 mg/kg (Becker, Plaa).

Caution: Hepatotoxic; may cause dermatitis.

6494. 2-(2-Naphthyloxy)ethanol. [93-20-9] 2-(2-Naphthalenyloxy)ethanol; β-naphthoxyethanol; betanaphthoxyethanol; β-hydroxyethyl 2-naphthyl ether; 2-(β-hydroxyethoxy)naphthalene; ethylene glycol mono-2-naphthyl ether. $C_{12}H_{12}O_2$; mol wt 188.23. C 76.57%, H 6.43%, O 17.00%. Prepd by the condensation of β-naphthol with ethylene chlorohydrin: Rindfusz *et al.*, *J. Am. Chem. Soc.* **42**, 157, 164, 165 (1920); Kirner, Richter, *ibid.* **51**, 3409 (1929); by heating β-naphthol with ethylene oxide and sodium ethoxide in alcohol: Boyd, Marle, *J. Chem. Soc.* **105**, 2117 (1914).

Crystals from benzene + petr ether, mp 76.7°. Insoluble in water. One gram dissolves in 4 g of 95% alcohol, in 2 g acetone; also sol in ether, chloroform. uv max (0.004% in chloroform): 273, 328 nm (E ~1.00, ~0.395).

THERAP CAT (VET): Formerly as anesthetic.

6495. 1-Naphthyl Salicylate. [550-97-0] 2-Hydroxybenzoic acid 1-naphthalenyl ester; α-naphthyl salicylate; α-naphthol salicylate; salicylic acid 1-naphthyl ester; Alphol. $C_{17}H_{12}O_3$; mol wt 264.28. C 77.26%, H 4.58%, O 18.16%.

Crystalline powder. mp 83°. Insol in water. Freely sol in alcohol, ether, oils.

THERAP CAT: Anti-infective; anti-inflammatory.

6496. 2-Naphthyl Salicylate. [613-78-5] 2-Hydroxybenzoic acid 2-naphthalenyl ester; β-naphthol salicylate; salicyclic acid 2-naphthyl ester; Betol; Naphthalol; Naphthosalol; Salinaphthol. C_{17}-$H_{12}O_3$; mol wt 264.28. C 77.26%, H 4.58%, O 18.16%.

White, odorless, tasteless, crystalline powder. mp 95°. Insol in water or glycerol. Sparingly sol in cold alcohol; freely sol in boiling alcohol, in benzene, ether.

THERAP CAT: Antiseptic.

6497. Napropamide. [15299-99-7] *N,N*-Diethyl-2-(1-naphthalenyloxy)propanamide; *N,N*-diethyl-2-(1-naphthyloxy)propionamide; 2-(α-naphthoxy)-*N,N*-diethylpropionamide; R-7465; Devrinol. $C_{17}H_{21}NO_2$; mol wt 271.36. C 75.25%, H 7.80%, N 5.16%, O 11.79%. Prepn: H. Filles *et al.*, US 3480671 (1969 to Stauffer). Characteristics in the soil: C. H. Wu *et al.*, *Weed Sci.* **23**, 54 (1975). Toxicity study: T. Kawada *et al.*, *Niigata Igakkai Zasshi* **87**, 289 (1973), *C.A.* **80**, 56257h (1974).

Light brown solid from *n*-pentane, mp 63-64° (tech, 69.5°). Solubility in water at 20°: 70 ppm. LD_{50} in mice (g/kg): >5 orally, >1 i.p., >1 s.c. (Kawada).

USE: Herbicide.

6498. Naproxcinod. [163133-43-5] (α*S*)-6-Methoxy-α-methyl-2-naphthaleneacetic acid 4-(nitrooxy)butyl ester; 4-(nitrooxy)butyl (2*S*)-2-(6-methoxy-2-naphthyl) propanoate; NO-naproxen; AR-P900758XX; AZD-3582; HCT-3012. $C_{18}H_{21}NO_6$; mol wt 347.37. C 62.24%, H 6.09%, N 4.03%, O 27.63%. Cyclooxygenase-inhibiting nitric oxide donator (CINOD). Consists of naproxen, *q.v.*, and a nitric oxide donating component linked together by a butyl moiety. Prepn: P. Del Soldato *et al.*, WO 9509831; *eidem*, US 5700947 (1995, 1997 both to NicOx). Preclinical pharmacokinetics: U. Fagerholm *et al.*, *J. Pharm. Pharmacol.* **57**, 587 (2005). Clinical pharmacokinetics and determn in plasma: U. Fagerholm, M. A. Björnsson, *ibid.* **57**, 1539 (2005). Clinical comparison with naproxen evaluating gastrointestinal impact in osteoarthritis patients: L. S. Lohmander *et al.*, *Ann. Rheum. Dis.* **64**, 449 (2005). Clinical trial in osteoarthritis: T. J. Schnitzer *et al.*, *Osteoarthr. Cartil.* **18**, 629 (2010). Review of development and clinical experience: P. Geusens, *Expert Opin. Biol. Ther.* **9**, 649-657 (2009).

Lipophilic, viscous oil. Low aq soly.
THERAP CAT: Anti-inflammatory; analgesic.

6499. Naproxen. [22204-53-1] (α*S*)-6-Methoxy-α-methyl-2-naphthaleneacetic acid; *d*-2-(6-methoxy-2-naphthyl)propionic acid; MNPA; RS-3540; Bonyl; Dysmenalgit; Equiproxen; Floginax; Laraflex; Laser; Malexin; Naixan; Napren; Naprius; Naprosyn; Naprosyne; Naprux; Naxen; Nycopren; Pranoxen; Prexan; Proxen; Proxine; Veradol; Xenar. $C_{14}H_{14}O_3$; mol wt 230.26. C 73.03%, H 6.13%, O 20.84%. Nonsteroidal anti-inflammatory. Prepn: J. H. Fried, I. T. Harrison, ZA 6707597; *eidem*, US 3904682; *eidem*, US 4009197 (1968, 1975, 1977 all to Syntex); I. T. Harrison *et al.*, *J. Med. Chem.* **13**, 203 (1970). Pharmacology: Roszkowski *et al.*, *J. Pharmacol. Exp. Ther.* **179**, 114 (1971). Activity may be due to the ability to inhibit prostaglandin biosynthesis. Mode of action studies: Tomlinson *et al.*, *Biochem. Biophys. Res. Commun.* **46**, 552 (1972). Metabolism: Runkel *et al.*, *J. Pharm. Sci.* **61**, 703 (1972). HPLC determn in plasma and serum: P. J. Streete, *J. Chromatogr.* **495**, 179 (1989). Stereoselective synthesis: K. T. Wan, M. E. Davis, *Nature* **370**, 449 (1994). Clinical studies: Katona *et al.*, *Clin. Trials J.* **8**, 3 (1972); Runkel, *Chem. Pharm. Bull.* **20**, 1457 (1972). *Review: Arzneim.-Forsch.* **25**, 278-332 (1975). Review of pharmacology and therapeutic efficacy: R. N. Brogden *et al.*, *Drugs* **18**, 241-277 (1979). Comprehensive description: F. J. Al-Shammary *et al.*, *Anal. Profiles Drug Subs. Excip.* **21**, 345-373 (1992).

Crystals from acetone-hexane, mp 152-154°. $[\alpha]_D$ +66° (in chloroform). Sol in 25 parts ethanol (96%), 20 parts methanol, 15 parts chloroform, 40 parts ether; sol in dehydrated alc. Practically insol in water. LD_{50} in mice (mg/kg): 435 i.v.; 1234 orally; in rats (mg/kg): 575 i.p.; 534 orally (Roszkowski).

Piperazine salt. [70981-66-7] Piproxen; Numidan. ($C_{14}H_{14}$-$O_3)_2.C_4H_{10}N_2$; mol wt 546.66.

Sodium salt. [26159-34-2] RS-3650; Aleve; Anaprox; Antalgin; Apranax; Axer Alfa; Flanax; Gynestrel; Miranax; Naprelan; Primeral; Synflex. $C_{14}H_{13}NaO_3$; mol wt 252.24. Crystals from acetone, mp 244-246°. $[\alpha]_D$ $-11°$ (in methanol). Sol in water, methanol; sparingly sol in alc; very slightly sol in acetone. Practically insol in chloroform, toluene.

THERAP CAT: Anti-inflammatory; analgesic; antipyretic.

THERAP CAT (VET): Anti-inflammatory.

6500. Naptalam. [132-66-1] 2-[(1-Naphthalenylamino)carbonyl]benzoic acid; *N*-1-naphthylphthalamic acid; α-naphthylphthalamic acid. $C_{18}H_{13}NO_3$; mol wt 291.31. C 74.22%, H 4.50%, N 4.81%, O 16.48%. Selective pre-emergence herbicide. Prepn: A. E. Smith, O. L. Hoffmann, **US 2556665** (1951 to U.S. Rubber). Activity: O. L. Hoffmann, A. E. Smith, *Science* **109**, 588 (1949). Mobility in soil: A. E. Smith *et al.*, *J. Agric. Food Chem.* **5**, 748 (1957).

Crystals from ethanol, d_4^{20} 1.40. mp 203° (technical grade, mp 175-180°). Soly: <0.02 g/100 ml of water. Sol in alkaline solns, but dec above pH 9.5. Slightly sol in ethanol, acetone, benzene. Hydrolyzed by strong acids and bases.

Sodium salt. ACP-332; NPA-3; Alanap. $C_{18}H_{12}NNaO_3$; mol wt 313.29.

USE: Herbicide; as analytical reagent for thorium and zirconium.

6501. Narasin. [55134-13-9] (4*S*)-4-Methylsalinomycin; narasin A; Compd 79891; Antibiotic A-28086 factor A; C-7819B; Monteban. $C_{43}H_{72}O_{11}$; mol wt 765.04. C 67.51%, H 9.49%, O 23.00%. Main component of a polyether antibiotic complex produced by *Streptomyces aureofaciens* NRRL 5758 & NRRL 8092. Production: D. H. Berg *et al.*, **DE 2525095** corresp to **US 4038384** (1975, 1977 to Lilly); L. D. Boeck *et al.*, *Dev. Ind. Microbiol.* **18**, 471 (1976). Isoln and characterization: D. H. Berg, R. L. Hamill, *J. Antibiot.* **31**, 1 (1978). Biosynthetic studies using ^{13}C-NMR study: D. E. Dorman *et al.*, *Helv. Chim. Acta* **59**, 2625 (1976). Structure: J. L. Occolowitz *et al.*, *Biomed. Mass Spectrom.* **3**, 272 (1976); H. Seto *et al.*, *J. Antibiot.* **30**, 530 (1977). Anticoccidial activity: M. D. Ruff *et al.*, *Poult. Sci.* **59**, 2008 (1980). Total synthesis: Y. Kishi *et al.*, *Front. Chem., Plenary Keynote Lect. IUPAC Congr., 28th* **1981**, K. J. Laidler, Ed. (Pergamon, Oxford, 1982) pp 287-304. HPLC determn in animal feeds: M. R. LaPointe, H. Cohen, *J. Assoc. Off. Anal. Chem.* **71**, 480 (1988).

Crystals from acetone-water, mp 98-100°; resolidif and remelts at 198-200°. uv max (ethanol): 285 nm (ε 58). $[\alpha]_D^{25}$ $-54°$ (c = 0.2 in methanol). pKa 7.9 (80% aq DMF). Sol in water, alcohols, acetone, methanol, DMF, DMSO, benzene, chloroform, ethyl acetate. LD_{50} i.p. in mice: 7.15 mg/kg (Berg).

THERAP CAT (VET): Coccidiostat; growth stimulant.

6502. Naratriptan. [121679-13-8] *N*-Methyl-3-(1-methyl-4-piperidinyl)-1*H*-indole-5-ethanesulfonamide. $C_{17}H_{25}N_3O_2S$; mol wt 335.47. C 60.87%, H 7.51%, N 12.53%, O 9.54%, S 9.56%. Serotonin 5-HT$_{1B/1D}$ receptor agonist. Prepn: A. W. Oxford *et al.*, **EP 303507**; *eidem*, **US 4997841** (1989, 1991 both to Glaxo). Pharmacology and receptor binding study: H. E. Connor *et al.*, *Cephalalgia* **17**, 145 (1997). Determn in plasma and pharmacokinetics in rabbits: B. D. Duléry *et al.*, *J. Pharm. Biomed. Anal.* **15**, 1009 (1997). Clinical trial in migraine: H. Havanka *et al.*, *Clin. Ther.* **22**, 970 (2000).

Crystals from ethyl acetate, mp 170-171°.

Hydrochloride. [143388-64-1] GR-85548A; Amerge; Naramig. $C_{17}H_{25}N_3O_2S.HCl$; mol wt 371.92. Microcrystals, mp 237-239°. Sol in water.

THERAP CAT: Antimigraine.

6503. Narbomycin. [6036-25-5] (10*E*)-3-De[(2,6-dideoxy-3-*C*-methyl-3-*O*-methyl-α-L-ribohexopyranosyl)oxy]-10,11-didehydro-10-demethyl-6,11,12-trideoxy-3-oxoerythromycin; 12-deoxypicromycin. $C_{28}H_{47}NO_7$; mol wt 509.68. C 65.98%, H 9.30%, N 2.75%, O 21.97%. Antibiotic substance produced by *Streptomyces narbonensis* from soil near Cannes, France: Corbaz *et al.*, *Helv. Chim. Acta* **38**, 935 (1955). Structure: Prelog *et al.*, *ibid.* **45**, 4 (1962). Stereochemical studies: Rickards, Smith, *Tetrahedron Lett.* **1970**, 1025; H. Ogura *et al.*, *J. Am. Chem. Soc.* **97**, 1930 (1975); *eidem*, *Tetrahedron* **37**, Suppl. 1, 165 (1981). Isoln and structure of the aglycone **narbonolide**: Hori *et al.*, *Chem. Commun.* **1971**, 304. Synthesis of narbonolide: T. Kaiho *et al.*, *J. Org. Chem.* **47**, 1612 (1982).

Crystals from ether + petr ether, mp 113.5-115°. $[\alpha]_D^{20}$ +68.5° (c = 1.35 in chloroform). uv max (abs ethanol): 225, 286 nm (log ε 4.06, 2.23). LD_{50} s.c. in mice: 500 mg/kg (Corbaz).

6504. Narceine. [131-28-2] 6-[2-[6-[2-(Dimethylamino)ethyl]-4-methoxy-1,3-benzodioxol-5-yl]acetyl]-2,3-dimethoxybenzoic acid; 6-[[6-[2-(dimethylamino)ethyl]-2-methoxy-3,4-(methylenedioxy)phenyl]acetyl]-*o*-veratric acid. $C_{23}H_{27}NO_8$; mol wt 445.47. C 62.01%, H 6.11%, N 3.14%, O 28.73%. Opioid analgesic. Occurs in opium to the extent of 0.1-0.5%. The separation from morphine mother liquors is tedious: Merck, *Chem. Ztg.* **13**, 525 (1889). Prepn from narcotine or gnoscopine: Roser, *Ann.* **247**, 167 (1888); Frankforter, Keller, *Ann. Chem. J.* **22**, 61 (1899); Frerichs, *Arch. Pharm.* **241**, 259 (1903); Hope, Robinson, *J. Chem. Soc.* **105**, 2100 (1914). Structure: Freund, Frankforter, *Ann.* **277**, 20 (1893); Freund, Michaels, *Ann.* **286**, 248 (1895); Freund, *Ber.* **40**, 194 (1907); Freund,

Oppenheim, *Ber.* **42**, 1084 (1909); Addinall, Major, *J. Am. Chem. Soc.* **55**, 1202, 2153 (1933).

The anhydr material is very hygroscopic, mp 138°. uv max (ethanol): 270 nm (log ε 3.98). Usually the alkaloid is obtained as the trihydrate. Clusters of silky, prismatic needles from water, mp 176°. pKb (20°) = 10.7, pKa = 9.3. pH of satd soln = 5.8. One gram dissolves in 770 ml water, 220 ml boiling water. Moderately sol in hot alcohol, nearly insol in benzene, chloroform, ether, petr ether; sol in alkali hydroxide solns forming salts, also in dil mineral acids.

Ethylnarceine hydrochloride. Narcyl. $C_{25}H_{32}ClNO_8$. Plates from water, mp 208-210°. Slightly sol in cold water, sol in hot water, alcohol, chloroform; insol in ether.

THERAP CAT: Antitussive.

6505. Narcotoline. [521-40-4] (3*S*)-6,7-Dimethoxy-3-[(5*R*)-5,6,7,8-tetrahydro-4-hydroxy-6-methyl-1,3-dioxolo[4,5-*g*]-isoquinolin-5-yl]-1(3*H*)-isobenzofuranone; desmethylnarcotine. $C_{21}H_{21}NO_7$; mol wt 399.40. C 63.15%, H 5.30%, N 3.51%, O 28.04%. Found in the shell of ripe poppyseed capsules: Wrede, *Arch. Pharmacol.* **184**, 331 (1937); Baumgarten, Christ, *Pharmazie* **5**, 80 (1950); Pfeifer, Weiss, *ibid.* **10**, 658 (1955); from opium prepns: Pfeifer, *Arch. Pharm.* **290**, 209 (1957). Preliminary stereochemical studies: Battersby, Spenser, *J. Chem. Soc.* **1965**, 1087. Revised stereochemistry: Blaha *et al.*, *Collect. Czech. Chem. Commun.* **29**, 2328 (1964); Snatzke *et al.*, *Tetrahedron* **25**, 5059 (1969).

Rectangular rods from dil methanol, mp 202°. $[\alpha]_D^{20}$ −189° (0.1 g in 25 ml chloroform, 20 cm tube). $[\alpha]_D^{20}$ +5.8° (0.065 g in 5 ml 0.1*N* HCl in 20 cm tube). Very sparingly sol in water. Moderately sol in warm alc and ether. Freely sol in chloroform, in dil acids and in dil aq solns of KOH and NaOH. Sparingly sol in dil aq solns of Na_2CO_3.

6506. Naringenin. [480-41-1] (2*S*)-2,3-Dihydro-5,7-dihydroxy-2-(4-hydroxyphenyl)-4*H*-1-benzopyran-4-one; 4′,5,7-trihydroxyflavanone; naringetol; salipurpol; pelargidanon 1602. $C_{15}H_{12}O_5$; mol wt 272.26. C 66.17%, H 4.44%, O 29.38%. The aglucon of naringin. Prepn by hydrolysis of naringin: Asahina, Inubuse, *Ber.* **61**, 1514 (1928); Haley, Bassin, *J. Am. Pharm. Assoc.* **40**, 111 (1951). From kino of *Eucalyptus maculata* Hook, *Myrtaceae:* Gell *et al.*, *Aust. J. Chem.* **11**, 372 (1958). Synthesis: Rosenmund, Rosenmund, *Ber.* **61**, 2608 (1928); Zemplén, Bognár, *Ber.* **75**, 648 (1942). Absolute configuration: Gaffield, Waiss, *Chem. Commun.* **1968**, 29.

Needles from dilute alc, mp 251°. uv max: 226, 292 nm. Sol in alcohol, ether, benzene; almost insol in water.

Triacetylnaringenin. $C_{15}H_9(CH_3CO)_3O_5$. Crystals, mp 55°.

6507. Naringin. [10236-47-2] (2*S*)-7-[[2-*O*-(6-Deoxy-α-L-mannopyranosyl)-β-D-glucopyranosyl]oxy]-2,3-dihydro-5-hydroxy-2-(4-hydroxyphenyl)-4*H*-1-benzopyran-4-one; 4′,5,7-trihydroxyflavanone 7-rhamnoglucoside; naringenin-7-rhamnoglucoside; aurantiin. $C_{27}H_{32}O_{14}$; mol wt 580.54. C 55.86%, H 5.56%, O 38.58%. In the flowers of grapefruit trees *(Citrus paradisi* Macfad, *Rutaceae)*, also in fruit and rind. Most abundant in immature fruit; main bitter component of grapefruit juice. Extraction from grapefruit peel: Zoller, *Ind. Eng. Chem.* **10**, 364 (1918); Pulley, von Loesecke, *J. Am. Chem. Soc.* **61**, 175 (1939); US 2421062; US 2421063 (both 1947 to California Fruit Growers Exchange). Structure: Asahina, Inubuse, *C.A.* **23**, 3475 (1929); *Ber.* **61**, 1514 (1928); Horowitz, Gentili, *Tetrahedron* **19**, 773 (1963). Solubility data: G. N. Pulley, *Ind. Eng. Chem. Anal. Ed.* **8**, 360 (1936). *Review:* Kesterson, Hendrickson, *Naringin, A Bitter Principle of Grapefruit* **Bulletin no. 511** (Univ. Florida Agric. Expt. Station, Jan. 1953).

When crystallized from water, it contains 6 to 8 mols H_2O. mp ~83°. After drying at 110° to constant weight, it contains 2 mols H_2O, mp 171°. Bitter taste (1:10,000 H_2O can be tasted). $[\alpha]_D^{19}$ −82° (alcohol). One gram dissolves in 1000 ml water at 40°. At 75° one gram dissolves in 10 ml water. Sol in acetone, alcohol, warm acetic acid.

6508. Natalizumab. [189261-10-7] Anti-(human integrin α4) immunoglobulin G4 (human-mouse monoclonal AN100226 γ_4-chain) disulfide with human-mouse monoclonal AN100226 light chain, dimer; Antegren; Tysabri. Recombinant humanized IgG4 monoclonal antibody directed against the leukocyte adhesion molecule VLA-4, also known as α4β1 integrin. α4-Integrin inhibitor designed to block lymphocyte activation, migration and adhesion to endothelial cells. Prepn: M. M. Bendig *et al.*, WO 9519790; *eidem*, US 5840299 (1995, 1998 both to Athena Neurosciences); O. J. P. Léger *et al.*, *Hum. Antibod.* **8**, 3 (1997). Clinical pharmacokinetics: W. A. Sheremata *et al.*, *Neurology* **52**, 1072 (1999). Clinical trial in multiple sclerosis: D. H. Miller *et al.*, *N. Engl. J. Med.* **348**, 15 (2003); in Crohn's disease: W. J. Sandborn *et al.*, *ibid.* **353**, 1912 (2005). Review of pharmacology and therapeutic potential: M. Elices, *Curr. Opin. Anti-Inflam. Immunomod. Invest. Drugs* **2**, 228-235 (2000); K. A. Keeley *et al.*, *Ann. Pharmacother.* **39**, 1833-1843 (2005).

THERAP CAT: Anti-inflammatory.

6509. Natamycin. [7681-93-8] (1*R*,3*S*,5*R*,7*R*,8*E*,12*R*,14*E*,-16*E*,18*E*,20*E*,22*R*,24*S*,25*R*,26*S*)-22-[(3-Amino-3,6-dideoxy-β-D-mannopyranosyl)oxy]-1,3,26-trihydroxy-12-methyl-10-oxo-6,-11,28-trioxatricyclo[22.3.1.0^{5,7}]octacosa-8,14,16,18,20-pentaene-25-carboxylic acid; antibiotic A 5283; pimaricin; tennecetin; CL-12625; Mycophyt; Myprozine; Natacyn; Pimafucin; Synogil. $C_{33}H_{47}NO_{13}$; mol wt 665.73. C 59.54%, H 7.12%, N 2.10%, O 31.24%. Polyene antifungal antibiotic produced by *Streptomyces natalensis* from soil near Pietermaritzburg, South Africa and by *S. chattanoogensis.* Isoln: Struyk *et al.*, *Antibiot. Annu.* **1957-1958**, 878. Prodn: GB 844289 (1960 to Koninklijke Nederlandsche Gist en Spiritus-Fabriek); GB 846933 (1960 to Am. Cyanamid). Identity with tennecetin: Divekar *et al.*, *Antibiot. Chemother.* **11**, 377 (1961). Structure: Golding *et al.*, *Tetrahedron Lett.* **1966**, 3551; Meyer, *Chem. Commun.* **1968**, 470. Configuration: Gaudiano *et al.*, *Chim. Ind. (Milan)* **48**, 1327 (1966). Revised structure: R. C. Pandey, K. L. Rinehart, *J. Antibiot.* **29**, 1035 (1976). Soly data: Marsh, Weiss, *J.*

Assoc. Off. Anal. Chem. **50**, 457 (1967). Toxicology: G. J. Levin-skas *et al., Toxicol. Appl. Pharmacol.* **8**, 97 (1966). Comprehensive description: H. Brik, *Anal. Profiles Drug Subs. Excip.* **23**, 399-419 (1994).

Crystals from methanol + water, darkens ~200°, dec 280-300°. $[\alpha]_D^{20}$ +278° (c = 1 in CH$_3$COOH). Sensitive to light, but otherwise very stable in the dry state. uv max (methanol + 0.1% CH$_3$COOH): 220, 280, 290, 303, 318 nm (ε 21300, 26630, 52930, 83220, 76230). Soly at 20° (g/ml): 87% glycerol 0.18. Sol in glacial acetic acid, DMF; slightly sol in methanol. Practically insol in water, higher alcohols, ether, esters, aromatic or aliphatic hydrocarbons, chlori-nated hydrocarbons, ketones, dioxane, cyclohexanol, oils. LD$_{50}$ orally in male, female rats (g/kg): 2.73, 4.67 (Levinskas).

THERAP CAT: Antifungal (topical).

6510. Nateglinide. [105816-04-4] *N*-[[*trans*-4-(1-Methyleth-yl)cyclohexyl]carbonyl]-D-phenylalanine; (−)-*N*-(*trans*-4-isopro-pylcyclohexyl-1-carbonyl)-D-phenylalanine; A-4166; AY-4166; SDZ-DJN-608; YM-026; Fastic; Starlix; Starsis. C$_{19}$H$_{27}$NO$_3$; mol wt 317.43. C 71.89%, H 8.57%, N 4.41%, O 15.12%. Amino acid derivative that stimulates insulin secretion. Prepn: S. Toyoshima *et al., EP 196222; eidem, US 4816484* (1986, 1989 both to Ajino-moto); H. Shinkai *et al., J. Med. Chem.* **32**, 1436 (1989). Effect on insulin secretion: Y. Sato *et al., Diabetes Res. Clin. Pract.* **12**, 53 (1991). HPLC determn in plasma and urine: I. Ono *et al., J. Chro-matogr. B* **692**, 397 (1997). Series of articles on pharmacology, pharmacokinetics and toxicology: *Yakuri to Chiryo* **25**, Suppl. 1, S9-S239 (1997). Toxicity data: K. Hasegawa *et al., ibid.* S5, *C.A.* **126**, 152622 (1997). Clinical trial in type 2 diabetes: A. Bellomo Damato *et al., Diabet. Med.* **28**, 560-566 (2011). Review of clinical experience: G. Grunberger, *Expert Opin. Pharmacother.* **12**, 2097-2106 (2011).

Crystals from methanol-water, mp 129-130°. $[\alpha]_D^{20}$ −9.4° (c = 1 in methanol). LD$_{50}$ orally in rats: >2.0 g/kg (Hasegawa).

THERAP CAT: Antidiabetic.

6511. Natural Gas. American natural gas consists of about 85% methane, 9% ethane, 3% propane, 2% nitrogen, 1% butane. Northern Texas gas contains more nitrogen, also sufficient helium to warrant commercial extraction.

Caution: Narcotic in high concns. Incomplete combustion can result in production of carbon monoxide.

USE: Fuel gas, in the manuf of hydrogen, methane, ammonia.

6512. Navitoclax. [923564-51-6] 4-[4-[[2-(4-Chlorophenyl)-5,5-dimethyl-1-cyclohexen-1-yl]methyl]-1-piperazinyl]-*N*-[[4-[[(1*R*)-3-(4-morpholinyl)-1-[(phenylthio)methyl]propyl]amino]-3-[(trifluoromethyl)sulfonyl]phenyl]sulfonyl]benzamide; 4-[4-[[2-(4-chlorophenyl)-5,5-dimethylcyclohex-1-en-1-yl]methyl]piperazin-1-yl]-*N*-[[4-[[(2*R*)-4-(morpholin-4-yl)-1-(phenylsulfanyl)butan-2-yl]-amino]-3-[(trifluoromethyl)sulfonyl]phenyl]sulfonyl]benzamide;

ABT-263. C$_{47}$H$_{55}$ClF$_3$N$_5$O$_6$S$_3$; mol wt 974.61. C 57.92%, H 5.69%, Cl 3.64%, F 5.85%, N 7.19%, O 9.85%, S 9.87%. Inhibitor of Bcl-2 family proteins; designed to restore apoptosis in tumor cells. Prepn: M. Bruncko *et al.,* **WO 07040650**; *eidem,* **US 7390799** (2007, 2008 both to Abbott); C.-M. Park *et al., J. Med. Chem.* **51**, 6902 (2008). Synthesis: G. Wang *et al., Synthesis* **15**, 2398 (2008). HPLC-MS/MS determn in urine: R. Rodila *et al., J. Chromatogr. B* **872**, 128 (2008). Biological properties: C. Tse *et al., Cancer Res.* **68**, 3421 (2008); and antitumor activity: A. R. Shoemaker *et al., Clin. Cancer Res.* **14**, 3268 (2008). Clinical pharmacokinetics and evaluation in lymphoid malignancies: W. H. Wilson *et al., Lancet Oncol.* **11**, 1149 (2010); in solid tumors: L. Gandhi *et al., J. Clin. Oncol.* **29**, 909 (2011). Review of structure-based design and tar-geted protein-protein interaction: M. D. Wendt, *Expert Opin. Drug Discov.* **3**, 1123-1143 (2008).

Pale yellow solid. $[\alpha]_D^{20}$ −59° (c = 0.28 in CHCl$_3$). Sol in meth-anol. Poorly sol in water.

THERAP CAT: Antineoplastic.

6513. NB-Enantride. [81572-37-4] (*T*-4)-1,5-Cyclooctane-diyl[(1*S*,2*S*,3*S*,5*R*)-6,6-dimethyl-2-[2-(phenylmethoxy)ethyl]bicy-clo[3.1.1]hept-3-yl]hydroborate(1−) lithium; lithium hydrido(9-BBN-nopol benzyl ether adduct). C$_{26}$H$_{40}$BLiO; mol wt 386.36. C 80.83%, H 10.44%, B 2.80%, Li 1.80%, O 4.14%. Lithium trialkyl-borohydride reagent for the asymmetric reduction of ketones. Prepn and synthetic applications: M. M. Midland, A. Kazubski, *J. Org. Chem.* **47**, 2495 (1982). Reactivity comparison with other asymmet-ric reducing agents: H. C. Brown *et al., ibid.* **52**, 5406 (1987); M. M. Midland *et al., ibid.* **56**, 1068 (1991); S. A. Weissman, P. V. Ramachandran, *Tetrahedron Lett.* **37**, 3791 (1996).

Sol in tetrahydrofuran. *Flammable. Irritant.* Reacts violently with water.

NB-Enantrane. [81971-15-5] 9-[(1*S*,2*R*,3*S*,5*S*)-6,6-Dimethyl-2-[2-(phenylmethoxy)ethyl]bicyclo[3.1.1]hept-3-yl]-9-borabicyclo-[3.3.1]nonane; 9-BBN-nopol benzyl ether adduct. C$_{26}$H$_{39}$BO; mol wt 378.41. Precursor to NB-enantride; also utilized for asymmetric ketone reductions. Prepn from nopol benzyl ether and synthetic ap-plications: M. M. Midland, A. Kazubski, *J. Org. Chem.* **47**, 2814 (1982). Synthetic limitations: H. C. Brown *et al., ibid.* **55**, 6328 (1990). Sol in tetrahydrofuran. *Flammable. Irritant.*

USE: Reagent in synthetic organic chemistry.

6514. Neamine. [3947-65-7] 2-Deoxy-4-*O*-(2,6-diamino-2,6-dideoxy-α-D-glucopyranosyl)-D-streptamine; neomycin A. C$_{12}$H$_{26}$-N$_4$O$_6$; mol wt 322.36. C 44.71%, H 8.13%, N 17.38%, O 29.78%. Component of the antibiotic complex neomycin, *q.v.* Isoln: Dutcher *et al., J. Am. Chem. Soc.* **73**, 1384 (1951). Identity of neomycin A and neamine: Leach, Teeters, *ibid.* **74**, 3187 (1952). Prepn from neomycin B: Peck *et al.,* **US 2691675** (1954 to Merck & Co.). Structure: Carter *et al., J. Am. Chem. Soc.* **83**, 3723 (1961); Hichens, Rinehart, *ibid.* **85**, 1547 (1963). Synthesis: Umezawa *et al., J. An-*

tibiot. **20A**, 53 (1967); H. Kohno *et al.*, *Agric. Biol. Chem.* **39**, 1091 (1975); A. Harayama *et al.*, *Bull. Chem. Soc. Jpn.* **52**, 3626 (1979).

Crystals from water or aq alc, dec 225-226°, $[\alpha]_D^{25}$ +112.8° (c = 1). IR spectrum: Leach, Teeters, *J. Am. Chem. Soc.* **73**, 2794 (1951).

Hydrochloride. $C_{12}H_{26}N_4O_6$.4HCl. Amorphous, dec 250-260°, $[\alpha]_D^{25}$ +83° (c = 1).

***N*-Acetyl derivative.** $C_{12}H_{26}N_4O_6$.$(CH_3CO)_4$. Crystals from methanol, mp 334-336°. $[\alpha]_D^{25}$ +87° (c = 1).

6515. Neatsfoot Oil. Fixed oil from feet of neat (bovine) cattle.

Pale yellow liquid; peculiar odor. d 0.915. Solidif 0° to −10°. n_D^{20} 1.4695-1.4708. Sapon no. 192-203. Iodine no. 44-73.2.

USE: Waterproofing and softening leather; lubricant; oiling wool, etc.

6516. Nebivolol. [118457-14-0]; [99200-09-6] (unspecified stereo). *rel*-($\alpha R,\alpha' R,2R,2'S$)-$\alpha,\alpha'$-[Iminobis(methylene)]bis[6-fluoro-3,4-dihydro-2*H*-1-benzopyran-2-methanol]; α,α'-(iminodimethylene)bis[6-fluoro-2-chromanmethanol]; *dl*-nebivolol; narbivolol; R-65824. $C_{22}H_{25}F_2NO_4$; mol wt 405.44. C 65.17%, H 6.22%, F 9.37%, N 3.45%, O 15.78%. β_1-Adrenergic blocker; nitric oxide-mediated vasodilator. Prepd as racemic mixture of (R,S,S,S)- and (S,R,R,R)-enantiomers: G. R. E. Van Lommen *et al.*, **EP 145067**; *eidem*, **US 4654362** (1985, 1987 both to Janssen); *eidem*, *J. Pharm. Belg.* **45**, 355 (1990). HPLC fluorescence determn in plasma: R. Woestenborghs *et al.*, *Methodol. Surv. Biochem. Anal.* **18**, 215 (1988). Receptor binding study: P. J. Pauwels *et al.*, *Mol. Pharmacol.* **34**, 843 (1988). Clinical study of vasodilating effects: J. R. Cockcroft *et al.*, *J. Pharmacol. Exp. Ther.* **274**, 1067 (1995). Clinical trial in hypertension: T. J. Cleophas *et al.*, *Clin. Ther. Res.* **62**, 451 (2001); in endothelial dysfunction with hypertension: N. Tzemos *et al.*, *Circulation* **104**, 511 (2001). Review of pharmacology: M. Mangrella *et al.*, *Pharmacol. Res.* **38**, 419-431 (1998); of clinical development: J. Cockcroft, *Expert Opin. Pharmacother.* **5**, 893-899 (2004).

Relative stereochemistry

Hydrochloride. [152520-56-4] R-67555; Bystolic; Lobivon; Nebilet; Nebilox. $C_{22}H_{25}F_2NO_4$.HCl; mol wt 441.90. White to almost white powder. Sol in methanol, dimethylsulfoxide, *N,N*-dimethylformamide; sparingly sol in ethanol, propylene glycol, polyethylene glycol; very slightly sol in hexane, dichloromethane, methylbenzene.

THERAP CAT: Antihypertensive.

6517. Nebularine. [550-33-4] 9-β-D-Ribofuranosyl-9*H*-purine. $C_{10}H_{12}N_4O_4$; mol wt 252.23. C 47.62%, H 4.80%, N 22.21%, O 25.37%. Isoln from the mushroom *Clitocybe nebularis* (Batsch.) Quel., *Agaricaceae:* Ehrenberg *et al.*, *Sven. Kem. Tidskr.* **58**, 269 (1946); Löfgren *et al.*, *Acta Chem. Scand.* **8**, 670 (1954); from a streptomycete: Isono, Suzuki, *J. Antibiot.* **13A**, 270 (1960). *In vitro* toxicity towards sarcoma 180 cells, mouse embryonic fibroblasts and epithelial cells: J. J. Biescle *et al.*, *Cancer* **8**, 87 (1955). Synthesis: Brown, Weliky, *J. Biol. Chem.* **204**, 1019 (1953); Fox *et al.*, *J. Am. Chem. Soc.* **80**, 1669 (1958); Hashizume, Iwamura, *Tetrahedron Lett.* **1966**, 643; *eidem*, *J. Org. Chem.* **33**, 1796 (1968). Crystal structure: T. Takeda, *Acta Crystallogr.* **30B**, 825 (1974). Alternate

syntheses: V. Nair, S. G. Richardson, *Tetrahedron Lett.* **1979**, 1181; P. K. Gupta, D. S. Bhakuni, *Indian J. Chem.* **20B**, 534 (1981). Toxicity studies: Truant, D'Amato, *Fed. Proc.* **14**, 391 (1955).

Small rhombohedra from ethyl methyl ketone + methanol, mp 181-182°; needles from methanol, mp 182-183°. $[\alpha]_D^{25}$ −48.6° (H_2-O). uv max (0.1*N* HCl): 262 nm ($E_{1cm}^{1\%}$ 232); (0.1*N* NaOH): 263 nm ($E_{1cm}^{1\%}$ 361). Considerably sol in water (about 10%). Slightly sol in cold ethanol. Very slightly sol in acetone, ether, chloroform. Aq solns may be sterilized by boiling without decompn. LD_{50} in rats, guinea pigs (mg/kg): 220, 15 s.c. (Truant, D'Amato).

6518. Nebulin. Family of giant, actin-binding proteins, varying in size from 700-900 kDa, found in the sarcomeres of vertebrate skeletal muscles; comprises ∼3% of the total myofibrillar protein. Composed of small repeats of ∼35 amino acid residues arranged in a 7-fold super-repeating pattern. Integral component of the cytoskeletal lattice of muscle cells (*see also* Titin). Forms inextensible filaments that are attached to the Z-line and span the length of the thin filaments. Thought to act as a "molecular ruler" that stabilizes the actin polymer and regulates thin filament length. Name derived from its histological localization in the N_2-line of the sarcomere, a nebulous striation within the I-band. Identification: K. Wang, C. L. Williamson, *Proc. Natl. Acad. Sci. USA* **77**, 3254 (1980). Distribution and organization in cytoskeletal matrix of striated muscle: K. Wang, *Adv. Exp. Med. Biol.* **170**, 285-305 (1984). Physiological role: M.-J. G. Chen *et al.*, *J. Biol. Chem.* **268**, 20327 (1993). Quantitative determn by gel electrophoresis: H. L. M. Granzier, K. Wang, *Electrophoresis* **14**, 56 (1993). Structural study: M. Pfuhl *et al.*, *EMBO J.* **13**, 1782 (1994). Complete primary structure: S. Labeit, B. Kolmerer, *J. Mol. Biol.* **248**, 308 (1995). *Review:* J. Trinick, *Trends Biochem. Sci.* **19**, 405-409 (1994).

6519. Neburon. [555-37-3] *N*-Butyl-*N'*-(3,4-dichlorophenyl)-*N*-methylurea; 3-(3,4-dichlorophenyl)-1-methyl-1-*n*-butylurea; Kloben Neburon. $C_{12}H_{16}Cl_2N_2O$; mol wt 275.17. C 52.38%, H 5.86%, Cl 25.77%, N 10.18%, O 5.81%. Selective pre-emergence herbicide. Prepn: Todd, **US 2655444; US 2655445; US 2655446; US 2655447** (all 1953 to du Pont). Toxicity study: G. W. Bailey, J. L. White, *Residue Rev.* **10**, 97 (1965).

Crystals from dioxane + water, mp 101.5-103°. Soly in water at 24°: 48 ppm; sparingly sol in hydrocarbon solvents. Stable toward oxidation and water at ordinary temperatures. LD_{50} orally in rats: >11000 mg/kg (Bailey, White).

USE: Herbicide.

6520. Necitumumab. [906805-06-9] Anti-(human epidermal growth factor receptor) immunoglobulin G1 (human monoclonal IMC-11F8 γ_1-chain) disulfide with human monoclonal IMC-11F8 κ-chain, dimer; IMC-11F8. Fully human IgG1 monoclonal antibody directed against epidermal growth factor receptor (EGFR). Prepn: M. Liu, Z. Zhu, **WO 05090407**; *eidem*, **US 7598350** (2005, 2009 both to ImClone Systems). EGFR binding studies and crystal structure: S. Li *et al.*, *Structure* **16**, 216 (2008). Clinical pharmacokinetics and evaluation in solid tumors: B. Keunen *et al.*, *Clin. Cancer Res.* **16**, 1915 (2010). Review of development and therapeutic potential: R. Dienstmann, E. Felip, *Expert Opin. Biol. Ther.* **11**, 1223-1231 (2011).

THERAP CAT: Antineoplastic.

6521. Nedocromil. [69049-73-6] 9-Ethyl-6,9-dihydro-4,6-di-oxo-10-propyl-4*H*-pyrano[3,2-*g*]quinoline-2,8-dicarboxylic acid; 4,6-dioxo-1-ethyl-10-propyl-4*H*,6*H*-pyrano[3,2-*g*]quinoline-2,8-dicarboxylic acid; FPL-59002. $C_{19}H_{17}NO_7$; mol wt 371.35. C 61.45%, H 4.61%, N 3.77%, O 30.16%. Mast cell stabilizer; also modulates production of cytokines by stimulated epithelial cells. Prepn and pharmacology: H. Cairns, D. Cox, **BE 866622**; *eidem*, **US 4474787** (1978, 1984 both to Fisons); H. Cairns *et al.*, *J. Med. Chem.* **28**, 1832 (1985). Series of articles on pharmacology, pharmacokinetics and clinical profile in allergic rhinitis: *Allergy* **51**, Suppl. 28, 8-34 (1996). Comparative clinical trial with cromolyn sodium, *q.v.*, in asthma: H. J. Schwartz *et al.*, *Chest* **109**, 945 (1996).

Yellow powder, mp 298-300° (dec).

Disodium salt. [69049-74-7] FPL-59002KP; Halamid; Irtan; Rapitil; Tilade; Tilarin; Tilavist. $C_{19}H_{15}NNa_2O_7$; mol wt 415.31. Pale yellow powder. Sol in water.

THERAP CAT: Antiallergic; antiasthmatic.

6522. Neem. Nim; Margosa tree; Indian lilac. Tropical evergreen tree, *Azadirachta indica*, A. Juss. (*Melia azadirachta* L.) *Meliaceae. Habit.* India, Southeast Asia, also in Indonesia, Australia, Western Africa. Key active constituent is the limonoid, azadirachtin, *q.v.* Medicinal portions are the bark, leaves, fruits, and seeds; traditionally used as anti-inflammatory, antipyretic, and anthelmintic. Series of articles on chemistry and activity of neem extracts: *Natural Pesticides from the Neem Tree*, Proc. 1st Int. Neem Conf., 1980, H. Schmutterer *et al.*, Eds. (German Agency for Technical Cooperation, Eschborn, 1981) 291 pp. Antifeedant activity of extracts against pea aphid: J. D. Stark, J. F. Walter, *J. Agric. Food Chem.* **43**, 507 (1995). Use in traditional medicinal remedies: R. N. Chopra *et al.*, *Indigenous Drugs of India* (U. N. Dhur & Sons, Calcutta, 2nd ed., 1958) pp 360-363. Cytogenetic toxicity of leaf extract: P. K. Khan, K. S. Awasthy, *Food Chem. Toxicol.* **41**, 1325 (2003). Review of biological activity of isolated compounds, medicinal use and safety: K. Biswas *et al.*, *Curr. Sci.* **82**, 1336-1345 (2002); of component and pesticide toxicity in humans and animals: S. J. Boeke *et al.*, *J. Ethnopharmacol.* **94**, 25-41 (2004); of pharmacology and use in pest management: G. Brahmachari, *Chembiochem* **5**, 408-421 (2004); of medicinal and pharmacological activities of leaf extracts: R. Subapriya, S. Nagini, *Curr. Med. Chem. Anticancer Agents* **5**, 149-156 (2005).

Fixed oil. [8002-65-1] Oil of neem. Obtained by extraction from seeds. *Constit.* Limonoids including azadirachtin, nimbin, nimbiol, salannin; fatty acids including oleic, stearic, palmitic. Properties and constituents: J. Kumar, B. S. Parmar, *J. Agric. Food Chem.* **44**, 2137 (1996). Determn by HPLC of azadirachtin variation among seeds: O. P. Sidhu *et al.*, *J. Agric. Food Chem.* **51**, 910 (2003); by LC-MS of oils: S. Barrek, P. Paisse, *Anal. Bioanal. Chem.* **378**, 753 (2004). Toxicology of debitterized oil: N. Chinnasamy *et al.*, *Food Chem. Toxicol.* **31**, 297 (1993). Yellow oil with bitter taste and odor of garlic. d_4^{25} 0.83-0.98. n_D^{27} 1.4575-1.4675. Sol in ether, chloroform. Practically insol in alcohol, water

USE: Insect repellent, antifeedant.

6523. Nefazodone. [83366-66-9] 2-[3-[4-(3-Chlorophenyl)-1-piperazinyl]propyl]-5-ethyl-2,4-dihydro-4-(2-phenoxyethyl)-3*H*-1,2,4-triazol-3-one; 2-[3-[4-(3-chlorophenyl)-1-piperazinyl]propyl]-5-ethyl-4-(2-phenoxyethyl)-2*H*-1,2,4-triazol-3(4*H*)-one. $C_{25}H_{32}$-ClN_5O_2; mol wt 470.01. C 63.89%, H 6.86%, Cl 7.54%, N 14.90%, O 6.81%. Selective serotonin 5-HT$_2$ receptor antagonist. Prepn: D. L. Temple *et al.*, W. G. Lobeck, Jr., **US 4338317** (1982 to Mead Johnson). Synthesis and x-ray crystal structure: G. D. Madding *et al.*, *J. Heterocycl. Chem.* **22**, 1121 (1985). Pharmacology: A. S. Eison *et al.*, *Psychopharmacol. Bull.* **26**, 311 (1990). HPLC determn in plasma: J. E. Franc *et al.*, *J. Chromatogr.* **570**, 129 (1991). Clinical

trial in depression: M. F. D'Amico *et al.*, *Psychopharmacol. Bull.* **26**, 147 (1990); in combination with psychotherapy for chronic depression: M. B. Keller *et al.*, *N. Engl. J. Med.* **342**, 1462 (2000). *Review:* W. E. Heydorn, *Expert Opin. Invest. Drugs* **4**, 131-137 (1995).

Crystals from 2-propanol/heptane, mp 83-84°.

Hydrochloride. [82752-99-6] BMY-13754; MJ-13754-1; Dutonin; Serzone. $C_{25}H_{32}ClN_5O_2$.HCl; mol wt 506.47. Crystals from 2-propanol, with slow cooling yields a polymorph, mp 186.0-187.0°; with rapid cooling yields a polymorph, mp 181.0-182.0° (Madding). Also reported as crystals from ethanol. Nonhygroscopic. mp 175-177° (Temple). Freely sol in chloroform; sol in propylene glycol; slightly sol in polyethylene glycol, water.

THERAP CAT: Antidepressant.

6524. Nefiracetam. [77191-36-7] *N*-(2,6-Dimethylphenyl)-2-oxo-1-pyrrolidineacetamide; 2-(2-oxo-1-pyrrolidinyl)-*N*-(2,6-dimethylphenyl)acetamide; 2-oxo-1-pyrrolidineaceto-2′,6′-xylidide; 2-oxo-1-pyrrolidinylacetic acid 2,6-dimethylanilide; DMMPA; DM-9384; DZL-221; Translon. $C_{14}H_{18}N_2O_2$; mol wt 246.31. C 68.27%, H 7.37%, N 11.37%, O 12.99%. Cyclic deriv of γ-aminobutyric acid. Prepn: H. Betzing *et al.*, **DE 2924011**; *eidem*, **US 4341790** (1980, 1982 both to Nattermann). Cognition enhancing effects in rats: T. Sakurai *et al.*, *Jpn. J. Pharmacol.* **50**, 47 (1989). HPLC determn in serum and urine: Y. Fujimaki *et al.*, *J. Chromatogr.* **575**, 261 (1992). Clinical pharmacokinetics: *eidem*, *Xenobiotica* **23**, 61 (1993). Series of articles on pharmacology and toxicology: *Arzneim.-Forsch.* **44**, 193-259 (1994).

Crystals from water, mp 153°. LD$_{50}$ in mice (mg/kg): 421 i.v.; 1766 orally (Betzing, 1982).

THERAP CAT: Nootropic.

6525. Nefopam. [13669-70-0] 3,4,5,6-Tetrahydro-5-methyl-1-phenyl-1*H*-2,5-benzoxazocine; 5-methyl-1-phenyl-1,3,4,6-tetrahydro-5*H*-benz[*f*]-2,5-oxazocine. $C_{17}H_{19}NO$; mol wt 253.35. C 80.59%, H 7.56%, N 5.53%, O 6.31%. A cyclized analog of orphenadrine and diphenhydramine, *q.q.v.*; representative of the benzoxazocine class of centrally acting analgesics. Prepn: **NL 6606390** (1966 to Rexall); M. W. Klohs *et al.*, **US 3830803** (1974 to Riker). LC-MS-MS determn in plasma: G. Hoizey *et al.*, *J. Pharm. Biomed. Anal.* **42**, 593 (2006). Pharmacology: Bassett *et al.*, *Br. J. Pharmacol.* **37**, 69 (1969); Klohs *et al.*, *Arzneim.-Forsch.* **22**, 132 (1972). Clinical trial as post-operative analgesic: B. Du Manoir *et al.*, *Br.J. Anaesth.* **91**, 836 (2003). Review of pharmacology and therapeutic efficacy: R. C. Heel *et al.*, *Drugs* **19**, 249-267 (1980).

Hydrochloride. [23327-57-3] Fenazoxine; R-738; Acupan; Ajan. $C_{17}H_{19}NO.HCl$; mol wt 289.80. mp 238-242°. LD_{50} in mice, rats (mg/kg): 119, 178 orally; 44.5, 28 i.v. (Baltes).

THERAP CAT: Analgesic.

6526. Negamycin. [33404-78-3] 3,6-Diamino-2,3,4,6-tetra-deoxy-L-*threo*-hexonic acid 2-(carboxymethyl)-2-methylhydrazide; 3,6-diamino-5-hydroxyhexanoic acid 2-(carboxymethyl)-2-methyl-hydrazide; [2-(3,6-diamino-5-hydroxy-1-oxohexyl)-1-methylhydra-zino]acetic acid. $C_9H_{20}N_4O_4$; mol wt 248.28. C 43.54%, H 8.12%, N 22.57%, O 25.78%. Hydrazide antibiotic isolated from strains related to *Streptomyces purpeofuscus:* Hamada *et al., J. Antibiot.* **23**, 170 (1970). Prepn: H. Umezawa *et al.*, **ZA 7002933**; *idem*, **US 3679742** (1970, 1972 both to Microbiochem. Res. Found.). Structure and partial synthesis: Kondo *et al., J. Am. Chem. Soc.* **93**, 6305 (1971). Total synthesis of negamycin and its antipode: Shibahara *et al., ibid.* **94**, 4353 (1972). Synthesis of racemic negamycin: W. Streicher *et al., J. Antibiot.* **31**, 725 (1978); G. Pasquet *et al., Tetrahedron Lett.* **21**, 931 (1980); A. Pierdet *et al., Tetrahedron* **36**, 1763 (1980). Stereocontrolled synthesis of (+)-negamycin: Y. F. Wang *et al., J. Am. Chem. Soc.* **104**, 6465 (1982); S. DeBernardo *et al., Tetrahedron Lett.* **29**, 4077 (1988). Mechanism of action: Mizuno *et al., J. Antibiot.* **23**, 581 (1970); Y. Uehara *et al., Biochim. Biophys. Acta* **442**, 251 (1976).

Colorless powder, mp 110-120° (dec). $[\alpha]_D^{29}$ +2.5° (c = 2). Amphoteric compound. pKa values after treatment with HCl-methanol: 3.55, 8.10, 9.75. Sol in water. Practically insol in methanol, ethanol, butanol, ethyl acetate, chloroform and benzene. LD_{50} i.v. in mice: 400-500 mg/kg (Hameda).

THERAP CAT: Antibacterial.

6527. Nelarabine. [121032-29-9] 9-β-D-Arabinofuranosyl-6-methoxy-9*H*-purin-2-amine; 2-amino-9-β-D-arabinofuranosyl-6-methoxy-9*H*-purine; GW-506U78; 506U; Arranon; Atriance. $C_{11}H_{15}N_5O_5$; mol wt 297.27. C 44.44%, H 5.09%, N 23.56%, O 26.91%. Purine nucleoside; prodrug of guanine arabinoside. Prepn: T. A. Krenitsky *et al.*, **EP 294114** (1988 to Wellcome Foundation); *idem*, D. J. T. Porter, **US 5424295** (1995 to Burroughs Wellcome); D. R. Averett *et al., Antimicrob. Agents Chemother.* **35**, 851 (1991). HPLC determn of active metabolites in leukemia cells: C. O. Rodriguez Jr. *et al., J. Chromatogr. B* **745**, 421 (2000). Mechanism of action study: *idem, et al., Blood* **102**, 1842 (2003). Clinical pharmacokinetics: D. F. Kisor *et al., J. Clin. Oncol.* **18**, 995 (2000). Clinical evaluation in T-cell malignancies: S. L. Berg *et al., ibid.* **23**, 3376 (2005). Review of pharmacology and clinical experience: D. F. Kisor, *Ann. Pharmacother.* **39**, 1056-1063 (2005).

mp 209-217° (dec). uv max (50 mM potassium phosphate buffer (pH 7.0)-ethanol (9:1, v/v)): 247.5, 279 nm (ε 9100, 9300). $[\alpha]_D^{20}$ +55.9° (c = 0.27 in DMF). Slightly sol to sol in water.

THERAP CAT: Antineoplastic.

6528. Nelfinavir. [159989-64-7] (3*S*,4a*S*,8a*S*)-*N*-(1,1-Di-methylethyl)decahydro-2-[(2*R*,3*R*)-2-hydroxy-3-[(3-hydroxy-2-methylbenzoyl)amino]-4-(phenylthio)butyl]-3-isoquinolinecar-boxamide; [3S-(3R*,4aR*,8aR*,2'S*,3'S*)]-2-[2'-hydroxy-3'-phen-ylthiomethyl-4'-aza-5'-oxo-5'-(2''-methyl-3''-hydroxyphenyl)pen-tyl]decahydroisoquinoline-3-*N*-t-butylcarboxamide; AG-1346. C_{32}-

$H_{45}N_3O_4S$; mol wt 567.79. C 67.69%, H 7.99%, N 7.40%, O 11.27%, S 5.65%. Peptidomimetic HIV protease inhibitor. Prepn: B. A. Dressman *et al.*, **WO 9509843**; *eidem*, **US 5484926** (1995, 1996 both to Agouron). Physical properties: M. Longer *et al., J. Pharm. Sci.* **84**, 1090 (1995). HIV protease inhibition and antiviral activity: A. K. Patick *et al., Antimicrob. Agents Chemother.* **40**, 292 (1996). Preclinical pharmacokinetics and tissue distribution: B. V. Shetty *et al., ibid.* 110. Review of pharmacology and clinical efficacy in HIV infection: A. Bardsley-Elliot, G. L. Plosker *et al., Drugs* **59**, 581-620 (2000); of clinical experience: M. Olmo, D. Podzam-czer, *Expert Opin. Drug Metab. Toxicol.* **2**, 285-300 (2006).

White foam. $[\alpha]_D$ −119.23° (c = 0.26 in methanol). pKa_1 6.0; pKa_2 11.06. Log P (octanol/water): 4.1. Practically insol in water.

Methanesulfonate. [159989-65-8] Nelfinavir mesylate; AG-1343; Viracept. $C_{32}H_{45}N_3O_4S.CH_3SO_3H$; mol wt 663.89. White powder. pKa −1.20. Soly (mg/ml): 4.5 in water; 2.6 in 0.1*N* HCl. Soly (mg/g): 70 in glycerin; >100 in propylene glycol; >200 in PEG 400. Very sol in methanol, ethanol, acetonitrile. Practically insol in soybean oil, mineral oil.

THERAP CAT: Antiretroviral.

6529. Nemadectin. [102130-84-7] (6*R*,23*S*)-5-*O*-Demethyl-28-deoxy-25-[(1*E*)-1,3-dimethyl-1-buten-1-yl]-6,28-epoxy-23-hy-droxymilbemycin B; antibiotic S-541A; LL-F-28249α; F-28249α; CL-287088. $C_{36}H_{52}O_8$; mol wt 612.80. C 70.56%, H 8.55%, O 20.89%. Macrocyclic antibiotic produced by *Streptomyces cyaneo-griseus* ssp. *noncyanogenus* and *Streptomyces thermoachaensis*. Structurally related to the milbemycins and the avermectins, *q.q.v.* Isoln: I. B. Wood *et al.*, **EP 170006**; I. B. Wood, J. A. Pankavich, **US 4869901** (1986, 1989 both to Am. Cyanamid). Chemical and structural characterization: G. T. Carter *et al., Chem. Commun.* **1987**, 402; G. T. Carter *et al., J. Antibiot.* **41**, 519 (1988). NMR spectroscopy: S. Rajan, G. W. Stockton, *Magn. Reson. Chem.* **27**, 437 (1989). Antiparasitic activity in horses: E. T. Lyons *et al., Am. J. Vet. Res.* **50**, 970 (1989); in dogs: M. E. Doscher *et al., Vet. Parasitol.* **34**, 255 (1989); in sheep: J. A. Pankavich *et al, Vet. Rec.* **130**, 241, (1992).

White fluffy solid from *tert*-butanol. $[\alpha]_D^{26}$ +133° (c = 0.3 in acetone). uv max (methanol): 244 nm, (log ε 4.47). Readily sol in common organic solvents. Practically insol in water.

THERAP CAT (VET): Antiparasitic.

6530. Nemifitide. [173240-15-8] 4-Fluoro-L-phenylalanyl-(4R)-4-hydroxy-L-prolyl-L-arginylglycyl-L-tryptophanamide; 4-F-Phe-4-OH-Pro-Arg-Gly-Trp-NH₂; netamifide. $C_{33}H_{43}FN_{10}O_6$; mol wt 694.77. C 57.05%, H 6.24%, F 2.73%, N 20.16%, O 13.82%. Synthetic pentapeptide produced by modification of melanocyte stimulating inhibitory factor. Prepn: H. B. Abajian, *et al.*, **WO 9530430**; *eidem*, **US 5589460** (1995, 1996 both to Innapharma). Structure activity study: J. J. Hlavka, *J. Appl. Res.* **2**, 58 (2002). Clinical pharmacokinetics: J. P. Feighner *et al.*, *Biopharm. Drug Dispos..* **23**, 33 (2002); G. Nicolau *et al.*, *ibid.* **26**, 379 (2005). Clinical evaluation: *eidem*, *Int. J. Neuropsychopharmacol.* **6**, 207 (2003). Review of development and therapeutic potential: J. Dingemanse, *Curr. Opin. Investig. Drugs* **4**, 859-862 (2003).

Bis(trifluoroacetate). [204992-09-6] Nemifitide ditriflutate; INN-00835. $C_{33}H_{43}FN_{10}O_6 \cdot 2C_2HF_3O_2$; mol wt 922.82.
THERAP CAT: Antidepressant.

6531. Nemonapride. [75272-39-8] *rel*-5-Chloro-2-methoxy-4-(methylamino)-*N*-[(2*R*,3*R*)-2-methyl-1-(phenylmethyl)-3-pyrrolidinyl]benzamide; (±)-*cis*-*N*-(1-benzyl-2-methyl-3-pyrrolidinyl)-5-chloro-4-(methylamino)-*o*-anisamide; *cis*-*N*-(1-benzyl-2-methylpyrrolidin-3-yl)-5-chloro-2-methoxy-4-(methylamino)benzamide; emonapride; YM-09151-2; Emilace. $C_{21}H_{26}ClN_3O_2$; mol wt 387.91. C 65.02%, H 6.76%, Cl 9.14%, N 10.83%, O 8.25%. Selective dopamine D₂ receptor antagonist. Prepn: M. Takashima *et al.*, **JP Kokai 79 14965**; *eidem*, **US 4210660** (1979, 1980 both to Yamanouchi); S. Iwanami *et al.*, *J. Med. Chem.* **24**, 1224 (1981). Pharmacology: M. Yamamoto *et al.*, *Neuropharmacology* **21**, 945 (1982). Dopamine receptor binding study: C. W. Grewe *et al.*, *Eur. J. Pharmacol.* **81**, 149 (1982). Use as a label for D₂ receptors: A. S. Unis *et al.*, *Life Sci.* **47**, PL151 (1990). HPLC determn in plasma: T. Nagasaki *et al.*, *J. Chromatogr. B* **714**, 293 (1998). Clinical evaluation in schizophrenia: K. Satoh *et al.*, *Int. Clin. Psychopharmacol.* **11**, 279 (1996).

Relative stereochemistry

Colorless crystals from isopropanol, mp 152-153° (Takashima, 1980). Also reported as mp 150° (Iwanami).
THERAP CAT: Antipsychotic.

6532. Neoarsphenamine. [457-60-3] Sulfoxylic acid mono-[[[5-[(3-amino-4-hydroxyphenyl)diarsenyl]-2-hydroxyphenyl]amino]methyl] ester sodium salt (1:1); [5-[(3-amino-4-hydroxyphenyl)-arseno]-2-hydroxyanilino]methanol sulfoxylate sodium; arsphenamine methylenesulfoxylic acid sodium salt; 3,3′-diamino-4,4′-dihydroxyarsenobenzenemethylenesulfoxylate sodium; Neosalvarsan; Collunovar; N.A.B.; Neo-Arsoluin; Vetarsenobillon; Novarsenobillon; Arsevan; Novarsan; Novarsenobenzol; Miarsenol. $C_{13}H_{13}As_2N_2NaO_4S$; mol wt 466.15. C 33.50%, H 2.81%, As 32.14%, N 6.01%, Na 4.93%, O 13.73%, S 6.88%. Medicinal grade contains a small amount of inert inorganic salts and some solvent. *The National Formulary* requires 19+% As. Prepn from arsphenamine + sodium formaldehydesulfoxylate: Krumwiede, *J. Am. Pharm. Assoc.* **8**, 795 (1919); Heyl, Miller, *ibid.* **11**, 432 (1922); Dohr, **US 1549465** (1925); Kober, **US 1564859** (1926); Kraft *et al.*, **SU 158388** (1962).

Physicochemical properties and toxicity: H. N. Wright *et al.*, *J. Pharmacol. Exp. Ther.* **73**, 12 (1941).

Yellow powder; odorless or slight odor. Oxidizes in air, becoming darker and more toxic; higher temps accelerate the oxidation; hence marketed in air-evacuated ampuls or filled with a nonoxidizing gas. Very sol in water; sol in glycerol. Slightly sol in alcohol or acetone. Practically insol in chloroform, ether. Its aq soln is practically neutral, unlike arsphenamine, which is acid.
THERAP CAT (VET): Has been used in contagious pleuropneumonia, babesiasis, equine petechial fever, eperythrozoonosis.

6533. Neocembrene. [31570-39-5] (1*E*,5*E*,9*E*,12*R*)-1,5,9-Trimethyl-12-(1-methylethyl)-1,5,9-cyclotetradecatriene; cembrene A; neocembrene A. $C_{20}H_{32}$; mol wt 272.48. C 88.16%, H 11.84%. Termite trail pheromone with all-*trans* configuration isolated from *Nasutitermes exitiosus* (Hill): B. P. Moore, *Nature* **211**, 746 (1966). Structure: A. J. Birch *et al.*, *J. Chem. Soc. Perkin Trans. 1* **1972**, 2653; V. D. Patil *et al.*, *Tetrahedron* **29**, 341 (1973). Synthesis: M. Kodama *et al.*, *Tetrahedron Lett.* **1975**, 3065; Y. Kitahara *et al.*, *Chem. Lett.* **1976**, 219; H. Takayanagi *et al.*, *Chem. Commun.* **1978**, 359. Synthesis of neocembrene and isomers: T. Kato *et al.*, *J. Org. Chem.* **45**, 1126 (1980).

Colorless oil, faint wax-like odor. bp₀.₈ 150-152°. n_D^{30} 1.5102.

6534. Neocuproine. [484-11-7] 2,9-Dimethyl-1,10-phenanthroline; 2,9-dimethyl-*o*-phenanthroline. $C_{14}H_{12}N_2$; mol wt 208.26. C 80.74%, H 5.81%, N 13.45%. Prepn: Case, *J. Am. Chem. Soc.* **70**, 3994 (1948); O'Reilly, Plowman, *Aust. J. Chem.* **13**, 145 (1960).

Hemihydrate. [34302-69-7] $C_{14}H_{12}N_2 \cdot \frac{1}{2}H_2O$; mol wt 217.27. Crystals from water or ligroin, mp 159-160°.
Dihydrate. [34302-70-0] $C_{14}H_{12}N_2 \cdot 2H_2O$; mol wt 244.29. Needles from water. Loses water over P_2O_5 or at 80°.
USE: Clinical reagent (blood glucose assay). In spectrophotometric determination of copper, *see:* Nebesar, *Anal. Chem.* **36**, 1961 (1964).

6535. Neodymium. [7440-00-8] Nd; at. wt 144.242; at. no. 60; valences 2, 3, 4. A lanthanide; belongs to the cerium group of rare earth metals. Naturally occurring isotopes (mass numbers): 142 (27.13%); 143 (12.18%); 144 (23.80%), radioactive, $T_{1/2}$ 2.1 × 10¹⁵ years, α-emitter; 145 (8.30%); 146 (17.19%); 148 (5.76%); 150 (5.64%). Known artificial radioactive isotopes: 127-141; 147; 149; 151; 152. Abundance in earth's crust: 24-40 ppm. Commercially important sources are the rare earth minerals monazite and bastnaesite; also found in cerite and gadolinite. Discovered by von Welsbach in 1885. Sepn from Pr: F. H. Spedding *et al.*, *J. Am. Chem. Soc.* **69**, 2786 (1947); **76**, 2557 (1954); Kauffman, Blank, *J. Chem. Educ.* **37**, 156 (1960); from other rare earths: Spedding *et al.*, *J. Am. Chem.*

Soc. **69**, 2812 (1947). Metal prepd by electrolysis of the chloride: Matignon, *Compt. Rend.* **131**, 891 (1900); Sieverts, Roell, *Z. Anorg. Chem.* **150**, 261 (1926); by reduction of the chloride: Bommer, Hohmann, *ibid.* **248**, 357 (1941). Spectrum: Alberston *et al.*, *Phys. Rev.* **61**, 167 (1942). Reviews of prepn, properties and compds: *The Rare Earths*, F. H. Spedding, A. H. Daane, Eds. (Krieger, Huntington, N.Y., 1971, reprint of 1961 ed.) 641 pp; Hulet, Bode, "Separation Chemistry of the Lanthanides and Transplutonium Actinides" in *MTP Int. Rev. Sci.: Inorg. Chem., Ser. One* vol. **7**, K. W. Bagnall, Ed. (University Park Press, Baltimore, 1972) pp 1-45; Moeller, "The Lanthanides" in *Comprehensive Inorganic Chemistry* vol. **4**, J. C. Bailar Jr. *et al.*, Eds. (Pergamon Press, Oxford, 1973) pp 1-101; F. H. Spedding in *Kirk-Othmer Encyclopedia of Chemical Technology* vol. **19** (John Wiley & Sons, New York, 3rd ed., 1982) pp 833-854; *Chemistry of the Elements*, N. N. Greenwood, A. Earnshaw, Eds. (Pergamon Press, New York, 1984) pp 1423-1449. Review of toxicology: T. J. Haley, *J. Pharm. Sci.* **54**, 663-670 (1965). Brief review of properties: G. T. Seaborg, *Radiochim. Acta* **61**, 115-122 (1993).

Silver-white metal, becomes yellowish on exposure to air. Crystalline forms: hexagonal α-form, d 7.003, transforms to β-form at 868°; body-centered cubic β-form, d 6.80, exists at >868°. mp 1021°. bp 3074°. Heat of fusion: 7.134 kJ/mol. Heat of sublimation (25°): 327.6 kJ/mol. E°(aq) Nd^{3+}/Nd −2.44 V (calc). Experimental reduction potentials (referred to a normal calomel electrode): −1.870, −1.960 V: Noddack, Brukl, *Angew. Chem.* **50**, 362 (1937). Reacts slowly with cold water; rapidly on heating.

Oxide. Nd_2O_3. Commercial uses: Pings, *Colo. Sch. Mines, Miner. Ind. Bull.* **12**, no. 2 (1969) 19 pp. Blue powder, hexagonal structure, exhibits a slightly red fluorescence. Prepd by heating the hydroxide, carbonate, nitrate or oxalate. Very stable. Sol in dil acids; soly in water: $5.7×10^{-6}$ g-mol/l at 29°.

Hydroxide. $Nd(OH)_3$. Bluish or pink precipitate; on heating at 300-350° is converted into $2Nd_2O_3.3H_2O$, grayish-brown; on further increase in temp is converted into $Nd_2O_3.H_2O$.

Chloride. $NdCl_3$. Large purple prisms. Sol in water, in alcohol. Forms addition compds with ammonia. A hexahydrate is obtained from the aq soln; the hydrate is very sol in water (2.46 parts per 1 part of water); mp 124°. LD_{50} in mice (mg/kg): 4000 s.c., 348.3 i.p.; in male mice (mg/kg): 600 i.p., 5250 orally; in rats (mg/kg): 150-250 i.p.; in guinea pigs (mg/kg): 70 i.v., 139.6 i.p.; in rabbits (mg/kg): 200-250 i.v. (Haley).

Sulfate. $Nd_2(SO_4)_3$. Pinkish needles; prepd by heating the oxide with concd sulfuric acid. Heat of formation 57.2 kcal. Sol in water; heat of soln 36.5 kcal. Is dec at 700-800°. Forms several double salts. A penta-, an octa-, and a pentadecahydrate have been prepared.

Nitrate. $Nd(NO_3)_3$. Penta- and hexahydrates; prepd by adding the oxide to nitric acid. LD_{50} (hexahydrate) in female mice (mg/kg): 270 i.p.; in female rats (mg/kg): 270 i.p., 2750 orally, 6.4 i.v.; in male rats (mg/kg): 66.8 i.v. (Haley).

USE: Oxide as glass coloring agent; in glass filter plate laminated on color TV tubes to improve contrast and brightness; in lasers. Alloyed with iron and boron to produce powerful permanent magnets.

6536. Neoergosterol. [516-98-3] (3β,22E)-19-Norergosta-5,-7,9,22-tetraen-3-ol. $C_{27}H_{40}O$; mol wt 380.62. C 85.20%, H 10.59%, O 4.20%. Prepd from ergosterol: Inhoffen, *Ann.* **497**, 130 (1932); Windaus, Deppe, *Ber.* **70**, 76 (1937); Mosettig, Scheer, *J. Org. Chem.* **17**, 764 (1952). Stereochemistry: Steele *et al.*, *J. Am. Chem. Soc.* **85**, 1134 (1963).

Needles from methanol, mp 152-154°. $[α]_D^{17}$ −11° (c = 2 in chloroform). Practically insol in water. Sol in organic solvents. Precipitated by digitonin.

Methyl ether. $C_{28}H_{42}O$. mp 94°. $[α]_D^{20}$ −5° (c = 1.2 in chloroform).

6537. Neohesperidin Dihydrochalcone. [20702-77-6] 1-[4-[[2-O-(6-Deoxy-α-L-mannopyranosyl)-β-D-glucopyranosyl]oxy]-2,6-dihydroxyphenyl]-3-(3-hydroxy-4-methoxyphenyl)-1-propanone; 3,5-dihydroxy-4-(3-hydroxy-4-methoxyhydrocinnamoyl)-phenyl-2-O-(6-deoxy-α-L-mannopyranosyl)-β-D-glucopyranoside; neohesperidin DHC; NHDC; Sukor. $C_{28}H_{36}O_{15}$; mol wt 612.58. C 54.90%, H 5.92%, O 39.18%. Prepn from naringen, a flavanone glycoside occurring naturally in grapefruit: Horowitz, Gentile, US **3087821** and US **3375242** (1963, 1968, both to U.S. Secy. Agr.); Robertson *et al.*, *Ind. Eng. Chem. Prod. Res. Dev.* **13**, 125 (1974). Sweetening effect: Inglett *et al.*, *J. Food Sci.* **34**, 101 (1969). Chromatography: Gentile, Horowitz, *J. Chromatogr.* **63**, 467 (1971); J. F. Fisher, *J. Agric. Food Chem.* **25**, 682 (1977). Subchronic oral toxicity study: B. A. R. Lina *et al.*, *Food Chem. Toxicol.* **28**, 507 (1990). *Review:* P. J. Pratter, *Perfum. Flavor.* **5**, 12-18 (1981).

Crystals from acetone, mp 156-158°. 1000 to 1500 times sweeter than sucrose; 20 times sweeter than saccharin.

USE: Sweetening agent, especially in chewing gum and dentifrices.

6538. Neomethymycin. [497-73-4] (3R,4S,5S,7R,9E,11R,-12S)-12-[(1R)-1-Hydroxyethyl]-3,5,7,11-tetramethyl-4-[[3,4,6-tri-deoxy-3-(dimethylamino)-β-D-xylohexopyranosyl]oxy]oxacyclododec-9-ene-2,8-dione; (12R)-10-deoxy-12-hydroxymethymycin. $C_{25}H_{43}NO_7$; mol wt 469.62. C 63.94%, H 9.23%, N 2.98%, O 23.85%. Macrolide antibiotic found in the mother liquors from methymycin, *q.v.* Differs from methymycin only in the location of one hydroxyl group, which results in marked changes in chemical behavior: Djerassi, Halpern, *J. Am. Chem. Soc.* **79**, 2022, 3926 (1957); *Tetrahedron* **3**, 255 (1958). Production from *Streptomyces venezuelae* cultures: Dutcher *et al.*, US **2916486** (1959 to Olin Mathieson). Synthesis of the aglycone, *neomethynolide*: J. Inanago *et al.*, *Chem. Lett.* **1981**, 1415.

Crystals from ether + hexane, mp 156-158°. $[α]_D^{25}$ +93°. uv max (ethanol): 227.5 nm (log ε 4.10). Solvates easily and when crystallized from dilute acetone yields 1 cm long crystals of the acetone solvate. Soluble in alcohol, acetone, chloroform, benzene, ethyl acetate, ether. Slightly sol in water, dibutyl ether. Practically insol in hexane, aliphatic hydrocarbons. *Compare also* Picromycin.

6539. Neomycin. [1404-04-2] Mycifradin; Fradiomycin; Neomin; Neolate; Neomas; Pimavecort; Vonamycin Powder V. Antibiotic complex composed of neomycins A, B and C. Produced by *Streptomyces fradiae*: Waksman, Lechevalier, *Science* **109**, 305 (1949); Waksman *et al.*, *J. Clin. Invest.* **28**, 934 (1949); Swart *et al.*, *ibid.* 1045; Waksman, Lechevalier, US **2799620** (1957). Purification: Jackson, **US 2848365** (1958 to Upjohn); Haak, **US 3108996** (1963 to Upjohn). Recovery: Miller, **US 3005815** (1961 to Merck & Co.); Moses, **US 3022228** (1962 to Penick); **GB 945475** (1964 to O.W.G. Chemie). Characterization: Peck *et al.*, *J. Am. Chem. Soc.* **71**, 2590 (1949); Regna, Murphy, *ibid.* **72**, 1045 (1950); Dutcher *et al.*, *ibid.* **73**, 1384 (1951). Structure of neomycins B and C: K. L. Rinehart *et al.*, *ibid.* **79**, 4567, 4568 (1957); *ibid.* **84**, 3216, 3218 (1962). Abs config of neomycin C: M. Hichens, K. L. Rinehart, *ibid.* **85**, 1547 (1963). Total synthesis of neomycin C: S. Umewaza, Y. Nishimura, *J. Antibiot.* **30**, 189 (1977); S. Umewaza *et al.*, *Bull. Chem. Soc. Jpn.* **53**, 3259 (1980); of neomycin B: T. Usui, S. Umezawa, *J. Antibiot.* **40**, 1464 (1987). Monographs: S. A. Waksman, *Neomycin* (Rutgers Univ. Press, New Brunswick, N. J., 1953) 219 pp; K. L. Rinehart, Jr., *The Neomycins and Related Antibiotics* (Wiley, New York, 1964) 137 pp. Comprehensive description: W. F. Heyes, *Anal. Profiles Drug Subs.* **8**, 399-488 (1979).

Neomycin complex is an amorphous base sol in water, methanol and acidified alcohol. Practically insol in common organic solvents. Solns up to 250 mg/ml H$_2$O may be prepared.

Neomycin A *see* Neamine.

Neomycin B. [119-04-0] Antibiotique EF 185; framycetin; Enterfram; Framygen; Actilin. C$_{23}$H$_{46}$N$_6$O$_{13}$; mol wt 614.65. Identity of neomycin B and framycetin: K. L. Rinehart *et al.*, *ibid.* **82**, 3938 (1960). Yields on hydrolysis neamine and *neobiosamine B*. Structure of neobiosamine B: K. L. Rinehart *et al.*, *J. Am. Chem. Soc.* **82**, 2970 (1960).

Neomycin B hydrochloride. [25389-99-5] Amorphous white powder. [α]$_D^{20}$ +57° (H$_2$O). Soly in mg/ml at ~28°: water 15.0; methanol 5.7; ethanol 0.65; isopropanol 0.05; isoamyl alcohol 0.33; cyclohexane 0.06; benzene 0.03. Practically insol in acetone, ether, other organic solvents. For additional soly data *see:* Weiss *et al.*, *Antibiot. Chemother.* **7**, 374 (1957).

Neomycin B sulfate. [1405-10-3] Biosol; Bykomycin; Endomixin; Fraquinol; Myacine; Neosulf; Neomix; Neobrettin; Nivemycin; Soframycin; Tuttomycin. Amorphous, hygroscopic white powder. Practically tasteless. [α]$_D^{20}$ +54° (c = 2 in H$_2$O). Soly in mg/ml at ~28°: water 6.3; methanol 0.225; ethanol 0.095; isopropanol 0.082; isoamyl alc, 0.247; cyclohexane 0.08; benzene 0.05. Insol in acetone, ether, chloroform. Aq solns are fairly stable at pH 2 to 9. Highly purified prepns are very stable to alkali and unstable to acids. Refluxing with barium hydroxide for 18 hrs showed no loss of activity. Boiling with mineral acids yields an aldehyde, characterized as furfural, and an organic base.

Neomycin C. [66-86-4] C$_{23}$H$_{46}$N$_6$O$_{13}$; mol wt 614.65. Yields on hydrolysis neamine and *neobiosamine C*. Structure of neobiosamine C: K. L. Rinehart, P. Woo, *J. Am. Chem. Soc.* **80**, 6463 (1958).

THERAP CAT: Antibacterial.

THERAP CAT (VET): Antibacterial.

6540. Neon. [7440-01-9] Ne; at. wt 20.1797; at. no. 10. Group VIIIA (18), also known as Group 0. A noble gas characterized by an electronic structure in which the outer *p* subshell is entirely filled. Naturally occurring stable isotopes (mass numbers): 20 (90.51%); 21 (0.27%); 22 (9.92%); known artificial radioactive isotopes: 16-19; 23-25. Longest-lived known isotope: 24 (T$_{1/2}$ 3.38 min, β-emitter). Abundance in igneous rock of the earth's crust: 7

× 10^{-5} ppm by wt; concentration in air: 18.2 ppm by vol. Obtained commercially from the atmosphere by distillation-liquefaction process. Discovered in 1898 by Travers and Ramsay: *The Discovery of the Rare Gases*, London, 1928. Monograph: *Argon, Helium and the Rare Gases* **vols. 1 2,**, G. A. Cook, Ed. (Interscience, New York, 1961) 818 pp. *Reviews:* Cockett, Smith, "The Monatomic Gases" in *Comprehensive Inorganic Chemistry* **vol. 1**, J. C. Bailar, Jr. *et al.*, Eds. (Pergamon Press, Oxford, 1973) pp 139-211; S.-C. Hwang, W. R. Weltmer, Jr. in *Kirk-Othmer Encyclopedia of Chemical Technology* **vol. 13** (John Wiley & Sons, 4th ed., 1995) p 1-38; *Chemistry of the Elements*, N. N. Greenwood, A. Earnshaw, Eds. (Pergamon Press, New York, 1984) pp 1042-1059. Review of use in inductively coupled plamsa-mass spectrometry: S. F. Durrant, *Fresenius J. Anal. Chem.* **347**, 389-392 (1993).

Colorless, odorless, tasteless, monatomic, inert gas; non-flammable. Will form compds with highly electronegative elements such as O, F, Cl. Does not condense at the temp of liquid air; solid at the temp of liquid hydrogen; the solid form exists as face-centered cubic crystals at normal pressure. Soly of gas in water (20°): 10.5 cm^3/kg water. Triple pt temp 24.562 K, press 43.37 kPa. Critical temp 44.40 K, critical press 2654 kPa, critical d 483 kg/m^3. Gas: d^0 (101.3 kPa) 0.89994 kg/m^3, d (normal bp) 9.552 kg/m^3. Liquid: normal bp −246.048°, d (normal bp) 1207 kg/m^3, d (triple pt) 1247 kg/m^3, heat of vaporization (normal bp) 1.741 kJ/mol. Solid: d (triple pt) 1444 kg/m^3, heat of vaporization (triple pt) 2.139 kJ/mol, heat of fusion (triple pt) 0.335 kJ/mol. Emission spectra: T. Jacksier, R. M. Barnes, *Appl. Spectrosc.* **48**, 65 (1994).

Caution: Can act as a simple asphyxiant by displacing air. *See: Matheson Gas Data Book* (Matheson Co., Inc., 4th ed., East Rutherford, NJ, 1966) pp 359-361.

USE: Gas in neon light tubes; ingredient of gaseous fillers for antifog devices, warning signals, electrical current detectors, high-voltage indicators for high-tension electric lines, lightning arresters, wave-meter tubes; in Ne-He lasers; in mixtures with He and Ar in Geiger counters. Liquid as cryogen to produce low temps.

6541. Neopentane. [463-82-1] 2,2-Dimethylpropane; tetramethylmethane. C$_5$H$_{12}$; mol wt 72.15. C 83.24%, H 16.77%. Found in petr naphtha. Prepn from *tert*-butyl iodide and dimethylzinc: Lwow, *Z. Chem.* **1870**, 520; *cf. J. Am. Chem. Soc.* **54**, 3460 (1932); from butylmagnesium iodide and dimethyl sulfate in ether: Ferrario, Fagetti, *Gaz. Chim.* **38, II**, 663 (1908); from amylene vapor and hydrogen in presence of an electric discharge: *ibid.* **62**, 621 (1932); by passing pentanes over aluminum chloride at 140°: *J. Gen. Chem. USSR* **14**, 343 (1944), *C.A.* **39**, 3783 (1945); from neopentylmagnesium chloride and water: Whitmore *et al.*, *J. Am. Chem. Soc.* **56**, 749 (1934).

Liquid or gas. d$_0^0$ 0.613 (liq). Solidifies to form tetragonal crystals, mp −19.8°. bp 9.5°. Insol in water.

6542. Neopentyl Alcohol. [75-84-3] 2,2-Dimethyl-1-propanol; *tert*-butyl carbinol; neoamyl alcohol; neopentanol. C$_5$H$_{12}$O; mol wt 88.15. C 68.13%, H 13.72%, O 18.15%. Prepn: Conant *et al.*, *J. Am. Chem. Soc.* **51**, 1246 (1929). Review of manuf (by fractionation of fusel oil and via chlorination of pentanes): *Industrial Chemicals*, W. L. Faith *et al.*, Eds. (Wiley & Sons, New York, 2nd ed., 1957) pp 107-114.

Volatile crystals, peppermint odor. bp 114°. mp 53°. d$_4^{20}$ 0.812. Vapor pressure: 16 mm at 20°; 50.5 mm at 46°; 533.5 mm at 100°. Slightly sol in water (~3.5% at 25°); miscible with alcohol, ether.

6543. Neopentyl Glycol. [126-30-7] 2,2-Dimethyl-1,3-propanediol; dimethyltrimethylene glycol. C$_5$H$_{12}$O$_2$; mol wt 104.15. C 57.66%, H 11.61%, O 30.72%. Prepd from isobutyraldehyde.

Needles from benzene, mp 127°. bp$_{760}$ 208°. Solubility in water ~65% w/w. Freely sol in alcohol, ether.

USE: In the manuf of plasticizers, polyesters, as modifier of alkyd resins.

6544. Neophyl Chloride. [515-40-2] (2-Chloro-1,1-dimeth-ylethyl)benzene; 1-chloro-2-methyl-2-phenylpropane; (β-chloro-*tert*-butyl)benzene. C$_{10}$H$_{13}$Cl; mol wt 168.66. C 71.21%, H 7.77%, Cl 21.02%. Prepn (68% yield) from benzene and methallyl chloride in the presence of concd sulfuric acid: Whitmore *et al.*, *J. Am. Chem. Soc.* **65**, 1469 (1943); Smith, Sellas, *Org. Synth.* **coll. vol. IV**, 702 (1963).

Liquid. d$_4^{25}$ 1.0379. bp$_{741}$ 221° (dec); bp$_{90}$ 111°; bp$_{30}$ 120°; bp$_{20}$ 105°; bp$_{18}$ 104°; bp$_{13}$ 97°; bp$_{10}$ 95°; bp$_{1.0}$ 53°. n_D^{20} 1.5249; n_D^{25} 1.5228. Considerably less reactive with sodium than neopentyl chloride. Forms a Grignard reagent readily and in good yield. Salt effect in acetolysis: Fainberg, Winstein, *J. Am. Chem. Soc.* **78**, 2763 (1956).

6545. Neopine. [467-14-1] (5α,6α)-8,14-Didehydro-4,5-ep-oxy-3-methoxy-17-methylmorphinan-6-ol; β-codeine. C$_{18}$H$_{21}$NO$_3$; mol wt 299.37. C 72.22%, H 7.07%, N 4.68%, O 16.03%. An alkaloid from opium, isomeric with codeine. Prepn from 14-bromo-codeinone: Conroy, **US 2797222** (1957 to Merck & Co.); from the-baine: S. Makleit *et al.*, *Acta Chim. Acad. Sci. Hung.* **94**, 161 (1977).

Long needles from petr ether, mp 127.5°. [α]$_D^{23}$ −28° (c = 7.5 in chloroform). The absorption spectrum seems identical with that of codeine: Dobbie, Lauder, *J. Chem. Soc.* **99**, 34 (1911). Solubilities about the same as those of codeine, *q.v.*

Hydrobromide. C$_{18}$H$_{21}$NO$_3$.HBr. Prismatic crystals from wa-ter, darkens about 240°, dec 283°. [α]$_D^{23}$ +17° (c = 3.7). Relatively insol in water, making possible a separation from other more sol hydrobromides of various opium alkaloids.

Note: Incorrectly called **hydroxycodeine.** *See:* L. J. Sargent, U. Weiss, *J. Org. Chem.* **25**, 987 (1960).

6546. Neoprene. Duprene; GR-M; polychloroprene; poly(2-chloro-1,3-butadiene). Mol wt range of 100,000-300,000. An oil-resistant synthetic rubber, with predominantly *trans* configuration made by polymerization of chloroprene. Prepn: W. H. Carothers *et al.*, *J. Am. Chem. Soc.* **53**, 4203 (1931). *Review:* P. J. Johnson in *Kirk-Othmer Encyclopedia of Chemical Technology* **vol. 8** (Wiley-Interscience, New York, 3rd ed., 1979) pp 515-534.

d^{20} approx 1.23. n_D^{20} approx 1.5512. Brittle point: −35°. Softens *ca.* 80°. Has high tensile strength, resilience and abrasion resistance.

USE: In mechanical and automotive products; as wire and cable jackets, gaskets, roof coatings, binder for fibers.

6547. Neopterin. [670-65-5] 2-Amino-6-(1,2,3-trihydroxy-propyl)-4(3H)-pteridinone; 1-(2-amino-4-hydroxy-6-pteridinyl)-1,-2,3-propanetriol; 6-(1′,2′,3′-trihydroxy)pterin; Crithidia factor. C$_9$-H$_{11}$N$_5$O$_4$; mol wt 253.22. C 42.69%, H 4.38%, N 27.66%, O 25.27%. Precursor in the biosynthesis of biopterin, *q.v.* Of the four possible isomers, D-*erythro*, L-*erythro*, D-*threo*, L-*threo*, two have been found in nature: the D-*erythro* form, to which the term neo-pterin originally referred; first isolated from the pupae of bees: Rem-bold, Buschmann, *Ann.* **662**, 72 (1963); the L-*threo* form, found to be the growth factor for the protozoan *Crithidia fasciculata* and iso-latable from cell-free extracts of *Serratia indica:* Kobashi, Iwai, *Agric. Biol. Chem.* **35**, 47 (1971); **36**, 1685, 1695 (1972). Both nat-ural forms found in human urine: Fukushima, Shiota, *J. Biol. Chem.* **247**, 4549 (1972). Early synthetic studies and structure: Rembold, Buschmann, *Ber.* **96**, 1406 (1963). Synthesis of L-(−)-form: Vis-contini, Provenzale, *Helv. Chim. Acta* **51**, 1495 (1968). Synthesis of natural D-neopterin: Viscontini *et al.*, *ibid.* **53**, 1202 (1970).

D-*erythro*-Neopterin

D-*erythro*-Form. [2009-64-5] [α]$_D^{25}$ +45 ± 3° (c = 0.3 in 0.1N HCl).

L-*erythro*-Form. [α]$_D^{25}$ −44 ± 3° (c = 0.3 in 0.1N HCl).

D-*threo*-Form. [α]$_D^{25}$ −92 ± 3° (c = 0.3 in 0.1N HCl).

L-*threo*-Form. Pale yellow, spiny crystals. [α]$_D^{25}$ +97 ± 3° (c = 0.3 in 0.1N HCl). uv max in 0.1N NaOH: 255, 363 nm (ε 20,900, 7050); in 0.1N HCl: 248, 323 nm (ε 11,200, 7880). Strong blue fluorescence in neutral or alkaline soln; weak fluorescence in acidic soln.

6548. Neoquassin. [76-77-7] 16-Hydroxy-2,12-dimethoxy-picrasa-2,12-diene-1,11-dione; 3a,4,5,6a,7,7a,8,11a,11b,11c-deca-hydro-5-hydroxy-2,10-dimethoxy-3,8,11a,11c-tetramethylphenan-thro[10,1-*bc*]pyran-1,11-dione. C$_{22}$H$_{30}$O$_6$; mol wt 390.48. C 67.67%, H 7.74%, O 24.58%. Found together with the quassin in the mixture of bitter constituents of the wood of *Quassia amara* L., *Simaroubaceae*, known in commerce as Surinam quassia. Forms quassin on oxidation. Isoln: London *et al.*, *J. Chem. Soc.* **1950**, 3431. Structure: Valenta *et al.*, *Tetrahedron Lett.* **1960** (20), 25; Carman, Ward, *ibid.* **1961**, 317; Valenta *et al.*, *Tetrahedron* **15**, 100 (1961).

Very bitter polymorphous crystals. Stable form, thick prisms, mp 227.5-228.5°; unstable form, long narrow plates, mp 213°. [α]$_D^{20}$ +41.0° (c = 4.98 in chloroform). uv max ~255 nm (ε ~11,650). Sol in cold acetone, chloroform, pyridine, acetic acid, in warm ethyl acetate, benzene, alcohol. Sparingly sol in ether, petr ether.

6549. Neostigmine. [59-99-4] 3-[[(Dimethylamino)carbon-yl]oxy]-N,N,N-trimethylbenzenaminium; (3-dimethylcarbamoxy-phenyl) trimethylammonium; synstigmin; proserine. [C$_{12}$H$_{19}$N$_2$-O$_2$]$^+$. Reversible cholinesterase inhibitor. Synthesis: Aeschlimann, **US 1905990** (1933 to Hoffmann-La Roche); Yanagisawa, **JP 51 3071** (1951), *C.A.* **47**, 4908g (1953). Pharmacology and pharmaco-kinetics: T. N. Calvey *et al.*, *Br. J. Clin. Pharmacol.* **7**, 149 (1979).

Comprehensive description: A. A. Al-Badr, M. Tariq, *Anal. Profiles Drug Subs.* **16**, 403-444 (1987). HPLC determn in plasma and CSF: F. Varin *et al.*, *J. Chromatogr. B* **723**, 319 (1999). Clinical evaluation in treatment of acute colonic pseudo-obstruction: R. J. Ponec *et al.*, *N. Engl. J. Med.* **341**, 137 (1999). Clinical comparison with edrophonium for neuromuscular block reversal: M. Naguib, W. Riad, *Can. J. Anaesth.* **47**, 1074 (2000).

Bromide. [114-80-7] Prostigmin (tabl.). $C_{12}H_{19}BrN_2O_2$; mol wt 303.20. White, crystalline powder with bitter taste. Crystals from alcohol + ether, dec ~167°. One gram dissolves in ~1 ml water. Sol in alc.

Methyl sulfate. [51-60-5] Intrastigmina; Normastigmin; Prostigmin (amp.). $C_{13}H_{22}N_2O_6S$; mol wt 334.39. Crystals from alc, mp 142-145°. One gram dissolves in ~10 ml water, less sol in alcohol. LD_{50} in mice (mg/kg): 0.16 i.v.; 0.42 s.c.; 7.5 orally. *See:* L. O. Randall, G. Lehmann, *J. Pharmacol. Exp. Ther.* **99**, 16 (1950).

THERAP CAT: Cholinergic; antidote (curare). In treatment of myasthenia gravis. Reversal agent for neuromuscular blocking drugs.

THERAP CAT (VET): Cholinergic; antidote (curare). In treatment of myasthenia gravis. Reversal agent for neuromuscular blocking drugs.

6550. Neotame. [165450-17-9] *N*-(3,3-Dimethylbutyl)-L-α-aspartyl-L-phenylalanine 2-methyl ester; *N*-[*N*-(3,3-dimethylbutyl)-L-α-aspartyl]-L-phenylalanine 1-methyl ester. $C_{20}H_{30}N_2O_5$; mol wt 378.47. C 63.47%, H 7.99%, N 7.40%, O 21.14%. Alkylated dipeptide reported to be 6000 to 10000 times sweeter than sucrose; structurally related to aspartame, *q.v.* Prepn: C. Nofre, J.-M. Tinti, **FR 2697844**; *eidem*, **US 5480668** (1994, 1996); and sweetness evaluation: I. Prakash *et al.*, *Synth. Commun.* **29**, 4461 (1999). Crystal structure: D. J. Wink *et al.*, *Acta Crystallogr.* **C55**, 1365 (1999). Analysis of polymorphic forms: B. E. Padden *et al.*, *Anal. Chem.* **71**, 3325 (1999). Review of discovery, properties, and utility: C. Nofre, J.-M. Tinti, *Food Chem.* **69**, 245-257 (2000).

Occurs as the monohydrate. White crystalline powder from ethyl acetate/hexane, mp 80-83°; also reported as mp 80.9-83.4° from aq MeOH. $[\alpha]_D$ −54.84° (c = 1 in methanol); $[\alpha]_D^{20}$ −39.8° (c = 0.5 in water). pKa_1 3.01; pKa_2 8.02. pI: 5.5. Caloric value: <1.2 kJ/g. Soly in water (g/l): 10.6 (15°), 12.6 (25°), 47.5 (60°); in ethyl acetate (g/l): 43.6 (15°), 77 (25°), >1000 (60°); in abs ethanol (g/l): ~950 (25°).

USE: Non-nutritive sweetener.

6551. Neotetrazolium Chloride. [298-95-3] 3,3′-[1,1′-Biphenyl]-4,4′-diylbis[2,5-diphenyl-2*H*-tetrazolium] dichloride; 3,3′-[4,4′-biphenylene]bis[2,5-diphenyl-2*H*-tetrazolium] dichloride; 2,2′,5,5′-tetraphenyl-3,3′-(*p*-diphenylene)ditetrazolium chloride; 2,2′-(*p*-diphenylene)-bis[3,5-diphenyl]ditetrazolium chloride; neotetrazolium blue; neo-T; TP; NTC. $C_{38}H_{28}Cl_2N_8$; mol wt 667.60. C 68.37%, H 4.23%, Cl 10.62%, N 16.78%. Prepn: L. J. Pannone, J. B. Rust, **US 2713581** (1955 to Montclair Res. Corp. and Ellis-Foster Co.). Purification: G. R. N. Jones, *Histochem. J.* **1**, 59 (1968). *See also* Triphenyltetrazolium chloride.

Light yellow powder, mp 297° (dec).

USE: Staining of microorganisms and plant and animal tissues. In determn of dehydrogenases in histochemical and cytochemical studies: K. Neumann, G. Koch, *Z. Physiol. Chem.* **295**, 35 (1953). In amylase determn.

6552. Neovitamin A. [2052-63-3] 13-*cis*-Retinol; 2-*cis*-vitamin A. $C_{20}H_{30}O$; mol wt 286.46. C 83.86%, H 10.56%, O 5.59%. Naturally occurring, bioactive isomer of vitamin A, *q.v.* The 13-*cis*- and all-*trans*-forms come to equilibrium in solution with ~33% as 13-*cis*. Isoln from shark liver oil: C. D. Robeson, J. G. Baxter, *J. Am. Chem. Soc.* **69**, 136 (1947). Synthesis: M. Matsui *et al.*, *J. Vitaminol.* **4**, 178 (1958). Stereoselective synthesis: J. Otera *et al.*, *Chem. Lett.* **1985**, 1883. HPLC determn in food samples: E. Brinkmann *et al.*, *J. Chromatogr. A* **693**, 271 (1995). Review of vitamin A compounds: W. Friedrich in *Vitamins* (Walter de Gruyter, Berlin, 1988) pp 63-140.

Pale yellow needles from ethyl formate, mp 58-60°. uv max: 328 nm (ε 41800).

6553. Nepafenac. [78281-72-8] 2-Amino-3-benzoylbenzeneacetamide; AHR-9434; AL-6515; Nevanac. $C_{15}H_{14}N_2O_2$; mol wt 254.29. C 70.85%, H 5.55%, N 11.02%, O 12.58%. Nonsteroidal anti-inflammatory with analgesic activity; selective COX-2 inhibitor. Prodrug of amfenac, *q.v.* Prepn: D. A. Walsh *et al.*, **DE 3035688** (1981 to A. H. Robins); *see also*: J. R. Shanklin, Jr. *et al.*, **US 4313949** (1982 to A. H. Robins); D. A. Walsh *et al.*, *J. Med. Chem.* **33**, 2296 (1990). Ocular pharmacology: D. A. Gamache *et al.*, *Inflammation* **24**, 357 (2000); T.-L. Ke *et al.*, *ibid.* 371. Inhibition of ocular neovascularization: K. Takahashi *et al.*, *Invest. Ophthalmol. Visual Sci.* **44**, 409 (2003). Review of ocular bioavailability and anti-inflammatory activity: R. Lindstrom, T. Kim, *Curr. Med. Res. Opin.* **22**, 397-404 (2006). Clinical trial for pain and inflammation following cataract surgery: W. A. Maxwell *et al.*, *J. Ocul. Pharmacol. Ther.* **24**, 593 (2008).

Yellow needles from isopropyl alcohol, mp 178.5-180.0°.

THERAP CAT: Anti-inflammatory; analgesic.

6554. Nepetalactone. [490-10-8] 5,6,7,7a-Tetrahydro-4,7-dimethylcyclopenta[*c*]pyran-1(4a*H*)-one. $C_{10}H_{14}O_2$; mol wt 166.22. C 72.26%, H 8.49%, O 19.25%. The first fully characterized methylcyclopentane monoterpenoid. Isolated from the volatile oil of catnip produced by *Nepeta cataria* L., *Labiatae*: S. M. McElvain *et al.*, *J. Am. Chem. Soc.* **63**, 1558 (1941); J. Meinwald, *ibid.* **76**, 4571 (1954). First believed to be a single entity, it was later shown to be a mixture of the *cis-trans* and *trans-cis* isomers, the *cis-trans* isomer comprising 70-99% of the oil. Structure: S. M. McElvain, E. J. Eisenbraun, *J. Am. Chem. Soc.* **77**, 1599 (1955). Configuration of the *cis-trans* and *trans-cis* isomers: R. B. Bates *et al.*, *ibid.* **80**, 3420 (1958). Separation and properties: R. B. Bates, C. W. Siegel, *Ex-*

perientia **19**, 564 (1963). Biosynthesis: F. E. Regnier *et al.*, *Phytochemistry* **7**, 221 (1968). Metabolism in cats: G. R. Waller *et al.*, *Science* **164**, 1281 (1969). Behavioral and toxicological study: J. W. Harney *et al.*, *Lloydia* **41**, 367 (1978). Structural study of isomeric nepetalactones: E. J. Eisenbraun *et al.*, *J. Org. Chem.* **45**, 3811 (1980).

cis-trans-Nepetalactone

Oil. Odor very attractive to cats. d_4^{25} 1.0663. $bp_{0.05}$ 71-72°. $[\alpha]_D^{23}$+3.6°. n_D^{25} 1.4859. Soluble in ether, carbon tetrachloride. When shaken with 10% NaOH yields **nepetalic acid**, crystals, mp 74°, $[\alpha]_D^{30}$ +47.6° (chloroform).

cis-trans-Form. [21651-62-7] 2-(2-Hydroxy-1-methylethenyl)-5-methylcyclopentanecarboxylic acid delta lactone. Oil, $[\alpha]_D^{21}$ +3.7° (c = 27 in chloroform). $[\alpha]_D^{27.5}$ +11.11° (chloroform). n_D^{25} 1.4878.

trans-cis-Form. mp 37-39°. $[\alpha]_D^{21}$ −24.4° (c = 6.15 in chloroform) (Eisenbraun); also reported as $[\alpha]_D^{27.5}$ +21.9° (chloroform) (Bates, Siegel). n_D^{25} 1.4878.

6555. Neptunium. [7439-99-8] Np; at. no. 93; valence 3, 4, 5, 6, 7. First man-made transuranium element; no stable nuclides. Known isotopes (mass numbers): 227-242. Discovery of isotope 239 ($T_{\frac{1}{2}}$ 2.355 days, β-decay, rel. at. mass 239.0529): E. McMillan, P. Abelson, *Phys. Rev.* **57**, 1185 (1940); of isotope 237 (α-emitter, $T_{\frac{1}{2}}$ 2.14 × 10^6 years, longest-lived known isotope, rel. at. mass 237.0482): A. C. Wahl, G. T. Seaborg, *ibid.* **73**, 940 (1948). Prepn of metal: S. Fried, N. Davidson, *J. Am. Chem. Soc.* **70**, 3539 (1948); L. B. Magnusson, T. J. LaChapelle, *ibid.* 3534. Presence in nature: Seaborg, Perlman, *ibid.* **70**, 1571 (1948). Chemical properties: Seaborg, Wahl, *ibid.* 1128. *Reviews:* C. Keller, *The Chemistry of the Transactinide Elements* (Verlag Chemie, Weinheim, English Ed., 1971) pp 253-332; W. W. Schulz, G. E. Benedict, *Neptunium-237; Production and Recovery*, AEC Critical Review Series (USAEC, Washington, D.C., 1972) 85 pp; *Comprehensive Inorganic Chemistry* **vol. 5**, J. C. Bailar, Jr. *et al.*, Eds. (Pergamon Press, Oxford, 1973) *passim*; J. A. Fahey in *The Chemistry of the Actinide Elements* **vol. 1**, J. J. Katz *et al.*, Eds. (Chapman and Hall, New York, 1986) pp 443-498; G. T. Seaborg in *Kirk-Othmer Encyclopedia of Chemical Technology* **vol. 1** (Wiley-Interscience, New York, 4th ed., 1991) pp 412-444.

Silvery metal; develops a thin oxide layer upon exposure to air for short periods. Reacts with air at high temperatures to form NpO_2. Exhibits 3 allotropic modifications: orthorhombic α-form, d 20.45, transforms to β-form at 280°; tetragonal β-form, d 19.36, transforms to γ-form at 577°; cubic γ-form transforms to liquid at mp 637°. Extrapolated bp 4174°. Np has been obtained in its five oxidation states in soln; the most stable is the pentavalent state. Tetravalent Np is readily oxidized to the hexavalent state by permanganate in the cold, or by strong oxidizing agents; on electrolytic reduction in an atmosphere of nitrogen, the trivalent form is obtained.

Caution: Radiation hazard; handling requires special equipment and shielding facilities (Katz *et al.*, *loc. cit.* **vol. 2**, p. 1128).

USE: Source material for prodn of ^{238}U (power source).

6556. Nequinate. [13997-19-8] 6-Butyl-1,4-dihydro-4-oxo-7-(phenylmethoxy)-3-quinolinecarboxylic acid methyl ester; 7-(benzyloxy)-6-*n*-butyl-1,4-dihydro-4-oxo-3-quinolinecarboxylic acid methyl ester; 3-methoxycarbonyl-6-*n*-butyl-7-benzyloxy-4-oxoquinoline; 7-(benzyloxy)-6-*n*-butyl-4-hydroxy-3-quinolinecarboxylic acid methyl ester; 7-(benzyloxy)-6-*n*-butyl-3-methoxycarbonyl-4-quinolone; methyl benzoquate; ICI-55052; Statyl. $C_{22}H_{23}NO_4$; mol wt 365.43. C 72.31%, H 6.34%, N 3.83%, O 17.51%. Prepn: **BE 677592** and **GB 1070223** (1966 to I.C.I.), *C.A.* **68**, 68899j (1968). Coccidiostatic activity: Bowie *et al.*, *Nature* **214**, 1349 (1967). HPLC determn in poultry feedstuffs: G. H. J. Merson *et al.*, *Analyst* **110**, 761 (1985).

Crystals, mp 287-288°.
THERAP CAT (VET): Coccidiostat.

6557. Neratinib. [698387-09-6] (2*E*)-*N*-[4-[[3-Chloro-4-(2-pyridinylmethoxy)phenyl]amino]-3-cyano-7-ethoxy-6-quinolinyl]-4-(dimethylamino)-2-butenamide; (*E*)-*N*-[4-[3-chloro-4-(2-pyridinylmethoxy)anilino]-3-cyano-7-ethoxy-6-quinolinyl]-4-(dimethylamino)-2-butenamide; 6-(4-*N*,*N*-dimethylaminocrotonyl)amido-4-(4-(2-pyridylmethoxy)anilino)-3-cyano-7-ethoxyquinoline; HKI-272; WAY-179272. $C_{30}H_{29}ClN_6O_3$; mol wt 557.05. C 64.69%, H 5.25%, Cl 6.36%, N 15.09%, O 8.62%. Pan-ErbB receptor tyrosine kinase inhibitor. Prepn: A. Wissner *et al.*, **WO 05028443**; *eidem*, **US 7399865** (2005, 2008 both to Wyeth); and receptor binding studies: H.-R. Tsou *et al.*, *J. Med. Chem.* **48**, 1107 (2005). Large-scale synthesis: J. L. Considine *et al.*, **US 7126025** (2006 to Wyeth). *In vitro* antitumor activity: S. K. Rabindran *et al.*, *Cancer Res.* **64**, 3958 (2004). Pharmacokinetics and efficacy in patients with ErbB2-positive breast cancer: H. J. Burstein *et al.*, *J. Clin. Oncol.* **28**, 1301 (2010). Review of development and therapeutic potential in breast and non-small cell lung cancer: P. Bose, H. Ozer, *Expert Opin. Invest. Drugs* **18**, 1-17 (2009).

Crystals from acetonitrile + THF.
Maleate. [915942-22-2] $C_{30}H_{29}ClN_6O_3 \cdot C_4H_4O_4$; mol wt 673.12. mp 178-183°.
THERAP CAT: Antineoplastic.

6558. Neridronic Acid. [79778-41-9] (6-Amino-1-hydroxyhexylidene)bisphosphonic acid. $C_6H_{17}NO_7P_2$; mol wt 277.15. C 26.00%, H 6.18%, N 5.05%, O 40.41%, P 22.35%. Bisphosphonate antiresorptive agent. Prepn: J. Jary *et al.*, **BE 885139**; *eidem*, **US 4304734** (both 1981 to Vysoka Skola Chem.-Tech.); and characterization of monosodium salt: M. Neves *et al.*, *Nucl. Med. Biol.* **29**, 329 (2002). Crystal structure: V. M. Coiro, D. Lamba, *Acta Crystallogr.* **C45**, 446 (1989). Clinical study in Paget's disease: S. Adami *et al.*, *Clin. Exp. Rheumatol.* **20**, 55 (2002); in osteogenesis imperfecta: *idem et al.*, *J. Bone Miner. Res.* **18**, 126 (2003); in osteoporosis: V. Braga *et al.*, *Bone* **33**, 342 (2003).

mp 245°.
Monosodium salt. [80729-79-9] Neridronate sodium; Nerixia. $C_6H_{16}NNaO_7P_2$; mol wt 299.13. White solid, mp 263-266° (hemihydrate).
THERAP CAT: Bone resorption inhibitor.

6559. Neriifolin. [466-07-9] (3β,5β)-3-[(6-Deoxy-3-*O*-methyl-α-L-glucopyranosyl)oxy]-14-hydroxycard-20(22)-enolide. C_{30}-

$H_{46}O_8$; mol wt 534.69. C 67.39%, H 8.67%, O 23.94%. Cardiac glycoside isolated from *Thevetia neriifolia* Juss.: M. Frèrejacque, *Compt. Rend.* **221**, 645 (1945); from *Cerbera odollam* Gaertn.: S. Rangaswami, E. V. Rao, *J. Sci. Ind. Res.* **16B**, 209 (1957); from *Thevetia thevetioides* seeds: J. L. McLaughlin *et al.*, *J. Econ. Entomol.* **73**, 39 (1980). Pharmacology: K. Mezey, *Arch. Int. Pharmacodyn. Ther.* **84**, 367 (1950); K. K. Chen *et al.*, *ibid.* 81.

Rhombic plates from methanol, mp 218-225°. Also reported as mp 208° (Frèrejacque). $[\alpha]_D^{23}$ −50.2° (CH_3OH). uv max (CH_3OH): 217 nm (log ε 4.1).

2'-Acetate. Cerberin; veneniferin; monoacetylneriifolin. $C_{32}H_{48}O_9$. Identity of cerberin with monoacetylneriifolin: M. Frèrejacque, *Compt. Rend.* **226**, 835 (1948); of cerberin with veneniferin: M. Frèrejacque, Durgeat, *ibid.* **228**, 1310 (1949). Coarse prisms from methanol + water, mp 212-215°. $[\alpha]_D^{19}$ −82° ($CHCl_3$). Sol in alc, $CHCl_3$, acetic acid, ether. Practically insol in water.

THERAP CAT: Cardiotonic.

6560. Nerol. [106-25-2] (2Z)-3,7-Dimethyl-2,6-octadien-1-ol; *cis*-2,6-dimethyl-2,6-octadien-8-ol. $C_{10}H_{18}O$; mol wt 154.25. C 77.87%, H 11.76%, O 10.37%. The *cis*-isomer of geraniol, *q.v.*; found in many essential oils. Readily loses water and cyclizes forming dipentene. Isoln from neroli oil: Hesse, Zeitschel, *J. Prakt. Chem.* **66**, 502 (1902). Structure: Verley, *Bull. Soc. Chim. Fr.* **25**, 68 (1919); J. L. Simonsen, *The Terpenes* vol. **I** (University Press, Cambridge, 2nd ed., 1947) pp 52-54. Stereochemistry: Burrell *et al.*, *Proc. Chem. Soc. London* **1959**, 263; Bates *et al.*, *J. Org. Chem.* **28**, 1086 (1963). Synthesis: Yukawa *et al.*, *Bull. Chem. Soc. Jpn.* **37**, 158 (1964). Stereochemistry and synthesis: Burrell *et al.*, *J. Chem. Soc. C* **1966**, 2144; K. Takabe *et al.*, *Chem. Lett.* **1977**, 1025.

Liquid. Odor of sweet rose. bp_{745} 224-225°; bp_{25} 125°. d^{15} 0.8813. Optically inactive. uv max: 189-194 nm (ε 18000). Sol in abs alc.

Tetrabromide. $C_{10}H_{18}Br_4O$. Crystals, mp 118°.

Allophanate. $C_{12}H_{20}N_2O_3$. Needles from petr ether, mp 84-86°.

USE: Base for manuf of perfumes.

6561. Nerolidol. [7212-44-4] 3,7,11-Trimethyl-1,6,10-dodecatrien-3-ol; peruviol. $C_{15}H_{26}O$; mol wt 222.37. C 81.02%, H 11.79%, O 7.19%. Found in essential oils from many flowers such as *Melaleuca viridiflora* Soland., *Myrtaceae* and *Myroxylon pereirae* (Royle) Klotzsch, *Leguminosae:* Naves, *Compt. Rend.* **251**, 900 (1960). Isoln: Hellyer, McKern, *J. Proc. R. Soc. N.S.W.* **89**, (Pt. 4), 188 (1955); Sutherland *et al.*, *Aust. J. Chem.* **13**, 357 (1960). Synthesis: Nazarov *et al.*, *Zh. Obshch. Khim.* **28**, 1444 (1958); Ofner *et al.*, *Helv. Chim. Acta* **42**, 2577 (1959); Shvarts, Petrov, *Zh. Obshch. Khim.* **30**, 3598 (1960). Exists in two stereoisomeric forms: Bates *et al.*, *J. Org. Chem.* **28**, 1086 (1963).

trans-form

trans-Form. [40716-66-3] Liquid. $bp_{0.15}$ 78°. n_D^{25} 1.4792.
cis-Form. [3790-78-1] Liquid. $bp_{0.10}$ 70°. n_D^{25} 1.4775.
cis/trans-Form. Liquid. bp_3 122°. n_D^{25} 1.4769. d_4^{25} 0.8720. Sol in abs alc (still clearly sol in 3 parts of 70% alc).

6562. Nerve Growth Factor. [9061-61-4] NGF. Polypeptide factor which promotes the growth and survival of peripheral sympathetic and sensory neurons and basal forebrain cholinergic neurons. Prototype of the neurotrophins, *q.v.* Synthesized in limiting amounts by target tissues of NGF-responsive cells; anomalous sources of large amounts of NGF include mouse submaxillary gland, snake venom gland, and guinea pig prostate. Bioactive form, also referred to as *β-NGF*, is a tightly associated dimer of 2 identical polypeptide chains. Mol wt of the dimer ~26,500 Da. Initial identification as a nerve growth stimulating factor from mouse sarcoma: R. Levi-Montalcini, V. Hamburger, *J. Exp. Zool.* **116**, 321 (1951); R. Levi-Montalcini, *Ann. N.Y. Acad. Sci.* **55**, 330 (1952). Identification in snake venom: S. Cohen, *J. Biol. Chem.* **234**, 1129 (1959); in mouse submaxillary gland: *idem, Proc. Natl. Acad. Sci. USA* **46**, 302 (1960); in guinea pig prostate: G. P. Harper *et al.*, *Nature* **279**, 160 (1979). Occurs in mouse submaxillary gland as a complex (known as 7S NGF) composed of α-, β-, and γ-subunits, of which only the β-subunit is bioactive. Characterization of subunits: S. Varon *et al.*, *Biochemistry* **7**, 1296 (1968); A. C. Server, E. M. Shooter, *Adv. Protein Chem.* **31**, 339 (1977). Amino acid sequence of mouse NGF: R. H. Angeletti, R. A. Bradshaw, *Proc. Natl. Acad. Sci. USA* **68**, 2417 (1971). Cloning of mouse NGF: J. Scott *et al.*, *Nature* **302**, 538 (1983); of human: A. Ullrich *et al.*, *ibid.* **303**, 821 (1983). Crystal structure: N. Q. McDonald *et al.*, *ibid.* **354**, 411 (1991). Review of NGF receptor: P. A. Barker, R. A. Murphy, *Mol. Cell. Biochem.* **110**, 1-15 (1992). Review of pharmacology and potential role in neurodegenerative disease: F. Hefti, P. A. Lapchak, *Adv. Pharmacol.* **24**, 239-273 (1993); in peripheral nerve regeneration: G. Raivich, G. W. Kreutzberg, *Clin. Neurol. Neurosurg.* **95** Suppl., S84-S88 (1993). *Reviews:* R. Levi-Montalcini, R. H. Angeletti, *Physiol. Rev.* **48**, 534-569 (1968); H. Thoenen, Y. A. Barde, *ibid.* **60**, 1284-1335 (1980); R. Levi-Montalcini, *Science* **237**, 1154-1162 (1987); R. A. Bradshaw *et al.*, *Trends Biochem. Sci.* **18**, 48-52 (1993). *Book: Neurotrophic Factors* S. E. Loughlin, J. H. Fallon, Eds. (Academic Press, San Diego, 1993) 607 pp.

6563. Neticonazole. [130726-68-0] 1-[(1E)-2-(Methylthio)-1-[2-(pentyloxy)phenyl]ethenyl]-1*H*-imidazole. $C_{17}H_{22}N_2OS$; mol wt 302.44. C 67.51%, H 7.33%; N 9.26%, O 5.29%, S 10.60%. Antifungal that inhibits cell wall synthesis. Prepn: M. Ogawa *et al.*, **EP 227011**; *eidem*, **US 4740601** (1987, 1988 both to SS Pharm.); and antifungal activity: M. Ogawa *et al.*, *Chem. Pharm. Bull.* **39**, 2301 (1991). Clinical study in tinea pedis: R. Tsuboi *et al.*, *Int. J. Dermatol.* **35**, 371 (1996).

Colorless crystals from isopropyl ether, mp 37-39°.

Hydrochloride. [130773-02-3] SS-717; Atolant. $C_{17}H_{22}N_2OS.HCl$; mol wt 338.89. Colorless columns from acetonitrile, mp 145-147°.

THERAP CAT: Antifungal (synthetic).

6564. Netilmicin. [56391-56-1] *O*-3-Deoxy-4-*C*-methyl-3-(methylamino)-β-L-arabinopyranosyl-(1 → 6)-*O*-[2,6-diamino-2,3,-4,6-tetradeoxy-α-D-*glycero*-hex-4-enopyranosyl-(1 → 4)]-2-deoxy-

N^1-ethyl-D-streptamine; (2*S-cis*)-4-*O*-[3-amino-6-(aminomethyl)-3,4-dihydro-2*H*-pyran-2-yl]-2-deoxy-6-*O*-[3-deoxy-4-*C*-methyl-3-(methylamino)-β-L-arabinopyranosyl]-N^1-ethyl-D-streptamine; 1-*N*-ethylsisomicin; Sch-20569. $C_{21}H_{41}N_5O_7$; mol wt 475.59. C 53.04%, H 8.69%, N 14.73%, O 23.55%. Broad spectrum semisynthetic aminoglycoside antibiotic, related to sisomicin, *q.v.* Prepn: M. J. Weinstein *et al.*, **DE 2437160** (1975 to Sherico); J. J. Wright, **US 4029882** (1977 to Schering); J. J. Wright *et al.*, **US 4002742** (1977 to Schering). Structure and synthesis: J. J. Wright, *Chem. Commun.* **1976**, 206. Biological activity: G. H. Miller *et al.*, *Antimicrob. Agents Chemother.* **10**, 827 (1976). Radioimmunoassay: A. Broughton *et al.*, *Clin. Chem.* **24**, 717 (1978). Pharmacology: W. Raab, *Adv. Clin. Pharmacol.* **15**, 91 (1978); I. Trestman *et al.*, *Antimicrob. Agents Chemother.* **13**, 832 (1978). Metabolism and pharmacokinetics: J. C. Pechere *et al.*, *Clin. Pharmacol. Ther.* **23**, 677 (1978); R. E. Brummett *et al.*, *Arch. Otolaryngol.* **104**, 579 (1978). Clinical studies: J. Klastersky *et al.*, *Antimicrob. Agents Chemother.* **12**, 503 (1977); A. P. Panwalker *et al.*, *ibid.* **13**, 170 (1978). Toxicity studies: L. Albiero *et al.*, *Arch. Int. Pharmacodyn. Ther.* **233**, 343 (1978); F. C. Lust, *J. Int. Med. Res.* **6**, 286 (1978). *Review:* P. Noone, *Drugs* **27**, 548-578 (1984).

$[\alpha]_D^{26}$ +164° (c = 3 in water). LD_{50} in mice (mg/kg): 40 i.v.; 125 i.p.; 175 s.c. (Miller).

Sulfate. [56391-57-2] Certomycin; Netillin; Netilyn; Netromicine; Netromycin; Nettacin; Vectacin; Zetamicin. $(C_{21}H_{41}N_5O_7)_2.$-$5H_2SO_4$; mol wt 1441.53. White to pale yellowish-white powder. Freely sol in water. Practically insol in dehydrated alc, ether.

THERAP CAT: Antibacterial.

6565. Netobimin. [88255-01-0] 2-[[[(Methoxycarbonyl)amino][2-nitro-5-(propylthio)phenyl]amino]methylene]amino]ethanesulfonic acid; 2-[[[(methoxycarbonyl)amino][2-nitro-5-(propylthio)phenyl]amino]methylene]amino]ethanesulfonic acid; methyl [*N'*-[2-nitro-5-(propylthio)phenyl]-*N*-(2-sulfoethyl)amidino]carbamate; *N*-methoxycarbonyl-*N'*-[2-nitro-5-(propylthio)phenyl]-*N''*-2-(ethylsulfonic acid)guanidine; totabin; Sch-32481; Hapadex. $C_{14}H_{20}N_4$-O_7S_2; mol wt 420.46. C 39.99%, H 4.79%, N 13.33%, S 15.25%. Phenylguanidine anthelmintic. Prepn: M. M. Nafissi-Varchei, **EP 50286**; *idem*, **US 4406893** (1982, 1983 both to Schering). Metabolism in animals: P. Delatour *et al.*, *J. Vet. Pharmacol. Ther.* **9**, 230 (1986). Anthelmintic efficacy in cattle: J. C. Williams *et al.*, *Am. J. Vet. Res.* **46**, 2188 (1985).

mp 215° (dec).

Sodium salt. $C_{14}H_{20}N_4NaO_7S_2$. Crystals, mp 150° (dec) (Nafissi-Varchei); also reported as yellow powder, mp 160° (Delatour). uv max (methanol): 225, 347 nm. Sol in water, acetone, methanol, dimethylformamide, DMSO. Insol in ether.

THERAP CAT (VET): Anthelmintic.

6566. Netropsin. [1438-30-8] 4-[[2-[(Aminoiminomethyl)-amino]acetyl]amino]-*N*-[5-[[(3-amino-3-iminopropyl)amino]carbonyl]-1-methyl-1*H*-pyrrol-3-yl]-1-methyl-1*H*-pyrrole-2-carboxamide; *N'*-(2-amidinoethyl)-4-(2-guanidinoacetamido)-1,1'-dimeth-

yl-*N*,4'-bi[pyrrole-2-carboxamide]; sinanomycin; congocidine; T-1384. $C_{18}H_{26}N_{10}O_3$; mol wt 430.47. C 50.22%, H 6.09%, N 32.54%, O 11.15%. Basic oligopeptide antibiotic produced by *Streptomyces netropsis* with wide range of antimicrobial activity. Prepn and antibacterial spectrum: A. C. Finlay *et al.*, *J. Am. Chem. Soc.* **73**, 341 (1951); and trypanocidal activity: C. Cosar *et al.*, *Compt. Rend.* **234**, 1498 (1952). Antiviral activity: F. M. Schabel *et al.*, *Proc. Soc. Exp. Biol. Med.* **83**, 1 (1953). Identity with sinanomycin: K. Watanabe, *J. Antibiot.* **9A**, 102 (1956). Structural studies: M. Julia, N. Joseph, *Compt. Rend.* **243**, 961 (1956). Structure: C. W. Waller *et al.*, *J. Am. Chem. Soc.* **79**, 1265 (1957). Revised structure: M. Julia, N. Préau-Joseph, *Compt. Rend.* **257**, 1115 (1963); *eidem*, *Bull. Soc. Chim. Fr.* **1967**, 4348. Specific binding to DNA: Ch. Zimmer *et al.*, *J. Mol. Biol.* **58**, 329 (1971); K. Zakrzewska *et al.*, *Nucleic Acids Res.* **11**, 8825, 8841 (1983). Use in gradient separation of DNA: K. M. Tatti *et al.*, *Anal. Biochem.* **89**, 561 (1978). Selective blocking of DNAase I cleavage: Ch. Zimmer *et al.*, *Nucleic Acids Res.* **8**, 2999 (1980). Review of activity: F. E. Hahn in *Antibiotics* **vol. 3**, J. W. Corcoran, F. E. Hahn, Eds. (Springer-Verlag, New York, 1975) pp 79-100. Review of effect on DNA structure and function: Ch. Zimmer, *Prog. Nucleic Acid Res. Mol. Biol.* **15**, 285-318 (1975).

Disulfate. $C_{18}H_{26}N_{10}O_3.2H_2SO_4$. Needles, mp 288°. Soly in water ~30 mg/ml at 80°, <0.5 mg/ml at 25°. Practically insol in the common organic solvents.

Dihydrochloride. $C_{18}H_{26}N_{10}O_3.2HCl$. Prisms, mp 228°. LD_{50} in mice (mg/kg): 17 i.v.; 70 s.c.; >300 orally (Finlay).

USE: Non-intercalative DNA binding agent.

6567. Nettle. Stinging nettle. Flowering perennial herb, *Urtica dioica* L., *Uricaceae*. Leaves and stems are covered with bristly stinging hairs that cause a characteristic urticaria upon contact. Used in traditional medicine as a diuretic, hemostatic, and in treatment of rheumatic conditions. Medicinal portions include leaves, fruits and root. *Habit.* Widely scattered throughout the Northern hemisphere, especially in moist woods, road sides and wasteland. *Constit.* Flavonoids: glycosides of quercetin, kaempferol, isorhamnetin in the flowers; histamine, acetylcholine and serotonin in the stinging hairs; β-sitosterol, coumarin, lignans and a lectin, *U. dioica* agglutinin, in the rhizome. Botanical description: T. W. McGovern, T. M. Barkley, *Cutis* **62**, 63 (1998). Clinical evaluation of nettle sting in osteoarthritic pain: C. Randall *et al.*, *J. R. Soc. Med.* **93**, 305 (2000). Reviews of medicinal preparations: M. Wichtl, N. G. Bisset, *Herbal Drugs and Phytopharmaceuticals*, English Ed. (CRC Press, Boca Raton, 1994) pp 502-509; J. Barnes *et al.*, *Herbal Medicines* (Pharmaceutical Press, London, 2nd ed., 2002) pp 360-364.

Root extract. Bazoton; Prostaherb. Antiproliferative effects on human prostatic epithelium: J. J. Lichius *et al.*, *Pharmazie* **54**, 768 (1999). Review of clinical experience: E. Koch, *Planta Med.* **67**, 489-500 (2001).

THERAP CAT: Root extract in treatment of micturitional disorders due to benign prostatic hypertrophy.

6568. Neuraminic Acid. [114-04-5] Prehemataminic acid. $C_9H_{17}NO_8$; mol wt 267.23. C 40.45%, H 6.41%, N 5.24%, O 47.90%. Parent acid of a family of amino sugars contg 9 or more carbon atoms (sialic acids or nonulosaminic acids). May be regarded as the aldol condensation product of pyruvic acid and *N*-acetyl-D-mannosamine. Neuraminic acid has not been isolated and characterized as such, but several *N*- and *O*-substituted derivatives are widely distributed in nature. *Review:* Whelan, *Annu. Rep. Prog. Chem.* **54**, 319 (1958); A. Gottschalk, *The Chemistry and Biology of Sialic Acids and Related Substances* (Cambridge Univ. Press, 1960). *See also* Sialic Acids.

6569. Neurine. [463-88-7] N,N,N-Trimethylethenaminium hydroxide; trimethylvinylammonium hydroxide. $C_5H_{13}NO$; mol wt 103.17. C 58.21%, H 12.70%, N 13.58%, O 15.51%. Found in egg yolk, brain, bile, in cadavers. Formed during putrefaction by dehydration of choline: Hofmann, *Compt. Rend.* **47**, 559 (1858); Renshaw, Ware, *J. Am. Chem. Soc.* **47**, 2992 (1925). Prepn: Meyer, Hopff, *Ber.* **54**, 2277 (1921); from aq trimethylamine and acetylene: Gardner *et al.*, *J. Chem. Soc.* **1949**, 789. Separation and identification in biological fluids: E. Merlevede *et al.*, *Arch. Int. Pharmacodyn.* **122**, 474 (1959). Stereoelectronic config: *Intern. Kongr. Entomol. Verhandl., 11th,* Vienna, **1960**, No. 3, pp 87-93, *C.A.* **57**, 7749b (1962). Toxicity study: Hunt, *J. Pharmacol. Exp. Ther.* **28**, 267 (1926).

Syrupy liq. Fishy odor. *Poisonous.* Forms a crystalline trihydrate. Alkaline reaction. Readily absorbs CO_2 from the air. Sol in water, alcohol. Dec readily, forming trimethylamine. Forms an HCl salt. LD s.c. in mice: 46 mg/kg (Hunt).

6570. Neuropeptide Y. [82785-45-3] Neuropeptide tyrosine; NPY. Neuroendocrine peptide that stimulates ingestive behavior. Named for its five tyrosine (Y) residues. Member of a structurally related peptide family which includes peptide YY and pancreatic polypeptide. Most abundant known peptide in mammalian brain; widely distributed in the neurons of the central and peripheral nervous system and adrenal medullary cells. Composed of a highly conserved sequence of 36 amino acids. Exhibits wide range of physiological effects including anxiolysis, sedation, vasoconstriction, and the stimulation of food intake. Isoln from porcine brain: K. Tatemoto *et al.*, *Nature* **296**, 659 (1982). Amino acid sequence: *idem, Proc. Natl. Acad. Sci. USA* **79**, 5485 (1982). Distribution in human brain: T. E. Adrian *et al.*, *Nature* **306**, 584 (1983). Role of NPY in the antiobesity effect of the *obese* gene product (*see also* Leptin): T. W. Stephens *et al.*, *Nature* **377**, 530 (1995). Conference proceedings: *Neuropeptide Y: Karolinska Inst. Nobel Conf. Series*, V. Mutt *et al.*, Eds. (Raven Press, N.Y., 1989) 353 p. Book: *The Biology of Neuropeptide Y and Related Peptides*, W. F. Colmers, C. Wahlestedt, Eds. (Humana Press, Totowa, N. J., 1993) 564 p. Review of pharmacology: J. G. Wettstein *et al.*, *Pharmacol. Ther.* **65**, 397-414 (1995); of role as potential therapeutic target: L. Grundemar, R. Hakanson, *Trends Pharmacol. Sci.* **15**, 153-159 (1994).

6571. Neurotensin. [39379-15-2] A basic tridecapeptide, found in mammalian brain and gut, having a wide variety of hormone-like activities. Neurotensin has been shown to induce hypotension in the rat, to stimulate contraction of guinea pig ileum and rat uterus, and to cause relaxation of rat duodenum. There is also evidence that it acts as a CNS neurotransmitter. Isoln from bovine hypothalamus: R. Carraway, S. Leeman, *J. Biol. Chem.* **248**, 6854 (1973). Amino acid sequence of bovine neurotensin: *eidem, ibid.* **250**, 1907 (1975). Synthesis: *eidem, ibid.* 1912; K. Kitagawa *et al.*, *Chem. Pharm. Bull.* **24**, 2692 (1976); H. Yajima *et al.*, *Int. J. Pept. Protein Res.* **14**, 169 (1979). Isoln, amino acid sequence of chicken neurotensin: R. Carraway, Y. M. Bhatnagar, *Peptides* **1**, 167 (1980). Synthesis: H. Yajima *et al.*, *Chem. Pharm. Bull.* **29**, 2587 (1981). Radioimmunoassay and differential distribution of bovine neurotensin in CNS, small intestine, and stomach: R. Carraway, S. Leeman, *J. Biol. Chem.* **251**, 7035, 7045 (1976). CNS effects: C. B. Nemeroff *et al.*, *Brain Res.* **128**, 485 (1977). Receptor binding in brain: G. R. Uhl *et al.*, *ibid.* **130**, 299 (1977). Pharmacokinetics and effect on gastrointestinal and pituitary hormones in man: A. M. Blackburn *et al.*, *J. Clin. Endocrinol. Metab.* **51**, 1257 (1980). Reviews of biolog-

ical activity: G. R. Uhl, S. H. Snyder, *Adv. Biochem. Psychopharmacol.* **28**, 87-106 (1981); D. R. Brown, R. J. Miller, *Annu. Rep. Med. Chem.* **17**, 271-280 (1982). Book: *Ann. N.Y. Acad. Sci.* **400**, entitled "Neurotensin, a Brain and Gastrointestinal Peptide", C. B. Nemeroff, A. J. Prange, Eds. (1982) pp 1-444.

5-oxoPro–Leu–Tyr–Glu–Asn–Lys–Pro–Arg–Arg–Pro–Tyr–Ile–LeuOH

Bovine Neurotensin

Bovine neurotensin triacetate hexahydrate. Neurotensin (ox) triacetate (salt). $C_{84}H_{133}N_{21}O_{26}\cdot 6H_2O$. Fluffy white powder. $[\alpha]_D^{23}$ $-65.6°$ (c = 0.5 in water).

6572. Neurotrophins. Family of highly conserved, target-derived, secretory proteins that promote the differentiation, growth, and survival of neurons in the peripheral and central nervous systems. Produced in limiting amounts by innervated tissues to enhance neuronal differentiation and survival. Five closely related factors have been identified: nerve growth factor (NGF), *q.v.*, brain-derived neurotrophic factor (BDNF), neurotrophin-3 (NT-3), NT-4, and NT-5. Neurotrophins differ in their sites of synthesis, developmental patterns of expression, and neuronal targets. Amino acid sequences are highly homologous. Most of the sequence differences are contained in 4 specific regions which are presumed to be responsible for the biological specificities of the factors. Purification of BDNF from pig brain: Y.-A. Barde *et al.*, *EMBO J.* **1**, 549 (1982). Cloning and expression of BDNF: J. Leibrock *et al.*, *Nature* **341**, 149 (1989). Identification of rat NT-3: P. C. Maisonpierre *et al.*, *Science* **247**, 1446 (1990); of human NT-3: Y. Kaisho *et al.*, *FEBS Lett.* **266**, 187 (1990); A. Rosenthal *et al.*, *Neuron* **4**, 767 (1990). Isolation of NT-4 from *Xenopus laevis*: F. Hallböök *et al.*, *ibid.* **6**, 845 (1991). Identification of NT-5 in rat and human: L. R. Berkemeier *et al.*, *ibid.* **7**, 857 (1991). Proposed identity of NT-5 as the mammalian homolog of Xenopus NT-4: N. Y. Ip *et al.*, *Proc. Natl. Acad. Sci. USA* **89**, 3060 (1992). Comparison of neurotrophins and other neurotrophic factors: H. Thoenen, *Trends Neurosci.* **14**, 165 (1991). Discussion of neurotrophin receptors: S. O. Meakin, E. M. Shooter, *ibid.* **15**, 323 (1992). Review of amino acid sequences and molecular structure: T. Ebendal, *J. Neurosci. Res.* **32**, 461-470 (1992). Review of bioactivity and distribution in the nervous system: G. Vantini, *Psychoneuroendocrinology* **17**, 401-410 (1992); of bioactivity and therapeutic potential in neurologic disease: F. F. Eide *et al.*, *Exp. Neurol.* **121**, 200-214 (1993).

Brain-derived neurotrophic factor. BDNF. 119 Amino acid peptide; mol wt ~13.5 kDa. pI ~9.99. Localized primarily, but not exclusively, in CNS. Amino acid sequence identical in human, pig, mouse and rat. Exhibits ~50% sequence homology with NGF.

Neurotrophin-3. [130939-66-1] NT-3; HDNF; hippocampal-derived neurotrophic factor; nerve growth factor-2; NGF-2; neuronotrophin-3. 119 Amino acid peptide; mol wt ~13.5 kDa. pI ~9.5. Produced primarily in peripheral tissues; also produced in the CNS. Amino acid sequence identical in human, mouse and rat. Exhibits ~57% sequence homology with NGF; ~58% homology with BDNF.

6573. Neutral Red. [553-24-2]; [366-13-2] (free base). N^8-N^8,3-Trimethyl-2,8-phenazinediamine hydrochloride (1:1); C.I. Basic Red 5; 3-amino-7-dimethylamino-2-methylphenazine hydrochloride; toluylene red; neutral red chloride; nuclear fast red (basic dye); aminodimethylaminotoluaminozine hydrochloride; C.I. 50040; Kernechtrot; Michrome no. 226. $C_{15}H_{17}ClN_4$; mol wt 288.78. C 62.39%, H 5.93%, Cl 12.28%, N 19.40%. Prepn from N,N-dimethyl-*p*-nitrosoaniline hydrochloride + toluene-2,4-diamine: Witt, *Ber.* **12**, 931 (1879); **DE 15272** (1880), *Frdl.* 1, 274; Bernthsen, Schweitzer, *Ann.* **236**, 332 (1886); Pokorny, *J. Soc. Dyers Colour.* **42**, 347 (1926). Toxicity data: F. Stolarsky, T. J. Haley, *Fed. Proc.* **10**, 337 (1951). Brief review: *H. J. Conn's Biological Stains*, R. W. Horobin, J. A. Kiernan Eds. (BIOSIS Scientific, Oxford, 10th ed., 2002) pp 271-272. *See also: Colour Index* vol. 4 (3rd ed., 1971) p 4446.

Dark green powder. Absorption max (50% alc): 533 nm. pKa 6.7. pH: 6.8 red; 8.0 yellow. Soluble in water or alcohol with red color. Soly in water 4.0%, in abs alcohol 1.8%, in Cellosolve 3.75%, in ethylene glycol 3.0%. Practically insol in xylene. LD$_{50}$ in mice, rats, rabbits (mg/kg): 142, 112, 97 i.v. (Stolarsky, Haley).

Note: Not to be confused with *Nuclear Fast Red (acid dye)* which is *C.I. 60760*.

USE: pH indicator. Indicator for organic nitrogen determn. Also used for preparing neutral-red paper. Stain used in histological and flow cytometry applications; assessment of cellular viability.

6574. Neutral Spirits. Distilled from suitable raw materials, is 95% ethanol (v/v) meaning that it is at least 190 proof when distilled. Used for blending with straight whisky (*see* Whisky) and for making gin, cordials, liqueurs and vodka. It may be made from grain or molasses or by redistillation of other beverages, such as brandy or rum, and is almost colorless and contains no extraneous flavoring.

6575. Nevirapine. [129618-40-2] 11-Cyclopropyl-5,11-dihydro-4-methyl-6*H*-dipyrido[3,2-*b*:2′,3′-*e*][1,4]diazepin-6-one; BI-RG-587; Viramune. C$_{15}$H$_{14}$N$_4$O; mol wt 266.30. C 67.65%, H 5.30%, N 21.04%, O 6.01%. Dipyridodiazepinone non-nucleoside reverse transcriptase inhibitor (NNRTI). Prepn: K. D. Hargrave *et al.*, EP 429987 (1991 to Boehringer, Ing; Thomae); *eidem*, US 5366972 (1994 to Boehringer, Ing.); *eidem*, *J. Med. Chem.* 34, 2231 (1991). Inhibition of HIV-1 reverse transcriptase: V. J. Merluzzi *et al.*, *Science* 250, 1411 (1990). Studies of structure-activity and binding site: P. M. Grob *et al.*, *AIDS Res. Hum. Retroviruses* 8, 145 (1992). HPLC determn in serum: R. M. Lopez *et al.*, *J. Chromatogr. B* 751, 371 (2001). Clinical pharmacokinetics: S. H. Cheeseman *et al.*, *Antimicrob. Agents Chemother.* 37, 178 (1993). Review of clinical experience: A. Carr, D. A. Cooper, *Adv. Exp. Med. Biol.* 394, 299-304 (1996); of safety profile: R. B. Pollard *et al.*, *Clin. Ther.* 20, 1071-1092 (1998). Clinical trial in prevention of perinatal HIV-1 transmission: L. A. Guay *et al.*, *Lancet* 354, 795 (1999).

Crystals from pyridine + water, mp 247-249°. pKa 2.8. Lipophilic. Soly in water ~0.1 mg/ml at neutral pH; highly sol at pH<3. Slightly sol in alc, methanol; hydrous form also slightly sol in propylene glycol.

THERAP CAT: Antiretroviral.

6576. Nialamide. [51-12-7] 4-Pyridinecarboxylic acid 2-[3-oxo-3-[(phenylmethyl)amino]propyl]hydrazide; 1-[2-(benzylcarbamoyl)ethyl]-2-isonicotinoylhydrazide; *N*-[2-(benzylcarbamyl)ethylamino]isonicotinamide; *N*-benzyl-β-(isonicotinoylhydrazino)propionamide; *N*-isonicotinoyl-*N*′-[β-(*N*-benzylcarboxamido)ethyl]hydrazide; isonicotinic acid 2-[2-(benzylcarbamoyl)ethyl]hydrazide; Espril; Niamid; Niamidal; Niaquitil; Nuredal; Nyazin. C$_{16}$H$_{18}$N$_4$O$_2$; mol wt 298.35. C 64.41%, H 6.08%, N 18.78%, O 10.72%. Monoamine oxidase (MAO) inhibitor. Prepn: Bloom, Carnahan, US 2894972; US 3040061 (1959, 1962 both to Pfizer). Toxicology study: C. S. Delahunt, J. M. Pepin, *Toxicol. Appl. Pharmacol.* 1, 524 (1959).

Slightly bitter crystals from ethyl acetate, mp 151.1-152.1°. Sparingly sol in water; freely sol in acidic solvents. LD$_{50}$ in mice, rats (mg/kg): 1000, 1700 orally; 742, 760 i.p. (Delahunt).

THERAP CAT: Antidepressant.

6577. Niaprazine. [27367-90-4] *N*-[3-[4-(4-Fluorophenyl)-1-piperazinyl]-1-methylpropyl]-3-pyridinecarboxamide; *N*-[3-[4-(*p*-fluorophenyl)-1-piperazinyl]-1-methylpropyl]nicotinamide; Nopron. C$_{20}$H$_{25}$FN$_4$O; mol wt 356.45. C 67.39%, H 7.07%, F 5.33%, N 15.72%, O 4.49%. Prepn: R. Y. Mauvernay, DE 1957371 corresp to US 3712893 (1970, 1973 both to CERM). Pharmacology: P. Duchene-Marrullaz *et al.*, *Therapie* 26, 1203 (1971). Psycholeptic effects in mice: J. Hache, J. Tachon, *J. Pharmacol.* 7, 469 (1976). Brain catecholamine depletion: P. E. Keane, M. S. Benedetti, *Neuropharmacology* 18, 595 (1979).

Cryst from ethyl acetate, mp 131°. Photosensitive. LD$_{50}$ in mice (mg/kg): 890 orally, 145 i.v. (Mauvernay).

THERAP CAT: Sedative; hypnotic.

6578. Nicaraven. [79455-30-4] *N*,*N*′-(1-Methyl-1,2-ethanediyl)bis-3-pyridinecarboxamide; 1,2-bis(nicotinamido)propane; (±)-*N*,*N*′-propylenedinicotinamide; AVS; Antevas. C$_{15}$H$_{16}$N$_4$O$_2$; mol wt 284.32. C 63.37%, H 5.67%, N 19.71%, O 11.25%. Hydroxyl radical scavenger. Prepn: T. Mori *et al.*, EP 29602; *eidem*, US 4366161 (1981, 1982 both to Chugai); T. Doi *et al.*, *Synlett* 1999, 1751. Effect on human platelet aggregation: T. Komiya *et al.*, *Clin. Neuropharmacol.* 22, 11 (1999). Clinical trial in subarachnoid hemorrhage: T. Asano *et al.*, *J. Neurosurg.* 84, 792 (1996). *Review*: K. K. Jain, *Expert Opin. Invest. Drugs* 9, 859-870 (2000).

Crystals from ethyl acetate, mp 156-157°. Sol in water.

THERAP CAT: In treatment of vasospasm following subarachnoid hemorrhage.

6579. Nicarbazin. [330-95-0] *N*,*N*′-Bis(4-nitrophenyl)urea, compd with 4,6-dimethyl-2(1*H*)-pyrimidinone (1:1); 4,4′-dinitrocarbanilide compd with 4,6-dimethyl-2-pyrimidinol (1:1); Nicarb; Nicoxin; Nicrazin. C$_{19}$H$_{18}$N$_6$O$_6$; mol wt 426.39. C 53.52%, H 4.26%, N 19.71%, O 22.51%. Prepn: Cuckler *et al.*, *Science* 122, 244 (1955); Basso, O'Neill, US 2731382; US 2731383; US 2731384 (all 1956 to Merck & Co.). Effect on nutritional encephalopathy in chicks: I. Bartov, P. Budowski, *Poult. Sci.* 58, 597 (1979).

Crystals, dec 265-275°. uv max (concd H$_2$SO$_4$): 298 nm (A$_{1cm}^{1\%}$ 670). Practically insol in water. Complex is slowly dec by trituration with water, or more rapidly by dil aq acids. The dry crystals are strongly electrostatic and present some dry-mixing problems.

THERAP CAT (VET): Coccidiostat.

6580. Nicardipine. [55985-32-5] 1,4-Dihydro-2,6-dimethyl-4-(3-nitrophenyl)-3,5-pyridinedicarboxylic acid 3-methyl 5-[2-[methyl(phenylmethyl)amino]ethyl] ester. C$_{26}$H$_{29}$N$_3$O$_6$; mol wt 479.53. C 65.12%, H 6.10%, N 8.76%, O 20.02%. Dihydropyridine calcium channel blocker. Prepn: M. Murakami *et al.*, BE 811324; *eidem*, US 3985758 (1974, 1976 both to Yamanouchi); M. Iwanami *et al.*, *Chem. Pharm. Bull.* 27, 1426 (1979). Vasodilator profile: T. Takenaka *et al.*, *Arzneim.-Forsch.* 26, 2172 (1976). Absorption, excretion, metabolism: S. Higuchi *et al.*, *Xenobiotica* **7**, 469 (1977). Clinical pharmacology: T. Seki, T. Takenaka, *Int. J. Clin. Pharmacol. Biopharm.* **15**, 267 (1977). Mechanism of action: K. Satoh *et al.*, *Clin. Exp. Pharmacol. Physiol.* **7**, 249 (1980). Acute toxicity study: Y. Odani, T. Sado, *Oyo Yakuri* **18**, 301 (1979), *C.A.* **92**, 104384u (1980). Symposium: *Br. J. Clin. Pharmacol.* **20**, Suppl. 1, 1S-208S (1985). Clinical synergy with captopril, *q.v.*: K. A. Conrad *et al.*, *Clin. Pharmacol. Ther.* **42**, 113 (1987). Review of pharmacodynamics, pharmacokinetics and therapeutic efficacy: E. M. Sorkin, S. P. Clissold, *Drugs* **33**, 296-345 (1987).

Hydrochloride. [54527-84-3] RS-69216; YC-93; Antagonil; Barizin; Bionicard; Cardene; Dacarel; Lecibral; Lescodil; Loxen; Nerdipina; Nicant; Nicardal; Nicarpin; Nicapress; Nicodel; Nimicor; Perdipina; Perdipine; Ranvil; Ridene; Rycarden; Rydene; Vasodin; Vasonase. $C_{26}H_{29}N_3O_6 \cdot HCl$; mol wt 515.99. Isolated in two crystalline forms from acetone. α-form: mp 179-181°; β-form: mp 168-170°. The forms also have different IR and x-ray diffraction patterns. LD_{50} in male, female rats (mg/kg): 634, 557 orally; 18.1, 25.0 i.v.; in male, female mice: 634, 650 orally; 20.7, 19.9 i.v. (Odani, Sado).

THERAP CAT: Antianginal; antihypertensive.

6581. Nicergoline. [27848-84-6] (8β)-10-Methoxy-1,6-dimethylergoline-8-methanol 5-bromo-3-pyridinecarboxylate (ester); 1-methyllumilysergol 8-(5-bromonicotinate) 10-methyl ether; 4,6,-6a,7,8,9,10,10a-octahydro-10aα-methoxy-4,7-dimethylindolo[4,3-*fg*]quinoline-9-methanol 5-bromonicotinate; 8β-[(5-bromonicotinyloxy)methyl]-1,6-dimethyl-10α-methoxyergoline; nicotergoline; nimergoline; MNE; FI-6714; Cergodum; Circo-Maren; Dilasenil; Duracebrol; Ergotop; Ergobel; Memoq; Nicergolent; Sermion; Vasospan. $C_{24}H_{26}BrN_3O_3$; mol wt 484.39. C 59.51%, H 5.41%, Br 16.50%, N 8.68%, O 9.91%. Prepn: Bernardi *et al.*, *US 3228943*; Temperilli, **DE 2112273** (1966, 1971, both to Farmitalia); Arcari *et al.*, *Experientia* **28**, 819 (1972). Series of articles on pharmacology, clinical studies, tolerability: *Arzneim.-Forsch.* **29**, 1213-1316 (1979). Toxicity study: B. W. Neumann, F. Lauschner, *ibid.* 1206. Hemodynamic effects in the dog: *ibid.* **31**, 1693 (1981). Use in acute myocardial infarction with diastolic hypertension: E. Triulzi *et al.*, *Farmaco Ed. Prat.* **36**, 449 (1981).

mp 136-138°. LD_{50} in male mice, rats (mg/kg): 860, 2800 orally; 46, 43 i.v. (Neumann, Lauschner).

THERAP CAT: Vasodilator (cerebral, peripheral).

6582. Niceritrol. [5868-05-3] 3-Pyridinecarboxylic acid 2,2-bis[[(3-pyridinylcarbonyl)oxy]methyl]-1,3-propanediyl ester; nicotinic acid neopentanetetrayl ester; pentaerythritol tetranicotinate; 8-AL; Perycit; Bufor. $C_{29}H_{24}N_4O_8$; mol wt 556.53. C 62.59%, H 4.35%, N 10.07%, O 23.00%. Prepn: **FR M2046** (1963 to Soc. d'Etudes et de Recherches Pharmacotechniques), *C.A.* **60**, 10656e

(1964); **GB 1022880**; **GB 1053689** (1966, 1967 to Aktiebolag Bofors), *C.A.* **64**, 19708d (1966), *C.A.* **66**, 104907e (1967); Kuriyama, Kudo, **JP 67 2359** (1967 to Yoshitomi), *C.A.* **66**, 115935p (1967). Acute toxicity data: T. Sugawara *et al.*, *Oyo Yakuri* **14**, 741 (1977), *C.A.* **88**, 131005v (1977).

Crystals, mp reported as 160-162° and as 163-164°. LD_{50} in mice, rats (g/kg): >20, >20 orally; >5, >5 s.c.; >5, >5 i.p.; in rabbits (g/kg): >10 orally; >5 i.p. (Sugawara).

THERAP CAT: Antilipemic.

6583. Nickel. [7440-02-0] Ni; at. wt 58.6934; at. no. 28; valence 2; seldom 1, 3, 4. Group VIII (10). Naturally occurring isotopes (mass numbers): 58 (68.27%), 60 (26.10%), 61 (1.13%), 62 (3.59%), 64 (0.91%); known artificial radioactive isotopes: 53, 55-57, 59 (longest-lived known isotope, $T_{1/2}$ 7.5 × 10^4 years, decay by electron capture), 63, 65-67. Abundance in earth's crust 99 ppm. Discovered by Cronstedt in 1754: Cronstedt, *Mineralogie* (Stockholm, 1758) p 218. Isoln: Berthier, *Ann. Chim. Phys.* [2] **14**, 52 (1820); **25**, 94 (1824). Occurs free in meteorites. Found in many ores as sulfides, arsenides, antimonides and oxides or silicates; chief sources include chalcopyrite, *q.v.*, pyrrhotite, **pentlandite** [(Fe,Ni)$_9$-S$_8$] and **garnierite** [3(Mg,Ni)O.2SiO$_2$.2H$_2$O]; other ores include **niccolite** (NiAs) and **millerite** (NiS). Methods of extraction and purification: Mackiw, *Can. J. Chem. Eng.* **46**, 3 (1968); Houot, *Ann. Mines* **1969** (April), 9; Queneau, *J. Met.* **22**, 44-48 (1970). Prepn of high purity nickel: Wise, Schaefer, *Metals Alloys* **16**, 424 (1924); from NiO and H$_2$: Glemser in *Handbook of Preparative Inorganic Chemistry* vol. 2, G. Brauer, Ed. (Academic Press, New York, 2nd ed., 1965) pp 1543-1544; by electrolysis: Vu Quang Kinh, Nardin, *C. R. Seances Acad. Sci. Ser. C* **266**, 307 (1968). Comprehensive reviews: *Gmelins, Nickel* (8th ed.) **57**, 5 vols, about 3500 pp (1965-1967); Nicholls in *Comprehensive Inorganic Chemistry* vol. 3, J. C. Bailar, Jr. *et al.*, Eds. (Pergamon Press, Oxford, 1973) pp 1109-1161; J. K. Tien, T. E. Howson in *Kirk-Othmer Encyclopedia of Chemical Technology* vol. 15 (Wiley-Interscience, New York, 3rd ed., 1981) pp 787-801; *Chemistry of the Elements* N. N. Greenwood, A. Earnshaw, Eds. (Pergamon Press, New York, 1984) pp 1328-1363. Book: *Nickel Toxicology*, S. S. Brown, F. W. Sunderman, Eds. (Academic Press, New York, 1980) 193 pp. Review of carcinogenic risk: *IARC Monographs* **11**, 75-112 (1976); of toxicology and human exposure: *Toxicological Profile for Nickel* (PB2006-100005, 2005) 397 pp.

Lustrous white, hard, ferromagnetic metal; face-centered cubic crystals. mp 1453°. bp (calc) 2732°. Also reported as bp (calc) 2837° (3110 K): D. R. Stull, G. C. Sinke, in *Adv. in Chem. Ser.* **18**, entitled "Thermodynamic Properties of the Elements" (ACS, Washington, 1956). d 8.908. Heat capacity (25°) 6.23 cal/g-atom/°C. Mohs' hardness 3.8. Latent heat of fusion 73 cal/g. Electrical resistivity (20°): 6.844 μohms-cm. E°(aq) Ni/Ni^{2+} 0.250 V. Stable in air at ordinary temp; burns in oxygen, forming NiO; not affected by water; decomposes steam at a red heat. Slowly attacked by dil hydrochloric or sulfuric acid; readily attacked by nitric acid. Not attacked by fused alkali hydroxides.

Caution: Potential symptoms of overexposure are sensitization dermatitis, allergic asthma, pneumonitis. *See NIOSH Pocket Guide to Chemical Hazards* (DHHS/NIOSH 97-140, 1997) p 224. *See also Patty's Industrial Hygiene and Toxicology* vol. 2C, G. D. Clayton, F. E. Clayton, Eds. (John Wiley & Sons, New York, 4th ed., 1994) 2157-2173. Metallic nickel is reasonably anticipated to be a human carcinogen; nickel compounds are listed as known human carcinogens: *Report on Carcinogens, Twelfth Edition* (PB2011-111646, 2011) p 280.

USE: Nickel-plating; for various alloys such as new silver, Chinese silver, German silver; for coins, electrotypes, storage batteries; magnets, lightning-rod tips, electrical contacts and electrodes, spark plugs, machinery parts; catalyst for hydrogenation of oils and other organic substances. *See also* Raney nickel. Manuf of Monel metal, stainless steels, heat resistant steels, heat and corrosion resistant alloys, nickel-chrome resistance wire; in alloys for electronic and space applications.

6584. Nickel Acetate. [373-02-4] Acetic acid nickel(2+) salt (2:1); nickel diacetate; nickelous acetate. $C_4H_6NiO_4$; mol wt 176.78. C 27.18%, H 3.42%, Ni 33.20%, O 36.20%. $Ni(CH_3CO_2)_2$. Review of toxicology and human exposure: *Toxicological Profile for Nickel* (PB2006-100005, 2005) 397 pp.
Tetrahydrate. Green, cryst mass or powder; acetic odor. d 1.744. Sol in 6 parts water, in alc. *Keep well closed.*
Caution: Nickel compounds are listed as known human carcinogens: *Report on Carcinogens, Twelfth Edition* (PB2011-111646, 2011) p 280.
USE: Catalyst; mordant for textiles.

6585. Nickel Acetylacetonate. [3264-82-2] (*SP*-4-1)-Bis-(2,4-pentanedionato-κO^2,κO^4)nickel; bisacetylacetonatonickel(II); bis(2,4-pentanediono)nickel(II); 2,4-pentanedione nickel complex. $C_{10}H_{14}NiO_4$; mol wt 256.91. C 46.75%, H 5.49%, Ni 22.85%, O 24.91%. $Ni(CH_3COCHCOCH_3)_2$. Also Ni(acac)$_2$ or Ni(AA)$_2$. Prepn from acetylacetone and Ni(OH)$_2$: Gach, *Monatsh. Chem.* **21**, 103 (1900); from acetylacetone and NiCl$_2$.6H$_2$O: Charles, Pawlikowski, *J. Phys. Chem.* **62**, 440 (1958); from 4-diethylamino-3-pentene-2-one and NiCl$_2$.6H$_2$O: Gash, *Can. J. Chem.* **45**, 2109 (1967). *See also:* Fernelius, Bryant, *Inorg. Synth.* **5**, 105 (1957). Exists as a trimer in the solid state: Bullen, *Nature* **177**, 537 (1956); Bullen *et al.*, *Inorg. Chem.* **4**, 456 (1965); as a monomer in the vapor phase: Fackler *et al.*, *J. Phys. Chem.* **72**, 4631 (1972). Structure of dihydrate: Montgomery, Lingafelter, *Acta Crystallogr.* **17**, 1481 (1964).
Emerald-green orthorhombic crystals. mp 229-230°. bp$_{11}$ 220-235°. d^{17} 1.455. uv max ($10^{-4}M$ in CHCl$_3$): 298, 265 nm (log ε 4.34, 4.44). Sol in water, alcohol, chloroform, benzene. Insol in ether, ligroin.
USE: Catalyst.

6586. Nickel Ammonium Sulfate. [15699-18-0] Sulfuric acid ammonium nickel(2+) salt (2:2:1); ammonium nickel sulfate; diammonium nickel disulfate. $H_8N_2NiO_8S_2$; mol wt 286.88. H 2.81%, N 9.77%, Ni 20.46%, O 44.62%, S 22.35%. Ni(NH$_4$)$_2$(SO$_4$)$_2$. Physical properties of aqueous solns: J. W. Mullin, M. M. Osman, *J. Chem. Eng. Data* **12**, 516 (1967); and crystal growth rates: *eidem, ibid.* **18**, 353 (1973). Toxicity study: E. L. Reagan, *J. Am. Coll. Toxicol.* **11**, 694 (1992). Crystal structure of hexahydrate: T. H. Tahirov, T.-H. Lu, *Acta Crystallogr.* **C50**, 668 (1994); and thermo-stability and optical transmission properties: G. Su *et al.*, *J. Phys. D: Appl. Phys.* **35**, 2652 (2002). Review of toxicology and human exposure: *Toxicological Profile for Nickel* (PB2006-100005, 2005) 397 pp.
Hexahydrate. [7785-20-8] Deep-green crystalline powder. d 1.923. d^{20} (satd soln): 1.050. Viscosity of satd soln: 1.2×10^{-3} N sec/m^2. Soly in water (g anhydrous salt/100 g water) at 10°: 5.19; at 25°: 7.52; at 40°: 10.9. Insol in alcohol. LD$_{50}$ orally in rats: 418 mg/kg (Reagan).
USE: Formerly in electroplating metals.

6587. Nickel Bromide. [13462-88-9] Nickel dibromide; nickelous bromide. Br$_2$Ni; mol wt 218.50. Br 73.14%, Ni 26.86%. NiBr$_2$.
Trihydrate. Yellowish-green, very deliquesc crystals; loses its water at about 200°, the anhydr salt is a golden-yellow color and sublimable in absence of air. Sol in one part water, in alcohol. *Keep well closed.*

6588. Nickel Carbonate Hydroxide. [12607-70-4] Nickel carbonate hydroxide (Ni$_3$(CO$_3$)(OH)$_4$); basic nickel carbonate. CH$_4$Ni$_3$O$_7$; mol wt 304.12. C 3.95%, H 1.33%, Ni 57.90%, O 36.83%. NiCO$_3$.2Ni(OH)$_2$. Tetrahydrate occurs in nature as the mineral *zaratite*. Review of toxicology and human exposure: *Toxicological Profile for Nickel* (PB2006-100005, 2005) 397 pp.
Tetrahydrate. Green, odorless powder. Sol in ammonia and in dil acids with effervescence. Insol in water.

Caution: Nickel compounds are listed as known human carcinogens: *Report on Carcinogens, Twelfth Edition* (PB2011-111646, 2011) p 280.
USE: Nickel-plating; catalyst for hardening of fats; in ceramic colors and glazes.

6589. Nickel Carbonyl. [13463-39-3] (*T*-4)-Nickel carbonyl (Ni(CO)$_4$); nickel tetracarbonyl. C$_4$NiO$_4$; mol wt 170.73. C 28.14%, Ni 34.38%, O 37.48%. Ni(CO)$_4$. Intermediate in nickel refining. Made by passing carbon monoxide over finely divided nickel: Mond *et al.*, *J. Chem. Soc.* **57**, 749 (1890); Gilliland, Blanchard, *Inorg. Synth.* **2**, 234 (1946). Use of nickel carbonyl in organic synthesis: G. Wilke *et al.*, *Angew. Chem. Int. Ed.* **5**, 151 (1966); M. F. Semmelhack in *Organic Reactions* vol. 19 (Wiley, New York, 1972) p 115; E. J. Corey, H. A. Kirst, *J. Am. Chem. Soc.* **94**, 667 (1972). Kinetic studies: D. H. Stedman *et al.*, *Science* **208**, 1029 (1980). Toxicity study: Hackett, Sunderman, *Arch. Environ. Health* **14**, 604 (1967). *Review:* Nicholls in *Comprehensive Inorganic Chemistry* vol. 3, J. C. Bailar, Jr. *et al.*, Eds. (Pergamon Press, Oxford, 1973) pp 1115-1119.
Colorless, volatile liquid. *Poisonous; flammable.* Oxidizes in the air: explodes at about 60°. d^{17} 1.318. bp 43°. mp −19.3°. Crit temp about 200°. Crit pressure about 30 atm. Sol in about 5000 parts water free from air; sol in alcohol, benzene, chloroform, acetone, carbon tetrachloride. LD$_{50}$ in rats (mg/kg): 39 i.p.; 63 s.c.; 66 i.v. (Hackett, Sunderman).
Caution: Potential symptoms of overexposure are headache, vertigo; nausea, vomiting, epigastric pain; substernal pain; cough, hyperpnea; cyanosis; weakness; leukocytosis; pneumonitis; delirium; convulsions. *See NIOSH Pocket Guide to Chemical Hazards* (DHHS/NIOSH 97-140, 1997) p 222. *See also Clinical Toxicology of Commercial Products*, R. E. Gosselin *et al.*, Eds. (Williams & Wilkins, Baltimore, 5th ed., 1984) Section II, p. 145. Nickel compounds are listed as known human carcinogens: *Report on Carcinogens, Twelfth Edition* (PB2011-111646, 2011) p 280.
USE: In organic synthesis; production of high-purity nickel powder and continuous nickel coatings on steel and other metals.

6590. Nickel Chloride. [7718-54-9] Nickel chloride (NiCl$_2$); nickel dichloride. Cl$_2$Ni; mol wt 129.59. Cl 54.71%, Ni 45.29%. NiCl$_2$. Evaluation of carcinogenic risk: *IARC Monographs* **49**, 257-445 (1990). Review of toxicology and human exposure: *Toxicological Profile for Nickel* (PB2006-100005, 2005) 397 pp.
Yellow deliquescent scales. Soly in water (g/l): 642 (20°), 876 (100°). Sol in ethanol, ammonium hydroxide. Insol in nitric acid. Sublimable in absence of air and readily absorbs NH$_3$. The aq soln is acid; pH about 4. *Keep well closed.* LD$_{50}$ in mice, rats (mg/kg): 48, 11 i.p. (IARC).
Hexahydrate. [7791-20-0] Cl$_2$Ni.6H$_2$O; mol wt 237.68. Green, deliquesc crystals or cryst powder. Monoclinic. Structure reported to be *trans*-[NiCl$_2$(H$_2$O)$_4$].2H$_2$O: Mizuno, *J. Phys. Soc. Jpn.* **16**, 1574 (1960), *C.A.* **55**, 26605g (1961). Sol in about one part water, in alcohol.
Caution: Nickel compounds are listed as known human carcinogens: *Report on Carcinogens, Twelfth Edition* (PB2011-111646, 2011) p 280.
USE: Anhydr salt as absorbent for NH$_3$ in gas masks. Hexahydrate for nickel electroplating; manuf nickel catalysts.

6591. Nickel Cyanide. [557-19-7] Dicyanonickel; nickel dicyanide. C$_2$N$_2$Ni; mol wt 110.73. C 21.69%, N 25.30%, Ni 53.01%. Ni(CN)$_2$. Prepn of yellow-brown anhydr salt: Aynsley, Campbell, *J. Chem. Soc.* **1958**, 1723. The commercial salt usually contains 20-25% water. Review of toxicology and human exposure: *Toxicological Profile for Nickel* (PB2006-100005, 2005) 397 pp.
Tetrahydrate. Apple-green powder. *Poisonous.* Insol in water. Slightly sol in dil acids, freely in alkali cyanides, in ammonia, and in ammonium carbonate.
USE: In nickel-plating.

6592. Nickel Dimethylglyoxime. [13478-93-8] (*SP*-4-1)-Bis[[2,3-butanedione 2,3-di(oximato-κN)](1−)]-nickel; bis(dimethylglyoximato)nickel. C$_8$H$_{14}$N$_4$NiO$_4$; mol wt 288.92. C 33.26%, H 4.88%, N 19.39%, Ni 20.31%, O 22.15%. Prepn: Banks *et al.*, *J. Am. Chem. Soc.* **77**, 324 (1955); F. J. Welcher, *Organic Analytical Reagents* vol. 3 (Van Nostrand, New York, 1947) pp 165-179; Thabet *et al.*, *Inorg. Nucl. Chem. Lett.* **8**, 211 (1972). Structure: Godycki, Rundle, *Acta Crystallogr.* **6**, 487 (1953); Merritt, *Anal. Chem.* **25**, 718 (1953).

Scarlet-red, cryst powder. Sublimes at 250°. Insol in water, acetic acid, ammonia. Sol in dil mineral acids and appreciably sol in abs alcohol.

USE: As sun-fast pigment in paints, lacquers, cellulose compounds and cosmetics.

6593. Nickel Fluoride. [10028-18-9] Nickel difluoride; nickelous fluoride. F_2Ni; mol wt 96.69. F 39.30%, Ni 60.70%. NiF_2. Prepn: Henkel, Klemm, *Z. Anorg. Allg. Chem.* **222**, 74 (1935); Priest, *Inorg. Synth.* **3**, 173 (1950); Rochow, Kukin, *J. Am. Chem. Soc.* **74**, 1615 (1952); Haendler *et al.*, *ibid.* 3167. Book: *Medical and Biologic Effects of Environmental Pollutants: Nickel* (National Acad. Sci., Washington DC, 1975) 277 pp.

Yellowish to green tetragonal crystals (rutile type). d 4.72. Sublimes in HF stream above 1000°. Slightly sol in water (4 g/100 ml at 25°). Aq solns are dec by boiling. Insol in alcohol, ether. LD_{50} i.v. in mice: 130 mg/kg (Nat. Acad. Sci.).

Caution: Chronic exposure may cause mottling of teeth, changes in bones.

6594. Nickel Formate. [3349-06-2] Formic acid nickel(2+) salt (2:1); nickel diformate; nickelous formate. $C_2H_2NiO_4$; mol wt 148.73. C 16.15%, H 1.36%, Ni 39.46%, O 43.03%. $Ni(HCOO)_2$. Prepd by reaction of formic acid with Ni: Johnson, US 2576072 (1951 to Harshaw Chemical); with $NiCO_3$: Bircumshaw, Edwards, *J. Chem. Soc.* **1950**, 1800.

Dihydrate. Fine, green, monoclinic crystals. Becomes anhydr on careful heating to 130-140°; decomposes at 180-200° yielding Ni, CO, CO_2, H_2, H_2O, CH_4. $d^{20.2}$ 2.154. Moderately sol in water. Practically insol in alc, formic acid.

USE: Manuf of Ni; prepn of Ni catalysts for organic reactions, particularly hydrogenation catalysts.

6595. Nickel Hydroxide. [12054-48-7] "Green nickel oxide"; nickel dihydroxide; nickelous hydroxide. H_2NiO_2; mol wt 92.71. H 2.17%, Ni 63.31%, O 34.51%. $Ni(OH)_2$.

Monohydrate. Apple-green powder. Decomp above 200° to form NiO and H_2O. Insol in water. Sol in dil acids, in ammonia.

Caution: Nickel compounds are listed as known human carcinogens: *Report on Carcinogens, Twelfth Edition* (PB2011-111646, 2011) p 280.

6596. Nickel Iodide. [13462-90-3] Nickel diiodide; nickelous iodide. I_2Ni; mol wt 312.50. I 81.22%, Ni 18.78%. NiI_2.

Iron-black color. Sublimes in absence of air. Hexahydrate, bluish-green very deliquesc crystals. Very sol in water or alcohol. *Keep well closed.*

6597. Nickel Monoxide. [1313-99-1] Nickel oxide (NiO); nickel(II) oxide; nickelous oxide; nickel protoxide. NiO; mol wt 74.69. Ni 78.58%, O 21.42%. Occurs as the mineral *bunsenite*. Review of toxicology and human exposure: *Toxicological Profile for Nickel* (PB2006-100005, 2005) 397 pp.

Green powder; yellow when hot. Insol in water. Sol in acids.

Caution: Nickel compounds are listed as known human carcinogens: *Report on Carcinogens, Twelfth Edition* (PB2011-111646, 2011) p 280.

USE: Painting on porcelain.

6598. Nickel Nitrate. [13138-45-9] Nitric acid nickel(2+) salt (2:1). N_2NiO_6; mol wt 182.70. N 15.33%, Ni 32.13%, O 52.54%. $Ni(NO_3)_2$. Toxicity study: H. F. Smyth *et al.*, *Am. Ind. Hyg. Assoc.*

J. **30**, 470 (1969). Review of toxicology and human exposure: *Toxicological Profile for Nickel* (PB2006-100005, 2005) 397 pp.

Hexahydrate. [13478-00-7] $N_2NiO_6.6H_2O$; mol wt 290.79. Green, deliquesc crystals. d 2.05. mp 56.7°. bp 137°. *Oxidizer; keep well closed.* Sol in 0.4 part water, in alcohol. The aq soln is acid; pH about 4. LD_{50} orally in rats: 1.62 g/kg (Smyth).

USE: Nickel-plating; manuf brown ceramic colors.

6599. Nickel Oxalate. [547-67-1] [Ethanedioato(2−)-κO^1,-κO^2]nickel; oxalic acid nickel(2+) salt (1:1). C_2NiO_4; mol wt 146.71. C 16.37%, Ni 40.01%, O 43.62%. NiC_2O_4.

Dihydrate, light green powder. Insol in water; sol in mineral acids, in solns of ammonium chloride, nitrate, or sulfate.

6600. Nickel Phosphate. [10381-36-9] Phosphoric acid nickel(2+) salt (2:3); nickelous phosphate; trinickel diphosphate. $Ni_3O_8P_2$; mol wt 366.02. Ni 48.11%, O 34.97%, P 16.92%. $Ni_3(PO_4)_2$.

Octahydrate, light green powder. Insol in water; sol in acids, ammonia.

USE: On ignition yields "nickel yellow", a pigment used in oil and water colors.

6601. Nickel Sesquioxide. [1314-06-3] Nickel oxide (Ni_2O_3); nickelic oxide; black nickel oxide; dinickel trioxide. Ni_2O_3; mol wt 165.38. Ni 70.98%, O 29.02%. Contains a variable quantity of water.

Gray-black powder. Dec at ~600° into NiO and oxygen. Insol in water; very slightly sol in cold acid; dissolved by hot HCl with evolution of Cl, and by hot H_2SO_4 or HNO_3 with evolution of oxygen.

6602. Nickel Sulfate. [7786-81-4] Sulfuric acid nickel(2+) salt (1:1). NiO_4S; mol wt 154.75. Ni 37.93%, O 41.35%, S 20.72%. $NiSO_4$. Soly study: F. C. Vilbrandt, J. A. Bender, *Ind. Eng. Chem.* **15**, 967 (1923). Specific gravity, viscosity, and soly of aqueous solns of α-hexahydrate: V. R. Phillips, *J. Chem. Eng. Data* **17**, 357 (1972). Thermal dehydration study of α-hexahydrate: N. Koga, H. Tanaka, *J. Phys. Chem.* **98**, 10521 (1994). Acute toxicity data: E. L. Reagan, *J. Am. Coll. Toxicol.* **1**, 685 (1992). Evaluation of carcinogenic risk: *IARC Monographs* **49**, 257-445 (1990); *Toxicology and Carcinogenesis Studies* (NTP TR-454, NIH 96-3370, 1996) 379 pp. Use in patch testing for nickel allergies: V. Simonetti *et al.*, *Contact Dermatitis* **39**, 187 (1998). Review of toxicology and human exposure: *Toxicological Profile for Nickel* (PB2006-100005, 2005) 397 pp.

Pale green to yellow crystals; dec at 848°. Soly at 20° (g/l): water 293. Insol in ethanol, diethyl ether.

Hexahydrate. [10101-97-0] $NiO_4S.6H_2O$; mol wt 262.84. Two known phases. α-Form, blue to blue-green tetragonal crystals; transition to β-form at 53.3°. β-Form, green transparent crystals. α-Form is more stable; commercially available. d 2.07. Soly in water of α-form (g anhydrous salt/100 g water) at 50.00°: 52.267; of β-form at 60.11°: 55.557. Sol in ethanol. pH of 5% soln at 25°: 3.8. LD_{50} in male, female rats (mg/kg): 335, 264 orally (Reagan).

Heptahydrate. [10101-98-1] $NiO_4S.7H_2O$; mol wt 280.85. Green, orthorhombic crystals; efflorescent. Soly in water (g anhydrous salt/100 g water) at 0.00°: 26.189; at 25.00°: 40.469. Sol in ethanol.

Caution: Nickel compounds are listed as known human carcinogens: *Report on Carcinogens, Twelfth Edition* (PB2011-111646, 2011) p 280.

USE: In electrolytic and electroless nickel-plating; as mordant in dyeing; in catalyst mfr; in prepn of other nickel compds.

THERAP CAT: Diagnostic aid (contact allergen).

6603. Niclofolan. [10331-57-4] 5,5′-Dichloro-3,3′-dinitro-[1,1′-biphenyl]-2,2′-diol; 4,4′-dichloro-6,6′-dinitro-*o,o′*-biphenol; 5,5′-dichloro-2,2′-dihydroxy-3,3′-dinitrobiphenyl; menichlopholan; Bayer 9015; ME-3625; Bilevon-M. $C_{12}H_6Cl_2N_2O_6$; mol wt 345.09. C 41.77%, H 1.75%, Cl 20.55%, N 8.12%, O 27.82%. Anthelmintic activity: Meiser, Federmann, US 3082151 (1963 to Bayer); P. J. Lane, J. M. Stewart, *Vet. Rec.* **80**, 702 (1967). Pharmacokinetics in desert sheep: B. H. Ali *et al.*, *J. Vet. Pharmacol. Ther.* **13**, 217 (1990).

THERAP CAT (VET): Anthelmintic (fasciolicide).

6604. Niclosamide. [50-65-7] 5-Chloro-*N*-(2-chloro-4-nitro-phenyl)-2-hydroxybenzamide; 2',5-dichloro-4'-nitrosalicylanilide; 5-chloro-*N*-(2'-chloro-4'-nitrophenyl)salicylamide; 5-chlorosalicyl-oyl-(*o*-chloro-*p*-nitranilide); *N*-(2'-chloro-4'-nitrophenyl)-5-chloro-salicylamide; Bayer 2353; Cestocide; Niclocide; Ruby; Trédémine; Yomesan. $C_{13}H_8Cl_2N_2O_4$; mol wt 327.12. C 47.73%, H 2.47%, Cl 21.67%, N 8.56%, O 19.56%. Prepn: **GB 824345** (1959 to Bayer), *C.A.* **54**, 15822b (1960). *See also:* E. Schraufstätter, R. Gönnert, **US 3079297**; R. Strufe *et al.*, **US 3113067** (both 1963 to Bayer); Bekhli *et al.*, *Med. Prom. SSSR* **1965**, 25.

Pale yellow crystals, mp 225-230°. Practically insol in water. Sparingly sol in ethanol, chloroform, ether.

Ethanolamine salt. [1420-04-8] Clonitrilide; Bayluscid. C_{13}-$H_8Cl_2N_2O_4\cdot C_2H_7NO$; mol wt 388.20. Yellow-brown solid, mp 204°.

USE: The ethanolamine salt as a molluscicide.

THERAP CAT: Anthelmintic (Cestodes).

THERAP CAT (VET): Anthelmintic (Cestodes).

6605. Nicomorphine. [639-48-5] (5α,6α)-7,8-Didehydro-4,5-epoxy-17-methylmorphinan-3,6-diol 3,6-di-3-pyridinecarboxyl-ate; 3,6-dinicotinoylmorphine; morphine dinicotinate; morphine bis-(pyridine-3-carboxylate). $C_{29}H_{25}N_3O_5$; mol wt 495.54. C 70.29%, H 5.09%, N 8.48%, O 16.14%. Semisynthetic opioid analgesic; morphine diester with nicotinic acid. Prepn: A. Pongratz, K. L. Zirm, *Monatsh. Chem.* **88**, 330 (1957); **GB 807115** (1959 to Lannacher Heilmittel). Pharmacology: K. L. Zirm, A. Pongratz, *Arz-neim.-Forsch.* **9**, 511 (1959). HPLC determn in serum: P. M. Koop-man-Kimenai *et al.*, *J. Chromatogr.* **416**, 382 (1987). Clinical phar-macokinetics: *idem et al.*, *Eur. J. Anaesthesiol.* **10**, 125 (1993). Clinical trial for post-operative analgesia: M. Hasenbos *et al.*, *Acta Anaesthesiol. Scand.* **29**, 572, 577 (1985).

Crystals, mp 178-178.5° (corr.). Practically insol in water. Sol in ethanol.

Hydrochloride. [12040-41-4] Vilan. $C_{29}H_{25}N_3O_5\cdot HCl$. Dec 248°. Sol in water with neutral reaction.

Note: This is a controlled substance (opium derivative): **21 CFR**, 1308.11.

THERAP CAT: Analgesic.

6606. Nicorandil. [65141-46-0] *N*-[2-(Nitrooxy)ethyl]-3-pyridinecarboxamide; *N*-(2-hydroxyethyl)nicotinamide nitrate (es-ter); SG-75; Adancor; Ikorel; Perisalol; Sigmart. $C_8H_9N_3O_4$; mol wt 211.18. C 45.50%, H 4.30%, N 19.90%, O 30.30%. Nicotinam-ide derivative; exhibits dual mechanism of action as both nitrovaso-dilator and potassium channel activator. Prepn: H. Nagano *et al.*, **DE 2714713**; *eidem*, **US 4200640** (1977, 1980 both to Chugai). Pharmacology: N. Taira *et al.*, *Clin. Exp. Pharmacol. Physiol.* **6**, 301 (1979). Mechanism of action: F. Yoneyama *et al.*, *Cardiovasc. Drugs Ther.* **4**, 1119 (1990). Symposia on pharmacology and clini-cal efficacy: *Am. J. Cardiol.* **63**, Suppl, 1J-85J (1989); *J. Cardio-vasc. Pharmacol.* **20**, Suppl. 3, S1-S108 (1992). Review of clinical potential in ischemic heart disease: H. Purcell, K. Fox, *Br. J. Clin. Pract.* **47**, 150-154 (1993). Clinical prevention of coronary events in patients with angina: IONA Study Group, *Lancet* **359**, 1269 (2002).

Colorless needles from ether/ethanol, mp 92-93°. LD_{50} in rats (mg/kg): 1200-1300 orally; 800-1000 i.v. (Nagano).

THERAP CAT: Antianginal.

6607. Nicosulfuron. [111991-09-4] 2-[[[[(4,6-Dimethoxy-2-pyrimidinyl)amino]carbonyl]amino]sulfonyl]-*N*,*N*-dimethyl-3-pyri-dinecarboxamide; SL-950; MU-495; DPX-V9360; Accent; Nostoc. $C_{15}H_{18}N_6O_6S$; mol wt 410.41. C 43.90%, H 4.42%, N 20.48%, O 23.39%, S 7.81%. Prepn: F. Kimura *et al.*, **EP 232067** (1987 to Ishihara Sangyo Kaisha); M. A. Hanagan, **US 4789393** (1988 to DuPont); and structure-activity study: S. Murai *et al.*, *ACS Symp. Ser.* **504**, 43 (1992). Activity in soil: T. C. Mueller *et al.*, *Tenn. Farm Home Sci.* **167**, 17 (1993). Adsorption on clay minerals: L. Ukrainczyk, N. Rashid, *J. Agric. Food Chem.* **43**, 855 (1995). Field trial for johnsongrass in corn and effect on yield: N. G. Gubbiga *et al.*, *Weed Technol.* **9**, 574 (1995).

Colorless white solid, mp 169-173°. Vapor pressure at 25°: 1.2 × 10^{-16} mm Hg. pKa (25°): 4.6. Soly in water (ppm): 44 (pH 3.5); 22000 (pH 7).

USE: Herbicide.

6608. Nicotinamide. [98-92-0] 3-Pyridinecarboxamide; nia-cinamide; nicotinic acid amide; Nicobion; Papulex. $C_6H_6N_2O$; mol wt 122.13. C 59.01%, H 4.95%, N 22.94%, O 13.10%. Precursor of the coenzymes NAD and NADP, *q.q.v.* The terms **niacin**, *Vitamin B$_3$*, and *Vitamin PP* have been used to refer to both nicotinamide and nicotinic acid, *q.v.* Isolation: H. v. Euler *et al.*, *Z. Physiol. Chem.* **258**, 212 (1939); P. Karrer, H. Keller, *Helv. Chim. Acta* **22**, 1292 (1939). Prepn from 3-cyanopyridine: E. J. Gasson, D. J. Had-ley, **US 2904552** (1959 to Distillers). Alternately prepd by passing NH_3 gas into molten nicotinic acid: A. Truchan, J. B. Davidson, **US 2993051** (1961 to Cowles Chem.). Isoln and analysis by liquid chro-matography: C. Bernofsky, *Methods Enzymol.* **66**, 23 (1980). Ex-traction from cereal products: R. B. Roy, J. J. Merten, *J. Assoc. Off. Anal. Chem.* **66**, 291 (1983). Toxicity data: F. G. Brazda, R. A. Coulson, *Proc. Soc. Exp. Biol. Med.* **62**, 19 (1946). Clinical trial in acne: A. R. Shalita *et al.*, *Int. J. Dermatol.* **34**, 434 (1995); to pre-serve beta cell function in diabetes: P. Pozzilli *et al.*, *Diabetologia*

38, 848 (1995). *Review:* "Niacin: Nicotinic Acid, Nicotinamide, NAD(P)" in *Vitamins*, W. Friedrich, Ed. (Walter de Gruyter, Berlin, 1988) pp 473-542. Comprehensive description: E. M. Abdel Moety *et al.*, *Anal. Profiles Drug Subs.* 20, 475-555 (1991).

Needles from benzene, mp 128-131°. Distills at 150-160° at 5×10^{-4} mm Hg. Absorption spectrum: Kuhn, Vetter, *Ber.* 68, 2374 (1935). uv max: 261 nm ($A_{1cm}^{1\%}$ 451). pK (20°C): 3.3. One gram dissolves in about one ml water, in about 1.5 ml alcohol, in 10 ml glycerol. A 10% w/v soln in water is neutral to litmus. Forms cryst salts with acids. LD_{50} s.c. in rats: 1.68 g/kg (Brazda, Coulson).

Compd with ascorbic acid. [1987-71-9] Nicotinamide ascorbate; nicoscorbine. $C_{12}H_{14}N_2O_7$. Prepn: W. Wenner, *J. Org. Chem.* 14, 22 (1949). Yellow crystals, mp 141-145°. $[\alpha]_D^{20}$ +27.5° (c = 8 in H_2O). pH of 5% aq soln ~4.0. Soly (20°): water 40%; abs ethanol 2.4%; methanol 10%. Sparingly sol in acetone. Practically insol in benzene, ether.

THERAP CAT: Antiacne; vitamin (enzyme cofactor).

THERAP CAT (VET): Vitamin (enzyme cofactor).

6609. Nicotine. [54-11-5] 3-[(2S)-1-Methyl-2-pyrrolidinyl]-pyridine; 1-methyl-2-(3-pyridyl)pyrrolidine; β-pyridyl-α-N-methylpyrrolidine; Habitrol; Nicabate; Nicoderm CQ; Nicolan; Nicopatch; Nicotinell; NiQuitin CQ; Tabazur. $C_{10}H_{14}N_2$; mol wt 162.24. C 74.03%, H 8.70%, N 17.27%. From the dried leaves of *Nicotiana tabacum* and *N. rustica* where it occurs to the extent of 2 to 8%, combined with citric and malic acids. Extraction procedure: Gattermann, Wieland, *Laboratory Methods of Organic Chemistry* (New York, 24th ed., 1937); Schwyzer, *Die Fabrikation pharmazeutischer und chemisch-technischer Produkte* (Berlin, 1931). Purification: Ratz, *Monatsh. Chem.* 26, 1241 (1905). Structure and synthesis: Pinner, *Ber.* 26, 294 (1893); Pictet, Rotschy, *Ber.* 37, 1225 (1904); Craig, *J. Am. Chem. Soc.* 55, 2854 (1933); M. Nakane. C. R. Hutchinson, *J. Org. Chem.* 43, 3922 (1978). Conformation in soln: T. P. Pitner *et al.*, *J. Am. Chem. Soc.* 100, 246 (1978). HPLC determn in plasma: M. Harlharan *et al.*, *Clin. Chem.* 34, 724 (1988). Toxicity data: R. B. Barlow, L. J. McLeod, *Br. J. Pharmacol.* 35, 161 (1969). Review and bibliography: Jackson, *Chem. Rev.* 29, 123 (1941). Review of pharmacology: R. W. Ryall in *Neuropoisons: Their Pathophysiological Actions* vol. 2, L. L. Simpson, D. R. Curtis, Eds. (Plenum, New York, 1974) pp 61-97; N. L. Benowitz, *Annu. Rev. Med.* 37, 21-32 (1986). Clinical trial of long-term efficacy in smoking cessation: T. Blondal *et al.*, *Br. Med. J.* 318, 285 (1999). Review of clinical experience in Tourette syndrome: A. A. Silver *et al.*, *CNS Spectr.* 4, 68-76 (1999).

Colorless to pale yellow, oily liq; very hygroscopic; turns brown on exposure to air or light. Acrid burning taste. Develops odor of pyridine. bp_{745} 247° (partial decompn); bp_{17} 123-125°. Volatile with steam. n_D^{20} 1.5282. d_4^{20} 1.00925. $[\alpha]_D^{20}$ −169.3° (neat); $[\alpha]_{5461}$ −204.1°. pK_1 (15°) 6.16; pK_2 10.96. pH of 0.05*M* soln: 10.2. Forms salts with almost any acid and double salts with many metals and acids. Absorption spectrum: Purvis, *J. Chem. Soc.* 97, 1035 (1910); Dobbie, Fox, *ibid.* 103, 1194 (1913). Optical rotatory properties: T. M. Lowry, W. V. Lloyd, *J. Chem. Soc.* 1929, 1771. *Poisonous.* Misc with water below 60°; on mixing nicotine with water the volume contracts. Very sol in alc, chloroform, ether, petr ether, kerosene, oils. Distribution of nicotine between water and petroleum oils: Norton, *Ind. Eng. Chem. Ind. Ed.* 32, 241 (1940). LD_{50} in mice (mg/kg): 0.3 i.v.; 9.5 i.p.; 230 orally (Barlow, McLeod).

Hydrochloride. [21361-93-3] $C_{10}H_{14}N_2 \cdot HCl$. Deliquesc crystals. $[\alpha]_D^{20}$ +104° (p = 10).

Dihydrochloride. [6019-02-9] $C_{10}H_{14}N_2 \cdot 2HCl$. Deliquesc crystals, very sol in water and alcohol. Nearly insol in ether.

Sulfate. [65-30-5] Nicotine neutral sulfate. $(C_{10}H_{14}N_2)_2 \cdot H_2SO_4$. Six-sided tablets. $[\alpha]_D^{20}$ +88° (p = 70). Sol in water, alcohol.

Bitartrate. [65-31-6] Nicotine tartrate. $C_{10}H_{14}N_2 \cdot 2C_4H_6O_6$. Dihydrate, crystals. mp 90°. $[\alpha]_D^{20}$ +26° (c = 10). Very sol in water or alcohol.

Zinc chloride double salt monohydrate. $C_{10}H_{16}Cl_4N_2Zn \cdot H_2O$. Also with $4H_2O$. Very sol in water; sparingly sol in abs alcohol and ether.

Salicylate. [29790-52-1] Eudermol. $C_{17}H_{20}N_2O_3$; mol wt 300.36. Six-sided plates, mp 118°. $[\alpha]_D^{20}$ +13° (c = 9). Freely sol in water or alcohol.

Polacrilex. [96055-45-7] Nicorette. Prepn: S. Lichtneckert *et al.*, **DE 2136119**; *eidem*, **US 3901248** (1972, 1975 both to AB Leo). Review of efficacy in smoking cessation: K. O. Fagerström, *Prog. Clin. Biol. Res.* 261, 109-128 (1988).

Caution: Nicotine can be absorbed through the alimentary canal, respiratory tract and intact skin. Potential symptoms of overexposure are nausea, salivation, abdominal pain, vomiting, diarrhea; headache, dizziness, auditory and visual disturbances; confusion, weakness, incoordination; paroxysmal atrial fibrillation; convulsions, dyspnea. Death may result from paralysis of respiratory muscles. *See NIOSH Pocket Guide to Chemical Hazards* (DHHS/NIOSH 97-140, 1997) p 224; *Clinical Toxicology of Commercial Products*, R. E. Gosselin *et al.*, Eds. (Williams & Wilkins, Baltimore, 5th ed., 1984) Section III, pp 311-314.

USE: Insecticide; fumigant. In the U.S. a 40% soln of nicotine sulfate, **Black Leaf 40**, was the commonly used form. As a contact poison it is most effective as soap, *i.e.*, as the laurate, oleate, or naphthenate. As a stomach poison a combination with bentonite has come into use.

THERAP CAT: Treatment of smoking withdrawal syndrome.

THERAP CAT (VET): Ectoparasiticide. Has been used as an anthelmintic.

6610. Nicotinic Acid. [59-67-6] 3-Pyridinecarboxylic acid; pyridine-β-carboxylic acid; P. P. factor; pellagra preventive factor; antipellagra vitamin; niacin; Niacor; Niaspan; Nicangin; Nicobid; Nicolar; Niconacid. $C_6H_5NO_2$; mol wt 123.11. C 58.54%, H 4.09%, N 11.38%, O 25.99%. Precursor of the coenzymes NAD and NADP, *q.q.v.* Widely distributed in nature; appreciable amounts are found in liver, fish, yeast and cereal grains. Dietary deficiency is associated with pellagra. The term "niacin" has also been applied to nicotinamide, *q.v.*, or to other derivatives exhibiting the biological activity of nicotinic acid. Prepn by oxidation of alkyl β-substituted pyridines: A. Ladenburg, *Ann.* 301, 152 (1898). Synthesis from pyridine: S. M. McElvain, M. A. Goese, *J. Am. Chem. Soc.* 63, 2283 (1941). Prepn by oxidation of nicotine: S. M. McElvain, *Org. Synth. coll. vol. I*, 385 (1941); C. F. Woodward *et al.*, *Ind. Eng. Chem.* 36, 540, 544 (1944). Toxicity study: F. G. Brazda, R. A. Coulson, *Proc. Soc. Exp. Biol. Med.* 62, 19 (1946). LC determn in foods: G. W. Chase, Jr. *et al.*, *J. AOAC Int.* 76, 390 (1993). Lipid-modifying effect in diabetic patients: M. B. Elam *et al.*, *J. Am. Med. Assoc.* 284, 1263 (2000). Review of nutritional requirements, bioavailability and relationship with tryptophan: W. J. Darby *et al.*, *Nutr. Rev.* 33, 289-297 (1975); of biosynthesis, metabolism and physiological effects: L. M. Henderson, *Annu. Rev. Nutr.* 3, 289-307 (1983); of pharmacology and therapeutic uses: J. R. DiPalma, W. S. Thayer, *ibid.* 11, 169-187 (1991). *Review:* "Niacin: Nicotinic Acid, Nicotinamide, NAD(P)" in *Vitamins*, W. Friedrich, Ed. (Walter de Gruyter, Berlin, 1988) pp 473-542.

Needles from alc or water, mp 236.6°. Nonhygroscopic and stable in air. Sublimes without decompn. pKa 4.85. pH of satd aq soln 2.7. uv max: 263 nm: *see:* Hünecke, *Ber.* 60, 1451 (1927). One gram dissolves in 60 ml water. Freely sol in boiling water, boiling alc, alkali hydroxides and carbonates; sol in propylene glycol; sparingly sol in water. Practically insol in ether. LD_{50} s.c. in rats: 5 g/kg (Brazda, Coulson).

N-Oxide *see* Oxiniacic Acid.

Sodium salt sesquihydrate. Direktan. $C_6H_4NNaO_2.1\frac{1}{2}H_2O$; mol wt 172.12. White crystals or crystalline powder; stable in air. One gram dissolves in ~1.4 ml water, in ~60 ml alc, in 10 ml glycerol. Insol in ether. pH of aq soln: ~7.

THERAP CAT: Antilipemic; vitamin (enzyme cofactor).

THERAP CAT (VET): Vitamin (enzyme cofactor).

6611. Nicotinic Acid Benzyl Ester. [94-44-0] 3-Pyridinecarboxylic acid phenylmethyl ester; pyridine-β-carboxylic acid benzyl ester; benzyl nicotinate; Rubriment; Pycaril; Pykaryl. $C_{13}H_{11}NO_2$; mol wt 213.24. C 73.22%, H 5.20%, N 6.57%, O 15.01%. Prepn: **GB 817103** (1959 to Nordmark-Werke).

Liquid, bp_{3-4} 170°.

THERAP CAT: Rubefacient.

6612. Nicotinyl Alcohol. [100-55-0] 3-Pyridinemethanol; β-pyridylcarbinol; 3-hydroxymethylpyridine; nicotinic alcohol; Nu-2121; Roniacol; Ronicol. C_6H_7NO; mol wt 109.13. C 66.04%, H 6.47%, N 12.84%, O 14.66%. Prepd by catalytic hydrogenation of 3-pyridinecarboxaldehyde: Panizzon, *Helv. Chim. Acta* **24**, supplemental issue in honor of Gadient Engi, p 26E (1941); by lithium aluminum hydride reduction of ethyl nicotinate: Rosenmund, Zymalkowski, *Ber.* **85**, 156 (1952); Cohen, **US 2520037** (1950); Mosher, Tessieri, *J. Am. Chem. Soc.* **73**, 4926 (1951); of methyl nicotinate: Bohlmann, *Ber.* **86**, 1423 (1953). From 3-cyanopyridine: Chase, **US 2615896** (1952 to Hoffmann-La Roche); from 3-aminomethylpyridine: Schläpfer, **US 2547048** (1951 to Hoffmann-La Roche); from thionicotinic acid S-methyl ester: Ruzicka, Prelog, **US 2509171** (1950 to Ciba).

Very hygroscopic liquid. bp_{28} 154°; bp_{16} 144-145°; bp_{12} 114°; $bp_{0.1}$ 110°. Freely sol in water, ether. Sparingly sol in petr ether.

d-Tartrate. [6164-87-0] Roniacol Tartrate; Radecol; Niltuvin. $C_6H_7NO.C_4H_6O_6$; mol wt 259.21. Crystals, sour taste, mp 147-148°. Sol in ether, freely sol in water, alcohol.

USE: Solubilizer for riboflavin: **US 2458430** (1949).

THERAP CAT: Vasodilator (peripheral).

6613. Nifedipine. [21829-25-4] 1,4-Dihydro-2,6-dimethyl-4-(2-nitrophenyl)-3,5-pyridinedicarboxylic acid dimethyl ester; 4-(2'-nitrophenyl)-2,6-dimethyl-3,5-dicarbomethoxy-1,4-dihydropyridine; Bay a 1040; Adalat(e); Adipine; Aldipin; Anifed; Aprical; Cardilate; Chronadalate; Coracten; Cordicant; Corotrend; Duranifin; Ecodipin; Fortipine; Hexadilat; Nifedicor; Nifelan; Nifelat; Nifensar; Orix; Pidilat; Procardia; Sepamit; Tensipine; Zenusin. $C_{17}H_{18}N_2O_6$; mol wt 346.34. C 58.96%, H 5.24%, N 8.09%, O 27.72%. Dihydropyridine calcium channel blocker. Prepn: Bossert, Vater, **ZA 6801482**; *eidem*, **US 3485847** (1968, 1969 both to Bayer). Series of articles on pharmacology, pharmacokinetics, biotransformation, clinical studies: *Arzneim.-Forsch.* **22**, 1-56, 330-388 (1972). Toxicity data: Vater *et al.*, *ibid.* 1. Comprehensive description: S. L. Ali, *Anal. Profiles Drug Subs.* **18**, 221-288 (1989). Review of analytical methods in biological fluids: P. A. Soons *et al.*, *J. Pharm. Biomed. Anal.* **9**, 475-484 (1991). Review of clinical efficacy in hypertension and Raynaud's phenomenon: W. Kiowski *et al.*, *Cardiovasc. Drugs Ther.* **4**, 935-940 (1990); in hypertension and angina: J. G. Tijssen, P. G. Hugenholtz, *Eur. Heart J.* **17**, 1152-1157 (1996). Comparative clinical trial in prevention of complications from hypertension: M. J. Brown *et al.*, *Lancet* **356**, 366 (2000).

Yellow crystals, mp 172-174°. uv max (methanol): 340, 235 nm (ε 5010, 21590); (0.1N HCl): 338, 238 nm (ε 5740, 20600); (0.1N NaOH): 340, 238 nm (ε 5740, 20510). Soly at 20° (g/L): acetone 250, methylene chloride 160, chloroform 140, ethyl acetate 50, methanol 26, ethanol 17. Practically insol in water. Very light sensitive in soln. LD_{50} in mice, rats (mg/kg): 494, 1022 orally; 4.2, 15.5 i.v. (Vater).

THERAP CAT: Antianginal; antihypertensive.

6614. Nifekalant. [130636-43-0] 6-[[2-[(2-Hydroxyethyl)][3-(4-nitrophenyl)propyl]amino]ethyl]amino]-1,3-dimethyl-2,4(1*H*,-3*H*)-pyrimidinedione. $C_{19}H_{27}N_5O_5$; mol wt 405.46. C 56.28%, H 6.71%, N 17.27%, O 19.73%. Nonselective potassium channel blocker. Prepn: T. Katakami *et al.*, **EP 369627**; *eidem*, **US 5008267** (1990, 1991 both to Mitsui Toatsu Chem.); *eidem*, *J. Med. Chem.* **35**, 3325 (1992). Pharmacology: J. Kamiya *et al.*, *Drug Dev. Res.* **30**, 373 (1993). Clinical experience: M. Igawa *et al.*, *J. Cardiovasc. Pharmacol.* **40**, 735 (2002); T. Minami *et al.*, *Pacing Clin. Electrophysiol.* **27**, 212 (2004). Review of pharmacology: H. Nakaya, H. Uemura, *Cardiovasc. Drug Rev.* **16**, 133-144 (1998); of development and therapeutic potential: A. Zaza, *Curr. Opin. Cardiovasc. Pulm. Renal Invest. Drugs* **2**, 86-94 (2000). Clinical evaluation in ventricular tachyarrhythmias: T. Katoh *et al.*, *Circ. J.* **69**, 1237 (2005); T. Washizuka *et al.*, *ibid.* 1508; in shock-refractory ventricular fibrillation: Y. Tahara *et al.*, *ibid.* **70**, 442 (2006).

Crystals from ethanol, mp 117.5-118.5°. Also reported as mp 125-126° (Katakami, 1992).

Hydrochloride. [130656-51-8] MS-551; Shinbit. $C_{19}H_{27}N_5-O_5.HCl$; mol wt 441.91. Light yellow to yellow crystals from methanol, mp 172-174°. Very sol in water, slightly sol in methanol, very slightly sol in ethanol. Practically insol in ether. pH 4.0 - 5.5 (5 mg/ml saline soln). pKa 7.05.

THERAP CAT: Antiarrhythmic (class III).

6615. Niflumic Acid. [4394-00-7] 2-[[3-(Trifluoromethyl)-phenyl]amino]-3-pyridinecarboxylic acid; 2-(α,α,α-trifluoro-*m*-toluidino)nicotinic acid; 2-[3-(trifluoromethyl)anilino]nicotinic acid; UP-83; Actol; Forenol; Landruma; Nifluril. $C_{13}H_9F_3N_2O_2$; mol wt 282.22. C 55.33%, H 3.21%, F 20.20%, N 9.93%, O 11.34%. Prepn: **NL 6414717**; C. Hoffmann, A. Faure, **US 3415834** (1965, 1968 both to Lab. U.P.S.A.); *eidem*, *Bull. Soc. Chim. Fr.* **1966**, 2316. Pharmacological and metabolic studies: Glasson *et al.*, *Biochem. Pharmacol.* **18**, 633 (1969); Boissier *et al.*, *Therapie* **26**, 211 (1971). Toxicity data: J. R. J. Sorenson, *J. Med. Chem.* **19**, 135 (1976). Determn in human plasma by GLC: G. Houin *et al.*, *J. Chromatogr.* **223**, 351 (1981).

Crystals from ethanol, mp 204°. LD_{50} in rats (mg/kg): 370 orally; 155 i.p. (Sorenson).

β-Morpholinoethyl ester. [65847-85-0] Morniflumate; Flomax; Nifluril (suppositories). $C_{19}H_{20}F_3N_3O_3$; mol wt 395.38. Prepn of the ester HCl: Hoffmann, **DE 1802777**; *idem*, **US 3708481** (1969, 1973 both to Hexachimie).

Phthalidyl ester *see* Talniflumate.

THERAP CAT: Anti-inflammatory.

6616. Nifuratel. [4936-47-4] 5-[(Methylthio)methyl]-3-[[(5-nitro-2-furanyl)methylene]amino]-2-oxazolidinone; 5-[(methyl-thio)methyl]-3-[(5-nitrofurfurylidene)amino]-2-oxazolidinone; 5-(methylmercaptomethyl)-3-(5-nitro-2-furfurylideneamino)-2-oxazolidinone; methylmercadone; Inimur; Macmiror; Magmilor; Omnes; Polmiror; Tydantil. $C_{10}H_{11}N_3O_5S$; mol wt 285.27. C 42.10%, H 3.89%, N 14.73%, O 28.04%, S 11.24%. Prepn: **BE 635608** (1963 to Polichimica SAP), *C.A.* **61**, 16069c (1964), corresp to **GB 969126**.

Crystals from acetic acid, mp 182°.

THERAP CAT: Antibacterial; antifungal; antiprotozoal (Trichomonas).

6617. Nifurfoline. [3363-58-4] 3-(4-Morpholinylmethyl)-1-[[(5-nitro-2-furanyl)methylene]amino]-2,4-imidazolidinedione; 3-(morpholinomethyl)-1-[(5-nitrofurfurylidene)amino]hydantoin; Urbac. $C_{13}H_{15}N_5O_6$; mol wt 337.29. C 46.29%, H 4.48%, N 20.76%, O 28.46%. Prepn: J. A. Bofill-Auge, J. M. Espinos-Taya, **ES 297087** (1964), *C.A.* **63**, 11572e (1966); **GB 1245095** (1971 to Esteve), *C.A.* **75**, 121382h (1971).

Yellow cryst, mp 206°; also reported as mp 194-196°, J. Klosa, H. Starke, **DD 508615** (1966), *C.A.* **66**, 65501w (1967).

THERAP CAT: Antibacterial.

6618. Nifuroquine. [57474-29-0] 4-(5-Nitro-2-furanyl)-2-quinolinecarboxylic acid 1-oxide; 2-carboxy-4-[2′-(5′-nitrofuryl)]-quinoline 1-oxide; 4-(5-nitro-2-furyl)quinaldic acid 1-oxide; quinaldofur; Abimasten 100. $C_{14}H_8N_2O_6$; mol wt 300.23. C 56.01%, H 2.69%, N 9.33%, O 31.97%. Prepn: R. R. G. Haber, E. Schoenberger, **ZA 6703320** corresp to **US 4217456** and **US 4224448** (1967, 1980, 1980). Use in bovine mastitis: R. R. G. Haber *et al.*, **DE 2612250** corresp to **US 4070469** (1977, 1978 both to ABIC). *In vitro* study: G. Ziv *et al.*, *Zentralbl. Veterinaermed. (B)* **23**, 301 (1976). Pharmacology: G. Ziv, A. Saran, *ibid.* 310. Activity under anaerobic conditions: G. M. Maluszynska, L. Bassalik-Chabielska, *Pr. Mater. Zootech.* **23**, 63 (1980), *C.A.* **93**, 143793 (1980).

Yellow cryst powder, mp 190° (dec). Practically insol in water.

THERAP CAT (VET): Antibacterial.

6619. Nifuroxazide. [965-52-6] 4-Hydroxybenzoic acid 2-[(5-nitro-2-furanyl)methylene]hydrazide; *p*-hydroxybenzoic acid (5-nitrofurfurylidene)hydrazide; 1-(*p*-hydroxybenzoyl)-2-(5-nitrofurfurylidene)hydrazine; 5-nitro-2-furaldehyde *p*-hydroxybenzoylhydrazone; RC-27109; Adral; Bacifurane; Diarlidan; Dicoferin; Ercefurol; Ercefuryl; Pentofuryl. $C_{12}H_9N_3O_5$; mol wt 275.22. C 52.37%, H 3.30%, N 15.27%, O 29.07%. Prepn: **FR 1327840** (1963 to Robert et Carriere), *C.A.* **59**, 12763b (1963); M. C. E. Carron, **GB 962706**; *idem*, **US 3290213** (1964, 1966 both to Robert et Carriere). Antiseptic and antibacterial properties: M. C. E. Carron *et al.*, *Ann. Pharm. Fr.* **21**, 287 (1963). *In vitro* study of activity spectrum: A. Thabaut, J. L. Durosoir, *Gaz. Med. Fr.* **85**, 4516 (1978), *C.A.* **90**, 1487 (1979).

Crystals from pyridine, mp 298°. Practically insol in water.

THERAP CAT: Intestinal antiseptic.

6620. Nifuroxime. [6236-05-1] [*C*(*Z*)]-5-Nitro-2-furancarboxaldehyde oxime; *anti*-5-nitro-2-furaldoxime; Micofur. $C_5H_4N_2O_4$; mol wt 156.10. C 38.47%, H 2.58%, N 17.95%, O 41.00%. Prepd by treating 5-nitrofurfural with hydroxylamine in alcohol: Ikeda, *C.A.* **50**, 10701 (1956). Claimed in Stillman *et al.*, **US 2319481** (1943 to Norwich Pharmacal).

Tasteless, pale yellow or greenish crystals from ethanol. Darkens on exposure to light. mp 156° (Ikeda); mp 163-164° (Stillman *et al.*). Soly at 25° in water: ~1 g/l, in methanol: 89.0 g/l, in 95% ethanol: 39.0 g/l.

THERAP CAT: Topical anti-infective; antiprotozoal (Trichomonas).

6621. Nifurpirinol. [13411-16-0] 6-[2-(5-Nitro-2-furanyl)-ethenyl]-2-pyridinemethanol; 6-[2-(5-nitro-2-furyl)vinyl]-2-pyridinemethanol; 6-(hydroxymethyl)-2-[2-(5-nitro-2-furyl)vinyl]pyridine; furpirinol; furpyrinol; P-7138; Furanace. $C_{12}H_{10}N_2O_4$; mol wt 246.22. C 58.54%, H 4.09%, N 11.38%, O 25.99%. Prepn: Fujita *et al.*, *J. Pharm. Soc. Jpn.* **86**, 1014 (1966); *C.A.* **115605f** (1967); Murakami, Iwanami, **JP 70 14072** (1970 to Yamanouchi), *C.A.* **73**, 45364v (1970). Antibacterial activity: Shimizu, Takase, *Prog. Antimicrob. Anticancer Chemother., Proc. 6th Int. Congr. Chemother.* **2**, 388 (1970); D. F. Amend, A. J. Ross, *Prog. Fish Cult.* **32**, 19 (1970); A. J. Ross, *ibid.* **34**, 18 (1972); Y. Takase *et al.*, *Chem. Pharm. Bull.* **21**, 144 (1973).

Yellow needles from acetone or methanol, mp 170-171°.

THERAP CAT: Antibacterial.

THERAP CAT (VET): Antibacterial in fish diseases.

6622. Nifurtimox. [23256-30-6] 3-Methyl-*N*-[(5-nitro-2-furanyl)methylene]-4-thiomorpholinamine 1,1-dioxide; 4-[(5-nitrofurfurylidene)amino]-3-methylthiomorpholine-1,1-dioxide; tetrahydro-3-methyl-4-[(5-nitrofurfurylidene)amino]-2*H*-1,4-thiazine 1,1-dioxide; 1-[(5-nitrofurfurylidene)amino]-2-methyltetrahydro-1,4-thiazine 4,4-dioxide; Bay 2502; Lampit. $C_{10}H_{13}N_3O_5S$; mol wt 287.29. C 41.81%, H 4.56%, N 14.63%, O 27.84%, S 11.16%. Prepn from 5-nitrofurfural and 4-amino-3-methyltetrahydro-1,4-thiazine 1,1-dioxide: H. Herlinger *et al.*, **DE 1170957**; *eidem*, **US 3262930** (1964, 1966 both to Bayer). Series of articles on pharmacology and clinical findings: *Arzneim.-Forsch.* **22**, 1563-1642 (1972). Toxicity data:

K. Hoffmann, *ibid.* 1590. Spectrophotometric determn in pharmaceutical formulations: C. S. P. Sastry *et al.*, *Talanta* **41**, 1957 (1994). Clinical trial of combination with eflornithine in trypanosomiasis: G. Priotto *et al.*, *Clin. Infect. Dis.* **45**, 1435 (2007).

Orange-red crystals from dil acetic acid, mp 180-182°. LD_{50} in mice, rats (mg/kg): 3720, 4050 by gavage (Hoffmann).

THERAP CAT: Antiprotozoal (Trypanosoma).

6623. Nifurtoinol. [1088-92-2] 3-(Hydroxymethyl)-1-[[(5-nitro-2-furanyl)methylene]amino]-2,4-imidazolidinedione; 3-(hydroxymethyl)-1-[(5-nitrofurfurylidene)amino]hydantoin; 1-[(5-nitrofurfurylidene)amino]-3-(hydroxymethyl)hydantoin; Urfadyn. $C_9H_8N_4O_6$; mol wt 268.19. C 40.31%, H 3.01%, N 20.89%, O 35.79%. Prepn: Michels, **BE 611941** corresp to **US 3446802** (1962, 1969 to Norwich Pharmacal); Spencer, Michels, *J. Org. Chem.* **29**, 3416 (1964).

Yellow crystals from aq formaldehyde. When heated on a melting-point block, loses formaldehyde, then further decomposes at same temp as nitrofurantoin. uv max (2% in dimethylformamide): 367.5, 265 nm (ε 17900, 12800).

THERAP CAT: Antibacterial.

6624. Nifurzide. [39978-42-2] 5-Nitro-2-thiophenecarboxylic acid 2-[3-(5-nitro-2-furanyl)-2-propen-1-ylidene]hydrazide; N^1-(5-nitro-2-furylacrylidene)-N^2-(5-nitro-2-thenoyl)hydrazine; 5-nitro-2-thiophenecarboxylic acid [3-(5-nitro-2-furyl)allylidene]hydrazide; Ricridene. $C_{12}H_8N_4O_6S$; mol wt 336.28. C 42.86%, H 2.40%, N 16.66%, O 28.55%, S 9.53%. Bactericidal agent related to nitrofurazone, *q.v.* Prepn: E. Szarvasi, L. Fontaine, **DE 2200375**; *eidem*, **US 3847911** (1972, 1974 both to Lipha); E. Szarvasi *et al.*, *J. Med. Chem.* **16**, 281 (1973). Mode of action study: A. Delsarte *et al.*, *Antimicrob. Agents Chemother.* **19**, 477 (1981).

trans-form

Yellow crystals from DMF/ether, mp 235-236°. LD_{50} orally in mice: 3200 mg/kg (Szarvasi, Fontaine, 1972).

THERAP CAT: Anti-infective.

6625. Nigericin. [28380-24-7] Antibiotic K 178; antibiotic X-464; azalomycin M; helixin C; polyetherin A. $C_{40}H_{68}O_{11}$; mol wt 724.97. C 66.27%, H 9.45%, O 24.28%. Polyether antibiotic which affects ion transport and ATPase activity in mitochondria; produced by *Streptomyces hygroscopicus* E-749 and structurally related to monensin, *q.v.* Isoln, characterization, production: R. L. Harned *et al.*, *Antibiot. Chemother.* **1**, 594 (1951); J. Berger *et al.*, *J. Am. Chem. Soc.* **73**, 5295 (1951); J. Shoji *et al.*, *J. Antibiot.* **21**, 402 (1968); T. Kubota *et al.*, *Chem. Commun.* **1968**, 1541; T. Kubota, S. Matsutani, *J. Chem. Soc. C* **1970**, 695. Use in coccidiosis: M. Gorman, R. L. Hamill, **US 3555150** (1971 to Lilly). Effect on calcium uptake and membrane potential in mitochondria: H. Rottenberg, A. Scarpa, *Biochemistry* **13**, 4811 (1974). Stimulation of ATPase activity: H. Sze, *Proc. Natl. Acad. Sci. USA* **77**, 5904 (1980). Approach to synthesis: C. P. Holmes, P. A. Bartlett, *J. Org. Chem.* **54**, 98 (1989).

Colorless needles, mp 183.5-185°. $[\alpha]_D^{24}$ +36.2° (c = 0.842 in $CHCl_3$). Sol in alcohols, acetone, ethyl acetate, chloroform, benzene, ether; slightly sol in satd hydrocarbons. Practically insol in water. LD_{50} in mice (mg/kg): 10-15 i.p. (Shoji); also reported as 2.5 i.p. (Harned).

Sodium salt. $C_{40}H_{67}NaO_{11}$. Crystals, mp 245-255° (dec). Sol in chloroform. Practically insol in water.

6626. Nihydrazone. [67-28-7] Acetic acid 2-[(5-nitro-2-furanyl)methylene]hydrazide; acetic acid 5-(nitrofurfurylidene)hydrazide; 5-nitro-2-furaldehyde acetylhydrazone; *N*-acetyl-5-nitro-2-furaldehyde hydrazide; 1-(5-nitro-2-furfurylidene)-2-acetylhydrazine; 1-acetyl-2-(5-nitro-2-furfurylidene)hydrazine; NF-64; HC-064; Furiton; Nidrafur. $C_7H_7N_3O_4$; mol wt 197.15. C 42.65%, H 3.58%, N 21.31%, O 32.46%. Prepn: Stillman, Scott, **US 2416234** and **US 2416236** (1947 to Eaton Labs).

Yellow crystals from acetic acid + ethanol, dec 230-235°. uv max (water): 253, 364 nm (log ε 4.11, 4.23). Soly in water 1:20,000.

THERAP CAT (VET): Antibacterial, antiprotozoal.

6627. Nikethamide. [59-26-7] *N,N*-Diethyl-3-pyridinecarboxamide; *N,N*-diethylnicotinamide; pyridine-3-carboxylic acid diethylamide; nicotinic acid diethylamide; Anacardone; Astrocar; Carbamidal; Cardamine; Cardiamid; Cardimon; Coracon; Coractiv N; Coramine; Cordiamin; Corediol; Cormed; Cormid; Corvitol; Corvotone; Dynacoryl; Eucoran; Inicardio; Niamine; Nicamide; Nicor; Nicorine; Nikardin; Pyricardyl; Salvacard; Stimulin; Ventramine. $C_{10}H_{14}N_2O$; mol wt 178.24. C 67.39%, H 7.92%, N 15.72%, O 8.98%. Prepd by the action of thionyl chloride on nicotinic acid, followed by treatment with diethylamine hydrochloride; also formed by heating nicotinic acid or quinolinic acid anhydride with diethylamine: Hartmann, Seiberth, **US 1403117** (1922); from nicotinic acid + benzenesulfonyldiethyl amide: Oxley *et al.*, *J. Chem. Soc.* **1946**, 763. Toxicity data: E. I. Goldenthal, *Toxicol. Appl. Pharmacol.* **18**, 185 (1971).

Slightly viscous liquid or cryst solid. Faintly bitter taste followed by a faint sensation of warmth. d_4^{25} 1.058-1.066 (liq). mp 24-26°. bp_{760} 296-300° (some decompn); bp_{10} 158-159°; bp_3 128-129°; $bp_{0.4}$ 115°. n_D^{20} 1.525-1.526; n_D^{25} 1.522-1.524. Misc with water, ether, chloroform, acetone, alc. Usually marketed as a 25% w/v aq soln, pH 6.0-6.5. Incompatible with Na_2CO_3 solns which cause precipitation; tannates produce an amorph ppt. LD_{50} i.p. in rats: 272 mg/kg (Goldenthal).

THERAP CAT: CNS and respiratory stimulant.

THERAP CAT (VET): Respiratory stimulant.

6628. Nile Red. [7385-67-3] 9-(Diethylamino)-5*H*-benzo[*a*]-phenoxazin-5-one; Nile Blue A oxazone. $C_{20}H_{18}N_2O_2$; mol wt 318.38. C 75.45%, H 5.70%, N 8.80%, O 10.05%. Hydrophobic fluorescent and solvatochromic dye. Prepn: R. Möhlau, K. Uhlmann, *Ann.* **289**, 90 (1896); J. F. Thorpe, *J. Chem. Soc.* **1907**, 324. Solvent-dependent spectrophotometric properties: M. M. Davis, H. B. Hetzer, *Anal. Chem.* **38**, 451 (1966). Spectrofluorometric studies of lipid interactions: P. Greenspan, S. D. Fowler, *J. Lipid Res.* **26**,

781 (1985). Solvatochromic and thermochromic properties: C. M. Golini *et al., J. Fluoresc.* **8**, 395 (1998); N. Ghoneim, *Spectrochim. Acta A* **56**, 1003 (2000). Use as a fluorescent probe of hydrophobic protein sites: D. L. Sackett, J. Wolff, *Anal. Biochem.* **167**, 228 (1987); as a stain for low density lipoproteins: P. Greenspan, R. L. Gutman, *Electrophoresis* **14**, 65 (1993); in detection of protein aggregates: M. Sutter *et al., J. Fluoresc.* **17**, 181 (2007); as an optochemical humidity sensor: I. Pellejero *et al., Ind. Eng. Chem. Res.* **46**, 2335 (2007).

Brown glistening plates from light petroleum, with faint green metallic reflex, mp 205° (Thorpe). Also reported as mp 192-193° (Greenspan, Fowler). Soly in acetone: 1 mg/ml; in *n*-heptane: 62 μg/ml; in water: <1 μg/ml. Partition coefficient at 4° (xylene/water): 210; (chloroform/water): 196; (*n*-heptane/water): 58. Absorption max in cyclohexane: 490, 512.5 nm; in acetone: 533 nm; in DMF: 541 nm; in ethanol: 550 nm; in formamide: 564 nm. Emission max in acetone: 615 nm; in DMF: 625 nm; in ethanol: 635 nm; in water: 665 nm.

USE: Polarity-sensitive fluorescent probe for lipids and hydrophobic protein domains.

6629. Nilutamide. [63612-50-0] 5,5-Dimethyl-3-[4-nitro-3-(trifluoromethyl)phenyl]-2,4-imidazolidinedione; 1-(3′-trifluoromethyl-4′-nitrophenyl)-4,4-dimethylimidazoline-2,5-dione; RU-23908; Anandron; Nilandron. $C_{12}H_{10}F_3N_3O_4$; mol wt 317.22. C 45.44%, H 3.18%, F 17.97%, N 13.25%, O 20.17%. Nonsteroidal antiandrogen. Prepn: J. Perronnet *et al.,* **DE 2649925**; *eidem,* **US 4097578** (1977, 1978 both to Roussel-UCLAF). Pharmacology: J.-P. Raynaud *et al., J. Steroid Biochem.* **11**, 93 (1979). Review of kinetics, metabolism and clinical studies: *idem et al., Prog. Clin. Biol. Res.* **185A**, 99-120 (1985); of therapeutic efficacy in prostate cancer: M. G. Harris *et al., Drugs Aging* **3**, 9-25 (1993).

Crystals from ethanol, mp 149°. Freely sol in ethyl acetate, acetone, chloroform, ethanol, dichloromethane, methanol. Soly in water at 25°: < 0.1%.

THERAP CAT: Antineoplastic (hormonal).

6630. Nilvadipine. [75530-68-6] 2-Cyano-1,4-dihydro-6-methyl-4-(3-nitrophenyl)-3,5-pyridinedicarboxylic acid 3-methyl 5-(1-methylethyl) ester; 5-isopropyl-3-methyl-2-cyano-1,4-dihydro-6-methyl-4-(*m*-nitrophenyl)-3,5-pyridinedicarboxylate; isopropyl 6-cyano-5-methoxycarbonyl-2-methyl-4-(3-nitrophenyl)-1,4-dihydropyridine-3-carboxylate; nivadipine; nivaldipine; FR-34235; FK-235; SKF-102362; Escor; Nivadil. $C_{19}H_{19}N_3O_6$; mol wt 385.38. C 59.22%, H 4.97%, N 10.90%, O 24.91%. Dihydropyridine calcium channel blocker. Prepn: Y. Sato, **BE 879263**; *idem,* **US 4338322** (1980, 1982 both to Fujisawa). Structural studies: A. Miyamae *et al., Chem. Pharm. Bull.* **34**, 3071 (1986). Determn in plasma and urine: Y. Tokuma *et al., J. Chromatogr.* **415**, 156 (1987). Preliminary pharmacokinetics and resolution of enantiomers: *eidem, J. Pharm. Sci.* **76**, 310 (1987). Pharmacokinetics in rabbits: Y. Nezasa *et al., Kankyo Igaku Kenkyusho Nenpo* **38**, 200 (1987), *C.A.* **107**, 89273q (1987). Mode of action: P. A. Molyvdas, N. Sperelakis, *J. Cardiovasc. Pharmacol.* **8**, 449 (1986). Cardiovascular effects: G. J. Gross *et al., Gen. Pharmacol.* **14**, 677 (1983). Clinical evaluation in hypertension: K. Mizuno *et al., Res. Commun. Chem. Pathol. Pharmacol.* **52**, 3 (1986).

Yellow prisms from ethanol, mp 148-150°.
(+)-Form. $[\alpha]_D^{20}$ +222.42° (c = 1 in methanol).
(−)-Form. $[\alpha]_D^{20}$ −219.62° (c = 1 in methanol).
THERAP CAT: Antihypertensive; antianginal.

6631. Nimbin. [5945-86-8] (2*R*,3a*R*,4a*S*,5*R*,5a*R*,6*R*,9a*R*,-10*S*,10a*R*)-5-(Acetyloxy)-2-(3-furanyl)-3,3a,4a,5,5a,6,9,9a,10,10a-decahydro-6-(methoxycarbonyl)-1,6,9a,10a-tetramethyl-9-oxo-2*H*-cyclopenta[*b*]naphthol[2,3-*d*]furan-10-acetic acid methyl ester; (4α,-5α,6α,7α,15β,17α)-6-(acetyloxy)-7,15:21,23-diepoxy-4,8-dimethyl-1-oxo-18,24-dinor-11,12-secochola-2,13,20,22-tetraene-4,11-dicarboxylic acid dimethyl ester; 5-(acetyloxy)-2-(3-furanyl)-3,3a,4a,-5,5a,6,9,9a,10,10a-decahydro-6-(methoxycarbonyl)-1,6,9a,10a-tetramethyl-9-oxo-2*H*-cyclopenta[*b*]naphtho[2,3-*d*]furan-10-acetic acid. $C_{30}H_{36}O_9$; mol wt 540.61. C 66.65%, H 6.71%, O 26.63%. Bitter principle from various parts of the neem tree, *Azadirachta indica* A. Juss. *(Melia azadirachta* L.), Meliaceae: Siddiqui, *Curr. Sci.* **11**, 278 (1942). Structure: C. R. Narayanan *et al., Chem. Ind. (London)* **1964**, 322; C. R. Narayanan, R. V. Pachapurkar, *Tetrahedron Lett.* **1965**, 4333. Stereochemistry: *eidem, J. Org. Chem.* **31**, 2691 (1966); Harris *et al., Tetrahedron* **24**, 1517 (1968). Crystal structure of dihydronimbin: C. R. Narayanan *et al., Acta Crystallogr.* **B36**, 486 (1980).

Crystals from methanol, mp 205°. $[\alpha]_D^{24}$ +170° (abs ethanol). uv max (95% ethanol): 210, 330 nm (ε 32700; 66). Practically insol in water. Sol in ether, abs alcohol.

Dihydronimbin. $C_{30}H_{38}O_9$. Crystals from methanol, mp 215-216°. $[\alpha]_D$ +167.5° (chloroform). uv max (95% ethanol): 210, 298 nm (ε 16200; 29).

6632. Nimbiol. [561-95-5] (4a*S*,10a*S*)-2,3,4,4a,10,10a-Hexahydro-6-hydroxy-1,1,4a,7-tetramethyl-9(1*H*)-phenanthrenone; 6-hydroxy-7-methyl-9-oxopodocarpane. $C_{18}H_{24}O_2$; mol wt 272.39. C 79.37%, H 8.88%, O 11.75%. (+)-Form found in the bark of the neem tree, *Azadirachta indica* A. Juss. *(Melia azadirachta* L.), Meliaceae: P. Sengupta *et al., Chem. Ind. (London)* **1958**, 861. Structure: P. Sengupta *et al., Tetrahedron* **10**, 45 (1960). Also obtained by conversion of podocarpic acid: Bible, *ibid.* **11**, 22 (1960); Wenkert *et al., J. Am. Chem. Soc.* **83**, 2320 (1961). Total synthesis of (±)-form: W. L. Meyer *et al., J. Org. Chem.* **40**, 3686 (1975). Synthesis of (±)-methyl ether: Ramachandran, Dutta, *J. Chem. Soc.* **1960**, 4766; Delobelle, Fétizon, *Bull. Soc. Chim. Fr.* **1961**, 1900; Nasipuri, Roy, *J. Indian Chem. Soc.* **40**, 327 (1963).

Crystals from dilute methanol or platelets by high vac sublimation. mp 248-252°. $[\alpha]_D^{25}$ +33° (chloroform). uv max (abs alcohol): 234, 283 nm (log ε 4.13, 4.10).

Methyl ether. $C_{19}H_{26}O_2$. Crystals, mp 143°. $[\alpha]_D^{25}$ +43.7° (chloroform). uv max (alcohol): 207, 232, 279 nm (log ε 4.18, 4.15, 4.12).

(±)-**Form.** White needles from CH_3OH, mp 237.0-237.5°.

(±)-**Methyl ether.** $C_{19}H_{26}O_2$. Crystals from hexane. mp 112-113°. uv max (alcohol): 207, 232, 279 nm (log ε 4.13, 4.13, 4.11) (Delobelle, Fétizon). Also reported as needles from CH_3OH, mp 117-118° (Nasipuri, Roy).

6633. Nimesulide. [51803-78-2] N-(4-Nitro-2-phenoxyphenyl)methanesulfonamide; 4-nitro-2-phenoxymethanesulfonanilide; R-805; Aulin; Fansidol; Flogovital; Mesulid; Nide; Nidol; Nisulid. $C_{13}H_{12}N_2O_5S$; mol wt 308.31. C 50.64%, H 3.92%, N 9.09%, O 25.95%, S 10.40%. Prostaglandin synthetase and platelet aggregation inhibitor. Prepn and anti-inflammatory activity: **BE 801812**; G. G. I. Moore, J. K. Harrington, **US 3840597** (both 1974 to Riker). Activity, comparison with other non-steroidal anti-inflammatories: K. F. Swingle et al., Arch. Int. Pharmacodyn. Ther. **221**, 132 (1976). Mechanism of action: R. L. Vigdahl, R. H. Tukey, Biochem. Pharmacol. **26**, 307 (1977). Chromatographic determn in plasma: S. F. Chang et al., J. Pharm. Sci. **66**, 1700 (1977). Pharmacology: K. F. Swingle, G. G. I. Moore, Drugs Exp. Clin. Res. **10**, 587 (1984). Clinical trials in rheumatic disorders: R. Weissenbach, J. Int. Med. Res. **9**, 349 (1981); in osteoarthritis: M. Reiner, ibid. **10**, 92 (1982); in acute inflammation: J. M. Pais, F. M. Rosteiro, ibid. **11**, 149 (1983); C. Milvio, ibid. **12**, 327 (1984); M. E. Nouri, Clin. Ther. **6**, 142 (1984).

Light tan crystals from ethanol, mp 143-144.5°. LD_{50} orally in rats: 324 mg/kg (Swingle, Moore).

THERAP CAT: Anti-inflammatory.

THERAP CAT (VET): Anti-inflammatory.

6634. Nimetazepam. [2011-67-8] 1,3-Dihydro-1-methyl-7-nitro-5-phenyl-2H-1,4-benzodiazepin-2-one; 1-methyl-5-phenyl-7-nitro-1,3-dihydro-2H-1,4-benzodiazepin-2-one; 1-methylnitrazepam; S-1530; Erimin. $C_{16}H_{13}N_3O_3$; mol wt 295.30. C 65.08%, H 4.44%, N 14.23%, O 16.25%. The desmethyl derivative of nitrazepam, q.v. Prepn: L. H. Sternbach et al., J. Med. Chem. **6**, 261 (1963); E. Reeder, L. H. Sternbach, **US 3109843**; **US 3144439**; **US 3141890**; **US 3243427** (1963, 1964, 1964, 1966 all to Hoffmann-La Roche); O. Keller et al., **US 3121114**; **US 3203990** (1964, 1965 to Hoffmann-La Roche); Sorrentino, **ZA 6706791** (1968 to Dumex), C.A. **70**, 57916c (1969); Yamamoto et al., **DE 1816046** (1970 to Sumitomo), C.A. **73**, 120690d (1970); Inaba et al., Chem. Pharm. Bull. **19**, 722 (1971). Pharmacology: S. Sakai et al., Arzneim.-Forsch. **22**, 534 (1972).

Pale yellow plates from ethanol, mp 156.5-157.5°. uv max (methanol): 259, 308 nm (ε 15800, 9600). LD_{50} in male, female mice,

rats (mg/kg): 910, 750, 1150, 970 orally; 970, 840, 970, 980 i.p.; 1500, 1500, 1000, 1000 s.c. (Sakai).

Note: This is a controlled substance (depressant): **21 CFR**, 1308.14.

THERAP CAT: Sedative, hypnotic.

6635. Nimidane. [50435-25-1] 4-Chloro-N-1,3-dithietan-2-ylidene-2-methylbenzenamine; cyclic methylene (4-chloro-o-tolyl)-dithioimidocarbonate; AC-84633; ENT-29106; Abequito. C_9H_8-$ClNS_2$; mol wt 229.74. C 47.05%, H 3.51%, Cl 15.43%, N 6.10%, S 27.91%. Prepn: R. W. Addor, S. Kantor, **DE 2305517** (1973 to Am. Cyanamid), C.A. **79**, 115553f (1973); W. W. Brand, M. W. Bullock, **US 3842096** (1974 to Am. Cyanamid). Tickicidal activity: N. K. Amaral et al., J. Econ. Entomol. **67**, 387 (1974).

White solid, mp 43-46°.

THERAP CAT (VET): Acaricide.

6636. Nimodipine. [66085-59-4] 1,4-Dihydro-2,6-dimethyl-4-(3-nitrophenyl)-3,5-pyridinedicarboxylic acid 3-(2-methoxyethyl) 5-(1-methylethyl) ester; 2-methoxyethyl 1,4-dihydro-5-(isopropoxy-carbonyl)-2,6-dimethyl-4-(3-nitrophenyl)-3-pyridinecarboxylate; isopropyl 2-methoxyethyl 1,4-dihydro-2,6-dimethyl-4-(3-nitrophenyl)-3,5-pyridinedicarboxylate; 2,6-dimethyl-4-(3′-nitrophenyl)-1,4-dihydropyridine-3,5-dicarboxylic acid 3-β-methoxyethyl ester 5-iso-propyl ester; Bay e 9736; Admon; Nimotop; Periplum. $C_{21}H_{26}$-N_2O_7; mol wt 418.45. C 60.28%, H 6.26%, N 6.69%, O 26.76%. Dihydropyridine calcium channel blocker. Prepn: H. Meyer et al., **DE 2117571**; eidem, **US 3799934** (1972, 1974 to Bayer). Pharmacology: R. Towart, S. Kazda, Br. J. Pharmacol. **67**, 409P (1979); K. Tanaka et al., Arzneim.-Forsch. **30**, 1494 (1980); L. M. Auer, ibid. **31**, 1423 (1981). Prepn of isomers and pharmacological comparison with racemate: R. Towart et al., ibid. **32**, 338 (1982). Use as cerebral vasodilator: H. Meyer et al., **GB 2018134**; eidem, **US 4406906** (1979, 1983 to Bayer). Effect on associative learning in aging rabbits: R. A. Deyo et al., Science **243**, 809 (1989). GC and LC determns in biological fluids: G. J. Krol et al., J. Chromatogr. **305**, 105 (1984). Clinical trial in prophylaxis of cerebral vasospasm: G. S. Allen et al., N. Engl. J. Med. **308**, 619 (1983). Toxicology: H. Schlüter, Arzneim.-Forsch. **36**, 1733 (1986). Series of articles on clinical pharmacology and therapeutic use: Am. J. Cardiol. **55**(3), 139B-153B (1985).

Crystals from petr ether/acetic ester, mp 125°. Affected by light. Exhibits polymorphism. Freely sol in ethyl acetate; sparingly sol in alc. Practically insol in water. LD_{50} in mice, rats (mg/kg): 3562, 6599 orally; 33, 16 i.v. (Schlüter).

(+)-**Form.** $[\alpha]_D^{20}$ +7.9° (c = 0.439 in dioxane).

(−)-**Form.** $[\alpha]_D^{20}$ −7.93° (c = 0.374 in dioxane).

THERAP CAT: Vasodilator (cerebral).

6637. Nimorazole. [6506-37-2] 4-[2-(5-Nitro-1H-imidazol-1-yl)ethyl]morpholine; N-2-morpholinoethyl-5-nitroimidazole; 1-(2-N-morpholinylethyl)-5-nitroimidazole; nitrimidazine; K-1900; Esclama; Naxofem; Naxogin. $C_9H_{14}N_4O_3$; mol wt 226.24. C 47.78%, H 6.24%, N 24.76%, O 21.22%. Prepn: **BE 667262**; Giraldi, Mariotti, **US 3399193** (1965, 1968 both to Carlo Erba); **NL 6609552**; **NL 6609553**; Gal and Carlson et al., **US 3458528** and **US 3646027** (1967, 1967, 1969, 1972, all to Merck & Co.). Synthesis and biological activity studies: Giraldi et al., Arzneim.-Forsch. **20**, 52 (1970).

Pharmacology and toxicology: de Carneri *et al.*, *Progr. Antimicrob. Anticancer Chemother., Proc. 6th Int. Congr. Chemother.* **vol. I**, Tokyo, 1969 (Univ. of Tokyo Press, 1970, Tokyo) pp 149-154. Clinical results: Emanueli, de Carneri, *ibid.* **vol. II**, pp 369-372; Evans, Catterall, *Br. Med. J.* **IV**, 146 (1971). Metabolic studies: Giraldi, *Biochem. Pharmacol.* **20**, 339 (1971). Acute toxicity data: B. Cavalleri *et al.*, *J. Med. Chem.* **21**, 781 (1978).

Crystals from water, mp 110-111°. Slightly sol in water at room temp; sol in alcohols, acetone, chloroform. LD_{50} orally in mice: 1530 mg/kg (Cavalleri).

THERAP CAT: Antiprotozoal (Trichomonas).

6638. Nimotuzumab. [780758-10-3] Anti-(human epidermal growth factor receptor) immunoglobulin G1 (human-mouse h-R3 heavy chain) disulfide with human-mouse h-R3 κ-chain, dimer; h-R3; Theraloc. Humanized monoclonal antibody directed against the external domain of epidermal growth factor receptor (EGRF). Prepn: C. M. Mateo del Acosta del Rio *et al.*, **EP 712863**; *eidem*, **US 5891996** (1996, 1999 both to Centr. Inmunol. Molec.). Prepn and characterization: C. Mateo *et al.*, *Immunotechnology* **3**, 71 (1997). Antiproliferative and antiangiogenic activity: T. Crombet-Ramos *et al.*, *Int. J. Cancer* **101**, 567 (2002). Crystal structure and binding study: A. Talavera *et al.*, *Cancer Res.* **69**, 5851 (2009). Clinical pharmacokinetics: T. Crombet *et al.*, *J. Immunother.* **26**, 139 (2003). Clinical evaluation in high-grade glioma: T. C. Ramos *et al.*, *Cancer Biol. Ther.* **5**, 375 (2006); in squamous cell carcinoma: M. O. Rodríguez *et al.*, *ibid.* **9**, 343 (2010). Review of clinical use in non-small cell lung cancer: W. Boland, G. Bebb, *Biologics Targets Therap.* **4**, 289-298 (2010); in pontine gliomas: M. Massimino *et al.*, *Expert Opin. Biol. Ther.* **11**, 247-256 (2011).

THERAP CAT: Antineoplastic.

6639. Nimustine. [42471-28-3] N'-[(4-Amino-2-methyl-5-pyrimidinyl)methyl]-N-(2-chloroethyl)-N-nitrosourea. C_9H_{13}-ClN_6O_2; mol wt 272.69. C 39.64%, H 4.81%, Cl 13.00%, N 30.82%, O 11.73%. Chloroethylnitrosourea derivative with antitumor activity. Prepn: H. Nakao *et al.*, **DE 2257360**; *eidem*, **US 4003901** (1973, 1977 both to Sankyo); *eidem*, *Yakugaku Zasshi* **94**, 1932 (1974), *C.A.* **82**, 43263y (1975). Effects on macrophage cytostatic activity in rats: N. Saijo *et al.*, *Br. J. Cancer* **42**, 162 (1980). Distribution, excretion, metabolism in mice: M. Tanaka *et al.*, *Cancer Treat. Rep.* **64**, 575 (1980). Exptl and clinical effect: N. Saijo *et al.*, *Cancer Chemother. Pharmacol.* **4**, 165 (1980). Acute toxicity: H. Masuda *et al.*, *Sankyo Kenkyusho Nempo* **29**, 118 (1977), *C.A.* **88**, 146298s (1978).

Pale yellow crystals from ethanol, dec 125°.
Hydrochloride. [55661-38-6] NSC-245382; ACNU; Nidran. $C_9H_{13}ClN_6O_2$·HCl; mol wt 309.15. White to light yellow cryst powder. uv max (0.04N HCl): 245 nm ($E_{1cm}^{1\%}$ 480-510). Sol in methanol, slightly sol in abs ethanol, n-butanol. Practically insol in ethyl acetate, ether, chloroform, benzene, n-hexane. Gradually develops greenish yellow color in light; decomposes slowly in humid air. LD_{50} in mice, rats (mg/kg): 62, 46 i.v. (Masuda).

THERAP CAT: Antineoplastic.

6640. Ninhydrin. [485-47-2] 2,2-Dihydroxy-1H-indene-1,3-(2H)-dione; 2,2-dihydroxy-1,3-indanedione; 1,2,3-indantrione monohydrate; triketohydrinden hydrate. $C_9H_6O_4$; mol wt 178.14. C 60.68%, H 3.40%, O 35.92%. Prepn: Ruhemann, *J. Chem. Soc.* **97**,

1438 (1910); Teeters, Shriner, *J. Am. Chem. Soc.* **55**, 3026 (1933); Wanag, Lode, *Ber.* **71**, 1267 (1938). Improved syntheses: L. F. Fieser, *Experiments in Organic Chemistry* (Boston, 3rd ed., 1955) p 123; Becker, Russell, *J. Org. Chem.* **28**, 1896 (1963). Absorption spectrum: Polonovski *et al.*, *Bull. Soc. Chim.* [5] **6**, 1557 (1939). Poisonous action on bacteria, mice, guinea pigs: O. Loew, *Biochem. Z.* **69**, 111 (1915). Acute toxicity study: J. C. Breton *et al.*, *C. R. Seances Soc. Biol. Ses Fil.* **151**, 719 (1957). Monograph: Henri Plagnol, *Influence de la Structure des Composés Aminés dans leur Reaction avec la Ninhydrine* (Bordeaux, 1962) 156 pp. *Reviews:* J. C. Breton, *Etudes Chimiques et Biologiques sur la Ninhydrine, Réactif des Aminoacides* (Imprimerie E. Drouillard, Bordeaux, 1958) 285 pp; McCaldin, *Chem. Rev.* **60**, 39 (1960); A. Schoenberg, E. Singer, *Tetrahedron* **34**, 1285-1300 (1978).

Monohydrate. Pale yellow prisms from water or alcohol. Reddens at 125°, swells at 139°, dec 241°. Freely sol in water. LD_{50} i.p. in mice: 78 mg/kg (Breton, 1957).

USE: Reagent for the detection of free amino and carboxyl groups in proteins and peptides, yielding a blue color under the proper conditions.

6641. Ninopterin. [2179-16-0] N-[4-[[1-(2-Amino-3,4-dihydro-4-oxo-6-pteridinyl)ethyl]amino]benzoyl]-L-glutamic acid; 9-methylpteroylglutamic acid; 9-methylfolic acid; Bremfol. $C_{20}H_{21}$-N_7O_6; mol wt 455.43. C 52.75%, H 4.65%, N 21.53%, O 21.08%. Folic acid analog. Prepn: Hultquist *et al.*, *J. Am. Chem. Soc.* **71**, 619 (1949); Hultquist, Smith, **GB 667098** (1952 to Am. Cyanamid). Clinical study in neoplastic diseases: J. C. Wright *et al.*, *J. Natl. Med. Assoc.* **43**, 211 (1951). *In vitro* evaluation as folic acid antagonist: S. Waxman *et al.*, *Chemotherapy (Basel)* **28**, 402 (1982).

uv max in 0.1N NaOH: 255, 283, 365 nm (ε 26, 30, 8 × 10^3); in 0.1N HCl: 297 nm (ε 18 × 10^3).

THERAP CAT: Experimental antineoplastic.

6642. Niobium. [7440-03-1] Columbium. Nb; at. wt 92.90638; at. no. 41; valence 2, 3, 4, 5; usually pentavalent. Group VB (5). One naturally occurring isotope: ^{93}Nb; artificial, radioactive isotopes: 88-92; 94-101. Approximately as abundant as nickel. Occurs in nature together with tantalum in the minerals *columbite* [(Fe,Mn)(Nb,Ta)$_2O_6$], *pyrochlore* (NaCaNb$_2O_6F$) and *tantalite* [(Fe,Mn)(Ta,Nb)$_2O_6$]. Discovered by Hatchett in 1801, isolated by Blomstrand in 1866, named after Niobe, daughter of Tantalos. Extracted from columbite which is mined largely in Nigeria and Zaire. Less than 10% of niobium-bearing ores come from the US, Canada, and Norway. Reviews of niobium and its compds: *Technology of Columbium (Niobium)* B. W. Gonser, E. M. Sherwood, Eds. (Wiley, New York, 1958); G. L. Miller, *Tantalum and Niobium* (Academic Press, New York, 1959) 767 pp; Brown, "The Chemistry of Niobium and Tantalum" in *Comprehensive Inorganic Chemistry* **Vol. 3**, J. C. Bailar, Jr. *et al.*, Eds. (Pergamon Press, Oxford, 1973) pp 553-622; P. H. Payton in *Kirk-Othmer Encyclopedia of Chemical Technology* **vol. 15** (Wiley-Interscience, New York, 3rd ed., 1981) pp 820-840.

Steel-gray, lustrous metal. Ductile and malleable when pure. Lattice structure: body-centered cube, lattice constant: 3.294 Å. d 8.57. mp 2468°. bp 4927°. Sp ht: 6.012 cal/g-atom/°C. Heat of sublimation: 170.9 kcal/g-atom; heat of combustion: 2379 cal/g.

Coefficient of linear expansion per °C: 7.1×10^{-6}. Electrical resistivity (20°): 13.2 μohm-cm. Temp coefficient of electrical resistivity per °C: 0.00395. Electron work function: 4.01 ev. Ionization potential: 6.77 V. Inert toward HCl, HNO$_3$ or aqua regia, but attacked by fusion with akali hydroxides or oxidizing agents.

USE: In ferrous metallurgy: Ferroniobium (produced by silicon reduction of columbite) is used to alloy stainless steels and metals for welding rods. In niobium base alloys for high temps and nuclear reactions. Niobium has some use as a getter in electronic vacuum tubes.

6643. Niobium Pentachloride. [10026-12-7] Niobium chloride (NbCl$_5$); columbium pentachloride; niobium(V) chloride. Cl$_5$-Nb; mol wt 270.16. Cl 65.61%, Nb 34.39%. NbCl$_5$. Exists as dimeric units in the solid state. Prepn: Epperson *et al., Inorg. Synth.* **7**, 163 (1963); *cf.* Rolsten, *J. Am. Chem. Soc.* **80**, 2952 (1958). Toxicity studies: T. J. Haley *et al., Toxicol. Appl. Pharmacol.* **4**, 385 (1962); N. A. Zhilova, A. A. Kasparov, *Hyg. Sanit.* **31**, 328 (1966). Review of niobium halides: Fairbrother in *Halogen Chemistry* **Vol. 3**, V. Gutmann, Ed. (Academic Press, New York, 1967) pp 123-178. Review of applications in organic synthesis: C. K. Z. Andrade, *Curr. Org. Synth.* **1**, 333-353 (2004).

Yellow, very deliquesc, monoclinic crystals; dec in moist air with evolution of HCl. d 2.75. mp 204.7-209.5°; bp ~250°, but begins to sublime at 125°. Sol in HCl, carbon tetrachloride. LD$_{50}$ in mice (mg/kg): 61 i.p., 940 orally (Haley). LD$_{50}$ in mice, rats (mg/kg): 829, 1400 orally (Zhilova, Kasparov).

USE: Lewis acid catalyst in organic synthesis.

6644. Niobium Pentafluoride. [7783-68-8] *(TB*-5-11)-Niobium fluoride (NbF$_5$); columbium pentafluoride. F$_5$Nb; mol wt 187.90. F 50.55%, Nb 49.44%. NbF$_5$. Prepn from NbCl$_5$ + HF: Ruff, Zedner, *Ber.* **42**, 492 (1909); Ruff, Schiller, *Z. Anorg. Allg. Chem.* **72**, 329 (1911); Kwasnik in *Handbook of Preparative Inorganic Chemistry* **Vol. 1**, G. Brauer, Ed. (Academic Press, New York, 2nd ed., 1963) p 254. Prepn from the elements: Fairbrother, Frith, *J. Chem. Soc.* **1951**, 3051; Junkins *et al., J. Am. Chem. Soc.* **74**, 3464 (1952). Review of transition metal pentafluorides: Peacock, *Adv. Fluorine Chem.* **7**, 113-145 (1973).

Strongly refractive, deliquescent, monoclinic crystals, d$_4^{80}$ 2.6955. mp 80.0°; bp 234.9°. Appreciable vapor pressure at 50°. Hydrolyzes in water, alc, caustic solns. Sparingly sol in carbon disulfide, chloroform. More sol in concd H$_2$SO$_4$ than TaF$_5$. Forms complexes with Lewis bases.

6645. Niobium Pentoxide. [1313-96-8] Niobium oxide (Nb$_2$O$_5$); columbium pentoxide; niobium(V) oxide. Nb$_2$O$_5$; mol wt 265.81. Nb 69.90%, O 30.09%. Separation: Münchow, *Chem. Ztg.* **84**, 490, 527 (1960). Review of polymorphism: Schäfer *et al., Angew. Chem. Int. Ed.* **5**, 40-52 (1966).

White, orthorhombic crystals. Becomes yellow on heating. d 4.6. mp 1520°. Insol in water; sol in HF, hot H$_2$SO$_4$.

6646. Nioxime. [492-99-9] 1,2-Cyclohexanedione 1,2-dioxime. C$_6$H$_{10}$N$_2$O$_2$; mol wt 142.16. C 50.69%, H 7.09%, N 19.71%, O 22.51%. Prepn: Hach *et al., Org. Synth.* **32**, 35 (1952); Boyer, Ellzey, *J. Am. Chem. Soc.* **82**, 2525 (1960); Borger *et al., Bull. Soc. Chim. Belg.* **73**, 73 (1964); Arthur, *FR 1339224* (1963 to du Pont).

Crystals, mp 185-188° (darkening at 170°).
USE: Analytical reagent for nickel and palladium.

6647. Nipecotic Acid. [498-95-3] 3-Piperidinecarboxylic acid; hexahydronicotinic acid. C$_6$H$_{11}$NO$_2$; mol wt 129.16. C 55.80%, H 8.58%, N 10.84%, O 24.77%. Prepd by hydrogenation of nicotinic acid with platinum oxide catalyst: McElvain, Adams, *J. Am. Chem. Soc.* **45**, 2738 (1923); with rhodium on alumina catalyst: Freifelder, *J. Org. Chem.* **28**, 1135 (1963); *idem, US 3159639* (1964 to Abbott).

Crystals, mp 261° (dec). Freely sol in water. Practically insol in abs alcohol and ether.
Hydrochloride. C$_6$H$_{11}$NO$_2$.HCl. mp 240-242°. Very sol in water; slightly sol in alc (also reported as freely sol in alc); slightly sol in chloroform. Practically insol in ether, benzene, and acetone.

6648. Nipradilol. [81486-22-8] 3,4-Dihydro-8-[2-hydroxy-3-[(1-methylethyl)amino]propoxy]-2*H*-1-benzopyran-3-ol 3-nitrate; 3,4-dihydro-8-(2-hydroxy-3-isopropylamino)propoxy-3-nitroxy-2*H*-1-benzopyran; 8-[2-hydroxy-3-(isopropylamino)propoxy]-3-chromanol 3-nitrate; nipradilol; K-351; Hypadil. C$_{15}$H$_{22}$N$_2$O$_6$; mol wt 326.35. C 55.21%, H 6.80%, N 8.58%, O 29.41%. β-Adrenergic blocker with vasodilating activity. Prepn: M. Shiratsuchi *et al.*, **EP 42299**; *eidem*, **US 4394382** (1981, 1983 both to Kowa). Synthesis: M. Shiratsuchi *et al., Chem. Pharm. Bull.* **35**, 632 (1987). Resolution of isomers: M. Shiratsuchi *et al.*, **EP 154511**; *eidem*, **US 4727085** (1985, 1988 both to Kowa). Synthesis and activity of isomers: M. Shiratsuchi *et al., Chem. Pharm. Bull.* **35**, 3691 (1987). Pharmacology: Y. Uchida *et al., Arch. Int. Pharmacodyn.* **262**, 132 (1983). Hemodynamic effects in dogs: H. Hisa *et al., ibid.* **271**, 169 (1984). Effects on cardiac function in dogs: M. Sakanashi *et al., ibid.* **274**, 47 (1985); M. Fujii *et al., Jpn. Heart J.* **27**, 233 (1986). Pharmacokinetics and metabolism in rats: S. Kabuto *et al., Arzneim.-Forsch.* **35**, 1674 (1985); H. Kimata *et al., ibid.* 1680. Metabolic fate in dog and man: M. Yoshimura *et al., Chem. Pharm. Bull.* **33**, 3456 (1985). Clinical evaluation in angina: H. Kishida *et al., Jpn. Heart J.* **29**, 309 (1988).

Colorless needles, mp 107-116° (Shiratsuchi, 1983); also reported as mp 110-122° (Shiratsuchi, 1987). LD$_{50}$ i.v. in mice, rats: 74.0, 73.0 mg/kg; orally in mice: 540 mg/kg (Shiratsuchi, 1988).
THERAP CAT: Antianginal; antihypertensive.

6649. Niridazole. [61-57-4] 1-(5-Nitro-2-thiazolyl)-2-imidazolidinone; 1-(5-nitro-2-thiazolyl)-2-oxotetrahydroimidazole; nitrothiamidazol; Ba-32644; Ciba 32644-Ba; Ambilhar. C$_6$H$_6$N$_4$O$_3$S; mol wt 214.20. C 33.64%, H 2.82%, N 26.16%, O 22.41%, S 14.97%. Prepn: **BE 632989** (1963 to Ciba), *C.A.* **61**, 1873d (1964), corresp to **GB 986562**; Lambert *et al., Experientia* **20**, 452 (1964).

Yellow crystals from dimethylformamide, mp 260-262°.
THERAP CAT: Anthelmintic (Schistosoma).

6650. Nisin. [1414-45-5] Nisin A; Nisaplin; Novasin. C$_{143}$-H$_{230}$N$_{42}$O$_{37}$S$_7$; mol wt 3354.09. C 51.21%, H 6.91%, N 17.54%, O 17.65%, S 6.69%. Ribosomally synthesized antimicrobial peptide produced by *Lactococcus lactis* (formerly *Streptococcus lactis*); name derived from "group N inhibitory substance". Member of the class I bacteriocins known as lantibiotics; structure contains 34 amino acid residues, eight of which are rarely found in nature, including lanthionine, *q.v.*, and β-methyllanthionine. Isoln and antibacterial activity: A. T. R. Mattick, A. Hirsch, *Nature* **154**, 551 (1944); *eidem, Lancet* **250**, 417 (1946); **253**, 5 (1947). Purification and nature of nisin: N. J. Berridge *et al.*, *Biochem. J.* **52**, 529 (1952). Production: H. B. Hawley, R. H. Hall, **US 2935503** (1960 to Aplin & Barrett). Structure: E. Gross, J. L. Morrell, *J. Am. Chem. Soc.* **93**, 4634 (1971). Confirmation of structure of nisin and its major degradation product: M. Barber *et al.*, *Experientia* **44**, 266 (1988). Partial synthesis: K. Fukase *et al.*, *Bull. Chem. Soc. Jpn.* **59**, 2505 (1986). Total synthesis: K. Fukase *et al.*, *Tetrahedron Lett.* **29**, 795 (1988). Biosynthetic study: G. W. Buchman *et al.*, *J. Biol. Chem.* **263**, 16260 (1988). Review of properties and commercial applications: A. Hurst, D. G. Hoover, *Food Sci. Technol.* **57**, 369-394 (1993); J. Delves-Broughton *et al.*, *Antonie van Leeuwenhoek* **69**, 193-202 (1996); of biosynthesis: C.-I. Cheigh, Y.-R. Pyun, *Biotechnol. Lett.* **27**, 1641-1648 (2005); of biosynthesis, mode of action and comparison with other lantibiotics: C. Chatterjee *et al.*, *Chem. Rev.* **105**, 633-683 (2005).

Abu = α-aminobutyric acid
Dha = dehydroalanine
Dhb = dehydrobutyrine

Crystals from ethanol. Soly in dilute HCl: 12% (pH 2.5); 4% (pH 5.0). Stable to boiling in acid soln.

Nisin Z. [137061-46-2] 27-L-Asparagine-nisin. $C_{141}H_{229}N_{41}O_{38}S_7$; mol wt 3331.05. Naturally occurring nisin variant. Identification: J. W. Mulders *et al.*, *Eur. J. Biochem.* **201**, 581 (1991).

USE: Food preservative, esp. for cheese and other dairy products, canned vegetables, and some baked goods.

6651. Nisoldipine. [63675-72-9] 1,4-Dihydro-2,6-dimethyl-4-(2-nitrophenyl)-3,5-pyridinedicarboxylic acid 3-methyl 5-(2-methylpropyl) ester; isobutyl methyl 1,4-dihydro-2,6-dimethyl-4-(*o*-nitrophenyl)-3,5-pyridinedicarboxylate; isobutyl 1,4-dihydro-5-methoxycarbonyl-2,6-dimethyl-4-(2-nitrophenyl)-3-pyridinecarboxylate; 2,6-dimethyl-3-carbomethoxy-4-(2-nitrophenyl)-5-carbisobutoxy-1,4-dihydropyridine; Bay k 5552; Baymycard; Sular; Syscor; Zadipina. $C_{20}H_{24}N_2O_6$; mol wt 388.42. C 61.85%, H 6.23%, N 7.21%, O 24.71%. Dihydropyridine calcium channel blocker. Prepn: E. Wehinger *et al.*, **DE 2549568**; *eidem*, **US 4154839** (1977, 1979 both to Bayer). Pharmacology: S. Kazda *et al.*, *Arzneim.-Forsch.* **30**, 2144 (1980). Hemodynamics: A. Vogt *et al.*, *ibid.* 2162. Series of articles on pharmacokinetics: *ibid.* **38**, 1093-1110 (1988). Reviews of pharmacology and clinical efficacy: H. A. Friedel, E. M. Sorkin, *Drugs* **36**, 682-731 (1988); J. Mitchell *et al.*, *J. Clin. Pharmacol.* **33**, 46-52 (1993).

Crystals from ethanol, mp 151-152°.
THERAP CAT: Antihypertensive; antianginal.

6652. Nitarsone. [98-72-6] *As*-(4-Nitrophenyl)arsonic acid; *p*-nitrobenzenearsonic acid. $C_6H_6AsNO_5$; mol wt 247.04. C 29.17%, H 2.45%, As 30.33%, N 5.67%, O 32.38%. Prepn: Jacobs *et al.*, *J. Am. Chem. Soc.* **40**, 1580 (1918); Doak, Freedman, **US**

2653160 (1953 to the U.S.A. as represented by the Administrator of the Federal Security Agency); Ruddy, Starkey, *Org. Synth.* **coll. vol. III**, 665 (1955).

Pale yellow leaflets from water, dec 298-300°. Very slightly sol in cold water, cold alcohol; readily sol in warm water, warm alcohol.
THERAP CAT (VET): Antihistomonad.

6653. Nitazoxanide. [55981-09-4] 2-(Acetyloxy)-*N*-(5-nitro-2-thiazolyl)benzamide; *N*-(5-nitro-2-thiazolyl)salicylamide acetate (ester); 2-(2'-acetoxy)benzamido-5-nitrothiazole; PH-5776; Alinia; Daxon; Dexidex; Kidonax; Navigator; Paramix. $C_{12}H_9N_3O_5S$; mol wt 307.28. C 46.91%, H 2.95%, N 13.68%, O 26.03%, S 10.43%. Broad spectrum antiparasitic agent; inhibits pyruvate ferredoxin oxidoreductase. Prepn: J. F. Rossignol, R. Cavier, **DE 2438037**; *eidem*, **US 3950351** (1975, 1976 both to S.P.R.L. Phavic); and antiparasitic activity: R. Cavier *et al.*, *Eur. J. Med. Chem. - Chim. Ther.* **13**, 539 (1978). Antibacterial spectrum *in vitro*: L Dubreuil *et al.*, *Antimicrob. Agents Chemother.* **40**, 2266 (1996). Toxicology: J. R. Murphy, J.-C. Friedmann, *J. Appl. Toxicol.* **5**, 49 (1985). Clinical pharmacokinetics: A. Stockis *et al.*, *Int. J. Clin. Pharmacol. Ther.* **34**, 349 (1996). Clinical trial in intestinal protozoan and helminthic infections: H. Abaza *et al.*, *Curr. Ther. Res.* **59**, 116 (1998). Review of mechanism of action and clinical experience: H. M. Gilles, P. S. Hoffman, *Trends Parasitol.* **18**, 95-97 (2002); of therapeutic potential in gastrointestinal infections: A. Hemphill *et al.*, *Expert Opin. Pharmacother.* **7**, 953-964 (2006).

Light yellow crystalline powder. Crystals from methanol, mp 202°. Sol in DMSO (>50 mg/ml). Poorly sol in ethanol. Practically insol in water. LD_{50} orally in male, female mice: 1350, 1380 mg/kg; in rats: >10 g/kg (Murphy, Friedmann).
THERAP CAT: Anthelmintic (cestodes); antiprotozoal (Cryptosporidium).
THERAP CAT (VET): In treatment of equine protozoal myeloencephalitis.

6654. Nitenpyram. [150824-47-8] (1*E*)-*N*-[(6-Chloro-3-pyridinyl)methyl]-*N*-ethyl-*N'*-methyl-2-nitro-1,1-ethenediamine; 1-[*N*-(6-chloro-3-pyridylmethyl)-*N*-ethyl]amino-1-methylamino-2-nitroethylene; TI-304; Bestguard; Capstar. $C_{11}H_{15}ClN_4O_2$; mol wt 270.72. C 48.80%, H 5.59%, Cl 13.09%, N 20.70%, O 11.82%. Chloronicotinyl insecticide; orally administered flea adulticide. Prepn: I. Minamida *et al.*, **EP 302389**; *eidem*, **US 5849768** (1989, 1998 both to Takeda); and insecticidal activity: *eidem*, *J. Pestic. Sci.* **18**, 41 (1993). Physical and biological properties: Y. Kashiwada, *Agrochem. Jpn.* **68**, 18 (1996). Effects on neuronal nicotinic acetylcholine receptor channel in rat cells: K. Nagata *et al.*, *J. Pestic. Sci.* **24**, 143 (1999). Efficacy study in flea infested dogs: M.-C. Cadiergues *et al.*, *Am. J. Vet. Res.* **60**, 1122 (1999). LC-MS determn in fruits and vegetables: D. Zywitz *et al.*, *Dtsch. Lebensm. Rundsch.* **99**, 188 (2003).

White crystals, mp 83-84°; also reported as pale yellow solid. Vapor pressure: 8.3×10^{-12} mmHg. Soly at 25° (g/l): water 840; chloroform 700; acetone 290; methanol 670; ethanol 89. Log P (octanol/water): −0.64. LD_{50} in male, female rats, male, female mice (mg/kg): 1680, 1575, 867, 1281 orally; in male, female rats (mg/kg): >2000, >2000 dermally. LC_{50} (48 hr) in carp, water fleas (ppm): >1000, >10000 (Kashiwada).

THERAP CAT (VET): Ectoparasiticide.

6655. Nithiazide. [139-94-6] N-Ethyl-N'-(5-nitro-2-thiazolyl)urea; Hepzide. $C_6H_8N_4O_3S$; mol wt 216.22. C 33.33%, H 3.73%, N 25.91%, O 22.20%, S 14.83%. Prepd by the condensation of ethyl isocyanate with 2-amino-5-nitrothiazole in toluene: Cuckler et al., Proc. Soc. Exp. Biol. Med. **92**, 483 (1956); O'Neill et al., US 2755285 (1956 to Merck & Co.).

Crystals, dec 228°. pKa 7.3. Practically insol in water: 3 mg/100 ml.

Sodium salt. Orange plates. Soly in water: 8 g/100 ml. Aq solns are alkaline and unstable.

Potassium salt. Crystals. Soly in water: 3 g/l00 ml. Aq solns are more stable than those of the Na salt.

THERAP CAT (VET): Antiprotozoal.

6656. Nitisinone. [104206-65-7] 2-[2-Nitro-4-(trifluoromethyl)benzoyl]-1,3-cyclohexanedione; NTBC; Orfadin. $C_{14}H_{10}F_3NO_5$; mol wt 329.23. C 51.07%, H 3.06%, F 17.31%, N 4.25%, O 24.30%. Herbicidal triketone that inhibits 4-hydroxyphenylpyruvate dioxygenase (HPPD), an enzyme involved in plastoquinone biosynthesis in plants and in tyrosine catabolism in mammals. Prepn: C. G. Carter, EP 186118 (1986 to Stauffer); idem, US 5006158 (1991 to ICI). Inhibition of HPPD in plants: M. P. Prisbylla et al., Brighton Crop Prot. Conf. - Weeds **1993**, 731; in rats: M. K. Ellis et al., Toxicol. Appl. Pharmacol. **133**, 12 (1995). LC determn in plasma: M. Bielenstein et al., J. Chromatogr. B **730**, 177 (1999). Clinical evaluation in hereditary tyrosinemia type I: S. Lindstedt et al., Lancet **340**, 813 (1992). Review of toxicology and therapeutic development: E. A. Lock et al, J. Inherit. Metab. Dis. **21**, 498-506 (1998); of clinical experience: E. Holme, S. Lindstedt, ibid. 507-517.

Solid, mp 88-94°.
THERAP CAT: In treatment of inherited tyrosinemia type I.

6657. Nitracrine. [4533-39-5] N^1,N^1-Dimethyl-N^3-(1-nitro-9-acridinyl)-1,3-propanediamine; 9-[[3-(dimethylamino)propyl]amino]-1-nitroacridine. $C_{18}H_{20}N_4O_2$; mol wt 324.38. C 66.65%, H 6.21%, N 17.27%, O 9.86%. Deriv of acridine, q.v., with cytostatic and cytotoxic properties. Prepn: FR 1458183 (1966 to Polfa), C.A. **68**, 39493s (1968); A. Ledochowski, B. Stefanska, Rocz. Chem. **40**, 301 (1966), C.A. **65**, 2219b (1966). Pharmacological studies: J. Gieldanowski et al., Arch. Immunol. Ther. Exp. **20**, 399 (1972); eidem, ibid. 419. Mechanism of action: J. Konopa et al., Mater. Med. Pol. **8**, 258 (1976). Cytotoxicity study: I. Szumiel, M. Walicka, Neoplasma **27**, 697 (1980). DNA binding activity: L. Szmigiero, M. Gniazdowski, Arzneim.-Forsch. **31**, 1875 (1981). Comprehensive review: M. Gniazdowski et al. in Antibiotics **Vol. V**, part 2, F. E. Hahn, Ed. (Springer-Verlag, New York, 1979) pp 275-297.

Crystals from benzene/petr ether, mp 134-135°. Practically insol in water. Sol in most organic solvents. pKa_1 6.45; pKa_2 8.8.

Dihydrochloride monohydrate. [55429-45-3] C-283; Ledakrin. $C_{18}H_{20}N_4O_2.2HCl.H_2O$; mol wt 415.32. Orange crystals, mp 223-224°. Sol in water, methanol, ethanol, slightly sol in benzene, diethyl ether. Conc water solns are acidic (pH 4). LD_{50} in rats, mice (mg/kg): 1, 0.72 i.v.; 34, 26 i.g. (Gniazdowski).

THERAP CAT: Antineoplastic.

6658. Nitramide. [7782-94-7] $H_2N_2O_2$; mol wt 62.03. H 3.25%, N 45.16%, O 51.58%. H_2NNO_2. Prepd from potassium nitrocarbamate: Thiele, Lachmann, Ber. **27**, 1909 (1894); Ann. **288**, 267 (1895); Marlies, La Mer, J. Am. Chem. Soc. **57**, 2008 (1935); Marlies et al., Inorg. Synth. **1**, 68 (1939).

Unstable, shiny, white leaflets from ether + petr ether, mp 72-75° (dec). Dipole moment 3.7. Should be freshly prepd when needed. Does not explode under ordinary conditions. Handle in glass or platinum. Sol in ether, alcohol, acetone, water. Slightly sol in benzene. Practically insol in petr ether, chloroform.

6659. Nitramine. [479-45-8] N-Methyl-N,2,4,6-tetranitrobenzenamine; N-methyl-N,2,4,6-tetranitroaniline; picrylmethylnitramine; picrylnitromethylamine; Tetralite; Tetryl. $C_7H_5N_5O_8$; mol wt 287.14. C 29.28%, H 1.76%, N 24.39%, O 44.57%. Review of toxicology and human exposure: Toxicological Profile for Tetryl (PB95-264271, 1995) 133 pp.

Yellow crystals. d 1.57. mp 130-132°; explodes at about 180-190°. Insol in water. Sol in alc, ether, benzene, glacial acetic acid. Keep soln in the dark. Causes yellow staining to skin and hair.

Caution: Potential symptoms of overexposure are sensitization dermatitis, itching, erythema; edema on nasal folds, cheeks and neck; keratitis; sneezing; anemia; fatigue; coughing, coryza; irritability; malaise, headache, lassitude and insomnia; nausea, vomiting; liver and kidney damage. See NIOSH Pocket Guide to Chemical Hazards (DHHS/NIOSH 97-140, 1997) p 304.

USE: As indicator, 0.1 g in 60 ml alcohol with water to make 100 ml. pH: 10.8 colorless, 13.0 reddish-brown. One to five drops of soln required for 10 ml liquid. Salt error said to be small. Also used in explosives.

6660. Nitrapyrin. [1929-82-4] 2-Chloro-6-(trichloromethyl)-pyridine; $\alpha,\alpha,\alpha,6$-tetrachloro-2-picoline; N-Serve. $C_6H_3Cl_4N$; mol wt 230.90. C 31.21%, H 1.31%, Cl 61.41%, N 6.07%. Prepn: Johnston et al., GB 957276, BE 624800 (1964, 1963 to Dow), C.A. **61**, 1841b (1964).

Crystals, mp 62.5-62.9° (CH_2Cl_2-pentane). bp_{11} 136-137.5°.
USE: Fertilizer additive to control nitrification and prevent loss of soil nitrogen.

6661. Nitrazepam. [146-22-5] 1,3-Dihydro-7-nitro-5-phenyl-$2H$-1,4-benzodiazepin-2-one; LA-1; Ro-4-5360; Benzalin; Calsmin; Eatan; Eunoctin; Imeson; Mogadan; Mogadon; Nelbon; Nitrados; Radedorm; Remnos; Somnased; Somnibel; Somnite; Surem; Unisomnia. $C_{15}H_{11}N_3O_3$; mol wt 281.27. C 64.05%, H 3.94%, N 14.94%, O 17.06%. Prepn: Sternbach et al., J. Med. Chem. **6**, 261 (1963); O. Keller et al., US 3121076 (1964 to Hoffmann-La Roche). Crystal structure: G. Gilli et al., Acta Crystallogr. **33B**, 2664 (1977). Pharmacokinetic studies: Rieder, Arzneim.-Forsch. **23**, 212 (1973); L. Kangas, Acta Pharmacol. Toxicol. **45**, 16 (1979). Toxicity data: Randall et al., Schweiz. Med. Wochenschr. **95**, 334 (1965). Review:

Rieder, Wendt in *Benzodiazepines*, S. Garattini *et al.*, Eds. (Raven Press, New York, 1973) pp 99-127. Comprehensive description: H. Y. Aboul-Enein *et al.*, *Anal. Profiles Drug Subs.* **9**, 487-517 (1980).

Yellow crystals from ethanol, mp 224-226°. uv max (0.1*N* H_2SO_4): 277.5 nm ($E_{1cm}^{1\%}$ 1500). Sol in alc, acetone, chloroform, ethyl acetate. Practically insol in water, ether, benzene, hexane. LD_{50} orally in rats: 825 ±80 mg/kg (Randall).

Note: This is a controlled substance (depressant): **21 CFR**, 1308.14

THER AT: Anticonvulsant; hypnotic.

6662. Nitrendipine. [39562-70-4] 1,4-Dihydro-2,6-dimethyl-4-(3-nitrophenyl)-3,5-pyridinedicarboxylic acid 3-ethyl 5-methyl ester; ethyl 1,4-dihydro-5-(acetoxycarbonyl)-2,6-dimethyl-4-(3-nitrophenyl)-3-pyridinecarboxylate; 3-ethyl-5-methyl-1,4-dihydro-2,6-dimethyl-4-(3-nitrophenyl)-3,5-pyridinedicarboxylate; Bay e 5009; Bayotensin; Baypress; Bylotensin; Deiten; Nidrel. $C_{18}H_{20}N_2O_6$; mol wt 360.37. C 59.99%, H 5.59%, N 7.77%, O 26.64%. Dihydropyridine calcium channel blocker. Prepn: H. Meyer *et al.*, **DE 2117571**; *eidem*, **US 3799934** (1972, 1974 both to Bayer); *eidem*, *Arzneim.-Forsch.* **31**, 407 (1981). Series of articles on pharmacology: *ibid.* 2056-2067. Hemodynamic effects: H. O. Ventura *et al.*, *Am. J. Cardiol.* **51**, 783 (1983). Pharmacokinetics: L. Hansson *et al.*, *Hypertension* **5**, Suppl. II, II-25 (1983). HPLC determn in human serum: R. A. Janis *et al.*, *J. Clin. Pharmacol.* **23**, 266 (1983). Double-blind, controlled clinical trial: U. Brugmann *et al.*, *Herz* **10**, 53 (1985). Symposium on pharmacology and clinical efficacy: *J. Cardiovasc. Pharmacol.* **6** Suppl. 7, S929-S1113 (1984). Toxicology studies: K. Hoffmann, "Toxicological Studies with Nitrendipine" in *Nitrendipine*, A. Scriabine *et al.*, Eds. (Urban & Schwarzenberg, Baltimore, 1984) pp 25-32. *Review:* A. Scriabine *et al.*, "Nitrendipine" in *New Drugs Annual: Cardiovascular Drugs* Vol. 2, A. Scriabine, Ed. (Raven Press, New York, 1984) pp 37-49.

Crystals from ethanol, mp 158°. Relatively insol in water. LD_{50} in mice, rats (mg/kg): 39, 12.6 i.v.; 2540, >10000 orally (Hoffmann).

THERAP CAT: Antihypertensive.

6663. Nitric Acid. [7697-37-2] Aqua fortis; azotic acid; Salpetersäure (German). HNO_3; mol wt 63.01. H 1.60%, N 22.23%, O 76.17%. Strong monobasic acid and oxidizing agent. Usually produced by the catalytic oxidation of ammonia. Reviews of industrial processes: F. D. Miles, *Nitric Acid, Manufacture and Uses* (Oxford Univ. Press, 1961, 1963); W. Sommer in *Ullmanns Encyklopädie der technischen Chemie* **vol. 15**, pp 3-67 (3rd ed., 1964). Pure acid prepd by distillation of concd nitric acid with concd sulfuric acid; by treating sodium or potassium nitrate with 100% H_2SO_4 and removing HNO_3 by distillation; by fractional crystallization of concd HNO_3. Review of preparation and properties of pure HNO_3: S. A. Stern *et al.*, *Chem. Rev.* **60**, 185-207 (1960); S. S. Pannu, *J. Chem. Educ.* **61**, 174-176 (1984). *Reviews:* Mellor's **Vol. VIII**, supplement II, *Nitro-*

gen (part 2), 278-352 (1967); Jones in *Comprehensive Inorganic Chemistry* **vol. 2**, J. C. Bailar, Jr. *et al.*, Eds. (Pergamon Press, Oxford, 1973) pp 375-388; D. J. Newman in *Kirk-Othmer Encyclopedia of Chemical Technology* **vol. 15** (Wiley-Interscience, New York, 1981) pp 853-871. Review of photochemistry: J. R. Huber, *Chem. Phys. Chem.* **5**, 1663-1669 (2004).

Colorless liquid. Fumes in moist air. Characteristic choking odor. d_4^{25} 1.51. mp −41.59°. bp_{760} 82.6°. Forms white, monoclinic crystals; $d^{−41.6}$ (solid) 1.895. n_D^{25} 1.3920. Heat of fusion: 2.503 kcal/mole. Heat of infinite dilution (298.1 K): −7971 cal/mole. Forms an azeotrope with water, so-called constant boiling acid at 68% HNO_3, bp 120.5°, d_4^{20} 1.41. Develops yellow color in the presence of light due to nitrogen oxide formation. Stains woolen fabrics and animal tissue a bright yellow. *Corrosive; oxidizer.* Reacts violently with combustible or readily oxidizable materials such as alcohols, turpentine, charcoal, organic refuse. Reacts with most metals to release hydrogen gas.

Nitric acid, concentrated. Defined as an aqueous solution containing approx 70% HNO_3. Density of aq solns: d_4^{20} 1.0036 (1% HNO_3 w/w); 1.0543 (10%); 1.1150 (20%); 1.1800 (30%); 1.2463 (40%); 1.3100 (50%); 1.3667 (60%); 1.4134 (70%); 1.4521 (80%); 1.4826 (90%); 1.5129 (100%): *International Critical Tables* **III**, 58 (1928).

Nitric acid, fuming. Concentrated nitric acid containing dissolved nitrogen dioxide. May be prepared from concd nitric acid by passing nitrogen dioxide into it or by adding a small amount of organic reducing agent, such as formaldehyde. Yellow to brownish-red, clear, strongly fuming, very corrosive liq; evolves suffocating, poisonous, yellowish-red fumes of nitrogen dioxide and nitrogen tetroxide. The density increases as the free NO_2 content increases: concd HNO_3 with 7.5% NO_2 added d_4^{20} = 1.526; with 12.7% NO_2 d_4^{20} = 1.544. Miscible with water. *Corrosive; oxidizer; poisonous.*

Caution: Potential symptoms of overexposure are irritation of eyes, mucous membranes and skin; delayed pulmonary edema, pneumonitis and bronchitis; dental erosion. *See NIOSH Pocket Guide to Chemical Hazards* (DHHS/NIOSH 97-140, 1997) p 224. *See also Clinical Toxicology of Commercial Products*, R. E. Gosselin *et al.*, Eds. (Williams & Wilkins, Baltimore, 5th ed., 1984) Section III, pp 8-11.

USE: Manufacture of inorganic and organic nitrates and nitro compounds for fertilizers, dye intermediates, explosives, rocket fuels. As laboratory reagent, in photoengraving, metal etching. Pharmaceutic aid (acidifier). Oxidizer. Solvent. Trace metal content analysis.

6664. Nitric Oxide. [10102-43-9] Nitrogen oxide (NO); mononitrogen monoxide; nitrogen monoxide; endothium-derived relaxing factor; EDRF. NO; mol wt 30.01. N 46.67%, O 53.31%. Highly reactive, potentially toxic gas produced by the partial oxidation of atmospheric nitrogen. Major air pollutant along with its oxidative by-products (NO_x) resulting from incomplete combustion of fossil fuels. Also found ubiquitously in animals; generated *in vivo* from arginine, *q.v.*, by *nitric oxide synthase* (NOS). Increases guanylate cyclase activity to produce cyclic GMP, *q.v.* Involved in a wide range of physiological functions, including vasodilation, neurotransmission, cytotoxicity of macrophages, and inhibition of platelet aggregation. Laboratory prepn: Blanchard, *Inorg. Synth.* **2**, 126 (1946); Schenk in *Handbook of Preparative Inorganic Chemistry* **vol. 1**, G. Brauer, Ed. (Academic Press, New York, 2nd ed., 1963) pp 485-487. General reviews: Beattie, "Nitric Oxide" in *Mellor's* **Vol. VIII**, supplement II, *Nitrogen* (part 2) 216-240 (1967); Jones in *Comprehensive Inorganic Chemistry* **vol. 2**, J. C. Bailar, Jr. *et al.*, Eds. (Pergamon Press, Oxford, 1973) pp 323-334. Role in cGMP production: W. P. Arnold *et al.*, *Proc. Natl. Acad. Sci. USA* **74**, 3203 (1977). Identification of endothelium dependent vasorelaxation: R. F. Furchgott, J. V. Zawadzki, *Nature* **288**, 373 (1980). Identity of NO and EDRF: L. J. Ignarro *et al.*, *Proc. Natl. Acad. Sci. USA* **84**, 9265 (1987). Reviews of physiological role: S. Moncada *et al.*, *Pharmacol. Rev.* **43**, 109-142 (1991); A. R. Butler, D. L. H. Williams, *Chem. Soc. Rev.* **22**, 233-242 (1993). Role in airway function: R. J. Martin *et al.*, *Semin. Perinatol.* **26**, 432 (2002). Clinical trial of inhaled NO in neonatal respiratory failure: G. M. Sokol, R. A. Ehrenkranz, *ibid.* **27**, 311 (2003). Review of pharmacology and therapeutic potential of inhaled NO: F. Ichinose *et al.*, *Circulation* **109**, 3106-3111 (2004). Review of chemistry: D. L. H. Williams, *Org.*

Biomol. Chem. **1**, 441-449 (2003); of coordination chemistry in bio-inorganic systems: J. A. McCleverty, *Chem. Rev.* **104**, 403-418 (2004).

Colorless gas. Burns only when heated with hydrogen. triple pt −163.6°. bp −151.8°. d$^{-150.2}$ (liq) 1.27. Relative d (gas) 1.036 (air = 1). Absolute d (gas) 1.227 (air = 1). n_D^{25} 1.0002697. Trouton constant 27.1. Contains an odd number of electrons and is paramagnetic. Crit temp −92.9°. Crit press. 64.6 atm. Heat of formation (18°): −21.5 kcal/mole. Heat of vaporization (bp): 3.293 kcal/mole. Ionization potential: 9.26 eV. Electron affinity: 0.024 eV. Solubility in water (ml/100 ml; 1 atm): 7.38 (0°); 4.6 (20°); 2.37 (60°). Combines with oxygen to form NO_2 and with chlorine and bromine to form the nitrosyl halides, such as NOCl, *see* N. V. Sidgwick, *Chemical Elements and Their Compounds* **vol. I** (Oxford, 1950) p 683.

Caution: Potential symptoms of overexposure are irritation of eyes, wet skin, nose and throat; drowsiness; unconsciousness; methemoglobinemia. *See NIOSH Pocket Guide to Chemical Hazards* (DHHS/NIOSH 97-140, 1997) p 224. On contact with air, nitric oxide is converted to the highly poisonous nitrogen dioxide, *q.v.* Respiratory protection and adequate ventilation should be used to avoid overexposure. *See Patty's Industrial Hygiene and Toxicology* **vol. 2F**, G. D Clayton, F. E. Clayton, Eds. (John Wiley & Sons, New York, 4th ed., 1994) pp 4566-4591.

USE: Manuf of nitric acid; in the bleaching of rayon; as stabilizer (to prevent free-radical decompn) for propylene, methyl ether, etc.

THERAP CAT: Vasodilator (pulmonary); in treatment of neonatal cardiorespiratory failure.

6665. Nitrilotriacetic Acid. [139-13-9] *N,N*-Bis(carboxymethyl)glycine; triglycollamic acid; α,α′,α″-trimethylaminetricarboxylic acid; tri(carboxymethyl)amine; triglycine; NTA. $C_6H_9NO_6$; mol wt 191.14. C 37.70%, H 4.75%, N 7.33%, O 50.22%. Prepn: Heintz, *Ann.* **122**, 260 (1862); Michaelis, Schubert, *J. Biol. Chem.* **106**, 331 (1934); Martell, Bersworth, *J. Org. Chem.* **15**, 46 (1950); Singer, Weisberg, **US 2855428** and **US 3061628** (1958 and 1962, both to Hampshire Chem.). IR studies: Nakamoto *et al.*, *J. Am. Chem. Soc.* **84**, 2081 (1962); Chapman *et al.*, *Proc. Chem. Soc. London* **1962**, 336. Solubility data: Bird, *J. Soc. Dyers Colour.* **56**, 473 (1940). Toxicity data: *Soap Chem. Spec.* **42**, 58 (1966). pK data: Schwarzenbach *et al.*, *Helv. Chim. Acta* **28**, 828 (1945). *Review:* Souchay, Graizon, *Bull. Soc. Chim. Fr.* **1952**, 34.

Prismatic crystals from hot water, mp 230-235° dec (Michaelis, Schubert). mp 241.5° (dec). 1.28 g dissolves in 1 liter of water at 22.5°, pH of satd aq soln is 2.3 (Bird). At 20° pK 3.03, pK$_2$ 3.07, pK$_3$ 10.70.

Trisodium salt. [5064-31-3] NTANa$_3$; Trilon A. $C_6H_6NNa_3O_6$; mol wt 257.08. MLD orally in rats >4,000 mg/kg (*Soap Chem. Spec.*).

Caution: Nitrilotriacetic acid is reasonably anticipated to be a human carcinogen: *Report on Carcinogens, Twelfth Edition* (PB2011-111646, 2011) p 284.

USE: Chelating and sequestering agent; builder in synthetic detergents.

6666. Nitrin. [553-74-2] 2-Aminobenzaldehyde phenylhydrazone. $C_{13}H_{13}N_3$; mol wt 211.27. C 73.91%, H 6.20%, N 19.89%. Prepd by refluxing 2-nitrobenzaldehyde with phenylhydrazine: Knöpfer, *Monatsh. Chem.* **31**, 97 (1910).

Needles from acetone, mp 227-229° (dec). Sol in acetone. Sparingly sol in cold alcohol, ether, chloroform, benzene. A soln in

alcohol or acetone develops a red color with nitrites upon addn of acid.

USE: Detection of nitrites, colibacilli in urine: Pfeiffer, *Muench. Med. Wochenschr.* **92**, 1315 (1950).

6667. Nitroacetanilide. $C_8H_8N_2O_3$; mol wt 180.16. C 53.33%, H 4.48%, N 15.55%, O 26.64%.

m-**Nitroacetanilide.** [122-28-1] *N*-(3-Nitrophenyl)acetamide. Leaflets. mp 151-153°. Sparingly sol in hot water; freely sol in chloroform, nitrobenzene. Practically insol in ether.

o-**Nitroacetanilide.** [552-32-9] Yellow leaflets, d 1.42. mp 93-94°. Moderately sol in cold water, freely in boiling water, in cold fixed alkali hydroxide solns; sol in chloroform, alcohol, ether.

p-**Nitroacetanilide.** [104-04-1] Prisms. mp 214-216°. Almost insol in cold water. Sol in hot water, alcohol, ether; sol in KOH with orange color.

6668. *m*-Nitroaniline. [99-09-2] 3-Nitrobenzenamine; *m*-nitraniline. $C_6H_6N_2O_2$; mol wt 138.13. C 52.17%, H 4.38%, N 20.28%, O 23.17%. Prepn by nitration of aniline: Holleman *et al.*, *Ber.* **44**, 704 (1911); by reduction of *m*-dinitrobenzene: Brady *et al.*, *J. Chem. Soc.* **1929**, 2266; Kubota *et al.*, *J. Pharm. Soc. Jpn.* **76**, 801 (1956); Kuhn, **US 2768209** (1956 to Ringwood); from *m*-nitrobenzoic acid: Snyder *et al.*, *J. Am. Chem. Soc.* **75**, 2014 (1953).

Yellow crystals from water, mp 114°. Calc bp 306°. d$_4^{25}$ 0.9011. One gram dissolves in 880 ml water, about 20 ml alcohol, 18 ml ether, 11.5 ml methanol. Forms water-sol salts with mineral acids.

Caution: Highly toxic; absorbed through skin. Avoid breathing dust; avoid contact with skin, eyes, clothing. In case of contact, immediately flush skin or eyes with plenty of water for at least 15 minutes; for eyes, get medical attention. Wash clothing before reusing. Acute exposure can cause methemoglobinemia, cyanosis. Chronic exposure may cause liver damage.

USE: Dyestuff intermediate.

6669. *o*-Nitroaniline. [88-74-4] 2-Nitrobenzenamine; *o*-nitraniline. $C_6H_6N_2O_2$; mol wt 138.13. C 52.17%, H 4.38%, N 20.28%, O 23.17%. Prepn from *o*-dinitrobenzene: Meisenheimer, Hesse, *Ber.* **52**, 1166 (1919); from *o*-nitroaniline-*p*-sulfonic acid: Ehrenfeld, Puterbaugh, *Org. Synth.* **coll. vol. I**, 388 (1941); from ψ-*o*-dinitrosobenzene: Boyer *et al.*, *J. Am. Chem. Soc.* **77**, 5688 (1955).

Orange-yellow crystals from boiling water, mp 69-71°. Calc bp 284°. d$_4^{25}$ 0.9015. Slightly sol in cold water; sol in hot water, alcohol, chloroform. Forms water soluble salts with mineral acids.

Caution: See m-Nitroaniline.

USE: Dyestuff intermediate.

6670. *p*-Nitroaniline. [100-01-6] 4-Nitrobenzenamine; *p*-nitraniline. $C_6H_6N_2O_2$; mol wt 138.13. C 52.17%, H 4.38%, N

20.28%, O 23.17%. Prepn from acetanilide: Bart, **US 2406578** (1946 to Am. Cyanamid); by Schmidt reaction: Stockel, Hall, *Nature* **197**, 787 (1963).

Bright yellow powder, mp 146°. Calc bp 332°. One gram dissolves in 1250 ml water, 45 ml boiling water, 25 ml alcohol, 30 ml ether; sol in benzene, methanol. Forms water soluble salts with mineral acids.

Caution: Potential symptoms of overexposure are irritation of nose and throat; cyanosis, ataxia; tachycardia, tachypnea; dyspnea; irritability; vomiting, diarrhea; convulsions; respiratory arrest; anemia; methemoglobinemia; jaundice. *See NIOSH Pocket Guide to Chemical Hazards* (DHHS/NIOSH 97-140, 1997) p 226.

USE: Dyestuff intermediate.

6671. Nitroanisole. Methoxynitrobenzene; nitrophenyl methyl ether. $C_7H_7NO_3$; mol wt 153.14. C 54.90%, H 4.61%, N 9.15%, O 31.34%.

m-**Nitroanisole.** [555-03-3] Crystals. d 1.373. mp 38-39°. bp 258°. Volatile with steam. Insol in water; sol in alc.

o-**Nitroanisole.** [91-23-6] Colorless to yellowish liquid. d_4^{20} 1.254. mp 9.4°. bp 277°; also stated as 272-273°. n_D^{20} 1.5620. Volatile with steam. Insol in water; sol in alc, ether.

p-**Nitroanisole.** [100-17-4] Crystals. d 1.233. mp 54°. bp 260°; also stated as 274°. Insol in water; freely sol in alc, ether, boiling petr ether; slightly sol in cold petr ether.

Caution: *o*-Nitroanisole is reasonably anticipated to be a human carcinogen: *Report on Carcinogens, Twelfth Edition* (PB2011-111646, 2011) p 285.

USE: *o*-Isomer as dye intermediate; in organic syntheses.

6672. 5-Nitrobarbituric Acid. [480-68-2] 5-Nitro-2,4,6(1*H*,-3*H*,5*H*)-pyrimidinetrione; dilituric acid. $C_4H_3N_3O_5$; mol wt 173.08. C 27.76%, H 1.75%, N 24.28%, O 46.22%.

Trihydrate. Prisms and leaflets from water. mp 176° with decompn, when anhydr. Soluble in ~1200 parts cold water; more sol in hot water; sol in alcohol, in sodium hydroxide soln. Insol in ether.

USE: As a microreagent for potassium with which it forms a characteristic precipitate.

6673. Nitrobenzaldehyde. $C_7H_5NO_3$; mol wt 151.12. C 55.64%, H 3.34%, N 9.27%, O 31.76%.

m-**Nitrobenzaldehyde.** Yellowish, cryst powder. mp 58°. bp$_{23}$ 164°. Volatile with steam. Almost insol in water; sol in alcohol, chloroform, ether.

o-**Nitrobenzaldehyde.** Light yellow needles. mp 42-44°. bp$_{23}$ 153°. Volatile with steam. Slightly sol in water, freely in alcohol, benzene, ether.

p-**Nitrobenzaldehyde.** White to yellow crystals, mp 106-107°. Sublimes; slightly volatile with steam. Slightly sol in water or ether; sol in alc, benzene, glacial acetic acid.

USE: *o*-Nitrobenzaldehyde as reagent for isopropyl alcohol and acetone.

6674. Nitrobenzene. [98-95-3] Nitrobenzol; essence of mirbane; oil of mirbane. $C_6H_5NO_2$; mol wt 123.11. C 58.54%, H 4.09%, N 11.38%, O 25.99%. Industrial prepn: *Faith, Keyes & Clark's Industrial Chemicals*, F. A. Lowenheim, M. K. Moran, Eds. (Wiley-Interscience, New York, 4th ed., 1975) p 571. Crystal structure: R. Boese *et al.*, *Struct. Chem.* **3**, 363 (1992). Vibrational spectra: V. A. Shlyapochnikov *et al.*, *J. Mol. Struct.* **326**, 1 (1994). HPLC determn of metabolite in urine: A. Astier, *J. Chromatogr.* **573**, 318 (1992). Review of toxicology and human exposure: *Toxicological Profile for Nitrobenzene* (PB91-180398, 1990) 117 pp; of carcinogenic risk: *IARC Monographs* **65**, 381-408 (1996). *Review:* R. L. Adkins in *Kirk-Othmer Encyclopedia of Chemical Technology* vol. **17** (Wiley-Interscience, New York, 4th ed., 1996) pp 133-152.

Colorless to pale yellow, oily liquid; odor of volatile oil almond. *Poisonous.* Use only with adequate ventilation. d_4^{15} 1.205. d_4^{25} 1.19864. mp 6°. bp 210-211°. Flash pt, closed cup: 190°F (88°C). n_D^{20} 1.5529. Volatile with steam. Sol in ~500 parts water; freely sol in alcohol, benzene, ether, oils. LD$_{50}$ orally in rats: 600 mg/kg (PB91-108398).

Caution: Potential symptoms of overexposure are irritation of skin and eyes; anoxia; dermatitis; anemia; methemoglobinemia. *See NIOSH Pocket Guide to Chemical Hazards* (DHHS/NIOSH 97-140, 1997) p 226. This substance is reasonably anticipated to be a human carcinogen: *Report on Carcinogens, Twelfth Edition* (PB2005-104914) p 294.

USE: For the manuf of aniline; in soaps, shoe polishes; for refining lubricating oils; manuf pyroxylin compds. Solvent. Anticoagulant for titrations.

6675. 4-Nitrobenzenesulfonamide. [6325-93-5] *p*-Nitrophenylsulfonamide; nosylamide; NsNH$_2$. $C_6H_6N_2O_4S$; mol wt 202.18. C 35.64%, H 2.99%, N 13.86%, O 31.65%, S 15.86%. Versatile reagent in synthetic chemistry, often utilized to introduce nitrogen into a more complex organic compnd. Prepn from 4-nitrobenzenesulfonyl chloride: J. J. Blanksma, *Recl. Trav. Chim. Pays-Bas* **20**, 121 (1901); from 4-nitrobenzenesulfinic acid: P. R. Carter, D. H. Hey, *J. Chem. Soc.* **1948**, 147. ^{13}C, ^{15}N, and ^{17}O NMR studies: P. Ruostesuo *et al.*, *Magn. Reson. Chem.* **25**, 189 (1987). Crystal structure: M. Tremayne *et al.*, *Acta Crystallogr.* **B58**, 823 (2002). Dissociation constant determn studies: A. V. Willi, *Helv. Chim. Acta* **39**, 46 (1956); M. Ludwig *et al.*, *Collect. Czech. Chem. Commun.* **49**, 1182 (1984). Carbonic anhydrase inhibition studies: A. Innocenti *et al.*, *Bioorg. Med. Chem. Lett.* **14**, 5703 (2004). Synthetic applications: H.-L. Kwong *et al.*, *Tetrahedron Lett.* **45**, 3965 (2004); G. Y. Cho, C. Bolm, *ibid.* **46**, 8007 (2005); J. G. Taylor *et al.*, *Org. Lett.* **8**, 3561 (2006); M. Abid *et al.*, *Tetrahedron Lett.* **48**, 4047 (2007). Review: A. C. Mayer, *Synlett* **2008**, 945-946.

Pale yellow needles from water, mp 177°. *Irritant.* Crystal density 1.618. pKa 9.14 (water, 20°); 8.63±0.06 (water, 25°); 13.92±0.06 (methanol, 25°); 14.68±0.07 (ethanol, 25°).

USE: Reagent in organic chemistry.

6676. Nitrobenzoic Acid. $C_7H_5NO_4$; mol wt 167.12. C 50.31%, H 3.02%, N 8.38%, O 38.29%. Prepn of *m*-isomer: O. Kamm, J. B. Segur, *Org. Synth.* **coll. vol. I** (2nd ed., 1964) p 391; of *p*-isomer: O. Kamm, A. O. Matthews, *ibid.* p 392.

m-**Nitrobenzoic acid.** Monoclinic leaflets from water. Bitter taste. d 1.494. mp 142°. Melts in hot water. K (25°) = 3.48 × 10^{-4}. Absorption spectrum: Purvis, *J. Chem. Soc.* **107**, 971 (1915). One gram dissolves in 320 ml water, 3 ml alc, 4 ml ether, 18 ml chloroform, ~2 ml methanol, 2.5 ml acetone; very slightly sol in benzene, carbon disulfide, petr ether.

o-**Nitrobenzoic acid.** Yellowish-white, intensely sweet crystals. d 1.58. mp 147-148°. One gram dissolves in 146 ml water, 3 ml alcohol, 220 ml chloroform, 4.5 ml ether, 2.5 ml acetone, 2.5 ml methanol; very slightly sol in benzene, carbon disulfide, petr ether.

p-**Nitrobenzoic acid.** Monoclinic leaflets, plates from benzene. d 1.58. mp 242.4°. Sublimes. One gram dissolves in 2380 ml water, 110 ml alcohol, 12 ml methanol, 150 ml chloroform, 45 ml ether, 20 ml acetone; slightly sol in benzene, carbon disulfide; insol in petr ether.

USE: Manuf intermediates; also as a reagent for alkaloids and thorium.

6677. 4-Nitrobenzoyl Chloride. [122-04-3] 4-Nitrobenzoic acid chloride; *p*-nitrobenzoyl chloride. $C_7H_4ClNO_3$; mol wt 185.56. C 45.31%, H 2.17%, Cl 19.10%, N 7.55%, O 25.87%. Prepd by the action of phosphorus pentachloride on *p*-nitrobenzoic acid: Adams, Jenkins, *Org. Synth.* **3**, 75 (1923).

Bright yellow needles from petr ether. Pungent odor. mp 75°. bp$_{105}$ 205°; bp$_{12}$ 154°. Decomposed by water and alcohol; sol in ether. *Keep well closed.*

6678. 4-Nitrobenzyl Cyanide. [555-21-5] 4-Nitrobenzene-acetonitrile; *p*-nitrophenylacetonitrile; *p*-nitro-α-toluinitrile. $C_8H_6N_2O_2$; mol wt 162.15. C 59.26%, H 3.73%, N 17.28%, O 19.73%. Prepd by the action of concd nitric acid on benzyl cyanide: Robertson, *Org. Synth.* **2**, 57 (1922).

Elongated prisms from alcohol. mp 117°. Insol in water; sol in alcohol, ether.

USE: In the prepn of 4-nitrophenylacetic acid.

6679. *o*-Nitrobiphenyl. [86-00-0] 2-Nitro-1,1'-biphenyl; *o*-nitrodiphenyl; ONB. $C_{12}H_9NO_2$; mol wt 199.21. C 72.35%, H 4.55%, N 7.03%, O 16.06%. Prepn from diazotized *o*-nitroaniline and benzene by a modified Gomberg reaction: Elks *et al., J. Chem. Soc.* **1940**, 1284. Prepd industrially by direct nitration of biphenyl (yields ~50% para, 35% ortho).

Orthorhombic bipyramidal plates from ethanol. Characteristic sweetish odor. mp 36.7°. d$_4^{25}$ 1.44. d$_{15.5}^{40}$ (liq) 1.189. The technical

liquid weighs ~10 lbs/gal. bp$_{760}$ 325°; bp$_{30}$ 205°; bp$_{13}$ 170°; bp$_4$ 166°. n$_D^{25}$ (tech liq) 1.613. Flash pt 179°C (354°F). uv max: 325 nm. Dipole moment: 3.79. Viscosity of tech liquid (cP): 12 (45°); 38 (25°). Insol in water. Sol in methanol, ethanol, tetrahydrofurfuryl alcohol, acetone, DMF, carbon tetrachloride, perchlorethylene, mineral spirits, turpentine, glacial acetic acid.

USE: Plasticizer for resins, cellulose acetate and nitrate, polystyrenes; fungicide for textiles; wood preservative; dye intermediate.

6680. *p*-Nitrobiphenyl. [92-93-3] 4-Nitro-1,1'-biphenyl; 4-nitrobiphenyl; *p*-nitrodiphenyl; PNB. $C_{12}H_9NO_2$; mol wt 199.21. C 72.35%, H 4.55%, N 7.03%, O 16.06%. Prepn: G. Schultz, *Ann.* **174**, 210 (1874); O. Kühling, *Ber.* **28**, 42 (1895); E. Bamberger, *ibid.* 404; R. L. Jenkins *et al., Ind. Eng. Chem.* **22**, (1) 31 (1930); K. Dimroth *et al., Ber.* **101**, 2215 (1968). Crystal structure: M. Prasad *et al., J. Indian Chem. Soc.* **13**, 519 (1936); *C.A.* **31**, 1271^5 (1937). Review of carcinogenic risk: *IARC Monographs* **vol. 4**, 113-117 (1974).

Needles from alcohol, mp 113.7°. bp$_{760}$ 340°; bp$_{30}$ 223.7-224.1°. Insol in water. Slightly sol in cold alc; more readily sol in hot alc; sol in chloroform, ether. uv spectrum: D. F. DeTor, H. J. Scheifele, *J. Am. Chem. Soc.* **73**, 1442 (1951).

Caution: Potential symptoms of overexposure are headache, drowsiness and dizziness; dyspnea; ataxia, lassitude; methemoglobinemia; urinary burning; acute hemorrhagic cystitis. Potential occupational carcinogen. *See NIOSH Pocket Guide to Chemical Hazards* (DHHS/NIOSH PB2003-100121, 2003) p 226.

USE: Formerly in prepn of *p*-biphenylamine, *q.v.*

6681. Nitrocefin. [41906-86-9] (6*R*,7*R*)-3-[(1*E*)-2-(2,4-Dinitrophenyl)ethenyl]-8-oxo-7-[(2-thienylacetyl)amino]-5-thia-1-azabicyclo[4.2.0]oct-2-ene-2-carboxylic acid; 3-(2,4-dinitrostyryl)-(6*R*,7*R*)-7-(2-thienylacetamido)-ceph-3-em-4-carboxylic acid, *E*-isomer; 87/312. $C_{21}H_{16}N_4O_8S_2$; mol wt 516.50. C 48.83%, H 3.12%, N 10.85%, O 24.78%, S 12.41%. Chromogenic cephalosporin. Prepd (not claimed): C. H. O'Callaghan *et al.*, US 3830700 (1974 to Glaxo). Characterization and use in detection of β-lactamases: *idem et al., Antimicrob. Agents Chemother.* **1**, 283 (1972). Synthesis: N. C. M. Barendse *et al., Synthesis* **1998**, 145. Improved synthesis: M. Lee *et al., J. Org. Chem.* **70**, 367 (2005). Review of commercially available nitrocefin-based β-lactamase tests: A. T. Meszaros *et al., Am. Clin. Lab.* **14**, 20-22 (1995).

Orange solid, mp 103-113° (dec) (O'Callaghan, 1974); also reported as crystals from methanol, mp 167-169° (dec) (Lee). [α]$_D^{20}$ −224° (c = 1.0 in dioxane). Absorption max (ethanol): 231, 289, 386 nm (ε 24300, 10300, 18000). Absorption max (0.1*M* pH 6 phosphate buffer): 233, 290, 391 nm (ε 22200, 12200, 17400). Undergoes a color change from yellow to pink or red upon exposure to β-lactamases.

USE: In determn of β-lactamase activity in biological samples.

6682. 3-Nitrocinnamic Acid. [555-68-0] 3-(3-Nitrophenyl)-2-propenoic acid; 3-nitrobenzenepropenoic acid; 3-(3-nitrophenyl)-acrylic acid. $C_9H_7NO_4$; mol wt 193.16. C 55.96%, H 3.65%, N 7.25%, O 33.13%. Prepd by heating a mixture of *m*-nitrobenzaldehyde, sodium acetate and acetic anhydride: Thayer, *Org. Synth.* **5**, 83 (1925). From *m*-nitrobenzaldehyde and malonic acid in the presence of glycine in a little water at 100°: Dakin, *J. Biol. Chem.* **7**, 54 (1909-1910).

(*E*)-form

(E)-Form. [1772-76-5] Prepd by the above methods. White needles from benzene or alcohol. mp 200-201°. Can be sublimed or distilled at atmospheric press. with very little decompn. Absorption spectrum: Purvis, *J. Chem. Soc.* **107**, 971 (1906). One gram dissolves in ~100 ml alcohol at 25°.

(Z)-Form. [5676-61-9] Prepd by ultraviolet irradiation of a soln in water-alcohol-ammonia giving a 22% yield: Wollring, *Ber.* **47**, 112 (1914). Minute needles from toluene + methanol, mp 158°. By reirradiation of a chloroform soln containing a trace of bromine, the more stable *trans*-form is regenerated.

6683. Nitrodan. [962-02-7] 3-Methyl-5-[2-(4-nitrophenyl)-diazenyl]-2-thioxo-4-thiazolidinone; 3-methyl-5-[(*p*-nitrophenyl)-azo]rhodanine; 3-methyl-5-[(*p*-nitrophenyl)azo]-2-thio-2,4-thiazoli-dinedione; CTR-6110; Nidanthel; Everfree. $C_{10}H_8N_4O_3S_2$; mol wt 296.32. C 40.53%, H 2.72%, N 18.91%, O 16.20%, S 21.64%. Prepn: Benghiat, Howard, **US 2952673** (1960 to Stauffer); **BE 637282** (1964 to Cooper, Tinsley Labs.).

Yellow powder, mp 267-268°. Sparingly sol in water.
THERAP CAT (VET): Anthelmintic.

6684. Nitroethane. [79-24-3] $C_2H_5NO_2$; mol wt 75.07. C 32.00%, H 6.71%, N 18.66%, O 42.62%. $CH_3CH_2NO_2$. Obtained by vapor phase nitration of ethane with HNO_3: Reidel, *Oil Gas J.* **54**, no. 36, 110-114 (1956). Laboratory prepn by treating 1.5 moles sodium nitrite with 1 mole sodium ethyl sulfate at 125-130° in the presence of 0.0625 moles potassium carbonate: Desseigne, Giral, *Mem. Poudres* **34**, 49-53 (1952), *C.A.* **49**, 836 (1955). Similar procedure according to the equation $EtOSO_2OEt + NaNO_2 \rightarrow Et-OSO_2ONa + EtNO_2$: McCombie *et al.*, *J. Chem. Soc.* **1944**, 24.

Oily liquid; pleasant odor. d_{25}^{25} 1.041; d_{20}^{20} 1.052. Flash pt, open cup: 106°F (41.11°C). mp ~ −50°. bp 114-115°. Undergoes thermal decompn at 335-382°. Heating value (liquid): 7,720 Btu/lb. Lower limit of flammability in air = 4.0% by volume. Viscosity (25°): 0.661 cP. n_D^{20} 3.3917; $n_D^{24.3}$ 3.39007. Soly in water: 4.5 ml/100 ml H_2O at 20°. Miscible with methanol, ethanol, ether. Sol in chloroform, aq solns of alkalies. Has high heat of absorption. Sudden absorption of the anhydr liquid or gas on activated carbon or Hopcalite may result in flames: *Chem. Eng. News* **30**, 2344 (1952).

Caution: Potential symptom of overexposure by direct contact is dermatitis. See *NIOSH Pocket Guide to Chemical Hazards* (DHHS/NIOSH 97-140, 1997) p 228. Potential symptom of overexposure by ingestion is methemoglobinemia. See G. Shepherd *et al.*, *Clin. Toxicol.* **36**, 613 (1998).

USE: Solvent, artificial fingernail glue remover; in organic syntheses. Experimentally as liq propellant.

6685. Nitrofen. [1836-75-5] 2,4-Dichloro-1-(4-nitrophenoxy)benzene; 2,4-dichlorophenyl-*p*-nitrophenyl ether; nitraphen; nitrophen; nitrofene; FW-925; TOK. $C_{12}H_7Cl_2NO_3$; mol wt 284.09. C 50.73%, H 2.48%, Cl 24.96%, N 4.93%, O 16.89%. Selective pre- and post-emergence herbicide. Prepn: H. F. Wilson, H. M. Dougal, **US 3080225** (1963 to Rohm & Haas). Toxicity studies: A. M. Ambrose *et al.*, *Toxicol. Appl. Pharmacol.* **19**, 263 (1971). Carcinogenicity studies: J. F. Robens, *Vet. Hum. Toxicol.* **22**, 328 (1980). Review of carcinogenic risk: *IARC Monographs* **30**, 271-282 (1983).

Crystalline solid, mp 70-71°. Vapor pressure at 40°: 8×10^{-6} mm Hg. Soly in water at 22°: 0.7-1.2 ppm. LD_{50} orally in rats: 3.58 g/kg (Ambrose).

Caution: This substance is reasonably anticipated to be a human carcinogen: *Report on Carcinogens, Twelfth Edition* (PB2011-111646, 2011) p 296.

USE: Formerly as herbicide.

6686. Nitrofurantoin. [67-20-9] 1-[[(5-Nitro-2-furanyl)-methylene]amino]-2,4-imidazolidinedione; *N*-(5-nitro-2-furfurylidene)-1-aminohydantoin; 1-(5-nitro-2-furfurylideneamino)hydantoin; Berkfurin; Chemiofuran; Cyantin; Cystit; Fua-Med; Furachel; Furalan; Furadantin; Furadantine MC; Furadoin; Furantoina; Furobactina; Furophen T-Caps; Ituran; Macrodantin; Parfuran; Trantoin; Urantoin; Urizept; Urodin; Urolong; Uro-Tablinen; Welfurin. $C_8H_6N_4O_5$; mol wt 238.16. C 40.35%, H 2.54%, N 23.53%, O 33.59%. Prepn from 1-aminohydantoin sulfate and 5-nitro-2-furaldehyde diacetate: K. J. Hayes, **US 2610181** (1952 to Eaton Labs.); Swirska *et al.*, *Przem. Chem.* **11** (34), 306 (1955), *C.A.* **52**, 14079b (1958). Alternate routes: J. G. Michels, **US 2898335**; G. Gever, C. J. O'Keefe, **US 2927110** (1959, 1960 both to Norwich Pharmacal). LC-MS/MS determn of residues in meat: P. Mottier *et al.*, *J. Chromatogr. A* **1067**, 85 (2005). Clinical trial in urinary tract infection: R. G. Rogers *et al.*, *Am. J. Obstet. Gynecol.* **191**, 182 (2004). Comprehensive description: D. E. Cadwallader, H. W. Jun, *Anal. Profiles Drug Subs.* **5**, 345-373 (1976).

Orange-yellow needles from dil acetic acid, dec 270-272°. pKa 7.2. uv max: 370 nm ($E_{1cm}^{1\%}$ 776). Soly (mg/100 ml): water (pH 7) 19.0; 95% ethanol 51.0; acetone 510; DMF 8000; peanut oil 2.1; glycerol 60; polyethylene glycol 1500.

Monohydrate. [17140-81-7] Macrobid.
THERAP CAT: Antibacterial.
THERAP CAT (VET): Antibacterial.

6687. Nitrofurazone. [59-87-0] 2-[(5-Nitro-2-furanyl)methylene]hydrazinecarboxamide; 5-nitro-2-furaldehyde semicarbazone; Amifur; Furacin; Chemofuran; Furesol; Nifuzon; Nitrofural; Nitrozone; Furacinetten; Furacoccid; Furazol W; Mammex; Furaplast; Coxistat; Aldomycin; Nefco; Vabrocid. $C_6H_6N_4O_4$; mol wt 198.14. C 36.37%, H 3.05%, N 28.28%, O 32.30%. Prepd from 2-formyl-5-nitrofuran and semicarbazide hydrochloride: Stillman, Scott, **US 2416234** (1947 to Eaton Labs.). Alternate route: Gever, O'Keefe, **US 2927110** (1960 to Norwich Pharmacal). LC-MS/MS determn of residues in meat: P. Mottier *et al.*, *J. Chromatogr.* **1067**, 85 (2005). Toxicity: J. Godwin *et al.*, *J. Am. Med. Assoc.* **133**, 299 (1947). Tumorigenic activity: Morris *et al.*, *Cancer Res.* **29**, 2145 (1969); Ertürk *et al.*, *ibid.* **30**, 1409 (1970).

Pale yellow needles, dec 236-240°. Bitter aftertaste. Darkens on prolonged exposure to light. uv max: 260, 375 nm. Sol in DMF, alkaline solns with dark orange color; slightly sol in alcohol (1:590), propylene glycol (1:350); very slightly sol in water (1:4200). pH of satd water soln 6.0 to 6.5. Practically insol in chloroform, ether. LD_{50} in rats (g/kg): 0.59 orally; 3.0 s.c. (Godwin).

THERAP CAT: Anti-infective (topical).
THERAP CAT (VET): Antimicrobial.

6688. Nitrogen. [7727-37-9] N; at. wt [14.00643; 14.00728]; conventional wt 14.007; at. no. 7; valences 3, 5; elemental state: N_2. Group VA (15). Two naturally occurring isotopes: 14 (99.635%); 15 (0.365%); five short-lived, artificial, radioactive isotopes: 12; 13; 16-18. Discovered in 1772 by Daniel Rutherford and independently by Scheele and Cavendish. Constitutes about 75.5%

by weight or 78.06% by volume of the atmosphere; found frequently in volcanic or mine gases, gases from springs and gases occluded in minerals and rocks; an essential constituent of all living organisms; fixed or combined nitrogen is present in many mineral deposits. Prepn from sodium (and alkaline earth) azides by heating the azide: Tiede, *Ber.* **46**, 4100 (1913); **49**, 1745 (1916); Justi, *Ann. Phys.* [5] **10**, 985 (1931). Prepd industrially by fractional distln of liquid air; by removal of oxygen by combustion; by reduction of ammonia. Purification of nitrogen furnished in steel cylinders: Kautsky, Thiele, *Z. Anorg. Allg. Chem.* **152**, 342 (1926); Kendall, *Science* **73**, 395 (1931); Schenk in *Handbook of Preparative Inorganic Chemistry* vol. 1, G. Brauer, Ed. (Academic Press, New York, 2nd ed., 1963) pp 458-460. Review of nitrogen and nitrogen compounds: Jones in *Comprehensive Inorganic Chemistry* vol. 2, J. C. Bailar, Jr. *et al.*, Eds. (Pergamon Press, Oxford, 1973) pp 147-388; R. W. Schroeder in *Kirk-Othmer Encyclopedia of Chemical Technology* vol. 15 (Wiley-Interscience, New York, 3rd ed., 1981) pp 932-941. Books: W. L. Jolly, *The Inorganic Chemistry of Nitrogen* (Benjamin, New York, 1964) 124 pp; *Mellor's* **Vol. VIII**, Supplements I, II, *Nitrogen*, part 1 (1964) 619 pp; part 2 (1967) 676 pp; M. Sittig, *Nitrogen in Industry* (Van Nostrand, Princeton, 1965) 278 pp.

Odorless gas; condenses to a liquid, bp $-195.79°$ (77.36 K); solidifies to a snow-white mass, mp $-210.01°$ (63.14 K). d^{gas} (0°, 1 atm) 1.25046 g/l. Critical temp: $-147.1°$; critical press: 33.5 atm; critical density: 0.311 g/cm^3. Sparingly sol in water: 100 volumes of water absorbs 2.4 volumes of gas at 0°, 1.6 volumes at 20°. Soly in water at 50, 75 and 100° from 25 to 1000 atmospheres: Wiebe *et al.*, *J. Am. Chem. Soc.* **55**, 947 (1933). Soly in liquid ammonia: Wiebe *et al.*, *ibid.* 975. Soly in alc: one volume of alcohol dissolves 0.1124 volume of nitrogen at 20°. Liquid oxygen at $-195.5°$ absorbs 50.7% of its weight of gaseous nitrogen. Heat of dissociation of the nitrogen molecule (N_2): 225.1 kcal/mole. Combines with oxygen and hydrogen on sparking, forming nitric oxide and ammonia, respectively. Combines directly with lithium, and at a red heat with calcium, strontium, and barium to form nitrides. Forms cyanides when heated with carbon in presence of alkalies or barium oxide.

Caution: In high concns it is a simple asphyxiant.

USE: In manuf of ammonia, nitric acid, nitrates, cyanides, etc.; in manuf explosives; in filling high-temp thermometers, incandescent bulbs; to form an inert atm for preservation of materials, for use in dry boxes or glove bags. Liquid nitrogen in food-freezing processes; in the laboratory as a coolant. Pharmaceutic aid (air displacement).

6689. Nitrogen Chloride. [10025-85-1] Nitrogen trichloride; chlorine nitride; trichlorine nitride; Agene. Cl_3N; mol wt 120.36. Cl 88.36%, N 11.64%. NCl_3. Prepared by the action of chlorine gas or hypochlorous acid on ammonium salts: Dulong, *Ann. Chim. Phys.* [1] **86**, 37 (1813); Balard, *ibid.* [2] **57**, 258 (1834); Hentschel, *Ber.* **30**, 1792, 2642 (1897); Noyes, *J. Am. Chem. Soc.* **42**, 2173 (1920); by reaction of anhydr ammonia and anhydr chlorine: Noyes, Haw, *ibid.* 2167; industrially by electrolyzing an acidified soln of ammonium chloride: **US 2118903** (1938), **DE 641816** (1937); **GB 494188** (1938); for bleaching flour by blowing an air stream containing hydrogen chloride gas through a bed of finely powdered ammonium chloride and potassium hypochlorite distributed on an inert carrier to form a gas mixture of chlorine dioxide and nitrogen chloride: **GB 597199** (1948). *Review:* Richards in *Mellor's* **Vol. VIII**, supplement II, *Nitrogen* (part 2), 411-415 (1967).

Yellow, thick, oily liquid, d 1.653, pungent odor, evaporates rapidly in air, very unstable. Dec in light, vapor pressure at room temp is ~150 mm Hg. Explodes when heated to 93°, subjected to a flash of direct sunlight or magnesium light, sealed in glass containers at 60° after 13 seconds, frozen in liquid air and thawed *in vacuo*, in contact with ozone, nitric oxide, grease, and several organic substances. Insol in water. Dec in water after 24 hours, sol in carbon disulfide, phosphorus trichloride, benzene, carbon tetrachloride, chloroform.

USE: Bleaching of flour; wastage control of citrus fruit.

6690. Nitrogen Dioxide. [10102-44-0] Nitrogen oxide (NO_2). NO_2; mol wt 46.01. N 30.44%, O 69.55%. Oxidizing free radical; damaging component of photochemical smog. Prepd industrially from nitric oxide and air. Convenient lab prepn from lead nitrate: Schenk in *Handbook of Preparative Inorganic Chemistry* vol. 1, G. Brauer, Ed. (Academic Press, New York, 1963) pp 488-489. Ultrapure NO_2 from N_2O_5: Hackspill, Besson, *Bull. Soc.*

Chim. Fr. Mem. [5] **16**, 479 (1949). Equilibrium constant with N_2O_4 and determn methods in the polar stratosphere: H. K. Roscoe, A. K. Hind, *J. Atmos. Chem.* **16**, 257 (1993). Review of structure and reactivity: P. Gray, A. D. Yoffe, *Chem. Rev.* **55**, 1069-1154 (1955). Symposium on chemistry and toxicology: *Toxicology* **89**, 1-312 (1994). Review of pathobiochemistry: M. Kirsch *et al.*, *Biol. Chem.* **383**, 389-399 (2002).

Reddish-brown paramagnetic gas. Irritating odor. *Poisonous.* Exists in equilibrium with dinitrogen tetroxide (N_2O_4), *q.v.* At high temperatures, the gas phase is predominantly NO_2. Liquid and solid phases are almost entirely pure N_2O_4. Colorless solid at low temp; pale lemon-yellow from $-20°$ to $-30°$; honey colored at mp $-11.2°$. bp 21.15°. d_4^{20} (liq) 1.448; d (gas) 1.58 (air = 1); $d_{gas}^{21.3}$ 3.3 g/liter. d^{-40} (solid) 1.95. Crit temp 158.2°. Crit press. 99.96 atm. Heat of vaporization (bp) 9.110 kcal/mole. Does not burn, but supports the combustion of carbon, phosphorus, sulfur. Sol in concd sulfuric and nitric acids. Dec in water forming nitric acid and nitric oxide, reacts with alkalies to form nitrates and nitrites. Corrosive to steel when wet, but may be stored in steel cylinders when moisture content is 0.1% or less.

Caution: Potential symptoms of overexposure are eye, nose and throat irritation, cough, mucoid frothy sputum, decreased pulmonary function, chronic bronchitis, dyspnea, chest pain, pulmonary edema, cyanosis, tachypnea and tachycardia. *See NIOSH Pocket Guide to Chemical Hazards* (DHHS/NIOSH 2005-149) p 228. One of the most insidious gases. Inflammation of lungs may cause only slight pain or pass unnoticed, but the resulting edema several days later may cause death. 100 ppm is dangerous for even a short exposure, and 200 ppm may be fatal: Y. Henderson, H. W. Haggard, *Noxious Gases*, A.C.S. Monograph Series, no. 35 (Reinhold, New York, 2nd ed., 1943) pp 134-137, 141.

USE: Intermediate in nitric and sulfuric acid production. Used in the nitration of organic compds and explosives, in the manuf of oxidized cellulose compds (hemostatic cotton). Has been used to bleach flour. Oxidizing agent in rocket propulsion.

6691. Nitrogen Fluoride. [7783-54-2] Nitrogen fluoride (NF_3); nitrogen trifluoride. F_3N; mol wt 71.00. F 80.27%, N 19.73%. NF_3. Prepd by electrolysis of melted ammonium acid fluoride, NH_4F_2H: Ruff *et al.*, *Z. Anorg. Allg. Chem.* **172**, 417 (1928); Ruff, *ibid.* **197**, 273 (1931); Ruff, Staub, *ibid.* **198**, 32 (1931); Kwasnik in *Handbook of Preparative Inorganic Chemistry* vol. 1, G. Brauer, Ed. (Academic Press, New York, 2nd ed., 1963) pp 181-183. Reviews of prepn and chemistry: Hoffman, Neville, *Chem. Rev.* **62**, 1-18 (1962); Kemmitt, Sharp, *Adv. Fluorine Chem.* **4**, 189-190 (1965).

Colorless, nonflammable gas. Moldy odor. mp $-208.5°$. bp $-129°$. d (liq at bp) 1.885. Trouton const 19.9. Insoluble in water. Rather inert chemically. Does not attack glass, mercury. Decomposed by electric sparks.

Caution: Potential symptoms of overexposure in exptl animals are anoxia, cyanosis; methemoglobinemia; weakness, dizziness, headache; liver, kidney injury. *See NIOSH Pocket Guide to Chemical Hazards* (DHHS/NIOSH 97-140, 1997) p 228.

USE: Chamber-cleaning gas in the manufacture of electronics.

6692. Nitrogen Pentoxide. [10102-03-1] Nitrogen oxide (N_2O_5); dinitrogen pentoxide; nitric anhydride. N_2O_5; mol wt 108.01. N 25.94%, O 74.06%. Usually prepd by dehydration of nitric acid by means of phosphorus pentoxide: Gruenhut *et al.*, *Inorg. Synth.* **3**, 78 (1950). *Reviews:* Beattie, "Dinitrogen Pentoxide" in *Mellor's* **Vol. VIII**, supplement II, *Nitrogen* (part 2), 269-277 (1967); Jones in *Comprehensive Inorganic Chemistry* vol. 2, J. C. Bailar, Jr. *et al.*, Eds. (Pergamon Press, Oxford, 1973) pp 356-360.

Colorless hexagonal crystals. mp 30°. Sublimes at 32.4° but undergoes moderately fast decompn into O_2 and the NO_2/N_2O_4 equilibrium mixture at temps above $-10°$. d^{15} 2.05. bp_{760} 47.0°; bp_{100} 7.0°; bp_{10} $-20°$. Dipole moment 1.39. Freely sol in chloroform without appreciable decomposition. Less sol in carbon tetrachloride. Chloroform solns may be stored at $-20°$ for as long as one week without excessive decompn.

USE: As nitrating agent in chloroform soln.

6693. Nitrogen Selenide. [12033-88-4] Selenium nitride. N_4Se_4; mol wt 371.87. N 15.07%, Se 84.93%. Prepd by passing dry ammonia over selenium tetrachloride: Wöhler, *Bull. Soc. Chim.* [1]

1, 25 (1859); by treating a dil soln of selenium oxychloride in benzene with dry ammonia and washing the precipitate with water and potassium cyanide: Lenher, Wolesensky, *J. Am. Chem. Soc.* **29**, 215 (1907); by the action of dry ammonia on a dilute soln of selenium monochloride in carbon disulfide: van Valkenburg, Bailar, *ibid.* **47**, 2134 (1925); by the action of dry ammonia on diethyl selenite or dimethyl selenite dissolved in benzene and washing with potassium cyanide: Strecker, Schwarzkopf, *Z. Anorg. Chem.* **221**, 193 (1934); by reacting anhyd, liquid ammonia with selenium dioxide: Jander, Doetsch, *Ber.* **93**, 561 (1960). Crystal structure: Baernighausen *et al.,* *Acta Crystallogr.* **15**, 615 (1962); **21**, 571 (1966).

Orange-red, amorphous powder or monoclinic crystals. d 4.2. Very hygroscopic. *Explosive.* Sparingly sol in carbon disulfide, benzene, acetic acid; insol in water, ether, abs alc.

6694. Nitroglycerin. [55-63-0] 1,2,3-Propanetriol 1,2,3-trinitrate; glyceryl trinitrate; glycerol nitric acid triester; nitroglycerol; trinitroglycerol; glonoin; trinitrin; blasting gelatin; blasting oil; S.N.G.; Adesitrin; Anginine; Aquo-Trinitrosan; Cordipatch; Corditrine; Deponit; Diafusor; Discotrine; Lenitral; Millisrol; Minitran; Nitradisc; Nitro-Bid; Nitrocine; Nitrocontin; Nitroderm; Nitro-Dur; Nitrogard; Nitroglin; Nitroglyn; Nitrolingual; Nitromex; Nitronal; Nitrong; Nitrostat; Nitrosylon; Percutol; Perlinganit; Rectogesic; Suscard; Sustac; Transderm-Nitro; Transiderm-Nitro; Tridil; Trinitrosan. $C_3H_5N_3O_9$; mol wt 227.09. C 15.87%, H 2.22%, N 18.50%, O 63.41%. Nitric oxide donor that induces vasodilation. Prepn: Sobrero, *Ann.* **64**, 398 (1847); Williamson, *Ann.* **92**, 305 (1854). Review of the early literature: J. W. Lawrie, *Glycerol and the Glycols* (New York, 1928). Review of chemistry and biochemistry: F. J. DiCarlo, *Drug Metab. Rev.* **4**, 1-38 (1975). Review of mechanism of action: S. F. Vatner, G. R. Heyndrickx, *Handb. Exp. Pharmakol.* **40**, 131-161 (1975). Molecular mechanism of nitric oxide release: Z. Chen *et al., Proc. Natl. Acad. Sci. USA* **99**, 8306 (2002). Review of the first hundred years: J. R. Parratt, *J. Pharm. Pharmacol.* **31**, 801-809 (1979). Comprehensive description: E. F. McNiff *et al., Anal. Profiles Drug Subs.* **9**, 519-541 (1980). Symposium on nitroglycerin therapy, perspectives and mechanisms: *Am. J. Med.* **74**, no. 6B, 1-94 (1983). Review of pharmacology and clinical studies of intravenous administration in heart disease: E. M. Sorkin *et al., Drugs* **27**, 45 (1984). Clinical trial in treatment of anal fissures: J. H. Scholefield *et al., Gut* **52**, 264 (2005).

Pale yellow, oily liquid. Sweet burning taste. Produces headache on tasting. *Explosive; poisonous.* Crystallizes in 2 forms: labile form, mp 2.8°; stable form, mp 13.5°. d_{15}^{15} 1.599; d_4^4 1.6144; d_4^{15} 1.6009; d_4^{25} 1.5918. n_D^{15} 1.474. Begins to dec at 50-60°, appreciably volatile at 100°, evolves nitrous yellow vapors at 135°, explodes at 218°. Vapor pressure at 20°: 0.00026 mm; at 93°: 0.31 mm. One gram dissolves in 800 ml water, in 4 g ethanol, in 18 g methanol, in 120 g carbon disulfide. Misc with ether, acetone, glacial acetic acid, ethyl acetate, benzene, nitrobenzene, pyridine, chloroform, ethylene bromide, dichloroethylene. Sparingly sol in petr ether, liquid petrolatum, glycerol. Heat of combustion: 1580 cal/g. Explodes on rapid heating or on concussion. On explosion harmless gases are produced: $4C_3H_5(ONO_2)_3 \rightarrow 12CO_2 + 10H_2O + 6N_2 + O_2$.

Spirit of Glyceryl Trinitrate. Spirit of nitroglycerin; spirit of glonoin. An alcoholic soln contg 1.0-1.1% glyceryl trinitrate. Colorless, clear liquid. d 0.814-0.820. Miscible with alcohol, chloroform, ether; 1 ml dissolves in 6 ml almond oil; very slightly sol in water; miscible with chloroform, ether.

Caution: Potential symptoms of overexposure by percutaneous or respiratory absorption are throbbing headache, dizziness, nausea, vomiting, abdominal pain; hypotension; flushing; palpitations; methemoglobinemia; delerium, CNS depression; angina; cyanosis; coma. Direct contact may cause skin irritation. *See NIOSH Pocket Guide to Chemical Hazards* (DHHS/NIOSH 97-140, 1997) p 228; *Patty's Industrial Hygiene and Toxicology* vol. 2B, G. D. Clayton, F. E. Clayton, Eds. (Wiley-Interscience, New York, 4th ed., 1994) pp 1064-1067.

USE: Manuf of **dynamite** (75% nitroglycerol, 24.5% diatomaceous earth, 0.5% sodium carbonate), smokeless powders and blasting gelatin; in rocket propellants.

THERAP CAT: Antianginal; vasodilator (coronary). Topically in treatment of anal fissures.

THERAP CAT (VET): Topical vasodilator (coronary); antianginal.

6695. N-Nitroguanidine. [556-88-7] $CH_4N_4O_2$; mol wt 104.07. C 11.54%, H 3.87%, N 53.84%, O 30.75%. Exists in two forms. The α-form is the usual stable form. The β-form is converted into the α-form by dissolving in concd H_2SO_4 and pouring into ice water. Prepn of the α-form from guanidine nitrate and the β-form from guanidine sulfate: T. L. Davis *et al., J. Am. Chem. Soc.* **47**, 1063 (1925). Modified prepn of the α-form: T. L. Davis, *Org. Synth.* **7**, 68 (1927). Absorption spectrum: E. R. Riegel, K. W. Buchwald, *J. Am. Chem. Soc.* **51**, 484 (1929).

Needles, prisms from water, dec 225-250° (depending on speed of heating) giving off NH_3 fumes. Slightly soluble in methanol (<0.5%), alcohol; practically insol in ether. Sol in concd acids from which it is precipitated by water. Sol in cold solns of alkalies (not of carbonates) with slow decompn.

USE: Explosive of moderate power. Can be exploded only with a detonator. Intermediate in the synthesis of pharmaceuticals.

6696. Nitrohydrochloric Acid. [8007-56-5] Aqua regia; nitromuriatic acid; chloronitrous acid; chloroazotic acid. Made by mixing 18 ml HNO_3, 82 ml HCl. In addition to the free acids, it contains nitrosyl chloride (NOCl) and some free chlorine.

Yellow, fuming, corrosive, volatile liq; suffocating odor; attacks all metals including gold and platinum. Miscible with water. *Keep in well-closed, glass-stoppered bottles in a cool place and protected from light.*

Caution: Undiluted form is a strong irritant, corrosive.

6697. Nitromersol. [133-58-4] [2-Methyl-5-nitrophenolato-(2−)-$\kappa C^6,\kappa O^1$]mercury; 5-methyl-2-nitro-7-oxa-8-mercurabicyclo-[4.2.0]octa-1,3,5-triene; 3-(hydroxymercuri)-4-nitro-o-cresol inner salt; Metaphen. $C_7H_5HgNO_3$; mol wt 351.71. C 23.91%, H 1.43%, Hg 57.03%, N 3.98%, O 13.65%. According to the National Formulary this compd is the anhydride of 4-nitro-3-hydroxymercuri-o-cresol. Prepd by treating 4-nitro-o-cresol with mercuric acetate: US 1544293; US RE 17563 (1925).

Yellow, odorless, tasteless powder or granules affected by light. Sol in boiling glacial acetic acid, solns of alkalies, ammonia by opening the anhydride ring and the formation of a salt; very slightly sol in water, alc, acetone, ether. Almost insol in aq soln of sodium carbonate.

THERAP CAT: Antiseptic.

THERAP CAT (VET): Disinfectant (topical).

6698. Nitromethane. [75-52-5] Nitrocarbol. CH_3NO_2; mol wt 61.04. C 19.68%, H 4.95%, N 22.95%, O 52.42%. Prepd by the interaction of sodium nitrite and sodium chloroacetate: Whitmore, *Org. Synth.* **3**, 83 (1923). Monograph: *Am. Ind. Hyg. Assoc. J.* **22**, 518 (1961). Toxicity study: Weatherby, *Arch. Ind. Health* **11**, 102 (1955).

Oily liquid with a moderately strong, somewhat disagreeable odor. d_4^{25} 1.1322. One gallon weighs 9.5 lbs. Flash pt 112°F. mp −29°. bp_{760} 101.2°; bp_{100} 46.6°; bp_{40} 27.5°; bp_{20} 14.1°; bp_{10} 2.8°; bp_5 −7.9°; $bp_{1.0}$ −29°. n_D^{32} 1.38056. Absorption spectrum: Hantzsch, Voigt, *Ber.* **45**, 106 (1912). *Flammable.* Slightly sol in water (9.5% by vol at 20°); sol in alc, ether, DMF. Water solns are acid to litmus. pH of $0.01M$ aq soln 6.12. Forms an explosive sodium salt which bursts into flame on contact with water. LC (in air) in guinea pigs: 1000 ppm; LD_{50} orally in mice: 1.44 g/kg (Weatherby).

Caution: Potential symptom of overexposure is dermatitis. *See NIOSH Pocket Guide to Chemical Hazards* (DHHS/NIOSH 97-140, 1997) p 230. This substance is reasonably anticipated to be a human carcinogen: *Report on Carcinogens, Twelfth Edition* (PB2011-111646, 2011) p 299.

USE: Rocket fuel; solvent for zein. Used in the coating industry.

6699. Nitromide. [121-81-3] 3,5-Dinitrobenzamide. C_7H_5-N_3O_5; mol wt 211.13. C 39.82%, H 2.39%, N 19.90%, O 37.89%. Prepd from 3,5-dinitrobenzoyl chloride and ammonium acetate: Finan, Fothergill, *J. Chem. Soc.* **1962**, 2824.

Leaflets from water, mp 183°. Slightly sol in cold water, somewhat more sol in hot water.

Note: An ingredient of *Tristat, Unistat, Unistat-3*.

THERAP CAT (VET): Antibacterial, coccidiostat (for poultry).

6700. Nitron. [2218-94-2] 1,4-Diphenyl-3-(phenylamino)-$4H$-1,2,4-triazolium inner salt. $C_{20}H_{16}N_4$; mol wt 312.38. C 76.90%, H 5.16%, N 17.94%. Orginally represented as *3,5,6-triphenyl-2,3,5,6-tetraazabicyclo[2.1.1]hex-1-ene*. Prepn: *DE* **159692** (1904 to E. Merck); M. Busch, *Ber.* **38**, 856 (1905). Corrected structure determn as meso-ionic compound: W. Baker, W. D. Ollis, *Q. Rev. Chem. Soc.* **11**, 15 (1957); G. A. Olah, *J. Inorg. Nucl. Chem.* **16**, 225 (1961). Mass spectrum: W. D. Ollis, C. A. Ramsden, *J. Chem. Soc. Perkin Trans. 1* **1974**, 856. NMR study: W. Bocian *et al.*, *Magn. Reson. Chem.* **33**, 134 (1995). Use in determn of metals: N. Gantchev, A. Dimitrova, *Mikrochim. Acta* **1971**, 476; T. J. Koralewski, G. A. Parker, *Anal. Chim. Acta* **113**, 389 (1980); H. Chen *et al.* *Acta Chim. Sin.* **1989**, 412. In determn of nitrate: A. Hulanicki, M. Maj, *Talanta* **22**, 767 (1975); A. Hioki *et al.* *Anal. Sci.* **6**, 757 (1990). Chemical sensor of stratospheric nitric acid: K. Meadows, V. R. Morris, *Proc. NOBCChE* **25**, 1 (1998).

Intensely yellow leaflets from alcohol, solvated needles from chloroform, dec ~189°. Insol in nitrates, water. Sol in alcohol, benzene, slightly sol acetone, chloroform, ethyl acetate; sparingly sol in ether. The alc soln undergoes partial decompn, indicated by a red color. Dipole moment in benzene: 7.2. pKa 10.34 ± 0.02

USE: Determination of boron, nitrate, perchlorate, rhenium, tungsten, molybdenum.

6701. 1-Nitronaphthalene. [86-57-7] α-Nitronaphthalene. $C_{10}H_7NO_2$; mol wt 173.17. C 69.36%, H 4.07%, N 8.09%, O 18.48%. Prepd from napthalene and a mixture of nitric and sulfuric acids at 50°.

Yellow crystals. d 1.331. mp 59-61°. bp 304°. Insol in water; sol in alcohol, freely in chloroform, ether, carbon disulfide. Gives a dark red soln with concd H_2SO_4.

USE: Deblooming petr oils, 2-3 parts suffice for 1000 parts oil; manuf dyes and intermediates.

6702. 1-Nitro-2-naphthol. [550-60-7] 1-Nitro-2-naphthalenol; α-nitro-β-naphthol; 1-nitro-2-hydroxynaphthalene. $C_{10}H_7$-NO_3; mol wt 189.17. C 63.49%, H 3.73%, N 7.40%, O 25.37%.

Yellow needles or platelets, mp 103°. Insol in water. Sol in alcohol, ether, glacial acetic acid, in alkali hydroxide solns.

USE: As a reagent for the determination of cobalt with which it gives a precipitate.

6703. 3-Nitropentane. [551-88-2] $C_5H_{11}NO_2$; mol wt 117.15. C 51.26%, H 9.46%, N 11.96%, O 27.31%.

Colorless liquid; fusel oil odor. d_4^0 0.957. bp 153-155°. Insol in water; sol in alcohol, ether.

6704. 5-Nitro-o-phenetidine. [136-79-8] 2-Ethoxy-5-nitrobenzenamine; 5-nitro-2-ethoxyaniline; 1-ethoxy-2-amino-4-nitrobenzene; 1-nitro-3-amino-4-phenyl ethyl ether; 4-nitro-2-aminophenetole; Neo-Douxan. $C_8H_{10}N_2O_3$; mol wt 182.18. C 52.74%, H 5.53%, N 15.38%, O 26.35%. Prepn: Verkade *et al.*, *Rec. Trav. Chim.* **65**, 346 (1946). Toxicology: Grupp, Bilger, *Arzneim.-Forsch.* **1**, 326 (1951).

Brown-yellow crystals from benzene, mp 97.5-98.5°. 950 times as sweet as cane sugar, claimed to have no aftertaste.

USE: Sweetening agent.

6705. Nitrophenide. [537-91-7] Bis(3-nitrophenyl)disulfide; *m,m'*-dinitrodiphenyl disulfide; NP; Megasul. $C_{12}H_8N_2O_4S_2$; mol wt 308.33. C 46.75%, H 2.62%, N 9.09%, O 20.76%, S 20.80%. Prepd by the reduction of *m*-nitrobenzenesulfonylchloride with hydriodic acid: Ekbom, *Ber.* **24**, 336 (1891); Foss *et al.*, *J. Am. Chem. Soc.* **60**, 2729 (1938). Anticoccidial activity: Waletzky *et al.*, *Ann. N.Y. Acad. Sci.* **52**, 543 (1949).

Yellow rhomboid crystals, mp 83°. Freely sol in ether, less sol in alcohol. Insol in water.

THERAP CAT (VET): Coccidiostat.

6706. *m*-Nitrophenol. [554-84-7] 3-Nitrophenol. $C_6H_5NO_3$; mol wt 139.11. C 51.81%, H 3.62%, N 10.07%, O 34.50%. Prepn by boiling diazotized *m*-nitroaniline with water and H_2SO_4: Adams, Wilson, *Org. Synth.* **3**, 87 (1923); Bachmann, Rottschaefer, *ibid.* **18**, 88 (1938); Manske, *ibid.* **coll. vol. I**, 404 (2nd ed., 1941).

Monoclinic prisms from ether or dil HCl. d_4^{20} 1.485; d_4^{100} 1.2797. mp 97°. bp_{70} 194°. Dec when distilled at ordinary pressure. pK (18°) 8.34. Absorption spectrum: Baly *et al.*, *J. Chem. Soc.* **97**, 586 (1910); Marchlewski, Moroz, *Bull. Soc. Chim.* [4] **35**, 476 (1924). An aq soln satd at 40° contains 3.02 g in 100 g soln; at 98.7°, 40.9 g: Sidgwick *et al.*, *J. Chem. Soc.* **107**, 1202 (1915). Soly (g/100 g solvent) in acetone: 169.35 (0.2°); 1305.9 (84°); in alcohol: 116.9 (1°); 1105.25 (85°); in ether: 105.9 (0.2°); 1065.8 (83°): Carrick, *J. Phys. Chem.* **25**, 635 (1921). Sol in hot and dil acids, in caustic solns. Insol in petr ether. LD_{50} orally in mice, rats: 1414, 933 mg/kg, K. C. Back *et al.*, *Reclassification of Materials Listed as Transportation Health Hazards* (TSA-20-72-3; PB214-270, 1972).

USE: As indicator in 0.3% soln in 50% alc. pH: 6.8 colorless, 8.6 yellow.

6707. *o*-Nitrophenol. [88-75-5] 2-Nitrophenol. $C_6H_5NO_3$; mol wt 139.11. C 51.81%, H 3.62%, N 10.07%, O 34.50%. Prepd by the nitration of phenol; separated by steam distillation. Review of toxicology and human exposure: *Toxicological Profile for Nitrophenols: 2-Nitrophenol and 4-Nitrophenol* (PB93-110823, 1992) 131 pp.

Light yellow needles or prisms; peculiar, aromatic odor. d 1.495. mp 44-45°. bp 214-216°. Volatile with steam. Slightly sol in cold water, freely in hot water, in alcohol, benzene, ether, carbon disulfide, alkali hydroxides. LD_{50} orally in mice, rats: 1.297, 2.828 g/kg, K. C. Back *et al.*, *Reclassification of Materials Listed as Transportation Health Hazards* (TSA-20-72-3; PB214-270, 1972).

USE: Manuf dyes, paint colorings, rubber chemicals, and fungicides. Indicator in 2% alc soln. pH: 5.0 colorless, 7.0 yellow, but the color change is not sharp and cannot be used where CO_2 is present; as reagent for glucose.

6708. *p*-Nitrophenol. [100-02-7] 4-Nitrophenol. $C_6H_5NO_3$; mol wt 139.11. C 51.81%, H 3.62%, N 10.07%, O 34.50%. Review of toxicology and human exposure: *Toxicological Profile for Nitrophenols: 2-Nitrophenol and 4-Nitrophenol* (PB93-110823, 1992) 131 pp.

Colorless to slightly yellow, odorless crystals; sweetish, then burning taste. d_4^{120} 1.270. mp 113-114°. Sublimes; slightly volatile with steam. Moderately sol in cold water, freely in alcohol, chloroform, ether; also sol in solns of fixed alkali hydroxides and carbonates. LD_{50} orally in mice, rats: 467, 616 mg/kg, K.C. Back *et al.*, *Reclassification of Materials Listed as Transportation Health Hazards* (TSA-20-72-3; PB214-270, 1972).

USE: Manuf of pharmaceuticals, fungicides, dyes. Indicator in 0.1% alcohol soln. pH: 5.6 colorless, 7.6 yellow.

6709. 4-Nitrophenylacetic Acid. [104-03-0] 4-Nitrobenzeneacetic acid; *p*-nitro-α-toluic acid. $C_8H_7NO_4$; mol wt 181.15. C 53.04%, H 3.90%, N 7.73%, O 35.33%. Prepd by hydrolysis of *p*-nitrobenzyl cyanide with 50% H_2SO_4: Robertson, *Org. Synth.* **2**, 59 (1922); **coll. vol. I**, 406 (2nd ed., 1941).

Long, pale yellow needles from water. mp 153°. pK (25°) 3.98. Absorption spectrum: Hewitt *et al.*, *J. Chem. Soc.* **101**, 1774 (1912). Sparingly sol in cold water; sol in alcohol, ether, benzene.
Methyl ester. [2945-08-6] $C_9H_9NO_4$. mp 54°.
Benzyl ester. [6035-06-9] $C_{15}H_{13}NO_4$. mp 92°.
Nitrile *see* 4-Nitrobenzyl Cyanide.

6710. 4-Nitro-*o*-phenylenediamine. [99-56-9] 4-Nitro-1,2-benzenediamine; 1,2-diamino-4-nitrobenzene; 4-nitro-1,2-diaminobenzene; 4-nitro-1,2-phenylenediamine. $C_6H_7N_3O_2$; mol wt 153.14. C 47.06%, H 4.61%, N 27.44%, O 20.89%. Prepd by the reduction of 2,4-dinitroaniline using H_2S in ammonia water: Kehrmann, *Ber.* **28**, 1707 (1895); Griffin, Peterson, *Org. Synth.* **coll. vol. III**, 242 (1955). Use as reagent for α-keto acids: Hockenhull, Floodgate, *Biochem. J.* **52**, 38 (1952); Taylor, Smith, *Analyst* **80**, 607 (1955). Toxicity study: C. Burnett *et al.*, *J. Toxicol. Environ. Health* **2**, 657 (1977).

Dark red needles from hot water. mp 201°. Sparingly sol in water. Sol in aq solns of hydrochloric acid. LD_{50} in rats (mg/kg): 3720 orally; >1600 i.p. (Burnett).

USE: Reagent for α-keto acids.

6711. (4-Nitrophenyl)hydrazine. [100-16-3] $C_6H_7N_3O_2$; mol wt 153.14. C 47.06%, H 4.61%, N 27.44%, O 20.89%.

Orange-red leaflets or needles. mp ~157° with dec. Slightly sol in cold water; sol in hot water or hot benzene, alcohol, chloroform, ether, ethyl acetate.

USE: As reagent for ketones, aliphatic aldehydes.

6712. *o*-Nitrophenylpropiolic Acid. [530-85-8] 3-(2-Nitrophenyl)-2-propynoic acid. $C_9H_5NO_4$; mol wt 191.14. C 56.55%, H 2.64%, N 7.33%, O 33.48%.

Yellowish to light brown scales or crystals. mp ~157° with decompn and may explode. Moderately sol in cold water, more in hot water or alcohol, very slightly in chloroform. Almost insol in carbon disulfide and petr ether.

USE: As a reagent for alkaloids and glucose.

6713. (4-Nitrophenyl)urea. [556-10-5] *N*-(4-Nitrophenyl)-urea. $C_7H_7N_3O_3$; mol wt 181.15. C 46.41%, H 3.90%, N 23.20%, O 26.50%. Prepn from *N*-substituted *S*-phenylthiocarbamates: Crosby, Niemann, *J. Am. Chem. Soc.* **76**, 4458 (1954).

Prisms from abs ethanol, needles from dil ethanol, mp 238°. Practically insol in cold water. Sol in boiling water, methanol, ethanol, dimethylformamide. Sparingly sol in ether, benzene.

6714. 1-Nitropropane. [108-03-2] $C_3H_7NO_2$; mol wt 89.09. C 40.45%, H 7.92%, N 15.72%, O 35.92%. Prepd by vapor-phase nitration of propane: Hass *et al.*, **US 1967667** (1934 to Purdue Res. Found.); *Ind. Eng. Chem.* **28**, 339 (1936); from natural gas: Bachman, Pollack, *ibid.* **46**, 713 (1954).

$$H_3C\diagdown\diagup NO_2$$

Liquid. d_4^{25} 0.9934. wt 8.4 lb/U.S. gal. mp $-108°$. bp_{760} 131.6°; bp_{728} 130°; bp_{401} 110°; bp_{94} 70°; $bp_{7.5}$ 20°; bp_4 10°. n_D^{20} 1.4018. Flash pt 34°C (93°F). Heat of formation -40.05 kcal/mole at 25°; heat of combustion 481.33 kcal/mole at 25°. Latent heat of vaporization 10.37 kcal/mole at 25°. Slightly sol in water (1.4 ml/l00 ml); 0.5 ml water dissolves in 100 ml of 1-nitropropane. Miscible with many organic solvents.
Caution: Potential symptoms of overexposure are eye irritation, headache, nausea, vomiting and diarrhea. *See NIOSH Pocket Guide to Chemical Hazards* (DHHS/NIOSH 97-140, 1997) p 230.
USE: Solvent for cellulose acetate, vinyl resins, lacquers, synthetic rubbers, fats, oils, dyes, other organic materials; as intermediate, propellant.

6715. 2-Nitropropane. [79-46-9] 1-Methylnitroethane; dimethylnitromethane; isonitropropane. $C_3H_7NO_2$; mol wt 89.09. C 40.45%, H 7.92%, N 15.72%, O 35.92%. Prepn: Hass *et al.*, **US 1967667** (1934 to Purdue Res. Found.); *Ind. Eng. Chem.* **28**, 339 (1936); from natural gas: Bachman, Pollack, *ibid.* **46**, 713 (1954). Toxicity and carcinogenicity study: T. R. Lewis *et al.*, *J. Environ. Pathol. Toxicol.* **2**, 233 (1979). Health hazard alert: *Am. Ind. Hyg. Assoc. J.* **41**, 18 (1980).

$$\underset{H_3C}{\overset{NO_2}{\diagup\hspace{-4pt}\diagdown}}CH_3$$

Colorless, oily liquid. *Flammable.* d_4^{25} 0.9821. wt 8.4 lb/U.S. gal. mp $-93°$. bp_{760} 120.3°; bp_{564} 110°; bp_{300} 90°; bp_{95} 60°; bp_7 10°. n_D^{20} 1.3944. Flash pt (closed cup): 24°C (75°F). Heat of formn -43.78 kcal/mole at 25°; heat of combustion 477.60 kcal/mole at 25°. Latent heat of vaporization 9.88 kcal/mole at 25°. Slightly sol in water (1.7 ml/l00 ml); 0.6 ml water dissolves in 100 ml of 2-nitropropane. Misc with many organic solvents.
Caution: Potential symptoms of overexposure are headache, anorexia, nausea, vomiting, diarrhea; irritation of eyes, skin, nose and respiratory system; kidney and liver damage. *See NIOSH Pocket Guide to Chemical Hazards* (DHHS/NIOSH 97-140, 1997) p 230. This substance is reasonably anticipated to be a human carcinogen: *Report on Carcinogens, Twelfth Edition* (PB2011-111646, 2011) p 300.
USE: Solvent in inks, paints, adhesives, varnishes, polymers, synthetic material; as chemical intermediate.

6716. 5-Nitro-2-propoxyaniline. [553-79-7] 5-Nitro-2-propoxybenzenamine; 2-amino-4-nitro-1-propoxybenzene; 1-propoxy-2-amino-4-nitrobenzene; P-4000; Ultrasüss; $C_9H_{12}N_2O_3$; mol wt 196.21. C 55.09%, H 6.16%, N 14.28%, O 24.46%. A substance which is 4000 times sweeter than sugar, with no bitter aftertaste. Prepn: Verkade *et al.*, *Rec. Trav. Chim.* **65**, 346-360 (1946); **NL 6514769** (1967 to N. V. Koffie en Theehandel), *C.A.* **67**, 108430n (1967); A. J. de Koning, *J. Chem. Educ.* **53**, 521 (1976); J. McK. Woollard, *ibid.* **57**, 464 (1980). Toxicity studies: Fitzhugh *et al.*, *J. Am. Pharm. Assoc.* **46**, 583 (1951).

$$\text{structure: benzene ring with } NH_2, \text{ O-propyl } (CH_3), \text{ and } O_2N$$

Orange crystals from *n*-propanol + petr ether. mp 47.5-48.5°. Soly in water at 20°: 136 mg/liter. Stable in boiling water and in dil acids.

6717. 3-Nitrosalicylic Acid. [85-38-1] 2-Hydroxy-3-nitrobenzoic acid. $C_7H_5NO_5$; mol wt 183.12. C 45.91%, H 2.75%, N 7.65%, O 43.68%. The 3- and 5-nitrosalicylic acids are formed by the nitration of salicylic acid with H_2SO_4 and HNO_3.

$$\text{structure: benzene ring with } COOH, OH, NO_2$$

Monohydrate. Yellowish crystals. mp 123°; mp 148° when anhydr. Slightly sol in water; freely sol in alcohol, benzene, chloroform, ether.
USE: In dye chemistry.

6718. 5-Nitrosalicylic Acid. [96-97-9] 2-Hydroxy-5-nitrobenzoic acid; anilotic acid. $C_7H_5NO_5$; mol wt 183.12. C 45.91%, H 2.75%, N 7.65%, O 43.68%.
Yellowish crystals. d 1.65. mp 228-230°. One gram dissolves in 1475 ml water; more sol in hot water; freely sol in alcohol, ether.

6719. Nitroscanate. [19881-18-6] 1-Isothiocyanato-4-(4-nitrophenoxy)benzene; isothiocyanic acid *p*-(*p*-nitrophenoxy)phenyl ester; *p*-(*p*-nitrophenoxy)phenyl isothiocyanate; 1-(4-isothiocyanatophenoxy)-4-nitrobenzene; 4-nitro-4'-isothiocyanodiphenyl ether; cantrodifene (obsolete); CGA-23654; GS-23654; Lopatol; Skanitrol. $C_{13}H_8N_2O_3S$; mol wt 272.28. C 57.35%, H 2.96%, N 10.29%, O 17.63%, S 11.77%. Prepn: **FR 1491477** (1967 to Agripat); F. Paltauf *et al.*, **US 3697555** (1972 to Ciba-Geigy); A. Martvon, K. Antos, *Chem. Zvesti* **23**, 181 (1969). Activity in dogs: M. A. Gemmell, G. Oudemans, *Res. Vet. Sci.* **19**, 217 (1975).

$$S=C=N\text{---}\diagdown\diagdown\diagup\text{---O---}\diagdown\diagdown\diagup\text{---}NO_2$$

Crystals, mp 107-113° (Martvon, Antos); also reported as 124-125° (Paltauf). Insol in water. Sol in organic solvents.
THERAP CAT (VET): Anthelmintic.

6720. 3-Nitrosobenzamide. [144189-66-2] NOBA. $C_7H_6N_2O_2$; mol wt 150.14. C 56.00%, H 4.03%, N 18.66%, O 21.31%. Antiretroviral zinc ejecting agent. Prepn: E. Kun *et al.*, **WO 9108902** (1991 to Octamer). Tumoricidal activity: W. G. Rice *et al.*, *Proc. Natl. Acad. Sci.* **89**, 7703 (1992). Inhibition of HIV-1 infectivity: W. G. Rice *et al.*, *Nature* **361**, 473 (1993); and site of action: *eidem, Proc. Natl. Acad. Sci. USA* **90**, 9721 (1993); of SIV: A. J. Chuang *et al.*, *FEBS Lett.* **326**, 140 (1993).

$$\text{structure: benzene ring with } ON \text{ and } C(=O)NH_2$$

Light yellow solid. Darkens above 135°, softens with polymerization 150-160°; mp 240-250° (dec). In solution the compound is dark green. uv/vis max (absol. ethanol): 218, 304, 750nm (ε 1.5 × 10^4, 5.35 × 10^3, 37.6).

6721. Nitrosobenzene. [586-96-9] C_6H_5NO; mol wt 107.11. C 67.28%, H 4.71%, N 13.08%, O 14.94%. Versatile electrophilic reagent in organic synthesis; involved in a variety of reactions, including additions, oxidations, and reductions. Prepn: A. Baeyer, *Ber.* **7**, 1638 (1874); O. Schmidt, *ibid.* **36**, 2459 (1903); G. H. Coleman *et al.*, *Org. Synth.* **coll. vol. III**, 668 (1955). *Reviews:* P. Zuman, B. Shah, *Chem. Rev.* **94**, 1621-1641 (1994); H. Yamamoto, N.

Momiyama, *Chem. Commun.* **2005**, 3514-3525; F. L. Silva da Machado, *Synlett* **2008**, 3075-3076.

Light yellow solid. *Poisonous.* mp 64-67°. bp_{21} 60-61°; bp_{18} 57-59°. Sol in most organic solvents. Insol in water.

USE: Reagent in synthetic organic chemistry.

6722. *N*-Nitrosodiethanolamine. [1116-54-7] 2,2′-(Nitrosoimino)bisethanol; diethanolnitrosamine; 2,2′-nitrosiminodiethanol; di-(2-hydroxyethyl)nitrosamine; NDELA. $C_4H_{10}N_2O_3$; mol wt 134.14. C 35.82%, H 7.51%, N 20.88%, O 35.78%. Formed by the action of nitrites on di- or triethanolamine. Prepn: E. R. H. Jones, W. Wilson, *J. Chem. Soc.* **1949**, 547; R. Preussmann, *Ber.* **95**, 1571 (1962); W. Lijinsky *et al.*, *J. Natl. Cancer Inst.* **49**, 1239 (1972). Carcinogenicity study: H. Druckery *et al.*, *Z. Krebsforsch.* **69**, 103 (1967). Impurity in cutting fluids: T. Y. Fan *et al.*, *Science* **196**, 70 (1977); in cosmetics: *eidem, Food Cosmet. Toxicol.* **15**, 423 (1977). Mutagenicity study: A. Hesbert *et al.*, *Mutat. Res.* **68**, 207 (1979).

Light yellow oil, $bp_{0.01}$ 125°. n_D^{20} 1.4849.
Caution: This substance is reasonably anticipated to be a human carcinogen: *Report on Carcinogens, Twelfth Edition* (PB2011-111646, 2011) p 304.

6723. *N*-Nitrosodiethylamine. [55-18-5] *N*-Ethyl-*N*-nitrosoethanamine; diethylnitrosamine; DEN; DENA; NDEA. $C_4H_{10}N_2O$; mol wt 102.14. C 47.04%, H 9.87%, N 27.43%, O 15.66%. Detected in trace amounts in tobacco smoke: Druckney, Preussman, *Naturwissenschaften* **49**, 498 (1962); and in various processed foods: Hedler, Marquardt, *Food Cosmet. Toxicol.* **6**, 341 (1968); Friemuth, Glaeser, *Nahrung* **14**, 357 (1970). Formed by the interaction of nitrite with diethylamine and by the action of nitrate-reducing bacteria. Industrial prepn: Reilly, **DE 1085166** (1960 to du Pont), *C.A.* **56**, 4594h (1962); Levering, Maury, **US 3090786** (1963 to Hercules Powder); Minisci, Galli, *Chim. Ind. (Milan)* **46**, 173 (1964). Hepatotoxicity and carcinogenicity studies: Schmaehl *et al.*, *Naturwissenschaften* **54**, 341 (1967); Grover, Fischer, *Eur. J. Cancer* **7**, 77 (1971); Bader *et al.*, *Arch. Geschwulstforsch.* **37**, 327 (1971). General review: Magee, Barnes, *Adv. Cancer Res.* **10**, 163-246 (1967).

Slightly yellow liq. d_4^{20} 0.9422. bp 175-177°. bp_5 47°. n_D^{20} 1.4388. Sol in water, alc, ether.
Caution: This substance is reasonably anticipated to be a human carcinogen: *Report on Carcinogens, Twelfth Edition* (PB2011-111646, 2011) p 306.

USE: Gasoline and lubricant additive; antioxidant; stabilizer in plastics.

6724. *N*-Nitrosodimethylamine. [62-75-9] *N*-Methyl-*N*-nitrosomethanamine; dimethylnitrosamine; DMN; DMNA. $C_2H_6N_2O$; mol wt 74.08. C 32.43%, H 8.16%, N 37.82%, O 21.60%. Reportedly found in trace amounts in tobacco smoke condensates: Rhoades, Johnson, *Nature* **236**, 307 (1972); in cured meat products, notably bacon: Sen *et al.*, *ibid.* **241**, 473 (1973); in smoked and salted fish: Fazio *et al.*, *J. Agric. Food Chem.* **19**, 250 (1971); Fong, Chan, *Nature* **243**, 421 (1973). Formed by the interaction of nitrite with dimethylamine and by the action of nitrate-reducing bacteria. Industrial prepn: **GB 772331** (1957 to Olin Mathieson), *C.A.* **51**, 14783a (1957); Ioffe, *Zh. Obshch. Khim.* **28**, 1296 (1958); Norris, *J. Am. Chem. Soc.* **81**, 3346 (1959); Campbell, **US 2981752** (1961 to C.S.C.); Datin, Elliott, **US 3136821** (1964 to Allied Chem.). Chem-

istry: Layne *et al.*, *J. Am. Chem. Soc.* **85**, 435, 1816 (1963). Metabolism and toxicity: Magee, Vandekar; Magee, *Biochem. J.* **70**, 600, 606 (1958); Heath, *ibid.* **85**, 72 (1962). Carcinogenicity studies: Magee, Barnes, *Br. J. Cancer* **10**, 114 (1956); *see also eidem, Adv. Cancer Res.* **10**, 163-246 (1967). Review of toxicology and human exposure: *Toxicological Profile for N-Nitrosodimethylamine* (PB90-182130, 1989) 132 pp.

Yellow liquid. bp 151-153°. bp_{40} 67.1°. d_4^{20} 1.0048. n_D^{20} 1.4368. Very sol in water, alcohol, ether. LD_{50} i.p. in rats: 34 mg/kg (Heath).
Caution: Potential symptoms of overexposure are nausea, vomiting, diarrhea and abdominal cramps; headache; fever; enlarged liver, jaundice; reduced function of liver, kidneys and lungs. *See NIOSH Pocket Guide to Chemical Hazards* (DHHS/NIOSH 97-140, 1997) p 232. This substance is reasonably anticipated to be a human carcinogen: *Report on Carcinogens, Twelfth Edition* (PB2011-111646, 2011) p 308.

USE: Formerly in the prodn of rocket fuels; antioxidant; additive for lubricants; softener of copolymers.

6725. *p*-Nitroso-*N,N*-dimethylaniline. [138-89-6] *N,N*-Dimethyl-4-nitrosobenzenamine; Accelerine. $C_8H_{10}N_2O$; mol wt 150.18. C 63.98%, H 6.71%, N 18.65%, O 10.65%. Prepd in the cold from $NaNO_2$ and a soln of dimethylaniline in HCl.

Green plates or leaflets. mp 92.5-93.5°; also stated as 87-88°. Volatile with steam. Insol in water. Sol in alcohol, ether.

USE: Manuf organic compds; accelerator in vulcanizing; in printing fabrics.

6726. 4-Nitrosodiphenylamine. [156-10-5] 4-Nitroso-*N*-phenylbenzenamine; 4-nitroso-*N*-phenylaniline; *p*-phenylaminonitrosobenzene; *N*-phenyl-*p*-nitrosoaniline; $C_{12}H_{10}N_2O$; mol wt 198.23. C 72.71%, H 5.09%, N 14.13%, O 8.07%. Review of toxicology and human exposure: *Toxicological Profile for N-Nitrosodiphenylamine* (PB93-182509, 1993) 114 pp.

Green plates with bluish luster (from benzene) or steel-blue prisms or plates (from ether + water). mp 144-145°. Slightly sol in water or petr ether; freely sol in alcohol, ether, chloroform, benzene. Dissolves in H_2SO_4 with a red color, which suddenly changes to violet on warming.
Note: This substance was formerly anticipated to be a carcinogen: *Fifth Annual Report on Carcinogens* (NTP 89-239, 1989) p 212; delisted for insufficient evidence of carcinogenicity: *Sixth Annual Report on Carcinogens* (PB92-120666, 1991) p 13.

USE: Has been used as retardant in vulcanizing rubber.

6727. *N*-Nitrosomorpholine. [59-89-2] 4-Nitrosomorpholine; NMOR. $C_4H_8N_2O_2$; mol wt 116.12. C 41.37%, H 6.94%, N 24.13%, O 27.56%. Prepn: L. Knorr, *Ann.* **301**, 1 (1898); F. Chapman, *J. Chem. Soc.* **1949**, 1631; G. Oláh *et al.*, *Ber.* **89**, 2374 (1956). Metabolism: B. W. Stewart, P. N. Magee, *Biochem. J.* **126**, 21P (1972). Carcinogenicity studies: H. Druckrey *et al.*, *Naturwissenschaften* **48**, 134 (1961); P. Bannasch, H.-A. Müller, *Arzneim.-Forsch.* **14**, 805 (1964); *IARC Monographs* **17**, 263 (1978).

Yellow crystals, mp 29°. bp$_{25}$ 139-140°; bp$_{747}$ 224-224.5°. Sol in water. LD$_{50}$ orally in rats: 282 mg/kg (Druckrey).

Caution: This substance is reasonably anticipated to be a human carcinogen: *Report on Carcinogens, Twelfth Edition* (PB2011-111646, 2011) p 319.

6728. 1-Nitroso-2-naphthol. [131-91-9] 1-Nitroso-2-naphthalenol; nitroso-β-naphthol. C$_{10}$H$_7$NO$_2$; mol wt 173.17. C 69.36%, H 4.07%, N 8.09%, O 18.48%. Prepared by the addition of H$_2$SO$_4$ to a mixture of β-naphthol dissolved in aq NaOH and NaNO$_2$: C. S. Marvel, P. K. Porter, *Org. Synth.* **coll. vol. I,** 411 (2nd ed., 1941).

Yellowish-brown needles from petr ether, mp 109-110°. Sol in 1000 parts water, 35 parts alcohol; also sol in hot alcohol, benzene, ether, carbon disulfide, caustic alkali solns, glacial acetic acid; slightly sol in cold petr ether.

USE: To prevent gum formation in gasoline; in analytical chemistry in the determination of cobalt (to separate it from nickel).

6729. 4-Nitrosophenol. [104-91-6] Quinone oxime; quinone monoxime. C$_6$H$_5$NO$_2$; mol wt 123.11. C 58.54%, H 4.09%, N 11.38%, O 25.99%.

Pale yellow orthorhombic needles, browns at 126° (dec 144°). *Explodes on contact with concentrated acid, alkali, or fire.* Ka at 25° = 3.3 × 10^{-7}. Moderately sol in water. Sol in dil alkalies giving green to brownish green solns. Sol in alc, ether, acetone.

6730. *N*-Nitrosopyrrolidine. [930-55-2] 1-Nitrosopyrrolidine; NPYR; NO-PYR. C$_4$H$_8$N$_2$O; mol wt 100.12. C 47.99%, H 8.05%, N 27.98%, O 15.98%. Cyclic nitrosamine that occurs in food products and tobacco smoke. Prepn: F. C. Peterson, *Ber.* **21,** 290 (1888). Carcinogenicity studies: H. Druckrey *et al.*, *Z. Krebsforsch.* **69,** 103 (1967); M. Greenblatt *et al.*, *J. Natl. Cancer Inst.* **48,** 1687 (1972); *IARC Monographs* **17,** 313 (1978). Mechanism of mutagenicity: L. I. Hecker *et al.*, *Mutat. Res.* **62,** 213 (1979). Metabolism: R. C. Cottrell *et al.*, *Adv. Exp. Med. Biol.* **136B,** 1165 (1982).

Yellow liquid, bp$_{20}$ 104-106°. Sol in water. LD$_{50}$ orally in rats: 900 mg/kg (Druckrey).

Caution: This substance is reasonably anticipated to be a human carcinogen: *Report on Carcinogens, Twelfth Edition* (PB2011-111646, 2011) p 323.

6731. Nitroso-R Salt. [525-05-3] 3-Hydroxy-4-nitroso-2,7-naphthalenedisulfonic acid sodium salt (1:2); sodium salt of 1-nitroso-2-hydroxynaphthalene-3,6-disulfonic acid. C$_{10}$H$_5$NNa$_2$S$_2$; mol wt 377.25. C 31.84%, H 1.34%, N 3.71%, Na 12.19%, O 33.93%, S 17.00%.

Golden-yellow, fan-shaped crystals. Sol in ~40 parts water; more sol in hot water; slightly sol in methyl or ethyl alcohol.

USE: As a reagent for cobalt and potassium.

6732. 2-Nitro-4-sulfobenzoic Acid. [552-23-8] C$_7$H$_5$NO$_7$S; mol wt 247.18. C 34.01%, H 2.04%, N 5.67%, O 45.31%, S 12.97%. Prepn by sulfonation of *o*-nitrotoluene and oxidation of the resulting 2-nitro-4-toluenesulfonic acid with potassium permanganate: Hart, *Am. Chem. J.* **1,** 352 (1879-80).

Needles from hydrochloric acid. Stable in air under ordinary conditions.

USE: Alkalimetric standard.

6733. Nitrosyl Chloride. [2696-92-6] Nitrosyl chloride ((NO)Cl). ClNO; mol wt 65.46. Cl 54.16%, N 21.40%, O 24.44%. NOCl. Best prepared from nitrosylsulfuric acid and dry HCl: Coleman *et al.*, *Inorg. Synth.* **1,** 55 (1939).

Non-explosive, reddish-yellow gas; liquid at −5.5°; solid at −61.5°. *Poisonous; corrosive.* Decomposed by water. Sol in fuming H$_2$SO$_4$. Critical temp 167°; crit press. 92.4 atm. The orange color of aqua regia is produced by nitrosyl chloride.

Caution: Intensely irritating to eyes, skin, mucous membranes. Inhalation may cause pulmonary edema, hemorrhage.

6734. Nitrosyl Fluoride. [7789-25-5] Nitrogen oxyfluoride. FNO; mol wt 49.00. F 38.77%, N 28.59%, O 32.65%. Preparation: Ruff *et al.*, *Z. Anorg. Allg. Chem.* **208,** 293 (1932); Balz, Mailänder, *ibid.* **217,** 166 (1934); Faloon, Kenna, *J. Am. Chem. Soc.* **73,** 2937 (1951); Kwasnik in *Handbook of Preparative Inorganic Chemistry* **vol. 1,** G. Brauer, Ed. (Academic Press, New York, 2nd ed., 1963) pp 184-185. *Reviews:* Hoffman, Neville, *Chem. Rev.* **62,** 1-18 (1962); Kemmitt, Sharp, *Adv. Fluorine Chem.* **4,** 194-195 (1965); Woolf, *ibid.* **5,** 1-30 (1965); Schmutzler, *Angew. Chem. Int. Ed.* **7,** 440-455 (1968).

Colorless gas. Often bluish because of impurities. Vigorous reaction with glass, corroding action on quartz. May be kept in quartz ampuls if cooled in liquid oxygen. mp −132.5°. bp −59.9°. d (liq at bp) 1.326. d (solid) 1.719. Trouton const. 21.3. Reacts with water to form NO, HNO$_3$ and HF.

Caution: Highly irritating to skin, eyes, mucous membranes. *See also* Fluorine.

USE: Oxidizer in rocket propellants; stabilizing agent for liquid SO$_3$; fluorinating agent.

6735. Nitrosylsulfuric Acid. [7782-78-7] Sulfuric acid anhydride with nitrous acid (1:1); nitrosyl sulfate; chamber crystals; nitrososulfuric acid; nitroxylsulfuric acid; nitrosulfonic acid; nitrosyl hydrogen sulfate; nitro acid sulfite; Nitrose. HNO$_5$S; mol wt 127.07. H 0.79%, N 11.02%, O 62.95%, S 25.23%. Formed as an intermediate in the lead chamber process for sulfuric acid by the reaction of sulfur dioxide, nitrogen trioxide, oxygen, and water: Clément, Désormes, *Ann. Chim. Phys.* [1] **59,** 329 (1806); Lunge, *J. Chem. Soc.* **47,** 470 (1885). Prepd from sulfur trioxide, nitrogen oxides and water: Döbereiner, *Schweigger's Journ.* **8,** 239 (1812); de Claubry, *Ann. Chim. Phys.* [2] **45,** 284 (1832); Kuhlmann, *ibid.* [3] **1,** 116 (1843); from silver acid sulfate and nitrosyl bromide: Berl *et al.*, *Z. Anorg. Allg. Chem.* **209,** 264 (1932). *See also:* US 1909557; US 1909558. The formation of crystals of nitrosylsulfuric acid may be observed by igniting a mixture of 1 part sulfur and 2 or 3 parts potassium nitrate under a bell jar.

Prisms, dec 73.5°. In moist air the crystals dec with the formation of sulfuric and nitric acids and above 50° nitric oxide and nitrogen dioxide are evolved. *Corrosive.* Sol in sulfuric acid, dec in water.

USE: For bleaching cereal milling products.

6736. Nitrosyl Tetrafluoroborate. [14635-75-7] Nitrosyl tetrafluoroborate(1−) (1:1); nitrosonium tetrafluoroborate; nitrosyl borofluoride; nitrosyl fluoborate. BF_4NO; mol wt 116.81. B 9.25%, F 65.06%, N 11.99%, O 13.70%. $NOBF_4$. Prepd according to the equation: $2\ HBF_4 + N_2O_3 \rightarrow 2\ NOBF_4 + H_2O$: Wilke-Dörfurt, Balz, *Z. Anorg. Allg. Chem.* **159**, 219 (1927); Balz, Mailänder, *ibid.* **217**, 162 (1934); H. S. Booth, D. R. Martin, *Boron Trifluoride and Its Derivatives* (New York, 1949) p 133 sqq. Review of tetrafluoroborates: Sharp, *Adv. Fluorine Chem.* **1**, 68-128 (1960).

Birefringent, orthorhombic, hygroscopic platelets. d_4^{25} 2.185. Sublimes at 0.01 mm and 250° without decompn. Decomposed by water. May be stored in glass bottles if absolutely dry.

USE: In the prepn of diazonium fluoborates.

6737. Nitrotoluene. Methylnitrobenzene. $C_7H_7NO_2$; mol wt 137.14. C 61.31%, H 5.15%, N 10.21%, O 23.33%. Nitration of toluol by a mixture of HNO_3 and H_2SO_4 yields principally *o*- and *p*-nitrotoluol. Prepn of *m*-nitrotoluene from 3-nitro-4-amino-toluene and $NaNO_2$: Clark, Taylor, *Org. Synth.* **3**, 91 (1923).

m-Nitrotoluene. [99-08-1] 3-Nitrotoluene. Liquid. d_4^{15} 1.1630; d_4^{20} 1.1581; d_4^{59} 1.124; d_4^{121} 1.063. Solidifies in an ice and salt cooling mixture; mp 15°. bp_{760} 232°; bp_{100} 156.9°; bp_{40} 130.7°; bp_{20} 112.8°; bp_{10} 96.0°; bp_5 81.0°; $bp_{1.0}$ 50.2°. n_D^{30} 1.5426. Absorption spectrum: Marchlewski, Mayer, *Bull. Acad. Polon.* [A] **1929**, 188. Miscible with alcohol and ether. Sol in benzene. Soly in water at 30° (mg/l): 498. Log P (octanol/water): 2.40.

o-Nitrotoluene. [88-72-2] 2-Nitrotoluene. Technical report: *NTP Technical Report on the Toxicology and Carcinogenesis Studies of o-Nitrotoluene* (NTP TR-504/NIH 02-4438, 2002) 368 pp. Yellowish liquid at ordinary temp. d_{15}^{19} 1.1622. mp −9.3°. bp 220.4°. n_D^{20} 1.5472. Soly in water at 30° (mg/l): 652. Sol in alc, benzene, petr ether. Log P (octanol/water): 2.30.

p-Nitrotoluene. [99-99-0] 4-Nitrotoluene. Technical report: *NTP Technical Report on the Toxicology and Carcinogenesis Studies of p-Nitrotoluene* (NTP TR-498/NIH 02-4432, 2002) 277 pp. Yellowish crystals. d 1.286. mp 51.7°. bp 238.3°. Flash pt 106°C. Soly in water at 30° (mg/l): 442. Sol in alcohol, benzene, ether, chloroform, acetone. Log P (octanol/water): 2.37.

Caution: Potential symptoms of overexposure are anoxia, cyanosis; headache, lassitude, dizziness; ataxia; dyspnea; tachycardia; nausea, vomiting. *See NIOSH Pocket Guide to Chemical Hazards* (DHHS/NIOSH 97-140, 2003) p 232. *o*-Nitrotoluene is listed as reasonably anticipated to be a human carcinogen: *Report on Carcinogens, Twelfth Edition* (PB2011-111646, 2011) p 331.

USE: Manuf of dyes, toluidines, nitrobenzoic acids, agricultural and rubber chemicals.

6738. Nitrourea. [556-89-8] *N*-Nitrourea; *N*-nitrocarbamide. $CH_3N_3O_3$; mol wt 105.05. C 11.43%, H 2.88%, N 40.00%, O 45.69%. Prepd by the action of concd sulfuric acid upon urea nitrate: J. Thiele, A. Lachman, *Ann.* **288**, 267 (1895); A. W. Ingersoll, B. F. Armendt, *Org. Synth.* **5**, 85 (1925). By dropwise addition of HCl to a cooled mixture of silver cyanate and nitramide in water: T. L. Davis, K. C. Blanchard, *J. Am. Chem. Soc.* **51**, 1790 (1929).

Platelets from alcohol + petr ether dec 158.4-158.8°. pK (20°) 2.15. Absorption spectrum: E. C. C. Baly, C. H. Desch, *J. Chem. Soc.* **93**, 1747 (1908). *Explosive.* Soluble in hot water, but water solns are unstable. Decompn in aq alkaline solns is almost instantaneous. Freely sol in acetone, alcohol, acetic acid. Sparingly sol in petr ether, chloroform, benzene. Stable to oxidizing agents. Can be detonated, but is not sensitive to percussion or heating.

6739. Nitrous Acid. [7782-77-6] HNO_2; mol wt 47.01. H 2.14%, N 29.80%, O 68.07%. Formed by the action of strong acids on inorganic nitrites. *Review:* Block, "Nitrous Acid, Hyponitrous Acid and their Salts" in *Mellor's* Vol. VIII, supplement II, *Nitrogen* (part 2) 353-408 (1967).

Known only in soln (pale blue in color). Weak acid. pK (25°) 3.35. In water it changes quickly into nitric oxide and nitric acid. Forms stable, water-sol nitrites with Li, Na, K, Ca, Sr, Ba, Ag. Does not form salts with weak polyvalent cations like Al or Be. Forms stable esters with alcohols.

6740. Nitrous Oxide. [10024-97-2] Nitrogen oxide (N_2O); dinitrogen monoxide; laughing gas; hyponitrous acid anhydride; factitious air. N_2O; mol wt 44.01. N 63.65%, O 36.35%. Constituent of the earth's atm, about 0.00005% by volume: Slobod, Krogh, *J. Am. Chem. Soc.* **72**, 1175 (1950). Prepd by thermal decompn of ammonium nitrate: E. H. Archibald, *The Preparation of Pure Inorganic Substances* (Wiley, New York, 1932) p 246; Castner, Kirst, US 2111276 (1938 to du Pont). Preparation and purification: Schenk in *Handbook of Preparative Inorganic Chemistry* vol. 1, G. Brauer, Ed. (Academic Press, New York, 2nd ed., 1963) pp 484-485. The chief impurity of the commercial product is N_2, although NO_2, N, O_2, and CO_2 may also be present. Teratogenicity study: G. A. Lane *et al.*, *Science* **210**, 899 (1980). Reviews: Beattie, "Nitrous Oxide" in *Mellor's* Vol. VIII, suppl II, *Nitrogen* (part 2) 189-215 (1967); Jones in *Comprehensive Inorganic Chemistry* vol. 2, J. C. Bailar, Jr. *et al.*, Eds. (Pergamon Press, Oxford, 1973) pp 316-323.

Colorless gas; non-flammable. *Oxidizer; asphyxiant.* Slightly sweetish odor and taste. Supports combustion. Very stable and rather inert chemically at room temperatures. Dissociation begins above 300° when the gas becomes a strong oxidizing agent. mp −90.81°; bp_{760} −88.46°; Trouton constant 21.4: Hoge, *J. Res. Natl. Bur. Stand.* **34**, 281 (1945). Dipole moment 0.166. d^{-89} (liq) 1.226; d(S.T.P.) 1.967; d(gas) 1.53 (air = 1). Critical temp 36.5°; crit press. 71.7 atm. Heat of vaporization (bp): 3.956 kcal/mole. While in the steel cylinder nitrous oxide is compressed to the form of gas over liquid and has a pressure of ~800 lbs/sq. in. at room temp. At 20° and 2 atm one liter of the gas dissolves in 1.5 liters of water. Freely sol in sulfuric acid. Sol in alcohol, ether, oils.

Caution: Potential symptoms of overexposure are dyspnea; drowsiness, headache; asphyxia; reproductive effects; direct contact with liquid may cause frostbite. *See NIOSH Pocket Guide to Chemical Hazards* (DHHS/NIOSH 97-140, 1997) p 234.

USE: To oxidize organic compds at temps >300°; to make nitrites from alkali metals at their boiling points; in rocket fuel formulations (with carbon disulfide); in the prepn of whipped cream.

THERAP CAT: Anesthetic (inhalation); analgesic.

6741. Nitrovin. [804-36-4] 2-[3-(5-Nitro-2-furanyl)-1-[2-(5-nitro-2-furanyl)ethenyl]-2-propen-1-ylidene]hydrazinecarboximidamide; [[3-(5-nitro-2-furyl)-1-[2-(5-nitro-2-furyl)vinyl]allylidene]amino]guanidine; *sym*-bis(5-nitro-2-furfurylidene)acetone guanylhydrazone; 1,5-bis(5-nitro-2-furyl)-3-pentadienone guanylhydrazone; 1,5-bis(5-nitro-2-furyl)-3-pentadienone amidinohydrazone; Panazon; Payzone. $C_{14}H_{12}N_6O_6$; mol wt 360.29. C 46.67%, H 3.36%, N 23.33%, O 26.64%. Prepn: Uota *et al.*, JP 52 2673 (1952 to Toyama), *C.A.* **48**, 2115h (1954); Uoda, Tanizaki, JP 64 4479 (1964 to Fukuju Pharm.), *C.A.* **62**, 10412f (1965).

Blackish violet crystals from ethyl alcohol, mp 217° (dec).
Hydrochloride. mp 280° (dec).

THERAP CAT (VET): Growth promoter; antibacterial.

6742. Nitroxoline. [4008-48-4] 5-Nitro-8-quinolinol; 5-nitro-8-hydroxyquinoline; Enterocol; Nibiol; Noxibiol; Uritrol; Urocoli. $C_9H_6N_2O_3$; mol wt 190.16. C 56.85%, H 3.18%, N 14.73%, O 25.24%. Prepn: Kostanecki, *Ber.* **24**, 154 (1891); Petrow, Sturgeon, *J. Chem. Soc.* **1954**, 570; Pratt, Duke, *J. Am. Chem. Soc.* **82**, 1155 (1960). *In vitro* antibacterial and antifungal activity: A. Desvignes,

Consult the Name Index before using this section.

P. Leguen, *Ann. Pharm. Fr.* **21**, 803 (1963); M. Medic-Saric *et al.*, *Chemotherapy* **26**, 263 (1980). Toxicological study: O. Angelova *et al.*, *Adv. Antimicrob. Antineoplast. Chemother., Proc. 7th Int. Congr. Chemother.* **1**, 507 (1972). Clinical pharmacokinetics: A. Mrhar *et al.*, *Int. J. Clin. Pharmacol. Biopharm.* **17**, 476 (1979). HPLC determn in plasma and urine: R. H. A. Sorel *et al.*, *J. Chromatogr.* **222**, 241 (1981). Clinical evaluation in urinary tract infections: M. R. Jacobs *et al.*, *S. Afr. Med. J.* **54**, 959 (1978); B. Cancet, A. Amgar, *Pathol. Biol.* **35**, 879 (1987).

Yellow needles from alcohol or acetic acid, mp 179.5-181.5°. Freely sol in alkali and hot HCl; sparingly sol in alcohol, ether.

Hydrochloride. $C_9H_7ClN_2O_3$. Yellow needles from alcohol, mp 258°.

THERAP CAT: Antibacterial.

6743. Nitroxynil. [1689-89-0] 4-Hydroxy-3-iodo-5-nitrobenzonitrile; Dovenix. $C_7H_3IN_2O_3$; mol wt 290.02. C 28.99%, H 1.04%, I 43.76%, N 9.66%, O 16.55%. Prepn: **NL 6516359** corresp to Collins *et al.*, **US 3331738** (1966, 1967 both to May & Baker).

Yellow crystals from benzene, mp 137-138°. Sparingly sol in water; moderately sol in most organic solvents.

D-N-Methylglucamine salt. [36508-79-9] Nitroxynil meglumine; 4-hydroxy-3-iodo-5-nitrobenzonitrile compd with D-1-deoxy-1-(methylamino)glucitol (1:1). $C_{14}H_{20}IN_3O_8$. Solid, mp 85-90°.

N-Ethylglucamine. [27917-82-4] Nitroxynil eglumine; 4-hydroxy-3-iodo-5-nitrobenzonitrile compd with 1-deoxy-1-(ethylamino)glucitol (1:1); Trodax. $C_{15}H_{22}IN_3O_8$; mol wt 499.26. Readily sol in water with a yellow, odorless and substantially neutral soln. Aq soln is very stable but contamination with calcium and certain other salts can result in pptn of an insol salt of nitroxynil.

THERAP CAT (VET): Anthelmintic (fasciolicide).

6744. Nitryl Chloride. [13444-90-1] Nitryl chloride ((NO₂)-Cl); nitroxyl chloride. $ClNO_2$; mol wt 81.46. Cl 43.52%, N 17.19%, O 39.28%. NO₂Cl. Conveniently prepd by the addn of chlorosulfonic acid to nitric acid: Dachlauer, **DE 509405** (1929 to I. G. Farben); Kaplan, Schechter, *Inorg. Synth.* **4**, 52 (1953); Collis *et al.*, *J. Chem. Soc.* **1958**, 438.

Corrosive, toxic, colorless gas. Chlorine-like odor. Vapor density (100°): 2.81 g/l. Dec >120°. bp −14.3°. mp −145°. d_{liq}^0 1.37; d_{liq}^{16} 1.33. Even the purest liquid may have a pale yellow color. Solns in polar solvents are always yellow. The gas or liquid may attack organic matter with explosive violence.

Caution: Strong irritant, corrosive.

USE: Nitrating and chlorinating agent in organic syntheses.

6745. Nitryl Fluoride. [10022-50-1] Nitryl fluoride ((NO₂)-F). FNO₂; mol wt 65.00. F 29.23%, N 21.55%, O 49.23%. Credit for original prepn by the spontaneous combustion of nitric oxide in an atm of fluorine is given to Moissan, Lebeau, *Compt. Rend.* **140**, 1573, 1621 (1905); more easily prepd by mixing nitrogen dioxide and fluorine: Ruff *et al.*, *Z. Anorg. Allg. Chem.* **208**, 298 (1932); Faloon, Kenna, *J. Am. Chem. Soc.* **73**, 2937 (1951). Reviews of prepn and chemistry: Hoffman, Neville, *Chem. Rev.* **62**, 1-18 (1962); Kwasnik in *Handbook of Preparative Inorganic Chemistry* **vol. 1**, G. Brauer, Ed. (Academic Press, New York, 2nd ed., 1963) pp 186-187; Kemmitt, Sharp, *Adv. Fluorine Chem.* **4**, 195-196

(1965); Woolf, *ibid.* **5**, 1-30 (1965); Schmutzler, *Angew. Chem. Int. Ed.* **7**, 440-455 (1968).

Colorless gas. Pungent odor. Attacks mucous membranes. mp −166.0°. bp −72.4°. d (liq at bp) 1.796. d (solid) 1.924. Trouton const 21.2. May be stored in quartz ampuls if cooled in liquid oxygen. Purification can be accomplished by fractional distillation at reduced press. in dry glass or quartz apparatus. Rapidly hydrolyzed in water to form nitric and hydrofluoric acids. Powerful oxidizing agent, with fluorinating powers; slightly weaker than fluorine. Absorbs mercury completely. Spontaneously ignites iodine, selenium, phosphorus (red and white), arsenic, antimony, boron, silicon, thorium, molybdenum. On mild warming attacks lead, bismuth, chromium, manganese, iron, nickel, tungsten, sulfur, charcoal. Does not react readily with hydrogen in the cold. Converts ethanol to ethyl nitrate; benzene to nitrobenzene.

Caution: See Fluorine.

USE: Oxidizer in rocket propellants; fluorinating agent.

6746. Nivalenol. [23282-20-4] $(3\alpha,4\beta,7\alpha)$-12,13-Epoxy-3,-4,7,15-tetrahydroxytrichothec-9-en-8-one; $3\alpha,4\beta,7\alpha,15$-tetrahydroxyscirp-9-en-8-one. $C_{15}H_{20}O_7$; mol wt 312.32. C 57.69%, H 6.45%, O 35.86%. **Trichothecene mycotoxin** isolated from *Fusarium nivale*: T. Tatsuno *et al.*, *Chem. Pharm. Bull.* **16**, 2519 (1968). Structure: *eidem, Tetrahedron Lett.* **1969**, 2823. Toxicology: T. Tatsuno, *Cancer Res.* **28**, 2393 (1968). Implicated as a chemical warfare agent in Southeast Asia with T-2 toxin, *q.v.*: N. Wade, *Science* **214**, 34 (1981); R. T. Rosen, J. D. Rosen, *Biomed. Mass Spectrom.* **9**, 443 (1982).

Crystals, mp 222-223° (dec). $[\alpha]_D^{24}$ +21.54° (c = 1.3 in ethanol). uv max (methanol): 218 nm (ε 6300). Slightly sol in water; sol in polar organic solvents. LD_{50} i.p. in mice: 40 μg/10 g (Tatsuno).

Caution: Potential symptoms of overexposure include fever, nausea, vomiting, diarrhea, leukopenia, bleeding, sepsis; necrotic lesions of skin and mucosa. *See* M. J. Ellenhorn, D. G. Barceloux in *Medical Toxicology: Diagnosis and Treatment of Human Poisoning* (Elsevier, New York, 1988) pp. 1312-1314.

6747. Nizatidine. [76963-41-2] N'-[2-[[[2-[(Dimethylamino)methyl]-4-thiazolyl]methyl]thio]ethyl]-N-methyl-2-nitro-1,1-ethenediamine; N-[4-(6-methylamino-7-nitro-2-thia-5-aza-6-heptene-1-yl)-2-thiazolylmethyl]-N,N-dimethylamine; LY-139037; ZE-101; ZL-101; Axid; Calmaxid; Cronizat; Distaxid; Gastrax; Nizax; Nizaxid. $C_{12}H_{21}N_5O_2S_2$; mol wt 331.45. C 43.49%, H 6.39%, N 21.13%, O 9.65%, S 19.35%. Histamine H₂-receptor antagonist related to ranitidine, *q.v.* Prepn: R. P. Pioch, **EP 49618**; *idem*, **US 4375547** (1982, 1983 both to Eli Lilly). Crystal structure: G. A. Stephenson *et al.*, *J. Mol. Struct.* **380**, 93 (1996). General pharmacology in animals: K. Bemis *et al.*, *Arzneim.-Forsch.* **39**, 240 (1989). Pharmacokinetics and gastric acid suppression in humans: J. T. Callaghan *et al.*, *Clin. Pharmacol. Ther.* **37**, 162 (1985). Voltammetric determn in biological fluids: A. A. Al-Majed *et al.*, *J. Pharm. Biomed. Anal.* **21**, 319 (1999). Determn in pharmaceutical prepns: F. A. El-Yazbi *et al. ibid.*, **31**, 1027 (2003). Symposium on pharmacology and clinical studies: *Scand. J. Gastroenterol.* **22**, Suppl. 136, 1-88 (1987). Comprehensive description: T. J. Wozniak, *Anal. Profiles Drug Subs.* **19**, 397-427 (1990). Review of prokinetic activity and clinical use in gastroesophageal reflux disease: E. J. Zarling, *Clin. Ther.* **21**, 2038-2046 (1999).

Crystals from ethanol-ethyl acetate, mp 130-132°. Bitter taste; mild sulfur-like odor. Crystal density: 1.324 g/cm³. uv max (methanol): 240, 325 nm (ε 8400, 19600); (water): 260, 314 nm (ε 11820, 15790). pKa$_1$ 2.1; pKa$_2$ 6.8. Partition coefficient (octanol/water): 0.3 (pH 7.4). Soly (mg/ml): chloroform >100; methanol 50.0-100.0; water 10.0-33.3; isopropanol 3.33-5.0; ethyl acetate 1.0-2.0; benzene, diethyl ether, octanol <0.5. LD$_{50}$ in mice, rats (mg/kg): 265, >300 i.v.; 1685, 1680 orally (Pioch).

THERAP CAT: Antiulcerative; gastroprokinetic.

THERAP CAT (VET): Gastroprokinetic.

6748. Nizofenone. [54533-85-6] (2-Chlorophenyl)[2-[2-[(diethylamino)methyl]-1H-imidazol-1-yl]-5-nitrophenyl]methanone; 2'-chloro-2-[2-[(diethylamino)methyl]imidazol-1-yl]-5-nitrobenzophenone; 1-[2-(2-chlorobenzoyl)-4-nitrophenyl]-2-(diethylaminomethyl)imidazole. C$_{21}$H$_{21}$ClN$_4$O$_3$; mol wt 412.87. C 61.09%, H 5.13%, Cl 8.59%, N 13.57%, O 11.63%. Imidazole derivative exhibiting protective activity against cerebral anoxia or ischemia. Prepn: M. Nakanishi *et al.*, DE 2403416; *eidem*, US 3915981 (1974, 1975 both to Yoshitomi). Antianoxic effect in animal models: H. Yasuda *et al.*, *Arch. Int. Pharmacodyn.* **233**, 136 (1978). Mechanism of action: *eidem, ibid.* **242**, 77 (1979). Multicenter clinical studies: I. Saito *et al.*, *Neurol. Res.* **5**, 29 (1983); T. Ohta *et al.*, *J. Neurosurg.* **64**, 420 (1986). Toxicity studies: H. Horizoe *et al.*, *Oyo Yakuri* **30**, 627 (1985); *C.A.* **104**, 61754m (1986); K. Okumura *et al., ibid.* 633, *C.A.* **104**, 61755 (1986).

Pale yellow crystals from isopropyl ether, mp 75-76°.

Fumarate. [54533-86-7] Midafenone; Y-9179; Ekonal. C$_{21}$H$_{21}$ClN$_4$O$_3$.C$_4$H$_4$O$_4$; mol wt 528.95. Pale yellow crystals from isopropyl ether, mp 157-158°. LD$_{50}$ in male, female mice, male, female rats (mg/kg): 495, 504, 1711, 1580 orally; 62, 70, 63, 65 i.v.; 270, 278, 1830, 1629 s.c. (Horizoe).

THERAP CAT: Nootropic.

6749. NMDA. [6384-92-5] *N*-Methyl-D-aspartic acid; *N*-methyl-D-aspartate; *N*-methyl-D-asparaginsaure (German). C$_5$H$_9$-NO$_4$; mol wt 147.13. C 40.82%, H 6.17%, N 9.52%, O 43.50%. Excitotoxic amino acid used to identify a specific subset of excitatory amino acid receptors; consequently the receptors are known as NMDA receptors. *See also* quisqualic acid, kainic acid. Prepn: O. Lutz, *Z. Physiol. Chem.* **70**, 256 (1909); O. Lutz, Br. Jirgensons, *Ber.* **63**, 448 (1930). Isoln from the blood shell, *Scapharca broughtonii:* M. Sato *et al.*, *Biochem. J.* **241**, 309 (1987). Identification as a neuroexcitant: D. R. Curtis, J. C. Watkins, *J. Physiol.* **166**, 1 (1963); of neurotoxic properties: J. W. Olney *et al.*, *Life Sci.* **25**, 537 (1979); R. Zaczek *et al.*, *Neurosci. Lett.* **24**, 181 (1981); M. V. Sofroniew, R. C. A. Pearson, *Brain Res.* **339**, 186 (1985). Receptor binding studies: C. K. Mitchell, D. A. Redburn, *Neurosci. Lett.* **28**, 241 (1982); D. T. Monaghan, C. W. Cotman, *J. Neurosci.* **5**, 2909 (1985); J. M. M. Olson *et al.*, *Neuroscience* **22**, 913 (1987). Review of NMDA and its receptor: J. C. Watkin, R. H. Evans, *Annu. Rev. Pharmacol. Toxicol.* **21**, 165-204 (1981); T. W. Stone, N. R. Burton, *Prog. Neurobiol.* **30**, 333-368 (1988).

6750. NMN. [1094-61-7] 3-(Aminocarbonyl)-1-(5-*O*-phosphono-β-D-ribofuranosyl)pyridinium inner salt; 3-carbamoyl-1-β-D-ribofuranosylpyridinium hydroxide 5'-(dihydrogen phosphate) inner

salt; nicotinamide mononucleotide. C$_{11}$H$_{15}$N$_2$O$_8$P; mol wt 334.22. C 39.53%, H 4.52%, N 8.38%, O 38.30%, P 9.27%. Prepd by incubation of NAD with potato pyrophosphatase in a non-phosphate buffer and in the presence of fluoride: Plaut, Plaut, *Arch. Biochem. Biophys.* **48**, 189 (1954); *Biochem. Prep.* **5**, 55 (1957). Enzymatic synthesis from nicotinamide by human erythrocytes and hemolyzates: Preiss, Handler, *J. Biol. Chem.* **225**, 759 (1957). Prepn by hydrolysis of NAD by a yeast enzyme: Takei, *Agric. Biol. Chem.* **34**, 23 (1970); by cleavage of NAD with crude rattlesnake venom: Apps, *FEBS Lett.* **15**, 277 (1971); R. Jeck, C. Wönckhaus, *Methods Enzymol.* **66**, 62 (1980). NMR studies and conformation of α- and β-nicotinamide nucleotide: N. J. Oppenheimer, N. O. Kaplan, *Biochemistry* **15**, 3981 (1976).

Can be precipitated from slightly acidic aq soln by a large vol of acetone. Freely sol in water, practically insol in acetone. More stable when stored frozen in soln, than when in dry form.

6751. Nobelium. [10028-14-5] No; at. no. 102; valence 2, 3. Man-made radioactive element. No stable nuclides; known isotopes (mass numbers): 250-259. Longest-lived known isotope: 259 (T$_{1/2}$ 1.00 hours, α-emitter, rel. at. mass 259.1011). Discovery of element 102 first claimed by P. R. Fields *et al.*, *Phys. Rev.* **107**, 1460 (1957); later experiments failed to confirm these results. Prepn of isotope originally assigned mass 254, later identified as ^{252}No (T$_{1/2}$ 2.3 sec, α-emitter) by ^{244}Cm (^{12}C,4n) reaction: A. Ghiorso *et al.*, *Phys. Rev. Lett.* **1**, 18 (1958). Prepn of ^{259}No by bombardment of ^{248}Cm with ^{18}O ions: Silva *et al.*, *Nucl. Phys. A* **216**, 97 (1973), *C.A.* **80**, 43040g (1974). *Reviews*: G. T. Seaborg, *J. Chem. Educ.* **36**, 38-44 (1959); A. Ghiorso, T. Sikkeland, *Phys. Today* **20**, (9) 25-32 (1967); G. N. Flerov, *U.S. Atomic Energy Commission* JINR-D7-3444 (1967) 36 pp, *C.A.* **68**, 34658q (1968); C. Keller, *The Chemistry of the Transuranium Elements* (Verlag Chemie, Weinheim, English Ed., 1971) pp 601-607; Silva, "Trans-Curium Elements" in *MTP Int. Rev. Sci.: Inorg. Chem., Ser. One* vol. 8, A. G. Maddock, Ed. (University Park Press, Baltimore, 1972) pp 71-105; *Comprehensive Inorganic Chemistry* vol. 5, J. C. Bailar, Jr., *et al.*, Eds. (Pergamon Press, Oxford, 1973) *passim;* Ghiorso, *Handb. Exp. Pharmakol.* **36**, 691-715 (1973); Taylor, *ibid.* 717-738 (1973); R. J. Silva in *The Chemistry of the Actinide Elements* vol. 2, J. J. Katz *et al.*, Eds. (Chapman and Hall, New York, 1986) pp 1096-1099; *The Elements Beyond Uranium*, G. T. Seaborg, W. D. Loveland, Eds. (John Wiley & Sons, Inc., New York, 1990) pp 46-49.

Caution: Radiation hazard; handling requires special equipment and shielding facilities (Katz *et al.*, *loc. cit.* vol. 2, p. 1128).

6752. Nocardamin. [26605-16-3] 1,12,23-Trihydroxy-1,6,-12,17,23,28-hexaazacyclotritriacontane-2,5,13,16,24,27-hexone. C$_{27}$H$_{48}$N$_6$O$_9$; mol wt 600.71. C 53.99%, H 8.05%, N 13.99%, O 23.97%. Antibiotic substance produced by a *Nocardia* isolated from old bee-honeycombs: Stoll *et al.*, *Schweiz. Z. Pathol. Bakteriol.* **14**, 225 (1951); *Helv. Chim. Acta* **34**, 862 (1951). Structure: Keller-Schierlein, Prelog, *ibid.* **44**, 1981 (1961).

Needles from hot methanol, mp 192-195°. Optically inactive. Has no basic properties. Soluble in hot water, methanol. An 0.5% aq soln at room temp has a pH of 4.55.

Triacetate. $C_{33}H_{54}N_6O_{12}$. Crystals, mp 115-117°.

6753. Nocardicins. Monocyclic β-lactam antibiotics with antimicrobial activity which inhibit bacterial cell wall biosynthesis. Nocardicins A, B, C, D, E, F, G have been identified. All are produced by *Nocardia uniformis* subsp. *tsuyamenensis*, A being the most important component. Prodn: H. Aoki *et al.*, **DE 2242699** (1973 to Fujisawa), *C.A.* **78,** 134496k (1973); **US 3923977** (1975 to Fujisawa). Isoln and characterization of A: H. Aoki *et al.*, *J. Antibiot.* **29,** 492 (1976); of B: M. Kurita *et al.*, *ibid.* 1243. Nocardicin A has also been produced by *Actinosynnema mirum:* K. Watanabe *et al.*, *ibid.* **36,** 321 (1983). Structural determination of A and B: M. Hashimoto *et al.*, *J. Am. Chem. Soc.* **98,** 3023 (1976), *J. Antibiot.* **29,** 890 (1976). Isoln, characterization, and structures of C, D, E, F, G: J. Hosoda *et al.*, *Agric. Biol. Chem.* **41,** 2013 (1977). Total synthesis of A: G. A. Koppel *et al.*, *J. Am. Chem. Soc.* **100,** 3933 (1978); W. V. Curran *et al.*, *J. Antibiot.* **35,** 329 (1981); of A and D: T. Kamiya *et al.*, *Tetrahedron* **1979,** 323; of A and B: H. P. Isenring, W. Hofheinz, *Tetrahedron* **1983,** 2591. Biosynthetic studies of A: J. Hosada *et al.*, *Agric. Biol. Chem.* **41,** 2007 (1977); C. A. Townsend *et al.*, *J. Am. Chem. Soc.* **105,** 919 (1983). Series of articles on antimicrobial activity of A: *J. Antibiot.* **30,** 917-944 (1977).

Nocardicin A

Nocardicin A. [39391-39-4] (α*R*,3*S*)-3-[[(2*Z*)-[4-[(3*R*)-3-Amino-3-carboxypropoxy]phenyl](hydroxyimino)acetyl]amino]-α-(4-hydroxyphenyl)-2-oxo-1-azetidineacetic acid. $C_{23}H_{24}N_4O_9$; mol wt 500.46. Colorless needles from acidic water, mp 214-216° (dec). $[\alpha]_D^{25}$ −135° (for the sodium salt). uv max (1/15*M* phosphate buffer): 272 nm ($E_{1cm}^{1\%}$ 310); (0.1*N* NaOH): 244, 283 nm ($E_{1cm}^{1\%}$ 460, 270). Sol in alkaline solns, slightly sol in methanol. Insol in chloroform, ethyl acetate, ethyl ether. LD_{50} in male mice, rats (mg/kg): >8000 orally; >2000 i.v.; 2500, 2600 i.p. (Aoki).

Nocardicin B. [60134-71-6] The *E*-isomer of nocardicin A. Colorless needles, mp 262-264° (dec). $[\alpha]_D^{25}$ −162° (for the sodium salt). uv max (ethanol/water): 224, 270 nm (ε 24600, 9700); (ethanol/0.1*N* NaOH): 245, 280 nm (ε 26000, 11100).

6754. Nociceptin. [170713-75-4] Orphanin FQ (swine); orphanin FQ; OFQ. $C_{79}H_{129}N_{27}O_{22}$; mol wt 1809.07. C 52.45%, H 7.19%, N 20.91%, O 19.46%. Seventeen amino acid neuropeptide identified as the endogenous ligand of the orphan opioid receptor, ORL-1. Shows a wide variety of pharmacological effects. Structurally related to the opioids, but does not bind to opiate receptors. Isoln from rat brain: J.-C. Meunier *et al.*, *Nature* **377,** 532 (1995); from porcine hypothalmus: R. K. Reinscheid *et al.*, *Science* **270,** 792 (1995). Distribution and localization of binding with opioid peptides: S. Schulz *et al.*, *NeuroReport* **7,** 3021 (1996). Binding affinity: I. D. Adapa, L. Toll, *Neuropeptides* **31,** 403 (1997). *Review:* J.-C. Meunier, *Eur. J. Pharmacol.* **340,** 1-15 (1997). Review of pharmacology and structure-activity: G. Henderson, A. T. McKnight, *Trends Pharmacol. Sci.* **18,** 293-300 (1997).

Phe–Gly–Gly–Phe–Thr–Gly–Ala–Arg–Lys–Ser–Ala–Arg–Lys–Leu–Ala–Asn–Gln

6755. Nodakenin. [495-31-8] 2-[1-(β-D-Glucopyranosyloxy)-1-methylethyl]-2,3-dihydro-7*H*-furo[3,2-*g*][1]benzopyran-7-one; nodakenetin glucoside. $C_{20}H_{24}O_9$; mol wt 408.40. C 58.82%, H 5.92%, O 35.26%. From root of *Peucedanum decursivum* Maxim., *Umbelliferae:* Arima, *Bull. Chem. Soc. Jpn.* **4,** 16, 113 (1929); Späth, Kainrath, *Ber.* **69,** 2062 (1936). Structure: Späth, Tyray, *ibid.* **72,** 2089 (1939).

Thin leaflets from alc, dec 218-219°. From dil alc or water it crystallizes with one mol H_2O in form of yellow prisms, melting at 216°. $[\alpha]_D^{30}$ +56.6°. Sol in hot water or alcohol.

Tetraacetate. $C_{28}H_{32}O_{13}$. Crystals from methanol, mp 195-196°.

6756. Nodulisporic Acid. [163120-03-4] (2*E*,4*E*)-5-[(3*S*,4*S*,-4a*R*,6a*S*,12a*R*,13*S*,15*S*,16b*S*,16c*S*)-2,3,4,4a,5,6,6a,7,10,12,12a,13,-14,15,16b,16c-Hexadecahydro-3,13-dihydroxy-4,10,10,12,12,16b,-16c-heptamethyl-15-(1-methylethenyl)-14-oxo-1*H*-benz[6,7]-indeno[1,2-*b*]pyrano[3′,4′:4,5]cyclopenta[1,2-*f*]pyrrolo[3,2,1-*hi*]indol-4-yl]-2-methyl-2,4-pentadienoic acid; nodulisporic acid A; NsA A. $C_{43}H_{53}NO_6$; mol wt 679.90. C 75.96%, H 7.86%, N 2.06%, O 14.12%. Indole diterpene alkaloid which activates glutamate-gated chloride channels. Isolated from the fungus *Nodulisporium* sp.; it occurs naturally as the (+)-form. Isoln: A. W. Dombrowski *et al.*, **US 5399582** (1995 to Merck & Co.); structure determination and insecticidal activity: J. G. Ondeyka *et al.*, *J. Am. Chem. Soc.* **119,** 8809 (1997). Mode of action: M. M. Smith *et al.*, *Biochemistry* **39,** 5543 (2000). Synthetic studies: A. B. Smith, III *et al.*, *Org. Lett.* **3,** 3967, 3971 (2001). Biosynthesis: K. M. Byrne *et al.*, *J. Am. Chem. Soc.* **124,** 7055 (2002). Systemic efficacy against fleas on dogs: W. L. Shoop *et al.*, *J. Parasitol.* **87,** 419 (2001). Review of chemical and biological properties: P. T. Meinke *et al.*, *Curr. Top. Med. Chem.* **2,** 655-674 (2002).

Yellow powder, mp 250-255°. $[\alpha]_D^{22}$ +13° (c = 0.4 in $CHCl_3$). uv max (methanol): 241, 265, 385 nm (ε 43900, 49800, 7360).

USE: Insecticide.

6757. Noformicin. [155-38-4] (2*S*)-5-Amino-*N*-(3-amino-3-iminopropyl)-3,4-dihydro-2*H*-pyrrole-2-carboxamide; *N*-(2-amidinoethyl)-5-imino-2-pyrrolidinecarboxamide; 2-[*N*-(2-amidinoethyl)carbamoyl]-5-iminopyrrolidine; β-(5-imino-2-pyrrolidinecarboxamido)propamidine; noformycin. $C_8H_{15}N_5O$; mol wt 197.24. C 48.72%, H 7.67%, N 35.51%, O 8.11%. Antiviral agent produced by *Nocardia formica*, culture no. N.R.R.L. 2470: Peck *et al.*, **US 2804463** (1957 to Merck & Co.); Gray, *Phytopathology* **45,** 281 (1955). Activity studies: Furusawa *et al.*, *Proc. Soc. Exp. Biol. Med.* **116,** 938 (1964); *eidem*, *Med. Pharmacol. Exp.* **12,** 259 (1965). Synthesis of (+)- and (±)-forms: Diana, *J. Med. Chem.* **16,** 857 (1973).

(+)-Form dihydrochloride. $C_8H_{17}Cl_2N_5O$. The biologically active form. Crystals from methanol, dec 263-264°. $[\alpha]_D^{25}$ +8.8° (methanol). Sol in water. Somewhat sol in organic solvents. pKa 9.4.

(±)-Form dihydrochloride. mp 252-254°.

6758. Nogalamycin. [1404-15-5] (2R,3S,4R,5R,6R,11S,13S,-14R)-11-[(6-Deoxy-3-C-methyl-2,3,4-tri-O-methyl-α-L-manno-pyranosyl)oxy]-4-(dimethylamino)-3,4,5,6,9,11,12,13,14,16-deca-hydro-3,5,8,10,13-pentahydroxy-9,16-dioxo-2,6-ep-oxy-2H-naphthaceno[1,2-b]oxocin-14-carboxylic acid methyl ester; U-15167; NSC-70845. C₃₉H₄₉NO₁₆; mol wt 787.81. C 59.46%, H 6.27%, N 1.78%, O 32.49%. Antitumor antibiotic isolated from *Streptomyces nogalater* var. *nogalater:* B. K. Bhuyan, A. Dietz, *Antimicrob. Agents Chemother.* **1965**, 836; B. K. Bhuyan *et al.*, *US* **3183157** (1965 to Upjohn). Characterization: P. F. Wiley *et al.*, *Tetrahedron Lett.* **1968**, 663. Structure: *eidem, J. Am. Chem. Soc.* **99**, 542 (1977). Stereochemistry: *eidem, J. Org. Chem.* **44**, 4030 (1979). Biosynthesis: *eidem, ibid.* **43**, 3457 (1978). Synthesis of the sugar component, **L-nogalose:** L. Valente *et al.*, *Tetrahedron Lett.* **1979**, 1153; J. Yoshimura *et al.*, *Chem. Lett.* **1979**, 687. Partial stereospecific synthesis: R. P. Joyce *et al.*, *Tetrahedron Lett.* **27**, 4885 (1986). Molecular structure, absolute stereochemistry and interactions with DNA: S. K. Arora, *J. Am. Chem. Soc.* **105**, 1328 (1983). Toxicity data: Z. Hadidian, *PB Report* 173334 (1966). Review of pharmacology: B. K. Bhuyan, C. G. Smith, *Handb. Exp. Pharmacol.* **38**, (pt. 2) 623-632 (1975).

Orange-red solid from methanol, mp 195-196° (dec). [α]²⁵_D +425° (c = 0.11 in CHCl₃). pKa' 7.45 (60% ethanol). uv max (ethanol): 236, 258, 292 nm (ε 52360, 24755, 9890). Sol in methylene chloride, acetone, chloroform, ethyl acetate. Insol in water, methanol, ethanol. LD₅₀ in mice (mg/kg): 11.75 i.v., 4.79 i.p. (Hadidian).

6759. Nolatrexed. [147149-76-6] 2-Amino-6-methyl-5-(4-pyridinylthio)-4(1H)-quinazolinone; 2-amino-3,4-dihydro-6-methyl-4-oxo-5-(4-pyridylthio)quinazoline; 2-amino-6-methyl-5-(pyridin-4-ylsulfanyl)-3H-quinazolin-4-one. C₁₄H₁₂N₄OS; mol wt 284.34. C 59.14%, H 4.25%, N 19.70%, O 5.63%, S 11.28%. Antifolate thymidylate synthase inhibitor. Prepn: S. E. Webber *et al.*, *WO* **9320055**; *eidem, US* **5707992** (1993, 1998 both to Agouron); S. E. Webber *et al.*, *J. Med. Chem.* **36**, 733 (1993). Solid-state properties: A. K. Dash, P. Tyle, *J. Pharm. Sci.* **85**, 1123 (1996). Pharmacology: S. Webber *et al.*, *Cancer Chemother. Pharmacol.* **37**, 509 (1996). Clinical pharmacokinetics: I. Rafi *et al.*, *Clin. Cancer Res.* **1**, 1275 (1995). Clinical evaluation in hepatocellular carcinoma: K. Stuart *et al.*, *Cancer* **86**, 410 (1999).

Tan solid, mp 301-302°.
Dihydrochloride. [152946-68-4] AG-337; Thymitaq. C₁₄H₁₂-N₄OS.2HCl; mol wt 357.25. Hygroscopic; occurs as a dihydrate. Sol in water. pKa 4.1; 5.9; 9.8. Partition coefficient (octanol/water) at 25°: 1.9. Upon heating, dehydrates and forms metastable, plate-like crystals, mp 213°. Recrystallizes at 261° into stable, needle-like crystals, mp 312° (dec).
THERAP CAT: Antineoplastic.

6760. Nomifensine. [24526-64-5] 1,2,3,4-Tetrahydro-2-methyl-4-phenyl-8-isoquinolinamine; 8-amino-1,2,3,4-tetrahydro-2-methyl-4-phenylisoquinoline. C₁₆H₁₈N₂; mol wt 238.33. C 80.63%, H 7.61%, N 11.75%. Novel antidepressant distinguished from existing tricyclic and tetracyclic antidepressants by its bicyclic structure. Prepn: *GB* **1164192** corresp to G. Ehrhart *et al.*, *US* **3577424** (1969, 1971, both to Farbwerke Hoechst); I. Hoffmann *et al.*, *Arzneim.-Forsch.* **21**, 1045 (1971). Pharmacology: I. Hoffmann, *ibid.* **23**, 45 (1973); P. Hunt *et al.*, *J. Pharm. Pharmacol.* **26**, 370 (1974); B. Costall *et al.*, *Psychopharmacologia* **41**, 153 (1975). Pharmacokinetics: L. Vereczkey *et al.*, *ibid.* **45**, 225 (1975). Study on abuse potential in humans: C. Spyraki, H. C. Fibiger, *Science* **212**, 1167 (1981). *Review: Br. J. Clin. Pharmacol.* **4**, Suppl. 2 (1977) pp 1S-248S; *Int. Pharmacopsychiatry* **17**, Suppl. 1 (1982) pp 1-148.

mp 179-181°.
Maleate. [32795-47-4] HOE-984; Alival; Hostalival; Merital; Neurolene; Psicronizer. C₁₆H₁₈N₂.C₄H₄O₄; mol wt 354.41. Crystals from ethanol, mp 199-201°. LD₅₀ in mice, rats (mg/kg): 400, 430 orally; 90, 72 i.v.; in mice (mg/kg): 410 s.c. (Hoffman, 1973).
THERAP CAT: Antidepressant.

6761. Nomilin. [1063-77-0] (1S,3aS,4aR,4bR,6aR,11S,11aR,-11bR,13aS)-11-(Acetyloxy)-1-(3-furanyl)decahydro-4b,7,7,11a,-13a-pentamethyloxireno[4,4a]-2-benzopyran[6,5-g][2]benzoxepin-3,5,9(3aH,4bH,6H)-trione; 1-(acetyloxy)-1,2-dihydroobacunoic acid ε-lactone; 1-(3-furyl)decahydro-11-hydroxy-4b,7,7-11a,13a-penta-methyloxireno(4,4a)-2-benzopyrano[6,5-g](2)benzoxepin-3,5,9-(3aH,4bH,6H)-trione acetate. C₂₈H₃₄O₉; mol wt 514.57. C 65.36%, H 6.66%, O 27.98%. Bitter principle from citrus juice, seeds. Isoln: O. H. Emerson, *J. Am. Chem. Soc.* **70**, 545 (1948). Structure elucidation: D. L. Dreyer, *Tetrahedron* **21**, 75 (1965); *idem, ibid.* **24**, 3273 (1968). Distribution in citrus seeds: S. Hasegawa *et al.*, *J. Agric. Food Chem.* **28**, 922 (1980); R. L. Rouseff, S. Nagy, *Phytochemistry* **21**, 85 (1982). Identification in grapefruit juice: R. L. Rouseff, *J. Agric. Food Chem.* **30**, 504 (1982). Reduction of nomilin bitterness by *Arthrobacter globiformis* cells: S. Hasegawa, V. A. Pelton, *ibid.* **31**, 178 (1983); by *Corynebacterium fascians:* S. Hasegawa *et al.*, *ibid.* **32**, 457 (1984); *eidem, J. Food Sci.* **50**, 330 (1985). Antifeedant effect against *Spodoptera frugiperda:* M. A. Altieri *et al.*, *Prot. Ecol.* **6**, 91 (1984). Biosynthetic pathways: S. Hasegawa *et al.*, *Phytochemistry* **23**, 1601 (1984).

Needles from methanol, mp 278-279°. Slightly sol in 2-propanol, ethyl acetate. [α]²³_D −95.7°.

6762. Nonactin. [6833-84-7] (1R,2R,5R,7R,10S,11S,14S,-16S,19R,20R,23R,25R,28S,29S,32S,34S)-2,5,11,14,20,23,29,32-Octamethyl-4,13,22,31,37,38,39,40-octaoxapentacyclo-[32.2.1.1^{7,10}.1^{16,19}.1^{25,28}]tetracontane-3,12,21,30-tetrone. C_{40}H_{64}O_{12}; mol wt 736.94. C 65.19%, H 8.75%, O 26.05%. Macrotetrolide antibiotic. Produced by several *Streptomyces* spp.: Corbaz *et al.*, *Helv. Chim. Acta* **38**, 1445 (1955); Wallhäusser *et al.*, *Arzneim.-Forsch.* **14**, 356 (1964). Structure: Dominguez *et al.*, *Helv. Chim. Acta* **45**, 129 (1962). Crystal structure: Dobler, *ibid.* **55**, 1371 (1972). Steroselective syntheses of *nonactic acid*, building block of nonactin: Gerlach, Wetter, *ibid.* **57**, 2306 (1974); R. E. Ireland, J.-P. Vevert, *J. Org. Chem.* **45**, 4260 (1980). Total synthesis of (+)- and (−)-nonactic acids: *eidem*, *Can. J. Chem.* **59**, 572 (1981). Synthesis of nonactin: Gombos *et al.*, *Monatsh. Chem.* **106**, 1043 (1975); *eidem*, *Tetrahedron Lett.* **1975**, 3391; Gerlach *et al.*, *Helv. Chim. Acta* **58**, 2036 (1975); U. Schmidt *et al.*, *Ber.* **109**, 2628 (1976). Three homologs, *monactin*, C_{41}H_{66}O_{12}, *dinactin*, C_{42}H_{68}O_{12}, and *trinactin*, C_{43}H_{70}O_{12} are known: Beck *et al.*, *Helv. Chim. Acta* **45**, 620 (1962); Gerlach, Prelog, *Ann.* **669**, 121 (1963). Activity of nonactin and its homologues: Meyers *et al.*, *J. Antibiot.* **18A**, 128 (1965). Biosynthesis study: J. E. Cox, N. D. Priestley, *J. Am. Chem. Soc.* **127**, 7976 (2005).

Needles from methanol, mp 147-148°. uv max (ethanol): slight peak at 264 nm (log ε 1.5). Remarkably inert to chemical compds.

6763. 2-Nonenal. [18829-56-6] (2E)-2-Nonenal; *trans*-2-nonenaldehyde. C_9H_{16}O; mol wt 140.23. C 77.09%, H 11.50%, O 11.41%. Widely distributed in nature; found in beer, coffee, cucumbers, watermelon, palm oil, potatoes, carrots etc. Prepn: R. Delaby, S. Guillot-Allègre, *Bull. Soc. Chim. Fr.* **53**, 301 (1933); J. Braun *et al.*, *Ber.* **67**, 269 (1934); J. Ficini, H. Normant, *Bull. Soc. Chim. Fr.* **1964**, 1294; J. Ficini *et al.*, *Tetrahedron Lett.* **1977**, 3589; M. Wado *et al.*, *Chem. Lett.* **1977**, 345. Insecticidal activity: P. M. Guerin, M. F. Ryan, *Experientia* **36**, 1387 (1980). Cockroach repellant activity: T. H. Maugh, *Science* **218**, 278 (1982).

Liquid, bp_{16} 100-102°. n_D^{18} 1.4403.
USE: As flavoring agent.

6764. Nonoxynol. [26027-38-3] α-(4-Nonylphenyl)-ω-hydroxypoly(oxy-1,2-ethanediyl); polyethyleneglycols mono(nonylphenyl) ether; macrogol nonylphenyl ether; nonoxinol; polyoxyethylene(n)nonylphenyl ether; nonylphenyl polyethyleneglycol ether; nonylphenoxypolyethoxyethanol. Nonionic surfactant mixtures prepd by reacting nonylphenol with ethylene oxide. Prepn of polyoxyethylated alkyl phenols: A. Steindorff *et al.*, **US 2213477** (1940 to GAF). Average number of ethylene oxide units (n) per molecule is indicated by number following nonoxynol (e.g. nonoxynol-15 for n = 15). Trademarks for nonoxynol series include *Conco NI, Dowfax 9N, Igepal CO, Makon, Neutronyx 600's, Nonipol NO, Polytergent B, Renex 600's, Solar NP, Sterox, Surfonic N, T-DET-N, Tergitol NP, Triton N*. Individual members of series indicated by numerical suffixes. Toxicology study: H. F. Smyth Jr., J. C. Calandra, *Toxicol. Appl. Pharmacol.* **14**, 315 (1969). Inactivation of HIV *in vitro* by nonoxynol-9: M. Malkovsky *et al.*, *Lancet* **1**, 645 (1988). Clinical evaluation of nonoxynol-9 in sexually transmitted diseases: W. C. Louv *et al.*, *J. Infect. Dis.* **158**, 518 (1988). *Review*: C. R. Enyeart, "Polyoxyethylene Alkylphenols" in *Nonionic Surfactants*, M. J. Schick, Ed. (Dekker, New York, 1967) pp 44-85.

Nonoxynol-9

Highly stable compounds. Lower adducts (n <15) are yellow to almost colorless liquids; higher adducts (n >20) are pale yellow to off-white pastes or waxes. Lower adducts (n <6) sol in oil; higher ones sol in water. Review of physical properties: Enyeart, *loc. cit.*

Nonoxynol-9. C-Film; Conco NI-90; Dowfax 9N9; Encare; Gynol II; Igepal CO-630; Intercept; Neutronyx 611; Semicid; Staycept; Tergitol TP-9. Colorless to yellow, viscous liquid. Av. mol wt 617. A component of *Conceptrol, Delfen, Gentersal, Ortho-creme*. For Igepal CO-630: d_4^{25} 1.06; solidif pt 26 ±2°F; pour point 37 ±2°F; flash point 535-555°F; cloud point (1% aq soln) 126-133°F; viscosity (25°): 175-250 cP. Sol in water, ethanol, ethylene glycol, ethylene dichloride, xylene, corn oil. Insoluble in Stoddard solvent, deodorized kerosene, low viscosity white mineral oil.

Nonoxynol-11. Duragel; Duracreme.

USE: Nonionic surfactants used as detergents, emulsifiers, wetting agents, dispersants, stabilizers, intermediates in synthesis of anionic surfactants, defoaming agents. Nonoxynols-9, 11 as spermaticide. Nonoxynols-4, 15, 30 as pharmaceutic aids (surfactants).

6765. n-Nonyl Acetate. [143-13-5] Acetic acid nonyl ester; *n*-nonyl ethanoate; nonanol acetate; pelargonyl acetate; Acetate C-9. C_{11}H_{22}O_2; mol wt 186.30. C 70.92%, H 11.90%, O 17.18%.

Liquid. Pungent odor, suggestive of mushrooms, resembles odor of gardenias when dil. d_4^{15} 0.8785 (commercial grade, d_{25}^{25} 0.864-0.868). bp 208-212°. n_D^{20} 1.4328 (commercial grade, n_D^{20} 1.422-1.426). Insoluble in water. Freely sol in abs alcohol and ether; 33 ml dissolve in 100 ml of 80% ethanol.

Note: The nonyl acetate of commerce may be the above compound or any of its isomers such as diisobutylcarbinyl acetate.

USE: In perfumery.

6766. n-Nonyl Alcohol. [143-08-8] 1-Nonanol; 1-hydroxynonane; nonalol; pelargonic alcohol. C_9H_{20}O; mol wt 144.26. C 74.93%, H 13.97%, O 11.09%. Occurs in oil of orange. Prepn by reduction of ethyl Δ^1-nonylenate: Harding, Weizmann, *J. Chem. Soc.* **97**, 304 (1910); by reduction of pelargonic aldehyde: Tomecko, Adams, *J. Am. Chem. Soc.* **49**, 529 (1927); by oxo process: Russum, Hengstebeck, **US 2638487** (1953 to Standard Oil of Indiana); by reductive cleavage of oleic acid ozonide: Sousa, Bluhm, *J. Org. Chem.* **25**, 108 (1960); from diborane and propylene trimer: Marshall, Smith, **GB 879242** (1961 to I.C.I.).

Colorless to yellowish liquid; odor of citronella oil. bp_{760} 215°; bp_{15} 107.5°; bp_{7.5} 95.6°. d_4^{20} 0.8279. n_D^{20} 1.4338. Practically insol in water. Miscible with alcohol, ether.

USE: In manufacture of artificial lemon oil.

6767. 2-Nonyldioxolane. [4353-06-4] 2-Nonyl-1,3-dioxolane; SEPA. C_{12}H_{24}O_2; mol wt 200.32. C 71.95%, H 12.08%, O 15.97%. Dermal penetration enhancer. Prepn: S. S. Nigam, B. C. L. Weedon, *J. Chem. Soc.* **1956**, 4049; B. B. Michniak *et al.*, *Drug Delivery* **2**, 117 (1995). Use as transdermal enhancer: C. M. Samour, S. Daskalakis, **EP 268460**; *eidem*, **US 4861764** (1988, 1989 both to MacroChem). GC-MS determn in serum: W. Z. Zhong *et al.*, *J. Chromatogr. B* **705**, 39 (1998). Pharmacology: R. W. Pelham, C. M. Samour, *Proc. Int. Symp. Controlled Release Bioact. Mater.* **22**, 694 (1995). Clinical evaluation for topical delivery of prostaglandin E_1: K. T. McVary *et al.*, *J. Urol.* **162**, 726 (1999).

Clear oil, $bp_{0.55 torr}$ 90-93°. $bp_{0.01mm}$ 68-70°. n_D^{20} 1.4390.

USE: Pharmaceutic aid (excipient).

6768. Nonyl Phenol. [25154-52-3] Approx mol wt 215. A technical grade mixture of monoalkyl phenols, predominantly *para* substituted. The side chains are isomeric branched-alkyl radicals. Manuf of *p*-nonylphenol: *Faith, Keyes & Clark's Industrial Chemicals*, F. A. Lowenheim, M. K. Moran, Eds. (Wiley-Interscience, New York, 4th ed., 1975) pp 575-578.

Pale yellow liquid. Slight characteristic phenolic odor. Comparatively high viscosity. Hydroxyl no. 253. d_4^{20} 0.950. bp 293-297°. n_D^{20} 1.513. Flash pt (open cup) 300°F. Practically insol in water or dil aq NaOH. Sol in benzene, chlorinated solvents, aniline, heptane, aliphatic alcohols, ethylene glycol.

USE: In the prepn of lubricating oil additives, resins, plasticizers, surface active agents.

6769. (*p*-Nonylphenoxy)acetic Acid. [3115-49-9] 2-(4-Nonylphenoxy)acetic acid. $C_{17}H_{26}O_3$; mol wt 278.39. C 73.35%, H 9.41%, O 17.24%. Prepd by the reaction of sodium chloroacetate with nonyl phenol in alkaline medium: Preston, **GB 812938** and Preston, Taylor, **GB 831883** (1959, 1960, to I.C.I.).

Liquid. d_4^{20} 1.010-1.025. Viscosity (30°): 5200 cP. Flow pt 5°. Practically insol in water. Miscible with benzene, mineral oil, kerosene, petr ether.

USE: Corrosion inhibitor and antifoaming agent in gasoline and cutting oils.

6770. Nootkatone. [4674-50-4] (4R,4aS,6R)-4,4a,5,6,7,8-Hexahydro-4,4a-dimethyl-6-(1-methylethenyl)-2(3H)-naphthalenone; (4R,4aS)-dimethyl-6R-isopropenyl-4,4a,5,6,7,8-hexahydro-3H-naphthalen-2-one; (+)-nootkatone. $C_{15}H_{22}O$; mol wt 218.34. C 82.52%, H 10.16%, O 7.33%. Sesquiterpene ketone; major flavor constituent of grapefruit (*Citrus paradisi*). Exhibits strong repellant properties against Formosan subterranean termites and other insects. Isoln from heartwood of Alaska yellow cedar, *Chamaecyparis nootkatensis*: H. Erdtman, Y. Hirose, *Acta Chem. Scand.* **16**, 1311 (1962); from grapefruit peel and juice: W. D. MacLeod, Jr., N. M Buigues, *J. Food Sci.* **29**, 565 (1964). Structural characterization: *idem*, *Tetrahedron Lett.* **6**, 4779 (1965). Total synthesis of (±)-form: M. Pesaro *et al.*, *Chem. Commun.* **1968**, 1152; of (+)-form: Y. Takagi *et al.*, *Tetrahedron* **34**, 517 (1978). Odor threshold and characterization: K. L. Stevens *et al.*, *J. Sci. Food Agric.* **21**, 590 (1970). Aroma and flavor analysis in grapefruit oil and juice: P. E. Shaw, C. W. Wilson, III, *J. Agric. Food Chem.* **29**, 677 (1981). Crystal structure: A. M. Sauer *et al.*, *Acta Crystallogr.* **C59**, o254 (2003). Repellent activity vs Formosan subterranean termites: B. C. R. Zhu *et al.*, *J. Chem. Ecol.* **27**, 523 (2001). Fragrance monograph: C. S. Letizia *et al.*, *Food Chem. Toxicol.* **38**, Suppl. 3, S165-S167 (2000).

Commercial product is a colorless to yellowish liquid with grapefruit taste and odor. Crystals from light petroleum, mp 36-37°. $[\alpha]_D$ +195.5° (c = 1.5 in $CHCl_3$). $[\alpha]_D^{20}$ + 170° (c = 0.5 in $CHCl_3$). $bp_{0.5}$ 125°. d^{25} 0.9963. n_D^{20} 1.5253. Flash pt (closed cup): 160°F. uv max (ethanol): 237 nm (ϵ 17000). Sol in alcohol, oil. Practically insol in water. LD_{50} orally in rats: > 5 g/kg (Letizia).

(±)-**Form.** [28834-25-5] Crystals, mp 45-46°.

USE: In beverages to impart grapefruit flavor; in fragrance compositions with other citrus oils.

6771. Nopol. [35836-73-8]; [128-50-7] ((±)-form). (1R,5S)-6,6-Dimethylbicyclo[3.1.1]hept-2-ene-2-ethanol. $C_{11}H_{18}O$; mol wt 166.26. C 79.47%, H 10.91%, O 9.62%. Bicyclic primary alcohol found in carrot root oil; synthesized from β-pinene, *q.v.* Prepn: J. P. Bain, *J. Am. Chem. Soc.* **68**, 638 (1946). Thermal isomerization: *idem et al.*, *ibid.* **74**, 4292 (1952). Isoln from carrot root oil: D. M. Alabran *et al.*, *J. Agric. Food Chem.* **23**, 229 (1975). Applications as a polymer substrate: J. V. Crivello, S. S. Liu, *J. Polym. Sci. A* **37**, 1199 (1999). Improved syntheses: U. R. Pillai, E. Sahle-Demessie, *Chem. Commun.* **2004**, 826; A. L. Villa de P *et al.*, *Catal. Today* **107-108**, 942 (2005).

bp_{10} 110.5°. n_D^{25} 1.4920. d_4^{25} 0.9647. $[\alpha]_D^{25}$ −36.5°.

USE: In the synthesis of pesticides, soaps, detergents and other household products.

6772. Noprylsulfamide. [576-97-6] 1-[[4-(Aminosulfonyl)phenyl]amino]-3-phenyl-1,3-propanedisulfonic acid sodium salt (1:2); 1-phenyl-3-*p*-sulfamoylanilino-1,3-propanedisulfonic acid disodium salt; N^4-(disodium 1,3-disulfo-3-phenylpropyl)sulfanilamide; disodium 1-phenyl-3-*p*-sulfamoylanilino-1,3-propanedisulfonate; disodium *p*-(γ-phenylpropylamino)benzenesulfonamide-α,γ-disulfonate; RP-40; Solucin; Soluseptasine; Soluseptazine; Solusetazine; Sulphasolucin; Sulphasolutin. $C_{15}H_{16}N_2Na_2O_8S_3$; mol wt 494.46. C 36.44%, H 3.26%, N 5.67%, Na 9.30%, O 25.89%, S 19.45%. Prepd by the action of sodium bisulfite on N^4-cinnamylidenesulfanilamide (prepd from cinnamic aldehyde and sulfanilamide): Despois, **US 2262544** (1941 to Rhône-Poulenc).

Crystals. One gram dissolves in 5 ml water. Neutral reaction. Breaks down in the body with the liberation of free sulfanilamide.

THERAP CAT: Antibacterial.

6773. Noracymethadol. [1477-39-0] α-Ethyl-β-[2-(methylamino)propyl]-β-phenylbenzeneethanol 1-acetate; α-dl-6-(methylamino)-4,4-diphenyl-3-heptanol acetate; α-dl-3-acetoxy-4,4-diphenyl-6-methylaminoheptane; α-dl-3-acetoxy-6-methylamino-4,4-diphenylheptane; α-dl-4,4-diphenyl-6-methylamino-3-heptanol acetate. $C_{22}H_{29}NO_2$; mol wt 339.48. C 77.84%, H 8.61%, N 4.13%, O 9.43%. Metabolite of methadyl acetate, *q.v.* Prepn: A. Pohland, **US 3021360** (1962 to Lilly). Metabolism: G. L. Henderson *et al.*, *Drug Metab. Dispos.* **5**, 321 (1977); M. Man *et al.*, *ibid.* **8**, 55 (1980). HPLC determn: C.-H. Kiang *et al.*, *J. Chromatogr.* **222**, 81 (1981).

Hydrochloride. [5633-25-0] NIH-7667. $C_{22}H_{29}NO_2$·HCl; mol wt 375.94. Crystals from acetone + ether, mp ~216-217°.

Note: This is a controlled substance (opiate): **21 CFR**, 1308.11.

6774. Norbolethone. [1235-15-0] (17α)-(±)-13-Ethyl-17-hydroxy-18,19-dinorpregn-4-en-3-one; dl-13β,17α-diethyl-17β-hydroxygon-4-en-3-one; Wy-3475. $C_{21}H_{32}O_2$; mol wt 316.49. C 79.70%, H 10.19%, O 10.11%. 19-nor anabolic steroid. Prepn of *dl*-, *d*-, and *l*-forms: Smith *et al.*, *J. Chem. Soc.* **1964**, 4472. Biological activities of *dl*-, *d*-, and *l*-forms: Edgren *et al.*, *Steroids* **2**, 731 (1963). Toxicity study: E. I Goldenthal, *Toxicol. Appl. Pharmacol.* **18**, 185 (1971). GC/MS determn in athletes' urine: D. H. Catlin *et al.*, *Rapid Commun. Mass Spectrom.* **16**, 1273 (2002).

Relative stereochemistry

Crystals from alcohol, mp 144-145°. uv max: 241 nm (ε 16500). LD_{50} orally in mice: >5010 mg/kg (Goldenthal).

d-**Form.** [28439-19-2] Crystals from acetone + hexane, mp 175-176°. $[\alpha]_D$ +20.7° (in chloroform).

l-**Form.** Crystals from acetone + hexane, mp 172-175.5°. $[\alpha]_D$ −18.1° (in chloroform).

Note: This is a controlled substance (anabolic steroid): **21 CFR,** 1308.13, as defined in 1300.01.

6775. Norbormide. [991-42-4] 3a,4,7,7a-Tetrahydro-5-(hydroxyphenyl-2-pyridinylmethyl)-8-(phenyl-2-pyridinylmethylene)-4,7-methano-1*H*-isoindole-1,3(2*H*)-dione; 5-(α-hydroxy-α-2-pyridylbenzyl)-7-(α-2-pyridylbenzylidene)-5-norbornene-2,3-dicarboximide; McN-1025; Raticate; Shoxin. $C_{33}H_{25}N_3O_3$; mol wt 511.58. C 77.48%, H 4.93%, N 8.21%, O 9.38%. Prepn: A. P. Roszkowski *et al.*, *Science* **144**, 412 (1964); R. J. Mohrbacher *et al.*, **BE 659610**; *eidem*, **US 3378566** (1965, 1968 both to McNeil). Stereoisomerism: *idem et al.*, *J. Org. Chem.* **31**, 2141 (1966); S. Abrahamsson, B. Nilsson, *ibid.* 3631. Structure-activity studies: G. I. Poos *et al.*, *J. Med. Chem.* **9**, 537 (1966). Toxicity data: A. P. Roszkowski, *J. Pharmacol. Exp. Ther.* **149**, 288 (1965).

Crystals from methylene chloride + ether, mp 190-198°. uv max (methanol): 250 nm (ε 17500). LD_{50} in rats (mg/kg): 5.3 orally; 0.65 i.v. (Roszkowski). Practically insol in water unless pH is <4.
USE: Rodenticide.

6776. Norbornadiene. [121-46-0] Bicyclo[2.2.1]hepta-2,5-diene. C_7H_8; mol wt 92.14. C 91.25%, H 8.75%. Highly strained, bridged hydrocarbon; isomer of toluene and quadricyclane, *q.q.v.* Prepn: J. Hine *et al.*, *J. Am. Chem. Soc.* **77**, 594 (1955); M. A. Pryanishnikova *et al.*, *Russ. Chem. Bull.* **16**, 1085 (1967). Microwave structural studies: G. Knuchel *et al.*, *J. Am. Chem. Soc.* **115**, 10845 (1993). Crystal structure determn: J. Benet-Buchholz *et al.*, *Chem. Commun.* **1998**, 2003. Review of synthetic applications: B. Hill *et al.*, *Mol. Photochem.* **5**, 195-208 (1973); V. R. Flid *et al.*, *Eurasian Chem. Technol. J.* **3**, 73-90 (2001); of solar energy storage applications of isomerization reaction with quadricyclane: A. D. Dubonosov *et al.*, *Russ. Chem. Rev.* **71**, 917-927 (2002).

Malodorous liquid, bp 89-89.5°. mp −16.02°. n_D^{26} 1.4670. d_4^{26} 0.8992. *Flammable.*
USE: Reagent in organic, organometallic, and petrochemical synthesis, especially cycloaddition reactions and synthesis of polycyclic hydrocarbons. Intermediate in prepn of the insecticide, aldrin, *q.v.*

6777. Norcarane. [286-08-8] Bicyclo[4.1.0]heptane; 1,2-methylenecyclohexane. C_7H_{12}; mol wt 96.17. C 87.43%, H 12.58%. Prepd by distilling the barium salt of norcaranecarboxylic acid with ZnO + BaO: Ebel *et al.*, *Helv. Chim. Acta* **12**, 19 (1929); by the action of methylene iodide and zinc-copper couple on cyclohexene: Smith, Simmons, *Org. Synth.* **41**, 72 (1961); Simmons, **US 3074984** (1963 to du Pont); from cyclohexene + diazomethane + zinc iodide: Applequist, Badad, *J. Org. Chem.* **27**, 288 (1962); from methylene chloride + cyclohexyl lithium: Closs, *J. Am. Chem. Soc.* **84**, 809 (1962).

Liquid. bp 116-117°. n_D^{25} 1.4546.

6778. Norcholanic Acid. [511-18-2] (5β)-24-Norcholan-23-oic acid. $C_{23}H_{38}O_2$; mol wt 346.56. C 79.71%, H 11.05%, O 9.23%. Prepn from ethyl cholanate: Wieland *et al.*, *Z. Physiol. Chem.* **161**, 80 (1926); from 3,7-diketonorcholanic acid: Windaus, van Schoor, *ibid.* **173**, 312 (1928); from 12-ketonorcholanic acid: Cook, Haslewood, *J. Chem. Soc.* **1934**, 428; from 3α,23-dihydroxycholanic acid: Yanuka *et al.*, *Tetrahedron Lett.* **1968**, 1725.

Needles from acetic acid, mp 177°.
Methyl ester. $C_{24}H_{40}O_2$. Needles from methanol, mp 74°.
Ethyl ester. $C_{25}H_{42}O_2$. Prisms from alcohol, mp 67°.

6779. Norcodeine. [467-15-2] (5α,6α)-7,8-Didehydro-4,5-epoxy-3-methoxymorphinan-6-ol; *N*-desmethylcodeine; normorphine 3-methyl ether. $C_{17}H_{19}NO_3$; mol wt 285.34. C 71.56%, H 6.71%, N 4.91%, O 16.82%. Prepn from acetylcodeine: von Braun, *Ber.* **47**, 2312 (1914); **DE 286743** (1914); *Chem. Zentralbl.* **1915**, II, 862; *Frdl.* **12**, 741; *Houben* **4**, 588; *cf.* **DE 289273** (1914); from codeine *N*-oxide: Diels, Fischer, *Ber.* **49**, 1723 (1916); from codeine: Speyer, Walther, *Ber.* **63**, 822 (1930).

Plates or needles from acetone or ethyl acetate, mp 185°. Sparingly sol in water, ether. Moderately sol in acetone. Freely sol in hot methanol, ethanol.
Hydrochloride trihydrate. $C_{17}H_{19}NO_3.HCl.3H_2O$. Needles from water. When anhydr, dec 309°. Sparingly sol in cold water, more sol in hot water. Freely sol in methanol, ethanol. Almost insol in acetone.
Sulfate. $2C_{17}H_{19}NO_3.H_2SO_4$. Crystals, freely sol in water.
Acetate (salt). $C_{17}H_{19}NO_3.C_2H_4O_2$. Crystals, freely sol in water.

Consult the Name Index before using this section.

6780. Nordazepam. [1088-11-5] 7-Chloro-1,3-dihydro-5-phenyl-2*H*-1,4-benzodiazepin-2-one; desmethyldiazepam; nordiazepam; DMDZ; A-101; Ro-5-2180; Calmday; Madar; Nordaz; Praxadium; Stilny. $C_{15}H_{11}ClN_2O$; mol wt 270.72. C 66.55%, H 4.10%, Cl 13.09%, N 10.35%, O 5.91%. Desmethyl analog and principal metabolite of diazepam, *q.v.* Prepn: L. H. Sternbach, E. Reeder, *J. Org. Chem.* **26**, 4936 (1961); E. Reeder *et al.*, **DE 1136709**; E. Reeder, L. Sternbach, **US 3051701** (both 1962 to Hoffmann-La Roche); A. Stempel, **BE 620020**; *idem*, **US 3202699** (1963, 1965, both to Hoffmann-La Roche); S. C. Bell *et al.*, *J. Org. Chem.* **27**, 562 (1962); A. Stempel, G. W. Landgraf, *ibid.* 4675. Pharmacology: U. Traversa *et al.*, *J. Pharm. Pharmacol.* **29**, 504 (1977); M. Babbini *et al.*, *Pharmacology* **17**, 121 (1978). Metabolism: U. Klotz *et al.*, *Br. J. Clin. Pharmacol.* **7**, 119 (1979). Pharmacokinetics: M. Konishi, *J. Pharm. Sci.* **67**, 1777 (1978). Clinical study: V. Andreoli *et al.*, *Arzneim.-Forsch.* **27**, 436 (1977). Toxicity study: L. O. Randall *et al.*, *Curr. Ther. Res.* **7**, 590 (1965). Teratogenicity study: R. P. Miller *et al.*, *Toxicol. Appl. Pharmacol.* **25**, 453 (1973).

White or pale yellow crystalline powder from acetone, mp 216-217°. uv max (chloroform): 313 nm ($E_{1cm}^{1\%}$ 82). Slightly sol in alc, chloroform. Practically insol in water. LD_{50} in mice (mg/kg): 2750 orally, >400 i.p. (Randall). Also reported as LD_{50} in mice, rats (mg/kg): 1300, >5200 orally (company communication).

Note: This is a controlled substance (depressant): **21 CFR**, 1308.14.

THERAP CAT: Anxiolytic.

6781. Nordefrin. [74812-63-8]; [6539-57-7] (unspecified stereo). *rel*-4-[(1*R*,2*S*)-2-Amino-1-hydroxypropyl]-1,2-benzenediol; (*erythro*)-(±)-α-(1-aminoethyl)-3,4-dihydroxybenzyl alcohol; (±)-3,4-dihydroxynorephedrine; (±)-α-methylnoradrenaline; (±)-α-methylnorepinephrine. $C_9H_{13}NO_3$; mol wt 183.21. C 59.00%, H 7.15%, N 7.65%, O 26.20%. α-Adrenergic agonist. Prepn: **DE 254438** (1912 to Bayer); W. H. Hartung *et al.*, *J. Am. Chem. Soc.* **53**, 4149 (1931); M. Bockmühl *et al.*, **US 1948162** (1934 to Winthrop); Bruckner *et al.*, *Ber.* **76B**, 466 (1943). Configuration: Fodor *et al.*, *Monatsh. Chem.* **83**, 1146 (1952). Prepn of optical isomers: **DE 269327** (1913 to Bayer); Bockmühl, Gorr, **DE 639126** (1936 to I. G. Farbenind.). Pharmacology: F. P. Luduena *et al.*, *J. Dent. Res.* **37**, 206 (1958). HPLC determn in plasma: D. A. Jenner *et al.*, *J. Chromatogr.* **224**, 507 (1981). Effect on toxicity of local anesthetics: S. E. Taylor, R. L. Dorris, *Anesth. Prog.* **36**, 79 (1989). Review of use in local anesthesia: T. M. Hensley *et al.*, *J. Foot Surg.* **26**, 504-510 (1987).

Levonordefrin

Hydrochloride. [61-96-1]; [138-61-4] (unspecified stereo). Corbasil; Cobefrin. $C_9H_{13}NO_3$.HCl; mol wt 219.67. Crystals, dec 178-179°. One gram dissolves in about 1.5 ml water, 15 ml alcohol. Practically insol in ether. Solns are neutral and are easily destroyed by traces of alkali.

(−)-Form. [829-74-3] Levonordefrin; corbadrine; Neo-Cobefrin. Crystals, dec 212-215°. $[\alpha]_D^{25}$ −31.0° (c = 0.5 in 0.01*N* HCl). uv max (dil HCl): 278 nm. Practically insol in water. Freely sol in aq solns of mineral acids. Very slightly sol in acetone, chloroform, ethanol, ether. LD_{50} in mice (mg/kg): 12.6 i.v. (Luduena).

THERAP CAT: Vasoconstrictor.

6782. Nordihydroguaiaretic Acid. [27686-84-6] (*meso*-form); [500-38-9] (unspecified stereo). *rel*-4,4′-[(2*R*,3*S*)-2,3-Dimethyl-1,4-butanediyl]bis-1,2-benzenediol; *meso*-4,4′-(2,3-dimethyltetramethylene)dipyrocatechol; 2,3-bis(3,4-dihydroxybenzyl)butane; β,γ-dimethyl-α,δ-bis(3,4-dihydroxyphenyl)butane; NDGA; masoprocol; CHX-100; Actinex. $C_{18}H_{22}O_4$; mol wt 302.37. C 71.50%, H 7.33%, O 21.16%. Lipoxygenase inhibitor; occurs as the *meso*-form in the resinous exudate of the creosote bush, *Larrea divaricata, Zygophyllaceae (Covillea tridentata)*. Isoln: C. W. Waller, O. Gisvold, *J. Am. Pharm. Assoc.* **34**, 78 (1945). Prepn from guaiaretic acid dimethyl ether: G. Schroeter *et al.*, *Ber.* **51**, 1587 (1918); R. D. Haworth *et al.*, *J. Chem. Soc.* **1934**, 1423. Synthesis: S. V. Lieberman *et al.*, *J. Am. Chem. Soc.* **69**, 1540 (1947); and configuration of naturally occurring form: C. W. Perry *et al.*, *J. Org. Chem.* **37**, 4371 (1972). Use as antioxidant: W. M. Lauer, **US 2373192** (1945 to U. S. Secr'y of Agriculture). Antiproliferative effect on cultured keratinocytes: D. I. Wilkinson, E. K. Orenberg, *Int. J. Dermatol.* **26**, 660 (1987); on cultured glioma cells: D. E. Wilson *et al.*, *J. Neurosurg.* **71**, 551 (1989). Clinical trial in actinic keratoses: E. A. Olsen *et al.*, *J. Am. Acad. Dermatol.* **24**, 738 (1991).

Relative stereochemistry

Crystals from acetic acid, mp 185-186°. uv max (methanol): 283, 218 nm (ε 6660, 13400). Sol in ethanol, methanol, ether, concd H_2SO_4; slightly sol in hot water, chloroform; sol in dil alkalies, developing a deep red color. Practically insol in petr ether, benzene, toluene, dil HCl.

USE: As antioxidant for fats and oils in foods.

THERAP CAT: Antineoplastic.

6783. Norelgestromin. [53016-31-2] (17α)-13-Ethyl-17-hydroxy-18,19-dinorpregn-4-en-20-yn-3-one oxime; 17-deacetylnorgestimate; 18-methylnorethindrone oxime; levonorgestrel 3-oxime; RWJ-10553. $C_{21}H_{29}NO_2$; mol wt 327.47. C 77.02%, H 8.93%, N 4.28%, O 9.77%. Bioactive metabolite of norgestimate, *q.v.* Prepd as intermediate of norgestrel, *q.v.*: **NL 7306609** (1973 to Richter Gedeon). Prepn and identification as norgestimate metabolite: S. F. Sisenwine *et al.*, *Contraception* **15**, 25 (1977). Pharmacology: J. L. McGuire *et al.*, *Am. J. Obstet. Gynecol.* **163**, 2127 (1990). Use as contraceptive patch: J. Jona *et al.*, **WO 9640355**; *eidem*, **US 5876746** (1996, 1999 both to Cygnus). Clinical pharmacokinetics: L. S. Abrams *et al.*, *Br. J. Clin. Pharmacol.* **53**, 141 (2002). Review of clinical pharmacology and efficacy: R. T. Burkman, *Int. J. Fertil.* **47**, 69-76 (2002); B. L. Sicat, *Pharmacotherapy* **23**, 472-480 (2003).

Crystals, mp 112°.

Combination with ethinyl estradiol. [396715-57-4] Evra; Ortho Evra.

THERAP CAT: Progestogen; in combination with estrogen as transdermal contraceptive.

6784. Norepinephrine. [51-41-2] 4-[(1*R*)-2-Amino-1-hydroxyethyl]-1,2-benzenediol; (−)-α-(aminomethyl)-3,4-dihydroxybenzyl alcohol; *l*-3,4-dihydroxyphenylethanolamine; noradrenaline; levarterenol; Adrenor; Levophed. $C_8H_{11}NO_3$; mol wt 169.18. C 56.80%, H 6.55%, N 8.28%, O 28.37%. Demethylated precursor of epinephrine, *q.v.* Occurs in animals and man, and is a sympathomimetic hormone of both adrenal origin and adrenergic orthosympathetic postganglionic origin in man. Physiologic review: Malmejac, *Physiol. Rev.* **44**, 186 (1964). It has also been found in plants, e.g., *Portulaca olerocea* L., *Portulacaceae:* Fing *et al.*, *Nature* **191**, 1108 (1961). Synthesis of *dl*-form: Payne, *Ind. Chem.* **37**, 523 (1961). Historic review of synthesis: Loewe, *Arzneim.-Forsch.* **4**, 583 (1954). Resolution of *dl*-form: Tullar, *J. Am. Chem. Soc.* **70**, 2067 (1948); *idem*, US 2774789 (1956 to Sterling Drug). Configuration: Pratesi *et al.*, *J. Chem. Soc.* **1959**, 4062. Comprehensive description: C. F. Schwender, *Anal. Profiles Drug Subs.* **1**, 149-173 (1972); T. D. Wilson, *ibid.* **11**, 555-586 (1982).

Microcrystals, dec 216.5-218°. $[\alpha]_D^{25}$ −37.3° (c = 5 in water with 1 equiv HCl).

Hydrochloride. [329-56-6] Arterenol. $C_8H_{11}NO_3 \cdot HCl$; mol wt 205.64. Crystals, mp 145.2-146.4°. $[\alpha]_D^{25}$ −40° (c = 6). Freely sol in water. Solns slowly oxidize under the influence of light and oxygen in a manner comparable to epinephrine hydrochloride.

***d*-Bitartrate.** [69815-49-2] Levarterenol bitartrate; Aktamin; Binodrenal. $C_8H_{11}NO_3 \cdot C_4H_6O_6$; mol wt 319.27. Obtained as the monohydrate, crystals, mp 102-104°. $[\alpha]_D^{25}$ −10.7° (c = 1.6 in H$_2$O). When anhydr, mp 158-159° (some decompn). Freely sol in water; slightly sol in alc. Practically insol in chloroform, ether.

***dl*-Form.** Crystals, dec 191°. Sparingly sol in water; very slightly sol in alc, ether; readily sol in dilute acids, caustic.

THERAP CAT: Adrenergic (vasopressor); antihypotensive.

THERAP CAT (VET): Sympathomimetic; vasopressor in shock.

6785. Norethandrolone. [52-78-8] (17α)-17-Hydroxy-19-norpregn-4-en-3-one; 17α-ethyl-19-nortestosterone; 17α-ethyl-17-hydroxy-4-norandrosten-3-one; 17α-ethyl-17-hydroxy-19-norandrost-4-en-3-one; Nilevar; Solevar. $C_{20}H_{30}O_2$; mol wt 302.46. C 79.42%, H 10.00%, O 10.58%. Prepn by catalytic hydrogenation of 17α-ethynyl-19-nortestosterone: Colton, US 2721871 (1955 to Searle); *J. Am. Chem. Soc.* **79**, 1123 (1957).

Crystals from methanol, mp 140-141°. uv max: 240 nm (ε 16500). Sol in alcohol, benzene, ether, ethyl acetate. Insol in water.

Note: This is a controlled substance (anabolic steroid): **21 CFR,** 1308.13, as defined in 1300.01.

THERAP CAT: Androgen.

6786. Norethindrone. [68-22-4] (17α)-17-Hydroxy-19-norpregn-4-en-20-yn-3-one; 19-nor-17α-ethynyltestosterone; 17α-ethynyl-19-nortestosterone; 19-nor-17α-ethynyl-17β-hydroxy-4-androsten-3-one; 19-nor-17α-ethynylandrosten-17β-ol-3-one; anhydrohydroxynorprogesterone; 19-norethisterone; norpregneninolone; "mini-pill"; Camila; Conludag; Menzol; Micronor; Micronovum; Mini-Pe; Norcolut; Noriday; Primolut N; Utovlan. $C_{20}H_{26}O_2$; mol wt 298.43. C 80.49%, H 8.78%, O 10.72%. Prepn from 19-nor-4-androstene-3,17-dione: Djerassi *et al.*, *J. Am. Chem. Soc.* **76**, 4092 (1954); US 2744122 (1956 to Syntex); De Ruggieri, US 2849462 (1958). Pharmacokinetics: H. Singh *et al.*, *Am. J. Obstet. Gynecol.* **135**, 409 (1979); M. Humpel, *Contraception* **26**, 83 (1982). Double-blind, comparative clinical trial: S. Koetsawang *et al.*, *ibid.* **25**, 231 (1982); A. Sheth *et al.*, *ibid.* 243. Multicenter trial of combination with ethinyl estradiol: M. Toews *et al.*, *Curr. Ther. Res.* **41**, 509 (1987). Review of carcinogenicity studies: *IARC Monographs* **21**, 441-460 (1979). Comprehensive description: A. P. Schroff, E. S. Moyer, *Anal. Profiles Drug Subs.* **4**, 268-293 (1975).

Crystals from ethyl acetate, mp 203-204°. $[\alpha]_D^{20}$ −31.7° (chloroform). First reported as $[\alpha]_D^{20}$ −25° (chloroform). uv max (ethanol): 240 nm (log ε 4.24). Sol in chloroform, dioxane; sparingly sol in alc; slightly sol in ether. Practically insol in water.

Mixture with ethinyl estradiol. Binovum; Brevicon; Estrostep; Modicon; Neocon 1/35; Norimin; Norinyl 1+35; Ortho-Novum; Ovcon; Ovysmen; Tri-Norinyl; Trinovum.

Mixture with mestranol. Norinyl 1+50; Ortho-Novin 1/50; Ortho-Novum 1/50.

Acetate. [51-98-5] Aygestin; Milligynon; Norlutate; Primolut-Nor. $C_{22}H_{28}O_3$; mol wt 340.46. Prepn: O. Engelfried *et al.*, US 2964537 (1960 to Schering AG). Crystals from methylene chloride + hexane, mp 161-162°. uv max: 240 nm (ε 18690). Very sol in chloroform; freely sol in dioxane; sol in ether, alc. Practically insol in water.

Acetate, mixture with ethinyl estradiol. Etalontin; Primosiston; Anovlar; Gynovlar; Loestrin; Minovlar; Norlestrin.

Enanthate. Noristerat. $C_{27}H_{38}O_3$; mol wt 410.60. Clinical trial as injectable contraceptive: S. K. Banerjee *et al.*, *Contraception* **30**, 561 (1984).

Caution: Norethindrone is reasonably anticipated to be a human carcinogen: *Report on Carcinogens, Twelfth Edition* (PB2011-111646, 2011) p 333.

THERAP CAT: Progestogen. Norethindrone and acetate in combination with estrogen as contraceptive (oral). Enanthate as contraceptive (injectable).

6787. Norethynodrel. [68-23-5] (17α)-17-Hydroxy-19-norpregn-5(10)-en-20-yn-3-one; 17α-ethynyl-17-hydroxy-5(10)-estren-3-one; 13-methyl-17-ethynyl-17-hydroxy-1,2,3,4,6,7,8,9,11,12,13,-14,16,17-tetradecahydro-15H-cyclopenta[a]phenanthren-3-one. $C_{20}H_{26}O_2$; mol wt 298.43. C 80.49%, H 8.78%, O 10.72%. Prepn: Colton, US 2691028, US 2725389 (1954, 1955 to Searle). Metabolism studies in mice: Freudenthal *et al.*, *Toxicol. Appl. Pharmacol.* **24**, 125 (1973).

Crystals from aq methanol, mp 169-170°. $[\alpha]_D$ +108° (1% chloroform). Freely sol in chloroform; sol in acetone; sparingly sol in alc; very slightly sol in water, solvent hexane.

Mixture with mestranol. [8015-30-3] Conovid E; Enavid; Enovid.

THERAP CAT: Progestogen. In combination with estrogen as contraceptive (oral).

6788. Norfenefrine. [536-21-0] α-(Aminomethyl)-3-hydroxybenzenemethanol; α-(aminomethyl)-*m*-hydroxybenzyl alcohol; 1-(*m*-hydroxyphenyl)-2-aminoethanol; *m*-hydroxyphenylethanolamine; norphenylephrine. $C_8H_{11}NO_2$; mol wt 153.18. C 62.73%, H 7.24%, N 9.14%, O 20.89%. Chem synthesis: Sachs, FR 866569 (1941), *C.A.* **43**, 5043c (1949); Legerlotz, US 2312916 (1943 to Ciba); Credner, Neugebauer, *Arzneim.-Forsch.* **3**, 462 (1953); Bretschneider, Hörmann, *Monatsh. Chem.* **84**, 1021 (1953). Synthesis and resolution: D'Amico *et al.*, *Chim. Ind. (Milan)* **38**, 93 (1956). Biosynthesis: Hartman *et al.*, *J. Am. Chem. Soc.* **77**, 816 (1955). Activity studies: Gersmeyer, Albrecht, *Med. Welt* **1966**, 657. Metabolism: Barac, *C. R. Seances Soc. Biol. Ses Fil.* **155**, 1598 (1961).

dl-**Form hydrochloride.** [15308-34-6] Coritat; Depot-Novadral; Energona; Esbuphon; Molycor-R; Novadral; Stagural; Tonolift; Vingsal; Zondel. $C_8H_{11}NO_2.HCl$; mol wt 189.64. Crystals, mp 159-160°. uv max: 274 nm ($E_{1cm}^{1\%}$ 91.21). Freely soluble in water.
THERAP CAT: Adrenergic.

6789. Norfloxacin. [70458-96-7] 1-Ethyl-6-fluoro-1,4-dihydro-4-oxo-7-(1-piperazinyl)-3-quinolinecarboxylic acid; AM-715; MK-366; Baccidal; Barazan; Chibroxin; Chibroxol; Floxacin; Lexinor; Noflo; Nolicin; Noroxin; Sebercim; Utinor; Zoroxin. $C_{16}H_{18}FN_3O_3$; mol wt 319.34. C 60.18%, H 5.68%, F 5.95%, N 13.16%, O 15.03%. Fluorinated quinolone antibacterial. Prepn: T. Irikura, **BE 863429**; *eidem*, **US 4146719** (1978, 1979 both to Kyorin); M. Pesson, **DE 2840910**; *eidem*, **US 4292317** (1979, 1981 both to Roger Bellon/Dainippon); H. Koga *et al.*, *J. Med. Chem.* **23**, 1358 (1980). HPLC determn in plasma and urine: H. J. Mascher, C. Kikuta, *J. Chromatogr. A* **812**, 381 (1998). Comparative antibacterial activity: K. Hirai *et al.*, *Antimicrob. Agents Chemother.* **19**, 188 (1981); M. Y. Khan *et al.*, *ibid.* 265; H. H. Gadebusch *et al.*, *Infection* **10**, 41 (1982). Series of articles on *in vivo* and *in vitro* activity, pharmacology, pharmacokinetics, metabolism, early clinical studies, toxicology: *Chemotherapy (Tokyo)* **29**, Suppl. 4, 1-1000 (1981). Acute toxicity data: T. Irikura *et al.*, *ibid.* 783. *In vitro* activity vs gentamicin-resistant *P. aeruginosa:* J. Downs *et al.*, *Antimicrob. Agents Chemother.* **21**, 670 (1982). Treatment of lower urinary tract infection: H. Giamarellou *et al.*, *Eur. J. Clin. Microbiol.* **2**, 266 (1983); of penicillin-resistant gonorrhea: S. R. Crider *et al.*, *N. Engl. J. Med.* **311**, 137 (1984). Review of pharmacokinetics, antibacterial activity, clinical trials: B. Holmes *et al.*, *Drugs* **30**, 482-513 (1985); R. C. Rowen *et al.*, *Pharmacotherapy* **7**, 92-110 (1987). Symposium on clinical experience: *Scand. J. Infect. Dis.* Suppl. 48, 1-91 (1986). Comprehensive description: C. Mazuel, *Anal. Profiles Drug Subs.* **20**, 557-600 (1991).

White to light-yellow crystalline powder, mp 220-221°. uv max (0.1*N* NaOH): ~274, 325, 336 nm ($A_{1cm}^{1\%}$ ~1109, 437, 425). pKa$_1$ 6.34; pKa$_2$ 8.75. Partition coefficient (octanol/water): 0.46. Soly at 25° (mg/ml): water 0.28; methanol 0.98; ethanol 1.9; acetone 5.1; chloroform 5.5; diethyl ether 0.01; benzene 0.15; ethyl acetate 0.94; octyl alcohol 5.1; glacial acetic acid 340. Solubility in water is pH dependent, increasing sharply at pH <5 or at pH >10. Hygroscopic in air, forms a hemihydrate. LD$_{50}$ in mice, rats (mg/kg): >4000 orally (both species); 1500 s.c. (both species); 470, >500 i.m.; 220, 270 i.v. (Irikura).
THERAP CAT: Antibacterial.

6790. Norflurazon. [27314-13-2] 4-Chloro-5-(methylamino)-2-[3-(trifluoromethyl)phenyl]-3(2*H*)-pyridazinone; 4-chloro-5-(methylamino)-2-(α,α,α-trifluoro-*m*-tolyl)-3(2*H*)-pyridazinone; 1-(3-trifluoromethylphenyl)-4-methylamino-5-chloropyridazone; SAN-9789; Solicam; Zorial. $C_{12}H_9ClF_3N_3O$; mol wt 303.67. C 47.46%, H 2.99%, Cl 11.67%, F 18.77%, N 13.84%, O 5.27%. Selective pre-emergent herbicide which inhibits carotenoid biosynthesis in susceptible species. Prepn: C. Ebner, M. Schuler, **BE 712832**; *eidem*, **US 3644355** (1968, 1972 both to Sandoz). Absorption and metabolism: R. H. Strang, R. L. Rogers, *J. Agric. Food Chem.* **22**, 1119 (1974). Mechanism of action: G. Sandmann *et al.*, *Pestic. Biochem. Physiol.* **14**, 185 (1980). HPLC determn in edible crops: W. M. Draper, J. C. Street, *J. Agric. Food Chem.* **29**, 724 (1981). Comprehensive description: E. Ummel *et al.*, *Proc. Brit. Weed Control Conf.* **13**, 313 (1976).

Crystals from alcohol, mp 183-185°. Soly in water (25°): 28 ppm. LD$_{50}$ orally in rats: 9300 mg/kg (Ummel).
USE: Herbicide.

6791. Norgesterone. [13563-60-5] (17α)-17-Hydroxy-19-norpregna-5(10),20-dien-3-one; 17β-hydroxy-17α-vinylestr-5(10)-en-3-one; 17α-vinyl-5(10)-estren-17β-ol-3-one; norvinodrel; vinylestrenolone. $C_{20}H_{28}O_2$; mol wt 300.44. C 79.96%, H 9.39%, O 10.65%. Prepn: Ruggieri, Ferrari, **US 2983735** (1961 to Richter); *eidem*, **US 3062713** (1962). Synthesis of racemic form: Hiscock, Whitehurst, *J. Chem. Soc.* **1965**, 5772. Biological properties: Ruggieri *et al.*, *Steroids* **5**, 73 (1965).

Crystals from diethyl ether-hexane, mp 142-143°. [α]$_D$ +161° (chloroform).
Mixture with ethinyl estradiol. [8064-75-3] Vestalin.
THERAP CAT: Progestogen.

6792. Norgestimate. [35189-28-7] (17α)-17-(Acetyloxy)-13-ethyl-18,19-dinorpregn-4-en-20-yn-3-one 3-oxime; (+)-13-ethyl-17-hydroxy-18,19-dinor-17α-pregn-4-en-20-yn-3-one oxime acetate (ester); 17α-acetoxy-13-ethyl-17-ethynylgon-4-en-3-one oxime; dexnorgestrel acetime; D-138; ORF-10131. $C_{23}H_{31}NO_3$; mol wt 369.51. C 74.76%, H 8.46%, N 3.79%, O 12.99%. Acetate oxime of D-norgestrel, *q.v.* Prepn: A. P. Shroff, **DE 2633210**; *idem*, **US 4027019** (both 1977 to Ortho). HPLC determn: P. A. Lane *et al.*, *J. Pharm. Sci.* **76**, 44 (1987). Pharmacology: D. W. Hahn *et al.*, *Contraception* **16**, 541 (1977); J. Killinger *et al.*, *ibid.* **32**, 311 (1985). Pharmacodynamics: H. S. Weintraub *et al.*, *J. Pharm. Sci.* **67**, 1406 (1978). Metabolism in female monkeys: S. F. Sisenwine *et al.*, *Contraception* **15**, 25 (1977); in women: K. B. Alton *et al.*, *ibid.* **29**, 19 (1984). Clinical trial as a contraceptive in combination with ethinyl estradiol: B. Rubio-Lotvin, R. Gonzales-Ansorena, *Acta Eur. Fertil.* **9**, 1 (1978).

Crystals from methylene chloride, mp 214-218°. [α]$_D^{25}$ +110°. Freely to very sol in methylene chloride; sparingly sol in acetonitrile. Insol in water.
Mixture with ethinyl estradiol. [79871-54-8] Cilest; Ortho Cyclen; Ortho Tri-Cyclen; Ortrel; TriCilest.
THERAP CAT: Progestogen. In combination with estrogen as oral contraceptive.

6793. Norgestrel. [6533-00-2] (17α)-(±)-13-Ethyl-17-hydroxy-18,19-dinorpregn-4-en-20-yn-3-one; Wy-3707; Neogest; Ovrette. $C_{21}H_{28}O_2$; mol wt 312.45. C 80.73%, H 9.03%, O 10.24%. The bioactive enantiomer is levorotatory. Prepn: H. Smith, **BE 623844** (1963), *C.A.* **61**, 4427c (1964); G. A. Hughes, H. Smith, **US 3959322** (1976 to Herchel Smith); H. Smith *et al.*, *Experientia* **19**,

394 (1963); *eidem, J. Chem. Soc.* **1964**, 4472; M. Rosenberger *et al., Helv. Chim. Acta* **54**, 2857 (1971). Comprehensive description: A. M. Sopirak, L. F. Cullen, *Anal. Profiles Drug Subs.* **4**, 294-318 (1975).

Crystals from methanol, mp 205-207°. uv max (ethanol): 241 nm (ε 16700). Freely sol in chloroform; sparingly sol in alc. Insol in water.

Mixture with ethinyl estradiol. [8056-51-7] Lo/Ovral; Ovral; Stédiril. Clinical efficiacy as oral contraceptive: T. R. Dunson *et al., Contraception* **48**, 109 (1993)

(+)-**Form.** [797-64-8] $(8\alpha,9\beta,10\alpha,13\alpha,14\beta)$-13-Ethyl-17-hydroxy-18,19-dinorpregn-4-en-20-yn-3-one; dextronorgestrel. Crystal structure determn: N. J. DeAngelis *et al., Acta Crystallogr.* **31B**, 2040 (1975). Crystals, mp 238-242°. $[\alpha]_D^{25}$ +40.7° (CHCl$_3$).

(−)-**Form.** [797-63-7] 17-Ethynyl-18-methyl-19-nortestosterone; 13β-ethyl-17α-ethynyl-17β-hydroxygon-4-en-3-one; levonorgestrel; D-norgestrel; dexnorgestrel (obsolete); Levonelle; Levonova; Microlut; Microval; Mirena; Norgeston; Norlevo; Norplant. Prepn: C. Rufer *et al., Ann.* **702**, 141 (1967); H. Baier *et al., Helv. Chim. Acta* **68**, 1054 (1985). Metabolism: F. Z. Stanczyk, S. Roy, *Contraception* **42**, 67 (1990). Comparative clinical trial with mifepristone, *q.v.*, in emergency contraception: H. von Hertzen *et al., Lancet* **360**, 1803 (2002). Review of clinical pharmacokinetics: K. Fotherby, *Clin. Pharmacokinet.* **28**, 203-215 (1995); of clinical efficacy of subdermal implant: A. J. Coukell, J. A. Balfour, *Drugs* **55**, 861-887 (1998); of intrauterine implant: P. Lähteenmäki *et al., Steroids* **65**, 693-697 (2000); T. Luukkainen, *ibid.* 699-702. Crystals from methanol, mp 235-237°. $[\alpha]_D^{20}$ −32.4° (c = 0.496 in CHCl$_3$). uv max (methanol): 241 nm (ε 16770).

Mixture of levonorgestrel with ethinyl estradiol. [39366-37-5] Levlen; Logynon; Microgynon; Nordette; Ovran; Ovranette; Tetragynon; Tri-Levlen; Trinordiol; Triphasil. Clinical trial in acne treatment: L. J. Diaz-Sandoval *et al., Fertil. Steril.* **76**, 461 (2001).

THERAP CAT: Progestogen; oral contraceptive; as contraceptive implant.

6794. Norgestrienone. [848-21-5] (17α)-17-Hydroxy-19-norpregna-4,9,11-trien-20-yn-3-one; 17α-ethynyl-4,9,11-estratrien-17β-ol-3-one; 17α-ethynyl-17β-hydroxy-3-oxo-4,9,11-estratriene; 17α-ethynyl-13β-methyl-$\Delta^{4,9,11}$-gonatriene-17β-ol-3-one; Ogyline. C$_{20}$H$_{22}$O$_2$; mol wt 294.39. C 81.60%, H 7.53%, O 10.87%. Prepn: NL **6401555**, corresp to Nominé *et al.,* US **3257278** (1964 and 1966, both to Roussel-UCLAF); Nominé *et al., Compt. Rend.* **260**, 4545 (1965); FR **M3060** and NL **6517141** (1965 and 1966, both to Roussel-UCLAF), *C.A.* **63**, 8449b (1965) and **65**, 15470b (1966).

Pale yellow needles from diisopropyl ether, mp 169°. $[\alpha]_D^{20}$ +63° (c = 0.5 in alc). uv max: 342, 238 nm (ε 29100, 5920). Sol in alcohols, ether, acetone, benzene, chloroform; practically insol in water, dil aq acids and alkalies.

THERAP CAT: Progestogen.

6795. Norleucine. [327-57-1] L-Norleucine; α-aminocaproic acid; (S)-2-aminohexanoic acid; glycoleucine; caprine. C$_6$H$_{13}$NO$_2$; mol wt 131.18. C 54.94%, H 9.99%, N 10.68%, O 24.39%. An

amino acid classified as nonessential with respect to its growth effect in rats. Several syntheses; prepn from α-bromo-*n*-caproic acid by action of 25% ammonia at 50-55°: Marvel, du Vigneaud, *Org. Synth.* **4**, 3 (1925).

Slightly sweet, shiny leaflets from water. mp 301° (partial decompn). Sublimes partially at 275-280°. $[M]_D$ +32.1° (5*N* HCl); +47.9° (glacial acetic acid). $[\alpha]_D^{20}$ +21.3° (c = 4.25 in 6*N* HCl); +6.26° (c = 0.70 in water).

DL-**Form.** [616-06-3] Lustrous leaflets from water. d 1.172. Dec 327°. pK$_1$ 2.39; pK$_2$ 9.76. Soly in water: 11.49 g/l at 25°, 17.27 g/l at 50°, 28.61 g/l at 75°, 52.0 g/l at 100°. Sparingly sol in alcohol: 0.42 g/100 g at 25°. Sol in acids.

D(−)-**Form.** [327-56-0] Bitter, shiny leaflets from water. mp 301° (partial decompn). Sublimes partially at 275-280°. $[\alpha]_D^{20}$ −22.4° (c = 4.69 in 6*N* HCl); −4.49° (c = 0.96 in water).

6796. Norlevorphanol. [1531-12-0] Morphinan-3-ol; (−)-3-hydroxymorphinan; (−)-1,3,4,9,10,10a-hexahydro-6-hydroxy-2*H*-10,4a-iminoethanophenanthrene; NIH-7539. C$_{16}$H$_{21}$NO; mol wt 243.35. C 78.97%, H 8.70%, N 5.76%, O 6.57%. Opioid analgesic. Prepn of (−) and (+) isomers: O. Schnider, A. Grüssner, *Helv. Chim. Acta* **34**, 2211 (1951); J. Hellerbach, *et al., ibid.* **39**, 429 (1956); GB **765920** (1957 to Hoffmann-La Roche).

Crystals from acetone + methanol, mp 270-272°. $[\alpha]_D^{21}$ −42 ±2° (c = 1 in methanol).

Hydrobromide. [63732-85-4] C$_{16}$H$_{21}$NO.HBr. Crystals from water, mp 222-224°.

(+)-**Form.** [15676-23-0] $(9\alpha,13\alpha,14\alpha)$-Morphinan-3-ol; nordextrorphan. Secondary metabolite of dextromethorphan, *q.v.*: G. Pfaff *et al., Int. J. Pharm.* **14**, 173 (1983). Stereocontrolled synthesis: Y. Génisson *et al., J. Org. Chem.* **58**, 2052 (1993). Crystals from acetone, mp 258-259°. $[\alpha]_D$ +40° (c = 0.5 in methanol).

Note: Norlevorphanol is a controlled substance (opiate): **21 CFR,** 1308.11.

USE: Intermediate in synthesis of morphinan derivatives.

6797. Normetanephrine. [97-31-4] α-(Aminomethyl)-4-hydroxy-3-methoxybenzenemethanol; α-(aminomethyl)vanillyl alcohol; 4-hydroxy-3-methoxy-α-(aminomethyl)benzyl alcohol; 1-(4-hydroxy-3-methoxyphenyl)-2-aminoethanol; 3-*O*-methylnoradrenaline; 3-*O*-methylnorepinephrine. C$_9$H$_{13}$NO$_3$; mol wt 183.21. C 59.00%, H 7.15%, N 7.65%, O 26.20%. A naturally occurring derivative of epinephrine, found together with metanephrine in urine and in certain tissues. Prepn: Fodor *et al., Acta Chim. Acad. Sci. Hung.* **1**, 395 (1951), *C.A.* **49**, 897 (1955); Axelrod *et al., J. Biol. Chem.* **233**, 697 (1958); Heacock, Hutzinger, *Chem. Ind. (London)* **1961**, 595.

dl-**Form hydrochloride.** C$_9$H$_{13}$NO$_3$.HCl. Prisms from abs ethanol, dec 206-207°. uv max (abs ethanol): 232, 282 nm (ε 7100, 2970).

6798. Normethadone. [467-85-6] 6-(Dimethylamino)-4,4-diphenyl-3-hexanone; 1-dimethylamino-3,3-diphenyl-4-hexanone; 1,1-diphenyl-1-(2-dimethylaminoethyl)-2-butanone; isoamidone I; desmethylmethadone; phenyldimazone; Hoechst 10582. $C_{20}H_{25}$-NO; mol wt 295.43. C 81.31%, H 8.53%, N 4.74%, O 5.42%. Opioid analgesic used as a cough suppressant. Prepn from 4-bromo-2,2-diphenylbutanenitrile and dimethylamine: Bockmühl, Ehrhart, *Ann.* **561**, 72 (1948); Easton *et al., J. Am. Chem. Soc.* **74**, 5772 (1952). Toxicity data: Eddy *et al., J. Pharmacol. Exp. Ther.* **98**, 121 (1950).

Oily liquid, bp_3 164-167°. Alkaline reaction.
Hydrochloride. [847-84-7] Ticarda. $C_{20}H_{25}$NO.HCl; mol wt 331.88. Crystals from acetone, mp 174-175°. Sol in water, alcohol. pH of 1% aq soln ~5. LD_{50} s.c. in mice: 90 mg/kg (Eddy).
Note: This is a controlled substance (opiate): **21 CFR**, 1308.11.
THERAP CAT: Antitussive.

6799. Normethandrone. [514-61-4] (17β)-17-Hydroxy-17-methylestr-4-en-3-one; 17α-methyl-19-nortestosterone; methylestrenolone; normethandrolone; normetandrone; methylnortestosterone; Orgasteron; Metalutin; Methalutin. $C_{19}H_{28}O_2$; mol wt 288.43. C 79.12%, H 9.79%, O 11.09%. Prepn: Djerassi *et al.*, **US 2744122** and **US 2774777** (1956 to Syntex); *J. Am. Chem. Soc.* **76**, 4092 (1956); De Ruggieri, **US 2849461** (1958).

Crystals from ether-hexane, mp 156-158° (Kofler). $[\alpha]_D$ +33°. uv max (ethanol): 240 nm (log ε 4.23).
Note: This is a controlled substance (anabolic steroid): **21 CFR**, 1308.13, as defined in 1300.01.
THERAP CAT: Androgen.

6800. Normorphine. [466-97-7]; [6035-32-1] (sesquihydrate). (5α,6α)-7,8-Didehydro-4,5-epoxymorphinan-3,6-diol; desmethylmorphine. $C_{16}H_{17}NO_3$; mol wt 271.32. C 70.83%, H 6.32%, N 5.16%, O 17.69%. Opioid analgesic; bioactive, minor metabolite of morphine, *q.v.* Prepn: I. J. von Braun *et al., Ber.* **47**, 2312 (1914); **DE 286743** (1914 to Hoffmann-La Roche); E. Speyer, L. Walther, *Ber.* **63**, 852 (1930); J. Weijlard, A. E. Erickson, *J. Am. Chem. Soc.* **64**, 869 (1942); M. M. Abdel-Monem, P. S. Portoghese, *J. Med. Chem.* **15**, 208 (1975). Characterization as morphine metabolite: U. Boerner *et al., J. Pharm. Pharmacol.* **26**, 393 (1974). Formation by cytochrome P450 isoenzymes: D. Projean *et al., Xenobiotica* **33**, 841 (2003). GC-MS determn in urine: C. Meadway *et al., Forensic Sci. Int.* **127**, 136 (2002).

Crystals from chloroform + isopropanol, mp 277°. Sesquihydrate, crystals from water, mp 272-273°. Sparingly sol in hot water, alcohol. Insol in ether, chloroform.

Hydrochloride. [3372-02-9] $C_{16}H_{17}NO_3$.HCl; mol wt 307.77. Monohydrate, crystals from water, dec 305°.
Note: This is a controlled substance (opium derivative): **21 CFR**, 1308.11.

6801. Nornicotine. [494-97-3] 3-(2S)-2-Pyrrolidinylpyridine; 2-(3-pyridyl)pyrrolidine. $C_9H_{12}N_2$; mol wt 148.21. C 72.94%, H 8.16%, N 18.90%. Occurs in ordinary tobacco, in other species of *Nicotiana*, and also in *Duboisia hopwoodii* F. Muell., *Solanaceae*. Synthesis: Mizogruchi, *Chem. Pharm. Bull.* **9**, 818 (1961); M. Nakane, C. R. Hutchinson, *J. Org. Chem.* **43**, 3922 (1978). Pharmacology: P. S. Larson, H. B. Haag, *J. Pharmacol. Exp. Ther.* **77**, 343 (1943). Toxicity: P. S. Larson *et al., Proc. Soc. Exp. Biol. Med.* **58**, 231 (1945). *Review:* Markwood, *U.S. Dept. Agr. Bur. Entomol. Plant Quarantine* **E-561** (1942), Supplement by Roark, **E-645** (1945).

Hygroscopic, somewhat viscous liquid, develops a slight amine odor, less pungent than that of nicotine. bp 270°; bp_{11} 131°; bp_3 105-107°. Hardly volatile with steam (difference from nicotine which is readily volatile). d_4^{20} 1.0737. $[\alpha]_D$ −89° (c = 100). n_D^{20} 1.5378. Miscible with water. Very sol in alcohol, chloroform, ether, petr ether, kerosene, oils. Less volatile and less easily oxidized than nicotine. LD_{50} in mice, rabbits (mg/kg): 21.7, >13.7 i.p.; 3.4, 3.0 i.v. (Larson *et al.*).
USE: Agricultural or horticultural insecticide.

6802. Norphenazone. [89-25-8] 2,4-Dihydro-5-methyl-2-phenyl-3H-pyrazol-3-one; 3-methyl-1-phenyl-2-pyrazolin-5-one; 1-phenyl-3-methyl-5-pyrazolone; C.I. Developer 1; developer Z; norantipyrine; MCI-186. $C_{10}H_{10}N_2O$; mol wt 174.20. C 68.95%, H 5.79%, N 16.08%, O 9.18%. Free radical scavenger; metabolite of antipyrine, *q.v.* Prepn: L. Knorr, *Ann.* **238**, 137 (1887). Structural determn by NMR and IR: R. Jones *et al., Tetrahedron* **19**, 1497 (1963); F.-D. Höppner *et al., J. Prakt. Chem.* **318**, 555 (1976). Crystal structure: F. Bechtel *et al., Cryst. Struct. Commun.* **3**, 469 (1973). HPLC determn in urine: M. A. Mikati *et al., J. Chromatogr.* **433**, 305 (1988). Anti-ischemic and radical scavenging actions: T. Watanabe *et al., J. Pharmacol. Exp. Ther.* **268**, 1597 (1994).

White crystals from benzene, mp 129-130°. Also reported as mp 127° (Knorr). bp_{265} 287°.
THERAP CAT: In treatment of stroke.

6803. Norpseudoephedrine. [492-39-7] (αS)-α-[(1S)-1-Aminoethyl]benzenemethanol; D-*threo*-2-amino-1-hydroxy-1-phenylpropane; (1S,2S)-2-amino-1-phenylpropan-1-ol; nor-d-ψ-ephedrine; d-ψ-norephedrine; d-norisoephedrine; cathine; katine. C_9H_{13}-NO; mol wt 151.21. C 71.49%, H 8.67%, N 9.26%, O 10.58%. Sympathomimetic amine naturally occurring in the leaves of the khat plant, *Catha edulis* Forsk. (Celastraceae) and in several species of *Ephedra* (Ephedraceae) known in traditional medicine as Ma Huang. Isoln from *C. edulis*: A. Beitter, *Arch. Pharm.* **239**, 17 (1901); O. Wolfes, *ibid.* **268**, 81 (1930); from Ma Huang: S. Smith, *J. Chem. Soc.* **1928**, 51. Physical measurements: C. S. Gibson, B. Levin, *ibid.* **1929**, 2754. Synthesis: W. N. Nagai, S. Kanao, *Ann.* **470**, 157 (1929); J. Sicher, M. Pankova, *Collect. Czech. Chem. Commun.* **20**, 1409 (1955). Pharmacology: H. Hofmann *et al., Arzneim.-Forsch.* **5**, 367 (1955). Clinical pharmacokinetics: F. Frosch, *ibid.* **27**, 665,

1076 (1977). Clinical evaluation for weight reduction: H. Muth, G. Issmeier, *Ther. Ggw.* **119**, 1173 (1980). HPLC determn in urine: P. J. van der Merwe *et al.*, *J. Chromatogr. B* **661**, 357 (1994).

Prisms from benzene, mp 77.5-78°. $[\alpha]^{20}_{546}$ +37.9° (c = 3 in methanol). pKa 8.92. Sol in alcohol, chloroform, ether, dil acids.

Hydrochloride. [2153-98-2] Amorphan; Adiposetten; Exponcit N; Fasupond; Minusin. $C_9H_{13}NO.HCl$; mol wt 187.67. Prisms from alcohol, mp 180-181°. $[\alpha]^{20}_{D}$ +42.53°. Sol in water.

Note: This is a controlled substance (stimulant): **21 CFR,** 1308.14.

USE: In the optical resolution of externally compensated acids.

THERAP CAT: Anorexic.

6804. Nortriptyline. [72-69-5] 3-(10,11-Dihydro-5*H*-dibenzo[*a,d*]cyclohepten-5-ylidene)-*N*-methyl-1-propanamine; 10,11-dihydro-*N*-methyl-5*H*-dibenzo[*a,d*]cycloheptene-$\Delta^{5,\gamma}$-propylamine; 5-(α-methylaminopropylidene)dibenzo[*a,d*]cyclohepta[1,4]diene; 3-(10,11-dihydro-5*H*-dibenzo[*a,d*]cyclohepten-5-ylidene)-*N*-methylpropylamine; 10,11-dihydro-5-(3-methylaminopropylidene)-5*H*-dibenzo[*a,d*][1,4]cycloheptene; desitriptilina; desmethylamitriptyline. $C_{19}H_{21}N$; mol wt 263.38. C 86.65%, H 8.04%, N 5.32%. Tricyclic antidepressant. Prepn: Hoffsommer *et al.*, *J. Org. Chem.* **27**, 4134 (1962); **NL 6408512**; E. I. Engelhardt, **US 3922305** (1965, 1975 both to Merck & Co.). Alternative process: N. L. Wendler, **US 3442949** (1969 to Merck & Co.). GC determn in plasma: J. E. Burch, *J. Chromatogr.* **308**, 165 (1984). Determn in plasma or serum by radioimmunoassay: J. F. Sayegh, *Neurochem. Res.* **11**, 193 (1986). HPLC determn in serum: O. V. Olesen *et al.*, *J. Chromatogr. B* **746**, 233 (2000). Pharmacokinetics: E. H. Rubin *et al.*, *J. Clin. Psychiatry* **46**, 418 (1985). Clinical pharmacogenetics: E. E. Kvist *et al.*, *Clin. Pharmacokinet.* **40**, 869 (2001). Clinical trial in post-stroke depression: J. R. Lipsey *et al.*, *Lancet* **1**, 297 (1984); in treatment-resistant depression: A. A. Nierenberg *et al.*, *J. Clin. Psychiatry* **64**, 35 (2003). Anticholinergic effect and clinical efficacy in urinary dysfunction: I. Nissenkorn *et al.*, *Eur. Urol.* **12**, 109 (1986). Comprehensive description: J. L. Hale, *Anal. Profiles Drug Subs.* **1**, 233-247 (1972).

Hydrochloride. [894-71-3] Acetexa; Allegron; Aventyl; Noritren; Nortrilen; Pamelor; Sensaval; Vividyl. $C_{19}H_{21}N.HCl$; mol wt 299.84. Crystals from ether + ethanol, mp 213-215°. Sol in ethanol, water, chloroform; sparingly sol in methanol. Practically insol in ether, acetone, benzene and in most other organic solvents. uv max (methanol): 240 nm (ε 13900).

THERAP CAT: Antidepressant.

THERAP CAT (VET): In treatment of canine behavioral disorders.

6805. Norvaline. [6600-40-4] L-Norvaline; (*S*)-2-aminovaleric acid; L-α-aminovaleric acid; (*S*)-2-aminopentanoic acid. C_5H_{11}-NO_2; mol wt 117.15. C 51.26%, H 9.46%, N 11.96%, O 27.31%. Prepd by treating butyraldehyde ammonia with HCN and HCl: Slimmer, *Ber.* **35**, 404 (1902); from 2-acetylvaleric acid ethyl ester: Hamlin, Hartung, *J. Biol. Chem.* **145**, 349 (1942); from acetamidomalonic acid diethyl ester: Archer, Albertson, **US 2445817** (1948 to Winthrop-Stearns); from 1-nitrobutane: Stiles, Finkbeiner, *J. Am. Chem. Soc.* **85**, 616 (1963); **US 3055936** (1962 to Res. Corp.). Prepn of optically active forms: Abderhalden, Kurton, *Fermentf.* **4**, 328; *Chem. Zentralbl.* **1921**, III, 296.

Crystals from dil alc. mp ~305° (closed capillary). [M]$_D$ +29.2° (5*N* HCl); [M]$_D$ +41.0° (glacial acetic acid). $[\alpha]^{20}_{D}$ +23.0° (c = 10 in 20% HCl). Freely sol in hot water; insol in alcohol, ether, chloroform, ethyl acetate, petr ether.

DL-Form. [760-78-1] Minute leaflets from alcohol or water, mp 303° (closed capillary). pK$_1'$ 2.36; pK$_2'$ 9.72. Sublimes without decompn. One gram dissolves in 10 ml water at 18°. Freely sol in hot water; practically insol in alcohol, ether, chloroform, ethyl acetate, petr ether.

D(−)-Form. [2013-12-9] Minute leaflets. mp ~307°. $[\alpha]^{20}_{D}$ −24.2° (c = 10 in 20% HCl). Freely sol in hot water; insol in alcohol, ether, chloroform, ethyl acetate, petr ether.

6806. Norvinisterone. [6795-60-4] (17α)-17-Hydroxy-19-norpregna-4,20-dien-3-one; 17-hydroxy-17α-vinyl-4-estren-3-one; 17-hydroxy-13-methyl-17α-vinyl-1,2,3,6,7,8,9,10,11,12,13,14,-16,17-tetradecahydro-15*H*-cyclopenta[*a*]phenanthren-3-one; 17α-vinyl-19-nortestosterone; Nor-Progestelea. $C_{20}H_{28}O_2$; mol wt 300.44. C 79.96%, H 9.39%, O 10.65%. Prepn: Colton, **US 2655518**; **US 2802015** (both 1953 to Searle).

Crystals from ethyl acetate + petr ether, mp 169-171°. $[\alpha]_D$ +36°.

THERAP CAT: Progestogen.

6807. Noscapine. [128-62-1] (3*S*)-6,7-Dimethoxy-3-[(5*R*)-5,6,7,8-tetrahydro-4-methoxy-6-methyl-1,3-dioxolo[4,5-*g*]isoquinolin-5-yl]-1(3*H*)-isobenzofuranone; narcotine; *l*-α-narcotine; *l*-α-2-methyl-8-methoxy-6,7-methylenedioxy-1-(6,7-dimethoxy-3-phthalidyl)-1,2,3,4-tetrahydroisoquinoline; narcosine; methoxyhydrastine; opian; opianine; NSC-5366; Nipaxon. $C_{22}H_{23}NO_7$; mol wt 413.43. C 63.91%, H 5.61%, N 3.39%, O 27.09%. An opium alkaloid, isolated from the plant *Papaver somniferum* L. Papaveraceae. Present in amounts up to 11% depending on season and locality. First isoln: Robiquet, *Ann. Chim. Phys.* [2] **5**, 275 (1817). Extractable from the water-insoluble residue remaining from the processing of opium for the manufacture of morphine. Racemization to gnoscopine: Rabe, McMillian, *Ann.* **377**, 233 (1910); and structural studies: Perkin, Robinson, *J. Chem. Soc.* **99**, 775 (1911); Marshall *et al.*, *ibid.* **1934**, 1318. Preliminary stereochemical studies: Ohta *et al.*, *Tetrahedron Lett.* **1963**, 1857; Battersby, Spenser, *J. Chem. Soc.* **1965**, 1087. Revised stereochemistry: Blaha *et al.*, *Collect. Czech. Chem. Commun.* **29**, 2328 (1964); Snatzke *et al.*, *Tetrahedron* **25**, 5059 (1969). Synthesis of racemate: Kerekes, Bognar, *J. Prakt. Chem.* **313**, 923 (1971). Biosynthesis: Battersby, Hirst, *Tetrahedron Lett.* **1965**, 669. Metabolism: N. Tsunoda, Y. Yoshimura, *Xenobiotica* **9**, 181 (1979); *eidem, ibid.* **11**, 23 (1981). Pharmacokinetics: B. Dhalstroem *et al.*, *Eur. J. Clin. Pharmacol.* **22**, 535 (1982). Clinical evaluation as antitussive: D. W. Empey *et al.*, *ibid.* **16**, 393 (1979). HPLC determn in serum: K. M. Jensen, *J. Chromatogr.* **274**, 381 (1983). Comprehensive description: M. A. Al-Yahya, M. M. A. Hassan, *Anal. Profiles Drug Subs.* **11**, 407-461 (1982).

Orthorhombic bisphenoidal prisms, tablets from diacetone. Triboluminescent. d 1.395. mp 176°. Sublimes at 150-160° under 11 mm pressure at 2 mm distance. Very weak base forming unstable salts with acids and strong bases. pK 7.8. uv max (ethanol): 209, 291, 309-310 nm (log ε 4.86, 3.60, 3.69). Practically insol in vegetable oils. Slightly sol in NH$_4$OH, hot solns of KOH and NaOH, forming salts. Salts formed with acids are dextrorotatory and unstable in water.

Hydrochloride. [912-60-7] Capval. $C_{22}H_{23}NO_7 \cdot HCl$; mol wt 449.88. Hemihydrate to tetrahydrate, crystals, very sol in water forming basic salts.

Camphorsulfonate. [25333-79-3] Tulisan. $C_{22}H_{23}NO_7 \cdot C_{10}H_{16}O_4S$; mol wt 645.72. Contains 35.97% camphosulfonic acid. Prepn: Maillard, **US 3108106** (1963 to Jacques Logeais). Crystals, mp 188-191°. $[\alpha]_D^{33} +32.7°$ (c = 4.56 in water). Freely sol in water. Sol in methanol, ethanol. Slightly sol in ethyl acetate. Practically insol in ether.

dl-Form. [6035-40-1] *dl*-Narcotine; gnoscopine. Long needles from methanol, mp 232° (dec). pK 7.8. Freely sol in carbon disulfide, hot chloroform; sol in ~1500 parts alcohol; sparingly sol in benzene, water.

THERAP CAT: Antitussive.

6808. Nosiheptide. [56377-79-8] *N*-[1-(Aminocarbonyl)ethenyl]-2-[(11*S*,14*Z*,21*S*,23*S*,29*S*)-14-ethylidene-9,10,11,12,13,14,-19,20,21,22,23,24,26,33,35,36-hexadecahydro-3,23-dihydroxy-11-[(1*R*)-1-hydroxyethyl]-31-methyl-9,12,19,24,33,43-hexaoxo-30,32-imino-8,5:18,15:40,37-trinitrilo-21,26-([2,4]-*endo*-thiazolomethanimino)-5*H*,15*H*,37*H*-pyrido[3,2-*w*][2,11,21,27,31,7,14,17]benzoxatetrathiatriazacyclohexatriacontin-2-yl]-4-thiazolecarboxamide; Multhiomycin; RP-9671; Primofax. $C_{51}H_{43}N_{13}O_{12}S_6$; mol wt 1222.34. C 50.11%, H 3.55%, N 14.90%, O 15.71%, S 15.74%. Polythiazole antibiotic produced by *Streptomyces actuosus*. Isoln and characterization: S. Pinnert *et al.*, **FR 1392453**; *eidem*, **US 3155581** (1961, 1964 both to Rhône-Poulenc); F. Benazet *et al.*, *Experientia* **36**, 414 (1980). NMR determination of mol wt and elemental formula: H. Depaire *et al.*, *Tetrahedron Lett.* **1977**, 1397, 1401. Structure and configuration: T. Prange *et al.*, *Nature* **265**, 189 (1977); C. Pascard *et al.*, *J. Am. Chem. Soc.* **99**, 6418 (1977). Biosynthetic study: D. P. Houck *et al.*, *ibid.* **109**, 1250 (1987). Identity with multhiomycin: T. Endo, H. Yonehara, *J. Antibiot.* **31**, 623 (1978). Mode of action: E. Cundliffe, J. Thompson, *J. Gen. Microbiol.* **126**, 185 (1981). *Review:* F. Benazet *et al.*, *Experientia* **36**, 414-416 (1980).

Yellow needles, mp 310-320° (dec). $[\alpha]_D^{20} +38°$ (c = 1 in pyridine). uv max (water/DMF): 242, 322 nm ($E_{1cm}^{1\%}$ 525, 229). Sol in chloroform, dioxane, pyridine, DMF, DMSO; slightly sol in methanol, ethanol, ethyl acetate, benzene. Insol in water and petr ether.

THERAP CAT (VET): Antibacterial; growth promotant.

6809. Novaluron. [116714-46-6] *N*-[[[3-Chloro-4-[1,1,2-trifluoro-2-(trifluoromethoxy)ethoxy]phenyl]amino]carbonyl]-2,6-difluorobenzamide; 1-[3-chloro-4-(1,1,2-trifluoro-2-trifluoromethoxy-ethoxy)phenyl]-3-(2,6-difluorobenzoyl)urea; MCW-275; Pedestal;

Rimon. $C_{17}H_9ClF_8N_2O_4$; mol wt 492.71. C 41.44%, H 1.84%, Cl 7.19%, F 30.85%, N 5.69%, O 12.99%. Insect growth regulator. Prepn: P. Massardo *et al.*, **EP 271923**; *eidem*, **US 4980376** (1988, 1990 both to Ist. Guido Donegani); F. Rama *et al.*, *Pestic. Sci.* **35**, 145 (1992). Greenhouse trials vs beet armyworm and greenhouse whitefly: I. Ishaaya *et al.*, *Brighton Crop Prot. Conf. - Pests Dis.* **1998**, 49.

mp 172-174°.

USE: Insecticide.

6810. Noviflumuron. [121451-02-3] *N*-[[[3,5-Dichloro-2-fluoro-4-(1,1,2,3,3,3-hexafluoropropoxy)phenyl]amino]carbonyl]-2,6-difluorobenzamide; *N*-[3,5-dichloro-2-fluoro-4-(1,1,2,3,3-hexafluoropropoxy)phenyl]-*N'*-(2,6-difluorobenzoyl)urea; XDE-007; X-550007; Recruit III; Recruit IV. $C_{17}H_7Cl_2F_9N_2O_3$; mol wt 529.14. C 38.59%, H 1.33%, Cl 13.40%, F 32.31%, N 5.29%, O 9.07%. Chitin synthesis inhibitor. Prepn: J. Drabek, **DE 3827133** (1989 to Ciba-Geigy). *See also:* R. J. Sbragia *et al.*, **US 5886221** (1999 to Dow AgroSciences). Laboratory performance and pharmacokinetics in eastern subterranean termites: L. L. Karr *et al.*, *J. Econ. Entomol.* **97**, 593 (2004).

White crystalline powder, mp 153-154° (Drabek). Also reported as light tan solid, mp 156-157° (Sbragia). Vapor pressure (20°): 2.20×10^{-10} mm Hg. Log P (octanol/water): 4.94 (20°). Relative density: 1.88 g/cm³. Soly in water (20°): 0.194 mg/l. Soly at 19° (g/l): acetone 425.0, acetonitrile 44.9, 1,2-dichloroethane 20.7, ethyl acetate 290.0, heptane 0.068, methanol 48.9, 1-octanol 8.1, *p*-xylene 93.3.

USE: Insecticide for control of termites.

6811. Novobiocin. [303-81-1] *N*-[7-[[3-*O*-(Aminocarbonyl)-6-deoxy-5-*C*-methyl-4-*O*-methyl-α-L-*lyxo*-hexopyranosyl]oxy]-4-hydroxy-8-methyl-2-oxo-2*H*-1-benzopyran-3-yl]-4-hydroxy-3-(3-methyl-2-buten-1-yl)benzamide; crystallinic acid; streptonivicin; PA-93; U-6591. $C_{31}H_{36}N_2O_{11}$; mol wt 612.63. C 60.78%, H 5.92%, N 4.57%, O 28.73%. Antibiotic substance produced by *Streptomyces spheroides*: Kaczka *et al.*, *J. Am. Chem. Soc.* **77**, 6404 (1955); Wolf, **US 3000873** (1961 to Merck & Co.); Stammer, Miller; Miller; Wallick, **US 3049475**; **US 3049476**; **US 3049534** (all 1962 to Merck & Co.). By *Streptomyces niveus:* Hoeksema *et al.*, *J. Am. Chem. Soc.* **77**, 6710 (1955); *Antibiot. Chemother.* **6**, 143 (1956); French, **US 3068221** (1962 to Upjohn). Structure: Shunk *et al.*, *J. Am. Chem. Soc.* **78**, 1770 (1956); Hoeksema *et al.*, *ibid.* 2019; Walton *et al.*, *ibid.* **82**, 1489 (1960). Conformation: Golding, Richards, *Chem. Ind. (London)* **1963**, 1081. Revised configuration: O. Achmatowicz *et al.*, *Tetrahedron* **32**, 1051 (1976). Synthesis: Stammer, **US 2925411** (1960); Walton, Spencer, **US 2966484** (1960 to Merck & Co.); Vaterlaus *et al.*, *Helv. Chim. Acta* **47**, 390 (1964). Conversion of *isonovobiocin* to novobiocin: Caron *et al.*, **US 2983723** (1961 to Upjohn). Antiviral activity: Chang, Weinstein, *Antimicrob. Agents Chemother.* **1970**, 165. Efficacy in canine respiratory infections: B. W. Maxey, *Vet. Med. Small Anim. Clin.* **75**, 89 (1980). Mechanism of action studies: Smith, Davis, *J. Bacteriol.* **93**, 71 (1967); H. T. Wright *et al.*, *Science* **213**, 455 (1981); I. W. Althaus *et al.*, *J. Antibiot.* **41**, 373 (1988). *Review:* Brock in *Antibiotics* vol. 1, R. Gottlieb, P. Shaw, Eds. (Springer-Verlag, New York, 1967) pp 651-665; M. J. Ryan, *ibid.* **vol. 5** (pt. 1), F. E. Hahn, Ed. (1979) pp 214-234.

Pale yellow orthorhombic crystals from ethanol. *Sensitive to light.* d 1.3448. Dec at 152-156° (a rarer modification dec 174-178°). Acid reaction: pKa_1 4.3; pKa_2 9.1. $[\alpha]_D^{24}$ −63.0° (c = 1 in ethanol). uv max (0.1N NaOH; 0.1N methanolic HCl; pH 7 phosphate buffer): 307; 324; 390 nm ($E_{1cm}^{1\%}$ 600, 390, 350 resp.). Sol in aq soln above pH 7.5. Practically insol in more acidic solns. Sol in acetone, ethyl acetate, amyl acetate, lower alcohols, pyridine. Additional soly data: Weiss *et al.*, *Antibiot. Chemother.* **7**, 374 (1957).

Monosodium salt. [1476-53-5] Albamycin. $C_{31}H_{35}N_2NaO_{11}$; mol wt 634.61. Minute crystals, dec 220°. $[\alpha]_D^{24}$ −38° (c = 2.5 in 95% ethanol); $[\alpha]_D^{24}$ −33° (c = 2.5 in water). Freely sol in water. A 100 mg/ml soln has a pH of 7.5 and a half-life of ~30 days at 25° and several months at 4°. Soly data: Weiss *et al.*, *loc. cit.* Properties: Birlova, Traktenberg, *Antibiotiki* **13**, 997 (1968).

THERAP CAT: Antibacterial.

THERAP CAT (VET): Antimicrobial.

6812. Novoldiamine. [140-80-7] N^1,N^1-Diethyl-1,4-pentanediamine; 1-diethylamino-4-aminopentane; 4-amino-1-diethylaminopentane; 2-amino-5-diethylaminopentane; δ-diethylamino-α-methylbutylamine; δ-diethylaminoisopentylamine. $C_9H_{22}N_2$; mol wt 158.29. C 68.29%, H 14.01%, N 17.70%. Prepd commercially from 2-diethylaminoethanol and ethyl acetoacetate. 2-Chlorotriethylamine (formed by the action of thionyl chloride on the alcohol) is condensed with the sodium derivative of ethyl acetoacetate to yield an intermediate ester, which is hydrolyzed and decarboxylated to *novol ketone* (5-diethylamino-2-pentanone). This is hydrogenated in the presence of ammonia to yield novoldiamine. Several other prepns, i.e., from 1,4-pentanediol and diethylamine: Kyrides, **US 2365825** (1944 to Monsanto). Purification procedure: Jones, **US 2400934** (1946 to Lilly).

Liquid. Amine odor. d_{26}^{20} 0.819. bp_{753} 200-200.5°. n_D^{26} 1.4403. Sol in water, alcohol, ether.

USE: Manuf quinacrine and other antimalarials having the same basic side chain.

6813. Noxythiolin. [15599-39-0] N-(Hydroxymethyl)-N'-methylthiourea; 1-(hydroxymethyl)-3-methyl-2-thiourea; noxytiolin; Noxyflex-S. $C_3H_8N_2OS$; mol wt 120.17. C 29.99%, H 6.71%, N 23.31%, O 13.31%, S 26.68%. Prepn: A. Aebi, E. Hafstetter, **GB 970414** (1964 to Ed. Geistlich Söhne), *C.A.* **61**, 15981e (1964). Antibacterial activity *in vitro*: D. Horsfield, *Clin. Trials J.* **4**, 625 (1967); J. I. Blenkharn, *J. Pharm. Pharmacol.* **42**, 589 (1990). Inhibition of bacterial adherence *in vitro*: S. P. Gorman *et al.*, *J. Appl. Bacteriol.* **60**, 311 (1986). Clinical evaluation in peritonitis: F. Antos, *Clin. Trials J.* **17**, 159 (1980); in urinary tract sepsis: M. A. Jones, A. Hasan, *Br. J. Urol.* **62**, 311 (1988).

Crystals. mp 88-90°. Sol in water (10% w/v), ethanol (4% w/v). pH of 2.5% soln: 6.31-7.0. LD_{50} orally in mice: >3 g/kg (Aebi, Hafstetter).

THERAP CAT: Antiseptic.

6814. NPPB. [107254-86-4] 5-Nitro-2-[(3-phenylpropyl)-amino]benzoic acid; HOE-144. $C_{16}H_{16}N_2O_4$; mol wt 300.31. C

63.99%, H 5.37%, N 9.33%, O 21.31%. Inhibits efflux of ion channels; specificity is cell type dependent. Prepn: H. C. Englert *et al.*, **DE 3528048**; *eidem*, **US 4994493** (1987, 1991 both to Hoechst). Synthesis and characterization as potassium channel blocker: K. R. Giles *et al.*, *Bioorg. Med. Chem. Lett.* **13**, 293 (2003). Blocking kinetics of chloride channels: J. Dreinhöfer *et al.*, *Biochim. Biophys. Acta* **946**, 135 (1988). Effect on chloride ion transport: B. R. Branchini *et al.*, *Biochem. Biophys. Res. Commun.* **176**, 459 (1991); *eidem*, *Arch. Biochem. Biophys.* **318**, 221 (1995); J.-A. Kim *et al.*, *Biochem. Biophys. Res. Commun.* **309**, 291 (2003).

Yellow crystals from isopropanol, mp 178-180°. Soly (mg/ml): 25 DMSO, 10 ethanol.

USE: Biochemical probe primarily as a Cl^- channel blocker.

6815. Nucleocidin. [24751-69-7] 4′-C-Fluoroadenosine 5′-sulfamate; 9-(4-fluoro-5-O-sulfamoylpentofuranosyl)adenine; 4′-fluoro-5′-O-sulfamoyladenosine. $C_{10}H_{13}FN_6O_6S$; mol wt 364.31. C 32.97%, H 3.60%, F 5.21%, N 23.07%, O 26.35%, S 8.80%. Antitrypanosomal antibiotic produced by *Streptomyces calvus*. The first naturally occurring derivative of a fluoro sugar. Isoln and activity studies: Thomas *et al.*, *Antibiot. Annu.* **1956-1957**, 716; Hewitt *et al.*, *ibid.* 722. Partial structure: Waller *et al.*, *J. Am. Chem. Soc.* **79**, 1011 (1957). Revised formula and structure: Morton *et al.*, *ibid.* **91**, 1535 (1969). Synthesis: Jenkins *et al.*, *ibid.* **93**, 4323 (1971).

As the monohydrate, crystalline, weakly alkaline substance. mp >190° (dec). uv max (methanol): 259 nm (ε 15000). LD_{50} i.p. in mice: ~0.2 mg/kg (Thomas).

6816. Nupharidine. [468-89-3] (1R,4S,5R,7S,9aS)-4-(3-Furanyl)octahydro-1,7-dimethyl-2H-quinolizine 5-oxide. $C_{15}H_{23}NO_2$; mol wt 249.35. C 72.25%, H 9.30%, N 5.62%, O 12.83%. From rhizome of *Nuphar luteum* (L.) Sibth. & Sm. (*Nymphaea lutea* L.), *Nymphaeaceae*: Goris, Crété, *Bull. Sci. Pharmacol.* **17**, 13 (1910). Isoln from *N. japonicum* DC., *Nymphaeaceae* and structure: Kotake *et al.*, *Ann.* **606**, 148 (1957). Total synthesis: Kotake *et al.*, *Bull. Chem. Soc. Jpn.* **35**, 698 (1962). Abs configuration: Kawasaki *et al.*, *ibid.* **41**, 1264 (1968); La Londe *et al.*, *J. Am. Chem. Soc.* **93**, 2501 (1971). Determn of crystal structure, abs config and stereochemistry by x-ray diffraction methods: J. Ohrt *et al.*, *J. Cryst. Mol. Struct.* **3**, 3 (1973).

Crystals, mp 221°. $[\alpha]_D$ +15°. Soluble in alcohol, chloroform, ether, acetone, amyl alcohol, dil acids. The base is tasteless, the salts bitter.

6817. Nutgall. Galla; galls; Aleppo-galls; Turkey-galls; Mecca-galls. Excrescence from the young twigs of *Quercus infectoria* Oliv. and other allied species of *Quercus*, *Fagaceae*. *Habit.* Asia Minor (Levant). *Constit.* 50-60% tannic acid, 2-4% gallic acid, ellagic acids, resin.

Chinese nutgall. Is the excrescence on the leaf or leafstalk of *Rhus semialata* Murr., *Anacardiaceae*. Contains about 70% tannin. *Incompat:* Alkalies, alkaloids; salts of Cu, Fe, Pb, Zn; AgNO$_3$; opium in soln.

Caution: Irritating to mucous membranes.

USE: Manuf tannin and ink; dyeing; tanning.

THERAP CAT: Astringent.

THERAP CAT (VET): Has been used as a topical astringent.

6818. Nutmeg. Myristica. Large tropical tree, *Myristica fragrans* Houtt., *Myristicaceae*, with fleshy edible fruit. The dried ripe seeds are widely used in cooking and in traditional medicine. The spice prepared from the seed's lacy red covering (aril) is known as **mace**. *Habit.* Molucca Islands; cultivated in Indonesia, West Indies. *Constit.* Seeds: 25-40% fixed oil; 5-15% volatile oils; triterpene saponins; sterols. Description of constituents: A. T. Weil, *Econ. Bot.* **19**, 194 (1965). Extraction of oils with liquid CO$_2$: C. B. Spricigo *et al.*, *J. Supercrit. Fluids* **15**, 253 (1999). Toxicity data: P. M. Jenner *et al.*, *Food Cosmet. Toxicol.* **2**, 327 (1964). Review of ethnobotany: C. Van Gils, P. A. Cox, *J. Ethnopharmacol.* **42**, 117-124 (1994); of constituents and uses: A. Y. Leung, S. Foster, *Encyclopedia of Common Natural Ingredients*, (Wiley-Interscience, Hoboken, 2nd Ed., 2003) pp 385-387; J. Gruenwald *et al.*, *PDR for Herbal Medicines* (Thomson PDR, Montvale, 3rd Ed., 2004) pp 594-595. Comprehensive description: B. Krishnamoorthy, J. Rema in *Handbook of Herbs and Spices*, **Vol. 1**, K. V. Peter, Ed., (Woodhead Publishing, Cambridge, 2001) pp 238-243.

Fixed oil. Nutmeg butter. Obtained by expressing the crushed seeds or by extracting with solvents. *Constit.* Trimyristin (84%), oleic acid (3.5%), resins (2.3%), linolenic acid (0.6%), other fatty acids. Orange-red to reddish-brown, soft solid. mp 45-51°. Odor and taste of nutmeg. d 0.990-0.995. Sapon no. 172-179. Iodine no. 40-52. Acid no. 17-23. Partly sol in cold alcohol, almost completely sol in hot alcohol; freely sol in chloroform, ether.

Volatile oil. [8008-45-5] Oil of nutmeg. Obtained by steam distillation from the dried kernals of the ripe seed. *Constit.* Monoterpene hydrocarbons (~88%) including camphene, α- and β-pinene, sabinene; monoterpene alcohols; myristicin (4-8%); elemicin; safrole. Colorless or pale yellow liquid; odor and taste of nutmeg. East Indian oil: d$_{25}^{25}$ 0.880-0.910; n_D^{20} 1.4740-1.4880; angular rotation between +8° and +30°. Sol in 3 vols 90% alcohol. West Indian oil: d$_{25}^{25}$ 0.854-0.880; n_D^{20} 1.469-1.476; angular rotation between +25° and +45°. Sol in 4 vols 90% alcohol. *Keep well closed, cool and protected from light.* LD$_{50}$ orally in rats: 2620 mg/kg (Jenner).

Caution: Ingestion of large quantities may cause intense thirst, nausea, anxiety, drowsiness, hallucinations, stupor, death (Gruenwald).

USE: As a cooking spice, flavoring in food and beverages. Nutmeg oil as fragrance in soaps, detergents, creams, lotions, perfumes. Fixed oil in candles.

THERAP CAT: Carminative.

6819. Nux Vomica. Quaker buttons; bachelor's buttons; poison nut; dog buttons; vomit nut. Dried, ripe seeds of *Strychnos nux-vomica* L., *Loganiaceae*. *Habit.* Southern Asia, Northern Australia. *Constit.* 1-1.4% strychnine, about an equal amount of brucine; strychnicine, loganin, caffeotannic (igasuric) acid, proteins. Nux vomica from Saigon contains 1.6-2% strychnine.

Caution: Extremely poisonous.

THERAP CAT: Formerly as bitter tonic.

THERAP CAT (VET): Has been used as a bitter tonic.

6820. Nybomycin. [30408-30-1] 8-(Hydroxymethyl)-6,11-dimethyl-2*H*,4*H*-oxazolo[5,4,3-*ij*]pyrido[3,2-*g*]quinoline-4,10-(11*H*)-dione; 6,11-dimethyl-8-(hydroxymethyl)pyrido[3,2-*g*]-oxazolo[5,4,3-*ij*]quinoline-4,10(2*H*,11*H*)-dione. C$_{16}$H$_{14}$N$_2$O$_4$; mol wt 298.30. C 64.42%, H 4.73%, N 9.39%, O 21.45%. Antibiotic

substance produced by *Streptomycete* A 717 isolated from Missouri soil: Strelitz *et al.*, *Proc. Natl. Acad. Sci. USA* **41**, 620 (1955); Eble *et al.*, *Antibiot. Chemother.* **8**, 627 (1958); Brock, Sokolski, *ibid.* 631. Structure: Rinehart, Renfroe, *J. Am. Chem. Soc.* **83**, 3729 (1961). Revised structure: Rinehart *et al.*, *ibid.* **92**, 6994 (1970). Total synthesis of *deoxynybomycin*: Forbis, Rinehart, *ibid.* 6995. Total synthesis of nybomycin: *eidem*, *J. Antibiot.* **24**, 326 (1971); *eidem*, *J. Am. Chem. Soc.* **95**, 5003 (1973).

Needles from acetic acid, mp 325-330°. Sublimes at 250° (15 mm). Optically inactive. uv max (ethanol): 266, 285 nm. Soluble in concd acids. Very slightly sol in water, alkalies, and common organic solvents. Shows antiphage and antibacterial properties. LD$_{50}$ i.p. in mice: 650 mg/kg (Brock, Sokolski). ^{13}C NMR spectrum: A. M. Nadzan, K. L. Rinehart, *J. Am. Chem. Soc.* **99**, 4647 (1977).

Acetate. C$_{18}$H$_{16}$N$_2$O$_5$. Crystals from chloroform + ethanol, mp 236-237°.

Succinate. C$_{20}$H$_{19}$N$_2$O$_7$. Crystals from dimethylformamide. Practically insol in water.

6821. Nylidrin. [447-41-6] 4-Hydroxy-α-[1-[(1-methyl-3-phenylpropyl)amino]ethyl]benzenemethanol; *p*-hydroxy-α-[1-[(1-methyl-3-phenylpropyl)amino]ethyl]benzyl alcohol; *p*-hydroxy-*N*-(1-methyl-3-phenylpropyl)norephedrine; buphenine; 1-(*p*-hydroxyphenyl)-2-(1'-methyl-3'-phenylpropylamino)-1-propanol; phenyl-*sec*-butyl norsuprifen. C$_{19}$H$_{25}$NO$_2$; mol wt 299.41. C 76.22%, H 8.42%, N 4.68%, O 10.69%. Prepn: **FR 968273** (1950 to Troponwerke Dinklage); **GB 669574**; **GB 669575** (1952); *Chem. Eng. News* **33**, 2896 (1955); Külz, Schöpf, **US 2661372**; **US 2661373** (1953). Pharmacology: T. T. Yen, D. V. Pearson, *Res. Commun. Chem. Pathol. Pharmacol.* **23**, 11 (1979); B. Fichtl, W. Felix, *Eur. J. Pharmacol.* **65**, 333 (1980).

Crystals from methanol, mp 111-112°.

Hydrochloride. [849-55-8] SKF-1700-A; Arlidin; Bufedon; Buphedrin; Dilatal; Dilatol; Dilatropon; Dilydrin; Opino; Penitardon; Perdilatal; Rudilin; Rydrin; Tocodilydrin; Tocodrin. C$_{19}$H$_{25}$NO$_2$.-HCl; mol wt 335.87. Crystals. Sparingly sol in water; slightly sol in alcohol. Practically insol in ether, chloroform, benzene.

THERAP CAT: Vasodilator (peripheral).

6822. Nylon. Polyamide. Generic term describing a manufactured fiber in which fiber-forming substances are any long-chain synthetic polyamide having recurring polyamide groups (—CONH—) as an integral part of the polymer chain. Formed from various combinations of diacids, diamines, and amino acids. May be formed also by addition polymerization. The linear polyamides have achieved the greatest commercial success. Shorthand nomenclature of nylons involves the use of numbers: a single numeral indicating the number of carbon atoms in a monomer, e.g. nylon 6; two numbers indicating a polymer formed from diamines and dibasic acids, the first numeral indicating the number of carbon atoms separating the nitrogen atoms of the diamine, the second indicating the number of straight-chain carbon atoms in the dibasic acid, e.g. nylon 6,6. First produced by E. I. du Pont de Nemours & Co. according to patents of W. H. Carothers. The name **nylon** was dedicated to public domain on Oct. 27, 1938 at the Herald Tribune Forum where the product itself was announced. *Reviews:* R. W. Moncrieff, *Man-Made Fibres* (John

Wiley, New York, 1963) pp 335-355; Snider, Richardson, "Polyamide Fibers" in *Encyclopedia of Polymer Science and Technology* vol. 10 (Interscience, New York, 1969) pp 347-460; J. H. Saunders, "Polyamides (Fibers)" in *Kirk-Othmer Encyclopedia of Chemical Technology* vol. 18 (Wiley-Interscience, New York, 3rd ed., 1982) pp 372-405. *Book: Nylon Plastics*, M. I. Kohan, Ed. (Wiley-Interscience, New York, 1973).

Crystalline solids characterized by low specific gravity, high strength, durability, high flexibility, and high tensile strength. Soluble in phenol, cresols (especially *m*-cresol), xylenol, formic acid. Insoluble in alcohols, esters, ketones, hydrocarbons. Hydrolysis and degradation occur at higher temperatures, esp in the melt. Stable to aqueous alkali. Degrades rapidly in aqueous acids. Undergoes photodegradation.

USE: In production of synthetic fibers for various textile and domestic uses. Surgical aid (nonabsorbable suture).

6823. Nylon 6. [25038-54-4] Poly[imino(1-oxo-1,6-hexanediyl)]; poly(iminocarbonylpentamethylene); Caprolan; Enkalon; Grilon; Kapron; Mirlon; Perlon; Phrilon; Amilan. Linear polymer obtained by polymerization of ε-caprolactam, *q.v.*: Schlack, US 2241321 (1941 to I. G. Farbenind.). The importance of this fiber increased with the discovery that caprolactam can be produced by the nitrosation of cyclohexanecarboxylic acid: Muench *et al.*, US 3022291 and US 3108096 (1962, 1963, both to Snia Viscosa). *Review:* R. W. Moncrieff, *Man-Made Fibres* (John Wiley & Sons, New York, 1963) pp 335-355; H. K. Reimschuessel, *J. Polym. Sci. Macromol. Rev.* 12, 65-139 (1977).

n = approx 200

Softens at 210°. mp 223°. Can withstand a temp of 100° for long periods of time. d_4^{20} 1.14. Moisture regain is about 4%. Swelling is low; if steeped in water and then centrifuged its volume increases by about 13-14%. Immune to microbiological attack. Resistant to most org chemicals, but dissolved by phenol, cresol, and strong acids.

USE: Tire cord; fishing lines; tow ropes; hose manuf; woven fabrics.

6824. Nylon 46. [50327-22-5] Poly[imino-1,4-butanediylimino(1,6-dioxo-1,6-hexanediyl)]; poly(tetramethyleneadipamide); Stanyl. Symmetrical polyamide with higher melting point and greater tensile strength than nylon 66 or nylon 6, *q.v.* Prepn from 1,4-diaminobutane and adipic acid: W. H. Carothers, US 2130948 (1938 to Du Pont). Improved process: E. H. J. P. Bour, J. M. M. Warnier, EP 77106 corresp to US 4463166 (1983, 1984 both to Stamicarbon). Account of prepn by melt polymerization, physical properties: R. J. Gaymans *et al.*, *J. Polym. Sci. Polym. Chem. Ed.* 15, 537 (1977). Brief account: D. O'Sullivan, *Chem. Eng. News* 62, 33 (May 21, 1984).

Clear polymer, mol wt 22,000-45,000. mp 283-319°. d 1.20 (solution-cast film). Insol in most solvents. Sol in formic acid. Slightly sol in trifluoroacetic acid.

USE: Industrial fiber; fiber-reinforced rubber products.

6825. Nystatin. [1400-61-9] Fungicidin; Biofanal; Diastatin; Candex; Candio-Hermal; Mycostatin; Moronal; Nystan; O-V Statin. Polyene antifungal antibiotic complex containing 3 biologically active components, A_1, A_2, A_3. Produced by *Streptomyces noursei*, *S. aureus* and other *Streptomyces* spp: Hazen, Brown, *Science* 112, 423 (1950); *Proc. Soc. Exp. Biol. Med.* 76, 93 (1951); Raubitscheck *et al.*, *Antibiot. Chemother.* 2, 179 (1952); Cohen, Webb, *Arch. Pediatr.* 69, 414 (1952); Dutcher *et al.*, *Antibiot. Annu.* 1953-1954, 191; *eidem*, *Therapy of Fungus Diseases* (Little, Brown, Boston, 1955) p 168. Review of early literature: Brown, Hazen, *Trans. N.Y. Acad. Sci.* 19 (1956-1957) pp 447-456. Purification: Vandeputte, US 2832719 (1958 to Olin Mathieson); Renella, US 3517100 (1970 to Am. Cyanamid). Chemistry and partial structure: Birch *et al.*, *Tetrahedron Lett.* 1964, 1491; of nystatin A_1 and A_2: Shenin *et al.*, *Antibiotiki* 13, 387 (1968). Structure of the aglycone: Manwaring *et al.*, *ibid.* 1969, 5319. Complete structure of A_1: Chong, Rickards, *ibid.* 1970, 5145; Borowski *et al.*, *ibid.* 1971, 685. Revised structure: R. C. Pandey, K. L. Rinehart, *J. Antibiot.* 29, 1035 (1976). Stereochemical study of A_1: J. M. Lancelin *et al.*, *Tetrahedron Lett.* 29, 2827 (1988). Structure of A_3: J. Zielinski *et al.*, *J. Antibiot.* 41, 1289 (1988). Mechanism of action: R. W. Holz in *Antibiotics* vol. 5, pt. 2, F. E. Hahn, Ed. (Springer-Verlag, New York, 1979) pp 313-340. Toxicity study: H. Seneca, *Antibiot. Annu.* 1955-1956, 697. Comprehensive description: G. W. Michel, *Anal. Profiles Drug Subs.* 6, 341-421 (1977).

Nystatin A₁

Light yellow, hygroscopic powder. Gradually decomposes above 160° without melting by 250°. $[\alpha]_D^{25}$ −10° (glacial acetic acid); +21° (pyridine); +12° (DMF); −7° (0.1N HCl in methanol). uv max (ethanol): 290, 307, 322 nm. Exhibits strong reducing properties. Freely sol in DMF, dimethyl sulfoxide; slightly to sparingly sol in *n*-propyl alc, *n*-butyl alc. Insol in ether. Solubility data: Weiss *et al.*, *Antibiot. Chemother.* 7, 374 (1957). Soly (mg/ml) at about 28°: water 4.0; methanol 11.2; ethanol 1.2; carbon tetrachloride 1.23; chloroform 0.48; benzene 0.28; ethylene glycol 8.75. Solns and aq suspensions begin to lose activity soon after prepn. Aq suspensions are stable for 10 minutes on heating to 100° at pH 7.0; also stable in moderately alkaline media, but labile at pH 9 and pH 2. Heat, light, and oxygen accelerate decompn. Activity not diminished by blood or serum. LD_{50} i.p. in mice: ~200 mg/kg (Seneca).

Nystatin A₁. [34786-70-4] $C_{47}H_{75}NO_{17}$; mol wt 926.11.

THERAP CAT: Antifungal.

THERAP CAT (VET): Antifungal; growth promotant.

6826. Nysted Reagent. [41114-59-4] Cyclo-dibromodi-μ-methylene[μ-(tetrahydrofuran)]trizinc. $C_6H_{12}Br_2OZn_3$; mol wt 456.20. C 15.80%, H 2.65%, Br 35.03%, O 3.51%, Zn 43.01%. Olefination reagent with two *gem*-dimetallic subunits. Prepn: GB 1304465; L. N. Nysted, US 3865848 (1973, 1975 both to G. D. Searle). Methylenation of carbonyl compounds: S. Matsubara, *et al.*, *Synlett* 1998, 313.

Commercial product is a white suspension in THF. Low soly in ether, THF, and other aprotic solvents.

USE: Reagent for the methylenation of ketones and aldehydes.

O

6827. Obidoxime Chloride. [114-90-9] 1,1'-[Oxybis(methylene)]bis[4-(hydroxyimino)methyl]pyridinium chloride (1:2); 1,1'-(oxydimethylene)bis[4-formylpyridinium]dichloride dioxime; N,N-dimethyleneoxidebis(pyridinium-4-aldoxime) dichloride; bis(4-hydroxyiminomethylpyridinium-1-methyl) ether dichloride; bis(isonicotinaldoxime 1-methyl) ether dichloride; BH-6; LüH7; Toksobidin; Toxogonin. $C_{14}H_{16}Cl_2N_4O_3$; mol wt 359.21. C 46.81%, H 4.49%, Cl 19.74%, N 15.60%, O 13.36%. Cholinesterase reactivator. Prepn from pyridine aldoxime and α,α'-dichlorodimethyl ether: **GB 930040**; Lüttringhaus et al., **US 3137702** (1963, 1964 to E. Merck); Lüttringhaus, Hagedorn, Arzneim.-Forsch. **14**, 1 (1964). Pharmacology and toxicology: W. D. Erdman, H. Engelhard, ibid. 5; Mayer, Michalek, Biochem. Pharmacol. **20**, 3029 (1971); Bajgar et al., Eur. J. Pharmacol. **19**, 199 (1972). Hydrolysis studies: Christenson, Acta Pharm. Suec. **9**, 309 (1972).

Occurs in two interchangeable isomeric forms (syn and anti). Crystals from HCl contg 70% alcohol, dec 225°. Also reported as syn, mp 235-236°; anti, mp 218-220°: Leitis et al., C.A. **71**, 81098d (1969). Freely sol in water, stable in 1-10% aq solns. LD_{50} in mice, rats (mg/kg): 70, 133 i.v.; 150, 225 i.p.; >2240, >4000 orally; in mice (mg/kg): 172 i.m. (Erdman, Engelhard).

Dibromide. $C_{14}H_{16}Br_2N_4O_3$. Dec 202-203°.

THERAP CAT: Antidote (organophosphate insecticide poisoning).

6828. Oblimersen Sodium. [190977-41-4] d(P-thio)(T-C-T-C-C-C-A-G-C-G-T-G-C-G-C-C-A-T)DNA; augmerosen; G-3139; Genasense. Synthetic 18 base phosphorothioate antisense oligonucleotide (AS ON); designed to bind to human bcl-2 oncogene, resulting in the down-regulation of an apoptosis-inhibiting protein Bcl-2. Prepn: J. C. Reed, **WO 9508350** (1995); idem, **US 5831066** (1998 to Univ. of Penn.). LC/MS determn in plasma and urine: G. Dai et al., J. Chromatogr. B **825**, 201 (2005). Pharmacokinetics: F. I. Raynaud et al., J. Pharmacol. Exp. Ther. **281**, 420 (1997). Mechanism of action study: L. Benimetskaya et al., Oligonucleotides **15**, 206 (2005). Clinical evaluations in malignant melanoma: B. Jansen et al., Lancet **356**, 1728 (2000); in non-Hodgkin's lymphoma: J. S. Waters et al., J. Clin. Oncol. **18**, 1812 (2000); in prostate cancer: K. N. Chi et al., Clin. Cancer Res. **7**, 3920 (2001). Review of pharmacology, preclinical and clinical development: R. J. Klasa et al., Antisense Nucleic Acid Drug Dev. **12**, 193-213 (2002). Review of therapeutic potential in lung cancer: R. S. Herbst, S. R. Frankel, Clin. Cancer Res. **10**, 4245s-4248s (2004); in prostate cancer: K. N. Chi, World J. Urol. **23**, 33-37 (2005).

THERAP CAT: Antineoplastic.

6829. Ochratoxins. Toxic metabolites from Aspergillus ochraceus Wilh.: Scott, Mycopathol. Mycol. Appl. **25**, 213 (1965); A. sulphureus and A. melleus: Lai et al., Appl. Microbiol. **19**, 542 (1970); and Penicillium viridicatum Westling: van Walbeek et al., Can. J. Microbiol. **15**, 1281 (1969). As these molds occur widely, some toxins have been found as natural contaminants on corn, peanuts, storage grains, cottonseed and other decaying vegetation. Isolns: van Walbeek et al., ibid. **14**, 131 (1968); Shotwell et al., Appl. Microbiol. **17**, 765 (1969); Nesheim, J. Assoc. Off. Anal. Chem. **52**, 975 (1969); Natori et al., Chem. Pharm. Bull. **18**, 2259 (1970). Structure and stereochemistry: van der Merwe et al., J. Chem. Soc. **1965**, 7083 (1965); Nature **205**, 1112 (1965). Synthesis of ochratoxins A, B and ester deriv: Steyn, Holzapfel, Tetrahedron **23**, 4449 (1967); Nesheim, loc. cit.; Roberts, Woolven, J. Chem. Soc. C **1970**, 278. Facile synthesis of A: G. A. Kraus, J. Org. Chem. **46**,

201 (1981). Toxicity study: Purchase, Theron, Food Cosmet. Toxicol. **6**, 479 (1968). Physicochemical data: A. E. Pohland et al., Pure Appl. Chem. **54**, 2220 (1982). Comprehensive reviews: Steyn, "Ochratoxin and other Dihydroisocoumarins" in Microbial Toxins vol. VI, A. Ciegler, et al., Eds. (Academic Press, New York, 1971) p 179-205; F. S. Chu, Crit. Rev. Toxicol. **2**, 499-524 (1974); P. Krogh in Natural Toxins, D. Eaker, P. Wadström, Eds. (Pergamon, New York, 1980) pp 673-680; A. E. Pohland et al., Pure Appl. Chem. **64**, 1029-1046 (1992).

Ochratoxin A

Ochratoxin A. [303-47-9] (R)-N-[(5-Chloro-3,4-dihydro-8-hydroxy-3-methyl-1-oxo-1H-2-benzopyran-7-yl)carbonyl]-L-phenylalanine. $C_{20}H_{18}ClNO_6$; mol wt 403.82. The major ochratoxin component. Crystals from xylene, mp 169°. Exhibits green fluorescence. $[\alpha]_D$ $-118°$ (c = 1.1 in $CHCl_3$). uv max (ethanol): 215, 333 nm (ε 34000; 2400) (van der Merwe). Also frequently reported as mp 90° from benzene (one mole of benzene of crystallization). uv max: 213, 332 nm (ε 36800; 6400) (Steyn, Holzapfel). LD_{50} orally in rats: 20-22 mg/kg (Purchase, Theron).

Ochratoxin B. [4825-86-9] (R)-N-[(3,4-Dihydro-8-hydroxy-3-methyl-1-oxo-1H-2-benzopyran-7-yl)carbonyl]-L-phenylalanine. $C_{20}H_{19}NO_6$; mol wt 369.37. The less toxic dechloro deriv of ochratoxin A. Crystals from methanol, mp 221° (van der Merwe); also reported as 208-209° (Nesheim). Exhibits blue fluorescence. $[\alpha]_D$ $-35°$ (c = 0.15 in ethanol). uv max: 218, 318 nm (ε 37200, 6900).

Ochratoxin C. [4865-85-4] (R)-N-[(5-Chloro-3,4-dihydro-8-hydroxy-3-methyl-1-oxo-1H-2-benzopyran-7-yl)carbonyl]-L-phenylalanine ethyl ester. $C_{22}H_{22}ClNO_6$; mol wt 431.87. The equally toxic amorphous ethyl ester of ochratoxin A: Steyn, Holzapfel, J. S. Afr. Chem. Inst. **20**, 186 (1967). uv max (ethanol): 333 nm (ε 6500).

Caution: Ochratoxin A is reasonably anticipated to be a human carcinogen: Report on Carcinogens, Twelfth Edition (PB2011-111646, 2011) p 335.

USE: Exptlly as teratogen and carcinogen.

6830. Ocimene. [29714-87-2] 3,7-Dimethyl-1,3,?-octatriene. $C_{10}H_{16}$; mol wt 136.24. C 88.16%, H 11.84%. From the leaves of Ocimum basilicum L., Labiatae; Baronia dentigeroides Cheel, Rutaceae; Litsea zeylanica C. & T. Nees, Lauraceae; Homoranthus flavescens A. Cunn., Myrtaceae. From the fruits of Evodia rutaecarpa (Juss.) Hook. f. & Thoms., Rutaceae. Exists in two modifications: α-form, **3,7-dimethyl-1,3,7-octatriene**; and β-form, **3,7-dimethyl-1,3,6-octatriene**. Cis and trans refers to the stereochemistry at the double bond between positions 3 and 4. Isoln and structure: van Romburgh, Proc. K. Ned. Akad. Wet. **3**, 454 (1900); Enklaar, Rec. Trav. Chim. **26**, 157 (1907); **27**, 422 (1908); **36**, 215 (1917); **45**, 337 (1926). Structure: Sutherland, J. Am. Chem. Soc. **74**, 2688 (1952). Separation of isomers: Ohloff et al., Ann. **675**, 83 (1964). Synthesis of trans-α-form: O. P. Vig et al., Indian J. Chem. **7**, 1111 (1969). Stereospecific synthesis: O. P. Vig et al., ibid. **15B**, 25 (1977).

trans-β-Ocimene

Oil. Pleasant odor. Mixture of isomers. bp_{70} 100°. d_4^{20} 0.8006. n_D^{20} 1.4862. uv max (methanol): 233 nm (ε 26,200). Practically insol in water. Sol in alcohol, chloroform, ether, glacial acetic acid.

trans-β-Form. [3779-61-1] d_4^{20} 0.799. n_D^{20} 1.4893. uv max (ethanol): 232 nm (ε 27,600).

cis-β-Form. d_4^{20} 0.799. n_D^{20} 1.4877. uv max (ethanol): 237.5 nm (ε 21,000).

***trans*-α-Form.** d_4^{20} 0.793. n_D^{20} 1.4802. uv max (ethanol): 231 nm (ε 27,300).

***cis*-α-Form.** d_4^{20} 0.794. n_D^{20} 1.4789. uv max (ethanol): 234.5 nm (ε 21,600).

6831. Octabenzone. [1843-05-6] [2-Hydroxy-4-(octyloxy)-phenyl]phenylmethanone; 2-hydroxy-4-(octyloxy)benzophenone; benzophenone-12; Spectra-Sorb UV 531. $C_{21}H_{26}O_3$; mol wt 326.44. C 77.27%, H 8.03%, O 14.70%. Prepn: Armitage *et al.*, **US 3098842** (1963 to du Pont).

Crystals, mp 45-46°.

USE: To stabilize polyethylene against deterioration by ultraviolet light.

THERAP CAT: Ultraviolet screen.

6832. Octacaine. [13912-77-1] 3-(Diethylamino)-*N*-phenyl-butanamide; 3-diethylaminobutyranilide. $C_{14}H_{22}N_2O$; mol wt 234.34. C 71.76%, H 9.46%, N 11.95%, O 6.83%. Prepn: Hofstetter, Wilder Smith, *Helv. Chim. Acta* **36**, 1698 (1953); Hofstetter, **US 2851393** (1958 to E. Geistlich Söhne).

Crystals from petr ether, mp 46-47°. bp$_1$ 200°. Easily sol in stoichiometric amount of HCl, in ether, alc, benzene.

Hydrochloride. [59727-70-7] Amplicain. $C_{14}H_{22}N_2O\cdot HCl$; mol wt 270.80. mp 132-134°. Sol in water.

THERAP CAT: Anesthetic (local).

6833. Octacosanol. [557-61-9] 1-Octacosanol; cluytyl alcohol; montanyl alcohol; *n*-octacosanol; octacosyl alcohol. $C_{28}H_{58}O$; mol wt 410.77. C 81.87%, H 14.23%, O 3.89%. Constituent of vegetable waxes. Isoln from the wax found on green blades of wheat: Pollard *et al.*, *Biochem. J.* **27**, 1889 (1933); from carnauba wax: Koonce, Brown, *Oil Soap (Chicago)* **21**, 231 (1944). Synthesis starting with behenic acid: Bleyberg, Ulrich, *Ber.* **64**, 2504 (1931); Francis *et al.*, *Proc. R. Soc. London* **A 158**, 691 (1937).

Crystals from much acetone, mp 83.4°. Insol in water. Sol in carbon disulfide, other fat solvents, oils.

6834. Octafluorocyclobutane. [115-25-3] 1,1,2,2,3,3,4,4-Octafluorocyclobutane; perfluorocyclobutane; Freon C318. C_4F_8; mol wt 200.03. C 24.02%, F 75.98%. Prepd by pyrolysis of chlorodifluoromethane: Downing *et al.*, **US 2384821** (1945 to Kinetic Chem.); **US 2551573**; **US 2615926** (1951; 1952 to du Pont). Thermodynamic studies: Furukawa *et al.*, *J. Res. Natl. Bur. Stand.* **52**, 11 (1954); Duus, *Ind. Eng. Chem.* **47**, 1445 (1955).

Nonflammable, nontoxic gas. d_{vapor}^{27} 8.2. mp $-41.4°$. bp $-6.04°$ (also reported bp $-5°$). Heat of formation at 25° = 352 kcal. Heat of combustion 1359 $\pm$39 cal/g.

USE: Refrigerant; heat-transfer medium.

6835. 2,2,3,3,4,4,5,5-Octafluoro-1-pentanol. [355-80-6] 1*H*,1*H*,5*H*-Octafluoropentanol; 1,1,5-trihydroperfluoropentanol. $C_5H_4F_8O$; mol wt 232.07. C 25.88%, H 1.74%, F 65.49%, O 6.89%. Prepd by free radical telomerization of tetrafluoroethylene in methanol: Hanford, Joyce, **US 2562547** and Joyce, **US 2559625** (both 1951 to du Pont).

Liquid. d_4^{20} 1.6647. bp$_{760}$ 140-141°. n_D^{20} 1.3178. Surface tension at 20° = 27.6 dynes/cm.

USE: To introduce fluoroalkyl groups into an organic molecule: Kramer, Gilbert, **US 2963526** (1960 to Esso). Proposed intermediate for plastics, surface-active agents, lubricants, elastomers.

6836. Octamethylcyclotetrasiloxane. [556-67-2] 2,2,4,4,6,-6,8,8-Octamethylcyclotetrasiloxane. $C_8H_{24}O_4Si_4$; mol wt 296.62. C 32.39%, H 8.16%, O 21.58%, Si 37.87%. Isolated from the hydrolysis product of dimethyldichlorosilane: Patnode, Wilcock, *J. Am. Chem. Soc.* **68**, 358 (1946).

Oily liquid. mp 17.5°. bp 175°. bp$_{20}$ 74°. d 0.9558. n_D^{20} 1.3968.

USE: Preparation of methyl silicon oils.

6837. Octamethyltrisiloxane. [107-51-7] 1,1,1,3,3,5,5,5-Octamethyltrisiloxane. $C_8H_{24}O_2Si_3$; mol wt 236.53. C 40.62%, H 10.23%, O 13.53%, Si 35.62%. Prepn: Patnode, Wilcock, *J. Am. Chem. Soc.* **68**, 358 (1946).

Liquid. bp 153°. d 0.8200. n_D^{20} 1.3848. mp $\sim -80°$. Stable. Inert to most chemical reagents and rubber. Maintains about the same viscosity over a wide temp range. Sol in benzene and the lighter hydrocarbons; slightly sol in alcohol and heavy hydrocarbons.

USE: As a basis for silicone oils or fluids designed to withstand extremes of temp; as a foam suppressant in petroleum lubricating oil.

6838. Octane. [111-65-9] C_8H_{18}; mol wt 114.23. C 84.12%, H 15.88%. Found in petroleum.

Liquid, mp $-56.8°$. d_4^{20} 0.7028. bp$_{760}$ 125.6°. n_D^{20} 1.39764. Flash pt, open cup: 72°F (22°C). *Flammable.* Insol in water. Slightly sol in alcohol; sol in ether; miscible with benzene, petr ether, gasoline. The term "octane rating" is explained under Gasoline, *q.v.*

Caution: Potential symptoms of overexposure are irritation of eyes and nose; drowsiness; dermatitis; aspiration of liquid may cause chemical pneumonia. *See NIOSH Pocket Guide to Chemical Hazards* (DHHS/NIOSH 97-140, 1997) p 236.

6839. Octanohydroxamic Acid. [7377-03-9] *N*-Hydroxy-octanamide; caprylohydroxamic acid; Oct HA; Taselin. $C_8H_{17}NO_2$; mol wt 159.23. C 60.35%, H 10.76%, N 8.80%, O 20.10%. Synthesis and chemical study: Inoue, Yukawa, *J. Agric. Chem. Soc. Jpn.* **16**, 504 (1940); Ichim *et al.*, *Igiena* **9**, 319 (1960), *C.A.* **55**, 16662a (1961). Activity as a urease inhibitor: Kobashi *et al.*, *Biochim. Biophys. Acta* **65**, 380 (1962); **227**, 429 (1971); as an antimicrobial: Hase *et al.*, *Chem. Pharm. Bull.* **19**, 363 (1971).

White plates from benzene, mp 78.5-79°. Water soluble. Practically insol in petr ether.

THERAP CAT (VET): Antimicrobial; growth promotant.

6840. 1-Octanol. [111-87-5] Caprylic alcohol; 1-hydroxyoctane. $C_8H_{18}O$; mol wt 130.23. C 73.78%, H 13.93%, O 12.29%. Occurs in the form of esters in some essential oils. Prepd from the esterified products of coconut oil, the methyl caprylate being reduced by Na and alcohol.

Colorless liquid; penetrating, aromatic odor. d_4^{20} 0.827. mp −17 to −16°. bp 194-195°. n_D^{20} 1.430. Insol in water. Miscible with alcohol, chloroform, ether.

USE: Solvent. Determn of partition coefficients. Manuf of perfumes and esters.

6841. 2-Octanol. [123-96-6] Secondary caprylic alcohol; methyl hexyl carbinol; hexylmethylcarbinol; 1-methyl-1-heptanol; 2-hydroxyoctane. $C_8H_{18}O$; mol wt 130.23. C 73.78%, H 13.93%, O 12.29%. Prepd by heating sodium ricinoleate with caustic soda in a copper vessel and distilling: Adams, Marvel, *Org. Synth.* **1**, 61 (1921); Kenyon, *ibid.* **6**, 68 (1926); Ellis, Reid, *J. Am. Chem. Soc.* **54**, 1674 (1932).

dl-Form. Oily, refractive liquid. Aromatic, yet somewhat unpleasant odor, particularly on heating. d_4^{20} 0.8193. mp −38.6°. bp$_{760}$ 178.5°; bp$_{60}$ 107.4°; bp$_{20}$ 83.3°; bp$_{10}$ 70.0°; bp$_5$ 57.6°; bp$_{1.0}$ 32.8°. n_D^{20} 1.42025. Béhal, *Bull. Soc. Chim.* [4] **25**, 482 (1919). Flash pt ~140°F (60°C). Soly in water: 0.096 ml/100 ml. Miscible with alc, ether.

d-Form. [6169-06-8] (2*S*)-2-Octanol. bp$_{20}$ 86°. d_4^{20} 0.8216. $[\alpha]_D^{17}$ +9.9°.

l-Form. [5978-70-1] (2*R*)-2-Octanol. bp$_{20}$ 86°. $[\alpha]_D^{17}$ −9.9°.

USE: In the manuf of perfumes; in disinfectant soaps. To prevent foaming. Solvent for fats and waxes.

6842. 3-Octanone. [106-68-3] Ethyl *n*-amyl ketone. C_8H_{16}-O; mol wt 128.22. C 74.94%, H 12.58%, O 12.48%. Volatile chemical found in lavender and other essential oils; also found in mushrooms. Mediates interactions between plants, herbivores, and their natural enemies. Isoln from bituminous tar: J. Herzenberg, E. v. Winterfeld, *Ber.* **64**, 1025 (1931); from the essential oil of *Rosmarinus officinalis* L: A. Koedam, M. J. M. Gijbels, *Z. Naturforsch.* **33C**, 144 (1978). Prepn: Y. R. Naves, *Helv. Chim. Acta* **26**, 1034 (1943); P. G. Gassman, G. D. Richmond, *J. Org. Chem.* **31**, 2355 (1966). Oxidation potential and uv spectrum: R. A. Day, Jr. *et al.*, *J. Am. Chem. Soc.* **72**, 1379 (1950). Role as an alarm pheromone in carpenter ants: R. M. Duffield, M. S. Blum, *Comp. Biochem. Physiol.* **51B**, 281 (1975); in herbivore and parasitoid response profiles: R. Ramachandran *et al.*, *J. Agric. Food Chem.* **39**, 2310 (1991). Colorimetric determn: D. H. E. Tattje, C. Scholtens, *Pharm. Weekbl.* **94**, 137 (1959). GC determn in lavender oil: P. A. Stadler, *Helv. Chim. Acta* **43**, 1601 (1960). Extraction from king oyster mushrooms: J.-L. Mau *et al.*, *J. Agric. Food Chem.* **46**, 4587 (1998).

Colorless to slightly yellow liquid; sharp, pungent, somewhat fruity odor. bp 162-165°. bp$_{30}$ 76-77°. bp$_{1.5}$ 55-58°. n_D^{20} 1.4153-1.4154 (Naves); also reported as n_D^{21} 1.4179 (Stadler). d^{21} 0.8260. uv max (methanol): 278 nm (ε 23.7).

USE: Ingredient in lavender perfumes; mushroom and pungent cheese flavors.

6843. Octenidine. [71251-02-0] *N,N'*-(1,10-Decanediyldi-1(4*H*)-pyridinyl-4-ylidene)bis[1-octanamine]; 1,10-bis-[4-(octylamino)-1-pyridinium]decane; Win-41464. $C_{36}H_{62}N_4$; mol wt 550.92. C 78.49%, H 11.34%, N 10.17%. Prepn: D. M. Bailey, **BE 851807**; *idem*, **US 4206215** (1977, 1980 both to Sterling); D. M. Bailey *et al.*, *J. Med. Chem.* **27**, 1457 (1984). Structure-activity study: A. M. Slee *et al.*, *Antimicrob. Agents Chemother.* **23**, 531 (1983). Antiplaque activity: A. M. Slee, J. R. O'Connor, *ibid.* 379; microbiocidal activity: D. M. Sedlock, D. M. Bailey, *ibid.* **28**, 786 (1985). Antimycotic effects: M. A. Ghannoum *et al.*, *J. Antimicrob. Chemother.* **25**, 237 (1990). Clinical evaluation as topical disinfectant: B. Christiansen, *Z. Bakteriol. Mikrobiol. Hyg.* **186**, 368 (1988). Clinical trials in prevention of plaque and gingivitis: M. R. Patters *et al.*, *J. Periodontal Res.* **21**, 154 (1986); B. B. Beiswanger *et al.*, *J. Dent. Res.* **69**, 454 (1990).

Dihydrochloride. [70775-75-6] Win-41464-2; Neo Kodan; Octeniderm; Octenisept. $C_{36}H_{62}N_2 \cdot 2HCl$; mol wt 595.82. Solid from ether, mp 215-217°.

Disaccharin. [86767-75-1] Win-41464-6. $C_{36}H_{62}N_2 \cdot 2C_7H_5$-$NO_3S$; mol wt 889.27.

THERAP CAT: Antiseptic (topical).

6844. Octhilinone. [26530-20-1] 2-Octyl-3(2*H*)-isothiazolone; 2-octyl-4-isothiazolin-3-one; RH-893; Kathon. $C_{11}H_{19}NOS$; mol wt 213.34. C 61.93%, H 8.98%, N 6.57%, O 7.50%, S 15.03%. Prepn: S. N. Lewis *et al.*, **FR 1555416** corresp to **US 3761488** (1969, 1973 to Rohm & Haas); *eidem*, *J. Heterocycl. Chem.* **8**, 571 (1971).

Liquid, bp$_{0.01}$ 120°. uv max (methanol): 280 nm (log ε 3.88).

USE: Fungicide. Biocide in cooling-tower water, paints, cutting oils, cosmetics and shampoo; for leather preservation.

6845. Octocrylene. [6197-30-4] 2-Cyano-3,3-diphenyl-2-propenoic acid 2-ethylhexyl ester; 2-ethylhexyl-2-cyano-3,3-diphenylacrylate; Eusolex OCR; Neo Heliopan 303; Uvinul 3039. $C_{24}H_{27}$-NO_2; mol wt 361.49. C 79.74%, H 7.53%, N 3.87%, O 8.85%. Photostable UV-B filter in sunscreens and cosmetics. Prepn: A. F. Strobel, S. C. Catino, **BE 619809**; *eidem*, **US 3215724** (1962, 1965 both to General Aniline & Film Corp.). Toxicity study: M. R. Odio *et al.*, *Fundam. Appl. Toxicol.* **22**, 355 (1994). HPTLC determn in sun protection lotions: J. Fisher, J. Sherma, *J. Planar Chromatogr. Mod. TLC* **13**, 388 (2000).

Clear yellow viscous liquid, bp$_{0.9}$ 205-209°. fp −10°. d^{25} 1.05. uv max: 303 nm. Miscible in methanol, *n*-butanol, ethyl acetate, mineral oil, hexane, toluene. Immiscible in water.

USE: Solvent for solid sunscreens. UV absorber for plastics and paints.

THERAP CAT: Ultraviolet screen.

6846. Octodrine. [543-82-8] 6-Methyl-2-heptanamine; 6-methyl-2-heptylamine; 2-methyl-6-aminoheptane; 6-amino-2-methylheptane; 2-amino-6-methylheptane; α,ε-dimethylhexylamine; 1,5-dimethylhexylamine; SKF-51; Vaporpac. C$_8$H$_{19}$N; mol wt 129.25. C 74.34%, H 14.82%, N 10.84%. α-Adrenergic agonist. Prepd from the corresponding saturated ketone: Rohrmann, Shonle, *J. Am. Chem. Soc.* **66**, 1516 (1944). Pharmacology: E. J. Fellows, *J. Pharmacol. Exp. Ther.* **90**, 351 (1947).

dl-**Form.** Viscous liquid, fishy odor, bp 154-156°. n$_D^{24}$ 1.4200.
Hydrochloride. C$_8$H$_{19}$N.HCl. Crystals, sol in water. LD$_{50}$ in mice, rats (mg/kg): 59, 41.5 i.p. (Fellows).
Sulfate. (C$_8$H$_{19}$N)$_2$.H$_2$SO$_4$. Crystals, sol in water.
THERAP CAT: Decongestant.

6847. Octogen. [2691-41-0] Octahydro-1,3,5,7-tetranitro-1,-3,5,7-tetrazocine; 1,3,5,7-tetranitro-1,3,5,7-tetrazacyclooctane; cyclotetramethylene tetranitramine; HMX. C$_4$H$_8$N$_8$O$_8$; mol wt 296.16. C 16.22%, H 2.72%, N 37.84%, O 43.22%. Prepn: H. Fischer, *Ber.* **82**, 192 (1949); A. F. McKay *et al.*, *Can. J. Res.* **27**, 462 (1949). Determn in explosives by infrared: W. Selig, *U.S. Atomic Energy Commission* UCRL-7873 (part II), 29 (1965); by HPLC: P. Vouros *et al.*, *Anal. Chem.* **49**, 1039 (1977). Mass spec: E. P. Burrows, *Org. Mass Spectrom.* **26**, 1027 (1991). Thermal analysis of phases using x-ray diffraction: M. Herrmann *et al.*, *Z. Kristallogr.* **204**, 121 (1993). ^{14}N-NMR: M. Kony *et al.*, *J. Org. Chem.* **59**, 5623 (1994). Review of syntheses and properties: B. Singh *et al.*, *Def. Sci. J.* **28**, 41-50 (1978); of thermal behavior: T. L. Boggs, *Prog. Astronaut. Aeronaut.* **90**, 121-175 (1984). Review of toxicology and human exposure: *Toxicological Profile for HMX* (PB98-101058, 1997) 130 pp.

Highly explosive. White crystalline solid, mp 281°. Non-hygroscopic. Four polymorphic forms: α, β, γ and δ. Stability at 300 K: β>α>γ>δ. β is stable at room temp; α at 115-156°C; γ at 156°C; δ at ~170-279°C. Density (g/cm^3): β=1.903; α=1.87; γ=1.82; δ=1.78. Practically insol in water.

USE: In solid propellants and explosives.

6848. Octopamine. [104-14-3] α-(Aminomethyl)-4-hydroxybenzenemethanol; α-(aminomethyl)-*p*-hydroxybenzyl alcohol; 1-(*p*-hydroxyphenyl)-2-aminoethanol; norsympatol; norsynephrine; *p*-hydroxyphenylethanolamine; WV-569. C$_8$H$_{11}$NO$_2$; mol wt 153.18. C 62.73%, H 7.24%, N 9.14%, O 20.89%. A biogenic amine that is the phenol analog of noradrenaline (norepinephrine, *q.v.*). It is a neurosecretory product found in several vertebrates and invertebrates. Formed by β-hydroxylation of tyramine by the enzyme dopamine β-hydroxylase: Pisano *et al.*, *Biochim. Biophys. Acta* **43**, 566 (1960). Identification: Erspamer, *Nature* **169**, 375 (1952). Found in the salivary glands of *Octopus vulgaris*, *O. macropus*, and of *Eledone moschata*: idem, *Arzneim.-Forsch.* **2**, 253 (1952); in mammalian nerves: Molinoff, Axelrod, *Science* **164**, 428 (1969); in cockroach nervous system: Nathanson, Greengard, ibid.

180, 308 (1973). Prepd synthetically: Asscher, US **2585988** (1952). The natural D(−) form is 3 times more potent than the L(+) form in producing cardiovascular adrenergic responses in anesthetized dogs and cats: Korol, Soffer, *Pharmacologist* **5**, 247 (1963). Prepn of D- and L-forms: Kappe, Armstrong, *J. Med. Chem.* **7**, 569 (1964). In invertebrate nervous systems octopamine may function as a neurotransmitter: Saavedra *et al.*, *Science* **185**, 364 (1974). Effects on neuromuscular transmission in crustacean muscle: C. A. Breen, H. L. Atwood, *Nature* **303**, 716 (1983).

D(−)-**Form.** [876-04-0] Crystals from hot water which change at about 160° to a compd which melts above 250° (dec). [α]$_D^{25}$ −56.0° (0.1*N* HCl); −37.4° (H$_2$O).
DL-**Form hydrochloride.** [770-05-8] Epirenor; Norden; Norfen; Norphen (ampules). C$_8$H$_{11}$NO$_2$.HCl; mol wt 189.64. Crystals, dec 170°. Freely sol in water.
THERAP CAT: Adrenergic.

6849. Octotiamine. [137-86-0] 6-(Acetylthio)-8-[[2-[[(4-amino-2-methyl-5-pyrimidinyl)methyl]formylamino]-1-(2-hydroxyethyl)-1-propen-1-yl]dithio]octanoic acid methyl ester; 8-[[2-[*N*-(4-amino-2-methyl-5-pyrimidinyl)methyl]formamido]-1-(2-hydroxyethyl)propenyl]dithio]-6-mercaptooctanoic acid methyl ester *S*(or 6)-acetate; *S*-(3-acetylthio-7-carbomethoxyheptylthio)thiamine; thiamine 8-(methyl 6-acetyldihydrothioctate) disulfide; Gerostop; Neuvitan; TATD. C$_{23}$H$_{36}$N$_4$O$_5$S$_3$; mol wt 544.74. C 50.71%, H 6.66%, N 10.29%, O 14.68%, S 17.66%. Prepn: Ohara *et al.*, US **3098856** (1963 to Fujisawa).

E-form

Crystals, mp 106-109°. uv max: 234, 277 nm (ε 16200, 5820).
Hydrochloride. C$_{23}$H$_{36}$N$_4$O$_5$S$_3$.HCl. Crystals from ether + abs ethanol, mp 134.5-135°. uv max: 233 nm (ε 23000).
THERAP CAT: Vitamin (enzyme cofactor).

6850. Octoxynol. [9002-93-1] α-[4-(1,1,3,3,-Tetramethylbutyl)phenyl]-ω-hydroxypoly(oxy-1,2-ethanediyl); octylphenoxy polyethoxyethanol; polyethylene glycol *p*-isooctylphenyl ether. Prepd by reacting isooctylphenol with ethylene oxide. Average number of ethylene oxide units (n) per molecule is indicated by number following octoxynol. *See also* nonoxynol. Trademarks for series of octoxynols include *Igepal CA*, *Polytergent G*, *Triton X*.

Mixture in which n ranges from about 5 to 15. Pale yellow, viscous liquid. d$_4^{25}$ 1.0595. n$_D^{25}$ 1.4894. Miscible with water, alcohol, acetone. Sol in benzene, toluene. Insol in petr ether. pH of 5% aq soln: 7-9. *Octoxynol-9*, average comp. (n = 9): C$_{34}$H$_{62}$O$_{11}$; av mol wt 647. Trademarks of products where n = 9 to 10: *Conco NIX-100*, *Igepal CA-630*, *Neutronyx 605*, *Triton X-100*. Ingredient of *Preceptin*.

USE: Nonionic detergent, emulsifier, dispersing agent. Octoxynol-9 as spermaticide.

6851. Octreotide. [83150-76-9] D-Phenylalanyl-L-cysteinyl-L-phenylalanyl-D-tryptophyl-L-lysyl-L-threonyl-N-[(1R,2R)-2-hydroxy-1-(hydroxymethyl)propyl]-L-cysteinamide cyclic (2 → 7)-disulfide; 1,2-dithia-5,8,11,14,17-pentaazacycloeicosane cyclic peptide deriv; SMS-201-995. $C_{49}H_{66}N_{10}O_{10}S_2$; mol wt 1019.25. C 57.74%, H 6.53%, N 13.74%, O 15.70%, S 6.29%. Octapeptide analog of somatostatin, $q.v.$ Prepn: W. Bauer, J. Pless, **EP 29579**; *eidem*, **US 4395403** (1981, 1983 both to Sandoz); and pharmacology: W. Bauer *et al.*, *Life Sci.* **31**, 1133 (1982). Opiate antagonist properties: R. Maurer *et al.*, *Proc. Natl. Acad. Sci. USA* **79**, 4815 (1982). HPLC determn in pharmaceutical formulations: N. Kyaterekera *et al.*, *J. Pharm. Biomed. Anal.* **21**, 327 (1999). Inhibitory effect on human gastroenteropancreatic hormone secretion: M. E. Kraenzlin *et al.*, *Experientia* **41**, 738 (1985). Endocrine profile in humans: E. del Pozo *et al.*, *Acta Endocrinol.* **111**, 433 (1986). Clinical evaluation in acromegaly: G. Plewe *et al.*, *Lancet* **2**, 782 (1984); in autonomic neuropathy: R. D. Hoeldtke *et al.*, *ibid.* **2**, 602 (1986). Symposium on chemistry, pharmacology and clinical trials: *Scand. J. Gastroenterol.* **21**, Suppl. 119, 1-274 (1986); on clinical evaluation in gastrointestinal endocrine tumors: *Am. J. Med.* **82**, Suppl. 5B, 1-99 (1987). Review of therapeutic potential in liver tumors: E. A. Kouroumalis, *Chemotherapy (Basel)* **47**, Suppl. 2, 150-161 (2001); of clinical use of long-acting release formulation in acromegaly: K. McKeage *et al.*, *Drugs* **63**, 2473-2499 (2003).

D-Phe–Cys–Phe–D-Trp–Lys–Thr–Cys–NH

Acetate. [79517-01-4] SMS-201-995ac; Longastatin; Sandostatin. $C_{49}H_{66}N_{10}O_{10}S_2 \cdot xC_2H_4O_2$.
$[\alpha]_D^{20}$ −42° (c = 0.5 in 95% acetic acid).

THERAP CAT: Gastric antisecretory agent. Treatment of acromegaly.

6852. Octyl Acetate. [103-09-3] Acetic acid 2-ethylhexyl ester; 2-ethylhexyl acetate. $C_{10}H_{20}O_2$; mol wt 172.27. C 69.72%, H 11.70%, O 18.57%. Toxicity data: H. F. Smyth, C. P. Carpenter, *J. Ind. Hyg. Toxicol.* **26**, 269 (1944).

Liquid. d_{20}^{20} 0.873. bp 199°. mp ∼−80°. n_D^{20} 1.4204. Flash pt, open cup: 190°F (88°C); closed cup: 56°F (13°C). Very slightly sol in water; misc with alcohol, oils, and other organic liquids. LD_{50} orally in rats: 3.0 g/kg (Smyth, Carpenter).

USE: Solvent for nitrocellulose, some resins, waxes, and oils.

6853. n-Octyl Bromide. [111-83-1] 1-Bromooctane. $C_8H_{17}Br$; mol wt 193.13. C 49.75%, H 8.87%, Br 41.37%. Prepd from hydrobromic acid and n-octanol: O. Kamm, C. S. Marvel, *Org. Synth.* **coll. vol. I**, 30 (2nd ed., 1941); *cf.* Norris *et al.*, *J. Am. Chem. Soc.* **38**, 1076 (1916); Whitmore *et al.*, *ibid.* **67**, 2059 (1945); from PBr$_3$ and n-octanol: Coulson *et al.*, *J. Chem. Soc.* **1965**, 2364. Toxicity study: H. F. Smyth *et al.*, *Am. Ind. Hyg. Assoc. J.* **30**, 470 (1969).

Colorless liquid. d_4^{25} 1.108. bp 198-200°. mp −55°. n_D^{25} 1.4503. Insol in water. Miscible with alcohol, ether. LD_{50} orally in rats: 4.49 ml/kg (Smyth).

USE: In organic syntheses.

6854. sec-Octyl Bromide. [557-35-7] 2-Bromooctane; 1-methylheptyl bromide. $C_8H_{17}Br$; mol wt 193.13. C 49.75%, H

8.87%, Br 41.37%. Prepd by the action of PBr$_3$ on sec-octyl alcohol: Hsueh, Marvel, *J. Am. Chem. Soc.* **50**, 856 (1928); Reynolds, Adkins, *ibid.* **51**, 279 (1929); Shriner, Young, *ibid.* **52**, 3332 (1930); Ellis, Reid, *ibid.* **54**, 1680 (1932); E. J. Coulson *et al.*, *J. Chem. Soc.* **1965**, 2364. Review of synthesis of optically active alkyl halides: H. R. Hudson, *Synthesis* **1**, 112-119 (1969).

d_4^{25} 1.0878. bp$_{14}$ 72°. bp$_6$ 66°. bp$_3$ 61°. n_D^{25} 1.4442. Insol in water, miscible with alcohol, ether.

(R)-Form. [5978-55-2] d_4^{25} 1.0982. bp$_{18}$ 73°. bp$_{14}$ 71°. bp$_3$ 60°. n_D^{25} 1.4475. $[\alpha]_D^{20}$ −44.91°.

(S)-Form. [1191-24-8] d_4^{25} 1.0982. bp$_{18}$ 77°. bp$_{14}$ 71°. bp$_3$ 60°. n_D^{25} 1.4475. $[\alpha]_D^{20}$ +43.4°.

6855. Octyl Cyanoacrylate. [6701-17-3] 2-Cyano-2-propenoic acid octyl ester; ocrilate; ocrylate; n-octyl α-cyanoacrylate; capryl α-cyanoacrylate; octyl 2-cyanoacrylate; Dermabond; Nexaband; Soothe-N-Seal. $C_{12}H_{19}NO_2$; mol wt 209.29. C 68.87%, H 9.15%, N 6.69%, O 15.29%. Biodegradable, topical surgical glue; upon contact with tissue anions, monomer rapidly polymerizes to form a strong adhesive. Preparative method: A. E. Ardis, **US 2467927** (1949 to B. F. Goodrich). Prepn, polymerization and degradation: F. Leonard *et al.*, *J. Appl. Polym. Sci.* **10**, 259 (1966). Biocompatibility, histomorphology, and clinical characteristics: A. Nitsch *et al.*, *Aesthetic Plast. Surg.* **29**, 53 (2005). Clinical trial for closure of lacerations: J. Quinn *et al*, *J. Am. Med. Assoc.* **277**, 1527 (1997); of lacerations and incisions: A. J. Singer *et al.*, *Surgery* **131**, 270 (2002); for repair of inguinal hernia: G. Miyano *et al.*, *J. Pediatr. Surg.* **39**, 1867 (2004).

bp$_{1.8}$ 117°; d^{20} 0.931; n_D^{20} 1.4489. Surface tension: 29.18 dynes/cm.

Polymer. [26877-34-9] Poly(octyl cyanoacrylate).

THERAP CAT: Tissue adhesive.

THERAP CAT (VET): Tissue adhesive.

6856. Octyldodecanol. [5333-42-6] 2-Octyl-1-dodecanol; Eutanol G. $C_{20}H_{42}O$; mol wt 298.56. C 80.46%, H 14.18%, O 5.36%. Lipophilic long-chain saturated fatty alcohol. Prepn: P. Mastagli, *Ann. Chim. (Paris)* **10**, 281 (1938); T. Matsu-ura *et al.*, *J. Org. Chem.* **71**, 8306 (2006). Safety assessment: *J. Am. Coll. Toxicol.* **4**, 1 (1985). Total and partial cohesion parameters: A. Munafo *et al.*, *J. Pharm. Sci.* **77**, 169 (1988). Sensory properties: M. E. Parente *et al.*, *J. Cosmet. Sci.* **56**, 175 (2005). Lipophilic character: C. J. Mbah *et al.*, *Pharmazie* **62**, 351 (2007).

Clear, slightly yellow, odorless oil. bp$_{15}$ 215°. d$_4^{19}$ 0.8463. d^{25} 0.8354. n_D^{19} 1.4545. Sol in ethanol, chloroform, ether. Insol in water. Molar volume = 357.40 ±0.018 cm³/mol. Dielectric constant, ε 2.64 ±0.11. Dipole moment, μ 1.24 ±0.105 D.

USE: In cosmetics as emulsifying and opacifying agent, carrier for oil soluble ingredients, coupling agent for waxes, and superfatting agent in shampoos and soaps.

6857. n-Octyl-β-D-glucoside. [29836-26-8] Octyl-β-D-glucopyranoside; n-octylglucoside; OG. $C_{14}H_{28}O_6$; mol wt 292.37. C 57.51%, H 9.65%, O 32.83%. Nonionic detergent primarily used for solubilizing membrane-bound proteins. Prepn: C. R. Noller, C. W. Rockwell, *J. Am. Chem. Soc.* **60**, 2076 (1938). Partition behavior between water and membrane phases: M. Ueno, *Biochemistry* **28**,

5631 (1989); thermodynamics and structural impact: M. R. Wenk *et al.*, *Biophys. J.* **72**, 1719 (1997). Solubilzation and reconstitution of liposomes: O. López *et al.*, *J. Phys. Chem. B* **105**, 9879 (2001). Solubilization of lipid vesicles: A. Meister, A. Blume, *Phys. Chem. Chem. Phys.* **6**, 1551 (2004). Micelle formation: A. Walter *et al.*, *Biochim. Biophys. Acta* **1508**, 20 (2000); in mixed systems: A. Lainez *et al.*, *Langmuir* **20**, 5745 (2004).

White solid, mp 65-99°. $[\alpha]_D^{25}$ −30.3° (methanol). Critical micelle concentration: 20-25 m*M*.

USE: Detergent and surfactant for biological systems.

6858. sec-Octyl Iodide. [557-36-8] 2-Iodooctane. $C_8H_{17}I$; mol wt 240.13. C 40.01%, H 7.14%, I 52.85%. Prepn: E. J. Coulson *et al.*, *J. Chem. Soc.* **1965**, 2364. Review of synthesis of optically active alkyl halides: H. R. Hudson, *Synthesis* **1**, 112-119 (1969).

Oily liquid; discolors in light. d_{15}^{18} 1.318. bp about 210° with decompn; also stated as 190°. *Protect from air and light.*
(***R***)-Form. [29117-48-4] $[\alpha]_D^{20}$ −64.63°.
(***S***)-Form. [1809-04-7] $[\alpha]_D^{24}$ +62.6°.

6859. Octyl Methoxycinnamate. [5466-77-3] 3-(4-Methoxyphenyl)-2-propenoic acid 2-ethylhexyl ester; 2-ethylhexyl *p*-methoxycinnamate; octinoxate; Eusolex 2292; Neo Heliopan AV; Parsol MCX; Uvinul MC 80. $C_{18}H_{26}O_3$; mol wt 290.40. C 74.45%, H 9.02%, O 16.53%. UV-B blocker. Manuf process and purification: P. Schudel *et al.*, **US 4713473** (1987 to Givaudan). Use in sunscreen preparations: D. H. Liem, L. T. H. Hilderink, *Int. J. Cosmet. Sci.* **1**, 341 (1979). Assessment of photostability: R. Aberturas *et al.*, *Boll. Chim. Farm.* **126**, 208 (1987); A. Deflandre, G. Lang, *Int. J. Cosmet. Sci.* **10**, 53 (1988). HPLC determn in cosmetics: L. Gagliardi *et al.*, *J. Chromatogr.* **408**, 409 (1987). Effect of reflectance of light on protection of skin: G. J. Smith *et al.*, *Photochem. Photobiol.* **75**, 122 (2002).

Pale yellow oil. bp at 1 mbar: 185-195°; bp at 0.1 mbar: 140-150°. Insol in water.

THERAP CAT: Ultraviolet screen.

6860. Octyl Salicylate. [118-60-5] 2-Hydroxybenzoic acid 2-ethylhexyl ester; 2-ethylhexyl salicylate; octisalate; Eusolex OS; Neo Heliopan OS. $C_{15}H_{22}O_3$; mol wt 250.34. C 71.97%, H 8.86%, O 19.17%. UV-B filter in sunscreens and cosmetics. Prepn: F. H. McMillan, J. A. King, *J. Am. Chem. Soc.* **67**, 2271 (1945). *In vitro* skin penetration studies: K. A. Walters *et al.*, *Food Chem. Toxicol.* **35**, 1219 (1997); R. Jiang *et al.*, *J. Pharm. Sci.* **86**, 791 (1997). Determn in sun protection lotions by GC-MS: K. W. Ro *et al.*, *J. Chromatogr. A* **688**, 375 (1994); by HPTLC: E. Westgate, J. Sherma, *Am. Lab.* **32**, 13 (2000).

Colorless to pale yellow liquid, bp_{21} 189-190°. n_D^{25} 1.5018. Misc with cosmetic oils, silicone oils.

USE: Solvent for solid sunscreens.

THERAP CAT: Ultraviolet screen.

6861. Odanacatib. [603139-19-1] (2*S*)-*N*-(1-Cyanocyclopropyl)-4-fluoro-4-methyl-2-[[[(1*S*)-2,2,2-trifluoro-1-[4′-(methylsulfonyl)[1,1′-biphenyl]-4-yl]ethyl]amino]pentanamide; N^1-(1-cyanocyclopropyl)-4-fluoro-N^2-{(1*S*)-2,2,2-trifluoro-1-[4′-(methylsulfonyl)-1,1′-biphenyl-4-yl]ethyl}-L-leucinamide; MK-0822. $C_{25}H_{27}$-$F_4N_3O_3S$; mol wt 525.56. C 57.13%, H 5.18%, F 14.46%, N 8.00%, O 9.13%, S 6.10%. Cathepsin K inhibitor with activity against bone resorption. Prepn: C. Bayly *et al.*, **WO 03075836**; *eidem*, **US 7375134** (2003, 2008 both to Merck & Co.). Enantioselective prepn: P. D. O'Shea *et al.*, *J. Org. Chem.* **74**, 1605 (2009). Physical properties and nanoparticle formulation: V. Kumar *et al.*, *Mol. Pharm.* **6**, 1118 (2009). Discovery and *in vitro* activity: J. Y. Gauthier *et al.*, *Bioorg. Med. Chem. Lett.* **18**, 923 (2008). Clinical trial in postmenopausal women with low bone mineral density: J. A. Eisman *et al.*, *J. Bone Miner. Res.* **26**, 242 (2011).

Hydrophobic, white solid, mp 223-224°. d (solid): 1.40. Bulk soly (mol/m³) at 5°: water 5.7×10^{-4}; 4.7% THF 12.0×10^{-4}.

THERAP CAT: Antiosteoporotic.

6862. Ofatumumab. [679818-59-8] Anti-(human CD20 (antigen)) immunoglobulin G1 (human monoclonal HuMax-CD20 heavy chain) disulfide with human monoclonal HuMax-CD20 κ-chain, dimer; 2F2; GSK-1841157; Arzerra; HuMax-CD20. Human monoclonal antibody directed against a membrane-proximal epitope of the CD20 surface protein on B lymphocytes. Induces complement-dependent lysis of CD20 positive cells. Prepn: J. Teeling *et al.*, **WO 04035807** (2004 to Genmab, Medarex); *eidem*, **US 04167319** (2004); *eidem*, *Blood* **104**, 1793 (2004). Binding studies and complement-dependent cytotoxicity: A. W. Pawluczkowycz *et al.*, *J. Immunol.* **183**, 749 (2009). Crystal structure of Fab fragment: J. Du *et al.*, *Mol. Immunol.* **46**, 2419 (2009). Clinical evaluation in chronic lymphocytic leukemia: B. Coiffier *et al.*, *Blood* **111**, 1094 (2008); in follicular lymphoma: A. Hagenbeek, *ibid.* 5486. Reviews of development and therapeutic potential: T. Robak, *Curr. Opin. Mol. Ther.* **10**, 294-300 (2008); B. Zhang, *mAbs* **1**, 326-331 (2009).

THERAP CAT: Antineoplastic.

6863. Ofloxacin. [82419-36-1] 9-Fluoro-2,3-dihydro-3-methyl-10-(4-methyl-1-piperazinyl)-7-oxo-7*H*-pyrido[1,2,3-*de*]-1,4-benzoxazine-6-carboxylic acid; ofloxacino; DL-8280; HOE-280; Exocin; Flobacin; Floxil; Floxin; Monoflocet; Oculflox; Oflocet; Oflocin; Tarivid. $C_{18}H_{20}FN_3O_4$; mol wt 361.37. C 59.83%, H 5.58%, F 5.26%, N 11.63%, O 17.71%. Broad spectrum, fluorinated quinolone antibacterial. Prepn: I. Hayakawa *et al.*, **EP 47005**; *eidem*, **US 4382892** (1982, 1983 both to Daiichi). Total synthesis: H. Egawa *et al.*, *Chem. Pharm. Bull.* **34**, 4098 (1986). Synthesis and activity of optical isomers: S. Atarashi *et al.*, *ibid.* **35**, 1896 (1987). Antibacterial spectrum of racemate: K. Sato *et al.*, *Antimicrob. Agents Chemother.* **22**, 548 (1982). Mechanism of differential activity of enantiomers: I. Morrissey *et al.*, *ibid.* **40**, 1775 (1996). Toxicity data: H. Ohno *et al.*, *Chemotherapy (Tokyo)* **32**, Suppl. 1, 1084 (1984). Pharmacology and clinical efficacy: *Infection* **14**, Suppl. 1, S1-S109 (1986). Symposium on pharmacokinetics and therapeutic use: *Scand. J. Infect. Dis.* **Suppl. 68**, 1-69 (1990). Review of antibacterial spectrum, pharmacology, and clinical efficacy: J. P. Monk,

D. M. Campoli-Richards, *Drugs* **33**, 346-391 (1987); of mechanism of action: K. Drlica, *Curr. Opin. Microbiol.* **2**, 504-508 (1999).

Colorless needles from ethanol, mp 250-257° (dec). Sparingly sol in chloroform; slightly sol in alc, methanol, water. LD_{50} in male, female mice, male, female rats (mg/kg): 5450, 5290, 3590, 3750 orally; 208, 233, 273, 276 i.v.; >10000, >10000, 7070, 9000 s.c. (Ohno).

S-(−)-Form. [100986-85-4]; [138199-71-0] (hemihydrate). Levofloxacin; DR-3355; Cravit; Levaquin; Tavanic; Quixin. Toxicity study: M. Kato *et al.*, *Arzneim.-Forsch.* **42**, 365 (1992). Series of articles on pharmacology and toxicology: *ibid.*, 368-418. Clinical study in bacterial conjunctivitis: D. G. Hwang *et al.*, *Br. J. Ophthalmol.* **87**, 1004 (2003). *Review:* D. S. North *et al.*, *Pharmacotherapy* **18**, 915-935 (1998). Prepd as the hemihydrate; needles from ethanol + ethyl ether, mp 225-227° (dec). $[\alpha]_D^{23}$ −76.9° (c = 0.385 in 0.5*N* NaOH). Freely sol in glacial acetic acid, chloroform; sparingly sol in water. LD_{50} in male, female mice, male, female rats (mg/kg): 1881, 1803, 1478, 1507 orally (Kato).

THERAP CAT: Antibacterial.

6864. **Oil Anise, Japanese.** Volatile oil from fruit of *Illicium anisatum* L., *Magnoliaceae* (Japanese star anise). *Constit.* Chiefly anethole; also safrol, eugenol.

Colorless to slightly yellow liquid. d about 1.006. Solidifies at −10° to −15°. α_D^{20} about −8°.

THERAP CAT: Carminative; expectorant.

6865. **Oil Bergamot.** Volatile oil expressed from rind of fresh fruit of *Citrus aurantium* L., var. *bergamia* Wight & Arn., *Rutaceae*. *Constit.* Complex mixture of components including 36-45% *l*-linalyl acetate, about 6% *l*-linalool, *d*-limonene, and the furocoumarins, bergapten and bergamottin. GC-MS determn of constituents: E. L. Belsito *et al.*, *J. Agric. Food Chem.* **55**, 7847 (2007). Phototoxicity study: L. Dubertret *et al.*, *J. Photochem. Photobiol. B* **7**, 251 (1990). Review of composition, pharmacology, and commercial uses: I. A. Khan, E. A. Abourashed in *Leung's Encyclopedia of Common Natural Ingredients* (Wiley, Hoboken, 3rd Ed., 2010) pp 91-93.

Yellowish-green liquid; aromatic bitter flavor and agreeable odor. d_{25}^{25} 0.875-0.880. α_D^{25} +8 to +24°. n_D^{20} 1.464-1.467. Acid no. 1-4. Almost insol in water. Sol in 0.5 vol 95% alcohol, 2 vols 80% alcohol. *Keep well closed in a cool place, protected from light.*

USE: Flavoring agent in foods and beverages, particularly Earl Grey tea. In perfumery; fragrance component of personal care products; in aromatherapy.

6866. **Oil of Balm.** Oil of melissa balm; oil of lemon balm. Volatile oil from leaves and tops of *Melissa officinalis* L., *Labiatae*. Chiefly citral. Composition studies: Hefendehl, *Arch. Pharm.* **303**, 345 (1970).

Yellow to yellowish-green liquid. d_{15}^{15} 0.89-0.925. Practically insol in water. Sol in alcohol. *Keep well closed, cool, and protected from light.*

6867. **Oil of Basil.** Volatile oil from leaves of *Ocimum basilicum* L., *Labiatae* (sweet basil). *Constit.* Methylchavicol, eucalyptol, linalool, estragol.

Yellowish to greenish liquid; aromatic odor. d_{20}^{20} 0.905-0.930. α_D^{20} −6 to −22°. Almost insol in water. Sol in 2 vols 80% alc; miscible with ether, chloroform. *Keep well closed, cool and protected from light.*

6868. **Oil of Bay.** Oil of Myrcia. Volatile oil distilled from leaves of *Pimenta (Myrcia) acris* Kostel., *Myrtaceae*. *Constit.* 40-55% eugenol; myrcene, chavicol, methyleugenol, methylchavicol, citral, *l*-phellandrene; total phenols, 50-65% by volume.

Yellow to brownish-yellow liq; pleasant odor; sharp, spicy taste; becomes brown on exposure to air. d_{25}^{25} 0.962-0.990. α_D^{20} −3°. n_D^{20}

1.500-1.520. Insoluble in water. Very sol in alcohol, carbon disulfide, glacial acetic acid.

USE: Pharmaceutic aid (aromatic). Manuf bay rum.

6869. **Oil of Bitter Almond.** Volatile oil from dried ripe kernels of bitter almonds or from other kernels containing amygdalin, such as apricots, cherries, plums, and especially peaches. Obtained by macerating with water, then steam distilling. *Constit.* Not less than 95% benzaldehyde; 2-4% HCN and phenoxyacetonitrile.

Colorless to yellow, very refractive liq; characteristic odor and taste of benzaldehyde. d_{25}^{25} 1.038-1.060. n_D^{20} 1.5428-1.5439. Slightly sol in water; miscible with alcohol, ether, oils. *Keep cool and protected from light.*

Caution: Hydrogen cyanide, *q.v.*, component responsible for highly toxic properties. *Very poisonous!*

USE: Only the oil *free from HCN* may be used for liqueurs and foods.

THERAP CAT: Formerly as topical antipruritic.

6870. **Oil of Bitter Orange.** Volatile oil expressed from fresh peel of *Citrus aurantium* L., *Rutaceae*. About 90% *d*-limonene; citral, decyl aldehyde, methyl anthranilate, linalool, terpineol.

Pale yellow liquid; bitter taste. d_{25}^{25} 0.842-0.848. α_D^{25} +88 to +98°. Very slightly sol in water; miscible with abs alcohol; sol in 4 vols alcohol, in 1 vol glacial acetic acid. *Keep well closed, cool, and protected from light.*

USE: As flavoring; in perfumery.

6871. **Oil of Cajeput.** Cajuput oil; cajeputi oil. Volatile oil from fresh leaves and twigs of several varieties of *Melaleuca leucadendron* L., and other species of *Melaleuca*, *Myrtaceae*. *Constit.* 50-60% eucalyptol (cineol); *l*-pinene, terpineol; valeric, butyric, benzoic and other aldehydes. Toxicity: P. M. Jenner *et al.*, *Food Cosmet. Toxicol.* **2**, 327 (1964).

Colorless or yellowish liquid, agreeable camphor odor, and bitter aromatic taste. d 0.912-0.925. α_D^{20} <−4°. n_D^{20} 1.4660-1.4710. Very slightly sol in water; sol in 1 vol 80% alcohol. Misc with alcohol, chloroform, ether, carbon disulfide. *Keep well closed, cool, and protected from light.* LD_{50} orally in rats: 3870 mg/kg (Jenner).

THERAP CAT: Expectorant, counterirritant, scabicide.

THERAP CAT (VET): Rubefacient, topical antimycotic.

6872. **Oil of Camphor, Rectified.** Formosa oil of camphor; Japanese oil of camphor; white oil of camphor; light oil of camphor. Volatile oil from *Cinnamomum camphora* T. Nees & Eberm., *Lauraceae*. *Constit.* Safrol, acetaldehyde, camphor, terpineol, eugenol, cineol, *d*-pinene, phellandrene, dipentene, cadinene.

Colorless or yellowish liquid; camphor odor. d_{20}^{20} 0.875-0.900. α_D^{25} +9 to +24°. n_D^{20} 1.465-1.470. Insol in water. Sol in chloroform, ether, oils, in about 3 vols alcohol. *Keep well closed, cool, and protected from light.*

USE: As solvent in paint and lacquer industry; in perfuming of soaps and detergents; in technical odor masking.

THERAP CAT: Rubefacient.

6873. **Oil of Cashew Nut Shell.** From *Anacardium occidentale* L., *Anacardiaceae* (Southern India). Consists mostly (90%) of anacardic acid, *q.v.* Review and bibliography: Sanyal, Das, *Indian Pharm.* **10**, 272-283 (1955).

Wijs iodine no. over 250. Polymerizes easily when heated with acid.

USE: In insulating varnishes, typewriter rolls, coldsetting cements, floor tile, brake linings; molding compounds from polymers: Harvey, **US 2767150** (1956 to The Harvel Corp.).

6874. **Oil of Cedar Wood.** Volatile oil from wood of *Juniperus virginiana* L., *Cupressaceae*, and other species of cedar. *Constit.* Chiefly cedrene (a terpene), and cedral (cedar camphor).

Colorless or slightly yellow, somewhat viscid liquid. d_{20}^{20} 0.940-0.950. α_D^{20} −25 to −46°. n_D^{20} 1.495-1.510. Insol in water. Sol in 10-20 vols 90% alcohol; sol in ether. *Keep well closed, cool, and protected from light.*

USE: In perfumery; as insect repellent; the thickened oil is used in microscopy as a clearing agent and for use with immersion lenses.

6875. **Oil of Celery.** Volatile oil from celery seed, *Apium graveolens* L., *Umbelliferae*. *Constit.* *d*-Limonene, phenols, sedanolide, sedanoic acid.

Colorless liquid; celery odor. d 0.870-0.895. α_D^{20} +67 to +79°. Slightly sol in water; very sol in alcohol. *Keep well closed, cool, and protected from light.*

USE: As flavor for soft drinks, unpleasant medicaments.

6876. Oil of Champaca. Volatile oil from flowers of *Michelia champaca* L., *Magnoliaceae. Constit.* Esters of benzoic acid, benzaldehyde, benzyl alcohol, isoeugenol.

Light yellow or reddish-yellow to brownish liquid. d_{15}^{15} 0.906-0.935. α_D −12 to −52°. Sapon no. ~77. Slightly sol in water; sol in chloroform, ether; sparingly sol in alcohol. *Keep well closed, cool and protected from light.*

USE: In perfumes.

6877. Oil of Chenopodium. Oil of American wormseed; Chenopodiol; Chenoposan; Chenoposetten. Volatile oil from fresh, aboveground parts of flowering and fruiting plant of *American wormseed*, *Chenopodium ambrosioides* L. var. *anthelminticum* (L.) Aellen, *Chenopodiaceae* also known as *Mexican tea*, *Spanish tea*, *Jerusalem tea*, *ambrosia. Habit.* Central America, U.S., Canada. Oil contains 60-70% ascaridole, the principal active constituent; *p*-cymene, α-terpinene, *l*-limonene, methadiene.

Colorless or pale yellow liquid; characteristic disagreeable odor and taste. d_{25}^{25} 0.950-0.980. α_D^{20} −4 to −8°. n_D^{20} 1.4723-1.4790. Insoluble in water. Sol in 8 vols 70% alcohol; partly sol in glacial acetic acid. *Keep well closed, cool and protected from light.*

THERAP CAT: Anthelmintic.

THERAP CAT (VET): Anthelmintic.

6878. Oil of Cherry Laurel. Volatile oil from leaves of *Prunus laurocerasus* L., *Rosaceae. Constit.* HCN, benzaldehyde, benzaldehyde cyanhydrin, benzyl alcohol.

Pale yellow liquid; odor and taste similar to oil of bitter almond. *Poisonous.* d_{20}^{20} 1.054-1.066. Slightly sol in water; sol in 2 vols 70% alcohol, in benzene, chloroform, ether. *Keep well closed, cool and protected from light.*

Caution: Hydrogen cyanide component responsible for highly toxic properties.

6879. Oil of Citronella. Volatile oil from fresh grass of *Cymbopogon (Andropogon) nardus* (L.) Rendle, *Gramineae. Constit.* Ceylon: about 60% geraniol, about 15% citronellal, 10-15% camphene and dipentene, small quantities of linalool, borneol. Java: 25-50% citronellal, 25-45% geraniol.

Almost colorless to pale yellow liq; gradually becomes reddish; pleasant odor. d Ceylon, 0.897-0.912; Java, 0.885-0.900. α_D^{20}: Ceylon, −6 to −14°; Java, −2 to −5°. n_D^{20} Ceylon 1.479-1.485; Java, 1.468-1.473. Slightly sol in water; sol in 10 vols 80% alcohol. *Keep well closed, cool and protected from light.*

USE: As perfume; insect repellent.

6880. Oil of Cumin. Volatile oil from fruit of *Cuminum cyminum* L., *Umbelliferae. Constit.* 30-40% cuminaldehyde; *p*-cymene, β-pinene, dipentene.

Colorless to yellow liquid. d_{25}^{25} 0.900-0.935. α_D^{20} +4 to +8°. n_D^{20} 1.4950-1.5090. Almost insol in water; sol in 10 vols 80% alcohol; more sol in stronger alcohol; very sol in chloroform, ether. *Keep well closed, cool and protected from light.*

USE: Flavoring in Indian curry powder.

6881. Oil of Cypress. Volatile oil from leaves and young branches of *Cupressus sempervirens* L., *Pinaceae. Constit.* Furfural, *d*-pinene, *d*-camphene, cymene, *d*-terpineol, *l*-cadinene, sylvestrene, cypress camphor.

Yellowish liquid. d 0.88-0.89. α_D^{20} +4 to +18°. Slightly sol in water; sol in 2-6 vols 90% alcohol. *Keep well closed, cool and protected from light.*

USE: In perfumery.

6882. Oil of Dill. Dill seed oil; dill fruit oil. Volatile oil from dried ripe fruit of *Anethum graveolens* L., *Umbelliferae. Constit.* About 50% carvone *d*-limonene, phellandrene and other terpenes.

Colorless or pale yellow liquid; characteristic odor. d_{15}^{15} 0.900-0.915. α_D^{20} +70 to +80°. n_D^{20} 1.481-1.492. Insol in water; sol in 1 vol 90% alcohol. *Keep well closed, cool and protected from light.*

THERAP CAT: Aromatic carminative.

6883. Oil of Dwarf Pine Needles. Oil of mountain pine; Pinus Montana oil; Pinus pumilio oil. Volatile oil from fresh leaves of

Pinus montana Mill. (*P. pumilio* Haenke), *Pinaceae. Constit.* *l*-Pinene, *l*-phellandrene, sylvestrene, dipentene, cadinene, 5-7% bornyl acetate.

Colorless or faintly yellow liq; pleasant odor; bitter taste. d_{25}^{25} 0.853-0.869. α_D^{25} −5 to −12°. n_D^{20} 1.4750-1.4800. Insol in water. Sol in 4.5-8 vols 90% alcohol; very sol in chloroform, ether. *Keep well closed, cool and protected from light.*

USE: Pharmaceutic aid (flavor and perfume).

THERAP CAT: Expectorant.

6884. Oil of Fir. Oil of silver pine; oil of silver fir. Volatile oil from needles and young twigs of *Abies alba* Mill. (*A. picea* Lindl., *A. pectinata* DC.), *Pinaceae. Constit.* *l*-Pinene, *l*-limonene, *l*-bornyl acetate.

Colorless, clear liquid; balsamic odor; terebinthinate taste. d_{15}^{15} 0.869-0.875. α_D^{20} −20 to −59°. Insol in water. Sol in 5 vols 90% alc, in ether. *Keep well closed, cool and protected from light.*

USE: Pharmaceutic aid (flavor and perfume).

THERAP CAT: Expectorant.

6885. Oil of Fir—Siberian. Oil of pine; oleum abietis; Siberian pine needle oil. Volatile oil from fresh leaves of *Abies sibirica* Ledeb., *Pinaceae. Constit.* About 40% esters calculated as bornyl acetate; pinene, camphene, dipentene, and phellandrene.

Colorless or pale yellow liquid; aromatic odor; pungent taste. d_{15}^{15} 0.905-0.925. α_D^{20} −32 to −45°. n_D^{20} 1.466-1.476. Sol in an equal vol 90% alcohol. *Keep well closed, cool and protected from light.*

USE: Pharmaceutic aid (flavor and perfume).

THERAP CAT: Expectorant.

6886. Oil of Fleabane. Oil of Canada fleabane; oil of erigeron. Volatile oil from fresh flowering herb of *Conyza canadensis* (L.) Cron. (*Erigeron canadensis* L., *Leptilon canadense* (L.)) Britt., *Compositae. Constit.* *d*-Limonene, aldehydes.

Pale yellow liquid; becomes darker and thicker with age and on exposure to air; peculiar odor; aromatic, slightly pungent taste. d_{25}^{25} 0.845-0.865. α_D^{20} ~+45°. Slightly sol in water; sol in 1 vol alcohol; very sol in chloroform, ether. *Keep well closed, cool and protected from light.*

6887. Oil of Geranium. Oil of pelargonium geranium; oil of rose geranium. Volatile oil from leaves of *Pelargonium odoratissimum* Ait. and allied species, *Geraniaceae. Constit.* Geraniol esters, calculated as geranyl tiglate, 21-30% in Algerian or French oil, 25-35% in Bourbon oil; citronellol; some linalool.

Colorless, greenish or brownish liquid. d_{15}^{15} 0.894-0.905. α_D^{20} −7 to −11°. n_D^{20} 1.4650-1.470. Slightly sol in water; sol in 3 vols 10% alcohol; more sol in stronger alcohol; very sol in chloroform, ether. *Keep well closed, cool and protected from light.*

USE: In perfumery; as odorant for tooth and dusting powders, ointments, etc. In manuf of rhodinol, *q.v.*

6888. Oil of Geranium—East Indian. Turkish geranium oil; palmarosa oil; Indian grass oil; rusa oil. Volatile oil from *Andropogon schoenanthus* L., *Gramineae*, and allied species grown in India (not Turkey). *Constit.* 85-95% geraniol; citronellol, dipentene; it is practically devoid of esters.

Colorless or light yellow liquid; pleasant rose odor. d_{15}^{15} 0.885-0.896. α_D +1.67 to −2°. n_D^{20} 1.476-1.4085.

6889. Oil of Hyssop. Volatile oil from *Hyssopus officinalis* L., *Labiatae. Constit.* About 50% pinene, small quantities of aromatic alcohol, probably also sesquiterpenes.

Colorless or greenish-yellow liquid; sharp, camphor-like taste. d_{15}^{15} 0.925-0.940. α_D^{20} −17 to −23°. bp ~200°. Almost insol in water; sol in 2 to 4 vols 80% alcohol. *Keep well closed and protected from light.*

6890. Oil of Lemon. Cedro oil. Volatile oil expressed from fresh peel of *Citrus limonum* (L.) Risso (*C. medica* var. *limon* L.), *Rutaceae. Constit.* About 90% of limonene, terpinene, phellandrene, and pinene combined; 4-6% aldehydes calculated as citral, some citronellal, geranyl acetate, sesquiterpenes.

Pale yellow or greenish-yellow liquid. d_{25}^{25} 0.849-0.855. α_D^{20} +57 to +65.6°. n_D^{20} 1.4742-1.4755. Slightly sol in water; sol in 3 vols alcohol; miscible with carbon disulfide, glacial acetic acid. *Keep cool, in well-filled and well-closed bottles, protected from light.*

USE: For flavoring medicaments; as a flavor in liqueurs, pastry, foods, beverages; also in perfumes.

6891. Oil of Lemon Grass. Indian oil of verbena; Indian melissa oil. Volatile oil from *Cymbopogon (Andropogon) citratus* (DC.) Stapf, and of *C. flexuosus* (Nees) Stapf, *Gramineae. Constit.* 75-85% citral; methylheptenone, citronellal, geraniol, limonene, dipentene.

Reddish-yellow or brownish-red liquid; strong odor of verbena. d_{15}^{15} 0.895-0.908 (0.878-0.882 of West Indian oil). α_D −3°. n_D^{20} 1.483-1.489. Slightly sol in water; sol in 3 vols 70% alcohol; sol in chloroform, ether. *Keep well closed, cool and protected from light.*

USE: As source of citral (synthesis of vitamin A) and in perfumery.

6892. Oil of Levant Wormseed. Volatile oil from the flowers of *Artemisia maritima* var *stechmanniana* Bess (*A. pauciflora* Weber), *Compositae. Constit.* Largely eucalyptol (cineol).

Pale yellow to yellowish-green liquid. d_{20}^{20} 0.915-0.940. Insol in water; sol in alcohol, ether. *Keep well closed, cool and protected from light.*

THERAP CAT: Anthelmintic.

6893. Oil of Linaloe. Volatile oil distilled from a Mexican wood *(Bursera delpechiana* Poiss. and probably other species of *Bursera, Burseraceae). Constit.* Linalool, geraniol, methylheptenone.

Colorless to yellowish liquid; pleasant odor. d_{15}^{15} 0.875-0.890. α_D −5 to −12°. n_D^{20} 1.4638. Slightly sol in water; sol in 2 vols 70% alcohol; sol in ether, chloroform. *Keep well closed, cool and protected from light.*

USE: In perfumery.

6894. Oil of Marjoram. Volatile oil from *Origanum marjorana* L., *Labiatae. Constit.* About 40% terpenes, chiefly terpinene; *d*-terpineol.

Yellow or greenish-yellow liquid. d_{15}^{15} 0.888-0.912. α_D^{20} +13 to +18°. Insol in water; sol in 2 vols 80% alcohol; sol in chloroform, ether.

USE: In perfumes and microscopy.

6895. Oil of Myrtle. Volatile oil from leaves of *Myrtus communis* L., *Myrtaceae. Constit. d*-Pinene, eucalyptol, dipentene, camphor.

Yellow to greenish liquid; fragrant odor. d_{15}^{15} 0.890-0.915. α_D +10 to +30°. Insol in water; sol in alcohol, chloroform, ether.

6896. Oil of Niaouli. Volatile oil from leaves of *Melaleuca viridiflora* (Soland.) Gaertn., *Myrtaceae. Constit.* About 65% cineol, about 30% terpineol, limonene and *d*-pinene combined.

Slightly yellow liquid; aromatic odor; pungent, refreshing peppermint-like taste. d_{15}^{15} 0.908-0.932. Insol in water or glycerol; sol in alcohol, benzene, ether.

THERAP CAT: Anthelmintic.

6897. Oil of Orange. Oil sweet orange. Volatile oil expressed from fresh peel of ripe fruit of the orange *(Citrus aurantium* var *sinensis* L., *Rutaceae). Constit.* About 90% *d*-limonene; citral, decyl aldehyde, methyl anthranilate, linalool, terpineol.

Yellow to deep orange liquid; characteristic orange taste and odor. d_{25}^{25} 0.842-0.846. α_D^{20} +94 to +99°. n_D^{20} 1.4723-1.4737. Slightly sol in water; sol in 2 vols 90% alcohol, in 1 vol glacial acetic acid; miscible with abs alcohol, carbon disulfide. *Keep well closed, cool and protected from light.*

USE: Chiefly as flavor and perfume. Pharmaceutic aid (flavor).

THERAP CAT: Expectorant.

6898. Oil of Orange Flowers. Oil of neroli. Volatile oil distilled from fresh orange flowers. *Constit.* Limonene, *l*-linalool, geraniol, 7-18% linalyl acetate; some methyl anthranilate, nerol and neroli camphor.

Yellowish, fluorescent liquid; very intense and pleasant odor; becomes brown on exposure to light. d_{25}^{25} 0.86-0.88. $[\alpha]_D^{25}$ +1.50 to +9.13°. n_D^{20} 1.475. Slightly sol in water; sol in 1.5-2 vols 80% alcohol with fine violet fluorescence. *Keep well closed, cool and protected from light.*

USE: As perfume and flavor.

6899. Oil of Origanum. Oil of wild marjoram. Volatile oil from flowering tops of *Origanum vulgare* L., *Labiatae. Constit.* Carvacrol, terpenes.

Light yellow liquid. d_{15}^{15} 0.870-0.910. α_D ~ −34°. Very slightly sol in water; very sol in alcohol. *Keep well closed, cool and protected from light.*

USE: In perfumery.

6900. Oil of Patchouli. Patchouli oil. Essential oil from leaves of several *Labiatae* species. The commercial oil is obtained from the cultivated species, *Pogostemon cablin* (Blanco) Benth. (*P. patchouly* Pellet. var *suavis* Hook. f.), *Labiatae.* Major constituent is patchouli alcohol, *q.v.,* minor constituents include patchoulene, azulene, eugenol, and several unidentified sesquiterpenes: Pfau, Plattner, *Helv. Chim. Acta* **19,** 874 (1936); Naoko *et al., Bull. Chem. Soc. Jpn.* **40,** 597 (1967). Review: E. Guenther, *The Essential Oils* **vol. III** (Van Nostrand, New York, 1949) pp 552-575.

Yellowish or greenish to dark brown oil, intense and persistent fragrant odor. Can be stored indefinitely. Odor seems to improve with age. d_{15}^{15} 0.975-0.987. $[\alpha]_D^{20}$ −54 to −65.3°. n_D^{20} 1.5099 to 1.5111. Saponif no. 3.3 to 9.3. Ester no. after acetylation: 17.7 to 22.4. Practically insol in water. Soluble in ether.

USE: In perfume formulations to impart a lasting oriental fragrance, in incense, soaps, cosmetics. To scent fine Indian fabrics and shawls.

6901. Oil of Pettigrain. Volatile oil from leaves, twigs and unripe fruit of *Citrus vulgaris* Risso (*C. bigaradia* Loisel.), *Rutaceae. Constit.* 40-80% linalyl acetate; geraniol, geranyl acetate, limonene.

Yellow liquid. d_{15}^{15} 0.887-0.900. α_D^{20} +3.72 to −1.37°. n_D^{20} 1.4623. Slightly sol in water; sol in 2 vols 80% alcohol. *Keep well closed, cool and protected from light.*

USE: In perfumes.

6902. Oil of Pine Needles. Oil of Scotch fir; fir-wood oil. Volatile oil from *Pinus sylvestris* L., *Pinaceae. Constit.* Dipentene, pinene, sylvestrene, cadinene, 3-3.5% bornyl acetate.

Yellowish liquid. d_{15}^{15} 0.884-0.886. α_D^{20} +7.05 to +10°. Insol in water. Sol in 10 vols 90% alcohol. *Keep well closed, cool and protected from light.*

USE: Pharmaceutic aid (flavor and perfume).

THERAP CAT: Expectorant.

6903. Oil of Rose. Otto of rose; essence of rose; attar of rose. Volatile oil from fresh flowers of *Rosa gallica* L. and *R. damascena* Mill. and varieties of these species *(Rosaceae). Constit.* 70-75% free geraniol and citronellol; small amounts of their esters; terpenes.

Colorless or pale yellow liquid; viscous at 25°; highly fragrant, rose odor. d_{15}^{30} 0.848-0.863. Congeals at 18-22° to a translucent, cryst mass. α_D^{25} −1 to −4°. n_D^{30} 1.457-1.463. Very slightly sol in water, sparingly sol in alcohol; sol in fatty oils, chloroform. *Keep cool in well-closed and well-filled containers and protected from light.*

USE: Largely in perfumery; for flavoring lozenges, ointments, and toilet prepns.

6904. Oil of Rue. Volatile oil from *Ruta graveolens* L., *Rutaceae. Constit.* About 90% methyl nonyl ketone, methyl anthranilate.

Pale yellow liquid; characteristic, sharp, unpleasant odor, but odor is pleasant on dilution. d_{15}^{15} 0.832-0.845. Solidif +8° to +10°. Optically inactive or slightly dextrorotatory. n_D^{20} 1.430-1.440. Almost insol in water; sol in 3 vols 70% alcohol. *Keep well closed, cool and protected from light.*

Caution: Frequent dermal contact produces erythema, vesication. Ingestion of large quantities causes epigastric pain, nausea, vomiting, confusion, convulsions, death; may cause abortion.

USE: Flavoring agent in food.

6905. Oil of Sweet Almond. Expressed almond oil. Fixed oil from kernels of varieties of *Prunus amygdalus* Stokes (*Amygdalus communis* L.), *Rosaceae. Constit.* Chiefly glyceryl oleate with small amounts of glycerides of linolic, etc., acids. Stated to contain no stearic acid.

Colorless or pale yellow, almost odorless, oily liq; bland taste. d_{25}^{25} 0.910-0.915. It is clear at −10°; congeals near −20°. n_D^{40} 1.4593-

1.4646. Sapon no. 191-200. Iodine no. 93-100. Insol in water. Slightly sol in alcohol; miscible with benzene, chloroform, ether, petr ether. *Keep well closed, cool and protected from light.*

USE: In perfumery, manuf fine soaps; as lubricant for delicate mechanisms such as watches, firearms, etc. Emollient.

THERAP CAT: Cathartic.

6906. Oil of Tansy. Volatile oil from leaves and tops of *Tanacetum vulgare* L., *Compositae*. *Constit.* Thujone, borneol, camphor.

Yellow liq; becomes brown on exposure to air and light. *Poisonous.* d_{15}^{15} 0.925-0.950. α_D +30 to +45°. Almost insol in water. Sol in alcohol, chloroform, ether. *Keep cool in well-filled and well-closed containers, protected from light.*

6907. Oil of Thyme. Volatile oil distilled from flowering plant *Thymus vulgaris* L., *Labiatae*. *Constit.* 20-40% by vol of thymol and carvacrol; cymene, pinene, linalool, bornyl acetate.

Colorless to reddish-brown liq; pleasant thymol odor; sharp taste. d_{25}^{25} 0.894-0.930. α_D^{25} <−4°. n_D^{20} 1.4830-1.5100. Very slightly sol in water; sol in 2 vols 80% alcohol. *Keep well closed, cool and protected from light.*

Note: Often misnamed "oil of origanum."

THERAP CAT: Rubefacient, counterirritant, antiseptic, carminative.

6908. Oil of Vetiver. Vetiver oil; oleum Andropogonis muricati; vetyver oil; khas khas oil; khus oil; cus cus oil; vetivert oil; vetiver oil Java; vetiver oil Haiti; vetiver oil Reunion (Bourbon). Distilled from roots of vetiver grass *Vetiveria zizanioides* Stapf., (*Andropogon muricatus* Retz., *Anatherum zizanioides* (L.) Hitch. & Chase, *Gramineae*), grown chiefly in Java, India, Reunion Island and Haiti. The constituents of vetiver oil vary according to the place of origin. Java and Reunion vetiver oils may contain 8-35% (usually 15-27%) sesquiterpene ketones of which α- and β-vetivones, *q.v.*, have been isolated. The content of vetivenols (vetiverols, vetiver alcohols) varies from 45 to 65%. Other isolated components are vetivenyl vetivenate, vetivenic acid ($C_{15}H_{22}O_2$), palmitic acid, benzoic acid, and vetivene ($C_{15}H_{24}$). Major constituents of Indian vetiver oil are khusol, khusitol, khusinol. Review of chemical studies: Anh, Fetizon, *Am. Perfum. Cosmet.* **80**, 40 (1965). Biogenetically significant components: Kaiser, Naegeli, *Tetrahedron Lett.* **1972**, 2009; Paknikar *et al., ibid.* **1975**, 2973. Isoln and synthesis of *zizanal* and *epizizanal*, two insect repellent constituents, from Javanese vetiver oil: S. C. Jain *et al., Tetrahedron Lett.* **23**, 4639 (1982).

Brown to reddish-brown viscous oil. Aromatic to harsh, woody odor which improves on aging. d_4^{15} 0.990-1.040. n_D^{20} 1.5200-1.5280. $[\alpha]_D^{20}$ +15 to +45°. Ester value after acetylation 110-165. Soluble in 1-3 volumes of 80% alcohol, becoming slightly turbid upon further dilution. Sol in all proportions in most fixed oils, diethyl phthalate, benzyl benzoate, mineral oil (slight turbidity). Practically insol in glycerol and propylene glycol. Fairly stable to dilute acids and weak alkalies; unstable to strong acids and alkalies. Should be stored in a cool place and protected from light. *Refs:* Kretchmar, Pictet, *Chimia* **8**, 123 (1954); Pfau, Plattner, *Helv. Chim. Acta* **22**, 640 (1939); Guenther, *The Essential Oils* vol. **IV** (Van Nostrand, New York, 1950) p 156.

USE: In soap and perfumery formulations.

6909. Oil of Wine, "Heavy". An oily liq obtained by distilling alcohol or ether (or both) with sulfuric acid. *Constit.* Etherin and etherol ($C_2H_4)_n$, polymers of ethylene; ethyl sulfate and ethyl sulfovinate.

Colorless to slightly yellow liq; penetrating odor; sharp, bitter taste. d_{15}^{15} 1.095-1.130. Almost insol in water; miscible with alcohol, ether.

USE: For making ethereal oil which is used in making Hoffman's anodyne; sometimes for flavoring brandy.

6910. Oil Palm. African oil palm. Tropical tree, *Elaeis guineensis* Jacq., *Arecaceae* (alt. *Palmae*), bearing large bunches of plum-sized fruit. Grown primarily for the edible oils obtained from the fruit and seed. *Habit.* West Africa, Malaysia. The closely related **American oil palm**, *E. oleifera* (Kunth) Cortés, is grown in Central and South America and is also used. Review of cultivation, oil production, characteristics and uses: J. A. Cornelius, *Prog. Chem. Fats Other Lipids* **15**, 5-27 (1977); Y. Basiron, *J. Oleo Sci.* **50**, 295-303

(2001). Review of safety assessment of oils: W. Johnson, Jr., *Int. J. Toxicol.* **19**, Suppl. 2, 7-28 (2000).

Palm oil. [8002-75-3] Obtained from the fleshy orange-red mesocarp of the fruit. *Constit.* Fatty acids, primarily palmitic (44%), oleic (39%), and linoleic (10%); carotenoids, esp. α- and β-carotene; tocopherols, esp. γ-tocotrienol; sterols, esp. β-sitosterol. May be refined to remove color and odor components and separated into olein and stearin fractions. Review of biochemistry, toxicology and role in nutrition: D. O. Edem, *Plant Foods Hum. Nutr.* **57**, 319-341 (2002). Crude oil is deep orange in color with characteristic sweetish odor. Semisolid at 21-27°; mp 27-50°. d_{25}^{50} 0.89-0.92. n_D^{40} 1.453-1.459. Sapon. no. 196-209. Iodine no. 46-60.

Palm olein. Liquid oil, mp 21.6°. d^{50} 0.91-0.92. *n* 1.47. Sapon. no. 189-198.0. Iodine no. 55.0-61.54.

Palm stearin. Solid, mp 44.5-56.2°. d^{50} 0.88-0.89. *n* 1.45. Sapon. no. 193-206. Iodine no. 21.6-49.4.

Palm kernel oil. [8023-79-5] Obtained from seed by mechanical expression or solvent extraction. *Constit.* Fatty acids, primarily lauric (48%), myristic (16%) and oleic (15%); sterols, esp. β-sitosterol; triterpene alcohols; hydrocarbons. Review of properties and uses: T. P. Pantzaris, M. J. Ahmed, *Palm Oil Devel.* **35**, 1123 (2001). Colorless to brownish yellow oil, solidifying to a white to yellowish fat. Crude oil has a strong, characteristic taste and odor and is most often refined before use. mp 25-30°. d_{25}^{40} 0.900-0.913. n_D^{40} 1.4495-1.4515. Sapon. no. 244-254. Iodine no. 14-20. Insol in water.

USE: Cooking and frying oil. In margarines, shortenings, nondairy creamers, whipping creams; as cocoa butter substitute. In manuf of soaps, oleochemicals. In cosmetics, personal care products as skin conditioner, viscosity increasing agent, solvent. Pharmaceutic aid (coating agent; emulsifying and solubilizing agent).

6911. Okadaic Acid. [78111-17-8] (α*R*,2*S*,5*R*,6*R*,8*S*)-α,5-Dihydroxy-α,10-dimethyl-8-[(1*R*,2*E*)-1-methyl-3-[(2*R*,4′a*R*,5*R*,-6′*S*,8′*R*,8′a*S*)-octahydro-8′-hydroxy-6′-[(1*S*,3*S*)-1-hydroxy-3-[(2*S*,-3*R*,6*S*)-3-methyl-1,7-dioxaspiro[5.5]undec-2-yl]butyl]-7′-methylenespiro[furan-2(3*H*),2′(3′*H*)-pyrano[3,2-*b*]pyran]5-yl]2-propen-1-yl]-1,7-dioxaspiro[5.5]undec-10-ene-2-propanoic acid; halochondrine A; 9,10-deepithio-9,10-didehydroacanthifolicin. $C_{44}H_{68}O_{13}$; mol wt 805.02. C 65.65%, H 8.51%, O 25.84%. First ionophoric polyether identified in marine organisms; isolated from marine black sponges, *Halichondria (okadai* or *melanodocia*). Tumor promoting cytotoxin associated with diarrhetic seafood poisoning. Isoln: K. Tachibana *et al., J. Am. Chem. Soc.* **103**, 2469 (1981). Total synthesis: M. Isobe *et al., Tetrahedron* **43**, 4767 (1987). Contractile effects: S. Shibata *et al., J. Pharmacol. Exp. Ther.* **223**, 135 (1982). Inhibition of protein phosphatases: A. Takai *et al., FEBS Lett.* **217**, 81 (1987); C. Bialojan, A. Takai, *Biochem. J.* **256**, 283 (1988). Tumor promoting activity: M. Suganuma *et al., Proc. Natl. Acad. Sci. USA* **85**, 1768 (1988). Review of mechanism of action and use as a probe for cellular regulation: P. Cohen *et al., Trends Biochem. Sci.* **15**, 98-102 (1990); A. Schönthal, *New Biol.* **4**, 16-21 (1992). Review of tumor promoting activity: H. Fujiki, M. Suganuma, *J. Biochem.* **115**, 1-5 (1994).

Crystals from dichloromethane/hexane, mp 171-175°. $[\alpha]_D^{20}$ +21° (c = 0.33 in $CHCl_3$). Also reported as crystals from benzene-$CHCl_3$, mp 164-166°. $[\alpha]_D^{25}$ +25.4° (c = 0.24 in $CHCl_3$). LD_{50} i.p. in mice: 192 μg/kg (Shibata).

USE: Biochemical tool as tumor promoter and probe of cellular regulation.

6912. Olaflur. [6818-37-7] 2,2′-[[3-[(2-Hydroxyethyl)octadecylamino]propyl]imino]bisethanol hydrofluoride (1:2); AmF 297; amine fluoride 297; *N*-(diethanolaminopropyl)-*N*-(ethanol)octadecylamine dihydrofluoride; *N,N,N′*-tris(2-hydroxyethyl)-*N′*-octadecyl-1,3-diaminopropane dihydrofluoride; SKF-38095; RonaCare.

$C_{27}H_{58}N_2O_3 \cdot 2HF$; mol wt 498.78. C 65.02%, H 12.13%, N 5.62%, O 9.62%, F 7.62%. Anticariogenic organic amine fluoride; often used in combination with stannous fluoride or sodium fluoride, *q.q.v.* Reduces surface tension of saliva, forming a protective film on tooth surfaces. Prepn: H. Schmid, H. R. Mühlemann, US 3083143 (1963 to GABA). *In vitro* antimicrobial activity: H. M. Kay, M. Wilson, *J. Periodontol.* **59**, 266 (1988). Adsorption and desorption to hydroxyapatite: J. Sefton *et al., Biomaterials* **17**, 37 (1996). Clinical evaluation of caries inhibition in children: T. M. Marthaler, *Br. Dent. J.* **119**, 153 (1965); of effects on plaque and gingivitis in adults: L. Shapira *et al., J. Int. Acad. Periodontol.* **4**, 117 (1999); of reduction of erosion in enamel and dentine: C. Ganss *et al., Caries Res.* **38**, 561 (2004).

Sol in water, ethanol, methanol. Aq soln foams. Slightly acidic pH.

USE: Dental caries prophylactic.

6913. Olah's Reagent. [62778-11-4] Hydrofluoric acid homopolymer compd with pyridine; pyridinium poly(hydrogen fluoride); PPHF. $(C_5H_5N)_x(HF)_y$. Nucleophilic fluorinating agent; acts as a stabilized, less volatile form of hydrogen fluoride, *q.v.* Consists of approx 30% pyridine and 70% hydrogen fluoride (w/w) in equilibrium with a small amount of free hydrogen fluoride. Use in fluorination of steroids: C. G. Bergstrom *et al., J. Org. Chem.* **28**, 2633 (1963). Prepn and use as an organic fluorinating reagent: G. A. Olah *et al., Synthesis* **1973**, 779; *idem et al., J. Org. Chem.* **44**, 3872 (1979). Peptide deprotection: S. Matsuura *et al., J. Chem. Soc. Chem. Commun.* **1976**, 451. Review of use in organic synthesis: G. A. Olah *et al., J. Fluorine Chem.* **33**, 377-396 (1986).

Liquid, stable to ~55°.

USE: Fluorinating reagent; deprotecting reagent in peptide chemistry.

6914. Olanzapine. [132539-06-1] 2-Methyl-4-(4-methyl-1-piperazinyl)-10*H*-thieno[2,3-*b*][1,5]benzodiazepine; LY-170053; Zyprexa. $C_{17}H_{20}N_4S$; mol wt 312.44. C 65.35%, H 6.45%, N 17.93%, S 10.26%. Serotonin (5-HT$_2$) and dopamine (D$_1$/D$_2$) receptor antagonist with anticholinergic activity. Prepn: J. K. Chakrabarti *et al., EP 454436; eidem, US 5229382* (1991, 1993 both to Lilly). Behavioral pharmacology: N. A. Moore *et al., J. Pharmacol. Exp. Ther.* **262**, 545 (1992). HPLC determn in human plasma: J. T. Catlow *et al., J. Chromatogr. B* **668**, 85 (1995). Clinical evaluation in schizophrenia: D. S. Baldwin, S. A. Montgomery, *Int. Clin. Psychopharmacol.* **10**, 239 (1995); in mania of bipolar disorder: M. Tohen *et al., Am. J. Psychiatry* **156**, 702 (1999). Review of pharmacology and clinical experience: B. C. Lund, P. J. Perry, *Expert Opin. Pharmacother.* **1**, 305-323 (2000).

Crystals from acetonitrile, mp 195°. Practically insol in water.

Pamoate monohydrate. [221373-18-8] ZypAdhera. $C_{17}H_{20}$-$N_4S \cdot C_{23}H_{16}O_6 \cdot H_2O$; mol wt 718.83. Review of efficacy and safety of depot formulation: L. Citrome, *Int. J. Clin. Pract.* **63**, 140-150 (2009).

THERAP CAT: Antipsychotic.

6915. Olaparib. [763113-22-0] 4-[[3-[[4-(Cyclopropylcarbonyl)-1-piperazinyl]carbonyl]-4-fluorophenyl]methyl]-1(2*H*)-

phthalazinone; 1-(cyclopropylcarbonyl)-4-[5-[(3,4-dihydro-4-oxo-1-phthalazinyl)methyl]-2-fluorobenzoyl]piperazine; AZD-2281; KU-0059436. $C_{24}H_{23}FN_4O_3$; mol wt 434.47. C 66.35%, H 5.34%, F 4.37%, N 12.90%, O 11.05%. Inhibitor of poly(ADP-ribose) polymerase-1 (PARP-1), an enzyme which helps repair single-stranded DNA breaks characteristic of BRCA-deficient tumors of breast, ovary, and prostate. Prepn: N. M. B. Martin *et al., WO 04080976; eidem, US 7449464* (2004, 2008 both to Kudos Pharm.); and SAR: K. A. Menear *et al., J. Med. Chem.* **51**, 6581 (2008). Large scale synthesis of crystalline form: *idem et al., WO 08047082* (2008 to Kudos). Mechanism of action and effect on radiosensitivity of human glioma cells: F. A. Dungey *et al., Int. J. Radiat. Oncol. Biol. Phys.* **72**, 1188 (2008). Inhibition of BRCA1-deficient murine mammary tumors: S. Rottenberg *et al., Proc. Natl. Acad. Sci. USA* **105**, 17079 (2008). Clinical pharmacokinetics and efficacy in patients with BRCA associated tumors: P. C. Fong *et al., N. Engl. J. Med.* **361**, 123 (2009).

Crystalline white solid from aqueous ethanol, mp 210.1°.
THERAP CAT: Antineoplastic.

6916. Olaquindox. [23696-28-8] *N*-(2-Hydroxyethyl)-3-methyl-2-quinoxalinecarboxamide 1,4-dioxide; Bay Va 9391; Bayo-nox; Fedan. $C_{12}H_{13}N_3O_4$; mol wt 263.25. C 54.75%, H 4.98%, N 15.96%, O 24.31%. Prepn: **FR 1594628**; K. Ley *et al., US 3908008* (1970, 1975 to Bayer). Growth promotant activity for pigs: R. S. Barber *et al., Anim. Feed Sci. Technol.* **4**, 117 (1979). HPLC analysis: G. F. Bories, *J. Chromatogr.* **172**, 505 (1979). LC determn in animal feeds: F. J. dos Ramos *et al., ibid.* **558**, 125 (1991).

Pale yellow crystals, mp 209° (dec). Slightly sol in water. Insol in most organic solvents.

THERAP CAT (VET): Growth stimulant.

6917. Old Yellow Enzyme. [9001-68-7] Reduced nicotinamide adenine dinucleotide phosphate dehydrogenase; dihydronicotinamide adenine dinucleotide phosphate diaphorase; NADPH$_2$ diaphorase; OYE. Mol wt 102,000-106,000. Flavoprotein which catalyzes the oxidation of reduced triphosphopyridine nucleotide by oxygen or ferricytochrome c. When the oxidizing agent is molecular oxygen, OYE functions as a true hydrogen carrier, whereas when ferricytochrome c is the oxidizing agent, only electrons are passed on to the cytochrome, the protons being liberated as free hydrogen ions. Shows diaphorase activity as well as low cytochrome c reductase activity. Isoln from dried brewer's bottom yeast: Warburg, Christian, *Biochem. Z.* **266**, 377 (1933); Theorell, Akeson, *Biochem. Prep.* **6**, 54 (1958). Each molecule contains two flavin mononucleotide groups: Ehrenberg, *Acta Chem. Scand.* **11**, 1257 (1957). *Review:* Akeson *et al., The Enzymes* vol. **7**, P. D. Boyer *et al.*, Eds. (Academic Press, New York, 1963) pp 477-494.

6918. Oleandomycin. [3922-90-5] PA-105; Amimycin; Landomycin; Romicil. $C_{35}H_{61}NO_{12}$; mol wt 687.87. C 61.11%, H 8.94%, N 2.04%, O 27.91%. Antibiotic substance produced by *Streptomyces antibioticus* no. ATCC 11891: Sobin *et al.*; Ratajak, Nubel, **US 2757123; US 2842481** (1956, 1958 to Pfizer). Structure: Hochstein *et al., J. Am. Chem. Soc.* **82**, 3225 (1960). Absolute configuration: Celmer, *ibid.* **87**, 1797 (1965); Celmer, Hobbs, *Carbo-*

hydr. Res. **1**, 137 (1965); S. Omura et al., Tetrahedron Lett. **1975**, 2939. Synthetic study: K. Tatsuta et al., ibid. **29**, 3975 (1988). Activity: Hahn, Antibiotics **1**, 378, 755 (1967). Review of macrolide antibiotics: Keller-Schierlein, Fortschr. Chem. Org. Naturst. **30**, 313-460 (1973). Toxicity: H. Sous et al., Arzneim.-Forsch. **8**, 386 (1958).

White amorphous powder. uv max (methanol): 286-289 nm. Moderately sol in water. Sol in dil acids. Freely sol in methanol, ethanol, butanol, acetone. Practically insol in hexane, carbon tetrachloride, dibutyl ether.

Hydrochloride. [6696-47-5] $C_{35}H_{61}NO_{12}.HCl$. Long needles from ethyl acetate, mp 134-135°. $[\alpha]_D^{25}$ −54° (methanol). Freely sol in water. Forms various cryst hydrates. LD_{50} in mice, rats (mg/kg): 8200, >10000 orally; 600, 400 i.v. (Sous).

Phosphate. [7060-74-4] Matromycin. $C_{35}H_{61}NO_{12}.H_3PO_4$; mol wt 785.86.

Triacetyl derivative see Troleandomycin.

THERAP CAT: Antibacterial.

THERAP CAT (VET): Antibacterial.

6919. Oleandrin. [465-16-7] (3β,5β,16β)-16-(Acetyloxy)-3-[(2,6-dideoxy-3-O-methyl-α-L-arabino-hexopyranosyl)oxy]-14-hydroxycard-20(22)-enolide; neriolin; Corrigen; Folinerin. $C_{32}H_{48}O_9$; mol wt 576.73. C 66.64%, H 8.39%, O 24.97%. Cytotoxic glycoside from the leaves of Nerium oleander L., Apocynaceae (Laurier rose). Component of the commercial oleander extract, **Anvirzel**. Isoln: Tanret, Compt. Rend. **194**, 914 (1932); Neumann, Ber. **70**, 1547 (1937). Prepn by enzymic hydrolysis of urechitoxin: Hassall, J. Chem. Soc. **1951**, 3193. Structure: Tschesche, Ber. **70**, 1554 (1937); Krasso et al., Helv. Chim. Acta **46**, 1691 (1963). LC/MS/MS determn in tissues and biological fluids: E. R. Tor et al., J. Agric. Food Chem. **53**, 4322 (2005). Apoptotic activity study: Y. Sreenivasan et al., Biochem. Pharmacol. **66**, 2223 (2003). Pharmacology: D. Ni et al., J. Exp. Ther. Oncol. **2**, 278 (2002). Review of toxicity: S. D. Langford, P. J. Boor, Toxicology **109**, 1-13 (1996).

Crystals from dil methanol, mp 250°. $[\alpha]_D^{25}$ −48.0° (c = 1.3 in methanol). uv max: 220 nm (log ε 4.20). Practically insol in water. Sol in alcohol, chloroform.

Desacetyloleandrin. [36190-93-9] $C_{30}H_{46}O_8$. Leaflets from alcohol, mp 238-240°. $[\alpha]_D^{18}$ −24.9°.

THERAP CAT: Cardiotonic; diuretic.

6920. Oleanolic Acid. [508-02-1] (3β)-3-Hydroxyolean-12-en-28-oic acid; oleanol; caryophyllin. $C_{30}H_{48}O_3$; mol wt 456.71. C 78.90%, H 10.59%, O 10.51%. Occurs in the free state in leaves of Olea europaea, Oleaceae, in leaves of Viscum album L., Loranthaceae, in buds of Syzygium aromaticum (L.) Merr. & Perry, Myrtaceae (cloves), in Swertia japonica (Maxim.) Makino, and in Centaurium umbellatum Gilib. (Erythraea centaurium (L.) Pers.), Gentianaceae; as acetate in birch bark, as glycoside in many saponins. Isoln procedures (from cloves): Winterstein, Stein, Z. Physiol. Chem. **202**, 222 (1931); Ruzicka, Hofmann, Helv. Chim. Acta **19**, 114 (1936); Picard et al., J. Chem. Soc. **1939**, 1047. Structure: Ruzicka et al., Helv. Chim. Acta **29**, 210 (1946). Review: J. Simonsen, W. C. T. Ross, The Terpenes vol. 5 (University Press, Cambridge, 1957) pp 221-285. See also α- amyrin and β-amyrin.

Fine, solvated needles from alc. After drying, mp 310°. $[\alpha]_D^{20}$ +83.3° (c = 0.6 in chloroform). pK 2.52. Insol in water. Sol in 65 parts ether, 106 parts 95% alcohol, 35 parts boiling 95% alcohol, 118 parts chloroform, 180 parts acetone, 235 parts methanol.

Acetate. $C_{32}H_{50}O_4$. Needles from methanol, mp 268°. $[\alpha]_D^{17}$ +74.5° (c = 0.6 in $CHCl_3$).

Methyl ester. $C_{31}H_{50}O_3$. mp 201°. $[\alpha]_D^{20}$ +75° (c = 0.6 in $CHCl_3$).

Acetate of methyl ester. $C_{33}H_{52}O_4$. Needles from alcohol, mp 223°. $[\alpha]_D^{20}$ +70° (c = 0.6 in $CHCl_3$).

6921. Oleic Acid. [112-80-1] (Z)-9-Octadecenoic acid. $C_{18}H_{34}O_2$; mol wt 282.47. C 76.54%, H 12.13%, O 11.33%. Obtained by the hydrolysis of various animal and vegetable fats and oils. Prepn from olive oil: Biochem. Prep. **2**, 100 (1952). Separation from olive oil by double fractionation via urea adducts: Rubin, Paisley, Biochem. Prep. **9**, 113 (1962). Stereochemistry: Thieme, Ann. **343**, 354 (1905). Synthesis: Robinson, Robinson, J. Chem. Soc. **127**, 175 (1925). ^{13}C-NMR studies: W. Stoffel et al., Z. Physiol. Chem. **353**, 1962 (1972); J. G. Batchelor et al., J. Org. Chem. **39**, 1698 (1974). Toxicity data: L. Orö, A. Wretlind, Acta Pharmacol. Toxicol. **18**, 141 (1961). Exptl use of ^{131}I-labelled oleic acid in myocardial imaging: F. J. Bonte et al., Radiology **108**, 195 (1973). Review of diagnostic use of ^{3}H-oleic acid in pancreatic function: N. T. Pedersen, Digestion **37**, Suppl. 1, 25-34 (1987).

Pure oleic acid is a colorless or nearly colorless liquid (above 5-7°). d_{25}^{25} ~0.895. Solidifies to cryst mass, mp 4°. bp_{100} 286°. At atm pressure it dec when heated at 80-100°. n_D^{18} 1.463; n_D^{26} 1.4585. Iodine no. 89.9; acid value 198.6. On exposure to air, especially when impure, it oxidizes and acquires a yellow to brown color and rancid odor. Practically insol in water. Misc with alc, benzene, chloroform, ether, fixed and volatile oils. Keep well closed, protected from light. LD_{50} i.v. in mice: 230±18 mg/kg (Orö, Wretlind). Several grades of the acid are available in commerce, varying in color from pale yellow to red-brown and, depending on the amount of saturated acid present, becoming turbid at 8-16°. The acid of commerce usually contains 7-12% saturated acids, e.g., stearic, palmitic; also some linoleic, etc., unsaturated acids.

Methyl ester. Methyl oleate. $C_{19}H_{36}O_2$. Prepd by refluxing oleic acid with p-toluene sulfonic acid in methanol: Rubin, Paisley, *loc. cit.* Iodine no. 85.6. d_4^{18} 0.879. n_D^{26} 1.4510. bp$_2$ 168-170°. Miscible with anhydr ethanol, ether.

Ethyl ester. Ethyl oleate; (*Z*)-9-octadecenoic acid ethyl ester. $C_{20}H_{38}O_2$. Yellowish, oily liquid. d 0.87. bp 205-208° (some dec). Insol in water. Misc with alc, ether.

Barium salt. Barium oleate. $C_{36}H_{66}BaO_4$. Yellowish-white, granular masses. *Poisonous.* Practically insol in water. Slightly sol in boiling alc.

Sodium salt. [143-19-1] Eunatrol. $C_{18}H_{33}NaO_2$; mol wt 304.45. White powder, slight tallow-like odor. Sol in ~10 parts water, ~20 parts alc. Generally contains small quantities of the sodium salts of stearic, etc. acids. Alkaline in aq solns due to hydrolysis but not in alc solns.

Caution: Mildly irritating to skin, mucous membranes.

USE: Prepn of Turkey red oil, soft soap and other oleates; in polishing compds; waterproofing textiles, oiling wool; manuf driers; thickening lubricating oils. Pharmaceutic aid (solvent; emulsifying and solubilizing agent). The barium salt in rodent extermination.

THERAP CAT: Diagnostic aid (pancreatic function).

6922. Oleocanthal. [289030-99-5] (3*S*,4*E*)-4-Formyl-3-(2-oxoethyl)-4-hexenoic acid 2-(4-hydroxyphenyl)ethyl ester; deacetoxy ligstroside aglycon. $C_{17}H_{20}O_5$; mol wt 304.34. C 67.09%, H 6.62%, O 26.28%. Secoiridoid found in virgin olive oil; induces a characteristic burning sensation in the throat when tasted. Isoln and proposed formation from ligstroside: G. Montedoro *et al., J. Agric. Food Chem.* **41**, 2228 (1993). Antioxidant activity: R. W. Owen *et al., Food Chem. Toxicol.* **38**, 647 (2000). Identification as key contributor to pungency of olive oil: P. Andrewes *et al., J. Agric. Food Chem.* **51**, 1415 (2003). Inhibitory effect on cyclooxygenase: G. K. Beauchamp *et al., Nature* **437**, 45 (2005). Synthesis and abs config: A. B. Smith, III *et al., Org. Lett.* **7**, 5075 (2005).

$[\alpha]_D^{25}$ −0.78° (c = 0.9 in chloroform). uv max (chloroform): 242, 277, 283 sh, 325 nm.

6923. Oleuropein. [32619-42-4] (2*S*,3*E*,4*S*)-3-Ethylidene-2-(β-D-glucopyranosyloxy)-3,4-dihydro-5-(methoxycarbonyl)-2*H*-pyran-4-acetic acid 2-(3,4-dihydroxyphenyl)ethyl ester. $C_{25}H_{32}O_{13}$; mol wt 540.52. C 55.55%, H 5.97%, O 38.48%. Bitter glucoside; first *secoiridoid* to be isolated. Isolation from olives and the leaves and bark of the olive tree, *Olea europaea* L., Oleaceae and structural studies: Panizzi *et al., Gazz. Chim. Ital.* **90**, 1449 (1960); Beyerman *et al., Bull. Soc. Chim. Fr.* **1961**, 1821; Shasha, Leibowitz, *J. Org. Chem.* **26**, 1948 (1961). Isoln from the ripe fruits of *Ligustrum lucidum* and *L. japonicum* Thunb, Oleaceae: Inouye, Nishioka, *Tetrahedron* **28**, 4231 (1972). Revised structure and stereochemistry: Inouye *et al., Tetrahedron Lett.* **1970**, 2459. Pharmacology: Petkov, Manolov, *Arzneim.-Forsch.* **22**, 1476 (1972). Partial synthesis: A. Bianco *et al., J. Nat. Prod.* **55**, 760 (1992).

O—β-D-glucopyranose

Minute crystals from ethyl acetate, mp 87-89°. Hygroscopic. $[\alpha]_D^{20}$ −147° (H$_2$O, alcohol, or acetone). Shows mutarotation $[\alpha]_D^{20}$ −127° after 9 hrs (H$_2$O). Freely sol in acetone, ethanol, methanol, pyridine, glacial acetic acid, 5% aq NaOH soln. Moderately sol in water, dioxane, butanol, ethyl acetate, butyl acetate. Practically insol in ether, petr ether, chloroform, benzene, carbon tetrachloride.

6924. Oleyl Alcohol. [143-28-2] (9*Z*)-9-Octadecen-1-ol; *cis*-9-octadecen-1-ol; Ocenol. $C_{18}H_{36}O$; mol wt 268.49. C 80.52%, H 13.52%, O 5.96%. Found in fish oils. Usually obtained as a mixture of C_{16} and C_{18} unsaturated alcohols with C_{18} predominating. Prepd from butyl oleate by a Bouveault-Blanc reduction with sodium and butyl alcohol; or from triolein by hydrogenation in the presence of zinc chromite: Noller, Bannerot, *J. Am. Chem. Soc.* **56**, 1563 (1934); Reid *et al., Org. Synth.* **15**, 72 (1935); Adkins, Gillespie, *ibid.* **coll. vol. III**, 671 (1955). Purification by fractional crystallization at −40° from acetone, followed by distillation: Swern *et al., Oil Soap (Chicago)* **21**, 113 (1944); Loev, Dawson, *J. Am. Chem. Soc.* **78**, 1182 (1956).

Oily liquid. Usually pale yellow. Gives off acrid fumes when heated. d_4^{20} 0.850. mp 13-19°. bp$_8$ 195°. bp$_{1.5}$ 182-184°. (Distilling range at 760 mm: 305-370°.) $n_D^{27.5}$ 1.4582. Insol in water; sol in alcohol, ether.

USE: Chiefly in the manuf of its sulfuric esters which are detergents and wetting agents, as an antifoam agent; metal cutting lubricant; in carbon paper, stencil paper, printing ink; as a plasticizer; for softening and lubricating textile fabrics; carrier for medicaments.

6925. Oleyl Hydroxyethyl Imidazoline. [21652-27-7]; [95-38-5] (unspecified stereo). 2-(8*Z*)-(8-Heptadecen-1-yl)-4,5-dihydro-1*H*-imidazole-1-ethanol; (*Z*)-2-(8-heptadecenyl)-2-imidazoline-1-ethanol; 1-(2-hydroxyethyl)-2-(8-heptadecenyl)-2-imidazoline; Amine 220; Crodazoline O; Mackazoline O; Monazoline O; Schercozoline O. $C_{22}H_{42}N_2O$; mol wt 350.59. C 75.37%, H 12.08%, N 7.99%, O 4.56%. Prepn from oleic acid and β-hydroxyethyl ethylene diamine: A. L. Wilson, **US 2267965** (1941 to Carbide and Carbon Chem.). Effect on herbicide leaching in sandy soil: R. S. Chandran, M. Singh, *Bull. Environ. Contam. Toxicol.* **62**, 315 (1999). Determn in marine sediments by ESI-MS: S. J. W. Grigson *et al., Rapid Commun. Mass Spectrom.* **14**, 2210 (2000).

Dark amber liquid with ammoniacal odor. bp$_1$ 230-240°. d_{20}^{20} 0.935. Miscible with water at 10°. pH (10% aq): 10.5-12.0. Alkali value: 150-160 mg KOH/g. Sol in dil hydrochloric and acetic acids.

USE: Industrial surfactant; corrosion inhibitor; water-in-oil emulsifier.

6926. Olibanum. [8050-07-5] Frankincense; gum thus. Gum resin from *Boswellia carterii* Birdwood and other species of *Boswellia* (Burseraceae). *Habit.* Ethiopia, Egypt, Arabia, Somaliland. *Constit.* 3-8% volatile oil (pinene, dipentene, etc.); about 60% resins; 20% gum (polysaccharide fraction); 6-8% bassorin; bitter principle. Discussion of polysaccharides present: F. Smith, R. Montgomery, *The Chemistry of Plant Gums and Mucilages* (Reinhold, New York, 1959) pp 312-313.

6927. Oligomycins. Macrolide antibiotic complex produced by an actinomycete similar to *Streptomyces diastatochromogenes:* Smith *et al., Antibiot. Chemother.* **4**, 962 (1954); McCoy, Peterson, **US 2927057** (1960 to Wisconsin Alumni Res. Foundation). Complex of several closely related compounds: Masamune *et al., J. Am. Chem. Soc.* **80**, 6092 (1958). Separation of oligomycins A, B and C: Marty, McCoy, *Antibiot. Chemother.* **9**, 286 (1959). Isoln and activity of D from *Streptomyces rutgersensis:* R. Q. Thompson *et al., Antimicrob. Agents Chemother.* **1961**, 474. Partial structure of B: Prouty *et al., Biochem. Biophys. Res. Commun.* **44**, 619 (1971); total structure of B: Glehn *et al., FEBS Lett.* **20**, 267 (1972). Structures of A and C: G. T. Carter, *Diss. Abstr. Int. B* **37**, 766 (1976); of D: C. Merienne, T. Staron, *Chem. Commun.* **1978**, 318. *Review:* P. D. Shaw in *Antibiotics* **vol. I**, D. Gottlieb, P. D. Shaw, Eds. (Springer-Verlag, New York, 1967) pp 585-610.

Oligomycin A

Oligomycin A. [579-13-5] $C_{45}H_{74}O_{11}$; mol wt 791.08. Two crystalline modifications, mp 140-141° and mp 150-151° (hexagonal crystals, melting with rapid decompn). uv max (abs ethanol): 225 nm (ε about 20,000). Solys at 25° (g/100 ml solvent): water 0.002; ether 28; benzene 6; Skellysolve B 0.02; abs ethanol 25; glacial acetic acid 37.5; acetone 85.

Oligomycin B. [11050-94-5] 28-Oxooligomycin A. $C_{45}H_{72}$-O_{12}; mol wt 805.06. Potent inhibitor of oxidative phosphorylation: Lardy *et al.*, *Arch. Biochem. Biophys.* **78**, 587 (1958).

Oligomycin C. [11052-72-5] 12-Deoxyoligomycin A. $C_{45}H_{74}$-O_{10}; mol wt 775.08.

Oligomycin D. [1404-59-7] 26-Demethyloligomycin A; rutamycin; A-272; RR-32705. $C_{44}H_{72}O_{11}$; mol wt 777.05. mp 116-119°. $[\alpha]_D^{20}$ −62° (c = 1.36 in $CHCl_3$).

THERAP CAT: Oligomycin D as antifungal.

6928. Olivacine. [484-49-1] 1,5-Dimethyl-6H-pyrido[4,3-b]-carbazole; guatambuinine. $C_{17}H_{14}N_2$; mol wt 246.31. C 82.90%, H 5.73%, N 11.37%. Isolated from the bark and stem of *Aspidosperma olivaceum* Müll. Arg., *A. australe* Müll. Arg., *A. longepetiolatum* Kuhlm., *Apocynaceae:* Schmutz, Hunziger, *Pharm. Acta Helv.* **33**, 341 (1958). Identity with guatambuinine: Marini-Bettolo, Carvalho-Ferreira, *Ann. Chim.* **49**, 869 (1959). Structure: Marini-Bettolo, Schmutz, *Helv. Chim. Acta* **42**, 2146 (1959); Ondetti, Deulofeu, *Tetrahedron* **15**, 160 (1961). Synthesis: Schmutz, Wittwer, *Helv. Chim. Acta* **43**, 793 (1960); Wenkert, Dave, *J. Am. Chem. Soc.* **84**, 94 (1962); Mosher *et al.*, *J. Med. Chem.* **9**, 237 (1966); R. Besselievre, H. Husson, *Tetrahedron Lett.* **1976**, 1873; J. Bergman, R. Carlsson, *ibid.* **1978**, 4055; J. P. Kutney *et al.*, *Heterocycles* **16**, 1469 (1981).

Fine yellow needles from dilute methanol, yellow prisms from undiluted methanol, mp 317-325°. Fluoresces in dilute alcoholic soln. uv max (ethanol): 224, 238, 276, 287, 292, 314, 329, 375 nm (log ε 4.39, 4.33, 4.70, 4.85, 4.83, 3.66, 3.80, 3.66). Soly in methanol, acetone, chloroform, carbon tetrachloride, carbon disulfide, tetrahydrofuran, dioxane less than 1%.

6929. Olivanic Acids. A family of naturally occurring carbapenem β-lactamase inhibitors with antibacterial activity, produced by *Streptomyces olivaceus:* D. Butterworth, G. N. Rolinson, **DE 2146400**; *eidem* US 3919415 (1972, 1975 both to Beecham). Isoln: A. G. Brown *et al.*, *J. Antibiot.* **29**, 668 (1976). The family contains seven members, all containing the 7-oxo-1-azabicyclo[3.2.0]hept-2-ene ring system. Compounds *MM 4550*, *MM 13902*, and *MM 17880* are sulfated and are predominant when *S. olivaceus* is grown in sodium sulfate-containing media. *MM 22380*, *MM 22381*, *MM*

22382, *MM 22383* are non-sulfated hydroxy analogs and are produced in sulfate-free media or by mutants of *S. olivaceus* unable to complete the sulfation process. Detection, properties, fermentation of the sulfated members: D. Butterworth *et al.*, *ibid.* **32**, 287 (1979); isoln, characterization: J. D. Hood *et al.*, *ibid.* 295. Fermentation, isoln, characterization of the non-sulfated members: S. J. Box *et al.*, *ibid.* 1239. Structures of MM 4550, MM 13902: A. G. Brown *et al.*, *Chem. Commun.* **1977**, 523; of MM 17880: D. F. Corbett *et al.*, *ibid.* 953. Structures of MM 22380-3: A. G. Brown *et al.*, *J. Antibiot.* **32**, 961 (1979). Total synthesis of racemic MM 22383: R. J. Ponsford, R. Southgate, *Chem. Commun.* **1980**, 1085. Comparative antibacterial activity *in vitro:* M. J. Basker *et al.*, *J. Antibiot.* **33**, 878 (1980). MM 22380, MM 22382, MM 22381, MM 22383, MM 13902, and MM 17880 are identical to **epithienamycins** A through F, respectively: E. O. Stapley *et al.*, *ibid.* **34**, 628 (1981); P. J. Cassidy *et al.*, *ibid.* 637. Synthetic study on epithienamycins: T. Kametani *et al.*, *J. Chem. Soc. Perkin Trans. 1* **1981**, 3048.

MM 4550	R = −S−CH=CH−NHCOCH₃	(E)
MM 13902	R = −S−CH=CH−NHCOCH₃	(E)
MM 17880	R = −S−CH₂CH₂NHCOCH₃	

MM 4550, MM 13902, MM 17880 are isolated as disodium salts. Sol in water, methanol, DMF, DMSO. Practically insol in other organic solvents. uv max (water): of MM 4550: 240, 287 nm ($E_{1cm}^{1\%}$ 268); MM 13902: 227, 307 nm ($E_{1cm}^{1\%}$ 356); MM 17880: 298 nm ($E_{1cm}^{1\%}$ 192). The antibiotics are unstable in aq soln outside a narrow pH range; in acids, degradation leads to changes in uv spectra. Addition of hydroxylamine or cysteine to neutral solns results in rapid degradation.

6930. Olive Oil. A fixed oil obtained from ripe olives, the fruit of the cultivated olive tree *Olea europaea* L., *Oleaceae.* Produced almost exclusively in the countries adjoining the Mediterranean Sea, Spain being the largest producer. Whole olives are crushed in edge runner mills and the oil is expressed in open hydraulic presses. *Constit.* Mixed glycerides of fatty acids, primarily oleic acid, palmitic acid, linoleic acid, stearic acid, and arachidic acid. Minor constituents are squalene, phytosterols, α- and γ-tocopherols, pigments, terpenic acids, flavonoids, and the phenolic antioxidants, tyrosol and hydroxytyrosol, *q.q.v.* HPLC determn of phenols, flavones, and lignans in virgin olive oils: R. Mateos *et al.*, *J. Agric. Food Chem.* **49**, 2185 (2001). *Reviews and bibliographies:* José M. de Soroa y Pineda, *El aceite de oliva* (Dossat, Madrid, 1944); R. F. Simari, G. B. Martinenghi, *Olivicoltura e Oleificio* (Hoepli, Milano, 1950); P. G. Garoglio, *Technologia de los Aceites Vegetales* (Mendoza, Argentina, 1951); E. W. Eckey, *Vegetable Fats and Oils* (Reinhold, New York, 1954).

Pale yellow or light greenish-yellow oil with a pleasing delicate flavor. Becomes rancid on exposure to air. Begins to get turbid at +5 to +10°, below 0° it forms a whitish, granular mass. Flash pt 437°F (225°C). Ignition temp 650°F (343°C). d_{15}^{15} 0.914-0.919; d_{25}^{25} 0.909-0.915. n_D^{25} 1.466-1.468; n_D^{40} 1.460-1.464. Titer 17-26°. Acid value 0.2-2.8. Saponification value 187-196. Iodine value 79-90. Thiocyanogen value 75-83. Hydroxyl value 4-12. Reichert-Meissl value 0.2-1.0. Unsaponifiable 0.5-1.3%. Slightly sol in alcohol. Miscible with ether, chloroform, carbon disulfide.

USE: As food in salads, with sardines, for cooking and baking. In the manuf of soaps, textile lubricants, sulfonated oils, cosmetics and pharmaceutical preparations. In ear drops to soften ear wax. Emollient.

THERAP CAT (VET): Laxative, emollient.

6931. Olivil. [2955-23-9] (2S,3R,4S)-Tetrahydro-4-hydroxy-2-(4-hydroxy-3-methoxyphenyl)-4-[(4-hydroxy-3-methoxyphenyl)-methyl]-3-furanmethanol. $C_{20}H_{24}O_7$; mol wt 376.41. C 63.82%, H 6.43%, O 29.75%. From gum-resin of *Olea europaea* L., *Oleaceae.* Isoln: Pelletier, *Ann. Chim. Phys.* **51**, 196 (1833). Structure: Traverso, *Gazz. Chim. Ital.* **90**, 792 (1960); Smith, *Tetrahedron Lett.* **1963**, 991. Revised structure: Ayres, Mhasalkar, *ibid.* **1964**, 335; *eidem, J. Chem. Soc.* **1965**, 3586.

Monohydrate. Crystals. mp 118-120°; when anhydr, mp 142-143°. $[\alpha]_D^{12}$ $-127°$. Sol in hot water, alcohol, acetic acid, fatty oils.

6932. Olivomycins. A mixture of antibiotics produced by *Streptomyces olivoreticuli:* Gauze *et al., Antibiotiki* **7**, 34 (1962); Brazhnikova *et al., ibid.* 39. Similar to the chromomycins, *q.v.:* Berlin *et al., Tetrahedron Lett.* **1966**, 1643. Prepn: Gauze, **GB 1152748; FR 1554600** (1969, both to Sci. Res. Inst. Antibiot.), *C.A.* **71**, 37490b, 128693m (1969). Composed of olivomycins A (major component), B, C, and D. Separation: Berlin *et al., Khim. Prir. Soedin.* **1967**, 331, *C.A.* **68**, 40017w (1968). Structural elucidation of the aglycone, *olivin,* and the carbohydrate moieties: Berlin *et al., Tetrahedron Lett.* **1964**, 1323; **1966**, 1431; Berlin *et al., Khim. Prir. Soedin.* **1969**, 567, *C.A.* **73**, 25823r (1970). Stereochemistry: Bakhaeva *et al., Chem. Commun.* **1967**, 10; Berlin *et al., Khim. Prir. Soedin.* **1972**, 519. Revised structure: J. Thiem, B. Meyer, *Tetrahedron* **37**, 551 (1981). Synthetic studies: J. H. Dodd, S. M. Weinreb, *Tetrahedron Lett.* **1979**, 3593; J. H. Dodd *et al., J. Org. Chem.* **47**, 4045 (1982). Series of articles on pharmacology: *Antibiotiki* **17** (1972). Toxicity data: M. Slavik, S. K. Carter, *Adv. Pharmacol. Chemother.* **12**, 1 (1975). *Review:* J. D. Skarbek, M. K. Speedie, in *Antitumor Compounds of Natural Origin* vol. **1**, A. Aszalos, Ed. (CRC Press, Boca Raton, 1981) pp 191-235.

Olivomycin A

Olivomycin A. [6988-58-5] 3^D-O-[2,6-Dideoxy-3-C-methyl-4-O-(2-methyl-1-oxopropyl)-α-L-*arabino*-hexopyranosyl]olivomycin D. $C_{58}H_{84}O_{26}$; mol wt 1197.28. Formerly known as olivomycin and variant I. Yellow crystals from ethanol-hexane, mp 160-165°. $[\alpha]_D^{20}$ $-36°$ (c = 0.5 in ethanol). uv max (ethanol): 228, 277, 318, 406 nm (log ε 4.39, 4.67, 3.81, 4.05). Sol in alc, ether, chloroform. Insol in benzene, carbon tetrachloride, petr ether, water. LD_{50} i.v. in mice: 13.75 mg/kg (Slavik, Carter).

THERAP CAT: Antineoplastic.

6933. Olmesartan. [144689-63-4] 4-(1-Hydroxy-1-methylethyl)-2-propyl-1-[[2'-(1*H*-tetrazol-5-yl)[1,1'-biphenyl]-4-yl]methyl]-1*H*-imidazole-5-carboxylic acid (5-methyl-2-oxo-1,3-dioxol-4-yl)methyl ester; (5-methyl-2-oxo-1,3-dioxolen-4-yl)methyl 4-(1-hydroxy-1-methylethyl)-2-propyl-1-[4-[2-(tetrazol-5-yl)phenyl]phenyl]methylimidazole-5-carboxylate; olmesartan medoxomil; CS-866; Benicar; Olmetec. $C_{29}H_{30}N_6O_6$; mol wt 558.60. C 62.36%, H 5.41%, N 15.05%, O 17.18%. Angiotensin II receptor antagonist. Prepn: H. Yanagisawa *et al.,* **EP 503785**; *eidem,* **US 5616599** (1992, 1997 both to Sankyo); *eidem, J. Med. Chem.* **39**, 323 (1996). Clinical evaluation in hypertension: K. Püchler *et al., J. Hypertens.* **15**, 1809 (1997). *Review:* M. Elisaf, *Curr. Opin. Cardiovasc. Pulm. Renal Invest. Drugs* **2**, 378-383 (2000).

Monoclinic crystals from ethanol, mp 180-182° (dec). Practically insol in water. Sparingly sol in methanol.

THERAP CAT: Antihypertensive.

6934. Olopatadine. [113806-05-6] (11*Z*)-11-[3-(Dimethylamino)propylidene]-6,11-dihydrodibenz[*b,e*]oxepin-2-acetic acid. $C_{21}H_{23}NO_3$; mol wt 337.42. C 74.75%, H 6.87%, N 4.15%, O 14.22%. Dual acting histamine H_1-receptor antagonist and mast cell stabilizer. Prepn: E. Oshima *et al.,* **EP 235796**; *eidem,* **US 5116863** (1987, 1992 both to Kyowa); *eidem, J. Med. Chem.* **35**, 2074 (1992). Pharmacology: C. Kamei *et al., Arzneim.-Forsch.* **45**, 1005 (1995); J. M. Yanni *et al., J. Ocul. Pharmacol. Ther.* **12**, 389 (1996). Receptor binding profile: N. A. Sharif *et al., J. Pharmacol. Exp. Ther.* **278**, 1252 (1996). Clinical trial in allergic conjunctivitis: M. B. Abelson, L. Spitalny, *Am. J. Ophthalmol.* **125**, 797 (1998); in seasonal allergic rhinitis: S. R. Shah *et al., Clin. Ther.* **31** 99 (2009). Review of development, pharmacology, and clinical experience: M. B. Abelson, P. J. Gomes, *Expert Opin. Drug Metab. Toxicol.* **4**, 453-461 (2008).

Crystallized as the hemihydrate from 2-propanol + water, mp 188-189.5°.

Hydrochloride. [140462-76-6] AL-4943A; KW-4679; Opatanol; Pataday; Patanase; Patanol. $C_{21}H_{23}NO_3$.HCl; mol wt 373.88. White crystals from acetone-water, mp 248° (dec). Sol in water.

THERAP CAT: Antiallergic; antihistaminic.

6935. Olsalazine. [15722-48-2] 3,3'-Azobis(6-hydroxybenzoic acid); C.I. Mordant Yellow 5; 3,3'-dicarboxy-4,4'-dihydroxyazobenzene; 5,5'-azobis(salicylic acid); azodisal. $C_{14}H_{10}N_2O_6$; mol wt 302.24. C 55.64%, H 3.34%, N 9.27%, O 31.76%. Prodrug cleaved by colonic bacteria at the azo-position to give two molecules of mesalamine, *q.v.* Originally used as a mordant dye. Prepn: **DE 278613**; C. Mettler, **US 1157169** (1914, 1915 both to J. R. Geigy). Improved prepn: K. H. Agback, A. S. Nygren, **EP 36636**; *eidem,* **US 4528367** (1981, 1985 both to Pharmacia AB). HPLC determn in biological samples: R. A. van Hogezand *et al., J. Chromatogr.* **305**, 470 (1984). *In vitro* effect on fecal bacteria: H. Sandberg-Gertzen *et al., Scand. J. Gastroenterol.* **20**, 607 (1985). Clinical pharmacokinetics and metabolism: C. P. Willoughby *et al., Gut* **23**, 1081 (1982). Clinical trial in ulcerative colitis: H. Sandberg-Gertzen *et al., Gastroenterology* **90**, 1024 (1986). Review of clinical experience in inflammatory bowel disease: A. N. Wadworth, A. Fitton, *Drugs* **41**, 647-664 (1991).

Disodium salt. [6054-98-4] Azodisal sodium; disodium azodisalicylate; C.I. 14130; CJ-91B; Dipentum. $C_{14}H_8N_2Na_2O_6$; mol wt 346.21. Yellow crystalline powder, mp 240° (dec). Sol in water, DMSO. Practically insol in ethanol, chloroform, ether.

THERAP CAT: Anti-inflammatory (gastrointestinal).

THERAP CAT (VET): Anti-inflammatory (gastrointestinal).

6936. Omalizumab. [242138-07-4] Anti-(human immunoglobulin E Fc region) immunoglobulin G (human-mouse monoclonal E25 clone pSVIE26 γ-chain) disulfide with human-mouse monoclonal E25 clone pSVIE26 κ-chain, dimer; olizumab; rhuMAb-E25; CGP-51901; Xolair. Humanized monoclonal antibody directed against human immunoglobulin E (IgE). Complexes with circulating IgE to prevent the IgE-mediated degranulation of mast cells during an allergic reaction. Prepn: L. G. Presta *et al.*, *J. Immunol.* **151**, 2623 (1993). Pharmacokinetics and effect on circulating IgE levels in allergy patients: J. Corne *et al.*, *J. Clin. Invest.* **99**, 879 (1997). Clinical trial in allergic asthma: H. Milgrom *et al.*, *N. Engl. J. Med.* **341**, 1966 (1999); in allergic rhinitis: T. B. Casale *et al.*, *J. Am. Med. Assoc.* **286**, 2956 (2001). Reviews of clinical efficacy in allergic asthma: S. Holgate *et al.*, *Curr. Med. Res. Opin.* **17**, 233-240 (2001); G. D'Amato, *Expert Opin. Biol. Ther.* **3**, 371-376 (2003).

THERAP CAT: Antiallergic; antiasthmatic.

6937. Omapatrilat. [167305-00-2] (4*S*,7*S*,10a*S*)-Octahydro-4-[[(2*S*)-2-mercapto-1-oxo-3-phenylpropyl]amino]-5-oxo-7*H*-pyrido[2,1-*b*][1,3]thiazepine-7-carboxylic acid; BMS-186716; Vanlev. $C_{19}H_{24}N_2O_4S_2$; mol wt 408.53. C 55.86%, H 5.92%, N 6.86%, O 15.66%, S 15.70%. Dual metalloprotease inhibitor of neutral endopeptidase (NEP) and angiotensin converting enzyme (ACE). Prepn: J. A. Robl *et al.*, **EP 629627**; *idem*, **US 5508272** (1994, 1996 both to Bristol-Myers Squibb); and pharmacological evaluation: *idem et al.*, *J. Med. Chem.* **40**, 1570 (1997). HPLC-MS determn in rat plasma: M. Jemal, D. J. Hawthorne, *J. Chromatogr. B* **698**, 123 (1997). Antihypertensive activity in rats: N. C. Trippodo *et al.*, *Am. J. Hypertens.* **11**, 363 (1998). Clinical pharmacodynamics: C. Massien *et al.*, *Clin. Pharmacol. Ther.* **65**, 448 (1999). Review of pharmacology and clinical use: M. Weber, *Am. J. Hypertens.* **12**, 139S-147S (1999). Clinical trial in heart failure: J. L. Rouleau *et al.*, *Lancet* **356**, 615 (2000).

Fine, white solid, mp 218-220°. $[\alpha]_D$ −78.9° (c = 0.46 in DMF).

THERAP CAT: Antihypertensive.

6938. Omega-3 Acid Ethyl Esters. [308081-97-2] Omega-3 polyunsaturated fatty acids ethyl esters; Lovaza; Omacor. Mixture of ethyl esters of long-chain omega-3 fatty acids extracted from fish oil, primarily docosahexaenoic and eicosapentaenoic acids, *q.q.v.* Purification from crude fish oil: H. Breivik *et al.*, **US 5656667** (1997 to Norsk Hydro). Clinical effect on lipid profile in combined hyperlipidemia: L. Calabresi *et al.*, *Atherosclerosis* **148**, 387 (2000). Clinical trial to prevent sudden death following myocardial infarction: A. Macchia *et al.*, *Eur. J. Heart Fail.* **7**, 904 (2005). Clinical effect of combination with simvastatin in persistent hypertriglyceridemia: M. H. Davidson *et al.*, *Clin. Ther.* **29**, 1354 (2007). Review of clinical experience: H. Bays, *Am. J. Cardiol.* **98**, Suppl. 4A, 71i-76i (2006).

Light yellow liquid with slight fish-like odor. Practically insol in water. Very sol in acetone, dehydrated alcohol, heptane, methanol.

THERAP CAT: Antilipemic.

6939. Omeprazole. [73590-58-6] 6-Methoxy-2-[[(4-methoxy-3,5-dimethyl-2-pyridinyl)methyl]sulfinyl]-1*H*-benzimidazole; H-168/68; Gastrogard; Losec; Mopral; OmeLich; Omelind; Omepral; Omeprazen; Osiren; Parizac; Pepticum; Prilosec; UlcerGard;

Zegerid; Zoltum. $C_{17}H_{19}N_3O_3S$; mol wt 345.42. C 59.11%, H 5.54%, N 12.17%, O 13.90%, S 9.28%. Gastric proton-pump inhibitor. Prepn: U. K. Junggren, S. E. Sjostrand, **EP 5129**; *eidem*, **US 4255431** (1979, 1981 both to AB Hässle). Resolution and activity of enantiomers: P. Erlandsson *et al.*, *J. Chromatogr.* **532**, 305 (1990). Manuf process for optically pure salts: S. Von Unge, **US 5693818** (1997 to Astra). Pharmacology: P. Muller *et al.*, *Arzneim.-Forsch.* **33**, 1685 (1983). Mechanism of action study: B. Wallmark *et al.*, *Biochim. Biophys. Acta* **778**, 549 (1984). LC determn in plasma and urine: P. Lagerstrom, B. Persson, *J. Chromatogr.* **309**, 347 (1984). Survey of preclinical data: *Scand. J. Gastroenterol.* **20**, Suppl 108, 1-120 (1985). Toxicological studies: L. Ekman *et al.*, *ibid.* 53. Clinical trial in Zollinger-Ellison syndrome: C. B. H. W. Lamers *et al.*, *N. Engl. J. Med.* **310**, 758 (1984); in duodenal ulcer: K. Lauritsen *et al.*, *ibid.* **312**, 958 (1985). Veterinary trial in race horses: M. J. Murray *et al.*, *Equine Vet. J.* **29**, 425 (1997). Review of pharmacology and clinical efficacy: H. D. Langtry, M. I. Wilde, *Drugs* **56**, 447-486 (1998).

Crystals from acetonitrile, mp 156°. Sol in dichloromethane; sparingly sol ethanol, methanol; slightly sol in acetone, isopropanol; very slightly sol in water. LD_{50} in mice, rats (g/kg): 0.08, >0.05 i.v.; >4, >4 orally (Ekman).

Magnesium salt. [95382-33-5] Antra; Gastracid; Gastroloc; Omebeta; Omep; Ome-Puren. $C_{34}H_{36}MgN_6O_6S_2$; mol wt 713.12. White to off-white powder. Sparingly sol in methanol; slightly sol in alc; very slightly sol in water, dichloromethane.

S-Form. [119141-88-7] Esomeprazole; perprazole; H-199/18. LC-MS determn in plasma: H. Stenhoff *et al.*, *J. Chromatogr. B* **734**, 191 (1999). Colorless syrup. $[\alpha]_D^{20}$ −155° (c = 0.5 in chloroform).

S-Form magnesium salt. [161973-10-0] (*T*-4)-Bis[6-methoxy-2-[(*S*)-[(4-methoxy-3,5-dimethyl-2-pyridinyl)methyl]sulfinyl-κO]-1*H*-benzimidazolato-κN^3]magnesium; esomeprazole magnesium; Nexium. Review of clinical experience in acid disorders: D. A. Johnson, *Expert Opin. Pharmacother.* **4**, 253-264 (2003). White powder. $[\alpha]_D^{20}$ −128.2° (c = 1 in methanol). Soluble in methanol; slightly sol in water. Practically insol in heptane.

THERAP CAT: Antiulcerative; in treatment of Zollinger-Ellison syndrome.

THERAP CAT (VET): Antiulcerative.

6940. Omiganan. [204248-78-2] L-Isoleucyl-L-leucyl-L-arginyl-L-tryptophyl-L-prolyl-L-tryptophyl-L-tryprophyl-L-prolyl-L-tryptophyl-L-arginyl-L-arginyl-L-lysinamide. $C_{90}H_{127}N_{27}O_{12}$; mol wt 1779.18. C 60.76%, H 7.20%, N 21.26%, O 10.79%. Cationic antimicrobial peptide. Synthetic analog of indolicidin, *q.v.*, an antimicrobial peptide isolated from bovine neutrophils. Prepn: J. R. Fraser *et al.*, **WO 9807745**; *eidem*, **US 6538106** (1998, 2003 both to Micrologix). Comparative *in vitro* antibacterial spectrum: H. S. Sader *et al.*, *Antimicrob. Agents Chemother.* **48**, 3112 (2004). Interaction with bacterial membranes and cell wall models: M. N. Melo, M. A. R. B. Castanho, *Biochim. Biophys. Acta* **1768**, 1277 (2007). Review of clinical development: R. E. Isaacson, *Curr. Opin. Investig. Drugs* **4**, 999-1003 (2003).

Ile–Leu–Arg–Trp–Pro–Trp–Trp–Pro–Trp–Arg–Arg–Lys–NH2

Pentahydrochloride. [269062-93-3] MBI-226; Omigard. $C_{90}H_{127}N_{27}O_{12}.5HCl$; mol wt 1961.47. Molar extinction coefficient at 280 nm: 1.72×10^4.

THERAP CAT: Antibacterial.

6941. Omoconazole. [74512-12-2] 1-[(1*Z*)-2-[2-(4-Chlorophenoxy)ethoxy]-2-(2,4-dichlorophenyl)-1-methylethenyl]-1*H*-imidazole; (*Z*)-1-[2,4-dichloro-β-[2-(*p*-chlorophenoxy)ethoxy]-α-methylstyryl]imidazole; CM-8282. $C_{20}H_{17}Cl_3N_2O_2$; mol wt 423.72. C 56.69%, H 4.04%, Cl 25.10%, N 6.61%, O 7.55%. Prepn: L. Zirngibl *et al.*, **DE 2839388**; *eidem*, **US 4210657** (both 1980 to

Siegfried); and crystal structure: K. Thiele *et al.*, *Helv. Chim. Acta* **70**, 441 (1987). Stereospecific synthesis: L. Zirngibl, K. Thiele, **US 4554356** (1985 to Siegfried). *In vitro* fungistatic activity: M. Mosse *et al.*, *Pathol. Biol.* **34**, 684 (1986).

Crystals from ethyl acetate/hexane (1:4), mp 89-90°.

Nitrate. [83621-06-1] Sgd-12878; Afongan; Azameno; Fungisan; Fongamil. $C_{20}H_{17}Cl_3N_2O_2.HNO_3$; mol wt 486.73. Crystals from ethyl acetate/ethanol, mp 118-120° (Büchi), 122.5° (Mettler).

THERAP CAT: Antifungal (topical).

6942. Onapristone. [96346-61-1] (11β,13α,17α)-11-[4-(Dimethylamino)phenyl]-17-hydroxy-17-(3-hydroxypropyl)estra-4,9-dien-3-one; 11β-(4-dimethylaminophenyl)-17α-hydroxy-17β-(3-hydroxypropyl)-13α-methyl-4,9-gonadien-3-one; ZK-98299. $C_{29}H_{39}NO_3$; mol wt 449.64. C 77.47%, H 8.74%, N 3.12%, O 10.67%. Progesterone receptor antagonist. Prepn: G. Neef *et al.*, **EP 129499**; *eidem*, **US 4780461** (1984, 1988 both to Schering AG); *idem et al.*, *Steroids* **44**, 349 (1984). HPLC determn in plasma and serum: C. Zurth, F. Kagels, *J. Chromatogr.* **532**, 115 (1990). *In vitro* effect on breast cancer cells: S. Classen *et al.*, *J. Steroid Biochem. Mol. Biol.* **45**, 315 (1993). Mechanism of action study: D. P. Edwards *et al.*, *ibid.* **53**, 449 (1995). Clinical antifolliculotrophic effect: H. B. Croxatto *et al.*, *Hum. Reprod.* **9**, 1442 (1994).

Amorphous glass. $[\alpha]_D^{25}$ +446.2° (c = 0.51 in chloroform).

THERAP CAT: Antiprogestin; antineoplastic (hormonal).

6943. Ondansetron. [99614-02-5] 1,2,3,9-Tetrahydro-9-methyl-3-[(2-methyl-1*H*-imidazol-1-yl)methyl]-4*H*-carbazol-4-one. $C_{18}H_{19}N_3O$; mol wt 293.37. C 73.69%, H 6.53%, N 14.32%, O 5.45%. Specific serotonin (5HT$_3$) receptor antagonist. Prepn: I. H. Coates *et al.*, **EP 191562**; *eidem*, **US 4695578** (1986, 1987 both to Glaxo). Pharmacology: A. Butler *et al.*, *Br. J. Pharmacol.* **94**, 397 (1988). LC-MS/MS determn in plasma: Y. Dotsikas *et al.*, *J. Chromatogr. B* **836**, 79 (2006). Clinical trials in cancer chemotherapy-induced nausea and vomiting: M. G. Kris *et al.*, *J. Clin. Oncol.* **6**, 659 (1988); L. Cubeddu *et al.*, *N. Engl. J. Med.* **322**, 810 (1990); M. Marty *et al.*, *ibid.* 816. Clinical evaluation in treatment of bulimia: P. L. Faris *et al.*, *Lancet* **355**, 792 (2000); in early onset alcoholism: B. A. Johnson *et al.*, *J. Am. Med. Assoc.* **284**, 963 (2000); in radiotherapy-induced nausea and emesis: J. P. LeBourgeois *et al.*, *Clin. Oncol.* **11**, 340 (1999). Review of pharmacology and therapeutic use: A. Markham, E. M. Sorkin, *Drugs* **45**, 931-952 (1993); of clinical pharmacokinetics: K. H. Simpson, F. M. Hicks, *J. Pharm. Pharmacol.* **48**, 774-781 (1996).

Crystals from methanol, mp 231-232°. Very sol in acid solns; sparingly sol in water.

Hydrochloride dihydrate. [99614-01-4] GR-38032F; GR-C507/75; SN-307; Zofran; Zophren. $C_{18}H_{19}N_3O.HCl.2H_2O$; mol wt 365.86. White crystalline solid from water/isopropanol, mp 178.5-179.5°. pKa 7.4.

3S-Form. [99614-58-1] $[\alpha]_D^{25}$ −14° (c = 0.19 in methanol).
3R-Form. [99614-60-5] $[\alpha]_D^{24}$ +16° (c = 0.34 in methanol).

THERAP CAT: Antiemetic.
THERAP CAT (VET): Antiemetic.

6944. Onion Oil. Obtained from crushed onion seeds, the seeds of *Allium cepa* L., *Liliaceae:* Loew, *Ind. Quim.* **10**, 5 (1948); Phadnis *et al.*, *J. Univ. Bombay* **17A**, no. 24, 62 (1948). Contains allylpropyl bisulfide, *S*-(1-propenyl)cysteine sulfoxide and 1-propenylsulfenic acid which is thought to be the lacrimator in onions: Virtanen, Sprare, *Suom. Kemistil. B* **35**, 28, 29 (1962); *Chem. Ztg.* **86**, 816 (1962).

Pale yellow oil. Strong odor of onions. d_4^{15} 0.9289. Solidifies at −15°. n_D^{25} 1.4730. $[\alpha]_D^{20}$ −1.5°. Sol in ether, chloroform, carbon disulfide.

6945. Oosporein. [475-54-7] 2,2′,5,5′-Tetrahydroxy-4,4′-dimethyl[bi-1,4-cyclohexadien-1-yl]-3,3′,6,6′-tetrone; 3,3′,6,6′-tetrahydroxy-5,5′-dimethyl-2,2′-bi-*p*-benzoquinone; chaetomidin; iso-oosporein. $C_{14}H_{10}O_8$; mol wt 306.23. C 54.91%, H 3.29%, O 41.80%. Fungal pigment isolated from *Oospora colorans* van Beyma.: Kögl, van Wessem, *Rec. Trav. Chim.* **63**, 5 (1944); from *Chaetomium aureum* Chivers and identity with chaetomidin: Lloyd *et al.*, *J. Chem. Soc.* **1955**, 2163; from *Acremonium* spp: Divekar *et al.*, *Can. J. Chem.* **37**, 2097 (1959); from *Beauveria bassiana* (Bals.) Vuill.: Vining *et al.*, *Can. J. Microbiol.* **8**, 931 (1962). Identity with iso-oosporein: Smith, Thomson, *Tetrahedron* **10**, 148 (1960). Synthesis: J. Kalamar *et al.*, *Helv. Chim. Acta* **57**, 2368 (1974). Biosynthesis: E. Steiner *et al.*, *ibid.* 2377.

Bronze plates from aq methanol, mp 290-295°. uv max (ethanol): 216, 287 nm (log ε 3.51, 4.67).

Tetraacetate. $C_{22}H_{28}O_{12}$. Yellow needles from methanol, mp 190°. uv max (ethanol): 262 nm (log ε 4.41).

Tetramethyl ether. $C_{18}H_{18}O_8$. Orange needles from aq methanol, mp 123°. uv max (ethanol): 285.5, 394 nm (log ε 4.40, 2.98).

6946. Opianic Acid. [519-05-1] 6-Formyl-2,3-dimethoxybenzoic acid; 5,6-dimethoxyphthalaldehydic acid. $C_{10}H_{10}O_5$; mol wt 210.19. C 57.14%, H 4.80%, O 38.06%. Obtained (together with cotarnine) by heating narcotine with dil HNO$_3$. Prepn: Wilson *et al.*, *J. Org. Chem.* **16**, 792 (1951); Blair, *J. Chem. Soc.* **1955**, 708. NMR studies: Buu-Hoi *et al.*, *Bull. Soc. Chim. Fr.* **1970**, 137.

Needles from water, mp 150°. Sol in 400 parts cold, 60 parts boiling water; sol in alcohol, ether. uv max: 215, 284 nm (ε 20,700, 6500).

6947. Opiniazide. [2779-55-7] 4-Pyridinecarboxylic acid [(2-carboxy-3,4-dimethoxyphenyl)methylene]hydrazide; 5,6-dimethoxyphthalaldehydic acid isonicotinoylhydrazone; 1-(2-carboxy-3,4-dimethoxybenzylidene)-2-isonicotinoylhydrazine; 2-carboxy-3,4-dimethoxybenzal isonicotinylhydrazone; saluside; saluzide; saliuzid; saluzid. $C_{16}H_{15}N_3O_5$; mol wt 329.31. C 58.36%, H 4.59%, N 12.76%, O 24.29%. Description: G. N. Pershin, S. A. Vichkanova, *C.A.* **51**, 10747e (1957); *C.A.* **52**, 1485-1486 (1958); V. A. Buskina, *C.A.* **54**, 744f (1960). Bioavailability: A. F. Zaeko, V. G. Perkova, *Farmatsiya (Moscow)* **28**, 28 (1979), *C.A.* **92**, 185805g (1980). Toxicity study: R. A. Akhundov, *Mater. Nauchn. Konf. Azgosmedinstituta* **1975**, 137, *C.A.* **86**, 50678y (1977).

LD$_{50}$ i.v. in guinea pigs: 1.634 g/kg (Akhundov).
THERAP CAT: Antibacterial (tuberculostatic).

6948. Opipramol. [315-72-0] 4-[3-(5*H*-Dibenz[*b*, *f*]azepin-5-yl)propyl]-1-piperazineethanol; 5-[γ-[4-(β-hydroxyethyl)piperazino]propyl]dibenzo[*b*, *f*]azepine; *N*-[3-[4-(2-hydroxyethyl)piperazino]propyl]iminostilbene; 4-[3-(5*H*-dibenzo[*b*, *f*]azepin-5-yl)propyl]-1-(2-hydroxyethyl)piperazine. $C_{23}H_{29}N_3O$; mol wt 363.51. C 76.00%, H 8.04%, N 11.56%, O 4.40%. Prepn: **FR M209** (1961 to Rhône-Poulenc), *C.A.* **58**, 3442f (1963); Schindler, **DE 1133729**; **CH 359143**; **CH 360061** (all 1962 to Geigy), *C.A.* **58**, 10219a, 10218f, 10218h (1963).

Crystals, mp 100-101°.
Dihydrochloride. [909-39-7] Dinsidon; Ensidon; Insidon. C_{23}-$H_{29}N_3O.2HCl$; mol wt 436.42. Crystals from ethanol, mp 210° (**FR M209**); mp 228-230° (**DE 1133729**). Sol in water, alcohol; sparingly sol in acetone.
THERAP CAT: Antidepressant; antipsychotic.

6949. Opium. Gum opium; crude opium. Air-dried latex from incised, unripe seed capsules of *Papaver somniferum* L., or *P. album* Mill., *Papaveraceae*. *Constit.* complex mixture of more than 20 alkaloids (20-30%) present as salts of meconic, lactic and sulfuric acids; volatile pyrazine derivatives responsible for the characteristic odor; meconin, sugars, proteins, gums, wax, fats, water. Morphine is the most important alkaloid and occurs to the extent of 10-16%, noscapine 4-8%, codeine 0.8-2.5%, papaverine 0.5-2.5%, thebaine 0.5-2%. For medicinal purposes, raw opium is dried at not above 70°, granulated or powdered, and adjusted with lactose or other inert diluent to contain 10-10.5% anhydr morphine. Historical review: M. J. Brownstein, *Proc. Natl. Acad. Sci. USA* **90**, 5391-5393 (1993). Review of alkaloid extraction processes: R. J. Bryant, *Chem. Ind. (London)* **1988** (5), 146-153; of opium characterization methods: B. Remberg *et al.*, *Bull. Narc.* **46**, 79-108 (1994). Review of composition, pharmacology, and abuse: H. Kalant, *Addiction* **92**, 267-277 (1997). HPLC analysis of alkaloid content: K. Yoshimatsu *et al.*, *Chem. Pharm. Bull.* **53**, 1446 (2005). Clinical use of opium tincture in neonatal abstinence syndrome: M. G. Coyle *et al.*, *J. Pediatr.* **140**, 561 (2002).

Pale olive-brown or gray, rounded, oval, brick-shaped or elongated, flattened masses with coarse surface, becoming hard or tough on storage. Internally, reddish brown and coarsely granular. Characteristic odor and very bitter taste.
Deodorized opium. Denarcotized opium. Powdered opium freed from its odor and nauseating substances by treatment with petr ether. Contains 10-10.5% anhydr morphine.
Laudanum. Opium tincture. Boiling water extract of opium mixed with dil alcohol. Contains 1 g anhydr morphine per 100 ml.
Paregoric. Camphorated tincture of opium. Mixture of powdered opium or opium tincture with benzoic acid, anise or other

essential oil, glycerin, and alcohol. Contains 35-45 mg anhydr morphine per 100 ml.
Note: This is a controlled substance: **21 CFR**, 1308.12.
USE: Largely for the manuf of morphine, codeine and other opium alkaloids.
THERAP CAT: Analgesic; antidiarrheal.
THERAP CAT (VET): Analgesic; antidiarrheal; antitussive.

6950. Opromazine. [969-99-3] 2-Chloro-*N*,*N*-dimethyl-10*H*-phenothiazine-10-propanamine 5-oxide; 2-chloro-10-[3-(dimethylamino)propyl]phenothiazine 5-oxide; 10-(γ-dimethylaminopropyl)-3-chlorophenothiazine 9-oxide; 5-oxychlorpromazine; chlorpromazine sulfoxide; Secotil. $C_{17}H_{19}ClN_2OS$; mol wt 334.86. C 60.98%, H 5.72%, Cl 10.59%, N 8.37%, O 4.78%, S 9.57%. Metabolite of chlorpromazine, *q.v.*: Salzman, Brodie, *J. Pharmacol. Exp. Ther.* **118**, 46 (1956). Preparation by enzymatic oxidation of chlorpromazine: Gillette, Kamm, *ibid.* **130**, 262 (1960). Synthesis: **FR 1167653** (1958 to Rhône-Poulenc). Toxicity data: Minami, Yoshimoto, *Nara Igaku Zasshi* **8**, 50 (1957), *C.A.* **51**, 16966a (1957).

Crystals, mp 115°. LD$_{50}$ s.c. in mice: 102 mg/kg (Minami, Yoshimoto).

6951. Opsins. Broad class of species-specific proteins which form the basis of the visual pigments and of bacteriorhodopsin, *q.v.* Structurally integrated into the rods and cones of the retina of the eye or into the photoreceptor membranes of certain bacteria. Each of these cell types produces a genetically specified opsin which has been classified on the basis of cellular source: *scotopsins* (rods), *photopsins* (cones), and *bacteriopsins* (bacteria). Methods for purification, prepn and assay: R. Hubbard *et al.*, *Methods Enzymol.* **18**, 615-653 (1971). Series of articles on photoreceptor biosynthesis: *ibid.* **81**, 763-815 (1982). Review of biosynthetic process: D. S. Papermaster, B. G. Schneider in *Cell Biology of the Eye*, D. McDevitt, Ed. (Academic Press, New York, 1982) p 475. The photoreceptor activity of visual pigments is due to a carotenoid chromophore, retinal or 3-dehydroretinal, *q.q.v.*, bound as a protonated Schiff base to a lysine moiety in the opsin portion of the molecule: A. R. Osenoff, R. Callender, *Biochemistry* **13**, 4243 (1974). Each pigment has unique physicochemical properties. The most significant is the absorption spectrum which is regulated by electrostatic interactions between the chromophore and the charged or dipolar groups on the opsin: R. Hubbard, L. Sperling, *Exp. Eye Res.* **17**, 581 (1973); B. Honig *et al.*, *J. Am. Chem. Soc.* **101**, 7084 (1979). A visual system, a set of pigments spanning the light sensitivity range of a particular species, is generally based on one type of chromophore combined with various opsins. Visual systems based on retinal are the most widespread in nature. The 11-*cis* isomer is utilized by rhodopsin, *q.v.*, the most common pigment of rod cells, and by the corresponding trichromatic cone pigments (*see* Iodopsin). Bacteriorhodopsin is composed of bacterioopsin and *trans* retinal. Visual systems based on 3-dehydroretinal, exemplified by the pigments porphyropsin and cyanopsin, *q.q.v.*, have been found to occur only in certain fish and amphibians. Visual pigments utilizing both types of chromophore have been found to coexist in the retina of some of these species: T. E. Reuter *et al.*, *J. Gen. Physiol.* **58**, 351 (1971). Exposure to light initiates the bleaching of the pigment through a series of distinct intermediates involving the isomerization and ultimate dissociation of the chromophore from the opsin: R. Hubbard, A. Kropf, *Proc. Natl. Acad. Sci. USA* **44**, 140 (1958); B. Honig *et al.*, *ibid.* **76**, 2503 (1979). This process initiates the mechanism of energy transduction and visual excitation: T. G. Ebrey, B. Honig, *Q. Rev. Biophys.* **8**, 129-184 (1975); B. Honig, *Annu. Rev. Phys. Chem.* **29**, 31-57 (1978); R. Uhl, E. W. Abrahamson, *Chem. Rev.* **81**, 291 (1981). Review of energy transduction in invertebrate photoreceptors: P. Hillman *et al.*, *Physiol. Rev.* **63**, 668-772 (1983); in bacteriorhodopsin: H. V. Westerhoff, Z. Dancshazy, *Trends Biochem. Sci.* **9**, 112 (1984). General reviews: G. Wald, *Science* **162**, 230-239 (1968); D. F. O'Brien, *ibid.* **218**, 961-966 (1982); A. Maeda, T. Yoshizawa, *Photochem. Photobiol.* **35**, 891-898 (1983); P. S. Zurer, *Chem. Eng. News* **61**, 24-35 (Nov. 28, 1983). *See also: Methods Enzymol.* **88**, 1-836 (1982).

6952. Orange I. [523-44-4] 4-[2-(4-Hydroxy-1-naphthalen-yl)diazenyl]benzenesulfonic acid sodium salt (1:1); Tropaeolin OOO no. 1; C.I. Acid Orange 20; FD & C Orange I; Ext. D & C Orange 3; C.I. 14600; α-naphthol orange; sodium azo-α-naphtholsulfanilate. $C_{16}H_{11}N_2NaO_4S$; mol wt 350.32. C 54.86%, H 3.17%, N 8.00%, Na 6.56%, O 18.27%, S 9.15%. Prepared by coupling diazotized sulfanilic acid with α-naphthol: *Colour Index* **vol. 4** (3rd ed., 1971) p 4065.

Reddish-brown powder. Sol in water to orange-red soln, in alcohol to orange soln. Acids ppt the aq soln. Sodium hydroxide intensifies the red color of the aq soln. pH: 7.6 brownish-yellow; 8.9 purple.

6953. Orange II. [633-96-5] 4-[2-(2-Hydroxy-1-naphthalen-yl)diazenyl]benzenesulfonic acid sodium salt (1:1); C.I. Acid Orange 7; β-naphthol orange; D & C Orange No. 4; C.I. 15510; Mandarin G; Tropaeolin OOO no. 2. $C_{16}H_{11}N_2NaO_4S$; mol wt 350.32. C 54.86%, H 3.17%, N 8.00%, Na 6.56%, O 18.27%, S 9.15%. Anionic azo dye. Prepd by coupling β-naphthol, *q.v.*, with diazotized sulfanilic acid, *q.v.*, in alkaline soln, *cf. Org. Synth.* **coll. vol. II**, 36 (1943); *Colour Index* **vol. 4** (3rd ed., 1971) p 4078. *Review: H. J. Conn's Biological Stains*, R. D. Lillie, Ed. (Williams & Wilkins, Baltimore, 9th ed., 1977) pp 112-113, 573.

Orange-brown powder. Indicator color transitions: amber to orange pH 7.4-8.6; orange to red pH 10.2-11.8. Soly (mg/ml): water 130 mg/ml; ethanol 4 mg/ml. Absorption max (water): 483 nm.

Pentahydrate. Orange needles from water. One gram dissolves in 20 ml water. Sol in alc. Absorption max 484.4 nm.

USE: Dye; biological stain; external colorant. Limited use as an indicator.

6954. Orange B. [15139-76-1] 4,5-Dihydro-5-oxo-4-[2-(4-sulfo-1-naphthalenyl)diazenyl]-1-(4-sulfophenyl)-1*H*-pyrazole-3-carboxylic acid 3-ethyl ester sodium salt (1:2); 5-oxo-4-[(4-sulfo-1-naphthyl)azo]-1-(*p*-sulfophenyl)-2-pyrazoline-3-carboxylic acid 3-ethyl ester disodium salt; 5-hydroxy-4-[(4-sulfo-1-naphthalenyl)-azo]-1-(4-sulfophenyl)-1*H*-pyrazole-3-carboxylic acid 3-ethyl ester disodium salt; 1-(4-sulfophenyl)-3-ethylcarboxyl-4-(4-sulfonaphthylazo)-5-hydroxypyrazole disodium salt; C.I. Acid Orange 137; C.I. 19235. $C_{22}H_{16}N_4Na_2O_9S_2$; mol wt 590.49. C 44.75%, H 2.73%, N 9.49%, Na 7.79%, O 24.39%, S 10.86%. Prepn: W. H. Kretlow *et al.*, **US 3285906** (1966 to Stange Co.). *Review: Fed. Regist.* **43**, 45611 (1978). *See also Colour Index* **vol. 4** (3rd ed., 1971) p 4134.

Dull orange crystals. Absorption max (0.04*N* ammonium acetate): 442 nm. Soly in water at 77°: 220 g/l. Violet soln in conc H_2SO_4, changing to fuchsia then red on dilution. Red soln in conc HCl. Yellowish soln in 10% NaOH, changing to brownish yellow on dilution.

USE: In coloring sausage and frankfurter casings.

6955. Orange Peel. Rind of the fruit of *Citrus aurantium* L., *Rutaceae. Habit.* N. India; cultivated near Mediterranean Sea, Spain, W. Indies, Florida, California. Review: A. Y. Leung, *Encyclopedia of Common Natural Ingredients* (John Wiley & Sons, New York, 1980) pp 248-252.

Bitter orange peel. Dried rind of unripe fruit. *Constit.* Volatile oil, hesperidine, naringin, aurantiamarin, acrid resin, gum, tannin.

Sweet orange peel. Fresh rind of ripe fruit of *C. aurantium*, var *sinensis* (also known as *C. sinensis*). *Constit.* Volatile oil, hesperidine, fixed oil, resin, gum, tannin. Closely resembles bitter orange peel, but has an orange-yellow color; sweetish, fragrant odor; aromatic and only slightly bitter taste.

USE: Flavoring agent.

6956. Orazamide. [2574-78-9] 1,2,3,6-Tetrahydro-2,6-dioxo-4-pyrimidinecarboxylic acid compd with 5-amino-1*H*-imidazole-4-carboxamide (1:1); 5-aminoimidazole-4-carboxamide orotate; orotic acid compd with 5(or 4)-aminoimidazole-4(or 5)-carboxamide (1:1); 4-amino-5-imidazolecarboxamide orotate; AICA orotate; Aicamin; Aicorat. $C_9H_{10}N_6O_5$; mol wt 282.22. C 38.30%, H 3.57%, N 29.78%, O 28.34%. Prepn: **FR 1351141**; Haraoka, Kamiya, **US 3271398** (1964, 1966 both to Fujisawa). Pharmacology and toxicology: Tamura, Shibayama, *Yakugaku Kenkyu* **35**, 94 (1963), *C.A.* **64**, 2652h (1966); Hashimoto, *J. Vitaminol.* **13**, 9, 19 (1967).

Dihydrate. Crystals, dec 284-285°. LD_{50} in mice (g/kg): 0.6 i.p.; >4.0 orally (Tamura, Shibayama).

THERAP CAT: Hepatoprotectant.

6957. Orbifloxacin. [113617-63-3] *rel*-1-Cyclopropyl-7-[(3*R*,5*S*)-3,5-dimethyl-1-piperazinyl]-5,6,8-trifluoro-1,4-dihydro-4-oxo-3-quinolinecarboxylic acid; 1-cyclopropyl-5,6,8-trifluoro-(*cis*-3,5-dimethyl-1-piperazinyl)-1,4-dihydro-4-oxoquinoline-3-carboxylic acid; Orbax. CP-104594; mol wt 395.38. C 57.72%, H 5.10%, F 14.42%, N 10.63%, O 12.14%. Fluorinated quinoline antibacterial. Prepn: J. Matsumoto *et al.*, **EP 242789**; *eidem*, **US 4886810** (1987, 1989 both to Dainippon). *In vitro* activity vs *Staphylococcus intermedius*: J.-P. Ganière *et al.*, *Res. Vet. Sci.* **77**, 67 (2004). Photodegradation kinetics: T. Morimura *et al.*, *Chem. Pharm. Bull.* **43**, 1000 (1995). Pharmacokinetics in mares: G. R. Haines *et al.*, *Can. J. Vet. Res.* **65**, 181 (2001). HPLC determn in rabbit plasma: M. A. Garcia *et al.*, *J. Chromatogr. Sci.* **37**, 199 (1999); LC/fluorescence/MS determn in eggs: M. J. Schneider, D. J. Donoghue, *Anal. Chim. Acta* **483**, 39 (2003). Localization of drug to diseased skin: P. A. Kay-Mugford *et al.*, *Vet. Ther.* **3**, 1 (2002). Review of veterinary use in Japan: S. Nakamura, *Drugs* **49**, Suppl. 2, 152-158 (1995); of use in companion animals: R. D. Walker, *Aust. Vet. J.* **78**, 84-90 (2000).

Relative stereochemistry

Crystals from chlororform-ethanol, mp 259-260°. pKa_1: 5.60; pKa_2 8.90.

THERAP CAT (VET): Antibacterial.

6958. Orcein. [1400-62-0] A dye first prepared from lichens (cudbear, *q.v.*, or *archil*). Prepn by oxidation of orcinol with H_2O_2 in the presence of ammonia water: Zulkowski, Peters, *Monatsh. Chem.* **11**, 227 (1890). Can be separated into 14 dyes by distribution chromatography: Musso, *Ber.* **89**, 1659 (1956). The eight compds depicted are the major components of orcein; β- and γ-components are *cis-trans* isomers of the same compd. Structure studies: Beecken *et al.*, *Angew. Chem.* **73**, 665 (1961). Brief review: *H. J. Conn's Biological Stains*, R. D. Lillie, Ed. (Williams & Wilkins, Baltimore, 9th ed., 1977) pp 400-403. Review of use as textile dye and histological stain for elastin: H. Puchter, S. N. Meloan, *Histochemistry* **64**, 119 (1979).

	R_1	R_2	R_3
α-Aminoorcein	H	O	NH_2
α-Hydroxyorcein	H	O	OH
β- and γ-Aminoorcein	orcinol	O	NH_2
β- and γ-Hydroxyorcein	orcinol	O	OH
β- and γ-Aminoorceimine	orcinol	NH	NH_2

Brownish-red microcryst powder. Practically insol in water, benzene, chloroform, ether, carbon disulfide. Sol in alcohol, acetone or acetic acid with red color, in dil aq alkali with bluish-violet color.

USE: Biological stain.

6959. Orcinol. [504-15-4] 5-Methyl-1,3-benzenediol; 5-methylresorcinol; orcin; 3,5-dihydroxytoluene. $C_7H_8O_2$; mol wt 124.14. C 67.73%, H 6.50%, O 25.78%. Occurs in many species of lichens: Sastry, Rao, *Curr. Sci.* **10**, 437 (1941). Prepn: Anker, Cook, *J. Chem. Soc.* **1945**, 311; Kisteneva, Rozhdestvenskii, *Zh. Prikl. Khim.* **22**, 1108 (1949); Stevens, US 2603662 (1952 to Gulf); Zimmer, US 3028410 (1962 to Hooker Chem.). Toxicity studies: I. Veldre *et al.*, *Vopr. Gig. Tr. Prof. Patol. Est. SSR* **2**, 160 (1970), *C.A.* **74**, 51746h (1971).

Monohydrate. Crystals; sweet but unpleasant taste; reddens on exposure to air due to oxidation. mp about 58°; 107° when anhyd. bp 290°; bp_{14-20} 165-170°; bp_5 147°. Freely sol in water, alcohol, ether; less sol in benzene; slightly sol in chloroform or carbon disulfide. *Keep well closed and protected from light.* LD_{50} in mice, rats, rabbits, guinea pigs (mg/kg): 772, 844, 2400, 1678 orally (Veldre).

USE: As a reagent for pentoses, lignin, beet sugar, saccharoses, arabinose, diastase.

6960. Oregovomab. [213327-37-8] Anti-(human CA125-(carbohydrate antigen)) immunoglobulin G1 (mouse monoclonal B43.13 $γ_1$-chain) disulfide with mouse monoclonal B43.13 κ-chain, dimer; OvaRex. Murine monoclonal antibody directed against the ovarian tumor associated antigen CA-125; induces tumor specific cellular immune response. Prepn: M. J. Krantz *et al.*, *J. Cell. Biochem.* Suppl. 12E, 139 (1988). Preparative method as radioimmunodiagnostic: V. Capstick *et al.*, *Int. J. Biol. Markers* **6**, 129 (1991). Clinical experience in antiidiotype induction therapy: R. Madiyalakan *et al.*, *Hybridoma* **14**, 199 (1995). Clinical pharmacokinetics: S. A. McQuarrie *et al.*, *Nucl. Med. Commun.* **18**, 878 (1997). Immunopharmacology: B. C. Schultes *et al.*, *Hybridoma* **18**, 47

(1999); A. A. Noujaim *et al.*, *Cancer Biother. Radiopharm.* **16**, 187 (2001). Assay to monitor immune response: B. C. Schultes, T. L. Whiteside, *J. Immunol. Methods* **279**, 1 (2003). Clinical evaluation in ovarian cancer: J. S. Berek *et al.*, *J. Clin. Oncol.* **22**, 3507 (2004); combined with chemotherapy: A. N. Gordon *et al.*, *Gynecol. Oncol.* **94**, 340 (2004). Reviews of development and therapeutic potential: I. Bruckner, *IDrugs* **4**, 457-462 (2001); J. S. Berek *Expert Opin. Biol. Ther.* **4**, 1159-1165 (2004).

THERAP CAT: Antineoplastic, immunomodulator.

6961. Orexins. Hypocretins. Neuropeptides produced in the hypothalamus that are involved in the regulation of feeding behavior and the sleep-wake cycle. Name derived from the Greek word, orexis, meaning appetite. Fasting increases orexin production; administration of exogenous orexin stimulates feeding. Also implicated in the pathophysiology of narcolepsy. Two forms are known (orexins A and B); both are derived from the same precursor protein. Two orexin receptors (OX_1R and OX_2R) have been identified in mammalian brain. OX_1R binds only orexin A while OX_2R is a nonselective receptor for both A and B. Identification of hypothalamus-specific mRNA that encodes the orexin precursor: L. De Lecea *et al.*, *Proc. Natl. Acad. Sci. USA* **95**, 322 (1998). Purification and biological activity of orexins: T. Sakurai *et al.*, *Cell* **92**, 573 (1998). Immunohistochemical localization in rat brain: Y. Date *et al.*, *Proc. Natl. Acad. Sci. USA* **96**, 748 (1999). Association of canine narcolepsy with OX_2R defect: L. Lin *et al.*, *Cell* **98**, 365 (1999); of sleep disorder in orexin knockout mice: R. M. Chemelli *et al.*, *ibid.* 437. Clinical determn of orexin deficiency in CSF of narcoleptic patients: S. Nishino *et al.*, *Lancet* **355**, 39 (2000). Review of structure, tissue distribution and role in feeding behavior: T. Sakurai, *Regul. Pept.* **85**, 25-30 (1999). Review of role in narcolepsy and potential therapeutic application: J. M. Siegel, *Cell* **98**, 409-412 (1999).

Orexin A. [205599-75-3] Hypocretin-1. Cyclic, 33-amino acid peptide; mol wt 3562 Da. Sequence is completely conserved in human, rat, mouse, cow and pig.

Orexin B. [205599-76-4] Hypocretin-2. Linear, 28-amino acid peptide; mol wt 2937 Da. Sequence is 46% homologous with orexin A. Human and rodent forms differ by 2 amino acid substitutions.

6962. Oripavine. [467-04-9] (5α)-6,7,8,14-Tetrahydro-4,5-epoxy-6-methoxy-17-methylmorphinan-3-ol; O^3-demethylthebaine. $C_{18}H_{19}NO_3$; mol wt 297.35. C 72.71%, H 6.44%, N 4.71%, O 16.14%. From *Papaver orientale* L., and *P. bracteatum* Lindl., *Papaveraceae*: Junusov *et al.*, *Ber.* **68**, 2158 (1935); Kiselev, Konovalova, *J. Gen. Chem. USSR* **18**, 142 (1948). Identity of *O*-methyl derivative with thebaine: *eidem, Zh. Obshch. Khim.* **18**, 855 (1948). Biosynthesis: Gross, Dawson, *Biochemistry* **2**, 186 (1963). Reviews: H. L. Holmes in *The Alkaloids* vol. II, Manske, Holmes, Eds. (Academic Press, New York, 1952) p 167; K. W. Bentley, *The Chemistry of the Morphine Alkaloids* (Oxford, 1954) pp 192-196. Synthesis from morphine: Barber, Rapoport, *J. Med. Chem.* **18**, 1074 (1975).

Crystals, mp 200-201°. $[α]_D^{20}$ −211.8°.

Hydrochloride. $C_{18}H_{19}NO_3$·HCl. Crystals, dec 244-245°.

Methiodide. $C_{18}H_{19}NO_3$·CH_3I. Crystals, dec 207-208°.

Note: This is a controlled substance (opiate): **21 CFR,** 1308.12.

6963. Oritavancin. [171099-57-3] (4″*R*)-22-*O*-(3-Amino-2,-3,6-trideoxy-3-*C*-methyl-α-L-*arabino*-hexopyranosyl)-*N*^{3n}-[(4′-chloro[1,1′-biphenyl]-4-yl)methyl]vancomycin; LY-333328. C_{86}-$H_{97}Cl_3N_{10}O_{26}$; mol wt 1793.12. C 57.61%, H 5.45%, Cl 5.93%, N 7.81%, O 23.20%. Semisynthetic glycopeptide antibiotic. Prepn: R. D. G. Cooper *et al.*, **WO 9630401**; *eidem* **US 5840684** (1996, 1998 both to Eli Lilly). *In vitro* antibacterial activity: F. Biavasco *et al.*, *Antimicrob. Agents Chemother.* **41**, 2165 (1997). Mechanism of action: N. E. Allen, T. I. Nicas, *FEMS Microbiol. Rev.* **26**, 511 (2003). Clinical pharmacokinetics: S. M. Bhavnani *et al.*, *Diagn. Microbiol. Infect. Dis.* **50**, 95 (2004); G. J. Fetterly *et al.*, *Antimicrob. Agents Chemother.* **49**, 148 (2005). Clinical evaluation in *Staphylococcus aureus* bacteremia: S. M. Bhavnani *et al.*, *ibid.* **50**, 994 (2006). Review of development and clinical experience: K. E. Ward *et al.*, *Expert Opin. Invest. Drugs*, **15**, 417-429 (2006).

Diphosphate. [192564-14-0] $C_{86}H_{97}Cl_3N_{10}O_{26}.2H_3PO_4$; mol wt 1989.10. Sol in water.

THERAP CAT: Antibacterial.

6964. Orlistat. [96829-58-2] *N*-Formyl-L-leucine (1*S*)-1-[[(2*S*,3*S*)-3-hexyl-4-oxo-2-oxetanyl]methyl]dodecyl ester; *N*-formyl-L-leucine ester with (3*S*,4*S*)-3-hexyl-4-[(2*S*)-2-hydroxytridecyl]-2-oxetanone; (−)-tetrahydrolipstatin; orlipastat; Ro-18-0647; alli; Xenical. $C_{29}H_{53}NO_5$; mol wt 495.75. C 70.26%, H 10.78%, N 2.83%, O 16.14%. Pancreatic lipase inhibitor. Isoln from fermentation broth of *Streptomyces toxytricini*: P. Hadvary *et al.*, **EP 129748**; eidem, **US 4598089** (1985, 1986 both to Hoffmann-La-Roche). Structural elucidation: E. Hochuli *et al.*, *J. Antibiot.* **40**, 1086 (1987). Synthesis and absolute configuration: P. Barbier, F. Schneider, *Helv. Chim. Acta* **70**, 196 (1987). Total synthesis: S. Hanessian *et al.*, *J. Org. Chem.* **58**, 7768 (1993). Mechanism of action: B. Borgström, *Biochim. Biophys. Acta* **962**, 308 (1988). Clinical pharmacology: J. B. Hauptman *et al.*, *Am. J. Clin. Nutr.* **55**, 309S (1992). Clinical trial: M. H. Davidson *et al.*, *J. Am. Med. Assoc.* **281**, 235 (1999). Review of pharmacology and clinical efficacy in treatment of obesity: W. McNeely, P. Benfield, *Drugs* **56**, 241-249 (1998). Evaluation of long-term efficacy and tolerability: J. Hauptman *et al.*, *Arch. Fam. Med.* **9**, 160 (2000).

Crystals, mp 43°. $[\alpha]_D^{20}$ −32.0° (c = 1 in chloroform).

THERAP CAT: Antiobesity agent.

6965. Orlon®. Polyacrylonitrile; Fiber A. Obtained by polymerizing acrylonitrile. Review of prepn, properties, and uses: R. W. Moncrieff, *Man-Made Fibres* (John Wiley, New York, 1963) pp 446-467.

White fiber. Sticks at 235°. Ironing temps above 160° may cause yellowing. Sp gr 1.17. Flammability similar to that of rayon and cotton. Generally has very good resistance to mineral acids; excellent resistance to common solvents, oils, greases, neutral salts, sunlight; fairly good resistance to weak alkalies but is degraded by strong alkalies. Resists attack by molds, mildew, insects. 100% polyacrylonitrile fibers rarely used commercially due to difficulty in dyeing.

USE: Fiber suitable for outdoor furnishings (awnings, tents, outdoor furniture), indoor furnishings, anode bags in electroplating, knitwear, rugs.

6966. Ormosinine. [14350-67-5] 21-Ormosanin-20-yl panamine. $C_{40}H_{66}N_6$; mol wt 631.01. C 76.14%, H 10.54%, N 13.32%. From seed of *Ormosia dasycarpa* Jacks., *Leguminosae*: Hess, Merck, *Ber.* **52**, 1976 (1919); from *O. panamensis* Benth., *Leguminosae*: Lloyd, Horning, *J. Am. Chem. Soc.* **80**, 1506 (1958). Structural studies: Wilson, *Chem. Ind. (London)* **1965**, 472; *Tetrahedron* **21**, 2561 (1965). Composed of one molecule of *panamine* and one molecule of *ormosanine*; NMR elucidation of structure: N. S. Bhacca *et al.*, *J. Am. Chem. Soc.* **105**, 2538 (1983).

Needles from ethyl acetate, mp 219-220°. $[\alpha]_{436}^{25}$ +16.0°. $[\alpha]_{589}^{25}$ +8.9° (c = 1.29 in chloroform). Sol in chloroform; slightly sol in ether. Practically insol in water, alcohol. Sublimes to panamine.

6967. Ornidazole. [16773-42-5] α-(Chloromethyl)-2-methyl-5-nitro-1*H*-imidazole-1-ethanol; 1-(3-chloro-2-hydroxypropyl)-2-methyl-5-nitroimidazole; Ro-7-0207; Madelen; Ornidal; Tiberal. $C_7H_{10}ClN_3O_3$; mol wt 219.63. C 38.28%, H 4.59%, Cl 16.14%, N 19.13%, O 21.85%. Prepn: **NL 6606853**; M. Hoffer, **US 3435049** and **US 3493582** (1966, 1969 and 1970, all to Hoffmann-La Roche). Activity studies: E. Grunberg *et al.*, *Proc. Soc. Exp. Biol. Med.* **133**, 490 (1970). Synthesis and antiprotozoal activity: M. Hoffer, E. Grunberg, *J. Med. Chem.* **17**, 1019 (1974). Pharmacokinetics: D. E. Schwartz, F. Jeunet, *Chemotherapy* **22**, 19 (1976).

Crystals from toluene, mp 77-78°. uv max (2-propanol): 288, 312 nm (ε 3720, 9150). pKa: 2.4 ± 0.1. LD₅₀ in rats, mice (mg/kg): 1780, 1420 orally (Grunberg). Also reported as LD_{50} in mice (mg/kg): >2000 orally, >2000 i.p. (Hoffer, Grunberg).

THERAP CAT: Anti-infective.

6968. Ornipressin. [3397-23-7] 8-L-Ornithinevasopressin; ornithine-vasopressin; Orn⁸-vasopressin; POR 8. $C_{45}H_{63}N_{13}O_{12}S_2$; mol wt 1042.20. C 51.86%, H 6.09%, N 17.47%, O 18.42%, S 6.15%. Synthetic analog of vasopressin, *q.v.*, in which L-ornithine replaces L-arginine. Prepn: R. L. Huguenin, R. A. Boissonnas, *Helv. Chim. Acta* **46**, 1669 (1963); eidem, **FR 1396607**; eidem, **US 3299036** (1965, 1967 both to Sandoz). HPLC determn in pharmaceutical formulations: R. H. Buck *et al.*, *J. Chromatogr. A* **548**, 335 (1991). Pharmacologic properties: B. Berde *et al.*, *Experientia* **20**,

42 (1964). Pharmacodynamics: S. Keppens, H. de Wulf, *Biochim. Biophys. Acta* **588**, 63 (1979). Comparison of hemostatic and cardiovascular effects: L. Saarnivaara, P. Leander, *Anaesthesist* **26**, 144 (1977). Clinical trial to reduce bleeding during surgery: D. J. Adendorff, D. Davies, *S. Afr. Med. J.* **51**, 131 (1977); in treatment of hepatorenal syndrome: V. Gülberg *et al.*, *Hepatology* **30**, 870 (1999). Review of pharmacology and clinical uses in surgery and anesthesia: P. C. A. Kam, T. M. Tay, *Eur. J. Anaesthesiol.* **15**, 133-139 (1998).

Cys–Tyr–Phe–Gln–Asn–Cys–Pro–Orn–GlyNH$_2$

THERAP CAT: Vasoconstrictor.

6969. Ornithine. [70-26-8] L-Ornithine; α,δ-diaminovaleric acid; 2,5-diaminopentanoic acid. $C_5H_{12}N_2O_2$; mol wt 132.16. C 45.44%, H 9.15%, N 21.20%, O 24.21%. Non-essential amino acid for human development but is required intermediate in arginine biosynthesis. Found in virtually all vertebrate tissues as well as incorporated into proteins, such as tyrocidine, *q.v.* Isolation from chicken excreta: M. Jaffe, *Ber.* **10**, 1925 (1877). Early chemistry and biochemistry: J. P. Greenstein, M. Winitz, *Chemistry of the Amino Acids* (John Wiley and Sons, Inc., New York, 1961) pp. 2477-2491, *passim.* Review of synthesis: R. Yoshida, *Synth. Prod. Util. Amino Acids*, T. Kaneko *et al.*, Eds (Kodansha Ltd, Tokyo, Japan, 1974) pp 166-170; of biosynthesis in plants: P. D. Shargool *et al.*, *Phytochemistry* **27**, 1571-1574 (1988). Brief historical review: H. A. Akers, E. V. Dromgoole, *Trends Biochem. Sci.* **7**, 156-157 (1982). Review of metabolism: V. E. Shih, *Enzyme* **26**, 254-258 (1981); of role in CNS functions: N. Seiler, G. Daune-Anglard, *Metab. Brain Dis.* **8**, 151-179 (1993).

Colorless crystals. $[\alpha]_D^{25}$ +11.5° (c = 6.5). Aq solns are alkaline. pK$_1'$ 1.94; pK$_2'$ 8.65; pK$_3'$ 10.76. Freely sol in water, alc; sparingly sol in ether.

L-Form aspartate. [3230-94-2] L-Ornithine-L-aspartate; Hepa-Merz. Crystal structure: D. M. Salunke, M. Vijayan, *Int. J. Pept. Protein Res.* **22**, 154 (1983). Clinical trial in hyperammonemia: J. Fehér *et al.*, *Med. Sci. Monit.* **3**, 669 (1997); S. Stauch *et al.*, *J. Hepatol.* **28**, 856 (1998).

DL-Form. [616-07-9] Crystals from water. Sparingly sol in alcohol.

THERAP CAT: In treatment of hyperammonemia.

6970. Ornoprostil. [70667-26-4] (1*R*,2*R*,3*R*)-3-Hydroxy-2-[(1*E*,3*S*,5*S*)-3-hydroxy-5-methyl-1-nonen-1-yl]-ε,5-dioxocyclopentaneheptanoic acid methyl ester; (11α,13*E*,15*S*,17*S*)-11,15-dihydroxy-17,20-dimethyl-6,9-dioxoprost-13-en-1-oic acid methyl ester; 17*S*,20-dimethyl-6-oxoprostaglandin E$_1$ methyl ester; 6-oxo-17*S*,20-dimethyl-PGE$_1$ methyl ester; ronoprost; ONO-1308; OU-1308; Alloca; Ronok. $C_{23}H_{38}O_6$; mol wt 410.55. C 67.29%, H 9.33%, O 23.38%. Analog of prostaglandin E$_1$, *q.v.* Prepn: M. Hayashi *et al.*, **DE 2840032**; *eidem*, **US 4278688** (1979, 1981 both to Ono). Effect on gastric mucosal lesions in rats: H. Kuwata *et al.*, *Nippon Shokakibyo Gakkai Zasshi* **82**, 1858 (1985), *C.A.* **104**, 15709c (1986); *eidem, Curr. Clin. Pract. Ser.* **36**, 243 (1986); on experimental esophagitis: S. Inoue *et al.*, *Gendai Iyro* **19**, 1348 (1987), *C.A.* **107**, 626m (1987).

THERAP CAT: Antiulcerative.

6971. Oroidin. [34649-22-4] *N*-[(2*E*)-3-(2-Amino-1*H*-imidazol-4-yl)-2-propenyl]-4,5-dibromo-1*H*-pyrrole-2-carboxamide. $C_{11}H_{11}Br_2N_5O$; mol wt 389.05. C 33.96%, H 2.85%, Br 41.08%, N 18.00%, O 4.11%. Pyrrole-imidazole alkaloid isolated from marine sponges of the genus, *Agelas* (*Agelasidae*). Considered to be the biosynthetic precursor for more complex sponge alkaloids. Acts as a chemical defense against predation by reef fish and as an antifoulant by inhibiting barnacle larval metamorphosis and the formation of bacterial biofilms. Isoln from *A. oroides*: S. Forenza *et al.*, *J. Chem. Soc. D* **1971**, 1129. Structure: E. E. Garcia *et al.*, *J. Chem. Soc. Chem. Commun.* **1973**, 78. Synthesis: G. De Nanteuil *et al.*, *Bull. Soc. Chim. Fr.* **1986**, 813; T. L. Little, S. E. Webber, *J. Org. Chem.* **59**, 7299 (1994). Total synthesis: C. Schroif-Gregoire *et al.*, *Org. Lett.* **8**, 2961 (2006); N. Ando, S. Terashima, *Synlett* **2006**, 2836. Antifoulant activity: S. Tsukamoto *et al.*, *J. Nat. Prod.* **59**, 501 (1996). Fish antifeedant activity: B. Chanas *et al.*, *J. Exp. Mar. Biol. Ecol.* **208**, 185 (1997). Structure-activity study of fish feeding deterrency: T. Lindel *et al.*, *J. Chem. Ecol.* **26**, 1477 (2000). Inhibits yeast multidrug resistance enzyme, Pdr5p: F. R. da Silva *et al.*, *J. Nat. Prod.* **74**, 279 (2011).

Light yellow oil. uv max (methanol): 278 nm (ε 21000).

Hydrochloride. [203399-38-6] $C_{11}H_{11}Br_2N_5O$.HCl; mol wt 425.51. Yellow-orange solid, mp 202-205° (dec); begins to evolve gas at 80-90°.

Hymenidin. [107019-95-4] *N*-[(2*E*)-3-(2-Amino-1*H*-imidazol-5-yl)-2-propen-1-yl]-4-bromo-1*H*-pyrrole-2-carboxamide. $C_{11}H_{12}$-BrN$_5$O; mol wt 310.16. Naturally occuring monobromo analog of oroidin. Isoln: J. Kobayashi *et al.*, *Experientia* **42**, 1176 (1986). Colorless amorphous solid. uv max (methanol): 270 nm (ε 23000).

6972. Orotic Acid. [65-86-1] 1,2,3,6-Tetrahydro-2,6-dioxo-4-pyrimidinecarboxylic acid; uracil-6-carboxylic acid; whey factor; animal galactose factor; Oropur; Orotyl. $C_5H_4N_2O_4$; mol wt 156.10. C 38.47%, H 2.58%, N 17.95%, O 41.00%. A pyrimidine precursor in animal organisms, found in milk: Bachstez, *Ber.* **64**, 2683 (1931); Hilbert, *J. Am. Chem. Soc.* **54**, 2082 (1932); Johnson, Schroeder, *ibid.* 2942. Synthesis from aspartic acid: Nye, Mitchell, *ibid.* **69**, 1382 (1947). Prepn by condensation of urea with monoethyl ester of oxalacetic acid in methanol: Scriabine, **US 2937175** (1960 to Rhône-Poulenc). Older syntheses from urea and oxalacetic ester: Müller, *J. Prakt. Chem.* **56**, 488 (1897); Behrend, Struve, *Ann.* **378**, 165 (1910). Microbial process by a pyrimidine-requiring *Micrococcus glutamicus* mutant: Kinoshita *et al.*, **US 3086917** (1963 to Kyowa). Classed as a vitamin: Moruzzi *et al.*, *Biochem. Z.* **333**, 318 (1960). Relationship to *vitamin B$_{13}$*: Manna, Hauge, *J. Biol. Chem.* **202**, 91 (1953).

Crystals from water, mp 345-346°.

Monohydrate. [50887-69-9] Lactinium; Oroturic. Crystals, mp 334°. uv max: 282 nm; min: 255 nm. Sol in water ~1.7 mg/ml.

Methyl ester. [6153-44-2] $C_6H_6N_2O_4$. Crystals, mp 249°.

Ethyl ester. [1747-53-1] $C_7H_8N_2O_4$. Crystals, mp 188-189°.

Choline salt. [24381-49-5] Choline orotate; Cholergol. C_5H_4-N$_2$O$_4$.C$_5$H$_{14}$NO; mol wt 260.27.

USE: Has been proposed as feed supplement in combination with methionine to aid growth of calves.

THERAP CAT: Uricosuric.

6973. Orotidine. [314-50-1] 1,2,3,6-Tetrahydro-2,6-dioxo-3-β-D-ribofuranosyl-4-pyrimidinecarboxylic acid; 3-β-D-ribofuranosylorotic acid; 6-carboxyuridine. $C_{10}H_{12}N_2O_8$; mol wt 288.21. C 41.67%, H 4.20%, N 9.72%, O 44.41%. An orotic acid riboside obtained from cultures of *Neurospora crassa* mutants: Michelson *et al., Proc. Natl. Acad. Sci. USA* **37**, 396 (1951). Isoln: Mitchell, Michelson, **US 2788346** (1957 to California Inst. Res. Found.). Structure: Fox *et al., Biochim. Biophys. Acta* **23**, 295 (1957). Synthesis: Curran, Angier, *J. Org. Chem.* **31**, 201 (1966); **US 3282919** (1966 to Am. Cyanamid). Synthesis of orotidine 5′-phosphate: Moffatt, *J. Am. Chem. Soc.* **85**, 1118 (1963).

Needles from methanol + benzene. Turned brown near 200° but failed to melt at 400°. uv max (methanol): 268 nm (ε 8900); in 0.1N HCl: 267 nm (ε 9570); in 0.1N methanolic NaOH: 265 nm (ε 8960). Soluble in hot water, lower aliphatic alcohols and aq solns of such alcohols.

Cyclohexamine salt. $C_{16}H_{25}N_3O_8$. Crystals from ethanol + benzene, mp 183-184°. $[\alpha]_D$ +15° (c = 1). Also isolated as the lead salt.

Orotidine 5′-phosphate trisodium salt trihydrate. $C_{10}H_{10}N_2$-$O_{11}PNa_3.3H_2O$. uv max (0.1N HCl): 267 nm (ε 9430). Soluble in water.

N^3-**Methylorotidine methyl ester.** $C_{12}H_{16}N_2O_8$. Stout crystals from isopropanol, mp 135-137°. uv max (methanol): 271 nm (ε 7620).

6974. Oroxylin A. [480-11-5] 5,7-Dihydroxy-6-methoxy-2-phenyl-4H-1-benzopyran-4-one; 5,7-dihydroxy-6-methoxyflavone; oroxylin. $C_{16}H_{12}O_5$; mol wt 284.27. C 67.60%, H 4.26%, O 28.14%. From root bark of *Oroxylum indicum* Vent., *Bignoniaceae:* Naylor, Chaplin, *Pharm. J.* **20**, 257 (1890); Naylor, Dyer, *J. Chem. Soc.* **79**, 954 (1901); Row *et al., Proc. Indian Acad. Sci.* **28A**, 189 (1948). Structure: Shah *et al., J. Chem. Soc.* **1936**, 591; **1938**, 1555. Synthesis: Murti, Seshadri, *Proc. Indian Acad. Sci.* **29A**, 1 (1949); Sarin, Seshadri, *J. Sci. Ind. Res.* **19B**, 117 (1960); Molho, Gerphagnon, *Bull. Soc. Chim. Fr.* **1963**, 607; Varady, *Tetrahedron Lett.* **1965**, 4281.

Yellow plates from ethanol, mp 231-232°. Sol in alc, acetone, hot benzene, ether, alkalies, glacial acetic acid; sparingly sol in chloroform; practically insol in water.

Diacetyl oroxylin. $C_{20}H_{16}O_7$. Needles from alcohol, mp 131-132°.

6975. Orphenadrine. [83-98-7] N,N-Dimethyl-2-[(2-methylphenyl)phenylmethoxy]ethanamine; N,N-dimethyl-2-(o-methyl-α-phenylbenzyloxy)ethylamine; o-methyldiphenhydramine; o-monomethyldiphenhydramine; 2-(phenyl-o-tolylmethoxy)ethyldimethylamine; phenyl-o-tolylmethyl dimethylaminoethyl ether; β-dimethylaminoethyl 2-methylbenzhydryl ether; BS-5930; Biorphen; Brocasipal. $C_{18}H_{23}NO$; mol wt 269.39. C 80.25%, H 8.61%, N 5.20%, O 5.94%. Prepd by the action of 2-methylbenzhydryl chloride on dimethylaminoethanol: Bijlsma *et al., Arzneim.-Forsch.* **5**, 72 (1955); Harms, Nauta, *J. Med. Pharm. Chem.* **2**, 57 (1960). Covered, but not described in Rieveschl, **US 2567351** (1951 to Parke,

Davis). Resolution of optical isomers: van der Stelt *et al., Arzneim.-Forsch.* **19**, 2010 (1969). Synthesis and pharmacological study of metabolites: Den Besten *et al., ibid.* **20**, 538 (1970). Prepn of dosage forms: Harms, **US 2991225** (1961 to Brocades-Stheeman).

Liquid. bp$_{12}$ 195°.

Hydrochloride. Disipal; Mephenamin. $C_{18}H_{23}NO.HCl$; mol wt 305.85. Crystals, mp 156-157°. Sol in water, alcohol, chloroform. Sparingly sol in acetone, benzene. Practically insol in ether. pH of aq soln about 5.5.

Citrate. [4682-36-4] Banflex; Norflex; X-Otag. $C_{18}H_{23}NO.C_6$-H_8O_7; mol wt 461.51. White, crystalline powder. Sparingly sol in water; slightly sol in alc. Insol in chloroform, benzene, ether.

THERAP CAT: Muscle relaxant (skeletal); antihistaminic.

6976. Orris. White flag. Rhizome of *Iris florentina* L., *I. pallida* Lam. or *I. germanica* L., *Iridaceae. Habit.* Northern Italy, Germany, France. *Constit.* Iridin, irone, ionone, resin, starch, volatile oil (butter of orris). *Review:* W. A. Poucher, *Perfumes, Cosmetics and Soaps* vol. 1 (Chapman & Hall, London, 1974) pp 287-290.

Oil. Orris root oil. *Constit.* About 85% myristic acid; the odorous principle irone; methyl myristate, oleic aldehyde. Yellowish-white to yellow, semisolid fatty substance; violet-like odor. mp 44-50°. Slightly dextrorotatory. Acid no. 213-222. Sapon no. 2-6.

USE: In perfumes and soaps. As dusting powder.

6977. o-Orsellinic Acid. [480-64-8] 2,4-Dihydroxy-6-methylbenzoic acid; 6-methyl-β-resorcylic acid; 4,6-dihydroxy-o-toluic acid; 2,4-dihydroxy-6-methylbenzenecarboxylic acid; orcinolcarboxylic acid. $C_8H_8O_4$; mol wt 168.15. C 57.14%, H 4.80%, O 38.06%. Found in conjugated form or in depside form in *Roccella* and *Lecanora* lichens, postulated to arise by autocondensation of acetoacetic acid. Isoln from the fungus *Chaetomium cochliodes:* Mosback, *Z. Naturforsch.* **14B**, 69 (1959). Synthesis: Sonn, *Ber.* **61**, 926 (1928); Kloss, Clayton, *J. Org. Chem.* **30**, 3566 (1965).

Needles from acetone, mp 176° (effervescence). pK (25°) 3.90. uv max (0.1N HCl): 214, 260, 296 nm; in 0.1N NaOH: 272 nm. Sol in water, alcohol, glycerol. Soly in ether at 20° = 15.7%. Slightly sol in benzene.

Monohydrate. Needles from water, mp 186-189°.

Methyl ester. $C_9H_{10}O_4$. Crystals, mp 140°.

Ethyl ester. $C_{10}H_{12}O_4$. Crystals, mp 132°.

6978. Orthanilic Acid. [88-21-1] 2-Aminobenzenesulfonic acid; o-sulfanilic acid; o-anilinesulfonic acid. $C_6H_7NO_3S$; mol wt 173.19. C 41.61%, H 4.07%, N 8.09%, O 27.71%, S 18.51%. Prepn from o-nitrobenzenesulfonyl chloride: Wertheim, *Org. Synth.* coll. **vol. II**, 271 (1943).

Minute hexagonal plates, dec ~325°. Slow crystallization from water below 13.5° may yield a hemihydrate. pK (25°) 2.48. Slowly and sparingly sol in water.

6979. Orthocaine. [536-25-4] 3-Amino-4-hydroxybenzoic acid methyl ester; methyl 3-amino-4-hydroxybenzoate; Orthoform; Orthoform New. $C_8H_9NO_3$; mol wt 167.16. C 57.48%, H 5.43%, N 8.38%, O 28.71%. Prepd by dissolving 3-amino-4-hydroxybenzoic acid in methanol, saturating with HCl gas: Einhorn, Pfyl, *Ann.* **311**, 46 (1900); **DE 97333**; *Chem. Zentralbl.* **1898**, II, 525. Also prepd by reduction of methyl 3-nitro-4-hydroxybenzoate with aluminum amalgam: Auwers, Röhrig, *Ber.* **30**, 991 (1897); with tin and HCl or with stannous chloride and alcoholic HCl: **DE 97334**; *Chem. Zentralbl.* **1898**, II, 526.

Needles from benzene, mp 143°. When crystallized from chloroform it sometimes assumes an allotropic form, mp 111°, which on melting changes to the normal form and after solidifying always melts at 143°. Odorless and tasteless. Neutral reaction. Almost insol in cold water. Moderately sol in hot water with gradual decompn forming 3-amino-4-hydroxybenzoic acid and methanol. One gram dissolves in 6 ml alcohol, 50 ml ether; readily dissolves in aq NaOH. Forms water sol salts with HCl and HBr.

Note: The name Orthoform was applied originally to the methyl ester of 4-amino-3-hydroxybenzoic acid which was also used as a local anesthetic. This compd now is designated as Orthoform Old.

THERAP CAT: Anesthetic (topical).

THERAP CAT (VET): Anesthetic (topical).

6980. Orthoformic Acid. [463-78-5] Methanetriol; trihydroxymethane. CH_4O_3; mol wt 64.04. C 18.76%, H 6.30%, O 74.95%. Hypothetical compd, illustrated for nomenclature purposes only. The hydrogens of the hydroxyl groups are replaceable with alkyl groups giving rise to orthoformic esters, which are real compds (produced by various syntheses, e.g. from sodium alcoholates and chloroform). Monograph: H. W. Post, *The Chemistry of the Aliphatic Orthoesters* (Reinhold, New York, 1943), 188 pp.

Orthoformic Acid   Methyl Orthoformate

Trimethyl ester. [149-73-5] Trimethoxymethane; methyl orthoformate. $C_4H_{10}O_3$. Prepd from chloroform and methanol in presence of sodium: Sah, Ma, *J. Am. Chem. Soc.* **54**, 2965 (1932). Liquid. d_4^{20} 0.9676; d_4^{25} 0.9623. bp_{760} 100.6°. n_D^{25} 1.3773.

Triethyl ester. [122-51-0] 1,1′,1″-[Methylidynetris(oxy)]tris[ethane]; ethyl orthoformate; triethoxymethane; triethyl orthoformate; Aethone. $C_7H_{16}O_3$; mol wt 148.20. Prepd from chloroform and ethanol in presence of sodium: Chu, Shen, *C.A.* **38**, 2930⁷ (1944). Toxicity study: H. F. Smyth *et al.*, *Arch. Ind. Hyg. Occup. Med.* **4**, 119 (1951). Liquid, sweetish odor resembling that of pine needles. d_4^{20} 0.8909; d_4^{25} 0.8858. bp_{765} 143°. mp < −18°. n_D^{25} 1.3900. Slightly sol in water with decompn; misc with alcohol, ether. LD50 orally in rats: 7.06 g/kg (Smyth).

6981. Orthosulfamuron. [213464-77-8] 2-[[[[[(4,6-Dimethoxy-2-pyrimidinyl)amino]carbonyl]amino]sulfonyl]amino]-*N*,*N*-dimethylbenzamide; 2-[(4,6-dimethoxypyrimidin-2-yl)carbamoylsulfamoylamino]-*N*,*N*-dimethylbenzamide; IR-5878; Strada. $C_{16}H_{20}N_6O_6S$; mol wt 424.43. C 45.28%, H 4.75%, N 19.80%, O 22.62%, S 7.55%. Sulfamoylurea herbicide for use in rice; inhibits the plant enzyme, acetolactate synthase (ALS), which is involved in the synthesis of branched chain amino acids. Prepn: F. Bettarini *et al.*, **WO**

9840361; *eidem*, **US 6329323** (1998, 2001 both to Isagro). ALS inhibition and evaluation of cross-resistance in the aquatic weed, *Cyperus difformis* L.: A. Merotto, Jr. *et al.*, *J. Agric. Food Chem.* **57**, 1389 (2009). Field trial in rice: M. Hasanuzzaman *et al.*, *Aust. J. Crop Sci.* **2**, 18 (2008).

Fine white powder, mp 160-162°. d^{20} 1.45. Log P (octanol/water): 2.0 (pH 4); 1.3 (pH 7). Soly in water at 20° (g/l): 0.062 (pH 4); 0.63 (pH 7); 39 (pH 8.5). Soly at 20° (g/l): acetone 20; ethyl acetate 3.3; dichloromethane 56; methanol 8.3.

USE: Herbicide.

6982. Oryzalin. [19044-88-3] 4-(Dipropylamino)-3,5-dinitrobenzenesulfonamide; 3,5-dinitro-N^4,N^4-dipropylsulfanilamide; EL-119; Surflan. $C_{12}H_{18}N_4O_6S$; mol wt 346.36. C 41.61%, H 5.24%, N 16.18%, O 27.72%, S 9.26%. Selective pre-emergence herbicide. Prepn: Q. F. Soper, **US 3367949** (1968 to Lilly). Soil degradation: T. Golab *et al.*, *Pestic. Biochem. Physiol.* **5**, 196 (1975); E. W. Stoller, L. M. Wax, *J. Environ. Qual.* **6**, 124 (1977). *Review:* O. D. Decker, W. S. Johnson, *Anal. Methods Pestic. Plant Growth Regul.* **8**, 433 (1976).

Yellow-orange crystals, mp 137-138° (tech, 141-142°). Soly in water at 25°: 2.5 ppm. Sol in acetone, ethanol, methanol, acetonitrile; slightly sol in benzene. Practically insol in hexane. Vapor pressure at 30°: <1 × 10⁻⁷ mm Hg. LD50 orally in rats: >10 g/kg (Decker, Johnson).

USE: Herbicide.

6983. γ-Oryzanol. [11042-64-1] OZ; γ-OZ; γ-orizanol; Caclate; Gammajust 50; Gamma-OZ; Gammariza; Gammatsul; Guntrin; Hi-Z; Maspiron; Oliver; Oryvita; Oryzaal; Thiaminogen. Mixture of ferulic acid esters of sterols (campestrol, stigmasterol, β-sitosterol) and triterpene alcohols (cycloartanol, cycloartenol, 24-methylenecycloartanol, cyclobranol) extracted from rice bran, corn and barley oils. Isoln from rice bran oil: R. Kaneko, T. Tsuchiya, *J. Chem. Soc. Jpn. Ind. Chem. Sect.* **57**, 526 (1954); T. Tsuchiya *et al.*, **JP 56 7182** (1956 to Agcy Ind. Sci. Technol.); *C.A.* **52**, 15848b (1958); *eidem*, **JP 57 4895** (1957 to Bureau Ind. Technics); *C.A.* **52**, 5758i (1958); from corn oil: T. Tsuchiya, O. Okubo, **JP 60 2945** (1960 to Bureau Ind. Technics); *C.A.* **54**, 25609g (1960). Extraction from rice bran, corn and barley oils and purification: T. Yamamoto, **DE 1301002** (1969 to Toyo Koatsu); *C.A.* **71**, 128704r (1969). Separation of 3 major components designated oryzanols A, B and C: M. Shimizu *et al.*, *Chem. Pharm. Bull.* **5**, 36 (1957). Structure of A: G. Ohta, M. Shimizu, *ibid.* **40**, 7 (1958); of C: *eidem*, *ibid.* **6**, 325 (1958); G. Ohta, *ibid.* **8**, 5, 9 (1960). Oryzanol B was subsequently found to be a mixture of A and C: M. Shimizu, G. Ohta, *ibid.* 108. Separation and properties of the seven components of γ-oryzanol: T. Endo *et al.*, *Yukagaku* **17**, 344 (1968); *C.A.* **69**, 37319m (1968); *eidem*, *ibid.* **18**, 63 (1969); *C.A.* **70**, 89042f (1969). General pharmacology: Y. Yamaji *et al.*, *Oyo Yakuri* **25**, 947 (1983); *C.A.* **99**, 151730h (1983). Antioxidant activity in cytochrome model systems: K. Tajima *et al.*, *Biochem. Biophys. Res. Commun.* **115**, 1002 (1983). Effect on lipid metabolism in rats: M. Shinomiya, *Tohoku J. Exp. Med.* **141**, 191

(1983). Antiulcerative effect on gastric lesions in mice: Y. Ichimaru *et al.*, *Nippon Yakurigaku Zasshi* **84**, 537 (1984); *C.A.* **102**, 55922g (1985). Clinical evaluation in chronic gastritis: T. Arai, *Kitakanto Igaku* **30**, 71, 85 (1980); in hyperlipidemia: G. Yoshino *et al.*, *Curr. Ther. Res.* **45**, 543 (1989).

Oryzanol A

White or slightly yellowish, tasteless powder with little or no odor. Crystals from acetone, mp 135-137°. uv max (heptane): 216, 231, 291, 315 nm.

Oryzanol A. [21238-33-5] (3β)-9,19-Cyclolanost-24-en-3-ol 3-(4-hydroxy-3-methoxyphenyl)-2-propenoate; cycloartenyl ferulate. $C_{40}H_{58}O_4$; mol wt 602.90. Prepd as the monohydrate, mp 150-151.5°. $[\alpha]_D$ +40° (c = 0.68). uv max (heptane): 231, 290, 315 nm (log ε 4.15, 4.24, 4.34).

Oryzanol C. [469-36-3] (3β)-24-Methylene-9,19-cyclolanostan-3-ol 3-(4-hydroxy-3-methoxyphenyl)-2-propenoate; 24-methylenecycloartanyl ferulate. $C_{41}H_{60}O_4$; mol wt 616.93.

THERAP CAT: Antiulcerative; antilipemic. Also used in treatment of menopausal syndrome.

6984. Osalmid. [526-18-1] 2-Hydroxy-*N*-(4-hydroxyphenyl)benzamide; 4′-hydroxysalicylanilide; *N*-(*p*-hydroxyphenyl)salicylamide; *N*-salicoylaminophenol; oksafenamide; oxaphenamide; Driol; Jestmin; Kanochol; Saryuurin; Yoshicol. $C_{13}H_{11}NO_3$; mol wt 229.24. C 68.11%, H 4.84%, N 6.11%, O 20.94%. Prepn: Weizmann *et al.*, *J. Org. Chem.* **13**, 796 (1948). Description: *Subsidia Med.* **8**, 103 (1956). HPLC determn in plasma: T. Sadanaga *et al.*, *J. Chromatogr.* **223**, 243 (1981).

Crystals, mp 179°. Practically insol in cold water, acetic acid. Slightly sol in warm water, benzene, toluene. Freely sol in methanol, ethanol, ether, acetone.

Diacetate. $C_{17}H_{15}NO_5$. Needles from alcohol, mp 151°.

THERAP CAT: Choleretic.

6985. Osaterone. [105149-04-0] (4a*R*,4b*S*,6a*S*,7*R*,9a*S*,9b*R*)-7-Acetyl-11-chloro-4a,4b,5,6,6a,7,8,9,9a,9b-decahydro-7-hydroxy-4a,6a-dimethyl-cyclopenta[5,6]naphtho[1,2-*c*]pyran-2(4*H*)-one; 6-chloro-17-hydroxy-2-oxapregna-4,6-diene-3,20-dione. $C_{20}H_{25}$-ClO_4; mol wt 364.87. C 65.84%, H 6.91%, Cl 9.72%, O 17.54%. Androgen receptor antagonist. Prepn: K. Shibata *et al.*, **EP 193871**; *eidem*, **US 4785103** (1986, 1988 both to Teikoku Hormone); K. Shibata *et al.*, *Chem. Pharm. Bull.* **40**, 935 (1992). Pharmacology: M. Murakoshi *et al.*, *Acta Pathol. Jpn.* **42**, 151 (1992); T. Ichikawa *et al.*, *Endocr. J.* **40**, 425 (1993); M. Murakoshi *et al.*, *ibid.* 479.

Colorless prisms from acetone/hexane, mp 218-221°.

Acetate ester. [105149-00-6] 2-Oxachlormadinone acetate; TZP-4238. $C_{22}H_{27}ClO_5$; mol wt 406.90. Colorless prisms from acetone/hexane, mp 253-255°.

THERAP CAT: Treatment of benign prostatic hypertrophy; antiandrogen.

6986. Oseltamivir. [196618-13-0] (3*R*,4*R*,5*S*)-4-(Acetylamino)-5-amino-3-(1-ethylpropoxy)-1-cyclohexene-1-carboxylic acid ethyl ester. $C_{16}H_{28}N_2O_4$; mol wt 312.41. C 61.51%, H 9.03%, N 8.97%, O 20.48%. Orally active inhibitor of influenza virus neuraminidase; converted *in vivo* to the active acid metabolite, **GS-4071**. Prepn: N. W. Bischofberger *et al.*, **US 5763483** (1998 to Gilead Sci.). Improved prepn: Y.-Y. Yeung *et al.*, *J. Am. Chem. Soc.* **128**, 6310 (2006). Degradation kinetics: R. Oliyai *et al.*, *Pharm. Res.* **15**, 1300 (1998). LC-MS/MS determn in biological fluids: K. Heinig, F. Bucheli, *J. Chromatogr. B* **876**, 129 (2008). Bioavailability and pharmacokinetics: W. Li *et al.*, *Antimicrob. Agents Chemother.* **42**, 647 (1998). Review of structure-activity studies and clinical development: C. U. Kim, *Med. Chem. Res.* **8**, 392-399 (1998). Review of syntheses: J. Gong, W. Xu, *Curr. Med. Chem.* **15**, 3145-3159 (2008). Clinical trial in prevention of influenza: F. G. Hayden *et al.*, *N. Engl. J. Med.* **341**, 1336 (1999); in treatment of acute influenza: J. J. Treanor *et al.*, *J. Am. Med. Assoc.* **283**, 1016 (2000). Review of clinical experience: P. Schirmer, M. Holodniy, *Expert Opin. Drug Saf.* **8**, 357-371 (2009).

Pale solid. pKa 7.7 (25°); 6.6 (70°).

Phosphate. [204255-11-8] GS-4104; Tamiflu. $C_{16}H_{28}N_2O_4$.-H_3PO_4; mol wt 410.40. White crystalline solid.

THERAP CAT: Antiviral.

6987. Osmaron B. [8031-66-1] Composed of benzoates of primary aliphatic fatty amines obtained from palm kernel oil and corn oil as starting materials. Preponderantly dodecylammonium benzoate and tetradecylammonium benzoate. Total nitrogen 4.15% by analysis; average benzoic acid content 27.6% by analysis.

Yellowish-brown, thick oil. mp −4 to −3°. n_D^{40} 1.4855-1.4885. Vapors of Osmaron B turn wet litmus paper blue and turmeric paper brown. Soluble in methanol, ethanol, ethylene glycol, glycerol, acetone, ethyl acetate, carbon disulfide, formic acid, acetic acid, and lactic acid. Forms turbid solns with ether, gasoline, benzene and its homologs, and chlorinated hydrocarbons which become clear upon deposition of the fine precipitate on standing. *Ref:* E. Benk, *Chem. Ztg.* **75**, 351 (1951).

USE: Disinfectant; cationic surface-active agent; in udder ointments used with milking machines.

6988. Osmium. [7440-04-2] Os; at. wt 190.23; at. no. 76; valences 1-8; most common states 3, 4, 6. Group VIII (8). Seven naturally occurring isotopes: 184 (0.02%); 186 (1.6%); 187 (1.6%); 188 (13.3%); 189 (16.1%); 190 (26.4%); 192 (41.0%); artificial radioactive isotopes: 181-183; 185; 191; 193-195. Occurrence in earth's crust ~0.001 ppm. Found in the mineral osmiridium and in all platinum ores. Discovered by Tennant in 1804. Prepn: Berzelius *et al.*, cited by Mellor, *A Comprehensive Treatise on Inorganic and Theoretical Chemistry* **15**, 687 (1936). Reviews of prepn, properties and chemistry of osmium and other platinum metals: Gilchrist, *Chem. Rev.* **32**, 277-372 (1943); Beamish *et al.* in *Rare Metals Handbook*, C. A. Hampel, Ed. (Reinhold, New York, 1956) pp 291-328; Griffith, *Q. Rev. Chem. Soc.* **19**, 254-273 (1965); *idem*, *The Chemistry of the Rarer Platinum Metals* (John Wiley, New York, 1967) pp 1-125; Livingstone in *Comprehensive Inorganic Chemistry* **vol. 3**, J. C. Bailar, Jr. *et al.*, Eds. (Pergamon Press, Oxford, 1973) pp 1163-1189, 1209-1233.

Bluish-white, lustrous metal; close-packed hexagonal structure. d_4^{20} 22.61; long believed to be the densest element; x-ray data show it to be slightly less dense than iridium. mp ~2700°. bp ~5500°. Sp heat (0°) 0.0309 cal/g/°C. Hardness 7.0 on Mohs' scale. Electrical resistivity (0°) 8.12 μohms-cm. Stable in air in the cold; when finely divided, is slowly oxidized by air even at ordinary temp to form tetroxide. Attacked by fluorine above 100°; by dry chlorine on heating; not attacked by bromine or iodine. Attacked by aqua regia, by oxidizing acids over a long period of time; barely affected by HCl, H_2SO_4. Burns in vapor of phosphorus to form a phosphide, in vapor of sulfur to form a sulfide. Attacked by molten alkali hydrosulfates, by potassium hydroxide and oxidizing agents. Finely divided osmium absorbs a considerable amount of hydrogen.

Osmarins. High-molecular weight polymers of carbohydrate and osmium. Potential use in the treatment of arthritis: *Chem. Eng. News* **60**, 8 (April 5, 1982).

USE: As alloy with iridium for pen points and fine machine bearings; as catalyst in the synthesis of ammonia; as catalyst in hydrogenation of organic compounds.

6989. Osmium Hexafluoride. [13768-38-2] *(OC-6-11)*-Osmium fluoride (OsF_6). F_6Os; mol wt 304.22. F 37.47%, Os 62.53%. OsF_6. Prepd by fluorination of osmium metal: Weinstock, Malm, *J. Am. Chem. Soc.* **80**, 4466 (1958); Hargreaves, Peacock, *Proc. Chem. Soc. London* **1959**, 85. Previously thought to be *osmium octafluoride:* Ruff, Tschirch, *Ber.* **46**, 929 (1913).

Pale yellow, volatile solid. *Highly poisonous! Very corrosive to skin!* mp 32.1°; bp 45.9°. Also reported: mp 33.4°; bp 47.5°. *See:* Cady, Hargreaves, *J. Chem. Soc.* **1961**, 1563. Hydrolyzed on exposure to moisture; forms white corrosive fumes which soon turn bluish. May be stored in quartz ampuls.

6990. Osmium Tetroxide. [20816-12-0] *(T-4)*-Osmium oxide (OsO_4); osmic acid. O_4Os; mol wt 254.23. O 25.17%, Os 74.83%. OsO_4. Prepd by heating (at 300-400°) finely divided osmium metal in a stream of air or oxygen. Lab prepn: Grube in *Handbook of Preparative Inorganic Chemistry* vol. **2**, G. Brauer, Ed. (Academic Press, New York, 2nd ed., 1965) pp 1603-1604. Use in treatment of arthritis: M. Nissilä *et al., Scand. J. Rheumatol.* **5**, 111 (1977); A. S. Hendricson *et al., Acta Orthop. Scand.* **52**, 17 (1982). Review of chemistry and biochemistry: W. P. Griffith, *Platinum Met. Rev.* **18**, 94-96 (1974); of toxicity: E. Browning, *Toxicity of Industrial Metals* (Appleton-Century-Crofts, New York, 2nd ed., 1969) pp 261-266.

Pale yellow solid; monoclinic crystals. *Vapor poisonous. Safeguards necessary when opening container.* Acrid, chlorine-like odor. Minimum perceptible concn 0.02 mg/liter of air. mp 40.6°. d 5.10 (calc): Ueki *et al., Acta Crystallogr.* **19**, 157 (1965). bp_{760} 130.0°; bp_{400} 109.3°; bp_{200} 89.5°; bp_{100} 71.5°; bp_{60} 59.4°. Begins to sublime and distil well below the boiling point. Vapor press at 27°: 11 mm. Critical temp 405°; crit press. 170 atm. Sol in benzene. Soly at 25° (g/100 g): water 7.24; carbon tetrachloride 375: Anderson, Yost, *J. Am. Chem. Soc.* **60**, 1822 (1938). Also sol in alc, ether, ammonium hydroxide, phosphorus oxychloride.

Caution: Potential symptoms of overexposure are irritation of eyes and respiratory system; lacrimation, visual disturbance; conjunctivitis; headache; coughing; dyspnea; dermatitis. *See NIOSH Pocket Guide to Chemical Hazards* (DHHS/NIOSH 97-140, 1997) p 238.

USE: Oxidizing agent, particularly for converting olefins to glycols. Catalyzes chlorate, peroxide, periodate, and other oxidations: P. N. Rylander, *Organic Syntheses with Noble Metal Catalysts* (Academic Press, New York, 1973) pp 121-144. As fixing and staining agent for cell and tissue studies.

6991. Ospemifene. [128607-22-7] 2-[4-[(1Z)-4-Chloro-1,2-diphenyl-1-buten-1-yl]phenoxy]ethanol; (deaminohydroxy)toremifene; FC-1271a; Ophena. $C_{24}H_{23}ClO_2$; mol wt 378.90. C 76.08%, H 6.12%, Cl 9.36%, O 8.44%. Selective estrogen receptor modulator (SERM); metabolite of toremifene, *q.v.* Prepn: R. J. Toivola *et al.*, **EP 95875** (1986 to Farmos); and use in osteoporosis: M. Degregorio *et al.*, **WO 9607402** (1996 to Orion). Pharmacology: L. Kangas, *Cancer Chemother. Pharmacol.* **27**, 8 (1990). HPLC determn in plasma: T. L. Taras *et al., J. Chromatogr. B* **724**, 163 (1999). Clinical pharmacokinetics: M. W. DeGregorio *et al., Eur. J. Clin. Pharmacol.* **56**, 469 (2000). Clinical trial in postmenopausal vulvovaginal atrophy: G. A. Bachmann *et al., Menopause* **17**, 480 (2010). Review of clinical experience: L. Gennari *et al., Expert Opin. Invest. Drugs* **18**, 839-849 (2009).

Crystals from ethanol + water. Sol in methanol.

THERAP CAT: In treatment of postmenopausal vaginal atrophy.

6992. Osteocalcin. [75757-02-7] (human reduced). Bone Gla protein; BGP. Abundant noncollagenous, calcium-binding protein of bone; constitutes ~1-2% of the total bone matrix. Also found in tooth dentin and cementum. Contains 46-50 amino acid residues (human: 49); mol wt ~5800 daltons. Amino acid sequence is highly conserved across species. Characterized by 3 residues of the vitamin K-dependent amino acid, γ-carboxyglutamic acid (Gla), *q.v.* Synthesized by osteoblasts and odontoblasts; modulated by calcitriol, *q.v.* Detectable in serum as an indicator of bone metabolism. Identification in chicken bone: P. V. Hauschka *et al., Proc. Natl. Acad. Sci. USA* **72**, 3925 (1975). Characterization of calf BGP: P. A. Price *et al. ibid.* **73**, 1447 (1976). Biosynthesis in bone cell culture: S. K. Nishimoto, P. A. Price, *J. Biol. Chem.* **254**, 437 (1979). RIA determn in plasma: P. A. Price *et al., J. Clin. Invest.* **66**, 878 (1980). Review of isoln and characterization: C. M. Gundberg *et al., Methods Enzymol.* **107B**, 516-544 (1984). Review of hydroxyapatite binding characteristics: P. V. Hauschka, F. H. Wians, Jr. *Anat. Rec.* **224**, 180-188 (1989). Comprehensive review: P. V. Hauschka *et al., Physiol. Rev.* **69**, 990-1047 (1989). Review of diagnostic methods and clinical applications: M. J. Power, P. F. Fottrell, *Crit. Rev. Clin. Lab. Sci.* **28**, 287-335 (1991). Immunohistochemical study and potential role in bone mineralization and resorption: H. I. Roach, *Cell Biol. Int.* **18**, 617-628 (1994).

Very acidic, pI 4.0. Selective affinity for insoluble Ca^{2+} salts; binds tightly to hydroxyapatite.

6993. Osthole. [484-12-8] 7-Methoxy-8-(3-methyl-2-butenyl)-2H-1-benzopyran-2-one; 7-methoxy-8-(3-methyl-2-butenyl)coumarin; 8-(3-methyl-2-butenyl)herniarin. $C_{15}H_{16}O_3$; mol wt 244.29. C 73.75%, H 6.60%, O 19.65%. From rhizome of *Peucedanum ostruthium* (L.) Koch *(Imperatoria ostruthium L.) Umbelliferae:* Herzog, Krohn, *Arch. Pharm.* **247**, 553 (1909); Butenandt, Marten, *Ann.* **495**, 187 (1932); from *Prangos pabularia* Lindl., *Umbelliferae:* Pigulevskii, Kuznetsova, *Dokl. Akad. Nauk SSSR* **61**, 309 (1948), *C.A.* **43**, 3416d (1949); from *Flindersia bennettiana* F. Muell., *Rutaceae:* Galbraith *et al., Aust. J. Chem.* **13**, 427 (1960). Structure: Späth, Pesta, *Ber.* **66**, 754 (1933). Synthesis: Späth, Holzen, *ibid.* **67**, 264 (1934); Murayama *et al., Chem. Pharm. Bull.* **20**, 741 (1972).

Prisms from ether. mp 83-84°. uv max: 322, 258 nm (ε 8000, 4300). Practically insol in water; sol in aq alkalies, alcohol, chloroform, acetone, boiling petr ether.

6994. Ostruthin. [148-83-4] 6-[(2E)-3,7-Dimethyl-2,6-octadien-1-yl]-7-hydroxy-2H-1-benzopyran-2-one; (E)-6-(3,7-dimethyl-2,6-octadienyl)-7-hydroxycoumarin; 6-(3,7-dimethyl-2,6-octadien-yl)umbelliferone. $C_{19}H_{22}O_3$; mol wt 298.38. C 76.48%, H 7.43%, O 16.09%. From the root of *Peucedanum ostruthium* (L.) Koch *(Imperatoria ostruthium L.), Umbelliferae:* Butenandt, Marten, *Ann.* **495**, 187 (1932); from *Eriostemon tomentellus, Rutaceae:*

Duffield, Jeffries, *Aust. J. Chem.* **16**, 123 (1963). Structure: Späth, Klager, *Ber.* **67**, 859 (1934); Späth, Kainrath, *ibid.* **70**, 2272 (1937).

Crystals from alcohol, mp 117-119°. Practically insol in water, petr ether. Sol in chloroform, ethyl acetate, hot alc.

Acetate. $C_{21}H_{24}O_4$. Leaflets from alcohol, mp 81°. Practically insol in water. Sol in alc, ether, benzene, chloroform.

6995. Ostruthol. [642-08-0] (2*Z*)-2-Methyl-2-butenoic acid (1*R*)-2-hydroxy-2-methyl-1-[[(7-oxo-7*H*-furo[3,2-*g*][1]benzopyran-4-yl)oxy]methyl]propyl ester. $C_{21}H_{22}O_7$; mol wt 386.40. C 65.28%, H 5.74%, O 28.98%. From rhizome of *Peucedanum ostruthium* L. Koch (*Imperatoria ostruthium* L.), *Umbelliferae:* Herzog, Koch, *Arch. Pharm.* **247**, 553 (1909). Structure: Späth, Christiani, *Ber.* **66**, 1150 (1933). Synthesis: Chatterjee, Dutta, *Sci. Cult.* **34**, 460 (1968).

Silky needles from benzene, mp 136-137°. $[\alpha]_D^{15}$ −18.3° (pyridine). Slightly sol in water or in ether; sol in hot alcohol, benzene, toluene, pyridine.

6996. Otamixaban. [193153-04-7] (*αR*)-3-(Aminoiminomethyl)-*α*-[(1*R*)-1-[[4-(1-oxido-4-pyridinyl)benzoyl]amino]ethyl]-benzenepropanoic acid methyl ester; methyl (2*R*,3*R*)-2-[3-amidino-benzyl]-3-[[4-(1-oxido-4-pyridinyl)benzoyl]amino]butanoate; (2*R*)-(3-carbamoylimidoylbenzyl)-(3*R*)-[4-(1-oxypyridin-4-yl)benzoyl-amino]butryic acid methyl ester; FXV-673; RPR-130673; XRP-0673. $C_{25}H_{26}N_4O_4$; mol wt 446.51. C 67.25%, H 5.87%, N 12.55%, O 14.33%. Direct inhibitor of Factor Xa. Prepn: K. R. Guertin *et al.*, **WO 97 24118**; S. I. Klein *et al.*, **US 6080767** (1997, 2000 both to Rhone Poulenc Rorer); K. R. Guertin *et al.*, *Bioorg. Med. Chem. Lett.* **12**, 1671 (2002). Pharmacology: V. Chu *et al.*, *Thromb. Res.* **103**, 309 (2001). Clinical pharmacokinetics: A. Paccaly *et al.*, *J. Clin. Pharmacol.* **46**, 37 (2006); and evaluation in coronary artery disease: M. Hinder *et al.*, *Clin. Pharmacol. Ther.* **80**, 691 (2006). *Review*: E. A. Nutescu, K. Pater, *IDrugs* **9**, 854-865 (2006).

THERAP CAT: Antithrombotic.

6997. Otilonium Bromide. [26095-59-0] *N*,*N*-Diethyl-*N*-methyl-2-[[4-[[2-(octyloxy)benzoyl]amino]benzoyl]oxy]ethanaminium bromide (1:1); octylonium bromide; SP-63; Doralin; Menoctyl; Pasminox; Spasen; Spasmoctyl; Spasmomen. $C_{29}H_{43}BrN_2O_4$; mol wt 563.58. C 61.80%, H 7.69%, Br 14.18%, N 4.97%, O 11.36%. Quaternary ammonium salt with gastrointestinal spasmolytic activity. Prepn: **BE 704269**; M. Ghelardoni *et al.*, **US 3536723** (1968, 1970 both to Menarini); *idem et al.*, *J. Med. Chem.* **16**, 1063 (1973).

Determn of finished pharmaceutical forms by HPLC: C. Mannucci *et al.*, *J. Pharm. Sci.* **82**, 367 (1993); by CE: S. Furlanetto *et al.*, *Analyst* **126**, 1700 (2001). Receptor binding study: S. Lindqvust *et al.*, *Br. J. Pharmacol.* **137**, 1134 (2002). Clinical trial in irritable bowel syndrome: G. Battaglia *et al.*, *Aliment. Pharmacol. Ther.* **12**, 1003 (1998). Review of pharmacology and clinical efficacy: S. Evangelista, *J. Int. Med. Res.* **27**, 207-222 (1999); comparative review of clinical studies: *idem, Curr. Pharm. Des.* **10**, 3561-3568 (2004).

Crystals from ethanol, mp 166-168°. LD_{50} in mice (mg/kg) 87.0 i.p. (Ghelardoni, 1973).

THERAP CAT: Antispasmodic.

6998. Otobain. [3738-01-0] (7*S*,8*R*,9*R*)-9-(1,3-Benzodioxol-5-yl)-6,7,8,9-tetrahydro-7,8-dimethylnaphtho[1,2-*d*]-1,3-dioxole; 5,6-methylenedioxy-2,3-dimethyl-4-(3′,4′-methylenedioxyphenyl)-1,2,3,4-tetrahydronaphthalene; otobite. $C_{20}H_{20}O_4$; mol wt 324.38. C 74.06%, H 6.21%, O 19.73%. Lignan isolated from otoba butter, the fat expressed from the fruit of *Myristica otoba* Humb. & Bonpl., *Myristicaceae.* Isoln: W. F. Baughman *et al.*, *J. Am. Chem. Soc.* **43**, 199 (1921). Structure: T. Gilchrist *et al.*, *J. Chem. Soc.* **1962**, 1780; N. S. Bhacca, R. Stevenson, *J. Org. Chem.* **28**, 1638 (1963). Absolute configuration: W. Klyne *et al.*, *J. Chem. Soc.* **C 1966**, 893. Synthesis of (±)-form: I. Maclean, R. Stevenson, *ibid.* 1717; T. B. H. McMurry, H. K. Kennedy-Skipton, *Tetrahedron Lett.* **1966**, 975.

Needles from ethanol, mp 137-138°. $[\alpha]_D$ −40.5° (c = 3.2 in chloroform). uv max (ethanol): 234, 287 nm (ε 9300, 6700). Sol in ether, hot alcohol. Practically insol in water.

6999. Ouabagenin. [508-52-1] (1*β*,3*β*,5*β*,11*α*)-1,3,5,11,-14,19-Hexahydroxycard-20(22)-enolide; G-strophanthidin. C_{23}-$H_{34}O_8$; mol wt 438.52. C 63.00%, H 7.82%, O 29.19%. Prepn from ouabain with HCl in cold acetone: Mannich, Siewert, *Ber.* **75**, 737 (1942). Review of structure: Reichstein, Reich, *Annu. Rev. Biochem.* **15**, 155 (1946). Proof of this structure: Tamm *et al.*, *Helv. Chim. Acta* **40**, 1469 (1957); *Experientia* **13**, 185 (1957); Turner, Meschino, *J. Am. Chem. Soc.* **80**, 4862 (1958); Volpp, Tamm, *Helv. Chim. Acta* **42**, 1408, 1418 (1959).

Monohydrate. Clusters of needles from water, mp 235-238°. One gram dissolves in about 10 ml boiling water. At room temp the soly in water is <1%. Also sol in dil alcohol. Practically insol in abs alcohol, ether, and chloroform. Becomes anhydr at 100° *in vacuo* over P_2O_5. The anhydr compd is hygroscopic, mp 255-256°. $[\alpha]_D^{17}$ +11.3° (c = 1.27).

Tetraacetylouabagenin trihydrate. $C_{31}H_{42}O_{12} \cdot 3H_2O$. Needles from 10% alcohol, mp 282-285°.

Dihydroouabagenin. $C_{23}H_{36}O_8 \cdot CH_3OH$. Solvated crystals from methanol + ether contg 1 mol methanol, mp 261°.

7000. Ouabain. [630-60-4] $(1\beta,3\beta,5\beta,11\alpha)$-3-[(6-Deoxy-$\alpha$-L-mannopyranosyl)oxy]-1,5,11,14,19-pentahydroxycard-20(22)-enolide; G-strophanthin; Gratus strophanthin; acocantherin. $C_{29}H_{44}O_{12}$; mol wt 584.66. C 59.58%, H 7.59%, O 32.84%. Obtained from the seeds of *Strophanthus gratus* (Wall. & Hock.) Baill.; also occurs in *Acokanthera ouabaio* Cathel and other *A.* spp, *Apocynaceae*. Isoln: Schwartze *et al.*, *J. Pharmacol.* **36**, 481 (1929). Hydrolysis yields one mol ouabagenin and one mol rhamnose: Jacobs, Bigelow, *J. Biol. Chem.* **96**, 647 (1932). Toxicity study: Small *et al.*, *Toxicol. Appl. Pharmacol.* **20**, 44 (1971). *See also* Fieser, Fieser, *Steroids* (1959, Reinhold, New York; Chapman & Hall, London) pp 768, 772; Reichstein, Reich, *Annu. Rev. Biochem.* **15**, 155 (1946).

Octahydrate. Purostrophan; Strodival; Strophoperm. Shiny plates (from water) which give up their water of crystn at 130°. When anhydr dec about 190°. $[\alpha]_D^{25}$ −31 to −32.5° (c = 1 calcd as anhydr). Stable in air, but affected by light. One gram dissolves in about 75 ml water, in 5 ml boiling water, in 100 ml alcohol, in 8 ml boiling alcohol. Also sol in amyl alcohol, dioxane. Slightly sol in ether, chloroform, ethyl acetate. Aq solns are neutral to litmus. LD_{50} i.v. in rats: 14 mg/kg (Small).

THERAP CAT: Cardiotonic.

THERAP CAT (VET): Cardiotonic, diuretic.

7001. Ovalbumin. Egg albumin. The major protein constituent (75%) of egg white from hen's eggs. Mol wt about 45,000. Produced under hormonal control by the bird oviduct. May be isolated and crystallized readily from the filtrate of an acidified mixture of egg white and an equal volume of satd. ammonium sulfate: Sorensen, Hoyrup, *C. R. Trav. Lab. Carlsberg* **12**, 12 (1917). Alternate method: Kekwick, Cannan, *Biochem. J.* **30**, 227 (1936). Can be separated by electrophoresis and chromatography from about ten other minor components including avidin, lysozyme, conalbumin, *q.q.v.*, and ovomucoid. Structure is a complex protein consisting of a single polypeptide chain of about 400 residues (about half of which are hydrophobic), a maximum of two phosphate residues per mole, and an oligosaccharide side chain composed of only mannose and glucosamine residues. Sequences of *N*- and *C*-terminal segments: Narita, Ishii, *J. Biochem.* **52**, 367 (1962); Thompson *et al.*, *Aust. J. Biol. Sci.* **24**, 525 (1971). *Reviews:* R. C. Warner in Neurath-Bailey, *The Proteins* vol. II, part A (New York, 1954) p 443 sqq; Taborsky, *Adv. Protein Chem.* **28**, 34-50 (1974).

Needles or elongated prisms, frequently forming rosettes. The crystals usually contain 2 mols protein and 3 mols H_2SO_4. $[\alpha]_D^{20}$ −30.7°. Coagulation temp 56°. Sol in electrolyte-free water. Combines with salts, acids, and bases. Denaturation can be induced by heating to 56°, by vigorous shaking, by electric current, and by various chemicals, such as acids, ammonia salts, heavy metal salts, and alcohols. All methods produce complete and irreversible denaturation. Isoelectric point: 4.63.

7002. Oxaceprol. [33996-33-7] $(4R)$-1-Acetyl-4-hydroxy-L-proline; 4-hydroxy-*N*-acetylproline; 1-acetyl-4-hydroxy-2-pyrrolidinecarboxylic acid; CO-61; AHP-200; Jonctum. $C_7H_{11}NO_4$; mol wt 173.17. C 48.55%, H 6.40%, N 8.09%, O 36.96%. Derivative of hydroxyproline having anti-inflammatory activity. Prepn: R. L. M. Synge, *Biochem. J.* **33**, 1924 (1939); J. J. Kolb, G. Toennies, *J. Biol. Chem.* **144**, 193 (1942); O. Leonardo *et al.*, **DE 2301358** corresp to **US 3860607** (1973, 1975 to Richardson-Merrell). Elucidation of cyclic conformation and *cis/trans* isomerism about the amide bond: T. Prange *et al.*, *Biochem. Biophys. Res. Commun.* **61**, 104 (1974). Crystal structure: M. Hospital *et al.*, *Biopolymers* **18**, 1141 (1979). In crystals, the acetyl group is in the *trans* conformation and the ring is puckered. The *trans* form is more stable than the *cis* form in solutions: W. A. Thomas, M. K. Williams, *J. Chem. Soc. Chem. Commun.* **1972**, 994. Anti-inflammatory and wound-healing activity in animals: P. Coirre, B. Coirre, **GB 1246141** (1971) related to P. Coirre *et al.*, **US 3891765** and **US 3932638** (1975, 1976 both to Franco Chimie). Clinical studies of use in several rheumatic conditions: P. Grellat, *Rhumatologie* **27**, 223 (1975); in degenerative joint disease: R. Schubotz, L. Hausmann, *Therapiewoche* **27**, 4248 (1977); in treatment of burns, tumors and other wounds (applied locally): Y. Privat, *Gaz. Med. Fr.* **84**, 618 (1977).

Crystals from acetone, mp 133-134° (Synge); also reported as mp 126-128° (Leonardo). $[\alpha]_D^{20}$ −116.5° (c = 3.2); $[\alpha]_D^{18}$ −119.5° (c = 3.75). Very sol in alcohol. Sol in water, methanol. Insol in ether, chloroform.

Monohydrate. $C_7H_{13}NO_5$. Crystals from moist ethyl acetate or acetone, mp 74-76°.

Zinc salt. $C_{14}H_{20}N_2O_8Zn$. mp 120°.

THERAP CAT: Anti-inflammatory; vulnerary.

7003. Oxacillin. [66-79-5] $(2S,5R,6R)$-3,3-Dimethyl-6-[[(5-methyl-3-phenyl-4-isoxazolyl)carbonyl]amino]-7-oxo-4-thia-1-azabicyclo[3.2.0]heptane-2-carboxylic acid; 5-methyl-3-phenyl-4-isoxazolylpenicillin; 6-(5-methyl-3-phenyl-2-isoxazoline-4-carboxamido)penicillanic acid; oxazocilline. $C_{19}H_{19}N_3O_5S$; mol wt 401.44. C 56.85%, H 4.77%, N 10.47%, O 19.93%, S 7.99%. Semisynthetic antibiotic related to penicillin. Prepn: Doyle, Nayler, **US 2996501** (1961); Doyle *et al.*, *Nature* **192**, 1183 (1961). Toxicity: E. I. Goldenthal, *Toxicol. Appl. Pharmacol.* **18**, 185 (1971).

Sodium salt monohydrate. [7240-38-2] Penicillin P-12; sodium oxacillin; BRL-1400; Bactocill; Bristopen; Cryptocillin; Micropenin; Penstapho; Prostaphlin; Stapenor. $C_{19}H_{18}N_3NaO_5S \cdot H_2O$; mol wt 441.43. Crystals from isopropanol mp 188° (dec). $[\alpha]_D^{20}$ +201° (c = 1 in water). Freely sol in water, methanol, DMSO; slightly sol in absolute alc, chloroform, pyridine, methyl acetate. Insol in ethyl acetate, ether, benzene, ethylene chloride. LD_{50} orally in rats: >8000 mg/kg (Goldenthal).

THERAP CAT: Antibacterial.

THERAP CAT (VET): Antibacterial.

7004. Oxadiargyl. [39807-15-3] 3-[2,4-Dichloro-5-(2-propyn-1-yloxy)phenyl]-5-(1,1-dimethylethyl)-1,3,4-oxadiazol-2(3*H*)-

one; 5-*tert*-butyl-3-(2,4-dichloro-5-propargyloxyphenyl)-1,3,4-oxa-diazol-2-(3*H*)-one; Raft; Topstar. $C_{15}H_{14}Cl_2N_2O_3$; mol wt 341.19. C 52.80%, H 4.14%, Cl 20.78%, N 8.21%, O 14.07%. Protopor-phyrinogen IX oxidase inhibitor; pre-emergence herbicide for use in rice and other food crops. Prepn: R. Boesch, **DE 2227012**; *idem*, **US 3818026** (1972, 1974 both to Rhone-Poulenc). Properties and field trials: R. Dickmann *et al.*, *Brighton Crop Prot. Conf. - Weeds* **1997**, 51; G. Tracchi *et al.*, *ibid.* 885.

Odorless, white powder with little agglomerates, mp 131°. Soly in water (20°): 0.37 mg/l. LD_{50} in rats (mg/kg): >5000 orally; >2000 dermally; LC_{50} (4 hr) in rats: >5.16 mg/l (Dickmann).
USE: Herbicide.

7005. 1,3,4-Oxadiazole. [288-99-3] $C_2H_2N_2O$; mol wt 70.05. C 34.29%, H 2.88%, N 39.99%, O 22.84%. Prepn: Ains-worth, *J. Am. Chem. Soc.* **87**, 5800 (1965).

Liquid. bp 150°. Thermally stable.

7006. Oxadiazon. [19666-30-9] 3-[2,4-Dichloro-5-(1-meth-ylethoxy)phenyl]-5-(1,1-dimethylethyl)-1,3,4-oxadiazol-2(3*H*)-one; 2-*tert*-butyl-4-(2,4-dichloro-5-isopropoxyphenyl)-Δ²-1,3,4-oxadi-azolin-5-one; 5-*tert*-butyl-3-(2,4-dichloro-5-isopropoxyphenyl)-1,-3,4-oxadiazolin-2-one; RP-17623; Ronstar. $C_{15}H_{18}Cl_2N_2O_3$; mol wt 345.22. C 52.19%, H 5.26%, Cl 20.54%, N 8.11%, O 13.90%. Selective pre-emergence herbicide. Prepn: J. Metivier, R. Boesch, **US 3385862** (1968 to Rhône Poulenc). Metabolism: A. Guardigli *et al.*, *Arch. Environ. Contam. Toxicol.* **4**, 145 (1976). GC-MS de-termn in biological and environmental samples: A. Navalón *et al.*, *J. Chromatogr. A* **946**, 239 (2002). Persistence in soil: D. Ambrosi *et al.*, *J. Agric. Food Chem.* **25**, 868 (1977). Efficacy and phytotox-icity: G. W. Knox, *Proc. Fla. State Hort. Soc.* **99**, 270 (1986). Weed control: S. S. Lim, *Proc. Nat. Semin. Workshop Rice Field Weed Mgmt.* **1988**, 169.

White, odorless, non-hygroscopic crystals, mp 88-90°. *Irritant.* Vapor pressure at 20°: <1×10^{-6} mm Hg. Soly at 20° (g/l): water ~0.0007; ethanol, methanol ~100; cyclohexanone ~200; acetone, acetophenone, anisole carbon tetrachloride, isophorone, methyleth-ylketone ~600; benzene, chloroform, methylene chloride, toluene ~1000. LD_{50} orally in rats, bobwhite quail, mallard duck: 8000, 6000, 1000 mg/kg (Lim).
USE: Herbicide.

7007. Oxadixyl. [77732-09-3] *N*-(2,6-Dimethylphenyl)-2-methoxy-*N*-(2-oxo-3-oxazolidinyl)acetamide; 2-methoxy-*N*-(2-oxo-1,3-oxazolidin-3-yl)acet-2',6'-xylidide; SAN-371F; Sandofan. $C_{14}H_{18}N_2O_4$; mol wt 278.31. C 60.42%, H 6.52%, N 10.07%, O 22.99%. Systemic fungicide belonging to the oxazolidinones. Prepn: J. Harr *et al.*, **BE 884661**; R. Sandmeier, H. Schelling, **US 4457937** (1981, 1984 both to Sandoz). Description of physical prop-erties, biological properties, and activity: U. Gisi *et al.*, *Meded. Fac. Landbouwwet. Univ. Gent* **48**, 541-549 (1983). Field trial of syner-gistic mixture with cymoxanil and mancozeb: M. R. Redbond *et al.*, *Brighton Crop Prot. Conf. - Pests Dis.* **1990**, 1115. GC determn in

cucumber: W.-O. Lee, S.-K. Wong, *Analyst* **120**, 2475 (1995). Soil degradation: N. D. Anan'yeva *et al.*, *Eurasian Soil Sci.* **28**, 157 (1996).

Colorless crystals from ethanol, mp 104-105°. Soly (% w/w) at 25°: water 0.34, ethanol 11.2, ethanol 5.0, acetone 34.4, DMSO 39.0. LD_{50} in female rats (mg/kg): 1860 orally; >2000 dermally (Gisi).
USE: Fungicide.

7008. Oxalacetic Acid. [328-42-7] 2-Oxobutanedioic acid; ketosuccinic acid; oxaloacetic acid; oxosuccinic acid; OAA. $C_4H_4O_5$; mol wt 132.07. C 36.38%, H 3.05%, O 60.57%. Prepd by hydrolyzing sodium diethyloxalacetate with concd HCl: C. Heidel-berger, R. B. Hurlbert, *J. Am. Chem. Soc.* **72**, 4704 (1950); C. Hei-delberger, *Biochem. Prep.* **3**, 59 (1953). Structural studies of the enol form: D. H. Flint *et al.*, *J. Org. Chem.* **57**, 7270 (1992).

***trans*-(Z)-enol Form.** [1115-67-9] Hydroxyfumaric acid. Crys-tals from acetone + benzene, mp 184°.
Diethyl ester *see* Ethyl Oxalacetate.

7009. Oxalenediuramidoxime. [580-52-9] N^1,N^2-Bis(ami-nocarbonyl)-N^1,N^2-dihydroxyethanediimidamide. $C_4H_8N_6O_4$; mol wt 204.15. C 23.53%, H 3.95%, N 41.17%, O 31.35%. Prepn: Zinkeisen, *Ber.* **1889**, 2952.

Needles from dil alcohol. mp 191-192° (dec). Insol in cold water, in ether, chloroform, benzene, petr ether. Very sol in alcohol, acids, alkalies.
USE: As a reagent for the determination of nickel, with which it gives an orange ppt in dil ammoniacal soln.

7010. Oxalic Acid. [144-62-7] Ethanedioic acid. $C_2H_2O_4$; mol wt 90.03. C 26.68%, H 2.24%, O 71.08%. Present in many plants and vegetables, notably in those of the *Oxalis* and *Rumex* families, where it occurs in the cell sap of the plant as the potassium or calcium salt. It is a product of the metabolism of many molds. Several species of *Penicillium* and *Aspergillus* convert sugar into calcium oxalate with 90% yields under optimum conditions. Oxalic acid was formerly manuf by fusion of cellulose matter, e.g. sawdust, with NaOH or by oxidation with HNO_3; it is now made by passing carbon monoxide into concd NaOH or by heating sodium formate in the presence of NaOH or Na_2CO_3: Wallace, **US 1602802** (1926); Beckman, **US 2687433** (1951 to Allied Chem.). Efficient laboratory prepn of anhydrous form: H. T. Clarke, A. W. Davis, *Org. Synth.* **coll. vol. I**, 421 (2nd ed., 1941). Toxicity study: E. H. Vernot *et al.*, *Toxicol. Appl. Pharmacol.* **42**, 417 (1977). *Review:* Wilson, "Mis-cellaneous *Aspergillus* Toxins" in *Microbial Toxins* **vol. VI**, A. Cie-gler *et al.*, Eds. (Academic Press, New York, 1971) pp 268-273; C. A. Bernales in *Kirk-Othmer Encyclopedia of Chemical Technology* **vol. 16** (Wiley-Interscience, New York, 3rd ed., 1981) pp 618-636.

Dihydrate. [6153-56-6] $C_2H_2O_4.2H_2O$; mol wt 126.06. Monoclinic tablets, prisms, granules. *Poisonous.* $d_4^{18.5}$ 1.653. mp 101-102° giving off water of crystn and starting to sublime. pK_1 1.27; pK_2 4.28. pH of 0.1*M* soln 1.3. One gram dissolves in about 7 ml water, 2 ml boiling water; 2.5 ml alcohol, 1.8 ml boiling alcohol, 100 ml ether, 5.5 ml glycerol. Practically insol in benzene, chloroform, petr ether. $d_4^{17.5}$ of aq solns: 1% (w/w) 1.0035; 3% 1.0105; 5% 1.0175; 10% 1.0350; 13% 1.0455. Oxalic acid can be dehydrated by careful drying at 100°, but considerable loss occurs through sublimation; this, moreover, is harmful to the oven. LD_{50} orally in rats (5% solution): 9.5 ml/kg (Vernot).

Anhydr oxalic acid. Crystallized from glacial acetic acid, is orthorhombic, the crystals being pyramidal or elongated octahedra. Hygroscopic, mp 189.5° (dec). Sublimes best at 157°. At higher temps dec into CO_2, CO, formic acid, and H_2O. d_4^{17} 1.90. 100 g of aq soln satd at 15° contain 6.71 g, at 20° 8.34 g, at 25° 9.81 g.

Caution: Potential symptoms of overexposure by ingestion are corrosion of alimentary tract mucosa; localized pain; vomiting, shock, hypotension, cardiovascular collapse; headache, muscle cramps, tatany; convulsions, stupor, coma; kidney damage. Direct contact may cause eye, skin and mucous membrane irritation; eye burns. *See NIOSH Pocket Guide to Chemical Hazards* (DHHS/NIOSH 97-140, 1997) p 238; *Clinical Toxicology of Commercial Products*, R. E. Gosselin *et al.*, Ed. (Williams & Wilkins, Baltimore, 5th ed., 1984) Section III, pp 326-328.

USE: As analytical reagent; in calico printing and dyeing; for bleaching straw (hats) and leather; removing paint or varnish, rust or ink stains; cleaning wood; manuf oxalates; blue ink; celluloid; intermediates and dyes; in metal polishes; in indigo dyeing; in purifying methanol; for decolorizing crude glycerol; for stabilizing hydrocyanic acid. As a general reducing agent; in ceramics and pigments; in metallurgy as cleanser; in the paper industry; in photography; in process engraving; in the rubber mfg industry; in making glucose from starch; as condensing agent in organic chemistry. *In vitro* blood specimen anticoagulant.

THERAP CAT (VET): In 5% solution with 5% malonic acid as hemostatic agent.

7011. Oxaliplatin. [61825-94-3] (*SP*-4-2)-[(1*R*,2*R*)-1,2-Cyclohexanediamine-*κN,κN′*][ethanedioato(2−)-*κO¹,κO²*]platinum; [(1*R*,2*R*)-1,2-cyclohexanediamine-*N,N′*][oxalato(2−)-*O,O′*]-platinum; oxalato (1*R*,2*R*-cyclohexanediamine)platinum(II); Pt(oxalato)(*trans-l*-dach); oxalato (*trans-l*-1,2-diaminocyclohexane)platinum(II); oxalatoplatin; oxalatoplatinum; *l*-OHP; RP-54780; NSC-266046; Eloxatin; Elplat. $C_8H_{14}N_2O_4Pt$; mol wt 397.29. C 24.19%, H 3.55%, N 7.05%, O 16.11%, Pt 49.10%. Third generation platinum complex. Prepn and antitumor activity: Y. Kidani, K. Inagaki, *J. Med. Chem.* **21**, 1315 (1978); Y. Kidani *et al.*, *Gann* **71**, 637 (1980). Crystal structure and abs config: M. A. Bruck, R. Bau, *Inorg. Chim. Acta* **92**, 279 (1984). Mass spectral analysis: J. Claereboudt *et al.*, *J. Pharm. Biomed. Anal.* **7**, 1599 (1989). LC determn in plasma: H. Ehrsson, I. Wallin, *J. Chromatogr. B* **795**, 291 (2003). Review of antitumor activity, toxicology, and pharmacokinetics: G. Mathé *et al.*, *Biomed. Pharmacother.* **43**, 237-250 (1989); of pharmacology and clinical experience as single and combination therapy in colorectal cancer: A. Grothey, R. M. Goldberg, *Expert Opin. Pharmacother.* **5**, 2159-2170 (2004); in epithelial ovarian cancer: S. Fu *et al.*, *Int. J. Gynecol. Cancer* **16**, 1717-1732 (2006).

Colorless, thin triangular plates with truncated vertices. mp 260°. Soly (mg/ml): water 7.9; methanol 2.1; dimethylformamide 9.0. Practically insol in ethanol, acetone, hexane.

THERAP CAT: Antineoplastic.

7012. Oxalomolybdic Acid. [53450-33-2] [Ethanedioato-(2−)-*κO¹,κO²*]trioxomolybdate(2−) dihydrogen; hydrogen trioxo-oxalatomolybdate(VI). $C_2H_2MoO_7$; mol wt 233.98. C 10.27%, H 0.86%, Mo 41.01%, O 47.86%. $H_2[MoO_3(C_2O_4)]$. Usually contains 1 or 2 H_2O.

Colorless crystals, sol in water.

USE: In invisible inks.

7013. Oxalyl Chloride. [79-37-8] Ethanedioyl dichloride; oxalic acid chloride; oxaloyl chloride. $C_2Cl_2O_2$; mol wt 126.92. C 18.93%, Cl 55.86%, O 25.21%. Obtained from oxalic acid and PCl_5.

Colorless, fuming liquid; penetrating odor. *Poisonous.* d_4^{13} 1.488. mp −12°. bp 63-64°. n_D^{13} 1.4340. Violently dec by water, also by alcohol.

Caution: Severely irritating to skin, eyes, respiratory tract.

USE: Chlorinating reagent.

7014. Oxamarin. [15301-80-1] 6,7-Bis[2-(diethylamino)ethoxy]-4-methyl-2*H*-1-benzopyran-2-one; 6,7-bis[2-(diethylamino)-ethoxy]-4-methylcoumarin. $C_{22}H_{34}N_2O_4$; mol wt 390.52. C 67.66%, H 8.78%, N 7.17%, O 16.39%. Deriv of *α*-pyrone, related structurally to sulmarin, *q.v.* Prepn: E. Massarani, *Farmaco Ed. Sci.* **12**, 691 (1957); G. Cavallini, E. Massarani, **US 2895963** (1959 to Maggioni); E. Massarani *et al.*, *J. Med. Pharm. Chem.* **3**, 231 (1961). Comparative study in treatment of varicose syndrome: G. Frantoli *et al.*, *Panminerva Med.* **14**, 290 (1972), *C.A.* **78**, 92741 (1973). Transcutaneous absorption: G. Ciaceri, P. Marini, *Boll. Chim. Farm.* **111**, 321 (1972). Cardiovascular, intestinal smooth muscle effects: *eidem*, *G. Ital. Patol. Sci. Affini* **19**, 11 (1972), *C.A.* **81**, 130823 (1974). Anti-inflammatory action: *eidem, ibid.* 1, *C.A.* **81**, 130952 (1974).

Oil, $bp_{0.5}$ 195°.

Dihydrochloride. [6830-17-7] MG-652; Idro P_3. $C_{22}H_{34}N_2$-$O_4.2HCl$; mol wt 463.44. Crystals from ethanol, mp 224-226° (Massarani, 1957); also reported as mp 234-236° (Massarani *et al.*, 1961).

THERAP CAT: Hemostatic.

7015. Oxametacine. [27035-30-9] 1-(4-Chlorobenzoyl)-*N*-hydroxy-5-methoxy-2-methyl-1*H*-indole-3-acetamide; 1-(*p*-chlorobenzoyl)-5-methoxy-2-methylindole-3-acetohydroxamic acid; indoxamic acid; Dinulcid; Flogar. $C_{19}H_{17}ClN_2O_4$; mol wt 372.81. C 61.21%, H 4.60%, Cl 9.51%, N 7.51%, O 17.17%. Deriv of indomethacin, *q.v.* Prepn: R. Aries, **FR 1599495** (1969), *C.A.* **73**, 3797h (1970); F. De Martiis *et al.*, **DE 2008332** corresp to **US 3624103** (both 1971 to ABC); *eidem, Boll. Chim. Farm.* **114**, 309 (1975). Pharmaco-toxicological evaluation: L. F. Elsom *et al.*, *Arzneim.-Forsch.* **29**, 1155 (1979). Pharmacokinetics: P. Dittrich *et al.*, *ibid.* **31**, 518 (1981). Effect on prostaglandin biosynthesis: J. S. Franzone *et al.*, *Farmaco Ed. Sci.* **35**, 498 (1980). Clinical study: J. Polderman, M. Colon, *J. Int. Med. Res.* **8**, 156 (1980).

Cryst from dioxane, mp 181-182° (dec). Sol in most organic solvents at elevated temperatures. In strong alkali it is hydrolyzed quickly to the debenzoylated product. LD_{50} orally in rats: 96 mg/kg (De Martiis, U.S. patent).

THERAP CAT: Anti-inflammatory.

7016. Oxamic Acid. [471-47-6] 2-Amino-2-oxoacetic acid; oxalamic acid; oxamidic acid. $C_2H_3NO_3$; mol wt 89.05. C 26.98%, H 3.40%, N 15.73%, O 53.90%. Prepd by heating oxamide in water with ammonia, followed by neutralization of resulting ammonium salt: Toussaint, *Ann.* **120**, 237 (1861).

Cryst powder from water, or prisms from alcohol, dec about 210 or 214°. Sparingly sol in water. Practically insol in abs alcohol, ether.

Ammonium salt. [516-00-7] Ammonium oxamate. $C_2H_6N_2O_3$; mol wt 106.08. Monoclinic prisms, dec about 226°. Very slightly sol in cold water, alcohol.

7017. Oxamide. [471-46-5] Ethanediamide; oxalamide; oxalic acid diamide; diaminoglyoxal; ethanedioic acid diamide. $C_2H_4N_2O_2$; mol wt 88.07. C 27.28%, H 4.58%, N 31.81%, O 36.33%. Prepd from formamide by glow-discharge electrolysis: Brown *et al.*, *J. Org. Chem.* **27**, 3698 (1962). Crystal structure: G. De With, S. Harkema, *Acta Crystallogr.* **33B**, 2367 (1977). Metabolized *in vivo* to form oxalic acid.

Triclinic needles, dec 350°. d_4^{20} 1.667. Sparingly sol in hot water, alcohol.

7018. Oxamniquine. [21738-42-1] 1,2,3,4-Tetrahydro-2-[[(1-methylethyl)amino]methyl]-7-nitro-6-quinolinemethanol; 1,2,-3,4-tetrahydro-2-[(isopropylamino)methyl]-7-nitro-6-quinolinemethanol; 6-hydroxymethyl-2-isopropylaminomethyl-7-nitro-1,2,-3,4-tetrahydroquinoline; UK-4271; Mansil; Vansil. $C_{14}H_{21}N_3O_3$; mol wt 279.34. C 60.20%, H 7.58%, N 15.04%, O 17.18%. Prepn: H. C. Richards, *ZA* **6803636**; *idem, US* **3821228** (1967, 1974 both to Pfizer). Discovery of schistosomicidal activity: H. C. Richards, R. Foster, *Nature* **222**, 581 (1969). Metabolism: N. M. Woolhouse, B. Kaye, *Parasitology* **15**, 111 (1977). Genetic activity: T. M. Ong, *Mutat. Res.* **55**, 43 (1978). *In vitro* activity: C. J. Chavasse *et al.*, *Ann. Trop. Med. Parasitol.* **72**, 293 (1978). Mutagenicity study: R. P. Batzinger, E. Bueding, *J. Pharmacol. Exp. Ther.* **200**, 1 (1977). Pharmacology: R. Foster, *Rev. Inst. Med. Trop. Sao Paulo* **15**, Suppl. 1, 189 (1973), *C.A.* **83**, 188417g (1975). Clinical study: R. J. Pitchford, M. Lewis, *S. Afr. Med. J.* **53**, 677 (1978). HPLC determn in biological fluids: H. W. Jun, M. A. Radwan, *Anal. Lett.* **18**, 1345 (1985). Comprehensive description: I. Ahmad, *et al.*, *Anal. Profiles Drug Subs.* **20**, 601-625 (1991).

Pale yellow crystals from isopropanol, mp 147-149°. Sol in acetone, chloroform and methanol. Sol in about 3300 parts of water at 27°C. pH of a 1% soln 8.0-10.0. uv max (methanol) 205.5, 249.5, 389.5 nm ($A_{1cm}^{1\%}$ 486, 695, 62.5). LD_{50} in mice, rabbits (mg/kg): >2000, >1000 i.m., 1300, 800 orally (Foster).

THERAP CAT: Anthelmintic (Schistosoma).

7019. Oxamyl. [23135-22-0] 2-(Dimethylamino)-*N*-[[(methylamino)carbonyl]oxy]-2-oxoethanimidothioic acid methyl ester; *N'*,*N'*-dimethyl-*N*-[(methylcarbamoyl)oxy]-1-thiooxamimidic acid methyl ester; *N*,*N*-dimethyl-α-methylcarbamoyloxyimino-α-(methylthio)acetamide; methyl 1-(dimethylcarbamoyl)-*N*-(methylcarbamoyloxy)thioformimidate; thioxamyl; DPX-1410; Vydate. $C_7H_{13}N_3$-O_3S; mol wt 219.26. C 38.35%, H 5.98%, N 19.16%, O 21.89%, S 14.62%. Prepn: J. B. Buchanan, *ZA* **6803629**; *eidem, US* **3530220**, and *US* **3658870** (1968, 1970, 1972 all to du Pont). Decompn: J. Harvey, Jr., J. Han, *J. Agric. Food Chem.* **26**, 536 (1978). Metabolism: *eidem, ibid.* 902; J. Harvey, Jr. *et al., ibid.* 529. Toxicity study: M. Fahmy *et al., J. Agric. Food Chem.* **26**, 550 (1978).

Crystalline solid, slight sulfurous odor. mp 100-102°, changing to a different crystalline form, mp 108-110°. Soly in g/100 ml at 25°: water 28; acetone 67; ethanol 33; 2-propanol 11; methanol 144; toluene 1. LD_{50} orally in rats: 5 mg/kg (Fahmy).

Caution: Toxic to wildlife, fish, bees.

USE: Insecticide, nematocide, acaricide.

7020. Oxandrolone. [53-39-4] (4a*S*,4b*S*,6a*S*,7*S*,9a*S*,9b*R*,-11a*S*)-Tetradecahydro-7-hydroxy-4a,6a,7-trimethylcyclopenta[5,6]-naphtho[1,2-*c*]pyran-2(1*H*)-one; 17β-hydroxy-17α-methyl-2-oxa-5α-androstan-3-one; dodecahydro-3-hydroxy-6-(hydroxymethyl)-3,3a,6-trimethyl-1*H*-benz[*e*]indene-7-acetic acid δ-lactone; Anavar; Lonavar; Provitar; Vasorome. $C_{19}H_{30}O_3$; mol wt 306.45. C 74.47%, H 9.87%, O 15.66%. Prepn: R. Pappo, C. J. Jung, *Tetrahedron Lett.* **9**, 365 (1962); R. Pappo, *US* **3128283** (1964 to Searle).

White crystals, mp 235-238°. $[\alpha]_D^{25}$ −23° (c = approx 1% in chloroform). Freely sol in chloroform; sparingly sol in alc, acetone. Practically insol in water.

Note: This is a controlled substance (anabolic steroid): **21 CFR**, 1308.13, as defined in 1300.01.

THERAP CAT: Androgen.

7021. Oxantel. [36531-26-7] 3-[(1*E*)-2-(1,4,5,6-Tetrahydro-1-methyl-2-pyrimidinyl)ethenyl]phenol; (*E*)-*m*-[2-(1,4,5,6-tetrahydro-1-methyl-2-pyrimidinyl)vinyl]phenol; 1-methyl-1,4,5,6-tetrahydro-2-[2-(3-hydroxyphenyl)vinyl]pyrimidine; CP-14445. $C_{13}H_{16}$-N_2O; mol wt 216.28. C 72.19%, H 7.46%, N 12.95%, O 7.40%. Analog of pyrantel, *q.v.*, with activity vs whipworms (*Trichuris* spp.). Prepn: J. W. Mc Farland, *ZA* **6804589** corresp to *US* **3579510** and *US* **3708584** (1968, 1971, 1973 all to Pfizer). Synthesis and evaluation in whipworm control: J. W. Mc Farland, H. L. Howes *J. Med. Chem.* **15**, 365 (1972). Evaluation vs *Trichuris* in dogs: H. L. Howes, *Proc. Soc. Exp. Biol. Med.* **139**, 394 (1972). Efficacy vs *T. trichiuris* in humans: E. L. Lee *et al., Am. J. Trop. Med. Hyg.* **25**, 563 (1976); vs *T. suis* in swine: M. Robinson, *Vet. Parasitol.* **5**, 223 (1979). Anthelmintic effects of combination with pyrantel pamoate: B. Sinniah, D. Sinniah, *Ann. Trop. Med. Parasitol.* **75**, 315 (1981).

Consult the Name Index before using this section.

Hydrochloride. $C_{13}H_{16}N_2O.HCl$. Crystals from ethanol, mp 207-208°. uv max (water): 231, 274 nm (ε 12700, 20100).

Pamoate. [68813-55-8] Oxantel embonate; CP-14445-16; Telopar. $C_{13}H_{16}N_2O.C_{23}H_{16}O_6$; mol wt 604.66.

Mixture with pyrantel pamoate (1:1) *see* Pyrantel.

THERAP CAT: Anthelmintic (Nematodes).

THERAP CAT (VET): Anthelmintic (Nematodes).

7022. Oxapropanium Iodide. [541-66-2] N,N,N-Trimethyl-1,3-dioxolane-4-methanaminium iodide (1:1); (1,3-dioxolan-4-yl-methyl)trimethylammonium iodide; 4-(dimethylaminomethyl)-1,3-dioxacyclopentane methiodide; 1-dimethylamino-2,3-dioxamethyl-enepropane methiodide; vasodilatateur 2249F; 2249-F; Dilvasene. $C_7H_{16}INO_2$; mol wt 273.11. C 30.79%, H 5.91%, I 46.47%, N 5.13%, O 11.72%. Prepd by the reaction of a cyclic acetal of an α-monohalohydrin of glycerol with dimethylamine which on treatment with methyl iodide results in a quaternary ammonium salt: Fourneau, **US 2445393** (1948 to Rhône-Poulenc).

Crystals, mp 158-160° (the free base bp_{21} 68°). Freely sol in water. Sol in boiling alcohol (about 40% w/v). Sparingly sol in cold alcohol, ether, chloroform, benzene.

THERAP CAT: Cholinergic.

7023. Oxaprozin. [21256-18-8] 4,5-Diphenyl-2-oxazolepropanoic acid; β-(4,5-diphenyloxazol-2-yl)propionic acid; Wy-21743; Alvo; Daypro. $C_{18}H_{15}NO_3$; mol wt 293.32. C 73.71%, H 5.15%, N 4.78%, O 16.36%. First description of anti-inflammatory properties: K. Brown *et al.*, *Nature* **219**, 164 (1968). Prepn: **FR 2001036** (1969 to Inst. Farm. Serono), *C.A.* **72**, 66930w (1970); K. Brown, **GB 1206403** and **US 3578671** (1970, 1971 both to Wyeth). Biochemical properties: M. W. Whitehouse *et al.*, *Biochem. Pharmacol.* **20**, 2309 (1971). Metabolism: F. W. Janssen *et al.*, *Drug Metab. Dispos.* **6**, 465 (1978). Pharmacology: D. A. Shriver *et al.*, *Toxicol. Appl. Pharmacol.* **42**, 75 (1977); F. Awouters *et al.*, *J. Pharm. Pharmacol.* **30**, 41 (1978). Clinical studies: R. Jamar, J. Dequeker, *Curr. Med. Res. Opin.* **5**, 433 (1978); J. A. Hubsher *et al.*, *J. Int. Med. Res.* **7**, 69 (1979); *eidem*, *Arthritis Rheum.* **25**, S117 (1982). Protein binding and clearance: C. A. Homon *et al.*, *Agents Actions* **12**, 211 (1982). Symposium on pharmacology and clinical efficacy: *Semin. Arthritis Rheum.* **15**, Suppl. 3, 1-107 (1986). Review of mechanism of action and clinical efficacy: F. Dallegri *et. al.*, *Expert Opin. Pharmacother.* **6**, 777-785 (2005).

Crystals from methanol, mp 160.5-161.5°. Slightly sol in alcohol. Insol in water. Partition coefficient (octanol/water): 4.8 (pH 7.4). pKa 4.3.

THERAP CAT: Anti-inflammatory.

7024. Oxatomide. [60607-34-3] 1-[3-[4-(Diphenylmethyl)-1-piperazinyl]propyl]-1,3-dihydro-2H-benzimidazol-2-one; 1-[3-[4-(diphenylmethyl)-1-piperazinyl]propyl]-2-benzimidazolinone; R-35443; Celtect; Cobiona; Dasten; Tinset. $C_{27}H_{30}N_4O$; mol wt 426.56. C 76.03%, H 7.09%, N 13.13%, O 3.75%. Orally active anti-allergic agent, related structurally to cinnarizine, *q.v.*, and having a novel biphasic mode of action. Prepn: J. Vandenberk *et al.*, **DE 2714437**; *eidem*, **US 4250176** (1977, 1981 both to Janssen). Inhibition of release and effects of allergic mediators: F. Awouters *et al.*, *Experientia* **33**, 1657 (1977). *In vitro* study of inhibition and stimulation of histamine release: M. K. Church *et al.*, *Agents Actions* **10**, 4 (1980). *In vivo* study: S. Gatti *et al.*, *Br. J. Dermatol.* **103**, 671 (1980). Clinical studies: S. W. Barham, F. Moran, *Br. J. Clin. Pract.* **34**, 323 (1980); E. F. Juniper *et al.*, *Clin. Allergy* **11**, 61

(1981); W. Peremans *et al.*, *Dermatologica* **162**, 42 (1981). Review of pharmacological and clinical studies: D. M. Richards *et al.*, *Drugs* **27**, 210 (1984).

White powder, mp 153.6°. LD_{50} in guinea pigs, mice, rats (mg/kg): 320, >2560, >2560 orally; 23, 27, 30 i.v. (Richards).

THERAP CAT: Antiallergic; antiasthmatic.

7025. Oxazepam. [604-75-1] 7-Chloro-1,3-dihydro-3-hydroxy-5-phenyl-2H-1,4-benzodiazepin-2-one; 7-chloro-3-hydroxy-5-phenyl-1,3-dihydro-2H-1,4-benzodiazepin-2-one; Wy-3498; Adumbran; Azutranquil; Durazepam; Hilong; Limbial; Noctazepam; Praxiten; Serax; Serenid; Serepax; Séresta; Serpax; Sobril; Uskan. $C_{15}H_{11}ClN_2O_2$; mol wt 286.72. C 62.84%, H 3.87%, Cl 12.36%, N 9.77%, O 11.16%. Benzodiazepine anxiolytic; metabolite of several other benzodiazepines including diazepam and chlordiazepoxide, *q.q.v.* Prepn: Bell, Childress, *J. Org. Chem.* **27**, 1691 (1962); **NL 298071**; S. C. Bell, **US 3296249** (1965, 1967 both to American Home Prod.). Crystal and molecular structure: G. Gilli *et al.*, *Acta Crystallogr.* **B34**, 2826 (1978). Metabolic studies: Sisenwine *et al.*, *Arzneim.-Forsch.* **22**, 682 (1972); Knowles, Ruelius, *ibid.* 687. Pharmacokinetics: D. J. Greenblatt, *Clin. Pharmacokinet.* **6**, 89 (1981). HPLC determn in urine: L. A. Berrueta *et al.*, *J. Chromatogr.* **616**, 344 (1993); dynamic HPLC determn of enantiomerization barrier: O. Trapp *et al.*, *J. Biochem. Biophys. Methods* **54**, 301 (2002). Toxicity data: E. I. Goldenthal, *Toxicol. Appl. Pharmacol.* **18**, 185 (1971). Comprehensive description: C. M. Shearer, C. R. Pilla, *Anal. Profiles Drug Subs.* **3**, 441-464 (1974). *Review: Acta Psychiatr. Scand. Suppl.* **274**, R. Chinnery, A. Sundwall, Eds. (1978) 128 pp.

Crystals from alc, mp 205-206°. Sol in dioxane; slightly sol in alc, chloroform; very slightly sol in ether. Practically insol in water. LD_{50} in mice, rats (mg/kg): >5010, >5010 orally (Goldenthal).

Note: This is a controlled substance (depressant): **21 CFR**, 1308.14.

THERAP CAT: Anxiolytic.

THERAP CAT (VET): Appetite stimulant in cats.

7026. Oxazolam. [27167-30-2] 10-Chloro-2,3,7,11b-tetrahydro-2-methyl-11b-phenyloxazolo[3,2-d][1,4]benzodiazepin-6(5H)-one; 10-chloro-2,3,5,6,7,11b-hexahydro-2-methyl-11b-phenyl-benzo[6,7]-1,4-diazepino[5,4-b]oxazol-6-one; oxazolazepam; Hializan; Serenal; Tranquit. $C_{18}H_{17}ClN_2O_2$; mol wt 328.80. C 65.75%, H 5.21%, Cl 10.78%, N 8.52%, O 9.73%. Prepn: Tachikawa *et al.*, **DE 1812252**; *eidem*, **US 3772371** (1969, 1973 to Sankyo). Synthesis and pharmacology: Miyadera *et al.*, *J. Med. Chem.* **14**, 520 (1971). Metabolism studies: Shindo *et al.*, *Chem. Pharm. Bull.* **19**, 60 (1971); Yasumura *et al.*, *ibid.* 1929.

Crystals, mp 186-188°. Sol in chloroform; slightly sol in ethanol. Practically insol in water.

Note: This is a controlled substance (depressant): **21 CFR,** 1308.14.

THERAP CAT: Anxiolytic.

7027. Ox Bile Extract. Purified oxgall; sodium choleate; Bi-Ketolan; Bilein; Bilicholan; Cholatol; Crescefel; Desicol; Doxychol; Panoxolin; Plebilin. One gram = 8 g ox bile. *Constit.* Chiefly the sodium salts of glycocholic and taurocholic acids; also cholesterol, lecithin, glycocol and choline compounds.

Yellowish-green, soft solid; peculiar odor; partly sweet, partly bitter, disagreeable taste. Very sol in water or alc.

THERAP CAT: Choleretic.

THERAP CAT (VET): Has been used as a choleretic.

7028. Oxcarbazepine. [28721-07-5] 10,11-Dihydro-10-oxo-5*H*-dibenz[*b,f*]azepine-5-carboxamide; oxacarbazepine; GP-47680; Trileptal. $C_{15}H_{12}N_2O_2$; mol wt 252.27. C 71.42%, H 4.79%, N 11.10%, O 12.68%. Ketoderivative of carbamazepine, *q.v.* Prepn: W. Schindler, **DE 2011087** (1970 to Geigy); *idem,* **US 3642775** (1972 to Ciba-Geigy). Improved prepn: D. Kaufmann *et al., Tetrahedron Lett.* **45,** 5275 (2004). Metabolism: H. Schütz *et al., Xenobiotica* **16,** 769 (1986). Hyponatremic effects: O. A. Nielsen *et al., Epilepsy Res.* **2,** 269 (1988). Determn of oxcarbazepine and main metabolites by GC in plasma: G. E. Von Unruh, W. D. Paar, *J. Chromatogr.* **345,** 67 (1985); by HPLC: A. A. Elyas, V. D. Goldberg, *ibid.* **528,** 473 (1990). Clinical evaluation in treatment of epilepsy: M. Dam *et al., Epilepsy Res.* **3,** 70 (1989); in management of trigeminal neuralgia: J. M. Zakrzewská, P. N. Patsalos, *J. Neurol. Neurosurg. Psychiatry* **52,** 472 (1989). Review of pharmacology and therapeutic efficacy: A. Beydoun, E. Kutluay, *Expert Opin. Pharmacother.* **3,** 59-71 (2001).

Crystals from ethanol, mp 215-216°.

THERAP CAT: Anticonvulsant.

7029. Oxeladin. [468-61-1] α,α-Diethylbenzeneacetic acid 2-[2-(diethylamino)ethoxy]ethyl ester; 2-ethyl-2-phenylbutyric acid 2-(2-diethylaminoethoxy)ethyl ester; 2-(2-diethylaminoethoxy)ethyl α,α-diethylphenylacetate; α,α-diethylphenylacetic acid 2-(2-diethylaminoethoxy)ethyl ester. $C_{20}H_{33}NO_3$; mol wt 335.49. C 71.60%, H 9.92%, N 4.18%, O 14.31%. Prepn: V. Petrow *et al., J. Pharm. Pharmacol.* **10,** 40 (1958); *eidem,* **US 2885404** (1959 to Brit. Drug Houses).

Yellow oil with an acrid odor and bitter taste. $bp_{0.5}$ 150°; $bp_{0.1}$ 140°. Soluble in dil HCl, ethanol, acetone, ether, toluene; practically insol in water. Stable in acids, unstable in alkalies. Non-hygroscopic.

Citrate. [52432-72-1] Pectamol; Pectamon; Paxeladine. $C_{20}H_{33}NO_3 \cdot C_6H_8O_7$; mol wt 527.61. Small needles from ethyl acetate, mp 90-91°. Sol in water.

THERAP CAT: Antitussive.

7030. Oxendolone. [33765-68-3] (16β,17β)-16-Ethyl-17-hydroxyestr-4-en-3-one; 16β-ethyl-19-nortestosterone; TSAA-291; Prostetin. $C_{20}H_{30}O_2$; mol wt 302.46. C 79.42%, H 10.00%, O 10.58%. Prepn: K. Hiraga *et al., DE 2100319; eidem,* **US 3856829** (1971, 1974 both to Takeda); K. Yoshioka *et al., Chem. Pharm. Bull.* **23,** 3203 (1975). Stereoselective synthesis and NMR study: G. Goto *et al., ibid.* **25,** 1295 (1977). Synthesis and anti-androgenic activity: *eidem, ibid.* **26,** 1718 (1978). Physico-chemical properties and stabilities: K. Itakura *et al., Takeda Kenkyushoho* **37,** 297 (1978), *C.A.* **91,** 20879 (1979). Disposition and metabolism: S. Tanayama *et al., Steroids* **33,** 65 (1979). Series of articles on pharmacology, mechanism of action, anti-androgen effects: *Acta Endocrinol.* **92** (Suppl 2), 1-107 (1979).

Crystals from ether, mp 152-153°. $[\alpha]_D$ +41° (c = 1.0 in ethanol). uv max (ethanol): 240 nm (ε 15800). Stable against heat, humidity, indoor light. Converted to the 16α- and 17α-epimers in sunlight. LD_{50} in rats, mice (g/kg): >10 orally; 5-10 i.m. and i.p. (Hiraga, 1971).

THERAP CAT: Anti-androgen; in treatment of benign prostatic hypertrophy.

7031. Oxenin. [3230-75-9] 11,12-Didehydro-7,10-dihydro-10-hydroxyretinol; 3,7-dimethyl-9-(2,6,6-trimethyl-1-cyclohexen-1-yl)nona-2,7-dien-4-yne-1,6-diol. $C_{20}H_{30}O_2$; mol wt 302.46. C 79.42%, H 10.00%, O 10.58%. Intermediate in the vitamin A synthesis. Prepn: Isler *et al., Helv. Chim. Acta* **30,** 1911 (1947); Isler, **US 2451739** (1948 to Hoffmann-La Roche); Kardys, **US 3046310** (1962 to Pfizer).

Very viscous yellow oil. d_4^{27} 0.9984; n_D^{22} 1.5344. Has been crystallized, mp 58.5-59°.

7032. Oxethazaine. [126-27-2] 2,2′-[(2-Hydroxyethyl)imino]bis[*N*-(1,1-dimethyl-2-phenylethyl)-*N*-methylacetamide]; *N,N*-bis[*N*-methyl-*N*-phenyl-*tert*-butylacetamido]-β-hydroxyethylamine; oxetacaine; oxethazine; Wy-806; Storocain; Topicain. $C_{28}H_{41}N_3O_3$; mol wt 467.65. C 71.91%, H 8.84%, N 8.99%, O 10.26%. Prepn: Seifter *et al., US 2780646* (1957 to Am. Home Prods.); Salomon, **IL 10062** (1957), *C.A.* **52,** 15569 (1958). Pharmacology: Axerio Agnessetti, *An. R. Acad. Farm.* **35,** 183 (1969), *C.A.* **71,** 122082q (1969). Toxicity studies: J. M. Glassman *et al., Toxicol. Appl. Pharmacol.* **5,** 184 (1963). Clinical trial of combination with antacid in ulcer pain: F. T. M. Cordeiro *et al., Curr. Ther. Res.* **19,** 230 (1976); A. Zanni *et al., Curr. Med. Res. Opin.* **10,** 128 (1986). Evaluation as dental anesthetic: P. A. Brennan, J. D. Langdon, *Br. J. Oral Maxillofac. Surg.* **28,** 26 (1990).

Crystals from benzene + hexane, mp 104-104.5°. Insol in water. Sol in dil acids.

Hydrochloride. [13930-31-9] Emoren. $C_{28}H_{41}N_3O_3 \cdot HCl$; mol wt 504.11. mp 146-147°. Easily sol in water. LD_{50} (calculated as base) in mice, rats (mg/kg): 399.9, 625.9 orally; 247.2, 502.3 i.m.; 3.6, 1.3 i.v.; LD_{50} (calculated as base) in rabbits (mg/kg): 0.54 i.v. (Glassman).

Combination with aluminum and magnesium hydroxides. Mutesa; Muthesa; Oxaine M; Tepilta.

THERAP CAT: Anesthetic (local).

7033. Oxetorone. [26020-55-3] 3-Benzofuro[3,2-c][1]benzoxepin-6(12H)-ylidene-N,N-dimethyl-1-propanamine; N,N-dimethylbenzofuro[3,2-c][1]benzoxepin-$\Delta^{6(12H),\gamma}$-propylamine; 6-(3-dimethylaminopropylidene)benzo[b]benzofurano[2,3-e]oxepine. $C_{21}H_{21}NO_2$; mol wt 319.40. C 78.97%, H 6.63%, N 4.39%, O 10.02%. Novel serotonin and histamine antagonist with antimigraine activity. Prepn: F. Binon, M. Descamps, **DE 1963205** corresp to **US 3651051** (1970, 1972 both to Labaz). Metabolism *in vitro*: E. Rossi *et al.*, *J. Chromatogr.* **152**, 228 (1978). Use in treatment of migraine: J. J. Dufresne, *Praxis* **67**, 1148 (1978); J. Florence, *ibid.* 1323; in chronic headache: U. Thoden, *Therapiewoche* **30**, 492 (1980).

Fumarate. [34522-46-8] L-6257; Nocertone; Oxedix. $C_{25}H_{25}NO_6$; mol wt 435.48. Cryst from isopropanol, mp 160°.

THERAP CAT: Analgesic (specific in migraine).

7034. Oxfendazole. [53716-50-0] N-[6-(Phenylsulfinyl)-1H-benzimidazol-2-yl]carbamic acid methyl ester; methyl-5-(phenylsulfinyl)-2-benzimidazolecarbamate; 5-phenylsulfinyl-2-carbomethoxyaminobenzimidazole; RS-8858; Autoworm; Benzelmin; Synanthic; Systamex. $C_{15}H_{13}N_3O_3S$; mol wt 315.35. C 57.13%, H 4.16%, N 13.33%, O 15.22%, S 10.17%. Prepn: C. Beard *et al.*, **DE 2363351**; *eidem*, **US 3929821** (1974, 1975, both to Syntex); E. A. Averkin *et al.*, *J. Med. Chem.* **18**, 1164 (1975). Metabolism: J. P. Bell, R. V. Tomlinson, *Fed. Proc.* **35**, 487 (1976). Radioimmunoassay: C. Nerenberg *et al.*, *J. Pharm. Sci.* **67**, 1553 (1978). Efficacy evaluations: J. L. Duncan, J. G. Reid, *Vet. Rec.* **103**, 332 (1978); N. F. Baker *et al.*, *Am. J. Vet. Res.* **39**, 1258 (1978).

Crystals from chloroform-methanol, mp 253° (dec). Slightly sol in alc, methylene chloride. Practically insol in water. LD_{50} in dogs, rats, mice: >1600, >6400, >6400 mg/kg (Averkin).

THERAP CAT (VET): Anthelmintic.

7035. Oxibendazole. [20559-55-1] N-(6-Propoxy-1H-benzimidazol-2-yl)carbamic acid methyl ester; 5-propoxy-2-benzimidazolecarbamic acid methyl ester; 5-propoxy-2-(carbomethoxyamino)benzimidazole; SKF-30310; Anthelcide EQ; Equitac. $C_{12}H_{15}N_3O_3$; mol wt 249.27. C 57.82%, H 6.07%, N 16.86%, O 19.26%. Prepn: **GB 1123317** corresp to P. P. Actor, J. F. Pagano, US

3574845 (1968, 1971 both to SK & F). Comparative anthelmintic effects in mice: C. S. Karunakaron, D. A. Denham, *J. Parasitol.* **66**, 929 (1980). Clinical studies in foals and horses: J. H. Drudge *et al.*, *Am. J. Vet. Res.* **42**, 526 (1981); E. T. Lyons *et al.*, *ibid.* 685; D. K. Hass *et al.*, *ibid.* **43**, 534 (1982).

Cryst, mp 230-230.5°.

THERAP CAT (VET): Anthelmintic.

7036. Oxiconazole. [64211-45-6] (1Z)-1-(2,4-Dichlorophenyl)-2-(1H-imidazol-1-yl)ethanone O-[(2,4-dichlorophenyl)methyl]oxime; 2′,4′-dichloro-2-(imidazol-1-yl)acetophenone O-(2,4-dichlorobenzyl)oxime. $C_{18}H_{13}Cl_4N_3O$; mol wt 429.12. C 50.38%, H 3.05%, Cl 33.04%, N 9.79%, O 3.73%. Broad spectrum topical antimycotic agent. Prepn: G. Mixich *et al.*, **DE 2657578**; *eidem*, **US 4124767** (1977, 1978 both to Siegfried). Stereospecific synthesis of (E)-, (Z)-isomers, ^{1}H NMR study: G. Mixich, K. Thiele, *Arzneim.-Forsch.* **29**, 1510 (1979). GLC determn in plasma: M. Zell, L. Herzfeld, *J. Chromatogr.* **229**, 111 (1982). *In vitro* and *in vivo* antifungal activity: A. Polak, *Arzneim.-Forsch.* **32**, 17 (1982). Study of skin, nail penetration: G. Stueltgen, E. Bauer, *Mykosen* **25**, 74 (1982). Comparison of antifungal imidazoles: W. H. Beggs, *IRCS Med. Sci.* **11**, 677 (1983).

Nitrate. [64211-46-7] Sgd-301-76; Ro-13-8996; Gyno-Myfungar; Myfungar; Oceral; Oxistat. $C_{18}H_{13}Cl_4N_3O \cdot HNO_3$; mol wt 492.13. Crystals from ethanol, mp 137-138°.

THERAP CAT: Antifungal.

7037. Oxidimethiin. [55290-64-7] 2,3-Dihydro-5,6-dimethyl-1,4-dithiin 1,1,4,4-tetraoxide; tetrathiin; UBI-N252; Harvade. $C_6H_{10}O_4S_2$; mol wt 210.26. C 34.27%, H 4.79%, O 30.44%, S 30.50%. Prepn: A. D. Brewer *et al.*, **FR 2228429** corresp to **US 3920438** (1974, 1975 to Uniroyal). Activity: R. W. Neidermyer *et al.*, *Proc. 12th Br. Weed Control Conf.* **3**, 959 (1974). Crystal structure: S. K. Arora *et al.*, *Acta Crystallogr.* **B34**, 2918 (1978).

Long white needles from water, mp 166-168°.

USE: Plant growth regulator.

7038. Oxidized Cellulose. Absorbable cellulose; cellulosic acid; polyanhydroglucuronic acid; Oxycel; Hemo-Pak. A cellulose of varied carboxyl content retaining the fibrous structure. Prepd by oxidizing cellulose with nitrogen dioxide: Yackel, Kenyon, *J. Am. Chem. Soc.* **64**, 121 (1942).

The degree of oxidation is sufficiently high to make the product sol in dil alk solns. Insol in water or acidic solns.

THERAP CAT: Hemostatic (local).

THERAP CAT (VET): Hemostatic (local).

7039. Oxidronic Acid. [15468-10-7] P,P′-(Hydroxymethylene)bisphosphonic acid; (hydroxymethylene)diphosphonic acid;

HDP; HMDP. $CH_6O_7P_2$; mol wt 192.00. C 6.26%, H 3.15%, O 58.33%, P 32.26%. Bisphosphonate similar to etidronic acid, *q.v.* Prepn: O. T. Quimby, **FR 1506840**; *idem*, **US 3422137** (1967, 1969 both to Procter & Gamble); of sodium salt: *idem et al., J. Org. Chem.* **32**, 4111 (1967). Prepn of ^{99m}Tc-complex: J. A. Bevan, **EP 7676**; *idem*, **US 4247534** (1980, 1981 both to Procter & Gamble); and osteotropic properties: T. S. T. Wang *et al., J. Nucl. Med.* **21**, 767 (1980). Biodistribution: J. A. Bevan *et al., ibid.* 961. Radiochemical purity: W. Majewski *et al., J. Nucl. Med. Technol.* **11**, 23 (1983). Clinical evaluation as skeletal imaging agent: P. A. Domstad *et al., Radiology* **136**, 209 (1980). Diagnostic use in Paget's disease: A. Evans *et al., Eur. J. Nucl. Med.* **18**, 757 (1991); for localization of bone metastases: F. Tenenbaum *et al., ibid.* **20**, 1168 (1993).

Sodium salt. [14255-61-9] Disodium monohydroxy methylene diphosphonate; oxidronate sodium. $CH_4Na_2O_7P_2$. Crystals from methanol/water (6/1), mp 297-300°.
^{99m}**Tc-complex.** [72945-61-0] Technetium Tc 99m oxidronate; TechneScan HDP; Osteoscan-HDP.
THERAP CAT: ^{99m}Tc-complex as diagnostic aid (radioactive imaging agent).

7040. Oxiniacic Acid. [2398-81-4] 3-Pyridinecarboxylic acid 1-oxide; nicotinic acid 1-oxide; 3-carboxypyridine *N*-oxide. C_6H_5-NO_3; mol wt 139.11. C 51.81%, H 3.62%, N 10.07%, O 34.50%. Prepn: Clemo, Koenig, *J. Chem. Soc.* **1949**, S231; Taylor, Crovetti, *J. Org. Chem.* **19**, 1633 (1954). Mass spectrum: Bild, Hesse, *Helv. Chim. Acta* **50**, 1885 (1967). Therapeutic effect: Debay, Thery, **BE 618968** (1962), *C.A.* **58**, 12375e (1963).

Needles, mp 254-255° (dec). Slightly soluble in cold water, more sol in hot water, hot glacial acetic acid, hot methanol, less sol in ethanol. Insol in light petroleum, benzene, chloroform. uv max ($0.1N$ H_2SO_4): 220, 260 nm (ε 22400, 10200).
Ethanolamine salt. [36296-31-8] Ethanolamine oxiniacate; Novacyl. $C_6H_5NO_3.C_2H_7NO$; mol wt 200.19.
THERAP CAT: Antilipemic.

7041. Oxiracetam. [62613-82-5] 4-Hydroxy-2-oxo-1-pyrrolidineacetamide; 2-(4-hydroxypyrrolidin-2-on-1-yl)acetamide; hydroxypiracetam; CT-848; ISF-2522; Neuractiv; Neuromet. C_6H_{10}-N_2O_3; mol wt 158.16. C 45.57%, H 6.37%, N 17.71%, O 30.35%. Analog of piracetam, *q.v.*, with psychostimulant activity. Prepn: G. Pifferi, M. Pinza, **DE 2635853**; *idem*, **US 4118396** (1977, 1978 both to I.S.F.); *eidem, Farmaco Ed. Sci.* **32**, 602 (1977). Effect on learning and memory in animals and prepn of enantiomers: S. Banfi *et al., ibid.* **39**, 16 (1984). Effect in animals with cerebral impairment: S. Banfi *et al., Pharmacol. Res. Commun.* **16**, 67 (1984). Pharmacokinetics in humans: E. Perucca *et al., Eur. J. Drug Metab. Pharmacokinet.* **9**, 267 (1984). Clinical evaluation in organic brain syndrome: B. Saletu *et al., Neuropsychobiology* **13**, 44 (1985).

White, microcrystalline powder from methanol, mp 165-168°.
(R)-Form. Crystals from acetone + water, mp 135-136°. $[\alpha]_D$ +36.2° (c = 1.00 in water).
(S)-Form. Crystals from acetone + water, mp 135-136°. $[\alpha]_D$ −36.0° (c = 1.00 in water).
THERAP CAT: Nootropic.

7042. Oxitropium Bromide. [30286-75-0] $(1\alpha,2\beta,4\beta,5\alpha,$-$7\beta)$-9-Ethyl-7-[2(S)-3-hydroxy-1-oxo-2-phenylpropoxy]-9-methyl-3-oxa-9-azoniatricyclo[3.3.1.0^{2,4}]nonane bromide (1:1); (8R)-6β,-7β-epoxy-8-ethyl-3α-hydroxy-1αH,5αH-tropanium bromide (−)-tropate; (−)-N-ethylnorscopolamine methobromide; Ba-253; Ba-253-BR-L; Oxivent; Tersigat; Ventilat. $C_{19}H_{26}BrNO_4$; mol wt 412.32. C 55.35%, H 6.36%, Br 19.38%, N 3.40%, O 15.52%. Anticholinergic bronchodilating agent. Prepn: K. Zeile *et al.*, **US 3472861** (1969 to Boehringer, Ing.). Comparison with fenoterol: D. Nolte, *Respiration* **36**, 32 (1978); with ipratropium bromide: J. Lulling *et al., ibid.* **42**, 188 (1981). Efficacy in bronchial asthma: H. M. Beumer *et al., Int. J. Clin. Pharmacol. Ther. Toxicol.* **19**, 168 (1981); in exercise-induced asthma: K. Larsson, *Respiration* **43**, 57 (1982).

Crystals, mp 203-204° (dec). $[\alpha]_D^{21}$ −25° (c = 2.0 in water).
THERAP CAT: Bronchodilator.

7043. Oxolamine. [959-14-8] *N,N*-Diethyl-3-phenyl-1,2,4-oxadiazole-5-ethanamine; 5-(2-diethylaminoethyl)-3-phenyl-1,2,4-oxadiazole; 3-phenyl-5-(β-diethylaminoethyl)-1,2,4-oxadiazole; 683-M. $C_{14}H_{19}N_3O$; mol wt 245.33. C 68.54%, H 7.81%, N 17.13%, O 6.52%. Prepn: **DE 1097998** (1961 to Angelini Francesco), *C.A.* **56**, 11598h (1962). Teratogenicity studies of oxolamine citrate: L. Nilsson, *Arzneim.-Forsch.* **17**, 781 (1967).

Liquid, bp$_{0.4}$ 127°.
Hydrochloride. [1219-20-1] $C_{14}H_{19}N_3O.HCl$. Crystals, mp 153-154°.
Citrate. [1949-20-8] AF-438; Bredon; Broncatar; Perebron; Prilon; Flogobron; Oxarmin. $C_{14}H_{19}N_3O.C_6H_8O_7$; mol wt 437.45. Crystals, slightly sol in water, alcohol. uv max (aq soln): 239 nm (ε 260), 273, 283 nm.
THERAP CAT: In inflammatory conditions of respiratory tract.

7044. Oxolinic Acid. [14698-29-4] 5-Ethyl-5,8-dihydro-8-oxo-1,3-dioxolo[4,5-g]quinoline-7-carboxylic acid; 1-ethyl-1,4-dihydro-6,7-methylenedioxy-4-oxo-3-quinolinecarboxylic acid; 1-ethyl-6,7-methylenedioxy-4-quinolone-3-carboxylic acid; W-4565; Inoxyl; Nidantin; Uritrate; Urotrate. $C_{13}H_{11}NO_5$; mol wt 261.23. C 59.77%, H 4.24%, N 5.36%, O 30.62%. Quinolone antibacterial. Prepn: Kaminsky, Meltzer, **US 3287458** (1966 to Warner-Lambert); *eidem, J. Med. Chem.* **11**, 160 (1968). Pharmacology: Turner *et al., Antimicrob. Agents Chemother.* **1967**, 475 sqq. Metabolism studies: DiCarlo *et al., Arch. Biochem. Biophys.* **127**, 503 (1968); Crew *et al., Xenobiotica* **1**, 193 (1971). Use in urinary tract infections: S. Kalowski *et al., Med. J. Aust.* **21**, 345 (1979); E. Anza *et al., Minerva Med.* **70**, 2333 (1979). Mechanism of action study: H. T. Wright *et al., Science* **213**, 455 (1981). Brief review of therapeutic use: R. Glickman *et al., Am. J. Hosp. Pharm.* **36**, 1077-1079 (1979).

Crystals from DMF, mp 314-316° (dec). LD_{50} in mice, rats (mg/kg): >6000, >2000 orally (Turner).

THERAP CAT: Antibacterial.

THERAP CAT (VET): Antibacterial.

7045. Oxomemazine. [3689-50-7] N,N,β-Trimethyl-10H-phenothiazine-10-propanamine 5,5-dioxide; 10-(3-dimethylamino-2-methylpropyl)phenothiazine 5,5-dioxide; 3-(9,9-dioxo-10-phenothiazinyl)-2-methyl-1-dimethylaminopropane; RP-6847; Dysedon. $C_{18}H_{22}N_2O_2S$; mol wt 330.45. C 65.43%, H 6.71%, N 8.48%, O 9.68%, S 9.70%. Prepn: Jacob, Robert, US 2972612 (1961 to Rhône-Poulenc).

Crystals from heptane, mp 115°.

Hydrochloride. Doxergan; Imakol. $C_{18}H_{22}N_2O_2S$.HCl; mol wt 366.90. Crystals from ethanol + isopropanol, mp 250°.

7046. Oxonic Acid. [937-13-3] 1,4,5,6-Tetrahydro-4,6-dioxo-1,3,5-triazine-2-carboxylic acid; allantoxanic acid; 5-azaorotic acid; s-triazine-2,4-dione-6-carboxylic acid; 2,4-dioxo-1,2,3,4-tetrahydro-1,3,5-triazine-6-carboxylic acid. $C_4H_3N_3O_4$; mol wt 157.09. C 30.58%, H 1.93%, N 26.75%, O 40.74%. Formed during alkaline oxidation of uric acid with H_2O_2 or $KMnO_4$. Prepn and structure: Brandenberger, *Helv. Chim. Acta* **37**, 641 (1954); Brandenberger, *ibid.* **37**, 2207 (1954); Piskala, Gut, *Collect. Czech. Chem. Commun.* **27**, 1572 (1962). Metabolism: Chelbova *et al.*, *Biochem. Pharmacol.* **19**, 2785 (1970); Cihak, Sorm, *ibid.* **21**, 607 (1972).

Free acid is very unstable; can be isolated as the monopotassium salt. uv max (pH 9): 255 nm (ε 6800); (pH 12): 252 nm (ε 4600).

Oxonic acid amide. $C_4H_4N_4O_3$. Microcrystalline substance. Practically insol in organic solvents, sparingly sol in boiling water. Does not melt below 350°, darkens above 300°.

7047. Oxophenarsine. [538-03-4] 2-Amino-4-arsenosophenol; 3-amino-4-hydroxyphenylarsene oxide; arsenoxide. C_6H_6-$AsNO_2$; mol wt 199.04. C 36.21%, H 3.04%, As 37.64%, N 7.04%, O 16.08%. Prepn: P. Ehrlich, A. Bertheim, *Ber.* **45**, 756 (1912). Purif of free base: A. B. Scott, J. A. Sultzaberger, US 2280132 (1942 to Parke, Davis); of hydrochloride: A. B. Scott, B. F. Tullar, CA 405532 (1942 to Parke, Davis). *See also:* G. W. Raiziss, J. L. Gavron, *Organic Arsenical Compounds* (Chem. Catalog Co., New York, 1923), pp 136-137.

White amorphous powder. Readily sol in alcohols, glycerin, pyridine; sol in acetone; moderately sol in water.

Hydrochloride. [538-03-4] 2-Amino-4-arsinylphenol hydrochloride (1:1); Ehrlich 5; Mapharsen. $C_6H_6AsNO_2$.HCl; mol wt 235.50. Hygroscopic, white or nearly white, odorless powder. Readily sol in water, methanol, ethanol, in solns of alkali hydroxides and carbonates, in dil mineral acids. Sparingly sol in glacial acetic acid. Practically insol in acetone, ether.

THERAP CAT: Formerly as antisyphilitic.

7048. Oxophenylarsine. [637-03-6] Arsenosobenzene; phenyl arsenoxide; phenylarsine oxide; Arzene. C_6H_5AsO; mol wt 168.03. C 42.89%, H 3.00%, As 44.59%, O 9.52%. Prepn: Blicke, Smith, *J. Am. Chem. Soc.* **51**, 3479 (1929); **52**, 2946 (1930).

Crystals, mp 144-146° (also reported as mp 119-120°).

USE: Pesticide.

THERAP CAT (VET): Coccidiostat (poultry).

7049. L-2-Oxo-4-thiazolidinecarboxylic Acid. [19771-63-2] (4R)-2-Oxo-4-thiazolidinecarboxylic acid; OTCA; OTC; Procysteine. $C_4H_5NO_3S$; mol wt 147.15. C 32.65%, H 3.43%, N 9.52%, O 32.62%, S 21.79%. Cysteine prodrug. Prepn: J. A. Maclaren, *Aust. J. Chem.* **21**, 1891 (1968); T. Kömives, *Org. Prep. Proced. Int.* **21**, 251 (1989). Crystal structure: N. Ramasubbu, R. Parthasarathy, *Int. J. Pept. Protein Res.* **34**, 153 (1989). Precursor for cysteine and use as glutathione biosynthesis stimulant: J. M. Williamson *et al.*, *Proc. Natl. Acad. Sci. USA* **79**, 6246 (1982); in chicks: T. K. Chung *et al.*, *J. Nutr.* **120**, 158 (1990). Toxicity: R. D. White *et al.*, *Acute Toxic. Data* **1**, 164 (1992); R. D. White *et al.*, *Toxicol. Lett.* **69**, 15 (1993). Clinical evaluation in HIV infection: G. Giorgi *et al.*, *Curr. Ther. Res.* **52**, 461 (1992); in asymptomatic HIV infection: R. C. Kalayjian *et al.*, *J. Acquir. Immune Defic. Syndr.* **7**, 369 (1994). HPLC determn in plasma: L. W. Webb *et al.*, *J. Chromatogr. B* **654**, 257 (1994).

Colorless cubes, mp 171-173° (dec) (Maclaren). Also reported as white crystals, mp 173-174° (Kömives). pKa (22°): 3.32. $[\alpha]_D^{20}$ −59.4° (c = 2). Readily sol in alcohol. Insol in benzene and ethyl acetate.

7050. Oxotremorine. [70-22-4] 1-[4-(1-Pyrrolidinyl)-2-butyn-1-yl]-2-pyrrolidinone; 1-(2-oxo-1-pyrrolidinyl)-4-(1-pyrrolidinyl)-2-butyne. $C_{12}H_{18}N_2O$; mol wt 206.29. C 69.87%, H 8.80%, N 13.58%, O 7.76%. Active metabolite of tremorine, *q.v.*: Cho *et al.*, *Biochem. Biophys. Res. Commun.* **5**, 276 (1961). Synthesis: Bebbington, Shakeshaft, *J. Med. Chem.* **8**, 274 (1965); Nau, US 3444185 (1969 to Soc. Civile Auguil). Cholinergic agent, used experimentally as convulsant in study of parkinsonism: Cho *et al.*, *Proc. 2nd Int. Pharmacol. Meet.* **2**, 75 (1964).

Pale yellow liquid. $bp_{0.6}$ 150-155°; $bp_{0.1}$ 124°. n_D^{25} 1.5156.

Picrolonate. $C_{22}H_{26}N_6O_6$. Crystals from acetone, mp 157-159°. USE: Pharmacological tool.

7051. Oxprenolol. [6452-71-7] 1-[(1-Methylethyl)amino]-3-[2-(2-propen-1-yloxy)phenoxy]-2-propanol; 1-[*o*-(allyloxy)phenoxy]-3-(isopropylamino)-2-propanol; 1-(isopropylamino)-2-hydroxy-3-[*o*-(allyloxy)phenoxy]propane. $C_{15}H_{23}NO_3$; mol wt 265.35. C 67.90%, H 8.74%, N 5.28%, O 18.09%. β-Adrenergic blocker. Prepn: BE 669402 (1966 to Ciba), *C.A.* **65**, 5402d (1966), corresp to GB 1077603. Metabolic studies: Garteiz, *J. Pharmacol.*

Exp. Ther. **179**, 354 (1971). Crystal structure of hydrochloride: J. M. Leger *et al., Acta Crystallogr.* **33B**, 2156 (1977). Long-term prevention study in coronary heart disease: S. H. Taylor *et al., N. Engl. J. Med.* **307**, 1293 (1982).

Crystals from hexane, mp 78-80°.

Hydrochloride. [6452-73-9] Ba-39089; Coretal; Laracor; Paritane; Trasicor; Trasacor. $C_{15}H_{23}NO_3.HCl$; mol wt 301.81. White crystals, mp 107-109°. Freely sol in water, alc, chloroform; sparingly sol in acetone. Practically insol in ether.

THERAP CAT: Antihypertensive, antianginal, antiarrhythmic.

7052. Oxyacanthine. [548-40-3] (4a*R*,16a*S*)-3,4,4a,5,16a,-17,18,19-Octahydro-21,22,26-trimethoxy-4,17-dimethyl-2*H*-1-24:12,15-dietheno-6,10-metheno-16*H*-pyrido[2′,3′:17,18][1,10]-dioxacycloeicosino[2,3,4-*ij*]isoquinolin-9-ol; 6,6′,7-trimethoxy-2,2′-dimethyloxyacanthan-12′-ol; vinetine. $C_{37}H_{40}N_2O_6$; mol wt 608.74. C 73.00%, H 6.62%, N 4.60%, O 15.77%. From root of *Berberis vulgaris* L., *Berberidaceae:* Rüdel, *Arch. Pharm.* **229**, 631 (1891); Späth, Kolbe, *Ber.* **58**, 2280 (1925). Structure: v. Bruchhausen *et al., Ann.* **507**, 144 (1933); Fujita, *J. Pharm. Soc. Jpn.* **72**, 213 (1952), *C.A.* **47**, 6429b (1953).

White, cryst, bitter powder. mp 216-217°. $[\alpha]_D^{20}$ +131.5° (chloroform). Practically insol in water; sol in alcohol, chloroform, ether, dil acids.

Dihydrochloride. $C_{37}H_{40}N_2O_6.2HCl$. mp 270-271°. $[\alpha]_D$ +185.5°. Sol in water.

7053. Oxybenzone. [131-57-7] (2-Hydroxy-4-methoxyphenyl)phenylmethanone; 2-hydroxy-4-methoxybenzophenone; 4-methoxy-2-hydroxybenzophenone; benzophenone-3; MOB; Cyasorb UV 9; Eusolex 4360; Neo Heliopan BB; Uvinul M-40. $C_{14}H_{12}O_3$; mol wt 228.25. C 73.67%, H 5.30%, O 21.03%. UV-A/B absorber in cosmetics and sunscreens. Prepn: König, v. Kostanecki, *Ber.* **39**, 4027 (1906); Hardy, Forster, **US 2773903** (1956 to Am. Cyanamid); Stanley *et al.*, **US 2861104**; **US 2861105**; **US 3073866** (1958, 1958, 1963 all to General Aniline & Film). Toxicity study: H. -J. Lewerenz *et al., Food Cosmet. Toxicol.* **10**, 41 (1972).

Crystals from isopropanol, mp 66°. Freely sol in alc, toluene; readily sol in most other organic solvents. Practically insol in water. LD_{50} orally in rats: >12.8 g/kg (Lewerenz).

USE: Ultraviolet light absorber and stabilizer, esp in plastics and paints.

THERAP CAT: Ultraviolet screen.

7054. Oxybutynin. [5633-20-5] α-Cyclohexyl-α-hydroxybenzeneacetic acid 4-(diethylamino)-2-butyn-1-yl ester; α-phenylcyclohexaneglycolic acid 4-(diethylamino)-2-butynyl ester; 4-diethylamino-2-butynyl phenylcyclohexylglycolate; oxibutinina. $C_{22}H_{31}NO_3$; mol wt 357.49. C 73.92%, H 8.74%, N 3.92%, O 13.43%. Muscarinic receptor antagonist. Prepn: **GB 940540** (1963 to Mead Johnson). Physico-chemical properties: E. Miyamoto *et al., Analyst* **119**, 1489 (1994). Synthesis of *S*-form: S. Masumoto *et al., Tetrahedron Lett.* **43**, 8647 (2002). GC-MS determn in plasma: K. S. Patrick *et al., J. Chromatogr.* **487**, 91 (1989). Clinical pharmacokinetics and pharmacodynamics: J. L. Reiz *et al., J. Clin. Pharmacol.* **47**, 351 (2007). Clinical trial in overactive bladder: P. Sand *et al., BJU Int.* **99**, 836 (2006). Review of pharmacodynamics and therapeutic use: Y. E. Yarker *et al., Drugs Aging* **6**, 243-262 (1995).

pKa 8.04. Log P (*n*-octanol/water): 2.9 (pH 6). Soly in water (mg/ml): 77 (pH 1); 0.8 (pH 6); 0.012 (pH >9.6).

Hydrochloride. [1508-65-2] Oxybutynin chloride; MJ-4309-1; Cystrin; Ditropan; Dridase; Driptane; Kentera; Pollakisu; Tropax. $C_{22}H_{31}NO_3.HCl$; mol wt 393.95. Crystals, mp 129-130°. Sol in water, acids. Practically insol in alkali. LD_{50} orally in rats: 1220 mg/kg. *See:* E. I. Goldenthal, *Toxicol. Appl. Pharmacol.* **18**, 185 (1971).

THERAP CAT: In treatment of urinary incontinence.

THERAP CAT (VET): In treatment of urinary incontinence.

7055. Oxychlorosene. [8031-14-9] Monoxychlorosene; Clorpactin; Clorpactin XCB. $C_{20}H_{35}ClO_4S$; mol wt 407.01. C 59.02%, H 8.67%, Cl 8.71%, O 15.72%, S 7.88%. $C_{20}H_{34}O_3$-S.HOCl. Described as a buffered organic hypochlorous acid derivative with slightly acid pH. Contains a long chain hydrocarbon (surface-active agent). The chain (14 carbons) may be branched or straight but usually consists of a constant mixture having predominantly the straight chain. The hydrocarbon chain also has a phenyl substituent which in turn holds a sulfonic acid group. The complex corresponds to a formula: $HO_3S-C_6H_4-(C_{14}H_{29}).HOCl$.

Aq solns are unstable and should be freshly prepd; the solid may be stored in properly stoppered bottles.

Sodium salt. [52906-84-0] Clorpactin WCS.

THERAP CAT: Antiseptic.

7056. Oxycinchophen. [485-89-2] 3-Hydroxy-2-phenyl-4-quinolinecarboxylic acid; 3-hydroxy-2-phenylcinchoninic acid; 3-hydroxycinchophen; HPC; Fenidrone; Magnofenyl; Magnophenyl; Oxinofen; Reumalon. $C_{16}H_{11}NO_3$; mol wt 265.27. C 72.45%, H 4.18%, N 5.28%, O 18.09%. Prepn: Berlingozzi, Capuano, *Atti Accad. Lincei* [V] **33**, II, 91 (1924); John, Fränkel, *J. Prakt. Chem.* **133**, 259 (1932); improved procedure: Marshall, Blanchard, *J. Pharmacol.* **95**, 186 (1949); Kreysa, **US 2776290** (1957 to Chemo Puro).

Minute, deep yellow prisms from alc, dec 206-207°. Sol in acetic acid, alkalies, hot alc, benzene. Sparingly sol in water, ether. Forms an alkaline, water-sol sodium salt.

THERAP CAT: Antidiuretic; uricosuric.

7057. Oxyclozanide. [2277-92-1] 2,3,5-Trichloro-*N*-(3,5-dichloro-2-hydroxyphenyl)-6-hydroxybenzamide; 3,3′,5,5′,6-pentachloro-2'-hydroxysalicylanilide; 3,3',5,5',6-pentachloro-2,2'-di-hydroxybenzanilide; Zanil. $C_{13}H_6Cl_5NO_3$; mol wt 401.45. C 38.89%, H 1.51%, Cl 44.15%, N 3.49%, O 11.96%. Prepn: **NL 6409325** corresp to Broome *et al.*, **US 3349090** (1965, 1967, both to I.C.I.).

Crystals, mp 209-211°.

THERAP CAT (VET): Anthelmintic (Trematodes).

7058. Oxycodone. [76-42-6] (5α)-4,5-Epoxy-14-hydroxy-3-methoxy-17-methylmorphinan-6-one; 6-deoxy-7,8-dihydro-14-hydroxy-3-*O*-methyl-6-oxomorphine; dihydrohydroxycodeinone; 14-hydroxydihydrocodeinone; Dihydrone. $C_{18}H_{21}NO_4$; mol wt 315.37. C 68.55%, H 6.71%, N 4.44%, O 20.29%. Semisynthetic opioid analgesic. Prepn from thebaine: M. Freund, E. Speyer, *J. Prakt. Chem.* **94**, 135 (1916); R. Krassnig *et al.*, *Arch. Pharm.* **329**, 325 (1996); from codeine: A. J. Walker, N. C. Bruce, *Tetrahedron* **60**, 561 (2004). Bibliography: Small, Lutz, *"Chemistry of the Opium Alkaloids,"* Suppl. No. 103 to Public Health Reports, Washington (1932); K. W. Bentley, *The Chemistry of the Morphine Alkaloids* (Oxford, 1954). *In vitro* liposolubility and protein binding: R. Pöyhiä, T. Seppälä, *Pharmacol. Toxicol.* **74**, 23 (1994). HPLC determn in serum: A. W. E. Wright *et al.*, *J. Chromatogr. B* **712**, 169 (1998). Review of pharmacokinetics and clinical experience: E. Kalso, *J. Pain Symptom Manage.* **29**, S47-S56 (2005); of abuse potential: T. J. Cicero *et al.*, *J. Pain* **6**, 662-672 (2005).

Long rods from alc, mp 218-220°. pKa 8.53. Sol in alcohol, chloroform. Nearly insol in ether, water, KOH, NaOH, NH₄OH.

Hydrochloride. [124-90-3] Dinarkon; Eubine; Eukodal; Oxy-Contin; Oxygesic; Oxynorm; Supeudol. $C_{18}H_{21}NO_4 \cdot HCl$; mol wt 351.83. Component of **Percocet** and **Percodan.** Long rods from water, dec 270-272°. $[\alpha]_D^{20}$ −125° (c = 2.5). Partition coefficient (octanol/water): 0.7. One gram dissolves in 6-7 ml water. Slightly sol in alcohol.

Pectinate. [9012-92-4] Proladone. Used for prolonged action.

Note: This is a controlled substance (opiate): **21 CFR**, 1308.12.

THERAP CAT: Analgesic.

7059. 4,4'-Oxydi-2-butanol. [821-33-0] 4,4'-Oxybis-2-butanol; 3,3'-dihydroxydibutyl ether; bis[3-hydroxybutyl] ether; DHBE; Diskin; Dyskinébyl; Dis-Cinil; Colenormol. $C_8H_{18}O_3$; mol wt 162.23. C 59.23%, H 11.18%, O 29.59%. Prepn: Joulty, **FR 1267084** (1960). Action on bile flow: Bornmann, *Arzneim.-Forsch.* **2**, 122 (1952). Use as choleretic and antispasmodic: Albot, Toulet, *Presse Med.* **67**, 2053 (1959). General pharmacological properties: Fregnan, Porta, *Arzneim.-Forsch.* **26**, 2116 (1976).

Clear, almost colorless, bitter fluid. Miscible with water.

THERAP CAT: Choleretic.

7060. 10,10'-Oxydiphenoxarsine. [58-36-6] 10,10'-Oxybis-10*H*-phenoxarsine; bis(phenoxarsin-10-yl)ether; bis(10-phenoxarsyl)oxide; bis(10-phenoxarsinyl)oxide; OBPA; Vinyzene. $C_{24}H_{16}$-As_2O_3; mol wt 502.23. C 57.40%, H 3.21%, As 29.84%, O 9.56%. Prepn: W. L. Lewis *et al.*, *J. Am. Chem. Soc.* **43**, 891 (1921). Synthesis: K. D. Shvetsova-Shilovskaya *et al.*, *J. Gen. Chem. USSR* **31**, 776 (1961). Use in control of plant growth: M. L. Joseph, J. L. Hardy, **US 3069252** (1962 to Dow Chemical); as bactericide in plastics: C. C. Yeager, **US 3288674** (1966 to Scientific Chemicals). Toxicity study: D. G. Clark, *Toxicol. Appl. Pharmacol.* **21**, 315 (1972).

Colorless monoclinic prisms, mp 184-185°. dec. 380°. sp. gr. 1.40-1.42. Sol in alcohol, chloroform, methylene chloride. Practically insol in water (5 ppm at 20°) and alkali. LD₅₀ in male rats: 35-50 mg/kg (Ventron, company data sheet).

USE: Primarily for fungicidal and bactericidal protection of plastics.

7061. Oxyfedrine. [15687-41-9] 3-[[(1*S*,2*R*)-2-Hydroxy-1-methyl-2-phenylethyl]amino]-1-(3-methoxyphenyl)-1-propanone; L-3-[(β-hydroxy-α-methylphenethyl)amino]-3'-methoxypropiophenone; L-(1-hydroxy-1-phenyl-2-propylamino)-1-(*m*-methoxyphenyl)-1-propanone; oxyphedrine. $C_{19}H_{23}NO_3$; mol wt 313.40. C 72.82%, H 7.40%, N 4.47%, O 15.31%. Partial β-adrenergic agonist with coronary vasodilating and positive inotropic effects. Prepn: K. Thiele, **BE 630296**; *idem*, **US 3225095** (1963, 1965 both to Degussa); K. Thiele *et al.*, *Arzneim.-Forsch.* **16**, 1064 (1966). Absolute configuration determined by circular dichroism: J. Engel *et al.*, *Chem. Ztg.* **105**, 85 (1981). TLC determn: Musumarra, *J. Chromatogr.* **350**, 151 (1985). Pharmacology: H. Hueller *et al.*, *Pharmazie* **27**, 242 (1972). Mode of action: P. Mentz, W. Forster, *Arzneim.-Forsch.* **34**, 1739 (1984); N. Sternitzke, *Z. Kardiol.* **73**, 586 (1984). In prevention of experimental myocardial necrosis in rats: S. D. Seth *et al.*, *Arzneim.-Forsch.* **34**, 678 (1984). Comparison with atenolol, *q.v.*, in angina pectoris: L. Fananapazir, C. Bray, *Br. J. Clin. Pharmacol.* **20**, 405 (1985).

L-Form hydrochloride. [16777-42-7] D-563; Ildamen; Modacor. $C_{19}H_{23}NO_3 \cdot HCl$; mol wt 349.86. Crystals from methanol, mp 192-194°. LD₅₀ in mice (mg/kg): 29 i.v. (Hueller).

DL-Form hydrochloride. [16648-69-4] mp 173-175°. LD₅₀ in mice (mg/kg): 34 i.v. (Hueller).

THERAP CAT: Antianginal. Treatment of coronary insufficiency.

7062. Oxyfluorfen. [42874-03-3] 2-Chloro-1-(3-ethoxy-4-nitrophenoxy)-4-(trifluoromethyl)benzene; 2-chloro-α,α,α-trifluoro-*p*-tolyl-3-ethoxy-4-nitrophenyl ether; RH-2915; Delta Goal; Goal. $C_{15}H_{11}ClF_3NO_4$; mol wt 361.70. C 49.81%, H 3.07%, Cl 9.80%, F 15.76%, N 3.87%, O 17.69%. Selective pre- and post-emergence herbicide. Prepn: **NL 7303590**; H. O. Bayer *et al.*, **US 3798276** (1973, 1974 both to Rohm & Haas). Activity: R. Y. Yih, C. Swithenbank, *J. Agric. Food Chem.* **23**, 592 (1975). Metabolism: I. L. Adler *et al.*, *ibid.* **25**, 1339 (1977).

Orange crystalline solid, mp 83-84°. Soly in water: 0.1 ppm. Sol in most organic solvents. LD_{50} orally in male albino rats: >5000 mg/kg (Yih, Swithenbank).

USE: Herbicide.

7063. Oxygen. [7782-44-7] O; at. wt [15.99903; 15.99977]; conventional at. wt 15.999; at. no. 8; valence 2. Group VIA (16). Occurs normally as the diatomic gas O_2, also as ozone O_3. Atomic oxygen (O) can be prepd. Three naturally occurring isotopes: 16 (99.759%); 17 (0.037%); 18 (0.204%); artificial radioactive isotopes: 13-15; 19; 20. The most abundant element on earth; makes up 46.6% of earth's crust; 20.95% by vol of dry air. Obtained on a large scale by liquefaction of air. First obtained by Scheele in 1771 and independently by Priestley in 1774. Monograph: M. Ardon, *Oxygen: Elementary Forms and Hydrogen Peroxide* (Benjamin, New York, 1965) 106 pp. Review of oxygen and its compounds: Ebsworth *et al.*, in *Comprehensive Inorganic Chemistry* **vol. 2**, J. C. Bailar, Jr. *et al.*, Eds. (Pergamon Press, Oxford, 1973) pp 685-794; A. H. Taylor in *Kirk-Othmer Encyclopedia of Chemical Technology* **vol. 16** (Wiley-Interscience, New York, 3rd ed., 1981) pp 653-673.

Colorless, odorless, tasteless, neutral gas; non-flammable, supports combustion. *Oxidizer. Explosion hazard; avoid smoking, flames, electric sparks.* d^0 (gas) 1.429 g/l; d^{-183} (liquid) 1.14 g/ml. mp −218.4°. bp −182.96°. Critical temp −118.95°. Critical press. 50.14 atm. Heat of vaporization (−183°): 50.9 cal/g. Usually marketed under pressure in metal cylinders. One vol gas dissolves in 32 vols water at 20°; in 7 vols alcohol at 20°; also sol in other organic liquids and usually to a greater extent than in water.

USE: In oxyhydrogen or oxyacetylene flame for welding metals and for lighting (calcium light, etc); submarine work by divers, propellant for rockets. In the production of synthesis gas which can be used in the Fischer-Tropsch process for liquid fuels.

THERAP CAT: Medicinal gas to relieve hypoxia; at hyperbaric pressures in cardiac and other surgery, anaerobic infections, carbon monoxide poisoning; in cryotherapy (liq form).

THERAP CAT (VET): In hypoxia and in conjunction with volatile anesthetics.

7064. Oxymesterone. [145-12-0] (17β)-4,17-Dihydroxy-17-methylandrost-4-en-3-one; 4,17β-dihydroxy-17α-methyl-3-oxoandrost-4-ene; 4-hydroxy-17α-methyltestosterone; 17α-methyl-4-androstene-4,17β-diol-3-one; oxymestrone; Anamidol; Oranabol; Theranabol. $C_{20}H_{30}O_3$; mol wt 318.46. C 75.43%, H 9.50%, O 15.07%. Prepn: **GB 848288**; Camerino *et al.*, **US 3060201** (1960, 1962 to Farmitalia).

Crystals, mp 169-171°. $[\alpha]_D^{20}$ +69° (ethanol). uv max (ethanol): 278 nm ($E_{1cm}^{1\%}$ 406). Practically insol in water. Sol in chloroform, acetone, alcohol.

Note: This is a controlled substance (anabolic steroid): **21 CFR**, 1308.13, as defined in 1300.01.

THERAP CAT: Androgen; anabolic.

7065. Oxymetazoline. [1491-59-4] 3-[(4,5-Dihydro-1*H*-imidazol-2-yl)methyl]-6-(1,1-dimethylethyl)-2,4-dimethylphenol; 6-*tert*-butyl-3-(2-imidazolin-2-ylmethyl)-2,4-dimethylphenol; 2-(4-*tert*-butyl-2,6-dimethyl-3-hydroxybenzyl)-2-imidazoline; H-990; Navisin; Hazol; Rhinofrenol; Rhinolitan; Sinerol; Nezeril. $C_{16}H_{24}$-N_2O; mol wt 260.38. C 73.81%, H 9.29%, N 10.76%, O 6.14%. Prepd from (4-*tert*-butyl-2,6-dimethyl-3-hydroxyphenyl)acetonitrile and ethylenediamine: Fruhstorfer, Mueller-Calgan, **DE 1117588**

(1961 to E. Merck), *C.A.* **57**, 4674a (1962). Toxicity data: Hotovy *et al.*, *Arzneim.-Forsch.* **11**, 1016 (1961).

Crystals from benzene, mp 181-183°.

Hydrochloride. [2315-02-8] Afrazine; Afrin; Iliadin; Nafrine; Nasivin; Oxilin; Sinex. $C_{16}H_{24}N_2O \cdot HCl$; mol wt 296.84. White, hygroscopic crystals, dec 300-303°. Sol in water, alc. Practically insol in ether, chloroform, benzene. LD_{50} orally in mice: 10 mg/kg (Hotovy).

Note: Ingredient of **Drixin**.

THERAP CAT: Adrenergic (vasoconstrictor); nasal decongestant.

7066. Oxymetholone. [434-07-1] (5α,17β)-17-Hydroxy-2-(hydroxymethylene)-17-methylandrostan-3-one; 2-hydroxymethylene-17α-methyldihydrotestosterone; 4,5α-dihydro-2-hydroxymethylene-17α-methyltestosterone; 2-hydroxymethylene-17α-methyl-17β-hydroxy-5α-androstan-3-one; 2-hydroxymethylene-17α-methylandrostan-17β-ol-3-one; anasterone; Adroyd; Anapolon; Anadrol; Pardroyd; Plenastril; Protanabol; Nastenon; Synasteron. $C_{21}H_{32}O_3$; mol wt 332.48. C 75.86%, H 9.70%, O 14.44%. Anabolic steroid. Prepn: Ringold *et al.*, *J. Am. Chem. Soc.* **81**, 427 (1959); Ringold, Rosenkranz, **DE 1070632** (1959 to Syntex).

Crystals from ethyl acetate, mp 178-180°. $[\alpha]_D$ +38°. uv max: 285 nm (log ε 3.99). Freely sol in chloroform; sol in dioxane; sparingly sol in alc; slightly sol in ether. Practically insol in water.

Enol acetate. $C_{23}H_{34}O_4$. Crystals from hexane, mp 144-148°. $[\alpha]_D$ +27° (ethanol). uv max: 255 nm (log ε 4.09).

Enol propionate. $C_{24}H_{36}O_4$. Crystals from hexane, mp 135°. $[\alpha]_D$ +26° (ethanol). uv max: 257 nm (log ε 4.11).

Enol benzoate. $C_{28}H_{36}O_4$. Crystals from acetone + water, mp 188-190°. uv max: 230 nm (log ε 4.19).

Caution: Oxymetholone is reasonably anticipated to be a human carcinogen: *Report on Carcinogens, Twelfth Edition* (PB2011-111646, 2011) p 338. This is a controlled substance (anabolic steroid): **21 CFR**, 1308.13, as defined in 1300.01.

THERAP CAT: Androgen.

THERAP CAT (VET): Anabolic steroid for small animals.

7067. Oxymethurea. [140-95-4] *N*,*N*'-Bis(hydroxymethyl)-urea; *N*,*N*'-dihydroxymethylurea; dimethanol urea; Methural. C_3-$H_8N_2O_3$; mol wt 120.11. C 30.00%, H 6.71%, N 23.32%, O 39.96%. Prepn: Einhorn, Hamburger, *Ber.* **41**, 26 (1908); Walter, **US 1863426** (1927); **US 2436355** (1946 to du Pont).

Crystals from alcohol, mp 137-139°. Very sol in cold water, hot ethyl alcohol, methanol.

USE: In the textile industry in cotton crease- and shrink-proofing, finishing, drying, sizing; in tanning; pesticides; in photographic developers.

THERAP CAT: Antiseptic.

7068. Oxymorphone. [76-41-5] (5α)-4,5-Epoxy-3,14-dihydroxy-17-methylmorphinan-6-one; dihydrohydroxymorphinone; dihydro-14-hydroxymorphinone; 14-hydroxydihydromorphinone.

$C_{17}H_{19}NO_4$; mol wt 301.34. C 67.76%, H 6.36%, N 4.65%, O 21.24%. Semisynthetic opioid analgesic. Prepn from dihydrohydroxycodeinone: U. Weiss, *J. Am. Chem. Soc.* **77**, 5891 (1955); M. J. Lewenstein, U. Weiss, US 2806033 (1957); by demethylation of oxycodone: A. Coop *et al.*, *J. Org. Chem.* **63**, 4392 (1998). Comparative toxicity study: P. E. Tullar, *Toxicol. Appl. Pharmacol.* **3**, 261 (1961). GC-MS determn in blood: R. Meatherall, *J. Anal. Toxicol.* **29**, 301 (2005). Pharmacokinetics of extended release formulation: M. P. Adams, H. Ahdieh, *Pharmacotherapy* **24**, 468 (2004). Review of pharmacology and clinical efficacy: E. Prommer, *Support. Care Cancer* **14**, 109-115 (2006).

Crystals from boiling ethanol, ethyl acetate or benzene. mp 248-249° (dec). Levorotatory. Sol in boiling acetone and chloroform; readily sol in aq alkalies; moderately sol in boiling ethanol; sparingly sol in benzene. LD_{50} in mice (mg/g): 0.200 s.c. (Tullar).

Hydrochloride. [357-07-3] Numorphan; Opana. $C_{17}H_{19}NO_4$.HCl; mol wt 337.80. White or slightly off-white, odorless powder. Darkens on exposure to light. Freely sol in water; sparingly sol in ethanol, ether. pKa_1 8.17; pKa_2 9.54. Partition coefficient (octanol/water at 37° and pH 7.4): 0.98

Note: This is a controlled substance (opiate): **21 CFR**, 1308.12.

THERAP CAT: Analgesic.

THERAP CAT (VET): Analgesic.

7069. Oxypendyl. [5585-93-3] 4-[3-(10*H*-Pyrido[3,2-*b*][1,4]benzothiazin-10-yl)propyl]-1-piperazineethanol; 10-[3-[4-(2-hydroxyethyl)-1-piperazinyl]propyl]-10*H*-pyrido[3,2-*b*][1,4]benzothiazine; 10-[3-(1-hydroxyethyl-4-piperazinyl)propyl]-4-azaphenothiazine; oxipendyl; D-706. $C_{20}H_{26}N_4OS$; mol wt 370.52. C 64.83%, H 7.07%, N 15.12%, O 4.32%, S 8.65%. Prepn: Schuler, Klebe, *Ann.* **653**, 172 (1962).

bp_1 260-265°, bp_6 280-300°.

Dihydrochloride. [17297-82-4] Pervetral. $C_{20}H_{26}N_4OS.2HCl$; mol wt 443.43. Crystals, mp 218-220°.

THERAP CAT: Antiemetic.

7070. Oxypertine. [153-87-7] 5,6-Dimethoxy-2-methyl-3-[2-(4-phenyl-1-piperazinyl)ethyl]-1*H*-indole; 1-[2-(5,6-dimethoxy-2-methyl-3-indolyl)ethyl]-4-phenylpiperazine; Win-18501-2; Equipertine; Forit; Integrin (formerly). $C_{23}H_{29}N_3O_2$; mol wt 379.50. C 72.79%, H 7.70%, N 11.07%, O 8.43%. Prepn: Archer *et al.*, *J. Am. Chem. Soc.* **84**, 1306 (1962).

THERAP CAT: Antidepressant.

7071. Oxyphenbutazone. [129-20-4] 4-Butyl-1-(4-hydroxyphenyl)-2-phenyl-3,5-pyrazolidinedione; 4-butyl-2-(*p*-hydroxyphenyl)-1-phenyl-3,5-pyrazolidinedione; 1-phenyl-2-(*p*-hydroxyphenyl)-3,5-dioxo-4-*n*-butylpyrazolidin; 1-(*p*-hydroxyphenyl)-2-phenyl-4-butylpyrazolidine-3,5-dione; *p*-hydroxyphenylbutazone; G-27202; Californit; Crovaril; Flogitolo; Flogoril; Frabel; Neo-Farmadol; Oxalid; Rapostan; Tandearil; Visubutina. $C_{19}H_{20}N_2O_3$; mol wt 324.38. C 70.35%, H 6.21%, N 8.64%, O 14.80%. Prepn: Hafliger, US 2745783 (1956 to Geigy); Pfister, Hafliger, *Helv. Chim. Acta* **40**, 395 (1957).

Monohydrate. [7081-38-1] Imbun; Phlogistol; Phlogase; Phlogont. Crystals, mp 96°. Anhydr crystals from ether + petr ether, mp 124-125°. Acidic reaction. Soluble in ethanol, methanol, chloroform, benzene, ether. Forms a water-soluble sodium salt.

THERAP CAT: Anti-inflammatory.

7072. Oxyphencyclimine. [125-53-1] α-Cyclohexyl-α-hydroxybenzeneacetic acid (1,4,5,6-tetrahydro-1-methyl-2-pyrimidinyl)methyl ester; 1,4,5,6-tetrahydro-1-methyl-2-pyrimidinylmethyl α-phenylcyclohexaneglycolate; α-phenylcyclohexaneglycolic acid 1-methyl-2-tetrahydroxypyrimidylmethyl ester; 1-methyl-1,4,5,6-tetrahydro-2-pyrimidylmethyl α-cyclohexyl-α-phenylglycolate; Antulcus; Caridan; Daricol; Setrol; Vio-Thene; Daricon; Naridan; Zamanil. $C_{20}H_{28}N_2O_2$; mol wt 344.46. C 69.74%, H 8.19%, N 8.13%, O 13.93%. Anticholinergic. Prepn: **GB 795758** (1958 to Pfizer); Faust *et al.*, *J. Am. Chem. Soc.* **81**, 2214 (1959).

Hydrochloride. [125-52-0] $C_{20}H_{28}N_2O_3$.HCl. Crystals, dec 231-232°. Solubility in water: 1.2 g/100 ml.

THERAP CAT: Antispasmodic.

7073. Oxyphenisatin Acetate. [115-33-3] 3,3-Bis[4-(acetyloxy)phenyl]-1,3-dihydro-2*H*-indol-2-one; 3,3-bis(*p*-acetoxyphenyl)oxindole; acetphenolisatin; endophenolphthalein; diacetyldiphenolisatin; diacetyldihydroxydiphenylisatin; diacetylhydroxyphenylisatin; diacetoxydiphenylisatin; di(acetoxyphenyl)oxindole; diphesatin; diacetyldioxyphenylisatin; Isacen; Isocrin; Isaphen; Laxo-Isatin; Promassolax; Sanapert; Bydolax; Cirotyl; Lisagal; Contax; Prulet; Purgaceen; Bisatin. $C_{24}H_{19}NO_5$; mol wt 401.42. C 71.81%, H 4.77%, N 3.49%, O 19.93%. Prepd from isatin by treatment with phenol followed by acetylation in the presence of H_2SO_4 at temps below 100°: Baeyer, Lazarus, *Ber.* **18**, 2641 (1885); Hoffmann-La Roche, CH 100806; DE 406210 (1923); US 1624675 (1927); DE 482435 (1929); Mizuno, Turuga, JP 129200 (1939).

Tasteless crystals, mp 242°. Practically insol in water, ether, dilute HCl. Slightly sol in alcohol.

THERAP CAT: Cathartic.

7074. Oxyphenonium Bromide. [50-10-2] 2-[(2-Cyclohexyl-2-hydroxy-2-phenylacetyl)oxy]-*N*,*N*-diethyl-*N*-methylethanaminium bromide (1:1); diethyl(2-hydroxyethyl)methylammonium α-phenylcyclohexaneglycolate bromide; α-phenylcyclohexaneglycolic acid ester diethyl(2-hydroxyethyl)methylammonium bromide; diethylaminoethyl α-phenylcyclohexaneglycolate methylbromide; 2-diethylaminoethyl α-cyclohexyl-α-phenylglycolate methobromide; cyclohexylhydroxyphenylacetic acid diethylmethylaminoethyl ester bromide; phenylcyclohexyloxyacetic acid diethylaminoethyl ester bromomethylate; Ba-5473; C-5473; Antrenyl; Spasmophen. C$_{21}$H$_{34}$BrNO$_3$; mol wt 428.41. C 58.88%, H 8.00%, Br 18.65%, N 3.27%, O 11.20%. Anticholinergic. Prepn: **CH 259948** (1949 to Ciba).

Crystals, mp 189-194° from ethyl acetate + alc. Freely sol in water. Sparingly sol in alc. Aq solns are neutral.

THERAP CAT: Antispasmodic.

7075. Oxytetracycline. [79-57-2] (4*S*,4a*R*,5*S*,5a*R*,6*S*,-12a*S*)-4-(Dimethylamino)-1,4,4a,5,5a,6,11,12a-octahydro-3,5,6,-10,12,12a-hexahydroxy-6-methyl-1,11-dioxo-2-naphthacenecarboxamide; glomycin; riomitsin; hydroxytetracycline. C$_{22}$H$_{24}$N$_2$O$_9$; mol wt 460.44. C 57.39%, H 5.25%, N 6.08%, O 31.27%. Antibiotic substance isolated from the elaboration products of the actinomycete, *Streptomyces rimosus*, grown on a suitable medium: Finlay *et al.*, *Science* **111**, 85 (1950). Isoln: Regna, Solomons, *Ann. N.Y. Acad. Sci.* **53**, 221 (1950); Regna *et al.*, *J. Am. Chem. Soc.* **73**, 4211 (1951). Production from *Streptomyces rimosus*: Sobin *et al.*, **US 2516080** (1950 to Pfizer). Isoln from *S. xanthophaeus:* Brockmann, Musso, *Naturwissenschaften* **41**, 451 (1954); Brockmann *et al.*, **DE 913687** (1954 to Bayer), *C.A.* **53**, 4662h (1959). Solubility data: Weiss *et al.*, *Antibiot. Chemother.* **7**, 374 (1957). Structure: Hochstein *et al.*, *J. Am. Chem. Soc.* **74**, 3708 (1952). Abs config: Dobrynin *et al.*, *Tetrahedron Lett.* **1962**, 901. Stereochemistry: Schach von Wittenau *et al.*, *J. Am. Chem. Soc.* **87**, 134 (1965). Total synthesis of the *dl*-form: H. Muxfeldt *et al.*, *ibid.* **101**, 689 (1979).

Stability: Oxytetracycline crystals show no loss in potency on heating for 4 days at 100°, the hydrochloride crystals show <5% inactivation after 4 mos at 56°. Aq solns of the hydrochloride at pH 1.0 to 2.5 are stable for at least 30 days at 25°. Solns at pH 3.0 to 9.0 show no detectable loss in potency on storage at 5° for at least 30 days. Half-life in hours of aq oxytetracycline solns at 37°: pH 1.0 = 114; pH 2.5 = 134; pH 4.6 = 45; pH 5.5 = 45; pH 7.0 = 26; pH 8.5 = 33; pH 10.0 = 14. Freely sol in 3*N* hydrochloric acid, alkaline solns; sparingly sol in alc; very slightly sol in water.

Dihydrate. Alamycin; Imperacin; Liquamycin; Terralon; Terramycin; Tetradure. Needles from water or methanol, dec 181-182°. [α]$_D^{25}$ −196.6° (0.1*N* HCl); [α]$_D^{25}$ −2.1° (0.1*N* NaOH); [α]$_D^{25}$ +26.5° (methanol). uv max (pH 4.5 phosphate buffer 0.1*M*): 249, 276, 353 nm (E$_{1cm}^{1\%}$ 240, 322, 301). Soly in water at 23°: pH 1.2 = 31,400 μg/ml, pH 2.0 = 4600 μg/ml, pH 3.0 = 1400 μg/ml; pH 5.0 = 500 μg/ml, pH 6.0 = 700 μg/ml, pH 7.0 = 1100 μg/ml, pH 9.0 = 38,600 μg/ml. Soly in abs ethanol 12,000 μg/ml, in 95% ethanol 200 μg/ml.

Hydrochloride. [2058-46-0] Berkmycen; Duphacycline; Engemycin; Geomycin; Oxy-Mycin; Oxytetrin; Posicycline; Tetra-Tablinen; Toxinal. Yellow, hygroscopic platelets from water. Soly (g/ml): water 1; (μg/ml): abs ethanol 12000, 95% ethanol 33000. Freely sol in water, but crystals of oxytetracycline base separate as a result of partial hydrolysis of the hydrochloride. Sparingly sol in methanol; even less sol in dehydrated alc. Insol in chloroform, ether.

Disodium salt dihydrate. C$_{22}$H$_{22}$N$_2$Na$_2$O$_9$.2H$_2$O. Yellow crystals; darkens on standing. Soly in abs alc: 8,000 μg/ml, in methanol: 1500 μg/ml.

USE: To treat lethal yellowing in palm trees.

THERAP CAT: Antibacterial.

THERAP CAT (VET): Antibacterial.

7076. Oxythiamine. [582-36-5] 3-[(1,6-Dihydro-2-methyl-6-oxo-5-pyrimidinyl)methyl]-5-(2-hydroxyethyl)-4-methylthiazolium chloride (1:1); 5-(2-hydroxyethyl)-3-[(4-hydroxy-2-methyl-5-pyrimidinyl)methyl]-4-methylthiazolium chloride. C$_{12}$H$_{16}$ClN$_3$O$_2$S; mol wt 301.79. C 47.76%, H 5.34%, Cl 11.75%, N 13.92%, O 10.60%, S 10.62%. Prepn: F. Bergel, A. R. Todd, *J. Chem. Soc.* **1937**, 1504; M. Soodak, L. R. Cerecedo, *J. Am. Chem. Soc.* **66**, 1988 (1944). Improved prepn: H. N. Rydon, *Biochem. J.* **48**, 383 (1951). Thiamine antagonist activity: A. J. Eusebi, L. R. Cerecedo, *Science* **110**, 162 (1949); L. J. Daniel, L. C. Norris, *Proc. Soc. Exp. Biol. Med.* **72**, 165 (1949); and distribution in tissues: C. J. Gubler, D. S. Murdock, *J. Nutr. Sci. Vitaminol.* **28**, 217 (1982). Proposed mechanism of action: S. A. Strumilo *et al.*, *Biomed. Biochim. Acta* **43**, 159 (1984). Determn by HPLC: B. C. Hemming, C. J. Gubler, *J. Liq. Chromatogr.* **3**, 1697 (1980).

Hydrochloride. [614-05-1] C$_{12}$H$_{16}$ClN$_3$O$_2$S.HCl; mol wt 338.25. Flat needles grouped in rosettes, dec 195°. uv max (acid soln): 265, 258, 228, 223 nm; (alkaline soln): 268, 260, 228, 221 nm. Does not give the thiochrome reaction.

Diphosphate. Crystals, mp 127-129°.

Monophosphoric acid ester. Hygroscopic crystals, dec 185-186°.

Triphosphoric acid ester. Minute crystals, dec 245-255°.

7077. Oxythioquinox. [2439-01-2] 6-Methyl-1,3-dithiolo-[4,5-*b*]quinoxalin-2-one; dithiocarbonic acid cyclic *S*,*S*-(6-methyl-2,3-quinoxalinediyl) ester; 6-methyl-2,3-quinoxalinedithiol cyclic *S*,*S*-dithiocarbonate; 6-methyl-2-oxo-1,3-dithio[4,5-*b*]quinoxaline; chinomethionat(e); quinomethionate; Bayer 36205; Forstan; Morestan. C$_{10}$H$_6$N$_2$OS$_2$; mol wt 234.29. C 51.27%, H 2.58%, N 11.96%, O 6.83%, S 27.37%. Prepn and review of chemical and fungicidal properties and toxicology: F. Grewe, H. Kaspers, *Pflanzenschutz-Nachr. Bayer (Engl. Ed.)* **18** (1), 1-23 (1965). Toxicity study: T. B. Gaines, *Toxicol. Appl. Pharmacol.* **14**, 515 (1969).

Yellow crystals from benzene, mp 172°. Practically insol in water. Freely sol in DMF. Sol in hot benzene, toluene, dioxane. Slightly sol in methanol, ethanol, acetone. LD$_{50}$ in male, female rats (mg/kg): 1800, 1100 orally (Gaines).

USE: Acaricide; agricultural fungicide.

7078. Oxytocin. [50-56-6] Alpha-hypophamine; ocytocin; Intertocine-S; Perlacton; Pitocin; Syntocinon; Orasthin; Partocon; Synpitan; Uteracon. C$_{43}$H$_{66}$N$_{12}$O$_{12}$S$_2$; mol wt 1007.19. C 51.28%, H 6.61%, N 16.69%, O 19.06%, S 6.37%. The principal uterus-contracting and lactation-stimulating hormone of the posterior pituitary gland. Isoln: Pierce *et al.*, *J. Biol. Chem.* **199**, 929 (1952). Structure and synthesis: Tuppy, Michl, *Monatsh. Chem.* **84**, 1011

(1953); Tuppy, *Biochim. Biophys. Acta* **11**, 449 (1953); du Vigneaud *et al.*, *J. Am. Chem. Soc.* **75**, 4879 (1953); **76**, 3115 (1954); Bodanszky, du Vigneaud, *ibid.* **81**, 2504 (1959); Cash *et al.*, *J. Med. Pharm. Chem.* **5**, 413 (1962); Sakakibara *et al.*, *Bull. Chem. Soc. Jpn.* **38**, 120 (1965). Solid phase synthesis: Bayer, Hagenmaier, *Tetrahedron Lett.* **1968**, 2037; Ives, *Can. J. Chem.* **46**, 2318 (1968). Synthesis of D-oxytocin: Flouret, du Vigneaud, *J. Am. Chem. Soc.* **87**, 3775 (1965). Description of commercial process: Velluz *et al.*, **US 2938891** and **US 3076797** (1960, 1963, both to Roussel-UCLAF). Radioimmunoassay: T. Chard, *Clin. Biochem. Anal.* **5**, 209 (1977). *Review:* du Vigneaud, *Experientia Suppl.* **II** (14th Intl. Congr. Pure and Appl. Chem.), 9-26 (1955); R. Caldeyro-Barcia, H. Heller, *Proc. Intl. Symp. on Oxytocin* (Montevideo 1959) 443 pp; several authors, *Adv. Exp. Med. Biol.* **2**, 53-104 (1968); C. R. W. Edwards in *Hormones in Blood* **vol. 2**, C. H. Gray, V. James, Ed. (Academic Press, New York, 3rd ed., 1979) pp 401-421. Review of role in parturition: A.-R. Fuchs, F. Fuchs, *Adv. Exp. Med.* **1980**, 403-428. Comprehensive description: F. Nachtmann *et al.*, *Anal. Profiles Drug Subs.* **10**, 563-600 (1981). Book: *Oxytocin: Cellular and Molecular Approaches in Medicine and Research*, R. Ivell, J. A. Russell, Eds. (Plenum Press, New York, 1995) 673 pp.

Cys–Tyr–Ile–Gln–Asn–Cys–Pro–Leu–GlyNH₂

White powder. $[\alpha]_D^{22} -26.2°$ (c = 0.53). Sol in water, 1-butanol, 2-butanol.

THERAP CAT: Oxytocic.

THERAP CAT (VET): Stimulates milk let-down, uterine contraction.

7079. Ozagrel. [82571-53-7] (2*E*)-3-[4-(1*H*-Imidazol-1-ylmethyl)phenyl]-2-propenoic acid; (*E*)-4-(imidazol-1-ylmethyl)cinnamic acid; OKY-046. $C_{13}H_{12}N_2O_2$; mol wt 228.25. C 68.41%, H 5.30%, N 12.27%, O 14.02%. Prepn: K. Iizuka *et al.*, **DE 2923815**; *eidem*, **US 4226878** (both 1980 to Ono; Kissei). Synthesis and thromboxane synthetase inhibitory activity: K. Iizuka *et al.*, *J. Med. Chem.* **24**, 1139 (1981). Pharmacology: S. Hiraku *et al.*, *Jpn. J. Pharmacol.* **41**, 393 (1986). Pulmonary vascular effects: R. Garcia-Szabo *et al.*, *Prostaglandins* **28**, 851 (1984). Metabolism in animals: M. Shimizu *et al.*, *Iyakuhin Kenkyu* **17**, 289 (1986), *C.A.* **105**, 72012q (1986). HPLC determn in biological fluids: *eidem*, *ibid.* 298, *C.A.* **105**, 72013r (1986). Clinical pharmacology and evaluation in myocardial infarction: T. Ito *et al.*, *Biomed. Biochim. Acta* **43**, S125 (1984). Clinical evaluation in prevention of cerebral vasospasm: S. Suzuki *et al.*, *Acta Neurochir.* **77**, 133 (1985); in coronary artery disease: M. Shikano *et al.*, *Jpn. Heart J.* **28**, 663 (1987). Toxicity data: T. Nishigake *et al.*, *Clin. Rep.* **20**, 2671 (1986). Series of articles on pharmacology: *Pharmacometrics* **31**, 527-565 (1986), *C.A.* **105**, 35349-35352 (1986).

Prisms from ethanol-ether, mp 223-224°.

Hydrochloride. [78712-43-3] $C_{13}H_{12}N_2O_2$.HCl. Crystals from ethanol-ether, mp 214-217°.

Sodium salt. Cataclot; Xanbon. $C_{13}H_{11}N_2NaO_2$; mol wt 250.23. LD₅₀ in male, female mice, male, female rats (mg/kg): 1940, 1580, 1150, 1300 i.v.; 3800, 3600, 5900, 5700 orally; 2450, 2100, 2300, 2250 s.c. (Nishigake).

THERAP CAT: Antithrombotic; antianginal.

7080. Ozone. [10028-15-6] Triatomic oxygen. O₃; mol wt 48.00. Found in the atm in varying proportions (about 0.05 ppm at sea level), since it is produced continuously in the outer layers of the atm by the action of solar uv radiation on the oxygen of the air. So-called sterilizing lamps operate on the same principle. In the laboratory ozone is prepd by passing dry air between two plate electrodes connected to an alternating current source of several thousand volts. The reaction is reversible, and after a little ozone has been produced it is dec at the same rate as it is generated. Obtained in pure form by cooling ozonized air to −180° when it separates as a dark blue liquid. *See also* C. E. Thorp, *Bibliography of Ozone Technology* (Armour Res. Found., Chicago). Lab prepn: *Org. Synth.* **coll. vol. III**, 673 (1955). Conference proceedings: *Adv. Chem. Ser.* **21**, entitled "Ozone Chemistry and Technology," H. A. Leedy, Ed. (ACS, Washington D.C., 1959) 465 pp. *Review:* C. Nebel in *Kirk-Othmer Encyclopedia of Chemical Technology* **vol. 16** (Wiley-Interscience, New York, 3rd ed., 1981) pp 683-713.

Bluish, explosive gas or blue liquid. Pleasant, characteristic odor in concns of less than 2 ppm. Irritating and injurious in higher concns. Powerful oxidizing agent. d^0 (gas): 2.144 g/l; $d^{-195.4}$ (liq) 1.614 g/ml. mp −193°. bp −111.9°. Critical temp −12.1°. Critical press. 53.8 atm. Heat of formation 34.4 kcal/mole at 25°. Intense absorption band beginning at about 290 nm. Unstable. Solutions contg ozone explode on warming. Prepn of ozone solns in liquid oxygen: Cook, **US 3008902** (1961 to Union Carbide). Although the stability of ozone in aq solns decreases as alkalinity rises, this effect is reversed at high concns. For example, the half life of ozone is 2 min in 1*N* NaOH; it is increased to 83 hrs in 20*N* NaOH: Heidt, Landi, *J. Chem. Phys.* **41**, 176 (1964).

Caution: Potential symptoms of overexposure are irritation of eyes and mucous membranes; pulmonary edema; chronic respiratory disease. *See NIOSH Pocket Guide to Chemical Hazards* (DHHS/NIOSH 97-140, 1997) p 238.

USE: As disinfectant for air and water by virtue of its oxidizing power. For bleaching waxes, textiles, oils. In organic syntheses. Forms ozonides which are sometimes useful oxidizing compds.

P

7081. Paclitaxel. [33069-62-4] ($\alpha R,\beta S$)-β-(Benzoylamino)-α-hydroxybenzenepropanoic acid (2aR,4S,4aS,6R,9S,11S,12S,-12aR,12bS)-6,12b-bis(acetyloxy)-12-(benzoyloxy)-2a,3,4,4a,5,6,9,-10,11,12,12a,12b-dodecahydro-4,11-dihydroxy-4a,8,13,13-tetramethyl-5-oxo-7,11-methano-1H-cyclodeca[3,4]benz[1,2-b]oxet-9-yl ester; 5β,20-epoxy-1,2α,4,7β,10β,13α-hexahydroxytax-11-en-9-one 4,10-diacetate 2-benzoate 13-ester with (2R,3S)-N-benzoyl-3-phenylisoserine; taxol A; NSC-125973; Abraxane; Anzatax; Taxol. $C_{47}H_{51}NO_{14}$; mol wt 853.92. C 66.11%, H 6.02%, N 1.64%, O 26.23%. Antiproliferative agent first isolated from the bark of the Pacific yew tree, *Taxus brevifolia, Taxaceae;* promotes the assembly and inhibits the tubulin disassembly process. Isoln and structure: M. C. Wani *et al., J. Am. Chem. Soc.* **93**, 2325 (1971). Effect on microtubule assembly: P. B. Schiff *et al., Nature* **277**, 665 (1979). Semisynthetic prepn from the naturally occurring precursor, 10-deacetylbaccatin III: J.-N. Denis *et al., J. Am. Chem. Soc.* **110**, 5917 (1988); I. Ojima *et al., Tetrahedron* **48**, 6985 (1992). Total synthesis: K. C. Nicolaou *et al., Nature* **367**, 630 (1994); R. A. Holton *et al., J. Am. Chem. Soc.* **116**, 1597, 1599 (1994). Production by *Taxomyces andreanae*, an endophytic fungus associated with *T. brevifolia*: A. Stierle *et al., Science* **260**, 214 (1993). Review of mechanism of action: J. J. Manfredi, S. B. Horwitz, *Pharmacol. Ther.* **25**, 83-125 (1984). Review of syntheses and structure-activity studies: Y. Fu *et al., Curr. Med. Chem.* **16**, 3966-3985 (2009). Review of clinical experience in cancer therapy: T. M. Mekhail, M. Markman, *Expert Opin. Pharmacother.* **3**, 755-766 (2002); of toxicities and drug delivery strategies: N. I. Marupudi *et al., Expert Opin. Drug Saf.* **6**, 609-621 (2007). Review of safety and efficacy of paclitaxel-eluting stents and restenosis: G. W. Stone *et al., JACC Cardiovasc. Interv.* **4**, 530-542 (2011).

Needles from aq methanol, mp 213-216° (dec). $[\alpha]_D^{20}$ −49° (methanol). uv max (methanol): 227, 273 nm (ε 29800, 1700). Highly lipophilic. Sol in alc. Insol in water.
Conjugate with poly-L-glutamic acid. [263351-82-2] Paclitaxel poliglumex; L-polyglutamic paclitaxel; PG-TXL; CT-2103; Opaxio; Xyotax. Biodegradable polymer-drug conjugate resulting from the random condensation of poly-L-glutamic acid and paclitaxel; contains approx 1 paclitaxel ester linkage per 11 glutamic acid units. Mol wt ~48 kDa. Prepn: C. Li *et al., Cancer Res.* **58**, 2404 (1998). Review of pharmacology and pharmacokinetics: J. W. Singer, *J. Controlled Release* **109**, 120-126 (2005); of clinical experience in ovarian cancer: V. L. Galic *et al., Expert Opin. Invest. Drugs* **20**, 813-821 (2011). White to off-white powder. Insol in ether; practically insol in 0.1M HCl and acetonitrile. Sol in 0.1M NaOH, 0.1M Na$_2$HPO$_4$ (pH 6.5); slightly sol in methanol, DMSO, DMF.
USE: Tool in study of structure and function of microtubules.
THERAP CAT: Antineoplastic; antirestenotic.

7082. Paclobutrazol. [76738-62-0] *rel*-($\alpha R,\beta R$)-β-[(4-Chlorophenyl)methyl]-α-(1,1-dimethylethyl)-1H-1,2,4-triazole-1-ethanol; 1-*tert*-butyl-2-(*p*-chlorobenzyl)-2-(1,2,4-triazol-1-yl)ethanol; (2RS,3RS)-1-(4-chlorophenyl)-4,4-dimethyl-2-(1H-1,2,4-triazol-1-yl)pentan-3-ol; ICI-PP-333; PP-333; Bonzi; Cultar; Parlay; Trimmit. $C_{15}H_{20}ClN_3O$; mol wt 293.80. C 61.32%, H 6.86%, Cl 12.07%, N 14.30%, O 5.45%. Plant growth regulator with fungicidal activity. Prepn: B. C. Baldwin *et al.*, **DE 2734426**; S. Balasubramanyan, M.

C. Shephard, **US 4243405** (1978, 1981 both to ICI). Physical properties and biological activity: B. G. Lever *et al., Proc. Br. Crop Prot. Conf. - Weeds* **1982**, 3. Resolution and activity of isomers: B. Sugavanam, *Pestic. Sci.* **15**, 296 (1984). GC determn in plant tissue: E. A. Stahly, D. A. Buchanan, *HortScience* **21**, 534 (1986). Comparison with daminozide, *q.v.*, of effect on apple trees: G. R. Stinchcombe *et al., J. Hortic. Sci.* **59**, 323 (1984).

Relative stereochemistry

White crystalline solid, mp 165-166°. d 1.22. Vapor pressure at 20°: 1×10^{-6} Pa. Soly: water 35 mg/l, methanol 15%, propylene glycol 5%, acetone 11%, cyclohexanone 18%, methylene dichloride 10%, hexane 1%, xylene 6%.
USE: Plant growth regulator.

7083. Pactamycin. [23668-11-3] 2-Hydroxy-6-methylbenzoic acid [(1S,2R,3R,4S,5S)-5-[(3-acetylphenyl)amino]-4-amino-3-[[(dimethylamino)carbonyl]amino]-1,2-dihydroxy-3-[(1S)-1-hydroxyethyl]-2-methylcyclopentyl]methyl ester; NSC-52947; U-15800. $C_{28}H_{38}N_4O_8$; mol wt 558.63. C 60.20%, H 6.86%, N 10.03%, O 22.91%. Antitumor antibiotic produced by *Streptomyces pactum* var *pactum*. Discovery and biological properties: Bhuyan *et al., Antimicrob. Agents Chemother.* **1961**, 184. Isoln and characterization: Argoudelis *et al., ibid.* 191. Manuf: **GB 980346** (1965 to Upjohn), *C.A.* **62**, 11115f (1965). Structure: Wiley *et al., J. Org. Chem.* **35**, 1420 (1970). Revised structure: D. J. Duchamp, *Am. Crystallogr. Assn.* (Winter Mtg., Albuquerque, 1972) p 23. Mechanism of action: T. A. Beerman *et al., Adv. Enzyme Regul.* **14**, 207 (1976). ^{13}C-NMR study: D. D. Weller *et al., J. Antibiot.* **30**, 997 (1978). Biosynthesis: D. D. Weller, K. L. Rinehart, *J. Am. Chem. Soc.* **100**, 6757 (1978). *Review:* Goldberg in *Antibiotics* vol. 3, J. W. Corcoran, F. E. Hahn, Eds. (Springer-Verlag, New York, 1975) pp 498-515.

$[\alpha]_D^{25}$ +79° (ethanol) changing to +23° on standing. $[\alpha]$ changes in acetone on standing from +25° to +76° in 24 hours. Amphoteric. Sol in ethanol, chloroform, methylene chloride, benzene, ether, in solns <pH 5 and >9.5. Insol in Skellysolve B, cyclohexane; insol at isoelectric pt, pH 8.3. Unstable in solution. LD$_{50}$ in mice (mg/kg): 10.7 orally; 15.6 i.v.; in rats (mg/kg): 1.4 i.v. (Bhuyan).

7084. Pagoclone. [133737-32-3] (+)-2-(7-Chloro-1,8-naphthyridin-2-yl)-2,3-dihydro-3-(5-methyl-2-oxohexyl)-1H-isoindol-1-one; (+)-2-(7-chloro-1,8-naphthyridin-2-yl)-3-(5-methyl-2-oxohexyl)-isoindolinone; CI-1043; IP-456; RP-62955. $C_{23}H_{22}ClN_3O_2$; mol wt 407.90. C 67.73%, H 5.44%, Cl 8.69%, N 10.30%, O 7.84%. Cyclopyrrolone GABA$_A$-receptor modulator with anxiolyic activity. Prepn: J.-D. Bourzat *et al.*, **US 4960779** (1990 to Rhone-Poulenc). Efficient synthesis: T. L. Stuk *et al., Org. Process Res. Dev.* **7**, 851 (2003). Clinical effects on neuropsychological performance: A. F. Caveney *et al., Neuropsychiatr. Dis. Treat.* **4**, 277 (2008). Evaluation of abuse potential: H. de Wit *et al., J. Clin. Psychopharmacol.*

26, 268 (2006). Clinical study in persistent developmental stuttering: G. Maguire *et al.*, *ibid.* **30**, 48 (2010). Review of development and therapeutic potential: A. Bateson, *Curr. Opin. Investig. Drugs* **4**, 91-95 (2003).

White crystalline solid, mp 169-172°. $[\alpha]_D^{20}$ +135° (c = 1 in dichloromethane).

(±)-**Form.** [133737-48-1] Crystals from acetonitrile, mp 180°.
THERAP CAT: Anxiolytic. In treatment of stuttering.

7085. Palau'amine. [148717-58-2] (3a*S*,4'*R*,5'*S*,10a*S*,11*S*,-12*S*,13a*S*,13b*R*)-2,2'-Diamino-11-(aminomethyl)-12-chloro-1,1',-3a,5',10a,11,12,13a-octahydro-5'-hydroxyspiro[8*H*-cyclopenta[3,-4]pyrrolo[1,2-*a*]imidazo[4,5-*b*]pyrrolo[1,2-*d*]pyrazine-13(10*H*),4'-[4*H*]imidazol]-8-one. $C_{17}H_{22}ClN_9O_2$; mol wt 419.87. C 48.63%, H 5.28%, Cl 8.44%, N 30.02%, O 7.62%. Immunosuppressive, hexacyclic bisguanidine produced by the marine sponge, *Stylotella aurantium*, also known as *Stylissa massa*. Isoln: R. B. Kinnel *et al.*, *J. Am. Chem. Soc.* **115**, 3376 (1993); *eidem*, *J. Org. Chem.* **63**, 3281 (1998). Revised structure: A. Grube, M. Köck, *Angew. Chem. Int. Ed.* **46**, 2320 (2007); M. S. Buchanan *et al.*, *Tetrahedron Lett.* **48**, 4573 (2007). Overview of structure elucidation and synthetic approaches: M. Köck *et al.*, *Angew. Chem. Int. Ed.* **46**, 6586 (2007). Total synthesis: I. B. Seiple *et al.*, *ibid.* **49**, 1095 (2010).

Off-white amorphous solid, dec upon heating. $[\alpha]_D^{25}$ −45.2° (c = 3.0 in methanol). uv max (methanol): 224, 272 nm (ε 7800, 7900). LD_{50} in mice (mg/kg): 13 i.p. (Kinnel).

7086. Palifermin. [162394-19-6] 24-163-Keratinocyte growth factor (human); 24-163-fibroblast growth factor 7 (human); rHuKGF; Kepivance. Recombinant epithelial tissue growth factor of 141 amino acid residues; N-terminal truncated form of endogenous KGF. Mol wt approx 16.3 kDa. Prepn: D. J. Gospodarowicz, F. R. Masiarz, *WO 9501434*; *eidem*, *US 5677278* (1995, 1997 both to Chiron). *In vivo* radioprotective effect: C. L. Farrell *et al.*, *Cancer Res.* **58**, 933 (1998). Clinical study in mucositis associated with myelosuppressive therapy: R. Spielberger *et al.*, *N. Engl. J. Med.* **351**, 2590 (2004). Review of therapeutic potential in oral mucositis: M. A. A. Siddiqui, K. Wellington, *Drugs* **65**, 2139-2146 (2005); M. L. Radtke, J. M. Kolesar, *J. Oncol. Pharm. Pract.* **11**, 121-125 (2005).
THERAP CAT: Radioprotective agent.

7087. Paliperidone. [144598-75-4] 3-[2-[4-(6-Fluoro-1,2-benzisoxazol-3-yl)-1-piperidinyl]ethyl]-6,7,8,9-tetrahydro-9-hydroxy-2-methyl-4*H*-pyrido[1,2-*a*]pyrimidin-4-one; 9-hydroxyrisperidone; RO-76477; R-76477; Invega. $C_{23}H_{27}FN_4O_3$; mol wt 426.49. C 64.77%, H 6.38%, F 4.45%, N 13.14%, O 11.25%. Dual antagonist of central dopamine D2 and serotonin type 2 ($5HT_{2a}$) receptors; also active as an antagonist at α_1 and α_2 adrenergic receptors and H_1 histamine receptors. Major active metabolite of risperidone, *q.v.* Prepn: C. G. M. Janssen *et al.*, *EP 368388*; *eidem*, *US 5158952* (1990, 1992 both to Janssen). Simultaneous LC/MS/MS determn

with risperidone in plasma: M. Aravagiri, S. R. Marder, *J. Mass Spectrom.* **35**, 718 (2000). Clinical trial in schizophrenia: J. Kane *et al.*, *Schizophr. Res.* **90**, 147 (2007). Review of development and clinical experience: L. Citrome, *Int. J. Clin. Pract.* **61**, 653-662 (2007).

Crystals from 2-propanol, mp 179.8°. Slightly sol in DMF; sparingly sol in 0.1*N* HCl, methylene chloride. Practically insol in water, hexane, 0.1*N* NaOH.

Palmitate. [199739-10-1] Hexadecanoic acid 3-[2-[4-(6-fluoro-1,2-benzisoxazol-3-yl)-1-piperidinyl]ethyl]-6,7,8,9-tetrahydro-2-methyl-4-oxo-4*H*-pyrido[1,2-*a*]pyrimidin-9-yl ester; 9-hydroxyrisperidone palmitate; RO-92670; Xeplion. $C_{39}H_{57}FN_4O_4$; mol wt 664.91. Prepn: M. K. J. François *et al.*, *WO 9744039*; *eidem*, *US 6077843* (1997, 2000 both to Janssen). Crystals from isopropanol.
THERAP CAT: Antipsychotic.

7088. Palitantin. [15265-28-8] *rel*-(2*R*,3*S*,5*R*,6*R*)-3-(1*E*,3*E*)-1,3-Heptadien-1-yl-5,6-dihydroxy-2-(hydroxymethyl)cyclohexanone. $C_{14}H_{22}O_4$; mol wt 254.33. C 66.12%, H 8.72%, O 25.16%. Metabolic product of *Penicillium palitans* Westling. Isoln: J. H. Birkinshaw, H. Raistrick, *Biochem. J.* **30**, 801 (1936). Derivs and degradation products: J. H. Birkinshaw, *ibid.* **51**, 271 (1952). Structure: K. Bowden *et al.*, *J. Chem. Soc.* **1959**, 1662. Biosynthesis: P. Chaplen, R. Thomas, *Biochem. J.* **77**, 91 (1960); A. J. Birch, M. Kocor, *J. Chem. Soc.* **1960**, 866. Reactivity studies: A. T. Austin, B. Pearson, *Chem. Ind. (London)* **1966**, 1228. Stereoselective synthesis of (±)-form: A. Ichihara *et al.*, *Tetrahedron Lett.* **1977**, 3473; *eidem*, *Tetrahedron* **36**, 1547 (1980).

Relative stereochemistry

Needles from hot water, mp 165°. $[\alpha]_{5461}^{23}$ +4.4° (c = 0.8 in $CHCl_3$). uv max (ethanol): 323 nm (ε 34,000). Sol in hot water, alc, chloroform; slightly sol in cold water, ether.

7089. Palivizumab. [188039-54-5] Anti-(respiratory syncytial virus protein F) immunoglobulin G1 (human-mouse monoclonal MEDI-493 γ_1-chain) disulfide with human-mouse monoclonal MEDI-493 κ-chain, dimer; MEDI-493; Synagis. Humanized monoclonal antibody directed to a neutralizing epitope on the F glycoprotein of respiratory syncytial virus (RSV). Mol wt ~148 kDa. Consists of ~95% human and ~5% murine antibody sequences. Prepn and neutralizing activity: L. S. Johnson, *WO 9605229*; *idem*, *US 5824307* (1996, 1998 both to MedImmune); S. Johnson *et al.*, *J. Infect. Dis.* **176**, 1215 (1997). Clinical pharmacokinetics and safety: X. Sáez-Llorens *et al.*, *Pediatr. Infect. Dis. J.* **17**, 787 (1998). Clinical trial in high risk infants: Impact-RSV Study Group, *Pediatrics* **103**, 531 (1998). *Review:* H. C. Meissner *et al.*, *Pediatr. Infect. Dis. J.* **18**, 223-231 (1999).
THERAP CAT: Antiviral (in prophylaxis of RSV infection).

7090. Palladium. [7440-05-3] Pd; at. wt 106.42; at. no. 46; valences 2, 4. Group VIII (10). Six naturally occurring isotopes: 102 (1.0%); 104 (11.0%); 105 (22.2%); 106 (27.3%); 108 (26.7%); 110 (11.8%); artificial, radioactive isotopes: 98-101, 103; 107; 109; 111-115. Abundance in earth's crust 0.001-0.01 ppm. Discovered in 1803 by Wollaston. Belongs to the platinum group of metals.

Occurs in nature alloyed with platinum or gold and as a selenide; found in nickel sulfide ores; found in the minerals stibiopalladinite, braggite, porpezite. Isoln: Vauguelin *et al.*, cited by Mellor, *A Comprehensive Treatise on Inorganic and Theoretical Chemistry* **15**, 595 (1936). Reviews of prepn, properties and chemistry of palladium and other platinum metals: Gilchrist, *Chem. Rev.* **32**, 277-372 (1943); Beamish *et al.*, in *Rare Metals Handbook*, C. A. Hampel, Ed. (Reinhold, New York, 1956) pp 291-328; Livingstone in *Comprehensive Inorganic Chemistry* vol. 3, J. C. Bailar, Jr. *et al.*, Eds. (Pergamon Press, Oxford, 1973) pp 1163-1189, 1274-1329. Review of uses: E. M. Wise, *Palladium, Recovery, Properties, Uses* (Academic Press, New York, 1968) 187 pp. Use of organic palladium derivatives in synthesis: P. M. Maitlis, *The Organic Chemistry of Palladium* (Academic Press, New York, 1971); J. Tsuji, *Top. Curr. Chem.* **91**, 29 (1980); B. M. Trost, *Tetrahedron* **33**, 2615 (1977); idem, *Acc. Chem. Res.* **13**, 385 (1980).

Silver-white metal, face-centered cubic structure; occurs also as black powder and as spongy masses which can be compressed to a compact mass. mp 1555°. bp 3167°. d_4^{20} 12.02. Hardness on Mohs' scale 4.8, Brinell hardness 61.0. Spec heat 0.0584 cal/g at 0°C. Electrical resistivity at 0° = 10.0 microohms-cm. Appreciably volatile at high temps. At a red heat is converted into the monoxide. Forms dihalides with fluorine or chlorine at a red heat. Reacts with nitric acid, sulfuric acid, a mixture of hydrochloric and chloric acids. Reacts slightly with concd HCl; more readily in the presence of air or free chlorine. Forms a sulfide when heated with sulfur, a phosphide when heated with phosphorus. Absorbs a considerable amount of hydrogen.

USE: In form of gold, silver, and copper alloys in dentistry; for alloy bearings, springs, balance wheels of watches; for mirrors in astronomical instruments; as catalyzer in manuf of sulfuric acid and in other oxidizing processes; in powder form as catalyst in hydrogenation and in ignition of hydrogen or hydrocarbons with oxygen; the spongy form is used in gas analysis for separating hydrogen from mixtures of gases.

7091. Palladium Chloride. [7647-10-1] Palladium chloride (PdCl$_2$); palladous chloride. Cl$_2$Pd; mol wt 177.32. Cl 39.98%, Pd 60.02%. PdCl$_2$. Prepn: Krustinsons, *Z. Elektrochem.* **44**, 537 (1938). Toxicity study: Orestano, *Boll. Soc. Ital. Biol. Sper.* **8**, 1154 (1933). Review of PdCl$_2$ and other halides: J. H. Canterford, R. Colton, *Halides of the Second and Third Row Transition Metals* (John Wiley, New York, 1968) pp 358-389.

Red crystals. mp 678-680°. Dec at high temp to palladium and chlorine. MLD i.v. in rabbits: 0.0186 g/kg (Orestano).

Dihydrate. [10038-97-8] Cl$_2$Pd.2H$_2$O; mol wt 213.35. Dark brown crystals. Sol in water, alcohol, acetone. Reduced in soln by hydrogen or CO to metal. *Keep tightly closed.*

USE: In photography, for preparing pictures to be transferred to porcelain; toning solutions; electroplating parts of clocks and watches; manuf indelible ink; for the prepn of the metal for use as a catalyst; PdCl$_2$ paper is used for detecting CO, to find leaks in buried gas pipes. Prepn of palladium catalysts using PdCl$_2$: Mozingo, *Org. Synth.* **coll. vol. III**, 685 (1955).

7092. Palladium Diacetate. [3375-31-3] Acetic acid palladium(2+) salt (2:1); bis(acetato)palladium; diacetatopalladium(II); palladium(II) acetate; Pd(OAc)$_2$. C$_4$H$_6$O$_4$Pd; mol wt 224.51. C 21.40%, H 2.69%, O 28.50%, Pd 47.40%. (CH$_3$COO)$_2$Pd. Palladium-based catalyst for a variety of reactions; often used in conjunction with a ligand or other reagent to form the reactive species. Prepn from palladium sponge or palladium nitrate and glacial acetic acid: S. M. Morehouse *et al.*, *Chem. Ind. (London)* **1964**, 544; T. A. Stephenson *et al.*, *J. Chem. Soc.* **1965**, 3632; from hydrated palladium oxide and acetic acid: E. A. Hausman *et al.*, **FR 1403398**; eidem, **US 3318891** (1965, 1967 to Engelhard). Crystal structure: A. C. Skapski, M. L. Smart, *J. Chem. Soc. Chem. Commun.* **1970**, 658. Synthetic utility in Heck reactions: R. F. Heck, J. P. Nolley, Jr., *J. Org. Chem.* **37**, 2320 (1972); in carbonylations: S. Cacchi *et al.*, *Tetrahedron Lett.* **26**, 1109 (1985); in Buchwald-Hartwig cross-coupling reactions: J. P. Wolfe, S. L. Buchwald, *J. Org. Chem.* **62**, 1264 (1997); in reductive aminations: B. Basu *et al.*, *Synlett* **2003**, 555; in alcohol oxidations: M. J. Schultz *et al.*, *J. Org. Chem.* **70**, 3343 (2005).

Orange-brown crystals, mp 205° (dec) (Stephenson). Also reported as red-brown monoclinic plates (Skapski). d 2.19. *Irritant.*

Sol in chloroform, methylene dichloride, acetone, acetonitrile, diethyl ether, benzene; sol with dec in aq HCl, aq KI solutions. Insol in water, aq NaCl, NaNO$_3$, NaOAc solns, petroleum, alcohols. Dec when warmed with alcohols. Air stable. Exists as trimer.

USE: Reagent and catalyst in synthetic organic chemistry.

7093. Palladium Nitrate. [10102-05-3] Nitric acid palladium(2+) salt (2:1); palladous nitrate. N$_2$O$_6$Pd; mol wt 230.43. N 12.16%, O 41.66%, Pd 46.18%. Pd(NO$_3$)$_2$. Prepn from palladium and nitric acid: *Gmelins, Palladium* (8th ed.) **65**, 269 (1942). Reported to be the dihydrate: Gatehouse *et al.*, *J. Chem. Soc.* **1957**, 4222.

Brown, deliquesc crystals. Sol in water with turbidity; with much water a brown basic salt precipitates; completely sol in dil HNO$_3$. *Keep well closed, protected from light.*

USE: Sepn of Cl$_2$ and I$_2$; catalyst in organic syntheses.

7094. Palladium Oxide. [1314-08-5] Palladium monoxide; palladous oxide. OPd; mol wt 122.42. O 13.07%, Pd 86.93%. PdO. Prepn: Shriner, Adams, *J. Am. Chem. Soc.* **46**, 1685 (1924).

Black powder. d 8.3. Insol in water, acids; slightly sol in aqua regia; sol in 48% HBr. Dec when strongly heated; also in the presence of H$_2$.

USE: Reduction catalyst in synthesis of organic compds.

7095. Palmatine. [3486-67-7] 5,6-Dihydro-2,3,9,10-tetramethoxydibenzo[*a,g*]quinolizinium; 7,8,13,13a-tetrahydro-2,3,-9,10-tetramethoxyberbinium; calystigine. [C$_{21}$H$_{22}$NO$_4$]$^+$. Obtained only in form of its salts. First isolated from Calumba root (*Jateorhiza palmata* (DC.) Miers, *Menispermaceae*). Now found in many other genera. Palmatine and tetrahydropalmatine, *q.v.* are probably the most widely distributed Berberis alkaloids. Extraction procedure: Feist, Dschu, *Arch. Pharm.* **263**, 301 (1925). Structure: Feist, Sandstede, ibid. **256**, 2, 5 (1918); Späth, Lang, *Ber.* **54**, 3064, 3068 (1921); Späth, Böhm, *Ber.* **55**, 2988 (1922); Späth, Meinhard, *Ber.* **75**, 400 (1942). Synthesis: Späth, Quientensky, *Ber.* **58**, 2267 (1925); R. D. Haworth *et al.*, *J. Chem. Soc.* **1927**, 548; Z. Kiparissides *et al.*, *Can. J. Chem.* **58**, 2770 (1980). Identity with calystigine, alkaloid of the Chinese drug Chi-Kuo-Lan: Huang, Chen, *C.A.* **52**, 15827i (1958).

Palmatine forms addn products with acetone and chloroform, as does berberine.

Iodide. (C$_{21}$H$_{22}$NO$_4$)I. Yellow needles from water, dec 239°; also a dihydrate. Sparingly sol in hot water and alc.

Nitrate. (C$_{21}$H$_{22}$NO$_4$)NO$_3$. Sesqui- or dihydrate, yellow needles, dec 239°. Freely sol in most solvents.

Chloride trihydrate. (C$_{21}$H$_{22}$NO$_4$)Cl.3H$_2$O. Yellowish-green needles from water. Freely sol in hot water and alc.

Sulfate pentahydrate. (C$_{21}$H$_{22}$NO$_4$)$_2$SO$_4$.5H$_2$O. Yellow needles, mp 250°. Very sol in alc, sol in water.

7096. Palmidrol. [544-31-0] N-(2-Hydroxyethyl)hexadecanamide; N-(2-hydroxyethyl)palmitamide. C$_{18}$H$_{37}$NO$_2$; mol wt 299.50. C 72.19%, H 12.45%, N 4.68%, O 10.68%. CH$_3$(CH$_2$)$_{14}$-CONHCH$_2$CH$_2$OH. A naturally occurring anti-inflammatory agent. Isolated from soybean lecithin, egg yolk, and peanut meal: Kuehl *et al.*, *J. Am. Chem. Soc.* **79**, 5577 (1957). Synthesis by refluxing ethanolamine with palmitic acid: Roe *et al.*, ibid. **74**, 3442 (1952).

Crystals from 95% ethanol or cyclohexane, mp 98-99°.

7097. Palmitic Acid. [57-10-3] Hexadecanoic acid; hexadecylic acid; cetylic acid. C$_{16}$H$_{32}$O$_2$; mol wt 256.43. C 74.94%, H 12.58%, O 12.48%. Occurs as the glyceryl ester in many oils and fats. Obtained from palm oil, Japan wax, or Chinese vegetable tallow. Purification: Magne *et al.*, **US 2791596** (1957 to Secretary of

Agriculture). Toxicity study: L. Orö, A. Wretlind, *Acta Pharmacol. Toxicol.* **18**, 141 (1961).

White, crystalline scales. d_4^{62} 0.853. mp 63-64°. bp_{15} 215°. n_D^{80} 1.4273. Freely sol in hot alc, propyl alc; sol in alc, ether, chloroform; sparingly sol in cold alc, petr ether. Practically insol in water. LD_{50} i.v. in mice: 57 ± 3.4 mg/kg (Orö, Wretlind).

7098. Palonosetron. [135729-61-2] (3aS)-2-(3S)-1-Azabicyclo[2.2.2]oct-3-yl-2,3,3a,4,5,6-hexahydro-1H-benz[de]isoquinolin-1-one. $C_{19}H_{24}N_2O$; mol wt 296.41. C 76.99%, H 8.16%, N 9.45%, O 5.40%. Serotonin 5-HT$_3$ receptor antagonist. Prepn: J. Berger *et al.*, **EP 430190**; *eidem*, **US 5202333** (1991, 1993 both to Syntex); R. D. Clark *et al.*, *J. Med. Chem.* **36**, 2645 (1993). Improved synthesis: B. A. Kowalczyk, *Heterocycles* **43**, 1439 (1996). Receptor binding study: E. H. F. Wong *et al.*, *Br. J. Pharmacol.* **114**, 851 (1995). Pharmacology: R. M. Eglen *et al.*, *ibid.* 860. Clinical efficacy and pharmacokinetics: P. Eisenberg *et al.*, *Ann. Oncol.* **15**, 330 (2004). HPLC determn in plasma: L. Ding *et al.*, *J. Pharm. Biomed. Anal.* **44**, 575 (2007). Clinical trial in prevention of chemotherapy-induced nausea and vomiting: R. Gralla *et al.*, *Ann. Oncol.* **14**, 1570 (2003). Review of pharmacology and clinical experience: M. A. A. Siddiqui, L. J. Scott, *Drugs* **64**, 1125-1132 (2004); M. Saito, M. Tsukuda, *Expert Opin. Pharmacother.* **11**, 1003-1014 (2010)

mp 87-88°. $[\alpha]_D$ −136° (c = 0.25 in chloroform).
Hydrochloride. [135729-62-3] RS-25259-197; Aloxi; Onicit. $C_{19}H_{24}N_2O \cdot HCl$; mol wt 332.87. Crystals from ethanol, mp >290°. $[\alpha]_D^{25}$ −94.1° (c = 0.4 in water). Freely sol in water; sol in propylene glycol; slightly sol in ethanol, 2-propanol.

THERAP CAT: Antiemetic.

7099. Palustric Acid. [1945-53-5] (1R,4aS,10aR)-1,2,3,4,-4a,5,6,9,10,10a-Decahydro-1,4a-dimethyl-7-(1-methylethyl)-1-phenanthrenecarboxylic acid; 13-isopropylpodocarpa-8,13-dien-15-oic acid. $C_{20}H_{30}O_2$; mol wt 302.46. C 79.42%, H 10.00%, O 10.58%. Isoln from gum rosin: Loeblich *et al.*, *J. Am. Chem. Soc.* **77**, 2823 (1955); Joye *et al.*, *J. Org. Chem.* **30**, 654 (1965). Structure: Shuller *et al.*, *J. Am. Chem. Soc.* **82**, 1734 (1960). Prepn of the 4-*epi*-form: Tabacik, Poisson, *Bull. Soc. Chim. Fr.* **1969**, 3264.

Crystals from methanol, mp 162-167°. $[\alpha]_D$ +71.6°. uv max (0.01N NaOH): 265-266 nm.
Methyl ester. $C_{21}H_{32}O_2$. Crystals from methanol, mp 25-27°. $[\alpha]_D$ +67.7°. uv max: 265-266 nm.

7100. Palytoxin. [77734-91-9] Palytoxin (C51-55 hemiacetal); PTX. $C_{129}H_{223}N_3O_{54}$; mol wt 2680.17. C 57.81%, H 8.39%, N 1.57%, O 32.23%. Potent toxin isolated from zoanthid coral of the genus *Palythoa* that is the most poisonous non-proteinaceous substance known. Isoln from "Limu-make-o-Hana", the Hawaiian name for the highly toxic coelenterate *Palythoa toxica*: R. E. Moore,

P. J. Scheuer, *Science* **172**, 495 (1971). Structure: R. E. Moore, G. Bartolini, *J. Am. Chem. Soc.* **103**, 2491 (1981). Structure of palytoxin from *P. tuberculosa* of Okinawa (differs from palytoxin from *P. toxica* at two positions): D. Uemura *et al.*, *Tetrahedron Lett.* **22**, 2781 (1981). Proposed absolute configuration of 60 of the 64 chiral centers: R. E. Moore *et al.*, *J. Am. Chem. Soc.* **104**, 3776 (1982). Structure and stereochemistry: J. K. Cha *et al.*, *ibid.* 7369. Discussion of the structural elucidation: Y. Shimizu, *Nature* **302**, 212 (1983); review: R. E. Moore, *Prog. Chem. Org. Nat. Prod.* **48**, 82-202 (1985). Synthetic studies: Y. Kishi, *Chem. Scr.* **27**, 573 (1987). Pharmacological study: P. N. Kaul *et al.*, *Proc. West. Pharmacol. Soc.* **17**, 294 (1974). Mechanism of action and treatment of palytoxin poisoning: J. A. Vick *et al.*, *Toxicol. Appl. Pharmacol.* **34**, 214 (1975). Mode of contractile action on vascular smooth muscle: K. Ito *et al.*, *Eur. J. Pharmacol.* **46**, 9 (1977). Depolarizing action on frog spinal cord: Y. Kudo, S. Shibata, *Br. J. Pharmacol.* **71**, 575 (1980). Toxicology and toxicity studies: J. S. Wiles *et al.*, *Toxicon* **12**, 427 (1974); K. Ito *et al.*, *Arch. Int. Pharmacodyn. Ther.* **258**, 146 (1982). Brief review of biology: P. J. Scheuer, *Acc. Chem. Res.* **10**, 33-39 (1977). Review of synthetic studies and conformational analysis: Y. Kishi, *Pure Appl. Chem.* **61**, 313-324 (1989).

White amorphous hygroscopic solid. No definite mp; chars when heated to 300°. $[\alpha]_D^{25}$ +26° (water). Insol in chloroform, ether, acetone. Sparingly sol in methanol, ethanol. Sol in pyridine, DMSO, water. LD_{50} in mice: 0.45 μg/kg i.v. (Wiles); 50-100 ng/kg i.p. (Kaul).

Caution: Palytoxin is an intense vasoconstrictor. In intact dogs, doses of >0.06 μg/kg i.v. caused a transient rise in arterial pressure followed by rapid hypotension and resulted in death within 5 minutes (Ito, 1982).

USE: As a physiological tool to evaluate anti-anginal chemotherapeutic agents.

7101. Pamabrom. [606-04-2] 8-Bromo-3,9-dihydro-1,3-dimethyl-1H-purine-2,6-dione compd with 2-amino-2-methyl-1-propanol (1:1); 8-bromotheophylline compd with 2-amino-2-methyl-1-propanol (1:1); 2-amino-2-methyl-2-propanol 8-bromotheophyllinate. $C_{11}H_{18}BrN_5O_3$; mol wt 348.20. C 37.94%, H 5.21%, Br 22.95%, N 20.11%, O 13.78%. Prepn from 2-amino-2-methyl-1-propanol and 8-bromotheophylline: J. M. Holbert, I. W. Grote, **US 2711411** (1955 to Chattanooga Med.). Prepn and diuretic activity: J. M. Holbert *et al.*, *J. Am. Pharm. Assoc. Sci. Ed.* **44**, 355 (1955).

Fine white powder, dec 300°. Soly in water >30 g/100 ml at 25°. pH of satd aq soln 8.0-8.5.

Mixture with acetaminophen and pyrilamine maleate. Midol PMS; Premsyn PMS; Sunril.

THERAP CAT: Diuretic.

7102. Pamaquine. [491-92-9] N^1,N^1-Diethyl-N^4-(6-methoxy-8-quinolinyl)-1,4-pentanediamine; 8-[[4-(diethylamino)-1-methylbutyl]amino]-6-methoxyquinoline. $C_{19}H_{29}N_3O$; mol wt 315.46. C 72.34%, H 9.27%, N 13.32%, O 5.07%. Prepn: **GB 295656** (1927 to I. G. Farben); K. H. Slotta, *Grundriss der modernen Arzneistoff-Synthese* (Stuttgart, 1931); Elderfield *et al.*, *J. Am. Chem. Soc.* **70**, 40 (1948). *Review:* Cooper, *Public Health Rep.* **64**, 717 (1949). HPLC determn in plasma: Y. S. Endoh *et al.*, *J. Chromatogr.* **579**, 123 (1992). Historical review: D. Greenwood, *J. Antimicrob. Chemother.* **36**, 857-872 (1995).

Dark yellow oil. $bp_{0.3}$ 175-180°; bp_1 182-194°.

Pamoate. [635-05-2] Pamaquine embonate; pamaquine naphthoate; Aminoquin; Plasmochin; Plasmoquine; Praequine. $C_{19}H_{29}$-$N_3O·C_{23}H_{16}O_6$; mol wt 703.84. Yellow to orange-yellow odorless, almost tasteless powder. Numbs the tongue. Insol in water. Sol in alcohol, acetone. *Protect from light.*

THERAP CAT: Antimalarial.

7103. Pamicogrel. [101001-34-7] 2-[4,5-Bis(4-methoxyphenyl)-2-thiazolyl]-1*H*-pyrrole-1-acetic acid ethyl ester; ethyl 2-[4,5-bis(*p*-methoxyphenyl)-2-thiazolyl]pyrrole-1-acetate; KB-3022; KBT-3022. $C_{25}H_{24}N_2O_4S$; mol wt 448.54. C 66.94%, H 5.39%, N 6.25%, O 14.27%, S 7.15%. Cyclooxygenase inhibitor. Prepn: K. Yoshino *et al.*, **EP 159677**; *eidem*, **US 4659726** (1985, 1987 both to Kanebo). HPLC determn in plasma and urine: Y. Nakada *et al.*, *Chem. Pharm. Bull.* **38**, 1093 (1990). Pharmacokinetics: Y. Nakada *et al.*, *Yakuzaigaku* **53**, 210 (1993), *C.A.* **120**, 315100 (1993). Activity as antithrombotic: K. Yokoto *et al.*, *Jpn. J. Pharmacol.* **68**, 201 (1995); as platelet aggregation inhibitor: K. Yokoto *et al.*, *J. Pharm. Pharmacol.* **47**, 768 (1995). Evaluation of cerebral protective effects: N. Yamamoto *et al.*, *Jpn. J. Pharmacol.* **69**, 421 (1995); *eidem*, *Eur. J. Pharmacol.* **297**, 225 (1996).

Crystals from ligroin, mp 132.5-135.5°. LD_{50} orally in male mice: >3000 mg/kg (Yoshino).

THERAP CAT: Antithrombotic.

7104. Pamidronic Acid. [40391-99-9] P,P'-(3-Amino-1-hydroxypropylidene)bisphosphonic acid; 3-amino-1-hydroxypropane-1,1-diphosphonic acid; ADP; AHPrBP. $C_3H_{11}NO_7P_2$; mol wt 235.07. C 15.33%, H 4.72%, N 5.96%, O 47.64%, P 26.35%. Bisphosphonate antiresorptive agent. Prepn: F. Krueger *et al.*, **DE 2130794** (1973 to Benckiser), *C.A.* **78**, 84528z (1973); K.-H. Worms *et al.*, *Z. Anorg. Allg. Chem.* **457**, 214 (1979). Improved process: H. Blum, K.-H. Worms, **US 4327039** (1982 to Henkel). Mechanism of action study: P. M. Boonekamp *et al.*, *Bone Miner.* **1**, 27 (1986). HPLC determn in serum and plasma: R. W. Sparidans *et al.*, *J. Chromatogr. B* **705**, 331 (1998). Clinical trial to prevent postmenopausal bone loss: B. Lees *et al.*, *Osteoporos. Int.* **6**, 480 (1996). Review of pharmacology and therapeutic use in resorptive bone disease: A. Fitton, D. McTavish, *Drugs* **41**, 289-318 (1991); A. J. Coukell, A. Markham, *Drugs Aging* **12**, 149-168 (1998); of therapeutic potential in Paget's disease: P. L. Selby, *Bone* **24**, 57S-58S (1999).

Disodium salt. [57248-88-1] Pamidronate disodium; GCP-23339A; Aredia; Ostepam; Pamifos; Pamisol. $C_3H_9NNa_2O_7P_2$; mol wt 279.03. White, crystalline powder. Sol in water, $2N$ sodium hydroxide; sparingly sol in $0.1N$ hydrochloric acid, $0.1N$ acetic acid. Practically insol in organic solvents.

THERAP CAT: Bone resorption inhibitor.

THERAP CAT (VET): Antidote (vitamin-D analog hypercalcemia).

7105. Pamiteplase. 275-L-Glutamic acid (1-91)-(174-527)-plasminogen activator (human tissue-type protein moiety); YM-866; Solinase. Genetically engineered variant of human tissue plasminogen activator (t-PA), *q.v.* Derived by deletion of the first kringle domain and the substitution of glutamic acid for arginine at position 275. Prepn: Y. Kawauchi *et al.*, **WO 8903874**; *eidem*, **US 5556621** (1989, 1996 both to Yamanouchi). Thrombolytic activity: T. Kawasaki *et al.*, *Jpn. J. Pharmacol.* **63**, 9 (1993); M. Suzuki *et al.*, *Curr. Ther. Res.* **61**, 7 (2000). Toxicology: A. Ishikawa *et al.*, *J. Toxicol. Sci.* **22**, 117 (1997). Review of thrombolytic agents: M. Verstraete, *Am. J. Med.* **109**, 52-58 (2000).

THERAP CAT: Thrombolytic.

7106. Pamoic Acid. [130-85-8] 4,4'-Methylenebis[3-hydroxy-2-naphthalenecarboxylic acid]; 4,4'-methylenebis(3-hydroxy-2-naphthoic acid); 4,4'-methylenedi(3-hydroxy-2-naphthoic acid); 2,2'-dihydroxy-1,1'-dinaphthylmethane-3,3'-dicarboxylic acid; embonic acid. $C_{23}H_{16}O_6$; mol wt 388.38. C 71.13%, H 4.15%, O 24.72%. Prepn from 2-hydroxy-3-naphthoic acid + formaldehyde: Strohback, *Ber.* **34**, 4162 (1901); Brass, Sommer, *ibid.* **61**, 993 (1928); Barber, Gaimster, *J. Appl. Chem.* **2**, 565 (1952). Prepn of salts: Puetzer, **US 2397903** (1946 to Vick Chem.); Barber, **US 2641610** (1953 to May & Baker).

Crystals from dil pyridine, dec above 280° without melting. Practically insoluble in water, alcohol, ether, benzene, acetic acid. Sparingly sol in chloroform; sol in nitrobenzene, pyridine.

Methyl ether. $C_{24}H_{18}O_6$. Yellow crystals from dil methanol, dec 277-282°.

7107. PAN. [85-85-8] 1-[2-(2-Pyridinyl)diazenyl]-2-naph-thalenol; 1-(2-pyridinylazo)-2-naphthalenol; 1-(2-pyridylazo)-2-naphthol. $C_{15}H_{11}N_3O$; mol wt 249.27. C 72.28%, H 4.45%, N 16.86%, O 6.42%. Cation chelant; reacts with metal ions to form stable, intensely colored complexes. Prepn: A. Tschitschibabin, M. Rjasanzew, *Chem. Zentralbl.* **1916**, III, 228. Improved synthesis: S. Wiejak *et al.*, *Pol. J. Chem.* **58**, 895 (1984). Spectroscopy studies of PAN and its metal chelates: D. Betteridge, D. John, *Analyst* **98**, 377, 390 (1973). Electronic absorption spectra: M. S. Masoud, H. H. Hammud, *Spectrochim. Acta A* **57**, 977 (2001). Metal ion extraction applications: J. Gao *et al.*, *Talanta* **40**, 195 (1993); I. Narin, M. Soylak, *ibid.* **60**, 215 (2003); M. Tuzen *et al.*, *Anal. Lett.* **37**, 473 (2004).

Orange-red solid, mp 140-141°. *Irritant.* Absorption max (diethyl ether): 458 nm; (acetone): 463 nm; (DMSO): 471 nm; (acetonitrile): 462 nm; (ethanol): 465 nm; (methanol): 465 nm. Metal complexes insol in water.

USE: Chromogenic indicator for complexometric and spectrophotometric detection of metal ions. In extraction of metal ions. In ion chromatography applications.

7108. Pancreatic Extract. Pancreas powder; Creon; Ku-Zyme; Nutrizym; Pancrease; Pancrex-V; Panzytrat; Ultrase; Vio-kase. Substance prepd from fresh or frozen mammalian pancreas, usually from swine, that contains various enzymes having proteolytic, lipolytic and amylolytic activities. Pharmacopeial specifications differ on the quantity of amylase, lipase and protease required; most commercial formulations are of higher digestive power. Method of production from cow or pig pancreas: S. Hoek, US **3223594** (1965 to N. American Philips). Clinical pharmacology and comparison of formulations: Y. W. Cho, D. M. Aviado, *J. Clin. Pharmacol.* **21**, 224 (1981). Analysis of pharmaceutical preparations: C. L. Case *et al.*, *Pancreas* **30**, 180 (2005). Clinical trial in weight loss prevention in pancreatic cancer: M. J. Bruno *et al.*, *Gut* **42**, 92 (1998); in cystic fibrosis: R. C. Stern *et al.*, *Am. J. Gastroenterol.* **95**, 1932 (2000); C. J. Patchell *et al.*, *J. Cyst. Fibros.* **1**, 287 (2002); M. S. Brady *et al.*, *J. Am. Diet. Assoc.* **106**, 1181-1186 (2006).

Slightly brown, amorphous powder. Partly sol in water. Practically insol in alcohol, ether.

Pancreatin. [8049-47-6] Pancreatic enzyme concentrate containing not less than 2 USP units of lipase, 25 of amylase and 25 of protease activities per mg. British Pharmacopeia specifies not less than 20 FIP units of lipase, 24 of amylase and 1.4 of protease activities per mg.

Pancrelipase. [53608-75-6] Porcine pancreatic enzyme concentrate containing not less than 24 USP units of lipase, 100 of amylase and 100 of protease activities per mg.

THERAP CAT: Enzyme replacement therapy (pancreatic exocrine insufficiency).

THERAP CAT (VET): Enzyme replacement therapy in pancreatic enzyme deficiency.

7109. Pancuronium Bromide. [15500-66-0] 1,1'-[(2β,3α,-5α,16β,17β)-3,17-Bis(acetyloxy)androstane-2,16-diyl]bis[1-methylpiperidinium] dibromide; 1,1'-(3α,17β-dihydroxy-5α-androstan-2β,16β-ylene)bis[1-methylpiperidinium] dibromide diacetate; 3α,-17β-diacetoxy-2β,16β-dipiperidino-5α-androstane dimethobromide; 2β,16β-dipiperidino-5α-androstane-3α,17β-diol diacetate dimethobromide; poncuronium bromide (rescinded USAN); NA-97; Org-NA-97; Mioblock; Pavulon. $C_{35}H_{60}Br_2N_2O_4$; mol wt 732.68. C 57.38%, H 8.25%, Br 21.81%, N 3.82%, O 8.73%. Aminosteroid, competitive neuromuscular blocker. Prepn: W. R. Buckett *et al.*, *Chim. Ther.* **2**, 186 (1967); *eidem*, *J. Med. Chem.* **16**, 1116 (1973). Structural studies: Savage *et al.*, *J. Chem. Soc. B* **1971**, 410. Structure-activity correlation: Waser, *Anaesthesist* **20**, 23 (1971). Phar-

macology: W. R. Buckett, I. L. Bonta, *Fed. Proc.* **25**, 718 (1966); W. R. Buckett *et al.*, *Br. J. Pharmacol. Chemother.* **32**, 671 (1968); I. L. Bonta *et al.*, *Eur. J. Pharmacol.* **4**, 83, 303 (1968). Comparative study of neuromuscular blocking and vagolytic effect: S. L. Son *et al.*, *Anesthesiology* **55**, 12 (1981). *Review:* Speight, Avery, *Drugs* **4**, 163-226 (1972).

White, yellowish-white or slightly pink, odorless, hygroscopic crystals with bitter taste, mp 215°. One gram dissolves in 30 parts chloroform, one part water (20°). Freely sol in alc, methylene chloride. LD_{50} in mice (mg/kg): 0.047 i.v.; 0.152 i.p.; 0.167 s.c.; 21.9 orally; in rats, rabbits: 0.153, 0.016 i.v. (Buckett, 1968).

THERAP CAT: Neuromuscular blocking agent.

THERAP CAT (VET): Neuromuscular blocking agent.

7110. Pangamic Acid. Controversial mixture of compounds erroneously labeled as *vitamin B₁₅*. Allegedly isolated from apricot kernel: E. T. Krebs, Sr. *et al.*, *Int. Rec. Med.* **164**, 18 (1951). Originally named pangamic acid because of its supposed ubiquity in seeds. There is no clear chemical identity for pangamic acid. Historical review of structural studies, analyses, syntheses, and components: J. C. Micheau *et al.*, *Chim. Ther.* **7**, 103 (1972). Products sold as pangamic acid in the U.S. vary considerably in their composition. Some are mixtures of calcium gluconate and N,N-dimethylglycine, q.q.v.; others contain diisopropylamine dichloroacetate, q.v.: *FDA Drug Bulletin* vol. **8**(6), Dec. 1978-Jan. 1979. *Ref.:* V. Herbert, *Am. J. Clin. Nutr.* **32**, 1534 (1979); V. Herbert, R. Herbert, in *Controversies in Nutrition*, L. Ellenbogen, Ed. (Churchill-Livingstone, New York, 1981) pp 159-170.

7111. Panipenem. [87726-17-8] (5R,6S)-6-[(1R)-1-Hydroxyethyl]-3-[[(3S)-1-(1-iminoethyl)-3-pyrrolidinyl]thio]-7-oxo-1-azabicyclo[3.2.0]hept-2-ene-2-carboxylic acid; (5R,6S)-6-[(R)-1-hydroxyethyl]-2-[(S)-1-acetimidoylpyrrolidin-3-ylthio]-1-carbapen-2-em-3-carboxylic acid; (+)-(5R,6S)-3-[[(S)-1-acetimidoyl-3-pyrrolidinyl]thio]-6-[(R)-1-hydroxyethyl]-7-oxo-1-azabicyclo[3.2.0]hept-2-ene-2-carboxylic acid; CS-533; RS-533. $C_{15}H_{21}N_3O_4S$; mol wt 339.41. C 53.08%, H 6.24%, N 12.38%, O 18.86%, S 9.45%. Carbapenem antibiotic. Prepn: T. Miyadera *et al.*, *J. Antibiot.* **36**, 1034 (1983). Manufacturing process: A. Yoshida, K. Oda, EP **587436** (1994 to Sankyo). Antibacterial spectrum *in vitro*: H. C. Neu *et al.*, *Antimicrob. Agents Chemother.* **30**, 828 (1986). Series of articles on pharmacology and clinical efficacy in combination with betamipron, q.v: *Chemotherapy (Tokyo)* **39** Suppl 3, 1-813 (1991). Toxicology: K. Kimura *et al.*, *ibid.* 140.

Prepd as the hemihydrate; colorless fine prisms, mp 198-200° (dec). uv max (water): 298 nm (ε 10400). Approx LD_{50} in male, female mice (mg/kg): 1700-2200, 1300-1700 i.v. (Kimura).

Mixture with betamipron. [138240-65-0] CS-976; Carbenin.

THERAP CAT: Antibacterial.

7112. Panitumumab. [339177-26-3] Anti-(human epidermal growth factor receptor) immunoglobulin (human monoclonal ABX-EGF heavy chain) disulfide with human monoclonal ABX-EGF light chain, dimer; ABX-EGF; Vectibix. Fully human monoclonal antibody secreted by the E7.6.3 hybridoma; directed against human epidermal growth factor receptor (EGFr), a transmembrane cell-surface glycoprotein overexpressed in a variety of cancers. Prepn: A. Jakobovits *et al.*, **WO 9850433**; *eidem*, **US 6235883** (1998, 2001 both to Abgenix); and *in vivo* antitumor activity: X.-D. Yang *et al.*, *Cancer Res.* **59**, 1236 (1999). Clinical pharmacokinetics: E. K. Rowinsky *et al.*, *J. Clin. Oncol.* **22**, 3003 (2004). Clinical trial in metastatic colorectal cancer: E. VanCustem *et al.*, *Ann. Oncol.* **19**, 92 (2008). Review of development and pharmacology: K. A. Foon *et al.*, *Int. J. Radiat. Oncol. Biol. Phys.* **58**, 984-990 (2004); of pharmacology and clinical experience with solid tumors: M. Wu *et al.*, *Clin. Ther.* **30**, 14-30 (2008).

THERAP CAT: Antineoplastic.

7113. Panobinostat. [404950-80-7] (2*E*)-*N*-Hydroxy-3-[4-[[[2-(2-methyl-1*H*-indol-3-yl)ethyl]amino]methyl]phenyl]-2-propenamide; LBH-589; Faridak. $C_{21}H_{23}N_3O_2$; mol wt 349.43. C 72.18%, H 6.63%, N 12.03%, O 9.16%. Hydroxamate type histone deacetylase (HDAC) inhibitor. Prepn: K. W. Bair, *et al.*, **WO 0222577**; S. W. Remiszewski *et al.*, **US 6552065** (2002, 2003 both to Novartis). Inhibition of tumor angiogenesis: D. Z. Qian *et al.*, *Clin. Cancer Res.* **12**, 634 (2006). *In vitro* and *in vivo* antitumor spectrum: M. C. Crisanti *et al.*, *Mol. Cancer Ther.* **8**, 2221 (2009). Review of mechanism of action: P. Atadja, *Cancer Lett.* **280**, 233-241 (2009); and of preclinical and clinical studies: H. M. Prince *et al.*, *Future Oncol.* **5** 601-612 (2009).

THERAP CAT: Antineoplastic.

7114. Pantetheine. [496-65-1] (2*R*)-2,4-Dihydroxy-*N*-[3-[(2-mercaptoethyl)amino]-3-oxopropyl]-3,3-dimethylbutanamide; 2,4-dihydroxy-*N*-[2-[(2-mercaptoethyl)carbamoyl]ethyl]-3,3-dimethylbutyramide; *N*-(pantothenyl)-β-aminoethanethiol; α,γ-dihydroxy-β,β-dimethylbutyryl-β-alanyl-β-aminoethanethiol. $C_{11}H_{22}N_2O_4S$; mol wt 278.37. C 47.46%, H 7.97%, N 10.06%, O 22.99%, S 11.52%. Intermediate in the biosynthesis of coenzyme A, *q.v.*, in mammalian liver and in some microorganisms. First identified as a growth factor for *Lactobacillus bulgaricus:* Williams *et al.*, *J. Biol. Chem.* **177**, 933 (1949). Synthesis: Schwyzer, *Helv. Chim. Acta* **35**, 1903 (1952); Baddiley, Thain, *J. Chem. Soc.* **1952**, 800; King *et al.*, *J. Am. Chem. Soc.* **75**, 1290 (1953); Walton *et al.*, *ibid.* **76**, 1146 (1954); Walton, **US 2744119** and **US 2835704** (1956, 1958, both to Merck & Co.). *Reviews:* Snell, Brown, *Adv. Enzymol.* **14**, 49 (1953); Snell, Wittle, *Methods Enzymol.* **3**, 918 (1957).

Syrup or glass. $[\alpha]_D^{20}$ +12.9° (c = 4.5 in water). Microbiological activity: 20,000 LBF units/mg. An LBF unit is that amount of the growth factor contained in one mg of Basamine-Busch, a standard yeast extract manuf by Anheuser-Busch, Inc.

Silver mercaptide. $C_{11}H_{21}AgN_2O_4S$. Yellow noncryst solid. $[\alpha]_D^{25}$ +8° (c = 4.27 in 0.9*N* NaCl). Very sol in water. 11,000 LBF units/mg.

Mercuric mercaptide. $C_{22}H_{42}HgN_4O_8S_2 \cdot C_3H_6O$. Crystals from acetone, mp 96-98°. $[\alpha]_D^{27}$ +9.6° (c = 4 in water). 15,500 LBF units/mg. uv max: 260-265 nm (ε 1000). Sol in water, methanol.

S-Benzoylpantetheine. $C_{18}H_{26}N_2O_5S$. Crystals from ethyl acetate, mp 116°. $[\alpha]_D^{27}$ +31° (ethanol). Sol in water, ethyl acetate, chloroform.

S-Acetylpantetheine. $C_{13}H_{24}N_2O_5S$. Thick syrup. $[\alpha]_D^{27}$ +39° (c = 0.8 in ethanol).

7115. Pantethine. [16816-67-4] (2*R*,2'*R*)-*N*,*N*'-[Dithiobis[2,1-ethanediylimino(3-oxo-3,1-propanediyl)]]bis[2,4-dihydroxy-3,3-dimethylbutanamide]; *N*,*N*'-[dithiobis(ethyleneiminocarbonylethylene)]bis(2,4-dihydroxy-3,3-dimethylbutyramide); D-bis(*N*-pantothenyl-β-aminoethyl) disulfide; Lipodel; Pantetina; Panthecin; Pantomin; Pantosin. $C_{22}H_{42}N_4O_8S_2$; mol wt 554.72. C 47.64%, H 7.63%, N 10.10%, O 23.07%, S 11.56%. Disulfide dimer of pantetheine, *q.v.* Growth factor for *Lactobacillus bulgaricus:* Williams *et al.*, *J. Biol. Chem.* **177**, 933 (1949). Formed by oxidation of pantetheine: Brown, Snell, *ibid.* **198**, 375 (1952). Structure: Snell *et al.*, *J. Am. Chem. Soc.* **72**, 5349 (1950). Synthesis: Wieland, Bokelmann, *Naturwissenschaften* **38**, 384 (1951); Wittle *et al.*, *J. Am. Chem. Soc.* **75**, 1694 (1953); Viscontini *et al.*, *Helv. Chim. Acta* **37**, 375 (1954); Bowman, Cavalla, *J. Chem. Soc.* **1954**, 1171; Shimizu *et al.*, *Chem. Pharm. Bull.* **13**, 180 (1965). Clinical trial in hyperlipoproteinemia: A. Gaddi *et al.*, *Atherosclerosis* **50**, 73 (1984).

Glassy, colorless to light yellow substance. $[\alpha]_D^{27}$ +13.5° (c = 3.75 in water). Freely sol in water; less sol in ethanol. Practically insol in ether, acetone, ethyl acetate, benzene, chloroform.

THERAP CAT: Antilipemic.

7116. Pantolactone. [599-04-2] (3*R*)-Dihydro-3-hydroxy-4,4-dimethyl-2(3*H*)-furanone; pantoic acid γ-lactone; pantoyl lactone; pantoic lactone; 2,4-dihydroxy-3,3-dimethylbutyric acid γ-lactone; α-hydroxy-β,β-dimethyl-γ-butyrolactone. $C_6H_{10}O_3$; mol wt 130.14. C 55.38%, H 7.75%, O 36.88%. A degradation product of pantothenic acid from liver: Williams, Major, *Science* **91**, 246 (1940). Important intermediate in the synthesis of pantothenic acid. May be prepd by condensing isobutyraldehyde with formaldehyde yielding α,α-dimethyl-β-hydroxypropionaldehyde which is condensed with hydrocyanic acid in the presence of calcium chloride to form racemic pantolactone. Various modifications of this procedure exist: Glaser, *Monatsh. Chem.* **25**, 46 (1904); Stiller *et al.*, *J. Am. Chem. Soc.* **62**, 1785 (1940); Reichstein, Grüssner, *Helv. Chim. Acta* **23**, 650 (1940); Carter, Ney, *J. Am. Chem. Soc.* **63**, 312 (1941). Vast patent literature, *e.g.*, Beckmann *et al.*; Klein, **US 2967869** and **US 3024250** (1961, 1962, both to Nopco).

D(−)-Form. Hygroscopic crystals from benzene + petr ether, mp 92°. $[\alpha]_D^{25}$ −50.7° (c = 2.05 in H_2O). Can be purified by microsublimation.

L(+)-Form. Hygroscopic crystals from benzene, mp 91°. $[\alpha]_D^{25}$ +50.1° (c = 2 in H_2O).

DL-Form. Hygroscopic rosettes or prisms, mp 80°, bp_{18} 130°. Freely sol in water. Sol in ether, benzene, chloroform, alcohol, carbon disulfide.

7117. Pantoprazole. [102625-70-7] 6-(Difluoromethoxy)-2-[[(3,4-dimethoxy-2-pyridinyl)methyl]sulfinyl]-1*H*-benzimidazole; SKF-96022; BY-1023. $C_{16}H_{15}F_2N_3O_4S$; mol wt 383.37. C 50.13%, H 3.94%, F 9.91%, N 10.96%, O 16.69%, S 8.36%. Gastric proton pump inhibitor. Prepn: B. Kohl *et al.*, **EP 166287**; *eidem*, **US 4758579** (1986, 1988 both to Byk Gulden); *eidem*, *J. Med. Chem.* **35**, 1049 (1992). Mechanism of action: W. A. Simon *et al.*, *Biochem. Pharmacol.* **39**, 1799 (1990). HPLC determn in plasma and serum: R. Huber *et al.*, *J. Chromatogr.* **529**, 389 (1990). Clinical pharmacokinetics: B. Simon *et al.*, *Z. Gastroenterol.* **28**, 443

(1990). Clinical trial in duodenal ulcer: P. Müller *et al., ibid.* **30**, 771 (1992); in Zollinger-Ellison syndrome: E. A. Lew *et al., Gastroenterology* **118**, 696 (2000). Review of pharmacology and clinical experience: P. Poole, *Am. J. Health Syst. Pharm.* **58**, 999-1008 (2001).

Off-white solid, mp 139-140° (dec). pKa$_1$ 3.92; pKa$_2$ 8.19.

Sodium salt. [138786-67-1] Eupantol; Pantecta; Pantozol; Pantopan; Pantorc; Peptazol; Protium; Protonix; Somac. $C_{16}H_{14}F_2N_3$-NaO$_4$S; mol wt 405.35. Prepd as the sesquihydrate. White to off-white crystalline powder, dec >130°. uv max (methanol): 289 (ε 1.64 × 10^4). Freely sol in water, methanol, dehydrated alc; very slightly sol in phosphate buffer (pH 7.4). Practically insol in *n*-hexane, dichloromethane.

THERAP CAT: Antiulcerative; in treatment of Zollinger-Ellison syndrome.

7118. Pantothenic Acid. [79-83-4] *N*-[(2*R*)-2,4-Dihydroxy-3,3-dimethyl-1-oxobutyl]-β-alanine; D(+)-*N*-(2,4-dihydroxy-3,3-dimethylbutyryl)-β-alanine; chick antidermatitis factor; vitamin B$_5$. $C_9H_{17}NO_5$; mol wt 219.24. C 49.31%, H 7.82%, N 6.39%, O 36.49%. A member of the B complex vitamins; essential vitamin for the biosynthesis of coenzyme A in mammalian cells. Occurs ubiquitously in all animal and plant tissue. The richest common source is liver, but jelly of the queen bee contains 6 times as much as liver. Rice bran and molasses are other good sources. Isoln from liver: R. J. Williams *et al., J. Am. Chem. Soc.* **60**, 2719 (1938). Synthesis: E. T. Stiller *et al., ibid.* **62**, 1785 (1940); Reichstein, Grüssner, *Helv. Chim. Acta* **23**, 650 (1940); Grüssner *et al., ibid.* 1276. Absolute configuration: Hill, Chan, *Biochem. Biophys. Res. Commun.* **38**, 181 (1970). Only the natural, dextrorotatory form has vitamin activity. Review of chemistry, biochemistry and pharmacology: W. Friedrich, *Vitamins* (de Gruyter, Berlin, 1988) pp 809-835; of metabolism: A. G. Tahiliani, C. J. Beinlich, *Vitam. Horm.* **46**, 165-228 (1991).

Unstable, viscous oil. Extremely hygroscopic. Easily destroyed by acids, bases, heat. $[\alpha]_D^{25}$ +37.5°. Freely sol in water, ethyl acetate, dioxane, glacial acetic acid; moderately sol in ether, amyl alcohol. Practically insol in benzene, chloroform.

Sodium salt. [867-81-2] $C_9H_{16}NNaO_5$. Very hygroscopic crystals. $[\alpha]_D^{25}$ +27.1° (c = 2). Can be stored in sealed ampuls only.

Calcium salt. [137-08-6] Calpanate; Pantholin. $C_{18}H_{32}CaN_2$-O_{10}; mol wt 476.54. Prepn: Wehrmeister, **US 2780645** (1957 to Commercial Solvents); Kagan, **US 2845456** (1958 to Upjohn). Minute needles from CH$_3$OH. Sweetish taste with slightly bitter aftertaste. Dec 195-196°. Moderately hygroscopic. $[\alpha]_D^{20}$ +28.2° (c = 5). One gram dissolves in 2.8 ml H$_2$O. Sol in glycerol; slightly sol in alcohol, acetone. pH of 5% aq soln: 7.2-8.0; pH in CO$_2$-free water: 8.7.

THERAP CAT: Vitamin.

THERAP CAT (VET): Nutritional factor: dietary essential except in horses, ruminants.

7119. Papain. [9001-73-4] Papayotin; vegetable pepsin; Arbuz; Nematolyt; Summetrin; Tromasin; Velardon; Vermizym. First recognized member of the class of proteolytic enzymes that needs a free sulfhydryl group for activity. Isolated from the latex of the green fruit and leaves of *Carica papaya* L., Caricaceae. Initial isolation and crystallization: Balls *et al., Science* **86**, 379 (1937); Balls, Lineweaver, *J. Biol. Chem.* **130**, 669 (1939). Prepn from commercial dried papaya latex and physical properties: Kimmel, Smith, *ibid.* **207**, 515 (1954); *ibid.* 533-573; Becker, *Econ. Bot.* **12**, 62 (1958).

Purification: Gibian, Bratfisch, **US 2950227** (1960 to Schering AG); Lesuk, **US 3011952** (1961 to Sterling Drug); Blumberg *et al., Eur. J. Biochem.* **15**, 97 (1970). The papain molecule consists of one folded polypeptide chain of 212 residues, mol wt ~23,400. Complete amino acid sequence: Drenth *et al., Nature* **218**, 929 (1968); Mitchel *et al., J. Biol. Chem.* **245**, 3485 (1970). Mechanism of action studies: Morihara, *J. Biochem.* **62**, 250 (1967). Use in treatment of contact lenses to prolong wearing time in keratoconic patients with papillary conjunctivitis: D. R. Korb *et al., Arch. Ophthalmol.* **101**, 48 (1983). *Reviews:* Kimmel, Smith, *Adv. Enzymol. Relat. Subj. Biochem.* **19**, 267-334 (1957); Glazer, Smith in *The Enzymes* **vol. III**, P. D. Boyer, Ed. (Academic Press, New York, 3rd ed., 1971) pp 501-537. Comprehensive review of the structural elucidation: Drenth *et al., ibid.* 485-498; *eidem, Adv. Protein Chem.* **25**, 79-11 (1971).

White or grayish-white, slightly hygroscopic powder. uv max: 278 nm (A$_{1cm}^{1\%}$ 25.0). Sol in water; incompletely sol in glycerol. Practically insol in alc, ether, chloroform, and most organic solvents. *Keep well closed.* Potency varies according to process of prepn, etc. with the usual grade digesting ~35 times its wt of lean meat. Best grades render sol 200-300 times their wt of coagulated egg albumin in alkaline media. A temp. range of 60-90° is favorable for the digestive process with 65° the optimum point. Best pH is 5.0, but it functions also in neutral or alkaline media. Activated by reduction (HCN, H$_2$S etc.) and inactivated by oxidation (H$_2$O$_2$, iodoacetate).

Note: The term papain is currently applied to both the crude dried latex and the crystalline proteolytic enzyme.

USE: For tenderizing meats; for clearing beverages; for bating skins.

THERAP CAT: Enzyme (proteolytic). Debriding agent; digestive aid. Has been used to prevent adhesions; as anthelmintic (Nematodes).

7120. Papaveraldine. [522-57-6] (6,7-Dimethoxy-1-isoquinolinyl)(3,4-dimethoxyphenyl)methanone; 6,7-dimethoxy-1-isoquinolyl 3,4-dimethoxyphenyl ketone; 6,7-dimethoxy-1-veratroylisoquinoline; xanthaline. $C_{20}H_{19}NO_5$; mol wt 353.37. C 67.98%, H 5.42%, N 3.96%, O 22.64%. The old name papaveraldine is retained to avoid confusion. From opium; whether papaveraldine occurs as such in the poppy plant, or is formed during the process of extraction, has not been investigated. Oxidation of papaverine to papaveraldine by SeO$_2$: Menon, *Proc. Indian Acad. Sci.* **19A**, 21 (1944). For older references *see:* Small, Lutz, "Chemistry of the Opium Alkaloids," in *U.S. Public Health Reports* **Suppl. No. 103** (Washington, 1932). Synthesis from Reissert compds: Popp, McEwen, *J. Am. Chem. Soc.* **79**, 3773 (1957).

Crystals from abs ethanol, mp 208-209°. Sol in benzene, chloroform; slightly sol in alcohol, ether, petr ether; nearly insol in water, alkalies or carbonates.

Hydrochloride. $C_{20}H_{19}NO_5$.HCl. Yellow crystals from abs alcohol, mp 200°.

7121. Papaveretum. [8002-76-4] Opium alkaloids hydrochlorides; opium active principles; Omnopon; Pantopon. Mixture of the hydrochlorides of opium alkaloids containing 253 parts morphine hydrochloride, 23 parts papaverine hydrochloride, 20 parts codeine hydrochloride. Earlier prepns described as approx 50% morphine, 3% codeine, 5% papaverine, and 20% noscapine. Exhibits biological action of morphine and other alkaloids present in opium. GLC determn of alkaloids: G. Fisher, R. Gillard, *J. Pharm. Sci.* **66**, 421 (1977). Clinical evaluation in intravenous analgesia: J. A. Catling *et al., Br. Med. J.* **281**, 478 (1980); in combination with buprenorphine: D. W. Green *et al., Br. J. Anaesth.* **70**, 626 (1993).

White or almost white crystalline powder. Sol in water; sparingly sol in alcohol. pH of 1.5% aq sol: 3.7-4.7. *Protect from light.*

Note: This is a controlled substance (opium derivative): **21 CFR,** 1308.12.

THERAP CAT: Analgesic.

7122. Papaverine. [58-74-2] 1-[(3,4-Dimethoxyphenyl)-methyl]-6,7-dimethoxyisoquinoline; 6,7-dimethoxy-1-veratrylisoquinoline. $C_{20}H_{21}NO_4$; mol wt 339.39. C 70.78%, H 6.24%, N 4.13%, O 18.86%. Smooth muscle relaxant found in opium (0.8-1.0%). Synthesis: Pictet, Gams, *Compt. Rend.* **149**, 210 (1909); *Ber.* **42**, 2943 (1909). Review of commercial syntheses: Goldberg, *Chem. Prod. Chem. News* **17**, 371 (1954). Improved synthetic procedures: Braz, Chizhov, *Soviet Pharmaceutical Research* **3**, 90-93 (New York, 1958). Biosynthetic studies: Battersby, Harper, *J. Chem. Soc.* **1962**, 3526; Brochmann-Hanssen *et al., J. Pharm. Sci.* **60**, 1672 (1971). Pharmacology and toxicology: Preininger in *The Alkaloids* **vol. 15**, R. H. F. Manske, Ed. (Academic Press, New York, 1975) pp 209-223. Toxicity: S. Levis *et al., Arch. Int. Pharmacodyn.* **123**, 264 (1960). Clinical effect on cerebral blood flow: H. L. Karpman, J. J. Sheppard, *Angiology* **26**, 592 (1975). Clinical evaluation in intermittent claudication: Y. Sheino *et al., ibid.* **34**, 257 (1983). Comprehensive description: M. S. Hifnawy, F. J. Muhtadi, *Anal. Profiles Drug Subs.* **17**, 367-447 (1988).

OCH₃ / H₃CO / structure

Triboluminescent, orthorhombic prisms from alc + ether, mp 147°. Sublimes at 135-140° at 11 mm pressure and 2 mm distance. d_4^{20} 1.337. pK (25°) 8.07. uv max (ethanol): 239, 278-280, 314, 327 nm (log ε 4.83, 3.86, 3.60, 3.67). Sol in hot benzene, glacial acetic acid, acetone; slightly sol in chloroform, carbon tetrachloride, petr ether. Almost insol in water. Optimal pH for storage of papaverine solns: 2.0-2.8.

Hydrochloride. [61-25-6] Artegodan; Cepaverin; Cerebid; Cerespan; Dynovas; Optenyl; Pameion; Panergon; Papital T.R.; Pavabid; Pavacen; Pavadel; Pavagen; Pavakey; Pavased; Spasmo-Nit; Therapav; Vasal; Vasospan. $C_{20}H_{21}NO_4$·HCl; mol wt 375.85. Monoclinic rods from water, mp 220-225°. uv max (ethanol): 249-250, 280-282, 311 nm (log ε 4.69, 3.80, 3.82). One gram dissolves in about 40 ml water. Sol in chloroform; slightly sol in alc. Practically insol in ether. pH of 0.05 molar soln 3.9; pH of 2% aq soln 3.3. LD_{50} in mice, rats (mg/kg): 27.5, 20 i.v.; 150, 370 s.c. (Levis).

Nitrite. [132-40-1] $C_{20}H_{21}NO_4$·HNO₂. Light yellow, crystalline powder. Slightly soluble in water and alcohol; freely sol in chloroform, acetone.

THERAP CAT: Vasodilator (cerebral).

7123. Papaya. Papaw; Carica; melon tree. Fruit of *Carica papaya* L., *Caricaceae. Habit.* Tropical America and Asia, Florida. *Constit.* Papain, the dried and purified latex of the fruit; carpaine; carposide (a glucoside).

USE: Manufacture of papain.

7124. Papuamine. [112455-84-2] (4a*S*,5a*R*,10a*R*,11a*S*,-15a*R*,15b*S*,16*E*,18*E*,19a*S*,19b*R*)-2,3,4,4a,5,5a,6,7,8,9,10,10a,11,-11a,12,13,14,15,15a,15b,19a,19b-Docosahydro-1*H*-diindeno[2,1-*f*:1′,2′-1][1,5]diazacyclotridecine. $C_{25}H_{40}N_2$; mol wt 368.61. C 81.46%, H 10.94%, N 7.60%. Alkaloid originally isolated from *Haliclona sp.*, a marine sponge found in Papua, New Guinea. Exhibits antifungal and antimicrobial activity. Isoln and characterization: B. J. Baker *et al., J. Am. Chem. Soc.* **110**, 965 (1988). Total synthesis: R. M. Borzilleri *et al., ibid.* **116**, 9789 (1994).

structure

White solid, mp 167.5-169°. $[\alpha]_D$ −150° (c = 1.5 in methanol).

Dihydrochloride. [112347-74-7] White solid, mp 230° (dec). $[\alpha]_D$ −140° (c = 1.3 in CH₃OH). uv max (methanol): 241 nm (ε 3000).

7125. PAR. [1141-59-9]; [113964-55-9] (*E*-form). 4-[2-(2-Pyridinyl)diazenyl]-1,3-benzenediol; 4-(2-pyridinylazo)-1,3-benzenediol; 4-(2-pyridinylazo)resorcinol; 4-(2-pyridylazo)resorcinol. $C_{11}H_9N_3O_2$; mol wt 215.21. C 61.39%, H 4.22%, N 19.53%, O 14.87%. Terdentate ligand for cation chelation; reacts with metal ions to form colored complexes. Prepn: A. E. Tschitschibabin, *Chem. Zentralbl.* **1923**, III, 1022. Improved synthesis: S. Wiejak *et al., Pol. J. Chem.* **58**, 895 (1984). ¹H NMR structural studies: K. Mochizuki *et al., Bull. Chem. Soc. Jpn.* **52**, 441 (1979). Acid-base properties and metal chelate formation: T. Iwamoto, *ibid.* **34**, 605 (1961). Absorption spectroscopy of PAR and its metal chelates: W. J. Geary *et al., Anal. Chim. Acta* **26**, 575 (1962). Stability constants of metal chelates: *eidem, ibid.* **27**, 71 (1962). Spectrophotometric determn of metal ions: T. M. Florence, Y. Farrar, *Anal. Chem.* **35**, 1613 (1963); E. Gómez *et al., Fresenius J. Anal. Chem.* **342**, 318 (1992); M. N. Abbas *et al., Anal. Chim. Acta* **436**, 223 (2001). Review of analytical uses: M. N. Desai, M. H. Gandhi, *Rec. Chem. Prog.* **30**, 223-231 (1969).

structure

Fine crystalline orange-red solid, mp 196-198° (Wiejak). Also reported as amorphous brown solid from methanol, mp 182° (Geary). *Irritant.* pKa₁ (25°) 5.83. pKa₂ (25°) 12.5. Absorption max (free base, pH 3.60): 383 nm (ε 15700); (monoionic form, pH 5.98-12.50): 415 nm (ε 25900). Readily sol in aq alkaline soln. Soly in water (10°): 5 mg/ml.

Sodium salt. [13311-52-9]; [16593-81-0] (monohydrate). $C_{11}H_8N_3NaO_2$; mol wt 237.19. Sol in acids, alkalies, alcohols. Insol in ether.

USE: Chromogenic indicator for complexometric, spectrophotometric, and spot test detection and determn of metal ions. In ion chromatography applications.

7126. Parabanic Acid. [120-89-8] 2,4,5-Imidazolidinetrione; imidazoletrione; oxalylurea. $C_3H_2N_2O_3$; mol wt 114.06. C 31.59%, H 1.77%, N 24.56%, O 42.08%. Prepd by the condensation of urea with diethyl oxalate in a methanol soln of sodium methoxide: Murray, *Org. Synth.* **37**, 71 (1957).

structure

Crystals, mp about 230°; also stated as 243° with decomposition. Sublimes at 100°. Soluble in about 20 parts water, in alcohol. Its salts are unstable.

7127. Paraffin. Paraffin wax; hard paraffin. A mixture of solid hydrocarbons having the general formula C_nH_{2n+2}, obtained from petroleum.

Colorless or white, somewhat translucent, odorless mass; greasy feel; burns with a luminous flame. d about 0.90. mp 50-57°; also available with higher and lower melting ranges. Miscible when melted with wax, spermaceti, fats. Freely sol in chloroform, ether, in volatile oils and most warm fixed oils; sol in benzene, carbon disulfide; slightly sol in dehydrated alc. Insol in water, alc.

Caution: Potential symptoms of overexposure to fumes are irritation of eyes, skin, respiratory system; discomfort and nausea. *See NIOSH Pocket Guide to Chemical Hazards* (DHHS/NIOSH 97-140, 1997) p 240.

USE: For raising mp of ointments. Manuf paraffin paper and candles (so-called wax paper or candles); for fixing drawings, etc., on muslin; water-proofing wood, cork, paper, leather; manuf varnishes; to render wooden vessels impermeable to water or alc; in lubricants;

to cover food products; in floor polishes, cosmetics, electrical insulators; for extracting perfumes from flowers. Pharmaceutic aid (stiffening agent).

7128. Paraffin Chlorinated. Chlorcosane; Cereclor. Prepared by chlorinating a liquid paraffin. Contains about 50% Cl.

Light yellow to amber, thick, oily liq; odorless and stable in air. d 1.00-1.07. Insol in water. Slightly sol in alc; miscible with benzene, chloroform, ether, carbon tetrachloride.

Caution: Chlorinated paraffins (C_{12}, 60% Cl) are reasonably anticipated to be human carcinogens: *Report on Carcinogens, Twelfth Edition* (PB2011-111646, 2011) p 95.

USE: As solvent for dichloramine-T, dissolving about 8%.

7129. Paraformaldehyde. [30525-89-4] Polyoxymethylene; Paraform; Formagene. Also erroneously referred to as *Triformol* or as *"trioxymethylene".* Polymerized formaldehyde, $(CH_2O)_n$. Obtained by concentrating formaldehyde soln. Use in mummifying dental pulp: I. Curson, *Br. Dent. J.* **121**, 519 (1966); P. Hobson, *ibid.* **128**, 275 (1970).

White, cryst powder, having an odor of formaldehyde. *Flammable. Keep tightly closed.* Slowly sol in cold, more readily in hot water, with evolution of formaldehyde; sol in fixed alkali hydroxide solns. Insol in alcohol, ether.

USE: For disinfecting sickrooms, clothing, linen, and sickroom utensils. Active ingredient of contraceptive creams. Also used as fumigant; in dentistry; in manuf synthetic resins and artificial horn or ivory.

7130. Paraherquamide. [77392-58-6] (1′*R*,5′a*S*,7′*R*,8′a*S*,-9′a*R*)-2′,3′,8′a,9′-Tetrahydro-1′-hydroxy-1′,4,4,8′,8′,11′-hexamethylspiro[4*H*,8*H*-[1,4]dioxepino[2,3-*g*]indole-8,7′(8′*H*)-[5*H*,6*H*-5a,9a](iminomethano)[1*H*]cyclopent[*f*]indolizine]-9,10′(10*H*)-dione; (−)-paraherquamide. $C_{28}H_{35}N_3O_5$; mol wt 493.60. C 68.13%, H 7.15%, N 8.51%, O 16.21%. Oxindole alkaloid fungal metabolite. Isolation from *Penicillium paraherquei* and structure: M. Yamazaki *et al.*, *Maikotokishin* **10**, 27 (1980), *C.A.* **95**, 19321p (1981); M. Yamazaki *et al.*, *Tetrahedron Lett.* **22**, 135 (1981); from *P. charlesii* and nematocidal activity: J. G. Ondeyka *et al.*, *J. Antibiot.* **43**, 1375 (1990). Absolute stereochemistry: T. A. Blizzard *et al.*, *J. Org. Chem.* **54**, 2657 (1989). Approach to synthesis: R. M. Williams, T. D. Cushing, *Tetrahedron Lett.* **31**, 6325 (1990). Anthelmintic activity in sheep: W. L. Shoop *et al.*, *J. Parasitol.* **76**, 349 (1990); in dogs: W. L. Shoop *et al.*, *Vet. Parasitol.* **40**, 339 (1991). Toxicity: *eidem*, *Am. J. Vet. Res.* **53**, 2032 (1992). Mode of action: J. M. Schaeffer *et al.*, *Biochem. Pharmacol.* **43**, 679 (1992).

Colorless prisms, mp 244-247° (dec). $[\alpha]_D^{22}$ −28° (c = 0.43 in methanol). uv max (ethanol): 226, 260, 290nm (ε 32400, 6100, 1600). Sol in methanol, ethyl acetate, acetone, DMSO. Practically insol in water.

THERAP CAT (VET): Anthelmintic.

7131. Paraldehyde. [123-63-7] 2,4,6-Trimethyl-1,3,5-trioxane; paracetaldehyde; Paral. $C_6H_{12}O_3$; mol wt 132.16. C 54.53%, H 9.15%, O 36.32%. A polymer of acetaldehyde. Prepd by the polymerization of acetaldehyde catalyzed by HCl and H_2SO_4 at medium to high temp: Kekulé, Zincke, *Ann.* **162**, 125 (1872); Baer, Mahan, *US 2864827* (1958 to Phillips). Toxicity data: Figot *et al.*, *Acta Pharmacol. Toxicol.* **8**, 290 (1952).

Liquid, characteristic aromatic odor and warm, but disagreeable taste. d_{25}^{25} ~0.994. bp ~124°. mp 12°. n_D^{20} 1.4049. *Flammable.* Sol in 8 parts water at 25°, in 17 parts boiling water; miscible with alc, chloroform, ether, oils. Gives acetaldehyde on heating with dil HCl or on warming with several drops concd H_2SO_4. LD_{50} orally in rats: 1.65 g/kg (Figot).

Note: This is a controlled substance (depressant): **21 CFR, 1308.14.**

USE: Manuf organic compounds.

THERAP CAT: Sedative; hypnotic.

7132. Paramethadione. [115-67-3] 5-Ethyl-3,5-dimethyl-2,4-oxazolidinedione; 3,5-dimethyl-5-ethyloxazolidine-2,4-dione; Paradione. $C_7H_{11}NO_3$; mol wt 157.17. C 53.49%, H 7.05%, N 8.91%, O 30.54%. Prepn: Spielman, *US 2575693* (1951 to Abbott).

Liquid. Fruity, esterlike odor. d_4^{25} 1.1180-1.1240. n_D^{25} 1.449. Slightly sol in water. Freely sol in alcohol, benzene, chloroform, ether.

THERAP CAT: Anticonvulsant.

7133. Paramethasone. [53-33-8] (6α,11β,16α)-6-Fluoro-11,17,21-trihydroxy-16-methylpregna-1,4-diene-3,20-dione; 6α-fluoro-16α-methylprednisolone; 16α-methyl-6α-fluoroprednisolone. $C_{22}H_{29}FO_5$; mol wt 392.47. C 67.33%, H 7.45%, F 4.84%, O 20.38%. Prepn: Edwards *et al.*, *J. Am. Chem. Soc.* **82**, 2318 (1960). Toxicity study: E. I. Goldenthal, *Toxicol. Appl. Pharmacol.* **18**, 185 (1971).

21-Acetate. [1597-82-6] Cortidene; Dilar; Dillar; Haldrate; Haldrone; Metilar; Monocortin; Paramezone; Syntecort; Stemex. $C_{24}H_{31}FO_6$; mol wt 434.50. mp 228-241° (dec). $[\alpha]_D$ +85°. uv max (ethanol): 243 nm (log ε 4.16). Sol in chloroform, ether, methanol, ethanol, acetone. Insol in water. LD_{50} i.p. in female rats: 392±23 mg/kg (Goldenthal).

Disodium phosphate. Soludillar.

Mixture of 21-acetate and disodium phosphate. Triniol.

THERAP CAT: Glucocorticoid.

7134. Paraoxon. [311-45-5] Phosphoric acid diethyl 4-nitrophenyl ester; diethyl *p*-nitrophenyl phosphate; phosphacol; E-600; Ester 25; Eticol; Mintacol. $C_{10}H_{14}NO_6P$; mol wt 275.20. C 43.64%, H 5.13%, N 5.09%, O 34.88%, P 11.26%. Organophosphate insecticide; cholinesterase inhibitor. Toxic oxidation product of parathion, *q.v.* Prepd by the action of diethyl chlorophosphate on sodium *p*-nitrophenolate or by nitration of diethyl phenyl phosphate: G. Schrader, *BIOS Final Report* **No. 714**, 52 (1947); *idem*, *Angew. Chem.* **62**, 471 (1950); *CH 257649* (1949 to Sandoz). Physical properties: E. F. Williams, *Ind. Eng. Chem.* **43**, 950 (1951). Acute toxicity: W. R. Pickering, J. C. Malone, *Biochem. Pharmacol.* **16**, 1183 (1967). Formation from parathion on citrus foliage and soil surface:

R. C. Spear *et al.*, *J. Agric. Food Chem.* **23**, 808 (1975); W. F. Spencer *et al.*, *Bull. Environ. Contam. Toxicol.* **14**, 265 (1975). GC determn in fish plasma and tissues: R. Abbas, W. L. Hayton, *J. Anal. Toxicol.* **20**, 151 (1996).

Oily liq. Slight odor. *Poisonous.* bp$_{1.0}$ 169-170°. d$_4^{25}$ 1.2683. n_D^{20} 1.50959. uv max: 274 nm (ε 8.9×10^3). Soly in water (25°): 2400 μg/mL. Freely sol in ether, other organic solvents. Aq solns are stable up to pH 7. LD$_{50}$ orally in rats: 1.8 mg/kg (Pickering, Malone).

USE: Insecticide.

7135. Paraquat. [4685-14-7] 1,1'-Dimethyl-4,4'-bipyridinium; *N,N'*-dimethyl-γ,γ'-dipyridylium; methyl viologen (2+). [C$_{12}$H$_{14}$N$_2$]$^{2+}$. Non-selective contact herbicide. Prepn of dichloride and bismethyl sulfate derivs: L. Michaelis, E. S. Hill, *J. Am. Chem. Soc.* **55**, 1481 (1933); R. C. Brian *et al.*, **GB 813531** (1959 to ICI). Activity: R. F. Homer *et al.*, *J. Sci. Food Agric.* **11**, 309 (1960); A. D. Dodge, *Endeavour* **30**, 130 (1971). Degradation: A. Calderbank, P. Slade, *Outlook Agric.* **5**, 55 (1966); A. Calderbank, T. E. Tomlinson, *ibid.* 252 (1968); A. Calderbank, *ibid.* **6**, 128 (1970). Toxicity studies: D. G. Clark *et al.*, *Br. J. Ind. Med.* **23**, 126 (1966); D. M. Conning *et al.*, *Br. Med. Bull.* **25**, 245 (1969); R. D. Kimbrough, T. B. Gaines, *Toxicol. Appl. Pharmacol.* **17**, 679 (1970); J. F. Dasta, *Am. J. Hosp. Pharm.* **35**, 1368 (1978). Controversial use on marijuana plants: R. J. Smith, *Science* **199**, 861 (1978). *Review:* A. A. Akhavein, D. L. Linscott, *Residue Rev.* **23**, 97-145 (1968); A. Calderbank, P. Slade in *Herbicides: Chemistry, Degradation and Mode of Action*, P. C. Kearney, D. Kaufman, Eds. (Dekker, New York, 2nd ed., 1976) pp 501-540.

$$H_3C-\overset{+}{N} \qquad \overset{+}{N}-CH_3$$

Dichloride. [1910-42-5] PP-148; Gramoxone. C$_{12}$H$_{14}$Cl$_2$N$_2$; mol wt 257.16. Colorless crystals, mp 300° (dec). Very sol in water, slightly sol in lower alcohols. Insol in hydrocarbons. Hydrolyzed by alkali. Inactivated by inert clays and anionic surfactants. Corrosive to metal. Non-volatile. Normal potential at 30°: −0.446 volts. LD$_{50}$ orally in rats: 125 mg/kg (Conning).
Bismethyl sulfate. [2074-50-2] Paraquat I; PP-910. C$_{14}$H$_{20}$N$_2$-O$_8$S$_2$; mol wt 408.44. Yellow solid. LD$_{50}$ orally in male rats: 100 mg/kg (Kimbrough, Gaines).
Caution: Potential symptoms of overexposure to paraquat dichloride are irritation of eyes, skin, nose, throat, respiratory system; epistaxis; dermatitis; fingernail damage; irritation of GI tract; heart, liver, kidney damage. *See NIOSH Pocket Guide to Chemical Hazards* (DHHS/NIOSH 97-140, 1997) p 240.

USE: Herbicide. Dichloride as biological oxidation-reduction indicator.

7136. Parasorbic Acid. [10048-32-5] (6*S*)-5,6-Dihydro-6-methyl-2*H*-pyran-2-one; 5-hydroxy-2-hexenoic acid lactone; δ-Δ^{α,β}-hexenolactone; 2-hexen-5,1-olide; sorbic oil. C$_6$H$_8$O$_2$; mol wt 112.13. C 64.27%, H 7.19%, O 28.54%. The sole constituent of "Vogelbeeröl", an oil obtained by steam distillation of the acidified juice of the ripe berries of the mountain ash, *Sorbus aucuparia* L., *Rosaceae:* Hofmann, *Ann.* **110**, 129 (1859); Doebner, *Ber.* **27**, 344 (1894); Kuhn, Jerchel, *Ber.* **76**, 413 (1943). Structure: *eidem, ibid.* Synthesis: Haynes, Jones, *J. Chem. Soc.* **1946**, 954; Lamberti *et al.*, *Recl. Trav. Chim. Pays-Bas* **86**, 504 (1967). Pharmacology and acute toxicity: H. J. Meyer, R. Kretzschmar, *Arzneim.-Forsch.* **19**, 617 (1969).

Oily liquid, sweet aromatic odor. bp$_{14}$ 104-105°; bp$_{22}$ 119-123°. n_D^{25} 1.4682. d$_4^{18}$ 1.079. [α]$_D^{18}$ +49.3°; [α]$_D^{19}$ +210° (c = 2 in alc). Soluble in water; freely sol in alcohol, ether. Aq solns are neutral and turn acid on storage. LD$_{50}$ in mice (mg/kg): 420 ±6.3 i.p.; 195 ±13.6 i.v. (Meyer, Kretzschmar).

7137. Parathion. [56-38-2] Phosphorothioic acid *O,O*-diethyl *O*-(4-nitrophenyl) ester; *O,O*-diethyl *O-p*-nitrophenyl phosphorothioate; diethyl-*p*-nitrophenyl monothiophosphate; DNTP; S.N.P.; E-605; AC-3422; ENT-15108; Alkron; Folidol; Fostox E; Rhodiatox; Thiophos. C$_{10}$H$_{14}$NO$_5$PS; mol wt 291.26. C 41.24%, H 4.85%, N 4.81%, O 27.47%, P 10.63%, S 11.01%. Non-systemic contact and stomach insecticide and acaricide; cholinesterase inhibitor. Prepn: Thurston, *FIAT Report* **949** (1946); Coates, Topley, *BIOS Final Report* **1808** (1947). See also: Fletcher *et al.*, *J. Am. Chem. Soc.* **70**, 3943 (1948). Conversion to toxic oxygen analogs: *See* Paraoxon. Toxicity study: T. B. Gaines, *Toxicol. Appl. Pharmacol.* **14**, 515 (1969). *Review:* Hall in *Adv. Chem. Ser.* **1**, entitled "Agricultural Control Chemicals" (ACS, Washington DC, 1950) p 150. Review of industrial syntheses: Chadwick, Watt, "Thiophosphates" in *Phosphorus and Its Compounds* vol. 2, J. R. Van Wazer, Ed. (Interscience, New York, 1961) pp 1257-1262. Review of distribution, transport and fate in the environment: M. S. Mulla *et al.*, *Residue Rev.* **81**, 1-159 (1981); of carcinogenic risk: *IARC Monographs* **30**, 153-181 (1983).

Pale yellow liquid. bp$_{760}$ 375°; bp$_{0.6}$ 157-162°. mp 6°. n_D^{25} 1.5370. d$_4^{25}$ 1.26. Vapor pressure at 20°: 3.78×10^{-5} mm Hg. Surface tension at 25°: 39.2 dynes/cm. Viscosity (25°): 15.30 cP. Absorption spectra: Williams, *Ind. Eng. Chem.* **43**, 950 (1951). Freely sol in alcohols, esters, ethers, ketones, aromatic hydrocarbons. Practically insol in water (20 ppm), petr ether, kerosene, and the usual spray oils. Incompatible with substances having a pH higher than 7.5. LD$_{50}$ in female, male rats (mg/kg): 3.6, 13 orally; 6.8, 21 dermally (Gaines).
Caution: Potential symptoms of overexposure are miosis; rhinorrhea; headache; tight chest, wheezing, laryngeal spasm, salivation and cyanosis; anorexia, nausea, vomiting, abdominal cramps and diarrhea; sweating; muscle fasciculation, weakness and paralysis; giddiness, confusion and ataxia; convulsions, coma; low blood pressure; cardiac irregularities; skin, eye and respiratory system irritation. *See NIOSH Pocket Guide to Chemical Hazards* (DHHS/NIOSH 97-140, 1997) p 240; *Clinical Toxicology of Commercial Products*, R. E. Gosselin *et al.*, Eds. (Williams & Wilkins, Baltimore, 5th ed., 1984) Section III, pp 336-343.

USE: Insecticide; acaricide.

7138. Parathyroid Hormone. [9002-64-6] Parathormone; PTH. Regulatory factor in the homeostatic control of calcium and phosphate metabolism, its principal sites of activity being the skeleton, kidneys, and gastrointestinal tract. Prime function is to raise plasma calcium concns. Acts synergistically with vitamin D$_3$, *q.v.* except in the kidneys where the latter causes phosphate retention. Secretion from the parathyroid gland varies inversely with serum Ca^{2+} concentrations, unlike calcitonin, *q.v.*, which is secreted in direct proportion to serum calcium levels. Structure consists of a single-chain polypeptide of 84 amino acid residues. Sequence varies slightly among mammalian species. Sequence of bovine PTH: Niall *et al.*, *Z. Physiol. Chem.* **351**, 1586 (1970); Brewer, Ronan, *Proc. Natl. Acad. Sci. USA* **67**, 1862 (1970). Sequence of porcine PTH: O'Riordan *et al.*, *Proc. R. Soc. Med.* **64**, 1263 (1971). Isoln of human PTH from parathyroid adenomas: O'Riordan *et al.*, *Endocrinology* **89**, 234 (1971). Fragment exhibiting full biological activity consists of about 35 amino acid residues from the *N*-terminal: Potts *et al.* in *Parathyroid Hormone and Thyrocalcitonin (Calcitonin)*, R. V. Talmage, L. F. Belanger, Eds. (Excerpta Medica, New York, 1968) p 44; *see* in entirety for review and special studies. Synthesis of active bovine fragment: Potts *et al.*, *Proc. Natl. Acad. Sci. USA*

68, 63 (1971); of human PTH (1-38): S. Funakoshi *et al.*, *Pept. Chem.* **18**, 223 (1980). Reviews of early literature: Potts *et al.*, *Recent Prog. Horm. Res.* **22**, 101 (1966); Arnaud *et al.*, *Annu. Rev. Physiol.* **29**, 349 (1967). *Reviews:* Behrens, Grinnan, *Annu. Rev. Biochem.* **38**, 83 (1969); Auerbach *et al.*, *Recent Prog. Horm. Res.* **28**, 353 (1972); Parsons, Potts, "Physiology and Chemistry of Parathyroid Hormone" in *Clinics in Endocrinology and Metabolism* I. MacIntyre, Ed. (Saunders, Philadelphia, 1972) pp 33-78. Biosynthetic review: J. F. Habener *et al.*, *Recent Prog. Horm. Res.* **33**, 249 (1977).

Note: Aqueous solns of the active principles of bovine parathyroid gland have been used under the names: *Parathorm, Para-thormone, Paroidin*.

THERAP CAT: Blood calcium regulator.

7139. Parbendazole. [14255-87-9] *N*-(6-Butyl-1*H*-benzimidazol-2-yl)carbamic acid methyl ester; methyl 5-butyl-2-benzimidazolecarbamate; 5-butyl-2-(carbomethoxyamino)benzimidazole; SKF-29044; Helmatac; Verminum; Worm Guard. $C_{13}H_{17}$-N_3O_2; mol wt 247.30. C 63.14%, H 6.93%, N 16.99%, O 12.94%. Prepd from 4-butyl-*o*-phenylenediamine and carbomethoxycyanamide: Actor *et al.*, *Nature* **215**, 321 (1967); **GB 1123317** and Stedman; Actor, Pagano, **US 3480642**; **US 3574845** (1968, 1969, 1971 all to SKF). Activity: Ostmann, Scheidy, *Prog. Antimicrob. Anticancer Chemother., Proc. 6th Int. Congr. Chemother.* **1**, 159 (1970). Identification of metabolites: Dunn *et al.*, *J. Med. Chem.* **16**, 996 (1973).

Crystals from aq ethanol, mp 225-227° (dec). uv max (95% ethanol/1*N* HCl): 282, 288 nm (ε 16200, 20000). Practically insol in water. LD_{50} in mice, rats (g/kg): >4 orally, both species (Actor).

THERAP CAT (VET): Anthelmintic.

7140. Parecoxib. [198470-84-7] *N*-[[4-(5-Methyl-3-phenyl-4-isoxazolyl)phenyl]sulfonyl]propanamide. $C_{19}H_{18}N_2O_4S$; mol wt 370.42. C 61.61%, H 4.90%, N 7.56%, O 17.28%, S 8.66%. Injectable prodrug of valdecoxib, *q.v.*, a selective cyclooxygenase-2 (COX-2) inhibitor. Prepn: J. J. Talley *et al.*, **WO 9738986**; *eidem*, **US 5932598** (1997, 1999 both to Searle); *eidem*, *J. Med. Chem.* **43**, 1661 (2000). Clinical pharmacokinetics: A. Karim *et al.*, *J. Clin. Pharmacol.* **41**, 1111 (2001). Pharmacological profile: S. S. V. Padi *et al.*, *Eur. J. Pharmacol.* **491**, 69 (2004). Clinical evaluation in post-surgical pain: N. A. Nussmeier *et al.*, *Anesthesiology* **104**, 518 (2006). Comparison of GI effects with ketorolac: R. R. Stoltz *et al.*, *Am. J. Gastroenterol.* **97**, 65 (2002). Review of pharmacology and clinical experience: S. M. Cheer, K. L. Goa, *Drugs* **61**, 1133-1141 (2001).

mp 148.9-151°.

Sodium salt. [198470-85-8] SC-69124A; Dynastat. $C_{19}H_{17}N_2$-NaO_4S; mol wt 392.40. mp 271.5-272.7°.

THERAP CAT: Anti-inflammatory; analgesic.

7141. Pareira. Pareira brava. Dried root of *Chondodendron platiphyllum* (A. St. Hil.) Miers, *C. microphyllum* (Eichl.) Mold., and *C. tomentosum* Ruiz et Pavon, *Menispermaceae*. *Habit.* Brazil. *Constit.* Bebeerine, chondrodine, fatty acids, tannin.

THERAP CAT: Diuretic, anti-infective (urinary).

7142. Pargyline. [555-57-7] *N*-Methyl-*N*-2-propyn-1-ylbenzenemethanamine; *N*-methyl-*N*-2-propynylbenzylamine; *N*-benzyl-*N*-methyl-2-propynylamine; *N*-methyl-*N*-propargylbenzylamine; MO-911; A-19120; Eudatin; Supirdyl. $C_{11}H_{13}N$; mol wt 159.23. C 82.97%, H 8.23%, N 8.80%. Monoamine oxidase inhibitor. Prepd from propargyl bromide and benzylmethylamine: **GB 906245**; Martin, **US 3155584** (1962 and 1964, both to Abbott). Activity as a glucuronyl transferase inducer: Yeh, Mitchell, *Experientia* **28**, 298 (1972).

Free base, bp_{11} 96-97°.

Hydrochloride. [306-07-0] Eutonyl. $C_{11}H_{13}N$.HCl; mol wt 195.69. Crystals from ethanol + ether, mp 154-155°. Readily sol in water. Aq solns are unstable.

THERAP CAT: Antihypertensive.

7143. Paricalcitol. [131918-61-1] (1*R*,3*R*,5*Z*)-5-[(2*E*)-2-[(1*R*,3a*S*,7a*R*)-Octahydro-1-[(1*R*,2*E*,4*S*)-5-hydroxy-1,4,5-trimethyl-2-hexen-1-yl]-7a-methyl-4*H*-inden-4-ylidene]ethylidene]-1,3-cyclohexanediol; (1α,3β,7*E*,22*E*)-19-nor-9,10-secoergosta-5,7,22-triene-1,3,25-triol; 19-nor-1α,25-dihydroxyvitamin D_2; Zemplar. C_{27}-$H_{44}O_3$; mol wt 416.65. C 77.83%, H 10.64%, O 11.52%. Synthetic analog of vitamin D. Prepn: H. F. DeLuca *et al.*, **EP 387077**; *eidem*, **US 5587497** (1990, 1996 both to Wisconsin Alum. Res.). Series of articles on pharmacology and clinical experience in secondary hyperparathyroidism: *Am. J. Kidney Dis.* **32**, Suppl. 2, S40-S66 (1998). *Review:* M. M. Goldenberg, *Clin. Ther.* **21**, 432-441 (1999).

White to almost white powder. Sol in alc. Insol in water.

THERAP CAT: Antihyperparathyroid.

7144. Parinaric Acid. [18427-44-6] 9,11,13,15-Octadecatetraenoic acid. $C_{18}H_{28}O_2$; mol wt 276.42. C 78.21%, H 10.21%, O 11.58%. Close structural analog of intrinsic membrane lipids. The naturally occuring *cis*-form is found in the seeds of the tropical rainforest Makita tree, indigenous to Fiji. Isoln from the kernel oil of *Parinarium laurinum* A. Gray (Rosaceae): M. Tsujimoto, H. Koyanagi, *Kogyo Kagaku Zasshi* **36**, 110 (1933). Stereochemical synthesis of isomers: H. P. Kaufmann, R. K. Sud, *Ber.* **92**, 2797 (1959); D. V. Kuklev, W. L. Smith, *Chem. Phys. Lipids* **131**, 215 (2004). Identification of α-parinaric acid as the 9*Z*,11*E*,13*E*,15*Z*-isomer: F. D. Gunstone, R. Subbarao, *ibid.* **1**, 349 (1967). Characterization as a fluorescent probe for phospholipid membrane studies: L. A. Sklar *et al.*, *Proc. Natl. Acad. Sci. USA* **72**, 1649 (1975); *idem et al.*, *Biochemistry* **16**, 813, 819, 829 (1977). Review of lipid partition behavior: *idem, Mol. Cell. Biochem.* **32**, 169-177 (1980). Use as a fluorescent probe for lipid peroxidation: F. A. Kuypers *et al.*, *Biochim. Biophys. Acta* **921**, 266 (1987); for lipase activity: A. M. Rogel *et al.*, *Lipids* **24**, 518 (1989).

cis-Parinaric Acid

***cis*-Form.** [593-38-4] (9*Z*,11*E*,13*E*,15*Z*)-9,11,13,15-Octadeca-tetraenoic acid; α-parinaric acid. Crystals from petroleum ether, mp 84-85°. uv max (methanol): 291, 304 (ε 70000), 319 nm. Fluorescence max: 432 nm (excitation λ=320 nm). Sol in benzene, chloroform, ether. Insol in water. *Protect from light and air. Store at ≤ −20°.*

***trans*-Form.** [18841-21-9] (9*E*,11*E*,13*E*,15*E*)-9,11,13,15-Octadecatetraenoic acid; β-parinaric acid. White needle-like crystals from hexane, mp 95-96°. uv max (methanol): 286, 299 (ε 73000), 313 nm.

USE: In measurement of phospholipase activity, lipase activity; indicator of lipid peroxidation.

7145. Park Nucleotide. [18836-50-5] (L-lysyl form); [120010-32-4] (pimelate form). *N*-(*N*-Acetyl-α-muramoyl)-L-alanyl-D-γ-glutamyl-L-lysyl-D-alanyl-D-alanine (1' → P')-ester with uridine 5'-(trihydrogendiphosphate); UDP-MurNAc-L-Ala-γ-D-Glu-X-D-Ala-D-Ala; stem peptide. Glyco-pentapeptide; key intermediate in bacterial cell-wall synthesis. Residue 3 varies from species to species; it is usually L-Lys in Gram positive bacteria and mesodiaminopimelate in Gram negative bacteria. Identification and isolation from *Staphylococcus aureus*: J. T. Park, *J. Biol. Chem.* **194**, 885 (1952). Total synthesis: S. A. Hitchcock *et al.*, *J. Am. Chem. Soc.* **120**, 1916 (1998); of radioiodinated analog: C. N. Eid *et al.*, *J. Labelled Compd. Radiopharm.* **41**, 705 (1998).

L-Lysyl Form

USE: Tool for antibacterial screen development.

7146. Paromomycin. [7542-37-2] *O*-2-Amino-2-deoxy-α-D-glucopyranosyl-(1 → 4)-*O*-[*O*-2,6-diamino-2,6-dideoxy-β-L-idopyranosyl-(1 → 3)-β-D-ribofuranosyl-(1 → 5)]-2-deoxy-D-streptamine; *O*-2,6-diamino-2,6-dideoxy-β-L-idopyranosyl-(1 → 3)-*O*-β-D-ribofuranosyl-(1 → 5)-*O*-[2-amino-2-deoxy-α-D-glucopyranosyl-(1 → 4)]-2-deoxystreptamine; paromomycin I; amminosidin; catenulin; crestomycin; estomycin; hydroxymycin; monomycin A; neomycin E; paucimycin; R-400. $C_{23}H_{45}N_5O_{14}$; mol wt 615.63. C 44.87%, H 7.37%, N 11.38%, O 36.38%. Oligosaccharide-type antibiotic isolated from various *Streptomyces*. From *S. rimosus* forma *paromomycinus*: Frohardt *et al.*, US 2916485 (1959 to Parke, Davis); from *S. catenulae*: Davisson, Finlay, US 2895876 (1959 to Pfizer); from *S. chrestomyceticus*: Canevazzi, Scotti, *G. Microbiol.* **7**, 242 (1959); Arcamone *et al.*, *ibid.* 251. Identity of paromomycin, catenulin, hydroxymycin and aminosidine: Schillings, Schaffner, *Antimicrob. Agents Chemother.* **1961**, 274. Structure: Haskell *et al.*, *J. Am. Chem. Soc.* **81**, 3480, 3482 (1959); Rinehart *et al.*, *ibid.* **84**, 3218 (1962); Hichens, Rinehart, *ibid.* **85**, 1547 (1963). Probable identity with **zygomycin A**: Horii, *J. Antibiot.* **15A**, 187 (1962). Identity with monomycin A: Konstantinova, Brazhnikova, *Antibiotiki* **10**(1), 34 (1965); with neomycin E: Hessler *et al.*, *J. Antibiot.* **23**, 464 (1970). Toxicity data: A. DiMarco, C. Bertazzoli, *Antibiot. Chemother.* **11**, 2 (1963). Review of antimicrobial activity: G. L. Coffey *et al.*, *ibid.* **9**, 730 (1959). Review of pharmacology: Gasparini, Pignatelli, *Veterinaria (Milan)* **21**, 7 (1972).

Amorphous white powder. $[α]_D^{25}$ +65 ±3°. Soluble in water; moderately sol in methanol; sparingly sol in abs ethanol. LD_{50} in rats, mice (mg/kg): >1625, >2275 orally; >650, 423 s.c.; 156, 90 i.v. (Coffey).

Sulfate. [1263-89-4] 1600 Antibiotic; FI-5853; Aminoxidin; Aminosidine; Farmiglucin; Farminosidin; Gabbromicina; Gabbromycin; Gabbroral; Humagel; Humatin; Pargonyl; Paramicina; Paricina; Sinosid. $C_{23}H_{45}N_5O_{14}·H_2SO_4$; mol wt 713.71. Creamy white to light yellow powder. Very hygroscopic. $[α]_D^{25}$ +50.5° (c = 1.5 in water pH 6). Very sol in water. Insol in alc, chloroform, ether. LD_{50} in mice (mg/kg): ~15,000 orally; 700 s.c.; 110 i.v. (Di Marco, Bertazzoli).

THERAP CAT: Antibacterial; antiamebic.

THERAP CAT (VET): Antiamebic.

7147. Parotin. [1392-81-0] Salivary gland hormone; protein of globulin nature having an isoelectric point of pH 5.7. Produced by the parotid gland. Structure studies: Y. Ito *et al.*, *Endocrinol. Jpn.* **12**, 249 (1966); H. Shimasaki *et al.*, *ibid.* **14**, 11 (1967). Generally acts on the mesenchymal tissues, esp the hard and connective tissues, to promote their development and growth. Also has a protein-anabolic function. Hypocalcemic and leukocytosis-promoting activities: Y. Ito *et al.*, *ibid.* **12**, 298 (1966). *Reviews:* Y. Ito, *J. Jpn. Biochem. Soc.* **25**, 143-164 (1953); *idem, Ann. N.Y. Acad. Sci.* **85**, 228-310 (1960).

7148. Paroxetine. [61869-08-7] (3*S*,4*R*)-3-[(1,3-Benzodioxol-5-yloxy)methyl]-4-(4-fluorophenyl)piperidine; (−)-*trans*-4-(*p*-fluorophenyl)-3-[[3,4-(methylenedioxy)phenoxy]methyl]piperidine; FG-7051; BRL-29060. $C_{19}H_{20}FNO_3$; mol wt 329.37. C 69.29%, H 6.12%, F 5.77%, N 4.25%, O 14.57%. Selective serotonin (5-HT) reuptake inhibitor (SSRI). Prepn: J. A. Christensen, R. F. Squires, DE 2404113; *eidem*, US 3912743; US 4007196 (1974, 1975, 1977 all to Ferrosan); of crystalline hydrochloride hemihydrate: R. D. B. Barnes *et al.*, EP 223403; US 4721723 (1987, 1988 both to Beecham). Characterization of serotonin inhibition: J. Buus Lassen, *Eur. J. Pharmacol.* **47**, 351 (1978). Binding to serotonin transporter complex: E. Habert *et al.*, *ibid.* **118**, 107 (1985). Clinical pharmacokinetics: J. Lund *et al.*, *Acta Pharmacol. Toxicol.* **51**, 351 (1982). HPLC determn in plasma: M. A. Brett *et al.*, *J. Chromatogr.* **419**, 438 (1987). Electroanalytical determn in pharmaceuticals: H. P. A. Nouws *et al.*, *J. Pharm. Biomed. Anal.* **42**, 341 (2006). Clinical trial in obsessive-compulsive disorder: J. Zohar *et al.*, *Br. J. Psychiatry* **169**, 468 (1996); in social phobia: M. B. Stein *et al.*, *J. Am. Med. Assoc.* **280**, 708 (1998). Review of pharmacology and clinical use in depression: K. L. Dechant, S. P. Clissold, *Drugs* **41**, 225-253 (1991); of clinical experience: D. Dunner, R. Kumar, *Pharmacopsychiatry* **31**, 89-101 (1998); of use in generalized social anxiety disorder: M. Van Ameringen *et al.*, *Expert Opin. Pharmacother.* **6**, 819-830 (2005); of clinical efficacy of controlled-release formulation: C.-U. Pae, A. A. Patkar, *Expert Rev. Neurother.* **7**, 107-120 (2007).

Hydrochloride hemihydrate. [110429-35-1]; [78246-49-8] (hydrochloride). Aropax; Deroxat; ParoLich; Paroxat; Paxil; Seroxat; Tagonis. $C_{19}H_{20}FNO_3.HCl.\frac{1}{2}H_2O$; mol wt 374.84. Crystals, mp 129-131°. Soly in water: 5.4 mg/ml.

Maleate. [64006-44-6] Crystals from ethanol-ether, mp 136-138°. $[\alpha]_D$ −87° (c = 5 in ethanol). LD_{50} in mice (mg/kg): 845 s.c.; 500 orally (Christensen, Squires, 1977).

Methanesulfonate. [217797-14-3] Paroxetine mesylate; Divarius; Euplix; Pexeva. $C_{19}H_{20}FNO_3.CH_4O_3S$; mol wt 425.47. Odorless, off-white powder, mp 147-150°. Soly in water: >1 g/ml.

THERAP CAT: Antidepressant; antiobsessional.

THERAP CAT (VET): In treatment of canine and feline behavioral disorders.

7149. Paroxypropione. [70-70-2] 1-(4-Hydroxyphenyl)-1-propanone; 4'-hydroxypropiophenone; p-hydroxypropiophenone; paraoxypropiophenone; p-propionylphenol; ethyl p-hydroxyphenyl ketone; P.O.P.; B-360; H-365; Profenone; Frenantol; Frenohypon; Paroxon; Possipione; Hypostat. $C_9H_{10}O_2$; mol wt 150.18. C 71.98%, H 6.71%, O 21.31%. Prepn: Perkin, *J. Chem. Soc.* **55**, 546 (1889); Goldzweig, Kaiser, *J. Prakt. Chem.* [2] **43**, 86 (1891); Cox, *J. Am. Chem. Soc.* **49**, 1028 (1927); Hartung *et al.*, *ibid.* **53**, 4153 (1931); Farinholt, *ibid.* **55**, 3386 (1933); Miller, Hartung, *Org. Synth.* **coll. vol. II** (1943) p 543. Derivatives: Buu-Hoi, *Rec. Trav. Chim.* **68**, 759 (1949).

Needles or prisms from water. mp 149°. One part dissolves in 2896 parts of water at 15°, in 30 parts at 100°. Freely sol in alcohol or ether.

THERAP CAT: Pituitary gonadotropic hormone inhibitor.

7150. Parsley. Biennial herb, *Petroselinum crispum* (Mill.) Nyman, *Umbelliferae*; also known as *P. sativum* Hoffm. *Habit.* Mediterranean region; cultivated worldwide. *Constit.* Flavonoids, esp. apiin, luteolin; furocoumarins, incl. bergapten (0.02%), oxypeucedanin (0.01%); volatile oils (0.05% in leaf, 2-7% in seed); fixed oil, protein, carbohydrate, vitamins, esp. A and C. GC-MS determn of volatile components: M. G. López *et al.*, *J. Agric. Food Chem.* **47**, 3292 (1999). Comprehensive description and medicinal uses: J. Barnes *et al.*, *Herbal Medicines* (Pharmaceutical Press, London, 2nd Ed., 2002) pp 365-367.

Parsley seed oil. [8000-68-8] Obtained by steam distillation of the ripe seed. *Constit.* Apiole, myristicin, tetramethoxyallylbenzene. Review of toxicology: *Food Cosmet. Toxicol.* **15**, 897-898 (1975). Yellow to light brown liquid with herbal-like odor. d_{25}^{25} 1.040-1.080. n_D^{20} 1.513-1.522. Saponification value between 2 and 10. Acid value not more than 4.0. Very slightly sol in water; slightly sol in propylene glycol; sol in 6 vols 80% alcohol. Insol in glycerin.

USE: Flavoring in foods. Pharmaceutic aid (flavor).

THERAP CAT: In treatment of urinary tract inflammation and kidney stones; carminative.

7151. Parthenin. [508-59-8] (3aS,6S,6aS,9aS,9bR)-3,3a,4,5,-6,6a,9a,9b-Octahydro-6a-hydroxy-6,9a-dimethyl-3-methyleneazuleno[4,5-b]furan-2,9-dione; 1,6β-dihydroxy-4-oxo-10αH-ambrosa-2,11(13)-dien-12-oic acid γ-lactone; parthenicin. $C_{15}H_{18}O_4$; mol wt 262.31. C 68.68%, H 6.92%, O 24.40%. From herb of *Parthenium hysterophorus* L., *Compositae*. Parthenin is the substance largely responsible for the allergic contact dermatitis caused by *P. hysterophorus*. Isoln: Arny, *Am. J. Pharm.* **69**, 169 (1897). Isoln and structure: Herz *et al.*, *Tetrahedron Lett.* **1961**, 82; Herz *et al.*, *J. Am. Chem. Soc.* **84**, 2601 (1962). Abs config: Emerson *et al.*, *Tetrahedron Lett.* **1966**, 6151. Total synthesis of (±)-form: P. Kok *et al.*, *Bull. Soc. Chim. Belg.* **87**, 615 (1978); C. H. Heathcock *et al.*, *J. Am. Chem. Soc.* **104**, 6081 (1982).

Crystals from water, mp 163-166°. $[\alpha]_D^{25}$ +7.02° (c = 2.71 in chloroform). uv max: 215, 340 nm (ε 15,100; 22). Practically insol in water. Sol in alcohol, chloroform, ether, ethyl acetate.

7152. Parthenolide. [20554-84-1] (1aR,4E,7aS,10aS,10bR)-2,3,6,7,7a,8,10a,10b-Octahydro-1a,5-dimethyl-8-methyleneoxireno[9,10]cyclodeca[1,2-b]furan-9(1aH)-one; 4,5α-epoxy-6β-hydroxygermacra-1(10),11(13)-dien-12-oic acid γ-lactone. $C_{15}H_{20}O_3$; mol wt 248.32. C 72.55%, H 8.12%, O 19.33%. Sesquiterpene lactone found in feverfew, q.v., and in other plants. Isolation from *Chrysanthemum parthenium* (L.) Bernh. *Compositae* and characterization: V. Herout *et al.*, *Chem. Ind. (London)* **1959**, 1069; M. Soucek *et al.*, *Collect. Czech. Chem. Commun.* **26**, 803 (1961); from *Magnolia grandiflora* L., *Magnoliaceae:* F. S. El-Feraly, Y.-M. Chan, *J. Pharm. Sci.* **67**, 347 (1978). Revised structure and spectral analysis: T. R. Govindachari *et al.*, *Tetrahedron* **21**, 1509 (1965). Absolute configuration: A. S. Bawdekar *et al.*, *Tetrahedron Lett.* **1966**, 1225. Crystal structure: A. Quick, D. Rogers, *J. Chem. Soc. Perkin Trans.* 2 **4**, 465 (1976). HPLC determn: D. Strack *et al.*, *Z. Naturforsch.* **35**, 915 (1980). Cytotoxicity: K.-H. Lee *et al.*, *Cancer Res.* **31**, 1649 (1971); L. A. J. O'Neill *et al.*, *Br. J. Clin. Pharmacol.* **23**, 81 (1987).

Colorless plates, mp 115-116°. $[\alpha]_D^{20}$ −81.4° (c = 1.04 in chloroform); $[\alpha]_D^{22}$ −71.4° (c = 0.220 in CH_2Cl_2). uv max: 214 nm (log ε 4.22).

7153. Partricin. [11096-49-4] Ayfactin; SPA-S-132. Heptaene macrolide antibiotic complex produced by *Streptomyces aureofaciens* NRRL 3878. Isoln: T. Bruzzese, R. Ferrari, US 3773925 (1973 to SPA). Recovery and purification process: S. Magnaghi *et al.*, GB 1462442 (1977 to SPA), *C.A.* **87**, 66588a (1977). In vitro activity: W. Ritzerfeld, *Farmaco Ed. Sci.* **27**, 235 (1972); G. A. Meloni *et al.*, *ibid.* **34**, 183 (1979). Use in treatment of benign prostatic hypertrophy: T. Bruzzese, L. Ferrari, US 4237117 (1980 to SPA). Separation, characterization and structure of components A and B: R. C. Tweit *et al.*, *J. Antibiot.* **35**, 997 (1982).

Partricin A R' = CH_3
Partricin B R' = H

Partricin A. $C_{59}H_{86}N_2O_{19}.4H_2O$; mol wt 1199.39. Greenish-yellow powder, mp >300° (dec). uv max (75% methanol in DMF): 232, 240, 247, 288, 342, 358, 378, 400 nm (ε 33476, 32237, 22936, 15493, 58282, 76883, 102308, 89280). pKa's (70% aq DMF): 6.07, 8.91.

Partricin B. $C_{58}H_{84}N_2O_{19}.2\frac{1}{2}H_2O$; mol wt 1158.34. Brownish-yellow powder, mp >300° (dec). uv max (75% methanol in DMF): 232, 240, 247, 288, 342, 358, 378, 400 nm (ε 34761, 32826, 23174, 20594, 50207, 73392, 100425, 87558). pKa's (70% aq DMF): 6.31, 8.95. $[\alpha]_D^{26}$ +87.2° (c = 0.06 in DMF).

Complex. Amphoteric yellow crystals. uv max (ethanol): 401, 379, 359, 341 nm. Sol in DMF, DMSO, dimethyl acetamide, pyridine. Practically insol in water, common organic solvents. LD_{50} in mice (mg/kg): 300 orally; 0.5 i.p. (Bruzzese).

Methyl ester see Mepartricin.

7154. Parvaquone. [4042-30-2] 2-Cyclohexyl-3-hydroxy-1,4-naphthalenedione; 2-cyclohexyl-3-hydroxy-1,4-naphthoquinone; 2-hydroxy-3-cyclohexyl-1,4-naphthoquinone; BW-993C; Clexon. $C_{16}H_{16}O_3$; mol wt 256.30. C 74.98%, H 6.29%, O 18.73%. Prepn: L. F. Fieser, *J. Am. Chem. Soc.* **70**, 3165 (1948); L. F. Fieser, M. T. Leffler, US 2553648 (1951 to Research Corp.). Antimalarial activity: L. F. Fieser, A. P. Richardson, *J. Am. Chem. Soc.* **70**, 3156 (1948). Electron transport inhibition: A. L. Tappel, *Biochem. Pharmacol.* **3**, 289 (1960); by uncoupling: J. L. Howland, *Biochim. Biophys. Acta* **131**, 247 (1967). *In vitro* and *in vivo* antiprotozoal activity: A. T. Hudson *et al.*, *Parasitology* **90**, 45 (1985). Bioassay in serum: N. McHardy, J. Mercer, *Kenya Vet.* **8**(2), 9 (1984). Treatment of East Coast fever in cattle: T. T. Dolan *et al.*, *Vet. Parasitol.* **15**, 103 (1984); of *Theileria annulata* in calves: N. McHardy, D. W. Morgan, *Res. Vet. Sci.* **39**, 1 (1985). Field comparison of antitheilarial activity with buparvaquone, *q.v.:* N. McHardy *et al.*, *ibid.* 39.

Bright yellow needles, mp 135-136°.

THERAP CAT (VET): Antiprotozoal (Theileria).

7155. Pasireotide. [396091-73-9] Cyclo[(2*S*)-2-phenylglycyl-D-tryptophyl-L-lysyl-*O*-(phenylmethyl)-L-tyrosyl-L-phenylalanyl-(4*R*)-4-[[[(2-aminoethyl)amino]carbonyl]oxy]-L-prolyl]; cyclo-[(4*R*)-4-(2-aminoethylcarbamoyloxy)-L-prolyl-L-phenylglycyl-D-tryptophyl-L-lysyl-4-*O*-benzyl-L-tyrosyl-L-phenylalanyl-]; SOM-230. $C_{58}H_{66}N_{10}O_9$; mol wt 1047.23. C 66.52%, H 6.35%, N 13.38%, O 13.75%. Cyclic hexapeptide somatostatin analog that binds to multiple somatostatin receptor subtypes. Prepn: R. Albert *et al.*, WO 0210192; *eidem*, US 7473761 (2002, 2009 both to Novartis); I. Lewis *et al.*, *J. Med. Chem.* **46**, 2334 (2003). Receptor binding, pharmacokinetics, and hormone inhibitory profile: C. Bruns *et al.*, *Eur. J. Endocrinol.* **146**, 707 (2002). Pharmacology: G. Weckbecker *et al.*, *Endocrinology* **143**, 4123 (2002). Clinical pharmacokinetics: P. Ma *et al.*, *Clin. Pharmacol. Ther.* **78**, 69 (2005). Clinical evaluation in acromegaly: J. van der Hoek *et al.*, *Clin. Endocrinol.* **63**, 176 (2005); in Cushing's disease: M. Boscaro *et al.*, *J. Clin. Endocrinol. Metab.* **94**, 115 (2009). Review of pharmacology and clinical experience: A. Ben-Shlomo, S. Melmed, *IDrugs* **10**, 885-895 (2007); H. A. Schmid, *Mol. Cell. Endocrinol.* **286**, 69-74 (2008).

White powder. Sol in water.

Acetate salt. [396091-76-2] $C_{58}H_{66}N_{10}O_9 \cdot C_2H_4O_2$; mol wt 1107.28. $[\alpha]_D^{20}$ −42° (c = 0.26 in 95% acetic acid).

THERAP CAT: In treatment of acromegaly and Cushing's disease.

7156. Passiflora. Passion flower; passion vine; May pops. Dried flowering and fruiting tops of *Passiflora incarnata* L., *Passifloraceae*. *Habit.* Southeastern U.S. *Constit.* Harman.

THERAP CAT: Sedative, analgesic.

7157. Patchouli Alcohol. [5986-55-0] (1*R*,4*S*,4a*S*,6*R*,8a*S*)-Octahydro-4,8a,9,9-tetramethyl-1,6-methanonaphthalen-1(2*H*)-ol; patchouli camphor. $C_{15}H_{26}O$; mol wt 222.37. C 81.02%, H 11.79%, O 7.19%. A tricyclic sesquiterpene alcohol isolated from oil of patchouli: Gadamer, Amenomiya, *Arch. Pharm.* **241**, 39 (1903). Proposed structure: Treibs, *Ann.* **564**, 141 (1949). Revised structure: Dobler *et al.*, *Proc. Chem. Soc. London* **1963**, 383. Structural studies: Büchi *et al.*, *J. Am. Chem. Soc.* **78**, 1262 (1956); **83**, 927 (1961); **84**, 3205 (1962); **86**, 4438 (1964). Synthesis of *dl*-form: Danishevsky, Dumas, *Chem. Commun.* **1968**, 1287; Mirrington, Schmalzl, *J. Org. Chem.* **37**, 2871 (1972); K. Yamada *et al.*, *Tetrahedron* **35**, 293 (1979). Stereoselective total synthesis of racemic form: F. Näf, G. Ohloff, *Helv. Chim. Acta* **57**, 1868 (1974); of natural, racemic and (+)-forms: F. Näf *et al.*, *ibid.* **64**, 1387 (1981). *Review:* Walker, *Manuf. Chem. Aerosol News* **39**, no. 7, 27 (1968).

Large crystals (hexagonal-trapezohedral) from the higher boiling fractions of oil of patchouli or from petr ether, mp 56°. mp (racemate) 39-40° (Danishevsky, Dumas); also reported as mp 46-47° (Mirrington, Schmalzl). bp$_8$ 140°. d_4^{20} 1.0284. $[\alpha]_D^{20}$ −97.4° (c = 24 in chloroform). n_D^{65} 1.5029. Practically insol in water. Sol in alcohol, ether, common organic solvents.

7158. Patulin. [149-29-1] 4-Hydroxy-4*H*-furo[3,2-*c*]pyran-2(6*H*)-one; clavacin; clavatin; claviformin; expansine; mycoin C$_3$; penicidin. $C_7H_6O_4$; mol wt 154.12. C 54.55%, H 3.92%, O 41.52%. An antibiotic derived from the metabolites of a number of fungi, e.g., *Aspergillus clavatus*, *A. claviforme*, *A. giganteus*, *A. terreus*, *Penicillium patulum*, *P. expansum*, *P. melinii*, *P. leucopus*, *P. urticae* and *Gymnoascus* spp. Isoln and antibacterial activity: Birkinshaw *et al.*, *Lancet* **245**, 625 (1943); Birkinshaw, Michael, US 2417584 (1947 to Therap. Res. Corp. of Great Britain); *cf* Waksman *et al.*, *Science* **96**, 202 (1942); Brack, *Helv. Chim. Acta* **30**, 1 (1947); Norstadt, McCalla, *Appl. Microbiol.* **17**, 193 (1969). Structure: Birkinshaw, *loc. cit.*; Bergel *et al.*, *J. Chem. Soc.* **1944**, 415; Woodward, Singh, *J. Am. Chem. Soc.* **71**, 758 (1949); Shemyakin, Khokhlov, *Dokl. Akad. Nauk SSSR* **75**, 47 (1950). Synthesis: Woodward, Singh, *J. Am. Chem. Soc.* **72**, 1428 (1950). Biosynthesis: Bu'Lock, Ryan, *Proc. Chem. Soc. London* **1958**, 222; Tanenbaum, Bassett, *J. Biol. Chem.* **234**, 1861 (1959). Inhibition of K$^+$ ion uptake in erythrocytes: Kahn, *J. Pharmacol. Exp. Ther.* **121**, 234 (1957). Physicochemical data: A. E. Pohland *et al.*, *Pure Appl. Chem.* **54**, 2219 (1982). Has carcinogenic activity attributable to α,β-unsaturation together with an external conjugated double bond attached to 4 position of the γ-lactone ring: Dickens, *Br. Med. Bull.* **20**, 96 (1964). Toxicity: R. Kinosita, T. Shikata, "On Toxic Moldy Rice" in *Mycotoxins in Foodstuffs*, G. N. Wogan, Ed. (The M.I.T. Press, Cambridge, 1965) pp 117-119. Review and evaluation of studies of carcinogenic action in laboratory animals: *IARC Monographs* **10**, 205-210 (1976). *Review:* Ciegler *et al.*, "Patulin, Penicillic Acid and Other Carcinogenic Lactones" in *Microbial Toxins* **vol. VI**, A. Ciegler *et al.*, Eds. (Academic Press, New York, 1971) p 409-414.

Compact prisms or thick plates from ether or chloroform, mp 111.0°; also reported as 105-108° (Pohland). $[\alpha]_D^{21}$ −6.2° (chloroform). uv max: 276.5 nm (Bergel). Sol in water and the common organic solvents except petr ether; very sol in ethyl or amyl acetate.

Unstable in alkali with loss of biological activity. LD_{50} s.c. in mice: 10-15 mg/kg (Kinosita, Shikata).

Acetylpatulin. $C_9H_8O_5$. Prisms from 50% alcohol, mp 118-120°.

7159. Pavoninins. Shark repellent compounds isolated from the defense secretion of the Pacific sole, *Pardachirus pavoninus*. Series of ichthyotoxic and hemolytic steroid amine glycosides; pavoninin-5 is the most abundant. Isoln, structure determn and evaluation of pavoninins 1-6 as shark repellents: K. Tachibana *et al.*, *Science* **226**, 703 (1984); and ^{13}C- and ^{1}H-NMR studies: *eidem*, *Tetrahedron* **41**, 1027 (1985). Synthesis of pavoninin-1: Y. Ohnishi, K. Tachibana, *Bioorg. Med. Chem.* **5**, 2251 (1997); of pavoninin-4: J. R. Williams *et al.*, *J. Org. Chem.* **70**, 10732 (2005).

Pavoninin-5

Pavoninin-1. [94426-01-4] $(7\alpha,25R)$-7-[[2-(Acetylamino)-2-deoxy-β-D-glucopyranosyl]oxy]-26-(acetyloxy)cholest-4-en-3-one. $C_{37}H_{59}NO_9$; mol wt 661.88. $[\alpha]_D^{20}$ +19° (c = 1.1 in chloroform). uv max (methanol): 244 nm (Log ε 4.1)

Pavoninin-4. [94359-66-7] $(3\alpha,5\alpha,15\alpha,25R)$-26-(Acetyloxy)-3-hydroxycholestan-15-yl 2-(acetylamino)-2-deoxy-β-D-glucopyranoside. $C_{37}H_{63}NO_9$; mol wt 665.91. White solid, mp 134-136°. $[\alpha]_D^{20}$ +28° (c = 0.4 in ethanol).

Pavoninin-5. [94480-49-6] $(3\beta,15\alpha,25R)$-26-(Acetyloxy)-3-hydroxycholest-5-en-15-yl 2-(acetylamino)-2-deoxy-β-D-glucopyranoside. $C_{37}H_{61}NO_9$; mol wt 663.89. Sol in ethyl acetate. $[\alpha]_D^{29}$ +21° (c = 0.7 in ethanol).

7160. Pazopanib. [444731-52-6] 5-[[4-[(2,3-Dimethyl-2*H*-indazol-6-yl)methylamino]-2-pyrimidinyl]amino]-2-methylbenzenesulfonamide; N^4(2,3-dimethyl-2*H*-indazol-6-yl)-N^4-methyl-N^2-(4-methyl-3-sulfonamidophenyl)-2,4-pyrimidinediamine; GW-786034. $C_{21}H_{23}N_7O_2S$; mol wt 437.52. C 57.65%, H 5.30%, N 22.41%, O 7.31%, S 7.33%. Multitargeted tyrosine kinase inhibitor with antiangiogenic activity. Prepn: A. Boloor *et al.*, **WO 02059110** (2002 to Glaxo); *eidem*, **US 7105530** (2006 to SKB); P. A. Harris *et al.*, *J. Med. Chem.* **51**, 4632 (2008). Pharmacology: R. Kumar *et al.*, *Mol. Cancer Ther.* **6**, 2012 (2007). Clinical trial in renal cell carcinoma: C. N. Sternberg *et al.*, *J. Clin. Oncol.* **28**, 1061 (2010). Review of pharmacology and clinical experience: B. Rini, M. Y. Al-Marrawi, *Expert Opin. Pharmacother.* **12**, 1171-1189 (2011).

Hydrochloride. [635702-64-6] GW-786034B; Votrient. $C_{21}H_{23}N_7O_2S \cdot HCl$; mol wt 473.98. White to slightly yellow solid. Practically insol above pH 4 in aqueous media.

THERAP CAT: Antineoplastic.

7161. Pazufloxacin. [127045-41-4] (3*S*)-10-(1-Aminocyclopropyl)-9-fluoro-2,3-dihydro-3-methyl-7-oxo-7*H*-pyrido[1,2,3-*de*]-1,4-benzoxazine-6-carboxylic acid; T-3761. $C_{16}H_{15}FN_2O_4$; mol wt 318.30. C 60.38%, H 4.75%, F 5.97%, N 8.80%, O 20.11%. Fluorinated quinolone. Prepn: H. Narita *et al.*, **DE 3913245**; *eidem*, **US**

4990508 (1989, 1991 both to Toyama). Synthesis: Y. Todo *et al.*, *Chem. Pharm. Bull.* **42**, 2569 (1994). Antimicrobial activity: Y. Fukuoka *et al.*, *Antimicrob. Agents Chemother.* **37**, 384 (1993).

Crystals, mp 269-271.5°. $[\alpha]_D^{25}$ -88.0° (c = 0.5 in 0.05*N* aq NaOH). LD_{50} i.v. in male mice: >500 mg/kg (Todo).

Methanesulfonate. [163680-77-1] Pazufloxacin mesylate; T-3762. $C_{16}H_{15}FN_2O_4 \cdot CH_3SO_3H$; mol wt 414.40. Colorless prisms, mp 258-259°(dec). $[\alpha]_D^{20}$ -64.2° (c = 1 in 1.0*N* NaOH). Soly (25°) in water >200 mg/ml.

THERAP CAT: Antibacterial.

7162. PBN. [3376-24-7] 2-Methyl-*N*-(phenylmethylene)-2-propanamine *N*-oxide; benzylidene-*tert*-butylamine oxide; *N*-*tert*-butyl-α-phenylnitrone. $C_{11}H_{15}NO$; mol wt 177.25. C 74.54%, H 8.53%, N 7.90%, O 9.03%. Spin trapping agent for short-lived radicals in biological systems. Prepn: W. D. Emmons, *J. Am. Chem. Soc.* **79**, 5739 (1957); R. W. Murray *et al.*, *J. Org. Chem.* **61**, 8099 (1996). Effects on ischemia in brain: J. W. Phillis, C. Clough-Helfman, *Med. Sci. Res.* **18**, 403 (1990); K. Pahlmark, B. K. Siesjö, *Acta Physiol. Scand.* **157**, 41 (1996). Review of chemical properties and biological applications: G. Chen *et al.* in *The Oxygen Paradox, PBN and Its Applications in Biology*, K. J. A. Davies, F. Ursini, Eds. (Cleup Press, Italy, 1995) pp. 789-800.

Crystals, mp 72-74°. Stored at ambient temperature, it is stable for one year unopened. Sensitive to light, especially in solution. Easily sol in medium polar organic solvents such as methanol, ethanol, acetonitrile, toluene, benzene and chloroform. Soly: 2.9% (water); 2.3% (saline); >5% (chloroform). uv max (ethanol): 293.5, 224 nm (ε 17700, 7240); (CH_3CN): 298 nm (ε 17716).

USE: Spin trap reagent for carbon- or oxygen- or nitrogen-centered free radicals.

7163. Peach Oil, Expressed. Persic oil. Fixed oil from seed of *Prunus persica (L.)* Stokes and allied spp., *Rosaceae*.

Light yellow liquid; closely resembles expressed oil of almond. d_{15}^{15} 0.917-0.921. Solidif not above -15°. n_D^{40} 1.464-1.465. Sapon no. 189-193. Iodine no. 100-110. Acid no. about 8. Sol in chloroform, ether, petr ether, slightly sol in alcohol.

USE: Emollient.

7164. Peanut. Groundnut; earthnut. Ripe, underground pods with seeds of *Arachis hypogaea* L., *Leguminosae*. *Habit.* Parana River Valley in Paraguay, Brazil and Argentina. Cultivated in the reasonably warm regions of all continents, *e.g.*, through the southern U.S.A. *Composition:* pericarp (shell) 21-29%, episperm (skin) 1.95-3.2%, kernel (+ germ) 71-75%. Botanically the peanut is kin to peas and beans, but its constituents are more like those of true nuts. *Constit.* of kernels (roasted with skin): proteins 26.2%, oil 48.7%, water 1.8%, carbohydrates 20.6%, ash 2.7%. The chief proteins are arachin (25% in oil-free meal) and conarchin (8%). Both are globulins of different soly: Johns, Jones, *J. Biol. Chem.* **28**, 77 (1916). The vitamin content of peanuts is moderate, the largest portion being in the episperm. Trace mineral content of kernels: iron 20 mg/kg; manganese 8.51 mg/kg; copper 6.8 mg/kg; zinc 16 mg/kg. *Reviews:* J. Adam, *Les Plantes a Matiere Grasse*, **vol. 3**, *L'Arachide* (Paris, 1947); N. J. Morris, F. G. Dollear, *Abstract Bibliography of the Chemistry and Technology of Peanuts*, Southern Regional Res. Lab. (New Orleans, 1949); J. G. Woodruff, C. T. Young in *Kirk-*

Othmer Encyclopedia of Chemical Technology **vol. 14** (Interscience, New York, 1967) p 122.

USE: In peanut butter, candy; as salted peanuts; for fodder and seeding; crushed for oil. Peanut proteins have been used to produce a fiber, **Sarelon**. The shells are used in the manuf of furfural, xylose, cellulose, plastics, mucilage, also in fertilizers and cattle feed.

7165. Peanut Oil. Arachis oil; groundnut oil; earthnut oil; katchung oil. Prepd by pressing the shelled and skinned seeds of *Arachis hypogaea* L., *Leguminosae:* N. J. Morris, F. G. Dollear, *Abstract Bibliography of the Chemistry and Technology of Peanuts* (Southern Regional Res. Lab., New Orleans, 1949); E. W. Eckey, *Vegetable Fats and Oils* (Reinhold, New York, 1954). *Constits.* of cold pressed oil: glycerides of the following fatty acids: palmitic 8.3%, stearic 3.1%, arachidic 2.4%, behenic 3.1%, lignoceric 1.1%, oleic 56.0%, linoleic 26.0%. Traces of capric and lauric acids have been reported. Unsaponifiable matter 0.8% (includes tocopherols 0.022 to 0.059%, sterols 0.19 to 0.25%, squalene 0.027% and other hydrocarbons).

Greenish-yellow or almost colorless oil. Mild, pleasant odor; bland taste. d_{15}^{15} 0.917-0.921. d_{25}^{25} 0.910-0.915. Clouds at low room temp. Solidifies at ~−5°. A.S.T.M. cloud point +4.5°; A.S.T.M. pour point +1°. Titer 26 to 32°. Flash pt 540°F (283°C); ignition temp 833°F (443°C); n_D^{25} 1.466-1.470; n_D^{40} 1.4605-1.4645. Acid value 0.08-6; saponification value 188-195; iodine value 84-102; thiocyanogen value 67-73; hydroxyl value 2.5-9.5; Reichert-Meissl value 0.2-1.0; Polenske value 0.2-0.7. Very slowly thickens and becomes rancid on prolonged exposure to air. Sol in benzene, carbon tetrachloride, oils; very slightly sol in alc. Miscible with ether, petr ether, chloroform, carbon disulfide.

USE: Edible oil: for salad oil as is; in hydrogenated state as shortening, in mayonnaise, in confections. For the manuf of margarine, soaps, paints. In pharmacy as vehicle for i.m. medication, in the laboratory as heat transfer medium in melting point apparatus. Pharmaceutic aid (solvent).

THERAP CAT (VET): Has been used in control of pasture bloat.

7166. Pearlman's Catalyst. [12135-22-7] Palladium hydroxide (Pd(OH)₂); palladium hydroxide on carbon. H_2O_2Pd; mol wt 140.43. H 1.44%, O 22.79%, Pd 75.78%. Pd(OH)₂/C. Commercial product contains 20 wt % Pd(OH)₂ on carbon. Prepn: W. M. Pearlman, *Tetrahedron Lett.* **8**, 1663 (1967); and catalytic activity study: R. J. Card *et al.*, *J. Catal.* **79**, 13 (1983). Catalytic applications in debenzylating amines: R. C. Bernotas, R. V. Cube, *Synth. Commun.* **20**, 1209 (1990); in Nef-type reactions: T. Capecchi *et al.*, *J. Chem. Soc. Perkin Trans. 1* **2000**, 2681; in Pd-catalyzed coupling reactions: Y. Mori, M. Seki, *J. Org. Chem.* **68**, 1571 (2003); in oxidation of silyl enol ethers: J.-Q. Yu *et al.*, *Org. Lett.* **7**, 1415 (2005).

USE: Nonpyrophoric catalyst for hydrogenolysis of benzyl-nitrogen and benzyl-oxygen bonds.

7167. Pebulate. [1114-71-2] *N*-Butyl-*N*-ethylcarbamothioic acid *S*-propyl ester; *S*-propyl butylethylthiocarbamate; propyl ethyl-*n*-butylthiolcarbamate; PEBC; Stauffer 2061; Tillam. $C_{10}H_{21}NOS$; mol wt 203.34. C 59.07%, H 10.41%, N 6.89%, O 7.87%, S 15.77%. Prepn: Campbell, Klingman, **US 2983747** (1961 to Stauffer). Toxicity data: G. W. Bailey, J. L. White, *Residue Rev.* **10**, 97 (1965).

Liquid, bp₂₀ 142°. n_D^{30} 1.4752. d_4^{30} 0.9458. Slightly sol in water; miscible with acetone, benzene, isopropanol, methanol, xylene. LD₅₀ orally in rats: 1.12 g/kg (Bailey, White).

USE: Selective herbicide.

7168. Pecilocin. [19504-77-9] 1-[(2*E*,4*E*,6*E*,8*R*)-8-Hydroxy-6-methyl-1-oxo-2,4,6-dodecatrien-1-yl]-2-pyrrolidinone; Supral; Variotin. $C_{17}H_{25}NO_3$; mol wt 291.39. C 70.07%, H 8.65%, N 4.81%, O 16.47%. Antifungal antibiotic isolated from *Paecilomyces varioti* Bainier var. *antibioticus:* S. Takeuchi *et al.*, *J. Antibiot.* **12A**, 109, 195 (1959); Sumiki *et al.*, **GB 866425** (1961 to Japan. Antibiot.

Res. Assoc. and Nippon Kayaku). Structure: S. Takeuchi *et al.*, *J. Antibiot.* **17A**, 267 (1964). Stereochemistry: S. Takeuchi, H. Yonehara, *Tetrahedron Lett.* **1966**, 5197. Revised structure and stereochemistry (*E,Z,E* to *E,E,E*): *eidem, J. Antibiot.* **22**, 179 (1969). Synthesis of the *dl*-form: A. Ishida, T. Mukaiyama, *Chem. Lett.* **1977**, 467; *eidem, Bull. Chem. Soc. Jpn.* **51**, 2077 (1978).

Neutral oil with ester-like odor. Does not show definite boiling or dec pt. $[\alpha]_D^{28}$ −5.68° (methanol). Freely sol in methanol, ethanol, acetone, ethyl acetate, benzene, ether, chloroform, pyridine, dioxane, acetic acid; slightly sol in water, petr ether, ligroin. uv max (methanol): *ca.* 318, 324 nm ($E_{1cm}^{1\%}$ 1198). Unstable and gradually loses antifungal activity in desiccator, though it is fairly stable in organic solvents. Unstable under alkaline conditions.

Monohydrate. Needles from ethyl acetate + petr ether, mp 41.5-42.5°. uv max: 320 nm (ε 46,000).

THERAP CAT: Antifungal.

7169. Pectin. [9000-69-5] Polysaccharide substance present in cell walls of all plant tissues which functions as an intercellular cementing material. One of the richest sources of pectin is lemon or orange rind which contains about 30% of this polysaccharide. Occurs naturally as the partial methyl ester of α-(1 → 4) linked D-polygalacturonate sequences interrupted with (1 → 2)-L-rhamnose residues. Neutral sugars: D-galactose, L-arabinose, D-xylose and L-fucose form side chains on the pectin molecule. Structure studies: D. A. Rees, A. W. Wight, *J. Chem. Soc. B* **1971**, 1366. Secondary and tertiary structure in solution and in gels: D. A. Rees, E. J. Welsh, *Angew. Chem. Int. Ed.* **16**, 214 (1977). Review and bibliography: Towle, Christensen, in *Industrial Gums*, R. L. Whistler, Ed. (Academic Press, New York, 2nd ed., 1973) p 429-461. Book: Z. I. Kertesz, *The Pectic Substances* (Interscience, New York, 1951).

Occurs as a coarse or fine powder, yellowish-white in color, practically odorless, and with a mucilaginous taste. Almost completely sol in 20 parts water, forming a viscous soln contg negatively charged, very much hydrated particles. Acid to litmus. Insol in alcohol or in diluted alcohol, and in other organic solvents. Dissolves more readily in water, if first moistened with alcohol, glycerol or sugar syrup, or if first mixed with 3 or more parts of sucrose. Stable under mildly acidic conditions; more strongly acidic or basic conditions cause depolymerization.

USE: In the preparation of jellies and similar food products: Owens *et al.*, "Factors Influencing Gelation with Pectin" in *Natural Plant Hydrocolloids, Advances in Chemistry Series* (ACS, Washington, 1954) pp 10-15.

THERAP CAT (VET): Antidiarrheal.

7170. Pectolinarigenin. [520-12-7] 5,7-Dihydroxy-6-methoxy-2-(4-methoxyphenyl)-4*H*-1-benzopyran-4-one; 5,7-dihydroxy-4',6-dimethoxyflavone; 6-methoxyacacetin; 6-hydroxypelargidenon-6,4'-dimethyl ether 1467. $C_{17}H_{14}O_6$; mol wt 314.29. C 64.97%, H 4.49%, O 30.54%. From leaves of *Linaria vulgaris* Mill., *Scrophulariaceae:* Schmid, Rumpel, *Monatsh. Chem.* **57**, 421 (1931); Merz, Wu, *Arch. Pharm.* **274**, 126 (1936). Structure: Schmid, Rumpel, *Monatsh. Chem.* **60**, 8 (1932). Synthesis: Wessely, Moser, *ibid.* **56**, 97 (1930); Zemplén, Farkas, *Ber.* **76**, 937 (1943); Murti, Seshadri, *Proc. Indian Acad. Sci.* **30A**, 78 (1949), *C.A.* **44**, 3987d (1950); Farkas, Strelisky, *Tetrahedron Lett.* **1970**, 187.

Yellow needles from methanol, mp 220-223°. uv max (methanol): 275, 335 nm. Sol in alcohol, acetone, ether, ethyl acetate. Practically insol in water, benzene, chloroform, petr ether.

Diacetate. $C_{21}H_{18}O_8$. Needles from 96% alcohol, mp 151°.

7-Rutinoside. Pectolinarin; neolinarin. $C_{29}H_{34}O_{15}$. From leaves of *L. vulgaris* Mill., *Scrophulariaceae:* Klobb, *Compt. Rend.* **145**, 331 (1907); Zemplén *et al.*, *Ber.* **75**, 489 (1942); from *Cirsium oleraceum* Scop., *Compositae:* Wagner *et al.*, *Arch. Pharm.* **293**, 1053 (1960). Structure: Zemplén, Bognár, *Ber.* **74**, 1818 (1941). Yellow crystals from methanol, mp 275°. uv max (methanol): 275, 330 nm (log ε 4.256, 4.365).

7171. Pederin. [27973-72-4] (1*S*)-2,6-Anhydro-3,5,7-trideoxy-1-*C*-[[(2*S*)-2-hydroxy-2-[(2*R*,5*R*,6*R*)-tetrahydro-2-methoxy-5,6-dimethyl-4-methylene-2*H*-pyran-2-yl]acetyl]amino]-5,5-dimethyl-1,8,9-tri-*O*-methyl-D-*manno*-nonitol; *N*-[[6-(2,3-dimethoxypropyl)tetrahydro-4-hydroxy-5,5-dimethyl-2*H*-pyran-2-yl]methoxymethyl]tetrahydro-2-methoxy-5,6-dimethyl-4-methylene-2*H*-pyran-2-glycolamide; pederine; paederine. $C_{25}H_{45}NO_9$; mol wt 503.63. C 59.62%, H 9.01%, N 2.78%, O 28.59%. The toxic principle isolated from blister beetles, *Paederus fuscipes* Curt.: Pavan, Bo, *Physiol. Comp. Oecol.* **3**, 307 (1953), *C.A.* **48**, 10217g (1954); **GB 932875** (1963 to Farmitalia). Powerful inhibitor of protein biosynthesis and mitosis. Structure: Cardani *et al.*, *Gazz. Chim. Ital.* **96**, 3 (1966). Revised structure: Matsumoto *et al.*, *Tetrahedron Lett.* **1968**, 6297. Biosynthesis: Cardani *et al.*, *ibid.* **1973**, 2815. Total synthesis: F. Matsuda *et al.*, *Tetrahedron Lett.* **23**, 4043 (1982); **24**, 1277 (1983).

Crystals from hexane, benzene + hexane, ether + hexane. mp 112-112.5°. Slightly sol in water, hexane. Sol in methanol, ethanol, carbon disulfide, chloroform, carbon tetrachloride, benzene, and acids. Practically insol in petr ether, NH_4OH, NaOH.

Pseudopederin. [10352-73-5] *N*-[[6-(2,3-Dimethoxypropyl)-tetrahydro-4-hydroxy-5,5-dimethyl-2*H*-pyran-2-yl]methoxymethyl]tetrahydro-α,2-dihydroxy-5,6-dimethyl-4-methylene-2*H*-pyran-2-acetamide; ψ-paederine; ψ-pederine. $C_{24}H_{43}NO_9$; mol wt 489.61. Isoln from blister beetles, *Paederus fuscipes:* Quilico *et al.*, *Chem. Ind. (Milan)* **43**, 1434 (1961); Cardani *et al.*, *Tetrahedron Lett.* **1965**, 2537; **GB 932875** (1963 to Farmitalia). Crystals from benzene, mp 133°.

7172. Pefloxacin. [70458-92-3] 1-Ethyl-6-fluoro-1,4-dihydro-7-(4-methyl-1-piperazinyl)-4-oxo-3-quinolinecarboxylic acid; pefloxacine; EU-5306; 1589RB; AM-725. $C_{17}H_{20}FN_3O_3$; mol wt 333.36. C 61.25%, H 6.05%, F 5.70%, N 12.61%, O 14.40%. Fluorinated quinolone antibacterial; analog of norfloxacin, *q.v.* Prepn: M. Pesson, **DE 2840910**; *idem*, **US 4292317** (1979, 1981 to Roger Bellon/Dainippon). Pharmacology and antibacterial spectrum: Y. Goueffon *et al.*, *C. R. Seances Acad. Sci. Ser. 3* **292**, 37 (1981). Pharmacokinetics: J. Barre *et al.*, *J. Pharm. Sci.* **73**, 1379 (1984). Bioavailability and metabolism: A. Contrepois *et al.*, *J. Antimicrob. Chemother.* **14**, 51 (1984); G. Montay *et al.*, *Antimicrob. Agents Chemother.* **25**, 463 (1984). HPLC determn in urine and plasma: *eidem*, *J. Chromatogr.* **272**, 359 (1983). Adsorptive stripping voltammetry determn in bulk form, tablets and serum: A. M. Beltagi, *J. Pharm. Biomed. Anal.* **31**, 1079 (2003). Symposium on pharmacokinetics, clinical efficacy and safety: *J. Antimicrob. Chemother.* **26**, Suppl. B, 1-229 (1990).

Crystals from DMF, mp 270-272° (dec). Slightly sol in water; sol in alkaline and acidic solutions. LD_{50} in mice (mg/kg): 225 i.v., 1000 orally; in rats (g/kg): 1.5 i.p., 2.5 orally (Goueffon).

Methanesulfonate dihydrate. [149676-40-4]; [70458-95-6] (methanesulfonate). Pefloxacin mesylate; 1589mRB; Peflacine; Peflox. $C_{17}H_{20}FN_3O_3.CH_3SO_3H.2H_2O$; mol wt 465.49.

THERAP CAT: Antibacterial.

7173. Pefurazoate. [101903-30-4] 2-[(2-Furanylmethyl)(1*H*-imidazol-1-ylcarbonyl)amino]butanoic acid 4-penten-1-yl ester; pent-4-enyl-*N*-furfuryl-*N*-imidazol-1-ylcarbonyl-DL-homoalaninate; UR-0003; UHF-8615; Healthied. $C_{18}H_{23}N_3O_4$; mol wt 345.40. C 62.59%, H 6.71%, N 12.17%, O 18.53%. Ergosterol biosynthesis inhibitor which acts as a rice seed disinfectant. Prepn: Y. Hirota *et al.*, **JP 60260572**; *eidem*, **CA 1324608** (1985, 1993 both to Ube); and fungicidal activity: M. Takenaka *et al.*, *J. Pestic. Sci.* **18**, 15 (1993). Soil degradation and adsorption: *idem et al.*, **16**, 631 (1991). Metabolism in rice seedlings: S. Sakai *et al.*, *ibid.* **18**, 217 (1993). Review of properties and use as disinfectant: M. Takenaka, I. Yamane, *Jpn. Pestic. Inf.* **57**, 33-36 (1990); T. Wada *et al.*, *J. Pestic. Sci.* **24**, 238-240 (1999).

Pale brownish liquid, dec at 235°. $n_D^{24.4}$ 1.5140. d_4^{20} 1.152. Vapor pressure (23°): 6.48 × 10^{-4} Pa. Soly at 25° (g/l): water 0.443; *n*-hexane 12.0; cyclohexane 36.9; DMSO >1000; ethanol >1000; acetone >1000; acetonitrile >1000; chloroform >1000; ethyl acetate >1000; toluene >1000. Log P (octanol/water): 3.00. LD_{50} in male, female rats, male, female mice (mg/kg): 981, 1051, 1299, 946 orally; in rats (mg/kg): >2000 dermally. LC_{50} in rats (mg/m³): >3450 by inhalation. LC_{50} (48 hr) in carp, rainbow trout, bluegill sunfish (ppm): 16.9, 4.0, 12.0 (Takenaka, 1990).

USE: Fungicide.

7174. Pegaptanib Sodium. [222716-86-1] PEG-t44-OMe; EYE-001; NX-1838; Macugen. 2′-Fluoropyrimidine RNA aptamer to human $VEGF_{165}$; composed of 28-mer oligonucleotide conjugated to 40 kDa polyethylene glycol moiety. Prepn: N. Janjic *et al.*, **WO 9818480**; *eidem*, **US 6168778** (1998, 2001 both to NeXstar Pharm.); and binding affinity studies: J. Ruckman *et al.*, *J. Biol. Chem.* **273**, 20556 (1998). Inhibition of *in vitro* VEGF-mediated responses: C. Bell *et al.*, *In Vitro Cell. Dev. Biol. Anim.* **35**, 533 (1999). HPLC determn in plasma: C. E. Tucker *et al.*, *J. Chromatogr. B* **732**, 203 (1999). Pharmacokinetics: D. W. Drolet *et al.*, *Pharm. Res.* **17**, 1503 (2000). Pharmacology and clinical safety study: Eyetech Study Group, *Retina* **22**, 143 (2002). Clinical study in age-related macular degeneration: E. S. Gragoudas *et al.*, *N. Engl. J. Med.* **351**, 2805 (2004). Review of development and therapeutic potential: S. A. Vinores, *Curr. Opin. Mol. Ther.* **5**, 673-679 (2003); of clinical experience and safety in macular degeneration: A. A. Moshfeghi, C. A. Puliafito, *Expert Opin. Invest. Drugs* **14**, 671-682 (2005).

THERAP CAT: In treatment of age-related macular degeneration and diabetic macular edema.

7175. Peginterferon Alfa-2a. [198153-51-4] Interferon αA (human leukocyte) mono(N^2,N^6-dicarboxy-L-lysyl) deriv. diester with α-methyl-ω-hydroxypoly(oxy-1,2-ethanediyl); peginterferon α2a; PEG-IFN-α2a; Ro-25-8310/000; Pegasys. Covalent conjugate of recombinant alfa-2a interferon with a branched 40 kDa monomethoxy polyethylene glycol (PEG), composed of 2 linked 20 kDa chains, attached to the interferon moiety via a stable amide bond to lysine. Mol wt ~60 kDa. Prepn: P. S. Bailon, A. V. Palleroni, **EP 0809996**; *eidem*, **US 7201987** (1997, 2007 both to Hoffmann-La Roche). Clinical trial in treatment of chronic hepatitis B: H. L.-Y. Chan *et al.*, *Antivir. Ther.* **13**, 555 (2008); in combination with ribavirin in hepatitis C: F. Mecenate *et al.*, *BMC Gastroenterol.* **10**, 21

(2010). Review of pharmacology, pharmacokinetics, and clinical experience in hepatitis B: G. M. Keating, *Drugs* **69**, 2633-2660 (2009).

THERAP CAT: Antiviral.

7176. Peginterferon Alfa-2b. [215647-85-1] Pegylated interferon α-2b (human); SCH-54031; PegIntron; Sylatron. Covalent conjugate of recombinant alfa-2b interferon with a linear, 12 kDa monomethoxy polyethylene glycol (PEG). Mol wt ~31 kDa. Manuf process: C. Gilbert, M.-O. Cho, **WO 9513090**; *eidem*, **US 5951974** (1995, 1999 both to Enzon). Description, pharmacology and clinical pharmacokinetics: R. M. Bukowski *et al.*, *Cancer* **95**, 389 (2002). Clinical trial in chronic hepatitis C: M. P. Manns *et al.*, *Lancet* **358**, 958 (2001); in melanoma: A. M. M. Eggermont *et al.*, ibid. **372**, 117 (2008). Review of clinical experience: S. Okuyama *et al.*, *Core Evid.* **5**, 39-48 (2010).

THERAP CAT: Antiviral; antineoplastic.

7177. Pegloticase. [885051-90-1] Urate oxidase (synthetic *Sus scrofa* variant pig*KS*-ΔN subunit) homotetramer amide with α-carboxy-ω-methoxypoly(oxy-1,2-ethanediyl); des-(1-6)-[7-threonine,46-threonine,291-lysine,301-serine]uricase (EC 1.7.3.3, urate oxidase) *Sus scrofa* (pig) tetramer, non acetylated, carbamates with α-carboxy-ω-methoxypoly(oxyethylene); Krystexxa; Puricase. Recombinant mammalian uricase, *q.v.*, produced by genetically modified strain of *Escherichia coli* and covalently conjugated to monomethoxypolyethylene glycol (mPEG). Catalyzes the oxidation of uric acid to allantoin, *q.v.*, and reduces serum uric acid. Average mol wt ~540 kDa. Prepn: J. Hartman, S. Mendelovitz, **WO 06110761**; *eidem*, **US 7811800** (2006, 2010 both to Savient). Clinical pharmacokinetics and pharmacodynamics in patients with hyperuricemia and treatment-failure gout: C. S. Yue, *et al.*, *J. Clin. Pharmacol.* **48**, 708 (2008). Clinical trial in chronic gout: J. S. Sundy *et al.*, *J. Am. Med. Assoc.* **306**, 711 (2011). Review of structure and development: M. R. Sherman *et al.*, *Adv. Drug Delivery Rev.* **60**, 59-68 (2008); of clinical experience in gout: M. K. Reinders, T. L. Jansen, *Ther. Clin. Risk Management* **6**, 543-550 (2010).

THERAP CAT: In treatment of gout.

7178. Pegvisomant. [218620-50-9] Pegylated somatotropin [18-aspartic acid, 21-asparagine, 120-lysine, 167-asparagine, 168-alanine, 171-serine, 172-arginine, 174-serine, 179-threonine] (human); B2036-PEG; Somavert. Growth hormone receptor antagonist. Genetically engineered analog of human growth hormone conjugated to polyethylene glycol (PEG), *q.v.*, to improve bioavailability. Prepn: B. C. Cunningham *et al.*, **WO 9711178**; *eidem*, **US 5849535** (1997, 1998 both to Genentech). Receptor binding and effects of pegylation on structure and function: R. J. M. Ross *et al. J. Clin. Endocrinol. Metab.* **86**, 1716 (2001). Clinical pharmacology: M. O. Thorner *et al.*, ibid. **84**, 2098 (1999). Clinical trial in acromegaly: P. J. Trainer *et al.*, *N. Engl. J. Med.* **342**, 1171 (2000); A. J. van der Lely *et al.*, *Lancet* **358**, 1754 (2001). Review of clinical experience: C. Parkinson, P. J. Trainer, *Expert Opin. Invest. Drugs* **10**, 1725-1735 (2001); of structure and function: S. Pradhananga *et al.*, *J. Mol. Endocrinol.* **29**, 11-14 (2002); of preclinical and clinical development: V. Goffin, P. Touraine, *Curr. Opin. Investig. Drugs* **5**, 463-468 (2003).

THERAP CAT: In treatment of acromegaly.

7179. Pelargonic Acid. [112-05-0] Nonanoic acid; nonylic acid; nonoic acid; 1-octanecarboxylic acid. $C_9H_{18}O_2$; mol wt 158.24. C 68.31%, H 11.47%, O 20.22%. Occurs as an ester in oil of pelargonium: Redtenbacher, *Ann.* **59**, 41, 52, 54 (1846). Prepn from unsaturated hydrocarbons by the oxo process: Hill, **US 2815355** (1957 to Standard Oil of Indiana); from tall oil unsaturated fatty acids: Maggiolo, **US 2865937** (1958 to Welsbach); by oxidation of oleic acid: Mackenzie, Morgan, **US 2820046** (1958 to Celanese); from rice bran oil fatty acid: Mihara *et al.*, **US 3060211** (1962 to Toya Koatsu Ind.). Purification: Port, Riser, **US 2890230** (1959 to U.S.D.A.). Toxicity study: L. Orö, A. Wretlind, *Acta Pharmacol. Toxicol.* **18**, 141 (1961).

H₃C~~~~~~~~~COOH

Colorless, oily liquid at ordinary temp; crystallizes when cooled; characteristic odor. d_4^{20} 0.907. mp 12.5°. bp$_{756}$ 252-253°; bp$_{14}$ 143-

145°; bp$_{6.3}$ 132-133°. n_D^{20} 1.4330; n_D^{40} 1.4245. Practically insol in water. Sol in alcohol, chloroform, ether. LD$_{50}$ i.v. in mice: 224±4.6 mg/kg (Orö, Wretlind).

Caution: Strong irritant.

USE: In the production of hydrotropic salts (hydrotropic salts form aq solns which dissolve sparingly sol substances to a greater extent than water); in the manuf of lacquers, plastics.

7180. Pelargonidin. [134-04-3] 3,5,7-Trihydroxy-2-(4-hydroxyphenyl)-1-benzopyrylium chloride (1:1); 3,4',5,7-tetrahydroxyflavylium chloride; 3,4',5,7-tetrahydroxy-2-phenylbenzopyrylium chloride. $C_{15}H_{11}ClO_5$; mol wt 306.70. C 58.74%, H 3.62%, Cl 11.56%, O 26.08%. The aglucone of pelargonin: Willstätter, Bolton, *Ann.* **408**, 42 (1914). Synthesis: Malkin, Robinson, *J. Chem. Soc.* **127**, 1190 (1925); Robertson *et al.*, **1928**, 1533. Prepn from kaempferol: Mirza, Robinson, *Nature* **166**, 997 (1950); King, White, *J. Chem. Soc.* **1957**, 3901.

Reddish-brown prisms from 2% HCl or from alcoholic HCl. Not melted at 350°. Absorption max (ethanol + 0.01% HCl): 530 nm (ε 32,000). Sol in alcohol, methanol; moderately sol in water; slightly sol in chloroform.

3,5-Diglucoside. [17334-58-6] 3,5-Bis(β-D-glucopyranosyloxy)-7-hydroxy-2-(4-hydroxyphenyl)-1-benzopyrylium chloride; pelargonin; salvinin; punicin. $C_{27}H_{31}ClO_{15}$; mol wt 630.98. From flowers of *Pelargonium zonale* Ait. var. meteor, *Geraniaceae:* Willstätter, Bolton, *loc. cit.*; from scarlet roses: Harborne, *Experientia* **17**, 72 (1961). Identity with monardin and salvinin: Robinson, Todd, *J. Chem. Soc.* **1932**, 2488. Identity with punicin: Karrer, Widmer, *Helv. Chim. Acta* **10**, 67 (1927). Structure: Leon *et al.*, *J. Chem. Soc.* **1931**, 2672. Red needles with green luster from methanol + HCl, dec 175-180°. [α]$_D$ −291°. Absorption max (methanol + HCl): 269, 505 nm. Sol in water, alc.

3-Glucoside. [18466-51-8] 3-(β-D-Glucopyranosyloxy)-5,7-dihydroxy-2-(4-hydroxyphenyl)-1-benzopyrylium chloride; callistephin. $C_{21}H_{21}ClO_{10}$; mol wt 468.84. From purple-red aster *(Callistephus chinensis* (L.) Nees, *Compositae):* Willstätter, Burdick, *Ann.* **412**, 149 (1916); from strawberries: Sondheimer, Kertesz, *J. Am. Chem. Soc.* **70**, 3476 (1948). Structure and synthesis: Robertson, Robinson, *J. Chem. Soc.* **1928**, 1460. Dark brownish-red needles with bronze luster. Absorption max (ethanol + HCl): 515 nm (ε 13,000). Sol in water, methanol, ethanol, 0.5-7% aq HCl; moderately sol in 10% HCl.

7181. Pelletierine. [2858-66-4] 1-(2R)-2-Piperidinyl)-2-propanone; punicine. $C_8H_{15}NO$; mol wt 141.21. C 68.05%, H 10.71%, N 9.92%, O 11.33%. From the rootbark of the pomegranate tree, *Punica granatum* L., *Punicaceae:* Isoln: Tanret, *Compt. Rend.* **86**, 1270 (1878). Structure: Gilman, Marion, *Bull. Soc. Chim. Fr.* **1961**, 1993; Drillien, Viel, ibid. **1963**, 2093. Abs config: H. C. Beyerman *et al.*, *Rec. Trav. Chim.* **86**, 80 (1967). Synthesis: Anet *et al.*, *Nature* **164**, 501 (1949); T. Nagasaka *et al.*, *Heterocycles* **29**, 155 (1989).

Isopelletierine. [4396-01-4] (±)-Pelletierine; 2-acetonylpiperidine. Slightly colored oily liq; bp 195°; bp$_{11}$ 102-107°. d$_4^{20}$ 0.988. Sol in alc, ether, chloroform. One gram dissolves in 20 ml water. Should be stored under nitrogen.

Isopelletierine hydrochloride. [5984-61-2] $C_8H_{15}NO$·HCl. Crystals from alcohol + ether, mp 145°. Soluble in water and alcohol.

N-methylisopelletierine. $C_9H_{17}NO$. Oily liquid. d_4^{20} 0.948. bp_{13} 96-98°. n_D^{20} 1.46737. Sol in water, petr ether.

Note: In the early literature the names pelletierine and isopelletierine were used interchangeably.

THERAP CAT: Anthelmintic (Cestodes).

7182. Pellitorine. [18836-52-7] (2E,4E)-N-(2-Methylpropyl)-2,4-decadienamide; (E,E)-N-isobutyl-2,4-decadienamide. $C_{14}H_{25}NO$; mol wt 223.36. C 75.28%, H 11.28%, N 6.27%, O 7.16%. Pungent principle isolated from *Anacyclus pyrethrum* DC., *Compositae:* J. M. Gulland, G. U. Hopton, *J. Chem. Soc.* **1930**, 6. Structure: M. Jacobson, *J. Am. Chem. Soc.* **71**, 366 (1949). Synthesis: *idem, ibid.* **75**, 2584 (1953); L. Crombie, *J. Chem. Soc.* **1955**, 1007; J. Tsuji *et al., Tetrahedron Lett.* **1977**, 1917; J. Nokami *et al., ibid.* **21**, 4455 (1980); T. Mandai *et al., Chem. Lett.* **1980**, 313. Stereoselective synthesis: R. Bloch, D. Hassangonzales, *Tetrahedron* **42**, 4975 (1986). Insecticidal activity: R. T. Lalonde *et al., J. Chem. Ecol.* **6**, 35 (1980).

Needles from petr ether, mp 90°. uv max (abs ethanol): 258 nm ($E_{1cm}^{1\%}$ 1330). Sol in organic solvents; sparingly sol in water. Practically insol in dil acid or alkalies.

7183. Pellotine. [83-14-7] (1S)-1,2,3,4-Tetrahydro-6,7-dimethoxy-1,2-dimethyl-8-isoquinolinol; N-methylanhalonidine; 8-hydroxy-6,7-dimethoxy-1,2-dimethyl-1,2,3,4-tetrahydroisoquinoline. $C_{13}H_{19}NO_3$; mol wt 237.30. C 65.80%, H 8.07%, N 5.90%, O 20.23%. From the mescal buttons (pellote) of *Lophophora williamsii* (Lemaire) Coult., *Cactaceae.* Structure: Späth *et al., Ber.* **65**, 1771 (1932). Synthesis: Brossi *et al., Helv. Chim. Acta* **47**, 2089 (1964); **49**, 403 (1966); Takido *et al., J. Pharm. Sci.* **59**, 271 (1970). Biosynthetic studies: Battersby *et al., Tetrahedron Lett.* **1967**, 563; **1968**, 6111.

Plates from petr ether, mp 112°. Alkaline reaction. Freely sol in alc, acetone, ether, chloroform; sparingly sol in water.

Hydrochloride. $C_{13}H_{19}NO_3 \cdot HCl$. Prisms, freely sol in water; sparingly in alcohol.

Hydriodide. $C_{13}H_{19}NO_3 \cdot HI$. Prisms, mp 130°, sol in water, alcohol, almost insol in ether.

7184. α-Peltatin. [568-53-6] (5R,5aR,8aR)-5,8,8a,9-Tetrahydro-10-hydroxy-5-(4-hydroxy-3,5-dimethoxyphenyl)furo[3',4':6,7]-naphtho[2,3-d]-1,3-dioxol-6(5aH)-one; 8-hydroxy-2-hydroxymethyl-6,7-methylenedioxy-4-(4'-hydroxy-3',5'-dimethoxyphenyl)-1,2,-3,4-tetrahydronaphthalene-3-carboxylic acid lactone. $C_{21}H_{20}O_8$; mol wt 400.38. C 63.00%, H 5.04%, O 31.97%. Exists as a glucoside in the rhizomes of *Podophyllum peltatum* L., *Berberidaceae:* von Wartburg *et al., Helv. Chim. Acta* **40**, 1331 (1957). Isoln from resin podophyllum: Hartwell, Detty, *J. Am. Chem. Soc.* **72**, 246 (1950).

Prismatic leaflets from abs ethanol. Begins to sinter at 236°. Dec 242-246°. $[\alpha]_D^{20}$ −124.8° (c = 0.5 in chloroform). Soly in water at 20°: ∼30 mg/liter. Fairly sol in chloroform, hot ethanol, acetic acid, acetone, dilute caustic; less sol in benzene, ether, carbon tetrachloride, propylene glycol. Practically insol in petr ether.

α-Peltatin-β-D-glucoside. $C_{27}H_{30}O_{13}$. Prismatic needles from acetone, mp 168-171°. $[\alpha]_D^{20}$ −128.9° (c = 0.590 in methanol); $[\alpha]_D^{20}$ −174.4° (c = 0.579 in pyridine).

Caution: Irritates the skin.

7185. β-Peltatin. [518-29-6] (5R,5aR,8aR)-5,8,8a,9-Tetrahydro-10-hydroxy-5-(3,4,5-trimethoxyphenyl)furo[3',4':6,7]-naphtho[2,3-d]-1,3-dioxol-6(5aH)-one; 8-hydroxy-2-hydroxymethyl-6,7-methylenedioxy-4-(3',4',5'-trimethoxyphenyl)-1,2,3,4-tetrahydronaphthalene-3-carboxylic acid lactone; β-peltatin A. $C_{22}H_{22}O_8$; mol wt 414.41. C 63.76%, H 5.35%, O 30.89%. Exists as a glucoside in the rhizomes of *Podophyllum peltatum* L., *Berberidaceae:* von Wartburg *et al., Helv. Chim. Acta* **40**, 1331 (1957). Isoln from resin podophyllum: Hartwell, Detty, *J. Am. Chem. Soc.* **70**, 2833 (1948); **72**, 246 (1950); from *Hyptis verticillata* Jacq., *Labiatae:* German, *J. Pharm. Sci.* **60**, 649 (1971).

Prisms from abs ethanol. Dec 238-241° (slight sintering begins at 234°). $[\alpha]_D^{20}$ −122.9° (c = 0.578 in chloroform). Somewhat less sol than α-peltatin. Soly in water at 23°: 13 mg/liter. Fairly sol in chloroform, hot ethanol, acetic acid, acetone, dil caustic; less sol in benzene, ether, carbon tetrachloride, propylene glycol. Practically insol in petr ether.

β-Peltatin-β-D-glucoside. $C_{28}H_{32}O_{13}$. White amorph powder from acetone + ether, dec 156-159°. $[\alpha]_D^{20}$ −122.7° (c = 0.587 in methanol); $[\alpha]_D^{20}$ −169.2° (c = 0.556 in pyridine).

Caution: Irritates the skin.

7186. Pemetrexed. [137281-23-3] N-[4-[2-(2-Amino-4,7-dihydro-4-oxo-3H-pyrrolo[2,3-d]pyrimidin-5-yl)ethyl]benzoyl]-L-glutamic acid. $C_{20}H_{21}N_5O_6$; mol wt 427.42. C 56.20%, H 4.95%, N 16.39%, O 22.46%. Multitargeted antifolate; inhibits thymidylate synthase as well as other folate dependent enzymes. Prepn: E. C. Taylor *et al., EP 432677; idem, US 5344932* (1991, 1994 both to Trustees Princeton Univ.); *idem et al., J. Med. Chem.* **35**, 4450 (1992). Profile of enzyme inhibition: C. Shih *et al., Adv. Enzyme Regul.* **38**, 135 (1998). HPLC determn in plasma: C. L. Hamilton, J. A. Kirkwood, *J. Chromatogr. B* **654**, 297 (1994). Clinical pharmacokinetics in cancer patients: A. C. McDonald *et al., Clin. Cancer Res.* **4**, 605 (1998). Clinical evaluation in colorectal carcinoma: W. John *et al., Cancer* **88**, 1807 (2000); in pancreatic cancer: K. D. Miller *et al., Ann. Oncol.* **11**, 101 (2000). Clinical trial as maintenance treatment in non-small-cell lung cancer: T. Ciuleanu *et al., Lancet* **374**, 1432 (2009).

Crystals from 50% methanol/methylene chloride.

Sodium salt. [150399-23-8] Pemetrexed disodium; LY-231514; Alimta. $C_{20}H_{19}N_5Na_2O_6$; mol wt 471.38.

THERAP CAT: Antineoplastic.

7187. Pemirolast. [69372-19-6] 9-Methyl-3-(2H-tetrazol-5-yl)-4H-pyrido[1,2-a]pyrimidin-4-one. $C_{10}H_8N_6O$; mol wt 228.22. C 52.63%, H 3.53%, N 36.82%, O 7.01%. Inhibitor of chemical mediator release from tissue mast cells. Prepn: P. F. Juby, **US 4122274** (1978 to Bristol-Myers). LC determn in plasma: H. Cheng et al., J. Pharm. Sci. **76**, 918 (1987). Immunopharmacology and mechanism of action studies: Y. Yanagihara et al., Jpn. J. Pharmacol. **48**, 91, 103 (1988). Clinical evaluation in allergic rhinitis: D. G. Tinkelman, R. B. Berkowitz, Ann. Allergy **66**, 162 (1991).

Crystals from DMF, mp 310-311° (dec).

Potassium salt. [100299-08-9] BMY-26517; TBX; Pemilaston. $C_{10}H_7KN_6O$; mol wt 266.31. Yellowish-white crystalline powder. Highly soluble in water.

THERAP CAT: Antiallergic.

7188. Pemoline. [2152-34-3] 2-Amino-5-phenyl-4(5H)-oxazolone; phenoxazole; phenylisohydantoin; phenylpseudohydantoin; azoxodone; PIO; LA-956; YH-1; Cylert; Tradon. $C_9H_8N_2O_2$; mol wt 176.18. C 61.36%, H 4.58%, N 15.90%, O 18.16%. Prepn: W. Traube, R. Ascher, Ber. **46**, 2077 (1913); L. Schmidt, H. Scheffler, **US 2892753** (1959 to Boehringer, Ing.). Prepn of magnesium hydroxide mixture, originally thought to be a chelate complex: W. E. Lange et al., J. Pharm. Sci. **51**, 477 (1962); B. H. Candon, M. Chessin, **US 3108045** (1963 to Purdue Frederick). CNS stimulant activity: L. Schmidt, Arzneim.-Forsch. **6**, 423 (1956). Efficacy in minimal brain dysfunction in hyperkinetic children: C. K. Connors et al., Psychopharmacologia **26**, 321 (1972); J. G. Page et al., J. Learn. Disabil. **7**, 498 (1974). Toxicity data: E. W. Schafer, Toxicol. Appl. Pharmacol. **21**, 315 (1972). Review: A. T. Dren, R. S. Janicki, in Pharmacological and Biochemical Properties of Drug Substances vol. 1, M. E. Goldberg, Ed. (Am. Pharm. Assoc., Washington, DC, 1977) pp 33-65.

Crystals, mp 256-257° (dec). Practically insol in water, ether, acetone, dil hydrochloric acid. Sol in propylene glycol (1%); in hot alcohol. LD_{50} orally in rats: 500 mg/kg (Schafer).

Magnesium hydroxide mixture. [18968-99-5] Magnesium pemoline; Abbott 30400

Note: This is a controlled substance (stimulant): **21 CFR, 1308.14.**

THERAP CAT: CNS stimulant.

7189. Pempidine. [79-55-0] 1,2,2,6,6-Pentamethylpiperidine. $C_{10}H_{21}N$; mol wt 155.29. C 77.35%, H 13.63%, N 9.02%. Ganglion blocking agent. Prepn by methylation of 2,2,6,6-tetramethylpiperidine: Leonard, Hauck, J. Am. Chem. Soc. **79**, 5289 (1957); Hall, ibid. 5447. Description: Spinks, Young, Nature **181**, 1397 (1958); Lee et al., ibid. 1717.

Liquid. pK at 30° = 11.25. Extremely basic reaction for a tertiary amine. bp_{760} 147°. n_D^{21} 1.4550.

p-Toluenesulfonate. $C_{17}H_{19}NO_3S$. Crystals from ethyl acetate + ethanol, mp 162-163°.

Tartrate. [546-48-5] M & B 4486; Pempidil; Pempiten; Perolysen; Tenormal; Tensinol; Tensoral. $C_{10}H_{21}N.C_4H_6O_6$; mol wt 305.37. Crystals, mp 160°. Sol in alcohol, moderately sol in water.

Hydrochloride. $C_{10}H_{21}N.HCl$. Crystals, sol in water, alc.

THERAP CAT: Tartrate as antihypertensive.

7190. Penamecillin. [983-85-7] (2S,5R,6R)-3,3-Dimethyl-7-oxo-6-[(2-phenylacetyl)amino]-4-thia-1-azabicyclo[3.2.0]heptane-2-carboxylic acid (acetyloxy)methyl ester; penicillin G hydroxymethyl ester acetate; acetoxymethyl benzylpenicillinate; benzylpenicillin acetoxymethyl ester; Wy-20788; Havapen; Maripen. $C_{19}H_{22}N_2O_6S$; mol wt 406.45. C 56.15%, H 5.46%, N 6.89%, O 23.62%, S 7.89%. Semi-synthetic antibiotic related to penicillin. Prepn: A. B. A. Jansen, T. J. Russell, J. Chem. Soc. **1965**, 2127; eidem, **GB 1003479** (1965 to John Wyeth & Brother); eidem, **US 3250679** (1966 to Am. Home).

Crystals from isopropanol + ethanol, mp 106-108°. $[\alpha]_D^{20}$ +154°. Not inactivated by gastric acid.

THERAP CAT: Antibacterial.

7191. Penbutolol. [38363-40-5] (2S)-1-(2-Cyclopentylphenoxy)-3-[(1,1-dimethylethyl)amino]-2-propanol; (S)-1-(tert-butylamino)-3-(o-cyclopentylphenoxy)-2-propanol; (−)-1-tert-butylamino-2-hydroxy-3-(2'-cyclopentylphenoxy)propane. $C_{18}H_{29}NO_2$; mol wt 291.44. C 74.18%, H 10.03%, N 4.81%, O 10.98%. β-Adrenergic blocker. Prepn: H. Ruschig et al., **ZA 6807915**; eidem, **US 3551493** (1969, 1970 both to Hoechst). Preliminary chemistry and pharmacology: G. Härtfelder et al., Arzneim.-Forsch. **22**, 930 (1972). Physicochemical and analytical study: P. Hajdu, D. Damm, ibid. **29**, 602 (1979). Pharmacology: J. Kaiser et al., ibid. **30**, 420 (1980). Action specificity: J. Kaiser ibid. 427. Crystallographic study: J. M. Leger et al., Mol. Pharmacol. **17**, 339 (1980). Hemodynamic effects: P. Lund-Johansen, Eur. J. Clin. Pharmacol. **16**, 149 (1979). Clinical study in hypertension: J. L. Cangiano et al., J. Clin. Pharmacol. **19**, 384 (1979). HPLC determn in plasma: R. K. Bhamra et al., Biomed. Chromatogr. **1**, 140 (1986). Review of pharmacology and therapeutic efficacy: R. C. Heel et al., Drugs **22**, 1-25 (1981).

Crystals, mp 68-72°. $[\alpha]_D^{20}$ −11.5° (c = 1 in methanol). pKa 9.3 (1.5 mmol/l in 25% ethanol). Sol in methanol, ethanol, chloroform.

Sulfate. [38363-32-5] HOE-893d; HOE-39-893d; Betapressin; Levatol; Paginol. $(C_{18}H_{29}NO_2)_2.H_2SO_4$; mol wt 680.94. White to off-white crystals, mp 216-218° (dec). $[\alpha]_D^{20}$ −24.6° (c = 1 in methanol). Sol in water, methanol.

THERAP CAT: Antihypertensive; antianginal; antiarrhythmic.

7192. Penciclovir. [39809-25-1] 2-Amino-1,9-dihydro-9-[4-hydroxy-3-(hydroxymethyl)butyl]-6H-purin-6-one; 9-[4-hydroxy-3-(hydroxymethyl)but-1-yl]guanine; PCV; BRL-39123; Denavir; Vectavir. $C_{10}H_{15}N_5O_3$; mol wt 253.26. C 47.43%, H 5.97%, N 27.65%, O 18.95%. Carba analog of ganciclovir, q.v., active against several herpes viruses. Prepn: U. K. Pandit et al., Synth. Commun. **2**, 345 (1972); R. L. Jarvest, M. R. Harnden, **US 5075445** (1991 to Beecham). Synthesis: M. R. Harnden et al., J. Med. Chem. **30**, 1636 (1987); J. Hannah et al., J. Heterocycl. Chem. **26**, 1261 (1989). Crystal and molecular structures: M. R. Harnden et al., Nucleosides

Nucleotides **9**, 499 (1990). *In vitro* activity of enantiomers in comparison with acyclovir, *q.v.*: G. Abele *et al.*, *Antivir. Chem. Chemother.* **2**, 163 (1991); against herpes simplex viruses: A. Weinberg *et al.*, *Antimicrob. Agents Chemother.* **36**, 2037 (1992). Clinical pharmacokinetics: S. E. Fowles *et al.*, *Eur. J. Clin. Pharmacol.* **43**, 513 (1992). HPLC determn in plasma and urine: J. R. McMeekin *et al.*, *Anal. Proc.* **29**, 178 (1992). Review of development and antiviral activity: M. R. Harnden, *Drugs Future* **14**, 347-358 (1989).

White crystalline solid from water, (monohydrate), mp 275-277°; also reported as colorless matted needles, mp 272-275°. uv max (in water): 253 nm (ε 11500). uv max (aq 0.01N NaOH): 215, 268 nm (ε 18140, 10710). Sol in water (20°): 1.7 mg/ml, pH 7.

Sodium salt. BRL-39123A. Occurs as monohydrate, stable crystalline solid. Sol in water (20°): >200 mg/ml. 30 mg/ml soln has pH 11.

THERAP CAT: Antiviral.

7193. Pendimethalin. [40487-42-1] *N*-(1-Ethylpropyl)-3,4-dimethyl-2,6-dinitrobenzenamine; *N*-(1-ethylpropyl)-2,6-dinitro-3,4-xylidine; *N*-(1-ethylpropyl)-3,4-dimethyl-2,6-dinitroaniline; penoxalin; AC-92553; Herbadox; Pendimax; Prowl; Stomp. $C_{13}H_{19}N_3O_4$; mol wt 281.31. C 55.51%, H 6.81%, N 14.94%, O 22.75%. Pre-emergence and pre-planting herbicide. Prepn: R. H. Kupelian, DE **2232263**, *C.A.* **78**, 84010z (1973); and A. W. Lutz, R. E. Diehl, DE **2241408** (both 1973 to Am. Cyanamid), *C.A.* **78**, 135858s (1973). Activity: H. A. Roberts, W. Bond, *Proc. 12th Br. Weed Control Conf.* **1**, 387 (1974). Persistence in soil: A. Walker, W. Bond, *Pestic. Sci.* **8**, 359 (1977).

Orange-yellow cryst solid, mp 56-57°. Vapor press at 25°: 3 × 10^{-5} mm Hg. Soly in water at 20°: 0.3 mg/l. Sol in most organic solvents.

USE: Herbicide.

7194. Penethamate Hydriodide. [808-71-9] (2*S*,5*R*,6*R*)-3,3-Dimethyl-7-oxo-6-[(phenylacetyl)amino]-4-thia-1-azabicyclo-[3.2.0]heptane-2-carboxylic acid 2-(diethylamino)ethyl ester hydriodide (1:1); penicillin G 2-diethylaminoethyl ester hydriodide; benzylpenicillin β-diethylaminoethyl ester hydriodide; β-diethylaminoethyl benzylpenicillinate hydriodide; ephicillin hydriodide; penethecillin; Bronchocillin; Estopen; Leocillin; Mamyzin; Neo-Penil. $C_{22}H_{32}IN_3O_4S$; mol wt 561.48. C 47.06%, H 5.74%, I 22.60%, N 7.48%, O 11.40%, S 5.71%. Prepn: K. A. Jensen *et al.*, *Ugeskr. Laeg.* **112**, 1043, 1075 (1950); E. K. Frederiksen, E. J. Nielsen, US **2694061**; A. B. A. Jansen, J. C. Hamlet, US **2880203** (1954, 1959 both to Lövens). Selective concentration in lung: A. G. S. Heathcote, E. Nassau, *Lancet* **260**, 1255 (1951). Properties: W. A. Woodard, *J. Pharm. Pharmacol.* **4**, 1009 (1952). Veterinary trial in cows: G. Ziv, M. Storper, *J. Vet. Pharmacol. Ther.* **8**, 276 (1985).

Crystals, mp 178-179°. Slightly soluble in water (0.96% at 20°). pH of satd aq soln 4.5-5.2. Aq solns are unstable, the ester is hydrolyzed to free penicillin and diethylaminoethanol, the velocity of the reaction increasing with rise in temperature and pH.

THERAP CAT: Antibacterial.

THERAP CAT (VET): Antibacterial.

7195. Penetratin. [188842-14-0] Penetratin-1. $C_{104}H_{168}N_{34}O_{20}S$; mol wt 2246.77. C 55.60%, H 7.54%, N 21.20%, O 14.24%, S 1.43%. Nuclear transport protein capable of delivering bioactive molecules to the cytoplasm and nucleus of living cells by an energy-independent mechanism. Consists of a 16 amino acid segment of the DNA-binding domain (homeodomain) of Antennapedia, a transcription factor found in *Drosophila*. Ref: D. Derossi *et al.*, *J. Biol. Chem.* **269**, 10444 (1994). Study of translocation mechanism: *idem et al.*, *ibid.* **271**, 18188 (1996). Use to deliver antigens to target cells: M.-P. Schutze-Redelmeier *et al.*, *J. Immunol.* **157**, 650 (1996). *Review:* A. Prochiantz, *Curr. Opin. Neurobiol.* **6**, 629-634 (1996); D. Derossi *et al.*, *Trends Cell Biol.* **8**, 84-87 (1998).

Arg–Gln–Ile–Lys–Ile–Trp–Phe–Gln–Asn–Arg–Arg–Met–Lys–Trp–Lys–Lys

USE: Internalization vector for peptides and oligonucleotides into living cells.

7196. Penfluridol. [26864-56-2] 1-[4,4-Bis(4-fluorophenyl)-butyl]-4-[4-chloro-3-(trifluoromethyl)phenyl]-4-piperidinol; 1-[4,4-bis(*p*-fluorophenyl)butyl]-4-(4-chloro-α,α,α-trifluoro-*m*-tolyl)-4-piperidinol; 1-(4,4-bis(4-fluorophenyl)butyl)-4-hydroxy-4-(3-trifluoromethyl-4-chlorophenyl)piperidine; R-16341; Semap. $C_{28}H_{27}ClF_5NO$; mol wt 523.97. C 64.18%, H 5.19%, Cl 6.77%, F 18.13%, N 2.67%, O 3.05%. Prepn: H. K. F. Hermans, C. J. E. Niemegeers, DE **2040231**; *eidem*, US **3575990** (both 1971 to Janssen); Sindelár *et al.*, *Collect. Czech. Chem. Commun.* **38**, 3879 (1973). Pharmacology and toxicology: Janssen *et al.*, *Eur. J. Pharmacol.* **11**, 139 (1970). Crystal structure: Koch, *Acta Crystallogr.* **29B**, 1538 (1973).

White, microcrystals, mp 105-107°. Slightly sol in water, dil HCl (<0.5 mg/ml). LD$_{50}$ orally in mice (day 7): 86.8 mg/kg (Janssen).

THERAP CAT: Antipsychotic.

7197. Penicillamine. [52-67-5] 3-Mercapto-D-valine; (*S*)-3,3-dimethylcysteine; α-amino-β-methyl-β-mercaptobutyric acid; DMC; β-thiovaline; D-penicillamine; Cuprimine; Cupripen; Depen; D-Penamine; Distamine; Mercaptyl; Metalcaptase; Trisorcin; Trolovol. $C_5H_{11}NO_2S$; mol wt 149.21. C 40.25%, H 7.43%, N 9.39%, O 21.44%, S 21.49%. Characteristic degradation product of penicillin type antibiotics. Active as copper chelating agent and as a disease modifying antirheumatic drug (DMARD). Prepn by hydrolysis of penicillins: E. P. Abraham *et al.*, *Nature* **151**, 107 (1943). Review of syntheses of DL-form and enantiomers: H. M. Crooks in *The Chemistry of Penicillin*, H. T. Clarke *et al.*, Eds. (Princeton Univ. Press, 1949) pp 455-472; W. M. Weigert *et al.*, *Angew. Chem. Int. Ed.* **14**, 330 (1975). Polymorphism of D-form: J. A. G. Vidler, *J. Pharm. Pharmacol.* **28**, 662 (1976). Toxicity data: Veis *et al.*, *Antibiotiki* **14**, 837 (1969). CE determn in pharmaceuticals: R. Gotti *et al.*, *J. Chromatogr. A* **844**, 361 (1999). General reviews: I. A. Jaffe in *Pharmacological and Biochemical Properties of Drug Substances* Vol. 2, M. E. Goldberg, Ed. (Am. Pharm. Assoc., Washington, DC, 1979) pp 465-478; C. C. Chiu, L. T. Grady, *Anal. Profiles Drug Subs.* **10**, 601-637 (1981). Review of assay methods: N. Kucharczyk, S. Shahiniam, *J. Rheumatol.* **8**, Suppl. 7, 28-34 (1981); of metabolism and pharmacology: D. Perrett, *ibid.* 41-50; of clinical pharmacokinetics: D. A. Joyce, *Pharmacol. Ther.* **42**, 405-427

(1989). Clinical trial in Wilson's disease: A. Czlonkowska *et al.*, *J. Neurol.* **243**, 269 (1996). Review of clinical experience in rheumatoid arthritis: R. Munro, H. A. Capell, *Br. J. Rheumatol.* **36**, 104-109 (1997).

White or practically white, crystalline powder. mp 202-206° (Weigert). $[\alpha]_D^{25}$ −63° (c = 0.1 in pyridine). Freely sol in water; slightly sol in alc. Insol in ether, chloroform, acetone, benzene, carbon tetrachloride. LD_{50} in rats (mg/kg): >10000 orally, >660 i.p. (Jaffe).

Hydrochloride. [2219-30-9] Pemine. $C_5H_{11}NO_2S.HCl$; mol wt 185.67. Hygroscopic crystals, dec 177.5°. $[\alpha]_D^{25}$ −63° (1*N* NaOH). Freely sol in water, sol in ethanol. Aq solns are comparatively stable at pH 2-4. LD_{50} i.v. in mice: 2289 mg/kg (Veis).

DL-Form. [52-66-4] Crystals, dec 201°. pK: 1.8 (carboxyl); 7.9 (α-amino); 10.5 (β-thiol). LD_{50} orally in rats: 365 mg/kg (Jaffe).

DL-Form hydrochloride. [22572-05-0] Crystals, dec 145-148°.

L-Form. [1113-41-3] Crystals, mp 190-194°. $[\alpha]_D^{25}$ +63° (in 1*N* NaOH). LD_{50} i.p. in rats: 350 mg/kg (Jaffe).

THERAP CAT: Antirheumatic. Chelating agent (copper); Wilson's Disease treatment.

THERAP CAT (VET): Chelating agent to control hepatic copper levels in dogs.

7198. Penicillamine Cysteine Disulfide. [18840-45-4] 3-[[(2*R*)-2-Amino-2-carboxyethyl]dithio]-D-valine; 3,3-dimethyl-3-3′-dithiodialanine; 1,6-diamino-5-5-dimethyl-3,4-dithiahexane-1,6-dicarboxylic acid. $C_8H_{16}N_2O_4S_2$; mol wt 268.35. C 35.81%, H 6.01%, N 10.44%, O 23.85%, S 23.89%. Prepn: Tabachnick *et al.*, *Nature* **174**, 701 (1954); Schöberl *et al.*, *Angew. Chem.* **68**, 213 (1956); Schöberl, Grafje, *Ann.* **617**, 71 (1958); Levine, *Nature* **187**, 940 (1960).

Crystals, mp 195°.

7199. Penicillamine Disulfide. [312-10-7] 3,3′-Dithiobisvaline; 3,3′-dithiodivaline; 3,3,3′,3′-tetramethylcystine. $C_{10}H_{20}N_2$-O_4S_2; mol wt 296.40. C 40.52%, H 6.80%, N 9.45%, O 21.59%, S 21.63%. Prepn: Süs, *Ann.* **561**, 31 (1948); Berg *et al.*, **GB 621915** (1949 to Merck & Co.); Butenandt *et al.*, *Z. Physiol. Chem.* **285**, 238 (1950).

DL-Form. [21174-80-1] Crystals, mp 181-183°.

D-Form. [20902-45-8] Crystals, mp 204-205°. $[\alpha]_D^{23}$ +27° (c = 1.46 in 1*N* HCl).

L-Form. [113626-33-8] Crystals, mp 207°. $[\alpha]_D^{22}$ −26° (1*N* HCl).

7200. Penicillanic Acid. [87-53-6] (2*S*,5*R*)-3,3-Dimethyl-7-oxo-4-thia-1-azabicyclo[3.2.0]heptane-2-carboxylic acid. Building block of penicillin, devoid of significant antibacterial activity: Sheehan *et al.*, *J. Am. Chem. Soc.* **75**, 3292 (1953); **81**, 5838 (1959). Separation from its esters and other penicillins by gas chromatog: Evard *et al.*, *Nature* **201**, 1124 (1964). *See also* 6-Aminopenicillanic Acid.

7201. Penicillic Acid. [90-65-3] 3-Methoxy-5-methyl-4-oxo-2,5-hexadienoic acid; γ-keto-β-methoxy-δ-methylene-Δ^α-hexenoic acid. $C_8H_{10}O_4$; mol wt 170.16. C 56.47%, H 5.92%, O 37.61%. Antibiotic mycotoxin produced by the following fungi: *Penicillium puberulum*, *P. cyclopium*, *P. thomii*, *P. suaveolens*, *P. baarnense*, *Aspergillus ochraceus*, *A. melleus*. Isoln: Alsberg, Black, *USDA Bur. Plant Ind. Bull.* **270**, (1913); Birkinshaw *et al.*, *Biochem. J.* **30**, 394 (1936); Oxford *et al.*, *Chem. Ind. (London)* **20**, 22 (1942); Karow *et al.*, *Arch. Biochem.* **5**, 279 (1944); Burton, *Nature* **165**, 274 (1950); Natori *et al.*, *Chem. Pharm. Bull.* **18**, 2259 (1970). Activity studies: Suzuki *et al.*, *Agric. Biol. Chem.* **35**, 287 (1971). Acid in tautomeric equilibrium with its lactone. Structure: Birkinshaw *et al.*, *loc. cit.* Physical properties: Kovac, Solcaniova, *Tetrahedron* **25**, 3617 (1969). Synthesis: Raphael, *Nature* **160**, 261 (1947); *J. Chem. Soc.* **1948**, 1508; C. L. Yeh *et al.*, *Tetrahedron Lett.* **1978**, 3987. Biosynthesis: Birch *et al.*, *J. Chem. Soc.* **1958**, 4582; Bentley, Keil, *J. Biol. Chem.* **237**, 867 (1962). Physicochemical data: A. E. Pohland *et al.*, *Pure Appl. Chem.* **54**, 2219 (1982). Toxicity studies: P. K. Chan *et al.*, *Toxicol. Appl. Pharmacol.* **52**, 1 (1980); P. K. Chan, A. W. Hayes, *J. Toxicol. Environ. Health* **7**, 169 (1981). Evaluation of carcinogenic risk: *IARC Monographs* **10**, 211-216 (1976). *Review:* Ciegler *et al.*, "Patulin, Penicillic Acid and Other Carcinogenic Lactones" in *Microbial Toxins* **vol. VI**, A. Ciegler *et al.*, Eds. (Academic Press, New York, 1971) p 414.

Needles from petr ether, mp 83-84°. uv max: about 220 nm. Acid reaction, turns Congo red paper blue. Moderately sol in cold water (2 g/100 ml); freely sol in hot water, alcohol, ether, benzene, chloroform; slightly sol in hot petr ether. Practically insol in pentanehexane. LD_{50} i.p. in mice: 90.00 mg/kg (Chan *et al.*).

Monohydrate. Large transparent monoclinic or triclinic, rhombic crystals from water, mp 58-64°.

7202. Penicillinase. [9073-60-3] β-Lactamase; Neutrapen. Mol wt about 50,000. Enzymes found in many bacteria which destroy penicillins and cephalosporins by catalyzing the hydrolysis of the amide bond in the β-lactam ring. Good penicillinase producers are *Bact. coli*, the *Bacillus subtilis-mesentericus* group, *Bacillus anthracis* and *Staphylococci*. Both intra- and extracellular penicillinase are of protein nature. There are probably as many different penicillinases as there are bacteria producing them. Ion-exchange procedures for the purification of penicillinase: Puetzer, Boschetti, **US 2982696** (1961 to Schenley). Amino acid sequence studies: Ambler, Meadway, *Nature* **222**, 24 (1969). Therapeutic use in penicillin allergy: Y. P. Borodin, *Allerg. Asthma* **14**, 43 (1968). *Reviews:* Chain *et al.* in *Antibiotics* **vol. 2**, Flory *et al.*, Eds. (Oxford, 1949) p 1090; Rothe, *Pharmazie* **5**, 25 (1950); Citri, Pollock, *Adv. Enzymol.* **28**, 237 (1966); Citri in *The Enzymes* P. D. Boyer, Ed. (Academic Press, New York, 3rd ed., 1971) pp 23-46.

USE: In culture media to antagonize antibacterial activity of penicillin.

THERAP CAT: Has been used in the treatment of allergic reactions to penicillin.

7203. Penicillin G. [61-33-6] (2*S*,5*R*,6*R*)-3,3-Dimethyl-7-oxo-6-[(2-phenylacetyl)amino]-4-thia-1-azabicyclo[3.2.0]heptane-2-carboxylic acid; benzylpenicillin; benzylpenicillinic acid; penicillin II. $C_{16}H_{18}N_2O_4S$; mol wt 334.39. C 57.47%, H 5.43%, N 8.38%, O 19.14%, S 9.59%. Discovery of antibiotic substance produced by *Penicillium* sp: A. Fleming, *Br. J. Exp. Pathol.* **10**, 226

(1929). Preliminary isoln: P. W. Clutterbuck *et al.*, *Biochem. J.* **26**, 1907 (1932); and chemotherapeutic properties: E. Chain *et al.*, *Lancet* **2**, 226 (1940); E. P. Abraham *et al.*, *ibid.* **2**, 177 (1941). Review of early studies: E. Chain, *Annu. Rev. Biochem.* **17**, 657-704 (1948); H. T. Clarke *et al.*, *The Chemistry of Penicillin* (Princeton Univ. Press, 1949) 1094 pp. Crystal structure: D. C. Hodgkin, *Adv. Sci.* **6**, 85 (1949). Fermentation process: A. L. Demain, N. L. Somerson, **US 3024169** (1962 to Merck & Co.). Total synthesis: J. C. Sheehan, K. R. Henery-Logan, *J. Am. Chem. Soc.* **81**, 5838 (1959); R. A. Firestone *et al.*, *J. Org. Chem.* **39**, 437 (1974). Review of clinical pharmacokinetics of penicillins: M. Barza, L. Weinstein, *Clin. Pharmacokinet.* **1**, 297 (1976). Comprehensive description of the potassium salt: J. Kirschbaum, *Anal. Profiles Drug Subs.* **15**, 427-507 (1987).

Amorphous white powder. $[\alpha]_D$ +282° (ethanol). Sparingly sol in water. Sol in methanol, ethanol, ether, ethyl acetate, benzene, chloroform, acetone. Insol in petr ether.

Sodium salt. [69-57-8] Crystapen; Penilevel. $C_{16}H_{17}N_2NaO_4S$; mol wt 356.37. Crystals from methanol + ethyl acetate. Moderately hygroscopic. $[\alpha]_D^{24.8}$ +301° (c = 2.0 in water). uv max (water): 252, 257.5, 264 nm (E_M about 300, 240, 180). Freely sol in water, isotonic saline, glucose solns. Sol in methanol; less sol in ethanol. Practically insol in acetone, chloroform, ether, ethyl acetate, fixed oils, liquid paraffin.

Potassium salt. [113-98-4] Crystapen; Falapen; Megacillin (tabl.); Pentids; Pfizerpen. $C_{16}H_{17}KN_2O_4S$; mol wt 372.48. Crystals from aq butanol, mp 214-217° (dec). Moderately hygroscopic. $[\alpha]_D^{22}$ +285° (c = 0.748 in water). Very sol in water, isotonic saline, dextrose solns; freely sol in glucose solns; sparingly sol in ethanol. Practically insol in chloroform, ether, fixed oils, liquid paraffin. pH of 6% aq soln 5.0 to 7.5.

Mixture with clemizole. [6011-39-8] Clemizole-penicillin; Neopenyl. $C_{19}H_{20}ClN_3$·$C_{16}H_{18}N_2O_4S$; mol wt 660.23. Repository form of penicillin. Prepn: H. Mückter *et al.*, *Arzneim.-Forsch.* **4**, 487 (1954). White powder, mp 144-145°. $[\alpha]_D^{24}$ +144.5° (c = 10 in DMF). Sol in methanol, ethanol, DMF; slightly sol in water, acetone, dioxane.

THERAP CAT: Antibacterial.
THERAP CAT (VET): Antibacterial.

7204. Penicillin G Benethamine. [751-84-8] (2*S*,5*R*,6*R*)-3,3-Dimethyl-7-oxo-6-[(2-phenylacetyl)amino]-4-thia-1-azabicyclo-[3.2.0]heptane-2-carboxylic acid compd with *N*-(phenylmethyl)benzeneethanamine (1:1); benzylpenicillinic acid *N*-benzyl-β-phenylethylamine salt; benethamine penicillin G. $C_{31}H_{35}N_3O_4S$; mol wt 545.70. C 68.23%, H 6.47%, N 7.70%, O 11.73%, S 5.88%. Repository form of penicillin G. Prepn: Jansen, Hems, **GB 732559** (1955 to Glaxo). Clinical studies: Nelson *et al.*, *Br. Med. J.* **II**, 339 (1954); Boger *et al.*, *Antibiot. Annu.* **1954-55**, 123.

Crystals, mp 146-147°. Slight, characteristic, amine taste. Very slightly sol in water (0.1 w/v at 40°).
THERAP CAT (VET): Antibacterial.

7205. Penicillin G Benzathine. [1538-09-6] (2*S*,5*R*,6*R*)-3,3-Dimethyl-7-oxo-6-[(2-phenylacetyl)amino]-4-thia-1-azabicyclo-[3.2.0]heptane-2-carboxylic acid compd with N^1,N^2-bis(phenylmethyl)-1,2-ethanediamine (2:1); penicillin G *N,N'*-dibenzylethylenediamine salt; *N,N'*-dibenzylethylenediamine bis[benzylpenicillin]; dibenzylethylenediamine dipenicillin G; benzethacil; benzathine penicillin G; DBED-penicillin; Beacillin; Bicillin L-A; Cepacilina; Extencilline; Lentopenil; Megacillin (susp.); Penidural; Permapen; Tardocillin. $C_{48}H_{56}N_6O_8S_2$; mol wt 909.13. C 63.42%, H 6.21%, N 9.24%, O 14.08%, S 7.05%. Repository form of penicillin. Prepn: J. L. Szabo *et al.*, *Antibiot. Chemother.* **1**, 499 (1951); J. L. Szabo, W. F. Bruce, **US 2627491** (1953 to Wyeth). Comprehensive

description: F. Kreuzig, *Anal. Profiles Drug Subs.* **11**, 463-482 (1982). Review of clinical efficacy in syphilis: E. W. Hook, III, *Rev. Infect. Dis.* **11**, Suppl. 6, S1511-S1517 (1989); in prophylaxis of rheumatic fever: B. J. Currie, *Pediatrics* **97**, 989 (1996).

Crystals from formamide, mp 123-124°. $[\alpha]_D^{25}$ +206° (c = 0.105 in formamide). Soly at 23° (mg/ml): water 0.15; benzene 0.38; alc 5.2; acetone 1.5; formamide 28.0. pH of satd aq soln about 6. Soly (mg/ml): water 0.315; methanol 16.9; ethanol 15.4; determined by Weiss *et al.*, *Antibiot. Chemother.* **7**, 374 (1957)

THERAP CAT: Antibacterial.
THERAP CAT (VET): Antibacterial.

7206. Penicillin G Procaine. [6130-64-9] (2*S*,5*R*,6*R*)-3,3-Dimethyl-7-oxo-6-[(phenylacetyl)amino]-4-thia-1-azabicyclo[3.2.0]-heptane-2-carboxylic acid compd with 2-(diethylamino)ethyl 4-aminobenzoate hydrate (1:1:1); penicillin G compd with 2-(diethylamino)ethyl *p*-aminobenzoate monohydrate; benzylpenicillin procaine; procaine benzylpenicillinate; procaine penicillin G; Abbocillin; Cilicaine; Crysticillin; Duracillin; Farmaproina; Mammacillin; Monocillin; Pfizerpen-AS; Wycillin. $C_{29}H_{38}N_4O_6S.H_2O$; mol wt 588.72. C 59.17%, H 6.85%, N 9.52%, O 19.02%, S 5.45%. Semi-synthetic antibiotic. Prepn: N. P. Sullivan *et al.*, *Science* **107**, 169 (1948); C. J. Sullivan *et al.*, *J. Am. Chem. Soc.* **70**, 1287 (1948); H. W. Rhodehamel, Jr., **US 2515898** (1950 to Eli Lilly). Crystal structure: Rose, *Anal. Chem.* **27**, 1841 (1955). Toxicity: K. Soehring *et al.*, *Arzneim.-Forsch.* **1**, 28 (1951). Soly profile: P. J. Weiss *et al.*, *Antibiot. Chemother.* **7**, 374 (1957). Pharmacokinetics in horses: S. M. Stover *et al.*, *Am. J. Vet. Res.* **42**, 629 (1981); in humans: B. T. Goh *et al.*, *Br. J. Vener. Dis.* **60**, 371 (1984). Review of use in syphilis: M. W. Adler, *Br. Med. J.* **288**, 551-553 (1984).

Monoclinic hemimorphic crystals from methanol-water, mp 106-110° (with decompn). d 1.255-1.256. Not appreciably affected by air or light. Aq solns are dextrorotatory. The pH of a satd aq soln is between 5 and 7.5. Soly in mg/ml at about 28°: water 6.8; methanol >20; isopropanol 6.5; benzene 0.075; toluene 1.05; petr ether 0.12; isooctane 0.0; carbon tetrachloride 0.12; ethyl acetate 3.35. LD_{50} s.c. in mice: 2.3 g/kg (Soehring).

THERAP CAT: Antibacterial.
THERAP CAT (VET): Antibacterial.

7207. Penicillin N. [525-94-0] (2*S*,5*R*,6*R*)-6-[[(5*R*)-5-Amino-5-carboxy-1-oxopentyl]amino]-3,3-dimethyl-7-oxo-4-thia-1-azabicyclo[3.2.0]heptane-2-carboxylic acid; 6-(D-5-amino-5-carboxyvaleramido)-3,3-dimethyl-7-oxo-4-thia-1-azabicyclo[3.2.0]heptane-2-carboxylic acid; (D-4-amino-4-carboxybutyl)penicillinic acid; cephalosporin N; adicillin; Synnematin B. $C_{14}H_{21}N_3O_6S$; mol wt 359.40. C 46.79%, H 5.89%, N 11.69%, O 26.71%, S 8.92%. Antibiotic substance produced by *Cephalosporium* spp found in sewage outpours: Gottshall *et al.*, *Proc. Soc. Exp. Biol. Med.* **76**, 307 (1951); Abraham *et al.*, *Nature* **171**, 343 (1953); **176**, 551 (1955). Produced also by *Paecilomyces persicimus:* Pisano *et al.*, *Antimicrob. Agents Annu.* **1960** (Plenum Press, New York, 1961) pp 41, 48; by *Penicillium chrysogenum:* Flynn *et al.*, *J. Am. Chem. Soc.* **84**, 4594 (1962). Structure: Abraham, Newton, *Biochem. J.* **58**, 103 (1954). Production: Miller *et al.*, **US 2831797** (1958). Purification: Goodall, Sutcliffe, **US 2899425** (1959 to ICI).

Soluble in water. Dextrorotatory. Inactivated by penicillinase as is penicillin G, but differs from the common penicillin by its antibacterial activity and hydrophilic character. When an aq soln is kept at pH 2.7 and 37° for 2 hrs, there is a loss of antibacterial activity and an increase in dextrorotation. Active against *Sarcina lutea, Proteus vulgaris, Salmonella typhimurium, Diplococcus pneumoniae.* Shows practically no activity against *B. subtilis* and *Staph. aureus.* The toxicity is somewhat less than that of penicillin G, although penicillin N is excreted more slowly.

Barium salt. White powder. $[\alpha]_D^{20}$ +187° (c = 0.6). Freely sol in water, sparingly sol in methanol. Practically insol in ethanol.

THERAP CAT: Antibacterial.

7208. Penicillin O. [87-09-2] (2*S*,5*R*,6*R*)-3,3-Dimethyl-7-oxo-6-[[2-(2-propenylthio)acetyl]amino]-4-thia-1-azabicyclo-[3.2.0]heptane-2-carboxylic acid; [(allylthio)methyl] penicillin; allylmercaptomethylpenicillin; allylmercaptomethylpenicillinic acid; penicillin AT. $C_{13}H_{18}N_2O_4S_2$; mol wt 330.42. C 47.26%, H 5.49%, N 8.48%, O 19.37%, S 19.41%. Antibiotic produced by *Penicillium chrysogenum*. Biosynthesis of salts: Behrens *et al.*, *J. Biol. Chem.* **175**, 793 (1948); Rhodehamel, Behrens *et al.*, US 2528175 and US 2623876 (1950, 1952, both to Lilly); Ford *et al.*, *Antibiot. Chemother.* **3**, 1149 (1953); Ford, US 2647894 (1953 to Upjohn); Palecková, Slechta, *C.A.* **50**, 17309g (1956).

2-Chloroprocaine salt monohydrate. Chloroprocaine penicillin O; penicillin O 2-chloroprocaine; Depo-Cer-O-Cillin Chloroprocaine. $C_{26}H_{37}ClN_4O_6S_2 \cdot H_2O$; mol wt 619.19. Slender needles from hot water, mp 79-81°. Practically insol in cold water. Stable in dry form at room temp. Aq suspensions are stable at room temp for 1 week, at refrigerator temps for 3 weeks. Calculated activity: 949 units/mg. Solubilities: Weiss *et al.*, *Antibiot. Chemother.* **7**, 374 (1957).

Potassium salt. [897-61-0] Potassium penicillin O; penicillin O potassium. $C_{13}H_{17}KN_2O_4S_2$. Crystals from acetone. Soluble in water. Stable in dry form at room temp for at least 3 years. Requires no refrigeration when dry. Aq solns may be kept for 3 days at +10° without significant loss of activity. Behrens' prepn assayed 1630 units/mg. Less toxic than benzylpenicillin in exptl animals.

Procaine salt. $C_{26}H_{38}N_4O_6S_2$. Crystals from water.

Sodium salt. [7177-54-0] Cer-O-Cillin Sodium. $C_{13}H_{17}N_2NaO_4S_2$; mol wt 352.40. Crystals from acetone.

THERAP CAT: Antibacterial.

7209. Penicillin V. [87-08-1] (2*S*,5*R*,6*R*)-3,3-Dimethyl-7-oxo-6-[(phenoxyacetyl)amino]-4-thia-1-azabicyclo[3.2.0]heptane-2-carboxylic acid; 6-phenoxyacetamidopenicillanic acid; penicillin phenoxymethyl; phenoxymethylpenicillin; phenoxymethylpenicillinic acid; Fenospen; Oracilline; V-Cillin. $C_{16}H_{18}N_2O_5S$; mol wt 350.39. C 54.85%, H 5.18%, N 8.00%, O 22.83%, S 9.15%. Obtained by adding 2-phenoxyethanol to the *Penicillium* culture using yeast autolyzate as source of nitrogen: Brandl *et al.*, *Wien. Med. Wochenschr.* **1953**, 602; Brandl, Margreiter, *Oesterr. Chem.-Ztg.* **55**, 11-21 (1954), *C.A.* **48**, 10296 (1954). Purification: Parker *et al.*, *J. Pharm. Pharmacol.* **7**, 683 (1953). Total synthesis: Sheehan, Henery-Logan, *J. Am. Chem. Soc.* **79**, 1262 (1957); *ibid.* **81**, 3089 (1959); *ibid.* **84**, 2983 (1962). Prepn from 6-aminopenicillanic acid: Glambitza, *Ann.* **673**, 166 (1964). Soly data: Weiss *et al.*, *Antibiot. Chemother.* **7**, 374 (1957). The biologically active form is the dextrorotatory D-form; the DL-form is half as active. L-Penicillin V has little, if any, antibiotic activity. Toxicity data: E. I. Goldenthal, *Toxicol. Appl. Pharmacol.* **18**, 185 (1971). Comprehensive description of the potassium salt: J. M. Dunham in *Anal. Profiles Drug Subs.* **1**, 249-300 (1972); D. H. Sieh, *ibid.* **17**, 677-748 (1988).

Crystals, dec 120-128°. Stable in air up to 37°; relatively stable to acid. uv max: 268, 274 nm (ε 1330, 1100). Soly in water at pH 1.8 (acidified with HCl): 25 mg/100 ml. Freely sol in alc, acetone; sol

in polar organic solvents. Practically insol in liquid petrolatum; insol in fixed oils.

Potassium salt. [132-98-9] Antibiocin; Calciopen; Cliacil; Fenoxypen; Milcopen; Ospen; Pen-Vee K; Primcillin; Veetids; Vepicombin; V-Pen; V-Tablopen. $C_{16}H_{17}KN_2O_5S$; mol wt 388.48. White, crystalline powder. $[\alpha]_D^{25}$ +223° (c = 0.2). Very soluble in water; slightly sol in alc. Insol in acetone. LD_{50} orally in rats: >1040 mg/kg (Goldenthal).

Calcium salt. [147-48-8] Arcasin; Calcipen; Isocillin; Ispenoral; Megacillin. $C_{32}H_{34}CaN_4O_{10}S_2$; mol wt 738.84.

Compd with dibenzylethylenediamine. [5928-84-7] Penicillin V benzathine; penicillin V DBED; benzathine penicillin V; phenoxymethylpenicillin benzathine; benzathine benzylpenicillin. ($C_{16}H_{18}$-$N_2O_5S)_2 \cdot C_{16}H_{20}N_2$; mol wt 941.13. Prepn: R. Brunner *et al.*, US 2820789 (1958 to American Home Products). Practically white powder having a characteristic odor, mp 105-109°. Soly at ~28° (mg/ml): water 0.321; ethanol 14.6. Sparingly sol in chloroform; slightly sol in ether.

THERAP CAT: Antibacterial.

THERAP CAT (VET): Antibacterial.

7210. Penicilloyl Polylysine. [27307-30-8] Poly[imino[1-[4-[[2-(4-carboxy-5,5-dimethyl-2-thiazolidinyl)-1-oxo-2-[(phenylacetyl)amino]ethyl]amino]butyl]-2-oxo-1,2-ethanediyl]]; benzylpenicilloyl polylysine; PPL; Cilligen; Pre-Pen; Testarpen. Prepn from polylysine and a penicillenic acid: Parker *et al.*, *J. Exp. Med.* **115**, 803 (1962). Prepn and use as diagnostic aid: M. A. Stahmann, S. S. Wagle, GB 1226773; *eidem*, US 3979508 (1971, 1976 both to Kremers-Urban Co.). Intradermal test for penicillin sensitivity: Brown *et al.*, *J. Am. Med. Assoc.* **189**, 599 (1964); Van Arsdale, *ibid.* **191**, 238 (1965); T. J. Sullivan, *J. Allergy Clin. Immunol.* **68**, 171 (1981).

THERAP CAT: Diagnostic aid (penicillin sensitivity).

7211. Pennyroyal. European pennyroyal. Perennial herb, *Mentha pulegium*, L., *Labiatae*. Medicinal portions are the fresh or dried leaves and flowering tops and the essential oil; traditionally used as an emmenagogue and abortifacient. *Habit*. Mediterranean region. *Constit*. Volatile oil (1-2%); tannins such as rosmaric acid; flavonoids incl. diosmin, hesperidin. Also used is the closely related *American pennyroyal*, Hedeoma pulegioides (L.) Pers., *Labiatae*. Fragrance monograph: D. L. J. Opdyke, *Food Cosmet. Toxicol.* **12**, 949 (1974). Extraction processes and composition of oil: N. Aghel *et al.*, *Talanta* **62**, 407 (2004). Review of toxicology: I. B. Anderson *et al.*, *Ann. Intern. Med.* **124**, 726-734 (1996); of components and pharmacology: J. Barnes *et al.*, *Herbal Medicines* (Pharmaceutical Press, London, 2nd Ed., 2002) pp 372-373; J. Gruenwald *et al.*, *PDR for Herbal Medicines* (Thomson PDR, Montvale, 3rd Ed., 2004) pp 627-628.

Pennyroyal oil. [8013-99-8] Oil of pennyroyal; oil of pulegium. Volatile oil obtained by steam distillation of the fresh or partially dried plant. *Constit*. Pulegone (60-90%), menthone, isomenthone, piperitone. Light yellow to yellow liquid; aromatic mint-like odor; aromatic taste. d_{25}^{25} 0.928-0.940. n_D^{20} 1.483-1.488. Rotation: +18° to +25°. Sol in most fixed oils, propylene glycol, in 2 vols 70% alcohol, with cloudiness in mineral oil. Practically insol in glycerin. LD_{50} orally in rats: 0.4 g/kg; dermally in rabbits: 4.2 g/kg (Opdyke). *Keep well closed, cool and protected from light.*

American pennyroyal oil. [8007-44-1] Oil of hedeoma. Volatile oil from leaves and flowering tops of *H. pulegioides*. *Constit*. Chiefly pulegone, menthone, isomenthone, acetic, formic, butyric, salicylic acids. Pale yellow liquid; herbaceous, mint-like odor; bitter, slightly burning taste. d_{15}^{15} 0.925-0.940. Rotation: +18° to +35°. n_D^{20} 1.482. Slightly sol in water; sol in 2 vols 70% alcohol; very sol in chloroform, ether.

Caution: Hepatotoxic. Ingestion of large quantities may cause vomiting, blood pressure elevation, anesthetic-like paralysis, death through respiratory failure (Gruenwald).

USE: Fragrance component in soaps, perfumes; flavoring agent; insect repellent.

7212. Penoxsulam. [219714-96-2] 2-(2,2-Difluoroethoxy)-*N*-(5,8-dimethoxy[1,2,4]triazolo[1,5-*c*]pyrimidin-2-yl)-6-(trifluoromethyl)benzenesulfonamide; DE-638; Granite; Grasp; Viper. C_{16}-$H_{14}F_5N_5O_5S$; mol wt 483.37. C 39.76%, H 2.92%, F 19.65%, N 14.49%, O 16.55%, S 6.63%. Triazolopyrimidine sulfonamide herbicide for weed control in rice; inhibits acetolactate synthase (ALS).

Prepn: T. C. Johnson *et al.*, **US 5858924** (1999 to Dow Agro-Sciences). Dissipation study in rice fields: D. W. Roberts *et al.*, *Proc. 12th Symp. Pestic. Chem.* **2003**, 349. Comprehensive review: D. Larelle *et al.*, *BCPC Int. Congr. - Crop Sci. Technol.* **2003**, 75-80.

Tan solid, mp 223-224°. Vapor pressure (mm Hg): 7.16×10^{-16} at 25°, 1.87×10^{-16} at 20°. pKa 5.1. Soly in water at 20° (mg/l): 5.7 pH 5, 408 pH 7, 1460 pH 9. Log P (octanol/water): -0.354 (19°, unbuffered water). LD_{50} (mg/kg): >5000 orally in rats; >5000 dermally in rabbits. LC_{50} (8 day) in bobwhite quail, mallard duck (ppm): >4411, >4310. LC_{50} (96 hr) in rainbow trout, bluegill sunfish, common carp (mg/l): >102, >103, >101 (Larelle).

USE: Herbicide.

7213. Pentaborane(9). [19624-22-7] Pentaboron nonahydride; nonahydropentaborane. B_5H_9; mol wt 63.12. B 85.63%, H 14.37%. Prepd from diborane: Stock, Mathing, *Ber.* **69B**, 1456 (1936); Schlesinger, Burg, *J. Am. Chem. Soc.* **53**, 4321 (1931); **55**, 4009 (1933). Molecular structure determination by rotational spectroscopy: D. Schwoch *et al.*, *Inorg. Chem.* **16**, 3219 (1977). Review of toxicity: *see* Decaborane(14). *Review:* Greenwood in *Comprehensive Inorganic Chemistry* **vol. 1**, J. C. Bailar, Jr. *et al.*, Eds. (Pergamon Press, Oxford, 1973) pp 792-801.

Liquid. mp $-46.6°$; bp 60°. d_4^0 0.61; vp 66 mm Hg at 0°. Dec very slowly at 150°. Ignites spontaneously in air. Hydrolyzes in water after long heating. Reacts with ammonia to form a diammoniate.

Caution: Potential symptoms of overexposure are dizziness, headache, drowsiness and lightheadedness; incoordination, tremor, convulsions, behavioral changes; tonic spasms of face, neck, abdomen and limbs; irritation of eyes and skin. *See NIOSH Pocket Guide to Chemical Hazards* (DHHS/NIOSH 97-140, 1997) p 242.

7214. Pentaborane(11). [18433-84-6] Dihydropentaborane-(9); pentaboron undecahydride; undecahydropentaborane. B_5H_{11}; mol wt 65.14. B 82.98%, H 17.02%. Prepd from diborane: Burg, Schlesinger, *J. Am. Chem. Soc.* **55**, 4009 (1933).

Liquid. mp $-123°$. bp 63°; bp_{53} 0°. Unstable. When heated or allowed to stand for long periods of time, it produces diborane, tetraborane, hydrogen, pentaborane, decaborane and brown nonvolatile liqs and solids. Ignites spontaneously in air. Hydrolyzes in water to boric acid and hydrogen. Reacts with ammonia to form a tetraammoniate.

7215. Pentabromoacetone. [79-49-2] 1,1,1,3,3-Pentabromo-2-propanone. C_3HBr_5O; mol wt 452.56. C 7.96%, H 0.22%, Br 88.28%, O 3.54%. Prepd by the addition of 12 parts bromine to 1 part acetone: Mulder, *Jahresber. Fortschr. Chem.* **1864**, 330.

Orthorhombic needles or prisms from alcohol or ether. Penetrating odor. mp 76° (sublimes above mp). Volatile with steam. Practically insol in water. Freely sol in organic solvents. Forms bromoform under the influence of alkalies.

7216. Pentacene. [135-48-8] Benzo[*b*]naphthacene; 2,3,6,7-dibenzoanthracene; *lin*-naphthoanthracene. $C_{22}H_{14}$; mol wt 278.35. C 94.93%, H 5.07%. Synthesis from *m*-xylene or from 4-benzyl-1,3-dimethylbenzene and benzoyl chloride in presence of aluminum chloride or from terephthalyl chloride and *o*-tolylmagnesium bromide: Clar, John, *Ber.* **62**, 940 (1929); **63**, 2967 (1930); **64**, 981

(1931). Synthesis by reduction of 6,13-pentacenequinone: Bruckner *et al.*, *Tetrahedron Lett.* **1960**, no. 1, 5; Bruckner, Tomasz, *Acta Chim. Acad. Sci. Hung.* **28**(4), 405 (1961). Structure: Campbell *et al.*, *Acta Crystallogr.* **14**, 705 (1961).

Deep blue needles with violet luster from hot nitrobenzene. Sublimes in CO_2 stream under reduced pressure at ~300° (Clar, John, *loc. cit.*). In presence of air dec >300°. Practically insol in water; sparingly sol in organic solvents.

Note: Anthracene (3 linear rings) is colorless; naphthacene (4 linear rings) is orange; pentacene (5 linear rings) is blue; hexacene (6 linear rings) is green.

7217. Pentachloroethane. [76-01-7] 1,1,1,2,2-Pentachloroethane; pentalin. C_2HCl_5; mol wt 202.28. C 11.88%, H 0.50%, Cl 87.63%. CCl_3CHCl_2. Toxicity data: G. S. Barsoum, K. Saad, *Q. J. Pharm. Pharmacol.* **7**, 205 (1934).

Liquid; chloroform-like odor. d_4^{25} 1.6712; bp 161-162°. mp $-29°$. n_D^{15} 1.5054. *Poisonous.* Insol in water. Miscible with alcohol, ether. MLD (mg/kg) in dogs: 1750 orally; 100 i.v.; in rabbits: 700 s.c. (Barsoum, Saad).

Caution: Potential symptoms of overexposure in exptl animals are irritation of eyes, skin; weakness, restlessness, irregular respiration, muscle incoordination; liver, kidney, lung changes. *See NIOSH Pocket Guide to Chemical Hazards* (DHHS/NIOSH 97-140, 1997) p 242.

7218. Pentachlorophenol. [87-86-5] 2,3,4,5,6-Pentachlorophenol; 1-hydroxypentachlorobenzene; penchlorol; PCP. C_6HCl_5O; mol wt 266.32. C 27.06%, H 0.38%, Cl 66.56%, O 6.01%. Broad spectrum biocide; prepd by the chlorination of phenol in the presence of a catalyst. GC-MS determn in plasma: Y. Zhou *et al.*, *Chemosphere* **70**, 256 (2007). Review of properties, uses, and environmental chemistry: D. G. Crosby *et al.*, *Pure Appl. Chem.* **53**, 1051-1080 (1981); of toxicology and human exposure: *Toxicological Profile for Pentachlorophenol* (PB2001-109106, 2001) 316 pp.

Needle-like crystals, mp 190-191°. *Poisonous.* bp ~309-310° (dec). d_4^{22} 1.978. pKa (25°): 4.70. Log P (octanol/water): 2.15; Log P (hexane/water): 5.01. Vapor pressure (torr): 1.7×10^{-5} at 0°; 1.7×10^{-4} at 20°; 0.14 at 100°; 758.4 at 300°. Soly at 25° (g/l): methanol 180; acetone 50; benzene 15. Soly in water (g/l): 0.005 at 0°; 0.014 at 20°; 0.085 at 70°. uv max: 303 nm (ε 2900). LD_{50} in male, female rats (mg/kg): 146, 175 orally; *see* T. B. Gaines, *Toxicol. Appl. Pharmacol.* **14**, 515 (1969).

Sodium salt. [131-52-2] Sodium pentachlorophenate; sodium pentachlorophenoxide; NaPCP. C_6Cl_5NaO; mol wt 288.30. White crystalline solid. Soly at 25° (g/l): methanol 22; acetone 37. Soly in water (g/l): 22.4 at 20°; 33 at 30°. Insol in benzene. LC_{50} in goldfish: 0.22 mg/l (Crosby).

Caution: Potential symptoms of overexposure are irritation of eyes, nose, throat; sneezing, cough; weakness, anorexia, weight loss; sweating; headache, dizziness; nausea, vomiting; dyspnea, chest pain; high fever. Direct contact may cause dermatitis. *See NIOSH Pocket Guide to Chemical Hazards* (DHHS/NIOSH 97-140, 1997) p 242; *Patty's Industrial Hygiene and Toxicology* **vol. 2B**, G. D. Clayton, F. E. Clayton, Eds. (Wiley-Interscience, New York, 1994) pp 1603-1613.

USE: Insecticide for termite control; molluscicide; general herbicide. Antimicrobial preservative and fungicide for wood, wood products, starches, textiles, paints, adhesives, leather, pulp, paper, industrial waste systems, building materials.

7219. 3-Pentadecylcatechol. [492-89-7] 3-Pentadecyl-1,2-benzenediol; 3-pentadecylpyrocatechol; tetrahydrourushiol; hydrourushiol; dihydrorhengol; 3-PDC. $C_{21}H_{36}O_2$; mol wt 320.52. C 78.69%, H 11.32%, O 9.98%. Constituent of the irritant oil of poison ivy (*Toxicodendron radicans* (L.) Kuntze) and other *Toxicodendron* spp. (*Anacardiaceae*). Prepn by hydrogenation of extracts from fruits of *Semecarpus heterophylla*: Backer, Haack, *Rec. Trav. Chim.* **57**, 225 (1938). Synthesis from 2,3-dimethoxybenzaldehyde and tetradecyl chloride: Mason, *J. Am. Chem. Soc.* **67**, 1538 (1945); from *o*-veratraldehyde: Backer, Haack, *loc. cit.;* Dawson *et al., J. Am. Chem. Soc.* **68**, 534 (1946); Keil *et al.*, US 2451955 (1948); from 2,3-dibenzyloxybenzaldehyde: Loev, Dawson, *J. Org. Chem.* **24**, 980 (1959); from furan derivs: Boehme, *J. Am. Chem. Soc.* **82**, 499 (1960); from catechol: Hanafusa, Yukawa, *Chem. Ind. (London)* **1961**, 23. Criticism of reported syntheses and synthesis of dimethyl deriv: Byck, Dawson, *J. Org. Chem.* **32**, 1084 (1967). Total synthesis: E. Wenkert *et al., J. Am. Chem. Soc.* **105**, 2021 (1983). Evaluation of diagnostic patch test: M. V. Dahl *et al., Arch. Dermatol.* **120**, 1022 (1984).

Short needles from toluene or petr ether, mp 59-60°. Can be purified by molecular distn. uv max: 277 nm. Sol in alc, ether, benzene, toluene. Sparingly sol in petr ether.

THERAP CAT: Diagnostic aid (contact allergen).

7220. 1,3-Pentadiene. [504-60-9] 1-Methyl-1,3-butadiene; piperylene. C_5H_8; mol wt 68.12. C 88.16%, H 11.84%. Synthetically useful diene that is a component of pyrolysis gasoline, formed as a by-product of ethylene manufacture from crude oil. Prepn: A. W. Hoffmann, *Ber.* **14**, 659 (1881); and sepn of isomers: D. Craig, *J. Am. Chem. Soc.* **65**, 1006 (1943); R. L. Frank *et al., ibid.* **69**, 2313 (1947). Isoln from pyrolysis gasoline: M. Morgan, *Int. J. Hydrocarbon Eng.* **4**, 35 (1999). Reactivity in Diels-Alder cycloadditions: C. A. Stewart, Jr., *J. Org. Chem.* **28**, 3320 (1963); T. Inukai, T. Kojima, *ibid.* **32**, 869 (1967); T. J. Brocksom, M. G. Constantino, *ibid.* **47**, 3450 (1982). Utility in polymer applications: V. Jankauskaite, R. Barauskas, *Pigment Resin Tech.* **28**, 75 (1999).

(3*E*)-Form

Clear, light yellow liquid. *Flammable.* bp 42-44°. $d_4^{16.5}$ 0.6957; d_4^{15} 0.6827; d_4^{10} 0.688; d_4^{0} 0.697. $n_D^{16.5}$ 1.44020; n_D^{15} 1.43398; n_D^{10} 1.4366. Flash pt, closed cup: −18°F (−28°C). Sol in acetone, alcohol, benzene, diethyl ether, heptane. Insol in water. Recommended storage temperature: 2-8°C.

(3*E*)-Form. [2004-70-8] *trans*-Piperylene. Colorless liquid. *Flammable, irritant.* bp 41.9°; bp_{745} 41.7°. d_4^{26} 0.673; d_4^{20} 0.6771. n_D^{20} 1.4320. Flash pt, closed cup: 5°F (−15°C). Recommended storage temperature: 2-8°C.

(3*Z*)-Form. [1574-41-0] *cis*-Piperylene. Clear, light yellow liquid. *Flammable.* bp 43.5°. fp −140.92. d_{20}^{20} 0.6916. n_D^{20} 1.4360. Flash pt, closed cup: 5°F (−15°C). Recommended storage temperature: 2-8°C.

USE: Reagent in synthetic organic chemistry; intermediate monomer in the manufacture of plastics, adhesives, and resins.

7221. Pentaerythritol. [115-77-5] 2,2-Bis(hydroxymethyl)-1,3-propanediol; tetrakis(hydroxymethyl)methane; tetramethylolmethane; Metab-Auxil; Penetek; Pentek. $C_5H_{12}O_4$; mol wt 136.15. C 44.11%, H 8.88%, O 47.00%. Prepd by treating acetaldehyde with formaldehyde in an aq soln of calcium hydroxide: H. B. J. Schurink, *Org. Synth.* **coll. vol. I**, 425 (2nd ed., 1941); Fieser, Fieser, *Organic Chemistry* (2nd ed, 1950) p 133. Review of mfg processes: P. W. Sherwood, *Petroleum Refiner* **Nov. 1956**, p 171-179; *Faith, Keyes & Clark's Industrial Chemicals*, F. A. Lowenheim, M. K. Moran, Eds. (Wiley-Interscience, New York, 4th ed., 1975) pp 598-603. Monograph: E. Berlow *et al.*, "The Pentaerythritols" in *ACS Monograph Series* no. **136** (Reinhold, New York, 1958).

Ditetragonal crystals from dil HCl, mp 260°. One gram dissolves in 18 ml water at 15°. Sol in ethanol, glycerol, ethylene glycol, formamide. Insol in acetone, benzene, paraffin, ether, carbon tetrachloride.

Tetraacetate. [597-71-7] 2,2-Bis[(acetyloxy)methyl]-1,3-propanediol diacetate; tetra-*O*-acetylpentaerythritol; pentaerythrityl tetraacetate. $C_{13}H_{20}O_8$; mol wt 304.30. Prepn: Wolfrom *et al., J. Am. Chem. Soc.* **73**, 874 (1951); Bonner *et al., J. Chem. Soc.* **1960**, 2914. Crystals, mp 83-84°.

Trinitrate see Pentrinitrol.

Caution: Potential symptoms of overexposure to pentaerythritol are irritation of eyes and respiratory system. *See NIOSH Pocket Guide to Chemical Hazards* (DHHS/NIOSH 97-140, 1997) p 244.

USE: In synthetic resins, in paint and varnish industries.

7222. Pentaerythritol Tetranitrate. [78-11-5] 2,2-Bis[(nitrooxy)methyl]-1,3-propanediol 1,3-dinitrate; pentaerythrityl tetranitrate; 2,2-bisdihydroxymethyl-1,3-propanediol tetranitrate; PETN; nitropentaerythritol; penthrit; niperyt; Lentrat; Hasethrol; Peritrate; Mycardol; Nitropenton; Pentral 80; Dilcoran-80; Terpate; Pentrite; Perityl; Pentanitrine; Prevangor; Subicard; Pentryate; Vasodiatol; Neo-Corovas; Pentafin; Quintrate; Pergitral; Pentitrate; Metranil; Cardiacap; Angitet; Nitropenta. $C_5H_8N_4O_{12}$; mol wt 316.14. C 19.00%, H 2.55%, N 17.72%, O 60.73%. Prepn: Acken, Vyverberg, US 2370437 (1945 to Du Pont). GC/MS analysis in post-explosion residues: T. Tamiri *et al.* in *Adv. Anal. Detect. Explosives*, J. Yinon, Ed. (Kluwer, Netherlands, 1993) pp 323-334.

Tetragonal holohedra from acetone + alcohol, mp 140°. d_4^{20} 1.773. Soluble in acetone. Practically insoluble in water (1.5 γ/ml). Sparingly sol in alcohol, ether. Does not reduce Fehling's soln (difference from erythrityl tetranitrate). *Explodes on percussion.* More sensitive to shock than TNT. For medicinal purposes it is dil with an inert ingredient, usually lactose, to prevent accidental explosions.

USE: Mainly in the manuf of detonating fuse (Primacord), a waterproof textile filled with powdered PETN.

THERAP CAT: Vasodilator (coronary).

7223. Pentafluorophenol. [771-61-9] 2,3,4,5,6-Pentafluorophenol; pentafluorohydroxybenzene; perfluorophenol. C_6HF_5O; mol wt 184.07. C 39.15%, H 0.55%, F 51.61%, O 8.69%. Reagent for the prepn of activated pentafluorophenyl ester intermediates in peptide synthesis. Prepn: W. J. Pummer, L. A. Wall, *Science* **127**, 643 (1958); and infrared spectroscopy studies: J. M. Birchall, R. N. Haszeldine, *J. Chem. Soc.* **1959**, 13; and characterization by derivative formation: E. J. Forbes *et al., ibid.* 2019. Crystal structure: D. Das *et al., Chem. Commun.* **2006**, 555. Use in peptide synthesis: Kisfaludy *et al., J. Org. Chem.* **35**, 3563 (1970). Prepn and synthetic applications of pentafluorophenyl derivatives: V. P. Rajappan, R. S. Hosmane, *Synth. Commun.* **28**, 753 (1998); S. Caddick *et al., Org. Lett.* **4**, 2549 (2002); E. Papavassilopoulou *et al., Tetrahedron Lett.* **48**, 8323 (2007).

Colorless, cylindrical monoclinic crystals. mp 34-36°. bp 143°; bp_{48} 72-73°. $d^{18.84}$ 1.956. n_D^{20} 1.4270; n_D^{26} 1.4263. Flash pt, closed cup: 162°F (72°C). *Irritant.* Hygroscopic. Phenolic odor. uv max (ethanol): 227, 269, 350 nm (ε 2640, 990, 198). Sol in most organic solvents.

USE: Reagent in synthetic organic chemistry.

7224. Pentagastrin. [5534-95-2] N-[(1,1-Dimethylethoxy)-carbonyl]-β-alanyl-L-tryptophyl-L-methionyl-L-α-aspartyl-L-phenylalaninamide; N-carboxy-β-alanyl-L-tryptophyl-L-methionyl-L-aspartylphenyl-L-alaninamide N-*tert*-butyl ester; N-(α-carbamoylphenethyl)-3-[2-[2-[3-(carboxyamino)propionamido]-3-indol-3-yl-propionamido]-4-(methylthio)butyramido]succinamic acid N-*tert*-butyl ester; N-[N-[N-[N-(N-*tert*-butoxycarbonyl-β-alanyl)-L-tryptophanyl]-L-methionyl]-L-aspartyl]-L-phenylalaninamide; Boc-β-Ala-Trp-Met-Asp-Phe(NH$_2$); AY-6608; ICI-50123; Gastrodiagnost; Peptavlon. $C_{37}H_{49}N_7O_9S$; mol wt 767.90. C 57.87%, H 6.43%, N 12.77%, O 18.75%, S 4.18%. Prepn: P. M. Hardy *et al.*, **BE 665591**; *eidem*, **US 3896103** (1965, 1975 to I.C.I.); Davey *et al.*, *J. Chem. Soc. C* **1966**, 555; Sakakibara *et al.*, *Bull. Chem. Soc. Jpn.* **41**, 438 (1968). Structure-function relationship studies: Morley *et al.*, *Nature* **207**, 1356 (1965). Formulation and pharmacological studies: Wai *et al.*, *J. Pharm. Pharmacol.* **22**, 923 (1970). *Reviews:* Makhlouf, *Fed. Proc.* **27**, 1322 (1968); Sanders, Schimmel, *Am. J. Med.* **49**, 380 (1970).

Fine, colorless needle-shaped crystals from 2-ethoxyethanol + water; mp 229-230° (dec). $[\alpha]_D^{22}$ −28.8 ±0.5° (DMF). uv max (2N NH$_4$OH): 280, 289 nm (ε 5340, 4590). Sol in DMF, DMSO. Almost insol in water, ethanol, ether, benzene.

THERAP CAT: Diagnostic aid (gastric secretion stimulant).

7225. Pentagestrone. [7001-56-1] 3-(Cyclopentyloxy)-17-hydroxypregna-3,5-dien-20-one; 17α-hydroxyprogesterone 3-cyclopentyl enol ether. $C_{26}H_{38}O_3$; mol wt 398.59. C 78.35%, H 9.61%, O 12.04%. Prepn of acetate: Ercoli, Gardi, *J. Am. Chem. Soc.* **82**, 746 (1960); of free alcohol and acetate: Ercoli, **US 3019241** (1962).

Solid, mp 184.5-186.5°. $[\alpha]_D$ −115° (dioxane).
Acetate. [1178-60-5] 17α-Acetoxyprogesterone 3-cyclopentyl enol ether; Gestovis. $C_{28}H_{40}O_4$; mol wt 440.62. Solid, mp 137-138°; also reported as 157-158°, Ercoli, **GB 893315** (1962 to Vismara). $[\alpha]_D$ −147° (dioxane).

THERAP CAT: Progestogen.

7226. Pentamethylcyclopentadienyliridium(III) Dichloride Dimer. [12354-84-6] Di-μ-chlorodichlorobis[(1,2,3,4,5-η)-1,2,3,-4,5-pentamethyl-2,4-cyclopentadien-1-yl]diiridium; bis(dichloro(η^5-pentamethylcyclopentadienyl)iridium); di-μ-chloro-dichlorobis(pentamethylcyclopentadienyl)diiridium(III); dichloropentamethylcyclopentadienyliridium dimer; iridium pentamethylcyclopentadienyl dichloride dimer; [Cp*IrCl$_2$]$_2$; [Ir(Cp*)Cl$_2$]$_2$; [(η-C$_5$Me$_5$)IrCl$_2$]$_2$. $C_{20}H_{30}Cl_4Ir_2$; mol wt 796.69. C 30.15%, H 3.80%, Cl 17.80%, Ir 48.25%. Iridium complex; homogenous catalyst for a variety of synthetic transformations. Prepn: J. W. Kang *et al.*, *Chem. Commun.* **1968**, 1304; *eidem*, *J. Am. Chem. Soc.* **91**, 5970 (1969); B. L. Booth *et al.*, *J. Organomet. Chem.* **16**, 491 (1969). Review of discovery and chemistry: P. M. Maitlis, *Acc. Chem. Res.* **11**, 301-307 (1978). Crystal structure: M. R. Churchill, S. A. Julius, *ibid.* **16**, 1488 (1977); and improved prepn: R. G. Ball *et al.*, *Inorg. Chem.* **29**, 2023 (1990). Synthetic applications in catalysis: K. Fujita, R. Yamaguchi, *Synlett* **2005**, 560; N. A. Owston *et al.*, *Org. Lett.* **9**, 73 (2007); S. Whitney *et al.*, *ibid.* 3299; V. S. Sridevi *et al.*, *Organometallics* **26**, 1157 (2007).

Orange crystals from chloroform-benzene, mp >230° (dec) (Kang); also reported as crystalline peach solid from dichloromethane-hexane, mp >300° (Booth). Crystal density: 2.259 g/cm³. Air and moisture stable.

USE: Versatile catalyst in organic chemistry.

7227. Pentamidine. [100-33-4] 4,4′-[1,5-Pentanediylbis-(oxy)]bisbenzenecarboximidamide; 4,4′-(pentamethylenedioxy)dibenzamidine; 4,4′-diamidino-α,ω-diphenoxypentane. $C_{19}H_{24}$-N_4O_2; mol wt 340.43. C 67.04%, H 7.11%, N 16.46%, O 9.40%. Prepn: A. J. Ewins, **GB 507565** (1939); J. N. Ashley *et al.*, *J. Chem. Soc.* **1942**, 103; of isethionate: G. Newbery, A. P. T. Easson, **US 2394003** (1946 to May & Baker). Trypanocidal activity: E. M. Lourie, W. Yorke, *Ann. Trop. Med. Parasitol.* **33**, 289 (1939). Preliminary pharmacological studies in animals: R. Wien, *ibid.* **37**, 1 (1943). Activity in fibrinolytic systems: J. D. Geratz, *Thromb. Diath. Haemorrh.* **29**, 154 (1973). Pharmacodynamics in men and mice: T. P. Waalkes *et al.*, *Clin. Pharmacol. Ther.* **11**, 505 (1970). *In vitro* activity against *Pneumocystis carinii:* E. L. Pesanti, C. Cox, *Infect. Immun.* **34**, 908 (1981). Uptake and distribution of aerosolized form in animals: R. J. Debs *et al.*, *Am. Rev. Respir. Dis.* **135**, 731 (1987). *In vivo* efficacy of aerosolized form in rats: *eidem*, *Antimicrob. Agents Chemother.* **31**, 37 (1987). Determn in plasma, urine and tissues: T. P. Waalkes, V. T. DeVita, *J. Lab. Clin. Med.* **75**, 871 (1970); by HPLC: C. M. Dickinson *et al.*, *J. Chromatogr.* **345**, 91 (1985). Preliminary clinical evaluation in *P. carinii* pneumonia: V. T. DeVita *et al.*, *N. Engl. J. Med.* **280**, 287 (1968). Comparison with sulfamethoxazole-trimethoprim mixture in *P. carinii* pneumonia in AIDS: J. M. Wharton *et al.*, *Ann. Intern. Med.* **105**, 37 (1986). Early review of pharmacology, mode of action and clinical applications: E. B. Schoenbach, E. M. Greenspan, *Medicine* **27**, 327-377 (1948). *Review:* S. Drake *et al.*, *Clin. Pharm.* **4**, 507-516 (1985); M. Sands *et al.*, *Rev. Infect. Dis.* **7**, 625-634 (1985); of activity, pharmacokinetics and therapeutic use: K. L. Goa, D. M. Campoli-Richards, *Drugs* **33**, 242-258 (1987).

Crystallizes as colorless plates from water. Dec 186°.
Dihydrochloride. [50357-45-4] $C_{19}H_{24}N_4O_2$.2HCl. Fine needles from dil HCl, mp 232-234°. LD$_{50}$ in mice (mg/g): 0.028 i.v.; 0.064 s.c. (Wein).
Isethionate. [140-64-7] M & B 800; RP-2512; NebuPent; Pentacarinat; Pentam. $C_{19}H_{24}N_4O_2$.2C$_2H_6O_4S$; mol wt 592.68. Hygroscopic, very bitter crystals, mp ~180°. Slight butyric odor. Sol in water (~1 in 10 at 25°, ~1 in 4 at 100°); sol in glycerol, more readily on warming; slightly sol in alcohol. Insol in ether, acetone, chloroform, liquid petr. pH of a 5% w/v soln in water: 4.5 to 6.5.
Dimethanesulfonate. [6823-79-6] Pentamidine mesylate. C_{19}-$H_{24}N_4O_2$.2CH$_3$SO$_3$H; mol wt 532.63. White powder.

THERAP CAT: Antiprotozoal (Trypanosoma, Leishmania); antipneumocystic.

THERAP CAT (VET): Antiprotozoal (Babesia, Leishmania).

7228. Pentane. [109-66-0] n-Pentane. C_5H_{12}; mol wt 72.15. C 83.24%, H 16.77%. Occurs in petroleum; it is a constituent of petr ether. Sepn from natural gasoline: Love, *Pet. Eng.* **12**, no. 10, 130 sqq (1941), *C.A.* **35**, 7162 (1941); from virgin naphthas: Tongberg *et al., Ind. Eng. Chem.* **30**, 166 (1938). Prepd by dehydration and subsequent hydrogenation of 2- and 3-pentanol: Mair, *Bur. Stand. J. Res.* **9**, 457 (1932); from 2-bromopentane by Grignard reaction: Noller, *Org. Synth.* **11**, 84 (1931). Toxicity data: Fühner, *Biochem. Z.* **115**, 235 (1921).

H3C⌃⌄CH3

Liquid. d_4^0 0.64529; d_4^{20} 0.62638; d_4^{30} 0.6163. mp −129.7°. bp$_{760}$ 36.1°; bp$_{400}$ 18.5°; bp$_{200}$ 1.9°; bp$_{100}$ −12.6°; bp$_{60}$ −22.2°; bp$_{40}$ −29.2°; bp$_{20}$ −40.2°; bp$_{10}$ −50.1°; bp$_5$ −62.5°; bp$_{1.0}$ −76.6°. n_D^{20} 1.35768. Flash pt, closed cup: <−40°F (−40°C). *Flammable.* Explosive limits, % by vol in air: lower 1.4; upper 8.0. Autoignition temp +588°F (+309°C). Soly in water at 16°: 0.36 g/l. Miscible with alc, ether, many organic solvents. LC (in air) in mice: 377 mg/l (Fühner).

Caution: Potential symptoms of overexposure are drowsiness; irritation of eyes, skin, nose. Direct contact may cause dermatitis; aspiration of liquid may cause chemical pneumonia. *See NIOSH Pocket Guide to Chemical Hazards* (DHHS/NIOSH 97-140, 1997) p 244.

7229. 1,5-Pentanediol. [111-29-5] Pentamethylene glycol; 1,5-dihydroxypentane. $C_5H_{12}O_2$; mol wt 104.15. C 57.66%, H 11.61%, O 30.72%. Prepd by hydrogenolysis of tetrahydrofurfuryl alcohol in the presence of copper chromite: Connor, Adkins, *J. Am. Chem. Soc.* **54**, 4678 (1932); D. Kaufman, W. Reeve, *Org. Synth.* **coll. vol. III**, 693 (1955). Toxicity study: H. F. Smyth *et al., Am. Ind. Hyg. Assoc. J.* **23**, 95 (1962).

HO⌃⌄⌃⌄OH

Viscous, oily liquid. Bitter taste. d^{20} 0.9941. mp −18°. bp$_{760}$ 239°, bp$_{3.0}$ 120°. n_D^{20} 1.4499. Flash pt 125°C (275°F). Miscible with water, methanol, alc, acetone, ethyl acetate. Soly in ether (25°): 11% w/w. Limited soly in benzene, trichloroethylene, methylene chloride, petr ether, heptane. LD$_{50}$ orally in rats: 5.89 g/kg (Smyth).

USE: As plasticizer in cellulose products and adhesives, in brake fluid compositions. Forms esters and polyesters which can be used as plasticizers, emulsifying agents and resin intermediates.

7230. 1-Pentanol. [71-41-0] Pentyl alcohol; *n*-amyl alcohol; *n*-butyl carbinol. $C_5H_{12}O$; mol wt 88.15. C 68.13%, H 13.72%, O 18.15%. Prepn from 1-pentene: Brown, Rao, *J. Am. Chem. Soc.* **81**, 6434 (1959); Brown, **US 2925437** (1960). Toxicity study: P. M. Jenner *et al., Food Cosmet. Toxicol.* **2**, 327 (1964). Review of manuf by fractionation of fusel oil and via chlorination of pentanes, and properties: *Industrial Chemicals*, W. L. Faith *et al.*, Eds. (John Wiley, New York, 2nd ed., 1957) pp 107-114.

HO⌃⌄⌃CH3

Liquid, mild characteristic odor. *Flammable. Irritant.* bp 137.5°. mp −79°. d_4^{22} 0.8146; d_4^{25} 0.8110. n_D^{20} 1.4103: Mumford, Phillips, *J. Chem. Soc.* **1950**, 75. Flash pt, closed cup: 100°F (38°C). Slightly sol in water (2.7 g/100 ml at 22°); misc with alc, ether. LD$_{50}$ orally in rats: 3030 mg/kg (Jenner).

USE: In organic syntheses; as solvent.

7231. 2-Pentanol. [6032-29-7] *dl-sec*-Amyl alcohol; methyl propyl carbinol; 2-hydroxypentane. $C_5H_{12}O$; mol wt 88.15. C 68.13%, H 13.72%, O 18.15%. Prepn: Brown, Nakagawa, *J. Am. Chem. Soc.* **77**, 3614 (1955); Brown, Wheeler, *ibid.* **78**, 2199 (1956). Sepn of optical isomers: Adembri, *Ann. Chim. (Rome)* **46**, 62 (1956). Review of manuf by fractionation of fusel oil and via chlorination of pentanes, and properties: *Industrial Chemicals*, W. L. Faith *et al.*, Eds. (John Wiley, New York, 1957) pp 107-114.

H3C⌃⌄⌃CH3
     |
     OH

Liquid, characteristic odor. *Flammable. Irritant.* bp 119.3°. bp$_{745}$ 118°. d_4^{20} 0.8098. n_D^{25} 1.4041, n_D^{20} 1.406. Flash pt, closed cup: 93°F (34°C). Slightly sol in water (16.6 g/100 ml at 20°). Miscible with alcohol, ether.

7232. 3-Pentanol. [584-02-1] Diethyl carbinol; 1-ethyl-1-propanol; *sec*-pentanol. $C_5H_{12}O$; mol wt 88.15. C 68.13%, H 13.72%, O 18.15%. Prepn: Brown, Nakagawa, *J. Am. Chem. Soc.* **77**, 3614 (1955). Toxicity study: H. F. Smyth *et al., Arch. Ind. Hyg. Occup. Med.* **10**, 61 (1954). Review of manuf and properties: *Industrial Chemicals*, W. L. Faith *et al.*, Eds. (John Wiley, New York, 2nd ed., 1957) pp 107-114.

         OH
          |
H3C⌃⌄⌃⌄CH3

Liquid, characteristic odor. *Flammable. Irritant.* bp 115.6°, bp$_{738}$ 113.5-113.7°. d_4^{25} 0.815; n_D^{25} 1.4077, n_D^{20} 1.4097. Flash pt, closed cup: 93°F (34°C). Slightly sol in water (5.5 g/100 g at 30°); sol in alcohol, ether. LD$_{50}$ orally in rats: 1.87 g/kg (Smyth).

USE: Flotation agent; solvent in organic synthesis.

7233. Pentazocine. [359-83-1] *rel*-(2*R*,6*R*,11*R*)-1,2,3,4,5,6-Hexahydro-6,11-dimethyl-3-(3-methyl-2-buten-1-yl)-2,6-methano-3-benzazocin-8-ol; (±)-*cis*-1,2,3,4,5,6-hexahydro-3-(3-methyl-2-butenyl)-6,11-dimethyl-8-hydroxy-2,6-methano-3-benzazocine; (±)-*cis*-2-dimethylallyl-5,9-dimethyl-2′-hydroxybenzomorphan; NSC-107430; Win-20228. $C_{19}H_{27}NO$; mol wt 285.43. C 79.95%, H 9.54%, N 4.91%, O 5.61%. Mixed opioid agonist-antagonist. Prepn: **BE 611000**; S. Archer, **US 4105659** (1962, 1978 both to Sterling Drug); *idem et al., J. Med. Chem.* **7**, 123 (1964). Prepn and activity of isomers: B. F. Tullar *et al., ibid.* **10**, 383 (1967). Enantioselective synthesis of (−)-form: B. M. Trost, W. Tang, *J. Am. Chem. Soc.* **125**, 8744 (2003). GC/MS determn in blood and urine: H. Seno *et al., J. Mass Spectrom.* **35**, 33 (2000). Review of pharmacology and clinical experience: G. Goldstein, *Drug Alcohol Depend.* **14**, 313-324 (1985).

cis-(−)-Form

White or very pale, tan-colored powder. Crystals from methanol + water, mp 145.4-148.6°. Freely sol in chloroform; sol in alcohol, acetone, ether; sparingly sol in benzene, ethyl actate. Practically insol in water. LD$_{50}$ s.c. in male rats: 175±36 mg/kg. *See:* E. I. Goldenthal, *Toxicol. Appl. Pharmacol.* **18**, 185 (1971).

Hydrochloride. [64024-15-3] Fortalgesic (tabl.); Fortral (cap.); Talwin (tabl.). $C_{19}H_{27}NO.HCl$; mol wt 321.89. Component of *Talacen*. White crystalline powder. mp 245-247° (dec). Freely sol in chloroform; sol in alcohol; sparingly sol in water; very slightly sol in acetone, ether. Practically insol in benzene.

Lactate. [17146-95-1] Fortalgesic (inj.); Fortralin; Pentagin; Sosegon; Talwin (inj.). $C_{19}H_{27}NO.C_3H_6O_3$; mol wt 375.51. White crystalline solid. Sol in acidic aq solns.

2*R*,6*R*,11*R*-Form. [7488-49-5] α-(−)-Pentazocine; *cis*-(−)-pentazocine. mp 180.6-182.2°. $[\alpha]_D^{25}$ −138.0° (c = 1.0 in CHCl$_3$).

2*S*,6*S*,11*S*-Form. [7361-76-4] α-(+)-Pentazocine; *cis*-(+)-pentazocine. mp 180.4-182.0°. $[\alpha]_D^{25}$ +135.5° (c = 1.0 in CHCl$_3$).

Note: This is a controlled substance: **21 CFR**, 1308.14.

THERAP CAT: Analgesic.

THERAP CAT (VET): Analgesic.

7234. 1-Pentene. [109-67-1] Propylethylene; α-*n*-amylene. C_5H_{10}; mol wt 70.14. C 85.62%, H 14.37%. Occurs in coal tar. Prepd from allyl bromide and ethylmagnesium bromide in ether or better in dipropyl ether: Meisenheimer, Casper, *Ber.* **54**, 1663

(1921); Norris, Joubert, *J. Am. Chem. Soc.* **49**, 885 (1927); Kirrmann, *Compt. Rend.* **184**, 1178 (1927).

H_2C ━━ CH_3

Liquid. d_4^{20} 0.6429; bp_{760} 30.1°; n_D^{20} 1.3714. Insol in water. Miscible with alcohol, ether, benzene.

7235. 2-Pentene. [109-68-2] β-*n*-Amylene; *sym*-methylethylethylene. C_5H_{10}; mol wt 70.14. C 85.62%, H 14.37%. Prepn of mixture of isomers by dehydration of 2-pentanol: Norris, *Org. Synth. coll. vol.* **I**, (2nd ed., 1941) p 430. Prepn of the geometrical isomers from *cis*- and *trans*-α-methyl-β-ethylacrylic acids: Lucas, Prater, *J. Am. Chem. Soc.* **59**, 1682 (1937). Prepn of the (*Z*)-form from 1-ethyl-2-iodobutyric acid with quinoline, of the (*E*)-form from 1-ethyl-2-iodobutyric acid with sodium carbonate: Sherrill, Matlak, *ibid.* 2134; prepn of both isomers from *sec*-amyl alcohol with H_2SO_4 and diatomaceous earth at 90-110° for 3 hrs: Lucas *et al.*, *ibid.* **63**, 22 (1941). Absorption spectra: Carr, Stücklen, *ibid.* **59**, 2138 (1937).

H_3C ━━ CH_3
(*E*)-form

(*E*)**-Form.** [646-04-8] Liquid. d_4^{20} 0.6482; d_4^{80} 0.5814; d_4^{30} 0.6381; d_4^0 0.6675. mp −136 to −135°. bp_{760} 35.85°. n_D^{20} 1.37921.
(*Z*)**-Form.** [627-20-3] Liquid. d_4^{20} 0.6503; d_4^{80} 0.5824; d_4^{30} 0.6392; d_4^0 0.6710. mp −180 to −178°. bp_{760} 37.0°. n_D^{20} 1.38130.

7236. Pentetate Calcium Trisodium. [12111-24-9] [*N*-[2-[Bis(carboxy-κ*O*)methyl]amino-κ*N*]ethyl]-*N*-[2-[[(carboxy-κ*O*)-methyl](carboxymethyl)amino-κ*N*]ethyl]glycinato(5−)-κ*N*-calciate(3−) sodium (1:3); [*N*,*N*-bis[2-[bis(carboxymethyl)amino]ethyl]glycinato(5−)]calciate(3−) trisodium; calcium trisodium pentetate; sodium[[[(carboxymethyl)imino]bis(ethylenenitrilo)]tetraacetato]calciate; [[(carboxymethyl)imino]bis(ethylenenitrilo)]tetraacetic acid calcium complex trisodium salt; trisodium calcium diethylenetriaminepentaacetate; pentacin; Calcium Chel 330; Ditripentat; Penthamil. $C_{14}H_{18}CaN_3Na_3O_{10}$; mol wt 497.36. C 33.81%, H 3.65%, Ca 8.06%, N 8.45%, Na 13.87%, O 32.17%. Prepn: Rubin, Dexter, **US 3062719** (1962 to Geigy). Effects in uranium poisoning in rats: R. Dagirmanjian *et al.*, *J. Pharmacol. Exp. Ther.* **117**, 20 (1956); in mixed fusion products poisoning: K. Kostial *et al.*, *J. Appl. Toxicol.* **3**, 291 (1983).

Solid. Sol in water. Practically insol in alcohol. Used in aq soln. LD_{50} i.p. in rats: 3.8 g/kg (Dagirmanjian).
THERAP CAT: Chelating agent (plutonium and other transuranium elements).

7237. Pentetic Acid. [67-43-6] *N*,*N*-Bis[2-[bis(carboxymethyl)amino]ethyl]glycine; [[(carboxymethyl)imino]bis(ethylenenitrilo)]tetraacetic acid; pentacarboxymethyl diethylenetriamine; diethylenetriamine pentaacetic acid; DTPA. $C_{14}H_{23}N_3O_{10}$; mol wt 393.35. C 42.75%, H 5.89%, N 10.68%, O 40.67%. Prepn of the acid: Curme *et al.*, **US 2384816** (1945 to Carbide and Carbon Chem.); Samoilova, Yashunskii, **SU 144479** (1962), *C.A.* **57**, 12326d (1962); of Na salt: **GB 601816**; **GB 601817** (both 1948 to Carbide and Carbon Chem.). Structure and infrared spectrum: Nakamoto *et al.*, *J. Am. Chem. Soc.* **85**, 309 (1963).

Trisodium calcium salt *see* Pentetate Calcium Trisodium.
USE: Chelating agent.
THERAP CAT: Chelating agent (iron).

7238. Pentetreotide. [138661-02-6] *N*-[2-[[2-[Bis(carboxymethyl)amino]ethyl](carboxymethyl)amino]ethyl]-*N*-(carboxymethyl)glycyl-D-phenylalanyl-L-cysteinyl-L-phenylalanyl-D-tryptophyl-L-lysyl-L-threonyl-*N*-[(1*R*,2*R*)-2-hydroxy-1-(hydroxymethyl)propyl]-L-cysteinamide cyclic (3 → 8)-disulfide; [DTPA-D-Phe¹]-octreotide; DTPA-SMS; SDZ-215-811. $C_{63}H_{87}N_{13}O_{19}S_2$; mol wt 1394.58. C 54.26%, H 6.29%, N 13.06%, O 21.80%, S 4.60%. Octapeptide analog of somatostatin, *q.v.* Prepn of compd and chelate, and binding characteristics: W. H. Bakker *et al.*, *Life Sci.* **49**, 1583 (1991). Synthesis, biological activity and somatostatin receptor-positive tumor imaging: R. Albert *et al.*, *Actual. Chim. Ther.* **21**, 111 (1994). Clinical metabolism and scintigraphic studies: E. P. Krenning *et al.*, *J. Nucl. Med.* **33**, 652 (1992). Diagnostic imaging: E. Bombardieri *et al.*, *Eur. J. Cancer* **31A**, 184 (1995); F. Jamar *et al.*, *J. Nucl. Med.* **36**, 542 (1995); G. Kahaly *et al.*, *ibid.* 550. Review of diagnostic use in neuroendocrine tumors: C. A. Hoefnagel, *Eur. J. Nucl. Med.* **21**, 561-581 (1994); of mechanism of action: E. P. Krenning *et al.*, *Semin. Oncol.* **21**, Suppl. 13, 6-14 (1994).

[¹¹¹]**In chelate.** [139096-04-1] Indium In 111 pentetreotide; [¹¹¹In-DTPA-D-Phe¹]-octreotide; MP-1727; OctreoScan. $C_{63}H_{84}{}^{111}InN_{13}O_{19}S_2$.
THERAP CAT: ¹¹¹In chelate as diagnostic aid (radioactive imaging agent).

7239. Penthiopyrad. [183675-82-3] *N*-[2-(1,3-Dimethylbutyl)-3-thienyl]-1-methyl-3-(trifluoromethyl)-1*H*-pyrazole-4-carboxamide; MTF-752; Aphet; Fontelis; Vertisan. $C_{16}H_{20}F_3N_3OS$; mol wt 359.41. C 53.47%, H 5.61%, F 15.86%, N 11.69%, O 4.45%, S 8.92%. Succinate dehydrogenase inhibitor; disrupts mitochondrial electron transport. Prepn: Y. Yoshikawa *et al.*, **EP 0737682**; *eidem*, **US 5747518** (1996, 1998 both to Mitsui Toatsu). Properties and activity: K. Tomiya, Y. Yanase, *BCPC Int. Congr. - Crop Sci. Technol.* **2003**, 99. Field trial in peanut crops: A. K. Culbreath *et al.*, *Pest Manag. Sci.* **65**, 66 (2009).

White powder, mp 103-105°. Vapor pressure (25°): 6.43×10^{-6} Pa. Soly in water at 20°: 7.53 mg/L. LD_{50} in rats (mg/kg): >2000 orally; >2000 dermally; LC_{50} (4 hr) in rats (mg/kg): >5669 by inhalation. LC_{50} in carp (96 hr): 1.17 ppm; in *Daphnia* (24 hr): 40 ppm (Tomiya).
USE: Agricultural fungicide.

7240. Pentifylline. [1028-33-7] 1-Hexyl-3,7-dihydro-3,7-dimethyl-1*H*-purine-2,6-dione; 1-hexyltheobromine; 1-hexyl-3,7-dimethylxanthine; 3,7-dimethyl-1-hexyl-1*H*,3*H*-purin-2,6-dione; SK-7; Cosaldon. $C_{13}H_{20}N_4O_2$; mol wt 264.33. C 59.07%, H 7.63%, N 21.20%, O 12.11%. Prepn from theobromine and hexyl halide: Eidebenz, Schuh, **DE 860217** (1952 to Chem. Werke Albert); Serchi,

Chimica **40**, 451 (1964); Chkhikvadze *et al.*, **SU 202152** (1967), *C.A.* **69**, 19368x (1968). Pharmacology: Cugurra, Echinard-Garin, *Arch. Int. Pharmacodyn. Ther.* **123**, 481 (1960); Ramos *et al.*, *ibid.* **153**, 430 (1965); Mohler *et al.*, *Arzneim.-Forsch.* **16**, 1524 (1966). Metabolic studies: Mohler *et al.*, *Arch. Pharm.* **299**, 448 (1966).

Crystals, mp 82-83°.

USE: Stabilizer of vitamin preparations: Nook, Eidebenz, **DE 1810705** (1970 to Chem. Werke Albert).

THERAP CAT: Vasodilator.

7241. Pentigetide. [62087-72-3] L-α-Aspartyl-L-seryl-L-α-aspartyl-L-prolyl-L-arginine; human IgE pentapeptide; HEPP; Pentyde. $C_{22}H_{36}N_8O_{11}$; mol wt 588.58. C 44.89%, H 6.17%, N 19.04%, O 29.90%. Synthetic pentapeptide with antiallergic activity. Structure corresponds to amino acids 320-324 of the epsilon chain of human immunoglobulin E (IgE). Prepn: R. N. Hamburger, **DE 2602443**; *idem*, **US 4171299** (1976, 1979 both to Univ. California). Inhibition of IgE-mediated immediate hypersensitivity response: *idem*, *Science* **189**, 389 (1975). Stimulation of peritoneal macrophages: E. Tzehoval *et al.*, *Proc. Natl. Acad. Sci. USA* **75**, 3400 (1978). Mechanism of action study: R. N. Hamburger, *Immunology* **38**, 781 (1979). Clinical evaluations in allergic rhinitis: G. A. Cohen *et al.*, *Ann. Allergy* **52**, 83 (1984); B. M. Prenner, *ibid.* **58**, 332 (1987).

Asp–Ser–Asp–Pro–Arg

$[\alpha]_D^{20}$ −78.6° (c = 1 in water).

THERAP CAT: Antiallergic.

7242. Pentisomide. [78833-03-1] α-[2-[Bis(1-methylethyl)-amino]ethyl]-α-(2-methylpropyl)-2-pyridineacetamide; 2-[2-(diisopropylamino)ethyl]-4-methyl-2-(2-pyridyl)pentanamide; propisomide; penticainide; CM-7857; ME-3202. $C_{19}H_{33}N_3O$; mol wt 319.49. C 71.43%, H 10.41%, N 13.15%, O 5.01%. Sodium channel blocker; derivative of disopyramide, *q.v.* Prepn: H. Demarne *et al.*, **EP 27412**; *eidem*, **US 4356177** (1981, 1982 both to C. M. Industries); C. A. Bernhart *et al.*, *J. Med. Chem.* **26**, 451 (1983). Electrophysiological study: V. Kühlkamp *et al.*, *Int. J. Cardiol.* **36**, 69 (1992). HPLC determn in plasma, urine, and tissues: T. A. Plomp, M. J. H. Buijs, *J. Chromatogr.* **612**, 123 (1993). Clinical evaluation: S. G. Priori *et al.*, *Am. J. Cardiol.* **60**, 1068 (1987); S. G. Mangini *et al.*, *Eur. Heart J.* **12**, 712 (1991). *Review:* S. Yuan, S. B. Olsson, *Cardiovasc. Drug Rev.* **11**, 74-93 (1993).

White crystals from diisopropyl ether, mp 108-109°. Sol in water.

THERAP CAT: Antiarrhythmic (class I).

7243. Pentobarbital. [76-74-4] 5-Ethyl-5-(1-methylbutyl)-2,4,6(1H,3H,5H)-pyrimidinetrione; 5-ethyl-5-(1-methylbutyl)barbituric acid; mebumal; pentobarbitone; Neodorm. $C_{11}H_{18}N_2O_3$; mol wt 226.28. C 58.39%, H 8.02%, N 12.38%, O 21.21%. Prepn: **DE 293163** (1916 to Bayer), *Frdl.* **13**, 799 (1923); and hypnotic activity: E. H. Volwiler, D. L. Tabern, *J. Am. Chem. Soc.* **52**, 1676 (1930). Toxicity data: E. W. Schafer, *Toxicol. Appl. Pharmacol.* **21**, 315 (1972). HPLC determn in plasma: M. J. Avram, T. C. Krejcie, *J. Chromatogr.* **414**, 484 (1986). Use in clinical management of elevated intracranial pressure: P. S. Woster, K. L. LeBlanc,

Clin. Pharm. **9**, 762 (1990); for euthanasia: A. T. Evans *et al.*, *J. Am. Vet. Med. Assoc.* **203**, 664 (1993).

White crystals from alc. Exhibits polymorphism. mp 129-130°. Very sol in alc, methanol, ether, chloroform, acetone; sol in benzene; very slightly sol in water, carbon tetrachloride.

Sodium salt. [57-33-0] Pentobarbital sodium; soluble pentobarbital; Carbrital; Narcoren; Nembutal; Pentone; Praecicalm; Sopental. $C_{11}H_{17}N_2NaO_3$; mol wt 248.26. White powder; slightly bitter taste. Dec ~127°. Very sol in water; freely sol in alc. Practically insol in benzene, ether. Aq solns are unstable. LD_{50} orally in rats: 118 mg/kg (Schafer).

Calcium salt. [7563-42-0] Pentobarbital calcium; Repocal. $C_{22}H_{34}CaN_4O_6$; mol wt 490.61.

Note: This is a controlled substance (depressant): **21 CFR,** 1308.12 and 1308.13.

THERAP CAT: Sedative, hypnotic.

THERAP CAT (VET): Anesthetic (intravenous); for euthanasia.

7244. 1-Pentol. [105-29-3] 3-Methyl-2-penten-4-yn-1-ol; 1-hydroxy-3-methyl-2-penten-4-yne. C_6H_8O; mol wt 96.13. C 74.97%, H 8.39%, O 16.64%. Prepd by allylic rearrangement of methylvinylethynyl carbinol with moderately strong acid at 80°: Oroshnik, *J. Am. Chem. Soc.* **78**, 2651 (1956).

trans - Isomer

cis-Form. 1′-Pentol. Oily liquid. $bp_{9.4}$ 65°. n_D^{20} 1.4820. uv max (ethanol): 223 nm (ε 11,000).

cis-Form p-nitrobenzoate. mp 61-62°.

trans-Form. 1″-Pentol. Oily liquid. $bp_{9.4}$ 73°. n_D^{20} 1.4934. uv max (ethanol): 224 nm (ε 13,100).

trans-Form p-nitrobenzoate. mp 63-64°.

Caution: Both isomers tend to polymerize and will explode when heated above 120° in a sealed bomb tube.

USE: Intermediate in vitamin A synthesis.

7245. Pentolinium Tartrate. [52-62-0] 1,1′-(1,5-Pentanediyl)bis[1-methylpyrrolidinium] (2R,3R)-2,3-dihydroxybutanedioate (1:2); pentamethylene-1,5-bis(1-methylpyrrolidinium) hydrogen tartrate; pentapyrrolidinium bitartrate; pentolonium bitartrate; M & B 2050A; Ansolysen Tartrate; Ansolysen Bitartrate; Pentilium. $C_{23}H_{42}N_2O_{12}$; mol wt 538.59. C 51.29%, H 7.86%, N 5.20%, O 35.65%. General procedure of prepn: Libman *et al.*, *J. Chem. Soc.* **1952**, 2305. Ganglionic blocking agent.

Crystals, dec 203°. Non-hygroscopic. Acid taste. Freely sol in water or in 25% aq polyvinylpyrrolidone soln. One gram dissolves in 0.4 ml water, in 810 ml ethanol. Insol in ether, chloroform. A 10% aq soln has a pH of about 3.5.

THERAP CAT: Antihypertensive.

7246. Pentorex. [434-43-5] α,α,β-Trimethylbenzeneethanamine; α,α,β-trimethylphenethylamine; 1,1-dimethyl-2-phenylpropylamine; DL-2-phenyl-3-methyl-3-butylamine; phenpentermine. $C_{11}H_{17}N$; mol wt 163.26. C 80.93%, H 10.50%, N 8.58%. Prepn:

FR M2594 (1964 to Nordmark), *C.A.* **61**, 16013f (1964). GC determn in plasma and urine: F. T. Delbeke, M. Debackere, *J. Chromatogr.* **273**, 141 (1983).

bp$_{20}$ 109-111°.

Hydrochloride. [5585-52-4] $C_{11}H_{17}N.HCl$. Crystals, mp 164-166°.

Hydrogen D-tartrate. [22232-55-9] Liprodène; Modatrop. C_{11}-$H_{17}N.C_4H_6O_6$; mol wt 313.35. Crystals, mp 160-162°. $[\alpha]_D^{20}$ +13.4° (c = 0.8 in water).

D-Form hydrogen D-tartrate. Crystals, mp 167-169°. $[\alpha]_D^{20}$ +17.9°.

L-Form hydrogen D-tartrate. Crystals, mp 164-166°. $[\alpha]_D^{20}$ +3.45°.

THERAP CAT: Anorexic.

7247. Pentosan Polysulfate. [37300-21-3] Xylan hydrogen sulfate; xylan polysulfate; CB-8061. Semi-synthetic sulfated polyanion composed of β-D-xylopyranose residues with properties similar to heparin, *q.v.* Mol wt ranges from 1500 to 5000. Prepn: **CH 293566** (1953 to Wander), *C.A.* **49**, 1787h (1955). Anticoagulant activity: B. Paramelle, *Therapie* **17**, 719 (1962); J.-B., Dureux *et al.*, *ibid.* **19**, 879 (1964). Pharmacokinetics: R. Taugner *et al.*, *Arch. Int. Pharmacodyn.* **189**, 250 (1971). Effects on aggregation of human blood platelets: G. Kindness *et al.*, *Thromb. Res.* **16**, 97 (1979). Comparison of pentosan polysulfate with heparin: C. Soria *et al.*, *ibid.* **19**, 455 (1980); M. Ryde *et al.*, *ibid.* **23**, 435 (1981); A.-M. Fischer *et al.*, *Thromb. Haemostasis* **47**, 104, 109 (1982). Mechanism of antiadherence activity in urinary bladder: C. L. Parsons *et al.*, *Science* **208**, 605 (1980); *eidem*, *Infect. Immun.* **27**, 876 (1980). Clinical trials in interstitial cystitis: A. Fritjofsson *et al.*, *J. Urol.* **138**, 508 (1987); S. G. Mulholland *et al.*, *Urology* **6**, 552 (1990).

Sodium salt. [37319-17-8] Sodium pentosan polysulfate; sodium xylan polysulfate; Cartrophen; Elmiron; Fibrase; Fibrezym; Hémoclar; SP-54; Thrombocid. White odorless powder, slightly hygroscopic. $[\alpha]_D^{20}$ −57°. pH of 10% aq soln ~6.0. n_D^{20} of 10% aq soln: 1.344. Sol in water.

THERAP CAT: In treatment of interstitial cystitis; antithrombotic.

THERAP CAT (VET): Anti-inflammatory.

7248. Pentostatin. [53910-25-1] (8R)-3-(2-Deoxy-β-D-*erythro*-pentofuranosyl)-3,4,7,8-tetrahydroimidazo[4,5-*d*][1,3]-diazepin-8-ol; 2'-deoxycoformycin; DCF; 2'-dCF; CL-67310465; NSC-218321; CI-825; Nipent. $C_{11}H_{16}N_4O_4$; mol wt 268.27. C 49.25%, H 6.01%, N 20.88%, O 23.86%. Adenosine deaminase inhibitor. Isoln from *Streptomyces antibioticus* and structure determn: P. W. K. Woo *et al.*, *J. Heterocycl. Chem.* **11**, 641 (1974). Isoln and purification: A. Ryder *et al.*, **DE 2517596**; *eidem*, **US 3923785** (both 1975 to Parke-Davis). Total synthesis: E. Chan *et al.*, *J. Org. Chem.* **47**, 3457 (1982). Preliminary biosynthetic study: J. C. Hanvey *et al.*, *Biochemistry* **27**, 5790 (1988). Inhibition of deaminases *in vitro*: T. Rogler-Brown *et al.*, *Biochem. Pharmacol.* **27**, 2289 (1978); C. Frieden *et al.*, *Biochem. Biophys. Res. Commun.* **91**, 278 (1979); *in vivo*: W. Plunkett *et al.*, *Biochem. Pharmacol.* **28**, 201 (1979). Toxicology study: J. F. Smyth *et al.*, *Cancer Chemother. Pharmacol.* **1**, 49 (1978). Clinical pharmacology: J. F. Smyth *et al.*, *ibid.* **5**, 93 (1980); P. P. Major *et al.*, *Blood* **58**, 91 (1981); F. J. Cummings *et al.*, *Clin. Pharmacol. Ther.* **44**, 501 (1988). Enzymatic determn in biological fluids: M. M. Chassin *et al.*, *Biochem. Pharmacol.* **28**, 1849 (1979); A. E. Staubus *et al.*, *ibid.* **33**, 1633 (1984). Clinical evaluation in hairy cell leukemia: J. B.

Johnston *et al.*, *J. Natl. Cancer Inst.* **80**, 765 (1988). Brief review: P. O'Dwyer *et al.*, *Ann. Intern. Med.* **108**, 733 (1988). Review of clinical pharmacology and immunosuppressive effects: M. Higman *et al.*, *Expert Opin. Pharmacother.* **5**, 2605-2613 (2004).

White crystals from methanol/water, mp 220-225° (Woo), also reported as 204-209.5° with darkening at >150° (Chan). uv max (water, pH 7): 282 nm (ε 8000); (pH 11): 283 nm (ε 7970); (pH 2): 273 nm (ε 7570 initially, 3143 after 6.5 hrs). $[\alpha]_D^{25}$ +76.4° (c = 1 in water); $[\alpha]_D^{23}$ +73.0° (c = 1, pH 7 buffer). pKa 5.2 in water. Freely sol in distilled water.

THERAP CAT: Antineoplastic.

7249. Pentoxifylline. [6493-05-6] 3,7-Dihydro-3,7-dimethyl-1-(5-oxohexyl)-1H-purine-2,6-dione; 1-(5-oxohexyl)theobromine; 1-(5-oxohexyl)-3,7-dimethylxanthine; 3,7-dimethyl-1-(5-oxohexyl)-1H,3H-purin-2,6-dione; oxpentifylline; vazofirin; BL-191; Durapental; Rentylin; Torental; Trental. $C_{13}H_{18}N_4O_3$; mol wt 278.31. C 56.10%, H 6.52%, N 20.13%, O 17.25%. Methylxanthine derivative that improves blood flow by decreasing blood viscosity. Identity as a metabolite of pentifylline, *q.v.*: Mohler *et al.*, *Arch. Pharm.* **299**, 448 (1966); Mohler *et al.*, *Arzneim.-Forsch.* **16**, 1524 (1966). Prepn: **NL 6511581**; Mohler *et al.*, **DE 1235320**; *eidem*, **US 3422107** (1966, 1967, 1969 all to Chem. Werke Albert). Series of articles on chemistry, pharmacology and clinical trials: *Arzneim.-Forsch.* **21**, 1159-1177 (1971). Toxicity data: K. Popendiker *et al.*, *ibid.* 1160. Review of hemorheologic properties and use in cerebrovascular disease: R. Müller, F. Lehrach, *Curr. Med. Res. Opin.* **7**, 253-263 (1981); of pharmacology and therapeutic uses: A. Ward, S. P. Clissold, *Drugs* **34**, 50-97 (1987); C. P. Samlaska, E. A. Winfield, *J. Am. Acad. Dermatol.* **30**, 603-621 (1994).

White or colorless needles from methanol. Bitter tasting. mp 105°. uv max: 273, 208 nm ($E_{1cm}^{1\%}$ 365, 935). Soly in water: 77 mg/ml at 25°, 191 mg/ml at 37°; in benzene: 11 g/100 ml. Freely sol in chloroform, methanol; sparingly sol in alc; slightly sol in ether. LD_{50} orally in mice: 1385 mg/kg (Popendiker).

THERAP CAT: Hemorheologic agent.

THERAP CAT (VET): In treatment of canine immune-mediated dermatologic conditions.

7250. Pentoxyl. [147-61-5] 5-(Hydroxymethyl)-6-methyl-2,-4(1H,3H)-pyrimidinedione; 5-hydroxymethyl-6-methyluracil; 5-hydroxymethyl-4-methyluracil; 4-methyl-5-hydroxymethyluracil. C_6-$H_8N_2O_3$; mol wt 156.14. C 46.15%, H 5.16%, N 17.94%, O 30.74%. Prepn: Endicott, Johnson, *J. Am. Chem. Soc.* **63**, 2063 (1941).

Crystals from boiling water, dec 314-315°.

THERAP CAT: Leukopoietic stimulant.

7251. Pentrinitrol. [1607-17-6] 2,2-Bis[(nitrooxy)methyl]-1,3-propanediol 1-nitrate; pentaerythritol trinitrate; W-2197; Petrin. $C_5H_9N_3O_{10}$; mol wt 271.14. C 22.15%, H 3.35%, N 15.50%, O 59.01%. Coronary vasodilator related to pentaerythritol tetranitrate, *q.v.* Prepn (no data given): DE 638422; DE 638423 (both 1936 to Westfälisch-Anhaltisch Sprengstoff), *C.A.* **31**, 1212²(1937); N. J. Marans *et al., J. Am. Chem. Soc.* **76**, 1304 (1954); A. T. Camp *et al., ibid.* **77**, 751 (1955); J. Simecek, *Collect. Czech. Chem. Commun.* **27**, 362 (1962). Use in coronary insufficiency: F. J. DiCarlo, US 3419571 (1968 to Warner-Lambert). Electrical moment and molecular rotation: A. R. Lawrence, A. J. Matuszko, *J. Phys. Chem.* **65**, 1903 (1961). Chromatographic determn: I. W. F. Davidson *et al., J. Chromatogr.* **57**, 345 (1971). Metabolism: F. J. DiCarlo *et al., Clin. Pharmacol. Ther.* **22**, 309 (1977). Comparative duration of action: J. Vohra *et al., Aust. N.Z. J. Med.* **9**, 554 (1979).

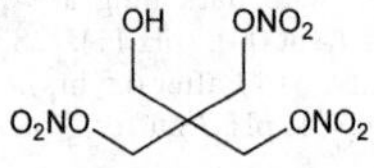

Viscous liq. n_D^{20} 1.4941. d_4^{20} 1.554. Viscosity (cSt): 166.8 (40°); 77.8 (50°). Soly in water: 0.705 g/100 ml at 20°; in benzene: 21.40 g/100 ml at 20°. Very sol in ethanol, ether. Forms a cryst hydrate, mp 32°, when washed with water and allowed to stand overnight at 20°; returns to unhydrated form when allowed to stand at 60° for 2 hrs. When used medicinally it is usually diluted with lactose or mannitol to reduce explosive liability, *see* pentaerythritol tetranitrate.

THERAP CAT: Vasodilator (coronary).

7252. Pentryl. [4481-55-4] 2-[Nitro(2,4,6-trinitrophenyl)-amino]ethanol 1-nitrate; 2-(N,2,4,6-tetranitroanilino)ethanol nitrate; *sym*-trinitrophenylnitroaminoethyl nitrate; $C_8H_6N_6O_{11}$; mol wt 362.17. C 26.53%, H 1.67%, N 23.21%, O 48.59%. Prepd by nitration of 2,4-dinitrophenylaminoethanol: Clark, *Ind. Eng. Chem.* **25**, 1385 (1933); Desseigne, *Mem. Poudres* **33**, 255 (1951); B. T. Fedoroff *et al., Encyclopedia of Explosives and Related Items* vol. I (Picatinny Arsenal, Dover, New Jersey, 1960) pp A425-429.

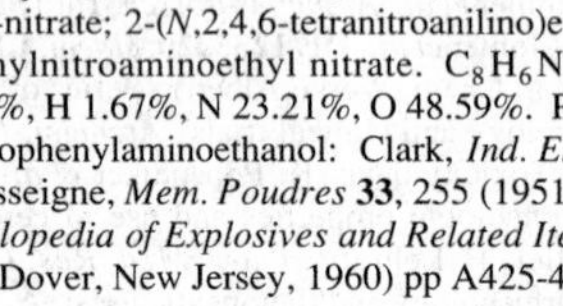

Small, cream-colored crystals from chloroform, mp 129° (slight decompn). d 1.82. Explodes when heated to 235°. Heat of combustion: 911.1 kcal/mole; heat of explosion: 372.4 kcal/mole; heat of formation: 43.4 kcal/mole. Soly (w/w) at 25° in benzene: 0.70; carbon tetrachloride: trace; chloroform: 0.07; ethanol: 0.11; ether: 0.16; ethylene dichloride: 0.72; methanol: 0.67; nitroglycerol: freely sol; toluene: 0.63; water: trace.

USE: High explosive. Base charge in detonators.

7253. *tert*-Pentyl Alcohol. [75-85-4] 2-Methyl-2-butanol; *tert*-amyl alcohol; dimethyl ethyl carbinol; ethyl dimethyl carbinol; *tert*-pentanol; amylene hydrate. $C_5H_{12}O$; mol wt 88.15. C 68.13%, H 13.72%, O 18.15%. Prepd from 2-methyl-2-butene alone and mixed with 2-methyl-1-butene: Fenske, Jones, US 2858331 (1958 to Esso); Odioso *et al., Ind. Eng. Chem.* **53**, 209 (1961). Review of manuf by fractionation of fusel oil and *via* chlorination of pentanes, and properties: *Industrial Chemicals*, W. L. Faith *et al.*, Eds. (John Wiley, New York, 2nd ed., 1957) pp 107-114. Toxicity: Schaffarzick, Brown, *Science* **116**, 663 (1952).

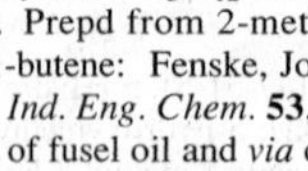

Volatile liquid; characteristic odor, burning taste. bp_{765} 102.5°. mp −9.0°. d^{20} 0.8084: Costello, Bowden, *Rec. Trav. Chim.* **77**, 36 (1958). n_D^{20} 1.4052. Flash pt, closed cup: 67°F (19°C); open cup:

70°F (21°C). Sol in 8 parts water; miscible with alcohol, ether, benzene, chloroform, glycerol, oils. *Keep tightly closed and protected from light.* LD_{50} orally in rats: 1.0 g/kg (Schaffarzick, Brown).

Caution: Moderately irritating to mucous membranes. Narcotic in high concns. *See also* 1-Pentanol.

THERAP CAT: Hypnotic.

7254. Pentylenetetrazole. [54-95-5] 6,7,8,9-Tetrahydro-5*H*-tetrazolo[1,5-*a*]azepine; α,β-cyclopentamethylenetetrazole; 1,5-pentamethylenetetrazole; 6,7,8,9-tetrahydro-5-azepotetrazole; 1,2,-3,3a-tetrazacyclohepta-8a,2-cyclopentadiene; 7,8,9,10-tetrazabicyclo[5.3.0]-8,10-decadiene; pentetrazol; Cardiazol; Cenalene-M; Cenazol; Coranormol; Corazole; Corvasol; Deumacard; Gewazol; Korazol; Metrazol; Phrenazol; Ventrazol. $C_6H_{10}N_4$; mol wt 138.17. C 52.16%, H 7.30%, N 40.55%. Prepd by the addition of cyclohexanone to a benzene or Tetralin soln of hydrazoic acid: Schmidt, *Ber.* **57**, 704 (1924); US 1564631 (1925); US 1599493 (1926); Prochazka, CS 92456 (1959), *C.A.* **56**, 4776e (1962). Synthesis from caprolactam: Glushkov, Golovchinskaya, *Zh. Prikl. Khim.* **32**, 920 (1959); *Med. Prom. SSSR* **14**, no. 1, 12 (1960). Toxicity study: E. I. Goldenthal, *Toxicol. Appl. Pharmacol.* **18**, 185 (1971).

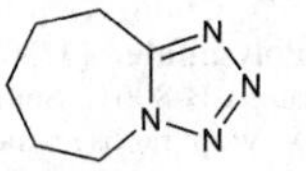

Slightly pungent, bitter crystals, mp 57-60°. Freely sol in water, most organic solvents. Aq solns are neutral to litmus. Very stable, not easily attacked by other substances. LD_{50} in rats (mg/kg): 85±2 s.c., 62 i.p. (Goldenthal).

THERAP CAT: CNS stimulant.

THERAP CAT (VET): CNS stimulant; narcotic antagonist.

7255. *p-tert*-Pentylphenol. [80-46-6] 4-(1,1-Dimethylpropyl)phenol; *p-tert*-amylphenol; 2-methyl-2-*p*-hydroxyphenylbutane; Pentaphen. $C_{11}H_{16}O$; mol wt 164.25. C 80.44%, H 9.82%, O 9.74%. Prepd by condensation of *tert*-pentanol or 2-methyl-3-butanol with phenol in the presence of aluminum chloride: Huston, Hsieh, *J. Am. Chem. Soc.* **58**, 439 (1936); Huston *et al., ibid.* **67**, 899 (1945). Physical properties: Pardee, Weinrich, *Ind. Eng. Chem.* **36**, 595 (1944). Toxicity study: H. F. Smyth *et al., Am. Ind. Hyg. Assoc. J.* **23**, 95 (1962).

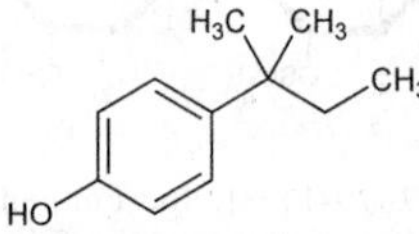

Crystals, mp 94-95°. bp 262.5°. bp_{740} 248-250°, bp_{15} 138.5°, bp_3 112-120°. d_4^{20} 0.962. Practically insol in water. Sol in alcohol, ether, benzene, chloroform. LD_{50} orally in rats: 3.08 g/kg (Smyth).

USE: In the manuf of oil-soluble resins; has been recommended as a germicide and fumigant; intermediate for organic mercury germicides, for pesticides, for chemicals used in rubber and petroleum industries.

7256. Peonidin. [134-01-0] 3,5,7-Trihydroxy-2-(4-hydroxy-3-methoxyphenyl)-1-benzopyrylium chloride (1:1); 3,4′,5,7-tetrahydroxy-3′-methoxyflavylium chloride; 3,4′,5,7-tetrahydroxy-3′-methoxy-2-phenylbenzopyrylium chloride. $C_{16}H_{13}ClO_6$; mol wt 336.72. C 57.07%, H 3.89%, Cl 10.53%, O 28.51%. The aglucon of peonin: R. Willstätter, T. J. Nolan, *Ann.* **408**, 136 (1915). Structure and synthesis: T. J. Nolan *et al., J. Chem. Soc.* **1926**, 1968; S. Murakami, R. Robinson, *ibid.* **1928**, 1537.

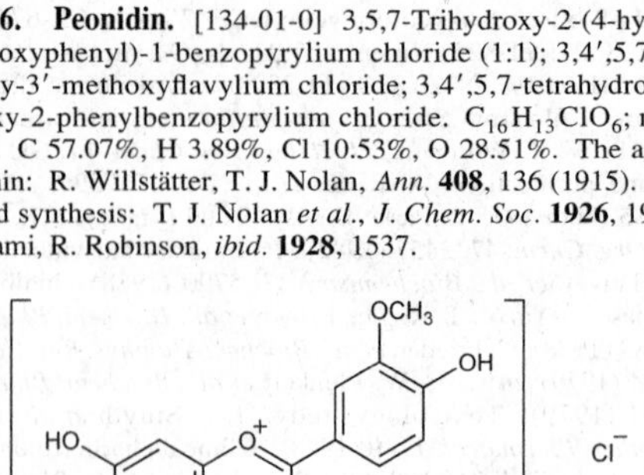

Monohydrate. Reddish-brown needles from aq 20% hydrochloric acid soln. Moderately sol in water giving a reddish-brown soln. Sol in alc giving a purplish-red soln.

3,5-Diglucoside. 3,5-Bis(β-D-glucopyranosyloxy)-7-hydroxy-2-(4-hydroxy-3-methoxyphenyl)-1-benzopyrylium chloride; peonin. $C_{28}H_{33}ClO_{16}$. From deep violet-red peonies: R. Willstätter, T. J. Nolan, *loc. cit.* Synthesis: R. Robinson, A. R. Todd, *J. Chem. Soc.* **1932**, 2488. Deep purple needles with water of crystn from dil HCl, mp 165-167° (dec).

7257. Peplomycin. [68247-85-8] N^1-[3-[[(1S)-1-Phenylethyl]amino]propyl]bleomycinamide; N^1-[[(S)-α-methylbenzyl]amino]propyl]bleomycinamide; pepleomycin; NK-631. $C_{61}H_{88}N_{18}$- $O_{21}S_2$; mol wt 1473.60. C 49.72%, H 6.02%, N 17.11%, O 22.80%, S 4.35%. Deriv of bleomycin, *q.v.* with cytostatic activity and less pulmonary toxicity than the natural bleomycin mixture. Prepn of the (R,S)-form: H. Umezawa *et al.*, *US 3846400* (1974 to Microbiochem. Res. Found.); of the (S)-form: T. Takita *et al.*, *DE 2828933*; *eidem*, *US 4195018* (1979, 1980 both to Nippon Kayaku); W. Tanaka *et al.*, *Heterocycles* **13**, 469 (1979). Biological study of degradation products: K. Takahashi *et al.*, *J. Antibiot.* **32**, 36 (1979). General pharmacology: Y. Ishii *et al.*, *Jpn. J. Antibiot.* **31**, 886 (1978), *C.A.* **91**, 215c (1979). Absorption, distribution, excretion, metabolism: H. Takayama *et al.*, *ibid.* 895, *C.A.* **91**, 32580j (1979). Properties and stability: A. Fuji *et al.*, *Iyakuhin Kenkyu* **10**, 197 (1979), *C.A.* **91**, 9408a (1979). Effect in prostatic cancer: T. Niijima, K. Koiso, *Scand. J. Respir. Dis.* **Suppl. 55**, 177 (1980). Relative pulmonary toxicity: B. I. Sikic *et al.*, *Cancer Treat. Rep.* **64**, 659 (1980). Acute toxicity study: K. Ito *et al.*, *Jpn. J. Antibiot.* **31**, 719 (1978), *C.A.* **91**, 68461k (1979). Review of clinical studies: S. Oko, *Recent Results Cancer Res.* **74**, 163 (1980).

Sulfate salt. [70384-29-1] Pepleo Injection. $C_{61}H_{88}N_{18}O_{21}S_2$.- H_2SO_4; mol wt 1571.67. Pale yellow amorphous powder, mp 196-198°. $[\alpha]_{436}^{25}$ −2.0° (c = 1 in water). pKa 2.9, 4.8, 7.4, 9.0. Sol in water, methanol, acetic acid, DMSO, DMF. Slightly sol in dioxane. Insol in ethyl acetate, acetone, ether, benzene. Stable at 37° for 3 months, 50° for 6 weeks, room temp for 30 months in a sealed container. LD$_{50}$ in male rats, mice (mg/kg): 234, 88 s.c.; 208, 85 i.p.; 245, 51 i.v. (Ito).

THERAP CAT: Antineoplastic.

7258. Peppermint. Brandy mint; lamb mint. Perennial herb, *Mentha piperita* L., *Labiatae*. *Habit.* Asia, Europe, North America; cultivated in gardens. Medicinal parts are the essential oil, dried leaves and flowering tops, or the fresh flowering tops. *Constit.* Volatile oil (1-3%), flavonoids incl. luteolin, rutin, hesperidin; phenolic acids incl. caffeic, chlorogenic and rosmarinic acids; triterpenes. GC determn of menthol components in oil: J. P. Sang, *J. Chromatogr.* **253**, 109 (1982); in pharmaceutical preparations: D. Y.-H. Yeung *et al.*, *J. Pharm. Biomed. Anal.* **30**, 1469 (2003). HPLC determn of phenolic compounds in leaves: F. M. Areias *et al.*, *Food Chem.* **73**, 307 (2001). Review of constituents and medicinal uses: P. R. Bradley, *British Herbal Compendium* (British Herbal Medicines Association, Dorset, 1992) pp 174-176. Series of articles on pharmacology and clinical experience in irritable bowel

syndrome: H.-G. Grigoleit, P. Grigoleit, *Phytomedicine* **12**, 601-616 (2005).

Volatile oil. [8006-90-4] Oil of peppermint; Colpermin; Mintec. Steam-distilled from fresh leaves and flowering tops. *Constit.* Chiefly menthol (35-55%), menthone (10-40%), menthyl acetate (1-10%), menthofuran (1-10%), cineol (2-13%), limonene, (+)-isomenthone, viridoflorol, dimethyl sulfide. Colorless to pale yellow liquid; strong, penetrating peppermint odor and pungent taste. d$_{25}^{25}$ 0.896-0.908. n_D^{20} 1.459-1.465. Rotation between −18 and −32. Very slightly sol in water; sol in 3 vols 70% alcohol. Insol in propylene glycol.

Spirit of Peppermint. An alcoholic soln contg per liter 100 ml oil of peppermint and the alcohol-soluble principles from 10 g of powdered peppermint previously macerated with water. Colorless liquid. Completely sol in water; easily sol in methanol, diethyl ether.

USE: Flavoring agent in foods, liqueurs, confectionary. Pharmaceutic aid (flavor). In hair tonics, stimulating shampoos, conditioners, perfumes. Leaves in herbal teas.

THERAP CAT: Carminative; antispasmodic.

7259. Pepsin. [9001-75-6] Pepsin A; E.C. 3.4.23.1; Puerzym. Principle digestive enzyme of gastric juice; controls the degradation of proteins to proteoses and peptones. Distinctive among enzymes for having a very low isoelectric point and a very low pH optimum. Hydrolyzes only peptide linkages. The N.F. grade digests not less than 3000 or more than 3500 times its wt of freshly coagulated and disintegrated egg albumin in 2½ hours at 52° in water acidulated with HCl. Isoln from gastric juice of swine and beef: Northrup, *J. Gen. Physiol.* **13**, 739 (1930); **16**, 615 (1933); from salmon and tuna: Norris, Edam, *J. Biol. Chem.* **134**, 443 (1940); **204**, 673 (1953). Amino acid composition: Brand in *Crystalline Enzymes*, J. H. Northrup *et al.* (Columbia Univ. Press, New York, 2nd ed., 1948) p 26. Active site: Herriott, *J. Gen. Physiol.* **45**, Suppl., 57 (1962). *Review:* Bovey, Yanari in *The Enzymes* vol. **4**, P. D. Boyer *et al.*, Eds. (Academic Press, New York, 1960) pp 63-92.

White or yellowish-white translucent scales or granules, or an amorphous slightly hygroscopic powder, or spongy masses. It has a slightly acid or saline taste. $[\alpha]_D^{26}$ −64.5° (water pH 4.6). Isoelectric point <pH 1.0. Freely sol in water with more or less opalescence. Practically insol in alc, chloroform, ether. Very stable to acid. The activity of pepsin in solns is destroyed by heating above 70°, or by alkalies; dry pepsin is not injured by heating to 100°. *Incompat.* Alc, tannin, alkaline substances and salts of heavy metals. *Note:* Pepsin of an activity of 4000, 5000, 6000, 10,000, 15,000, and 20,000 is likewise available, and it has also been obtained in pure crystalline form.

THERAP CAT: Enzyme (digestive).

THERAP CAT (VET): Has been used as a digestive aid in deficiency of pepsin secretion.

7260. Pepstatin. [26305-03-3] N-(3-Methyl-1-oxobutyl)-L-valyl-L-valyl-(3S,4S)-4-amino-3-hydroxy-6-methylheptanoyl-N-[(1S)-1-[(1S)-2-carboxy-1-hydroxyethyl]-3-methylbutyl]-L-alaninamide; pepstatin A; N-isovaleryl-L-valyl-L-valyl-3-hydroxy-6-methyl-γ-aminoheptanoyl-L-alanyl-3-hydroxy-6-methyl-γ-aminoheptanoic acid. $C_{34}H_{63}N_5O_9$; mol wt 685.90. C 59.54%, H 9.26%, N 10.21%, O 20.99%. A pentapeptide pepsin inhibitor, isolated from cultured broths of *Streptomyces testaceus* Hamada *et* Okami and *Streptomyces argenteolus* var. *toyonakensis*: H. Umezawa *et al.*, *J. Antibiot.* **23**, 259 (1970). Prepn: H. Umezawa *et al.*, *DE 2028403* (1971 to Microbiochemical Res. Found.), *C.A.* **74**, 75201c (1971). Structure: H. Morishima *et al.*, *J. Antibiot.* **23**, 263 (1970). Synthesis: H. Morishima *et al.*, *ibid.* **25**, 551 (1972). Biological properties: T. Aoyagi *et al.*, *ibid.* **24**, 687 (1971). N-n-Caproyl and N-*iso*-caproyl derivatives, **pepstatin B** and **pepstatin C**, also isolated from pepstatin-producing *Streptomyces*, as minor components of crude preparations: T. Miyano *et al.*, *ibid.* **25**, 489 (1972). Mechanism of pepsin inhibition: S. Kunimoto *et al.*, *ibid.* 251; *eidem, ibid.* **27**, 413 (1974); J. Marciniszyn *et al.*, *Adv. Exp. Med. Biol.* **95**, 199 (1977); D. H. Rich, E. T. O. Sun, *Biochem. Pharmacol.* **29**, 2205 (1980); of other acid protease inhibition: D. H. Rich *et al.*, *Biochemistry* **24**, 3165 (1985). Tissue distribution in rats: D. A. W. Grant *et al.*, *Biochem. Pharmacol.* **31**, 2302 (1982). Effect on gastric ulcers in man: O. Bonnevie *et al.*, *Gut* **20**, 624 (1979); L. B. Svendsen *et al.*, *Scand. J. Gastroenterol.* **14**, 929 (1979).

Colorless needles, mp 228-229° (dec). $[\alpha]_D^{27}$ −90.3° (c = 0.288 in methanol). Sol in methanol, ethanol, acetic acid DMSO. Practically insol in benzene, chloroform, ether, and water. LD_{50} in mice, rats, rabbits, dogs (mg/kg): 1090, 875, 820, 450 i.p.; all >2000 orally (Umezawa, 1970).

7261. Peptide T. [106362-32-7] L-Alanyl-L-seryl-L-threonyl-L-threonyl-L-threonyl-L-asparaginyl-L-tyrosyl-L-threonine. $C_{35}H_{55}$-N_9O_{16}; mol wt 857.87. C 49.00%, H 6.46%, N 14.69%, O 29.84%. Octapeptide segment of the human immunodeficiency virus (HIV) envelope glycoprotein (gp 120); named peptide T because of its high threonine content. Has been reported to block the *in vitro* binding of HIV envelope to human leukocyte receptor CD4. Isolation, neuropharmacology and anti-HIV activity of peptide and analogs: C. B. Pert *et al., Proc. Natl. Acad. Sci. USA* **83**, 9254 (1986). Characterization of active core structure, T[4-8], and chemotactic effects: C. B. Pert, M. R. Ruff, *Clin. Neuropharmacol.* **9**, Suppl. 4, 482 (1986). Chemotactic effects and structural homology with vasoactive intestinal peptide (VIP), *q.v.:* M. R. Ruff *et al., FEBS Lett.* **211**, 17 (1987). Competitive binding studies at VIP receptor: T. D. Nguyen, *Peptides* **9**, 425 (1988). Structural homology with thymosin α_1, *q.v.:* T. D. Nguyen, L. A. Scheving, *Biochem. Biophys. Res. Commun.* **145**, 884 (1987). Evaluation of anti-HIV activity: J. Sodroski *et al., Lancet* **1**, 1428 (1987). Clinical evaluation in treatment of AIDS: L. Wetterberg *et al., ibid.* 159. Chromatographic purification of *[D-ala¹] peptide T amide*, a metabolically stable and more potent analog of peptide T: T. R. Burke, M. Knight, *J. Chromatogr.* **411**, 431 (1987). Blood to brain transport of the amide in mice: C. M. Barrera *et al., Brain Res. Bull.* **19**, 629 (1987). Brief review of debate over effectiveness of peptide T: D. M. Barnes, *Science* **237**, 128 (1987).

Ala–Ser–Thr–Thr–Thr–Asn–Tyr–Thr

7262. Peptide YY. [106388-42-5] (unspecified source); [118997-30-1] (human); [81858-94-8] (porcine). Peptide tyrosine-tyrosine; PYY_{1-36}. 36 amino acid peptide hormone named for its N- and C-terminal tyrosine (Y) residues. Evolutionarily related to pancreatic polypeptide and neuropeptide Y, *q.v.* Regulates gut motility and secretion of gastric fluids. The two main endogenous circulating forms, PYY_{1-36} and PYY_{3-36}, are released from endocrine L-cells in response to caloric intake. Abnormalities in intestinal PYY are linked to many gastrointestinal disorders including obesity and diabetes. Isoln and characterization of porcine form: K. Tatemoto, V. Mutt, *Nature* **285**, 417 (1980); K. Tatemoto, *Proc. Natl. Acad. Sci. USA* **79**, 2514 (1982); of human form: idem *et al., Biochem. Biophys. Res. Commun.* **157**, 713 (1988). Structural characterization of human PYY_{3-36} and PYY_{1-36}: G. A. Eberlein *et al., Peptides* **10**, 797 (1989). CD and NMR soln structure studies: D. A. Keire *et al., Biochemistry* **39**, 9935 (2000). Review of physiologic role in digestion: F. L. C. Hill *et al., Steroids* **56**, 77-82 (1991). Series of articles on physiological roles and clinical potential: *Peptides* **23**, 249-407 (2002). Review of potential role in appetite regulation and obesity therapy: B. M. C. McGowan, S. R. Bloom, *Curr. Opin. Pharmacol.* **4**, 583-588 (2004); D. Renshaw, R. L. Batterham, *Curr. Drug Targets* **6**, 171-179 (2005).

Tyr–Pro–Ile–Lys–Pro–Glu–Ala–Pro–Gly–Glu–Asp–Ala–Ser–Pro–Glu–Glu–Leu–Asn
                                  |
                                   Arg
NH₂–Tyr–Arg–Gln–Arg–Thr–Val–Leu–Asn–Leu–Tyr–His–Arg–Leu–Ser–Ala–Tyr–Tyr

Human Form

Peptide YY (3-36). [126339-09-1] (porcine). PYY_{3-36}. Produced by cleavage of the two N-terminal amino acids by dipeptidyl peptidase IV. Inhibition of food intake in rats: R. L. Batterham *et*

al., Nature **418**, 650 (2002); in mice: B. G. Challis *et al., Biochem. Biophys. Res. Commun.* **311**, 915 (2003).

7263. Peracetic Acid. [79-21-0] Ethaneperoxoic acid; peroxyacetic acid; acetyl hydroperoxide; Proxitane. $C_2H_4O_3$; mol wt 76.05. C 31.59%, H 5.30%, O 63.11%. Commercial product is an equilib soln of peracetic acid, hydrogen peroxide, acetic acid and water, *q.q.v.* Prepn: J. D'Ans, **DE 251802** (1911); J. D'Ans, W. Frey, *Ber.* **45**, 1845 (1912); using solid superacid catalysts: M. S. Saha *et al., Tetrahedron Lett.* **44**, 5535 (2003). Decomposition study: E. Koubek *et al., J. Am. Chem. Soc.* **85**, 2263 (1963). HPLC Determn: U. Pinkernell *et al., Anal. Chem.* **66**, 2599 (1994); idem et al., *J. Chromatogr. A* **730**, 203 (1996). Biocidal properties and water treatment applications: J. F. Kramer, *Mater. Perform.* **36**, 42 (1997). Review of of use in pharmaceutical synthesis: H. Feigenbaum, *Spec. Chem.* **17**, 80-84 (1997); of prepn and properties: B. B. Klopotek, *Chim. Oggi* **16**, 33-37 (1998); of use as a bleaching agent: R. B. Chavan *et al., Colourage* **47**, 15-20 (2000).

Liquid, acrid odor. mp 0°. $bp_{1.6\ hPa}$ 25°. d_4^{20} 1.0375. n_D^{20} 1.3974. Flash pt: 133°F (56°C). pKa 8.2. Explodes violently on heating to 110°. Freely sol in water, alcohol, ether, H_2SO_4. Stable in dil aq soln. Strong oxidizing agent. LD_{50} (mg/kg) in rats: 1540 orally; in rabbits: 1410 dermally. LC_{50} in rats (mg/m³): 450 by inhalation (Klopotek).

Caution: Strongly irritating to skin and eyes.

USE: Environmentally friendly biocide; disinfectant in the food and beverage industry; bleaching agent for textiles and paper. Oxidizing agent in organic synthesis.

7264. Peramivir. [330600-85-6] (1S,2S,3R,4R)-3-[(1S)-1-(Acetylamino)-2-ethylbutyl]-4-[(aminoiminomethyl)amino]-2-hydroxycyclopentanecarboxylic acid. $C_{15}H_{28}N_4O_4$; mol wt 328.41. C 54.86%, H 8.59%, N 17.06%, O 19.49%. Selective inhibitor of influenza virus neuraminidase. Prepn: Y. S. Babu *et al., WO 9933781; eidem, US 6562861* (1999, 2003 both to BioCryst Pharm.); and structure activity studies: P. Chand *et al., J. Med. Chem.* **44**, 4379 (2001). Total synthesis of (±)-form: T. Mineno, M. J. Miller, *J. Org. Chem.* **68**, 6591 (2003). Crystal structure of the trihydrate: E. Keller, V. Krämer, *Z. Naturforsch.* **62**, 983 (2007). HILIC determn in plasma: Y. Li *et al., J. Chromatogr. B* **877**, 933 (2009). Clinical pharmacokinetics: L. Barroso *et al., Antivir. Ther.* **10**, 901 (2005). Review of pharmacology and clinical experience in H1N1 infection: C. E. Mancuso *et al., Ann. Pharmacother.* **44**, 1240-1249 (2010).

Trihydrate. [1041434-82-5] BCX-1812; RWJ-270201; Rapiacta. $C_{15}H_{28}N_4O_4.3H_2O$; mol wt 382.46. Colorless, octahedral crystals from methanol/water.

(±)-Form. [229614-56-6] Crystals from water/methanol, mp >250°.

THERAP CAT: Antiviral.

7265. Perazine. [84-97-9] 10-[3-(4-Methyl-1-piperazinyl)-propyl]-10*H*-phenothiazine; *N*-methylpiperazinyl-*N*'-propylphenothiazine; 10-(γ-methylpiperazinopropyl)phenothiazine; P-725; Taxilan. $C_{20}H_{25}N_3S$; mol wt 339.50. C 70.76%, H 7.42%, N 12.38%, S 9.44%. Prepn: Hromatka *et al., Monatsh. Chem.* **88**, 56, 193

(1957); **91**, 107 (1960); Horclois, **GB 780193** (1957 to Rhône-Poulenc).

Crystals, mp 51-53°. $bp_{0.001}$ 160-170° (air bath temp).
Dihydrochloride. $C_{20}H_{27}Cl_2N_3S$. Hygroscopic needles, dec 228-230°.
Dihydrochloride hemihydrate. Platelets from ethanol, mp 225-227°.
Dimaleate. $C_{28}H_{33}N_3O_8S$. Crystals from water, mp 210°.
THERAP CAT: Antipsychotic.

7266. Perbenzoic Acid. [93-59-4] Benzenecarboperoxoic acid; peroxybenzoic acid; benzoyl hydroperoxide. $C_7H_6O_3$; mol wt 138.12. C 60.87%, H 4.38%, O 34.75%. Prepd from dibenzoyl peroxide by treatment with a soln of sodium methoxide in methanol at 0°: Braun, *Org. Synth.* **13**, 86 (1933); *cf.* Bergmann, Witte, **DE 409779**; *Chem. Zentralbl.* **1925**, I, 1911.

Leaflets from benzene. Acrid odor. mp 41-43°. Very volatile. Sublimes in desiccator. bp_{15} 100-110° (partial decomposition). Volatile with steam. Very sparingly sol in water, but turns liquid upon contact with water; sparingly sol in petr ether; freely sol in other organic solvents. Stability of solns with chloroform, carbon tetrachloride, ether, benzene: Prileshajew, *Chem. Zentralbl.* **1911**, I, 1280; Kolthoff *et al.*, *J. Polym. Sci.* **2**, 199 (1947), *C.A.* **41**, 4960 (1947). Forms an unstable acid sodium salt, and a somewhat more stable neutral sodium salt.
USE: To convert ethylenic compounds into oxides; in analysis of unsatd compounds, to determine the number of double bonds.

7267. Perchloric Acid. [7601-90-3] $ClHO_4$; mol wt 100.45. Cl 35.29%, H 1.00%, O 63.71%. $HClO_4$. Prepd from potassium perchlorate and sulfuric acid: Schmeisser in *Handbook of Preparative Inorganic Chemistry* vol. **1**, G. Brauer, Ed. (Academic Press, New York, 2nd ed., 1963) pp 318-320. Comprehensive monograph: J. C. Schumacher, *Perchlorates* (Reinhold, New York, 1960).

The anhyd acid is a colorless, volatile, very hygroscopic liquid. d^{22} 1.768; bp_{11} 19°. *Oxidizer; corrosive.* Dec when distilled at atmospheric pressure, sometimes with explosive violence. mp −112°. Combines vigorously with water with evolution of heat. Undergoes spontaneous and explosive decompn, hence it is marketed only in mixture with water contg 60-70% $HClO_4$, density 1.5 and 1.6, respectively. The aq acid is very caustic and may deflagrate in contact with oxidizable substances. Density of aq solns at 15°: 1% = 1.0050; 10% = 1.0597; 20% = 1.1279; 30% = 1.2067; 40% = 1.2991; 50% = 1.4103; 60% = 1.5389; 70% = 1.6736. Density of aq solns at 25°: 65.0% = 1.597; 70.0% = 1.664; 75.0% = 1.731.
Caution: Corrosive to skin, mucous membranes.
USE: The acid in analytical chemistry as an oxidizer and for separation of potassium from sodium. Solvent for metals and ores. In nonaqueous titrimetry. In wet ashing. Its salts for explosives and for plating of metals.

7268. Perchloryl Fluoride. [7616-94-6] Trioxychlorofluoride. $ClFO_3$; mol wt 102.45. Cl 34.60%, F 18.54%, O 46.85%. *Reviews:* Downs, Adams in *Comprehensive Inorganic Chemistry* vol. **2**, J. C. Bailar, Jr. *et al.*, Eds. (Pergamon Press, Oxford, 1973) pp 1391-1393; Christe, Schack in *Adv. Inorg. Chem. Radiochem.* **18**, 319-398.

Characteristic sweet odor. Usually stored in cylinders as liquid under pressure. mp −147.7°. bp −46.7°. d^{20} (liq) 1.434. Critical temp 95.2°. Crit pressure 53 atm. Crit density: 0.637. Heat of vaporization 4.6 kcal/mol. Trouton constant: 20.4. Dipole moment 0.03. Heat of formation of gas at 25° −5.12 kcal/mol. Sp heat of liquid: 0.229 cal/g/°C at −40°; 0.290 at +50°. Does not corrode base metals when anhyd. Shows the greatest resistance to electrical breakdown known for any gas. *Handle with caution.* Do not bring in direct contact with reducing agents, alcohols, etc.: *Chem. Eng. News* **37**, 60 (1959). Explosions have occurred.
Caution: Potential symptom of overexposure is respiratory system irritation; direct contact with liquid may cause frostbite. *See NIOSH Pocket Guide to Chemical Hazards* (DHHS/NIOSH 97-140, 1997) p 246. May cause methemoglobinemia, cyanosis, alveolar edema, bronchopneumonia. *See Patty's Industrial Hygiene and Toxicology* vol. 2F, G. D. Clayton, F. E. Clayton, Eds. (John Wiley & Sons, New York, 4th ed., 1994) p 4478.
USE: In organic synthesis to introduce fluorine atoms into organic molecules. As oxidizing agent; insulator for high voltage systems.

7269. Perezone. [3600-95-1] 2-[(1R)-1,5-Dimethyl-4-hexen-1-yl]-3-hydroxy-5-methyl-2,5-cyclohexadiene-1,4-dione; 2-(1,5-dimethyl-4-hexenyl)-3-hydroxy-5-methyl-*p*-benzoquinone; pipitzahoic acid. $C_{15}H_{20}O_3$; mol wt 248.32. C 72.55%, H 8.12%, O 19.33%. From roots of *Trixis pipitzahuac* Shaffner *(Perezia adnata* Gr., *Compositae):* Weld, *Ann.* **95**, 188 (1855); from roots of *Radix pereziae:* Mylius, *Ber.* **18**, 463, 480, 937 (1885); Anschütz, *Ber.* **18**, 709 (1885); Fichter *et al.*, *Ann.* **395**, 1, 15 (1913). Structure: Archer, Thomson, *Chem. Commun.* **1965**, 354; Bates *et al.*, *Chem. Ind. (London)* **1965**, 1793.

Yellow plates from water or alcohol, mp 103-104°. $[\alpha]_D^{20}$ −17° (ether).

7270. Perflubron. [423-55-2] 1-Bromo-1,1,2,2,3,3,4,4,5,5,-6,6,7,7,8,8,8-heptadecafluorooctane; bromoperfluorooctane; perfluorooctyl bromide; PFOB; L-1913; Imagent; LiquiVent; Oxygent. C_8BrF_{17}; mol wt 498.96. C 19.26%, Br 16.01%, F 64.73%. Synthetic second generation brominated fluorocarbon. Prepd not claimed: **FR 1512068**; L. A. Loree, **US 3456024** (1968, 1969 both to Dow Corning). Radiopacity and toxicology: D. M. Long *et al.*, *Radiology* **105**, 323 (1972). Biodistribution and determn in plasma and tissue: F. H. Lee *et al.*, *J. Pharm. Sci.* **67**, 1038 (1978). Clinical trial in abdominal imaging: R. F. Mattrey, *Radiology* **191**, 841 (1994). Toxicity studies: A. R. Burgan *et al.*, *Biomater. Artif. Cells Artif. Organs* **16**, 681 (1988). Review of physicochemical properties and therapeutic potential: D. M. Long *et al.*, *ibid.* 411-420; of oxygen delivery characteristics: J. G. Riess, *ibid.* **20**, 183-202 (1992).

Liquid, bp_{760} 141°. d^{25} 1.928. Vapor pressure (37.5°) 14 torr. Surface tension 18.2 dynes/cm. Insol in water. LD_{50} i.v. in female rats: 41 g/kg (Burgan).
THERAP CAT: Diagnostic aid (contrast agent).

7271. Perfluorohexane. [355-42-0] 1,1,1,2,2,3,3,4,4,5,5,6,6,-6,6-Tetradecafluorohexane; perfluoro-*n*-hexane; Imavist. C_6F_{14}; mol wt 338.04. C 21.32%, F 78.68%. Inert perfluorocarbon with applications as a coolant and medical imaging agent. Discovery: J. H. Simons, L. P. Block, *J. Am. Chem. Soc.* **59**, 1407 (1937). Prepn and physical properties: V. E. Stiles, G. H. Cady, *ibid.* **74**, 3771 (1952); R. D. Dunlap *et al.*, *ibid.* **80**, 83 (1958). Critical properties and virial coefficients: Z. L. Taylor, Jr., T. M. Reed, III, *AIChE J.* **16**, 738 (1970). Conformational analysis: P. Piaggio *et al.*, *J. Mol. Struct.* **26**, 421 (1975). ^{19}F-NMR characterization of commercial product: T. A. Kestner, *J. Fluorine Chem.* **36**, 77 (1987). Use as a solvent in bromination reactions: S. M. Pereira *et al.*, *Synth. Commun.* **25**, 1023 (1995). Evaluation in treatment of acute lung injury in rabbits: M. Hübler *et al.*, *Crit. Care Med.* **30**, 422 (2002).

Coarse, transparent, needle-like crystals, mp $-82.26 \pm 0.01°$. bp 57.23°. d^0 1.7560. d^{25} 1.66970 (air satd, 1 atm). d^{25} 1.6717 (degassed, equilib vapor pressure). n_D^{22} 1.2515. Viscosity at 0°: 9.79 mP. Heat of fusion: 1580 $\pm$90 cal/mol. Molar heat of vaporization (cal) at bp: 7307; at 20°: 7793.

USE: In the electronics industry as a coolant and test bath medium. Non-toxic, non-ozone-depleting, inert reaction medium.

THERAP CAT: Diagnostic aid (ultrasound contrast agent).

7272. Perfluorooctanoic Acid. [335-67-1] 2,2,3,3,4,4,5,5,6,-6,7,7,8,8,8-Pentadecafluorooctanoic acid; perfluorocaprylic acid; PFOA. $C_8HF_{15}O_2$; mol wt 414.07. C 23.21%, H 0.24%, F 68.82%, O 7.73%. Environmentally persistent perfluorocarboxylate (PFCA) used as a surfactant and synthetic intermediate. Prepn by electrochemical fluorination: A. R. Diesslin *et al.*, US 2567011 (1951 to Minnesota Mining & Manuf.); E. A. Kauck, A. R. Diesslin, *Ind. Eng. Chem.* **43**, 2332 (1951); H. W. Prokop *et al.*, *J. Fluorine Chem.* **43**, 277 (1989). LC-MS/MS determn in food and drinking water: Y.-C. Chang *et al.*, *Anal. Bioanal. Chem.* **402**, 1315 (2012); in human hair and urine: F. Perez *et al.*, *ibid.*, 2369. Review of toxicology: N. Kudo, Y. Kawashima, *J. Toxicol. Sci.* **28**, 49-57 (2003); of industrial uses and environmental fate: K. Prevedouros *et al.*, *Environ. Sci. Technol.* **40**, 32-44 (2006). Review of analytical methods: P. de Voogt, M. Sáez, *Trends Anal. Chem.* **25**, 326-342 (2006); of environmental occurrence and biological monitoring studies: M. Houde *et al.*, *Environ. Sci. Technol.* **40**, 3463-3473 (2006). Review of health effects: K. Steenland *et al.*, *Environ. Health Perspect.* **118**, 1100-1108 (2010).

White crystals, mp 55°. bp$_{736}$ 189°. pKa 2-3. Soly in water (22°): 0.414 g/100ml. Sol in acetone.

Ammonium salt. [3825-26-1] Ammonium perfluorooctanoate; ammonium perfluorocaprylate; APFO. $C_8H_4F_{15}NO_2$; mol wt 431.10. Prepn: D. Lines, H. Sutcliffe, *J. Fluorine Chem.* **25**, 505 (1984). Crystals from ether, mp 157-165°.

USE: Polymerization aid in production of fluoroelastomers and fluoropolymers, such as polytetrafluoroethylene. As water repellent; surfactant in fire fighting foams, Intermediate in the synthesis of fluoroacrylic esters.

7273. Perfluoropropane. [76-19-7] 1,1,1,2,2,3,3,3-Octafluoropropane; octafluoropropane; perflutren; K-218; PFC-218; R-218; Ispan. C_3F_8; mol wt 188.02. C 19.16%, F 80.84%. Non-ozone depleting fluorocarbon; formulated for surgical and diagnostic use. Prepn: J. H. Simons, L. P. Block, *J. Am. Chem. Soc.* **61**, 2962 (1939). *See also:* J. H. Simons, US 2456027 (1948 to Minnesota Mining Manuf.). Physical properties: J. A. Brown, *J. Chem. Eng. Data* **8**, 106 (1963). Thermophysical properties: D. R. Defibaugh, M. R. Moldover, *ibid.* **42**, 160 (1997). Clinical experience in repair of retinal detachment: N. R. Sabates *et al.*, *Retina* **16**, 7 (1996). Review of properties and description: *Matheson Gas Data Book*, W.

Braker, A. L. Mossman, Eds. (Matheson, Lyndhurst, NJ, 6th ed., 1980) p 590-595; of electron interactions for modeling behavior: L. G. Christophorou, J. K. Oltoff, *J. Phys. Chem. Ref. Data* **27**, 889-913 (1998).

Colorless, nonflammable gas. mp $-183°$. bp $-36.7°$. d^{20} (liq) 1.352. Critical temp 71.9°. Critical press 25.45 atm. Critical density 0.629.

Human albumin microsphere suspension. FS-069; Optison. Clinical study as echocardiographic contrast agent: J. L. Cohen *et al.*, *J. Am. Coll. Cardiol.* **32**, 746 (1998). Toxicology and safety study: Y. Greener *et al.*, *Int. J. Toxicol.* **17**, 631 (1998). Physical and biochemical stability: S. Podell *et al.*, *Biotechnol. Appl. Biochem.* **30**, 213 (1999). Review of diagnostic use: L. N. Clark, H. C. Dittrich, *Am. J. Cardiol.* **86**, Suppl., 14G-18G (2000). Mean microsphere particle diameter: 3.0 - 4.5 μm.

Lipid microsphere suspension. DMP-115; DMX-115; MRX-115; Definity. Preparative method: E. C. Unger, US 5874062 (1999 to ImaRx Pharm.). Clinical evaluation in echocardiography: D. W. Kitzman *et al.*, *Am. J. Cardiol.* **86**, 669 (2000). Mean microsphere particle diameter: 1.1 - 3.3 μm.

USE: High voltage insulating gas; refrigerant when combined with chlorofluoro hydrocarbons; plasma processing gas; fire protection agent.

THERAP CAT: Diagnostic aid (ultrasound contrast agent). Adjunct in repair of retinal detachment.

7274. Perforin. Cytolysin; pore-forming protein; PFP; C9-related protein. Complement-like protein produced by cytotoxic T-lymphocytes (CTLs) and natural killer (NK) cells; released from cytoplasmic granules during the cytolytic process. Perforates the membrane of target cells by forming transmembrane pores that perturb membrane permeability and result in cell lysis. Pore-formation is mediated by the Ca^{2+}-dependent polymerization of perforin within the target cell membrane. Also thought to act as a conduit for the endocytosis of **granzymes**, serine proteinases produced by cytotoxic lymphocytes, that initiate nuclear disintegration. Glycoprotein; mol wt 66-70 kDa. Identification of pore-forming tubular complexes produced by NK cells: E. R. Podack, G. Dennert, *Nature* **302**, 442 (1983). Partial purification of protein: E. R. Podack, P. J. Konigsberg, *J. Exp. Med.* **160**, 695 (1984). Cloning and amino acid sequence of human perforin: M. G. Lichtenheld *et al.*, *Nature* **335**, 448 (1988); of murine: D. M. Lowrey *et al.*, *Proc. Natl. Acad. Sci. USA* **86**, 247 (1989). Confirmation of biological activity: D. Kägi *et al.*, *Nature* **369**, 31 (1994). Review of structure and function: E. R. Podack, *Curr. Top. Microbiol. Immunol.* **178**, 175-184 (1992); C-C. Liu *et al.*, *Immunol. Today* **16**, 194-201 (1995).

7275. Performic Acid. [107-32-4] Methaneperoxoic acid; peroxyformic acid; permethanoic acid; formyl hydroperoxide. CH_2O_3; mol wt 62.02. C 19.37%, H 3.25%, O 77.39%. HCOOOH. A strong oxidizing agent. A 90% soln is obtained when a mixture of 20 g formic acid, 25 g 100% H_2O_2 and 6.5 g H_2SO_4 is allowed to interact for 2 hrs and is then distilled: D'Ans, Kneip, *Ber.* **48**, 1137 (1915); Greenspan, *J. Am. Chem. Soc.* **68**, 907 (1946).

The 90% soln is a colorless liq. *Prone to explode on contact with metals, their oxides, reducing substances, or on distillation.* Has lower vapor pressure than formic acid. Miscible with water, alc, ether. Sol in benzene, chloroform. Solns are unstable, gassing being noticeable after a few hours, and the effective concn showing a definite decline in 2 hrs.

Caution: Irritant.

USE: For oxidation, epoxidation and hydroxylation reactions.

7276. Perfosfamide. [62435-42-1]; [39800-16-3] (unspecified stereo). *rel*-(2R,4R)-N,N-Bis(2-chloroethyl)tetrahydro-4-hydroperoxy-2H-1,3,2-oxazaphosphorin-2-amine 2-oxide; *rel*-(2R,4R)-2-[bis(2-chloroethyl)amino]tetrahydro-2H-1,3,2-oxazaphosphorin-4-yl hydroperoxide P-oxide; *cis*-($\pm$)-2-[bis(2-chloroethyl)amino]-4-

hydroperoxytetrahydro-2*H*-1,3,2-oxazaphosphorine 2-oxide; *cis*-4-hydroperoxycyclophosphamide; 4-HC; NSC-181815; Pergamid. $C_7H_{15}Cl_2N_2O_4P$; mol wt 293.08. C 28.69%, H 5.16%, Cl 24.19%, N 9.56%, O 21.84%, P 10.57%. Prepn (stereochem. unspecified): **NL 7208895**; A. Takamizawa, T. Iwata, **US 3808297** (1972, 1974 both to Shionogi); *idem et al., J. Am. Chem. Soc.* **95**, 985 (1973). Structural study: A. Camerman *et al., Acta Crystallogr.* **B33**, 678 (1977). Cytostatic activity: A. Takamizawa *et al., J. Med. Chem.* **18**, 376 (1975). Clinical use: D. Biggs *et al., Prog. Clin. Biol. Res.* **389**, 9 (1994). Review of use as purging agent in autologous bone marrow transplantation: A. M. Yeager *et al., Am. J. Pediatr. Hematol. Oncol.* **12**, 245-256 (1990); R. J. Jones, *J. Hematother.* **1**, 343-348 (1992).

Relative stereochemistry

White crystals from acetone/ether (unspecified form). mp 107-108°. LD_{50} in rats, mice (mg/kg): 115, 235 i.v.; 131, 181 i.p. (Takamizawa, 1975).

THERAP CAT: Antineoplastic; bone marrow purging agent.

7277. Pergolide. [66104-22-1] (8β)-8-[(Methylthio)methyl]-6-propylergoline; D-6-*n*-propyl-8β-methylmercaptomethylergoline; LY-141B. $C_{19}H_{26}N_2S$; mol wt 314.49. C 72.56%, H 8.33%, N 8.91%, S 10.19%. Dopamine D_1/D_2 receptor agonist. Prepn: E. C. Kornfeld, N. J. Bach, **US 4166182** (1979 to Lilly). Industrial synthesis: W. Cabri *et al., Org. Process Res. Dev.* **10**, 198 (2006). Dopaminergic effects in rats: R. W. Fuller *et al., Life Sci.* **24**, 375 (1979); T. T. Yen *et al., ibid.* **25**, 209 (1979). Clinical pharmacology: L. Lemberger, R. E. Crabtree, *Science* **205**, 1151 (1979). Pharmacological evaluation as antiparkinson agent: W. C. Koller, *Neuropharmacology* **19**, 831 (1980). Comprehensive description: D. J. Sprankle, E. C. Jensen, *Anal. Profiles Drug Subs. Excip.* **21**, 375-413 (1992). Review of clinical experience in Parkinson's disease: U. Bonuccelli *et al., Clin. Neuropharmacol.* **25**, 1-10 (2002).

Solid, mp 217.5°.

Methanesulfonate. [66104-23-2] Pergolide mesylate; LY-127809; Celance; Nopar; Permax; Prascend. $C_{19}H_{26}N_2S \cdot CH_3SO_3H$; mol wt 410.59. Crystals, mp 258-260° (dec) (Sprankle, Jensen); also reported as mp 267.4° (Cabri). uv max (water): 279 nm (ε 6385); (methanol): 280 nm (ε 6980); (dehydrated ethanol): 281 nm (ε 6993). α_D^{20} between −18.0° and −23.0° (c = 10 mg/ml in DMF). pKa (66% DMF) 7.8. Sparingly sol in DMF, methanol; slightly sol in water, 0.01*N* HCl, chloroform, acetonitrile, dichloromethane, dehydrated ethanol; very slightly sol in acetone. Practically insol in 0.1*N* NaOH, 0.1*N* HCl, ether. Partition coefficient at 25° (chloroform/water): 6.14 (pH 2.19); 119.6 (pH 4.02).

THERAP CAT: Antiparkinsonian.

THERAP CAT (VET): In treatment of equine Cushing's disease.

7278. Perhexiline. [6621-47-2] 2-(2,2-Dicyclohexylethyl)-piperidine; 1,1-dicyclohexyl-2-(2-piperidyl)ethane; perhexilene.

$C_{19}H_{35}N$; mol wt 277.50. C 82.24%, H 12.71%, N 5.05%. Calcium blocking agent. Prepn: **GB 1025578** (1966 to Richardson-Merrell), *C.A.* **65**, 2229f (1966). Pharmacology: Hudak *et al., J. Pharmacol. Exp. Ther.* **173**, 371 (1970). Clinical studies: Winsor, *Clin. Pharmacol. Ther.* **11**, 85 (1970). Toxicity study: Causa, Perri, *Arzneim.-Forsch.* **21**, 114 (1971). Series of articles: *Postgrad. Med. J. Suppl.* **49**, 8-132 (1973). Book: *Antiarrhythmic Action and the Puzzle of Perhexiline*, E. M. Vaughan Williams, Ed. (Academic Press, New York, 1980) 143 pp.

Maleate. [6724-53-4] Pexid. $C_{19}H_{35}N \cdot C_4H_4O_4$; mol wt 393.57. mp 188.5-191°. LD_{50} in rats, mice (g/kg): >7, 4.37 orally (Causa, Perri).

Hydrochloride. $C_{19}H_{35}N \cdot HCl$. White crystalline powder, mp 243-245.5° (dec).

THERAP CAT: Vasodilator (coronary); diuretic.

7279. Pericyazine. [2622-26-6] 10-[3-(4-Hydroxy-1-piperidinyl)propyl]-10*H*-phenothiazine-2-carbonitrile; 2-cyano-10-[3-(4-hydroxypiperidino)propyl]phenothiazine; periciazine; propericiazine; RP-8908; Aolept; Neulactil; Neuleptil. $C_{21}H_{23}N_3OS$; mol wt 365.50. C 69.01%, H 6.34%, N 11.50%, O 4.38%, S 8.77%. Psychotherapeutic phenothiazine. Prepn: Robert, Jacques, **FR 1212031** (1960 to Rhône-Poulenc). Hypotensive action in dogs: K. P. Singh *et al., Indian J. Med. Res.* **58**, 1467 (1970). GLC determn in human urine: A. P. De Leenheer, *J. Pharm. Sci.* **63**, 389 (1974). Clinical studies in psychiatric patients: U. Spiegelberg, G. Kleu, *Arzneim.-Forsch.* **17**, 159 (1967); J. C. Barker, M. Miller, *Br. J. Psychiatry* **115**, 169 (1969). Toxicity and metabolism: L. Julou *et al., Proc. Eur. Soc. Study Drug Toxic.* **9**, 11 (1968). Use in spectrophotometry: H. Gowda *et al., Anal. Chem.* **55**, 1816 (1983); A. T. Gowda *et al., Anal. Chim. Acta* **154**, 347 (1983).

Crystals, mp 116-117°. uv max: 232.5, 271.5 nm (log ε 4319, 4503). LD_{50} orally in rats: 395 mg/kg (Schafer).

USE: Spectrophotometric reagent for palladium and ruthenium.

THERAP CAT: Antipsychotic.

7280. Perifosine. [157716-52-4] 4-[[Hydroxy(octadecyloxy)phosphinyl]oxy]-1,1-dimethylpiperidinium inner salt; octadecyl-(1,1-dimethylpiperidinio-4-yl)phosphate; D-21266; KRX-0401; NSC-639966. $C_{25}H_{52}NO_4P$; mol wt 461.67. C 65.04%, H 11.35%, N 3.03%, O 13.86%, P 6.71%. Synthetic alkylphosphocholine anticancer agent. Inhibits the phosphatidylinositol 3-kinase regulated Akt/PKB survival pathway and induces apoptosis. Prepn: G. Nössner *et al.,* **DE 4222910**; *eidem,* **US 6172050** (1994, 2001 both to Asta Medica). Mechanism of action study: G. A. Ruiter *et al., Anti-Cancer Drugs* **14**, 167 (2003). LC-ESI-MS determn in plasma: E. W. Woo *et al., J. Chromatogr. B* **759**, 247 (2001). Biodistribution and tumor uptake: S. R. Vink *et al., Invest. New Drugs* **23**, 279 (2005). Clinical pharmacokinetics: M. Crul *et al., Eur. J. Cancer* **38**, 1615 (2002). Clinical evaluation in colorectal cancer: J. C. Bendell *et al., J. Clin. Oncol.* **29**, 4394 (2011).

Crystals from methyl ethyl ketone, mp 271-272° (dec).
THERAP CAT: Antineoplastic.

7281. Perilla Ketone. [553-84-4] 1-(3-Furanyl)-4-methyl-1-pentanone; 1-(3-furyl)-4-methyl-1-pentanone; β-furyl isoamyl ketone. $C_{10}H_{14}O_2$; mol wt 166.22. C 72.26%, H 8.49%, O 19.25%. Potent pulmonary edematogenic agent. Isoln from the mint plant, *Perilla frutescens* Britton, *Labiatae*, and structure: R. Goto, *J. Pharm. Soc. Jpn.* **57**, 77 (1937). Synthesis: T. Matsuura, *Bull. Chem. Soc. Jpn.* **30**, 430 (1957); K. Kondo, M. Matsumoto, *Tetrahedron Lett.* **1976**, 4363; T. Kitamura *et al.*, *Synth. Commun.* **7**, 521 (1977); K. Inomata *et al.*, *Chem. Lett.* **1979**, 709. Potent lung toxin implicated in emphysema of grazing cattle: B. J. Wilson *et al.*, *Science* **197**, 573 (1977).

Colorless oil, bp 196°. n_D^{20} 1.4781; d_{15}^{15} 0.9920. uv max (ethanol): 207, 253 nm (ε 14100, 5800). Sensitive to oxygen; becomes reddish-orange on standing. LD$_{50}$ in female, male mice, male rats (mg/kg): 2.5, 6, 10 i.p. (Wilson).

7282. Perillaldehyde. [2111-75-3] 4-(1-Methylethenyl)-1-cyclohexene-1-carboxaldehyde; 4-isopropenyl-1-cyclohexene-1-carboxaldehyde. $C_{10}H_{14}O$; mol wt 150.22. C 79.96%, H 9.39%, O 10.65%. Isoln from *Perilla arguta* Benth., *Labiatae*: Semmler, Zaar, *Ber.* **44**, 52, 815 (1911); from essential oil of *Sium latifolium* L., *Umbelliferae*: Parczewski, *Diss. Pharm.* **12**, 223 (1960), *C.A.* **55**, 7765c (1961); from mandarin peel oil *(Citrus reticulata* Blanco, *Rutaceae)*: Kugler, Kováts, *Helv. Chim. Acta* **46**, 1480 (1963). Prepn by chromic oxidation of perilla alcohol: Naves, *ibid.* **29**, 553 (1946); Ritter, Ginsburg, *J. Am. Chem. Soc.* **72**, 2381 (1950); Kergomard, Philibert-Bigou, *Bull. Soc. Chim. Fr.* **1958**, 393, 1174; Naves, Grampoloff, *ibid.* **1960**, 37. Brief review: *Food Chem. Toxicol.* **20**, Suppl. I-IV, 799-800 (1982).

LD$_{50}$ in mice (g/kg): 1.72 orally; in guinea pigs (g/kg): >5 dermally *(Food Chem. Toxicol.).*

***d*-Form.** Liquid. bp$_{745}$ 237°; bp$_7$ 98-100°. d_4^{20} 0.953. n_D^{20} 1.5058. $[\alpha]_D^{20}$ +127° (c = 13.1 in carbon tetrachloride).

***l*-Form.** Liquid. bp$_{10}$ 104-105°. d_4^{20} 0.9645. n_D^{20} 1.5069. $[\alpha]_D^{20}$ -146°.

Oxime. [138-91-0] *l*-Perillaldehyde α-*syn*-oxime; perillartine; "perilla sugar". $C_{10}H_{15}NO$. Previously referred to as *l-perillaldehyde α-anti-oxime.* Synthesis: Andô *et al.*, *Science (Tokyo)* **17**, 241 (1947), *C.A.* **45**, 1976d (1951). Clarification of structure: Acton *et al.*, *Experientia* **26**, 473 (1970). Needles, mp 102°. uv max (alc): 232 nm (ε 21800). About 2000 times as sweet as sucrose: Furukawa, *Koryo* No. **11**, 11, 40 (1950), *C.A.* **44**, 6083g (1950).

USE: The oxime is used as sweetening agent in Japan.

7283. Perimethazine. [13093-88-4] 1-[3-(2-Methoxy-10*H*-phenothiazin-10-yl)-2-methylpropyl]-4-piperidinol; 2-methoxy-10-

[2-methyl-3-(4-hydroxypiperidino)propyl]phenothiazine; 3-methoxy-10-[3-(4-hydroxypiperidyl)-2-methylpropyl]phenothiazine; perimetazine; RP-9159; AN-1317; Leptryl. $C_{22}H_{28}N_2O_2S$; mol wt 384.54. C 68.72%, H 7.34%, N 7.29%, O 8.32%, S 8.34%. Prepn: **GB 904210**; Jacob, Robert, **US 3075976** (1962, 1963 both to Rhône-Poulenc). Pharmacology: L. Julon *et al.*, *C. R. Seances Soc. Biol. Ses Fil.* **160**, 1852 (1966). Clinical evaluation as antipsychotic: M. Bourgeois, A.-M. Sicart, *Bordeaux Med.* **3**, 515 (1970).

Crystalline powder from benzene:cyclohexane (15:85), mp 137-138°.

Hydrochloride. $C_{22}H_{28}N_2O_2S.HCl$. LD$_{50}$ in mice (mg/kg): 115 i.v.; 140 i.p.; 330 s.c.; 310 orally (Julon).

THERAP CAT: Antipsychotic.

7284. Perimycin. [11016-07-2] Aminomycin; fungimycin; WX-2412; NC-1968. Polyene antifungal antibiotic complex produced by *Streptomyces coelicolor* var. *aminophilus* NRRL 2390: W. E. Wooldridge, **US 2956925** (1960); R. R. Mohan *et al.*, *Antimicrob. Agents Chemother.* **1963**, 462; L. E. McDaniel *et al.*, **US 3182004** (1965 to Warner-Lambert). Purification: E. Borowski *et al.*, *Antimicrob. Agents Annu.* **1960**, 532. Found to be a mixture of three active components, perimycin A (major), B, and C. Mechanism of action with succinyl adduct: E. Borowski, B. Cybulska, *Nature* **213**, 1034 (1967). Ionophoric and hemolytic activities: B. Cybulska *et al.*, *Biochem. Pharmacol.* **38**, 1755 (1989). Isoln of perimycins A, B, C and structural elucidation of A: P. Kolodziejczyk *et al.*, *Tetrahedron Lett.* **1976**, 3603. Stereochemistry and revised structure: J. Pawlak *et al.*, *J. Antibiot.* **48**, 1034 (1995).

Perimycin A

Amorphous, golden-yellow solid. Has no definite mp but dec slowly with darkening upon heating. uv max (methanol): 383 nm ($E_{1cm}^{1\%}$ 1000). Sol in the following solvents in the presence of water: lower alcohols, pyridine, tetrahydrofuran, acetone, dioxane. Sol in warm methanol, DMF, dimethylsulfoxide, and in the lower fatty acids. Practically insol in water, petr ether, ethyl acetate, benzene.

Perimycin A. [62327-61-1] 4'-Amino-3'-deamino-18-decarboxy-40-demethyl-4'-deoxy-3,7-dideoxo-3,3',7-trihydroxy-N^{47},18-dimethyl-5-oxocandicidin D cyclic 15,19-hemiacetal. $C_{59}H_{88}N_2$-O_{17}; mol wt 1097.35. uv max: 380 nm (E_1^1 1000).

THERAP CAT: Antifungal.

7285. Perindopril. [82834-16-0] (2*S*,3a*S*,7a*S*)-1-[(2*S*)-2-[[(1*S*)-1-(Ethoxycarbonyl)butyl]amino]-1-oxopropyl]octahydro-1*H*-indole-2-carboxylic acid; (2*S*,3a*S*,7a*S*)-1-[(*S*)-*N*-[(*S*)-1-carboxybutyl]alanyl]hexahydro-2-indolinecarboxylic acid 1-ethyl ester; (2*S*)-2-[(1*S*)-1-carbethoxybutylamino]-1-oxopropyl]-(2*S*,3a*S*,7a*S*)-perhydroindole-2-carboxylic acid; S-9490; McN-A-2833. $C_{19}H_{32}N_2O_5$; mol wt 368.47. C 61.93%, H 8.75%, N 7.60%, O 21.71%. Angiotensin-converting enzyme (ACE) inhibitor. Hydrolyzed *in vivo* to the active diacid metabolite. Prepn: G. Remond *et al.*, **EP 49658** (1982 to Sci. Union et Cie. - Soc. Franc. Rech. Med.); M. Vincent *et* *al.*, **US 4508729** (1985 to ADIR). Stereoselective synthesis: M. Vincent *et al.*, *Tetrahedron Lett.* **23**, 1677 (1982). NMR study: N. Platzer *et al.*, *Magn. Reson. Chem.* **26**, 296 (1988). Hemodynamic effects in humans: K. R. Lees, J. L. Reid, *Br. J. Clin. Pharmacol.* **23**, 159 (1987). Clinical evaluation in essential hypertension: T. Morgan *et al.*, *J. Cardiovasc. Pharmacol.* **10**, Suppl. 7, S116 (1987). Pharmacokinetics, pharmacodynamics of the diacid: K. R. Lees, J. L. Reid, *ibid.* **10**, 129 (1987). Review of pharmacology and clinical use: P. A. Todd, A. Fitton, *Drugs* **42**, 90-114 (1991). Symposium on clinical efficacy in cardiovascular disease: *Am. J. Cardiol.* **88**, Suppl. 1, 1-40 (2001).

tert-**Butylamine.** [107133-36-8] Perindopril erbumine; S-9490-3; McN-A-2833-109; Aceon; Coversum; Coversyl; Procaptan. $C_{19}H_{32}N_2O_5 \cdot C_4H_{11}N$; mol wt 441.61.

Diacid form. [95153-31-4] Perindoprilat; S-9780. $C_{17}H_{28}N_2O_5$; mol wt 340.42.

THERAP CAT: Antihypertensive.

7286. Periodic Acid. [10450-60-9] H_5IO_6; mol wt 227.94. H 2.21%, I 55.67%, O 42.11%. Prepd by electrolytic oxidation of iodic acid or from barium periodate and nitric acid: Willard, *Inorg. Synth.* **1**, 172 (1939). Chemistry of periodic acid and periodates: H. Siebert, *Fortschr. Chem. Forsch.* **8**, 470 (1967). Periodic acid and periodates in organic and bioorganic chemistry: A. J. Fatiadi, *Synthesis* **1974**, 229. Book: G. Dryhurst, *Periodate Oxidation of Diol and Other Functional Groups* (Pergamon Press, New York, 1970).

Monoclinic, hygroscopic crystals. mp 122°; dec 130-140° forming I_2O_5, H_2O, and O_2. Freely sol in water. A 38% w/w soln had d_4^{17} 1.3875. Sol in alcohol, slightly in ether. Soly in nitric acid (d 1.42) at 26° = 7.82 g/100 ml. When aq solns are evapd at room temp the "ortho" acid H_5IO_6 crystallizes out. If this is heated in a vacuum at 100° and 12 mm it loses water and is converted to the "meta" acid HIO_4, an intermediate $H_4I_2O_9$ being formed at 80°. Periodic acid is a dibasic acid, much weaker than perchloric acid. K_1 at 25° = 2.3 × 10^{-2}; $K_2 = 2 × 10^{-6}$. Very easily reduced to iodic acid by nitrous or sulfurous acids and even by hydrochloric and sulfuric acids. Oxidizes organic material.

USE: In organic synthesis. Oxidizer. Determination of manganese.

7287. Periodyl. [53586-99-5] 12-Hydroxy-9,10-diiodo-9-octadecenoic acid; diiodoricinstearolic acid; ricinstearolic acid diiodide; 8,9-diiodo-11-hydroxy-8-heptadecene-1-carboxylic acid; Diiodyl; Joristen. $C_{18}H_{32}I_2O_3$; mol wt 550.26. C 39.29%, H 5.86%, I 46.13%, O 8.72%. $CH_3(CH_2)_5CHOHCH_2CI=CI(CH_2)_7COOH$. Prepn: Mühle, *Ber.* **46**, 2091 (1913); **DE 296495** (to Riedel).

Tasteless needles from dil alcohol, mp 62°. Practically insol in water, acids; sol in dil alkalies, alcohol, ether, chloroform; slightly sol in benzene.

THERAP CAT: Iodine source.

7288. Periplanones. Sex pheromones of the American cockroach, *Periplaneta americana* L., found primarily in the alimentary tract and excreta of the insect. They act as close proximity sex-excitants and function over relatively short distances compared to long range insect sex-attractants. Isoln: D. R. A. Wharton *et al.*, *Science* **137**, 1062 (1962). Improved isolns, identification and proposed structures of periplanone B: C. J. Persoons *et al.*, *Tetrahedron Lett.* **1976**, 2055; E. Talman *et al.*, *Isr. J. Chem.* **17**, 227 (1978); C. J. Persoons *et al.*, *J. Chem. Ecol.* **5**, 221 (1979). Total synthesis and structure of (±) periplanone B: W. C. Still, *J. Am. Chem. Soc.* **101**, 2493 (1979); H. Hauptmann, G. Muhlbauer, *Tetrahedron Lett.* **27**, 1315 (1986); of the (−)-form: T. Kihara *et al.*, *ibid.* 1343. Absolute configuration of B: M. A. Adams *et al.*, *J. Am. Chem. Soc.* **101**, 2495 (1979). Structural and stereochemical studies of periplanone A: C. J. Persoons *et al.*, *J. Chem. Ecol.* **8**, 439 (1982); H. Hauptmann, G. Muhlbauer, *Tetrahedron Lett.* **27**, 6189 (1986); Y. Shizuri

et al., *ibid.* **28**, 1791, 1795 (1987); L. MacDonald *et al.*, *Heterocycles* **25**, 305 (1987). Short synthesis of periplanone A: Y. Shizuri *et al.*, *Tetrahedron Lett.* **29**, 1971 (1988). Use in insecticidal formulations: W. J. Bell *et al.*, *Environ. Entomol.* **13**, 448 (1984); W. J. Bell *et al.*, *Pest Control* **54**, 40 (1986).

Periplanone B

Periplanone A. [112709-47-4] (3*S*,6*S*,7*E*,11*Z*)-9-Methylene-6-(1-methylethyl)-1-oxaspiro[2.9]dodeca-7,11-dien-4-one. $C_{15}H_{20}O_2$; mol wt 232.32. Found primarily in excreta of *P. americana*, it is unstable and gradually rearranges to a biologically inactive compound. uv max (hexane): 220 nm (log ε 4.18).

Periplanone B. [61228-92-0] (1*R*,2*R*,5*S*,6*E*,10*R*)-8-Methylene-5-(1-methylethyl)spiro[11-oxabicyclo[8.1.0]undec-6-ene-2,2'-oxiran]-3-one; (−)-periplanone B. $C_{15}H_{20}O_3$; mol wt 248.32. More stable and more active than periplanone A, it is found in both the alimentary tract and excreta of *P. americana*. The ratio of A to B in excreta is 1:10. Crystals, mp 47-50°. $[α]_D^{22}$ −667° (c = 0.13 in *n*-hexane).

(±)-Periplanone B. [70613-99-5] Crystals, mp 48-50°. uv max (hexane): 226 nm.

7289. Periplocin. [13137-64-9] (3*β*,5*β*)-3-[(2,6-Dideoxy-4-*O*-β-D-glucopyranosyl-3-*O*-methyl-β-D-*ribo*-hexopyranosyl)oxy]-5,14-dihydroxycard-20(22)-enolide; glucoperiplocymarin; periplocoside. $C_{36}H_{56}O_{13}$; mol wt 696.83. C 62.05%, H 8.10%, O 29.85%. Isoln from *Periploca graeca* L., *Asclepiadaceae*: Lehmann, *Arch. Pharm.* **235**, 157 (1897); W. A. Jacobs, A. Hoffmann, *J. Biol. Chem.* **79**, 519 (1928); from *Strophanthus preussii* Engl. and Pax., *Apocynaceae*: Ruppol, Trukovic, *J. Pharm. Belg.* **10**, 221 (1955), *C.A.* **50**, 12089d (1956). Structure: Stoll, Renz, *Helv. Chim. Acta* **22**, 1193 (1939). Pharmacology and toxicity: M. H. MacKeith, *J. Pharmacol. Exp. Ther.* **27**, 449 (1926).

Amorphous yellowish powder. Readily sol in water. Acid hydrolysis yields periplogenin, *q.v.*, and periplobiose (cymarose + glucose, $C_{13}H_{24}O_9$, $[α]_D^{20}$ +32°). Enzymatic hydrolysis with strophanthobiase splits off glucose, yielding periplocymarin. The biose is attached to the hydroxyl group at C-3 of the aglycon. LD in rabbits, rats (mg/kg): 10, 480 s.c. (MacKeith).

Dihydrate. Fine needles from water. Becomes anhyd after drying for one hour in high vacuum at 105°. mp 224° when bath is preheated to 200°. $[α]_D^{20}$ +23° (c = 0.7 in alcohol). One gram dissolves in ~20 ml boiling water, at 25° the soly is ~1:2500. Freely sol in alcohol. Almost insol in ether, chloroform.

Tetraacetylperiplocin. $C_{44}H_{64}O_{17}$. Six-sided prisms from alcohol, mp 195°. $[α]_D^{20}$ +20° (c = 0.5 in alcohol). Sol in alcohol, chloroform; very slightly sol in water.

7290. Periplocymarin. [32476-67-8] (3*β*,5*β*)-3-[(2,6-Dideoxy-3-*O*-methyl-β-D-*ribo*-hexopyranosyl)oxy]-5,14-dihydroxycard-20(22)-enolide. $C_{30}H_{46}O_8$; mol wt 534.69. C 67.39%, H 8.67%, O 23.94%. By extracting bark and wood of *Periploca graeca* L., *Asclepiadaceae* with 70% alcohol and treating with the enzyme strophanthobiase from seeds of *Strophanthus courmonti* Sacleux, *Apocynaceae:* Jacobs, Hoffmann, *J. Biol. Chem.* **79**, 519 (1928); Katz, Reichstein, *Pharm. Acta Helv.* **19**, 231 (1944); from *S. hypoleucus* Stapf: von Euw, Reichstein, *Helv. Chim. Acta* **33**, 544 (1950); from *S. ledienii* Stein: Lichti *et al.*, *ibid.* **39**, 1914 (1956); from *S. eminii* Asch. et Pax.: Zelnik, Schindler, *ibid.* **40**, 2110 (1957). Structure: von Euw, Reichstein, *ibid.* **31**, 883 (1948). Acid hydrolysis yields one mol periplogenin and one mol cymarose, *q.q.v.* The cymarose is attached to the hydroxyl group at C-3 of the aglycon.

Lustrous needles from methanol contg approx one mol CH_3OH of crystn. Bitter taste, but not nearly as marked as that of cymarin. Becomes solvent-free at 100° *in vacuo.* Sinters at 138°. mp 148°. $[\alpha]_D^{27}$ +29° (c = 0.94 in 95% alcohol for the anhydr substance). Readily sol in alcohol, chloroform, acetone; less readily in methanol; very sparingly sol in water and practically insol in ether. Readily hydrolyzed in the cold.

7291. Periplogenin. [514-39-6] (3β,5β)-3,5,14-Trihydroxy-card-20(22)-enolide; desoxostrophanthidin. $C_{23}H_{34}O_5$; mol wt 390.52. C 70.74%, H 8.78%, O 20.48%. The aglycon of periplocin and periplocymarin; accompanies strophanthidin in *Strophanthus eminii* Asch. et Pax., *Apocynaceae.* Isoln: Lehmann, *Arch. Pharm.* **235**, 157 (1897); Jacobs, Hoffmann, *J. Biol. Chem.* **79**, 519 (1928); Stoll, Renz, *Helv. Chim. Acta* **22**, 1193 (1939); Lardon, *ibid.* **33**, 639 (1950). Structure: Speiser, Reichstein, *Experientia* **3**, 323 (1947); *eidem*, *Helv. Chim. Acta* **30**, 2143 (1947); **31**, 622 (1948). Synthesis: Deghenghi, Gaudry, *Tetrahedron Lett.* **1963**, 2045; Kamano *et al.*, *J. Org. Chem.* **39**, 2319 (1974).

Solvated prisms from methanol which contain an undetermined amount of methanol, sinter at 140°, mp 235°. $[\alpha]_D^{27}$ +31.5° (c = 1.04 in alcohol). Sol in alcohol and chloroform; slightly sol in ether, water (1:2500); practically insol in benzene, petr ether. pH of aq solns ~7. Strong positive Legal test.

Monobenzoate. $C_{30}H_{38}O_6$. Stout glistening wedges from 95% alcohol, mp 235°.

Dihydroperiplogenin. $C_{23}H_{36}O_5$. Stout prisms from 25% alcohol, mp 204° (slight preliminary softening).

3-O-Acetylperiplogenin. $C_{25}H_{36}O_6$. Crystals, mp 242-244°, $[\alpha]_D^{22}$ +46.9° (c = 0.32 in chloroform).

7292. Perivine. [2673-40-7] 4-Demethyl-3-oxovobasan-17-oic acid methyl ester. $C_{20}H_{22}N_2O_3$; mol wt 338.41. C 70.98%, H

6.55%, N 8.28%, O 14.18%. Indole alkaloid from *Vinca rosea* Linn., *Apocynaceae:* G. Svoboda, *J. Am. Pharm. Assoc.* **47**, 834 (1958); G. Svoboda *et al.*, *ibid.* **48**, 659 (1959). Structure: Gorman, Sweeny, *Tetrahedron Lett.* **1964**, 3105. Cytotoxicity: D. G. I. Kingston, *J. Pharm. Sci.* **67**, 272 (1978).

Prisms from methanol, dec 218-221°. Also reported as 180-181° (Svoboda 1959). $[\alpha]_D^{26}$ −121.4° (chloroform). pKa in 66% DMF: 7.5. uv max (ethanol): 314 nm ($E_{1cm}^{1\%}$ 2.67).

7293. Perlapine. [1977-11-3] 6-(4-Methyl-1-piperazinyl)-11*H*-dibenz[*b,e*]azepine; 6-(4-methyl-1-piperazinyl)morphanthridine; AW-14'2333; HF-2333; MP-11; Hypnodin. $C_{19}H_{21}N_3$; mol wt 291.40. C 78.31%, H 7.26%, N 14.42%. Prepn: **GB 1006156**; J. Schmutz, F. Hunziker, **US 3389139** (1965, 1968 both to Wander); Hunziker *et al.*, *Helv. Chim. Acta* **49**, 1433 (1966). Pharmacology: Y. Take *et al.*, *J. Takeda Res. Lab.* **29**, 416 (1970); G. Stille *et al.*, *Psychopharmacologia* **28**, 325 (1973). Toxicology study: H. Yokotani *et al.*, *J. Takeda Res. Lab.* **29**, 441 (1970).

Yellow, prismatic crystals from acetone-petrol ether, mp 136-138°. LD_{50} in male, female mice, rats (mg/kg): 61, 61, 60, 66 i.v.; 250, 300, 480, 420 s.c., 270, 280, 660, 720 orally (Yokotani).

THERAP CAT: Hypnotic.

7294. Permethrin. [52645-53-1] 3-(2,2-Dichloroethenyl)-2,2-dimethylcyclopropanecarboxylic acid (3-phenoxyphenyl)methyl ester; 3-(phenoxyphenyl)methyl (±)-*cis,trans*-3-(2,2-dichloroethenyl)-2,2-dimethylcyclopropanecarboxylate; *m*-phenoxybenzyl (±)-*cis,trans*-3-(2,2-dichlorovinyl)-2,2-dimethylcyclopropanecarboxylate; FMC-33297; NIA-33297; NRDC-143; PP-557; SBP-1513; S-3151; Ambush; Coopex; Dragnet; Ectiban; Eksmin; Elimite; Nix; Perigen; Pounce; Permasect; Ridect. $C_{21}H_{20}Cl_2O_3$; mol wt 391.29. C 64.46%, H 5.15%, Cl 18.12%, O 12.27%. Synthetic pyrethroid insecticide, more stable to light and at least as active as the natural pyrethrins and with low mammalian toxicity: M. Elliott *et al.*, *Nature* **246**, 169 (1973). Of the four possible isomers, the (1*R,trans*)- and the (1*R,cis*)-isomers are the two esters primarily responsible for insecticidal activity: P. E. Burt *et al.*, *Pestic. Sci.* **5**, 791 (1974). Prepn of the racemic mixture: T. Mizutani *et al.*, **DE 2437882** (1975 to Sumitomo); F. Mori *et al.*, **DE 2544150**; *eidem*, **US 4113968** (1976, 1978, both to Kuraray). Metabolism: M. Elliott *et al.*, *J. Agric. Food Chem.* **24**, 270 (1976); L. C. Gaughan *et al.*, *ibid.* **26**, 613 (1978). Photodecompn: R. L. Holmstead *et al.*, *ibid.* 590. Analysis: H. Swaine, M. J. Tandy, *Anal. Methods Pestic. Plant Growth Regul.* **13**, 103-120 (1984). Field evaluation to control flies and ticks in cows: S. S. Quisenberry, D. R. Strohbehn, *J. Econ. Entomol.* **77**, 422 (1984); R. B. Davey, E. H. Ahrens, *Am. J. Vet. Res.* **45**, 1008 (1984). Clinical trial in pediatric pediculosis: E. Ares Mazas *et al.*, *Int. J. Dermatol.* **24**, 603 (1985). Toxicity studies in mammals: L. Metker *et al.*, *U.S. NTIS AD Rep.* **1977**, AD-AO47284, 70 pp; F. Cantalamessa, *Arch. Toxicol.* **67**, 510 (1993). *Review:* C.

N. E. Ruscoe, *Pestic. Sci.* **8**, 236 (1977). Review of toxicology and human exposure: *Toxicological Profile for Pyrethrins and Pyrethroids* (PB2004-100004, 2003) 332 pp.

Technical material is a mixture of ~60% *trans*- and 40% *cis*-isomers: Colorless crystals to a pale yellow viscous liquid, mp ~35°. $bp_{0.05}$ 220°. d^{20} 1.190-1.272. Vapor pressure at 50° <1 × 10^{-6} mm Hg. Soly in water: <1 ppm. Sol or miscible with org solvents except ethylene glycol. LD_{50} orally in female rats: 3801 mg/kg (Metker). LD_{50} in 8 day old rats, male adult rats (mg/kg): 340.5, 1500.0 orally (Cantalamessa). Toxic to bees and fish.

Caution: Mild irritant to skin and eyes (Metker).

USE: Insecticide.

THERAP CAT: Ectoparasiticide.

THERAP CAT (VET): Ectoparasiticide.

7295. Pernambuco. Fernambuco; Brazil wood; Nicaragua wood; Lima wood; redwood. Wood of *Caesalpinia echinata* Lam., *Leguminosae.* *Habit. Constit.* Brazilin.

USE: Dyeing red; manuf of a red lake pigment. With alkalies = purplish-red; with acids = yellow.

7296. Perospirone. [150915-41-6] *rel*-(3a*R*,7a*S*)-2-[4-[4-(1,2-Benzisothiazol-3-yl)-1-piperazinyl]butyl]hexahydro-1*H*-isoindole-1,3(2*H*)-dione; *cis-N*-[4-[4-(1,2-benzisothiazol-3-yl)-1-piperazinyl]butyl]-1,2-cyclohexanedicarboximide. $C_{23}H_{30}N_4O_2S$; mol wt 426.58. C 64.76%, H 7.09%, N 13.13%, O 7.50%, S 7.52%. Serotonin (5-HT$_2$) and dopamine (D$_2$) antagonist (SDA). Prepn (stereochem unspec): K. Ishizumi *et al.*, **EP 196096**; *eidem*, **US 4745117** (1986, 1988 both to Sumitomo). Prepn of *cis*-form: *eidem*, *Chem. Pharm. Bull.* **43**, 2139 (1995). Receptor binding profile in rat brain: T. Kato *et al.*, *Jpn. J. Pharmacol.* **54**, 478 (1990). Pharmacology: H. Sakamoto *et al.*, *Pharmacol. Biochem. Behav.* **60**, 873 (1998). Clinical trial in schizophrenia: M. Murasaki *et al.*, *Clin. Eval.* **24**, 159 (1997). Review of pharmacological profile: Y. Ohno *et al.*, *Pol. J. Pharmacol.* **49**, 213-219 (1997); and clinical experience: S. V. Onrust, K. McClellan, *CNS Drugs* **15**, 329-337 (2001).

Relative stereochemistry

Hydrochloride. [129273-38-7] SM-9018; Lullan. $C_{23}H_{30}N_4$- $O_2S \cdot HCl$; mol wt 463.04. Prepd as dihydrate, mp 192-193°.

THERAP CAT: Antipsychotic.

7297. Perphenazine. [58-39-9] 4-[3-(2-Chloro-10*H*-phenothiazin-10-yl)propyl]-1-piperazineethanol; 2-chloro-10-[3-[1-(2-hydroxyethyl)-4-piperazinyl]propyl]phenothiazine; 1-(2-hydroxyethyl)-4-[3-(2-chloro-10-phenothiazinyl)propyl]piperazine; chlorpiprazine; chlorpiprozine; PZC; Sch-3940; Trilafon; Trilifan; Decentan; Fentazin; Perphenan. $C_{21}H_{26}ClN_3OS$; mol wt 403.97. C 62.44%, H 6.49%, Cl 8.78%, N 10.40%, O 3.96%, S 7.94%. Prepn: Cusic, **US 2766235** (1956); Sherlock, Sperber, **US 2860138** (1958 to Schering). Metabolism: U. Breyer, H. J. Gaertner, *Adv. Biochem. Psychopharmacol.* **9**, 167 (1974); H. J. Gaertner *et al.*, *Drug Metab. Dispos.* **3**, 437 (1975). Crystal structure: J. J. H. McDowell, *Acta Crystallogr.* **B34**, 686 (1978).

White crystals. Sensitive to light. mp 94-100°. $bp_{0.15}$ 214-218°; bp_1 278-281°. Freely sol in chloroform. Soly (mg/ml): ethanol 153; acetone 82. Practically insol in water, sesame oil.

Dihydrochloride. $C_{21}H_{28}Cl_3N_3OS$. Crystals from alcohol, mp 225-226°.

THERAP CAT: Antipsychotic.

THERAP CAT (VET): Tranquilizer, pre-anesthetic agent.

7298. Pertuzumab. [380610-27-5] Anti-(human neu (receptor)) immunoglobulin G1 (human-mouse monoclonal 2C4 heavy chain) disulfide with human-mouse monoclonal 2C4 κ-chain, dimer; rhuMAb 2C4; omnitarg; R-1273. Humanized monoclonal antibody directed against human epidermal growth factor receptor 2, HER2/Erb2, acting as HER dimerization inhibitor (HDI). Description and use in cancer treatment: C. W. Adams *et al.*, **WO 0100245**; M. Sliwkowski, **US 6949245** (2001, 2005 both to Genentech). Prepn and pharmacology: C. W. Adams *et al.*, *Cancer Immunol. Immunother.* **55**, 717 (2006). Co-crystal structure of receptor complex and dimerization inhibition studies: M. C. Franklin *et al.*, *Cancer Cell* **5**, 317 (2004). Clinical pharmacokinetics and antitumor activity: D. B. Agus *et al.*, *J. Clin. Oncol.* **23**, 2534 (2005). Clinical evaluation in breast cancer: L. Gianni *et al.*, *ibid.* **28**, 1131 (2010); with trastuzumab: J. Baselga *et al.*, *ibid.* 1138. Review of development and therapeutic potential: J. Spicer, *Curr. Opin. Mol. Ther.* **6**, 337-343 (2004); K. Kristjansdottir, D. Dizon, *Expert Opin. Biol. Ther.* **10**, 243-250 (2010).

THERAP CAT: Antineoplastic.

7299. Perylene. [198-55-0] Dibenz[*de,kl*]anthracene; *peri*-dinaphthalene. $C_{20}H_{12}$; mol wt 252.32. C 95.20%, H 4.79%. Occurs in coal tar. Isoln from pitch distillate: Cook *et al.*, *J. Chem. Soc.* **1933**, 395. From 2,2′-dihydroxy-1,1′-dinaphthyl: Zinke *et al.*, *Monatsh. Chem.* **64**, 415 (1934). From the reaction of phenanthrene with acrolein in anhydr HF: Weinmayr, **US 2145905** (1939); *cf.* *J. Am. Chem. Soc.* **61**, 949 (1939).

Yellow to colorless crystals from toluene. mp 273-274°. Sublimes 350-400°. d 1.35. Absorption spectrum: Clar, *Ber.* **65**, 848 (1932). Freely sol in CS_2, chloroform; moderately sol in benzene; slightly in ether, alcohol, acetone; very sparingly sol in petr ether. Insol in water.

Monopicrate. $C_{26}H_{15}N_3O_7$. Dark violet-blue needles, mp 223-225°.

7300. Petrolatum. Petroleum jelly; paraffin jelly; vasoliment; Cosmoline; Saxoline; Stanolene; Vaseline. Purified mixture of semisolid hydrocarbons, chiefly of the methane series of the general formula C_nH_{2n+2}. A colloidal system of nonstraight-chain solid hydrocarbons and high-boiling liq hydrocarbons, in which most of the liq hydrocarbons are held inside the micelles. Detailed historical account including chemistry and modern mfg methods: Schindler, *Drug Cosmet. Ind.* **89**, 36-37, 76, 78-80, 82 (1961).

Yellowish to light amber or white, semisolid, unctuous mass; practically odorless and tasteless. d_{25}^{60} 0.820-0.865. mp 38-54°. n_D^{60} 1.460-1.474. White petrolatum is transparent in thin layers even at

0°. Freely sol benzene, carbon disulfide, chloroform, turpentine oil; sol in solvent hexane, ether, petr ether, in most fixed and volatile oils. Practically insol in glycerol, alc; insol in water.

USE: As ointment base in pharmaceuticals and cosmetics. Lubricating firearms and machinery, leather grease, shoe polish, rust preventives, modeling clays.

7301. Petrolatum, Liquid. Liquid paraffin; mineral oil; white mineral oil; paraffin oil; Clearteck; Drakeol; Hevyteck; Kremol; Kaydol; Alboline; Nujol; Paroleine; Saxol; Adepsine oil; Glymol. A mixture of liquid hydrocarbons from petroleum.

Colorless, oily liquid; practically tasteless and odorless even when warmed. The density of the "light" oil is usually 0.83-0.860; the "heavy" 0.875-0.905. Surface tension at 25° slightly below 35 dynes/cm. Insol in water, alc. Sol in benzene, chloroform, ether, carbon disulfide, petr ether, oils.

USE: Lubricant. Pharmaceutic aid (vehicle, solvent). As formulation aid in foods. In cosmetics as emollient.

THERAP CAT: Cathartic.

THERAP CAT (VET): Laxative, externally as a protectant, lubricant.

7302. Petroleum. Crude oil; mineral oil; rock oil; coal oil; seneca oil. Consists of a mixture of hydrocarbons from C_2H_6 and up—chiefly of the paraffins, cycloparaffins, or of cyclic aromatic hydrocarbons, with small amounts of benzene hydrocarbons, sulfur, and oxygenated compounds. *Occurrence:* U.S., Mexico, Iran, Russia, Roumania, Poland, Dutch East Indies, etc.

Dark yellow to brown or greenish-black, oily liquid. Insol in water and only a small portion of it may dissolve in alcohol; sol in benzene, chloroform, ether.

USE: Source of gasoline, petr ether, liq and solid petrolatum, fuel and lubricating oils, butane, isopropyl alcohol, etc.

7303. Petroleum Benzin. [8030-30-6] Naphtha; benzin; petroleum naphtha. Term that has been applied to low boiling fractions of petroleum, consisting chiefly of hydrocarbons of the methane series, principally pentanes and hexanes.

Clear, colorless, nonfluorescent, *highly flammable*, volatile liq; characteristic odor; does not solidify in the cold. The vapors mixed with air explode if ignited. *Keep tightly closed in a cool place and away from fire.* d 0.625-0.660; bp between 35-80°. Insol in water. Miscible with abs alc, benzene, chloroform, ether, carbon disulfide, carbon tetrachloride, and oils except castor oil.

Caution: Potential symptoms of overexposure are lightheadedness, drowsiness; irritation of eyes, nose and skin; dermatitis. *See NIOSH Pocket Guide to Chemical Hazards* (DHHS/NIOSH 97-140, 1997) p 220.

USE: Pharmaceutic aid (solvent).

THERAP CAT: Counterirritant.

7304. Petroselinic Acid. [593-39-5] (6Z)-6-Octadecenoic acid; *cis*-6-Octadecenoic acid; petroselic acid; *cis*-5-heptadecylene-1-carboxylic acid; *cis*-Δ⁶-octadecylenic acid. $C_{18}H_{34}O_2$; mol wt 282.47. C 76.54%, H 12.13%, O 11.33%. Isoln from parsley seed oil, the oil extracted from dried ripe seed of *Petroselinum hortense* Hoffm., *Umbelliferae:* Fore *et al., J. Am. Oil Chem. Soc.* **37**, 490 (1960).

Leaflets from petr ether, mp 29.5-30.1°. bp$_{18}$ 237-238°; d$_4^{40}$ 0.8700. n_D^{40} 1.4533. Low temp solubilities: Heptane at −10° = 0.50 g/100 g solution; methanol at −20° = 0.48 g/100 g; ethyl acetate at −20° = 0.73 g/100 g; ether at −20° = 3.52 g/100 g. Ozonolysis yields 85% of adipic acid. Neutralization equivalent: 282.45; iodine value 89.87%.

Methyl ester. [2777-58-4] $C_{19}H_{36}O_2$; mol wt 296.50. Liq, d$_4^{20}$ 0.8767; n_D^{20} 1.4501; bp$_{10}$ 208-210°.

Glyceryl triester. [3296-43-3] Glyceryl tripetroselinate; tripetroselin. $C_{57}H_{104}O_6$; mol wt 885.45. Solidifies at 16.5°. n_D^{40} 1.4619.

Amide. [24222-02-4] Petroselinamide. $C_{18}H_{35}NO$; mol wt 281.48. Needles, mp 76°.

7305. Petunidin. [1429-30-7] 2-(3,4-Dihydroxy-5-methoxyphenyl)-3,5,7-trihydroxy-1-benzopyrylium chloride (1:1); 3,3′,4′,-

5,7-pentahydroxy-5′-methoxyflavylium chloride; petunidol. C_{16}-$H_{13}ClO_7$; mol wt 352.72. C 54.48%, H 3.72%, Cl 10.05%, O 31.75%. The aglucone of petunin: Willstätter, Burdick, *Ann.* **412**, 217 (1917). Synthesis: Bradley *et al., J. Chem. Soc.* **1930**, 793; Robinson, Robinson, *Biochem. J.* **25**, 1687 (1931). Chromatographic separation: Spaeth, Rosenblatt, *Anal. Chem.* **22**, 1321 (1950).

Gray-brown leaflets or prisms from dil HCl.

3,5-Diglucoside. [25846-73-5] 2-(3,4-Dihydroxy-5-methoxyphenyl)-3,5-bis(β-D-glucopyranosyloxy)-7-hydroxy-1-benzopyrylium chloride; petunin. $C_{28}H_{33}ClO_{17}$; mol wt 677.01. From *Petunia hybrida* Hort., *Solanaceae:* Willstätter, Burdick, *loc. cit.* Synthesis: Bell, Robinson, *J. Chem. Soc.* **1934**, 1604. Violet plates with a coppery luster from dil HCl, mp ~178°. Absorption max (methanolic HCl): 540 nm.

7306. Peucedanin. [133-26-6] 3-Methoxy-2-(1-methylethyl)-7H-furo[3,2-g][1]benzopyran-7-one; 6-hydroxy-2-isopropyl-3-methoxy-5-benzofuranacrylic acid δ-lactone; 4-methoxy-5-isopropylfuro[2,3:6,7]coumarin; oreoselone methyl ether. $C_{15}H_{14}O_4$; mol wt 258.27. C 69.76%, H 5.46%, O 24.78%. Coumarin deriv obtained from rhizome of *Peucedanum officinale* L., *Umbelliferae:* Schlatter, *Ann.* **5**, 201 (1833); Hlasiwetz, Weidel, *ibid.* **174**, 67 (1874); A. Jassoy, P. Haensel, *Arch. Pharm.* **236**, 662 (1898); Popper, *Monatsh. Chem.* **19**, 268 (1898); from *Peucedanum morisonii*, Bess., *Umbelliferae:* G. K. Nikonov, A. A. Ivashenko, *Zh. Obshch. Khim.* **33**, 2740 (1963). Fluorescence spectrum: R. H. Goodwin, F. Kavanagh, *Arch. Biochem.* **27**, 182 (1950). Antitumor activity and toxicity studies: E. M. Vermel, S. A. Kruglyak-Syrkina, *Vopr. Onkol.* **5**, 43 (1959), *C.A.* **53**, 19162h (1959). Structure: E. Späth *et al., Ber.* **64**, 2203 (1931); E. Späth, K. Klager, *ibid.* **66**, 749 (1933). Synthesis: H. Schmid, A. Ebnöther, *Helv. Chim. Acta* **34**, 1982 (1951).

Colorless needles from ether/petr ether, mp 84-87°; also reported as mp 95-97° (Schmid, Ebnöther); from ligroin, mp 102.5° (Nikonov, Ivashenko). uv max (methanol): 255, 295, 340 nm (log ε 4.40, 4.05, 3.70). Practically insol in water. Freely sol in chloroform, CS_2; sol in hot alcohol, ether, acetic acid; sparingly sol in benzene, petr ether. LD_{50} orally in mice: 315 mg/kg (Vermel, Kruglyak-Syrkina).

7307. Pexelizumab. [219685-93-5] Anti-(human complement C5 α-chain) immunoglobulin (human-mouse monoclonal 5G1.1-SC single chain); h5G1.1-scFv. Single-chain fragment of humanized monoclonal antibody directed against complement component C5; designed to prevent complement-mediated inflammation and tissue injury. Prepn: M. J. Evans *et al.,* **WO 9529697**; *eidem,* **US 6355245** (1995, 2002 both to Alexion); and complement inhibition study: T. C. Thomas *et al., Mol. Immunol.* **33**, 1389 (1996). Clinical pharmacology: J. C. K. Fitch *et al., Circulation* **100**, 2499 (1999). Clinical studies in acute myocardial infarction: K. W. Mahaffey *et al., ibid.* **108**, 1176 (2003), C. B. Granger *et al., ibid.* 1184; in coronary artery bypass graft surgery: E. D. Verrier *et al., J. Am. Med. Assoc.* **291**, 2319 (2004). Review of development and therapeutic potential: P. A. Whiss, *Curr. Opin. Investig. Drugs* **3**, 870-

877 (2002); A. J. Fleisig, E. D. Verrier, *Expert Opin. Biol. Ther.* **5**, 833-839 (2005).

THERAP CAT: Anti-inflammatory.

7308. Pexiganan. [147664-63-9] Glycyl-L-isoleucylglycyl-L-lysyl-L-phenylalanyl-L-leucyl-L-lysyl-L-lysyl-L-alanyl-L-lysyl-L-lysyl-L-phenylalanylglycyl-L-lysyl-L-alanyl-L-phenylalanyl-Lvalyl-L-lysyl-Lisoleucyl-L-leucyl-L-lysyl-L-lysinamide; 7-L-lysine-8-L-lysine-10-L-lysine-18-L-lysine-19-*de*-L-glutamic acid-21-L-leucine-23-L-lysinamide magainin I. $C_{122}H_{210}N_{32}O_{22}$; mol wt 2477.22. C 59.15%, H 8.55%, N 18.09%, O 14.21%. Synthetic, 22 amino acid peptide; analog of the magainins, *q.v.* Prepn: B. Berkowitz, L. Jacob, **WO 9301723** (1993 to Magainin). *In vitro* antibacterial spectrum: Y. Ge *et al.*, *Diagn. Microbiol. Infect. Dis.* **35**, 45 (1999). Review of development: L. Jacob, M. Zasloff, *Ciba Found. Symp.* **186**, 197-216 (1994); of structure-activity studies: W. L. Maloy, U. P. Kari, *Biopolymers* **37**, 105-122 (1995); of pharmacology and clinical trials in diabetic foot ulcers: H. M. Lamb, L. R. Wiseman, *Drugs* **56**, 1047-1052 (1998).

Gly–Ile–Gly–Lys–Phe–Leu–Lys–Lys–Ala–Lys–Lys
|
H₂N–Lys–Lys–Leu–Ile–Lys–Val–Phe–Ala–Lys–Gly–Phe

Acetate. [172820-23-4] MSI-78; Locilex. $C_{122}H_{210}N_{32}O_{22}.xC_2H_4O_2$

THERAP CAT: Antibacterial.

7309. Pfeiffer's Substance. [8015-18-7] 5,5-Diethyl-2,4,6-(1*H*,3*H*,5*H*)-pyrimidinetrione mixture with 4-(dimethylamino)-1,2-dihydro-1,5-dimethyl-2-phenyl-3*H*-pyrazol-3-one; 4-(dimethylamino)antipyrine compd with 5,5-diethylbarbituric acid (1:1); corps de Pfeiffer. $C_{21}H_{29}N_5O_4$; mol wt 415.49. C 60.71%, H 7.04%, N 16.86%, O 15.40%. A molecular compd of aminopyrine and barbital, prepd by repeated crystn of stoichiometric amounts of the components from a minimum of hot water: Pfeiffer, *Z. Physiol. Chem.* **146**, 98 (1925).

Silky needles, mp 113-115°. At 115° the melt is turbid, but turns into a clear liquid at 140°. Freely sol in water. The aminopyrine moiety dissolves in benzene, while the barbital moiety precipitates out.

7310. P-Glycoprotein. Glycoprotein-P; P-gp; P-170; permeability glycoprotein. Highly conserved cell-surface glycoprotein of M_r 170,000 daltons; protein product of the mammalian multidrug resistance (*mdr*) gene. Present in selective human tissues where it functions as an energy-dependent efflux pump. Increased expression in mammalian tumor cells is associated with multidrug resistance (MDR) to cancer chemotherapy agents. Identification and isolation from colchicine-resistant Chinese hamster ovary (CHO) cells: R. J. Juliano, V. Ling, *Biochim. Biophys. Acta* **455**, 152 (1976). Purification from plasma membrane of CHO cells: J. R. Riordan, V. Ling, *J. Biol. Chem.* **254**, 12701 (1979). MDR associated with P-glycoprotein in mammalian cell lines: N. Kartner *et al.*, *Science* **221**, 1285 (1983); N. Kartner *et al.*, *Cancer Res.* **43**, 4413 (1983). Deduced amino acid sequence from mouse *mdr* cDNA: P. Gros *et al.*, *Cell* **47**, 371 (1986); from human *mdr*1 cDNA: C.-j. Chen *et al.*, *ibid.* 381. Transfection and expression of human *mdr*1 gene in cell culture: K. Ueda *et al.*, *Proc. Natl. Acad. Sci. USA* **84**, 3004 (1987). Characterization of ATPase activity: H. Hamada, T. Tsuruo, *Cancer Res.* **48**, 4926 (1988). Detection in normal human tissues: F. Thiebaut *et al.*, *Proc. Natl. Acad. Sci. USA* **84**, 7735 (1987); R. N. Hitchins *et al.*, *Eur. J. Cancer Clin. Oncol.* **24**, 449 (1988); and in human tumor cells: D. R. Bell *et al.*, *J. Clin. Oncol.* **3**, 311 (1985); W. S. Dalton *et al.*, *Blood* **73**, 747 (1989). Review of role in MDR in human cancer: I. Pastan, M. Gottesman, *N. Engl. J. Med.* **316**, 1388 (1987); of biochemistry of MDR: J. A. Endicott, V. Ling, *Annu. Rev. Biochem.* **58**, 137-171 (1989); of mechanism of action: S. Ruetz, P. Gros, *Trends Pharmacol. Sci.* **15**, 260-263 (1994).

7311. Phalloidin. [17466-45-4] Phalloidine. $C_{35}H_{48}N_8O_{11}$S; mol wt 788.87. C 53.29%, H 6.13%, N 14.20%, O 22.31%, S 4.06%. Best known of the toxins isolated from the poisonous green fungus *Amanita phalloides* (Fr.) Seer., *Agaricaceae*, known as "the green death cap" or "deadly agaric": Lynen, Wieland, *Ann.* **533**, 93

(1938). Structure: Wieland, Schön, *ibid.* **593**, 157 (1955); Wieland, Schöpf, *ibid.* **626**, 174 (1959); Wieland, Schnabel, *ibid.* **657**, 225 (1962). Differs from amanitin in rapidity of action; at high dose levels, death of mice or rats occurs within 1 or 2 hours. Phalloidin acts by binding actin, *q.v.*, an essential internal structural protein. Ultrastructural pathology: M. A. Russo *et al.*, *Am. J. Pathol.* **109**, 133 (1982). Toxicity study: Vogt, *Arch. Exp. Pathol. Pharmakol.* **190**, 406 (1938). Review of the chemistry and toxicology of the toxins of *Amanita phalloides:* Wieland, Wieland, *Pharmacol. Rev.* **11**, 87-107 (1959); *see also* T. Wieland, *Fortschr. Chem. Org. Naturst.* **25**, 214-250 (1967); T. Wieland, H. Faulstich, *Crit. Rev. Biochem.* **5**, 185-260 (1978).

Hexahydrate, needles from water, mp 280-282°. uv max (water): 295 nm ($E_{1cm}^{1\%}$ 0.597). Soly in water (0°): 0.5%; much more sol in hot water; freely sol in methanol, ethanol, butanol, pyridine. LD_{50} i.m. in albino mice: 3.3 μg/g (Vogt). LD_{50} i.p. in mice: 2 mg/kg (Wieland, Wieland).

Caution: See Amanitin.

7312. Phanquinone. [84-12-8] 4,7-Phenanthroline-5,6-dione; 4,7-phenanthroline-5,6-quinone; 5,6-dioxo-5,6-dihydro-4,7-phenanthroline; phanchinone; phanquone; Ciba 11925; Entobex. $C_{12}H_6N_2O_2$; mol wt 210.19. C 68.57%, H 2.88%, N 13.33%, O 15.22%. Prepd from 5,(6)-methoxy-4,7-phenanthroline: Druey, Schmidt, *Helv. Chim. Acta* **33**, 1080 (1950); **GB 688802** (1953 to Ciba).

Crystals from methanol, mp 295° (dec). Sparingly sol in water, alcohol; sol in dil mineral acids.

THERAP CAT: Antiamebic.

7313. Phaseolin. [13401-40-6] (6b*R*,12b*R*)-6b,12b-Dihydro-3,3-dimethyl-3*H*,7*H*-furo[3,2-*c*:5,4-*f* ']bis[1]benzopyran-10-ol; phaseollin. $C_{20}H_{18}O_4$; mol wt 322.36. C 74.52%, H 5.63%, O 19.85%. Antifungal phytoalexin isolated from French bean *(Phaseolus vulgaris* L., *Leguminosae):* I. A. M. Cruickshank, D. R. Perrin, *Life Sci.* **2**, 680 (1963). Structure: D. R. Perrin, *Tetrahedron Lett.* **1964**, 29; D. R. Perrin *et al.*, *ibid.* **1972**, 1673. Crystal structure: C. DeMartinis *et al.*, *Tetrahedron* **34**, 1849 (1978). Biosynthesis: S. L. Hess *et al.*, *Phytopathology* **61**, 79 (1971); P. M. Dewick, M. J. Steele, *Phytochemistry* **21**, 1599 (1982). Total synthesis: S. E. N. Mohamed *et al.*, *J. Chem. Soc. Perkin Trans. 1* **1987**, 431. Mode of action: F. D. Van Etten, D. F. Bateman, *Phytopathology* **61**, 1363 (1971). Antifungal properties: I. A. M. Cruickshank, D. R. Perrin, *Phytopathol. Z.* **70**, 209 (1971); M. A. Gordon *et al.*, *Antimicrob. Agents Chemother.* **17**, 120 (1980).

Consult the Name Index before using this section.

Crystals, mp 177-178°. $[\alpha]_{578}$ −145°. pKa 9.13. uv max (ethanol): 207, 230, 280, 286 nm (log ε 4.68, 4.40, 3.97, 3.90).

7314. Phasin. [1392-87-6] Poisonous agglutinin from beans. A polypeptide composed of glutamic acid, aspartic acid, serine, alanine, tyrosine, lysine and arginine: Piekarski, *Diss. Pharm.* **9**, 255 (1957), *C.A.* **52**, 6456i (1958). Prepn: Wienhaus, *Biochem. Z.* **18**, 228 (1909), *C.A.* **3**, 2471 (1909).

Amorphous powder. Its toxicity and agglutination properties are destroyed by heating.

7315. α-Phellandrene. [99-83-2] 2-Methyl-5-(1-methylethyl)-1,3-cyclohexadiene; *p*-mentha-1,5-diene; 5-isopropyl-2-methyl-1,3-cyclohexadiene; 4-isopropyl-1-methyl-1,5-cyclohexadiene. $C_{10}H_{16}$; mol wt 136.24. C 88.16%, H 11.84%. Isoln of *l*-form from essential oils of *Eucalyptus dives* Schau. and *E. phellandra* Baker & Smith, *Myrtaceae*: Smith *et al.*, *J. Chem. Soc.* **123**, 1657 (1923). Isoln of *d*-form from oil of bitter fennel *(Foeniculum vulgare* Mill., *Umbelliferae)*: Wallach, *Ann.* **336**, 9 (1904). Structure: Semmler, *Ber.* **36**, 1749 (1903). Synthesis of *dl*-form: Read, Storey, *J. Chem. Soc.* **1930**, 2770; B. Singaram, J. Verghese, *Indian J. Chem.* **14B**, 1003 (1976). Configuration: Burgstahler *et al.*, *J. Am. Chem. Soc.* **83**, 4660 (1961). Circular dichroism: G. Snatzke *et al.*, *Tetrahedron Lett.* **1966**, 4551. Review: J. L. Simonsen, *The Terpenes* **vol. 1** (Univ. Press, Cambridge, 2nd ed., 1947) pp 193-204; B. Singaram, J. Verghese, *Perfum. Flavor.* **2**, 33-38 (1978).

l-Form. [4221-98-1] Mobile oil. bp_{758} 171-172°; bp_{16} 58-59°. d_4^{20} 0.8410. n_D^{20} 1.4709. $[\alpha]_D^{20}$ −217°. Practically insol in water; sol in ether.

d-Form. [2243-33-6] Mobile oil. bp_{16} 66-68°. d_4^{25} 0.8463. n_D^{25} 1.4777. $[\alpha]_D^{16}$ +86.4°. Practically insol in water; sol in ether.

Caution: Can be irritating to, and absorbed through, skin. Ingestion can cause vomiting, diarrhea.

USE: In fragrances.

7316. β-Phellandrene. [555-10-2] 3-Methylene-6-(1-methylethyl)cyclohexene; *p*-mentha-1(7),2-diene; 3-isopropyl-6-methylene-1-cyclohexene; 4-isopropyl-1-methylene-2-cyclohexene. $C_{10}H_{16}$; mol wt 136.24. C 88.16%, H 11.84%. Isoln of *d*-form from oil of water fennel *(Phellandrium aquaticum* L., *Umbelliferae)*: Berry *et al.*, *J. Chem. Soc.* **1937**, 1448. Isoln of *l*-form from Canada balsam oil: Macbeth *et al.*, *ibid.* **1938**, 119. Structure: Wallach, *Ann.* **343**, 28 (1905). Synthesis of *dl*-form: Deorha, Sareen, *Rec. Trav. Chim.* **82**, 137 (1965); B. Singaram, J. Verghese, *Indian J. Chem.* **14B**, 1003 (1976).

d-Form. [6153-16-8] Mobile oil. bp_{760} 171-172°; bp_{11} 57°. d_4^{20} 0.8520. n_D^{20} 1.4788. $[\alpha]_D^{20}$ +65.2°. Practically insol in water; sol in ether.

l-Form. [6153-17-9] Mobile oil. bp_{758} 178-179°; bp_{12} 53°. d_{15}^{15} 0.8497. n_D^{20} 1.4800. $[\alpha]_D^{20}$ −51.9°. Practically insol in water, alcohol; sol in ether.

7317. Phenacaine Hydrochloride. [620-99-5] *N,N'*-Bis(4-ethoxyphenyl)ethanimidamide hydrochloride (1:1); N^1,N^2-bis(*p*-ethoxyphenyl)acetamidine hydrochloride; Holocaine Hydrochloride. $C_{18}H_{23}ClN_2O_2$; mol wt 334.84. C 64.57%, H 6.92%, Cl 10.59%, N 8.37%, O 9.56%. Prepn: **DE 79868**; **DE 80568**.

Monohydrate. [6153-19-1] Faintly bitter crystals producing transient numbness of the tongue. Stable in air. When anhydr, mp 190-192°. One gram dissolves in 50 ml water. Freely sol in boiling water, alcohol, chloroform. Insol in ether. Incompatible with alkalies and their carbonates and the usual alkaloidal reagents. Aq solns are stable and are not dec by boiling.

THERAP CAT: Topical anesthetic (ophthalmic).

THERAP CAT (VET): Ocular anesthetic.

7318. Phenacemide. [63-98-9] *N*-(Aminocarbonyl)benzeneacetamide; (phenylacetyl)urea; phenacetylurea; Epiclase; Phacetur; Phenurone; Phetylureum. $C_9H_{10}N_2O_2$; mol wt 178.19. C 60.67%, H 5.66%, N 15.72%, O 17.96%. Prepd by the action of aq NH_3 on phenacetylurethan: Basterfield, Greig, *Can. J. Res.* **8**, 454 (1933); of phenacetyl chloride on urea: Spielman *et al.*, *J. Am. Chem. Soc.* **70**, 4189 (1948). Toxicity study: K. Nakamura *et al.*, *Arzneim.-Forsch.* **18**, 524 (1968).

Crystals from alcohol, mp 212-216°. Very slightly sol in water; slightly sol in alcohol, benzene, chloroform, ether. LD_{50} in mice, rats (mmol/kg): 5.54, >10 orally (Nakamura).

THERAP CAT: Anticonvulsant.

7319. Phenacetin. [62-44-2] *N*-(4-Ethoxyphenyl)acetamide; *p*-acetophenetidide; *p*-ethoxyacetanilide; acetophenetidin; *para*-acetophenetidin; *p*-acetophenetide. $C_{10}H_{13}NO_2$; mol wt 179.22. C 67.02%, H 7.31%, N 7.82%, O 17.85%. Prepn: *Beilstein* **vol. XIII**, 461. Improved process: Eaker, Campbell, **US 2887513** (1959 to Monsanto). *Monograph*: P. K. Smith, *Acetophenetidin* (Interscience, New York, 1958) 180 pp. Toxicity: Boyd, *Toxicol. Appl. Pharmacol.* **1**, 240 (1959). Review of toxicity and metabolism studies: L. Fishbein, *IARC Sci. Publ.* **40**, 287-310 (1981). Epidemiologic study of renal morbidity and mortality: U. C. Dubach *et al.*, *N. Engl. J. Med.* **308**, 357 (1983). Evaluation of renal effects: D. P. Sandler *et al.*, *ibid.* **320**, 1238 (1989).

Slightly bitter, cryst scales or powder. mp 134-135°. One gram dissolves in 1310 ml cold water, 82 ml boiling water; 15 ml cold alcohol, 2.8 ml boiling alcohol; 14 ml chloroform, 90 ml ether; sol in glycerol. Gives a pasty mass with phenol, chloral hydrate or pyrocatechol, strong acids or alkalies, salicylic acid, oxidizers, iodine, spirit nitrous ether. LD_{50} orally in rats: 1.65 g/kg (Boyd).

Note: Component of *APC* tablets, analgesic mixture also containing aspirin and caffeine.

Caution: Phenacetin is reasonably anticipated to be a human carcinogen; analgesic mixtures containing phenacetin are listed as known human carcinogens: *Report on Carcinogens, Twelfth Edition* (PB2011-111646, 2011) p 340.

THERAP CAT: Analgesic, antipyretic.

THERAP CAT (VET): Analgesic, antipyretic.

7320. Phenacetolin. [1340-26-7] Degener's indicator. Reaction product of concd H_2SO_4 and glacial acetic acid on phenol. Yellowish-brown powder. Slightly sol in water; sol in alc.

USE: Has been employed as indicator, particularly for mixtures of alkali hydroxides and carbonates; the yellow color changes to red when hydroxide is neutralized, and again to yellow when carbonate is fully dec by acid; was also used for determining alkalinity of water. The indicator soln is prepd by digesting 1 g with 100 ml warm alcohol and filtering after cooling; 2-3 drops used for 100 ml liquid.

7321. Phenacylamine. [613-89-8] 2-Amino-1-phenylethanone; 2-aminoacetophenone; α-aminoacetophenone; ω-aminoacetophenone. C_8H_9NO; mol wt 135.17. C 71.09%, H 6.71%, N 10.36%, O 11.84%. Prepn from α-phenylethylamine: H. E. Baumgarten, J. M. Petersen, *J. Am. Chem. Soc.* **82**, 459 (1960); *eidem, Org. Synth.* **coll. vol. V**, 909 (1973).

Hydrochloride. $C_8H_9NO.HCl$. Crystals from isopropanol + HCl, dec 188.5°.

7322. Phenamidine. [101-62-2] 4,4'-Oxybisbenzenecarboximidamide; 4,4'-oxydibenzamidine; 4,4'-diamidinodiphenyl ether. $C_{14}H_{14}N_4O$; mol wt 254.29. C 66.13%, H 5.55%, N 22.03%, O 6.29%. Prepn: A. J. Ewins *et al.,* **GB 507565** (1939 to May & Baker); J. N. Ashley *et al., J. Chem. Soc.* **1942**, 103; of isethionate: G. Newbery, A. P. T. Easson, **US 2410796** (1946 to May & Baker). Trypanocidal activity: E. M. Lourie, W. Yorke, *Ann. Trop. Med. Parasitol.* **33**, 289 (1939). Preliminary pharmacology: R. Wien, *ibid.* **37**, 1 (1943). Use in treatment of Babesia infections in dogs: M. D. Ruff *et al., Am. J. Vet. Res.* **34**, 641 (1973); R. Gothe *et al., Kleintier-Praxis* **32**, 97 (1987). Early review of pharmacology, mode of action and clinical applications: E. M. Schoenbach, E. M. Greenspan, *Medicine* **27**, 327-377 (1948).

Irregular plates from water, mp 215-216°.

Isethionate. [620-90-6] M & B 736; Lomadine. $C_{14}H_{14}N_4O.$-$2C_2H_6O_4S$; mol wt 506.55. Crystals, mp 225° (dec). One part is sol in 1.4 parts water; 300 parts alcohol. Practically insol in ether, chloroform.

THERAP CAT (VET): Antiprotozoal (Babesia).

7323. Phenampromide. [129-83-9] N-[1-Methyl-2-(1-piperidinyl)ethyl]-N-phenylpropanamide; N-(1-methyl-2-piperidinoethyl)propionanilide; 1-piperidino-2-(N-propionylanilino)propane; phenampromid. $C_{17}H_{26}N_2O$; mol wt 274.41. C 74.41%, H 9.55%, N 10.21%, O 5.83%. Prepn and analgesic activity: W. B. Wright, Jr. *et al., J. Am. Chem. Soc.* **81**, 1518 (1959); W. B. Wright, Jr., H. J. Brabander, **US 3016382** (1962 to Am. Cyanamid). (−)-Phenampromide, the more active enantiomer, has the R-configuration. Absolute configuration: Portoghese, *J. Med. Chem.* **8**, 147 (1965). Pharmacology: Kikuchi *et al., Nippon Yakurigaku Zasshi* **57**, 585 (1961), *C.A.* **59**, 2082b (1963). Determn by HPLC: I. Jane, A. McKinnon, *J. Chromatogr.* **323**, 191 (1985).

(−)-Phenampromid(e)

Liquid, $bp_{0.2}$ 124-128°. n_D^{28} 1.518.

(−)-Form hydrochloride. $C_{17}H_{27}ClN_2O$. Crystals from alcoholic HCl + ether, mp 201-202°.

Note: This is a controlled substance (opiate): **21 CFR**, 1308.11.

7324. Phenazopyridine. [94-78-0] 3-(2-Phenyldiazenyl)-2,6-pyridinediamine; 2,6-diamino-3-phenylazopyridine; β-phenylazo-α,α'-diaminopyridine. $C_{11}H_{11}N_5$; mol wt 213.24. C 61.96%, H 5.20%, N 32.84%. Azo dye used in treatment of urinary tract infections. Prepn: A. E. Chichibabin, O. A. Zeide, *J. Russ. Phys. Chem. Soc.* **46**, 1216 (1914); *Chem. Zentralbl.* **1915**, I, 1064; I. Ostromislensky, **US 1680109** (1928 to Pyridium); and bacteriostatic properties: R. N. Shreve *et al., J. Am. Chem. Soc.* **65**, 2241 (1943). Physical characteristics: G. W. Collins, *J. Am. Pharm. Assoc.* **20**, 455 (1931). Toxicity data: B. A. Becker, J. G. Swift, *Toxicol. Appl. Pharmacol.* **1**, 42 (1959). Comprehensive description: K. W. Blessel *et al., Anal. Profiles Drug Subs.* **3**, 465-482 (1974). LC-MS determn in plasma and pharmacokinetics: E. Shang *et al., Anal. Bioanal. Chem.* **382**, 216 (2005). Review of clinical experience in urinary tract infections: S. A. Zelenitsky, G. G. Zhanel, *Ann. Pharmacother.* **30**, 866-868 (1996).

mp 136-137°.

Hydrochloride. [136-40-3] Prodium; Pyridium; Pyridacil; Sedural. $C_{11}H_{11}N_5.HCl$; mol wt 249.70. Brick-red microcrystals, slight violet luster. Slightly bitter taste. mp ~235°. Sol in boiling water, about 1 part in 20; one part is sol in about 100 parts glycerol; also sol in ethylene and propylene glycols, acetic acid; slightly sol in cold water, about 1 part in 300; also slightly sol in alc, chloroform, lanolin. Insol in acetone, benzene, ether, toluene. Aq solns are yellow to brick-red and slightly acid. Forms supersatd solns easily. Will precipitate out of a 2% soln at 25° after about 2 days, out of a 1% soln only after months. Aq solns may be stabilized by the addn of 10% glucose. LD_{50} orally in rats: 403 mg/kg (Becker, Swift).

Caution: Phenazopyridine hydrochloride is reasonably anticipated to be a human carcinogen: *Report on Carcinogens, Twelfth Edition* (PB2011-111646, 2011) p 341.

THERAP CAT: Analgesic (urinary tract).

THERAP CAT (VET): Analgesic (urinary tract).

7325. Phenothrin. [26002-80-2] 2,2-Dimethyl-3-(2-methyl-1-propen-1-yl)cyclopropanecarboxylic acid (3-phenoxyphenyl)-methyl ester; 2,2-dimethyl-3-(2-methylpropenyl)cyclopropanecarboxylic acid m-phenoxybenzyl ester; 3-phenoxybenzyl *cis,trans*-chrysanthemate; S-2539; Sumithrin. $C_{23}H_{26}O_3$; mol wt 350.46. C 78.83%, H 7.48%, O 13.70%. Synthetic pyrethroid insecticide. Prepn of racemic mixture: N. Itaya *et al.,* **DE 1926433** corresp to **US 3666789** (1969, 1972 to Sumitomo). Comparative activity of isomers: K. Fujiwo *et al., Agric. Biol. Chem.* **37**, 2681 (1973); Y. Okuno *et al.,* **DE 2348930** corresp to **US 3934023** (1973, 1976 to Sumitomo). Analysis: Y. Takimoto *et al., Anal. Methods Pestic. Plant Growth Regul.* **13**, 133-146 (1984). Review of toxicology and human exposure: *Toxicological Profile for Pyrethrins and Pyrethroids* (PB2004-100004, 2003) 332 pp.

(1R-trans)-form

The commercial product is a mixture of isomers. Colorless liquid. d_{25}^{25} 1.06; n_D^{25} 1.5483. Sol in acetone, xylene. Insol in water.

USE: Insecticide.

7326. Phenanthrene. [85-01-8] $C_{14}H_{10}$; mol wt 178.23. C 94.35%, H 5.66%. Isomeric with anthracene. Occurs in coal tar, *q.v.,* and in products of incomplete combustion. Isoln: Ostermayer,

Fittig, *Ber.* **5**, 933 (1872); Glaser, *ibid.* 982. Purification (from contaminating carbazole and anthracene): Clar, *Ber.* **65**, 852 (1932). Formation from toluene, bibenzil, 9-methylfluorene or stilbene by passage through red-hot tube: Graebe, *Ber.* **7**, 48 (1874); *Ann.* **167**, 161 (1879); *Ber.* **37**, 4145 (1904). Also from coumarone and benzene: Kraemer, Spilker, *Ber.* **23**, 85 (1890). Pschorr synthesis from *o*-nitrobenzaldehyde and phenylacetic acid: *Ber.* **29**, 500 (1896). From diphenylethylene: Cook, Hewett, *J. Chem. Soc.* **1933**, 1098. Diene synthesis from 1-vinylnaphthalene and maleic anhydride: Cohen, Warren, *ibid.* **1937**, 1315. From *o*-phenylbenzoic acid: Schönberg, Warren, *Chem. Ind. (London)* **58**, 199 (1939). By irradiation of stilbene: Mallory *et al.*, *J. Am. Chem. Soc.* **84**, 4361 (1962). Synthesis by double succinoylation of benzene: Rahman *et al.*, *J. Org. Chem.* **28**, 3571 (1963). Structure: Trotter, *Acta Crystallogr.* **16**, 605 (1963). Review of carcinogenic risk: *IARC Monographs* **32**, 419-430 (1983); of toxicology and human exposure: *Toxicological Profile for Polycyclic Aromatic Hydrocarbons* (PB95-264370, 1995) 487 pp.

Monoclinic plates from alcohol. d^{25} 1.179; mp 100°; bp 340°. Sublimes in high vacuum. Absorption spectrum: Clar, Lombardi, *Ber.* **65**, 1412 (1932); Mayneord, Roe, *Proc. R. Soc. London* **A 152**, 317 (1935). Practically insol in water. Sol in organic solvents, especially in aromatic hydrocarbons. One gram dissolves in 60 ml cold, 10 ml boiling 95% alcohol, 25 ml abs alcohol, 2.4 ml toluene or carbon tetrachloride, 2 ml benzene, 1 ml carbon disulfide, 3.3 ml anhyd ether. Soluble in glacial acetic acid. Solns exhibit a blue fluorescence. Forms molecular compds with picric acid, picryl chloride, dinitrobenzene and similar nitro compounds. LD_{50} i.p. in mice: 700 mg/kg (IARC).

Caution: Can cause photosensitization of skin.

7327. Phenanthrenequinone. [84-11-7] 9,10-Phenanthrenedione; 9,10-phenanthraquinone. $C_{14}H_8O_2$; mol wt 208.22. C 80.76%, H 3.87%, O 15.37%. Prepn by oxidation of phenanthrene: Curtis *et al.*, **US 2956065** (1960 to U.S. Steel); Binder, Koch, **DE 1166176** (1964).

Orange-red crysts. mp 206-207°. bp ~360°. d 1.405. Sublimes. Practically insol in water; sol in benzene, ether, glacial acetic acid, hot alc. With concd H_2SO_4 gives a dark green color.

7328. *o*-Phenanthroline. [66-71-7] 1,10-Phenanthroline; 4,5-phenanthroline. $C_{12}H_8N_2$; mol wt 180.21. C 79.98%, H 4.47%, N 15.55%. Metal chelator. Prepn: F. Blau, *Monatsh. Chem.* **19**, 666 (1898); K. Madeja, *J. Prakt. Chem.* **17**, 104 (1962). Solubilities: J. Burgess, R. I. Haines, *J. Chem. Eng. Data* **23**, 196 (1978). Crystal and molecular structure: S. Nishigaki *et al.*, *Acta Crystallogr.* **B34**, 875 (1978). Review of analytical uses: F. Vydra, M. Kopanica, *Chemist-Analyst* **52**, 88-94 (1963); of chemistry and reactivity: L. A. Summers, *Adv. Heterocycl. Chem.* **22**, 1-69 (1978); W. Sliwa, *Heterocycles* **12**, 1207-1237 (1979); P. G. Sammes, G. Yahioglu, *Chem. Soc. Rev.* **1994**, 327-334; of chiral derivatives used in asymmetric catalysis: E. Schoffers, *Eur. J. Org. Chem.* **2003**, 1145-1152.

White, cryst powder. mp 117°. pKa_1: 4.8-5.2. uv max (water): 265 nm (ε 29510). Soly at 298.2 K (mol/dm^3): water 0.0149; ethanol 2.78.

Monohydrate. [5144-89-8] $C_{12}H_8N_2 \cdot H_2O$; mol wt 198.23.

USE: As an analytical reagent for determn of metals in chemical and biological systems through complex formation. Colorimetric indicator; indicator (*"Ferroin"*) in combination with ferrous ions for oxidation/reduction reactions. In organic syntheses as an activator.

7329. Phenarsazine Chloride. [578-94-9] 10-Chloro-5,10-dihydrophenarsazine; 5-aza-10-arsenaanthracene chloride; 10-chloro-5,10-dihydroarsacridine; diphenylaminechlorarsine; phenazarsine chloride; adamsite; DM. $C_{12}H_9AsClN$; mol wt 277.58. C 51.92%, H 3.27%, As 26.99%, Cl 12.77%, N 5.05%. Prepd by heating diphenylamine with arsenic trichloride: **DE 281049** (1914 to I. G. Farben.); Wieland, Reinheimer, *Ann.* **423**, 12 (1921); Lewis, Hamilton, *J. Am. Chem. Soc.* **43**, 2222 (1921); Burton, Gibson, *J. Chem. Soc.* **1926**, 450.

Canary-yellow crystals from carbon tetrachloride. *Poisonous.* Dimorphous. The stable form occurs as orthorhombic crystals; d 1.65; mp 195°; bp 410° (decompn). Sublimes readily. Vapor press. at 20° = 2×10^{-13} mm; volatility 0.02 mg/cu m; heat of volatilization 54.8 cal; spec heat 0.268 cal. Practically insol in water. Slightly sol in benzene, xylene, carbon tetrachloride. Corrodes iron, bronze, brass. (The metastable form melts at 186° if monoclinic, and at 182° if triclinic.)

Caution: Irritating to skin and respiratory tract. Causes profuse watery nasal discharge; severe pain in nose, sinuses, chest; sneezing, coughing, nausea, vomiting, marked depression, weakness. Sensory disturbances may occur later.

USE: As war gas, dispersed in air in the form of minute particles. For riots in combination with tear gas (chloroacetophenone). In the formulation of wood-treating solns, against marine borers and similar pests.

7330. Phenatine. [139-68-4] *N*-(1-Methyl-2-phenylethyl)-3-pyridinecarboxamide; *N*-(α-methylphenethyl)nicotinamide; nicotinic acid β-phenylisopropylamide; nicotinoyl-β-phenylisopropylamine; fenatin; perviton. $C_{15}H_{16}N_2O$; mol wt 240.31. C 74.97%, H 6.71%, N 11.66%, O 6.66%. Synthesis from nicotinoyl chloride and $C_6H_5CH_2CH(CH_3)NH_2$: Arbuzov *et al.*, *Sb. Statei Obshch. Khim.* **1**, 714 (1953), *C.A.* **49**, 1072 (1955). Pharmacology: Arbuzov, *Farmakol. Toksikol.* **31**, 373 (1968). Analysis: P. P. Suprun, *Farm. Zh. (Kiev)* **30**, 49 (1975); O. N. Shcherbina *et al.*, *ibid.* **1979**, 49.

Crystals from benzene, mp 99-100°.

Phosphate. $C_{15}H_{16}N_2O \cdot 2H_3PO_4$. Crystals from alcohol + ether, mp 162°. Sol in water, warm alcohol. Insol in ether. Aq solns may be boiled for 3 hrs without decompn.

7331. Phenazine. [92-82-0] Dibenzopyrazine; dibenzoparadiazine; azophenylene. $C_{12}H_8N_2$; mol wt 180.21. C 79.98%, H 4.47%, N 15.55%. Obtained (with other products) by passing aniline vapor through a red-hot tube: Bernthsen, *Ber.* **19**, 3257 (1886); by heating aniline with nitrobenzene and sodium hydroxide to 140°: Wohl, Aue, *Ber.* **34**, 2446 (1901); Wohl, *Ber.* **36**, 4135 (1903); by heating *o*-phenylenediamine with pyrocatechol in sealed tube: Ris, *Ber.* **19**, 2206 (1886); Hinsberg, Garfunkel, *Ann.* **292**, 258 (1896); upon distilling 2-aminodiphenylamine with lead monoxide: Fischer,

Heiler, *Ber.* **26**, 383 (1893); by heating 2-aminodiphenylamine with 2-nitrodiphenylamine in the presence of sodium acetate: Kehrmann, Havas, *Ber.* **46**, 342 (1913); by heating nitrobenzene with barium oxide: Zerewitinoff, Ostromisslensky, *Ber.* **44**, 2402 (1911); by boiling 2,2'-dinitrodiphenylamine with stannous chloride in hydrochloric and acetic acids, followed by oxidation with hydrogen peroxide: Eckert, Steiner, *Monatsh. Chem.* **35**, 1154 (1914).

Pale yellow needles from alcohol or by sublimation. Colorless needles from dilute alcohol. mp 171°; bp above 360°. Practically insol in water. One part dissolves in 50 parts alcohol. Moderately sol in ether, benzene; sol in mineral acids giving yellow to red solns.

7332. Phenazocine. [127-35-5] 1,2,3,4,5,6-Hexahydro-6,11-dimethyl-3-(2-phenylethyl)-2,6-methano-3-benzazocin-8-ol; 2'-hydroxy-5,9-dimethyl-2-phenethyl-6,7-benzmorphan; phenethylazocine; phenobenzorphan. $C_{22}H_{27}NO$; mol wt 321.46. C 82.20%, H 8.47%, N 4.36%, O 4.98%. Synthetic opioid analgesic. Prepn of ($\pm$)-*cis*-form from metazocine, *q.v.*: E. L. May, N. B. Eddy, *J. Org. Chem.* **24**, 294 (1959). Optical resolution and pharmacology: *eidem, ibid.* 1435. Improved syntheses: J. H. Ager, E. L. May, *ibid.* **25**, 984 (1960); G. A. Brine *et al., J. Heterocycl. Chem.* **27**, 2139 (1990).

cis-(–)-Form

cis-(**±**)-**Form.** [58073-76-0] (±)-α-Phenazocine. Rods from methanol, mp 181-182°.
cis-(**±**)-**Form hydrobromide.** [70878-79-4]; [1239-04-9] (unspecified stereo). NIH-7519; Narphen; Prinadol. $C_{22}H_{27}NO.HBr$; mol wt 402.38. Rods from acetone or abs alc + ether, mp 166-170°. LD$_{50}$ s.c. in mice: 332 mg/kg (May, Eddy).
cis-(**–**)-**Form.** [58640-87-2] α-(–)-Phenazocine. Needles from methanol, mp 159-159.5°. $[\alpha]_D^{20}$ – 122° (c = 0.74 in 95% ethanol).
cis-(**–**)-**Form hydrobromide.** Crystals, mp 284-287°. $[\alpha]_D^{20}$ –84.1° (c = 1.12 in 95% ethanol). LD$_{50}$ s.c. in mice: 147 mg/kg (May, Eddy).
Note: This is a controlled substance (opiate): **21 CFR**, 1308.12.
THERAP CAT: Analgesic.

7333. Phencyclidine. [77-10-1] 1-(1-Phenylcyclohexyl)-piperidine; angel dust; HOG; PCP; CI-395. $C_{17}H_{25}N$; mol wt 243.39. C 83.89%, H 10.35%, N 5.75%. Prepn: **GB 836083** and Godefroi *et al.,* **US 3097136** (1960, 1963 to Parke, Davis); V. H. Maddox *et al., J. Med. Chem.* **8**, 230 (1965). Pharmacology: G. Chen *et al., J. Pharmacol. Exp. Ther.* **127**, 241 (1959); J. C. Munch, *Bull. Narc.* **26**, 9 (1974). Human metabolism: L. K. Wong, K. Biemann, *Biomed. Mass Spectrom.* **2**, 204 (1975). Toxicity: K. Bailey *et al., J. Pharm. Pharmacol.* **28**, 713 (1976). Extensive bibliography: R. L. Balster, R. S. Pross, *J. Psychedelic Drugs* **10**, 1-15 (1978). *Review:* R. E. Garey, *ibid.* **11**, 265-275 (1979). Review of neuropharmacology: K. M. Johnson, S. M. Jones, *Annu. Rev. Pharmacol. Toxicol.* **30**, 707-750 (1990).

Colorless crystals, mp 46-46.5°. bp$_{1.0}$ 135-137°. uv max (0.1*N* HCl): 252, 257.5, 262, 268.5 nm (E$_{1cm}^{1\%}$ 7.9, 11.2, 13.0, 9.7).
Hydrochloride. [956-90-1] Sernyl; Sernylan. $C_{17}H_{25}N.HCl$; mol wt 279.85. Crystals from 2-propanol, mp 233-235°. uv max (ethanol): 254, 258, 262.5, 269 nm (E$_{1cm}^{1\%}$ 7.9, 10.8, 12.7, 10.0). LD$_{50}$ orally in mice: 76.5 mg/kg (Bailey).
Hydrobromide. Crystals, mp 214-218°.
Caution: This is a controlled substance (depressant): **21 CFR**, 1308.12. The ethylamine, pyrrolidine and thiophene analogs are controlled substances (hallucinogens): **21 CFR**, 1308.11.
THERAP CAT: Anesthetic (intravenous).
THERAP CAT (VET): Analgesic; anesthetic.

7334. Phendimetrazine. [634-03-7] (2*S*,3*S*)-3,4-Dimethyl-2-phenylmorpholine; *d*-2-phenyl-3,4-dimethylmorpholine; 3,4-dimethyl-2-phenyltetrahydro-1,4-oxazine. $C_{12}H_{17}NO$; mol wt 191.27. C 75.36%, H 8.96%, N 7.32%, O 8.36%. Prepn from ethylene chlorohydrin and *l*-ephedrine: Otto, *Angew. Chem.* **68**, 181 (1956); from 2-phenyl-2-hydroxy-3,4-dimethylmorpholine: Boehringer *et al.,* **GB 791416** (1958); **GB 862198** (1961). Stereochemistry: Dvornik, Schilling, *J. Med. Chem.* **8**, 466 (1965). Toxicity data: Stegen *et al., Toxicol. Appl. Pharmacol.* **2**, 589 (1960).

bp$_8$ 122-124°; bp$_{12}$ 134-135°.
Bitartrate. [50-58-8] Adphen; Bacarate; Neo-Nilorex; Obepar; Phenazine; Plegine; Statobex; Symetra; Trimstat; Trimtabs.
Hydrochloride. [7635-51-0] Antapentan. $C_{12}H_{17}NO.HCl$; mol wt 227.73. mp 191° (Otto), 208° (Boehringer). $[\alpha]_D^{20}$ +35.7°. LD$_{50}$ in rats (mg/kg): 455 orally; 245 i.p. (Stegen).
Pamoate. [27922-80-1] Fringanor. $(C_{12}H_{17}NO)_2.C_{23}H_{16}O_6$; mol wt 770.92.
Note: This is a controlled substance (stimulant): **21 CFR**, 1308.13.
THERAP CAT: Anorexic.

7335. Phenelzine. [51-71-8] (2-Phenethyl)hydrazine; β-phenylethylhydrazine; phenalzine. $C_8H_{12}N_2$; mol wt 136.20. C 70.55%, H 8.88%, N 20.57%. Monoamine oxidase inhibitor. Prepn: Votocek, Leminger, *Collect. Czech. Chem. Commun.* **4**, 271 (1932), *C.A.* **26**, 5294 (1932); Biel *et al., J. Am. Chem. Soc.* **81**, 2805 (1959); Biel, **US 3000903** (1959 to Lakeside). Toxicity: J. A. Gylys *et al., Ann. N.Y. Acad. Sci.* **107**, 899 (1963). Comprehensive description: R. E. Daly, *Anal. Profiles Drug Subs.* **2**, 383-407 (1973).

Liquid. bp$_{0.1}$ 74°. n$_D^{20}$ 1.5494.
Acid sulfate. [156-51-4] W-1544a; Nardelzine; Nardil. $C_8H_{12}N_2.H_2SO_4$; mol wt 234.27. White powder, sol in water. LD$_{50}$ orally in mice: 156 mg/kg (Gylys).
Hydrochloride. [5470-36-0] $C_8H_{12}N_2.HCl$; mol wt 172.66. Crystals, mp 174°.
THERAP CAT: Antidepressant.

7336. Phenetharbital. [357-67-5] 5,5-Diethyl-1-phenyl-2,4,6(1*H*,3*H*,5*H*)-pyrimidinetrione; 5,5-diethyl-1-phenylbarbituric acid; 5,5-diethyl-2,4,6-trioxo-1-phenylhexahydropyrimidine; 1-phenyl-5,5-diethylbarbituric acid; *N*-phenylbarbital; phenidiemal; Fedibaretta; Pyrictal. $C_{14}H_{16}N_2O_3$; mol wt 260.29. C 64.60%, H 6.20%, N 10.76%, O 18.44%. Prepd from ethyl diethylmalonate and phenylurea: Fischer, Dilthey, *Ann.* **335**, 334 (1904); Buck, *J. Am. Chem. Soc.* **58**, 1284 (1936).

Small, thick, glittering plates, mp 178°. Freely sol in hot alcohol, alkali.

Note: This is a controlled substance (depressant): **21 CFR, 1308.13.**

THERAP CAT: Anticonvulsant.

7337. Phenethicillin Potassium. [132-93-4] 3,3-Dimethyl-7-oxo-6-[(1-oxo-2-phenoxypropyl)amino]-4-thia-1-azabicyclo[3.2.0]-heptane-2-carboxylic acid potassium salt (1:1); (α-phenoxyethyl)-penicillin potassium; α-phenoxyethylpenicillinic acid potassium salt; 6-(α-phenoxypropionamido)penicillanic acid potassium salt; penicillin-152; penicillin MV; penicillin-152 potassium; potassium phenethicillin; Maxipen; Syncillin. $C_{17}H_{19}KN_2O_5S$; mol wt 402.51. C 50.73%, H 4.76%, K 9.71%, N 6.96%, O 19.87%, S 7.97%. Semisynthetic antibiotic related to penicillin. Prepn from 6-aminopenicillanic acid: Perron *et al.*, *Antibiot. Annu.* **1959-1960**, 107; *J. Am. Chem. Soc.* **82**, 3934 (1960); Glombitza, *Ann.* **673**, 166 (1964).

***dl*-Form.** Crystals from acetone, dec 230-232°. Much less hygroscopic than benzylpenicillin sodium. Freely sol in water.

***l*-Form.** Epiphenethicillin potassium. Crystals from 20% butanol, dec 238-239°. $[\alpha]_D^{24}$ +218° (c = 0.01 in water).

THERAP CAT: Antibacterial.

7338. Phenethyl Alcohol. [60-12-8] Benzeneethanol; 2-phenylethanol; β-phenylethyl alcohol; benzyl carbinol; β-hydroxyethylbenzene. $C_8H_{10}O$; mol wt 122.17. C 78.65%, H 8.25%, O 13.10%. Found in a number of natural essential oils, such as rose, carnation, hyacinth, Aleppo pine, orange blossom, geranium Bourbon, neroli and in the essential oil of champaca. Prepd by reduction of ethyl phenylacetate with sodium in abs alcohol: Bouveault, Blanc, *Bull. Soc. Chim.* [3] **31**, 672 (1904); Leonard, *J. Am. Chem. Soc.* **47**, 1778 (1925); by hydrogenation of phenylacetaldehyde in the presence of nickel catalyst: Skita, Ritter, *Ber.* **43**, 3398 (1910); v. Braun, Kochendörfer, *Ber.* **56**, 2176 (1923); Milligan, Reid, *J. Am. Chem. Soc.* **44**, 204 (1922). Isoln from the fungus *Gibberella fujikuroi:* Cross *et al.*, *J. Chem. Soc.* **1963**, 2937. Antibacterial activity: R. M. E. Richards *et al.*, *J. Pharm. Pharmacol.* **21**, 681 (1969). Use as preservative in ophthalmic solutions: R. M. E. Richards, R. J. McBride, *ibid.* **24**, 145 (1972). Toxicity: Jenner *et al.*, *Food Cosmet. Toxicol.* **2**, 327 (1964).

Liquid. Floral odor, rose character. mp −27°. d_{25}^{25} 1.017 to 1.019. bp_{750} 219-221°; bp_{14} 104°; bp_{12} 98-100°. n_D^{20} 1.530 to 1.533. Two ml dissolve in 100 ml water after thorough shaking. One part is clearly sol in 1 part of 50% alc. Miscible with alcohol, ether. LD_{50} orally in rats: 1790 mg/kg (Jenner).

USE: Pharmaceutic aid (antimicrobial). In flavors and perfumery (esp rose perfumes).

7339. Phenethylamine. [64-04-0] Benzeneethanamine; β-phenylethylamine; 1-amino-2-phenylethane; β-aminoethylbenzene;

PEA. $C_8H_{11}N$; mol wt 121.18. C 79.29%, H 9.15%, N 11.56%. Endogenous amine related structurally and pharmacologically to amphetamine: P. Mantegazza, M. Riva, *J. Pharm. Pharmacol.* **15**, 472 (1963). Present in oil of bitter almonds. Found in normal human urine (about 30 μg/liter). Prepn: Johnson, Guest, *Am. Chem. J.* **42**, 346 (1909); Robinson, Snyder, *Org. Synth.* coll. vol. **III**, 720 (1955). Elevated levels in urine due to paranoid chronic schizophrenia: S. G. Potkin *et al.*, *Science* **206**, 470 (1979); due to stress: M. A. Paulos, R. E. Tessel, *ibid.* **215**, 1127 (1982).

Liquid. Fishy odor. Absorbs CO_2 from air. Does not solidify when cooled in an ice-salt mixture. Strong base. d_4^{25} 0.9640; bp 194.5-195°. Sol in water. Freely sol in alc, ether.

Hydrochloride. [156-28-5] $C_8H_{11}N.HCl$; mol wt 157.64. Orthorhombic bipyramidal platelets from abs alcohol, mp 217°. Freely sol in water (100 parts H_2O will dissolve 80 parts at 15°). Sol in alc. Insol in ether. LD_{50} s.c. in mice: 470 mg/kg (Mantegazza, Riva).

Caution: Skin irritant and possible sensitizer.

7340. Phenethyl Isothiocyanate. [2257-09-2] (2-Isothiocyanatoethyl)benzene; 2-phenylethyl isothiocyanate; PEITC. C_9H_9NS; mol wt 163.24. C 66.22%, H 5.56%, N 8.58%, S 19.64%. Chemopreventive agent found in cruciferous vegetables; produced by hydrolysis of *gluconasturtiin*. Prepn: J. v. Braun, H. Deutsch, *Ber.* **45**, 2188 (1912); E. Schmidt *et al.*, *Ann.* **612**, 11 (1958); and isoln from turnips: E. P. Lichtenstein *et al.*, *J. Agric. Food Chem.* **10**, 30 (1962). Isoln from cabbage and watercress seeds: N. Kaoulla *et al.*, *Phytochemistry* **19**, 1053 (1980). LC/MS determn in plasma and urine: Y. Ji, M. E. Morris, *Anal. Biochem.* **323**, 39 (2003). Chemopreventive studies in animal models: G. D. Stoner, *Curr. Top. Plant Physiol.* **15**, 78 (1995); A. Nishikawa *et al.*, *Curr. Cancer Drug Targets* **4**, 373 (2004). Mechanistic studies of anticancer activity and induction of apoptosis: Y.-R. Chen *et al.*, *J. Biol. Chem.* **277**, 39334 (2002); R. Hu *et al.*, *Carcinogenesis* **24**, 1361 (2003); D. Xiao *et al.*, *Clin. Cancer Res.* **11**, 2670 (2005).

Colorless to pale yellow liquid, bp_{11} 139-140°. $bp_{1.4-1.5}$ 102-103°. n_D^{20} 1.5904. d_4^{20} 1.0942. LD_{50} in mice (mg/kg): 700 orally; 150 s.c.; 50 i.v. (Lichtenstein).

USE: Flavor ingredient; provides the hot or burning sensation in horseradish.

7341. *o*-Phenetidine. [94-70-2] 2-Ethoxybenzenamine; 2-aminophenetole; 2-ethoxyaniline. $C_8H_{11}NO$; mol wt 137.18. C 70.05%, H 8.08%, N 10.21%, O 11.66%.

Oily liquid; rapidly becomes brown on exposure to light and air. mp < −20°. bp 228-230°. *Poisonous.* Sol in alcohol. Insol in water.

Caution: Absorbed through skin; vapor hazardous.

USE: Manuf of dyes.

7342. *p*-Phenetidine. [156-43-4] 4-Ethoxybenzenamine; 4-aminophenetole; 4-ethoxyaniline; *p*-aminophenyl ethyl ether. $C_8H_{11}NO$; mol wt 137.18. C 70.05%, H 8.08%, N 10.21%, O 11.66%. Prepn by reduction of *p*-nitrophenetole: West, *J. Chem. Soc.* **127**, 494 (1925). Prepn from nitrobenzene, ethanol, Mg, and sulfuric acid: Yukawa, *J. Chem. Soc. Jpn.* **71**, 547 (1950).

Colorless liquid; becomes red to brown on exposure to air and light. d_4^{16} 1.0652; mp about 3°; bp 253-255°. Practically insol in water. Sol in alcohol. *Keep well closed and protected from light.*

Citrate. Citrophen. $C_8H_{11}NO.C_6H_8O_7$; mol wt 329.31. Crystalline powder, mp 186-188°. Slightly acid taste. Sol in about 40 parts water.

Caution: Absorbed through skin; vapor hazardous.

USE: Manuf acetophenetidine, phenocoll, dulcin, dyes.

7343. Phenetole. [103-73-1] Ethoxybenzene; ethyl phenyl ether; phenyl ethyl ether. $C_8H_{10}O$; mol wt 122.17. C 78.65%, H 8.25%, O 13.10%. Prepd from phenol or its salts by the use of the ethylating agent ethyl chloride: Wohl, *Ber.* **39**, 1951 (1906); ethyl bromide: White *et al., J. Am. Chem. Soc.* **46**, 965 (1924); ethyl *p*-toluenesulfonate: Finzi, *Ann. Chim. Appl.* **15**, 41 (1925); diethyl sulfate or triethyl phosphate: Noller, Dutton, *J. Am. Chem. Soc.* **55**, 424 (1933). Toxicity data: Binet, *Rev. Med. Suisse Romande* **15**, 561 (1895).

Oily liquid. d_4^{20} 0.967; mp −30°; bp 171-173°; n_D^{20} 1.507. Practically insol in water. Freely sol in alcohol, ether. MLD in rats (mg/kg): 3500-4000 s.c. (Binet).

7344. Pheneturide. [90-49-3] *N*-(Aminocarbonyl)-α-ethylbenzeneacetamide; (2-phenylbutyryl)urea; α-phenyl-α-ethylacetylurea; *N*-(α-phenylbutyryl)urea; α-ethyl-α-phenylacetylurea; ethylphenacemide; EPA; PBU; Benuride. $C_{11}H_{14}N_2O_2$; mol wt 206.25. C 64.06%, H 6.84%, N 13.58%, O 15.51%. Prepd by heating α-phenylbutyric acid chloride with urea: **DE 249241** (1912 to Bayer); *Chem. Zentralbl.* **1912**, II, 396; *Frdl.* **10**, 1165. From α-phenylbutyric acid chloride, urea, and antipyrine: Gold-Aubert, *Helv. Chim. Acta* **41**, 1512 (1958); **CH 374644** (1964 to Labs. Sapos). Pharmacokinetics: R. L. Galeazzi *et al., J. Pharmacokinet. Biopharm.* **7**, 453 (1979). Toxicity: M. J. Orloff *et al., Neurology* **1**, 377 (1951). Clinical studies: J. C. Bowe, *Br. J. Clin. Pract.* **27**, 174 (1973); F. B. Gibberd *et al., J. Neurol. Neurosurg. Psychiatry* **45**, 1113 (1982).

***dl*-Form.** Needles from ethanol, mp 149-150°.

***d*-Form.** Needles from ethanol, mp 168-169°. $[\alpha]_D^{17}$ +54.0° (c = 1 in ethanol); $[\alpha]_D^{22}$ +53.8° (c = 1 in acetone); +48.2° (c = 1 in dioxane).

***l*-Form.** Crystals from 50% ethanol, mp 162-163°. $[\alpha]_D^{30}$ −51.6° (c = 1 in ethanol).

THERAP CAT: Anticonvulsant.

7345. Phenformin. [114-86-3] *N*-(2-Phenylethyl)imidodicarbonimidic diamide; 1-phenethylbiguanide; phenethyldiguanide; *N'*-β-phenethylformamidinyliminourea; fenformin; fenormin; β-PEBG; PEDG. $C_{10}H_{15}N_5$; mol wt 205.27. C 58.51%, H 7.37%, N 34.12%. Prepn: Shapiro *et al., J. Am. Chem. Soc.* **81**, 2220 (1959); Shapiro, Freedman, **US 2961377; US 3057780** (1960, 1962 both to U.S.V.). Metabolism: R. Beckmann, *Diabetologia* **3**, 368 (1967). Association with lactic acidosis in diabetic patients: R. I. Misbin, *Ann. Intern. Med.* **87**, 591 (1977). Toxicity study: G. Proske *et al., Arzneim.-Forsch.* **12**, 314 (1962). Comprehensive description: J. E. Moody, *Anal. Profiles Drug Subs.* **4**, 319-332 (1975). Review of phenformin-induced lactic acidosis: M. E. McGuinness, R. L. Talbert, *Ann. Pharmacother.* **27**, 1183-1187 (1993).

Hydrochloride. [834-28-6] Insoral. $C_{10}H_{15}N_5.HCl$; mol wt 241.72. Crystals from isopropanol, mp 175-178°. Sol in water; pH (0.1*M* aq soln): 6.7. LD_{50} in mice (mg/kg): 19 i.v., 450 orally; in rats: 1050 orally; in guinea pigs: 47 orally, 19 s.c. (Proske).

THERAP CAT: Antidiabetic.

7346. Phenicin. [128-68-7] 2,2′-Dihydroxy-4,4′-dimethyl-[bi-1,4-cyclohexadien-1-yl]-3,3′,6,6′-tetrone; 3,3′-dihydroxy-5,5′-dimethyl-2,2′-bi-*p*-benzoquinone; phoenicin. $C_{14}H_{10}O_6$; mol wt 274.23. C 61.32%, H 3.68%, O 35.00%. Fungal pigment produced by *Penicillium phoeniceum* and *P. rubrum:* Friedheim, *Helv. Chim. Acta* **21**, 1464 (1938); Charollais *et al., Arch. Sci.* **16**, 474 (1963). Synthesis: Posternak *et al., Helv. Chim. Acta* **26**, 2031 (1943); Musso, Beecken, *Ber.* **92**, 1416 (1959); J. Kalamar *et al., Helv. Chim. Acta* **57**, 2368 (1974). Biosynthesis: E. Steiner *et al., ibid.* 2377.

Yellowish-brown crystals from alc, mp 230-231°. uv max (chloroform): 268, 406 nm (log ε 4.52, 3.36). Sparingly sol in water; freely sol in chloroform, acetic acid, hot alcohol. Very acidic solns are yellow, turning red at pH 1.6-3.6 and violet at pH 4.9-6.

7347. Phenindamine. [82-88-2] 2,3,4,9-Tetrahydro-2-methyl-9-phenyl-1*H*-indeno[2,1-*c*]pyridine; 2-methyl-9-phenyl-2,3,4,9-tetrahydro-1-pyridinene; 1,2,3,4-tetrahydro-2-methyl-9-phenyl-2-azafluorene; Nu-1504. $C_{19}H_{19}N$; mol wt 261.37. C 87.31%, H 7.33%, N 5.36%. Prepn: J. T. Plati, W. Wenner, **US 2470108** (1949 to Hoffmann-La Roche). Pharmacology: G. Lehmann, *J. Pharmacol. Exp. Ther.* **92**, 249 (1948); W. Schallek, *ibid.* **105**, 291 (1952).

Crystals from ether, mp 90-91°, d 1.17.

Hydrogen tartrate. [569-59-5] Pernovin; Thephorin. $C_{19}H_{19}$-N.$C_4H_6O_6$; mol wt 411.45. Crystals, mp 160°. Soly: about 2.5% in water; sparingly sol in propylene glycol. Insol in 95% ethanol, glycerol, ether. LD_{50} in mice (mg/kg): 27 i.v., 170 i.p., 265 orally (Schallek).

Hydrochloride. [5503-08-2] $C_{19}H_{19}N.HCl$. Crystals, sol in water. LD_{50} in mice (mg/kg): 88 ±11 i.p., 22.5 ±2.5 i.v., 255 ±21 orally, 270 ±60 s.c.; in rats (mg/kg): 280 ±50 orally; in guinea pigs (mg/kg): 140 ±42 i.p.; in rabbits (mg/kg): 15 ±2.4 i.v., 500 orally; in dogs (mg/kg): 33 i.v. (Lehmann).

Note: Thephorin was formerly used to designate theobromine sodium formate.

THERAP CAT: Tartrate as antihistaminic.

7348. Phenindione. [83-12-5] 2-Phenyl-1*H*-indene-1,3(2*H*)-dione; 2-phenyl-1,3-indandione; 2-phenyl-1,3-diketohydrindene; PID; Pindione; Bindan; Dindevan; Dineval; Hemolidione; Indon; Indema; Fenilin; Fenhydren; Rectadione; Cronodione; Thrombasal. $C_{15}H_{10}O_2$; mol wt 222.24. C 81.07%, H 4.54%, O 14.40%. By heating phthalide with benzaldehyde and sodium ethylate soln: Dieckmann, *Ber.* **47**, 1439 (1914).

Leaflets from alcohol, mp 149-151°. Practically insol in cold water. Slightly sol in warm water. pH of satd soln at 25° = 4.5. Readily sol in alkaline solns. Soluble in methanol, ethanol, ether, acetone, benzene, chloroform. Solns in alkalies are red, in concd H_2SO_4 blue.

THERAP CAT: Anticoagulant.

7349. Pheniramine. [86-21-5] N,N-Dimethyl-γ-phenyl-2-pyridinepropanamine; 2-[α-(2-dimethylaminoethyl)benzyl]pyridine; 1-phenyl-1-(2-pyridyl)-3-dimethylaminopropane; 3-phenyl-3-(2-pyridyl)-N,N-dimethylpropylamine; prophenpyridamine; propheniramine. $C_{16}H_{20}N_2$; mol wt 240.35. C 79.96%, H 8.39%, N 11.66%. Synthesis: Sperber *et al.*, **US 2567245** and **US 2676964** (1951, 1954, both to Schering).

Oily liq. Slightly yellow color. Characteristic amine-like odor. d 1.0081; bp_{13} 181°; bp_2 142°; $bp_{0.5}$ 135°; n_D^{25} 1.5519 to 1.5521. Sol in dil acids, alc, benzene, chloroform, ether. Insol in water.

Maleate. [132-20-7] Avil; Daneral; Inhiston; Trimeton. $C_{16}H_{20}$-N_2.$C_4H_4O_4$; mol wt 356.42. Crystals from amyl alcohol, faint amine-like odor, mp 107°. Sol in water, alc; slightly sol in ether, benzene. pH of 1% aq soln between 4.3 and 4.9.

p-Aminosalicylate. [3269-83-8] $C_{16}H_{20}N_2$.$C_7H_7NO_3$. Small needles from acetone + ethyl acetate, large octahedra from water, dec 142°. One gram dissolves in 10 ml water. Freely sol in alc; sparingly sol in ethyl acetate, ether, acetone.

THERAP CAT: Antihistaminic.

7350. Phenmedipham. [13684-63-4] N-(3-Methylphenyl)-carbamic acid 3-[(methoxycarbonyl)amino]phenyl ester; m-hydroxycarbanilic acid methyl ester m-methylcarbanilate; methyl 3-(m-tolylcarbamoyloxy)phenylcarbamate; Schering 38584; Betanal. $C_{16}H_{16}N_2O_4$; mol wt 300.31. C 63.99%, H 5.37%, N 9.33%, O 21.31%. Prepn: **NL 6604363** (1966 to Schering AG), *C.A.* **66**, 104813w (1967). Persistence in the soil: K. Kossmann, *Weed Res.* **10**, 349 (1970).

Colorless crystals, mp 139-142° (tech, 143-144°). Soly in water at room temp: <10 ppm.

USE: Herbicide.

7351. Phenmetrazine. [134-49-6] 3-Methyl-2-phenylmorpholine; 3-methyl-2-phenyltetrahydro-2H-1,4-oxazine; 2-phenyl-3-methyltetrahydro-1,4-oxazine; A-66. $C_{11}H_{15}NO$; mol wt 177.25. C 74.54%, H 8.53%, N 7.90%, O 9.03%. Prepn: Thomae, Wick, **US 2835669** (1958 to Boehringer, Ing.); Siemer, Hengen, **US 3018222** (1962 to Ravensberg Chem.); Clark, *J. Org. Chem.* **27**, 3251 (1962); Klosa, *J. Prakt. Chem.* **21**, 12 (1963).

Liquid. bp_{12} 138-140°; bp_1 104°.

Hydrochloride. [1707-14-8] Marsin; Neo-Zine; Preludin. C_{11}-$H_{15}NO.HCl$; mol wt 213.71. Crystals from ethanol + ether, mp 182°. One gram dissolves in 0.4 ml water, in 2.0 ml 95% alc, in 2.0 ml chloroform. Sparingly sol in ether.

Note: This is a controlled substance (stimulant): **21 CFR**, 1308.12.

THERAP CAT: Anorexic.

7352. Phenobarbital. [50-06-6] 5-Ethyl-5-phenyl-2,4,6(1H,-3H,5H)-pyrimidinetrione; 5-ethyl-5-phenylbarbituric acid; phenylethylmalonylurea; phenobarbitone; Agrypnal; Barbiphenyl; Barbipil; Gardenal; Luminal; Phenobal. $C_{12}H_{12}N_2O_3$; mol wt 232.24. C 62.06%, H 5.21%, N 12.06%, O 20.67%. Prepn: **DE 247952** (1911 to Bayer); Hoerlein, **US 1025872** (1912), *Frdl.* **11**, 926; *Chem. Zentralbl.* **1912**, II, 212; Chamberlain *et al.*, *J. Am. Chem. Soc.* **57**, 352 (1935); Inman, Bitler, **US 2358072** (1944 to Kay-Fries Chem.); J. T. Pinhey, B. A. Rowe, *Tetrahedron Lett.* **21**, 965 (1980). Toxicity data: Schaffarzick, Brown, *Science* **116**, 663 (1952); E. I. Goldenthal, *Toxicol. Appl. Pharmacol.* **18**, 185 (1971). Series of articles on pharmacology and mechanism of antiepileptic action: *Adv. Neurol.* **27**, 473-562 (1980). Clinical evaluation in febrile convulsions: R. W. Newton, *Arch. Dis. Child.* **63**, 1189 (1988). Comprehensive description: M. K. C. Chao *et al.*, *Anal. Profiles Drug Subs.* **7**, 359-399 (1978).

White crystals (3 different phases); may exhibit polymorphism. mp 174-178°. uv max (pH 10 buffer): 240 nm ($A_{1cm}^{1\%}$ 431); (0.1N NaOH): 256 nm ($A_{1cm}^{1\%}$ 314). Slightly bitter taste. One gram dissolves in about one liter of water, 8 ml alc, 40 ml chloroform, 13 ml ether, about 700 ml benzene. Soluble in alkali hydroxides and carbonates. A satd aq soln is acid to litmus. pK_1 7.3, pK_2 11.8. LD_{50} orally in rats: 162 ±14 mg/kg (Goldenthal).

Sodium salt. [57-30-7] Sol phenobarbital; sol phenobarbitone. $C_{12}H_{11}N_2NaO_3$. Bitter, slightly hygroscopic crystals or white powder. One gram dissolves in about 1 ml water, about 10 ml alc. Practically insol in ether, chloroform. Aq solns are alkaline to litmus and phenolphthalein, pH ~9.3. LD_{50} orally in rats: 660 mg/kg (Schaffarzick, Brown).

Note: This is a controlled substance (depressant): **21 CFR**, 1308.14.

THERAP CAT: Anticonvulsant; sedative; hypnotic.

THERAP CAT (VET): Anticonvulsant; sedative.

7353. Phenoctide. [78-05-7] N,N-Diethyl-N-[2-[4-(1,1,3,3-tetramethylbutyl)phenoxy]ethyl]benzenemethanaminium chloride (1:1); benzyldiethyl-2-[(p-1,1,3,3-tetramethylbutyl)phenoxy]ethylammonium chloride; β-p-$tert$-octylphenoxyethyldiethylbenzylammonium chloride; Octaphen. $C_{27}H_{42}ClNO$; mol wt 432.09. C 75.05%, H 9.80%, Cl 8.20%, N 3.24%, O 3.70%. Prepn: Goldberg, Besly, **GB 703477** (1954 to Ward Blenkinsop); Erekaev, *C.A.* **54**, 12483i (1960).

Crystals from ethyl acetate. mp 112-114° (Goldberg, Besly); mp 95° (Erekaev).

USE: Orthophosphate as lubricant: Semmens, Summers-Smith, **GB 790056** (1958 to I.C.I.).

THERAP CAT: Anti-infective (topical).

7354. Phenol. [108-95-2] Carbolic acid; phenic acid; phenylic acid; phenyl hydroxide; hydroxybenzene; oxybenzene. C_6H_6-O; mol wt 94.11. C 76.58%, H 6.43%, O 17.00%. Obtained from coal tar, or made by fusing sodium benzenesulfonate with NaOH, or by heating monochlorobenzene with aq NaOH under high pressure. The crystalline article of commerce contains at least 98% phenol.

Review of mfg processes: A. Dierichs, R. Kubicka, *Phenole und Basen, Vorkommen und Gewinnung* (Akademie-Verlag, Berlin, 1958) 472 pp; *Faith, Keyes & Clark's Industrial Chemicals*, F. A. Lowenheim, M. K. Moran, Eds. (Wiley-Interscience, New York, 4th ed., 1975) pp 612-623. Use in treatment of spasticity: D. E. Garland *et al.*, *Clin. Orthop.* **165**, 217 (1982); *eidem*, *Arch. Phys. Med. Rehabil.* **65**, 243 (1984). Toxicity study: W. B. Deichmann, S. Witherup, *J. Pharmacol. Exp. Ther.* **80**, 233 (1944). Review of use in pain relief: K. M. Wood, *Pain* **5**, 205-229 (1978). *Review*: C. Thurman in *Kirk-Othmer Encyclopedia of Chemical Technology* vol. 17 (Wiley-Interscience, New York, 3rd ed., 1982) pp 373-384. Review of toxicology: H. Babich, D. L. Davis, *Regul. Toxicol. Pharmacol.* **1**, 90-109 (1981); and human exposure: *Toxicological Profile for Phenol* (PB2009-100007, 2008) 269 pp.

Colorless, acicular crystals or white, crystalline mass. Characteristic odor, somewhat sickeningly sweet and acrid with a sharp and burning taste. *Poisonous and caustic.* Prone to redden on exposure to air and light, hastened by presence of alkalinity. d 1.071. When free from water and cresols it congeals at 41° and melts at 43°. Ultrapure material mp 40.85°. The commercial product contains an impurity which raises the mp. bp 182°. Flash pt, closed cup: 175°F (79°C). n_D^{41} 1.5425. pKa at 25° = 10.0. pH of aq solns ~6.0. It is liquefied by mixing with ~8% water. One gram dissolves in ~15 ml water, 12 ml benzene; very sol in alc, chloroform, ether, glycerol, carbon disulfide, petrolatum, aq alkali hydroxides, volatile and fixed oils; sparingly sol in mineral oil. Almost insol in petr ether. LD_{50} orally in rats: 530 mg/kg (Deichmann, Witherup). *Keep well closed and protected from light. Do not handle with bare hands.*

Ammonium salt. [5973-17-1] Ammonium phenate; ammonium carbolate. $C_6H_6O.NH_3$; mol wt 111.14. White to pink crystalline masses. Sol in water.

Sodium salt. [139-02-6] Phenol sodium; sodium carbolate; sodium phenolate; sodium phenoxide. C_6H_5NaO; mol wt 116.09. Crystal structure and reactivity study: M. Kunert *et al.*, *Ber./Recl.* **130**, 1461 (1997). Colorless crystals, mp 380°. Very sol in water; sol in alc. The aq soln is caustic.

Caution: Potential symptoms of acute overexposure are steatorous breathing, mucous rales, froth at mouth and nose, frank pulmonary edema; cyanosis; tremor, convulsions, twitching; death due to respiratory failure. Chronic overexposure may result in vomiting, difficulty swallowing, excess salivation, diarrhea, anorexia, weight loss; headache, fainting, vertigo, mental disturbances; muscle aches and pain, weakness; damage to liver and kidney, dark urine. Ingestion may cause burning of mouth and throat; white necrotic lesions in mouth, esophagus and stomach; abdominal pain. Direct contact may cause irritation of eyes, nose and throat; skin burns; dermatitis; ochronosis. *See NIOSH Pocket Guide to Chemical Hazards* (DHHS/NIOSH 2003-100121, 2003) p 248; *Clinical Toxicology of Commercial Products*, R. E. Gosselin *et al.*, Eds. (Williams & Wilkins, Baltimore, 5th ed., 1984) Section III, p344-348; *Patty's Industrial Hygiene and Toxicology* vol. 2B, G. D. Clayton, F. E. Clayton, Eds. (Wiley-Interscience, New York, 4th ed., 1994) p 1567-1584.

USE: As a general disinfectant, either in soln or mixed with slaked lime, etc., for toilets, stables, cesspools, floors, drains, etc.; for the manuf of colorless or light-colored artificial resins, many medical and industrial organic compds and dyes; as a reagent in chemical analysis. pH indicator. Pharmaceutic aid (preservative).

THERAP CAT: Aqueous soln as topical anesthetic; topical antiseptic; topical antipruritic.

THERAP CAT (VET): Antiseptic caustic. Topical anesthetic in pruritic skin conditions. Has been used internally and externally as an antiseptic.

7355. Phenoldisulfonic Acid. [96-77-5] 4-Hydroxy-1,3-benzenedisulfonic acid; 4-hydroxy-*m*-benzenedisulfonic acid; 1-phenol-2,4-disulfonic acid. $C_6H_6O_7S_2$; mol wt 254.23. C 28.35%, H 2.38%, O 44.05%, S 25.22%. Conveniently prepd by hydrolysis of the dichloride which is obtained by the action of chlorosulfonic acid

on phenol at room temp: Pollak *et al.*, *Monatsh. Chem.* **46**, 395 (1925). Monograph: E. E. Gilbert, *Sulfonation and Related Reactions* (Interscience, New York, 1965) 529 pp.

Deliquescent needles, vague melting range from 89° to 100°. Decomp above 100°. *Corrosive.* Freely sol in water, methanol. Practically insol in ether, petr ether.

USE: In the manuf of aminophenoldisulfonic acids which are intermediates in the dye industry.

7356. Phenolphthalein. [77-09-8] 3,3-Bis(4-hydroxyphenyl)-1(3H)-isobenzofuranone; 3,3-bis(p-hydroxyphenyl)phthalide; α-(p-hydroxyphenyl)-α-(4-oxo-2,5-cyclohexadien-1-ylidine)-o-toluic acid; white phenolphthalein. $C_{20}H_{14}O_4$; mol wt 318.33. C 75.46%, H 4.43%, O 20.10%. Prepn: A. Baeyer, *Ber.* **4**, 658 (1871); *idem*, *Ann.* **202**, 36 (1880); M. Hubacher, US 2192485 (1940 to Ex Lax). Reactions at various pH values: G. Wittke, *J. Chem. Educ.* **60**, 239 (1983). Polarographic analysis in aqueous soln: M. M. Ellaithy *et al.*, *Farmaco Ed. Prat.* **41**, 326 (1986). Subchronic toxicity studies: D. D. Dietz *et al.*, *Fundam. Appl. Toxicol.* **18**, 48 (1992). Toxicology and carcinogenicity studies: *NTP Technical Report 465* (PB97-3390) 348 pp. Comprehensive description of properties, clinical and laboratory uses: F. J. Al-Shammary *et al.*, *Anal. Profiles Drug Subs.* **20**, 627-664 (1991).

White or yellowish white minute, triclinic crystals, often twinned. mp 258-262°. d 1.299. Sol in 95% alcohol and ether, slightly sol in chloroform. One gram dissolves in 12 ml alcohol, in ~100 ml ether. Almost insol in water. Sol in dil solns of alkali hydroxides and hot solns of alkali carbonates forming a red soln. pKa (25°C): 9.7. uv max (methanol): 205, 229, 276 nm (ε 27261.147, 14692.144, 2006.369). Log P (octanol/pH 7.4): 2.4. Shows color change from colorless in acid range to purple at pH 8-9.

Yellow Phenolphthalein. [8053-05-2] Produced in the mfg process prior to final purification of white phenolphthalein. Contains yellow bodies which impart characteristic color. Review of chemistry and use as a laxative: M. H. Hubacher, S. Doernberg, *J. Am. Pharm. Assoc. Sci. Ed.* **37**, 261 (1948). d_4^{20} 1.290-1.296. mp 255-260°. One gram dissolves in 12 ml alcohol, in 102 ml ether. Solns show a slight greenish fluorescence.

Caution: Phenolphthalein is reasonably anticipated to be a human carcinogen: *Report on Carcinogens, Twelfth Edition* (PB2011-111646, 2011) p 342.

USE: A 1% alcoholic soln as an indicator in titrations of mineral and organic acids and most alkalies.

THERAP CAT: Cathartic.

THERAP CAT (VET): Has been used as a laxative.

7357. Phenolphthalin. [81-90-3] 2-[Bis(4-hydroxyphenyl)-methyl]benzoic acid; decolorized phenolphthalein; phthalin; 4',4''-dihydroxytriphenylmethane-2-carboxylic acid. $C_{20}H_{16}O_4$; mol wt 320.34. C 74.99%, H 5.03%, O 19.98%. Made by boiling phenolphthalein with zinc dust in alkaline soln. Called "*Kastle-Meyer reagent*" when in soln. Prepd by dissolving 2 g phenolphthalin + 20 g KOH in a desired amt of doubly distd H_2O and diluting with an equal vol of 95% ethanol. Detailed directions for prepn of test soln

Phenolphthalol

and its use: Lecoq, *Bull. Soc. Chim. Belg.* **54**, 186-202 (1945), *C.A.* **41**, 2346i (1947).

Colorless crystals, mp 237°. Insol in water; sol in alcohol, ether, aq alkali. The alkaline solns gradually become pink on exposure to air or other oxidizing substances.

USE: As a reagent for oxidases, blood, HCN, peroxides, copper.

7358. Phenolphthalol. [81-92-5] 2-[Bis(4-hydroxyphenyl)-methyl]benzenemethanol; *o*-[bis(*p*-hydroxyphenyl)methyl]benzyl alcohol; dihydroxyphenylmethenylbenzyl alcohol; 2-(4,4'-dihydroxybenzhydryl)benzyl alcohol; bis(4-hydroxyphenyl)-(2-hydroxymethylphenyl)methane; Egmol; Regolax. $C_{20}H_{18}O_3$; mol wt 306.36. C 78.41%, H 5.92%, O 15.67%. Prepn: Hubacher, *J. Am. Chem. Soc.* **74**, 5216 (1952); Schultz, Geller, *Arch. Pharm.* **288**, 234 (1955); Bulcsu, **DE 1141293** (1962 to Iromedica), *C.A.* **59**, 1535b (1963).

Crystals from dil alcohol, mp 201-202°.

Monoacetate. $C_{22}H_{20}O_4$. Crystals from benzene or chloroform, mp 171-174°. Soluble in dil NaOH.

Triacetate. $C_{26}H_{24}O_6$. Crystals from methanol or ethanol, mp 104-106°. Practically insol in dil alkali.

THERAP CAT: Cathartic.

7359. *p*-Phenolsulfonic Acid. [98-67-9] 4-Hydroxybenzenesulfonic acid; sulfocarbolic acid. $C_6H_6O_4S$; mol wt 174.17. C 41.38%, H 3.47%, O 36.74%, S 18.41%. Commercially available as a 65% soln. Prepn by hydrolysis of *p*-chloro- or *p*-bromobenzenesulfonic acid: Zollinger, Roehling, **US 1321271** (1920); by sulfonation of phenol: Davidson, Byrne, **GB 820659** (1959 to Hardman & Holden). Prepn of ammonium salt: Oxley *et al.*, *J. Chem. Soc.* **1948**, 303.

Deliquescent needles. Miscible with water, alcohol.

Ammonium salt. Ammonium *p*-phenolsulfonate. $C_6H_9NO_4S$. Plates from water, dec 270-271°.

Barium salt. Barium *p*-phenolsulfonate. $C_{12}H_{10}BaO_8S_2$. Monohydrate, powder. *Poisonous*. Sol in water; slightly sol in alcohol.

Caution: Irritating to skin.

USE: Intermediate in mfg of pharmaceuticals, dyestuffs. In the Ferrostan process of tin plating (U.S. Steel).

7360. Phenolsulfonphthalein. [143-74-8] 4,4'-(1,1-Dioxido-3*H*-2,1-benzoxathiol-3-ylidene)bisphenol; 4,4'-(3*H*-2,1-benzoxathiol-3-ylidene)bisphenol *S,S*-dioxide; α-hydroxy-α,α-bis(*p*-hydroxy-

phenyl)-*o*-toluenesulfonic acid γ-sultone; 3,3-bis(*p*-hydroxyphenyl)-3*H*-2,1-benzoxathiole 1,1-dioxide; phenol red; P.S.P.; Sulfonphthal. $C_{19}H_{14}O_5S$; mol wt 354.38. C 64.40%, H 3.98%, O 22.57%, S 9.05%. Prepd by the action of *o*-sulfobenzoic anhydride or of *o*-sulfobenzoyl chloride on phenol: Kekulé, Barbaglia, *Ber.* **5**, 876 (1872); Kekulé, *Ber.* **6**, 943 (1873); Heumann, Kochlin, *Ber.* **15**, 1118 (1882); **DE 142116**; *Chem. Zentralbl.* **II**, 79 (1903); Orndorff, Sherwood, *J. Am. Chem. Soc.* **45**, 486 (1923). Diagnostic use: G. Dunea, P. Freedman, *J. Am. Med. Assoc.* **204**, 159 (1968); R. D. Wilkes *et al.*, *Vet. Med. Small Anim. Clin.* **76**, 289 (1981). Molecular structure: K. Yamaguchi *et al.*, *Anal. Sci.* **13**, 521 (1997).

Bright red to dark red crystals. Stable in air. pK = 7.9. d 1.445 Mg m^{-3}. One gram dissolves in ~1300 ml water, in ~350 ml alc, in 500 ml acetone. Almost insol in chloroform, ether. Readily sol in aq alkali hydroxides or carbonates with red color, which is discharged by boiling with zinc dust.

Sodium salt. [34487-61-1] $C_{19}H_{13}NaO_5S$; mol wt 376.36.

USE: As indicator in 0.02-0.05% alcohol soln. pH 6.8 yellow, 8.4 red.

THERAP CAT: Diagnostic aid (renal function).

THERAP CAT (VET): Diagnostic aid (renal function).

7361. Phenoltetrachlorophthalein. [639-44-1] 4,5,6,7-Tetrachloro-3,3-bis(4-hydroxyphenyl)-1(3*H*)-isobenzofuranone; 3,-4,5,6-tetrachlorophenolphthalein. $C_{20}H_{10}Cl_4O_4$; mol wt 456.10. C 52.67%, H 2.21%, Cl 31.09%, O 14.03%. Prepd by condensation of phenol and tetrachlorophthalic acid or its anhydride: W. R. Orndorff, J. A. Black, *Am. Chem. J.* **41**, 349 (1909); Zalkind, Belikova, *Zh. Prikl. Khim.* **8**, 1210 (1935). Metabolism and hepatic chromosecretion of phthalein dyes: P. Hykes, J. Jirsa, *Acta Univ. Carol. Med.* **25**, 135 (1979), *C.A.* **95**, 111282z (1981).

White powder, dec >300°. Almost insol in water, chloroform, benzene. Sol in alcohol, ether, acetone, glacial acetic acid; sol in aq alkali hydroxides or carbonates with deep purple color when concd, violet-red when dil, and bluish when very dil.

Disodium salt. [128-71-2] Chlor-Tetragnost. $C_{20}H_8Cl_4Na_2O_4$; mol wt 500.06. Dihydrate, violet crystals. Freely sol in water. Dec on exposure to air.

THERAP CAT: Cathartic; disodium salt as diagnostic aid (hepatic function).

7362. Phenoperidine. [562-26-5] 1-(3-Hydroxy-3-phenylpropyl)-4-phenyl-4-piperidinecarboxylic acid ethyl ester; 1-(3-hydroxy-3-phenylpropyl)-4-phenylisonipecotic acid ethyl ester; 1-[γ-hydroxy-γ-phenylpropyl]-4-phenyl-4-carbethoxypiperidine; 3-(4-carbethoxy-4-phenylpiperidino)-1-phenyl-1-propanol; 1-phenyl-3-[(4'-phenyl-4'-carbethoxy)piperidino]-1-propanol; phenoperidin. $C_{23}H_{29}NO_3$; mol wt 367.49. C 75.17%, H 7.95%, N 3.81%, O 13.06%. Synthetic opioid analgesic; metabolized *in vivo* to meperidine, *q.v.* Prepn: P. A. J. Janssen, **BE 576331** (1959); F. A. Cutler,

Consult the Name Index before using this section.

Jr., J. F. Fisher, US 2962501 (1960 to Merck & Co.). Structure-activity study: P. A. J. Janssen, N. B. Eddy, *J. Med. Pharm. Chem.* **2**, 31 (1960). Resolution, config, and activity of isomers: R. H. Mazur, *J. Org. Chem.* **26**, 962 (1961). Crystal structure: C. Humblet *et al.*, *Acta Crystallogr.* **B34**, 1389 (1978). GC determn in plasma: P. Kintz *et al.*, *Forensic Sci. Int.* **43**, 267 (1989).

Crystals from hot ethanol, mp 84-87°.

Hydrochloride. [3627-49-4] R-1406; Lealgin; Operidine. C_{23}-$H_{29}NO_3$.HCl; mol wt 403.95. Crystals from ethyl acetate + methanol, or ethanol, mp 200-202°. Soluble in water.

l-**Form.** Glistening plates from aq methanol, mp 86-86.5°. $[\alpha]_D$ $-21°$.

l-**Form hydrochloride.** Irregular prisms from isopropyl alcohol, mp 187-188°. $[\alpha]_D$ $-23°$.

d-**Form hydrochloride.** mp 186-187°. $[\alpha]_D$ $+25°$.

Note: This is a controlled substance (opiate): 21 CFR, 1308.11.

THERAP CAT: Analgesic.

7363. Phenosafranin. [81-93-6] 2,8-Diamino-10-phenyl-phenazinium chloride (1:1); C.I. 50200; Safranin B Extra. $C_{18}H_{15}$-ClN_4; mol wt 322.80. C 66.98%, H 4.68%, Cl 10.98%, N 17.36%. Prepd by dichromate oxidation of 1 mol p-phenylenediamine hydrochloride and 2 mols aniline hydrochloride: Witt, *Ber.* **12**, 939 (1879); **19**, 3121 (1886). *See also Colour Index* **vol. 4** (3rd ed., 1971) p 4449.

Green, lustrous needles from dil HCl. Freely sol in water, also sol in alc. Aq and alcoholic solns are purplish-red and have a greenish-yellow fluorescence. Absorption max ~530 nm.

USE: Biological stain.

7364. Phenothiazine. [92-84-2] 10H-Phenothiazine; thiodiphenylamine; dibenzothiazine; AFI-Tiazin; Antiverm; Fentiazin; Helmetina; Lethelmin; Nemazine; Orimon; Phénégic; Phenoverm; Phenovis; Phenoxur; Reconox; Souframine; Vermitin. $C_{12}H_9NS$; mol wt 199.27. C 72.33%, H 4.55%, N 7.03%, S 16.09%. Prepd by fusing diphenylamine with sulfur: Bernthsen, *Ber.* **16**, 2896 (1883); *Ann.* **230**, 73 (1885); DE 25150 (1883), *Frdl.* **1**, 252. Improved yields with iodine as catalyst: Knoevenagel, *J. Prakt. Chem.* [2] **89**, 11 (1914); Mitchell, Webb, US 2415363 (1947 to Koppers). Purification: Vierling, US 2887482 (1959); Rigby, US 3000887 (1961 to Shell Oil). Crystal structure: J. D. Bell *et al.*, *Chem. Commun.* **1968**, 1656.

Yellow, rhombic leaflets or diamond-shaped plates from toluene or butanol, mp 185.1°. Sublimes at 130° at 1 mm. bp_{760} 371°; bp_{40} 290°. Freely sol in benzene; sol in ether, in hot acetic acid; slightly sol in alcohol and in mineral oils. Practically insol in petr ether, chloroform, water. Readily oxidized by sunlight or when in presence of a finely divided inert carrier, acquiring a greenish-brown tint. This can be prevented by the admixture of 0.3-1.0% methenamine.

Caution: Potential symptoms of overexposure are itching, irritation and reddening skin; hepatitis, hemolytic anemia, abdominal cramps, tachycardia; kidney damage; skin sensitization and photophobia. *See NIOSH Pocket Guide to Chemical Hazards* (DHHS/NIOSH 97-140, 1997) p 248.

USE: Insecticide; manuf pharmaceuticals.

THERAP CAT (VET): Anthelmintic.

7365. Phenoxazine. [135-67-1] 10H-Phenoxazine; phenazoxine. $C_{12}H_9NO$; mol wt 183.21. C 78.67%, H 4.95%, N 7.65%, O 8.73%. Prepn from 2-aminophenol and 2-aminophenol-HCl: Kehrmann, Neil, *Ber.* **47**, 3107 (1903); de Antoni, *Bull. Soc. Chim. Fr.* **1963**, 2871; by heating 2-amino-2'-hydroxydiphenyl ether in a sealed tube at 270-280° for 40 hrs: Cullinane *et al.*, *J. Chem. Soc.* **1934**, 718.

Leaflets from alc, mp 156°. Freely sol in abs methanol, ethanol, ether, CHCl3, benzene. Sparingly sol in petr ether.

10-Acetylphenoxazine. $C_{14}H_{11}NO_2$. Prisms, mp 142°. Sparingly sol in hot water; sol in alc; freely sol in glacial acetic acid.

7366. Phenoxyacetic Acid. [122-59-8] 2-Phenoxyacetic acid; O-phenylglycolic acid; phenyl ether glycolic acid; Phenylium. C_8-H_8O_3; mol wt 152.15. C 63.15%, H 5.30%, O 31.55%. Prepd from phenol and monochloroacetic acid: Giacosa, *J. Prakt. Chem.* [2] **19**, 396 (1879); van Alphen, *Rec. Trav. Chim.* **46**, 148 (1927).

Needles from water, mp 98°; bp 285° (some decompn). pK (25°) 3.12. One gram dissolves in ~75 ml water. Freely sol in alcohol, ether, benzene, carbon disulfide, glacial acetic acid.

Caution: Mild irritant.

USE: Fungicide; keratin exfoliative (to relieve and to soften calluses, corns, and other hard skin surfaces; applied as plasters, pads or in liquids).

7367. Phenoxyacetyl Cellulose. [68332-77-4] 2-Phenoxyacetate cellulose; Enzorb-A. Support used to immobilize proteins, enzymes and microsomes by hydrophobic adsorption. The attractive forces between the hydrophobic phenoxyacetate groups of the support and the hydrophobic surface areas of protein molecules arise from the common repulsion of the aqueous medium. Prepn and properties: L. G. Butler, *Arch. Biochem. Biophys.* **171**, 645 (1975). Applications: J. Dixon *et al.*, *Biotechnol. Bioeng.* **21**, 2113 (1979).

Stable in dry form; slowly hydrolyzed in aq media above pH 8. Resistant to enzymatic hydrolysis.

USE: In enzyme immobilization.

7368. Phenoxybenzamine. [59-96-1] N-(2-Chloroethyl)-N-(1-methyl-2-phenoxyethyl)benzenemethanamine; N-(2-chloroethyl)-N-(1-methyl-2-phenoxyethyl)benzylamine; N-phenoxyisopropyl-N-benzyl-β-chloroethylamine; bensylyt; 688-A. $C_{18}H_{22}ClNO$; mol wt 303.83. C 71.16%, H 7.30%, Cl 11.67%, N 4.61%, O 5.27%. Irreversible α-adrenergic antagonist. Prepn: Kerwin, Ullyot, US 2599000 (1952 to SK&F). Review of clinical use in urinary tract disorders: A. E. Te, *Clin. Ther.* **24**, 851-861 (2002).

Crystals from petr ether, mp 38-40°. Sol in benzene.

Hydrochloride. [63-92-3] Dibenyline; Dibenzyline; Dibenzyran. $C_{18}H_{22}ClNO.HCl$; mol wt 340.29. Crystals from alcohol + ether, mp 137.5-140°. Sol in alcohol, propylene glycol, water, chloroform. Insol in ether. Solutions in propylene glycol should be sterilized by filtration.

Caution: Phenoxybenzamine hydrochloride is reasonably anticipated to be a human carcinogen: *Report on Carcinogens, Twelfth Edition* (PB2011-111646, 2011) p 344.

THERAP CAT: Antihypertensive.

THERAP CAT (VET): In small animals to reduce internal urethral sphincter tone.

7369. 2-Phenoxyethanol. [122-99-6] 1-Hydroxy-2-phenoxyethane; ethylene glycol monophenyl ether; β-hydroxyethyl phenyl ether; Phenoxethol; Phenoxetol; Phenyl Cellosolve. $C_8H_{10}O_2$; mol wt 138.17. C 69.54%, H 7.30%, O 23.16%. Obtained by treating phenol with ethylene oxide in an alkaline medium: Becker, Barthell, *Monatsh. Chem.* **77**, 80 (1947); *see also* Roithner, *ibid.* **15**, 674, 678 (1894); Rindfusz, *J. Am. Chem. Soc.* **41**, 669 (1919). Toxicity study: H. F. Smyth *et al., J. Ind. Hyg. Toxicol.* **23**, 259 (1941).

Oily liquid. Faint aromatic odor. Burning taste. d_{20}^{20} 1.1094; d_4^{22} 1.102; mp 14°; bp_{760} 245.2°; bp_{80} 165°; bp_{25} 137°; bp_{20} 128-130°. n_D^{20} 1.534. Flash pt 250°F. Soly in water: 2.67 g/100 ml. Freely sol in alcohol, ether, NaOH solns. LD_{50} orally in rats: 1.26 g/kg (Smyth).

Acetate. [6192-44-5] $C_{10}H_{12}O_3$; mol wt 180.20. Liquid, bp 243°.

USE: Fixative for perfumes, in org synthesis; as bactericide in conjunction with quaternary ammonium compds; as insect repellent.

THERAP CAT: Antiseptic (topical).

7370. Phenprobamate. [673-31-4] Benzenepropanol 1-carbamate; carbamic acid 3-phenylpropyl ester; 1-carbamoyloxy-3-phenylpropane; γ-phenylpropyl carbamate; proformiphen; MH-532; Extacol; Spantol; Gamaquil. $C_{10}H_{13}NO_2$; mol wt 179.22. C 67.02%, H 7.31%, N 7.82%, O 17.85%. Prepn: **GB 837718** (1960 to Siegfried AG). Pharmacology: G. Stille, *Arzneim.-Forsch.* **12**, 340 (1962). HPLC determn in plasma: J. X. S. Sun *et al., Biopharm. Drug Dispos.* **8**, 341 (1987). Clinical pharmacokinetics: F. C. Tulunay *et al., Arzneim.-Forsch.* **48**, 1068 (1998).

Shiny leaflets from aq ethanol, mp 101-104°. Slightly bitter, somewhat burning taste. Numbs the tongue slightly afterwards. Soluble in abs ethanol, chloroform, propylene glycol, ethylenediamine, dimethylformamide; sparingly sol in ether, 50% ethanol. Practically insol in water. LD_{50} orally in mice: 840 mg/kg (Stille).

THERAP CAT: Muscle relaxant (skeletal).

7371. Phenprocoumon. [435-97-2] 4-Hydroxy-3-(1-phenylpropyl)-2*H*-1-benzopyran-2-one; 3-(α-ethylbenzyl)-4-hydroxycoumarin; 3-(1-phenylpropyl)-4-hydroxycoumarin; Falithrom; Marcoumar; Marcumar; Liquamar. $C_{18}H_{16}O_3$; mol wt 280.32. C 77.13%, H 5.75%, O 17.12%. Prepn: Grüssner, Balthasar, **US 2723276** (1955 to Hoffmann-La Roche); Junek, Ziegler, *Monatsh. Chem.* **87**, 218 (1956); Schroeder, Link, *J. Am. Chem. Soc.* **79**, 3291 (1957); **US 2872457** (1959 to Wisconsin Alumni Res. Found.); **GB 805748** (1958 to Geigy); L. R. Pohl, *J. Med. Chem.* **18**, 513 (1975). Resolution: Preis *et al.,* **US 3239529** (1966 to Wisconsin Alumni Res. Found.). Conformation in soln: E. J. Valente *et al., J. Med. Chem.* **21**, 141, 231 (1978).

Crystals or prisms from dil methanol, mp 179-180°.

THERAP CAT: Anticoagulant.

7372. Phenserine. [101246-66-6]; [159652-53-6] (racemate). (3a*S*,8a*R*)-1,2,3,3a,8,8a-Hexahydro-1,3a,8-trimethylpyrrolo[2,3-*b*]-indol-5-ol 5-(*N*-phenylcarbamate); (−)-eseroline phenylcarbamate. $C_{20}H_{23}N_3O_2$; mol wt 337.42. C 71.19%, H 6.87%, N 12.45%, O 9.48%. Acetylcholinesterase inhibitor that reduces the formation of β-amyloid precursor protein (β-APP); analog of physostigmine, *q.v.* Prepn: M. Polonovski, *Bull. Soc. Chim. Fr.* **19**, 46 (1916). *See also:* M. Pomponi *et al.,* **EP 154864**; M. Brufani *et al.,* **US 5306825** (1985, 1994 both to Consig. Naz. Delle Ricerche). Anticholinesterase activity: M. Brzostowska *et al., Med. Chem. Res.* **2**, 238 (1992). Pharmacology: S. Iijima *et al., Psychopharmacology* **112**, 415 (1993). Pharmacokinetics: N. H. Greig *et al., Acta Neurol. Scand.* Suppl. **176**, 74 (2000). Mode of action study for regulating β-APP expression: K. T. Y. Shaw *et al., Proc. Natl. Acad. Sci. USA* **98**, 7605 (2001). Review of total synthesis and biological activity: N. H. Greig *et al., Med. Res. Rev.* **15**, 3-31 (1995); A. Brossi *et al., Aust. J. Chem.* **49**, 171-181 (1996). Review of development and therapeutic potential: U. Thatte, *IDrugs* **3**, 1222-1228 (2000).

mp 150°. $[\alpha]_D$ −80° (in ethanol).

Tartrate. [156910-61-1] $C_{20}H_{23}N_3O_2.C_4H_6O_6$; mol wt 487.51. mp 142-145°. $[\alpha]_D$ −58.7° (c = 0.75 in methanol). Highly sol in water.

THERAP CAT: In treatment of Alzheimer's disease. Cholinesterase inhibitor.

7373. Phensuximide. [86-34-0] 1-Methyl-3-phenyl-2,5-pyrrolidinedione; *N*-methyl-2-phenylsuccinimide; *N*-methyl-α-phenylsuccinimide; Milontin. $C_{11}H_{11}NO_2$; mol wt 189.21. C 69.83%, H 5.86%, N 7.40%, O 16.91%. Prepd by the action of methylamine on phenylsuccinic acid: Miller, Long, *J. Am. Chem. Soc.* **73**, 4895 (1951); **US 2643258** (1953 to Parke, Davis). Toxicity study: Chen, Ensor, *J. Lab. Clin. Med.* **41**, 78 (1953).

Fine crystals from hot 95% ethanol, mp 71-73°. Very sol in chloroform; readily sol in methanol; sol in alc; slightly sol in water (about 4.2 mg/ml at 25°). Aq solns are fairly stable at pH 2-8, but hydrolysis sets in under more alkaline conditions. LD_{50} orally in mice: 960 mg/kg (Chen, Ensor).

THERAP CAT: Anticonvulsant.

7374. Phentermine. [122-09-8] α,α-Dimethylbenzeneethanamine; α,α-dimethylphenethylamine; phenyl-*tert*-butylamine; α-benzylisopropylamine. $C_{10}H_{15}N$; mol wt 149.24. C 80.48%, H 10.13%, N 9.39%. Prepn: Shelton, Van Campen, **US 2408345** (1946 to Wm. S. Merrell); Abell *et al.,* **US 2590079** (1952 to Wyeth).

Oily liquid. bp$_{750}$ 205°; bp$_{21}$ 100°.

Hydrochloride. [1197-21-3] Adipex-P; Fastin; Wilpo. C$_{10}$H$_{15}$-N.HCl; mol wt 185.70. White, hygroscopic crystals, mp 198°. Sol in water and in lower alcs; slightly sol in chloroform. Insol in ether.

Ion exchange resin complex. Duromine; Ionamin; Linyl; Mirapront; Omnibex.

Note: This is a controlled substance (stimulant): **21 CFR,** 1308.14.

THERAP CAT: Anorexic.

7375. Phentetiothalein Sodium. [18265-54-8] 3,3-Bis(4-hydroxyphenyl)-4,5,6,7-tetraiodo-1(3H)-isobenzofuranone sodium salt (1:2); 4,5,6,7-tetraiodophenolphthalein sodium; 3,4,5,6-tetraiodophenolphthalein disodium salt; phenoltetraiodophthalein sodium; Iso-Iodeikon. C$_{20}$H$_8$I$_4$Na$_2$O$_4$; mol wt 865.88. C 27.74%, H 0.93%, I 58.62%, Na 5.31%, O 7.39%. Prepd by the condensation of phenol and tetraiodophthalic acid or its anhydride: Orndorff, Black, *Am. Chem. J.* **41**, 349 (1909); Zalkind, Belikova, *Zh. Prikl. Khim.* **8**, 1210 (1935). Isomeric with iodophthalein sodium, *q.v.*

Bronze-purple, odorless, slightly hygroscopic granules. Dec on exposure becoming incompletely sol. Sol in water, alcohol. *Keep tightly closed.*

THERAP CAT: Diagnostic aid (radiopaque medium—cholecystography and hepatic function).

7376. Phentolamine. [50-60-2] 3-[[(4,5-Dihydro-1H-imidazol-2-yl)methyl](4-methylphenyl)amino]phenol; 2-[N-(m-hydroxyphenyl)-p-toluidinomethyl]imidazoline; 2-(m-hydroxy-N-p-tolylanilinomethyl)-2-imidazoline; C-7337. C$_{17}$H$_{19}$N$_3$O; mol wt 281.36. C 72.57%, H 6.81%, N 14.93%, O 5.69%. α-Adrenergic blocker. Prepn: K. Miescher *et al.*, US 2503059 (1950 to Ciba); E. Urech *et al.*, *Helv. Chim. Acta* **33**, 1386 (1950). Pharmacology and toxicity: R. Meier *et al.*, *Proc. Soc. Exp. Biol. Med.* **71**, 70 (1949). HPLC determn in biological samples: B. D. Kerger *et al.*, *Anal. Biochem.* **170**, 145 (1988). Diagnostic use: G. Spergel *et al.*, *J. Am. Med. Assoc.* **211**, 266 (1970). Review of pharmacology and therapeutic applications: L. Gould, C. V. R. Reddy, *Am. Heart J.* **92**, 397-402 (1976). Clinical efficacy of combination with papaverine in impotence: A. Bechara *et al.*, *J. Urol.* **157**, 2132 (1997).

Crystals, mp 174-175°.

Hydrochloride. [73-05-2] C$_{17}$H$_{19}$N$_3$O.HCl. Bitter crystals, mp 239-240°. One gram dissolves in 50 ml water, in 70 ml alcohol. Very slightly sol in chloroform. Practically insol in acetone, ethyl acetate. pH of 1% aq soln 4.5-5.5. Aq solns cannot be stored. LD$_{50}$ in rats (mg/kg): 75 i.v.; 275 s.c.; 1250 orally (Meier).

Methanesulfonate. [65-28-1] Phentolamine mesylate; Regitine; Rogitine. C$_{17}$H$_{19}$N$_3$O.CH$_3$SO$_3$H; mol wt 377.46. Crystals, mp 177-181°. One gram dissolves in 50 ml water, 23 ml alcohol, 660 ml chloroform. pH of 1% aq soln 4.5-5.5. Aq solns cannot be stored.

THERAP CAT: Antihypertensive in treatment of pheochromocytoma; diagnostic aid (pheochromocytoma). In treatment of male erectile dysfunction.

7377. Phenylacetaldehyde. [122-78-1] Benzeneacetaldehyde; phenylethanal; α-toluic aldehyde; Hyacinthin. C$_8$H$_8$O; mol wt 120.15. C 79.97%, H 6.71%, O 13.32%. Prepd by oxidizing phenylethyl alcohol with chromic acid. High-yield synthesis from styrene oxide or styrene glycol: G. Paparatto, G. Gregorio, *Tetrahedron Lett.* **29**, 1471 (1988).

Oily, colorless liquid which polymerizes and grows more viscous on standing. Odor reminiscent of lilac and hyacinth. (Has been crystallized, mp 33-34°.) d$_{25}^{25}$ 1.023-1.030; bp$_{760}$ 195°; bp$_{18}$ 88°; bp$_{10}$ 78°; n$_D^{20}$ 1.524-1.528. Slightly sol in water. Soluble in alcohol, ether. One part is sol in 2 parts of 80% alc forming a clear solution.

USE: In perfumery; intermediate in organic synthesis.

7378. α-Phenylacetamide. [103-81-1] Benzeneacetamide; α-toluamide. C$_8$H$_9$NO; mol wt 135.17. C 71.09%, H 6.71%, N 10.36%, O 11.84%. The amide of phenylacetic acid, not to be confused with N-phenylacetamide which is acetanilide. Prepd by Willgerodt reaction from acetophenone or from styrene. Review and lab procedures: Carmack, Spielman in *Org. React.* **II** (New York, 1946) p 83. Has also been prepd by heating the ammonium salt of phenylacetic acid: Menschutkin, *Ber.* **31**, 1429 (1898). Lab procedure from benzyl cyanide: Wenner, *Org. Synth.* **coll. vol. IV**, 760 (1963).

Bimorphous plates, leaflets. mp 155°. Distils *in vacuo*. Freely sol in alc. Slightly sol in water, ether, benzene.

USE: In manuf of penicillin G.

7379. Phenyl Acetate. [122-79-2] Acetic acid phenyl ester; acetoxybenzene; acetylphenol. C$_8$H$_8$O$_2$; mol wt 136.15. C 70.58%, H 5.92%, O 23.50%. Prepd from phenol and acetyl chloride.

Colorless, mobile, highly refractive liquid; phenolic odor. d$_4^{20}$ 1.073; bp 195-196°. n$_D^{20}$ 1.5030 (almost the same as glass). Practically insol in water. Miscible with alc, chloroform, ether; sol in glacial acetic acid. LD$_{50}$ orally in rats: 1.63 ml/kg; *see:* H. F. Smyth *et al.*, *Am. Ind. Hyg. Assoc. J.* **30**, 470 (1969).

7380. Phenylacetic Acid. [103-82-2] Benzeneacetic acid; phenylethanoic acid; α-toluic acid. C$_8$H$_8$O$_2$; mol wt 136.15. C 70.58%, H 5.92%, O 23.50%. Made by refluxing benzyl cyanide with dil H$_2$SO$_4$ or HCl: Adams, Thal, *Org. Synth.* **2**, 63 (1922). Absorption spectrum: Baly, Tryhorn, *J. Chem. Soc.* **107**, 1065 (1915).

Leaflets on distillation *in vacuo;* plates, tablets from petr ether; mp 76.5°. bp$_{760}$ 265.5°; bp$_{100}$ 198.2°; bp$_{40}$ 173.6°; bp$_5$ 127°; bp$_{1.0}$ 97°. d$_4^{77}$ 1.091. pK (25°): 4.25. Slightly sol in cold, freely in hot water. The aq soln satd at 25° is 0.131*N*. Sol in alcohol, ether. Soly at 25° in chloroform (moles/l): 4.422; in carbon tetrachloride: 1.842; in acetylene tetrachloride: 4.513; in trichlorethylene: 3.299; in tetrachlorethylene: 1.558; in pentachloroethane: 3.252.

Methyl ester. [101-41-7] C$_9$H$_{10}$O$_2$; mol wt 150.18. Liquid, bp 215°.

Ethyl ester *see* Ethyl phenylacetate.

USE: Starting material in manuf synthetic perfumes, condensation products with aldehydes.

7381. Phenylacetone. [103-79-7] 1-Phenyl-2-propanone; benzyl methyl ketone. C$_9$H$_{10}$O; mol wt 134.18. C 80.56%, H 7.51%, O 11.92%. Prepn from phenylacetic and acetic acids: R. H. Pickard, J. Kenyon, *J. Chem. Soc.* **105**, 1124 (1914); R. M. Herbst, R. H. Manske, *Org. Synth.* **coll. vol. II,** 389 (1943); from α-phenylacetoacetonitrile: P. L. Julian, J. J. Oliver, *ibid.* 391; from α-methyl-α-phenylethylene oxide: S. Danilow, E. Venus-Danilowa, *Ber.* **60,** 1050 (1927); from diethyl malonate: H. G. Walker, C. R. Hauser, *J. Am. Chem. Soc.* **68,** 1386 (1946). Conformational calculations: M. Hirota *et al., Tetrahedron* **39,** 3091 (1983). Use as prochiral ketone in enantioselective hydrosilylation: H. Brenner *et al., Ber.* **117,** 1330 (1984).

Oil, mp −16 to −15°. bp$_{760}$ 214°; bp$_{14}$ 100-101°. d$_4^{20}$ 1.0157. n$_D$ 1.5174. uv max (ethanol): 258, 283 nm (ε 255, 150).

Note: This is a controlled substance: (immediate precursor to amphetamine and methamphetamine) **21 CFR,** 1308.12.

USE: In organic synthesis; production of benzyl radicals by photolysis.

7382. Phenylalanine. [63-91-2] L-Phenylalanine; Phe; F; β-phenylalanine; α-aminohydrocinnamic acid; (*S*)-2-amino-3-phenylpropanoic acid; α-amino-β-phenylpropionic acid. C$_9$H$_{11}$NO$_2$; mol wt 165.19. C 65.44%, H 6.71%, N 8.48%, O 19.37%. Essential amino acid for human development. Originally isolated from the sprouts of lupine: E. Schulze, J. Barbieri, *Ber.* **12,** 1924 (1879). Early chemistry and biochemistry: *Amino Acids and Proteins*, D. M. Greenberg, Ed. (Charles C. Thomas, Springfield, IL, 1951) 950 pp. *passim*; J. P. Greenstein, M. Winitz, *Chemistry of the Amino Acids* **vols 1-3** (John Wiley and Sons, Inc., New York, 1961) pp. 2156-2177, *passim*. Synthesis from L-tyrosine: V. Viswanatha, V. J. Hruby, *J. Org. Chem.* **45,** 2010 (1980). Colorimetric determn in blood: R. S. Campbell *et al., Ann. Clin. Biochem.* **31,** 140 (1994). Effects on food intake: G. H. Anderson, L. A. Leiter, *Appetite* **11,** Suppl. 48 (1988); P. J. Rogers, J. E. Blundell, *Physiol. Behav.* **56,** 247 (1994). Clinical trial in vitiligo: C. Antoniou *et al., Int. J. Dermatol.* **28,** 545 (1989); A. H. Siddiqui *et al., Dermatology* **188,** 215 (1994). Review of metabolism: H. N. Munro, *J. Toxicol. Environ. Health* **2,** 189-206 (1976). Review of role in phenylketonuria: S. P. Bessman, *Nutr. Rev.* **37,** 209-220 (1979); F. Güttler, *Acta Paediatr. Scand.* **73,** 705-716 (1984). Review of microbial production: T. K. Maiti, S. P. Chatterjee, *Hind. Antibiot. Bull.* **32,** 3-26 (1990).

Monoclinic plates, leaflets from warm concd aq solns. Hydrated needles from dil solns. Dec 283°. Sublimes *in vacuo*. [α]$_D^{20}$ −35.1° (c = 1.94). pK$_1$ 1.83; pK$_2$ 9.13. Soly in water (g/l): 19.8 at 0°; 29.6 at 25°; 44.3 at 50°; 66.2 at 75°; 99.0 at 100°. Very slightly sol in methanol, ethanol, dilute mineral acids.

D-Form. [673-06-3] (*R*)-Phenylalanine; Sabiden. Occurs naturally in microbial products such as tyrocidine, *q.v.*: A. H. Gordon *et al., Biochem. J.* **37,** 313 (1943). May be used as a source for L-form

in man. Biotransformation: S. Tokuhisa *et al., Chem. Pharm. Bull.* **29,** 514 (1981). Evaluation of D-form in multiple sclerosis: A. Winter, *Neurol. Orthopaed. J. Med. Surg.* **5,** 39 (1984). Leaflets from water, dec 285°. [α]$_D^{20}$ +35.0° (c = 2.04); [α]$_D^{20}$ +7.1° (c = 3.8 in 18% HCl). One gram dissolves in 35.5 ml water at 16°. Sparingly sol in methanol.

DL-Form. [150-30-1] Monoclinic leaflets or prisms from water or alc, sweetish taste. Dec 271-273°. Sublimes *in vacuo*. pK$_1$ 2.58; pK$_2$ 9.24. Soly in water (g/l): 9.97 at 0°; 14.11 at 25°; 21.87 at 50°; 37.08 at 75°; 68.9 at 100°.

USE: Component of the artificial sweetner aspartame, *q.v.*; nutrient.

7383. Phenyl Aminosalicylate. [133-11-9] 4-Amino-2-hydroxybenzoic acid phenyl ester; *p*-aminosalicylic acid phenyl ester; phenyl *p*-aminosalicylic acid; *p*-aminosalol; fenamisal; Phenyl PAS; Pheny-PAS-Tebamin; Tebamin; Tebanyl. C$_{13}$H$_{11}$NO$_3$; mol wt 229.24. C 68.11%, H 4.84%, N 6.11%, O 20.94%. Description of properties: Meyer, *Antibiot. Annu.* **1957-1958,** 614. Prepd by Raney nickel reduction of the corresp nitro ester in ethyl acetate: Friere, **US 2604488** (1952 to Rhône-Poulenc).

Crystals from isopropanol, mp 153°. Soly in water: 0.7 mg/100 ml; in serum: 12 mg/100 ml. One gram of phenyl PAS is equivalent to 0.67 g PAS.

THERAP CAT: Antibacterial (tuberculostatic).

7384. *N*-Phenylanthranilic Acid. [91-40-7] 2-(Phenylamino)benzoic acid; 2-anilinobenzoic acid; diphenylamine-2-carboxylic acid. C$_{13}$H$_{11}$NO$_2$; mol wt 213.24. C 73.22%, H 5.20%, N 6.57%, O 15.01%. Prepd from *o*-chlorobenzoic acid and aniline: Ullmann, Dieterle, *Ann.* **365,** 322 (1907); Ullmann, *Ber.* **36,** 2382 (1907); from 2-iodobenzoic acid and phenylhydroxylamine: Wieland, Roseeu, *Ber.* **48,** 1120 (1915). Use: Adamovich, Zagurlko, *Zavod. Lab.* **9,** 465 (1940), *C.A.* **37,** 1346 (1943).

Leaflets from alcohol, dec 183-184°. Sol in hot alcohol. Very slightly sol in hot water, hot benzene, ether.

USE: Detection of vanadium in steel.

7385. Phenyl Azide. [622-37-7] Azidobenzene; triazobenzene. C$_6$H$_5$N$_3$; mol wt 119.13. C 60.49%, H 4.23%, N 35.27%. Reagent used to prepare nitrogen heterocycles. Prepn: P. Griess, *Ann.* **137,** 39 (1866); R. O. Lindsay, C. F. H. Allen, *Org. Synth.* **coll. vol. III,** 710 (1955). Synthetic utility in dipolar cycloadditions: A. Michael, *J. Prakt. Chem.* **48,** 94 (1893); K. R. Henery-Logan, R. A. Clark, *Tetrahedron Lett.* **9,** 801 (1968); L. K. Rasmussen *et al., Org. Lett.* **9,** 5337 (2007); in photochemical reactions: W. E. Doering, R. A. Odum, *Tetrahedron* **22,** 81 (1966); A. K. Schrock, G. B. Schuster, *J. Am. Chem. Soc.* **106,** 5228 (1984); in Staudinger reactions: J. E. Leffler, R. D. Temple, *ibid.* **89,** 5235 (1967).

Pungent, pale yellow oil. *Explosive.* mp −27.5 to −27.1°. bp$_{13}$ 53-54°; bp$_5$ 49-50°; bp$_{0.7}$ 35°. n$_D^{25.5}$ 1.5591; n$_D^{20}$ 1.5598. Freely sol in common organic solvents. Insol in water, aq acidic and alkali solns. Light sensitive. May be stored in cool, dark place for up to one month.

USE: Reagent in synthetic organic chemistry.

7386. Phenyl Benzenethiosulfonate. [1212-08-4] Benzenesulfonothioic acid *S*-phenyl ester; benzenethiol benzenesulfonate; phenyl phenylthio sulfone; PhSSO$_2$Ph. C$_{12}$H$_{10}$O$_2$S$_2$; mol wt 250.33. C 57.58%, H 4.03%, O 12.78%, S 25.61%. Sulfenylating agent for the introduction of the phenylthio functional group. Prepn: C. Pauly, R. Otto, *Ber.* **9**, 1639 (1876); D. Barnard, *J. Chem. Soc.* **1957**, 4673; and utility in sulfenylations: B. M. Trost, G. S. Massiot, *J. Am. Chem. Soc.* **99**, 4405 (1977). Use in bissulfenylations: Y. K. Yee, A. G. Schultz, *J. Org. Chem.* **44**, 719 (1979); B. M. Trost, M. K. T. Mao, *Tetrahedron Lett.* **21**, 3523 (1980). Carbonylation reaction applications: S. Kim *et al.*, *Angew. Chem. Int. Ed.* **44**, 6183 (2005). Crystal structure: R. Caputo *et al.*, *Gazz. Chim. Ital.* **114**, 421 (1984).

Colorless crystals from methanol, mp 36-37° (Trost); also reported as monoclinic plates, mp 45° (Pauly). bp$_{0.1}$ 125°. d 1.386. Flash pt, closed cup: 235.40°F (113.00°C).

USE: Reagent in synthetic organic chemistry.

7387. 2-Phenyl-1*H*-benzimidazole. [716-79-0] *N,N'*-Benzenyl-*o*-phenylenediamine; phenzidole; Gainex. C$_{13}$H$_{10}$N$_2$; mol wt 194.24. C 80.39%, H 5.19%, N 14.42%. Prepn from *N*-benzoyl-*o*-phenylenediamine: v. Auwers, Frese, *Ber.* **59**, 548 (1926); from *o*-phenylenediamine and benzonitrile: Hölljes, Wagner, *J. Org. Chem.* **9**, 31 (1944). Anthelmintic in sheep: Forsyth, *Aust. Vet. J.* **38**, 398 (1962).

Needles from benzene, plates from water, mp 291°. Sol in methanol, abs ethanol. Sparingly sol in water, benzene, chloroform.

Hydrochloride. C$_{13}$H$_{10}$N$_2$.HCl. Needles from alcohol, dec 328°. Freely sol in water.

THERAP CAT (VET): Has been used as an anthelmintic.

7388. Phenyl Benzoate. [93-99-2] Benzoic acid phenyl ester. C$_{13}$H$_{10}$O$_2$; mol wt 198.22. C 78.77%, H 5.09%, O 16.14%.

Monoclinic prisms; geranium odor. d 1.235; mp 70°; bp 314°. Insoluble in water. Freely sol in hot alcohol; slightly sol in cold alcohol or ether.

7389. Phenyl Biguanide. [102-02-3] *N*-Phenylimidodicarbonimidic diamide; phenyl diguanide; 1-phenylbiguanide; *N*-phenyl-*N'*-guanylguanidine. C$_8$H$_{11}$N$_5$; mol wt 177.21. C 54.22%, H 6.26%, N 39.52%. Prepd by reacting aniline hydrochloride and dicyandiamide in water: Cohn, *J. Prakt. Chem.* [2] **84**, 396; from aniline, dicyandiamide in pyridine contg HCl: Jacobs, Jolles, **GB 587907** (1947 to ICI).

Crystals from water or toluene. Sharp, somewhat bitter taste. mp 144-146°. pK$_1$ 10.76, pK$_2$ 2.13. Freely sol in water and in alcohol.

Hydrochloride. C$_8$H$_{11}$N$_5$.HCl. Prisms, mp 237°. Sol in about 115 parts of water at 32°.

Nitrate. C$_8$H$_{11}$N$_5$.HNO$_3$. Crystals, mp 208-209°.

7390. Phenylbutazone. [50-33-9] 4-Butyl-1,2-diphenyl-3,5-pyrazolidinedione; 4-butyl-1,2-diphenyl-3,5-dioxopyrazolidine; 3,5-dioxo-1,2-diphenyl-4-*n*-butylpyrazolidine; flexazone; G-13871; R-3-ZON; Ambene; Artrizin; Azolid; Bizolin; Butacote; Butadion; Butapirazol; Butadiona; Butatron; Butoz; Butazolidin; Buzon; Ecobutazone; Equipalazone; Exrheudon N; Fenibutol; Intrabutazone; Intrazone; Mepha-Butazon; Phenyzene; Robizone-V; Tevcodyne; Uzone. C$_{19}$H$_{20}$N$_2$O$_2$; mol wt 308.38. C 74.00%, H 6.54%, N 9.08%, O 10.38%. Prepn: Stenzl, **US 2562830** (1951 to Geigy); *cf.* **GB 812449** (1959 to Geigy). Review of synthesis: *Ullmanns Encyklopädie der technischen Chemie* **vol. 13**, 298 (1962). Physical properties and pharmacology: v. Rechenberg, *Phenylbutazone* (Edward Arnold, London, 1962) 197 pp. Acute toxicity: T. B. Gaines, R. E. Linder, *Fundam. Appl. Toxicol.* **7**, 299 (1986). Soly data: Pulver *et al.*, *Schweiz. Med. Wochenschr.* **86**, 1080 (1956). Comprehensive description: S. L. Ali, *Anal. Profiles Drug Subs.* **11**, 483-521 (1982). Review of hematological effects: G. A. Faich, *Pharmacotherapy* **7**, 25 (1987).

Crystals from ethanol, mp 105°. Soly in water at 22.5°: 0.7 mg/ml (also reported as 2.2 mg/ml). Freely sol in acetone, ether; sol in alc. pK 4.5 (from uv in water), pK 4.89 (titration in 50% ethanol), pK 5.25 (titration in 80% 2-methoxyethanol). uv max (acid methanol): 239.5 nm (log ε 4.19).

Sodium salt. [129-18-0] GP-26872; Elmedal. C$_{19}$H$_{19}$N$_2$NaO$_2$; mol wt 330.36. Crystals, freely sol in water. pH of aq solns ~8.2.

Calcium salt. Ticinil Calcico. C$_{38}$H$_{36}$CaN$_4$O$_4$; mol wt 652.81. LD$_{50}$ in adult male, female rats (mg/kg): 1311, 647 orally (Gaines, Linder).

Piperazine salt. Pyrazinobutazone; pyrasanone; Carudol. C$_{19}$H$_{20}$N$_2$O$_2$.C$_4$H$_{10}$N$_2$; mol wt 394.52. mp 140-141° (solidifies and remelts at ~180°).

2-Amino-2-thiazoline salt. [54749-86-9] 4-Butyl-1,2-diphenyl-3,5-pyrazolidinedione compd with 4,5-dihydro-2-thiazolamine (1:1); thiazolinobutazone; TZB; LAS-11871; Fordonal. C$_{22}$H$_{26}$N$_4$O$_2$S; mol wt 410.54. Prepn: J. Moragues *et al.*, *Arzneim.-Forsch.* **24**, 1785 (1974). Pharmacology: M. Márquez, D. J. Roberts, *ibid.* 1786, 1790; M. Colombo *et al.*, *ibid.* **26**, 1347 (1976). White crystals, mp 164-166°. LD$_{50}$ orally in rats, mice: 1425, 1650 mg/kg (Colombo).

THERAP CAT: Anti-inflammatory.

THERAP CAT (VET): Analgesic; anti-inflammatory.

7391. 2-Phenyl-6-chlorophenol. [85-97-2] 3-Chloro[1,1'-biphenyl]-2-ol; 6-chlorothoxenol. C$_{12}$H$_9$ClO; mol wt 204.65. C 70.43%, H 4.43%, Cl 17.32%, O 7.82%.

Pale yellow, viscous liquid. Slight, characteristic odor. d$_4^{25}$ about 1.24; mp 6°; bp 317-318° (dec). n_D^{30} 1.6237. Insoluble in water. Sol in fixed alkali hydroxide solns and in most organic solvents.

USE: Germicide; fungicide.

7392. 4-Phenyl-2-chlorophenol. [92-04-6] 3-Chloro-(1,1'-biphenyl)-4-ol. C$_{12}$H$_9$ClO; mol wt 204.65. C 70.43%, H 4.43%, Cl 17.32%, O 7.82%.

Pale yellow crystals. mp ~77°. bp₇ ~160-162°. Soly and use as for 2-phenyl-6-chlorophenol, *q.v.*

7393. α-Phenylcinnamic Acid. [3368-16-9] α-(Phenylmethylene)benzeneacetic acid; 2,3-diphenylacrylic acid; 2,3-diphenylpropenoic acid; stilbene-α-carboxylic acid. C₁₅H₁₂O₂; mol wt 224.26. C 80.34%, H 5.39%, O 14.27%. Prepn of *cis*-form: Buckles, Hausman, *J. Am. Chem. Soc.* **70**, 415 (1948); Buckles, *J. Chem. Educ.* **27**, 210 (1950); Buckles, Bremer, *Org. Synth.* **coll. vol. IV**, 777 (1963); B. H. Patwardhan, G. Bagavant, *Indian J. Chem.* **10**, 59 (1972). Prepn of *cis*- and *trans*-forms: L. F. Fieser, *Experiments in Organic Chemistry* (Heath, Boston, 3rd ed., 1955) p 182. Isomerization studies: S. V. Kessar *et al.*, *Indian J. Chem.* **20B**, 4 (1981).

cis-form

cis-Form. [91-47-4] Silky needles from ether + petr ether, mp 174°. pKa 6.1 in 60% ethanol. uv max (ethanol): 223, 280 nm (ε 32100, 19500). Sol in hot water, methanol, ethanol, isopropanol, ether, benzene.

trans-Form. [91-48-5] Stout prisms from ether + petr ether, mp 138-139°. Less stable than the *cis*-form. pKa 4.8 in 60% ethanol. uv max (ethanol): 222, 289 nm (ε 14500, 22500). More sol than the *cis*-form.

7394. Phenyl Dichlorophosphate. [770-12-7] Phosphorodichloridic acid phenyl ester; PCDP. C₆H₅Cl₂O₂P; mol wt 210.98. C 34.16%, H 2.39%, Cl 33.61%, O 15.17%, P 14.68%. Phosphorylation reagent. Prepn: G. Jacobsen, *Ber.* **8**, 1519 (1875); R. Anschütz, W. O Emery, *Ann.* **253**, 105 (1889). Activation of carboxyl groups: H.-J. Liu, S. I. Sabesan, *Can. J. Chem.* **58**, 2645 (1980); for lactam synthesis: J. M. Aizpurua *et al.*, *Tetrahedron Lett.* **25**, 3905 (1984); L. Banfi, G. Guanti, *ibid.* **43**, 7427 (2002). Oxidative rearrangements: H.-J. Liu, J. M. Nyangulu, *ibid.* **30**, 5097 (1989); J. P. Jeyadevan *et al.*, *J. Med. Chem.* **47**, 1290 (2004). Brief review: A. K. Adak, *Synlett* **2004**, 1651-1652.

Liquid, bp₉ₘₘHg 103-106°; bp₁₃₋₁₄ₘₘ 116°. d₄²⁰ 1.41214.
USE: In organic synthesis for preparation of lactams, phosphate diesters and for oxidation reactions.

7395. m-Phenylenediamine. [108-45-2] 1,3-Benzenediamine; *m*-diaminobenzene. C₆H₈N₂; mol wt 108.14. C 66.64%, H 7.46%, N 25.91%. Prepd by the reduction of *m*-dinitrobenzene: Kuhn, *US 2768209* (1956 to Ringwood); Faust, *J. Prakt. Chem.* **6**, 14 (1958); Neilson *et al.*, *J. Chem. Soc.* **1962**, 371; Tallee, Peltier, *Compt. Rend.* **259**, 400 (1964). Toxicity study: C. Burnett *et al.*, *J. Toxicol. Environ. Health* **2**, 657 (1977).

White crystals becoming red on exposure to air. d 1.139; mp 62-63°; bp 284-287°; dipole moment 1.79. Fire point: 175°. Sol in water, methanol, ethanol, chloroform, acetone, dimethylformamide, methyl ethyl ketone, dioxane. Slightly sol in ether, carbon tetrachloride, dibutyl phthalate, isopropanol. Very slightly sol in benzene,

toluene, xylene, butanol. *Keep well closed and protected from light.* LD₅₀ in rats (mg/kg): 650 orally; 283 i.p. (Burnett).

Hydrochloride. White or slightly red, crystalline powder; becomes darker on exposure to air. Freely sol in water; sol in alcohol.

USE: Manuf dyes; rubber curing agents, ion exchange resins, decoloring resins, formaldehyde condensates, resinous polyamides, block polymers, textile fibers, urethanes, petroleum additives, rubber chemicals, corrosion inhibitors; in photography; as reagent for gold and bromine. The hydrochloride chiefly as a reagent for nitrite.

7396. o-Phenylenediamine. [95-54-5] 1,2-Benzenediamine; *o*-diaminobenzene. C₆H₈N₂; mol wt 108.14. C 66.64%, H 7.46%, N 25.91%. Made by reducing *o*-nitroaniline with Zn and NaOH.

Brownish-yellow crystals. mp 103-104°. bp 256-258°. Slightly sol in water; freely sol in alcohol, chloroform, ether. LD₅₀ in rats (mg/kg): 1070 orally; 516 i.p., C. Burnett *et al.*, *J. Toxicol. Environ. Health* **2**, 657 (1977).

USE: Manufacture of dyes.

7397. p-Phenylenediamine. [106-50-3] 1,4-Benzenediamine; *p*-diaminobenzene; *p*-aminoaniline; orsin; C.I. 76076; Ursol D. C₆H₈N₂; mol wt 108.14. C 66.64%, H 7.46%, N 25.91%. Prepn: A. Rinne, T. Zincke, *Ber.* **7**, 869 (1874); **DE 202170** (1907 to BASF), *C.A.* **3**, 382 (1909); A. J. Quick, *J. Am. Chem. Soc.* **42**, 1033 (1920); J. F. Norris, E. O. Cummings, *Beilstein* **17**, 305 (1925). *See also: Beilstein* **XIII**, 61 (1930). Crystal structure: A. Domenicano *et al.*, *Acta Crystallogr.* **B33**, 1664 (1977). Mutagenicity studies: B. N. Ames *et al.*, *Proc. Natl. Acad. Sci. USA* **72**, 2423 (1975); W. G. H. Blijleven, *Mutat. Res.* **48**, 181 (1977). Toxicity study: C. Burnett *et al.*, *J. Toxicol. Environ. Health* **2**, 657 (1977).

White to slightly red crystals; darkens on exposure to air. mp 145-147°. bp 267°. Sol in 100 parts cold water; sol in alcohol, chloroform, ether. A black color is developed with 3% H₂O₂; brown with 5% FeCl₃ soln. *Keep well closed and protected from light.* LD₅₀ in rats (mg/kg): 80 orally, 37 i.p. (Burnett).

Hydrochloride. White to slightly reddish crystals. Freely sol in water, slightly in alcohol, ether.

Caution: Potential symptoms of overexposure are irritation of pharynx and larynx; bronchial asthma; sensitization dermatitis. *See NIOSH Pocket Guide to Chemical Hazards* (DHHS/NIOSH 97-140, 1997) p 248.

USE: Dyeing furs; also in photochemical measurements, accelerating vulcanization; manuf azo dyes, etc. The hydrochloride as reagent for blood, H₂S, amyl alcohol; in testing of milk.

7398. Phenylephrine. [59-42-7] (αR)-3-Hydroxy-α-[(methylamino)methyl]benzenemethanol; (−)-*m*-hydroxy-α-[(methylamino)methyl]benzyl alcohol; *l*-1-(*m*-hydroxyphenyl)-2-methylaminoethanol; *l*-α-hydroxy-β-methylamino-3-hydroxy-1-ethylbenzene; *m*-methylaminoethanolphenol; metaoxedrin. C₉H₁₃NO₂; mol wt 167.21. C 64.65%, H 7.84%, N 8.38%, O 19.14%. α-Adrenergic agonist. Prepn: H. Legerlotz, *US 1932347* (1933 to Frederick Stearns); E. D. Bergmann, M. Sulzbacher, *J. Org. Chem.* **16**, 84 (1951); M. K. Gurjar *et al.*, *Org. Process Res. Dev.* **2**, 422 (1998). Spectrophotometric determn: S. A. Shama, *J. Pharm. Biomed. Anal.* **30**, 1385 (2002). Toxicity data: M. R. Warren, H. W. Werner, *J. Pharmacol. Exp. Ther.* **86**, 284 (1946). Clinical efficacy in mydriasis: V. Tanner, A. G. Casswell, *Eye* **10**, 95 (1996). Clinical evaluation in fecal incontinence: M. J. Cheetham *et al.*, *Gut* **48**, 356 (2001). Comprehensive description: C. A. Gaglia, Jr., *Anal. Profiles Drug Subs.* **3**, 483-512 (1974). Review of ophthalmologic uses: S. M. Meyer, F. T. Fraunfelder, *Ophthalmology* **87**, 1177-1180 (1980); of clinical trials vs ephedrine, *q.v.*, in treatment of hypotension during cesarean delivery: A. Lee *et al.*, *Anesth. Analg.* **94**, 920-926 (2002).

mp 170-171°.

Hydrochloride. [61-76-7] Ak-Dilate; Ak-Nefrin; Alconefrin; Incostop; Mydfrin; Neo-Synephrine; Nostril; Prefrin; Rhinall. C_9H_{13}-NO_2.HCl; mol wt 203.67. White, bitter-tasting crystals, mp 140-145°. $[\alpha]_D$ −44.0° (c = 2.16 in H_2O). pK_1 8.77; pK_2 9.84. uv max (0.05 N HCl): 216, 274, 279 nm ($\varepsilon \times 10^{-3}$ 5.91, 1.81, 1.65). uv max (0.05 N NaOH): 239, 292.5 ($\varepsilon \times 10^{-3}$ 8.95, 3.04). Freely sol in water, alc. LD_{50} in rats (mg/kg): 17 ±1.1 i.p.; 33 ±2.0 s.c. (Warren, Werner).

THERAP CAT: Mydriatic; decongestant; in treatment of fecal incontinence.

7399. Phenylethanolamine. [7568-93-6] α-(Aminomethyl)-benzenemethanol; α-(aminomethyl)benzyl alcohol; β-hydroxyphen-ethylamine. $C_8H_{11}NO$; mol wt 137.18. C 70.05%, H 8.08%, N 10.21%, O 11.66%. Prepn: Dornow, Theidel, *Ber.* **88**, 1267 (1955).

Pale yellow crystals, mp 56-57°. bp_{17} 157-160°. Freely sol in water. Gives an alkaline reaction in water and forms salts with acids under mild conditions.

Sulfate. [613-82-1] *dl*-α-Phenyl-β-aminoethanol sulfate; Apophedrin. $(C_8H_{11}NO)_2.H_2SO_4$; mol wt 372.44. Prepn: Roth, *Arch. Pharm.* **292**, 76 (1959). Crystals, mp 275-276° (Kofler). Soluble in water.

USE: Free base as a stopping agent during polymerization of styrene-butadiene rubber; in the hardening of waxes. Intermediate in the manuf of pressor amines.

THERAP CAT: Sulfate as topical vasoconstrictor.

7400. 5-(α-Phenylethyl)semioxamazide. [93-95-8] 2-Oxo-2-[(1-phenylethyl)amino]acetic acid hydrazide; 5-(α-methylbenzyl)-semioxamazide. $C_{10}H_{13}N_3O_2$; mol wt 207.23. C 57.96%, H 6.32%, N 20.28%, O 15.44%. Prepd from α-phenylethylamine, ethyl oxalate, and hydrazine: Leonard, Boyer, *J. Org. Chem.* **15**, 42 (1950).

dl-Form. Fine needles from ethanol, mp 157°.
d-Form. Crystals, mp 167-168°. $[\alpha]_D^{25}$ +102.0° (c = 1.04 in chloroform).
l-Form. Crystals, mp 167-168°. $[\alpha]_D^{25}$ −102.5° (c = 0.625 in chloroform).

USE: The racemic form as reagent for the characterization of aldehydes and ketones. The optically active forms as resolving agents for carbonyl compds possessing asymmetric carbon atoms.

7401. Phenylglyceryl Ether. [538-43-2] 3-Phenoxy-1,2-propanediol; 1,2-dihydroxy-3-phenoxypropane; glycerol α-monophenyl ether; Antodyne. $C_9H_{12}O_3$; mol wt 168.19. C 64.27%, H 7.19%, O 28.54%. Prepd from phenol and 3-chloro-1,2-propylene carbonate: Smith, *US 2967892* (1961 to Dow). Toxicity study: Hine *et al.*, *J. Pharmacol. Exp. Ther.* **97**, 414 (1949).

Crystals, mp 50-52°, $bp_{0.6}$ 129-142°. Sol in water, alcohol. LD_{50} i.p. in mice: 1280 mg/kg (Hine).

7402. α-Phenylglycine. [2835-06-5] α-Aminobenzeneacetic acid; α-aminophenylacetic acid; α-amino-α-toluic acid; C-phenyl-glycine. $C_8H_9NO_2$; mol wt 151.17. C 63.56%, H 6.00%, N 9.27%, O 21.17%. Prepd by hydrolysis of α-aminophenylacetonitrile with dil hydrochloric acid: Marvel, Noyes, *J. Am. Chem. Soc.* **42**, 2264 (1920); R. E. Steiger, *Org. Synth.* **coll. vol. III**, 84 (1955).

Lustrous platelets. Sublimes without melting at ~255°. Slightly sol in the usual organic solvents. Sol in alkalies.

(R)-Form. [875-74-1] Needles from dil alc, mp 305-310°. $[\alpha]_D^{20}$ −157.8° (dil HCl).
(S)-Form. [2935-35-5] Crystals. $[\alpha]_D^{20}$ +156° (dil HCl).
Methyl ester. [26682-99-5] Methyl DL-phenylglycinate. C_9H_{11}-NO_2; mol wt 165.19. Needles from petr ether, mp 32°. Sol in alcohol, ether, benzene.
Ethyl ester. [6097-58-1] Ethyl DL-phenylglycinate. $C_{10}H_{13}$-NO_2; mol wt 179.22. Liquid, bp 257°; bp_{16} 149°; bp_5 114-115°; n_D^{25} 1.500.

7403. N-Phenylglycine. [103-01-5] Anilinoacetic acid; (phenylamino)acetic acid. $C_8H_9NO_2$; mol wt 151.17. C 63.56%, H 6.00%, N 9.27%, O 21.17%. Prepd from aniline and chloroacetic acid: Curtius, *J. Prakt. Chem.* **38**, 436 (1888); Thorpe, Wood, *J. Chem. Soc.* **103**, 1606 (1913).

Crystals, mp 127-128°. pK (25°): 5.42. Moderately sol in water. Less sol in alc. Sparingly sol in ether. Forms water-sol salts with alkali hydroxides.

Ethyl ester. [2216-92-4] Ethyl phenylglycinate. $C_{10}H_{13}NO_2$; mol wt 179.22. Leaflets, mp 58°. Sol in alc, ether.

USE: In the Heumann synthesis of indigo.

7404. Phenylhydrazine. [100-63-0] Hydrazinobenzene. C_6-H_8N_2; mol wt 108.14. C 66.64%, H 7.46%, N 25.91%. Prepd by diazotizing aniline with $NaNO_2$ and HCl, then treating the soln with Na_2SO_3 followed by NaOH: Fischer, *Anleitung zur Darstellung organischer Präparate* (Braunschweig, 10th ed., 1922) p 23; Coleman, *Org. Synth.* **2**, 71 (1922), **coll. vol. I** (2nd ed., 1941) p 442. Mechanism of oxidative hemolysis of erythrocytes: B. Goldberg, A. Stern, *Mol. Pharmacol.* **13**, 832 (1977); B. Vilsen, H. Nielsen, *Biochem. Pharmacol.* **33**, 2739 (1984). Review of interaction with hemoglobin: M. D. Shetlar, H. A. O. Hill, *Environ. Health Perspect.* **64**, 265-281 (1985).

Monoclinic prisms or oil. Turns yellow to dark red on exposure to air and light. Faint aromatic odor. d_4^{20} 1.0978; mp 19.5°; bp_{760} 243.5° (dec); bp_{100} 173.5°; bp_{40} 148.2°; bp_{20} 131.5°; bp_{10} 115.8°; bp_5 101.6°; $bp_{1.0}$ 71.8°; $n_D^{20.3}$ 1.60813. Weak base, pK (15°): 8.79. *Poi-*

sonous. Keep well closed and protected from light. Miscible with alcohol, ether, chloroform, benzene. Sparingly sol in water, petr ether; sol in dil acids. Forms a hemihydrate, mp 24°.

Hydrochloride. [59-88-1] Phenylhydrazinium chloride. C_6H_8-N_2.HCl; mol wt 144.60. Prepn: Fischer, *Ann.* **190**, 83 (1878); Brunner, Eiermann, *Ber.* **31**, 1406 (1898); Rüetschi, Trümpler, *Helv. Chim. Acta* **36**, 1649 (1953); Hupfer, **DE 1143825** (1963 to Hoechst). Leaflets from alc, mp 243-246° (slight browning). Sublimes. Freely sol in water; sol in alcohol. Practically insol in ether.

Caution: Potential symptoms of overexposure are skin sensitization, hemolytic anemia, dyspnea and cyanosis; jaundice; kidney damage; vascular thrombosis. Potential occupational carcinogen. *See NIOSH Pocket Guide to Chemical Hazards* (DHHS/NIOSH 97-140, 1997) p 250. *See also Patty's Industrial Hygiene and Toxicology* vol. **2A**, G. D. Clayton, R. E. Clayton, Eds. (Wiley-Interscience, New York, 3rd ed., 1981) pp 2792-2795, 2804-2805.

USE: Manuf dyes, antipyrine, nitron (a stabilizer for explosives); reagent for sugars, aldehydes, ketones.

THERAP CAT: Hemolytic.

7405. Phenylhydroxylamine. [100-65-2] *N*-Hydroxybenzenamine; *N*-hydroxyaniline; β-phenylhydroxylamine; *N*-phenylhydroxylamine. C_6H_7NO; mol wt 109.13. C 66.04%, H 6.47%, N 12.84%, O 14.66%. Prepd by zinc reduction of nitrobenzene in ammonium chloride soln: Kamm, *Org. Synth.* **4**, 57 (1925).

Needles from satd NaCl soln. mp 82°. *Deteriorates on storage and should be used promptly.* Sol in 50 parts cold, in 10 parts hot water. Freely sol in alcohol, ether, carbon disulfide, chloroform, hot benzene, dil mineral acids, acetic acid. Slightly sol in petr ether. The oxalate is more stable.

USE: Manufacture of cupferron.

7406. Phenyliodine(III) Bis(trifluoroacetate). [2712-78-9] Phenylbis(2,2,2-trifluoroacetato-κ*O*)iodine; bis(trifluoroacetoxy)iodobenzene; iodobenzene ditrifluoroacetate; PIFA. $C_{10}H_5F_6IO_4$; mol wt 430.04. C 27.93%, H 1.17%, F 26.51%, I 29.51%, O 14.88%. Hypervalent iodine(III) reagent. Prepn: N. W. Alcock, T. C. Waddington, *J. Chem. Soc.* **1963**, 4103; I. I. Maletina *et al.*, *J. Org. Chem. USSR* **10**, 294 (1974). Structural study: N. W. Alcock *et al.*, *J. Chem. Soc. Dalton Trans.* **1984**, 1709. Use as a one-electron oxidant: L. Eberson *et al.*, *Acta Chem. Scand.* **49**, 640 (1995). Review of applications in organic synthesis: G. Pohnert, *J. Prakt. Chem.* **342**, 731-734 (2000).

Crystals from chloroform + petroleum ether, mp 120-121°.

USE: In oxidation, cyclization, dehydrogenation, and dearomatization reactions.

7407. Phenyl Isocyanate. [103-71-9] Isocyanatobenzene; carbanil; phenylcarbimide. C_7H_5NO; mol wt 119.12. C 70.58%, H 4.23%, N 11.76%, O 13.43%. Prepd by passing carbonyl chloride into a hot soln of aniline in toluene, saturated with HCl: Hardy, *J. Chem. Soc.* **1934**, 2011.

Liquid, with acrid odor. $d_4^{11.6}$ 1.101; d_4^{15} 1.092; $d_4^{19.6}$ 1.0956; $d_4^{25.9}$ 1.08870. bp 158-168°; bp$_{18-20}$ 58.2-59.5°; bp$_{13}$ 55°. $n_D^{19.6}$ 1.53684; $n_D^{25.9}$ 1.53412. *Poisonous; flammable.*

Caution: Irritating to eyes.

7408. Phenyl Isothiocyanate. [103-72-0] Isothiocyanatobenzene; isothiocyanic acid phenyl ester; phenyl mustard oil; thiocarbanil. C_7H_5NS; mol wt 135.18. C 62.20%, H 3.73%, N 10.36%, S 23.72%. Prepd from ammonium phenyldithiocarbamate by the action of lead nitrate: Dains *et al.*, *Org. Synth.* **coll. vol. I** (2nd ed., 1941) p 447.

Liquid. mp −21°. d_4^{25} 1.1288; d_4^{35} 1.1202; d_4^{50} 1.1061. bp$_{760}$ 221°; bp$_{33}$ 117.1°; bp$_{12}$ 95°. Distills with water without dec. $n_D^{23.4}$ 1.64918. Insol in water. Sol in alcohol, ether.

USE: Derivatizing agent for primary, secondary amines. In sequencing of peptides by Edman degradation. In amino acid analyses by HPLC (Pico-Tag).

7409. Phenylmagnesium Chloride. [100-59-4] Chlorophenylmagnesium. C_6H_5ClMg; mol wt 136.86. C 52.66%, H 3.68%, Cl 25.90%, Mg 17.76%. C_6H_5MgCl. One of the Grignard reagents. Prepd by reacting chlorobenzene with magnesium at reflux temperatures in the presence of catalytic amounts of an organic nitrate: Ramsden, **US 2816937** (1957 to M. & T. Corp.).

Sol in ether (ethyl ether, other ethers may be used as solvents). A solution which is about 3 molar has an approx strength of 48%, and a d_4^{20} of ~1.15. Reacts with water, steam or acids to produce toxic and flammable vapors with evolution of heat.

USE: In organic synthesis, especially in the production of hydrocarbons, alcohols, ketones, organic acids, amines, silicones, boranes.

7410. *N*-Phenylmaleimide. [941-69-5] 1-Phenyl-1*H*-pyrrole-2,5-dione. $C_{10}H_7NO_2$; mol wt 173.17. C 69.36%, H 4.07%, N 8.09%, O 18.48%. Prepn from maleanilic acid: Cava *et al.*, *Org. Synth.* **41**, 93 (1961).

Canary-yellow needles from cyclohexane, mp 89-89.8°.

USE: Dienophile in the Diels-Alder reaction. Usually gives cryst adducts.

7411. Phenylmercuric Acetate. [62-38-4] (Acetato-κ*O*)-phenylmercury; acetoxyphenylmercury; phenylmercury acetate; PMA; PMAC; PMAS; Agrosan; Ceresan; Mergamma; Unisan. $C_8H_8HgO_2$; mol wt 336.74. C 28.53%, H 2.39%, Hg 59.57%, O 9.50%. Prepn: E. Dreher, R. Otto, *Ann.* **154**, 93 (1870); F. C. Whitmore, *Organic Compounds of Mercury* (Chemical Catalog Co., New York, 1921) pp 175; Grave *et al.*, *J. Am. Pharm. Assoc.* **25**, 752 (1936). Control of crabgrass: J. A. DeFrance, *Greenkeepers Rep.* **15**, 30 (1947). Metabolism in animals: V. L. Miller *et al.*, *Toxicol. Appl. Pharmacol.* **2**, 344 (1960). GC determn in air: M. G. Rosell *et al.*, *Pergamon Ser. Environ. Sci.* **7**, 299 (1982).

Small, lustrous prisms from ethanol, mp 149°. *Poisonous.* Sol in ~600 parts water, in alc, benzene, acetone.

USE: Herbicide; fungicide. Pharmaceutic aid (antimicrobial preservative).

7412. Phenylmercuric Chloride. [100-56-1] Chlorophenylmercury; phenylmercury chloride. C_6H_5ClHg; mol wt 313.15. C 23.01%, H 1.61%, Cl 11.32%, Hg 64.06%. Antibacterial activity: H. L. Friedman, *Ann. N.Y. Acad. Sci.* **65**, 461 (1957). Use as fungicide: U. Prota, *Phytopathol. Mediterr.* **8**, 87 (1970), *C.A.* **72**, 42138t

(1970); M. Makes, Z. Lokaj, *Agrochemia* **16**, 181 (1976), *C.A.* **85**, 187602t (1976). Toxicity: M. Umeda *et al.*, *Jpn. J. Exp. Med.* **39**, 47 (1969).

White satiny leaflets, mp 250-252°. Sol in ~20,000 parts cold water, in benzene, ether, pyridine; slightly sol in hot alcohol. LD₅₀ s.c. in rats: 30 mg/kg (Umeda).

USE: Has been used as an agricultural fungicide.

7413. Phenylmercuric Nitrate. [55-68-5] (Nitrato-κO)-phenylmercury; merphenyl nitrate; Phe-Mer-Nite; Phenmerzyl nitrate. C₆H₅HgNO₃; mol wt 339.70. C 21.21%, H 1.48%, Hg 59.05%, N 4.12%, O 14.13%. Prepn: R. Otto *J. Prakt. Chem.* **1**, 179 (1870); F. L. Pyman, H. A. Stevenson, *Pharm. J.* **133**, 269 (1934); T. B. Grave *et al.*, *J. Am. Pharm. Assoc.* **25**, 752 (1936). Toxicity of basic salt: R. Wien, *Q. J. Pharm. Pharmacol.* **12**, 212 (1939). HPLC determn in pharmaceuticals: M. Larroque, L. Vian, *J. Pharm. Biomed. Anal.* **11**, 173 (1993).

Needles or plates from dry chloroform, mp 130-132°. *Poisonous.* Slightly sol in alc, glycerin; very slightly sol in water. More sol in presence of nitric acid or alkali hydroxides.

Basic salt. [8003-05-2] Hydroxyphenylmercury mixt with (nitrato-*O*)phenylmercury; basic phenylmercuric nitrate. C₆H₅-HgOH.C₆H₅HgNO₃; mol wt 634.40. Pearly lustrous scales, mp 187-190° (dec). Soluble in ~1250 parts water; slightly sol in alc; moderately sol in glycerol. Practically insol in other organic solvents. LD₅₀ in mice (mg/g): 0.045 s.c.; 0.027 i.v. (Wien).

USE: Pharmaceutic aid (antimicrobial preservative). Tree wound dressing.

THERAP CAT (VET): Antiseptic.

7414. Phenylmercury Borate. [102-98-7] [Orthoborato-(3−)-κO]phenylmercurate(2−) hydrogen (1:2); (dihydrogen borato)phenylmercury; phenylmercuric borate; Famosept; Merfen. C₆-H₇BHgO₃; mol wt 338.52. C 21.29%, H 2.08%, B 3.19%, Hg 59.25%, O 14.18%. Prepn: Christiansen, US 2196384 (1950 to Lever Bros.).

Crystalline powder, mp 112-113°. Sol in water, alcohol, glycerol.
Note: Product of commerce may be an equimolar composition of phenylmercuriborate and phenylmercurihydroxide.

THERAP CAT: Antiseptic (topical).

7415. *o*-Phenylphenol. [90-43-7] (1,1'-Biphenyl)-2-ol; 2-biphenylol; orthoxenol; *o*-hydroxydiphenyl; 2-hydroxydiphenyl; Dowicide 1. C₁₂H₁₀O; mol wt 170.21. C 84.68%, H 5.92%, O 9.40%. Prepn from phenyl ether: Lüttringhaus, Sääf, *Ann.* **542**, 241 (1939); from dibenzofuran: Gilman, Esmay, *J. Am. Chem. Soc.* **75**, 2947 (1953); Müller, US 2862035 (1958 to Bayer). Purification: Widiger, US 3087969 (1963 to Dow). Toxicity data: Hodge *et al.*, *J. Pharmacol. Exp. Ther.* **104**, 202 (1952). Review of carcinogenic risk: *IARC Monographs* **30**, 329-344 (1983).

White, flaky crystals. Mild, characteristic odor. mp 55.5-57.5°; bp 280-284°; bp₁₅ 152-154°. Practically insol in water. Sol in fixed alkali hydroxide solns and most organic solvents. LD₅₀ orally in rats: 2.48 g/kg (Hodge).

Sodium salt. [132-27-4] Natriphene; sodium *o*-phenylphenate; sodium 2-phenylphenoxide; Dowicide A. C₁₂H₉NaO; mol wt 192.19. White flakes. Soly at 25° (g/100 g solvent): water 120; acetone 330; methanol 468; propylene glycol >200. Practically insol

in petroleum fractions, pine oil. pH of satd water soln at 25° = 12.0-13.5.
Caution: Toxic symptoms similar to phenol, *q.v. See Clinical Toxicology of Commercial Products*, R. E. Gosselin *et al.*, Eds. (Williams & Wilkins, Baltimore, 5th ed., 1984) Section II, p 316.

USE: Antimicrobial additive in the mfr of adhesives, leather, metalworking fluids, and textiles; preservative in automotive polishes, ceramic glazes, laundry starch, inks, floor wax emulsions; agricultural fungicide.

7416. *p*-Phenylphenol. [92-69-3] [1,1'-Biphenyl]-4-ol; 4-hydroxydiphenyl; Paraxenol. C₁₂H₁₀O; mol wt 170.21. C 84.68%, H 5.92%, O 9.40%.

mp 164-165°; bp 305-308°. *Irritant.* Possesses approx the soly of the ortho compd.

USE: As intermediate in the manuf of resins; also in the rubber industry.

7417. Phenylpropanolamine. [14838-15-4] *rel*-(αS)-α-[(1R)-1-Aminoethyl]benzenemethanol; (1RS,2SR)-2-amino-1-phenyl-1-propanol; *erythro*-2-amino-1-phenyl-1-propanol; *dl*-norephedrine. C₉H₁₃NO; mol wt 151.21. C 71.49%, H 8.67%, N 9.26%, O 10.58%. Sympathomimetic amine; racemic mixture of *d*- and *l*-norephedrine. Prepn: F. W. Calliess, *Arch. Pharm.* **250**, 141 (1912); P. Rabe, *Ber.* **45**, 2163 (1912); F. W. Hoover, H. B. Hass, *J. Org. Chem.* **12**, 506 (1947); and prepn of enantiomers: W. N. Nagai, S. Kanao, *Ann.* **470**, 157 (1929); C. Jarowski, W. H. Hartung, *J. Org. Chem.* **8**, 564 (1943). Crystal structure: A. Podder *et al.*, *Indian J. Phys.* **53A**, 652 (1979). HPLC determn in plasma and urine: K. Yamashita *et al.*, *J. Chromatogr.* **527**, 103 (1990). Clinical studies of decongestant activity: O. K. Haugeto *et al.*, *J. Otolaryngol.* **10**, 359 (1981); M. Bende *et al.*, *Rhinology* **23**, 43 (1985); in stress incontinence: E. Fossberg *et al.*, *Urol. Int.* **38**, 293 (1983); in weight loss: D. E. Schteingart, *Int. J. Obes.* **16**, 487 (1992). Toxicity: E. I. Goldenthal, *Toxicol. Appl. Pharmacol.* **18**, 185 (1971). Comprehensive description: I. Kanfer *et al.*, *Anal. Profiles Drug Subs.* **12**, 357-383 (1983). Review of anorexic effects: P. J. Wellman, *Neurosci. Biobehav. Rev.* **14**, 339-355 (1990); of toxicology: R. Hanzlick, G. Davis, *Am. J. Forensic Med. Pathol.* **13**, 37-41 (1992). Evaluation of risk of hemorrhagic stroke: W. N. Kernan *et al.*, *N. Engl. J. Med.* **343**, 1826 (2000).

Relative stereochemistry

mp 101-101.5°.
Hydrochloride. [154-41-6] Boxogetten; Kontexin; Monydrin; Proin; Propalin; UriCon. White crystals; affected by light. mp 190-194°. Odor resembling that of crude benzoic acid. pKa 9.44±0.04. Freely sol in water, alc. Practically insol in chloroform, benzene; insol in ether. The aq soln is neutral to litmus. LD₅₀ orally in rats: 1490 mg/kg (Goldenthal).

(+)-Form hydrochloride. [40626-29-7] [α]²⁵_D +32° (water). mp 171-172°.
D-*threo*-Form see Norpseudoephedrine.

THERAP CAT: Decongestant (nasal); anorexic; in treatment of urinary incontinence.

THERAP CAT (VET): In treatment of urinary incontinence in dogs and cats; decongestant (nasal).

7418. Phenylpropylmethylamine. [93-88-9] *N*,β-Dimethylbenzeneethanamine; *dl-N*,β-dimethylphenethylamine; *dl-N*-methyl-

2-phenylpropylamine; 1-methylamino-2-phenylpropane; phenpromethamine; 1-methylamino-2-methyl-2-phenylethane; Vonedrine. $C_{10}H_{15}N$; mol wt 149.24. C 80.48%, H 10.13%, N 9.39%. Synthesis starting with the condensation of chlorobenzene with allyl chloride, followed by ammonolysis: Patrick et al., J. Am. Chem. Soc. **68**, 1009 (1946).

Volatile liquid. d_4^{25} 0.915-0.925; bp_{760} 205-210°; bp_{15} 95-96°; n_D^{20} 1.5102. Slightly sol in water (1.2 g/100 ml). Freely sol in alcohol, ether, benzene. Aq solns are strongly alkaline; pH of a soln of 2 drops (~0.1 ml) dil with 10 ml H_2O is ~10.5.

Hydrochloride. $C_{10}H_{15}N.HCl$. mp 144-148°.

THERAP CAT: Adrenergic.

7419. 1-Phenyl-3-pyrazolidinone. [92-43-3] 1-Phenyl-3-pyrazolidone; Phenidone. $C_9H_{10}N_2O$; mol wt 162.19. C 66.65%, H 6.21%, N 17.27%, O 9.86%. Prepd by heating phenylhydrazine with β-chloropropionic acid: **DE 53834** (1889); by acid hydrolysis of 3-amino-1-phenylpyrazoline: Kendall, Duffin, **GB 650911** and **GB 669591** (1951, 1952, both to Ilford). Review: Kendall, Br. J. Photogr. **100**, 56 (1953).

Leaflets or needles from benzene, mp 121°. One gram dissolves in 10 ml boiling water, in 10 ml hot alcohol, in 37.5 ml boiling benzene. Practically insol in ether, petr ether. Freely sol in dil aq solns of acids and alkalies.

USE: Non-staining, high contrast photographic developer. The amount required is about one-fifth to one-tenth that of Metol.

7420. Phenyl Salicylate. [118-55-8] 2-Hydroxybenzoic acid phenyl ester; salol. $C_{13}H_{10}O_3$; mol wt 214.22. C 72.89%, H 4.71%, O 22.41%. Made by the action of phosphorus oxychloride on a mixture of phenol and salicylic acid.

White, small crystals or crystalline powder; pleasant aromatic odor and taste. d 1.25; mp 41-43°; bp_{12} 173°. One gram dissolves in 6670 ml water, 6 ml alcohol, 1.5 ml benzene, 5 ml amyl alcohol, 10 ml liquid paraffin, 4 ml almond oil. Sol in acetone, chloroform, ether, oils; very slightly sol in glycerol. Solubility at 25° in g/100 g: absolute ethanol 53; ethyl acetate 470; methyl ethyl ketone 620; toluene 460; Stoddard solvent 88; water less than 0.1%. Incompat. Bromine water, ferric salts; camphor, monobromated camphor, phenol, chloral hydrate, thymol or urethan in trituration.

USE: In the manuf of various polymers for the plastics industry, also in lacquers, adhesives, waxes, polishes. In suntan oils and creams. As light absorber to prevent discoloration of plastics. Has some plasticizer properties.

THERAP CAT: Ultraviolet screen.

7421. Phenylselenotrimethylsilane. [33861-17-5] [(Trimethylsilyl)seleno]benzene; trimethylsilyl phenyl selenide; PSTMS.

$C_9H_{14}SeSi$; mol wt 229.26. C 47.15%, H 6.16%, Se 34.44%, Si 12.25%. Organoselenium compound for the introduction of selenium into molecules. Prepn: N. Y. Derkach et al., Zh. Org. Khim. **7**, 1543 (1971); and use: D. Liotta et al., Tetrahedron Lett. **1978**, 5091. Improved synthesis: N. Miyoshi et al., Synthesis **1979**, 300. Synthetic applications: S. D. Rychnovsky, D. J. Skalitzky, Synlett **1995**, 555; H. Abe et al., Chem. Pharm. Bull. **46**, 1311 (1998); M. W. DeGroot, J. F. Corrigan, Dalton **2000**, 1235. Brief review: M. David, Synlett **2001**, 445.

Colorless mobile liquid with a very unpleasant odor, slowly decomposes on exposure to air. $bp_{9.10\ mm}$ 93-95°; $bp_{5\ mm}$ 86.5°. mp −14-−9°. d_{20}^{20} 1.1960. n_D^{20} 1.5525.

USE: Selenating reagent.

7422. 4-Phenylsemicarbazide. [537-47-5] N-Phenylhydrazinecarboxamide; anilinoformylhydrazine; hydrazinecarboxanilide. $C_7H_9N_3O$; mol wt 151.17. C 55.62%, H 6.00%, N 27.80%, O 10.58%. Prepd by the action of hydrazine hydrate on phenylurea: Curtius, J. Prakt. Chem. [2] **58**, 216 (1898); Wheeler, Org. Synth. **coll. vol. I** (2nd ed., 1941) p 450.

Orthorhombic plates from water, mp 122°. Insol in ether, difficultly sol in hot water. Freely sol in alcohol, chloroform, dil acids and alkalies.

Hydrochloride. [5441-14-5] $C_7H_9N_3O.HCl$; mol wt 187.63. Prisms, mp 215°; freely sol in water and alcohol.

7423. N-Phenylsulfanilic Acid. [101-57-5] 4-(Phenylamino)benzenesulfonic acid; 4-diphenylaminesulfonic acid; p-anilinobenzenesulfonic acid. $C_{12}H_{11}NO_3S$; mol wt 249.28. C 57.82%, H 4.45%, N 5.62%, O 19.25%, S 12.86%. Prepd by acetylation followed by sulfonation of diphenylamine: Merz, Weith, Ber. **6**, 1512 (1873); Gnehm, Werdenberg, Z. Angew. Chem. **12**, 1027 (1899); Sarver, Kolthoff, J. Am. Chem. Soc. **53**, 1902 (1931).

Leaves. Becomes blue on exposure to light. Soluble in water, in alcohol. Insol in ether. Dec into diphenylamine and sulfuric acid when heated above 200° with water contg hydrochloric acid.

Sodium salt. [6152-67-6] $C_{12}H_{10}NNaO_3S$; mol wt 271.27. Very sol in water.

Potassium salt. [6152-42-7] $C_{12}H_{10}KNO_3S$; mol wt 287.37. Leaves; slightly sol in alcohol, very sol in water.

Barium salt. [6211-24-1] $(C_{12}H_{10}NO_3S)_2Ba$; mol wt 633.88. Leaflets; slightly sol in water. Poisonous.

USE: Colorimetric determination of nitrates; oxidation-reduction indicator; detection of oxidizing substances.

7424. N-Phenylthiourea. [103-85-5] Phenylthiocarbamide. $C_7H_8N_2S$; mol wt 152.22. C 55.23%, H 5.30%, N 18.40%, S 21.06%. Prepd by evaporating an aq soln of aniline hydrochloride and ammonium thiocyanate and carefully heating the residue. Toxicity data: Scheline et al., J. Med. Pharm. Chem. **4**, 109 (1961).

Bitter or tasteless needles, depending upon heredity of taster. d 1.3. mp 154°. Soluble in 400 parts cold water, 17 parts boiling water; soluble in alcohol. LD$_{50}$ in rats, rabbits (mg/kg): 3, 40 orally (Scheline).

USE: In medical genetics.

7425. Phenyltoloxamine. [92-12-6] N,N-Dimethyl-2-[2-(phenylmethyl)phenoxy]ethanamine; N,N-dimethyl-2-(α-phenyl-o-tolyloxy)ethylamine; N,N-dimethyl-2-(α-phenyl-o-toloxy)ethylamine; 2-(2-dimethylaminoethoxy)diphenylmethane; 2-benzhydryl β-dimethylaminoethyl ether; 2-benzylphenyl β-dimethylaminoethyl ether; PRN; bistrimin; C-5581H; Antin; Phenoxadrine. C$_{17}$H$_{21}$NO; mol wt 255.36. C 79.96%, H 8.29%, N 5.49%, O 6.27%. Prepn: Cheney et al., J. Am. Chem. Soc. **71**, 60 (1949); Binkley, Cheney, US 2703324 (1955 to Bristol Labs.).

Oily liquid, bp 141-144° (at <0.1 mm Hg).

Dihydrogen citrate. C$_{17}$H$_{21}$NO.C$_6$H$_8$O$_7$. Crystals from water or methanol, mp 138-140°. Soluble in water.

Hydrochloride. C$_{17}$H$_{21}$NO.HCl. Crystals from methyl isobutyl ketone, mp 119-121°. Soluble in water.

THERAP CAT: Dihydrogen citrate as antihistaminic.

7426. Phenyl Tolyl Ketone. C$_{14}$H$_{12}$O; mol wt 196.25. C 85.68%, H 6.16%, O 8.15%. o-Isomer made from benzene and o-toluic acid chloride in the presence of AlCl$_3$. p-Isomer made from benzoyl chloride and toluene in presence of AlCl$_3$.

o-**Phenyl tolyl ketone.** (2-Methylphenyl)phenylmethanone; 2-methylbenzophenone. Oily liquid. mp below −18°. bp 309-311°. Insol in water. Freely sol in alcohol, oils, most organic solvents.

p-**Phenyl tolyl ketone.** (4-Methylphenyl)phenylmethanone; 4-methylbenzophenone. Crystals, mp 59-60°. bp$_{720}$ 311-312°. Insol in water. Sol in alcohol; easily sol in benzene, ether, oils.

USE: As fixative in perfumery.

7427. Phenyl Triflimide. [37595-74-7] 1,1,1-Trifluoro-N-phenyl-N-[(trifluoromethyl)sulfonyl]methanesulfonamide; N,N-bis-(trifluoromethanesulfonyl)aniline; N-phenylbis(trifluoromethanesulfonimide); N-phenyltriflimide; N-phenyltrifluoromethanesulfonimide; Tf$_2$NPh. C$_8$H$_5$F$_6$NO$_4$S$_2$; mol wt 357.24. C 26.90%, H 1.41%, F 31.91%, N 3.92%, O 17.91%, S 17.95%. Reagent used to introduce the trifluoromethanesulfonyl (triflate) group under mild conditions. Prepn: L. Z. Gandel'sman et al., J. Org. Chem. USSR (Engl. Transl.) **8**, 1696 (1972); and use as triflating reagent: J. B. Hendrickson, R. Bergeron, Tetrahedron Lett. **14**, 4607 (1973). Application for the prepn of enol triflates: J. E. McMurry, W. J. Scott, ibid. **24**, 979 (1983); of triflones: J. B. Hendrickson, K. W. Bair, J. Org. Chem. **42**, 3875 (1977). Utility in dehydration reactions: I. Torrini et al., Synth. Commun. **19**, 695 (1989). Incorporation into a solid phase triflating reagent: A. D. Wentworth et al., Org. Lett. **2**, 477 (2000).

Crystalline solid from petr ether, mp 99-100° (Gandel'sman); also reported as mp 93-94° (Hendrickson). *Irritant.* uv max: 218, 262, 269 nm. Stable and nonhygroscopic.

USE: Reagent in synthetic organic chemistry.

7428. Phenyltrimethylammonium Iodide. [98-04-4] N,N,N-Trimethylbenzaminium iodide (1:1); N,N-dimethylaniline methiodide; trimethylanilinium iodide; trimethylphenylammonium iodide. C$_9$H$_{14}$IN; mol wt 263.12. C 41.08%, H 5.36%, I 48.23%, N 5.32%. Ref: Pass, Ward, Analyst **58**, 667 (1933).

White, cryst powder. mp 175°. Sol in water or alcohol.

USE: For the detection and determination of cadmium.

7429. N-Phenylurea. [64-10-8] Phenylcarbamide. C$_7$H$_8$N$_2$O; mol wt 136.15. C 61.75%, H 5.92%, N 20.58%, O 11.75%. Obtained with carbanilide as byproduct by refluxing aniline hydrochloride and urea in water: Davis, Blanchard, Org. Synth. **3**, 95 (1923).

Monoclinic prisms from water or alcohol. d 1.302; mp 147° (dec); bp 238°. On slow cooling separates in needles several centimeters in length. Soluble in hot water, hot alcohol, ether, ethyl acetate, glacial acetic acid.

7430. Phenylurethane. [101-99-5] N-Phenylcarbamic acid ethyl ester; ethyl phenylcarbamate; ethyl carbanilate; phenylurethan. C$_9$H$_{11}$NO$_2$; mol wt 165.19. C 65.44%, H 6.71%, N 8.48%, O 19.37%. Made by the action of aniline on ethyl chloroformate.

White, acicular crystals. d 1.106; mp 52-53°; bp 238° with slight decompn; n$_D^{20}$ 1.5376. Slightly sol in water, freely in alcohol, ether; scarcely attacked by boiling for a short time with HCl or NaOH.

7431. Phenyl Vinyl Sulfone. [5535-48-8] (Ethenylsulfonyl)-benzene; phenylsulfonylethylene. C$_8$H$_8$O$_2$S; mol wt 168.21. C 57.12%, H 4.79%, O 19.02%, S 19.06%. Reagent that serves as a dienophile in cycloaddition reactions and as a nucleophilic acceptor in Michael reactions. Prepn: A. H. Ford-Moore et al., J. Chem. Soc. **1949**, 1754; L. A. Paquette, R. V. C. Carr, Org. Synth. coll. vol. VII, 453 (1990). Improved prepn: N. O. Brace, J. Org. Chem. **58**, 4506 (1993). Crystal structure: A. D. Briggs et al., Acta Crystallogr. C **54**, 1335 (1998). Use as ethylene equivalent in cycloaddition reactions: R. V. C. Carr, L. A. Paquette, J. Am. Chem. Soc. **102**, 853 (1980); R. V. C. Carr et al., J. Org. Chem. **48**, 4976 (1983). Synthetic application as Michael acceptor in thiol protections: Y. Kuroki, R. Lett, Tetrahedron Lett. **25**, 197 (1984). Review: O. De Lucchi, S. Cossu in Encyclopedia of Reagents for Organic Synthesis **6**, L. A. Paquette, Ed. (Wiley, New York, 1995) pp 4041-4045.

Colorless crystals from hexane, mp 66-67°. d 1.396. *Irritant, skin sensitizer.* Sol in common organic solvents.

USE: Reagent in synthetic organic chemistry.

7432. Phenyramidol. [553-69-5] α-[(2-Pyridinylamino)-methyl]benzenemethanol; 2-(β-hydroxyphenethylamino)pyridine; fenyramidol. $C_{13}H_{14}N_2O$; mol wt 214.27. C 72.87%, H 6.59%, N 13.07%, O 7.47%. Prepn from 2-mandelamidopyridine or 2-amino-pyridine: Gray et al., J. Am. Chem. Soc. **81**, 4347, 4351 (1959).

Crystals from dil methanol, mp 82-85°. uv max (95% ethanol): 243, 303 nm (log ε 4.24, 3.63). pKa 5.85.

Hydrochloride. [326-43-2] IN-511; MJ-505; NSC-17777; Cabral. $C_{13}H_{14}N_2O.HCl$; mol wt 250.73. Crystals from ethanol + ether, mp 140-142°. Soluble in water. Aq solns are slightly acidic and stable in ampuls.

Methiodide. $C_{13}H_{14}N_2O.CH_3I$. Crystals from ethanol + ether, mp 164-166°.

THERAP CAT: Analgesic.

7433. Phenytoin. [57-41-0] 5,5-Diphenyl-2,4-imidazolidine-dione; diphenylhydantoin; Di-Hydan; Dilantin; Hydantin; Hydantol; Lehydan; Zentropil. $C_{15}H_{12}N_2O_2$; mol wt 252.27. C 71.42%, H 4.79%, N 11.10%, O 12.68%. Prepn from benzophenone: H. R. Henze, US **2409754** (1946 to Parke, Davis); of sodium salt from benzil, urea and NaOH: Biltz, Ber. **41**, 1391 (1908); **44**, 411 (1911). Pharmacology: Gillis et al., J. Pharmacol. Exp. Ther. **179**, 599 (1971). Toxicity of base: G. Stille, I. Brunckow, Arzneim.-Forsch. **4**, 723 (1954); of sodium salt: G. B. Fink, E. A. Swinyard, J. Pharmacol. Exp. Ther. **127**, 318 (1959). Reviews: Damato, Prog. Cardiovasc. Dis. **12**, 1-15 (1969); Dreifus, Watanabe, Am. Heart J. **80**, 709-713 (1970). Review of carcinogenicity studies: IARC Monographs **13**, 201-225 (1977). Comprehensive description: J. Philip et al., Anal. Profiles Drug Subs. **13**, 417-445 (1984).

White powder, mp 295-298°. One gram dissolves in about 60 ml alc, about 30 ml acetone. Sol in alkali hydroxides, hot alc; slightly sol in cold alc, chloroform, ether. Practically insol in water. LD_{50} in mice (mg/kg): 92 i.v.; 110 s.c. (Stille, Brunckow).

Sodium salt. [630-93-3] Phenytoin soluble; sodium 5,5-diphenyl hydantoinate; Aurantin; Dintoina; Epanutin; Phenhydan; Pyoredol; Tacosal. $C_{15}H_{11}N_2NaO_2$; mol wt 274.25. White powder; bitter, soapy taste. Somewhat hygroscopic. Easily dissociated even by weak acids (incl CO_2 absorbed on exposure to air) regenerating phenytoin. One gram dissolves in 10.5 ml alc; in ~66 ml water (aq soln turbid unless pH adjusted to >11.7, the pH of a satd soln). Practically insol in ether, chloroform. LD_{50} orally in mice: 490 mg/kg (Fink, Swinyard).

Caution: Phenytoin is reasonably anticipated to be a human carcinogen: Report on Carcinogens, Twelfth Edition (PB2011-111646, 2011) p 345.

THERAP CAT: Anticonvulsant; antiepileptic.

THERAP CAT (VET): Anticonvulsant.

7434. Phillyrin. [487-41-2] 4-[(1S,3aR,4R,6aR)]-4-[4-(3,4-Dimethoxyphenyl)tetrahydro-1H,3H-furo[3,4-c]furan-1-yl]-2-methoxyphenyl-β-D-glucopyranoside; phyllyrin; phillyroside; forsythin. $C_{27}H_{34}O_{11}$; mol wt 534.56. C 60.67%, H 6.41%, O 32.92%. From bark of Phillyrea latifolia L., and allied Oleaceae: Campona, Ann. **24**, 242 (1837); Eijkman, Rec. Trav. Chim. **5**, 127 (1886); Kramer, Compt. Rend. **196**, 814 (1933); Sosa, Bull. Soc. Chim. Biol. **29**, 918 (1947). Structure: Kaku et al., J. Pharm. Soc. Jpn. **59**, 248 (1939); M. Chiba et al., Chem. Pharm. Bull. **25**, 3435 (1977).

α-Form. Needles from dil alc, mp 154-155°. $[α]_D$ +48.4° (alcohol).

β-Form. Needles, mp 184-185°. $[α]_D$ +48.5° (alcohol).

7435. PhIP. [105650-23-5] 1-Methyl-6-phenyl-1H-imidazo-[4,5-b]pyridin-2-amine; 2-amino-1-methyl-6-phenylimidazo[4,5-b]-pyridine. $C_{13}H_{12}N_4$; mol wt 224.27. C 69.62%, H 5.39%, N 24.98%. Most abundant of the mutagenic heterocyclic amines found in well-cooked meat and fish. Isoln from fried ground beef: J. S. Felton et al., Carcinogenesis **7**, 1081 (1986). Synthesis: M. G. Knize, J. S. Felton, Heterocycles **24**, 1815 (1986); T. Choshi et al., J. Org. Chem. **58**, 7952 (1993); F. S. Bavetta et al., Tetrahedron Lett. **44**, 7793 (1997). Mutagenicity in Chinese hamster ovary cells: L. H. Thompson et al., Mutagenesis **2**, 483 (1987). Evaluation of carcinogenic risk: IARC Monographs **56**, 229-242 (1993). Review of bioavailability, content in foods, metabolism: N. J. Gooderham et al., Br. J. Clin. Pharmacol. **42**, 91-98 (1996). Correlation of dietary intake with breast cancer risk: R. Sinha et al., J. Natl. Cancer Inst. **92**, 1352 (2000).

Gray-white crystals, mp 327-328°. uv max (methanol): 225, 273, 316 nm (log ε 4.46, 4.00, 4.46). Sol in methanol, DMSO.

Caution: This substance is reasonably anticipated to be a human carcinogen: Report on Carcinogens, Twelfth Edition (PB2011-111646, 2011) p 222.

7436. Phloionic Acid. [23843-52-9] rel-(9R,10R)-9,10-Dihydroxyoctadecanedioic acid; floionic acid. $C_{18}H_{34}O_6$; mol wt 346.46. C 62.40%, H 9.89%, O 27.71%. Isoln from cork: Guillemonat, Cesaire, Bull. Soc. Chim. Fr. **1949**, 792; Ribas, Seoane, An. R. Soc. Esp. Fis. Quim. **50B**, 963 (1954), C.A. **50**, 806f (1956); Brown, Rosen, US **2872464** (1959 to Crown Cork). Structure: Duhamel, Bull. Soc. Chim. Fr. **1965**, 399. Synthesis: Ruzicka et al., Helv. Chim. Acta **25**, 1086 (1942); Gensler, Schlein, J. Am. Chem. Soc. **77**, 4846 (1955). Resolution of isomers: Gender, Mahadevan, ibid. **78**, 169 (1956); Alvarez-Varquez, Ribas-Marques, An. Quim. **64**, 783 (1968). Synthesis of (±), (+), and (−) forms: McGhie et al., Chem. Ind. (London) **1972**, 536.

Relative stereochemistry

Crystals from ethanol + water, mp 126°.

Dimethyl ester. $C_{20}H_{38}O_6$. Crystals from ethanol, mp 77.5-78°.

7437. Phloretin. [60-82-2] 3-(4-Hydroxyphenyl)-1-(2,4,6-trihydroxyphenyl)-1-propanone; 2′,4′,6′-trihydroxy-3-(p-hydroxy-phenyl)propiophenone; β-(p-hydroxyphenyl)phloropropiophenone; β-(p-hydroxyphenyl)-2,4,6-trihydroxypropiophenone. $C_{15}H_{14}O_5$; mol wt 274.27. C 65.69%, H 5.15%, O 29.17%. The aglucon of phloridzin, q.v. From root bark of apple trees: Rochleder, J. Prakt. Chem. **98**, 205 (1866). Structure: Seshadri, Annu. Rev. Biochem. **20**, 495 (1951). Synthesis: Fischer, Nouri, Ber. **50**, 611 (1917);

Rosenmund, Rosenmund, *ibid.* **61B**, 2608 (1928); Shinoda *et al.*, *J. Pharm. Soc. Jpn.* **49**, 797 (1929); Johnston, **US 2789995** (1957 to Union Carbide).

Needles from dil alcohol, dec 262°. Absorption spectrum: Lambrechts, *Arch. Int. Physiol.* **44**, Suppl., 1-39 (1937). Freely sol in alc, methanol, acetone; sol in alkalies in hot glacial acetic acid; very sparingly sol in benzene, chloroform. Practically insol in water, ether, the soly being increased by addition of alcohol.

7438. Phloridzin. [60-81-1] 1-[2-(β-D-Glucopyranosyloxy)-4,6-dihydroxyphenyl]-3-(4-hydroxyphenyl)-1-propanone; phlorhizin; phlorizin; phlorrhizen; phloretin-2′-β-glucoside; 4,6-dihydroxy-2-(β-D-glucosido)-β-(*p*-hydroxyphenyl)propiophenone. $C_{21}H_{24}O_{10}$; mol wt 436.41. C 57.80%, H 5.54%, O 36.66%. Dihydrochalcone originally isolated from apple tree root bark, subsequently found in several other species. Competitive inhibitor of sodium glucose co-transporter (SGLT) -1 and -2. Isoln from root bark: De Koninck, *Ann.* **15**, 75, 258 (1835); Stass, *Ann.* **30**, 192 (1839); Bridel, Kramer, *Bull. Soc. Chim. Biol.* **15**, 544 (1933). Hydrolysis by dil mineral acids yields phloretin and glucose. Procedure for acid hydrolysis: Wessely, Sturm, *Monatsh. Chem.* **53-54**, 557 (1929); Müller, Robertson, *J. Chem. Soc.* **1933**, 1170. Synthesis: Zemplen, Bognár, *Ber.* **75B**, 1040 (1942); *ibid.* 645; *ibid.* **76B**, 386 (1943). Review of biosynthesis, distribution and physiological relevence in plants: C. Gosch *et al.*, *Phytochemistry* **71**, 838-843 (2010). Review of discovery, pharmacology and effect on glucose transport: J. R. L. Ehrenkranz *et al.*, *Diabetes Metab. Res. Rev.* **21**, 31-38 (2005); as prototype for SGLT2 inhibitors in diabetes: J. R. White, *Clin. Diabetes* **28**, 5-10 (2010).

Dihydrate. Long needles from water, mp 110°. Sweet, with bitter aftertaste. $[\alpha]_D^{25}$ -52° (0.16 g in 5 ml of 96% alcohol). One gram dissolves in about one liter of water at 22°, in 64 ml at 60°, in 22 ml at 70°. Freely sol in boiling water; in about 4 parts alcohol, in methanol, amyl alcohol, acetone, ethyl acetate, pyridine, aniline, quinoline and other organic bases; in aq alkaline solns and in glacial acetic acid. Practically insol in ether, chloroform, benzene.

USE: Experimentally to produce glycosuria in animals.

7439. Phloroglucinol. [108-73-6] 1,3,5-Benzenetriol; 1,3,5-trihydroxybenzene; phloroglucin. $C_6H_6O_3$; mol wt 126.11. C 57.15%, H 4.80%, O 38.06%. Isolation as dimer from phloretin, *q.v.*: H. Hlasiwetz, *Ann.* **96**, 118 (1855). Prepn: J. R. Hwu, S.-C. Tsay, *J. Org. Chem.* **55**, 5987 (1990). GC-MS determn in human plasma: C. Lartigue-Mattei *et al.*, *J. Chromatogr.* **617**, 140 (1993). Clinical evaluation: G. Cargill *et al.*, *Presse Med.* **21**, 19 (1992). Toxicology: R. Cahen *et al.*, *Therapie* **17**, 1349 (1962); *J. Am. Coll. Toxicol.* **14**, 468 (1995). Review: G. Leston in *Kirk-Othmer Encyclopedia of Chemical Technology* vol. 19 (Wiley-Interscience, New York, 4th ed., 1996) pp 792-812. Biosynthesis study: J. Achkar *et al.*, *J. Am. Chem. Soc.* **127**, 5332 (2005).

White micro crystals, mp 217-219° with quick heating; 200-209° with slow heating. Sweet taste. Discolors in light. Sol in ether, benzene, pyrimidine, pyridine. Soluble in 10 parts methanol, 10 parts ethanol, 100 parts water (25°). LD_{50} in mice, rats (g/kg): 4.7, 4.0 i.g. (Cahen).

Dihydrate. [6099-90-7] Spasfon-Lyoc; Spassirex. Colorless, odorless white rhombic crystals, mp 113-116° with quick heating. Loses water of crystallization at 110°.

USE: In diazo-type printing, textile dyeing. In cosmetics, as antioxidant; hair colorant. As reagent for detection of aldehydes, carbohydrates including lignin, and HCl. Cloud seeding. Rooting medium for woody plants.

THERAP CAT: Antispasmodic.

7440. Phloxine B. [18472-87-2] 2′,4′,5′,7′-Tetrabromo-4,5,-6,7-tetrachloro-3′,6′-dihydroxyspiro[isobenzofuran-1(3*H*),9′-[9*H*]-xanthen]-3-one sodium salt (1:2); 2′,4′,5′,7′-tetrabromo-4,5,6,7-tetrachlorofluorescein disodium salt; cyanosin; eosin blue; C.I. 45410; C.I. Acid Red 92; D & C Red No. 28. $C_{20}H_2Br_4Cl_4Na_2O_5$; mol wt 829.63. C 28.96%, H 0.24%, Br 38.53%, Cl 17.09%, Na 5.54%, O 9.64%. Photosensitive xanthene based dye; structural analog of fluorescein and eosin Y, *q.q.v.* Prepn: W. R. Orndorff, E. F. Hitch, *J. Am. Chem. Soc.* **36**, 680 (1914). Purification by counter-current chromatography: A. Weisz *et al.*, *J. Chromatogr. A* **678**, 77 (1994). Vibrational spectra: V. A. Narayanan *et al.*, *Analusis* **24**, 1 (1996). Isoln from lipstick matrices by supercritical fluid extraction: S. Scalia, S. Simeoni, *Chromatographia* **53**, 490 (2001). Teratogenicity study: M. Seno *et al.*, *Food Chem. Toxicol.* **22**, 55 (1984). Use as histological stain: E. P. C. Tock, *Am. J. Med. Technol.* **35**, 302 (1969); E. H. Oldmixon, *Stain Technol.* **63**, 165 (1988). Review of use as insecticide: D. Bergsten, *Pestic. Outlook* **8**, 20-23 (1997); as bacterial stain: R. Rasooly, *FEMS Immunol. Med. Microbiol.* **49**, 261-265 (2007).

Brick red to brown powder. Absorption max: 548 nm in 50% aq ethanol. Soly: 11% in water; 5% in ethanol.

Lactone. [13473-26-2] C.I. 45410:1; C.I. Solvent Red 48; D & C Red No. 27; tetrachloroeosin; phloxine O. $C_{20}H_4Br_4Cl_4O_5$; mol wt 785.66. White crystals from acetic acid. Sol as hydrate in alcohol, acetone, ether, and ethyl acetate; difficultly sol in chloroform and benzene. Insol in petr ether and ligroin.

USE: Histological and bacterial stain. Colorant for cosmetics and drugs; inks and laquers; wool, nylon, jute, leather, silk, and acid dyeable acrylics.

7441. Pholcodine. [509-67-1] (5α,6α)-7,8-Didehydro-4,5-epoxy-17-methyl-3-[2-(4-morpholinyl)ethoxy]morphinan-6-ol; 3-[2-(4-morpholinyl)ethyl]morphine; tetrahydro-1,4-oxazinylmethylcodeine; 3-(2-morpholinoethyl)morphine; β-morpholinylethylmorphine; homocodeine; Ethnine; Galenphol; Galphol; Memine; Codylin; Pectolin; Weifacodine. $C_{23}H_{30}N_2O_4$; mol wt 398.50. C 69.32%, H 7.59%, N 7.03%, O 16.06%. Prepn: Chabrier *et al.*, **US 2619485** (1952 to Dausse). Description: *Ann. Pharm. Fr.* **8**, 261 (1950). Toxicity data: B. Kelentey *et al.*, *Arzneim.-Forsch.* **8**, 325 (1958).

Monohydrate. Crystals, mp 91°. Bitter taste. $[\alpha]_D^{20}$ −95.3° (c = 2 in ethanol). Slightly sol in water (2% w/v), ether. Sol in alc (1:3), chloroform, benzene. pH of 2% aq soln 9.5 to 9.8. LD_{50} s.c. in mice: 0.540 g/kg (Kelentey).

Note: This is a controlled substance (opium derivative): **21 CFR, 1308.11.**

THERAP CAT: Antitussive.

7442. Pholedrine. [370-14-9] 4-[2-(Methylamino)propyl]-phenol; *p*-hydroxy-*N*,α-dimethylphenethylamine; β-(*p*-hydroxyphenyl)isopropylmethylamine; α-(*p*-hydroxyphenyl)-β-methylaminopropane; *p*-hydroxy-*N*-methylbenzedrine; Knoll H_{75}. $C_{10}H_{15}NO$; mol wt 165.24. C 72.69%, H 9.15%, N 8.48%, O 9.68%. α-Adrenergic agonist. Prepd from *p*-methoxybenzyl methyl ketone: **FR 822422, GB 482414, DE 674753** (1937, 1938 and 1939 to Knoll); Hildebrandt and Hildebrandt, Freese, **DE 665793** and **DE 767161** (1938 and 1951 to Knoll). Synthesis: Savitskii, Makhnenko, *J. Gen. Chem. USSR* **10**, 1819 (1940); Buzas, Dufour, *Bull. Soc. Chim. Fr.* **1950**, 139. Toxicity data: Lindner, *Arch. Exp. Pathol. Pharmakol.* **188**, 675 (1938).

Crystals from methanol, mp 162-163°. Acrid, burning taste. Alkaline reactions. Slightly sol in water. Sol in alcohol, ether; readily sol in dil acids.

Sulfate. [6114-26-7] Paredrinol; Pulsotyl; Veritol. $(C_{10}H_{15}-NO)_2.H_2SO_4$; mol wt 428.54. Crystals, dec 320-323°. Sol in water. LD s.c. in rats: 500 mg/kg (Lindner).

THERAP CAT: Antihypotensive.

THERAP CAT (VET): Circulatory stimulant.

7443. Phorate. [298-02-2] Phosphorodithioic acid *O,O*-diethyl *S*-[(ethylthio)methyl] ester; *O,O*-diethyl *S*-(ethylthio)methyl phosphorodithioate; *O,O*-diethyl *S*-ethylmercaptomethyl dithiophosphate; AC-3911; EI-3911; ENT-24042; Thimet. $C_7H_{17}O_2PS_3$; mol wt 260.36. C 32.29%, H 6.58%, O 12.29%, P 11.90%, S 36.94%. Organophosphate insecticide; cholinesterase inhibitor. Prepn: Schrader, Lorenz, **US 2759010; GB 797307** (1956, 1958 both to Bayer). Metabolism: J. B. Bowman, J. E. Casida, *J. Agric. Food Chem.* **5**, 192 (1957). Persistence in soil: D. L. Suett, *Pestic. Sci.* **6**, 385 (1975). Toxicity study: T. B. Gaines, *Toxicol. Appl. Pharmacol.* **14**, 515 (1969).

Clear liquid, $bp_{0.1}$ 75-78°, $bp_{0.8}$ 118-120°, $bp_{2.0}$ 125-127°. d_4^{25} 1.156. n_D^{25} 1.5329. Vapor pressure at 20°: 8.4 × 10^{-4} mm Hg. Soly: 50 ppm in water. Misc with xylene, carbon tetrachloride, dioxane, methyl cellosolve, dibutyl phthalate, vegetable oils. Stable at room temp. Hydrolyzed in the presence of water and alkali. *Poisonous.* LD_{50} in female, male rats (mg/kg): 1.1, 2.3 orally; 2.5, 6.2 dermally (Gaines).

Caution: Potential symptoms of overexposure are irritation of eyes, skin, respiratory system; miosis; rhinorrhea; headache, chest tightness, wheezing, laryngeal spasms, salivation, cyanosis; anorexia, nausea, vomiting, abdominal cramps, diarrhea; sweating; muscle fasciculations, weakness, paralysis; giddiness, confusion, ataxia; convulsions, coma; low blood pressure; cardiac irregularities. *See NIOSH Pocket Guide to Chemical Hazards* (DHHS/NIOSH 97-140, 1997) p 252.

USE: Insecticide.

7444. Phorbol. [17673-25-5] (1a*R*,1b*S*,4a*R*,7a*S*,7b*S*,8*R*,9*R*,-9a*S*)-1,1a,1b,4,4a,7a,7b,8,9,9a-Decahydro-4a,7b,9,9a-tetrahydroxy-3-(hydroxymethyl)-1,1,6,8-tetramethyl-5*H*-cyclopropa[3,4]benz-[1,2-*e*]azulen-5-one. $C_{20}H_{28}O_6$; mol wt 364.44. C 65.91%, H 7.74%, O 26.34%. Parent alcohol of the tumor promoting com-

pounds in croton oil, *q.v.*, the oil expressed from the seeds of *Croton tiglium* L., *Euphorbiaceae.* Phorbol has a structural skeleton based on cyclopropabenzazulene. Isoln: Flaschenträger, Wigner, *Helv. Chim. Acta* **25**, 569 (1942); Kauffmann, Neumann, *Ber.* **92**, 1715 (1959); S. Tseng *et al., J. Org. Chem.* **42**, 3645 (1977). Structure and stereochemistry: Hecker *et al., Tetrahedron Lett.* **1967**, 3165; Pettersen, Ferguson, *Chem. Commun.* **1967**, 716. Unlike its diesters, phorbol does not appear to be co-carcinogenic or to enhance chemically-induced mutagenesis: C. J. Soper, F. J. Evans, *Cancer Res.* **37**, 2487 (1977). Mechanism of action study on phorbol esters: A. S. Kraft, W. B. Anderson, *Nature* **301**, 621 (1983). Comprehensive review of phorbol and its esters: Hecker, Schmidt, *Fortschr. Chem. Org. Naturst.* **31**, 377-467 (1974); P. M. Blumberg, *Crit. Rev. Toxicol.* **8**, 199-234 (1981).

Anhyd crystals, dec 250-251°. Two forms of solvated crystals from ethyl acetate: mp 162-163° and 233-234°. Solvated crystals from methanol or ethanol; mp 249-250°. $[\alpha]_D^{24}$ +102° (water). $[\alpha]_D^{20}$ +118° (c = 0.4 in dioxane). uv max (ethanol): 235, 334 nm (ε 5200, 70). Quite sol in polar solvents, including water.

12-Myristate 13-acetate diester. 12-*O*-Tetradecanoylphorbol-13-acetate; TPA; croton oil factor A_1. $C_{36}H_{56}O_8$. uv max (ethanol): 232, 333 nm (ε 5400, 73). Differentiation of human leukemia cells: J. B. Weinberg, *Science* **213**, 655 (1981).

7445. Phorone. [504-20-1] 2,6-Dimethyl-2,5-heptadien-4-one; diisobutenyl ketone; diisopropylidene acetone. $C_9H_{14}O$; mol wt 138.21. C 78.21%, H 10.21%, O 11.58%. Prepn from isobutenyllithium + CO_2: Braude, Timmons, *J. Chem. Soc.* **1950**, 2000; from acetone: Dolgov, Samsonova, *Zh. Obshch. Khim.* **22**, 632 (1952); Joseph, Blumenthal, *J. Org. Chem.* **24**, 1371 (1959); Tsmur, *Zh. Prikl. Khim.* **34**, 1628 (1961); M. Konieczny, G. Sosnovsky, *Z. Naturforsch.* **33B**, 454 (1978).

Yellow liquid or yellowish-green prisms, mp 28°. bp 198-199°; bp_{17} 88°. d_4^{20} 0.885. n_D^{21} 1.4968.

7446. Phosalone. [2310-17-0] Phosphorodithioic acid *S*-[(6-chloro-2-oxo-3(2*H*)-benzoxazolyl)methyl] *O,O*-diethyl ester; phosphorodithioic acid *O,O*-diethyl ester *S*-ester with 6-chloro-3-(mercaptomethyl)-2-benzoxazolinone; 3-(*O,O*-diethyldithiophosphorylmethyl)-6-chlorobenzoxazolinone; 6-chloro-3-(*O,O*-diethyldithiophosphorylmethyl)benzoxazolone; *S*-(6-chloro-2-oxobenzoxazolin-3-yl)methyl diethyl phosphorothiolothionate; RP-11974; Zolone. $C_{12}H_{15}ClNO_4PS_2$; mol wt 367.80. C 39.19%, H 4.11%, Cl 9.64%, N 3.81%, O 17.40%, P 8.42%, S 17.43%. Organophosphate insecticide; cholinesterase inhibitor. Prepn: **GB 1005372** (1965 to Rhône-Poulenc). Multiresidue determn in serum and blood: E. Lacassie *et al., J. Chromatogr. B* **759**, 109 (2001); in fruit samples: K. S. Liapis *et al., J. Chromatogr. A* **996**, 181 (2003). Toxicology, metabolism, insecticidal properties: D. L. Colinese, H. J. Terry, *Chem. Ind. (London)* **1968**, 1507.

Crystals, mp 47.5-48°. Sol in ketones, alcohols and most aromatic solvents. Practically insol in water and aliphatic hydrocarbons. LD_{50} in mice, female, male rats, guinea pigs (mg/kg): 180-205, 135-170, 120, 82-150 orally; female rats (mg/kg): 390 dermally (Colinese, Terry).

USE: Insecticide, acaricide.

7447. Phosgene. [75-44-5] Carbonic dichloride; carbonyl chloride; chloroformyl chloride. CCl_2O; mol wt 98.91. C 12.14%, Cl 71.68%, O 16.18%. $Cl_2C=O$. Prepn from chlorine + carbon monoxide: Whitehouse, **US 1231226** (1917); Peacock, **US 1360312** (1921); Bradner, **US 1457493** (1923); Douthitt, **US 2847470** (1958 to Texas Co.); from carbon monoxide + nitrosyl chloride: Williams, **US 1746506** (1930 to du Pont Ammonia Corp.); from carbon tetrachloride + oleum: Murphy, Reuter, *Aust. Chem. Inst. J. Proc.* **15**, 144 (1948). Toxicology: S. A. Cucinell, *Arch. Environ. Health* **28**, 272 (1974). *Review:* E. E. Hardy in *Kirk-Othmer Encyclopedia of Chemical Technology* **vol. 17** (Wiley-Interscience, New York, 3rd ed., 1982) pp 416-425.

Colorless, highly toxic gas; suffocating odor; when much diluted with air there is an odor reminiscent of moldy hay. Condenses at ~0° to a clear, colorless, fuming liquid. d_4^0 1.432. mp −118°. bp_{760} 8.2°. Vapor press at 20°: 1215 mm. *Poisonous; corrosive; spontaneously combustible.* Slightly sol in water and slowly hydrolyzed by it; freely sol in benzene, toluene, glacial acetic acid and most liquid hydrocarbons.

Caution: Severe pulmonary irritant, although not immediately irritating even in potentially lethal exposures. Potential symptoms of overexposure by inhalation may initially be mild and transient and include burning of eyes, cough, dry burning throat, dyspnea, foamy sputum, chest pain, vomiting. Delayed symptoms include peribronchial edema, pulmonary congestion, alveolar edema, cyanosis, anoxia. Direct contact with liquid may cause frostbite. *See Clinical Toxicology of Commercial Products*, R. E. Gosselin *et al.*, Eds. (Williams & Wilkins, Baltimore, 5th ed., 1984) Section II, p 96; *NIOSH Pocket Guide to Chemical Hazards* (DHHS/NIOSH, 97-140, 1997) p 252; *Patty's Industrial Hygiene and Toxicology* **vol. 2F**, G. D. Clayton, F. E. Clayton, Eds. (Wiley-Interscience, New York, 4th ed., 1994) pp 4557-4563.

Note: Paper soaked in alcoholic or carbon tetrachloride soln contg 10% of a mixture of equal parts of *p*-dimethylaminobenzaldehyde and colorless diphenylamine, then dried, will turn from yellow to deep orange in the presence of approx the max allowable concn of phosgene, and should always be used where the generation of this gas is possible or suspected.

USE: For the prepn of many organic chemicals; as a war gas.

7448. Phosmet. [732-11-6] Phosphorodithioic acid S-[(1,3-dihydro-1,3-dioxo-2*H*-isoindol-2-yl)methyl] *O,O*-dimethyl ester; phosphorodithioic acid *O,O*-dimethyl ester *S*-ester with *N*-(mercaptomethyl)phthalimide; *O,O*-dimethyl *S*-phthalimidomethyl phosphorothionate; *N*-(mercaptomethyl)phthalimide *S*-(*O,O*-dimethyl phosphorodithioate); phthalophos (USSR); ENT-25705; R-1504; Fosdan; Imidan. $C_{11}H_{12}NO_4PS_2$; mol wt 317.31. C 41.64%, H 3.81%, N 4.41%, O 20.17%, P 9.76%, S 20.21%. Organophosphate insecticide; cholinesterase inhibitor. Prepn: Fancher, **US 2767194** (1956 to Stauffer). Activity: B. A. Butt, J. C. Keller, *J. Econ. Entomol.* **54**, 813 (1961). Multiresidue determ in serum and blood: E. Lacassie *et al.*, *J. Chromatogr. B* **759**, 109 (2001). GC-MS determn in olives: S. C. Cunha *et al.*, *Talanta* **73**, 514 (2007). Post-harvest degradation in blueberries: K. M. Crowe *et al.*, *J. Agric. Food Chem.* **54**, 9608 (2006). Toxicity study: T. B. Gaines, *Toxicol. Appl. Pharmacol.* **14**, 515 (1969).

Off-white, crystalline solid, mp 71.9°. Tech product (95-98% pure), mp 66.5-69.5°. Dec below its boiling point. Vapor pressure at 50°: 1×10^{-3} mm Hg. Soly in water at 25°: 25 ppm. LD_{50} in male, female rats (mg/kg): 113, 160 orally (Gaines).

USE: Insecticide, acaricide.

7449. Phosphamidon. [13171-21-6] Phosphoric acid 2-chloro-3-(diethylamino)-1-methyl-3-oxo-1-propen-1-yl dimethyl ester; phosphoric acid dimethyl ester, ester with 2-chloro-*N,N*-diethyl-3-hydroxycrotonamide; 2-chloro-2-diethylcarbamoyl-1-methylvinyl dimethyl phosphate; Ciba 570; ENT-25515; Kinadon. $C_{10}H_{19}$-$ClNO_5P$; mol wt 299.69. C 40.08%, H 6.39%, Cl 11.83%, N 4.67%, O 26.69%, P 10.34%. Organophosphate insecticide; cholinesterase inhibitor. Commercial product consists of 73/27 mixture of *cis/trans* isomers. Prepn: Beriger, Sallmann, **US 2908605** (1959 to Ciba); Anliker *et al.*, *Helv. Chim. Acta* **44**, 1622 (1961). GC determn: W. E. Westlake *et al.*, *J. Agric. Food Chem.* **21**, 846 (1973). Toxicity data: T. B. Gaines, *Toxicol. Appl. Pharmacol.* **14**, 515 (1969). *Review: Residue Rev.* **37**, 1-202 (1971).

Z-isomer

Oil. *Poisonous.* d_4^{25} 1.2132. n_D^{25} 1.4718. $bp_{1.5}$ 162°; $bp_{0.001}$ 120°. mp −45°. Vapor pressure at 20°: 2.5×10^{-5} mm Hg. Misc with water and most organic solvents except saturated hydrocarbons. One gram dissolves in ~30 g hexane. Stable in neutral or acid media; hydrolyzed by alkali. LD_{50} orally in rats: 24 mg/kg (Gaines).

USE: Insecticide.

7450. Phosphine. [7803-51-2] H_3P; mol wt 34.00. H 8.89%, P 91.10%. PH_3. Formed in small quantity in the putrefaction of organic matter contg phosphorus. Prepd from white phosphorus and aq alkali hydroxide; also by treatment of PH_4I with KOH: Klement in *Handbook of Preparative Inorganic Chemistry* **vol. 1**, G. Brauer, Ed. (Academic Press, New York, 2nd ed., 1963) pp 525-530; by pyrolysis of phosphorous acid: Gokhale, Jolly, *Inorg. Synth.* **9**, 56 (1967); by hydrolysis of a metal phosphide such as calcium phosphide: Klement, *loc. cit.*; Baudler *et al.*, *Z. Anorg. Allg. Chem.* **353**, 122 (1967). Review of human exposure and toxicity: N. Brautbar, J. Howard, *Toxicol. Ind. Health* **18**, 71-75 (2002).

Gas; odor of decaying fish. *Poisonous; flammable.* bp −87.7°. mp −133°. Spontaneously flammable in air if there is a trace of P_2H_4 present; burns with a luminous flame. Slightly sol in water (0.26 vol. at 20°). Combines violently with oxygen and the halogens. Liberates hydrogen and forms the phosphide when passed over heated metal. Forms phosphonium salts when brought in contact with the halogen acids.

Caution: Potential symptoms of overexposure are nausea, vomiting, abdominal pain, diarrhea; thirst; chest tightness, dyspnea; muscle pain, chills; stupor or syncope; pulmonary edema. Direct contact with liquid may cause frostbite. *See NIOSH Pocket Guide to Chemical Hazards* (DHHS/NIOSH 97-140, 1997) p 254. *See also Clinical Toxicology of Commercial Products*, R. E. Gosselin *et al.*, Eds. (Williams & Wilkins, Baltimore, 5th ed., 1984) Section II, p 119.

7451. Phosphinothricin. [35597-44-5] (2*S*)-2-Amino-4-(hydroxymethylphosphinyl)butanoic acid; 2-ammonio-4-methylphosphinicobutyrate; L-PPT. $C_5H_{12}NO_4P$; mol wt 181.13. C 33.16%, H 6.68%, N 7.73%, O 35.33%, P 17.10%. First natural L-amino acid known to contain a phosphinic acid group. Isolated from the tripeptide antibiotic, **bialaphos**; inhibits glutamine synthetase. Isoln from *S. viridochromogenes* and synthesis of DL-form: E. Bayer *et al.*, *Helv. Chim. Acta* **55**, 224 (1972). Synthesis of DL-form: H. Gross, T. Gnauk, *J. Prakt. Chem.* **318**, 157 (1976); of enantiomers: N. Minowa *et al.*, *Tetrahedron Lett.* **25**, 1147 (1984). Crystal structure: E. F. Paulus, S. Grabley, *Z. Kristallogr.* **160**, 63 (1982). Determn in biological fluids: A. Suzuki, M. Kawana, *Bull. Environ. Contam. Toxicol.* **43**, 17 (1989). Environmental impact in lakes: M. J. Faber *et al.*, *Environ. Toxicol. Chem.* **17**, 1291 (1998). Field trial in orchards and vineyards as herbicide: P. Langelüddeke *et al.*, *Meded. Fac. Landbouwwet. Rijksuniv. Gent* **47**, 95 (1982); on seed potato as desiccant: H. M. Lawson, J. S. Wiseman, *Brighton Crop Prot. Conf. - Weeds* **1991**, 233. Review of syntheses, activities and uses: G. Hoerlein, *Rev. Environ. Contam. Toxicol.* **138**, 73-145 (1994).

$[\alpha]_D^{25}$ +13.4° (c = 1 in water).
D-Form. [73679-07-9] $[\alpha]_D^{25}$ −12.4° (c = 1 in water).
DL-Form. [51276-47-2] Glufosinate. mp 229-231° (dec).
DL-Form monoammonium salt. [77182-82-2] Glufosinate-ammonium; ammonium DL-homoalanine-4-yl(methyl)phosphinate; HOE-661; HOE-39866; Basta; Liberty; Rely. $C_5H_{15}N_2O_4P$; mol wt 198.16. Sol in water. LD_{50} in male, female mice, male, female rats (mg/kg): 431, 416, 2000, 1620 orally (Langelüddeke).
USE: Postemergent herbicide; desiccant.

7452. Phosphocreatine. [67-07-2] *N*-[Imino(phosphonoamino)methyl]-*N*-methylglycine; *N*-(phosphonoamidino)sarcosine; creatine phosphate; creatinephosphoric acid; PC. $C_4H_{10}N_3O_5P$; mol wt 211.11. C 22.76%, H 4.77%, N 19.90%, O 37.89%, P 14.67%. Phosphorylated form of creatine, *q.v.* Occurs primarily in skeletal muscle; also found in heart and brain. Involved in the transfer of high energy phosphate to ADP. Isoln from frog muscle: Eggleton, *Biochem. J.* **21**, 190 (1927); from cat muscle: Fiske, Subbarow, *J. Biol. Chem.* **81**, 629 (1929). Synthesis by phosphorylation of creatine: Zeile, Fawaz, *Z. Physiol. Chem.* **256**, 193 (1938); Ennor, Stocken, *Biochem. J.* **43**, 190 (1948). Enzymic determrn in skeletal muscle: K. Yoshikawa *et al.*, *Anal. Biochem.* **159**, 303 (1986). HPLC determrn in cardiac muscle: T. Teerlink *et al.*, *Anal. Biochem.* **214**, 278 (1993). Clinical trial as cardioprotective: M. L. Semenovsky *et al.*, *J. Thorac. Cardiovasc. Surg.* **94**, 762 (1987); as antiarrhythmic: M. Y. Ruda *et al.*, *Am. Heart J.* **116**, 393 (1988). Review of role in energy metabolism: S. P. Bessman, C. L. Carpenter, *Annu. Rev. Biochem.* **54**, 831-862 (1985).

Sodium salt. Creatergyl; Neoton. $C_4H_8N_3Na_2O_5P$; mol wt 255.08. Occurs as the hexahydrate, platelets from water + ethanol. Very sol in water. (Free acid, pKa_2 is 4.6).
Calcium salt tetrahydrate. $C_4H_8CaN_3O_5P.4H_2O$. Hygroscopic crystals. Sol in water; sparingly sol in alcohol.
Note: The term **phosphagen**, originally a synonym for phosphocreatine, has since been used to describe any naturally occurring phosphorylated guanidine compd.
THERAP CAT: Cardioprotective.

7453. Phosphocysteamine. [5746-40-7] 2-Aminoethanethiol 1-dihydrogen phosphate; phosphorothioic acid *S*-(2-aminoethyl) ester; cysteamine-*S*-phosphate; MEAP. $C_2H_8NO_3PS$; mol wt 157.12. C 15.29%, H 5.13%, N 8.91%, O 30.55%, P 19.71%, S 20.40%. Radioprotectant. Prepn of sodium salt: S. Åkerfeldt, *Acta Chem. Scand.* **13**, 1479 (1959); of free acid: *idem, ibid.* **14**, 1980 (1960). Radioprotective effect: B. Hansen, B. Sörbo, *Acta Radiol.* **56**, 141 (1961); A. Vacek *et al.*, *Folia Biol. (Prague)* **17**, 340 (1971). Mechanism of action: B. Shapiro *et al.*, *Radiat. Res.* **44**, 421 (1970). Radiobiological and biochemical profile: J. W. Harris, T. L. Phillips, *ibid.* **46**, 362 (1971). Photoreactivity: S. A. Grachev *et al.*, *Radiochem. Radioanal. Lett.* **19**, 283 (1974); H. Neumann, M. Sokolovsky, *Biochim. Biophys. Acta* **381**, 292 (1975). Clinical evaluation in cystinosis: L. A. Smolin *et al.*, *Pediatr. Res.* **23**, 616 (1988); D. S. Theodoropoulos *et al.*, *J. Am. Med. Assoc.* **270**, 2200 (1993).

Crystals from glacial acetic acid, mp 156°. pKa: 2.2, 5.0, 10.3.
Sodium salt. [3724-89-8] Sodium hydrogen-*S*-(2-aminoethyl)-phosphorothioate; cystafos; cystaphos; WR-638. $C_2H_7NNaO_3PS$; mol wt 179.11. White crystalline water soluble powder. LD_{50} i.p. in male mice: 0.93 ±0.03 g/kg (Hansen).

USE: Research radioprotective agent.
THERAP CAT: Treatment of cystinosis.

7454. Phosphomolybdic Acid. [11104-88-4] Molybdenumphosphorus hydroxide oxide; molybdophosphoric acid; dodecamolybdophosphoric acid. Formula approx $24MoO_3.P_2O_5.xH_2O$. Prepn: Wu, *J. Biol. Chem.* **43**, 189 (1920); Hastings, Frediani, *Anal. Chem.* **20**, 382 (1948). Formula also reported to be $20MoO_3.P_2O_5.51H_2O$: *USP* **XXI**, 1398. Review of phosphomolybdic acids: *Mellor's* **Vol. XI**, pp 659-672 (1931).
Bright yellow crystals. Sol in less than 0.4 part water; very sol in alcohol, ether.
USE: Weighting silks; as reagent for alkaloids, uric acid, xanthine, creatinine, some metals, with hematoxylin as nerve stain in microscopy.

7455. Phosphonium Iodide. [12125-09-6] H_4IP; mol wt 161.91. H 2.49%, I 78.38%, P 19.13%. PH_4I. Prepd by hydrolysis of a mixture of diphosphorus tetraiodide and white phosphorus: Work, *Inorg. Synth.* **2**, 141 (1946). Improved apparatus for its prepn: Beredjick, *ibid.* **6**, 91 (1960).
Large, transparent, colorless crystals (usually cubes). Tetragonal system. Sublimes at room temp. *Store in sealed ampuls in refrigerator.* Vapor pressure: 50 mm at 20°; 760 mm at 62.5°. mp 18.5° under its own vapor pressure. Heat of fusion 12,680 cal/mol. *Caution:* Heat or traces of moisture or alcohol cause decompn into PH_3 and HI. Will detonate if heated rapidly.
USE: In the laboratory prepn of phosphine.

7456. Phosphoric Acid. [7664-38-2] Orthophosphoric acid. H_3O_4P; mol wt 97.99. H 3.09%, O 65.31%, P 31.61%. H_3PO_4. Obtained commercially from phosphate rock deposits in Florida, Tennessee, and the Western United States. Phosphate rock is essentially tricalcium phosphate and one of the large scale processes is based on the equation: $Ca_3(PO_4)_2 + 3H_2SO_4 + 6H_2O \rightarrow 2H_3PO_4 + 3(CaSO_4.2H_2O)$. Description of various processes: W. H. Waggaman, *Phosphoric Acid, Phosphates and Phosphatic Fertilizers* (Reinhold, New York, 1952); *Phosphoric Acid* **Vol. 1**, parts I, II, A, V. Slack, Ed. (Dekker, New York, 1968) 1159 pp; *Faith, Keyes & Clark's Industrial Chemicals*, F. A. Lowenheim, M. K. Moran, Eds. (Wiley-Interscience, New York, 4th ed., 1975) pp 628-639. Prepn of ultrapure, cryst H_3PO_4: Simon, Schulze, *Z. Anorg. Allg. Chem.* **242**, 322 (1939); Weber, King, *Inorg. Synth.* **1**, 101 (1939). Reviews: J. R. Van Wazer, *Phosphorus and Its Compounds* **Vol. 1**, *Chemistry* (Interscience, New York, 1958) pp 479-491; R. B. Hudson *et al.*, "Phosphoric Acids and Phosphates" in *Kirk-Othmer Encyclopedia of Chemical Technology* **Vol. 17** (Wiley-Interscience, New York, 3rd ed., 1982) pp 426-472.

Unstable, orthorhombic crystals, mp 42.35°, or clear, syrupy liquid; easily supercooled into a glass. Pleasing acid taste when suitably diluted. An acid containing ~88% H_3PO_4 will frequently crystallize on prolonged cooling; forms hemihydrate, mp 29.32°. Becomes anhydr at 150°, gradually changes to pyrophosphoric acid at ~200°, and changes to metaphosphoric acid when heated above 300°. The hot concd acid attacks porcelain and granite ware. May be stored in suitable stainless steel containers. Tribasic acid: pK_1: 2.15; pK_2: 7.09; pK_3: 12.32. Other reported values of dissociation constants and viscosity data reviewed by Van Wazer. The pH of a 0.1N aq soln is 1.5. Heat of formation (crystals): −306.2 kcal/mole. Heat of soln: +2.79 kcal/mole. *Corrosive.* Misc with water, alc; sol in 8 vols of a 3:1 ether: alc mixture. Properties of phosphoric acid solns. d^{25} 1.8741 (100% soln); 1.6850 (85% soln), 1.3334 (50% soln); 1.0523 (10% soln). Density measurements: Christensen, Reed, *Ind. Eng. Chem.* **47**, 1277 (1950); Egan, Luff, *ibid.* 1280. $n_D^{17.5}$ 1.34203 (10% soln); 1.35032 (20% soln), 1.35846 (30% soln). Spec heat (21.3°): 0.4359 (88% soln).
Caution: Potential symptoms of overexposure are irritation of upper respiratory system, eyes and skin; burns skin and eyes; dermatitis. *See NIOSH Pocket Guide to Chemical Hazards* (DHHS/NIOSH 97-140, 1997) p 254.

USE: In the manuf of superphosphates for fertilizers, other phosphate salts, polyphosphates, detergents. Acid catalyst in making ethylene, purifying hydrogen peroxide. As acidulant and flavor, synergistic antioxidant and sequestrant in food. Pharmaceutic aid (solvent; acidifying and buffering agent). In dental cements; process engraving; rustproofing of metals before painting; coagulating rubber latex; as analytical reagent for trace metal analysis. In buffers.

THERAP CAT (VET): Has been used in lead poisoning.

7457. Phosphoric Acid, Meta. [37267-86-0] Metaphosphoric acid (HPO$_3$); glacial phosphoric acid. HO$_3$P; mol wt 79.98. H 1.26%, O 60.01%, P 38.73%.

Transparent, glass-like solid or soft silky masses; hygroscopic. Volatilizes at red heat. Very slowly sol in cold water, slowly changing to H$_3$PO$_4$, the change is hastened by boiling; sol in alcohol. *Keep tightly closed.* For greater convenience it is also marketed in form of rods made by the addition of sodium phosphate.

USE: In dentistry for making zinc oxyphosphate cement. Reagent in chemical analysis; acid digestion.

7458. Phosphorous Acid. [13598-36-2] Phosphonic acid. H$_3$O$_3$P; mol wt 81.99. H 3.69%, O 58.54%, P 37.78%. Prepd by hydrolysis of PCl$_3$ according to the equation PCl$_3$ + 3H$_2$O $\rightarrow$ H$_3$PO$_3$ + 3HCl, the rather violent reaction can be slowed down by the initial presence of concd HCl: Milobedzki, Friedmann, *Chem. Pol.* **15**, 76 (1917); *Chem. Z.* **1918**, I, 933; Simon, Fehér, *Z. Anorg. Allg. Chem.* **230**, 298 (1937). Alternate procedure carrying out the reaction in carbon tetrachloride: Voight, Gallais, *Inorg. Synth.* **4**, 55 (1953). *Review:* Ohashi, "Lower Oxo Acids of Phosphorus and Their Salts" in *Topics in Phosphorus Chemistry* Vol. 1, M. Grayson, E. J. Griffith, Eds. (Interscience, New York, 1964) pp 113-187.

White, very hygroscopic and deliquesc, crystalline mass; garliclike taste; slowly oxidized by oxygen (air) to H$_3$PO$_4$. d$_4^{21}$ 1.65; d$_4^{76}$ liq 1.597. mp ~73°; above 180° is dec into PH$_3$ and H$_3$PO$_4$. Very sol in water, alcohol. pK$_1$ 1.29; pK$_2$ 6.74. Usually marketed as a 20% aq soln.

7459. Phosphorus. [7723-14-0] P; at. wt 30.973762; at. no. 15; valences 3, 5. Group VA (15). One naturally occurring isotope: ^{31}P; artificial, radioactive isotopes: 28-30; 32-34. Abundance in earth's crust: about 0.12%. Does not occur free in nature; found in the form of phosphates in the minerals *chlorapatite* [3Ca$_2$-(PO$_4$)$_2$.CaCl$_2$], *fluorapatite* [3Ca(PO$_4$)$_2$.CaF$_2$], vivianite, wavellite and "phosphate rock" or phosphorite; occurs in small quantities in granite rocks; occurs in all fertile soil; an essential constituent of protoplasm, nervous tissue and bones. Discovered in 1669 by Brandt. Prepn: Ullmann, *Enzyklopädie der Technischen Chemie* **8**, 362 (1931); DeWitt, Skolnik, *J. Am. Chem. Soc.* **68**, 2305 (1946); Skolnik et al., *ibid.* 2310. Lab prepn and purification: Klement in *Handbook of Preparative Inorganic Chemistry* vol. 1, G. Brauer, Ed. (Academic Press, New York, 2nd ed., 1963) pp 518-525. *Reviews:* J. R. Van Waser, *Phosphorus and Its Compounds* 2 vols. (Interscience, New York, 1958, 1961) 2046 pp; Corbridge, "The Structural Chemistry of Phosphorus Compounds" in *Topics in Phosphorus Chemistry* Vol. 3, E. J. Griffith, M. Grayson, Eds. (Interscience, New York, 1966) pp 57-394; Toy, "Phosphorus" in *Comprehensive Inorganic Chemistry* Vol. 2, J. C. Bailar, Jr. et al., Eds. (Pergamon Press, Oxford, 1973) pp 389-545; J. R. Van Wazer in *Kirk-Othmer Encyclopedia of Chemical Technology* vol. 17 (Wiley-Interscience, New York, 3rd ed., 1982) pp 473-490. Review of toxicology and human exposure: *Toxicological Profile for White Phosphorous* (PB98-101090, 1997) 248 pp.

Phosphorus exists in three main allotropic forms: white, black, and red. The same liquid is obtained on melting. *Caution:* Avoid contact with KClO$_3$, KMnO$_4$, peroxides and other oxidizing agents; explosions may result on contact or friction. *Flammable.* Phosphorus atoms exist as symmetrical, tetrahedral P$_4$ molecules in the liquid phase and in the vapor phase below 800°; molecules dissociate to P$_2$ above 800°.

White phosphorus. Colorless or white, transparent, cryst solid; waxy appearance; darkens on exposure to light. Sometimes called yellow phosphorus; color due to impurities. Two allotropic modifications: α-form exists at room temp; cubic crystals containing P$_4$ molecules; d 1.83. β-Form prepd by conversion of α-form at −79.6°; hexagonal crystals; d 1.88. mp 44.1° (vapor press. 0.181 mm); bp 280°. Volatile; sublimes *in vacuo* at ordinary temp when exposed to light. When exposed to air in the dark, emits a greenish light and gives off white fumes. Solubilities in water: one part/300,000 parts water; in abs alc: one g/400 ml; in abs ether: one g/102 ml; in CHCl$_3$: one g/40 ml; in benzene: one g/35 ml; in CS$_2$: one g/0.8 ml. Soly in oils: one gram phosphorus dissolves in 80 ml olive oil, 60 ml oil of turpentine, ~100 ml almond oil. Ignites at about 30° in moist air; the ignition temp is higher when the air is dry. *Handle with forceps. Keep under water. Poisonous.* Combines directly with the halogens to form tri- or pentahalides; combines with sulfur to form sulfides. Reacts with several metals to form phosphides. Yields orthophosphoric acid when treated with nitric acid. Reacts with alkali hydroxides with formation of phosphine and sodium hypophosphite.

Black Phosphorus. Polymorphic. Orthorhombic crystalline form: stable in air; resembles graphite in texture; produced from the white modification under high pressures: Bridgman, *J. Am. Chem. Soc.* **36**, 1344 (1914); Jacobs, *J. Chem. Phys.* **5**, 945 (1937); Krebs, *Inorg. Synth.* **7**, 60 (1963). d 2.691. Does not catch fire spontaneously. Insol in organic solvents. Amorphous form prepd at lower pressures: Jacobs, *loc. cit.* At higher pressure the orthorhombic form undergoes reversible transition to a rhombohedral structure, d 3.56, and a cubic structure, d 3.83: Jamieson, *Science* **139**, 1291 (1963).

Red phosphorus. Polymorphism: Roth *et al.*, *J. Am. Chem. Soc.* **69**, 2881 (1947); Corbridge, *loc. cit.* Crystal structure of one form, *Hittorf's phosphorus*: Thurn, Krebs, *Acta Crystallogr.* **25B**, 125 (1969). Red to violet powder. The properties of red phosphorus are intermediate between those of the white and black forms. Sublimes at 416°, triple point 589.5° under 43.1 atm. d 2.34. Insol in organic solvents. Sol in phosphorus tribromide. Less active than the white form; reacts only at high temp. Yields the white modification when distilled at 290°. Catches fire when heated in air to about 260° and burns with formation of the pentoxide. Burns when heated in an atmosphere of chlorine.

Caution: Potential symptoms of acute poisoning by ingestion of white (or yellow) phosphorous are burning pain in throat and abdomen; intense thirst; nausea, vomiting, diarrhea; garlic breath; luminescent vomitus and feces. Direct contact may cause eye and skin burns. May be followed by symptoms of systemic poisoning, including hematemesis; hepatomegaly, jaundice; hemorrhages into skin, mucous membranes, viscera; oliguria, hematuria, albuminuria; cardiovascular collapse; convulsions, confusion, coma. Potential symptoms of chronic poisoning from inhalation or ingestion are cachexia, anemia, bronchitis. general debility, necrosis of mandible. *See Clinical Toxicology of Commercial Products*, R. E. Gosselin *et al.*, Eds. (Williams & Wilkins, Baltimore, 5th ed., 1984) Section III, pp 348-352; *NIOSH Pocket Guide to Chemical Hazards* (DHHS/NIOSH 97-140, 1997) p 254.

USE: White phosphorus: manuf rat poisons; for smoke screens, gas analysis. Red phosphorus: pyrotechnics; manuf safety matches; in organic synthesis; manuf phosphoric acid, phosphine, phosphoric anhydride, phosphorus pentachloride, phosphorus trichloride; manuf fertilizers, pesticides, incendiary shells, smoke bombs, tracer bullets.

7460. Phosphorus Oxybromide. [7789-59-5] Phosphoric tribromide; phosphoryl tribromide. Br$_3$OP; mol wt 286.68. Br 83.62%, O 5.58%, P 10.80%. POBr$_3$. Prepd according to the equation: 3PBr$_5$ + P$_2$O$_5$ $\rightarrow$ 5POBr$_3$: Hönigschmid, Hirschbold-Wittner, *Z. Anorg. Allg. Chem.* **243**, 355 (1940); Johnson, Nunn, *J. Am. Chem. Soc.* **63**, 141 (1941); Booth, Seegmiller, *Inorg. Synth.* **2**, 151 (1946).

Thin plates, faint orange tint, d 2.822. mp 56° (in hot water). bp$_{758}$ 193° (dec). Slowly hydrolyzes in water forming H$_3$PO$_4$ and HBr. Sol in ether, benzene, chloroform, carbon disulfide, concd H$_2$SO$_4$. *Corrosive; store in sealed glass ampuls.*

7461. Phosphorus Oxychloride. [10025-87-3] Phosphoric trichloride; phosphorus chloride; phosphoryl chloride. Cl$_3$OP; mol wt 153.32. Cl 69.36%, O 10.44%, P 20.20%. POCl$_3$. Manuf:

Faith, Keyes & Clark's Industrial Chemicals, F. A. Lowenheim, M. K. Moran, Eds. (Wiley-Interscience, New York, 4th ed., 1975) pp 646-649.

Colorless, clear, strongly fuming liquid; pungent odor. d^{25} 1.645. bp 105.8°; mp 1.25°: J. R. Van Wazer, *Phosphorus and Its Compounds* vol. 1 (Interscience, New York, 1958) p 254. Reacts exothermically with water, alc. *Corrosive; poisonous. Keep in tightly closed containers.*

Caution: Potential symptoms of overexposure are irritation of eyes, skin, respiratory system; eye, skin burns; dyspnea, cough, lung edema; dizziness, headache, weakness; abdominal pain, nausea, vomiting; nephritis. *See NIOSH Pocket Guide to Chemical Hazards* (DHHS/NIOSH 97-140, 1997) p 254.

USE: As chlorinating agent, especially to replace oxygen in organic compounds; as solvent in cryoscopy.

7462. Phosphorus Pentabromide. [7789-69-7] Pentabromophosphorane; phosphoric bromide; phosphorus perbromide. Br_5P; mol wt 430.49. Br 92.81%, P 7.20%. PBr_5. *Review:* Payne, "The Chemistry of Phosphorus Halides" in *Topics in Phosphorus Chemistry* Vol. 4, M. Grayson, E. J. Griffith, Eds. (Interscience, New York, 1967) pp 85-155.

Yellow, cryst mass. mp >100° with decompn. Dec by water or alcohol; sol in carbon disulfide or tetrachloride. *Keep tightly closed. Corrosive.*

USE: Brominating agent for converting organic acids to acyl bromides.

7463. Phosphorus Pentachloride. [10026-13-8] Pentachlorophosphorane; phosphoric chloride; phosphorus perchloride. Cl_5P; mol wt 208.22. Cl 85.13%, P 14.88%. PCl_5. Prepn: Maxson, *Inorg. Synth.* **1**, 99 (1939). *Review:* Payne, "The Chemistry of Phosphorus Halides" in *Topics in Phosphorus Chemistry* Vol. 4, M. Grayson, E. J. Griffith, Eds. (Interscience, New York, 1967) pp 85-155.

White to pale yellow, fuming, deliquesc, cryst mass; pungent, unpleasant odor; attacks the eyes and mucous membranes. mp 148° under pressure; sublimes at about 100° without melting. bp 160°. Hydrolyzed by water to form phosphoric acid and hydrogen chloride. Reacts with alcohols (ROH) to form the corresponding chloride (RCl). Sol in carbon disulfide or tetrachloride. *Keep in tightly closed containers and handle with caution. Corrosive.*

Caution: Potential symptoms of overexposure are irritation of eyes, skin and respiratory system; bronchitis; dermatitis. *See NIOSH Pocket Guide to Chemical Hazards* (DHHS/NIOSH 97-140, 1997) p 256.

USE: As catalyst in manuf acetylcellulose; for replacing hydroxyl groups by Cl, particularly for converting acids into acid chlorides.

7464. Phosphorus Pentafluoride. [7647-19-0] Pentafluorophosphorane. F_5P; mol wt 125.97. F 75.41%, P 24.59%. PF_5. Prepd by treating PF_3 with bromine to form PF_3Br which then disproportionates to PF_5 and PBr_5: Moissan, *C. R. Hebd. Seances Acad. Sci.* **100**, 1348; **101**, 1490 (1885); by heating P_2O_5 with CaF_2: Lucas, Ewing, *J. Am. Chem. Soc.* **49**, 1270 (1927); Booth, Bozarth, *ibid.* **55**, 3890 (1933); from PCl_5 and AsF_3: Thorpe, *Ann.* **182**, 201 (1876); *Proc. Roy. Soc.* **25**, 122 (1877); O. Ruff, *Die Chemie des Fluors* (Berlin, 1920), p 29; Kwasnik in *Handbook of Preparative Inorganic Chemistry* vol. 1, G. Brauer, Ed. (Academic Press, New York, 2nd ed., 1963) pp 190-191; from PCl_5 and CaF_2: Muetterties *et al.*, *J. Inorg. Nucl. Chem.* **16**, 52 (1960); from phosphoryl fluoride, hydrogen fluoride and sulfur trioxide: Wiesboeck, US 3584999; US 3592594 (1971 to U.S. Steel). *Reviews:* Burg in *Fluorine Chemistry* vol. 1, J. H. Simons, Ed. (Academic Press, New York, 1950) pp 97-98; Kemmitt, Sharp, *Adv. Fluorine Chem.* **4**, 197-198 (1965); Schmutzler, *ibid.* **5**, 32-285 (1965).

Colorless gas. Fumes strongly in air. d (gas) 5.805 g/l. mp −93.8°. bp −84.6°. Trouton constant 21.8. Dipole moment: zero. High thermal stability. *Poisonous; corrosive.* Does not attack dry glass even at 250°, but a slight trace of moisture leads to formation of POF_3 and HF. Water hydrolysis ultimately yields phosphoric acid; intermediates are oxyfluophosphates. Lewis acid; forms complexes with amines, ethers, nitrates, sulfoxides, organic bases. Forms a cryst addn product $PF_5.NO_2$ at −10° which dissociates on warming. May be stored in steel cylinders.

Caution: Intensely irritating to skin, eyes, mucous membranes. Inhalation may cause pulmonary edema.

USE: Catalyst in ionic polymerization reactions.

7465. Phosphorus Pentaselenide. [1314-82-5] Phosphorus selenide (P_2Se_5). P_2Se_5; mol wt 456.75. P 13.56%, Se 86.44%. Prepd by melting a mixture of phosphorus and selenium in an atm of CO_2 or N_2: Carius, Bogen, *Ann.* **124**, 57 (1862); Kudchadker *et al.*, *Can. J. Chem.* **46**, 1415 (1968).

Amorphous, glass, black-purple solid. Dec in steam and boiling water. Reacts with CCl_4; insol in carbon disulfide.

7466. Phosphorus Pentasulfide. [1314-80-3] Phosphorus sulfide (P_2S_5); phosphoric sulfide; thiophosphoric anhydride; phosphorus persulfide. P_2S_5; mol wt 222.25. P 27.87%, S 72.13%. Exists as P_4S_{10}. Conveniently prepd by fusing red phosphorus with sulfur: Stock, Herscovici, *Ber.* **43**, 1223 (1910); Klement in *Handbook of Preparative Inorganic Chemistry* vol. 1, G. Brauer, Ed. (Academic Press, New York, 2nd ed., 1963) p 567. Manuf: *Faith, Keyes & Clark's Industrial Chemicals*, F. A. Lowenheim, M. K. Moran, Eds. (Wiley-Interscience, New York, 4th ed., 1975) pp 650-653.

Light yellow, triclinic crystals; peculiar odor. d 2.09. mp 286-290°; bp 513-515°. Dec by water forming H_3PO_4 and H_2S; sol in carbon disulfide, in aq soln of alkali hydroxides. *Dangerous when wet; flammable. Keep tightly closed.*

Caution: Potential symptoms of overexposure are irritation of eyes, skin and respiratory system; apnea, coma and convulsions; conjunctival pain, lacrimation, photophobia, kerato-conjunctivitis and corneal vesiculation; dizziness; headache; fatigue; irritability, insomnia; GI disturbance. *See NIOSH Pocket Guide to Chemical Hazards* (DHHS/NIOSH 97-140, 1997) p 256.

USE: In manuf of lube oil additives and pesticides. Manuf safety matches, ignition compds, and for introducing sulfur into organic compds.

7467. Phosphorus Pentoxide. [1314-56-3] Phosphorous oxide (P_2O_5); phosphoric anhydride; diphosphorus pentoxide. O_5P_2; mol wt 141.94. O 56.36%, P 43.64%. P_2O_5. Exists as P_4O_{10}. Prepd commercially by burning phosphorus in a current of dry air. Purification: Manley, *J. Chem. Soc.* **121**, 331 (1922); de Decker, McGillavry, *Rec. Trav. Chim.* **60**, 153, 413 (1941). Review of phosphorus oxides: J. R. Van Waser, *Phosphorus and Its Compounds* vol. 1 (Interscience, New York, 1958) pp 267-286.

Very deliquescent crystals. *Corrosive.* Several crystalline and amorphous modifications. Commercial form; hexagonal; d 2.30. mp 340°. Sublimation temperature 360°. Nonflammable. Does not support combustion. Heat of formation: −365.83 kcal/mole. Sp heat: 0.170 cal/g/°C. Heat of fusion: 8.2 kcal/mole. Heat of volatilization: 22.7 kcal/mole of P_4O_{10}. Readily absorbs moisture from the air. Exothermic hydrolysis by water to form phosphoric acid. The reaction with alcohol is similar.

Caution: Strong irritant; corrosive to skin, mucous membranes and eyes.

USE: Drying and dehydrating agent. Condensing agent in organic synthesis.

7468. Phosphorus Sulfochloride. [3982-91-0] Phosphorothioic trichloride; thiophosphoryl chloride. Cl_3PS; mol wt 169.38. Cl 62.79%, P 18.29%, S 18.93%. $PSCl_3$. Prepd from P_2S_5 + PCl_5: Martin, Duvall, *Inorg. Synth.* **4**, 73 (1953). Alternate procedure from PCl_3, $AlCl_3$ and S: Moeller *et al.*, *ibid.* 71.

Fuming liq, crystallizes as α-form at −40.8° or as β-form at −36.2°. bp$_{760}$ 125°. n_D^{25} 1.635. Sol in benzene, carbon tetrachloride, carbon disulfide, chloroform. Hydrolyzes slowly in water, rapidly in alkaline solns. In water the hydrolysis products are orthophosphoric acid, hydrochloric acid, and hydrogen sulfide.

Caution: Strong irritant.

7469. Phosphorus Tribromide. [7789-60-8] Phosphorous bromide. Br_3P; mol wt 270.69. Br 88.56%, P 11.44%. PBr_3. Prepn: Gray, Maxson, *Inorg. Synth.* **2**, 147 (1946). *Review:* Payne, "The Chemistry of Phosphorus Halides" in *Topics in Phosphorus Chemistry* vol. 4, M. Grayson, E. J. Griffith, Eds. (Interscience, New York, 1967) pp 85-155.

Colorless, fuming liquid; very penetrating odor. d^{15} 2.85. mp −41.5°. bp 173.2°. Vapor pressure: 10 mm (48°). Dissolved and dec by water or alc; sol in acetone, carbon disulfide. *Corrosive. Keep tightly closed.*

7470. Phosphorus Trichloride. [7719-12-2] Phosphorous chloride. Cl_3P; mol wt 137.32. Cl 77.45%, P 22.56%. PCl_3. Prepd from red phosphorus and dry chlorine in the presence of refluxing PCl₃: Forbes *et al., Inorg. Synth.* **2**, 145 (1946). Manuf: *Faith, Keyes & Clark's Industrial Chemicals,* F. A. Lowenheim, M. K. Moran, Eds. (Wiley-Interscience, New York, 4th ed., 1975) pp 654-657. *Review:* Payne, "Chemistry of Phosphorus Halides" in *Topics in Phosphorus Chemistry* vol. **4**, M. Grayson, E. J. Griffith, Eds. (Interscience, New York, 1967) pp 85-155.

Colorless, clear, fuming liquid. d_4^{21} 1.574. mp −112°. bp 76°. Vapor pressure: 100 mm (21°). Decomposed by water or alc. Sol in benzene, chloroform, ether, carbon disulfide. *Poisonous; corrosive. Keep in tightly closed containers and handle with caution.*

Caution: Potential symptoms of overexposure are irritation of eyes, skin, nose and throat; pulmonary edema; burns eyes and skin. *See NIOSH Pocket Guide to Chemical Hazards* (DHHS/NIOSH 97-140, 1997) p 256.

USE: As of phosphorus oxychloride; manuf POCl₃, PCl₅; producing iridescent metallic deposits.

7471. Phosphorus Trifluoride. [7783-55-3] Phosphorus fluoride. F₃P; mol wt 87.97. F 64.79%, P 35.21%. PF₃. Prepd by halogen exchange between PCl₃ and AsF₃: Moissan, *Compt. Rend.* **100**, 272 (1885); Hoffman, *Inorg. Synth.* **4**, 149 (1953); between PCl₃ and ZnF₂: Williams, *ibid.* **5**, 95 (1957); between PCl₃ and CaF₂ or SbF₃: Booth, Bozarth, *J. Am. Chem. Soc.* **61**, 2927 (1939); Muetterties *et al., J. Inorg. Nucl. Chem.* **16**, 52 (1960); between PCl₃ and HF: Kwasnik in *Handbook of Preparative Inorganic Chemistry* vol. **1**, G. Brauer, Ed. (Academic Press, New York, 2nd ed., 1963) pp 189-190. *Reviews:* Burg in *Fluorine Chemistry* vol. **1**, J. H. Simons, Ed. (Academic Press, New York, 1950) pp 98-100; Kemmitt, Sharp, *Adv. Fluorine Chem.* **4**, 198-199 (1965); Schmutzler, *ibid.* **5**, 32-285 (1965).

Colorless gas. Does not fume in air. Does not attack glass except at high temps. *Poisonous.* d (gas) 3.907 g/l. mp −151.30°. bp −101.38°. Critical temp −2.05°; critical press. 42.69 atm. May be stored in steel cylinders or in glass, also in a gasometer over Hg. Slowly hydrolyzed by water. Absorbed by aq bases at a rate increasing with pH, producing a fluophosphite which boiling aq HNO₃ does not convert to phosphate. Anhydr KOH may be used to dry PF₃ with little loss. Dry NH₃ forms a solid addn product. Aq oxidizers such as chromic acid, permanganate, or bromine rapidly destroy PF₃, and alcohols convert it to alkyl phosphite. Phosphides and fluorides are formed upon reaction with hot metals.

Caution: See Phosphorus Pentafluoride.

7472. Phosphorus Trioxide. [1314-24-5] Phosphorus oxide (P₂O₃); diphosphorus trioxide. O₃P₂; mol wt 109.94. O 43.66%, P 56.35%. P₂O₃. Exists as P₄O₆. Prepd by treating PCl₃ with tetramethylammonium sulfite in liq SO₂: Jander *et al., Ber.* **77**, 689 (1944). Alternate prepn from the elements: Thorpe, Tutton, *J. Chem. Soc.* **57**, 545 (1890); Miller, *ibid.* **1928**, 1847; **1929**, 1823; Wolf, Schmager, *Ber.* **62**, 779 (1929). Chemistry: Riess, Van Wazer, *Inorg. Chem.* **5**, 178 (1966).

Transparent monoclinic crystals or colorless liquid. d_4^{21} 2.135. mp 23.8°. bp 173.1° in nitrogen atm. *Corrosive; poisonous.* Disproportionates into red P and P₂O₄ when heated above 210°. Sol in benzene, carbon disulfide. When placed in cold water, H₃PO₃ is formed slowly. Hot water produces a violent reaction with the formation of red phosphorus, phosphine, and H₃PO₄.

7473. Phosphorus Triselenide. [1314-86-9] 3,5,7-Triselena-1,2,4,6-tetraphosphatricyclo[2.2.1.0²,⁶]heptane; tetraphosphorus triselenide. P₄Se₃; mol wt 360.78. P 34.34%, Se 65.66%. Prepn: Meyer, *Z. Anorg. Chem.* **30**, 258 (1902); **61B**, 1807 (1928); Irgolic *et al., Inorg. Chem.* **4**, 1421 (1965).

Orange-red crystals. Irritating odor. mp 245-246°. d 1.31. bp 360-400°. Phosphoresces at 160°. Flammable when heated in air. Dec in moist air. Sol in carbon tetrachloride, carbon disulfide, chloroform, benzene, toluene, acetone, acetylene dichloride, acetylene trichloride.

7474. Phosphorylcholine. [107-73-3] N,N,N-Trimethyl-2-(phosphonooxy)ethanaminium chloride (1:1); choline chloride dihydrogen phosphate; (2-hydroxyethyl)trimethylammonium chloride phosphate; phosphorylcholine chloride; choline phosphate chloride; choline chloride phosphate; choline phosphoric acid ester (chloride). C₅H₁₅ClNO₄P; mol wt 219.60. C 27.35%, H 6.89%, Cl 16.14%, N

6.38%, O 29.14%, P 14.10%. Prepd by phosphorylation of choline chloride with diphenylphosphoryl chloride: Baer, McArthur, *J. Biol. Chem.* **154**, 451 (1944); Baer, *J. Am. Chem. Soc.* **69**, 1253 (1947); *Biochem. Prep.* **2**, 96 (1952); by heating choline or choline chloride with pyrophosphoric or polyphosphoric acid: Cherbuliez, Rabinowitz, *Helv. Chim. Acta* **42**, 1154 (1959).

Barium salt. [6484-71-5] C₅H₁₃BaClNO₄P; mol wt 354.91. Hydrated glistening leaflets from water + alcohol. Must be dried *in vacuo* (0.5 mm) over P₂O₅ at 100° for 16 hrs. Freely sol in water, practically insol in abs ethanol, ether.

Calcium salt. [4826-71-5] Colifos; Epafosforil; Fosfocolina; Isocolin. C₅H₁₃CaClNO₄P; mol wt 257.66.

Magnesium salt. [17032-39-2] Heparexine. C₅H₁₃ClMgNO₄P; mol wt 241.89.

THERAP CAT: Hepatobiliary dysfunction.

7475. Phosphoserine. [407-41-0] O-Phosphono-L-serine; serine dihydrogen phosphate (ester); serine phosphate. C₃H₈NO₆P; mol wt 185.07. C 19.47%, H 4.36%, N 7.57%, O 51.87%, P 16.74%. Prepn: Neuhaus, Korkes, *Biochem. Prep.* **6**, 75 (1958); D. M. Theodoropoulos, *J. Chem. Soc.* **1960**, 5257. Other methods of prepn and isoln from protein hydrolysates: Greenstein, Winitz, *Chemistry of the Amino Acids* (Wiley, New York, 1961), *passim.* Crystal structure: G. H. McCallum, *Nature* **184**, 1863 (1959).

Crystals from ethanol + ether, mp 166-167° (dec).

Note: Used in combination with L-glutamine and vitamin B₁₂ as a nutritional supplement: *Fosforina B₁₂, Vitasprint B₁₂.*

7476. Phosphotungstic Acid. [12067-99-1] Tungsten hydroxide oxide phosphate; tungstophosphoric acid. Approx composition 24WO₃.2H₃PO₄.48H₂O. The H₂O content may vary appreciably. Prepn: Bailar, *Inorg. Synth.* **1**, 132 (1939).

White or slightly yellowish-green, slightly efflorescent crystals or cryst powder. Sol in about 0.5 part water. Also sol in alcohol and in ether.

USE: As reagent for alkaloids and many other nitrogen bases, for phenols, albumin, peptone, amino acids, uric acid, urea, blood, carbohydrates; as biological stain.

7477. Phosvitins. [9008-96-2] A phosphoprotein of mol wt about 40,000. Isoln from egg yolk: Mecham, Olcott, *J. Am. Chem. Soc.* **71**, 3670 (1949); *Biochem. Prep.* **2**, 15 (1952); Joubert, Cook, *Can. J. Biochem. Physiol.* **36**, 399 (1958); from roe: Mano, Yoshida, *J. Biochem. (Tokyo)* **66**, 105 (1969). Represents about 7% of the total yolk protein; high in phosphorus content and a ratio of serine to all other amino acid residues of 6:4. *In vitro* synthesis in hens liver: Heald, McLachlan, *Biochem. J.* **94**, 32 (1965); Rudack, Wallace, *Biochim. Biophys. Acta* **155**, 299 (1968). Estradiol-induced synthesis in roosters: Veuving, Gruber, *ibid.* **232**, 524, 529 (1971); Jailkhani, Talwar, *Nature New Biol.* **236**, 239 (1972). Amino acid sequence: Belitz, *Angew. Chem.* **76**, 574 (1964); Shainkin, Perlmann, *J. Biol. Chem.* **246**, 2278 (1971). Conformation: Grizzuti, Perlmann, *ibid.* **245**, 2573 (1970). Anticlotting activity studies: Sato *et al., Am. J. Physiol.* **203**, 1170 (1962). *Review:* Taborsky, *Adv. Protein Chem.* **28**, 50-78 (1974).

THERAP CAT: Anticoagulant.

7478. Phoxim. [14816-18-3] 5-Ethoxy-2-phenyl-4,6-dioxa-3-aza-5-phosphaoct-2-enenitrile 5-sulfide; 4-ethoxy-7-phenyl-3,5-dioxa-6-aza-4-phosphaoct-6-ene-8-nitrile 4-sulfide; phenylglyoxylonitrile oxime O,O-diethyl phosphorothioate; O,O-diethyl O-(α-cyanobenzylideneamino)phosphorothioate; α-[[(diethoxyphosphinothioyl)oxy]imino]benzeneacetonitrile; Bay 5621; Bay 77488;

Baythion; Byemite; Sebacil. $C_{12}H_{15}N_2O_3PS$; mol wt 298.30. C 48.32%, H 5.07%, N 9.39%, O 16.09%, P 10.38%, S 10.75%. Organophosphate insecticide; cholinesterase inhibitor. Prepn: **NL 6605907**; W. Lorenz et al., **US 3591662** (1966, 1971 both to Bayer). Activity: C. R. Harris, J. Econ. Entomol. **63**, 782 (1970); D. C. Read, ibid. **69**, 429 (1976). Toxicity studies: J. H. Vinopal, T. R. Fukuto, Pestic. Biochem. Physiol. **1**, 44 (1971). Degradation in soil: G. Dräger, Pflanzenschutz-Nachr. **30**, 28 (1977). HPLC determn in water samples: P. Liang et al., J. Sep. Sci. **29**, 366 (2006). Field trials in chickens: B. Meyer-Kühling et al., Vet. Parasitol. **147**, 289 (2007).

Pale yellow oil, $bp_{0.01}$ 102°. n_D^{20} 1.5405. d_4^{20} 1.176. Sol in alc, ketones, aromatic hydrocarbons. LD_{50} orally in mice: >2000 mg/kg (Vinopal, Fukuto).

USE: Insecticide.

THERAP CAT (VET): Ectoparasiticide.

7479. Phrenosin. [586-02-7] (2R)-N-[(1S,2R,3E)-1-[(β-D-Galactopyranosyloxy)methyl]-2-hydroxy-3-heptadecen-1-yl]-2-hydroxytetracosanamide. $C_{48}H_{93}NO_9$; mol wt 828.27. C 69.61%, H 11.32%, N 1.69%, O 17.38%. Cerebroside easily separated from a commercially available beef spinal cord lipid concentrate or from nerve tissue. A large proportion of its fatty acid content (25%) is 2-hydroxystearic acid, the rest is cerebronic (2-hydroxylignoceric) acid. Isoln: Radin et al., J. Biol. Chem. **219**, 977 (1956); Skipski et al., Arch. Biochem. Biophys. **82**, 487 (1959). Synthesis: Shapiro, Flowers, J. Am. Chem. Soc. **83**, 3327 (1961). IR spectrum: H. P. Schwarz et al., Ann. N.Y. Acad. Sci. **69**, 116 (1957).

Cryst from pyridine + acetone, methanol + chloroform, toluene + alcohol (1:1) or chloroform + alcohol + water, [α] +4.5° (c = 2 in pyridine). Sol in hot dioxane, 1,1-dichloro-1-nitroethane, butanol and acetonitrile + chloroform.

7480. o-Phthalaldehyde. [643-79-8] 1,2-Benzenedicarboxaldehyde; o-phthaldialdehyde; Cidex OPA. $C_8H_6O_2$; mol wt 134.13, C 71.64%, H 4.51%, O 23.86%. Alternative to glutaraldehyde, q.v. for high-level disinfection. Prepn: A. Colson, H. Gautier, Bull. Soc. Chim. **45**, 509 (1886); J. Thiele, O. Günther, Ann. **347**, 106 (1906); T. C. Chaudhuri, J. Am. Chem. Soc. **64**, 315 (1942). Polarographic determn: N. H. Furman, D. R. Norton, Anal. Chem. **26**, 1111 (1954). Use in fluorogenic determn of peptides: M. Roth, ibid. **43**, 880 (1971); T. M. Joys, H. Kim, Anal. Biochem. **94**, 371 (1979); of biological thiols: K. Mopper, D. Delmas, Anal. Chem. **56**, 2557 (1984). Biocidal properties: S. E. Walsh et al., J. Appl. Microbiol. **86**, 1039 (1999). Mechanism of antibacterial action: idem et al., ibid. **87**, 702 (1999). Bactericidal efficacy study: J. C. N. Shackelford et al., J. Antimicrob. Chemother. **57**, 335 (2006). Hospital disinfection of endoscopes; S. M. Hession, Gastroenterol. Nurs. **26**, 110 (2003). Review of reactivity with nucleophiles: P. Zuman, Chem. Rev. **104**, 3217-3238 (2004).

Long, pale yellow needles from petroleum ether mp 56-56.5° (Thiele, Günther). Also reported as colorless powder, mp 54° (Chaudhuri).

USE: Disinfectant. Reagent in fluorometric determn of primary amines and thiols.

7481. Phthalamide. [88-96-0] 1,2-Benzenedicarboxamide; phthalic acid diamide. $C_8H_8N_2O_2$; mol wt 164.16. C 58.53%, H 4.91%, N 17.07%, O 19.49%. Prepd from phthalic anhydride and anhydr ammonia in benzene: Dominikiewicz, Arch. Chem. Farm. **3**, 141 (1937).

Minute crystals, mp 228° (commercial grades mp 221- 223°). Melts with decomposition to NH_3 and phthalimide. Slightly sol in cold water, methanol. More sol in hot solvents. Heat of combustion 921.7 cal/mol. Boiling of aq or alcoholic solns produces NH_3 and phthalimide. The presence of acids accelerates decompn.

7482. Phthalazine. [253-52-1] 2,3-Benzodiazine; benzo[d]-pyridazine; β-phenodiazine. $C_8H_6N_2$; mol wt 130.15. C 73.83%, H 4.65%, N 21.52%. Prepd by reduction of 1-chlorophthalazine with hydriodic acid and red phosphorus: Gabriel, Eschenbach, Ber. **30**, 3024 (1897); Paul, Ber. **32**, 2015 (1899); by redn of 1-hydrazinophthalazine: Armarego, J. Appl. Chem. **11**, 70 (1961); from o-phthalaldehyde + hydrazine sulfate: Smith, Otremba, J. Org. Chem. **27**, 879 (1962); from 1,2-benzodinitrile: Carter, Cheeseman, Org. Prep. Proced. Int. **6**, 67 (1974).

Pale yellow needles from ether, mp 90-91°. bp 315-317° (dec); bp_{29} 189°; bp_{17} 175°. uv max (water): 218, 261, 292, 305 nm (log ε 4.83, 3.53, 3.18, 3.11). Freely sol in water; sol in ethanol, methanol, benzene, ethyl acetate; less sol in ether. Practically insol in ligroin.

Hydrochloride. $C_8H_6N_2$·HCl. Needles from ether, mp 231° (effervescence).

7483. Phthalic Acid. [88-99-3] 1,2-Benzenedicarboxylic acid. $C_8H_6O_4$; mol wt 166.13. C 57.84%, H 3.64%, O 38.52%. Manufactured by catalytic oxidation of o-toluic acid and oxidation of xylene: Taylor, Dean, **US 3064046** (1962 to I.C.I.); Cier, **US 3088974** (1963 to Esso Res. & Eng.). Isoln from the fungus, Gibberella fujikuroi: Cross et al., J. Chem. Soc. **1963**, 2937. Toxicity studies: C. B. Shaffer et al., J. Ind. Hyg. Toxicol. **27**, 130 (1945); A. R. Singh et al., J. Pharm. Sci. **61**, 59 (1972). Review of toxicology: A. M. Api, Food Chem. Toxicol. **39**, 97-108 (2001).

Crystals, mp ~230° when rapidly heated, forming phthalic anhydride and water. One gram dissolves in 160 ml water, 10 ml alc, 205 ml ether, 5.3 ml methanol. Practically insol in chloroform. LD_{50} orally in rats: 7.9 g/kg (Shaffer).

Ethyl ester. [84-66-2] Ethyl phthalate; diethyl phthalate; Neantine; Palatinol A. $C_{12}H_{14}O_4$; mol wt 222.24. Review of toxicology and human exposure: *Toxicological Profile for Diethyl Phthalate* (PB95-264214, 1995) 159 pp. Colorless, practically odorless, oily liq; bitter disagreeable taste. d_4^{14} 1.232. bp 295°. Flash pt 140°C. n_D^{14} 1.5049. Insol in water. Misc with alcohol, ether and many other organic solvents. LD_{50} i.p. in rats: 5.06 ml/kg (Singh).

Caution: Potential symptoms of overexposure to diethyl phthalate are irritation of eyes, skin, nose, throat; headache, dizziness, nausea; lacrimation; polyneuritis, vestibular dysfunction; pain, numbness, weakness, spasms in arms and legs. *See NIOSH Pocket Guide to Chemical Hazards* (DHHS/NIOSH 97-140, 1997) p 108.

USE: Phthalic acid used in buffers. Ethyl phthalate used in manuf celluloid; solvent for cellulose acetate in manuf varnishes and dopes; denaturing alc. Vehicle for fragrance and cosmetic ingredients.

7484. Phthalic Anhydride. [85-44-9] 1,3-Isobenzofurandione; 1,2-benzenedicarboxylic anhydride. $C_8H_4O_3$; mol wt 148.12. C 64.87%, H 2.72%, O 32.40%. Prepd from naphthalene by oxidation with a mixture of $HgSO_4$ and $CuSO_4$ in presence of H_2SO_4; by passing naphthalene and oxygen over a suitable catalyst at 400-500°. Review of mfg processes: *Faith, Keyes & Clark's Industrial Chemicals*, F. A. Lowenheim, M. K. Moran, Eds (Wiley-Interscience, New York, 4th ed., 1975) pp 658-665.

White, lustrous needles. d 1.53; mp 130.8°; bp 295°. Sublimes. *Corrosive.* Sol in 162 parts water, more in hot water with conversion into phthalic acid, in 125 parts carbon disulfide; sol in alcohol, sparingly in ether.

Caution: Potential symptoms of overexposure are irritation of eyes, skin and upper respiratory system; conjunctivitis; nasal ulcer bleeding; bronchitis, bronchial asthma; dermatitis. *See NIOSH Pocket Guide to Chemical Hazards* (DHHS/NIOSH 97-140, 1997) p 256.

USE: Manuf phthaleins, phthalates, benzoic acid, synthetic indigo, artificial resins (glyptal). Buffer.

7485. Phthalimide. [85-41-6] 1*H*-Isoindole-1,3(2*H*)-dione. $C_8H_5NO_2$; mol wt 147.13. C 65.31%, H 3.43%, N 9.52%, O 21.75%. Prepd from phthalic anhydride and NH_4OH or $(NH_4)_2CO_3$: W. A. Noyes, P. K. Porter, *Org. Synth.* **coll. vol. I**, 457 (2nd ed., 1941). Teratogenicity study: K. Fickentscher *et al.*, *Pharmazie* **31**, 172 (1976).

Monoclinic prisms from water or by sublimation. mp 238°. Has slightly acidic properties, Ka = 5 × 10^{-9}. Absorption spectrum: Hartley, Hedley, *J. Chem. Soc.* **91**, 317 (1907). Slightly soluble in water. A satd aq soln contains 0.036 g at 25°; 0.07 g at 40°; about 0.4 g at the bp. 100 g of boiling alcohol dissolves 5 g phthalimide. Almost insol in benzene, petr ether; fairly sol in boiling acetic acid; freely sol in aq alkali hydroxides.

7486. Phthalofyne. [131-67-9] 1,2-Benzenedicarboxylic acid 1-(1-ethyl-1-methyl-2-propyn-1-yl) ester; phthalic acid 1-ethyl-1-methyl-2-propynyl ester; 1-ethyl-1-methyl-2-propynyl acid phthalate; 3-methyl-1-pentyn-3-yl acid phthalate; ftalofyne; NSC-25614; Whipcide. $C_{14}H_{14}O_4$; mol wt 246.26. C 68.28%, H 5.73%, O 25.99%. Prepn: Sugimoto, Okumura, **JP 54 1833** (1954 to Tanabe), *C.A.* **49**, 11711e (1955); **GB 736993** (1955 to Schering).

Crystals from benzene or hexane, mp 96-98°. Weak acid. Slightly sol in water. Unstable in strong alkali.

THERAP CAT (VET): Anthelmintic.

7487. Phthaloyl Chloride. [88-95-9] 1,2-Benzenedicarbonyl dichloride. $C_8H_4Cl_2O_2$; mol wt 203.02. C 47.33%, H 1.99%, Cl 34.92%, O 15.76%. Obtained by the action of PCl_5 on phthalic anhydride.

Colorless, oily liquid. d^{20} 1.409. Solidif +12°; mp 15-16°; bp 280-282°; n_D^{20} 1.5692. Dec by water or alcohol; sol in ether. *Keep tightly closed.*

7488. Phthalylsulfacetamide. [131-69-1] 2-[[[4-[(Acetylamino)sulfonyl]phenyl]amino]carbonyl]benzoic acid; 4′-(acetylsulfamyl)phthalanilic acid; phthalsulfacetimide; N^1-acetyl-N^4-phthaloylsulfanilamide; N^1-acetyl-N^4-phthalylsulfanilamide; phthaloylsulfacetamide; ftalicetimida; N-[p-(o-carboxybenzamido)benzenesulfonyl]acetamide; N-(o-carboxybenzoyl)sulfacetamide; Enterocid; Enterosulfamid; Enterosulfon; Talecid; Thalamyd; Rabalan; Sterathal. $C_{16}H_{14}N_2O_6S$; mol wt 362.36. C 53.03%, H 3.89%, N 7.73%, O 26.49%, S 8.85%. Prepn: Basu, *J. Indian Chem. Soc.* **26**, 130 (1949).

Needles from dil alcohol, mp 196°. Soluble in alcohol. Very sparingly sol in water. Forms a water-sol sodium salt.

THERAP CAT: Antibacterial.

THERAP CAT (VET): Antibacterial.

7489. Phthalylsulfathiazole. [85-73-4] 2-[[[4-[(2-Thiazolylamino)sulfonyl]phenyl]amino]carbonyl]benzoic acid; 4′-(2-thiazolylsulfamyl)phthalanilic acid; 2-(N^4-phthalylaminobenzenesulfonamido)thiazole; 2-(N^4-phthalylsulfanilamido)thiazole; phthalylsulfonazole; Estreptocarbocaftiazol; Thalazole. $C_{17}H_{13}N_3O_5S_2$; mol wt 403.43. C 50.61%, H 3.25%, N 10.42%, O 19.83%, S 15.89%. May be prepd by condensing sulfathiazole with phthalic anhydride: M. L. Moore, **US 2324013**; **US 2324015** (1943 to Sharp & Dohme). Prepn of 8-hydroxyquinoline salt: **DE 1008296** (1957 to Geistlich Söhne). Toxicity data: P. A. Mattis *et al.*, *J. Pharmacol. Exp. Ther.* **81**, 116 (1944).

Crystals. Slightly bitter taste. Darkens on prolonged exposure to light. Effervesces at 244° to 250°. mp 272-277° (dec) when the melting point bath is preheated to 220-225°. Slightly sol in alcohol; very slightly sol in ether; readily sol in NaOH or KOH soln, ammonia water and concd HCl. Practically insol in chloroform and water. LD$_{50}$ i.p. in mice: 920 mg/kg (Mattis).

8-Hydroxyquinoline salt. Yellow crystals from dil methanol, dec above 220°. Split into its components by dil acids.

THERAP CAT: Antibacterial.

THERAP CAT (VET): Antibacterial.

7490. Phthiocol. [483-55-6] 2-Hydroxy-3-methyl-1,4-naphthalenedione; 2-hydroxy-3-methyl-1,4-naphthoquinone. C$_{11}$H$_8$O$_3$; mol wt 188.18. C 70.21%, H 4.29%, O 25.51%. Antibiotic substance produced by *Mycobacterium tuberculosis:* Terni, *Boll. Soc. Ital. Biol. Sper.* **25**, 60 (1949), *C.A.* **45**, 2054f (1951). Possesses some vitamin K activity. Isoln and synthesis: Anderson, Newman, *J. Biol. Chem.* **101**, 773 (1933); **103**, 197, 405 (1933); L. F. Fieser, *ibid.* **133**, 391 (1940); Tarbell *et al.*, *J. Am. Chem. Soc.* **72**, 379 (1950); Burton, Praill, *J. Chem. Soc.* **1952**, 755; Eistert, Müller, *Ber.* **92**, 2071 (1959); H. Kallmayer, *Arch. Pharm.* **307**, 806 (1974); K. Maruyama, S. Arakawa, *J. Org. Chem.* **42**, 3793 (1977).

Yellow prisms from ether-petr ether, mp 173-174°. Sublimes. Volatile with steam. Slightly sol in water; sol in the usual organic solvents except petr ether. Forms deep red water-sol salts. E$_0^{alc}$ 0.299 volts.

Acetate. C$_{13}$H$_{10}$O$_4$. Pale yellow crystals, mp 101-102°.

7491. Phycobiliproteins. Deeply colored, highly fluorescent photoreceptor pigments found in blue-green, red and cryptomonad algae that contain a linear tetrapyrrole as the prosthetic group. They are composed of a bile pigment or **phycobilin** and an apoprotein. Phycobiliproteins are classified according to uv-vis absorption maxima as **phycocyanins** (blue pigment), **phycoerythrins** (red pigment), and **allophycocyanins** (pale blue pigment). Phycocyanins and phycoerythrins occur as large mol wt aggregates called **phycobilisomes** that are attached to the photosynthetic membranes. They are closely linked to the chlorophyll containing system for efficient energy transfer. **Phytochrome** is a similar biliprotein that functions in plant photomorphogenesis. It is widely distributed in plants but only in trace amounts. It exists in two forms that are interconverted when alternate exposure to red and far-red light. *Reviews:* H. W. Siegelman *et al.*, *Biochem. Soc. Symp.* **28**, 107-120 (1968); P. O'Carra, C. O'hEocha in *Chemistry and Biochemistry of Plant Pigments* vol. **1**, T. W. Goodwin, Ed. (Academic, New York, 2nd ed., 1976) pp 328-376; A. Bennett, H. W. Siegelman in *Porphyrins* vol. **6**, D. Dolphin, Ed. (Academic, New York, 1979) pp 493-520.

7492. Phylloquinone. [84-80-0] 2-Methyl-3-[(2*E*,7*R*,11*R*)-3,7,11,15-tetramethyl-2-hexadecen-1-yl]-1,4-naphthalenedione; 2-methyl-3-phytyl-1,4-naphthoquinone; 3-phytylmenadione; phytomenadione; phytonadione; vitamin K$_1$; AquaMephyton; Konakion; Mephyton; Mono-Kay; Veda-K$_1$; Veta-K$_1$. C$_{31}$H$_{46}$O$_2$; mol wt 450.71. C 82.61%, H 10.29%, O 7.10%. Photosynthetic electron carrier; occurs widely in green plants, algae, photosynthetic bacteria. Major dietary source of vitamin K, *q.v.* Isoln from alfalfa: H. Dam *et al.*, *Helv. Chim. Acta* **22**, 310 (1939). Structure: D. W. MacCorquodale *et al.*, *J. Biol. Chem.* **131**, 357 (1939); L. F. Fieser, *J. Am. Chem. Soc.* **61**, 3467 (1939). Partial syntheses from menadione and phytol: H. J. Almquist, A. A. Klose, *ibid.* 2557; S. B. Binkley *et al.*, *ibid.* 2558; L. F. Fieser, *ibid.* 2559. Stereochemistry and total synthesis: H. Mayer *et al.*, *Helv. Chim. Acta* **47**, 221 (1964); L. M. Jackman *et al.*, *ibid.* **48**, 1332 (1965). Synthesis using a π-allylic nickel(I) complex: Sato *et al.*, *J. Chem. Soc. Perkin Trans. I* **1973**, 2289. Alternate synthesis: Y. Tachibana, *Chem. Lett.* **1977**, 901. Metabolic studies: M. J. Shearer *et al.*, *Br. J. Haematol.* **18**, 297 (1970); **22**, 579 (1972). The *cis* isomer is not bioactive: J. T. Matschiner *et al.*, *J. Nutr.* **102**, 625 (1972). Isoln from chloroplasts: E.

Interschick-Niebler, H. K. Lichtenthaler, *Z. Naturforsch.* **36C**, 276 (1981). Role in photosystem I: K. Brettel *et al.*, *FEBS Lett.* **203**, 220 (1986). Conversion *in vivo* to menaquinone-4, *q.v.*: H. H. W. Thijssen, M. J. Drittij-Reijnders, *Br. J. Nutr.* **72**, 415 (1994). HPLC determn in foods: S. L. Booth *et al.*, *J. Agric. Food Chem.* **42**, 295 (1994). Clinical efficacy in hemorrhagic disease of newborn: P. M. Loughnan, P. N. McDougall, *J. Paediatr. Child Health* **29**, 171 (1993). Comprehensive description: M. M. A. Hassan *et al.*, *Anal. Profiles Drug Subs.* **17**, 449-531 (1988).

Yellow viscous oil. [α]$_D^{25}$ −0.28° (dioxane). n$_D^{20}$ 1.5263. uv max (petr ether): 242, 248, 260, 269, 325 nm (E$_{1cm}^{1\%}$ 396, 419, 383, 387, 68). Sol in dehydrated alc, acetone, benzene, petr ether, hexane, dioxane, chloroform, ether, in other fat solvents and in vegetable oils; sparingly sol in methanol; slightly sol in alc. Insol in water. Stable to air and moisture, but dec in sunlight. Unaffected by dil acids, but destroyed by solns of alkali hydroxides and by reducing agents. *Keep well closed and protected from light.*

Dihydro form. [572-96-3] Phytonadiol; dihydrovitamin K$_1$; α-phyllohydroquinone. Waxy mass. Freely sol in ether; sparingly sol in petr ether. Insol in water.

Dihydro form sodium diphosphate. [5988-22-7] Phytonadiol sodium diphosphate; Kayhydrin. C$_{31}$H$_{48}$Na$_2$O$_8$P$_2$. mp 138°. Sol in water, methanol.

2,3-Epoxide. [25486-55-9] Vitamin K$_1$ epoxide; vitamin K$_1$ oxide. C$_{31}$H$_{46}$O$_3$. Prepn: Fieser *et al.*, *J. Am. Chem. Soc.* **61**, 3216 (1939). Colorless oil. uv max (95% alc): 259, 305 nm (log E$_M$ 3.79, 3.31). Insol in water.

THERAP CAT: Vitamin (prothrombogenic).

THERAP CAT (VET): Vitamin (prothrombogenic); antidote for dicoumarol poisoning.

7493. Physalaemin. [2507-24-6] 5-Oxo-L-prolyl-L-alanyl-L-α-aspartyl-L-prolyl-L-asparaginyl-L-lysyl-L-phenylalanyl-L-tyrosylglycyl-L-leucyl-L-methioninamide; physalemin. C$_{58}$H$_{84}$N$_{14}$O$_{16}$S; mol wt 1265.45. C 55.05%, H 6.69%, N 15.50%, O 20.23%, S 2.53%. An undecapeptide belonging to the group of proteins named tachykinins. Found in skin of the amphibian *Physalaemus fuscumaculatus:* Erspamer *et al.*, *Experientia* **18**, 562 (1962). Structure: Erspamer *et al.*, *ibid.* **20**, 489 (1964); Anastasi *et al.*, *Arch. Biochem. Biophys.* **108**, 341 (1964). Synthesis: Bernardi *et al.*, *Experientia* **20**, 490 (1964); Nakamura *et al.*, **JP 71 25384** (1971 to Dainippon), *C.A.* **75**, 152083r (1971). Solid-phase synthesis: W. Voelter *et al.*, *Tetrahedron* **28**, 5963 (1972). Biological activities similar to the tachykinins eledoisin and substance P, *q.q.v.* Exerts a powerful hypotensive action, stimulates salivary secretion, intestinal contraction, and vasodilation. Occurrence in other *Physalaemus* spp. and pharmacology: G. Bertaccini *et al.*, *Br. J. Pharmacol.* **25**, 363 (1965); G. Bertaccini, *Pharmacol. Rev.* **28**, 127 (1976). Differentiation of physalaemin and substance P: Geipert *et al.*, *Arch. Pharmacol.* **265**, 225 (1969). Immunoreactivity study in human lung small-cell carcinoma: L. H. Lazarus *et al.*, *Science* **219**, 79 (1983).

5-oxo-Pro–Ala–Asp–Pro–Asn–Lys–Phe–Tyr–Gly–Leu–MetNH$_2$

Trifluoroacetate. [4705-64-0] mp 180° (dec). [α]$_D^{20}$ −57° (c = 0.3 in ethanol). uv max: 278 nm (ε 1780).

7494. Physodic Acid. [84-24-2] 3,8-Dihydroxy-11-oxo-1-(2-oxoheptyl)-6-pentyl-11*H*-dibenzo[*b*,*e*][1,4]dioxepin-7-carboxylic acid; 4,4′,6′-trihydroxy-6-(2-oxoheptyl)-2′-pentyl-2,3′-oxydibenzoic acid-ε-lactone; physodalin. C$_{26}$H$_{30}$O$_8$; mol wt 470.52. C 66.37%, H 6.43%, O 27.20%. From the lichens *Parmelia physodes* (L.) Ach.: Hesse, *Ber.* **30**, 1983 (1897); Klosa, *Pharm. Ind.* **15**, 46 (1953); and *Cetraria ciliaris* Ach.: Culberson, *Science* **143**, 255 (1964). Structure: Asahina, Nogami, *Ber.* **67**, 805 (1934); **68**, 77 (1935).

Needles from methanol, mp 205°. uv max (95% ethanol): 256 nm (log ε 4.2). Sol in ether, acetone, hot methanol; somewhat sol in cold ethanol, methanol, hot chloroform; practically insol in benzene, petr ether, carbon disulfide.

Diacetate. $C_{30}H_{34}O_{10}$. Plates from acetone + carbon disulfide, mp 153-155.5°.

Methyl ester. $C_{27}H_{32}O_8$. Prisms from 80% alcohol, mp 156-157°.

7495. Physostigma. Calabar bean; ordeal bean; chop nut; split nut. Dried ripe seed of *Physostigma venenosum* Balf., *Leguminosae. Poisonous! Habit.* West Africa (near mouths of Niger and Old Calabar Rivers); introduced into India and Brazil. *Constit.* 0.15-0.3% alkaloids consisting of physostigmine (eserine), physovenine, geneserine, eseramine. Review of isoln, structures, and properties of constituents: B. Robinson, *Alkaloids*, vol. X (Academic Press, New York, 1968) pp 383-400. Review of historical uses, pharmacology, and constituents: A. Proudfoot, *Toxicol. Rev.* **25**, 99-138 (2006).

7496. Physostigmine. [57-47-6] (3a*S*,8a*R*)-1,2,3,3a,8,8a-Hexahydro-1,3a,8-trimethylpyrrolo[2,3-*b*]indol-5-ol 5-(*N*-methyl-carbamate); eserine; Synapton. $C_{15}H_{21}N_3O_2$; mol wt 275.35. C 65.43%, H 7.69%, N 15.26%, O 11.62%. Acetylcholinesterase inhibitor. From Calabar beans, the seeds of *Physostigma venenosum* Balf., *Leguminosae:* Jobst, Hesse, *Ann.* **129**, 115 (1864); Hesse, *ibid.* **141**, 82 (1867). Extraction procedure: Schwyzer, *Die Fabrikation pharmazeutischer und chemisch-technischer Produkte* (Berlin, 1931) p 338; Chemnitius, *J. Prakt. Chem.* **116**, 59 (1927). Structure: Stedman, Barger, *J. Chem. Soc.* **127**, 247 (1925). Absolute configuration: R. B. Longmore, B. Robinson, *Chem. Ind. (London)* **1969**, 622. Crystal structure: Petcher, Pauling, *Nature* **241**, 277 (1973). Synthesis: Julian, Pikl, *J. Am. Chem. Soc.* **57**, 755 (1935); Harley-Mason, Jackson, *J. Chem. Soc.* **1954**, 3651; J. Wijnberg, W. N. Speckamp, *Tetrahedron* **34**, 2399 (1978). Cholinesterase inhibitor: Engelhart, Loewi, *Arch. Exp. Pathol. Pharmakol.* **150**, 1 (1930); Matthes, *J. Physiol.* **70**, 338 (1930). Toxicity data: W. T. Lynch, J. M. Coon, *Toxicol. Appl. Pharmacol.* **21**, 53 (1972); R D. Sofia, L. C. Knabloch, *ibid.* **28**, 227 (1974). Improvement of long-term memory: K. L. Davis *et al., Science* **201**, 272 (1978). Memory enhancement in Alzheimer's disease: L. J. Thal *et al., N. Engl. J. Med.* **308**, 720 (1983); *eidem, Ann. Neurol.* **13**, 491 (1983); L. Gustafson *et al., Psychopharmacology* **93**, 31 (1987). *Review:* B. Robinson in *The Alkaloids* **Vol. XIII**, R. H. F. Manske, Ed. (Academic Press, New York, 1971) pp 213-226. Comprehensive description: F. J. Muhtadi, S. S. El-Hawary, *Anal. Profiles Drug Subs.* **18**, 289-350 (1989).

Orthorhombic sphenoidal prisms or clusters of leaflets from ether or benzene. mp 105-106° (also an unstable, low-melting form, mp 86-87°). $[\alpha]_D^{17}$ −76° (c = 1.3 in chloroform); $[\alpha]_D^{25}$ −120° (benzene). pKa₁ 6.12; pKa₂ 12.24. Very sol in chloroform, dichloromethane; freely sol in alc; sol in benzene, fixed oils; slightly sol in water. Solid and solns turn red on exposure to heat, light, air, and on contact with traces of metals. Under certain conditions the oxidation may pro-

ceed to yield *eserine blue*, $C_{26}H_{31}N_5O_2$. LD₅₀ orally in mice: 4.5 mg/kg (Lynch, Coon).

Salicylate. [57-64-7] Antilirium. $C_{22}H_{27}N_3O_5$; mol wt 413.47. Acicular crystals, mp 185-187°. uv max (methanol): 239, 252, 303 nm (log ε 4.09, 4.04, 3.78). One gram dissolves in 75 ml water at 25° (pH of 0.5% aq soln 5.8); in 16 ml water at 80°; in 16 ml alc, 5 ml boiling alc, in 6 ml chloroform, 250 ml ether. *Solns should be kept well closed in light-resistant, alkali-free glass containers and used within a week.* Turns red and loses effectiveness on exposure to heat, light, air. LD₅₀ i.p. in mice: 0.64 mg/kg (Sofia, Knabloch).

Sulfate. [64-47-1] $(C_{15}H_{21}N_3O_2)_2 \cdot H_2SO_4$. Deliquescent scales, mp 140° (after drying at 100°). One gram dissolves in 0.4 ml alc, 4 ml water (pH of 0.05*M* soln 4.7), 1200 ml ether. The solns are more prone to change color than those of the salicylate.

Sulfite. $(C_{15}H_{21}N_3O_2)_2 \cdot H_2SO_3$. White powder, freely sol in water, alc. The aq soln is stated to remain colorless for a long time.

THERAP CAT: Cholinergic; miotic.

THERAP CAT (VET): Cholinergic; miotic.

7497. Physovenine. [6091-05-0] (3a*S*,8a*S*)-3,3a,8,8a-Tetrahydro-3a,8-dimethyl-2*H*-furo[2,3-*b*]indol-5-ol 2-(*N*-methylcarbamate). $C_{14}H_{18}N_2O_3$; mol wt 262.31. C 64.11%, H 6.92%, N 10.68%, O 18.30%. From seed of *Physostigma venenosum* Balf., *Leguminosae:* Salway, *J. Chem. Soc.* **99**, 2148 (1911). Structure: Robinson, *ibid.* **1964**, 1503. Synthesis: Longmore, Robinson, *Chem. Ind. (London)* **1965**, 1297. Abs config: *eidem, ibid.* **1969**, 622.

Plates from ether, mp 124-125°. $[\alpha]_D^{22.5}$ −92° (ethanol). uv max (ethanol): 252, 310 nm (ε 13200, 3300). Sol in alcohol, benzene, chloroform; slightly sol in ether. Practically insol in water, petr ether.

7498. Phytantriol. [74563-64-7] 3,7,11,15-Tetramethyl-1,-2,3-hexadecanetriol; phytane-1,2,3-triol. $C_{20}H_{42}O_3$; mol wt 330.55. C 72.67%, H 12.81%, O 14.52%. Cosmetic ingredient: improves moisture retention properties of hair and skin; acts as a penetration enhancer to increase the effects of panthenol, vitamins, and amino acids. Prepn: W. Guex *et al.,* DE 1149700 (1963 to F. Hoffmann-La Roche); T. C. Jain, R. J. Striha, *Can. J. Chem.* **47**, 4359 (1969). Cosmetic uses: E. Wagner, *Parfuem. Kosmet.* **75**, 260 (1994). Aqueous phase behavior: J. Barauskas, T. Landh, *Langmuir* **19**, 9562 (2003).

Light yellow oil, bp₀.₀₁ 130° (Guex); also reported as white crystals from acetone-acetonitrile, mp 56-57° (Jain).

USE: Penetration enhancer in hair and skin care products.

7499. Phytic Acid. [83-86-3] *myo*-Inositol 1,2,3,4,5,6-hexakis(dihydrogen phosphate); inositolhexaphosphoric acid; 1,2,3,4,-5,6-cyclohexanehexolphosphoric acid; cyclohexanehexyl hexaphosphate; Alkalovert. $C_6H_{18}O_{24}P_6$; mol wt 660.03. C 10.92%, H 2.75%, O 58.18%, P 28.16%. $C_6H_6[OPO(OH)_2]_6$. Major phosphorus compound in plants; particularly abundant in oil seeds, legumes and cereal grains. Forms insoluble complexes with di- and trivalent cations. Prepn: S. Posternak, US 1313014 (1919 to Ciba); *idem, Helv. Chim. Acta* **4**, 155 (1921); M. J. Thomas, US 2718523 (1955 to Staley Mfg.); D. S. Bolley *et al.,* US 2732395 (1956 to Natl. Lead Co.); A. R. Baldwin *et al.,* US 2815360 (1957 to Corn Prods. Ref.).

HPLC determn in plants: E. Graf, F. R. Dintzis, *Anal. Biochem.* **119**, 413 (1982); in foods and biological samples: B.E. Knuckles *et al.*, *J. Food Sci.* **47**, 1257 (1982). Review of effect of dietary phytate on zinc metabolism: *Nutr. Rev.* **41**, 64-66 (1983); on mineral bioavailability: J. L. Kelsay, *Am. J. Gastroenterol.* **82**, 983-986 (1987).

Syrupy, straw-colored liquid. Dec on heating. Acid reaction; the pH of a 10% aq soln has been reported as 0.86. Miscible with water, 95% alc, glycerol; sol in water contg alcohol-ether mixtures; very slightly sol in abs alc, methanol. Practically insol in anhydr ether, benzene, chloroform.

Calcium iron salt. Obtained in the processing of corn. Grayish powder contg 20% Ca, 14% P, and 2% Fe. Sparingly sol in water, dil mineral acids.

Calcium magnesium salt. [3615-82-5] Phytin. White, odorless powder. Poor soly in water. Sol in dil acids.

Sodium salt. [7205-52-9] Sodium phytate; Phytat D.B.; Rencal. $C_6H_9Na_9O_{24}P_6$; mol wt 857.86. Soluble in water with neutral reaction.

USE: Complexing agent for the removal of traces of heavy metal ions (also employed as the sodium salt). Starting material in manufacture of inositol. Fermentation nutrient.

THERAP CAT: Sodium salt as hypocalcemic.

7500. Phytochlorin. [19660-77-6] (7*S*,8*S*)-3-Carboxy-5-(carboxymethyl)13-ethenyl-18-ethyl-7,8-dihydro-2,8,12,17-tetramethyl-21*H*,23*H*-porphine-7-propanoic acid; (2*S*-*trans*)-18-carboxy-20-(carboxymethyl)-8-ethenyl-13-ethyl-2,3-dihydro-3,7,-12,17-tetramethyl-21*H*,23*H*-porphine-2-propanoic acid; chlorin e₆; phytochlorin e. $C_{34}H_{36}N_4O_6$; mol wt 596.68. C 68.44%, H 6.08%, N 9.39%, O 16.09%. Prepn from broccoli-leaf extract: Wall, *US 2555583* (1951 to U.S.D.A.). *Review:* H. Fischer, A. Stern, *Die Chemie des Pyrrols* vol. **II**, 2 (Leipzig, 1940) pp 91-94; *The Chlorophylls*, L. P. Vernon, G. R. Seely, Eds. (Academic Press, New York, 1966) pp 9, 12, 71, 90, 95, 98, 142.

Greenish-brown rectangular plates. May contain one mol water. $[\alpha]_D^{20}$ −141° in acetone. Ether solns are olive-green with deep red fluorescence. Sparingly sol in ethanol, ether, acetone. Freely sol in pyridine.

7501. Phytofluene. [540-05-6] 7,7′,8,8′,11,12-Hexahydro-ψ,ψ-carotene; 5,6,7,8,9,10,10′,9′,8′,7′,6′,5′-dodecahydrolycopene; *all-trans*-phytofluene. $C_{40}H_{68}$; mol wt 548.98. C 87.52%, H 12.49%. Polyene hydrocarbon widespread in the vegetable kingdom where it has been observed in chlorophyll-free tissues which contain considerable amounts of carotenoid pigments. Extracted from *Diospyros kaki* L.f., *Ebenaceae*, *Arbutus unedo* L., *Ericaceae*, *Pyracantha angustifolia* (Franch.) Schneid., *Rosaceae*, and tomatoes: Zechmeister, Sandoval, *J. Am. Chem. Soc.* **68**, 197 (1946); Wallace, Porter, *Arch. Biochem. Biophys.* **36**, 468 (1952); from the basidiomycete, *Dacromyces stillatus:* Goodwin, *Biochem. J.* **53**, 538 (1953). Structure: Porter, Lincoln, *Arch. Biochem. Biophys.* **27**, 390 (1950); Zechmeister, *Experientia* **10**, 1 (1954).

Pale orange, viscous oil which solidifies upon cooling, forming a glassy mass without apparent crystal structure. bp₀.₀₀₀₁ 140-185° (bath temp). uv max (petr ether): 367, 348, 332 nm. Freely sol in petr ether, ether, benzene. Practically insol in water, methanol, ethanol. Strong green fluorescence in ultraviolet spectra.

7502. Phytol. [150-86-7] (2*E*,7*R*,11*R*)-3,7,11,15-Tetramethyl-2-hexadecen-1-ol; 2,6,10,14-tetramethylhexadec-14-en-16-ol. $C_{20}H_{40}O$; mol wt 296.54. C 81.01%, H 13.60%, O 5.40%. Decompn product of chlorophyll: Willstätter, *Ann.* **354**, 205 (1907); **371**, 1 (1909); **378**, 1, 73 (1911); **418**, 121 (1918). Synthesis: Fischer, Löwenberg, *Ann.* **475**, 183 (1929); Karrer, Ringier, *Helv. Chim. Acta* **22**, 610 (1939); Karrer *et al.*, *ibid.* **26**, 1741 (1943); from ethyl levulinate: Lukes, Zobacova, *Chem. Listy* **51**, 330 (1957); from acetone: Sato *et al.*, *J. Org. Chem.* **32**, 177 (1967). Stereochemistry: Burrell *et al.*, *Proc. Chem. Soc. London* **1959**, 263. Abs config: Crabbe *et al.*, *ibid.* **1959**, 264. Stereochemistry and synthesis: Burrell *et al.*, *J. Chem. Soc. C* **1966**, 2144. Stereoselective total synthesis of natural phytol: T. Fujisawa *et al.*, *Tetrahedron Lett.* **22**, 4823 (1981); M. Schmid *et al.*, *Helv. Chim. Acta* **65**, 684 (1982). *Review:* J. Simonsen, D. H. R. Barton, *The Terpenes* vol. **III** (Cambridge University Press, Cambridge, 1952) pp 345-349.

Oily liquid. d_4^{25} 0.8497. n_D^{25} 1.4595. bp₁₀ 203-204°; bp₀.₀₃ 145°. uv max (abs alcohol): 212 nm (log ε 3.04); *see:* Bader, *Helv. Chim. Acta* **34**, 1632 (1951). Practically insol in water; sol in the usual organic solvents.

USE: Preparation of vitamins E and K₁.

7503. Phytolacca. Poke root; garget; pocan; American nightshade root; scoke. Dried root of *Phytolacca americana* L. (*P. decandra* L.), *Phytolaccaceae* (berries also used). *Habit.* North America; naturalized in Southern Europe. *Constit.* Resin, tannin, about 10% sugar; phytolaccine, phytolaccic acid, asparagine.

THERAP CAT: Antirheumatic; emetic; ectoparasiticide.

7504. Piberaline. [39640-15-8] [4-(Phenylmethyl)-1-piperazinyl]-2-pyridinylmethanone; 1-(phenylmethyl)-4-(2-pyridinylcarbonyl)piperazine; 1-benzyl-4-picolinoylpiperazine; EGYT-475; Trelibet. $C_{17}H_{19}N_3O$; mol wt 281.36. C 72.57%, H 6.81%, N 14.93%, O 5.69%. Prepn: J. Körösi *et al.*, *BE 781494*; *eidem*, *US 3865828* (1972, 1975 both to EGYT). Alternate prepns: Z. Budai *et al.*, *DE 2828888*, *C.A.* **90**, 168643u (1979); *eidem*, *HU 17182*, *C.A.* **92**, 215463p (1980) (both 1979 to EGYT); *eidem*, *Acta Chim. Acad. Sci. Hung.* **105**, 241 (1980), *C.A.* **94**, 192275d (1981). HPLC separation: J. Borda *et al.*, *J. Chromatogr.* **258**, 271 (1983). Effect on avoidance behavior in rats: G. Telegdy *et al.*, *Arch. Int. Pharmacodyn. Ther.* **266**, 50 (1983). Metabolism in animals and humans: K. Magyar, *Pol. J. Pharmacol. Pharm.* **39**, 107 (1987). Mechanism of action study: K. Tekes *et al.*, *ibid.* 203.

Dihydrochloride. $C_{17}H_{19}N_3O.2HCl$. Crystals from ethanol, mp 214-215°.

Maleate. $C_{17}H_{19}N_3O.C_4H_4O_4$. Crystals, mp 169-170°.

Fumarate. $C_{17}H_{19}N_3O.C_4H_4O_4$. Crystals, mp 165°.

THERAP CAT: Antidepressant.

7505. Piboserod. [152811-62-6] *N*-[(1-Butyl-4-piperidinyl)methyl]-3,4-dihydro-2*H*-[1,3]oxazino[3,2-*a*]indole-10-carboxamide; SB-207266. $C_{22}H_{31}N_3O_2$; mol wt 369.51. C 71.51%, H 8.46%, N 11.37%, O 8.66%. Selective 5-HT₄ receptor antagonist. Prepn: L. M. Gaster, P. A. Wyman, *WO 9318036*; *eidem*, *US 5852014* (1993, 1998 both to SKB). *In vitro* and *in vivo* studies of bioavailability, pharmacokinetics, and duration: K. A. Wardle *et al.*, *Br. J. Pharmacol.* **118**, 665 (1996). Effect on gastrointestinal function: G. J. Sanger *et al.*, *Neurogastroenterol. Motil.* **10**, 271 (1998); *idem et al.*, *Br. J. Pharmacol.* **130**, 706 (2000). Clinical evaluation in irritable bowel syndrome: L. A. Houghton *et al.*, *Aliment. Pharmacol. Ther.* **13**, 1437 (1999).

White solid from ether, mp 110-113°.
Hydrochloride. [178273-87-5] SB-207266A. $C_{22}H_{31}N_3O_2$·HCl; mol wt 405.97. White crystals from ETOH/petrol, mp 254-256° (dec).

THERAP CAT: Gastroprokinetic

7506. Picaridin. [119515-38-7] 2-(2-Hydroxyethyl)-1-piperidinecarboxylic acid 1-methylpropyl ester; 1-methylpropyl 2-(2-hydroxyethyl)piperidine-1-carboxylate; hydroxyethyl isobutyl piperidine carboxylate; icaridin; KBR 3023; Bayrepel; Saltidin. $C_{12}H_{23}NO_3$; mol wt 229.32. C 62.85%, H 10.11%, N 6.11%, O 20.93%. Acts on certain olfactory receptor cell types to reduce the activating or attracting effect of odor sources. Prepn: B.-W. Krüger *et al.*, **EP 289842**; *eidem*, **US 4900834** (1988, 1990 both to Bayer). Synthesis and mode of action: J. Boeckh *et al.*, *Pestic. Sci.* **48**, 359 (1996). Field trials in mosquito repellency: H. H. Yap *et al.*, *J. Vector Ecol.* **23**, 62 (1998); A. Badolo *et al.*, *Trop. Med. Int. Health* **9**, 330 (2004). Physical, chemical and toxicological properties: *WHO Interim Specification for Icaridin* (WHO/IS/TC/ 668/2001) 16 pp. MS characterization: T. P. Knepper, *J. Chromatogr. A* **1046**, 159 (2004).

Colorless, nearly odorless liquid. mp <−170°. $bp_{1013\ hPa}$ 280°. d^{20} 1.07. n_D^{20} 1.4717. Flash pt: 142°C. Log P (octanol/water) at 20°: 2.11 (unbuffered), 2.23 (pH 4-9). Vapor pressure (hPa): 3.4×10^{-4} at 20°; 5.9×10^{-4} at 25°; 7.1×10^{-3} at 50°. Soly at 20° (g/l): water 8.2 (pH 4-9), 8.6 (unbuffered); acetone 7250. Viscosity at 20°: 135.5 mPa·s. LD_{50} in rats (mg/kg): 4743 orally; >2000 dermally (Yap). LC_{50} in rainbow trout (96 hr): 173 mg/l; in bobwhite quail (5 day diet): >5000 ppm a.i. in diet (WHO).

USE: Insect repellent.

7507. Picein. [530-14-3] 1-[4-(β-D-Glucopyranosyloxy)-phenyl]ethanone; *p*-(acetylphenyl)-β-D-glucopyranoside; 4′-(β-D-glucopyranosyloxy)acetophenone; *p*-hydroxyacetophenone-D-glucoside; piceoside; salinigrin; salicinerein; ameliaroside. $C_{14}H_{18}O_7$; mol wt 298.29. C 56.37%, H 6.08%, O 37.55%. In needles and sprouts of *Pinus picea* L., *Picea excelsa* Link., *Picea glehnii* Mast., *Coniferae*: Tanret, *Bull. Soc. Chim.* [3] **11**, 944 (1894); Kariyone *et al.*, *Yakugaku Zasshi* **79**, 394 (1959), *C.A.* **53**, 14096i (1953). In various willow barks, especially in bark of *Salix discolor* Muhl., *S. nigra* Marsh., *Salicaceae*: Jowett, *J. Chem. Soc.* **77**, 707 (1900); Nonomura, *J. Pharm. Soc. Jpn.* **75**, 80 (1955). In English mistletoe, *Amelanchier vulgaris* Moench., *Rosaceae*: Bridel *et al.*, *Compt. Rend.* **187**, 56 (1928). Synthesis: Mauthner, *J. Prakt. Chem.* [2] **85**, 564 (1912).

Needles or prisms from methanol; mp 195-196°. $[\alpha]_D^{23}$ −88° (c = 1). One gram dissolves in 50 ml water at 15°, in 1 ml boiling water, in ~650 ml abs alc at 15°, in ~40 ml boiling abs alc, in 140 ml ethyl acetate at 15°. Sol in hot glacial acetic acid. Practically insol in ether, chloroform. Hydrolysis with dil mineral acids or with emulsin yields D-glucose and *p*-hydroxyacetophenone (*Piceol*). Alkaline hydrolysis yields levoglucosan and *p*-hydroxyacetophenone: Montgomery *et al.*, *J. Org. Chem.* **10**, 194 (1945).

7508. Picene. [213-46-7] 3,4-Benzchrysene; 1,2,7,8-dibenzphenanthrene; dibenzo[*a,i*]phenanthrene; β,β-binaphthyleneethene. $C_{22}H_{14}$; mol wt 278.35. C 94.93%, H 5.07%. Found in tar oils from soft coal: Burg, *Ber.* **13**, 1834 (1880); Lang *et al.*, *ibid.* **97**, 494 (1964); in petroleum residues from the cracking process: Meyer, Hofmann, *Monatsh. Chem.* **37**, 681 (1916). Synthesis from cholic acid by selenium dehydrogenation: Ruzicka *et al.*, *Helv. Chim. Acta* **17**, 200 (1934); from *o*-xylylenedicyanide with *o*-nitrobenzaldehyde by a double Pschorr synthesis: Waldmann, Pitschak, *Ann.* **527**, 183 (1937); by a diene synthesis with tetrahydrodinaphthyl and maleic anhydride: Weidlich, *Ber.* **71**, 1203 (1938); from 9,10-dihydrophenanthrene: Phillips, *J. Am. Chem. Soc.* **75**, 3223 (1953); from ethyl 4,6-dioxoheptane-1,5-dicarboxylate: Nasipuri, *Chem. Ind. (London)* **1956**, 795.

Fluorescent plates from ethyl acetate, mp 366-367°. bp 518-520°. Absorption spectrum: Mayneord, Roe, *Proc. R. Soc. London* **A152**, 319 (1935). Difficultly sol in most solvents. Somewhat sol in boiling benzene, chloroform, glacial acetic acid; more sol in cumene (isopropylbenzene).

7509. Picloram. [1918-02-1] 4-Amino-3,5,6-trichloro-2-pyridinecarboxylic acid; 4-amino-3,5,6-trichloropicolinic acid; Pinene; Tordon. $C_6H_3Cl_3N_2O_2$; mol wt 241.45. C 29.85%, H 1.25%, Cl 44.05%, N 11.60%, O 13.25%. Plant growth regulator. Prepn: H. Johnston, M. S. Tomita, **BE 628487**; *eidem*, **US 3285925** (1963, 1966 both to Dow). Activity: J. W. Hamaker *et al.*, *Science* **141**, 363 (1963). Carcinogenicity study: M. D. Reuber, *J. Toxicol. Environ. Health* **7**, 207 (1981). Review of properties and analytical methods: J. R. Ramsey, *Anal. Methods Pestic. Plant Growth Regul. Food Addit.* **5**, 507-525 (1967); of toxicology and environmental fate: W. R. Mullison, *Proc. West. Soc. Weed Sci.* **38**, 21-92 (1985).

White powder, mp 218-219° with decomp starting at 190°. Readily sublimes at 190° under 12 mm Hg. Vapor pressure at 45°: 1.07 $\times 10^{-6}$ mm Hg. pKa 4.1. Soly at 25° (ppm): acetone 19800; ethanol 10500; acetonitrile 1600; ethyl ether 1200; water 430; benzene 200; carbon disulfide 50. LD_{50} in rats, mice, rabbits, guinea pig, chickens, sheep, cattle (mg/kg): 8200, 2000-4000, 2000, 3000, 6000, >1000, >750 orally (Mullison).

Caution: Potential symptoms of overexposure are irritation of eyes, skin, respiratory system; nausea. *See NIOSH Pocket Guide to Chemical Hazards* (DHHS/NIOSH 97-140, 1997) p 258.

USE: Herbicide.

7510. Picloxydine. [5636-92-0] N^1,N^4-Bis[[(4-chlorophenyl)amino]iminomethyl]-1,4-piperazinedicarboximidamide; *N,N″*-bis[(*p*-chlorophenyl)amidino]-1,4-piperazinedicarboxamidine; 1,4-bis(*N*4-*p*-chlorophenylamidinoamidinyl)piperazine; 1,1′-[1,4-piperazinediylbis(imidocarbonyl)]bis[3-(*p*-chlorophenyl)guanidine]. $C_{20}H_{24}Cl_2N_{10}$; mol wt 475.38. C 50.53%, H 5.09%, Cl 14.91%, N 29.46%. Heterocyclic biguanide with antibacterial activity. Prepn: J. W. James, L. F. Wiggins, **GB 855017**; *eidem*, **US 3101336** (1960, 1963 both to Aspro-Nicholas). Prepn and antibacterial spectrum: J.

W. James *et al.*, *J. Med. Chem.* **11**, 942 (1968). Bactericidal effect in disinfectant formulations: A. M. Gordon, *J. Clin. Pathol.* **22**, 496 (1969). Mode of action study: B. D. Rawal, J. V. Hardy, *Microbios* **14**, 135 (1974). *In vitro* activity vs *Chlamydia trachomatis:* D. Thomas *et al.*, *Pathol. Biol.* **32**, 544 (1984).

Dihydrochloride. [19803-62-4] Vitabact. $C_{20}H_{24}Cl_2N_{10}\cdot2HCl$; mol wt 548.30. Crystals from water, mp 274°. LD_{50} in mice (mg/kg): 150 i.p. (James).

THERAP CAT: Antibacterial (topical).

7511. Picolinafen. [137641-05-5] *N*-(4-Fluorophenyl)-6-[3-(trifluoromethyl)phenoxy]-2-pyridinecarboxamide; 4′-fluoro-6-[(α,-α,α-trifluoro-*m*-tolyl)oxy]picolinanilide; AC-900001; Pico. C_{19}-$H_{12}F_4N_2O_2$; mol wt 376.31. C 60.64%, H 3.21%, F 20.19%, N 7.44%, O 8.50%. Aryloxypicolinamide herbicide for use in barley and wheat. Inhibits phytoene desaturase, an enzyme involved in carotenoid biosynthesis. Prepn: C. J. Foster *et al.*, **EP 447004**; *eidem*, **US 5294597** (1991, 1994 both to Shell). Comprehensive description of properties and field trials: R. H. White *et al.*, *Brighton Crop Prot. Conf. - Weeds* **1999**, 47-52.

White to chalky-white, finely crystalline solid, mp 107.2-107.6°. Vapor pressure (20°): 1.66×10^{-7} Pa (est.). Log P (octanol/water): 5.36-5.43 (pH 5-9). Soly in water (20°): $3.8\text{-}4.7 \times 10^{-5}$ g/l. LD_{50} in rats (mg/kg): >5000 orally; >4000 dermally. LC_{50} in rats (mg/l): >5.9 by inhalation. LC_{50} in rainbow trout (96 hr): 0.281 mg/l (White).

USE: Herbicide.

7512. α-Picoline. [109-06-8] 2-Methylpyridine. C_6H_7N; mol wt 93.13. C 77.38%, H 7.58%, N 15.04%. Found in coal tar and in bone oil. Synthesis (40-50% yield) from cyclohexylamine with excess ammonia and zinc chloride at 350°: Nordt, *PB Report 704* (1941). Prepn from ethylene-mercuric acetate adduct and ammonia water (70% yield): Gumboldt, Feichtinger, *Z. Naturforsch.* **4b**, 123 (1949). Toxicity study: H. F. Smyth *et al.*, *Arch. Ind. Hyg. Occup. Med.* **4**, 119 (1951). Physical properties: Biddiscombe *et al.*, *J. Chem. Soc.* **1954**, 1957. Comprehensive review of methods for condensing aldehydes, ketones, ethylene, butadiene, etc., with ammonia: F. Brody, P. R. Ruby in A. Weissberger, *The Chemistry of Heterocyclic Compounds* **vol. 14**, part I (New York, 1960) pp 99-589.

Colorless liquid; strong unpleasant odor. mp −70°. bp 128-129°. d_4^{15} 0.950; n_D^{20} 1.501. Freely sol in water; miscible with alcohol, ether. LD_{50} orally in rats: 1.41 g/kg (Smyth).

Caution: Irritating to respiratory tract.

USE: Solvent; intermediate in the dye and resins industries.

7513. β-Picoline. [108-99-6] 3-Methylpyridine. C_6H_7N; mol wt 93.13. C 77.38%, H 7.58%, N 15.04%.

Colorless liq; sweetish, not unpleasant odor. d_4^{15} 0.9613; bp 143-144°; n_D^{24} 1.5043. Miscible with water, alcohol, ether.

USE: Solvent; intermediate in the dye and resins industries; in the manufacture of insecticides, waterproofing agents, niacin, and niacinamide.

7514. γ-Picoline. [108-89-4] 4-Methylpyridine; 4-picoline. C_6H_7N; mol wt 93.13. C 77.38%, H 7.58%, N 15.04%. Found in coal tar, in bone oil, in urine of horses. Isoln from technical picolines: Flaschner, *J. Chem. Soc.* **95**, 669 (1909). Toxicity study: H. F. Smyth *et al.*, *Arch. Ind. Hyg. Occup. Med.* **10**, 61 (1954).

Liquid. Obnoxious, sweetish odor. *Flammable.* Turns brown if not very pure. d_4^{15} 0.9571. bp_{760} 145°. n_D^{17} 1.5064. Kb at 25° = 1.1×10^{-8}. Sol in water, alcohol, ether. LD_{50} orally in rats: 1.29 g/kg (Smyth).

USE: Manuf isonicotinic acid and derivatives. In waterproofing agents for fabrics. As solvent for resins.

7515. Picolinic Acid. [98-98-6] 2-Pyridinecarboxylic acid; *o*-pyridinecarboxylic acid. $C_6H_5NO_2$; mol wt 123.11. C 58.54%, H 4.09%, N 11.38%, O 25.99%. Naturally occuring isomer of nicotinic acid; by-product of tryptophan catabolism. Prepn: Singer, McElvain, *Org. Synth.* **coll. vol. III**, 740 (1955). *Review:* Oliveto, "Pyridinecarboxylic Acids" in *Pyridine and Its Derivatives* **Part 3**, E. Klingsberg, Ed. (Interscience, New York, 1962) p 179. Review of immunobiology and effect on macrophage activation: G. Melillo *et al.*, *Adv. Exp. Med. Biol.* **398**, 135-141 (1996). HPLC determn in biological fluids: C. Dazzi *et al.*, *J. Chromatogr. B* **751**, 61 (2001).

Needles from water, alcohol or benzene, mp 134-136°. pK 5.4. Sublimes. Very sol in glacial acetic acid; practically insol in ether, chloroform, carbon disulfide.

Hydrochloride. Crystals from abs ethanol/ether, mp 210-212° (slow heating).

7516. Picoperine. [21755-66-8] *N*-Phenyl-*N*-[2-(1-piperidinyl)ethyl]-2-pyridinemethanamine; 1-[2-[*N*-(2-pyridylmethyl)anilino]ethyl]piperidine; *N*-(2-pyridylmethyl)-*N*-phenyl-*N*-2-(piperidinoethyl)amine; 1-[2-[phenyl(2-pyridylmethyl)amino]ethyl]piperidine; *N*-(2-picolyl)-*N*-phenyl-*N*-(2-piperidinoethyl)amine; *N*-(2-piperidinoethyl)-*N*-(2-pyridylmethyl)aniline; *N*-phenyl-*N*-(2-pyridylmethyl)-2-piperidinoethylamine; picoperidamine; TAT-3. C_{19}-$H_{25}N_3$; mol wt 295.43. C 77.25%, H 8.53%, N 14.22%. Prepn: **FR 1511398** and **US 3471501** (1968 and 1969 to Takeda); Miyano *et al.*, *J. Med. Chem.* **13**, 704 (1970). Prepn of salts: Sawa *et al.*, *Takeda Kenkyusho* **29**(2), 275 (1970), *C.A.* **73**, 98758h (1970); Masuda *et al.*, **DE 1935172** (1970 to Takeda), *C.A.* **73**, 98818c (1970). Pharmacology: Y. Kasé *et al.*, *Arzneim.-Forsch.* **19**, 1916 (1969); *eidem, ibid.* **20**, 37 (1970). *Review:* *Jpn. Med. Gaz.* **8**(11), 9 (1971).

Pale yellow liquid, bp$_4$ 195-196°.

Hydrochloride. [24699-40-9] Coben. C$_{19}$H$_{25}$N$_3$.HCl; mol wt 331.89. Water-soluble white crystalline powder, odorless with slightly bitter taste, mp 183-185°. Freely sol in ethanol and chloroform. Slightly sol in acetone, dioxane, benzene. Practically insol in hexane, ether. LD$_{50}$ in mice (mg/kg): 210 s.c.; 85 i.p.; 17 i.v.; 240 orally (Kasé, 1969).

Tripalmitate. [24656-22-2] Coben P. C$_{19}$H$_{25}$N$_3$.3C$_{16}$H$_{32}$O$_2$; mol wt 1064.72. White, tasteless, odorless crystalline powder, mp 57-60°. Freely sol in ethanol, chloroform; slightly sol in dioxane, acetone, benzene. Practically insol in hexane, water. LD$_{50}$ orally in mice: 1900 mg/kg (Kasé, 1969).

THERAP CAT: Antitussive.

7517. Picoplatin. [181630-15-9] (*SP*-4-3)-Amminedichloro(2-methylpyridine)platinum; *cis*-amminedichloro(2-methylpyridine)platinum(II); AMD-473; JM-473; NX-473; ZD-473. C$_6$H$_{10}$-Cl$_2$N$_2$Pt; mol wt 376.14. C 19.16%, H 2.68%, Cl 18.85%, N 7.45%, Pt 51.86%. Third generation, sterically hindered Pt-complex that exhibits reduced reactivity with sulfur ligands such as glutathione, *q.v.*; designed to overcome acquired Pt-chemotherapy resistance. Prepn: B. A. Murrer, **EP 727430** (1996 to AnorMED; Inst. of Cancer Research); *idem*, **US 5665771** (1997 to Johnson Matthey). Pharmacology and pharmacokinetics: F. I. Raynaud *et al.*, *Clin. Cancer Res.* **3**, 2063 (1997). Cytotoxicity and DNA binding activity: J. Holford *et al.*, *Anti-Cancer Drug Des.* **13**, 1 (1998). *In vitro* circumvention of cisplatin resistance: *idem, et al,. Br. J. Cancer* **77**, 366 (1998); of radiosensitizing activity: G. P. Raaphorst *et al.*, *Anticancer Res.* **24**, 613 (2004). LC/MS determn in plasma: T. Oe *et al.*, *Anal. Chem.* **74**, 591 (2002); in urine: *eidem, J. Chromatogr. B* **792**, 217 (2003). Series of articles on clinical evaluations: *Eur. J. Cancer* **38**, Suppl. S1-S31 (2002). Review of early development: L. R. Kelland *et al.*, *J. Inorg. Biochem.* **77**, 111-115 (1999); and therapeutic potential: M. P. Hay, *Curr. Opin. Investig. Drugs* **1**, 263-266 (2000).

Pale yellow solid. *Sensitive to light.* uv max (0.15M sodium chloride): 268 nm. A$_{250}$/A$_{268}$ = 0.62. LD$_{50}$ in mice (mg/kg): 43 i.p., 560 orally (Murrer).

THERAP CAT: Antineoplastic.

7518. Picosulfate Sodium. [10040-45-6] 4,4′-(2-Pyridinylmethylene)bisphenol 1,1′-bis(hydrogen sulfate) sodium salt (1:2); 4,4′-(2-pyridylmethylene)diphenolbis(hydrogen sulfate) (ester) disodium salt; 4,4′-(2-picolylidene)bis(phenylsulfuric acid) disodium salt; 2-picolylidenebis(*p*-phenyl sodium sulfate); disodium 4,4′-disulfoxydiphenyl-(2-pyridyl)methane; picosulfol; sodium picosulfate; Guttalax-Fher; Laxoberal; Laxoberon; Neopax; Pico-Salax. C$_{18}$H$_{13}$-NNa$_2$O$_8$S$_2$; mol wt 481.40. C 44.91%, H 2.72%, N 2.91%, Na 9.55%, O 26.59%, S 13.32%. Prepn: Pala, **FR M5832** and **US 3528986, US 3558643** (1968, 1970, 1971 all to Ist. de Angeli); Seeger, Machleidt, **DE 1904322** (1970 to Thomae). Synthesis and physical-chemical data: Pala *et al.*, *Helv. Chim. Acta* **51**, 1164 (1968). Pharmacology: Pala *et al.*, *Arch. Int. Pharmacodyn. Ther.* **164**, 356 (1966). Metabolism: Perego *et al.*, *Arzneim.-Forsch.* **19**, 1889 (1969).

White crystalline solid from ethanol or methanol, mp 272- 275° (dec). uv max (H$_2$O): 218, 262 nm (ε 20450, 6075). Readily sol in

water; slightly sol in alcohol. Practically insol in most organic solvents.

THERAP CAT: Cathartic.

7519. Picotamide. [32828-81-2] 4-Methoxy-N^1,N^3-bis(3-pyridinylmethyl)-1,3-benzenedicarboxamide; 4-methoxy-N,N'-bis(3-pyridylmethyl)isophthalamide; N,N'-bis(3-picolyl)-4-methoxyisophthalamide; G-137. C$_{21}$H$_{20}$N$_4$O$_3$; mol wt 376.42. C 67.01%, H 5.36%, N 14.88%, O 12.75%. Prepn: R. Selleri *et al.*, *Chim. Ther.* **6**, 203 (1971); G. Orzalesi, R. Selleri, **FR 2100850**, *C.A.* **77**, 164502f (1972); and use as inhibitor of blood platelet aggregation: *eidem*, **US 3973026** (1972, 1976 both to Soc. Italo-Brit. L. Manetti-H. Roberts). Crystal structure of monohydrate: E. Foresti *et al.*, *Acta Crystallogr.* **C42**, 220 (1986). Mechanism of action study: M. Berrettini *et al.*, *Boll. Soc. Ital. Biol. Sper.* **59**, 309 (1983). Clinical studies: R. Schmutzler *et al.*, *Age Ageing* **7**, 246 (1978); V. Coto *et al.*, *Minerva Cardioangiol.* **34**, 601 (1986).

Crystalline powder from benzene, mp 124°. LD$_{50}$ i.p. in male mice: 1205 mg/kg (Orzalesi, Selleri, 1976).

Monohydrate. [80530-63-8] Plactidil. C$_{21}$H$_{20}$N$_4$O$_3$.H$_2$O; mol wt 394.43.

THERAP CAT: Antithrombotic; fibrinolytic; anticoagulant.

7520. Picoxystrobin. [117428-22-5] ($αE$)-$α$-(Methoxymethylene)-2-[[[6-(trifluoromethyl)-2-pyridinyl]oxy]methyl]benzeneacetic acid methyl ester; methyl (*E*)-2-[2-[6-(trifluoromethyl)pyridin-2-yloxymethyl]phenyl]-3-methoxyacrylate; ZA-1963; Acanto. C$_{18}$-H$_{16}$F$_3$NO$_4$; mol wt 367.32. C 58.86%, H 4.39%, F 15.52%, N 3.81%, O 17.42%. Strobilurin fungicide for use in cereal crops. Prepn: J. M. Clough *et al.*, **EP 278595**; *eidem*, **US 5021581** (1988, 1991 both to ICI). Comprehensive description: J. R. Godwin *et al.*, *Brighton Crop Prot. Conf. - Pests Dis.* **2000**, 533-540; M. Hiemer *et al.*, *Gesunde Pflanz.* **53**, 191-195 (2001).

Solid, mp 75°. d^{20} 1.4. Vapor pressure at 20°: 5.5 × 10^{-9} kPa. Log P (*n*-octanol/water) at 20°: 3.6. Soly in water (20°): 3.1 mg/l. MLD in rats (mg/kg): >5000 orally; >2000 dermally; MLC in rats (mg/l): 2.12 by inhalation (Godwin).

USE: Agricultural fungicide.

7521. Picramic Acid. [96-91-3] 2-Amino-4,6-dinitrophenol; 4,6-dinitro-2-aminophenol. C$_6$H$_5$N$_3$O$_5$; mol wt 199.12. C 36.19%, H 2.53%, N 21.10%, O 40.17%. Prepd from picric acid, concd NH$_4$-OH, and H$_2$S followed by acetic acid neutralization of ammonium salt: Egerer, *J. Biol. Chem.* **35**, 565 (1918). Prepn of ammonium salt from picric acid, aqueous NH$_3$-soln, and ammonium sulfide: Dehn, **US 1472791** (1924).

Dark red needles from alcohol, prisms from chloroform. mp 169-170°. At 22-25°, 0.065 g dissolves in 100 ml H$_2$O; not much more sol in hot water. Sparingly sol in ether, chloroform; moderately sol in alcohol; sol in benzene, glacial acetic acid, aniline. Flashes at 210°; in contact with open flame in glass tube or beaker, ignites rapidly and burns relatively fast: Blinov, *Khim. Prom.* **1959**, 419, *C.A.* **55**, 6866d (1961).

Caution: Toxic symptoms similar to 2,4-dinitrophenol, *q.v.*

USE: Manuf of azo dyes; rarely as indicator (yellow with acids, red with alkalies); reagent for albumin.

7522. Picric Acid. [88-89-1] 2,4,6-Trinitrophenol; 1-hydroxy-2,4,6-trinitrobenzene; picronitric acid; carbazotic acid; nitroxanthic acid. $C_6H_3N_3O_7$; mol wt 229.10. C 31.46%, H 1.32%, N 18.34%, O 48.88%. Prepd by sulfonating phenol then treating with nitric acid: Olsen, Goldstein, *Ind. Eng. Chem.* **16**, 66 (1924); by treating benzene with nitric acid and mercuric nitrate: Teeters, Mueller, US 2455322 (1948 to Allied Chem.); by nitration of 2-*tert*-butyl-4,6-dinitrophenol: Ley, Müller, *Ber.* **89**, 1402 (1956). Crystal structure: E. N. Duesler *et al.*, *Cryst. Struct. Commun.* **7**, 449 (1978).

Pale yellow, odorless, intensely bitter crystals. d 1.763. mp 122-123°. Explodes above 300°. One gram dissolves in 78 ml water, 15 ml boiling water, 12 ml alc, 10 ml benzene, 35 ml chloroform, 65 ml ether. *Keep in a cool place and remote from fire. Explodes when rapidly heated or by percussion. Note:* For safety in transportation, 10-20% water is usually added.

Caution: Potential symptoms of overexposure are irritation of eyes, skin; sensitization dermatitis; yellow stained hair, skin; weakness, myalgia, anuria, polyuria; bitter taste, GI disturbances; hepatitis; hematuria, albuminuria, nephritis. *See NIOSH Pocket Guide to Chemical Hazards* (DHHS/NIOSH 97-140, 1997) p 258; *Clinical Toxicology of Commercial Products*, R. E. Gosselin *et al.*, Eds. (Williams & Wilkins, Baltimore, 5th ed., 1984) Section II, p 197.

USE: Explosives; matches; in leather industry; electric batteries; etching copper; manuf colored glass; textile mordant. Reagent for prepn of organic derivatives for identification.

7523. Picrocrocin. [138-55-6] (4*R*)-4-(β-D-Glucopyranosyloxy)-2,6,6-trimethyl-1-cyclohexene-1-carboxaldehyde; saffron-bitter. $C_{16}H_{26}O_7$; mol wt 330.38. C 58.17%, H 7.93%, O 33.90%. From stigmas of *Crocus sativus* L., *Iridaceae.* Isoln: Kayser, *Ber.* **17**, 2228 (1884). Structure: Kuhn, Winterstein, *Ber.* **67**, 344 (1934). Exerts sex-determining influences in the plant organism: Kuhn, *Angew. Chem.* **53**, 1 (1940). Its moieties are glucose and safranal, *q.q.v.* Abs config: Buchecker, Eugster, *Helv. Chim. Acta* **56**, 1121 (1973). Synthesis: H. Mayer, J.-M. Santer, *Helv. Chim. Acta* **63**, 1463 (1980).

Crystals, mp 154-156°. $[\alpha]_D^{20}$ −58° (c = 0.6). Bitter taste. Alkali unstable. Sol in water, alcohol; slightly sol in chloroform, ether. Practically insol in petr ether, benzene.

7524. Picrolonic Acid. [550-74-3] 2,4-Dihydro-5-methyl-4-nitro-2-(4-nitrophenyl)-3*H*-pyrazol-3-one; 3-methyl-4-nitro-1-(*p*-nitrophenyl)-2-pyrazolin-5-one. $C_{10}H_8N_4O_5$; mol wt 264.20. C 45.46%, H 3.05%, N 21.21%, O 30.28%.

Yellow leaflets, mp 116-117°; dec at 125°. Sparingly sol in water; sol in alcohol.

USE: As reagent for alkaloids, tryptophan, phenylalanine, and for the detection and estimation of calcium.

7525. Picromycin. [19721-56-3] (3*R*,5*R*,6*S*,7*S*,9*R*,11*E*,13*S*,-14*R*)-14-Ethyl-13-hydroxy-3,5,7,9,13-pentamethyl-6-[[3,4,6-trideoxy-3-(dimethylamino)-β-D-*xylo*-hexopyranosyl]oxy]oxacyclotetradec-11-ene-2,4,10-trione; pikromycin; albomycetin; amaromycin. $C_{28}H_{47}NO_8$; mol wt 525.68. C 63.98%, H 9.01%, N 2.66%, O 24.35%. First macrolide antibiotic isolated. Isoln from *Actinomyces* spp by Lindenbein and Bauer: Brockmann, Henkel, *Naturwissenschaften* **37**, 138 (1950); *Ber.* **84**, 284 (1951); Brockmann, Bohne, US 2693433 (1954 to Schenley). Structure: Brockmann, Oster, *Ber.* **90**, 605 (1957); Anliker, Gubler, *Helv. Chim. Acta* **40**, 119, 1768 (1957). Studies in stereochemistry: Djerassi, Halpern, *J. Am. Chem. Soc.* **79**, 3926 (1957); Ogura *et al.*, *ibid.* **97**, 1930 (1975); *eidem*, *Tetrahedron* **37**, Suppl. 1, 165 (1981). Revised structure: Rickards, Smith, *Chem. Commun.* **1968**, 1049; Muxfeldt *et al.*, *J. Am. Chem. Soc.* **90**, 4748 (1968). Synthesis from narbonolide: Maezawa *et al.*, *J. Antibiot.* **26**, 771 (1973).

Very bitter, rectangular platelets from methanol, mp 169.5-170°. Stable to heat. $[\alpha]_D^{24}$ +8.2° (c = 3.5 in ethanol); $[\alpha]_D^{20}$ −33.5° (c = 2.07 in chloroform); $[\alpha]_D^{24}$ −50.2° (c = 6.3 in chloroform). uv max (ethanol): 225 nm (log ε 3.97). Rotary dispersion data: Djerassi, Halpern, *Tetrahedron* **3**, 268 (1958). Very sparingly sol in water, petr ether, carbon disulfide. Soly in ethanol: 3.5 g/100 ml at 20°. Freely sol in acetone, benzene, chloroform, ethyl acetate, dioxane. Moderately sol in ether, methanol.

7526. Picrotin. [21416-53-5] (1a*R*,2a*R*,3*S*,6*R*,6a*S*,8a*S*,8b*R*,-9*S*)-Hexahydro-2a-hydroxy-9-(1-hydroxy-1-methylethyl)-8b-methyl-3,6-methano-8*H*-1,5,7-trioxacyclopenta[*ij*]cycloprop[*a*]azulene-4,8(3*H*)-dione. $C_{15}H_{18}O_7$; mol wt 310.30. C 58.06%, H 5.85%, O 36.09%. Nontoxic component of picrotoxin. Prepn from picrotoxin: Meyer, Bruges, *Ber.* **31**, 2958 (1898). Structure studies: Bensted *et al.*, *J. Chem. Soc.* **1952**, 1042; Slater, Wilson, *ibid.* 1597; Holker *et al.*, *ibid.* **1958**, 2987. Total synthesis: E. J. Corey, H. L. Pearce, *Tetrahedron Lett.* **21**, 1823 (1980). Toxicity data: C. H. Jarboe *et al.*, *J. Med. Chem.* **11**, 729 (1968). *Review:* Porter, *Chem. Rev.* **67**, 441 (1967).

Fine needles from water; thick, shiny prisms from aq alc soln, mp 248-250°. $[\alpha]_D$ −64.7° (c = 2.31 in abs alc). Freely sol in abs alc, acetic acid, boiling water. Slightly sol in cold water. Practically insol in ether, chloroform, benzene. LD$_{50}$ i.p. in mice: 135 mg/kg (Jarboe).

7527. Picrotoxin. [124-87-8] (1a*R*,2a*R*,3*S*,6*R*,6a*S*,8a*S*,8b*R*,-9*S*)-Hexahydro-2a-hydroxy-9-(1-hydroxy-1-methylethyl)-8b-methyl-3,6-methano-8*H*-1,5,7-trioxacyclopenta[*ij*]cycloprop[*a*]azulene-4,8(3*H*)-dione compd. with (1a*R*,2a*R*,3*S*,6*R*,6a*S*,8a*S*,8b*R*,9*R*)-hexahydro-2a-hydroxy-8b-methyl-9-(1-methylethenyl)-3,6-methano-8*H*-1,5,7-trioxacyclopenta[*ij*]cycloprop[*a*]azulene-4,8(3*H*)-dione (1:1); Cocculin. $C_{30}H_{34}O_{13}$; mol wt 602.59. C 59.80%, H 5.69%, O 34.52%. Bitter principle isolated from the seed of *Anamirta cocculus* L. Wight & Arn., *Menispermaceae*, also found in *Tinomiscium philippinense* Diels. Molecular compd of one mole picrotoxinin and one mole picrotin, *q.q.v.*, into which it is readily separated. Extraction procedure: Clark, *J. Am. Chem. Soc.* **57**, 1111 (1935). Chemical bibliography: *Helv. Chim. Acta* **32**, 1859 (1949). Crystal and molecular structure: L. Dupont *et al., Acta Crystallogr.* **B32**, 2987 (1976). Toxicity study: I. Setnikar *et al., J. Pharmacol. Exp. Ther.* **128**, 176 (1960).

Intensely bitter shiny rhomboid leaflets, mp 203°. *Poisonous.* $[\alpha]_D^{16} -29.3°$ (c = 4 in abs ethanol). One gram dissolves in ~350 ml water, in ~5 ml boiling water, in 13.5 ml 95% ethanol, in ~3 ml boiling alcohol. Sparingly sol in ether, chloroform. Readily sol in strong ammonia water, in aq solns of NaOH. Highly toxic to fish. LD_{50} i.p. in mice: 7.2 mg/kg (Setnikar).

THERAP CAT: CNS and respiratory stimulant.

THERAP CAT (VET): CNS stimulant, antidote to barbiturates.

7528. Picrotoxinin. [17617-45-7] (1a*R*,2a*R*,3*S*,6*R*,6a*S*,8a*S*,-8b*R*,9*R*)-Hexahydro-2a-hydroxy-8b-methyl-9-(1-methylethenyl)-3,6-methano-8*H*-1,5,7-trioxacyclopenta[*ij*]cycloprop[*a*]azulene-4,8-(3*H*)-dione. $C_{15}H_{16}O_6$; mol wt 292.29. C 61.64%, H 5.52%, O 32.84%. Toxic component of picrotoxin, *q.v.* Prepn from picrotoxin: Horrmann, *Ber.* **45**, 2090 (1912). Structure: Conroy, *J. Am. Chem. Soc.* **73**, 1889 (1951); **79**, 5550 (1957); Craven, *Tetrahedron Lett.* no. 19, 21 (1960). Total synthesis: E. J. Corey, H. L. Pearce, *J. Am. Chem. Soc.* **101**, 5841 (1979). Structure-activity relationship: C. H. Jarboe *et al., J. Med. Chem.* **11**, 729 (1968). *Review:* Porter, *Chem. Rev.* **67**, 441 (1967).

Very bitter large prisms or small crystals contg water, mp 209.5°. $[\alpha]_D^{17} +4.4°$ (c = 4.28 in abs alc), +3.49° (c = 7.57 in acetone). Soluble in hot common organic solvents, in cold alcohol and chloroform. LD_{50} i.p. in mice: 3 mg/kg (Jarboe).

7529. Picryl Chloride. [88-88-0] 2-Chloro-1,3,5-trinitrobenzene. $C_6H_2ClN_3O_6$; mol wt 247.55. C 29.11%, H 0.81%, Cl 14.32%, N 16.97%, O 38.78%.

Almost white needles. d 1.797. mp 83°. Insol in water; freely sol in benzene, hot chloroform, boiling alcohol, slightly in ether or petr ether.

7530. Picumast. [39577-19-0] 7-[3-[4-[(4-Chlorophenyl)methyl]-1-piperazinyl]propoxy]-3,4-dimethyl-2*H*-1-benzopyran-2-one; 7-[3-[4-(*p*-chlorobenzyl)-1-piperazinyl]propoxy]-3,4-dimethylcoumarin; 1-(4-chlorobenzyl)-4-[3-(3,4-dimethylcoumarin-7-yloxy)propyl]piperazine. $C_{25}H_{29}ClN_2O_3$; mol wt 440.97. C 68.09%, H 6.63%, Cl 8.04%, N 6.35%, O 10.88%. Mast cell degranulation inhibitor with histamine antagonist activity. Prepn: E. C. Witte *et al.*, **DE 2123924**; *eidem*, **US 3810898** (1972, 1974 both to Boehringer, Mann.); E. C. Witte, *Arzneim.-Forsch.* **39**, 1309 (1989). Mechanism of action studies: O. H. Wilhelms, *Int. Arch. Allergy Appl. Immunol.* **82**, 544, 547 (1987). Series of articles on pharmacology, pharmacokinetics, and clinical trials: *Arzneim.-Forsch.* **39**, 1307-1376 (1989).

Crystals from isopropanol, mp 112-114° (Witte, 1974); also reported as mp 115-117° (Witte, 1989).

Dihydrochloride. [39577-20-3] BM-15100; Auteral. $C_{25}H_{29}$-$ClN_2O_3 \cdot 2HCl$; mol wt 513.88. Crystals from aq ethanol, mp 266-268°.

THERAP CAT: Antiallergic.

7531. Pidotimod. [121808-62-6] (4*R*)-3-[[(2*S*)-5-Oxo-2-pyrrolidinyl]carbonyl]-4-thiazolidinecarboxylic acid; 3-L-pyroglutamyl-L-thiazolidine-4-carboxylic acid; (*R*)-3-[(*S*)-5-oxoprolyl]-4-thiazolidinecarboxylic acid; PGT/1A; Axil; Onaka; Pigitil; Polimod. $C_9H_{12}N_2O_4S$; mol wt 244.27. C 44.25%, H 4.95%, N 11.47%, O 26.20%, S 13.12%. Peptide-like biological response modifier. Prepn: S. Poli, **EP 276752**; *idem*, **US 4839387** (1988, 1989 both to Poli). Immunostimulant effect in nude mice: A. Vacca *et al., Int. J. Immunother.* **9**, 85 (1993); on human peripheral blood monocytes: M. O. Borghi *et al., ibid.* **10**, 35 (1994). HPLC determn in plasma and urine: G. Coppi, M. Barchielli, *J. Chromatogr.* **563**, 385 (1991). Clinical trial in recurrent acute tonsillitis: P. Careddu *et al., Adv. Otorhinolaryngol.* **47**, 328 (1992). Toxicity study: G. Coppi *et al., Arzneim.-Forsch.* **44**, 1448 (1994). Series of articles on chemistry, pharmacology, toxicology and clinical experience: *ibid.* 1399-1530.

White or slightly ivory, odorless microcrystalline powder, mp 194-198° (dec). Also reported as white crystals from water, mp 192-194° (Poli). $[\alpha]_D^{25} -150°$ (c = 2 in 5*N* HCl). pKa 3.03. Soly (g/l): water 37.8; methanol 13.8; ethanol 4.4; DMF 72.4. Practically insol in chloroform, hexane. LD_{50} in mice and rats (mg/kg): >4000 i.v.; >4000 i.m.; >8000 i.p.; >8000 orally (Coppi).

THERAP CAT: Immunomodulator.

7532. Pifarnine. [56208-01-6] 1-(1,3-Benzodioxol-5-ylmethyl)-4-(3,7,11-trimethyl-2,6,10-dodecatrien-1-yl)piperazine; 1-piperonyl-4-(3,7,11-trimethyl-2,6,10-dodecatrienyl)piperazine; U-27; Pifazin. $C_{27}H_{40}N_2O_2$; mol wt 424.63. C 76.37%, H 9.50%, N 6.60%, O 7.54%. Non-anticholinergic gastric anti-secretory agent. Prepn: S. Tricerri *et al.*, **DE 2310044** corresp to **US 3875163** (1973, 1975 both to Pierrel); *eidem*, *Eur. J. Med. Chem.* **9**, 555 (1974). Pifarnine is a mixture of four stereoisomers: *ZZ, EZ, ZE, EE*. Separation does not give compounds with significantly different activity/toxicity ratios from the mixture. Stereochemistry and pharmacology of the four isomers: G. Guadagnini *et al., ibid.* **10**, 585 (1975). Pharmacological study of pifarnine: A. Bianchetti *et al., Arzneim.-Forsch.* **25**, 580 (1975). Physical-chemical and analytical studies: G. Guadagnini *et al., Pharm. Ind.* **38**, 296 (1976). Absorption, distribution, excretion: M. Riva *et al., Farmaco Ed. Prat.* **34**, 542 (1979). Clinical studies: A. Porro *et al., ibid.* 85; M. Petrillo *et al., Curr. Ther. Res.* **25**, 457 (1979).

(*E,E*)-isomer

Light yellow viscous liquid with a slight odor and bitter taste. uv max (ethanol): 287 nm ($E_{1cm}^{1\%}$ 94.6). n_D^{20} 1.5235. Relative density at 20°: 1.013-1.015. pK_1 4.10; pK_2 3.25. Readily sol in most org solvents. Slightly sol in aq solns of organic acids. Practically insol in alkali, water. LD_{50} in mice, rats (mg/kg): 2175, 2610 orally; 40.6,

33.3 i.v. (Bianchetti). Approx LD$_{50}$ i.p. in mice: 500 mg/kg (Tricerri, 1974).

THERAP CAT: Antiulcerative.

7533. Pifithrin-α. [63208-82-2] 1-(4-Methylphenyl)-2-(4,5,-6,7-tetrahydro-2-imino-3(2H)-benzothiazolyl)ethanone hydrobromide (1:1); 2-(2-imino-4,5,6,7-tetrahydrobenzothiazol-3-yl)-1-p-tolylethanone hydrobromide; PFT-α. C$_{16}$H$_{18}$N$_2$OS.HBr; mol wt 367.31. C 52.32%, H 5.21%, N 7.63%, O 4.36%, S 8.73%, Br 21.75%. Small molecule inhibitor of p53-mediated gene activation and apoptosis; the name pifithrin is an abbreviation for p-fifty three inhibitor. Converted *in vitro* to the planar tricyclic condensation product, pifithrin-β. Prepn: A. Singh *et al.*, *Indian J. Chem.* **14B**, 997 (1976). Identification of p53 inhibition: P. G. Komarov *et al.*, *Science* **285**, 1733 (1999). Synthesis and neuroprotective activity: X. Zhu *et al.*, *J. Med. Chem.* **45**, 5090 (2002). Inhibition of heat shock and glucocorticoid signal transduction pathways: E. A. Komarova *et al.*, *J. Biol. Chem.* **278**, 15465 (2003). Characterization as an aryl hydrocarbon receptor agonist: M. S. Hoagland *et al.*, *J. Pharmacol. Exp. Ther.* **314**, 603 (2005). Conversion to pifithrin-β: R. K. Gary, D. A. Jensen, *Mol. Pharm.* **2**, 462 (2005).

mp 182°. pKa 9.11. Sol in water, DMSO.

Pifithrin-β. [60477-34-1]; [511296-88-1] (hydrobromide). 5,6,7,8-Tetrahydro-2-(4-methylphenyl)imidazo[2,1-b]benzothiazole; 2-(4-methylphenyl)imidazo[2,1b]-5,6,7,8-tetrahydrobenzothiazole; 2-p-tolyl-5,6,7,8-tetrahydrobenzo[d]imidazo[2,1-b]thiazole; cyclic pifithrin-α. C$_{16}$H$_{16}$N$_2$S; mol wt 268.38. Crystal structure: W. Clegg, C. Jamieson, *Acta Crystallogr.* **E61**, o1486 (2005). Crystals from methanol. mp 185°. pKa 4.36. Log P (octanol/water): 4.26. Crystal density: 1.318 Mg/m^3.

USE: Biochemical tool in studies of p53 function and apoptosis.

7534. Piketoprofen. [60576-13-8] 3-Benzoyl-α-methyl-N-(4-methyl-2-pyridinyl)benzeneacetamide; m-benzoyl-N-(4-methyl-2-pyridyl)hydratropamide; 2-(3-benzoylphenyl)-N-(4-methyl-2-pyridyl)propionamide; Calmatel (aerosol). C$_{22}$H$_{20}$N$_2$O$_2$; mol wt 344.41. C 76.72%, H 5.85%, N 8.13%, O 9.29%. Derivative of ketoprofen, *q.v.* used topically as cream or aerosol. Prepn: R. G. W. Spickett *et al.*, *GB 1436502* (1976 to Antonio Gallardo, S.A.). Pharmacology and clinical efficacy: E. Tarrus *et al.*, *2nd Eur. Congr. Biopharm. Pharmacokinet.* **1**, 483 (1984).

Oil, sol in methylene chloride, ethanol. Insol in water.

Hydrochloride. [59512-37-9] Calmatel (cream). C$_{22}$H$_{20}$N$_2$O$_2$.HCl; mol wt 380.87. mp 180-182°.

THERAP CAT: Anti-inflammatory (topical).

7535. Pildralazine. [64000-73-3] 1-[(6-Hydrazinyl-3-pyridazinyl)methylamino]-2-propanol; 6-[(2-hydroxypropyl)methylamino]-3(2H)-pyridazinone hydrazone; 3-hydrazino-6-[(2-hydroxypropyl)methylamino]pyridazine; propyldazine; propildazine. C$_8$H$_{15}$N$_5$O; mol wt 197.24. C 48.72%, H 7.67%, N 35.51%, O 8.11%. Peripheral vasodilator with hypotensive activity; related to hydralazine, *q.v.* Prepn: G. Pifferi, *DE 2154245*; *idem*, *US 3769278* (1972, 1973 both to I.S.F.); G. Pifferi *et al.*, *J. Med. Chem.* **18**, 741 (1975). Pharmacology in animals: L. Dorigotti *et al.*, *Pharmacol. Res. Commun.* **8**, 295 (1976); L. Dorigotti *et al.*, *Arzneim.-Forsch.* **34**, 876

(1984); in humans: L. Terzoli *et al.*, *Boll. Soc. Ital. Cardiol.* **22**, 1053 (1977). GLC determn in plasma: P. Ventura *et al.*, *J. Chromatogr.* **161**, 237 (1978). Clinical evaluation in essential hypertension: R. Pellegrini, G. Abbondati, *Farmaco Ed. Prat.* **32**, 19 (1977). Pharmacokinetics and tissue binding: E. Noack *et al.*, *Arzneim.-Forsch.* **37**, 407 (1987).

Dihydrochloride. [56393-22-7] ISF-2123; Atensil. C$_8$H$_{15}$N$_5$-O.2HCl; mol wt 270.16. Crystals from ethanol, mp 206-209° (dec). LD$_{50}$ in mice, rats (mg/kg): 357, 355 i.p.; 1170, 1230 orally (Dorigotti, 1976).

THERAP CAT: Antihypertensive.

7536. Pilocarpine. [92-13-7] (3S,4R)-3-Ethyldihydro-4-[(1-methyl-1H-imidazol-5-yl)methyl]-2(3H)-furanone; Ocusert Pilo. C$_{11}$H$_{16}$N$_2$O$_2$; mol wt 208.26. C 63.44%, H 7.74%, N 13.45%, O 15.36%. Cholinergic principle from *Pilocarpus jaborandi* Holmes, *Rutaceae*. Isoln: Petit, Polanovski, *Bull. Soc. Chim.* [3] **17**, 557, 702 (1897). Structure: Jowett, *J. Chem. Soc.* **77**, 473, 851 (1900); **83**, 438 (1903). Stereoisomeric with isopilocarpine: Polonovski, Polonovski, *Bull. Soc. Chim.* [4] **31**, 1314 (1922). Has the *cis* configuration; isopilocarpine is *trans*: Zav'yalov, *Dokl. Akad. Nauk SSSR* **82**, 257 (1952). Absolute configuration: Hill, Barcza, *Tetrahedron* **22**, 2889 (1966). Synthesis: Preobrashenski *et al.*, *Ber.* **66**, 1187 (1933); Samokhvalov, *Med. Prom. SSSR* **11**, no. 2, 10 (1957); DeGraw, *Tetrahedron* **28**, 967 (1972); Link, Bernauer, *Helv. Chim. Acta* **55**, 1053 (1972). Stereoselective synthesis: A. Noordam *et al.*, *Rec. Trav. Chim.* **98**, 467 (1979). *Review:* Langenbeck, *Angew. Chem.* **60**, 297 (1948); van Rossum *et al.*, *Experientia* **16**, 373 (1960). Toxicity studies: Beccari, *Boll. Chim. Farm.* **106**, 8 (1967). Comprehensive description: A. A. Al-Badr, H. Y. Aboul-Enein, *Anal. Profiles Drug Subs.* **12**, 385-432 (1983). Clinical trial in Sjögren's syndrome: F. B. Vivino *et al.*, *Arch. Intern. Med.* **159**, 174 (1999); in radiation-induced xerostomia: J.-C. Horiot *et al.*, *Radiother. Oncol.* **55**, 233 (2000).

Oil or crystals. Hygroscopic. mp 34°. bp$_5$ 260° (partial conversion to isopilocarpine). [α]$_D^{18}$ +106° (c = 2). pK$_1$ (20°) 7.15; pK$_2$ (20°) 12.57. Sol in water, alc, chloroform; sparingly sol in ether, benzene. Practically insol in petr ether.

Hydrochloride. [54-71-7] Akarpine; Almocarpine; Isopto Carpine; Pilogel; Pilopine HS; Pilostat; Salagen. C$_{11}$H$_{16}$N$_2$O$_2$.HCl; mol wt 244.72. Hygroscopic crystals from alc, mp 204-205°. [α]$_D^{18}$ +91° (c = 2). Very sol in water; freely sol in alc; slightly sol in chloroform. Insol in ether. *Keep well closed and protected from light.*

Nitrate. [148-72-1] Chibro Pilocarpine; Licarpin; Pilo; Pilofrin; Pilagan. C$_{11}$H$_{16}$N$_2$O$_2$.HNO$_3$; mol wt 271.27. White crystals; affected by light. mp 173.5-174.0° (dec). *Poisonous.* [α]$_D$ +77 to +83° (c = 10). One gram dissolves in 4 ml water, 75 ml alc. Insol in chloroform, ether. *Incompat.* Silver nitrate, mercury bichloride, iodides, gold salts, tannin, calomel, KMnO$_4$, alkalies.

Isopilocarpine. β-Pilocarpine. Hygroscopic oily liquid or prisms. bp$_{10}$ 261°. [α]$_D^{18}$ +50° (c = 2). pK$_1$ (18°) 7.17. Miscible with water and alc; very sol in chloroform; less sol in benzene, ether. Almost insol in petr ether.

Isopilocarpine hydrochloride hemihydrate. C$_{11}$H$_{16}$N$_2$O$_2$.-HCl.½H$_2$O. Scales from alc + ether, mp 127°; when anhydr, mp 161°. [α]$_D^{18}$ +39° (c = 5). Sol in 0.27 part water; 2.1 parts alc.

Isopilocarpine nitrate. C$_{11}$H$_{16}$N$_2$O$_2$.HNO$_3$. Prisms from water, scales from alc, mp 159°. [α]$_D^{18}$ +39° (c = 2). Sol in 8.4 parts water, in 350 parts abs alc.

THERAP CAT: Antiglaucoma agent; miotic; sialogogue.

THERAP CAT (VET): Parasympathomimetic; miotic; gastric secretory stimulant.

7537. Pilocarpus. Jaborandi. Leaves of *Pilocarpus jaborandi* Holmes (Pernambuco jaborandi), or of *P. microphyllus* Stapf (Maranhao jaborandi), *Rutaceae. Habit.* Brazil, Paraguay. *Constit.* About 1% alkaloids of which about 0.5% is pilocarpine; pilocarpidine, isopilocarpine, jaborine, jaboridine, jaboric acid, pilocarpic acid.

THERAP CAT: Sudorific; miotic.

7538. Piloty's Acid. [599-71-3] *N*-Hydroxybenzenesulfonamide; benzenesulfohydroxamic acid; *N*-(phenylsulfonyl)hydroxylamine. $C_6H_7NO_3S$; mol wt 173.19. C 41.61%, H 4.07%, N 8.09%, O 27.71%, S 18.51%. Decomposes to produce nitroxyl (HNO). Prepn: O. Piloty, *Ber.* **29**, 1559 (1896). Reaction with aldehydes: A. Hassner *et al., J. Org. Chem.* **35**, 1962 (1970). Crystal structure: J. N. Scholz *et al., Tetrahedron* **45**, 7695 (1989). Decompn study: F. T. Bonner, Y. Ko, *Inorg. Chem.* **31**, 2514 (1992).

mp 126° (Piloty). Also reported as crystals from water, mp 109-110° (Scholz). pKa 9.29. Sol in alcohol, ether, ethyl acetate, acetone, warm water. Slightly sol in toluene, benzene, chloroform.

USE: HNO/NO⁻ donor in biochemical model studies. In the Angeli-Rimini test for aldehydes.

7539. Pilsicainide. [88069-67-4] *N*-(2,6-Dimethylphenyl)-tetrahydro-1*H*-pyrrolizine-7a(5*H*)-acetamide; tetrahydro-1*H*-pyrrolizine-7a(5*H*)-aceto-2′,6′-xylidide; *N*-(2,6-dimethylphenyl)-8-pyrrolizidineacetamide. $C_{17}H_{24}N_2O$; mol wt 272.39. C 74.96%, H 8.88%, N 10.28%, O 5.87%. Structural analog of lidocaine, *q.v.* Prepn: S. Miyano *et al.,* **EP 89061**; *eidem,* **US 4564624** (1983, 1986 both to Suntory); *eidem, J. Med. Chem.* **28**, 714 (1985). Pharmacology: K. Aisaka *et al., Arzneim.-Forsch.* **35**, 1239 (1985); T. Hidaka *et al., ibid.* 1381. Clinical evaluation in supraventricular tachycardia: T. Terazawa *et al., Am. Heart J.* **121**, 1437 (1991).

Hydrochloride hemihydrate. [88069-49-2] SUN-1165; Sunrythm. $C_{17}H_{24}N_2O \cdot HCl \cdot \frac{1}{2}H_2O$; mol wt 317.86. Crystals from ethanol-ether, mp 212-214°. LD_{50} in mice (mg/kg): 410 s.c. (Miyano, 1985).

THERAP CAT: Antiarrhythmic.

7540. Pimaric Acid. [127-27-5] (1*R*,4a*R*,4b*S*,7*S*,10a*R*)-7-Ethenyl-1,2,3,4,4a,4b,5,6,7,9,10,10a-dodecahydro-1,4a,7-trimethyl-1-phenanthrenecarboxylic acid; 13α-methyl-13-vinylpodocarp-8(14)-ene-15-oic acid; dextropimaric acid; *d*-pimaric acid; α-pimaric acid. $C_{20}H_{30}O_2$; mol wt 302.46. C 79.42%, H 10.00%, O 10.58%. Isoln from American rosin: Rimbach, *Ber. Dtsch. Pharm. Ges.* **6**, 61 (1896); from French galipot from *Pinus maritima* Mill, *Pinaceae:* Ruzicka, Balas, *Helv. Chim. Acta* **6**, 677 (1923); Ruzicka *et al., ibid.* **15**, 915 (1932). Structure: Ruzicka, Sternbach, *ibid.* **23**, 124 (1940); Fleck, *J. Am. Chem. Soc.* **62**, 2044 (1940); Harris, Sanderson, *ibid.* **70**, 2081 (1948). Stereochemistry: Wenkert, Chamberlin, *ibid.* **81**, 688 (1959).

Orthorhombic, bisphenoidal crystals from acetone or acetic acid, mp 217-219°; bp_{18} 282°; $[\alpha]_D^{18}$ +74.7° (c = 0.4 in chloroform); $[\alpha]_D^{20}$ +87.3° (in chloroform).

Hydrochloride. $C_{20}H_{31}O_2Cl$. Dec 184°. $[\alpha]_D^{20}$ +13.6° (c = 0.5 in alcohol). Can be reconverted by heating with quinoline at 250°.

Quinidine salt. $C_{40}H_{54}N_2O_4$. mp 90°.

Methyl ester. $C_{21}H_{32}O_2$. mp 69°.

7541. Pimecrolimus. [137071-32-0] (3*S*,4*R*,5*S*,8*R*,9*E*,12*S*,-14*S*,15*R*,16*S*,18*R*,19*R*,26a*S*)-3-[(1*E*)-2-[(1*R*,3*R*,4*S*)-4-Chloro-3-methoxycyclohexyl]-1-methylethenyl]-8-ethyl-5,6,8,11,12,13,14,-15,16,17,18,19,24,25,26,26a-hexadecahydro-5,19-dihydroxy-14,16-dimethoxy-4,10,12,18-tetramethyl-15,19-epoxy-3*H*-pyrido[2,1-*c*]-[1,4]oxaazacyclotricosine-1,7,20,21(4*H*,23*H*)-tetrone; 33-*epi*-chloro-33-desoxyascomycin; SDZ-ASM-981; Elidel. $C_{43}H_{68}ClNO_{11}$; mol wt 810.46. C 63.73%, H 8.46%, Cl 4.37%, N 1.73%, O 21.71%. Macrolactam ascomycin derivative. Specific calcineurin inhibitor that blocks production of pro-inflammatory cytokines by T cells and mast cells. Prepn: K. Baumann, G. Emmer, **EP 427680** (1991 to Sandoz); *eidem,* **US 5912238** (1999 to Novartis). *In vitro* pharmacology: M. Grassberger *et al., Br. J. Dermatol.* **141**, 264 (1999). Clinical evaluation in psoriasis: U. Mrowietz *et al., ibid.* **139**, 992 (1998); in contact dermatitis: C. Queille-Roussel *et al., Contact Dermatitis* **42**, 349 (2000). Review of clinical experience in inflammatory skin conditions: U. Wollina *et al., Expert Opin. Pharmacother.* **7**, 1967-1975 (2006).

White to off-white fine crystalline powder. Highly lipophilic. Sol in methanol and ethanol. Insol in water.

THERAP CAT: Immunosuppressant; dermatological in treatment of atopic eczema.

7542. Pimefylline. [10001-43-1] 3,7-Dihydro-1,3-dimethyl-7-[2-[(3-pyridinylmethyl)amino]ethyl]-1*H*-purine-2,6-dione; 7-[2-[(3-pyridylmethyl)amino]ethyl]theophylline; 7-(β-3′-picolylaminoethyl)theophylline; pimephylline; ES-771. $C_{15}H_{18}N_6O_2$; mol wt 314.35. C 57.31%, H 5.77%, N 26.74%, O 10.18%. Prepn: **NL 6600250**; Suter, Zutter, **US 3350400** (1966, 1967 both to Eprova). Structure determn, properties and chemistry: *eidem, Pharm. Acta Helv.* **48**, 133 (1973). Pharmacology: Ciaceri, Attaguile, *Gazz. Med. Ital.* **132**, 36, 108 (1973). Metabolism: Pitrè, *Farmaco Ed. Prat.* **29**, 46 (1974).

Crystals from isopropyl acetate, mp 111-112°. uv max (water): 270 nm. Readily sol in cold water, chloroform; sol in acetone, ethanol. LD$_{50}$ in mice (mg/kg): 1900 orally, 402 i.v. (Suter, Zutter, 1973).

Nicotinate. [10058-07-8] ES-902; Teonicon. C$_{15}$H$_{18}$N$_6$O$_2$·C$_6$-H$_5$NO$_2$; mol wt 437.46. Colorless, odorless powder from ethanol, mp 159-160°. uv max (water): 267 nm. Very sol in water (40% soln can be prepd); slightly sol in methanol; very slightly sol in acetone, chloroform. LD$_{50}$ in mice (mg/kg): 2530 orally, 470 i.v. (Suter, Zutter, 1973).

THERAP CAT: Vasodilator (coronary).

7543. Pimelic Acid. [111-16-0] Heptanedioic acid; 1,5-pentanedicarboxylic acid. C$_7$H$_{12}$O$_4$; mol wt 160.17. C 52.49%, H 7.55%, O 39.96%. Prepn starting with cyclohexanone and ethyl oxalate: Snyder *et al.*, *Org. Synth.* **coll. vol. II**, p 531 (1943); by the action of sodium and isoamyl alcohol on salicylic acid: Müller, *ibid.* p 535.

HOOC⌒⌒⌒COOH

Monoclinic prisms from benzene, mp 105.7-105.8°. Tends to sublime. bp$_{100}$ 272°; bp$_{50}$ 251.5°; bp$_{15}$ 223°; bp$_{10}$ 212°. 2.52 parts dissolve in 100 parts water at 13.5°; 5 parts dissolve in 100 parts water at 20°. Freely sol in alcohol, ether. Practically insol in cold benzene.

Diethyl ester. [2050-20-6] Diethyl pimelate. C$_{11}$H$_{20}$O$_4$; mol wt 216.28. Liquid, d$_4^{20}$ 0.99448; bp$_{748}$ 254°; bp$_{24}$ 155°; bp$_{15}$ 140°. Sol in alcohol, ether, ethyl acetate.

7544. Pimenta. Jamaica pepper. Small, evergreen tree, *Pimenta dioica* L. (syn. *P. officinalis* Lindl.), *Myrtaceae*, bearing deep purple or glossy black berries. The dried unripe fruits resemble peppercorns and are known as *allspice*. Flavor similar to a combination of clove, cinnamon, and nutmeg. *Habit.* West Indies, Central America, Mexico. *Constit.* Volatile oil (4%), quercetin glycosides, catechins, proanthocyanidins, protein, lipid, fixed oil, sugar, vitamins, minerals. *Reviews:* A. Y. Leung, S. Foster, *Encyclopedia of Common Natural Ingredients*, (Wiley-Interscience, Hoboken, 2nd Ed., 2003) pp 20-22; B. Krishnamoorthy, J. Rema in *Handbook of Herbs and Spices*, **Vol. 2**, K. V. Peter, Ed., (Woodhead Publishing, Cambridge, 2004) pp 117-139.

Pimenta berry oil. [8006-77-7] Oil of pimenta; pimento oil; allspice berry oil. Volatile oil obtained by distillation of dried immature berries. *Constit.* Complex mixture of components, chiefly eugenol (60-80%), methyleugenol, β-caryophyllene, humulene, terpinen-4-ol, α-phellandrene, eucalyptol. Yellow to brownish-yellow liquid with warm spicy sweet odor. d$_{25}^{25}$ 1.018-1.048. n$_D^{20}$ 1.527-1.540. Rotation between −4° and 0°. Sol in propylene glycol, most vegetable oils, in 2 vols 70% alcohol. *Keep well closed, cool and protected from light.*

Pimenta leaf oil. Volatile oil obtained by distillation of fresh or dried leaves. *Constit.* Eugenol (up to 96%) and other terpenes. Brownish-yellow liquid becoming darker with age; woody, spicy aromatic odor. d$_{25}^{25}$ 1.037-1.050. n$_D^{20}$ 1.531-1.536. Rotation between −2° and +0.5°. Sol in propylene glycol, in most fixed oils with slight opalescence, in 2 vols 70% alcohol. Insol in glycerin, mineral oil. *Keep well closed, cool and protected from light.*

USE: Spice in cooking; component of Caribbean jerk seasoning. Fragrance component of cosmetics, perfumes, soaps.

THERAP CAT: Carminative.

7545. Pimobendan. [74150-27-9] 4,5-Dihydro-6-[2-(4-methoxyphenyl)-1*H*-benzimidazol-6-yl]-5-methyl-3(2*H*)-pyridazinone; UD-CG 115; UD-CG 115 BS; 2-(4-methoxyphenyl)-5(6)-(5-methyl-3-oxo-4,5-dihydro-2*H*-6-pyridazinyl)benzimidazole; Acardi; Vetmedin. C$_{19}$H$_{18}$N$_4$O$_2$; mol wt 334.38. C 68.25%, H 5.43%, N 16.76%, O 9.57%. Positive inotropic agent. Prepn: V. Austel *et al.*, **DE 2837161**; *eidem*, **US 4361563** (1980, 1982 both to Thomae). Cardiovascular effects: J. C. A. von Meel, *Arzneim.-Forsch.* **35**, 284 (1985); P. D. Verdouw *et al.*, *Eur. J. Pharmacol.* **126**, 21 (1986). Comparison of enantiomer activity: K. Fujino *et al.*, *J. Pharmacol. Exp. Ther.* **247**, 519 (1988). HPLC determn of enantiomers in plasma: M. Asakura *et al.*, *J. Chromatogr.* **614**, 135 (1993). Clinical trial in congestive heart failure: F. Hagemeijer, *Am. Heart J.* **122**, 517 (1991); S. D. Katz *et al.*, *ibid.* **123**, 95 (1992). Clinical pharmacokinetics and pharmacodynamics: K.-M. Chu *et al.*, *Drug Metab. Dispos.* **27**, 701 (1999). Veterinary use in heart failure: V.

L. Fuentes, *Vet. Clin. North Am. Small Anim. Pract.* **34**, 1145 (2004). Review of pharmacology and therapeutic potential: A. Fitton, R. N. Brogden, *Drugs Aging* **4**, 417-441 (1994).

Hydrochloride. [77469-98-8] C$_{19}$H$_{18}$N$_4$O$_2$·HCl. Crystals from methanol and ethereal HCl, mp 311°(dec). LD$_{50}$ orally in mice: ~600 mg/kg (Austel, 1982).

THERAP CAT: Cardiotonic.

THERAP CAT (VET): Cardiotonic.

7546. Pimozide. [2062-78-4] 1-[1-[4,4-Bis(4-fluorophenyl)-butyl]-4-piperidinyl]-1,3-dihydro-2*H*-benzimidazol-2-one; 1-[1-[4,4-bis(*p*-fluorophenyl)butyl]-4-piperidyl]-2-benzimidazolinone; 1-[4,4-di-(4-fluorophenyl)butyl]-4-(2-oxo-1-benzimidazolinyl)-piperidine; R-6238; Orap; Opiran. C$_{28}$H$_{29}$F$_2$N$_3$O; mol wt 461.56. C 72.86%, H 6.33%, F 8.23%, N 9.10%, O 3.47%. Prepn: Janssen, **FR M3695** (1965 to Janssen), *C.A.* **66**, 115709t (1967). Pharmacological studies: Janssen *et al.*, *Arzneim.-Forsch.* **18**, 261, 279, 282 (1968). Metabolism in rats: Soudijn, Wijngaarden, *Life Sci.* **8**, 291 (1969). Clinical experience: Poldinger, *Curr. Ther. Res.* **13**, 23 (1971). Pimozide blocks establishment but not expression of amphetamine-produced environment-specific conditioning: R. J. Beninger, B. L. Hahn, *Science* **220**, 1304 (1983). Use in treatment of acute schizophrenia: B. Shopsin, G. Selzer, *Curr. Ther. Res.* **21**, 755 (1977); in treatment of chronic schizophrenia: F. Kline *et al.*, *ibid.* 768. In management of Gilles de la Tourette's syndrome: M. S. Ross, H. Moldofsky, *Lancet* **1**, 103 (1977); A. K. Shapiro, E. Shapiro, *Am. J. Psychiatry* **140**, 1235 (1983); *eidem*, *J. Am. Acad. Child Psychiatry* **23**, 161 (1984).

White microcrystals, mp 214-218°. Weak base, pKa 7.32. Freely sol in chloroform; slightly sol in ether, alc; very slightly sol in dil aq solns of organic and mineral acids (<5 mg/ml). Insol in water (<0.01 mg/ml).

THERAP CAT: Antipsychotic.

7547. Pimpinella. Pimpernel; brunet saxifrage; small saxifrage. Root of *Pimpinella saxifraga* L. or *P. magna* L. *Umbelliferae.* *Habit.* Europe; adventitious in U.S. *Constit.* Volatile oil, resin, benzoic acid, pimpinellin.

THERAP CAT: Aromatic carminative.

7548. Pimpinellin. [131-12-4] 5,6-Dimethoxy-2*H*-furo[2,3-*h*]-1-benzopyran-2-one; 4-hydroxy-6,7-dimethoxy-5-benzofuran-acrylic acid δ-lactone. C$_{13}$H$_{10}$O$_5$; mol wt 246.22. C 63.42%, H 4.09%, O 32.49%. Found in *Pimpinella saxifraga* L., *Heracleum spondylium* L., *H. lanatum* Michx., and *H. panaces* L., *Umbelliferae.* Isoln: Heut, *Arch. Pharm.* **236**, 162 (1898); Herzog, Hancu, *ibid.* **246**, 402 (1908); Wessely, Kallab, *Monatsh. Chem.* **59**, 161 (1932); Späth, Simon, *ibid.* **67**, 344 (1936); Fujita, Furuya, *J. Pharm. Soc. Jpn.* **74**, 795 (1954); *ibid.* **76**, 535 (1956); Svendsen, Ottestad, *Pharm. Acta Helv.* **32**, 457 (1957); Svendsen, *C.A.* **52**, 2173g (1958); Svendsen *et al.*, *Planta Med.* **7**, 113 (1959); *Pharm. Acta Helv.* **34**, 33 (1959); Jastrzebski, *C.A.* **54**, 1205b (1960). Synthesis: M. W. Reed, H. W. Moore, *J. Org. Chem.* **53**, 4166 (1988).

Off-white needles from methylene chloride/hexane, mp 119°. Practically insol in water. Sol in alcohol. Absorption spectrum: Wessely, Koltan, *Monatsh. Chem.* **86**, 430 (1955).

7549. Pinacidil. [60560-33-0] *N*-Cyano-*N'*-4-pyridinyl-*N''*-(1,2,2-trimethylpropyl)guanidine; P-1134. $C_{13}H_{19}N_5$; mol wt 245.33. C 63.65%, H 7.81%, N 28.55%. Potassium channel opening vasodilator. Prepn: H. J. Petersen, **DE 2557438**; *idem*, **US 4057636** (1976, 1977 both to Leo Pharm.); H. J. Petersen *et al.*, *J. Med. Chem.* **21**, 773 (1978). Mechanism of action: E. Arrigoni-Martelli *et al.*, *Experientia* **36**, 445 (1980); K. M. Bray *et al.*, *Br. J. Pharmacol.* **91**, 421 (1987). Metabolism: E. Eilertsen *et al.*, *Xenobiotica* **12**, 187 (1982). Bioavailability: *eidem, ibid.* 177. Determn in plasma: M. Hamilton *et al.*, *J. Chromatogr.* **375**, 359 (1986). Pharmacokinetics and hypotensive effects: J. W. Ward *et al.*, *Eur. J. Clin. Pharmacol.* **26**, 603 (1984). Clinical comparison with hydralazine, *q.v.*: R. L. Byyny *et al.*, *Clin. Pharmacol. Ther.* **42**, 50 (1987). Review of pharmacology and mechanism of action: M. L. Cohen: *Drug Dev. Res.* **9**, 249-258 (1986).

Monohydrate. [85371-64-8] Pindac. $C_{13}H_{19}N_5 \cdot H_2O$; mol wt 263.35. Crystals, mp 164-165°. LD_{50} in mice, rats (mg/kg): 600, 570 orally (Petersen, 1978).

THERAP CAT: Antihypertensive.

7550. Pinacol. [76-09-5] 2,3-Dimethyl-2,3-butanediol; pinacone; tetramethylethylene glycol. $C_6H_{14}O_2$; mol wt 118.18. C 60.98%, H 11.94%, O 27.08%. Prepd by the reduction of acetone: Holleman, *Rec. Trav. Chim.* **25**, 206 (1906); R. Adams, E. W. Adams, *Org. Synth.* **coll. vol. I**, 459 (2nd ed., 1941); **DE 233894**, *Frdl.* **10**, 1000. Convenient lab procedure: L. F. Fieser, *Experiments in Organic Chemistry* (Boston, 3rd ed., 1955) p 101. Crystal structure: G. A. Jeffrey, A. Robbins, *Acta Crystallogr.* **B34**, 3817 (1978).

Hexahydrate. $C_6H_{14}O_2 \cdot 6H_2O$. Four-sided plates from water. mp 45.4°; d^{15} 0.967 (supercooled liquid). The anhyd compd crystallizes in needles from alc or ether, mp 41.1°; bp_{760} 174.4°. Freely sol in hot water, in alc, in ether; slightly sol in cold water, carbon disulfide.

Dimethyl ether. $C_8H_{18}O_2$. Liquid. Agreeable odor. bp 144°.

7551. Pinacolone. [75-97-8] 3,3-Dimethyl-2-butanone; *tert*-butyl methyl ketone; pinacolin. $C_6H_{12}O$; mol wt 100.16. C 71.95%, H 12.08%, O 15.97%. Prepd by distillation of pinacol hydrate with dil H_2SO_4: Fittig, *Ann.* **114**, 56 (1860); Hill, Flosdorf, *Org. Synth.* **coll. vol. I** (2nd ed., 1941) p 462; L. F. Fieser, K. L. Williamson, *Organic Experiments* (D. C. Heath and Co., Lexington, Mass., 5th ed., 1983) p 332. Almost quantitative yields are obtained when 56.5 g pinacol hydrate is boiled for 3 hrs with 0.5 liter 25% H_2SO_4 and the product is steam distilled: Boeseken, van Tonningen, *Rec. Trav. Chim.* **39**, 189 (1920).

Liquid. Peppermint or camphor-like odor. d_{25}^{25} 0.7250. mp $-52.5°$. bp_{760} 106.2°. n_D^{25} 1.3939. Volatile in steam. Moderately sol in water (2.44% at 15°). Soluble in alcohol, ether, acetone.

Oxime. $C_6H_{13}NO$. Needles from aq alcohol. mp 78°. bp_{748} 171.6°. Sol in alc, ether, petr ether, benzene, chloroform.

7552. Pinaverium Bromide. [53251-94-8] 4-[(2-Bromo-4,5-dimethoxyphenyl)methyl]-4-[2-[2-(6,6-dimethylbicyclo[3.1.1]hept-2-yl)ethoxy]ethyl]morpholinium bromide (1:1); 4-(6-bromoveratryl)-4-[2-[2-(6,6-dimethyl-2-norpinyl)ethoxy]ethyl]morpholinium bromide; LAT-1717; Dicetel. $C_{26}H_{41}Br_2NO_4$; mol wt 591.43. C 52.80%, H 6.99%, Br 27.02%, N 2.37%, O 10.82%. Spasmolytic agent with low incidence of anticholinergic effects. Prepn: **BE 769469**; R. Baronnet, **US 3845048** (1971, 1974 both to Societe Berri-Balzac). Synthesis and pharmacology: R. Baronnet *et al.*, *Eur. J. Med. Chem.* **9**, 182 (1974). Pharmacodynamics: J. Bretaudeau *et al.*, *Therapie* **30**, 919 (1975). Synthesis of ^{14}C pinaverium bromide and pharmacokinetics: C. Jacquot *et al.*, *Eur. J. Med. Chem.* **13**, 61 (1978). Mechanism of action: J. Bretaudeau, O. Foussard-Blanpin, *J. Pharmacol.* **11**, 233 (1980). Inhibition of gastrointestinal contractile activity in dogs: Z. Itoh, T. Takahashi, *Arzneim.-Forsch.* **31**, 1450 (1981). Effects in humans on gastric emptying and transit time: J. Bertrand *et al.*, *Therapie* **36**, 555 (1981).

Crystals from methyl ethyl ketone, mp 181°; also reported as mp ~170° (Baronnet, 1974). LD_{50} in mice (mg/kg): 1400 orally; 66 i.v. (Baronnet, 1974); also reported as LD_{50} i.v. in mice: 37 ±2.4 mg/kg (Baronnet *et al.*).

THERAP CAT: Antispasmodic.

7553. Pinazepam. [52463-83-9] 7-Chloro-1,3-dihydro-5-phenyl-1-(2-propyn-1-yl)-2*H*-1,4-benzodiazepin-2-one; 7-chloro-1-propargyl-5-phenyl-3*H*-1,4-benzodiazepin-2(1*H*)-one; Z-905; Domar; Duna. $C_{18}H_{13}ClN_2O$; mol wt 308.77. C 70.02%, H 4.24%, Cl 11.48%, N 9.07%, O 5.18%. Prepn: F. Tenconi *et al.*, **DE 2339790** (1974 to Zambeletti), *C.A.* **80**, 133492k (1974); C. Podesva, K. Vagi, **US 3842094** (1974 to Delmar). Metabolism study: A. Trebbi *et al.*, *J. Chromatogr.* **110**, 309 (1975). Pharmacological and toxicological studies: F. Scrollini *et al.*, *Arzneim.-Forsch.* **25**, 934 (1975). Pharmacokinetics: P. M. Boselli, F. Scrollini, *Boll. Chim. Farm.* **116**, 363 (1977). Psychopharmacological properties: F. Scrollini *et al.*, *Arzneim.-Forsch.* **28**, 423 (1978). Physico-chemical profile: G. Filipi, A Trebbi, *Boll. Chim. Farm.* **118**, 105 (1979). Clinical study: V. Bertoncelli *et al.*, *Clin. Ter.* **94**, 641 (1980).

Cryst from methanol/water, mp 140-142°. LD_{50} in mice, rats (mg/kg): 1355, 5819 orally; 266, 622 i.p. (Scrollini, 1975). Also reported as LD_{50} in mice (mg/kg): 670 orally (Podesva, Vagi).

Note: This is a controlled substance (depressant): **21 CFR,** 1308.14.

THERAP CAT: Anxiolytic.

7554. Pindolol. [13523-86-9] 1-(1*H*-Indol-4-yloxy)-3-[(1-methylethyl)amino]-2-propanol; 4-[2-hydroxy-3-(isopropylamino)-propoxy]indole; prinodolol; LB-46; Betapindol; Blocklin L; Decreten; Pectobloc; Pinbetol; Pindoptan; Visken. $C_{14}H_{20}N_2O_2$; mol wt 248.33. C 67.71%, H 8.12%, N 11.28%, O 12.89%. Mixed β-adrenergic blocker and serotonin $5HT_{1A}$-receptor antagonist. Prepn: **NL 6601040**; F. Troxler, A. Hofmann, **US 3471515** (1966, 1969 both to Sandoz). HPLC determn in plasma: H. Smith, *J. Chromatogr.* **415,** 95 (1987). Symposia on cardiovascular pharmacology and clinical studies: *Am. Heart J.* **104,** Suppl. 2, pt. 2, 333-520 (1982); *Br. J. Clin. Pharmacol.* **13,** Suppl. 2, 143S-450S (1982). Effect on serotonin levels in rat brain: L. J. Dreshfield *et al., Neurochem. Res.* **21,** 557 (1996). Clinical trial with fluoxetine, *q.v.,* in depression: V. Pérez *et al., Lancet* **349,** 1594 (1997).

Crystals from ethanol, mp 171-173°. Slightly sol in methanol; very slightly sol in chloroform. Practically insol in water.

THERAP CAT: Antihypertensive; antianginal; antiarrhythmic; antiglaucoma.

7555. Pindone. [83-26-1] 2-(2,2-Dimethyl-1-oxopropyl)-1*H*-indene-1,3(2*H*)-dione; 2-pivaloyl-1,3-indandione; 2-pivalyl-1,3-indandione; 2-trimethylacetyl-1,3-indandione; pivalyl indandione; pivaldione; Pival; Pivalyl Valone; Tri-Ban. $C_{14}H_{14}O_3$; mol wt 230.26. C 73.03%, H 6.13%, O 20.84%. Prepn: Kilgore *et al., Ind. Eng. Chem.* **34,** 494 (1942); Zelmens, Vanags, *C.A.* **53,** 21830d (1959). Toxicity study: T. B. Gaines, *Toxicol. Appl. Pharmacol.* **14,** 515 (1969).

Bright yellow crystals from ethanol, mp 108-110°. LD_{50} orally in male rats: 280 mg/kg (Gaines).

Sodium salt. Pivalyn. Bright yellow crystals, mp 205-210°. Schwarz, **US 2880132** (1959 to Morton Chem.). Sol in water.

Caution: Potential symptoms of overexposure are epistaxis, excess bleeding from minor cuts and bruises; smoky urine, black tarry stools; abdominal and back pain. *See NIOSH Pocket Guide to Chemical Hazards* (DHHS/NIOSH 97-140, 1997) p 258.

USE: Rodenticide; insecticide.

7556. α-Pinene. [80-56-8] 2,6,6-Trimethylbicyclo[3.1.1]-hept-2-ene; 2-pinene; pinene. $C_{10}H_{16}$; mol wt 136.24. C 88.16%, H 11.84%. Obtained from oil of turpentine which contains 58-65% α-pinene along with 30% β-pinene, *q.v.:* E. Gildemeister, F. Hoffmann, *Die ätherischen Ole* **Band IV** (Akademie-Verlag, Berlin, 4th ed., 1956) p 39. α-Pinene in North American oils is dextrorotatory, in most European oils it is levorotatory. Constituent of many volatile oils. Isoln of *d*-α-pinene from Port Oxford cedar wood oil *(Chamaecyparis lawsoniana* Parl., *Pinaceae):* Thurber, Roll, *Ind. Eng. Chem.* **19,** 739 (1927). Isoln of *l*-α-pinene from mandarin peel oil *(Citrus reticulata* Blando, *Rutaceae):* Kugler, Kováts, *Helv. Chim. Acta* **46,** 1480 (1963). Total synthesis of α- and β-forms: Komppa, *Ann. Acad. Sci. Fenn.* **A59,** 3 (1943), *C.A.* **41,** 425 (1947); Thomas, Fallis, *Tetrahedron Lett.* **1973,** 4687; *eidem, J. Am. Chem. Soc.* **98,** 1227 (1976). *Review:* Palmer, *Ind. Eng. Chem.* **34,** 1028 (1942); J.

L. Simonsen, *The Terpenes* **vol. II** (Cambridge Univ. Press, 2nd ed., 1949) pp 105-191; D. V. Banthorpe, D. Whittaker, *Chem. Rev.* **66,** 643-654 (1966); *Food Cosmet. Toxicol.* **16,** Suppl. I, 853-857 (1978).

dl-**Form.** Liquid, characteristic odor of turpentine. bp_{760} 155-156°; bp_{20} 52.5°. d_4^{20} 0.8592. n_D^{20} 1.4664. Practically insol in water. Sol in alc, chloroform, ether, glacial acetic acid.

Hydrochloride. $C_{10}H_{17}Cl$. mp 132°.

d-**Form.** bp_{760} 155-156°. d_4^{20} 0.8591. n_D^{20} 1.4663. $[\alpha]_D^{20}$ +51.14°.

Hydrochloride. mp 132°. $[\alpha]_D^{20}$ +33.52° (alcohol).

l-**Form.** bp_{760} 155-156°. d_4^{20} 0.8590. n_D^{20} 1.4662. $[\alpha]_D^{20}$ −51.28°.

Hydrochloride. mp 132°. $[\alpha]_D^{20}$ −33.24° (alcohol).

Caution: Toxic effects similar to turpentine, *q.v.*

USE: Manufacture of camphor, insecticides, solvents, plasticizers, perfume bases, synthetic pine oil.

7557. β-Pinene. [127-91-3] 6,6-Dimethyl-2-methylenebicyclo[3.1.1]heptane; nopinene. $C_{10}H_{16}$; mol wt 136.24. C 88.16%, H 11.84%. Found in most essential oils which contain α-pinene, but in far smaller proportions; the *l*-form occurs most commonly. Initial identification: A. von Baeyer, *Ber.* **29,** 25 (1896). Isoln of the *d*-form from *Ferula galbaniflua* Boiss. et Buhse, *Umbelliferae:* B. N. Rutovski, I. V. Vinogradova, *J. Prakt. Chem.* **120,** 41 (1928); from *Cynomarathrum nuttallii* A. Gray, *Umbelliferae:* E. K. Nelson, *J. Am. Chem. Soc.* **55,** 3400 (1933). Irreversible isomerization of β-pinene to α-pinene occurs on shaking with platinum black satd with hydrogen: F. Richter, W. Wolff, *Ber.* **59,** 1733 (1926). Synthesis: G. Bonnet, *Bull. Inst. Pin* **1938,** 217; *ibid.* **1939,** 1, *C.A.* **33,** 4223⁴ (1939); K. J. Crowley, *Proc. Chem. Soc. London* **1962,** 245; *Tetrahedron* **21,** 1001 (1965); L. M. Harwood, M. Julia, *Synthesis* **1980,** 456. For general refs *see* α-pinene.

dl-**Form.** bp_{760} 165-166°.

d-**Form.** bp_{760} 164-166°. d_{20}^{20} 0.8654. n_D^{20} 1.4739. $[\alpha]_D$ +28.59° (Nelson). Also reported as bp_{760} 162-163°. d_{20}^{20} 0.8662. n_D^{20} 1.4745. $[\alpha]_D$ +20.75° (Rutovski, Vinogradova).

l-**Form.** bp_{760} 162-163°. d^{15} 0.874. n_D^{15} 1.4872, $[\alpha]_D$ −22.4°.

7558. Pine Oil. Yarmor. An oil from *Pinus palustris* Mill. and certain other species of pines, *Pinaceae.* It is obtained from pitch-soaked pine wood by steam distillation or solvent extraction followed by steam distillation and also by destructive distillation. It consists mainly of isomeric tertiary and secondary, cyclic terpene alcohols.

Colorless to pale yellow liquid, turpentine-like odor. d about 0.9; bp 200-220°. *Flammable.* Insol in water. Sol in the usual organic solvents.

Caution: Irritating to skin, mucous membranes. Large doses may cause CNS depression.

USE: Pharmaceutic aid (flavor and perfume). Manuf terpin hydrate and other terpin products; as a solvent, disinfectant and deodorant; in textile scouring; for flotation of lead and zinc ores.

7559. Pine Tar. A product obtained by destructive distillation of wood of *Pinus palustris* Mill., or other species of pine, *Pinaceae.*

Blackish-brown, viscous liquid; heavier than water; empyreumatic odor and sharp taste. Slightly sol in water; sol in alc, chloroform, ether, acetone, glacial acetic acid, fixed and volatile oils, and in solns of caustic alkalies. Principal constituents: turpentine, resin, guaiacol, creosol, methylcreosol, phenol, phlorol, toluene, xylene, and other hydrocarbons.

THERAP CAT: Topical antieczematic; rubefacient.

THERAP CAT (VET): Mild irritant, antiseptic in chronic skin conditions. Expectorant.

7560. Pinguinain. [37288-97-4] A proteolytic enzyme obtained from the juice of the fruit of *Bromelia pinguin* Plum. ex L., *Bromeliaceae* (pineapple family): Asenjo, *Science* **95**, 48 (1942); Bloch, Messing, US 2977287 (1961 to Ethicon). Immunochemical studies: E. Toro-Goyco, I. Rodriguez-Costas, *Arch. Biochem. Biophys.* **175**, 359 (1976). Structure studies: E. Toro-Goyco *et al.*, *Biochim. Biophys. Acta* **622**, 151 (1980).

Aq solns are capable of digesting necrotic tissue, but do not attack viable tissue. Optimum proteolytic activity is at pH 5.2 to 5.5 and also at pH 7.3. Inactivated at temps above 80°.

7561. Pinitol. [10284-63-6]; [484-68-4] (DL-form). 3-*O*-Methyl-D-*chiro*-inositol; D-pinitol; pinite; Inzitol. $C_7H_{14}O_6$; mol wt 194.18. C 43.30%, H 7.27%, O 49.44%. Naturally occurring cyclitol discovered in sugar pine, *Pinus lambertiana* Dougl., *Pinaceae*; subsequently found in other conifers, soybean and other legumes, and the medicinal plants, sutherlandia and bougainvillea. Isoln from sugar pine: M. Berthelot, *C. R. Acad. Sci.* **41**, 392 (1855); H. W. Wiley, *J. Am. Chem. Soc.* **13**, 228 (1891); E. G. Griffin, J. M. Nelson, *ibid.* **37**, 1552 (1915). Large scale prodn from pine stump wood: A. B. Anderson, *Ind. Eng. Chem.* **45**, 593 (1953). Isoln from soybean: D. V. Phillips *et al.*, *J. Agric. Food Chem.* **30**, 456 (1982). Structure: A. B. Anderson *et al.*, *J. Am. Chem. Soc.* **74**, 1479 (1952). Total synthesis: S. V. Ley *et al.*, *Tetrahedron Lett.* **28**, 225 (1987); J. L. Aceña *et al.*, *Tetrahedron: Asymmetry* **7**, 3535 (1996). Determn in plant extracts by ^{13}C-NMR spectroscopy: E. Duquesnoy *et al.*, *Carbohydr. Res.* **343**, 893 (2008). Exptl insulin-like activity: S. H. Bates *et al.*, *Br. J. Pharmacol.* **130**, 1944 (2000). Clinical effect on blood glucose in diabetic patients: M.-J. Kang *et al.*, *J. Med. Food* **9**, 182 (2006).

Small white prisms from methanol, mp 186°. $[\alpha]_D^{22}$ +66.1° (c = 1 in water). Insol in abs alcohol, ether.

USE: Dietary supplement.

7562. Pinolenic Acid. [16833-54-8] (5*Z*,9*Z*,12*Z*)-5,9,12-Octadecatrienoic acid. $C_{18}H_{30}O_2$; mol wt 278.44. C 77.65%, H 10.86%, O 11.49%. Characteristic fatty acid of the seed oils of *Pinus* spp., *Pinaceae*. Positional isomer of γ-linolenic acid, *q.v.* Isoln from tall oil: E. Elomaa *et al.*, *Suom. Kemistil. B* **36**, 52 (1963); A. Hase *et al.*, *J. Am. Oil Chem. Soc.* **69**, 832 (1992). Review of occurrence in the genus *Pinus*: R. L. Wolff *et al.*, *Lipids* **35**, 1-22 (2000). Effect of pinolenic acid-enriched pine nut oil on fatty acid metabolism in rats: M. Sugano *et al.*, *Br. J. Nutr.* **72**, 775 (1994); on LDL-receptor activity: J.-W. Lee *et al.*, *Lipids* **39**, 383 (2004).

n_D^{20} 1.4804. Acid no. 201.8. Iodine no. 268.

THERAP CAT: Nutritional supplement.

7563. Pinosylvin. [22139-77-1] 5-[(1*E*)-2-Phenylethenyl]-1,3-benzenediol; *E*-3,5-stilbenediol; 5-styrylresorcinol; *trans*-3,5-dihydroxystilbene. $C_{14}H_{12}O_2$; mol wt 212.25. C 79.22%, H 5.70%, O 15.08%. Occurs together with its monomethyl and dimethyl ethers in the heartwood of pine and other woody plants. Naturally occurring pinosylvins have the *trans* configuration. Isoln from *Pinus sylvestris* L., *Pinaceae*: H. Erdtman, *Ann.* **539**, 116 (1939); from other *Pinus* species: G. Lindstedt, *Acta Chem. Scand.* **3**, 755-772 (1949); J. C. Alvarez-Novoa *et al.*, *ibid.* **4**, 444 (1950); from *Alnus sieboldiana*, *Betulaceae*: Y. Asakawa, *Bull. Chem. Soc. Jpn.* **44**, 2761 (1971); from *Polygonum nodosum*, *Polygonaceae*: M. Kuro-

yanagi *et al.*, *Chem. Pharm. Bull.* **30**, 1602 (1982). Synthesis of pinosylvin: E. Späth, F. Liebherr, *Ber.* **74**, 869 (1941); of monomethyl ether: E. Späth, K. Kromp, *ibid.* 1424; of dimethyl ether: G. Aulin-Erdtman, H. Erdtman, *ibid.* 50; of pinosylvin and derivatives: A. A. Loman, L. R. Snowdon, *Can. J. Chem.* **48**, 1554 (1970). Biosynthesis: Birch, *Fortschr. Chem. Org. Naturst.* **14**, 186 (1957). Toxicological study: K. O. Frykholm, *Nature* **155**, 454 (1945). Use as antimicrobial agent: E. H. Sheers, DE 1952451; *idem*, US 3577230 (1970, 1971 both to Arizona Chem. Co.). Deterrent to feeding behavior of snowshoe hare: J. P. Bryant *et al.*, *Science* **222**, 1023 (1983).

Fine needles from glacial acetic acid, mp 155.5-156°. uv max (ethanol): 305 nm (log ε 4.49). Practically insol in water. Sol in benzene, acetone, chloroform, glacial acetic acid.

Monomethyl ether. $C_{15}H_{14}O_2$; mol wt 226.28. Crystals, mp 122-123°. uv max (ethanol): 303 nm (log ε 4.26). More sol in benzene than pinosylvin. Also sol in methanol, glacial acetic acid.

Dimethyl ether. $C_{16}H_{16}O_2$; mol wt 240.30. Crystals from methanol-water, mp 55-56°. uv max (ethanol): 305 nm (log ε 4.39).

7564. Pinoxaden. [243973-20-8] 2,2-Dimethyl propanoic acid 8-(2,6-diethyl-4-methylphenyl)-1,2,4,5-tetrahydro-7-oxo-7*H*-pyrazolo[1,2-*d*][1,4,5]oxadiazepin-9-yl ester; SYN-407855; Axial. $C_{23}H_{32}N_2O_4$; mol wt 400.52. C 68.97%, H 8.05%, N 6.99%, O 15.98%. Broad spectrum graminicide for use in cereal crops. Interrupts fatty acid biosynthesis by inhibiting the enzyme, acetyl CoA carboxylase. Prepn: M. Mühlebach *et al.*, WO 9947525 (1999 to Novartis); *eidem*, US 6410480 (2002 to Syngenta); M. Muehlebach *et al.*, *Bioorg. Med. Chem.* **17**, 4241 (2009). Properties, mechanism of action, and activity in wheat and barley: U. Hofer *et al.*, *J. Plant Dis. Prot.* **2006**, Sp. Iss. 20, 989.

Crystalline white solid, mp 122-123°. *Irritant*. d 1.16. Soly in water: 200 mg/l. Log P (octanol/water): 3.2. Vapor pressure: 4.6×10^{-7} Pa. LD_{50} in rats (mg/kg): >5000 orally, >2000 dermally; LC_{50} in rats (mg/l): 5.2 by inhalation (Hofer).

USE: Herbicide.

7565. Pioglitazone. [111025-46-8] 5-[[4-[2-(5-Ethyl-2-pyridinyl)ethoxy]phenyl]methyl]-2,4-thiazolidinedione; (±)-5-[*p*-[2-(ethyl-2-pyridyl)ethoxy]benzyl]-2,4-thiazolidinedione; AD-4833. $C_{19}H_{20}N_2O_3S$; mol wt 356.44. C 64.02%, H 5.66%, N 7.86%, O 13.47%, S 8.99%. Insulin sensitizer. Prepn: K. Meguro, T. Fujita, EP 193256; *eidem*, US 4687777 (1986, 1987 both to Takeda); T. Sohda *et al.*, *Arzneim.-Forsch.* **40**, 37 (1990). Pharmacology: H. Ikeda *et al.*, *ibid.* 156. HPLC determn in serum: W. Z. Zhong, D. B. Lakings, *J. Chromatogr.* **490**, 377 (1989). Mechanism of action: C. Hofmann *et al.*, *Endocrinology* **129**, 1915 (1991); M. Kobayashi *et al.*, *Diabetes* **41**, 476 (1992). Effect on adipocyte differentiation: T. Sandouk *et al.*, *Am. J. Physiol.* **264**, C1600 (1993). Clinical evaluation in noninsulin-dependent diabetes: R. Kawamori *et al.*, *Diabetes Res. Clin. Pract.* **41**, 35 (1998).

Colorless needles from DMF + water, mp 183-184°.

Hydrochloride. [112529-15-4] U-72107A; Actos. $C_{19}H_{20}N_2$-$O_3S.HCl$; mol wt 392.90. Colorless prisms from ethanol, mp 193-194°. Sol in DMF; slightly sol in ethanol; very slightly sol in acetone, acetonitrile. Practically insol in water; insol in ether.

THERAP CAT: Antidiabetic.

7566. Pipacycline. [1110-80-1] (4S,4αS,5aS,6S,12aS)-4-(Dimethylamino)-1,4,4a,5,5a,6,11,12a-octahydro-3,6,10,12,12a-pentahydroxy-N-[[4-(2-hydroxyethyl)-1-piperazinyl]methyl]-6-methyl-1,11-dioxo-2-naphthacenecarboxamide; N-[[4-(2-hydroxyethyl)-1-piperazinyl]methyl]tetracycline; N-[4-(β-hydroxyethyl)diethylenediamino-1-methyl]tetracycline; mepicycline; mepiciclina; AmbraVena; Sieromicin; Valtomicina. $C_{29}H_{38}N_4O_9$; mol wt 586.64. C 59.38%, H 6.53%, N 9.55%, O 24.55%. Semi-synthetic broad spectrum antibiotic related to tetracycline. Prepn: Pedrazzoli *et al., Boll. Chim. Farm.* **98**, 516 (1959), *C.A.* **54**, 3856a (1960); Gradnik *et al.,* **GB 888968** corresp to **US 3149114** (1962 and 1964 to E.R.A.S.M.E.). Properties: *eidem, Pharm. Acta Helv.* **35**, 529 (1960). Pharmacokinetic studies: A. Scalvini, A. Delmonte, *Gazz. Med. Ital.* **131**, 1 (1972).

Yellow cryst powder, dec 162-163°. $[\alpha]_D^{20}$ −195° (c = 0.5). $[\alpha]_D^{20}$ −175° (c = 0.5 in methanol). uv max (10 γ/ml 0.1N HCl): 286, 355 nm. pH of 2% aq soln, 7.2-7.4. Freely sol in water, methanol, formamide; slightly sol in ethanol, isopropanol. Practically insol in ether, benzene, chloroform. Sensitive to light, heat, and air. LD_{50} i.v. in white mice: 188 mg/kg (Scalvini, Delmonte).

THERAP CAT: Antibacterial.

7567. Pipamazine. [84-04-8] 1-[3-(2-Chloro-10H-phenothiazin-10-yl)propyl]-4-piperidinecarboxamide; 1-[3-(2-chlorophenothiazin-10-yl)propyl]isonipecotamide; 10-[3-(4-carbamoylpiperidin-1-yl)propyl]-2-chlorophenothiazine; 2-chloro-10-[3-(4-carbamoyl-piperidinyl)propyl]phenothiazine; 10-[3-(4-carbamoylpiperidino)-propyl]-2-chlorophenothiazine; SC-9387; Nausidol; Mornidine. $C_{21}H_{24}ClN_3OS$; mol wt 401.95. C 62.75%, H 6.02%, Cl 8.82%, N 10.45%, O 3.98%, S 7.98%. Prepn: Cusic *et al.,* **US 2957870** (1960 to Searle).

Crystals from 2-propanol + petr ether, mp about 139°.

Hydrochloride. Crystals, mp about 196-197° with formation of bubbles.

THERAP CAT: Antiemetic.

7568. Pipamperone. [1893-33-0] 1'-[4-(4-Fluorophenyl)-4-oxobutyl]-[1,4'-bipiperidine]-4'-carboxamide; 1'-[3-(p-fluorobenzoyl)propyl]-[1,4'-bipiperidine]-4'-carboxamide; 1-(p-fluorophen-yl)-4-(4-piperidino-4-carbamoylpiperidino)-1-butanone; 1-[γ-(4-fluorobenzoyl)propyl]-4-piperidinopiperidine-4-carboxamide; 4'-fluoro-4-[N-[4-(N-piperidino)-4-carbamido]piperidino]butyrophenone; floropipamide; R-3345. $C_{21}H_{30}FN_3O_2$; mol wt 375.49. C 67.17%, H 8.05%, F 5.06%, N 11.19%, O 8.52%. Prepn of the dihydrochloride by reaction of γ-chloro-4-fluorobutyrophenone and 4-piperidinopiperidine-4-carboxamide: Janssen, **BE 610830** (1962 to Janssen), *C.A.* **57**, 13740b (1962).

Dihydrochloride. [2448-68-2] Dipiperon; Piperonil; Propitan. $C_{21}H_{30}FN_3O_2.2HCl$; mol wt 448.40. Crystals, mp 124.5-126.0°.

THERAP CAT: Antipsychotic.

7569. Pipazethate. [2167-85-3] 10H-Pyrido[3,2-b][1,4]-benzothiadiazine-10-carboxylic acid 2-[2-(1-piperidinyl)ethoxy]-ethyl ester; 2-(2-piperidinoethoxy)ethyl 10H-pyrido[3,2-b][1,4]ben-zothiadiazine-10-carboxylate; thiophenylpyridylamino-10-carbox-ylic acid piperidinoethoxyethyl ester; 2-(2-piperidinoethoxy)ethyl 10-thia-1,9-diazaanthracene-10-carboxylate; 1-azaphenothiazine-10-carboxylic acid 2-(2-piperidinoethoxy)ethyl ester; D-254. $C_{21}H_{25}N_3O_3S$; mol wt 399.51. C 63.14%, H 6.31%, N 10.52%, O 12.01%, S 8.02%. Prepn: Schuler *et al., Ann.* **673**, 102 (1964); Schuler, **US 2989529** (1961 to Degussa).

Hydrochloride. [6056-11-7] Lenopect; Selvigon; Selvjgon; Theratuss. $C_{21}H_{25}N_3O_3S.HCl$; mol wt 435.97. Pale yellow crystals from isopropanol, mp 160-161°. Soluble in water, methanol. Practically insol in acetone, petr ether. LD_{50} orally in rats: 560 mg/kg (Schuler).

THERAP CAT: Antitussive.

7570. Pipecolic Acid. [535-75-1] 2-Piperidinecarboxylic acid; pipecolinic acid; hexahydropicolinic acid; homoproline; dihy-drobaikiaine. $C_6H_{11}NO_2$; mol wt 129.16. C 55.80%, H 8.58%, N 10.84%, O 24.77%. The *l*-form occurs in plants: Phillips, *Chem. Ind. (London)* **1953**, 127. Prepn: A. Ladenburg, *Ber.* **24**, 640 (1891); Stevens, Ellman, *J. Biol. Chem.* **182**, 75 (1950); V. Asher *et al., Tetrahedron Lett.* **22**, 141 (1981). Synthesis of L-pipecolic acid from L-lysine: Fujii, Miyoshi, *Bull. Chem. Soc. Jpn.* **48**, 1341 (1975). Synthesis of racemate: R. T. Shuman *et al., J. Org. Chem.* **55**, 738 (1990).

l-**Form.** Needles by sublimation, mp 270°. $[\alpha]_D^{25}$ −34.9°. Sol in water, dil alcohol. Sparingly sol in abs alcohol, acetone, chloroform. Insol in ether.

d-**Form.** Platelets from alcohol, mp 270°. $[\alpha]_D^{25}$ +35.7°. Sol in water. Somewhat sol in alcohol.

dl-**Form.** Leaflets from water, mp 264°. Sol in water, boiling alcohol.

dl-**Hydrochloride.** $C_6H_{11}NO_2.HCl$. Crystals (warts) from alcohol + benzene, mp 258-262°.

7571. Pipecurium Bromide. [52212-02-9] 4,4′-[(2β,3α,5α,-16β,17β)-3,17-Bis(acetyloxy)androstane-2,16-diyl]bis(1,1-dimethylpiperazinium)bromide (1:2); 4,4′-(3α,17β-dihydroxy-5α-androstan-2β,16β-ylene)bis(1,1-dimethylpiperazinium)dibromide diacetate; pipecuronium bromide; RGH-1106; Arduan. $C_{35}H_{62}Br_2N_4O_4$; mol wt 762.71. C 55.12%, H 8.19%, Br 20.95%, N 7.35%, O 8.39%. Aminosteroid, competitive neuromuscular blocker. Prepn: Z. Tuba et al., **DE 2337882** (1974 to Gedeon Richter), C.A. **80**, 121210d (1974); Z. Tuba, Arzneim.-Forsch. **30**, 342 (1980). Toxicity study: E. Kárpáti, K. Biro, ibid. 346. Series of articles on pharmacology, properties, disposition, pharmacokinetics, safety tests, clinical studies: ibid. 346-394. Determn in human serum: G. Szabo, E. Tasonyi, ibid. **31**, 1013 (1981).

Crystals from methylene dichloride/acetone, mp 262-264° (dec). $[\alpha]_D^{25}$ +8.1° (c = 1 in water). LD_{50} in mice, rats (mg/kg): 29.7, 172.6 i.v.; 70.6, 449.6 i.p.; 60.5, 455.8 s.c. (Kárpáti, Biro).
THERAP CAT: Neuromuscular blocking agent.

7572. Pipemidic Acid. [51940-44-4] 8-Ethyl-5,8-dihydro-5-oxo-2-(1-piperazinyl)pyrido[2,3-d]pyrimidine-6-carboxylic acid; piperamic acid; 1489-RB; Filtrax; Memento 400; Pi-Coli; Pipecid; Pipedac; Pipemid; Pipurin; Tractur; Uropimid; Urosten; Uroval. $C_{14}H_{17}N_5O_3$; mol wt 303.32. C 55.44%, H 5.65%, N 23.09%, O 15.82%. Quinolone antibacterial. Prepn: S. Minami et al., **DE 2341146**; eidem, **US 3887557** (1974, 1975 both to Dainippon Pharm.); Pesson et al., C. R. Seances Acad. Sci. Ser. C **278**, 1169 (1974); De Lajudie et al., C. R. Seances Acad. Sci. Ser. D **279**, 1931 (1974); Matsumoto, Minami, J. Med. Chem. **18**, 74 (1975). Characterization: Ficicchia, Farmaco Ed. Prat. **30**, 207 (1974). Antibacterial activity: Ficicchia, De Lajudie, ibid. 252. Pharmacology: Shimizu et al., Antimicrob. Agents Chemother. **7**, 441 (1975).

Yellowish-white, odorless, bitter-tasting crystals, mp 253-255°. Hygroscopic. Yellows slowly in light. Sol in acid soln and alkaline. Very slightly sol in water, alcohol; slightly sol in chloroform (0.5%), methanol (0.4%). Practically insol in ether, benzene. LD_{50} in mice (mg/kg): 4000 orally; 1000 i.p., 50 i.v. (Ficicchia).
Trihydrate. [72571-82-5] Deblaston; Dolcol; Pipram; Solupemid. Nearly colorless needles, mp 253-255°.
THERAP CAT: Antibacterial.
THERAP CAT (VET): Antibacterial.

7573. Pipenzolate Bromide. [125-51-9] 1-Ethyl-3-[(2-hydroxy-2,2-diphenylacetyl)oxy]-1-methylpiperidinium bromide (1:1); 1-ethyl-3-hydroxy-1-methylpiperidinium bromide benzilate; benzilic acid 1-ethyl-3-piperidyl ester methyl bromide; N-ethyl-3-piperidyl benzilate methobromide; pipenzolate methylbromide; pipenzolone bromide; JB-323; Piptal. $C_{22}H_{28}BrNO_3$; mol wt 434.37. C 60.83%, H 6.50%, Br 18.40%, N 3.22%, O 11.05%. Anticholinergic. Prepn from N-ethyl-3-chloropiperidine and benzilic acid: Biel et al., J. Am. Chem. Soc. **74**, 1485 (1952); **US 2918406** (1959 to Lakeside).

Crystals from methyl ethyl ketone, mp 179-180°. Soluble in water.
Note: N-Ethyl-3-piperidyl benzilate is a controlled substance (hallucinogen): **21 CFR, 1308.11.**
THERAP CAT: Antispasmodic.

7574. Piperacetazine. [3819-00-9] 1-[10-[3-[4-(2-Hydroxyethyl)-1-piperidinyl]propyl]-10H-phenothiazin-2-yl]ethanone; 10-[3-[4-(2-hydroxyethyl)piperidino]propyl]phenothiazin-2-yl methyl ketone; 2-acetyl-10-[3-[4-(β-hydroxyethyl)piperidino]propyl]phenothiazine; 2-acetyl-10-[3-[γ-(2-hydroxyethyl)piperidino]propyl]phenothiazine; PC-1421; Quide. $C_{24}H_{30}N_2O_2S$; mol wt 410.58. C 70.21%, H 7.37%, N 6.82%, O 7.79%, S 7.81%. Prepn: **GB 861807** (1961 to Searle). Clinical trial in schizophrenia: R. T. Rada, P. T. Donlon, Curr. Ther. Res. **16**, 124 (1974).

Hydrochloride. mp 100-110°.
THERAP CAT: Antipsychotic.

7575. Piperacillin. [61477-96-1] (2S,5R,6R)-6-[[(2R)-2-[[(4-Ethyl-2,3-dioxo-1-piperazinyl)carbonyl]amino]-2-phenylacetyl]-amino]-3,3-dimethyl-7-oxo-4-thia-1-azabicyclo[3.2.0]heptane-2-carboxylic acid; (2S,5R,6R)-6-[(R)-2-(4-ethyl-2,3-dioxo-1-piperazinecarboxamido)-2-phenylacetamido]-3,3-dimethyl-7-oxo-4-thia-1-azabicyclo[3.2.0]heptane-2-carboxylic acid; 6-D(−)-α-(4-ethyl-2,3-dioxo-1-piperazinylcarbonylamino)-α-phenylacetamidopenicillanic acid; 4-ethyl-2,3-dioxopiperazine carbonyl ampicillin. $C_{23}H_{27}N_5O_7S$; mol wt 517.56. C 53.38%, H 5.26%, N 13.53%, O 21.64%, S 6.19%. Broad spectrum semi-synthetic antibiotic related to pencillin. Prepn: I. Saikawa et al., **DE 2519400**; eidem, **US 4087424** (1976, 1978 both to Toyama). In vitro studies: G. P. Bodey, B. LeBlanc, Antimicrob. Agents Chemother. **14**, 78 (1978); R. Wise et al., ibid. 549. In vitro and in vivo: N. A. Kuck, G. S. Redin, J. Antibiot. **31**, 1175 (1978). Pharmacokinetics: M. A. Evans et al., J. Antimicrob. Chemother. **4**, 255 (1978). Metabolism: K. Iida et al., Antimicrob. Agents Chemother. **14**, 257 (1978). Clinical study: T. Saito, Y. Yamada, Jpn. J. Antibiot. **30**, 835 (1977). Toxicity study: A. Takai et al., Chemotherapy (Tokyo) **25**, 816 (1977). Review of antibacterial activity, pharmacokinetics: G. L. Mandell, Clin. Ther. **7**, Suppl. B, 28-35 (1985). Review of clinical experience in urinary tract infections: S. J. Childs, ibid. 36-45. Series of articles on antibacterial activity and clinical use: Chemotherapy (Tokyo) **36**, Suppl. 7, 1-85 (1988).

White to off-white, crystalline powder. Very sol in methanol; sparingly sol in isopropyl alc; slightly sol in ethyl acetate; very slightly sol in water.

Sodium salt. [59703-84-3] CL-227193; T-1220; Isipen; Pentcillin; Pipracil; Pipril. $C_{23}H_{26}N_5NaO_7S$; mol wt 539.54. White to off-white solid. mp 183-185° (dec). Freely sol in water, alc. LD_{50} in mice, rats, dogs, monkeys (g/kg): 5, 2.7, >6, >4 i.v. (Takai).

THERAP CAT: Antibacterial.

7576. Piperazine. [110-85-0] 1,4-Diazacyclohexane; hexahydropyrazine; piperazidine; diethylenediamine. $C_4H_{10}N_2$; mol wt 86.14. C 55.77%, H 11.70%, N 32.52%. Prepn: Cloez, *Jahresber.* **1853**, 468; Wolff, *Ber.* **26**, 724 (1893); Kyrides, US 2267686 (1941); Martin, Martell, *J. Am. Chem. Soc.* **70**, 1817 (1948); MacKenzie, Turbin, US 2901482 (1959 to Dow); Moss, Godfrey, US 3037023 (1962 to Jefferson Chem.). Prepn of salts: Hefferren *et al.*, *J. Am. Pharm. Assoc.* **44**, 678 (1955). Determn in animal feeds: J. D. McLean, O. L. Daniels, *J. Assoc. Off. Anal. Chem.* **54**, 555 (1971). Veterinary trial in combination with levamisole: J. H. Drudge *et al.*, *Am. J. Vet. Res.* **35**, 67 (1974). Clinical efficacy in roundworm infection: A. S. Nanivadekar *et al.*, *J. Postgrad. Med.* **30**, 144 (1984).

Leaflets from alc; ammoniacal odor. mp 106°. Salty taste. bp_{760} 146°. Strong base: pKa: 4.19, absorbs water and CO_2 from air. Absorption spectrum: Purvis, *J. Chem. Soc.* **103**, 2286, 2293 (1913). Freely sol in glycerol, glycols; one gram dissolves in 2 ml of 95% alcohol. Sol in water. Insol in ether. pH of a 10% aq soln 10.8-11.8. Forms a sol compd with theophylline. *Corrosive. Keep tightly closed and protect from light.*

Hexahydrate. Dietelmin; Helmifren; Parid; Paravermin; Tasnon (elixir); Upixon; Uvilon. $C_4H_{10}N_2.6H_2O$; mol wt 194.23. Crystals from water (contg 44.34% anhydr piperazine), mp 44°. bp 125-130°. The piperazine of commerce is usually this hydrate. Freely sol in water; sol in alc (~1:2). Practically insol in ether. pH of a 10% aq soln 10.8-11.8.

Citrate. [144-29-6] Tripiperazine dicitrate; Helmezine; Oxucide; Parazine; Pinozan; Pipizan Citrate; Pipracid (syrup); Rhomex; Ta-Verm; Worm Away. $3C_4H_{10}N_2.2C_6H_8O_7$; mol wt 642.66. White crystals, dec 182-187°. Sol in water. Practically insol in chloroform. Insol in alc, ether. pH of a 10% aq soln 5.0-6.0.

Edetate calcium. [12002-30-1] Piperazine calcium edathamil; Perin. $C_{14}H_{24}CaN_4O_8$; mol wt 416.44. Prepn: Schlesinger *et al.*, US 2834782 (1958 to Endo). Occurs as the dihydrate. Crystals; slightly salty taste. Freely sol in water. Very slightly sol in alc, chloroform. Practically insol in ether. pH of 20% aq soln 4.3-5.4.

Phosphate. Antepar; Pinrou; Piperverm; Piperazate; Tasnon (tabl.). $C_4H_{10}N_2.H_3PO_4$; mol wt 184.13. White, minute crystals. Sparingly sol in water. Practically insol in alc. pH of saturated aq soln, 6.5.

Tartrate. [133-36-8] Noxiurotan; Piperate. $C_4H_{10}N_2.C_4H_6O_6$; mol wt 236.22. Crystals, dec 258-263°. Soly in g/100 ml at 25°: water 26; alcohol 0.01; chloroform 0.01. pH of 1% soln: 4.8.

N^1,N^4**-Dibenzoylpiperazine.** $C_{18}H_{18}N_2O_2$. Prepd from piperazine and benzoyl chloride in dil NaOH. Crystals from alc, mp 191°.

THERAP CAT: Anthelmintic (Nematodes).

THERAP CAT (VET): Anthelmintic (Nematodes).

7577. Piperazine Adipate. [142-88-1] Hexanedioic acid compd with piperazine (1:1); Entacyl; Oxyzin (tabl.); Vermicompren (tabl.); Nometan; Oxypaat; Pipadox; Oxurasin. $C_{10}H_{20}N_2O_4$; mol wt 232.28. C 51.71%, H 8.68%, N 12.06%, O 27.55%. $C_4H_{10}N_2.C_6H_{10}O_4$. The neutral salt of piperazine and adipic acid: Davies *et al.*, *J. Pharm. Pharmacol.* **6**, 707 (1954). Prepn in methanolic medium: Forrest, Petrow, US 2799617 (1957 to British Drug Houses); GB 767826. Pharmacology: B. G. Cross *et al.*, *J. Pharm. Pharmacol.* **6**, 711 (1954).

Prisms, mp 256-257°. Stable to heat and air. Pleasant, slightly acid taste. Dissolves slowly. Soly in 100 ml water at 20°: 5.53 g, at 30°: 6.61 g, at 56.3°: 10.14 g; in 100 g methanol at 25°: 0.02 g.

Practically insol in abs ethanol, isopropanol, dioxane. Aq solns of 0.2-0.01M have pH 5.45 which is only slightly affected by increases in ionic strength caused by the addition of simple neutral salts. LD_{50} in mice, rats (g/kg): 11.4, 7.9 orally (Cross).

THERAP CAT: Anthelmintic (Nematodes).

THERAP CAT (VET): Anthelmintic (Nematodes).

7578. 2,5-Piperazinedione. [106-57-0] Glycine anhydride; cycloglycylglycine; α,γ-diacipiperazine; glycylglycine lactam; diglycolyldiamide; 2,5-diketopiperazine; 2,5-dioxopiperazine. $C_4H_6N_2O_2$; mol wt 114.10. C 42.11%, H 5.30%, N 24.55%, O 28.04%. Can exist in 5 enol forms: Richardson *et al.*, *J. Am. Chem. Soc.* **51**, 3074 (1929). Prepn from glycine ethyl ester hydrochloride: Fischer, *Ber.* **39**, 2930 (1906).

Platelets from water, needles by sublimation. Dec 311-312° (begins to sublime at 260° and sinters at 305°). Sparingly sol in water; sol in HCl (d 1.19) from which it can be precipitated by the addition of alcohol. Acts as a weak base. Hydrolyzed to glycylglycine by acids and alkalies.

Hydrochloride. $C_4H_6N_2O_2.HCl$. Crystals, mp 129-130°.

7579. Piperic Acid. [5285-18-7] 5-(1,3-Benzodioxol-5-yl)-2,4-pentadienoic acid; 5-(3,4-methylenedioxyphenyl)-2,4-pentadienoic acid. $C_{12}H_{10}O_4$; mol wt 218.21. C 66.05%, H 4.62%, O 29.33%. Early literature treats the title compound without specifying stereochemistry; however, four isomers exist. Physical constants given in early references for piperic acid agree with those of piperinic acid, the (E,E)-isomer. *See* Beilstein **19**, 281; suppl. II, 300. Synthesis of (E,E)-form: L. von Babo, E. Keller, *J. Prakt. Chem.* **72**, 53 (1857); A. Ladenburg, M. Sholtz, *Ber.* **27**, 2958 (1894); of (E,Z)-form: H. Lohaus, H. Gall, *Ann.* **517**, 278 (1935); of (Z,E)-form: E. Ott, F. Eichler, *Ber.* **55**, 2653 (1922); of (Z,Z)-form: *idem, loc. cit.*; H. Lohaus, H. Gall, *loc. cit.;* of all isomers: R. Grewe *et al.*, *Ber.* **103**, 3752 (1970); R. De Cleyn, A. Verzele, *Bull. Soc. Chim. Belg.* **81**, 529 (1972).

(E,E)-form

(E,E)-Form. [136-72-1] Piperinic acid; piperinsäure (German). Needles from alc. Colorless when freshly prepd, rapidly turns yellow on exposure to light. mp 216-217°. Sublimes as yellow needles with partial decomp. uv max (methanol): 340 nm (ε 28800). Sol in 50 parts boiling alcohol, 275 parts abs alcohol at 25°. Practically insol in water, ether, benzene, carbon disulfide.

(E,Z)-Form. [495-87-4] Isochavicinic acid. Yellow crystals from methanol/water, sublimes as needles. mp 134-136° (Lohaus, Gall); 143° (Grewe). uv max (methanol): 335 nm (ε 14500). Sol in methanol, benzene.

(Z,E)-Form. [495-88-5] Isopiperinic acid. Needles from benzene. mp 145° (Ott, Eichler); 153° (Grewe); 138° (DeCleyn, Verzele). uv max (methanol): 328 nm (ε 22000). Sol in benzene.

(Z,Z)-Form. [495-89-6] Chavicinic acid. Amorphous yellow granules from benzene. Wide disparities in mp have been reported: 200-202° (Ott, Eichler); 130° (Grewe); 120° (De Cleyn, Verzele). uv max (methanol): 335 nm (ε 17500). 0.55 g sol in 16 g boiling 95% alcohol, and in 65 g boiling benzene.

7580. Piperidine. [110-89-4] Azacyclohexane; hexahydropyridine. $C_5H_{11}N$; mol wt 85.15. C 70.53%, H 13.02%, N 16.45%. Found in small quantities in *Piper nigrum* L., Piperaceae (black pepper). May be obtained from piperine by heating with alcoholic KOH, or from 1,5-diaminopentane hydrochloride by cyclization. Usually prepd by electrolytic reduction of pyridine. Forms complexes with salts of heavy metals. Because of its reactivity, piperi-

Consult the Name Index before using this section.

dine is useful in the prepn of cryst derivatives of aromatic nitro compds contg nuclear halogen atoms: Seikel, *J. Am. Chem. Soc.* **62**, 750 (1940). Toxicity study: H. F. Smyth *et al.*, *Am. Ind. Hyg. Assoc. J.* **23**, 95 (1962). Review of physical constants of piperidine and *N*-alkyl piperidines: Magnusson, Schierz, *Univ. Wyo. Publ.* **7**, 1 (1940).

Liquid. Characteristic odor. Soapy feel. Solidifies −13° to −17°; mp −7°; bp_{760} 106°; bp_{20} 18°; d_4^{20} 0.8622; n_D^{20} 1.4534. Infrared absorption spectrum: Freymann, *Compt. Rend.* **205**, 852 (1937); *Ann. Chim.* **11**, 11 (1939). Ultraviolet and Raman spectra: Lecomte, *Compt. Rend.* **207**, 395 (1938). *Corrosive; flammable.* Strong base: pK (25°): 2.80. Misc with water. Sol in alcohol, benzene, chloroform. LD_{50} orally in rats: 0.52 ml/kg (Smyth).
Hydrochloride. $C_5H_{11}N.HCl$. Orthorhombic prisms from alcohol, mp 247°. Freely sol in water, alcohol.
Nitrate. $C_5H_{11}N.HNO_3$. Hygroscopic plates, sublimes 75° (10 mm), mp 110°. Freely sol in water, alcohol, ether. Absorption spectrum: Harper, Macbech, *J. Chem. Soc.* **107**, 91 (1915).
Bitartrate. $C_5H_{11}N.C_4H_6O_6$. Crystals, freely sol in water.
Aurichloride. $C_5H_{11}N.HAuCl_4$. Yellow crystals, mp 219°.
Platinichloride. $2C_5H_{11}N.H_2PtCl_6$. Yellow monoclinic prisms from water, mp 202°. Freely sol in water, slightly in alcohol.
***N*-Benzoylpiperidine.** Long needles, mp 44-48°. *Ref: Org. Synth.* **coll. vol. I** (2nd ed., 1941) p 99.
Phosphate. [767-21-5] Prepn from piperidine + phosphoric acid: Abood, *US* **3035977** (1962). Crystals, mp 204-206°.

7581. Piperidione. [77-03-2] 3,3-Diethyl-2,4-piperidinedione; 3,3-diethyl-2,4-dioxopiperidine; dihyprylone; Sedulon; Tusseval. $C_9H_{15}NO_2$; mol wt 169.22. C 63.88%, H 8.94%, N 8.28%, O 18.91%. Prepn: Tsukita, *J. Pharm. Soc. Jpn.* **69**, 194 (1949).

Crystals, mp 102-107°. Bitter taste. Freely sol in water, alcohol, chloroform.
THERAP CAT: Sedative, antitussive.

7582. Piperidolate. [82-98-4] α-Phenylbenzeneacetic acid 1-ethyl-3-piperidinyl ester; diphenylacetic acid 1-ethyl-3-piperidyl ester; *N*-ethyl-3-piperidyl diphenylacetate; JB-305. $C_{21}H_{25}NO_2$; mol wt 323.44. C 77.98%, H 7.79%, N 4.33%, O 9.89%. Anticholinergic. Prepd from diphenylacetyl chloride and *N*-ethyl-3-hydroxypiperidine: Biel *et al.*, *J. Am. Chem. Soc.* **74**, 1485 (1952); *US* **2918407** (1959 to Lakeside Labs.).

Liquid. $bp_{0.18}$ 191-192°.
Hydrochloride. [129-77-1] Crapinon; Dactil. $C_{21}H_{25}NO_2.HCl$; mol wt 359.89. Crystals, mp 195-196°. Soluble in water.
THERAP CAT: Antispasmodic.

7583. Piperilate. [4546-39-8] α-Hydroxy-α-phenylbenzeneacetic acid 2-(1-piperidinyl)ethyl ester; β-piperidylethyl benzilate;

pipethanate; benzilic acid 1-piperidineethanol ester; 1-piperidineethanol benzilate; 2-(1-piperidino)ethyl benzilate. $C_{21}H_{25}NO_3$; mol wt 339.44. C 74.31%, H 7.42%, N 4.13%, O 14.14%. Anticholinergic. Prepn: Ford-Moore, Ing, *J. Chem. Soc.* **1947**, 55.

Hydrochloride. [4544-15-4] Daipisate; Norticon; Pensanate; Pipenale. $C_{21}H_{25}NO_3.HCl$; mol wt 375.89. Crystals from acetone or ethanol, mp 170-171°.
Ethyl bromide. Panpurol. $C_{23}H_{30}BrNO_3$; mol wt 448.40.
THERAP CAT: Antispasmodic.

7584. Piperine. [94-62-2] (2*E*,4*E*)-5-(1,3-Benzodioxol-5-yl)-1-(1-piperidinyl)-2,4-pentadien-1-one; 1-[(2*E*,4*E*)-5-(1,3-benzodioxol-5-yl)-1-oxo-2,4-pentadienyl]piperidine; (*E*,*E*)-1-piperoylpiperidine. $C_{17}H_{19}NO_3$; mol wt 285.34. C 71.56%, H 6.71%, N 4.91%, O 16.82%. Isolated from black pepper (*Piper nigrum* L.); also in *P. longum* L., *P. retrofractum* Vahl. (*P. officinarum* C.D.C.), and *P. clusii* C.D.C.; in root bark of *Piper geniculatum* Sw., *Piperaceae*. Extraction procedure: Cazeneuve, Caillot, *Bull. Soc. Chim.* [2] **27**, 291 (1877). Synthesis: Rugheimer, *Ber.* **15**, 1390 (1882); Newman, *Chem. Prod.* **16**, 379 (1953); Normant, Feugeas, *Compt. Rend.* **258**, 2846 (1964). Spectroscopic structural elucidation and preparative separation of piperine and its stereoisomers isopiperine, isochavicine and chavicine, *q.v.*: R. De Cleyn, M. Verzele, *Bull. Soc. Chim. Belg.* **84**, 435 (1975). Synthesis of isomers: R. Grewe *et al.*, *Ber.* **103**, 3752 (1970); of piperine and isochavicine: S. Tsuboi *et al.*, *Tetrahedron Lett.* **1979**, 1043. Stereoselective synthesis of piperine: R. A. Olsen, G. O. Spessard, *J. Agric. Food Chem.* **29**, 942 (1981). More toxic to houseflies than pyrethrum: Harvill *et al.*, *Contrib. Boyce Thompson Inst.* **13**, 87 (1943).

Monoclinic prisms from alcohol, mp 130°. Tasteless at first, but burning aftertaste. Neutral to litmus. pK (18°): 12.22. Almost insol in water (40 mg/liter at 18°), in petr ether. One gram dissolves in 15 ml alcohol, 1.7 ml chloroform, 36 ml ether. Sol in benzene, acetic acid.
(*E*,*Z*)-Form. Isochavicine. Crystals from chloroform + hexane, mp 89° (Grewe), 103° (De Cleyn). uv max (methanol): 333 nm (ε 16300).
(*Z*,*E*)-Form. Isopiperine. Crystals from chloroform + hexane, mp 110° (Grewe), 86° (De Cleyn). uv max (methanol): 332 nm (ε 21800).
USE: To impart pungent taste to brandy. As insecticide.

7585. Piperitone. [89-81-6] 3-Methyl-6-(1-methylethyl)-2-cyclohexen-1-one; *p*-menth-1-en-3-one; 4-isopropyl-1-methyl-1-cyclohexen-3-one. $C_{10}H_{16}O$; mol wt 152.24. C 78.90%, H 10.59%, O 10.51%. Isoln of *d*-form from oil of *Cymbopogon sennaarensis* Chiov., *Gramineae:* Roberts, *J. Chem. Soc.* **107**, 1465 (1915); from oil of *Andropogon iwarancusa* Jones: Simonsen, *ibid.* **119**, 1644 (1921); from oil of *Mentha* spp., *Labiatae:* Reitsma, *J. Am. Pharm. Assoc.* **47**, 265, 267 (1958). Isoln of *l*-form from Sitka spruce oil: von Rudloff, *Can. J. Chem.* **42**, 1057 (1964). Isoln of *dl*-form from oil of *Eucalyptus dives* Schau., *Myrtaceae:* Read, Smith, *J. Chem. Soc.* **119**, 779 (1921); from peppermint oil: Katsuragi, *Koryo* **No. 24**, 16 (1953). Synthesis of *dl*-form: Misrock, Church, *Ind. Eng. Chem.* **49**, 822 (1957); Bain *et al.*, *US* **2972632** (1961 to Glidden); Wiemann, Dubois, *Bull. Soc. Chim. Fr.* **1962**, 1813; Stepanov, Myrsina, *Zh. Obshch. Khim.* **34**, 3092 (1964).

Piperlongumine

d-**Form.** Liquid. Peppermint odor. bp 232-235°; bp$_{20}$ 116-118.5°. d$_4^{20}$ 0.9344. [α]$_D^{20}$ +49.13°. n$_D^{20}$ 1.4848.

l-**Form.** Liquid. bp$_{15}$ 109.5-110.5°. d$_4^{20}$ 0.9324. [α]$_D^{20}$ −15.9°. n$_D^{20}$ 1.4823. Practically insol in water. Sol in alcohol, oils.

dl-**Form.** Liquid. bp$_{769}$ 232-233°; bp$_{16}$ 116-118°. d$_4^{20}$ 0.9331. n$_D^{24}$ 1.4823. uv max (ethanol): 232.5 nm (ε 13350).

USE: In masking odors in dentifrices.

7586. Piperlongumine. [20069-09-4] 5,6-Dihydro-1-[(2*E*)-1-oxo-3-(3,4,5-trimethoxyphenyl)-2-propen-1-yl]-2(1*H*)-pyridinone; 5,6-dihydro-1-(3,4,5-trimethoxycinnamoyl)-2(1*H*)-pyridone; piplartine. C$_{17}$H$_{19}$NO$_5$; mol wt 317.34. C 64.34%, H 6.04%, N 4.41%, O 25.21%. Pyridone alkaloid isolated from Indian long pepper, *Piper longum* L., *Piperaceae*, an Ayurvedic medicinal plant primarily used in the treatment of asthma and bronchitis. Selectively induces apoptosis in cancer cells by blocking the stress response to reactive oxygen species (ROS). Isoln: C. K. Atal, S. S. Banga, *Indian J. Pharm.* **24**, 105 (1962). Structure: A. Chatterjee, C. P. Dutta, *Sci. Cult.* **29**, 568 (1963). Synthesis: *eidem, Tetrahedron* **23**, 1769 (1967). Crystal structure: T. Banerjee, S. Chaudhuri, *Can. J. Chem.* **64**, 876 (1986). Biosynthesis: B. R. Prabhu, N. B. Mulchandani, *Phytochemistry* **24**, 2589 (1985). Inhibition of platelet aggregation: M. Iwashita *et al., Eur. J. Pharmacol.* **570**, 38 (2007). Selective effect on ROS in cancer cells: L. Raj *et al., Nature* **475**, 231 (2011).

White acicular crystals from alc, mp 124-125°. uv max in ethanol: 220, 328 nm (log ε 4.99, 4.50); in 0.01*N* HCl: 328 nm (log ε 4.28); in 0.01*N* alkali: 232, 304 nm (log ε 4.37, 4.28). Sol in chloroform, conc acids and alkalis. Sparingly sol in ethanol, methanol, benzene. Insol in water.

7587. Piperocaine. [136-82-3] 2-Methyl-1-piperidinepropanol benzoate; γ-(2-methylpiperidyl)propyl benzoate; (2-methylpiperidino)propyl benzoate; 3-benzoxy-1-(2-methylpiperidino)propane; benzoyl-γ-(2-methylpiperidino)propanol. C$_{16}$H$_{23}$NO$_2$; mol wt 261.37. C 73.53%, H 8.87%, N 5.36%, O 12.24%. Prepn of the hydrochloride: S. M. McElvain, **US 1784903** (1930). Pharmacology: K. H. Beyer, A. R. Latven, *J. Pharmacol. Exp. Ther.* **106**, 37 (1952). Structure activity study: P. P. Koelzer, K. H. Wehr, *Arzneim.-Forsch.* **8**, 708 (1958).

LD$_{50}$ i.v. in rabbits: 18 ±6 mg/kg (Beyer, Latven).

Hydrochloride. [533-28-8] C$_{16}$H$_{23}$NO$_2$.HCl. Crystals from alc-ether, mp 167-169°. Stable in air. One gram dissolves in 1.5 ml water, 4.5 ml alcohol. Sol in alc, chloroform. Practically insol in ether, fixed oils. LD$_{50}$ in mice (mg/20g): 9 s.c.; in rats (mg/kg): 129 i.p. (Koelzer).

THERAP CAT: Anesthetic (local).

THERAP CAT (VET): Anesthetic (local).

7588. Piperonal. [120-57-0] 1,3-Benzodioxole-5-carboxaldehyde; 3,4-(methylenedioxy)benzaldehyde; heliotropin; piperonylaldehyde; dioxymethyleneprotocatechuic aldehyde. C$_8$H$_6$O$_3$; mol wt 150.13. C 64.00%, H 4.03%, O 31.97%. Prepn: Blair, **US 2916499** (1959 to Welsbach Corp.); Holum, *J. Org. Chem.* **26**, 4814 (1961); Feugeas, *Bull. Soc. Chim. Fr.* **1964**, 1892. Toxicity: Hagan *et al., Toxicol. Appl. Pharmacol.* **7**, 18 (1965).

Colorless, lustrous crystals, mp 37°. Heliotrope odor. bp ~263°; bp$_{0.5}$ 88°. Sol in 500 parts water; freely sol in alcohol, ether. *Keep in cool place protected from light.* LD$_{50}$ orally in rats: 2700 mg/kg (Hagan).

USE: In perfumery, in cherry and vanilla flavors, in organic syntheses.

THERAP CAT: Has been used as pediculicide.

7589. Piperonyl Butoxide. [51-03-6] 5-[[2-(2-Butoxyethoxy)ethoxy]methyl]-6-propyl-1,3-benzodioxole; α-[2-(2-butoxyethoxy)ethoxy]-4,5-methylenedioxy-2-propyltoluene; [3,4-(methylenedioxy)-6-propylbenzyl] butyl diethyleneglycol ether; 6-propylpiperonyl butyl diethylene glycol ether; butylcarbityl (6-propylpiperonyl) ether; ENT-14250; Butacide. C$_{19}$H$_{30}$O$_5$; mol wt 338.44. C 67.43%, H 8.94%, O 23.64%. Prepn: H. Wachs, **US 2485681**; **US 2550737** (1949, 1951 both to U.S. Industrial Chemicals); *idem, Science* **105**, 530 (1947). Toxicity: T. B. Gaines, *Toxicol. Appl. Pharmacol.* **14**, 515 (1969). Review of carcinogenic risk: *IARC Monographs* **30**, 183-195 (1983).

Liquid. d 1.04-1.07. bp$_{1.0}$ 180°. n$_D^{20}$ 1.50. Flash pt 340°F. Miscible with methanol, ethanol, benzene, Freons, Geons, other organic solvents, oils. LD$_{50}$ orally in female, male rats: 6150, 7500 mg/kg (Gaines).

Mixture with synthetic pyrethroids. Derringer; Duracide; Grovex; Prentox; Scourge.

In combination with rotenone. PB-NOX; Chem-Fish; Rotacide. *Caution:* Large doses have caused vomiting, diarrhea: Sarles *et al., Am. J. Trop. Med.* **29**, 151 (1949).

USE: Insecticide synergist, especially for pyrethroids and rotenone.

7590. Piperonylic Acid. [94-53-1] 1,3-Benzodioxole-5-carboxylic acid; 3,4-methylenedioxybenzoic acid; protocatechuic acid methylene ether. C$_8$H$_6$O$_4$; mol wt 166.13. C 57.84%, H 3.64%, O 38.52%. Occurs in Paracoto bark. Prepd by permanganate oxidation of piperonal: Shriner, Kleiderer, *Org. Synth.* **10**, 82 (1930).

Prisms (by sublimation), needles from alc, feathery crystals from water. mp 229°. Sublimes around 210°. Slightly sol in water, chloroform, cold alcohol, ether. Absorption spectrum: Dobbie, Lauder, *J. Chem. Soc.* **83**, 621 (1903).

Methyl ester. C$_9$H$_8$O$_4$. mp 53°. Sublimes easily. Volatile in steam. Freely sol in alcohol, ether.

7591. Piperoxan. [59-39-2] 1-[(2,3-Dihydro-1,4-benzodioxin-2-yl)methyl]piperidine; 2-piperidinomethyl-1,4-benzodioxan; 2-(1-piperidylmethyl)-1,4-benzodioxan; benzodioxane; Benodaine. C$_{14}$H$_{19}$NO$_2$; mol wt 233.31. C 72.07%, H 8.21%, N 6.00%, O 13.71%. α-Adrenergic blocker. Prepn: Fourneau, **US 2056046** (1936 to Rhône-Poulenc).

Consult the Name Index before using this section.

bp$_{17}$ 193°.

dl-Form hydrochloride. Compd 933F; Fourneau 933. C$_{14}$H$_{19}$-NO$_2$.HCl; mol wt 269.77. Crystals, mp 232-234° (darkens at 220°). uv max: 275 nm (E$^{1\%}_{1cm}$ 82); min: 240 nm (E$^{1\%}_{1cm}$ <5). Freely sol in water, pH of 1% soln ~5. Alkalies liberate the water-insol base. Soluble in acid solns. Soly in isopropanol: ~10.8 mg/g at 25°. Crystals are not hygroscopic and are stable to light, air, and normal storage temps. An aq soln at its own pH (5) is stable to autoclaving and to many months of storage at room temp.

THERAP CAT: Antihypertensive. Diagnostic aid (pheochromocytoma).

7592. PIPES. [5625-37-6] 1,4-Piperazinediethanesulfonic acid; piperazine-N,N'-bis(2-ethanesulfonic acid); 1,4-piperazinebis-(ethanesulfonic acid). C$_8$H$_{18}$N$_2$O$_6$S$_2$; mol wt 302.36. C 31.78%, H 6.00%, N 9.27%, O 31.75%, S 21.21%. One of the zwitterionic N-substituted aminosulfonic acids known as "Good" buffers; active in the pH range 6-8.5. Prepn: N. Good et al., Biochemistry 5, 467 (1966). Interference with Lowry protein determn: H. M. Himmel, W. Heller, J. Clin. Chem. Clin. Biochem. 25, 909 (1987). Use as biological buffer: R. Salema, I. Brandao, J. Submicrosc. Cytol. 5, 79 (1973); S. Haviernick et al., J. Microsc. 135, 83 (1984).

Monosodium salt. [10010-67-0] C$_8$H$_{17}$N$_2$NaO$_6$S$_2$. Crystals from water and alcohol, mp >300° (dec). pKa$_1$ ~3, pKa$_2$ (20°): 6.82 (0.1M); 6.82 (0.2M); 6.96 (0.01M). ΔpKa/°C −0.0085.

Disodium salt. [76836-02-7] Sodium pipesate. C$_8$H$_{16}$N$_2$Na$_2$-O$_6$S$_2$.

USE: Biological buffer.

7593. Pipobroman. [54-91-1] 1,1'-(1,4-Piperzinediyl)bis-[3-bromo-1-propanone]; 1,4-bis-(3-bromo-1-oxopropyl)piperazine; 1,4-bis-(3-bromopropionyl)piperazine; A-8103; NSC-25154; Amedel; Vercyte. C$_{10}$H$_{16}$Br$_2$N$_2$O$_2$; mol wt 356.06. C 33.73%, H 4.53%, Br 44.88%, N 7.87%, O 8.99%. Prepn: Horrom, Carbon, DE 1138781 (1962 to Abbott), C.A. 58, 7955c (1963); Groszkowski, Rocz. Chem. 38, 229 (1964), C.A. 60, 14506a (1964). Structure-cytostatic effect studies: Groszkowski et al., J. Med. Chem. 11, 621 (1968); Oteleanu, Retezeanu, Farmacia (Bucharest) 16, 279 (1968), C.A. 69, 33591w (1968).

Crystals, mp 106-107°.

THERAP CAT: Antineoplastic.

7594. Piposulfan. [2608-24-4] 1,1'-(1,4-Piperazinediyl)-bis[3-[(methylsulfonyl)oxy]-1-propanone]; 1,4-bis[3-[(methylsulfonyl)oxy]-1-oxopropyl]piperazine; 1,4-dihydracryloylpiperazine dimethanesulfonate; 1,4-bis(3-hydroxypropionyl)piperazine dimethanesulfonate; N,N'-bis(3-methanesulfonyloxypropionyl)piperazine; N,N'-bis(3-methylsulfonyloxypropionyl)piperazine; A-20968; NSC-47774; Ancyte. C$_{12}$H$_{22}$N$_2$O$_8$S$_2$; mol wt 386.43. C 37.30%, H 5.74%, N 7.25%, O 33.12%, S 16.59%. Prepn: Horrom, Carbon, DE 1177162 (1964 to Abbott), C.A. 61, 13329a (1964).

Crystals from water, mp 175-177°.

THERAP CAT: Antineoplastic.

7595. Pipotiazine. [39860-99-6] 10-[3-[4-(2-Hydroxyethyl)-1-piperidinyl]propyl]-N,N-dimethyl-10H-phenothiazine-2-sulfonamide; 2-[1-[3-[2-[(dimethylamino)sulfonyl]-10H-phenothiazin-10-yl]propyl]-4-piperidinyl]ethanol; pipothiazine; RP-19366; Piportil. C$_{24}$H$_{33}$N$_3$O$_3$S$_2$; mol wt 475.67. C 60.60%, H 6.99%, N 8.83%, O 10.09%, S 13.48%. Prepn: FR M7835 (1970 to Rhône-Poulenc), C.A. 78, 43499x (1973). Pharmacokinetics: P. J. De Schepper et al., Arzneim.-Forsch. 29, 1056 (1979); and HPLC determn in plasma: D. A. Ogden et al., J. Pharm. Biomed. Anal. 7, 1273 (1989). Series of articles on pharmacology and clinical use: Acta Psychiat. Scand., Suppl. 241, 9-138 (1973). Toxicity: L. Julou et al., ibid. 9.

LD$_{50}$ in mice (mg/kg): 108 i.p.; 360 s.c.; 440 orally (Julou).

Palmitic ester. [37517-26-3] Pipotiazine palmitate; RP-19552; Piportil L4. C$_{40}$H$_{63}$N$_3$O$_4$S$_2$; mol wt 714.08.

Undecylenic ester. [22178-11-6] Pipotiazine undecylenate; RP-19551; Piportil M2. C$_{35}$H$_{51}$N$_3$O$_4$S$_2$; mol wt 641.93.

THERAP CAT: Antipsychotic.

7596. Pipoxolan Hydrochloride. [18174-58-8] 5,5-Diphenyl-2-[2-(1-piperidinyl)ethyl]-1,3-dioxolan-4-one hydrochloride; 2-(β-N-piperidylethyl)-4,4-diphenyl-1,3-dioxolan-5-one hydrochloride; BR-18; Rowapraxin. C$_{22}$H$_{26}$ClNO$_3$; mol wt 387.90. C 68.12%, H 6.76%, Cl 9.14%, N 3.61%, O 12.37%. Prepn: M. Pailer et al., Monatsh. Chem. 99, 891 (1968); GB 1109959 and BE 719230 (1968, 1969 to Rowa-Wagner). Pharmacology and acute toxicity: K. Morsdorf, H. Wengenroth, Pharmacology 3, 193 (1970).

Crystals from isopropyl alc, mp 207-209°. Soluble in water; stable to mild attack by acids or bases. LD$_{50}$ in rats, mice (mg/kg): 1500, 700 orally; 60, 35 i.v.; 130, 130 i.p.; in rats (mg/kg): >300 s.c. (Morsdorf).

THERAP CAT: Antispasmodic.

7597. Pipradrol. [467-60-7] α,α-Diphenyl-2-piperidine-methanol; α-(2-piperidyl)benzhydrol; alpha-pipradrol; pipradrol; MRD-108. C$_{18}$H$_{21}$NO; mol wt 267.37. C 80.86%, H 7.92%, N 5.24%, O 5.98%. Prepd by hydrogenation of α,α-diphenyl-2-pyridinemethanol: Tilford et al., J. Am. Chem. Soc. 70, 4001 (1948); Werner, Tilford, US 2624739 (1953 to Merrell).

Hydrochloride. [71-78-3] Meratran; Stimolag. C$_{18}$H$_{21}$NO.-HCl; mol wt 303.83. Crystals from butanone, dec 308-309°. Slightly bitter taste. One gram dissolves in 60 ml of hot water.

Note: This is a controlled substance (stimulant): 21 CFR, 1308.14.

THERAP CAT: CNS stimulant.

7598. Piprozolin. [17243-64-0] 2-[3-Ethyl-4-oxo-5-(1-piperidinyl)-2-thiazolidinylidene]acetic acid ethyl ester; 3-ethyl-4-oxo-5-piperidino-Δ²,ᵅ-thiazolidineacetic acid ethyl ester; Gö-919; W-3699; Coleflux; Epsyl; Probilin; Secrebil. $C_{14}H_{22}N_2O_3S$; mol wt 298.40. C 56.35%, H 7.43%, N 9.39%, O 16.08%, S 10.74%. Prepn: G. Satzinger, *Ann.* **665**, 150 (1963); G. Satzinger *et al.*, **DE 2414345** (1975 to Goedecke), corresp to US 3971794 (1976 to Warner-Lambert). Metabolism and pharmacokinetics: K. O. Vollmer, F. W. Koss, *Arch. Int. Pharmacodyn. Ther.* **198**, 312 (1972). Mechanism of action: F. W. Koss *et al.*, *ibid.* 333. Series of articles on synthesis, pharmacology, toxicology and clinical studies: *Arzneim.-Forsch.* **27**, 463-526 (1977). Toxicity data: M. Herrmann *et al.*, *ibid.* 467.

Colorless cryst, mp 86-87°. Practically insol in water. Sol in dil aq acids and most organic solvents. uv max (methanol): 245, 285 nm (ε 8200, 20000) (Vollmer). LD₅₀ in mice, rats (mg/kg): 1070, 3256 orally (Herrmann).

THERAP CAT: Choleretic.

7599. Pipsyl Chloride. [98-61-3] 4-Iodobenzenesulfonyl chloride; *p*-iodophenyl sulfonyl chloride. $C_6H_4ClIO_2S$; mol wt 302.51. C 23.82%, H 1.33%, Cl 11.72%, I 41.95%, O 10.58%, S 10.60%. Usually prepd from iodide ion and *p*-diazobenzenesulfonic acid, followed by treatment with phosphorus pentachloride. A 5-10-fold excess reacts quantitatively with amino acids as indicated by the disappearance of amino nitrogen. Prepn and use of radioactive pipsyl chloride: A. S. Keston *et al.*, *J. Am. Chem. Soc.* **68**, 1390 (1946).

USE: Radioactive iodine form in the analysis of proteins.

7600. Piracetam. [7491-74-9] 2-Oxo-1-pyrrolidineacetamide; 2-pyrrolidoneacetamide; 2-pyrrolidinoneacetamide; 2-ketopyrrolidine-1-ylacetamide; 1-acetamido-2-pyrrolidinone; UCB-6215; Avigilen; Axonyl; Cerebroforte; Encetrop; Gabacet; Geram; Nootrop; Nootropil; Nootropyl; Norzetam; Normabraïn; Piracebral; Piracetrop; Sinapsan. $C_6H_{10}N_2O_2$; mol wt 142.16. C 50.69%, H 7.09%, N 19.71%, O 22.51%. Prepn: H. Morren, **NL 6509994**; *eidem*, **US 3459738** (1966, 1969 both to U.C.B.). Pharmacology: Giurgea *et al.*, *Arch. Int. Pharmacodyn. Ther.* **166**, 238 (1967); Giurgea, Moyersoons, *ibid.* **188**, 401 (1970); Giurgea *et al.*, *Psychopharmacologia* **20**, 160 (1971). Metabolism and biochemical studies: Gobert, *J. Pharm. Belg.* **27**, 281 (1972). Clinical studies: W. J. Oosterveld, *Arzneim.-Forsch.* **30**, 1947 (1980); G. Chouinard *et al.*, *Psychopharmacol. Bull.* **17**, 129 (1981); in dyslexia: M. Di Ianni *et al.*, *J. Clin. Psychopharmacol.* **5**, 272 (1985).

Crystals from isopropanol, mp 151.5-152.5°.
THERAP CAT: Nootropic.

7601. Pirarubicin. [72496-41-4] (8*S*,10*S*)-10-[[3-Amino-2,3,6-trideoxy-4-*O*-[(2*R*)-tetrahydro-2*H*-pyran-2-yl]-α-L-*lyxo*-hexopyranosyl]oxy]-7,8,9,10-tetrahydro-6,8,11-trihydroxy-8-(2-hydroxyacetyl)-1-methoxy-5,12-naphthacenedione; 4′-*O*-tetrahydropyranyl doxorubicin; (2″*R*)-4′-*O*-tetrahydropyranyladriamycin; tepirubicin; THP; THP-ADM; THP-adriamycin; 1609-RB; Theprubicin. $C_{32}H_{37}NO_{12}$; mol wt 627.64. C 61.24%, H 5.94%, N 2.23%, O

30.59%. Structural analog of doxorubicin, *q.v.* Prepn of (2″*R*) and (2″*S*)-diastereomers: H. Umezawa *et al.*, *J. Antibiot.* **32**, 1082 (1979); *idem et al.*, **EP 14853**; *eidem*, US 4303785 (1980, 1981 both to Microbiochem. Res. Found. Japan). Absolute configuration: *idem et al.*, *J. Antibiot.* **37**, 1094 (1984). HPLC determn in biological fluids: Y. Matsushita *et al.*, *ibid.* **36**, 880 (1983). Cellular uptake and inhibition of DNA synthesis: S. Kunimoto *et al.*, *ibid.*. 312 (1983). Mechanism of action study: K.-I. Kiyomiya *et al.*, *Int. J. Oncol.* **21**, 1081 (2002). Clinical pharmacokinetics and toxicity: A. A. Miller, C. G. Schmidt, *Cancer Res.* **47**, 1461 (1987). Overview of clinical experience: H. Majima, K. Ohta, *Biomed. Pharmacother.* **41**, 237-243 (1987). Clinical evaluation in metastatic colon cancer: D. Fallik *et al.*, *Ann. Oncol.* **14**, 856 (2003).

Red solid, mp 188-192° (dec). [α]$_D^{25}$ +175 ±25° (c = 0.2 in CHCl₃). uv and visible max (methanol): 234, 252, 290, 498, 531.5, 580 nm (E$_{1cm}^{1\%}$ 480, 350, 110, 140, 100, 45). Sol in ethyl acetate, chloroform, and ethanol; slightly sol in water, *n*-hexane, petr ether. Ethanolic and acidic solutions are red in color; give a positive ninhydrin reaction and do not reduce Fehling's solution. LD₅₀ i.v. in mice: 27.8 mg/kg (Umezawa, 1979).

Hydrochloride. [95343-20-7] Pinorubin; Therarubicin. $C_{32}H_{37}NO_{12}$·HCl; mol wt 664.10. Red crystalline solid. Sol in water and methanol.

THERAP CAT: Antineoplastic.

7602. Pirbuterol. [38677-81-5] α⁶-[[(1,1-Dimethylethyl)amino]methyl]-3-hydroxy-2,6-pyridinedimethanol; 2-hydroxymethyl-3-hydroxy-6-(1-hydroxy-2-*tert*-butylaminoethyl)pyridine. $C_{12}H_{20}N_2O_3$; mol wt 240.30. C 59.98%, H 8.39%, N 11.66%, O 19.97%. Analog of albuterol, *q.v.*, with β₂-adrenergic stimulating activity. Prepn: W. E. Barth, **DE 2204195**; *idem*, US 3700681 (both 1972 to Pfizer). Stability study: P. C. Bansal, D. C. Monkhouse, *J. Pharm. Sci.* **66**, 819 (1977). Biotransformation: H. M. McIlhenny, M. S. D. Ghaly, *Fed. Proc.* **38**, 1130 (1979). Pharmacokinetics and cardiopulmonary effects in dogs: J. W. Constantine *et al.*, *J. Pharmacol. Exp. Ther.* **208**, 371 (1979). Comparative study in respiratory disease: A. J. Dyson, A. D. Mackay, *Br. J. Dis. Chest* **74**, 70 (1980). Use in treatment of cardiac failure: N. A. Awan *et al.*, *Clin. Res.* **28**, 17A (1980); W. S. Colucci *et al.*, *N. Engl. J. Med.* **305**, 185 (1981); G. I. Nelson *et al.*, *Eur. Heart J.* **3**, 238 (1982); K. T. Weber *et al.*, *Circulation* **66**, 1262 (1982). Review of pharmacology and efficacy in bronchospastic disease: D. M. Richards, R. N. Brogden, *Drugs* **30**, 6-21 (1985).

Dihydrochloride. [38029-10-6] CP-24314-1. $C_{12}H_{22}Cl_2N_2O_3$; mol wt 313.22. Crystals from ethanol/isopropyl ether, mp 182° (dec).

Monoacetate. Maxair; Spirolair. $C_{14}H_{24}N_2O_5$; mol wt 300.36. Freely sol in water.

THERAP CAT: Bronchodilator.

7603. Pirenoxine. [1043-21-6] 1-Hydroxy-5-oxo-5*H*-pyrido[3,2-*a*]phenoxazine-3-carboxylic acid; 1-hydroxy-5*H*-pyrido[3,2-*a*]phenoxazin-5-one-3-carboxylic acid; 1-hydroxy-3-carboxy-5*H*-pyrido[3,2-*a*]phenoxazin-5-one; pirfenoxone; Catalin. $C_{16}H_8N_2O_5$;

mol wt 308.25. C 62.34%, H 2.62%, N 9.09%, O 25.95%. Prepn: S. Ogino, **JP 59 2227**; S. Ishii, **JP 61 1782** (1959, 1961 both to Chizu Drug), *C.A.* **54**, 11058i (1960), *C.A.* **55**, 21494e (1961); of the sodium salt: S. Ishii, K. Ogata, **JP 73 2672** (1973 to Senju), *C.A.* **80**, 6959t (1974). Pharmacological studies: F. Ikemoto *et al.*, *Oyo Yakuri* **8**, 937 (1974), *C.A.* **83**, 71663t (1975). Toxicological studies: *eidem*, *ibid.* 911, 923, *C.A.* **82**, 51617g, 68262k (1975). Influence on carbohydrate metabolism in the lens: I. Korte *et al.*, *Ophthalmic Res.* **7**, 282 (1975). Effect on NADH, NADPH: *eidem*, *ibid.* 440. Effect on senile cataracts: T. Murata, *Folia Ophthalmol. Jpn.* **31**, 1217 (1980). Clinical trial in treatment of cataracts: S. K. Angra *et al.*, *Indian J. Ophthalmol.* **31**, 5 (1983).

Orange-yellow powder, mp 247-248° (dec).
Sodium salt. [51410-30-1] Clarvisan. $C_{16}H_7N_2NaO_5$; mol wt 330.23. Very sol in water. LD_{50} in mice (mg/kg): >10000 orally; >5000 s.c.; 2120-2250 i.p.; LD_{50} i.p. in rats (mg/kg): 2400 (males); 1460 (females) (Ikemoto).
THERAP CAT: Treatment of cataracts.

7604. Pirenzepine. [28797-61-7] 5,11-Dihydro-11-[2-(4-methyl-1-piperazinyl)acetyl]-6H-pyrido[2,3-b][1,4]benzodiazepin-6-one; LS-519. $C_{19}H_{21}N_5O_2$; mol wt 351.41. C 64.94%, H 6.02%, N 19.93%, O 9.11%. Tricyclic gastric-acid inhibitor. Prepn: **FR 1505795** (1967 to Thomae), *C.A.* **70**, 4154w (1969). Pharmacology: W. Eberlein *et al.*, *Arzneim.-Forsch.* **27**, 356 (1977). Pharmacokinetics: R. Hammes *et al.*, *ibid.* 928. Mechanism of action: G. Heller *et al.*, *Verh. Dtsch. Ges. Inn. Med.* **84**, 991 (1978), *C.A.* **90**, 132984s (1979). Human pharmacology: H. Brunner *et al.*, *Arzneim.-Forsch.* **27**, 684 (1977). Multicenter controlled clinical trial: *Scand. J. Gastroenterol.* **17**, Suppl. 81, 1-42 (1982). Radioimmunoassay determn in human plasma and urine: C. A. Homon *et al.*, *Ther. Drug Monit.* **9**, 236 (1987). Symposium: *ibid.* Suppl. 72, 1-273. Review of pharmacology and therapeutic efficacy: A. A. Carmine, R. N. Brogden, *Drugs* **30**, 85-126 (1985). Comprehensive description: H. A. El-Obeid *et al.*, *Anal. Profiles Drug Subs.* **16**, 445-506 (1987).

Dihydrochloride. [29868-97-1] LS-519-Cl2; Duogastral; Durapirenz; Gasteril; Gastrozepin; Leblon; Maghen; Renzepin; Tabe; Ulcuforton; Ulcosan. $C_{19}H_{21}N_5O_2 \cdot 2HCl$; mol wt 424.33. Sol in water, slightly sol in methanol. Practically insol in ether.
THERAP CAT: Antiulcerative.

7605. Piretanide. [55837-27-9] 3-(Aminosulfonyl)-4-phenoxy-5-(1-pyrrolidinyl)benzoic acid; 4-phenoxy-3-(1-pyrrolidinyl)-5-sulfamoylbenzoic acid; HOE-118; S-73-4118; Diumax; Eurelix; Tauliz. $C_{17}H_{18}N_2O_5S$; mol wt 362.40. C 56.34%, H 5.01%, N 7.73%, O 22.07%, S 8.85%. High-ceiling loop diuretic, structurally related to bumetanide, *q.v.* Prepn: D. Bormann *et al.*, **DE 2419970**; *eidem*, **US 4010273** (1975, 1977 both to Hoechst). Chemistry and pharmacology: W. Merkel *et al.*, *Eur. J. Med. Chem.* **11**, 399 (1976).

Diuretic activity in man: N. Pozet *et al.*, *Br. J. Clin. Pharmacol.* **9**, 577 (1980); T. Saruta, E. Kato, *Arzneim.-Forsch.* **30**, 1807 (1980). Vascular effects of piretanide: E. Klaus *et al.*, *ibid.* **33**, 1273 (1983); renal effects: M. Omosu *et al.*, *ibid.* 1277. Review of pharmacology and therapeutic efficacy: S. P. Clissold, R. N. Brogden, *Drugs* **29**, 489-530 (1985).

Pale yellow platelets from methanol/water, mp 225-227°. Exhibits intense light blue fluorescence at 366 nm. LD_{50} in rats, mice (mg/kg): 5601, 3672 orally (Merkel).
Monosodium salt. [112132-09-9] Arelix. $C_{17}H_{17}N_2NaO_5S$; mol wt 384.38.
THERAP CAT: Diuretic.

7606. Pirfenidone. [53179-13-8] 5-Methyl-1-phenyl-2(1H)-pyridinone; AMR-69; Deskar; Esbriet; Pirespa. $C_{12}H_{11}NO$; mol wt 185.23. C 77.81%, H 5.99%, N 7.56%, O 8.64%. Anti-inflammatory and antifibrotic agent. Prepd by phenylation of 5-methyl-2-pyridinone: S. M. Gadekar, **US 3839346** (1974 to Affiliated Med. Res.); by photochemical rearrangement of N-phenyl-6-methylpyridinium-3-olate: T. Laerum, K. Undheim, *Acta Chem. Scand. B* **32**, 68 (1978). HPLC determn in plasma: Y. Wang *et al.*, *Biomed. Chromatogr.* **20**, 1375 (2006). Clinical pharmacokinetics: S. Shi *et al.*, *J. Clin. Pharmacol.* **47**, 1268 (2007). Clinical trial in idiopathic pulmonary fibrosis: P. W. Noble *et al.*, *Lancet* **377**, 1760 (2011). Review of clinical experience: L. Richeldi, R. M. duBois, *Expert Rev. Respir. Med.* **5**, 473-481 (2011).

White solid from hot water, mp 102-104° (Gadekar); also reported as mp 103-105° from benzene-light petroleum (Laerum, Undheim). Sol in methanol.
THERAP CAT: Antifibrotic; in treatment of idiopathic pulmonary fibrosis.

7607. Piribedil. [3605-01-4] 2-[4-(1,3-Benzodioxol-5-yl-methyl)-1-piperazinyl]pyrimidine; 2-(4-piperonyl-1-piperazinyl)pyrimidine; 2-[4-(3,4-methylenedioxybenzyl)piperazino]pyrimidine; 1-(2-pyrimidyl)-4-piperonylpiperazine; 1-(2″-pyrimidyl)-4-(methylene-3′,4′-dioxybenzyl)piperazine; ET-495; EU-4200; Trivastal. $C_{16}H_{18}N_4O_2$; mol wt 298.35. C 64.41%, H 6.08%, N 18.78%, O 10.72%. Central dopaminergic agonist. Prepn: **NL 6413349**; G. Regnier *et al.*, **US 3299067**; *eidem*, **GB 1101425** (1965, 1967, 1968 all to Sci. Union et Cie-Soc. Franc. Rech. Med.); G. Regnier *et al.*, *J. Med. Chem.* **11**, 1151 (1968). Activity in man: R. Royer, *Proc. 3rd Int. Pharmacol. Meet.* **3**, R. K. Richards, Ed. (Pergamon Press, New York, 1968) pp 45-55. Pharmacology: M. Laubie *et al.*, *Eur. J. Pharmacol.* **6**, 75 (1969). Metabolism: D. B. Campbell *et al.*, *Adv. Neurol.* **3**, 199 (1973). Review of mechanism of action: P. Jenner, *J. Neurol.* **239**, Suppl. 1, S2-S8 (1992).

Crystals from anhydr ethanol, mp 98°. LD$_{50}$ in mice (mg/kg): 88 i.v., 690 i.p., 1460 orally (Laubie).

THERAP CAT: Vasodilator (peripheral).

7608. Pirifibrate. [55285-45-5] 2-(4-Chlorophenoxy)-2-methylpropanoic acid [6-(hydroxymethyl)-2-pyridinyl]methyl ester; 2,6-pyridinedimethanol mono-p-chlorophenoxyisobutyrate; EL-466; Bratenol. C$_{17}$H$_{18}$ClNO$_4$; mol wt 335.78. C 60.81%, H 5.40%, Cl 10.56%, N 4.17%, O 19.06%. Hypolipemic agent related structurally to clofibrate, q.v. Prepn: D. Humbert, R. Ratouis, **DE 2432322**; eidem, **US 3971798** (1975, 1976 both to Roussel). Multicenter study in hyperlipoproteinemias: A. J. Domingo et al., Clin. Ther. **3**, 219 (1980).

Crystals from isopropyl ether, mp 46°. LD$_{50}$ in mice (mg/kg): 915-1098 i.p. (Humbert, Ratouis, 1975).

Hydrochloride. C$_{17}$H$_{18}$ClNO$_4$.HCl. Crystals, mp 110°. LD$_{50}$ (calculated as base) in mice (mg/kg): 1098-1281 i.p. (Humbert, Ratouis, 1975).

THERAP CAT: Antilipemic.

7609. Pirimicarb. [23103-98-2] N,N-Dimethylcarbamic acid 2-(dimethylamino)-5,6-dimethyl-4-pyrimidinyl ester; 2-(dimethylamino)-5,6-dimethyl-4-pyrimidinyl dimethylcarbamate; 5,6-dimethyl-2-dimethylamino-4-dimethylcarbamoyloxypyrimidine; PP-062; ENT-27766; Aphox; Fernos; Pirimor. C$_{11}$H$_{18}$N$_4$O$_2$; mol wt 238.29. C 55.45%, H 7.61%, N 23.51%, O 13.43%. Selective aphicide. Prepn: F. L. C. Baranyovits et al., **ZA 6701588**; eidem, **US 3493574** (1968, 1970 to ICI). Activity: F. L. C. Baranyovits, R. Ghosh, Chem. Ind. (London) **1969**, 1018. Degradn in soil: I. R. Hill, ACS Symp. Ser. **29**, 358 (1976). GLC analysis: P. D. Bland, J. Assoc. Off. Anal. Chem. **64**, 1315 (1981).

Crystalline solid, mp 90.5°. Vapor pressure at 30°: 3×10^{-5} mm Hg. Soly in water at 25°: 2.7 g/l. Sol in most organic solvents. Dec by prolonged boiling with acids or alkali. Aq solns are unstable to light. LD$_{50}$ orally in female rats: 147 (mg/kg) (Baranyovits, Ghosh).

USE: Insecticide.

7610. Pirimiphos-methyl. [29232-93-7] O-[2-(Diethylamino)-6-methyl-4-pyrimidinyl]phosphorothioic acid O,O-dimethyl ester; O,O-dimethyl O-[2-(diethylamino)-6-methyl-4-pyrimidinyl]phosphorothioate; PP-511; Actellic; Blex; Dominator; Silosan. C$_{11}$H$_{20}$N$_3$O$_3$PS; mol wt 305.33. C 43.27%, H 6.60%, N 13.76%, O 15.72%, P 10.14%, S 10.50%. Organophosphate insecticide; cholinesterase inhibitor. Prepn: S. P. Sharpe, B. K. Snell, **GB 1204552**; eidem, **US 3651224** (1970, 1972 both to ICI). Description of properties and analytical methods: D. J. W. Bullock, Anal. Methods Pestic. Plant Growth Regul. **8**, 185-206 (1976). Toxicity study: P. S. Rajini, M. K. Krishnakumari, J. Environ. Sci. Health **B23**, 127 (1988). Photodegradation study: J. M. Herrmann et al., Catal. Today **54**, 353 (1999). Efficacy in stored wheat: F. Huang, B. Subramanyam, Pest Manag. Sci. **61**, 356 (2005).

Straw-colored liquid. mp 15°. d^{30} 1,157. n$_D^{24}$ 1.5291. Vapor pressure at 30°: 1.1×10^{-4} Torr. Soly in water at 30°: 5 mg/l. uv max: 248 nm (ε 20761). LD$_{50}$ in male, female rats (mg/kg): 1861, 1667 orally (Rajini, Krishnakumari).

O,O-Diethyl analog. [23505-41-1] Pirimiphos-ethyl; PP-211; Primicid. C$_{13}$H$_{24}$N$_3$O$_3$PS; mol wt 333.39. Prepn: B. K. Snell, S. P. Sharpe, **GB 1205000** (1970 to ICI). Persistence and degradn in soil: D. L. Suett, Pestic. Sci. **6**, 385 (1975). Review of properties and analytical methods: D. J. W. Bullock, Anal. Methods Pestic. Plant Growth Regul. **8**, 171-184 (1976). Straw-colored liquid, dec >130°. d^{20} 1.14. n$_D^{20}$ 1.520. Vapor pressure at 25°: 2.9×10^{-4} mm Hg. Miscible with most organic solvents. Soly in water (30°): 1 ppm. Weakly basic. uv max (methanol): 248 nm (ε 21300). LD$_{50}$ in rats, mice, guinea pigs (mg/kg); 140-200, 105, 50-150 orally; dermally in rats: 1000-2000 mg/kg (Bullock).

USE: Insecticide.

THERAP CAT (VET): Ectoparasiticide.

7611. Piritramide. [302-41-0] 1'-(3-Cyano-3,3-diphenylpropyl)-[1,4'-bipiperidine]-4'-carboxamide; 1-(3,3-diphenyl-3-cyanopropyl)-4-piperidino-4-piperidinecarboxamide; 2,2-diphenyl-4-(4-piperidino-4-carbamoylpiperidino)butyronitrile; pirinitramide; A-65; R-3365; Dipidolor. C$_{27}$H$_{34}$N$_4$O; mol wt 430.60. C 75.31%, H 7.96%, N 13.01%, O 3.72%. Opioid analgesic. Prepn: **BE 606850**; P. A. J. Janssen, **US 3080366** (1961, 1963 both to Janssen); C. van de Westeringh et al., J. Med. Chem. **7**, 619 (1964). Crystal structure: C. Humblet et al., Acta Crystallogr. **B33**, 1615 (1977). LC/MS/MS determn in plasma and urine: R. Kahlich et al., Rapid Commun. Mass Spectrom. **20**, 275 (2006). Pharmacology: N. Kumar, D. J. Rowbotham, Br. J. Anaesth. **82**, 3 (1999). Clinical pharmacokinetics: T. Bouillon et al., Anesthesiology **90**, 7 (1999). Clinical comparison with morphine: U. R. Döpfmer et al., Eur. J. Anaesthesiol. **18**, 389 (2001).

Crystals from acetone, mp 149-150°.

Note: This is a controlled substance (opiate): **21 CFR**, 1308.11.

THERAP CAT: Analgesic.

7612. Piritrexim. [72732-56-0] 6-[(2,5-Dimethoxyphenyl)-methyl]-5-methylpyrido[2,3-d]pyrimidine-2,4-diamine; 2,4-diamino-6-(2,5-dimethoxybenzyl)-5-methylpyrido[2,3-d]pyrimidine; BW-301U; NSC-351521. C$_{17}$H$_{19}$N$_5$O$_2$; mol wt 325.37. C 62.76%, H 5.89%, N 21.52%, O 9.83%. Lipid soluble dihydrofolate reductase inhibitor. Prepn: E. M. Grivsky et al., J. Med. Chem. **23**, 327 (1980); E. M. Grivsky et al., **EP 21292** (1981 to Wellcome); E. M. Grivsky, **US 4959474** (1990 to Burroughs Wellcome). Biochemistry and antitumor activity: D. S. Duch et al., Cancer Res. **42**, 3987 (1982). Cytotoxicity: I. W. Taylor et al., ibid. **45**, 978 (1985). Pharmacology and toxicology: C. W. Sigel et al., NCI Monogr. **5**, 111 (1987). Antipneumocystis and antitoxoplasma activities: J. A. Kovacs et al., Antimicrob. Agents Chemother. **32**, 430 (1988). HPLC determn in plasma: R. G. Foss, C. W. Sigel, J. Pharm. Sci. **71**, 1176 (1982). Determn by protein binding assay in plasma: J. L. Woolley, Jr., et al., ibid. **78**, 749 (1989). Crystal structure and conformational analysis: P. A. Sutton, V. Cody, J. Am. Chem. Soc. **110**, 6219 (1988). Clinical pharmacology in children: P. C. Adamson et al., Cancer Res. **50**, 4464 (1990). Clinical evaluation in lung cancer: M. G. Kris et al., Cancer Treat. Rep. **71**, 763 (1987); in advanced squamous head and neck cancer: W.-C. Uen et al., Cancer **69**, 1008 (1992).

Yellow powder from ethanol/water, mp 252-254°. log P (octanol/water): 1.74.

Isethionate. [79483-69-5] $C_{17}H_{19}N_5O_2.C_2H_6O_4S$. LD_{50} orally in rats: 1168 mg/kg (Sigel).

THERAP CAT: Antineoplastic.

7613. Pirlimycin. [79548-73-5] Methyl 7-chloro-6,7,8-trideoxy-6-[[[(2S,4R)-4-ethyl-2-piperidinyl]carbonyl]amino]-1-thio-L-*threo*-α-D-*galacto*-octopyranoside. $C_{17}H_{31}ClN_2O_5S$; mol wt 410.95. C 49.69%, H 7.60%, Cl 8.63%, N 6.82%, O 19.47%, S 7.80%. Semi-synthetic lincosaminide antibiotic; structural analog of clindamycin, *q.v.* Prepn: R. D. Birkenmeyer, **DE 3043502**; *idem*, **US 4278789** (both 1981 to Upjohn); R. D. Birkenmeyer *et al.*, *J. Med. Chem.* **27**, 216 (1984). *In vitro* activity vs aerobic bacteria: V. I. Ahonkhai *et al.*, *Antimicrob. Agents Chemother.* **21**, 902 (1982); vs anaerobic bacteria: S. M. H. Qadri *et al.*, *J. Antibiot.* **36**, 42 (1983). HPLC determn in biological fluids: J. A. Shah, D. J. Weber, *J. Chromatogr.* **309**, 95 (1984). Ion-pairing LC determn in pharmaceutic prepns: D. L. Theis, *ibid.* **402**, 335 (1987). Metabolism and residue studies in dairy cows: R. E. Hornish *et al.*, *ACS Symp. Ser.* **503**, 132-147 (1992). Evaluation in bovine mastitis: W. E. Owens *et al.*, *Agri-Pract.* **15**, 19 (1994).

Hydrochloride. [78822-40-9] U-57903E; Pirsue. Crystals from water, mp 222-224°. $[\alpha]_D^{25}$ +176° (**US 4278789**). Also reported as mp 210-212°. $[\alpha]_D^{25}$ +181°. LD_{50} i.p. in mice: 600 mg/kg (Birkenmeyer, 1984).

THERAP CAT (VET): Antibacterial.

7614. Pirmenol. [68252-19-7] *rel*-α-[3-[(2R,6S)-2,6-Dimethyl-1-piperidinyl]propyl]-α-phenyl-2-pyridinemethanol; (±)-*cis*-2,6-dimethyl-α-phenyl-α-2-pyridyl-1-piperidinebutanol; (±)-1-phenyl-1-(2-pyridyl)-4-(*cis*-2,6-dimethyl-1-piperidyl)butanol. $C_{22}H_{30}N_2O$; mol wt 338.50. C 78.06%, H 8.93%, N 8.28%, O 4.73%. Prepn: R. W. Fleming, **DE 2806654**; *idem*, **US 4112103** (both 1978 to Parke, Davis). Anti-arrhythmic profile in dogs: T. E. Mertz, T. J. Steffe, *J. Cardiovasc. Pharmacol.* **2**, 527 (1980). Toxicology study: J. L. Schardein *et al.*, *Toxicol. Appl. Pharmacol.* **56**, 294 (1980). HPLC determn in biological fluids: E. L. Johnson, L. A. Pachla, *J. Pharm. Sci.* **73**, 754 (1984). Pharmacokinetics in humans: S. C. Hammill *et al.*, *Circulation* **65**, 369 (1982); S. W. Sanders *et al.*, *J. Clin. Pharmacol.* **23**, 113 (1983). Hemodynamic effects in cardiac patients: M. S. Nieminen *et al.*, *Eur. Heart J.* **7**, 150 (1986). Clinical evaluations in ventricular arrhythmias: L. K. Toivonen *et al.*, *J. Cardiovasc. Pharmacol.* **8**, 156 (1986); E. M. Hampton *et al.*, *Eur. J. Clin. Pharmacol.* **31**, 15 (1986). Chronic toxicity study: J. R. Watkins *et al.*, *Drug Invest.* **3**, 141 (1991). Evaluation of carcinogenicity: G. E. Macallum *et al.*, *ibid.* 278. Symposium on pharmacology and clinical efficacy: *Am. J. Cardiol.* **59**, Suppl., 1H-57H (1987).

Relative stereochemistry

Crystals from petroleum ether, mp 70-71°.

Monohydrochloride. CI-845. $C_{22}H_{30}N_2O.HCl$; mol wt 374.95. mp 171-172°. LD_{50} in mice, rats, dogs (mg/kg): 20.8, 23.6, >7.0 i.v.; 215.5, 359.9, >40.0 orally (Schardein).

THERAP CAT: Antiarrhythmic (class IA).

7615. Piroctone. [50650-76-5] 1-Hydroxy-4-methyl-6-(2,4,4-trimethylpentyl)-2(1H)-pyridinone. $C_{14}H_{23}NO_2$; mol wt 237.34. C 70.85%, H 9.77%, N 5.90%, O 13.48%. Pyridone deriv related structurally to ciclopirox, *q.v.* Prepn: G. Lohaus, W. Dittmar, **DE 2, DE 214608** corresp to **US 3972888** (1973, 1976 both to Hoechst); *eidem*, *Arzneim.-Forsch.* **31**, 1311 (1981). Evaluation of efficacy as anti-dandruff agent: E. Futterer, *J. Soc. Cosmet. Chem.* **32**, 327 (1981).

Crystals, mp 108°.

Ethanolamine salt (1:1). [68890-66-4] Octopirox. $C_{16}H_{30}N_2O_3$; mol wt 298.43.

THERAP CAT: Antiseborrheic.

7616. Piroheptine. [16378-21-5] 3-(10,11-Dihydro-5H-dibenzo[*a,d*]cyclohepten-5-ylidene)-1-ethyl-2-methylpyrrolidine. $C_{22}H_{25}N$; mol wt 303.45. C 87.08%, H 8.30%, N 4.62%. Prepn: **NL 6609280**; Y. Deguchi *et al.*, **US 3454595** (1967, 1969 both to Fujisawa). Crystal structure: Y. Tokuma *et al.*, *Bull. Chem. Soc. Jpn.* **44**, 2665 (1971). Pharmacological studies: M. Hitomi *et al.*, *Arzneim.-Forsch.* **22**, 953, 961 (1972). *In vitro* study: T. Ohashi *et al.*, *ibid.* 966. Metabolism: Y. Tokuma *et al.*, *Bull. Chem. Soc. Jpn.* **48**, 294 (1975). Clinical pharmacology: A. Barbeau, *Annu. Rev. Pharmacol.* **14**, 91 (1974). Toxicity study: M. Hitomi *et al.*, *Arzneim.-Forsch.* **22**, 961 (1972).

Liquid, bp_4 167°. uv max (95% ethanol): 240 nm (ε 12100).

Hydrochloride. [16378-22-6] Trimol. $C_{22}H_{25}N.HCl$; mol wt 339.91. Crystals, mp 250-253°. LD_{50} in male mice, rats (mg/kg): 153, 600 orally; 19, 17 i.v.; 95, 110 i.p.; 109, 330 s.c. (Hitomi).

THERAP CAT: Antiparkinsonian.

7617. Piromen. [9008-99-5] Desacchromin dispersion; Pyromen. Pyrogenic, pseudomonal polysaccharide-nucleic acid complex. Prepd by proteolyzing the bacterial organism and separating the complex from inactive material by dialyzing: N. M. Nesset, L. G. Ginger, **GB 699663** (1953 to Baxter Labs.), *C.A.* **48**, 6083 (1954). Contains deoxyribonucleic acid, ribonucleic acid, and hexosamine, the pyrogenic reducing sugar: N. M. Nesset *et al.*, *J. Am. Pharm. Assoc.* **39**, 456 (1950). Use in determn of bone marrow granulocyte reserves: B. C. Korbitz *et al.*, *Curr. Ther. Res.* **11**, 491 (1969). Effects on spinal cord regeneration: M. A. Matthews *et al.*, *Neuropathol. Appl. Neurobiol.* **5**, 161 (1979); *eidem*, *Acta Neurobiol. Exp.* **40**, 489 (1980).

7618. Piromidic Acid. [19562-30-2] 8-Ethyl-5,8-dihydro-5-oxo-2-(1-pyrrolidinyl)pyrido[2,3-*d*]pyrimidine-6-carboxylic acid; 5,8-dihydro-8-ethyl-5-oxo-2-pyrrolidinopyrido[2,3-*d*]pyrimidine-6-carboxylic acid; PD-93; Bactramyl; Enterol; Gastrurol; Panacid; Pirodal; Purim; Reelon; Septural; Uropir. $C_{14}H_{16}N_4O_3$; mol wt 288.31. C 58.32%, H 5.59%, N 19.43%, O 16.65%. Quinolone antibacterial. Prepn: S. Minami *et al.*, **JP 67 25912**; *eidem*, **GB 1129358** (1967, 1968 both to Dainippon Pharm.); *eidem*, *Chem.*

Pharm. Bull. **19**, 1426 (1971). Activity studies: M. Shimizu *et al.*, *Antimicrob. Agents Chemother.* **1970**, 117. Metabolism: *eidem, ibid.* 123.

Crystals from ethanol-chloroform, mp 314-316°. LD_{50} in male, female mice, male, female rats (mg/kg): 287, 268, 177, 158 i.v.; all >4000 orally, s.c. and i.p. (Shimizu).

THERAP CAT: Antibacterial.

7619. Piroxicam. [36322-90-4] 4-Hydroxy-2-methyl-*N*-2-pyridinyl-2*H*-1,2-benzothiazine-3-carboxamide 1,1-dioxide; 3,4-dihydro-2-methyl-4-oxo-*N*-2-pyridyl-2*H*-1,2-benzothiazine-3-carboxamide 1,1-dioxide; CP-16171; Artroxicam; Baxo; Bruxicam; Durapirox; Erazon; Feldene; Felden; Flexase; Geldene; Improntal; Inflaced; Larapam; Pirkam; Piroflex; Proxalyoc; Riacen; Roxiden; Sasulen; Solocalm; Zunden. $C_{15}H_{13}N_3O_4S$; mol wt 331.35. C 54.37%, H 3.95%, N 12.68%, O 19.31%, S 9.68%. Non-steroidal anti-inflammatory with long half-life. Prepn (keto form): J. Lombardino, **DE 1943265**; *idem*, **US 3591584** (1970, 1971 to Pfizer). Synthesis and biological properties: J. Lombardino, E. Wiseman, *J. Med. Chem.* **15**, 848 (1972); J. Lombardino *et al.*, *ibid.* **16**, 493 (1973). Characterization of crystal modifications: F. Vrecer *et al.*, *Int. J. Pharm.* **256**, 3 (2003). HPLC determn in plasma: S. Dadashzadeh *et al.*, *J. Pharm. Biomed. Anal.* **28**, 1201 (2002). Pharmacology: E. Wiseman *et al.*, *Arzneim.-Forsch.* **26**, 1300 (1976). Evaluation of ulcerogenic effects: G. Palacios *et al.*, *Methods Find. Exp. Clin. Pharmacol.* **9**, 353 (1987). Antitumor effects in canine squamous cell carcinoma: B. R. Schmidt *et al.*, *J. Am. Vet. Med. Assoc.* **218**, 1783 (2001). Review of pharmacology and therapeutic efficacy: R. N. Brogden *et al.*, *Drugs* **22**, 165-187 (1981); *eidem, ibid.* **28**, 292-323 (1984). Symposium on clinical efficacy and safety: *Am. J. Med.* **81**, Suppl. 5B, 1-55 (1986). Comprehensive description: M. Mihalic *et al.*, *Anal. Profiles Drug Subs.* **15**, 509-531 (1986). Review of therapeutic potential of β-cyclodextrin complex: C. R. Lee, J. A. Balfour, *Drugs* **48**, 907-929 (1994).

Crystals from methanol. mp 198-200°. Slightly sol in alc, aq alkaline solns; very slightly sol in water, dilute acid, most organic solvents. Pyridyl nitrogen pKa 1.8. 4-Hydroxy pKa 5.1. LD_{50} orally in mice: 360 mg/kg (Wiseman).

Cinnamic acid ester. [87234-24-0] Piroxicam cinnamate; cinnoxicam; SPA-S-510; Sinartrol; Zelis; Zen. $C_{24}H_{19}N_3O_5S$; mol wt 461.49.

Compd with β-cyclodextrin. [121696-62-6] Brexidol; Brexin; Cicladol; Cycladol. $C_{57}H_{83}N_3O_{39}S$; mol wt 1466.33.

THERAP CAT: Anti-inflammatory.

THERAP CAT (VET): Treatment of carcinoma in dogs; anti-inflammatory.

7620. Pirozadil. [54110-25-7] 3,4,5-Trimethoxybenzoic acid 1,1'-[2,6-pyridinediylbis(methylene)]ester; 2,6-pyridinedimethanol bis(3,4,5-trimethoxybenzoate); 722 D; Pemix. $C_{27}H_{29}NO_{10}$; mol wt 527.53. C 61.47%, H 5.54%, N 2.66%, O 30.33%. Hypolipidemic agent that inhibits platelet aggregation. Prepn: J. P. Cochs, **DE 2411902** (1974 to Inst. Int. Ter.), *C.A.* **82**, 4136q (1975). Effect on cerebral metabolic blood flow in rabbits: J. Balasch, L. Palacios, *Arch. Farmacol. Toxicol.* **3**, 137 (1977). Efficacy in exptl athero-

sclerosis: M. R. Parwaresch *et al.*, *Atherosclerosis* **31**, 395 (1978). Toxicological and histopathological study: J. Roca *et al.*, *Arch. Farmacol. Toxicol.* **6**, 41 (1980). Clinical trials in hyperlipoproteinemia: M. Shinomiya *et al.*, *Arzneim.-Forsch.* **37**, 1069 (1987); R. Tapounet, I. Marti Ragué, *Drugs Exp. Clin. Res.* **13**, 447 (1987).

White cryst powder, mp 119-126°. Very sol in chloroform; sol in dioxane, acetonitrile. Practically insol in ether, water.

THERAP CAT: Antilipemic.

7621. Pitavastatin. [147511-69-1] (3*R*,5*S*,6*E*)-7-[2-Cyclopropyl-4-(4-fluorophenyl)-3-quinolinyl]-3,5-dihydroxy-6-heptenoic acid; itavastatin; nisvastatin. $C_{25}H_{24}FNO_4$; mol wt 421.47. C 71.24%, H 5.74%, F 4.51%, N 3.32%, O 15.18%. HMG CoA reductase inhibitor. Prepn: Y. Fujikawa *et al.*, **EP 304063**; *eidem*, **US 5011930** (1989, 1991 both to Nissan Chem. Ind.). Prepn of lactone: S. Takano *et al.*, *Tetrahedron: Asymmetry* **4**, 201 (1993). Chiral synthesis: M. Suzuki *et al.*, *Bioorg. Med. Chem. Lett.* **9**, 2977 (1999). Structure-activity study: *idem et al.*, *Bioorg. Med. Chem.* **9**, 2727 (2001). Pharmacology: T. Aoki *et al.*, *Arzneim.-Forsch.* **47**, 904 (1997). Determn in plasma by HPLC: J. Kojima *et al.*, *J. Chromatogr. B* **724**, 173 (1999); in plasma and urine by LC-MS/MS: L. Tian *et al.*, *ibid.* **865**, 127 (2008). Metabolism: I. Yamada *et al.*, *Xenobiotica* **33**, 789 (2003). Review of pharmacology and safety assessment: K. Kajinami *et al.*, *Cardiovasc. Drug Rev.* **21**, 199-215 (2003); of pharmacokinetics and metabolism: T. Teramoto *et al.*, *Expert Opin. Pharmacother.* **11**, 817-828 (2010). Review of clinical experience: R. Y. A. Mukhtar *et al.*, *Int. J. Clin. Pract.* **59**, 239-252 (2005).

Calcium salt. [147526-32-7] Monocalcium bis[(3*R*,5*S*,6*E*)-7-[2-cyclopropyl-4-(4-fluorophenyl)-3-quinolinyl]-3,5-dihydroxy-6-heptenoate]; NK-104; Livalo. $C_{50}H_{46}CaF_2N_2O_8$; mol wt 881.00. White to pale-yellow, odorless powder. Lipophilic. $[\alpha]_D^{20}$ +23.1° (c = 1.00 in acetonitrile/water). Freely sol in pyridine, chloroform, dilute hydrochloric acid, tetrahydrofuran; sol in ethylene glycol; sparingly sol in octanol; slightly sol in methanol; very slightly sol in water, ethanol. Practically insol in acetonitrile, diethyl ether. Hygroscopic; slightly unstable in light.

Lactone. [141750-63-2] (4*R*,6*S*)-6-[(1*E*)-2-[2-Cyclopropyl-4-(4-fluorophenyl)-3-quinolinyl]ethenyl]tetrahydro-4-hydroxy-2*H*-pyran-2-one. $C_{25}H_{22}FNO_3$; mol wt 403.45. mp 138-139°. $[\alpha]_D^{32}$ +8.84° (c = 0.92 in chloroform). Sol in acetonitrile.

THERAP CAT: Antilipemic.

7622. Pithecolobine. [22368-82-7] 19-Heptyl-10-hydroxy-1,5,10,14-tetraazacyclononadecan-15-one. $C_{22}H_{46}N_4O_2$; mol wt 398.64. C 66.29%, H 11.63%, N 14.05%, O 8.03%. Occurs in the bark of *Samanea saman* Merr. (formerly *Pithecolobium saman* Benth.), bark and seed of *P. bigeminum* Mart. and *P. lobatum* Benth., Leguminosae. Isoln: Greshoff, *Ber.* **23**, 3541 (1890); K. Wiesner *et al.*, *Can. J. Chem.* **30**, 761 (1952). Structure: *eidem, J. Am. Chem. Soc.* **75**, 6348 (1953); D. E. Orr, K. Wiesner, *Chem. Ind. (London)*

1959, 672; K. Wiesner, D. E. Orr, *Tetrahedron Letters* no. 16, 11 (1960); K. Wiesner *et al.*, *Can. J. Chem.* **46**, 1886, 3617 (1968).

Crystals, mp 67-69° or oily liquid, bp$_{0.007}$ 230°. Sublimes at 135° at 0.007 mm pressure. Sol in water, alcohol, chloroform, ether, petr ether.

7623. Pituitary, Posterior. Pituamin; Di-Sipidin; Pituitrin. Desiccated hypophysis. The cleaned, dried, and powdered posterior lobe of pituitary body of domesticated animals used for food by man. Contains both oxytocin and vasopressin, *q.q.v.*

Yellowish or grayish, amorphous powder; characteristic odor. Partially sol in water. *Keep well closed and in a cool place.*

THERAP CAT: Oxytocic; antidiuretic.

THERAP CAT (VET): Oxytocic; antidiuretic.

7624. Pivalaldehyde. [630-19-3] 2,2-Dimethylpropanal; 2,2-dimethylpropionaldehyde; neopentaldehyde; trimethylacetaldehyde; *tert*-pentanal. $C_5H_{10}O$; mol wt 86.13. C 69.73%, H 11.70%, O 18.58%. Sterically hindered aldehyde that serves as an electrophilic precursor and reagent in organic synthesis. Prepn: L. Tissier, *Ann. Chim. (Paris)* **6**, 321 (1893); L. Bouveault, *Compt. Rend.* **138**, 1108 (1904); K. N. Campbell, *J. Am. Chem. Soc.* **59**, 1980 (1937). Improved prepn: R. B. Nazarski *et al.*, *Bull. Soc. Chim. Belg.* **101**, 817 (1992). Thermodynamic studies: M. A. White, A. Perrott, *Can. J. Chem.* **66**, 729 (1988). Structural studies: A. P. Cox *et al.*, *J. Chem. Soc. Faraday Trans.* **87**, 2689 (1991). Synthetic applications: D. Seebach *et al.*, *J. Am. Chem. Soc.* **105**, 5390 (1983); K. Kaneda *et al.*, *Tetrahedron Lett.* **33**, 6827 (1992). *Review:* J. P. Konopelski in *Encyclopedia of Reagents for Organic Synthesis* **6**, L. A. Paquette, Ed. (Wiley, New York, 1995) pp 4154-4158.

Colorless liquid. *Flammable. Irritant.* bp$_{760}$ 78°; bp$_{730}$ 71-74°. n_D^{20} 1.3791. Flash pt, closed cup: 4°F (−16°C). Store at 2-8°C. Solidies upon refrigeration. Sol in ethanol, diethyl ether.

USE: Reagent in synthetic organic chemistry.

7625. Pivalic Acid. [75-98-9] 2,2-Dimethylpropanoic acid; α,α-dimethylpropionic acid; trimethylacetic acid; *tert*-pentanoic acid. $C_5H_{10}O_2$; mol wt 102.13. C 58.80%, H 9.87%, O 31.33%. Prepd by the reaction of *tert*-butylmagnesium chloride and carbon dioxide: Bouveault, *Compt. Rend.* **138**, 1108 (1904); Puntambeker, Zoellner, *Org. Synth.* **8**, 104 (1928); other methods: *ibid.* 108. Forms higher esters (e.g., isobutyl ester) only with difficulty.

Needles, mp 35.5°. bp$_{760}$ 163.8°; d^{50} 0.905; $n_D^{36.5}$ 1.3931. pKa (25°): 5.01. One gram dissolves in 40 ml water. Freely sol in alcohol, ether.

Ethyl ester. [3938-95-2] $C_7H_{14}O_2$; mol wt 130.19. Liquid. d$_4^{18}$ 0.8580; bp 118.2°; n_D^{18} 1.3922.

7626. Pivampicillin. [33817-20-8] (2S,5R,6R)-6-[[(2R)-Aminophenylacetyl]amino]-3,3-dimethyl-7-oxo-4-thia-1-azabicyclo-[3.2.0]heptane-2-carboxylic acid (2,2-dimethyl-1-oxopropoxy)-methyl ester; hydroxymethyl 6-(2-amino-2-phenylacetamido)-3,3-dimethyl-7-oxo-4-thia-1-azabicyclo[3.2.0]heptane-2-carboxylate pi-

valate (ester); 6-[D-α-aminophenylacetamido]penicillanic acid pivaloyloxymethyl ester; pivaloyloxymethyl D-α-aminöbenzylpenicillinate; ampicillin pivaloyloxymethyl ester; pivaloyloxymethyl ampicillinate; MK-191. $C_{22}H_{29}N_3O_6S$; mol wt 463.55. C 57.00%, H 6.31%, N 9.07%, O 20.71%, S 6.92%. Semi-synthetic antibiotic related to penicillin. Prepn: E. K. Frederiksen, W. O. Godtfredsen, **ZA 6805952**; *eidem*, **US 3660575** (1969, 1972 to Lövens Kemiske Fabrik); von Daehne *et al.*, *J. Med. Chem.* **13**, 607 (1970). Pharmacology: *eidem, loc. cit.*; Jordan *et al.*, *Antimicrob. Agents Chemother.* **1970**, 438; Foltz *et al.*, *ibid.* 442. Pharmacokinetics in man: M. Ehrnebo *et al.*, *J. Pharmacokinet. Biopharm.* **7**, 429 (1979). Toxicity: von Daehne *et al.*, *Antimicrob. Agents Chemother.* **1970**, 430.

Hydrochloride. [26309-95-5] Pondocil; Pondocillin; Pondocillina; Sanguicillin. $C_{22}H_{29}N_3O_6S.HCl$; mol wt 500.01. Microcrystalline powder, mp 155-156° (decomp). $[\alpha]_D^{20}$ +196° (c = 1 in water). Weak uv max (water): 268, 262, 256 nm (E$_{1cm}^{1\%}$ ~3.9, 5.7, 6.3). pKa ~7.0. Relatively stable in acid soln; ester hydrolyzes slowly in neutral soln. pH of 0.5 g/100 ml water: ~4.5. Very sol in water and chloroform; freely sol in ethanol; sparingly sol in *n*-propanol, *tert*-butanol, and ethyl ether. LD$_{50}$ in mice, rats (g/kg): 3.34, 5.00 orally; 3.60, 4.50 s.c. (von Daehne).

THERAP CAT: Antibacterial.

7627. Pivcefalexin. [63836-75-9] (6R,7R)-7-[[(2R)-Aminophenylacetyl]amino]-3-methyl-8-oxo-5-thia-1-azabicyclo[4.2.0]oct-2-ene-2-carboxylic acid (2,2-dimethyl-1-oxopropoxy)methyl ester; 7-(D-2-amino-2-phenylacetamido)desacetoxycephalosporanic acid pivaloyloxymethyl ester; pivcephalexin. $C_{22}H_{27}N_3O_6S$; mol wt 461.53. C 57.25%, H 5.90%, N 9.10%, O 20.80%, S 6.95%. Orally active semi-synthetic cephalosporin antibiotic; pivaloyloxymethyl ester of cephalexin, *q.v.* Prepn: W. O. Godtfredsen, E. T. Binderup, **DE 1951012** (1970 to Lövens Kemiske Fabrik), *C.A.* **72**, 132761v (1972). Pharmacology: P. Foresta *et al.*, *Arzneim.-Forsch.* **27**, 819 (1977). Absorption and excretion: E. Trabucchi *et al.*, *Clin. Ther.* **81**, 299 (1977). Clinical trial: C. Vittorini *et al.*, *Arzneim.-Forsch.* **31**, 1163 (1981).

Hydrochloride. [27726-31-4] ST-21; Cefalen; Pivacef. $C_{22}H_{27}$-$N_3O_6S.HCl$; mol wt 497.99.

THERAP CAT: Antibacterial.

7628. Pixantrone. [144510-96-3] 6,9-Bis[(2-aminoethyl)-amino]benz[g]isoquinoline-5,10-dione. $C_{17}H_{19}N_5O_2$; mol wt 325.37. C 62.76%, H 5.89%, N 21.52%, O 9.83%. Second generation aza-anthracenedione analog which intercalates DNA and inhibits topoisomerase II; structurally similar to mitoxantrone, *q.v.* Prepn: P. A. Krapcho, **EP 503537** (1992 to Univ. of Vermont); *idem et al.*, *J. Med. Chem.* **37**, 828 (1994). Solid-state characterization: A. Marini *et al.*, *J. Pharm. Sci.* **92**, 577 (2003). Antitumor spectrum and evaluation of cardiotoxic potential: G. Beggiolin *et al.*, *Tumori* **87**, 407 (2001). Immunosuppressant activity: G. Cavaletti *et al.*, *J. Neuroimmunol.* **151**, 55 (2004). Clinical pharmacokinetics: S. Faivre *et al.*, *Clin. Cancer Res.* **7**, 43 (2001). Review of clinical experience in non-Hodgkins lymphoma: D. Mukherji, R. Pettengell, *Expert Opin. Pharmacother.* **11**, 1915-1923 (2010).

Dimaleate salt. [144675-97-8] BBR-2778; Pixuvri. $C_{17}H_{19}N_5O_2.2C_4H_{14}O_4$; mol wt 577.68. Blue solid from ethanol and ether, mp 192° (dec). Moderately sol in physiological saline.

Hydrochloride. [175989-38-5] $C_{17}H_{19}N_5O_2$.HCl; mol wt 361.83. Dark blue hygroscopic solid from chloroform, mp 209-212°.

THERAP CAT: Antineoplastic.

7629. Pizotyline. [15574-96-6] 4-(9,10-Dihydro-4*H*-benzo[4,5]cyclohepta[1,2-*b*]thien-4-ylidene)-1-methylpiperidine; 4-(1-methyl-4-piperidylidene)-9,10-dihydro-4*H*-benzo[4,5]cyclohepta[1,2-*b*]thiophene; pizotifen; pizotifan; BC-105. $C_{19}H_{21}NS$; mol wt 295.44. C 77.24%, H 7.16%, N 4.74%, S 10.85%. Serotonin antagonist structurally related to cyproheptadine, *q.v.* Prepn: Jucker *et al.*, **BE 636717** and **US 3272826** (1964, 1966 to Sandoz); Bastian *et al.*, *Helv. Chim. Acta* **49**, 214 (1966). Toxicity data: Bagdon, Dorado, *Pharmacologist* **12**, No. 2, 297 (1970). Mechanism of action study: E. Müller-Schweinitzer, *J. Cardiovasc. Pharmacol.* **8**, 805 (1986). Clinical trial in migraine: D. Crowder, W. P. Maclay, *Curr. Med. Res. Opin.* **9**, 280 (1984). Review of pharmacology and therapeutic efficacy: Speight, Avery, *Drugs* **3**, 159-203 (1972).

Hydrochloride. [73391-87-4] Crystals from isopropanol-ether, mp 261-263° (dec).

Malate. [5189-11-7] Litec; Sandomigran; Sanmigran; Sanomigran; Mosegor. $C_{19}H_{21}NS.C_4H_6O_5$; mol wt 429.53. Crystals from methanol, mp 185-186° (dec).

THERAP CAT: Antimigraine; appetite stimulant.

7630. Plafibride. [63394-05-8] 2-(4-Chlorophenoxy)-2-methyl-*N*-[[(4-morpholinylmethyl)amino]carbonyl]propanamide; *N*-2-(*p*-chlorophenoxy)isobutyryl-*N*′-morpholinomethylurea; ITA-104; Idonor; Perifunal. $C_{16}H_{22}ClN_3O_4$; mol wt 355.82. C 54.01%, H 6.23%, Cl 9.96%, N 11.81%, O 17.99%. Analog of clofibrate, *q.v.* Prepn: J. Iniesta-Pons, **DE 2716374** (1977 to Investigacion Tecnica y Aplicada), *C.A.* **88**, 37810g (1978). Toxicity data: J. Zapatero, L. Brugeghini, *Arch. Farmacol. Toxicol.* **4**, 137 (1978). Series of articles on antiplatelet aggregation activity, hypolipemic activity, pharmacology, and toxicology: *ibid.* 132-142.

Crystals, mp 100-102°. Sol in acetone. Slightly sol in alcohol. Practically insol in water, petr ether. LD_{50} in mice, rats, guinea pigs (mg/kg): 3569, >4000, 2168 orally (Zapatero, Brugeghini).

THERAP CAT: Antithrombotic.

7631. Plantago Seed. Psyllium seed; plantain seed; flea seed. Seed from *Plantago ovata* Forsk., *Plantaginaceae*, known as blond or Indian plantago seed, *P. psyllium* L., known as Spanish psyllium seed, or *P. indica* L. (*P. arenaria* Waldst. & Kit.), known as French psyllium seed. Habit. Mediterranean countries, India. Cultivated in India, Pakistan, France. *Psyllium husk* is the epidermis from seeds of any of the above plantago species. *Ispaghula husk* refers specifically to the epidermis of *P. ovata* seeds. The husk contains a water-soluble mucilage consisting of a highly branched, acidic arabinoxylan: J. F. Kennedy *et al.*, *Carbohydr. Res.* **75**, 265 (1979). Use in chronic constipation: M. Borgia *et al.*, *J. Int. Med. Res.* **11**, 124 (1983). *Review:* J. N. BeMiller in *Industrial Gums*, R. L. Whistler, Ed. (Academic Press, New York, 2nd ed., 1973) pp 345-354. Review of therapeutic potential in reducing blood cholesterol: H. Lipsky *et al.*, *J. Clin. Pharmacol.* **30**, 699-703 (1990).

Small, dark reddish-brown, odorless, almost tasteless seeds. Mixed with an equal bulk of water, forms a mucilaginous mass.

Ispaghula husk. Fibrolax; Fybozest; Fybogel; Isogel; Regulan.

Psyllium hydrophilic mucilloid. [8063-16-9] Psyllium hydrocolloid; psyllium seed gum; Effer-Syllium; Fiberall; Metamucil; Perdiem Fiber; Serutan. Obtained from epidermis of *P. ovata* seeds. White to cream-colored, slightly granular powder. Slightly acid taste; little or no odor.

THERAP CAT: Laxative.

7632. Plantisul. [1407-93-8] Banabins. Orally effective antidiabetic principle contained in aq extracts from the leaves and fruits of the banaba tree, *Lagerstroemia speciosa* (L.) Pers., *Lythraceae*. Extraction procedure: Faustino Garcia, *Philipp. J. Sci.* **76**, no. 3, 3-21 (1944); *J. Philipp. Med. Assoc.* **31**, 216-224 and 276-282 (1955). *Constit.* of banaba extracts: Carew, Chin, *Nature* **190**, 1108 (1961).

7633. Plasmalogens. Aldehydogenic lipids characteristic of the animal kingdom; references to their existence in plants and bacteria are rare. Plasmalogens contain an aldehydogenic chain linked to glycerol as an α,β-unsaturated ether. Although most are aldehydogenic phosphatides, nonphosphatide or neutral plasmalogens have been detected in animal tissues. "Native plasmalogens", when deacylated, yield lysoderivatives, *lysoplasmalogens*, contg the aldehydogenic chain linked to glycerol as an α,β-unsaturated ether. The nomenclature *"phosphatidal ethanolamine"*, *"phosphatidal choline"*, etc. has been proposed for native plasmalogen phosphatides, and *"lysophosphatidal ethanolamine"*, *"lysophosphatidal choline"*, etc. for corresponding deacylated derivatives. This nomenclature minimizes the confusion arising from less precise terms, such as "plasmalogen" or *"ethanolamine plasmalogen"*, *"choline plasmalogen"*, etc. to designate either native compds, lysoderivatives, or other structures which may occur (molecules with two α,β-unsaturated ether chains or with one saturated ether and one α,β-unsaturated ether chain, and true cyclic glyceryl acetal derivatives). First isoln of a pure native plasmalogen (phosphatidal choline): Gottfried, Rapport, *Fed. Proc.* **20**, 278 (1961); *eidem*, *J. Biol. Chem.* **237**, 329 (1962). *Ref:* Rapport, Norton, *Annu. Rev. Biochem.* **31**, 103 (1962). Synthesis: Piantadosi *et al.*, *J. Org. Chem.* **28**, 2425 (1963); Chacko *et al.*, *ibid.* **32**, 3698 (1967); Slotboom *et al.*, *Chem. Phys. Lipids* **1**, 192 (1967); Gigg, Gigg, *J. Chem. Soc.* **C 1968**, 16, 2030; Vtorov *et al.*, *Tetrahedron Lett.* **1971**, 4605. *Reviews:* E. Klenk, H. Debuch, "Plasmalogens" in Holman *et al.*, *Progr. Chem. Fats Lipids* **vol. 6**, (Macmillan, New York, 1963) pp 1-29; Piantadosi, Snyder, *J. Pharm. Sci.* **59**, 283-297 (1970).

7634. Plasmin. [9001-90-5] Fibrinolysin; serum tryptase; E.C. 3.4.21.7; Actase; Thrombolysin. Mol wt about 90,000. Trypsin-like proteolytic enzyme which cleaves fibrin, fibrinogen, *q.q.v.*, and other plasma proteins. Component of the mammalian fibrinolytic system specifically responsible for the dissolution of fibrin clots. Exists in plasma as an inactive precursor, plasminogen, *q.v.* Converted to the active enzyme at the clot site by tissue plasminogen activator, *q.v.* Also activated by streptokinase and urokinase, *q.q.v.* Rapidly inactivated in plasma by α_2-*antiplasmin*, a glyceroprotein with high specific binding affinity for plasmin. Prepn: Christensen, MacLeod, *J. Gen. Physiol.* **28**, 599 (1945); E. C. Loomis, **US 2624691** (1953 to Parke, Davis); J. H. Hink, J. K. McDonald, **US 3234106** (1966 to Cutter Labs); K. C. Robbins, L. Summaria, *J. Biol. Chem.* **238**, 952 (1963). Converted from plasminogen by proteolysis of a single arg-val bond: K. C. Robbins *et al.*, *ibid.* **242**,

2333, 4279 (1967). Structure consists of two polypeptide chains connected by two disulfide bonds. The heavy (A) chain (mol wt ~65000) originates from the amino-terminal portion of plasminogen and contains the binding sites. The light (B) chain (mol wt ~25000) originates from the carboxy-terminus and contains the active site. Isolation and characterization of heavy and light chains: L. Summaria et al., J. Biol. Chem. **242**, 5046 (1967); eidem, ibid. **246**, 2143 (1971). Amino acid sequence studies: W. R. Groskopf et al., ibid. **244**, 3590 (1969); Hartley, Philos. Trans. R. Soc. London Ser. B **257**, 77 (1970); S. Nagasawa, T. Suzuki, Biochem. Biophys. Res. Commun. **41**, 562 (1970). Amino acid sequence of active site: K. C. Robbins et al., J. Biol. Chem. **248**, 1631 (1973). Review of structural studies: F. J. Castellino, Semin. Thromb. Hemostasis **10**, 18-23 (1984). Enzyme specificity for lysine and arginine peptide bonds: W. Troll et al., J. Biol. Chem. **208**, 85 (1954). Review of biochemistry: D. Ogston, J. Clin. Pathol. **33**, Suppl. 14, 5-9 (1980). Reviews: D. Collen, Thromb. Haemostasis **43**, 77-89 (1980); K. C. Robbins et al., Methods Enzymol. **80**, 379-387 (1981); K. K. Kane, Ann. Clin. Lab. Sci. **14**, 443-449 (1984).

THERAP CAT: Thrombolytic enzyme.

7635. Plasminogen. [9001-91-6] Profibrinolysin; plasma trypsinogen. Mol wt ~90,000. The circulating plasma precursor (zymogen) or inactive form of plasmin, q.v. Prepn from blood plasma: Loomis et al., Arch. Biochem. **12**, 1 (1947); Oncley et al., J. Am. Chem. Soc. **71**, 541 (1949). Purification: Christensen, Smith, Proc. Soc. Exp. Biol. Med. **74**, 840 (1950); Kline, J. Biol. Chem. **204**, 949 (1953); Hagan et al., ibid. **235**, 1005 (1960); Hagan et al., US **3066079** (1962 to Am. Cyanamid); Derechin et al., Biochem. J. **84**, 336 (1962); Mertz, Chan, Can. J. Biochem. Physiol. **41**, 1811 (1963); D. G. Deutsch, E. T. Mertz, Science **170**, 1095 (1970). Physical properties: Davies, Englert, J. Biol. Chem. **235**, 1011 (1960). Converted to plasmin by natural activators such as streptokinase, urokinase, or tissue plasminogen activator, q.q.v., through the cleavage of a single arg-val bond. Mechanism of activation: K. C. Robbins et al., ibid. **242**, 2333, 4279 (1967). Native plasminogen is a single chain glycopeptide. Two major molecular forms exist, differing in carbohydrate content and separable by affinity chromatography. Mol wt studies: Barlow et al., ibid. **244**, 1138 (1969). Sequence studies: K. C. Robbins et al., ibid. **247**, 6757 (1972); F. J. Castellino et al., Biochem. Biophys. Res. Commun. **53**, 845 (1973). Carbohydrate comp: M. L. Hayes, F. J. Castellino, J. Biol. Chem. **254**, 8768, 8772, 8777 (1979). Amino acid sequence and biochemistry: F. J. Castellino, Semin. Thromb. Hemostasis **10**, 18 (1984). Biosynthesis and secretion by cultured hepatocytes: J. F. Bohmfalk, G. M. Fuller, Science **209**, 408 (1980). Therapeutic use with streptokinase: I. D. Walker et al., Thromb. Haemostasis **51**, 204 (1984). Review of fibrinolytic system: D. Collen, ibid. **43**, 77-89 (1980). Reviews: Fibrinolysis, D. L. Kline, K. N. N. Reddy, Eds. (CRC, Boca Raton, 1980) 256 pp; F. J. Castellino, Chem. Rev. **81**, 431-446 (1981).

Soluble below pH 5 and above pH 9 and only sparingly sol at intermediate pH values. Resistant to heat below pH 5. Displays maximum stability at pH 2-3.

7636. Plasmocid. [551-01-9] N^1,N^1-Diethyl-N^3-(6-methoxy-8-quinolinyl)-1,3-propanediamine; 8-(3-diethylaminopropyl-amino)-6-methoxyquinoline; 6-methoxy-8-(3-diethylaminopropyl-amino)quinoline; 710-F; SN-3115; Fourneau 710; Antimalarine; Rhodoquine. $C_{17}H_{25}N_3O$; mol wt 287.41. C 71.04%, H 8.77%, N 14.62%, O 5.57%. Prepn: Fourneau et al., Ann. Inst. Pasteur **44**, 503 (1930); eidem, ibid. **46**, 514 (1931); eidem, ibid. **50**, 731 (1933); Magidson, Strukov, Arch. Pharm. **271**, 359 (1933); Strukov, SU **39105** (1934), C.A. **30**, 3445 (1936); Magidson et al., J. Appl. Chem. USSR **9**, 304 (1936); Giral, Ciencia (Mexico City) **6**, 253 (1945), C.A. **40**, 3563 (1946); Yanko et al., J. Am. Chem. Soc. **67**, 664 (1945).

Oily liquid, d_4^{24} 1.0569. $bp_{1.0}$ 182°. n_D^{24} 1.5855.

Dihydrochloride. $C_{17}H_{27}Cl_2N_3O$. Yellow crystals, mp 218-220°. Slightly sol in water, alcohol.

Diphosphate. Yellow crystals, mp 169-171°. Slightly sol in water, alcohol.

Note: The name Rhodoquine was first applied to 8-(3-dimethylaminopropylamino)-6-methoxyquinoline.

THERAP CAT: Antimalarial.

7637. Plastoquinones. PQs. Family of lipid-soluble benzoquinone derivatives involved in photosynthetic electron transport. Structurally and functionally analogous to the ubiquinones, q.v. Structures are characterized by a side chain of repeating isoprenoid units with 9 units being the most common. Originally discovered by Kofler in 1946. Isoln of PQ-9 from alfalfa: F. L. Crane, Plant Physiol. **34**, 546 (1959). Localization in chloroplasts and role in photosynthesis: N. I. Bishop, Proc. Natl. Acad. Sci. USA **45**, 1696 (1959). Structure: M. Kofler et al., Helv. Chim. Acta **42**, 2252 (1959); N. R. Trenner et al., J. Am. Chem. Soc. **81**, 2026 (1959). Synthesis: C. H. Shunk et al., ibid. 5000. Identification of plastoquinone analogs B and C: L. P. Kegel et al., Biochem. Biophys. Res. Commun. **8**, 294 (1962). Review of role in photosynthetic electron flow: U. Siggel, Bioelectrochem. Bioenerg. **3**, 302-318 (1976); A. Trebst in Coenzyme Q, G. Lenaz, Ed. (John Wiley & Sons, Chichester, 1985) pp 257-284.

Plastoquinone 9

Plastoquinone 9. [4299-57-4] (all-E)-2,3-Dimethyl-5-(3,7,11,-15,19,23,27,31,35-nonamethyl-2,6,10,14,18,22,26,30,34-hexatria-contanonaenyl)-2,5-cyclohexadiene-1,4-dione; 2,3-dimethyl-5-solanesyl-1,4-benzoquinone; Kofler's quinone; plastoquinone A; plastoquinone 45; Q-254; PQ-9. $C_{53}H_{80}O_2$; mol wt 749.22. Yellow crystals, mp 48-49°. uv max (isooctane): 254, 262 nm (E $_{1cm}^{1\%}$ 247, 226).

7638. Platelet Activating Factor. [65154-06-5] Blood platelet-activating factor; PAF; 1-O-alkyl-2-acetyl-sn-glycero-3-phosphorylcholine; acetyl glyceryl ether phosphorylcholine; AGEPC; PAF-acether; antihypertensive polar renomedullary lipid; APRL. Phospholipid mediator of platelet aggregation, inflammation, and anaphylaxis. Produced in response to specific stimuli by a variety of cell types, including neutrophils, basophils, platelets, and endothelial cells. Several molecular species of PAF have been identified which vary in the length of the O-alkyl side chain. In vitro identification of a soluble platelet activating factor produced by antigen stimulated leukocytes: P. M. Henson, J. Exp. Med. **131**, 287 (1970). Isoln from rabbit basophils, characterization of PAF response and potential role in immune complex disease: J. Benveniste et al., ibid. **136**, 1356 (1972). Isoln from human leukocytes: J. Benveniste, Nature **249**, 581 (1974). Release of PAF in vivo during anaphylaxis: R. N. Pinckard et al., J. Immunol. **123**, 1847 (1979). Structural study: J. Benveniste et al., Nature **269**, 170 (1977); and synthetic approaches: J. Benveniste et al., C. R. Seances Acad. Sci. Ser. D **289**, 1037 (1979); C. A. Demopoulos et al., J. Biol. Chem. **254**, 9355 (1979). Identity with APRL and antihypertensive activity: M. L. Blank et al., Biochem. Biophys. Res. Commun. **90**, 1194 (1979). Structures of 1-O-octadecyl- and 1-O-hexadecyl-PAF, two of the predominant molecular forms: D. J. Hanahan et al., ibid. **255**, 5514 (1980). Total synthesis of 1-O-octadecyl-PAF: J. Godfroid et al., FEBS Lett. **116**, 161 (1980). Enantiomeric synthesis of C_{16}- and C_{18}-PAF: M. Ohno et al., Chem. Pharm. Bull. **33**, 572 (1985). Molecular heterogeneity of naturally occurring PAF: R. N. Pinckard et al., Biochem. Biophys. Res. Commun. **122**, 325 (1984); H. W. Mueller et al., J. Biol. Chem. **259**, 14554 (1984). Ulcerogenic effects: A.-C. Rosam et al., Nature **319**, 54 (1986). Effects on human pulmonary and cardiovascular function: F. M. Cuss et al., Lancet **2**, 189 (1986). Review of isoln and analytical methods: D. J. Hanahan, S. T. Weintraub, Methods Biochem. Anal. **31**, 195-219 (1985); of chemical and biochemical characteristics: D. J. Hanahan, R. Kumar, Prog. Lipid Res. **26**, 1-28 (1987). Reviews of biosynthesis, biological activities and PAF antagonists: P. Braquet *et al.*, *Pharmacol. Rev.* **39**, 97-145 (1987); P. J. Barnes *et al.*, *J. Allergy Clin. Immunol.* **81**, 919-934 (1988). Books: *Platelet-Activating Factor and Related Lipid Mediators*, F. Snyder, ed. (Plenum Press, New York, 1987) 471 pp; *Prog. Biochem. Pharmacol.* **22** entitled "Biologically Active Ether Lipids" by P. Braquet *et al.*, Eds. (Karger, Basel, 1988).

n usually represents 15 or 17

1-O-Hexadecyl PAF. (*R*)-7-(Acetyloxy)-4-hydroxy-*N,N,N*-trimethyl-3,5,9-trioxa-4-phosphapentacosan-1-aminium hydroxide inner salt 4-oxide; C_{16}-PAF. $C_{26}H_{54}NO_7P$. White, amorphous solid, mp 247° (dec). $[\alpha]_D^{21}$ −3.66° (c = 0.71 in chloroform).
1-O-Octadecyl-PAF. (*R*)-7-(Acetyloxy)-4-hydroxy-*N,N,N*-trimethyl-3,5,9-trioxa-4-phosphaheptacosan-1-aminium hydroxide inner salt 4-oxide; C_{18}-PAF. $C_{28}H_{58}NO_7P$. mp 212-215° (dec). $[\alpha]_D^{20}$ −4.00° (c = 0.71 in chloroform).

7639. Platelet-Derived Growth Factor. PDGF. Cationic glycoprotein; mol wt ~30 kDa. Chemoattractant and mitogen for fibroblasts, smooth muscle cells and glial cells. Involved in both normal and pathologic processes such as wound healing, atherosclerosis, and neoplasia. Composed of two disulfide-linked polypeptide chains, A and B, which may be assembled as a heterodimer or as AA or BB homodimers. PDGF-AB is the most abundant. Originally isolated from platelets, subsequently found to be secreted by a number of cell types including macrophages, fibroblasts, and endothelial cells. Identification: R. Ross *et al.*, *Proc. Natl. Acad. Sci. USA* **71**, 1207 (1974); N. Kohler, A. Lipton, *Exp. Cell Res.* **87**, 297 (1974). Isoln: H. N. Antoniades *et al.*, *Proc. Natl. Acad. Sci. USA* **72**, 2635 (1975). Purification: H. N. Antoniades *et al.*, *ibid.* **76**, 1809 (1979); C.-H. Heldin *et al.*, *ibid.* 3722. Structure: A. Johnsson *et al.*, *Biochem. Biophys. Res. Commun.* **104**, 66 (1982); A. Hammacher *et al.*, *J. Biol. Chem.* **263**, 16493 (1988). PDGF-B is structurally homologous to *p28*^{sis}, the transforming protein of the simian sarcoma virus: M. D. Waterfield *et al.*, *Nature* **304**, 35 (1983); R. F. Doolittle *et al.*, *Science* **221**, 275 (1983). Comparison of biological properties and transforming potential of A and B chains: M. P. Beckmann *et al.*, *Science* **241**, 1346 (1988). Identification of PDGF receptor: C.-H. Heldin *et al.*, *Proc. Natl. Acad. Sci. USA* **78**, 3664 (1981). Molecular mechanisms of action: L. T. Williams, *Science* **243**, 1564 (1989). Review of role in atherosclerosis: R. Ross, *N. Engl. J. Med.* **314**, 488-500 (1986); in wound healing: G. F. Pierce *et al.*, *J. Cell. Biochem.* **45**, 319-326 (1991). *Reviews:* R. Ross, *Annu. Rev. Med.* **38**, 71-79 (1987); B. Westermark *et al.*, *Ciba Found. Symp.* **150**, 6-22 (1990); K. C. Hart *et al.*, *Genet. Eng. (NY)* **17**, 181-208 (1995); W. Meyer-Ingold, W. Eichner, *Cell Biol. Int.* **19**, 389-398 (1995).
Becaplermin. [165101-51-9] rhPDGF-BB; RWJ-60235; Regranex. Recombinant protein corresponding to endogenous human PDGF-BB homodimer. Clinical trial in treatment of diabetic foot ulcers: D. L. Steed *et al.*, *J. Am. Coll. Surg.* **183**, 61 (1996).
THERAP CAT: Vulnerary.

7640. Platensimycin. [835876-32-9] 3-[[3-[(1*S*,3*S*,4*S*,5a*S*,9*S*,9a*R*)-1,4,5,8,9,9a-Hexahydro-3,9-dimethyl-8-oxo-3*H*-1,4:3,5a-dimethano-2-benzoxepin-9-yl]-1-oxopropyl]amino]-2,4-dihydroxybenzoic acid. $C_{24}H_{27}NO_7$; mol wt 441.48. C 65.29%, H 6.16%, N 3.17%, O 25.37%. Broad spectrum antibiotic produced by *Streptomyces platensis*. Inhibits gram-positive bacterial growth by selective inhibition of the condensing enzyme, FabF in type II bacterial fatty acid biosynthesis. Discovery and characterization: J. Wang *et al.*, *Nature* **441**, 358 (2006). Isoln, structure determn and stereochemistry: S. B. Singh *et al.*, *J. Am. Chem. Soc.* **128**, 11916 (2006). Total synthesis of racemic form: K. C. Nicolaou *et al.*, *Angew. Chem. Int. Ed.* **45**, 7086 (2006). Review of mechanism of action: D. Häbich, F. von Nussbaum, *ChemMedChem* **1**, 951-954 (2006).

Buff colored prisms from nitromethane, mp 220-222°. $[\alpha]_D^{23}$ −51.1° (c = 0.135 in CH_3OH). uv max (CH_3OH): 227, 240(sh), 296 nm (ε 28167, 4825).

7641. Platinic Chloride. [16941-12-1] Hydrogen (*OC*-6-11)-hexachloroplatinate(2−) (2:1); hydrogen hexachloroplatinate(IV); acid platinic chloride; hexachloroplatinic(IV) acid; chloroplatinic acid. Cl_6H_2Pt; mol wt 409.80. Cl 51.90%, H 0.49%, Pt 47.60%. H_2PtCl_6. Exists as hexahydrate.
Hexahydrate. $Cl_6H_2Pt.6H_2O$; mol wt 517.89. Brownish-yellow, very deliquesc, cryst mass. d 2.431; mp 60°. Easily sol in water, alcohol. *Corrosive. Keep tightly closed, protected from light.*
Caution: May cause asthma or dermatitis, E. Browning, *Toxicity of Industrial Metals* (Appleton-Century-Crofts, New York, 2nd ed., 1969) pp 270-275.
USE: In platinum plating, photography, platinum mirrors, platinum luster on glass and porcelain, platinized carbon for acetic acid manuf; platinizing pumice stone or asbestos, as catalyst in manuf of SO_3; indelible ink; relief etching of zinc for artistic and commercial purposes; fixing microscopic preps, etc.

7642. Platinic Oxide. [1314-15-4] Platinum oxide (PtO_2); platinum dioxide; Adams' catalyst. O_2Pt; mol wt 227.08. O 14.09%, Pt 85.91%. PtO_2. Prepd by the interaction of platinic chloride and excess NaOH: Wöhler, *Z. Anorg. Chem.* **40**, 434 (1904); Bellucci, *ibid.* **44**, 171 (1905); by the reduction of chloroplatinic acid with formaldehyde: Willstätter, Waldschmidt-Leitz, *Ber.* **54**, 113 (1921); Feulgen, *ibid.* 360; by fusing chloroplatinic acid with sodium nitrate: Adams *et al.*, *Org. Synth.* **8**, 92 (1928). Determination of orthorhombic ($CaCl_2$) crystal structure of β-PtO_2: Siegel *et al.*, *J. Inorg. Nucl. Chem.* **31**, 3803 (1969).
Several hydrates have been reported; the compd generally used for catalysis is the monohydrate. Black powder. When freshly pptd it is sol in concd acids, also in dil H_3PO_4, esp when warmed. Easily sol in dilute solns of potassium hydroxide.
USE: As catalyst in hydrogenations. The actual catalyst is platinum black which is formed *in situ* by reduction of the PtO_2 by the hydrogen used for the hydrogenation. Especially useful for reduction at room temp and hydrogen pressures up to 4 atmospheres. Suitable for the reduction of double and triple bonds, aromatic nuclei, carbonyl groups, nitro groups, and nitriles.

7643. Platinous Chloride. [10025-65-7] Platinum chloride ($PtCl_2$); platinous dichloride; platinum dichloride. Cl_2Pt; mol wt 265.98. Cl 26.66%, Pt 73.35%. $PtCl_2$. Prepn from hexachloroplatinic acid, $H_2PtCl_6.6H_2O$: Cohen, *Inorg. Synth.* **6**, 209 (1960).
Grayish-green to brown powder. d 5.87. Insol in water, alcohol, ether. Sol in hydrochloric acid. Combines with PCl_3 to form a compd sol in benzene or chloroform.

7644. Platinum. [7440-06-4] Pt; at. wt 195.084; at. no. 78; valences 2,4; seldom 1, 5, 6. Group VIII (10). Six naturally occurring ring isotopes: 190 (0.01%); 192 (0.8%); 194 (32.9%); 195 (33.8%); 196 (25.2%); 198 (7.2%); 190 is radioactive: $T_{1/2}$ 6.9 × 10^{11} years. Artificial, radioactive isotopes: 173-189; 191; 193; 197; 199-201. Abundance in earth's crust about 0.01 ppm. Believed to be mentioned by Pliny under the name *"alutiae"*. Has been known and used in South America as "platina del Pinto". Reported by Ulloa in 1735; brought to Europe by Wood, and described by Watson in 1741. Occurs native alloyed with one or more members of its group (iridium, osmium, palladium, rhodium, and ruthenium) in gravels and sands. Prepn: Wichers *et al.*, *Trans. Am. Inst. Min. Metall. Eng.* **76**, 602 (1928). Reviews of prepn, properties and chemistry of platinum and other platinum metals: Gilchrist, *Chem. Rev.* **32**, 277-372 (1943); Beamish *et al.*, in *Rare Metals Handbook*, C. A. Hampel, Ed. (Reinhold, New York, 1956) pp 291-328; Livingstone in *Comprehensive Inorganic Chemistry* Vol. 3, J. C. Bailar, Jr. *et al.*, Eds. (Pergamon Press, Oxford, 1973) pp 1163-1189, 1330-1370; F. R.

Hartley, *The Chemistry of Platinum and Palladium with Particular Reference to Complexes of the Elements* (Halsted Press, New York, 1973).

Silver-gray, lustrous, malleable and ductile metal; face-centered cubic structure. Also prepd in the form of a black powder (platinum black) and as spongy masses (platinum sponge). mp 1773.5 ± 1°. *See:* Roeser *et al.*, *Natl. Bur. Stand. J. Res.* **6**, 1119 (1931); bp about 3827°. d 21.447 (calcd). Brinell hardness: 55. Sp heat 0.0314 cal/ g at 0°. Electrical resistivity (20°) 10.6 μohm-cm. Does not tarnish on exposure to air. Absorbs hydrogen at a red heat and retains it tenaciously at ord. temp, gives off the gas at a red heat *in vacuo.* Occludes carbon monoxide, carbon dioxide, nitrogen. Volatilizes considerably when heated in air at 1500°. The heated metal absorbs oxygen; gives it off on cooling. Not affected by water or by single mineral acids. Reacts with boiling aqua regia with formation of chloroplatinic acid, also with molten alkali cyanides. Attacked by halogens, by fusion with caustic alkalies, alkali nitrates, alkali peroxides; by arsenates and phosphates in presence of reducing agents.

Caution: Potential symptoms of overexposure to Pt metal are irritation of eyes, skin, respiratory system; dermatitis. Potential symptoms of overexposure to soluble Pt salts are irritation of eyes, nose; cough, dyspnea, wheezing, cyanosis; dermatitis, skin sensitization; lymphocytosis. *See NIOSH Pocket Guide to Chemical Hazards* (DHHS/NIOSH 97-140, 1997) p 260. *See also* E. Browning, *Toxicity of Industrial Metals* (Appleton-Century-Crofts, New York, 2nd ed., 1969) pp 270-275.

USE: Manuf apparatus for laboratory and industrial use, thermocouples, platinum resistance thermometers, acidproof containers, electrodes, etc. In dentistry; jewelry; electroplating. As oxidation catalyst in manuf acetic acid, nitric acid from ammonia, manuf sulfuric acid; control of automotive emissions.

7645. Platonin. [3571-88-8] 2,2'-[3-[2-(3-Heptyl-4-methyl-2(3*H*)-thiazolylidene)ethylidene]-1-propene-1,3-diyl]bis(3-heptyl-4-methylthiazolium) iodide (1:2); 4,4',4''-trimethyl-3,3',3'-triheptyl-8-(2''-thiazolyl)-2,2'-pentamethinethiazolocyanine 3,3''-diiodide; 3,-3',3''-triheptyl-4,4',4''-trimethyl-7-(2''-methylthiazolyl-2,2'-trimethine)thiazolcyanine 3,3''-diiodide; platonin J; NK-19; Kankohso 101; Photosensitizer 101. $C_{38}H_{61}I_2N_3S_3$; mol wt 909.92. C 50.16%, H 6.76%, I 27.89%, N 4.62%, S 10.57%. Photosensitizing cyanine dye with antimicrobial and immunomodulating activities. General prepn and use in rheumatoid arthritis: I. Yamamoto, **BE 894635**; *idem*, **US 4464383** (1983, 1984). Antifungal properties: K. Ito, P. K. Kuroda, *Bull. Pharm. Res. Inst. Osaka Med. Coll.* **3**, 20 (1952), *C.A.* **47**, 11335f (1953). Effect on experimentally induced leukopenia: H. Iwata *et al.*, *Folia Pharmacol. Jpn.* **50**, 169 (1954), *C.A.* **49**, 10518b (1955). Immunomodulation effect in animals: H. Ichihashi, T. Kondo, *Gann* **58**, 529 (1967); Y. Oyanagui, *Arch. Int. Pharmacodyn. Ther.* **266**, 162 (1983). Enzyme immunoassay in plasma: I. Yamamoto, K. Morishita, *J. Immunoassay* **7**, 17 (1986). Toxicity study: T. Kimoto, K. Nishitani, *Kanko Shikiso* **85**, 43 (1977), *C.A.* **86**, 84374m (1977).

Brilliant green crystalline powder, mp 204°. Sol in water, alcohol. LD_{50} in male, female mice (mg/kg): 46.9, 50.5 i.p.; in male, female rats (mg/kg): 1539, 1571 orally (Kimoto, Nishitani).

THERAP CAT: Immunomodulator.

7646. Platyphylline. [480-78-4] (3*Z*,5*R*,6*R*,9a*S*,14a*R*,14b*R*)-3-Ethylidenedodecahydro-6-hydroxy-5,6-methyl[1,6]dioxacylcodo-decino[2,3,4-*gh*]pyrrolizine-2,7-dione; (1α)-1,2-dihydro-12-hydroxysenecionan-11,16-dione; platifillin. $C_{18}H_{27}NO_5$; mol wt 337.42. C 64.07%, H 8.07%, N 4.15%, O 23.71%. Toxic pyrrolizidine alkaloid from *Senecio platyphyllus* DC. and other *Senecio* spp., *Compositae:* Orékhov, Tiedebel, *Ber.* **68**, 650 (1935); Orékhov *et*

al., ibid. 1886; Konovalova, Orékhov, *ibid.* **69**, 1908 (1936); *Bull. Soc. Chim. Fr.* [5] **4**, 2037 (1937); Allen, *Am. J. Pharm.* **117**, 110 (1945). Structure: Dry *et al.*, *J. Chem. Soc.* **1955**, 63. Comprehensive reviews of platyphylline and other pyrrolizidine alkaloids: L. Bull *et al.*, *The Pyrrolizidine Alkaloids* (North-Holland, Amsterdam, 1968) 293 pp; F. L. Warren in *The Alkaloids* **vol. 12**, R. H. F. Manske, Ed. (Academic Press, New York, 1970) pp 245-331.

Crystals, mp 129°. $[\alpha]_D^{20}$ −56.4° (c = 0.7 in chloroform). Practically insol in water. Sol in alcohol, chloroform, ether, dil acids.
Bitartrate. $C_{18}H_{27}NO_5 \cdot C_4H_6O_6$. Crystals, mp 199°. $[\alpha]_D^{25}$ −40° (c = 2). Sol in 10 parts of water, in 5 parts of hot water. Sparingly sol in alc, in 42 parts of boiling alc. Nearly insol in chloroform, ether.

7647. Plaunotol. [64218-02-6] (2*Z*,6*E*)-2-[(3*E*)-4,8-Dimethyl-3,7-nonadienyl]-6-methyl-2,6-octadiene-1,8-diol; (*E,Z,E*)-7-hydroxymethyl-3,11,15-trimethyl-2,6,10,14-hexadecatetraen-1-ol; CS-684; Kelnac. $C_{20}H_{34}O_2$; mol wt 306.49. C 78.38%, H 11.18%, O 10.44%. Acyclic diterpene alcohol isolated from a Thai medicinal plant identified as *Croton sublyratus* Kurz, *Euphorbiaceae.* Isoln, synthesis and anti-ulcer activity: H. Mishima *et al.*, **JP Kokai 77 62213**; *eidem*, **US 4059041** (both 1977 to Sankyo); A. Ogiso *et al.*, *Chem. Pharm. Bull.* **26**, 3117 (1978). Pharmacology: S. Kobayashi *et al.*, *Oyo Yakuri* **24**, 599 (1982).

Light yellow oil. Aromatic odor, bitter taste. Sol in benzene, acetone, alcohols, ethers. Insol in water. LD_{50} in male, female mice, rats (μl/kg): 8800, 8100, 10900, 11200 orally (Kobayashi).

THERAP CAT: Antiulcerative.

7648. Pleconaril. [153168-05-9] 3-[3,5-Dimethyl-4-[3-(3-methyl-5-isoxazolyl)propoxy]phenyl]-5-(trifluoromethyl)-1,2,4-oxadiazole; 5-[3-[2,6-dimethyl-4-[5-(trifluoromethyl)-1,2,4-oxadiazol-3-yl]phenoxy]propyl]-3-methylisoxazole; VP-63843; Win-63843. $C_{18}H_{18}F_3N_3O_3$; mol wt 381.36. C 56.69%, H 4.76%, F 14.95%, N 11.02%, O 12.59%. Picornavirus replication inhibitor. Prepn: G. D. Diana, T. J. Nitz, **EP 566199**; *eidem*, **US 5464848** (1993, 1995 both to Sterling Winthrop). Stability and HPLC determn of degradation products: A. B. C. Yu *et al.*, *Drug Dev. Ind. Pharm.* **21**, 1827 (1995). Activity spectrum vs enteroviruses: D. C. Pevear *et al.*, *Antimicrob. Agents Chemother.* **43**, 2109 (1999). Clinical pharmacokinetics in children: G. L. Kearns *et al.*, *ibid.* 634; in adults: S. M. Abdel-Rahman, G. L. Kearns, *J. Clin. Pharmacol.* **39**, 613 (1999). Clinical trial in induced respiratory infection: G. M. Schiff, J. R. Sherwood, *J. Infect. Dis.* **181**, 20 (2000).

White solid, mp 61-62°. Soly at 25° (ng/ml): water < 20. Sol in safflower seed oil, corn oil or corn oil-ethanol solutions.

THERAP CAT: Antiviral.

7649.　Plerixafor. [110078-46-1] 1,1′-[1,4-Phenylenebis-(methylene)]bis-1,4,8,11-tetraazacyclotetradecane. $C_{28}H_{54}N_8$; mol wt 502.80. C 66.89%, H 10.83%, N 22.29%. Symmetric bicyclam; specific CXCR4 chemokine receptor antagonist. Induces mobilization of CD34$^+$ hematopoietic progenitor cells from bone marrow to peripheral blood. Prepn: M. Ciampolini *et al., Inorg. Chem.* **26**, 3527 (1987); M. Achmatowicz, L. S. Hegedus, *J. Org. Chem.* **68**, 6435 (2003); F. Boschetti *et al., ibid.* **70**, 7042 (2005). Large scale synthesis: W. Yang *et al., Tetrahedron Lett.* **44**, 2481 (2003). Receptor binding studies: M. M. Rosenkilde *et al., J. Biol. Chem.* **279**, 3033 (2004); S. P. Fricker *et al., Biochem. Pharmacol.* **72**, 588 (2006). Mechanism of action: H. E. Broxmeyer *et al., J. Exp. Med.* **201**, 1307 (2005). Clinical pharmacokinetics and pharmacodynamics: N. A. Lack *et al., Clin. Pharmacol. Ther.* **77**, 427 (2005). Clinical evaluation in mobilization of peripheral blood stem cells in multiple myeloma and non-Hodgkin's lymphoma: S. M. Devine *et al., J. Clin. Oncol.* **22**, 1095 (2004). Review of discovery and clinical development: E. De Clercq, *Nat. Rev. Drug Discovery* **2**, 581-587 (2003).

Colorless oil or foam.
　Octahydrobromide. [155148-32-6] $C_{28}H_{54}N_8 \cdot 8HBr$; mol wt 1150.09. Prepn: G. J. Bridger *et al., J. Med. Chem.* **38**, 366 (1995). Dihydrate as white solid, mp 239-241° (dec).
　Octahydrochloride. [155148-31-5] AMD-3100; JM-3100; Mozobil. $C_{28}H_{54}N_8 \cdot 8HCl$; mol wt 794.46. Trihydrate as fine colorless crystals.
　THERAP CAT: Adjunct in bone marrow transplantation (stem cell mobilizer).

7650.　Pleuromutilin. [125-65-5] 2-Hydroxyacetic acid (3a*S*,-4*R*,5*S*,6*S*,8*R*,9*R*,9a*R*,10*R*)-6-ethenyldecahydro-5-hydroxy-4,6,9,10-tetramethyl-1-oxo-3a,9-propano-3a*H*-cyclopentacycloocten-8-yl ester; glycolic acid 8-ester with octahydro-5,8-dihydroxy-4,6,9,10-tetramethyl-6-vinyl-3a,9-propano-3a*H*-cyclopentacycloocten-1(4*H*)-one; 14-deoxy-14-[(hydroxyacetyl)oxy]mutilin; drosophilin B; BC-757. $C_{22}H_{34}O_5$; mol wt 378.51. C 69.81%, H 9.05%, O 21.13%. Antibiotic substance produced by the basidiomycetes *Pleurotus mutilus* (Fr.) Sacc. and *P. passeckerianus* Pilat.: F. Kavanagh *et al., Proc. Natl. Acad. Sci. USA* **37**, 570 (1951); **38**, 555 (1952). Isoln procedure: M. Anchel, *J. Biol. Chem.* **199**, 133 (1952). Structure: D. Arigoni, *Gazz. Chim. Ital.* **92**, 884 (1962); A. J. Birch *et al., Chem. Ind. (London)* **1963**, 374; *Tetrahedron* **22**, Suppl. 8, 359 (1966). Stereochemistry: D. Arigoni, *Pure Appl. Chem.* **17**, 331 (1968). Fermentation and biosynthetic study: F. Knauseder, E. Brandl, *J. Antibiot.* **29**, 125 (1976). Mechanism of action: G. Hoegenauer, *Antibiotics* vol. **5**, pt. 1, F. E. Hahn, Ed. (Springer-Verlag, New York, 1979) p 344. Synthetic studies: E. G. Gibbons, *J. Org. Chem.* **45**, 1540 (1980); M. Kahn, *Tetrahedron Lett.* **21**, 4547 (1980). Total synthesis of (±)-form: E. G. Gibbons, *J. Am. Chem. Soc.* **104**, 1767 (1982). Review of discovery, pharmacology, and therapeutic potential of derivatives: R. Novak, D. M. Shlaes, *Curr. Opin. Investig. Drugs* **11**, 182-191 (2010).

Crystals from ethyl acetate + Skellysolve B, mp 170-171°. $[\alpha]_D^{24}$ +20° (c = 3 in abs ethanol). uv max (5 mg/ml in 95% ethanol): 290 nm. Soluble in methanol, ethanol, ethyl acetate, chloroform. LD_{50} i.v. in mice: >60 mg/kg (Kavanagh, 1951).
　Diacetate. $C_{26}H_{38}O_7$. Crystals from abs ethanol, mp 145.5°.
　USE: Starting material in the production of semi-synthetic antibiotics.

7651.　Pleurotine. [1404-23-5] (2aα,4aβ,5β,6β,8aα,12bβ,-12cβ,12dβ)-(−)-2a,3,4,4a,5,6,7,8a,12b,12c-Decahydro-6-methyl-2*H*-5,12d-ethanofuro[4′,3′,2′:4,10]anthra[9,1-*bc*]oxepin-2,9,12-trione; pleurotin. $C_{21}H_{22}O_5$; mol wt 354.40. C 71.17%, H 6.26%, O 22.57%. Antibiotic substance produced by the fungus *Pleurotus griseus*: Robbins *et al., Bull. Torrey Bot. Club* **72**, 165 (1945). Isoln: Robbins *et al., Proc. Natl. Acad. Sci. USA* **33**, 171 (1947). Partial structure: Huls, *Publ. Univ. Congo Elisabethville* **4**, 109 (1962), *C.A.* 3894f (1965). Structure and stereochemistry: J. Grandjean, R. Huls, *Tetrahedron Lett.* **1974**, 1893. Synthesis of racemate: D. J. Hart, H.-C. Huang, *J. Am. Chem. Soc.* **110**, 1634 (1988).

Amber crystals from ether + chloroform, mp 200-215°. $[\alpha]_D^{23}$ −20° (chloroform). Neutral reaction. Sparingly sol in water which inactivates it. More stable at acid pH, but still loses ~50% of its activity when boiled for 10 min at pH 3. Aq solns also are inactivated on exposure to light. Moderately sol in alcohol, ether. Freely sol in chloroform.

7652.　Plicamycin. [18378-89-7] Mithramycin; aureolic acid; mithracin; mitramycin; aurelic acid; antibiotic LA-7017. $C_{52}H_{76}$-O_{24}; mol wt 1085.16. C 57.56%, H 7.06%, O 35.38%. Oligosaccharide antibiotic produced by *Streptomyces argillaceus* n. sp. and *S. tanashiensis*. Isoln: Grundy *et al., Antibiot. Chemother.* **3**, 1215 (1953); Philip, Schenck, *ibid.* 1218; Rao *et al., ibid.* **12**, 182 (1962). Identity of aureolic acid with mithramycin: Berlin *et al., Nature* **218**, 193 (1968). Prepn: Gado *et al.,* **HU 155679** (1969 to Gyogyszerkutato Intezet), *C.A.* **70**, 118093f (1969). Similar to the chromomycins and olivomycins, *q.q.v.* Structural elucidation of the aglycone, *chromomycinone*, and the carbohydrate moieties: Bakhaeva *et al., Tetrahedron Lett.* **1968**, 3595; Berlin *et al., Khim. Prir. Soedin.* **1972**, 537, 542, *C.A.* **78**, 16436t, 16443t (1973). Revised structure: J. Thiem, B. Meyer, *Tetrahedron* **37**, 551 (1981). NMR and fluorometric characterization: N. R. Krishna *et al., J. Antibiot.* **43**, 1543 (1990). Metabolic studies: Kennedy *et al., Cancer Res.* **27**, 1534 (1967). Mode of action studies: Kushch *et al., Antibiotiki* **17**, 504 (1972). Toxicity study: M. Slavik, S. K. Carter, *Adv. Pharmacol. Chemother.* **12**, 1 (1975). *Review:* J. D. Skarbek, M. K. Speedie in *Antitumor Compounds of Natural Origin* vol. **1**, A. Aszalos, Ed. (CRC Press, Boca Raton, 1981) pp 191-235.

Yellow solid, mp 180-183°. $[\alpha]_D^{20}$ −51° (c = 0.4 in ethanol). Sol in lower alcohols, acetone, ethyl acetate, water. Moderately sol in chloroform. Slightly sol in ether, benzene. LD_{50} in mice, rats (mg/kg): 2.14, 1.74 i.v. (Slavik, Carter).

THERAP CAT: Antineoplastic.

7653. Plicatic Acid. [16462-65-0] (1S,2S,3R)-1-(3,4-Dihydroxy-5-methoxyphenyl)-1,2,3,4-tetrahydro-2,3,7-trihydroxy-3-(hydroxymethyl)-6-methoxy-2-naphthalenecarboxylic acid. $C_{20}H_{22}$-O_{10}; mol wt 422.39. C 56.87%, H 5.25%, O 37.88%. Coloring constituent and agent associated with western red cedar asthma. Low-molecular weight lignan which occurs as the (−)-form. Isolation from the heartwood of western red cedar, *Thuja plicata* Donn: J. A. F. Gardner *et al.*, *Can. J. Chem.* **37**, 1703 (1959). Structure determn: *idem et al.*, *ibid.* **44**, 52 (1966). Absolute configuration: R. J. Swan *et al.*, *ibid.* **45**, 319 (1967). Chromatographic analysis: B. F. MacDonald, E. P. Swan, *J. Chromatogr.* **51**, 553 (1970). Identification as coloring material: Y. Kai, E. P. Swan, *Mokuzai Gakkaishi* **36**, 218 (1990). Role in asthma: M. Chan-Yeung *et al.*, *Am. Rev. Respir. Dis.* **108**, 1094 (1973); *eidem, J. Allergy Clin. Immunol.* **79**, 792 (1987). Mechanism of action study: M. Chan-Yeung, *Am. J. Ind. Med.* **25**, 13 (1994).

Light tan powder. $[\alpha]_D^{21}$ −9.99° (water). uv max (water): 281 nm (log ε 3.58). pKa: 3.0. Very sol in water; sol in absolute acetone and ethanol; very slightly sol in ethyl acetate and ether.

7654. Plinabulin. [714272-27-2] (3Z,6Z)-3-[[5-(1,1-Dimethylethyl)-1H-imidazol-4-yl]methylene]-6-(phenylmethylene)-2,5-piperazinedione; KPU-2; NPI-2358. $C_{19}H_{20}N_4O_2$; mol wt 336.40. C 67.84%, H 5.99%, N 16.66%, O 9.51%. Tubulin polymerization inhibitor that disrupts tumor endothelial vasculature and reduces blood flow to the tumor site. Structural analog of halimide, a cytotoxic antibiotic produced by a marine *Aspergillus* sp. Prepn: Y. Hayashi *et al.*, **WO 04054498**; *eidem*, **US 7064201** (2004, 2006 both to Nereus). Cytotoxic activity as vascular disrupting agent: B. Nicholson *et al.*, *Anti-Cancer Drugs* **17**, 25 (2006); A. V. Singh *et al.*, *Blood* **117**, 5692 (2011). Clinical pharmacokinetics and efficacy vs solid tumors: M. M. Mita *et al.*, *Clin. Cancer Res.* **16**, 5892 (2010).

Yellow solid from aq methanol or chloroform. Protect from light. THERAP CAT: Antineoplastic.

7655. Plumbagin. [481-42-5] 5-Hydroxy-2-methyl-1,4-naphthalenedione; 5-hydroxy-2-methyl-1,4-naphthoquinone. C_{11}-H_8O_3; mol wt 188.18. C 70.21%, H 4.29%, O 25.51%. Found in the roots of *Plumbago europaea* L., *P. rosea* L., *Plumbaginaceae.* Isoln: Dulong d'Astafort, *J. Pharm. Chim.* **14**, 441 (1828); Wefers-Bettink, *Rec. Trav. Chim.* **8**, 319 (1889); Ray, Dutt, *J. Indian Chem. Soc.* **5**, 419 (1928). Structure and synthesis: Fieser, Dunn, *J. Am. Chem. Soc.* **58**, 572 (1936). Biosynthesis: Durand, Zenk, *Tetrahedron Lett.* **1971**, 3009. Chemotherapeutic properties: Vichkanova *et al.*, *C.A.* **78**, 66906s (1973). Efficient syntheses: A. Ichihara *et al.*, *Agric. Biol. Chem.* **44**, 211 (1980); G. Wurm *et al.*, *Arch. Pharm.* **314**, 861, 1055 (1981); H. Möhrle, H. Foltmann, *ibid.*

321, 259 (1988). Toxicity data: M. Debray *et al.*, *Plant Med. Phytother.* **7**, 77 (1973), *C.A.* **79**, 14255e (1973).

Yellow needles from dil alcohol, mp 78-79°. Sublimes. Volatile with steam. Slightly sol in hot water; sol in alc, acetone, chloroform, benzene, acetic acid. LD_{50} i.p. in mice: ~0.015 g/kg (Debray).

7656. Plumericin. [77-16-7] (3E,3aS,4aR,7aS,9aS,9bS)-3-Ethylidene-3,3a,7a,9b-tetrahydro-2-oxo-2H,4aH-1,4,5-trioxadicyclopent[a,hi]indene-7-carboxylic acid methyl ester. $C_{15}H_{14}O_6$; mol wt 290.27. C 62.07%, H 4.86%, O 33.07%. A sesquiterpenoid, exhibiting *in vitro* activity against fungi, some bacteria including *Mycobacterium tuberculosis* 607. Isoln from roots of *Plumeria multiflora* Muell.-Arg., *Apocynaceae*, also from roots of *P. rubra* var. *alba*: J. E. Little, D. B. Johnstone, *Arch. Biochem.* **30**, 445 (1951); from *Allamanda cathartica* Linn., *Apocynaceae*: B. R. Pai *et al.*, *Indian J. Chem.* **8**, 851 (1970). Structure, physical properties and IR, ^{1}HNMR spectra: G. Albers-Schönberg, H. Schmid, *Helv. Chim. Acta* **44**, 1447 (1961). Synthetic approach: J. K. Whitesell *et al.*, *Synth. Commun.* **7**, 355 (1977). Biomimetic total synthesis of (±)-form: B. M. Trost *et al.*, *J. Am. Chem. Soc.* **105**, 6755 (1983); *eidem*, *ibid.* **108**, 4974 (1986). Alternate synthesis of (±)-form: *eidem*, *ibid.* 4965; K. E. B. Parkes, G. Pattenden, *J. Chem. Soc. Perkin Trans. 1* **1988**, 1119.

Narrow, rectangular plates from alc, toluene or methylene chloride + ether, dec 211.5-212.5°. Sublimes in high vacuum at 160-180°. $[\alpha]_D^{25}$ +197.5 ±2° (c = 0.982 in chloroform). uv max (ethanol): 214-215 nm (log ε 4.24). Sol in chloroform; slightly sol in methanol, alc, ether, acetone, benzene. Practically insol in petr ether, water.

7657. Plumieride. [511-89-7] (1S,2'R,4aS,7aS)-1-(β-D-Glucopyranosyloxy)-4a,7a-dihydro-4'-[(1S)-1-hydroxyethyl]-5'-oxospiro[cyclopenta[c]pyran-7(1H),2'(5'H)-furan]-4-carboxylic acid methyl ester; agoniadin. $C_{21}H_{26}O_{12}$; mol wt 470.43. C 53.62%, H 5.57%, O 40.81%. Found in bark of *Plumeria lancifolia* Muell.-Arg., *Apocynaceae*, also in *P. acutifolia* and *P. rubra* var. *alba*: Peckolt, *Arch. Pharm.* [2] **142**, 40 (1870); E. Merck's *Jahresber.* **1895**, 11; Boorsma, *Mededeel. Lands Plant.* **13**, 27 (1894); **31**, 132 (1899); Franchimont, *Rec. Trav. Chim.* **18**, 334, 477 (1899); **19**, 350 (1900); Halpern, Schmid, *Helv. Chim. Acta* **41**, 1109 (1958). Structure: Albers-Schönberg, Schmid, *ibid.* **44**, 1447 (1961). Biosynthesis: D. A. Yeowell, H. Schmid, *Experientia* **20**, 250 (1964); K. Inoue *et al.*, *Chem. Pharm. Bull.* **27**, 3115 (1979).

Monohydrate, bitter crystals, mp 156-158°. When anhydrous, mp 224-225°. $[\alpha]_D^{20}$ −114°; $[\alpha]_D^{20}$ −80° (methanol). Broad uv absorption: 217-238 nm. Sol in water, alcohol, ethyl acetate.

7658. Plutonium. [7440-07-5] Pu; at. no. 94; valences 3, 4, 5, 6, 7. No stable nuclides, known isotopes (mass numbers): 232-246. Longest-lived known isotopes: ^{242}Pu (T$_{1/2}$ 3.76 × 10^5 years, α-emitter, rel. at. mass 242.0587), 244 (T$_{1/2}$ 8.26 × 10^7 years, α-emitter, rel. at. mass 244.0642). Commercially useful isotopes: ^{238}Pu (T$_{1/2}$ 87.74 years; α-emitter, rel. at. mass 238.0496); ^{239}Pu (T$_{1/2}$ 2.41 × 10^4 years; α-emitter, rel. at. mass 239.0522). Occurrence in the earth's crust: 10^{-22}%. Discovery of isotope ^{238}Pu: G. T. Seaborg *et al.*, *Phys. Rev.* **69**, 366, 367 (1946); of isotope ^{239}Pu: J. W. Kennedy *et al.*, *ibid.* **70**, 555 (1946). Isoln of ^{239}Pu from pitchblende: G. T. Seaborg, M. L. Perlman, *J. Am. Chem. Soc.* **70**, 1571 (1948). Prepn of metal: B. B. Cunningham, L. B. Werner, *ibid.* **71**, 1521 (1949). Chemical properties: Seaborg, Wahl, *ibid.* 1128; Harvey *et al.*, *J. Chem. Soc.* **1947**, 1010. *Reviews:* J. M. Cleveland, *The Chemistry of Plutonium* (Gordon & Breach, New York, 1970) 653 pp; C. Keller, *The Chemistry of the Transuranium Elements* (Verlag Chemie, Weinheim, English Ed., 1971) pp 333-484; *Comprehensive Inorganic Chemistry* vol. **5**, J. C. Bailar, Jr. *et al.*, Eds. (Pergamon Press, Oxford, 1973) *passim; Handb. Exp. Pharmakol.* **36**, 307-688 (1973); F. Weigel in *Kirk-Othmer Encyclopedia of Chemical Technology* vol. **18** (Wiley-Interscience, New York, 3rd ed., 1982) pp 278-301; *Plutonium Chemistry*, W. T. Carnall, G. R. Choppin, Eds. (Am. Chem. Soc., Washington, D.C., 1983) 484 pp; F. Weigel *et al.* in *The Chemistry of the Actinide Elements* vol. **1**, J. J. Katz *et al.*, Eds. (Chapman and Hall, New York, 1986) pp 499-886. Review of toxicology: W. J. Bair, R. C. Thompson, *Science* **183**, 715-722 (1974); and health effects: *Toxicological Profile for Plutonium* (PB2010-100006, 2010) 320 pp.

Silvery-white metal. Highly reactive. Oxidizes readily in dry air and oxygen, rate increases in presence of moisture. Six allotropic forms: simple monoclinic α-form, d^{21} 19.86, transforms to β-form at 122 ±4°; body-centered monoclinic β-form, d^{190} 17.70, transforms to γ-form at 207 ±5°; face-centered orthorhombic γ-form, d^{235} 17.14, transforms to δ-form at 315 ±3°; face-centered cubic δ-form, d^{320} 15.92, transforms to δ'-form at 457 ±2°; body-centered tetragonal δ'-form, d^{405} 16.00, transforms to ε-form at 479 ±4°; body-centered cubic ε-form, d^{490} 16.51, transforms to liquid at mp 640 ±2°.

Trivalent plutonium. Weak reducing agent. Stable in soln in absence of air. Slowly oxidized to the tetravalent plutonium by atmospheric oxygen, by permanganate in acid soln in the cold; oxidized to the hexavalent form by permanganate at 60°. Trivalent salts are blue; form complexes very readily; form a series of double sulfates. Crystal structure of the complex and double salts: Zachariasen, *J. Am. Chem. Soc.* **70**, 2147 (1948).

Tetravalent plutonium. Reduced in aq soln to the trivalent form by sulfur dioxide, hydroxylamine hydrochloride, hydrazine hydrochloride, the uranous ion, the iodide ion; by shaking with mercury in chloride soln; electrolytically at a platinum cathode. Tetravalent salts are pink or greenish; form complexes very readily.

Hexavalent plutonium. Obtained by the action of strong oxidizing agents (ceric salts, dichromates, permanganates, or hot bromate soln contg nitric acid) on the tri- or tetravalent form. Reduced to tri- or tetravalent plutonium by sulfur dioxide or ferrocyanide.

Caution: Radiation hazard; handling requires special equipment and shielding facilities. Animal studies have indicated that inhaled ^{239}Pu as particulate matter may remain in the lungs, or move to the bones, liver or other body organs, and over a period of time may give rise to neoplasms due to its α-emitting radioactive decay. *See* Katz *et al.*, *loc. cit.* vol. **2**, p. 1128, 1182-1188. Max permissible concn of ^{238}Pu in air: 7 × 10^{-13} μCurie/cc; of ^{239}Pu in air: 6 × 10^{-13} μCurie/cc: *Natl. Bur. Stand. Handb.* **69**, 17 (1959).

USE: ^{238}Pu as heat source; as radioisotope thermoelectric generator; in radionuclide batteries for pacemakers. ^{239}Pu as fuel in atomic weapons and nuclear power reactors.

7659. PMSF. [329-98-6] Benzenemethanesulfonyl fluoride; phenylmethanesulfonyl fluoride; α-toluenesulfonyl fluoride. C$_7$H$_7$-FO$_2$S; mol wt 174.19. C 48.27%, H 4.05%, F 10.91%, O 18.37%, S 18.41%. Serine and cysteine protease inhibitor. Prepn: W. Davies, J. H. Dick, *J. Chem. Soc.* **1932**, 483; and protease inhibition study: D. E. Fahrney, A. M. Gold, *J. Am. Chem. Soc.* **85**, 997 (1963).

Stability studies: G. T. James, *Anal. Biochem.* **86**, 574 (1978). Analgesic effects and toxicity in rodents: C. Pinsky *et al.*, *Life Sci.* **31**, 1193 (1982). Neuropathologic effects in organophosphorus ester-induced delayed neuropathy in hens: C. Massicotte *et al.*, *Neurotoxicology* **20**, 749 (1999).

Colorless needles from light petroleum, mp 90-91°. LD$_{50}$ (24 hr) in mice (mg/kg): 215 ±55 i.p. (Pinsky).

USE: Standard protease inhibitor in biological research; in protein purification to prevent proteolytic degradation.

7660. Podocarpic Acid. [5947-49-9] (1S,4aS,10aR)-1,2,3,4,-4a,9,10,10a-Octahydro-6-hydroxy-1,4a-dimethyl-1-phenanthrene-carboxylic acid; 12-hydroxypodocarpa-8,11,13-trien-16-oic acid. C$_{17}$H$_{22}$O$_3$; mol wt 274.36. C 74.42%, H 8.08%, O 17.49%. Chief acidic constituent of the resin of the Javanese *Podocarpus cupressina* (L'Hérit.) Pers., *Coniferae*, esp. in *var. imbricata*. Also in New Zealand kahikatea resin (from *Podocarpus dacrydioides*) and in rimu resin (*Dacrydium cupressinum*). A related component, nimbiol, *q.v.*, occurs in the nim tree, *Azadirachta indica* Juss. (*Melia azadirachta* L.), *Meliaceae*. Isoln from *Podocarpus cupressina:* A. C. Oudemans, *Ber.* **6**, 1122 (1873); from *Dacrydium cupressinum:* I. R. Sherwood, W. F. Short, *J. Chem. Soc.* **1938**, 1008. Structure: Campbell, Todd, *J. Am. Chem. Soc.* **64**, 928 (1942). Conversion to nimbiol as proof of structure: Bible, *Tetrahedron* **11**, 22 (1960). Synthesis from desoxypodocarpic acid: Wenkert, Jackson, *J. Am. Chem. Soc.* **80**, 217 (1958). Total synthesis of *dl*-form: Meyer, Maheshwari, *Tetrahedron Lett.* **1964**, 2175; P. R. Kanjilal *et al.*, *Synth. Commun.* **11**, 795 (1981). Synthesis: Pelletier *et al.*, *Tetrahedron Lett.* **1971**, 4179.

Platelets from dil alcohol, mp 193.5°. $[\alpha]_{546}^{20}$ +165° (c = 4 in abs ethanol). Sol in methanol, ethanol, ether, acetic acid. Practically insol in water, chloroform, benzene, carbon disulfide.

Methyl ester. C$_{18}$H$_{24}$O$_3$. Crystals from alcohol, mp 208°.

7661. Pododacric Acid. [32630-75-4] (1S,4aS,10aR)-1,2,3,-4,4a,9,10,10a-Octahydro-6-hydroxy-7-[2-hydroxy-1-(hydroxymethyl)ethyl]-1,4a-dimethyl-1-phenanthrenecarboxylic acid; 12-hydroxy-13-[2-hydroxy-1-(hydroxymethyl)ethyl]podocarpa-8,11,13-trien-16-oic acid. C$_{20}$H$_{28}$O$_5$; mol wt 348.44. C 68.94%, H 8.10%, O 22.96%. From heartwood of *Podocarpus dacrydioides; P. totara:* Briggs *et al.*, *Tetrahedron* **7**, 270 (1959); Cambie, Mander, *ibid.* **18**, 465 (1962). Structure: Cambie, Mathai, *Chem. Commun.* **1971**, 154.

Colorless needles from 30% methanol, mp 213-214°. $[\alpha]_D^{25}$ +118° (c = 0.9 in 5:1 chloroform-methanol). uv max: 225, 284 nm (ε 5500, 3100). pK 8.44 (in methyl cellosolve system).

7662. Podophyllic Acids. C$_{22}$H$_{24}$O$_9$; mol wt 432.43. C 61.11%, H 5.59%, O 33.30%. Prepn of a "podophyllic acid", mp 145-150° (crude), 163-165° (pure): Borsche, Niemann, *Ann.* **494**, 126 (1932). Prepn of the 2,3-*trans* acid: Kuhn, Wartburg, *Experientia* **19**, 391 (1963); *NL 6405480* (1964 to Sandoz), *C.A.* **62**, 9083c (1965); and 2,3-*cis* acid: Renz *et al.*, *Ann.* **681**, 207 (1965). Prepn of a DL-stereoisomer: Gensler *et al., J. Am. Chem. Soc.* **76**, 315 (1954). Configuration and nomenclature: Rutschmann, Renz, *Helv. Chim. Acta* **42**, 890 (1959).

Podophyllinic Acid

2,3-*trans*-Hydroxy acid. [1853-37-8] Podophyllinic acid. Crystals from acetone + ether, mp 164-168°. $[\alpha]_D^{20}$ −199.9° (c = 0.463 in ethanol), −292.6° (c = 0.671 in pyridine).

2,3-*cis*-Hydroxy acid. [477-67-8] Picropodophyllic acid. Needles from methanol + ether, double mp 150-155° and 200-232°. $[\alpha]_D^{21}$ −100.4° (c = 0.615 in ethanol), −185° (c = 0.746 in pyridine).

2,3-*trans*-Hydroxy acid hydrazide. [78178-41-3] Podophyllinic acid hydrazide. Prepn and properties: Rutschmann, **US 2977359** (1961 to Sandoz). Prisms from methanol, mp 198-199°. $[\alpha]_D$ −202° (c = 0.4 in ethanol).

***trans*-2-Ethylhydrazide.** [1508-45-8] Podophyllinic acid 2-ethylhydrazide; SP-I. Prepn: Rutschmann, **US 3054802** (1962 to Sandoz). Amorphous powder, pptd from chloroform + petr ether. $[\alpha]_D$ −154° (c = 0.5 in chloroform).

7663. Podophyllotoxin. [518-28-5] (5*R*,5a*R*,8a*R*,9*R*)-5,8,-8a,9-Tetrahydro-9-hydroxy-5-(3,4,5-trimethoxyphenyl)furo[3′,-4′:6,7]naphtho[2,3-*d*]-1,3-dioxol-6(5a*H*)-one; 1-hydroxy-2-hydroxymethyl-6,7-methylenedioxy-4-(3′,4′,5′-trimethoxyphenyl)-1,-2,3,4-tetrahydronaphthalene-3-carboxylic acid lactone; podophyllinic acid lactone; podofilox; Condyline; Condylox; Wartec; Warticon. $C_{22}H_{22}O_8$; mol wt 414.41. C 63.76%, H 5.35%, O 30.89%. Precursor of antineoplastics etoposide, teniposide, *q.q.v.* Found in the rhizomes of North American *Podophyllum peltatum* L., *Podophyllaceae:* V. Podwyssotski, *Arch. Exp. Pathol. Pharmakol.* **13**, 29 (1880); also in *P. emodi* Wall.: von Wartburg *et al., Helv. Chim. Acta* **40**, 1331 (1957); in *Juniperus virginiana* L., *Cupressaceae:* Kupchan *et al., J. Pharm. Sci.* **54**, 659 (1965). Structure and absolute configuration: Schrecker, Hartwell, *J. Org. Chem.* **21**, 381 (1956). Crystal structure: T. J. Petcher *et al., J. Chem. Soc. Perkin Trans. 2* **1973**, 288. Synthesis: Gensler, Gatsonis, *J. Am. Chem. Soc.* **84**, 1748 (1962); T. Kaneko, H. Wong, *Tetrahedron Lett.* **28**, 517 (1987). Asymmetric total synthesis: R. C. Andrews *et al., J. Am. Chem. Soc.* **110**, 7854 (1988); E. J. Bush, D. W. Jones, *Chem. Commun.* **1993**, 1200. Toxicity data: F. S. Phillips *et al., Fed. Proc.* **7**, 249 (1948). ELISA determn in plants: K.-J. Yoo, J. R. Porter, *J. Nat. Prod.* **56**, 715 (1993). Clinical evaluation for treatment of genital warts: A. Lassus, *Lancet* **2**, 513 (1987); K. R. Beutner *et al., ibid.* **1**, 831 (1989). Review of early literature: Hartwell, Schrecker in *Fortschr. Chem. Org. Naturst.* **15**, 98-121 (1958). Review of chemistry and antineoplastic activity: I. Jardine, *Anticancer Agents Based on Natural Product Models*, J. M. Cassady, J. D. Douros, Eds. (Academic Press, New York, 1980) pp 319-351. Review of syntheses: R. S. Ward, *Synthesis* **1992**, 719-730. Review: D. L. Sackett, *Pharmacol. Ther.* **59**, 163-228 (1993).

Solvated crystals. mp 114-118° (effervescence). Several polymorphic modifications. After drying: mp 183.3-184.0°. $[\alpha]_D^{20}$ −132.7° (chloroform). Soly in water at 23°: 120 mg/l. Sol in alcohol, chloroform, acetone, warm benzene, glacial acetic acid. LD_{50} in rats (mg/kg): 8.7 i.v.; 15 i.p. (Phillips).

Podophyllotoxin-β-D-glucoside. $C_{28}H_{32}O_{13}$. Hygroscopic amorphous white flakes, mp 152-154°. $[\alpha]_D^{20}$ −76.4° (c = 0.576 in methanol); $[\alpha]_D^{20}$ −117.0° (c = 0.668 in pyridine).

Picropodophyllin. [477-47-4] Picropodophyllinic acid lactone. $C_{22}H_{22}O_8$; mol wt 414.41. Found in resin podophyllum, a dried alcoholic extract of *Podophyllum peltatum* L., *Berberidaceae* but not in the fresh plant. 5a*S*-Isomer of podophyllotoxin. Isoln from resin podophyllum: Späth *et al., Ber.* **65**, 1545 (1932). Structure: Schrecker, Hartwell, *J. Am. Chem. Soc.* **76**, 752 (1954). Synthesis: Gensler, Wang, *J. Am. Chem. Soc.* **76**, 5890 (1954); Gensler *et al., ibid.* **82**, 1714 (1960). Crystals from acetone. Irritates the skin. mp 214°. Higher-melting modification from abs ethanol, mp 228°. $[\alpha]_D^{20}$ +9.4° (c = 0.7 in chloroform). Soly in water at 25° ~100 mg/l. Fairly sol in chloroform, hot ethanol, acetic acid, acetone, dil caustic; less sol in benzene, ether, carbon tetrachloride, propylene glycol. Practically insol in petr ether.

THERAP CAT: Antiviral (topical).

7664. Podophyllum. Genus of perennial flowering plants of the family *Podophyllaceae*, formerly, *Berberidaceae*; represented in traditional medicine by two species: *Podophyllum peltatum* L., known as American podophyllum, and *P. hexandrum* R., syn. *P. emodi* Wall., known as Indian podophyllum. Medicinal portions are the dried rhizome and the resin extracted from it. Has been used as a cathartic, emetic and cholagogue and externally as a caustic. Constituents are similar in both species, but vary in concentration. The antimitotic lignin, podophyllotoxin, *q.v.*, is the principal active component. History, isoln procedures, structures: J. L. Hartwell, A. W. Schrecker, *Fortschr. Chem. Org. Naturst.* **15**, 83-166 (1958); and review of medicinal uses: J. Singh, N. C. Shah, *Curr. Res. Med. Aromat. Plants* **16**, 53-83 (1994). HPLC determn of constituents in plant extracts: J. K. Bastos *et al., Phytochem. Anal.* **6**, 101 (1995). Production of cytotoxic lignins by cell culture: M. Petersen, A. W. Alfermann, *Appl. Microbiol. Biotechnol.* **55**, 135 (2001).

American podophyllum. Mayapple; American mandrake; Indian apple; vegetable calomel. *Habit.* North America. *Constit.* Chiefly podophyllotoxin (0.25-1.0%), α- and β-peltatins, quercetin, kaempferol.

Indian podophyllum. Papra; ban-kakari. *Habit.* Himalayas, east Asia. *Constit.* Chiefly podophyllotoxin (1.0-4.0%), 4′-demethylpodophyllotoxin, podophyllotoxone, quercetin, kaempferol.

Podophyllin. [9000-55-9] Podophyllum resin. Dried ethanolic extract obtained from the dried rhizomes and roots. Description, pharmacology and comparison with podophyllotoxin in treatment of anogenital warts: E. Longstaff, G. von Krogh, *Regul. Toxicol. Pharmacol.* **33**, 117-137 (2001). Light brown to greenish-yellow powder, darkens on exposure to light. *Irritating to skin and mucous membranes.*

THERAP CAT: Caustic.

7665. Poi. The root of the taro plant, *Colocasia esculenta* (L.) Schott, *Araceae*. Used as a cereal substitute, particularly in allergy cases and potentially allergic infants.

THERAP CAT: Nutrient.

7666. Poison Ivy. Poison vine; markweed. *Toxicodendron radicans* (L.) Kuntze, *Anacardiaceae*. Erroneously called poison oak. Vigorous woody vine, shrub, or subshrub with trifoliate, alternate leaves. *Habit.* All states of the United States east of the Cascade Mountains, the Great Basin, and the Mojave Desert (absent in Nevada); all states of Mexico except in the Yucatan Peninsula and northern Baja California. Its southern limit is Huehuetenango Department of Guatemala and its northern limit is the 52nd parallel of latitude. It is found in Bermuda and several islands in the Bahamas, in Japan, and in the middle elevations in the mountains of Taiwan and of central and western China. *Review:* Gillis in J. M. Kingsbury, *Poisonous Plants of the United States and Canada* (Prentice-Hall, Englewood Cliffs, N. J., 1964) pp 209-214. *Constit.* Toxic constituent: urushiol, *q.v.:* Dawson, *Trans. N.Y. Acad. Sci.* **18**, 427 (1956).

Caution: Can cause severe allergic dermatitis.
THERAP CAT: Extract as antiallergic (hyposensitization therapy).

7667. Poison Oak. Western poison oak. *Toxicodendron diversilobum* (T. & G.) Greene, *Anacardiaceae.* Similar to poison ivy in that it has three leaflets, grows as a shrub or vine and produces dermatitis in man. Interbreeds with poison ivy. *Habit.* Western North America from southern British Columbia to northern Baja California. *Constit.* Poisons presumably closely related if not identical to those in poison ivy.

Eastern poison oak. *Toxicodendron quercifolium* (Michx.) Greene, *Anacardiaceae.* Differs significantly from poison ivy. Never climbs or produces aerial roots and rarely hybridizes with poison ivy. *Habit.* Southern New Jersey to Florida, west to eastern Texas and Kansas. *Constit.* Poisons may be the same as in poison ivy or Western poison oak.

Extract of *T. quercifolium.* Anergex.

Caution: Direct contact can cause severe allergic dermatitis. *See* M.J. Ellenhorn, D.G. Barceloux, *Medical Toxicology: Diagnosis and Treatment of Human Poisoning* (Elsevier, New York, 1988) pp 1301-1306.

THERAP CAT: Extract as antiallergic (hyposensitization therapy).

7668. Poison Sumac. Poison elder. *Toxicodendron vernix* (L.) Kuntze, *Anacardiaceae.* Tall, rangy shrub with leaves compounded with seven to eleven leaflets. Rachis is bright red and leaflets have no teeth on their margins, thus differing from nonpoisonous sumacs. *Habit.* Southern Quebec to central Florida predominantly east of Mississippi River. Found only in bogs, swamps and wet bottom lands.

Caution: Can cause severe allergic dermatitis.

THERAP CAT: Extract as antiallergic (hyposensitization therapy).

7669. Polaprezinc. [107667-60-7] [*N*-β-Alanyl-L-histidinato(2−)-*N*,*N*^{N},*O*^{α}]zinc; *catena-(S)-[μ-[N^α-(3-aminopropionyl)-L-histidinato(2−)-N^1,N^2,O:N^τ]zinc]; zinc L-carnosine; β-alanyl-L-histidinato zinc; *N*-(3-aminopropionyl)-L-histidinato zinc; Z-103; Promac. Polymeric zinc(II) complex with L-carnosine, corresponding to the formula, $(C_9H_{12}N_4O_3Zn)_n$. Exhibits antioxidant and gastroprotective properties. Prepn of crystalline complex: T. Matsukura *et al.*, *EP* 303380; *eidem*, *US* 4981846 (1989, 1991 both to Hamari). Prepn and structure elucidation: T. Matsukura *et al.*, *Chem. Pharm. Bull.* **38**, 3140 (1990). Pharmacology: T. Arakawa *et al.*, *Dig. Dis. Sci.* **35**, 559 (1990). Mechanism of action study: T. Yoshikawa *et al.*, *Biochim. Biophys. Acta* **1115**, 15 (1991). Series of articles on metabolism: *Arzneim.-Forsch.* **41**, 965-995 (1991); on toxicology: *ibid.* 1033-1057. Acute toxicity: K. Matsuda *et al.*, *ibid.* 1033.

White or pale yellow crystalline powder. Insol in water and common organic solvents. LD_{50} in male, female mice, rats (mg/kg): 220, 165, 405, 422 i.p.; 758, 874, >5000, >5000 s.c.; 1269, 1331, 8441, 7375 orally (Matsuda).

THERAP CAT: Antiulcerative.

7670. Polar® Yellow. [6372-96-9] 5-Chloro-2-[4,5-dihydro-3-methyl-4-[2-[4-[[(4-methylphenyl)sulfonyl]oxy]phenyl]diazenyl]-5-oxo-1*H*-pyrazol-1-yl]benzenesulfonic acid sodium salt (1:1); 4-*p*-hydroxybenzeneazo-1-*p*-chloro-*o*-sulfophenyl-3-methyl-5-hydroxypyrazole toluene-*p*-sulfonyl ester sodium salt; C.I. Acid Yellow 40; C.I. 18950. $C_{23}H_{18}ClN_4NaO_7S_2$; mol wt 584.98. C 47.22%, H 3.10%, Cl 6.06%, N 9.58%, Na 3.93%, O 19.14%, S 10.96%. Prepn from diazotized *p*-aminophenol and 1-(*o*-sulfo-*p*-chlorophenyl)-3-methyl-5-pyrazolone: B. Richard, *US* 1067881 (Geigy). *See also: Colour Index* vol. 4 (3rd ed., 1971) p 4128.

Yellowish-brown powder. Sol in water giving a pure yellow soln; in alcohol (yellow color); in alkalies; in concd sulfuric acid. Slightly sol in ethanol, acetone. Insol in other org solvents. Precipitated by dil acids.

USE: To dye wool. In the isoln of polymyxin.

7671. Poldine Methylsulfate. [545-80-2] 2-[[(2-Hydroxy-2,2-diphenylacetyl)oxy]methyl]-1,1-dimethylpyrrolidinium methyl sulfate (1:1); 2-hydroxymethyl-1,1-dimethylpyrrolidinium methyl sulfate benzilate; 2-benziloyloxymethyl-1,1-dimethylpyrrolidinium methyl sulfate; (1-methyl-2-pyrrolidinyl)methyl benzilate methylmethosulfate; IS-499; McN-R-726-47; Nacton; Nactate. $C_{22}H_{29}NO_7S$; mol wt 451.53. C 58.52%, H 6.47%, N 3.10%, O 24.80%, S 7.10%. Anticholinergic. Prepn: Doyle *et al.*, *J. Chem. Soc.* **1958**, 4458. Prepn of the base: Blicke, Lu, *J. Am. Chem. Soc.* **79**, 29 (1955).

Needles from methyl ethyl ketone + ethanol + ether, mp 154-155°. Soluble in water.

THERAP CAT: Antispasmodic.

7672. Policosanol. [142583-61-7] Sugar cane wax alcohols; Ateromixol; Lipex; Mercol; Phytocor; PPG. Mixture of higher (C_{24} to C_{34}) primary aliphatic alcohols purified from wax of sugar cane, *Saccharum officinarum* L., *Gramineae*. Composed of octacosanol (60-70%), triacontanol (10-15%), dotriacontanol (5-10%), hexacosanol (3-8%), heptacosanol, nonacosanol, tetratriacontanol and tetracosanol. Prepn: L. Granja *et al.*, *WO* 9407830; *eidem*, *US* 5856316 (1994, 1999 both to Dalmar). Clinical pharmacology: I. Gouni-Berthold, H. K. Berthold, *Am. Heart J.* **143**, 356 (2002). Antiplatelet effects: M. L. Arruzazabala *et al.*, *Clin. Exp. Pharmacol. Physiol.* **29**, 891 (2002). Review of clinical studies: M. Janikula, *Altern. Med. Rev.* **7**, 203-217 (2002); and comparison with plant sterols and stanols: J. T. Chen *et al.*, *Pharmacotherapy* **25**, 171-183 (2005).

Off-white, odorless crystalline powder. mp 78-82°. Insol in water, hexane, methanol, ethanol. Slightly sol in chloroform.

THERAP CAT: Antilipemic.

7673. Policresulen. [9011-02-3] Hydroxymethylbenzenesulfonic acid polymer with formaldehyde; methylenebis(hydroxytoluenesulfonic acid) polymer; dihydroxydimethyldiphenylmethanedisulfonic acid polymer; Albocresil; Albothyl; Negatol. Condensation product prepd by reacting *m*-cresolsulfonic acid and formaldehyde, e.g. the product obtained by Thuau, *CA* 417272 (1943 to Lilly), *C.A.* **38**, 1077 (1944).

Sol in water, forming colloidal solns.

THERAP CAT: Antiseptic.

THERAP CAT (VET): Antiseptic.

7674. Polidexide. [9064-92-0] Sephadex 2-(diethylamino)-ethyl ether; dextran 2-(diethylamino)ethyl 2-[[2-(diethylamino)ethyl]diethylammonio]ethyl ether chloride hydrochloride epichlorohydrin crosslinked; poly[2-(diethylamino)ethyl] polyglycerylene dextran hydrochloride; PDX chloride; Secholex; DEAE-Sephadex. Anion exchange resin containing quaternary ammonium groups. Bile acid sequestrant that has been used to reduce serum cholesterol. Prepn: *GB* 1013585 (1965 to Pharmacia AB). Clinical evaluation in hyperlipidemia: R. J. C. Evans *et al.*, *Angiology* **24**, 22 (1973); S. Ritland *et al.*, *Scand. J. Gastroenterol.* **10**, 791 (1975). Review of use in chromatographic purification of proteins: D. M. Bollag, *Methods Mol. Biol.* **36**, 11-22 (1994).

Tasteless powder; forms gel in water.
USE: Ion exchange resin.
THERAP CAT: Antilipemic.

7675. Polidocanol. [9002-92-0] α-Dodecyl-ω-hydroxypoly-(oxy-1,2-ethanediyl); polyethylene glycol (9) monododecyl ether; dodecyl alcohol polyoxyethylene ether; hydroxypolyethoxydecane; laureth 9; polyoxyethylene lauryl ether; Aetoxisclerol; Aethoxysklerol; Asclera; Atlas G-4829; Hetoxol L-9. Contains an average of nine ethylene oxide units and has an average mol wt ~600. Prepd by reaction of ethylene oxide and dodecyl alcohol: Pertsemlides, Soehring, *Arzneim.-Forsch.* **10**, 990 (1960). Toxicology: H. S. Zipf *et al.*, *ibid.* **7**, 162 (1957). Review of clinical experience for sclerotherapy of telangiectasias: P. M. Goldman, *J. Dermatol. Surg. Oncol.* **15**, 204-209 (1989); for varicose veins: D. M. Eckmann, *Expert Opin. Invest. Drugs* **18**, 1919-1927 (2009).

average, *n* = 8

Viscous liquid, mp 15-21°. d 0.97. Sol in water, ethanol, toluene. Miscible with hot mineral, natural and synthetic oils; with fats and fatty alcohols. pH of 1% aq soln: 6.0-8.0. LD_{50} in mice (mg/kg): 1170 orally, 125 i.v. (Zipf).
USE: Solvent; nonionic emulsifier; pharmaceutic aid (surfactant).
THERAP CAT: Sclerosing agent.

7676. Polifeprosan. [90409-78-2] Decanedioic acid polymer with 4,4′-[1,3-propanediylbis(oxy)]bis[benzoic acid]; 1,3-bis(*p*-carboxyphenoxy)propane-sebacic acid copolymer; poly[1,3-bis(*p*-carboxyphenoxy)propane-sebacic acid]; PCPP-SA. Biodegradable, hydrophobic polymers developed for use as implantable drug delivery systems. Prepn of various copolymers: K. W. Leong *et al.*, *Polym. Prepr.* **25**, 201 (1984); *eidem*, *J. Biomed. Mater. Res.* **19**, 941 (1985). Toxicology and biocompatibility studies: C. T. Laurencin *et al.*, *Clin. Lab. Med.* **10**, 549 (1990). Review of polymer design, synthesis and physical properties: J. Tamada, R. Langer, *J. Biomater. Sci. Polym. Ed.* **3**, 315-353 (1992).

Polifeprosan 20   *m* = 20   *n* = 80

Polifeprosan 20. High mol wt polymer consisting of bis(*p*-carboxyphenoxy)propane and sebacic acid in a 20:80 ratio. Mol wt 20,000-200,000. Prepn: A. J. Domb, R. S. Langer, **EP 266603**; *eidem*, **US 4757128** (both 1988 to MIT); *eidem*, *J. Polym. Sci.* **25**, 3373 (1987). White solid, mp 68°. Viscosity (in chloroform at 23°): 0.92 dl/g. Soly (g/100ml): >20 in chloroform; <1 in hexane.
Gliadel. Polifeprosan 20 combined with carmustine, *q.v.*, in an implantable wafer. Prepn and characterization of polymer-drug matrix: W. Dang *et al.*, *Pharm. Res.* **13**, 683 (1996); A. J. Domb *et al.*, *ibid.* **16**, 762 (1999). Clinical trial of surgically implanted wafer for malignant gliomas: H. Brem *et al.*, *Lancet* **345**, 1008 (1995); S. Valtonen *et al.*, *Neurosurgery* **41**, 44 (1997).
USE: Pharmaceutic aid for controlled release drug delivery.

7677. Polihexanide. [32289-58-0]; [28757-47-3] (base). Poly(iminocarbonimidoyliminocarbonimidoylimino-1,6-hexanediyl) hydrochloride (1:?); polyaminopropylbiguanide; polyhexamethylene biguanide hydrochloride; polyhexanide; PHMB; Cosmocil CQ; Lavasept; Vantocil. $(C_8H_{17}N_5)_n$.xHCl. Broad spectrum antimicrobial polymer; binds and disrupts cytoplasmic membranes. Prepn: F. L. Rose, G. Swain, **US 2643232** (1953 to ICI). Antibacterial action: A. Davies *et al.*, *J. Appl. Bacteriol.* **31**, 448 (1968). Synthesis and characterization of solution behavior: G. C. East *et al.*, *Polymer* **38**, 3973 (1997). Interactions with phospholipid membranes: P. Broxton *et al.*, *J. Appl. Bacteriol.* **57**, 115 (1984); T. Ikeda *et al.*, *Bull. Chem. Soc. Jpn.* **58**, 705 (1985). Mechanism of action study: W. Khunkitti *et al.*, *J. Appl. Microbiol.* **82**, 107 (1997). Titrimetric determn of biguanide groups: T. Hattori *et al.*, *Anal. Sci.* **19**, 1525 (2003). Clinical evaluation of mouthrinse formulations on bacterial count and plaque growth: M. Rosin *et al.*, *J. Clin. Periodontol.* **28**, 1121 (2001); *eidem*, *ibid.* **29**, 392 (2002).

Sol in water.
USE: Biocide in personal hygiene products, contact lens solutions, industrial disinfectants, textiles, and household products; preservative in cosmetics.
THERAP CAT: Antiseptic; disinfectant.

7678. Polonium. [7440-08-6] Po; at. no. 84; valence 4, occasionally 2, rarely 6. No stable nuclides. Isotopes range in mass number from 193-218; all are radioactive; 210 is naturally occurring. The first radioactive substance discovered by Mme. Curie in 1898. A product of disintegration of radium; one gram is contained in ~25,000 tons of pitchblende or in 7.5 kg radium that is more than 30 years old. Separated in form of a deposit on a bismuth plate immersed in a soln of the chloride: Marckwald, *Ber.* **35**, 2285 (1902); using a silver, gold or nickel plate: Curie, Joliot, *J. Chim. Phys.* **28**, 201 (1931); Haissinsky, *ibid.* **33**, 97 (1936); Rollier, *Gazz. Chim. Ital.* **66**, 797 (1936); Ziv, *C. R. Acad. Sci. USSR* **25**, 743 (1939). Obtained in the metallic form by volatilization from nickel on a collodion film: Rollier *et al.*, *J. Chim. Phys.* **4**, 648 (1936). The only readily accessible isotope is the penultimate member of the radium decay series, $^{210}_{84}$Po, also called **Radium F** (Ra-F). Decays by α-emission ($T_{1/2}$ 138.4 days, rel. at. mass 209.9828) to $^{206}_{82}$Pb. Physical properties: Maxwell, *J. Chem. Phys.* **17**, 1288 (1949). Comprehensive reviews: K. W. Bagnall, *Chemistry of the Rare Radioelements* (Butterworths, London, 1957); *idem*, *Endeavour* **22**(86), 61 (May 1963); *idem*, "Selenium, Tellurium and Polonium" in *Comprehensive Inorganic Chemistry* vol. 2, J. C. Bailar, Jr., et al., Eds. (Pergamon Press, Oxford, 1973) pp 935-1008. Review of radiation toxicology: J. Harrison *et al.*, *J. Radiol. Prot.* **27**, 17-40 (2007). Simple cubic crystal structure. Two allotropic forms; coexist between 18° and 54°; d (α-form) 9.196; d (β-form) 9.398. mp 254°; bp 962°. Latent heat of vaporization: 24.597 kcal/mole. Resistivity: α-Po = 42 μohm-cm at 0°; β-Po = 44 μohm-cm at 0°. Chemically resembles tellurium and bismuth. Forms a volatile, unstable hydride, PoH_2. Forms a polonide, Na_2Po; a carbonyl PoCO; a hydroxide $Po(OH)_4$.
Caution: Radiation hazard; alpha emitter. Sol compds hazard to kidneys, spleen; insol, airborne compds hazard to lungs. Max per-

missible concn of insol $_{84}^{210}$Po in air: 7×10^{-11} μ-Curie/cc, *Natl. Bur. Stand. Handb.* **69**, 79 (1959).

7679. Poloxamers. [106392-12-5] 2-Methyloxirane block polymer with oxirane; polyethylene glycol-polypropylene glycol block copolymer; α-hydro-ω-hydroxypoly(oxyethylene)poly(oxypropylene)poly(oxyethylene) block copolymer. Series of nonionic surfactants with the structure HO(CH$_2$CH$_2$O)$_a$(CH(CH$_3$)CH$_2$O)$_b$-(CH$_2$CH$_2$O)$_c$H where b is at least 15, a and c are approximately equal and the polyoxyethylene content is 10 to 80% of the total polymer weight. Mol wt ranges from 1000 to >16,000. The polyoxypropylene segment is hydrophobic; the polyoxyethylene segment hydrophilic. Prepn: L. G. Lundsted, **US 2674619** (1954 to Wyandotte). Comprehensive description of manufacture and properties: I. R. Schmolka in *Nonionic Surfactants*, M. Schick, Ed. (Dekker, New York, 1967) pp 300-371; I. F. Paterson *et al.* in *Handbook of Engineering Polymeric Materials*, N. P. Cheremisinoff, Ed. (Marcel Dekker, New York, 1997) pp 765-774. Review of chemistry and medical applications: L. E. Reeve in *Handbook of Biodegradable Polymers*, A. J. Domb *et al.*, Eds. (Gordon & Breach, New York, 1997) pp 231-249.

Mobile liquids, pastes or flakeable solids. Relatively nonhygroscopic. Vary from water-insol to very water-sol compds; more sol in cold than hot water. Low-foaming properties. Stable to acids, alkalies, metallic ions.

Poloxalene. [9003-11-6] SKF-18667; Bloat Guard; Therabloat. a = 12, b = 34, c = 12; av. mol wt 3000. Use in treatment of ruminant bloat: E. E. Bartley, G. C. Scott, **US 3465083** (1969 to Kansas State Univ. and SKF). Comparative field trial in cattle: W. Majak *et al.*, *J. Anim. Sci.* **73**, 1493 (1995). Colorless or pale yellow liquid. Sol in water, chloroform, ethylene dichloride.

Poloxamer 182. Pluronic L62. a = 8, b = 30, c = 8; av. mol wt 2500. Liquid, mp $-4°$. d$_{25}^{25}$ 1.03. Brookfield viscosity (25°): 450 cP; cloud pt (10% aq soln): 24°. Sol in water, ethanol, toluene. Insol in kerosene, ethylene glycol.

Poloxamer 188. Exocorpol; Pluronic F68; Synperonic PE/F68. a = 75, b = 30, c = 75; av. mol wt 8400. White, flakeable solid. mp 52°. Brookfield viscosity (77°): 1000 cP; cloud pt (10% aq soln): >100°. Freely sol in water, ethanol; partially sol in toluene. Insol in kerosene, ethylene glycol.

Poloxamer 331. Pluronic L101. a = 7, b = 54, c = 7; av. mol wt 3800. Liquid, mp $-23°$. d$_{25}^{25}$ 1.018; Brookfield viscosity (25°): 800 cP; cloud pt (10% aq soln): 11°. Sol in ethanol, toluene. Insol in water, kerosene, ethylene glycol.

Purified Poloxamer 188. Flocor. Highly purified pharmaceutical grade for intravenous use; lowers blood viscosity and improves blood flow. Pharmacology: R. M. Emanuele, *Expert Opin. Invest. Drugs* **7**, 1193-1200 (1998). Clinical trial in sickle cell disease: E. P. Orringer *et al.*, *J. Am. Med. Assoc.* **286**, 2099 (2001).

USE: Pharmaceutic aids; food additives; defoamers; antistatic agents; demulsifiers; detergents; wetting agents; gelling agents; emulsifiers; foam controllers; dispersants; dye levelers.

THERAP CAT: Poloxamer 188 as hemorheologic agent.

THERAP CAT (VET): Poloxalene in prevention of bloat in cattle.

7680. Polybrominated Biphenyls. PBBs; brominated biphenyls; polybromobiphenyls. Mixtures of isomers and congeners with structures similar to polychlorinated biphenyls, *q.v.*, where each X = H or Br. Once widely used commercially. Prepn: H. Hahn *et al.*, **DE 1161547** (1964 to Chem. Fabrik Kalb); G. A. Burk, **US 3733366** (1973 to Dow); L. C. Mitchell, D. R. Breckenridge, **US 3763248** and **US 3833674** (1973, 1974 both to Ethyl Corp.). Persistence in soils: L. W. Jacobs *et al.*, *J. Agric. Food Chem.* **24**, 1198 (1976). Photodegradation: L. O. Ruzo *et al.*, *ibid.* 1062. Reviews of environmental hazards: K. Kay, *Environ. Res.* **13**, 74-93 (1977); F. J. DiCarlo *et al.*, *Environ. Health Perspect.* **23**, 351-365 (1978); of toxicology and mechanism of action: S. Safe, *Crit. Rev. Toxicol.* **13**, 319-395 (1984); of carcinogenicity: E. M. Silberhorn *et al.*, *ibid.* **20**, 440-496 (1990); of toxicology and human exposure: *Toxicological Profile for Polybrominated Biphenyls and Polybrominated Diphenyl Ethers* (PB2004-107334, 2004) 624 pp.

Firemaster BP-6. [59536-65-1] Major component is *2,2',4,4',-5,5'-hexabromobiphenyl*. Softens at 72°, dec >300°. Low vapor pressure; degraded by uv light. Very sol in benzene, toluene. Insol in water.

Caution: The 1973 "Michigan Incident" in which BP-6 was accidentally added to animal feed, and resulted in the contamination of the food chain and wide-spread destruction of contaminated farm animals, led to the removal of BP-6 from the market: L. J. Carter, *Science* **192**, 240 (1976). Reported symptoms of overexposure include impaired immune system; hypothyroidism; neurological effects, headache, joint stiffness, memory loss; chloracne-like lesions (Safe). *See also Patty's Industrial Hygiene and Toxicology* **vol. 2D**, G. D. Clayton, F. E. Clayton, Eds. (John Wiley & Sons, New York, 4th ed., 1994) p 2433-2504. These substances are reasonably anticipated to be human carcinogens: *Report on Carcinogens, Twelfth Edition* (PB2011-111646, 2011) p 347.

USE: Formerly as flame retardant.

7681. Polybrominated Diphenyl Ethers. PBDEs. Group of variably brominated phenolic compounds with flame retardant properties. There are 209 possible congeners, depending on the number and position of the bromine atoms. Technical formulations are mixtures of congeners, typically the penta-, octa-, or decabromodiphenyl ether, *q.v.* Manufacturing processes: H. Jenkner, **US 3285965** (1966 to Chem. Fabrik Kalk); D. R. Brackenridge, **US 3959387** (1976 to Ethyl Corp.). Overview of congener composition of technical flame retardant mixtures: M. J. La Guardia *et al.*, *Environ. Sci. Technol.* **40**, 6247 (2006). Review of manufacture, uses, and toxicology: F. Rahman *et al.*, *Sci. Total Environ.* **275**, 1-17 (2001). Review of analytical methods in environmental and human samples: A. Covaci *et al.*, *Environ. Int.* **29**, 735-756 (2003); of toxicology and human exposure: A. Sjödin *et al.*, *ibid.*, 829-839; *Toxicological Profile for Polybrominated Biphenyls and Polybrominated Diphenyl Ethers* (PB2004-107334, 2004) 624 pp.

X = H or Br

PBDEs are generally resistant to acids and bases, to heat and light, and to reducing or oxidizing compounds.

2,2',4,4'-Tetrabromodiphenyl ether. [5436-43-1] 2,4-Dibromo-1-(2,4-dibromophenoxy)benzene; BDE-47. C$_{12}$H$_6$Br$_4$O; mol wt 485.80. Most abundant congener found in environmental samples; major impurity in penta-BDE formulations. Vapor pressure: 1.40×10^{-6} mm Hg. Soly in water: 15 μg/l. Log P (octanol/water): 6.81.

Pentabromodiphenyl ether. [32534-81-9] Pentabromodiphenyl oxide; PeBDE; penta-BDE. Technical mixture contains *2,2',4,4',5-pentabromodiphenyl ether* as the major component. Amber to pale yellow, highly viscous liquid. mp -7 to $-3°$. bp >300°; starts to dec above 200°. Soly in water: 13.3 μg/l; in methanol: 10 g/kg. Miscible in toluene. Log P (octanol/water): 6.57. Vapor pressure: 3.5×10^{-7} mm Hg.

Note: Tetrabromodiphenyl ether and pentabromodiphenyl ether are listed as persistent organic pollutants (POPs) in Annex A of the *Stockholm Convention on Persistent Organic Pollutants* (United Nations, Stockholm, 2001) 43 pp; amended (Geneva, 2009) 63 pp.

USE: Additive flame retardant for resins and polymers.

7682. Polychlorinated Biphenyls. PCBs; chlorinated biphenyls; chlorobiphenyls; Aroclor; Clophen; Fenclor; Kanechlor; Pyralene. Once widely used industrial chemicals whose high stability contributed to both their commercial usefulness and their long-term deleterious environmental and health effects. Synthesis: H. Schmidt, G. Schulz, *Ann.* **207**, 338 (1881). Commercially available since 1930: C. Penning, *Ind. Eng. Chem.* **22**, 1180 (1930). Commercial PCBs are mixtures of various isomers and congeners. The Aroclors are characterized by four digit numbers. The first two digits indicate that the mixture contains biphenyls (12), triphenyls (54) or both (25, 44); the last two digits give the weight percent of chlorine in the mixture (e.g. Aroclor 1242 contains biphenyls with ~42% chlorine). Accumulation of airborne PCBs in foliage: E. H. Buckley, *Science* **216**, 520 (1982). *Reviews:* H. L. Hubbard in *Kirk-Othmer Encyclopedia of Chemical Technology* **vol. 5** (Interscience, New York, 2nd ed., 1964) pp 289-297; O. Hutzinger *et al.*, *The Chemistry of PCBs* (CRC Press, Cleveland, Ohio, 1974) 269 pp; J. W. Lloyd *et al.*, *J. Occup. Med.* **18**, 109-113 (1976). Reviews of environmental impact and toxicity: L. Fishbein, *Annu. Rev. Pharmacol.* **14**, 139-156 (1974); *National Conference on Polychlorinated Biphenyls*, Nov. 19-21, 1975 (EPA-560/6-75-004, 1976) 487 pp; R. D. Kimbrough, *Crit. Rev. Toxicol.* **2**, 445-498 (1974); S. H. Safe, *ibid.* **24**, 87-149 (1994). Reviews of carcinogenicity: *IARC Monographs* **18**, 43-103 (1978); E. M. Silberhorn *et al.*, *Crit. Rev. Toxicol.* **20**, 440-496 (1990); of toxicology and mechanism of action: S. Safe, *ibid.* **13**, 319-395 (1984); of toxicology and human exposure: *Toxicological Profile for Polychlorinated Biphenyls* (PB2000-108027, 2000) 948 pp.

X = H or Cl

Aroclor 1242. [53469-21-9] PCB 1242. Clear, mobile liquid; av. number Cl/molecule: 3.10. d_4^{25} 1.381, $d_4^{15.5}$ 1.392. Distillation range 325-366°. Flash point (open cup) 348-356°F. n_D^{20} 1.627-1.629. Dielectric constant (1000 cycles) 5.6 (25°), 4.9 (100°).

Aroclor 1254. [11097-69-1] PCB 1254. Light yellow, viscous liquid; av. number Cl/molecule: 4.96. d_4^{65} 1.495; $d_4^{15.5}$ 1.505. Distillation range 365-390°. No open cup flash point to boiling. n_D^{20} 1.629-1.641. Dielectric constant (1000 cycles) 5.0 (25°), 4.3 (100°). LD$_{50}$ orally in weanling rats: 1295 mg/kg (Kimbrough).

Aroclor 1260. [11096-82-5] PCB 1260. Light yellow, soft, sticky resin; av. number Cl/molecule: 6.30. d_4^{90} 1.555; $d_4^{15.5}$ 1.566. Distillation range 385-420. No open cup flash point to boiling. n_D^{20} 1.647-1.649. Dielectric constant (1000 cycles) 4.3 (25°); 3.7 (100°). LD$_{50}$ orally in weanling rats: 1315 mg/kg (Kimbrough).

Caution: In Japan, 1968, oral intoxication to humans due to accidental contamination of rice bran oil with Kanechlor 400 led to an outbreak of what became known as "Yusho disease". Symptoms of oral intoxication in humans included nausea, lethargy, chloracne, brown pigmentation of skin and nails, subcutaneous edema of the face, distinctive hair follicles, excessive eye discharge, swelling of eyelids, visual disturbances, GI disturbances and jaundice. *See* M. Kuratsune *et al.*, (EPA-560/6-75-004, 1976) p 14. Potential symptoms of occupational overexposure are chloracne, dermal lesions; hepatic injury; decreased pulmonary function; decreased birth weight in offspring of exposed mothers; eye irritation (Safe, 1994). *See also Patty's Industrial Hygiene and Toxicology* **vol. 2D**, G. D. Clayton, F. E. Clayton, Eds. (John Wiley & Sons, Inc., New York, 4th ed., 1994) 2433-2504. These substances are reasonably anticipated to be human carcinogens: *Report on Carcinogens, Twelfth Edition* (PB2011-111646, 2011) p 349.

Note: Polychlorinated biphenyls are listed as persistent organic pollutants (POPs) in Annex A and Annex C of the *Stockholm Convention on Persistent Organic Pollutants* (United Nations, Stockholm, 2001) 43 pp; amended (Geneva, 2009) 63 pp.

USE: In electrical capacitors, electrical transformers, gas-transmission turbines, vacuum pumps. Formerly used in U.S. as hydraulic fluids, plasticizers, adhesives, fire retardants, wax extenders, dedusting agents, pesticide extenders, inks, lubricants, cutting oils, in heat transfer systems, carbonless reproducing paper.

7683. Polychlorinated Dibenzo-*p*-dioxins. Polychlorinated dibenzodioxins; PCDDs; CDDs. Class of hazardous and persistent environmental pollutants; 75 congeners are possible depending on the chlorine substitution pattern of the aromatic rings. Chlorinated dioxin congeners are produced predominately as the by-products of industrial processes, such as paper pulp bleaching or in the commercial synthesis of chlorinated aromatic compds. The most well known compd of the class, TCDD, *q.v.*, is a carcinogenic contaminant in the

manufacture of the defoliant, Agent Orange. Structurally related to polychlorinated dibenzofurans and polychlorinated biphenyls, *q.q.v.*, PCDDs possess comparable physical properties and cause similar adverse effects. Elicit biochemical responses upon binding to the arylhydrocarbon receptor (AhR). Toxic potencies of individual compds vary considerably; congeners with chlorine in the 2,3,7, and 8 positions are considered to be toxic to organisms. Toxicity differences among relevant congeners are distinguished on the basis of toxic equivalency factors (TEFs), a conceptual scale that estimates the toxicity of a compd relative to TCDD. Prepn of polychlorinated congeners: A. P. Gray *et al.*, *J. Org. Chem.* **41**, 2435 (1976). Review of mechanism of formation from potential industrial sources: K. C. M. Ree *et al.*, *Toxicol. Environ. Chem.* **17**, 171-195 (1988); of environmental photochemistry: G. G. Choudhry, G. R. B. Webster, *ibid.* **14**, 43-61 (1987). Review of analytical techniques and procedures for isoln and determn: E. J. Reiner *et al.*, *Anal. Bioanal. Chem.* **386**, 791-806 (2006); of bioanalytical determn methods: J. Díaz-Ferrero *et al.*, *Trends Anal. Chem.* **16**, 563-573 (1997); of congener distribution in environmental and human samples: K. Srogi, *Environ. Chem. Lett.* **6**, 1-28 (2008). Review of environmental toxicology: J. P. Vanden Heuvel, G. Lucier, *Environ. Health Perspect.* **100**, 189-200 (1993); of congener toxic equivalency factors (TEFs): M. Van den Berg *et al.*, *ibid.* **106**, 775-792 (1998); of risks to animal and human health: S. A. Skene *et al.*, *Hum. Toxicol.* **8**, 173-203 (1989).

X = H or Cl

Generally stable upon heating to 700°C. Highly lipophilic with low water solubilities and low vapor pressures; solubility in polar solvents and volatility decrease with additional chlorine atoms. Resistant to chemical and biological degradation. Long half-lives in animals; bioaccumulate in animal and human adipose tissue. High sorbent potential for particulate matter.

1,2,3,7,8-Pentachlorodibenzo[*b,e*][1,4]dioxin. [40321-76-4] 1,-2,3,7,8-PeCDD. $C_{12}H_3Cl_5O_2$; mol wt 356.40. C 40.44%, H 0.85%, Cl 49.73%, O 8.98%. Dibenzo-*p*-dioxin congener with the same TEF profile as TCDD. Solid from chloroform-methanol, mp 240-241°. uv max (chloroform): 308 nm (ε 6100). TEF in humans/mammals, fish, birds: 1, 1, 1 (Van den Berg).

1,2,3,4,7,8-Hexachlorodibenzo[*b,e*][1,4]dioxin. [39227-28-6] 1,2,3,4,7,8-HxCDD. $C_{12}H_2Cl_6O_2$; mol wt 390.85. C 36.88%, H 0.52%, Cl 54.42%, O 8.19%. Most toxic hexachlorinated dibenzo-*p*-dioxin congener toward fish. Solid from chloroform, mp 272.5-273°. uv max (chloroform): 313 nm (ε 4100). TEF in humans/mammals, fish, birds: 0.1, 0.5, 0.05 (Van den Berg).

Caution: Potential symptoms of overexposure in experimental animals are: enzyme induction, wasting syndrome, lymphoid involution, hepatic damage, hepatic porphyria, chloracne, gastric lesions, urinary tract hyperplasia, edema, hyperlipidemia, reproductive toxicity, teratogenicity, increase in tumor incidence, and lethality. *See* A. Poland, J. C. Knutson, *Annu. Rev. Pharmacol. Toxicol.* **22**, 517 (1982).

Note: Polychlorinated dibenzo-*p*-dioxins are listed as persistent organic pollutants (POPs) in Annex C of the *Stockholm Convention on Persistent Organic Pollutants* (United Nations, Stockholm, 2001) 43 pp; amended (Geneva, 2009) 63 pp.

7684. Polychlorinated Dibenzofurans. PCDFs; CDFs. Class of hazardous and persistent environmental pollutants; 135 congeners are possible depending on the chlorine substitution pattern of the aromatic rings. Produced predominately as by-products of industrial processes, such as paper pulp bleaching, or in the commercial synthesis of chlorinated aromatic compds. Structurally related to polychlorinated dibenzo-*p*-dioxins and polychlorinated biphenyls, *q.q.v.*; possess comparable physical properties and cause similar adverse effects. Elicit biochemical responses upon binding to the arylhydrocarbon receptor (AhR). Toxic potencies of individual compds vary considerably; congeners with chlorine in the 2,3,7, and 8 positions are considered to be toxic to organisms. Toxicity differences among relevant congeners are distinguished on the basis of toxic equivalency factors (TEFs), a conceptual scale that estimates the toxicity of a compd relative to TCDD, *q.v.*, a known human carcinogen. Prepn and characterization of polychlorinated congeners: H. Kuroki *et al.*, *Chemosphere* **13**, 561 (1984). Review of mechanism of formation from potential industrial sources: K. C. M. Ree *et al.*, *Toxicol. Environ. Chem.* **17**, 171-195 (1988); of environmental photochemistry: G. G. Choudhry, G. R. B. Webster, *ibid.* **14**, 43-61 (1987). Review of analytical techniques and procedures for isoln and determn: E. J. Reiner *et al.*, *Anal. Bioanal. Chem.* **386**, 791-806 (2006); of bioanalytical determn methods: J. Díaz-Ferrero *et al.*, *Trends Anal. Chem.* **16**, 563-573 (1997); of congener distribution in environmental and human samples: K. Srogi, *Environ. Chem. Lett.* **6**, 1-28 (2008). Review of environmental toxicology: J. P. Vanden Heuvel, G. Lucier, *Environ. Health Perspect.* **100**, 189-200 (1993); of congener toxic equivalency factors (TEFs): M. Van den Berg *et al.*, *ibid.* **106**, 775-792 (1998); of risks to animal and human health: S. A. Skene *et al.*, *Hum. Toxicol.* **8**, 173-203 (1989).

X = H or Cl

Generally stable upon heating to 700°C. Highly lipophilic with low water solubilities and low vapor pressures; solubility in polar solvents and volatility decreases with additional chlorine atoms. Resistant to chemical and biological degradation. Long half-lives in animals; bioaccumulate in animal and human adipose tissue. High sorbent potential for particulate matter.

2,3,7,8-Tetrachlorodibenzofuran. [51207-31-9] 2,3,7,8-TCDF. $C_{12}H_4Cl_4O$; mol wt 305.96. C 47.11%, H 1.32%, Cl 46.35%, O 5.23%. Structural analog of TCDD. Solid, mp 219-221°. TEF in humans/mammals, fish, birds: 0.1, 0.05, 1 (Van den Berg).

2,3,4,7,8-Pentachlorodibenzofuran. [57117-31-4] 2,3,4,7,8-PeCDF. $C_{12}H_3Cl_5O$; mol wt 340.41. C 42.34%, H 0.89%, Cl 52.07%, O 4.70%. Most toxic dibenzofuran congener across taxa. Solid, mp 196-196.5°. TEF in humans/mammals, fish, birds: 0.5, 0.5, 1 (Van den Berg).

Caution: Potential symptoms of overexposure in experimental animals are: enzyme induction, wasting syndrome, lymphoid involution, hepatic damage, hepatic porphyria, chloracne, gastric lesions, urinary tract hyperplasia, edema, hyperlipidemia, reproductive toxicity, teratogenicity, increase in tumor incidence, and lethality. *See* A. Poland, J. C. Knutson, *Annu. Rev. Pharmacol. Toxicol.* **22**, 517 (1982).

Note: Polychlorinated dibenzofurans are listed as persistent organic pollutants (POPs) in Annex C of the *Stockholm Convention on Persistent Organic Pollutants* (United Nations, Stockholm, 2001) 43 pp; amended (Geneva, 2009) 63 pp.

7685. Polydextrose. [68424-04-4] Litesse. Randomly bonded condensation polymer of dextrose with small amounts of bound sorbitol and citric acid. Functions to replace the bulk and mouthfeel of sugar and/or fat in reduced calorie foods. Prepn: H. H. Rennhard, US 3766165 (1973 to Pfizer). Improved process: D. B. Guzek *et al.*, EP 473333 (1992 to Pfizer), *C.A.* **116**, 237725s (1992). Reviews of physical properties and applications in foods: A. Torres, R. D. Thomas, *Food Technol.* **35**, 44-49 (1981); F. K. Moppett in *Food Sci. Technol.* **48**, entitled "Alternative Sweeteners", L. O. Nabors, R. C. Gelardi, Eds. (1991) pp 401-421.

White to light tan, amorphous powder, mp >130°. Bland, nonsweet taste. Hygroscopic. pH of 10% w/w aq soln: 2.5-3.5. Viscosity of 50% aq soln: 35 cP. Very sol in water (to ~80%). Partially sol in glycerin, propylene glycol. Insol in ethanol. Caloric utilization value in humans: 1 kcal/g.

USE: Bulking agent for reduced calorie foods.

7686. Polyestradiol Phosphate. [28014-46-2] (17β)-Estra-1,3,5(10)-triene-3,17 diol polymer with phosphoric acid; estradiol phosphate polymer; PEP; Estradurin. Polymeric ester of phosphoric acid and estradiol. Mol wt ~26,000. Prepn: Diczfalusy, *Endocrinology* **54**, 471 (1954); Fernö *et al.*, *Acta Chem. Scand.* **12**, 1675 (1958); Diczfalusy *et al.*, US 2928849 (1960 to AB Leo). Clinical pharmacology: P. O. Gunnarsson, B. J. Norlén, *Prostate* **13**, 299 (1988). Clinical trial in prostatic carcinoma: J. Aro, *ibid.* **18**, 131 (1991).

n is approx. 80

Solid, mp 195-202°. Very sol in aq pyridine; sol in aq alkali; very slightly sol in ethanol, ethanol + water (1:1), water, dioxane, acetone, chloroform. Intrinsic viscosity [η] in 0.25M NaCl soln at pH 7.5 = 0.04.

THERAP CAT: Antineoplastic (hormonal).

7687. Polyethylene. [9002-88-4] Ethene homopolymer; Agilene; Alathon; Alkathene; Courlene; Lupolen; Platilon; Polythene; Pylen; Reevon. Mol wt about 1500-100,000. C 85.7%, H 14.3%. Prepd by polymerization of liq ethylene at high temps and high or low pressure. *Reviews:* Aggarwal, Sweeting, *Chem. Rev.* **57**, 665-742 (1957); Raff, Allison, *Polyethylene* vol. XI of High Polymers series (Interscience, New York, 1956); Faith *et al.*, *Industrial Chemicals* (Wiley, New York, 3rd ed., 1965) pp 624-630.

Plastic solid of milky transparency. d_4^{20} 0.92. Tough and flexible at room temps, mp 85-110°. Breaks with cryst fracture at −50°. Good electrical insulator. Surface resistivity: 10^{14} ohms. Will burn, but hardly supports combustion. Stable to water, non-oxidizing acids and alkalies, alcohols, ethers, ketones, esters at ordinary temps. Attacked by oxidizing acids such as nitric acid and perchloric acid, free halogens, benzene, petr ether, gasoline and lubricating oils, aromatic and chlorinated hydrocarbons.

USE: Laboratory tubing; in making prostheses; electrical insulation; packaging materials; kitchenware; tank and pipe linings; paper coatings; textile stiffeners.

7688. Polyethylene Glycol. [25322-68-3] α-Hydro-ω-hydroxypoly(oxy-1,2-ethanediyl); macrogol; PEG; Carbowax; Pluracol E; Poly-G; Polyglycol E. Liquid and solid polymers of the general formula $H(OCH_2CH_2)_nOH$, where *n* is greater than or equal to 4. In general, each PEG is followed by a number which corresponds to its average mol wt. Synthesis: Fordyce, Hibbert, *J. Am. Chem. Soc.* **61**, 1905, 1910 (1939). Polyethylene glycols are compds of low toxicity: Smyth *et al.*, *J. Am. Pharm. Assoc. Sci. Ed.* **39**, 349 (1950). Toxicity data (PEG 400): W. Bartsch *et al.*, *Arzneim.-Forsch.* **26**, 1581 (1976). *Reviews:* Glycols, G. O. Curme, F. Johnston, Eds. (Reinhold, New York, 1952) pp 176-202; Kastens in *High Polymers*, H. Mark *et al.*, Eds., vol. **13** entitled *Polyethers*, part 1 (Interscience, New York, 1963) pp 169-189, 274-291; G. M. Powell, III in *Handbook of Water-Soluble Gums & Resins*, R. L. Davidson, Ed. (McGraw-Hill, New York, 1980) pp 18/1-18/31. *Book: Poly(Ethylene Glycol) Chemistry: Biotechnical and Biomedical Applications*, J. M. Harris, Ed. (Plenum Press, New York, 1992) 385 pp. Series of articles on pegylation to enhance delivery of protein drugs: *Adv. Drug*

Delivery Rev. **54**, 453-609 (2002). Clinical evaluation as laxative: S. Chaussade, M. Minic, *Aliment. Pharmacol. Ther.* **17**, 165 (2003).

Clear, viscous liquids or white solids which dissolve in water forming transparent solns. Sol in many organic solvents. Readily sol in aromatic hydrocarbons. Only slightly sol in aliphatic hydrocarbons. Do not hydrolyze or deteriorate on storage, will not support mold growth. Solvent action on some plastics.

Polyethylene glycol 200. Average value of n is 4, mol wt range 190-210. Viscous, hygroscopic liq; slight characteristic odor; d_{25}^{25} 1.127. Viscosity (210°F): 4.3 cSt. Supercools upon freezing.

Polyethylene glycol 400. Average value of n between 8.2 and 9.1, mol wt range 380-420. Viscous, slightly hygroscopic liq; slight characteristic odor; d_{25}^{25} 1.128. mp 4-8°. Viscosity (210°F): 7.3 cSt. LD$_{50}$ orally in rats: 30 ml/kg (Bartsch).

Polyethylene glycol 600. Average value of n between 12.5 and 13.9, mol wt range 570-630. Viscous, slightly hygroscopic liq; characteristic odor; d_{25}^{25} 1.128. mp 20-25°. Viscosity (210°F): 10.5 cSt.

Polyethylene glycol 1500. Average value of n between 29 and 36, mol wt range 1300-1600. White, free-flowing powder; d_{25}^{25} 1.210. mp 44-48°. Viscosity (210°F): 25-32 cSt.

Polyethylene glycol 4000. Forlax; Idrolax. Average value of n between 68 and 84, mol wt range 3000-3700. White, free-flowing powder or creamy-white flakes; d_{25}^{25} 1.212. mp 54-58°. Viscosity (210°F): 76-110 cSt. LD$_{50}$ orally in rats (divided doses): 59 g/kg (Smyth).

Polyethylene glycol 6000. Average value of n between 158 and 204, mol wt range 7000-9000. Powder or creamy-white flakes; d_{25}^{25} 1.21. mp 56-63°. Viscosity (210°F): 470-900 cSt. LD$_{50}$ orally in rats: >50 g/kg (Smyth).

USE: Pharmaceutic aid (ointment and suppository base; tablet excipient). As water-soluble lubricants for rubber molds, textile fibers, and metal-forming operations. In food and food packaging. In hair prepns, in cosmetics in general. As a stationary phase in gas chromatography. Also in water paints, paper coatings, polishes and in the ceramics industry.

THERAP CAT: Laxative.

7689. Polyethylene Terephthalates. PET. Fiber forming polyesters prepd from terephthalic acid, *q.v.*, or its esters and ethylene glycol: Whinfield, Dickson, **US 2465319** (1949 to du Pont). Review of structures, definition of trade names: R. W. Moncrieff, *Man-Made Fibres* (John Wiley & Sons, New York, 4th ed., 1963) pp 361-389, 707-723.

R = OH, ***Dacron, Amilar, Fiber V***. Solid, dec at approx 250°. Sp gr 1.38. Sol in hot *m*-cresol, trifluoroacetic acid, *o*-chlorophenol, a mixture of 7 parts of trichlorophenol and 10 parts (by wt) of phenol, a mixture of 2 parts of tetrachloroethane and 3 parts (by wt) of phenol. Fiber has good resistance to weak acids even at boiling temp, to strong acids in the cold, to weak alkalies, to bleaches, to most alcohols, ketones, soaps, detergents, and dry cleaning agents. Fabric has good resistance to creasing, abrasion, heat aging, and sunlight when behind glass. When "heat-set", fabric will not shrink in either boiling water or boiling drycleaning solvent. Fabric burns, but local melting generally prevents spread of fire. Insects cannot thrive on the fiber, but some can cut through it. Molds, mildew, and fungi may grow on applied finishes, but do not attack fiber. R = OCH$_3$, **Terylene**. For physical properties, *see* Dacron above. Other similar products: ***Diolen, Enkalene, Fortrel, Tergal, Terital, Terlenka, Trevira, Mylar***.

USE: In fabric manufacture; as films; as base for magnetic coatings. Surgical aid (arterial grafts).

7690. Polygodial. [6754-20-7] (1*R*,4a*S*,8a*S*)-1,4,4a,5,6,7,8,-8a-Octahydro-5,5,8a-trimethyl-1,2-naphthalenedicarboxaldehyde; tadeonal. C$_{15}$H$_{22}$O$_2$; mol wt 234.34. C 76.88%, H 9.46%, O 13.65%. Widely distributed drimane sesquiterpene with insect anti-

feedant properties; naturally occurring as the (−)-form. Isoln from *Polygonum hydropiper* L., *Polygonaceae* (Australia) and structure: C. S. Barnes, J. W. Loder, *Aust. J. Chem.* **15**, 322 (1962); from the bark of *Warburgia stuhlmanni* Engl. or *W. ugandensis, Canellacceae* (E. Africa): I. Kubo *et al., Chem. Commun.* **1976**, 1013; from nudibranch *Dendrodons limbata* (Mediterranean): G. Cimino *et al., Science* **219**, 1237 (1983); from nudibranchs *D. nigra, D. tuberculosa* (Hawaii) and *D. krebsii* (Mexico): R. K. Okuda *et al., J. Org. Chem.* **48**, 1866 (1983). Relationship between structure and antifeedant-activity: K. Nakanishi, I. Kubo, *Isr. J. Chem.* **16**, 28 (1977); M. D'Ischia *et al., Tetrahedron Lett.* **1982**, 3295. Synthesis of racemate: T. Kato *et al., Tetrahedron Lett.* **1971**, 1961; S. C. Howell *et al., Chem. Commun.* **1981**, 507. Synthesis of (−)-form: I. Razmilic *et al., Synth. Commun.* **17**, 95 (1987).

Colorless needles from petroleum (40-60°), mp 57° (Barnes, Loder). [α]$_D^{24}$ −131° (c = 0.96 in ethanol). uv max (ethanol): 231, 295 nm (ε 11800, 76).

(±)-Form. mp 93-94° (Tanis, Nakanishi).

7691. Polylysine. [25104-18-1] L-Lysine homopolymer. A lysine polypeptide or homopolymer, the chain length of which varies with the method of prepn. Prepn: Katchalski *et al., J. Am. Chem. Soc.* **69**, 2564 (1947); **70**, 2094 (1948); Fasman *et al., ibid.* **83**, 709 (1961); Sela *et al., Biopolymers* **1**, 517 (1963); Strojny, White, **US 3215684** (1965 to Dow). For structure *see* Lysine.

L-Form hydriodide. Average dp (or n) = 32. Transparent, solid, film-like polymer. Readily sol in water; practically insol in the usual organic solvents. Transition of high-mol-wt poly-L-lysine (dp 1500) in aq soln from a helical to a randomly coiled conformation under the influence of decreasing pH or increasing temp: Applequist, Doty, *C.A.* **58**, 6925b (1963).

7692. Polymerized Pyridoxylated Hemoglobin. Poly SFH-P; PolyHeme. Acellular oxygen carrier consisting of pyridoxylated, stroma-free hemoglobin polymerized with glutaraldehyde. Average mol wt 150 kDa. Prepn and oxygen-carrying capacity: L. R. Sehgal *et al., Prog. Clin. Biol. Res.* **122**, 19 (1983); *eidem, Surgery* **95**, 433 (1984). Pharmacology: S. A. Gould *et al., Ann. Emerg. Med.* **15**, 1416 (1986). Clinical trial in acute blood loss: *idem et al., J. Am. Coll. Surg.* **187**, 113 (1998). Review of clinical development: S. A. Gould, G. S. Moss, *World J. Surg.* **20**, 1200-1207 (1996).

Prepd as solution containing 12-14 g hemoglobin/dl. Oxygen-carrying capacity: 16-19 vol%. Binding coefficient (ml O$_2$/g Hb): 1.30. Colloid osmotic pressure: 20-25 mm Hg.

THERAP CAT: Blood substitute.

7693. Polymyxin. [1406-11-7] Antibiotic complex produced by *Bacillus polymyxa:* Brownlee, Jones, *Biochem. J.* **43**, XXV (1948). Prepn: Ainsworth, Pope, **US 2565057** (1951 to Burroughs Wellcome); Petty, **US 2595605** (1952 to Am. Cyanamid); Benedict, Stodola, **US 2771397** (1956 to USDA). Purification: Hastings *et al.*, **GB 782926** (1957 to Distillers Co.). Polymyxins A, B, C, D, E, F, K, M, P, S and T have been identified. Isoln of polymyxins B, C and E: Few, Schulman, *Biochem. J.* **54**, 171 (1953); of *polymyxin D*: Stansley *et al., Bull. Johns Hopkins Hosp.* **81**, 43 (1947); of *polymyxin F*: W. L. Parker *et al, J. Antibiot.* **30**, 767 (1977); of *polymyxin K*: Kimura, **JP 71 16152** (1971), *C.A.* **75**, 62180r (1971); of *polymyxin M*: Khokhlov *et al., C.A.* **55**, 5653h (1961); of *polymyxin P*: Kimura *et al., J. Antibiot.* **22**, 449 (1969); of *polymyxin S$_1$* and *polymyxin T$_1$*: J. Shoji *et al., ibid.* **30**, 1029 (1977). Resolution of polymyxin B into B$_1$ and B$_2$: Hausmann, Craig, *J. Am. Chem. Soc.* **76**, 4892 (1954). Structure and synthesis of polymyxin B$_1$: Wilkinson, Lowe, *Nature* **202**, 1211 (1964); Vogler *et al., Helv. Chim. Acta* **48**, 1161 (1965). Structure of polymyxin B$_2$: Wilkinson, Lowe, *Nature* **204**, 993 (1964). Separation of polymyxin D into D$_1$ and D$_2$ and structures: Hayashi *et al., Experientia* **22**, 354 (1966).

Structure of polymyxin S$_1$: J. Shoji *et al.*, *J. Antibiot.* **30**, 1035 (1977); of polymyxin T$_1$: *eidem, ibid.* 1042. *Review:* Vogler, Studer, *Experientia* **22**, 345-354 (1966); Paulus, "Polymyxins" in *Antibiotics* **II**, D. Gottlieb, P. Shaw, Eds. (Springer-Verlag, New York, 1967) pp 254-267.

DAB = L-α,γ-diaminobutyric acid

Polymyxin B$_1$	R = (+)-6-methyloctanoyl	X = Phe	Y = Leu	Z = DAB
B$_2$	R = 6-methylheptanoyl	X = Phe	Y = Leu	Z = DAB
D$_1$	R = (+)-6-methyloctanoyl	X = Leu	Y = Thr	Z = D-Ser
D$_2$	R = 6-methylheptanoyl	X = Leu	Y = Thr	Z = D-Ser

Obtained as the hydrochloride. Nearly colorless powder, dec 228-230°. $[\alpha]_D^{23}$ −40° (c = 1.05). The hydrochloride is very sol (more than 40%) in water and methanol. The soly decreases in the higher alcs. Practically insol in the usual ethers, esters, ketones, hydrocarbons, and the chlorinated solvents. Forms water insol salts with a number of ppts such as picric acid, helianthic acid, Reinecke salt. The free base is slightly sol in water; almost insol in alc.

Polymyxin B. [1404-26-8] Mixture of polymyxins B$_1$ and B$_2$. $[\alpha]_{5461}$ −106.3° (1*N* HCl).

Polymyxin B sulfate. [1405-20-5] Aerosporin; Mastimyxin. Solubilities: Weiss *et al.*, *Antibiot. Chemother.* **7**, 374 (1957). White to buff-colored powder. Freely sol in water; slightly sol in alc.

Polymyxin B sulfate mixture with trimethoprim. Polytrim.

Polymyxin B$_1$. [4135-11-9] C$_{56}$H$_{98}$N$_{16}$O$_{13}$; mol wt 1203.50.

Polymyxin B$_1$ pentahydrochloride. C$_{56}$H$_{98}$N$_{16}$O$_{13}$.5HCl; mol wt 1385.79. White powder. $[\alpha]_D^{25}$ −85.11° (c = 2.33 in 75% ethanol).

Polymyxin B$_2$. [34503-87-2] C$_{55}$H$_{96}$N$_{16}$O$_{13}$; mol wt 1189.47. $[\alpha]_{5461}^{22}$ −112.4° (2% Acetic acid).

Polymyxin D$_1$. [10072-50-1] C$_{50}$H$_{93}$N$_{15}$O$_{15}$; mol wt 1144.38.

Polymyxin D$_2$. [34167-45-8] C$_{49}$H$_{91}$N$_{15}$O$_{15}$; mol wt 1130.36.

Polymyxin E *see* Colistin.

THERAP CAT: Antibacterial.

THERAP CAT (VET): Antibacterial.

7694. Polynoxylin. Anaflex; Larex; Ponoxylan. Alkali condensation product of formaldehyde and urea. Prepn: D. Haler, A. Aebi, *Nature* **190**, 734 (1961); GB 905195 (1962 to Ed. Geistlich Soehne). Antimicrobial activity and toxicity study: H. Brodhage, A. R. Stofer, *Antibiot. Chemother.* **11**, 205 (1961). Mechanism of action study: J. I. Blenkharn, *J. Clin. Hosp. Pharm.* **10**, 367 (1985).

Amorphous powder, dec 200° (without melting). Soly in water: 0.28-0.31%.

THERAP CAT: Antiseptic.

7695. Polyoxidonium. Synpol. Immunostimulant copolymer of *N*-oxidized 1,4-ethylenepiperazine and (*N*-carboxyethyl)-1,4-ethylenepiperazine bromide; mol wt 60-100 kDa. Prepn: R. V. Petrov *et al.*, WO 9507100; A. V. Nekrasov *et al.*, US 5503830 (1995, 1996 both to Petrovax). Adjuvant activity in mice: K. S. Nandakumar, V. R. Muthukkaruppan, *Scand. J. Immunol.* **50**, 188 (1999). Effect on

human peripheral blood leukocytes: V. A. Dyakonova *et al.*, *Int. Immunopharmacol.* **4**, 1615 (2004). Review of development and use in influenza vaccine: V. A. Kabanov, *Pure Appl. Chem.* **76**, 1659-1677 (2004).

Sol in water.

USE: Pharmaceutic aid (adjuvant).

THERAP CAT: Immunomodulator.

7696. Polyoxins. [11113-80-7] Agricultural antifungal antibiotic complex produced by *Streptomyces cacaoi* var *asoensis* and *S. piomogenus.* Polyoxins A through O are known, all except C and I having specific activity against phytopathogenic fungi by inhibiting cell wall chitin synthesis. Isoln and characterization of polyoxin A: S. Suzuki *et al.*, *J. Antibiot.* **18A**, 131 (1965); of A and B: K. Isono *et al.*, *Agric. Biol. Chem.* **29**, 848 (1965); S. Suzuki *et al.*, JP 66 15520; of C through L: K. Isono *et al.*, *Agric. Biol. Chem.* **30**, 817 (1966); *ibid.* **31**, 190 (1967); *ibid.* **32**, 792 (1968); of M: K. Isono, S. Suzuki, *Tetrahedron Lett.* **1970**, 425. Isoln of polyoxins N and O: S. Suzuki *et al.*, JP Kokai 72 23596 (to Inst. Phys. Chem. Res.), and Hokko Chem. Ind.), *C.A.* **78**, 41566t (1973); production of N and O: *eidem*, JP 77 20555 (1977 to Inst. Phys. Chem. Res.), *C.A.* **87**, 150183x (1977). Isoln and structure of N: M. Uramoto *et al.*, *Nucleic Acids Res. Spec. Publ.* **5**, 327 (1978). Structural elucidation of polyoxins A through L: K. Isono *et al.*, *J. Am. Chem. Soc.* **91**, 7490 (1969). Revised structure of A: S. Hanessian *et al.*, *Tetrahedron Lett.* **34**, 4153 (1993). Total synthesis of J: H. Kuzuhara *et al.*, *ibid.* **1973**, 5055. Biosynthetic studies: K. Isono *et al.*, *Biochemistry* **14**, 2992 (1975); S. Funayama, K. Isono, *ibid.* 5568; K. Isono, R. J. Suhadolnik, *Arch. Biochem. Biophys.* **173**, 141 (1976); S. Funayama, K. Isono, *Biochemistry* **16**, 3121 (1977); K. Isono *et al.*, *J. Am. Chem. Soc.* **100**, 3937 (1978). Mode of action: N. Ohta *et al.*, *Agric. Biol. Chem.* **34**, 1224 (1970); M. Hori *et al.*, *ibid.* **35**, 1280 (1971); **38**, 691, 699 (1974). *Review:* R. J. Suhadolnik, *Nucleoside Antibiotics* (Wiley-Interscience, New York, 1970) pp 218-234; K. Isono, S. Suzuki, *Heterocycles* **13**, 333-351 (1979).

Polyoxin A

Polyoxin A. [19396-03-3] [*S*-(*Z*)]-1-[5-[[2-Amino-5-*O*-(aminocarbonyl)-2-deoxy-L-xylonoyl]amino]-1,5-dideoxy-1-[3,4-dihydro-5-(hydroxymethyl)2,4-dioxo-1(2*H*)-pyrimidinyl]-β-D-allofuranuronoyl]-3-ethylidene-2-azetidinecarboxylic acid. C$_{23}$H$_{32}$N$_6$O$_{14}$; mol wt 616.54. Colorless needles from aq ethanol, dec >180°. $[\alpha]_D^{20}$ −30°. uv max (0.05*N* HCl): 262 nm (log ε 3.94); (0.05*N* NaOH): 264 nm (log ε 3.80).

Polyoxin B. [19396-06-6] 5-[[2-Amino-5-*O*-(aminocarbonyl)-2-deoxy-L-xylonoyl]amino]-1,5-dideoxy-1-[3,4-dihydro-5-(hydroxy-methyl)-2,4-dioxo-1(2*H*)-pyrimidinyl]-β-D-allofuranuronic acid; Polyoxin AL. C$_{17}$H$_{25}$N$_5$O$_{13}$; mol wt 507.41. Amorphous powder from aq ethanol. $[\alpha]_D^{20}$ +34°. uv max (0.05*N* HCl): 262 nm (log ε 3.94); (0.05*N* NaOH): 264 nm (log ε 3.82).

USE: Fungicide, esp for *Alternaria* leaf spot of many agricultural products.

7697. Polyoxyethylene Alcohols. Polyethylene glycol fatty alcohol ethers; ethoxylated fatty alcohols; macrogol fatty alcohol ethers; alcohol ethoxylates; POE alcohol ethers. Nonionic surfactants prepared by ethoxylation of fatty alcohols with ethylene oxide. Of the general structure $R(OCH_2CH_2)_nOH$ where R is a long chain alkyl group or mixture of alkyl groups. Prepn and reaction mechanism: J. D. Malkemus, *J. Am. Oil Chem. Soc.* **33**, 571 (1956); W. Satkowski, C. G. Hsu, *Ind. Eng. Chem.* **49**, 1875 (1957); R. D. Fine, *J. Am. Oil Chem. Soc.* **35**, 542 (1958). Compounds with a broad range of properties can be prepared by varying the fatty alcohol (lipophile) used and the degree of polymerization of the polyethylene glycol (hydrophile) segment. CFTA-assigned names based on fatty alcohol segment include *ceteth* (cetyl alc), *laureth* (lauryl alc), *myreth* (myristyl alc), *oleth* (oleyl alc), *steareth* (stearyl alc), *trideth* (tridecyl alc). The average number of ethylene oxide units in the polyethylene glycol segment is indicated by an appended number (e.g. ceteth-20). Additional products prepared from fatty alcohol mixtures include *ceteareth* (cetyl/stearyl alcs), *laneth* (lanolin alcs). Trademarks for some commercially available series of compounds: *Alfonic*, *Bio Soft* (E, EA, EN), *Brij*, *Dehydol*, *Ethosperse*, *Eumulgin* (B, O), *Hetoxol*, *Lipocol*, *Macol* (LA, SA, TD), *Polychol*, *Rhodasurf*, *Tergitol* (S, TMN), *Trycol*, *Volpo*. Comprehensive description: W. B. Satkowski *et al.*, "Polyoxyethylene Alcohols" in *Nonionic Surfactants* M. J. Schick, Ed. (Dekker, New York, 1967) pp 86-141. Effects of structure on properties: T. Kuwamura, *ACS Symp. Ser.* **253**, 27-47 (1984). Review of properties and uses in household cleaning products: K. W. Dillan *et al.*, *Household Pers. Prod. Ind.* **23**, 32 (1986).

Liquids to waxy solids. Compds with one to five moles ethylene oxide are sol in oil and many hydrocarbons. Water soly increases with increasing ethylene oxide content. Properties of Brij surfactants: G. King, *Drug Cosmet. Ind.* **90**, 24 (1962).

Laureth 9 *see* Polidocanol.

Ceteth 20. Polyoxyethylene (20) cetyl ether; Brij 58; Hetoxol CS-20; Lipocol C-20. $C_{16}H_{33}(OCH_2CH_2)_nOH$ where the average value of n is 20. Waxy solid. Pour point $\sim39°$; cloud point $\sim45°$. Sol in water; 2% ethanol.

USE: Used as emulsifiers, wetting agents, antistats, solubilizers, defoamers, detergents, lubricants in pharmaceutical, cosmetic and other industrial applications.

7698. Polyoxyethylene Fatty Acid Esters. Polyethylene glycol esters of fatty acids; PEG fatty acid esters; POE fatty acid esters; ethoxylated fatty acid esters; macrogol fatty acid esters. Nonionic surfactants prepared commercially by esterification of fatty acids with ethylene oxide or with polyethylene glycol. Of the general structure $RCOO(CH_2CH_2O)_nH$ or $RCOO(CH_2CH_2O)_nOCR$ where R is a long chain alkyl group or mixture of alkyl groups. Wide range of properties achieved by changing the hydrophobic fatty acid segment and/or varying the degree of polymerization of the hydrophilic polyoxyethylene segment. Trademarks for some commercially available series of compounds include *Emerest* (2600, 2700 series), *Emulsynt*, *Ethofat*, *Lipopeg*, *Mapeg*, *Myrj*, *Nopalcol*, *Pegosperse*, *Renex 20*, *Varonic LI*. Review of prepn, properties and uses: W. B. Satkowski *et al.*, "Polyoxyethylene Esters of Fatty Acids" in *Nonionic Surfactants*, M. J. Schick, Ed. (Dekker, New York, 1967) pp 142-174.

Liquids, soft solids, solids or flakes. Solubility properties depend on length of polyoxyethylene (POE) segment added to a specific fatty acid. Sol in oil and hydrocarbon solvents when <8 ethylene oxide units are added. Soly in water begins when 12-15 ethylene oxide units are added. Specific gravity and viscosity increase with increasing ethylene oxide content.

Polyoxyl 8 stearate. Polyethylene glycol 400 monostearate; PEG-400 monostearate; POE (8) stearate; macrogol ester 400; Emerest 2711; Lipopeg 4-S; Myrj 45; Mapeg 400 MS; Pegosperse 400-MS. $C_{17}H_{35}COO(CH_2CH_2)_nH$ where the average value of n is 8. Prepn and properties: R. L. Birkmeier, J. D. Brandner, *J. Agric. Food Chem.* **6**, 471 (1958).

Polyoxyl 40 stearate. Polyoxyethylene (40) stearate; POE (40) monostearate; Myrj 52; Myrj 52S; Pegosperse 1750-MS. The average number of oxyethylene units is 40. Waxy, white to light tan solid; odorless or faint fat-like odor. Sol in water, alc, ether, acetone. Insol in mineral oil, vegetable oils.

USE: As antistats, emulsifiers, defoamers, wetting agents, solubilizers, conditioning agents, lubricants, detergents. Have wide range of cosmetic, pharmaceutical and other industrial applications. Polyoxyl 8 and 40 stearates as pharmaceutic aid (surfactant).

7699. Polyphosphazenes. Polymers containing an inorganic backbone of alternating nitrogen and phosphorus atoms substituted with very long chains of skeletal atoms. They exist as glasses, flexible solids, or rubbery solids with a low tendency for crystallization; non-flammable and more elastic than carbon-backbone polymers. Prepn of the first polyphosphazene, hexachlorocyclotriphosphazene: J. Liebig, *Ann.* **11**, 139 (1834). Improved synthesis and basis of modern mfg methods: R. Schenck, G. Römer, *Ber.* **57B**, 1343 (1924). Review of syntheses, properties, chemistry and applications: H. R. Allcock, *Science* **193**, 1214-1219 (1976); *idem*, *Angew. Chem. Int. Ed.* **16**, 147-156 (1977); E. N. Peters, "Inorganic High Polymers", in *Kirk-Othmer Encyclopedia of Chemical Technology* **vol. 13** (Wiley-Interscience, New York, 3rd ed., 1981) pp 398-413.

USE: In waterproofing; as flame retardants; in gaskets, o-rings, hydrocarbon fuel hoses.

7700. Polyphosphoric Acid. Pholeum; tetraphosphoric acid. May be prepd by heating H_3PO_4 with sufficient phosphoric anhydride to give the resulting product an 82-85% P_2O_5 content: Bell, *Ind. Eng. Chem.* **40**, 1464 (1949); Van Wazer, Holst, *J. Am. Chem. Soc.* **72**, 639 (1950); Kennard, *Org. Chem. Bull.* **29**, no. 1 (1957). Consists of about 55% tripolyphosphoric acid, the remainder being H_3PO_4 and other polyphosphoric acids. Typical analysis: 83.0% P_2O_5; ortho equivalent 115.0%.

Viscous liquid at room temps. Conveniently fluid at 60°. Solidifies to a glass at low temps. Sol in water with evolution of heat and hydrolysis to H_3PO_4.

Caution: In strong concns moderately irritating to skin, mucous membranes.

USE: In organic synthesis for cyclizations and acylations.

7701. Polypropylene. [9003-07-0] 1-Propene homopolymer; propylene polymer. Class of compounds also known as olefins. Three forms are possible. *Isotactic* (fiber-forming): methyl groups are all on same side of plane of zig-zag carbon atom chain. *Syndiotactic:* methyl groups are on alternate sides of plane of carbon atom chain. *Atactic* (not fiber-forming, amorphous): methyl groups are in a random arrangement with respect to plane of carbon atom chain. Early synthesis of isotactic form with Ziegler catalyst and comparison with atactic form: Natta *et al.*, *J. Chem. Soc.* **77**, 1708 (1955); Natta, *J. Polym. Sci.* **16**, 143 (1955). *Reviews:* N. G. Gaylord, H. F. Mark, *Linear and Stereoregular Addition Polymers* (Interscience, New York, 1959) pp 54-65; R. W. Moncrieff, *Man-Made Fibres* (Wiley, New York, 4th ed., 1963) pp 500-510; J. G. Cook, *Handbook of Textile Fibres* (Merrow Publishing Co., England, 3rd ed., 1964) pp 369-379; G. Crespi, L. Luciani, "Olefin Polymers (Polypropylene)" in *Kirk-Othmer Encyclopedia of Chemical Technology* **Vol. 16** (Wiley-Interscience, New York, 3rd ed., 1981) pp 453-469. Review of thermal decomposition and toxicity: V. Purohit, R. A. Orzel, *J. Am. Coll. Toxicol.* **7**, 221-242 (1988).

Isotactic form. Amco; Amerfil; Beamette; Courlene PY; DLP; Gerfil; Herculon; Lambeth; Meraklon; Moplen; Olane; Prolene; Tuff-Lite; Ulstron. Solid material, softens at $\sim155°$, mp $\sim165°$. Low flammability comparable to that of wool. Keeps strength down to $-100°$. d 0.90-0.92. Practically insol in cold org solvents. Sol in hot decalin, hot tetralin, boiling tetrachloroethane. Shrinks in boiling trichloroethylene. Resistant to acids, alkalies; attacked by strong oxidizing agents, *e.g.*, hydrogen peroxide. Good resistance to abrasion ("pilling"). Tendency to develop static charges. Unstabilized material has poor resistance to sunlight. Difficult to dye, lacks dye-attracting polar groups in structure.

USE: Isotactic form: for fishing gear, ropes, filter cloths, laundry bags, protective clothing, blankets, fabrics, carpets, yarns, etc.

7702. Polysaccharide-K. [134192-05-5] Krestin; PSK. Protein-bound polysaccharide composed of a glucan with a β1-β4 bond in the main chain, and β1-β3 and β1-β6 bonds in the side chain. Biological response modifier having immunostimulant and anticancer properties; Toll-like receptor (TLR) 2 agonist. Isolated from a basidiomycete, *Coriolus versicolor* (Fr.) Quel.: S. Otsuka *et al.*, **JP 73 8489** (1973 to Kureha), *C.A.* **80**, 41025g (1974). Antitumor activity: S. Tsukagoshi *et al.*, *Prog. Chemother. (Antibacterial, Antiviral, Antineoplast.), Proc. Int. Congr. Chemother., 8th, Athens 1973* **vol. 3**, G. K. Daikos, Ed. (Hellen. Soc. Chemother., Athens, 1974) pp 799-803. Clinical studies as postoperative immunotherapy: M. Torisu *et al.*, *Cancer Immunol. Immunother.* **31**, 261 (1990); Y. Kuroda *et al.*, *Int. J. Clin. Oncol.* **3**, 311 (1998). Mechanism of action study: H. Lu *et al.*, *Clin. Cancer Res.* **17**, 67 (2010). *Review:* S. Tsukagoshi *et al.*, *Cancer Treat. Rev.* **11**, 131-155 (1984). Review of antimicrobial and biological activity: H. Sakagami, M. Takeda in *Mushroom Biology and Mushroom Products* (Chinese Univ. Press, Hong Kong, 1993) p 237-245; antimetastatic effects: H. Kobayashi *et al.*, *Cancer Epidemiol. Biomarkers Prev.* **4**, 275-281 (1995).

Brownish powder. Tasteless, but has a slight odor. Sol in water. Practically insol in methanol, pyridine, chloroform, benzene, hexane. pH of 1% soln is 6.6-7.2.

THERAP CAT: Antineoplastic.

7703. Polysorbates. Polyoxyethylene sorbitan esters; POE sorbitan esters. Nonionic surfactants derived from sorbitan esters, *q.v.* Comprehensive description: P. Becher, "Polyol Surfactants" in *Nonionic Surfactants*, M. J. Schick, Ed. (Dekker, New York, 1967) pp 247-299. Description of prepn and uses: L. R. Chislett, J. Walford, *Int. Flavours Food Addit.* **7**, 61 (1976). Pharmacology of polysorbate 80: R. K. Varma *et al.*, *Arzneim.-Forsch.* **35**, 804 (1985). Determn in foods: H. Kato *et al.*, *J. Assoc. Off. Anal. Chem.* **72**, 27 (1989).

Polysorbate 80
(Sum of w, x, y, and z is 20)

Polysorbate 80. [9005-65-6] Polyoxyethylene (20) sorbitan monooleate; POE (20) sorbitan monooleate; Emsorb 6900; Liposorb O-20; Monitan; Sorlate; T-Maz 80; Tween 80. Lemon- to amber-colored, oily liquid. d 1.06-1.09. Viscosity (25°): 300-500 cSt. Very sol in water; sol in alcohol, cottonseed oil, corn oil, ethyl acetate, methanol, toluene. Insol in mineral oil. pH of 5% aq soln between 6 and 8. LD_{50} in mice, rats (ml/kg): 7.5, 6.3 i.p. (Varma).

USE: As emulsifiers and dispersing agents in medicinal products; as defoamers and emulsifiers in foods. Pharmaceutic aid (surfactant).

7704. Polytetrafluoroethylene. [9002-84-0] 1,1,2,2-Tetrafluoroethene homopolymer; tetrafluoroethylene polymer; polytetrafluoroethylene resin; polytef; PTFE; Fluon; Teflon; Tetran. A highly stable thermoplastic tetrafluoroethylene, *q.v.*, homopolymer. Composed of at least 20,000 C_2F_4 monomer units linked into very long unbranched chains. Prepd by polymerization of tetrafluoroethylene: Plunkett, **US 2230654** (1941 to Kinetic Chem.); Brubaker, **US 2393967**; Joyce, **US 2394243** (both 1946 to du Pont); Hanford, Joyce, *J. Am. Chem. Soc.* **68**, 2082 (1946); Renfrew, **US 2534058** (1950 to du Pont); C. E. Schildknecht, *Vinyl and Related Polymers* (Wiley, New York, 1952) pp 483-494. Account of discovery by Roy J. Plunkett: A. B. Garrett, *J. Chem. Educ.* **39**, 288 (1962). *Reviews:* M. M. Renfrew, E. E. Lewis, *Ind. Eng. Chem.* **38**, 870-877 (1946); R. W. Moncrieff, *Man-Made Fibres* (John Wiley, New York, 4th ed., 1963) pp 512-517; McCane in *Encyclopedia of Polymer Science and Technology* **vol. 13**, N. M. Bikales, Ed. (Interscience, New York, 1970) pp 623-654; S. V. Gangal in *Kirk-Othmer Encyclopedia of Chemical Technology* **vol. 11** (Wiley-Interscience, New York, 4th ed., 1994) pp 621-644.

Nonflammable, high polymer. White translucent to opaque solid (depending on thickness). Very inert chemically. Useful temp range from cryogenic to +260°. Melts to an extremely viscous gel at 327° and reverts to the gaseous monomer at temperatures above 400°. d 2.2. Shore hardness 55-56. Tensile strength 3500-4500 psi. Flexural modulus ~80,000-90,000 psi at room temp. Brittle point below −80°. Dielectric constant (at 60 to 3×10^9 cycles) 2.0-2.05. Not affected by water, aqua regia, chlorosulfonic acid, acetyl chloride, boron fluoride, hot nitric acid, boiling solns of sodium hydroxide, and organic solvents. Not wetted by water. No substance has been found which will dissolve the polymer at moderate temperatures, but prolonged contact with fluorine, hot plasticizers and polymeric waxes is not recommended. Is subject to cold flow at high pressure. Because of its high melt viscosity molding and sintering techniques similar to those used in powder metallurgy are normally used for fabrication.

Caution: Potential symptom of overexposure by inhalation to the heated polymer is polymer fume fever, characterized by dizziness, headache, nausea, chills, weakness, cough, chest tightness, sore throat, pyrexia. *See Clinical Toxicology of Commercial Products*, R. E. Gosselin *et al.*, Eds. (Williams & Wilkins, Baltimore, 5th ed., 1984) Section II, p 412; *Patty's Industrial Hygiene and Toxicology* **vol. 2E**, G. D. Clayton, F. E. Clayton, Eds. (John Wiley & Sons, New York, 4th ed., 1994) 3791-3793.

USE: For hookup and hookup-type wire in electronic equipment; in computer wire, electrical tape, electrical components, spaghetti tubing. Seals and piston rings, basic shapes, bearings, mechanical tapes, coated glass fabrics. As tubing and sheets for chemical laboratory and process work; for lining reaction vessels; for gaskets and pump packings, sometimes mixed with graphite or glass fibers; as electrical insulator esp in high frequency applications; filtration fabrics; protective clothing. Prosthetic aid.

7705. Polythiazide. [346-18-9] 6-Chloro-3,4-dihydro-2-methyl-3-[[(2,2,2-trifluoroethyl)thio]methyl]-2*H*-1,2,4-benzothiadiazine-7-sulfonamide 1,1-dioxide; 2-methyl-3-(β,β,β-trifluoroethylthiomethyl)-6-chloro-7-sulfamyl-3,4-dihydro-1,2,4-benzothiadiazine 1,1-dioxide; 6-chloro-3,4-dihydro-2-methyl-7-sulphamoyl-3-(2,2,2-trifluoroethylthiomethyl)-2*H*-benzo-1,2,4-thiadiazine 1,1-dioxide; Drenusil; Nephril; Renese. $C_{11}H_{13}ClF_3N_3O_4S_3$; mol wt 439.87. C 30.04%, H 2.98%, Cl 8.06%, F 12.96%, N 9.55%, O 14.55%, S 21.87%. Prepn: J. M. McManus, **US 3009911** (1961 to Pfizer). Comprehensive description: T. Negendra Varo Prasad, *Anal. Profiles Drug Subs.* **20**, 665-692 (1991).

Crystals from isopropanol, mp 202.5°. Sol in methanol and acetone. Practically insol in water and chloroform. Sol in aq solns made alkaline with carbonates or hydroxides of the alkali metals. Rate of decompn increases with increase in pH.

THERAP CAT: Diuretic, antihypertensive.

7706. Polyvinyl Alcohol. [9002-89-5] Ethenol homopolymer; PVA; Akwa Tears; Elvanol; Gelvatol; Liquifilm; Mowiol; Polyviol; Sno Tears; Vinarol; Vinol. A polymer prepd from polyvinyl acetates by replacement of the acetate groups with hydroxyl groups. The alcoholysis proceeds most rapidly in a methanol + methyl acetate mixture in the presence of catalytic amounts of alkali or mineral acids: Hermann, Haehnel, *Ber.* **60**, 1658 (1927). *Monograph:* C. E. Schildknecht, *Vinyl and Related Polymers* (Wiley, New York, 1952). The head-to-tail or 1,3-glycol structure is favored: Staudinger *et al.*, *Ber.* **60**, 1782 (1927); *J. Prakt. Chem.* **155**,

261 (1940); Marvel, Denoon, *J. Am. Chem. Soc.* **60**, 1045 (1938); McDowell, Kenyon, *ibid.* **62**, 415 (1940); Marvel, Inskeep, *ibid.* **65**, 1710 (1943). *Reviews:* M. Leeds in *Kirk-Othmer Encyclopedia of Chemical Technology* **vol. 21** (Wiley-Interscience, New York, 2nd ed., 1970) pp 353-368; *Polyvinyl Alcohol*, A. C. Finch, Ed. (Wiley, New York, 1973) 640 pp; A. S. Dunn, *Chem. Ind. (London)* **1980**, 801-806. Vinyl alcohol monomer has not been isolated. *Review:* R. B. Seymour, G. B. Kauffman, *J. Chem. Educ.* **71**, 582 (1994). Comprehensive description: D. Wong, J. Parasrampuria, *Anal. Profiles Drug Subs. Excip.* **24**, 397-441 (1996).

Dry, unplasticized polyvinyl alcohol powders are white to cream colored, soften at about 200° with decompn. Commercial polyvinyl alcohols have different contents of residual acetyl groups and therefore different viscosity characteristics. The first code number following the trade name indicates the degree of hydrolysis, while the second set of numbers indicates the approx viscosity in cP (4% aq soln at 20°). Polyvinyl alcohols are essentially sol in hot and cold water, but those coded 20-105 require alcohol-water mixtures. Aq solns are colloidal and compatible with lower alcohols. Pure aq solns are neutral or faintly acid and subject to mold growth. Insol in petroleum solvents.

USE: In the plastics industry in molding compds, surface coatings, films resistant to gasoline, textile sizes and finishing compositions; can be compounded to yield elastomers to be used in manuf artificial sponges, fuel hoses, etc., also in printing inks for plastics and glass, in pharmaceutical finishing, cosmetics, water-sol film and sheeting. Pharmaceutic aid (viscosity increasing agent); ophthalmic lubricant.

7707. Polyvinyl Chloride. [9002-86-2] Chloroethene homopolymer; chloroethylene polymer; PVC; Geon; Breon; Welvic; Movyl; Tevilon; Koroseal; Marvinol. Polyvinyl chloride fibers are marketed under the names: *Rhovyl, Fibravyl, Thermovyl, Isovyl, Retractyl, Crinovyl, Envilon, Nip.* Average mol wt ~60,000 to 150,000. Prepn: Baumann, *Ann.* **163**, 308 (1872); Schoenfeld, **US 2168808** (1937 to B. F. Goodrich). Structure: Natta, Rigamonti, *Atti Accad. Naz. Lincei Cl. Sci. Fis. Mat. Nat. Rend.* **24**, 381 (1936); *C.A.* **31**, 4563⁹ (1937); Marvel *et al.*, *J. Am. Chem. Soc.* **61**, 3241 (1939). *Reviews:* C. E. Schildknecht, *Vinyl and Related Polymers* (Wiley, New York, 1952) pp 392-442. Technology: W. S. Penn, *PVC Technology* (Maclaren, London, 1962); J. A. Davidson, K. L. Gardner, "Vinyl Polymers (PVC)" in *Kirk-Othmer Encyclopedia of Chemical Technology* **vol. 23** (Wiley-Interscience, New York, 3rd ed., 1983) pp 886-936. Book: Sarvetnick, *Polyvinyl Chloride* (Van Nostrand, Reinhold, New York, 1969).

Plastic solid. d 1.406; *n* 1.54. Stabilizers are necessary to prevent discoloration from exposure to light or heat. Solvents for unmodified polyvinyl chloride of high mol wt: cyclohexanone, methyl cyclohexanone, dimethyl formamide, nitrobenzene, tetrahydrofuran, isophorone, mesityl oxide. Solvents for lower polymers: dipropyl ketone, methyl amyl ketone, methyl isobutyl ketone, acetonylacetone, methyl ethyl ketone, dioxane, methylene chloride.

USE: Rubber substitutes, electric wire and cable-coverings, pliable thin sheeting, film finishes for textiles, non-flammable upholstery, raincoats, tubing, belting, gaskets, shoe soles.

7708. Pomalidomide. [19171-19-8] 4-Amino-2-(2,6-dioxo-3-piperidinyl)-1*H*-isoindole-1,3(2*H*)-dione; 3-amino-*N*-(2,6-dioxo-3-piperidyl)phthalamide; 1,3-dioxo-2-(2,6-dioxopiperidin-3-yl)-4-aminoisoindoline; CC-4047; Actimid. $C_{13}H_{11}N_3O_4$; mol wt 273.25. C 57.14%, H 4.06%, N 15.38%, O 23.42%. Second generation immunomodulatory drug (IMiD) and TNF-α inhibitor; amino analog of

thalidomide, *q.v.* Prepn: G. W. Muller *et al.*, **WO 9803502**; *eidem*, **US 6335349** (1998, 2002 both to Celgene); *eidem*, *Bioorg. Med. Chem. Lett.* **9**, 1625 (1999). Structure-embryotoxicity study of thalidomide analogs: R. L. Smith *et al.*, in *Symposium on Embryopathic Activity of Drugs* (J. & A. Churchill, London, 1965) pp 194-209. Antiangiogenic activity: R. J. D'Amato *et al.*, *Semin. Oncol.* **28**, 597 (2001). HPLC determn in plasma and racemization of the *S*-isomer: S. K. Teo *et al.*, *Chirality* **15**, 348 (2003). Effect on T cell differentiation: W. Xu *et al.*, *Clin. Immunol.* **128**, 392 (2008); on erythropoiesis and hemoglobulin synthesis: S. E. Meiler *et al.*, *Blood* **118**, 1109 (2011). Review of pharmacology and clinical experience in myeloma: S. Schey, K. Ramasamy, *Expert Opin. Invest. Drugs* **20**, 691-700 (2011); and myelofibrosis: M. Q. Lacy, A. Tefferi, *Leuk. Lymphoma* **52**, 560-566 (2011).

Yellow solid from dioxane + ethyl acetate.
THERAP CAT: Antineoplastic.

7709. Pomegranate. Granatum. Dried bark of stem or root of *Punica granatum* L., *Punicaceae. Habit.* Mediterranean region; Eastern, Western, and Southern Asia; cultivated in subtropical countries. *Constit.* 0.5-1% alkaloids consisting of pelletierine, methylpelletierine, pseudopelletierine (granatonine), and isopelletierine; mannite, about 20% tannin. Rind of fruit contains about 30% tannin.
THERAP CAT: Formerly as teniafuge.

7710. Ponasterone A. [13408-56-5] (2β,3β,5β,22R)-2,3,14,-20,22-Pentahydroxycholest-7-en-6-one; 25-deoxyecdysterone; 25-deoxy-20-hydroxyecdysone. $C_{27}H_{44}O_6$; mol wt 464.64. C 69.80%, H 9.55%, O 20.66%. Polyhydroxylated steroid with strong moulting hormone activity; first phytoecdysteroid to be isolated. Isoln with ponasterones B, C, D, from plant source, *Podocarpus nakaii* Hay., *Podocarpaceae*, and structure determn: K. Nakanishi *et al.*, *Chem. Commun.* **1966**, 915; isoln from various Japanese ferns: T. Takemoto *et al.*, *Chem. Pharm. Bull.* **21**, 2336 (1973). Isoln from crustaceans: J. F. McCarthy, *Steroids* **34**, 799 (1979). Stereochemical elucidation: H. Moriyama, K. Nakanishi, *Tetrahedron Lett.* **1968**, 1111; using ¹³C-NMR: H. Hikino *et al.*, *Chem. Pharm. Bull.* **23**, 125 (1975). Chromatographic sepn: M. Hori, *Steroids* **14**, 33 (1969). HPLC analysis: I. D. Wilson *et al.*, *J. Chromatogr.* **238**, 97 (1982); R. E. Isaac *et al.*, *ibid.* **246**, 317 (1982). Synthesis: G. Hüppi, J. B. Siddall, *Tetrahedron Lett.* **1968**, 1113. Moulting hormone activity on houseflies, silkworms: M. Kobayashi *et al.*, *Steroids* **9**, 529 (1967); on various insects: W. E. Robbins *et al.*, *ibid.* **16**, 105 (1970). Hormonal regulation to increase yield from silkworms: T. Okauchi *et al.*, **US 3941879** (1976 to Takeda). Used to characterize ecdysteroid receptors in *Drosophilia* cells: P. Maroy *et al.*, *Proc. Natl. Acad. Sci. USA* **75**, 6035 (1978); B. A. Sage *et al.*, *J. Biol. Chem.* **257**, 6373 (1982). *Review* of *ponasterones*: K. Nakanishi, *Bull. Soc. Chim. Fr.* **1969**, 3475. *See also* Ecdysteroids.

Crystals from ethanol, mp 259-260° (dec). $[\alpha]_D^{15}$ +90° (methanol). uv max (methanol): 244, 326 nm (ε 12400, 130).

7711. Ponazuril. [69004-04-2] 1-Methyl-3-[3-methyl-4-[4-[(trifluoromethyl)sulfonyl]phenoxy]phenyl]-1,3,5-triazine-2,4,6-(1*H*,3*H*,5*H*)-trione; toltrazuril sulfone; Bay Vi 9143; Marquis. C_{18}-$H_{14}F_3N_3O_6S$; mol wt 457.38. C 47.27%, H 3.09%, F 12.46%, N 9.19%, O 20.99%, S 7.01%. Triazine trione antiprotozoal; metabolite of toltrazuril, *q.v.* Prepn: A. Haberkorn *et al.*, **DE 2718799**; *eidem*, **US 4219552** (1978, 1980 both to Bayer). Stereoselective generation by microsomes: E. Benoit *et al.*, *Biochem. Pharmacol.* **46**, 2337 (1993). LC/MS determn in meat and eggs: V. Hormazábal *et al.*, *J. Liq. Chromatogr. Relat. Technol.* **26**, 791 (2003). Activity vs *Sarcocystis neurona*, the causative agent of EPM: D. S. Lindsay *et al.*, *Vet. Parasitol.* **92**, 165 (2000); R. P. Franklin *et al.*, *ibid.* **114**, 123 (2003); vs *Neospora canium*: A. K. Darius *et al.*, *Parasitol. Res.* **92**, 453, 520 (2004). Pharmacokinetics in cattle: L. Dirikolu *et al.*, *J. Vet. Pharmacol. Ther.* **32**, 280 (2008).

mp 242°.

THERAP CAT (VET): Antiprotozoal. In treatment of equine protozoal myeloencephalitis (EPM).

7712. Ponceau 3R. [3564-09-8] 3-Hydroxy-4-[2-(2,4,5-trimethylphenyl)diazenyl]-2,7-naphthalenedisulfonic acid sodium salt (1:2); sodium cumeneazo-β-naphthol disulfonate; C.I. Food Red 6; C.I. 16155; FD & C Red no. 1; Ext. D & C Red no. 15. $C_{19}H_{16}N_2$-$Na_2O_7S_2$; mol wt 494.44. C 46.16%, H 3.26%, N 5.67%, Na 9.30%, O 22.65%, S 12.97%. Prepn: *Colour Index* **vol. 4** (3rd ed., 1971) p 4092. Toxicity studies: Hansen *et al.*, *Toxicol. Appl. Pharmacol.* **5**, 105 (1963).

Dark red powder. Soluble in water with cherry-red color, slightly in alcohol. Addition of HCl to its aq soln does not change the color, but NaOH produces a yellow ppt. It dissolves in concd H_2SO_4 to a cherry-red soln which does not change color on dilution.

USE: Dyeing wool. Colorant in food, drugs and cosmetics.

7713. Ponceau SX. [4548-53-2] 3-[2-(2,4-Dimethyl-5-sulfophenyl)diazenyl]-4-hydroxy-1-naphthalenesulfonic acid sodium salt (1:2); FD & C Red No. 4; C.I. Food Red 1; C.I. 14700. $C_{18}H_{14}N_2$-$Na_2O_7S_2$; mol wt 480.42. C 45.00%, H 2.94%, N 5.83%, Na 9.57%, O 23.31%, S 13.35%. Azo dye. Prepn: E. Nölting, O. Kohn, *Ber.* **19**, 137 (1886). Metabolism: J. L. Radomski, T. J. Mellinger, *J. Pharmacol. Exp. Ther.* **136**, 259 (1962). Toxicity studies: F. C. Lu, A. Lavalle, *Can. Pharm. J.* **97**, 30 (1964); K. J. Davis *et al.*, *Toxicol. Appl. Pharmacol.* **8**, 306 (1966). Determn in maraschino cherries: R. E. Draper, *J. Assoc. Off. Anal. Chem.* **56**, 703 (1973); HPLC determn in lipstick: L. Gagliardi *et al.*, *J. Chromatogr.* **394**, 345 (1987). *Review: IARC Monographs* **8**, 207-214 (1975). *See also: Colour Index* **vol. 4** (3rd ed., 1971) p 4068.

Red crystals. Absorption max (0.02 *N* CH_3COONH_4): 500 nm. Sol in water; slightly sol in ethanol. Insol in vegetable oils. Deep red soln in conc H_2SO_4, changing to red ppt on dilution. Orange soln in conc HNO_3, turning yellow. LD_{50} orally in rats: >2 g/kg (Lu, Lavalle).

USE: Dye in externally applied drugs, cosmetics; food colorant in maraschino cherries.

7714. Poppy Capsules. Poppy heads. Fully grown, dried capsules of *Papaver somniferum* L., *Papaveraceae*. *Habit*. Europe, Asia. *Constit*. Capsules: 0.15-0.5% morphine and small amounts of other opium alkaloids. Seed: Fixed oil (poppy oil), albuminoids.

USE: For preparing emulsions—only the white seeds should be used. Seeds are chiefly used for making the oil, and in baking; the bluish-black variety is generally used in the US for baking.

7715. Poppy Oil. Poppy-seed oil. Expressed from poppy seeds. Contains no morphine or other opium alkaloids. *See also* Ethiodized Oil; Iodized Oil.

Pale yellow, drying oil; pleasant odor and taste. d 0.924-0.927. Solidif about −18°. n_D^{20} 1.4766-1.4774. Sapon. no. 189-197. Iodine no. 133-158.

USE: Manuf paints, varnishes, and soaps. Edible grades are marketed in Europe and Asia.

7716. Populin. [99-17-2] 2-(Hydroxymethyl)phenyl-β-D-glucopyranoside 6-benzoate; populoside; salicin benzoate. C_{20}-$H_{22}O_8$; mol wt 390.39. C 61.53%, H 5.68%, O 32.79%. In bark and leaves of *Populus tremula* L., *P. nigra* L., *P. nigra* L. var. *italica* Duroi, *P. canadensis* Moench., *P. grandidentate* Michx., and *P. tremuloides* Michx., *Salicaceae*, perhaps also in *Salix helix*, *Salix purpureae* L. var *helix* (L.) Koch. Isoln: Pearl *et al.*, *J. Org. Chem.* **27**, 2685 (1962). May be made from salicin by melting with benzoic anhydride, or from salicin and benzoyl chloride in presence of potassium hydroxide: Richtmyer, Yeakel, *J. Am. Chem. Soc.* **56**, 2495 (1934). Alkaline cleavage produces benzoic acid and salicin. Enzymatic hydrolysis with taka-diastase, *q.v.*, gives salicyl alcohol and benzoyl glucose: Kitasato, *Biochem. Z.* **190**, 109 (1927).

Dihydrate. Needles from water, sweet taste, like licorice. Becomes anhydr at 100°. mp 179°. $[\alpha]_D^{20}$ −2.0° (c = 5 in pyridine); $[\alpha]_D^{25}$ −29.7° (c = 5 in acetone). One gram dissolves in about 2 liters water, in 42 ml boiling water, in about 100 ml alcohol. Practically insol in ether.

7717. Poractant Alfa. [129069-19-8] PLS; Curosurf. Porcine lung extract composed of 99% polar lipids and 1% apoproteins. The phospholipid component consists primarily of phosphatidylcholine and its derivatives. Differs from other pulmonary surfactants in its 9.9% sphingomyelin content. Isolation: T. Curstedt *et al.*, *Eur. J. Biochem.* **168**, 255 (1987). Characterization of phospholipid content: E. Redenti *et al.*, *Farmaco* **49**, 285 (1994). Evaluation as surfactant replacement in rabbit: B. Robertson *et al.*, *Eur. J. Pediatr.* **147**, 168 (1988); in comparison with synthetic surfactants: S. Bongrani *et al.*, *Biol. Neonate* **65**, 406 (1994). Multicenter clinical trial in RDS in infants: Collab. Eur. Multicent. Study Group, *Pediatrics* **82**, 683 (1988); H. L. Halliday *et al.*, *Arch. Dis. Child.* **69**, 276 (1993).

THERAP CAT: Pulmonary surfactant; in treatment of respiratory distress syndrome.

7718. Porfimer Sodium. [87806-31-3] Photofrin porfimer sodium; CL-184116; Photofrin. Light sensitive polyporphyrin oligomer linked via ethers and esters; purification product of hematoporphyrin derivative. Forms aggregates with combined mol wt ~10000. Prepn: T. J. Dougherty *et al.*, "Photoradiation Therapy - Clinical and Drug Advances" in *Porphyrin Photosensitization*, D.

Kessel, T. J. Dougherty, Eds. (Plenum Press, New York, 1983) pp 3-13; T. J. Dougherty *et al.*, **US 4649151** (1987 to Health Res.). Structural studies: *idem, Photochem. Photobiol.* **46**, 569 (1987); R. K. Pandey *et al., Biomed. Environ. Mass Spectrom.* **19**, 405 (1990). Photobleaching: J. Moan, D. Kessel, *J. Photochem. Photobiol. B* **1**, 429 (1988); J. D. Spikes, *Photochem. Photobiol.* **55**, 797 (1992). Pharmacokinetics in mice: D. A. Bellnier *et al., ibid.* **50**, 221 (1989). Clinical evaluation in treatment of esophageal cancer: T. Okunaka *et al., Surg. Endosc.* **4**, 150 (1990). Imaging study for premalignant lesions: C. Liebow *et al., Proc. Natl. Acad. Sci. USA* **90**, 1897 (1993). Clinical use as imaging agent for lung cancer: S. Lam *et al., Chest* **97**, 333 (1990).

Orange red color.

USE: Fluorescent imaging of malignancies.

THERAP CAT: Antineoplastic (photosensitizer).

7719. Porfiromycin. [801-52-5] (1a*S*,8*S*,8a*R*,8b*S*)-6-Amino-8-[[(aminocarbonyl)oxy]methyl]-1,1a,2,8,8a,8b-hexahydro-8a-methoxy-1,5-dimethylazirino[2′,3′:3,4]pyrrolo[1,2-*a*]indole-4,7-dione; *N*-methylmitomycin C; U-14743. $C_{16}H_{20}N_4O_5$; mol wt 348.36. C 55.17%, H 5.79%, N 16.08%, O 22.96%. Antibiotic substance isolated from a *Streptomyces ardus* fermentation broth: Herr *et al., Antimicrob. Agents Annu.* **1960**, 23. Isoln from *S. verticillatus* and structure: Webb *et al., J. Am. Chem. Soc.* **84**, 3185, 3187 (1962). Production process: Bohonos *et al.,* **US 3219530** (1965 to Am. Cyanamid). Synthesis of (±)-form: F. Nakatsubo *et al., J. Am. Chem. Soc.* **99**, 8115 (1977). For stereochemistry and other synthetic studies, *see* Mitomycins.

Dark purple triclinic crystals, dec 201-201.5°. $[\alpha]_D^{25}$ +275 ±55° (c = 0.1% in methanol); $[\alpha]_D^{25}$ +242 ±100° (c = 0.045% in methanol). uv max (methanol): 217, 360, 555 nm (ε 24600, 23000, 209). Slightly sol in water, moderately sol in polar organic solvents. Practically insol in hydrocarbon solvents.

THERAP CAT: Antibacterial; antineoplastic.

7720. Porphine. [101-60-0] 21*H*,23*H*-Porphine; porphin. $C_{20}H_{14}N_4$; mol wt 310.36. C 77.40%, H 4.55%, N 18.05%. Parent substance of the *porphyrins*, a group of compounds found in all living matter which are the basis of respiratory pigments in animals and plants. In porphyrins, side chains are substituted for the hydrogens in the porphine pyrrole rings. *Chlorins* are dihydroporphyrins. *See also:* chlorophyll, hemoglobin, vitamin B₁₂, hematin. Prepn of porphine: Fischer, Gleim, *Ann.* **521**, 157 (1935); Rothemund, *J. Am. Chem. Soc.* **58**, 625 (1936); Krol, *J. Org. Chem.* **24**, 2065 (1959). Study of porphyrin analogs: C. L. Honeybourne *et al., Tetrahedron* **36**, 1833 (1980). Review of biosyntheses of porphyrins and chlorins: A. R. Battersby, E. McDonald in *Porphyrins Metalloporphyrins*, K. M. Smith, Ed. (Elsevier, New York, 1975) pp 61-122. Review of porphyrin syntheses: R. P. Evstigneeva, *Pure Appl. Chem.* **53**, 1129-1140 (1981). Comprehensive seven volume treatise: *The Porphyrins*, D. Dolphin, Ed. (Academic Press, New York, 1978).

Dark red, shiny leaflets from chloroform-methanol. Darkens at 360° but does not melt. The absorption bands are those characteristic of the substituted porphyrins, details in Fischer-Orth, *Die Chemie des Pyrrols* **vol. II**, 1, 175 (1937). Sol in pyridine, dioxane, and phenol; slightly sol in chloroform, bromoform, glacial acetic acid. Almost insol in acetone, methanol, ether. HCl number: 1.7 (Fischer); 3.3 (Rothemund).

Iron salt. $C_{20}H_{12}N_4$.FeCl. Brown cubes from ether.
Magnesium salt. $C_{20}H_{12}MgN_4$. Red needles.
Copper salt. $C_{20}H_{12}N_4Cu$. Brown needles.

7721. Porphobilinogen. [487-90-1] 5-(Aminomethyl)-4-(carboxymethyl)-1*H*-pyrrole-3-propanoic acid; 2-aminomethylpyrrol-3-acetic acid 4-propionic acid. $C_{10}H_{14}N_2O_4$; mol wt 226.23. C 53.09%, H 6.24%, N 12.38%, O 28.29%. An intermediate in the biosynthesis of heme, *q.v.*, found in the urine of patients with acute porphyria: Westall, *Nature* **170**, 614 (1952). Isoln and structure: Cookson, Rimington, *Biochem. J.* **57**, 476 (1954). Synthesis: Jackson, MacDonald, *Can. J. Chem.* **35**, 715 (1957). Chemistry of conversion into porphyrins: Mathewson, Corwin, *J. Am. Chem. Soc.* **83**, 135 (1961).

Monohydrate. Minute pink crystals from dil ammonium acetate soln at pH 4. Dec 172-175°. p*K*′ = 3.70; 4.95; 10.1. Slightly sol in water. Converted to uroporphyrins by hot dil HCl or by blood hemolyzates.

Hydrochloride monohydrate. Fine triclinic needles from dil HCl, dec 165-170°. Soluble in water.

7722. Porphyropsin. [9009-58-9] Photoreceptor protein found in the retinal rod cells of fresh water and migrating fish, lampreys, and certain amphibians. Absorption maximum approx 520-530 nm. Composed of the chromophore, 11-*cis*-3-dehydroretinal, *q.v.*, bound to scotopsin, the specific protein component of rod pigments (see Opsins). Biological activity is similar to that of rhodopsin, *q.v.* Isoln from retinas: G. Wald, *Nature* **139**, 1017 (1937); *idem, J. Gen. Physiol.* **22**, 775 (1939). Prepn from 3-dehydroretinal and opsin: *idem, Annu. Rev. Biochem.* **22**, 497 (1953). Methods of purification, prepn and assay: R. Hubbard *et al., Methods Enzymol.* **18**, 615-653 (1971). Exposure to light initiates the conversion of porphyropsin through a series of distinct intermediates to yield opsin and *trans*-3-dehydroretinal. Bleaching kinetics and photochemistry: T. Yoshizawa in *Handbook of Sensory Physiology* **Vol. VII**(2), H. J. A. Dartnall, Ed. (Springer-Verlag, New York, 1972) pp 146-179. Porphyropsin may co-exist with rhodopsin in the retinas of certain amphibians and fish: T. E. Reuter *et al., J. Gen. Physiol.* **58**, 351 (1971). Environmental effect on visual pigment formation and interconversion: P. Witkovsky *et al., ibid.* **72**, 821 (1978); A. T. C. Tsin, D. D. Beatty, *Exp. Eye Res.* **30**, 143 (1980). *Review:* G. Wald, *Science* **162**, 230-239 (1968).

7723. Posaconazole. [171228-49-2] 2,5-Anhydro-1,3,4-tri-deoxy-2-*C*-(2,4-difluorophenyl)-4-[[4-[4-[4-[1-[(1*S*,2*S*)-1-ethyl-2-hydroxypropyl]-1,5-dihydro-5-oxo-4*H*-1,2,4-triazol-4-yl]phenyl]-1-piperazinyl]phenoxy]methyl]-1-(1*H*-1,2,4-triazol-1-yl)-D-*threo*-pentitol; (3*R-cis*)-4-[4-[4-[4-[5-(2,4-difluorophenyl)-5-(1,2,4-triazol-1-ylmethyl)tetrahydrofuran-3-ylmethoxy]phenyl]piperazin-1-yl]phenyl]-2-[1(*S*)-ethyl-2(*S*)-hydroxypropyl]-2,4-dihydro-2*H*-1,2,4-triazol-3-one; Sch-56592; Noxafil. $C_{37}H_{42}F_2N_8O_4$; mol wt 700.79. C 63.42%, H 6.04%, F 5.42%, N 15.99%, O 9.13%. Orally active triazole antifungal. Prepn: A. K. Saksena *et al.*, **WO 9517407**; *eidem*, **US 5661151** (1995, 1997 both to Schering); *eidem, Tetrahedron Lett.* **37**, 5657 (1996). Comparative antifungal spectrum: A. Cacciapuoti *et al., Antimicrob. Agents Chemother.* **44**, 2017 (2000). Pharmacokinetics, safety and tolerability: R. Courtney *et al., ibid.* **47**, 2788 (2003). LC-MS/MS determn in plasma: B. Rochat *et al., ibid.* **54**, 5074 (2010). Review of development: A. K. Saksena *et al.* in *Anti-Infectives: Recent Advances in Chemistry and Structure Activity Relationships* (Royal Soc. Chem., Cambridge, 1997) pp 180-199; and clinical efficacy in fungal infections: R. Herbrecht, *Int. J. Clin. Pract.* **58**, 612-624 (2004).

White solid, mp 170-172°.
THERAP CAT: Antifungal.

7724. Potasan. [299-45-6] Phosphorothioic acid *O,O*-diethyl *O*-(4-methyl-2-oxo-2*H*-1-benzopyran-7-yl) ester; *O,O*-diethyl *O*-(4-methyl-7-coumarinyl) thiophosphate; 4-methylumbelliferone *O,O*-diethyl phosphorothioate; hymecromone *O,O*-diethyl phosphorothioate; E-838. $C_{14}H_{17}O_5PS$; mol wt 328.32. C 51.22%, H 5.22%, O 24.36%, P 9.43%, S 9.76%. Organophosphate insecticide; cholinesterase inhibitor. Dechlorination product of coumaphos, *q.v.* Prepn: Schrader, Kükenthal, **US 2583744** (1952 to Bayer). Synthesis of deuterium labeled compd: J. Kochansky, *J. Agric. Food Chem.* **48**, 2826 (2000). Conversion from coumaphos by anaerobic bacte[ria]. R. Shelton, J. S. Karns, *ibid.* **36**, 831 (1988). Pharmacolog[y,] toxicology: J. P. Frawley *et al., J. Pharmacol. Exp. Ther.* **105**, 156 (1952). Water partition and molecular interactions: A. Lopes *et al., Environ. Sci. Technol.* **29**, 562 (1995). *Review:* G. Schrader, *Die Entwicklung neuer insektizider Phosphorsäure-Ester* (Verlag Chemie, 1963) p 187.

Long needles from petr ether. Weak aromatic odor. mp 39.5-41.3°. $bp_{1.0}$ 210° (dec). d_4^{38} 1.260 (liq); n_D^{37} 1.5685 (liq). Log P (1-octanol/water): 14. Very sparingly sol in water. Slightly sol in petr ether. Sol in most other organic solvents. Aq solns of emulsions adjusted to pH 7 to 8 show a blue fluorescence. LD_{50} in mice, guinea pigs, male, female rats (mg/kg): 98.5 ±5.0, 25.0 ±2.3, 42.0 ±3.1, 19.0 ±2.5 orally (Frawley).
USE: Insecticide.

7725. Potassium. [7440-09-7] Kalium. K; at. wt 39.0983; at. no. 19; valence 1. Group IA (1). Alkali metal. Occurrence in earth's crust: 2.59% by wt. Naturally occurring isotopes: 39 (93.26%); 40 (0.012%); 41 (6.73%); ^{40}K is radioactive: $T_{1/2}$ 1.27 × 10^9 years; known isotopes range in mass number from 35 to 54. Found mainly as the chloride (sylvite); also in the aluminosilicates *orthoclase*, and *microcline* ($KAlSi_3O_8$), and as *carnallite* (KCl.-$MgCl_2.6H_2O$). Major essential element for plant growth. First prepd in free form by Davy in 1807 by electrolysis of fused potassium hydroxide. Produced industrially by chemical reduction. Prepns: Hackspill, *Helv. Chim. Acta* **11**, 1003 (1928); Jackson, Werner, **US 2480655** (1949 to Mine Safety Appliances Co.). NMR spectrum of potassium anion (K⁻): P. P. Edwards *et al., Nature* **317**, 242 (1985). *Reviews:* Whaley, "Sodium, Potassium, Rubidium, Cesium, and Francium" in *Comprehensive Inorganic Chemistry* **vol. 1**, J. C. Bailar, Jr. *et al.*, Eds. (Pergamon Press, Oxford, 1973) pp 369-529; *Chemistry of the Elements* N. N. Greenwood, A. Earnshaw, Eds. (Pergamon Press, New York, 1984) pp 75-116; K.-W. Chiu in *Kirk-Othmer Encyclopedia of Chemical Technology* **vol. 19** (Wiley-Interscience, New York, 4th ed., 1996) pp 1047-1057.

Soft, silvery-white metal; body-centered cubic structure; tarnishes on exposure to air; becomes brittle at low temps; mp 63.2°. bp 765.5°. d^{20} 0.856. Specific heat (0°): 0.176 cal/g deg. Thermal conductivity (cal/sec °C cm): 0.23 (21°); 0.956 (400°). Sol in liquid ammonia, ethylenediamine, aniline; sol in several metals; forms liquid alloys with other alkali metals. Emits characteristic violet color (766.5 nm) in flame. One of the most active metals; E^0 (aq) K/K⁺ 2.922 V. Reacts vigorously with oxygen; with water even at −100°;

with acids; with the halogens, igniting with bromine and iodine. Molten metal reacts with sulfur; with hydrogen sulfide. Reacts with hydrogen slowly at 200°, rapidly at 350-400°. Reacts slowly with anhyd hydrogen halides at room temp; molten metal ignites in the reaction. Reduces silicates, sulfates, nitrates, carbonates, phosphates, oxides and hydroxides of the heavy metals, often with the separation of the metal. Reacts with organic compds containing active groups. Inert to saturated aliphatic and to aromatic hydrocarbons. *Dangerous when wet. Keep under liquid containing no oxygen, e.g., liquid petrolatum, petroleum, etc.*

Caution: Direct contact with metal may be corrosive and cause skin and eye burns. *See: Fire Protection Guide to Hazardous Materials* (National Fire Protection Assoc., Quincy, MA, 12th ed., 1997) Section 49, p 109.

USE: In synthesis of inorganic potassium compds; in organic syntheses involving condensation, dehalogenation, reduction, and polymerization reactions. As heat transfer medium together with sodium: *Chem. Eng. News* **33**, 648 (1955). Radioactive decay of ^{40}K to ^{40}Ar used as tool for geological dating.

7726. Potassium Acetate. [127-08-2] Acetic acid potassium salt (1:1). $C_2H_3KO_2$; mol wt 98.14. C 24.48%, H 3.08%, K 39.84%, O 32.60%. CH_3COOK.
Colorless, lustrous, rapidly deliquesc crystals or white cryst powder or flakes. d^{25} 1.57. mp 292°. One gram dissolves in 0.5 ml water, 0.2 ml boiling water, 2.9 ml alcohol. The aq soln is alkaline to litmus. pH of 0.1 molar aq soln 9.7. *Keep tightly closed.* LD_{50} orally in rats: 3.25 g/kg; *see:* H. F. Smyth *et al., Am. Ind. Hyg. Assoc. J.* **30**, 470 (1969).
USE: Buffer.
THERAP CAT: Electrolyte replenisher; potassium supplement.

7727. Potassium Aminobenzoate. [138-84-1] 4-Aminobenzoic acid potassium salt (1:1); aminobenzoate potassium; *p*-aminobenzoate; KPABA; Potaba. $C_7H_6KNO_2$; mol wt 175.23. C 47.98%, H 3.45%, K 22.31%, N 7.99%, O 18.26%. Prepn: E. A. Meyers *et al., J. Am. Chem. Soc.* **89**, 3565 (1967). Crystal structure: G. S. Ciminago, *C. R. Seances Acad. Sci. Ser. C* **267**, 1402 (1968). Antifibrotic efficacy in idiopathic pulmonary fibrosis: U. H. Cegla *et al., Pneumonologie* **152**, 75 (1975); in Peyronie's disease: A. J. Riley, *Br. J. Sex. Med.* **6**, 29 (1970); G. Williams, N. A. Green, *Br. J. Urol.* **52**, 392 (1980).

Crystals from alcohol. Saline taste. Slightly alkaline to litmus. pH of 1% soln about 7. Very freely sol in water, less sol in alc. Practically insol in ether. Reported to cause less gastric irritation than the free acid or the sodium salt.
USE: Catalyst in the manuf of condensation polymers of polyglycol ethers.
THERAP CAT: Antifibrotic.

7728. Potassium Arsenate. [7784-41-0] Arsenic acid (H_3-AsO_4) potassium salt (1:1); potassium acid arsenate; potassium dihydrogen arsenate; Macquer's salt. AsH_2KO_4; mol wt 180.03. As 41.62%, H 1.12%, K 21.72%, O 35.55%. KH_2AsO_4.
Colorless crystals or white, cryst mass or powder. *Poisonous.* d 2.8. Sol in 5.5 parts cold, more sol in hot water, slowly in 1.6 parts glycerol. Insol in alcohol.
USE: In the textile, tanning, and paper industries. In insecticidal formulations (especially fly paper).

7729. Potassium Arsenite. [13464-35-2] Arsenenous acid potassium salt (1:1); potassium meta-arsenite. $AsKO_2$; mol wt 146.02. As 51.31%, K 26.78%, O 21.91%. $KAsO_2$. Prepd by the reaction of arsenic trioxide with potassium hydroxide: R. Cernatesco, A. Mayer, *Z. Phys. Chem.* **160**, 305 (1932). Infrared spectra and molecular characterization: J. S. Ogden, S. J. Williams, *J. Chem. Soc. Dalton Trans.* **1982**, 825. Toxicology: H. Tinwell *et al., Environ. Health Perspect.* **95**, 205 (1991). Toxicity data: A. J. Lehman, *Q. Bull. Assoc. Food Drug Off. U.S.* **15**, 122 (1951). Evaluation of carcinogenic risk: *IARC Monographs* **2**, 48-73 (1973); *Comm. Eur. Communities, Toxicol. Chem.* **3**, 53-58 (1991).

White, hygroscopic powder. *Poisonous.* Commercial material is of variable composition, approx corresponding to $KAsO_2.HAsO_2$. Sol in water. Gradually decomposes on exposure to air. *Keep well closed.* LD_{50} orally in rats: 14 mg/kg (Lehman).

Fowler's solution. [1332-10-1] Potassium arsenite solution; arsenical solution. Prepd by dissolving arsenic trioxide in aq potassium bicarbonate. Formerly used in treatment of chronic myelogenous leukemia and as a dermatologic and tonic in human and veterinary medicine.

Caution: Arsenic and inorganic arsenic compounds are listed as known human carcinogens: *Report on Carcinogens, Twelfth Edition* (PB2011-111646, 2011) p 50.

USE: In manuf of mirrors to reduce the silver salt to metallic silver.

7730. Potassium Bicarbonate. [298-14-6] Carbonic acid potassium salt (1:1); potassium acid carbonate; K-Lyte. $CHKO_3$; mol wt 100.11. C 12.00%, H 1.01%, K 39.06%, O 47.94%. $KHCO_3$. Contains not less than 99% $KHCO_3$.

Colorless, transparent crystals, white granules or powder. Sol in 2.8 parts water, 2 parts water at 50°. Almost insol in alcohol. pH: 8.2 (in 0.1 molar concn).

USE: In baking powders, effervescent salts. Buffer.

THERAP CAT: Potassium supplement.

7731. Potassium Bifluoride. [7789-29-9] Potassium fluoride ($K(HF_2)$); potassium acid fluoride; potassium hydrogen fluoride. F_2HK; mol wt 78.10. F 48.65%, H 1.29%, K 50.06%. KF.HF. Prepd according to the eq $KOH + 2HF = KHF_2 + H_2O$: Lange, Eichler, *Z. Phys. Chem.* **129**, 285 (1927); Kwasnik in *Handbook of Preparative Inorganic Chemistry* vol. 1, G. Brauer, Ed. (Academic Press, New York, 2nd ed., 1963) p 237. Made commercially from potassium carbonate and hydrofluoric acid.

Tetragonal crystals. *Poisonous.* d 2.37. mp 238.7°. Transformation pt 195°. Soly in water (g/100 ml): 30.1 (10°); 39.2 (20°); 114.0 (19°). Sol in dil alc. Insol in abs alc.

Caution: Corrosive and irritating to skin, mucous membranes.

USE: In the prepn of pure potassium fluoride; as an electrolyte in the manuf of fluorine; frosting glass; treating coal to prevent slag formation; flux for silver solders; catalyst in the alkylation of benzene with olefins.

7732. Potassium Binoxalate. [127-95-7] Ethanedioic acid potassium salt (1:1); potassium acid oxalate; potassium hydrogen oxalate; monopotassium oxalate; oxalic acid monopotassium salt. C_2HKO_4; mol wt 128.12. C 18.75%, H 0.79%, K 30.52%, O 49.95%. KOOCCOOH. Isolated from the leaves of the wood sorrel, *Oxalis acetosella*, and originally known as **salt of sorrel**, **sal acetosella**, or by the somewhat misleading name, **salt of lemon**. *See also* potassium tetroxalate. Identification method: J. Liebig, *Lancet* **44**, 149 (1844). Prepn from oxalic acid and potassium fluoride: M. A. Borodin, *C. R. Hebd. Seances Acad. Sci.* **55**, 553 (1862). Identification in leaves of the African sorrel, *Rumex abyssinicus*: G. P. Walton, *Bot. Gazz.* **74**, 158 (1922). Crystal structure: H. Einspahr *et al., Acta Crystallogr. B* **28**, 2194 (1972).

Monohydrate. [6100-03-4] $C_2HKO_4.H_2O$; mol wt 146.14. White, odorless crystals. *Poisonous; irritant.* d 2.0. Sol in 40 parts cold, 6 parts boiling water, slightly in alcohol. pH of 0.1 molar aq soln: 2.7.

USE: Removing ink stains, scouring metals; in photography.

7733. Potassium Biphthalate. [877-24-7] 1,2-Benzenedicarboxylic acid potassium salt (1:1); phthalic acid potassium salt; potassium acid phthalate; potassium hydrogen phthalate; acid potassium phthalate. $C_8H_5KO_4$; mol wt 204.22. C 47.05%, H 2.47%, K 19.15%, O 31.34%. Prepd by half-neutralization of a phthalic anhydride soln: F. J. Welcher, *Organic Analytical Reagents* vol. 2 (Van Nostrand, New York, 1947) pp 75-79. Crystal structure studies: N. G. Furmanova *et al., J. Struct. Chem.* **35**, 697 (1994). Thermal stability studies: E. R. Caley, R. H. Brundin, *Anal. Chem.* **25**, 142 (1953).

Orthorhombic crystals, stable in air. d_4^{25} 1.636. Acid reaction; pH of $0.05M$ aq soln at 25° = 4.005 (glass electrode). Sol in about 12 parts cold water, 3 parts boiling water; slightly sol in alcohol.

USE: As primary standard for preparing volumetric alkali solns, also as a buffer in pH determinations.

7734. Potassium Bisulfate. [7646-93-7] Sulfuric acid potassium salt (1:1); potassium acid sulfate; potassium hydrogen sulfate; sal enixum. HKO_4S; mol wt 136.16. H 0.74%, K 28.71%, O 47.00%, S 23.55%. $KHSO_4$.

White, deliquesc crystals, pieces, or granules. d 2.24. mp 197°; at higher temp loses water and is converted into pyrosulfate. Sol in 1.8 parts water, 0.85 part boiling water. *Keep well closed.*

USE: As flux in analysis of ores and siliceous compds.

THERAP CAT: Cathartic.

7735. Potassium Bisulfide. [1310-61-8] Potassium sulfide ($K(SH)$); potassium hydrosulfide; potassium hydrogen sulfide; potassium sulfhydrate. HKS; mol wt 72.17. H 1.40%, K 54.18%, S 44.42%. KHS. Prepd industrially from $Ca(HS)_2$ and K_2SO_4: Hene, **DE 380385** (1922); from H_2S and K_2S: Bassett, **US 1662735** (1925); Strosacker, Jones, **US 1771384** (1926 to Dow). Prepn of pure material by the action of dry H_2S upon potassium metal dissolved in abs ethanol: Rule, *J. Chem. Soc.* **99**, 558, 564 (1911); West, *Z. Kristallogr.* **88**, 102 (1934).

Colorless, deliquescent crystals or white, strongly hygroscopic, cryst mass. Usually present as the hemihydrate. Trigonal system. d 1.70. Rapidly becomes yellow on exposure to air with formation of polysulfides and H_2S. Becomes anhyd at 175-200°. mp 450-510° forming a dark red liquid. Heat of formation: +62.5 kcal. Heat of soln at 17°: +0.77 kcal, for hemihydrate at 16°: +0.62 kcal. Freely sol in water, alcohol.

7736. Potassium Bitartrate. [868-14-4] (2*R*,3*R*)-2,3-Dihydroxybutanedioic acid potassium salt (1:1); potassium acid tartrate; acid potassium tartrate; potassium hydrogen tartrate; cream of tartar; cremor tartari; faecula; faecla. $C_4H_5KO_6$; mol wt 188.18. C 25.53%, H 2.68%, K 20.78%, O 51.01%. $KHC_4H_4O_6$. Obtained from the sediments in the manuf of wine, known as argols or wine lees. The salt is at least 99.5% pure. *See also* Argol and Tartaric Acid.

Colorless crystals or white, cryst powder; pleasant acidulous taste. One gram dissolves in 162 ml water, in 16 ml boiling water, 8820 ml alcohol; readily sol in dil mineral acids, in solns of alkalies or borax. Soly in water also given as about 0.4% at 10° to about 6% at 100°.

USE: Largely in baking powders; coloring metals, galvanic tinning of metals; reducer of CrO_3 in mordants for wool. Buffering agent.

THERAP CAT: Cathartic.

THERAP CAT (VET): Laxative, diuretic.

7737. Potassium Borohydride. [13762-51-1] Tetrahydroborate(1−) potassium (1:1); potassium tetrahydroborate. BH_4K; mol wt 53.94. B 20.04%, H 7.47%, K 72.48%. KBH_4. Prepn: H. I. Schlesinger *et al., J. Am. Chem. Soc.* **75**, 199 (1953). Commercial process: M. D. Banus *et al., ibid.* **76**, 3848 (1954). NMR relaxation study: T. Tsang, T. C. Farrar, *J. Chem. Phys.* **50**, 3498 (1969); IR and Raman spectra: K. B. Harvey, N. R. McQuaker, *Can. J. Chem.* **49**, 3272 (1971). Use as reducing agent in protein labelling: E. K, J. Pauwels *et al., Nucl. Med. Biol.* **20**, 825 (1993); in simple reductions: C. Than *et al., J. Labelled Compd. Radiopharm.* **38**, 693 (1996); J. C. Briggs *et al., Tetrahedron* **53**, 3943 (1997). Review of potassium and other metal tetrahydroborates: B. D. James, M. G. H. Wallbridge, *Prog. Inorg. Chem.* **11**, 99-231 (1970).

Non-hygroscopic crystals. Stable to air. d 1.11. n_D 1.490. Dec commences at about 500°. Supports combustion. Negative heat of soln in H_2O = 6.3 kcal/mol. *Dangerous when wet.* Soly (w/w) in water at 25°: 19%; liquid ammonia at 25°: 20%; ethylenediamine at

75°: 3.9%; methanol at 20°: 0.7%; DMF at 20°: 15.0%. 0.25 g dissolves in 100 g of 95% ethanol at 25°. Soly in a 4:1 water-methanol mixture: 13 g/100 g. Insol (< 0.01%) in isopropylamine, benzene, hexane, ether, dioxane, tetrahydrofuran and acetonitrile. Alkaline aq solutions are stable.

USE: Reducing agent; source of H⁻.

7738. Potassium Bromate. [7758-01-2] Bromic acid potassium salt (1:1). $BrKO_3$; mol wt 167.00. Br 47.85%, K 23.41%, O 28.74%. $KBrO_3$.

White crystals or granules. d 3.27. mp about 350°, decomposing at about 370° with evolution of oxygen. *Oxidizer.* Sol in 12.5 parts water, 2 parts boiling water. Almost insol in alc.

Caution: Ingestion may cause vomiting, diarrhea, methemoglobinemia, renal injury.

USE: Bread- and flour-improving agent. In analytical chemistry as oxidizing agent in acid solutions.

7739. Potassium Bromide. [7758-02-3] Potassium bromide (KBr). BrK; mol wt 119.00. Br 67.15%, K 32.86%. KBr. Continuous electrolytic process of prepn: Maylott, Elkins, *US 2989450* (1961 to Dow).

Colorless crystals or white granules or powder. d 2.75. mp 730°. One gram dissolves in 1.5 ml water, 1 ml boiling water, 250 ml alc, 4.6 ml glycerol. The aq soln is neutral.

Caution: Large doses cause CNS depression. Prolonged intake may cause mental deterioration, acneform skin eruptions.

USE: Manuf photographic papers and plates; process engraving. Analytical standard. Redox reagent.

THERAP CAT: Sedative, anticonvulsant.

THERAP CAT (VET): Sedative.

7740. Potassium Carbonate. [584-08-7] Carbonic acid potassium salt (1:2); salt of tartar; pearl ash; potash. CK_2O_3; mol wt 138.20. C 8.69%, K 56.58%, O 34.73%. K_2CO_3.

Hygroscopic, odorless granules or granular powder. d 2.29; mp 891°. Sol in 1 part cold, 0.7 part boiling water. Practically insol in alcohol. Its aq soln is strongly alkaline. pH 11.6. *Keep tightly closed.* LD_{50} orally in rats: 1.87 g/kg. *See:* H. F. Smyth *et al., Am. Ind. Hyg. Assoc. J.* **30**, 470 (1969).

Sesquihydrate. [6381-79-9] $CK_2O_3 \cdot 1\frac{1}{2}H_2O$; mol wt 165.23. Small granular crystals. When it contains the full amount of water (16.36%) it is not hygroscopic. Sol in less than 1 part water. Practically insol in alcohol. The aq soln is strongly alkaline.

Caution: Irritant, caustic.

USE: Manuf soap, glass, pottery, smalts and many potassium salts; in process engraving and lithography; tanning and finishing leather; liq shampoos. Base. Anhydrous form as drying agent for organic solvents. Sesquihydrate used in invert sugar test, in buffers. Pharmaceutic aid (alkalizer).

7741. Potassium Chlorate. [3811-04-9] Chloric acid potassium salt (1:1); potcrate. $ClKO_3$; mol wt 122.55. Cl 28.93%, K 31.90%, O 39.17%. $KClO_3$. Contains at least 99% $KClO_3$. Thermal decompn studies: H. Iwakura *et al., Ind. Eng. Chem. Res.* **30**, 778 (1991).

Colorless, lustrous crystals, or white granules or powder. d 2.32. mp 350°; dec 635° into perchlorate and oxygen. One gram dissolves slowly in 16.5 ml water, 1.8 ml boiling water, about 50 ml glycerol. Almost insol in alcohol. *Oxidizer. Keep out of contact with organic matter or other oxidizable substances. Caution:* Explodes with sulfuric acid; inflames with explosion if triturated with any organic substances, sulfur, phosphorus, sulfite, hypophosphite, and other oxidizable substances. *Incompat.* Iodides, tartaric acid.

Caution: Irritating to G.I. tract, kidney; can cause hemolysis of red blood cells and methemoglobinemia: *Clinical Toxicology of Commercial Products*, R. E. Gosselin *et al.*, Eds. (Williams & Wilkins, Baltimore, 5th ed., 1984) Section II, p 112; Section III, pp 74-77.

USE: Explosives; fireworks; matches; printing and dyeing cotton and wool black; manuf aniline black and other dyes; source of oxygen; in chemical analyses.

THERAP CAT: Formerly as topical antiseptic.

THERAP CAT (VET): In dilute soln as antiseptic mouthwash.

7742. Potassium Chloride. [7447-40-7] Potassium chloride (KCl); Diffu-K; Kaleorid; Kalitabs; Kaon-Cl; Kay-Cee-L; K-Contin;

K-Dur; Lento-Kalium; Potassiject; Slow-K; Span-K. ClK; mol wt 74.55. Cl 47.55%, K 52.45%. KCl. Occurs in nature as the mineral *sylvine* or *sylvite*. Industrial prepns: *Faith, Keyes & Clark's Industrial Chemicals*, F. A. Lowenheim, M. K. Moran, Eds. (Wiley-Interscience, New York, 4th ed., 1975) pp 666-673.

White crystals or crystalline powder. d 1.98. mp 773°. One gram dissolves in 2.8 ml water, 1.8 ml boiling water, 14 ml glycerol, about 250 ml alcohol. Insol in ether, acetone. Hydrochloric acid, sodium or magnesium chlorides diminish its soly in water. d of saturated aq soln at 15° is 1.172. pH: about 7.

Caution: Large doses by mouth can cause G.I. irritation, purging, weakness and circulatory disturbances.

USE: In photography. In buffer solns, electrode cells.

THERAP CAT: Electrolyte replenisher.

THERAP CAT (VET): Potassium supplement.

7743. Potassium Chromate(VI). [7789-00-6] Chromic acid (H_2CrO_4) potassium salt (1:2); neutral potassium chromate. CrK_2O_4; mol wt 194.19. Cr 26.78%, K 40.27%, O 32.96%. K_2CrO_4.

Lemon-yellow crystals; d 2.73; mp 975°. Sol in 1.6 parts cold, 1.2 parts boiling water. Insol in alc. The aq soln is alkaline to litmus or phenolphthalein.

Caution: Chromium hexavalent (VI) compounds are listed as known human carcinogens: *Report on Carcinogens, Twelfth Edition* (PB2011-111646, 2011) p 106.

USE: Has a limited application in enamels, finishing leather, rust-proofing of metals, being replaced by the sodium salt. Oxidizing agent in analytical chemistry.

7744. Potassium Citrate. [866-84-2] 2-Hydroxy-1,2,3-propanetricarboxylic acid potassium salt (1:3); tripotassium citrate. $C_6H_5K_3O_7$; mol wt 306.39. C 23.52%, H 1.64%, K 38.28%, O 36.55%. $K_3C_6H_5O_7$. Manufacturing process: G. Kominek, *US 3819696* (1974 to Jungbunzlauer). Effect on urinary chemistry and crystallization of stone forming salts: K. Sakhaee *et al., Kidney Int.* **24**, 348 (1983). Use in oral rehydration solutions: M. R. Islam, *Arch. Dis. Child.* **60**, 852 (1985). Clinical trial in nephrolithiasis: C. Y. C. Pak *et al., J. Urol.* **134**, 11 (1985); following shockwave lithotripsy: T. Soygür *et al., J. Endourol.* **16**, 149 (2002).

Monohydrate. [6100-05-6] K-CIT-V; Urocit-K. $C_6H_5K_3O_7 \cdot H_2O$; mol wt 324.41. White crystals, granules or powder; odorless with cooling, salty taste. Loses its water at 180°. One gram dissolves in 0.65 ml water; very slowly in 2.5 ml glycerol. Practically insol in alcohol. The aq soln is alkaline to litmus; pH about 8.5.

USE: In foods and beverages as buffering, sequestering or emulsifying agent.

THERAP CAT: Alkalinizing agent; antiurolithic.

THERAP CAT (VET): Alkalinizing agent; antiurolithic.

7745. Potassium Citrate, Monobasic. [866-83-1] 2-Hydroxy-1,2,3-propanetricarboxylic acid potasium salt (1:1); monopotassium citrate. $C_6H_7KO_7$; mol wt 230.21. C 31.30%, H 3.07%, K 16.98%, O 48.65%. $KH_2C_6H_5O_7$.

White, cryst powder. Sol in water; the soln is subject to molding.

USE: A 0.05 molal solution as standard for pH scale (pH at 25° 3.776): Staples, Bates, *J. Res. Natl. Bur. Stand.* **73A**, 37 (1969).

7746. Potassium Cyanate. [590-28-3] Cyanic acid potassium salt (1:1); potassium isocyanate. CKNO; mol wt 81.12. C 14.81%, K 48.20%, N 17.27%, O 19.72%. KOCN. Inhibitor of sickling of erythrocytes *in vitro:* Cerami, Manning, *Proc. Natl. Acad. Sci. USA* **68**, 1180 (1971). Pharmacology: A. Cerami *et al., J. Pharmacol. Exp. Ther.* **185**, 653 (1973). Brief review: *Dangerous Prop. Ind. Mater. Rep.* **13**, 408-415 (1993).

White, cryst powder. mp 315°. Highly hygroscopic. d 2.05. Sol in water, very slightly in alcohol. LD_{50} i.p. in mice: 320 mg/kg (Cerami). *Keep well closed in cool, dry place.*

USE: As starting material in organic synthesis.

7747. Potassium Cyanide. [151-50-8] Potassium cyanide (K(CN)). CKN; mol wt 65.12. C 18.44%, K 60.04%, N 21.51%. KCN. Article of commerce contains about 95% KCN. Toxicity study: Hayes, *Toxicol. Appl. Pharmacol.* **11**, 327 (1967). Review of toxicology and human exposure: *Toxicological Profile for Cyanide* (PB2007-100674, 2006) 341 pp.

White, deliquesc, granular powder or fused pieces; odor of HCN. *Poisonous.* On exposure to air it is gradually dec by CO_2 and moisture. d 1.52; mp 634°. Sol in 2 parts cold, 1 part boiling water, 2 parts glycerol, 100 parts alc, 25 parts methanol. The aq soln is strongly alkaline and rapidly dec. pH of $0.1N$ aq soln: 11.0. *Keep tightly closed and protected from light.* LD_{50} orally in rats: 10 mg/kg (Hayes).

Caution: Potential symptoms of overexposure are irritation of eyes, skin and upper respiratory system; weakness, headache and confusion; nausea, vomiting; increased rate of respiration; slow gasping respiration; asphyxia; thyroid and blood changes. *See NIOSH Pocket Guide to Chemical Hazards* (DHHS/NIOSH 97-140, 1997) p 262.

USE: Extracting gold and silver from ores; electroplating baths; silver plating; case hardening steel by liquid nitriding. Complexing agent in alkaline solutions.

7748. Potassium Dichromate(VI). [7778-50-9] Chromic acid ($H_2Cr_2O_7$) potassium salt (1:2); potassium bichromate. Cr_2-K_2O_7; mol wt 294.18. Cr 35.35%, K 26.58%, O 38.07%. $K_2Cr_2O_7$. In the U.S.A. it is usually prepared by the reaction of potassium chloride on sodium dichromate: Vetter in *Kirk-Othmer Encyclopedia of Chemical Technology* vol. 3 (Interscience, New York, 1949) p 951; Hartford, Copson, *ibid.* vol. 5 (2nd ed., 1964) pp 484-488. In Germany it is obtained from potassium chromate produced by roasting the chrome ore with KOH. *Ref:* Müller, Glissmann in *Ullmanns Encyklopädie der technischen Chemie* vol. 5 (Munich, 3rd ed., 1954) p 580.

Bright orange-red crystals. Not hygroscopic or deliquescent (difference from sodium dichromate). Crystal habit: prismatic. Crystal system: triclinic pinacoidal, transition to monoclinic at 241.6°. d_4^{25} 2.676. Bulk density: 100 lbs/cu ft. mp 398°. Dec at about 500°. Heat of fusion 29.8 cal/g. Heat of soln -62.5 cal/g. Specific heat 0.186 at 16° -98°. Soluble in water. A satd aq soln contains at 0°: 4.3%, at 20°: 11.7%, at 40°: 20.9%, at 60°: 31.3%, at 80°: 42.0%, at 100°: 50.2%. Acid reaction: A 1% aq soln has a pH of 4.04 and a 10% soln has a pH of 3.57.

Caution: Corrosive poison if ingested. Industrial contact may result in ulceration of hands, destruction of mucous membranes and perforation of nasal septum. *See* E. Browning, *Toxicity of Industrial Metals* (Appleton-Century Crofts, New York, 2nd ed., 1969) pp 119-131. *See also* Chromium. Chromium hexavalent (VI) compounds are listed as known human carcinogens: *Report on Carcinogens, Twelfth Edition* (PB2011-111646, 2011) p 106.

USE: In tanning leather, dyeing, painting, decorating porcelain, printing, photolithography, pigment-prints, staining wood, pyrotechnics, safety matches; for bleaching palm oil, wax, and sponges; waterproofing fabrics; as oxidizer in the manuf of organic chemicals; in electric batteries; as depolarizer for dry cells. As corrosion inhibitor in preference to sodium dichromate where lower soly is advantageous. Pharmaceutic aid (oxidizing agent). Redox reagent.

THERAP CAT (VET): Caustic.

7749. Potassium Dicyanoaurate(I). [13967-50-5] Bis(cyano-κC)aurate(1$-$) potassium (1:1); gold potassium cyanide; potassium aurocyanide; potassium gold cyanide. C_2AuKN_2; mol wt 288.10. C 8.34%, Au 68.37%, K 13.57%, N 9.72%. KAu(CN)$_2$. Prepd by electrolysis of Au in KCN: Glassford, Napier, *Philos. Mag.* **25**, 61 (1844).

White powder. *Poisonous.* d^{25} 3.45. Sol in water. *Store in cool, dry conditions in well sealed container.*

Dihydrate. [6227-35-6] Cryst powder. One gram dissolves in 7 ml water, 0.5 ml boiling water; slightly sol in alcohol. Practically insol in ether.

USE: For electroplating.

7750. Potassium Diformate. [20642-05-1] Formic acid potassium salt (2:1); potassium hydrogen diformate; Formi. $C_2H_3KO_4$; mol wt 130.14. C 18.46%, H 2.32%, K 30.04%, O 49.17%. KH-(HCOO)$_2$. Conjugated salt of potassium formate and formic acid. Feed additive for pigs; promotes growth and controls microbial activity by decreasing the pH of the digestive tract. Prepn: E. Groschuff, *Ber.* **36**, 1783 (1903). X-ray crystal structure: G. Larsson, I. Nahringbauer, *Acta Crystallogr.* **B24**, 666 (1968). Electron density study: H. Hermansson, R. Tellgren, *ibid.* **B45**, 252 (1989). Growth-promoting effects in pigs: B. R. Paulicks *et al.*, *Agribiol. Res.* **49**,

318 (1996); M. Kirchgessner *et al.*, *ibid.* **50**, 1 (1997). Antimicrobial effects in pig digestive tracts: N. Canibe *et al.*, *J. Anim. Sci.* **79**, 2123 (2001).

Crystals from ethanol, mp 108.6°.

THERAP CAT (VET): Antimicrobial; growth promotant.

7751. Potassium Ferricyanide. [13746-66-2] Potassium (OC-6-11)-hexakis(cyano-κC)ferrate(3$-$) (3:1); potassium hexacyanoferrate(III); red prussiate of potash. $C_6FeK_3N_6$; mol wt 329.25. C 21.89%, Fe 16.96%, K 35.62%, N 25.53%. $K_3Fe(CN)_6$.

Ruby-red crystals. d 1.89. Slowly sol in 2.5 parts cold water, in 1.3 parts boiling water; slightly sol in alc; dec by acids. The aq soln dec slowly on standing. *Protect from light.*

USE: Chiefly for blueprints; in photography; also for staining wood, dyeing wool, calico printing, as etching liquid (Mercer's liquor), tempering iron and steel; in electroplating; as a mild oxidizing agent in organic synthesis; in analytical chemistry.

7752. Potassium Ferrocyanide. [13943-58-3] Potassium (OC-6-11)-hexakis(cyano-κC)ferrate(4$-$) (4:1); tetrapotassium hexakis(cyano-C)ferrate(4$-$); potassium hexacyanoferrate(II); yellow prussiate of potash. $C_6FeK_4N_6$; mol wt 368.35. C 19.56%, Fe 15.16%, K 42.46%, N 22.82%. $K_4Fe(CN)_6$. Review of properties, chemistry and syntheses: *The Chemistry of Ferrocyanides*, American Cyanamid Co. (Beacon Press, New York, 1953) 112 p.

Trihydrate. [14459-95-1] $K_4Fe(CN)_6.3H_2O$; mol wt 422.39. Soft, slightly efflorescent crystals; begins to lose water at 60°, becomes anhyd at 100°. d 1.85.

USE: Moderately strong oxidizer when coupled with ferricyanide.

7753. Potassium Fluoride. [7789-23-3] Potassium fluoride (KF). FK; mol wt 58.10. F 32.70%, K 67.29%. KF. Prepd by thermal decompn of KHF_2 or by neutralizing HF with K_2CO_3: Lange, Eichler, *Z. Phys. Chem.* **129**, 285, 286 (1927); Kwasnik in *Handbook of Preparative Inorganic Chemistry* vol. **1**, G. Brauer, Ed. (Academic Press, New York, 2nd ed., 1963) p 236. Toxicity data: Simonin, *C. R. Seances Soc. Biol. Ses Fil.* **124**, 133 (1937).

Cubic crystals (NaCl lattice). Usually obtained as white, deliquesc powder or solid. *Poisonous.* d 2.481. mp 859.9°. bp 1505°. Soly in water (g/100 ml): 92.3 (18°); 96.4 (21°). Very freely sol in boiling water. Also sol in aq HF, liquid NH_3. Insol in alcohol unless water is present. May be stored in aluminum containers. Attracts moisture from the air. Aq solns corrode glass and porcelain. MLD in guinea pigs (mg/kg): 250 orally, 350 s.c.; in frogs (mg/kg): 375 s.c. (Simonin).

Dihydrate. [13455-21-5] $FK.2H_2O$; mol wt 94.13. Monoclinic crystals, mp 41°. Soly in water (18°): 349.3 g/100 ml.

Tetrahydrate. [34341-58-7] $FK.4H_2O$; mol wt 130.16. Crystals, mp 19.3°.

Caution: Irritating to skin, eyes, mucous membranes.

USE: In the fluorination of organic compds; in flux for hard solder; to prevent unwanted fermentations; in insecticide formulations; for frosting glass. Complexing agent.

7754. Potassium Formate. [590-29-4] Formic acid potassium salt (1:1). CHKO$_2$; mol wt 84.12. C 14.28%, H 1.20%, K 46.48%, O 38.04%. HCOOK. Environmentally friendly, biodegradable source of potassium ion. Prepn: E. Groschuff, *Ber.* **36**, 1783 (1903); and complex formation with formic acid: J. Kendall, H. Adler, *J. Am. Chem. Soc.* **43**, 1470 (1921). Crystal structure: B. F. Mentzen, Y. Oddon, *Inorg. Chim. Acta* **43**, 237 (1980); J. W. Bats, H. Fuess, *Acta Crystallogr.* **B36**, 1940 (1980). Properties and applications as a secondary refrigerant: A. Aittomäki, A. Lahti, *Int. J. Refrig.* **20**, 276 (1997).

mp 157° (Groschuff); also reported as mp 167.5 ± 0.5° (Kendall). d^{18} 1.95. Very sol in water. The aq soln is practically neutral. Exceedingly hygroscopic.

USE: Reducing agent; secondary refrigerant; anti-oxidant in oil drilling applications.

7755. Potassium Guaiacolsulfonate. [1321-14-8] Hydroxymethoxybenzenesulfonic acid potassium salt (1:1); sulfogaiacol. C_7-H_7KO_5S; mol wt 242.29. C 34.70%, H 2.91%, K 16.14%, O 33.02%, S 13.23%. Expectorant used in cough and cold preparations. Commercial formulations may contain a variable mixture of the potassium salts of 4- and 5-guaiacolsulfonic acid with the 4-isomer predominating. Prepn of isomers: A. Rising, *Ber.* **39**, 3685

(1906). Analysis of constituents in pharmaceutical formulations: P. Tangtatsawasdi, S. E. Krikorian, *J. Pharm. Sci.* **73**, 1238 (1984); and physical properties: K. Kawamura *et al.*, *J. Assoc. Off. Anal. Chem.* **70**, 673 (1987). HPLC determn in cough syrups: O. A. Dönmez *et al.*, *Talanta* **83**, 1601 (2011).

Potassium 4-guaiacolsulfonate

White, odorless crystals or cryst powder; faint bitter taste. Gradually turns pink on exposure to air and light. Soluble in 7.5 parts water. Almost insol in alcohol. Insol in ether. The aq soln is neutral to litmus. *Keep well closed and protected from light.*

Potassium 4-guaiacolsulfonate. [16241-25-1] 4-Hydroxy-3-methoxybenzenesulfonic acid potassium salt (1:1). Crystals from water or ethanol. pKa 8.74. Soly in water (20°): 13.6 parts per 100. uv max: 279 nm ($E_{1\ cm}^{1\%}$ 116.17 at pH 6; 239.44 at pH 12).

Potassium 5-guaiacolsulfonate. [5011-21-2] 3-Hydroxy-4-methoxybenzenesulfonic acid potassium salt (1:1). Crystals from water as the dihydrate. pKa 9.16. Soly in water (20°): 65.8 parts per 100. uv max: 279 nm ($E_{1\ cm}^{1\%}$ 125.35 at pH 6; 89.55 at pH 12).

THERAP CAT: Expectorant.

7756. Potassium Hexachloroosmate(IV). [16871-60-6] (*OC*-6-11)-Hexachloroosmate(2−) potassium (1:2); dipotassium hexachloroosmate; osmium potassium chloride. Cl_6K_2Os; mol wt 481.13. Cl 44.21%, K 16.25%, Os 39.54%. $K_2[OsCl_6]$. Preparation: Turner *et al.*, *Anal. Chem.* **30**, 1708 (1958).

Dark red to almost black, cubic crystals. *Poisonous; corrosive.* Freely sol in water; sparingly sol in alcohol.

7757. Potassium Hexachloroplatinate(IV). [16921-30-5] (*OC*-6-11)-Hexachloroplatinate(2−) potassium (1:2); dipotassium hexachloroplatinate; platinic potassium chloride; potassium platinic chloride. Cl_6K_2Pt; mol wt 485.98. Cl 43.77%, K 16.09%, Pt 40.14%. $K_2[PtCl_6]$.

Orange-yellow crystals or yellow powder. *Poisonous; corrosive.* d 3.50. Slightly sol in cold water; sol in hot water. Almost insol in alcohol.

USE: In photography.

7758. Potassium Hexacyanocobaltate(III). [13963-58-1] (*OC*-6-11)-Hexakis(cyano-κC)cobaltate(3−) potassium (1:3); potassium cyanocobaltate(III); potassium cobalticyanide; potassium cobaltihexacyanide; cobalt potassium cyanide; tripotassium hexacyanocobaltate. $C_6CoK_3N_6$; mol wt 332.34. C 21.68%, Co 17.73%, K 35.29%, N 25.29%. $K_3[Co(CN)_6]$. Prepd by reacting cobalt acetate or cyanide with potassium cyanide, followed by air-oxidation of the cobaltocyanide formed: Biltz, Biltz, *Z. Anorg. Allg. Chem.* **50**, 108 (1906); Grube, *Z. Elektrochem.* **32**, 561 (1926); Benedetti-Pichler, *Z. Anal. Chem.* **70**, 258 (1927); Bigelow, *Inorg. Synth.* **2**, 225 (1946).

Faintly yellow, monoclinic crystals from water. *Poisonous.* d 1.906. Melts with decompn forming an olive-green mass. Freely sol in water, acetic acid solns. Insol in alcohol. Very slightly sol in liquid ammonia. Dec by strong mineral acids.

7759. Potassium Hexafluorosilicate. [16871-90-2] Hexafluorosilicate(2−) potassium (1:2); potassium fluosilicate; potassium silicofluoride. F_6K_2Si; mol wt 220.27. F 51.75%, K 35.50%, Si 12.75%. $K_2[SiF_6]$.

White, fine powder or crystals. d 2.27. Slightly sol in cold water. Insol in alcohol. In hot water it hydrolyzes to KF, HF, and silicic acid. pH of 1% aq soln 3.4.

Caution: Strong irritant. Ingestion can cause vomiting, diarrhea.

USE: In the manuf of opalescent glass, in porcelain enamels, in insecticides. Also used in aluminum metallurgy.

7760. Potassium Hexafluorozirconate. [16923-95-8] (*OC*-6-11)-Hexafluorozirconate(2−) potassium (1:2); dipotassium hexafluorozirconate; potassium zirconium fluoride; zirconium potassium fluoride. F_6K_2Zr; mol wt 283.41. F 40.22%, K 27.59%, Zr 32.19%. $K_2[ZrF_6]$.

Crystals. *Poisonous; irritant.* Slightly sol in cold water, sol in hot water.

USE: For manuf of metallic zirconium.

7761. Potassium Hydroxide. [1310-58-3] Potassium hydroxide (K(OH)); potassium hydrate; caustic potash; potassa. HKO; mol wt 56.11. H 1.80%, K 69.68%, O 28.51%. KOH. Prepd industrially by electrolysis of potassium chloride: *Faith, Keyes & Clark's Industrial Chemicals*, F. A. Lowenheim, M. K. Moran, Eds. (Wiley-Interscience, New York, 4th ed., 1975) pp 674-678. Toxicity: H. F. Smyth *et al.*, *Am. Ind. Hyg. Assoc. J.* **30**, 470 (1969).

White or slightly yellow lumps, rods, pellets. Rapidly absorbs moisture and CO_2 from the air and deliquesces. mp about 360°; mp 380° when anhydr. *Corrosive. Keep tightly closed and do not handle with bare hands.* Sol in 0.9 part water, about 0.6 part boiling water, 3 parts alcohol, 2.5 parts glycerol. When dissolved in water or alcohol or when the soln is treated with an acid, much heat is generated. A 0.1*M* aq soln has a pH of 13.5. LD_{50} orally in rats: 1.23 g/kg (Smyth).

Caution: Extremely corrosive. Potential symptoms of overexposure are irritation of eyes, skin, respiratory system; eye and skin burns; cough, sneezing; vomiting, diarrhea. Direct contact may cause painful burns and liquefaction necrosis. *See NIOSH Pocket Guide to Chemical Hazards* (DHHS/NIOSH 97-140, 1997) p 262; *Clinical Toxicology of Commercial Products*, R. E. Gosselin *et al.*, Eds. (Williams & Wilkins, Baltimore, 5th ed., 1984) Section III, pp 245-252.

USE: Manuf liq soap; mordant for wood; absorbing CO_2; mercerizing cotton; paint and varnish removers; electroplating, photoengraving and lithography; printing inks; in analytical chemistry for alkalimetric titrations; in organic synthesis. Pharmaceutic aid (alkalizer).

THERAP CAT (VET): Caustic. In disbudding calves' horns. In aq solution to dissolve scales and hair in skin scrapings.

7762. Potassium Hypophosphite. [7782-87-8] Phosphinic acid potassium salt (1:1); potassium phosphinate. H_2KO_2P; mol wt 104.09. H 1.94%, K 37.56%, O 30.74%, P 29.76%. KH_2PO_2.

White crystals or granular, deliquesc powder. Odorless; pungent, saline taste. When strongly heated it dec with evolution of phosphine which ignites spontaneously in air. *It explodes when triturated with chlorates or other oxidizing agents.* One gram dissolves in 0.6 ml water, 9 ml alc, 5 ml boiling alc. The aq soln is neutral or slightly alkaline. *Keep well closed.*

7763. Potassium Iodate. [7758-05-6] Iodic acid (HIO_3) potassium salt (1:1). IKO_3; mol wt 214.00. I 59.30%, K 18.27%, O 22.43%. KIO_3.

White, odorless crystals or cryst powder. d 3.89. mp 560° with partial decompn. Slowly sol in 12 parts water, in 3.1 parts boiling water. Insol in alcohol.

USE: Oxidizing agent in volumetric chemical analysis. Fast-acting oxidant in dough conditioning; promotes disulfide bond formation in gluten and improves dough-forming properties.

THERAP CAT: Iodine supplement.

THERAP CAT (VET): In feeds as a source of iodine.

7764. Potassium Iodide. [7681-11-0] Potassium iodide (KI); Jodid; Thyroblock; Thyrojod. IK; mol wt 166.00. I 76.45%, K 23.55%. KI. Potassium iodide of commerce contains about 99.5% KI. Prepd from HI and $KHCO_3$. Purification by melting in dry hydrogen: Lingane, Kolthoff, *Inorg. Synth.* **1**, 163 (1939). Continuous electrolytic process for large scale industrial prepn: Morylott, Elkins, US 2989450 (1961 to Dow). Toxicity data: Hildebrandt, *Arch. Exp. Pathol. Pharmakol.* **96**, 292 (1923). Use in the treatment of radiation poisoning resulting from a nuclear accident: W. K. Waterfall, *Br. Med. J.* **281**, 988 (1980); *Bull. N.Y. Acad. Med.* **57**, 395 (1981).

Colorless or white, cubical crystals, white granules, or powder. Slightly deliquescent in moist air; on long exposure to air becomes yellow due to liberation of iodine, and small quantities of iodate may be formed; light and moisture accelerate the decompn. Aq solns also become yellow in time due to oxidation, but a small amount of alkali prevents it. d 3.12. mp 680° (volatilizes at higher temp). One gram dissolves in 0.7 ml water, 0.5 ml boiling water, 22 ml alcohol, 8 ml boiling alcohol, 51 ml abs alcohol, 8 ml methanol, 75 ml acetone, 2 ml glycerol, about 2.5 ml glycol. Potassium iodide solns readily dissolve elemental iodine. The aq soln is neutral or, usually, slightly alkaline. pH: 7-9. 30 g KI with 21 ml water gives 30 ml of a saturated soln at 25°. Approx LD i.v. in rats: 285 mg/kg (Hildebrandt). *Incompat.* Alkaloidal salts, chloral hydrate, tartaric and other acids, calomel, potassium chlorate, metallic salts.

USE: Manuf photographic emulsions; in animal and poultry feeds to the extent of 10-30 parts per million; in table salt as a source of iodine and in some drinking water; also in analytical chemistry for iodometric titrations. Reducing agent.

THERAP CAT: Antifungal; expectorant; iodine supplement.

THERAP CAT (VET): In actinobacillosis, actinomycosis. For simple goiter. As expectorant. In iodine deficiency and in chronic poisoning with lead or mercury. Orally only, not by injection. Externally for treatment of bursal enlargements.

7765. Potassium Manganate. [10294-64-1] Manganic acid (H_2MnO_4) potassium salt (1:2); dipotassium manganate. K_2MnO_4; mol wt 197.13. K 39.67%, Mn 27.87%, O 32.46%. Prepn: Scholder, Waterstradt, *Z. Anorg. Allg. Chem.* **277**, 172 (1954).

Dark green crystals; dec at 190°. *Irritant.* Sol in water. Sol and stable in KOH solns. *Oxidizer.* With HCl it gives free chlorine.

7766. Potassium Metabisulfite. [16731-55-8] Disulfurous acid dipotassium salt (1:2); dipotassium disulfite. $K_2O_5S_2$; mol wt 222.31. K 35.17%, O 35.98%, S 28.84%. $K_2S_2O_5$. The article of commerce contains ~95% $K_2S_2O_5$.

White crystals or cryst powder; sulfur dioxide odor; acid reaction; liberates SO_2 with acids; oxidizes in air to sulfate, more readily in presence of moisture. It may catch fire if much heat develops in powdering it. Sol in water. Insol in alc. *Keep dry and well closed.*

USE: As antifermentative in breweries and wineries; bleaching straw; preservative for fruits and vegetables.

7767. Potassium Metaphosphate. [7790-53-6] Metaphosphoric acid (HPO_3) potassium salt (1:1); potassium Kurrol's salt; potassium polymetaphosphate; potassium polyphosphate. $(KPO_3)_x$. High mol wt polymer; degree of polymerization dependent upon preparative conditions. Prepd by dehydration of KH_2PO_4: Pfansteil, Iler, *J. Am. Chem. Soc.* **74**, 6059 (1952). Structural studies: Jost, *Acta Crystallogr.* **16**, 623 (1963); Jost, Schulze, *ibid.* **25B**, 1110 (1969); *eidem, ibid.* **27B**, 1345 (1971). Reviews of metaphosphates: J. R. Van Wazer, *Phosphorus and Its Compounds* vol 1 (Interscience, New York, 1958) pp 601-678; Thilo, *Adv. Inorg. Chem. Radiochem.* **4**, 1-75 (1962).

White, monoclinic crystals. d^{20} 2.45. Insol in pure water. Sol in aq solns of alkali metal (except potassium) salts.

7768. Potassium Molybdate. [13446-49-6] (*T*-4)-Molybdate (MoO_4^{2-}) potassium (1:2); dipotassium molybdate; dipotassium tetraoxomolybdate; potassium molybdenum oxide. K_2MoO_4; mol wt 238.14. K 32.84%, Mo 40.29%, O 26.87%.

Odorless white powder. *Irritant.* d 2.3; mp 919°. Sol in 0.6 part water. Insol in alc. *Keep well closed.*

7769. Potassium Nitrate. [7757-79-1] Nitric acid potassium salt (1:1). KNO_3; mol wt 101.10. K 38.67%, N 13.85%, O 47.47%. Occurs in nature as the mineral **saltpeter** or **niter.** Polymorphism: F. C. Kracek, *J. Phys. Chem.* **34**, 225 (1930). Prepn from potassium chloride: D. L. Reed, K. G. Clark, *Ind. Eng. Chem.* **29**, 333 (1937); by molten salt technique: R. W. Pfeiffer *et al., J. Agric. Food Chem.* **15**, 949 (1967); by solvent separation of strong acids: A. Eyal *et al., Ind. Eng. Chem. Process Des. Dev.* **24**, 387 (1985). Viscosity: R. E. Wellman *et al., J. Chem. Eng. Data* **11**, 156 (1966). Heat capacity: E. W. Dewing, *ibid.* **20**, 221 (1975). Clinical evaluations in tooth hypersensitivity: M. Hodosh, *J. Am. Dent. Assoc.* **88**, 831 (1974); T. Nagata *et al., J. Clin. Periodontol.* **21**, 217 (1994); S. C. Frechoso *et al., ibid.* **30**, (2003). Toxicity data: Dollahite, Rowe, *Southwest. Vet.* **27**, 246 (1974).

Colorless transparent prisms, white granular or cryst powder; cooling, saline, pungent taste. d 2.11; mp 333°; dec at 400° with evolution of O_2. *Oxidizer.* One gram dissolves in 2.8 ml water, 0.5 ml boiling water, 620 ml alc. Sol in glycerol; insol in abs alc. Dissolves in water with a lowering of the temp. pH ~7. LD_{50} orally in rabbits: 1.166 g anion/kg (Dollahite, Rowe).

Caution: Ingestion of large quantities may cause violent gastroenteritis. Prolonged exposure to small amts may produce anemia, methemoglobinemia, nephritis.

USE: In fireworks, fluxes, pickling meats; production of nitric acid; manuf glass, matches, gunpowder; freezing mixtures. Agricultural fertilizer. Preservative in foods. In dentrifices to reduce tooth hypersensitivity.

7770. Potassium Nitrite. [7758-09-0] Nitrous acid potassium salt (1:1). KNO_2; mol wt 85.10. K 45.94%, N 16.46%, O 37.60%. The nitrite of commerce usually contains ~85% KNO_2, the remainder consisting chiefly of nitrate.

White or slightly yellow, deliquesc granules or rods. Dec even by weak acids with evolution of brown fumes of nitrous anhydride. d 1.915; mp 441° (decompn starts at 350°). Sol in 0.35 part water, slightly in alc. The aq soln is alkaline. *Oxidizer. Keep well closed.* LD_{50} orally in rabbits: 108 mg anion/kg. *See:* Dollahite, Rowe, *Southwest. Vet.* **27**, 246 (1974).

USE: Complexing agent.

THERAP CAT: Vasodilator; antidote (cyanide poisoning).

7771. Potassium Oleate. [143-18-0] (9Z)-9-Octadecenoic acid potassium salt (1:1); oleic acid potassium salt. Approx $C_{18}H_{33}$-KO_2.

Yellowish or brownish, soft mass. Sol in water, alc. The aq soln is alkaline to phenolphthalein.

USE: Detergent.

7772. Potassium Oxalate. [583-52-8] Ethanedioic acid potassium salt (1:2); oxalic acid dipotassium salt. $C_2K_2O_4$; mol wt 166.21. C 14.45%, K 47.05%, O 38.50%. $K_2C_2O_4$.

Occurs as the monohydrate, colorless, odorless crystals; effloresent in warm dry air. *Poisonous.* d 2.13. Loses its water at ~160°; when ignited is converted into carbonate without appreciable charring. Sol in 3 parts water.

Monohydrate. [6487-48-5] $C_2K_2O_4 \cdot H_2O$; mol wt 184.23.

USE: Cleaning and bleaching straw, removing stains in photography; *in vitro* blood anticoagulant; also as reducing agent in analytical chemistry.

7773. Potassium Percarbonate. [589-97-9] Peroxydicarbonic acid potassium salt (1:2); dipotassium peroxydicarbonate. C_2-K_2O_6; mol wt 198.21. C 12.12%, K 39.45%, O 48.43%. $K_2C_2O_6$. Prepn of practically anhydr compd: Partington, Fathallah, *J. Chem. Soc.* **1950**, 1934.

Monohydrate. White, granular mass. Sol in water with evolution of oxygen. One part potassium percarbonate is sol in 15 parts of cold water; dec in boiling water; 100 parts water dissolve 6.5 parts potassium percarbonate at ordinary temp. *Keep dry and protected from light.*

Caution: Strong irritant. Causes vomiting if swallowed. Large quantities can be fatal.

USE: Has been used in microscopy for detecting tubercle bacilli stained with fuchsin in smears; in photography under the name **Antihypo,** to remove last traces of sodium thiosulfate; also as oxidizing agent in chem analyses, but is no longer favored.

7774. Potassium Perchlorate. [7778-74-7] Perchloric acid potassium salt (1:1); peroidin; Perchloracap. $ClKO_4$; mol wt 138.54. Cl 25.59%, K 28.22%, O 46.19%. $KClO_4$.

Colorless crystals or white, cryst powder. *Oxidizer.* Dec at 400°; also dec by organic matter, oxidizable substances and on concussion, but is less reactive than the chlorate. d 2.52. Sol in 65 parts cold water, 15 parts boiling water. Practically insol in alcohol.

USE: In explosives, pyrotechnics and photography. Oxidizing agent.

7775. Potassium Periodate. [7790-21-8] Periodic acid (HIO_4) potassium salt (1:1); potassium metaperiodate. IKO_4; mol wt 230.00. I 55.18%, K 17.00%, O 27.82%. KIO_4. Prepd by oxidizing potassium iodate with chlorine in alkaline soln: Hill, *J. Am. Chem. Soc.* **50**, 2678 (1928); *Inorg. Synth.* **1**, 171 (1939).

Colorless tetragonal crystals, d_4^{15} 3.618. mp 582°. Soly in water (g/100 g H_2O): 0.168 at 0°; 0.42 at 20°; 0.93 at 40°; 2.16 at 60°; 4.44 at 80°; 7.87 at 100°; also given as 0.66 at 13°. Sparingly sol in aq KOH.

Caution: Highly irritating to skin, eyes, mucous membranes.

USE: Powerful oxidizer in acid soln, oxidizing manganese compds to permanganate; used for this purpose in analytical chemistry (colorimetric estimation of Mn), also for the oxidation of some organic compds.

7776. Potassium Permanganate. [7722-64-7] Permanganic acid ($HMnO_4$) potassium salt (1:1); chameleon mineral. $KMnO_4$; mol wt 158.03. K 24.74%, Mn 34.76%, O 40.50%. Prepn from manganese ore by electrolytic oxidation: *Faith, Keyes & Clark's Industrial Chemicals*, F. A. Lowenheim, M. K. Moran, Eds. (Wiley-Interscience, New York, 4th ed., 1975) pp 679-683. Toxicity study: H. F. Smyth *et al.*, *Am. Ind. Hyg. Assoc. J.* **30**, 470 (1969).

Dark purple or bronze-like, odorless crystals. Almost opaque by transmitted light and of a blue metallic luster by reflected light. Sweet with astringent aftertaste; stable in air. Dec ~240° with evolution of oxygen. d 2.7. Soluble in 14.2 parts cold, 3.5 parts boiling water. Dec by alc and many other organic solvents, also by concd acids with liberation of oxygen; with HCl, chlorine is liberated. Readily dec by many reducing substances, such as ferrous salts, iodides, oxalates, etc., especially in the presence of an acid. *Oxidizer. Take great care in handling as explosions may occur if it is brought into contact with organic or other readily oxidizable substances, either in soln or in the dry state. Incompat.* Alcohol, arsenites, bromides, iodides, hydrochloric acid, charcoal; organic substances generally; ferrous or mercurous salts, hypophosphites, hyposulfites, sulfites, peroxides, oxalates. LD_{50} orally in rats: 1.09 g/kg (Smyth).

Caution: Dilute solns are mildly irritating and high concns are caustic.

USE: Bleaching resins, waxes, fats, oils, straw, cotton, silk and other fibers and chamois skins; dyeing wood brown; printing fabrics; washing CO_2 in manuf mineral waters; exterminating *Oidium tuckeri;* photography; tanning leathers; purifying water; with formaldehyde soln to expel formaldehyde gas for disinfecting; as an important oxidizing reagent in analytical and synthetic organic chemistry.

THERAP CAT: Anti-infective (topical).

THERAP CAT (VET): Antiseptic (topical), astringent, deodorant.

7777. Potassium Persulfate. [7727-21-1] Peroxydisulfuric acid ([(HO)S(O)$_2$]$_2$O$_2$) potassium salt (1:2); potassium peroxydisulfate. $K_2O_8S_2$; mol wt 270.31. K 28.93%, O 47.35%, S 23.72%. $K_2S_2O_8$. The article of commerce contains 93-97% $K_2S_2O_8$.

Colorless or white, odorless crystals. *Oxidizer. Keep well closed in a cool place.* Gradually dec, losing available oxygen; dec more quickly at higher temps; completely dec ~100°. Sol in ~50 parts water, 25 parts water at 40°. Aq soln dec at ordinary temp and more rapidly on warming; aq soln is acid. Insol in alc.

USE: Bleaching fabrics, soaps; in photography under the name **Anthion** to remove last traces of thiosulfate from plates and paper; as an oxidizing agent in analytical chemistry.

7778. Potassium Phenoxide. [100-67-4] Phenol potassium salt (1:1); potassium phenate; potassium phenylate; potassium carbolate. C_6H_5KO; mol wt 132.20. C 54.51%, H 3.81%, K 29.58%, O 12.10%. C_6H_5OK. Prepd from phenol and KOH in dil methanol: Kornblum, Lurie, *J. Am. Chem. Soc.* **81**, 2710 (1959).

White to reddish, hygroscopic, cryst lumps. Very sol in water; sol in alcohol. The aq soln is strongly alkaline. *Keep tightly closed.*

7779. Potassium Phosphate, Dibasic. [7758-11-4] Phosphoric acid potassium salt (1:1); dipotassium phosphate; dikalium phosphate; DKP; dipotassium hydrogen phosphate. HK_2O_4P; mol wt 174.17. H 0.58%, K 44.90%, O 36.74%, P 17.78%. K_2HPO_4.

White, somewhat hygroscopic granules. Very sol in water, slightly in alcohol. 100 g will dissolve rapidly and completely in 67 g of cold water. Converted into pyrophosphate by ignition. The aq soln is slightly alkaline to phenolphthalein. *Keep well closed.*

USE: Buffering agent in antifreeze solns; nutrient in the culturing of antibiotics; ingredient of instant fertilizers; as sequestrant in the prepn of non-dairy powdered coffee creams.

THERAP CAT: Cathartic.

7780. Potassium Phosphate, Monobasic. [7778-77-0] Phosphoric acid potassium salt (1:1); potassium biphosphate; potassium acid phosphate; potassium dihydrogen phosphate; monopotassium phosphate; Sörensen's potassium phosphate. H_2KO_4P; mol wt 136.08. H 1.48%, K 28.73%, O 47.03%, P 22.76%. KH_2PO_4.

Colorless crystals or white, granular powder; permanent in air; at 400° loses H_2O, forming metaphosphate. d 2.34. pH 4.4-4.7. Sol in ~4.5 parts water. Practically insol in alc.

USE: In buffers for determination of pH. Pharmaceutic aid (buffering agent).

7781. Potassium Phosphate, Tribasic. [7778-53-2] Phosphoric acid potassium salt (1:3); tripotassium phosphate. K_3O_4P;

mol wt 212.26. K 55.26%, O 30.15%, P 14.59%. K_3PO_4. Purification: Jänecke, *Z. Phys. Chem.* **127**, 75 (1927); Simon, Schulze, *Z. Anorg. Allg. Chem.* **242**, 331 (1939).

Deliquescent, orthorhombic crystals. d_4^{17} 2.564. mp 1340°. Soly in water: 43.7% at 0°; 50.8% at 25°; 59.7% at 45.1°. Insol in alcohol. Aq solns are strongly alkaline.

Heptahydrate. [22763-02-6] $K_3O_4P.7H_2O$; mol wt 338.37.

Octahydrate. $K_3O_4P.8H_2O$; mol wt 356.38. Flat, rectangular platelets, mp 45.1°.

USE: Buffering agent.

7782. Potassium Phosphite. [13492-26-7] Phosphonic acid potassium salt (1:2); dipotassium hydrogen phosphite; dipotassium phosphite; dipotassium phosphonate. HK_2O_3P; mol wt 158.18. H 0.64%, K 49.44%, O 30.34%, P 19.58%. K_2HPO_3.

White, deliquesc powder. Slowly oxidizes in the air to phosphate; dec by heat. Very sol in water; insol in alcohol. *Keep well closed.*

7783. Potassium Pyrophosphate. [7320-34-5] Diphosphoric acid potassium salt (1:4); diphosphoric acid tetrapotassium salt; tetrapotassium diphosphate; tetrapotassium pyrophosphate. $K_4O_7P_2$; mol wt 330.33. K 47.34%, O 33.90%, P 18.75%. $K_4P_2O_7$. Manuf: *Faith, Keyes & Clark's Industrial Chemicals*, F. A. Lowenheim, M. K. Moran, Eds. (Wiley-Interscience, New York, 4th ed., 1975) pp 684-687.

Trihydrate. Colorless, deliquesc granules or cryst mass; freely sol in water; insol in alcohol. The aq soln is alkaline.

USE: In detergents and surfactants; in water treatment; in drilling muds as a clay thinner.

7784. Potassium Pyrosulfate. [7790-62-7] Disulfuric acid potassium salt (1:2); "anhydrous" potassium acid sulfate. $K_2O_7S_2$; mol wt 254.31. K 30.75%, O 44.04%, S 25.21%. $K_2S_2O_7$.

Colorless, fused pieces. d 2.28. mp ~325°. Sol in water. The aq soln is strongly acid.

7785. Potassium Salicylate. [578-36-9] 2-Hydroxybenzoic acid potassium salt (1:1); potassium 2-hydroxybenzoate. $C_7H_5KO_3$; mol wt 176.21. C 47.71%, H 2.86%, K 22.19%, O 27.24%. HOC$_6$-H$_4$COOK.

White, odorless powder. Becomes pink on exposure to light. Very sol in water, alcohol. One gram dissolves in 0.85 ml H_2O. A satd aq soln contains 55.82% w/w at 28.5°. *See:* Sidgwick, Ewbank, *J. Chem. Soc.* **121**, 1847, 1850 (1922). The aq soln is neutral or slightly acid to litmus. *Keep well closed and protected from light.*

7786. Potassium Selenate. [7790-59-2] Selenic acid potassium salt (1:2); dipotassium selenate. K_2O_4Se; mol wt 221.15. K 35.36%, O 28.94%, Se 35.70%. K_2SeO_4.

Colorless crystals or white powder. d 3.07. Sol in about one part water.

USE: Reagent.

7787. Potassium Selenide. [1312-74-9] Potassium selenide (K_2Se); dipotassium monoselenide; dipotassium selenide. K_2Se; mol wt 157.16. K 49.76%, Se 50.24%. Prepd by heating selenium with excess potassium: Fonzes-Diacon, *Contribution a l'Etudes des Séléniures Métalliques*, Montpellier (1901); by adding selenium to a soln of potassium in liq ammonia: Hugot, *Recherches sur l'Action du Sodammonium et du Potassammonium sur Quelque Métalloides*, Paris (1900); Feher in *Handbook of Preparative Inorganic Chemistry* vol. 1, G. Brauer, Ed. (Academic Press, New York, 2nd ed., 1963) p 421; by igniting potassium selenite or selenate in an atm of hydrogen: Berzelius, cited in *Mellor's* vol. 10, 767 (1930).

Crystalline mass. d 2.29. Reddens on exposure to air. Turns brownish-black when heated. Deliquescent. Forms potassium selenite with selenious acid. Forms potassium hydroselenide with hydrogen selenide. Sol in water; insol in ammonia.

Nonahydrate. Needle-like cryst. Freely sol in water.

7788. Potassium Silicate. [1312-76-1] Silicic acid potassium salt; soluble potash glass; soluble potash water glass. Variable composition: $K_2Si_2O_5$ to $K_2Si_3O_7$; may also contain water.

Colorless or yellowish, translucent to transparent, hygroscopic, glass-like pieces; strong alkaline reaction. Usually very slowly sol in cold water, or depending on the composition, almost insol. More readily sol in water when heated with it under pressure. Insol in alcohol; dec by acids with precipitation of silica. *Keep well closed.*

USE: As binder (*e.g.*, in carbon electrodes, lead pencils, protective coatings, insol pigments); detergent, in glass and ceramics manuf.

7789. Potassium Silver Cyanide. [506-61-6] Bis(cyano-κC)-argentate(1−) potassium (1:1); potassium bis(cyano-κC)argentate-(1−); potassium dicyanoargentate. C_2AgKN_2; mol wt 199.00. C 12.07%, Ag 54.21%, K 19.65%, N 14.08%. $KAg(CN)_2$. Review of toxicology and human exposure: *Toxicological Profile for Cyanide* (PB2007-100674, 2006) 341 pp.

White crystals; sensitive to light. *Poisonous.* Sol in water; acids ppt silver cyanide from the soln. Slightly sol in ethanol. *Protect from light.*

USE: In silver plating.

7790. Potassium Sodium Tartrate. [304-59-6] (2*R*,3*R*)-2,3-Dihydroxybutanedioic acid potassium sodium salt (1:1:1); Rochelle salt; Seignette salt. $C_4H_4KNaO_6$; mol wt 210.16. C 22.86%, H 1.92%, K 18.60%, Na 10.94%, O 45.68%. $KNaC_4H_4O_6$.

Tetrahydrate. [6381-59-5] $C_4H_4KNaO_6 \cdot 4H_2O$; mol wt 282.22. Translucent crystals or white, cryst powder; cooling saline taste. Slightly effloresces in warm air. d 1.79; mp 70-80°; at 100° loses $3H_2O$; becomes anhydr at 130-140°; at 220° begins to dec. Sol in 0.9 part water. Practically insol in alc. The aq soln is slightly alkaline to litmus. pH 7-8. *Incompat.* Acids, calcium or lead salts, magnesium sulfate, silver nitrate.

USE: Manuf of mirrors; as a constituent of Fehling's soln. For the control of radio frequencies, and wherever piezoelectric crystals are used. In aqueous extractions as a chelator of metal ions, especially aluminum.

THERAP CAT: Cathartic.

7791. Potassium Stannate. [12142-33-5] Stannate (SnO_3^{2-}) potassium (1:2); dipotassium trioxostannate; potassium tin(IV) oxide; potassium metastannate. K_2O_3Sn; mol wt 244.90. K 31.93%, O 19.60%, Sn 48.47%. K_2SnO_3.

Trihydrate. [12125-03-0] $K_2O_3Sn \cdot 3H_2O$; mol wt 298.95. Colorless crystals. *Irritant.* d 3.197. Sol in one part water. Insol in alcohol. The aq soln is alkaline.

USE: In textile dyeing and printing, in tin plating.

7792. Potassium Stearate. [593-29-3] Octadecanoic acid potassium salt (1:1); stearic acid potassium salt; potassium octadecanoate. $C_{18}H_{35}KO_2$; mol wt 322.57. C 67.02%, H 10.94%, K 12.12%, O 9.92%. Prepd from alcoholic solution of stearic acid with potassium hydroxide. Commercial grades prepd from plant or animal fats may also contain other fatty acid potassium salts, especially the palmitate. Prepn and phase transition behavior: A. S. C. Lawrence, *Trans. Faraday Soc.* **1938**, 660; T. R. Lomer, *Acta Crystallogr.* **5**, 11 (1952). Safety assessment for use in cosmetics: *Int. J. Toxicol.* **1**, 143-177 (1982).

White to pale yellow powder; usually with slight odor of fat. d^{25} 1.110; d^{75} 1.037. Slowly sol in cold, readily in hot water or alcohol. The aq soln is strongly alkaline to litmus or phenolphthalein, but the alcoholic soln is only slightly alkaline to phenolphthalein.

USE: Cleansing agent, emulsifier, or lubricant in cosmetics; antitack or release agent for elastomers; binder, emulsifier, or anticaking agent in foods.

7793. Potassium Sulfate. [7778-80-5] Sulfuric acid potassium salt (1:2); sal polychrestum; arcanum duplicatum; tartarus vitriolatus. K_2O_4S; mol wt 174.25. K 44.88%, O 36.73%, S 18.40%. K_2SO_4. Contains at least 99% K_2SO_4.

Colorless or white, odorless, hard, bitter crystals, or white granules or powder; permanent in air. n_D^{20} 1.49333-1.49733. One gram dissolves in 8.3 ml water, 4 ml boiling water, 75 ml glycerol. Insol in alcohol. Its soly in water is decreased by KCl or $(NH_4)_2SO_4$ and is practically insol in a saturated soln of the latter. The aq soln is neutral. pH about 7.

Caution: Swallowing large doses causes severe G.I. irritation.

USE: Technical grades are used in fertilizers for manuf of potassium alum, potassium carbonate and glass; the reagent grade is used in the Kjeldahl determination of nitrogen.

THERAP CAT: Cathartic.

7794. Potassium Sulfide. [1312-73-8] Potassium sulfide (K_2S); dipotassium monosulfide; potassium monosulfide. K_2S; mol wt 110.26. K 70.92%, S 29.08%. Best prepd from the elements in liq ammonia: Klemm *et al.*, *Z. Anorg. Allg. Chem.* **241**, 281 (1939); Feher in *Handbook of Preparative Inorganic Chemistry* vol. 1, G. Brauer, Ed. (Academic Press, New York, 2nd ed., 1963) pp 360-361.

White, cubic crystals or fused plates. Discolors in air. Very hygroscopic. Unstable. *Spontaneously combustible on percussion or rapid heating.* d 1.74. mp 912°. Freely sol in water. Aq solns are strongly alkaline.

Pentahydrate. Colorless rhombs. Odor of hydrogen sulfide. Discolors upon exposure to light and air (yellow to yellowish red). mp 60°. *Spontaneously combustible.* Freely sol in water, alc, glycerol. Insol in ether. Densities of aq solutions at 25° (calcd as % $K_2S.5H_2O$): 1.82%: 1.009; 10.90%: 1.049; 21.80%: 1.100; 29.07%: 1.136; 39.97%: 1.192; 50.88%: 1.250; 81.77%: 1.432. Aq solns are very caustic and prone to decompn.

7795. Potassium Sulfite. [10117-38-1] Sulfurous acid potassium salt (1:2); dipotassium sulfite. K_2O_3S; mol wt 158.25. K 49.41%, O 30.33%, S 20.26%. K_2SO_3. Reducing agent. Prepd from sulfur dioxide and potassium hydroxide: H. F. Johnstone, *Inorg. Synth.* **2**, 166 (1946). Manufacturing process: R. L. Zeller, III, D. L. Johnson, **US 5567406** (1996). Crystal structure: L. Andersen, D. Strömberg, *Acta Chem. Scand. A* **40**, 479 (1986). Measurement of oxidation kinetics: G. C. Mishra, R. D. Srivastava, *Chem. Eng. Sci.* **31**, 969 (1976). Safety evaluation for use in cosmetics: *Int. J. Toxicol.* **22**, Suppl. 2, 63 (2003).

White, odorless granular powder. *Irritant.* d 2.49. Gradually oxidizes in air to sulfate. Soluble in about 3.5 parts water, slightly sol in alcohol. Dec by dil acids with evolution of sulfur dioxide. Moisture sensitive. *Keep well closed in a cool place.*

USE: Antioxidant and preservative in food and cosmetics; in hairwaving preparations. In developing soln for photographic negatives.

7796. Potassium Tartrate. [921-53-9] (2*R*,3*R*)-2,3-Dihydroxybutanedioic acid potassium salt (1:2); tartaric acid potassium salt; dipotassium tartrate; soluble tartar. $C_4H_4K_2O_6$; mol wt 226.27. C 21.23%, H 1.78%, K 34.56%, O 42.42%.

Hemihydrate. White crystals or granular powder; loses its water at about 150°. d 1.98. Sol in about 0.5 part water. Almost insol in alc. The aq soln is slightly alkaline to litmus; pH 7-8.

THERAP CAT: Cathartic.

7797. Potassium Tellurate. [15571-91-2] Telluric acid (H_2TeO_4) potassium salt (1:2). K_2O_4Te; mol wt 269.79. K 28.98%, O 23.72%, Te 47.30%. K_2TeO_4. Prepn: E. B. Hutchins, Jr., *J. Am. Chem. Soc.* **27**, 1157 (1905).

Hydrate. [339091-77-9] White, cryst powder. *Irritant.*

7798. Potassium Tellurite. [7790-58-1] Telluric acid (H_2TeO_3) potassium salt (1:2); dipotassium tellurite. K_2O_3Te; mol wt 253.79. K 30.81%, O 18.91%, Te 50.28%. K_2TeO_3. Prepn: V. Lenher, E. Wolensensky, *J. Am. Chem. Soc.* **35**, 718 (1913). Crystal structure: L. Andersen *et al.*, *Acta Crystallogr. B* **45**, 344 (1989). Use in selective medium for staphylocci: A. G. Innes, *J. Appl. Bacteriol.* **23**, 108 (1960); A. C. Baird-Parker, *ibid.* **25**, 12 (1962); for enteropathogenic *E. coli*: R. Hiramatsu *et al.*, *J. Clin. Microbiol.* **40**, 922 (2002).

White, granular powder. d 3.492. *Poisonous; irritant.* Sol in water on an alkaline soln.

USE: In differential microbiological media to detect pathogenic microorganisms which reduce tellurite to insoluble, black, elemental tellurium.

7799. Potassium Tetraborate. [1332-77-0] Boron potassium oxide ($B_4K_2O_7$); boric acid ($H_2B_4O_7$) dipotassium salt; potassium biborate; potassium borate. $B_4K_2O_7$; mol wt 233.43. B 18.52%, K 33.50%, O 47.98%. $K_2B_4O_7$.

Pentahydrate. White, cryst powder. Sol in 4 parts water, slightly in alc.

7800. Potassium Tetrachloroaurate(III). [13682-61-6] (*SP*-4-1)-Tetrachloroaurate(1−) potassium (1:1); gold potassium chloride; potassium aurichloride; potassium gold(III) chloride; potassium gold tetrachloride. $AuCl_4K$; mol wt 377.86. Au 52.13%, Cl 37.53%, K 10.35%.

Dihydrate. [13005-39-5] Yellow, monoclinic crystals. *Irritant.* Sol in water. *Protect from light.*

USE: Photography, painting on porcelain and glass; prepn of several other gold compds.

7801. Potassium Tetrachloroplatinate(II). [10025-99-7] (*SP*-4-1)-Tetrachloroplatinate(2−) potassium (1:2); dipotassium tetrachloroplatinate; platinous potassium chloride; potassium platinochloride; potassium chloroplatinate. Cl_4K_2Pt; mol wt 415.08. Cl 34.16%, K 18.84%, Pt 47.00%. $K_2[PtCl_4]$.
Ruby-red crystals. *Poisonous; irritant.* Sol in water.
USE: In photography, in acid toning baths.

7802. Potassium Tetracyanonickelate(II). [14220-17-8] (*SP*-4-1)-Tetrakis(cyano-*κC*)nickelate(2−) potassium (1:2); dipotassium nickel tetracyanide; dipotassium tetracyanonickelate; nickel potassium cyanide. $C_4K_2N_4Ni$; mol wt 240.96. C 19.94%, K 32.45%, N 23.25%, Ni 24.36%. $K_2[Ni(CN)_4]$. Prepn: W. C. Fernelius *et al., Inorg. Synth.* **2**, 227 (1946).
Hydrate. [339527-86-5] Orange-yellow, cryst powder. *Poisonous.* Loses H_2O at about 100°. Sol in water.

7803. Potassium Tetracyanoplatinate(II). [562-76-5] (*SP*-4-1)-Tetrakis(cyano-*κC*)platinate(2−) potassium; dipotassium platinum tetracyanide; dipotassium tetracyanoplatinate. $C_4K_2N_4Pt$; mol wt 377.35. C 12.73%, K 20.72%, N 14.85%, Pt 51.70%. K_2-$[Pt(CN)_4]$. Crystallizes as a tri- or a dodecahydrate.
Trihydrate. [14323-36-5] $C_4K_2N_4Pt.3H_2O$; mol wt 431.40. Almost colorless, rhombic prisms; blue in direction of principal axis. *Poisonous.* Sol in hot water.

7804. Potassium Tetrafluoroborate. [14075-53-7] Tetrafluoroborate(1−) potassium (1:1); potassium borofluoride; potassium fluoborate; avogadrite. BF_4K; mol wt 125.90. B 8.59%, F 60.36%, K 31.06%. KBF_4. Prepared according to the eq H_3BO_3 + $4HF + KOH = KBF_4 + 4H_2O$: Vorländer *et al., Ber.* **65**, 535 (1932); Kwasnik in *Handbook of Preparative Inorganic Chemistry* **vol. 1**, G. Brauer, Ed. (Academic Press, New York, 2nd ed., 1963) p 223. For other methods of prepn *see* H. S. Booth, D. R. Martin, *Boron Trifluoride and Its Derivatives* (New York, 1949) pp 99-106.
Orthorhombic bipyramidal or cubic crystals. d_4^{20} 2.505. mp 530°. Soly in water (g/100 g): 0.3 (3°); 0.448 (20°); 0.55 (25°); 1.4 (40°); 6.27 (100°). Index of refraction of solns is lower than that of water. A satd soln (0.6%) on heating gives color effects, if excess crystals are added. With a 10% aq soln a transparent blue color appears at 100°, turning green at 90°, and yellow at 60°. In a concd fluoboric acid soln these phenomena occur with variations in room temp. Aq solns of KBF_4 are at first neutral to litmus, but upon standing, diluting, or heating become acidic without etching their glass containers. Slightly sol in boiling alc.
USE: Has been proposed as a flux for soldering and brazing; filler in resin-bonded grinding wheels.

7805. Potassium Tetraiodomercurate(II). [7783-33-7] (*T*-4)-Tetraiodomercurate(2−) potassium (1:2); mercuric potassium iodide; potassium mercuric iodide. HgI_4K_2; mol wt 786.40. Hg 25.51%, I 64.55%, K 9.94%. K_2HgI_4. Bactericidal properties: R. A. Lambert, *J. Exp. Med.* **24**, 683 (1916). Analysis of reaction with ammonia in Nessler's reagent: A. P. Vanselow, *Ind. Eng. Chem. Anal. Ed.* **12**, 516 (1940).
Sulfur-yellow crystals; deliquesc in moist air. *Poisonous.* Very sol in water; sol in alcohol, ether, acetone. *Keep well closed.*
USE: In Nessler's reagent for the detection of ammonia.

7806. Potassium Tetroxalate. [127-96-8] Ethanedioic acid potassium salt (2:1); oxalic acid hemipotassium salt; potassium quadroxalate; potassium trihydrogen dioxalate. $C_4H_3KO_8$; mol wt 218.16. C 22.02%, H 1.39%, K 17.92%, O 58.67%. $KHC_2O_4.H_2$-C_2O_4. Prepn and use in volumetric titrations: R. Ulbricht, E. Meissl, *Z. Anal. Chem.* **26**, 350 (1887); O. Kühling, *Z. Angew. Chem.* **16**, 1030 (1903). Identification as primary component in commercial *salt of sorrel* preparations: H. Thoms, T. Sabalitschka, *Ber. Dtsch. Pharm. Ges.* **28**, 137 (1918). *See also* potassium binoxalate. Crystal structure: D. J. Haas, *Acta Crystallogr.* **17**, 1511 (1964). Calculation of pH of standard solution: P. M. Juusola *et al., J. Chem. Eng. Data* **52**, 973 (2007).
Dihydrate. [6100-20-5] $C_4H_3KO_8.2H_2O$; mol wt 254.19. Colorless or white crystals. d 1.85. Sol in 60 parts cold, 12 parts boiling

water; slightly sol in alcohol. pH of 0.05 mol/kg solution: 1.68 (25°); 1.71 (50°).
USE: Removing rust and ink spots; in metal polishes; as standard reference buffer to measure pH.

7807. Potassium Thiocyanate. [333-20-0] Thiocyanic acid potassium salt (1:1); potassium sulfocyanate; potassium rhodanide; Rhocya. CKNS; mol wt 97.18. C 12.36%, K 40.23%, N 14.41%, S 32.99%. KSCN.
Colorless, deliquesc crystals. d 1.89; mp ∼173°, the fused salt turning brown, then green, blue, and white again on cooling. One gram dissolves in 0.5 ml acetone, 12 ml alcohol, 8 ml boiling alcohol. When dissolved in its own wt of water, the temp drops about 30°. The aq soln is neutral. *Poisonous. Keep well closed.* LD_{50} orally in mice, rats: 594, 854 mg/kg; *see:* Andersen, Chen, *J. Am. Pharm. Assoc.* **29**, 152 (1940).
Caution: Potential symptoms of overexposure are skin eruptions, psychosis, and collapse.
USE: Manuf artificial mustard oil; printing and dyeing textiles; in photography as intensifier; in analytical chemistry for analysis of the silver ion and indirect determn of chloride, bromide, and iodide.

7808. Potassium Thiosulfate. [10294-66-3] Thiosulfuric acid ($H_2S_2O_3$) potassium salt (1:2); potassium hyposulfite. $K_2O_3S_2$; mol wt 190.31. K 41.09%, O 25.22%, S 33.69%. $K_2S_2O_3$.
Colorless, hygroscopic crystals. Sol in water. Insol in alcohol. May crystallize with 0.33 to 1.5 mol H_2O. *Keep well closed.*

7809. Potassium Titanyl Oxalate. [14481-26-6] (*SP*-5-21)-Bis[ethanedioato(2−)-*κO^1,κO^2*]oxotitanate(2−) potassium (1:2); potassium oxalatotitanate(IV); potassium oxodioxalatotitanate(IV); titanium potassium oxalate; titanyl potassium oxalate. $C_4K_2O_9Ti$; mol wt 318.10. C 15.10%, K 24.58%, O 45.27%, Ti 15.05%. K_2-$TiO(C_2O_4)_2$.
Crystals or cryst powder. Very sol in water.
Dihydrate. [14402-67-6] $C_4K_2O_9Ti.2H_2O$; mol wt 354.13.
USE: As mordant in dyeing cotton and leather.

7810. Potassium Trithiocarbonate. [584-10-1] Carbonotrithioic acid dipotassium salt; potassium sulfocarbonate; potassium thiocarbonate. CK_2S_3; mol wt 186.39. C 6.44%, K 41.95%, S 51.60%. K_2CS_3. Prepn: E. W. Yeoman, *J. Chem. Soc. Trans.* **119**, 38 (1921). Monograph: K. N. Johri, *Chemical Analysis without H$_2$S using Potassium Trithiocarbonate* (Asia Pub., New York, 1963) 107 pp. Crystal structure of monohydrate: E. Philippot, O. Lindqvist, *Acta Crystallogr. B* **26**, 877 (1970). Use as complexing agent and precipitant for metal ions: K. N. Johri *et al., Talanta* **16**, 432 (1969); N. K. Kaushik, K. N. Johri, *ibid.* **18**, 1061 (1971).
Yellowish-red, deliquesc granules or crystals. Very sol in water. The aq soln is strongly alkaline. *Keep tightly closed.*
Monohydrate. [19086-12-5] $CK_2S_3.H_2O$; mol wt 204.40. Orange needles from diethyl ether. d^{20} 1.82. *Irritant.*
USE: In analytical chemistry.

7811. Potassium Tungstate. [7790-60-5] (*T*-4)-Tungstate ($WO_4^{2−}$) potassium (1:2); potassium tungstate(VI); potassium tungsten oxide; dipotassium tetraoxotungstate. K_2O_4W; mol wt 326.03. K 23.98%, O 19.63%, W 56.39%. K_2WO_4. Crystallizes also with $2H_2O$.
Heavy, deliquesc, cryst powder. *Irritant.* d 3.12. mp 921°. Sol in about 2 parts cold, about 0.7 part boiling water. Insol in alcohol. *Keep well closed.*

7812. Potassium Uranate(VI). [7790-63-8] Potassium uranium oxide ($K_2U_2O_7$); potassium diuranate; uranium oxide orange. $K_2O_7U_2$; mol wt 666.25. K 11.74%, O 16.81%, U 71.45%. K_2U_2-O_7.
Orange powder. Insol in water; sol in acids.
USE: Painting on porcelain.

7813. Potassium Xanthogenate. [140-89-6] Carbonodithioic acid *O*-ethyl ester potassium salt (1:1); ethylxanthic acid potassium salt; potassium ethyldithiocarbonate; potassium ethylxanthogenate; potassium ethylxanthate. $C_3H_5KOS_2$; mol wt 160.29. C 22.48%, H 3.14%, K 24.39%, O 9.98%, S 40.00%. $C_2H_5OCS_2K$. Made by treating an alcoholic soln of CS_2 with alcoholic KOH. Usually contains 8-10% H_2O.

White to pale yellow crystals or cryst powder. Very sol in water; sol in alcohol. The aq soln is strongly alkaline. *Keep well closed and protected from light.*

USE: As reagent in analytical chemistry.

7814. Povidone. [9003-39-8] 1-Ethenyl-2-pyrrolidinone homopolymer; 1-vinyl-2-pyrrolidinone polymers; poly[1-(2-oxo-1-pyrrolidinyl)ethylene]; polyvinylpyrrolidone; polyvidone; PVP; Kollidon; Luviskol; Periston; Plasdone; Protagent. Homopolymer of *N*-vinyl-2-pyrrolidone, produced commercially as a series of products having mean mol wts ranging from 2,500 to 1,000,000. Prepd by free radical polymerization of the monomer. See J. W. Reppe, *Acetylene Chemistry* (PB Report 18852-s, U.S. Dept. Commerce, 1949) pp 68-72. Review of clinical use and early literature: W. Wessel *et al., Arzneim.-Forsch.* **21**, 1468-1482 (1971); of synthesis and physical properties: H. Warson, *Polym. Paint Colour J.* **161**, 637-644 (1972); F. Haaf *et al., Polym. J.* **17**, 143-152 (1985); of use in cosmetics: F. G. M. Vogel, *Soap Cosmet. Chem. Spec.* **65**, 42-47, 128 (1989). Book: *PVP: A Critical Review of the Kinetics and Toxicology of Polyvinylpyrrolidone (Povidone)*, B. V. Robinson *et al.*, Eds. (Lewis Publishers, Chelsea, MI, 1990) 209 pp. Comprehensive description: C. M. Adeyeye, E. Barabas, *Anal. Profiles Drug Subs. Excip.* **22**, 555-685 (1993). Size exclusion chromatography: C. Wu *et al., Chromatogr. Sci. Ser.* **69**, 311 (1995).

White, hygroscopic powder. Freely sol in water, methanol, alc; sol in chloroform, formic acid, acetic acid, *N*-methylpyrrolidone, methylcyclohexanone, dichloromethane, ethylenediamine, glycerol, diethyleneglycol, PEG 400; slightly sol in acetone. Practically insol in ether; insol in xylene, toluene, diethylether, ethylacetate, cyclohexanone, chlorobenzene, dioxane, carbon tetrachloride, mineral oil.

Monomer. [88-12-0] 1-Ethenyl-2-pyrrolidinone; *N*-vinyl-2-pyrrolidinone; NVP. C_6H_9NO; mol wt 111.14. Clear to light straw colored liquid. bp_{14} 96°, bp_{400} 193°. Freezing pt 13.5°. d_4^{24} 1.04. n_D^{25} 1.511. Flash pt (open cup) 100.5°C (213°F). Viscosity (25°): 2.07 cP. Sol in water and many organic solvents.

Crospovidone. Polyvinylpolypyrrolidone; PVPP; Divergan; Kollidon CL; Polyclar; Polyplasdone XL. Crosslinked insoluble homopolymer of NVP. Review of properties and applications: A. H. Bronnsack in *Proc. Intl. Symp. Povidone*, G. A. Digenis, J. Ansell, Eds. (Univ. Kentucky, Coll. Pharmacy, Lexington, 1983) pp 471-490. Comprehensive description: E. S. Barabas, C. M. Adeyeye, *Anal. Profiles Drug Subs. Excip.* **24**, 87-163 (1996). Free flowing, white, almost tasteless powder. Hygroscopic; swells on contact with water. Insol in water, strong mineral acids, caustic solns, and common organic solvents.

Complex with iodine see Povidone-Iodine.

USE: Povidone as pharmaceutic aid (dispersing, suspending and viscosity-increasing agent; tablet coating and binder). Thickener, dispersant, lubricant, film-forming agent and binder in cosmetics. Stabilizer, diluent, and dye dispersant in food. Dye dispersant in paper and textiles. Adhesive; paper coating. Coating and processing aid in photographic products. Manuf of plastics and rubber. Cryoprotectant for biological samples. Crospovidone as pharmaceutic aid (tablet binder and disintegrant); clarifying and stabilizing agent in beverages. Monomer as dispersant and wetting agent in pigments.

THERAP CAT: Povidone formerly as a synthetic blood plasma expander. Crospovidone as antidiarrheal.

7815. Povidone-Iodine. [25655-41-8] 1-Ethenyl-2-pyrrolidinone homopolymer compd with iodine; 1-vinyl-2-pyrrolidinone polymers, iodine complex; iodine-polyvinylpyrrolidone complex; polyvinylpyrrolidone-iodine complex; PVP-I; Betadine; Betaisodona; Braunol; Braunosan H; Disadine D.P.; Efodine; Inadine; Isodine; Proviodine; Traumasept. An iodophor, *q.v.*, prepd by Beller, Hosmer, **US 2706701**; Hosmer, **US 2826532**; Siggia, **US 2900305** (1955, 1958, and 1959, all to GAF). Prepn, history and use: Shelanski, Shelanski, *J. Int. Coll. Surg.* **25**, 727 (1956).

Yellowish-brown, amorphous powder with slight characteristic odor. Aq solns have a pH near 2 and may be made more neutral (but less stable) by the addition of sodium bicarbonate. Sol in alc, water. Practically insol in chloroform, carbon tetrachloride, ether, solvent hexane, acetone. Solns do not give the familiar starch test when freshly prepared.

THERAP CAT: Anti-infective (topical).

THERAP CAT (VET): Anti-infective (topical).

7816. PPACK. [71142-71-7] D-Phenylalanyl-*N*-[(1*S*)-4-[(aminoiminomethyl)amino]-1-(2-chloroacetyl)butyl]-L-prolinamide; D-phenylalanyl-prolyl-arginine chloromethyl ketone; D-Phe-Pro-Arg-chloromethyl ketone; FPRMeCl. $C_{21}H_{31}ClN_6O_3$; mol wt 450.97. C 55.93%, H 6.93%, Cl 7.86%, N 18.64%, O 10.64%. Selective irreversible inhibitor of thrombin: C. A. Kettner, E. N. Shaw, *Thromb. Res.* **14**, 969 (1979). Synthesis: E. N. Shaw, C. A. Kettner, **US 4318904** (1982 to Research Corp.). Anticoagulant and antithrombotic activity: J. Hauptman, F. Markwardt, *Thromb. Res.* **20**, 347 (1980). *In vivo* studies: D. Collen *et al., J. Lab. Clin. Med.* **99**, 76 (1982).

Dihydrochloride. [82188-90-7] $C_{21}H_{31}ClN_6O_3 \cdot 2HCl$; mol wt 523.88.

Trifluoroacetate salt. [157379-44-7] $C_{21}H_{31}ClN_6O_3 \cdot C_2HF_3O_2$; mol wt 564.99.

USE: As a research tool; as a clinical reagent to measure the level of thrombin in blood.

7817. P-Phos Ligands. Family of chiral dipyridylphosphine compds; metal complexes of the atropisomers serve as air stable catalysts for a variety of asymmetric transformations. Ligands are named for the Ar-substituent. Prepn: A. S.-C. Chan, C.-C. Pai, **US 5886182** (1999 to Hong Kong Polytechnic University); of use of Ru complexes in asymmetric hydrogenations: C.-C. Pai *et al., J. Am. Chem. Soc.* **122**, 11513 (2000); J. Wu *et al., Synlett* **2001**, 1050; *eidem, Tetrahedron Lett.* **43**, 1539 (2002). Additional asymmetric hydrogenations: *eidem, J. Org. Chem.* **67**, 7908 (2002); *eidem, ibid.* **68**, 2490 (2003); *eidem, Tetrahedron: Asymmetry* **14**, 987 (2003). Asymmetric catalysis applications: Q. Shi *et al., Tetrahedron Lett.* **44**, 6505 (2003); L. Wang *et al., J. Mol. Catal. A: Chem.* **196**, 171 (2003); J. Wu *et al., Proc. Natl. Acad. Sci. USA* **102**, 3570 (2005). *Review*: J. Wu, A. S. C. Chan, *Acc. Chem. Res.* **39**, 711-720 (2006).

(*R*)-P-Phos. [221012-82-4] (3*R*)-4,4'-Bis(diphenylphosphino)-2,2',6,6'-tetramethoxy-3,3'-bipyridine; (*R*)-(+)-2,2',6,6'-tetramethoxy-4,4'-bis(diphenylphosphino)-3,3'-bipyridine. $C_{38}H_{34}N_2O_4P_2$; mol wt 644.65. C 70.80%, H 5.32%, N 4.35%, O 9.93%, P 9.61%. White solid, mp 261-265°. $[\alpha]^D + 36.4°$ (c = 1.1 in dichloromethane).

(S)-P-Phos. [362524-23-0] Light yellow powder, mp 261-265°. *Irritant.*

(R)-Tol-P-Phos. [358622-76-1] (3R)-4,4'-Bis[bis(4-methylphenyl)phosphino]-2,2',6,6'-tetramethoxy-3,3'-bipyridine; (R)-2,2',6,6'-tetramethoxy-4,4'-bis[di(p-tolyl)phosphino]-3,3'-bipyridine. $C_{42}H_{42}N_2O_4P_2$; mol wt 700.76. C 71.99%, H 6.04%, N 4.00%, O 9.13%, P 8.84%.

(S)-Tol-P-Phos. [358622-79-4]

(R)-Xyl-P-Phos. [442905-33-1] (3R)-4,4'-Bis[bis(3,5-dimethylphenyl)phosphino]-2,2',6,6'-tetramethoxy-3,3'-bipyridine; (R)-(+)-2,2',6,6'-tetramethoxy-4,4'-bis[di(3,5-xylyl)phosphino]-3,3'-bipyridine. $C_{46}H_{50}N_2O_4P_2$; mol wt 756.86. C 73.00%, H 6.66%, N 3.70%, O 8.46%, P 8.18%. mp 190-194°. *Irritant.*

(S)-Xyl-P-Phos. [443347-10-2] mp 158-162°. *Irritant.*

USE: Ligands in asymmetric synthesis.

7818. Practolol. [6673-35-4] N-[4-[2-Hydroxy-3-[(1-methylethyl)amino]propoxy]phenyl]acetamide; 4'-[2-hydroxy-3-(isopropylamino)propoxy]acetanilide; 1-(4-acetamidophenoxy)-3-isopropylamino-2-propanol; AY-21011; ICI-50172; Dalzic; Eraldin. $C_{14}H_{22}N_2O_3$; mol wt 266.34. C 63.14%, H 8.33%, N 10.52%, O 18.02%. β-Adrenergic blocker. Prepn: Howe, Smith, **NL 6512676**; *eidem,* **US 3408387** (1966, 1968, to I.C.I.). Synthesis of R(+)-form: Danilewicz, Kemp, *J. Med. Chem.* **16**, 168 (1973). Pharmacological studies: Dunlop, Shanks, *Br. J. Pharmacol. Chemother.* **32**, 201 (1968); Barrett, *Postgrad. Med. J. Suppl.* **47**, 7 (1971). Metabolic studies: Scales, Cosgrove, *J. Pharmacol. Exp. Ther.* **175**, 338 (1970).

mp 134-136° (BuOAc). Sol in warm isopropanol.
Hydrochloride monohydrate. mp 140-142°.
R(+)-Form. Crystals from dioxane, mp 130-131.5°. $[\alpha]^{25}_{365}$ +4.3°, $[\alpha]^{25}_{578}$ +3.5° (ethanol).
R(+)-Hydrochloride. $[\alpha]^{25}_{436}$+26.0°, $[\alpha]^{25}_{578}$ +14.0°.
THERAP CAT: Antiarrhythmic.

7819. Pradofloxacin. [195532-12-8] 8-Cyano-1-cyclopropyl-6-fluoro-1,4-dihydro-7-[(4aS,7aS)-octahydro-6H-pyrrolo[3,4-b]pyridin-6-yl]-4-oxo-3-quinolinecarboxylic acid; 8-cyano-1-cyclopropyl-7-((1S,6S)-2,8-diazabicyclo[4.3.0]nonan-8-yl)-6-fluoro-1,4-dihydro-4-oxo-3-quinolinecarboxylic acid; Veraflox. $C_{21}H_{21}FN_4O_3$; mol wt 396.42. C 63.63%, H 5.34%, F 4.79%, N 14.13%, O 12.11%. Cyanofluorinated quinoline antibiotic. Prepn: S. Bartel *et al.,* **WO 9731001**; *eidem,* **US 6323213** (1997, 2001 both to Bayer). Comparative activity vs anaerobic bacteria: P. Silley *et al., J. Antimicrob. Chemother.* **60**, 999 (2007). Pharmacokinetics in cats: A. Hartmann *et al., J. Vet. Pharmacol. Ther.* **31**, 87 (2008). Clinical efficacy in urinary tract infection in cats: A. Litster *et al., J. Vet. Intern. Med.* **21**, 990 (2007); in canine pyoderma: R. S. Mueller, B. Stephan, *Vet. Dermatol.* **18**, 144 (2007).

mp 246-248° (dec).
THERAP CAT (VET): Antibacterial.

7820. Prajmaline. [35080-11-6] (17R,21α)-17,21-Dihydroxy-4-propylajmalanium; N^4-propylajmalinium; prajmalium; N-propylajmaline. $[C_{23}H_{33}N_2O_2]^+$. Prepn: J. Keck, Z. *Naturforsch.* **18b**, 177 (1963); A. Bonati, E. Bombardelli, *Farmaco Ed. Sci.* **18**, 851 (1963); J. Keck, **DE 1154120** (1963 to Thomae); *idem,* **US 3414577** (1968 to Boehringer, Ing.). Pharmacology: R. Koch, *Arzneim.-Forsch.* **22**, 2079 (1972); H. M. Mertens *et al., ibid.* **23**, 642 (1973); K.-W. Diederich, D. Boyksen, *ibid.* 1302. Toxicology: G. Von Philipsborn, B. Stalder, *ibid.* **22**, 2085 (1972). Pharmacokinetics: A. T. Trompler *et al., ibid.* **33**, 436 (1983). Mechanism of action: H. Langenfeld *et al., J. Cardiovasc. Pharmacol.* **15**, 338 (1990). GC/MS determn in urine: H. H. Maurer, *Arch. Toxicol.* **64**, 218 (1990). Clinical comparison with tocainide, *q.v.*: B. Schwartzkopff *et al., Arzneim.-Forsch.* **33**, 153 (1983).

Hydrogen tartrate. [2589-47-1] Prajmaline bitartrate; NPAB; GT-1012; Neo-Gilurytmal. $C_{23}H_{33}N_2O_2 \cdot C_4H_5O_6$; mol wt 518.61. White crystals from ethanol + ether, mp 149-152°. LD_{50} in mice (mg/kg): 43 orally; 1.7 i.v. (Von Philipsborn, Stalder).
THERAP CAT: Antiarrhythmic.

7821. Pralatrexate. [146464-95-1] N-[4-[1-[(2,4-Diamino-6-pteridinyl)methyl]-3-butyn-1-yl]benzoyl]-L-glutamic acid; (2S)-2-[[4-[(1RS)-1-[(2,4-diaminopteridin-6-yl)methyl]but-3-ynyl]benzoyl]amino]pentanedioic acid; 10-propargyl-10-deazaaminopterin; PDX; Folotyn. $C_{23}H_{23}N_7O_5$; mol wt 477.48. C 57.86%, H 4.86%, N 20.53%, O 16.75%. Analog of folic acid, *q.v.*, that inhibits dihydrofolate reductase (DHFR) and polyglutamylation by folylpolyglutamyl synthetase. Synthesis and pharmacology: J. I DeGraw *et al., J. Med. Chem.* **36**, 2228 (1993). Prepd (not claimed): J. I. DeGraw *et al.,* **US 5354751** (1994 to SRI). Improved prepn: F. M. Sirotnak *et al.,* **US 6028071** (2000 to Sloan-Kettering Inst. Cancer Res.). Mechanism of action: E. Izbicka *et al., Cancer Chemother. Pharmacol.* **64**, 993 (2009). Clinical pharmacokinetics and pharmacodynamics: D. R. Mould *et al., Clin. Pharmacol. Ther.* **86**, 190 (2009). Clinical evaluation in malignant pleural mesothelioma: L. M. Krug *et al., J. Thorac. Oncol.* **2**, 317 (2007); in lymphoma: O. A. O'Connor *et al., J. Clin. Oncol.* **27**, 4357 (2009). Review of development and clinical experience: J. Zain, O. O'Connor, *Expert Opin. Pharmacother.* **11**, 1705-1714 (2010); of clinical evaluation in peripheral T-cell lymphoma: S. M. Malik *et al., Clin. Cancer Res.* **16**, 4921-4927 (2010)..

Off-white to yellow solid. pKa 3.25, 4.76, 6.17. uv max (0.1 N NaOH): 256, 372 nm (ε 29800, 7000). Sol in aq solns (pH ≥ 6.5). Practically insol in chloroform, ethanol.
THERAP CAT: Antineoplastic.

7822. Pralidoxime Chloride. [51-15-0] 2-[(Hydroxyimino)methyl]-1-methylpyridinium chloride (1:1); 2-formyl-1-methylpyridinium chloride oxime; 1-methyl-2-formylpyridinium chloride oxime; N-methylpyridinium-2-aldoxime chloride; 2-pyridine aldoxime methyl chloride; 2-PAM chloride; Protopam. $C_7H_9ClN_2O$; mol wt 172.61. C 48.71%, H 5.26%, Cl 20.54%, N 16.23%, O 9.27%. Cholinesterase reactivator. Prepn of salts: I. B. Wilson *et al.,* **US 2816113** (1957 to U.S. Sec'y. of Army); A. A. Kondritzer *et al., J.*

Pharm. Sci. **50**, 109 (1961). Manufacturing processes: R. I. Ellin *et al.*, **US 3140289** (1964 to U.S. Dept. of the Army); McDowell, **US 3155674** (1964 to Olin Mathieson). Commercial prepn: R. I. Ellin, *Ind. Eng. Chem. Prod. Res. Dev.* **3** (1), 20 (1964). HPLC determn in serum: P. Houzé *et al.*, *J. Chromatogr. B* **814**, 149 (2005). Pharmacokinetics: D. Jovanovic, *Arch. Toxicol.* **63**, 416 (1989). Toxicology: Christensen, Richter, *Arch. Environ. Health* **15**, 599 (1967). Toxicity data: Fleisher *et al.*, *Toxicol. Appl. Pharmacol.* **16**, 40 (1970); R. I. Ellin, J. H. Wills, *J. Pharm. Sci.* **53**, 1143 (1964). Comprehensive description: U. V. Banakar, U. N. Patel, *Anal. Profiles Drug Subs.* **17**, 533-569 (1988). Clinical trial in organophosphorus pesticide poisoning: K. S. Pawar *et al.*, *Lancet* **368**, 2136 (2006).

Crystals from alcohol + ether, mp 235-238° (dec). Soly (g/100 ml) 25°: acetone 0, isopropanol 0.09, ethanol 0.89, methanol 8.5, water 65.5. LD_{50} in rats (mg/kg): 96 i.v. (Fleisher). LD_{50} in rabbits (mg/kg): 95 i.v.; LD_{50} in mice (mg/kg): 115 i.v., 205 i.p., 4100 orally (Ellin, Wills).

Pralidoxime iodide. [94-63-3] 2-Pyridine aldoxime methiodide; 2-PAM. $C_7H_9IN_2O$; mol wt 264.07. Yellow crystals from alcohol, mp 225-226°. Soly at 25°: 48 mg/ml. Very sol in water, fairly sol in hot alcohol, poorly sol in cold alcohol. Insol in ether, acetone. LD_{50} in mice (mg/kg): 140-178 i.v., 136-260 i.p., 290-340 s.c., 1500-4000 orally (Ellin, Wills).

Pralidoxime mesylate. [154-97-2] Contrathion. $C_8H_{12}N_2O_4S$; mol wt 232.25. Very hygroscopic crystals from ethanol, mp 155°. pKa 8.0. Soly in water: 1 g in 2 ml. LD_{50} in mice (mg/kg): 118-122 i.v., 216 i.p., 3700 orally; in rats (mg/kg): 109 i.v., 262 i.p. (Ellis, Wills).

THERAP CAT: Antidote (nerve gases and organophosphate insecticide poisoning).

THERAP CAT (VET): Antidote (organophosphate poisoning).

7823. Prallethrin. [23031-36-9] 2,2-Dimethyl-3-(2-methyl-1-propenyl)cyclopropanecarboxylic acid 2-methyl-4-oxo-3-(2-propynyl)-2-cyclopenten-1-yl ester. $C_{19}H_{24}O_3$; mol wt 300.40. C 75.97%, H 8.05%, O 15.98%. Synthetic pyrethroid; propynyl analog of allethrin. Of the 8 possible stereoisomers, the (4*S*,1*R*)-*trans* isomer has the most insecticidal activity. Commercially formulated as a mixture of (4*S*,1*R*)-*trans* and (4*S*,1*R*)-*cis* isomers in a ratio of approximately 4:1. Initial report: W. A. Gersdorff, P. G. Piquett, *J. Econ. Entomol.* **54**, 1250 (1961). Synthesis and resolution of labelled commercial isomers: H. Kanamaru *et al.*, *J. Labelled Compd. Radiopharm.* **23**, 995 (1986). Comparative insecticidal activity of all 8 isomers: T. Matsunaga *et al.*, *Jpn. J. Sanit. Zool.* **38**, 219 (1987). Structure-activity: N. Matsuo, *Pestic. Sci.* **52**, 21-28 (1998). Description of properties and activity of commercial formulation: Y. Abe *et al.*, *Brighton Crop Prot. Conf. - Pests Dis.* **1994**, 1023. Efficacy against mosquitoes: R. L. Groves *et al.*, *J. Am. Mosq. Control Assoc.* **13**, 184 (1997).

(4*S*,1*R*)-*trans* isomer

Commercial formulation. (*S*)-2-Methyl-4-oxo-3-(2-propynyl)-cyclopent-2-enyl (1*R*)-*cis*,*trans*-chrysanthemate; S-4068; S-4068SF; Etoc. Yellow to yellow brown liquid. d_4^{20} 1.03. Log P (octanol/water): 4.49 (20°, pH 5.6-5.9). Vapor pressure (20°): 3.5×10^{-5};

(30°) 1.0×10^{-4} mm Hg. Sol in most organic solvents. Soly in water (25°): 8.51 ppm.

USE: Insecticide.

7824. Pralmorelin. [158861-67-7] D-Alanyl-3-(2-naphthalenyl)-D-alanyl-L-alanyl-L-tryptophyl-D-phenylalanyl-L-lysinamide; D-Ala-D-(β-naphthyl)-Ala-Ala-Trp-D-Phe-Lys-NH₂; growth hormone-releasing peptide-2; GHRP-2. $C_{45}H_{55}N_9O_6$; mol wt 817.99. C 66.08%, H 6.78%, N 15.41%, O 11.74%. Synthetic hexapeptide growth hormone secretagogue. Prepn: C. Y. Bowers, D. Coy, **WO 9304081**; *eidem*, **US 5663146** (1993, 1997 both to Admin. Tulane Ed. Fund). LC-MS/MS determn in urine for doping detection: M. Okano *et al.*, *Rapid Commun. Mass Spectrom.* **24**, 2046 (2010). Growth hormone releasing activity: N. Doi *et al.*, *Arzneim.-Forsch.* **54**, 857 (2004). General pharmacology: S. Furata *et al.*, *ibid.*, 868. Clinical trial as diagnostic aid for adult growth hormone deficiency: K. Chilhara *et al. Eur. J. Endocrinol.* **157**, 19 (2007).

Dihydrochloride. [158827-34-0] KP-102; WAY-GPA-748. $C_{45}H_{55}N_9O_6$·2HCl; mol wt 890.91.

THERAP CAT: Diagnostic aid (growth hormone deficiency).

7825. Pramipexole. [104632-26-0] (6*S*)-4,5,6,7-Tetrahydro-N^6-propyl-2,6-benzothiazolediamine; (*S*)-2-amino-4,5,6,7-tetrahydro-6-(propylamino)benzothiazole. $C_{10}H_{17}N_3S$; mol wt 211.33. C 56.84%, H 8.11%, N 19.88%, S 15.17%. Dopamine agonist active at D_3 and D_2 receptor subtypes. Prepn of racemate: G. Griss *et al.*, **EP 186087**; *eidem*, **US 4886812** (1986, 1989 both to Thomae); of enantiomers: C. S. Schneider, J. Mierau, *J. Med. Chem.* **30**, 494 (1987). Pharmacology: J. Mierau, G. Schingnitz, *Eur. J. Pharmacol.* **215**, 161 (1992). Clinical safety and pharmacodynamics: J. C. Schilling *et al.*, *Clin. Pharmacol. Ther.* **51**, 541 (1992). Clinical trial in Parkinson's disease: K. Kieburtz *et al.*, *J. Am. Med. Assoc.* **278**, 125 (1997). Review of clinical trials in restless legs syndrome: C. A. Kushida, *Expert Opin. Pharmacother.* **7**, 441-451 (2006).

Dihydrochloride monohydrate. [191217-81-9]; [104632-25-9] (dihydrochloride). SND-919; Mirapex; Mirapexin; Sifrol. $C_{10}H_{17}N_3S$·2HCl·H₂O; mol wt 302.26. White to off-white powder, mp 296-301° (dec). $[\alpha]_D^{20}$ −67.2° (c = 1 in methanol). Soly: water >20%; methanol ~8%; ethanol ~0.5%. Practically insol in dichloromethane.

THERAP CAT: Antiparkinsonian. In treatment of restless legs syndrome.

7826. Pramiracetam. [68497-62-1] *N*-[2-[Bis(1-methylethyl)amino]ethyl]-2-oxo-1-pyrrolidineacetamide; *N*-[2-(diisopropylamino)ethyl]-2-oxo-1-pyrrolidineacetamide; amacetam. $C_{14}H_{27}N_3O_2$; mol wt 269.39. C 62.42%, H 10.10%, N 15.60%, O 11.88%. Cognition enhancer structurally related to piracetam, *q.v.* Prepn: Y. J. L'Italien, I. C. Nordin, **DE 2808067**; *eidem*, **US 4145347** (1978, 1979 both to Parke, Davis); D. E. Butler *et al.*, *J. Med. Chem.* **27**, 684 (1984). Pharmacology: B. P. H. Poschel *et al.*, *Drugs Exp. Clin. Res.* **9**, 853 (1983). GC determn in plasma: T. Chang, R. M. Young, *J. Chromatogr.* **274**, 346 (1983). Clinical pharmacokinetics: T. Chang *et al.*, *J. Clin. Pharmacol.* **25**, 291 (1985). Clinical evaluation in Alzheimer's disease: R. J. Branconnier *et al.*, *Psychopharmacol. Bull.* **19**, 726 (1983); in various cognitive disorders: R. Dejong, *Curr. Ther. Res.* **41**, 254 (1987).

bp$_{0.15}$ 162-164°. Monohydrate, mp 47-48°.

Sulfate. [72869-16-0] CI-879. $C_{14}H_{27}N_3O_2.H_2SO_4$; mol wt 367.46. LD$_{50}$ in male, female mice (mg/kg): 5434, 4355 orally (Poschel).

THERAP CAT: Nootropic.

7827. Pramiverin. [14334-40-8] N-(1-Methylethyl)-4,4-diphenylcyclohexanamine; N-isopropyl-4,4-diphenylcyclohexylamine; pramiverine; primaverine; propaminodiphen. $C_{21}H_{27}N$; mol wt 293.45. C 85.95%, H 9.27%, N 4.77%. Prepn: **NL 6515046**; R. Unger *et al.*, **US 3376312** (1966, 1968 both to E. Merck). Pharmacology: H. J. Enenkel *et al.*, *Arzneim.-Forsch.* **26**, 690 (1976). Toxicological study: M. Von Eberstein *et al.*, *ibid.* 703. Series of articles on pharmacology, pharmacokinetics, GC determn, clinical studies: *ibid.* 686-752. Use in management of labor: G. LoDico *et al.*, *Minerva Ginecol.* **31**, 683 (1979).

Liq, bp$_{0.05}$ 164-165°; also reported as a solid, mp 70° (Enenkel).

Hydrochloride. [14334-41-9] EMD-9806; HSP-2986; Monoverin; Sistalgin. $C_{21}H_{27}N.HCl$; mol wt 329.91. Off-white crystalline powder, mp 230°; also reported as 234-237° (Enenkel). Soly at 25° (g/100 ml): in water 0.3; in ethanol 4; in chloroform 5. Practically insol in ether. LD$_{50}$ (after 14 days) in mice, rats (mg/kg): 346, 623 orally; 25, 26 i.v. (Von Eberstein).

THERAP CAT: Antispasmodic.

7828. Pramlintide. [151126-32-8] L-Lysyl-L-cysteinyl-L-asparaginyl-L-threonyl-L-alanyl-L-threonyl-L-cysteinyl-L-alanyl-L-threonyl-L-glutaminyl-L-arginyl-L-leucyl-L-alanyl-L-asparaginyl-L-phenylalanyl-L-leucyl-L-valyl-L-histidyl-L-seryl-L-seryl-L-asparaginyl-L-asparaginyl-L-phenylalanylglycyl-L-prolyl-L-isoleucyl-L-leucyl-L-prolyl-L-prolyl-L-threonyl-L-asparaginyl-L-valylglycyl-L-seryl-L-asparaginyl-L-threonyl-L-tyrosinamide, cyclic-(2 → 7)-disulfide; 25-L-proline-28-L-proline-29-L-proline-amylin (human); tripro-amylin. $C_{171}H_{267}N_{51}O_{53}S_2$; mol wt 3949.44. C 52.00%, H 6.81%, N 18.09%, O 21.47%, S 1.62%. Analog of human amylin, *q.v.* Prepn: L. S. L. Gaeta *et al.*, **WO 9310146**; *eidem*, **US 5686411** (1993, 1997 both to Amylin Pharm.). Pharmacology: A. A. Young *et al.*, *Drug Dev. Res.* **37**, 231 (1996). Clinical pharmacokinetics, pharmacodynamics: W. A. Colburn *et al.*, *J. Clin. Pharmacol.* **36**, 13 (1996). Clinical evaluation in insulin-dependent diabetes: O. G. Kolterman *et al.*, *Diabetologia* **39**, 492 (1996). Review of clinical effect on glycemic control and weight: D. Singh-Franco *et al.*, *Diabetes Obes. Metab.* **13**, 169-180 (2011).

Lys—Cys—Asn—Thr-Ala-Thr-Cys-Ala-Thr-Gln-Arg-Leu-Ala-Asn-Phe-Leu-Val-His

Ser

Ser

H$_2$N-Tyr—Thr-Asn—Ser-Gly-Val-Asn-Thr-Pro-Pro-Leu-Ile-Pro-Gly-Phe-Asn-Asn

Acetate. [187887-46-3] AC-137; Symlin. $C_{171}H_{267}N_{51}O_{53}S_2.xC_2H_4O_2$, ($x$ = 3-8). White powder. Sol in water.

THERAP CAT: In treatment of insulin-dependent diabetes.

7829. Pramoxine. [140-65-8] 4-[3-(4-Butoxyphenoxy)propyl]morpholine; p-butoxyphenyl γ-morpholinopropyl ether; γ-mor-

pholinopropyl 4-n-butoxyphenyl ether; pramocaine; proxazocain. $C_{17}H_{27}NO_3$; mol wt 293.41. C 69.59%, H 9.28%, N 4.77%, O 16.36%. Prepn: Wright, Moore, *J. Am. Chem. Soc.* **73**, 2281 (1951); **76**, 4396 (1954); **US 2870151** (1959 to Abbott). Pharmacology: O. Blanpin, *Anesth. Analg. Reanimat.* **14**, 225 (1957). Toxicity studies: J. L. Schmidt *et al.*, *Toxicol. Appl. Pharmacol.* **1**, 454 (1959); S. Monash, D. Gibbs, *Anesth. Analg.* **38**, 265 (1959).

bp$_6$ 196°; bp$_{2.8}$ 183-184°. LD$_{50}$ in mice (mg/kg): 300 i.p., 900 s.c. (Monash, Gibbs).

Hydrochloride. [637-58-1] Proctofoam-NS; Tronolane; Tronothane. $C_{17}H_{27}NO_3.HCl$; mol wt 329.87. Crystals, mp 181-183°. Sol in water. LD$_{50}$ i.v. in mice: 79.5 ±2.7 mg/kg (Schmidt). LD$_{50}$ in mice (mg/kg): 300 i.p., 750 s.c. (Monash, Gibbs).

THERAP CAT: Anesthetic (local).

7830. Pranlukast. [103177-37-3] N-[4-Oxo-2-(2H-tetrazol-5-yl)-4H-1-benzopyran-8-yl]-4-(4-phenylbutoxy)benzamide; 8-[4-(4-phenylbutoxy)benzamido]-2-(tetrazol-5-yl)-4H-1-benzopyran-4-one; 8-[p-(4-phenylbutoxy)benzoyl]amino-2-(5-tetrazolyl)-4-oxo-4H-1-benzopyran; ONO-1078; ONO-RS-411; Azlaire; Onon. $C_{27}H_{23}N_5O_4$; mol wt 481.51. C 67.35%, H 4.81%, N 14.54%, O 13.29%. Leukotriene antagonist. Prepn: M. Toda *et al.*, **EP 173516**; *eidem*, **US 4780469** (1986, 1988 both to Ono); H. Nakai *et al.*, *J. Med. Chem.* **31**, 84 (1988). Pharmacology: T. Obata *et al.*, *Adv. Prostaglandin Thromboxane Leukotriene Res.* **15**, 229 (1985); *idem et al.*, *ibid.* **17**, 540 (1987). Clinical evaluations in asthma: Y. Taniguchi *et al.*, *J. Allergy Clin. Immunol.* **92**, 507 (1993); H. Yamamoto *et al. Am. J. Respir. Crit. Care Med.* **150**, 254 (1994).

Occurs as the hemihydrate. Crystals, mp 244-245°. LD$_{50}$ in male mice (mg/kg): >1000 i.v. (Toda).

THERAP CAT: Antiasthmatic.

7831. Pranoprofen. [52549-17-4] α-Methyl-5H-[1]benzopyrano[2,3-b]pyridine-7-acetic acid; 2-(5H-[1]benzopyrano[2,3-b]pyridin-7-yl)propionic acid; Y-8004; Niflan; Oftalar; Pranoflog. $C_{15}H_{13}NO_3$; mol wt 255.27. C 70.58%, H 5.13%, N 5.49%, O 18.80%. Prepn: M. Nakanishi *et al.*, **DE 2337052**; *eidem*, **US 3931205** (1974, 1976 both to Yoshitomi); *eidem*, *Yakugaku Zasshi* **96**, 99 (1976), *C.A.* **84**, 135515b (1976). Physicochemical properties: M. Nobutoki, Y. Ota, *Iyakuhin Kenkyu* **7**, 200 (1976), *C.A.* **88**, 110468h (1978). Toxicity study: M. Edanaga *et al.*, *Iyakuhin Kenkyu* **7**, 211 (1976), *C.A.* **88**, 131015y (1978). LC-MS/MS determn in plasma and urine: J. Yu *et al.*, *J. Chromatogr. B* **878**, 3249 (2010). Evaluation of formulation for transdermal delivery: S.-C. Shin, C.-W. Cho, *Arch. Pharmacal Res.* **29**, 928 (2006). Clinical trial for post-surgical pain and inflammation: I. Akyol-Salman *et al.*, *J. Ocul. Pharmacol. Ther.* **23**, 280 (2007).

Cryst from aq dioxane, mp 182-183°. LD_{50} in male mice, rats (mg/kg): 447.3, 87.3 orally (Edanaga).

THERAP CAT: Anti-inflammatory.

7832. Praseodymium. [7440-10-0] Pr; at. wt 140.90765; at. no. 59; valences 3, 4. A lanthanide; belongs to the cerium group of rare earth metals. Naturally occurring isotope (mass number): 141; known artificial radioactive isotopes: 126, 129-140; 142-152. Prepn of radioactive isotopes by bombardment: Seaborg, *Chem. Rev.* **27**, 199 (1940); Huber *et al., Helv. Phys. Acta* **18**, 221 (1945). Abundance in earth's crust 5.5-9.1 ppm; found in rare earth minerals. Reported in 1843 by Mosander as *didymium* which was a mixture of Pr and Nd; separated in 1885 by von Welsbach. Separated from other rare earth elements by fractional crystn. Also by ion exchange: Spedding *et al., J. Am. Chem. Soc.* **69**, 2786, 2812 (1947). Prepn of metal by electrolysis: Muthmann, Weiss, *Ann.* **331**, 1 (1904); Canneri, Rossi, *Gazz. Chim. Ital.* **62**, 1160 (1932); Mazza, *Atti X Congr. Int. Chim.* **3**, 604 (1939); Trombe, Mahn, *Compt. Rend.* **220**, 778 (1945). Toxicity study: Haley *et al., Toxicol. Appl. Pharmacol.* **6**, 614 (1964). Reviews of prepn, properties and compds: *The Rare Earths*, F. H. Spedding, A. H. Daane, Eds. (Krieger, Huntington, N.Y., 1971, reprint of 1961 ed.) 641 pp; Hulet, Bode, "Separation Chemistry of the Lanthanides and Transplutonium Actinides" in *MTP Int. Rev. Sci.: Inorg. Chem., Ser. One* Vol. 7, K. W. Bagnall, Ed. (Univ. Park Press, Baltimore, 1972) pp 1-45; Moeller, "The Lanthanides", *Comprehensive Inorganic Chemistry* Vol. 4, J. C. Bailar, Jr. *et al.*, Eds. (Pergamon Press, Oxford, 1973) pp 1-101; F. H. Spedding in *Kirk-Othmer Encyclopedia of Chemical Technology* vol. 19 (John Wiley & Sons, New York, 3rd ed., 1982) pp 833-854; *Chemistry of the Elements*, N. N. Greenwood, A. Earnshaw, Eds. (Pergamon Press, New York, 1984) pp 1423-1449. Brief review of properties: G. T. Seaborg, *Radiochim. Acta* **61**, 115-122 (1993).

Yellowish metal, forms oxide film on exposure to moist air. Two crystalline forms: hexagonal α-form, d 6.475, transforms to β-form at 798°; body-centered cubic β-form, d 6.64, exists at >798°. mp 935°. bp 3020°; also reported as bp 3520° (Spedding). E^0 (aq) Pr^{3+}/Pr −2.47 V (calc). Experimental reduction potentials (referred to a normal calomel electrode): −1.875, −1.990 V: Noddack, Brukl, *Angew. Chem.* **50**, 362 (1937). Heat of sublimation (25°): 355.6 kJ/mol. Heat of fusion: 6.912 kJ/mol (Spedding, 1982). Also reported as heat of fusion: 11.3 ±2.1 kJ/mol (Greenwood, Earnshaw).

Oxide. Pr_2O_3. d 7.07, prepd by reducing with hydrogen the oxides PrO_2 or Pr_6O_{11}. Oxidizes to Pr_6O_{11} on heating in air; forms the dioxide, PrO_2, on fusing with potassium chlorate.

Hydroxide. $Pr(OH)_3$. A gelatinous pale green precipitate, obtained by the action of alkali hydroxide on a soln of a praseodymium salt; a purple powder when obtained by the action of water on praseodymium carbide.

Chloride. $PrCl_3$. Heptahydrate, green crystals; mp in its water of crystn at 111°; on partial dehydration yields the hexa-, tri-, or monohydrate; on heating to 180-200° in a stream of hydrogen chloride yields the anhyd chloride, green needles, mp 769-782°, sol in water, alcohol. LD_{50} in mice: 600 mg/kg i.p.; 4.5 g/kg orally (Haley).

Sulfate. $Pr_2(SO_4)_3$. Light green crystals, prepd by evaporating a soln of the oxide in H_2SO_4 and cooling the resulting product over phosphorus pentoxide. Several hydrates and double sulfates are known.

7833. Prasugrel. [150322-43-3] 2-[2-(Acetyloxy)-6,7-dihydrothieno[3,2-*c*]pyridin-5(4*H*)-yl]-1-cyclopropyl-2-(2-fluorophenyl)ethanone; 2-acetoxy-5-(α-cyclopropylcarbonyl-2-fluorobenzyl)-4,5,6,7-tetrahydrothieno[3,2-*c*]pyridine; 5-[(1*RS*)-2-cyclopropyl-1-(2-fluorophenyl)-2-oxoethyl]-4,5,6,7-tetrahydrothieno[3,2-*c*]pyridin-2-yl acetate; CS-747. $C_{20}H_{20}FNO_3S$; mol wt 373.44. C 64.33%, H 5.40%, F 5.09%, N 3.75%, O 12.85%, S 8.59%. Thienopyridine antiplatelet prodrug. Deacetylated *in vivo* to an inactive, short-lived thiolactone metabolite; further metabolized by hepatic cytochrome P isoenzymes leading to the open ring, thiol-containing active metabolite, R-138727, which binds irreversibly to platelet $P2Y_{12}$ adenosine diphosphate (ADP) receptors. Prepn: H. Koike *et al.*, **CA 2077695**; *eidem*, **US 5288726** (1993, 1994 to Sankyo). LC-MS/MS determn of metabolites in plasma: N. A. Farid *et al., Rapid Commun. Mass Spectrom.* **21**, 169 (2007). Stereoselective antagonism in receptor-binding and platelet-aggregation assays: M. Hasegawa *et al., Thromb. Haemostasis* **94**, 593 (2005). *In vivo* pharmacology: A. Sugidachi *et al., Br. J. Pharmacol.* **129**, 1439 (2000).

Clinical comparison with clopidogrel: M. Mariani *et al., Expert Rev. Cardiovasc. Ther.* **7**, 17 (2009). Review of chemistry, metabolism, and clinical pharmacology: F. Asai *et al., Annu. Rep. Sankyo Res. Lab.* **51**, 1-44 (1999); of clinical experience and therapeutic potential: U. S. Tantry *et al., Expert Opin. Invest. Drugs* **15**, 1627-1633 (2006).

White to pale yellow crystalline powder containing masses. mp 121-122°. pKa (25°) 5.40±0.05. Log P (*n*-octanol/phosphate buffer): 3.7 (pH 7). Practically insol in water; becomes slightly sol with decreasing pH. Soly at 20° (mg/ml): benzene >100; chloroform >100; acetone >100; ethyl acetate 82.5; acetonitrile 67.0; methanol 9.01; ethanol 5.91; *n*-hexane 0.65. Soly at 20° (μg/ml): aq soln 59.8 (pH 3.57), 14.2 (pH 4.85), 4.01 (pH 6.00), 0.15 (pH 7.12).

Hydrochloride. [389574-19-0] LY-640315; Efient. $C_{20}H_{20}FNO_3S \cdot HCl$; mol wt 409.90.

R-138727. [204204-73-9] (2*Z*)-2-[1-[2-Cyclopropyl-1-(2-fluorophenyl)-2-oxoethyl]-4-mercapto-3-piperidinylidene] acetic acid. $C_{18}H_{20}FNO_3S$; mol wt 349.42. Active metabolite. Mixture of four stereoisomers; (4*R*,1*S*)-isomer identified as most active.

THERAP CAT: Antithrombotic.

7834. Pratensein. [2284-31-3] 5,7-Dihydroxy-3-(3-hydroxy-4-methoxyphenyl)-4*H*-1-benzopyran-4-one; 3′,5,7-trihydroxy-4′-methoxyisoflavone. $C_{16}H_{12}O_6$; mol wt 300.27. C 64.00%, H 4.03%, O 31.97%. Isoln from red clover (*Trifolium pratense* L., *Leguminosae*), structure and synthesis: E. Wong, *J. Org. Chem.* **28**, 2336 (1963). Alternate syntheses: A. C. Jain, P. K. Bambab, *Indian J. Chem.* **26B**, 488 (1987).

Needles from ethanol, mp 272-273°. uv max (ethanol): 263 nm (log ε 4.53).

Triacetylpratensein. $C_{22}H_{18}O_9$. White needles from chloroform, mp 175-177°.

7835. Pravastatin Sodium. [81131-70-6] (β*R*,δ*R*,1*S*,2*S*,6*S*,-8*S*,8a*R*)-1,2,6,7,8,8a-Hexahydro-β,δ,6-trihydroxy-2-methyl-8-[(2*S*)-2-methyl-1-oxobutoxy]-1-naphthaleneheptanoic acid sodium salt (1:1); sodium (+)-(3*R*,5*R*)-3,5-dihydroxy-7-[(1*S*,2*S*,6*S*,8*S*,8a*R*)-6-hydroxy-2-methyl-8-[(*S*)-2-methylbutyryloxy]-1,2,6,7,8,8a-hexahydro-1-naphthyl]heptanoate; eptastatin sodium; 3β-hydroxycompactin sodium salt; CS-514; SQ-31000; Elisor; Lipostat; Liprevil; Mevalotin; Oliprevin; Pravachol; Pravaselect; Pravasin; Selectin; Selipran; Vasten. $C_{23}H_{35}NaO_7$; mol wt 446.52. C 61.87%, H 7.90%, Na 5.15%, O 25.08%. HMG-CoA reductase inhibitor; bioactive metabolite of mevastatin, *q.v.* Prepn by microbial hydroxylation: A. Terahara, M. Tanaka, **DE 3122499**; *eidem*, **US 4346227** (1981, 1982 both to Sankyo); N. Serizawa *et al., J. Antibiot.* **36**, 604 (1983). Structure elucidation: H. Haruyama *et al., Chem. Pharm. Bull.* **34**, 1459 (1986). HPLC determn in biological fluids: S. Bauer *et al., J. Chromatogr. B* **818**, 257 (2005). Effect on serum lipid concentration: N. Nakaya *et al., Atherosclerosis* **61**, 125 (1986); on hepatic metabolism of cholesterol: E. Reihnér *et al., N. Engl. J. Med.* **323**, 224 (1990). Clinical comparison with probucol, *q.v.*: G. Yoshino *et al., Lancet* **2**, 740 (1986). Clinical reduction of risk of major cardiovascular events in patients with coronary heart disease: LIPID Study Group, *N. Engl. J. Med.* **339**, 1349 (1998). Clinical effect on risk of stroke: H. D. White *et al., ibid.* **343**, 317 (2000).

Odorless, white to off-white, fine or crystalline powder. Hygroscopic. uv max (methanol): 230, 237, 245 nm. Freely sol in water, methanol; sol in alc; slightly sol in isopropanol; very slightly sol in acetonitrile. Practically insol in acetone, ethyl acetate, chloroform, ether.

Lactone. $C_{23}H_{34}O_6$. Colorless plate crystals, mp 138-142°. $[\alpha]_D^{22}$ +194.0° (c = 0.51 in methanol). uv max (methanol): 230, 237, 245 nm.

THERAP CAT: Antilipemic.

7836. Prazepam. [2955-38-6] 7-Chloro-1-(cyclopropylmethyl)-1,3-dihydro-5-phenyl-2H-1,4-benzodiazepin-2-one; 1-(cyclopropylmethyl)-5-phenyl-7-chloro-1H-1,4-benzodiazepin-2(3H)-one; W-4020; Centrax; Demetrin; Lysanxia; Prazene; Sedapran; Trepidan. $C_{19}H_{17}ClN_2O$; mol wt 324.81. C 70.26%, H 5.28%, Cl 10.91%, N 8.62%, O 4.93%. Prepn: McMillan, Pattison, **FR 1394287**, corresp to **US 3192199** (both 1965 to Warner-Lambert); Wuest, **US 3192200** (1965); Inaba et al., Chem. Pharm. Bull. **17**, 1263 (1969). Pharmacology: Robichaud et al., Arch. Int. Pharmacodyn. Ther. **185**, 213 (1970). Metabolism: DiCarlo et al., J. Pharm. Sci. **58**, 960 (1969).

Crystals from methanol, mp 145-146°.

Note: This is a controlled substance (depressant): **21 CFR**, 1308.14.

THERAP CAT: Anxiolytic.

7837. Praziquantel. [55268-74-1] 2-(Cyclohexylcarbonyl)-1,2,3,6,7,11b-hexahydro-4H-pyrazino[2,1-a]isoquinolin-4-one; EMBAY 8440; Biltricide; Cesol; Droncit. $C_{19}H_{24}N_2O_2$; mol wt 312.41. C 73.05%, H 7.74%, N 8.97%, O 10.24%. Prepn: J. Seubert et al., **DE 2362539**; eidem, **US 4001411** (1975, 1977, both to E. Merck, W. Ger.); D. Frehel et al., Heterocycles **20**, 1731 (1983). Properties: J. Seubert et al., Experientia **33**, 1036 (1977). In vitro study: C. J. Chavasse et al., Z. Parasitenkd. **58**, 169 (1979). Effect on ultrastructure of trematodes: H. Mehlhorn et al., Arzneim.-Forsch. **33**, 91 (1983). Pharmacodynamics: G. C. Coles, J. Helminthol. **53**, 31 (1979). Ovicidal activity in Echinococcus: A. S. Thakur et al., Exp. Parasitol. **47**, 131 (1979). Clinical pharmacology: G. Leopold et al., Eur. J. Clin. Pharmacol. **14**, 281 (1978). Efficacy in dogs: F. L. Anderson et al., Am. J. Vet. Res. **40**, 700 (1979). Mutagenicity study: H. Bartsch et al., Mutat. Res. **58**, 133 (1978). Tolerance study: P. Muermann et al., Vet. Med. Rev. **1976** (2), 142. Symposium on African schistosomiasis: Arzneim.-Forsch. **31**, 535-618 (1981). Clinical studies (human schistosomiasis): T. E. Nash et al., Am. J. Trop. Med. Hyg. **31**, 977 (1982); R. N. H. Pugh, C. H. Teesdale, Br. Med. J. **286**, 429 (1983); C. H. Schutte et al., S. Afr. Med. J. **64**, 7 (1983). Clinical trial in human parenchymal brain cysticercosis: J. Sotelo et al., N. Engl. J. Med. **310**, 1001 (1984).

Crystals, mp 136-138°. Soly (g/100 ml): ethanol 9.7; chloroform 56.7; water 0.04. LD_{50} in mice, rats (mg/kg): 2000-3000 orally; >3000 s.c. (Muermann).

THERAP CAT: Anthelmintic (Schistosoma).
THERAP CAT (VET): Anthelmintic.

7838. Prazosin. [19216-56-9] [4-(4-Amino-6,7-dimethoxy-2-quinazolinyl)-1-piperazinyl]-2-furanylmethanone; 1-(4-amino-6,7-dimethoxy-2-quinazolinyl)-4-(2-furanylcarbonyl)piperazine; 2-[4-(2-furoyl)piperazin-1-yl]-4-amino-6,7-dimethoxyquinazoline; furazosin. $C_{19}H_{21}N_5O_4$; mol wt 383.41. C 59.52%, H 5.52%, N 18.27%, O 16.69%. α_1-Adrenergic blocker. Prepn: **GB 1156973**; Hess, **US 3511836** (1969, 1970 both to Pfizer); **NL 7206067** (1972 to Brocades-Stheeman), C.A. **78**, 72180s (1973); E. Honkanen et al., J. Heterocycl. Chem. **17**, 797 (1980). Crystal structure determn: V. V. Chernyshev et al., J. Pharm. Sci. **93**, 3090 (2004). Pharmacology and clinical data: Scriabine et al., Experientia **24**, 1150 (1968); Cohen, J. Clin. Pharmacol. J. New Drugs **10**, 408 (1970). Pharmacokinetics: P. Jaillon, Clin. Pharmacokinet. **5**, 365 (1980). HPLC determn in biological fluids: R. K. Bhamra et al., J. Chromatogr. **380**, 216 (1986). Voltammetic and spectrophotometric determn in urine and pharmaceutical formulations: A. Arranz et al., J. Pharm. Biomed. Anal. **21**, 797 (1999). Clinical evaluation in Raynaud's phenomenon: I. J. Russell, J. A Lessard, J. Rheumatol. **12**, 94 (1985); in dysuria with benign prostatic hypertrophy: A. LeDuc et al., Urol. Int. **45**, Suppl. 1, 56 (1990). Book: Prazosin: Pharmacology, Hypertension and Congestive Heart Failure, M. D. Rawlins, Ed. (Grune & Stratton, New York, 1981) 143 pp. Review of pharmacology and therapeutic use: W. F. Stanaszek et al., Drugs **25**, 339-384 (1983); J. L. Reid, J. Vincent, Cardiology **73**, 164-174 (1986). Comprehensive description: L. J. Kostek, Anal. Profiles Drug Subs. **18**, 351-378 (1989).

Crystals, mp 278-280°. d 1.362 g/cm³.

Hydrochloride. [19237-84-4] CP-12299-1; Alpress LP; Duramipress; Eurex; Hypovase; Minipress; Peripress. $C_{19}H_{21}N_5O_4$·HCl; mol wt 419.87. d 1.449 g/cm³. uv max (methanolic 0.01N HCl): 246, 329 nm (a_M 137 ±3, 27.6 ±0.3). Soly at ambient temp (mg/ml): acetone 0.0072, methanol 6.4, ethanol 0.84, dimethylformamide 1.3, dimethylacetamide 1.2, water (pH~3.5) 1.4, chloroform 0.041.

THERAP CAT: Antihypertensive.
THERAP CAT (VET): Antihypertensive.

7839. Precocenes. Anti-JH. Anti-juvenile hormones found in plants that induce reversible precocious metamorphosis and sterilization in insects by suppressing the function of the corpora allata gland: W. S. Bowers et al., Science **193**, 542 (1976); W. S. Bowers, R. Martinez-Pardo, ibid. **197**, 1369 (1977). Isoln from Ageratum mexicanum Sw., Compositae and structure: A. R. Alertsen, Acta Chem. Scand. **9**, 1725 (1955). Synthesis of precocene I: R. Livingstone, R. B. Watson, J. Chem. Soc. **1957**, 1509; of precocene II: R. Huls, Bull. Soc. Chim. Belg. **66**, 409 (1957); **67**, 22 (1958); of I and

II: J. R. Hlubucek *et al.*, *Aust. J. Chem.* **24**, 2347 (1971); A. Banerji, N. C. Goomer, *Indian J. Chem.* **20B**, 144 (1981); V. K. Ahluwalia *et al.*, *Chem. Ind. (London)* **1982**, 369. Biosynthesis: A. V. Vyas, N. B. Mulchandani, *Phytochemistry* **19**, 2597 (1980). Metabolism studies: D. M. Soderlund *et al.*, *J. Agric. Food Chem.* **1980**, 724; B. J. Bergot *et al.*, *Pestic. Biochem. Physiol.* **13**, 95 (1980). Mechanism of action: W. S. Bowers, *Am. Zool.* **21**, 737 (1981).

Precocene I   R = H
Precocene II   R = OCH₃

Precocene I. 7-Methoxy-2,2-dimethyl-2*H*-1-benzopyran; 7-methoxy-2,2-dimethylchromene; 6-demethoxyageratochromene. $C_{12}H_{14}O_2$; mol wt 190.24. Oil, bp₆ 120°. n_D^{21} 1.5548. uv max (ethanol): 280, 304 nm (ε 6644, 5844).

Precocene II. 6,7-Dimethoxy-2,2-dimethyl-2*H*-1-benzopyran; 6,7-dimethoxy-2,2-dimethylchromene; ageratochromene. $C_{13}H_{16}O_3$; mol wt 220.27. Crystals, mp 47.5°, bp₆ 136°. uv max (ethanol): 280, 323 nm (ε 5500, 9200).

7840. Prednicarbate. [73771-04-7] (11β)-17-[(Ethoxycarbonyl)oxy]-11-hydroxy-21-(1-oxopropoxy)pregna-1,4-diene-3,20-dione; 11β,17,21-trihydroxypregna-1,4-diene-3,20-dione 17-(ethyl carbonate) 21-propionate; prednisolone 17-ethylcarbonate 21-propionate; HOE-777; S-77-0777; Dermatop; Prednitop. $C_{27}H_{36}O_8$; mol wt 488.58. C 66.38%, H 7.43%, O 26.20%. Topical anti-inflammatory agent. Prepn: U. Stache *et al.*, **DE 2735110**; *eidem*, **US 4242334** (1979, 1980 both to Hoechst AG); U. Stache *et al.*, *Arzneim.-Forsch.* **35**, 1753 (1985). Topical and systemic activity in rats: H. G. Alpermann *et al.*, *ibid.* **32**, 633 (1982). Series of articles on pharmacology, pharmacokinetics and clinical efficacy: *Z. Hautkrankhr.* **61**, Suppl. 1, 1-96 (1986).

Crystals from ethanol + diethylether, mp 110-112°; second crystalline form, mp 183°. uv max (ethanol): 241 nm (ε 15000). $[\alpha]_D^{20}$ +63° (c = 0.1 in ethanol). Freely sol in acetone, ethanol; sparingly sol in propylene glycol. Practically insol in water.

THERAP CAT: Glucocorticoid.

7841. Prednisolone. [50-24-8] (11β)-11,17,21-Trihydroxypregna-1,4-diene-3,20-dione; 1,4-pregnadiene-3,20-dione-11β,-17α,21-triol; 1,4-pregnadiene-11β,17α,21-triol-3,20-dione; 3,20-dioxo-11β,17α,21-trihydroxy-1,4-pregnadiene; metacortandralone; delta F; Δ¹-dehydrocortisol; Δ¹-hydrocortisone; Δ¹-dehydrohydrocortisone; hydroretrocortine; Codelcortone; Decaprednil; Decortin H; Deltacortril; Klismacort; Meticortelone; Precortisyl; Prednicen; Prednisolone; Solone. $C_{21}H_{28}O_5$; mol wt 360.45. C 69.98%, H 7.83%, O 22.19%. Synthetic corticosteroid; metabolically interconvertible with prednisone, *q.v.* Prepn: A. Nobile *et al.*, *J. Am. Chem. Soc.* **77**, 4184 (1955); A. Nobile, **US 2837464**; **US 3134718** (1958, 1964, both to Schering); Herzog *et al.*, *Tetrahedron* **18**, 581 (1962). HPLC determn in urine: B. M. Frey, F. J. Frey, *J. Chromatogr.* **229**, 283 (1982). Clinical use in advanced cancer: R. G. Twycross, D. Guppy, *Practitioner* **229**, 57 (1985); in inflammatory bowel disease: C. J. Hawkey, *Neth. J. Med.* **35** Suppl 1, S21-S26 (1989). Review of clinical pharmacology: W. J. Jusko, J. Q. Rose, *Ther. Drug*

Monit. **2**, 169-176 (1980); of pharmacokinetics and metabolism: B. M. Frey, F. J. Frey, *Clin. Pharmacokinet.* **19**, 126-146 (1990). Comprehensive description: S. L. Ali, *Anal. Profiles Drug Subs. Excip.* **21**, 415-500 (1992).

Crystals, dec 240-241°. $[\alpha]_D^{25}$ +102° (dioxane). uv max (methanol): 242 nm (ε 15000; $A_{1cm}^{1\%}$ 414). One gram dissolves in about 30 ml of alc, in about 180 ml of chloroform, in about 50 ml of acetone. Sol in methanol, dioxane; very slightly sol in water.

21-Acetate. [52-21-1] Ak-Tate; Deltastab; Inflanefran; Pred Forte; Pred Mild; Scherisolon; Inflamase; Orapred; Predsol. $C_{23}H_{30}O_6$; mol wt 402.49. Crystals, decomp 237-239°. $[\alpha]_D^{25}$ +116° (dioxane). Slightly sol in acetone, alc, chloroform. Practically insol in water.

21-*tert*-Butylacetate. [7681-14-3] Prednisolone tebutate; Codelcortone TBA. $C_{27}H_{38}O_6$; mol wt 458.60. Prepn: Sarett, **US 2736734** (1956 to Merck & Co.). Crystals from ethanol, mp 266-273°.

21-Disodium Phosphate. [125-02-0] Prednisolone sodium phosphate; Codelsol; Colicort; Hefasolon; Inflamase; Orapred; Predsol; Solucort. $C_{21}H_{27}Na_2O_8P$; mol wt 484.39. Prepn: Sarett, **US 2789117** (1957 to Merck & Co.). Alternate synthesis: Poos *et al.*, *Chem. Ind. (London)* **1958**, 1260; Elks, Phillips, **US 2936313** (1960 to Glaxo). White powder. Slightly hygroscopic. Stable at room temp. $[\alpha]_D^{25}$ +102.5°. uv max (methanol): 243 nm ($A_{1cm}^{1\%}$ 308). Freely sol in water; sol in methanol; slightly sol in alc, chloroform; very slightly sol in acetone, dioxane. pH of 1% aq soln 7.5 to 8.5.

21-Succinate sodium salt. [1715-33-9] Prednisolone sodium succinate; Meticortelone Soluble; Solu-Decortin-H. $C_{25}H_{31}NaO_8$; mol wt 482.50. Prepn: Shull, Kita, **DE 1045400** (1958 to Pfizer), *C.A.* **55**, 2746f (1961).

21-Stearoylglycolate. [5060-55-9] Prednisolone steaglate; Sintisone. $C_{41}H_{64}O_8$; mol wt 684.96. Prepn: Giraldi, Nannini, **US 3171846** (1965 to Carlo Erba); Giraldi *et al.*, *Arzneim.-Forsch.* **16**, 162 (1966). Crystals from dilute alcohol or butyl ether, mp 105-107°. uv max (methanol): 242 nm ($E_{1cm}^{1\%}$ 212 ±10). $[\alpha]_D^{20}$ +57-63°.

21-*m*-Sulfobenzoate sodium salt. [630-67-1] 11,17-Dihydroxy-21-[(3-sulfobenzoyl)oxy]pregna-1,4-diene-3,20-dione monosodium salt; prednisolone sodium metasulfobenzoate; Predenema; Predfoam; Solupred. $C_{28}H_{31}NaO_9S$; mol wt 566.60. Prepn: Allais, Girault, **US 3032568**; Joly, Warrant, **US 3037034** (both 1962 to Roussel-UCLAF). Crystals from water, dec 293-295°. $[\alpha]_D^{20}$ +170° (water).

21-Trimethylacetate. [1107-99-9] Prednisolone 21-pivalate; Ultracortenol. $C_{26}H_{36}O_6$; mol wt 444.57. Prepn: Vischer *et al.*, *Helv. Chim. Acta* **38**, 1502 (1955); Joly, Warrant, **US 3037034** (1962 to Roussel-UCLAF). Crystals from acetone, mp 233-236°. $[\alpha]_D^{26}$ +103° (c = 1.208 in chloroform); $[\alpha]_D^{20}$ +97.5° (c = 1 in chloroform). uv max: 244 nm (ε 14700).

THERAP CAT: Glucocorticoid.

THERAP CAT (VET): Glucocorticoid.

7842. Prednisone. [53-03-2] 17,21-Dihydroxypregna-1,4-diene-3,11,20-trione; 1,4-pregnadiene-17α,21-diol-3,11,20-trione; Δ¹-dehydrocortisone; Δ¹-cortisone; deltacortisone; delta E; metacortandracin; retrocortine; NSC-10023; Colisone; Cortancyl; Dacortin; Decortin; Deltacortene; Deltacortone; Deltasone; Deltison; Di-Adreson; Encorton; Hostacortin; Meticorten; Orasone; Rectodelt; Sone; Ultracorten. $C_{21}H_{26}O_5$; mol wt 358.43. C 70.37%, H 7.31%, O 22.32%. Prepn: Oliveto, Gould, **US 2897216** (1959 to Schering). Microbiological prepn: Nobile *et al.*, *J. Am. Chem. Soc.* **77**, 4184 (1955); Nobile, **US 2837464** and **US 3134718** (1958, 1964 both to Schering); Herzog *et al.*, *Tetrahedron* **18**, 581 (1962). Structure: Herzog *et al.*, *Science* **121**, 176 (1955); *cf.* Djerassi *et al.*, **US 2579479** (1951 to Syntex).

Crystals, dec 233-235°. $[\alpha]_D^{25}$ +172° (dioxane). uv max (methanol): 238 nm (ε 15500). One gram dissolves in about 150 ml alcohol, in about 200 ml chloroform. Slightly sol in methanol, dioxane. Very slightly sol in water.

21-Acetate. [125-10-0] $C_{23}H_{28}O_6$. Crystals, dec 226-232°. $[\alpha]_D^{25}$ +186° (dioxane). uv max (ethanol): 238 nm (ε 16100).

THERAP CAT: Glucocorticoid.

THERAP CAT (VET): Adrenocortical steroid. Glucocorticoid, antiinflammatory.

7843. Pregabalin. [148553-50-8] (3S)-3-(Aminomethyl)-5-methylhexanoic acid; (S)-(+)-4-amino-3-(2-methylpropyl)butanoic acid; (S)-(+)-3-isobutyl-γ-aminobutyric acid; CI-1008; PD-144723; Lyrica. $C_8H_{17}NO_2$; mol wt 159.23. C 60.35%, H 10.76%, N 8.80%, O 20.10%. Structural analog of γ-aminobutyric acid, q.v.; ligand at $\alpha_2\delta$ subunit of voltage-gated calcium channels. Prepn of racemate: R. Andruszkiewicz, R. B. Silverman, *Synthesis* **1989**, 953. Enantioselective synthesis: P. Yuen *et al.*, *Bioorg. Med. Chem. Lett.* **4**, 823 (1994). Manufacturing process: M. S. Hoekstra *et al.*, *Org. Process Res. Dev.* **1**, 26 (1997). HPLC determn in biological fluids: B. L. Windsor, L. L. Radulovic, *J. Chromatogr. B* **674**, 143 (1995). Clinical trial in post-herpetic neuralgia: R. Sabatowski *et al.*, *Pain* **109**, 26 (2004); in fibromyalgia syndrome: L. J. Crofford *et al.*, *Arthritis Rheum.* **52**, 1264 (2005). Overview of mechanism and pharmacology: S. M. Stahl, *J. Clin. Psychiatry* **65**, 596, 1033 (2004). Review of pharmacology and clinical experience: R. Huckle, *Curr. Opin. Investig. Drugs* **5**, 82-89 (2004); of clinical experience in generalized anxiety disorder: S. A. Montgomery, *Expert Opin. Pharmacother.* **7**, 2139-2154 (2006).

White to off-white crystalline solid, mp 186-188°. pKa_1 4.2, pKa_2 10.6. $[\alpha]_D^{23}$ +10.52° (c = 1.06 in water). Log P (*n*-octanol/0.05M phosphate buffer): −1.35 (pH 7.4). Freely sol in water and both basic and acidic aqueous solns.

Note: This is a controlled substance (depressant): **21 CFR, 1308.15**

THERAP CAT: Anticonvulsant; anxiolytic; analgesic in treatment of peripheral neuropathic pain and fibromyalgia syndrome.

7844. Pregnane. [481-26-5] 5β-Pregnane; 17β-ethyletiocholane. $C_{21}H_{36}$; mol wt 288.52. C 87.42%, H 12.58%. Prepd by reduction of etiocholyl methyl ketone or of pregnanedione: Butenandt, *Ber.* **64**, 2529 (1931); Marker *et al.*, *J. Am. Chem. Soc.* **60**, 1067 (1938); Steiger, Reichstein, *Helv. Chim. Acta* **21**, 161 (1938).

Monoclinic scales, plates from methanol; mp 83.5°. d_4^{15} 1.032. $[\alpha]_D^{19}$ +20° (c = 2 in chloroform).

7845. Pregnanediol. [80-92-2] (3α,5β,20S)-Pregnane-3,20-diol. $C_{21}H_{36}O_2$; mol wt 320.52. C 78.69%, H 11.32%, O 9.98%. A metabolite of progesterone, present in large amounts in pregnancy urine. Isoln from pregnancy urine of women: Marrian, *Biochem. J.* **23**, 1090 (1929); Butenandt, *Ber.* **63**, 659 (1930); of cows, mares, and chimpanzees: Fish *et al.*, *J. Biol. Chem.* **143**, 716 (1942). Prepn by reduction of pregn-16-ene-3,20-dione: Marker *et al.*, **US 2352852** (1944 to Parke, Davis). Conversion to progesterone: Butenandt, Schmidt, *Ber.* **67**, 1893, 1901 (1934). Conversion to 3α-hydroxypregn-20-one: Marker, **US 2223377** (1940 to Parke, Davis). Prepn of the 3-acetate: Hirschmann, *J. Biol. Chem.* **140**, 797 (1941); Ralls *et al.*, *ibid.* **210**, 709 (1954). Prepn of the 20-acetate: Hirschmann, *loc. cit.* Prepn of the diacetate: Johnson *et al.*, *J. Chem. Soc.* **1954**, 1302. Crystal structure: Haner, Norton, *Acta Crystallogr.* **16**, 707 (1963). Review of metabolism, bioactivity and assay during pregnancy: P. J. Keller, *Contrib. Gynecol. Obstet.* **2**, 75-91 (1976).

Crystals from acetone or ethanol, mp 239°. $[\alpha]_D^{20}$ +27.4° (c = 0.7 in alc). Sparingly sol in organic solvents. Not precipitated by digitonin.

3-Acetate. 3α-Acetoxypregnan-20α-ol. $C_{23}H_{38}O_3$. Crystals, mp 132°. $[\alpha]_D^{25}$ +45° (CHCl$_3$).

20-Acetate. 20α-Acetoxypregnan-3α-ol. $C_{23}H_{38}O_3$. Crystals, mp 174°.

Diacetate. 3α,20α-Diacetoxypregnane. $C_{25}H_{40}O_4$. Crystals from light petroleum, mp 180°, also reported as mp 182-183°. $[\alpha]_D^{15}$ +35° (c = 1.1 in CHCl$_3$).

USE: In manuf of progesterone.

7846. 3,20-Pregnanedione. [128-23-4] (5β)-Pregnane-3,20-dione. $C_{21}H_{32}O_2$; mol wt 316.49. C 79.70%, H 10.19%, O 10.11%. From pregnancy urine of mares. Prepn from other steroids: Butenandt, *Ber.* **63**, 659 (1930); Butenandt, Fleischer, *Ber.* **68**, 2094 (1935); Marker *et al.*, *J. Am. Chem. Soc.* **59**, 1595 (1937); Shoppee, Reichstein, *Helv. Chim. Acta* **24**, 356 (1941); **US 2160719; US 2352852; US 2323276; US 2229818.**

Needles from dil alc, mp 123°. Insol in water. Freely sol in the usual organic solvents.

Dioxime. $C_{21}H_{34}N_2O_2$. Dec above 250°.

7847. Pregnan-3α-ol-20-one. [128-20-1] (3α,5β)-3-Hydroxypregnan-20-one; 3α-hydroxy-5β-pregnan-20-one; pregnanolone; eltanolone; KABI 2213. $C_{21}H_{34}O_2$; mol wt 318.50. C 79.19%, H 10.76%, O 10.05%. Naturally occurring metabolite of progesterone, q.v. Isoln from urine of pregnant women: R. E. Marker, O. Kamm, *J. Am. Chem. Soc.* **59**, 1373 (1937); from bile of pregnant cows: W. H. Pearlman, E. Cerceo, *J. Biol. Chem.* **176**, 847 (1948). Prepn: R. E. Marker *et al.*, *J. Am. Chem. Soc.* **59**, 1841 (1937); L. Gyermek *et al.*, *J. Med. Chem.* **11**, 117 (1968); T. L. G. Lemos, J. D. McChesney, *J. Nat. Prod.* **53**, 152 (1990). Pharmacokinetics and pharmacodynamics: P. Carl *et al.*, *Acta Anaesthesiol.*

Scand. **38**, 734 (1994). Comparative clinical evaluation: J. Van Hemelrijck *et al.*, *Anesthesiology* **80**, 36 (1994); H. Eriksson *et al.*, *Acta Anaesthesiol. Scand.* **39**, 479 (1995). Hemodynamic effects in humans: J. W. Sear *et al.*, *J. Clin. Anesth.* **7**, 126 (1995).

Needles from aq methanol, mp 148-148.5° (Pearlman, Cerceo); also reported as cryst from hexane, mp 131-132° (Lemos, Mc-Chesney). $[\alpha]_D^{23}$ +59.6° (c = 0.3 in chloroform). $[\alpha]_D^{26}$ +108.5 ±1° (c = 9.23 mg/1.23 ml abs ethanol). LD$_{50}$ in mice, rats (mg/kg): 66 ± 10, 27.5 ± 2.4 i.v. (Gyermek).

THERAP CAT: Anesthetic (local).

7848. 4-Pregnene-11β,17α,20β,21-tetrol-3-one. [116-58-5] (11β,20R)-11,17,20,21-Tetrahydroxypregn-4-en-3-one; 17-(1,2-di-hydroxyethyl)androsten-3-one-11,17-diol; Reichstein's substance E; 11β,17,20β,21-tetrahydroxypregn-4-en-3-one. C$_{21}$H$_{32}$O$_5$; mol wt 364.48. C 69.20%, H 8.85%, O 21.95%. Occurs in adrenal cortex. Isoln: Reichstein, *Helv. Chim. Acta* **19**, 29 (1936); **20**, 953 (1937); Reichstein, von Euw, *ibid.* **24**, 247E (1941).

Hydrated crystals from dil acetone, dec 125°. $[\alpha]_D^{20}$ +87° (alc). [M]$_D$ +317°. uv max: 240 nm.

20,21-Diacetate. Crystals, dec 229-230°. $[\alpha]_D^{20}$ +162.7°. [M]$_D$ +730° (acetone).

7849. 4-Pregnene-17α,20β,21-triol-3,11-dione. [116-59-6] (20R)-17,20,21-Trihydroxypregn-4-ene-3,11-dione; Reichstein's substance U. C$_{21}$H$_{30}$O$_5$; mol wt 362.47. C 69.59%, H 8.34%, O 22.07%. Isoln from adrenal glands: Reichstein, von Euw, *Helv. Chim. Acta* **24**, 247E (1941).

Clusters of needles from acetone + ether, mp 208-209°.

20,21-Diacetate. C$_{25}$H$_{34}$O$_7$. Pointed needles from acetone + ether or chloroform + ether, mp 253°. $[\alpha]_D^{21}$ +178.5° (c = 0.924 in acetone). uv max (alc): 239 nm (log ε 4.1). Less sol than the 20,21-diacetate of 4-pregnene-11β,17α,20,21-tetrol-3-one.

7850. 4-Pregnene-17α,20β,21-triol-3-one. [128-19-8] (20R)-17,20,21-Trihydroxypregn-4-en-3-one; 17-(1,2-dihydroxy-ethyl)-Δ^4-androsten-3-on-17α-ol; 17α-pregnenetriolone. C$_{21}$-H$_{32}$O$_4$; mol wt 348.48. C 72.38%, H 9.26%, O 18.36%. Prepd from 17-ethynyltestosterone by hydrogenation with palladium in pyridine, allylic rearrangement, and hydroxylation with osmium tetroxide: Ruzicka, Müller, *Helv. Chim. Acta* **22**, 755 (1939); Logemann, *Naturwissenschaften* **27**, 196 (1939); from a 3-enol ester of a 17,21-

diacyloxyprogesterone by reduction followed by saponification: **CH 207496** (1940), *C.A.* **36**, 3636 (1942).

Crystals from methanol. mp 190°. Sol in dioxane, chloroform, methanol. $[\alpha]_D$ +63° (c = 1 in dioxane). uv max: 240 nm (log ε 4.1).

20,21-Diacetate. C$_{25}$H$_{36}$O$_6$. Crystals from acetone + ether. Polymorphic; mp 170° and 194°; $[\alpha]_D^{20}$ +125° (dioxane). Reaction with zinc in toluene yields 17-isodesoxycorticosterone acetate.

7851. Pregnenolone. [145-13-1] (3β)-3-Hydroxypregn-5-en-20-one; Δ^5-pregnen-3β-ol-20-one; 17β-(1-ketoethyl)-Δ^5-androsten-3β-ol. C$_{21}$H$_{32}$O$_2$; mol wt 316.49. C 79.70%, H 10.19%, O 10.11%. Prepn from stigmasterol: Butenandt *et al.*, *Ber.* **67**, 1611 (1934); Butenandt, Fleischer, *Ber.* **70**, 96 (1937); from androstenolone: H. Butenandt, J. Schmidt-thome, *Ber.* **72**, 182 (1939); S. Danishefsky *et al.*, *J. Org. Chem.* **40**, 1989 (1975); from Δ^5-3-acetoxyetiocholenic acid chloride: Wettstein, *Helv. Chim. Acta* **23**, 1373 (1940); from diosgenin: Marker, Krueger, *J. Am. Chem. Soc.* **62**, 3349 (1940); Marker *et al.*, *ibid.* **69**, 2173 (1947); from nologenin: *eidem, ibid.* 2395; by treating a 21-halo-Δ^5-pregnen-3-ol-20-one with a reducing agent: **CH 215139** (1941), *C.A.* **42**, 3144 (1948). Crystal structure: J. Bordner *et al.*, *Cryst. Struct. Commun.* **7**, 513 (1978).

Needles from dil alc, mp 193°. $[\alpha]_D^{20}$ +28° (alc). Very sparingly sol in water. Soly (g/100 ml of soln): carbon tetrachloride 0.5; petr ether 0.1; ethyl acetate 1.1; acetone 0.6; chloroform 17.0; ethanol 1.9; benzene 0.9; isopropanol 1.5. Soly (g/100 ml of solvent): propylene glycol 0.1; dioxane 3.1; benzyl alcohol 8.1. On refluxing with methyl alcohol yields the 17-isopregnenolone, mp 172-173°, $[\alpha]_D^{20}$ −140.5° (alcohol).

Acetate. C$_{23}$H$_{34}$O$_3$. Needles from alcohol, mp 149-151°. $[\alpha]_D^{20}$ +22° (alcohol). Soly (g/100 ml of soln): carbon tetrachloride 5.0; petr ether 1.0; ethyl acetate 7.9; acetone 2.7; chloroform 55.0; ethanol 2.5; benzene 26.0; isopropanol 2.0. Soly (g/100 ml of solvent): propylene glycol 0.1; dioxane 20.2; benzyl alcohol 11.1; benzyl benzoate 9.1.

Methyl ether. [511-26-2] C$_{22}$H$_{34}$O$_2$. Crystals from abs or dil methanol, mp 123.5°. $[\alpha]_D^{18}$ +18° (c = 1.085 in chloroform).

7852. Prelog-Djerassi Lactone. [69056-12-8] (αR,2S,3S,-5R)-Tetrahydro-α,3,5-trimethyl-6-oxo-2H-pyran-2-acetic acid; (+)-Prelog-Djerassi lactonic acid; 6-(1-carboxyethyl)-3,4,5,6-tetrahydro-3,5-dimethyl-2-pyranone. C$_{10}$H$_{16}$O$_4$; mol wt 200.23. C 59.99%, H 8.05%, O 31.96%. Intermediate in the synthesis of macrolide antibiotics. Isoln as degradation product of narbomycin, *q.v.*: R. Anliker *et al.*, *Helv. Chim. Acta* **39**, 1785 (1956); of methymycin, *q.v.*: C. Djerassi, J. A. Zderic, *J. Am. Chem. Soc.* **78**, 6390 (1956). Abs config: R. W. Rickards, R. M. Smith, *Tetrahedron Lett.* **1970**, 1025. Synthesis: P. A. Grieco *et al.*, *J. Am. Chem. Soc.* **101**, 4749 (1979); D. A. Evans, J. Bartroli, *Tetrahedron Lett.* **23**, 807 (1982). Synthesis of (±)-form: S. Masamune *et al.*, *J. Am. Chem. Soc.* **97**, 3512 (1975); C. Santelli-Rouvier *et al.*, *Tetrahedron Lett.* **35**, 6101 (1994). Review of stereoselective syntheses: S. F. Martin, D. E. Guinn, *Synthesis* **1991**, 245.

CH₃

$$O = \quad O. \quad \overset{CH_3}{\underset{COOH}{}}$$

H₃C   CH₃

Crystals, mp 124-125°. $[\alpha]_D$ +33° (c = 0.797 in CHCl₃).

7853. Prenalterol. [57526-81-5] 4-[(2S)-2-Hydroxy-3-[(1-methylethyl)amino]propoxy]phenol; (−)-(S)-1-(p-hydroxyphenoxy)-3-(isopropylamino)-2-propanol. C₁₂H₁₉NO₃; mol wt 225.29. C 63.98%, H 8.50%, N 6.22%, O 21.30%. A β₁-adrenergic agonist. Prepn: K. A. Jaeggi et al., **DE 2503968**; **US 3978041**; **US 4049797** (1974, 1976, 1977, all to Ciba-Geigy). Pharmacologic study: E. Carlsson et al., Arch. Pharmacol. **300**, 101 (1977). Metabolism, hemodynamic effects, pharmacokinetics in man: O. Rönn et al., Eur. J. Clin. Pharmacol. **17**, 81 (1980). Cardiovascular effects: D. H. Scott et al., Br. J. Clin. Pharmacol. **7**, 365 (1979). Clinical study in coronary heart disease: I. Hutton et al., Br. Heart J. **43**, 134 (1980). Selectivity of action: T. P. Kenakin, D. Beek, J. Pharmacol. Exp. Ther. **213**, 406 (1980). Prepns of the racemic mixture: **NL 6409883**; H. Köppe et al., **US 3637852** (1965, 1972 both to Boehringer, Ing.); **NL 301580**; A. F. Crowther, L. H. Smith, **US 3501769** (1965, 1970 both to ICI); A. F. Crowther et al., J. Med. Chem. **12**, 638 (1969). Symposium: Acta Med. Scand. Suppl. **659**, 1-325 (1982).

OH   H
                                      |
HO⟍   O⟍⟍⟍⟍   N⟍CH₃
                                          |
                                          CH₃

Crystals from ethyl acetate, mp 127-128°. $[\alpha]_D^{20}$ −1 ±1°; $[\alpha]_{Hg}^{20}$ +2 ±1° (c = 0.940 in methanol).

Hydrochloride. [61260-05-7] H-133/22; (−)-H-80/62; CGP-7760B; Hyprenan. C₁₂H₁₉NO₃.HCl; mol wt 261.75.

THERAP CAT: Cardiotonic.

7854. Prenoxdiazine. [47543-65-7] 1-[2-[3-(2,2-Diphenylethyl)-1,2,4-oxadiazol-5-yl]ethyl]piperidine; 3-(β,β-diphenylethyl)-5-(β-piperidinoethyl)-1,2,4-oxadiazole. C₂₃H₂₇N₃O; mol wt 361.49. C 76.42%, H 7.53%, N 11.62%, O 4.43%. Prepn: K. Harsanyi et al., **HU 151748**: eidem, **US 3280122** (1964, 1966 both to Chinoin); of the hibenzate: G. Vita, P. Melloni, **GB 1347172** (1974 to Chinoin). Pharmacology: L. Tardos, I. Erdély, Arzneim.-Forsch. **16**, 617 (1966). Stability study: E. Pandula et al., Acta Pharm. Hung. **38**, 68 (1968). Review of pharmacology and clinical studies: K. Harsanyi et al., Boll. Chim. Farm. **112**, 691 (1973).

                                                    N
                                          N |
                         ⟋⟍   ⟍   ∥   ⟍ |
                        | | | N O   N
                         ⟍ |

Hydrochloride. [982-43-4] HK-256; Libexin; Prenoxid; Rhinathiol Tusso. C₂₃H₂₇N₃O.HCl; mol wt 397.95. Crystals from ethanol, mp 192-193°. LD₅₀ in mice, rats (mg/kg): 920, >2000 orally; 34, 32 i.v. (Tardos, Erdély).

Compd with 2-(4-hydroxybenzoyl)benzoic acid. [37671-82-2] Prenoxdiazine hibenzate. C₂₃H₂₇N₃O.C₁₄H₁₀O₄; mol wt 603.72. Crystals from ethanol, mp 164-166°.

THERAP CAT: Antitussive.

7855. Prenylamine. [390-64-7] N-(1-Methyl-2-phenylethyl)-γ-phenylbenzenepropanamine; N-(3,3-diphenylpropyl)-α-methylphenethylamine; N-(3′-phenyl-3′-propyl)-1,1-diphenyl-3-propylamine; 1-phenyl-2-[1′,1′-diphenylpropyl-3′-amino]propane; B-436; Elecor. C₂₄H₂₇N; mol wt 329.49. C 87.49%, H 8.26%, N 4.25%. Prepn: G. Ehrhart et al., **DE 1100031**; C.A. **56**, 3413h (1962) and **DE 1111642** corresp to **US 3152173** (1961, 1961, 1964 all to Hoechst); G. Erhart, Arch. Pharm. **295**, 196 (1962). Series of arti-

cles on pharmacology and chemistry: Arzneim.-Forsch. **10**, 569-588 (1960). Metabolism: M. Volz, ibid. **21**, 1320 (1971). Review: J. E. Murphy, J. Int. Med. Res. **1**, 204-209 (1973).

                              CH₃
                              |
              ⟋⟍   ⟋⟍ |
             | | | | CH
                    ⟍   ⟍   N⟍   ⟋⟍
                                            H | |
                                                      ⟍

mp 36.5-37.5°.

Lactate. [69-43-2] Angormin; Bismetin; Carditin-Same; Coredamin; Corontin; Crepasin; Daxauten; Hostaginan; Incoran; Irrorin; Lactamin; Plactamin; Reocorin; Roinin; Seccidin; Sedolatan; Segontin; Synadrin. C₂₇H₃₃NO₃; mol wt 419.57. mp 140-142°. Sparingly sol in water (~0.5%); sol in organic solvents. uv max (chloroform): 260 nm (E₁ₐₘ¹ᐟ⁸ 170).

THERAP CAT: Vasodilator (coronary).

7856. Prephenic Acid. [126-49-8] 1-Carboxy-4-hydroxy-α-oxo-2,5-cyclohexadiene-1-propanoic acid; 1-carboxy-4-hydroxy-2,5-cyclohexadiene-1-pyruvic acid. C₁₀H₁₀O₆; mol wt 226.18. C 53.10%, H 4.46%, O 42.44%. Non-aromatic biosynthetic intermediate that represents a secondary branch-point in the pathway from chorismic acid to phenylalanine and tyrosine, q.q.v., in many organisms. Isoln from cultures of mutant Escherichia coli: B. D. Davis, Science **118**, 251 (1953). Characterized as cryst Ba salt: U. Weiss et al., ibid. **119**, 774 (1954); R. L. Metzenberg, H. K. Mitchell, Arch. Biochem. Biophys. **64**, 51 (1956). Structure: H. Plieninger, G. Keilich, Z. Naturforsch. **16b**, 81 (1961). Synthetic approaches: H. Plieninger, Angew. Chem. Int. Ed. **1**, 367 (1962). Total synthesis of disodium salt: S. Danishevsky, M. Hirama, J. Am. Chem. Soc. **99**, 7740 (1977); S. Danishevsky et al., ibid. **101**, 7013 (1979); W. Gramlich, H. Plieninger, Ber. **112**, 1571 (1979). Review: U. Weiss, J. M. Edwards, The Biosynthesis of Aromatic Compounds (Wiley, New York, 1980) pp 144-184. See also shikimic acid.

        HOOC⟍   ⟍   ⟋COOH
                    ⟍   ⟍ ⟋
                     ⟍   ⟍⟋
                      | | O
                       ⟍ ⟋
              HO⟋ ⟍H

Barium salt monohydrate. C₁₀H₈BaO₆.H₂O. Crystals from water + methanol or water + pyridine.

7857. Presenilins. Integral membrane proteins required for the proteolytic processing of the *amyloid precursor protein* (APP) to yield amyloid-β peptide (Aβ), q.v. Missense mutations in the presenilin genes result in the production of an abnormal form of Aβ which is the major component of the amyloid plaques deposited in the brains of patients with Alzheimer's disease (AD). Highly hydrophobic, *presenilin 1* (*PS1*) and its homolog *presenilin 2* (*PS2*), consist of 463 and 448 amino acid residues, respectively. PS1 is believed to be a catalytic subunit of γ-*secretase*. Thought to have 7 or 8 transmembrane domains that span the membranes of the endoplasmic reticulum, nuclear envelope and Golgi apparatus. Identification of the gene for PS1: R. Sherrington et al., Nature **375**, 754 (1995); for PS2: E. I. Rogaev et al., ibid. **376**, 775 (1995). Characterization of PS1 and distribution in human and rodent brain: G. A. Elder et al., J. Neurosci. Res. **45**, 308 (1996). Structural study: N. N. Dewji et al., Proc. Natl. Acad. Sci. USA **101**, 1057 (2004). Review of structure, possible physiological functions, and role in Alzheimer's disease: C. Haass, B. De Strooper, Science **286**, 916-919 (1999); P. E. Fraser et al., Biochim. Biophys. Acta **1502**, 1-15 (2000).

7858. Pretilachlor. [51218-49-6] 2-Chloro-N-(2,6-diethylphenyl)-N-(2-propoxyethyl)acetamide; N-propoxyethyl-N-chloroacetyl-2,6-diethylaniline; 2,6-diethyl-N-(2′-n-propoxyethyl)chloroacetanilide; CGA-26423; CG-113; Rifit. C₁₇H₂₆ClNO₂; mol wt 311.85. C 65.48%, H 8.40%, Cl 11.37%, N 4.49%, O 10.26%. Selective herbicide for control of perennial weeds in transplanted rice.

Prepn, herbicidal properties: **NL 7307584**; C. Vogel, R. Aebi, **US 4168965** (1973, 1979 both to Ciba-Geigy). Residue analysis, stability: H. Egli, *J. Agric. Food Chem.* **30**, 861 (1982). Effect on leaf elongation in rice, use in combination with a safening agent (CGA 123,407): R. A. Christ, *Weed Res.* **25**, 193 (1985). Brief account: J. Rufener, M. Quadranti, *Proc. 10th Conf. Int. Congr. Plant Prot.* **1**, 332 (1983).

Colorless liquid, $bp_{0.001}$ 135°. Vapor pressure at 20°: 1×10^{-6} mm Hg. Almost insol in water (50 mg/l at 20°). Sol in most organic solvents. n_D^{20} 1.5204. LD_{50} in rats (mg/kg): 6099 orally; >3100 dermally (Rufener, Quadranti). LC_{50} in rainbow trout, carp, catfish: 3.0, 3.0, 2.6 ppm (Vogel, Aebi).

USE: Herbicide for use in rice paddys.

7859. Pridinol. [511-45-5] α,α-Diphenyl-1-piperidinepropanol; 1,1-diphenyl-3-piperidino-1-propanol; 3-piperidino-1,1-diphenyl-1-propanol; 1,1-diphenyl-3-(1-piperidyl)-1-propanol; 3-(N-piperidyl)-1,1-diphenyl-1-propanol; ridinol; C-238. $C_{20}H_{25}NO$; mol wt 295.43. C 81.31%, H 8.53%, N 4.74%, O 5.42%. May be prepd from ethyl 1-piperidinepropionate and phenylmagnesium bromide: Adamson, **GB 624118** (1949 to Wellcome Found.). Toxicity study: R. W. Cunningham *et al.*, *J. Pharmacol. Exp. Ther.* **96**, 151 (1949).

Crystals, mp 120-121°. Soluble in acetone.
Hydrochloride. [968-58-1] Parks 12. $C_{20}H_{25}NO \cdot HCl$; mol wt 331.88. Crystals, dec 238°. Sol in alc. LD_{50} in mice, rats (mg/kg): 35, 33 i.v.; 131, 91 i.p. (Cunningham).
Methanesulfonate. [6856-31-1] Pridinol mesylate; Konlax; Loxeen; Lyseen; Myoson. $C_{20}H_{25}NO \cdot CH_3SO_3H$; mol wt 391.53. Crystals, mp 152.5-155.0°. Sparingly sol in water.
THERAP CAT: Antiparkinsonian; muscle relaxant (skeletal).

7860. Pridopidine. [346688-38-8] 4-[3-(Methylsulfonyl)-phenyl]-1-propylpiperidine; ACR-16; ASP-2314; FR-310826. $C_{15}H_{23}NO_2S$; mol wt 281.41. C 64.02%, H 8.24%, N 4.98%, O 11.37%, S 11.39%. Dopamine D_2-receptor antagonist that preferentially binds to activated receptors; stabilizes the dopaminergic system by either enhancing or inhibiting dopaminergic dependent functions in the brain depending on the initial level of dopaminergic activity. Prepn: B. Andersson *et al.*, **WO 0146145**; C. Sonesson *et al.*, **US 6903120** (2001, 2005 both to A. Carlsson Res.); and dopaminergic activity: F. Pettersson *et al.*, *J. Med. Chem.* **53**, 2510 (2010). Receptor occupancy studies: S. Natesan *et al.*, *J. Pharmacol. Exp. Ther.* **318**, 810 (2006); T. Dyhring *et al.*, *Eur. J. Pharmacol.* **628**, 19 (2010). Behavioral pharmacology: H. Ponten *et al.*, *ibid.* **644**, 88 (2010). Clinical evaluation in Huntington's disease: A. Lundin *et al.*, *Clin. Neuropharmacol.* **33**, 260 (2010).

Hydrochloride. [882737-42-0] Huntexil. $C_{15}H_{23}NO_2S \cdot HCl$; mol wt 317.87. Crystals from ethanol + diethyl ether, mp 212-214°.
THERAP CAT: In treatment of Huntington's disease.

7861. Prifinium Bromide. [4630-95-9] 3-(Diphenylmethylene)-1,1-diethyl-2-methylpyrrolidinium bromide (1:1); 3-(diphenyl-

methylene)-1-ethyl-2-methylpyrrolidine ethyl bromide; pyrodifenium bromide; Padrin; Riabal. $C_{22}H_{28}BrN$; mol wt 386.38. C 68.39%, H 7.30%, Br 20.68%, N 3.63%. Prepn: Sadao Oki, **JP 65 22462** (1965), *C.A.* **64**, 3489a (1966); of the free base: *idem*, **JP 65 17015** (1965), *C.A.* **63**, 18034b (1965). Ganglion blocking agent. Metabolism: Nakai *et al.*, *Arzneim.-Forsch.* **20**, 1112 (1970). Toxicity studies: Kumada *et al.*, *ibid.* **22**, 706 (1972).

Crystals, mp 216-218°. (Free base, $bp_{0.15}$ 183-185°). LD_{50} in male mice (mg/kg): 11 i.v.; 43 i.p.; 30 s.c.; 330 orally (Kumada).
THERAP CAT: Antispasmodic.

7862. Prilocaine. [721-50-6] N-(2-Methylphenyl)-2-(propylamino)propanamide; 2-(propylamino)-o-propionotoluidide; N-(α-propylaminopropionyl)-o-toluidine; α-propylamino-2-methylpropionanilide; propitocaine. $C_{13}H_{20}N_2O$; mol wt 220.32. C 70.87%, H 9.15%, N 12.72%, O 7.26%. Prepn: N. Löfgren, C. Tegner, *Acta Chem. Scand.* **14**, 486 (1960); **GB 839943**; N. Löfgren, C. Tegner, **US 3160662** (1960, 1964 both to Astra).

Needles, mp 37-38°. $bp_{0.1}$ 159-162°. n_D^{20} 1.5298. Very sol in alc, acetone; slightly sol in water.
Hydrochloride. [1786-81-8] L-67; Citanest; Xylonest. $C_{13}H_{20}N_2O \cdot HCl$; mol wt 256.77. Crystals from ethanol + isopropyl ether, mp 167-168°. Freely sol in water, alc; slightly sol in chloroform; very slightly sol in acetone. Practically insol in ether.
THERAP CAT: Anesthetic (local).

7863. Primaquine. [90-34-6] N^4-(6-Methoxy-8-quinolinyl)-1,4-pentanediamine; 8-(4-amino-1-methylbutylamino)-6-methoxyquinoline; SN-13272. $C_{15}H_{21}N_3O$; mol wt 259.35. C 69.47%, H 8.16%, N 16.20%, O 6.17%. Prepn following the synthesis of pamaquine: Elderfield *et al.*, *J. Am. Chem. Soc.* **68**, 1525 (1946); improved procedure: Elderfield *et al.*, *ibid.* **77**, 4816 (1955). *Review:* Olenick in *Antibiotics* vol. 3, J. W. Corcoran, F. E. Hahn, Eds. (Springer-Verlag, New York, 1975) pp 516-520.

Viscous liquid, $bp_{0.2}$ 175-179°. Soluble in ether.
Diphosphate. [63-45-6] $C_{15}H_{21}N_3O \cdot 2H_3PO_4$. Yellow crystals from 90% ethanol, mp 197-198°. Sol in water. Insol in chloroform, ether.
Oxalate. $C_{15}H_{21}N_3O \cdot C_2H_2O_4$. Yellow crystals from 80% ethanol, mp 182.5-185°.
THERAP CAT: Antimalarial.

7864. Primeverose. [26531-85-1] 6-O-β-D-Xylopyranosyl-D-glucose; 6-(β-D-xylosido)-D-glucose. $C_{11}H_{20}O_{10}$; mol wt 312.27. C 42.31%, H 6.46%, O 51.23%. By enzymatic hydrolysis of gaultherin (monotropitoside), primeverin, rhamnicoside and other glycosides: Goris, Vischniac, *Compt. Rend.* **169**, 871 (1919); Bridel, *ibid.* **179**, 991 (1924); Bridel, Charaux, *ibid.* **180**, 857 (1925); Richter, *J.*

Chem. Soc. **1936**, 1701. Synthesis: Helferich, Rauch, *Ber.* **59**, 2655 (1926); *Ann.* **455**, 168 (1927); McCloskey, Coleman, *J. Am. Chem. Soc.* **65**, 1778 (1943).

Crystals from methanol or 80% alcohol. Darkens at 190°, mp 209-210° (copper block). Shows mutarotation. $[\alpha]_D^{20}$ +23° → −3.2° (c = 5 in water). Sol in water, methanol, 80% alcohol. Is hydrolyzed to 1 mol D-glucose and 1 mol D-xylose by boiling 2% H_2SO_4 for 5 hrs. Reduces Fehling's soln slowly in the cold, instantly when hot.

7865. Primidone. [125-33-7] 5-Ethyldihydro-5-phenyl-4,6-(1*H*,5*H*)-pyrimidinedione; 5-ethyl-5-phenylhexahydropyrimidine-4,6-dione; 2-desoxyphenobarbital; Liskantin; Mylepsinum; Mysoline; Resimatil; Sertan. $C_{12}H_{14}N_2O_2$; mol wt 218.26. C 66.04%, H 6.47%, N 12.84%, O 14.66%. Prepd by electrolytic reduction of phenobarbital or by catalytic desulfuration of the corresp 2-thiobarbituric acid: Boon *et al.*, **GB 666027** (1952 to I.C.I.); Bogue, Carrington, *Br. J. Pharmacol.* **8**, 230 (1953). Comprehensive description: R. D. Daley in *Anal. Profiles Drug Subs.* **2**, 409-437 (1973); A. A. Al-Badr, H. A. El-Obeid, *ibid.* **17**, 749-795 (1988). HPLC determn in urine: V. Ferranti *et al.*, *J. Chromatogr. B* **718**, 199 (1998). Clinical trial in essential tremor: M. Serrano-Dueñas, *Parkinsonism Relat. Disord.* **10**, 29 (2003).

Crystals, mp 281-282°. Practically tasteless. Has no acidic properties. Slightly sol in alc; very slightly sol in water and in most organic solvents.

THERAP CAT: Anticonvulsant.

THERAP CAT (VET): Anticonvulsant; chiefly to control epileptiform seizures.

7866. Primisulfuron-methyl. [86209-51-0] 2-[[[[[4,6-Bis-(difluoromethoxy)-2-pyrimidinyl]amino]carbonyl]amino]sulfonyl]-benzoic acid methyl ester; 2-[3-(4,6-bis(difluoromethoxy)pyrimidin-2-yl)ureidosulfonyl]benzoic acid methyl ester; *N*-(2-methoxycarbonylphenylsulfonyl)-*N'*-[4,6-bis(difluoromethoxy)pyrimidin-2-yl]-urea; CGA-136872; Beacon; Tell. $C_{15}H_{12}F_4N_4O_7S$; mol wt 468.34. C 38.47%, H 2.58%, F 16.23%, N 11.96%, O 23.91%, S 6.85%. Post-emergence sulfonylurea herbicide. Prepn: W. Meyer *et al.*, **EP 84020**; *eidem*, **US 4478635** (1983, 1984 both to Ciba-Geigy). Comprehensive description: W. Maurer *et al.*, *Proc. Br. Crop Prot. Conf. - Weeds* **1987**, 41-48. Field trial in corn: C. L. Foy, H. L. Witt, *Weed Technol.* **4**, 615 (1990).

Colorless crystals, mp 203.1° (Maurer); also reported as mp 186-188° (Meyer). Vapor pressure (20°): <7.5 × 10⁻¹² mm Hg. Soly in water (pH 7): 0.07 g/l at 20°. LD_{50} in rats (mg/kg): >4000 orally; >2000 dermally (Maurer).

USE: Herbicide.

7867. Primulaverin. [154-61-0] 5-Methoxy-2-[(6-*O*-β-D-xylopyranosyl-β-D-glucopyranosyl)oxy]-benzoic acid methyl ester; 2-hydroxy-5-methoxymethyl benzoate-2-primeveroside. $C_{20}H_{28}O_{13}$; mol wt 476.43. C 50.42%, H 5.92%, O 43.66%. Isoln from species of *Primulaceae*: Thieme, Winkler, *Pharmazie* **26**, 434 (1971). Synthesis: Chaudhury *et al.*, *J. Chem. Soc.* **1948**, 2220.

Dihydrate. Star-shaped clusters from ethyl acetate, needles from alc, mp 163°. $[\alpha]_D^{20}$ −67°. Sol in water, alc, acetone.

7868. Primycin. Macrolide antibiotic complex of more than 20 components produced by actinomycetes found in the intestinal tract of the wax moth *(Galeria melonella)*. Nine primary components, in 3 major groups designated A, B, and C, represent 90% of the total material. Originally thought to be a single entity which was subseqently identified as the major component, primycin A_1. Isoln from cultures of *Streptomyces primycini*: Vályi-Nagy *et al.*, *Nature* **174**, 1105 (1954); **HU 146332** and **HU 151197** (1962, 1964 both to Hung. Acad. Sci.). Production by *Thermopolyspora galeriensis* (also designated *Micromonospora galeriensis)*: T. Vályi-Nagy *et al.*, **HU 153593**; *eidem*, **US 3498884** (1967, 1970 both to Chinoin). Structure of A_1: J. Aberhart *et al.*, *J. Am. Chem. Soc.* **92**, 5816 (1970); *eidem*, *J. Chem. Soc. Perkin Trans. 1* **1974**, 816, 836. TLC sepn of components: I. Szilagyi *et al.*, *J. Chromatogr.* **295**, 141 (1984). Structures of the nine primary components: J. Frank *et al.*, *Tetrahedron Lett.* **28**, 2759 (1987). Antibacterial activity and toxicity: T. Vályi-Nagy *et al.*, *Pharmazie* **11**, 304 (1956). Antifungal activity *in vitro*: J. V. Uri, P. Actor, *J. Antibiot.* **32**, 1207 (1979). Mode of action: I. Horvath *et al.*, *Arch. Microbiol.* **121**, 135 (1979). Clinical studies in dermatological infections: C. Mészáros, K. Vezekényi, *Ther. Hung.* **35**, 77 (1987); J. Biro, V. Várkonyi, *ibid.* 136; G. Bálint, *ibid.* 140. *Review:* J. V. Uri, *Acta Microbiol. Hung.* **33**, 141 (1986).

Primycin A_1

Sulfate. mp 192-195° (dec) (Aberhart, 1974); also reported as 202-206° (dec) (Uri, 1986). Infrared and ultraviolet spectra: Szilágyi *et al.*, *Nature* **193**, 243 (1962). Fairly soluble in methanol (2.5%), less sol in the higher alcohols, sparingly sol in pyridine, glacial acetic acid, water. LD_{50} in mice, guinea pigs, rats, rabbits (mg/kg): 2.5, 5.0, 10.0, 10.0 i.p. (Vályi-Nagy, 1956).

THERAP CAT: Antibacterial (topical).

7869. Prinomastat. [192329-42-3] (3*S*)-*N*-Hydroxy-2,2-dimethyl-4-[[4-(4-pyridinyloxy)phenyl]sulfonyl]-3-thiomorpholine-carboxamide; 3(*S*)-*N*-hydroxy-4-((4-((pyrid-4-yl)oxy)benzenesulfonyl)-2,2-dimethyl)tetrahydro-2*H*-1,4-thiazine-3-carboxamide; AG-3340. $C_{18}H_{21}N_3O_5S_2$; mol wt 423.50. C 51.05%, H 5.00%, N 9.92%, O 18.89%, S 15.14%. Selective matrix metalloproteinase inhibitor with antiangiogenic activity. Prepn: S. E. Zook *et al.*, **WO 9720824**; S. L. Bender, M. J. Melnick, **US 5753653** (1997, 1998 both to Agouron). Pharmacokinetic and antitumor efficacy in rats: O. Santos *et al.*, *Clin. Exp. Metastasis* **15**, 499 (1997). Review of pharmacology: D. R. Shalinsky *et al.*, *Ann. N.Y. Acad. Sci.* **878**, 236-270 (1999); and clinical experiences: R. Scatena, *Expert Opin. Invest. Drugs* **9**, 2159-2165 (2000).

White powder, mp 149.8°. Also reported as 184-186° with gas evolution. Water soluble.

THERAP CAT: Antineoplastic.

7870. Pristane. [1921-70-6] 2,6,10,14-Tetramethylpentadecane; norphytane; Robuoy. $C_{19}H_{40}$; mol wt 268.53. C 84.98%, H 15.02%. Isoprenoid alkane obtained from the unsaponifiable fraction of shark liver oil where it occurs to an extent of 14%: Tsujimoto, *J. Soc. Chem. Ind.* **51**, 317T (1932); Sörensen, Mehllum, *Acta Chem. Scand.* **2**, 140 (1948). Identity with norphytane: Pliva, Sörensen, *ibid.* **4**, 846 (1950). Isoln from petroleum crude oils: Bendoraitis *et al.*, *Anal. Chem.* **34**, 49 (1962); from wool wax: Mold *et al.*, *Nature* **199**, 283 (1963). Synthesis from phytol: Sörensen, Sörensen, *Acta Chem. Scand.* **3**, 939 (1949). Metabolism: McKenna, Kallio, *Proc. Natl. Acad. Sci. USA* **68**, 1552 (1971). Use in inducing murine plasmacytomas: P. N. Anderson, M. Potter, *Nature* **222**, 994 (1969). Effect on ascites tumor formation and monoclonal antibody production: N. J. Hoogenraad, C. J. Wraight, *Methods Enzymol.* **121**, 375 (1986).

Mobile, transparent, stable liq, d_4^{20} 0.78267. Congealing point $-100°$. bp_{760} 296°; bp_{12} 158°; $bp_{0.001}$ 68° (bath temp); n_D^{20} 1.43848. Acid no. 0-5. Iodine no. 0-7.5. Sapon no. 0-5. Viscosity at 25°: 5 cP. Soluble in ether, petr ether, benzene, chloroform, carbon tetrachloride.

USE: Lubricant; transformer oil. Anti-corrosion agent. Biological marker. In experimental systems to induce plasmacytomas; in production of monoclonal antibodies.

7871. Pristinamycin. [270076-60-3] RP-7293; Pyostacine. One of the streptogramins, *q.v.*, produced by *Streptomyces pristinaespiralis* (NRRL 2958). The two major components are IA and IIA. Isoln: W. D. Celmer, B. A. Sobin, *Antibiot. Annu.* **1955-1956**, 437; D. I. Mancy *et al.*, **FR 1301857**; *eidem*, **US 3154475** (1962, 1964 both to Rhône-Poulenc); J. Preud'homme *et al.*, *Compt. Rend.* **260**, 1309 (1965). Isoln and characterization of constituents: *idem et al.*, *Bull. Soc. Chim. Fr.* **1968**, 585. HPLC determn of major components in plasma: C. Koechlin *et al.*, *J. Chromatogr.* **425**, 197 (1988). Total synthesis of IIB: P. Breuilles, D. Uguen, *Tetrahedron Lett.* **39**, 3149 (1998). Pharmacokinetics: C. Koechlin *et al.*, *J. Antimicrob. Chemother.* **25**, 651 (1990). Clinical trial in pneumonia: R. Poirier, *Presse Med.* **28**, Suppl. 1, 13 (1999).

Pristinamycin IA

Pristinamycin IA. [3131-03-1] 4-[4-(Dimethylamino)-N-methyl-L-phenylalanine]virginiamycin S₁; streptogramin B; mikamycin IA; ostreogrycin B; vernamycin Bα. $C_{45}H_{54}N_8O_{10}$; mol wt 866.97. Acid-base properties: M. Largeron, M. B. Fleury, *J. Pharm. Sci.* **81**, 565 (1992). Biosynthesis: V. de Crècy-Lagard *et al.*, *Antimicrob. Agents Chemother.* **41**, 1904 (1997). White microcrystalline powder from methanol, mp 198°. $[\alpha]_D^{22}$ $-57.5°$ (c = 0.25 in ethanol). uv max (ethanol): 243, 260, 281, 303 nm ($E_{1cm}^{1\%}$ 140, 213, 60, 104).

Pristinamycin IIA *see* Virginiamycin M₁.

THERAP CAT: Antibacterial.

7872. Probenecid. [57-66-9] 4-[(Dipropylamino)sulfonyl]benzoic acid; p-(dipropylsulfamoyl)benzoic acid; p-(dipropylsulfamyl)benzoic acid; Benemid; Probecid; Proben. $C_{13}H_{19}NO_4S$; mol wt 285.36. C 54.72%, H 6.71%, N 4.91%, O 22.43%, S 11.23%. Prepn from p-carboxybenzenesulfonyl chloride and dipropylamine: C. S. Miller, **US 2608507** (1952 to Sharp & Dohme). Study of metabolites: Z. H. Israili *et al.*, *J. Med. Chem.* **15**, 709 (1972). Clinical pharmacokinetics: R. F. Cunningham *et al.*, *Clin. Pharmacokinet.* **6**, 135 (1981). Toxicology studies: S. E. McKinney *et al.*, *J. Pharmacol. Exp. Ther.* **102**, 208 (1951); *NTP Technical Report 395* (PB92-129584, 1991) 220 pp. Comprehensive description: A. A. Al-Badr, H. A. El-Obeid, *Anal. Profiles Drug Subs.* **10**, 639-663 (1981).

Crystals from dilute alcohol, mp 194-196°. uv max (0.1N NaOH): 242.5 nm. pKa 5.8. Slightly bitter taste, pleasant aftertaste. Sol in alc, acetone, chloroform, and in dil solns of NaOH buffered to pH 7.4 and other dilute alkalis. Practically insol in water and in dilute acids.

Sodium salt. [23795-03-1] $C_{13}H_{18}NNaO_4S$. LD_{50} in mice (mg/kg): 458 i.v., 1156 s.c., 1666 orally; in rats (mg/kg): 394 i.p., 611 s.c., 1604 orally (McKinney).

THERAP CAT: Uricosuric.

7873. Probucol. [23288-49-5] 4,4'-[(1-Methylethylidene)bis(thio)]bis[2,6-bis(1,1-dimethylethyl)phenol]; 4,4'-(isopropylidenedithio)bis[2,6-di-*tert*-butylphenol]; acetone bis(3,5-di-*tert*-butyl-4-hydroxyphenyl)mercaptole; DH-581; Lorelco; Lurselle; Sinlestal. $C_{31}H_{48}O_2S_2$; mol wt 516.84. C 72.04%, H 9.36%, O 6.19%, S 12.41%. Lipid-lowering antioxidant. Prepn: M. B. Neuworth, **FR 1561853**; *eidem*, **US 3576883** (1969, 1971 both to Consolidation Coal Co.); and use as a cholesterol-lowering agent: J. W. Barnhart, P. J. Shea, **US 3862332** (1975 to Dow). Prepn and activity studies: M. B. Neuworth *et al.*, *J. Med. Chem.* **13**, 722 (1970). Review of pharmacology and therapeutic use in hyperlipidemia: R. C. Heel *et al.*, *Drugs* **15**, 409-428 (1978). Symposium on mechanism of action, clinical efficacy and safety: *Am. J. Cardiol.* **57**, 1H-54H (1986). Clinical trial to prevent restenosis following angioplasty: J.-C. Tardif *et al.*, *N. Engl. J. Med.* **337**, 365 (1997).

White crystalline solid from ethanol, mp 124.5-126°; fine, yellow crystals from isopropanol, mp 125-126.5°. Freely sol in chloroform, *n*-propyl alc; sol in alc and in solvent hexane. Insol in water.

THERAP CAT: Antilipemic.

7874. Procainamide Hydrochloride. [614-39-1] 4-Amino-N-[2-(diethylamino)ethyl]benzamide hydrochloride (1:1); Amisalin; Novocamid; Procamide; Procanbid; Procan-SR; Procapan; Pronestyl. $C_{13}H_{22}ClN_3O$; mol wt 271.79. C 57.45%, H 8.16%, Cl 13.04%, N 15.46%, O 5.89%. Prepn: M. Yamazaki *et al.*, *J. Pharm. Soc. Jpn.* **73**, 294 (1953); Y. Tashika, M. Kuranari, *ibid.* 1069. Comprehensive description: R. B. Poet, H. Kadin, *Anal. Profiles Drug Subs.* **4**, 333-383 (1975).

Crystals, mp 165-169°. uv max: 278 nm. Very sol in water; sol in alc; slightly sol in chloroform; very slightly sol in benzene, ether. The pH of a 10% aq soln is 5.5.

THERAP CAT: Antiarrhythmic (class IA).

7875. Procaine. [59-46-1] 4-Aminobenzoic acid 2-(diethylamino)ethyl ester; *p*-aminobenzoyldiethylaminoethanol; 2-diethylaminoethyl *p*-aminobenzoate. $C_{13}H_{20}N_2O_2$; mol wt 236.32. C 66.07%, H 8.53%, N 11.85%, O 13.54%. Benzoic acid derivative with anesthetic activity. Prepn: A. Einhorn, **US 812554** (1906); *idem, Ann.* **371**, 125 (1909); A. Einhorn, E. Uhlfelder, *ibid.* 131. CNS effects: C. G. Peterson, *Anesthesiology* **16**, 976 (1955). Intravenous pharmacokinetics in humans: A. B. Seifen *et al.*, *Anesth. Analg.* **58**, 382 (1979). Clinical evaluation as anti-arrhythmic and cough suppressant during anesthesia: D. S. Thompson *et al.*, *Am. J. Surg.* **138**, 798 (1979). Stabilization of vascular smooth muscle *in vitro*: K. Kitamura *et al.*, *Drugs Exp. Clin. Res.* **12**, 773 (1986). Toxicity data: W. C. North, K. F. Urbach, *J. Am. Pharm. Assoc. Sci. Ed.* **45**, 382 (1956); E. I. Goldenthal, *Toxicol. Appl. Pharmacol.* **18**, 185 (1971).

Hygroscopic, anhydr plates, tablets from ligroin or ether, mp 61°. When freshly precipitated, one gram dissolves in 200 ml water. Sol in alc, ether, benzene, chloroform. LD_{50} in mice (mg/kg): 195 i.p.; 45 i.v. (North, Urbach).

Dihydrate. [6192-89-8] Needles from aq alc, mp 51°. Slightly bitter taste; applied to the tongue causes transitory numbing sensation.

Nitrate. [6192-92-3] $C_{13}H_{20}N_2O_2 \cdot HNO_3$. Crystals, mp 100-102°. Sol in water, alc. The aq soln is neutral. Particularly useful with silver nitrate because no precipitate forms.

Butyrate. [136-55-0] Probutylin. $C_{13}H_{20}N_2O_2 \cdot C_4H_8O_2$; mol wt 324.42. Hygroscopic crystals. Soluble in water, alc, vegetable oils.

Hydrochloride. [51-05-8] Anestil; Enpro; Gero; Jenacaine; Medaject; Naucaine; Neocaine; Novocain; Omnicain; Planocaine; Rocain; Syntocain. $C_{13}H_{20}N_2O_2 \cdot HCl$; mol wt 272.77. Crystals. Six-sided plates, monoclinic or triclinic. mp 153-156°. Numbing taste. Stable in air. One gram dissolves in 1 ml water and in 30 ml alcohol. Slightly sol in chloroform. Practically insol in ether. The pH of a $0.1M$ aq soln is 6.0. Aq solns may be sterilized by boiling. LD_{50} in mice (mg/kg): 660 ± 60 s.c. (Goldenthal).

THERAP CAT: Anesthetic (local).

THERAP CAT (VET): Anesthetic (local).

7876. Procarbazine. [671-16-9] *N*-(1-Methylethyl)-4-[(2-methylhydrazinyl)methyl]benzamide; *N*-isopropyl-α-(2-methylhydrazino)-*p*-toluamide; *N*-4-isopropylcarbamoylbenzyl-*N'*-methylhydrazine; *p*-(*N*¹-methylhydrazinomethyl)-*N*-isopropylbenzamide; ibenzmethyzin; MIH; Ro-4-6467. $C_{12}H_{19}N_3O$; mol wt 221.30. C 65.13%, H 8.65%, N 18.99%, O 7.23%. Alkylating agent. Prepn: **BE 618638**; W. Bollag *et al.*, **US 3520926** (1962, 1970 to Hoffmann-La Roche). Comprehensive description: R. J. Rucki, *Anal. Profiles Drug Subs.* **5**, 403-427 (1976). HPLC determn in plasma: X. He *et al.*, *J. Chromatogr. B* **799**, 281 (2004). Veterinary use in treatment of granulomatous meningoencephalomyelitis: J. R. Coates *et al.*, *J.*

Vet. Intern. Med. **21**, 100 (2007). Review of clinical experience in hematological malignancies: M. Massoud *et al.*, *Eur. J. Cancer* **40**, 1924-1927 (2004).

Hydrochloride. [366-70-1] Matulane; Natulan. $C_{12}H_{19}N_3O \cdot HCl$; mol wt 257.76. Crystals from methanol, mp 223-226°. Soluble but unstable in water or aq. solutions. LD_{50} orally in rats: 785 ±34 mg/kg. *See*: E. I. Goldenthal, *Toxicol. Appl. Pharmacol.* **18**, 185 (1971).

Hydrobromide. [18969-59-0] $C_{12}H_{19}N_3O \cdot HBr$. Crystals from methanol + ether, dec 216-217°.

Caution: Procarbazine hydrochloride is reasonably anticipated to be a human carcinogen: *Report on Carcinogens, Twelfth Edition* (PB2011-111646, 2011) p 361.

THERAP CAT: Antineoplastic.

THERAP CAT (VET): Antineoplastic. In treatment of granulomatous meningoencephalitis in dogs.

7877. Procaterol. [72332-33-3] *rel*-8-Hydroxy-5-[(1*R*,2*S*)-1-hydroxy-2-[(1-methylethyl)amino]butyl]-2(1*H*)-quinolinone; (±)-*erythro*-8-hydroxy-5-[1-hydroxy-2-(isopropylamino)butyl]carbostyril. $C_{16}H_{22}N_2O_3$; mol wt 290.36. C 66.19%, H 7.64%, N 9.65%, O 16.53%. Sympathomimetic amine with selective β_2-adrenergic agonist activity. Prepn: K. Nakagawa *et al.*, **BE 823841**; *eidem*, **US 4026897** (1975, 1977 both to Otsuka); S. Yoshizaki *et al.*, *J. Med. Chem.* **19**, 1138 (1976). Prepn of optical isomers: *eidem, ibid.* **20**, 1103 (1977). Assessment of selective action: H. Himori, N. Taira, *Br. J. Pharmacol.* **61**, 9 (1977). Metabolism: Y. Yasuda *et al.*, *Arzneim.-Forsch.* **29**, 261 (1979). Pharmacokinetics: M. Ishigami *et al.*, *ibid.* 266. Series of reproduction studies: *Iyakuhin Kenkyu* **10**, 68-111 (1979), *C.A.* **90**, 197728f-30a (1979). Antigenicity test: N. Nakagiri, S. Tei, *Oyo Yakuri* **17**, 363 (1979), *C.A.* **91**, 186577a (1979).

Relative stereochemistry

Hydrochloride hemihydrate. [81262-93-3] OPC-2009; Lontermin; Masacin; Meptin; Onsukil; Procadil; Promaxol; Propulm. $C_{16}H_{22}N_2O_3 \cdot HCl \cdot \frac{1}{2}H_2O$; mol wt 335.83. Off-white crystalline powder, mp 193-197° (dec). Sol in methanol; slightly sol in ethanol. Practically insol in acetone, ether, ethyl acetate, chloroform, benzene. Colors on exposure to light. LD_{50} of the hydrochloride in male rats (mg/kg): 2600 orally, 80 i.v. (Nakagawa).

THERAP CAT: Bronchodilator.

7878. Prochloraz. [67747-09-5] *N*-Propyl-*N*-[2-(2,4,6-trichlorophenoxy)ethyl]-1*H*-imidazole-1-carboxamide; 1-[*N*-propyl-*N*-[2-(2,4,6-trichlorophenoxy)ethyl]carbamoyl]imidazole; BTS-40542; Sportak. $C_{15}H_{16}Cl_3N_3O_2$; mol wt 376.66. C 47.83%, H 4.28%, Cl 28.24%, N 11.16%, O 8.50%. Broad spectrum fungicide for use on cereal crops, fruit and vegetables. Prepn from imidazole: R. F. Brookes *et al.*, **DE 2429523**; *eidem*, **US 3991071** (1975, 1976 both to Boots). Fungicidal properties: R. J. Birchmore *et al.*, *Proc. Br. Crop Prot. Conf. - Pests Dis.* **1977**, 593; R. G. Harris *et al.*, *ibid.* **1979**, 53. Efficacy against cereal powdery mildew: D. M. Weighton *et al.*, *ibid.* **1977**, 25. Protective effect against blotch in winter wheat and barley: R. G. Harris, G. Barnes, *ibid.* **1981**, 267. Effect on cytochrome P-450 and sterol biosynthesis in Japanese quail: J.-L. Riviere *et al.*, *Pestic. Sci.* **15**, 317 (1984). HPLC/MS determn in rainbow trout: L. Debrauwer *et al.*, *J. Agric. Food Chem.* **49**, 3821 (2001). *Review:* A. de Saint-Blanquat, J. My, *Def. Veg.* **37**, 121 (1983).

White crystalline solid, mp 38.5-41°. Technical product is pale yellow viscous oil. Vapor pressure at 20°: 0.57×10^{-9} torr. Almost insol in water (0.0055 g/l). Soly in chloroform, diethyl ether, toluene, xylene: 2500 g/l; in acetone: 3500 g/l. LD_{50} in rats (mg/kg): 1600 orally; >5000 s.c.; 400-800 i.p.; LC_{50} (96 hour) in rainbow trout, bluegill (mg/l): 1, 2.2 (de Saint-Blanquat, My). Relatively non-toxic to bees.

USE: Fungicide.

7879. Prochlorperazine. [58-38-8] 2-Chloro-10-[3-(4-methyl-1-piperazinyl)propyl]-10*H*-phenothiazine; 3-chloro-10-[3-(4-methyl-1-piperazinyl)propyl]phenothiazine; 2-chloro-10-[3-(1-methyl-4-piperazinyl)propyl]phenothiazine; *N*-[γ-(4′-methylpiperazinyl-1′)propyl]-3-chlorophenothiazine; chlormeprazine; prochlorpemazine; proclorperazine; Bayer A 173; RP-6140; SKF-4657. $C_{20}H_{24}ClN_3S$; mol wt 373.94. C 64.24%, H 6.47%, Cl 9.48%, N 11.24%, S 8.57%. Prepn: R. J. Horclois, **GB 780193**; **FR 1167627**; **US 2902484** (1957, 1958, 1959 all to Rhône-Poulenc). Pharmacology and toxicity: S. Courvoisier *et al.*, *C. R. Seances Soc. Biol. Ses Fil.* **152**, 1371 (1958). Pharmacokinetics: W. B. Taylor, D. N. Bateman, *Br. J. Clin. Pharmacol.* **23**, 137 (1987). Clinical efficacy as antiemetic and antipsychotic: J. Lapierre *et al.*, *Can. Psychiatr. Assoc. J.* **14**, 267 (1969). Clinical trial in treatment of nausea, dizziness and vertigo: A. E. Ward, *Br. J. Clin. Pract.* **42**, 228 (1988).

Clear to pale yellow, viscous liquid. Sensitive to light. Freely sol in alc, chloroform, ether. Very slightly sol in water.

Dimaleate. [84-02-6] Buccastem; Compazine; Meterazine; Stemetil; Vertigon. $C_{20}H_{24}ClN_3S.2C_4H_4O_4$; mol wt 606.09. Minute crystals, mp 228°. Practically insol in water, alc, ether, benzene. Slightly sol in methanol, warm chloroform. LD_{50} in mice (mg/kg): 400 s.c.; 120 i.p.; 90 i.v., 400 orally (Courvoisier).

Dimethanesulfonate. [51888-09-6] Prochlorperazine mesylate; Novamin; Tementil. $C_{20}H_{24}ClN_3S.2CH_4O_3S$; mol wt 566.14.

THERAP CAT: Antiemetic; antipsychotic. In treatment of vertigo.

THERAP CAT (VET): Antiemetic.

7880. Procodazole. [23249-97-0] 1*H*-Benzimidazole-2-propanoic acid; β-(2-benzimidazolyl)propionic acid; 2-(2-carboxyethyl)benzimidazole; propazol; AL-1241; Estimulocel. $C_{10}H_{10}N_2O_2$; mol wt 190.20. C 63.15%, H 5.30%, N 14.73%, O 16.82%. Non-specific, active immunoprotective agent against viral and bacterial infections. Prepn: J. Maier, *Ann.* **327**, 17 (1903); R. Meyers, H. Lüders, *ibid.* **415**, 29 (1918); B. Chatterjee, *J. Chem. Soc.* **1929**, 2966; **ES 407882** (1972 to Lafarquim). Pharmacological studies: C. Fernández *et al.*, *Rev. Clin. Esp.* **135**, 539 (1974); *eidem, ibid.* **141**, 51 (1976). Clinical studies: M. Pérez Tascon, J. M. Monturio, *Med. Klin.* **181**, 78 (1976).

Silky white needles from water, mp 228° (dec). Sol in alc, warm water. Practically insol in ether, benzene. The water soln is extremely sweet-tasting.

THERAP CAT: Immunopotentiator (non-specific).

7881. Procyclidine. [77-37-2] α-Cyclohexyl-α-phenyl-1-pyrrolidinepropanol; 1-cyclohexyl-1-phenyl-3-(1-pyrrolidinyl)-1-propanol; 1-cyclohexyl-1-phenyl-3-pyrrolidino-1-propanol. $C_{19}H_{29}NO$; mol wt 287.45. C 79.39%, H 10.17%, N 4.87%, O 5.57%. Anticholinergic. Prepn of free base and hydrochloride: Adamson *et al.*, *J. Chem. Soc.* **1951**, 52; Adamson, **US 2891890** (1959 to Burroughs Wellcome). Prepn of methochloride: Bottorff, **US 2826590** (1958 to Lilly); Harfenist, Magnien, **US 2842555** (1958 to Burroughs Wellcome).

Crystals from petr ether, mp 85.5-86.5°. uv max (0.17% in ethanol): 258.5 nm (ε 233).

Hydrochloride. [1508-76-5] Arpicolin; Kemadrin; Osnervan. $C_{19}H_{29}NO.HCl$; mol wt 323.91. Crystals from ethanol + ethyl acetate, dec 226-227°. Sol in water (about 3.0 g/100 ml), alc, chloroform. Insol in ether, acetone.

Methochloride. 1-(3-Cyclohexyl-3-hydroxy-3-phenylpropyl)-1-methylpyrrolidinium chloride; tricyclamol chloride; Lergine. $C_{19}H_{29}NO.CH_3Cl$; mol wt 337.93. Crystals from nitroethane, mp 159-164°. Moderately sol in water, alc. Practically insol in ether.

Methosulfate. $C_{21}H_{35}NO_5S$. Crystals, mp ~100°. Soluble in water (about 2% at 25°), alc. Practically insol in ether.

THERAP CAT: Antiparkinsonian.

7882. Procymidone. [32809-16-8] 3-(3,5-Dichlorophenyl)-1,5-dimethyl-3-azabicyclo[3.1.0]hexane-2,4-dione; *N*-(3,5-dichlorophenyl)-1,2-dimethyl-1,2-cyclopropanedicarboximide; dicyclidine; S-7131; Sumisclex; Sumilex. $C_{13}H_{11}Cl_2NO_2$; mol wt 284.14. C 54.95%, H 3.90%, Cl 24.95%, N 4.93%, O 11.26%. Dicarboximide fungicide systemically active against *Botrytis* and *Sclerotinia spp.* on fruits and vegetables. Prepn: **NL 7003836**; A. Fujinami *et al.*, **US 3903090** (1970, 1975 both to Sumitomo). Antimicrobial spectrum and systemic activity: Y. Hisada *et al.*, *J. Pestic. Sci.* **1**, 145 (1976). Mechanism of action: A. C. Pappas, D. J. Fisher, *Pestic. Sci.* **10**, 239 (1979). Field evaluation: I. F. Jackson, B. N. Smith, *Proc. 32nd N.Z. Weed Pest Control Conf.* 278 (1979). HPLC determn in wine musts and extracts: P. Cabras *et al.*, *J. Chromatogr.* **256**, 176 (1983). *Review:* N. Mikami, J. Miyamoto, *Rev. Plant Prot. Res.* **14**, 85-95 (1981).

Crystalline solid, mp 165-167°. d^{25} 1.42-1.46. Vapor pressure: 1.32×10^{-4} mm Hg. uv max: 207.5, 275 nm (ε 4.2×10^4, 4.1×10^2). Soly in water (25°): 4.5 ppm. Highly sol in acetonitrile, acetone, ether, chloroform. Moderately sol in benzene, toluene. Stable in solvents. Unstable in alkaline media. LD_{50} in male rats (mg/kg): 6800 orally, >10000 dermally (Jackson). Also reported as LD_{50} in male, female rats (g/kg): 7.8, 9.1 orally (Mikami).

USE: Systemic agricultural fungicide.

7883. Prodiamine. [29091-21-2] 2,6-Dinitro-N^1,N^1-dipropyl-4-(trifluoromethyl)-1,3-benzenediamine; α,α,α-trifluoro-3,5-dinitro-N^4,N^4-dipropyltoluene-2,4-diamine; 5-dipropylamino-α,α,α-trifluoro-4,6-dinitro-o-toluidine; 2,4-dinitro-N^3,N^3-dipropyl-6-(trifluoromethyl)-1,3-benzenediamine; N^3,N^3-di-n-propyl-2,4-dinitro-6-trifluoromethyl-1,3-phenylenediamine; USB-3153; CN-11-2936; Barricade; Endurance; Factor. $C_{13}H_{17}F_3N_4O_4$; mol wt 350.30. C 44.57%, H 4.89%, F 16.27%, N 15.99%, O 18.27%. Broad-spectrum, pre-emergence dinitroaniline herbicide. Prepn: D. L. Hunter et al., **DE 2013510**; eidem, **US 3764623** (1970, 1973 both to U.S. Borax and Chemical Corp.). Weed control in container grown plants: T. A. Fretz, W. J. Sheppard, HortScience **15**, 489 (1980); S. A. Duray, F. T. Davies, J. Environ. Hort. **5**, 82 (1987). Field studies: M. G. Sybouts, Proc. West. Soc. Weed Sci. **40**, 169 (1987). Evaluation of herbicidal activity: W. Bond, Crop Prot. **7**, 75 (1988). Review of herbicidal activity: S. J. Bowe, Proc. West. Soc. Weed Sci. **39**, 216-218 (1986).

Orange needles from 95% ethanol, mp 124-125°.
USE: Herbicide.

7884. Prodigiosin. [82-89-3] 4-Methoxy-5-[(5-methyl-4-pentyl-2H-pyrrol-2-ylidene)methyl]-2,2'-bi-1H-pyrrole; 2,2'-[3-methoxy-4'-amyl-5'-methyl-5-(2''-pyrryl)]dipyrrylmethene; prodigiosine. $C_{20}H_{25}N_3O$; mol wt 323.44. C 74.27%, H 7.79%, N 12.99%, O 4.95%. Antibiotic pigment produced by Chromobacterium prodigiosum (Serratia marcescens). Exhibits antimicrobial and cytotoxic properties. Isoln: Wrede, Hettche, Ber. **62**, 2678 (1929); Lasseur, Georges, Trav. Lab. Microbiol. Fac. Pharm. Nancy **9**, 47 (1936); Lasseur, Melcion, ibid. **13**, 192 (1944). Purification: Morgan, Tanner, J. Chem. Soc. **1955**, 3305. Structure: Wrede, Rothhaas, Z. Physiol. Chem. **226**, 95 (1934). Revised structure and synthesis: H. H. Wasserman et al., J. Am. Chem. Soc. **82**, 506 (1960); H. Rapoport, K. G. Holden, ibid. 5510; ibid. **84**, 635 (1962); A. J. Castro et al., J. Org. Chem. **28**, 857 (1963). Total synthesis and in vitro cytotoxic activity: D. L. Boger, M. Patel, Tetrahedron Lett. **28**, 2499 (1987); eidem, J. Org. Chem. **53**, 1405 (1988). NMR studies: R. J. Cushley et al., Can. J. Chem. **53**, 148 (1975). Antimalarial activity: A. J. Castro, Nature **213**, 903 (1967). Review: R. P. Williams, W. R. Hearn in Antibiotics vol. **2**, D. Gottlieb, P. D. Shaw, Eds. (Springer-Verlag, New York, 1967) pp 410-432, 449-451. Review of synthesis: A. H. Jackson, K. M. Smith, in The Total Synthesis of Natural Products vol. **1**, J. ApSimon, Ed. (Wiley-Interscience, New York, 1973) pp 227-232; of biological activity and production: V. Alonzo, Ig. Mod. **81**, 557-564 (1984), C.A. **101**, 126398h (1984).

Lustrous square pyramids (dark red with green reflex) from petr ether, mp 151-152°. Almost insol in water. Moderately sol in alcohol, ether; freely sol in chloroform, bromoform, benzene. Alkaline or neutral solns are orange-yellow, acid solns are red. Absorption max (isopropanol): 466 nm (ε 43000); 336, 280 nm.
Hydrochloride. [112373-40-7] $C_{20}H_{26}ClN_3O$. Magenta crystals from benzene + petr ether, dec 148.5-150°. Absorption max (isopropanol): 540, 294 nm (ε 70700, 10800).

7885. Prodlure. [50767-79-8] (9Z,11E)-9,11-Tetradecadien-1-ol 1-acetate; cis-9,trans-11-tetradecadienyl acetate; cis-9,trans-11-TDDA; litlure A. $C_{16}H_{28}O_2$; mol wt 252.40. C 76.14%, H 11.18%, O 12.68%. Major component of sex pheromone of female cotton leafworm, Spodoptera litura (F.) and Egyptian cotton leafworm, S. littoralis (Boisd.). Isoln and prepn: Y. Tamaki et al., Appl. Entomol.

Zool. **8**, 200 (1973); B. F. Nesbitt et al., Nature New Biol. **244**, 208 (1973). Prepn: T. Yushimo et al., **DE 2406259** (1974 to Natl. Inst. Agr. Sci.), C.A. **82**, 97681b (1975). Stereoselective synthesis: D. R. Hall et al., Chem. Ind. (London) **1975**, 216; G. Goto et al., Chem. Lett. **1975**, 103; G. Decodts et al., Synthesis **7**, 510 (1979). Activity studies in presence of minor component, (Z,E)-9,12-tetradecadien-1-ol acetate: S. Neumark et al., Environ. Lett. **6**, 219 (1974); Y. Tamaki, T. Yushima et al., J. Insect Physiol. **20**, 1005 (1974); M. Kehat et al., Appl. Entomol. Zool. **11**, 45 (1976).

Colorless liquid, bp$_{0.2}$ 147-148°, bp$_{0.003}$ 85-86°. uv max (hexane): 232 nm (ε 27300).
USE: Insect attractant.

7886. Profenofos. [41198-08-7] Phosphorothioic acid O-(4-bromo-2-chlorophenyl) O-ethyl S-propyl ester; O-(4-bromo-2-chlorophenyl) O-ethyl S-propyl phosphorothioate; CGA-15324; OMS-2004; Curacron; Selecron. $C_{11}H_{15}BrClO_3PS$; mol wt 373.63. C 35.36%, H 4.05%, Br 21.39%, Cl 9.49%, O 12.85%, P 8.29%, S 8.58%. Organophosphate insecticide for use in cotton and food crops; cholinesterase inhibitor. Prepn: E. Beriger, J. Drabek, **BE 789937**; eidem, **US 3992533** (1973, 1976 both to Ciba-Geigy). Comprehensive description: F. Buholzer, Proc. 8th Br. Insectic. Fungic. Conf. **1975**, 659-665. Resolution and biological activity of chiral isomers: H. Leader, J. E. Casida, J. Agric. Food Chem. **30**, 546 (1982). Chromatographic determn of residue in potatoes: R. A. Habiba et al., ibid. **40**, 1852 (1992); in tomatoes: S. M. M. Ismail et al., ibid. **41**, 610 (1993); in tea: S. K. Pramanik et al., Bull. Environ. Contam. Toxicol. **74**, 645 (2005). Effect on microbial flora of soil: A. Y. Abdel-Mallek et al., Microbiol. Res. **149**, 167 (1994). Toxicity to fish: A Kumar, J. C. Chapman, Environ. Toxicol. Chem. **17**, 1799 (1998); M. Ismail et al., Bull. Environ. Contam. Toxicol. **82**, 569 (2009).

Slightly yellowish liquid. bp$_{0.001}$ 110°. d^{20} 1.455 g/cc. Soly in water at 20°: 20 ppm. Misc with methanol, dichloromethane, benzene, hexane. Vapor pressure at 20°: ~10^{-5} mm Hg. n_D^{20} 1.5466. LD$_{50}$ in rat (mg/kg): 358-400 orally; ~3300 dermally; LC$_{50}$ (4 hr) in rat (mg/m^3): ~3000 by inhalation (Buholzer). LC$_{50}$ (96 hr) in rainbow fish: 0.91 mg/l (Kumar, Chapman).
USE: Insecticide; acaricide.

7887. Proflavine. [92-62-6] 3,6-Acridinediamine; 3,6-diaminoacridine; 2,8-diaminoacridine. $C_{13}H_{11}N_3$; mol wt 209.25. C 74.62%, H 5.30%, N 20.08%. Prepn: M. Schöpff, Ber. **27**, 2320 (1894); **DE 230412** (1910 to Cassella), C.A. **5**, 2734 (1911); W. P. Thompson, **GB 137214** (1919 to Poulenc Frères), C.A. **14**, 1445 (1920); A. Albert, J. Chem. Soc. **1941**, 121, 484; ibid. **1947**, 244. Toxicity study: S. D. Rubbo, Br. J. Exp. Pathol. **28**, 1 (1947). Review: The Acridines, A. Albert, Ed. (St. Martin's Press, New York, 2nd ed., 1966) pp 300-302; Acridines, R. M. Acheson, Ed. (Interscience, New York, 1956) pp 341-344.

Yellow needles from alc, mp 281° (Schöpff), mp 288° (Albert). Sol in water, ethanol. Practically insol in benzene, ether. pKa 9.7. Solns are brownish and when diluted are fluorescent. Solns should be discarded when they become turbid. LD$_{50}$ s.c. in mice: 0.14 g/kg (Rubbo).

Dihydrochloride. [531-73-7] $C_{13}H_{11}N_3.2HCl$. Orange-yellow needles. pH of 0.1% soln 2.5-3.0.

Sulfate. [553-30-0] Proflavine hemisulfate; neutral proflavine sulfate. $C_{13}H_{11}N_3.H_2SO_4$. Occurs as the hydrate. Red needles, hygroscopic. Sol in 300 parts cold water, in 1 part boiling water; slightly sol in alc. Practically insol in ether, CHCl$_3$. pH of satd soln is 6-8. pH of 0.1% soln is 2.5.

Mixture with 3,6-diamino-10-methylacridinium chloride *see* Acriflavine.

THERAP CAT: Topical antiseptic.

THERAP CAT (VET): Topical antiseptic.

7888. Progabide. [62666-20-0] 4-[[(4-Chlorophenyl)(5-fluoro-2-hydroxyphenyl)methylene]amino]butanamide; 4-[[α-(*p*-chlorophenyl)-5-fluorosalicylidene]amino]butyramide; 4-[[α-(*p*-chlorophenyl)-5-fluoro-2-hydroxybenzylidene]amino]butyramide; halogabide; SL-76.002; Gabren(e). $C_{17}H_{16}ClFN_2O_2$; mol wt 334.78. C 60.99%, H 4.82%, Cl 10.59%, F 5.67%, N 8.37%, O 9.56%. Gamma-aminobutyric acid (GABA) antagonist with anti-epileptic activity. Prepns: J.-P. Kaplan *et al.*, **DE 2634288**; *eidem*, **US 4094992** (1977, 1978 to Synthelabo); *eidem*, *J. Med. Chem.* **23**, 702 (1980). Use as analgesic: J.-P. Kaplan, **US 4361583** (1982 to Synthelabo). Pharmacokinetics: I. Johno *et al.*, *J. Pharm. Sci.* **71**, 633 (1982). HPLC determn in biological fluids: P. Padovani *et al.*, *J. Chromatogr.* **308**, 229 (1984). Double-blind clinical trial in therapy resistant epilepsy: P. Loiseau *et al.*, *Epilepsia* **24**, 703 (1983); in spasticity: K. Mondrup, E. Pedersen, *Acta Neurol. Scand.* **69**, 200 (1984).

Crystals from cyclohexane and toluene, mp 133-135°; possible second crystalline form, mp 142.5° (Kaplan, *J. Med. Chem.*). uv max in methanol: 332, 250, 210 (ε 4200, 10800, 24000). LD$_{50}$ i.p. in mice: 900 mg/kg (Kaplan).

THERAP CAT: Anticonvulsant.

7889. Progesterone. [57-83-0] Pregn-4-ene-3,20-dione; Δ^4-pregnene-3,20-dione; corpus luteum hormone; luteohormone; Crinone; Cyclogest; Estima; Gestone; Menaelle; Progestasert; Progestogel; Progeston; Proluton; Utrogestan. $C_{21}H_{30}O_2$; mol wt 314.47. C 80.21%, H 9.62%, O 10.18%. Active principle of the corpus luteum, secreted during the latter half of the menstrual cycle. If pregnancy ensues, secretion continues. Exerts an antiovulatory effect when administered during days 5 to 25 of the normal menstrual cycle. Isoln from corpus luteum of pregnant sows: Butenandt, Westphal, *Ber.* **67**, 1440 (1934); Wintersteiner, Allen, *J. Biol. Chem.* **107**, 321 (1934). Structure: Butenandt *et al.*, *Ber.* **67**, 1611 (1934). Synthesis of DL-form: Johnson *et al.*, *J. Am. Chem. Soc.* **93**, 4332 (1971). Prepn from other steroids in review by W. H. Strain in Gilman's *Organic Chemistry* vol. II (Wiley, New York, 2nd ed., 1943) pp 1487-1489. Numerous patents, e.g., **US 2379832**; **US 2232438**; **US 2314185**. Anesthetic effect and toxicity: H. Selye, *Proc. Soc. Exp. Biol. Med.* **46**, 116 (1941). Review of physiology: Csapo, *Sci. Am.* **198**, 40-46 (April, 1958); Rothchild, *Vitam. Horm.* **23**, 209-327 (1965). Book: *Progesterone and Progestins*, C. W. Bardin *et al.*, Eds. (Raven Press, New York, 1982) 462 pp.

Exists in two cryst forms of equal physiologic activity and which are readily interconverted. The α-form is orthorhombic (prisms from

dil alc) with a:b:c = 0.750:1:0.905. Crystals show (011), (110), (010). Poor (011) cleavage. d^{23} 1.166. mp 127-131°. The β-form is orthorhombic (needles) with a:b:c = 0.563:1:0.275. Crystals acicular with parallel extinction and negative elongation. Cleavage on (001) and (110). d^{20} 1.171. mp 121°. [α]$_D^{20}$ +172 to +182° (c = 2 in dioxane). uv max: 240 nm. Sol in alc, acetone, dioxane, concd H$_2$SO$_4$; sparingly sol in vegetable oils. Practically insol in water. 1 mg = 1 I.U.

Caution: This substance is reasonably anticipated to be a human carcinogen: *Report on Carcinogens, Twelfth Edition* (PB2011-111646, 2011) p 362.

THERAP CAT: Progestogen.

THERAP CAT (VET): Progestogen. Has been used to control habitual abortion, to suppress or synchronize estrus.

7890. Proglumetacin. [57132-53-3] 1-(4-Chlorobenzoyl)-5-methoxy-2-methyl-1*H*-indole-3-acetic acid 2-[4-[3-[[4-(benzoylamino)-5-(dipropylamino)-1,5-dioxopentyl]oxy]propyl]-1-piperazinyl]-ethyl ester; (±)-*N*-[2-[1-(*p*-chlorobenzoyl)-5-methoxy-2-methyl-3-indolylacetoxy]ethyl]-*N'*-[3-(*N*-benzoyl-*N'*,*N'*-di-*n*-propyl-DL-isoglutaminoyl)oxypropyl]piperazine. $C_{46}H_{58}ClN_5O_8$; mol wt 844.45. C 65.43%, H 6.92%, Cl 4.20%, N 8.29%, O 15.16%. Deriv of indomethacin, *q.v.* Prepn: F. Makovec *et al.*, **DE 2535799** corresp to **US 3985878** (both 1976 to Rotta). Series of articles on pharmacology, mechanism of action, safety: *Arzneim.-Forsch.* **29**, 1116-1129 (1979). Bioavailability study: A. A. Bignamini, P. L. Casula, *Curr. Med. Res. Opin.* **6**, 299 (1979). Clinical evaluation: J. Münzenberg, S. Tachibana, *Pharmatherapeutica* **2**, 279 (1980); P. Loizzi *et al.*, *ibid.* 285. Mutagenicity study: R. Vidal y Plana *et al.*, *Farmaco Ed. Prat.* **33**, 543 (1978). Toxicity data for dimaleate: A. L. Rovati *et al.*, *Arzneim.-Forsch.* **29**, 1116 (1979).

Dimaleate. [59209-40-4] Protacine; CR-604; Afloxan; Miridacin; Protaxon; Proxil. $C_{46}H_{58}ClN_5O_8.2C_4H_4O_4$; mol wt 1076.59. Crystals from ethanol, mp 146-148°. LD$_{50}$ in male mice, rats (mg/kg): 262, 450 orally (Rovati).

THERAP CAT: Anti-inflammatory.

7891. Proglumide. [6620-60-6] 4-(Benzoylamino)-5-(dipropylamino)-5-oxopentanoic acid; DL-4-benzamido-*N*,*N*-dipropylglutaramic acid; *N*-benzoyl-*N'*,*N'*-di-*n*-propyl-DL-isoglutamine; xylamide; CR-242; W-5219; Milid; Milide; Promid. $C_{18}H_{26}N_2O_4$; mol wt 334.42. C 64.65%, H 7.84%, N 8.38%, O 19.14%. Prepn: **NL 6510006**; **ZA 6504065** (both 1966 to Rotta), *C.A.* **65**, 3793b (1966). Pharmacological activity: A. L. Rovati *et al.*, *Minerva Med.* **58**, 3653 (1967); T. Umetzu *et al.*, *Eur. J. Pharmacol.* **64**, 69 (1980). Pharmacokinetic study: A. A. Bignamini *et al.*, *Arzneim.-Forsch.* **29**, 639 (1979). Clinical study in duodenal ulcer: W. Bergemann *et al.*, *Med. Klin.* **76**, 226 (1981). Cholecystokinin receptor antagonist activity: W. F. Hahne *et al.*, *Proc. Natl. Acad. Sci. USA* **78**, 6304 (1981). Selective blockade of cholecystokinin CNS effects: L. A. Chiodo, B. S. Bunney, *Science* **219**, 1449 (1983).

Crystals, mp 142-145°. LD$_{50}$ in mice (mg/kg): 2211-2649 i.v.; 7350-8861 orally (Rovati).

THERAP CAT: Anticholinergic.

7892. Prohexadione. [88805-35-0] 3,5-Dioxo-4-(1-oxopropyl)cyclohexanecarboxylic acid; 3,5-dioxo-4-propionylcyclohexanecarboxylic acid. $C_{10}H_{12}O_5$; mol wt 212.20. C 56.60%, H 5.70%, O 37.70%. Gibberellin biosynthesis inhibitor. Prepn: K. Motojima *et al.*, **EP 123001**; *eidem*, **US 4678496** (1984, 1987 both to Kumiai). Review of properties and field trials: T. Miyazawa *et al.*, *Brighton Crop Prot. Conf. - Weeds* **1991**, 967-972.

mp 98-99°.

Calcium salt. [127277-53-6] BX-112; KIM-112; KUH-833; Medax; Viviful. $C_{10}H_{10}CaO_5$; mol wt 250.26. Fine white powder, mp >360°. Vapor pressure (20°): 1.335×10^{-5} Pa. Soly at 20° (mg/l): water 174.2; methanol 1.11. LD_{50} in rats (mg/kg): >5000 orally; >2000 dermally; LC_{50} in carp, bluegill sunfish, rainbow trout (mg/l): >150, >100, >100 (Miyazawa).

USE: Plant growth regulator.

7893. Proinsulin. [9035-68-1] Single chain insulin precursor consisting of the insulin A and B chains and a connecting polypeptide (C-peptide), which contains 30-35 amino acids; the number and sequence of these amino acids are species dependent. Its presence was discovered in a human islet cell adenoma: D. F. Steiner, P. E. Oyer, *Proc. Natl. Acad. Sci. USA* **57**, 473 (1967). Conversion of proinsulin to insulin has a half-time of about 1 hour in rat islets *in vitro;* it is postulated that proteolytic enzymes cleave proinsulin at the sites where two amino acids connect the C-peptide to the insulin chain.This cleavage results in the production of insulin and the C-peptide, both of which are retained in the secretory granules of the beta cells and discharged in equimolar amounts during exocytosis of the granules. *See:* W. Kemmler, D. F. Steiner, *Biochem. Biophys. Res. Commun.* **41**, 1223 (1970). *In vitro* conversion of proinsulin to insulin with trypsin and carboxypeptidase B: W. Kemmler *et al.*, *J. Biol. Chem.* **246**, 6786 (1971); *ibid.* **248**, 4544 (1973). Isoln of bovine proinsulin: D. F. Steiner *et al.*, *Diabetes* **17**, 725 (1968). Structural studies on porcine mammalian proinsulin: R. E. Chance *et al.*, *Science* **161**, 165 (1968); on bovine: C. Nolan *et al.*, *J. Biol. Chem.* **246**, 2780 (1971); on human and monkey: P. E. Oyer *et al.*, *ibid.* 1365; on monkey, sheep and dog: J. D. Petersen *et al.*, *ibid.* **247**, 4866 (1972); on rat and horse: H. S. Tager, D. F. Steiner, *ibid.* 7936; on guinea pig: D. E. Massey, D. G. Smyth, *ibid.* **250**, 6288 (1975). Proposed three-dimensional structure: C. R. Snell, D. G. Smyth, *ibid.* 6291. Syntheses of C-peptides and human proinsulin: N. Yanaihara *et al.*, *Diabetes* **27** (Suppl. 1), 149 (1978). Synthesis of rat proinsulin in bacteria: L. Villa-Komaroff *et al.*, in *Polypeptide Hormones*, R. F. Beers, E. G. Bassett, Eds. (Raven Press, New York, 1980) pp 49-65. *Reviews:* D. F. Steiner *et al.* in *Diabetes, 8th Proc. Congr. Int. Diabetes Fed.*, W. J. Malaisse, J. Pirart, Eds. (Excerpta Med., Amsterdam, 1974) pp 119-133; A. H. Rubenstein *et al.*, in *Recent Prog. Horm. Res.* **33**, R. O. Greep, Ed. (Academic Press, New York, 1977) pp 435-475; A. E. Kitabshi, *Metabolism* **26**, 547-587 (1977). Book: *Proinsulin, Insulin, C-Peptides,* S. Baba *et al.*, Eds. (Excerpta Medica, Amsterdam, 1979) 468 pp.

7894. Prolactin. [9002-62-4] Adenohypophysial luteotropin; anterior pituitary luteotropin; galactin; lactogen; LTH; luteotropic hormone; luteotropin; mammotropin; pituitary lactogenic hormone; Ferolactan. Polypeptide hormone of mol wt about 23,000; active principle of adenohypophysial gland essential in the induction of lactation in mammals at parturition. Its synergistic action with estrogen promotes mammary gland proliferation. Also brings about the release of progesterone, *q.v.* from lutein cells which renders the uterine mucosa suited for the imbedding of the ovum, should fertilization occur. Isoln procedures from adenohypophyseal tissue or whole pituitary glands of ox, sheep, and swine: Lyons, *Cold Spring Harbor Symp. Quant. Biol.* **5**, 198 (1937); Li *et al.*, *J. Biol. Chem.* **146**, 627 (1942); White *et al.*, *ibid.* **143**, 447 (1942). Isoln from sheep pituitaries: Reisfeld *et al.*, *J. Am. Chem. Soc.* **83**, 3719 (1961); from other mammalian pituitaries: Nelson; Eppstein, **US 3265580** and **US 3317392** (1966, 1967, both to Upjohn); from human pituitaries: Lewis *et al.*, *Biochem. Biophys. Res. Commun.* **44**, 1169 (1971).

Amino acid sequence of ovine prolactin: Li *et al.*, *Nature* **224**, 695 (1969); Li, Dixon, *Arch. Biochem. Biophys.* **146**, 233 (1971). Review of structural studies of human prolactin and relationship to somatotropin, *q.v.*: H. D. Niall *et al.*, *Recent Prog. Horm. Res.* **29**, 387 (1973). Amino acid sequence of human prolactin: B. Shome, A. F. Parlow, *J. Clin. Endocrinol. Metab.* **45**, 1112 (1977). Effects of prolactin on the murine immune system: E. Nagy *et al.*, *Acta Endocrinol.* **102**, 351 (1983). Symposium on clinical endocrinology: *Horm. Res.* **22**, 129-252 (1985). General reviews: Li, Evans, *Hormones* **1**, 631 (1948); White, *Vitam. Horm.* **7**, 253 (1949); Voss, *Arzneim.-Forsch.* **4**, 467 (1954); Apostalakis, *Vitam. Horm.* **26**, 197 (1968). Books: *Prolactin* vols. **1-8**, D. F. Horrobin, Ed. (Eden Press, Quebec, 1973-1981); *Prolactin: Physiology, Pharmacology and Clinical Findings,* O. Hutzinger *et al.*, Eds. (Springer-Verlag, New York, 1982) 224 pp; *Prolactin and Prolactinomas,* G. Tolis *et al.*, Eds. (Raven Press, New York, 1983) 478 pp.

Crystals. Isoelectric point 5.73. $[\alpha]_D^{25}$ −40.5° (c = 1 in phosphate buffer of pH 7). Practically insol in water (0.102 g/l) unless electrolytes are present. Forms a water-soluble hydrochloride. Sol in abs methanol or ethanol, if a small amount of acid is present. These data apply to prolactin obtained from ox glands. Prolactin from sheep glands is slightly different: In 0.357*M* NaCl at pH 2.25 the sheep hormone has a soly of 0.506 g/l, while the soly of the ox hormone is only 0.316 g. In citrate buffer (1*M*, pH 6.36) and in alcohol the ox protein is more sol than the sheep protein. In the absence of salt, prolactin shows little loss of potency after boiling for one hour at pH 8.0 or at 60° for 5 hours; in the presence of salts complete destruction may occur. An 0.04% soln was stable in a boiling water bath for 15 minutes at pH 1 to 9, but lost activity rapidly at pH 11 and 13. As a rule the hormone is more stable in acid than in alkaline solns. 1 mg = 30 I.U.

THERAP CAT: Lactation stimulating hormone.

7895. Proline. [147-85-3] L-Proline; Pro; P; (*S*)-2-pyrrolidinecarboxylic acid. $C_5H_9NO_2$; mol wt 115.13. C 52.16%, H 7.88%, N 12.17%, O 27.79%. Non-essential amino acid for human development. Only imino acid of the 20 amino acids commonly found in proteins. Shows *cis-trans* isomerism. First synthesized: R. Willstätter, *Ber.* **33**, 1160 (1900); prior to identification in casein: E. Fischer, *ibid.* **34**, 454 (1901). Named in 1904 by Fischer. Early chemistry and biochemistry: *Amino Acids and Proteins,* D. M. Greenberg, Ed. (Charles C. Thomas, Springfield, IL, 1951) 950 pp., *passim*; J. P. Greenstein, M. Winitz, *Chemistry of the Amino Acids* vols 1-3 (John Wiley and Sons, Inc., New York, 1961) pp. 2178-2201, *passim*. Stereospecific synthesis: S. L. Titouani *et al.*, *Tetrahedron* **36**, 2961 (1980). HPLC determn: A. Carisano, *J. Chromatogr.* **318**, 132 (1985). HPLC separation of *cis-trans* isomers of peptides: S. Friebe *et al.*, *J. Chromatogr. A* **661**, 7 (1994). Review of biosynthesis and degradation: E. Adams, L. Frank, *Annu. Rev. Biochem.* **49**, 1005-1061 (1980). Review of effects on protein structure and biological function: A. Yaron, F. Naider, *Crit. Rev. Biochem. Mol. Biol.* **28**, 31-81 (1993); G. Vanhoof *et al.*, *FASEB J.* **9**, 736-744 (1995).

Flat needles from alcohol + ether, prisms from water, dec 220-222°. $[\alpha]_D^{23.4}$ −85.0°; $[\alpha]_D^{20}$ −52.6° (c = 0.57 in 0.50*N* HCl); $[\alpha]_D^{20}$ −93.0° (c = 2.42 in 0.6*N* KOH). pI 6.30. pK_1 1.99; pK_2 10.60. Soly in 100 ml water: 127.4 g at 0°; 162.3 g at 25°; 206.7 g at 50°; 239 g at 65°. Sol in alcohol 1.55% at 35°. Insol in ether, butanol, isopropanol.

DL-Form. [609-36-9] Monohydrate, crystals, mp 190° (when anhydr, dec 205°). Sol in water, alc; sparingly sol in acetone, chloroform, benzene; insol in ether.

7896. Prolintane. [493-92-5] 1-[1-(Phenylmethyl)butyl]pyrrolidine; 1-(α-propylphenethyl)pyrrolidine; 1-phenyl-2-pyrrolidylpentane; phenylpyrrolidinopentane; SP-732. $C_{15}H_{23}N$; mol wt 217.36. C 82.89%, H 10.67%, N 6.44%. Prepn: **GB 807835**; Seeger, Kottler, **DE 1093799** (1959, 1960 both to Thomae), *C.A.* **55**,

19950c (1961). Pharmacological studies: R. Kadatz, E. Pötzsch, *Arzneim.-Forsch.* **7**, 344 (1957); K. Takagi *et al., Oyo Yakuri* **5**, 5 (1971), *C.A.* **76**, 94575k (1972).

bp$_{0.5}$ 105°; bp$_{16}$ 153°.
Hydrochloride. [1211-28-5] Katovit; Promotil. C$_{15}$H$_{23}$N.HCl; mol wt 253.81. Crystals from methyl ether, mp 133-134°. LD$_{50}$ orally in mice: 257 mg/kg (Kadatz, Pötzsch).

THERAP CAT: CNS stimulant; antidepressant.

7897. Prolonium Iodide. [123-47-7] 2-Hydroxy-$N^1,N^1,N^1,$-N^3,N^3,N^3-hexamethyl-1,3-propanediaminium iodide (1:2); (2-hydroxytrimethylene)bis[trimethylammonium] iodide; 1,3-bis(trimethylamino)-2-propanol diiodide; hexamethyldiaminoisopropanol diiodide; di(iodohexamethyl)diaminoisopropanol; iodisan; Endojodin. C$_9$H$_{24}$I$_2$N$_2$O; mol wt 430.11. C 25.13%, H 5.62%, I 59.01%, N 6.51%, O 3.72%. Prepn: Callsen, **US 1526627** (1925).

White, crystalline powder. mp ~275° with decompn, but becomes brown at 240°. Freely sol in water, slightly in alcohol. Practically insol in ether, acetone.

THERAP CAT: Iodine source.

7898. Promazine. [58-40-2] *N,N*-Dimethyl-10*H*-phenothiazine-10-propanamine; 10-(3-dimethylaminopropyl)phenothiazine; RP-3276; Wy-1094; A-145. C$_{17}$H$_{20}$N$_2$S; mol wt 284.42. C 71.79%, H 7.09%, N 9.85%, S 11.27%. Prepd by heating a xylene soln of phenothiazine and 3-dimethylamino-1-chloropropane in the presence of sodamide: Charpentier, **US 2519886** (1950 to Rhône-Poulenc).

Oily liq. Amine odor. Alkaline reaction. bp$_{0.3}$ 203-210°.
Hydrochloride. [53-60-1] Liranol; Promwill; Prazine; Protactyl; Sparine; Talofen. C$_{17}$H$_{20}$N$_2$S.HCl; mol wt 320.88. White to slightly yellow crystals, dec 181° (microblock). Oxidizes upon prolonged exposure to air and acquires a blue or pink color. Hygroscopic. One gram dissolves in about 3 ml water. Freely sol in chloroform; sol in methanol, ethanol. Practically insol in ether, benzene. Aq solns are slightly acid to litmus. Incompatible with alkalies, oxidizing agents, heavy metals.

THERAP CAT: Antipsychotic.
THERAP CAT (VET): Tranquilizer.

7899. Promedol. [64-39-1] 1,2,5-Trimethyl-4-phenyl-4-piperidinol 4-propanoate; 1,2,5-trimethyl-4-phenyl-4-propionyloxypiperidine; 1,2,5-trimethyl-4-phenyl-4-piperidyl propionate; dimethylmeperidine. C$_{17}$H$_{25}$NO$_2$; mol wt 275.39. C 74.14%, H 9.15%, N 5.09%, O 11.62%. Prepn: Nazarov *et al., J. Gen. Chem. USSR* **26**, 3117 (1956); Nazarov, Shvestov, *ibid.* 3533. Conformation studies: Prostakov, Mikheeva, *ibid.* **31**, 108 (1961); **33**, 2931 (1963); *eidem, Russ. Chem. Rev.* **31**, 556 (1962). Of the four possible isomers α and γ are shown.

α-Isomer hydrochloride. α-Promedol. C$_{17}$H$_{25}$NO$_2$.HCl. Crystals from benzene, mp 153-154°. Has been also reported as mp 106-107° or 126-131° (Prostakov, Mikheeva, 1961).
β-Isomer hydrochloride. Isopromedol. Crystals, mp 183-184°. See: Nazarov, Shvestov, *Bull. Acad. Sci. USSR Phys. Ser.* **1959**, 2059.
γ-Isomer hydrochloride. Trimeperidine; γ-promedol. Crystals from acetone, mp 222-223°.
Note: This is a controlled substance (opiate): **21 CFR,** 1308.11.

THERAP CAT: Analgesic.

7900. Promegestone. [34184-77-5] (17β)-17-Methyl-17-(1-oxopropyl)estra-4,9-dien-3-one; 17α-methyl-17-propionylestra-4,9-dien-3-one; 17α-methyl-17β-propionyl-19-nor-4,9-androstadien-3-one; 17α,21-dimethyl-19-norpregna-4,9-diene-3,20-dione; R-5020; RU-5020; Surgestone. C$_{22}$H$_{30}$O$_2$; mol wt 326.48. C 80.94%, H 9.26%, O 9.80%. Synthetic progestin with no androgenic activity and with high affinity for the progesterone receptor. Prepn: J. Warnant, A. Farcilli, **BE 763099**; *eidem,* **US 3679714, US 3761591** (1971, 1972, 1973 all to Roussel-UCLAF). Inhibition of gonadotropin secretion and lack of androgenic activity: F. Labrie *et al., Fertil. Steril.* **28**, 1104 (1977). Binding studies in mouse uterus: D. Philibert, J.-P. Raynaud, *Steroids* **22**, 89 (1973); *eidem, Endocrinology* **94**, 627 (1974). Binding to human endometrium: *eidem, Contraception* **10**, 457 (1974); M. Haukkamaa, T. Luukkainen, *J. Steroid Biochem.* **5**, 447 (1974). Review and possible use in treatment of hormone-dependent breast cancer: J.-P. Raynaud, T. Ojasoo, *J. Gynecol. Obstet. Biol. Reprod.* **12**, 697 (1983).

Colorless crystals from isopropyl ether, mp 152°. Sol in acetone, benzene. Insol in water. [α]$_D^{20}$ −262° (c = 0.5 in ethanol). uv max in ethanol: 215, 305 nm (E$_{1cm}^{1\%}$ 202, 648).

USE: As radioligand for the progestin receptor.
THERAP CAT: Progestogen.

7901. Promethazine. [60-87-7] *N,N,α*-Trimethyl-10*H*-phenothiazine-10-ethanamine; 10-[2-(dimethylamino)propyl]phenothiazine; *N*-(2'-dimethylamino-2'-methyl)ethylphenothiazine; proazamine; RP-3277; Fargan. C$_{17}$H$_{20}$N$_2$S; mol wt 284.42. C 71.79%, H 7.09%, N 9.85%, S 11.27%. Antihistamine with sedating and antiemetic effects. Prepn from 10-phenothiazinepropyl chloride and dimethylamine in presence of Cu: P. Charpentier, *Compt. Rend.* **225**, 306 (1947); *idem,* **US 2530451** (1950 to Rhone-Poulenc); from Grignard complexes of dimethylaminopropyl halide and phenothiazine: S. S. Berg, J. N. Ashley, **US 2607773** (1950, 1952 both to Rhone-Poulenc). Acute toxicity: M. Rajsner, *Collect. Czech. Chem. Commun.* **34**, 1019 (1969). Comprehensive description of properties: C. M. Shearer, S. M. Miller, *Anal. Profiles Drug Subs.* **5**, 429-465 (1976). HPLC determn of enantiomers in serum: J. Liu, J. T. Stewart, *J. Pharm. Biomed. Anal.* **16**, 303 (1997). Spectrophotometric determn in pharmaceutical formulations: M. J. Saif, J. Anwar, *Talanta* **67**, 869 (2005). Clinical trial for post-operative nausea and vomiting: A. S. Habib *et al., Anesth. Analg.* **104**, 548 (2007); for prevention of motion sickness: A. Estrada *et al., Aviat. Space Environ. Med.* **78**, 408 (2007).

Crystals, mp 60°. bp$_3$ 190-192°.

Hydrochloride. [58-33-3] RP-3389; Atosil; Farganesse; Fenazil; Hiberna; Lergigan. C$_{17}$H$_{20}$N$_2$S.HCl; mol wt 320.88. White to faintly yellow crystalline powder, mp 203-204° (Charpentier); also reported as white prisms from alcohol-ether, mp 218-220° (Berg, Ashley). Turns blue on prolonged exposure to air and moisture. uv max (water): 249, 297 nm (ε 28770, 3400). pH of 10% aq soln 5.3. Freely sol in water, chloroform, in hot dehydrated alc. Sol in methylene chloride. Practically insol in acetone, ether, ethyl acetate. LD$_{50}$ i.v. in mice: 55.0 mg/kg (Rajsner).

Compd with 8-chlorotheophylline. [17693-51-5] Promethazine teoclate; Avomine. C$_{17}$H$_{20}$N$_2$S.C$_7$H$_7$ClN$_4$O$_2$; mol wt 499.03.

THERAP CAT: Antihistaminic; antiemetic.

THERAP CAT (VET): Antiemetic.

7902. Promethium. [7440-12-2] Pm; at. no. 61; valence 3. No stable nuclides, known radioactive isotopes (mass numbers): 132-155. Longest-lived known isotope: 145 (rel. at. mass 144.9127, T$_{1/2}$ 17.7 years, α-emitter). Most available isotope: 147 (rel. at. mass 146.9151, T$_{1/2}$ 2.6234 years, β-emitter). Abundance in the earth's crust reported as 4.5×10^{-20} ppm; present at low levels in uranium ores. Independent discovery in rare earth concentrate claimed by: J. A. Harris, B. S. Hopkins, *J. Am. Chem. Soc.* **48**, 1585 (1926); J. A. Harris *et al.*, *ibid.* 1594; and by: L. Rolla, L. Fernandez, *Gazz. Chim. Ital.* **56**, 435 (1926). Synthetic prepn: H. B. Law *et al.*, *Phys. Rev.* **59**, 936 (1941). Positive identification from fission products of ^{235}U by ion-exchange chromatography: J. A. Marinsky *et al.*, *J. Am. Chem. Soc.* **69**, 2781 (1947). Metal prepd by reduction of halides: Weigel, *Angew. Chem. Int. Ed.* **2**, 326 (1963); E. J. Wheelwright, *J. Phys. Chem.* **73**, 2867 (1969). Review of prepn, properties and compds: Boyd, *J. Chem. Educ.* **36**, 3-14 (1959); Weigel, *Fortschr. Chem. Forsch.* **12**, 539-621 (1969); F. H. Spedding in *Kirk-Othmer Encyclopedia of Chemical Technology* vol. **19** (John Wiley & Sons, New York, 3rd ed., 1982) pp 833-854; *Chemistry of the Elements*, N. N. Greenwood, A. Earnshaw, Eds. (Pergamon Press, New York, 1984) pp 1423-1449. Review of chemistry, toxicology and industrial uses: *Promethium Technology*, E. J. Wheelwright, Ed. (American Nuclear Society, Hindsdale, IL, 1973) 395 pp. Brief review of properties: G. T. Seaborg, *Radiochim. Acta* **61**, 115-122 (1993).

^{147}Pm, metallic solid; d 7.22. mp 1080° (Weigel); also reported as 1169° (Wheelwright). A number of salts have been prepared including the trihalides (PmX$_3$), the sesquioxide (Pm$_2$O$_3$), the hydroxide [Pm(OH)$_3$] and the nitrate [Pm(NO$_3$)$_3$.xH$_2$O].

USE: ^{147}Pm as energy source for nuclear powered batteries, β-particle source for thickness gauges; in the prepn of self-luminous compds; as portable x-ray sources.

7903. Prometon. [1610-18-0] 6-Methoxy-N^2,N^4-bis(1-methylethyl)-1,3,5-triazine-2,4-diamine; 2,4-bis(isopropylamino)-6-methoxy-s-triazine; 2-methoxy-4,6-bis(isopropylamino)-s-triazine; methoxypropazine; prometone; G-31435; Gesafram; Pramitol. C$_{10}$H$_{19}$N$_5$O; mol wt 225.30. C 53.31%, H 8.50%, N 31.09%, O 7.10%. Nonselective pre- and post-emergent herbicide for use on noncrop land. General preparative information and use as herbicide: H. Gysin, E. Knüsli, **US 2909420** (1959 to Geigy); *see also:* M. A. Priola, **US 3713806** (1973 to Ciba-Geigy). Alternate process: H. V. Lemaster, **US 3663542** (1972 to Geigy). Comparison with other triazine herbicides: H. Gysin, E. Knüsli, *Adv. Pest Contr. Res.* **3**, 289 (1960). Metabolism in rats: J. E. Bakke *et al.*, *J. Agric. Food Chem.* **15**, 628 (1967). GC determn in human urine: *ibid.* **27**, 740 (1979). Toxicology study in sheep and cattle: A. E. Johnson *et al.*, *Am. J. Vet. Res.* **33**, 1433 (1972). Review of properties and analytical methods: B. G. Tweedy, R. A. Kahrs, *Anal. Methods Pestic. Plant Growth Regul.* **10**, 493 (1978).

Crystalline solid, mp 91-92°. Soly in water (20°): 750 ppm. Very sol in organic solvents. Vapor pressure at 20°: 2.3×10^{-6} Torr. LD$_{50}$ in mice, rats (mg/kg): 2160, 2980 orally (Gysin, Knüsli, 1960). LC$_{50}$ (96 hour) in bluegill sunfish, rainbow trout (ppm): >32, 20 (Tweedy, Kahrs).

USE: Nonselective herbicide.

7904. Prometryn. [7287-19-6] N^2,N^4-Bis(1-methylethyl)-6-methylthio-1,3,5-triazine-2,4-diamine; 2,4-bis(isopropylamino)-6-(methylthio)-s-triazine; 2-methylthio-4,6-bis(isopropylamino)-s-triazine; G-34161; Gesagard; Caparol. C$_{10}$H$_{19}$N$_5$S; mol wt 241.36. C 49.76%, H 7.94%, N 29.02%, S 13.28%. Selective pre- and post-emergence herbicide. General preparative information and use as herbicide: H. Gysin, E. Knüsli, **CH 337019**; *eidem*, **US 2909420** (both 1959 to Geigy). See also: M. A. Priola, **US 3713806** (1973 to Ciba-Geigy). Alternative process: **FR 1372089**; E. Knüsli, W. Stammbach, **US 3207756** (1964, 1965 both to Geigy). Activity: H. Gysin, *Chem. Ind. (London)* **1962**, 1393. Metabolism in rats and rabbits: C. Böhme, F. Bär, *Food Cosmet. Toxicol.* **5**, 23 (1967). Toxicity: Geigy, *Toxicology Data on Prometryne* July, 1964. Review of properties and analytical methods: B. G. Tweedy, R. A. Kahrs, *Anal. Methods Pestic. Plant Growth Regul.* **10**, 493 (1978).

Crystals, mp 118-120°. Vapor pressure at 20°: 1×10^{-6} mm Hg. Soly in water at 20°: 48 ppm. Readily sol in org solvents. Stable in neutral or slightly acid or alkaline media. Hydrolyzed under stronger acidic or basic conditions. LD$_{50}$ orally in rats: 3.75 g/kg (Geigy). LC$_{50}$ (96 hour) in bluegill sunfish, rainbow trout (ppm): 10.0, 2.5 (Tweedy, Kahrs).

USE: Herbicide.

7905. Pro-Opiomelanocortin. [66796-54-1] ACTH-β-lipotropin common precursor; precursor to ACTH-LPH-β-endorphin; 31K-precursor; pro-opiocortin; POMC. A precursor protein of mol wt ~30,000, synthesized in the hypothalamus, pituitary gland, brain, and several peripheral tissues that incorporates the amino acid sequences of the pituitary hormones ACTH and β-lipotropin. These two hormones, in turn, contain biologically active component peptides: α-MSH, *corticotropin-like intermediate lobe peptide* (*CLIP*), α-LPH, β-MSH, endorphins, and met-enkephalin. First description of the presence of ACTH and LPH in the same molecule: P. J. Lowry *et al.*, *Int. Congr. Ser.* **402**, 71 (1976). Confirmation of the common precursor by radioimmunoassay: R. E. Mains *et al.*, *Proc. Natl. Acad. Sci. USA* **74**, 3014 (1977); J. L. Roberts, E. Herbert, *ibid.* 4826, 5300. Isoln of rat pro-opiomelanocortin: M. Rubenstein *et al.*, *ibid.* **75**, 669 (1978). Partial sequence analysis using cDNA from mouse: J. L. Roberts *et al.*, *ibid.* **76**, 2153 (1979). Complete sequence of bovine POMC using cloned cDNA: S. Nakanishi *et al.*, *Nature* **278**, 423 (1979). The precursor has also been found in human ACTH ectopic tumor: X. Bertagna *et al.*, *Proc. Natl. Acad. Sci. USA* **75**, 5160 (1978). Primary structure of the NH$_2$-terminal glycopeptide of human pituitary POMC: N. G. Seidah *et al.*, *J. Biol. Chem.* **256**, 7977 (1981). *Reviews:* M. Chrétien, N. G. Seidah, *Mol. Cell. Biochem.* **34**, 101-127 (1981); D. De Wied, J. Jolles, *Physiol.*

Rev. **62**, 976-1059 (1982); Y. P. Loh *et al.*, *Peptides* **3**, 397-404 (1982).

7906. Propacetamol. [66532-85-2] *N,N*-Diethylglycine 4-(acetylamino)phenyl ester; *N,N*-diethylglycine ester with 4′-hydroxyacetanilide; 4-acetamidophenyl (diethylamino)acetate. $C_{14}H_{20}N_2O_3$; mol wt 264.33. C 63.62%, H 7.63%, N 10.60%, O 18.16%. Injectable prodrug of acetaminophen, *q.v.* Prepn and pharmacology: J. C. Cognacq, **BE 854376**; *idem*, **US 4127671** (1977, 1978 both to Hexachimie). Clinical comparison with injectable aspirin: R. De Marneffe, L. Mokassa, *C. R. Ther. Pharmacol. Clin.* **3**, 23 (1985). Clinical studies: P. Delacroix *et al.*, *Sem. Hop.* **61**, 2739 (1985); J. Modai, *ibid.* **62**, 587 (1986).

Thick oil.

Hydrochloride. [66532-86-3] UP-34101; Pro-Dafalgan. $C_{14}H_{20}N_2O_3 \cdot HCl$; mol wt 300.78. mp 228°. Sol in water.

THERAP CAT: Analgesic; antipyretic.

7907. Propachlor. [1918-16-7] 2-Chloro-*N*-(1-methylethyl)-*N*-phenylacetamide; 2-chloro-*N*-isopropylacetanilide; *N*-isopropyl-α-chloroacetanilide; CP-31393; Ramrod. $C_{11}H_{14}ClNO$; mol wt 211.69. C 62.41%, H 6.67%, Cl 16.75%, N 6.62%, O 7.56%. Selective pre-emergence herbicide. Prepn: P. C. Hamm, A. J. Speziale, **US 2863752** (1958 to Monsanto). Activity studies: Duke, *Diss. Abstr. B* **28**, 1315 (1967); Jaworski, *J. Agric. Food Chem.* **17**, 165 (1969); Lamoureaux *et al.*, *ibid.* **19**, 346 (1971); Dhillon, Anderson, *Weed Res.* **12**, 182 (1972). Metabolism: J. E. Bakke *et al.*, *Science* **210**, 433 (1980). Toxicity data: E. E. Kenaga, *Down Earth* **35**, 25 (1979).

Light tan solid, mp 67-76°. Vapor pressure at 110°: 0.03 mm Hg. Soly in water at 20°: 700 mg/l. Sol in common organic solvents except aliphatic hydrocarbons. LD_{50} orally in rats: 710 mg/kg (Kenaga).

USE: Herbicide.

7908. Propafenone. [54063-53-5] 1-[2-[2-Hydroxy-3-(propylamino)propoxy]phenyl]-3-phenyl-1-propanone; 2′-[2-hydroxy-3-(propylamino)propoxy]-3-phenylpropiophenone; SA-79. $C_{21}H_{27}NO_3$; mol wt 341.45. C 73.87%, H 7.97%, N 4.10%, O 14.06%. Sodium channel blocker. Prepn: R. Sachse, **DE 2001431** (1971 to Helopharm), *C.A.* **75**, 151538f (1971). Pharmacology: H.-J. Hapke, E. Prigge, *Arzneim.-Forsch.* **26**, 1849 (1976). Pharmacokinetics: M. Hollmann *et al.*, *ibid.* **33**, 763 (1983). HPLC determn in plasma and urine: U. Hofmann *et al.*, *J. Chromatogr. B* **748**, 113 (2000). Review of pharmacology and therapeutic efficacy: H. M. Bryson *et al.*, *Drugs* **45**, 85-130 (1993). Review of clinical trials in supraventricular arrhythmias and atrial fibrillation: A. P. Rae, *Am. J. Cardiol.* **82**, 59N-65N (1998); S. C. Reimold *et al.*, *ibid.* 66N-71N.

Hydrochloride. [34183-22-7] Arythmol; Pronon; Rythmol; Rytmonorm. $C_{21}H_{27}NO_3 \cdot HCl$; mol wt 377.91. Fine white crystals. Slightly bitter taste. Sol in methanol, CCl_4, hot water; slightly sol in

alc, chloroform, cold water; very slightly sol in acetone. Insol in diethyl ether, toluene. LD_{50} in rats (mg/kg): 18.8 i.v.; 700 orally (Hapke, Prigge).

THERAP CAT: Antiarrhythmic (class IC).

7909. Propagermanium. [12758-40-6] 3,3′-(1,3-Dioxo-1,3-digermoxanediyl)bispropanioic acid; 2-(carboxyethyl)germanium sesquioxide; 3-oxygermylpropionic acid polymer; poly-*trans*-[(2-carboxyethyl)germasesquioxane]; proxigermanium; repagermanium; GE-132; SK-818; Serocion. $C_6H_{10}Ge_2O_7$; mol wt 339.40. C 21.23%, H 2.97%, Ge 42.80%, O 33.00%. Polymeric, organogermanium biologic response modifier. Orginally synthesized at Asai Germanium Institute in 1967. Prepn and crystal structure: M. Tsutsui *et al.*, *J. Am. Chem. Soc.* **98**, 8287 (1976). Toxicology and clinical pharmacokinetics: K. Miyao *et al.*, *Curr. Chemother. Infect. Dis.* **2**, 1527 (1979). TLC quantitation: X. Peishan *et al.*, *J. Planar Chromatogr. Mod. TLC* **3**, 141 (1990). Electrochemical behavior and determn: K. Hasebe *et al.*, *Electroanalysis* **6**, 779 (1994). Antiviral activity: Y. Ishiwata *et al.*, *Arzneim.-Forsch.* **44**, 357 (1994). Review of biological activities and mechanism of antitumor action: R. R. Brutkiewicz, F. Suzuki, *In Vivo* **1**, 189-204 (1987). *Review:* K. Miyao, N. Tanaka, *Drugs Future* **13**, 441-453 (1988). Review of clinical use as an immunostimulant: H. Fukazawa *et al.*, *Head Neck* **16**, 30-38 (1994).

Colorless, monoclinic crystals or crystalline powder having no odor but a slightly acidic taste. mp 270° (dec). pKa 3.6. Sol in water at 1.09% at 20°. Insol or very slightly sol in almost all organic solvents. Very sol in water under alkaline conditions; >10% at pH 7.4. Aq solns exhibit acid reactions. Stable at pH 2-12, especially at pH 7.4 especially where decomposition does not occur at 110° for 5 min. LD_{50} i.p. in mice: 2.8 g/kg (Brutkiewicz).

THERAP CAT: Antineoplastic; immunostimulant.

7910. Propallylonal. [545-93-7] 5-(2-Bromo-2-propen-1-yl)-5-(1-methylethyl)-2,4,6(1*H*,3*H*,5*H*)-pyrimidinetrione; 5-(2-bromoallyl)-5-isopropylbarbituric acid; Noctal. $C_{10}H_{13}BrN_2O_3$; mol wt 289.13. C 41.54%, H 4.53%, Br 27.64%, N 9.69%, O 16.60%. Prepn: **US 1622129** (1927); *cf.* Herzog, *Arch. Pharm.* **263**, 216 (1925). Acute toxicity: Maloney, *J. Pharmacol. Exp. Ther.* **42**, 267 (1931).

Crystals. mp 177-179°. Slightly bitter taste. Slightly sol in water; freely sol in alcohol, glacial acetic acid, acetone, alkalies. Sparingly sol in ether, chloroform, benzene. MLD orally in rabbits: 300-350 mg/kg (Maloney).

Note: This is a controlled substance (depressant): **21 CFR,** 1308.13.

THERAP CAT: Sedative, hypnotic.

7911. Propamidine. [104-32-5] 4,4′-[1,3-Propanediylbis(oxy)]bisbenzenecarboximidamide; 4,4′-(trimethylenedioxy)dibenzamidine; 4,4′-diamidino-α,ω-diphenoxypropane. $C_{17}H_{20}N_4O_2$; mol wt 312.37. C 65.37%, H 6.45%, N 17.94%, O 10.24%. Prepn: A. J. Ewins *et al.*, **GB 507565** (1939 to May & Baker); J. N. Ashley *et al.*, *J. Chem. Soc.* **1942**, 103; of isethionate: G. Newbery, A. P. T. Easson, **US 2394003** (1946 to May & Baker). Trypanocidal activity: E. M. Lourie, W. Yorke, *Ann. Trop. Med. Parasitol.* **33**, 289 (1939). Preliminary pharmacology: R. Wien, *ibid.* **37**, 1 (1943). Determn in biological fluids: D. P. Jackson *et al.*, *J. Biol. Chem.* **167**, 377 (1947). Mode of action: M. J. Pine, *Biochem. Pharmacol.* **17**, 75 (1968). Activity in fibrinolytic systems: J. D. Geratz, *Thromb. Diath. Haemorrh.* **29**, 154 (1973). Clinical use in treatment of *Acanthamoeba* keratitis: D. L. Easty, *Br. Med. J.* **296**, 228

(1988); J. J. Wiens, W. B. Jackson, *Can. J. Ophthalmol.* **23**, 107 (1988). Early review of pharmacology, mode of action and clinical applications: E. B. Schoenbach, E. M. Greenspan, *Medicine* **27**, 327-377 (1948).

Isethionate. [140-63-6] M & B 782; Brolene Drops. $C_{21}H_{32}N_4O_{10}S_2$; mol wt 564.63. Hygroscopic, very bitter crystals or granular powder, mp ~235°. Soluble in water (~1 in 5), glycerol, 95% alcohol (~1 in 33). Practically insol in ether, chloroform, fixed oils, liquid petrolatum. pH of a 5% w/v soln in water = 4.5 to 6.5.

THERAP CAT: Antiprotozoal (Trypanosoma); antiamebic.

THERAP CAT (VET): Anti-infective (topical). Formerly used as antiprotozoal (Trypanosoma, Babesia).

7912. Propamocarb. [24579-73-5] *N*-[3-(Dimethylamino)-propyl]carbamic acid propyl ester; propyl-(3-dimethylaminopropyl) carbamate. $C_9H_{20}N_2O_2$; mol wt 188.27. C 57.42%, H 10.71%, N 14.88%, O 17.00%. Systemic carbamate fungicide. Prepn: **BE 708057** (1968 to Schering AG). Prepn of salts: G.-A. Hoyer, E. A. Pieroh, **US 3649674** (1972 to Schering AG). Properties and activity: E. A. Pieroh *et al.*, *Meded. Fac. Landbouwwet. Rijksuniv. Gent* **43**, 933 (1978). Metabolic fate in bluegills and catfish: C. Gray, C. O. Knowles, *Chemosphere* **10**, 469 (1981). Efficacy vs late blight fungus on potato: Y. Samoucha, Y. Cohen, *Phytoparasitica* **18**, 27 (1990). GC determn in food products: T. Nagayama *et al.*, *J. AOAC Int.* **79**, 769 (1996). Uptake and redistribution in potato and grapevine: R. I. Harris, *Brighton Crop Prot. Conf. - Pests Dis.* **1996**, 281.

bp$_{18mm}$ 139-141°. n_D^{20} 1.4490.

Hydrochloride. [25606-41-1] SN-66752; AE-B066752; HOE-102791; Banol; Previcur; Proplant. $C_9H_{20}N_2O_2$·HCl; mol wt 224.73. Colorless, odorless crystals, mp 45-55°. Soly at 25° (g/100ml): water >70; dichloromethane >43; methanol >50. Vapor pressure at 25°: 6×10^{-6} torr. LD$_{50}$ orally in rats: 8600 mg/kg. LC$_{50}$ (96 hr) in rainbow trout, sunfish, carp (ppm): 616, 415, 234 (Pieroh).

USE: Agricultural fungicide.

7913. Propane. [74-98-6] Dimethylmethane; propyl hydride. C_3H_8; mol wt 44.10. C 81.71%, H 18.29%. Constituent of natural gas and of crude petroleum. Obtained by the so-called "stabilization process" using fractional distillation under pressure: Francis, Robbins, *J. Am. Chem. Soc.* **55**, 4339 (1933). Many syntheses, e.g., by using butyronitrile and sodium: Timmermans, *J. Chim. Phys.* **18**, 133 (1920).

Gas. Odorless when pure. *Flammable.* Burns with a luminous, smoky flame. Explosive limits, % by vol in air: 2.37-9.5. Heavier than air. One liter weighs 2.0200 g at 0° and 760 mm; 1.8324 g at 25° and 760 mm. Liquefies at −42°; solid at −187.7°. bp (1 atm) −42.1°; bp (2 atm) −25.6°; bp (5 atm) 1.4°; bp (10 atm) 26.9°; bp (20 atm) 58.1°; bp (30 atm) 78.7°; bp (40 atm) 94.8°. Crit temp 96.81°; crit press. 42.01 atm. Heat of combustion (const vol) 528.4 cal, (const pressure) 553.5 cal. 100 vols water dissolve 6.5 vols at 17.8° and 753 mm pressure; 100 vols abs alc dissolve 790 vols at 16.6° and 754 mm pressure; 100 vols ether dissolve 926 vols at 16.6° and 757 mm pressure; 100 vols chloroform dissolve 1299 vols at 21.6° and 757 mm pressure; 100 vols benzene dissolve 1452 vols at 21.5° and 757 mm pressure; 100 vols turpentine dissolve 1587 vols at 17.7° and 757 mm pressure.

Caution: Potential symptoms of overexposure are dizziness, confusion, excitation, asphyxia; direct contact with liquid may cause frostbite. *See NIOSH Pocket Guide to Chemical Hazards* (DHHS/NIOSH 97-140, 1997) p 262.

USE: As fuel gas, sometimes mixed with butane. In organic syntheses. As refrigerant.

7914. 1-Propanearsonic Acid. [107-34-6] *As*-propylarsonic acid. $C_3H_9AsO_3$; mol wt 168.02. C 21.45%, H 5.40%, As 44.59%, O 28.57%.

White needles, mp 125°. Freely sol in water; sol in alcohol. Insol in ether.

USE: For the determination of zirconium.

7915. 1,3-Propanedithiol. [109-80-8] 1,3-Dimercaptopropane; dithiotrimethyleneglycol; trimethylenedimercaptan; trimethylenedithioglycol. $C_3H_8S_2$; mol wt 108.22. C 33.30%, H 7.45%, S 59.25%. Prepd by alkaline hydrolysis of propylene-1,3-diisothiuronium dihydrochloride: Grogan *et al.*, *J. Org. Chem.* **20**, 50 (1955).

Oil. Disagreeable odor. d_4^{20} 1.0772. bp$_{760}$ 169-170°; bp$_{759}$ 170-171°; bp$_{56}$ 92-98°. n_D^{20} 1.5392. Volatile with steam. Slightly sol in water. Miscible with alcohol, ether, chloroform and benzene.

7916. Propanethial S-Oxide. [32157-29-2]; [70565-74-1] (*Z*-form). Thiopropanal *S*-oxide; thiopropionaldehyde *S*-oxide. C_3H_6OS; mol wt 90.14. C 39.97%, H 6.71%, O 17.75%, S 35.57%. Lachrymatory factor of the onion, *Allium cepa L.*, found as a 95% (*Z*)- and 5% (*E*)- mixture. Early structure studies: W. D. Niegisch, W. H. Stahl, *Food Res.* **21**, 657 (1956); C. G. Spare, A. I. Virtanen, *Acta Chem. Scand.* **17**, 641 (1963). Structure: M. H. Brodnitz, J. V. Pascale, *J. Agric. Food Chem.* **19**, 269 (1971). Stereochemistry: E. Block *et al.*, *J. Am. Chem. Soc.* **101**, 2200 (1979); *eidem, Tetrahedron Lett.* **21**, 1277 (1980). Chemistry: *eidem, J. Am. Chem. Soc.* **102**, 2490 (1980).

(Z)-form

7917. Propanidid. [1421-14-3] 4-[2-(Diethylamino)-2-oxoethoxy]-3-methoxybenzeneacetic acid propyl ester; [4-[(diethylcarbamoyl)methoxy]-3-methoxyphenyl]acetic acid propyl ester; [3-methoxy-4-[(*N*,*N*-diethylcarbamido)methoxy]phenyl]acetic acid *n*-propyl ester; propyl [4-[(diethylcarbamoyl)methoxy]-3-methoxyphenyl]acetate; Bayer 1420; FBA-1420; Epontol; Sombrevin. $C_{18}H_{27}NO_5$; mol wt 337.42. C 64.07%, H 8.07%, N 4.15%, O 23.71%. Prepn: R. Hiltmann *et al.*, **DE 1134981**; **US 3086978** (1962, 1963 both to Bayer). Toxicity study: E. I. Goldenthal, *Toxicol. Appl. Pharmacol.* **18**, 185 (1971).

Pale yellow oil, bp$_{0.7}$ 210-212°. Practically insol in water; sol in alcohol, chloroform. LD$_{50}$ orally in rats: >10,000 mg/kg (Goldenthal).

THERAP CAT: Anesthetic (intravenous).

7918. Propanil. [709-98-8] *N*-(3,4-Dichlorophenyl)propanamide; 3',4'-dichloropropionanilide; *N*-(3,4-dichlorophenyl)propionamide; DPA; FW-734; Stam; Stampede. $C_9H_9Cl_2NO$; mol wt 218.08. C 49.57%, H 4.16%, Cl 32.51%, N 6.42%, O 7.34%. Selective contact herbicide. Prepn: W. Schäfer *et al.*, **DE 1039779** (1958 to Bayer), *C.A.* **54**, 20060i (1960); Huffman, Allen, *J. Agric. Food Chem.* **8**, 298 (1960). Use in nematocide formulations: Fielding, Stoddard, **US 3108038** (1963 to du Pont). Toxicity study: G. W. Bailey, J. L. White, *Residue Rev.* **10**, 97 (1965).

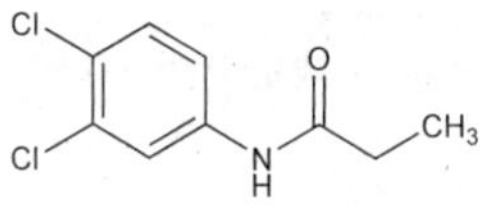

White crystalline solid, mp 91-93°. Soly in water at room temp: 225 ppm. LD$_{50}$ orally in rats: 1384 mg/kg (Bailey, White).

USE: Herbicide.

7919. Propantheline Bromide. [50-34-0] *N*-Methyl-*N*-(1-methylethyl)-*N*-[2-[(9*H*-xanthen-9-ylcarbonyl)oxy]ethyl]-2-propanaminium bromide (1:1); (2-hydroxyethyl)diisopropylmethyl-ammonium bromide xanthene-9-carboxylate; β-diisopropylaminoethyl 9-xanthenecarboxylate methobromide; Corrigast; Pro-Banthine. C$_{23}$H$_{30}$BrNO$_3$; mol wt 448.40. C 61.61%, H 6.74%, Br 17.82%, N 3.12%, O 10.70%. Synthetic, quaternary ammonium anticholinergic. Prepn: Cusic, Robinson, *J. Org. Chem.* **16**, 1921 (1951); **US 2659732** (1953 to Searle). Metabolic studies: Beermann *et al.*, *Clin. Pharmacol. Ther.* **13**, 212 (1972). Clinical pharmacokinetics: C. W. Vose *et al.*, *Br. J. Clin. Pharmacol.* **7**, 89 (1979); HPLC determn in serum: B. G. Charles, P. J. Ravenscroft, *J. Chromatogr.* **306**, 424 (1984).

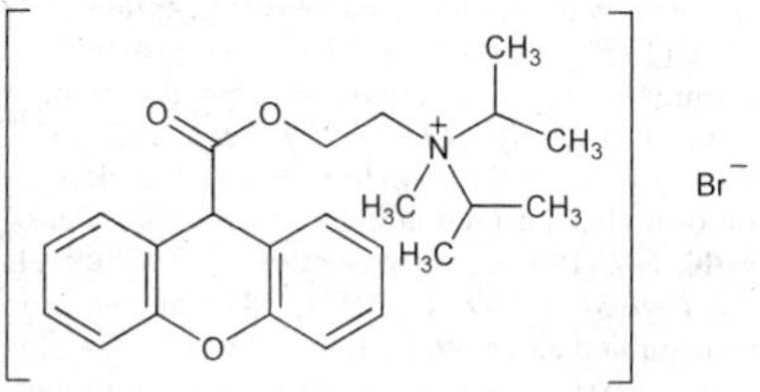

Crystals from isopropanol + ether, mp 159-161°. Very soluble in water, alcohol, chloroform. Practically insol in ether, benzene.

THERAP CAT: Antispasmodic; in treatment of urinary incontinence.

THERAP CAT (VET): Antispasmodic.

7920. Propaquizafop. [111479-05-1] (2*R*)-2-[4-[(6-Chloro-2-quinoxalinyl)oxy]phenoxy]propanoic acid 2-[[(1-methylethylidene)amino]oxy]ethyl ester; 2-isopropylideneaminooxyethyl (*R*)-2-[4-(6-chloroquinoxalin-2-yloxy)phenoxy]propionate; Ro-17-3664; Prilan. C$_{22}$H$_{22}$ClN$_3$O$_5$; mol wt 443.88. C 59.53%, H 5.00%, Cl 7.99%, N 9.47%, O 18.02%. Post-emergence graminicide for use in broadleaf crops. Prepn: G. Frater *et al.*, **US 4687849** (1987 to Hoffmann-La Roche); R. Klaus *et al.*, *ACS Symp. Ser.* **443**, 226-235 (1991). Comprehensive description: P. F. Bocion *et al.*, *Proc. Br. Crop Prot. Conf. - Weeds* **1987**, 55-62. Review of comparative field trials: J. D. A. Wevers, *Meded. Fac. Landbouwwet. Rijksuniv. Gent* **56**, 611-615 (1991).

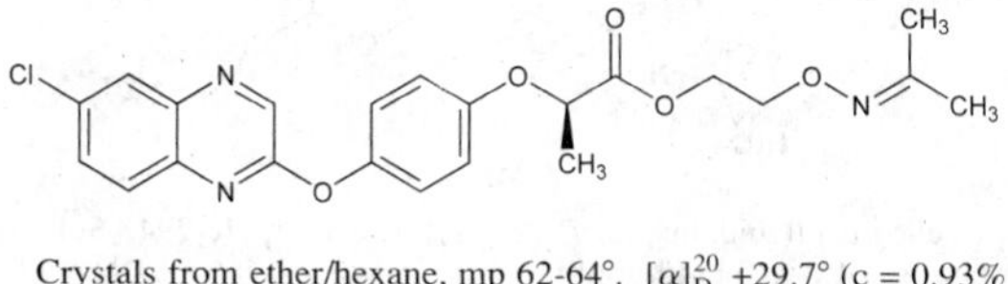

Crystals from ether/hexane, mp 62-64°. [α]$_D^{20}$ +29.7° (c = 0.93% in chloroform). d^{20} 1.29. Vapor pressure (20°): 1.3 × 10^{-8} Pa. Log P (*n*-octanol/water at pH 7): 3.5. Soly in water (25°): 2 ppm (pH 7). LD$_{50}$ in rats (mg/kg): >5000 orally; >2000 dermally (Bocion).

USE: Herbicide.

7921. Proparacaine. [499-67-2] 3-Amino-4-propoxybenzoic acid 2-(diethylamino)ethyl ester; 2-(diethylamino)ethyl 3-amino-4-propoxybenzoate; proxymetacaine. C$_{16}$H$_{26}$N$_2$O$_3$; mol wt 294.40. C 65.28%, H 8.90%, N 9.52%, O 16.30%. Prepn of the hydrochloride: Clinton *et al.*, *J. Am. Chem. Soc.* **74**, 592 (1952). Pharmacology:

McIntyre, Sievers, *J. Pharmacol. Exp. Ther.* **63**, 369 (1938). Comprehensive description: D. B. Whigan, *Anal. Profiles Drug Subs.* **6**, 423-456 (1977).

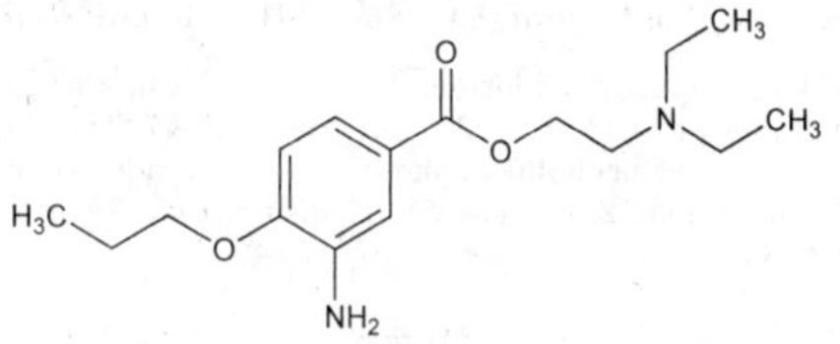

Hydrochloride. [5875-06-9] Ak-Taine; Alcaine; Ophthaine; Ophthetic. C$_{16}$H$_{26}$N$_2$O$_3$.HCl; mol wt 330.85. Prisms from abs alcohol + ethyl acetate, mp 182.0-183.3°. uv max (methanol): 225, 270, 300 nm. Sol in water, warm alcohol, methanol. Insol in ether, benzene. Solns are neutral to litmus. pKa 3.2.

THERAP CAT: Topical anesthetic (ophthalmic).

THERAP CAT (VET): Anesthetic (local).

7922. Propargite. [2312-35-8] Sulfurous acid 2-[4-(1,1-dimethylethyl)phenoxy]cyclohexyl 2-propyn-1-yl ester; sulfurous acid 2-(*p-tert*-butylphenoxy)cyclohexyl 2-propynyl ester; 2-(*p-tert*-butylphenoxy)cyclohexyl propargyl sulfite; cyclosulfyne; propargil; BPPS; ENT-27226; DO-14; Omite; Comite. C$_{19}$H$_{26}$O$_4$S; mol wt 350.47. C 65.12%, H 7.48%, O 18.26%, S 9.15%. Acaricide for control of mites on crops. Prepn and insecticidal properties: R. A. Covey *et al.*, **US 3272854** (1966 to U.S. Rubber). Field tests, comparison with other acaricides in control of Pacific spider mite: E. M. Stafford, *J. Econ. Entomol.* **61**, 1641 (1968). Efficacy against citrus red mite: L. R. Jeppson *et al.*, *ibid.* **62**, 531 (1969). Residue determn: J. M. Devine, H. R. Sisken, *J. Agric. Food Chem.* **20**, 59 (1972). Brief account of properties, analysis: G. M. Stone, *Anal. Methods Pestic. Plant Growth Regul.* **7**, 355 (1973). Toxicity studies: T. B. Gaines, *Toxicol. Appl. Pharmacol.* **14**, 515 (1969).

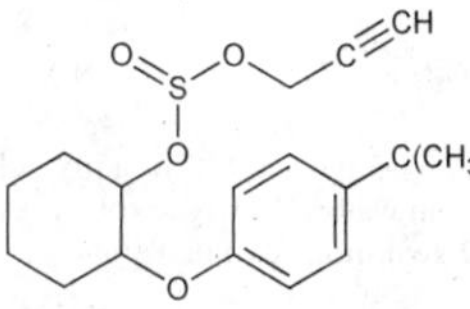

Viscous liquid, distils with decomp, stabilized by adding 0.5-1.0% propylene oxide. Soluble in most organic solvents. Practically insoluble in water (10.5 ppm). LD$_{50}$ in male, female rats (mg/kg): 1480, 1480 orally; 250, 680 dermally (Gaines). LC$_{50}$ in rainbow trout, bluegill sunfish (ppb): >100, 31 (Stone).

USE: Acaricide.

7923. Propargyl Alcohol. [107-19-7] 2-Propyn-1-ol. C$_3$H$_4$O; mol wt 56.06. C 64.28%, H 7.19%, O 28.54%. Prepd by heating β-bromoallyl alcohol with conc KOH: Henry, *Ber.* **5**, 456, 569 (1872); **6**, 729 (1873); **14**, 404 (1881); from formaldehyde and sodium acetylide: Henne, Greenlee, *J. Am. Chem. Soc.* **67**, 484 (1945); from epichlorohydrin + sodium: Eglinton *et al.*, *J. Chem. Soc.* **1952**, 2873; from acetylene and formaldehyde: Reppe, *Ann.* **596**, 1 (1955). Toxicity study: W. Guilian, B. Naibin, *Chemosphere* **36**, 1475 (1998).

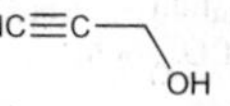

Moderately volatile liquid. Mild geranium odor. d$_4^{20}$ 0.9715. mp −52 to −48°. bp$_{760}$ 114-115°; bp$_{490.3}$ 100°; bp$_{147.6}$ 70°; bp$_{35.4}$ 40°; bp$_{20.6}$ 30°; bp$_{11.6}$ 20°. n$_D^{20}$ 1.43064. Viscosity at 20° = 1.68 cP. Flash pt 33°C. Spec heat: 0.616 cal/g. Misc with water, benzene, chloroform, 1,2-dichloroethane, ethanol, ether, acetone, dioxane, tetrahydrofuran, pyridine. Appreciable heat is evolved on mixing with pyridine. Moderately sol in carbon tetrachloride. Immiscible with aliphatic hydrocarbons. With water, propargyl alcohol forms an azeotrope, bp 97.5°. This mixture has a composition of 39.5 parts by weight of propargyl alcohol and 60.5 parts of water. Polymerized by heat or caustic. Acidified aq solns are resistant to polymerization. LD_{50} in rats, mice (mg/kg): 20, 50 orally (Guilian, Naibin).

Caution: Potential symptoms of overexposure are irritation of skin, mucous membranes; CNS depression. *See NIOSH Pocket Guide to Chemical Hazards* (DHHS/NIOSH 97-140, 1997) p 264.

7924. Propargyl Chloride. [624-65-7] 3-Chloro-1-propyne. C_3H_3Cl; mol wt 74.51. C 48.36%, H 4.06%, Cl 47.58%. Prepd by treating propargyl alcohol with phosphorus trichloride: Henry, *Ber.* **7**, 761 (1874); *ibid.* **8**, 398 (1875); Kirsmann, *Bull. Soc. Chim. Fr.* [4] **39**, 698 (1926).

$$HC\equiv C-CH_2Cl$$

Liquid. d_4^{25} 1.0306. mp $-78°$. bp_{760} 57°. Flash pt <60°F. Practically insol in water, glycerol. Miscible with benzene, carbon tetrachloride, ethanol, ethylene glycol, ether, ethyl acetate. Reacts with hydroxy compds to form ethers; with sulfides, ammonia, amines or metal hypoiodites to give the corresp propargyl compds; with aldehydes and ketones to give β-acetylenic alcohols. Undergoes isomerization.

USE: Intermediate in organic synthesis.

7925. Propatyl Nitrate. [2921-92-8] 2-Ethyl-2-[(nitrooxy)-methyl]-1,3-propanediol 1,3-dinitrate; 2-ethyl-2-(hydroxymethyl)-1,3-propanediol trinitrate; 1,1,1-tris(nitratomethyl)propane; ethyltrimethylolmethane trinitrate; trimethylolethylmethane trinitrate; 2,2-bis(hydroxymethyl)-1-butanol trinitrate; 1-hydroxy-2,2-bis(hydroxymethyl)butane trinitrate; ettriol trinitrate; ETTN; Win-9317; Atrilon 5; Etrynit; Gina; Ginapect; Vasangor. $C_6H_{11}N_3O_9$; mol wt 269.17. C 26.77%, H 4.12%, N 15.61%, O 53.49%. Prepn: Médard, *Mem. Poudres* **35**, 113 (1953); Bourjol, *ibid.* **36**, 79 (1954); Hensinger, **FR 1103113** (1955).

White powder, mp 51-52°. d 1.49. Readily sol in acetone, alcohol. Practically insol in water. Lowest explosive temp: 220°. Heat of combustion 829.2 kcal/mol. Explosive but only slightly sensitive to shock.

THERAP CAT: Vasodilator (coronary).

7926. Propazine. [139-40-2] 6-Chloro-N^2,N^4-bis(1-methylethyl)-1,3,5-triazine-2,4-diamine; 2-chloro-4,6-bis(isopropylamino)-*s*-triazine; 2,4-bis(isopropylamino)-6-chloro-*s*-triazine; G-30028; Prozinex. $C_9H_{16}ClN_5$; mol wt 229.71. C 47.06%, H 7.02%, Cl 15.43%, N 30.49%. Selective pre-emergence herbicide. Prepn: Gysin, Knüsli, **CH 342784**; **CH 342785** (both 1960 to Geigy). Toxicity study: G. W. Bailey, J. L. White, *Residue Rev.* **10**, 97 (1965).

Crystals, mp 213°. Solubility in water at 20° = 8.6 ppm; difficultly sol in organic solvents. LD_{50} orally in rats: >5000 mg/kg (Bailey, White).

USE: Herbicide.

7927. Propentofylline. [55242-55-2] 3,7-Dihydro-3-methyl-1-(5-oxohexyl)-7-propyl-1H-purine-2,6-dione; 3-methyl-1-(5-oxohexyl)-7-propylxanthine; 1-(5'-oxohexyl)-3-methyl-7-propylxanthine; HWA-285; Albert-285; HOE-285; Hextol; Karsivan. $C_{15}H_{22}N_4O_3$; mol wt 306.37. C 58.81%, H 7.24%, N 18.29%, O 15.67%. Peripheral vasodilator which inhibits cyclic AMP phosphodiesterase. Prepn: W. Mohler *et al.*, **DE 2330742**; *eidem*, **US 4289776** (1975, 1981 both to Hoechst). Cardiovascular effects in animals: O. Hudlicka *et al.*, *Br. J. Pharmacol.* **72**, 723 (1981). Effect on cAMP

phosphodiesterase: K. Nagata *et al.*, *Arzneim.-Forsch.* **35**, 1034 (1985). Inhibition of adenosine uptake: B. B. Fredholm, K. Lindström, *Acta Pharmacol. Toxicol.* **58**, 187 (1986). Cerebrovascular effects in animals: J. J. Grome *et al.*, *Drug Dev. Res.* **5**, 111 (1985). Effect on postischemic brain edema: B. B. Mrsulja *et al.*, *ibid.* **6**, 339 (1985). Effect on cognitive function in humans: I. Hindmarch, Z. Subhan, *ibid.* **5**, 379 (1985). Acute toxicity: M. Inazu *et al.*, *Oyo Yakuri* **31**, 357 (1986), *C.A.* **104**, 218960a (1986).

Crystals from diisopropyl ether, mp 69-70°. Soly in water at 25°: 3.2%; in ethanol: >10%; in DMSO: >10%. LD_{50} in male, female mice, male, female rats (mg/kg): 900, 780, 1150, 940 orally; 168, 170, 180, 195 i.v.; 375, 346, 199, 196 i.p.; 450, 508, 400, 338 s.c. (Inazu).

THERAP CAT: Nootropic.

THERAP CAT (VET): Vasodilator (peripheral and cerebral).

7928. Properdin. [11016-39-0] Complement factor P. A highly basic serum protein believed to be a factor in natural immunity against germ and virus diseases, perhaps also against cancer: L. Pillemer *et al.*, *Science* **120**, 279 (1954). Combines with zymosan, *q.v.*, the insol cell wall residue from yeast: L. Pillemer, O. A. Ross, *ibid.* **121**, 732 (1955). Participates also in a nonspecific manner in a variety of immunological reactions of normal serum. Isoln from human serum: L. Pillemer *et al.*, *J. Exp. Med.* **103**, 1 (1956). Purification: Spicer *et al.*, **US 3038838** (1962 to Merck & Co.). Characterization of highly purified human properdin: J. Pensky *et al.*, *J. Immunol.* **100**, 142 (1968). Activity studies: O. Götze, H. J. Müller-Eberhard, *J. Exp. Med.* **139**, 44 (1974). The human properdin system has been studied extensively; it was the subject of long years of scientific controversy. Components of the human properdin system that have been identified and characterized are: properdin, factor B, factor D and C3. Comprehensive review of L. Pillemer's work and history of the properdin system discovery: I. H. Lepow, *J. Immunol.* **125**, 471 (1980). The rabbit properdin system: G. B. Naff, *ibid.* **124**, 2625 (1980).

7929. Propetamphos. [31218-83-4] (2E)-3-[[(Ethylamino)-methoxyphosphinothioyl]oxy]-2-butenoic acid 1-methylethyl ester; (E)-3-hydroxycrotonic acid isopropyl ester O-ester with O-methyl ethylphosphoramidothioate; (E)-1-methylethyl 3-[[(ethylamino)-methoxyphosphinothioyl]oxy]-2-butenoate; (E)-O-2-isopropoxycarbonyl-1-methylvinyl O-methyl ethylphosphoramidothioate; SAN-322I; Catalyst. $C_{10}H_{20}NO_4PS$; mol wt 281.31. C 42.70%, H 7.17%, N 4.98%, O 22.75%, P 11.01%, S 11.40%. Prepn: J. P. Leber, K. Lutz, **DE 2035103** corresp to **US 3758645** (1971, 1973 to Sandoz). Activity: W. Berg, R. Gothe, *Proc. Br. Crop Prot. Conf. - Pests Dis.* **1979**, 517.

Yellowish liquid, $bp_{0.005}$ 87-89°. n_D^{20} 1.495. d_4^{20} 1.1294. Soly in water at 24°: 110 mg/l. Sol in most organic solvents. LD_{50} orally in male rats: 82 mg/kg (Berg, Gothe).

USE: Insecticide.

THERAP CAT (VET): Ectoparasiticide.

7930. Propham. [122-42-9] N-Phenylcarbamic acid 1-methylethyl ester; carbanilic acid isopropyl ester; N-phenyl isopropyl carbamate; isopropyl carbanilate; O-isopropyl N-phenyl carbamate; INPC; IPC; isoPPC. $C_{10}H_{13}NO_2$; mol wt 179.22. C 67.02%, H 7.31%, N 7.82%, O 17.85%. Prepn: Allen, **US 2615916** (1952 to Columbia-Southern Chem. Corp.); Kovalenko, *Zh. Obshch. Khim.*

24, 1041 (1954); J. L. Hermanson, R. Olson, *Trans. Kans. Acad. Sci.* 64, 231 (1961). Acute toxicity: T. B. Gaines, R. E. Linder, *Fundam. Appl. Toxicol.* 7, 299 (1986).

Crystals, mp 90°. Practically insol in water. Sol in most organic solvents. LD_{50} in male, female rats (mg/kg): 3724, 4315 orally (Gaines, Linder).

USE: Herbicide, applied as a spray to the soil.

7931. Propicillin. [551-27-9] (2S,5R,6R)-3,3-Dimethyl-7-oxo-6-[(1-oxo-2-phenoxybutyl)amino]-4-thia-1-azabicyclo[3.2.0]-heptane-2-carboxylic acid; 6-(α-phenoxybutyramido)penicillanic acid; α-phenoxypropylpenicillin; levopropylcillin. $C_{18}H_{22}N_2O_5S$; mol wt 378.44. C 57.13%, H 5.86%, N 7.40%, O 21.14%, S 8.47%. Semi-synthetic antibiotic related to penicillin. Prepn from 6-amino-penicillanic acid: **GB 877120** (1961 to Beecham); of salts: Perron *et al., J. Am. Chem. Soc.* 82, 3934 (1960); Glombitza, *Ann.* 673, 166 (1964). Metabolism in humans: M. Cole *et al., Antimicrob. Agents Chemother.* 3, 463 (1973). TLC determn: S. Hendrickx *et al., J. Chromatogr.* 291, 211 (1984).

Potassium salt. [1245-44-9] BRL-284; PA-248; Baycillin; Bro-cillin; Cetacillin; Oricillin; Trescillin; Ultrapen. $C_{18}H_{21}KN_2O_5S$; mol wt 416.53. Crystals, dec 195-197°. Soluble at 20° in 1.2 parts water, 23 parts 95% (w/v) alcohol. pH of 1% w/v soln: 5-7.5.

THERAP CAT: Antibacterial.

7932. Propiconazole. [60207-90-1] 1-[[2-(2,4-Dichloro-phenyl)-4-propyl-1,3-dioxolan-2-yl]methyl]-1H-1,2,4-triazole; pro-conazole; CGA-64250; Banner; Orbit; PropiMax; Tilt. $C_{15}H_{17}Cl_2$-N_3O_2; mol wt 342.22. C 52.65%, H 5.01%, Cl 20.72%, N 12.28%, O 9.35%. Systemic foliar fungicide. Prepn: G. Van Reet *et al.*, **DE 2551560**; *eidem*, **US 4079062** (1976, 1978 both to Janssen). Physi-cochemical properties, toxicity and antifungal activity: P. A. Urech *et al., Proc. Br. Crop Prot. Conf. - Pests Dis.* **1979**, 508. Efficacy vs cereal diseases: G. Eyries, *Phytiatr.-Phytopharm.* 30, 37 (1981). GC determn in soil, water, plant material: B. Büttler, *J. Agric. Food Chem.* 31, 762 (1983).

Yellowish, viscous liquid. $bp_{0.1mm}$ 180°. Vapor pressure at 20°: $<3 \times 10^{-6}$ mm Hg. Soly in water at 20°: 110 mg/l. Sol in most organic solvents. LD_{50} orally in rats: 1517 mg/kg (Urech).

USE: Agricultural fungicide.

7933. Propineb. [12071-83-9] (monomer); [9016-72-2] (homopolymer). [N-[2-[(Dithiocarboxy)amino]-1methylethyl]car-bamodithioato(2−)-κS,κS']zinc; [[(1-methyl-1,2-ethanediyl)bis[car-bamodithioato]](2−)]zinc; [propylenebis(dithiocarbamato)]zinc; zinc 1,2-propylene bisdithiocarbamate; mezineb; methylzineb; Bayer 46131; Antracol. $C_5H_8N_2S_4Zn$; mol wt 289.78. C 20.72%, H 2.78%, N 9.67%, S 44.25%, Zn 22.57%. Prepn and use as fungi-cide: **IT 611046** (1960 to Montecatini); Lehmann *et al.*, **BE 611960** and **BE 628114** (1962 and 1963 to Bayer). Chemistry, biological

activity and toxicology: Grewe, *Pflanzenschutz-Nachr.* 20, 581 (1967). Fungicidal activity: A. G. Channon, *Ann. Appl. Biol.* 65, 481 (1965); E. J. S. Reddy *et al., Pesticides* 19, 67 (1985). Distri-bution and degradation in soil: W. Mittelstaedt, F. Fuhr, *Land-wirtsch. Forsch.* 30, 221 (1977); in fruit: K. Vogeler *et al., Pflanzen-schutz-Nachr.* 30, 72 (1977).

White to yellowish powder, practically odorless. Readily poly-merizes. Dec and turns brown above 160°. Dec in strongly acid or alkaline media. Practically insol in all conventional solvents. LD_{50} in male rats, rabbits, cats (mg/kg): 8500, >2500, >2500 orally (Grewe).

USE: Agricultural fungicide.

7934. β-Propiolactone. [57-57-8] 2-Oxetanone; hydracrylic acid β-lactone; β-propionolactone; propanolide; NSC-21626; Beta-prone. $C_3H_4O_2$; mol wt 72.06. C 50.00%, H 5.60%, O 44.40%. Prepd by the condensation of ketene with formaldehyde: Küng, **US 2356459** (1941). Purification: Gresham, Jansen, **US 2602802** (1952 to B. F. Goodrich). Use in Diels-Alder diene synthesis: Gresham *et al., J. Am. Chem. Soc.* 76, 609 (1954). Review of carcinogenic risk: *IARC Monographs* 4, 259-269 (1974).

Liquid. d_4^{20} 1.1460; d_4^{25} 1.1420; d_{20}^{20} 1.1490. mp −33.4°. bp_{760} 162° (dec); bp_{750} 150° (dec); bp_{20} 61°; bp_{10} 51°. Flash pt 70°C (158°F). n_D^{20} 1.4131; n_D^{25} 1.4110. Dipole moment 3.8. Slowly hy-drolyzed to hydracrylic acid. Stable when stored at 5° in glass con-tainers. Soly in water: 37% v/v. Misc with alcohol, acetone, ether, chloroform.

Caution: Potential symptoms of overexposure are skin irritation, blistering and burns; corneal opacity; frequent urination; dysuria; hematuria. *See NIOSH Pocket Guide to Chemical Hazards* (DHHS/ NIOSH 97-140, 1997) p 264. This substance is reasonably antici-pated to be a human carcinogen: *Report on Carcinogens, Twelfth Edition* (PB2011-111646, 2011) p 366.

USE: Versatile intermediate in organic synthesis.

THERAP CAT: Disinfectant.

7935. Propiolic Acid. [471-25-0] 2-Propynoic acid; acet-ylenecarboxylic acid; propargylic acid. $C_3H_2O_2$; mol wt 70.05. C 51.44%, H 2.88%, O 45.68%. Prepn: Wilson, Wenzke, *J. Am. Chem. Soc.* 57, 1265 (1935); Owen, Sultanbawa, *J. Chem. Soc.* **1949**, 3109; Wolf, *Ber.* 86, 735 (1953); 87, 668 (1954). Manuf: Wolf, **US 2786022** (1957); Pachter, **US 2799703** (1957 to Ethyl Corp.). Review on its occurrence and biological activity: Reisch, *Pharmazie* 20, 194 (1965).

Liquid at room temp (crystals from carbon disulfide, mp 9°). bp 144° (dec), $bp_{10.5}$ 54-55°, bp_{50} 70-75°. d_4^{15} 1.1435, d_4^{20} 1.1380, d_4^{25} 1.1325. $n_D^{20.4}$ 1.4302. Dipole moment at 25°, 2.08D (in dioxane): Wilson, Wenzke, *loc. cit.*

7936. Propiomazine. [362-29-8] 1-[10-[2-(Dimethylamino)-propyl]-10H-phenothiazin-2-yl]-1-propanone; 10-dimethylamino-isopropyl-2-propionylphenothiazine; propionylpromethazine; Wy-1359. $C_{20}H_{24}N_2OS$; mol wt 340.49. C 70.55%, H 7.11%, N 8.23%, O 4.70%, S 9.42%. Prepn: Schmitt *et al., Bull. Soc. Chim. Fr.* **1957**, 1474; Schmitt, **FR addn. 71342** (addn. to **FR 1176919** to Clin-Byla). Comprehensive description of the hydrochloride: K. B. Crombie, L. F. Cullen, *Anal. Profiles Drug Subs.* 2, 439-466 (1973). GC determn in plasma and urine: U. Ahs, G. Wickström, *J. Chro-*

matogr. **183**, 229 (1980). Clinical use as sedative: W. F. Powell *et al.*, *Anesth. Analg.* **49**, 132 (1970); as hypnotic: M. Viukari, P. Miettinen, *Neuropsychobiology* **12**, 134 (1984).

bp$_{0.5}$ 235-245°.

Maleate. CB-1678; Indorm; Propavan. $C_{20}H_{24}N_2OS.C_4H_4O_4$; mol wt 456.56. Crystals from isopropanol, mp 160-161°.

Hydrochloride. [1240-15-9] Largon. $C_{20}H_{24}N_2OS.HCl$; mol wt 376.94. Yellow, practically odorless powder. Very soluble in water; freely sol in alcohol. Insol in benzene.

THERAP CAT: Sedative, hypnotic.

THERAP CAT (VET): Tranquilizer.

7937. Propionaldehyde. [123-38-6] Propanal; methylacetaldehyde; propylaldehyde. C_3H_6O; mol wt 58.08. C 62.04%, H 10.41%, O 27.55%. Prepd by treating propyl alcohol with a bichromate oxidizing mixture: C. D. Hurd, R. N. Meinert, *Org. Synth.* **coll. vol. II**, 541 (1943); by passing propyl alcohol vapor over copper at high temp: Sabatier, Senderens, *Compt. Rend.* **136**, 923 (1903); Willimott, *Analyst* **50**, 13; *Chem. Zentralbl.* **1925**, I, 2097. Absorption spectrum: Kwiecinski, Marchlewski, *Bull. Soc. Chim.* [4] **45**, 608 (1929). Toxicity study: H. F. Smyth *et al.*, *Arch. Ind. Hyg. Occup. Med.* **4**, 119 (1951).

Liquid. Suffocating odor. d_4^0 0.8432; $d_4^{9.7}$ 0.8192; d_4^{20} 0.8071; d_4^{33} 0.7898. mp −81°. bp$_{760}$ 49°; bp$_{740}$ 47°; bp$_{687}$ 45°. $n_{580}^{16.6}$ 1.3695; n_D^{19} 1.36460. Flash pt, open cup: <20°F (<−6°C). *Flammable.* Sol in 5 vols water at 20°. Misc with alc and ether. LD$_{50}$ orally in rats: 1.4 g/kg; LC for rats in air: 8000 ppm (Smyth).

7938. Propionamide. [79-05-0] Propanamide; propionic acid amide. C_3H_7NO; mol wt 73.10. C 49.29%, H 9.65%, O 21.89%. Prepd by heating ammonium propionate under pressure: Hofmann, *Ber.* **15**, 981 (1882); by dropping propionyl chloride into cooled ammonia water: Aschan, *Ber.* **31**, 2347 (1898).

Orthorhombic platelets from benzene. d_4^{20} 1.0335; d_4^{80} 0.9597; mp 79°; bp 222.2°. Volatile with steam. n_D^{110} 1.4160. Freely sol in water, alcohol, ether, chloroform.

7939. Propionic Acid. [79-09-4] Propanoic acid; methylacetic acid; ethylformic acid. $C_3H_6O_2$; mol wt 74.08. C 48.64%, H 8.16%, O 43.19%. Occurs in dairy products in small amounts. Can be obtained from wood pulp waste liquor by a fermentation process using bacteria of the genus *Propionibacterium:* Wayman *et al.*, US 3067107 (1962 to Columbia Cellulose). Prepn from ethylene, carbon monoxide and steam: Reppe, *Angew. Chem.* **1956**, 46; Larson, US 2448375 (1948 to du Pont); from ethanol and carbon monoxide using a boron trifluoride catalyst: Loder, US 2135448; US 2135451; US 2135453 (all 1939 to du Pont); by oxidation of propionaldehyde: Hasche, US 2294984 (1942 to Kodak); from natural gas by the Fischer-Tropsch process; as a byproduct in the pyrolysis of wood; by the action of microorganisms on a variety of materials in small yields. Very pure propionic acid can be obtained from propionitrile. Toxicity study: H. F. Smyth *et al.*, *Am. Ind. Hyg. Assoc. J.* **23**, 95 (1962).

Oily liquid. Slightly pungent, disagreeable, rancid odor. d_4^{20} 0.99336. mp −21.5°. bp$_{760}$ 141.1°; bp$_{400}$ 122.0°; bp$_{100}$ 85.8°; bp$_{1.0}$ 4.6°. n_D^{25} 1.3848. Flash pt, open cup: 136°F (58°C). Viscosity (cP) at 15°: 1.175; at 25°: 1.020; at 30°: 0.956; at 60°: 0.668; at 90°: 0.495. Surface tension in dynes/cm at 15°: 27.21. Ka at 25°: 1.34 × 10^{-5}. *Corrosive.* Misc with water, alc, other organic solvents. Can be salted out of water solns by the addn of $CaCl_2$ or other salts. Sol in ether, chloroform. Azeotrope with water, bp 99.98°, contains 17.7% acid; with toluene, bp 110.45°, contains 3% acid; with *o*-xylene, bp 135.4°, contains 43% acid; with ethylbenzene, bp 131.1°, contains 28% acid. LD$_{50}$ orally in rats: 4.29 g/kg (Smyth).

Barium salt monohydrate. [5915-88-8] Barium propionate. $C_6H_{10}BaO_4.H_2O$. Powder, usually with slight odor. *Poisonous.* Freely sol in water; slightly sol in alcohol.

Caution: Potential symptoms of overexposure to propionic acid are irritation of eyes, skin, nose, throat; blurred vision, corneal burns; skin burns; abdominal pain, nausea, vomiting. *See NIOSH Pocket Guide to Chemical Hazards* (DHHS/NIOSH 97-140, 1997) p 266.

USE: Esterifying agent; acidifying agent; in the production of cellulose propionate (thermoplastic) and other propionates, e.g., calcium propionate, used as mold inhibitors and preservatives; in the manuf of ester solvents, fruit flavors, and perfume bases.

THERAP CAT: Antifungal.

7940. Propionic Anhydride. [123-62-6] Propanoic acid 1,1'-anhydride; propanoic anhydride; propionyl oxide; methylacetic anhydride. $C_6H_{10}O_3$; mol wt 130.14. C 55.38%, H 7.75%, O 36.88%. Obtained by dehydration of the acid or by carbonylation of esters: Reppe, Friederich, US 2730546 (1956 to Badische Anilin- & Soda-Fabrik); from propionaldehyde by air oxidation in the presence of cobalt and copper acetate catalysts: McFarlane, US 2491572 (1949 to Celanese). Other syntheses, e.g., from ethanol and carbon monoxide. Toxicity study: H. F. Smyth *et al.*, *Arch. Ind. Hyg. Occup. Med.* **10**, 61 (1954).

Liquid. Odor more pungent than that of the acid. d_4^0 1.0336; d_4^{15} 1.0169; d_4^{20} 1.0125; d_4^{40} 0.98974; d_4^{50} 0.97913. 8.4 lbs/gal at 20°. mp −45°. bp$_{760}$ 167.0°; bp$_{400}$ 146.0°; bp$_{200}$ 127.8°; bp$_{100}$ 107.2°; bp$_{60}$ 94.5°; bp$_{40}$ 85.6°; bp$_{20}$ 70.4°; bp$_{10}$ 57.7°; bp$_5$ 45.3°; bp$_{1.0}$ 20.6°. Flash pt, open cup: 165°F (74°C). n_D^{17} 1.4041; n_D^{20} 1.4038. Viscosity (cP): 1.144 at 20°; 0.978 at 30°; 0.853 at 40°; 0.7511 at 50°. Dec by water. Sol in methanol, ethanol, ether, chloroform. LD$_{50}$ orally in rats: 2.36 g/kg (Smyth).

USE: Esterifying agent for certain perfume oils, fats, oils, and especially cellulose. In the production of alkyd resins, dyestuffs and drugs. Has been used as a dehydrating agent in some sulfonations and nitrations.

7941. Propionitrile. [107-12-0] Propanenitrile; ethyl cyanide. C_3H_5N; mol wt 55.08. C 65.42%, H 9.15%, N 25.43%. Prepd by dehydration of propionamide (or propionic acid + NH_3) or by distilling ethyl sulfate and concd aq KCN, also by reduction of acrylonitrile. Studies as duodenal ulcerogen in rats: S. Szabo *et al.*, *Res. Commun. Chem. Pathol. Pharmacol.* **16**, 311 (1977); L. M. Lichtenberger *et al.*, *Gastroenterology* **73**, 1305 (1977). Toxicity study: H. F. Smyth *et al.*, *Arch. Ind. Hyg. Occup. Med.* **4**, 119 (1951).

Liquid. *Flammable. Poisonous when heated to decompn or on contact with acids.* Pleasant, ethereal, sweetish odor. d_4^0 0.8020; d_4^{20} 0.7818; d_4^{30} 0.7716; d_4^{56} 0.7515; $d_4^{70.2}$ 0.7291. mp −91.8°. bp$_{760}$ 97.2°; bp$_{400}$ 77.7°; bp$_{200}$ 58.2°; bp$_{100}$ 41.4°; bp$_{60}$ 30.1°; bp$_{40}$ 22.0°; bp$_{20}$ 8.8°; bp$_{10}$ −3.0°; bp$_5$ −13.6°; bp$_{1.0}$ −35.0°. n_D^{15} 1.36812; n_D^{20} 1.36585; n_D^{30} 1.36132. Flash point, closed cup: 43°F (6°C). Soly in water at 40° = 11.9 g/100 g H_2O; at 100° = 29 g/100 g H_2O. Misc with alcohol, ether, DMF. LD_{50} orally in rats: 39 mg/kg (Smyth).

Caution: Readily absorbed through skin; irritating to eyes and respiratory system. Potential symptoms of overexposure are nausea, vomiting, chest pains, weakness, stupor, convulsions. *See NIOSH Pocket Guide to Chemical Hazards* (DHHS/NIOSH 97-140, 1997) p 266.

USE: Reagent and solvent in organic synthesis.

7942. Propionyl Chloride. [79-03-8] Propanoyl chloride. C_3H_5ClO; mol wt 92.52. C 38.95%, H 5.45%, Cl 38.32%, O 17.29%.

Liquid, pungent odor. d_4^{20} 1.065; mp −94°; bp 80°; n_D^{20} 1.4051. *Flammable; corrosive.* Vigorously dec and dissolved by water or alcohol.

7943. Propionylpromazine. [3568-24-9] 1-[10-[3-(Dimethylamino)propyl]-10*H*-phenothiazin-2-yl]-1-propanone; 3-propionyl-10-(γ-dimethylaminopropyl)phenothiazine; 10-(3-dimethylaminopropyl)-2-propionylphenothiazine; propiopromazine; 1497-CB. $C_{20}H_{24}N_2OS$; mol wt 340.49. C 70.55%, H 7.11%, N 8.23%, O 4.70%, S 9.42%. Prepn: Schmitt *et al., Bull. Soc. Chim. Fr.* **1957**, 938, 1474; Schmitt, **FR addn. 71342** (1956 to Clin-Byla), *C.A.* **56**, 3486h (1962).

Crystals, mp 69-70°.
Hydrochloride. Tranvet. $C_{20}H_{24}N_2OS.HCl$; mol wt 376.94.
Maleate. $C_{20}H_{24}N_2OS.C_4H_4O_4$. Crystals from acetone, mp 135°.
Methiodide. $C_{20}H_{24}N_2OS.CH_3I$. Yellow crystals from isopropanol, mp 79-80°.
THERAP CAT (VET): Tranquilizer.

7944. Propiophenone. [93-55-0] 1-Phenyl-1-propanone; ethyl phenyl ketone; propionylbenzene; phenyl ethyl ketone. C_9H_{10}-O; mol wt 134.18. C 80.56%, H 7.51%, O 11.92%. Prepared from propionyl chloride and benzene in the presence of anhydr aluminum chloride: Read, *J. Am. Chem. Soc.* **44**, 1751 (1922).

Leaflets or tabular crystals. Strong, persistent, agreeable, flowery odor. Usually supplied as a liquid. d_4^0 (solid) 1.157; d_4^{20} (liq) 1.0105; d_{20}^{20} 1.0118; d_{25}^{25} 1.0087; $d_4^{41.8}$ 0.9934; $d_4^{61.2}$ 0.9776; $d_4^{85.5}$ 0.9572. mp 21°. Typical freezing point: 18.6°. bp_{760} 218.0°; bp_{400} 194.2°; bp_{200} 170.2°; bp_{100} 149.3°; bp_{60} 135.0°; bp_{40} 124.3°; bp_{20} 107.6°; bp_{10} 92.2°; bp_5 77.9°. $n_D^{15.9}$ 1.5290; n_D^{20} 1.5269. uv max (hexane): 250, 280, 323 nm. Dipole moment: 2.7. Parachor 328.7. Flash pt 99°C (210°F). Misc with methanol, anhydr ethanol, ether, benzene, toluene. Insol in water, glycerol, ethylene glycol, propylene glycol.

USE: In perfumery; in the synthesis of ephedrine and related compds.

7945. Propiverine. [60569-19-9] α-Phenyl-α-propoxybenzeneacetic acid 1-methyl-4-piperidinyl ester; α,α-diphenyl-α-propoxyacetic acid 1-methyl-4-piperidyl ester; 1-methyl-4-piperidyl diphenylpropoxyacetate. $C_{23}H_{29}NO_3$; mol wt 367.49. C 75.17%, H

7.95%, N 3.81%, O 13.06%. Prepn: C. Starke *et al.,* **DD 106643** (1974), *C.A.* **82**, 155841s (1975). GC-MS determn in plasma and urine: T. Marunaka *et al., J. Chromatogr.* **420**, 43 (1987). Pharmacology: A. Haruno, *Arzneim.-Forsch.* **42**, 815, 1459 (1992). Acute toxicity: K. Yamashita *et al., J. Toxicol. Sci.* **14**, Suppl. 2, 1 (1989), *C.A.* **112**, 172060 (1990). Clinical pharmacokinetics: C. Müller *et al., Eur. J. Drug Metab. Pharmacokinet.* **18**, 265 (1993). Clinical trial in urinary incontinence: D. Mazur *et al., Urologe* A33, 447 (1994).

Hydrochloride. [54556-98-8] P-4; Detrunorm; Mictonorm; Mictonetten. $C_{23}H_{29}NO_3.HCl$; mol wt 403.95. mp 216-218°. Sparingly sol in water and ethanol. LD_{50} in male, female mice, male, female rats (mg/kg): 36, 36, 22, 25 i.v.; 223, 283, 1632, 1411 s.c.; 410, 323, 1000, 1092 orally (Yamashita).
THERAP CAT: In treatment of urinary incontinence.

7946. Propizepine. [10321-12-7] 6-[2-(Dimethylamino)propyl]-6,11-dihydro-5*H*-pyrido[2,3-*b*][1,5]benzodiazepin-5-one; 6,11-dihydro-6-[2-(dimethylamino)-2-methylethyl]-5*H*-pyrido[2,3-*b*][1,-5]benzodiazepin-5-one. $C_{17}H_{20}N_4O$; mol wt 296.37. C 68.90%, H 6.80%, N 18.90%, O 5.40%. Prepn: **NL 6600065** (1966 to Labs. U.P.S.A.); Hoffmann, Faure, *Bull. Soc. Chim. Fr.* **1966**, 2316. Psychopharmacology: Lwoff *et al., Therapie* **26**, 451 (1971).

mp 122°.
Hydrochloride. [14559-79-6] UP-106; Vagran. $C_{17}H_{20}N_4O$.-HCl; mol wt 332.83. mp 235°.
THERAP CAT: Antidepressant.

7947. Propofol. [2078-54-8] 2,6-Bis(1-methylethyl)phenol; 2,6-diisopropylphenol; disoprofol; ICI-35868; Ansiven; Diprivan; Disoprivan; Rapinovet. $C_{12}H_{18}O$; mol wt 178.28. C 80.85%, H 10.18%, O 8.97%. Prepn: A. J. Kolka *et al., J. Org. Chem.* **21**, 712 (1956); **22**, 642 (1957); G. G. Ecke, A. J. Kolka, **US 2831898** (1958 to Ethyl Corp.); T. J. Kealy, D. D. Coffman, *J. Org. Chem.* **26**, 987 (1961); B. E. Firth, T. J. Rosen, **US 4447657** (1984 to Universal Oil Products). Chromatographic study: J. K. Carlton, W. C. Bradbury, *J. Am. Chem. Soc.* **78**, 1069 (1956). Animal studies: J. B. Glen, *Br. J. Anaesth.* **52**, 731 (1980). Pharmacokinetics: H. K. Adam *et al., ibid.* 743; idem, *ibid.* **55**, 97 (1983). Determn in blood: *eidem, J. Chromatogr.* **223**, 232 (1981). Comparative studies vs other injectable anesthetics: B. Kay, D. K. Stephenson, *Anaesthesia* **35**, 1182 (1980); D. V. Rutter *et al., ibid.* 1188. Use in i.v. anesthesia: E. Major *et al., ibid.* **37**, 541 (1982). Cardiovascular effects: D. Al-Khudhairi *et al., ibid.* 1007. Pharmacology of emulsion formulation: J. B. Glen, S. C. Hunter, *Br. J. Anaesth.* **56**, 617 (1984). Series of articles on pharmacology and clinical experience: *Postgrad. Med. J.* **61**, Suppl. 3, 1-169 (1985).

Clear to slightly yellowish liquid.bp$_{30}$ 136°. bp$_{17}$ 126°. mp 19°. n_D^{20} 1.5134. n_D^{25} 1.5111. d$_{20}$ 0.955. Very sol in methanol, ethanol; slightly sol in cyclohexane, isopropyl alc; very slightly sol in water.

THERAP CAT: Anesthetic (intravenous).

THERAP CAT (VET): Intravenous anesthetic (dogs and cats).

7948. Propolis. Bee glue. Resinous substance of complex composition produced by bees for use in construction and protection of hives. Used in traditional folk medicine for its broad spectrum biological activities. *Constit.* 50% resin and vegetable balsam, 30% wax, 10% essential and aromatic oils, 5% pollen, and 5% other substances, including organic debris. Analysis by LC-MS: K. Midorikawa *et al.*, *Phytochem. Anal.* **12**, 366 (2001). Review of the origin, chemical constituents and therapeutic activity: M. H. Haydak, *State of Iowa*, *Repts. State Apiarist 1953*, p 74-87; M. Vanhaelen, R.Vanhaelen-Fastre, *J. Pharm. Belg.* **34**, 253 (1979); of composition and toxicology: G. A. Burdock, *Food Chem. Toxicol.* **36**, 347-363 (1998); of plant sources and chemical constituents: V. S. Bankova *et al.*, *Apidologie* **31**, 3-15 (2000). Review of biological activities and therapeutic use: S. L. De Castro, *ARBS Annu. Rev. Biomed. Sci.* **3**, 49-83 (2001); of pharmacology and mechanism of action: A. H. Banskota *et al.*, *Phytother. Res.* **15**, 561-571 (2001).

Yellow-green to dark brown material. Aromatic odor. Hard and brittle when cold; soft and sticky when warm. Extraction with alcohol yields the insol *propolis wax* and alcohol sol *propolis balsam*.

7949. Propoxur. [114-26-1] 2-(1-Methylethoxy)phenol 1-(N-methylcarbamate); *o*-isopropoxyphenyl *N*-methylcarbamate; aprocarb; Bay 39007; Bay 9010; Baygon; Bifex; Blattanex; Propyon; Suncide; Unden. C$_{11}$H$_{15}$NO$_3$; mol wt 209.25. C 63.14%, H 7.23%, N 6.69%, O 22.94%. Prepn: **US 3111539** (1963 to Bayer; Chemagro Corp.). Properties: W. Behrenz, E. Boecker, *Pflanzenschutz-Nachr. Bayer (Ger. Ed.)* **18**, 53 (1965). Toxicity data: T. B. Gaines, *Toxicol. Appl. Pharmacol.* **14**, 515 (1969). Efficacy vs fleas and ticks in domestic animal collars: R. G. Hughes, *Vet. Med. Rev.* **1**, 80 (1985). Teratogenicity study: K. D. Courtney *et al.*, *J. Environ. Sci. Health* **B20**, 373 (1985). Environmental fate and LC/MS determn: L. Sun, H. K. Lee, *J. Chromatogr. A* **1014**, 153 (2003).

Minute crystals, mp 91.5°. Dec at high temp forming methyl isocyanate. Sol in methanol, acetone and many organic solvents, but only slightly sol in cold hydrocarbons. Water soly about 0.2% at 20°. Unstable in highly alkaline media. LD$_{50}$ orally in male, female rats: 83, 86 mg/kg (Gaines).

Caution: Potential symptoms of overexposure are miosis, blurred vision; sweating, salivation; abdominal cramps, nausea, diarrhea, vomiting, headache, weakness, muscle twitching. *See NIOSH Pocket Guide to Chemical Hazards* (DHHS/NIOSH 97-140, 1997) p 266.

USE: Insecticide.

7950. Propoxycaine Hydrochloride. [550-83-4] 4-Amino-2-propoxybenzoic acid 2-(diethylamino) ethyl ester hydrochloride (1:1); 2-diethylaminoethyl 4-amino-2-propoxybenzoate hydrochloride; 2-diethylaminoethyl 2-propoxy-4-aminobenzoate hydrochloride; Ravocaine Hydrochloride; Blockain Hydrochloride. C$_{16}$H$_{26}$ClN$_2$O$_3$; mol wt 330.85. C 58.09%, H 8.23%, Cl 10.71%, N 8.47%, O 14.51%. Prepn: Clinton, Laskowski, **US 2689248** (1954 to Sterling Drug).

White, odorless crystals, mp 148-150°. Discolors upon prolonged exposure to light and to air. Freely sol in water; sol in alc; sparingly sol in ether. Practically insol in acetone, chloroform. pH of a 2% aq soln 5.4.

THERAP CAT: Anesthetic (local).

7951. Propoxycarbazone. [145026-81-9] 2-[[[(4,5-Dihydro-4-methyl-5-oxo-3-propoxy-1H-1,2,4-triazol-1-yl)carbonyl]amino]-sulfonyl]benzoic acid methyl ester; methyl 2-[[[(4-methyl-5-oxo-3-propoxy-4,5-dihydro-1H-1,2,4-triazol-1-yl)carbonyl]amino]sulfonyl]benzoate. C$_{15}$H$_{18}$N$_4$O$_7$S; mol wt 398.39. C 45.22%, H 4.55%, N 14.06%, S 8.05%. Triazolinone graminicide for use in cereal crops; acetolactate synthase inhibitor. Prepn: K.-H. Müller *et al.*, **EP 507171**; *eidem*, **US 5534486** (1992, 1996 both to Bayer). Review of properties and field trials: D. Feucht *et al.*, *Brighton Crop Prot. Conf. - Weeds 1999*, 53-58. Absorption and fate in weeds: L. Fandrich *et al.*, *Weed Sci.* **49**, 717 (2001).

mp 141°. pKa: 2.1.

Sodium salt. [181274-15-7] BAY MKH 6561; Attribut; Olympus. C$_{15}$H$_{17}$N$_4$NaO$_7$S; mol wt 420.37. Colorless crystalline powder, mp 230-240° (dec). Vapor pressure (20°): <1 × 10^{-8} Pa. Log P (octanol/water) at 20°: −0.30 (pH 4); −1.55 (pH 7); −1.59 (pH 9). Soly in water at 20° (g/l): 2.9 (pH 4); 42.0 (pH 7, pH 9). LD$_{50}$ in rats (mg/kg): >5000 orally; >5000 dermally. LC$_{50}$ in rats (mg/m^3): >5030 in air by inhalation. LC$_{50}$ (96 hr) in bluegill sunfish, rainbow trout (mg/l): >94.7; >77.6 (Feucht).

USE: Herbicide.

7952. Propoxyphene. [469-62-5] (αS)-α-[(1R)-2-(Dimethylamino)-1-methylethyl]-α-phenylbenzeneethanol 1-propanoate; α-*d*-4-dimethylamino-3-methyl-1,2-diphenyl-2-butanol propionate; (+)-1,2-diphenyl-2-propionoxy-3-methyl-4-dimethylaminobutane; (+)-4-dimethylamino-1,2-diphenyl-3-methyl-2-propionyloxybutane; α-*d*-propoxyphene; dextropropoxyphene. C$_{22}$H$_{29}$NO$_2$; mol wt 339.48. C 77.84%, H 8.61%, N 4.13%, O 9.43%. Opioid analgesic. Prepn of racemate: Pohland, Sullivan, *J. Am. Chem. Soc.* **75**, 4458 (1953); Pohland, **US 2728779** (1955 to Lilly). Prepn of (+)-form: Pohland, Sullivan, *J. Am. Chem. Soc.* **77**, 3400 (1955). Stereochemistry: Sullivan *et al.*, *J. Org. Chem.* **28**, 2381 (1963); Casy, Myers, *J. Pharm. Pharmacol.* **16**, 455 (1964). Stereospecific synthesis: Pohland *et al.*, *J. Org. Chem.* **28**, 2483 (1963). Metabolism: S. L. Due *et al.*, *Biomed. Mass Spectrom.* **3**, 217 (1976). The α-*dl*- and *d*-diastereoisomers possess marked analgesic activity in contrast to the β-diastereoisomers which are substantially inactive. Toxicity: E. I. Goldenthal, *Toxicol. Appl. Pharmacol.* **18**, 185 (1971); J. L. Emerson *et al.*, *ibid.* **19**, 445 (1971). Comprehensive description: B. McEwan, *Anal. Profiles Drug Subs.* **1**, 301-318 (1972). Symposium on pharmacology, toxicology, and clinical efficacy of propoxyphene alone and in combination with acetaminophen: *Hum. Toxicol.* **3**, Suppl., 1S-238S (1984).

Crystals from petr ether, mp 75-76°. [α]$_D^{25}$ +67.3° (c = 0.6 in chloroform).

Hydrochloride. [1639-60-7] Darvon; Deprancol; Develin. C$_{22}$H$_{29}$NO$_2$.HCl; mol wt 375.94. Bitter crystals from methanol + ethyl acetate, mp 163-168.5°. [α]$_D^{25}$ +59.8° (c = 0.6 in water). Freely sol in water; sol in alc, chloroform, acetone. Practically insol in benzene, ether. LD_{50} in mice, rats (mg/kg): 28, 15 i.v.; 111, 58 i.p.; 211, 134 s.c.; 282, 230 orally (Emerson).

Napsylate monohydrate. [26570-10-5] Darvon-N. $C_{22}H_{29}$-$NO_2.C_{10}H_8O_3S.H_2O$; mol wt 565.73. Odorless, white crystalline powder; bitter taste. Sol in methanol, ethanol, chloroform, acetone; very slightly sol in water. LD_{50} orally in female rats: 990 mg/kg (Goldenthal).

α-l-Form *see* Levopropoxyphene.

α-dl-Form. Racemic propoxyphene; diméprotane.

α-dl-Form hydrochloride. Crystals from methanol + ethyl acetate, mp 170-171°. Soluble in water, alc, chloroform. Practically insol in benzene, ether.

β-dl-Form. Crystals from acetone + ether. mp 187-188°. More soluble than the α-form.

Note: Bulk dextropropoxyphene (non-dosage forms) is a controlled substance (opiate): **21 CFR**, 1308.12; dextropropoxyphene is a controlled substance (narcotic): **21 CFR**, 1308.14.

THERAP CAT: Analgesic.

7953. Propranolol. [525-66-6] 1-[(1-Methylethyl)amino]-3-(1-naphthalenyloxy)-2-propanol; 1-(isopropylamino)-3-(1-naphthyloxy)-2-propanol. $C_{16}H_{21}NO_2$; mol wt 259.35. C 74.10%, H 8.16%, N 5.40%, O 12.34%. β-Adrenergic blocker. Prepn: **BE 640312** and **BE 640313**; Crowther, Smith, **US 3337628** and **US 3520919** (1964, 1964, 1967, 1970 all to I.C.I.). Description of optical isomers: Howe, Shanks, *Nature* **210**, 1336 (1966). Biological studies: Barrett, Cullum, *Br. J. Pharmacol.* **34**, 43 (1968). Metabolism: Bond, *Nature* **213**, 721 (1967). Multi-center clinical trial in myocardial infarction: V. Hansteen *et al.*, *Br. Med. J.* **284**, 155 (1982). CZE determn in urine: J. J. B. Nevado *et al.*, *Anal. Chim. Acta* **559**, 9 (2006). Use as migraine prophylactic: S. Diamond, E. Millstein, *J. Clin. Pharmacol.* **28**, 193 (1988); S. Diamond *et al.*, *Headache* **27**, 70 (1987). Toxicity data: M. Martin, P. Linee, *Eur. J. Med. Chem.* **9**, 563 (1974). Clinical evaluation in hypertension: D. A. Sica *et al.*, *J. Clin. Hypertens.* **6**, 231 (2004). Review of pharmacokinetics: P. A. Routledge, D. G. Shand, *Appl. Pharmacokinet.* **1980**, 464-485; of pharmacology: J. D. Fitzgerald in *Pharmacology of Antihypertensive Drugs*, A. Scriabine, Ed. (Raven Press, New York, 1980) pp 195-208. Series of articles on use in hypertension: *Drugs* **37**, Suppl. 2, 42-76 (1989).

Crystals from cyclohexane, mp 96°.

Hydrochloride. [318-98-9] AY-64043; ICI-45520; NSC-91523; Apsolol; Avlocardyl; Bedranol; Berkolol; Beta-Tablinen; Cardinol; Deralin; Dociton; Efektolol; Elbrol; Hémipralon; Inderal; InnoPran; Obsidan; Propabloc; Prophylux; Propranur; Rapynogen; Sumial; Syprol. $C_{16}H_{21}NO_2.HCl$; mol wt 295.81. Crystals from *n*-propanol, mp 163-164°. Sol in water, alc; slightly sol in chloroform. Practically insol in ether, benzene, ethyl acetate. LD_{50} in mice (mg/kg): 565 orally; 22 i.v.; 107 i.p. (Martin, Linee).

THERAP CAT: Antihypertensive; antianginal; antiarrhythmic (class II).

THERAP CAT (VET): Antiarrhythmic (class II).

7954. Propyl Acetate. [109-60-4] Acetic acid propyl ester; 1-acetoxypropane; propyl ethanoate. $C_5H_{10}O_2$; mol wt 102.13. C 58.80%, H 9.87%, O 31.33%. Prepn from acetic acid and *n*-propyl alcohol: Wagner, *J. Chem. Educ.* **27**, 245 (1950). Manuf from acetic acid and mixture of propene + propane in the presence of $ZnCl_2$ catalyst: Biller, **GB 872876** (1961). Toxicity study: P. M. Jenner *et al.*, *Food Cosmet. Toxicol.* **2**, 327 (1964).

Liquid, odor of pears, bp 101.6°. mp −92°. d_4^{20} 0.836, d_{20}^{20} 0.887. n_D^{20} 1.3844. Flash pt, closed cup: 58°F (14°C). Soly in water at 16°: 1.6:100. Misc with alcohol, ether. LD_{50} in rats, mice (mg/kg): 9370, 8300 orally (Jenner).

Caution: Potential symptoms of overexposure in exptl animals are irritation of eyes, nose and throat; narcosis; dermatitis. *See NIOSH Pocket Guide to Chemical Hazards* (DHHS/NIOSH 97-140, 1997) p 266.

USE: Mfg flavors, perfumes. Solvent for resins, cellulose derivatives, plastics.

7955. Propyl Alcohol. [71-23-8] 1-Propanol; 1-hydroxypropane; propylic alcohol; Optal. C_3H_8O; mol wt 60.10. C 59.96%, H 13.42%, O 26.62%. Discovered by Chancel in 1853 in crude fusel oil and obtained therefrom by fractionation; available as a byproduct of the reaction between carbon monoxide and hydrogen. Toxicity study: Smyth *et al.*, *Arch. Ind. Hyg. Occup. Med.* **10**, 61 (1954).

Liquid; alcoholic and slightly stupefying odor. mp −127°. bp 97.2°. d_4^{20} 0.8053; d_4^{25} 0.8016: Mumford, Phillips, *J. Chem. Soc.* **1950**, 75. Flash pt 22°C. n_D^{20} 1.3862. Miscible with water, alcohol, ether. LD_{50} orally in rats: 1.87 g/kg (Smyth).

Caution: Potential symptoms of overexposure are irritation of eyes, nose and throat; dry cracking skin; drowsiness, headache; ataxia, GI pain; abdominable cramps; nausea, vomiting and diarrhea. *See NIOSH Pocket Guide to Chemical Hazards* (DHHS/NIOSH 97-140, 1997) p 268.

USE: As a solvent for resins and cellulose esters, etc.

7956. Propylamine. [107-10-8] 1-Propanamine; 1-aminopropane. C_3H_9N; mol wt 59.11. C 60.96%, H 15.35%, N 23.70%. Prepn from propionaldehyde + ammonia with a Raney nickel catalyst: Olin, Schwoegler, **US 2373705** (1945 to Sharples); by low pressure catalytic hydrogenation of nitropropane: Iffland, Cassis, *J. Am. Chem. Soc.* **74**, 6284 (1952); by catalytic hydrogenation of propionitrile: Rylander, Kaplan, **US 3117162** (1964 to Englehard). Toxicity study: H. F. Smyth *et al.*, *Am. Ind. Hyg. Assoc. J.* **23**, 95 (1962).

Colorless, alkaline liq; strong ammonia odor. d_{20}^{20} 0.719; mp −83°; bp 48-49°; n_D^{20} 1.389. Flash pt, closed cup: 10°F (−12°C). *Flammable; corrosive. Keep tightly closed.* Misc with water, alcohol, ether. LD_{50} orally in rats: 0.57 g/kg (Smyth).

Hydrochloride. [556-53-6] $C_3H_9N.HCl$; mol wt 95.57. Deliquesc crystals, mp 157-158°. Soluble in 0.4 part water, 1.5 parts chloroform. *Keep well closed.*

Caution: Strong irritant, possible skin sensitizer.

7957. Propylbenzene. [103-65-1] Isocumene; 1-phenylpropane. C_9H_{12}; mol wt 120.20. C 89.93%, H 10.06%. Prepd by the action of diethyl sulfate on benzylmagnesium chloride: H. Gilman, W. E. Catlin, *Org. Synth.* **coll. vol. I**, 471 (2nd ed., 1941). Toxicity study: P. M. Jenner *et al.*, *Food Cosmet. Toxicol.* **2**, 327 (1964).

Liquid. d_4^{20} 0.8621. mp −99.2°. bp_{760} 159.2°; bp_{400} 135.7°; bp_{200} 113.5°; bp_{100} 94°; bp_{40} 71.6°; bp_{20} 56.8°; bp_{10} 43.4°; bp_5 31.3°; $bp_{1.0}$ 6.3°. n_D^{20} 1.4919. Flash pt, closed cup: 107.6°F (42°C). *Flammable. Irritant.* Very slightly sol in water (0.06 g/l); sol in alcohol, ether. LD_{50} orally in rats: 6040 mg/kg (Jenner).

USE: In textile dyeing and printing; as solvent for cellulose acetate.

7958. Propyl Bromide. [106-94-5] 1-Bromopropane. C_3H_7-Br; mol wt 122.99. C 29.30%, H 5.74%, Br 64.97%. Review of human reproductive risks: *NTP-CERHR Monograph* (NIH Publ. No. 04-4479, 2003) 88 pp.

Colorless liquid. d_{20}^{20} 1.353; mp $-110°$; bp 71°; n_D^{20} 1.4341. Sol in 400 parts water; miscible with alcohol, etc.

USE: Solvent for fats, waxes and resins; intermediate in synthesis of pharmaceuticals, insecticides and other compds; in spray adhesives; degreaser; cleaner for metal and precision electronic components.

7959. Propyl Butyrate. [105-66-8] Butanoic acid propyl ester; propyl butanoate. $C_7H_{14}O_2$; mol wt 130.19. C 64.58%, H 10.84%, O 24.58%.

Colorless liquid. d_4^{15} 0.879; mp $-95°$; bp 143°; n_D^{20} 1.4005. Slightly sol in water; misc with alcohol, ether. LD_{50} orally in rats: 15,000 mg/kg; *see:* P. M. Jenner *et al., Food Cosmet. Toxicol.* **2**, 327 (1964).

7960. Propyl Chloride. [540-54-5] 1-Chloropropane. C_3H_7Cl; mol wt 78.54. C 45.88%, H 8.98%, Cl 45.14%. Prepd from propyl alc and PCl_3 in the presence of $ZnCl_2$.

Colorless liquid. d_{20}^{20} 0.890; mp -123 to $-122°$; bp 46-47°; n_D^{20} 1.3886. Sol in ~300 parts water; miscible with alcohol, ether.

7961. Propyl Chloroformate. [109-61-5] Carbonochloridic acid propyl ester; chloroformic acid propyl ester; propyl chlorocarbonate. $C_4H_7ClO_2$; mol wt 122.55. C 39.20%, H 5.76%, Cl 28.93%, O 26.11%.

Colorless liquid. *Flammable. Poisonous.* d^{20} 1.090; bp 114-116°. Flash pt, closed cup: 84°F (29°C). Gradually dec by water or alc; miscible with benzene, chloroform, ether.

Caution: Vapors strongly irritating to eyes and mucous membranes.

7962. Propylene. [115-07-1] 1-Propene; methylethylene; methylethene. C_3H_6; mol wt 42.08. C 85.63%, H 14.37%. Obtained from petr oils during the refining of gasoline. Catalytic or thermal cracking of hydrocarbons always yields propylene. Can be obtained by catalytic dehydrogenation of propane. Chronic toxicity studies: J. A. Quest *et al., Toxicol. Appl. Pharmacol.* **76**, 288 (1984). *Reviews:* R. F. Goldstein, *The Petroleum Chemicals Industry* (New York-London, 1949) p 114 sqq.; Sherwood, *Ind. Chem.* **1960**, 542-546; *Chim. Ind. (Paris)* **1961**, 576-587; Haney, "Ethylene, Propylene and 1-Butene" in *Vinyl and Diene Monomers*, E. C. Leonard, Ed. (Interscience, New York, 1971) pp 577-689; M. R. Schoenberg *et al.,* in *Kirk-Othmer Encyclopedia of Chemical Technology* **vol. 19** (Wiley-Interscience, New York, 3rd ed., 1982) pp 228-246. Review of carcinogenic risk: *IARC Monographs* **60**, 161-180 (1994).

Gas. *Flammable.* Burns with yellow sooty flame. d 1.49 (air = 1.0). mp (triple pt) $-185°$. bp_{760} $-48°$. Critical temp 91.8°. Critical pressure 45.6 atm. Heat of fusion 717.6 cal/mol. Liquefies at 7-8 atm. d_4^{20} (liq) 0.5139. Flash pt, closed cup: $-162.0°F$ ($-108.0°C$). Flammable limits in air: 2.4-10.3% (by volume). Latent heat of vaporization at bp: 104.62 cal/g. Dipole moment 0.35. n_D^{-40} 1.3567. Surface tension at 90°: 16.70 dynes/cm. Shipped as a liquefied gas in low pressure steel cylinders under its own vapor pressure of ~136 pounds per square inch. Contaminants are propane, ethane, carbon dioxide.

Caution: Gas may act as a simple asphyxiant and mild anesthetic. Direct contact with liquid may cause skin burns. *See Patty's Industrial Hygiene and Toxicology* **Vol. 2B**, G. D. Clayton, F. E. Clayton, Eds. (John Wiley & Sons, Inc., New York, 4th ed., 1994) p 1244-1245.

USE: In polymerized form as polypropylene for plastics and carpet fibers. Chemical intermediate in the manuf of acetone, isopropylbenzene, isopropanol, isopropyl halides, propylene oxide, acrylonitrile, cumene.

7963. Propylene Chlorohydrin. [78-89-7] 2-Chloro-1-propanol; 2-chloropropyl alcohol; 1-hydroxy-2-chloropropane. C_3H_7ClO; mol wt 94.54. C 38.11%, H 7.46%, Cl 37.50%, O 16.92%.

Colorless liquid; pleasant odor. d^{20} 1.103; bp 133-134°; n_D^{20} 1.4362. *Poisonous; flammable.* Sol in water, alcohol, etc. LD_{50} orally in rats: 0.22 ml/kg; by skin penetration in rabbits: 0.48 ml/kg; *see:* Smyth *et al., Am. Ind. Hyg. Assoc. J.* **30**, 470 (1969).

USE: In prepn of propylene oxide, *q.v.*

7964. *sec*-Propylene Chlorohydrin. [127-00-4] 1-Chloro-2-propanol; 1-chloro-2-hydroxypropane; 1-chloroisopropyl alcohol. C_3H_7ClO; mol wt 94.54. C 38.11%, H 7.46%, Cl 37.50%, O 16.92%.

Colorless liquid. d^{20} 1.115; bp 126-127°; n_D^{20} 1.4392. Sol in water, alcohol, etc.

7965. Propylenediamine. [78-90-0] 1,2-Propanediamine; 1,2-diaminopropane. $C_3H_{10}N_2$; mol wt 74.13. C 48.61%, H 13.60%, N 37.79%. Prepd from propylene dibromide and alcoholic ammonia at 100°.

Extremely hygroscopic, strongly alkaline liq. Rapidly absorbs moisture to form a hemihydrate. d^{15} 0.878 in anhydr form. bp 119-120°. Very sol in water. *Keep tightly closed.*

USE: In conjunction with cupric sulfate as a very sensitive reagent for mercury.

7966. Propylene Dibromide. [78-75-1] 1,2-Dibromopropane. $C_3H_6Br_2$; mol wt 201.89. C 17.85%, H 3.00%, Br 79.16%. Prepd from propyl bromide and Br_2 in the presence of $AlCl_3$ or $AlBr_3$.

Colorless liquid. mp $-55°$; bp 140-142°; n_D^{20} 1.5203; d^{20} 1.933. Slightly sol in water; miscible with organic solvents.

7967. Propylene Dichloride. [78-87-5] 1,2-Dichloropropane. $C_3H_6Cl_2$; mol wt 112.98. C 31.89%, H 5.35%, Cl 62.75%. Prepd from propyl chloride and Sb_2Cl_5. Toxicity data: H. F. Smyth *et al., Am. Ind. Hyg. Assoc. J.* **30**, 470 (1969). Review of toxicology and human exposure: *Toxicological Profile for 1,2-Dichloropropane* (PB90-182122, 1989) 131 pp.

Flammable, mobile liq. Odor of chloroform. d_{25}^{25} 1.159; bp 95-96°. Solidifies below −70°. n_D^{20} 1.4388. Flash point (ASTM open cup) 21°C (70°F). Despite the low flash pt it does not catch fire readily in industrial applications. Fire pt 38°. Slightly sol in water; miscible with organic solvents. LD$_{50}$ orally in rats: 1.19 ml/kg (Smyth).

Caution: Potential symptoms of overexposure are eye, skin and respiratory system irritation; drowsiness, lightheadedness; liver and kidney damage. Potential occupational carcinogen. *See NIOSH Pocket Guide to Chemical Hazards* (DHHS/NIOSH 97-140, 1997) p 268.

USE: Solvent for oils, fats, resins, waxes, rubber; in ion exchange manuf; in photographic film prodn; paper coating; petroleum catalyst regeneration. Has been used as a soil fumigant.

7968. Propylene Glycol. [57-55-6] 1,2-Propanediol; methyl glycol; 1,2-dihydroxypropane. $C_3H_8O_2$; mol wt 76.10. C 47.35%, H 10.60%, O 42.05%. Prepn from glycerol: Raschig, Prahl, *Ber.* **61**, 185 (1928). Prepn of levorotatory propylene glycol from hydroxyacetone by yeast reduction: Levene, Walti, *Org. Synth.* **coll. vol. II**, 545 (1943). Prepn of S-(+)-form: E. Baer, H. O. L. Fischer, *J. Am. Chem. Soc.* **70**, 609 (1948); C. Melchiorre, *Chem. Ind. (London)* **1976**, 218. Manuf from propylene oxide by hydration: *Faith, Keyes & Clark's Industrial Chemicals*, F. A. Lowenheim, M. K. Moran, Eds. (Wiley-Interscience, New York, 4th ed., 1975) pp 688-691. Toxicity data: W. Bartsch *et al.*, *Arzneim.-Forsch.* **26**, 1581 (1976). GC/MS determn in plasma: C. Giachetti *et al.*, *Biomed. Environ. Mass Spectrom.* **18**, 592 (1989). Review of toxicity, metabolism and biochemistry: Ruddick, *Toxicol. Appl. Pharmacol.* **21**, 102 (1972); of toxicology and human exposure: *Toxicological Profile for Ethylene Glycol and Propylene Glycol* (US DHHS, PB98-101109, 1997) 250 pp.

dl-**Form.** Hygroscopic, viscous liquid. Slightly acrid taste. mp −59°. bp$_{760}$ 188.2°; bp$_{400}$ 168.1°; bp$_{200}$ 149.7°; bp$_{100}$ 132.0°; bp$_{60}$ 119.9°; bp$_{40}$ 111.2°; bp$_{20}$ 96.4°; bp$_{10}$ 83.2°; bp$_5$ 70.8°; bp$_{1.0}$ 45.5°. d_4^{25} 1.036. Flash pt, open cup: 210°F (99°C). n_D^{20} 1.4324. Miscible with water, acetone, chloroform. Sol in ether. Will dissolve many essential oils, but is immiscible with fixed oils. It is a good solvent for rosin. Under ordinary conditions propylene glycol is stable, but at high temps it tends to oxidize giving rise to products such as propionaldehyde, lactic acid, pyruvic acid and acetic acid. LD$_{50}$ orally in rats: 25 ml/kg (Bartsch).

l-**Form.** [4254-14-2] (*R*)-1,2-Propanediol. bp$_{12}$ 88-90°, bp$_{760}$ 187-189°. $[\alpha]_D^{20}$ −15.0°.

d-**Form.** [4254-15-3] (*S*)-1,2-Propanediol. bp$_{14}$ 94-96°. bp$_{765}$ 186-188°. d^{25} 1.04. $n_D^{23.5}$ 1.4312. $[\alpha]_D^{23}$ +20.1° (c = 7.5 in water). $[\alpha]_D^{22}$ +4.2° (c = 6.6 in dry ethanol). $[\alpha]_D^{20}$ +15.84° (neat).

USE: Nontoxic antifreeze in breweries and dairy establishments. Substitute for ethylene glycol and glycerol. In the manuf of synthetic resins and de-icing solutions. Emulsifier in foods; solvent for food colors and flavors. Pharmaceutic aid (humectant, solvent). As mist to disinfect air; to create artificial smoke and mist for theatrical use.

THERAP CAT (VET): Glucogenic (orally) in ruminants.

7969. Propylene Oxide. [75-56-9] 2-Methyloxirane; propene oxide. C_3H_6O; mol wt 58.08. C 62.04%, H 10.41%, O 27.55%. Results from the action of KOH (aq) on propylene chlorohydrin. Toxicity: H. F. Smyth *et al.*, *J. Ind. Hyg. Toxicol.* **23**, 259 (1941). Toxicological comparison with ethylene oxide, *q.v.*: E. Agurell *et al.*, *Mutat. Res.* **250**, 229 (1991). *Reviews:* Holden in *Glycols*, G. O. Curme, F. Johnston, Eds. (Reinhold, New York, 1952) pp 250-261; *Faith, Keyes & Clark's Industrial Chemicals*, F. A. Lowenheim, M. K. Moran, Eds. (Wiley-Interscience, New York, 4th ed., 1975) pp 692-697; R. O. Kirk, T. J. Dempsey in *Kirk-Othmer Encyclopedia of Chemical Technology* vol. 19 (Wiley-Interscience, New York, 3rd ed., 1982) pp 246-274. Review of manufacturing processes: *Chem. Eng. News* **70**, 9-12 (Mar. 2, 1992); of carcinogenic risk: *IARC Monographs* **60**, 181-213 (1994).

Colorless ethereal liquid. *Flammable.* d_4^0 0.859. mp −112.13°. bp 34.23°. Flash pt, closed cup: −31°F (−35°C). Soly in water (20°): 40.5% by wt; soly of water in propylene oxide: 12.8% by wt; miscible with alcohol, ether. LD$_{50}$ orally in rats: 1.14 g/kg (Smyth).

Caution: Potential symptoms of overexposure are irritation of eyes, skin and respiratory system; blistering and burns. *See NIOSH Pocket Guide to Chemical Hazards* (DHHS/NIOSH 97-140, 1997) p 270. This substance is reasonably anticipated to be a human carcinogen: *Report on Carcinogens, Twelfth Edition* (PB2011-111646, 2011) p 367.

USE: Chemical intermediate in prepn of polyethers to form polyurethanes; in prepn of urethane polyols and propylene and dipropylene glycols; in prepn of lubricants, surfactants, oil demulsifiers. As solvent; fumigant; soil sterilant.

7970. Propyl Formate. [110-74-7] Formic acid propyl ester; propyl methanoate. $C_4H_8O_2$; mol wt 88.11. C 54.53%, H 9.15%, O 36.32%.

Colorless liquid; pleasant odor. d^{20} 0.901; mp −93°; bp 81-82°. Flash pt, closed cup: 27°F (−3°C). n_D^{20} 1.3771. *Flammable.* Sol in 45 parts water; misc with alcohol, ether. LD$_{50}$ orally in rats: 3980 mg/kg; *see:* P. M. Jenner *et al.*, *Food Cosmet. Toxicol.* **2**, 327 (1964).

7971. Propyl Gallate. [121-79-9] 3,4,5-Trihydroxybenzoic acid propyl ester; *n*-propyl gallate; gallic acid propyl ester; PG; Progallin P; Tenox PG. $C_{10}H_{12}O_5$; mol wt 212.20. C 56.60%, H 5.70%, O 37.70%. Spectrophotometric determn: C. S. Sastry *et al.*, *Talanta* **29**, 917 (1982). Effects on survival of *Saccharomyces cerevisiae*: V. L. Eubanks, L. R. Beuchat, *J. Food Prot.* **46**, 29 (1983). Antioxidant effectiveness: M. A. Augustin, S. K. Berry, *J. Am. Oil Chem. Soc.* **60**, 105 (1983). Comprehensive review of biological effects and toxicology: *J. Am. Coll. Toxicol.* **4**, 23-64 (1985).

Crystals, mp 150°. Soly at 25° in water = 0.35 g/100 ml; in alcohol = 103 g/100 g; in ether = 83 g/100 g. Soly in cottonseed oil at 30° = 1.23 g/100 g; in lard at 45° = 1.14 g/100 g. Darkens in the presence of iron and iron salts. Synergic with acids, BHA, BHT. pKa 8.11. Partition coefficient (oleyl alcohol:water) 17. Partition coefficient (octanol:water) 32. pH (0.05% aq soln) 6.3; (0.1% aq soln) 5.9; (0.2% aq soln) 5.7. LD$_{50}$ in mice, rats, hamsters, rabbits (g/kg): 1.70-3.50, 2.1-7, 2.48, 2.75 orally; LD$_{50}$ i.p. in rats: 0.38 g/kg (*J. Am. Coll. Toxicol.*).

USE: Antioxidant for cosmetics, foods, fats, oils, ethers, emulsions, waxes, transformer oils.

7972. Propylhexedrine. [101-40-6] *N*,α-Dimethylcyclohexaneethanamine; 1-cyclohexyl-2-methylaminopropane; hexahydrodesoxyephedrine; Benzedrex. $C_{10}H_{21}N$; mol wt 155.29. C 77.35%, H 13.63%, N 9.02%. Prepd by catalytic hydrogenation of the phenyl analog using Adams platinum catalyst and glacial acetic acid as the solvent: Zenitz *et al.*, *J. Am. Chem. Soc.* **69**, 1117 (1947); Ullyot, *US 2454746* (1948 to SK & F).

dl-**Form.** [3595-11-7] Oily liq, d_4^{25} 0.8501, amine odor, bp_{760} 205°; bp_{20} 92-93°; volatilizes slowly at room temp. n_D^{20} 1.4600. Absorbs carbon dioxide from air and its aq solns are alkaline to litmus. Miscible with alc, chloroform, ether; very slightly sol in water.

dl-**Form hydrochloride.** [6192-95-6] $C_{10}H_{21}N.HCl$. Crystals, dec 127-128°. Sol in water.

d-**Form.** [6556-29-2] Oily liq, bp_{10} 82-83°; n_D^{20} 1.4588.

d-**Form hydrochloride.** [6192-96-7] Crystals, dec 138-139°; $[\alpha]_D^{26}$ +14.73°. Sol in water.

l-**Form.** [6192-97-8] Oily liq, bp_9 80-81°. n_D^{20} 1.4590.

l-**Form ethylphenylbarbiturate.** [4388-82-3] *l*-1-Cyclohexyl-2-(methylamino)propane ethylphenylbarbiturate; phenobarbital compd with 1-propylhexedrine; barbexaclone; Maliasin. $C_{22}H_{33}N_3O_3$; mol wt 387.52.

l-**Form hydrochloride.** [6192-98-9] Crystals, dec 138-139°; $[\alpha]_D^{26}$ −14.74°. Sol in water.

THERAP CAT: Adrenergic (vasoconstrictor); decongestant (nasal).

7973. Propylidene Chloride. [78-99-9] 1,1-Dichloropropane. $C_3H_6Cl_2$; mol wt 112.98. C 31.89%, H 5.35%, Cl 62.75%. Obtained by the action of PCl_5 on propionaldehyde. Toxicity data: Smyth *et al.*, *Arch. Ind. Hyg. Occup. Med.* **10**, 61 (1954).

Liquid. d^{10} 1.143; bp 87°. Very slightly sol in water; sol in many organic solvents. LD_{50} orally in rats: 6.5 g/kg (Smyth *et al.*).

7974. Propyl Iodide. [107-08-4] 1-Iodopropane. C_3H_7I; mol wt 169.99. C 21.20%, H 4.15%, I 74.65%. Prepd by heating propyl alcohol with iodine and red phosphorus.

Colorless or slightly yellow liquid. d_4^{20} 1.747; mp ~−98°; bp 102-103°; n_D^{20} 1.5051. Sol in 575 parts water; miscible with alcohol, ether.

7975. Propyl Nitrate. [627-13-4] Nitric acid propyl ester. $C_3H_7NO_3$; mol wt 105.09. C 34.29%, H 6.71%, N 13.33%, O 45.67%. Prepd by nitration of propanol with nitric acid, usually in the presence of urea and ammonium nitrate or sulfuric acid: Vogel, *J. Chem. Soc.* **1948**, 1847. Physical data concerning combustion and propulsion: Penner, Ducarme, *The Chemistry of Propellants* (Pergamon Press, 1960).

Pale yellow liquid. Sweet, sickly odor. d_4^{20} 1.0538. bp_{762} 110°. *Flammable. Explosive.* n_D^{20} 1.3979. Dipole moment; 2.98. Azeotrope with water contg 75% $C_3H_7NO_3$, bp_{760} 84.8°. Very slightly sol in water. Sol in alcohol, ether.

Caution: Potential symptoms of overexposure in exptl animals are irritation of eyes and skin; methemoglobinemia, anoxia, cyanosis; dyspnea, weakness, dizziness, headache. *See NIOSH Pocket Guide to Chemical Hazards* (DHHS/NIOSH 97-140, 1997) p 270.

USE: Fuel ignition promoter, in rocket fuel formulations, as organic intermediate.

7976. Propyl Nitrite. [543-67-9] Nitrous acid propyl ester. $C_3H_7NO_2$; mol wt 89.09. C 40.45%, H 7.92%, N 15.72%, O 35.92%. Prepd from silver nitrite and *n*-propyl bromide: Reynolds, Adkins, *J. Am. Chem. Soc.* **51**, 279 (1929); from *n*-propanol, sodium nitrite, and dil sulfuric acid: Cowley, Partington, *J. Chem. Soc.* **1933**, 1252.

Liquid. d_4^{20} 0.8864. bp_{760} 46-48°. n_D^{20} 1.3592; also reported as n_D^{20} 1.3613. Soluble in alcohol, ether.

Caution: Inhalation causes vasodilation, smooth muscle relaxation, hypotension.

USE: Jet propellant.

7977. Propylparaben. [94-13-3] 4-Hydroxybenzoic acid propyl ester; propyl *p*-hydroxybenzoate; Nipasol M; Solbrol P; Propyl Parasept. $C_{10}H_{12}O_3$; mol wt 180.20. C 66.65%, H 6.71%, O 26.64%. Prepn: Stohmann, *J. Prakt. Chem.* **36**, 368 (1887); L. Nobli, *Giorn. Farm. Chim.* **84**, 168 (1935), *C.A.* **30**, 3423⁹ (1936). Review of safety assessment: M. G. Soni *et al.*, *Food Chem. Toxicol.* **39**, 513-532 (2002).

White crystals, mp 96-97°. Sol in 2000 parts water; freely sol in alcohol, ether; slightly sol in boiling water.

USE: Pharmaceutic aid (antifungal). Antimicrobial preservative in foods and cosmetics.

7978. Propylphosphonic Anhydride. [68957-94-8] 2,4,6-Tripropyl-1,3,5,2,4,6-trioxatriphosphorinane 2,4,6-trioxide; *n*-propane phosphonic acid anhydride; *n*-propylphosphonic cyclic anhydride; PPAA; T3P. $C_9H_{21}O_6P_3$; mol wt 318.18. C 33.97%, H 6.65%, O 30.17%, P 29.20%. Coupling and water removal reagent often used in peptide synthesis. Prepn from *n*-propylphosphonic dichloride: H. Wissmann, H.-J. Kleiner, *Angew. Chem. Int. Ed. Engl.* **19**, 133 (1980). Alternate synthetic method: M. Wehner *et al.*, US 060264654 (2006). Synthetic applications: M. Ueda, T. Honma, *Polym. J.* **20**, 477 (1988); K. S. Crichfield *et al.*, *Synth. Commun.* **30**, 3737 (2000); J. Hartund, M. Schwarz, *Synlett* **2000**, 371. Reviews: M. Schwarz, *ibid.* 1369; A. L. L. García, *ibid.* **2007**, 1328-1329.

Colorless liquid, $bp_{0.3}$ 200°. Sol in ethyl acetate, dioxane, dichloromethane, THF, DMF.

USE: Reagent in synthetic organic chemistry.

7979. Propyl Propanoate. [106-36-5] Propanoic acid propyl ester; propyl propionate. $C_6H_{12}O_2$; mol wt 116.16. C 62.04%, H 10.41%, O 27.55%.

Liquid. d^{20} 0.883; mp −76°; bp 122-124°; n_D^{20} 1.3935. Sol in 200 parts water; miscible with alcohol, ether.

7980. Propylthiouracil. [51-52-5] 2,3-Dihydro-6-propyl-2-thioxo-4(1*H*)pyrimidinone; 6-propyl-2-thiouracil; 2-thio-4-oxo-6-propyl-1,3-pyrimidine; 2-thio-6-propyl-1,3-pyrimidin-4-one; Propacil; Propycil; Propyl-Thyracil; Thyreostat II. $C_7H_{10}N_2OS$; mol wt 170.23. C 49.39%, H 5.92%, N 16.46%, O 9.40%, S 18.83%. Prepd by the condensation of ethyl β-oxocaproate with thiourea: Anderson *et al.*, *J. Am. Chem. Soc.* **67**, 2197 (1945). Comprehensive description: H. Y. Aboul-Enein, *Anal. Profiles Drug Subs.* **6**, 457-486 (1977). Review of pharmacology and clinical experience: D. S. Cooper *et al.*, *N. Engl. J. Med.* **311**, 1353-1362 (1984).

White, bitter cryst powder of starch-like appearance to the eye and to the touch, mp 219-221°. uv max (methanol): 275, 214 nm (ε 15800, 15600); (methanolic KOH): 315.5, 260, 207.5 nm (ε 10900, 10700, 15400). One part dissolves in ~900 parts water at 20°, in 100 parts boiling water, in 60 parts ethyl alcohol, in 60 parts acetone. Practically insol in ether, chloroform, benzene. Freely sol in aq solns of ammonia and alkali hydroxides. A satd aq soln is neutral or slightly acid to litmus.

Caution: This substance is reasonably anticipated to be a human carcinogen: *Report on Carcinogens, Twelfth Edition* (PB2011-111646, 2011) p 369.

THERAP CAT: Antihyperthyroid.

THERAP CAT (VET): Antihyperthyroid. Has been used to promote fattening.

7981. Propylure. [10297-61-7] (*E*)-10-Propyl-5,9-tridecadien-1-ol acetate; *trans*-1-acetoxy-10-(*n*-propyl)trideca-5,9-diene; 10-propyl-*trans*-5,9-tridecadienyl acetate. $C_{18}H_{32}O_2$; mol wt 280.45. C 77.09%, H 11.50%, O 11.41%. Once thought to be the sex pheromone of the pink bollworm moth, *Pectinophora gossypiella* (Saunders), a destructive cotton pest: H. E. Hummel *et al.*, *Science* **181**, 873 (1973), *cf.* Gossyplure. Reported as the first known natural product possessing di-*n*-propyl branching. Isoln, structure, and synthesis: Jones *et al.*, *Science* **152**, 1516 (1966). Improved total syntheses: Pattenden, *J. Chem. Soc. C* **1968**, 2385; Meyers, Collington, *Tetrahedron* **27**, 5979 (1971); Vig *et al.*, *J. Indian Chem. Soc.* **50**, 39 (1973); K. Utimoto *et al.*, *Tetrahedron Lett.* **1975**, 4233. Photochemical synthesis: Kossanyi *et al.*, *ibid.* **1973**, 3459. Stereoselective synthesis: A. Alexakis *et al.*, *ibid.* **1978**, 2027. *trans*-Propylure is rendered inactive by the presence of >15% of the *cis*-isomer: Jacobson, *Science* **163**, 190 (1969).

Colorless liquid having no detectable odor. bp$_{0.1}$ 135°. n_D^{25} 1.4635. Strong IR band at 970 cm^{-1}.

USE: Insect attractant.

7982. Propyphenazone. [479-92-5] 1,2-Dihydro-1,5-dimethyl-4-(1-methylethyl)-2-phenyl-3*H*-pyrazol-3-one; 4-isopropylantipyrine; 4-isopropyl-2,3-dimethyl-1-phenyl-3-pyrazolin-5-one; 2,3-dimethyl-1-phenyl-4-isopropylpyrazolone; isopropylphenazone; Budirol; Causyth; Eufibron; Isoprochin P. $C_{14}H_{18}N_2O$; mol wt 230.31. C 73.01%, H 7.88%, N 12.16%, O 6.95%. Prepn: Stenz, **US 1972036** (1934 to Hoffmann-La Roche); Sawa, *J. Pharm. Soc. Jpn.* **57**, 953 (1937), *C.A.* **32**, 2533 (1938).

Slightly bitter crystals, mp 103°. Readily sol in alcohol, ether. Soly in water: 0.24 g/100 ml at 16.5°.

THERAP CAT: Analgesic; antipyretic; anti-inflammatory.

7983. Propyzamide. [23950-58-5] 3,5-Dichloro-*N*-(1,1-dimethyl-2-propyn-1-yl)benzamide; pronamid; RH-315; Kerb. $C_{12}H_{11}Cl_2NO$; mol wt 256.13. C 56.27%, H 4.33%, Cl 27.68%, N 5.47%, O 6.25%. Selective pre-emergence herbicide. Prepn: B. W. Horrom *et al.*, **ZA 6800090**; *eidem*, **US 3534098**; **US 3640699** (1969, 1970, 1972 all to Rohm & Haas). Activity: K. L. Viste *et al.*, *Science* **167**, 280 (1970); C. Swithenbank *et al.*, *J. Agric. Food Chem.* **19**, 417 (1971). Metabolism: R. Y. Yih *et al.*, *Weed Sci.* **18**, 604 (1970); R. Y. Yih, C. Swithenbank, *J. Agric. Food Chem.* **19**, 314, 320 (1971); J. D. Fisher, *ibid.* **22**, 606 (1974). Carcinogenicity study: M. D. Reuber, *Environ. Res.* **23**, 1 (1980).

White solid, mp 155-156°. Soly in water at 25°: 15 ppm. Sol in aliphatic, aromatic solvents. Vapor pressure at 25°: 8.5×10^{-5} mm Hg. LD$_{50}$ in male, female rats (mg/kg): 8350, 5620 orally (Viste).

USE: Herbicide.

7984. Proquazone. [22760-18-5] 7-Methyl-1-(1-methylethyl)-4-phenyl-2(1*H*)-quinazolinone; 1-isopropyl-7-methyl-4-phenyl-2(1*H*)-quinazolinone; RU-43-715; Sandoz 43-715; Biarison. $C_{18}H_{18}N_2O$; mol wt 278.36. C 77.67%, H 6.52%, N 10.06%, O 5.75%. Prepn: H. Ott, M. Denzer, **DE 1805501** corresp to **US 3845128** and **US 3925548** (1969, 1974 and 1975, all to Sandoz); R. V. Coombs *et al.*, *J. Med. Chem.* **16**, 1237 (1973). Metabolism: M. B. Zucker, *Proc. Soc. Exp. Biol. Med.* **156**, 209 (1977); H. Ott, J. Meier, *Scand. J. Rheumatol. Suppl.* **21**, 12 (1978). Pharmacology: H. U. Gubler, M. Baggiolini, *ibid.* 8; H. Ott, *ibid.* 5. Clinical studies: P. Sfikakis, P. Tsachalos, *Ther. Umsch.* **34**, 730 (1977); series of articles in *Scand. J. Rheumatol. Suppl.* **21**, 15-39 (1978). Review of pharmacodynamics, pharmacokinetics and therapeutic efficacy: S. P. Clissold, R. Beresford, *Drugs* **33**, 478-502 (1987).

Yellow crystals from ethyl acetate, mp 137-138°. Sol in chloroform. Insol in water.

THERAP CAT: Anti-inflammatory.

7985. Proscillaridin. [466-06-8] (3β)-3-[(6-Deoxy-α-L-mannopyranosyl)oxy]-14-hydroxybufa-4,20,22-trienolide; 14-hydroxy-3β-(rhamnosyloxy)bufa-4,20,22-trienolide; 3β-rhamnosido-14β-hydroxy-$\Delta^{4,20,22}$-bufatrienolide; proscillaridin A; desglucotransvaaline; scillarenin 3β-rhamnoside; Caradrin; Proscillan; Purosin-TC; Stellarid; Talusin. $C_{30}H_{42}O_8$; mol wt 530.66. C 67.90%, H 7.98%, O 24.12%. Prepn by acid cleavage of scillaren A: Stoll *et al.*, *Helv. Chim. Acta* **16**, 703 (1933); by enzymic decompn of glucoscillaren A with strophanthobiase: Stoll *et al.*, *ibid.* **35**, 2495 (1952); from *Urginea burkei* Baker, *Liliaceae:* Zoller, Tamm, *ibid.* **36**, 1744 (1953); from *U. (Scilla) maritima* (L.) Baker, *Liliaceae:* Görlich, *Arzneim.-Forsch.* **10**, 770 (1960). Structure: Stoll *et al.*, *Helv. Chim. Acta* **35**, 1934 (1952). Pharmacology: Lenke, Brock, *Arzneim.-Forsch.* **20**, 1 (1970). Metabolic studies: Davis *et al.*, *Arch. Int. Pharmacodyn.* **177**, 231 (1969); Nakano *et al.*, *ibid.* **183**, 199 (1970). Clinical studies: Several authors, *Minerva Med.* **58**, 4243-4322 (1967). Toxicity study: E. I. Goldenthal, *Toxicol. Appl. Pharmacol.* **18**, 185 (1971). Series of articles on prepn, pharmacology, toxicology, pharmacokinetics, metabolism of 4'-methyl ether: *Arzneim.-Forsch.* **28**, 493-573 (1978).

Prisms from methanol, mp 219-222°. $[\alpha]_D^{20}$ −91.5° (CH$_3$OH). LD$_{50}$ in male, female rats (mg/kg): 56, 76 orally (Goldenthal).

Proscillaridin-4-methyl ether. [33396-37-1] Meproscillarin; Clift. C$_{31}$H$_{44}$O$_8$; mol wt 544.69. mp 213-217°. $[\alpha]_D^{20}$ −94° (CH$_3$-OH). uv max (CH$_3$OH): 297 nm (log ε 3.79), (1*N* KOH/CH$_3$OH): 355 nm (log ε 4.65). Sol in methanol, ethanol, THF, dioxane; slightly sol in CHCl$_3$, CH$_2$Cl$_2$, acetone. Practically insol in water, nonpolar organics.

THERAP CAT: Cardiotonic.

7986. Prostacyclin. [35121-78-9] (5*Z*,9α,11α,13*E*,15*S*)-6,9-Epoxy-11,15-dihydroxyprosta-5,13-dien-1-oic acid; (5*Z*)-9-deoxy-6,9α-epoxy-Δ^5-PGF$_{1α}$; epoprostenol; prostaglandin I$_2$; prostaglandin X; PGI$_2$; PGX; U-53217. C$_{20}$H$_{32}$O$_5$; mol wt 352.47. C 68.15%, H 9.15%, O 22.70%. A prostaglandin produced by enzymatic transformation of prostaglandin endoperoxides (*PGG$_2$*, *PGH$_2$*), which dilates blood vessels and is a potent platelet aggregation inhibitor. Evidence for its occurrence during biosynthetic conversion of arachidonic acid by rat stomach homogenates: C. Pace-Asciak, L. S. Wolfe, *Biochemistry* **10**, 3657 (1971). Isoln from microsomes of pig and rabbit aorta by J. R. Vane and co-workers: S. Moncada *et al.*, *Nature* **263**, 663 (1976). Prepn: S. Moncada, N. Whittaker, **DE 2720999**; S. Moncada, **US 4539333** (1977, 1985 both to Wellcome Found.). PGI$_2$ is also synthesized in bovine coronary arteries as well as human arteries and veins: *eidem*, *Lancet* **1**, 18 (1977); G. J. Dusting *et al.*, *Prostaglandins* **13**, 3 (1977); by cultured human and bovine endothelial cells: B. B. Weksler *et al.*, *Proc. Natl. Acad. Sci. USA* **74**, 3922 (1977); by pig aortic endothelial cells: D. E. MacIntyre *et al.*, *Nature* **271**, 549 (1978). It has been suggested that endoperoxides released by platelets can be converted to PGI$_2$ by vascular tissue and that a balance between formation of PGI$_2$ and release of thromboxane A$_2$, *q.v.*, which induces platelet aggregation, controls the formation of thrombi in blood vessels. It has also been postulated that PGI$_2$ acts to stimulate platelet adenylate cyclase and to prevent the action of thrombi on phospholipid breakdown as well as platelet aggregation. Structure: R. A. Johnson *et al.*, *Prostaglandins* **12**, 915 (1976). Synthesis: E. J. Corey *et al.*, *J. Am. Chem. Soc.* **99**, 2006 (1977); of sodium salt and stereochemistry: R. A. Johnson *et al.*, *ibid.* 4182. Additional syntheses: I. Tomoskozi *et al.*, *Tetrahedron Lett.* **18**, 2627 (1977); N. Whittaker, *ibid.* 2805; K. Nicolaou, *Chem. Commun.* **1977**, 630. Synthesis of the 5*E*-isomer: E. J. Corey *et al.*, *Tetrahedron Lett.* **18**, 35293529 (1977). Chemical stability in aq solns: M. J. Cho, M. A. Allen, *Prostaglandins* **15**, 943 (1978). Biosynthetic study: V. Tomasi *et al.*, *Nature* **273**, 670 (1978). Biological properties: R. J. Gryglewski *et al.*, *Prostaglandins* **12**, 685 (1976). Preliminary clinical study: A. E. S. Gimson *et al.*, *Lancet* **1**, 173 (1980). Antimetastatic effects: K. V. Honn *et al.*, *Science* **212**, 1270 (1981); *eidem*, *ibid.* **217**, 542 (1982). Preliminary study of effect of PGX infusion in patients with acute myocardial infarction: O. Edhag *et al.*, *N. Engl. J. Med.* **308**, 1032 (1983). Review of biological properties: S. Moncada, J. R. Vane, *Clin. Sci.* **61**, 369-372 (1981); of therapeutic potential: *eidem*, *Adv. Pharmacol. Ther.* **4**, 215-233 (1982); of physiological role: J. R. Vane *et al.*, *Int. Rev. Exp. Pathol.* **23**, 161-207 (1982). General reviews: S. Mon-

cada, J. R. Vane, *Fed. Proc.* **38**, 66-71 (1979); J. C. McGiff, *Annu. Rev. Pharmacol. Toxicol.* **21**, 479-509 (1981); S. Moncada *et al.*, *Adv. Pharmacol. Ther.* **6**, 39-47 (1982). Review of clinical efficacy in pulmonary arterial hypertension: W. Jacobs, A. Vonk-Noordegraaf, *Expert Opin. Drug Metab. Toxicol.* **5**, 83-90 (2009).

Chemically unstable in aq soln. Hydrolyzes to 6-oxo-PGF$_{1α}$. Half-life at 4° is approx 14.5 min when total phosphate is 0.165 *M*. Anti-aggregating activity disappears within 0.25 min on boiling or within 10 min at 37°.

Sodium salt. [61849-14-7] Epoprostenol sodium; U-53217A; Flolan. C$_{20}$H$_{31}$NaO$_5$; mol wt 374.45. Hygroscopic, free-flowing white powder. Stable for 2 months if kept dry at −30°.

THERAP CAT: Platelet aggregation inhibitor. In treatment of pulmonary arterial hypertension.

7987. Prostaglandin E$_1$. [745-65-3] (11α,13*E*,15*S*)-11,15-Dihydroxy-9-oxoprost-13-en-1-oic acid; 3-hydroxy-2-(3-hydroxy-1-octenyl)-5-oxocyclopentaneheptanoic acid; alprostadil; PGE$_1$; U-10136; Caverject; Edex; Liple; Liprostin; Minprog; Muse; Palux; Prostandin; Prostin VR; Prostivas. C$_{20}$H$_{34}$O$_5$; mol wt 354.49. C 67.76%, H 9.67%, O 22.57%. A primary prostaglandin; easily crystallized from purified biological extracts. Isoln from sheep seminal vesicle tissue, and structure: Bergstrom *et al.*, *Acta Chem. Scand.* **16**, 501 (1962); *eidem*, *J. Biol. Chem.* **238**, 3555 (1963). Enzymic conversion from 8,11,14-eicosatrienoic acid: Nugteren *et al.*, *Rec. Trav. Chim.* **85**, 405 (1966). Synthesis of the *dl*-form: Corey *et al.*, *J. Am. Chem. Soc.* **90**, 3245, 3247 (1968); Schneider *et al.*, *ibid.* **5895**; *eidem*, *ibid.* **91**, 5372 (1969); Axen *et al.*, *Chem. Commun.* **1969**, 303; Taub *et al.*, *ibid.* **1970**, 1258; Slates *et al.*, *ibid.* **1972**, 304; Kuo *et al.*, *Tetrahedron Lett.* **1972**, 5317; Taub *et al.*, *Tetrahedron* **29**, 1447 (1973); Miyano, Stealey, *Chem. Commun.* **1973**, 180; Finch *et al.*, *J. Org. Chem.* **38**, 4412 (1973). Synthesis of natural form: Corey *et al.*, *J. Am. Chem. Soc.* **91**, 535 (1969); **92**, 2586 (1970); Sih *et al.*, *ibid.* **94**, 3643 (1972); **95**, 1676 (1973); Schaaf, Corey, *J. Org. Chem.* **37**, 2921 (1974); Slates *et al.*, *Tetrahedron* **30**, 819 (1974). Metabolism in guinea pigs: Anggard, Samuelsson, *J. Biol. Chem.* **239**, 4097 (1964). Metabolism in humans: Hamberg, Samuelsson, *ibid.* **246**, 6713 (1971). Review of biological activities: Berti *et al.*, *Prog. Biochem. Pharmacol.* **3**, 110 (1967). Comparative pharmacology with respect to other prostaglandins: Weeks, *Annu. Rev. Pharmacol.* **12**, 317 (1972). Clinical use in neonates with cyanotic congenital heart disease: P. M. Olley *et al.*, *Adv. Prostaglandin Thromboxane Res.* **7**, 913 (1980). Use in non-atherosclerotic vasculopathy: D. L. Wooster *et al.*, *J. Am. Med. Assoc.* **245**, 1846 (1981). Clinical trials in impotence: O. I. Linet, F. G. Ogrinc, *N. Engl. J. Med.* **334**, 873 (1996); H. Padma-Nathan *et al.*, *ibid.* **336**, 1 (1997).

Crystals from ethyl acetate + heptane, mp 115-116°. $[\alpha]_{578}$ −61.6° (c = 0.56 in tetrahydrofuran). Soly at 35°: 8000 μg/100 ml double distilled water. Easily dehydrated in soln at pHs <4 or >8.

THERAP CAT: Vasodilator (peripheral). In treatment of male erectile dysfunction.

7988. Prostaglandin E$_2$. [363-24-6] (5*Z*,11α,13*E*,15*S*)-11,15-Dihydroxy-9-oxoprosta-5,13-dien-1-oic acid; 7-[3-hydroxy-

2-(3-hydroxy-1-octenyl)-5-oxocyclopentyl]-5-heptenoic acid; dinoprostone; PGE₂; U-12062; Minprostin E₂; Prepidil; Propess; Prostin E₂. $C_{20}H_{32}O_5$; mol wt 352.47. C 68.15%, H 9.15%, O 22.70%. The most common and most biologically potent of mammalian prostaglandins. Isoln from sheep prostate: S. Bergström, J. Sjövall, **GB 851827**; *idem*, **US 3598858** (1960, 1971); from sheep seminal vesicle tissue: S. Bergström *et al.*, *Acta Chem. Scand.* **16**, 501 (1962). Total synthesis of the *dl*-form: W. P. Schneider, *Chem. Commun.* **1969**, 304; E. J. Corey *et al.*, *J. Am. Chem. Soc.* **91**, 5675 (1969); E. J. Corey *et al.*, *Tetrahedron Lett.* **1970**, 307; W. P. Schneider, **DE 2011969** (1970 to Upjohn), *C.A.* **74**, 87486n (1971); J. Fried *et al.*, *J. Am. Chem. Soc.* **94**, 4342 (1972). Synthesis of naturally occurring form: E. J. Corey *et al.*, *ibid.* **92**, 397, 2586 (1970); J. B. Heather *et al.*, *Tetrahedron Lett.* **1973**, 2313; from *Plexaura homomalla* prostaglandin intermediates: G. L. Bundy *et al.*, *J. Am. Chem. Soc.* **94**, 2123 (1972); W. P. Schneider *et al.*, *Chem. Commun.* **1973**, 254. Biosynthesis: D. A. Van Dorp *et al.*, *Biochim. Biophys. Acta* **90**, 204 (1964); S. Bergström *et al.*, *ibid.* 207; **NL 6505799** (1965 to Unilever), *C.A.* **65**, 7584h (1966). Metabolism: E. Anggard, B. Samuelsson, *Mem. Soc. Endocrinol.*, **14**, 107 (1966); M. Hamberg, B. Samuelsson, *J. Biol. Chem.* **246**, 6713 (1971). Several reviews in *Prostaglandin Symp. Worcester Found. Exp. Biol.*, P. Ramwell, Ed. (Interscience, New York, 1968). For general refs *see* Prostaglandins.

Natural form, colorless crystals. mp 66-68°. $[\alpha]_D^{26}$ −61° (c = 1 in tetrahydrofuran). Easily dehydrated in soln at pHs <4 or >8. Freely sol in acetone, alc, ether, ethyl acetate, isopropyl alc, methanol, methylene chloride; sol in toluene, diisopropyl ether. Practically insol in hexanes.

THERAP CAT: Oxytocic; abortifacient.

7989. Prostaglandin F₂α. [551-11-1] (5Z,9α,11α,13E,15S)-9,11,15-Trihydroxyprosta-5,13-dien-1-oic acid; 7-[3,5-dihydroxy-2-(3-hydroxy-1-octenyl)cyclopentyl]-5-heptenoic acid; dinoprost; PGF₂α; U-14583; Enzaprost F; Prostarmon F. $C_{20}H_{34}O_5$; mol wt 354.49. C 67.76%, H 9.67%, O 22.57%. One of the most biologically studied of the primary prostaglandins. Closely related to prostaglandin E₂ (PGE₂) in that both prostaglandins are biosynthesized from the same precursors and that PGF₂α is the synthetic reduction product of PGE₂. For refs to synthesis of *dl* and natural forms *see* Prostaglandin E₂. Prepn of the tromethamine salt: W. Morozowich, **DE 2126127**; *idem*, **US 3657327** (1971, 1972 both to Upjohn). Alternate synthesis of natural PGF₂α: Schneider, Murray, *J. Org. Chem.* **38**, 397 (1973); R. B. Woodward *et al.*, *J. Am. Chem. Soc.* **95**, 6853 (1973); G. Stork *et al.*, *ibid.* **100**, 8272 (1978); K. Kondo *et al.*, *Tetrahedron Lett.* **1978**, 3927; N. R. A. Beeley *et al.*, *Tetrahedron* **37**, Suppl. 9, 411 (1981); R. J. Cave *et al.*, *J. Chem. Soc. Perkin Trans. 1* **1981**, 646. Causes vasocontraction and exhibits luteolytic activity; is most commonly associated with its role in pregnancy: Karim *et al.*, *J. Obstet. Gynaecol. Br. Commonw.* **78**, 172 (1971). Metabolism in female subjects: Granstrom, Samuelsson, *J. Biol. Chem.* **246**, 5254 (1971). Toxicity data: T. Fujita *et al.*, *Iyakuhin Kenkyu* **9**, 261 (1978), *C.A.* **89**, 71399k (1978). For general refs *see* Prostaglandins.

Natural form, crystals, mp 25-35°. $[\alpha]_D^{25}$ +23.5° (c = 1 in tetrahydrofuran). Freely sol in methanol, abs ethanol, ethyl acetate, chloroform; slightly sol in water. Stable for two years in light resistant containers at 5-15°. Degrades in one week when exposed to sunlight

or in three months at 40°. LD₅₀ in rabbits (mg/kg): 2.5-5.0 i.v.; 2.5-5.0 i.m. (Fujita).

Tromethamine salt. [38562-01-5] In-Synch; Lutalyse. $C_{20}H_{34}O_5 \cdot C_4H_{11}NO_3$; mol wt 475.62. White or off-white cystalline powder, mp 100-101°. Readily sol in water to at least 200 mg/ml.

THERAP CAT: Oxytocic; abortifacient.

THERAP CAT (VET): Oxytocic.

7990. Prostaglandins. A family of biologically potent lipid acids first discovered in seminal fluid and extracts of accessory genital glands of man and sheep: von Euler, *Arch. Exp. Pathol. Pharmakol.* **175**, 78 (1934); *Klin. Wochenschr.* **14**, 1182 (1935). Isoln: Bergstrom, Sjovall, **US 3069322** and **US 3598858** (1962, 1971); Samuelsson, *J. Biol. Chem.* **238**, 3229 (1963). Also found in lower concns in other organs: *idem*, *Biochim. Biophys. Acta* **84**, 707 (1964). The single non-mammalian source of prostaglandin intermediates, or *syntons*, is the gorgonian sea whip or sea fan, *Plexaura homomalla:* Weinheimer, Spraggins, *Tetrahedron Lett.* **1969**, 5185; Schneider *et al.*, *J. Am. Chem. Soc.* **94**, 2122 (1972). Prostaglandins are named as derivatives of *prostanoic acid*. Prostaglandins are divided into the types E, F, A, B, C, and D based on functions in the cyclopentane ring. Numerical subscripts refer to the number of unsaturations in the side chains; α or β subscripts refer to the configuration of substituents in the ring. Six naturally occurring prostaglandins, E₁, E₂, E₃, F₁α, F₂α, F₃α, are considered primary in that no one is derived from another in the living organism. First structural and stereochemical elucidations: Bergström *et al.*, *Acta Chem. Scand.* **16**, 501 (1962); *idem*, *J. Biol. Chem.* **238**, 3555 (1963). Absolute config: Nugteren *et al.*, *Nature* **212**, 38 (1966). First total synthesis of racemic PGE₁ and PGF₁α: Corey *et al.*, *J. Am. Chem. Soc.* **90**, 3245 (1968). Review of synthetic studies: Pike, *Fortschr. Chem. Org. Naturst.* **28**, 313 (1970); Axen *et al.*, in *The Total Synthesis of Natural Products* vol. 1, J. ApSimon, Ed. (Wiley-Interscience, New York, 1973) pp 81-143; Clarkson in *Progress in Organic Chemistry* vol. 8, W. Carruthers, J. K. Sutherland, Eds. (Wiley, New York, 1973) pp 1-28. Book: J. S. Bindra, R. Bindra, *Prostaglandin Synthesis* (Academic Press, New York, 1977). Biosynthesis occurs by enzymatic conversion of unsaturated twenty-carbon fatty acids. Review of biosynthetic studies: Samuelsson, *Prog. Biochem. Pharmacol.* **5**, 109 (1969). Review of metabolism: Samuelsson *et al.*, *Ann. N.Y. Acad. Sci.* **180**, 138 (1971). Biological activities include stimulation of smooth muscle, dilation of small arteries, bronchial dilation, lowering of blood pressure, inhibition of gastric secretion, of lipolysis, and of platelet aggregation, induction of labor, abortion, and menstruation, and increase in ocular pressure. Implicated also in dysmenorrhea, inflammatory reactions, nasal vasoconstriction, kidney function, and in autonomic neurotransmission. Reviews of pharmacological and biochemical aspects: Horton, *Experientia* **21**, 113 (1965); Weeks, *Annu. Rev. Pharmacol.* **12**, 317 (1972); Hinman, *Annu. Rev. Biochem.* **41**, 161 (1972). Review of biological activities of synthetic prostaglandins: Ramwell *et al.*, *Nature* **221**, 1251 (1969). Review of analytical and preparative techniques: "Prostaglandins and Arachidonate Metabolites" in *Methods Enzymol.* **86**, 705, (1982). General reviews: Bergstrom, *Science* **157**, 382 (1967); Bergstrom *et al.*, *Pharmacol. Rev.* **20**, 1 (1968); Ramwell, Shaw, *Recent Prog. Horm. Res.* **26**, 139 (1970); Pike, *Sci. Am.* **225**, 84 (Nov., 1971); Bindra, Bindra, *Progress in Drug Research* vol. 17 (Birkhäuser Verlag, Basel, 1973) pp 410-487. Books: *The Prostaglandins* vols. 1 2,, P. Ramwell, Ed. (Plenum Press, New York, 1973, 1974); *Prostaglandins in Cardiovascular and Renal Function*, A. Scriabine *et al.*, Eds. (Spectrum Publications, New York, 1980) 498 pp; *Cardiovascular Pharmacology of the Prostaglandins*, A. Herman, Ed. (Raven Press, New York, 1982) 472 pp.

Prostanoic acid

7991. Prostalene. [54120-61-5] (9α,11α,13E,15R)-(±)-9,-11,15-Trihydroxy-15-methylprosta-4,5,13-trien-1-oic acid methyl ester; (±)-methyl-7-[3,5-dihydroxy-2-[(E)-3-hydroxy-3-methyl-1-octenyl]cyclopentyl]-4,5-heptadienoate; RS-9390; Synchrocept. $C_{22}H_{36}O_5$; mol wt 380.53. C 69.44%, H 9.54%, O 21.02%. Synthetic analog of prostaglandin $F_{2\alpha}$, q.v. Prepn: P. Crabbé, J. H. Fried, **DE 2258668** corresp to **US 3879438** (1973, 1975 both to Syntex). Use in induction of parturition in sows: W. Holtz et al., J. Anim. Sci. **49**, 367 (1979). Pharmacodynamics in mares: R. G. Loy et al., J. Reprod. Fertil. Suppl. **27**, 229 (1979).

Relative stereochemistry

THERAP CAT (VET): Luteolytic.

7992. Prosulfuron. [94125-34-5] N-[[(4-Methoxy-6-methyl-1,3,5-triazin-2-yl)amino]carbonyl]-2-(3,3,3-trifluoropropyl)benzenesulfonamide; 1-(4-methoxy-6-methyl-1,3,5-triazin-2-yl)-3-[2-(3,3,3-trifluoropropyl)phenylsulfonyl]urea; 1-[2-(3,3,3-trifluoropropyl)phenylsulfonyl]-3-(4-methoxy-6-methyl-1,3,5-triazin-2-yl)urea; CGA-152005; Peak. $C_{15}H_{16}F_3N_5O_4S$; mol wt 419.38. C 42.96%, H 3.85%, F 13.59%, N 16.70%, O 15.26%, S 7.64%. Post-emergence sulfonylurea herbicide for use in maize; acetolactate synthase inhibitor. Prepn: W. Meyer, K. Oertle, **EP 120814**; eidem, **US 4671819** (1984, 1987 both to Ciba-Geigy); J. G. Dingwall, Pestic. Sci. **41**, 259 (1994). Comprehensive description: M. Schulte et al., Brighton Crop Prot. Conf. - Weeds **1993**, 53-59.

Colorless, odorless crystals, mp 155° (dec). Vapor pressure (25°): $<3.5 \times 10^{-6}$ Pa. Soly in water (25°): 4000 mg/l (pH 6.8). Log P (n-octanol/water): −0.21 (pH 6.9). LD_{50} orally in rats: 986 mg/kg; dermally in rabbits: >2000 mg/kg. LC_{50} (4 hr) by inhalation in rats: >5000 mg/m³ (Schulte).

USE: Herbicide.

7993. Prosultiamine. [59-58-5] N-[(4-Amino-2-methyl-5-pyrimidinyl)methyl]-N-[4-hydroxy-1-methyl-2-(propyldithio)-1-buten-1-yl]formamide; 2-(2-methyl-4-aminopyrimidin-5-yl)methylformamido-5-hydroxy-2-penten-3-yl propyl disulfide; vitamin B_1 propyl disulfide; thiamine propyl disulfide; dithiopropylthiamine; DTPT; TPD; Alinamin; Binova. $C_{15}H_{24}N_4O_2S_2$; mol wt 356.50. C 50.54%, H 6.79%, N 15.72%, O 8.98%, S 17.99%. Synthesis: Matsukawa, Kawasuki, J. Pharm. Soc. Jpn. **73**, 216 (1953), C.A. **48**, 2071 (1954); Matsukawa et al., J. Vitaminol. **1**, 13 (1954); Fujiwara et al., **US 2833768** (1958 to Takeda); **FR 1068459** (1954 to Takeda). Structural studies: Nishikawa et al., Chem. Pharm. Bull. **17**, 932 (1969). Metabolism: Suzuoki-Ziro et al., J. Biochem. **58**, 279 (1965); Nishikawa et al., J. Pharmacol. Exp. Ther. **157**, 589 (1967).

E-form

Prisms from benzene, dec 128-129°. Sparingly soluble in water. Sol in organic solvents and lipids. Better absorbed upon oral ingestion by man, than thiamine hydrochloride.

Hydrochloride. [973-99-9] $C_{15}H_{24}N_4O_2S_2$.HCl. Crystals, dec 160-161°.

THERAP CAT: Vitamin (enzyme co-factor).

7994. Protactinium. [7440-13-3] Protoactinium; eka-tantalum. Pa; at. no. 91; at. wt 231.03588 (characteristic naturally occurring isotopic mixture); valences 3, 4, 5. No stable nuclides; known isotopes (mass numbers): 216, 222-238; naturally occurring isotopes: 231, 234m, 234. First isotope discovered, ^{234m}Pa ($T_{1/2}$ 1.17 minutes); called **brevium**, **uranium X_2** or **UX_2**, natural decay product of ^{238}U. Modes of decay: γ by isomeric transition to yield ^{234}Pa; β⁻ to yield ^{234}U. Longest-lived isotope, ^{231}Pa ($T_{1/2}$ 3.276 × 10⁴ years, rel. at. mass 231.0359); natural decay product of ^{235}U. Decays by α emission; parent of ^{227}Ac. ^{234}Pa ($T_{1/2}$ 6.75 hrs); called **Uranium Z** or **UZ**; natural decay product of ^{238}U. Decays by β⁻ emission. Discovery of ^{m234}Pa: K. Fajans, O. H. Göhring, Naturwissenschaften **1**, 339 (1913), C.A. **7**, 3916 (1913); eidem, Phys. Z. **14**, 877 (1913), C.A. **7**, 297 (1913); of ^{231}Pa: O. Hahn, L. Meitner, ibid. **19**, 208 (1918); F. Soddy, J. A. Cranston, Proc. Roy. Soc. **94A**, 384 (1918); of ^{234}Pa: O. Hahn, Ber. **54B**, 1131 (1921). Prepn of metal: A. V. Grosse, M. Agrass, J. Am. Chem. Soc. **56**, 2200 (1934). Review of discovery: K. Fajans, D. F. C. Morris, Nature **244**, 137-138 (1973); of use in nuclear fuel cycle: O. L. Keller, Radiochim. Acta **25**, 211-223 (1978). Reviews: Comprehensive Inorganic Chemistry **vol. 5**, J. C. Bailar, Jr. et al., Eds. (Pergamon Press, Oxford, 1973) passim; H. W. Kirby in The Chemistry of the Actinide Elements **vol. 1**, J. J. Katz et al., Eds. (Chapman and Hall, New York, 1986) pp 102-168.

Shiny, silvery, malleable, ductile metal. Body-centered tetragonal crystal structure. Easily tarnished in air to an undetermined oxide. mp 1560°; also reported as 1575° (Bailar). bp 4227°. d^{25} 15.37. Reacts with H_2 at 250-300° to form PaH_3. In dil solns of HF, is deposited (10-96%) on Be, Al, Mn, Zn, and Pl; gives small deposits on Cr, Ta, Fe, Cd, Ni, Cu, Hg, W: Camarcat et al., J. Chim. Phys. **46**, 153-157 (1949).

Caution: Radiation hazard; handling requires special equipment and shielding facilities (Katz et al., loc. cit. **vol 2**, p 1128). Inhalation hazard in insol form; general hazard if absorbed systemically. Max permissible concn of insoluble form in air: 4×10^{-11} μCi/cc; of sol form in air: 4×10^{-13} μCi/cc; Natl. Bur. Stand. Handb. **69**, 83 (1959).

USE: ^{233}Pa as intermediate in production of fissile ^{233}U in thorium breeder reactors.

7995. Protein C. [60202-16-6] Blood-coagulation factor XIV. Vitamin K-dependent plasma protein involved in the regulation of hemostasis. Two-chain, multi-domain glycoprotein, mol wt ~62,000. Circulates as a zymogen; activated by thrombin complexed with thrombomodulin, q.v. Activated protein C is a serine protease that controls coagulation by the selective inactivation of factors Va and VIIIa. This activity is potentiated by factor V and by **Protein S**, a vitamin K-dependent protein unrelated to the serine proteases. Identification of **autoprothrombin II-A**, an anticoagulant obtained from purified prothrombin complex: E. F. Mammen et al., Thromb. Diath. Haemorrh. **5**, 218 (1960). Purification: W. H. Seegers et al., Thromb. Res. **1**, 443 (1972). Characterization: J. Stenflo, J. Biol. Chem. **251**, 355 (1976). Identity with autoprothrombin II-A: W. H. Seegers et al., Thromb. Res. **8**, 543 (1976). Isoln from human plasma: W. Kisiel, J. Clin. Invest. **64**, 761 (1979). Amino acid sequence of bovine protein C: P. Fernlund, J. Stenflo, J. Biol. Chem. **257**, 12170, 12180 (1982). Review of role in hemostasis: L. H. Clouse, P. C. Comp, N. Engl. J. Med. **314**, 1298-1304 (1986); C. T. Esmon, J. Biol. Chem. **264**, 4743-4746 (1989); of preparative methods and mechanism of activation: C. T. Esmon et al., Methods Enzymol. **222**, 359-385 (1993). Review of protein C system and role in thromboembolic disease: B. Dahlbäck, Thromb. Res. **77**, 1-43 (1995).

Drotrecogin alfa. [98530-76-8] Recombinant human activated protein C; rhAPC; Xigris; Zovant. Prepn: S. C. B. Yan et al., Biotechnology **8**, 655-661 (1990). Review of development and clinical potential: S. B. Yan, B. W. Grinnell, Perspect. Drug Discovery Des. **1**, 503-520 (1993). Clinical trial in severe sepsis: G. R. Bernard et al., N. Engl. J. Med. **344**, 699 (2001).

THERAP CAT: In treatment of severe sepsis.

7996. Protein Hydrolysates. Aminosol; Dekamin; Parentamin; Travamin. Sterile solution of amino acids and short-chain peptides which represent the approx nutritive equivalent of the casein, lactalbumin, plasma, fibrin or other suitable protein from which it is derived by acid, enzymatic or other method of hydrolysis. It may be modified by partial removal and restoration or addn of one or more amino acids. It may contain alcohol, dextrose, or other carbohydrate suitable for i.v. infusion. Not <50% of the total nitrogen present is in the form of α-amino nitrogen.

THERAP CAT: Parenteral nutrient.

THERAP CAT (VET): Parenteral nutrient.

7997. Prothioconazole. [178928-70-6] 2-[2-(1-Chlorocyclopropyl)-3-(2-chlorophenyl)-2-hydroxypropyl]-1,2-dihydro-3H-1,-2,4-triazole-3-thione; 2-(1-chlorocyclopropyl)-1-(2-chlorophenyl)-3-(5-mercapto-1,2,4-triazol-1-yl)propan-2-ol; JAU-6476; Proline; Redigo. $C_{14}H_{15}Cl_2N_3OS$; mol wt 344.25. C 48.85%, H 4.39%, Cl 20.60%, N 12.21%, O 4.65%, S 9.31%. Fungal sterol demethylation inhibitor. Prepn: M. Jautelat *et al.*, **DE 19528046**; *eidem*, **US 5789430** (1996, 1998 both to Bayer). Comprehensive description: A. Mauler-Machnik *et al.*, *BCPC Conf. - Pests Dis.* **2002**, 389-394.

White to light beige crystalline powder. mp 139.1-144.5°. Vapor pressure (20°): $<4 \times 10^{-7}$ Pa. pKa: 6.9. Log P (octanol/water): 4.05 (unbuffered at 20°). Soly in water (20°): 0.3 g/l. LD_{50} in rats (mg/kg): >6200 orally; >2000 dermally. LC_{50} in rats (mg/m³): >4990 by inhalation. LC_{50} (96 hr) in rainbow trout: 1.83 mg/l (Mauler-Machnik).

USE: Agricultural fungicide.

7998. Prothipendyl. [303-69-5] N,N-Dimethyl-10H-pyrido-[3,2-b][1,4]benzothiazine-10-propanamine; 10-(3-dimethylaminopropyl)-10H-pyrido[3,2-b][1,4]benzothiazine; 10-(γ-dimethylaminopropyl)-1-azaphenothiazine; N-(3-dimethylaminopropyl)thiophenylpyridylamine; 2,3-pyridino-(5',6')-5,6-benzo-4-(3″-dimethylaminopropyl)-1,4-thiazine. $C_{16}H_{19}N_3S$; mol wt 285.41. C 67.33%, H 6.71%, N 14.72%, S 11.23%. Prepn: Yale, Bernstein, **US 2943086** (1960 to Olin Mathieson); von Schlichtegroll, *Proc. 1st Int. Congr. Neuro-Pharmacol.* **1958**, 408 (1959), *C.A.* **54**, 13400g (1960); **FR 1173134** (1959 to Rhône-Poulenc).

Liquid, $bp_{0.7}$ 217-219°; $bp_{0.5}$ 195-198°.

Hydrochloride. [1225-65-6] D-206; Dominal; Tolnate. $C_{16}H_{19}N_3S.HCl$; mol wt 321.87. mp 177-178° with sintering ~176°. Monohydrate as crystals, mp 108-112°. Freely sol in water, methanol. Practically insol in ether, petr ether.

Dihydrochloride. $C_{16}H_{19}N_3S.2HCl$. Crystals from acetonitrile, mp 205-207°.

THERAP CAT: Antipsychotic.

7999. Prothrombin. [9001-26-7] Blood-coagulation factor II; Factor II; prothrombase; serozyme; thrombogen. Mol wt 69,000-74,000. Coagulation proenzyme present in highest concentration in blood. Prothrombin is one of the vitamin-K dependent blood coagulation factors. It is converted to thrombin by the action of factor X_a, factor V and phospholipid in the presence of Ca^{2+} ions. Accounts for < 0.2% of total plasma protein. Prepn of human and bovine prothrombin: Goldstein *et al.*, *J. Biol. Chem.* **234**, 2857 (1959); Lanchantin *et al.*, *ibid.* **238**, 238 (1963). Chemistry of activation: Magnussen, *Biochem. J.* **115**, 2P (1969). Synthesized normally by liver parenchymal cells in a cyclic asynchronous manner. Dicou-

marol derivatives halt the synthesis, while vitamin K_1 stimulates synchronized activity of all the liver parenchymal cells: Barnhart, Anderson, *Biochem. Pharmacol.* **9**, 23 (1962). The glycoprotein structure probably consists of a single polypeptide chain containing between 8 and 10% carbohydrate: Magnussen, *Thromb. Diath. Haemorrh.* **suppl. 54**, 31 (1973). *Reviews:* W. H. Seegers, *Prothrombin* (Harvard Univ. Press, 1962) 728 pp; Magnussen "Prothrombin and Thrombin," in *The Enzymes* **Vol. III**, P. D. Boyer, Ed. (Academic Press, New York, 3rd ed., 1971) pp 277-321; K. G. Mann, *Methods Enzymol.* **45B**, 123-156 (1976). Review on prothrombin activation: several authors in *Ann. N.Y. Acad. Sci.* **370**, 336-528 (1981).

Most stable within the range pH 4-9.5. Isoelec pt pH 4.2. Very sol in water but pptd at pH 4.2-4.5. Solns tend to activate spontaneously. There is little loss of activity when drying from the frozen state but the dry material alters in a few months. Drying with organic solvents destroys activity.

8000. Protionamide. [14222-60-7] 2-Propyl-4-pyridinecarbothioamide; 2-propylthioisonicotinamide; prothionamide; 2-propyl-4-thiocarbamoylpyridine; Ektebin; Peteha; Trevintix. $C_9H_{12}N_2S$; mol wt 180.27. C 59.97%, H 6.71%, N 15.54%, S 17.78%. Prepn: D. Liberman *et al.*, *Compt. Rend.* **242**, 2409 (1956); **GB 800250**, D. Liberman, **US 2901488** (1958, 1959 both to Chimie et Atomistique); L. N. Yakhontov *et al.*, *Khim. Farm. Zh.* **10**, 96 (1976), *C.A.* **85**, 32787h (1976). Pharmacology: A. M. Il'in, *Farmakol. Toksikol.* **38**, 471 (1975).

Crystals, mp 142°. Sol in ethanol, methanol; slightly sol in ether, chloroform. Practically insol in water. LD_{50} in mice, rats (g/kg): 1.0, 1.32 orally (Il'in).

THERAP CAT: Antibacterial (tuberculostatic).

8001. Protoanemonin. [108-28-1] 5-Methylene-2(5H)-furanone; 4-hydroxy-2,4-pentadienoic acid γ-lactone; 5-methylene-2-oxodihydrofuran; Isomycin. $C_5H_4O_2$; mol wt 96.09. C 62.50%, H 4.20%, O 33.30%. An antibacterial principle of *Anemone pulsatilla* L., *Ranunculaceae*: Seegal, Holden, *Science* **101**, 413 (1945). Exists as a glucoside, *ranunculin*, in the intact plant, and is released by an enzymatic process on maceration of the plant tissue: R. Hill, R. van Heyningen, *Biochem. J.* **49**, 332 (1951). Isoln: Asahina, Fujita, *Acta Phytochim.* **1**, 1 (1922); Baer *et al.*, *J. Biol. Chem.* **162**, 65 (1946). Structure and synthesis: Asahina, Fujita, *loc. cit.*; Muskat *et al.*, *J. Am. Chem. Soc.* **52**, 326 (1930); Fox, Jr., *Proc. Soc. Exp. Biol.* **51**, 102 (1942); Shaw, *J. Am. Chem. Soc.* **68**, 2510 (1946). Industrial prepns: **GB 759999** (1956 to Olin Mathieson), *C.A.* **51**, 9678f (1957); Reicheneder *et al.*, **DE 1088047** (BASF), *C.A.* **56**, 14086i (1962); Sakuma, Hirano, **US 3203863** (1965 to Lion Dentifrice).

Pale yellow oil; volatile with steam. $bp_{1.5}$ 45°. Sol in ethylene dichloride, chloroform. Solubility in water ~1%. Stable in water. When pure the compd turns to a hard polymer which, when ground and extracted with boiling ethyl acetate, yields anemonin, *q.v.*

THERAP CAT: Antibacterial.

8002. Protocatechualdehyde. [139-85-5] 3,4-Dihydroxybenzaldehyde; 3,4-dihydroxybenzenecarbonal; protocatechuic aldehyde; rancinamycin IV. $C_7H_6O_3$; mol wt 138.12. C 60.87%, H 4.38%, O 34.75%. Prepn from catechol: Reimer, Tiemann, *Ber.* **9**, 1268 (1876); Tiemann, Koppe, *ibid.* **14**, 2015 (1881); from vanillin: Tiemann, Haarmann, *ibid.* **7**, 620 (1874); from veratric aldehyde: Dreyfus, **DE 193958**, *Frdl.* **9**, 161 (1908-10); from piperonal: Hoering, Baum, *Ber.* **41**, 1914 (1908); Barger, *J. Chem. Soc.* **93**, 563 (1908); Buck, Zimmerman, *Org. Synth.* **coll. vol. II**, 549 (1943).

Platelets from water or toluene. Dimorphic. Dec 153-154°. pK (25°) 7.55. Soly in water (g/100 ml): 5 (20°); 33 (99°); in ethanol: 79 (78°). Freely sol in ether.

8003. Protocatechuic Acid. [99-50-3] 3,4-Dihydroxybenzoic acid. $C_7H_6O_4$; mol wt 154.12. C 54.55%, H 3.92%, O 41.52%. Minute amounts are found in wheat grains, in wheat seedlings, and in many other plants: L. Hörhammer, A. Scherm, *Arch. Pharm.* **288**, 441 (1955). Prepd by the alkaline fusion of vanillin: Tiemann, Haarmann, *Ber.* **7**, 617 (1874); Pearl, *J. Am. Chem. Soc.* **68**, 2180 (1946); *Org. Synth.* **coll. vol. III**, 745 (1955).

White to brownish, cryst powder; discolors in air. d 1.54. mp ~200° with decompn. Sol in 50 parts water; sol in alcohol, ether. *Keep well closed.*

8004. Protokylol. [136-70-9] 4-[2-[[2-(1,3-Benzodioxol-5-yl)-1-methylethyl]amino]-1-hydroxyethyl]-1,2-benzenediol; α-[(α-methyl-3,4-methylenedioxyphenethylamino)methyl]protocatechuyl alcohol; α-(3,4-dihydroxyphenyl)-β-[2-(3,4-methylenedioxyphenyl)isopropylamino]ethanol; 1-(3,4-dihydroxyphenyl)-2-(α-methyl-3,4-methylenedioxyphenethylamino)ethanol; *N*-[β-(3,4-methylene-dioxyphenyl)isopropyl]-β-(3,4-dihydroxyphenyl)-β-hydroxyethylamine; *N*-[2-(3,4-methylenedioxyphenylisopropyl)]norepinephrine. $C_{18}H_{21}NO_5$; mol wt 331.37. C 65.24%, H 6.39%, N 4.23%, O 24.14%. β-Adrenergic agonist. Prepn: Biel *et al., J. Am. Chem. Soc.* **76**, 3149 (1954); Biel, **US 2900415** (1959 to Lakeside Labs.). Toxicity: E. I. Goldenthal, *Toxicol. Appl. Pharmacol.* **18**, 185 (1971).

Hydrochloride. [136-69-6] JB-251; Caytine; Ventaire. $C_{18}H_{21}$-NO_5.HCl; mol wt 367.83. Crystals from isopropanol, mp 126-127°. Sol in water. LD_{50} orally in rats: 938 ±96 mg/kg (Goldenthal).

THERAP CAT: Bronchodilator.

THERAP CAT (VET): Bronchodilator.

8005. Protopine. [130-86-9] 4,6,7,14-Tetrahydro-5-methyl-bis[1,3]benzodioxolo[4,5-*c*:5',6'-*g*]azecin-13(5*H*)-one; 7-methyl-2,-3:9,10-bis(methylenedioxy)-7,13a-secoberbin-13a-one; fumarine; macleyine. $C_{20}H_{19}NO_5$; mol wt 353.37. C 67.98%, H 5.42%, N 3.96%, O 22.64%. From opium: Hesse, *Ber.* **4**, 693 (1871); also from the herb *Fumaria officinalis* L., *Chelidonium majus* L., and many other *Papaveraceae* and *Fumariaceae:* Manske in *The Alkaloids* **vol. IV**, R. H. F. Manske, H. L. Holmes, Eds. (Academic Press, New York, 1954) pp 157-159. Structure: Perkin, Jr., *J. Chem. Soc.* **109**, 815 (1916); Gadamer, Bruchhausen, *Arch. Pharm.* **260**, 97 (1922); Mottus *et al., Can. J. Chem.* **31**, 1144 (1953); Manet, Marion, *ibid.* **32**, 452 (1954). Synthesis: Haworth, Perkin, *J. Chem. Soc.* **1926**, 1769. Crystal structure: Hall, Ahmed, *Acta Crystallogr.* **24B**, 337 (1968).

Monoclinic prisms from alcohol + chloroform, mp 208°. d 1.399 (calc). uv max (95% ethanol): 293 nm (log ε 3.93). Sol in 15 parts chloroform, 900 parts alc, 1000 parts ether. Slightly sol in ethyl acetate, carbon disulfide, benzene, petr ether. Practically insol in water.

Hydrochloride. $C_{20}H_{19}NO_5$.HCl. Prisms from alcohol, sol in 143 parts water at 13°, sol in alcohol. Also a hexahydrate, needles from water.

Methiodide. $C_{20}H_{19}NO_5$.CH_3I. Twinned crystals from methanol, dec 217°.

8006. Protoporphyrin IX. [553-12-8] 7,12-Diethenyl-3,8,-13,17-tetramethyl-21*H*,23*H*-porphine-2,18-dipropanoic acid; 3,7,-12,17-tetramethyl-8,13-divinyl-2,18-porphinedipropionic acid; 1,3,-5,8-tetramethyl-2,4-divinylporphine-6,7-dipropionic acid; ooporphyrin; Kämmerer's porphyrin. $C_{34}H_{34}N_4O_4$; mol wt 562.67. C 72.58%, H 6.09%, N 9.96%, O 11.37%. Biological precursor of blood and plant pigments. Prepd from hemin: Fischer-Orth, *Die Chemie des Pyrrols* **II**, 1, 396 (Leipzig, 1937); Ramsey, *Biochem. Prep.* **3**, 39 (1953). Structure: Sparatore, Mauzerall, *J. Org. Chem.* **25**, 1073 (1960). Synthesis: Carr *et al., J. Chem. Soc. C* **1971**, 487. Crystal and molecular structure: W. S. Caughey, J. A. Ibers, *J. Am. Chem. Soc.* **99**, 6639 (1977). Chelates with metals, esp iron, in the ferrous state to form heme, *q.v.,* in the ferric state to form hematin, *q.v.* Review: Rimington, Kennedy, in M. Florkin, H. S. Mason, *Comparative Biochemistry* (Academic Press, New York, 1962) pp 557-614.

Monoclinic, brownish-yellow prisms from ether. Absorption max (25% HCl): 602.4, 582.2, 557.2 nm. Freely sol in chloroform, glacial acetic acid, alcohol contg HCl, ether contg some glacial acetic acid, hydrochloric acid. Somewhat sol in dil alkalies, aniline, pyridine. Forms sparingly sol disodium and dipotassium salts.

Disodium salt. [50865-01-5] Depocolin-S. $C_{34}H_{32}N_4Na_2O_4$; mol wt 606.63.

Dimethyl ester. [5522-66-7] $C_{36}H_{38}N_4O_4$. Crystals from chloroform + methanol, mp 228-230°. Absorption max (25% HCl): 601, 556, 409 nm. Soluble in chloroform, slightly sol in methanol. Insol in sodium carbonate solns.

THERAP CAT: In liver disease.

8007. Protostephanine. [549-28-0] 6,7,8,9-Tetrahydro-2,3,-10,12-tetramethoxy-7-methyl-5*H*-dibenz[*d*,*f*]azonine. $C_{21}H_{27}$-NO_4; mol wt 357.45. C 70.56%, H 7.61%, N 3.92%, O 17.90%. A member of the hasubanan alkaloids. First alkaloid known to possess the unique dibenz[*d*,*f*]azonine structure. Isoln from *Stephania japonica*, Miers, *Menispermaceae:* H. Kondo, T. Sanada, *J. Pharm. Soc. Jpn.* **541**, 177 (1927), *C.A.* **21**, 2700[4] (1927); H. Kondo, T. Watanabe, *ibid.* **58**, 268 (1938), *C.A.* **32**, 5403[5] (1938). Structure: K. Takeda, *Bull. Agric. Chem. Soc. Jpn.* **20**, 165 (1956). Synthesis:

idem, C.A. **60**, 5570f (1964); B. Pecherer, A. Brossi, *J. Org. Chem.* **32**, 1053 (1967); A. R. Battersby *et al., J. Chem. Soc. Perkin Trans. 1* **1981**, 2002. Biosynthesis: *eidem, ibid.* 2010, 2016, 2030.

White crystalline solid from benzene, mp 84-86° (Pecherer). Also reported as mp 75° (Kondo).

8008. Protoveratrines. From the rhizome of *Veratrum album* L., *Liliaceae:* Salzberger, *Arch. Pharm.* **228**, 462 (1890); Poethke, *ibid.* **275**, 357 (1937); Craig, Jacobs, *J. Biol. Chem.* **143**, 427 (1942); **149**, 271 (1943). Mixture of protoveratrines A and B: Glen *et al., Nature* **170**, 932 (1952); Klohs *et al., J. Am. Chem. Soc.* **74**, 5107 (1952); Nash, Brooker, *ibid.* **75**, 1942 (1953); Stoll, Seebeck, *Helv. Chim. Acta* **36**, 718 (1953); Nash, Brooker, US 2929812 (1960 to Allied Labs). Structure of protoveratrines A and B: Kupchan, Ayres, *J. Am. Chem. Soc.* **82**, 2252 (1960). Comparative toxicity: O. Krayer *et al., J. Pharmacol. Exp. Ther.* **82**, 167 (1944); K. Tanaka, *ibid.* **113**, 89 (1955).

Protoveratrine A R = H
Protoveratrine B R = OH

Slightly bitter, sternutative crystals from alc, dec 266-267°. $[\alpha]_D^{25}$ −38.6° (pyridine). $[\alpha]_D^{25}$ −8.5° (c = 1.99 in chloroform). Soluble in chloroform; dil aq acidic solns; slightly sol in ether. Practically insol in water, petr ether. LD$_{50}$ i.v. in mice: 0.048 mg/kg (Krayer).
Protoveratrine A. [143-57-7] [$3\beta(S),4\alpha,6\alpha,7\alpha,15\alpha(R),16\beta$]-4,9-Epoxycevane-3,4,6,7,14,15,16,20-octol 6,7-diacetate 3-(2-hydroxy-2-methylbutanoate) 15-(2-methylbutanoate); Protalba. C$_{41}$H$_{63}$NO$_{14}$; mol wt 793.95. Crystals from acetone, dec 267-269°. $[\alpha]_D^{25}$ −40.5° (pyridine); $[\alpha]_D^{25}$ −10.5° (chloroform). Soluble in chloroform, pyridine, hot alcohol. LD$_{50}$ s.c. in male mice: 0.29 mg/kg (Tanaka).
Protoveratrine B. [124-97-0] [$3\beta(2R,3R),4\alpha,6\alpha,7\alpha,15\alpha$-$(R),16\beta$]-4,9-Epoxycevane-3,4,6,7,14,15,16,20-octol 6,7-diacetate 3-(2,3-dihydroxy-2-methylbutanoate) 15-(2-methylbutanoate); veratetrine; neoprotoveratrine. C$_{41}$H$_{63}$NO$_{15}$; mol wt 809.95. Crystals from acetone, dec 268-270°. $[\alpha]_D^{25}$ −37° (pyridine); $[\alpha]_D^{25}$ −3.5° (chloroform). Soluble in chloroform, pyridine, hot alcohol. LD$_{50}$ s.c. in male mice: 0.21 mg/kg (Tanaka).
Mixture of protoveratrine A and B. Provell; Tensatrin.
THERAP CAT: Antihypertensive.

8009. Protoverine. [76-45-9] ($3\beta,4\alpha,6\alpha,7\alpha,15\alpha,16\beta$)-4,9-Epoxycevane-3,4,6,7,14,15,16,20-octol. C$_{27}$H$_{43}$NO$_9$; mol wt 525.64. C 61.70%, H 8.25%, N 2.66%, O 27.39%. Obtained by

alkaline hydrolysis of protoveratrine A: Poethke, *Arch. Pharm.* **275**, 571 (1937); Stoll, Seebeck, *Helv. Chim. Acta* **36**, 718 (1953). Structure: Kupchan *et al., Chem. Ind. (London)* **1958**, 1626; *J. Am. Chem. Soc.* **81**, 1009 (1959); **82**, 2242 (1960). Comparative toxicity: O. Krayer *et al., J. Pharmacol. Exp. Ther.* **82**, 167 (1944); K. Tanaka, *ibid.* **113**, 89 (1955).

Fine needles from methanol, dec 220-222°. $[\alpha]_D^{20}$ −15.7° (c = 1.1 in pyridine). Sol in CHCl$_3$, methanol, pyridine. LD$_{50}$ i.v. in mice: 194.0 mg/kg (Krayer). LD$_{50}$ s.c. in male mice: 520 mg/kg (Tanaka).

8010. Protriptyline. [438-60-8] *N*-Methyl-5*H*-dibenzo[*a,d*]-cycloheptene-5-propanamine; 5-(3-methylaminopropyl)-5*H*-dibenzo[*a,d*]cycloheptene; 7-(3-methylaminopropyl)-1,2:5,6-dibenzocycloheptatriene; amimetilina. C$_{19}$H$_{21}$N; mol wt 263.38. C 86.65%, H 8.04%, N 5.32%. Prepn: Engelhardt, Christy, **BE 617967** (1962 to Merck & Co.), *C.A.* **59**, 517f (1963); Tishler *et al.,* US 3244748 and US 3271451 (both 1966 to Merck & Co.); Engelhardt *et al., J. Med. Chem.* **11**, 325 (1968). Metabolism in man, pig and dog: Sisenwine *et al., J. Pharmacol. Exp. Ther.* **175**, 51 (1970). Use in treatment of sleep apnea: R. W. Clark *et al., Neurology* **29**, 1287 (1979); L. G. Brownell *et al., N. Engl. J. Med.* **307**, 1037 (1982).

Hydrochloride. [1225-55-4] MK-240; Concordin; Triptil; Vivactil. C$_{19}$H$_{21}$N.HCl; mol wt 299.84. Crystals from isopropanol-ethyl ether, mp 169-171°. uv max: 290 nm (ε 13311). Freely sol in water, alc, chloroform. Practically insol in ether. pKa 8.2.
THERAP CAT: Antidepressant.

8011. Pro-Urokinase. [82657-92-9] Prourokinase (enzyme-activating); single-chain urokinase-type plasminogen activator; single-chain pro-urokinase; scu-PA; pro-UK; pro u-PA; PUK; Sandolase; Thombolyse; Tomieze. Single-chain proenzyme form of urokinase, *q.v.* Produced by the kidney; present in urine and blood. Consists of 411 amino acid residues, mol wt ~54,000 daltons. Converted by plasmin or kallikrein to active two-chain urokinase by proteolytic cleavage of the Lys 158 - Ile 159 peptide bond. Identification of proenzyme activity from cultured human kidney cells: M. B. Bernik, *J. Clin. Invest.* **52**, 823 (1973). Isoln from culture media: C. Nolan *et al., Biochim. Biophys. Acta* **496**, 384 (1977). Purification from human plasma: T.-C. Wun *et al., J. Biol. Chem.* **257**, 3276 (1982); from human urine: S. S. Husain *et al., Arch. Biochem. Biophys.* **220**, 31 (1983). Amino acid sequence and cleavage site: S. Kasai *et al., J. Biol. Chem.* **260**, 12382 (1985). Clot-specific activity: V. Gurewich *et al., J. Clin. Invest.* **73**, 1731 (1984). Clinical pharmacology: G. Trübestein *et al., Haemostasis* **17**, 238 (1987). Clinical trials in acute myocardial infarction: C. Bode *et al., Am. J. Cardiol.* **61**, 971 (1988); in combination with t-PA: V. Gurewich, *J. Am. Coll. Cardiol.* **10**, 16B (1987). Review of mechanism of action studies: V. Gurewich, R. Pannell, *Semin. Thromb. Hemostasis* **13**, 146-151 (1987); and physicochemical properties: V. Gurewich, *ibid.* **14**, 110-115 (1988).

Saruplase. [99149-95-8] Prourokinase (enzyme-activating) (human clone pUK4/pUK18 protein moiety reduced); CG-4509. Nonglycosylated human prourokinase produced in *E. coli* by recombinant DNA technology. Prepn: H. L. Heyneker *et al.*, **EP 92182**; *eidem*, **US 5112755** (1983, 1992 both to Genentech); W. E. Holmes *et al.*, *Biotechnology* **3**, 923 (1985). Series of articles on thrombolytic activity: *Thromb. Haemostasis* **52**, 19-33 (1984). Pharmacokinetics: A. de Boer *et al.*, *ibid.* **70**, 320 (1993). Clinical trial: PRIMI Trial Study Group, *Lancet* **1**, 863 (1989); J. Schofer *et al.*, *Eur. Heart J.* **14**, 958 (1993).

THERAP CAT: Thrombolytic.

8012. Proxazole. [5696-09-3] *N,N*-Diethyl-3-(1-phenylpropyl)-1,2,4-oxadiazole-5-ethanamine; 5-[2-(diethylamino)ethyl]-3-(α-ethylbenzyl)-1,2,4-oxadiazole; propaxoline. $C_{17}H_{25}N_3O$; mol wt 287.41. C 71.04%, H 8.77%, N 14.62%, O 5.57%. Smooth muscle relaxant. Prepn: **GB 924608**; Palazzo, Silvestrini, **US 3141019** (1963, 1964 both to Angelini Francesco). Pharmacology: Silvestrini, Pozzatti, *Arzneim.-Forsch.* **13**, 798 (1963). Separation and pharmacology of the enantiomers: De Feo *et al.*, *Farmaco Ed. Sci.* **26**, 370 (1971). Clinical trial in cerebrovascular insufficiency: G. Esposito, M. De Gregorio, *Arzneim.-Forsch.* **24**, 1692 (1974).

$bp_{0.2}$ 132°.

Citrate. [132-35-4] AF-634; Flou; Pirecin; Toness. $C_{17}H_{25}N_3$-$O.C_6H_8O_7$; mol wt 479.53. LD_{50} in rats (mg/kg): 39 i.p., 60 orally (Silvestrini, Pozzatti).

Nitrate. Crystals, mp 127-128°.

THERAP CAT: Antispasmodic.
THERAP CAT (VET): Antispasmodic.

8013. Proxibarbal. [2537-29-3] 5-(2-Hydroxypropyl)-5-(2-propen-1-yl)-2,4,6(1*H*,3*H*,5*H*)-pyrimidinetrione; 5-allyl-5-(2-hydroxypropyl)barbituric acid; 5-allyl-5-(β-hydroxypropyl)barbituric acid; 5-allyl-5-(β-hydroxypropyl)malonylurea; proxibarbital; HH-184; Axeen; Centralgol; Centralgyl; Ipronal. $C_{10}H_{14}N_2O_4$; mol wt 226.23. C 53.09%, H 6.24%, N 12.38%, O 28.29%. Prepn: Bobranski *et al.*, *Rocz. Chem.* **30**, 157 (1956), *C.A.* **51**, 438f (1957); **GB 953387** (1964 to Hommel A.G.), *C.A.* **61**, 3123a (1964); Smissman *et al.*, *J. Med. Chem.* **14**, 853 (1971). Metabolism: Bobranski *et al.*, *Arch. Immunol. Ther. Exp.* **10**, 895 (1962); B. Lambrey *et al.*, *Eur. J. Med. Chem. - Chim. Ther.* **15**, 463 (1980). Pharmacokinetics: *eidem, ibid.* **12**, 565 (1977).

Crystals from benzene + ethanol, mp 157-158°; also reported as mp 166.5-168.5° from acetone + chloroform (Smissman). Moderately sol in water.

Note: This is a controlled substance (depressant): **21 CFR,** 1308.13.

THERAP CAT: Sedative, hypnotic.

8014. Proxyphylline. [603-00-9] 3,7-Dihydro-7-(2-hydroxypropyl)-1,3-dimethyl-1*H*-purine-2,6-dione; 7-(2-hydroxypropyl)theophylline; Brontyl; Spasmolysin; Theon. $C_{10}H_{14}N_4O_3$; mol wt

238.25. C 50.41%, H 5.92%, N 23.52%, O 20.15%. Smooth muscle relaxant. Prepd by refluxing 1-chloro-2-propanol with theophylline in an alkaline medium: Rice, **US 2715125** (1955 to Gane's Chem. Works).

Crystals from abs ethanol, mp 135-136°. One gram dissolves in about 1 ml water, in 14 ml abs ethanol. More sol in boiling ethanol. pH of a 5% aq soln 5.5 to 7.0. Solns may be sterilized by heating.

THERAP CAT: Bronchodilator; vasodilator.

8015. Prucalopride. [179474-81-8] 4-Amino-5-chloro-2,3-dihydro-*N*-[1-(3-methoxypropyl)-4-piperidinyl]-7-benzofurancarboxamide; R-93877; Resolor. $C_{18}H_{26}ClN_3O_3$; mol wt 367.87. C 58.77%, H 7.12%, Cl 9.64%, N 11.42%, O 13.05%. Benzofuran derivative; specific serotonin 5-HT$_4$ receptor agonist. Prepn: G. H. P. Van Daele *et al.*, **WO 9616060**; *eidem*, **US 5854260** (1996, 1998 both to Janssen). Effect on isolated gastrointestinal muscle: N. H. Prins *et al.*, *Br. J. Pharmacol.* **127**, 1431 (1999). Clinical stimulation of colonic transit: E. P. Bouras *et al.*, *Gut* **44**, 682 (1999). Clinical trial in severe chronic constipation: J. Tack *et al.*, *ibid.* **58**, 357 (2009).

Monohydrate, mp 90.7°.

THERAP CAT: Gastroprokinetic.

8016. Prulifloxacin. [123447-62-1] 6-Fluoro-1-methyl-7-[4-[(5-methyl-2-oxo-1,3-dioxol-4-yl)methyl]-1-piperazinyl]-4-oxo-1*H*,4*H*-[1,3]thiazeto[3,2-*a*]quinoline-3-carboxylic acid; NM-441; Pruvel; Sword; Unidrox. $C_{21}H_{20}FN_3O_6S$; mol wt 461.46. C 54.66%, H 4.37%, F 4.12%, N 9.11%, O 20.80%, S 6.95%. Fluoroquinolone antibacterial; prodrug for active metabolite, *ulifloxacin.* Prepn: M. Kise *et al.*, **EP 315828**; *eidem*, **US 5086049** (1989, 1992 both to Nippon Shinyaku); J. Segawa *et al.*, *J. Med. Chem.* **35**, 4727 (1992). Comparative *in vivo* activity: M. Ozaki *et al.*, *Antimicrob. Agents Chemother.* **35**, 2496 (1991). LC-MS/MS determn of active metabolite in plasma: L. Guo *et al.*, *J. Chromatogr. B* **832**, 280 (2006). Clinical pharmacokinetics: M. Nakashima *et al.*, *J. Clin. Pharmacol.* **34**, 930 (1994). Clinical comparison with ciprofloxacin in chronic bronchitis: C. Grassi *et al.*, *Respiration* **69**, 217 (2002). Review of pharmacology: M. G. Matera, *Pulm. Pharmacol. Ther.* **19**, 20-29 (2006); of clinical development: M. Cazzola *et al.*, *ibid.* 30-37.

Pale yellow powder from acetonitrile, mp 220° (dec).

THERAP CAT: Antibacterial.

8017. Prunetin. [552-59-0] 5-Hydroxy-3-(4-hydroxyphenyl)-7-methoxy-4*H*-1-benzopyran-4-one; 4′,5-dihydroxy-7-methoxyisoflavone; prunusetin. $C_{16}H_{12}O_5$; mol wt 284.27. C 67.60%, H 4.26%, O 28.14%. Isoln from *Prunus* spp., *Rosaceae:* Hasegawa, Shirato, *J. Am. Chem. Soc.* **79**, 450 (1957); Hasegawa, *ibid.* 1738; Goel, Seshadri, *Tetrahedron* **5**, 91 (1959); Plouvier, *Compt. Rend.* **250**, 594 (1960). Identity with prunusetin: King, Jurd, *J. Chem. Soc.* **1952**, 3211. Structure: Shrimer, Hull, *J. Org. Chem.* **10**, 288 (1945). Synthesis: Bradbury, White, *J. Chem. Soc.* **1953**, 871.

Needles from ethanol, mp 240°.
Diacetate. $C_{20}H_{16}O_7$. Rods from methanol, mp 222.5°.
4'-Glucoside. Prunitrin. $C_{22}H_{22}O_{10}$. Isoln: Finnemore, *Pharm. J.* **31**, 604 (1910). Structure and synthesis: Zemplén, Farkas, *Ber.* **90**, 836 (1957). Needles, mp 235-236°. Sol in hot water, ethyl acetate.

8018. Prussian Blue. [14038-43-8]; [12240-15-2] (unspecified formula). *(OC*-6-11)-Iron(3+) (3:4) hexakis(cyano-κ*C*) ferrate-(4−); ferric ferrocyanide; ferric hexacyanoferrate(II); iron(III) hexacyanoferrate(4−); C.I. 77510; Berlin blue; Chinese blue; Hamburg blue; insoluble Prussian blue; Iron blue; Milori blue; Paris blue; Radiogardase. $C_{18}Fe_7N_{18}$; mol wt 859.24. C 25.16%, Fe 45.50%, N 29.34%. $Fe_4[Fe(CN)_6]_3$. Discovered by Diesbach in 1704; one of the first synthetic coordination compounds and modern colors. The structure consists of alternating ferric and ferrous ions on face centered cubic lattice sites; the remaining charge is balanced by potassium or ferric ions. Early characterization: J. Brown, *Philos. Trans.* **33**, 17 (1724). Prepn: *Colour Index* **vol. 4** (3rd ed, 1971) p 4673. Structural study: J. F. Keggin, F. D. Miles, *Nature* **137**, 577 (1936). Crystal structure: H. J. Buser *et al., Inorg. Chem.* **16**, 2704 (1977). Physicochemical properties and efficacy in thallium poisoning: J. Kravzov *et al., J. Appl. Toxicol.* **13**, 213 (1993). Review of toxicology: J. Pearce, *Food Chem. Toxicol.* **32**, 577-582 (1994); of clinical efficacy in radiocesium poisoning: D. F. Thompson, C. O. Church, *Pharmacotherapy* **21**, 1364-1367 (2001); of use in chemical and biological sensors: R. Koncki, *Crit. Rev. Anal. Chem.* **32**, 79-96 (2002).
Purple powder. d 1.75-1.81. Soly (mg/ml): water 6; ethanol 20. Slightly sol in most organic solvents. Sol in oxalic acid; precipitates on exposure to light. Absorption max (water): 694 nm.
Ammonium Ferric Hexacyanoferrate. [25869-00-5] Ferric ammonium ferrocyanide; $C_6H_4Fe_2N_7$; mol wt 285.84. $NH_4Fe[Fe(CN)_6]$.
LD_{50} orally in mice: >5 g/kg (Pearce).
Potassium Ferric Ferrocyanide. [25869-98-1] C.I. 77520; C.I. Pigment Blue 27; iron(III) potassium hexacyanoferrate(4−); soluble Prussian blue. $C_6Fe_2KN_6$; mol wt 306.90. $KFe[Fe(CN)_6]$.
USE: As pigment in printing inks, paints, alkyd resin enamels, linoleum, leathercloth, carbon papers, typewriter ribbons, plastics, artists' colors; in chemical and biological sensors.
THERAP CAT: Antidote (radioactive cesium and thallium poisoning).

8019. Pseudobaptigenin. [90-29-9] 3-(1,3-Benzodioxol-5-yl)-7-hydroxy-4*H*-1-benzopyran-4-one; 7-hydroxy-3',4'-(methylenedioxy)isoflavone. $C_{16}H_{10}O_5$; mol wt 282.25. C 68.09%, H 3.57%, O 28.34%. The aglycon of pseudobaptisin. Prepn by hydrolysis of natural pseudobaptisin: Gorter, *Arch. Pharm.* **235**, 494 (1897). Structure: Späth, Schmidt, *Monatsh. Chem.* **53**, 454 (1929). Syntheses: Späth, Lederer, *Ber.* **63**, 743 (1930); Mahal *et al., J. Chem. Soc.* **1934**, 1771; Baker *et al., ibid.* **1937**, 805; **1953**, 1852; Farkas, *Ber.* **91**, 2858 (1958); Dhoubhadel, Joshi, *J. Indian Chem. Soc.* **52**, 440 (1975).

Long, felted needles from alc, dec 296-298°. Sublimes in high vacuum at ~220°. Sparingly sol in the usual solvents.
7-Rhamnoglucoside trihydrate. Pseudobaptisin. $C_{28}H_{30}O_{14}$.·$3H_2O$. Crystals, becomes anhydr at 120°. mp 148-150° (evac tube). If heating is continued, it resolidif at 180-210° and melts again at 249-251°. $[\alpha]_D^{14}$ −98.1° (c = 1.1 in methanol). Freely sol in methanol, hot water, hot acetone.

8020. Pseudococaine. [478-73-9] (1*R*,2*S*,3*S*,5*S*)-3-(Benzoyloxy)-8-methyl-8-azabicyclo[3.2.1]octane-2-carboxylic acid methyl ester; 3β-hydroxy-1α*H*,5α*H*-tropane-2α-carboxylic acid methyl ester benzoate; 2α-carbomethoxy-3β-benzoxytropane; depsococaine; dextrocaine; isococaine; Delcaine. $C_{17}H_{21}NO_4$; mol wt 303.36. C 67.31%, H 6.98%, N 4.62%, O 21.10%. Cocaine diastereomer with greater local anesthetic activity than the natural substance. Synthesis: Einhorn, Marquardt, *Ber.* **23**, 468, 981 (1890); Willstätter, Bode, *Ann.* **326**, 42 (1903); Willstätter, Bommer, *ibid.* **422**, 34 (1921); Willstätter *et al., ibid.* **434**, 138 (1923); **GB 210050** (1923 to E. Merck). Configuration: S. P. Findlay, *J. Am. Chem. Soc.* **76**, 2855 (1954). Conformational analysis and ^1H, ^{13}C NMR studies: F. I. Carroll *et al., J. Org. Chem.* **1982**, 13. Pharmacokinetics: A. L. Misra *et al., Experientia* **32**, 895 (1976). Interaction with sodium channels: J. C. Matthews, A. Collins, *Biochem. Pharmacol.* **32**, 455 (1983). Pharmacology of cocaine and pseudococaine: G. Schmidt *et al., Arch. Exp. Pathol. Pharmakol.* **240**, 523 (1961).

Prisms, mp 47°. $[\alpha]_D^{20}$ +42° (c = 5 in chloroform). Freely sol in ether, chloroform, benzene, petr ether. Slightly sol in water.
Hydrochloride. [6363-57-1] $C_{17}H_{21}NO_4$·HCl. Crystals from alcohol, mp 210°. $[\alpha]_D^{20}$ +41° (c = 5), less sol in water than cocaine hydrochloride.
Tartrate. [1176-03-0] Psicaine. $C_{17}H_{21}NO_4$·$C_4H_6O_6$; mol wt 453.44. Crystals, mp 139°. $[\alpha]_D^{20}$ +43° (c = 5 in water). Sol in 4 parts water, in alcohol. The aq soln is stable and may be sterilized by boiling, without decompn. pH about 3.7 (2% soln).
Sodium tartrate. Psicaine N. More sol in water than the tartrate; esp useful where a neutral soln is desired.
n-**Propyl ester analog.** [55608-72-5] Neopsicaine. $C_{19}H_{25}NO_4$; mol wt 331.41.
THERAP CAT: Formerly as anesthetic (local).

8021. Pseudocodeine. [466-96-6] (5α,8β)-6,7-Didehydro-4,5-epoxy-3-methoxy-17-methylmorphinan-8-ol; ψ-codeine; neoisocodeine. $C_{18}H_{21}NO_3$; mol wt 299.37. C 72.22%, H 7.07%, N 4.68%, O 16.03%. An isomer of codeine. Review and bibliography: K. W. Bentley, *The Chemistry of the Morphine Alkaloids* (Oxford, 1954).

White needles, mp 181-182°. $[\alpha]_D$ −96.6° (alc). Slightly sol in water; sol in alcohol.

8022. Pseudoconhydrine. [140-55-6] (3*S*,6*S*)-6-Propyl-3-piperidinol; (3*S-trans*)-6-propyl-3-piperidinol; 5-hydroxy-2-propylpiperidine; 5-hydroxyconiine. $C_8H_{17}NO$; mol wt 143.23. C 67.09%, H 11.96%, N 9.78%, O 11.17%. In *Conium maculatum* L., *Umbelliferae* (hemlock). Extraction procedure: Ladenburg *et al.*,

Ber. **24**, 1671 (1891); Braun, *Ber.* **38**, 3108 (1905); Löffler, *Ber.* **42**, 116 (1909). Structure: E. Späth *et al.*, *Ber.* **66**, 591 (1933). Stereochemistry: R. K. Hill, *J. Am. Chem. Soc.* **80**, 1611 (1958); K. Tadano *et al.*, *J. Carbohydr. Chem.* **4**, 129 (1985).

Hygroscopic needles from abs ether, mp 106° (monohydrate, scales, mp 60° from moist ether). bp 236°. $[\alpha]_D^{20}$ +11° (c = 10 in alc). pK (18°): 3.70. Soluble in water and most organic solvents.

Hydrochloride. $C_8H_{17}NO.HCl$. Crystals from alcohol, mp 213°. Freely sol in water, sparingly sol in alcohol, acetone.

8023. Pseudocumene. [95-63-6] 1,2,4-Trimethylbenzene; pseudocumol; asymmetrical trimethylbenzene. C_9H_{12}; mol wt 120.20. C 89.93%, H 10.06%. Occurs in coal tar and in many petroleums. Physical properties: Hirschler, Falconer, *J. Am. Chem. Soc.* **68**, 210 (1946). Metabolism study: J. Huo *et al.*, *Xenobiotica* **19**, 161 (1989). GC/MS determn in serum: E. Kenndler *et al.*, *J. Anal. Toxicol.* **13**, 211 (1989). Neurotoxic effects of inhalation exposure: Z. Korsak, K. Rydzynski, *Int. J. Occup. Med. Environ. Health* **9**, 341 (1996). Genotoxicity study: E. Janik-Spiechowicz *et al.*, *Mutat. Res.* **412**, 299 (1998).

Liquid. d_4^{20} 0.8761. bp 169-171°. mp −43.78° (Hirschler, Falconer). n_D^{21} 1.5044. Practically insol in water. Sol in alc, benzene, ether. Oxidation yields trimellitic anhydride. LD_{50} i.p. in male, female mice (mg/kg): 5000, 4100 (Janik-Spiechowicz).

Caution: Potential symptoms of overexposure are irritation of eyes, skin, nose, throat, respiratory system; bronchitis; hypochromic anemia; headache, drowsiness, fatigue, dizziness, nausea, incoordination; vomiting, confusion; aspiration of liquid may cause chemical pneumonia. *See NIOSH Pocket Guide to Chemical Hazards* (DHHS/NIOSH 97-140, 1997) p 320.

USE: Sterilizing catgut by heating one hour at 160°; solvent in manuf dyes, perfumes and resins. Solvent for liquid scintillation counting solns.

8024. Pseudoephedrine. [90-82-4] (αS)-α-[(1S)-1-(Methylamino)ethyl]benzenemethanol; (1S,2S)-2-methylamino-1-phenylpropan-1-ol; *d-ψ*-ephedrine; *d*-isoephedrine. $C_{10}H_{15}NO$; mol wt 165.24. C 72.69%, H 9.15%, N 8.48%, O 9.68%. Sympathomimetic amine found in plants of the genus *Ephedra (Ephedraceae)* known in traditional medicine as Ma Huang. α-Adrenergic agonist; (+)-*threo*-isomer of ephedrine, *q.v.* Isoln from *E. vulgaris:* A. Ladenburg, C. Oelschlägel, *Ber.* **22**, 1823 (1889). Synthesis: E. Späth, R. Göhring, *Monatsh. Chem.* **41**, 319 (1920). HPLC determn in Ma Huang: B. J. Gurley *et al.*, *J. Pharm. Sci.* **87**, 1547 (1998); in plasma: P. Guo *et al.*, *Biomed. Chromatogr.* **13**, 61 (1999). Toxicity: M. D. Fairchild, G. A. Alles, *J. Pharmacol. Exp. Ther.* **158**, 135 (1967). Comprehensive description: S. A. Benezra, J. W. McRae, *Anal. Profiles Drug Subs.* **8**, 489-507 (1979). Review of clinical pharmacology: D. T. D. Hughes *et al.*, *J. Clin. Hosp. Pharm.* **8**, 315-321 (1983). Clinical trial for prevention of otic barotrauma: J. S. Jones *et al.*, *Am. J. Emerg. Med.* **16**, 262 (1998); in nasal congestion: D. Taverner *et al.*, *Clin. Otolaryngol.* **24**, 47 (1999).

Crystals from ether, mp 118-118.7°. $[\alpha]_D^{20}$ +51.2° (in ethanol). Sparingly sol in water. Freely sol in alc or ether.

Hydrochloride. [345-78-8] Efidac 24; Galpseud; Galsud; Rhinalair; Otrinol; Sudafed. $C_{10}H_{15}NO.HCl$; mol wt 201.69. Needles,

mp 182.5-183.5°. $[\alpha]_D^{20}$ +62.05°. uv max (ethanol): 208, 251, 257, 264 nm (ε 8300, 161, 201, 161). pKa 9.22. Soly at 25° (g/ml): water 2.0; chloroform 0.011; ethanol 0.278. Practically insol in ether. Partition coefficient (25°): 0.010 (*n*-octanol/water, pH 1.2); 0.049 (*n*-octanol/water, pH 6.0). LD_{50} i.p. in mice: 1.0 mmole/kg (Fairchild, Alles).

Sulfate. [7460-12-0] $C_{20}H_{30}N_2O_2.H_2SO_4$; mol wt 428.54. White crystalline powder, bitter taste. mp 174-179°. $[\alpha]_D^{20}$ + 56.0-58.0°. Very freely sol in water; freely sol in ethanol. Practically insol in ether.

THERAP CAT: Decongestant (nasal).

THERAP CAT (VET): In treatment of urinary incontinence; decongestant.

8025. Pseudoionone. [141-10-6] 6,10-Dimethyl-3,5,9-undecatriene-2-one; citrylideneacetone; 2,6-dimethylhendeca-2,6,8-trien-10-one. $C_{13}H_{20}O$; mol wt 192.30. C 81.20%, H 10.48%, O 8.32%. Intermediate in the synthesis of α- and β-ionone. Prepn from citral and acetone: Tiemann, Krüger, *Ber.* **26**, 2692 (1893); Stiehl, *J. Prakt. Chem.* [2] **58**, 84 (1898); Tiemann, *Ber.* **32**, 115 (1899); Hibbert, Cannon, *J. Am. Chem. Soc.* **46**, 119 (1924); A. Russel, R. L. Kenyon, *Org. Synth.* **coll. vol. III**, 747 (1955). A less pure product is obtained by the treatment of oil of lemon grass and acetone with bleaching powder, cobalt nitrate, and alcohol: Ziegler, *J. Prakt. Chem.* [2] **57**, 493 (1898); Tiemann, *Ber.* **31**, 2313 (1898); Haarmann, Reimer & Co., **DE 73089**, *Frdl.* **3**, 889. Synthesis: T. Onishi *et al.*, *Synthesis* **1980**, 651.

Pale yellow oil, bp_2 114-116°; bp_4 124-126°; bp_{12} 143-145°; d^{20} 0.8984; n_D^{20} 1.53346.

8026. Pseudomonic Acids. A group of antibacterial antibiotics produced by *Pseudomonas fluorescens* NCIB 10586 that have unusual structural features. Four members of the group are known: *pseudomonic acid A,* the major component; *pseudomonic acid B,* the 3,4,5-trihydroxy analog of A (also referred to as *pseudomonic acid I*); *pseudomonic acid D,* the 4-nonenoic acid analog of A; and *pseudomonic acid C,* in which the epoxide oxygen is replaced by a double bond. Isoln and characterization of A and B: A. T. Fuller *et al.*, *Nature* **234**, 416 (1971); K. D. Barrow, G. Mellows, **DE 2227739**; *eidem*, **US 3977943** (1973, 1976 both to Beecham). Structure of B: E. B. Chain, G. Mellows, *J. Chem. Soc. Perkin Trans. 1* **1977**, 318. Prepn of C: N. H. Rogers, P. J. O'Hanlon, **EP 3069**; *eidem*, **US 4205002** (1979, 1980 both to Beecham). Isoln, structure, configuration of C: J. P. Clayton *et al.*, *Tetrahedron Lett.* **21**, 881 (1980). Total syntheses of naturally occurring (+)-form of C: A. P. Kozikowski *et al.*, *J. Am. Chem. Soc.* **102**, 6577 (1980); in high yield: C. Mckay *et al.*, *Chem. Commun.* **2000**, 1109; of racemic A or C: B. B. Snider, G. B. Phillips, *J. Am. Chem. Soc.* **104**, 1113 (1982); B. B. Snider *et al.*, *J. Org. Chem.* **48**, 3003 (1983). Prepn of D: P. J. O'Hanlon, **EP 68680** (1983 to Beecham), *C.A.* **98**, 159135t (1983); P. J. O'Hanlon *et al.*, *J. Chem. Soc. Perkin Trans. 1* **1983**, 2655. Antimycoplasmal activity in vitro: R. M. Banks *et al.*, *J. Antibiot.* **41**, 609 (1988).

Pseudomonic Acid C

Isolated as a mixture of sodium salts. Can be stored at 0° for several months with no activity loss. Stable within pH 4-9 at 37° for 24 hrs. Hemolytic; inactivated by serum at conc >50%.

Pseudomonic Acid A *see* Mupirocin.

Pseudomonic Acid C. [71980-98-8] (2E)-5,9-Anhydro-2,3,4,8-tetradeoxy-8-[(2E,4R,5S)-5-hydroxy-4-methyl-2-hexenyl]-3-methyl-L-*talo*-non-2-enonic acid 8-carboxyoctyl ester. $C_{26}H_{44}O_8$; mol wt 484.63. $[\alpha]_D^{25}$ +7.64° (c = 0.78 in chloroform). uv max (ethanol): 222 nm (ε 14100).

Pseudomonic Acid D. [85248-93-7] (2E)-5,9-Anhydro-2,3,4,8-tetradeoxy-8-[[(2S,3S)-3-[(1S,2S)-2-hydroxy-1-methylpropyl]oxiranyl]methyl]-L-*talo*-non-2-enoic acid (5E)-8-carboxy-5-octenyl ester. $C_{26}H_{42}O_9$; mol wt 498.61. Oil. uv max (ethanol): 220 nm (ε 15499).

8027. Pseudomorphine. [125-24-6] (5α,6α)-(5'α,6'α)-7,7',8,8'-Tetradehydro-4,5:4',5'-diepoxy-17,17'-dimethyl-[2,2'-bimorphinan]-3,3',6,6'-tetrol; 2,2'-bimorphine; pseudomorphine (C34 alkaloid). $C_{34}H_{36}N_2O_6$; mol wt 568.67. C 71.81%, H 6.38%, N 4.93%, O 16.88%. A dimolecular base formed by the gentle oxidation of morphine in alkaline soln. *Refs:* Broockmann, Polstorff, *Ber.* **13**, 88 (1880); Bentley, Dyke, *J. Chem. Soc.* **1959**, 2574. Review and bibliography: K. W. Bentley, *The Chemistry of the Morphine Alkaloids* (Oxford, 1954) p 54.

Trihydrate. Crystalline powder; levorotatory in acid soln. Becomes anhydr at 150° and dec at about 327°. Insol in cold water, alcohol, ether, dil H_2SO_4. Slightly sol in cold, more sol in hot ammonia; sol in fixed alkali hydroxide solns, in pyridine, aniline, benzyl alcohol, guaiacol.

Note: Name pseudomorphine is also used for the C17 alkaloid base.

8028. Pseudopelletierine. [552-70-5] 9-Methyl-9-azabicyclo[3.3.1]nonan-3-one; 9-methyl-3-granatanone; pseudopunicine. $C_9H_{15}NO$; mol wt 153.23. C 70.55%, H 9.87%, N 9.14%, O 10.44%. In root bark of *Punica granatum* L., *Punicaceae.* Extraction procedure: Hess, Eichel, *Ber.* **50**, 380, 1391, 1395 (1917); *ibid.* **52**, 1012 (1919). Synthesis under physiological conditions from calcium acetonedicarboxylate, glutaraldehyde, and methylamine: Menzies, Robinson, *J. Chem. Soc.* **1924**, 2163; *cf.* Schöpf, *Angew. Chem.* **50**, 779, 797 (1937). Alternate syntheses: Cope *et al., Org. Synth.* **37**, 73 (1957); Robinson, Hunt, *J. Pharm. Pharmacol.* **22**, 29S (1970); Bottini, Gal, *J. Org. Chem.* **36**, 1718 (1971). Conformation: Chen, LeFevre, *J. Chem. Soc. B* **1966**, 539.

Orthorhombic prisms from petr ether, mp 54°; bp 246°. Volatile. Strong base. One gram dissolves in about 2.5 ml water, 10 ml ether. Freely sol in alcohol and chloroform; sparingly sol in petr ether.

Dihydrate. Plates from water.

Hydrochloride. $C_9H_{15}NO.HCl$. Rhombohedra. One gram dissolves in about 1 ml water.

Sulfate tetrahydrate. $2C_9H_{15}NO.H_2SO_4.4H_2O$. Crystals, sol in water.

8029. Pseudopterosins. Family of anti-inflammatory, diterpene glycosides originally isolated from the marine gorgonian coral, *Pseudopterogorgia elisabethae.* Isoln of A-D: S. A. Look *et al., Proc. Natl. Acad. Sci. USA* **83**, 6238 (1986); *idem et al., J. Org. Chem.* **51**, 5140 (1986); of E-L: V. Roussis *et al., ibid.* **55**, 4916

(1990). Total synthesis of A: C. A. Broka *et al., ibid.* **53**, 1584 (1988); of A and E: E. J. Corey, P. Carpino, *J. Am. Chem. Soc.* **111**, 5472 (1989). Pharmacological characterization of A and E: A. M. S. Mayer *et al., Life Sci.* **62**, 401 (1998). Biosynthesis: R. G. Kerr *et al., J. Ind. Microbiol. Biotechnol.* **33**, 532 (2006). *Review:* T. J. Heckrodt, J. Mulzer, *Top. Curr. Chem.* **244**, 1-41 (2005).

Pseudopterosin A

Pseudopterosin A. [104855-20-1] (3S,7R,9S,9aR)-2,3,7,8,9,9a-Hexahydro-5-hydroxy-3,6,9-trimethyl-7-(2-methyl-1-propenyl)-1H-phenalen-4-yl-β-D-xylopyranoside. $C_{25}H_{36}O_6$; mol wt 432.56. Amorphous solid. $[\alpha]_D^{20}$ -85° (c = 0.69 in chloroform). uv max (methanol): 230, 278, 283 nm (ε 11200, 2060, 2200).

Pseudopterosin B. [104855-21-2] (3S,7R,9S,9aR)-2,3,7,8,9,9a-Hexahydro-5-hydroxy-3,6,9-trimethyl-7-(2-methyl-1-propenyl)-1H-phenalen-4-yl-β-D-xylopyranoside 2-acetate. $C_{27}H_{38}O_7$; mol wt 474.59. Oil. $[\alpha]_D^{20}$ -55.2° (c = 2.1 in chloroform). uv max (methanol): 230-235, 274, 285 nm (ε 6000, 1400, 1600).

Pseudopterosin C. [104881-78-9] (3S,7R,9S,9aR)-2,3,7,8,9,9a-Hexahydro-5-hydroxy-3,6,9-trimethyl-7-(2-methyl-1-propen-1-yl)-1H-phenalen-4-yl-β-D-xylopyranoside 3-acetate. $C_{27}H_{38}O_7$; mol wt 474.59. Crystals from ethanol + ethyl acetate, mp 113.5-115°. $[\alpha]_D^{20}$ -77° (c = 1.09 in chloroform). uv max (methanol): 229, 275, 282 nm (ε 9600, 1500, 1700).

Pseudopterosin E. [121011-80-1] (1S,3R,7S,9aR)-2,3,7,8,9,9a-Hexahydro-6-hydroxy-1,4,7-trimethyl-3-(2-methyl-1-propenyl)-1H-phenalen-5-yl 6-deoxy-α-L-galactopyranoside. $C_{26}H_{38}O_6$; mol wt 446.58. Amorphous white solid. $[\alpha]_D$ -255.0° (c = 0.4 in methanol). uv max (methanol): 226, 275, 285 nm (ε 20800, 2770, 3350).

USE: In cosmetics as anti-irritant and anti-inflammatory ingredient.

8030. Pseudotropine. [135-97-7] (3-*exo*-)-8-Methyl-8-azabicyclo[3.2.1]octan-3-ol; 1αH,5αH-tropan-3β-ol; 3β-tropanol; ψ-tropine; 3-pseudotropanol. $C_8H_{15}NO$; mol wt 141.21. C 68.05%, H 10.71%, N 9.92%, O 11.33%. Stereoisomeric with tropine. Prepn, structure, separation from tropine, and stereochemical configuration: *See* Tropine. Stereochemical synthesis: J. J. Tufariello *et al., J. Am. Chem. Soc.* **101**, 2435 (1979).

Orthorhombic, bipyramidal crystals from petr ether + benzene. mp 109°; bp 241°; pK (15°): 3.80. pH of 0.05M soln 11.5. Freely sol in water, alcohol, benzene.

Hydrochloride. $C_8H_{15}NO.HCl$. Needles from alcohol, dec 282°. Soluble in water, hot alcohol.

Tropate. $C_8H_{15}NO.C_9H_8O_2$. Cryst from benzene, mp 132°.

8031. Pseudoyohimbine. [84-37-7] (3β,16α,17α)-17-Hydroxyyohimban-16-carboxylic acid methyl ester. $C_{21}H_{26}N_2O_3$; mol wt 354.45. C 71.16%, H 7.39%, N 7.90%, O 13.54%. Present in the bark of *Corynanthe johimbe* K. Schum., *Rubiaceae.* Obtained from the residues of commercial isoln procedures for the manuf of yohimbine: Karrer, Salomon, *Helv. Chim. Acta* **9**, 1059 (1926). Structure and stereochemistry: Janot *et al., Bull. Soc. Chim. Fr.* **1952**, 1085; **1961**, 637. Synthesis: van Tamelen *et al., J. Am. Chem. Soc.* **80**, 5006 (1958); **91**, 7315 (1969); Stork, Guthikonda, *ibid.* **94**, 5109 (1972).

Rhombic platelets, mp 293° (corr, Maquenne block), mp 268° (open capillary), browns at 250°. Also reported as crystals from methanol, mp 252-256° (van Tamelen). $[\alpha]_D^{19}$ +27° (pyridine). uv max (methanol): 225, 281, 290 nm (log ε 4.54, 3.86, 3.80).

Hydrochloride. $C_{21}H_{26}N_2O_3 \cdot HCl$. Needles from alcohol + ether, dec 258°. $[\alpha]_D^{20}$ −10° (c = 1 in water).

8032. Psicofuranine. [1874-54-0] 1'-C-(Hydroxymethyl)-adenosine; 9-β-D-psicofuranosyl-9H-purin-6-amine; 9β-D-psicofuranosyladenine; 6-amino-9-D-psicofuranosylpurine; angustmycin C; U-9586. $C_{11}H_{15}N_5O_5$; mol wt 297.27. C 44.44%, H 5.09%, N 23.56%, O 26.91%. Nucleoside antibiotic produced by *Streptomyces hygroscopicus* var. *decoyicus:* Eble *et al., Antibiot. Chemother.* **9**, 419 (1959); Yüntsen, *J. Antibiot.* **11A**, 244 (1958); Eble, Lewis, **US 3020274** (1962 to Upjohn). Antibacterial and antitumor activity: N. Tanaka *et al., J. Antibiot.* **14A**, 98 (1961). Structure: Schroeder, Hoeksema, *J. Am. Chem. Soc.* **81**, 1767 (1959); Garrett, *ibid.* **82**, 827 (1960). Synthesis: Farkas, Sorm, *Collect. Czech. Chem. Commun.* **28**, 882 (1963); L. A. Aleksandrova, F. W. Lichtenthaler, *Nucleic Acids Symp. Ser.* **9**, 263 (1981). Biosynthesis: Sugimori, Suhadolnik, *J. Am. Chem. Soc.* **87**, 1136 (1965). Inhibits the conversion of xanthosine-5'-phosphate to guanosine-5'-phosphate: Slechta, *Biochem. Pharmacol.* **5**, 96 (1960).

Crystals, dec 212-214°. $[\alpha]_D^{25}$ −68° (dimethylformamide). uv max (0.01N acid): 259 nm ($E_{1cm}^{1\%}$ 508); in 0.01N base: 261 nm ($E_{1cm}^{1\%}$ 527). Soly at 25° (mg/ml): water 8; methanol 8; ethanol 6; butanol 2; ethyl acetate 0.23.

8033. D-Psicose. [551-68-8] D-*ribo*-2-Ketohexose; D-ribohexulose; D-allulose; D-erythrohexulose; pseudofructose. $C_6H_{12}O_6$; mol wt 180.16. C 40.00%, H 6.71%, O 53.28%. Occurrence and identification as a non-fermentable substance in cane molasses: Zerban, Sattler, *Ind. Eng. Chem.* **34**, 1180 (1942). Structure: Ohle, Just, *Ber.* **68**, 601 (1935). Prepn starting with D-ribono-γ-lactone: Wolfrom *et al., J. Am. Chem. Soc.* **67**, 1793 (1945); from D-allose: Steiger, Reichstein, *Helv. Chim. Acta* **19**, 184 (1937); from D-glucose: Hough *et al., J. Chem. Soc.* **1953**, 2005.

Sweet syrupy liquid. $[\alpha]_D^{25}$ +4.7° (c = 4.3 in water; no detectable mutarotation). Soluble in water, methanol, ethanol. Practically insol in acetone.

Phenylosazone. $C_{18}H_{22}N_4O_4$. Yellow crystals from water, mp about 162-163° (dec); also reported mp 178°.

8034. Psilocin. [520-53-6] 3-[2-(Dimethylamino)ethyl]-1H-indol-4-ol; 4-hydroxy-N,N-dimethyltryptamine; psilocyn. $C_{12}H_{16}N_2O$; mol wt 204.27. C 70.56%, H 7.90%, N 13.71%, O 7.83%. The minor hallucinogenic component of Teonanácatl, the sacred mushroom of Mexico. Isolated in trace amounts from the fruiting bodies of *Psilocybe mexicana* Heim, *Agaricaceae:* Hofmann *et al., Experientia* **14**, 107 (1958); Heim *et al., Helv. Chim. Acta* **42**, 1557 (1959). Prepn: Heim *et al.,* **DE 1087321** (1960 to Sandoz). Synthetic precursor of psilocybin: Hofmann, Troxler, **US 3075992** (1963 to Sandoz). Psilocin, the 4-hydroxy analog of psilocybin, is formed by metabolic dephosphorylation of psilocybin and is the active species in the central nervous system: Horita, Weber, *Toxicol. Appl. Pharmacol.* **4**, 730 (1962). Crystal structure: T. J. Petcher, H. P. Weber, *J. Chem. Soc. Perkin Trans.* 2 **1974**, 946. *Review:* Hofmann, *Bull. Narc.* **23**, 3 (1971).

Plates from methanol, mp 173-176°. Amphoteric substance. Unstable in soln, esp. alkaline soln. Very slightly sol in water. uv max: 222, 260, 267, 283, 293 nm (log ε 4.6, 3.7, 3.8, 3.7, 3.6).

Note: This is a controlled substance (hallucinogen): **21 CFR,** 1308.11.

8035. Psilocybin. [520-52-5] 3-[2-(Dimethylamino)ethyl]-1H-indol-4-ol 4-(dihydrogen phosphate); O-phosphoryl-4-hydroxy-N,N-dimethyltryptamine; Indocybin. $C_{12}H_{17}N_2O_4P$; mol wt 284.25. C 50.71%, H 6.03%, N 9.86%, O 22.51%, P 10.90%. The major of two hallucinogenic components of Teonanácatl, the sacred mushroom of Mexico, the other component being psilocin, *q.v.* from the fruiting bodies of *Psilocybe mexicana* Heim, *Agaricaceae:* Hofmann *et al., Experientia* **14**, 107 (1958); Heim *et al., Helv. Chim. Acta* **42**, 1557 (1959); Heim *et al.,* **DE 1087321** (1960 to Sandoz). Structure: Hofmann *et al., Experientia* **14**, 397 (1958). Synthesis: Hofmann, Troxler, **US 3075992** (1963 to Sandoz). Crystal structure: H. P. Weber, T. J. Petcher, *J. Chem. Soc. Perkin Trans.* 2 **1974**, 942. Converted to psilocin *in vivo.* Toxicity data: E. Usdin, D. H. Efron, *Psychotropic Drugs and Related Compounds* (National Institute of Mental Health, Rockville, Md., 2nd ed., 1972) p 138. *Reviews:* Hofmann, *Proc. 1st Int. Congr. Neuro-Pharmacol.* **1958**, 446; Cerletti, *Dtsch. Med. Wochenschr.* **84**, 2317 (1959); Hofmann, *Bull. Narc.* **23**, 3 (1971).

Crystals from boiling water, mp 220-228°; from boiling methanol, mp 185-195°. uv max (methanol): 220, 267, 290 nm (log ε 4.6, 3.8, 3.6). pH 5.2 in 50% aq ethanol. Sol in 20 parts boiling water, 120 parts boiling methanol; difficultly sol in ethanol. Practically insol in chloroform, benzene. LD$_{50}$ in mice, rats, rabbits (mg/kg): 285, 280, 12.5 i.v. (Usdin, Efron).

Note: This is a controlled substance (hallucinogen): **21 CFR,** 1308.11.

THERAP CAT: Psychomimetic.

8036. Psitticofulvins. Family of endogenously synthesized, lipid soluble pigments produced only by parrots (Order Psittaciformes) and responsible for the characteristic coloration of their plumage. Structures consist of unmethylated polyene aldehydes of varying length. Biosynthesized within the maturing feather follicle and keratinized into feathers preferentially over carotenoids. Color is modulated by combination with keratin; green feathers also contain melanin. Discovery in parrot feathers: C. F. W. Krukenberg, *Ver-*

gleichend-physiologische Studien Reihe 2, Abtlg. 2, 29-36 (1882). Characterization: O. Völker, *J. Ornithol.* **84**, 618 (1936); *idem, ibid.* **85**, 136 (1937). Synthesis of polyene aldehydes: R. Kuhn, *Angew. Chem.* **50**, 703 (1937); J. Schmitt, A. Obermeit, *Ann.* **547**, 285 (1941). Absorption spectra: E. R. Blout, M. Fields, *J. Am. Chem. Soc.* **70**, 189 (1948). Structural study of natural pigments: M. Veronelli *et al., J. Raman Spectrosc.* **26**, 683 (1995). Purification and confirmation of structures: R. Stradi *et al., Comp. Biochem. Physiol. B* **130**, 57 (2001). HPLC determn in feathers and distribution in parrot species: K. J. McGraw, M. C. Nogare, *Biol. Lett.* **1**, 38 (2005). Radical scavenging ability: A. Martínez, *J. Phys. Chem. B* **113**, 4915 (2009). Role in promoting resistance to bacterial degradation: E. H. Burtt, Jr. *et al., Biol. Lett.* **7**, 214 (2011). *Review:* M. L. Berg, A. T. D. Bennett, *Emu* **110**, 10-20 (2010).

Hexadecaheptaenal. [64512-29-4]; [328385-64-4] (*all-E*). 2,4,-6,8,10,12,14-Hexadecaheptaenal. $C_{16}H_{18}O$; mol wt 226.32. Red needles, mp 218° (dec) (Kuhn); also reported as deep red crystals from chloroform, mp 219-220° (dec) (Blout). Abs max in dioxane: 305, 415 nm (log ε 3.89, 4.80); in chloroform: 310, 424 nm (log ε 4.03, 4.85); in acetonitrile/methanol: 420 nm.

Tetradecahexaenal. [64512-28-3]; [774604-49-8] (*all-E*). 2,4,-6,8,10,12-Tetradecahexaenal. $C_{14}H_{16}O$; mol wt 200.28. Orange needles from alcohol, mp 192° (Schmitt); also reported as red-orange crystals from chloroform, mp 195-196° (Blout). Abs max in chloroform: 285, 393, 399 nm (log ε 4.05, 4.81 4.71); in acetonitrile/methanol: 403 nm

8037. Psoralen. [66-97-7] 7*H*-Furo[3,2-*g*][1]benzopyran-7-one; 6-hydroxy-5-benzofuranacrylic acid δ-lactone; furo[3,2-*g*]coumarin; ficusin. $C_{11}H_6O_3$; mol wt 186.17. C 70.97%, H 3.25%, O 25.78%. One of a group of furocoumarins occurring naturally in more than two dozen plant sources, including *Rutaceae* (*e.g.* bergamot, limes, cloves), *Umbelliferae* (*e.g.* celery, parsnips), *Leguminosae* (*e.g. Psoralen coryfolia*), and *Moraceae* (*e.g.* figs). Isoln: H. S. Jois *et al., J. Indian Chem. Soc.* **10**, 41 (1933); A. Stoll *et al., Helv. Chim. Acta* **33**, 1637 (1950); F. E. King *et al., J. Chem. Soc.* **1954**, 1392. Synthesis: E. Späth *et al., Ber.* **69**, 1087 (1936); R. C. Esse, B. E. Christensen, *J. Org. Chem.* **25**, 1565 (1960); O. Dann, D. Volz, *Arch. Pharm.* **308**, 121 (1975); V. K. Ahluwalia *et al., Monatsh. Chem.* **111**, 877 (1980). Psoralens are *phytoalexins*; they are used by plants in a defensive response to attacks by fungi and insects: M. Berenbaum, P. Feeny, *Science* **212**, 927 (1981). They have also shown photosensitizing and phototoxic effects in animals and humans and have been used in photochemotherapy for management of vitiligo, psoriasis, and mycosis fungoides: T. F. Anderson, J. J. Voorhees, *Annu. Rev. Pharmacol. Toxicol.* **20**, 235 (1980); A. Kornhauser *et al., Science* **217**, 733 (1982). Review of psoralen photochemistry: B. J. Parsons, *Photochem. Photobiol.* **32**, 813-821 (1980). Review of genetic toxicity of psoralen and uv radiation in human cells: *Acta Derm. Venereol.* **Suppl. 104**, 4-40 (1982). *See* Methoxsalen, Trioxsalen, Bergapten for additional refs.

Crystals from ether, mp 163-164°; 169-179° (Späth). Absorption spectra: Wessely, Kaltan, *Monatsh. Chem.* **86**, 430 (1955).

USE: As photochemical probe in biological systems: P.-S. Song, C.-N. Ou, *Ann. N.Y. Acad. Sci.* **346**, 355 (1980).

8038. Psychotrine. [7633-29-6] 1-[[(2*R*,3*R*,11b*S*)-3-Ethyl-1,3,4,6,7,11b-hexahydro-9,10-dimethoxy-2*H*-benzo[*a*]quinolizin-2-yl]methyl]-3,4-dihydro-7-methoxy-6-isoquinolinol; 1′,2′-didehydro-7′,10,11-trimethoxyemetan-6′-ol. $C_{28}H_{36}N_2O_4$; mol wt 464.61. C 72.39%, H 7.81%, N 6.03%, O 13.77%. Minor alkaloid found in ipecac, the ground roots of *Uragoga ipecacuanha* (Brot.) Baill. [*Cephaelis ipecacuanha* (Brot.) A. Rich.], *Rubiaceae*: F. H. Carr, F. L.

Pyman, *J. Chem. Soc.* **105**, 1591 (1914); O. Hesse, *Ann.* **405**, 1 (1914). Structure and stereochemistry: A. R. Battersby *et al., J. Chem. Soc.* **1959**, 2704, 3512. Mass spectrum: H. Budzikiewicz *et al., Tetrahedron* **20**, 399 (1964); ^{1}H,^{13}C-NMR of *O*-methylpsychotrine: T. Fujii *et al., Chem. Pharm. Bull.* **30**, 598 (1982); ^{13}C-NMR of psychotrine: *eidem, ibid.* **31**, 2583 (1983). Proposed alternate structure with exocyclic double bond: E. E. van Tamelen *et al., J. Am. Chem. Soc.* **79**, 4817 (1957); C. Schuij *et al., J. Chem. Soc. Perkin Trans. 1* **1979**, 970. *Review:* M. Janot in Manske, Holmes, *The Alkaloids* vol. 3 (Academic Press, New York, 1953) pp 363-394.

Tetrahydrate. Yellow prisms with blue fluorescence from dil acetone or alcohol. *Very bitter taste, produces nausea instantly.* The anhydr material sinters at 120°, becomes transparent at 120-126° and melts completely at 128°. $[\alpha]_D^{15}$ +69.3° (c = 2 in alcohol, calcd as the tetrahydrate). Sparingly sol in water, benzene, petr ether, ether. More sol in alcohol, acetone, chloroform. uv max (0.1*N* HCl): 240, 288, 306, 356 nm (ε 13900, 5700, 6250, 6800).

Sulfate trihydrate. $C_{28}H_{36}N_2O_4 \cdot H_2SO_4 \cdot 3H_2O$. Pale yellow scales from water, dec 214-217° when anhydr. $[\alpha]_D^{20}$ +39.2°.

O-Methylpsychotrine. Has a methoxy group in place of the hydroxyl group. mp 123-124°. $[\alpha]_D^{20}$ +43.2° (alc). uv max in water, pH 1: 241.5, 288.5, 305, 354 (log ε 4.26, 3.86, 3.92, 3.91); in water, pH 13: 226, 278.5, 307 (log ε 4.43, 3.96, 3.77).

8039. Ptaquiloside. [87625-62-5] (2′*R*,3′a*R*,4′*S*,7′a*R*)-7′a-(β-D-Glucopyranosyloxy)-1′,3′a,4′,7′a-tetrahydro-4′-hydroxy-2′,-4′,6′-trimethylspiro[cyclopropane-1,5′-[5*H*]inden]-3′(2′*H*)-one; braxin C; PT. $C_{20}H_{30}O_8$; mol wt 398.45. C 60.29%, H 7.59%, O 32.12%. Norsequiterpene glucoside identified as the carcinogenic prinicple of the bracken ferns, *Pteridium spp.* Isoln from *P. aquilinum* var. *latiusculum*: H. Niwa *et al., Tetrahedron Lett.* **24**, 4117 (1983). Structure elucidation: *eidem, ibid.* 5371. HPLC determn in ferns: M. P. Agnew, D. R. Lauren, *J. Chromatogr.* **538**, 462 (1991). Mechanism of carcinogenesis: M. Ojika *et al., Tetrahedron* **43**, 5261 (1987); A. S. Prakash *et al., Nat. Toxins* **4**, 221 (1996). Levels in cow's milk and health implications: M. E. Alonso-Amelot *et al., Lait* **78**, 413 (1998). Review of chemistry and carcinogenic mechanisms: K. Yamada *et al., Angew. Chem. Int. Ed.* **37**, 1818-1826 (1998); of biological activity: M. Shahin *et al., Mutat. Res.* **443**, 69-79 (1999).

Colorless, hygroscopic, amorphous compound. $[\alpha]_D^{22}$ −188° (c = 1 in methanol). Sol in water, methanol and ethyl acetate.

8040. Pteridine. [91-18-9] Pyrazino[2,3-*d*]pyrimidine; pyrimido[4,5-*b*]pyrazine; pyrimidine-4′,5′:2,3-pyrazine; 1,3,5,8-tetraazanaphthalene; azinepurine; benzotetrazine. $C_6H_4N_4$; mol wt

132.13. C 54.54%, H 3.05%, N 42.40%. Prepn from 4,5-diamino-pyrimidine and polyglyoxal or glyoxal sodium bisulfite: Albert *et al.*, *J. Chem. Soc.* **1951**, 474.

Yellow crystals from benzene, mp 138-138.5°. uv max (water): 298, 309 nm (log ε 3.87, 3.83). Sol in 7.2 parts water at 20-25°.
Caution: Related to folic acid antagonists.

8041. Pterocarpin. [524-97-0] (6aR,12aR)-6a,12a-Dihydro-3-methoxy-6H-[1,3]dioxolo[5,6]benzofuro[3,2-c][1]benzopyran. $C_{17}H_{14}O_5$; mol wt 298.29. C 68.45%, H 4.73%, O 26.82%. From red sandalwood *(Pterocarpus santalinus L.f., Leguminosae):* Cazeneuve, Hugouneng, *Compt. Rend.* **104**, 1722 (1887); **107**, 737 (1888); Leonhardt, Fay, *Arch. Pharm.* **273**, 53 (1935); McGookin *et al.*, *J. Chem. Soc.* **1940**, 787; from *Pterocarpus* spp: King *et al.*, *ibid.* **1953**, 3693. Structure: Bredenberg, Shoolery, *Tetrahedron Lett.* **1966**, 1805; *eidem*, *Bull. Chem. Soc. Jpn.* **42**, 1408 (1969); Uchiyama, Matsui, *Agric. Biol. Chem.* **31**, 1490 (1967). Stereochemistry: Ito *et al.*, *Chem. Commun.* **1965**, 595; Pachler, Underwood, *Tetrahedron* **23**, 1817 (1967).

Crystalline plates from petr ether or from ethanol, mp 164.5°. $[\alpha]_D$ −214.5° and $[\alpha]_{5461}^{20.5}$ −207.5° (c = 0.53 in chloroform). Practically insol in water, cold alcohol or ether. Sol in hot alcohol, chloroform.

8042. Pteroic Acid. [119-24-4] 4-[[(2-Amino-3,4-dihydro-4-oxo-6-pteridinyl)methyl]amino]benzoic acid; p-[(2-amino-4-hydroxy-6-pteridylmethyl)amino]benzoic acid. $C_{14}H_{12}N_6O_3$; mol wt 312.29. C 53.85%, H 3.87%, N 26.91%, O 15.37%. Intermediate in the synthesis of folic acids. Prepn: Waller *et al.*, *J. Am. Chem. Soc.* **70**, 19 (1948); *cf.* Hultquist *et al.*, *ibid.* 23; Boothe, **US 2472462** (1949 to Am. Cyanamid); C. Temple *et al.*, *J. Org. Chem.* **46**, 3666 (1981). Synthesis: M. G. Nair *et al.*, *ibid.* 3152.

Crystals from dil HCl. Sol in aq NaOH solns.

8043. Pteropterin. [89-38-3] N-[4-[[(2-Amino-3,4-dihydro-4-oxo-6-pteridinyl)methyl]amino]benzoyl]-L-γ-glutamyl-L-γ-glutamyl-L-glutamic acid; N-[N-(N-pteroyl-γ-glutamyl)-γ-glutamyl]glutamic acid; fermentation L. casei Factor; pteroyl-γ-glutamyl-γ-glutamylglutamic acid; pteroyldi-γ-glutamylglutamic acid; pteroyltriglutamic acid; PTGA; Teropterin. $C_{29}H_{33}N_9O_{12}$; mol wt 699.63. C 49.79%, H 4.75%, N 18.02%, O 27.44%. Isoln from aerobic culture of *Corynebacterium:* Hutchings *et al.*, *J. Am. Chem. Soc.* **70**, 1 (1948). Identification, structure and synthesis: Boothe *et al.*, *ibid.* 1099; *ibid.* **71**, 2304 (1949). Prepn: Cosulich, **US 2563707** (1951 to Am. Cyanamid).

Forms a crystalline barium salt.
Monohydrate. Crystals from water adjusted to pH 2.8 with HCl and contg some NaCl. Has the general properties of a polypeptide. Absorption spectrum, Hutchings, *loc. cit.* Soly in water at 5°: 0.10 mg/ml; at 80°: 3.00 mg/ml. Sol in NaOH solns.
Methyl ester. Crystals from methanol contg NaCl. Soly in water at 5°: 0.12 mg/ml; at 80°: 1.00 mg/ml; in methanol at −5°: 0.30 mg/ml; at 60°: 5.00 mg/ml.
THERAP CAT: Formerly as antineoplastic.

8044. Pteroylhexaglutamylglutamic Acid. [6484-74-8] N-[4-[[(2-Amino-1,4-dihydro-4-oxo-6-pteridinyl)methyl]amino]benzoyl]-L-γ-glutamyl-L-γ-glutamyl-L-γ-glutamyl-L-γ-glutamyl-L-γ-glutamyl-L-γ-glutamyl-L-glutamic acid; pteroylheptaglutamic acid; vitamin Bc conjugate; Bc conjugate; PHGA. $C_{49}H_{61}N_{13}O_{24}$; mol wt 1216.09. C 48.40%, H 5.06%, N 14.97%, O 31.57%. One of the naturally occurring polyglutamates of folic acid. Isoln from yeast: Pfiffner *et al.*, *Science* **102**, 228 (1945); *J. Am. Chem. Soc.* **68**, 1392 (1946). Solid phase synthesis: C. L. Krumdieck, C. M. Baugh, *Biochemistry* **8**, 1568 (1969); *eidem*, *Methods Enzymol.* **66**, 523 (1980). Chain length characterization of pteroylpolyglutamates: D. G. Priest *et al.*, *Anal. Biochem.* **115**, 163 (1981).

Rosettes of minute needles from 5% NaCl soln. Has no definite mp. When heated on the hot stage it begins to darken from about 200° and partially melts at 230-260°, and remains partially melted up to 360°. The ultraviolet absorption curves are practically identical with those of folic acid. Sparingly sol in water contg NaCl. Sol in dil NaOH soln.

8045. Pukateine. [81-67-4] (7aR)-6,7,7a,8-Tetrahydro-7-methyl-5H-benzo[g]-1,3-benzodioxolo[6,5,4-de]quinolin-12-ol; 1,2-(methylenedioxy)-6aβ-aporphin-11-ol. $C_{18}H_{17}NO_3$; mol wt 295.34. C 73.20%, H 5.80%, N 4.74%, O 16.25%. From bark of *Laurelia novaezelandiae* A. Cunn, *Lauraceae.* Isoln: Aston, *J. Chem. Soc.* **97**, 1381 (1910); Bernauer, *Helv. Chim. Acta* **50**, 1583 (1967). Structure: Barger, Giradet, *ibid.* **14**, 481, 504 (1931); Barger, Schlittler, *ibid.* **15**, 381 (1932). Total synthesis: Zymalkowski, Happel, *Tetrahedron Lett.* **1969**, 219; *Ber.* **102**, 2959 (1969); Kametani *et al.*, *J. Chem. Soc. Perkin Trans. 1* **1972**, 1435.

Crystals from alcohol, mp 208-212°. $[\alpha]_D^{25}$ −240° (c = 0.097 in alcohol). Practically insol in water. Sol in alc, ether, chloroform, pyridine; slightly sol in petr ether.
dl-Form. Crystals from absolute alcohol, mp 232-233°.

8046. Pulegone. [89-82-7] (5R)-5-Methyl-2-(1-methylethylidene)cyclohexanone; R-(+)-p-menth-4(8)-en-3-one; 1-methyl-4-isopropylidene-3-cyclohexanone. $C_{10}H_{16}O$; mol wt 152.24. C 78.90%, H 10.59%, O 10.51%. Found in oils derived from plants of the *Labiatae* family as (+)-form. Readily isolated in quantity from the pennyroyal oils from *Mentha pulegium* L., *M. longifolia* (L.) Huds., and *Hedeoma pulegioides* (L.) Pers., *Labiatae:* Gildemeister,

Hoffmann, *Die ätherischen Öle* **Vol. I,** p 560 (1928); Simonsen, *The Terpenes* **Vol. I** (2nd ed, 1947) p 370. Synthesis: Kuhn, Schinz, *Helv. Chim. Acta* **36,** 161 (1953). Synthesis of the (±)-form: Black *et al., J. Chem. Soc.* **1956,** 2971; Wolinsky *et al., J. Org. Chem.* **30,** 3207 (1965); of the *S*(−)-form: E. J. Corey *et al., ibid.* **41,** 380 (1976). Improved synthesis of the *R*(+)-form: T. Sato *et al., Tetrahedron Lett.* **1980,** 3377. Conversion of the *R*(+) to the *S*(−)-form: H. E. Ensley, R. V. C. Carr, *ibid.* **1977,** 513. Biosynthetic study: A. Akhila, D. V. Banthorpe, *Z. Pflanzenphysiol.* **99,** 277 (1980), *C.A.* **94,** 2073r (1981).

Oil. Pleasant odor, midway between peppermint and camphor. d_4^{15} 0.9346. bp$_{760}$ 224°; bp$_{100}$ 151-153°; bp$_{17}$ 103°; bp$_6$ 84°. $[\alpha]_D^{20}$ +21°; $[\alpha]_{546}^{20}$ +28.2°. n_D^{20} 1.4894. Practically insol in water. Miscible with alcohol, ether, chloroform.

(±)-**Form.** Liquid. bp$_7$ 78-80°. n_D^{16} 1.4856. uv max (alc): 253.3 nm (log ε 3.86).

(−)-**Form.** Liquid. bp$_{20}$ 104-108°. $[\alpha]_D^{23}$ −22.5° (neat).

8047. Pullulan. [9057-02-7] High molecular weight, linear polysaccharide of α-(1 → 4) and α-(1 → 6) linked D-glucose produced and secreted by the yeast-like fungus, *Aureobasidium pullulans*. Usually described as a repeating polymer of α-(1 → 6) linked maltotriose units. Isoln: B. Bernier, *Can. J. Microbiol.* **4,** 195 (1958); and characterization: H. Bender *et al., Biochim. Biophys. Acta* **36,** 309 (1959). Rheological properties in aqueous solutions: Y. Hemar, D. N. Pinder, *Biomacromolecules* **7,** 674 (2006). Mol wt and polydispersity values: G. J. Doucet *et al., Macromolecules* **39,** 9446 (2006). Review of structure, biosynthesis, and chemically modified derivatives: K. I. Shingel, *Carbohydr. Res.* **339,** 447-460 (2004); of production methods and commercial applications: R. S. Singh *et al., Carbohydr. Polym.* **73,** 515-531 (2008).

White to off-white odorless powder. $[\alpha]_D^{20}$ +168° (water). Freely sol in water (25°). Sol in DMF, DMSO. Practically insol in ethanol.

USE: Low calorie food additive and source of fiber; stabilizer, thickener, and glazing agent; bulking agent for freeze-drying; plasticizer; wetting and solubilizing agent. As edible film in food and health care products. Pharmaceutic aid (tablet binder; encapsulating agent).

8048. Pulsatilla. Pasque flower; wind flower; meadow anemone; Easter flower. Dried herb of *Anemone pulsatilla* L. *(Pulsatilla vulgaris* Mill.), *A. pratensis* L. *(P. pratensis* (L.) Mill.), or *A. patens* L. *(P. patens* (L.) Mill.), *Ranunculaceae. Habit.* Europe, Asia. *Constit.* Anemone camphor, volatile oil, tannin.

8049. Pumactant. ALEC; Zofac. Artificial lung expanding compound composed of *dipalmitoylphosphatidylcholine* (DPPC) and egg *phosphatidylglycerol* in a 7:3 (w/w) ratio. Description of prepn and clinical evaluation in premature babies: C. J. Morley *et al., Lancet* **1,** 64 (1981). Clinical trials in premature babies: Ten Centre Study Group, *Br. Med. J.* **294,** 991 (1987). Review of phys-

ical properties and design of mixture: A. D. Bangham *et al.,* "Easy Breathing and the Design of Artificial Lung Expanding Compounds (ALEC)" in *Topics in Lipid Research,* R. Klein, B. Schmitz, Eds (Royal Society of Chemistry, London, Great Britain, 1986) 268-275; of various surfactant treatments; C. J. Morley, in *Surfactant Replacement Therapy,* D. L. Shapiro, R. H. Notter, Eds (Alan R. Liss, Inc., NY, NY, 1989) 219-234. Clinical trial in treatment of asthma: K. S. Babu *et al., Eur. Respir. J.* **21,** 1046 (2003).

Fine crystalline powder.

THERAP CAT: Pulmonary surfactant.

8050. Pumice. Light, hard, rough, porous, gray masses or a gritty, gray-colored powder of volcanic origin. Found chiefly in the Lipari islands and in the Greek archipelagos. Consists mainly of complex silicates of Al, K, and Na. It is insol in water and is not attacked by acids.

Very light, hard, rough, porous, grayish masses or gritty, grayish powder. Practically insol in water.

USE: Abrasive in metal polishes; in fireproofing and insulating compds; also as a carrier for metal catalysts. In pharmacy as filtering medium and dispersant. In cosmetics for removing rough skin.

8051. Pumpkin Seed. Pepo. Dried ripe seeds of cultivated varities of *Cucurbita pepo* L., *Cucurbitaceae. Habit.* Southern Asia, Europe, America. *Constit.* Fixed oil, acrid resin, myosin, vitellin, sugar.

THERAP CAT: Anthelmintic.

8052. 1*H*-Purine. [120-73-0] 9*H*-Purine; 7*H*-imidazo[4,5-*d*]-pyrimidine. $C_5H_4N_4$; mol wt 120.12. C 50.00%, H 3.36%, N 46.64%. Prepd by heating 4,5-diaminopyrimidine with anhydr formic acid in a current of carbon dioxide: Isay, *Ber.* **39,** 251 (1906); by boiling 2,6-diiodopurine with zinc dust and water under carbon dioxide: Fischer, *ibid.* **31,** 2551, 2564 (1898).

Needles from toluene or alcohol, mp 216-217° (partial sublimation on rapid heating). Freely sol in water, hot alcohol. Slightly sol in hot ethyl acetate, acetone. Practically insol in ether, chloroform. Aq solns are neutral to litmus. Forms salts with acids and bases.

8053. Purmorphamine. [483367-10-8] 9-Cyclohexyl-*N*-[4-(4-morpholinyl)phenyl]-2-(1-naphthalenyloxy)-9*H*-purin-6-amine. $C_{31}H_{32}N_6O_2$; mol wt 520.64. C 71.52%, H 6.20%, N 16.14%, O 6.15%. Agonist of the Sonic Hedgehog (Shh) signaling pathway. Binds directly to the Smoothened (Smo) transmembrane receptor and activates Gli-mediated transcription in a variety of cell types. Induces differentiation of multipotent mesenchymal progenitor cells into osteoblasts. Prepn and osteogenesis-inducing activity: X. Wu *et al., J. Am. Chem. Soc.* **124,** 14520 (2002). Effect on Hedgehog signaling: *eidem, Chem. Biol.* **11,** 1229 (2004); and activation of Smo: S. Sinha, J. K. Chen, *Nat. Chem. Biol.* **2,** 29 (2006). Use to generate motor neurons from embryonic stem cells: B.-Y. Hu, S.-C. Zhang, *Methods Mol. Biol.* **636,** 123 (2010). Review of effect on Shh signaling: B. Stanton, L. Peng, *Mol. Biosyst.* **6,** 44-54 (2010).

Off-white to pale beige crystalline solid. Soly (mg/ml): 10 in DMSO; 20 in DMF. Sparingly sol in aqueous buffers. *Protect from light.*

USE: Reagent in cell signaling research and to induce stem cell differentiation.

8054. Puromycin. [53-79-2] 3'-[[(2S)-2-Amino-3-(4-methoxyphenyl)-1-oxopropyl]amino]-3'-deoxy-N,N-dimethyladenosine; L-3'-(α-amino-p-methoxyhydrocinnamamido)-3'-deoxy-N,N-dimethyladenosine; 6-dimethylamino-9-[3-deoxy-3-(p-methoxy-L-phenylalanylamino)-β-D-ribofuranosyl]-β-purine; 6-dimethylamino-9-[3'-(p-methoxy-L-phenylalanylamino)-3'-deoxy-β-D-ribofuranosyl]purine; stylomycin; CL-13900; P-638; 3123-L. $C_{22}H_{29}$-N_7O_5; mol wt 471.52. C 56.04%, H 6.20%, N 20.79%, O 16.97%. Aminonucleoside antibiotic produced by the soil actinomycete *Streptomyces alboniger.* Structural analog of aminoacyl-tRNA; inhibits protein synthesis. Isoln: J. N. Porter *et al., Antibiot. Chemother.* **2**, 409 (1952); *eidem,* US 2763642 (1956 to Am. Cyanamid). Structure: C. W. Waller *et al., J. Am. Chem. Soc.* **75**, 2025 (1953); P. W. Fryth *et al., ibid.* **80**, 2736 (1958). Synthesis: B. R. Baker *et al., ibid.* **76**, 4044 (1954); **77**, 12 (1955); M. J. Robins *et al., J. Org. Chem.* **66**, 8204 (2001). Conformation: O. Jardetzky, *J. Am. Chem. Soc.* **85**, 1823 (1963). Effect on protein synthesis: D. Nathans, *Proc. Natl. Acad. Sci. USA* **51**, 585 (1964). Use in gene transfer studies: S. de la Luna *et al., Gene* **62**, 121 (1988). *Review:* D. Nathans in *Antibiotics* **vol. I**, D. Gottlieb, P. D. Shaw, Eds. (Springer-Verlag, New York, 1967) pp 259-277.

Crystals from water, mp 175.5-177°. $[\alpha]_D^{25} -11°$ (ethanol). uv max (0.1N NaOH): 275 nm (ε 20300); in 0.1N HCl: 267.5 nm (ε 19500). pKa: 6.8, 7.2. Sparingly sol in water, organic solvents. LD_{50} in mice (mg/kg): 350 i.v.; 525 i.p.; 675 orally (Porter, 1952).

Dihydrochloride. [58-58-2] $C_{22}H_{29}N_7O_5.2HCl$; mol wt 544.43. White crystals or powder. Prepd as the dihydrate, mp 174° (dec). Sol in water.

USE: Research tool for studying protein synthesis; cell line selection agent in gene transfer experiments.

8055. Purothionin. [9009-72-7] A low mol wt wheat protein composed of α_1-, α_2- and β-purothionins. Contains about 20% cystine, 10% arginine and 10% lysine. Isoln from unbleached patent flour: A. K. Balls, W. S. Hale, *Cereal Chem.* **17**, 243 (1940); A. K. Balls *et al., ibid.* **19**, 279 (1942). Inhibits the growth of bacteria and yeasts: L. S. Stuart, T. H. Harris, *ibid.* 288. Toxic effects in animals: E. J. Coulson *et al., ibid.* 301. Sepn and characterization of α- and β-purothionin: D. G. Redman, N. Fisher, *J. Sci. Food Agric.* **19**, 651 (1968); C. C. Nimmo *et al., ibid.* **25**, 607 (1974). Amino acid sequences of α_1- and α_2-purothionin: S. Ohtani *et al., Agric. Biol. Chem.* **39**, 2269 (1975); *eidem, J. Biochem.* **82**, 753 (1977); of β-purothionin: A. S. Mak, B. L. Jones, *Can. J. Chem.* **54**, 835 (1976); of α_1 and α_2-purothionin: B. L. Jones, A. S. Mak, *Cereal Chem.* **54**, 511 (1977). *Review:* D. D. Kasarda *et al.,* "Wheat Proteins" in *Advances in Cereal Science and Technology* **vol. 1** (Am. Cereal Chem., St. Paul, 1976) pp 208-210.

Forms a cryst hydrochloride. Freely sol in water. Easily digested by proteolytic enzymes including cryst chymotrypsin, chymopapain, papain.

α_1-Purothionin. 5-L-Arginine-27-glycine-33-L-isoleucine-34-L-serine-42-glycine-purothionin A I (reduced); purothionin A II. C_{198}-$H_{340}N_{68}O_{56}S_8$; mol wt 4825.80.

α_2-Purothionin. 5-L-Arginine-6-L-threonine-18-L-serine-26-L-serine-27-L-threonine-42-glycine-purothionin A I (reduced). C_{202}-$H_{348}N_{68}O_{59}S_8$; mol wt 4929.90.

β-Purothionin. Purothionin A I. $C_{203}H_{339}N_{67}O_{59}S_8$; mol wt 4918.84.

8056. Purpurin. [81-54-9] 1,2,4-Trihydroxy-9,10-anthracenedione; 1,2,4-trihydroxyanthraquinone; C.I. Natural Red 8; C.I. Natural Red 16; C.I. 58205; C.I. 75410. $C_{14}H_8O_5$; mol wt 256.21. C 65.63%, H 3.15%, O 31.22%. Occurs as glycoside in the madder root *(Rubia tinctorum* L., *Rubiaceae)* of commerce. Is formed during storage; no appreciable amount in the fresh root: Hill, Richter, *J. Chem. Soc.* **1936**, 1714. Although a dye itself, it is usually considered as an undesirable contaminant of alizarin extracted from madder. May be prepd from alizarin by oxidation with ammonium persulfate: Wacker, *J. Prakt. Chem.* [2] **54**, 90 (1896); also by Friedel-Crafts condensation of hydroxyhydroquinone with phthalic anhydride: Dimroth, Fick, *Ann.* **411**, 321 (1916).

Long orange needles with 1 H_2O from dil alcohol, anhydr at 100°. Anhydr red needles from abs alcohol or by sublimation around 150° in high vacuum (less than 2 mm Hg). mp 257°. Absorption spectrum: Meek, *J. Chem. Soc.* **111**, 969 (1917); Ezaby, *J. Chem. Soc. B* **1970**, 1293. More sol in boiling water than alizarin (yellow color with yellowish hue). Freely sol in alcohol (red), in ether (intensely yellow with fluorescence). Soluble in benzene, toluene, xylene (dark yellow), in boiling alum soln (red).

2-Methyl ether. $C_{15}H_{10}O_5$. Orange crystals from benzene, mp 240°.

2,4-Dimethyl ether. $C_{16}H_{12}O_5$. Orange needles, mp 186-189°.

USE: Forms colored "lakes" with various metal salts and is a fast dye for cotton printing. Now used mostly in the manuf of acid and chrome dyes. Reagent for the detection of boron; for detection of insol calcium salts in the cell contents of histological material and as a nuclear stain.

8057. Purpurogallin. [569-77-7] 2,3,4,6-Tetrahydroxy-5H-benzocyclohepten-5-one. $C_{11}H_8O_5$; mol wt 220.18. C 60.01%, H 3.66%, O 36.33%. The aglycone of several glycosides from various nutgalls. Prepn by oxidation of pyrogallol: Perkin, Steven, *J. Chem. Soc.* **83**, 192 (1903); Perkin, *ibid.* **101**, 803 (1912); Nierenstein, Spiers, *Ber.* **46**, 3151 (1913); Evans, Dehn, *J. Am. Chem. Soc.* **52**, 3647 (1930). Structure: Willstätter, Heiss, *Ann.* **433**, 17 (1923); Barltrop, Nicholson, *J. Chem. Soc.* **1948**, 116. Synthesis: Caunt *et al., ibid.* **1950**, 1631; **1951**, 1313.

Deep red needles from glacial acetic acid, dec 274-275°. Sparingly sol in most solvents.

Diglucoside. Dryophantin. $C_{23}H_{28}O_{15}$. Coloring matter of "red pea" gall produced by *Dryophanta divisa* Ald. on *Quercus pedunculata* Ehrh., *Fagaceae:* Nierenstein, *J. Chem. Soc.* **115**, 1328 (1919); Nierenstein, Swanton, *Biochem. J.* **38**, 373 (1944). Dark red needles with bronze luster, mp 220-221°. Slightly sol in water, cold alcohol; sol in boiling alcohol; in methanol, acetone.

USE: As an additive to edible or inedible fats or oils, hydrocarbon fuels or lubricants, retards oxidation or metal contamination: Thompson, US 2770545 (1956 to Universal Oil Prod.).

8058. Putrescine. [110-60-1] 1,4-Butanediamine; 1,4-diaminobutane; tetramethylenediamine. $C_4H_{12}N_2$; mol wt 88.15. C

54.50%, H 13.72%, N 31.78%. Biogenic polyamine and precursor of spermidine, *q.v.*, initially detected in decaying animal tissues, but now known to be present in all cells and certain bacterial cultures. It is essential for both normal and neoplastic tissue growth. Formed via decarboxylation of ornithine or by decarboxylation of arginine, followed by hydrolysis. Prepn: A. Ladenburg, *Ber.* **19**, 780 (1886); G. Ciamician, C. U. Zanetti, *ibid.* **22**, 1970 (1889); R. Willstätter, W. Heubner, *ibid.* **40**, 3871 (1907); of the dihydrochloride: *Org. Synth.* **coll. vol. IV** (1963) p 819. Role in cell growth processes: C. W. Tabor, H. Tabor, *Annu. Rev. Biochem.* **45**, 285 (1976); J. Janne *et al., Biochim. Biophys. Acta* **473**, 241 (1978). Formation and interconversion of putrescine and spermidine in mammalian cells: A. E. Pegg *et al., Adv. Enzyme Regul.* **19**, 427 (1980). Biosynthetic study in fungi: L. Stevens, *Med. Biol.* **59**, 308 (1981). Regulation of tRNA methyl transferase activity: M. Mach *et al., Biochem. J.* **202**, 153 (1982). Alteration of DNA conformation in rat brain tumor cells by depletion of intracellular putrescine: D. T. Hung *et al., Science* **221**, 368 (1983). Use of labeled putrescine as a positron-emission tomographic tracer in brain tumors: N. Volkow *et al., ibid.* 673. Reviews of early literature: M. Guggenheim, *Die biogenen Amine* (S. Karger, Basel, 1951, 4th ed.) 619 pp; H. Tabor *et al., Annu. Rev. Biochem.* **30**, 579-604 (1961). Review of formation of GABA, the major inhibitory neurotransmitter in vertebrate brains, from putrescine: N. Seiler, *Physiol. Chem. Phys.* **12**, 411-429 (1980). Review of metabolism: T. L. Sourkes, K. Missala, *Agents Actions* **11**, 20-27 (1981). Book: *Polyamines in Biology and Medicine*, D. R. Morris, L. J. Marton, Eds. (Dekker, New York, 1981) 512 pp.

Colorless oil, bp 158-160°. Cryst on cooling, mp 23-24°. Strong piperidine-like odor. Very sol in water.

Dihydrochloride. [333-93-7] $C_4H_{12}N_2.2HCl$. Cryst from 85% alc, mp >275°.

USE: As a tool in biochemical research.

8059. PVNO. [26715-00-4] 4-Ethenylpyridine 1-oxide homopolymer; poly(4-vinylpyridine-*N*-oxide); PVPNO. $(C_7H_7NO)_x$. Prepn and thermal properties: P. Ranganathan *et al., J. Polym. Sci. A* **28**, 2711 (1990). Viscosity and light scattering: C.-M. Lee *et al., Polymer* **37**, 4283 (1996); fluorescence and metal-ion complexation: Y. Okamoto *et al., Macromolecules* **31**, 9201 (1998). Interaction with sodium dodecyl sulfate, *q.v.*: P. Bahadur *et al., Langmuir* **11**, 1951 (1995). Use as a dye transfer inhibitor: R. K. Panandiker *et al., WO 9513354; eidem, US 5466802* (both 1995 to Proctor & Gamble); as adsorbent stabilizer for metal catalysis: A. B. R. Mayer, J. E. Mark, *J. Macromol. Sci. Pure Appl. Chem.* **A34**, 2151 (1997); T. Balakrishnan, V. Rajendran, *J. Appl. Polym. Sci.* **78**, 2075 (2000). GC/MS determn in laundry detergents: T. Uchiyama *et al., J. Anal. Appl. Pyrolysis* **45**, 111 (1998).

Crystallized as white solid monomer from benzene-hexane. mp 122°. Polymerizes at 100-125° under vacuum in a sealed tube; decomposes 290°. uv max (water): 257 nm (ε 19844). Sol in water and acidic media. Insol in most organic solvents including DMSO, dimethylacetamide, THF, and acetone.

USE: Dye transfer inhibitor used in laundry detergents. Support for metal catalysts.

8060. Pygeum Africanum Extract. [85865-74-3] Tadenan. Extract from the bark of the tree *Pygeum africanum* Hook, also known as *Prunus africana* (Hook) F. Kalkm (Rosaceae-Amygdaleae), which is native to tropical Asia and Africa. Prepn: J. Debat,

FR **1578711**; *idem*, FR **M6703**; *idem*, US **3856946** (1969, 1969, 1974 all to Labs. du Debat). Pharmacology: L. Thieblot *et al., Therapie* **26**, 575 (1971). Identification of 1-docosanol, *q.v.*, as active principle: *idem et al., ibid.* **32**, 99 (1977). Characterization of components: R. Longo, S. Tira, *Farmaco Ed. Prat.* **38**, 287 (1983); E. M. Martinelli *et al., J. High Resolut. Chromatogr. Chromatogr. Commun.* **9**, 106 (1986). Clinical study in BPH: A. Barlet *et al., Wien. Klin. Wochenschr.* **102**, 667 (1990). Review of clinical experience: A. Ishani *et al., Am. J. Med.* **109**, 654-664 (2000).

THERAP CAT: In treatment of micturitional disorders due to benign prostatic hypertrophy.

8061. Pymetrozine. [123312-89-0] 4,5-Dihydro-6-methyl-4-[(*E*)-(3-pyridinylmethylene)amino]-1,2,4-triazin-3(2*H*)-one; CGA-215944; Plenum; Fulfill. $C_{10}H_{11}N_5O$; mol wt 217.23. C 55.29%, H 5.10%, N 32.24%, O 7.37%. Pyridine azomethine antifeedant. Prepn: H. Kristinsson, *EP 314615; idem, US 4931439* (1989, 1990 both to Ciba-Geigy). Comprehensive description: C. R. Flückiger *et al., Brighton Crop Prot. Conf. - Pests Dis.* **1992**, 43-50. Field trials against aphids and white flies: *idem et al., ibid.* 1187. Historical account of discovery, synthesis, and development: H. Kristinsson, *Spec. Publ. - R. Soc. Chem.* **147**, 85-102 (1994).

Crystalline, mp 234.4°. Vapor pressure at 20°: $< 9.7 \times 10^{-8}$ Pa. Soly at 20° (g/l): water 0.270; ethanol 2.25; hexane <0.001. Partition coefficient (*n*-octanol/water): 0.2. LD_{50} orally in rats: 5820 mg/kg (Flückiger).

USE: Insecticide.

8062. Pyocyanine. [85-66-5] 5-Methyl-1(5*H*)-phenazinone; sanazin. $C_{13}H_{10}N_2O$; mol wt 210.24. C 74.27%, H 4.79%, N 13.32%, O 7.61%. Blue redox pigment produced by *Pseudomonas aeruginosa*. Important virulence factor in pseudomonal infection. Isoln: Wrede, Strack, *Z. Physiol. Chem.* **140**, 1 (1924); Schoental, *Br. J. Exp. Pathol.* **22**, 137 (1941). Synthesis: Wrede, Strack, *Z. Physiol. Chem.* **181**, 58 (1929); Surrey, *Org. Synth.* **26**, 86 (1946); **coll. vol. III**, 753 (1955). HPLC determn in sputum: R. Wilson *et al., Infect. Immun.* **56**, 2515 (1988). Review of biosynthesis and role in pathogenesis: G. W. Lau *et al., Trends Mol. Med.* **10**, 599-606 (2004).

Dark blue needles from water (usually with 1 H_2O which is lost at 50° over P_2O_5 *in vacuo*). mp 133°. Upon further heating it sublimes with decompn. Absorption spectrum: Nitzsche, *Ber.* **77**, 337 (1944). Freely sol in chloroform. Sol in nitrobenzene, pyridine, phenol, acetic acid, hot water, hot alcohol. Slightly sol in cold water and benzene. The blue water soln, made alkaline with Na_2CO_3, is easily rendered colorless by reduction with glucose or sodium hydrosulfite. Acidic $KMnO_4$ solns are decolorized by pyocyanine. An alkaline water soln of pyocyanine acquires a maroon color upon heating.

8063. Py-Phe. [199612-75-4] *N*-[1-Oxo-4-(1-pyrenyl)butyl]-L-phenylalanine; *N*-(1-phenylalanine)-4-(1-pyrene)butyramide. $C_{29}H_{25}NO_3$; mol wt 435.52. C 79.98%, H 5.79%, N 3.22%, O 11.02%. Photoactivated chemical protease. Prepn: C. V. Kumar, A. Buranaprapuk, *Angew. Chem. Int. Ed.* **36**, 2085 (1997). Site specificity: *idem et al., Proc. Natl. Acad. Sci. USA* **95**, 10361 (1998).

ε 42600 M^{-1}cm^{-1} at 343 nm.

USE: Photocleavage probe.

8064. Pyraclostrobin. [175013-18-0] *N*-[2-[[[1-(4-Chloro-phenyl)-1*H*-pyrazol-3-yl]oxy]methyl]phenyl]-*N*-methoxycarbamic acid methyl ester; methyl *N*-[2-[[1-(4-chlorophenyl)-1*H*-pyrazol-3-yl]oxymethyl]phenyl]-*N*-methoxycarbamate; BAS-500F; Cabrio; Comet; Headline. C$_{19}$H$_{18}$ClN$_3$O$_4$; mol wt 387.82. C 58.84%, H 4.68%, Cl 9.14%, N 10.84%, O 16.50%. Strobilurin fungicide for use in seed grass and food crops. Prepn: B. Müller *et al.*, **DE 4423612**; *eidem*, **US 5869517** (1996, 1999 both to BASF). Comprehensive description: E. Ammermann *et al.*, *Brighton Crop Prot. Conf. - Pests Dis.* **2000**, 541-548. Review of activity and field trials: R. Stierl *et al.*, *ibid.* 859-864.

White or light beige crystalline solid, mp 63.7-65.2°. Vapor pressure at 20°: 2.6 × 10^{-8} Pa. Log P (*n*-octanol/water) at 22°: 3.99. Soly in water (20°): 1.9 mg/l. LD$_{50}$ in rats (mg/kg): >5000 orally; >2000 dermally; LC$_{50}$ in rainbow trout: 0.006 mg/l (Ammermann).

USE: Agricultural fungicide.

8065. Pyranine. [6358-69-6]; [27928-00-3] (free acid). 8-Hy-droxy-1,3,6-pyrenetrisulfonic acid sodium salt (1:3); C.I. 59040; C.I. Solvent Green 7; D & C green no. 8; HPTS. C$_{16}$H$_7$Na$_3$O$_{10}$S$_3$; mol wt 524.37. C 36.65%, H 1.35%, Na 13.15%, O 30.51%, S 18.34%. Fluorescent dye with pH-sensitive spectral properties. Prepn: E. Tietze, O. Bayer, *Ann.* **540**, 189 (1939). Use as a pH probe for liposome interiors: K. Kano, J. H. Fendler, *Biochim. Biophys. Acta* **509**, 289 (1978); for phospholipid vesicles: N. R. Clement, J. M. Gould, *Biochemistry* **20**, 1534 (1981); in yeast cells: A. Peña *et al.*, *J. Bacteriol.* **177**, 1017 (1995); in organelles: C. C. Overly *et al.*, *Proc. Natl. Acad. Sci. USA* **92**, 3156 (1995); in cytosol: B. S. Gan *et al.*, *Am. J. Physiol. Cell Physiol.* **275**, 1158 (1998). Effects of salt and polyethylene glycol concentration on pKa: Y. Avnir, Y. Barenholz, *Anal. Biochem.* **347**, 34 (2005). Review of use as a biological stain: *Conn's Biological Stains*, R. W. Horobin, J. A. Kiernan, Eds. (BIOS Scientific Publishers Ltd, Oxford, UK, 10th ed., 2002) 399-400.

Crystals from aq acetone. pKa 7.22 ± 0.04. Changes from blue to green at pH 6.5-7.5. Highly sol in water; slightly sol in glacial acetic acid. Absorption max (acidic aq soln): 403 nm. Absorption max (aq alkali soln): 454 nm. Emission max (aq alkali soln): 511 nm.

USE: Fluorescent pH indicator for biological systems. Biological stain; dye for solvents, cosmetics and pharmaceuticals.

8066. Pyrantel. [15686-83-6] 1,4,5,6-Tetrahydro-1-methyl-2-[(1*E*)-2-(2-thienyl)ethenyl]pyrimidine. C$_{11}$H$_{14}$N$_2$S; mol wt 206.31. C 64.04%, H 6.84%, N 13.58%, S 15.54%. Prepn: **BE 658987** (1965 to Pfizer), *C.A.* **64**, 8192c (1966); Austin *et al.*, *Nature*

212, 1273 (1966); Kasubrick, McFarland, **ZA 6800516**; *eidem*, **US 3502661** (1968, 1970 both to Pfizer). Structure-activity studies: McFarland *et al.*, *J. Med. Chem.* **12**, 1066 (1969). Pharmacology and animal trials: Cornwell, *Vet. Rec.* **79**, 590 (1966); Howes, Lynch, *J. Parasitol.* **53**, 1085 (1967); Eyre, *J. Pharm. Pharmacol.* **22**, 26 (1970).

Crystals from methanol, mp 178-179°.

Tartrate. [33401-94-4] CP-10423-18; Banminth; Strongid. C$_{11}$H$_{14}$N$_2$S.C$_4$H$_6$O$_6$; mol wt 356.39. White crystals from hot methanol, mp 148-150°. uv max (water): 312 nm (log ε 4.27).

Pamoate. [22204-24-6] Pyrantel embonate; CP-10423-16; Cobantril; Combantrin; Helmex; Helmintox; Nemex. C$_{11}$H$_{14}$N$_2$S.C$_{23}$H$_{16}$O$_6$; mol wt 594.68. Tasteless, yellow crystalline powder. Sol in dimethyl sulfoxide; slightly sol in DMF. Practically insol in water, methanol.

THERAP CAT: Anthelmintic (Nematodes).

THERAP CAT (VET): Anthelmintic (Nematodes).

8067. Pyrazinamide. [98-96-4] 2-Pyrazinecarboxamide; pyrazinoic acid amide; pyrazine carboxylamide; D-50; Pezetamid; Pyrafat; Pirilène; Piraldina; Tebrazid; Unipyranamide; Zinamide. C$_5$H$_5$N$_3$O; mol wt 123.12. C 48.78%, H 4.09%, N 34.13%, O 12.99%. Prepn: O. Dalmer, E. Walter, **DE 632257** (1936 to E. Merck); *eidem*, **US 2149279** (1939 to Merck & Co.); S. A. Hall, P. E. Spoerri, *J. Am. Chem. Soc.* **62**, 664 (1940). Structure-activity study: S. Kushner *et al.*, *ibid.* **74**, 3617 (1952). Pharmacology: I. M. Weiner, J. P. Tinker, *J. Pharmacol. Exp. Ther.* **180**, 411 (1972). HPLC determn in biological samples: J. E. Conte, Jr. *et al.*, *J. Chromatogr. Sci.* **38**, 33 (2000). Clinical pharmacokinetics: C. A. Peloquin *et al.*, *Pharmacotherapy* **18**, 1205 (1998). Clinical trial in combination with rifampicin and isoniazid, *q.q.v.*: S. K. Teo, *Int. J. Tuberc. Lung Dis.* **3**, 126 (1999); in prevention of tuberculosis: F. Gordin *et al.*, *J. Am. Med. Assoc.* **283**, 1445 (2000). Comprehensive description: E. Feldner, D. Pitre, *Anal. Profiles Drug Subs.* **12**, 433-462 (1983). Review of clinical experience: M. A. Steele, R. M. Des Prez, *Chest* **94**, 845-850 (1988).

Crystals from water or alcohol, mp 189-191°. Begins to sublime at 60°. pKa 0.5. uv max: 269 nm (E$_{1cm}^{1\%}$ 660). Soly (mg/ml): water 15; methanol 13.8; abs ethanol 5.7; isopropanol 3.8; ether 1.0; isooctane 0.01; chloroform 7.4. Aq solns are neutral.

THERAP CAT: Antibacterial (tuberculostatic).

8068. Pyrazine. [290-37-9] 1,4-Diazine; paradiazine. C$_4$H$_4$N$_2$; mol wt 80.09. C 59.99%, H 5.03%, N 34.98%. Prepn: S. Gabriel, G. Pinkus, *Ber.* **26**, 2197 (1893); P. Brandes, C. Stoehr, *J. Prakt. Chem.* [2] **54**, 481 (1896). *See also*: *Beilstein* **XXIII**, 91 (1936). Crystal structure: G. De With *et al.*, *Acta Crystallogr.* **B32**, 3178 (1976).

Crystals or wax-like solid. Strong pyridine-like odor. d$_4^{61}$ 1.031; mp 53°; bp 115-118°; n$_D^{61}$ 1.4953. Freely sol in water, alcohol, ether. Volatile with steam.

8069. 2,3-Pyrazinedicarboxylic Acid. [89-01-0] C$_6$H$_4$N$_2$O$_4$; mol wt 168.11. C 42.87%, H 2.40%, N 16.66%, O 38.07%. Prepd by permanganate oxidation of quinoxaline: Gabriel, Sonn, *Ber.* **40**, 4850 (1907); Sausville, Spoerri, *J. Am. Chem. Soc.* **63**, 3153 (1941); Jones, McLaughlin, *Org. Synth.* **30**, 86 (1950).

Dihydrate. Prisms from water. Becomes anhydr at 100°, then melts at 183-185° (with evolution of CO_2). The dry dust provokes sneezing. Freely sol in water. Soluble in methanol, acetone, ethyl acetate. Slightly sol in ethanol, ether, chloroform, benzene, petr ether.

Dimethyl ester. $C_8H_8N_2O_4$. Crystals, mp 50°. Sol in water, acetone. Insol in benzene, petr ether.

Diamide. $C_6H_6N_4O_2$. Crystals, dec 240°. Sol in boiling water.

8070. Pyrazinoic Acid. [98-97-5] 2-Pyrazinecarboxylic acid; pyrazinemonocarboxylic acid. $C_5H_4N_2O_2$; mol wt 124.10. C 48.39%, H 3.25%, N 22.57%, O 25.78%. Prepn by thermal decompn of 2,3-pyrazinedicarboxylic acid: Hall, Spoerri, *J. Am. Chem. Soc.* **62**, 664 (1940); by oxidation of alkylpyrazines: Gainer, *J. Org. Chem.* **24**, 691 (1959).

Fine white needles from water, dec 225-229°. Sublimes. pK (25°) 2.92. Slightly sol in cold water, more sol in hot water. One gram dissolves in 120 g abs ethanol at 25°. Practically insol in ether, chloroform, benzene.

Methyl ester. Crystals from ether, mp 59°.

Amide see Pyrazinamide.

8071. Pyrazole. [288-13-1] 1*H*-Pyrazole; 1,2-diazole. $C_3H_4N_2$; mol wt 68.08. C 52.93%, H 5.92%, N 41.15%. Prepn from acetylene and diazomethane: H. v. Pechmann, *Ber.* **31**, 2950 (1898). Two tautomeric forms. ^{1}H-NMR: J. F. K. Wilshire, *Aust. J. Chem.* **19**, 1935 (1966); ^{13}C-NMR: W. M. Litchman, *J. Am. Chem. Soc.* **101**, 545 (1979); ^{15}N-NMR: I. I. Schuster *et al.*, *J. Org. Chem.* **44**, 1765 (1979). Spectral studies: D. Dumanovic *et al.*, *Talanta* **22**, 819 (1975). Inhibition of alcohol dehydrogenase: W. K. Lelbach, *Experientia* **25**, 816 (1969). Induction of thyroid necrosis: S. Szabo *et al.*, *Science* **199**, 1209 (1979). Toxicity: G. Magnussen *et al.*, *Experientia* **28**, 1198 (1972).

Needles or prisms from petr ether; pyridine-like odor, bitter taste. mp 69.5-70°. bp$_{757.9}$ 186-188°. pK (25°) 11.52. uv max (12*M* H_2-SO_4): 215 nm (log ε 3.76); at pH 6.1: 310 nm (log ε 3.61). Sol in water, alcohol, ether, benzene. LD$_{50}$ (24 hr) in rats, mice (mmol/kg): 19, 21 i.v.; 21, 22 orally (Magnussen).

USE: Chelating agent; in organic synthesis.

8072. 2-Pyrazoline. [109-98-8] 4,5-Dihydro-1*H*-pyrazole. $C_3H_6N_2$; mol wt 70.10. C 51.40%, H 8.63%, N 39.96%. Prepn from ethylene and diazomethane: Azzarello, *Atti Accad. Lincei* [5] **14**, II, 285 (1905).

Liquid. Faint amine odor. An odor resembling chocolate has been detected in some prepns. d$_4^{17}$ 1.0200. bp$_{760}$ 144°. Volatile with steam. $n_{587}^{17.2}$ 1.4796. Miscible with water, alc.

8073. Pyrazophos. [13457-18-6] 2-[(Diethoxyphosphinothi-oyl)oxy]-5-methylpyrazolo[1,5-*a*]pyrimidine-6-carboxylic acid ethyl ester; *O*,*O*-diethyl *O*-6-ethoxycarbonyl-5-methylpyrazolo[1,5-*a*]pyrimidin-2-yl phosphorothioate; 2-hydroxy-5-methylpyrazolo-[1,5-*a*]pyrimidine-6-carboxylic acid ethyl ester, *O*-ester with *O*,*O*-

diethyl phosphorothioate; HOE-2873; Afugan; Curamil. $C_{14}H_{20}$-N_3O_5PS; mol wt 373.36. C 45.04%, H 5.40%, N 11.25%, O 21.43%, P 8.30%, S 8.59%. Organophosphate fungicide. Prepn: **NL 6602131**; O. Scherer, H. Mildenberger, **US 3632757** (1966, 1972 both to Hoechst). Fungicidal properties and toxicity: F. M. Smit, *Meded. Rijksfac. Landbouwwet. Gent* **34**, 763 (1969). Degradation study in wines: P. Stavropoulos *et al.*, *Food Chem.* **72**, 473 (2001). Multiresidue determn in blood by GC-MS: E. Lacassie *et al.*, *J. Chromatogr. B* **759**, 109 (2001). Effectiveness against powdery mildew: D. J. Butt *et al.*, *Ann. Appl. Biol.* **75**, 217 (1973); R. T. Burchill, M. E. Cook, *Plant Pathol.* **24**, 194 (1975). Absorption and metabolism by plants: S. Gorbach *et al.*, *Environ. Qual. Saf. Suppl.* **3**, 840 (1975).

Yellow oil, mp 38-40°C. LD$_{50}$ in rats, marmots, rabbits (mg/kg): 140, 184, 435 orally (Smit).

USE: Fungicide.

8074. Pyrene. [129-00-0] Benzo[*def*]phenanthrene. $C_{16}H_{10}$; mol wt 202.26. C 95.01%, H 4.98%. Occurs in coal tar, *q.v.* Isoln: Kruber, *Ber.* **64**, 84 (1931). Also obtained by the destructive hydrogenation of hard coal: I. G. Farben., **DE 639240**; **DE 640580**; **DE 654201**. Purification by chromatography: Winterstein *et al.*, *Z. Physiol. Chem.* **230**, 162 (1934). Synthesis from *o*,*o*'-ditolyl: Weitzenböck, *Monatsh. Chem.* **34**, 193 (1913). From *peri*-trimethylene-naphthalene and malonyl chloride with AlCl$_3$: Fleischer, Retze, *Ber.* **55**, 3280 (1922). From α-tetralone: Braun, Rath, *Ber.* **61**, 956 (1928). From 4-keto-1,2,3,4-tetrahydrophenanthrene by a Reformatsky reaction: Cook, Hewett, *J. Chem. Soc.* **1934**, 366. Review of toxicology and human exposure: *Toxicological Profile for Polycyclic Aromatic Hydrocarbons* (PB95-264370, 1995) 487 pp.

Monoclinic prismatic tablets from alcohol or by sublimation. d^{23} 1.271. Pure pyrene is colorless, the usual contaminant which gives it a yellow color is tetracene. Solid and solns have slight blue fluorescence. mp 156°. bp 404°. Absorption spectrum: Clar, *Ber.* **69**, 1677 (1936); Seshan, *Proc. Indian Acad. Sci.* **A3**, 148 (1936). Fluorescence maxima: Sannié, Poremski, *Bull. Soc. Chim.* [5] **3**, 1139 (1936). Insol in water. Fairly sol in organic solvents.

8075. Pyrethrins. Active insecticidal constituents of pyrethrum flowers. Isoln: Staudinger, Ruzicka, *Helv. Chim. Acta* **7**, 177 (1924). Prepn by reconstitution from pyrethrolone and chrysanthemic acid: Elliott, Janes, *Chem. Ind. (London)* **1969**, 270. Structure: Crombie *et al.*, *J. Chem. Soc.* **1956**, 3963; Godin *et al.*, *J. Chem. Soc. C* **1966**, 332. Stereochemistry: Begley *et al.*, *Chem. Commun.* **1972**, 1276. Biosynthesis: Crowley *et al.*, *Biochim. Biophys. Acta* **60**, 312 (1962). *Review:* Crombie, Elliott, *Fortschr. Chem. Org. Naturst.* **19**, 120 (1961). Review of toxicology and human exposure: *Toxicological Profile for Pyrethrins and Pyrethroids* (PB2004-100004, 2003) 332 pp.

Pyrethrin I   R = CH_3
Pyrethrin II   R = $COOCH_3$

Pyrethrin I. [121-21-1] (1R,3R)-2,2-Dimethyl-3-(2-methyl-1-propenyl)cyclopropanecarboxylic acid (1S)-2-methyl-4-oxo-3-(2Z)-2,4-pentadienyl-2-cyclopenten-1-yl ester; chrysanthemummonocarboxylic acid pyrethrolone ester. $C_{21}H_{28}O_3$; mol wt 328.45. Viscous liquid. bp$_{0.0005}$ 146-150°. n_D^{20} 1.5242. $[\alpha]_D^{20}$ −14° (isooctane). uv max (95% ethanol): 225 nm (ε 36400). Oxidizes readily and becomes inactive in air. *Should be refrigerated and stored in darkness.* Stability data of pyrethrins I and II: Godin, *Pyrethrum Post* **9**, 17 (1968), *C.A.* **71**, 12031q (1969). Practically insol in water. Sol in alcohol, petr ether, kerosene, carbon tetrachloride, ethylene dichloride, nitromethane.

Pyrethrin II. [121-29-9] (1R,3R)-3-[(1E)-3-Methoxy-2-methyl-3-oxo-1-propenyl]-2,2-dimethylcyclopropanecarboxylic acid (1S)-2-methyl-4-oxo-3-(2Z)-2,4-pentadienyl-2-cyclopenten-1-yl ester; chrysanthemumdicarboxylic acid monomethyl ester pyrethrolone ester. $C_{22}H_{28}O_5$; mol wt 372.46. Viscous liquid. Oxidizes rapidly and becomes inactive in air. bp$_{0.007}$ 192-193°. n_D^{20} 1.5355. $[\alpha]_D^{19}$ +14.7° (isooctane-ether). uv max (95% ethanol): 229 nm (ε 45850). Practically insol in water. Sol in alc, petr ether (less sol than pyrethrin I), kerosene, carbon tetrachloride, ethylene dichloride, nitromethane.

Caution: Direct skin contact or inhalation may cause severe allergic attacks in sensitive people. Potential symptoms of overexposure are erythema, dermatitis, papules, pruritus; asthma, sneezing, vasomotor rhinitis, anaphylactic reactions; numbness of lips and tongue, sneezing, vomiting, diarrhea; tinnitus, headache, restlessness, incoordination, clonic convulsions, stupor, prostration; death due to respiratory paralysis. *See: Clinical Toxicology of Commercial Products*, R. E. Gosselin *et al.*, Eds. (Williams & Wilkins, Baltimore, 5th ed., 1984) Section III, pp 352-355; *NIOSH Pocket Guide to Chemical Hazards* (DHHS/NIOSH 97-140, 1997) p 270.

USE: Insecticide.

8076. Pyrethrosin. [28272-18-6] (1aR,4E,6R,6aR,9aS,10aR)-6-(Acetyloxy)-1a,3,6,6a,7,9a,10,10a-octahydro-4,10a-dimethyl-7-methyleneoxireno[8,9]cyclodeca[1,2-b]furan-8(2H)-one; 1,10-epoxy-6,8-dihydroxygermacra-4,11(13)-dien-12-oic acid 12,8-lactone acetate. $C_{17}H_{22}O_5$; mol wt 306.36. C 66.65%, H 7.24%, O 26.11%. From flowers of *Chrysanthemum cinerariaefolium* Vis., Compositae. Isoln: Thoms, *Pharm. Ztg.* **1891**, 503. Structure: Barton, de Mayo, *J. Chem. Soc.* **1957**, 150; Barton *et al.*, *ibid.* **1960**, 2263. Revised structure: S. Iriuchijima, S. Tamura, *Agric. Biol. Chem.* **34**, 204 (1970); E. J. Gabe *et al.*, *Chem. Commun.* **1971**, 559.

Crystals from ethanol, ethyl acetate or benzene + light petroleum, mp 198-200°. $[\alpha]_D$ −31° (c = 1.73 in chloroform). uv max: 204, 210, 220, 230 nm (ε 14,600; 12,200; 6000; 1700). Practically insol in water. Sol in hot alcohol, chloroform; slightly sol in ether or petr ether.

8077. Pyrethrum Flowers. Dalmatian insect powder; Persian insect powder. Flowers of *Chrysanthemum (Pyrethrum) cinerariaefolium* Vis., and of *C. coccineum* Willd. (*C. roseum* Adam), Compositae. Habit. Dalmatia, Montenegro, Western Asia. Constit. About 1-1.5% volatile oil; pyrethrin, pyretol, pyrethrotoxic acid, pyrethrosin, chrysanthemine, chrysathemumic acid.

USE: In insecticide prepns.

THERAP CAT: Scabicide.

THERAP CAT (VET): Ectoparaciticide.

8078. Pyridaben. [96489-71-3] 4-Chloro-2-(1,1-dimethylethyl)-5-[[[4-(1,1-dimethylethyl)phenyl]methyl]thio]-3(2H)-pyridazinone; 2-*tert*-butyl-5-(4-*tert*-butylbenzylthio)-4-chloropyridazin-3(2H)-one; NC-129; NCI-129; Sanmite; Nexter. $C_{19}H_{25}ClN_2OS$; mol wt 364.93. C 62.54%, H 6.91%, Cl 9.71%, N 7.68%, O 4.38%, S 8.79%. Pyridazinone derivative. Prepn: **JP Kokai 85 4173**; M. Taniguchi *et al.*, **US 4877787** (1985, 1989 both to Nissan). Chemi-

cal and biological properties: K. Hirata *et al.*, *Brighton Crop Prot. Conf. - Pests Dis.* **1988**, 41; K. Hirata *et al.*, *J. Pestic. Sci.* **20**, 177 (1995). HPLC determn in soil: C. Y. Deng *et al.*, *Fenxi Huaxue* **23**, 1357 (1995); *C.A.* **124**, 7749 (1995). Field trials against mites: G. Sterk, C. Versmissen, *Meded. Fac. Landbouwwet. Univ. Gent* **57**, 941 (1992). Review of field trials: G. Sterk, *Brighton Crop Prot. Conf. - Pests Dis.* **1994**, 559-568.

Odorless, white crystalline solid, mp 111-112°. d_4^{20}: 1.2. Vapor pressure (20°): 1.9×10^{-6} mm Hg. Soly at 20° (g/100ml): acetone 46, corn oil 4.2, ethanol 5.7, methyl cellosolve 11, xylene 39, benzene 11, cyclohexane 32, hexane 1.0, *n*-octanol 6.3, water 1.2×10^{-6}. Stable in water at pH 4, 7, 9. Stable in most organic solvents. Stable to heat at 50° for 3 months. LD$_{50}$ in male, female rats, bobwhite quail, mallard ducks (mg/kg): 435, 358, >2250, >2500 orally (Hirata). LD$_{50}$ in male, female rabbits (mg/kg): >2000, >2000 dermally (Hirata).

USE: Acaricide.

8079. Pyridate. [55512-33-9] O-(6-Chloro-3-phenyl-4-pyridazinyl)carbonothioic acid S-octyl ester; fenpyrate; CL-11344; Lentagran. $C_{19}H_{23}ClN_2O_2S$; mol wt 378.92. C 60.23%, H 6.12%, Cl 9.36%, N 7.39%, O 8.44%, S 8.46%. Contact herbicide for control of weeds in cereals. Prepn: R. Schonbeck *et al.*, **US 3953445** (1974, 1976 to Chemie Linz). Description: A. Diskus *et al.*, *Proc. Br. Crop Prot. Conf. - Weeds* **1976**, 717; M. Chéroux, J.-M. Moncorge, *Def. Veg.* **38**, 310 (1984). Residue analysis by HPLC: W. Lindner, H. Ruckendorfer, *Int. J. Environ. Anal. Chem.* **16**, 205 (1983).

Oily brown liquid, bp$_{0.1}$ 220°. mp 27°. n_D^{20} 1.568. d^{20} 1.555. Flash pt (open cup): 200°C. Almost insol in water (90 ppm). Sol in organic solvents. LD$_{50}$ in male rats, female rats, mice (mg/kg): 1970, 2400, >10,000 orally; in rabbits (mg/kg): >3450 dermally. LC$_{50}$ (96 hr) in carp, trout: >100, 81 mg/l (Chéroux, Moncorge).

USE: Herbicide.

8080. Pyridazine. [289-80-5] 1,2-Diazine; orthodiazine; oizine. $C_4H_4N_2$; mol wt 80.09. C 59.99%, H 5.03%, N 34.98%. Prepd from maleic hydrazide: Mizzoni, Spoerri, *J. Am. Chem. Soc.* **73**, 1873 (1951). Resonance energy and position of double bonds: Maccoll, *J. Chem. Soc.* **1946**, 671.

Liquid. d_4^{18} 1.107; $d_4^{23.5}$ 1.1035. mp −8°. bp$_{760}$ 208°; bp$_{14}$ 87°; bp$_{1.0}$ 48°. $n_D^{23.5}$ 1.52311. Dipole moment in dioxane at 35° = 3.94. uv max: 338 nm. Resonance energy: 22 kcal. Miscible with water, benzene, DMF. Freely sol in methanol, ethanol, ether. Practically insol in petr ether.

Hydrochloride. $C_4H_4N_2$·HCl. Yellow solid, mp 161-163°.

8081. Pyridine. [110-86-1] C_5H_5N; mol wt 79.10. C 75.92%, H 6.37%, N 17.71%. Discovered in coal tar: T. Anderson, *Ann.* **60**, 86 (1846). Physical data: B. A. Middleton, J. R. Partington, *Nature* **141**, 516 (1938). Acute toxicity study: H. F. Smyth Jr. *et al.*, *Arch. Ind. Hyg. Occup. Med.* **4**, 119 (1951). Books: H. Maier-Bode, *Das Pyridin und Seine Derivate in Wissenschaft und Technik* (Edwards Bros., Ann Arbor, Mich., 1943); Hill in *Chemistry of Coal*

Utilization **vol 2**, H. H. Lowry, Ed. (Wiley, New York, 1945), Chapter 27; *The Chemistry of Heterocyclic Compounds* **vol. 14**, a series of books entitled "Pyridine and its Derivatives": Parts 1-4, E. Klingsberg, Ed. (1960-1964); Suppl. Parts 1-4, R. A. Abramovitch, Ed. (1974-1975); Part 5, G. R. Newkome, Ed. (1984) (Wiley-Interscience, New York). *Review*: G. L. Goe in *Kirk-Othmer Encyclopedia of Chemical Technology* **vol. 19** (Wiley-Interscience, New York, 3rd ed., 1982) pp 454-483. Review of commercial synthesis and uses of pyridine and derivatives: D. J. Berry, *Spec. Chem.* **3**, 13 (1983); of occurrence of pyridine and derivatives in foods, tobacco and essential oils: G. Vernin, *Perfum. Flavor.* **7**, 23-26 (1982); of use in the study of surface properties of transition metal oxides: M. C. Kung, H. H. Kung, *Catal. Rev. Sci. Eng.* **27**, 425-460 (1985). Review of toxicology and environmental fate: A. Jori *et al.*, *Ecotoxicol. Environ. Saf.* **7**, 251 (1983); and human exposure: *Toxicological Profile for Pyridine* (PB93-110831, 1992) 108 pp.

Colorless liq; characteristic disagreeable odor; sharp taste. d_4^{20} 0.98272. Flash pt, closed cup: 68°F (20°C). mp −41.6°. bp 115.2-115.3°. n_D^{20} 1.50920. *Flammable*. Dipole moment in benzene: 2.26. Forms an azeotropic mixture with 3 mols water, boiling at 92-93°. Volatile with steam. Miscible with water, alcohol, ether, petr ether, oils and many other organic liquids. Good solvent for many organic and inorganic compds. Weak base; forms salts with strong acids. pKa 5.19. pH of 0.2 molar soln in H_2O: 8.5. LD_{50} orally in rats: 1.58 g/kg (Smyth).

Caution: Potential symptoms of overexposure are headache, nervousness, dizziness and insomnia; nausea, anorexia; eye irritation; dermatitis; liver and kidney damage. *See NIOSH Pocket Guide to Chemical Hazards* (DHHS/NIOSH 97-140, 1997) p 272.

USE: As solvent for anhydr mineral salts. Synthetic intermediate in laboratory and industry. Complexing agent.

8082. 3-Pyridineacetic Acid. [501-81-5] Lioxone; 3PAA; Minedil. $C_7H_7NO_2$; mol wt 137.14. C 61.31%, H 5.15%, N 10.21%, O 23.33%. Metabolite of nicotine, *q.v.* Prepn: K. Miescher, H. Kägi, *Helv. Chim. Acta* **24**, 1471 (1941); S. Carboni, *Gazz. Chim. Ital.* **85**, 1194 (1955). Manuf. process: M. Hartmann *et al.*, **US 2408020** (1946 to Ciba). Pharmacology: E. Ginoulhiac, L. T. Tenconi, *Minerva Med.* **51**, 1166 (1960). Clinical evaluation as hypocholesteremic: F. Pupita, *ibid.* 1140. Thermal and photodecarboxylation: F. R. Stermitz, W. H. Huang, *J. Am. Chem. Soc.* **93**, 3427 (1971). Determn in biological fluids by HPLC: G. A. Kyerematen *et al.*, *J. Chromatogr.* **419**, 191 (1987); by thermospray LC-MS: K. T. McManus *et al.*, *J. Chromatogr. Sci.* **28**, 510 (1990).

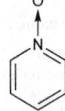

Crystals from ethyl acetate or alcohol, mp 144°.

Note: Often confused with homonicotinic acid, (4-methyl-3-pyridinecarboxylic acid), in early literature.

THERAP CAT: Antilipemic.

8083. Pyridine Bis(oxazoline) Ligands. Pybox ligands. C_2-symmetric ligands that coordinate with metal cations; the resulting tridentate complexes function as chiral catalysts. Structurally related to bis(oxazoline) ligands, *q.v.*, a similar group of compds also used in asymmetric catalysis when bound to metals. Prepn of *i*-Pr-pybox and evaluation of the rhodium complex as a catalyst for asymmetric ketone hydrosilylations: H. Nishiyama *et al.*, *Organometallics* **8**, 846 (1989); and design and prepn of Ph-pybox: *eidem, ibid.* **10**, 5090 (1991). Improved prepn of Ph-pybox: I. W. Davies *et al.*, *J. Org. Chem.* **61**, 9629 (1996). Large scale prepn of *i*-Pr-pybox: M. J. Totleben *et al.*, *ibid.* **66**, 1057 (2001). Synthetic utility of metal-pybox ligand complexes as chiral catalysts for olefin cyclopropanations: H. Nishiyama *et al.*, *J. Am. Chem. Soc.* **116**, 2223 (1994); for 1,3-dipolar cycloadditions: A. I. Sanchez-Blanco *et al.*, *Tetrahedron Lett.* **38**, 7923 (1997); for Mukaiyama aldol reactions: D. A. Evans *et al.*, *J. Am. Chem. Soc.* **121**, 669 (1999); for Diels-Alder reactions: S. Fukuzawa *et al.*, *Synlett* **2001**, 709. *Review*: G. Desimoni *et al.*, *Chem. Rev.* **103**, 3119-3154 (2003).

(S,S)-*i*-Pr-pybox: R = CH(CH₃)₂
(S,S)-Ph-pybox: R = phenyl

(S,S)-*i*-Pr-pybox. [118949-61-4] 2,6-Bis[(4S)-4,5-dihydro-4-(1-methylethyl)-2-oxazolyl]pyridine; 2,6-bis[(4S)-isopropyl-2-oxazolin-2-yl]pyridine. $C_{17}H_{23}N_3O_2$; mol wt 301.39. White needles from hexane + ethyl acetate, mp 152-153°. *Irritant*. $[\alpha]_D^{26}$ −116.8° (c = 1.0 in dichloromethane); $[\alpha]_D^{20}$ −110° (c = 0.70 in dichloromethane).

(R,R)-*i*-Pr-pybox. [131864-67-0] White needles, mp 152-153°. *Irritant*. $[\alpha]_D^{23}$ +118.4° (c = 1.00 in dichloromethane).

(S,S)-Ph-pybox. [174500-20-0] 2,6-Bis[(4S)-4,5-dihydro-4-phenyl-2-oxazolyl]pyridine; 2,6-bis[(4S)-4-phenyl-2-oxazolin-2-yl]pyridine. $C_{23}H_{19}N_3O_2$; mol wt 369.42. Opaque white solid from hexane + ethyl acetate. *Irritant*. $[\alpha]_D^{27}$ −182° (c = 0.91 in dichloromethane).

(R,R)-Ph-pybox. [128249-70-7] White solid, mp 170-172°. *Irritant*. $[\alpha]_D^{26}$ +183.5° (c = 1.03 in dichloromethane).

USE: Ligands in asymmetric synthesis.

8084. Pyridine 1-Oxide. [694-59-7] Pyridine *N*-oxide. C_5H_5NO; mol wt 95.10. C 63.15%, H 5.30%, N 14.73%, O 16.82%. Prepd industrially by heating pyridine in glacial acetic acid at 80° with 30% hydrogen peroxide. The acid is neutralized and the oxide recovered by distillation: Katritzky *et al.*, *J. Chem. Soc.* **1957**, 1769. Laboratory procedure using peracetic acid: H. S. Mosher *et al.*, *Org. Synth. coll. vol. IV*, 828 (1963).

Deliquescent crystals, mp 66°; $bp_{1.0}$ 100-105°.

Hydrochloride. $C_5H_5NO.HCl$. Crystals from isopropanol, mp 179.5-181°.

USE: Synthetic intermediate.

8085. Pyridinium Bromide Perbromide. [39416-48-3] Hydrogen tribromide compd with pyridine (1:1). $C_5H_6Br_3N$; mol wt 319.82. C 18.78%, H 1.89%, Br 74.95%, N 4.38%. $C_5H_5N^+HBr_3^-$. Prepd from pyridine, hydrobromic acid, and bromine: L. F. Fieser, M. Fieser, *Reagents for Organic Chemistry* **vol. 1** (John Wiley, New York, 1967) p 967.

Orange-needles from acetic acid. Odorless. Nonvolatile. Sparingly sol in acetic acid. Dissociates in the presence of a bromine acceptor, such as an alkene, to liberate one mole of bromine.

USE: In small-scale brominations, where it is much more convenient and agreeable to measure and use than elemental bromine.

8086. Pyridinium Chlorochromate. [26299-14-9] (*T*-4)-Chlorotrioxochromate(1−) hydrogen compd with pyridine (1:1:1); PCC. $C_5H_6ClCrNO_3$; mol wt 215.55. C 27.86%, H 2.81%, Cl 16.45%, Cr 24.12%, N 6.50%, O 22.27%. Oxidizing agent for the efficient conversion of primary or secondary alcohols to carbonyl compounds. Prepn: R. J. Meyer, H. Best, *Z. Anorg. Allg. Chem.* **22**, 192 (1899); J. Bernard, M. Camelot, *C. R. Hebd. Seances Acad. Sci.* **258**, 5881 (1964); and use as an oxidant: E. J. Corey, J. W. Suggs, *Tetrahedron Lett.* **1975**, 2647. Improved prepn: S. Agarwal *et al.*, *Tetrahedron* **46**, 4417 (1990). Oxidation, ring enlargement of furans to pyranones: G. Piancatelli *et al.*, *Tetrahedron Lett.* **1977**, 2199; oxidation of tertiary allylic alcohols: W. G. Dauben, D. M. Michno, *J. Org. Chem.* **42**, 682 (1977). Role in oxidative cationic cyclization: E. J. Corey, D. L. Boger, *Tetrahedron Lett.* **1978**, 2461. *Review*: G. Piancatelli *et al.*, *Synthesis* **1982**, 245-258.

Orange-yellow solid, mp 205° (dec). Insol in dichloromethane.
USE: Oxidizing agent in organic synthesis.

8087. Pyridinium Dichromate. [20039-37-6] Chromic acid ($H_2Cr_2O_7$) compound with pyridine (1:2); dipyridinium dichromate; pyridine dichromate; PDC. $C_{10}H_{12}Cr_2N_2O_7$; mol wt 376.21. C 31.93%, H 3.22%, Cr 27.64%, N 7.45%, O 29.77%. Mild and selective oxidizing agent for the efficient conversion of primary and secondary alcohols to carbonyl groups. Original prepn for use as fixing agent: P. Chevalier, *Rev. Cytol. Cytophysiol. Vegetales* **6**, 221 (1943). Prepn and use in oxidation reactions: W. M. Coates, J. R. Corrigan, *Chem. Ind. (London)* **1969**, 1594; J. R. Corrigan, W. M. Coates, US **3734917** (1973 to Mead Johnson); E. J. Corey, G. Schmidt, *Tetrahedron Lett.* **20**, 399 (1979). Oxidation procedure modifications with molecular sieves: J. Herscovici, K. Antonakis, *J. Chem. Soc. Chem. Commun.* **1980**, 561; with bis(trimethylsilyl) peroxide: S. Kanemoto *et al.*, *Tetrahedron Lett.* **24**, 2185 (1983); with acetic acid: S. Czernecki *et al.*, *ibid.* **26**, 1699 (1985); with chlorotrimethylsilane: F. P. Cossío *et al.*, *Can. J. Chem.* **64**, 225 (1986). Kinetic studies: S. Meenakshisundaram, M. Amutha, *Bull. Pol. Acad. Sci. Chem.* **49**, 165 (2001). Crystal structure: A. Lennartson, M. Hakansson, *Acta Crystallogr. C* **65**, m182 (2009).

Stable bright orange solid, mp 144-146°. *Oxidizer, skin sensitizer.* Very sol in water, DMF, DMSO, dimethylacetamide, acetonitrile. Sparingly sol in dichloromethane, chloroform, acetone. Insol in hexane, toluene, diethyl ether, ethyl acetate.
USE: Reagent in synthetic organic chemistry.

8088. Pyridinol Carbamate. [1882-26-4] 2,6-Pyridinedimethanol 2,6-bis(N-methylcarbamate); 2,6-pyridinylenebis[methyl-N-methylcarbamate]; pyricarbate; H-3749; Anginin; Angioxine; Aterosan; Atover; Cicloven; Colesterinex; Duvaline; Movecil; Prodectin; Ravenil; Sospitan; Vasagin; Vasocil; Vasoverin. $C_{11}H_{15}N_3O_4$; mol wt 253.26. C 52.17%, H 5.97%, N 16.59%, O 25.27%. Prepd by the reaction of methylamine with 2,6-pyridinedimethanol bis[phenylcarbonate] in methanol: Matsumoto, **JP 66 22185** (1966 to Banyu), *C.A.* **66**, 75907x (1967). Pharmacology and clinical evaluation: T. Shimamoto *et al.*, *Am. Heart J.* **71**, 297 (1966). Metabolism: Mallein *et al.*, *Therapie* **28**, 115 (1973). Effects on lipid metabolism and antiatherogenic effect in animals: V. Orbetzova *et al.*, *Artery* **8**, 560 (1980). HPLC determn in biological fluids: A. Suenaga *et al.*, *Acta Pharm. Suec.* **23**, 245 (1986).

Needles from methanol or acetone, mp 136-137°. uv max (methanol): 264-265 nm. Sparingly sol in cold water; freely sol in hot water. LD_{50} in mice, rats, rabbits, dogs (mg/kg): 4500, 3400, 5200, 1000 orally (Shimamoto).
THERAP CAT: Antiarteriosclerotic.

8089. Pyridomycin. [18791-21-4] 3-Hydroxy-N-[(2Z,5R,-6S,9S,10S,11R)-10-hydroxy-5,11-dimethyl-2-(1-methylpropylidene)-3,7,12-trioxo-9-(3-pyridinylmethyl)-1,4-dioxa-8-azacyclododec-6-yl]-2-pyridinecarboxamide. $C_{27}H_{32}N_4O_8$; mol wt 540.57. C 59.99%, H 5.97%, N 10.36%, O 23.68%. Antimycobacterial antibiotic substance produced by *Streptomyces albidofuscus* Okami et Umezawa, nov sp., renamed *S. pyridomyceticus*. Isoln: Maeda *et al.*, *J. Antibiot.* **6A**, 140 (1953); K. Yagishita, *ibid.* **7A**, 143 (1954). Degradation studies: Maeda, *ibid.* **10A**, 94 (1957). Structure: Koyama *et al.*, *Tetrahedron Lett.* **1967**, 3587; H. Ogawara *et al.*, *Chem.*

Pharm. Bull. **16**, 679 (1968). Synthetic study: M. Kinoshita, M. Awamura, *Bull. Chem. Soc. Jpn.* **51**, 869 (1978); M. Kinoshita *et al.*, *ibid.* 3595. Production: **JP 54 7048**; **JP 55 1349**; **JP 56 9566** (1954, 1955, 1956 all to Nippon Antibiotic Subst. Sci. Assoc.).

Crystals from ethanol, mp 222°. uv max (ethanol): 303 nm ($E_{1cm}^{1\%}$ 209). Soluble in lower alcohols, ethyl or butyl acetate, benzene, acetone, dioxane, tetrahydrofuran. Practically insol in water.
Hydrochloride. Crystals, mp 194-196°. Freely sol in water. $[\alpha]_D^{16}$ −53.2° (water). pH of aq solns about 2.0. LD_{50} i.p. in mice: >2100 mg/kg (Yagishita).

8090. Pyridostigmine Bromide. [101-26-8] 3-[[(Dimethylamino)carbonyl]oxy]-1-methylpyridinium bromide; 3-hydroxy-1-methylpyridinium bromide dimethylcarbamate; 1-methyl-3-hydroxypyridinium bromide dimethylcarbamate; 3-(dimethylcarbamyloxy)-1-methylpyridinium bromide; Ro-1-5130; Kalymin; Mestinon; Regonol. $C_9H_{13}BrN_2O_2$; mol wt 261.12. C 41.40%, H 5.02%, Br 30.60%, N 10.73%, O 12.25%. Reversible inhibitor of acetylcholinesterase. Prepn: Urban, US **2572579** (1951 to Hoffmann-La Roche). Mechanism of protective effect in soman poisoning: X. Deyi *et al.*, *Fundam. Appl. Toxicol.* **1**, 217 (1981). Evaluation of side effects profile under desert conditions: J. E. Cook *et al.*, *Mil. Med.* **157**, 250 (1992). Comparative clinical trial for reversal of neuromuscular block: L. Gyermek *et al.*, *Br. J. Anaesth.* **74**, 410 (1995). Review of determn methods in biological samples: B. Zhao *et al.*, *J. Pharm. Pharm. Sci.* **9**, 71-81 (2006).

Shiny, hygroscopic crystals from abs ethanol, mp 152-154°. Characteristic, agreeable odor. Freely sol in water, alcohol, chloroform; slightly sol in hexane. Practically insol in ether, acetone, benzene. Aq solns may be sterilized by autoclaving with steam.
THERAP CAT: Cholinergic; in treatment of myasthenia gravis. Reversal agent for neuromuscular blocking drugs. Pre-exposure antidote to chemical warfare agents.
THERAP CAT (VET): In treatment of myasthenia gravis.

8091. Pyridoxal. [66-72-8] 3-Hydroxy-5-(hydroxymethyl)-2-methyl-4-pyridinecarboxaldehyde; 3-hydroxy-5-(hydroxymethyl)-2-methylisonicotinaldehyde; 2-methyl-3-hydroxy-4-formyl-5-hydroxymethylpyridine. $C_8H_9NO_3$; mol wt 167.16. C 57.48%, H 5.43%, N 8.38%, O 28.71%. One of the naturally occurring forms of vitamin B_6; metabolically converted to the active coenzyme, pyridoxal 5-phosphate, *q.v.* Synthesis and structure: Harris *et al.*, *J. Biol. Chem.* **154**, 315 (1944); *J. Am. Chem. Soc.* **66**, 2088 (1944).

Hydrochloride. [65-22-5] $C_8H_9NO_3 \cdot HCl$. Rhombic crystals, mp approx 165° with decompn. Soluble in water (1 g/2 ml); sol in 95% ethanol (1.7 g/100 ml); pH of 1% water soln = 2.65. The water solns are sensitive to heat. uv max: 292.5 nm (E mol 7600). Can be reduced to pyridoxine hydrochloride (mp 206-208°).

Monoethylacetal hydrochloride. 1,3-Dihydro-1-ethoxy-6-methylfuro[3,4-c]pyridin-7-ol hydrochloride. $C_{10}H_{14}ClNO_3$. Crystals from alcohol and ether, mp 142-143°.

8092. Pyridoxal 5-Phosphate. [54-47-7] 3-Hydroxy-2-methyl-5-[(phosphonooxy)methyl]-4-pyridinecarboxaldehyde; pyridoxal 5-monophosphoric acid ester; 3-hydroxy-5-(hydroxymethyl)-2-methylisonicotinaldehyde 5-phosphate; codecarboxylase; PLP; Pyromijin; Vitazechs. $C_8H_{10}NO_6P$; mol wt 247.14. C 38.88%, H 4.08%, N 5.67%, O 38.84%, P 12.53%. Active coenzyme form of vitamin B_6. Prepn by the action of adenosine triphosphate on pyridoxal: Gunsalus et al., J. Biol. Chem. **155**, 685 (1944); by the action of phosphorus oxychloride on pyridoxal in aq soln: Gunsalus et al., ibid. **161**, 743 (1945); Umbreit et al., Arch. Biochem. **7**, 189 (1945); by phosphorylation of pyridoxamine with 100% H_3PO_4 followed by oxidation: Wilson, Harris, J. Am. Chem. Soc. **73**, 4693 (1951). Alternate route: Schorre, US 3124587 (1964 to E. Merck). Isoln in pure form as the oxime and as the O-methyl oxime and structure: Heyl et al., J. Am. Chem. Soc. **73**, 3430 (1951). Activity as a cotransaminase in the synthesis of amino acids: Lichstein et al., J. Biol. Chem. **161**, 311 (1945). Activity of natural and synthetic material: Umbreit et al., loc. cit. Review of biosynthesis: T. B. Fitzpatrick et al., Chembiochem **11**, 1185-1193 (2010).

Colorless in acid soln, bright-yellow in alkaline soln. uv max (alkaline soln): 390 nm (E_m 3.7); in acid soln: 295 nm (E_m 5.1). Gives a negative 2,6-dichloroquinone chlorimide test. On oxidation with H_2O_2 in alkaline solution yields [(2-methyl-3,4-dihydroxy-5-pyridyl)methyl]phosphoric acid.

Oxime. [634-25-3] $C_8H_{11}N_2O_6P$. Crystals, dec 229-230°. Practically insol in water, alcohol, ether.

O-Methyloxime. [20905-71-9] $C_9H_{13}N_2O_6P$. Crystals, dec 212-213°. Practically insol in water, alcohol, ether.

Calcium salt. [29956-24-9] Aderoxal. Yellow precipitate. Practically insol in water, alcohol or ether.

THERAP CAT: Vitamin (enzyme co-factor).

8093. Pyridoxamine. [85-87-0] 4-(Aminomethyl)-5-hydroxy-6-methyl-3-pyridinemethanol; 2-methyl-3-hydroxy-4-amino-methyl-5-hydroxymethylpyridine; pyridoxylamine. $C_8H_{12}N_2O_2$; mol wt 168.20. C 57.13%, H 7.19%, N 16.66%, O 19.02%. One of the naturally occurring forms of vitamin B_6; metabolically converted to the active coenzyme form, pyridoxal 5-phosphate, q.v. Has antioxidant activity; inhibits advanced glycation and lipoxidation reactions. Synthesis and structure: S. A. Harris et al., J. Biol. Chem. **154**, 315 (1944); eidem, J. Am. Chem. Soc. **66**, 2088 (1944). HPLC determn in foods: H. T. V. Do et al., Food Nutr. Res. **56**, 5409 (2012). Review of advanced glycation inhibition and therapeutic potential in diabetic nephropathy: T. O. Metz et al., Arch. Biochem. Biophys. **419**, 41-49 (2003).

Crystals from alc; mp 193-193.5°; sol in alc.

Dihydrochloride. [524-36-7] $C_8H_{12}N_2O_2 \cdot 2HCl$; mol wt 241.11. Platelets, mp 226-227° with dec. Soluble in water (approx 1 g/2 ml); sol in 95% alc (0.65 g/100 ml). Reasonably stable at room temp; shows no decompn in a few days at 60°. Liquefies on exposure to 80% relative humidity. pH of a 1% water soln 2.4. uv max (pH 1.94): 287.5 nm (E mol 91000).

5'-Phosphate. [529-96-4] 4-(Aminomethyl)-5-hydroxy-6-methyl-3-pyridinemethanol 3-(dihydrogen phosphate); 2-methyl-3-hydroxy-4-aminomethyl-5-pyridylmethylphosphoric acid; pyridoxamine phosphate; PMP. $C_8H_{13}N_2O_5P$; mol wt 248.17. Prepn in soln by autoclaving pyridoxal 5-phosphate with glutamic acid: McNutt, Snell, J. Biol. Chem. **182**, 557 (1950); prepn by direct phosphorylation of pyridoxamine in aq soln with phosphorus oxychloride: Heyl et al., J. Am. Chem. Soc. **73**, 3430 (1951).

8094. 4-Pyridoxic Acid. [82-82-6] 3-Hydroxy-5-(hydroxymethyl)-2-methyl-4-pyridinecarboxylic acid; 3-hydroxy-5-(hydroxymethyl)-2-methylisonicotinic acid; 2-methyl-3-hydroxy-4-carboxy-5-hydroxymethylpyridine. $C_8H_9NO_4$; mol wt 183.16. C 52.46%, H 4.95%, N 7.65%, O 34.94%. Occurs in urine. It is the chief metabolic product of pyridoxine, pyridoxal, and pyridoxamine. Isoln from human urine: Singal, Sydenstricker, Science **78**, 545 (1941). Isoln and synthesis: Huff, Perlzweig, J. Biol. Chem. **155**, 345 (1944).

Wedge-shaped crystals, mp 247-248°. Slightly sol in water, alcohol, pyridine. Insoluble in ether and in aq acid soln, but completely sol in aq alkaline soln. Possesses two acidic groups, one a phenolic and the other a carboxyl having pK values of 9.75 and 5.50, respectively. Characteristic blue fluorescence, max at pH 3 to 4. The fluorescence disappears on reduction with hydrosulfite and is restored to the original intensity with H_2O_2. Is adsorbed on zeolite from aq solns at pH 4 to 5 and can be eluted with 25% KCl; butanol extracts of neutral eluates also show characteristic blue fluorescence which is increased by a trace of acetic acid. Stable to boiling with dil alkali ($1N$); but upon being heated with $0.5N$ acid for a few minutes it is converted to the lactone.

Lactone. β-Pyracine. $C_8H_7NO_3$. mp 263-265°, exhibits much stronger fluorescence. Is more easily followed in the course of biochemical investigations.

8095. Pyridoxine. [65-23-6] 5-Hydroxy-6-methyl-3,4-pyridinedimethanol; pyridoxol; 2-methyl-3-hydroxy-4,5-bis(hydroxymethyl)pyridine; 5-hydroxy-6-methyl-3,4-pyridinedicarbinol; 3-hydroxy-4,5-dimethylol-α-picoline. $C_8H_{11}NO_3$; mol wt 169.18. C 56.80%, H 6.55%, N 8.28%, O 28.37%. One of the vitamins of the B_6 complex; see also pyridoxal and pyridoxamine. Present in many foodstuffs; esp good sources are yeast, liver and cereals. Metabolically converted to the active coenzyme form, pyridoxal 5-phosphate, q.v. Isoln: Keresztesy, Stevens, Proc. Soc. Exp. Biol. Med. **38**, 64 (1938); György, J. Am. Chem. Soc. **60**, 983 (1938); Kuhn, Wendt, Ber. **71**, 780, 1118 (1938). Structure: E. T. Stiller et al., J. Am. Chem. Soc. **61**, 1237 (1939); R. Kuhn et al., Ber. **72**, 305 sqq. (1939). Absorption spectrum: E. A. Peterson, H. A. Sober, J. Am. Chem. Soc. **76**, 169 (1954). Synthesis: Harris, Folkers, J. Am. Chem. Soc. **61**, 1242, 1245, 3307 (1939); P. G. Stevens, US 2680743 and US 2734063 (1954, 1956 both to Gen. Aniline); P. I. Pollak, US 2904551, US 3024244, US 3024245 (1959, 1962, 1962 all to Merck & Co.); E. E. Harris et al., J. Org. Chem. **27**, 2705 (1962); W. Böll, H. König, Ann. **1979**, 1657; T. Shono et al., Chem. Lett. **1981**, 1121. Biosynthesis: Hiel, Spenser, Science **169**, 773 (1970). HPLC determn in foods: H. T. V. Do et al., Food Nutr. Res. **56**, 5409 (2012). Review: The Vitamins vol. 2, W. H. Sebrell, R. S. Harris, Eds. (Academic, New York, 2nd ed., 1968) pp 1-117.

Hydrochloride. [58-56-0] Pyridoxinium chloride; Benadon; Bonasanit; Hexavibex; Hexobion; Pyridox. $C_8H_{11}NO_3 \cdot HCl$; mol wt 205.64. Platelets or thick, birefringent rods from alcohol + acetone. Reasonably stable to light and air. Dec 205-212°. Sublimes. (Free base, mp 160°). uv max (0.1N HCl): 290 nm (ε 8400); (phosphate buffer, pH 7): 253, 325 nm (ε 3700, 7100). One gram dissolves in about 4.5 ml water, 90 ml alcohol. Sol in propylene glycol. Sparingly sol in acetone. Insol in ether, chloroform. pH of a 10% w/v soln in water: 3.2. Acidic aq solns are stable and may be heated for 30 min at 120° without decompn.

THERAP CAT: Vitamin (enzyme cofactor).

THERAP CAT (VET): Nutritional factor.

8096. Pyrifenox. [88283-41-4] 1-(2,4-Dichlorophenyl)-2-(3-pyridinyl)ethanone O-methyloxime; 2′,4′-dichloro-2-(3-pyridyl)-acetophenone O-methyloxime; ACR-3453A; Ro-15-1297; Dorado. $C_{14}H_{12}Cl_2N_2O$; mol wt 295.16. C 56.97%, H 4.10%, Cl 24.02%, N 9.49%, O 5.42%. Ergosterol biosynthesis inhibitor for use on crops. Prepn: F. Dorn, DE 3310148; idem, US 4605656 (1983, 1986 both to Hoffmann-La Roche). Review of physical properties, toxicity and bioactivity: P. Zobrist et al., Proc. Br. Crop Prot. Conf. - Pests Dis. **1986**, 47-53; of field trials: G. Neumann et al., Meded. Fac. Landbouwwet. Rijksuniv. Gent **54**, 635-642 (1989).

Slightly viscous, tan liquid with mild aromatic odor. $bp_{0.1}$ >150°. vapor pressure at 25°: 1.4×10^{-5} torr. Soly (at 20°) in water at pH 7: 115 mg/l; in hexane: <1 g/l; in acetone, ethyl acetate, chloroform, diethyl ether, DMF, isopropyl alcohol, toluene: >250 g/l. LD_{50} in rats (mg/kg): 950 i.p.; 2900 orally; >5000 dermally; LC_{50} by inhalation in rats: >2.05 mg/l (Zobrist).

USE: Agricultural fungicide.

8097. Pyrilamine. [91-84-9] N^1-[(4-Methoxyphenyl)methyl]-N^2,N^2-dimethyl-N^1-2-pyridinyl-1,2-ethanediamine; 2-[(2-dimethylaminoethyl)(p-methoxybenzyl)amino]pyridine; N-p-methoxybenzyl-N',N'-dimethyl-N-α-pyridylethylenediamine; mepyramine; pyranisamine; RP-2786. $C_{17}H_{23}N_3O$; mol wt 285.39. C 71.55%, H 8.12%, N 14.72%, O 5.61%. Prepn: Bovet et al., C. R. Seances Soc. Biol. Ses Fil. **138**, 99 (1944); Huttrer et al., J. Am. Chem. Soc. **68**, 1999 (1946); Viaud, Prod. Pharm. **2**, 53 (1947); Horclois, US 2502151 (1950 to Rhône-Poulenc). Toxicity: A. von Schlichtergroll, Arzneim.-Forsch. **7**, 237 (1957); of the maleate: F. Hunziker et al., ibid. **13**, 324 (1963).

Oily liquid. bp_5 201°; $bp_{0.06}$ 168-172°. n_D^{25} 1.5760-1.5765. LD_{50} orally in mice: 312 mg/kg (Schlichtergroll).

Hydrochloride. $C_{17}H_{23}N_3O \cdot HCl$. Crystals, mp 143-143.5°. Very sol in water.

Maleate. [59-33-6] Antamine; Anthisan; Dorantamin; Enrumay; Histalon; Histan; Histapyran; Histatex; Neo-Antergan; Paraminyl; Parmal; Pyramal; Stamine; Stangen; Thylogen. Crystals. Bitter saline taste. Stable in air. mp 100-101°. uv max: 244 nm ($E_{1cm}^{1\%}$ 420). One gram dissolves in about 0.4 ml water, in about 15 ml abs alc. pH of 10% soln ~5.1. On raising the pH to 7.5 or 8.0 pptn of the oily, free base begins. Freely sol in chloroform; slightly sol in benzene, ether. LD_{50} orally in mice: 338 mg/kg (Hunziker).

THERAP CAT: Antihistaminic.

THERAP CAT (VET): Antihistaminic.

8098. Pyrimethamine. [58-14-0] 5-(4-Chlorophenyl)-6-ethyl-2,4-pyrimidinediamine; 2,4-diamino-5-(p-chlorophenyl)-6-ethylpyrimidine; RP-4753; Daraprim; Malocide. $C_{12}H_{13}ClN_4$; mol wt 248.71. C 57.95%, H 5.27%, Cl 14.25%, N 22.53%. Dihydrofolate

reductase inhibitor; generally used in combination with other antimicrobial agents. Prepn: P. B. Russell, G. H. Hitchings, J. Am. Chem. Soc. **73**, 3763 (1951); G. H. Hitchings et al., US 2576939 (1951 to Burroughs Wellcome); W. Logemann et al., Ber. **87**, 435 (1954); R. M. Jacob, US 2680740 (1954 to Rhône-Poulenc). Review of antimicrobial activity and mechanism of action: Burchall in Antibiotics vol. 3, J. W. Corcoran, F. E. Hahn, Eds. (Springer-Verlag, New York, 1975) pp 312-320. Comprehensive description: M. A. Loutfy, H. Y. Aboul-Enein, Anal. Profiles Drug Subs. **12**, 463-482 (1983). LC-MS determn in plasma: B. A. Sinnaeve et al., J. Chromatogr. A **1076**, 97 (2005). Clinical evaluations in toxoplasmosis in AIDS patients: C. Leport et al., Am. J. Med. **84**, 94 (1988); B. Dannemann et al., Ann. Intern. Med. **116**, 33 (1992). Review of clinical experience in malaria: H. M. McIntosh et al., Ann. Trop. Med. Parasitol. **93**, 265-270 (1998); C. V. Plowe et al., Br. Med. J. **328**, 545 (2004).

Crystals, mp 233-234° (capillary); mp 240-242° (copper block). Sol in boiling ethanol (about 25 g/l); slightly sol in acetone, chloroform; also slightly sol in ethanol, (about 9 g/l), in dil HCl (about 5 g/l). Very sparingly sol in propylene glycol and dimethylacetamide at 70°. Practically insol in water.

Combination with sulfadoxine. [37338-39-9] Fansidar.

THERAP CAT: Antiprotozoal (Toxoplasma); antimalarial.

THERAP CAT (VET): Antiprotozoal (Toxoplasma).

8099. Pyrimethanil. [53112-28-0] 4,6-Dimethyl-N-phenyl-2-pyrimidinamine; 2-anilino-4,6-dimethylpyrimidine; N-(4,6-dimethylpyrimidin-2-yl)aniline; Scala. $C_{12}H_{13}N_3$; mol wt 199.26. C 72.33%, H 6.58%, N 21.09%. Anilino-pyrimidine fungicide for control of Botrytis and Venturia spp. in food crops; inhibits secretion of cell wall degrading enzymes required for fungal infection. Prepn: S. Angerstein, Ber. **34**, 3956 (1901). Mode of action studies: A. Daniels et al., Brighton Crop Prot. Conf. - Pests Dis. **1994**, 525; A. Daniels, J. A. Lucas, Pestic. Sci. **45**, 33 (1995); R. J. Milling, C. J. Richardson, ibid. 43. Degradation study: A. Agüera et al., Environ. Sci. Technol. **34**, 1563 (2000). Determn in soil by headspace solid-phase microextraction and GC/MS: A. Navalón et al., Anal. Bioanal. Chem. **379**, 1100 (2004). Comprehensive review: G. L. Neumann et al., Brighton Crop Prot. Conf. - Pests Dis. **1992**, 395-402.

White crystalline solid, mp 96.3°. Soly in water (25°): 0.121 g/l. Sol in most organic solvents. Log P (n-octanol/water): 2.48. Vapor pressure (25°): 2.2×10^{-3} Pa. LD_{50} (mg/kg): 4061-5358 orally in mice; 4150-5971 orally in rats; >5000 dermally in rats. LC_{50} (96 hr) in mirror carp, rainbow trout (mg/l): 35.36, 10.56 (Neumann).

USE: Fungicide.

8100. Pyrimidine. [289-95-2] 1,3-Diazine; metadiazine; miazine. $C_4H_4N_2$; mol wt 80.09. C 59.99%, H 5.03%, N 34.98%. Prepd by reducing 2,4,6-trichloropyrimidine with zinc dust: Gabriel, Ber. **33**, 3666 (1900); by reducing tetrachloropyrimidine: Emery, ibid. **34**, 4180 (1901); by reducing 2,4-dichloropyrimidine with magnesium oxide and palladium-charcoal: Wittaker, J. Chem. Soc. **1953**, 1646.

Liquid or cryst mass; penetrating odor. mp 20-22°. bp$_{762}$ 123-124°. uv max (water): 240 nm (ε 2400). Soluble in water, alcohol, ether.

Methiodide. $C_4H_4N_2 \cdot CH_3I$. Yellow plates from ethanol, mp 136-137°.

8101. Pyriminil. [53558-25-1] *N*-(4-Nitrophenyl)-*N*'-(3-pyridinylmethyl)urea; *N*-3-pyridylmethyl-*N*'-*p*-nitrophenylurea; pyrinuron; RH-787; DLP-787; Vacor. $C_{13}H_{12}N_4O_3$; mol wt 272.26. C 57.35%, H 4.44%, N 20.58%, O 17.63%. Prepn: J. E. Ware *et al.*, **DE 2409686** (1974 to Rohm & Haas). Evaluation as rodenticide: D. L. Peardon, *Pest Control* **42**, 14, 16, 18, 27 (1974); J. E. Brooks, P. T. Htun, *J. Hyg.* **80**, 401 (1978). Mode of action: D. L. Peardon, J. E. Ware, *Pest Control* **45**, 49 (1977).

mp 223-225° (dec). LD$_{50}$ orally in male, female rats: 6.2, 7.2 mg/kg (Brooks, Htun).

USE: Rodenticide.

8102. Pyrinoline. [1740-22-3] α-[3-(Di-2-pyridinylmethylene)-2,4-cyclopentadien-1-yl]-α-2-pyridinyl-2-pyridinemethanol; 3-(di-2-pyridylmethylene)-α,α-di-2-pyridyl-1,4-cyclopentadiene-1-methanol; di-2-pyridyl-(6,6-di-2-pyridylfulven-2-yl)methanol; McN-1210; Surexin. $C_{27}H_{20}N_4O$; mol wt 416.48. C 77.87%, H 4.84%, N 13.45%, O 3.84%. Prepn: **GB 1009012** (1965 to McNeil).

Crystals, mp 146.5-147.5°.

THERAP CAT: Antiarrhythmic (cardiac depressant).

8103. Pyriproxyfen. [95737-68-1] 2-[1-Methyl-2-(4-phenoxyphenoxy)ethoxy]pyridine; 4-phenoxyphenyl (*RS*)-2-(2-pyridyloxy)propyl ether; S-9318; S-31183; Sumilarv. $C_{20}H_{19}NO_3$; mol wt 321.38. C 74.75%, H 5.96%, N 4.36%, O 14.93%. Juvenile hormone mimic. Prepn: S. Nishida *et al.*, **GB 2140010**; *eidem*, **US 4751225** (1984, 1988 both to Sumitomo). Ovicidal activity: K. R. S. Ascher, M. Eliyahu, *Phytoparasitica* **16**, 15 (1988). Environmental persistence study: C. H. Schaefer *et al.*, *Ecotoxicol. Environ. Saf.* **21**, 207 (1991). Field trial to control tsetse flies: J. W. Hargrove, P. A. Langley, *Bull. Entomol. Res.* **83**, 361 (1993).

Pale yellow liquid. $n_D^{20.5}$ 1.5823. Solidifies upon standing to give crystals, mp 49.7°.

USE: Insecticide.

THERAP CAT (VET): Ectoparasiticide.

8104. Pyrisuccideanol. [33605-94-6] Butanedioic acid 2-(dimethylamino)ethyl [5-hydroxy-4-(hydroxymethyl)-6-methyl-3-pyridinyl]methyl ester; pyridoxal 3-[2-(dimethylamino)ethyl succinate]; 2-(dimethylamino)ethyl [5-hydroxy-4-(hydroxymethyl)-6-methyl-3-pyridyl]succinic acid methyl ester; pirisudanol. $C_{16}H_{24}N_2O_6$; mol wt 340.38. C 56.46%, H 7.11%, N 8.23%, O 28.20%. Prepn: Esanu, **DE 2102831** (1971 to Soc. d'Etudes de Prods. Chim.) corresp to **US 3717636** (1973). Activity studies: Hugelin *et al.*, *C. R. Seances Soc. Biol. Ses Fil.* **166**, 1435 (1972).

Dimaleate. [53659-00-0] Mentium; Nadex; Stivane. $C_{16}H_{24}N_2O_6 \cdot 2C_4H_4O_4$; mol wt 572.52. mp 134°.

THERAP CAT: Antidepressant.

8105. Pyrithiamine. [534-64-5] 1-[(4-Amino-2-methyl-5-pyrimidinyl)methyl]-3-(2-hydroxyethyl)-2-methylpyridinium bromide hydrobromide (1:1:1); 1-(2-methyl-4-amino-5-pyrimidyl)methyl-2-methyl-3-hydroxyethylpyridinium bromide hydrobromide; neopyrithiamine. $C_{14}H_{20}Br_2N_4O$; mol wt 420.15. C 40.02%, H 4.80%, Br 38.04%, N 13.34%, O 3.81%. Thiamine antagonist prepd by condensation of 2-methyl-3-(β-hydroxyethyl)pyridine with the pyrimidine moiety of vitamin B$_1$: Tracy, Elderfield, *J. Org. Chem.* **6**, 54 (1941); improved procedure: Wilson, Harris, *J. Am. Chem. Soc.* **71**, 2231 (1949); **US 2587262** (1952 to Merck & Co.). *See also:* Woolley, *J. Am. Chem. Soc.* **72**, 5763 (1950).

Crystals from acetone, dec 218-220°. uv max (water): 238, 271 nm. Sol in water.

Chloride hydrochloride. $C_{14}H_{20}Cl_2N_4O$. Dec 234-236°. uv max (water): 235, 273 nm. Sol in water.

8106. Pyrithiobac. [123342-93-8] 2-Chloro-6-[(4,6-dimethoxy-2-pyrimidinyl)thio]benzoic acid; KIH-8921. $C_{13}H_{11}ClN_2O_4S$; mol wt 326.75. C 47.79%, H 3.39%, Cl 10.85%, N 8.57%, O 19.59%, S 9.81%. Pyrimidinylsalicylic acid herbicide for use in cotton; acetolactate synthase inhibitor. Prepn: Y. Saito *et al.*, **EP 315889**; *eidem*, **US 4923501** (1989, 1990 both to Kumiai Chem. Ind.; Ihara Chem. Ind.). Field trials in cotton: J. W. Keeling *et al.* *Weed Technol.* **7**, 930 (1993). Mechanism of action: T. Shimizu *et al.*, *J. Pestic. Sci.* **19**, 59 (1994). Comprehensive description: S. Takahashi *et al.*, *Brighton Crop Prot. Conf. - Weeds* **1991**, 57-62.

Ochre powder, mp 148-151°.

Sodium salt. [123343-16-8] KIH-2031; DPX-PE350; Staple. $C_{13}H_{10}ClN_2NaO_4S$; mol wt 348.73. White solid, mp 247.7° (dec). Soly in water (10°): 760 g/l. LD$_{50}$ in male, female rats (mg/kg): 1000-3000, 3000-5000 orally. LD$_{50}$ in rats (mg/kg): >2000 dermally (Takahashi).

USE: Herbicide.

8107. Pyrithione. [1121-30-8] 1-Hydroxy-2(1*H*)-pyridinethione; 2-pyridinethiol 1-oxide; 2-mercaptopyridine 1-oxide; PTO; Omadine. C_5H_5NOS; mol wt 127.16. C 47.23%, H 3.96%, N 11.02%, O 12.58%, S 25.21%. Prepn: Shaw *et al.*, *J. Am. Chem. Soc.* **72**, 4362 (1950); Semenoff, Dolliver, **US 2745826** (1956 to Olin Mathieson).

Sodium salt. Fonderma. C_5H_4NNaOS; mol wt 149.14.

Zinc derivative. [13463-41-7] Zinc pyrithione; zinc pyridinethione; bis(2-pyridylthio)zinc 1,1'-dioxide; bis-(1-hydroxy-2(1H)-pyridinethionato-O,S)zinc; Desquaman. $C_{10}H_8N_2O_2S_2Zn$; mol wt 317.72. Prepn: **GB 761171** (1956 to Olin Mathieson). Activity: Karsten *et al.*, **US 3236733**; Judge *et al.*, **US 3281366** (both 1966 to Procter and Gamble).

Dimer. Dipyrithione; 2,2'-dithiobispyridine 1,1'-dioxide; OMDS; Omadine Disulfide. $C_{10}H_8N_2O_2S_2$; mol wt 252.31.

USE: Fungicide, bactericide.

THERAP CAT: Antibacterial; antifungal. Zinc deriv also as antiseborrheic.

8108. Pyrithyldione. [77-04-3] 3,3-Diethyl-2,4-(1H,3H)pyridinedione; 3,3-diethyl-2,4-dioxotetrahydropyridine; 2,4-dioxo-3,3-diethyltetrahydropyridine; 2,4-dioxo-3,3-diethyl-1,2,3,4-tetrahydropyridine; Presidon; Persedon; Tetridin; Benedorm. $C_9H_{13}NO_2$; mol wt 167.21. C 64.65%, H 7.84%, N 8.38%, O 19.14%. Prepn: Schnider, *Festschr. Emil Barell* **1936**, 195; Preiswerk, Schnider, **US 2090068** (1937 to Hoffmann-La Roche); Strukov *et al.*, *Med. Prom. SSSR* **13**, no. 9, 9 (1959). Alternate procedure: Rechen *et al.*, **US 3019230** (1962 to Hoffmann-La Roche).

Exists in three cryst modifications: I, mp 92-93°; II, mp 97-98°; III, mp 81-86°. *See:* Scheibl, Wachter, *Arch. Toxikol.* **18**, 253 (1960). bp$_{14}$ 187-189°. Moderately sol in water; freely sol in the usual organic solvents except petr solvents. Aq solns are just acid to litmus.

THERAP CAT: Sedative, hypnotic.

8109. Pyritinol. [1098-97-1] 3,3'-[Dithiobis(methylene)]-bis[5-hydroxy-6-methyl-4-pyridinemethanol]; bis(4-hydroxymethyl-5-hydroxy-6-methyl-3-pyridylmethyl) disulfide; bis[(3-hydroxy-4-hydroxymethyl-2-methyl-5-pyridyl)methyl] disulfide; dipyridoxolyldisulfide; pyridoxine-5-disulfide; pyrithioxin. $C_{16}H_{20}N_2O_4S_2$; mol wt 368.47. C 52.16%, H 5.47%, N 7.60%, O 17.37%, S 17.40%. Prepn: Zima, Schorre, **US 3010966** (1961 to E. Merck); Iwanami *et al.*, *Bitamin* **36**, 122 (1967); *J. Vitaminol.* **14**, 321, 326 (1968). HPLC determn in urine: K. Kitao *et al.*, *Chem. Pharm. Bull.* **25**, 1335 (1977). Pharmacokinetics and metabolism: Darge *et al.*, *Arzneim.-Forsch.* **19**, 5, 9, (1969); Nowak, Schorre, *ibid.* 11. Clinical trial in dementia: S. Hoyer *et al.*, *ibid.* **27**, 671 (1977); A. J. Cooper, R. V. Magnus, *Pharmatherapeutica* **2**, 317 (1980); in cerebrovascular disorders: Y. Tazaki *et al.*, *J. Int. Med. Res.* **8**, 118 (1980).

Crystals, mp 218-220°.

Dihydrochloride monohydrate. Biocefalin; Bonifen; Enbol; Encephabol; Enerbol; Epocan; Life. $C_{16}H_{20}N_2O_4S_2.2HCl.H_2O$; mol wt 459.40. mp 184°. *Note:* Has no vitamin B$_6$ activity.

THERAP CAT: Nootropic.

8110. Pyrocalciferol. [128-27-8] (3β,10α,22E)-Ergosta-5,-7,22-trien-3-ol; 9α-lumisterol; 9α-lumista-5,7,22-trien-3β-ol. $C_{28}H_{44}O$; mol wt 396.66. C 84.78%, H 11.18%, O 4.03%. Differs from ergosterol in the steric configuration at C-9 and C-10. Formation via isopyrocalciferol: Askew *et al.*, *Proc. R. Soc. London* **B109**, 488 (1932); Windaus *et al.*, *Nachr. Ges. Wiss. Goettingen Math.-Phys. Kl.* **1932**, 150. Dehydrogenation with selenium gives Diel's hydrocarbon (3'-methyl-1,2-cyclopentenophenanthrene). Fieser, Fieser, *Steroids* (New York, 1959) pp 136-143, argue the configuration at C-9 and C-10. *See:* Castells *et al.*, *J. Chem. Soc.* **1959**, 1159. Crit-

ical examination and comparison with other unnatural steroids: Castells *et al.*, *ibid.* **1962**, 2907.

Needles from methanol, mp 93-95°. $[\alpha]_D^{20}$ +512°, $[\alpha]_{546}^{20}$ +624° (c = 0.15 in alcohol). uv max: 274, 294 nm (Askew).

Acetate. $C_{30}H_{46}O_2$. mp 81-82°, $[\alpha]_D$ +403° (chloroform).

3,5-Dinitrobenzoate. $C_{35}H_{46}N_2O_6$. mp 168-170°. $[\alpha]_D^{21}$ +195°, $[\alpha]_{546}^{20}$ +249° (both c = 2 in chloroform).

8111. Pyrocatechol. [120-80-9] 1,2-Benzenediol; pyrocatechin; catechol; 1,2-dihydroxybenzene. $C_6H_6O_2$; mol wt 110.11. C 65.45%, H 5.49%, O 29.06%. Prepd by treating salicylaldehyde with hydrogen peroxide, or from its monomethyl ether (guaiacol) by treatment with hydrobromic acid: Dakin, *Org. Synth.* **coll. vol. I**, 149 (2nd ed., 1941). Toxicity data: A. J. Lehman *et al.*, *Adv. Food Res.* **3**, 197 (1951). Carcinogenicity study: M. Hirose *et al.*, *Carcinogenesis* **14**, 525 (1993). *Review:* J. Varagnat, "Hydroquinone, Resorcinol, and Catechol", in *Kirk-Othmer Encyclopedia of Chemical Technology* **vol. 13** (Wiley-Interscience, New York, 3rd ed., 1981) pp 39-69.

Monoclinic tablets, prisms from toluene. Discolors in air and light. d 1.344; mp 105°; bp$_{760}$ 245.5°; bp$_{400}$ 221.5°; bp$_{200}$ 197.7°; bp$_{100}$ 176°; bp$_{60}$ 161.7°; bp$_{40}$ 150.6°; bp$_{20}$ 134°; bp$_{10}$ 118.3°; bp$_5$ 104°. Sublimes. Volatile with steam. pK (18°) 9.48. Sol in 2.3 parts water, in alcohol, benzene, chloroform, ether; very sol in pyridine, aq alkalies. Its aq solns soon turn brown. LD$_{50}$ in mice (mg/kg): 260 orally; 190 i.p. (Lehman).

Note: Catechol also refers to catechin, *q.v.*

Caution: Potential symptoms of overexposure are irritation of eyes, skin, respiratory system; lacrimation, eye burns; convulsions, increased blood pressure. Direct contact may cause skin sensitization and dermatitis. Systemic effects similar to phenol, *q.v. See NIOSH Pocket Guide to Chemical Hazards* (DHHS/NIOSH 97-140, 1997) p 56; *Patty's Industrial Hygiene and Toxicology* **vol. 2B**, G. D. Clayton, F. E. Clayton, Eds. (Wiley-Interscience, New York, 4th ed., 1994) pp 1584-1586.

USE: In photography; dyeing fur; as reagent.

THERAP CAT: Antiseptic.

8112. Pyrogallol. [87-66-1] 1,2,3-Benzenetriol; 1,2,3-trihydroxybenzene; pyrogallic acid. $C_6H_6O_3$; mol wt 126.11. C 57.15%, H 4.80%, O 38.06%. Observed by Scheele in 1786; prepd by Braconot in 1818. Prepn from gallic acid: Marsh, **GB 144897** (1919); Rinderknecht, Niemann, *J. Am. Chem. Soc.* **70**, 2605 (1948); from *p-tert*-butylphenol: Stevens, **US 2603662** (1952 to Gulf). Synthesis from aliphatic sources: Shipchandler *et al.*, *J. Chem. Soc. Perkin Trans. 1* **1975**, 1400. Isoln from *Penicillium patulum:* Tanenbaum, Bassett, *Biochim. Biophys. Acta* **28**, 21 (1958). Acute toxicity: J. W. Dollahite *et al.*, *Am. J. Vet. Res.* **23**, 1264 (1962).

White, odorless crystals, mp 131-133°. Becomes grayish on exposure to air and light. *Poisonous.* d 1.45; bp 309°. Sublimes when slowly heated. One gram dissolves in 1.7 ml water, 1.3 ml alc, 1.6 ml ether; slightly sol in benzene, chloroform, carbon disulfide. The aq soln darkens on exposure to air, quite rapidly when alkaline. *Keep well closed and protected from light. Incompat.* Alkalies, ammonium hydroxide, antipyrine, camphor, phenol, menthol. LD_{50} orally in rabbits: 1.6 g/kg (Dollahite).

Monoacetate. [1330-51-4] Eugallol. $C_8H_8O_4$; mol wt 168.15. White or brownish liquid, bp_{23} ~185°. Sol in water, alcohol, chloroform, ether, acetone and castor oil. Marketed as a 67% soln in acetone.

Triacetate. [525-52-0] Acetpyrogall; Lenigallol. $C_{12}H_{12}O_6$; mol wt 252.22. White, crystalline powder, mp 165°. Slightly sol in water; sol in alcohol. Dec by alkali hydroxide solns.

Caution: Potential symptoms of overexposure are vomiting, hypothermia, fine tremors, weakness, muscle incoordination, diarrhea, loss of reflexes, coma, asphyxia. Direct contact may cause skin irritation and sensitization. *See Patty's Industrial Hygiene and Toxicology* **Vol. 2B,** G. D. Clayton, F. E. Clayton, Eds. (Wiley Interscience, New York, 4th ed., 1994) p 1595-1597.

USE: Developer in photography; making colloidal solns of metals; as mordant for wool, staining leather, process engraving; manuf various dyes; dyeing furs, hair, etc. In analytical chemistry as a complexing agent; reducing agent; alkaline soln as an indicator of gaseous oxygen.

8113. L-Pyroglutamic Acid. [98-79-3] 5-Oxo-L-proline; 5-oxo-2-pyrrolidinecarboxylic acid; 2-pyrrolidone-5-carboxylic acid; glutimic acid; glutiminic acid; α-aminoglutaric acid lactam; glutamic acid lactam; pyroGlu. $C_5H_7NO_3$; mol wt 129.12. C 46.51%, H 5.46%, N 10.85%, O 37.17%. Cyclized internal amide of L-glutamic acid found in vegetables, fruits, grasses, and molasses. Easily prepd from L-glutamic acid by autoclaving with an equal wt of water at 135-140°: Dearborn, Stekol, **US 2528267**; purification: Blish, **US 2738353** (1950, 1956, both to International Minerals and Chem.). *Review:* Orlowski, Meister, in *The Enzymes* vol. 4, P. D. Boyer, Ed. (Academic Press, New York, 3rd ed., 1971) pp 123-151; C. Moret, M. Briley, *Trends Pharmacol. Sci.* **9,** 278-279 (1988).

Orthorhombic bisphenoidal crystals from alcohol + petr ether. mp 162-163°. $[\alpha]_D^{20}$ −11.9° (c = 2); $[\alpha]_D^{25}$ −23.6° (c = 5 at pH 7). Sol in water, alcohol, acetone.

USE: In the resolution of racemic amines.

8114. Pyrolan. [87-47-8] *N,N*-Dimethylcarbamic acid 3-methyl-1-phenyl-1*H*-pyrazol-5-yl ester; 1-phenyl-3-methyl-5-pyrazolyl dimethylcarbamate; 3-methyl-1-phenyl-5-pyrazolyl dimethylcarbamate; ENT-17588; G-22008. $C_{13}H_{15}N_3O_2$; mol wt 245.28. C 63.66%, H 6.16%, N 17.13%, O 13.05%. Prepn from 1-phenyl-3-methyl-5-pyrazolone and dimethylcarbamoyl chloride: **CH 279553** (1952 to Geigy); **GB 681376**. Anticholinesterase activity: Pulver, Domenjoz, *Experientia* **7,** 306 (1951); Ferguson, Alexander, *J. Agric. Food Chem.* **1,** 888 (1953); Müller, Spindler, *Experientia* **10,** 91 (1954). Toxicity study: E. W. Schafer, *Toxicol. Appl. Pharmacol.* **21,** 315 (1972).

Crystals, mp 50°. $bp_{0.2}$ 160-162°. Possesses water and lipid solubility. LD_{50} orally in rats: 62 mg/kg (Schafer).
Caution: Cholinesterase inhibitor.
USE: Insecticide.

8115. Pyroligneous Acids. [8030-97-5] Wood vinegar; pyroligneous vinegar. Obtained by destructive distillation of wood.

Yellowish, acid liquid; empyreumatic odor; contains about 6% acetic acid. Miscible with water, alcohol.
USE: Largely for smoking meats.

8116. Pyromellitic Acid. [89-05-4] 1,2,4,5-Benzenetetracarboxylic acid. $C_{10}H_6O_8$; mol wt 254.15. C 47.26%, H 2.38%, O 50.36%. Prepd by heating mellitic (benzenehexacarboxylic) acid with $KHSO_4$ and H_2SO_4: Silberrad, *J. Chem. Soc.* **89,** 1795 (1906) (actually the dianhydride, mp 286°, is obtained by this method). Laboratory prepn from pine or spruce charcoal: Philippi, Thelen, *Org. Synth.* coll. vol. **II,** 551 (1943); by oxidation of benzene derivatives contg substituents in the 1,2,4,5 positions: Jacobsen, *Ber.* **17,** 2516 (1884).

Dihydrate. Triclinic plates from water. When anhydr, mp 276°. Distills with anhydride formation. 1.5 g (anhydr) dissolves in 100 ml water. Freely sol in alc.

Tetramethyl ester. $C_{14}H_{14}O_8$. Leaflets from alc, mp 141.5°.

8117. Pyronaridine. [74847-35-1] 4-[(7-Chloro-2-methoxy-benzo[*b*]-1,5-naphthyridin-10-yl)amino]-2,6-bis(1-pyrrolidinyl-methyl)phenol; malaridine. $C_{29}H_{32}ClN_5O_2$; mol wt 518.06. C 67.24%, H 6.23%, Cl 6.84%, N 13.52%, O 6.18%. Prepn: X.-Y. Zheng *et al.*, *Yao Hsueh Hsueh Pao* **14,** 736 (1979), *C.A.* **93,** 132397 (1980). HPLC determn in blood and urine: S. A. Wages *et al.*, *J. Chromatogr.* **527,** 115 (1990). *In vitro* activity vs drug resistant malaria: E. I. Elueze *et al.*, *J. Antimicrob. Chemother.* **37,** 511 (1996). Clinical trial: P. Ringwald *et al.*, *Lancet* **347,** 24 (1996). *Reviews:* B. Shao, *Chin. Med. J.* **103,** 428-434 (1990); C. Chang *et al.*, *Trans. R. Soc. Trop. Med. Hyg.* **86,** 7-10 (1992).

Tetraphosphate. [76748-86-2] $C_{29}H_{32}ClN_5O_2 \cdot 4H_3PO_4$. Hygroscopic, yellow needles, mp 233-236° (dec). Odorless; slightly bitter taste. Sol in water; very sparingly sol in ethanol. Insol in choroform, ether, and other organic solvents. LD_{50} in mice (mg/kg): 251 ±33 i.m.; 1368 ±239 orally (Shao).

THERAP CAT: Antimalarial.

8118. Pyronine B. [2150-48-3] 3,6-Bis(diethylamino)xanthylium chloride (1:1); [6-(diethylamino)-3*H*-xanthen-3-ylidene]diethylammonium chloride; C.I. 45010. $C_{21}H_{27}ClN_2O$; mol wt 358.91. C 70.28%, H 7.58%, Cl 9.88%, N 7.81%, O 4.46%. Prepn of metallic complex: **DE 54190** (1889 to Bayer); **DE 59003** (1891 to A. Leonhardt); Biehringer, *Ber.* **27,** 3299 (1894); *J. Prakt. Chem.* **54,** 217 (1896); Albert, *J. Chem. Soc.* **1947,** 244. Structure: Chamberlin *et al.*, *J. Org. Chem.* **27,** 2263 (1962).

Ferric chloride complex. Green metallic needles, mp 176-178°. Absorption max (50% ethanol): 555 nm ($E_{1cm}^{1\%}$ 2324).

USE: Stain for bacteria, molds, ribonucleic acids.

8119. Pyronine Y. [92-32-0] 3,6-Bis(dimethylamino)xanthylium chloride (1:1); *N*-[6-(dimethylamino)-3*H*-xanthene-3-ylidene]-*N*-methylmethanaminium chloride; [6-(dimethylamino)-3*H*-xanthen-3-ylidene]dimethylammonium chloride; tetramethyldiaminoxanthylium chloride; C.I. 45005; pyronine G. $C_{17}H_{19}ClN_2O$; mol wt 302.80. C 67.43%, H 6.32%, Cl 11.71%, N 9.25%, O 5.28%. Prepn from dimethyl-*m*-aminophenol: Mohlau, Koch, *Ber.* **27**, 2887 (1894); **DE 58955**; **DE 59003**; **DE 63081** (1892 to Leonhardt); *Frdl.* **3**, 92-94. *See also: Colour Index* **vol. 4** (3rd ed., 1971) p 4417.

Ferric chloride complex. Lustrous green crystals from alc. Soly in water at 26°: 8.96% giving a red soln. Soly in alc at 26°: 0.60%. Solns show a yellowish fluorescence. Absorption max about 552 nm; curves; *see:* Stotz *et al.*, *Stain Technol.* **25**, 57 (1950).

USE: Bacterial and biological stain.

8120. Pyrophosphoric Acid. [2466-09-3] Diphosphoric acid. $H_4O_7P_2$; mol wt 177.97. H 2.27%, O 62.93%, P 34.81%. Prepd according to the equation $5H_3PO_4 + POCl_3 \rightarrow 3H_4P_2O_7 + 3HCl$: Geuther, *J. Prakt. Chem.* [2] **8**, 359 (1874); Partington, Wallsom, *Chem. News* **136**, 97 (1928); by heating H_3PO_4: Bell, *Ind. Eng. Chem.* **40**, 1464 (1948).

Hygroscopic glass, seldom acicular crystals, mp 61°. K_1 at 18° = 0.14; $K_2 = 0.011$; $K_3 = 2.1 \times 10^{-7}$; $K_4 = 4.1 \times 10^{-10}$. Forms normal salts, such as $Na_4P_2O_7$, and dihydrogen salts, such as $Na_2H_2P_2O_7$. Solubility in water (23°): 709 g/100 ml. Quickly converted to phosphoric acid when dissolved in hot water. Sol in alcohol and ether.

8121. Pyrosulfuric Acid. [7783-05-3] Disulfuric acid. $H_2O_7S_2$; mol wt 178.13. H 1.13%, O 62.87%, S 36.00%. $H_2S_2O_7$.

Colorless to slightly yellow, very hygroscopic crystals, fuming strongly in air. d 1.89; mp 35°. Very sol in water with violent hissing and liberation of much heat. *Keep tightly closed and handle with caution.*

8122. Pyrovalerone. [3563-49-3] 1-(4-Methylphenyl)-2-(1-pyrrolidinyl)-1-pentanone; 4′-methyl-2-(1-pyrrolidinyl)valerophenone; α-pyrrolidino-*p*-methylvalerophenone; 1-(1-pyrrolidinyl)butyl *p*-tolyl ketone; 1-(*p*-tolyl)-1-oxo-2-pyrrolidino-*n*-pentane; 1-(*p*-tolyl)-2-pyrrolidino-1-pentanone. $C_{16}H_{23}NO$; mol wt 245.37. C 78.32%, H 9.45%, N 5.71%, O 6.52%. Prepn: **GB 927475** (1963 to Wander); **GB 933507**; E. Seeger, **US 3314970** (1963, 1967 both to Thomae); Heffe, *Helv. Chim. Acta* **47**, 1289 (1964). Pharmacology: Stille *et al.*, *Arzneim.-Forsch.* **13**, 871 (1963). Metabolism: Michaelis *et al.*, *J. Med. Chem.* **13**, 497 (1970).

$bp_{0.08}$ 104°.

Hydrochloride. [1147-62-2] F-1983; Centroton; Thymergix. $C_{16}H_{23}NO.HCl$; mol wt 281.82. Crystals from 2-butanone or from methanol + acetone + diethyl ether, mp 178°. LD_{50} orally in mice: 350 mg/kg (Stille).

Note: This is a controlled substance (stimulant): **21 CFR**, 1308.15.

THERAP CAT: CNS stimulant.

8123. Pyroxasulfone. [447399-55-5] 3-[[[5-(Difluoromethoxy)-1-methyl-3-(trifluoromethyl)-1*H*-pyrazol-4-yl]methyl]sulfonyl]-4,5-dihydro-5,5-dimethylisoxazole; KIH-485; Sakura. $C_{12}H_{14}F_5N_3O_4S$; mol wt 391.31. C 36.83%, H 3.61%, F 24.28%, N 10.74%, O 16.35%, S 8.19%. Weed seedling growth inhibitor that reduces very long chain fatty acid (VLCFA) biosynthesis. Prepn: M. Nakatani *et al.*, **EP 1364946**; *eidem*, **US 7238689** (2003, 2007 both to Kumiai Chem. and Ihara Chem.). Tolerance of corn hybrids: S. R. Sikkema *et al.*, *HortScience* **43**, 170 (2008). Mechanism of action: Y. Tanetani *et al.*, *Pestic. Biochem. Physiol.* **95**, 47 (2009). Field studies: S. Z. Knezevic *et al.*, *Weed Technol.* **23**, 34 (2009).

White powder, mp 129-130°.

USE: Herbicide.

8124. Pyroxsulam. [422556-08-9] *N*-(5,7-Dimethoxy[1,2,4]triazolo[1,5-*a*]pyrimidin-2-yl)-2-methoxy-4-(trifluoromethyl)-3-pyridinesulfonamide; XDE-742; Admitt; Powerflex. $C_{14}H_{13}F_3N_6O_5S$; mol wt 434.35. C 38.71%, H 3.02%, F 13.12%, N 19.35%, O 18.42%, S 7.38%. Post-emergent herbicide for control of weeds in wheat and other cereal crops. Acetolactate synthase (ALS) inhibitor. Prepn: T. C. Johnson *et al.*, **WO 0236595**; *eidem*, **US 6559101** (2002, 2003 both to Dow AgroSci.); B. M. Bell *et al.*, *Org. Process Res. Dev.* **10**, 1167 (2006). Large scale process: M. A. Gonzalez *et al.*, *ibid.* **12**, 301 (2008). Review of activity and environmental fate: J. Becker *et al.*, *J. Plant Dis. Protect.* **XXI**, 623-628 (2008). Comprehensive description: *Pyroxsulam Pesticide Fact Sheet* (U.S. EPA, February, 2008) 55 pp.

White needles from methanol/acetonitrile. mp 208.3°. d^{20} 1.62. pKa 4.7. Vapor pressure (20°): $<7.5 \times 10^{-10}$ Pa. Log P (octanol/water): 1.08 (pH 4); −1.01 (pH 7). uv max: 297 nm (ε 8000). Soly at 20° (mg/l): 1,2-dichloromethane 39400; acetone 27900; ethyl acetate 21700; methanol 10100; xylene 352; octanol 730; heptane <10. Soly in water at 20° (g/l): 0.0164 (pH 4); 3.20 (pH 7). LD_{50} in rats (mg/kg): >2000 orally; >2000 dermally; LC_{50} in rainbow trout: >87 mg/l (Becker). LC_{50} in rats (mg/l): >5.12 (U.S. EPA).

USE: Herbicide.

8125. Pyroxylin. [9004-70-0] Cellulose nitrate; nitrocellulose; collodion cotton; soluble gun cotton; collodion wool; colloxylin; xyloidin; celloidin; Parlodion. Variable mixture which consists chiefly of cellulose tetranitrate. *Review:* R. T. Bogan *et al.* in *Kirk-Othmer Encyclopedia of Chemical Technology* **vol. 5** (Wiley-Interscience, New York, 3rd ed., 1979) pp 129-143.

Yellowish-white, matted mass of filaments, having the appearance of raw cotton. *Highly flammable;* pyroxlin with higher nitrogen content *may explode!* When kept in well-closed containers and exposed to light it dec. Sol in 25 parts of a mixture of 1 vol alcohol and 3 vols ether; also sol in methanol, acetone, glacial acetic acid, amyl acetate. *Keep loosely packed in cartons and protected from light and moisture.* Can be shipped with safety only when wet with 25-30% water or alcohol.

USE: In manuf of collodions; in lacquer coatings, inks, adhesives. Cellulose hexanitrate is used in explosives and propellants. Celloidin is used for embedding sections in microscopy; in electrotechnics, photography, galvanoplasty.

THERAP CAT: Topical protectant.

8126. Pyrrobutamine. [91-82-7] 1-[4-(4-Chlorophenyl)-3-phenyl-2-buten-1-yl]pyrrolidine; 1-[γ-*p*-(chlorobenzyl)cinnamyl]-pyrrolidine; 1-*p*-chlorophenyl-2-phenyl-4-pyrrolidyl-2-butene; Pyronil. $C_{20}H_{22}ClN$; mol wt 311.85. C 77.03%, H 7.11%, Cl 11.37%, N 4.49%. Prepn: J. Mills, US 2655509 (1953 to Eli Lilly). Pharmacology and toxicity: H. M. Lee *et al.*, *Proc. Soc. Exp. Biol. Med.* **80**, 458 (1952). Pharmacology and physical properties: N. G. Lordi, J. E. Christian, *J. Am. Pharm. Assoc.* **45**, 300 (1956). Structure-activity and stereochemical study: A. F. Casy, R. R. Ison, *J. Pharm. Pharmacol.* **22**, 270 (1970). GC-MS determn in urine: H. Maurer, K. Pfleger, *J. Chromatogr.* **430**, 31 (1988).

Oily liquid, $bp_{0.3}$ 190-195°. On standing gives crystals, mp 48-49°. d^{25} 1.1052. uv max (95% ethanol): 360, 243 nm (α 112, 9500), (c = 0.538, 0.00538 mg/ml).

Diphosphate. [135-31-9] $C_{20}H_{28}ClNO_8P_2$. Crystals from alcohol + ether, mp 129.5-130°. pKa_1 8.77, pka_2 5.23. Sol in warm water to the extent of 10%. Soly in alcohol at 25° about 5%. Soly at pH 7.4 (37.5°): 0.00087 M/l. Practically insol in chloroform, ether. LD_{50} in mice, guinea pigs (mg/kg): 1270 ±156, 1241 ±165 s.c.; 836.8 ±95, 625.6 ±41.9 i.m.; 1116 ±73, 992.6 ±107 orally; in mice (mg/kg): 53.53 ±1.61 i.v. (Lee).

Hydrochloride. $C_{20}H_{22}ClN.HCl$. Crystals from alcohol + ether, mp 227-228°.

Hydrobromide. $C_{20}H_{22}ClN.HBr$. Crystals from alcohol + ether, mp 228-229°.

THERAP CAT: Antihistaminic.

8127. Pyrrocaine. [2210-77-7] *N*-(2,6-Dimethylphenyl)-1-pyrrolidineacetamide; 1-pyrrolidineaceto-2′,6′-xylidide; 2-(1-pyrrolidinyl)-2′,6′-acetoxylidide; 1-pyrrolidinoaceto-2,6-dimethylanilide; EN-1010; NSC-52644; Endocaine; Dynacaine. $C_{14}H_{20}N_2O$; mol wt 232.33. C 72.38%, H 8.68%, N 12.06%, O 6.89%. Prepn: Schlesinger, Gordon, US 2813861 (1957 to Endo); Löfgren *et al.*, *Acta Chem. Scand.* **11**, 1724 (1957).

Crysts from hexane or petr ether + dibutyl ether, mp 83°.

Hydrochloride. [2210-64-2] $C_{14}H_{20}N_2O.HCl$. Crystals from isopropanol, mp 205°. Soluble in water, alcohol, isopropyl alcohol. Practically insol in chloroform, ether.

THERAP CAT: Anesthetic (local).

8128. 1*H*-Pyrrole. [109-97-7] Azole; imidole; divinylenimine. C_4H_5N; mol wt 67.09. C 71.61%, H 7.51%, N 20.88%. A constituent of coal tar and bone oil: Runge, *Ann. Phys.* **31**, 67 (1834). Prepd industrially by fractional distillation of bone oil, or by the thermal decompn of ammonium mucate with glycerol or mineral oil: McElvain, Bolliger, *Org. Synth.* **coll. vol. I** (2nd ed., 1941) p 473; Blicke, Powers, *Ind. Eng. Chem.* **19**, 1334 (1927). Also formed on heating of albumin; on heating sheep's wool with aq barium hydroxide soln; by pyrolysis of gelatin. Alternate prepns from acetaldehyde and ammonia: Tschitschibabin, *Chem. Zentralbl.* **1916**, I,

920; from succindialdehyde with ammonia and acetic acid: Harries, *Ber.* **34**, 1496 (1901); **35**, 1183 (1902); distilling succinimide with zinc or sodium: Bell, Bernthsen, *Ber.* **13**, 877, 1049 (1880). Purification and physical properties: R. V. Helm *et al.*, *J. Phys. Chem.* **62**, 858 (1958). *Review:* Fischer-Orth, *Die Chemie des Pyrrols* (Leipzig, 1934-1940); E. Vittort, L. R. Anderson in *Kirk-Othmer Encyclopedia of Chemical Technology* **vol. 19** (Wiley-Interscience, New York, 3rd ed., 1982) pp 499-520.

Liquid. Agreeable empyreumatic odor resembling that of chloroform. Colorless when freshly distilled, darkens unless every trace of oxygen is removed. d_4^{20} 0.9691. bp_{760} 129.8°. Best distilled *in vacuo.* n_D^{20} 1.5085. Flash pt, closed cup: 102°F (39.0°C). Absorption spectrum: Menczel, *Phys. Chem.* **125**, 161; *Chem. Zentralbl.* **1927**, I, 2510. Sparingly sol in water; freely sol in alcohol, benzene, ether. Insol in aq alkalies. Sol in dil acids with decompn. Solns in dil HCl yield pyrrole red, an amorphous, orange-colored substance; also polymerization takes place under the influence of acids and glycols.

8129. Pyrrolidine. [123-75-1] Tetrahydropyrrole. C_4H_9N; mol wt 71.12. C 67.55%, H 12.76%, N 19.69%. Found in tobacco and carrot leaves. Probable biosynthesis from ornithine and putrescine. Usually prepd by reduction of pyrrole.

Almost colorless liquid; unpleasant ammonia-like odor. Fumes in air. bp 88.5-89°. $d_4^{22.5}$ 0.8520. n_D^{28} 1.4402. Strong base. pK (25°) 2.89. *Flammable; corrosive.* Miscible with water. Soluble in alcohol, ether, chloroform.

8130. 2-Pyrrolidone. [616-45-5] 2-Pyrrolidinone; 2-oxopyrrolidine; α-pyrrolidone; 2-ketopyrrolidine. C_4H_7NO; mol wt 85.11. C 56.45%, H 8.29%, N 16.46%, O 18.80%. Prepd on a large scale from butyrolactone by a Reppe process: DE **1085525** (to BASF). Other prepns: Metzger, Seelert, *Angew. Chem.* **75**, 919 (1963); Copenhaver, Ney, US **3095423** (1963 to Minnesota Mining & Manuf); Lidov, US **3109005** (1963 to Halcon International).

Liquid above 25°. bp_{760} 245°; $bp_{9.2}$ 113-114°; $bp_{0.2}$ 76°. d_4^{25} 1.116. Flash pt, open cup: 265°F (129°C). Viscosity at 25° = 13.3 cP. Dipole moment: 2.3. In the presence of the stoichiometric amount of water a cryst monohydrate, mp 30°, can be formed. Non-corrosive. Good chemical stability. Miscible with water, ethanol, ether, chloroform, benzene, ethyl acetate, carbon disulfide.

USE: Intermediate in the manuf of polyvinylpyrrolidone and polypyrrolidone (a polymer, formed in the presence of alkaline catalysts). Also used as high-boiling solvent in petroleum processing, acrylonitrile manuf. Industrial solvent for polymers, chlordane, DDT, sorbitol, glycerol, iodine, sugars. In specialty printers inks. As plasticizer and coalescing agent for acrylic-styrene emulsion-type floor polishes.

8131. 3-Pyrroline. [109-96-6] 2,5-Dihydro-1*H*-pyrrole. C_4H_7N; mol wt 69.11. C 69.52%, H 10.21%, N 20.27%. Prepd by reduction of pyrrole with zinc and glacial acetic or hydrochloric acid.

Almost colorless liquid. Unpleasant, ammonia-like odor. Fumes in air. Very hygroscopic, also absorbs CO_2. bp_{748} 90-91°; d_4^{20}

0.9097; n_D^{20} 1.4664. Strong base. Miscible with water. Sol in alcohol, ether, chloroform.

8132. Pyrrolnitrin. [1018-71-9] 3-Chloro-4-(3-chloro-2-nitrophenyl)pyrrole; 3-chloro-4-(2′-nitro-3′-chlorophenyl)pyrrole; PN; Pyroace. $C_{10}H_6Cl_2N_2O_2$; mol wt 257.07. C 46.72%, H 2.35%, Cl 27.58%, N 10.90%, O 12.45%. Antifungal antibiotic isolated from *Pseudomonas pyrrocinia* n. sp.: K. Arima *et al.*, *Agric. Biol. Chem.* **28**, 575 (1964); *eidem*, *J. Antibiot.* **18**, 201 (1965). Structure: Imanaka *et al.*, *ibid.* 207. Synthesis: S. Umio *et al.*, **BE 670427**; *eidem*, **US 3428648** (1964, 1969 both to Fujisawa); Nakano *et al.*, *Tetrahedron Lett.* **1966**, 737; S. Umio *et al.* (10 publications) *Chem. Pharm. Bull.* **17**, 559-628 (1969); Gosteli, *Helv. Chim. Acta* **55**, 451 (1972). Pharmacological studies: M. Nishida *et al.*, *J. Antibiot.* **18**, 211 (1965).

Pale yellow crystals from hot cyclohexane, mp 124.5°. Gradually changes to a red or brown color on exposure to sunlight and loses antibiotic activity. uv max: 252 nm (ε 7500). Slightly sol in water, petr ether, cyclohexane; sol in methanol, ethanol, butanol, acetone, ethyl acetate, benzene, ether, chloroform, carbon tetrachloride, pyridine, acetic acid. LD_{50} in rats, rabbits (mg/kg): 68, 105 i.p. (Nishida).

THERAP CAT: Antifungal.

8133. Pyrrolysine. [448235-52-7] N^6-[(3R)-1,5-Didehydro-3-methyl-D-prolyl]-L-lysine; L-pyrrolysine. $C_{12}H_{21}N_3O_3$; mol wt 255.32. C 56.45%, H 8.29%, N 16.46%, O 18.80%. Naturally occurring, genetically encoded amino acid found in certain methanogenic Archaea; originally described as L-lysine ε-amide linked to (4R,5R)-4-substituted-1-pyrroline-5-carboxylate. Identification of pyrrolysine-specific tRNA in *Methanosarcina barkeri*: G. Srinivasan *et al.*, *Science* **296**, 1459 (2002). Identification as UAG-encoded residue in monomethylamine methyltransferase: B. Hao *et al.*, *ibid.* 1462. Identification of C_4-substituent as a methyl group, synthesis and properties: *idem et al.*, *Chem. Biol.* **11**, 1317 (2004). MS analysis: J. A. Soares *et al.*, *J. Biol. Chem.* **280**, 36962 (2005). *Reviews:* C. Fenske *et al.*, *Angew. Chem. Int. Ed.* **42**, 606-610 (2003); J. A. Krzycki, *Curr. Opin. Microbiol.* **8**, 706-712 (2005).

Lithium salt. $C_{12}H_{20}LiN_3O_3$; mol wt 261.25. White solid. $[\alpha]_D^{20}$ −2.1° (c = 0.25 in CH_3OH). uv max (methanol): 310 nm (sh) (ε 35.4).

8134. Pyruvate Decarboxylase. [9001-04-1] Pyruvic decarboxylase; E.C. 4.1.1.1; pyruvic carboxylase; α-carboxylase; 2-oxo-acid carboxylase. An enzyme found in yeast, plants, wheat germ and bacteria: Neuberg, Rosenthal, *Biochem. Z.* **51**, 142 (1913); Holzer *et al.*, *ibid.* **327**, 331 (1956); Singer, Pensky, *J. Biol. Chem.* **196**, 375 (1952); King, Chelelin, *ibid.* **208**, 821 (1954); Howells, Lindstrom, *J. Bacteriol.* **75**, 305 (1958). Cofactors: Diphosphothiamine, divalent metal ions. Catalyzes decarboxylation of α-ketoacids to aldehydes + CO_2: Juni, *J. Biol. Chem.* **236**, 2302 (1961). Structure and mechanism of action of the yeast enzyme: Schellenberger, *Angew. Chem. Int. Ed.* **6**, 1024 (1967). *Review:* Utter, *The Enzymes* vol. 5, P. D. Boyer *et al.*, Eds. (Academic Press, New York, 2nd ed., 1961) pp 320-323.

8135. Pyruvic Acid. [127-17-3] 2-Oxopropanoic acid; α-ketopropionic acid; acetylformic acid; pyroracemic acid; brenztraubensäure (German). $C_3H_4O_3$; mol wt 88.06. C 40.92%, H 4.58%, O 54.50%. Intermediate in sugar metabolism and in enzymatic carbohydrate degradation (alcoholic fermentation) where it is converted to acetaldehyde and CO_2 by carboxylase. *See:* Nord, *Chem. Rev.* **26**, 423 (1940). In muscle, pyruvic acid (derived from glycogen) is reduced to lactic acid during exertion, which is reoxidized and partially retransformed to glycogen during rest. The liver can convert pyruvic acid to alanine by amination. Pyruvic acid has been isolated from cane sugar fermentation broth by fixation with β-naphthylamine giving α-methyl-β-naphthocinchonic acid: Grab, *Biochem. Z.* **123**, 84 (1921). The most practical method of preparing pyruvic acid is by distillation of tartaric acid in presence of potassium acid sulfate as dehydrating agent. The distillate must be fractionated under reduced pressure: Erlenmeyer, *Ber.* **14**, 321 (1881); Döbner, *Ann.* **242**, 269 (1887); Howard, Fraser, *Org. Synth.* **coll. vol. I** (2nd ed., 1941) p 475. Absorption spectrum: Henri, Fromageot, *Bull. Soc. Chim.* [4] **37**, 846 (1925).

Liquid. Odor resembling that of acetic acid. d_4^{15} 1.267; mp 11.8°; bp_{760} 165° (dec); bp_{100} 106.5°; bp_{40} 85.3°; bp_{20} 70.8°; bp_{10} 57.9°; bp_5 45.8°; $bp_{1.0}$ 21.4°; n_D^{20} 1.4138. pK (25°) 2.49. Miscible with water, alcohol, ether. Polymerizes and dec on standing unless pure and kept in container with airtight closure.

Methyl ester. [600-22-6] Methyl pyruvate. $C_4H_6O_3$; mol wt 102.09. Liquid, bp_{760} 134-137°.

Ethyl ester. [617-35-6] Ethyl methylglyoxylate; ethyl pyruvate. $C_5H_8O_3$; mol wt 116.12. Liquid, bp_{760} 155°; bp_{750} 147.5°; bp_{42} 69-71°; bp_{20} 66° also 56°. Prepn from ethyl lactate by oxidation with $KMnO_4$: Cornforth, *Org. Synth.* **coll. vol. IV**, 467 (1963).

8136. Pyrvinium Pamoate. [3546-41-6] 6-(Dimethylamino)-2-[2-(2,5-dimethyl-1-phenyl-1H-pyrrol-3-yl)ethenyl]-1-methylquinolinium salt with 4,4′-methylenebis[3-hydroxy-2-naphthalenecarboxylic acid] (2:1); bis[6-(dimethylamino)-2-[2-(2,5-dimethyl-1-phenylpyrrol-3-yl)vinyl]-1-methylquinolinium] 4,4′-methylenebis(3-hydroxy-2-naphthoate); 6-(dimethylamino)-2-[2-(2,5-dimethyl-1-phenyl-3-pyrryl)vinyl]-1-methylquinolinium salt of 2,2′-dihydroxy-1,1′-dinaphthylmethane-3,3′-dicarboxylic acid; pyrvinium embonate; viprynium embonate; Molevac; Povanyl; Pyrcon; Vanquin. $C_{75}H_{70}N_6O_6$; mol wt 1151.42. C 78.24%, H 6.13%, N 7.30%, O 8.34%. Cyanine dye. Prepn: E. F. Elslager, D. F. Worth, **US 2925417** (1960 to Parke, Davis). Determn in pharmaceutical prepn: H. D. Beckstead, S. J. Smith, *J. Pharm. Sci.* **56**, 390 (1967); by HPLC: U. G. G. Hennig *et al.*, *Arch. Toxicol.* **60**, 278 (1987). Clinical trial: R. A. Buchanan *et al.*, *Clin. Pharmacol. Ther.* **16**, 716 (1974). Clinical pharmacology: T. C. Smith *et al.*, *ibid.* **19**, 802 (1976). Review of use in *Enterobius vermicularis* infections: G. C. Cook, *Gut* **35**, 1159-1162 (1994).

Bright orange or orange-red to almost black precipitate, mp 210-215° (softens at 190°). Absorption max: 236, 356, 503 nm. Stable to heat, light and air. Freely sol in glacial acetic acid; slightly sol in chloroform, methoxyethanol; very slightly sol in alc, methanol. Practically insol in water, ether.

Pyrvinium chloride. [548-84-5] Viprynium chloride; SN-4395. $C_{26}H_{28}ClN_3$; mol wt 417.98. Prepn: E. Van Lare, L. G. S. Brooker, **US 2515912** (1950 to Eastman Kodak). May occur as dihydrate, deep-red powder, dec 249-251°. Sparingly sol in water.

THERAP CAT: Anthelmintic (Nematodes).

Q

8137. Q-Enzyme. [9001-97-2] α-Glucan branching glycosyltransferase; potato branching enzyme; enzyme Q; branching factor; glucosan transglycosylase. A widely distributed enzyme which transfers part of a 1,4-glucosan chain from a 4- to a 6-position in polysaccharides. Isoln from potatoes: Barker *et al.*, *J. Chem. Soc.* **1949**, 1705. Crystallization: Gilbert, Patrick, *Biochem. J.* **51**, 181 (1952). Improved isoln procedure: Griffen, Wu, *Biochemistry* **7**, 3063 (1968). Although Q-enzyme is unstable in aq soln, it can be stored for long periods without appreciable loss in activity when it is freeze-dried. *Review:* Barker, Bourne, *Q. Rev. Chem. Soc.* **7**, 65-68 (1953); Dixon, Webb, *Enzymes* (Academic Press, New York, 1958) *passim.*

8138. Quadricyclane. [278-06-8] Tetracyclo[3.2.0.0^{2,7}.0^{4,6}]-heptane; quadricyclo[2.2.1.0^{2,6}.0^{3,5}]heptane. C_7H_8; mol wt 92.14. C 91.25%, H 8.75%. Highly strained, thermally stable, saturated hydrocarbon. Prepn by photochemical isomerization of norbornadiene: W. G. Dauben, R. L. Cargill, *Tetrahedron* **15**, 197 (1961); G. S. Hammond *et al.*, *J. Am. Chem. Soc.* **83**, 4674 (1961); C. D. Smith, *Org. Synth.* **coll. vol. VI**, 962 (1988). Cycloaddition reactions: M. E. Landis, J. C. Mitchell, *J. Org. Chem.* **44**, 2288 (1979). Solar energy storage applications of photoisomerization reactions with norbornadiene: C. Philippopoulos *et al.*, *Ind. Eng. Chem. Prod. Res. Dev.* **22**, 627 (1983); K. Maruyama *et al.*, *J. Org. Chem.* **50**, 4742 (1985). Gas phase reactivity and thermochemistry: H. S. Lee *et al.*, *J. Am. Chem. Soc.* **118**, 5068 (1996). Vibrational spectroscopy: X. Zhou, R. Liu, *Vib. Spectrosc.* **12**, 65 (1996). Toxicity data: E. R. Kinkead *et al.*, *J. Am. Coll. Toxicol.* **12**, 634 (1993). Review of isomerization reaction with norbornadiene and its energy storage applications: A. D. Dubonosov *et al.*, *Russ. Chem. Rev.* **71**, 917-927 (2002); of synthetic chemistry: V. A. Petrov, N. V. Vasil'ev, *Curr. Org. Synth.* **3**, 215-259 (2006).

Liquid, bp$_{740}$ 108°. $n_D^{26.5}$ 1.4830. Oxidation potential, E$_{\frac{1}{2}}$ = 0.91 V (SCE). LD$_{50}$ orally in rats: 1.2-3.5 g/kg (Kinkead).

USE: Versatile organic synthon in thermal [2+2+2] cycloadditions, photochemical transformations, ionic and redox reactions.

8139. Quadrone. [66550-08-1] (3a*S*,5a*R*,6*R*,8a*R*,8b*R*)-Octahydro-10,10-dimethyl-6-8b-ethano-8b*H*-cyclopenta[*de*]-2-benzopyran-1,4-dione. $C_{15}H_{20}O_3$; mol wt 248.32. C 72.55%, H 8.12%, O 19.33%. Cytotoxic sesquiterpene produced by *Aspergillus terreus*. First known member of the quadrane family of bioactive natural products bearing the distinctive tricyclic core structure. Isoln: R. L. Ranieri, G. J. Calton, *Tetrahedron Lett.* **19**, 499 (1978). Total synthesis of (±)-form: S. Danishefsky *et al.*, *J. Am. Chem. Soc.* **103**, 4136 (1981). Abs config: K. Kon *et al.*, *Tetrahedron Lett.* **25**, 3739 (1984). Synthesis of naturally occuring (−)-form and abs config of enantiomers: A. B. Smith, III *et al.*, *J. Am. Chem. Soc.* **113**, 3533 (1991). Biosynthesis: D. E. Cane *et al.*, *Bioorg. Chem.* **14**, 417 (1986). Review of syntheses: M. Presset *et al.*, *Eur. J. Org. Chem.* **2010**, 2247-2260.

Crystals from methanol, mp 185-186°. $[\alpha]_D^{25}$ −60.53° (c = 0.5 in chloroform).

(±)-Form. [74807-65-1] Crystals from methanol + water, mp 140-142°.

(+)-Form. [87480-01-1] mp 177-178°. $[\alpha]_D$ +49.8° (c = 0.62 in ethanol).

8140. Quantum Dye. $C_{38}H_{36}EuN_8S_2$; mol wt 820.85. C 55.60%, H 4.42%, Eu 18.51%, N 13.65%, S 7.81%. Macrocyclic europium-chelate fluorimetric tracer. Prepn and optical properties: L. M. Vallarino *et al.*, *Proc. SPIE Int. Soc. Opt. Eng.* **1885**, 376 (1993). Quantification of Eu^{3+} in dye: K. Saito *et al.*, *Anal. Biochem.* **258**, 311 (1998). Use as quantitative fluorescent label for glycoproteins: Y. C. Lee *et al.*, *Glycobiology* **8**, 849 (1998); I. L. Deras *et al.*, *Carbohydr. Res.* **306**, 469 (1998).

R is attached at (a or b) and (c or d).

White crystalline solid. Water soluble. Absorption max: 360 nm (excitation); 615 nm (emission).

USE: Fluorescent probe especially for glycoproteins.

8141. Quassia. Bitter wood; bitter ash. The wood of *Picrasma excelsa* (Sw.) Planch. or of *Quassia amara* L., Simaroubaceae. The first is known in commerce as Jamaica quassia, the second as Surinam quassia. *Habit.* *Picrasma escelsa* inhabits Jamaica and the Caribbean Islands; *Quassia amara* is a native of Brazil and Guiana and is cultivated in Colombia, Panama, and the West Indies. Quassin and neoquassin are the bitter principles of Surinam quassia; picrasmin, that of Jamaica quassia. These bitter pinciples are obtained in yields of 0.1-0.2% and appear commercially under the name of quassin.

Unground quassia occurs usually in chips, raspings, or shavings, occasionally in billets; yellowish-white to bright yellow with a few light gray pieces; coarsely grained, fibrous. Slight odor; very bitter taste. The powdered form is pale yellow in color.

USE: The extract is used for fly poison on flypaper; to imitate hops.

THERAP CAT: Anthelmintic.

THERAP CAT (VET): Has been used as a bitter, as an anthelmintic.

8142. Quassin. [76-78-8] 2,12-Dimethoxypicrasa-2,12-diene-1,11,16-trione; 3aβ,6aβ,7,7aα,8,11a,11bα,11c-octahydro-2,10-dimethoxy-3,8α,11aβ,11cβ-tetramethylphenanthro[10,1-*bc*]pyran-1,5,11(4*H*)-trione; nigakilactone D. $C_{22}H_{28}O_6$; mol wt 388.46. C 68.02%, H 7.27%, O 24.71%. One of the bitter constituents of the wood of *Quassia amara* L., Simaroubaceae known in commerce as Surinam quassia. Obtained by the resolution of the mixture of bitter constituents of quassia wood: E. P. Clark, *J. Am. Chem. Soc.* **59**, 927 (1937); London *et al.*, *J. Chem. Soc.* **1950**, 3431. Structure: Valenta *et al.*, *Tetrahedron Lett.* no. **20**, 25 (1960); Carman, Ward, *ibid.* **1961**, 317; Valenta *et al.*, *Tetrahedron* **15**, 100 (1961). Stereochemistry: Valenta *et al.*, *ibid.* **18**, 1433 (1962). Identity with nigakilactone D: Murae *et al.*, *ibid.* **27**, 1545 (1971). Synthetic approach: Stojanac *et al.*, *Can. J. Chem.* **53**, 619 (1975); P. A. Grieco *et al.*, *Tetrahedron Lett.* **21**, 1619 (1980). Total synthesis of *dl*-form: *eidem*, *J. Am. Chem. Soc.* **102**, 7586 (1980).

Very bitter rectangular plates from dilute methanol, mp 222°. $[\alpha]_D^{20}$ +34.5° (c = 5.09 in $CHCl_3$). uv max: ~255 nm (ε ~11,650). Sol in benzene, alc, acetone, chloroform, pyridine, acetic acid, hot ethyl acetate. Sparingly sol in ether, petr ether. Bitterness threshold 1:60,000.

8143. Quaternium-15. [4080-31-3] 1-(3-Chloro-2-propen-1-yl)-3,5,7-triaza-1-azoniatricyclo[3.3.1.1³⁷]decane chloride (1:1); 1-(3-chloroallyl)-3,5,7-triaza-1-azoniaadamantane chloride; N-(3-chloroallyl)hexaminium chloride; CTAC; Dowicil 75. $C_9H_{16}Cl_2N_4$; mol wt 251.16. C 43.04%, H 6.42%, Cl 28.23%, N 22.31%. Microbiocide. Prepn and antibacterial activity: C. R. Scott, P. A. Wolf, *Appl. Microbiol.* **10**, 211 (1962). Cosmetic ingredient review: *J. Am. Coll. Toxicol.* **5**, 61-101 (1986).

cis-Form

Cream colored powder with pungent odor. Dec above 60°. Soly at 25° (g/100 g): mineral oil <0.1, isopropanol <0.1, ethanol 2.4, methanol 20.8, glycerine 12.6, propylene glycol 18.7, water 127.2. LD_{50} (as 50% aq soln) in female rats, rabbits (mg/kg): 1552, 78.5 orally; in female rabbits (mg/kg): 565 dermally (cosmetic ingredient review), *Keep dry and cool.*

cis-**Form.** [51229-78-8] Dowicil 200; CoSept 200. Off-white powder with slight amine odor. Bulk density (25°): 0.44 g/cc. Vapor pressure (25°): 1×10^{-9} mm Hg. Miscible with water. *Flammable, keep dry and cool.*

USE: Preservative in cosmetics, hair care products, soaps, adhesives.

8144. Quatrimycin. [79-85-6] (4R,4aS,5aS,6S,12aS)-4-(Dimethylamino)-1,4,4a,5,5a,6,11,12a-octahydro-3,6,10,12,12a-pentahydroxy-6-methyl-1,11-dioxo-2-naphthacenecarboxamide; epitetracycline. $C_{22}H_{24}N_2O_8$; mol wt 444.44. C 59.46%, H 5.44%, N 6.30%, O 28.80%. Prepn by epimerization of tetracycline: McCormick *et al.*, *J. Am. Chem. Soc.* **79**, 2849 (1957); Kaplan *et al.*, *Antibiot. Chemother.* **7**, 569 (1957); Remmers *et al.*, *J. Pharm. Sci.* **52**, 752 (1963). Acute toxicity: L. Kung, H. Sun, *Yao Hsueh Hsueh Pao* **13**, 244 (1966); *C.A.* **65**, 7869e (1966).

Monohydrate. Crystals, mp 178° dec. $[\alpha]_D^{25}$ -335° (c = 0.5 in 0.03N HCl). uv max (0.01N H_2SO_4): 216, 255, 270, 355 (ε 13,900, 16,400, 15,200, 14,700). LD_{50} i.v. in mice: 85.8 mg/kg (Kung, Sun).

Ammonium salt monohydrate. $C_{22}H_{27}N_3O_8 \cdot H_2O$. Yellow crystals, mp 170° dec. $[\alpha]_D^{25}$ -321° (c = 0.5 in 0.03N HCl). Soly in water >50 mg/kg.

Hydrochloride. $C_{22}H_{25}ClN_2O_8$. Amorphous yellow solid. $[\alpha]_D^{25}$ -325° (c = 0.58 in 0.2N HCl). Soly in water >100 mg/ml.

Methiodide. $C_{23}H_{27}IN_2O_8$. Crystals, mp 161-162°. $[\alpha]_D^{25}$ -265° (c = 0.5 in 0.03N HCl).

8145. Quazepam. [36735-22-5] 7-Chloro-5-(2-fluorophenyl)-1,3-dihydro-1-(2,2,2-trifluoroethyl)-2H-1,4-benzodiazepine-2-thione; Sch-16134; Doral; Dormalin; Oniria; Prosedar; Quazium; Selepam. $C_{17}H_{11}ClF_4N_2S$; mol wt 386.79. C 52.79%, H 2.87%, Cl 9.17%, F 19.65%, N 7.24%, S 8.29%. Benzodiazepine hypnotic. Prepn: M. Steinman, **DE 2138773**; *idem*, **US 3845039** (1972, 1974 both to Schering Corp.); M. Steinman *et al.*, *J. Med. Chem.* **16**, 1354 (1973). Pharmacology: A. Barnett *et al.*, *Arzneim.-Forsch.* **32**, 1452 (1982); E. Ongini *et al.*, *ibid.* 1456. GC determn in plasma: J. M. Hilbert *et al.*, *J. Pharm. Sci.* **73**, 516 (1984). Pharmacokinetics: M. Chung *et al.*, *Clin. Pharmacol. Ther.* **35**, 520 (1984). Metabolism: N. Zampaglione *et al.*, *Drug Metab. Dispos.* **13**, 25 (1985). Clinical trials in insomnia: J. W. Goethe, G. Kader, *Curr. Ther. Res.* **32**, 150 (1982); M. Mamelak *et al.*, *J. Clin. Pharmacol.* **24**, 65 (1984). Toxicology: H. E. Black *et al.*, *Arzneim.-Forsch.* **37**, 906 (1987).

Crystals from methylene chloride-hexane, mp 137.5-139°. LD_{50} in mice (mg/kg): >1370 i.v., >5000 orally (Orgini). LD_{50} in male, female mice, rats (mg/kg): 845, 921, 3072, 2749 i.p. (Black).

Note: This is a controlled substance (depressant): **21 CFR**, 1308.14.

THERAP CAT: Sedative, hypnotic.

8146. Quebrachamine. [4850-21-9] 7-Ethyl-1,4,5,6,7,8,-9,10-octahydro-2H-3,7-methanoazacycloundecino[5,4-b]indole; kamassin. $C_{19}H_{26}N_2$; mol wt 282.43. C 80.80%, H 9.28%, N 9.92%. From bark of *Aspidosperma quebracho blanco* Schlecht., *Apocynaceae:* Hesse, *Ann.* **211**, 249 (1882); Field, *J. Chem. Soc.* **125**, 1444 (1924). Identity with kamassin: Gellert, Witkop, *Helv. Chim. Acta* **35**, 114 (1952). Structure: Witkop, *J. Am. Chem. Soc.* **79**, 3193 (1957); Kny, Witkop, *J. Org. Chem.* **25**, 635 (1960); Biemann, Spiteller, *J. Am. Chem. Soc.* **84**, 4578 (1962). Crystal structure: C. Puglisi *et al.*, *Acta Crystallogr.* **B32**, 1900 (1976). Total synthesis of *dl*-form: Stork, Dolfini, *J. Am. Chem. Soc.* **85**, 2872 (1963); Ziegler *et al.*, *ibid.* **91**, 2342 (1969); Kutney *et al.*, *ibid.* **92**, 1727 (1970); V. S. Giri *et al.*, *J. Heterocycl. Chem.* **17**, 1133 (1980); S. Takano *et al.*, *Heterocycles* **16**, 247 (1981). Enantioselective synthesis: *eidem*, *Chem. Commun.* **1980**, 616; **1981**, 1153.

Bitter leaflets, mp 145-147°. $[\alpha]_D^{20}$ -109 to -110° (acetone). uv max (methanol): 230, 287, 293 nm (log ε 4.55, 3.85, 3.84). Sol in acetone, alc, chloroform, ether, dil acids.

8147. Quebracho Colorado. Red quebracho. Wood of *Loxopterygium lorentzii* Griseb., *Anacardiaceae. Habit.* Argentine Republic. *Constit.* Tannin, coloring matter, loxopterygine.

USE: In dyeing and tanning.

8148. Queen Substance. [334-20-3] (2E)-9-Oxo-2-decenoic acid. $C_{10}H_{16}O_3$; mol wt 184.24. C 65.19%, H 8.75%, O 26.05%. Secreted in the mandibular gland of queen honey bees *(Apis mellifera, A. florea, A. cerana, A. dorsata)*; inhibits the development of ovaries in worker bees, prevents queen cell formation and attracts male bees (drones) to virgin queens for the purpose of mating: Butler, *Experientia* **13**, 256 (1957); Sannasi, Rajulu, *Life Sci.* **10**, part 2, 195 (1971). Similarity with the ovary inhibiting hormone of prawns *(Leander serratus):* Carlisle, Butler, *Nature* **177**, 276 (1956). Extraction and purification: Carlisle, Butler, *loc. cit.;* Butler *et al., Nature* **184**, 1871 (1959). Synthesis: Barbier *et al., Compt. Rend.* **251**, 1133 (1960); Jaeger, Robinson, *Tetrahedron* **14**, 320 (1961); Y. Naoshima *et al., Agric. Biol. Chem.* **48**, 2151 (1984); U. P. Dhokte, A. S. Rao, *Synth. Commun.* **17**, 355 (1987).

Transparent elongated plates from ether + petr ether or aq methanol, mp 54.5-55.5°. Stable to heat, acids, less stable to alkalies. Sol in acetone, alcohol. IR spectrum: Butler *et al., loc. cit.*

8149. Quercetagetin. [90-18-6] 2-(3,4-Dihydroxyphenyl)-3,- 5,6,7-tetrahydroxy-4*H*-1-benzopyran-4-one; 3,3′,4′,5,6,7-hexahydroxyflavone; 6-hydroxycyanidenolon 1555. $C_{15}H_{10}O_8$; mol wt 318.24. C 56.61%, H 3.17%, O 40.22%. From flowers of French marigold, *Tagetes patula* Linn., *Compositae:* Perkin, *J. Chem. Soc.* **103**, 209 (1913). Synthesis: Baker *et al., ibid.* **1929**, 74; Rao, Seshadri, *Proc. Indian Acad. Sci.* **23A**, 23 (1946), *C.A.* **40**, 5052[2] (1946).

Dihydrate. Pale yellow needles from dil alcohol, mp 318°. uv max (alc): 259, 361 nm (log ε 4.23, 4.34). Sol in hot alcohol; sparingly sol in boiling water.
Hexaacetate. $C_{27}H_{22}O_{14}$. Needles from alcohol + acetic acid, mp 209-211°. Sparingly sol in alc.
7-Glucoside. Quercetagitrin. $C_{21}H_{20}O_{13}$. From flowers of the African marigold, *Tagetes erecta* L., *Compositae:* Rao, *Proc. Indian Acad. Sci.* **14A**, 289 (1941), *C.A.* **36**, 2555[3] (1942); from *Chrysanthemum coronarium* L., *Compositae:* Anyas, Steelink, *Arch. Biochem. Biophys.* **90**, 63 (1960). Structure: Rajagopalan, Seshadri, *Proc. Indian Acad. Sci.* **28A**, 31 (1948), *C.A.* **43**, 4265b (1949). Crystals from aqueous pyridine, dec 236-238°. uv max (95% ethanol): 260, 272, 362 nm.

8150. Quercetin. [117-39-5] 2-(3,4-Dihydroxyphenyl)-3,5,7- trihydroxy-4*H*-1-benzopyran-4-one; 3,3′,4′,5,7-pentahydroxyflavone; meletin; sophoretin; cyanidenolon 1522. $C_{15}H_{10}O_7$; mol wt 302.24. C 59.61%, H 3.34%, O 37.05%. The aglucon of quercitrin, of rutin, and of other glycosides. Widely distributed in the plant kingdom, esp in rinds and barks, in clover blossoms and in ragweed pollen. Isoln from *Rhododendron cinnabarinum* Hook, *Ericaceae:* Rangaswami *et al., Proc. Indian Acad. Sci.* **56A**, 239 (1962), *C.A.* **58**, 9414a (1963). Structure: Underhill *et al., Can. J. Biochem. Physiol.* **35**, 219 (1957). Biosynthesis: Watkin *et al., ibid.* 229; Grisebach, *Biochem. J.* **85**, 3p (1962); Patschke *et al., Z. Naturforsch.* **21b**, 201 (1966). Synthesis: Shakhova *et al., Zh. Obshch. Khim.* **32**, 390 (1962), *C.A.* **58**, 1426f (1963). Metabolism: Nakagawa *et al., Biochim. Biophys. Acta* **97**, 233 (1965). Toxicity data: M. Sullivan *et al., Proc. Soc. Exp. Biol. Med.* **77**, 269 (1951). *See also* Bioflavonoids.

Dihydrate. [6151-25-3] Yellow needles from dil alcohol. Becomes anhydr at 95-97°. When anhydr dec 314°. uv max (alc): 258,

375 nm (log ε 2.75, 2.75). One gram dissolves in 290 ml abs alc, in 23 ml boiling alc. Soluble in glacial acetic acid; in aq alkaline solns with yellow color. Practically insol in water. Alcoholic solns taste very bitter. LD_{50} orally in mice: 160 mg/kg (Sullivan).
Pentabenzyl ether. [13157-90-9] 3,3′,4′,5,7-Pentakis(benzyloxy)flavone; penta-*O*-benzylquercetin; Parietrope. $C_{50}H_{40}O_7$; mol wt 752.86. Prepn: Chopin, Chadenson, *C. R. Seances Acad. Sci. Ser. C* **263**, 729 (1966); Binovic, **DE 2122514** (1972 to Biosedra), *C.A.* **76**, 113072n (1972). Crystals, mp 123-125°. uv max (chloroform): 249, 343 nm (log ε 4.43, 4.14).
3-β-D-Galactoside hemipentahydrate. Hyperin; hyperoside. $C_{21}H_{20}O_{12}.2\frac{1}{2}H_2O$. From *Acacia melanoxylon* R. Br., *Leguminosae:* Falco, de Vries, *Naturwissenschaften* **51**, 462 (1964). Yellow needles from ethanol, dec 227-230°. $[\alpha]_D^{20}$ -83° (c = 0.2 in pyridine). uv max: 259, 364 nm (log ε 4.31, 4.39).

THERAP CAT: Capillary protectant.

8151. Quercimeritrin. [491-50-9] 2-(3,4-Dihydroxyphenyl)- 7-(β-D-glucopyranosyloxy)-3,5-dihydroxy-4*H*-1-benzopyran-4-one; quercetin-7-D-glucoside; 3,3′,4′,5,7-pentahydroxyflavone-7-D-glucoside. $C_{21}H_{20}O_{12}$; mol wt 464.38. C 54.32%, H 4.34%, O 41.34%. Found in flowers of *Gossypium herbaceum* L., *Malvaceae:* Perkin, *J. Chem. Soc.* **95**, 2181 (1909); from leaves of *Chrysanthemum ségetum* L. and *C. coronarium* L., *Compositae:* Geissman, Steelink, *J. Org. Chem.* **22**, 946 (1957); Anyas, Steelink, *Arch. Biochem. Biophys.* **90**, 63 (1960). In the mother liquor from quercimeritrin the glucosides gossypitrin and isoquercitrin, *q.v.*, are also found. Structure: Attree, Perkin, *J. Chem. Soc.* **1927**, 234; Rao, Seshadri, *Proc. Indian Acad. Sci.* **9A**, 365 (1939), *C.A.* **34**, 107[1] (1940); Pacheco, Grouiller, *Compt. Rend.* **253**, 1178 (1961).

Trihydrate. Yellow plates from aq pyridine. The water of crystn is given up at 100°, the anhydr material is hygroscopic, mp 247-249°. uv max (ethanol): 372, 257 nm (log ε 4.33, 4.38). Practically insol in cold water, more sol in hot water; sol in methanol. Sol in aq alkaline solns with deep yellow color. Is hydrolyzed by 7% H_2SO_4 yielding 1 mol quercetin and 1 mol D-glucose.
Gossypitrin. $C_{21}H_{20}O_{13}$; mol wt 480.38. Orange-yellow needles melting at 200-202°; slightly sol in alcohol and acetic acid.

8152. *d*-Quercitol. [488-73-3] 2-Deoxy-D-*chiro*-inositol; D- 1-deoxy-*muco*-inositol; "acorn sugar"; (+)-protoquercitol; 1,2,3,4,5- cyclohexanepentol. $C_6H_{12}O_5$; mol wt 164.16. C 43.90%, H 7.37%, O 48.73%. Found in the acorns of various spp of *Quercus, Fagaceae:* Prunier, *Ann. Chim. Phys.* **15**, 1 (1878); from leaves of the European palm (*Chaemerops humilis*): Muller, *J. Chem. Soc.* **91**, 1766 (1907). Configuration: Posternak, *Helv. Chim. Acta* **19**, 1007 (1936). Synthesis: G. E. McCasland *et al., J. Org. Chem.* **33**, 4220 (1968).

Sweet crystals. mp 234-235°. $[\alpha]_D^{20}$ +24 to +26°. Sol in water; slightly sol in hot, almost insol in cold alcohol; practically insol in ether.

8153. Quercitrin. [522-12-3] 3-[(6-Deoxy-α-L-mannopyranosyl)oxy]-2-(3,4-dihydroxyphenyl)-5,7-dihydroxy-4*H*-1-benzopyran-4-one; quercitroside; quercimelin; quercetin-3-L-rhamnoside; thujin. $C_{21}H_{20}O_{11}$; mol wt 448.38. C 56.25%, H 4.50%, O 39.25%. From *Aesculus hippocastanum* L., *Hippocastanaceae:* Hörhammer *et al., Arch. Pharm.* **292**, 113 (1959). Structure: Zemplén *et al.,*

Ber. **61**, 2486 (1928); Marchlewski, Skarzynski, *Biochem. Z.* **297**, 56 (1938); Wolfrom, Thompson in R. L. Whistler, M. L. Wolfrom, *Methods in Carbohydrate Chemistry* **vol. 1** (Academic Press, New York, 1962) p 202.

Yellow crystals from dil methanol or ethanol, mp 176-179°; from water, mp 167°. uv max (ethanol): 350, 258 nm (log ε 4.18, 4.30). Practically insol in cold water, ether; sol in alc; moderately sol in hot water; sol in aq alkaline solns with intense yellow color which is oxidized by air to brown.

USE: Has been used as textile dye. *Flavine yellow shade* is prepared by extracting quercitron bark with high pressure steam and consists mainly of quercitrin: Tisdale, *Can. Text. J.* **57**, 44 (1941).

8154.   Quercus. White oak. Dried inner bark of trunk and branches of *Quercus alba* L., *Fagaceae.* *Habit.* Eastern U.S. and Canada. *Constit.* Tannic acid, oak-red, resin, pectin, levulin, quercitol.

THERAP CAT: Astringent.

8155.   Quetiapine. [111974-69-7] 2-[2-(4-Dibenzo[*b*,*f*][1,-4]thiazepin-11-yl-1-piperazinyl)ethoxy]ethanol; 11-[4-[2-(2-hydroxyethoxy)ethyl]-1-piperazinyl]dibenzo[*b*,*f*][1,4]thiazepine. $C_{21}H_{25}N_3O_2S$; mol wt 383.51. C 65.77%, H 6.57%, N 10.96%, O 8.34%, S 8.36%. Combined serotonin (5HT$_2$) and dopamine (D$_2$) receptor antagonist. Prepn: E. J. Warawa, B. M. Migler, **EP 240228**; *eidem*, **US 4879288** (1987, 1989 both to ICI). HPLC and GC/MS determn in plasma: R. H. Pullen *et al.*, *J. Chromatogr.* **573**, 49 (1992). Series of articles on pharmacology: *Psychopharmacology* **112**, 285-307 (1993). Review of clinical experience in schizophrenia: B. Green, *Curr. Med. Res. Opin.* **15**, 145-151 (1999); in major depressive disorder: M. Bauer *et al.*, *J. Affect. Disord.* **127**, 19-30 (2010); in bipolar depression: P. G. Janicak, J. T. Rado, *Expert Opin. Pharmacother.* **12**, 1643-1651 (2011).

Hemifumarate. [111974-72-2] ICI-204636; Seroquel. $(C_{21}H_{25}N_3O_2S)_2 \cdot C_4H_4O_4$; mol wt 883.09. Crystals from ethanol, mp 172-173°.

THERAP CAT: Antipsychotic.

8156.   Quillaja. Soap bark; quillay bark; Panama bark; China bark; Murillo bark. Inner dried bark of *Quillaja saponaria* Molina, *Rosaceae.* *Habit.* South America (Peru, Chile); cultivated in Northern Hindustan. *Constit.* Quillaic acid, quillajasaponin, sucrose, tannin.

USE: Manuf of saponin; in mineral water industry, shampoo liquids, etc.; as foam producer.

8157.   Quillaja Saponin. The saponin of quillay bark. Isolation: Kobert, *Arch. Exp. Pathol. Pharmakol.* **23**, 233 (1887); Pachorukow, *Arbeiten des Pharmakologischen Instituts zu Dorpat* **I**, 5

(1888); Cofman-Nicoresti, *Pharm. J.* **111**, 103 (1923), found the total saponin content of quillay bark to be ~9 or 10%. The saponin is built from quillaic acid, a sugar and possibly other substances.

Amorphous, deliquescent powder which causes sneezing when dispersed in the air. Freely sol in dil alc. Foams easily when shaken with water, the foam being relatively stable.

8158.   Quin2. [73630-23-6] (salt); [83014-44-2] (acid). *N*-[2-[[8-[bis(carboxymethyl)amino]-6-methoxy-2-quinolinyl]methoxy]-4-methylphenyl]-*N*-(carboxymethyl)glycine potassium salt (1:4); 2-[[2-[bis(carboxymethyl)amino]-5-methylphenoxy]methyl]-6-methoxy-8-[bis(carboxymethyl)amino]quinoline; quin-2. $C_{26}H_{23}K_4N_3O_{10}$; mol wt 693.87. C 45.01%, K 22.54%, N 6.06%, O 23.06%. Fluorescent indicator dye which is primarily used for measuring changes in intracellular Ca^{2+}. Increases in ion concn results in increased fluorescence without a shift in wavelengths. Prepn: R. Y. Tsien, *Biochemistry* **19**, 2396 (1980). Use in measuring Ca^{2+} flow: J. N. Crofts, G. J. Barritt, *Biochem. J.* **264**, 61 (1989); Zn^{2+}-protein interactions: J. R. Jefferson *et al.*, *Anal. Biochem.* **187**, 328 (1990); in fluorescent lifetime imaging of cells: J. R. Lakowicz *et al.*, *Cell Calcium* **15**, 7 (1994). Fluorescent properties of acetoxymethyl ester: N. Miyoshi *et al.*, *Appl. Fluoresc. Technol.* **4**, 3 (1992). Dissociation parameters for Ca^{2+} and Mg^{2+} complexes: D. T. W. Bryant, *Biochem. J.* **226**, 613 (1985); and photophysics: V. Van Den Bergh *et al.*, *Photochem. Photobiol.* **61**, 442 (1995). Review of use in measurement of Ca^{2+}: R. Y. Tsien, T. Pozzan, *Methods Enzymol.* **172**, 230-232 (1989).

uv max absorption (0.1*M* KCl): 261, 354 nm (ε 37000, 5000) free dye; 240, 332 nm (ε 36000, 5000) Ca^{2+} complex.

Acetoxymethyl ester. [83104-85-2] Quin-2AM. $C_{38}H_{43}N_3O_{18}$.

USE: Fluorescent probe for measurement of calcium in biological systems and for selected divalent cations.

8159.   Quinacillin. [1596-63-0] (2*S*,5*R*,6*R*)-6-[[(3-Carboxy-2-quinoxalinyl)carbonyl]amino]-3,3-dimethyl-7-oxo-4-thia-1-azabicyclo[3.2.0]heptane-2-carboxylic acid; 3-carboxy-2-quinoxalinylpenicillin. $C_{18}H_{16}N_4O_6S$; mol wt 416.41. C 51.92%, H 3.87%, N 13.46%, O 23.05%, S 7.70%. Semi-synthetic antibiotic related to penicillin. Prepd by condensation of quinoxaline-2,3-dicarboxylic anhydride with 6-aminopenicillanic acid: Richards *et al.*, *Nature* **199**, 354 (1963).

Disodium salt. [985-32-0] $C_{18}H_{14}N_4Na_2O_6S$. Crystals, dec 261-262°. $[\alpha]_D^{23}$ +183.5° (water). Very hygroscopic. uv max (containing 9.2% water): 242, 326 nm (ε 32,100; 7280). Acquires a bright yellow color on exposure to strong sunlight but is stable at 100° for at least 3 months. Freely sol in water; a 25% aq soln is stable for 2 months at 0°. Antimicrobial activity is highest against *Staphylococcus aureus*.

Bistriethylammonium salt monohydrate. $C_{30}H_{46}N_6O_6S \cdot H_2O$. Crystals from acetone, dec 135-137°. $[\alpha]_D^{20}$ +142° (c = 0.376 in water).

THERAP CAT: Antibacterial.

8160. Quinacrine. [83-89-6] N^4-(6-Chloro-2-methoxy-9-acridinyl)-N^1,N^1-diethyl-1,4-pentanediamine; 6-chloro-9-[[4-(diethylamino)-1-methylbutyl]amino]-2-methoxyacridine; 3-chloro-7-methoxy-9-(1-methyl-4-diethyl aminobutylamino)acridine; 2-chloro-5-(ω-diethylamino-α-methylbutylamino)-7-methoxyacridine; mepacrine. $C_{23}H_{30}ClN_3O$; mol wt 399.96. C 69.07%, H 7.56%, Cl 8.86%, N 10.51%, O 4.00%. Prepd by condensing 1-diethylamino-4-aminopentane with 3,9-dichloro-7-methoxyacridine: F. Mietzsch, H. Mauss, **DE 553072**; **DE 571499** (1934); *eidem*, **US 571499** (1938 to Winthrop Chemical); *cf.* Drosdov, Cherutzov, *J. Gen. Chem. USSR* **5**, 1576, 1736 (1935); **8**, 1192 (1938); Magidson, Grigorovskii, *Khim. Farm. Prom.* **1933**, 187; Jensch, Eisleb, **US 1782727** (1930); Schulemann *et al.*, **US 1889704** (1932). Review of quinacrine and other acridines: A. O. Wolfe in *Antibiotics* **vol. 3**, J. W. Corcoran, F. E. Hahn, Eds. (Springer-Verlag, New York, 1975) pp 203-233.

Dihydrochloride dihydrate. [6151-30-0] RP-866; SN-390; Atabrine hydrochloride. $C_{23}H_{30}ClN_3O.2HCl.2H_2O$; mol wt 508.91. Bitter, bright yellow crystals. Dec 248-250° (mp poorly discernible). One gram dissolves in ~35 ml water. Much more sol in hot water. Slightly sol in ethanol, somewhat more sol in methanol. Insol in ether, benzene, acetone. pH of a 1% aq soln ~4.5. Under uv light the yellow aq solns exhibit a vivid fluorescence which is detectable in a dilution of 1:5,000,000.

Methanesulfonate monohydrate. [6598-46-5] $C_{25}H_{38}ClN_3O_7S_2.H_2O$. Bitter, bright yellow crystals. One part dissolves in 3 parts water at 15.5°, in 36 parts 95% alcohol at 15.5°. pH of 2.0% w/v soln in water 3.0-5.0.

THERAP CAT: Anthelmintic (Cestodes); antimalarial.

THERAP CAT (VET): Antiprotozoal, teniacide.

8161. Quinagolide. [87056-78-8] *rel-N,N*-Diethyl-N'-[(3R,4aR,10aS)-1,2,3,4,4a,5,10,10a-octahydro-6-hydroxy-1-propylbenzo[g]quinolin-3-yl]sulfamide; ($\pm$)-1-*n*-propyl-3α-diethylsulfamoylamino-6-hydroxy-1,2,3,4,4aα,5,10,10aβ-octahydrobenzo[g]quinoline. $C_{20}H_{33}N_3O_3S$; mol wt 395.56. C 60.73%, H 8.41%, N 10.62%, O 12.13%, S 8.10%. Nonergot dopamine D_2-receptor agonist. Prepn: R. Nordmann, T. J. Petcher, **EP 77754**; *eidem*, **US 4565818** (1983, 1986 both to Sandoz); *eidem*, *J. Med. Chem.* **28**, 367 (1985). Receptor binding study *in vivo*: A. Closse *et al.*, *Brain Res.* **440**, 123 (1988). Clinical pharmacology: R. C. Gaillard, J. Brownell, *Life Sci.* **43**, 1355 (1988). Clinical evaluation in hyperprolactinemia: C. Rasmussen *et al.*, *Acta Endocrinol.* **125**, 170 (1991).

Relative stereochemistry

Beige powder, mp 122.5-124°.

Hydrochloride. [94424-50-7] CV-205-502; SDZ-205-502; Norprolac. $C_{20}H_{33}N_3O_3S.HCl$; mol wt 432.02. Crystals from methylene chloride + methanol, mp 234-236°.

THERAP CAT: Prolactin inhibitor.

8162. Quinaldic Acid. [93-10-7] 2-Quinolinecarboxylic acid; quinaldinic acid. $C_{10}H_7NO_2$; mol wt 173.17. C 69.36%, H 4.07%, N 8.09%, O 18.48%.

Dihydrate, crystals. mp 155-157° (anhydr). Moderately sol in water; sol in alcohol and in alkali solns.

USE: For the determination of copper, zinc and uranium with which it forms insol salts.

8163. Quinaldine. [91-63-4] 2-Methylquinoline. $C_{10}H_9N$; mol wt 143.19. C 83.88%, H 6.34%, N 9.78%. Occurs in coal tar; made from aniline, acetaldehyde and HCl. Lab procedure: A. I. Vogel, *Practical Organic Chemistry* (Longmans, London, 3rd ed., 1959) p 831; Gattermann-Wieland, *Praxis des Organischen Chemikers* (de Gruyter, Berlin, 40th ed., 1961) p 318. Pharmacology: Brown *et al.*, *Comp. Biochem. Physiol.* **42A**, 223 (1972). Toxicity data: Smyth *et al.*, *Arch. Ind. Hyg. Occup. Med.* **4**, 119 (1951); L. L. Marking, *Invest. Fish Control* **23**, 3 (1969).

Colorless, oily liquid; quinoline odor; becomes reddish-brown on exposure to air. d ~1.06. bp 246-247°. Practically insol in water. Sol in chloroform, ether. *Keep tightly closed and protected from light.* LD_{50} orally in rats: 1.23 g/kg (Smyth).

Sulfate. [655-76-5] $C_{10}H_9N.H_2SO_4$. Prepn: S. Hoogewerff, W. A. Van Dorp, *Rec. Trav. Chim.* **3**, 345 (1884); J. L. Allen, J. B. Sills, *Invest. Fish Control* **47**, 3 (1973). Efficacy as anesthetic in fish: P. A. Gilderhus *et al.*, *ibid.* **49**, 3 (1973); G. C. Blasiola Jr., *J. Fish Biol.* **10**, 113 (1977). Toxicity: L. L. Marking, V. K. Dawson, *Invest. Fish Control* **48**, 3 (1973). Light yellow crystalline powder, mp 211-214°. uv max (0.1N H_2SO_4): 236, 317 nm. Soly (g/100 ml) in water 104.05, in methanol 7.44, in ethanol 2.27, in acetone 0.08. Practically insol in ether, benzene, hexane. LC_{50} (96 hour) in largemouth bass: 6.80 mg/l; in carp: 72.5 mg/l (Marking, Dawson).

USE: As anesthetic in transport and handling of fish.

8164. Quinaldine Blue. [2768-90-3] 1-Ethyl-2-[3-(1-ethyl-2(1H)-quinolinylidene)-1-propen-1-yl]quinolinium chloride (1:1); bis[1-ethylquinoline-(2)]trimethinecyanine chloride; 1,1'-diethyl-2,2'-trimethinequinocyanine chloride; pinacyanol chloride; Vernitest Reagent. $C_{25}H_{25}ClN_2$; mol wt 388.94. C 77.20%, H 6.48%, Cl 9.11%, N 7.20%. Prepn: Fischer, *J. Prakt. Chem.* **98**, 204 (1918). Properties and use as histological stain: F. Proescher, *Proc. Soc. Exp. Biol. Med.* **31**, 79 (1933); H. P. Klinger *et al.*, *Stain Technol.* **46**, 43 (1971); R. K. J. Narayan, *ibid.* **55**, 9 (1980).

Bright blue-green alcohol-containing prisms or needles from alcohol. Drying at 110° expels alcohol of crystn. Dec at ~263°. Moderately sol in water with a violet-red color, in alcohol with a blue color. Solns are dichroic.

USE: Histological stain (chromosomes).

8165. Quinaldine Red. [117-92-0] 2-[2-[4-(Dimethylamino)phenyl]ethenyl]-1-ethylquinolinium iodide (1:1); 2-[p-(dimethylamino)styryl]-1-ethylquinolinium iodide; 2-(p-dimethylaminostyryl)quinoline ethiodide; α-(p-dimethylaminophenylethylene)quinoline ethiodide; Eastman no. 1361. $C_{21}H_{23}IN_2$; mol wt 430.33. C 58.61%, H 5.39%, I 29.49%, N 6.51%. Prepd by condensing quinaldine ethyliodide and p-dimethylaminobenzaldehyde in alc in the presence of piperidine: König, *J. Prakt. Chem.* [2] **86**, 172 (1912); **DE 294744**, *Frdl.* **14**, 735.

Very dark red powder. Sparingly sol in water. Freely sol in alcohol, giving a dark red soln. Solns are slowly dec by light. pKb 11.25. Colorless at pH 1.4; red at pH 3.2. In the presence of chlorides the transition interval is from pH 1.2 to pH 3.0.

USE: As an indicator.

8166. Quinalizarin. [81-61-8] 1,2,5,8-Tetrahydroxy-9,10-anthracenedione; 1,2,5,8-tetrahydroxyanthraquinone; alizarin bordeaux B; C.I. Mordant Violet 26; C.I. 58500. $C_{14}H_8O_6$; mol wt 272.21. C 61.77%, H 2.96%, O 35.26%. From hemipic (hemipinic) acid and hydroquinone with H_2SO_4: Liebermann, Kostanecki, *Ann.* **240**, 245 (1887); Liebermann, Wense, *Ber.* **20**, 862 (1887). From alizarin or quinizarin by treatment with 80% oleum, then boiling with caustic: Gattermann, *J. Prakt. Chem.* [2] **43**, 246 (1891); Schmidt, *ibid.* **237**, 242 (1891); *Bull. Soc. Ind. Mulhouse* **84**, 409 (1914). *See also: Colour Index* **vol. 4** (3rd ed., 1971) p 4519.

Red needles with green metallic luster from acetic acid or from nitrobenzene or by sublimation *in vacuo*. mp >275°. Absorption spectrum: Meek, Watson, *J. Chem. Soc.* **109**, 544 (1916); Meek, *ibid.* **111**, 969 (1917). Insol in water. Very slightly sol in most other solvents. Dissolves in aq solns of alkalies with reddish-violet, in acetic acid with yellow, in sulfuric acid with blue-violet color.

USE: Dyes cotton mordanted with Al salts dark red, cotton mordanted with Cr salts bluish-violet. No longer used as a dye, except occasionally in printing cotton.

8167. Quinapril. [85441-61-8] (3S)-2-[(2S)-2-[[(1S)-1-(Ethoxycarbonyl)-3-phenylpropyl]amino]-1-oxopropyl]-1,2,3,4-tetrahydro-3-isoquinolinecarboxylic acid; (S)-2-[(S)-N-[(S)-1-carboxy-3-phenylpropyl]alanyl]-1,2,3,4-tetrahydro-3-isoquinolinecarboxylic acid 1-ethyl ester. $C_{25}H_{30}N_2O_5$; mol wt 438.52. C 68.47%, H 6.90%, N 6.39%, O 18.24%. Angiotensin converting enzyme (ACE) inhibitor. Prepn: M. L. Hoefle, S. Klutchko, *EP 49605*; *eidem, US 4344949* (both 1982 to Warner-Lambert); S. Klutchko *et al., J. Med. Chem.* **29**, 1953 (1986). Synthesis of purified crystalline form: O. P. Goel, U. Krolls, *US 4761479* (1988 to Warner-Lambert). Pharmacology: H. R. Kaplan *et al., Fed. Proc.* **43**, 1326 (1984); M. J. Ryan *et al., ibid.* 1330. HPLC determn of quinapril and quinaprilat in plasma and urine: H. Hengy, M. Most, *J. Liq. Chromatogr.* **11**, 517 (1988). Toxicity data: H. R. Kaplan *et al., Angiology* **40**, 335 (1989). Review of pharmacology, pharmacokinetics and clinical experience in congestive heart failure and hypertension: C. R. Culy, B. Jarvis, *Drugs* **62**, 339-385 (2002).

Hydrochloride. [82586-55-8] CI-906; PD-109452-2; Accupril; Accupro; Acequin; Acuitel; Korec; Quinazil. $C_{25}H_{30}N_2O_5 \cdot HCl$; mol wt 474.98. Crystals from ethyl acetate-toluene, mp 120-130°. $[\alpha]_D^{23}$ +14.5° (c = 1.2 in ethanol) (Klutchko). Also reported as white crystalline solid from acetonitrile, mp 119-121.5°. $[\alpha]_D^{25}$ +15.4° (c =

2 in methanol) (Goel, Krolls). Freely sol in aq solvents. LD_{50} in male, female mice, rats (mg/kg): 1739, 1840, 4280, 3541 orally; 504, 523, 158, 107 i.v. (Kaplan, 1989).

Diacid. [82768-85-2] Quinaprilat; CI-928; Accuprin. $C_{23}H_{26}N_2O_5$; mol wt 410.47. Active metabolite. Hydrate, crystals from methanol-ethyl ether, mp 166-168°. $[\alpha]_D^{23}$ +20.9° (c = 1 in methanol).

THERAP CAT: Antihypertensive. In treatment of congestive heart failure.

8168. Quinapyramine. [20493-41-8] 4-Amino-6-[(2-amino-1,6-dimethyl-4(1H)-pyrimidinylidene)amino]-1,2-dimethylquinolinium conjugate monoacid; 4-amino-6-[(2-amino-6-methyl-4-pyrimidinyl)amino]-1-methylquinaldinium methosalts; 4-amino-6-[(2-amino-1,6-dimethyl-4-pyrimidinyl)amino]-1,2-dimethylquinoline salts; 4-amino-6-(2-amino-6-methyl-4-pyrimidylamino)quinaldine-1,1'-dimethosalts; M-7555; Antrycide. Prepn: F. H. S. Curd, D. G. Davey, *Br. J. Pharmacol.* **5**, 25 (1950); Curd, *US 2585917* (1952 to I.C.I.); Ainley *et al., J. Chem. Soc.* **1953**, 59.

Dimethosulfate. $C_{19}H_{28}N_6O_8S_2$. Creamy white crystals from aq methanol, mp 265-266°. Freely sol in water. LD_{50} i.v. in mice: 10-15 mg/kg (Curd, Davey).

Diiodide. Pale cream needles, mp 312-313° (dec). Sparingly sol in water.

Dichloride. Crystals from water, mp 316-317° (dec). Sparingly sol in water. LD_{50} i.v. in mice: 10-15 mg/kg (Curd, Davey).

THERAP CAT: Antiprotozoal (Trypanosoma).

8169. Quinazoline. [253-82-7] 1,3-Benzodiazine; benzo[a]-pyrimidine; 5,6-benzopyrimidine; phenmiazine. $C_8H_6N_2$; mol wt 130.15. C 73.83%, H 4.65%, N 21.52%. Prepn from 2-nitrobenzylidenebis(formamide): Riedel, *DE 174941*; *Chem. Zentralbl.* **1906**, II, 1372; *Frdl.* **8**, 1238; Bogert, McColm, *J. Am. Chem. Soc.* **49**, 2650 (1927).

Leaflets from petr ether; odor of quinoline. Slightly bitter taste. mp 48.0-48.5°. $bp_{772.5}$ 243°; bp_{764} 241.5°. Freely sol in water. Neutral reaction. Sol in many organic solvents.

8170. Quince Seed. Gum quince seed; semen cydonia; golden apple seed; cydonia seed. Seed of *Cydonia oblonga* Mill. (*C. vulgaris* Pers.), *Rosaceae*. *Habit.* Southern Asia, Europe; widely cultivated. *Constit.* Amygdalin, emulsin, about 15% fatty oil, about 20% of a mucilage named cydonin. Brief review of seed and gum uses: BeMiller in *Industrial Gums*, R. L. Whistler, Ed. (Academic Press, New York, 2nd ed., 1973) pp 339-345.

USE: Gum from the seeds as suspending agent, stabilizer; in hair and cosmetic prepns.

8171. Quinestrol. [152-43-2] (17α)-3-(Cyclopentyloxy)-19-norpregna-1,3,5(10)-trien-20-yn-17-ol; 17α-ethinylestradiol 3-cyclopentyl ether; W-3566; Estrovis. $C_{25}H_{32}O_2$; mol wt 364.53. C 82.37%, H 8.85%, O 8.78%. Prepn: Ercoli, Gardi, *Chem. Ind. (London)* **1961**, 1037; Ercoli, *US 3159543*; Ercoli *et al., US 3231567* (1964, 1966, both to Vismara).

Crystals, mp 107-108°. $[\alpha]_D^{25}$ +5° (c = 0.5 in dioxane).
THERAP CAT: Estrogen.

8172. Quinethazone. [73-49-4] 7-Chloro-2-ethyl-1,2,3,4-tetrahydro-4-oxo-6-quinazolinesulfonamide; 7-chloro-2-ethyl-6-sulfamoyl-1,2,3,4-tetrahydro-4-quinazolinone; 7-chloro-2-ethyl-1,2,-3,4-tetrahydro-4-oxo-6-sulfamoylquinazoline; CL-36010; Hydromox; Aquamox. $C_{10}H_{12}ClN_3O_3S$; mol wt 289.73. C 41.46%, H 4.17%, Cl 12.24%, N 14.50%, O 16.57%, S 11.07%. Prepn: Cohen et al., J. Am. Chem. Soc. **82**, 2731 (1960); Cohen, Vaughan, Jr., US 2976289 (1961 to Am. Cyanamid).

Fibrous crystals from 50% acetone, mp 250-252°. Sol in acetone, alcohol.
THERAP CAT: Diuretic, antihypertensive.

8173. Quinfamide. [62265-68-3] 2-Furancarboxylic acid 1-(dichloroacetyl)-1,2,3,4-tetrahydro-6-quinolinyl ester; 2-furoic acid ester with 1-(dichloroacetyl)-1,2,3,4-tetrahydro-6-quinolinol; 1-(dichloroacetyl)-6-(2-furoyloxy)-1,2,3,4-tetrahydroquinoline; Win-40014; Amenox; Amenide. $C_{16}H_{13}Cl_2NO_4$; mol wt 354.18. C 54.26%, H 3.70%, Cl 20.02%, N 3.95%, O 18.07%. Prepn: D. M. Bailey, US 3997542 (1976 to Sterling); D. M. Bailey et al., J. Med. Chem. **22**, 600 (1979). Amebicidal activity and toxicological evaluation: R. G. Slighter et al., Parasitology **81**, 157 (1980). Distribution and metabolism in rats: J. F. Baker, Arch. Int. Pharmacodyn. **258**, 29 (1982). HPLC determn in biological samples: J. M. Morales et al., J. Chromatogr. B **746**, 133 (2000). Clinical evaluation in adults with chronic amebiasis: L. Guevara et al., Clin. Ther. **6**, 43 (1983); in children: F. A. Rojas et al., ibid. 47.

Crystals from ethyl acetate, mp 150.5-151°.
THERAP CAT: Antiamebic.

8174. Quinhydrone. [106-34-3] 2,5-Cyclohexadiene-1,4-dione compd with 1,4-benzenediol (1:1); green hydroquinone. $C_{12}H_{10}O_4$; mol wt 218.21. C 66.05%, H 4.62%, O 29.33%. An addn compd of one mol hydroquinone and one mol quinone. Prepn from hydroquinone and quinone: Gattermann-Wieland, Praxis des Organischen Chemikers (de Gruyter, Berlin, 40th ed., 1961) p 270. Alternate prepn by the action of ferric ammonium sulfate on hydroquinone: A. I. Vogel, Practical Organic Chemistry (Longmans, London, 3rd ed., 1959) p 747. Toxicity study: Woodward et al., Fed. Proc. **8**, 348 (1949).

Green crystals with metallic luster; reddish-brown by transmitted light. d 1.40. mp 171°; sublimes with partial decompn. Slightly sol in cold water; sol in hot water, ammonia, alc, ether. Insol in petr ether. LD_{50} orally in rats: 225 mg/kg (Woodward).
USE: In pH determinations (quinhydrone electrode).

8175. Quinic Acid. [77-95-2] (1α,3R,4α,5R)-1,3,4,5-Tetrahydroxycyclohexanecarboxylic acid; chinic acid; kinic acid; (−)-hexahydro-1,3,4,5-tetrahydroxybenzoic acid. $C_7H_{12}O_6$; mol wt 192.17. C 43.75%, H 6.29%, O 49.95%. Found in cinchona bark, particularly in South American barks; also in many other plants, such as tobacco leaves, carrot leaves, apples, peaches, pears, plums, etc. Structure and configuration: Fischer, Dangschat, Ber. **65**, 1009 (1932). Total synthesis: Grewe et al., ibid. **87**, 793 (1954); Smissman, Oxman, J. Am. Chem. Soc. **85**, 2184 (1963). Stereospecific synthesis: Wolinsky et al., J. Org. Chem. **29**, 3596 (1964). Review: Bohm, Chem. Rev. **65**, 435 (1965).

White crystals; strong acid taste. d 1.64. mp 162-163°; at higher temps forms a lactone. $[\alpha]_D^{20}$ −42 to −44° in water. Sol in 2.5 parts water, in alcohol, glacial acetic acid.

8176. Quinidine. [56-54-2] (9S)-6'-Methoxycinchonan-9-ol; α-(6-methoxy-4-quinolyl)-5-vinyl-2-quinuclidinemethanol; conquinine; pitayine; β-quinine. $C_{20}H_{24}N_2O_2$; mol wt 324.42. C 74.05%, H 7.46%, N 8.64%, O 9.86%. A dextrorotatory stereoisomer of quinine, q.v. Present in cinchona barks to the extent of 0.25-3.0%. Found in quinine sulfate mother liquors. Review of structural elucidation and early synthetic studies: R. B. Turner, R. B. Woodward, in The Alkaloids vol. 3, 1-63 (1953). Configuration: Prelog, Zalán, Helv. Chim. Acta **27**, 535 (1944); Prelog, Häfliger, ibid. **33**, 2021 (1950); Roth, Pharmazie **16**, 257 (1961). Crystal and molecular structure: R. Doherty et al., J. Pharm. Sci. **67**, 1698 (1978). Rotatory dispersion studies: Lyle, Gaffield, Tetrahedron Lett. **1963**, 1371. Prepn by isomerization of quinine: W. E. Doering et al., J. Am. Chem. Soc. **69**, 1700 (1949). Total synthesis: J. Gutzwiller, M. Uskokovic, ibid. **92**, 204 (1970); eidem, Helv. Chim. Acta **56**, 1494 (1973); eidem, J. Am. Chem. Soc. **100**, 576 (1978). Toxicity data: C. Turba et al., Arzneim.-Forsch. **18**, 1127 (1968); K. Dietmann et al., ibid. **27**, 589 (1977). Comprehensive description of the sulfate: M. A. Loutfy et al., Anal. Profiles Drug Subs. **12**, 483-546 (1983). Clinical evaluation in severe malaria: R. E. Phillips et al., N. Engl. J. Med. **312**, 1273 (1985). Review of pharmacology and clinical efficacy in cardiac arrhythmias: J. W. Mason, L. M. Hondeghem, Ann. N.Y. Acad. Sci. **432**, 162-176 (1984); A. R. Leon, J. D. Merlino, Heart Dis. Stroke **2**, 407-413 (1993).

Triboluminescent. mp 174-175° after drying of solvated crystals. $[\alpha]_D^{15}$ +230° (c = 1.8 in chloroform), $[\alpha]_D^{17}$ +258° (alc), $[\alpha]_D^{17}$ +322° (c = 1.6 in 2M HCl). pK_1 (20°) 5.4; pK_2 10.0. Blue fluorescence in dil H_2SO_4. The uv absorption spectrum is identical with that of quinine. One gram dissolves in about 2000 ml cold, 800 ml boiling water, 36 ml alcohol, 56 ml ether, 1.6 ml chloroform; very sol in methanol. Practically insol in petr ether. LD_{50} in rats (mg/kg): 30 i.v., 263 orally (Dietmann).
Hemipentahydrate. Prisms from dil alcohol, loses ½ H_2O in air, mp ~168°.
Hydrogen sulfate tetrahydrate. [6151-39-9] Quinidine bisulfate; Chinidin-Duriles; Kiditard; Kinichron; Kinidin Durules; Quiniduran. $C_{20}H_{24}N_2O_2 \cdot H_2SO_4 \cdot 4H_2O$; mol wt 494.56. Rods, sol in 8 parts water with blue fluorescence.

Sulfate dihydrate. [6591-63-5]; [50-54-4] (anhydrous). Cin-Quin; Quinidex Extentabs; Quinicardine; Quinora. $(C_{20}H_{24}N_2O_2)_2.H_2SO_4.2H_2O$; mol wt 782.95. White, very bitter, odorless, fine crystals, frequently cohering in masses. Darkens on exposure to light. Does not lose all of its water below 120°. $[\alpha]_D^{25}$ ~+212° (95% alcohol); ~+260° (dil HCl). pKa 4.2, 8.8. pH (1% aq soln): 6.0-6.8. One gram dissolves in about 90 ml water, 15 ml boiling water, 10 ml alcohol, 3 ml methanol, 12 ml chloroform. Insol in ether, benzene. *Protect from light.* LD_{50} in mice, rats (mg/kg): 700, 455.8 orally; 83, 56 i.v. (Turba).

Gluconate. [7054-25-3] Gluconic acid quinidine salt; Duraquin; Quinaglute. $C_{26}H_{36}N_2O_9$; mol wt 520.58. Crystals, mp 175-176.5°. Sol in 9 parts water, 60 parts alcohol.

Polygalacturonate. [7681-28-9] Cardioquin; Galactoquin; Nati-cardina. $(C_{20}H_{24}N_2O_2.C_6H_{10}O_7.H_2O)_x$. Prepn: A. Halpern *et al., Am. J. Pharm.* **130**, 190 (1958). Pharmacology: A. Halpern *et al., Antibiot. Chemother.* **9**, 97 (1959). Amorphous powder, mp 180° (dec). Anhydr product is insol in methanol, ethanol, chloroform, ether, acetone, dioxane. Soly in hot 40% methanol or ethanol: 12%; in water at 25°: ~2%. LD_{50} in rats, mice (mg/kg): 3200 ± 350, 2680 ± 210 orally (Halpern, 1959).

THERAP CAT: Antiarrhythmic (class IA); antimalarial.

THERAP CAT (VET): Antiarrhythmic.

8177. Quinine. [130-95-0] $(8\alpha,9R)$-6'-Methoxycinchonan-9-ol. $C_{20}H_{24}N_2O_2$; mol wt 324.42. C 74.05%, H 7.46%, N 8.64%, O 9.86%. Primary alkaloid of various species of *Cinchona* (Rubiaceae), *see* Cinchona. Representative samples of dried bark contain ~0.8 to 4% quinine. Optical isomer of quinidine, q.v. Isoln: Pelletier, Caventau, *Ann. Chim. Phys.* [2] **15**, 291 (1820). Extraction procedure: Jucker, Stoll, in *Ullmanns Encyklopädie der technischen Chemie* vol. 3, 213-218 (1953). Configuration: Prelog, Zalán, *Helv. Chim. Acta* **27**, 535 (1944); Prelog, Häfliger, *ibid.* **33**, 2021 (1950); Roth, *Pharmazie* **16**, 257 (1961). Synthesis: Woodward, Doering, *J. Am. Chem. Soc.* **66**, 849 (1944); **67**, 860 (1945); Taylor, Martin, *ibid.* **94**, 6218 (1972); Gutzwiller, Uskokovic, *ibid.* **100**, 576 (1978); G. Grethe *et al., ibid.* 589; T. Imanishi *et al., Chem. Pharm. Bull.* **30**, 1925 (1982). Review of structural elucidation and early synthetic studies: R. B. Turner, R. B. Woodward in *The Alkaloids* vol. 3, 1-63 (1953); of bioactivity: F. E. Hahn, Ed. in *Antibiotics* vol. 5 (pt. 2) (Springer-Verlag, New York, 1979) pp 353-362. Comprehensive description of the hydrochloride: F. J. Muhtadi *et al., Anal. Profiles Drug Subs.* **12**, 547-621 (1983). LC determn in soft drinks: L. P. Valenti, *J. Assoc. Off. Anal. Chem.* **68**, 782 (1985). HPLC determn in blood: V. K. Dua *et al. J. Chromatogr.* **614**, 87 (1993). Clinical evaluation to relieve nocturnal leg cramps: P. S. Connolly *et al., Arch. Intern. Med.* **152**, 1877 (1992). Clinical efficacy in malaria: P. G. Kremsner *et al., J. Infect. Dis.* **169**, 467-470 (1994). Review of historical importance and total syntheses: T. S. Kaufman, E. A. Ruveda, *Angew. Chem. Int. Ed.* **44**, 854-885 (2005).

Triboluminescent, orthorhombic needles from abs alcohol, mp 177° (some decompn). Sublimes in high vacuum at 170-180°. $[\alpha]_D^{15}$ −169° (c = 2 in 97% alcohol), $[\alpha]_D^{17}$ −117° (c = 1.5 in chloroform), $[\alpha]_D^{15}$ −285° (c = 0.4M in 0.1N H_2SO_4). pK_1 (18°) 5.07; pK_2 9.7. pH of satd aq soln 8.8. Absorption spectra: Dobbie, Lauder, *J. Chem. Soc.* **99**, 1260 (1911); Dobbie, Fox, *ibid.* **101**, 78 (1912). Fluorescence: Rabe, Marschall, *Ann.* **382**, 362 (1911). The blue fluorescence is especially strong in dil H_2SO_4. One gram dissolves in 1900 ml water, 760 ml boiling water, 0.8 ml alcohol, 80 ml benzene (in 18 ml benzene at 50°), in 1.2 ml chloroform; 250 ml dry ether, 20 ml glycerol, 1900 ml of 10% ammonia water. Almost insol in petr ether.

Trihydrate. Microcrystalline powder, mp 57°, efflorescent, loses one H_2O in air, two H_2O over H_2SO_4, anhydr at 125°.

Bisulfate heptahydrate. [6183-68-2]; [549-56-4] (anhydrous). Biquinate. $C_{20}H_{24}N_2O_2.H_2SO_4.7H_2O$; mol wt 548.60. Very bitter crystals or cryst powder; efflorescent on exposure to air and darkens on exposure to light. One gram dissolves in 9 ml water, 0.7 ml boiling water, 23 ml alcohol, 0.7 ml alcohol at 60°, 625 ml chloroform, 2500 ml ether, 15 ml glycerol. pH: 3.5.

Dihydrochloride. [60-93-5] Quinine dichloride; quinine bimuriate; acid quinine hydrochloride. $C_{20}H_{24}N_2O_2.2HCl$. Very bitter powder or crystals. One gram dissolves in about 0.6 ml water, in about 12 ml alcohol. Slightly sol in chloroform, very slightly sol in ether. Aq solns are strongly acid to litmus paper (pH about 2.6).

Hydrochloride dihydrate. [6119-47-7]; [130-89-2] (anhydrous). $C_{20}H_{24}N_2O_2.HCl.2H_2O$. Bitter, silky needles. Effloresces on exposure to warm air. Does not lose all its water below 120°. One gram dissolves in 16 ml water, in 0.5 ml boiling water, in 1.0 ml alcohol, in about 7.0 ml glycerol, in about 1 ml chloroform, in about 350 ml ether. pH (1% aq soln): 6.0-7.0. Bitterness threshold 1:30000. *Protect from light.*

Sulfate dihydrate. [6119-70-6]; [804-63-7]. Quinamm; Quinate. $(C_{20}H_{24}N_2O_2)_2.H_2SO_4.2H_2O$; mol wt 782.95. Dull needles or rods, making a light and readily compressible mass. Becomes brownish on exposure to light. Loses its water of crystn at about 100°. $[\alpha]_D^{15}$ −220° (5% soln in about 0.5N HCl). One gram dissolves in 810 ml water, 32 ml boiling water, 120 ml alcohol, 10 ml alcohol at 78°. Slightly sol in chloroform, ether, but freely sol in a mixture of 2 vols chloroform and 1 vol abs alcohol. Aq solns are neutral to litmus, pH of satd soln 6.2.

USE: Flavor in carbonated beverages.

THERAP CAT: Antimalarial; muscle relaxant (skeletal).

THERAP CAT (VET): Antiprotozoal for fish.

8178. Quininic Acid. [86-68-0] 6-Methoxy-4-quinolinecarboxylic acid; 6-methoxycinchoninic acid. $C_{11}H_9NO_3$; mol wt 203.20. C 65.02%, H 4.46%, N 6.89%, O 23.62%.

Pale yellow crystals. mp about 280° with dec. Slightly sol in water, cold alcohol or ether; sol in about 80 parts boiling abs alcohol; sol in aq alkalies.

8179. Quininone. [84-31-1] (8α)-6'-Methoxycinchonan-9-one. $C_{20}H_{22}N_2O_2$; mol wt 322.41. C 74.51%, H 6.88%, N 8.69%, O 9.92%. Prepn by careful oxidation of quinine or quinidine: Rabe, Kuliga, *Ann.* **364**, 346, 349 (1909); Woodward *et al., J. Am. Chem. Soc.* **67**, 1425 (1945). Alternate procedure: Rabe, Kindler, *Ber.* **51**, 466 (1918). Prepn of amorphous epimeric mixture of quininone and quinidinone from quinotoxine: Gutzwiller, Uskokovic, *Helv. Chim. Acta* **56**, 1494 (1973).

Crystals from ether, mp 108° (rapid heating). Shows mutarotation, final $[\alpha]_D^{20}$ +76° (c = 2 in alcohol). Alkaline reaction to litmus. Freely sol in alcohol, ether, chloroform, benzene. Almost insol in water and petr ether.

Hydrochloride. $C_{20}H_{22}N_2O_2.HCl$. Hygroscopic crystals, mp 212°. Final $[\alpha]_D^{18}$ +59° (c = 2 in abs alcohol).

8180. Quinizarin. [81-64-1] 1,4-Dihydroxy-9,10-anthra-cenedione; 1,4-dihydroxyanthraquinone; C.I. 58050. $C_{14}H_8O_4$; mol wt 240.21. C 70.00%, H 3.36%, O 26.64%. Prepn from *p*-chloro-phenol and phthalic anhydride: Reynolds, Bigelow, *J. Am. Chem. Soc.* **48**, 420 (1926); US 1845632, *C.A.* **26**, 2203; *Org. Synth.* **coll. vol. I**, 476 (New York, 1941). Prepn from hydroquinone: Gatter-mann-Wieland, *Praxis des Organischen Chemikers* (de Gruyter, Berlin, 40th ed., 1961) p 299. Also prepd from diazotized *p*-chloroaniline and phthalic anhydride: **GB 373999**, *C.A.* **27**, 3946 (1933); by treating anthraquinone with ammonium persulfate in sulfuric acid: Wacker, *J. Prakt. Chem.* [2] **54**, 90 (1896). Purification from contaminating purpurin: *Org. Synth., loc. cit. See also: Colour Index* **vol. 4** (3rd ed., 1971) p 4515.

Orange crystals from acetic acid, mp 200-203° (*Org. Syn.*). Orange plates from ether. Deep red needles from alcohol, benzene, toluene, xylene. mp 196°. Sublimes in high vacuum. Absorption spectrum: Meek, Watson, *J. Chem. Soc.* **109**, 544 (1916); Meek, *ibid.* **111**, 969 (1917). pK (18°) 9.51. Moderately sol in alcohol with red color. Sol in ether with brown color and yellow fluorescence. Sol with violet color in aq alkalies and in ammonia. Black precipitate with CO_2. One gram dissolves in about 13 g of boiling glacial acetic acid.

Dimethyl ether. $C_{16}H_{12}O_4$. mp 177°.

8181. Quinizarin Green SS. [128-80-3] 1,4-Bis[(4-methyl-phenyl)amino]-9,10-anthracenedione; 1,4-di-*p*-toluidinoanthraqui-none; D & C Green No. 6; C.I. Solvent Green 3; C.I. 61565. $C_{28}H_{22}N_2O_2$; mol wt 418.50. C 80.36%, H 5.30%, N 6.69%, O 7.65%. Discovered by R. E. Schmidt in 1894: *Colour Index* **vol. 4** (3rd ed., 1971) p 4541.

Dark violet needles, mp 218°. Blue soln in conc H_2SO_4 giving a blue-green ppt on dilution.

USE: For sutures: *Fed. Regist.* **40**, 18167 (1975). Permitted for use in externally applied drugs and cosmetics: *ibid.* **47**, 14138 (1982).

8182. Quinmerac. [90717-03-6] 7-Chloro-3-methyl-8-quin-olinecarboxylic acid; BAS-518; BAS-518H. $C_{11}H_8ClNO_2$; mol wt 221.64. C 59.61%, H 3.64%, Cl 15.99%, N 6.32%, O 14.44%. Aux-in-type herbicide. Prepn: H. Hagen *et al.*, **DE 3233089**; *idem*, **US 4715889** (1984, 1987 both to BASF). ELISA determn in cereals: R. A. Baumann, V. M. 't Hart-De Kleijn, *Meded. Fac. Landbouwwet. Univ. Gent* **58**, 173 (1993). Mode of action study: K. Grossman, F. Scheltrup, *Pestic. Sci.* **52**, 111 (1998). Field trials in cereals, rapeseed and sugarbeets: W. Nuyken *et al.*, *Proc. Br. Crop Prot. Conf. - Weeds* **1985**, 71. *Review:* B. Wuerzer *et al., ibid.* 63-70.

Colorless, odorless crystalline solid from DMF, mp 244°. Soly at 20° (g/100g solvent): water 2.1×10^{-3}, olive oil <0.1, acetone 0.2. Low soly in other organic solvents. Vapor pressure (20°): $<1 \times 10^{-5}$ Pa. LD_{50} in rats (mg/kg): >5000 orally, >2000 dermally (Wuerzer).

USE: Herbicide.

8183. Quinocide. [525-61-1] N^1-(6-Methoxy-8-quinolinyl)-1,4-pentanediamine; 8-(4-aminopentylamino)-6-methoxyquinoline; 8-[(4-amino-4-methylbutyl)amino]-6-methoxyquinoline; 6-meth-oxy-8-(4-aminopentylamino)quinoline; chinocide; khinocyde. $C_{15}H_{21}N_3O$; mol wt 259.35. C 69.47%, H 8.16%, N 16.20%, O 6.17%. Prepn: Braude, Stavrovskaya, *J. Gen. Chem. USSR* **26**, 999 (1956); modified method: *idem, Med. Prom. SSSR* **11**(7), 19 (1957), *C.A.* **52**, 11043 (1958); B. Balkrishen *et al.*, *Chem. Ind. (London)* **1983**, 899.

bp$_1$ 183-186°. mp 46°.

Hydrochloride. $C_{15}H_{21}N_3O.HCl$. Crystals from abs alcohol, mp 224-224.5°.

Dihydrochloride. $C_{15}H_{21}N_3O.2HCl$. mp 227-227.5°.

Diphosphate. $C_{15}H_{27}N_3O_9P_2$. mp 174-176°.

THERAP CAT: Antimalarial.

8184. Quinoclamine. [2797-51-5] 2-Amino-3-chloro-1,4-naphthalenedione; 2-amino-3-chloro-1,4-naphthoquinone; ACN; ACNQ; Mogeton. $C_{10}H_6ClNO_2$; mol wt 207.61. C 57.85%, H 2.91%, Cl 17.08%, N 6.75%, O 15.41%. Surface algicide and herbicide for liverwort and moss control; inhibits photosynthesis at the photosystem I (PS I) receptor site. Prepn: F. Ullmann, M. Ettisch, *Ber.* **54**, 259 (1921). Mechanism of action study: S. Koura *et al.*, *Biosci. Biotechnol. Biochem.* **58**, 1210 (1994). Herbicidal activity in submerged paddy rice: H. Hagimoto, *Weed Res.* **9**, 296 (1969); *idem, ibid.* **10**, 11 (1970); *idem, ibid.* 187. GC-MS determn in river water: A. Tanabe *et al.*, *J. Chromatogr. A* **754**, 159 (1996). LC-MS/MS determn in agricultural products: S. Takatori *et al.*, *J. AOAC Int.* **91**, 871 (2008). Nursery studies vs. liverwort weed: J. E. Altland *et al.*, *Weed Technol.* **521**, 483 (2007).

Orange powder or yellow crystals, mp 202°. *Irritant.* pH 5.3. d 2.08. Log P (octanol/water): 1.58. Soly at 20° (g/l): water 0.02, methanol 6.57, toluene 3.14.

USE: Herbicide; algicide. Surface cleaner for removal of weathering deposits on outdoor paths and similar surfaces.

8185. Quinoline. [91-22-5] Leucoline; chinoleine; 1-benza-zine; benzo[*b*]pyridine. C_9H_7N; mol wt 129.16. C 83.69%, H 5.46%, N 10.84%. Occurs in small amounts in coal tar. Prepd by the Skraup synthesis by heating aniline with glycerol and nitrobenzene in presence of sulfuric acid: Clarke, Davis, *Org. Synth.* **coll. vol. I**, 478 (2nd ed., 1941) 478; alternate syntheses: Manske, *Chem. Rev.* **30**, 113 (1942); Bergstrom, *ibid.* **35**, 150 (1944). Process based on the interaction of aniline with acetaldehyde and a formaldehyde hemiacetal: Cislak, Wheeler, **US 3020281** (1962 to Reilly Tar & Chem.). Toxicity data: Smyth *et al.*, *Arch. Ind. Hyg. Occup. Med.* **4**, 119 (1951).

Hygroscopic liquid. Absorbs as much as 22% water. Darkens on storage in ordinary, stoppered bottle. Penetrating odor, not as offensive as pyridine. Volatile with steam. d_4^{25} 1.0900; mp −15°; bp_{760} 237.7°; bp_{100} 163.2°; bp_{40} 136.7°; bp_{20} 119.8°; bp_{10} 103.8°; bp_5 89.6°; $bp_{1.0}$ 59.7°. n_D^{20} 1.62683. Absorption spectrum: Hantzsch, *Ber.* **44**, 1824 (1911). Weak base (neutral to phenolphthalein), forms water-soluble salts with strong acids: pK 9.5. Difficultly sol in cold water, more easily in hot water. Miscible with alcohol, ether, carbon disulfide. Dissolves sulfur, phosphorus, arsenic trioxide. *Protect from light and moisture.* LD_{50} orally in rats: 460 mg/kg (Smyth).

Bisulfate. $C_9H_9NO_4S$. White to grayish-white, crystalline powder. mp 163-165°. Freely sol in water; one gram dissolves in 50 ml cold, in 9 ml boiling abs alcohol. *Protect from light.*

Hydrochloride. $C_9H_7N.HCl$. White deliquesc crystals. mp 93-94°. Freely sol in water, alcohol, hot benzene, chloroform, sparingly in cold ether. *Keep well closed and protected from light.*

Salicylate. $C_{16}H_{13}NO_3$. Reddish-gray crystalline powder. Sol in 80 parts water; freely sol in alcohol, benzene, ether, glycerol, oils. *Protect from light.*

Tartrate. $C_{43}H_{45}N_3O_{24}$. White crystals, pungent odor, sharp taste. mp 125° (dec). Sol in 80 parts water, 150 parts alcohol. Insol in ether.

USE: Manuf dyes; prepn hydroxyquinoline sulfate, niacin. As preservative for anatomical specimens. Solvent for resins, terpenes.

THERAP CAT: Antimalarial.

8186. 8-Quinolineboronic Acid. [86-58-8] *B*-8-Quinolinyl-boronic acid. $C_9H_8BNO_2$; mol wt 172.98. C 62.49%, H 4.66%, B 6.25%, N 8.10%, O 18.50%. Prepd by reacting 8-bromoquinoline with butyllithium and tributyl borate: Letsinger, Dandegaonker, *J. Am. Chem. Soc.* **81**, 498 (1959).

Crystals from alc, mp >300°.

Butyl diester. $C_{17}H_{24}BNO_2$. bp_4 180°. n_D^{25} 1.4840.

Chloroethyl diester. $C_{13}H_{14}BCl_2NO_2$. Crystals from toluene, mp 193-194°.

8187. 8-Quinolinecarboxylic Acid. [86-59-9] $C_{10}H_7NO_2$; mol wt 173.17. C 69.36%, H 4.07%, N 8.09%, O 18.48%. Prepd by heating 2-aminobenzoic acid with 2-nitrobenzoic acid, glycerol and sulfuric acid: Schlosser, Skraup, *Monatsh. Chem.* **2**, 530 (1881); by oxidizing 8-quinolylaldehyde with chromic acid: Howitz, *Ber.* **35**, 1275 (1902).

Needles from water. mp 186-187.5°. Sublimes above mp. Slightly sol in cold water, appreciably sol in hot water and in alcohol. Freely sol in acids and alkalies.

USE: Detection of cadmium, copper, iron, lead, mercury, silver, thallium; quantitative determination of copper.

8188. Quinoline Yellow. [8004-92-0] C.I. Acid Yellow 3; D & C Yellow No. 10; Acid Yellow 3; Food Yellow 13; C.I. 47005. Synthetic dye. A mixture of the sodium salts of the mono- and disulfonic acids of quinoline yellow spirit soluble, *q.v.* Principal constituents: sodium salts of *2-(2,3-dihydro-1,3-dioxo-1H-indene-2-yl)-6-quinolinesulfonic acid* and *2-(2,3-dihydro-1,3-dioxo-1H-indene-2-yl)-8-quinolinesulfonic acid.* Prepn of sulfonic acids: M. C. Traub, *Ber.* **16**, 297 (1883); *Colour Index* vol. 4 (3rd ed., 1971) p 4435. HPLC determn: J. Chudy *et al., J. Chromatogr.* **154**, 306 (1978). Purification: N. A. Ambrosiano *et al., US 4398916* (1983 to Sterling Drug). Stability: H. Delonca *et al., Pharm. Acta Helv.* **58**, 332 (1983). ^{13}C-NMR spectrum: M. Gelbcke *et al., Bull. Soc.*

Chim. Belg. **91**, 237 (1982). TLC identification: J. A. Steele, *J. Assoc. Off. Anal. Chem.* **67**, 540 (1984). Composition and skin sensitivity: Y. Sata *et al., Contact Dermatitis* **10**, 30 (1984). LC determn of unsulfonated impurity: A. L. Goldberg, *J. Assoc. Off. Anal. Chem.* **68**, 477 (1985). Toxicity: F. C. Lu, A. Lavalle, *Can. Pharm. J.* **97**, 30 (1964).

Bright greenish yellow. Orange in conc H_2SO_4. Sol in water, slightly sol in ethanol. Practically insol in vegetable oils. LD_{50} in rats (g/kg): >2 orally (Lu, Lavalle).

USE: Textile dye for wool, nylon, silk. Paper dye. Barium salt in printing inks. Color for food, drugs, cosmetics.

8189. Quinoline Yellow Spirit Soluble. [8003-22-3] C.I. Solvent Yellow 33; quinoline yellow base; quinoline yellow A; D & C Yellow No. 11; C.I. 47000. Synthetic dye consisting principally of *quinophthalone (2-(2-quinolinyl)-1H-indene-1,3-(2H)-dione).* Prepn: M. C. Traub, *Ber.* **16**, 297 (1883); E. Jacobsen, C. L. Reimer, *ibid.* 1082; *Colour Index* vol. 4 (3rd ed., 1971) p 4435. ^{13}C-NMR spectrum: M. Gelbcke *et al., Bull. Soc. Chim. Belg.* **91**, 237 (1982). Composition and skin sensitivity: Y. Sato *et al., Contact Dermatitis* **10**, 30 (1984); S. Kita *et al., ibid.* **11**, 210 (1984). Inhalation toxicity in animals: T. C. Marrs *et al., Hum. Toxicol.* **3**, 289 (1984).

Quinophthalone

Bright greenish yellow. Insol in water. Slightly sol in ethanol (yellow), linseed oil, mineral oil, oleic acid, paraffin wax, stearic acid, turpentine. Sol in acetone, chloroform, benzene, toluene. Yellow brown in conc H_2SO_4, yellow flocculent ppt in dilution.

USE: In spirit lacquers, polystyrenes, polycarbonates, polyamides, and acrylic resins. In colored smokes. Occasionally in hydrocarbon solvents. In externally applied drugs and cosmetics.

8190. Quinolinic Acid. [89-00-9] 2,3-Pyridinedicarboxylic acid. $C_7H_5NO_4$; mol wt 167.12. C 50.31%, H 3.02%, N 8.38%, O 38.29%. Metabolite of tryptophan, *q.v.* Prepn: S. Hoogewerff, W. A. van Dorp, *Ber.* **12**, 747 (1879); W. Koenigs, *ibid.* 983; O. Fischer, E. Renouf, *ibid.* **17**, 755 (1884); E. Sucharda, *ibid.* **58B**, 1727 (1925); V. Yu. Stiks, S. A. Bulgach, *ibid.* **65B**, 11 (1932); A. F. Lindenstruth, C. A. VanderWerf, *J. Am. Chem. Soc.* **71**, 3020 (1949). Metabolism studies: L. M. Henderson *et al., J. Biol. Chem.* **181**, 667, 677, 687, 731 (1949); H. P. Sarett, *ibid.* **193**, 627 (1951). Neuroexcitatory activity and possible role in neurodegenerative disorders: R. Schwarcz *et al., Science* **219**, 316 (1983).

Odorless crystals. mp 190° when rapidly heated, with decompn into CO_2 and nicotinic acid. Sol in 180 parts water, in alkalies, slightly in alcohol. Almost insol in ether or benzene.

8191. Quinone. [106-51-4] 2,5-Cyclohexadiene-1,4-dione; *p*-quinone; 1,4-benzoquinone; 1,4-cyclohexadienedione. $C_6H_4O_2$; mol wt 108.10. C 66.67%, H 3.73%, O 29.60%. Made by oxidation of aniline with sodium dichromate in presence of sulfuric acid. Laboratory prepn from hydroquinone: Vliet, *Org. Synth.* **2**, 85 (1922); also *Org. Synth.* **coll. vol. I** (2nd ed., 1941). Cf. Underwood, Walsh, *ibid.* **16**, 73 (1936). Toxicity study: Woodard *et al., Fed. Proc.* **8**, 348 (1949).

Yellow monoclinic prisms from water or petr ether. Penetrating odor resembling that of chlorine. Irritating vapors. d_4^{20} 1.318. mp 115.7°. Sublimes. Sublimation velocities *in vacuo:* Kempf, *J. Prakt. Chem.* [2] **78**, 236 (1908). Volatile with steam. Absorption spectrum: Hantzsch, *Ber.* **49**, 522 (1916). Dipole moment: 0.67. Polemic over correct values: Paoloni, *J. Am. Chem. Soc.* **80**, 3879 (1958). Slightly sol in water; sol in alcohol, ether, hot petr ether, alkalies. LD_{50} orally in rats: 130 mg/kg (Woodard).

Caution: Potential symptoms of overexposure are eye and skin irritation, conjunctivitis; keratitis. *See NIOSH Pocket Guide to Chemical Hazards* (DHHS/NIOSH 97-140, 1997) p 272.

USE: Oxidizing agent; in photography; manuf dyes; manuf hydroquinone; tanning hides; making gelatin insol; strengthening animal fibers; as reagent.

8192. Quinovic Acid. [465-74-7] (3β)-3-Hydroxyurs-12-ene-27,28-dioic acid; quinovaic acid; chinovic acid; chinova acid. $C_{30}H_{46}O_5$; mol wt 486.69. C 74.04%, H 9.53%, O 16.44%. Isoln from cinchona bark as the glycoside quinovin: Hlasiwetz, *Ann.* **79**, 129 (1851); **111**, 182 (1859); Votocek, Rác, *Collect. Czech. Chem. Commun.* **1**, 234 (1929); Tschesche *et al., Ann.* **667**, 151 (1963); from *Zygophyllum coccineum* L., *Zygophyllaceae:* Soliman, *J. Chem. Soc.* **1939**, 1760; from bark of *Mitragyna inermis* Kuntze and leaves of *M. ciliata* Aubrev & Pellegr. and *M. rubrostipulacea* Havil., *Rubiaceae:* Badger *et al., ibid.* **1950**, 867. Structure: Brossi *et al., Helv. Chim. Acta* **34**, 244 (1951); Barton, de Mayo, *J. Chem. Soc.* **1953**, 3111. *Review:* J. Simonsen, W. C. J. Ross, *The Terpenes* vol. 5 (University Press, Cambridge, 1957) pp 75-113.

Prisms from dil pyridine, decomp 297°. Very bitter taste. $[\alpha]_D^{20}$ +99° (c = 2.48 in pyridine). Practically insol in water and other solvents except pyridine.

O-**Acetylquinovic acid.** $C_{32}H_{48}O_6$. Needles from dilute acetone, mp 282-284°.

Dimethyl ester. $C_{32}H_{50}O_5$. Needles from dil acetone, mp 175°. $[\alpha]_D^{19}$ +117.4° (c = 2.2 in $CHCl_3$). Sol in organic solvents.

8193. Quinovin. [53516-73-7] Quinova-bitter; chinovin. Isoln from cinchona bark *(cortex chinae):* Hlasiwetz, *Ann.* **111**, 182 (1859); Liebermann, Giesel, *Ber.* **16**, 926 (1883). A mixture of three glycosides, 60% A, 5% B and 30% C: Tschesche *et al., Ann.* **667**, 151 (1963).

Glycoside A. Quinovic acid β-D-quinovoside. $C_{36}H_{56}O_9$. Crystals from methanol + water, mp 237-238°. $[\alpha]_D^{21}$ +57 ± 2° (ethanol).

Glycoside B. Cincholic acid β-D-quinovoside. $C_{36}H_{56}O_9$. Crystals from methanol + water, mp 193-195°. $[\alpha]_D^{19}$ +78 ± 3° (c = 0.78 in ethanol).

Glycoside C. Quinovic acid β-D-glucoside. $C_{36}H_{56}O_{10}$. Crystals from methanol + water, mp 247-250°. $[\alpha]_D^{20}$ +62 ±2° (ethanol).

8194. Quinovose. [7658-08-4] 6-Deoxy-D-glucose; D-glucomethylose; D-isorhamnose; isorhodeose; epifucose; chinovose. $C_6H_{12}O_5$; mol wt 164.16. C 43.90%, H 7.37%, O 48.73%. Isoln from cinchona bark: Freudenerg, *Ber.* **62**, 373 (1929). Identity with D-glucomethylose: Votocek, Rác, *Collect. Czech. Chem. Commun.* **1**, 239 (1929). Structure: Karrer, Boettcher, *Helv. Chim. Acta* **36**, 570 (1953). Synthesis from 6-*O-p*-tolylsulfonyl-D-glucose: Schmidt, "6-Deoxy-α-D-glucose" in *Methods in Carbohydrate Chemistry* vol. I, R. L. Whistler, M. L. Wolfrom, Eds. (Academic Press, New York, 1962) pp 198-201. Alter-

nate facile synthesis: V. K. Srivastava, L. M. Lerner, *Carbohydr. Res.* **64**, 263 (1978).

Crystals from ethyl acetate, mp 146°. $[\alpha]_D^{20}$ +73° (5 min) → +30° (3 hr, final) (c = 8.3 in water). Sol in water, ethanol. Practically insol in ether, acetone.

8195. Quinoxaline. [91-19-0] 1,4-Benzodiazine; benzo[a]-pyrazine; benzoparadiazine; phenpiazine. $C_8H_6N_2$; mol wt 130.15. C 73.83%, H 4.65%, N 21.52%. Prepd from *o*-phenylenediamine and glyoxal or glyoxal sodium bisulfite: Hinsberg, *Ber.* **17**, 320 (1884); *Ann.* **237**, 334 (1887); Jones, McLaughlin, *Org. Synth.* **30**, 86 (1950).

Crystals, mp 29-30°. Odor of quinoline when cool, odor of piperidine when hot. d_4^{48} (liq) 1.1334. $bp_{760.3}$ 229.5°. bp_{12} 108-111°. n_D^{48} 1.6231. Very freely sol in water, alc, ether, benzene.

Monohydrate. Crystals, mp 37°.

Sulfate. $C_8H_6N_2 \cdot H_2SO_4$. Leaflets, mp 186-187°. Freely sol in water. Less sol in alcohol.

8196. Quinoxyfen. [124495-18-7] 5,7-Dichloro-4-(4-fluorophenoxy)quinoline; DE-795; Fortress; Quintec. $C_{15}H_8Cl_2FNO$; mol wt 308.13. C 58.47%, H 2.62%, Cl 23.01%, F 6.17%, N 4.55%, O 5.19%. Quinoline fungicide for powdery mildew control in cereals, grapes and hops; inhibits formation of appressoria through which the fungus invades the plant. Prepn: W. R. Arnold *et al., EP 326330* (1989 to Eli Lilly); *eidem,* **US 5145843** (1992 to DowElanco). Properties and field trials: C. Longhurst *et al., Brighton Crop Prot. Conf. - Pests Dis.* **1996**, 27. Resistance profile: D. W. Hollomon *et al., ibid.* 701. Metabolism and environmental fate: G. L. Reeves *et al., ibid.* 1169. Determn and fate in grapes, raisins and wine: P. Cabras *et al., J. Agric. Food Chem.* **48**, 6128 (2000).

Crystals from heptane, mp 105-106°. LD_{50} orally in rats: >5000 mg/kg; dermally in rabbits: >2000 mg/kg; by inhalation in rats: >3.38 mg/l (Longhurst).

USE: Agricultural fungicide.

8197. Quintozene. [82-68-8] 1,2,3,4.5-Pentachloro-6-nitrobenzene; PCNB; terrachlor; PKhNB; Botrilex; Brassicol; Folosan; Pentagen; Terraclor; Tilcarex. $C_6Cl_5NO_2$; mol wt 295.32. C 24.40%, Cl 60.02%, N 4.74%, O 10.84%. Prepd by treating pentachlorobenzene with fuming nitric acid: Jungfleisch, *Ann. Chim.* [4] **15**, 286 (1868); A. Roedig, K. Kiepert, *Ann.* **593**, 71 (1955). Persistence of residues in foods: P. B. Baker, B. Flaherty, *Analyst* **97**, 378 (1972); A. R. P. Paxton, D. Purser, *Pestic. Sci.* **13**, 401 (1982); in soil: J. Beck, K. E. Hansen, *ibid.* **5**, 41 (1974). Metabolism by sheep and goats: P. W. Aschbacher, V. J. Feil, *J. Agric. Food Chem.* **31**, 1150 (1983). Toxicity data: J. K. Finnegan *et al., Arch. Int. Pharmacodyn. Ther.* **114**, 38 (1958). *Review: WHO Environmental Health Criteria* **41**, 1-38 (1984). Comparative review of the effects on soil organisms: E. R. Ingham, *Crop Prot.* **4**, 3-32 (1985).

Fine needles from alcohol, platelets from carbon disulfide. d_4^{25} 1.718. mp 144°. bp_{760} 328° (some dec). Practically insol in water, cold alcohol. Freely sol in carbon disulfide, benzene, chloroform. LD_{50} in male, female rats (g/kg): 1.71 ±0.20, 1.65 ±0.17 by gavage (Finnegan).

USE: Fungicide for seed and soil treatment.

8198. Quinuclidine. [100-76-5] 1-Azabicyclo[2.2.2]octane; 1,4-ethylenepiperidine. $C_7H_{13}N$; mol wt 111.19. C 75.62%, H 11.79%, N 12.60%. Prepn: Löffler, Stietzel, *Ber.* **42**, 124 (1909); Meisenheimer *et al.*, *Ann.* **420**, 191 (1920); Clemo, Metcalfe, *J. Chem. Soc.* **1937**, 1989; Prelog, US 2192840 (1940); Wawzonek *et al.*, *J. Am. Chem. Soc.* **73**, 2806 (1951); Leonard, Elkin, *J. Org. Chem.* **27**, 4635 (1962).

Prisms from petr ether. Sublimes. mp 156° (sealed tube). Very sol in water and the usual organic solvents.

Hydrochloride. $C_7H_{13}N\cdot HCl$. Crystals from abs ethanol, mp 364-365°.

8199. 3-Quinuclidinol. [1619-34-7] 1-Azabicyclo[2.2.2]octan-3-ol; 3-hydroxyquinuclidine. $C_7H_{13}NO$; mol wt 127.19. C 66.10%, H 10.30%, N 11.01%, O 12.58%. Prepn and resolution of *dl*-forms: Sternbach, Kaiser, *J. Am. Chem. Soc.* **74**, 2215 (1952); Sternbach, US 2648667 (1953 to Hoffmann-La Roche); of free alcohol, acetate and/or benzoate esters: Grob *et al.*, *Helv. Chim. Acta* **40**, 2170 (1957); Mikhalina, Rubtsov, *Zh. Obshch. Khim.* **30**, 163 (1960), *C.A.* **54**, 22632h (1960). Structure activity of esters: M. D. Mashkovsky in *Proc. 1st Int. Pharmacol. Mtg. Stockholm, 1961* **vol. 7** (Macmillan, NewYork, 1963) pp 359-366.

dl-**Form.** [3684-26-2] Crystals from benzene or acetone, mp 221-223° also reported as 225-227°. Sublimes at 120° and 20 mm Hg. Very sol in water.

dl-**Form hydrochloride.** [25333-43-1] Prisms from methanol + acetone, mp >300°.

dl-**Form acetate (ester).** Aceclidine; 3-acetoxyquinuclidine; 3-quinuclidinyl acetate. $C_9H_{15}NO_2$. Oily liq, $bp_{0.4}$ 73-74°, bp_{11} 113-115°. n_D^{25} 1.4675.

dl-**Form acetate hydrochloride.** [6109-70-2] Glaucostat; Glaucotat; Glaudin. $C_9H_{15}NO_2\cdot HCl$; mol wt 205.68. Crystals, mp 166°.

dl-**Form benzilate (ester).** 3-Quinuclidinyl benzilate; BZ; QNB; Ro-2-3308. $C_{21}H_{23}NO_3$; mol wt 337.42. Prepn: Sternbach, Kaiser, *loc. cit.*, 2219. Crystals from acetone-ether, mp 164-165°.

dl-**Form benzoate (ester).** [66-93-3] 3-Benzoyloxyquinuclidine; 3-quinuclidinyl benzoate. $C_{14}H_{17}NO_2$. Oily liquid, $bp_{0.3}$ 148-150°.

dl-**Form benzoate hydrochloride.** [25333-48-6] Oksilidin; oxylidine. $C_{14}H_{17}NO_2\cdot HCl$. Crystals, mp 238-240°.

l-**Form.** [25333-42-0] Prisms from acetone, mp 220-222°. $[\alpha]_D^{25}$ -2.0° (c = 6.5), -43.8° (c = 3 in 1N HCl).

USE: Benzilate (ester) has been used as an incapacitating agent in chemical warfare: *Health Aspects of Chemical and Biological Weapons* (WHO, Geneva, 1970) pp 49-51.

THERAP CAT: Hypotensive; acetate (ester) as cholinergic.

8200. Quinupramine. [31721-17-2] 5-(1-Azabicyclo[2.2.2]-oct-3-yl)-10,11-dihydro-5H-dibenz[b,f]azepine; 10,11-dihydro-5-(3-quinuclidinyl)-5H-dibenz[b,f]azepine; LM-208; Kinupril; Kevopril. $C_{21}H_{24}N_2$; mol wt 304.44. C 82.85%, H 7.95%, N 9.20%. Analog of imipramine, *q.v.* Prepn: C. Gueremy, P. C. Wirth, DE 2030492 (1971 to Sogeras), *C.A.* **74**, 141581e (1971). Animal studies: W. Van Dorsser, A. Dresse, *Arch. Int. Pharmacodyn. Ther.* **208**, 373 (1974); *eidem, ibid.* **220**, 164 (1976). Clinical study: R. Volmat *et al.*, *Clin. Neurol. Psychiat.* **239**, 445 (1978).

Crystals, mp 150°.
THERAP CAT: Antidepressant.

8201. Quinupristin. [120138-50-3] 4-[4-(Dimethylamino)-N-methyl-L-phenylalanine]-5-[(2S,5R)-5-[[[(3S)-1-azabicyclo[2.2.-2]oct-3-yl]thio]methyl]-4-oxo-2-piperidinecarboxylic acid]virginiamycin S1; 5δ-[(3S)-3-quinuclidinyl]thiomethylpristinamycin IA; RP-57669. $C_{53}H_{67}N_9O_{10}S$; mol wt 1022.23. C 62.27%, H 6.61%, N 12.33%, O 15.65%, S 3.14%. Semisynthetic depsipeptide type I streptogramin, *q.v.* Marketed in combination with dalfopristin, *q.v.* Prepn: J.-C. Barriere *et al.*, EP 248703, *eidem*, US 4798827 (1987, 1989 both to Rhone-Poulenc). *In vitro* activity: H. C. Neu *et al.*, *J. Antimicrob. Chemother.* **30**, Suppl. A, 83 (1992). HPLC determn in plasma: A. Le Liboux *et al.*, *J. Chromatogr. B* **708**, 161 (1998).

White crystals in combination with methanol, mp ~200°.
THERAP CAT: Antibacterial.

8202. Quisqualic Acid. [52809-07-1] (αS)-α-Amino-3,5-di-oxo-1,2,4-oxadiazolidine-2-propanoic acid; L-quisqualic acid; β-(3,5-dioxo-1,2,4-oxadiazolidin-2-yl)-L-alanine. $C_5H_7N_3O_5$; mol wt 189.13. C 31.75%, H 3.73%, N 22.22%, O 42.30%. Excitatory amino acid (EAA) used to identify a specific subset of EAA receptors; consequently, the receptors are known as quisqualate receptors. *See also* NMDA, kainic acid. Isoln from the seeds of *Quisqualis chinesis* and anthelmintic activity: Y.-C. Tuan *et al.*, *Yao Hsueh Hsueh Pao* **5**, 87 (1957), *C.A.* **56**, 14896b (1958); from *Q. indica*: S.-T. Fang, J.-H. Chu, *Hua Hsueh Hsueh Pao* **30**, 226 (1964), *C.A.* **61**, 7359f (1962); from *Q. fructus*: T. Takemoto *et al.*, *Yakugaku Zasshi* **95**, 176 (1975), *C.A.* **82**, 152211a (1975). Enzymic synthesis: I. Murakoshi *et al.*, *Chem. Pharm. Bull.* **22**, 473 (1974). Total synthesis: J. E. Baldwin *et al.*, *Chem. Commun.* **1985**, 256. Crystal structure: J. L. Flippen, R. D. Gilardi, *Acta Crystallogr.* **B32**, 951 (1976). Identification as a neuroexcitant: H. Shinozaki, I. Shibuya, *Neuropharmacology* **13**, 665 (1974). Receptor binding studies: K.

Koshiya, *Life Sci.* **37,** 1373 (1985); J. T. Greenamyre, *J. Pharmacol. Exp. Ther.* **233,** 254 (1985). Review of isolation of quisqualic acid and other EAAs: T. Takemoto, in *Kainic Acid as a Tool in Neurobiology,* R. G. McGeer *et al.,* Eds. (Raven Press, New York, 1978) pp 1-15.

Crystals from water-ethanol, mp 190-191°. $[\alpha]_D^{20}$ +17.0° (c = 2.0 in 6M HCl).

8203. Quizalofop-ethyl. [76578-14-8] 2-[4-[(6-Chloro-2-quinoxalinyl)oxy]phenoxy]propanoic acid ethyl ester; ethyl 2-[4-(6-chloro-2-quinoxalinyloxy)phenoxy]propionate; quinofop-ethyl; DPX-Y6202; NCI-96683; NC-302; Assure; Targa; Pilot. $C_{19}H_{17}$-ClN_2O_4; mol wt 372.81. C 61.21%, H 4.60%, Cl 9.51%, N 7.51%, O 17.17%. Post-emergence herbicide for control of grassy weeds in broad-leaved crops. Prepn and herbicidal activity: Y. Ura *et al.,* **DE 3004770** (1980 to Nissan), *C.A.* **94,** 103421h (1981). Prepn and structure-activity relationships: G. Sakata *et al., J. Pestic. Sci.* **10,** 61 (1985). Pharmacological effects in laboratory animals: K. Inokuchi *et al., Oyo Yakuri* **30,** 509 (1985), *C.A.* **104,** 30101s (1986). Site of action studies: T. Ikai *et al., Proc. Br. Crop Prot. Conf. - Weeds* **1985,** 163. Account of properties, toxicity, crop tolerance:

G. Sakata *et al., Proc. 10th Conf. Int. Congr. Plant Prot.* **1,** 315 (1983).

White crystals, mp 92-93°. $bp_{0.2}$ 220°. Practically insol in water (0.3 × 10^{-6} g/ml at 20°). Soly in acetone, ethanol, benzene, xylene at 20° (g/ml): 0.11, 0.009, 0.29, 0.12. LD_{50} in male, female rats, male, female mice (mg/kg): 1670, 1480, 2350, 2360 orally; all 10000 dermally. LC_{50} (96 hr) in rainbow trout: 10.7 mg/l (Sakata).

USE: Herbicide.

8204. Quorn®. Myco-protein. High protein, meat substitute derived from the biomass of the filamentous fungus, *Fusarium graminearum* (Schwabe). Description of production: J. Edelman *et al., Nutr. Abstr. Rev.* **53,** 471 (1983). See also: R. A. Marsh, **US 4555485** (1985 to Ranks Hovis McDougall). Nutritional profile: M. Sadler, *Nutr. Food Sci.* **112,** 9 (1988). Metabolic study in comparison with meat: W. H. Turnbull *et al., Am. J. Clin. Nutr.* **52,** 646 (1990). Effect on blood lipids: *eidem, ibid.* **55,** 415 (1992). *Review:* A. P. J. Trinci, *Mycol. Res.* **96,** 1-13 (1992).

Very mild, wheaty taste; slight aroma of mushrooms. Fibrous texture resembling that of meat.

USE: Meat substitute.

R

8205. **R-11.** [126-15-8] 1,5a,6,9,9a,9b-Hexahydro-4a(4H)-dibenzofurancarboxaldehyde; 2,3,4,5-bis(2-butenylene)tetrahydrofurfural; MGK-11; MGK Repellent 11. $C_{13}H_{16}O_2$; mol wt 204.27. C 76.44%, H 7.90%, O 15.66%. Prepd by heating furfuraldehyde with butadiene and water under pressure: Hillyer, Nicewander, **US 2683151** (1954 to Phillips Petroleum). Clinical efficacy vs sand flies: F. P. Fossati, M. Maroli, *Trans. R. Soc. Trop. Med. Hyg.* **80**, 771 (1986).

Liquid, bp 307°. d_4^{20} 1.10. mp −80°. n_D^{20} 1.5254. Practically insol in water.

Oxime. mp 97.2°.

Dinitrophenylhydrazone. mp 152.8°.

USE: Insect repellent.

8206. **Rabeprazole.** [117976-89-3] 2-[[[4-(3-Methoxypropoxy)-3-methyl-2-pyridinyl]methyl]sulfinyl]-1H-benzimidazole; pariprazole. $C_{18}H_{21}N_3O_3S$; mol wt 359.44. C 60.15%, H 5.89%, N 11.69%, O 13.35%, S 8.92%. Partially reversible gastric proton pump inhibitor. Prepn: S. Souda *et al.*, **EP 268956**; *eidem*, **US 5045552** (1988, 1991 both to Eisai). Pharmacology: H. Goto *et al.*, *Arzneim.-Forsch.* **41**, 635 (1991). Mode of action: M. Morii, N. Takeguchi, *J. Biol. Chem.* **268**, 21553 (1993). HPLC determn in plasma: H. Nakai *et al.*, *J. Chromatogr. B* **660**, 211 (1994). Clinical pharmacokinetics: S. Yasuda *et al.*, *Int. J. Clin. Pharmacol. Ther.* **32**, 466 (1994). Review of clinical trials in acid peptic diseases: M. L. Cloud *et al.*, *Dig. Dis. Sci.* **43**, 993-1000 (1998); of pharmacology and clinical experience in gastro-esophageal reflux disease: B. Thjodleifsson, *Expert Opin. Pharmacother.* **5**, 137-149 (2004).

White crystals from CH_2Cl_2/ether, mp 99-100° (dec).

Sodium salt. [117976-90-6] E-3810; Aciphex; Pariet. $C_{18}H_{20}N_3NaO_3S$; mol wt 381.43. White crystals from ether, mp 140-141° (dec). Very sol in water, methanol; freely sol in ethanol, chloroform, ethyl acetate. Insol in ether, n-hexane.

THERAP CAT: Antiulcerative.

8207. **Racecadotril.** [81110-73-8] N-[2-[(Acetylthio)methyl]-1-oxo-3-phenylpropyl]glycine phenylmethyl ester; N-[(R,S)-3-acetylthio-2-benzylpropanoyl]glycine benzyl ester; acetorphan; Hidrasec; Tiorfan. $C_{21}H_{23}NO_4S$; mol wt 385.48. C 65.43%, H 6.01%, N 3.63%, O 16.60%, S 8.32%. Antisecretory enkephalinase inhibitor. Prepn: B. Roques *et al.*, **EP 38758** (1981); *eidem*, **US 4513009** (1985 to Bioprojet). Pharmacology: J.-M. Lecomte *et al.*, *J. Pharmacol. Exp. Ther.* **237**, 937 (1986). Effect on intestinal transit: J. F. Bergmann *et al.*, *Aliment. Pharmacol. Ther.* **6**, 305 (1992). Clinical trial in acute diarrhea: P. Baumer *et al.*, *Gut* **33**, 753 (1992); in children: E. Salazar-Lindo *et al.*, *N. Engl. J. Med.* **343**, 463 (2000). Symposium on clinical pharmacology and clinical experience: *Aliment. Pharmacol. Ther.* **13**, Suppl. 6, 1-32 (1999). Review of clinical development: J.-C. Schwartz, *Int. J. Antimicrob. Agents* **14**, 75-79 (2000); J. M. Lecomte, *ibid.* 81-87.

White crystals from ether mp 89°.

(S)-**Form.** [112573-73-6] Ecadotril; sinorphan; Bay-y-7432. $C_{21}H_{23}NO_4S$; mol wt 385.48. Prepn: P. Duhamel *et al.*, **EP 318377**; *eidem*, **US 5208255** (1989, 1993 both to Bioprojet); and pharmacology: B. Giros *et al.*, *J. Pharmacol. Exp. Ther.* **243**, 666 (1987). Clinical effect on plasma ANP levels in CHF: J. C. Kahn *et al.*, *Lancet* **335**, 118 (1990); on renal function: F. Schmitt *et al.*, *Am. J. Physiol.* **267**, F20 (1994). Clinical trial in heart failure: C. M. O'Connor *et al.*, *Am. Heart J.* **138**, 1140 (1999); J. G. F. Cleland, K. Swedberg, *Lancet* **351**, 1657 (1998). mp 71°. $[\alpha]_D^{25}$ −24.1° (c = 1.3 in methanol). LD_{50} i.v. in mice: >100 mg/kg (Duhamel, 1993).

THERAP CAT: Antidiarrheal.

8208. **Ractopamine.** [97825-25-7] 4-Hydroxy-α-[[[3-(4-hydroxyphenyl)-1-methylpropyl]amino]methyl]benzenemethanol; 1-(4-hydroxyphenyl)-2-[1-methyl-3-(4-hydroxyphenyl)-propylamino]ethanol; N-[2-(4-hydroxyphenyl)-2-hydroxyethyl]-1-methyl-3-(4-hydroxyphenyl)propylamine. $C_{18}H_{23}NO_3$; mol wt 301.39. C 71.73%, H 7.69%, N 4.65%, O 15.93%. β-Adrenergic agonist; repartitioning agent. Prepn (stereochemistry unspecified): J. van Dijk, H. D. Moed, *Rec. Trav. Chim.* **92**, 1281 (1973). Prepn of R,R-isomer: J. Mills *et al.*, **EP 7205** (1980 to Lilly). Prepd (not claimed): D. B. Anderson *et al.*, **US 4690951** (1987 to Lilly). LC determn in animal feeds: M. P. Turberg *et al.*, *J. AOAC Int.* **77**, 840 (1994). Metabolism and tissue residue studies: J. E. Dalidowicz *et al.*, *ACS Symp. Ser.* **503**, 234 (1992). Effect on growth, carcass characteristics and meat quality in swine: B. E. Uttaro *et al.*, *J. Anim. Sci.* **71**, 2439 (1993). Effect on β-receptor affinity and density in pigs: M. E. Spurlock *et al.*, *ibid.* **72**, 75 (1994).

Hydrochloride. [90274-24-1] EL-737; LY-031537; Optaflexx; Paylean. $C_{18}H_{23}NO_3 \cdot HCl$; mol wt 337.84. Mixture of 4 stereoisomers in approx equal proportions. Product containing 51% RR,SS- and 49% RS,SR-diastereomers, mp 124-129°.

R,R-**Form Hydrochloride.** [74432-68-1] Butopamine hydrochloride. Crystals from ethanol + diethyl ether, mp 176-176.5° (dec). $[\alpha]_D$ −22.7°; $[\alpha]_{365}$ −71.2° (c = 3.7 mg/ml in methanol).

THERAP CAT (VET): Animal growth promotant.

8209. **Radicinin.** [10088-95-6] (2S,3S)-2,3-Dihydro-3-hydroxy-2-methyl-7-[(1E)-1-propen-1-yl]-4H,5H-pyrano[4,3-b]pyran-4,5-dione; stemphylone. $C_{12}H_{12}O_5$; mol wt 236.22. C 61.02%, H 5.12%, O 33.86%. A mold metabolite from the plant pathogen *Stemphylium radicinum*: Clark, Nord, *Arch. Biochem. Biophys.* **45**, 469 (1953); *eidem, ibid.* **59**, 269 (1955). Structure and identity with stemphylone: Grove, *J. Chem. Soc.* **1964**, 3234. Absolute configuration: M. Nukina, S. Marumo, *Tetrahedron Lett.* **1977**, 3271. Synthesis of dl-form: Kato *et al.*, *Chem. Commun.* **1969**, 95.

Needles from ethanol, dec 238-240°. $[\alpha]_D^{27}$ −217.4° (c = 2.37 in pyridine); $[\alpha]_D^{27}$ −175.7° (c = 0.2 in ethanol); $[\alpha]_D^{27}$ −208° (c = 1.25 in chloroform). uv max: 343, 280, 270 nm (log ε 4.27, 3.62, 3.79). Soluble in alkali. Practically insol in dil acids, sodium bicarbonate, sodium carbonate.

Monoacetate. $C_{14}H_{14}O_6$; mol wt 278.26. Crystals from methanol, mp 197°. $[\alpha]_D^{27}$ −267° (c = 0.69 in pyridine).

8210. Radium. [7440-14-4] Ra; at. no. 88; valence 2. Group IIA (2). Radioactive alkaline earth metal. Occurrence in earth's crust: approx $10^{-11}\%$ by wt. No stable nuclides. Naturally occurring isotopes: 223, *actinium X* ($T_{\frac{1}{2}}$ 11.435 days, α emitter, rel. at. mass 223.0185); 224, *thorium X* ($T_{\frac{1}{2}}$ 3.66 days, α emitter, rel. at. mass 224.0202); 226 (longest-lived isotope, $T_{\frac{1}{2}}$ 1599 years, α emitter, rel. at. mass 226.0254); 228, *mesothorium I* ($T_{\frac{1}{2}}$ 5.75 years, β^- emitter, rel. at. mass 228.0311). Known radioactive isotopes: 205-222, 225, 227, 229-234. ^{226}Ra is a product of disintegration of uranium and is present in all ores contg uranium. Separated in the form of a salt by P. and M. S. Curie from the pitchblende of Joachimsthal, Bohemia: Curie *et al.*, *Compt. Rend.* **127**, 1215 (1898). Isoln of the element by electrolysis of an aq soln of radium chloride: Curie, Debierne, *ibid.* **151**, 523 (1910). Production of ^{228}Ra by disintegration of thorium (^{232}Th) discovered in 1907 by O. Hahn in monazite residues from isolating thorium. Determn in environmental and monitoring samples: W. C. Lawrie *et al.*, *Appl. Radiat. Isot.* **53**, 133 (2000). Clinical evaluation in brachytherapy of tongue: J. Horiuchi *et al.*, *Int. J. Radiat. Oncol. Biol. Phys.* **8**, 82 (1982); M. Hoshina *et al.*, *Br. J. Radiol.* **62**, 59 (1989); in intracavitary radiation of uterus: D. A. Jones, R. Stout, *Clin. Radiol.* **37**, 169 (1986). Review of radiotherapy in cervical carcinoma: P. R. Reddi *et al.*, *Obstet. Gynecol.* **43**, 238-247 (1974); in rectal carcinoma: *Dis. Colon Rectum* **29**, 600-614 (1986), reprint of C. Gordon-Watson, *Br. J. Surg.* **17**, 649-669 (1930). Comprehensive reviews: K. W. Bagnall, *Chemistry of the Rare Radioelements* (New York, Academic Press, 1957); Goodenough, Stenger, "Magnesium, Calcium, Strontium, Barium, and Radium" in *Comprehensive Inorganic Chemistry* **vol. 1**, J. C. Bailar, Jr. *et al.*, Eds. (Pergamon Press, Oxford, 1973) pp 591-664. Review of toxicology and human exposure: *Toxicological Profile for Radium* (PB 91-180414, 1990) 138 pp; of discovery and medical use: J.-J. Mazeron, A. Gerbaulet, *Radiother. Oncol.* **49**, 205-216 (1998); of industrial use: D. I. Harvie, *Endeavour* **23**, 100-105 (1999).

Brilliant, white metal; body-centered cubic structure; blackens on exposure to air. mp 700°; bp 1737°; d 5.5. One gram of radium evolves about 1000 kcal per year. Undergoes spontaneous disintegration with formation of radon. One gram of radium produces about 0.0001 ml of radon per day at normal temp and pressure. Radium is produced and used in the form of its salts: the chloride, bromide, carbonate, sulfate. Its compds closely resemble those of barium; the element itself is more volatile than barium. Radium salts impart a carmine-red color to a flame.

Bromide. [10031-23-9] Br_2Ra. White or slightly brownish crystals. d 5.79. mp 728°. Sublimes at 900°. Sol in water. The salt of commerce is usually a mixture with barium bromide.

Chloride. [10025-66-8] Cl_2Ra. White or slightly brownish crystals. d 4.91. mp 1000°. Sol in water. The salt of commerce is usually a mixture with barium chloride.

Caution: Potential symptoms of overexposure to radioactive emissions are anemia, cataracts, fractured teeth and cancer, esp. bone cancer (*Toxicological Profile*).

USE: In physical research. In radiography of metals because the penetration of gamma rays is more pronounced than that of x-rays. As source of radon. No longer used to make luminous paints.

THERAP CAT: Antineoplastic (radiation source).

8211. Radium-223 Chloride. [444811-40-9] Radium chloride ($^{223}RaCl_2$); radium ^{223}Ra chloride; ATI-BC-1; Alpharadin. Cl_2^{223}Ra. Bone-seeking radiopharmaceutical for targeted radiotherapy of bone metastases. Prepn from ^{227}Ac: R. W. Atcher *et al.*, *J. Radioanal. Nucl. Chem. Lett.* **135**, 215 (1989); by neutron irradiation of ^{226}Ra: R. W. Howell *et al.*, *Radiat. Res.* **147**, 342 (1997). Improved process and use as radiopharmaceutical: R. H. Larsen, G. Henriksen, WO **0040275**; R. H. Larsen *et al.*, US **6635234** (2000, 2003 both to Anticancer Therapeutic Inventions); G. Henriksen *et al.*, *Radiochim. Acta* **89**, 661 (2001). Bone targeting activity: *idem*

et al., *J. Nucl. Med.* **44**, 252 (2003). Radiotoxicity studies: R. H. Larsen *et al.*, *In Vivo* **20**, 325 (2006). Clinical evaluation in prostate cancer metastases: S. Nilsson *et al.*, *Lancet Oncol.* **8**, 587 (2007). Review of development and therapeutic potential: O. S. Bruland *et al.*, *Clin. Cancer Res.* **12**, Suppl 20, 6250s-6257s (2006); K. Liepe, *Curr. Opin. Investig. Drugs* **10**, 1346-1358 (2009).

Prepd for use as a sterile isotonic solution. Contains ^{223}Ra which is an alpha emitter with a half-life of 11.4 days.

THERAP CAT: Antineoplastic (radiation source).

8212. Radon. [10043-92-2] Rn; at. no. 86. Obsolete synonyms: *niton*, symbol Nt, and *radium emanation* or *emanation*, symbol Em. Group VIIIA (18), also known as Group 0. A noble gas characterized by an electronic structure in which the outer p subshell is entirely filled. No stable nuclides; naturally occurring isotope (mass number): 222 (longest-lived known isotope, $T_{\frac{1}{2}}$ 3.825 days, rel. at. mass 222.0176, α-emitter, member of decay chain of ^{238}U). Known artificial radioactive isotopes: 200-221, 223-226. Isotope ^{220}Rn ($T_{\frac{1}{2}}$ 55.6 sec, α-emitter, rel. at. mass 220.0114, formed in decay chain of ^{232}Th), unofficial name *Thoron*, symbol Tn, was discovered by Owens and Rutherford in 1900. ^{222}Rn was discovered by Dorn in 1900 and isolated by Rutherford and Sody in 1902. Isotope ^{219}Rn ($T_{\frac{1}{2}}$ 3.96 sec, α-emitter, formed in decay chain of ^{235}U), unofficial name *actinon*, symbol An, was discovered by Debièrne and Giesel in 1902. Concentration in air: 6×10^{-14} ppm by vol; in igneous rock of the earth's crust: 1.7×10^{-10} ppm by wt. Prepn of fluoride: Fields *et al.*, *J. Am. Chem. Soc.* **84**, 4164 (1962); Stein, *Science* **168**, 362 (1970). Prepn of gas from radium salt: Jennings, Russ, *Radon: Its Technique and Use* (Murray, London, 1948) pp 79-95. Reviews of chemistry and compds: *Argon, Helium and the Rare Gases*, G. A. Cook, Ed. (Interscience, New York, 1961); Haissinsky, Adloff, *Radiochemical Survey of the Elements* (Elsevier, New York, 1965) pp 126-128; Cockett, Smith, "The Monatomic Gases" in *Comprehensive Inorganic Chemistry* **vol. 1**, J. C. Bailar, Jr. *et al.*, Eds. (Pergamon Press, Oxford, 1973) pp 139-211; Bartlett, Sladky, *ibid.* pp. 213-330; S.-C. Hwang, W. R. Weltmer, Jr. in *Kirk-Othmer Encyclopedia of Chemical Technology* **vol. 13** (John Wiley & Sons, 4th ed., 1995) pp 1-38; G. J. Schrobilgen, J. M. Whalen, *ibid.* pp 38-53; *Chemistry of the Elements*, N. N. Greenwood, A. Earnshaw, Eds. (Pergamon Press, New York, 1984) pp 1042-1059. Reviews of sources and health effects of indoor radon exposure: E. P. Radford, *Environ. Health Perspect.* **62**, 281-287 (1985); D. L. Henshaw, *Contemp. Phys.* **34**, 31-48 (1993).

Colorless, odorless, tasteless, inert, monatomic gas; will form compds with highly electronegative elemnts such as O, F, Cl. Soly of gas in water (20°) 230 cm^3/l water. Soluble in organic solvents. Strongly adsorbed on various surfaces. Heat capacity at 25° and 1 atm: 4.9860 cal/g-atom/K. Triple point temp 202 K, press 70 kPa. Critical temp 378 K, critical press 6280 kPa. Gas: d^0 (101.3 kPa) 9.73 kg/m^3. Liquid: normal bp −62°, d (normal bp) 4400 kg/m^3, heat of vaporization (normal bp) 18,100 J/mol. Heat of fusion (triple point) 3247 J/mol. Solid form exists as face-centered cubic crystals at normal pressure.

Caution: Toxicity due to ionizing radiation (α-emitter). Increased incidences of lung cancer have been reported due to occupational exposure to high doses of radon, especially in hard rock miners. Max permissible concn of ^{222}Rn in air: 10^{-8} μ-Curie/cc: *Natl. Bur. Stand. Handb.* **69**, 79 (1959). The U.S. EPA has established an action level of 4 pCi/l (0.02 working levels) for home indoor radon and radon decay product measurements. *See: Interim Protocols for Screening and Follow-up Radon and Radon Decay Product Measurements* (PB89-224265, EPA-520/1-86-014-1, 1987) 22 p. Radon and its isotopic forms, ^{222}Rn and ^{220}Rn, are listed as known human carcinogens: *Report on Carcinogens, Twelfth Edition* (PB2011-111646, 2011) p 242.

USE: To initiate and influence chemical reactions, as a surface label in the study of surface reactions; in the determination of radium or thorium; in the study of the behavior of filters; in combination with Be or other light materials as a source of neutrons.

THERAP CAT: Antineoplastic (radiation source).

8213. Raffinose. [512-69-6] β-D-Fructofuranosyl-*O*-α-D-galactopyranosyl-(1 → 6)-α-D-glucopyranoside; gossypose; melitose; melitriose. $C_{18}H_{32}O_{16}$; mol wt 504.44. C 42.86%, H 6.39%, O 50.75%. A trisaccharide built from 1 mol each of D-galactose, D-glucose, and D-fructose which are obtained from it by acid hydrolysis. Invertase splits it into melibiose and saccharose. Occurs in Australian manna (from *Eucalyptus* spp, *Myrtaceae*); in cottonseed meal. Prepn: C. S. Hudson, T. S. Harding, *J. Am. Chem. Soc.* **36**, 2110 (1914); E. P. Clark, *ibid.* **44**, 210 (1922); Harding, *Sugar* **25**, 82 (1923); Hungerford, Nees, *Ind. Eng. Chem.* **26**, 462 (1934). Configuration: Haworth *et al.*, *J. Chem. Soc.* **1923**, 3125; Charlton *et al.*; Haworth *et al.*, *ibid.* **1927**, 1527, 3146. Structure: Hassid, Ballou in W. Pigman, *The Carbohydrates* (Academic Press, New York, 1957) p 517. Synthesis: Suami *et al.*, *Carbohydr. Res.* **26**, 234 (1973). Review: E. B. Rathbone, *Dev. Food Carbohydr.* **2**, 145-185 (1980).

Pentahydrate. Crystals in clusters from dil alc. Indifferent taste. d 1.465. mp 80°. Loses water of crystn upon slow heating to 100°. The anhydrous form dec 118-119°. $[\alpha]_D^{20}$ +105.2° (c = 4). One gram dissolves in 7 ml water (soly table: Hungerford, Nees), in 10 ml methanol. Sol in pyridine, slightly sol in alc. Does not form an osazone and does not reduce Fehling's soln.

8214. Rafoxanide. [22662-39-1] *N*-[3-Chloro-4-(4-chlorophenoxy)phenyl]-2-hydroxy-3,5-diiodobenzamide; 3′-chloro-4′-(*p*-chlorophenoxy)-3,5-diiodosalicylanilide; MK-990; Flukanide; Ranide. $C_{19}H_{11}Cl_2I_2NO_3$; mol wt 626.01. C 36.45%, H 1.77%, Cl 11.33%, I 40.54%, N 2.24%, O 7.67%. Prepn: **NL 6815783**; **BE 724668** (both 1969 to Merck & Co.); Mrozik *et al.*, *Experientia* **25**, 883 (1969). Trials in sheep and cattle: Snijders, Horak, Louw, *J. S. Afr. Vet. Med. Assoc.* **43**, 397 (1972); **44**, 251 (1973); **46**, 265 (1975); Horak, Snijders, *Vet. Rec.* **94**, 12 (1974).

Crystals, mp 168-170°. Moderately sol in acetone and acetonitrile. Practically insol in water.

THERAP CAT (VET): Fasciolicide; anthelmintic.

8215. Raloxifene. [84449-90-1] [6-Hydroxy-2-(4-hydroxyphenyl)benzo[*b*]thien-3-yl][4-[2-(1-piperidinyl)ethoxy]phenyl]methanone; keoxifene; LY-139481. $C_{28}H_{27}NO_4S$; mol wt 473.59. C 71.01%, H 5.75%, N 2.96%, O 13.51%, S 6.77%. Nonsteroidal, selective estrogen receptor modulator (SERM). Prepn: C. D. Jones, **EP 62503**; *idem*, **US 4418068** (1982, 1983 both to Lilly); *idem et al.*, *J. Med. Chem.* **27**, 1057 (1984). Review of pharmacology and toxicology: J. Buelke-Sam *et al.*, *Reprod. Toxicol.* **12**, 217-221 (1998); of clinical pharmacology and pharmacokinetics: D. Hochner-Celnikier, *Eur. J. Obstet. Gynecol. Reprod. Biol.* **85**, 23-29 (1999); of clinical efficacy in osteoporosis: D. Agnusdei, *ibid.* 43-46. Clinical effect on risk of breast cancer: S. R. Cummings *et al.*, *J. Am. Med. Assoc.* **281**, 2189 (1999); on reduction of fracture risk: B. Ettinger *et al.*, *ibid.* **282**, 637 (1999).

Crystals from acetone, mp 143-147°. uv max (ethanol): 290 nm (ε 34000).

Hydrochloride. [82640-04-8] LY-156758; Evista. $C_{28}H_{27}NO_4S.HCl$; mol wt 510.05. Crystals from methanol/water, mp 258°. uv max (ethanol): 286 nm (ε 32800). Freely sol in dimethylsulfoxide; sparingly sol in methanol; slightly sol in alc; very slightly sol in water, isopropyl alc, octanol. Practically insol in ether, ethyl acetate.

THERAP CAT: Antiosteoporotic.

8216. Raltegravir. [518048-05-0] *N*-[(4-Fluorophenyl)methyl]-1,6-dihydro-5-hydroxy-1-methyl-2-[1-methyl-1-[[(5-methyl-1,-3,4-oxadiazol-2-yl)carbonyl]amino]ethyl]-6-oxo-4-pyrimidinecarboxamide; *N*-(4-fluorobenzyl)-5-hydroxy-1-methyl-2-[1-methyl-1-[[(5-methyl-1,3,4-oxadiazol-2-yl)carbonyl]amino]ethyl]-6-oxo-1,6-dihydropyrimidine-4-carboxamide. $C_{20}H_{21}FN_6O_5$; mol wt 444.42. C 54.05%, H 4.76%, F 4.27%, N 18.91%, O 18.00%. HIV-1 integrase strand transfer inhibitor. Prepn: B. Crescenzi *et al.*, **WO 03035077**; *eidem*, **US 7169780** (2003, 2007 both to IRBM); of the crystalline potassium salt: K. M. Belyk *et al.*, **US 060122205** (2006 to Merck & Co.). HPLC/MS/MS determn in plasma: S. A. Merschman *et al.*, *J. Chromatogr. B* **857**, 15 (2007). Metabolism and disposition: K. Kassahun *et al.*, *Drug Metab. Dispos.* **35**, 1657 (2007). Clinical evaluation in treatment-experienced patients: B. Grinsztejn *et al.*, *Lancet* **369**, 1261 (2007).

Sol in organic solvents such as acetonitrile, chloroform, DMSO, methylene chloride, THF. Limited soly in water, ethanol.

Monopotassium salt. [871038-72-1] MK-0518; Isentress. $C_{20}H_{20}FKN_6O_5$; mol wt 482.51. White to off-white powder. Sol in water; slightly sol in methanol; very slightly sol in ethanol, acetonitrile. Insol in isopropanol.

THERAP CAT: Antiretroviral.

8217. Raltitrexed. [112887-68-0] *N*-[[5-[[(3,4-Dihydro-2-methyl-4-oxo-6-quinazolinyl)methyl]methylamino]-2-thienyl]carbonyl]-L-glutamic acid; *N*-(5-[*N*-(3,4-dihydro-2-methyl-4-oxoquinazolin-6-ylmethyl)-*N*-methylamino]-2-thenoyl)-L-glutamic acid; ICI-D-1694; ZD-1694; Tomudex. $C_{21}H_{22}N_4O_6S$; mol wt 458.49. C 55.01%, H 4.84%, N 12.22%, O 20.94%, S 6.99%. Folate-based inhibitor of thymidylate synthase; rapidly and extensively metabolized to its more potent polyglutamate derivatives. Prepn: L. R. Hughes, **EP 239362** (1987 to ICI; Natl. Res. Dev.); *idem*, **US 4992550** (1991 to ICI); P. R. Marsham *et al.*, *J. Med. Chem.* **34**, 1594 (1991). Pharmacokinetics: D. I. Jodrell *et al.*, *Cancer Chemother. Pharmacol.* **28**, 331 (1991). HPLC determn of metabolites in cultured cells: W. Gibson *et al.*, *Biochem. Pharmacol.* **45**, 863 (1993). Review of clinical toxicology: S. J. Clarke *et al.*, *Adv. Exp. Med. Biol.* **338**, 601-604 (1993); of mechanism of action: A. L. Jackman *et al.*, *ibid.* **339**, 265-276. Historical development, pharmacology, and clinical evaluation: S. J. Clarke *et al.*, *ibid.* 277-287; A. L. Jackman *et al.*, *Eur. J. Cancer* **31A**, 1277-1282 (1995).

Pale yellow powder. Sol in water.

Monohydrate. $C_{21}H_{22}N_4O_6S.H_2O$. mp 180-184°.

THERAP CAT: Antineoplastic.

8218. Ramatroban. [116649-85-5] (3*R*)-3-[[(4-Fluorophenyl)sulfonyl]amino]-1,2,3,4-tetrahydro-9*H*-carbazole-9-propanoic acid; (+)-(3*R*)-3-(*p*-fluorobenzenesulfonamido)-1,2,3,4-tetrahydrocarbazole-9-propionic acid; (+)-3-(4-fluorophenylsulfonamido)-9-(2-carboxyethyl)-1,2,3,4-tetrahydrocarbazole; Bay u 3405; Baynas. $C_{21}H_{21}FN_2O_4S$; mol wt 416.47. C 60.56%, H 5.08%, F 4.56%, N

6.73%, O 15.37%, S 7.70%. Thromboxane A$_2$ receptor antagonist. Prepn: H. Böshagen *et al.*, **DE 3631824**; *eidem*, **US 4965258** (1988, 1990 both to Bayer); and absolute configuration: U. Rosentreter *et al.*, *Arzneim.-Forsch.* **39**, 1519 (1989). Series of articles on pharmacology: *ibid.* 1522-1530. Clinical evaluation in asthma: H. Aizawa *et al.*, *Chest* **109**, 338 (1996).

Crystals from ether, mp 134-135°. $[\alpha]_D$ +70.1° (c = 1.0 in methanol).

THERAP CAT: Antiasthmatic; antiallergic.

8219. Ramelteon. [196597-26-9] *N*-[2-[(8*S*)-1,6,7,8-Tetrahydro-2*H*-indeno[5,4-*b*]furan-8-yl]ethyl]propanamide; TAK-375; Rozerem. C$_{16}$H$_{21}$NO$_2$; mol wt 259.35. C 74.10%, H 8.16%, N 5.40%, O 12.34%. Melatonin MT$_1$/MT$_2$ receptor agonist. Prepn: S. Ohkawa *et al.*, **WO 9732871**; *eidem* **US 6034239** (1997, 2000 both to Takeda); O. Uchikawa *et al.*, *J. Med. Chem.* **45**, 4222 (2002). Pharmacology: N. Yukuhiro *et al.*, *Brain Res.* **1027**, 59 (2004). Receptor binding study: K. Kato *et al.*, *Neuropharmacology* **48**, 301 (2005). Reviews of development and therapeutic potential: C. Cajochen, *Curr. Opin. Investig. Drugs* **6**, 114-121 (2005); N. N. Nguyen *et al.*, *Formulary* **40**, 146-155 (2005).

Crystals from ethyl acetate, mp 113-115°. $[\alpha]_D^{20}$ −57.8° (c = 1.004 in chloroform).

THERAP CAT: Sedative, hypnotic.

8220. Ramifenazone. [3615-24-5] 1,2-Dihydro-1,5-dimethyl-4-[(1-methylethyl)amino]-2-phenyl-3*H*-pyrazol-3-one; 4-isopropylamino-2,3-dimethyl-1-phenyl-3-pyrazolin-5-one; 4-isopropylaminoantipyrine; isopropylaminophenazone; isopyrin. C$_{14}$H$_{19}$N$_3$O; mol wt 245.33. C 68.54%, H 7.81%, N 17.13%, O 6.52%. Prepn: E. Skita, W. Stühmer, **DE 930328**; **DE 932677** (both 1955), *C.A.* **52**, 16372i, 20200h (1958). Acute toxicity: E. Tubaro *et al.*, *Arzneim.-Forsch.* **20**, 1024 (1970). Clinical evaluation of combination with phenylbutazone, *q.v.*, in migraine: W. Sacks, *S. Afr. Med. J.* **58**, 444 (1980).

Crystals from acetone + glacial acetic acid, mp 80°. LD$_{50}$ in mice (mg/kg): 843 i.p.; 1070 orally (Tubaro).

Hydrochloride monohydrate. C$_{14}$H$_{19}$N$_3$O.HCl.H$_2$O.

Mixture with phenylbutazone. [76120-72-4] Tomanol.

THERAP CAT: Analgesic, antipyretic, anti-inflammatory.

8221. Ramipril. [87333-19-5] (2*S*,3a*S*,6a*S*)-1-[(2*S*)-2-[[(1*S*)-1-(Ethoxycarbonyl)-3-phenylpropyl]amino]-1-oxopropyl]octahydrocyclopenta[*b*]pyrrole-2-carboxylic acid; *N*-(1*S*-carboethoxy-3-phenylpropyl)-*S*-alanyl-*cis,endo*-2-azabicyclo[3.3.0]octane-3*S*-carboxylic acid; (2*S*,3a*S*,6a*S*)-1-[(*S*)-*N*-[(*S*)-1-carboxy-3-phenylpropyl]alanyl]octahydrocyclopenta[*b*]pyrrole-2-carboxylic acid 1-ethyl ester; HOE-498; Altace; Cardace; Delix; Pramace; Quark; Ramace; Triatec; Tritace; Unipril; Vesdil. C$_{23}$H$_{32}$N$_2$O$_5$; mol wt 416.52. C 66.32%, H 7.74%, N 6.73%, O 19.21%. Angiotensin converting enzyme (ACE) inhibitor; converted to active, diacid metabolite, *ramiprilat*. Prepn: V. Teetz *et al*, **EP 79022** (1983 to Hoechst A.G.), *C.A.* **100**, 52012h (1984); E. H. Gold *et al.*, **US 4587258** (1986 to Schering Corp.); V. Teetz *et al.*, *Arzneim.-Forsch.* **34**, 1399 (1984). Radioimmunoassay determn in human serum and plasma: H. G. Eckert *et al.*, *ibid.* **35**, 1251 (1985). Toxicology: H. H. Donaubauer, D. Mayer, *ibid.* **38**, 14 (1988). Symposium on pharmacology and clinical efficacy: *Am. J. Cardiol.* **59**, 1D-177D (1987). Clinical trial in chronic nephropathy: P. Ruggenenti *et al.*, *Lancet* **352**, 1252 (1998). Clinical prevention of cardiovascular events in high-risk patients: HOPE Study Investigators, *N. Engl. J. Med.* **342**, 145 (2000). Clinical trial in prevention of stroke: J. Bosch *et al.*, *Br. Med. J.* **324**, 1 (2002).

Felty needles from ether, mp 109°. $[\alpha]_D^{24}$ +33.2° (c = 1 in 0.1*N* ethanolic HCl). Freely sol in methanol; sparingly sol in water. LD$_{50}$ (14 day) in male, female mice, male, female rats (mg/kg): 1194, 1158, 687, 608 i.v.; 10933, 10048, >10000, >10000 orally (Donaubauer, Mayer).

THERAP CAT: Antihypertensive.

8222. Ramoplanin. [76168-82-6] A-16686; MDL-62198. Glycolipodepsipeptide antibiotic complex consisting of 6 related components (A$_1$, A$_2$, A$_3$, A'$_1$, A'$_2$, A'$_3$) of which A$_2$ is the most abundant. Isoln and inital characterization from *Actinoplanes* sp.: B. Cavalleri *et al.*, **DE 3013246**; *eidem*, **US 4303646** (1980, 1981 both to Gruppo Lepetit); *idem et al.*, *J. Antibiot.* **37**, 309 (1984). Bactericidal activity: R. Pallanza *et al.*, *ibid.* 318. Structure elucidation: R. Ciabatti *et al.*, *ibid.* **42**, 254 (1989); and sequence determn: J. K. Kettenring *et al.*, *ibid.* 268. 3D structure: M. Kurz, W. Guba, *Biochemistry* **35**, 12570 (1996). Isoln and characterization of the A' factors: L. Gastaldo *et al.*, *J. Ind. Microbiol.* **11**, 13 (1992). Determn in sera: M. T. Kenny *et al.*, *Diagn. Microbiol. Infect. Dis.* **18**, 117 (1994). Clinical evaluation against vancomycin resistant enterococci (VRE): M. T. Wong *et al.*, *Clin. Infect. Dis.* **33**, 1476 (2001). Review of discovery and chemistry: F. Parenti *et al.*, *Drugs Exp. Clin. Res.* **16**, 451-455 (1990); of chemistry and mechanism of action: S. Walker *et al.*, *Chem. Rev.* **105**, 449-475 (2005); of pharmacology and clinical potential: D. K. Farver *et al.*, *Ann. Pharmacother.* **39**, 863-868 (2005).

Ramoplanin A$_2$

White powder, slightly hygroscopic. mp 210-230°. $[\alpha]_D^{20}$ +78.3° (c = 1.04 in H_2O). Sol in DMF and lower alcohols. Insol in ethyl ether, petroleum ether and benzene. LD_{50} in mice (mg/kg): 328 i.p., 122 i.v.; orally in rats: 2000 mg/kg (Pallanza).

Ramoplanin A₂. [81988-88-7] $C_{119}H_{154}ClN_{21}O_{40}$; mol wt 2554.10. C 55.96%, H 6.08%, Cl 1.39%, N 11.52%, O 25.06%. Total synthesis: W. Jiang *et al.*, *J. Am. Chem. Soc.* **124**, 5288 (2002); *eidem*, *ibid.* **125**, 1877 (2003). White powder, mp 210-220°. $[\alpha]_D^{20}$ +73±4° (c = 0.49 in H_2O). uv max in methanol: 234, 268 nm ($E_{1cm}^{1\%}$ 206, 114).

THERAP CAT: Antibacterial.

8223. Ramosetron. [132036-88-5] (1-Methyl-1*H*-indol-3-yl)[(6*R*)-4,5,6,7-tetrahydro-1*H*-benzimidazol-6-yl]methanone; (*R*)-5-[(1-methylindol-3-yl)carbonyl]-4,5,6,7-tetrahydrobenzimidazole; Nor-YM-060. $C_{17}H_{17}N_3O$; mol wt 279.34. C 73.10%, H 6.13%, N 15.04%, O 5.73%. Selective serotonin 5HT₃ receptor antagonist; structurally different from ondansetron, granisetron, *q.q.v.* Prepn: M. Ohta *et al.*, **EP 381422**; *eidem*, **US 5344927** (1990, 1994 both to Yamanouchi). Pharmacology: K. Miyata *et al.*, *J. Pharmacol. Exp. Ther.* **259**, 15 (1991). Effects on gastric emptying in rodents: K. Miyata *et al.*, *Jpn. J. Pharmacol.* **69**, 205 (1995). Toxicity evaluation: H. Tabata *et al.*, *Arzneim.-Forsch.* **45**, 760 (1995). HPLC determn in plasma and urine: H. Miura *et al.*, *Biomed. Chromatogr.* **8**, 103 (1994). Clinical prevention of postoperative nausea and vomiting: Y. Fujii *et al.*, *Anesth. Analg.* **90**, 472 (2000).

Foam. $[\alpha]_D^{20}$ −16.5° (c 1.13 in methanol).

Hydrochloride. [132907-72-3] YM-060; Nasea. $C_{17}H_{17}N_3O\cdot HCl$; mol wt 315.80. Slightly yellowish-white, odorless crystalline powder. Easily sol in water (Tabata). Also as crystals from ethanol, mp 215-230°. $[\alpha]_D$ −42.9° (c 1.02 in methanol).

THERAP CAT: Antiemetic.

8224. Ranelic Acid. [135459-90-4] 5-[Bis(carboxymethyl)amino]-2-carboxy-4-cyano-3-thiopheneacetic acid; 2-[*N*,*N*-di(carboxymethyl)amino]-3-cyano-4-carboxymethylthiophene-5-carboxylic acid. $C_{12}H_{10}N_2O_8S$; mol wt 342.28. C 42.11%, H 2.94%, N 8.18%, O 37.39%, S 9.37%. Prepn: M. Wierzbicki *et al.*, **EP 415850**; *eidem*, **US 5128367** (1991, 1992 both to Adir & Co.).

Crystals from ethyl ether, contains 4% ether.

Strontium salt. [135459-87-9] Strontium ranelate; S-12911; Osseor; Protelos; Protos. $C_{12}H_6N_2O_8SSr_2$; mol wt 513.49. Bone metabolism modulator; inhibits bone resorption while maintaining bone formation. Effect on human chondrocytes *in vitro*: Y. Henrotin *et al.*, *J. Bone Miner. Res.* **16**, 299 (2001). Effect on bone remodeling in monkeys: J. Buehler *et al.*, *Bone* **29**, 176 (2001). Clinical trial in postmenopausal osteoporosis: P. J. Meunier *et al.*, *N. Engl. J. Med.* **350**, 459 (2004).

THERAP CAT: Strontium salt as antiosteoporotic.

8225. Raney® Nickel. [7440-02-0] Raney nickel catalyst. Prepd by fusing 50 parts nickel with 50 parts aluminum: M. Raney, **US 1628190** (1927); **US 1915473** (1933); pulverizing the alloy and dissolving out most of the aluminum with NaOH soln: Covert, Adkins, *J. Am. Chem. Soc.* **54**, 4116 (1932); Ruggli, Preiswerk, *Helv. Chim. Acta* **22**, 494 (1939); Mozingo, *Org. Synth.* **21**, 15 (1941).

Absorption studies: Kokes, Emmett, *J. Am. Chem. Soc.* **83**, 29 (1961). The residual aluminum, which amounts to several percent, appears to be necessary for proper catalytic activity. *Review:* J. S. Pizey, *Synthetic Reagents* **vol. 2** (John Wiley, New York, 1974) pp 175-311.

Grayish-black powder or cubic crystals. *Ignites on contact with air.* Contains hydrogen and has been attributed the formula Ni_2H. Generally stored under alcohol, although ether, water, methylcyclohexane, and dioxane may be used. Raney nickel loses its hydrogen slowly on storage and becomes inactive. Properly prepd and stored it should remain active for 6 months.

USE: Catalyst for the hydrogenation of organic compds with gaseous hydrogen. Usually from 1 to 10% of the substance to be reduced is employed. Is active at room temp and 1 atm pressure, but is also used at high temps and high pressures, *see* H. B. Adkins, *Reactions of Hydrogen* (Madison, 1937).

8226. Ranibizumab. [347396-82-1] Anti-(human vascular endothelial growth factor) immunoglobulin G1 Fab fragment (human-mouse monoclonal rhuFAB V2 γ₁-chain) disulfide with human-mouse monoclonal rhuFAB V2 light chain; rhuFabV2; Lucentis. Recombinant humanized monoclonal antibody fragment directed against vascular endothelial growth factor (VEGF); exhibits antiangiogenic activity. Mol wt approx 48 kDa. Prepn: M. Baca *et al.*, **WO 9845331**; *eidem*, **US 6884879** (1998, 2005 both to Genentech). Intraocular biodistribution study: J. Mordenti *et al.*, *Toxicol. Pathol.* **27**, 536 (1999). Inhibition of exptl choroidal neovascularization: M. G. Krzystolik *et al.*, *Arch. Ophthalmol.* **120**, 338 (2002). Pharmacokinetics: J. Gaudreault *et al.*, *Invest. Ophthalmol. Visual Sci.* **46**, 726 (2005). Determn in serum by electrochemiluminescence assay: J. Lowe *et al.*, *J. Pharm. Biomed. Anal.* **52**, 680 (2010). Review of clinical experience in treatment of neovascular age-related macular degeneration: H. Kourlas, P. Abrams, *Clin. Ther.* **29**, 1850-1861 (2007); U. Schmidt-Erfurth, *Expert Opin. Drug Saf.* **9**, 149-165 (2010). Clinical trial for macular edema following branch retinal vein occlusion: P. A. Campochiaro *et al.*, *Ophthalmology* **117**, 1102 (2010).

THERAP CAT: In treatment of age-related macular degeneration and macular edema following retinal vein occlusion.

8227. Ranimustine. [58994-96-0] Methyl 6-[[[(2-chloroethyl)nitrosoamino]carbonyl]amino]-6-deoxy-α-D-glucopyranoside; methyl *N*-carbamyl-*N′*-(2-chloroethyl)-*N′*-nitroso-6-amino-6-deoxy-α-D-glucopyranoside; 6-[3-(2-chloroethyl)-3-nitrosoureido]-6-deoxy-α-D-glucopyranoside; 3-(methyl-α-D-glucopyranos-6-yl)-1-(2-chloroethyl)-1-nitrosourea; ranomustine; MCNU; NSC-0270516; Cymerin; Thymerin. $C_{10}H_{18}ClN_3O_7$; mol wt 327.72. C 36.65%, H 5.54%, Cl 10.82%, N 12.82%, O 34.17%. Chloroethylnitrosourea derivative with antitumor activity. Similar to carmustine, chlorozotocin, lomustine, nimustine, *q.q.v.* Prepn: **NL 7507973**; G. Kimura, J. Sekine, **US 4057684**; **NL 7800920**; G. Kimura, **US 4156777** (1976, 1977, 1978, 1979, all to Tokyo Tanabe). Antitumor activity in rodents, comparison with other nitrosoureas: S. Sekido *et al.*, *Cancer Treat. Rep.* **63**, 961 (1979); S. Fujimoto, M. Ogawa, *Cancer Chemother. Pharmacol.* **9**, 134 (1982); S. Fujimoto *et al.*, *Gann* **75**, 937 (1984). Mechanism of action: R. Kanamaru *et al.*, *Curr. Chemother. Immunother.* **2**, 1377 (1982). Effect of sugar alcohols on toxicity in mice: T. Tashiro *et al.*, *Cancer Chemother. Pharmacol.* **8**, 183 (1982). Series of articles on metabolic fate in rats: Y. Esumi *et al.*, *Iyakuhin Kenkyu* **16**, 381-428 (1985), *C.A.* **103**, 115620q, 115621r, 134369f (1985). Clinical study in hematological malignant diseases: T. Masaoka *et al.*, *Chemotherapy (Tokyo)* **33**, 271 (1985).

Pale yellowish needles from anhydrous ethanol-ethyl ether (l:1), mp 111-112°. $[\alpha]_D^{20}$ +93.2° (c = 0.5 in methanol) (Kimura). Also reported as 101-103° (dec) from isopropanol. $[\alpha]_D^{25}$ +73.2° (c = 0.3 in methanol) (Kimura, Sekine). Solubility in water: 900 mg/ml at 25°. LD_{50} in male rats (mg/kg): 42 i.p., 42 i.v., 50 orally (Kimura).

THERAP CAT: Antineoplastic.

8228. Ranitidine. [66357-35-5] N'-[2-[[[-5-[(Dimethylamino)methyl]-2-furanyl]methyl]thio]ethyl]-N-methyl-2-nitro-1,1-ethenediamine. $C_{13}H_{22}N_4O_3S$; mol wt 314.40. C 49.66%, H 7.05%, N 17.82%, O 15.27%, S 10.20%. Histamine H_2-receptor antagonist which inhibits gastric acid secretion. Prepn: B. J. Price *et al.*, **FR 2384765**; *eidem*, **US 4128658** (both 1978 to Allen & Hanburys). HPLC determn in plasma: P. F. Carey, L. E. Martin, *J. Liq. Chromatogr.* **1979**, 1291. HPTLC determn in urine and bioequivalence study: S. A. Shah *et al.*, *J. Chromatogr. B* **767**, 83 (2002). CV determn in pharmaceutical formulations: P. Norouzi *et al.*, *J. Pharmacol. Toxicol. Methods* **55**, 289 (2007). Pharmacological studies: J. Bradshaw *et al.*, *Br. J. Pharmacol.* **66**, 464 (1979); M. J. Daly *et al.*, *Gut* **21**, 408 (1980). Characterization and soly studies of polymorphic forms: M. Mirmehrabi *et al.*, *Int. J. Pharm.* **282**, 73 (2004). Efficacy in treatment of duodenal ulcers: A. Berstad *et al.*, *Scand. J. Gastroenterol.* **15**, 637 (1980); R. P. Walt *et al.*, *Gut* **22**, 49 (1981); of premeal therapy in reduction of heartburn: K. A. Pappa *et al.*, *Curr. Ther. Res.* **59**, 454 (1998). Review of pharmacology and therapeutic use: R. N. Brogden *et al.*, *Drugs* **24**, 267-303 (1982). Comprehensive description: M. Hohnjec *et al.*, *Anal. Profiles Drug Subs.* **15**, 533-561 (1986).

Solid, mp 69-70°. Sulfurous odor. Low soly in water. *Protect from light.*

Hydrochloride. [66357-59-3] AH-19065; Azantac; Noctone; Raniben; Ranidil; Raniplex; Sostril; Taural; Terposen; Ulcex; Ultidine; Zantac; Zantic. $C_{13}H_{22}N_4O_3S$.HCl; mol wt 350.86. White to pale yellow solid, mp 133-134°. Slightly bitter taste; sulfur like odor. Very sol in water; freely sol in acetic acid; sol in methanol; sparingly sol in ethanol. Practically insol in chloroform. uv max (isopropanol): 228, 326 nm.

Bismuth citrate. [128345-62-0] Ranitidine bismutrex; GR-122311X; Pylorid; Tritec, $C_{13}H_{22}N_4O_3S.C_6H_5BiO_7$; mol wt 712.48. Pharmacology and activity vs *Helicobacter* sp: R. Stables *et al.*, *Aliment. Pharmacol. Ther.* **7**, 237 (1993).

THERAP CAT: Antiulcerative.

THERAP CAT (VET): Antiulcerative; gastroprokinetic.

8229. Ranolazine. [95635-55-5] N-(2,6-Dimethylphenyl)-4-[2-hydroxy-3-(2-methoxyphenoxy)propyl]-1-piperazineacetamide; (±)-4-[2-hydroxy-3-(*o*-methoxyphenoxy)propyl]-1-piperazineaceto-2',6'-xylidide; (±)-1-[3-(2-methoxyphenoxy)-2-hydroxypropyl]-4-[N-(2,6-dimethylphenyl)carbamoylmethyl]piperazine; Ranexa. $C_{24}H_{33}N_3O_4$; mol wt 427.55. C 67.42%, H 7.78%, N 9.83%, O 14.97%. Anti-ischemic agent which modulates myocardial metabolism. Prepn: A. F. Kluge *et al.*, **EP 126449**; *eidem*, **US 4567264** (1984, 1986 both to Syntex). HPLC resolution of enantiomers: E. Delée *et al.*, *Chromatographia* **24**, 357 (1987). Clinical trial in angina: B. R. Chaitman *et al.*, *J. Am. Coll. Cardiol.* **43**, 1375 (2004). Review of pharmacology and clinical development: J. G. McCormack *et al.*, *Gen. Pharmacol.* **30**, 639-645 (1998); R. S. Schofield, J. A. Hill, *Expert Opin. Invest. Drugs* **11**, 117-123 (2002).

Dihydrochloride. [95635-56-6] RS-43285. $C_{24}H_{33}N_3O_4$.2HCl; mol wt 500.46. White crystalline powder from methanol/ether, mp 164-166°. Readily sol in water.

THERAP CAT: Antianginal.

8230. Ranpirnase. [133737-96-9] Ribonuclease (*Rana pipiens* reduced); P-30 protein; Onconase. Antitumor ribonuclease isolated from oocytes and early embryos of *Rana pipiens*. Single chain protein containing 104 amino acid residues; mol wt ~12 kDa. Description of cytotoxic activity: Z. Darzynkiewicz *et al.*, *Cell Tissue Kinet.* **21**, 169 (1988). Amino acid sequence and identification as a ribonuclease: W. Ardelt *et al.*, *J. Biol. Chem.* **266**, 245 (1991). Crystallization: S. C. Mosimann *et al.*, *Proteins Struct. Funct. Genet.* **14**, 392 (1992). Mechanism of action: Y. Wu *et al.*, *J. Biol. Chem.* **268**, 10686 (1993). Prepn by recombinant technology: E. Notomista *et al.*, *FEBS Lett.* **463**, 211 (1999). Clinical trial in malignant mesothelioma: S. M. Mikulski *et al.*, *J. Clin. Oncol.* **20**, 274 (2002).

THERAP CAT: Antineoplastic.

8231. Rapacuronium Bromide. [156137-99-4] 1-[(2β,3α,-5α,16β,17β)-3-(Acetyloxy)-17-(1-oxopropoxy)-2-(1-piperidinyl)-androstan-16-yl]-1-(2-propen-1-yl)piperidinium bromide (1:1); 1-allyl-2-(3α,17β-dihydroxy-2β-piperidino-5α-androstan-16β-yl)piperidinium bromide 3-acetate 17-propionate; Org-9487; Raplon. $C_{37}H_{61}N_2O_4$.Br; mol wt 677.81. C 65.57%, H 9.07%, N 4.13%, O 9.44%, Br 11.79%. Aminosteroid, competitive neuromuscular blocker. Prepn: T. Sleigh *et al.*, **CA 2094457**; *eidem*, **US 5418226** (both to Akzo). Clinical pharmacodynamics: P. M. C. Wright *et al.*, *Anesthesiology* **90**, 16 (1999). Clinical trial in pediatric patients: R. F. Kaplan *et al.*, *Anesth. Analg.* **89**, 1172 (1999). Review of pharmacology and use in endotracheal intubation: S. V. Onrust, R. H. Foster, *Drugs* **58**, 887-918 (1999).

Crystals from diethyl ether-acetone, mp 184°. $[\alpha]_D^{20}$ −12.7° (c = 1.01 in CHCl₃).

THERAP CAT: Neuromuscular blocking agent.

8232. Rapamycin. [53123-88-9] Sirolimus; RAPA; RPM; AY-22989; NSC-226080; Rapamune. $C_{51}H_{79}NO_{13}$; mol wt 914.19. C 67.01%, H 8.71%, N 1.53%, O 22.75%. Triene macrolide antibiotic isolated from *Streptomyces hygroscopicus*. Name derived from the native word for Easter Island, Rapa Nui. Isoln: S. N. Sehgal *et al.*, **DE 2347682**; *eidem*, **US 3929992** (1974, 1975 both to Ayerst McKenna Harrison); purification and characterization: C. Vézina *et al.*, *J. Antibiot.* **28**, 721 (1975); S. N. Sehgal *et al.*, *ibid.* 727. Inhibition of immune response: R. R. Martel *et al.*, *Can. J. Physiol. Pharmacol.* **55**, 48 (1977); of graft rejection in mice: C. P. Eng *et al.*, *Transplant. Proc.* **23**, 868 (1991). Total synthesis: K. C. Nicolaou *et al.*, *J. Am. Chem. Soc.* **115**, 4419 (1993); D. Romo *et al.*, *ibid.* 7906. Series of articles on therapeutic monitoring and pharmacokinetics: *Clin. Ther.* **22**, Suppl. 2, B1-B132 (2000); on pharmacology and clinical experience in transplantation: *Transplant. Proc.* **35**, Suppl. 1, S1-S233 (2003). Clinical trial in prevention of coronary restenosis: D. R. Holmes, Jr. *et al.*, *Circulation* **109**, 634 (2004).

Colorless crystalline solid from ether, mp 183-185°. uv max (95% ethanol): 267, 277, 288 nm ($E_{1cm}^{1\%}$ 417, 541, 416). $[\alpha]_D^{25}$ −58.2° (methanol). Sol in ether, chloroform, acetone, methanol and DMF; very sparingly sol in hexane and petr ether. Substantially insol in water. LD_{50} in mice (mg/kg): ≈600 i.p.; >2,500 orally (Vézina).

USE: Tool for immunochemistry.

THERAP CAT: Immunosuppressant; antirestenotic.

8233. Rapeseed Oil. Colza oil. Oil expressed from seeds of *Brassica campestris* L., *Cruciferae*.

Pale yellow, rather viscid liq. d 0.913-0.917; n_D^{20} 1.4720-1.4752. Solidif −2° to −10°. Sapon no. 170-177. Iodine no. 97-105. Sol in chloroform, ether, CS_2.

USE: Lubricant; manuf rubber substitutes, margarine, soft soaps, blown oils; oiling woolens.

8234. Rasagiline. [136236-51-6] (1R)-2,3-Dihydro-N-2-propyn-1-yl-1H-inden-1-amine; (R)-N-2-propynyl-1-indanamine; N-propargyl-1-(R)-aminoindan; AGN-1135. $C_{12}H_{13}N$; mol wt 171.24. C 84.17%, H 7.65%, N 8.18%. Irreversible MAO-B inhibitor. Prepn: M. B. H. Youdim et al., **EP 436492**; eidem, **US 5457133** (1991, 1995 both to Teva). Pharmacology: J. P. M. Finberg, M. B. H. Youdim, Neuropharmacology **43**, 1110 (2002); of metabolites: S. Glezer, J. P. M. Finberg, Eur. J. Pharmacol. **472**, 173 (2003). Mechanism of action: M. B. H. Youdim et al., Biochem. Pharmacol. **66**, 1635 (2003). Double-blind clinical trial as adjunctive therapy to levodopa: J. M. Rabey et al., Clin. Neuropharmacol. **23**, 324 (2000); multicenter trial in early Parkinson's disease: Arch. Neurol. **59**, 1937 (2002); in levodopa-treated Parkinson's patients: Parkinson Study Group, ibid. **62**, 241 (2005); O. Rascol et al., Lancet **365**, 947 (2005). Review of pharmacology: J. J. Chen, D. M. Swope, J. Clin. Pharmacol. **45**, 878-894 (2005); of clinical development: O. Rascol, Expert Opin. Pharmacother. **6**, 2061-2075 (2005); F. Stocchi, Int. J. Clin. Pract. **60**, 215-221 (2006).

Methanesulfonate. [161735-79-1] Rasagiline mesylate; TVP-1012; Agilect; Azilect. $C_{12}H_{13}N.CH_3SO_3H$; mol wt 267.34.

THERAP CAT: Antiparkinsonian.

8235. Rasburicase. [134774-45-1] Urate oxidase (*Aspergillus flavus* clone 9C/9A reduced); SR-29142; Elitek; Fasturtec. Recombinant form of fungal urate oxidase expressed in *Saccharomycese cerevisiae*. Tetramer of identical subunits, each comprised of a single, 301 amino acid polypeptide chain with mol wt ~34 kDa. Converts uric acid into the more soluble compound, allantoin, which is readily excreted by the kidneys. The natural enzyme is endoge-

nous in most mammals but is absent in higher primates where uric acid is the end-product of purine metabolism. Prepn: D. Caput et al., **EP 408461**; eidem, **US 5382518** (1991, 1995 both to Sanofi); P. Leplatois et al., Gene **122**, 139 (1992). Comparison with natural enzyme: A. Bayol et al., Biotechnol. Appl. Biochem. **36**, 21 (2002). Radioimmunoassay in plasma: D. Dussossoy et al., J. Pharm. Sci. **85**, 955 (1996). Clinical pharmacokinetics in cancer patients: C.-H. Pui et al., J. Clin. Oncol. **19**, 697 (2001). Clinical comparison with allopurinol in pediatric hyperuricemia: S. C. Goldman et al., Blood **97**, 2998 (2001); in adults at risk for tumor lysis syndrome: J. Cortes et al., J. Clin. Oncol. **27**, 4207 (2010). Review: C.-H. Pui, Expert Opin. Pharmacother. **3**, 433-452 (2002); V. Oldfield, C. M. Perry, Drugs **66**, 529-545 (2006).

THERAP CAT: In treatment of malignancy associated hyperuricemia.

8236. Raspberry. Red raspberry. Deciduous, woody bush, *Rubus idaeus* L., *Rosaceae*, bearing edible fruit. The leaf has been used medicinally for diarrhea, nausea, vomiting, and to prepare the uterus for childbirth. Habit. Europe, Asia, cultivated in Canada and U.S. Constit. Anthocyanins, ellagitannins, terpenoids, flavonoids, sugar, malic and citric acids. HPLC determn of antioxidants in fruit: J. Beekwilder et al., J. Agric. Food Chem. **53**, 3313 (2005). Clinical trial of raspberry leaf in pregnancy: M. Simpson et al, J. Midwifery Womens Health **46**, 51 (2001). Review of therapeutic constituents: A. V. Patel et al., Curr. Med. Chem. **11**, 1501-1512 (2004); of pharmacology and medicinal uses: J. Barnes et al., Herbal Medicines (Pharmaceutical Press, London, 2nd Ed., 2002) pp 397-398; J. Gruenwald et al., PDR for Herbal Medicines (Thomson PDR, Montvale, 3rd Ed., 2004) pp 675-676.

USE: Fruit in jams, desserts, sauces, beverages. Pharmaceutic aid (flavor).

THERAP CAT: Astringent.

8237. Raubasine. [483-04-5] (19α)-16,17-Didehydro-19-methyloxayohimban-16-carboxylic acid methyl ester; δ-yohimbine; pytetrahydroserpentine; tetrahydroserpentine; ajmalicine; Circolene; Hydrosarpan; Isoarteril; Lamuran. $C_{21}H_{24}N_2O_3$; mol wt 352.43. C 71.57%, H 6.86%, N 7.95%, O 13.62%. α_1-Adrenergic blocker isolated from bark of *Corynanthe johimbe* K. Schum., *Rubiaceae*: H. Heinemann, Ber. **67**, 15 (1934); from roots of *Rauwolfia serpentina* (L.) Benth., *Apocynaceae*: S. Siddiqui, R. H. Siddiqui, J. Indian Chem. Soc. **8**, 667 (1931); A. H. Popelak et al., Naturwissenschaften **40**, 625 (1953); A. Hofmann, Helv. Chim. Acta **37**, 849 (1954); M. W. Klohs et al., J. Am. Chem. Soc. **76**, 1332 (1954). Review of early literature: R. E. Woodson et al., Rauwolfia: Botany, Pharmacognosy, Chemistry and Pharmacology (Little, Brown and Co., Boston, 1957) 147 pp. Structure: Goutarel, Le Hir, Bull. Soc. Chim. Fr. **18**, 909 (1951). Stereochemistry: Wenkert et al., J. Am. Chem. Soc. **83**, 5037 (1961); Shamma, Richey, ibid. **85**, 2507 (1963). Total synthesis of dl-form: van Tamelen, Placeway, ibid. **83**, 2594 (1961); van Tamelen et al., ibid. **91**, 7359 (1969); J. Gutzwiller et al., Helv. Chim. Acta **64**, 1663 (1981); T. Kametani et al., J. Chem. Soc. Perkin Trans. 1 **1981**, 3168. Biosynthesis: N. Nagakura et al., ibid. **1979**, 2308; M. Rueffer et al., Chem. Commun. **1979**, 1016. Pharmacokinetics: A. Marzo et al., Farmaco Ed. Prat. **36**, 173 (1981). Clinical evaluation of platelet anti-aggregant activity: J. Neuman et al., Arzneim.-Forsch. **36**, 1394 (1986). Evaluation of combination with almitrine, q.v., in cerebral ischemia in rats: M. G. Borzeix, J. Cahn, ibid. **37**, 491 (1987).

Prisms from methanol, dec 257°. $[\alpha]_D^{20}$ −60° (c = 0.5 in chloroform); $[\alpha]_D^{20}$ −45° (c = 0.5 in pyridine); $[\alpha]_D^{20}$ −39° (c = 0.25 in methanol). uv max (methanol): 227, 292 nm (log ε 4.61, 3.79).

Hydrochloride. $C_{21}H_{24}N_2O_3 \cdot HCl$. Leaflets from ethanol, mp 290° (dec). $[\alpha]_D^{20} -17°$ (c = 0.5 in methanol). Sparingly sol in water or dil HCl.

Hydrobromide. $C_{21}H_{24}N_2O_3 \cdot HBr$. Diamond-shaped platelets from methanol, mp 295-296°.

THERAP CAT: Antihypertensive; anti-ischemic (cerebral and peripheral).

8238. Rauwolfia serpentina. Hiwolfia; Koglucoid; Raudixin; Rauserpol; Rauverid; Rauwoldin; Rauwolfemms; Serfolia; Serpina; Wolfina. Powdered, dried root of *Rauwolfia serpentina* (L.) Benth. Apocynaceae, a small shrub native to the Orient from India to Sumatra. Contains many indole alkaloids of which reserpine, rescinnamine and deserpidine are the primary constitutents. Others include: reserpiline, yohimbine, ajmaline, serpentine, serpetinine. Review and bibliography: Monachino, *Econ. Bot.* **8**, 349-365 (1954). Clinical trial in hypertension: M. C. Srivastava *et al., Indian Pract.* **20**, 279 (1967); in combination with chlorothiazide: W. M. Smith, *Circ. Res.* **40**, Suppl I, I98 (1977). LC determn of reserpine-rescinnamine content: U. R. Cieri, *J. Assoc. Off. Anal. Chem.* **70**, 540 (1987).

Alseroxylon. [8001-95-4] Austrawolf; Egalin; Gendon; Rautensin; Rauwiloid; Ra-Valeas; Rivadescin. Alkaloidal extract of the dried root. Red, amorphous powder with characteristic odor. Contains 0.15-0.2% reserpine-rescinnamine alkaloids.

THERAP CAT: Antihypertensive.

8239. Ravuconazole. [182760-06-1] 4-[2-[(1R,2R)-2-(2,4-Difluorophenyl)-2-hydroxy-1-methyl-3-(1H-1,2,4-triazol-1-yl)propyl]-4-thiazolyl]benzonitrile; (2R,3R)-3-[4-(4-cyanophenyl)thiazol-2-yl]-2-(2,4-difluorophenyl)-1-(1H-1,2,4-triazol-1-yl)-2-butanol; BMS-207147; ER-30346. $C_{22}H_{17}F_2N_5OS$; mol wt 437.47. C 60.40%, H 3.92%, F 8.69%, N 16.01%, O 3.66%, S 7.33%. Ergosterol biosynthesis inhibitor. Prepn (stereochemistry unspecified): T. Naito *et al*, **EP 667346**; *eidem*, **US 5648372** (1995, 1997 both to Eisai); of optically acitve form: A. Tsuruoka *et al., Chem. Pharm. Bull.* **46**, 623 (1998). Chiral synthesis: Y. Kaku *et al., ibid.* 1125. *In vitro* comparative antifungal spectrum: J. C. Fung-Tomc *et al., Antimicrob. Agents Chemother.* **42**, 313 1998. Antifungal activity in candidosis: K. V. Clemons, D. A. Stevens, *ibid.* **45**, 3433 (2001); in aspergillosis: W. R. Kirkpatrick *et al., J. Antimicrob. Chemother.* **49**, 353 (2002). Clinical evaluation in onychomycosis: A. K. Gupta *et al., J. Eur. Acad. Dermatol. Venereol.* **19**, 437 (2005). Review of development and therapeutic potential: S. Arikan, J. H. Rex, *Curr. Opin. Investig. Drugs* **3**, 555-561 (2002).

Colorless prisms from diisopropyl ether/n-hexane, mp 164-166°. $[\alpha]_D^{24} -29.1°$ (c = 1.03 in methanol).

THERAP CAT: Antifungal.

8240. Rayon. Regenerated cellulose. The Code of Federal Regulations defines rayon as a manufactured fiber composed of regenerated cellulose, as well as manufactured fibers composed of regenerated cellulose in which substituents have replaced not more than 15 percent of the hydrogens of the hydroxyl groups. Now produced almost exclusively by the viscose process. *Review:* J. Lundberg, A. Turbak in *Kirk-Othmer Encyclopedia of Chemical Technology* **Vol. 19** (Wiley-Interscience, New York, 3rd ed., 1982) pp 855-880.

Purified rayon. Avisco Rayon. A medicinal grade of rayon.

USE: Purified rayon as surgical aid.

8241. Razoxane. [21416-67-1] 4,4'-(1-Methyl-1,2-ethanediyl)bis-2,6-piperazinedione; (±)-4,4'-propylenedi-2,6-piperazinedione; (±)-(3,5,3',5'-tetraoxo)-1,2-dipiperazinopropane; (±)-1,2-

bis(3,5-dioxopiperazinyl)propane; ICI-59118; ICRF-159; NSC-129943. $C_{11}H_{16}N_4O_4$; mol wt 268.27. C 49.25%, H 6.01%, N 20.88%, O 23.86%. Cyclized analog of EDTA, *q.v.*; exhibits intracellular iron chelating activity. Plays role in preventing the formation of doxorubicin-iron complexes which generate radical oxygen species that are associated with cardiotoxicity. Prepn: A. M. Creighton *et al., Nature* **222**, 384 (1969); *idem*, **GB 1234935**; *idem*, **US 3941790** (1971, 1976 both to Natl. Res. Dev. Corp.). Mode of action: H. B. A. Sharpe *et al., Nature* **226**, 524 (1970). Metabolism: R. E. Bellet *et al., Eur. J. Cancer* **13**, 1293 (1977). Pharmacology: A. Atherton, *ibid.* **11**, 383 (1975). Clinical studies: M. T. Bakowski *et al., Int. J. Radiat. Oncol. Biol. Phys.* **4**, 115 (1978); H. W. Bruckner *et al., Cancer Treat. Rep.* **66**, 1713 (1982). Toxicity study: E. Hassenstein, K. Renner, *Strahlentherapie* **154**, 122 (1978). Clinical evaluation in renal cell cancer: J. P. Braybrooke *et al., Clin. Cancer Res.* **6**, 4697 (2000). Review of antineoplastic activity and radiosensitization: M. T. Bakowski, *Cancer Treat. Rev.* **3**, 95-107 (1976); K. Hellmann, W. Rhomberg, *ibid.* **18**, 225-240 (1991).

Pale cream microcrystalline solid, mp 237-239°.

(+)-Form. [24584-09-6] Dexrazoxane; ADR-529; ICRF-187; NSC-169780; Savene; Totect; Zinecard. Clinical pharmacokinetics: R. H. Earhart *et al., Cancer Res.* **42**, 5255 (1982). HPLC determn in plasma and urine: R. C. Lewis *et al., Pharm. Res.* **9**, 101 (1992). Clinical study of protection vs doxorubicin-associated cardiomyopathy: J. L. Speyer *et al., J. Clin. Oncol.* **10**, 117 (1992); S. E. Lipshultz *et al., N. Engl. J. Med.* **351**, 145 (2004). Review of pharmacology and clinical studies: J. Koning *et al., Cancer Treat. Rev.* **18**, 1-19 (1991); of therapeutic role as a cardioprotectant: S. M. Swain, P. Vici, *J. Cancer Res. Clin. Oncol.* **130**, 1-7 (2004); of development and clinical experience: S. W. Langer, *Expert Rev. Anticancer Ther.* **7**, 1081-1088 (2007). Crystals from aq methanol/ether, mp 193°. $[\alpha]_D +11.35°$ (c = 5 in DMF). Soly (mg/ml): 10-12 water, 35-43 0.1N HCl, 25-34 0.1N NaOH, 6.7-10 10% ethanol, 1 methanol, 7.1-10 H_2O/DMA (1:1), 9.7-14.5 0.1M citrate buffer (pH 4), 8.7-13 0.1M borate buffer (pH 9). pKa 2.1. Partition coefficient (octanol/water): 0.025. Degrades rapidly above pH 7.0.

(+)-Form hydrochloride. [149003-01-0] Cardioxane. Clinical experience of cardioprotective effect in anthracycline treated breast cancer: M. Marty *et al., Ann. Oncol.* **17**, 614 (2006).

THERAP CAT: Racemate as antineoplastic. (+)-Form as cardioprotectant.

8242. Rebamipide. [90098-04-7] α-[(4-Chlorobenzoyl)amino]-1,2-dihydro-2-oxo-4-quinolinepropanoic acid; (±)-α-(p-chlorobenzamido)-1,2-dihydro-2-oxo-4-quinolinepropionic acid; 2-(4-chlorobenzoylamino)-3-[2(1H)-quinolinon-4-yl]propionic acid; proamipide; OPC-12759; Mucosta. $C_{19}H_{15}ClN_2O_4$; mol wt 370.79. C 61.55%, H 4.08%, Cl 9.56%, N 7.56%, O 17.26%. Gastric cytoprotectant. Prepn: M. Uchida *et al.*, **DE 3324034**; *eidem*, **US 4578381**; (1984, 1986 both to Otsuka). Synthesis and pharmacology: M. Uchida *et al., Chem. Pharm. Bull.* **33**, 3775 (1985); of enantiomers: *eidem, ibid.* **35**, 853 (1987). Antiulcer activity in rats: K. Yamasaki *et al., Eur. J. Pharmacol.* **142**, 23 (1987); K. Yamasaki *et al., Jpn. J. Pharmacol.* **49**, 441 (1989). HPLC determn in plasma and urine: Y. Shioya, T. Shimizu, *J. Chromatogr.* **434**, 283 (1988).

White powder from DMF-water, mp 288-290° (dec) as hemihydrate.

(−)-Form. Colorless needles from DMF, mp 305-306° (dec). $[\alpha]_D^{20}$ −116.7° (c = 1.0 in DMF).

(+)-Form. Colorless needles from DMF, mp 305-306° (dec). $[\alpha]_D^{20}$ +116.9° (c = 1.0 in DMF).

THERAP CAT: Antiulcerative.

8243. Rebaudiosides. Group of intensely sweet, steviol glycosides isolated from the stevia plant, *Stevia rebaudiana* Bert., *Asteraceae*. Rebaudiosides A and C are the most abundant. Structurally similar to stevioside, *q.v.*; differing in the number and type of sugars attached. Isoln of A and B: H. Kohda *et al.*, *Phytochemistry* **15**, 981 (1976). Structure determn by ^{13}C-NMR: K. Yamasaki *et al.*, *Tetrahedron Lett.* **17**, 1005 (1976). Revised structure of A: W. E. Steinmetz, A. Lin, *Carbohydr. Res.* **344**, 2533 (2009). Isoln of C D, E: I. Sakamoto *et al.*, *Chem. Pharm. Bull.* **25**, 844 (1977). Structures and separative analysis: O. Tanaka, *Trends Anal. Chem.* **1**, 246 (1982). Photostability of A in carbonated beverages: S. S. Chang, J. M. Cook, *J. Agric. Food Chem.* **31**, 409 (1983); J. F. Clos *et al.*, *ibid.* **56**, 8507 (2008). Series of articles on safety assessment and development of A as food additive: *Food Chem. Toxicol.* **46**, Suppl 1, S1-S92 (2008).

Rebaudioside A

Rebaudioside A. [58543-16-1] (4α)-13-[(O-β-D-Glucopyranosyl-(1 → 2)-O-[β-D-glucopyranosyl-(1 → 3)]-β-D-glucopyranosyl)oxy]kaur-16-en-18-oic acid β-D-glucopyranosyl ester; stevioside a3; rebiana; Truvia. $C_{44}H_{70}O_{23}$; mol wt 967.02. One of the sweetest natural products known. Trihydrate, colorless needles from methanol, mp 242-244°. $[\alpha]_D^{24}$ −20.8° (c = 0.84 in methanol). Sweetness potency: 170 (relative to 10% sucrose).

Rebaudioside B. [58543-17-2] (4α)-13-[(O-β-D-Glucopyranosyl-(1 → 2)-O-[β-D-glucopyranosyl-(1 → 3)]-β-D-glucopyranosyl)oxy]kaur-16-en-18-oic acid; stevioside a4. $C_{38}H_{60}O_{18}$; mol wt 804.88. Hydrolysis product of rebaudioside A. Dihydrate, colorless needles from methanol, mp 193-195°. $[\alpha]_D^{24}$ −45.4° (c = 0.96 in methanol).

Rebaudioside C. [63550-99-2] (4α)-13-[(O-6-Deoxy-α-L-mannopyranosyl-(1 → 2)-O-[β-D-glucopyranosyl-(1 → 3)]-β-D-glucopyranosyl)oxy]kaur-16-en-18-oic acid β-D-glucopyranosyl ester; dulcoside B. $C_{44}H_{70}O_{22}$; mol wt 951.02. Colorless needles from methanol, mp 215-217°. $[\alpha]_D^{25}$ −29.9° (methanol).

Rebaudioside D. [63279-13-0] $C_{50}H_{80}O_{28}$; mol wt 1129.16. Contains 5 glucopyransyl residues. Colorless needles from ethanol, mp 283-286°. $[\alpha]_D^{25}$ −22.7° (methanol).

Rebaudioside E. [63279-14-1] $C_{44}H_{70}O_{23}$; mol wt 967.02. Colorless needles from methanol, mp 205-207°. $[\alpha]_D^{25}$ −34.2° (methanol).

USE: Rebaudioside A as non-nutritive sweetener.

8244. Reboxetine. [71620-89-8] *rel*-(2R)-2-[(R)-(2-Ethoxyphenoxy)phenylmethyl]morpholine; (±)-(2R*)-2-[(αR*)-α-(o-ethoxyphenoxy)benzyl]morpholine. $C_{19}H_{23}NO_3$; mol wt 313.40. C

72.82%, H 7.40%, N 4.47%, O 15.31%. Selective noradrenaline reuptake inhibitor. Prepn: P. Melloni *et al.*, **DE 2901032**; *eidem*, **US 4229449** (1979, 1980 both to Farmitalia Carlo Erba); *eidem et al.*, *Eur. J. Med. Chem. - Chim. Ther.* **19**, 235 (1984). Configuration: P. Melloni *et al.*, *Tetrahedron* **41**, 1393 (1985). HPLC determn of enantiomers in plasma: E. Frigerio *et al.*, *J. Chromatogr. A* **660**, 351 (1994). Metabolism: G. Cocchiara *et al.*, *Eur. J. Drug Metab. Pharmacokinet.* **16**, 231 (1991). Stereoselective kinetics: M. S. Benedetti *et al.*, *Chirality* **7**, 285 (1995). Clinical pharmacokinetics: C. Pellizzoni *et al.*, *Biopharm. Drug Dispos.* **17**, 623 (1996). Clinical trial in depression and social phobia: S. A. Montgomery, *J. Psychopharmacol.* **11**, Suppl., S9 (1997); A. Dubini *et al.*, *ibid.* S17.

Relative stereochemistry

mp 170-171°.

Methanesulfonate. [98769-82-5] Reboxetine mesylate; FCE-20124; Edronax; Prolift; Vestra. $C_{19}H_{23}NO_3 \cdot CH_3SO_3H$; mol wt 409.50. mp 145-146°.

THERAP CAT: Antidepressant.

8245. Red Fluorescent Protein. RFP. Tetrameric, 28kDa, fluorescent protein responsible for the red coloration of the *Discosoma* coral. Named following the pattern of two lower case letters indicating the species, FP for fluorescent protein, and the emission maximum wavelength in nm. Identification and cloning from *Discosoma* by homology to green fluorescent protein, *q.v.*: M. V. Matz *et al.*, *Nat. Biotechnol.* **17**, 969 (1999); S. A. Lukyanov *et al.*, **WO 0034324** (2000 to Clontech). Structure and biochemical characterization: G. S. Baird *et al.*, *Proc. Natl. Acad. Sci. USA* **97**, 11984 (2000). Chromophore structure: L. A. Gross *et al.*, *ibid.* 11990; molecular spectroscopy and hydrodynamics: A. A. Heikal *et al.*, *ibid.* 11996. Brief review in measurement of signal transduction: D. A. Zacharias *et al.*, *Curr. Opin. Neurobiol.* **10**, 416-421 (2000).

drFP583. [251925-26-5] Fluorescent protein (Discosoma clone FP583); DsRed. Abs max (excitation aq. soln): 558 nm (ε 75000 $M^{-1}cm^{-1}$). Abs max (emission): 583 nm.

USE: Reagent for fluorescent imaging of cell signals.

8246. Reductic Acid. [80-72-8] 2,3-Dihydroxy-2-cyclopenten-1-one; 2-cyclopenten-2,3-diol-1-one. $C_5H_6O_3$; mol wt 114.10. C 52.63%, H 5.30%, O 42.07%. Antioxidant obtained from pectic substances: Reichstein, Oppenauer, *Helv. Chim. Acta* **16**, 988 (1933); **17**, 390 (1934); Goldstein, **US 2854484** (1958). Alternate syntheses: Hesse *et al.*, *Ann.* **563**, 31 (1949); **592**, 137 (1955); **736**, 134 (1970). Crystal structure: D. Semmingsen, *Acta Chem. Scand. B* **31**, 81 (1977). Similar to vitamin C in structure but not activity.

Pale yellow crystals (has been obtained colorless) from ethyl acetate, dec 213-213.5° (also reported as 207.5° and 211°). Soluble in water, methanol, ethanol. Sparingly sol in ether, ethyl acetate, acetone. Insol in benzene.

5-Methylreductic acid. mp 71°. Stronger reducing agent than reductic acid.

USE: As antioxidant, like isoascorbic or ascorbic acid.

8247. Regadenoson. [313348-27-5] 2-[4-[(Methylamino)carbonyl]-1H-pyrazol-1-yl]adenosine; 1-(6-amino-9-β-D-ribofuranosyl-9*H*-purin-2-yl)-*N*-methyl-1*H*-pyrazole-4-carboxamide; [1-[9-[(4*S*,2*R*,3*R*,5*R*)-3,4-dihydroxy-5-(hydroxymethyl)oxolan-2-yl]-6-aminopurin-2-yl]pyrazol-4-yl]-*N*-methylcarboxamide; CVT-3146. $C_{15}H_{18}N_8O_5$; mol wt 390.36. C 46.15%, H 4.65%, N 28.71%, O 20.49%. Adenosine A_{2A}-receptor agonist; pharmacologic stress agent for use in cardiac imaging by producing coronary vasodilation. Prepn: J. A. Zablocki *et al.*, **WO 0078779**; *eidem*, **US 6403567** (2000, 2002 both to CV Therapeutics). Receptor binding study and coronary vasodilatory effects: Z. Gao *et al.*, *J. Pharmacol. Exp. Ther.* **298**, 209 (2001). Structure-affinity relationship: V. P. Palle *et al.*, *Bioorg. Med. Chem. Lett.* **12**, 2935 (2002). Clinical pharmacokinetics in patients with impaired renal function: T. Gordi *et al.*, *J. Clin. Pharmacol.* **47**, 825 (2007). Clinical effect on coronary blood flow: H. D. Lieu *et al.*, *J. Nucl. Cardiol.* **14**, 514 (2007). Clinical comparison with adenosine in myocardial perfusion imaging: A. E. Iskandrian *et al.*, *ibid.* 645. Review of development and clinical experience: H. Eggebrecht, M. Gössl, *Curr. Opin. Investig. Drugs* **7**, 264-271 (2006).

Sol in methanol, phosphate-buffered saline.
Monohydrate. [875148-45-1] Lexiscan; Rapiscan. $C_{15}H_{18}N_8$-$O_5.H_2O$; mol wt 408.38.
THERAP CAT: Diagnostic aid (cardiac stress testing).

8248. Reinecke Salt. [13573-16-5] Ammonium (*OC*-6-11)-diamminetetrakis(thiocyanato-*κN*)chromate(1−) (1:1); ammonium reineckate; ammonium tetrathiocyanodiammonochromate. C_4H_{10}-CrN_7S_4; mol wt 336.41. C 14.28%, H 3.00%, Cr 15.46%, N 29.15%, S 38.12%. $NH_4[Cr(NH_3)_2(SCN)_4]$. Exists as monohydrate; made by fusing ammonium thiocyanate with ammonium dichromate. Prepn: Dakin, *Org. Synth.* **coll. vol. II**, 555 (1943).
Monohydrate. $C_4H_{10}CrN_7S_4.H_2O$; mol wt 354.42. Dark red crystals or red cryst powder. Sparingly sol in cold water; sol in hot water, alcohol. Dec in aq soln with formation of a blue color and free HCN; dec occurs in about 2 weeks at room temp, rapidly above 65°. A similar decompn takes place in boiling alcohol.
Note: The K-salt is also known as Reinecke salt.
USE: Precipitant for primary and secondary amines, proline, hydroxyproline, and certain amino acids; also a reagent for mercury with which it gives a red color or precipitate.

8249. Relaxin. [9002-69-1] Cervilaxin; Releasin (formerly). Polypeptide hormone secreted by the corpora lutea of many mammalian species during pregnancy; also produced in several non-mammalians, including the shark. Facilitates the birth process by causing a softening and lengthening of the pubic symphysis and cervix; also inhibits contraction of the uterus and may play a role in timing of parturition. First discovered in estrogen-primed guinea pigs: F. L. Hisaw, *Proc. Soc. Exp. Biol. Med.* **23**, 661 (1926). Extraction from corpora lutea: Fevold *et al.*, *J. Am. Chem. Soc.* **52**, 3340 (1930); Albert *et al.*, *Endocrinology* **40**, 370 (1947); from pregnant rabbit serum: Abramowitz *et al.*, *Anat. Rec.* **84**, 456 (1942); *cf.* Hall, Newton, *J. Physiol.* **106**, 18 (1947). Can be obtained commercially from pregnant sows' ovaries. Isoln procedure and purification by resin chromatography: Lehrman *et al.*, *J. Am. Pharm. Assoc.* **44**, 206 (1955); R. L. Kroc, G. E. Phillips, **US 2852431**; G. E. Phillips, **US 2852432** (both 1958 to Warner-Lambert). Isoln from ovarian tissue using glacial acetic acid: Cohen, **US 2930737** (1960 to Princeton Laboratories). Alternate method: Keck, **US 3008878** (1961 to Thomae). Prepn from hog ovaries: Doczi, **US 3096246** (1963 to Warner-Lambert). Porcine relaxin has a mol wt of approx 6000 and consists of two peptide chains, A and B, of 22 and 31 residues respectively, linked covalently by one intra- and two inter-chain disul-

fide bonds. Purification, characterization of porcine relaxin: O. D. Sherwood, E. O'Byrne, *Arch. Biochem. Biophys.* **160**, 185 (1974). Structure of the A chain: C. Schwabe *et al.*, *Biochem. Biophys. Res. Commun.* **70**, 397 (1976); of the B chain: *eidem*, *ibid.* **75**, 503 (1977). Relaxin is structurally homologous to insulin and related growth factors: R. A. Bradshaw, *Rev. Biochem.* **47**, 191 (1978); S. Bedarkar *et al.*, *Nature* **270**, 449 (1977); N. Isaacs *et al.*, *ibid.* **271**, 278 (1978). Complete amino acid sequence of porcine relaxin: R. James *et al.*, *ibid.* **267**, 544 (1977); of rat: M. J. John *et al.*, *Endocrinology* **108**, 726 (1981). Isoln, characterization of relaxin from the sand tiger shark (*Odontaspis taurus*): J. W. Reinig *et al.*, *ibid.* **109**, 537 (1981); structure: L. K. Gowan *et al.*, *FEBS Lett.* **129**, 80 (1981). Demonstration of synthesis of rat relaxin as a preprorelaxin molecule with a connecting peptide of 105 amino acid residues, using molecular cloning: P. Hudson *et al.*, *Nature* **291**, 127 (1981). Structure of a genomic clone from which the amino acid sequence of biologically active human relaxin was predicted: *eidem*, *ibid.* **301**, 628 (1983). *Reviews:* C. Schwabe *et al.*, *Recent Prog. Horm. Res.* **34**, 123-211 (1978); *Ann. N.Y. Acad. Sci.* **380**, entitled "Relaxin: Structure, Function, and Evolution", B. G. Steinetz *et al.*, Eds. (1982) pp 1-244.

Amorphous powder. Diffusion constant 12.6×10^{-7} cm²/sec. Sedimentation constant 1.6×10^{-13} sec. Isoelec pt around pH 7.0. uv max (water, pH 7.0): 277.5 nm ($E_{1cm}^{1\%}$ 7.20). Slightly sol in water and 95% alcohol. Sol in acid or alkaline solns. Insol in abs alcohol, ether, acetone, petr ether, benzene. Dec above pH 9. Neutral or acid solns which are free from oxidizing agents, are stable. Ampuled solns at pH 7.0 and 3.2 have been kept in refrigerator for one year without loss of activity.
THERAP CAT: Hormone (ovarian).

8250. Remacemide. [128298-28-2] 2-Amino-*N*-(1-methyl-1,2-diphenylethyl)acetamide. $C_{17}H_{20}N_2O$; mol wt 268.36. C 76.09%, H 7.51%, N 10.44%, O 5.96%. NMDA receptor antagonist. Prepn: R. C. Griffith, J. J. Napier, **EP 279937** (1988 to Pennwalt); *see also:* *eidem*, **US 5331007** (1994 to Fisons). Pharmacology, metabolism and pharmacokinetics: K. T. Muir, G. C. Palmer, *Epilepsy Res. Suppl.* **3**, 147 (1991). HPLC determn in plasma: J. W. Flynn, J. E. O'Brien, *J. Chromatogr.* **583**, 91 (1992). Mechanism of action study: S. Subramaniam *et al.*, *J. Pharmacol. Exp. Ther.* **276**, 161 (1996). Neuroprotective properties: G. C. Palmer *et al.*, *Ann. N.Y. Acad. Sci.* **765**, 236 (1995). Clinical evaluation in neuroprotection during cardiac surgery: J. E. Arrowsmith *et al.*, *Stroke* **29**, 2357 (1998); in Parkinson's disease: Parkinson Study Group, *Neurology* **56**, 455 (2001). Review of mechanism of action and clinical applications: S. C. Schachter, D. Tarsy, *Expert Opin. Invest. Drugs* **9**, 871-883 (2000).

Hydrochloride. [111686-79-4] FPL-12924AA; PR-934-423A. $C_{17}H_{20}N_2O.HCl$; mol wt 304.82. Crystals from isopropanol + methanol, mp 253-254°. Soly (g/l): water 40; normal saline 22; dilute HCl 56; ethanol 24. LD_{50} (calculated as base) in rats (mg/kg): ~50 i.v., ~900 orally (Muir, Palmer).
THERAP CAT: Neuroprotective.

8251. Remifentanil. [132875-61-7] 4-(Methoxycarbonyl)-4-[(1-oxopropyl)phenylamino]-1-piperidinepropanoic acid methyl ester; 4-carboxy-4-(*N*-phenylpropionamido)-1-piperidinepropionic acid dimethyl ester; remifentanyl; GI-87084. $C_{20}H_{28}N_2O_5$; mol wt 376.45. C 63.81%, H 7.50%, N 7.44%, O 21.25%. Synthetic *mu*-opioid agonist; derivative of fentanyl, *q.v.* Prepn: P. L. Feldman *et al.*, **EP 383579**; *eidem*, **US 5019583** (1990, 1991 both to Glaxo); *eidem et al.*, *J. Med. Chem.* **34**, 2202 (1991). Opioid receptor activity: M. K. James *et al.*, *J. Pharmacol. Exp. Ther.* **259**, 712 (1991). LC-MS determn in plasma: E. Bossu *et al.*, *J. Pharm. Biomed. Anal.* **42**, 367 (2006). Review of pharmacokinetics and therapeutic use: R. Beers, E. Camporesi, *CNS Drugs* **18**, 1085-1104 (2004); of pharmacology and clinical use in general anesthesia: L. J. Scott, C. M. Perry, *Drugs* **65**, 1793-1823 (2005); in analgesia-based sedation: A. J. Battershill, G. M. Keating, *ibid.* **66**, 365-385 (2006).

Oil. Sol in methanol, ethyl acetate. pKa 7.26.

Hydrochloride. [132539-07-2] GI-87084B; Ultiva. $C_{20}H_{28}N_2$-O_5.HCl; mol wt 412.91. Crystals from methanol + ether, mp 212-214°.

Note: This is a controlled substance (opiate): **21 CFR**, 1308.12.

THERAP CAT: Analgesic; anesthesia adjunct.

8252. Remoxipride. [80125-14-0] 3-Bromo-*N*-[[(2*S*)-1-eth-yl-2-pyrrolidinyl]methyl]-2,6-dimethoxybenzamide; (−)-*N*-ethyl-2-(3-bromo-2,6-dimethoxybenzamidomethyl)pyrrolidine. $C_{16}H_{23}$-BrN_2O_3; mol wt 371.28. C 51.76%, H 6.24%, Br 21.52%, N 7.55%, O 12.93%. Specific dopamine D_2-receptor antagonist. Prepn: G. L. Florvall, S. O. Ogren, **EP 4831**; *eidem*, **US 4232037** (1979, 1980 both to Astra); *eidem, J. Med. Chem.* **25**, 1280 (1982). Antidopaminergic activity: S. O. Ogren *et al., Eur. J. Pharmacol.* **102**, 459 (1984). Radiometric determn in urine: A. C. Veltkamp *et al., J. Chromatogr.* **384**, 357 (1987). Clinical pharmacology and pharmacokinetics: L. Farde *et al., Psychopharmacology* **95**, 157 (1988). Clinical evaluation in tardive dyskinesia: U. Andersson *et al., ibid.* **94**, 167 (1988); in schizophrenia: R. G. McCreadie *et al., Acta Psychiatr. Scand.* **78**, 49 (1988).

$[\alpha]_D^{20}$ −64° (c = 2 in ethanol).

Hydrochloride monohydrate. [117591-79-4]; [73220-03-8]. A-33547; FLA-731; Roxiam. $C_{16}H_{23}BrN_2O_3$.HCl.H_2O; mol wt 425.75. White to off-white crystalline solid. Loss of water at 105°; mp 173°. $[\alpha]_D^{20}$ −11° (c = 2 in water). pKa 8.9. uv max in water, ethanol, 0.1*N* HCl: 286, 287, 286 nm (ε 2280, 2260, 2280). Soly (mg/ml): water 300; ethanol 400; dichloromethane 80; acetone 20. LD_{50} i.p. in rats: 794 μmol/kg (Florvall, Ogren, 1982).

THERAP CAT: Antipsychotic.

8253. Renin. [9015-94-5] Highly specific aspartyl proteinase of mol wt about 40000 Da, produced and secreted by the kidney. Found also in amniotic fluid. Conc. of renin in the human kidney is about 20 times less than in hog kidney. Purification from hog kidneys: Haas *et al., Arch. Biochem. Biophys.* **42**, 368 (1953); from kidneys of various species: Haas *et al., ibid.* **48**, 256 (1954). Prepn of human renin: Haas *et al., ibid.* **110**, 534 (1965). Exists in α, β and γ configurations: Haas *et al., ibid.* **44**, 79 (1953). Purification on DEAE cellulose columns: Passananti, *Biochim. Biophys. Acta* **34**, 246 (1959); Maier, Morgan, *ibid.* **128**, 193 (1966). Has been observed to exist in high molecular weight inactive forms which upon acid treatment decrease in molecular size and increase in enzymic activity. *See* review of purification studies: E. Haber, E. E. Slater, *Circ. Res.* **40**, Suppl. I, 136 (1977). Renin is secreted by juxtaglomerular cells and acts on the plasma substrate, angiotensinogen, to split off the inactive decapeptide angiotensin I which is converted to the active pressor agent angiotensin II, *q.v.* Renin itself has no activity. Secretion of renin is stimulated by constriction of a renal artery, blood loss, low sodium levels, adrenal insufficiency. Increased levels of renin shown also in pregnancy. Cloning and sequence analysis of human renin cDNA: T. Imai *et al., Proc. Natl. Acad. Sci. USA* **80**, 7405 (1983). Crystal structure of human recombinant renin: A. R. Sielecki *et al., Science* **243**, 1346 (1989). Re-

view: W. S. Peart, "The Renin-Angiotensin System" in *Pharmacol. Rev.* **17**, 143 (1965); *idem, Proc. R. Soc. London Ser. B* **173**, 317 (1969); Smeby, Bumpus, "Renin" in *Renal Hypertension*, I. Page, J. McCubbin, Eds. (Year Book Medical Publishers, Chicago, 1968) pp 14-61. Books: M. R. Lee, *Renin and Hypertension* (Williams & Wilkins, Baltimore, 1969); *Renin* vol. 5, S. Oparil *et al.*, Eds. (Eden Press, Quebec, 1980) 368 pp. Review of the renin-angiotensin system: J. C. Romero, F. G. Knox, *Hypertension* **11**, 724-738 (1988).

8254. Rennin. [9001-98-3] Chymosin; rennase; lab; abomasal enzyme; Lab Ferment (German). Mol wt about 31,000. The predominant milk-clotting enzyme from the true stomach or abomasum of the suckling calf: Tauber, Kleiner, *J. Biol. Chem.* **96**, 745 (1932). Secreted as an inactive precursor called ***prorennin*** and converted in the acid environment of the stomach to the active enzyme. Obtained by extracting dried strips of calf stomach with 5 to 10% NaCl soln contg boric acid for 5 days; improved process: Keil, **US 2339931** (1942). Structure consists of a single polypeptide chain with internal disulfide bridges. Amino acid composition: Schwander *et al., Helv. Chim. Acta* **35**, 553 (1952). Determn of *N*-terminal amino acid sequence: Foltmann, *Philos. Trans. R. Soc. London* **257B**, 147 (1970). Homologies in amino acid sequences of rennin and pepsin: Tang, Hartley, *Biochem. J.* **118**, 611 (1970). *Reviews:* Berridge, *Adv. Enzymol.* **15**, 423 (1954); Foltmann, *Milk Proteins* **2**, 217 (1971).

Yellowish powder, grains or scales. Peculiar, not unpleasant odor. Characteristic, slightly salty taste. Slightly hygroscopic. Can be purified enough to crystallize. The dry substance (sometimes diluted with NaCl or lactose) is stable, aq solns are not. Partially sol in water, dil alcohol. Insol at pH values near its isoelectric point of pH 4.5. pH of aq soln 5.8 (acceptable range 5.3-6.3). Strongly affected by ultraviolet (sunlight). Causes milk to coagulate, optimum temp 37-43°, optimum pH of milk 5.8. Not active below 15° nor above 55°. When a commercial prepn is labeled 1:10,000 it means that one gram will curdle 10 liters of milk at 35° within 40 minutes. Other strengths are 1:100,000 to 1:7,600,000.

USE: **Rennet** (a dried extract contg rennin) is used in the manuf of cheese and rennet casein, also sold retail for the making of junket or rennet custards.

THERAP CAT: Digestive enzyme.

8255. Renzapride. [112727-80-7] *rel*-4-Amino-*N*-(1*R*,4*S*,-5*R*)-1-azabicyclo[3.3.1]non-4-yl-5-chloro-2-methoxybenzamide; (±)-*endo*-4-amino-5-chloro-2-methoxy-*N*-(1-azabicyclo[3.3.1]non-4-yl)benzamide; ATL-1251; BRL-24924. $C_{16}H_{22}ClN_3O_2$; mol wt 323.82. C 59.35%, H 6.85%, Cl 10.95%, N 12.98%, O 9.88%. Serotonin receptor 5-HT_4 agonist/5-HT_3 antagonist. Prepn: F. D. King, **EP 94742**; *idem*, **US 4612319** (1983, 1986 both to Beecham); *idem et al., J. Med. Chem.* **36**, 683 (1993). 5-HT_4 receptor selectivity as quaternary salt: G. S. Baxter *et al., Bioorg. Med. Chem. Lett.* **3**, 633 (1993). Clinical pharmacology: D. H. Staniforth, M. Pennick, *Eur. J. Clin. Pharmacol.* **38**, 161 (1990). Clinical evaluation in irritable bowel syndrome: M. Camilleri *et al., Clin. Gastroenterol. Hepatol.* **2**, 895 (2004). *Review:* P. Norman, *Curr. Opin. Cent. Peripher. Nerv. Syst. Invest. Drugs* **2**, 355-368 (2000).

Relative stereochemistry

mp >260°.

THERAP CAT: Gastroprokinetic; in treatment of irritable bowel syndrome.

8256. Repaglinide. [135062-02-1] 2-Ethoxy-4-[2-[[(1*S*)-3-methyl-1-[2-(1-piperidinyl)phenyl]butyl]amino]-2-oxoethyl]benzoic acid; (+)-2-ethoxy-α-[[(*S*)-α-isobutyl-*o*-piperidinobenzyl]carbamoyl]-*p*-toluic acid; (*S*)-(+)-2-ethoxy-4-[*N*-[1-(2-piperidinophenyl)-3-methyl-1-butyl]aminocarbonylmethyl]benzoic acid; AG-EE-623-

ZW; Novonorm; Prandin. $C_{27}H_{36}N_2O_4$; mol wt 452.60. C 71.65%, H 8.02%, N 6.19%, O 14.14%. Non-sulfonylurea oral hypoglycemic agent. Prepn: W. Grell *et al.*, **WO 9300337** (1993 to Thomae). HPLC determn of racemate in human plasma: A. Greischel *et al.*, *J. Chromatogr.* **568**, 246 (1991). Clinical pharmacokinetics: F. J. Ampudia-Blasco *et al.*, *Diabetologia* **37**, 703 (1994). Comparative clinical evaluation in NIDDM: B. H. R. Wolffenbuttel *et al.*, *Eur. J. Clin. Pharmacol.* **45**, 113 (1993).

White, odorless, crystalline powder. Crystals from ethanol/water (2:1), mp 126-128°. Crystals from neutralized water, mp 130-131°. $[\alpha]_D^{20}$ +6.97° (c = 0.975 in methanol); $[\alpha]_D^{20}$ +7.45° (c = 1.06 in methanol). Sol in methanol. LD_{50} orally in rats: >1 g/kg (Grell).

THERAP CAT: Antidiabetic.

8257. Repinotan. [144980-29-0] 2-[4-[[[(2R)-3,4-Dihydro-2H-1-benzopyran-2-yl]methyl]amino]butyl]-1,2-benzisothiazol-3(2H)-one 1,1-dioxide; R-(−)-2-[4-[(chroman-2-ylmethyl)amino]-butyl]-1,1-dioxide; (−)-2-[N-[4-(1,1-dioxido-3-oxo-2,3-dihydrobenzisothiazol-2-yl)butyl]aminomethyl]chroman. $C_{21}H_{24}N_2O_4S$; mol wt 400.49. C 62.98%, H 6.04%, N 6.99%, O 15.98%, S 8.01%. Serotonin 5-HT$_{1A}$-receptor agonist. Prepn: B. Junge *et al.*, **EP 352613**; *eidem*, **US 5137901** (1990, 1992 both to Bayer AG). Pharmacology and receptor binding study: J. De Vry *et al.*, *J. Pharmacol. Exp. Ther.* **284**, 1082 (1998). Neuroprotective effect *in vitro* and *in vivo*: I. Semkova *et al.*, *Eur. J. Pharmacol.* **359**, 251 (1998). *Review:* H. A. M. Mucke, *Curr. Opin. Cent. Peripher. Nerv. Syst. Invest. Drugs* **1**, 621-628 (1999).

Hydrochloride. [144980-77-8] Bay x 3702. $C_{21}H_{24}N_2O_4S$.-HCl; mol wt 436.95. mp 192-194°. $[\alpha]_D$ −42.2° (c = 1 in trichloromethane).

THERAP CAT: Neuroprotective.

8258. Repirinast. [73080-51-0] 5,6-Dihydro-7,8-dimethyl-4,5-dioxo-4H-pyrano[3,2-c]quinoline-2-carboxylic acid 3-methylbutyl ester; isopentyl 5,6-dihydro-7,8-dimethyl-4,5-dioxo-4H-pyrano[3,2-c]quinoline-2-carboxylate; MY-5116; Romet. $C_{20}H_{21}$-NO$_5$; mol wt 355.39. C 67.59%, H 5.96%, N 3.94%, O 22.51%. Prepn: **BE 876751**; Y. Morinaka, K. Takahashi, **US 4298610** (1979, 1981 both to Mitsubishi Yuka); Y. Morinaka *et al.*, *Eur. J. Med. Chem. - Chim. Ther.* **16**, 251 (1981). General pharmacology in animals: K. Takahashi *et al.*, *Oyo Yakuri* **32**, 233 (1986), *C.A.* **105**, 218691j (1986). Anti-allergic effects: S. Kono, K. Ohata, *Arerugi* **35**, 1105 (1986), *C.A.* **106**, 165870z (1987). Toxicity: M. Nagase *et al.*, *Iyakuhin Kenkyu* **17**, 545 (1986), *C.A.* **106**, 108121r (1987).

Crystals from chloroform + *n*-hexane, mp 236-241°. LD_{50} in rats, mice (mg/kg): >5000 orally, >5000 s.c. both species (Nagase).

THERAP CAT: Antiallergic.

8259. Reproterol. [54063-54-6] 7-[3-[[2-(3,5-Dihydroxyphenyl)-2-hydroxyethyl]amino]propyl]-3,7-dihydro-1,3-dimethyl-1H-purine-2,6-dione; 7-[3-[(β,3,5-trihydroxyphenethyl)amino]propyl]theophylline; D-1959. $C_{18}H_{23}N_5O_5$; mol wt 389.41. C 55.52%, H 5.95%, N 17.98%, O 20.54%. Theophylline deriv which has selective activity on β_2 receptors. Prepn: **FR M5969** (1968 to Degussa), *C.A.* **71**, 70644c (1969). *See also* **FR Addn. 0308** (1970 to Degussa), *C.A.* **78**, 72216h (1973). Synthesis: K. H. Klingler, *Arzneim.-Forsch.* **27**, 4 (1977). Toxicology: S. Habersang, *ibid.* 45. Series of articles on metabolism, pharmacology and clinical trials: *ibid.* 15-76. Other clinical studies: D. Nolte *et al.*, *Dtsch. Med. Wochenschr.* **102**, 619 (1977); H. Budmiger *et al.*, *Schweiz. Med. Wochenschr.* **108**, 1190 (1978).

Hydrochloride. [13055-82-8] W-2946M; Asmaterolo; Bronchodil; Bronchospasmin. $C_{18}H_{23}N_5O_5$.HCl; mol wt 425.87. Crystals, mp 249-250°. LD_{50} in mice (mg/kg): 148 i.v., >10,000 orally (Habersang).

THERAP CAT: Bronchodilator.

8260. Resacetophenone. [89-84-9] 1-(2,4-Dihydroxyphenyl)ethanone; 2′,4′-dihydroxyacetophenone. $C_8H_8O_3$; mol wt 152.15. C 63.15%, H 5.30%, O 31.55%. Made by heating resorcinol, glacial acetic acid, and anhydr zinc chloride at 145-150°: Cooper, *Org. Synth.* **21**, 103 (1941); from resorcinol, acetic anhydride and boron trifluoride: Killelea, Lindwall, *J. Am. Chem. Soc.* **70**, 428 (1948).

Needles or leaflets, mp 145-147°. Gradually dec by water; sol in pyridine, in warm alcohol, in glacial acetic acid; practically insol in benzene, ether, chloroform.

USE: Reagent for iron as a 10% alcoholic soln. A red color is obtained with ferric ions in slightly acid soln: Cooper, *Ind. Eng. Chem. Anal. Ed.* **9**, 334 (1937).

8261. Resazurin. [550-82-3] 7-Hydroxy-3H-phenoxazin-3-one 10-oxide; "diazoresorcinol"; resazoin. $C_{12}H_7NO_4$; mol wt 229.19. C 62.89%, H 3.08%, N 6.11%, O 27.92%. Prepn from resorcinol: Rast *et al.*, **DE 961829** (1957 to Bayer), *C.A.* **53**, 15098 (1959).

Dark red, small crystals with greenish luster. Insol in water, ether; sparingly sol in alcohol, glacial acetic acid; sol in dil alkali hydroxides.

USE: As indicator, using a soln of 0.1 g in 20 ml 0.1N NaOH, and water to make 500 ml. pH: 3.8 orange, 6.5 dark violet. In the detection of hyposulfite (sulfoxylate). In food research (reductase test).

8262. Rescinnamine. [24815-24-5] (3β,16β,17α,18β,20α)-11,17-Dimethoxy-18-[[[(2E)-1-oxo-3-(3,4,5-trimethoxyphenyl)-2-propen-1-yl]oxy]yohimban-16-carboxylic acid methyl ester; 3,4,5-

trimethoxycinnamic acid ester of methyl reserpate; methyl 3,4,5-trimethoxycinnamoyl reserpate; reserpinine; Cartric; Cinnaloid. C_{35}-$H_{42}N_2O_9$; mol wt 634.73. C 66.23%, H 6.67%, N 4.41%, O 22.69%. Found in *Rauwolfia serpentina* Benth., *Apocynaceae:* Haack *et al.,* *Naturwissenschaften* **41**, 214 (1954); Ordway, Guercio, **US 2876228** (1959 to Pfizer); Klohs *et al.,* **US 2974144** (1961 to Riker); from *Rauwolfia* spp: Banes *et al., J. Am. Pharm. Assoc.* **47**, 625 (1958). Identity with reserpinine: Haack *et al., Naturwissenschaften* **42**, 47 (1955). Structure: Klohs *et al., J. Am. Chem. Soc.* **77**, 2241 (1955). LC-ED determn in urine: J. Wang, M. Bonakdar, *J. Chromatogr.* **382**, 343 (1986). LC determn in rauwolfia preparations: U. R. Cieri, *J. AOAC Int.* **81**, 373 (1998).

Fine needles from benzene, mp 238-239° (vac). $[\alpha]_D^{24} -97°$ (c = 1 in chloroform). uv max (methanol): 228, 302 nm (log ε 4.79, 4.48). Practically insol in water. Moderately sol in methanol, benzene, chloroform, other organic solvents.

THERAP CAT: Antihypertensive.

8263.   **Reserpic Acid.** [83-60-3] (3β,16β,17α,18β,20α)-18-Hydroxy-11,17-dimethoxyyohimban-16-carboxylic acid; reserpinolic acid. $C_{22}H_{28}N_2O_5$; mol wt 400.48. C 65.98%, H 7.05%, N 7.00%, O 19.97%. Obtained by controlled alkaline hydrolysis of reserpine: Dorfman *et al., Helv. Chim. Acta* **37**, 59 (1954); Schlittler, **US 2824874** (1958).

Crystals from methanol, mp 241-243°.
Hydrochloride hemihydrate. $C_{22}H_{28}N_2O_5 \cdot HCl \cdot \frac{1}{2}H_2O$. Crystals, mp 257-259°. $[\alpha]_D^{23} -81°$.
Methyl ester. $C_{23}H_{30}N_2O_5$. Needles from methanol, mp 235-240°. $[\alpha]_D^{25} -106°$.
Methyl ester hydrochloride. mp ~228°.

8264.   **Reserpiline.** [131-02-2] (4S,4aS,13bR,14aS)-4a,5,7,8,-13,13b,14,14a-Octahydro-10,11-dimethoxy-4-methyl-4H-indolo-[2,3-a]pyrano[3,4-g]quinolizine-1-carboxylic acid methyl ester; (3β,19α,20α)-16,17-didehydro-10,11-dimethoxy-19-methyloxayohimban-16-carboxylic acid methyl ester. $C_{23}H_{28}N_2O_5$; mol wt 412.49. C 66.97%, H 6.84%, N 6.79%, O 19.39%. Isoln from *Rauwolfia serpentina* Benth., *Apocynaceae:* Klohs *et al., Chem. Ind. (London)* **1954**, 1264; from *R. canescens* L.: Stoll *et al., Helv. Chim. Acta* **38**, 270 (1955). Stereochemistry: Shamma, Richey, *J. Am. Chem. Soc.* **85**, 2507 (1963). Partial synthesis: S.-I. Sakai *et al., Heterocycles* **17**, 99 (1982).

Amorphous powder. $[\alpha]_D^{20} -38°$ (ethanol); −14° (c = 1.5 in pyridine); −12° (c = 1.7 in chloroform). uv max (ethanol): 229, 300-304 nm (log ε 4.57, 4.03). Freely sol in ethanol, acetone, chloroform, benzene.

Hydrochloride. $C_{23}H_{28}N_2O_5 \cdot HCl$. Crystals, mp 205-207°. $[\alpha]_D^{24} -40°$ (c = 0.44 in ethanol).

Reserpilic acid dimethylaminoethyl ester dihydrochloride. [3735-84-0] Paratensiol; Resporisan. $C_{26}H_{35}N_3O_5 \cdot 2HCl$; mol wt 542.50.

THERAP CAT: Antihypertensive.

8265.   **Reserpine.** [50-55-5] (3β,16β,17α,18β,20α)-11,17-Dimethoxy-18-[(3,4,5-trimethoxybenzoyl)oxy]yohimban-16-carboxylic acid methyl ester; 3,4,5-trimethoxybenzoyl methyl reserpate; Rivasin; Serpasil. $C_{33}H_{40}N_2O_9$; mol wt 608.69. C 65.12%, H 6.62%, N 4.60%, O 23.66%. Indole alkaloid found in *Rauwolfia* spp. Isoln from the roots of *Rauwolfia serpentina* L. Benth., *Apocynaceae:* J. M. Müller *et al., Experientia* **8**, 338 (1952); and structural studies: L. Dorfman *et al., Helv. Chim. Acta* **37**, 59 (1954); N. Neuss *et al., J. Am. Chem. Soc.* **76**, 2463 (1954). Stereochemistry: P. E. Aldrich *et al., ibid.* **81**, 2481 (1959); Y. Ban, O. Yonemitsu, *Tetrahedron* **20**, 2877 (1964). First total synthesis: R. B. Woodward *et al., J. Am. Chem. Soc.* **78**, 2023 (1956); *eidem, Tetrahedron* **2**, 1 (1958). Review of total syntheses: F.-E. Chen, J. Huang, *Chem. Rev.* **105**, 4671-4706 (2005). Comprehensive description: R. E. Schirmer, *Anal. Profiles Drug Subs.* **4**, 384-430 (1975). *Review:* A. Scriabine, Ed. in *Pharmacology of Antihypertensive Drugs* (Raven, New York, 1980) pp 119-125. Review of carcinogenicity studies: *IARC Monographs* **10**, 217-229 (1976); R. M. Diener *et al., Toxicol. Pathol.* **8**, 1-21 (1980).

Long prisms from dil acetone, dec 264-265° (dec 277-277.5° in evac tube). $[\alpha]_D^{23} -118°$ (CHCl$_3$); $[\alpha]_D^{26} -164°$ (c = 0.96 in pyridine); $[\alpha]_D^{26} -168°$ (c = 0.624 in DMF). uv max (CHCl$_3$): 216, 267, 295 nm (ε 61700, 17000, 10200). Weak base. pK 6.6. Freely sol in chloroform (~1 g/6 ml), methylene chloride, acetic acid; sol in ethyl acetate; slightly sol in benzene, acetone, methanol, and in aq solns of citric acids; very slightly sol in alc (1 g/1800 ml), ether. Insol in water.

Hydrochloride hydrate. $C_{33}H_{40}N_2O_9 \cdot HCl \cdot H_2O$. Crystals, dec 224°.

Caution: Reserpine is reasonably anticipated to be a human carcinogen: *Report on Carcinogens, Twelfth Edition* (PB2011-111646, 2011) p 370.

THERAP CAT: Antihypertensive.

THERAP CAT (VET): Hypotensive, tranquilizer. Has been used to prevent aortic rupture in turkeys.

8266.   **Resibufogenin.** [465-39-4] (3β,5β,15β)-14,15-Epoxy-3-hydroxy-bufa-20,22-dienolide; Respigon. $C_{24}H_{32}O_4$; mol wt 384.52. C 74.97%, H 8.39%, O 16.64%. Cytotoxic constituent of toad venom. Isoln: Meyer, *Helv. Chim. Acta* **35**, 2444 (1952); Linde, Meyer, *Pharm. Acta Helv.* **33**, 327 (1958). Structure: Thiessen, *Chem. Ind. (London)* **1958**, 440; Linde, Meyer, *Experientia* **14**, 238 (1958); *eidem, Helv. Chim. Acta* **42**, 807 (1959). Synthesis: Pettit *et al., J. Org. Chem.* **36**, 3736 (1971); Haede *et al., Ann.* **1973**, 5. Stereochemical study of cytotoxic properties: Y. Kamono *et al., J. Chem. Res. Synop.* **1977**, 78. Pharmacology: Leigh, Caldwell, *J. Pharm. Pharmacol.* **21**, 708 (1969).

Crystals from acetone + water, mp 113-140°/155-168°. $[\alpha]_D^{22}$ −7.1° (c = 1.259 in chloroform). Also obtained as an amorphous solid, $[\alpha]_D^{16}$ −5.4° (c = 2.030 in chloroform).

Hydrochloride. $C_{24}H_{33}ClO_4$. Crystals from acetone, dec 230-232°. $[\alpha]_D^{15}$ +15.1° (c = 0.5302 in chloroform). uv max (alc): 298 nm (log ε 3.74).

THERAP CAT: Cardiotonic.

8267. Resilin. Highly amorphous rubberlike protein found in insect cuticle. Isoln and characterization: Weis-Fogh, *J. Exp. Biol.* **37**, 889 (1960). The structure is a cross-linked random network of flexible protein chains, incorporating 15 of the more common amino acids and involving covalent bonds. Cross-links have been identified as di- and trityrosine. Biosynthesis: Coles, *J. Insect Physiol.* **12**, 679 (1966). Similar to elastin, *q.v.*, but does not form fibers; occurs as masses or typically as 2-5 μ layers. Elasticity approaches that of the ideal cross-linked rubber. Responsible for various spring-like actions of winged insects. In flea: Rothschild, *Nature* **239**, 45 (1972). *Review:* Andersen, *Acta Physiol. Scand.* **66**, Suppl. 263 (1966); *idem*, "Resilin" in *Comprehensive Biochemistry* **vol. 26C**, M. Florkin, E. H. Stotz, Eds. (Elsevier, New York, 1971) pp 633-657.

The colorless material is insol in water <140° and in all other solvents not rupturing peptide bonds. Swells in water and protein solvents; shrinks and becomes hard in absolute methanol, ethanol, dioxane, acetone. Shows no tendency to crystallize at all. Fluorescent in uv light.

8268. Resin Ipomea. [9000-34-4] Resin of Mexican scammony. Made by extracting the dried roots of *Ipomoea orizabensis* Ledenois, *Convolvulaceae* with alcohol and pptg with water.

Brown, translucent, brittle fragments or pale brown powder. Insol in water; sol in alc, chloroform, partially in ether.

THERAP CAT: Cathartic.

8269. Resin Jalap. [9000-35-5] Prepd by extracting the powdered root of *Exogonium purga* (Hayne) Lindl. (*E. jalapa* Baill., *Ipomoea purga* Hayne), *Convolvulaceae* with alcohol and pptg with water.

Yellow to brown, amorphous mass or powder. Insol in water, benzene, carbon disulfide, oils; freely sol in alc.

THERAP CAT: Cathartic.

8270. Resin Scammony. [9000-58-2] Made by extracting the tubers of *Convolvulus scammonia* L., *Convolvulaceae* with alcohol and pptg with water.

Brown, amorphous mass. Freely sol in alc; not less than 95% sol in ether and in caustic alkalies (with gentle heat).

THERAP CAT: Cathartic.

8271. Resiquimod. [144875-48-9] 4-Amino-2-(ethoxymethyl)-α,α-dimethyl-1*H*-imidazo[4,5-*c*]quinoline-1-ethanol; R-848; S-28463. $C_{17}H_{22}N_4O_2$; mol wt 314.39. C 64.95%, H 7.05%, N 17.82%, O 10.18%. Immune response modulator structurally similar to imiquimod, *q.v.* Induces cytokine secretion by monocytes and activates antibody secretion by B cells. Prepn: J. F. Gerster *et al.*, **WO 9215582**; *eidem*, **US 5389640** (1992, 1995 both to 3M). Immunomodulating and antiviral activities: M. A. Tomai *et al.*, *Antiviral Res.* **28**, 253 (1995). Mechanism of action study: G. A. Bishop *et al.*, *Cell. Immunol.* **208**, 9 (2001). Clinical evaluation in genital herpes: S. L. Spruance *et al.*, *J. Infect. Dis.* **184**, 196 (2001). Review of development: J. J. Wu *et al.*, *Antiviral Res.* **64**, 79-83 (2004).

Colorless, crystalline solid, mp 190-193°.

THERAP CAT: Antiviral.

8272. Resistomycin. [20004-62-0] 3,5,7,10-Tetrahydroxy-1,-1,9-trimethyl-2*H*-benzo[*cf*]pyrene-2,6(1*H*)-dione; X-340. C_{22}-$H_{16}O_6$; mol wt 376.36. C 70.21%, H 4.29%, O 25.51%. Unusually stable antibiotic substance produced by *Streptomyces resistomycificus:* Brockmann, Schmidt-Kastner, *Naturwissenschaften* **38**, 479 (1951); *Ber.* **87**, 1460 (1954); Brockmann, **DE 888918** (1953). Structure: Brockmann, *Angew. Chem.* **76**, 863 (1964). Revised structure: Bailey *et al.*, *Chem. Commun.* **1968**, 374; Brockmann *et al.*, *Ber.* **102**, 1224 (1969). Synthetic studies: J. F. Kingston, L. Weiler, *Can. J. Chem.* **55**, 785 (1977); K. James, R. A. Raphael, *Tetrahedron Lett.* **1979**, 3895. Total synthesis: B. A. Keay, R. Rodrigo, *J. Am. Chem. Soc.* **104**, 4725 (1982). Isoln from *Streptomyces griseoflavus* B71, ^{13}C-NMR and biosynthesis: G. Höfle, H. Wolf, *Ann.* **1983**, 835.

Yellow needles from dioxane, dec 315°. Sublimes at 200-205° (10^{-4} mm) without activity loss. Stable to hot conc H_2SO_4 or hot *N* KOH. Weakly acid. Absorption max: 268, 290, 320, 337, 366, 457 nm (ε 24000, 23000, 14400, 13900, 11000, 15400). Slight soly in water; fair in ether, benzene, alcohol, acetone, acetic acid. Orange-red in sodium hydroxide and piperidine; red in pyridine. Fluoresces in alcohol, benzene, and acetone solns, also in sulfuric acid.

8273. Reslizumab. [241473-69-8] Anti-(human interleukin 5) immunoglobulin G4 (human-rat monoclonal SCH 55700 γ_4-chain) disulfide with human-rat monoclonal SCH 55700 light chain, dimer; SCH-55700; Cinquil. Humanized monoclonal antibody (mAb) directed against the pro-inflammatory cytokine, interleukin-5 (IL-5). Consists of antigen recognition sites from a rat IgG$_{2a}$ antibody directed against human IL-5 (JES1-39D10) grafted into a consensus human IgG$_4$ constant region. Mol wt ~150 kDa. Prepn: C.-C. Chou *et al.*, **WO 9316184**; *eidem*, **US 6451982** (1993, 2002 both to Schering Corp); and pharmacology: R. W. Egan *et al.*, *Arzneim.-Forsch.* **49**, 779 (1999). Epitope mapping and binding study: J. Zhang *et al.*, *Int. Immunology* **11**, 1935 (1999). Manufacturing batch quality validation assay: R. A. DiGiacomo *et al.*, *Anal. Biochem.* **327**, 165 (2004). Clinical pharmacokinetics in patients with asthma: J. C. Kips *et al.*, *Am. J. Respir. Crit. Care Med.* **167**, 1655 (2003). Clinical trial in eosinophilic asthma: M. Castro *et al.*, *Am. J. Respir. Crit. Care Med.* **184**, 1125 (2011). Review of development and clinical experience in eosinophil-mediated inflammatory conditions: G. M. Walsh, *Curr. Opin. Mol. Ther.* **11**, 329-336 (2009); of therapeutic potential in pediatric eosinophilic esophagitis: *idem*, *Immunotherapy* **2**, 461-465 (2010).

THERAP CAT: Anti-inflammatory.

8274. Resmethrin. [10453-86-8] 2,2-Dimethyl-3-(2-methyl-1-propen-1-yl)cyclopropanecarboxylic acid [5-(phenylmethyl)-3-furanyl]methyl ester; 5-benzyl-3-furanmethanol-2,2-dimethyl-3-(2-methylpropenyl) cyclopropanecarboxylate; (5-benzyl-3-furyl)-methyl chrysanthemate; (5-benzyl-3-furyl)methyl *cis,trans*-(±)-2,2-dimethyl-3-(2-methylpropenyl)cyclopropanecarboxylate; FMC-17370; SBP-1382; Chryson. $C_{22}H_{26}O_3$; mol wt 338.45. C 78.07%, H 7.74%, O 14.18%. Synthetic pyrethroid insecticide. Prepn: M.

Elliott, **GB 1168797**; *idem*, **US 3465007** (both 1969 to Natl. Res. Dev. Corp.); *idem et al.*, *Nature* **213**, 493 (1967). GLC determn in corn, cornmeal, flour, wheat: R. A. Simonaitis, R. S. Cail, *J. AOAC* **58**, 1032 (1975); and infrared spec in nonaqueous aerosols: M. W. Law, *J. AOAC Int.* **59**, 745 (1976). LC determn in urine and water: B. L. Loper, K. A. Anderson, *ibid.* **86**, 1236 (2003). Photodecomposition: K. Ueda *et al.*, *J. Agric. Food Chem.* **22**, 212 (1974). Environmental fate and ectoxicological effects in aquatic systems: G. M. Rand, *Ecotoxicology* **11**, 101 (2002). Efficacy against greenhouse pests: R. E. Webb *et al.*, *J. Econ. Entomol.* **67**, 295 (1974); against *Anopheles quadrimaculatus*: C. A. Sandoski *et al. ibid.*, **76**, 646 (1983). Review of toxicology and human exposure: *Toxicological Profile for Pyrethrins and Pyrethroids* (PB2004-100004, 2003) 332 pp.

Prepd as a mixture of a 70-80% (±)-*trans* isomer and a 20-30% (±)-*cis* isomer. bp >180° (dec). d^{20} 0.96-0.97. Flash pt, closed cup: 264.2°F (129°C). Log P (octanol/water): 5.43. Vapor pressure: <10^{-5} mm Hg. Soly at 25° in water (μg/l): 35-40. Sol in organic solvents. LC$_{50}$ (96 hr) in rainbow trout, bluegill sunfish (μg/l): 3.14, 7.2 (Rand).

(+)-*trans*-Form see Bioresmethrin.

USE: Insecticide.

8275. Resorantel. [20788-07-2] *N*-(4-Bromophenyl)-2,6-dihydroxybenzamide; 4′-bromo-γ-resorcylanilide; 4′-bromo-2,6-dihydroxybenzanilide; 2,6-dihydroxybenzoic acid 4′-bromoanilide; resorcylam; HOE-296V; Terenol. C$_{13}$H$_{10}$BrNO$_3$; mol wt 308.13. C 50.67%, H 3.27%, Br 25.93%, N 4.55%, O 15.58%. Prepn: LeMaire *et al.*, *J. Pharm. Sci.* **50**, 831 (1961); **GB 1124613**; H. Ruschig *et al.*, **US 3449420** (1968, 1969 to Hoechst). Animal studies: Düwel; Behrens, Matschullat; Pfeiffer, *Dtsch. Tieraerztl. Wochenschr.* **77**, 97-107 (1970); Christ *et al.*, *Berl. Muench. Tieraerztl. Wochenschr.* **83**, 61 (1970).

Colorless powder, mp 229-230°. Also reported as mp 183-186° (LeMaire). Stable at room temp for at least two years. Very resistant to acids and alkalies. Highly sensitive to iron compds which stain it deep red or violet. Very soluble in DMF; slightly sol in lower alcohols. Insol in water, hydrocarbons, vegetable oils.

THERAP CAT (VET): Anthelmintic (sheep and other ruminants).

8276. Resorcinol. [108-46-3] 1,3-Benzenediol; *m*-dihydroxybenzene; resorcin. C$_6$H$_6$O$_2$; mol wt 110.11. C 65.45%, H 5.49%, O 29.06%. Prepd by fusing *m*-benzenedisulfonic acid with excess NaOH. *Review:* J. Varagnat in *Kirk-Othmer Encyclopedia of Chemical Technology* **vol. 13** (Wiley-Interscience, New York, 3rd ed., 1981) pp 39-69. Review of toxicology and risk assessment: B. S. Lynch *et al.*, *Regul. Toxicol. Pharmacol.* **36**, 198-210 (2002).

White, needle-like crystals; sweetish taste. Becomes pink on exposure to light and air, or by contact with iron. d 1.272. mp 109-111°. bp 280°, but volatilizes at lower temp and is slightly volatile with steam. pH: 5.2. One gram dissolves in 0.9 ml water, 0.2 ml

water at 80°, 0.9 ml alcohol; freely sol in ether, glycerol; slightly sol in chloroform. *Poisonous. Protect from light.*

Monoacetate. [102-29-4] Acetylresorcinol; Euresol. C$_8$H$_8$O$_3$; mol wt 152.15. Golden-yellow, thick, syrupy, oily liquid. bp ~283° with decompn. Insol in water. Miscible with alc, benzene, chloroform, acetone. Sol in solns of alkali hydroxides.

Caution: Potential symptoms of overexposure are irritation of eyes, skin, nose, throat, upper respiratory system; dermatitis; methemoglobinemia, cyanosis, convulsions; restlessness, bluish skin, increased heart rate, dyspnea; dizziness, drowsiness, hypothermia, hemoglobinuria; spleen, kidney, liver changes. See *NIOSH Pocket Guide to Chemical Hazards* (DHHS/NIOSH 97-140, 1997) p 272; *Patty's Industrial Hygiene and Toxicology* **vol. 2B**, G. D. Clayton, F. E. Clayton, Eds. (Wiley-Interscience, New York, 4th ed., 1994) pp 1586-1590.

USE: In tanning, photography; manuf of resins, resin adhesives, hexylresorcinol, *p*-aminosalicylic acid, explosives, dyes, tires, rubber products; in cosmetics; dyeing and printing textiles; as a reagent for zinc.

THERAP CAT: Keratolytic; antiseborrheic.

THERAP CAT (VET): Topical antipruritic and antiseptic. Has been used as an intestinal antiseptic.

8277. β-Resorcylaldehyde. [95-01-2] 2,4-Dihydroxybenzaldehyde; 2,4-dihydroxybenzenecarbonal. C$_7$H$_6$O$_3$; mol wt 138.12. C 60.87%, H 4.38%, O 34.75%. Prepd by saturating with HCl a soln of resorcinol and anhyd hydrocyanic acid in abs ether and boiling the precipitated resorcylaldimide-HCl with water: Gattermann, Köbner, *Ber.* **32**, 278 (1899); **DE 106508**; *Chem. Zentralbl.* **1900**, I, 742. By adding POCl$_3$ to a soln of resorcinol and formanilide in ether, then boiling with dil NaOH, acidifying with H$_2$SO$_4$ and extracting with ether: Dimrath, Zöppritz, *Ber.* **35**, 995 (1902); by passing HCl into a mixture of resorcinol, ether, anhyd zinc chloride and cyanogen bromide, then boiling the precipitate with water: Karrer, *Helv. Chim. Acta* **2**, 92 (1919).

Needles from water or ether + ligroin. mp 135-136°. bp$_{22}$ 226°. Freely sol in water, alcohol, ether, chloroform, glacial acetic acid. Difficultly sol in cold benzene. β-Resorcylaldehyde is easily dec by acids and alkalies. Prolonged exposure to moist air converts it to a brown, amorphous powder which is insol in ether.

8278. β-Resorcylic Acid. [89-86-1] 2,4-Dihydroxybenzoic acid; 2,4-dihydroxybenzenecarboxylic acid; BRA. C$_7$H$_6$O$_4$; mol wt 154.12. C 54.55%, H 3.92%, O 41.52%. Prepared from resorcinol and KHCO$_3$ in glycerol or water by heating and passing CO$_2$ through the mixture: Brunner, *Ann.* **351**, 320 (1907). *Cf.* Bistrycki, Kostanecki, *Ber.* **18**, 1984 (1885); Nierenstein, Clibbens, *Org. Synth.* **coll. vol. II**, 557 (1943).

Hydrated crystals from water. Becomes anhyd at 100°, mp 213° (rapid heating). pK (25°) 3.30. Sol in hot water, alcohol, ether, olive oil. Boiling with water, acids or salt solns results in loss of CO$_2$.

USE: Intermediate for dyestuffs and drugs; spot test reagent for iron.

8279. Resveratrol. [501-36-0] 5-[(1*E*)-2-(4-Hydroxyphenyl)ethenyl]-1,3-benzenediol; *trans*-resveratrol; (*E*)-5-(*p*-hydroxystyryl)resorcinol; 3,4′,5-stilbenetriol; 3,5,4′-trihydroxystilbene. C$_{14}$-

$H_{12}O_3$; mol wt 228.25. C 73.67%, H 5.30%, O 21.03%. Phytoalexin found in a variety of plants; active ingredient of Asian traditional medicine, Kojo-Kon, which is the powdered root of the Japanese knotweed. Isoln from white hellebore and structure: M. Takaoka, *J. Fac. Sci., Hokkaido Imp. Univ. Ser. 3* **3**, 1 (1940); *C.A.* **34**, 7887 (1940); from fescue grass: R. G. Powell *et al.*, *Phytochemistry* **35**, 335 (1994). Isoln of monomers, oligomers, isomers and glucosides from wine: F. Mattivi *et al.*, *J. Agric. Food Chem.* **43**, 1820 (1995). GC-MS determn: D. M. Goldberg *et al.*, *J. Am. Enol. Viticult.* **46**, 159 (1995). Produced as stress metabolite in response to fungal infection or injury: G. H. Dai *et al.*, *Physiol. Mol. Plant Pathol.* **46**, 177 (1995); P. Jeandet *et al.*, *J. Phytopathol.* **143**, 135 (1995). Antiplatelet aggregating activity: A. A. E. Bertelli *et al.*, *Int. J. Tissue React.* **17**, 1 (1995); and implications in coronary heart disease: C. R. Pace-Asciak *et al.*, *Clin. Chim. Acta* **235**, 207 (1995).

Off-white powder from MeOH, mp 253-255°.

3-β-mono-D-glucoside. [27208-80-6] Polydatin; piceid. $C_{20}H_{22}O_8$. Isoln from fresh root of *Polygonum cuspidatum* Sieb. & Zucc., *Polygonaceae*, and structure: Nonomura *et al.*, *Yakugaku Zasshi* **83**, 988 (1963).

(Z)-Form. [61434-67-1] *cis*-Resveratrol. GC-MS determn in wine: D. M. Goldberg *et al.*, *J. Agric. Food Chem.* **43**, 1245 (1995).

8280. Retamine. [2122-29-4] (1*S*,7*R*,7a*S*,14*R*,14a*S*)-Dodecahydro-7,14-methano-2*H*,6*H*-dipyrido[1,2-*a*:1′,2′-*e*][1,5]diazocin-1-ol. $C_{15}H_{26}N_2O$; mol wt 250.39. C 71.95%, H 10.47%, N 11.19%, O 6.39%. From bark and young branches of *Genista sphaerocarpa* Lam. (*Retama sphaerocarpa* (Lam.) Boiss.), *Leguminosae*. Isoln: Battandier, Malosse, *Compt. Rend.* **125**, 360, 450 (1897); Ribas *et al.*, *An. Fis. Quim.* **42**, 516 (1946), *C.A.* **41**, 4894c (1947). Structure: Bohlmann *et al.*, *Ber.* **98**, 659 (1965). Synthesis: Bohlmann *et al.*, *ibid.* 653. Stereochemistry: Ribas *et al.*, *Tetrahedron Lett.* **1965**, 3181.

Bitter crystals from alc, mp 165-166°. $[\alpha]_D^{20}$ +46.2° (c = 1.017 in abs ethanol). Practically insol in water; slightly sol in ether; sol in chloroform, alcohol, methanol; very sol in benzene.

Hydrochloride. mp 272-273°.

8281. Retapamulin. [224452-66-8] 2-[[(3-*exo*)-8-Methyl-8-azabicyclo[3.2.1]oct-3-yl]thio]acetic acid (3a*S*,4*R*,5*S*,6*S*,8*R*,9*R*,-9a*R*,10*R*)-6-ethenyldecahydro-5-hydroxy-4,6,9,10-tetramethyl-1-oxo-3a,9-propano-3a*H*-cyclopentacyclooocten-8-yl ester; mutilin 14-(*exo*-8-methyl-8-azabicyclo[3.2.1]oct-3-ylsulfanyl)acetate; SB-275833; Altabax; Altargo. $C_{30}H_{47}NO_4S$; mol wt 517.77. C 69.59%, H 9.15%, N 2.71%, O 12.36%, S 6.19%. Topical antibiotic; semi-synthetic derivative of pleuromutilin, *q.v.* Inhibits bacterial protein synthesis by binding to ribosomes. Prepn: V. Berry *et al.*, **WO 9921855**; *eidem*, **US 6281226** (1999, 2001 both to SKB). Antibacterial spectrum: R. N. Jones *et al.*, *Antimicrob. Agents Chemother.* **50**, 2583 (2006). Mechanism of action: K. Yan *et al.*, *ibid.*, 3875 (2006); W. S. Champney, W. K. Rodgers, *ibid.* **51**, 3385 (2007). Clinical trial in treatment of secondarily infected dermatitis: L. C. Parish *et al.*, *J. Am. Acad. Dermatol.* **55**, 1003 (2006); of impetigo: S. Koning *et al.*, *Br. J. Dermatol.* **158**, 1077 (2008). Review of pharmacology and clinical experience: L. P. H. Yang, S. J. Keam, *Drugs* **68**, 855-873 (2008); of microbiological profile: N. E. Scangarella-Oman *et al.*, *Expert Rev. Anti Infect. Ther.* **7**, 269-279 (2009).

White to pale-yellow crystalline solid.

THERAP CAT: Antibacterial.

8282. Retaspimycin. [857402-23-4] 18,21-Didehydro-17-demethoxy-18,21-dideoxo-18,21-dihydroxy-17-(2-propen-1-ylamino)geldanamycin; 17-allylamino-17-demethoxygeldanamycin hydroquinone; 17-AAGH$_2$. $C_{31}H_{45}N_3O_8$; mol wt 587.71. C 63.35%, H 7.72%, N 7.15%, O 21.78%. Semisynthetic ansamycin antibiotic; derivative of geldanamycin. Inhibitor of heat shock protein 90 (Hsp90), a molecular chaperone that regulates key proteins in cell growth, survival, and differentiation pathways. Prepn: J. Adams *et al.*, **WO 05063714**; *eidem*, **US 7282493** (2005, 2007 both to Infinity); and bioactivity: J. Ge *et al.*, *J. Med. Chem.* **49**, 4606 (2006). Pharmacology, metabolism, and biochemical characterization: J. R. Sydor *et al.*, *Proc. Natl. Acad. Sci. USA* **103**, 17408 (2006). Review of pharmacology and clinical experience in various cancers: B. E. Hanson, D. H. Vesole, *Expert Opin. Invest. Drugs* **18**, 1375-1383 (2009).

Hydrochloride. [857402-63-2] IPI-504. $C_{31}H_{45}N_3O_8$·HCl; mol wt 624.17. Pale yellow powder from methanol/ethyl acetate. Soly in water: 250-275 mg/ml. Stable in acidic aq buffer (pH 3). *Store under inert atmosphere at −20°.*

THERAP CAT: Antineoplastic.

8283. Retene. [483-65-8] 1-Methyl-7-(1-methylethyl)phenanthrene. $C_{18}H_{18}$; mol wt 234.34. C 92.26%, H 7.74%. Occurs in pine tar, in fossilized pine, in high-boiling tar oils. Formed by Se or Pd dehydrogenation of abietic acid: Ruzicka, Waldmann, *Helv. Chim. Acta* **16**, 842 (1933). Synthesis starting with methyl β-(6-isopropyl-2-naphthoyl)propionate: Bardhan, Sengupta, *J. Chem. Soc.* **1932**, 2520, 2798; *cf.* Bogert, *Science* **77**, 289 (1933). Review and bibliography: Fieser, Fieser, *Natural Products Related to Phenanthrene* (New York, 3rd ed., 1949).

Plates, scales from alc, mp 99°. bp$_{10}$ 208°; bp$_{760}$ 390-394°. Absorption spectrum: Askew, *J. Chem. Soc.* **1935**, 509. Insol in water. Sol in hot alcohol, benzene, hot ether, carbon disulfide. Upon treatment with CrO_3 in acetic acid forms retenequinone, $C_{18}H_{16}O_2$, mp 197.5°.

Reteplase

8284. Reteplase. [133652-38-7] 173-L-Serine-174-L-tyrosine-175-L-glutamine-173-527-plasminogen activator (human tissue-type); r-PA; BM-06022; Rapilysin; Retavase. Genetically engineered deletion variant of human tissue plasminogen activator (t-PA), *q.v.* Single-chain, nonglycosylated form containing only the protease and kringle-2 domains of native, human t-PA. Expression in *E. coli* cells and chromatographic purifn: A. Stern *et al.*, **EP 382174**; *eidem*, **US 5223256** (1990, 1993 both to Boehringer, Mann.). Clinical pharmacokinetics: U. Martin *et al.*, *Thromb. Haemostasis* **66**, 569 (1991). Review of pharmacology, toxicology and clinical evaluation: *idem et al.*, *Cardiovasc. Drug Rev.* **11**, 299-311 (1993). Clinical comparison with alteplase in acute myocardial infarction: C. Bode *et al.*, *Circulation* **94**, 891 (1996); E. J. Topol *et al.*, *N. Engl. J. Med.* **337**, 1118 (1997); in pulmonary embolism: U. Tebbe *et al.*, *Am. Heart J.* **138**, 39 (1999).

THERAP CAT: Thrombolytic.

8285. Reticulin. One of the connective tissue proteins occurring wherever connective tissue forms a boundary: Jacobson, *Nature and Structure of Collagen*, J. T. Randall, Ed. (Butterworths, London, 1953) pp 6-13, *C.A.* **48**, 10814e (1954). Found together with collagen and elastin. Particularly abundant in membranes of the glomeruli and tubules of the kidney and in supportive structure of the vasculature and lung alveoli. Isoln from kidney tissue slices: Kramer, Little, *ibid.* pp 33-43, *C.A.* **48**, 10815a (1954). Contains about 85% protein, the amino acid content of which is very similar to that of collagen. Also contains about 4.2% carbohydrate and approx 11% bound lipid which yields, on hydrolysis, 95% myristic acid and 5% palmitic acid: Windrum *et al.*, *Br. J. Exp. Pathol.* **36**, 49 (1955). Lacks hydroxyproline; glycine content is that of normal proteins: Pras, Glynn, *ibid.* **54**, 449 (1973).

Stable in boiling water and in 1*N* HCl. Can be dispersed in distilled water but precipitated by physiological saline (0.15 mol/l).

8286. Reticuline. [485-19-8] (1*S*)-1,2,3,4-Tetrahydro-1-[(3-hydroxy-4-methoxyphenyl)methyl]-6-methoxy-2-methyl-7-isoquinolinol; 1,2,3,4-tetrahydro-6-methoxy-2-methyl-1-vanillyl-7-isoquinolinol; 2-methyl-7-hydroxy-6-methoxy-1-(3-hydroxy-4-methoxybenzyl)-1,2,3,4-tetrahydroisoquinoline; coclanoline. $C_{19}H_{23}NO_4$; mol wt 329.40. C 69.28%, H 7.04%, N 4.25%, O 19.43%. Precursor of many aporphine and morphine-type alkaloids. Isoln of *d*-form from *Anona reticulata* Linn., *Anonaceae:* Gopinath *et al.*, *Ber.* **92**, 776 (1959); *dl*-form from opium: Brochmann-Hanssen, Furuya, *J. Pharm. Sci.* **53**, 575 (1964); *Planta Med.* **12**, 328 (1964). Structure: Brochmann-Hanssen, Nielson, *Tetrahedron Lett.* **1965**, 1271. Configuration: Battersby, Evans, *ibid.* 1275. Synthesis of *dl*-form: Gopinath *et al.*, *Ber.* **92**, 1657 (1959); P. Kerekes *et al.*, *Acta Chim. Acad. Sci. Hung.* **98**, 491 (1978). Synthesis of crystalline *dl*-form: Baxter *et al.*, *J. Chem. Soc. C* **1965**, 3645; Chan, Maitland, *ibid.* **1966**, 753; K. C. Rice, A. Brossi, *J. Org. Chem.* **45**, 592 (1980); G. Dornyei *et al.*, *Tetrahedron Lett.* **1982**, 2913. Asymmetric synthesis of (*S*)-reticuline: Konda *et al.*, *Chem. Pharm. Bull.* **23**, 1063 (1975). Biosynthesis: D. S. Dewan *et al.*, *J. Chem. Soc. Perkin Trans. 1* **1977**, 1662; A. Barrett *et al.*, *ibid.* **1979**, 652.

***dl*-Form.** Pink crystals, mp 146°. uv max: 284 nm (log ε 3.85). Sol in aqueous buffer of pH <7.5 or >11. Practically insol in water at pH 8-10.

(*S*)-Form perchlorate. $C_{19}H_{23}NO_4 \cdot HClO_4$. Colorless prisms from ethanol, mp 203-204°. $[\alpha]_D^{18}$ +88.3° (c = 0.21 in alc).

8287. Retinal. [116-31-4] (*all-E*)-3,7-Dimethyl-9-(2,6,6-trimethyl-1-cyclohexen-1-yl)-2,4,6,8-nonatetraenal; *trans*-retinal; retinaldehyde; retinene₁; vitamin A aldehyde; axerophthal. $C_{20}H_{28}O$; mol wt 284.44. C 84.45%, H 9.92%, O 5.62%. Carotenoid component of the visual pigments. The 11-*cis* isomer is the chromophore

of the majority of naturally occurring opsins, *q.v.* Exposure of opsin-bound retinal to light initiates the *cis-trans* isomerization and resultant dissociation of retinal from the apoprotein. This catalyzes an enzyme cascade which leads to visual excitation. Isoln from retinas: G. Wald, *J. Gen. Physiol.* **19**, 351 (1935); recognition as vitamin A aldehyde: R. A. Morton, *Nature* **153**, 69 (1944). Prepd by the oxidation of vitamin A: Ball *et al.*, *Biochem. J.* **42**, 516 (1948); by the oxidation of β-carotene: Wendler *et al.*, *J. Am. Chem. Soc.* **72**, 234 (1950); from β-ionone: Eiter, Truscheit, **US 3060229** (1962 to Bayer); Jacobs *et al.*, *Rec. Trav. Chim.* **84**, 1113 (1965). Synthesis *in vitro* from vitamin A₁: G. Wald, R. Hubbard, *Proc. Natl. Acad. Sci. USA* **36**, 92 (1950). Synthesis of 11-*cis* isomer: J. M. Dieterle, C. D. Robeson, *Science* **120**, 219 (1954). X-ray crystallography of retinals: R. Gilardi *et al.*, *Nature* **232**, 187 (1971); H. Matsumoto *et al.*, *J. Am. Chem. Soc.* **102**, 4259 (1980). HPLC determn in biological material: T. Suzuki, M. Makino-Tasaka, *Anal. Biochem.* **129**, 111 (1983). Role of photo-isomerization in visual excitation: G. Wald, *Science* **162**, 230 (1968). Review of syntheses: V. Balogh-Nair, K. Nakanishi, *Methods Enzymol.* **88**, 496-506 (1982); R. Liu, A. Asata, *ibid.* 506-516. Review of role in vision: D. F. O'Brien, *Science* **218**, 961-966 (1982); P. S. Zurer, *Chem. Eng. News* **61**, 24-35 (Nov. 28, 1983). Book: *The Retinoids* Vol. 1-2, M. B. Sporn *et al.*, Eds. (Academic Press, New York, 1984).

11-*cis* Retinal

all-*trans* Retinal

Orange colored crystals from petr ether, mp 61-64°. uv max (hexane): 368 nm (ε 48000). Sol in ethanol, chloroform, cyclohexane, petr ether, oils. Practically insol in water.

11-*cis*-Retinal. [564-87-4] Orange prisms from petr ether, mp 63.5-64.4°. uv max (hexane): 365 nm (ε 26360).

8288. Retinoic Acid. [302-79-4] (*all-E*)-3,7-Dimethyl-9-(2,-6,6-trimethyl-1-cyclohexen-1-yl)-2,4,6,8-nonatetraenoic acid; *all-trans*-retinoic acid; vitamin A acid; tretinoin; Aberel; Airol; Avita; Epi-Aberel; Eudyna; Kerlocal; Renova; Retin-A; Retinova; Vesanoid. $C_{20}H_{28}O_2$; mol wt 300.44. C 79.96%, H 9.39%, O 10.65%. Physiological metabolite of vitamin A, *q.v.* Effects gene expression via nuclear retinoic acid receptors (RAR); mediates cellular growth and differentiation. Prepn: D. A. van Dorp, J. R. Arens, *Rec. Trav. Chim.* **65**, 338 (1946); C. D. Robeson *et al.*, *J. Am. Chem. Soc.* **77**, 4111 (1955). Single-step process: R. Marbet, **DE 2061507**; **US 3746730** (1971, 1973 both to Hoffmann-La Roche). Crystal structure: C. H. Stam, C. H. MacGillavry, *Acta Crystallogr.* **16**, 62 (1963). Toxicology: J. J. Kamm, *J. Am. Acad. Dermatol.* **6**, 652 (1982). Clinical evaluation in treatment of photoaged skin: J. S. Weiss *et al.*, *J. Am. Med. Assoc.* **259**, 527 (1988); in promyelocytic leukemia: R. P. Warrell, Jr. *et al.*, *Leukemia* **8**, 929 (1994). Review of pharmacology and therapeutic potential: G. D. Goss *et al.*, *Crit. Rev. Clin. Lab. Sci.* **29**, 185-215 (1992). Book: *The Retinoids*, M. B. Sporn *et al.*, Eds. (Raven Press, New York, 2nd ed., 1994) 679 pp.

all-*trans* Retinoic Acid

Consult the Name Index before using this section.

Yellow to light orange crystalline powder with characteristic floral odor. Crystals from ethanol, mp 180-182°. uv max (methanol): 351 nm (ε 45000). Sol in DMSO; slightly sol in polyethylene glycol 400, octanol, ethanol. Practically insol in water, mineral oil, glycerin. LD_{50} (10 day) in mice, rats (mg/kg): 790, 790 i.p.; 2200, 2000 orally (Kamm).

9-*cis*-Form *see* Alitretinoin.

13-*cis*-Form *see* Isotretinoin.

α-Tocopheryl ester *see* Tocoretinate.

THERAP CAT: Antiacne; adjunct in treatment of photodamaged skin; antineoplastic (hormonal).

8289. Retronecine. [480-85-3] (1*R*,7a*R*)-2,3,5,7a-Tetrahydro-1-hydroxy-1*H*-pyrrolizine-7-methanol. $C_8H_{13}NO_2$; mol wt 155.20. C 61.91%, H 8.44%, N 9.03%, O 20.62%. The most common base portion of pyrrolizidine alkaloids. *"Necine" bases* are *1-methylpyrrolizidines* of different stereochemical configurations and degrees of hydroxylation that occur in the form of esters in alkaloids of *Senecio*, *Crotalaria* and a number of genera of the *Boraginaceae*. Retronecine occurs in nature in the (+)-form. It is the necine base of monocrotaline, senecionine, seneciphylline, retrorsine, *q.q.v.*, and numerous other hepatotoxic pyrrolizidine alkaloids. Structure: R. Adams, E. F. Rogers, *J. Am. Chem. Soc.* **63**, 228 (1941). Total synthesis: T. A. Geissman, A. C. Waiss, *J. Org. Chem.* **27**, 139 (1962). Stereospecific synthesis: E. Vedejs, G. R. Martinez, *J. Am. Chem. Soc.* **102**, 7993 (1980); H. Niwa *et al.*, *Tetrahedron Lett.* **27**, 4605 (1986). Total synthesis of (−)-form: J. Cooper *et al.*, *Chem. Commun.* **1988**, 509. Toxicity data: P. N. Harris *et al.*, *J. Pharmacol. Exp. Ther.* **75**, 78 (1942). Biosynthesis: D. J. Robins, J. R. Sweeney, *Chem. Commun.* **1979**, 120; G. Grue Sorensen, I. D. Spenser, *J. Am. Chem. Soc.* **103**, 3208 (1981). Comprehensive reviews on retronecine and other necine bases: F. L. Warren, *Fortschr. Chem. Org. Naturst.* **12**, 198-269 (1955); **24**, 329-406 (1966); D. J. Robins, *ibid.* **41**, 115-203 (1982); F. L. Warren in *Alkaloids* Vol. **XII**, R. H. F. Manske, Ed. (Academic Press, New York, 1970) pp 245-331.

Crystals from acetone, mp 119-120°. $[\alpha]_D^{20}$ +4.95° (c = 0.58 in alc). LD_{50} i.v. in mice: 634.0 ±26.0 mg/kg (Harris).

(±)-Form. Crystals from acetone, mp 130-131°.

8290. Retrorsine. [480-54-6] (3*Z*,5*R*,6*S*,14a*R*,14b*R*)-3-Ethylidene-3,4,5,6,9,11,13,14,14a,14b-decahydro-6-hydroxy-6-(hydroxymethyl)-5-methyl-[1,6]dioxacyclododecino[2,3,4-*gh*]pyrrolizine-2,7-dione; 12,18-dihydroxysenecionan-11,16-dione; β-longilobine. $C_{18}H_{25}NO_6$; mol wt 351.40. C 61.52%, H 7.17%, N 3.99%, O 27.32%. Hepatotoxic pyrrolizidine alkaloid; common constituent of *Senecio* species. Isoln from *Senecio retrorsus* DC, Compositae: R. H. F. Manske, *Can. J. Chem.* **5**, 651 (1931); G. Barger *et al.*, *J. Chem. Soc.* **1935**, 11; from *Crotalaria usaramoensis* E. G. Baker, *Leguminosae*: C. C. J. Culvenor, L. W. Smith, *Aust. J. Chem.* **19**, 2127 (1966); S. M. H. Christie *et al.*, *J. Chem. Soc.* **1949**, 1700; E. C. Leisegang, F. L. Warren, *ibid.* **1950**, 702. Identity with β-longilobine: F. L. Warren *et al.*, *J. Am. Chem. Soc.* **72**, 1421 (1950). Review and evaluation of toxicity and carcinogenicity studies: *IARC Monographs* **10**, 303-312, 333-342 (1976). Comprehensive reviews of pyrrolizidine alkaloids: L. B. Bull *et al.*, *The Pyrrolizidine Alkaloids* (North Holland, Amsterdam, 1968) 293 pp; F. L. Warren in *The Alkaloids* vol. **12**, R. H. F. Manske, Ed. (Academic Press, New York, 1970) pp 245-331.

Crystals from ethyl acetate, mp 212° (Barger *et al.*); 216-216.5° (Bull *et al.*). $[\alpha]_D^{18}$ −17.6° (c = 1.99 in ethanol). uv max (water): 217 nm (log ε 3.85). Readily sol in alcohol, chloroform; slightly sol in water, acetone, ethyl acetate; practically insol in ether.

N-Oxide. Isatidine; 12,18-dihydroxysenecionan-11,16-dione-4-oxide; retrorsine *N*-oxide. $C_{18}H_{25}NO_7$. Isolated from *Senecio* species. Review and evaluation of carcinogenicity and toxicity studies: *IARC Monographs* **10**, 269-273 (1976). Crystals from ethanol, mp 140.5-141.5° (Christie *et al.*).

8291. Revaprazan. [199463-33-7] 4-(3,4-Dihydro-1-methyl-2(1*H*)-isoquinolinyl)-*N*-(4-fluorophenyl)-5,6-dimethyl-2-pyrimidinamine; 5,6-dimethyl-2-(4-fluorophenylamino)-4-(1-methyl-1,2,3,4-tetrahydroisoquinolin-2-yl)pyrimidine. $C_{22}H_{23}FN_4$; mol wt 362.45. C 72.90%, H 6.40%, F 5.24%, N 15.46%. Gastric proton pump inhibitor. Prepn: J. W. Lee *et al.*, **WO 9605177**; *eidem*, **US 5750531** (1996, 1998 both to Yuhan). HPLC determn in plasma and urine: K. S. Han *et al.*, *J. Chromatogr. B* **696**, 312 (1997). Clinical pharmacokinetics: K.-S. Yu *et al.*, *J. Clin. Pharmacol.* **44**, 73 (2004).

Hydrochloride. [178307-42-1] YH-1885; Revanex. $C_{22}H_{23}$-$FN_4 \cdot HCl$; mol wt 398.91. mp 205-208°.

THERAP CAT: Antiulcerative.

8292. Reversine. [656820-32-5] N^6-Cyclohexyl-N^2-[4-(4-morpholinyl)phenyl]-9*H*-purine-2,6-diamine; 2-(4-morpholinoanilino)-6-cyclohexylaminopurine; 2-(4-morpholinoanilino)-N^6-cyclohexyladenine. $C_{21}H_{27}N_7O$; mol wt 393.50. C 64.10%, H 6.92%, N 24.92%, O 4.07%. Synthetic small molecule reported to induce murine myoblasts to dedifferentiate into multipotent progenitor cells. Discovery of induction of cell dedifferentiation: S. Chen *et al.*, *J. Am. Chem. Soc.* **126**, 410 (2004). Prepn: S. Chen *et al.*, **WO 05047524**; *eidem*, **US 050176707** (both 2005 to Scripps); *see also* combinatorial synthesis method: S. Ding *et al.*, *J. Am. Chem. Soc.* **124**, 1594 (2002). Adenosine A_3-receptor antagonist activity: M. Perreira *et al.*, *J. Med. Chem.* **48**, 4910 (2005).

Pale white solid.

8293. Reviparin Sodium. LU-47311; Clivarin(e). Low molecular weight heparin, prepared by controlled nitrous acid degradation of a standard heparin extracted from porcine intestinal mucosa. Mean molecular weight 3900 Da (range: 3500-4500 Da). Biochemical and pharmacological profile: W. Jeske *et al.*, *Semin. Thromb. Hemostasis* **19**, Suppl 1, 229 (1993). Clinical pharmacokinetics: K. Andrassy *et al.*, *Thromb. Res.* **73**, 95 (1994). Symposium on pharmacology and clinical efficacy in the prevention of thromboembolism and restenosis: *Blood Coagulation Fibrinolysis* **4**, S3-S60 (1993). Clinical trial in deep-vein thrombosis: H. K. Breddin *et al.*, *N. Engl. J. Med.* **344**, 626 (2001).

THERAP CAT: Antithrombotic.

8294. Rhamnetin. [90-19-7] 2-(3,4-Dihydroxyphenyl)-3,5-dihydroxy-7-methoxy-4*H*-1-benzopyran-4-one; 3,3′,4′,5-tetrahydroxy-7-methoxyflavone; 7-methylquercetin; β-rhamnocitrin; cyanidenolon-7-methyl ether 1537. $C_{16}H_{12}O_7$; mol wt 316.27. C 60.76%, H 3.82%, O 35.41%. The aglucon of xanthorhamnin. From fruit of *Rhamnus cathartica* L., *Rhamnaceae*: Krassowski, *J. Russ. Phys. Chem. Soc.* **40**, 1510 (1909); from commercial xanthorhamnin: Nystrom *et al.*, *J. Org. Chem.* **22**, 1272 (1957). Structure: Oesch, Perkin, *J. Chem. Soc.* **105**, 2354 (1914); Jurd, Horowitz, *J. Org. Chem.* **22**, 1618 (1957). Synthesis: Kuhn, Low, *Ber.* **77**, 211 (1944); Jurd, *J. Am. Chem. Soc.* **80**, 5531 (1958); US 2892845 (1959 to USDA); Anand *et al.*, *J. Sci. Ind. Res.* **21B**, 322 (1962), *C.A.* **57**, 13712f (1962); Kawano *et al.*, *Chem. Pharm. Bull.* **15**, 711 (1967).

Yellow needles from acetone + methanol, mp 292-294°. uv max (ethanol): 371, 256 nm (log ε 4.41, 4.40). Sol in hot phenol. Slightly sol in hot water, hot alcohol, hot glacial acetic acid, hot acetone. Freely sol in dil alkalies with intense yellow color.

Tetraacetate. $C_{24}H_{20}O_{11}$. Needles from acetone + methanol, mp 189-190°.

USE: Has been used for dyeing wool and cotton.

8295. Rhamnose. [3615-41-6] 6-Deoxy-L-mannose; L-rhamnose; L-mannomethylose; isodulcit. $C_6H_{12}O_5$; mol wt 164.16. C 43.90%, H 7.37%, O 48.73%. Occurs free in poison sumac *(Rhus toxicodendron* L., *Anacardiaceae)*, combined in the form of glycosides of many plants. Preparation: Clark, *J. Biol. Chem.* **38**, 255 (1919). Structure and configuration: Fischer, Morrell, *Ber.* **27**, 384 (1894); Fischer, Zach, *ibid.* **45**, 3762 (1912); Hirst, Macbeth, *J. Chem. Soc.* **1926**, 22; Avery, Hirst, *ibid.* **1929**, 2466. Isoln from walls of gram-negative bacteria: Salton, *Biochim. Biophys. Acta* **45**, 364 (1960); from *Afraegle paniculata* Engl., *Rutaceae:* Torto, *J. Chem. Soc.* **1961**, 5234; from rabbit skin: Malawista, Davidson, *Nature* **192**, 871 (1961); from leaves of *Solanum chacoense* Bitter, *Solanaceae:* Kuhn, Löw, *Ber.* **94**, 1088 (1961).

α-Form. Always obtained by crystn from H_2O or EtOH. Monohydrate, holohedric rods from water, hemihedric monoclinic columns from alcohol. Loses water of crystn on heating and partially changes to the β-modification. Very sweet taste. mp 82-92°: Ghosh, *Proc. R. Soc. Edinburgh* **36**, 216 (1915/16). Sublimes at 105° and 2 mm Hg. d_4^{20} 1.4708. Shows mutarotation. $[α]_D^{20}$ −7.7° → +8.9°: Hudson, Yanovsky, *J. Am. Chem. Soc.* **39**, 1013 (1917).

β-Form. Prepd by heating α-rhamnose monohydrate on a steam bath; crystallized from anhydr acetone + alcohol. Needles, mp 122-126° (rapid heating). $[α]_D^{20}$ +31.5° (1 min, p = 10). After a short time the rotation adjusts to the same final value as α-rhamnose. The β-form is hygroscopic and changes into crystals of the α-modification upon exposure to moist air.

8296. Rhamnus cathartica. Buckthorn bark. Dried, ripe fruit of *Rhamnus cathartica* L., *Rhamnaceae*. *Habit*. Europe, Northern Africa to Middle Asia. *Constit*. Rhamnocathartin, rhamnotannic acid, rhamnin, rhamnetin.

THERAP CAT: Cathartic.

8297. Rhapontin. [155-58-8] 3-Hydroxy-5-[(1E)-2-(3-hydroxy-4-methoxyphenyl)ethenyl]phenyl β-D-glucopyranoside; 4′-methoxy-3,3′,5-stilbenetriol 3-glucoside; rhaponticin; ponticin. $C_{21}H_{24}O_9$; mol wt 420.41. C 60.00%, H 5.75%, O 34.25%. From root of *Rheum rhaponticum* L., *Polygonaceae:* Hesse, *J. Prakt. Chem.* **77**, 321 (1908); Schürhoff, Plettner, *Arch. Pharm.* **275**, 281 (1937). On hydrolysis yields rhapontigenin and glucose: Kawamura, *J. Pharm. Soc. Jpn.* **58**, 405 (1938), *C.A.* **32**, 6655⁴ (1938). Constitu-

tion of rhapontigenin: Takaoka, *Proc. Imp. Acad. Tokyo* **16**, 408 (1940), *C.A.* **35**, 1399² (1941).

Crystals, dec 236-237°. $[α]_D^{32}$ −59.5° (acetone). Exhibits bright blue fluorescence. Sol in dil alcohol, hot acetone, hot water; slightly sol in ether, alc, acetone, cold water; practically insol in benzene, petr ether, chloroform.

8298. Rheadine. [2718-25-4] (5bR,13bR,15S)-5b,6,7,8,-13b,15-Hexahydro-15-methoxy-6-methyl-[1,3]dioxolo[4,5-h]-1,3-dioxolo[7,8][2]benzopyrano[3,4-a][3]benzazepine; (8β)-8-methoxy-16-methyl-2,3:10,11-bis[methylenebis(oxy)]rheadan; rhoeadine. $C_{21}H_{21}NO_6$; mol wt 383.40. C 65.79%, H 5.52%, N 3.65%, O 25.04%. Isoln from the seed capsules of corn poppy *(Papaver rhoeas* L., *Papaveraceae)*: Hesse, *Ann.* **140**, 145 (1866); *Arch. Pharm.* **228**, 7 (1890); Awe, *ibid.* **274**, 439 (1936); Awe, Winkler, *ibid.* **290**, 367 (1957); Nemecková, Santavy, *Collect. Czech. Chem. Commun.* **27**, 1210 (1962). Structure and stereochemistry: Santavy *et al.*, *ibid.* **30**, 3479 (1965); **32**, 4452 (1967). Total synthesis: Klotzer *et al.*, *Helv. Chim. Acta* **54**, 2057 (1971); **55**, 2228 (1972); Irie *et al.*, *J. Chem. Soc. Perkin Trans. 1* **1972**, 2986.

Crystals from ethanol, mp 252-254°. $[α]_D^{23}$ +235° (c = 1.01 in chloroform); $[α]_D^{22}$ +174° (c = 0.69 in pyridine). uv max (ethanol): 239, 290 nm (ε 9150, 9180). Slightly sol in ethyl acetate, methylene chloride. Practically insol in water, alcohol, chloroform, ether, benzene.

8299. Rhein. [478-43-3] 9,10-Dihydro-4,5-dihydroxy-9,10-dioxo-2-anthracenecarboxylic acid; 1,8-dihydroxyanthraquinone-3-carboxylic acid; 4,5-dihydroxyanthraquinone-2-carboxylic acid; chrysazin-3-carboxylic acid; monorhein; rheic acid; cassic acid; parietic acid; rhubarb yellow. $C_{15}H_8O_6$; mol wt 284.22. C 63.39%, H 2.84%, O 33.77%. Found in the free state and as glucoside in *Rheum* spp, *Polygonaceae* (rhubarb) and in Senna leaves; also in several spp of *Cassia (Leguminosae)*. Diacetate used as antirheumatic. Isoln from Chinese rhubarb: J. Schlossberger, O. Doepping, *Ann.* **50**, 196 (1844); O. Hesse, *ibid.* **309**, 32 (1899); F. Tutin, H. W. B. Clewer, *J. Chem. Soc.* **99**, 946 (1911); from aloe-emodin, *q.v.:* O. A. Oesterle, *Arch. Pharm.* **241**, 604 (1903); from *Cassia alata* L.: H. Hauptmann, L. L. Nazario, *J. Am. Chem. Soc.* **72**, 1492 (1950); from *C. fistula:* V. K. Murty *et al.*, *Tetrahedron* **23**, 515 (1967). Prepn from chrysophanic diacetate: Fischer *et al.*, *J. Prakt. Chem.* [2] **83**, 208; **84**, 369 (1911). Structure: H. Nawa *et al.*, *J. Org. Chem.* **26**, 979 (1961). Use in arthritis: C. A. Friedmann, DE 2711493; *idem*, US 4244968 (1977, 1981 both to Proter). Spectrophotometric study: A. A. Habib, N. A. El-Sebakhy, *J. Nat. Prod.* **43**, 452 (1980). Determn by gas chromatography, mass spectroscopy: G. W. Van Eijk, H. J. Roeijmans, *J. Chromatogr.* **295**, 497 (1984); HPLC: A. J. J. Van den Berg, R. P. Labadie, *ibid.* **329**, 311 (1985). Pharmacological effects on colonic mucosa: R. Wanitschke, *Pharmacology* **20**, Suppl. 1, 21 (1980); K. Ewe, *ibid.* 27. Metabolism study: J. Lemli, L. Lemmons, *ibid.* 50. In electron transport: P. Egerer *et al.*, *Z. Physiol. Chem.* **363**, 627 (1982). Inhibition of proteases: L. Raimondi *et al.*, *Pharmacol. Res. Commun.* **14**, 103 (1982). Effect on prostaglandin biosynthesis: S. Franchi-Micheli *et al.*, *J. Pharm. Pharmacol.* **35**, 262 (1983).

Yellow needles by sublimation, mp 321-322°, dec 330°. Absorption max (methanol): 229, 258, 435 nm (ε 36800, 20100, 11100). Practically insol in water. Sol in alkalies, pyridine; slightly sol in alc, benzene, chloroform, ether, petr ether. Forms a red potassium salt and a pink sodium salt. Red precips are also obtained with $Ca(OH)_2$ and $Ba(OH)_2$.

Diacetate see Diacerein.

8300. Rhenium. [7440-15-5] Re; at. wt 186.207; at. no. 75; valences 1-7; the heptavalent state is the most stable. Group VIIB (7). Two naturally occurring isotopes: 185 (37.07%); 187 (62.93%); the latter is radioactive, $T_{1/2} \sim 10^{11}$ years. Artificial, radioactive isotopes: 177-184; 186; 188-192. Occurs in gadolinite, molybdenite, columbite, rare earth minerals, and some sulfide ores. Average concn in earth's crust 1×10^{-9} (0.001 ppm). Discovered by Noddack *et al., Naturwissenschaften* **13**, 567, 571 (1925). Prepn of metallic rhenium by reduction of potassium perrhenate or ammonium perrhenate: Hurd, Brim, *Inorg. Synth.* **1**, 175 (1939). Prepn of high purity rhenium: Rosenbaum *et al., J. Electrochem. Soc.* **103**, 518 (1956). *Reviews:* Melaven in *Rare Metals Handbook,* C. A. Hampel, Ed. (Reinhold, New York, 1954) pp 347-364; Peacock in *Comprehensive Inorganic Chemistry* **vol. 3,** J. C. Bailar, Jr. *et al.,* Eds. (Pergamon Press, Oxford, 1973) pp 905-978; P. M. Treichel in *Kirk-Othmer Encyclopedia of Chemical Technology* **vol. 20** (Wiley-Interscience, New York, 3rd ed., 1982) pp 249-258.

Hexagonal close-packed crystals, black to silver-gray. d 21.02. mp 3180°. bp 5900° (estimated). Specific heat 0-20° 0.03263 cal/g/°C. Specific electrical resistance: 0.21×10^{-4} ohm/cm at 20°. Brinell hardness: 250. Latent heat of vaporization 152 kcal/mol. Reacts with oxidizing acids, nitric and concd sulfuric; not with HCl.

USE: Electron tube and semiconductor applications, in alloys for electrical contacts, as catalyst; possibly in high temp thermocouples and to improve the workability of tungsten and molybdenum alloys; plating jewelry, medical instruments, high vac equipment, mirror backings.

8301. Rhenium Heptoxide. [1314-68-7] Rhenium oxide (Re_2O_7); dirhenium heptoxide. O_7Re_2; mol wt 484.41. O 23.12%, Re 76.88%. Re_2O_7. Prepd from rhenium metal or its lower oxides in a stream of air or oxygen at 400-425°: Noddack, Noddack, *Naturwissenschaften* **17**, 93 (1929); *Z. Anorg. Allg. Chem.* **181**, 1 (1929); **215**, 129 (1933); Briscoe *et al., Nature* **129**, 618 (1932); Melaven *et al., Inorg. Synth.* **3**, 188 (1950). Crystal structure: Krebs *et al., Chem. Commun.* **1968**, 263.

Canary-yellow, very deliquescent crystals. Begins to sublime at 250°. mp 300.3°; bp 360.3°: Smith *et al., J. Am. Chem. Soc.* **74**, 4964 (1952). Freely sol in water, alcohol, ether, ethyl acetate, dioxane, pyridine. Readily absorbs water forming perrhenic acid, $HReO_4$.

8302. Rhenium Hexafluoride. [10049-17-9] (*OC*-6-11)-Rhenium fluoride (ReF_6). F_6Re; mol wt 300.20. F 37.97%, Re 62.03%. ReF_6. Prepd by direct fluorination of rhenium metal: Ruff, Kwasnik, *Z. Anorg. Allg. Chem.* **209**, 113 (1932); **219**, 65 (1934); Malm, Selig, *J. Inorg. Nucl. Chem.* **20**, 189 (1961). The product obtained by Ruff was probably a mixture of ReF_6 and ReF_7: *eidem, J. Am. Chem. Soc.* **82**, 1510 (1960).

Yellow, cubic crystals or liquid; colorless gas; extremely hygroscopic. Forms bluish vapors on contact with air, discolors to a dark purple liquid. mp 18.5°. bp 33.7°. Sol in nitric acid. Soly in anhydr HF: 1.75 moles/1000 g HF, Frlec, Hyman, *Inorg. Chem.* **6**, 1596 (1967). Does not attack dried Pyrex glass or silica. May be handled in copper up to 150°.

8303. Rhenium Trioxide. [1314-28-9] Rhenium oxide (ReO_3); rhenic anhydride. O_3Re; mol wt 234.20. O 20.49%, Re 79.51%. ReO_3. Prepd by reduction of Re_2O_7 with dioxane: Ne-

chamkin, Hiskey, *Inorg. Synth.* **3**, 186 (1950); with CO: Melaven *et al., ibid.* 187.

Red cubic crystals; green luster by transmitted light; the color of the powder may appear blue (**rhenium blue**). At 400° disproportionates *in vacuo* to Re_2O_7 and ReO_2. bp 750°. d 6.9-7.4. Practically insol in water, alkalies, non-oxidizing acids; oxidized by HNO_3 to $HReO_4$.

8304. Rhizoxin. [90996-54-6] WF-1360; NSC-332598. C_{35}-$H_{47}NO_9$; mol wt 625.76. C 67.18%, H 7.57%, N 2.24%, O 23.01%. Rice seedling blight pathogen; phytotoxin which inhibits tubulin polymerization. Isoln from *Rhizopus chinensis:* S. Iwasaki *et al., J. Antibiot.* **37**, 354 (1984). Absolute configuration: *eidem, ibid.* **39**, 424 (1986). Antitumor activity in comparison with vincristine, *q.v.:* T. Tsuruo *et al., Cancer Res.* **46**, 381 (1986). Mechanism of action: M. Takahashi *et al., Biochim. Biophys. Acta* **926**, 215 (1987); Y. Li *et al., Biochem. Biophys. Res. Commun.* **187**, 722 (1992). Angiogenesis inhibition: K. Aoki *et al., Eur. J. Pharmacol.* **459**, 131 (2003). Clinical pharmacology: A. W. Tolcher *et al., Ann. Oncol.* **11**, 333 (2000). Review of chemistry and biology: J. Hong, J. D. White, *Tetrahedron* **60**, 5653-5681 (2004).

Pale yellow powder, mp 131-135°. $[\alpha]_D^{24}$ +155.5° (c = 0.8 in methanol). uv max (methanol): 295, 308, 325 nm (ε 42300, 54000, 39000).

8305. Rhod-2. [132523-91-2] 9-[4-[Bis(carboxymethyl)amino]-3-[2-[2-[bis(carboxymethyl)amino]-5-methylphenoxy]ethoxy]-phenyl]-3,6-bis(dimethylamino)xanthylium chloride (1:1). $C_{40}H_{43}$-ClN_4O_{11}; mol wt 791.25. C 60.72%, H 5.48%, Cl 4.48%, N 7.08%, O 22.24%. Rhodamine-based calcium specific fluorescent dye. Prepn: R. Y. Tsien, A. Minta, **EP 314480**; *eidem,* **US 5049673** (1989, 1991 both to Univ. California); A. Minta *et al, J. Biol. Chem.* **264**, 8171 (1989). Use in monitoring intramitochondrial Ca^{2+}: M. Hoth, R. S. Lewis in *Imaging Neurons,* R. Yuste *et al.,* Eds. (Cold Spring Harbor Laboratory Press, Cold Spring Harbor, NY, 2000) pp 1-14; in free/intracellular Ca^{2+} levels: M.-P. Muriel *et al., J. Comp. Neurol.* **426**, 297 (2000); G. A. MacGowan *et al., J. Biomed. Opt.* **6**, 23 (2001).

Dark purplish solid. K_D for Ca^{2+}: 1 μM.

USE: Indicator for calcium measurement in biological systems.

8306. Rhodamine 123. [62669-70-9] 3,6-Diamino-9-[2-(methoxycarbonyl)phenyl]xanthylium chloride (1:1); 2-(6-amino-3-imino-3H-xanthen-9-yl)benzoic acid methyl ester hydrochloride; rhodamine 110 methyl ester; Rh 123. $C_{21}H_{17}ClN_2O_3$; mol wt 380.83. C 66.23%, H 4.50%, Cl 9.31%, N 7.36%, O 12.60%. Mitochondrial specific fluorescent dye for living cells. Substrate for P-glycoprotein. Prepd as methyl ester of rhodamine 110: P. Pal *et al.*, *Photochem. Photobiol.* **63**, 161 (1996); J. A. Ross *et al.*, *Synth. Commun.* **36**, 1745 (2006). Mitochondrial staining applications: L. V. Johnson *et al.*, *Proc. Natl. Acad. Sci. USA* **77**, 990 (1980); L. B. Chen *et al.*, *Cold Spring Harbor Symp. Quant. Biol.* **46**, 141 (1982); R. K. Emaus *et al.*, *Biochim. Biophys. Acta* **850**, 436 (1986). Photophysical and photochemical properties: A. Chow *et al.*, *Photobiochem. Photobiophys.* **11**, 139 (1986); M. W. Ferguson *et al.*, *Phys. Chem. Chem. Phys.* **1**, 261 (1999). Binding studies with P-glycoprotein: B. Nare *et al.*, *Mol. Pharmacol.* **45**, 1145 (1994); with multidrug resistance protein: R. Daoud *et al.*, *Biochemistry* **39**, 15344 (2000). Flow cytometric assessment of P-glycoprotein function: J. Petríz, J. García-López, *Leukemia* **11**, 1124 (1997). HPLC detection in cell lysate: T. Iqbal *et al.*, *J. Chromatogr. B* **814**, 259 (2005).

Brownish-red solid. Abs max: 500 nm (ε 75000, water); 507 nm (methanol). Excitation max: 511 nm (ethanol). Emission max: 529 nm (methanol); 530 nm (ethanol). Fluorescence emission: green (excitation at 485 nm); red (excitation at 546 nm). Sol in ethanol, ether, DMF, methanol, and water.

USE: Mitochondrial stain. Laser dye. In measurement of membrane potential. In flow cytometric analysis of P-glycoprotein function. In assays for assessment of multidrug resistance of cancer cells.

8307. Rhodamine B. [81-88-9] 9-(2-Carboxyphenyl)-3,6-bis(diethylamino)xanthylium chloride (1:1); *N*-[9-(2-carboxyphenyl)-6-(diethylamino)-3H-xanthen-3-ylidene]-*N*-ethylethanaminium chloride; tetraethylrhodamine; D & C Red No. 19; C.I. Basic Violet 10; C.I. 45170. $C_{28}H_{31}ClN_2O_3$; mol wt 479.02. C 70.21%, H 6.52%, Cl 7.40%, N 5.85%, O 10.02%. Prepn: *Colour Index* vol. **4** (3rd ed., 1971) p 4420. Toxicity study: J. M. Webb *et al.*, *Toxicol. Appl. Pharmacol.* **3**, 696 (1961).

Green crystals or reddish-violet powder. Very sol in water with a bluish-red color, dil solns being strongly fluorescent; very sol in alcohol; slightly sol in HCl, NaOH. LD_{50} i.v. in rats: 89.5 mg/kg (Webb).

USE: As a dye, especially for paper; as a reagent for antimony, bismuth, cobalt, niobium, gold, manganese, mercury, molybdenum, tantalum, thallium, tungsten; as biological stain.

8308. Rhodanine. [141-84-4] 2-Thioxo-4-thiazolidinone; 2-thio-4-ketothiazolidine; 4-oxo-2-thionothiazolidine; rhodanic acid. $C_3H_3NOS_2$; mol wt 133.18. C 27.06%, H 2.27%, N 10.52%, O 12.01%, S 48.15%. Prepd by the action of sodium chloroacetate on ammonium dithiocarbamate: Julian, Sturgis, *J. Am. Chem. Soc.* **57**, 1126 (1935); Redemann *et al.*, *Org. Synth.* **coll. vol. III**, 763 (1955).

Pale yellow crystals (long blades) from glacial acetic acid or water. d 0.868. mp 168.5° (capillary); mp 170.5-171° (hot stage). May explode on rapid heating. pK (25°) 5.52. Freely sol in boiling water, in methanol, ethanol, DMF, ether, alkalies, ammonia, hot acetic acid.

USE: In the synthesis of phenylalanine.

8309. Rhodinol. [6812-78-8] (3S)-3,7-Dimethyl-7-octen-1-ol; α-citronellol. $C_{10}H_{20}O$; mol wt 156.27. C 76.86%, H 12.90%, O 10.24%. Name that has been used to describe a group of related chemicals as well as a single chemical. The term, at times, was applied to what was considered the most important alcohol in geranium oil, *1-citronellol*. *See:* Barbier, Bouveault, *Ber.* **29**, 289 (1896). Prepn in pure form: Sutherland, *J. Am. Chem. Soc.* **73**, 2385 (1951). Synthesis from *cis*-pinane: Rienäcker, *Chimia* **27**, 97 (1973).

Oily liquid, odor of rose. d_4^{20} 0.8549. bp_{12} 114-115°. n_D^{20} 1.4556. $[\alpha]_D^{20}$ −2.88°. uv max: 186-189 nm (ε 9000). Very slightly sol in water, miscible with alcohol, ether.

USE: In perfumery. *See also* β-Citronellol.

8310. Rhodium. [7440-16-6] Rh; at. wt 102.90550; at. no. 45; valences 1-6; most common states 1, 3. Group VIII (9). One naturally occurring isotope: 103; artificial radioactive isotopes: 97-102; 104-110. Member of platinum group of metals. One of the rarest elements, constitutes about 1×10^{-7}% of the earth's crust; found in small quantities associated with all native platinum; in the minerals rhodite, sperrylite, iridosmine; in some nickel-copper ores. Discovered in 1803 by Wollaston, *Philos. Trans. R. Soc. London* **94**, 419 (1804). Prepn: Vauquelin, etc., cited by Mellor, *A Comprehensive Treatise on Inorganic and Theoretical Chemistry* **15**, 546 (1936). Practical methods of separation: Wichers, Gilchrist, *Trans. Am. Inst. Min. Metall. Eng.* **76**, 619 (1928). Reviews of prepn, properties and chemistry of rhodium and other platinum metals: Gilchrist, *Chem. Rev.* **32**, 277-372 (1943); Beamish *et al.*, in *Rare Metals Handbook*, C. A. Hampel, Ed. (Reinhold, New York, 1956) pp 291-328; W. P. Griffith, *The Chemistry of the Rarer Platinum Metals* (John Wiley, New York, 1967) pp 1-41, 313-430; Livingstone in *Comprehensive Inorganic Chemistry* **Vol. 3**, J. C. Bailar Jr. *et al.*, Eds. (Pergamon Press, Oxford, 1973) pp 1163-1189, 1233-1253.

Silvery-white, soft, ductile, malleable metal; face-centered cubic structure. mp 1966°: Roeser, Wensel, *Natl. Bur. Stand. J. Res.* **12**, 519 (1934); d^{20} 12.41. Electr resistivity (0°) 4.51 μ-ohms-cm. Brinell hardness: 100. Not attacked by acids even aqua regia when in compact form; the finely divided metal reacts with aqua regia. Absorbs oxygen when melted; at a red heat is slowly oxidized to the sesquioxide. Converted to the trihalide by chlorine or bromine at a red heat; not attacked by fluorine.

Caution: Potential symptom of overexposure to metal fumes and insoluble compds is respiratory sensitization. Potential symptoms of overexposure to soluble compds in exptl animals are eye irritation, CNS damage. *See NIOSH Pocket Guide to Chemical Hazards* (DHHS/NIOSH 97-140, 1997) p 272-275.

USE: As an alloy with platinum; as a corrosion-resistant electroplate for protecting silverware from tarnishing; for making high-reflectivity mirrors for cinema projectors, searchlights. Spongy or black rhodium is used as a catalyst in various organic hydrogenation and oxidation reactions.

8311. Rhodium(II) Acetate. [15956-28-2] Tetrakis[μ-(acetato-$\kappa O:\kappa O'$)]dirhodium (Rh-Rh); bis(rhodium diacetate); dirhodium tetraacetate; rhodium diacetate dimer; rhodium(II) acetate dimer; $Rh_2(OAc)_4$. $C_8H_{12}O_8Rh_2$; mol wt 441.99. C 21.74%, H

2.74%, O 28.96%, Rh 46.56%. [(CH₃COO)₂Rh]₂. Efficient catalyst for reactions of diazo-carbonyl compds that proceed through transient rhodium-carbenoid intermediates. Preliminary X-ray diffraction studies: M. A. Porai-Koshits, A. S. Antsyshkina, *Dokl. Akad. Nauk SSSR* **146**, 1102 (1962). Prepn: S. A. Johnson *et al.*, *Inorg. Chem.* **2**, 960 (1963); G. A. Rempel *et al.*, *Inorg. Synth.* **13**, 90 (1972). Crystal structure: E. V. Dikarev *et al.*, *Inorg. Chem.* **43**, 5558 (2004). Description of catalytic activiy: R. Paulissen *et al.*, *Tetrahedron Lett.* **14**, 2233 (1973); A. J. Hubert *et al.*, *Synthesis* **1976**, 600. Synthetic applications in ylide formation: M. P. Doyle *et al.*, *J. Org. Chem.* **46**, 5094 (1981); in heteroatom-hydrogen bond insertions: M. P. Moyer *et al.*, *ibid.* **50**, 5223 (1985); in carbon-hydrogen bond insertions: D. F. Taber, R. E. Ruckle, Jr., *J. Am. Chem. Soc.* **108**, 7686 (1986); in cyclopropanations: M. P. Doyle *et al.*, *ibid.* **112**, 1906 (1990). *Review*: J. Adams, D. M. Spero, *Tetrahedron* **47**, 1765-1808 (1991).

Dark emerald-green crystals. d 2.371. Abs max: 587, 447 nm (ε 230, 94, water); 590, 446 nm (ε 210, 98, ethanol); 603, 440 nm (ε 260, 115, acetone); 597, 441 nm (ε 235, 110, tetrahydrofuran); 592, 442 nm (ε 255, 100, acetic acid); 552, 437 nm (ε 235, 125, acetonitrile); 500 nm (ε 275, dimethyl sulfoxide). Sol in water, tetrahydrofuran, ethanol, methanol, acetone, acetic acid, acetonitrile, dimethyl sulfoxide, nitromethane, triethyl phosphate; slightly sol in toluene. Insol in diethyl ether, benzene, carbon tetrachloride, 1,2-dichloroethane, dichloromethane.

USE: Catalyst in synthetic organic chemistry.

8312. Rhodium Carbonyl Chloride. [14523-22-9] Tetracarbonyldi-μ-chlorodirhodium; dichlorotetracarbonyldirhodium; chlorodicarbonylrhodium(I) dimer. C₄Cl₂O₄Rh₂; mol wt 388.75. C 12.36%, Cl 18.24%, O 16.46%, Rh 52.94%. [Rh(CO)₂Cl]₂. Prepn: Hieber, Lagally, *Z. Anorg. Allg. Chem.* **251**, 96 (1943); McCleverty, Wilkinson, *Inorg. Synth.* **8**, 211 (1966); Colton *et al.*, *Aust. J. Chem.* **23**, 1351 (1970); Cramer, *Inorg. Synth.* **15**, 14 (1974). Structure: Dahl *et al.*, *J. Am. Chem. Soc.* **83**, 1761 (1961).

Orange-red crystalline solid or red crystalline sublimate. mp 124-125°. Sol in most organic solvents (except aliphatic hydrocarbons). Stable to dry air. Solutions in organic solvents decompose when exposed to air.

USE: Catalyst.

8313. Rhodium Chloride. [10049-07-7] Rhodium chloride (RhCl₃); rhodium trichloride. Cl₃Rh; mol wt 209.26. Cl 50.82%, Rh 49.18%. RhCl₃. Similar to the chlorides of other platinum group metals, rhodium chloride readily forms double salts with alkali chlorides. The chloride is obtainable also as a water-sol hydrate with variable amounts of water. Prepn of trihydrate: Anderson, Basolo, *Inorg. Synth.* **7**, 214 (1963). Toxicity data: Landolt *et al.*, *Toxicol. Appl. Pharmacol.* **21**, 589 (1972).

Red powder. Insol in water. Sol in alkali hydroxide or cyanide solns. LD₅₀ in rats (mg/kg): 198 i.v. (Landolt).

8314. Rhododendrin. [497-78-9] (1*R*)-3-(4-Hydroxyphenyl)-1-methylpropyl β-D-glucopyranoside; betuloside. C₁₆H₂₄O₇; mol wt 328.36. C 58.53%, H 7.37%, O 34.11%. From leaves of *Rhododendron* spp., *Ericaceae*: Archangelski, *Arch. Exp. Pathol. Pharmakol.* **46**, 313 (1901); Kawaguchi *et al.*, *J. Pharm. Soc. Jpn.* **62**, 4 (1942), *C.A.* **44**, 9634a (1950). Structure: Kim, *ibid.* 455, *C.A.* **45**, 4222h (1951). Identity with betuloside: Kim, *ibid.* **63**, 103 (1943), *C.A.* **45**, 4222i (1951). Chemotaxonomic significance: Thieme, Winkler, *Pharmazie* **24**, 703 (1969).

Bitter crystals, mp 187°. Sol in hot water, alcohol; slightly sol in chloroform, ether.

Pentaacetate. Needles, mp 96-97°.

Methyl ether. Prisms from warm water, mp 102-103°. [α]²¹_D −17.1° (96% alcohol).

USE: In systematic classification of *Rhododendron* species.

8315. Rhodomycins. Antibiotic substance produced by *Streptomyces purpurascens* nov. sp.: Brockmann, Bauer, *Naturwissen-*

schaften **37**, 492 (1950); Brockmann *et al.*, *Ber.* **84**, 700 (1951); **GB 708749** (1954 to Bayer). Separation of rhodomycin A and rhodomycin B: Brockmann, Patt, *Ber.* **88**, 1455 (1955). Isoln and structural elucidation of additional rhodomycins produced by *Streptomyces stammes:* Biedermann, Brauniger, *Pharmazie* **27**, 782 (1972); Brockmann *et al.*, *Tetrahedron Lett.* **1973**, 3699. Total synthesis of racemic rhodomycinones (the aglycones of the rhodomycins): A. S. Kende, Y.-G. Tsay, *Chem. Commun.* **1977**, 140.

Rhodomycin A

Rhodomycin A. [23666-50-4] [7*R*-(7α,8β,10β)]-8-Ethyl-7,8,-9,10-tetrahydro-1,6,8,11-tetrahydroxy-7,10-bis[[2,3,6-trideoxy-3-(dimethylamino)-α-L-lyxo-hexopyranosyl]oxy]-5,12-naphthacenedione; β-rhodomycin II. C₃₆H₄₈N₂O₁₂; mol wt 700.78. Dec on mild acidic hydrolysis into 1 mole β-rhodomycinone and 2 moles rhodosamine: Brockmann, Spohler, *Naturwissenschaften* **48**, 716 (1961). Structure of β-rhodomycinone: Brockmann, Niemeyer, *ibid.* 570. Structure of rhodosamine: Brockmann, Spohler, *ibid.* **42**, 154 (1955). Total structure: Brockmann *et al.*, *Tetrahedron Lett.* **1969**, 415.

Dihydrochloride. C₃₆H₄₈N₂O₁₂·2HCl. Red prisms from ethanol + isopropanol, mp 205°. Freely sol in water, lower alcohols; sparingly sol in benzene, chloroform; practically insol in ether, petr ether.

Perchlorate. C₃₆H₄₈N₂O₁₂·2HClO₄. Minute red needles, mp 188°. Freely sol in methanol, acetone; sparingly sol in water, butanol.

Rhodomycin B. [1404-52-0] [7*R*-(7α,8β,10β)]-8-Ethyl-7,8,-9,10-tetrahydro-1,6,7,8,11-pentahydroxy-10-[[2,3,6-trideoxy-3-(dimethylamino)-α-L-lyxo-hexopyranosyl]oxy]-5,12-naphthacenedione. C₂₈H₃₁NO₉; mol wt 525.55. Dec on mild acidic hydrolysis into 1 mole β-rhodomycinone and 1 mole rhodosamine: Brockmann, Spohler, *ibid.* **48**, 716 (1961).

Hydrochloride. Red prisms, mp 180°. Freely sol in lower alcohols; sparingly sol in water, acetone. Practically insol in chloroform, benzene.

8316. Rhodopin. [105-92-0] 1,2-Dihydro-1-hydroxy-ψ,ψ-carotene; 1,2-dihydro-1-hydroxylycopene. C₄₀H₅₈O; mol wt 554.90. C 86.58%, H 10.54%, O 2.88%. Pigment from *Rhodovibrio* and *Thiocystis* bacteria: Karrer, Solmssen, *Helv. Chim. Acta* **18**, 1306 (1935); **19**, 1019 (1936); Karrer *et al.*, *ibid.* **21**, 454 (1938); from *Rhodomicrobium vannielii:* Volk, Pennington, *J. Bacteriol.* **59**, 169 (1950); Ryvarden, Jensen, *Acta Chem. Scand.* **18**, 643 (1964). Structure: Jensen, *ibid.* **13**, 842, 2142 (1959). Synthesis: Bonnett *et al.*, *ibid.* **18**, 1739 (1964); Surmatis *et al.*, *J. Org. Chem.* **31**, 186 (1966). Synthesis of deuterated rhodopin: Johansen, Liaaen-Jensen, *Acta Chem. Scand.* **28B**, 301 (1974).

Aggregates of dark red crystals from carbon disulfide + petr ether, mp 171° (preliminary sintering). Absorption max (chloroform): 521, 486, 453 nm.

8317. Rhodopsin. [9009-81-8] Visual purple. Mol wt ~40000. A photoreceptor protein found in the rods of the retina of the eye. Composed of 11-*cis* retinal, *q.v.*, bound as a protonated Schiff base to a lysine moiety in the apoprotein opsin, *q.v.* Prepn from retinal and opsin: G. Wald, P. K. Brown, *Proc. Natl. Acad. Sci. USA* **36**, 84 (1950); R. Hubbard, G. Wald, *ibid.* **37**, 69 (1951). Purification and amino acid compn of bovine rhodopsin: Shields *et al.*, *Biochim. Biophys. Acta* **147**, 238 (1967). Synthesis of labelled rhodopsin and studies on the active site: Hirtenstein, Akhtar, *Biochem. J.* **119**, 359 (1970). Characterization in synthetic membranes: W. L. Hubbell, *Acc. Chem. Res.* **8**, 85 (1975). Amino acid sequence of bovine rhodopsin: P. A. Hargrave *et al.*, *Biophys. Struct. Mech.* **9**, 235 (1983). Studies on transmembrane molecular structure: E. A. Dratz, P. A. Hargrave, *Trends Biochem. Sci.* **8**, 128 (1983). Crystal structure: K. Palczewski *et al.*, *Science* **289**, 739 (2000). Isoln and sequence of gene encoding bovine rhodopsin: J. Nathans, D. S. Hogness, *Cell* **34**, 807 (1983); of gene encoding human rhodopsin: *eidem*, *Proc. Natl. Acad. Sci. USA* **81**, 4851 (1984). Exposure to light initiates the conversion of rhodopsin through a series of distinct intermediates to yield opsin + *trans*-retinal: R. Hubbard, A. Kropf, *ibid.* **44**, 130 (1958); R. Hubbard *et al.*, *Nature* **183**, 142 (1959); B. Honig *et al.*, *Proc. Natl. Acad. Sci. USA* **76**, 2503 (1979). Review of visual transduction and bioregulation of absorption maxima: K. Nakanishi, *Pure Appl. Chem.* **57**, 769-776 (1985). *Reviews:* T. G. Ebrey, B. Honig, *Q. Rev. Biophys.* **8**, 129-184 (1975); D. F. O'Brien, *Science* **218**, 961-966 (1982); A. Maeda, T. Yoshizawa, *Photochem. Photobiol.* **35**, 891-898 (1982); P. S. Zurer, *Chem. Eng. News* **61**, 24-35 (Nov. 28, 1983). *See also: Methods Enzymol.* **81** and **88**, Lester Packer, Ed. (Academic Press, New York, 1982).

8318. Rhodoquinone. [5591-74-2] 2-Amino-5-(3,7,11,15,- 19,23,27,31,35,39-decamethyl-2,6,10,14,18,22,26,30,34,38-tetra-contadecaen-1-yl)-3-methoxy-6-methyl-2,5-cyclohexadiene-1,4-dione; rhodoquinone-10. $C_{58}H_{89}NO_3$; mol wt 848.35. C 82.12%, H 10.57%, N 1.65%, O 5.66%. Isoln from *Rhodospirillum rubrum:* Gloves, Threlfall, *Biochem. J.* **85**, 14P (1962). Structure and synthesis: Moore, Folkers, *J. Am. Chem. Soc.* **87**, 1409 (1965); **88**, 567 (1966); Daves *et al.*, *ibid.* **90**, 5587 (1968).

Violet needles, mp 70-71°. uv max (cyclohexane): 280, 500 nm ($E^{1\%}_{1cm}$ 154, 14).

8319. Rhodoviolascin. [34255-08-8] 3,3′,4,4′-Tetradehydro-1,1′,2,2′-tetrahydro-1,1′-dimethoxy-ψ,ψ-carotene; 3,3′,4,4′-tetradehydro-1,1′,2,2′-tetrahydro-1,1′-dimethoxylycopene; spirilloxanthin. $C_{42}H_{60}O_2$; mol wt 596.94. C 84.51%, H 10.13%, O 5.36%. Carotenoid pigment isolated from *Rhodovibrio* and *Thioceptis* bacteria: Karrer, Solmssen, *Helv. Chim. Acta* **18**, 1306 (1935). Identity with spirilloxanthin: Zechmeister *et al.*, *Arch. Biochem.* **5**, 243 (1944). Structure: Karrer, Koenig, *Helv. Chim. Acta* **23**, 460 (1940); Barber *et al.*, *Proc. Chem. Soc. London* **1959**, 96. Synthesis: Surmatis, Ofner, *J. Org. Chem.* **28**, 2735 (1963); Surmatis, US 3160658 (1964 to Hoffmann-La Roche).

Dark red, spindle-shaped crystals from benzene, mp 218°. Absorption max (chloroform): 544, 507, 476 nm. Almost insol in petr ether, ligroin, methanol, somewhat more sol in hot benzene.

8320. Rhodoxanthin. [116-30-3] 4′,5′-Didehydro-4,5′-*retro*-β,β-carotene-3,3′-dione. $C_{40}H_{50}O_2$; mol wt 562.84. C 85.36%, H 8.95%, O 5.69%. Carotenoid pigment found widely distributed in

nature, but in very small amounts only. One of very few carotenoids with *retro* configuration. Best isolated from the arils of *Taxus baccata* L., *Taxaceae:* Kuhn, Brockmann, *Ber.* **66**, 828 (1933). Occurs also in the feathers of birds such as *Phoenicircus nigricollis*, *Megaloprepia magnifica.* Isoln: Volker, *Z. Physiol. Chem.* **292**, 75 (1953). Structure: Kuhn, Brockmann, *loc. cit.*; Karrer, Solmssen, *Helv. Chim. Acta* **18**, 477 (1935). Synthesis: Mayer *et al.*, *ibid.* **50**, 1606 (1967); Surmatis, Walser, US 3466331; US 3624105 (1969, 1971 to Hoffmann-La Roche); P. R. Ellis *et al.*, *Helv. Chim. Acta* **64**, 1092 (1981); E. Widmer *et al.*, *ibid.* **65**, 944 (1982). *Review:* Liaaen-Jensen in *Aspects Terpenoid Chem. Biochem.*, *Proc. Phytochem. Soc. Symp.*, 2nd, 1970, T. W. Goodwin, Ed. (Academic Press, London, 1971) p 223-254.

Rosettes of deep purple, lancet-shaped crystals from benzene:methanol (1:4). mp 219° (evacuated tube). Absorption max (chloroform): 546, 510, 482 nm. Freely sol in pyridine; sol in benzene, chloroform. Sparingly sol in ethanol and methanol. Practically insol in hexane, petr ether.

USE: Coloring material for food, beverages, pharmaceuticals and cosmetics.

8321. Rhubarb. Rhizome and roots of *Rheum officinale* Baill., *R. palmatum* L., or other *Rheum* spp, *Polygonaceae.* *Habit.* Central Asia; cultivated in Europe, Southern Siberia, North America. *Constit.* Chrysophanic acid, emodin, rhein, rheotannic acid, erythroretin, methylchrysophanic acid, rhabarberon, cinnamic and gallic acids, calcium oxalate. Comprehensive studies: Workman, Hiner, *J. Am. Pharm. Assoc.* **49**, 118 (1960). *Incompat.* Mineral acids, iron salts, lead acetate, zinc salts, lime water; infusions of catechu, cinchona, and nutgall; tartar emetic, tannic acid.

THERAP CAT: Cathartic.

THERAP CAT (VET): Has been used as a laxative.

8322. Rhynchophylline. [76-66-4] (αE,1′R,6′R,7′S,8′aS)-6′-Ethyl-1,2,2′,3′,6′,7′,8′,8′a-octahydro-α-(methoxymethylene)-2-oxo-spiro[3H-indole-3,1′(5′H)-indolizine]-7′-acetic acid methyl ester; (7β,16E,20α)-16,17-didehydro-17-methoxy-2-oxocorynoxan-16-carboxylic acid methyl ester; rhyncophylline; mitrinermine. $C_{22}H_{28}N_2O_4$; mol wt 384.48. C 68.73%, H 7.34%, N 7.29%, O 16.64%. From stems and roots of *Uncaria rynchophylla* Miq. [*Ourouparia rynchophylla* (Miq.)] Matsum, *Rubiaceae:* Kondo *et al.*, *J. Pharm. Soc. Jpn.* **48**, 54 (1928); from *Mitragyna rotundifolia* (Roxb.), *Rubiaceae:* Barger *et al.*, *J. Org. Chem.* **4**, 418 (1939). Structure: Seaton, Marion, *Can. J. Chem.* **35**, 1102 (1957); **38**, 1035 (1960). Stereochemistry: Nozoye, *Chem. Pharm. Bull.* **6**, 309 (1958); Ban, Oishi, *Tetrahedron Lett.* **1961**, 791; *eidem*, *Chem. Pharm. Bull.* **11**, 441, 446, 451 (1963). Partial synthesis: Finch, Taylor, *J. Am. Chem. Soc.* **84**, 1318, 3871 (1962). Total synthesis: Ban *et al.*, *Tetrahedron Lett.* **1972**, 2113.

Crystals from methanol, mp 216°; *dl*-form reported as colorless pillars from ethyl acetate-ether, mp 197-199° (Ban *et al.*, *loc. cit.*). $[\alpha]_D^{13}$ −14.7° (c = 2.5 in chloroform). pKa 6.4. uv max: 245, 280 nm (log ε 4.24, 3.15). Sol in chloroform; moderately sol in acetone, alcohol, benzene; sparingly sol in ether, ethyl acetate. Practically insol in petr ether.

8323. Ribavirin. [36791-04-5] 1-β-D-Ribofuranosyl-1H-1,-2,4-triazole-3-carboxamide; RTCA; tribavirin; ICN-1229; Copegus; Rebetol; Viramid; Virazide; Virazole. $C_8H_{12}N_4O_5$; mol wt 244.21.

C 39.35%, H 4.95%, N 22.94%, O 32.76%. Purine nucleoside analog; inhibits inosine monophosphate dehydrogenase (IMPDH). Prepn: J. T. Witkowski, R. Robins, **DE 2220246**; *eidem*, **US 3798209** (1972, 1974 both to ICN); J. T. Witkowski *et al.*, *J. Med. Chem.* **15**, 1150 (1972); *eidem*, *J. Carbohydr. Nucleosides Nucleotides* **5**, 363 (1978). Regioselective synthesis: Y. Ito *et al.*, *Tetrahedron Lett.* **1979**, 2521. NMR study: G. P. Kreishman *et al.*, *J. Am. Chem. Soc.* **94**, 5894 (1972). Crystal structure: P. Prusiner, M. Sundaralingam, *Nature New Biol.* **244**, 116 (1973). Teratogenicity study: D. M. Kochhar *et al.*, *Toxicol. Appl. Pharmacol.* **52**, 99 (1980). Review of antiviral activity: R. W. Sidwell *et al.*, *Pharmacol. Ther.* **6**, 123-146 (1979); of clinical experience: H. Fernandez *et al.*, *Eur. J. Epidemiol.* **2**, 1-14 (1986); of molecular modes of action: J. L. Patterson, R. Fernandez-Larsson, *Rev. Infect. Dis.* **12**, 1139-1146 (1990). Clinical trial in infants with respiratory syncytial viral infection: C. B. Hall *et al.*, *N. Engl. J. Med.* **308**, 1443 (1983); in combination with interferon alfa-2b in hepatitis C: O. Reichard *et al.*, *Lancet* **351**, 83 (1998).

White to off-white powder. Freely sol in water; slightly sol in anhydrous alcohol. Exists in two polymorphic forms: mp 166-168° (aq ethanol); mp 174-176° (ethanol) (Witkowski, 1972). Also reported as mp 180-182° (Ito). $[\alpha]_D^{25}$ −36.5° (c = 1 in water); $[\alpha]_D^{20}$ −38° (c = 1 in water). LD$_{50}$ i.p. in mice: 1.3 g/kg; orally in rats: 5.3 g/kg (Witkowski, 1972).

THERAP CAT: Antiviral.

8324. **α-Ribazole.** [132-13-8] 5,6-Dimethyl-1-α-D-ribofuranosyl-1*H*-benzimidazole. $C_{14}H_{18}N_2O_4$; mol wt 278.31. C 60.42%, H 6.52%, N 10.07%, O 22.99%. Nucleoside moiety of vitamin B$_{12}$, *q.v.* Isoln by acid hydrolysis of vitamin B$_{12}$: N. G. Brink *et al.*, *J. Am. Chem. Soc.* **72**, 1866 (1950); N. G. Brink, K. Folkers, *ibid.* **74**, 2856 (1952). Syntheses of α- and β-anomers: F. W. Holly *et al.*, *ibid.* 4521; R. S. Wright *et al.*, *ibid.* **80**, 2004 (1958); J. D. Stevens *et al.*, *J. Org. Chem.* **33**, 1806 (1968). Biosynthetic study: J. Hörig, P. Renz, *FEBS Lett.* **80**, 337 (1977). Spectroscopic characterization: K. L. Brown *et al.*, *Inorg. Chem.* **23**, 1463 (1984).

Crystals from water, mp 198-199°. $[\alpha]_D^{23}$ +14° (c = 0.9 in pyridine).

8325. **Riboflavin.** [83-88-5] Vitamin B$_2$; lactochrome; lactoflavine; vitamin G; 1-deoxy-1-(3,4-dihydro-7,8-dimethyl-2,4-dioxobenzo[*g*]pteridin-10(2*H*)-yl)-D-ribitol; 7,8-dimethyl-10-(D-*ribo*-2,3,4,5-tetrahydroxypentyl)isoalloxazine; 7,8-dimethyl-10-ribitylisoalloxazine; C.I. Food Yellow 15; C.I. 50900; Beflavin; Flavaxin. $C_{17}H_{20}N_4O_6$; mol wt 376.37. C 54.25%, H 5.36%, N 14.89%, O 25.51%. Nutritional factor found in milk, eggs, malted barley, liver, kidney, heart, leafy vegetables. Biologically active forms occurring in tissues and cells are flavin mononucleotide and flavin-adenine dinucleotide, *q.q.v.* Isoln from milk whey: A. W. Blyth, *J. Chem. Soc., Trans.* **35**, 530 (1879). Syntheses: Karrer *et al.*, *Helv. Chim. Acta* **18**, 426, 522 (1935); Kuhn *et al.*, *Naturwissenschaften* **23**, 260 (1935); Howe, **US 2807611** (1957 to Merck & Co.); F. Yoneda *et al.*, *J. Chem. Soc. Perkin Trans. 1* **1978**, 348. May also be produced by fermentation: Malzahn *et al.*, **US 2876169** (1959 to Grain Processing Corp.). Toxicity studies: Kuhn, Boulanger, *Z. Physiol.*

Chem. **241**, 233 (1936); K. Unna, J. G. Greslin, *J. Pharmacol. Exp. Ther.* **76**, 75 (1942). Comprehensive description: F. J. Al-Shammary *et al.*, *Anal. Profiles Drug Subs.* **19**, 429-476 (1990). Review of extraction and assay methods: Pearson, *The Vitamins* vol. VII, P. György, W. N. Pearson, Eds. (Academic Press, New York, 2nd ed., 1967) pp 99-136. Review of metabolism: Rivlin, *N. Engl. J. Med.* **283**, 463 (1970); of biosynthesis: M. Fischer, A. Bacher, *Arch. Biochem. Biophys.* **474**, 252-265 (2008). Review of discovery and bioactivity: V. Massey, *Biochem. Soc. Trans.* **28**, 283-296 (2000); of role in human nutrition: H. J. Powers, *Am. J. Clin. Nutr.* **77**, 1352-1360 (2003).

Fine orange-yellow needles from 2*N* acetic acid, alcohol, water, or pyridine. Dec at 278-282° (darkens at about 240°). Three different crystal forms having different solubilities in water: Dale, **US 2603633**; **US 2797215**. $[\alpha]_D^{25}$ −112 to −122° (50 mg in 2 ml 0.1*N* alcoholic NaOH dil to 10 ml with water). Absorption max: 220-225, 266, 371, 444, 475 nm. One gram dissolves in from 3000 to about 15,000 ml of water, the variations in the soly being due to differences in the crystal structure. Soly in alcohol (27.5°): 0.0045 g/100 ml of abs ethanol; sol in dilute solns of alkalies (with decompn); slightly sol in cyclohexanol, amyl acetate, benzyl alcohol, phenol; very slightly sol in isotonic sodium chloride solution. Insol in ether, chloroform, acetone, benzene. pKa 10.2; pKb 1.7. pH of satd aq soln: ∼6. Isoelectric pt: pH 6. Aq solns are yellow showing a green fluorescence with max at 565 nm. Sensitive to alkalies; stable to mineral acids in the dark. Visible or ultraviolet irradiation of alkaline solns causes the formation of lumiflavine, *q.v.*, whereas irradiation of acid or neutral solns gives rise to the production of lumichrome, *q.v.*, together with varying amounts of lumiflavine. LD$_{50}$ in rats (g/kg): >10 orally, 5.0 s.c., 0.56 i.p. (Unna, Greslin).

2′,3′,4′,5′-Tetrabutyrate. [752-56-7] Bituvitan; Eyekas; Hibon; Lacflavin; Ribolact; Wakaflavin L; Viras. $C_{33}H_{44}N_4O_{10}$; mol wt 656.73.

USE: Food coloring.

THERAP CAT: Vitamin (enzyme cofactor).

THERAP CAT (VET): Vitamin (enzyme cofactor).

8326. **Ribonuclease.** [9001-99-4] RNase. An enzyme that digests RNA. First isolated from beef pancreas in 1920. Crystallization from acid extracts of pancreas: Kunitz, *J. Gen. Physiol.* **24**, 15 (1940). Isoln: Kunitz, McDonald, *Biochem. Prep.* **3**, 9 (1953). Bovine pancreatic RNase A is a single-chain peptide of 124 amino acid residues. Primary structure: C. H. W. Hirs *et al.*, *J. Biol. Chem.* **235**, 633 (1960). Amino acid sequence: D. H. Spackman *et al.*, *ibid.* 648. Pictorial representation of entire structure: Stein, Moore, *Sci. Am.* **204**, no. 2, 81-92 (Feb. 1961). Ribonuclease from plant leaves has slightly different characteristics: Markham, Strominger, *Biochem. J.* **64**, 46P (1956). Can be obtained as a by-product of microbial erythromycin production: **JP 63 26938** (1963 to Shionogi). Chemical synthesis of materials possessing partial RNase enzyme activity: Gutte, Merrifield, *J. Am. Chem. Soc.* **91**, 501 (1969); Denkewalter, Hirschmann *et al.*, *ibid.* 502. Series of articles describing the total synthesis of a protein having the full enzymic activity of bovine pancreatic RNase: N. Fujii, H. Yajima, *J. Chem. Soc. Perkin Trans. 1* **1981**, 789-841. Specifically catalyzes the cleavage of the phosphodiester bond between the 3′ and 5′ positions of the ribose moieties in RNA with the formation of oligonucleotides terminating in 2′,3′-cyclic phosphate derivs: Roberts *et al.*, *Proc. Natl. Acad. Sci. USA* **62**, 1151 (1969). *Reviews:* Anfinsen, White, *The Enzymes* vol. 5, P. D. Boyer *et al.*, Eds. (Academic Press, New York, 2nd ed., 1961) pp 95-122; Richards, Wyckoff, *ibid.* **vol. 4** (3rd ed., 1971) pp 647-806.

Consult the Name Index before using this section.

Crystals. $[\alpha]_D^{25}$ (per mg N) $-0.47°$ (c = 5). uv max (0.1M KCl soln): 277.5 nm (ε 9700). Isoelec pt about pH 8.0. Freely sol in water. Aq solns of cryst ribonuclease are quite stable at temps below 25°. The region of maximum stability is between pH 2 and 4.5. Stable for years stored as refrigerated dry powder or in frozen soln. Aggregates on lyophilization and storage. Shows affinity for glass surfaces. The optimum temp for digestion of yeast ribonucleic acid is 65°. The optimum pH is 7.7. Inhibited by heavy metal ions; by magnesium ions at concns as low as 0.0005M. Competitively inhibited by DNA, the effect of denatured DNA being greater than that of the native nucleic acid. Ribonuclease is precipitated by trichloroacetic acid, and does not diffuse through collodion or cellophane.

8327. Ribonucleic Acid. RNA; yeast nucleic acid. Polynucleotide directly involved in protein synthesis; found in both the nucleus and the cytoplasm of cells. The four primary nucleosides are adenosine, guanosine, cytidine and uridine, $q.q.v.$; minor nucleosides are also found. The nucleosides are linked by phosphate diester bonds from the 3'-hydroxyl of one D-ribose to the 5'-hydroxyl of the next. The secondary structure of RNA is that of an incompletely organized single-stranded polynucleotide consisting of some areas with helical structure alternating with nonhelical lengths. *Compare* Deoxyribonucleic Acid (DNA). Structure: Brown, Todd in *The Nucleic Acids* **vol. 1**, E. Chargaff, J. N. Davidson, Eds. (Academic Press, New York, 1955) pp 409-439; Spirin, *Prog. Nucleic Acid Res.* **1**, 301-345 (1963); J. N. Davidson, *The Biochemistry of Nucleic Acids* (Academic Press, New York, 7th ed., 1972) pp 106-128. Review of NMR studies: B. R. Reid, *Annu. Rev. Biochem.* **50**, 969-996 (1981). Book: *The Ribonucleic Acids*, P. R. Stewart, D. S. Letham, Eds. (Springer-Verlag, New York, 2nd ed., 1977).

Ribosomal RNA. rRNA. Metabolically stable form; about 80% of the RNA found in cells; an important component of ribosomes, which play a central role in protein synthesis. Two rRNA species of high molecular weight have been isolated from ribosomes: mol wts approx 0.6×10^6 and 1.1×10^6 in bacterial cells; higher in eukaryotic cells. In addition, at least one more rRNA species of low mol wt, 5S RNA, has been identified; approx 120 nucleotides. *Reviews:* Attardi, Amaldi, *Annu. Rev. Biochem.* **39**, 183-226; Möller, Garrett, "The Ribonucleic Acids of the Bacterial Ribosome" in *Protein Synthesis* **vol. 1**, E. H. McConkey, Ed. (Dekker, New York, 1971) pp 229-272; Craig, "Ribosomal RNA Synthesis in Eukaryotes and its Regulation" in *MTP Int. Rev. Sci.: Biochem., Ser. One* **vol. 6**, K. Burton, Ed. (University Park Press, Baltimore, 1974) pp 255-288.

Messenger RNA. mRNA; informational RNA. A short-lived form of high molecular weight; acts as template for protein synthesis in the cell; complementary to one strand of DNA. A linear relationship exists between the sequence of amino acids in a polypeptide and the sequence of nucleotides in the corresponding mRNA and DNA. The four purine and pyrimidine bases are treated as letters which can be combined to form 3-letter words or *codons*; 4^3 or 64 codons can be formed. 61 have been found to code for specific amino acids; the remaining three are chain termination codons. Reviews of mRNA and the genetic code: Lipmann, "Messenger Ribonucleic Acid" in *Prog. Nucleic Acid Res.* **1**, 135-161 (1963); Crick, "The Recent Excitement in the Coding Problem" *ibid.* pp 163-217; Woese, "The Present Status of the Genetic Code" in *Prog. Nucleic Acid Res. Mol. Biol.* **7**, 107-172 (1967); Hadjiolov, "Ribonucleic Acids and Information Transfer in Animal Cells" *ibid.* 195-242; Jukes, Gatlin, "Recent Studies Concerning the Coding Mechanism" *ibid.* **11**, 303-350 (1971); J. N. Davidson, *The Biochemistry of the Nucleic Acids* (Academic Press, New York, 7th ed., 1972) pp 290-383; Mathews, "Mammalian Messenger RNA" in *Essays in Biochemistry* **vol. 9**, P. N. Campbell, F. Dickens, Eds. (Academic Press, New York, 1973) pp 59-102; Brawerman, *Annu. Rev. Biochem.* **43**, 621-642 (1974); Watts, Watts, "The Genetic Code" in *MTP Int. Rev. Sci.: Biochem., Ser. One* **vol. 7**, H. R. V. Arnstein, Ed. (University Park Press, Baltimore, 1975) pp 255-294.

Transfer RNA. tRNA; soluble RNA; sRNA. Low mol wt: 23,000-27,000; approx 75-85 nucleotides. Each tRNA is specific for and binds with a particular amino acid; more than one may exist for each amino acid. Performs three functions during protein synthesis: binds with its specific amino acid; recognizes the corresponding codon on mRNA and places the amino acid in the correct position for attachment to the polypeptide chain being formed; binds the polypeptide to the ribosome. First determination of total structure of a transfer RNA (yeast alanine tRNA): Holley *et al.*, *Science* **147**, 1462 (1965). Reviews of structure and function: Miura, "Specificity in the Structure of Transfer RNA" in *Prog. Nucleic Acid Res. Mol. Biol.* **6**, 39-82 (1967); Cramer, "Three-Dimensional Structure of tRNA", *ibid.* **11**, 391-421 (1971); *Nucleic Acid Sequence Analysis*, S. Mandeles (Columbia University Press, New York, 1972) pp 256-280; Nishimura, "Transfer RNA: Structure and Biosynthesis" in *MTP Int. Rev. Sci.: Biochem., Ser. One* **vol. 6**, K. Burton, Ed. (University Park Press, Baltimore, 1974) pp 289-322; A. Rich, V. L. Raj Bhandary, *Annu. Rev. Biochem.* **45**, 805-860 (1976); P. F. Agris, *The Modified Nucleosides of Transfer RNA* **II** (A. R. Liss, New York, 1983) 220 pp.

8328. D-Ribose. [50-69-1] $C_5H_{10}O_5$; mol wt 150.13. C 40.00%, H 6.71%, O 53.28%. Prepd by hydrolysis of yeast-nucleic acid: Levene, Jacobs, *Ber.* **42**, 1201, 3247 (1909); Levene, Clark, *J. Biol. Chem.* **46**, 19 (1921); Bredereck, *Ber.* **71**, 408 (1938); Bredereck *et al.*, *ibid.* **73**, 956 (1940); Phelps, **US 2152662** (1939 to U.S. Gov't); by ion-exchange resin chromatography: Cohn, *Science* **109**, 377 (1949); *J. Am. Chem. Soc.* **71**, 2275 (1949); **72**, 1471 (1950). From glucose: Karrer, *Helv. Chim. Acta* **18**, 1435 (1935); Austin, Humoller, *J. Am. Chem. Soc.* **56**, 1152 (1934); Kuhn *et al.*, *Ber.* **68**, 1765 (1935); from nucleosides: Laufer, Charney, **US 2379913**; **US 2379914** (both 1945 to Schwarz Labs.); from D-erythrose: Sowden, *J. Am. Chem. Soc.* **72**, 808 (1950); from L-glutamic acid: Koga *et al.*, *Tetrahedron Lett.* **1971**, 263. Reduction of D-ribonic acid: van Ekenstein, Blanksma, *Chem. Zentralbl.* **1913**, II, 1562; Steiger, *Helv. Chim. Acta* **19**, 189 (1936). *Review:* Overend, Stacey, in *Nucleic Acids* **vol. I**, E. Chargaff, J. N. Davidson, Eds. (Academic Press, New York, 1955) pp 1-80.

Plates from abs alcohol, mp 87°. Shows complex mutarotation: Phelps *et al.*, *J. Am. Chem. Soc.* **56**, 748 (1934). Final $[\alpha]_D^{24} -25°$ (water). Sol in water, slightly sol in alc.

Phenylosazone. $C_{17}H_{20}N_4O_3$. Yellow needles from pyridine + water, mp 163-164°.

Methyl-D-riboside. Crystals from ethyl acetate, mp 83-84°. $[\alpha]_D^{20} -113.6°$ (p = 3): Minsaas, *Ann.* **512**, 286 (1934).

8329. D-Ribose-5-phosphoric Acid. [4300-28-1] 5-(Dihydrogen phosphate)-D-ribose; D-ribofuranose-5-phosphoric acid; D-ribose-5-phosphate. $C_5H_{11}O_8P$; mol wt 230.11. C 26.10%, H 4.82%, O 55.62%, P 13.46%. From inosinic acid: Levene, Jacobs, *Ber.* **41**, 2703 (1908); **44**, 746 (1911); Levene, Mori, *J. Biol. Chem.* **81**, 215 (1929); Levene, Stiller, *ibid.* **104**, 299 (1934); LePage, Umbreit, *ibid.* **148**, 255 (1943); Marmur *et al.*, *Arch. Biochem. Biophys.* **34**, 209 (1951). Purification: Groth *et al.*, *J. Biol. Chem.* **199**, 389 (1952). Differentiation between ribose-5-phosphate and ribose-3-phosphate by means of the orcinol-pentose reaction: Albaum, Umbreit, *ibid.* **167**, 369 (1947).

Barium salt hemihendecahydrate. $C_5H_9BaO_8P \cdot 5\frac{1}{2}H_2O$. Hexagonal plates from water. Sparingly soluble in cold water. Reduces Fehling's soln.

8330. Ribostamycin. [25546-65-0] O-2,6-Diamino-2,6-dideoxy-α-D-glucopyranosyl-(1 → 4)-O-[β-D-ribofuranosyl-(1 → 5)]-2-deoxy-D-streptamine; SF-733 antibiotic. $C_{17}H_{34}N_4O_{10}$; mol wt 454.48. C 44.93%, H 7.54%, N 12.33%, O 35.20%. Aminocyclitol antibiotic belonging to the neomycin, $q.v.$, group of antibiotics. Produced by *Streptomyces ribosidificus* (formerly called *S. thermoflavus*). Taxonomy, isoln and toxicity data: T. Shomura *et al.*, *J. Antibiot.* **23**, 155 (1970). Prepn: *eidem*, **DE 1814735** corresp to **US 3661892**, **US 3799842** (1969, 1972, 1974 all to Meiji). Structure:

E. Akita *et al.*, *J. Antibiot.* **23**, 173 (1970). Synthesis from neamine, *q.v.*: Ito *et al.*, *Antimicrob. Agents Chemother.* **1970**, 33; *eidem*, *Agric. Biol. Chem.* **34**, 980 (1970); V. Kumar, W. A. Remers, *J. Org. Chem.* **43**, 3327 (1978); **46**, 4298 (1981). Total synthesis: H. Fukami *et al.*, *Tetrahedron Lett.* **1976**, 545; *eidem*, *Agric. Biol. Chem.* **41**, 1689 (1977). *In vitro* antibacterial activity: E. Yourassowsky, M. P. Vander Linden, *Arzneim.-Forsch.* **26**, 184 (1976).

Colorless needles from methanol, mp 192-195°. Also reported as monohydrate from water-methanol-ethanol, mp 175-180° (dec). $[\alpha]_D^{23}$ +42°. pKa' 7.70. Stable in neutral and alkaline solns; slightly unstable in acidic media. Sol in water; slightly sol in methanol. Practically insol in acetone, *n*-butanol, ethyl acetate, benzene, hexane, ether.

Sulfate. [53797-35-6] Ibistacin; Landamycine; Ribostamin; Ribomycine; Vistamycin. White or yellowish powder. $[\alpha]_D^{20}$ +39° (c = 1). LD$_{50}$ i.v. in mice: 225 mg/kg (Shomura).

THERAP CAT: Antibacterial.

8331. D-Ribulose. [488-84-6] D-*erythro*-2-Pentulose; D-adonose; D-*erythro*-2-ketopentose. $C_5H_{10}O_5$; mol wt 150.13. C 40.00%, H 6.71%, O 53.28%. Prepn from D-arabinose: Glatthaar, Reichstein, *Helv. Chim. Acta* **18**, 80 (1935); from 2-keto-D-gluconic acid: Hall *et al.*, *Biochem. J.* **60**, 271 (1955); from D-fructose: P. M. Collins *et al.*, *J. Chem. Soc. Perkin Trans. 1* **1980**, 277. Stereoselective synthesis: M. Yamaguchi, T. Mukaiyama, *Chem. Lett.* **1981**, 1005; K. Suzuki *et al.*, *ibid.* 1529.

Sweet syrup. $[\alpha]_D^{24}$ −15° (c = 0.5 in water).
Diphosphate. $C_5H_{12}O_{11}P_2$; mol wt 310.09. Prepn from D-ribose-5-phosphate: Horecker *et al.*, *Biochem. Prep.* **6**, 83 (1958).

8332. Rice Bran Oil. Oil of rice bran; rice oil. Extracted from bran obtained in making white or polished rice (which is actually the endosperm of the seed of *Oryza sativa* L., *Gramineae*). Rice bran separated by milling of hulled rice consists of bran, aleurone layer and germ, and contains about 15% oil. The oil is usually obtained by solvent extraction: J. R. Loeb, N. J. Morris, *Abstract Bibliography of the Chemistry, Processing, and Utilization of Rice Bran and Rice Bran Oil* (U.S. Dept. of Agriculture) Ms. no **AIC-328** (1952 and suppl.). Rice bran contains an unusually active lipase which raises the free fatty acid content on storage. A variety of methods to inhibit this rise can be found in the literature: Williams *et al.*, *J. Am. Oil Chem. Soc.* **42**, 151 (1965). The free fatty acid content of the oil should not exceed 4 to 5%, but may run as high as 70%. Acceptable rice bran oil contains 15-20% of satd and 80-85% unsatd fatty acids as glycerides: Myristic 0.4-1%, palmitic 12-18%, stearic 1-3%, C_{20}-C_{22} satd 1%, oleic 40-50%, linoleic 29-42%, linolenic trace to 1%, palmitoleic 0.2-0.4%. *See also:* Sreenivasan, *J. Am. Oil Chem. Soc.* **45**, 263 (1968).

Golden yellow oil, difficult to bleach. Not affected by temporary heating to 160°. d_{25}^{25} 0.916-0.921. n_D^{25} 1.470-1.473; n_D^{40} 1.465-1.468. Titer (solidification pt) 24-28. Saponification value 181-189. Acid value 4-120 (see above comments on the lipase content of the bran). Iodine value 99-108. Thiocyanogen value 69-76. Hydroxyl value 5-14. Reichert-Meissl value <0.5. Polenské value <0.5. Hehner value 95.3. Unsaponifiable matter 3-5%. The high refractive index is ascribed to the presence of squalene. Also contains other antioxidants such as tocopherols. Miscible with hexane and other fat solvents.

USE: Soaps, hydrogenated shortenings. Edible oil, also suitable for cosmetics and pharmaceuticals when low in free fatty acids and all the lipase has been deactivated.

8333. Ricin. [9009-86-3] A toxic lectin and hemagglutinin isolated from castor bean, *Ricinus communis* L., *Euphorbiaceae*: Stillmark, *Arb. Pharmak. Inst. Dorpat* **1889**, 59; T. Osborne *et al.*, *Am. J. Physiol.* **14**, 259 (1905); H. L. Craig *et al.*, US 3060165 (1962 to U.S. Dept. Army). Purification of toxic ricin protein: M. Ishiguro *et al.*, *J. Biochem.* **55**, 587 (1964). The hemagglutinating activity of ricin was initially believed to be the cause of its high toxicity, but later studies have shown that separate proteins are responsible for the toxicity and hemagglutination: S. Olsnes, A. Pihl, *Biochemistry* **12**, 3121 (1973). "Ricin", the original term for the mixed extract, is now used in various ways. Two ricin agglutinins and two toxins have been identified: *RCL I*, *RCL II* (agglutinins), *RCL III* or *Ricin D* and *RCL IV* (toxins). *See:* T. T. Lin, S. S. Li, *Eur. J. Biochem.* **105**, 453 (1980). All four lectins consist of two different polypeptide chains joined by a disulfide bond; the toxins are dimers of an A-chain (30,000 Da) and a B-chain (33,000 Da) and the agglutinins occur as a tetramer composed of two 30,000 and two 33,000 mol wt subunits. Ricin has been shown to have antitumor properties: J. Y. Lin *et al.*, *Nature* **227**, 92 (1970); P. E. Thorp *et al.*, *ibid.* **297**, 594 (1982). Synergistic effect with daunorubicin, cisplatin, and vincristine, *q.q.v.*, in systemic L 1210 leukemia: O. Fodstad, A. Pihl, *Cancer Res.* **42**, 2152 (1982). Conjugates of ricin and cell-binding antigens or antibodies, called immunotoxins, have been used in cancer therapy: E. S. Vitetta *et al.*, *Science* **219**, 644 (1983). Experimental study with ricin-A chain conjugate in AIDS: M. A. Till *et al.*, *Science* **242**, 1166 (1988). RIA determn in blood: A. Godal *et al.*, *J. Toxicol. Environ. Health* **8**, 409 (1981). *Reviews:* M. Funatsu, "The Structure and Toxic Function of Ricin" in *Proteins, Structure and Function* vol. 2, M. Funatsu *et al.*, Eds. (Kodansha, Tokyo, Wiley, New York, 1972) pp 103-139; G. A. Balint, *Toxicology* **2**, 77-122 (1974); C. Winder, *J. Toxicol. Toxin Rev.* **23**, 97-103 (2004). *See also* Abrin, Lectins.

Isoelectric pt 7.1. uv max: 280 nm (ε 85000). MLD i.p. in mice at 48 hrs: 0.001 μg ricin nitrogen/g body wt (Ishiguro).

Caution: Ricin is among the most toxic compounds known. Seeds of *R. communis*, if thoroughly masticated, can produce serious poisoning and death: J. M. Kingsbury, *Poisonous Plants of the United States and Canada* (Prentice-Hall, New Jersey, 1964) pp 194-197. Potential symptoms of overexposure by inhalation are weakness, fever, cough, pulmonary edema, respiratory distress, hypoxia; by ingestion are abdominal pain, GI hemorrhage, liver, spleen and kidney damage (Widner).

USE: As a tool in studies of cell-surface properties; exptly in cancer research.

8334. Ricinine. [524-40-3] 1,2-Dihydro-4-methoxy-1-methyl-2-oxo-3-pyridinecarbonitrile; 1,2-dihydro-4-methoxy-1-methyl-2-oxonicotinonitrile; ricidine. $C_8H_8N_2O_2$; mol wt 164.16. C 58.53%, H 4.91%, N 17.07%, O 19.49%. From seeds and leaves of the castor plant, *Ricinus communis* L., *Euphorbiaceae*. Extraction procedure: Böttcher, *Ber.* **51**, 673 (1918). Synthesis starting with polymerization of cyanoacetyl chloride: Schroeter *et al.*, *Ber.* **65**, 432 (1932). Biosynthesis: Waller, Henderson, *J. Biol. Chem.* **236**, 1186 (1961); Essery *et al.*, *Can. J. Chem.* **41**, 1142 (1963).

Prisms, needles from alc, mp 201.5°. Sublimes at 170-180° under 20 mm pressure. Sparingly sol in water, alc, chloroform, ether. Neutral to litmus. Does not form salts with acids.

Caution: Ingestion may cause nausea, vomiting, hemorrhagic gastroenteritis, hepatic and renal damage, convulsions, coma, hypotension, respiratory depression, death.

8335. Ricinoleic Acid. [141-22-0] (9Z,12R)-12-Hydroxy-9-octadecenoic acid; *d*-12-hydroxyoleic acid. $C_{18}H_{34}O_3$; mol wt

298.47. C 72.44%, H 11.48%, O 16.08%. Found primarily in oils from the seeds of *Ricinus* spp, *Euphorbiaceae*. Accounts for about 90% of the triglyceride fatty acids of castor oil, and up to about 40% of the glyceride fatty acids of ergot oil. Bibliography on its isoln: Ralston, *Fatty Acids* (New York, 1948) p 189. Also isolated from *Linum mucronatum* (flax), *Linaceae:* Kleiman, Spencer, *Lipids* **6**, 962 (1971). Structure: Goldsobel, *Ber.* **27**, 3121 (1894). Mechanism of biosynthesis: Morris, *Biochem. Biophys. Res. Commun.* **29**, 311 (1967).

Liquid. $d_4^{27.4}$ 0.940; mp 5.5°; bp_{10} 245°. $[\alpha]_D^{22}$ +6.67°; $[\alpha]_D^{26}$ +7.15° (c = 5 in acetone). n_D^{20} 1.4716. Neutralization value 187.98; iodine value 85.05. Sol in alcohol, acetone, ether, chloroform (*cf.* the solubilities of castor oil).

Acid sulfate. [36634-48-7] Ricinolsulfuric acid. $C_{18}H_{34}O_6S$; mol wt 378.52. Obtained by the action of chlorosulfuric acid. Viscous brown liquid with weak blue fluorescence. Sol in water (about 10%), alcohol, ether, chloroform.

Sodium salt. [5323-95-5] Sodium ricinoleate; Soricin; Colidosan. $C_{18}H_{33}NaO_3$; mol wt 320.45. Sodium salts of the fatty acids from castor oil. White or slightly yellow, odorless or almost odorless powder. Sol in water or alcohol. The aq soln is alkaline.

USE: In textile finishing; sometimes added to Turkey red oil, drycleaning soaps.

THERAP CAT: Has been used in contraceptive jellies. The sodium salt has been used as sclerosing agent.

8336. Ridaforolimus. [572924-54-0] Rapamycin 42-(dimethylphosphinate); (1*R*,9*S*,12*S*,15*R*,16*E*,18*R*,19*R*,21*R*,23*S*,24*E*,-26*E*,28*E*,30*S*,32*S*,35*R*)-12-[(1*R*)-2-[(1*S*,3*R*,4*R*)-4-[(dimethylphosphinoyl)oxy]-3-methoxycyclohexyl]-1-methylethyl]-1,18-dihydroxy-19,30-dimethoxy-15,17,21,23,29,35-hexamethyl-11,36-dioxa-4-azatricyclo[30.3.1.0⁴,⁹]hexatriaconta-16,24,26,28-tetraene-2,-3,10,14,20-pentone; deforolimus; AP-23573; MK-8669. $C_{53}H_{84}$-$NO_{14}P$; mol wt 990.22. C 64.29%, H 8.55%, N 1.41%, O 22.62%, P 3.13%. Dimethylphosphinic acid ester of rapamycin, *q.v.* Inhibits mammalian target of rapamycin (mTOR), a serine-threonine kinase involved in cell growth and regulation. Prepn: D. L. Berstein *et al.*, **WO 03064383**; C. A. Metcalf, III *et al.*, **US 7091213** (2003, 2006 both to Ariad). Clinical pharmacokinetics and safety: C. M. Hartford *et al.*, *Clin. Cancer Res.* **15**, 1428 (2009). Pharmacology and mechanism of action: V. M. Rivera *et al.*, *Mol. Cancer Ther.* **10**, 1059 (2011). Antitumor activity in human sarcoma and endometrial cancer cell lines: R. M. Squillace *et al.*, *ibid.*, 1959. Review of clinical experience: M. Mita *et al.*, *Expert Opin. Invest. Drugs* **17**, 1947-1954 (2008).

White solid. Sol in ethanol.
THERAP CAT: Antineoplastic.

8337. Ridogrel. [110140-89-1] 5-[[(*E*)-[3-Pyridinyl[3-(trifluoromethyl)phenyl]methylene]amino]oxy]pentanoic acid; R-68070. $C_{18}H_{17}F_3N_2O_3$; mol wt 366.34. C 59.02%, H 4.68%, F 15.56%, N 7.65%, O 13.10%. Combined thromboxane A_2 synthase inhibitor and thromboxane A_2/prostaglandin endoperoxide receptor antagonist. Prepn: E. J. E. Freyne *et al.*, **EP 221601**; *eidem*, **US 4963573** (1987, 1990 both to Janssen). Clinical pharmacology: B. Hoet *et al.*, *Thromb. Haemostasis* **64**, 87 (1990); C. Weber *et al.*, *ibid.* **68**, 214 (1992). Clinical study in peripheral arterial obstructive disease: J. De Cree *et al.*, *Int. Angiol.* **12**, 59 (1993); as adjunct to thrombolysis in acute myocardial infarction: RAPT Investigators, *Circulation* **89**, 588 (1994).

Crystals from diisopropyl ether/hexane (2:1), mp 70.3°.
THERAP CAT: Antithrombotic.

8338. Rifabutin. [72559-06-9] (9*S*,12*E*,14*S*,15*R*,16*S*,17*R*,-18*R*,19*R*,20*S*,21*S*,22*E*,24*Z*)-16-(Acetyloxy)-6,18,20-trihydroxy-14-methoxy-7,9,15,17,19,21,25-heptamethyl-1′-(2-methylpropyl)-spiro[9,4-(epoxypentadeca[1,11,13]trienimino)-2*H*-furo[2′,3′:7,8]-naphth[1,2-*d*]imidazole-2,4′-piperidine]-5,10,26(3*H*,9*H*)-trione; 1′,4-didehydro-1-deoxy-1,4-dihydro-5′-(2-methylpropyl)-1-oxo-rifamycin XIV; (9*S*,12*E*,14*S*,15*R*,16*S*,17*R*,18*R*,19*R*,20*S*,21*S*,22*E*,-24*Z*)-6,16,18,20-tetrahydroxy-1′-isobutyl-14-methoxy-7,9,15,17,-19,21,25-heptamethylspiro[9,4-(epoxypentadeca[1,11,13]trienimino)-2*H*-furo[2′,3′:7,8]naphth[1,2-*d*]imidazole-2,4′-piperidine]-5,-10,26-(3*H*,9*H*)-trione-16-acetate; 4-deoxo-3,4-[2-spiro-(*N*-isobutyl-4-piperidyl)]-(1*H*)-imidazo-(2,5-dihydro)rifamycin S; 4-*N*-isobutyl-spiropiperidylrifamycin S; LM-427; Ansatipine; Mycobutin. C_{46}-$H_{62}N_4O_{11}$; mol wt 847.02. C 65.23%, H 7.38%, N 6.61%, O 20.78%. Semisynthetic derivative of rifamycin S that inhibits nucleic acid synthesis. Prepn: L. Marsili *et al.*, **DE 2825445** (1979 to Farmitalia); *eidem*, **US 4219478** (1980 to Archifar Labs). *In vitro* and *in vivo* antibacterial activity: A. Sanfilippo *et al.*, *J. Antibiot.* **33**, 1193 (1980); C. Della Bruna *et al.*, *ibid.* **36**, 1502 (1983). Mechanism of action: D. Ungheri *et al.*, *Drugs Exp. Clin. Res.* **10**, 681 (1984). Comparative *in vitro* antimycobacterial spectrum: J. M. Dickinson, D. A. Mitchison, *Tubercle* **68**, 177 (1987). *In vitro* inhibition of HIV-1 replication: R. Anand *et al.*, *Antimicrob. Agents Chemother.* **32**, 684 (1988). Clinical pharmacokinetics: M. H. Skinner *et al.*, *ibid.* **33**, 1237 (1989). Pharmacology and clinical efficacy in mycobacterial infections: R. J. O'Brien *et al.*, *Rev. Infect. Dis.* **9**, 519 (1987).

Violet-red crystalline powder. Sol in chloroform, methanol; sparingly sol in alc; very slightly sol in water. uv max (methanol): 493, 315, 274, 238 nm.
THERAP CAT: Antibacterial (tuberculostatic).

Consult the Name Index before using this section.

8339. Rifalazil. [129791-92-0] 1',4-Didehydro-1-deoxy-1,4-dihydro-3'-hydroxy-5'-[4-(2-methylpropyl)-1-piperazinyl]-1-oxorifamycin VIII; (2S,16Z,18E,20S,21S,22R,23R,24R,25S,26R,27S,28E)-5,12,21,23,25-pentahydroxy-10-(4-isobutyl-1-piperazinyl)-27-methoxy-2,4,16,20,22,24,26-heptamethyl-2,7-(epoxypentadeca[1,11,13]trienimino)-6H-benzofuro[4,5-a]phenoxazine-1(2H),6,15-trione 25-acetate; 3'-hydroxy-5'-(4-isobutyl-1-piperazinyl)benzoxazinorifamycin; KRM-1648. $C_{51}H_{64}N_4O_{13}$; mol wt 941.09. C 65.09%, H 6.86%, N 5.95%, O 22.10%. Semisynthetic derivative of rifamycin S, q.v. Prepn: T. Yamane et al., **EP 366914**; eidem, **US 4983602** (1990, 1991 both to Kanegafuchi); eidem, Chem. Pharm. Bull. **41**, 148 (1993). Antimycobacterial efficacy in comparison with rifampin, q.v.: T. Yamamoto et al., Antimicrob. Agents Chemother. **40**, 426 (1996). Pharmacokinetics: K. Hosoe et al., ibid. 2749. HPLC determn in biological fluids: eidem, J. Chromatogr. B **653**, 177 (1994). Review of pharmacology and clinical studies: D. M. Rothstein et al., Expert Opin. Invest. Drugs **12**, 255-271 (2003).

Crystals from chloroform-hexane, mp 195-200° (dec).
THERAP CAT: Antibacterial (tuberculostatic).

8340. Rifamide. [2750-76-7] 4-O-[2-(Diethylamino)-2-oxoethyl]-rifamycin; 2-[(1,2-dihydro-5,6,17,19,21-pentahydroxy-23-methoxy-2,4,12,16,18,20,22-heptamethyl-1,11-dioxo-2,7-(epoxypentadeca[1,11,13]trienimino)naphtho[2,1-b]furan-9-yl)oxy]-N,N-diethylacetamide 21-acetate; N,N-diethylrifomycin B amide; rifamycin B diethylamide; M-14; Rifocin M. $C_{43}H_{58}N_2O_{13}$; mol wt 810.94. C 63.69%, H 7.21%, N 3.45%, O 25.65%. Prepn: **BE 632770**; Sensi, Maggi, **US 3313804** (1963, 1967 both to Lepetit); Sensi et al., J. Med. Chem. **7**, 596 (1964). Physical and chemical properties: Maggi et al., Farmaco Ed. Prat. **20**, 147 (1965). Activity data: Pallanza et al., Arzneim.-Forsch. **15**, 800 (1965); Monnier, Bourse, Pathol. Biol. **16**, 901 (1968). Metabolic studies: Fürész et al., Arzneim.-Forsch. **15**, 802 (1965); Maffii, Shiatti, Toxicol. Appl. Pharmacol. **8**, 138 (1966). Toxicology: Dezulian et al., ibid. 126.

Yellow-orange ppt, crystallized from benzene + hexane. No definite mp, begins to soften at 140°, melts completely at 170° (dec). $[\alpha]_D^{20}$ −48.7° (c = 0.4 in methanol). Absorption max in phosphate

buffer (pH 7.38): 222, 302, 421 nm (ε 42820, 20770, 16200). LD_{50} in mice, rats (mg/kg): 2450, >4000 orally; 640, 2500 s.c.; 320, 535 i.p.; 315, 380 i.v. (Dezulian).
THERAP CAT: Antibacterial.

8341. Rifampin. [13292-46-1] 3-[[(4-Methyl-1-piperazinyl)imino]methyl]rifamycin; 5,6,9,17,19,21-hexahydroxy-23-methoxy-2,4,12,16,18,20,22-heptamethyl-8-[N-(4-methyl-1-piperazinyl)formimidoyl]-2,7-(epoxypentadeca[1,11,13]trienimino)naphtho-[2,1-b]furan-1,11(2H)-dione 21-acetate; rifampicin; rifaldazine; rifamycin AMP; R/AMP; Abrifam; Eremfat; Rifa; Rifadin(e); Rifaldin; Rifapiam; Rifaprodin; Rifoldin; Rimactan(e). $C_{43}H_{58}N_4O_{12}$; mol wt 822.95. C 62.76%, H 7.10%, N 6.81%, O 23.33%. Semisynthetic antibiotic obtained by reacting 3-formylrifamycin SV with 1-amino-4-methylpiperazine in tetrahydrofuran. Prepn and structure: Maggi et al., Chemotherapia **11**, 285 (1966); **NL 6509961**; Maggi, Sensi, **US 3342810** (1966, 1967 both to Lepetit). Chemical and biological properties: Fürész, Antibiot. Chemother. **16**, 316 (1970). Activity studies and clinical survey: Arioli et al., Arzneim.-Forsch. **17**, 523 (1967); Pallanza et al., ibid. 529; Bergamini, ibid. **20**, 1546 (1970); Dans et al., Am. J. Med. Sci. **259**, 120 (1970). Metabolism: Meyer-Brunot et al., Int. Congr. Chemother. Proc., 5th, Vienna 1967 **1**(2), 763; Fürész et al., Arzneim.-Forsch. **17**, 534 (1967); Maggi et al., ibid. **19**, 651 (1969). Inhibition of protein synthesis in mammalian cells: W. C. Buss et al., Science **200**, 432 (1978). Comprehensive reviews: Binda et al., Arzneim.-Forsch. **21**, 1907-1978 (1971); Lester, Annu. Rev. Microbiol. **26**, 88-102 (1972). Comprehensive description: G. G. Gallo, P. Radaelli, Anal. Profiles Drug Subs. **5**, 467-513 (1976). Symposium on the use of rifampin in the treatment of nontuberculous infections: Rev. Infect. Dis. **5**, Suppl. 3, S399-S632 (1983).

Red to orange platelets from acetone, dec 183-188°. Absorption max (pH 7.38): 237, 255, 334, 475 nm (ε 33200, 32100, 27000, 15400). Rifampin is a "zwitterion" with pKa 1.7 related to the 4-hydroxy and pKa 7.9 related to the 3-piperazine nitrogen. Very stable in DMSO; rather stable in water. Freely sol in chloromethane, DMSO, chloroform; sol in ethyl acetate, methanol, tetrahydrofuran; slightly sol in acetone, carbon tetrachloride; very slightly sol in water (pH <6). LD_{50} in mice, rats (mg/kg): 885, 1720 orally; 260, 330 i.v.; 640, 550 i.p. (Fürész).
THERAP CAT: Antibacterial (tuberculostatic).
THERAP CAT (VET): Antibacterial.

8342. Rifamycins. Rifomycins. Group of antibiotics characterized by a natural ansa structure (chromophoric naphthohydroquinone group spanned by a long aliphatic bridge) not previously found in other known antibiotics. Isoln from the fermentation broths of Streptomyces mediterranei: P. Sensi et al., Antibiot. Annu. **1959-1960**, 262. Among rifamycins, rifamycin B, O, S, SV and X are the more studied members. Prepn of rifamycin B derivs: P. Sensi et al., **US 3313804** (1967 to Lepetit). Structure: Prelog, Pure Appl. Chem. **7**, 551 (1963); Chemotherapia **7**, 133 (1963); Oppolzer et al., Experientia **20**, 336 (1964); Oppolzer, Prelog, Helv. Chim. Acta **56**, 2287 (1973). Total synthesis of rifamycin S: H. Nagaoka et al., J. Am. Chem. Soc. **102**, 7962 (1980); H. Iio et al., ibid. 7965; H. Nagaoka, Y. Kishi, Tetrahedron **37**, 3873 (1981); S. Hanessian et al., J. Am. Chem. Soc. **104**, 6164 (1982). Review of chemistry of ansamycin antibiotics: K. L. Rinehart et al., Fortschr. Chem. Org. Naturst. **33**, 231-307 (1976). General reviews: P. Sensi, Research Progress in Organic-Biological & Medicinal Chemistry vol. 1, U. Gallo, L. Santamaria, Eds. (Società Editoriale Farmaceutica, Milan, Italy, 1964) pp 337-421; Wehrli, Staehelin, in *Antibiotics* **vol. 3**, J. W. Corcoran, F. E. Hahn, Eds. (Springer-Verlag, New York, 1975) pp 252-268; C. Gurgo *et al.*, "Rifamycins" in *Handbook of Experimental Pharmacology* **vol. 61**, G. V. R. Born *et al.*, Eds., entitled "Chemotherapy of Viral Infections", P. E. Came, L. A. Caligiuri, Eds. (Springer-Verlag, New York, 1982) pp 519-555.

	R	R'
Rifamycin B	—OCH$_2$COOH	—OH
Rifamycin O	—(1,3-dioxolan-4-on)-2-yl	=O
Rifamycin S	=O	=O
Rifamycin X	=N$\overset{+}{=}$N$^-$	=O

Rifamycin AMP *see* Rifampin.

Rifamycin B. [13929-35-6] 4-*O*-(Carboxymethyl)rifamycin; nancimycin. C$_{39}$H$_{49}$NO$_{14}$; mol wt 755.81. Yellow prismatic needles from benzene, mp 300° (dec 160-164°). $[\alpha]_D^{20}$ −11° (methanol). Absorption max (phosphate buffer soln pH 7.3): 223, 304, 425 nm (E$_{1cm}^{1\%}$ 555, 275, 220). Dibasic acid. Very stable. Solubilities: water 0.027% (w/w), methanol 2.62%, ethanol 0.44%. LD$_{50}$ in mice (mg/kg): 2040 i.v., >3000 i.p., s.c., and orally (Sensi, 1964).

Rifamycin O. [14487-05-9] 4-*O*-(Carboxymethyl)-1-deoxy-1,4-dihydro-4-hydroxy-1-oxorifamycin γ-lactone. C$_{39}$H$_{47}$NO$_{14}$; mol wt 753.80. Prepn: P. Sensi *et al.*, *Farmaco Ed. Sci.* **15**, 228 (1960); Umezawa, *JP 66 15518* (1966), *C.A.* **66**, 1583v (1967). Pale yellow crystals from methanol, mp 300° (dec 160°). Also reported as mp 180-185° (Umezawa). $[\alpha]_D^{20}$ +71.5° (c = 1 in dioxane). uv max (methanol contg 5% acetate buffer soln pH 4.62): 226, 273, 370 nm (E$_{1cm}^{1\%}$ 365, 440, 60). Weak acid. Slowly sol in alkaline soln with red-violet color. Sol in acetone, tetrahydrofuran; slightly sol in methanol, ethanol, ethyl acetate. Practically insol in ether, petr ether, water, dilute acids.

Rifamycin S. [13553-79-2] 1,4-Dideoxy-1,4-dihydro-1,4-dioxorifamycin. C$_{37}$H$_{45}$NO$_{12}$; mol wt 695.76. Activation product found in solns of rifamycin B and rifamycin O: P. Sensi *et al.*, *Experientia* **16**, 412 (1960). Yellow-orange crystals from methanol, dec 179-181°. $[\alpha]_D^{20}$ +476° (c = 0.1 in methanol). Absorption max (phosphate buffer soln pH 7.3): 317, 525 nm (E$_{1cm}^{1\%}$ 426, 62). LD$_{50}$ in mice (mg/kg): 122 i.v.; 258 i.p.; 3000 orally (Sensi, 1964).

Rifamycin SV *see separate entry.*

Rifamycin X. C$_{37}$H$_{45}$N$_3$O$_{11}$; mol wt 707.78. Prepn: Greco *et al.*, *Farmaco Ed. Sci.* **16**, 766 (1961). Yellow crystals, no definite mp, dec 135-140°. Unstable to light. $[\alpha]_D^{20}$ +491.8° (c = 0.981 in dioxane). Absorption max (acetate buffer soln): 286, 317, 402 nm (E$_{1cm}^{1\%}$ 400, 362, 195). Practically insol in water. Sol in methanol, ethanol, ethyl acetate, benzene.

Rifamycin AG. A condensation product of rifamycin O and aminoguanidine: P. Sensi *et al.*, *Antibiot. Chemother.* **12**, 448 (1962).

8343. Rifamycin SV. [6998-60-3] 5,6,9,17,19,21-Hexahydroxy-23-methoxy-2,4,12,16,18,20,22-heptamethyl-2,7-(epoxypentadeca[1,11,13]trienimino)naphtho[2,1-*b*]furan-1,11(2*H*)-dione 21-acetate; rifomycin SV; rifamicine SV. C$_{37}$H$_{47}$NO$_{12}$; mol wt 697.78. C 63.69%, H 6.79%, N 2.01%, O 27.51%. Semi-synthetic antibiotic derived from Rifamycin S. Prepn: Sensi *et al.*, *Experientia* **16**, 412 (1960); *Farmaco Ed. Sci.* **16**, 165 (1961). Comprehensive review: Bergamini, Fowst, *Arzneim.-Forsch.* **15**, 951-1002 (1965). For general refs *see* Rifamycins.

Yellow-orange crystals, mp 300° (dec 140°). $[\alpha]_D^{20}$ −4° (methanol). uv max (phosphate buffer soln pH 7.3): 223, 314, 445 nm (E$_{1cm}^{1\%}$ 586, 322, 204). Acid reaction. Slightly sol in water, petr ether; sol in ether, bicarbonate soln; very sol in methanol, ethanol, acetone, ethyl acetate. A reducing substitute, such as ascorbic acid, should be added to aq solns of rifamycin SV to prevent its transformation to rifamycin S. LD$_{50}$ in mice (mg/kg): 550 i.v.; 625 i.p.; 2120 orally (Bergamini, Fowst).

Sodium salt. [14897-39-3] Rifamastene; Rifocin. Orange-red crystals. Soly in water pH 7.2: ~5 g/100 ml.

THERAP CAT: Antibacterial.

THERAP CAT (VET): Antibacterial.

8344. Rifapentine. [61379-65-5] 3-[[(4-Cyclopentyl-1-piperazinyl)imino]methyl]rifamycin; MDL-473; DL-473; R-773; Priftin. C$_{47}$H$_{64}$N$_4$O$_{12}$; mol wt 877.05. C 64.37%, H 7.36%, N 6.39%, O 21.89%. Semi-synthetic rifamycin. Prepn: R. Cricchio, V. Arioli, *DE 2608218*; *eidem*, *US 4002752* (1976, 1977 both to Lepetit). Antibacterial activity: V. Arioli *et al.*, *J. Antibiot.* **34**, 1026 (1981); H. C. Neu, *Antimicrob. Agents Chemother.* **24**, 457 (1983). Pharmacokinetics: A. Assandri *et al.*, *J. Antibiot.* **37**, 1066 (1984). HPLC determn in plasma: E. Riva *et al.*, *J. Chromatogr.* **553**, 35 (1991). Clinical trial in tuberculosis: Natl. Co-op Group Clin. Study Rifapentine, *Chin. J. Antibiot.* **17**, 313 (1992), *Biological Abstracts* **95**, 70840 (1993). Review of pharmacology and clinical efficacy: B. Jarvis, H. M. Lamb, *Drugs* **56**, 607-616 (1998).

Crystals from ethyl acetate, mp 179-180°. uv max: 475, 334 nm (ε 15200, 26700). LD$_{50}$ in mice (mg/kg): >2000 orally; 750 i.p. (Cricchio). Also reported as LD$_{50}$ in mice (mg/kg): 3300 orally; 710 i.p. (Aroli).

THERAP CAT: Antibacterial (tuberculostatic).

8345. Rifaximin. [80621-81-4] (2*S*,16*Z*,18*E*,20*S*,21*S*,22*R*,-23*R*,24*R*,25*S*,26*R*,27*S*,28*E*)-25-(Acetyloxy)-5,6,21,23-tetrahydroxy-27-methoxy-2,4,11,16,20,22,24,26-octamethyl-2,7-(epoxypentadeca[1,11,13]trienimino)benzofuro[4,5-*e*]pyrido[1,2-*a*]benzimidazole-1,15(2*H*)-dione; 4-deoxy-4'-methylpyrido[1',2'-1,2]-imidazo[5,4-*c*]rifamycin SV; rifamycin L 105; rifaxidin; L-105; Flonorm; Normix; Redactiv; Rifacol; Xifaxan. C$_{43}$H$_{51}$N$_3$O$_{11}$; mol wt 785.89. C 65.72%, H 6.54%, N 5.35%, O 22.39%. Nonabsorbable semisynthetic rifamycin antibiotic. Prepn: *BE 888895*; E. Marchi, L. Montecchi, *US 4341785* (1981, 1982 both to Alfa); E. Marchi *et* *al., J. Med. Chem.* **28**, 960 (1985); and NMR study: M. Brufani *et al., J. Antibiot.* **37**, 1611 (1984). X-ray crystal structure: *idem et al., ibid.* 1623. *In vitro* and *in vivo* antibacterial activity: A. P. Venturini, E. Marchi, *Chemioterapia* **5**, 257 (1986). Toxicological study: G. Borelli, D. Bertoli, *ibid.* 263. Clinical trial in travelers' diarrhea: R. Steffen *et al., Am. J. Gastroenterol.* **98**, 1073 (2003). Review of activity, pharmacokinetics and clinical experience in gastrointestinal infections: J. C. Gillis, R. N. Brogden, *Drugs* **49**, 467-484 (1995); D. B. Huang, H. L. DuPont, *J. Infect.* **50**, 97-106 (2005). Review of clinical efficacy in irritable bowel syndrome: M. Pimentel, *Expert Opin. Invest. Drugs* **18**, 349-358 (2009).

Red orange powder, mp 200-205° (dec). uv max: 232, 260, 292, 320, 370, 450 nm ($E_{1cm}^{1\%}$ 489, 339, 295, 216, 119, 159). Sol in alcohols, ethyl acetate, chloroform, toluene. Insol in water. LD_{50} orally in rats: >2000 mg/kg (Borelli, Bertoli).

THERAP CAT: Antibacterial.

THERAP CAT (VET): Antibacterial.

8346. Rilmazafone. [99593-25-6] 5-[[(Aminoacetyl)amino]methyl]-1-[4-chloro-2-(2-chlorobenzoyl)phenyl]-*N,N*-dimethyl-1*H*-1,2,4-triazole-3-carboxamide; 1-(2-*o*-chlorobenzoyl-4-chlorophenyl)-5-glycylaminomethyl-3-dimethylcarbamoyl-1*H*-1,2,4-triazole; 2′,5-dichloro-2-(3-dimethylcarbamoyl-5-glycylaminomethyl-1*H*-1,-2,4-triazol-1-yl)benzophenone. $C_{21}H_{20}Cl_2N_6O_3$; mol wt 475.33. C 53.06%, H 4.24%, Cl 14.92%, N 17.68%, O 10.10%. Ring-opened triazolobenzodiazepine derivative. Prepn: K. Hirai, H. Sugimoto, **DE 2725164**; *eidem*, **US 4159374** (1977, 1979 both to Shionogi); and CNS activity: K. Hirai *et al., J. Heterocycl. Chem.* **19**, 1363 (1982). Receptor binding study of rilmazafone and its closed ring active metabolites: M. Fujimoto *et al., Biochem. Pharmacol.* **33**, 1645 (1984). Combined HPLC determn and enzyme immunoassay in plasma: G. Kominami *et al., J. Chromatogr.* **417**, 216 (1987). Series of articles on teratogenicity studies: *Oyo Yakuri* **30**, 749-833 (1985), *C.A.* **104**, 81885h-81887k; 102328w (1986).

Monohydrochloride dihydrate. 450191-S; Rhythmy. $C_{21}H_{20}$-$Cl_2N_6O_3$·HCl.$2H_2O$; mol wt 547.82. Solid from 95% ethanol, mp 107°. LD_{50} orally in mice: >1500 mg/kg (Hirai, 1979).

THERAP CAT: Sedative, hypnotic.

8347. Rilmenidine. [54187-04-1] *N*-(Dicyclopropylmethyl)-4,5-dihydro-2-oxazolamine; 2-[*N*-(dicyclopropylmethyl)amino]oxazoline; oxaminozoline; S-3341. $C_{10}H_{16}N_2O$; mol wt 180.25. C

66.64%, H 8.95%, N 15.54%, O 8.88%. α_2-Adrenoceptor agonist. Prepn: C. Malen *et al.*, **DE 2362754**; *eidem*, **US 4102890** (1974, 1978 both to Sci. Union et Cie. - Soc. Franc. Recher. Med.). Adrenoceptor binding study: P. Guicheney *et al., J. Pharmacol.* **12**, 255 (1981). Pharmacokinetics in hypertensive subjects: J. Velly *et al., ibid.* **13**, 413 (1982). Animal pharmacology: M. Laubie *et al., ibid.* **16**, 259 (1985); P. A. van Zwieten *et al., Arch. Int. Pharmacodyn.* **279**, 130 (1986). Hypotensive effect in humans: K. Weerasuriya *et al., Eur. J. Clin. Pharmacol.* **27**, 281 (1984). Mechanism of action study: R. E. Gomez *et al., Eur. J. Pharmacol.* **195**, 181 (1991). *Review*: T. J. Verbeuren *et al., Cardiovasc. Drug Rev.* **8**, 56-70 (1990).

Crystals from hexane, mp 106-107°.

Phosphate. S-3341-3; Hyperium. $C_{10}H_{16}N_2O.H_3PO_4$; mol wt 278.24. pKa 9.3. Soly in H_2O: ~19% w/v; in methanol: ~7% w/v; in chloroform and ethanol: ~0.7% w/v. LD_{50} orally in mice, rats: 375, 295 mg/kg (Verbeuren).

Fumarate. $C_{14}H_{20}N_2O_5$. Crystals from ethanol, mp 170°.

THERAP CAT: Antihypertensive.

8348. Rilonacept. [501081-76-1] Interleukin 1 receptor accessory protein (human extracellular domain fragment) fusion protein with type I interleukin 1 receptor (human extracellular domain fragment) fusion protein with immunoglobulin G1 (human Fc fragment), homodimer; [653-glycine][human interleukin-1 receptor accessory protein-(1-339)-peptide (extracellular domain fragment) fusion protein with human type I interleukin-1 receptor-(5-316)-peptide (extracellular domain fragment) fusion protein with human immunoglobulin G1-(229 *C*-terminal residues)-peptide (Fc fragment)] dimer; interleukin-1 trap; Arcalyst. Dimeric fusion protein consisting of the extracellular portions of the human interleukin-1 receptor component (IL-1R1) and the IL-1 receptor accessory protein (IL-1RAcP) linked to the Fc portion of human IgG1. Produced in Chinese hamster ovary cells (CHO) using recombinant technology; comprised of 880 amino acid residues. mol wt ~251 kDa. Blocks IL-1β signaling by acting as a soluble decoy receptor that binds IL-1β and prevents its interaction with cell surface receptors; also binds with reduced affinity to IL-1α and interleukin-1 receptor antagonist (IL-1ra). Prepn: N. Stahl, G. D. Yancopoulos, **WO 04039951**; *eidem*, **US 6927044** (2004, 2005 both to Regeneron). Clinical trial in patients with cryopyrin-associated periodic syndromes (CAPS): H. M. Hoffman *et al., Arthritis Rheum.* **58**, 2443 (2008); in treatment of gouty arthritis: R. Terkeltaub *et al., Ann. Rheum. Dis.* **68**, 1613 (2009). Review of pharmacology and clinical experience: H. M. Hoffman, *Expert Opin. Biol. Ther.* **9**, 519-531 (2009).

THERAP CAT: Anti-inflammatory in treatment of interleukin-1 modulated disease.

8349. Rilpivirine. [500287-72-9] 4-[[4-[[4-[(1*E*)-2-Cyanoethenyl]-2,6-dimethylphenyl]amino]-2-pyrimidinyl]amino]benzonitrile; R-278474; TMC-278. $C_{22}H_{18}N_6$; mol wt 366.43. C 72.11%, H 4.95%, N 22.94%. Diarylpyrimidine (DAPY) nonnucleoside reverse transcriptase inhibitor (NNRTI). Prepn: J. E. G. Guillemont *et al.*, **WO 03016306**; *eidem*, **US 7125879** (2003, 2006 both to Janssen). Crystal structure in complexes with wild-type and mutant HIV-1 reverse transcriptase: K. Das *et al., Proc. Natl. Acad. Sci. USA* **105**, 1466 (2008); and 2D-IR spectroscopy: C. Fang *et al., ibid.*, 1472. Clinical pharmacokinetics and antiviral activity: F. Goebel *et al., AIDS* **20**, 1721 (2006). Review of discovery and development: P. A. J. Janssen *et al., J. Med. Chem.* **48**, 1901-1909 (2005).

Slightly yellow crystalline powder, mp 242°. pKa 5.6. Readily sol in DMSO (>50 mg/ml); moderately sol in PEG 400 (40 mg/ml). Practically insol in water (20 ng/ml, pH 7.0).
Hydrochloride. [700361-47-3] Edurant. $C_{22}H_{18}N_6$·HCl; mol wt 402.89. White to almost white powder. Practically insol in water.

THERAP CAT: Antiretroviral.

8350. Riluzole. [1744-22-5] 6-(Trifluoromethoxy)-2-benzothiazolamine; 2-amino-6-(trifluoromethoxy)benzothiazole; PK-26124; RP-54274; Rilutek. $C_8H_5F_3N_2OS$; mol wt 234.20. C 41.03%, H 2.15%, F 24.34%, N 11.96%, O 6.83%, S 13.69%. Modulates glutamatergic transmission. Prepn: L. M. Yagupol'skii, L. Z. Gandel'sman, *Zh. Obshch. Khim.* **33**, 2301 (1963), *C.A.* **60**, 692a (1964). Prepd but not claimed: J. Mizoule, *EP 50551*; *idem*, *US 4370338* (1982, 1983 both to Pharmindustrie). Pharmacology: J. Mizoule *et al.*, *Neuropharmacology* **24**, 767 (1985); F. Wahl *et al.*, *Eur. J. Pharmacol.* **230**, 209 (1993). Mechanism of action studies: M.-W. Debono *et al.*, *ibid.* **235**, 283; D. Martin *et al.*, *ibid.* **250**, 473. Use in treatment of motor nerve diseases: E. Louvel, *EP 558861* (1993 to Rhône-Poulenc Rorer). Clinical study in amyotrophic lateral sclerosis: G. Bensimon *et al.*, *N. Engl. J. Med.* **330**, 585 (1994). Review: J. Wokke, *Lancet* **348**, 795-799 (1996).

Crystals from ethanol/water (1:1), mp 119°. LD_{50} in mice (mg/kg): 46 i.p.; 67 orally (Mizoule).

THERAP CAT: Neuroprotective.

8351. Rimantadine. [13392-28-4] α-Methyltricyclo[3.3.1.1^{3,7}]decane-1-methanamine; α-methyl-1-adamantanemethylamine; remantadin(e). $C_{12}H_{21}N$; mol wt 179.31. C 80.38%, H 11.81%, N 7.81%. Deriv of adamantane, *q.v.* Prepn: *NL 6408505*; W. W. Prichard, *US 3352912* (1965, 1967 both to du Pont). Antiviral activity: A. Tsunoda *et al.*, *Antimicrob. Agents Chemother.* **1965**, 553. Effects on influenza in mice: J. W. McGahen *et al.*, *Ann. N.Y. Acad. Sci.* **173**, 557 (1970). Mechanism of action study: A. Bukrinskaya *et al.*, *Arch. Virol.* **66**, 275 (1980); *eidem*, *J. Gen. Virol.* **60**, 49 (1982). Pharmacokinetics in humans: R. J. Wills *et al.*, *Antimicrob. Agents Chemother.* **31**, 826 (1987). Clinical trial in prophylaxis of influenza A infection: R. Dolin *et al.*, *N. Engl. J. Med.* **307**, 580 (1982). Comparative toxicity of rimantadine and amantadine in healthy adults: F. G. Hayden *et al.*, *Antimicrob. Agents Chemother.* **19**, 226 (1981). Controlled study of CNS effects: V. M. Millet *et al.*, *ibid.* **21**, 1 (1982). Review of studies in the USSR on exptl and clinical pharmacology: D. M. Zlydnikov *et al.*, *Rev. Infect. Dis.* **3**, 408-421 (1981).

Hydrochloride. [1501-84-4] EXP-126; Flumadine; Meradan(e); Roflual. $C_{12}H_{21}N$·HCl; mol wt 215.77. White crystals, mp 373-375° (sealed tube).

THERAP CAT: Antiviral.

8352. Rimexolone. [49697-38-3] (11β,16α,17β)-11-Hydroxy-16,17-dimethyl-17-(1-oxopropyl)androsta-1,4-dien-3-one; 11β-hydroxy-16α,17α-dimethyl-17-propionylandrosta-1,4-dien-3-one; 11β-hydroxy-16α,17α,21-trimethylpregna-1,4-diene-3,20-dione; trimexolone; Org-6216; Rimexel; Vexol. $C_{24}H_{34}O_3$; mol wt 370.53. C 77.80%, H 9.25%, O 12.95%. Prepn: *NL 7300313*; G. F. Woods *et al.*, *US 3947478* (1973, 1976 both to Akzo); J. Cairns *et al.*, *J. Chem. Soc. Perkin Trans. 1* **1981**, 2306. Pharmacology: P. K. Fox *et al.*, *Arzneim.-Forsch.* **30**, 55 (1980). Clinical trial in rheumatoid arthritis: E. van Vliet-Daskalopoulou *et al.*, *Br. J. Rheuma-*

tol. **26**, 450 (1987). Clinical pharmacokinetics: G. Gevers *et al.*, *Clin. Rheumatol.* **13**, 103 (1994).

White to off-white powder or crystals, mp 258-268°. $[\alpha]_D$ +100° (c = 0.92 in pyridine). uv max: 244 nm (ε 14600). Freely sol in chloroform; sparingly sol in methanol.

THERAP CAT: Anti-inflammatory (local).

8353. Rimiterol. [32953-89-2] *rel*-4-[(R)-Hydroxy-(2S)-2-piperidinylmethyl]-1,2-benzenediol; *erythro*-α-(3,4-dihydroxyphenyl)-2-piperidinemethanol; *erythro*-3,4-dihydroxyphenyl-2-piperidinylcarbinol. $C_{12}H_{17}NO_3$; mol wt 223.27. C 64.56%, H 7.68%, N 6.27%, O 21.50%. Prepn: G. H. Sankey, K. D. E. Whiting, *DE 2024049*; *idem*, *US 3910934* (1970, 1975 to 3M); C. Kaiser, S. T. Ross, *DE 2047937*; *idem*, *US 3705169* (1971, 1972 to SK & F); G. H. Sankey, K. D. E. Whiting, *J. Heterocycl. Chem.* **9**, 1049 (1972). Pharmacology: Carney *et al.*, *Arch. Int. Pharmacodyn. Ther.* **194**, 334 (1971); Griffin, Turner, *J. Clin. Pharmacol.* **11**, 280 (1971); Bowman, Rodger, *Br. J. Pharmacol.* **45**, 574 (1972).

Relative stereochemistry

Crystals from ethyl acetate, mp 203-204°.
Hydrobromide. [31842-61-2] R-798; WG-253; Asmaten; Pulmadil. $C_{12}H_{17}NO_3$·HBr; mol wt 304.18. White powder, mp 220° (dec).

THERAP CAT: Bronchodilator.

8354. Rimocidin. [1393-12-0] (1R,3S,9R,10S,13R,15E,17E,-19E,21E,23R,25S,26R,27S)-23-[(3-Amino-3,6-dideoxy-β-D-mannopyranosyl)oxy]-10-ethyl-1,3,9,27-tetrahydroxy-7,11-dioxo-13-propyl-12,29-dioxabicyclo[23.3.1]nonacosa-15,17,19,21-tetraene-26-carboxylic acid. $C_{39}H_{61}NO_{14}$; mol wt 767.91. C 61.00%, H 8.01%, N 1.82%, O 29.17%. Polyene antifungal antibiotic produced by *Streptomyces rimosus* along with oxytetracycline, *q.v.* Recovered from fermentation broth by extracting the mycelium with butanol. Isoln and antifungal activity: Davisson *et al.*, *Antibiot. Chemother.* **1**, 289 (1951); Seneca *et al.*, *ibid.* **2**, 435 (1952); Davisson *et al.*, *US 2963401* (1960 to Pfizer). Partial structure: Cope *et al.*, *J. Am. Chem. Soc.* **87**, 5452 (1965). Structure: L. Falkowski *et al.*, *J. Antibiot.* **29**, 197 (1976); R. Pandey, K. L. Rinehart, *ibid.* **30**, 146 (1977).

Dec above 110°. $[\alpha]_D^{25}$ +116° (pyridine). uv max (80% methanol): 279, 291, 304, 318 nm. Slightly sol in water, acetone, lower alcohols. A cryst sodium salt was prepd by reaction with NaOH in methanol.

Sulfate heptahydrate. Large plates from dil methanol, dec 151°. $[\alpha]_D^{25}$ +75.2° (methanol). Sol in water. LD_{50} i.v. in mice: 20 mg/kg (Seneca).

8355. Rimonabant. [168273-06-1] 5-(4-Chlorophenyl)-1-(2,4-dichlorophenyl)-4-methyl-N-1-piperidinyl-1H-pyrazole-3-carboxamide; N-(piperidin-1-yl)-5-(4-chlorophenyl)-1-(2,4-dichlorophenyl)-4-methyl-1H-pyrazole-3-carboxamide; SR-141716; Acomplia. $C_{22}H_{21}Cl_3N_4O$; mol wt 463.79. C 56.97%, H 4.56%, Cl 22.93%, N 12.08%, O 3.45%. Brain cannabinoid receptor (CB_1) antagonist. Prepn: F. Barth et al., **EP 576357**; eidem., **US 5624941** (1993, 1997 both to Sanofi); H. H. Seltzman et al., J. Chem. Soc. Chem. Commun. **1995**, 1549; and pharmacology: A. K. Dutta et al., Med. Chem. Res. **5**, 54 (1994). Receptor antagonist activity: M. Rinaldi-Carmona et al., FEBS Lett. **350**, 240 (1994). Clinical trial in blockade of effects of smoked marijuana: M. A. Huestis et al., Arch. Gen. Psychiatry **58**, 322 (2001). Clinical trials in obesity: L. F. Van Gaal et al., Lancet **365**, 1389 (2005); J.-P. Després et al., N. Engl. J. Med. **353**, 2121 (2005). Review of therapeutic potential in obesity and drug addiction: M. A. M. Carai et al., Life Sci. **77**, 2339-2350 (2005).

Hydrochloride. [158681-13-1] SR-141716A. $C_{22}H_{21}Cl_3N_4$-O.HCl; mol wt 500.25. White crystals from ether + petr ether, mp 154-156° (Dutta); also reported as mp 157-159° (Seltzman).

THERAP CAT: Antiobesity agent.

8356. Rimsulfuron. [122931-48-0] N-[[(4,6-Dimethoxy-2-pyrimidinyl)amino]carbonyl]-3-(ethylsulfonyl)-2-pyridinesulfonamide; DPX-E9636; Matrix; Titus. $C_{14}H_{17}N_5O_7S_2$; mol wt 431.44. C 38.98%, H 3.97%, N 16.23%, O 25.96%, S 14.86%. Sulfonylurea herbicide; blocks branched-chain amino acid synthesis by inhibiting the plant enzyme, acetolactate synthase. Prepd (not claimed): P. H. Liang, **US 4774337** (1988 to DuPont); and use in herbicidal mixtures: idem, **EP 341011** (1989 to DuPont), C.A. **113**, 6360 (1990). Comprehensive description: H. L. Palm et al., Brighton Crop Prot. Conf. - Weeds **1989**, 23-28. HPLC, TLC determn and environmental fate: G. E. Schneiders et al., J. Agric. Food Chem. **41**, 2402 (1993). Field trials in potatoes: C. V. Eberlein et al., Weed Technol. **8**, 428 (1994).

mp 176-178°. pKa 4.1. Vapor pressure at 25°: 1.1×10^{-8} Torr. Soly in water at 25° (ppm): 135 (pH 5); 7300 (pH 7); 5560 (pH 9). Partition coefficient (octanol/water): 0.034 (pH 7). LD_{50} orally in rats: >5000 mg/kg; dermally in rabbits: >2000 mg/kg (Palm).

USE: Herbicide.

8357. Rintatolimod. [38640-92-5] 5'-Inosinic acid homopolymer complex with 5'-cytidylic acid polymer with 5'-uridylic

acid (1:1); polyI:polyC12U; polyI.poly($C_{12}U$); poly(rI).poly[r(C_{12}-U)$_n$]; Ampligen. Synthetic, mismatched double-stranded RNA capable of inducing interferons. Consists of a strand of 5'-inosinic acid units hydrogen bonded to a second strand of 5'-cytidylic- 5'-uridylic-acid units in a 12C:U ratio. Prepn and pharmacology: P. O. P. Ts'o et al., Mol. Pharmacol. **12**, 299 (1976). In vitro activity vs HIV: D. C. Montefiori, W. M. Mitchell, Proc. Natl. Acad. Sci. USA **84**, 2985 (1987); D. C. Montefiori et al., Antiviral Res. **9**, 47 (1988). Induction of interferon: D. A. Stringfellow, S. D. Weed, Antimicrob. Agents Chemother. **17**, 988 (1980). HPLC determn in plasma: M. G. Rosenblum, L. Cheung, J. Liq. Chromatogr. **9**, 2869 (1986). Overview of preclinical studies: W. A. Carter et al., J. Biol. Response Modif. **4**, 495-502 (1985). Clinical pharmacology in AIDS patients: W. A. Carter et al., Lancet **329**, 1286 (1987). Clinical trial in chronic fatigue syndrome: D. R. Strayer et al., Clin. Infect. Dis. **18**, Suppl. 1, S88 (1994); in treatment of HIV: K. A. Thompson et al., Eur. J. Clin. Microbiol. Infect. Dis. **15**, 580 (1996).

THERAP CAT: Antiviral; immunomodulator.

8358. Riociguat. [625115-55-1] N-[4,6-Diamino-2-[1-[(2-fluorophenyl)methyl]-1H-pyrazolo[3,4-b]pyridin-3-yl]-5-pyrimidinyl]-N-methylcarbamic acid methyl ester; methyl [4,6-diamino-2-[1-(2-fluorobenzyl)-1H-pyrazolo[3,4-b]pyridin-3-yl]pyrimidin-5-yl]methylcarbamate; BAY 63-2521. $C_{20}H_{19}FN_8O_2$; mol wt 422.42. C 56.87%, H 4.53%, F 4.50%, N 26.53%, O 7.57%. Soluble guanylate cyclase (sGC) stimulator; increases cGMP production and induces pulmonary vasodilation. Prepn: C. Alonso-Alija et al., **WO 03095451**; eidem, **US 7173037** (2003, 2007 both to Bayer); J. Mittendorf et al., ChemMedChem **4**, 853 (2009). Clinical pharmacokinetics: R. Frey et al., J. Clin. Pharmacol. **48**, 926 (2008). Clinical evaluation in pulmonary hypertension: F. Grimminger et al., Eur. Respir. J. **33**, 785 (2009). Review of development and clinical experience: R. T. Schermuly et al., Expert Opin. Invest. Drugs **20**, 567-576 (2011).

White crystals from methanol.

THERAP CAT: In treatment of pulmonary hypertension.

8359. Rioprostil. [77287-05-9] (11α,13E)-1,11,16-Trihydroxy-16-methylprost-13-en-9-one; (2R,3R,4R)-4-hydroxy-2-(7-hydroxyheptyl)-3-[(E)-(4RS)-(4-hydroxy-4-methyl-1-octenyl)]cyclopentanone; 16-methyl-1,11α,16RS-trihydroxyprost-13E-en-9-one; Bay o 6893; ORF-15927; TR-4698; Rostil. $C_{21}H_{38}O_4$; mol wt 354.53. C 71.15%, H 10.80%, O 18.05%. Prostaglandin E_1 analog with cytoprotective and gastric antisecretory activity. Prepn: H. C. Kluender et al., **US 4132738** (1979 to Miles). Pharmacology: D. A. Shriver et al., Arzneim.-Forsch. **35**, 839 (1985). Effect on human gastric secretion: P. Demol et al., ibid. 861; B. Vaona et al., Adv. Prostaglandin Thromboxane Leukotriene Res. **17**, 328 (1987). Clinical evaluation in duodenal ulcer: H. G. Dammann et al., ibid. 303.

$[\alpha]_D$ −58.6° (c = 1 in chloroform).

THERAP CAT: Antiulcerative.

8360. Risedronic Acid. [105462-24-6] P,P'-[1-Hydroxy-2-(3-pyridinyl)ethylidene]bisphosphonic acid; 2-(3-pyridinyl)-1-hydroxyethane diphosphonic acid. $C_7H_{11}NO_7P_2$; mol wt 283.11. C 29.70%, H 3.92%, N 4.95%, O 39.56%, P 21.88%. Bisphosphonate antiresorptive agent. Prepn: J. J. Benedict, C. M. Perkins, **EP 186405**; *eidem*, **US 5583122** (1986, 1996 both to Procter & Gamble). Pharmacology: W. S. S. Jee *et al.*, *Bone* **14**, 493 (1993). Clinical trial in Paget's disease: E. S. Siris *et al.*, *J. Bone Miner. Res.* **13**, 1032 (1998); in postmenopausal bone loss: L. Mortensen *et al.*, *J. Clin. Endocrinol. Metab.* **83**, 396 (1998). Clinical effect on reduction of fracture risk: S. T. Harris *et al.*, *J. Am. Med. Assoc.* **282**, 1344 (1999). Review of pharmacology and clinical efficacy: K. L. Goa, J. A. Balfour, *Drugs Aging* **13**, 83-91 (1998).

Monosodium salt. [115436-72-1] Risedronate sodium; NE-58095; Actonel; Optinate. $C_7H_{10}NNaO_7P_2$; mol wt 305.09. Fine, white to off-white, odorless crystalline powder. Occurs as the hemipentahydrate. Sol in water. Essentially insol in common organic solvents.

THERAP CAT: Bone resorption inhibitor.

8361. Risperidone. [106266-06-2] 3-[2-[4-(6-Fluoro-1,2-benzisoxazol-3-yl)-1-piperidinyl]ethyl]-6,7,8,9-tetrahydro-2-methyl-4H-pyrido[1,2-a]pyrimidin-4-one; R-64766; Belivon; Risperdal. $C_{23}H_{27}FN_4O_2$; mol wt 410.49. C 67.30%, H 6.63%, F 4.63%, N 13.65%, O 7.80%. Combined serotonin (5-HT$_2$) and dopamine (D$_2$) receptor antagonist. Prepn: L. E. J. Kennis, J. Vandenberk, **EP 196132**; *eidem*, **US 4804663** (1986, 1989 both to Janssen). Pharmacology: P. A. J. Janssen *et al.*, *J. Pharmacol. Exp. Ther.* **244**, 685 (1988). Receptor binding studies: J. E. Leysen *et al.*, *ibid.* **247**, 661 (1988). HPLC determn in plasma: A. Avenoso *et al.*, *J. Chromatogr. B* **746**, 173 (2000). Clinical study in psychoses: Y. G. Gelders *et al.*, *Pharmacopsychiatry* **23**, 206 (1990); in autism: L. Scahill *et al.*, *N. Engl. J. Med.* **347**, 314 (2002). Brief review: M. G. Livingston, *Lancet* **343**, 457-460 (1994). Review of pharmacology and therapeutic potential: S. Grant, A. Fitton, *Drugs* **48**, 253-273 (1994); B. Green, *Curr. Med. Res. Opin.* **16**, 57-65 (2000); of clinical experience in schizophrenia: H.-J. Möller, *Expert Opin. Pharmacother.* **6**, 803-818 (2005).

Crystals from DMF + 2-propanol, mp 170.0°. Sol in methylene chloride; sparingly sol in alc. Practically insol in water. LD$_{50}$ in male, female mice, rats, dogs (mg/kg): 29.7, 26.9, 34.3, 35.4, 14.1, 18.3 i.v.; 82.1, 63.1, 113, 56.6, 18.3, 18.3 orally (Janssen, 1988).

THERAP CAT: Antipsychotic.

8362. Ristocetin. [1404-55-3] Ristomycin. Glycopeptide antibiotic complex produced by the actinomycete *Nocardia lurida*. Ristocetin A and the more active ristocetin B differ in the number of glucose, mannose, rhamnose and D-arabinose groups in their side chains. Isoln, crystallizn and chemical properties: J. E. Philip *et al.*, *Antibiot. Annu.* **1956-57**, p 699; *eidem*, **US 2990329** (1961 to Abbott). Total structure determn of ristocetin A: D. H. Williams *et al.*, *Chem. Commun.* **1979**, 906; J. R. Kalman, D. H. Williams, *J. Am. Chem. Soc.* **102**, 897 (1980). Identity of ristocetin A with ristomycin A: D. H. Williams *et al.*, *J. Chem. Soc. Perkin Trans. 1* **1979**, 787. ^{13}C-NMR studies: F. Sztaricskai *et al.*, *Tetrahedron Lett.* **21**, 2983 (1980); M. P. Williamson, D. H. Williams, *J. Chem. Soc. Perkin*

Trans. 1 **1981**, 1483. Revised configuration: C. M. Harris, T. M. Harris, *J. Am. Chem. Soc.* **104**, 363 (1982). Biosynthesis: S. J. Hammond *et al.*, *Chem. Commun.* **1983**, 116. Tool for investigation of platelet aggregation: Howard, Firkin, *Thromb. Diath. Haemorrh.* **26**, 362 (1971). Clinical use in diagnosis of von Willebrand disease: B. M. Ewenstein, *Haemophilia* **7**, Suppl 1, 10 (2001); A. E. Bowyer *et al.*, *Thromb. Res.* **127**, 341 (2011). *Review:* D. C. Jordan, *Antibiotics* **vol. 1**, D. Gottlieb, P. Shaw, Eds. (Springer-Verlag, New York, 1967) pp 84-89.

Ristocetin A

Crystalline sulfates. Mixture of ristocetin A and B. Sol in acidic aq solns; much less sol in the neutral pH range. Generally insol in organic solvents. Both components show good stability in aq acidic solns, but are readily inactivated above pH 7.0.

Ristocetin A. [11021-66-2] Ristomycin A. $C_{95}H_{110}N_8O_{44}$; mol wt 2067.94. C 55.18%, H 5.36%, N 5.42%, O 34.04%. Cryst sulfate, $[\alpha]_D$ −120 to −133° (water).

Ristocetin B. [1405-59-0] Ristomycin B. $C_{84}H_{92}N_8O_{35}$. C 56.88%, H 5.23%, N 6.32%, O 31.57%. Cryst sulfate, $[\alpha]_D$ −144 to −149° (water).

USE: Peptide-based probe for platelet aggregation.

THERAP CAT: Diagnostic aid (von Willebrand disease).

8363. Ritanserin. [87051-43-2] 6-[2-[4-[Bis(4-fluorophenyl)methylene]-1-piperidinyl]ethyl]-7-methyl-5H-thiazolo[3,2-a]pyrimidin-5-one; R-55667; Tiserton. $C_{27}H_{25}F_2N_3OS$; mol wt 477.57. C 67.91%, H 5.28%, F 7.96%, N 8.80%, O 3.35%, S 6.71%. Selective serotonin (5-HT$_2$) receptor antagonist. Prepn: L. E. J. Kennis *et al.*, **EP 110435**; *eidem*, **US 4533665** (1984, 1985 both to Janssen). Pharmacological profile: F. Awouters *et al.*, *Drug Dev. Res.* **15**, 61 (1988). GC/MS determn in plasma and pharmacokinetics: P. Timmerman *et al*, *Biomed. Environ. Mass Spectrom.* **18**, 498 (1989). Clinical studies: G. Nappi *et al.*, *Headache* **30**, 439 (1990); G. Bersani *et al.*, *Acta Psychiatr. Scand.* **83**, 244 (1991); J. M. Monti *et al.*, *Sleep* **16**, 647 (1993).

Crystals from acetonitrile, mp 145.5°. LD$_{50}$ in male, female mice, rats, dogs (mg/kg): 28.2, 28.2, 20.0, 22.2, 24.1, 33.2 i.v.; 626, 993, 956, 515, ~1280, 640-1280 orally (Awouters).

L-Tartrate. [93076-39-2] C$_{27}$H$_{25}$F$_2$N$_3$OS.C$_4$H$_6$O$_6$. Solid from 2-propanol, mp 198.7°.

THERAP CAT: Anxiolytic; antidepressant.

8364. Ritipenem. [84845-57-8] (5R,6S)-3-[[(Aminocarbonyl)oxy]methyl]-6-[(1R)-1-hydroxyethyl]-7-oxo-4-thia-1-azabicyclo-[3.2.0]hept-2-ene-2-carboxylic acid; (5R,6S,8R)-6α-hydroxyethyl-2-carbamoyloxymethyl-2-penem-3-carboxylic acid; (5R,6S)-6-[(1R)-1-hydroxyethyl]-3-(hydroxymethyl)-7-oxo-4-thia-1-azabicyclo-[3.2.0]hept-2-ene-2-carboxylic acid, 3-carbamate. C$_{10}$H$_{12}$N$_2$O$_6$S; mol wt 288.27. C 41.67%, H 4.20%, N 9.72%, O 33.30%, S 11.12%. Prepn: M. Alpegiani *et al.*, **DE 3245270**; M. Foglio *et al.*, **US 4482565** (1983, 1984 both to Carlo Erba). Synthesis: G. Franceschi *et al.*, *J. Antibiot.* **36**, 938 (1983). Total synthesis: W. Cabri *et al.*, *Tetrahedron Lett.* **34**, 3491 (1993). Toxicity study: M. Brughera *et al.*, *J. Antimicrob. Chemother.* **23**, Suppl. C, 129 (1989). Series of articles on synthesis, *in vitro* activity, metabolism: *ibid.* 1-204 (1989). Clinical pharmacokinetics of acid and ester forms: S. R. Norrby *et al.*, *ibid.* **25**, 371 (1990); A. M. Lovering *et al.*, *ibid.* **29**, 179 (1992). HPLC determn in serum and urine: R. Mendez *et al.*, *J. Chromatogr.* **579**, 115 (1992).

Sodium salt. [84845-58-9] FCE-22101. C$_{10}$H$_{11}$N$_2$NaO$_6$S; mol wt 310.26. [α]$_D^{20}$ +140°. uv max (H$_2$O): 258, 306 nm (ε 4150, 6030). LD$_{50}$ in male, female mice, male, female rats (mg/kg): 3872, 4393, 2000, 2201 i.v. (Brughera).

Acetoxymethyl ester. [87238-52-6] Ritipenem acoxil; FCE-22891. C$_{13}$H$_{16}$N$_2$O$_8$S; mol wt 360.34. LD$_{50}$ in male, female mice, male, female rats (mg/kg): 4363, 6167, >5000, >5000 orally (Brughera).

THERAP CAT: Antibacterial.

8365. Ritodrine. [26652-09-5] *rel*-(αS)-4-Hydroxy-α-[(1R)-1-[[2-(4-hydroxyphenyl)ethyl]amino]ethyl]benzenemethanol; *erythro*-p-hydroxy-α-[1-[(p-hydroxyphenethyl)amino]ethyl]benzyl alcohol; N-[2-(p-hydroxyphenyl)ethyl]-N-[2-(p-hydroxyphenyl)-2-hydroxy-1-methylethyl]amine; 1-(4-hydroxyphenyl)-2-[2-(4-hydroxyphenyl)ethylamino]propanol; N-(p-hydroxyphenylethyl)-4-hydroxynorephedrine. C$_{17}$H$_{21}$NO$_3$; mol wt 287.36. C 71.06%, H 7.37%, N 4.87%, O 16.70%. β$_2$-Adrenergic agonist. Prepn: **BE 660244** (1965 to N.V. Philips); Claassen *et al.*, **US 3410944** (1968 to Am. Philips). Clinical investigations: Coutinho *et al.*, *Am. J. Obstet. Gynecol.* **104**, 1053 (1969); Landesman *et al.*, *ibid.* **110**, 111 (1971); Wesselius-De Casparis *et al.*, *Br. Med. J.* **3**, 144 (1971). Clinical efficacy in treatment of preterm labor: J. F. Larsen *et al.*, *Obstet. Gynecol.* **67**, 607 (1986).

Relative stereochemistry

Base. Resinous mass, mp 88-90°.

Hydrochloride. [23239-51-2] DU-21220; Miolene; Prempar; Pre-Par; Utemerin; Utopar; Yutopar. C$_{17}$H$_{21}$NO$_3$.HCl; mol wt 323.82. White crystalline powder. mp 193-195° (dec) from ethanol-ether. uv max: 267.5 nm (ε 3310). Freely sol in water, alc; sol in n-propyl alc. Practically insol in ether.

THERAP CAT: Tocolytic.

8366. Ritonavir. [155213-67-5] (3S,4S,6S,9S)-4-Hydroxy-12-methyl-9-(1-methylethyl)-13-[2-(1-methylethyl)-4-thiazolyl]-8,11-dioxo-3,6-bis(phenylmethyl)-2,7,10,12-tetraazatridecanoic acid 5-thiazolymethyl ester; (5S,8S,10S,11S)-10-hydroxy-2-methyl-5-(1-methylethyl)-1-[2-(1-methylethyl)-4-thiazolyl]-3,6-dioxo-8,11-bis(phenylmethyl)-2,4,7,12-tetraazatridecan-13-oic acid 5-thiazolylmethyl ester; (2S,3S,5S)-5-[N-[N-[N-methyl-N-[(2-isopropyl-4-thiazolyl)methyl]amino]carbonyl]valinyl]amino]-2-[N-[(5-thiazolyl)methoxycarbonyl]amino]-1,6-diphenyl-3-hydroxyhexane; A-84538; Abbott 84538; ABT-538; Norvir. C$_{37}$H$_{48}$N$_6$O$_5$S$_2$; mol wt 720.95. C 61.64%, H 6.71%, N 11.66%, O 11.10%, S 8.89%. Peptidomimetic HIV-1 protease inhibitor. Also inhibits cytochrome P450 3A4 (CYP3A4) and enhances systemic availablity of coadministered drugs by blocking their metabolism by the enzyme. Prepn: D. J. Kempf *et al.*, **WO 9414436**; *eidem*, **US 5541206** (1994, 1996 both to Abbott). Antiretroviral spectrum, pharmacokinetics: *idem et al.*, *Proc. Natl. Acad. Sci. USA* **92**, 2484 (1995). Structural model for drug resistance: M. Markowitz *et al.*, *J. Virol.* **69**, 701 (1995). HPLC determn in biological fluids: R. M. W. Hoetelmans *et al.*, *J. Chromatogr. B* **705**, 119 (1998). Clinical pharmacokinetics and anti-CYP3A4 activity: A. Hsu *et al.*, *Clin. Pharmacokinet.* **35**, 275-291 (1998). Clinical trial with nucleoside analogs in HIV-infected children: S. A. Nachman *et al.*, *J. Am. Med. Assoc.* **283**, 492 (2000); as pharmacoenhancer for lopinavir in HIV infection: S. Walmsley *et al.*, *N. Engl. J. Med.* **346**, 2039 (2002). Review of clinical experience: A. P. Lea, D. Faulds, *Drugs* **52**, 541-546 (1996); as pharmacokinetic enhancer in HIV therapy: M. W. Hull, J. S. G. Montaner, *Ann. Med.* **43**, 375-388 (2011).

White to light tan powder with bitter, metallic taste. Freely sol in methanol, methylene chloride; very slightly sol in acetonitrile. Practically insol in water.

THERAP CAT: Antiretroviral. Pharmacokinetic enhancer for antiretroviral agents.

8367. Rituximab. [174722-31-7] Anti-(human CD20 (antigen)) immunoglobulin G1 (human-mouse monoclonal IDEC-C2B8 γ$_1$-chain) disulfide with human-mouse monoclonal IDEC-C2B8 κ-chain, dimer; IDEC-C2B8; Mabthera; Rituxan. Chimeric murine-human monoclonal antibody directed against the CD20 surface protein found on normal and malignant B lymphocytes. Contains murine light and heavy-chain variable regions and γ$_1$ heavy-chain and κ light chain human constant regions. Composed of 2 heavy chains of 451 amino acids and 2 light chains of 213 amino acids; approx mol wt 145 kDa. Biological response modifier that depletes mature B cells from blood and tissue. Prepn: D. R. Anderson *et al.*, **WO 9411026**; *eidem*, **US 5736137** (1994, 1998 both to IDEC); M. E. Reff *et al.*, *Blood* **83**, 435 (1994). HPLC determn of product purity: K. G. Moorhouse *et al.*, *J. Pharm. Biomed. Anal.* **16**, 593 (1997). Clinical pharmacokinetics: N. L. Berinstein *et al.*, *Ann. Oncol.* **9**, 995 (1998). Review of pharmacology and clinical experience in non-Hodgkin's lymphoma: S. V. Onrust *et al.*, *Drugs* **58**, 79-88 (1999); A. Molina, *Annu. Rev. Med.* **59**, 237-250 (2008); in rheumatoid arthritis: A. A. Schuna, *Pharmacotherapy* **27**, 1702-1710 (2007). Clinical effect on immunocompetency in patients with autoimmune disease: R. J. Looney *et al.*, *Arthritis Rheum.* **58**, 5-14 (2008).

THERAP CAT: Antineoplastic; anti-inflammatory.

8368. Rivaroxaban. [366789-02-8] 5-Chloro-N-[[(5S)-2-oxo-3-[4-(3-oxo-4-morpholinyl)phenyl]-5-oxazolidinyl]methyl]-2-thiophenecarboxamide; BAY-59-7939; Xarelto. C$_{19}$H$_{18}$ClN$_3$O$_5$S;

mol wt 435.88. C 52.36%, H 4.16%, Cl 8.13%, N 9.64%, O 18.35%, S 7.36%. Direct factor Xa inhibitor. Prepn: A. Straub *et al.*, **WO 0147919**; *eidem*, **US 7157456** (2001, 2007 both to Bayer). HPLC-MS/MS determn in plasma: G. Rohde, *J. Chromatogr. B* **872**, 43 (2008). Clinical pharmacokinetics and pharmacodynamics: W. Mueck *et al.*, *Thromb. Haemostasis* **100**, 453 (2008). Clinical evaluation in deep-vein thrombosis: G. Agnelli *et al.*, *Circulation* **116**, 180 (2007); for thromboprophylaxis after arthroplasty: B. I. Eriksson *et al.*, *N. Engl. J. Med.* **358**, 2765 (2008); M. R. Lassen *et al.*, *ibid.* 2776. Review of pharmacology: V. Laux *et al.*, *Semin. Thromb. Hemostasis* **33**, 515-523 (2007); of development and clinical experience: J. P. Piccini *et al.*, *Expert Opin. Invest. Drugs* **17**, 925-937 (2008). Review of clinical trials in acute coronary syndromes: D. Alexander, A. Jeremias, *Expert Opin. Invest. Drugs* **20**, 849-857 (2011).

mp 232-233°. $[\alpha]_D^{21}$ −38° (c = 0.2985 in DMSO).
THERAP CAT: Antithrombotic.

8369. Rivastigmine. [123441-03-2] *N*-Ethyl-*N*-methylcarbamic acid 3-[(1*S*)-1-(dimethylamino)ethyl]phenyl ester; (*S*)-*N*-ethyl-3-[(1-dimethylamino)ethyl]-*N*-methylphenylcarbamate; Exelon Patch. $C_{14}H_{22}N_2O_2$; mol wt 250.34. C 67.17%, H 8.86%, N 11.19%, O 12.78%. Brain selective acetylcholinesterase inhibitor. Prepn: A. Enz, **DE 3805744**; *idem*, **US 5602176** (1988, 1997 both to Sandoz); and activity: R. Amstutz *et al.*, *Helv. Chim. Acta* **73**, 739 (1990). Pharmacology: A. Enz *et al.*, *Prog. Brain Res.* **98**, 431 (1993). LC-MS/MS determn in plasma: J. Bhatt *et al.*, *J. Chromatogr. B* **852**, 115 (2007). Clinical trial in Alzheimer's disease: M. Rösler *et al.*, *Br. Med. J.* **318**, 633 (1999). Review of clinical experience with oral and transdermal formulations in Alzheimer's disease and Parkinson's disease dementia: T. Darreh-Shori, V. Jelic, *Expert Opin. Drug Saf.* **9**, 167-176 (2010).

Clear, colorless to yellow or very slightly brown, viscous liquid. Sparingly sol in water; very sol in ethanol, acetonitrile, *n*-octanol, ethyl acetate. Distribution coefficient (*n*-octanol/pH 7 phosphate buffer) at 37°: 4.27.

Hydrogen tartrate. [129101-54-8] ENA-713; SDZ-ENA-713; SDZ-212-713; Exelon. $C_{14}H_{22}N_2O_2 \cdot C_4H_6O_6$; mol wt 400.43. White to off-white crystals from ethanol, mp 123-125°. $[\alpha]_D^{20}$ +4.7° (c = 5 in ethanol). Very sol in water; sol in ethanol, acetonitrile; slightly sol in *n*-octanol; very slightly sol in ethyl acetate. Distribution coefficient (*n*-octanol/pH 7 phosphate buffer) at 37°: 3.0.
THERAP CAT: Nootropic.

8370. Rizatriptan. [144034-80-0] *N,N*-Dimethyl-5-(1*H*-1,2,4-triazol-1-ylmethyl)-1*H*-indole-3-ethanamine; 3-[2-(dimethylamino)ethyl]-5-(1*H*-1,2,4-triazol-1-ylmethyl)indole; *N,N*-dimethyl-2-[5-(1,2,4-triazol-1-ylmethyl)-1*H*-indol-3yl]ethylamine. $C_{15}H_{19}N_5$; mol wt 269.35. C 66.89%, H 7.11%, N 26.00%. Selective serotonin 5-HT$_{1B/1D}$ receptor agonist; structurally derived from tryptamine. Prepn: R. Baker *et al.*, **EP 497512**; *eidem*, **US 5298520** (1992, 1994 both to Merck Sharp & Dohme); and binding characteristics: L. J. Street *et al.*, *J. Med. Chem.* **38**, 1799 (1995). Synthesis: C. Chen *et al.*, *Tetrahedron Lett.* **35**, 6981 (1994). Clinical pharmacokinetics: H. Cheng *et al.*, *Biopharm. Drug Dispos.* **17**, 17 (1996). LC-MS determn in plasma: D. A. McLoughlin *et al.*, *J. Chromatogr. A* **726**, 115 (1996). Comparative clinical trial in treatment of migraine: J. U. Adelman *et al.*, *Neurology* **57**, 1377 (2001). Review of pharmacology and therapeutic efficacy: K. Wellington,

G. L. Plosker, *Drugs* **62**, 1539-1574 (2002); of development and clinical experience: A. V. Krymchantowski, M. E. Bigal, *Expert Rev. Neurother.* **5**, 597-603 (2005).

mp 120-121°.
Benzoate. [145202-66-0] MK-0462; Maxalt. $C_{15}H_{19}N_5 \cdot C_7H_6O_2$; mol wt 391.48. Crystal structure: K. Ravikumar *et al.*, *Acta Crystallogr.* **E63**, 1958 (2007). White to off-white, crystalline solid, mp 178-180°. Sol in methanol. Soly in water (25°C): ~42 mg/ml (expressed as free base).
THERAP CAT: Antimigraine.

8371. Robenidine. [25875-51-8] 2,2′-Bis[(4-chlorophenyl)methylene]carbonimidic dihydrazide; 1,3-bis[(*p*-chlorobenzylidene)amino]guanidine; $C_{15}H_{13}Cl_2N_5$; mol wt 334.20. C 53.91%, H 3.92%, Cl 21.21%, N 20.96%. Prepn: Tomcufcik, **DE 1933112** (1970 to Am. Cyanamid), *C.A.* **72**, 90113c (1970). Activity studies: Kantor *et al.*, *Science* **168**, 373 (1970); Wong *et al.*, *Biochem. Biophys. Res. Commun.* **46**, 621 (1972). Metabolism: Zulalian *et al.*, 163rd Am. Chem. Soc. Meeting (Boston, April 1972) *Abstracts of Papers*, PEST 12, 13. Animal studies: Millard, *Res. Vet. Sci.* **11**, 394 (1970); Joyner, Norton, *ibid.* **13**, 279 (1972).

Hydrochloride. [25875-50-7] Robenzidene; Cycostat; Robenz. $C_{15}H_{13}Cl_2N_5 \cdot HCl$; mol wt 370.66. Crystals from ethanol, mp 289-290°.
THERAP CAT (VET): Coccidiostat.

8372. Robinin. [301-19-9] 3-[[6-*O*-(6-Deoxy-α-L-mannopyranosyl)-β-D-galactopyranosyl]oxy]-7-[(6-deoxy-α-L-mannopyranosyl)oxy]-5-hydroxy-2-(4-hydroxyphenyl)-4*H*-1-benzopyran-4-one; kaempferol 3-robinoside 7-rhamnoside. $C_{33}H_{40}O_{19}$; mol wt 740.66. C 53.51%, H 5.44%, O 41.04%. Dimorphic flavanoid isolated from the leaves and flowers of *Robinia pseudoacacia* L., *Leguminosae*: C. Zwenger, F. Dronke, *Ann.* **suppl. 1**, 257 (1861); C. Sando, *J. Biol. Chem.* **94**, 675 (1932). Structure: Zemplén, Bognár, *Ber.* **74B**, 1783 (1941). Total synthesis and structure: L. Farkas *et al.*, *Phytochemistry* **15**, 215 (1976).

β-Form. Yellow crystals, mp 250-254° (Farkas); also reported as straw-yellow needles from alc, mp 249-250° (Sando). uv max (ethanol): 352, 368 nm (log ε 4.14, 4.18), Jurd, Horowitz, *J. Org. Chem.* **22**, 1619 (1957). Sol in hot water, hot alc; practically insol in ether. On hydrolysis yields kaempferol, *q.v.*

α-Form. Obtained by crystallization from water and dehydrating, mp 195-197° (Sando). Also reported as hydrate, yellow needles from aq methanol, mp 196-199° (Farkas).

8373. Roccellic Acid. [29838-46-8] (2*R*,3*S*)-2-Dodecyl-3-methylbutanedioic acid; (2*R*,3*S*)-2-dodecyl-3-methylsuccinic acid; *d*-α-dodecyl-β-methylsuccinic acid; *d*-α-methyl-α′-dodecylsuccinic acid. $C_{17}H_{32}O_4$; mol wt 300.44. C 67.96%, H 10.74%, O 21.30%. Occurs in lichens. Isoln from *Lecanora sordida* (Pers.) Th. Fries, *Parmeliaceae*: Hesse, *J. Prakt. Chem.* **58**, 497 (1898); Kennedy *et al.*, *Sci. Proc. R. Dublin Soc.* **21**, 557 (1937); from *Roccella montag-*

nei, Graphidaceae: Subbaraya, Seshadri, *Proc. Indian Acad. Sci.* **12A**, 466 (1940); from *Crocynia membranacae* (Dicks.) Zahlbr., *Chrysotrichaceae:* Akermark *et al., Acta Chem. Scand.* **13**, 1855 (1959). Structure: Kennedy *et al., loc. cit.* Absolute configuration: Akermark, *Acta Chem. Scand.* **16**, 599 (1962).

Rectangular rods from acetone, mp 132-133°. $[\alpha]_D^{20}$ +18° (c = 1.84 in ethanol). Practically insol in water. Freely sol in alcohol, ether; sol in aq sodium bicarbonate solns. Forms a water-sol sodium salt.

8374. Rociverine. [53716-44-2] *rel*-(1*R*,2*R*)-1-Hydroxy-[1,1′-bicyclohexyl]-2-carboxylic acid 2-(diethylamino)-1-methylethyl ester; 2-(diethylamino)-1-methylethyl *cis*-1-hydroxy[bicyclohexyl]-2-carboxylate; LG-30158; Rilaten. $C_{20}H_{37}NO_3$; mol wt 339.52. C 70.75%, H 10.98%, N 4.13%, O 14.14%. Spasmolytic agent with balanced neurotropic and myotropic properties. Prepn: L. Turbanti, **ZA 6705649** *C.A.* **70**, 47117d (1969) and **US 3700675** (1968, 1972 both to Guidotti). Antispasmodic activity *in vitro* and *in vivo:* G. Toson *et al., Arzneim.-Forsch.* **28**, 1130 (1978). Effect in cystitis or bladder spasm: A. Manganelli, *Farmaco Ed. Prat.* **34**, 384 (1979). Clinical studies: M. Petrillo *et al., Curr. Med. Res. Opin.* **7**, 73 (1980); R. Assisi, S. deStefano, *Acta Ther.* **6**, 353 (1980); F. Marsala, *Minerva Med.* **73**, 2179 (1982).

Relative stereochemistry

Oil, bp$_{0.1}$ 148-150°. n_D^{20} 1.4820. Sol in alc, ether, chloroform, benzene, dil mineral acids. Insol in water.
THERAP CAT: Antispasmodic.

8375. Rocuronium. [143558-00-3] 1-[(2β,3α,5α,16β,17β)-17-(Acetyloxy)-3-hydroxy-2-(4-morpholinyl)androstan-16-yl]-1-(2-propenyl)pyrrolidinium; 1-allyl-1-(3α,17β-dihydroxy-2β-morpholino-5α-androstan-16β-yl)pyrrolidinium 17-acetate. $[C_{32}H_{53}N_2O_4]^+$. Aminosteroid, competitive neuromuscular blocker. Prepn: D. S. Savage *et al.,* **EP 287150**; T. Sleigh *et al.,* **US 4894369** (1988, 1990 both to Akzo). Pharmacology: A. W. Muir *et al., Br. J. Anaesth.* **63**, 400 (1989); K. Khuenl-Brady *et al., Anesthesiology* **72**, 669 (1990). Clinical pharmacodynamics: T. J. Quill *et al., Anesth. Analg.* **72**, 203 (1991); and pharmacokinetics in the elderly: R. S. Matteo *et al., ibid.* **77**, 1193 (1993). Comparative clinical trial: T. Magorian *et al., Anesthesiology* **79**, 913 (1993). HPLC determn: U. W. Kleef *et al., J. Chromatogr.* **621**, 65 (1993). Review: T. C. Wicks, *J. Am. Assoc. Nurse Anesth.* **62**, 33-38 (1994).

Bromide. [119302-91-9] Org-9426; Esmeron; Zemuron. $C_{32}H_{53}BrN_2O_4$; mol wt 609.69. Crystals, mp 161-169°. $[\alpha]_D^{20}$ +18.7° (c = 1.03 in CHCl$_3$).
THERAP CAT: Neuromuscular blocking agent.

8376. Roentgenium. [54386-24-2] Element 111; unununium. Rg, Uuu; at. no. 111. Group IB (11). Transuranium element. No stable nuclides. Prepn and decay of isotope ^{272}111 (T$_{1/2}$ ~1.5 msec,

α decay) by ^{209}Bi(^{64}Ni,n): S. Hofmann *et al., Z. Phys. A* **350**, 281 (1995). Synthesis and decay chains: *idem, et al., ibid.* **14**, 147 (2002). Name and symbol approved: J. Corish, G. M. Rosenblatt, *Pure Appl. Chem.* **76**, 2101 (2004). Review of production and properties: *idem, Rep. Prog. Phys.* **61**, 639-689 (1998).

8377. Rofecoxib. [162011-90-7] 4-[4-(Methylsulfonyl)phenyl]-3-phenyl-2(5*H*)-furanone; MK-0966; Vioxx. $C_{17}H_{14}O_4S$; mol wt 314.36. C 64.95%, H 4.49%, O 20.36%, S 10.20%. Selective cyclooxygenase-2 (COX-2) inhibitor. Prepn: Y. Ducharme *et al.,* **WO 9500501**; *eidem,* **US 5474995** (both 1995 to Merck Frosst). HPLC determn in plasma: C. M. Chavez-Eng *et al., J. Chromatogr. B* **748**, 31 (2000). Enzyme inhibition and clinical evaluation in dental pain: E. W. Ehrich *et al., Clin. Pharmacol. Ther.* **65**, 336 (1999). Evaluation of risk of gastrointestinal effects in patients with osteoarthritis: M. J. Langman *et al., J. Am. Med. Assoc.* **282**, 1929 (1999); with rheumatoid arthritis: C. Bombardier *et al., N. Engl. J. Med.* **343**, 1520 (2000). Review of pharmacology and clinical experience: A. J. Matheson, D. P. Figgitt, *Drugs* **61**, 833-865 (2001).

White to off-white to light yellow powder. Sparingly sol in acetone; slightly sol in methanol, isopropyl acetate; very slightly sol in ethanol. Practically insol in octanol; insol in water.
THERAP CAT: Anti-inflammatory.

8378. Roflumilast. [162401-32-3] 3-(Cyclopropylmethoxy)-*N*-(3,5-dichloro-4-pyridinyl)-4-(difluoromethoxy)benzamide; BY-217; Daliresp; Daxas. $C_{17}H_{14}Cl_2F_2N_2O_3$; mol wt 403.21. C 50.64%, H 3.50%, Cl 17.58%, F 9.42%, N 6.95%, O 11.90%. Selective phosphodiesterase 4 (PDE4) inhibitor. Prepn: D. Flockerzi *et al.,* **WO 9501338**; H. Amschler, **US 5712298** (1995, 1998 both to Byk-Gulden). Pharmacology: A. Hatzelmann, C. Schudt, *J. Pharmacol. Exp. Ther.* **297**, 267 (2001); D. S. Bundschuh *et al., ibid.* 280. Clinical evaluation in allergic rhinitis: B. M. W. Schmidt *et al., J. Allergy Clin. Immunol.* **108**, 530 (2001); in exercise-induced asthma: W. Timmer *et al., J. Clin. Pharmacol.* **42**, 297 (2002). Clinical trial in chronic obstructive pulmonary disease (COPD): K. F. Rabe *et al., Lancet* **366**, 563 (2005). Review of pharmacology and clinical experience in inflammatory ariway disease: V. Boswell-Smith, C. P. Page, *Expert Opin. Invest. Drugs* **15**, 1105-1113 (2006); in COPD: M. Cazzola *et al., Expert Opin. Pharmacother.* **11**, 441-449 (2010).

Crystals from isopropanol, mp 158°. Practically insol in water, hexane. Sparingly sol in ethanol; freely sol in acetone.
THERAP CAT: Antiasthmatic; in treatment of chronic obstructive pulmonary disease.

8379. Rokitamycin. [74014-51-0] Leucomycin V 4B-butanoate 3B-propanoate; 3″-propionylleucomycin A5; 5-[*O*-(2,6-dideoxyhexopyranosyl)-(1 → 4)-3,6-dideoxy-3-dimethylamino-β-D-glucopyranosyloxy]-6-formylmethyl-3,9-dihydroxy-4-methoxy-8-methyl-10,12-hexadecadien-15-olide; rikamycin; M-19-Q; TMS-19Q; Ricamycin; Rokital. $C_{42}H_{69}NO_{15}$; mol wt 828.01. C 60.92%, H 8.40%, N 1.69%, O 28.98%. Macrolide antibiotic active against Mycoplasma and macrolide resistant strains of *Staphylococcus aureus* and *Streptococcus pyogenes.* Prepn: H. Sakakibara *et al.,* **DE 2918954** corresp to **US 4242504** (1979, 1980 both to Toyo Jozo);

eidem, J. Antibiot. **34**, 1001 (1981). Series of articles on antibacterial activity, pharmacology, metabolism, toxicology: *Chemotherapy (Tokyo)* **32**, Suppl. 6, 1-627 (1984). Toxicity data: K. Matsumoto *et al., ibid.* 138.

mp 116°. $[\alpha]_D^{20}$ −71° (c = 1.0 in chloroform). uv max (ethanol): 232 nm (ε 28000).

THERAP CAT: Antibacterial.

8380. Rolipram. [61413-54-5] 4-[3-(Cyclopentyloxy)-4-methoxyphenyl]-2-pyrrolidinone; ZK-62711. $C_{16}H_{21}NO_3$; mol wt 275.35. C 69.79%, H 7.69%, N 5.09%, O 17.43%. Prototypical selective phosphodiesterase 4 (PDE4) inhibitor. Prepn: **BE 826923**; R. Schmiechen *et al.*, **US 4193926** (1975, 1980 both to Schering AG). Selective enzyme inhibition and enhancement of cAMP accumulation in rat brain tissue: U. Schwabe *et al., Mol. Pharmacol.* **12**, 900 (1976). Antidepressant mechanism of action study: H. Wachtel, *Neuropharmacology* **22**, 267 (1983). Receptor binding study: H. H. Schneider *et al., Eur. J. Pharmacol.* **127**, 105 (1986). Pharmacokinetics: W. Krause, G. Kühne, *Xenobiotica* **18**, 561 (1988). Clinical evaluation in severe depression: F. Eckmann *et al., Curr. Ther. Res.* **43**, 291 (1988). Pharmacology and efficacy in exptl autoimmune encephalomyelitis: N. Sommer *et al., Nat. Med.* **1**, 244 (1995).

Crystals from ethyl acetate, mp 132°.

USE: Pharmacological tool for characterization of phosphodiesterase isoenzymes.

THERAP CAT: Antidepressant.

8381. Rolitetracycline. [751-97-3] (4S,4aS,5aS,6S,12aS)-4-(Dimethylamino)-1,4,4a,5,5a,6,11,12a-octahydro-3,6,10,12,12a-pentahydroxy-6-methyl-1,11-dioxo-N-(1-pyrrolidinylmethyl)-2-naphthacenecarboxamide; N-(pyrrolidinomethyl)tetracycline; N-(1-pyrrolidinylmethyl)tetracycline; Reverin; Syntetrin; Tetraverin; Transcycline. $C_{27}H_{33}N_3O_8$; mol wt 527.57. C 61.47%, H 6.31%, N 7.97%, O 24.26%. Semi-synthetic antibiotic prepd from tetracycline: Siedel *et al., Muench. Med. Wochenschr.* **100**, 661 (1958); Lindner *et al.*, **ZA 5703169**. Alternate procedure and structure: Gottstein *et al., J. Am. Chem. Soc.* **81**, 1198 (1959); Cheney *et al.*, **US 3104240** (1963 to Bristol-Myers). Toxicity: E. I. Goldenthal, *Toxicol. Appl. Pharmacol.* **18**, 185 (1971).

Fine, pale yellow needles, dec 162-165°. Amphoteric. More sol than tetracycline and tetracycline hydrochloride. Soly in water at 25°: 1.25 g/ml. Freely sol in alc. Sol in dil acids and alkalies.

Nitrate sesquihydrate. [26657-13-6] Bristacin; Pyrrocycline-N. $C_{27}H_{33}N_3O_8 \cdot HNO_3 \cdot 1\frac{1}{2}H_2O$; mol wt 617.61. LD_{50} i.v. in rats: 91 mg/kg (Goldenthal).

Compd with chloramphenicol succinate. [4154-10-3] Cafrolicycline; gradocycline; levocycline; senociclin; Clorociclin; Crovicina; Metilcaf; Proterciclina; Reicaf; Tecaf; Tetrafenicol. $C_{27}H_{33}N_3$-$O_8 \cdot C_{15}H_{16}Cl_2N_2O_8$; mol wt 950.77. Prepn: **BE 636234**; Scevola, **US 3218335** (1963, 1965 both to Lab. Pro-Ter). Bright yellow powder, dec 140-144°. Bitter taste. Very sol in water. Practically insol in ether, ligroin, hexane.

THERAP CAT: Antibacterial.

8382. Romidepsin. [128517-07-7] Cyclo[(2Z)-2-amino-2-butenoyl-L-valyl-(3S,4E)-3-hydroxy-7-mercapto-4-heptenoyl-D-valyl-D-cysteinyl] cyclic (3 → 5)-disulfide; [S-(E)]-N-(3-hydroxy-7-mercapto-1-oxo-4-heptenyl)-D-valyl-D-cysteinyl-(Z)-2,3-didehydro-2-aminobutanoyl-L-valine (4 → 1)-lactone cyclic (1 → 2)-disulfide; (1S,4S,7Z,10S,16E,21R)-7-ethylidene-4,21-bis(1-methylethyl)-2-oxa-12,13-dithia-5,8,20,23-tetraazabicyclo[8.7.6]tricos-16-ene-3,6,-9,19,22-pentone; FK-228; FR-901228; NSC-630176; Chromadax; Istodax. $C_{24}H_{36}N_4O_6S_2$; mol wt 540.69. C 53.31%, H 6.71%, N 10.36%, O 17.75%, S 11.86%. Bicyclic depsipeptide antibiotic produced by *Chromobacterium violaceum* WB968. Natural prodrug converted to glutathione-mediated reduction of the disulfide bond to the active histone deacetylase (HDAC) inhibitor. Isoln: M. Okuhara *et al.*, **EP 352646**; *eidem*, **US 4977138** (both 1990 to Fujisawa). Series of articles on isoln, structure, and activity: *J. Antibiot.* **47**, 301-323 (1994). Total synthesis: K. W. Li *et al., J. Am. Chem. Soc.* **118**, 7237 (1996). HPLC determn in plasma: C. Chassaing *et al., J. Chromatogr. B* **719**, 169 (1998). Mechanism of action studies: H. Nakajima *et al., Exp. Cell Res.* **241**, 126 (1998); R. Furumai *et al., Cancer Res.* **62**, 4916 (2002). Clinical pharmacokinetics and toxicological evaluation: V. Sandor *et al., Clin. Cancer Res.* **8**, 718 (2002). Clinical evaluation in leukemias: J. C. Byrd *et al., Blood* **105**, 959 (2005). Review of pharmacology and clinical experience: D. M. Vigushin, *Curr. Opin. Investig. Drugs* **3**, 1396-1402 (2002); P. A. Konstantinopoulos *et al., Cancer Chemother. Pharmacol.* **58**, 711-715 (2006).

Colorless prisms, mp 235-245° (dec). $[\alpha]_D^{23}$ +39° (c = 1.0 in $CHCl_3$). Sol in chloroform, ethyl acetate; sparingly sol in methanol, ethanol. Insol in water, hexane.

THERAP CAT: Antineoplastic.

8383. Romiplostim. [267639-76-9] AMG-531; Nplate; Romiplate. Thrombopoiesis stimulating recombinant fusion protein that binds to the thrombopoietin (TPO) receptor (mpl). Peptidyl structure consists of 2 disulfide-bonded human IgG_1 Fc domains each covalently linked to a peptide containing 2 TPO receptor binding sequences. Has no sequence homology to endogenous TPO. Mol wt 60 kDa. Prepn by cloning and expression in *E. coli*: C.-F. Liu *et al.*, **WO 0024770**; *eidem*, **US 6835809** (2000, 2004 both to Amgen). Stimulation of megakaryopoiesis *in vitro* by Mpl binding: V. C. Broudy, N. L. Lin, *Cytokine* **25**, 52 (2004). Clinical pharmacodynamics and pharmacokinetics: Y. Kumagai *et al., J. Clin. Pharmacol.* **47**, 1489 (2007). Clinical impact on quality of life in immune thrombocytopenic purpura (ITP): S. D. Mathias *et al., Clin. Ther.* **29**, 1574 (2007); on durable platelet response in chronic ITP: D. J. Kuter *et al., Lancet* **371**, 395 (2008). Review of development and clinical experience in thrombocytopenia in ITP and myelodyplastic syndromes: R. V. Tiu, M. A. Sekeres, *Expert Opin. Biol. Ther.* **8**, 1021-1030 (2008).

THERAP CAT: Antithrombocytopenic.

8384. Romurtide. [78113-36-7] N-(N-Acetylmuramoyl)-L-alanyl-D-α-glutaminyl-N^6-(1-oxooctadecyl)-L-lysine; N^2-(N-acetylmuramoyl-L-alanyl-D-isoglutaminyl)-N^6-stearoyl-L-lysine; MDP-Lys (L18); muroctasin; DJ-7041; Nopia. $C_{43}H_{78}N_6O_{13}$; mol wt 887.13. C 58.22%, H 8.86%, N 9.47%, O 23.44%. Derivative of muramyl dipeptide, q.v., with immunostimulating activity. Prepn: T. Shiba et al., **EP 21367**; eidem, **US 4317771** (1981, 1982 both to Daiichi Seiyaku). Immunopotentiating effect in vitro: Y. Osada et al., Infect. Immun. **38**, 848 (1982); in vivo: K. Matsumoto et al., ibid. **39**, 1029 (1983); M. Akasaki et al., Agents Actions **22**, 144 (1987). Enzyme immunoassay in plasma: H. Masayasu et al., Chem. Pharm. Bull. **33**, 5522 (1985). Supplement on pharmacology, toxicology and clinical studies: Arzneim.-Forsch. **38**, Suppl. 7A, 951-1074 (1988). Acute toxicity: Y. Ono et al., ibid. 1022.

Odorless, tasteless white powder, mp 176-178°. $[\alpha]_D^{20}$ +24.9° (c = 1 in methanol). pKa 5.70. Soly at 20°: DMF 1 g/72 ml; methanol 1 g/190 ml; ethanol 1 g/330 ml. Practically insol in acetone, acetonitrile, chloroform, water. LD_{50} in male, female mice, male, female rats, dogs (mg/kg): 436, 625, 761, 801, >200 s.c. (Ono).

THERAP CAT: Immunostimulant.

8385. Ronidazole. [7681-76-7] 1-Methyl-5-nitro-1H-imidazole-2-methanol 2-carbamate; carbamic acid (1-methyl-5-nitroimidazol-2-yl)methyl ester; 1-methyl-2-[(carbamoyloxy)methyl]-5-nitroimidazole; (1-methyl-5-nitroimidazole-2-yl)methyl carbamate; MCMN; Ridzol. $C_6H_8N_4O_4$; mol wt 200.15. C 36.01%, H 4.03%, N 27.99%, O 31.97%. Prepn and antiprotozoal activity: **NL 6609552**; **NL 6609553** (both 1967 to Merck & Co.), C.A. **67**, 54123u, 11487y (1967); Verdi; Kollonitsch, **US 3450710**; **US 3459764** (both 1969 to Merck & Co.).

Pale yellow crystals, mp 167-169°. Soly in water (at pH 6.5) ~2.9 mg/ml at room temp. More sol in acid solns. Unstable in alkaline solns. pKa 1.2. Freely sol in acetone. Sol in methanol, ethanol, chloroform and ethyl acetate.

THERAP CAT (VET): Antimicrobial.

8386. Ronifibrate. [42597-57-9] 3-Pyridinecarboxylic acid 3-[2-(4-chlorophenoxy)-2-methyl-1-oxopropoxy]propyl ester; 3-(nicotinoyloxy)propyl p-chlorophenoxyisobutyrate; 3-[α-(p-chlorophenoxy)isobutyryloxy]propyl nicotinate; I-612; Cloprane. $C_{19}H_{20}ClNO_5$; mol wt 377.82. C 60.40%, H 5.34%, Cl 9.38%, N 3.71%, O 21.17%. Diester of clofibric and nicotinic acids, q.q.v. Prepn: Y. Hirata et al., **JP Kokai 73 40777** (1973 to Yamanouchi), C.A. **79**, 66180w (1973); H. Shindo et al., **JP Kokai 74 30377** (1974 to Kowa Pharm.), C.A. **81**, 135984s (1974). Clinical trial in hyperdyslipidemia: A. Bucalossi et al., Clin. Ter. **89**, 127 (1979). Efficacy and tolerability: G. Buzzelli et al., ibid. 251.

LD_{50} orally in mice: 4.08 g/kg (Bucalossi).

THERAP CAT: Antilipemic.

8387. Ronnel. [299-84-3] Phosphorothioic acid O,O-dimethyl O-(2,4,5-trichlorophenyl)ester; fenchlorphos; Trolene; Etrolene; Nankor; Korlan; Viozene. $C_8H_8Cl_3O_3PS$; mol wt 321.53. C 29.88%, H 2.51%, Cl 33.08%, O 14.93%, P 9.63%, S 9.97%. Organophosphate insecticide; cholinesterase inhibitor. Prepn: Schrader, **DE 814152** (1948 to Bayer); Moyle, **US 2599516** (1952 to Dow). Description of mfg process: Can. Dept. Agr. Bull. (3rd ed., Oct. 1957) p 136. Toxicity study: T. B. Gaines, Toxicol. Appl. Pharmacol. **14**, 515 (1969). GC-MS analysis: L. Amendola et al., Anal. Chim. Acta **461**, 97 (2002).

White powder, mp 41°. d^{25} 1.48. Log P (octanol/water): 5.07. Vapor pressure at 25°: 8×10^{-4} mm Hg. Practically insol in water (0.004 g/100 ml at 25°). Freely sol in acetone, carbon tetrachloride, ether, methylene chloride, toluene, kerosene. LD_{50} in male, female rats (mg/kg): 1250, 2630 orally (Gaines).

Caution: Potential symptoms of overexposure in exptl animals are cholinesterase inhibition; eye irritation; liver and kidney damage. See NIOSH Pocket Guide to Chemical Hazards (DHHS/NIOSH 97-140, 1997) p 274.

USE: Insecticide.

8388. Ropinirole. [91374-21-9] 4-[2-(Dipropylamino)ethyl]-1,3-dihydro-2H-indol-2-one; 4-[2-(di-n-propylamino)ethyl]-2(3H)-indolone; SKF-101468. $C_{16}H_{24}N_2O$; mol wt 260.38. C 73.81%, H 9.29%, N 10.76%, O 6.14%. Selective dopamine D_2-receptor agonist. Prepn: G. Gallagher, Jr., **US 4452808** (1984 to Smithkline Beckman); idem et al., J. Med. Chem. **28**, 1533 (1985). HPLC determn in plasma: J. E. Swagzdis, B. A. Mico, J. Pharm. Sci. **75**, 90 (1986). Pharmacology and receptor binding study: R. J. Eden et al., Pharmacol. Biochem. Behav. **38**, 147 (1991). Clinical pharmacology, pharmacokinetics: C. de Mey et al., Br. J. Clin. Pharmacol. **32**, 483 (1991). Review of clinical experience in Parkinson's disease: R. Pahwa et al., Expert Rev. Neurother. **4**, 581-588 (2004); in restless legs syndrome: C. Trenkwalder, Eur. J. Neurol. **13**, Suppl. 3, 21-30 (2006).

Hydrochloride. [91374-20-8] SKF-101468A; Adartrel; Requip. $C_{16}H_{24}N_2O.HCl$; mol wt 296.84. Crystals from acetonitrile, mp 241-243°. Soly in water: 133 mg/ml.

THERAP CAT: Antiparkinsonian. In treatment of restless legs syndrome.

8389. Ropivacaine. [84057-95-4] (2S)-N-(2,6-Dimethylphenyl)-1-propyl-2-piperidinecarboxamide; (S)-(−)-1-propyl-2',6'-pipecoloxylidide; l-N-n-propylpipecolic acid-2,6-xylidide; LEA-103. $C_{17}H_{26}N_2O$; mol wt 274.41. C 74.41%, H 9.55%, N 10.21%, O 5.83%. Prepn: A. F. Thuresson, C. Bovin, **WO 8500599** (1985 to Apothekernes); H.-J. Federsel et al., Acta Chem. Scand. **B41**, 757 (1987). Physicochemical properties: G. R. Strichartz et al., Anesth. Analg. **71**, 158 (1990). HPLC determn in human plasma: Z. Yu et al., J. Chromatogr. B **654**, 221 (1994). In vitro metabolism: Y. Oda et al., Anesthesiology **82**, 214 (1995). Clinical pharmacokinetics: D. J. Kopacz et al., ibid. **81**, 1139 (1994). Toxicity study in sheep: A. C. Santos et al., ibid. **82**, 734 (1995). Clinical evaluation in relief of surgical pain: I. Cederholm et al., Reg. Anesth. **19**, 18 (1994); B. Johansson et al., Anesth. Analg. **78**, 210 (1994); labor pain: R. Stienstra et al., ibid. **80**, 285 (1995).

Crystals from toluene, mp 144-146°. $[\alpha]_D^{25}$ −82.0° (c = 2 in methanol). pKa 8.16. Distribution coefficient (1-octanol/aq buffer, pH 7.4): 115.0.

Hydrochloride. [98717-15-8] Naropin. $C_{17}H_{26}N_2O$.HCl; mol wt 310.87. Crystals from isopropyl alcohol, mp 260-262°. $[\alpha]_D^{25}$ −6.6° (c = 2 in water). Sol in water.

Hydrochloride monohydrate. [132112-35-7] Crystals from acetone + water, mp 269.5-270.6°. $[\alpha]_D^{20}$ −7.28° (c = 2 in water).

THERAP CAT: Anesthetic (local).

8390. Roquinimex. [84088-42-6] 1,2-Dihydro-4-hydroxy-N,1-dimethyl-2-oxo-N-phenyl-3-quinolinecarboxamide; N-phenyl-N-methyl-1,2-dihydro-4-hydroxy-1-methyl-2-oxoquinoline-3-carboxamide; 1,2-dihydro-4-hydroxy-N,1-dimethyl-2-oxo-3-quinolinecarboxanilide; LS-2616; Linomide. $C_{18}H_{16}N_2O_3$; mol wt 308.34. C 70.12%, H 5.23%, N 9.09%, O 15.57%. Biological response modifier. Prepn: E. Eriksoo et al., **EP 59698**; eidem, **US 4738971** (1982, 1988 both to AB Leo). Immunopharmacology: A. Tarkowski et al., Immunology **59**, 589 (1986). Mechanism of action study: E.-L. Larsson et al., Int. J. Immunopharmacol. **9**, 425 (1987). Clinical evaluation in cancer patients: J. C. S. Bergh et al., Cancer Invest. **15**, 204 (1997).

Crystals from pyridine, mp 200-204°.

THERAP CAT: Antineoplastic.

8391. Rosaprostol. [56695-65-9] 2-Hexyl-5-hydroxycyclo-pentaneheptanoic acid; 9-hydroxy-19,20-bisnorprostanoic acid; C-83; IBI-C83; Rosal. $C_{18}H_{34}O_3$; mol wt 298.47. C 72.44%, H 11.48%, O 16.08%. Prostaglandin analog. Prepn, hypolipemic, platelet aggregation inhibitory activity: U. Valcavi, **DE 2535343**, eidem, **US 4073938** (1976, 1978 both to Ist. Biochim. Ital.). Alternate process: V. Marotta, G. Zabban, **EP 155392** (1985 to Ist. Biochim. Ital.). Gastric antisecretory, cytoprotective activity: U. Valcavi et al., Arzneim.-Forsch. **32**, 657 (1982). Effect on mucus and gastrin secretion in duodenal ulcer: D. Foschi et al., Prostaglandins Leukotrienes Med. **15**, 147 (1984). Comparison with cimetidine, q.v.: eidem, Drugs Exp. Clin. Res. **10**, 427 (1984). Clinical evaluation in treatment of ulcers: G. P. Tincani et al., Minerva Med. **78**, 847 (1987).

Oil.

Sodium salt. [56695-66-0] $C_{18}H_{33}NaO_3$. White solid. LD_{50} orally in mice: ~3000 mg/kg (Valcavi, 1978); orally in rats: >5 g/kg (Valcavi, 1982).

THERAP CAT: Antiulcerative.

8392. Rosaramicin. [35834-26-5] 4′-Deoxycirramycin A₁; 3-ethyl-7-hydroxy-2,8,12,16-tetramethyl-5,13-dioxo-9-[[3,4,6-tri-deoxy-3-(dimethylamino)-β-D-xylo-hexopyranosyl]oxy]-4,17-di-oxabicyclo[14.1.0]heptadec-14-ene-10-acetaldehyde; antibiotic 67-694; juvenimicin A₃; rosamicin; M-4365A2; Sch-14947. $C_{31}H_{51}$-NO_9; mol wt 581.75. C 64.00%, H 8.84%, N 2.41%, O 24.75%.

Macrolide antibiotic isolated from fermentations of *Micromonospora rosaria* NRRL-3718. Isoln and properties: M. J. Weinstein et al., **ZA 7100402**; **FR 2081448** (1971, 1972 both to Sherico), C.A. **76**, 139065n (1972); **78**, 109310n (1973); G. H. Wagman et al., J. Antibiot. **25**, 641 (1972). Biological studies: J. A. Waitz et al., ibid. 647. Structure: H. Reimann, R. S. Jaret, Chem. Commun. **1972**, 1270. Crystal structure: A. K. Ganguly et al., Tetrahedron Lett. **21**, 4699 (1980). Biosynthesis: A. K. Ganguly et al., J. Antibiot. **29**, 976 (1976). Identity with juvenimicin A₃: T. Kishi et al., ibid. 1171. Isoln from *M. capillata* MCRL 0940 and identity with antibiotic M-4365A2: A. Kinumaki et al., ibid. **30**, 450 (1977). In vitro activity: S. Feltham et al., J. Antimicrob. Chemother. **5**, 731 (1979). Use in experimental pneumococcal meningitis: C. M. Nolan et al., Antimicrob. Agents Chemother. **16**, 776 (1979).

Crystals from chloroform, mp 119-122°. $[\alpha]_D^{26}$ −35° (ethanol). uv max (methanol): 240 nm (ε 14600). Very sol in methanol, acetone, chloroform, benzene. Sparingly sol in ether. Slightly sol in water. LD_{50} in mice (mg/kg): 625 s.c.; 350 i.p.; 155 i.v. (Wagman).

THERAP CAT: Antibacterial.

8393. Rose Bengal. [632-68-8] 4,5,6,7-Tetrachloro-3′,6′-di-hydroxy-2′,4′,5′,7′-tetraiodospiro[isobenzofuran-1(3H),9′-[9H]-xanthen]-3-one potassium salt (1:2); 4,5,6,7-tetrachloro-2′,4′,5′,7′-tetraiodofluorescein dipotassium salt; C.I. Acid Red 94; Rose Bengale B; C.I. 45440. $C_{20}H_2Cl_4I_4K_2O_5$; mol wt 1049.85. C 22.88%, H 0.19%, Cl 13.51%, I 48.35%, K 7.45%, O 7.62%. Discovered by Gnehm 1882; tetraiodinate of 4,5,6,7-tetrachlorofluorescein (from resorcinol and tetrachlorophthalic anhydride) converted to potassium salt: *Colour Index* vol. 4 (3rd ed., 1971) 4428. Labeling with ¹³¹I: Liebner, Andrysek, Nature **184**, 913 (1959). See also: H. J. Conn's Biological Stains, R. D. Lillie, Ed., (Williams & Wilkins, Baltimore, 9th ed., 1977) p 350. Diagnostic use in hepatic function: K. S. Nijran, D. C. Barber, Phys. Med. Biol. **31**, 563 (1986).

Bright bluish pink. Soluble in water (bluish red without fluorescence); brown soln in concd H_2SO_4, on dilution gives flesh pink ppt.

Disodium salt. [632-69-9] Rose Bengal Extra. $C_{20}H_2Cl_4I_4Na_2$-O_5.

¹³¹I-Labeled disodium salt. [50291-21-9] Sodium rose bengal I 131; rose bengal sodium I 131; Robenogatope I-131.

Note: C.I. Acid Red 94 and C.I. 45440 also used for disodium salt.

USE: As a dye, biological stain; in coloring straw, wood chips, and inks; for coloring edible products and cosmetics.

THERAP CAT: Diagnostic aid (corneal trauma indicator). Sodium Rose Bengal I 131 as diagnostic aid (hepatic function).

8394. Rose Hips. Hipberries. The fruits or berries from wild rose bushes, notably *Rosa canina* L., *R. gallica* L., *R. condita* Scop. and *R. rugosa* Thunb., Rosaceae. Rich in ascorbic acid: M. Szczyglowa *et al.*, *Rocz. Panstw. Zakl. Hig.* **1**, 523-532 (1950), *C.A.* **46**, 1177d. Ascorbic acid content of hips of *R. canina* reported as 19-27 mg/g dry weight: Å. Gustafsson, J. Schröderheim, *Nature* **153**, 196 (1944); as 33.1 mg/g dry weight: I. Roubani *et al.*, *J. Hortic. Sci.* **51**, 375 (1976). Comparison of methods for determn of ascorbic acid in hips: K. Gliniecki *et al.*, *Pharm. Ztg.* **127**, 823 (1982).

8395. Rosemary. Low, branched perennial shrub, *Rosmarinus officinalis* L., *Lamiaceae* (*Labiatae*). *Habit*. Mediterranean basin; widely cultivated in gardens. Leaves are used as a seasoning in cooking; extracts and essential oil are used in traditional medicine. *Constit.* Volatile oil (1.0-2.5%); antioxidant diterpenoids incl. carnosic acid, carnosol, rosmanol; rosmarinic acid; flavonoids incl. genkwanin, cirsimarin. HPLC determn of antioxidant components: K. Schwarz, W. Ternes, *Z. Lebensm.-Unters. Forsch.* **195**, 95 (1992). Extraction of volatile components: A. Basile *et al.*, *J. Agric. Food Chem.* **46**, 5205 (1998). Use as antioxidant in edible oils: N. V. Yanishlieva, E. M. Marinova, *Eur. J. Lipid Sci. Technol.* **103**, 752 (2001); in packaged beef: N. T. M. McBride *et al.*, *Int. J. Food Chem. Technol.* **42**, 1201 (2007). Botanical description, constituents, and uses: B. Sasikumar in *Handbook of Herbs and Spices*, **Vol. 2**, K. V. Peter, Ed., (Woodhead Publishing, Cambridge, 2004) pp 243-255. Review of pharmacology: J. Barnes *et al.*, *Herbal Medicines* (Pharmaceutical Press, London, 3rd Ed., 2007) pp 508-511.

Volatile oil. [8000-25-7] Oil of rosemary. Obtained from the fresh flowering tops. *Constit.* Complex mixture of monoterpenes, primarily α-Pinene (15-25%), eucalyptol (20-50%), camphor (10-25%), borneol, (16-20%) camphene, limonene, linalool. Colorless or pale yellow liquid; characteristic rosemary odor; camphoraceous taste. d_{25}^{25} 0.894-0.912. α_D^{25} −5 to +10°. n_D^{20} 1.464-1.476. Almost insol in water. Sol in 10 vols 80% alcohol. *Keep well closed, cool and protected from light.* LD$_{50}$ orally in rats: ~5 ml/kg; dermally in rabbits: >10 ml/kg. *See: Food Cosmet. Toxicol.* **12**, 977 (1974).

USE: Condiment and flavoring agent in foods; antioxidant in edible oils, meats, and other fat containing foods. Dietary supplement. Extracts and oil in hair conditioners, mouth rinses, fragrances.

THERAP CAT: Carminative; rubefacient; antimicrobial.

8396. Rosiglitazone. [122320-73-4] 5-[[4-[2-(Methyl-2-pyridinylamino)ethoxy]phenyl]methyl]-2,4-thiazolidinedione; 5-[4-[2-[*N*-methyl-*N*-(2-pyridinyl)amino]ethoxy]benzyl]-2,4-thiazolidinedione; BRL-49653. $C_{18}H_{19}N_3O_3S$; mol wt 357.43. C 60.49%, H 5.36%, N 11.76%, O 13.43%, S 8.97%. Insulin sensitizer; binds to peroxisome proliferator activated receptor gamma (PPAR-γ). Prepn: R. M. Hindley, **EP 306228**; *idem*, **US 5002953** (1989, 1991 both to Beecham); B. C. C. Cantello *et al.*, *J. Med. Chem.* **37**, 3977 (1994). Pharmacology: N. D. Oakes *et al.*, *Diabetes* **43**, 1203 (1994). Binding study: D. E. Mais *et al.*, *Med. Chem. Res.* **7**, 325 (1997). Clinical pharmacokinetics and bioavailability: M. I. Freed *et al.*, *Eur. J. Clin. Pharmacol.* **55**, 53 (1999). Review of pharmacology and clinical experience: J. A. Barman Balfour, G. L. Plosker, *Drugs* **57**, 921-930 (1999).

Colorless crystals from methanol, mp 153-155°.

Maleate. [155141-29-0] BRL-49653c; Avandia. $C_{18}H_{19}N_3O_3 \cdot C_4H_4O_4$; mol wt 473.50. White to off-white solid, mp 122-123°. pKa 6.1; 6.8. Readily sol in ethanol and in buffered aq soln with pH of 2.3; soly decreases with increasing pH in the physiological range.

THERAP CAT: Antidiabetic.

8397. Rosin. Colophony; yellow resin; abietic anhydride. Residue left after distilling off the volatile oil from the oleoresin obtained from *Pinus palustris* and other species of *Pinus*, *Pinaceae*. Offered as wood rosin (from Southern pine stumps), gum rosin (the exudate from incisions in the living tree, *P. palustris* and *P. caribaea*), and tall oil rosin, *see* Tall Oil. Rosin is chiefly produced in the U.S.A. *Constit.* About 90% resin acids and 10% neutral matter. Of the resin acids about 90% are isomeric with abietic acid ($C_{20}H_{30}$-

O_2); the other 10% is a mixture of dihydroabietic acid ($C_{20}H_{32}O_2$) and dehydroabietic acid ($C_{20}H_{28}O_2$).

Pale yellow to amber, translucent fragments; brittle fracture at ordinary temp; slight turpentine odor and taste. Readily fusible when heated. d 1.07-1.09. Acid no. not less than 150. Insol in water. Freely sol in alc, benzene, ether, glacial acetic acid, oils, carbon disulfide; also sol in dil solns or fixed alkali hydroxides.

USE: Manuf varnishes, varnish and paint driers, printing inks, cements, soap, sealing wax, wood polishes, floor coverings, paper, plastics, fireworks, tree wax, sizes, rosin oil; for water-proofing cardboard, walls, etc. Pharmaceutic aid (stiffening agent).

8398. Rosin Oil. Rosinol. Obtained by dry distillation of rosin.

Yellow, viscid, fluorescent, oily liquid. bp >280°. *Flammable.* Insol in water. Sol in ether, oil turpentine and other oils. It dissolves phosphorus, sulfur, camphor, phenols, and many other organic compds.

USE: Manuf of carbon black for lithography and printing inks; in varnishes, retinol colors, lacquers, brewers' pitch, axle greases.

8399. Rosmarinic Acid. [20283-92-5] (α*R*)-α-[[(2*E*)-3-(3,4-Dihydroxyphenyl)-1-oxo-2-propen-1-yl]oxy]-3,4-dihydroxybenzenepropanoic acid; 3,4-dihydroxycinnamic acid 2-ester with 3-(3,4-dihydroxyphenyl)lactic acid; α-*O*-caffeoyl-(3,4-dihydroxyphenyl)-lactic acid. $C_{18}H_{16}O_8$; mol wt 360.32. C 60.00%, H 4.48%, O 35.52%. Naturally occurring antioxidant principally found in rosemary, sage, and other aromatic herbs of the mint family; occurs in lesser amounts in wide variety of plants. Isoln from *Rosmarinus officinalis* L., *Lamiaceae*: M. L. Scarpati, G. Oriente, *Ricerca Sci.* **28**, 2329 (1958); from lemon balm, peppermint, and sage: K. Herrmann, *Arch. Pharm.* **293**, 1043 (1960). Synthesis: T. Eicher *et al.*, *Synthesis* **1996**, 755. Large scale production in plant cell suspensions: A. I. Pavlov *et al.*, *Biotechnol. Prog.* **21**, 394 (2005). HPLC determn in plant material: H. Wang *et al.*, *Food Chem.* **87**, 307 (2004). Biosynthesis in *Mentha* species: B. E. Ellis, G. H. N. Towers, *Biochem. J.* **118**, 291 (1970). Antioxidant activity in bulk corn oil: E. N. Frankel *et al.*, *J. Agric. Food Chem.* **44**, 131 (1996). Clinical evaluation as anti-allergic: N. Osakabe *et al.*, *BioFactors* **21**, 127 (2004). Review: M. Petersen, M. S. J. Simmonds, *Phytochemistry* **62**, 121-125 (2003).

White crystalline powder, mp 172-174°. $[\alpha]_D^{25}$ +39.2° (c = 1.0 in methanol). Soly (mg/ml): 25 in ethanol, DMSO, DMF; 15 in phosphate buffered saline (pH 7.2).

USE: Antioxidant in cosmetics, edible oils; dietary supplement.

8400. Rosoxacin. [40034-42-2] 1-Ethyl-1,4-dihydro-4-oxo-7-(4-pyridinyl)-3-quinolinecarboxylic acid; acrosoxacin; Win-35213; Eracine; Eradacil; Eradacin; Winuron. $C_{17}H_{14}N_2O_3$; mol wt 294.31. C 69.38%, H 4.79%, N 9.52%, O 16.31%. Quinolone antibacterial. Prepn: G. Y. Lesher, P. M. Carabateas, **DE 2224090**; *eidem*, **US 3753993** (1972, 1973 both to Sterling). HPLC determn in plasma and urine: M. P. Kullberg *et al.*, *J. Chromatogr.* **173**, 155 (1979). Pharmacological studies: S. Maigaard *et al.*, *Urol. Res.* **8**, 113 (1980); *eidem*, *Invest. Urol.* **17**, 149 (1979). Clinical study in gonorrhea: B. M. Limson *et al.*, *Curr. Ther. Res.* **26**, 842 (1979).

Yellow crystals from DMF, mp 290°. Stable in dry heat at 70°. Light sensitive. Solns and suspensions stable at pH 2.0-9.5.

THERAP CAT: Antibacterial.

8401. Rostaporfin. [284041-10-7] (OC-6-13)-Dichloro[rel-ethyl (18R,19S)-3,4,20,21-tetradehydro-4,9,14,19-tetraethyl-18,19-dihydro-3,8,13,18-tetramethyl-20-phorbinecarboxylato(2−)-κN^{23},-κN^{24},κN^{25},κN^{26}]tin; SnET$_2$; tin ethyl etiopurpurin dichloride; tin etiopurpurin dichloride; Photrex; Purlytin. $C_{37}H_{42}Cl_2N_4O_2Sn$; mol wt 764.38. C 58.14%, H 5.54%, Cl 9.28%, N 7.33%, O 4.19%, Sn 15.53%. Metallopurpurin photosensitizer for photodynamic therapy. Prepn: A. R. Morgan et al., SPIE **847**, 172 (1987); eidem, Cancer Res. **48**, 194 (1988). Pharmacokinetics in dogs and rats: D. L. Frazier et al., J. Vet. Pharmacol. Ther. **15**, 275 (1992). Intracranial tissue uptake and specificity: L. Lilge, B. C. Wilson, J. Clin. Laser Med. Surg. **16**, 81 (1998). Photophysical properties: B. W. Pogue et al., Photochem. Photobiol. **68**, 809 (1998). Clinical evaluation in cutaneous metastatic breast cancer: T. S. Mang et al., Cancer J. Sci. Am. **4**, 378 (1998). Review of clinical development in wet age-related macular degeneration: D. W. C. Hunt, IDrugs **5**, 180-186 (2002).

Relative stereochemistry

Crystals from dichloromethane:methanol(10:1). Absorption max (dichloromethane): 659 nm (ε 30347). Absorption max: 656 nm (methanol); 661 (phosphate-buffered saline); (ε 42800; 15200).

THERAP CAT: In treatment of age-related macular degeneration.

8402. Rosuvastatin. [287714-41-4] (3R,5S,6E)-7-[4-(4-Fluorophenyl)-6-(1-methylethyl)-2-[methyl(methylsulfonyl)amino]-5-pyrimidinyl]-3,5-dihydroxy-6-heptenoic acid. $C_{22}H_{28}FN_3O_6S$; mol wt 481.54. HMG-CoA reductase inhibitor. Prepn: K. Hirai et al., **EP 521471**; eidem, **US 5260440** (both 1993 to Shionogi); M. Watanabe et al., Bioorg. Med. Chem. **5**, 437 (1997). Pharmacology: F. McTaggart et al., Am. J. Cardiol. **87**, Suppl., 28B (2001). Comparative clinical trial with atorvastatin: M. Davidson et al., ibid. **89**, 268 (2002); in metabolic syndrome: A. F. H. Stalenhoef et al., Eur. Heart J. **26**, 2664 (2005); in African-American patients: K. C. Ferdinand et al., Am. J. Cardiol. **97**, 229 (2006). Pharmacokinetics and pharmacogenetics in Asian and white ethnic groups: E. Lee et al., Clin. Pharmacol. Ther. **78**, 330 (2005). Review of therapeutic potential: K. C. Ferdinand, Expert Opin. Pharmacother. **6**, 1897-1910 (2005).

Partition coefficient (octanol/water): 0.13 (pH 7.0).

Calcium salt. [147098-20-2] S-4522; ZD-4522; Crestor. C_{44}CaH$_{54}$F$_2$N$_6$O$_{12}$S$_2$; mol wt 1001.14. White powder from water as the monohydrate; begins to melt at 155° with no definitive mp.

$[\alpha]_D^{24}$ +14.8° (c = 1.012 in 50% methanol). Sparingly sol in water, methanol; slightly sol in ethanol.

THERAP CAT: Antilipemic.

8403. Rotenone. [83-79-4] (2R,6aS,12aS)-1,2,12,12a-Tetrahydro-8,9-dimethoxy-2-(1-methylethenyl)-[1]benzopyrano[3,4-b]-furo[2,3-h][1]benzopyran-6(6aH)-one; Canex; Noxfire. $C_{23}H_{22}O_6$; mol wt 394.42. C 70.04%, H 5.62%, O 24.34%. Principal insecticidal constituent of derris root, cubé, etc. Powerful inhibitor of mitochondrial electron transport. Isoln from Lonchocarpus nicou (Aubl.) DC., Leguminosae: Geoffrey, Ann. Inst. Colon. Marseille **2**, 1 (1895). Review of structure: La Forge et al., Chem. Rev. **12**, 181 (1933); Butenandt, McCartney, Ann. **494**, 17 (1932); King, Annu. Rep. Prog. Chem. **29**, 186 (1932). Absolute configuration: Büchi et al., J. Chem. Soc. **1961**, 2843; Nakazaki, Arakawa, Bull. Chem. Soc. Jpn. **34**, 1246 (1961); Begley et al., Chem. Commun. **1975**, 850. NMR spectrum: Crombie, Lown, ibid. **1962**, 775. Total synthesis: Miyano et al., Agric. Biol. Chem. **25**, 673 (1961); Miyano, J. Am. Chem. Soc. **87**, 3958 (1965). Alternate synthesis: Crombie et al., J. Chem. Soc. Perkin Trans. 1 **1973**, 1277; eidem, Chem. Commun. **1979**, 1142; I. Sasaki, K. Yamashita, Agric. Biol. Chem. **43**, 137 (1979). Synthesis of stereoisomers: Unai, Yamamoto, ibid. **37**, 897 (1973). Toxicology: Santi, Toth, Farmaco Ed. Sci. **20**, 270 (1965). Toxicity data: J.-I. Fukami et al., Science **155**, 713 (1967). Review: H. Fukami, M. Nakajima in Naturally Occurring Insecticides, M. Jacobson, D. G. Crosby, Eds. (Dekker, New York, 1971) pp 71-97; S. B. Soloway, Environ. Health Perspect. **14**, 109-117 (1976); T. J. Haley, J. Environ. Pathol. Toxicol. **1**, 315-337 (1978).

Orthorhombic, six-sided plates from trichloroethylene, mp 165-166° (dimorphic form, mp 185-186°). $[\alpha]_D^{20}$ −228° (c = 2.22 in benzene). uv spectra: Büchi et al., loc. cit. Practically insol in water. Sol in alcohol, acetone, carbon tetrachloride, chloroform, ether, and many other organic solvents. Dec upon exposure to light and air. Colorless solns in organic solvents oxidize upon exposure and become yellow, orange and then deep red and may deposit crystals of dehydrorotenone and rotenonone which are toxic to insects. LD$_{50}$ i.p. in mice: 2.8 mg/kg (Fukami); in rats (mg/kg): 132 orally; 6 i.v. (Soloway).

Caution: Potential symptoms of overexposure are irritation of eyes, skin, respiratory system; numbness of mucous membranes; nausea, vomiting, abdominal pain; muscle tremors, incontinence, clonic convulsions, stupor. See NIOSH Pocket Guide to Chemical Hazards (DHHS/NIOSH 97-140, 1997) p 274. Inhalation may cause severe pulmonary irritation. See Clinical Toxicology of Commercial Products, R. E. Gosselin et al., Eds. (Williams & Wilkins, Baltimore, 5th ed., 1984) Section III, pp 366-368.

USE: Pesticide.

THERAP CAT (VET): Ectoparasiticide.

8404. Rotigotine. [99755-59-6] (6S)-5,6,7,8-Tetrahydro-6-[propyl[2-(2-thienyl)ethyl]amino]-1-naphthalenol; (−)-2-(N-propyl-N-2-thienylethylamino)-5-hydroxytetralin; (−)-5-hydroxy-2[N-n-propyl-N-2-(2-thienyl)ethylamino]tetralin; N-0923; SPM-962; Neupro. $C_{19}H_{25}NOS$; mol wt 315.48. C 72.34%, H 7.99%, N 4.44%, O 5.07%, S 10.16%. Non-ergot dopamine receptor agonist. Prepn and activity of enantiomers: W. Timmerman et al., Eur. J. Pharmacol. **166**, 1 (1989). Improved synthesis: G. Minaskanian, K. Rippel, **US 6372920** (2002 to Aderis). Clinical trial of transdermal formulation in early Parkinson's disease: R. L. Watts et al., Neurology **68**, 272 (2007). Review of pharmacology and clinical experience of transdermal formulation in Parkinson's disease: Y. Naidu, K. Ray

Chaudhuri, *Expert Opin. Drug Deliv.* **4**, 111-118 (2007); and in restless legs syndrome: M. Y. Splinter, *Ann. Pharmacother.* **41**, 285-295 (2007).

White to off-white powder. Sensitive to oxidation. Non-hygroscopic. Lipophilic and poorly sol in water at neutral pH; soly increases at more acidic pH.

Hydrochloride. [125572-93-2] $C_{19}H_{25}NOS.HCl$; mol wt 351.93. Determn of enantiomeric purity by chiral HPLC: D. T. Witte *et al.*, *Chirality* **4**, 62 (1992). GC/MS determn in plasma: P. A. Zavitsanos *et al.*, *Rapid Commun. Mass Spectrom.* **7**, 1145 (1993). White crystalline solid. $[\alpha]_{578}^{25}$ $-55.79°$ (c = 0.99 in MeOH).

(±)-Form. [92206-54-7] N-0437. Prepn: A. S. Horn, **US 4564628** (1986 to Nelson).

THERAP CAT: Antiparkinsonian.

8405. Rottlerin. [82-08-6] (2E)-1-[6-[(3-Acetyl-2,4,6-trihydroxy-5-methylphenyl)methyl]-5,7-dihydroxy-2,2-dimethyl-2*H*-1-benzopyran-8-yl]-3-phenyl-2-propen-1-one; 5,7-dihydroxy-2,2-dimethyl-6-(2,4,6-trihydroxy-3-methyl-5-acetylbenzyl)-8-cinnamoyl-1,2-chromene; mallotoxin. $C_{30}H_{28}O_8$; mol wt 516.55. C 69.76%, H 5.46%, O 24.78%. Principal phenolic component of kamala, *q.v.*, an anthelmintic dye obtained from *Mallotus philippinensis* (Lam.) Muell. Arg. (also known as *Rottlera tinctoria* Roxb.), *Euphorbiaceae*. Isoln: H. Telle, *Arch. Pharm.* **244**, 446 (1906); S. Dutt, *J. Chem. Soc.* **127**, 2044 (1925); A. McGookin *et al.*, *ibid.* **1937**, 748. Structure: Brockmann, Maier, *Ann.* **535**, 149 (1938); A. McGookin *et al.*, *J. Chem. Soc.* **1939**, 1579. Synthesis of tetrahydrorottlerin: Backhouse *et al.*, *ibid.* **1948**, 113. Review of isoln and chemistry: M. Lounasmaa *et al.*, *Planta Med.* **28**, 16 (1975). Activity as protein kinase inhibitor: M. Gschwendt *et al.*, *Biochem. Biophys. Res. Commun.* **199**, 93 (1994). HPTLC determn in serum and pharmacokinetics: V. D. Mody *et al.*, *J. Pharm. Technol.* **10**, 71 (1994).

Light reddish-brown plates or needles with golden lustre from ethyl acetate, mp 212° (McGookin). Also reported as brownish-yellow plates from toluene, mp 206-207° (Dutt). Sol in ether, chloroform, benzene, ethyl acetate; sparingly sol in cold alc, acetic acid. Practically insol in water.

5,7-Dimethyl ether. $C_{32}H_{32}O_8$. Yellow crystals from ethyl acetate + acetone or chloroform + methanol, dec 245-246°. Sol in chloroform, pyridine, hot glacial acetic acid; slightly sol in cold methanol, ethyl acetate, acetone, benzene, ether.

Pentamethyl ether. $C_{35}H_{38}O_8$. Crystals from petr ether, methanol or 90% alcohol, mp 144°. Very sol in acetone, glacial acetic acid, benzene, ethyl acetate.

Pentaacetate. $C_{40}H_{38}O_{13}$. Leaflets from benzene + alcohol or acetone + alcohol, prisms from ethyl acetate, mp 212°.

Tetrahydrorottlerin. $C_{30}H_{32}O_8$. Yellow prisms from alcohol, mp 214°.

8406. Roxarsone. [121-19-7] *As*-(4-Hydroxy-3-nitrophenyl)-arsonic acid; 4-hydroxy-3-nitrobenzenearsonic acid; 3-nitro-4-hydroxyphenylarsonic acid; 2-nitro-1-hydroxybenzene-4-arsonic acid; nitrophenolarsonic acid; NSC-2101; Ren-o-sal. $C_6H_6AsNO_6$; mol

wt 263.04. C 27.40%, H 2.30%, As 28.48%, N 5.33%, O 36.49%. Prepd by treating sodium *p*-hydroxyphenylarsonate with a mixture of nitric and sulfuric acids at 0°: Benda, Bertheim, *Ber.* **44**, 3446 (1911); **DE 224953** *C.A.* **5**, 156 (1911). Toxicity study: Kerr *et al.*, *Toxicol. Appl. Pharmacol.* **5**, 507 (1963). Review of toxicology and human exposure: *Toxicological Profile for Arsenic* (PB2008-100002, 2007) 559 pp.

Tufts of pale-yellow needles or rhombohedral plates from water. Puffs up and deflagrates on heating. Freely sol in methanol, ethanol, acetic acid, acetone, alkalies; sol in ~30 parts boiling water; sparingly sol in dil mineral acids; slightly sol in cold water. Insol in ether, ethyl acetate. Forms mono-, di-, and trisodium salts. LD_{50} in rats, chickens (mg/kg): 155, 110-123 orally; 66, 34 i.p. (Kerr).

USE: Formerly in manuf arsphenamines; as reagent for zirconium.

THERAP CAT (VET): Control of enteric infections. To improve growth and feed efficiency.

8407. Roxatidine Acetate. [78628-28-1] 2-(Acetyloxy)-*N*-[3-[3-(1-piperidinylmethyl)phenoxy]propyl]acetamide; aceroxatidine. $C_{19}H_{28}N_2O_4$; mol wt 348.44. C 65.49%, H 8.10%, N 8.04%, O 18.37%. Histamine H_2-receptor antagonist. Prepn and anti-ulcerative activity: K. Shibata *et al.*, **EP 24510**; *eidem*, **US 4293557** (both 1981 to Teikoku Hormone). Pharmacology in animals: M. Tarutani *et al.*, *Arzneim.-Forsch.* **35**, 703 (1985). Comparison with cimetidine of effect on gastric acid secretion and ulcer formation in animals: *eidem, ibid.* 844. Metabolism in humans: S. Honma *et al.*, *Oyo Yakuri* **30**, 555 (1985); *C.A.* **104**, 61453n (1985).

mp 59-60°. LD_{50} orally in male mice: 1000 mg/kg (Shibata).

Hydrochloride. [93793-83-0] Pifatidine; HOE-760; TZU-0460; Altat; Gastralgin; Neo H2; Roxit. $C_{19}H_{28}N_2O_4.HCl$; mol wt 384.90. mp 145-146°.

THERAP CAT: Antiulcerative.

8408. Roxindole. [112192-04-8] 3-[4-(3,6-Dihydro-4-phenyl-1(2*H*)-pyridinyl)butyl]-1*H*-indol-5-ol; 5-hydroxy-3-[4-(1,2,3,6-tetrahydro-4-phenyl-1-pyridyl)butyl]indole. $C_{23}H_{26}N_2O$; mol wt 346.47. C 79.73%, H 7.56%, N 8.09%, O 4.62%. Dopamine D_2-receptor agonist. Synthesis: H.-H. Hausberg *et al.*, *Acta Pharm. Suec.* **1983**, Suppl 2, 213; H. Böttcher *et al.*, *J. Med. Chem.* **35**, 4020 (1992). Pharmacology: C. A. Seyfried *et al.*, *Eur. J. Pharmacol.* **160**, 31 (1989). Clinical pharmacology: K. Wiedemann, M. Kellner, *Exp. Clin. Endocrinol.* **102**, 284 (1994). Clinical evaluation in schizophrenia: H. Wetzel *et al.*, *Am. J. Psychiatry* **151**, 1499 (1994); in depression: G. Gründer *et al.*, *Psychopharmacology* **111**, 123 (1993). Neuroendocrinology in schizophrenics: G. Gründer *et al.*, *ibid.* **117**, 472 (1995).

Hydrochloride. [108050-82-4] EMD-38362. $C_{23}H_{26}N_2O.HCl$; mol wt 382.93. Monoclinic crystals, mp 274°.

Methanesulfonate. [119742-13-1] Roxindole mesylate; EMD-49980

THERAP CAT: Antidepressant.

8409. Roxithromycin. [80214-83-1] Erythromycin 9-[*O*-[(2-methoxyethoxy)methyl]oxime]; 9-(2′,5′-dioxahexyloxyimino)-erythromycin; RU-28965; RU-965; Assoral; Claramid; Forilin; Overal; Rossitrol; Rotramin; Rulid; Surlid. $C_{41}H_{76}N_2O_{15}$; mol wt 837.06. C 58.83%, H 9.15%, N 3.35%, O 28.67%. Semisynthetic erythromycin derivative. Prepn: S. Gouin d'Ambrieres *et al.*, **FR 2473525**; *eidem*, **US 4349545** (1981, 1982 both to Roussel-UCLAF). *In vitro* antibacterial spectrum: R. N. Jones *et al.*, *Antimicrob. Agents Chemother.* **24**, 209 (1983); Y. J. Drabu *et al.*, *Drugs Exp. Clin. Res.* **13**, 201 (1987). Antiprotozoal activity in mice: B. J. Luft, *Eur. J. Clin. Microbiol.* **6**, 479 (1987); H. R. Chang, J.-C. R. Pechere, *Antimicrob. Agents Chemother.* **31**, 1147 (1987). Pharmacokinetics: R. Wise *et al.*, *ibid.* 1051. Bioassay in biological fluids: A. L. Barry, R. R. Packer, *Eur. J. Clin. Microbiol.* **5**, 536 (1986). Clinical evaluations in respiratory tract infections: J. M. Lachat *et al.*, *Schweiz. Med. Wochenschr.* **116**, 1739 (1986); C. Grassi *et al.*, *Chemioterapia* **6**, 41 (1987). Series of articles on antimicrobial activity, pharmacokinetics and clinical efficacy: *J. Antimicrob. Chemother.* **20**, Suppl. B, 1-185 (1987).

$[\alpha]_D^{25}$ −77.5 ±2° (c = 0.45 in chloroform).

THERAP CAT: Antibacterial.

8410. Royal Jelly. [8031-67-2] Queen bee jelly; apilak; Weiselfuttersaft (German); Gelée royale (French). Secretion from the salivary glands of the worker honey bee which is essential for the development of queen bees. *See also* Queen Substance. Production from bee hives: Ritschel, *Oesterr. Drogisten-Ztg.* **12**, 4-7 (1958). Synthetic mixture fed to bee larvae maintains life, but does not produce queens. Identification of constituents by GC-MS: V. A. Isidorov *et al.*, *J. Chromatogr. B* **885-886**, 109 (2012). Review of composition and biological activity: A. D. Dayan, *J. Pharm. Pharmacol.* **12**, 377-383 (1960); H. Rembold, *Vitam. Horm.* **23**, 359-382 (1965); M. Viuda-Martos *et al.*, *J. Food Sci.* **73**, R117-R124 (2008).

Royal Jelly Acid

Milky white, highly viscous secretion. Analysis of fresh specimen (% wt at pH 5): moisture 65-70, protein 15-20, carbohydrate 10-15, lipid 1.7-6, ash 0.7-2.0; elemental analysis: P up to 0.5, S up to 0.6; trace elements present: Na, K, Fe, Cu, Mg, Mn, Ca. Vitamins (μg/g): thiamine 2, riboflavine 10, pyridoxine 2, nicotinic acid 75, biotin 2, folic acid 0.3, inositol 100, pantothenic acid 250, ascorbic acid 3-5, vitamin D trace, vitamin E trace. When stored at room temp, changes to a lightly yellow gum, and after some weeks, to a brittle amber solid.

Royal jelly acid. [14113-05-4] (2*E*)-10-Hydroxy-2-decenoic acid; *trans*-10-hydroxy-Δ²-decenoic acid. $C_{10}H_{18}O_3$; mol wt 186.25. Constitutes ~10% of the dried royal jelly. Isoln: Townsend, Lucas, *Biochem. J.* **34**, 1155 (1940); Butenandt, Rembold, *Z. Physiol. Chem.* **308**, 284 (1957). Synthesis: Fray *et al.*, *Tetrahedron Lett.* **4**, 15 (1960); Smissman *et al.*, *J. Org. Chem.* **29**, 3517 (1964);

Bestmann *et al.*, *Ann.* **699**, 33 (1966); J. Tsuji *et al.*, *Bull. Chem. Soc. Jpn.* **50**, 2507 (1977); T. Fujisawa *et al.*, *Chem. Lett.* **1982**, 219; R. Chiron, *J. Chem. Ecol.* **8**, 709 (1982). Leukemia prevention in mice: Townsend *et al.*, *Nature* **183**, 1270 (1959). Prisms from ether + petr ether or methanol + water, mp 64-65°. uv max: 211 nm (ε 12000).

8411. Rubber. Caoutchouc; India rubber. Primarily obtained by coagulating the milk juice (latex) of several tropical trees, chiefly *Hevea brasiliensis* Muell.-Arg., *Euphorbiaceae*. Habit. Brazil, East Indies, Java, etc. Rubber also occurs in a number of plants, among which *guayule* (*Parthenium argentatum* Gray), a shrub which grows primarily in the northern Mexican desert, represents a potentially useful renewable source. Guayule produces rubber which is chemically identical to *Hevea* rubber in amounts ranging from 10-20% (dry basis) distributed in the roots and stems. *See:* E. Campos-Lopez, *J. Polym. Sci. Polym. Lett. Ed.* **14**, 649 (1976); *J. Polym. Sci. Polym. Chem. Ed.* **14**, 1561 (1976); J. D. Johnson, C. W. Hinman, *Science* **208**, 460 (1980). Possible biosynthesis from acetate through mevalonic acid and isopentenyl pyrophosphate: Archer *et al.*, *Nature* **184**, 268 (1959). Comprehensive review of chemistry: Ellis, *Chem. Ind. (London)* **1962**, 1447. Book: L. G. Polhamus, *Rubber* (Wiley, New York, 1962). *Review:* Jirgenson's, *Natural Organic Macromolecules* (Pergamon Press, New York, 1962) pp 115-124; D. R. St. Cyr in *Kirk-Othmer Encyclopedia of Chemical Technology* vol. 20 (Wiley-Interscience, New York, 3rd ed., 1982) pp 468-491. Natural rubber is defined as a *cis*-1,4-polyisoprene with a molecular weight varying from 100,000 to one million.

The best grades of raw rubber (pale crepe or smoked sheet) contain ~95% rubber hydrocarbon. The rest consists of proteins (2-3%), acetone-sol resins and fatty acids (2%), small amounts of sugar and a little mineral matter. Vulcanization, which consists of heating rubber with 1-3% of sulfur, introduces cross links between chains to produce a three-dimensional lattice of improved elasticity, strength, and temp sensitivity. Accelerators such as zinc dimethyldithiocarbamate greatly decrease the time or lower the temp required for vulcanization. Pure rubber is nearly colorless and transparent in thin layers; odorless and tasteless. It is very elastic and lighter than water. Dec at about 120°. Burns with smoky flame. Emits characteristic offensive odor while burning. Practically insol in water, alcohol, dil acids, or alkali; sol in abs ether, chloroform, most fixed and volatile oils, petr ether, carbon disulfide, oil of turpentine.

8412. Rubeanic Acid. [79-40-3] Ethanedithioamide; dithiooxamide. $C_2H_4N_2S_2$; mol wt 120.19. C 19.99%, H 3.35%, N 23.31%, S 53.35%. Review of rubeanic acid and derivs as chelating ligands and analytical reagents: Ray, Xavier, *J. Indian Chem. Soc.* **38**, 535 (1961).

Red crystals or cryst powder; dec ~200°. Slightly sol in water; sol in alcohols. Insol in ether.

USE: As a reagent for copper, cobalt, and nickel. As a stabilizer of ascorbic acid solns: Smoczkiewicz, Grochmalicka, *Nature* **192**, 16 (1961).

8413. Ruberythric Acid. [152-84-1] 1-Hydroxy-2-[(6-*O*-β-D-xylopyranosyl-β-D-glucopyranosyl)oxy]-9,10-anthracenedione; 1-hydroxy-2-anthraquinonyl 6-*O*-β-D-xylopyranosyl-β-D-glucopyranoside; β-2-alizarin primeveroside; ruberythrinic acid; rubian; rubianic acid. $C_{25}H_{26}O_{13}$; mol wt 534.47. C 56.18%, H 4.90%, O 38.91%. Isoln from root of *Rubia tinctorium* L., *Rubiaceae*: Hill,

Richter, *Proc. R. Soc. London Ser. B* **121**, 547 (1937). Structure: Richter, *J. Chem. Soc.* **1936**, 1701. Synthesis: Zemplén, Bognár, *Ber.* **72B**, 913 (1939). Metabolism: Maehner, Dulce, *Z. Klin. Chem. Klin. Biochem.* **6**, 99 (1968).

Golden-yellow, silky, lustrous prisms or long needles. mp 259-261° (from water). Sol in hot water, alkalies; slightly sol in alc, ether. Practically insol in benzene.

8414. Rubiadin. [117-02-2] 1,3-Dihydroxy-2-methyl-9,10-anthracenedione; 1,3-dihydroxy-2-methylanthraquinone. $C_{15}H_{10}O_4$; mol wt 254.24. C 70.86%, H 3.96%, O 25.17%. From *Rubia tinctorum* L., *Coprosma* var., *Morinda citrifolia* Linn, *Rubiaceae*: Schunck, *Ann.* **87**, 344 (1853); Briggs, Nicholls, *J. Chem. Soc.* **1949**, 1241; Briggs *et al., ibid.* **1952**, 1718; Borvie, Cooke, *Aust. J. Chem.* **15**, 332 (1962). Structure: Marchlewski, *J. Chem. Soc.* **63**, 1137 (1893); Schunck, Marchlewski, *ibid.* **65**, 182 (1894); Jones, Robertson, *ibid.* **1930**, 1699. Synthesis: Joshi *et al., J. Sci. Ind. Res.* **14B**, 87 (1955); Hirase, *Chem. Pharm. Bull.* **8**, 417 (1960).

Yellow plates from glacial acetic acid, mp 302°; yellow slender plates from alc, mp 290°. Absorption max (ethanol): 246, 280, 415 nm (log ε 4.39, 4.52, 3.87). Sol in alc, ether. Practically insol in water, alkalies.
Diacetate. Yellow rods from acetic anhydride, mp 228°.
Dimethyl ether. Yellow needles from alc, mp 160.5°.

8415. Rubidium. [7440-17-7] Rb; at. wt 85.4678; at. no. 37; valence 1. Group IA (1). Alkali metal. Widely distributed in very small quantities in earth's crust: 0.0034% by wt. Naturally occurring isotopes: 85 (72.15%); 87 (27.85%); [87]Rb is radioactive, $T_{1/2}$ 4.88×10^{10} yr, β^- emitter. Artificial isotopes (mass nos.): 74-102. Found with other alkali metals in *rhodizite* (borate), *lepidolite* (aluminosilicate), *rubidium carnallite* (chloride); in sea water; in mineral springs and salt lakes. Discovered by Bunsen and Kirchhoff in 1861. Prepn: Hackspill, *Helv. Chim. Acta* **11**, 1003 (1928). Review: Whaley, "Sodium, Potassium, Rubidium, Cesium and Francium" in *Comprehensive Inorganic Chemistry* **vol. I**, J. C. Bailar Jr. *et al.*, Eds. (Pergamon Press, Oxford, 1973) pp 369-529; *Chemistry of the Elements* N. N. Greenwood, A. Earnshaw, Eds. (Pergamon Press, New York, 1984) pp 75-116; F. S. Wagner in *Kirk-Othmer Encyclopedia of Chemical Technology* **vol. 21** (Wiley-Interscience, New York, 4th ed., 1997) pp 591-600.

Lustrous, silvery-white, soft metal; body-centered cubic structure; rapidly tarnishes on exposure to air. mp 39°. bp 688°. d^{20} 1.532. Specific heat 0.0802 cal/g deg. One of the most active metals. E° (aq) Rb/Rb$^+$ 2.924 V. Emits characteristic red-violet color (780.0 nm) in flame. Chemical properties closely resemble potassium. *Dangerous when wet. Keep under benzene, petroleum or other liquid not containing oxygen.* Reacts violently with water, ice, steam, lower alcohols, chlorinated hydrocarbons. Ignites spontaneously in oxygen; when molten readily takes fire in the air. Reacts vigorously with the halogens. Forms a series of solid solns with potassium, cesium, sodium. Combines vigorously with mercury.

USE: In making rubidium salts; as a reagent in making zeolite catalysts; in photoelectric cells.

8416. Rubidium Bromide. [7789-39-1] Rubidium monobromide. BrRb; mol wt 165.37. Br 48.32%, Rb 51.68%. RbBr.

White, cryst powder; d 3.35; mp 682°; bp 1340°. Sol in about 1 part cold water, 0.5 part boiling water. The aq soln is neutral.

8417. Rubidium Chloride. [7791-11-9] Rubidium monochloride; Rubinorm. ClRb; mol wt 120.92. Cl 29.32%, Rb 70.68%. RbCl. Causes shock-induced aggression in rats: B. S. Eichelman, *Psychopharmacol. Bull.* **9**, 21 (1973). Biochemical, behavioral and metabolic studies in humans: R. R. Fieve *et al., Am. J. Psychiatry* **130**, 55 (1973). Clinical evaluations in depression: C. Paschalis *et al., J. R. Soc. Med.* **71**, 343 (1978); G. Placidi *et al., J. Clin. Psychopharmacol.* **8**, 184 (1988).

White, crystalline powder. d 2.76. mp 715°. bp 1390°. One gram dissolves in 1 ml cold, 0.7 ml boiling water, 90 ml methanol, 1650 ml alc. The aq soln is neutral.
USE: As catalyst; as gasoline octane number improver.
THERAP CAT: Antidepressant.

8418. Rubidium Hydroxide. [1310-82-3] HORb; mol wt 102.47. H 0.98%, O 15.61%, Rb 83.41%. RbOH. Book: L. S. Itkina, *Lithium, Rubidium and Cesium Hydroxide* (Nauka, Moscow, USSR, 1973) 136 pp. Toxicological study: G. T. Johnson *et al., Toxicol. Appl. Pharmacol.* **32**, 239 (1975).

Grayish-white, deliquesc mass. d 3.203. Melts at red heat. Sol in 0.6 part water; sol in alcohol. It is a stronger base than potassium hydroxide. *Corrosive. Keep well closed.*
USE: Catalyst in oxidative chlorination.

8419. Rubidium Iodide. [7790-29-6] Rubidium monoiodide. IRb; mol wt 212.37. I 59.76%, Rb 40.24%. RbI.

White crystals or cryst powder; discolors on exposure to air and light; d 3.55; mp 642°; bp 1300°. Sol in 0.66 part water; sol in alcohol. The aq soln is neutral or slightly alkaline. *Keep well closed and protected from light.*
THERAP CAT: Iodine source.

8420. Rubijervine. [79-58-3] (3β,12α)-Solanid-5-ene-3,12-diol; Δ^5-3β,12α-dihydroxysolanidene; rubigervine. $C_{27}H_{43}NO_2$; mol wt 413.65. C 78.40%, H 10.48%, N 3.39%, O 7.74%. From various species of *Veratrum, Liliaceae.* Isoln: Wright, Luff, *J. Chem. Soc.* **35**, 405 (1879); Salzberger, *Arch. Pharm.* **228**, 462 (1890); Jacobs, Craig, *J. Biol. Chem.* **148**, 41 (1943); Sato, Jacobs, *ibid.* **179**, 623 (1949); Pelletier, Locke, *J. Am. Chem. Soc.* **79**, 4531 (1957). Stereochemistry: Höhne *et al., Tetrahedron* **22**, 673 (1966). Comparative toxicity: O. Krayer *et al., J. Pharmacol. Exp. Ther.* **82**, 167 (1944).

Solvated needles from alc, mp 240-246°. $[α]_D^{25}$ +19.0° (ethanol). Very sparingly sol in water. Sol in ethanol, methanol, benzene, chloroform. Slightly sol in ether, petr ether. Red color in concd H_2SO_4. Precipitated by digitonin. LD_{50} i.v. in mice: 70 mg/kg (Krayer).
Hydrobromide. $C_{27}H_{43}NO_2 \cdot HBr$. Needles from methanol + acetone, dec 265-270°.
THERAP CAT: Antifungal.

8421. Rubitecan. [91421-42-0] (4S)-4-Ethyl-4-hydroxy-10-nitro-1H-pyrano[3',4':6,7]indolizino[1,2-b]quinoline-3,14(4H,-12H)-dione; 9-nitrocamptothecin; 9-nitro-(20S)-camptothecin; 9-NC; Orathecin. $C_{20}H_{15}N_3O_6$; mol wt 393.36. C 61.07%, H 3.84%, N 10.68%, O 24.40%. Semisynthetic camptothecin which inhibits DNA topoisomerase I. Prodrug of 9-aminocamptothecin, (9-AC), *q.v.* Prepn: **JP Kokai 59 51288** (1984 to Yakult Honsha), *C.A.* **101**, 91322 (1984); M. C. Wani *et al., J. Med. Chem.* **29**, 2358 (1986); and activity: S. Sawada *et al., Chem. Pharm. Bull.* **39**, 3183 (1991). Conversion to 9-AC by human cells: P. Pantazis *et al., Eur. J. Haematol.* **53**, 246 (1994); eidem, *ibid.* **55**, 211 (1995). Clinical trial in pancreatic cancer: J. S. Stehlin *et al., Int. J. Oncol.* **14**, 821 (1999).

Review of development and clinical experience: J. W. Clark, *Expert Opin. Invest. Drugs* **15**, 71-79 (2006).

Yellow amorphous powder from CH_3OH-$CHCl_3$ (13:87), mp 182-186°. $[\alpha]_D^{23}$ +27° (c = 0.2 in CH_3OH-$CHCl_3$, 1:4).

THERAP CAT: Antineoplastic.

8422. Rubixanthin. [3763-55-1] (3*R*)-β,ψ-Caroten-3-ol. $C_{40}H_{56}O$; mol wt 552.89. C 86.90%, H 10.21%, O 2.89%. Carotenoid isomeric with cryptoxanthin. Isolated from hipberries (rose hips, *Rosa rubiginosa* L., *Rosaceae*): Kuhn, Grundmann, *Ber.* **67**, 339, 1133 (1934); from the flowers of *Tagetes patula* L., *Compositae*. Found also in other *Rosa* varieties. Structural studies: Brown, Weedon, *Chem. Commun.* **1968**, 382. Abs config: L. Bartlett *et al.*, *J. Chem. Soc. C* **1969**, 2527. Biosynthesis studies: McDermott *et al.*, *Biochem. J.* **134**, 1115 (1973). Has no vitamin A activity.

Deep red needles with metallic luster from benzene + methanol. Orange crystals from benzene + petr ether, mp 160°. Absorption max (chloroform): 509, 474, 439 nm. Sol in benzene, chloroform; slightly sol in petr ether, alcohol.

8423. Ruboxistaurin. [169939-94-0] (9*S*)-9-[(Dimethylamino)methyl]-6,7,10,11-tetrahydro-9*H*,18*H*-5,21:12,17-dimethenodibenzo[*e*,*k*]pyrrolo[3,4*h*][1,4,13]oxadiazacyclohexadecine-18,-20(19*H*)-dione; (*S*)-3,4-[(*N*,*N*'-1,1'-((2″-ethoxy)-3‴(*O*)-4‴-(*N*,*N*-dimethylamino)butane)-bis-(3,3'-indolyl))]-1(1*H*)-pyrrole-2,3-dione; LY-333531. $C_{28}H_{28}N_4O_3$; mol wt 468.56. C 71.77%, H 6.02%, N 11.96%, O 10.24%. Protein kinase Cβ (PKCβ) inhibitor; macrocyclic bisindolylmaleimide. Prepn: W. F. Heath, Jr. *et al.*, **EP 657458**; *eidem*, **US 5552396** (1995, 1996 both to Eli Lilly); M. R. Jirousek *et al.*, *J. Med. Chem.* **39**, 2664 (1996); M. M. Faul *et al.*, *J. Org. Chem.* **63**, 1961 (1998). Characterization of salt forms: G. L. Engel *et al.*, *Int. J. Pharm.* **198**, 239 (2000). Enantioselective synthesis: B. M. Trost, W. Tang, *Org. Lett.* **3**, 3409 (2001). Antiangiogenic activity: R. P. Danis *et al.*, *Invest. Ophthalmol. Visual Sci.* **39**, 171 (1998). Clinical trial in diabetic retinopathy: PKC-DRS Study Group, *Diabetes* **54**, 2188 (2005); in diabetic neuropathy: A. I. Vinik *et al.*, *Clin. Ther.* **27**, 1164 (2005); in diabetic nephropathy: K. R. Tuttle *et al.*, *Diabetes Care* **28**, 2686 (2005). Review of pharmacology and therapeutic potential: S. V. Joy *et al.*, *Ann. Pharmacother.* **39**, 1693-1699 (2005); A. Vinik, *Expert Opin. Invest. Drugs* **14**, 1547-1559 (2005).

uv max (ethanol): 493, 382, 282, 232, 204 nm (ε 3944, 3796, 9946, 38382, 43887). Soly in water: <1 μg/ml.

Hydrochloride. [169939-93-9] $C_{28}H_{28}N_4O_3$.HCl; mol wt 505.02. Orange cake. $[\alpha]_D^{25}$ −28.7° (c = 1.0 in ethanol). Soly in water: 0.1 mg/ml.

Methanesulfonate monohydrate. [202260-21-7]; [192050-59-2] (anhydrous). Ruboxistaurin mesylate monohydrate; Arxxant. $C_{28}H_{28}N_4O_3$.CH_4O_3S.H_2O; mol wt 582.67. Thin red or red-orange plates from acetone/water. Dec 270°. Soly in water: 0.5 mg/ml.

THERAP CAT: In treatment of diabetic microvascular complications.

8424. Rubus. Blackberry bark. Dried bark of rhizome and roots of the section *Eubatus* Focke of the genus *Rubus* L. or *R. nigrobaccus* Bailey, *Rosaceae*. *Habit.* Eastern U.S. *Constit.* Tannin, gallic acid, villosin (a saponin).

THERAP CAT: Antidiarrheal.

8425. Rufigallol. [82-12-2] 1,2,3,5,6,7-Hexahydroxy-9,10-anthracenedione; 1,2,3,5,6,7-hexahydroxyanthraquinone; rufigallic acid; C.I. 58600. $C_{14}H_8O_8$; mol wt 304.21. C 55.28%, H 2.65%, O 42.07%. Prepd by the action of H_2SO_4 on gallic acid: Robiquet, *Ann.* **19**, 204 (1836); by dry distn of gallic acid: Kunz-Krause, Manicke, *Ber.* **53**, 190 (1920). *See also: Colour Index* **vol. 4** (3rd ed., 1971) p 4520.

Red needles. Does not melt, but sublimes with partial decompn upon heating. Absorption spectrum: Treibs, Steinmetz, *Ann.* **506**, 171, 191 (1933). Practically insol in water. Freely sol in acetone; slightly sol in alcohol, ether, with yellow color. Sol in alkali hydroxide solns with violet color, but is soon dec by oxidation.

Hexamethyl ether. $C_{20}H_{20}O_8$. Yellow needles, mp 240°.

USE: For color reactions with Zr and Hf.

8426. Rufinamide. [106308-44-5] 1-[(2,6-Difluorophenyl)methyl]-1*H*-1,2,3-triazole-4-carboxamide; 1-(2,6-difluorobenzyl)-1*H*-1,2,3-triazole-4-carboxamide; CGP-33101; Banzel; Inovelon. $C_{10}H_8F_2N_4O$; mol wt 238.20. C 50.42%, H 3.39%, F 15.95%, N 23.52%, O 6.72%. Antiepileptic triazole derivative which decreases firing by neurons at sodium channels. Prepn: R. Meier, **EP 199262**; *eidem*, **US 4789680** (1986, 1988 both to Ciba-Geigy). Determn in plasma by HPLC: L. A. Brunner, M. L. Powell, *Biomed. Chromatogr.* **6**, 278 (1992); in dried blood spot samples by LC-MS/MS: G. la Marca *et al.*, *J. Pharm. Biomed. Anal.* **54**, 192 (2011). Clinical trial for refractory partial-onset seizures: V. Biton *et al.*, *Epilepsia* **52**, 234 (2011). Review of clinical pharmacokinetics: E. Perucca *et al.*, *ibid.* **49**, 1123-1141 (2008); of pharmacology and clinical experience: J. W. Wheless, B. Vazquez, *Epilepsy Curr.* **10**, 1-6 (2010). Review of clinical trials in Lennox-Gastaut syndrome: F. M. C. Besag, *Expert Opin. Pharmacother.* **12**, 801-806 (2011).

Crystals from ethanol, mp 237-240°. Moderate soly in 0.1*N* HCl; slightly sol in tetrahydrofuran, methanol; very slightly sol in ethanol, acetonitrile. Practically insol in water.

THERAP CAT: Anticonvulsant.

8427. Rufloxacin. [101363-10-4] 9-Fluoro-2,3-dihydro-10-(4-methyl-1-piperazinyl)-7-oxo-7*H*-pyrido[1,2,3-*de*]-1,4-benzothiazine-6-carboxylic acid; MF-934. $C_{17}H_{18}FN_3O_3S$; mol wt 363.41. C 56.19%, H 4.99%, F 5.23%, N 11.56%, O 13.21%, S 8.82%. Fluorinated quinolone; structurally similar to ofloxacin, *q.v.* Prepn:

G. Mascellani *et al.*, **AU 85 39142**; *eidem*, **US 4684647** (1985, 1987 both to Mediolanum); V. Cecchetti *et al.*, *J. Med. Chem.* **30**, 465 (1987). *In vitro* activity: R. Wise *et al.*, *Antimicrob. Agents Chemother.* **29**, 649 (1992). Mechanism of action and resistance: L. J. V. Piddock *et al.*, *J. Antimicrob. Chemother.* **31**, 855 (1993). Pharmacokinetics and safety: J. C. Kisicki *et al.*, *ibid.* **36**, 1296 (1992). HPLC determn in biological fluids: F. Lombardi *et al.*, *J. Chromatogr.* **576**, 129 (1992). Clinical study in cystitis and urinary tract infections: R. Mattina *et al.*, *Infection* **21**, 106 (1993); in bronchitis: W. Klietmann *et al.*, *Antimicrob. Agents Chemother.* **37**, 2298 (1993).

Hydrochloride. [106017-08-7] ISF-09334; Qari; Monos; Tebraxin. $C_{17}H_{18}FN_3O_3S.HCl$; mol wt 399.87. Crystals from ethanol/water, mp 322-324° (Cecchetti). LD_{50} in rats, mice (mg/kg): 285, 224 i.v.; LD_{50} in rabbits, male, female rats (mg/kg): 660, 631, 501 orally (Cecchetti).

THERAP CAT: Antibacterial.

8428. Rugulovasines. Intercovertible, isomeric indole alkaloids produced by *Penicillium concavo-rugulosum* Abe and other *penicillium* strains. Isolated as 2 racemates, rugulovasine A (*cis*-isomers) and rugulovasine B (*trans*-isomers). Isoln: M. Abe *et al.*, *Agric. Biol. Chem.* **33**, 469 (1969); M. Abe, **DE 1919255**; *idem*, **US 3651220** (1969, 1972 both to Takeda); R. J. Cole *et al.*, *Can. J. Microbiol.* **22**, 741 (1976). Structure: S. Yamatodani *et al.*, *Agric. Biol. Chem.* **34**, 485 (1970). Configuration of the isomers and mechanism of interconversion: R. J. Cole *et al.*, *Tetrahedron Lett.* **17**, 3849 (1976). Total synthesis of (+)-A, racemization, and interconversion: J. Rebek, Y. K. Shue, *J. Am. Chem. Soc.* **102**, 5426 (1980); J. Rebek *et al.*, *J. Org. Chem.* **49**, 3540 (1984). Hypotensive and cardiovascular activity: A. Nagaoka *et al.*, *Arzneim.-Forsch.* **22**, 137, 143 (1972).

(+)-Rugulovasine A

Rugulovasine A. [26909-33-1] *rel*-(2'*R*,4*S*)-3,4-Dihydro-4'-methyl-4-(methylamino)spiro[benz[*cd*]indole-5(1*H*),2'(5'*H*)-furan]-5'-one. $C_{16}H_{16}N_2O_2$; mol wt 268.32. Colorless prisms from benzene, chloroform, or acetonitrile, mp 138° (dec). uv max (ethanol): 224, 277, 288, 295 nm (log ε 4.37, 3.70, 3.78, 3.77). Sparingly sol in water, petr ether. Moderately sol in ether, chloroform, benzene, acetonitrile. Readily sol in ethyl acetate, acetone, methanol, ethanol, pyridine, and dil acids.

Rugulovasine A hydrochloride. [28375-03-3] $C_{16}H_{16}N_2O_2.$-HCl; mol wt 304.77. Colorless prisms from water, mp 225° (dec).

Rugulovasine B. [26909-34-2] *rel*-(2'*R*,4*R*)-3,4-Dihydro-4'-methyl-4-(methylamino)spiro[benz[*cd*]indole-5(1*H*),2'(5'*H*)-furan]-5'-one. $C_{16}H_{16}N_2O_2$; mol wt 268.32. Difficult to crystallize from common solvents. Separated from benzene as colorless resinous oil. uv max (ethanol): 227, 278, 288, 295 nm (log ε 4.16, 3.68, 3.73, 3.72). Same solubilities as rugulovasine A.

Rugulovasine B hydrochloride. [28510-15-8] $C_{16}H_{16}N_2O_2.$-HCl; mol wt 304.77. White prisms from water, mp 187° (dec).

8429. Rumex. Yellow dock; curled dock. Dried root of *Rumex crispus* L., or of *R. obtusifolius* L., *Polygonaceae*. Habit. Eu-

rope, North America. *Constit.* Chrysophanic acid, emodin, tannin, calcium oxalate, lapathin.

THERAP CAT: Cathartic, astringent.

8430. Rupatadine. [158876-82-5] 8-Chloro-6,11-dihydro-11-[1-[(5-methyl-3-pyridinyl)methyl]-4-piperidinylidene]-5*H*-benzo[5,6]cyclohepta[1,2-*b*]pyridine; UR-12592. $C_{26}H_{26}ClN_3$; mol wt 415.97. C 75.07%, H 6.30%, Cl 8.52%, N 10.10%. Dual antagonist of histamine H_1 and platelet-activating factor receptors. Prepn: E. Carceller *et al.*, **ES 2042421**; *eidem*, **US 5407941** (1993, 1995 both to Uriach); *eidem*, *J. Med. Chem.* **37**, 2697 (1994). Mechanism of action: M. Merlos *et al.*, *J. Pharmacol. Exp. Ther.* **280**, 114 (1997). Clinical trial in seasonal allergic rhinitis: F. Saint-Martin *et al.*, *J. Investig. Allergol. Clin. Immunol.* **14**, 34 (2004); and comparison with ebastine: E. M. Guadaño *et al.*, *Allergy* **59**, 766 (2004). Review of pharmacology and clinical experience: C. Picado, *Expert Opin. Pharmacother.* **7**, 1989-2001 (2006).

Creamy solid, mp 58-61°.

Fumarate. [182349-12-8] Rupafin. $C_{26}H_{26}ClN_3.C_4H_4O_4$; mol wt 532.04.

Trihydrochloride. [156611-76-6] $C_{26}H_{26}ClN_3.3HCl$; mol wt 525.34. Crystals from ethyl acetate + ether, mp 213-217°.

THERAP CAT: Antihistaminic.

8431. Rusalatide. [497221-38-2] L-Alanylglycyl-L-tyrosyl-L-lysyl-L-prolyl-L-α-aspartyl-L-α-glutamylglycyl-L-lysyl-L-arginyl-glycyl-L-α-aspartyl-L-alanyl-L-cysteinyl-L-α-glutamylglycyl-L-α-aspartyl-L-serylglycylglycyl-L-prolyl-L-phenylalanyl-L-valinamide; TP508; TRAP-508. $C_{97}H_{147}N_{29}O_{35}S$; mol wt 2311.47. C 50.40%, H 6.41%, N 17.57%, O 24.23%, S 1.39%. Synthetic, 23 amino acid peptide corresponding to the nonproteolytic receptor binding domain of human thrombin (prothrombin amino acids 508-530). Stimulates angiogenesis, revascularization, osteogenesis, and tissue repair. Synthesis and identification as thrombin receptor binding site: D. H. Carney, K. C. Glenn, **WO 8803151** (1988 to Board of Regents, Univ. Texas); *eidem*, **US 5352664** (1994 to Board of Regents, Univ. Texas; Monsanto); K. C. Glenn *et al.*, *Pept. Res.* **1**, 65 (1988). Wound healing effects on experimental incisions: D. H. Carney *et al.*, *J. Clin. Invest.* **89**, 1469 (1992). Overview of bioactivities and clinical pharmacology: J. T. Ryaby *et al.*, *J. Bone Joint Surg.* **88**, Suppl. 3, 132-139 (2006). Clinical trial in treatment of diabetic foot ulcers: C. Fife *et al.*, *Wound Repair Regen.* **15**, 23 (2007).

Acetate (3:2). [875455-82-6] Chrysalin. $(C_{97}H_{147}N_{29}O_{35}S)_3.$-$2C_2H_4O_2$; mol wt 7054.52.

THERAP CAT: Vulnerary.

8432. Rutecarpine. [84-26-4] 8,13-Dihydroindolo[2',3':3,4]pyrido[2,1-*b*]quinazolin-5(7*H*)-one; rutaecarpine. $C_{18}H_{13}N_3O$; mol wt 287.32. C 75.25%, H 4.56%, N 14.63%, O 5.57%. From fruit of *Evodia rutaecarpa* Hook. & Thoms. and *Hortia arborea* Engl., *Rutaceae*: Asahina, Kashiwaki, *J. Pharm. Soc. Jpn.* **1915**, 1293; Pachter *et al.*, *J. Am. Chem. Soc.* **82**, 5187 (1960). Structure: Y. Asahina, *J. Pharm. Soc. Jpn.* **1924**, 1. Synthesis: Y. Asahina *et* *al., J. Chem. Soc.* **1927**, 1708; T. Kametani *et al., J. Am. Chem. Soc.* **99**, 2306 (1977); *eidem, Chem. Pharm. Bull.* **26**, 1922 (1978); H. Möhrle *et al., Arch. Pharm.* **313**, 990 (1980); J. Bergman, S. Bergman, *Heterocycles* **16**, 347 (1981). Simple synthesis: J. Kökösi *et al., Tetrahedron Lett.* **22**, 4861 (1981). Synthesis under physiological conditions: Schöpf, *Angew. Chem.* **50**, 779, 797 (1937). Biosynthesis: M. Yamazaki *et al., Tetrahedron Lett.* **1966**, 3221; **1967**, 3317. Mass spec.: J. Tamas *et al., Acta Chim. Acad. Sci. Hung.* **89**, 85 (1976).

Needles from ethyl acetate, mp 259.5-260°. uv max (ethanol): 278, 290, 332, 345, 364 nm (log ε 3.83, 3.88, 4.49, 4.54, 4.44). Sol in alc, benzene, chloroform, ether. Practically insol in water.

8433. Ruthenium. [7440-18-8] Ru; at. wt 101.07; at. no. 44; valences 1-8; most common states 2, 3, 4. Group VIII (8). Naturally occurring isotopes: 96 (5.46%); 98 (1.87%); 99 (12.63%); 100 (12.53%); 101 (17.02%); 102 (31.6%); 104 (18.87%); artificial radioactive isotopes: 93-95; 97; 103; 105-108. Belongs to the platinum group of metals. Found in the minerals osmiridium, laurite, in platinum ores, in some copper-nickel ores. Discovered in 1828 by Osann; prepd in the pure state in 1845 by Klaus. Constitutes about 0.0004 ppm of the crust of the earth. Prepn from osmiridium: Fremy *et al.*, cited by Mellor, *A Comprehensive Treatise on Inorganic and Theoretical Chemistry* **15**, 499 (1936). Reviews of prepn, properties and chemistry of ruthenium and other platinum metals: Gilchrist, *Chem. Rev.* **32**, 277-372 (1943); W. P. Griffith, *The Chemistry of the Rarer Platinum Metals* (John Wiley, New York, 1967) pp 1-41, 126-226; Livingstone in *Comprehensive Inorganic Chemistry* vol. **3**, J. C. Bailar, Jr. *et al.*, Eds. (Pergamon Press, Oxford, 1973) pp 1163-1209.

Lustrous, hard metal; hexagonal, close-packed structure. d_4^{20} 12.45. mp about 2450°; bp about 4150°. Sp heat (0°): 0.057 cal/g/°C. Does not react with acids, even aqua regia. Not oxidized by air in the cold; on heating combines readily with oxygen; the powdered metal forms the dioxide on igniting in air. Superficially attacked by concd alkaline hypochlorites. The powdered metal is attacked by chlorine above 200°; by bromine between 300-700°. Oxidized by fused alkali hydroxides. Forms alloys with platinum, palladium, cobalt, nickel, tungsten; forms definite compds with zinc and with tin.

USE: As substitute for platinum in jewelry; for pen nibs; as hardener in electrical contact alloys, electrical filaments; in ceramic colors; catalyst in synthesis of long chain hydrocarbons.

8434. Ruthenium Red. [11103-72-3] Ruthenium oxychloride ammoniated; C.I. 77800. $Cl_6H_{42}N_{14}O_2Ru_3$; mol wt 786.34. Cl 27.05%, H 5.38%, N 24.94%, O 4.07%, Ru 38.56%. [(NH₃)₅Ru—O—Ru(NH₃)₄—O—Ru(NH₃)₅]Cl₆. Exists as tetrahydrate. Identification: A. Joly, *C. R. Hebd. Seances Acad. Sci.* **115**, 1299 (1893). Prepn and structure: Fletcher *et al., J. Chem. Soc.* **1961**, 2000. Mössbauer study: Clausen *et al., Inorg. Nucl. Chem. Lett.* **7**, 485 (1971). Alternate structure and composition: Sterling, *Am. J. Bot.* **57**, 172 (1970). Use as microscopic stain: L. Mangin, *C. R. Hebd. Seances Acad. Sci.* **116**, 653 (1893); in electron microscopy: N. Dierichs, *Histochemistry* **64**, 171 (1979). Differential staining: P. R. Blanquet, *ibid.* **47**, 175 (1976); M. G. Gutierrez-Gonzalvez *et al., J. Microsc.* **145**, 333 (1987). Series of articles on chemistry, mechanism of action, specific uses: *Anat. Rec.* **171**, 347-442 (1971). Conversion *in situ* to **ruthenium violet:** J. R. Luft, *ibid.* 347; M. G. Guiterrez-Gonzalvez *et al., loc. cit.*

Brownish-red powder. Sol in water, ammonia.

USE: In microscopy as a general stain.

8435. Ruthenium Tetroxide. [20427-56-9] (*T*-4)-Ruthenium oxide (RuO₄); ruthenium(VIII) oxide. O₄Ru; mol wt 165.07. O 38.77%, Ru 61.23%. RuO₄. Prepd by fusing Ru with KMnO₄ and KOH in a proportion of 1:2:20. Lab procedures: Ruff, Vidic, Z.

Anorg. Allg. Chem. **136**, 49 (1924); Grube in *Handbook of Preparative Inorganic Chemistry* vol. **2**, G. Brauer, Ed. (Academic Press, New York, 2nd ed., 1965) p 1599. Crystal structure: Tréhoux *et al., C. R. Seances Acad. Sci. Ser. C* **268**, 246 (1969). Review of use as oxidizing agent: P. N. Rylander, *Organic Syntheses with Noble Metal Catalysts* (Academic Press, New York, 1973) pp 134-144.

Golden-yellow monoclinic prisms. Very volatile, sublimes at room temp. *Handle in hood only.* mp 25.4°. bp 40°. Sparingly sol in water, about 2.0% (w/v) at 20°. Freely sol in carbon tetrachloride, other chlorinated solvents. Also sol in bromine, liquid SO₂. Reacts explosively with alcohol and filter paper.

Caution: Direct contact with vapors may be irritating to eyes and respiratory tract.

USE: Oxidizing agent similar to osmium tetroxide, but more difficult to handle. The solvents normally employed in OsO₄ oxidations (ether, benzene, pyridine) cannot be used because of their violent reaction with RuO₄. Only CCl₄ is recommended: Djerassi, Engle, *J. Am. Chem. Soc.* **75**, 3838 (1953).

8436. Ruthenium Trichloride. [10049-08-8] Ruthenium chloride (RuCl₃). Cl₃Ru; mol wt 207.42. Cl 51.27%, Ru 48.73%. RuCl₃. Two forms, α and β, are known. Prepn: Fletcher *et al., Nature* **199**, 1089 (1963).

α-RuCl₃. Black lustrous crystals. Insol in alcohol, water.

β-RuCl₃. Dark-brown, fluffy, hexagonal crystals. Sol in alcohol.

8437. Rutherfordium. [53850-36-5] Element 104; kurchatovium; unnilquadium. Rf, Ku, Unq; at. no. 104. Group IVB (4). No stable nuclides. Initial production claimed by two groups. Prepn of isotope ^{260}104 (T$_½$ ~300 msec; spontaneous fission) by ^{242}Pu (^{22}Ne,-4n): G. N. Flerov *et al., At. Energ.* **17**, 310 (1964), *C.A.* **62**, 189d (1965); *eidem, Phys. Lett.* **13**, 73 (1964); I. Zvara *et al., Radiokhimiya* **11**, 161, 163 (1969). Prepn of two α-emitting isotopes, ^{257}104 (T$_½$ 4.5 seconds) by ^{249}Cf (^{12}C,4n) and ^{259}104 (T$_½$ 3 seconds) by ^{249}Cf (^{13}C,3n): A. Ghiorso *et al., Phys. Rev. Lett.* **22**, 1317 (1969). Ten isotopes reported; mass numbers 253-262. Longest-lived known isotope, ^{261}104, α-emitter, T$_½$ 65 ±10 seconds, rel. at. mass 261.1089; produced by ^{248}Cm (^{18}O,5n): A. Ghiorso *et al., Phys. Lett. B* **32**, 95 (1970). Chemical properties of isotopes and revised T$_½$ 21 ±1 msec for ^{260}104: L. P. Somerville *et al., Phys. Rev.* **C31**, 1801-1815 (1985). Solution chemistry: K. R. Czerwinski *et al., Radiochim. Acta* **64**, 23 (1994); *eidem, ibid.* 29. Reviews of history, prepn and properties: C. Keller, *The Chemistry of the Transuranium Elements* (Verlag Chemie, Weinheim, English Ed., 1971) pp 613-618; Silva, "Trans-Curium Elements" in *MTP Int. Rev. Sci.: Inorg. Chem., Ser. One* vol. **8**, A. G. Maddock, Ed. (University Park Press, Baltimore, 1972) pp 71-105; R. J. Silva in *The Chemistry of the Actinide Elements* vol. **2**, J. J. Katz *et al.*, Eds. (Chapman and Hall, New York, 1986) pp 1103-1106; E. K. Hyde *et al., Radiochim. Acta* **42**, 57-102 (1987); Transfermium Working Group, *Pure Appl. Chem.* **65**, 1757-1814 (1993); M. Schädel, *Inst. Phys. Conf. Ser.* **132**, 413-422 (1993). Review of chemistry: D. C. Hoffman, *Proc. Robert A. Welch Found. Conf. on Chem. Res., XXXIV, Fifty Years with Transuranium Elements* (Houston, Texas, 1990) pp 255-276.

8438. Rutin. [153-18-4] 3-[[6-*O*-(6-Deoxy-α-L-mannopyranosyl)-β-D-glucopyranosyl]oxy]-2-(3,4-dihydroxyphenyl)-5,7-dihydroxy-4*H*-1-benzopyran-4-one; rutoside; quercetin-3-rutinoside; 3,3',4',5,7-pentahydroxyflavone-3-rutinoside; melin; phytomelin; eldrin; ilixathin; sophorin; globularicitrin; paliuroside; osyritrin; osyritin; myrticolorin; violaquercitrin; Birutan; $C_{27}H_{30}O_{16}$; mol wt 610.52. C 53.12%, H 4.95%, O 41.93%. Identity with ilixanthin: Schindler, Herb, *Arch. Pharm.* **288**, 372 (1955). Found in many plants, especially the buckwheat plant (*Fagopyrum esculentum* Moench., *Polygonaceae*) which contains about 3% (dry basis): Couch *et al., Science* **103**, 197 (1946). From tobacco (*Nicotiana tabacum* L., *Solanaceae)*: Couch, Krewson, *U.S. Dept. Agr., Eastern Regional Res. Lab.* AIC-52 (1944). In forsythia (*Forsythia suspensa* (Thunb.) Vahl. var. *fortunei* (Lindl.) Rehd., *Oleaceae)*, in hydrangea (*Hydrangea paniculata* Sieb., *Saxifragaceae)*, in pansies (*Viola* sp., *Violaceae)*. General extraction procedure: *Beilstein* **XXXI**, 376. From leaves of *Eucalyptus macroryncha* F. v. Muell., *Myrtaceae*: Attree, Perkin, *J. Chem. Soc.* **1927**, 234. Industrial production from *Eucalyptus* spp.: Humphreys, *Econ. Bot.* **18**, 195 (1964). Structure: Zemplén, Gerecs, *Ber.* **68B**, 1318 (1935). Synthesis: Shakhova *et al.*, *Zh. Obshch. Khim.* **32**, 390 (1962), *C.A.* **58**, 1426e (1963). Rutin is hydrolyzed by rhamnodiastase from the seed of *Rhamnus utilis* Decne, *Rhamnaceae* (Chinese buckthorn); emulsin is not effective: Bridel, Charaux, *Compt. Rend.* **181**, 925 (1925). Toxicity data: Harrison *et al.*, *J. Am. Pharm. Assoc.* **39**, 557 (1950). *Book:* J. Q. Griffith, Jr., *Rutin and Related Flavonoids* (Mack, Easton, Pa., 1955). Comprehensive description: T. I. Khalifa *et al.*, *Anal. Profiles Drug Subs.* **12**, 623-681 (1983).

Pale yellow needles from water, gradual darkening on exposure to light. The crystals contain 3 H_2O and become anhyd after 12 hrs at 110° and 10 mm Hg. Anhyd rutin browns at 125°, becomes plastic at 195-197°, and dec 214-215° (with effervescence). $[\alpha]_D^{23}$ +13.82° (ethanol); $[\alpha]_D^{23}$ −39.43° (pyridine). Anhyd rutin is hygroscopic. One gram dissolves in about 8 liters water, about 200 ml boiling water, 7 ml boiling methanol. Sol in pyridine, formamide and alkaline solns; slightly sol in alcohol, acetone, ethyl acetate. Practically insol in chloroform, carbon bisulfide, ether, benzene, petr solvents. Dil solns give green color with ferric chloride. Rutin is colored brown by tobacco enzyme under experimental conditions: Neuberg, Kobel, *Naturwissenschaften* **23**, 800 (1935). LD_{50} i.v. in mice: 950 mg/kg (propylene glycol soln) (Harrison).

THERAP CAT: Capillary protectant.

8439. Rutinose. [90-74-4] 6-*O*-(6-Deoxy-α-L-mannopyranosyl)-D-glucose; 6-*O*-α-L-rhamnosyl-D-glucose. $C_{12}H_{22}O_{10}$; mol wt 326.30. C 44.17%, H 6.80%, O 49.03%. A disaccharide present in glycosides. Prepn from rutin by enzymatic hydrolysis using rhamnodiastase: Charaux, *Compt. Rend.* **178**, 1312 (1924); *Bull. Soc. Chim. Biol.* **6**, 634 (1924); Zemplén, Gerecs, *Ber.* **68B**, 1320 (1935). Synthesis: *eidem*, *Ber.* **67B**, 2049 (1934); **68B**, 1318 (1935). Structure: Gorin, Perlin, *Can. J. Chem.* **37**, 1930 (1959).

Very hygroscopic powder from alcohol and ether turning into a syrup on exposure to air. The freshly pptd material loses about 6% of its wt at 100°. The dried rutinose softens around 140°, dec 189-192°. $[\alpha]_D^{20}$ +3.2° → +0.8° (c = 4 in water). $[\alpha]_D^{20}$ −10° (95% alc). Very sol in water. Sol in alc. Hydrolysis with dil HCl yields 1 mol L-rhamnose and 1 mol D-glucose.

Heptaacetylrutinose. mp 169-170°. $[\alpha]_D^{20}$ −27.7° (chloroform).

8440. Ruxolitinib. [941678-49-5] (β*R*)-β-Cyclopentyl-4-(7*H*-pyrrolo[2,3-*d*]pyrimidin-4-yl)-1*H*-pyrazole-1-propanenitrile; INCB-018424. $C_{17}H_{18}N_6$; mol wt 306.37. C 66.65%, H 5.92%, N 27.43%. Selective JAK1 and JAK2 tyrosine kinase inhibitor. Prepn: J. D. Rodgers *et al.*, **WO 07070514**; J. D. Rodgers, S. Shepard, **US 7598257** (2007, 2009 both to Incyte). Enantioselective synthesis: Q. Lin *et al.*, *Org. Lett.* **11**, 1999 (2009). Preclinical characterization of JAK inhibition: A. Quintás-Cardama *et al.*, *Blood* **115**, 3109 (2010). Clinical pharmacokinetics: A. D. Shilling *et al.*, *Drug Metab. Dispos.* **38**, 2023 (2010). Clinical evaluation in myelofibrosis: S. Verstovsek *et al.*, *N. Engl. J. Med.* **363**, 1117 (2010). Review of development and therapeutic potential in myelofibrosis: A. Ostojic

et al., *Future Oncol.* **7**, 1035-1043 (2011); in leukemia: K. Naqvi *et al.*, *Expert Opin. Invest. Drugs* **20**, 1159-1166 (2011).

Colorless oil; solidifies upon standing at room temp under vacuum.

Phosphate. [1092939-17-7] Jakafi. $C_{17}H_{18}N_6 \cdot H_3O_4P$; mol wt 404.37. White to off-white to light pink powder. Sol in aq buffer at pH 1-8.

THERAP CAT: Antineoplastic

8441. Ryania. [8047-13-0] A genus of tropical American shrubs and trees belonging to the *Flacourtiaceae* family. The wood of various species is insecticidal. Ground stem wood of *Ryania speciosa* Vahl., *Flacourtiaceae* is employed in the commercial insecticide formulations **Ryanex, Ryanicide** (formerly). *See:* Folkers *et al.*, **US 2400295** (1946 to Merck & Co.); Pepper, Carruth, *J. Econ. Entomol.* **38**, 59 (1945); Heal, *Agric. Chem.* **4**, 37 (May, 1949). Insecticidal components such as ryanodine, *q.v.*, are extractable by water, chloroform, or methanol. Toxicity data: Kuna, Heal, *J. Pharmacol. Exp. Ther.* **93**, 407 (1948).

LD_{50} orally in rats, mice, rabbits, guinea pigs (mg/kg): 1200, 650, 650, 2500 (Kuna, Heal).

Caution: In exptl animals causes weakness, deep and slow respiration, vomiting, diarrhea, tremors, convulsions, coma; death may occur.

USE: Insecticide.

8442. Ryanodine. [15662-33-6] (3*S*,4*R*,4a*R*,6*S*,7*S*,8*R*,8a*S*,-8b*R*,9*S*,9a*S*)-1*H*-Pyrrole-2-carboxylic acid dodecahydro-4,6,7,8a,-8b,9a-hexahydroxy-3,6a,9-trimethyl-7-(1-methylethyl)-6,9-methanobenzo[1,2]pentaleno[1,6-*bc*]furan-8-yl ester; ryanodol 3-(1*H*-pyrrole-2-carboxylate). $C_{25}H_{35}NO_9$; mol wt 493.55. C 60.84%, H 7.15%, N 2.84%, O 29.17%. Insecticidal principle isolated from *Ryania speciosa* Vahl., *Flacourtiaceae*: E. F. Rogers *et al.*, *J. Am. Chem. Soc.* **70**, 3086 (1948); R. B. Kelly *et al.*, *Can. J. Chem.* **29**, 905 (1951). Structure: D. R. Babin *et al.*, *Experientia* **21**, 425 (1965); J. Santroch *et al.*, *ibid.* **21**, 730 (1965); K. Wiesner *et al.*, *Tetrahedron Lett.* **1967**, 221; K. Wiesner, *Adv. Org. Chem.* **8**, 295 (1972). Crystal structure: S. N. Srivastava, M. Przybylska, *Can. J. Chem.* **46**, 795 (1968). ¹H-NMR: A. L. Waterhouse *et al.*, *Chem. Commun.* **1984**, 1265. Synthetic studies: P. Deslongchamps, *Pure Appl. Chem.* **49**, 1329 (1977). Effect on calcium ion uptake by cardiac sarcoplasmic reticulum vesicles: L. R. Jones *et al.*, *J. Pharmacol. Exp. Ther.* **209**, 48 (1979); L. R. Jones, S. E. Cala, *J. Biol. Chem.* **256**, 11809 (1981).

Crystals, dec 219-220°. $[\alpha]_D^{25}$ +26° (methanol). uv max (alc): 268.5 nm (log ε 4.18). Sol in water, alcohol, acetone, ether, chloroform. Practically insol in benzene, petr ether.

USE: Insecticide.

S

8443. Sabadilla. Cevadilla; caustic barley. The dried ripe seeds of *Schoenocaulon officinale* (Schlecht. & Cham.) A. Gray (*Sabadilla officinarum* Brandt, *Asagraea officinalis* Lindl.), *Liliaceae. Habit.* The Andes of Mexico, Guatemala, Venezuela. *Constit.* About 3 to 6% total alkaloids of which cevadine (cryst veratrine) and veratridine (amorphous veratrine) are the most important; other alkaloids are sabadinine, sabadilline, and sabadine; also contains sabadillic and veratric acids; fatty oil, resin. Toxicity data: E. D. Swiss, R. O. Bauer, *Proc. Soc. Exp. Biol. Med.* **76**, 847 (1951).

LD$_{50}$ i.p. in mice: 7.5 mg/kg (Swiss, Bauer).

THERAP CAT: Pediculicide.

THERAP CAT (VET): Has been used as an insecticide.

8444. Sabcomeline. [159912-53-5] (αZ,3R)-α-(Methoxyimino)-1-azabicyclo[2.2.2]octane-3-acetonitrile; SB-202026. C$_{10}$H$_{15}$-N$_3$O; mol wt 193.25. C 62.15%, H 7.82%, N 21.74%, O 8.28%. Selective muscarinic M$_1$-receptor partial agonist. Prepn: B. S. Orlek *et al.*, **EP 392803**; *eidem*, **US 5278170**; (1990, 1994 both to Beecham); S. M. Bromidge *et al.*, *J. Med. Chem.* **40**, 4265 (1997). Pharmacology: J. M. Loudon *et al.*, *J. Pharmacol. Exp. Ther.* **283**, 1059 (1997). Cognition improvement in rats: H. Hodges *et al.*, *Behav. Brain Res.* **99**, 81 (1999).

White crystals from methanol-acetone, mp 154-156°. [α]$_D^{20}$ +14.4° (c = 0.424 in EtOH).

Hydrochloride. [159912-58-0] Memric. C$_{10}$H$_{15}$N$_3$O.HCl; mol wt 229.71. mp 218-219°. [α]$_D^{20}$ +25.3° (c = 1.00 in EtOH).

THERAP CAT: Nootropic.

8445. Saccharin. [81-07-2] 1,2-Benzisothiazol-3(2H)-one 1,1-dioxide; 2,3-dihydro-3-oxobenzisosulfonazole; 1,2-dihydro-2-ketobenzisosulfonazole; saccharin insoluble; benzosulfimide; o-sulfobenzimide; benzoic sulfimide; o-sulfobenzoic acid imide; Gluside; Glucid; Garantose; Saccharinol; Saccharinose; Saccharol; Saxin; Sykose; Hermesetas. C$_7$H$_5$NO$_3$S; mol wt 183.18. C 45.90%, H 2.75%, N 7.65%, O 26.20%, S 17.50%. Prepn from o-sulfamoylbenzoic acid: Remsen, Fahlberg, *Ber.* **12**, 470 (1879). History and comparison with other sweetening agents: R. W. Moncrieff, *The Chemical Senses* (Wiley, New York, 1946). Mfg processes: O. Beyer, *Handbuch der Saccharinfabrikation* (Rascher, Zürich, 1923); *FIAT Report* **PB 901** (1945); *Chem. Eng. (N.Y.)* **61**, 128, 150 (July 1954). Maumee Chemical's process: *Chem. Eng. News* **41**, 76-78 (Dec. 9, 1963). Stability studies: DeGarmo *et al.*, *J. Am. Pharm. Assoc. Sci. Ed.* **41**, 17 (1952). Toxicity data: J. D. Taylor *et al.*, *Food Cosmet. Toxicol.* **6**, 313 (1968). Series of articles on toxicology and metabolism: *Food Chem. Toxicol.* **23**, 419-546 (1985). Review of toxicology and carcinogenicity studies of saccharin and its salts: *IARC Monographs* **22**, 111-170 (1980); D. L. Arnold, *Fundam. Appl. Toxicol.* **4**, 674-685 (1984). *Review:* R. Mazur, "Sweeteners" in *Kirk-Othmer Encyclopedia of Chemical Technology* **vol. 22** (John Wiley & Sons, New York, 3rd ed., 1983) pp 448-464.

Monoclinic crystals, mp 228.8-229.7°. Twinning on (001). Perfect 100 cleavage. Acicular crystals by vacuum sublimation. In dil aq soln it is 500 times as sweet as sugar; the sweet taste is still detectable in 1:100,000 dilution. Bitter, metallic aftertaste. d 0.828. Heat of combustion at constant volume: 4753.1 cal/g. uv max (0.1N

NaOH): 267.3 nm (ε 1570). Soly: one gram dissolves in 290 ml water, 25 ml boiling water, 31 ml alcohol, 12 ml acetone, about 50 ml glycerol. Readily dissolved by dilute solns of ammonia, by solns of alkali hydroxides, by solns of alkali carbonates; slightly sol in chloroform, ether; sparingly sol in alc. pH of 0.35% aq soln 2.0. Hydrolysis products of saccharin are o-sulfamoylbenzoic acid (alkaline hydrolysis) and ammonium o-sulfobenzoic acid (acid hydrolysis).

Ammonium salt. Saccharin ammonium; Daramin; Sucline. C$_7$-H$_8$N$_2$O$_3$S; mol wt 200.21.

Sodium salt dihydrate. Soluble saccharin; saccharin sodium; Kristallose; Crystallose; Dagutan; Sucaryl; Sucromat. C$_7$H$_4$-NNaO$_3$S.2H$_2$O; mol wt 241.19. White, crystalline powder; effloresces in dry air. In dil aq soln it is 300-500 times as sweet as sugar. One gram dissolves in 1.2 ml water, in about 50 ml alcohol. Aq solns are neutral or alkaline to litmus, but not alkaline to phenolphthalein. LD$_{50}$ in mice, rats (g/kg): 6.3, 7.1 i.p.; 17.5, 17.0 orally (Taylor). Anhydrous form, *Sucrédulcor*.

Note: Saccharin was formerly listed as reasonably anticipated to be a human carcinogen: *Eighth Report on Carcinogens* (PB99-128746, 1998) p III-905; delisted because the cancer data are not sufficient to meet the current criteria for this listing: *Ninth Report on Carcinogens* (PB2000-107509, 2000) p B-3.

USE: Non-nutritive sweetener; pharmaceutic aid (flavor). In formulations for electroplating-bath brighteners.

8446. L-Saccharopine. [997-68-2] N-[(5S)-5-Amino-5-carboxypentyl]-L-glutamic acid; ε-N-(L-glutar-2-yl)-L-lysine. C$_{11}$H$_{20}$-N$_2$O$_6$; mol wt 276.29. C 47.82%, H 7.30%, N 10.14%, O 34.74%. A lysine precursor in the aminoadipic acid-lysine pathway in yeast: Darling, Larsen, *Acta Chem. Scand.* **15**, 743 (1961); Kjaer, Larsen, *ibid.* 750; Kuo *et al.*, *Biochem. Biophys. Res. Commun.* **8**, 227 (1962); Trupin, Broquist, *J. Biol. Chem.* **240**, 2524 (1965); Jones, Broquist, *ibid.* **241**, 3430 sqq. (1966). Isoln from mycelium of the yeast *Candida utilis*: Marimoto, Yamano, *Biochem. Z.* **340**, 155 (1964).

Hydrated crystals; loses water of crystn over P$_2$O$_5$ at 100°. Dec 240-248°, when anhydr. [α]$_D^{23}$ +33.6° (c = 1 in 0.5N HCl); [M]$_D$ +93°. pK$_2$ 2.6; pK$_3$ 4.1; pK$_4$ 9.2; pK$_5$ 10.3. Sparingly sol in water, ethanol. Readily sol in alkaline solns and in strong acids.

8447. Sacrosidase. [85897-35-4] β-Fructofuranosidase (*Saccharomyces cerevisiae* clone FI4 protein moiety reduced); β-D-fructofuranoside fructohydrolase; Sucraid. Form of sucrase (invertase) produced by baker's yeast, *Saccharomyces cerevesiae*. Glycosylated peptide containing 513 amino acid residues; mol wt 66-116 kDa. Hydrolyzes sucrose into glucose and fructose. Purification and properties of natural yeast invertase: N. P. Neumann, J. O. Lampen, *Biochemistry* **6**, 468 (1967). Prepn by clone FI4: C. N. Chang *et al.*, **US 5010003** (1991 to Genentech). Clinical trial in congenital sucrase-isomaltase deficiency: W. R. Treem *et al.*, *J. Pediatr. Gastroenterol. Nutr.* **28**, 137 (1999).

May exist in soln as monomer, dimer, tetramer, and octomer. pI 4.093. *Protect from heat and light.*

THERAP CAT: Enzyme replacement therapy (congenital sucrase-isomaltase deficiency).

8448. Safflower Oil. The oil from the seed of *Carthamus tinctorius* L., *Compositae.* Milling and extraction procedures: Winter, *J. Am. Oil Chem. Soc.* **27**, 82 (1950). Purification and stabilization: Freedman, Shapiro, **US 2978381** (1961). Monograph: R. E. Woodward, G. M. Severson, *Industrial Survey of Safflower*, Chemurgy Dept. Report no. S-3 (Agricult. Expt. Sta., Lincoln, Nebraska, 1951). Fatty acids present as glycerides: palmitic 6.4%, stearic 3.1%, arachidic 0.2%, oleic 13.4%, linoleic 76.6-79.0%, linolenic 0.04-0.13%. Edible drying oil, intermediate between soybean and linseed oil. d$_{25}^{25}$ 0.9211-0.9215. Titer: 15-18°. n_D^{25} 1.472-1.475. n_D^{40} 1.4690-1.4692. Acid value 1.0-9.7. Saponif value 188-194. Iodine value 140-150. Thiocyanogen value 82.5-86.0. Reichert-Meissl value below 0.5. Hydroxyl value 2.9-6.0. Unsaponifiable below 1.5%. Thickens and becomes rancid on prolonged exposure to air. Miscible with ether, chloroform. Sol in the usual oil and fat solvents. Insol in water.

USE: As linseed oil in paints. For salad oil blends, in hydrogenated state as shortening.

THERAP CAT: Dietary supplement in hypercholesteremia (and possible prophylaxis and treatment of atherosclerosis).

8449. Saffron. Crocus; Spanish saffron; French saffron. Stigmas of *Crocus sativus* L., *Iridaceae. Habit.* Western Asia, Southern Europe. *Constit.* About 1% volatile oil, picrocrocin—a bitter glucoside, crocin (polychroit), fixed oil, wax.

Flattish-tubular, almost thread-like stigmas ~1.25 inches (3 cm) long; orange-brown color; strong, peculiar, aromatic odor; bitterish, aromatic taste.

USE: Coloring and flavoring.

8450. Safinamide. [133865-89-1] (2*S*)-2-[[[4-[(3-Fluorophenyl)methoxy]phenyl]methyl]amino]propanamide; (*S*)-(+)-2-[[4-(3-fluorobenzoxy)benzyl]amino]propanamide; FCE-26743. $C_{17}H_{19}FN_2O_2$; mol wt 302.35. C 67.53%, H 6.33%, F 6.28%, N 9.27%, O 10.58%. Voltage-dependent sodium and calcium channel blocker; selective and reversible inhibitor of monoamine oxidase B (MAO-B). Prepn: P. Dostert *et al.*, **EP 400495**; *eidem*, **US 5236957** (1990, 1993 both to Farmitalia); and anticonvulsant activity: P. Pevarello *et al., J. Med. Chem.* **41**, 579 (1998). HPLC determn of enantiomeric purity in bulk drugs: K. Zhang *et al., J. Pharm. Biomed. Anal.* **55**, 226 (2011). Mechanism of action study: P. Salvati *et al., J. Pharmacol. Exp. Ther.* **288**, 1151 (1999). Clinical pharmacokinetics and pharmacodynamics: A. Marzo *et al., Pharmacol. Res.* **50**, 77 (2004). Review of pharmacology: R. G. Fariello, *Neurotherapeutics* **4**, 110-116 (2007); of clinical experience in Parkinson's disease: A. H. V. Schapira, *Expert Opin. Pharmacother.* **11**, 2261-2268 (2010).

mp 208-212°.

Methanesulfonate. [202825-46-5] Safinamide mesylate; NW-1015; PNU-151774E. $C_{17}H_{19}FN_2O_2 \cdot CH_4O_3S$; mol wt 398.45. mp 210° (dec). $[\alpha]_D^{25}$ +12.9° (c = 1.1% in 98% acetic acid).

THERAP CAT: Antiparkinsonian; anticonvulsant.

8451. Saflufenacil. [372137-35-4] 2-Chloro-5-[3,6-dihydro-3-methyl-2,6-dioxo-4-(trifluoromethyl)-1(2*H*)-pyrimidinyl]-4-fluoro-*N*-[[methyl(1-methylethyl)amino]sulfonyl]benzamide; *N'*-{2-chloro-4-fluoro-5-[1,2,3,6-tetrahydro-3-methyl-2,6-dioxo-4-(trifluoromethyl)pyrimidin-1-yl]benzoyl}-*N*-isopropyl-*N*-methylsulfamide; BAS-800H; Kixor; Sharpen; Treevix. $C_{17}H_{17}ClF_4N_4O_5S$; mol wt 500.85. C 40.77%, H 3.42%, Cl 7.08%, F 15.17%, N 11.19%, O 15.97%, S 6.40%. Protoporphyrinogen IX oxidase (PPO) inhibitor; designed to control broadleaf weeds. Prepn: M. Carlsen *et al.*, **WO 0183459**; *eidem*, **US 6534492** (2001, 2003 both to BASF). Field studies in crops: P. H. Sikkema *et al., Crop Prot.* **27**, 1495 (2008); N. Soltani *et al., Weed Technol.* **23**, 331 (2009). Greenhouse studies: P. W. Geier *et al., ibid.* **23**, 313 (2009). Mechanism of action study: K. Grossmann *et al., Weed Sci.* **58**, 1 (2010). Review of chemistry, activity and toxicology: *EPA Pesticide Fast Sheet: Saflufenacil* (2009).

mp 189.9°. d^{20} 1.595. pH (25°): 4.426. Vapor pressure (20°): 4.5 x 10^{-15} Pa. pKa 4.41. Log P (octanol/water): 2.6. Soly in water at 20° (g/100ml): 0.0014 pH 4; 0.0025 pH 5; 0.21 pH 7.

USE: Herbicide.

8452. Safranal. [116-26-7] 2,6,6-Trimethyl-1,3-cyclohexadiene-1-carboxaldehyde; 2,2,6-trimethyl-4,6-cyclohexadien-1-aldehyde. $C_{10}H_{14}O$; mol wt 150.22. C 79.96%, H 9.39%, O 10.65%. Prepn from picrocrocin: Kuhn, Winterstein, *Ber.* **67**, 354 (1934). Synthesis: Kuhn, Wendt, *Ber.* **69**, 1549 (1936); Könst *et al., Tetrahedron Lett.* **1974**, 3175. Biogenetic synthesis: T. Kametani *et al., Chem. Pharm. Bull.* **29**,105 (1981).

Liquid. d_4^{19} 0.9734. $bp_{1.0}$ 70° (bath temp). n_D^{19} 1.5281. Sol in methanol, ethanol, petr ether, glacial acetic acid.

8453. Safrole. [94-59-7] 5-(2-Propen-1-yl)-1,3-benzodioxole; 4-allyl-1,2-methylenedioxybenzene; allylcatechol methylene ether; allyldioxybenzene methylene ether; *m*-allylpyrocatechin methylene ether. $C_{10}H_{10}O_2$; mol wt 162.19. C 74.06%, H 6.21%, O 19.73%. Constituent of several essential oils, notably of sassafras in which it is present to the extent of about 75%. Has been used as topical antiseptic; pediculicide. Metabolism studies: Oswald *et al., Biochim. Biophys. Acta* **230**, 237 (1971). Toxicity data: Hagan *et al., Toxicol. Appl. Pharmacol.* **7**, 18 (1965). Review of carcinogenic risk: *IARC Monographs* **10**, 231-241 (1976).

Colorless or slightly yellow liq; sassafras odor. d^{20} 1.096. mp ~11°. bp 232-234°. n_D^{20} 1.5383. Insol in water. Very sol in alcohol. Miscible with chloroform, ether. LD_{50} in rats, mice (mg/kg): 1950, 2350 orally (Hagan).

Caution: This substance is reasonably anticipated to be a human carcinogen: *Report on Carcinogens, Twelfth Edition* (PB2011-111646, 2011) p 374.

USE: In perfumery; denaturing fats in soap manuf; also in manuf of heliotropin. Formerly as flavoring agent in foods, drugs and beverages.

8454. SAICAR. [3031-95-6] *N*-[[5-Amino-1-(5-*O*-phosphono-β-D-ribofuranosyl)-1*H*-imidazol-4-yl]carbonyl]-L-aspartic acid; *N*-[(5-amino-1-ribofuranosylimidazol-4-yl)carbonyl]aspartic acid 5'-phosphate; 5-amino-*N*-(1,2-dicarboxyethyl)-1-ribofuranosyl-imidazole-4-carboxamide 5'-phosphate; 5-amino-4-imidazole-*N*-succinocarboxamide ribonucleotide; 5-amino-4-imidazole-*N*-succinocarboxamide ribotide; *N*-[5-amino-1-(5'-phosphoribofuranosyl)-4-imidazolecarbonyl]aspartic acid; succino-AICAR. $C_{13}H_{19}N_4O_{12}P$; mol wt 454.28. C 34.37%, H 4.22%, N 12.33%, O 42.26%, P 6.82%. Important purine precursor. Synthesis by enzymic reaction according to the equation AIR + ATP + aspartic acid + $CO_2 \rightarrow$ SAICAR + ADP + orthophosphate: Lukens, Buchanan, *J. Biol. Chem.* **234**, 1791 (1959). Production by adenine-requiring mutants of *Bacillus subtilis:* Ohmura *et al.*, **US 3280007** (1966 to Takeda). Synthesis of SAICAR and derivatives: Burrows, Shaw, *J. Chem. Soc. C* **1967**, 1088; Burrows *et al., ibid.* **1968**, 40.

Dibarium salt. $C_{13}H_{15}Ba_2N_4O_{12}P$. uv max (pH 1): 269, 244 nm (ε 11850, 9580). If the Bratton-Marshall procedure is modified by keeping the sample in ice during, and for 10 minutes after the addition of the reagents, SAICAR produces a purple chromophore having max absorption at 550 nm. *See:* A. C. Bratton, E. K. Marshall Jr., *J. Biol. Chem.* **128**, 537 (1939). The diazonium salt of SAICAR is very unstable and dec before it can couple under the usual test conditions.

8455. Sakuranetin. [2957-21-3] (2S)-2,3-Dihydro-5-hydroxy-2-(4-hydroxyphenyl)-7-methoxy-4*H*-1-benzopyran-4-one; 4',5-dihydroxy-7-methoxyflavanone; naringen 7-methyl ether. C_{16}-$H_{14}O_5$; mol wt 286.28. C 67.13%, H 4.93%, O 27.94%. Isoln by hydrolysis of 5-glucoside obtained from bark of *Prunus yedoensis* Matsum., *Rosaceae:* Asahina, *Arch. Pharm.* **246**, 259 (1908); from wood of various *Prunus* species: Hasegawa, Shirato, *J. Am. Chem. Soc.* **76**, 5559 (1954); *eidem, ibid.* **77**, 3557 (1955); Hasegawa, *ibid.* **79**, 1738 (1957). Structure: Asahina *et al., J. Pharm. Soc. Jpn.* **No. 550**, 1007 (1927); Shinoda, Sato, *ibid.* **48**, 220 (1928). Synthesis: Zemplén *et al., Ber.* **75B**, 1432 (1942). Absolute configuration of (+)-isomer: Arakawa, Nakazaki, *Ann.* **636**, 111 (1960).

Needles, mp 152°. Also reported as dihydrate, mp 131-132° (Arakawa, Nakazaki, *loc. cit.*). (+)-Isomer, $[\alpha]_D^{16}$ +8.0° (c = 7.92 in acetone); $[\alpha]_D^{12}$ +8.4° (c = 6.28 in methanol). Sol in alcohol, ether, chloroform, benzene, ethyl acetate, pyridine; slightly sol in boiling water. Practically insol in cold water.

5-Glucoside. 4',5-Dihydroxy-7-methoxyflavanone 5-glucoside; sakuranin. $C_{22}H_{24}O_{10}$. Tetrahydrate, bitter needles from dil alcohol. Begins to lose water at 100°. mp 212-214°. $[\alpha]_D^{28}$ −106.6° (tetrahydrate in acetone); $[\alpha]_D^{28}$ −123.2° (anhydr compd in acetone); $[\alpha]_D^{16}$ −97.4° (c = 1.58 for anhydr compd in 75% acetone). Very sol in dil alcohol, pyridine; slightly sol in abs alcohol, hot water, ethyl acetate. Practically insol in ether, cold water.

8456. Salatrim. [177403-56-4] Family of low-calorie fats composed of mixtures of long-chain saturated fatty acids (predominantly stearic) and short-chain saturated fatty acids (acetic, propionic, and/or butyric) esterified to the glycerol backbone. Name is an acronym derived from *s*hort and *l*ong *a*cyl*tri*glyceride *m*olecule. Prepd but not claimed: T. A. Pelloso *et al., WO 9418290; eidem,* **US 5434278** (1994, 1995 both to Nabisco). Series of articles on prepn, characterization, HPLC determn, pharmacology, and clinical assessment: *J. Agric. Food Chem.* **42**, 432-604 (1994). GC/MS determn in foods: A. S. Huang *et al., ibid.* **43**, 1834 (1995). Review of applications in foods: R. Kosmark, *Food Technol.* **50**, 98-101 (1996); of analytical methods: M. Y. Khaled *et al., Semin. Food Anal.* **2** 131-143 (1997).

Average caloric availability in humans: 4.9 kcal/g.
USE: Alternative fat used in foods.

8457. Salazosulfadimidine. [2315-08-4] 5-[2-[4-[[(4,6-Dimethyl-2-pyrimidinyl)amino]sulfonyl]phenyl]diazenyl]-2-hydroxybenzoic acid; 5-[*p*-[(4,6-dimethyl-2-pyrimidinyl)sulfamoyl]phenylazo]salicylic acid; 4'-(4,6-dimethylpyrimidin-2-ylsulfamoyl)-4-hydroxyazobenzene-3-carboxylic acid; 5-[*p*-[(4,6-dimethyl-2-pyrimidinyl)aminosulfonyl]phenylazo]salicylic acid; salicylazosulfadimidine; salicylazosulfamethazine; Azudimidine. $C_{19}H_{17}N_5O_5S$; mol wt 427.44. C 53.39%, H 4.01%, N 16.38%, O 18.71%, S 7.50%. Prepn: Korkuczanski, *Przem. Chem.* **37**, 162 (1958), *C.A.* **52**, 13727 (1958); Malatesta, Lotti, *Ann. Chim. (Rome)* **50**, 114 (1960), *C.A.* **56**, 8613g (1962).

Brown crystals, mp 207°.
THERAP CAT: Antibacterial.

8458. Salen. [129409-01-4] (*E,E*-form); [94-93-9] (unspecified stereo). 2,2'-[1,2-Ethanediylbis[(E)-nitrilomethylidyne]]bisphenol; (*E,E*)-2,2'-[1,2-ethanediylbis(nitrilomethylidyne)]bisphenol; *N,N'*-bis[(2-hydroxyphenyl)methylene]-1,2-ethanediamine; *N,N'*-ethylenebis(salicylideneimine); H_2sal$_2$en. $C_{16}H_{16}N_2O_2$; mol wt 268.32. C 71.62%, H 6.01%, N 10.44%, O 11.93%. Tetradentate Schiff base chelating ligand. Prepn: A. T. Mason, *Ber.* **20**, 267 (1887); and spectral analysis of Cu(II) complex: G. V. Panova *et al., J. Gen. Chem. USSR* **53**, 1452 (1983). Prepn of transition metal complexes: P. Pfeiffer *et al., Ann.* **505**, 84 (1933); and their physicochemical properties: A. Z. El-Sonbati *et al., Transition Met. Chem.* **17**, 66 (1992); M. M. Bhadbhade, D. Srinivas, *Inorg. Chem.* **32**, 6122 (1993). Mass spectra of transition metal complexes: S. M. Schildcrout *et al., ibid.* **34**, 4117 (1995). Thermodynamic properties of salen and its metal complexes: F. Lloret *et al., Inorg. Chim. Acta* **189**, 195 (1991). Use of Mn(III) complex as catalyst in epoxidation of olefins: K. Srinivasan *et al., J. Am. Chem. Soc.* **108**, 2309 (1986); as specific DNA cleaving agent: D. J. Gravert, J. H. Griffin, *J. Org. Chem.* **58**, 820 (1993).

Crystals from alcohol, mp 125-126°. Insol in water.
USE: As stabilizing ligand for a variety of oxidation states of metal ions.

8459. Salep. Dried tubers of several species of *Orchis, Orchidaceae. Habit.* Europe, Asia Minor. *Constit.* Mucilage, starch.
USE: Nutrient, demulcent; as a vehicle for acrid remedies.

8460. Salicin. [138-52-3] 2-(Hydroxymethyl)phenyl-β-D-glucopyranoside; salicoside; salicyl alcohol glucoside; saligenin-β-D-glucopyranoside. $C_{13}H_{18}O_7$; mol wt 286.28. C 54.54%, H 6.34%, O 39.12%. Usually obtained by making hot water extracts from the ground bark of poplar *(Populus)* and willow *(Salix);* also found in the leaves and female flowers of the willow. Isoln from root bark of *Viburnum prunifolium* L., *Caprifoliaceae:* Evans *et al., J. Am. Pharm. Assoc.* **34**, 207 (1945). Structure and synthesis: Irvine, Rose, *J. Chem. Soc.* **89**, 814 (1906); Kunz, *J. Am. Chem. Soc.* **48**, 262 (1926). Hydrolysis by acids: Moelwyn-Hughes, *Trans. Faraday Soc.* **25**, 503 (1929). Enzymatic hydrolysis: Pigman, *J. Res. Natl. Bur. Stand.* **27**, 6 (1941). Enzymatic hydrolysis in heavy water: Stacie, *Z. Phys. Chem.* **28B**, 236 (1935).

Orthorhombic crystals from water, mp 199-202°. $[\alpha]_D^{25}$ −62 to −67° (c = 3). $[\alpha]_D^{20}$ −45.6° (c = 0.6 in abs alc). *See:* Brauns, *J. Am. Chem. Soc.* **47**, 1292 (1925). One gram dissolves in 23 ml water, 3 ml boiling water, 90 ml alcohol, 30 ml alcohol at 60°. Soluble in alkalies, pyridine, glacial acetic acid. Practically insol in ether, chloroform. Aq solns are neutral to litmus and have a bitter taste.
USE: Standard substrate in evaluating enzyme prepns contg β-glucosidase: Weidenhagen, *Z. Ver. Zucker-Ind.* **79**, 591 (1929).
THERAP CAT: Analgesic.
THERAP CAT (VET): Has been used as a bitter stomachic, as an antirheumatic and as an analgesic.

8461. Salicyl Alcohol. [90-01-7] 2-Hydroxybenzenemethanol; saligenin; saligenol; *o*-hydroxybenzyl alcohol. $C_7H_8O_2$; mol wt 124.14. C 67.73%, H 6.50%, O 25.78%. Prepd by the action of emulsin on salicin; by heating phenol with methylene chloride and aq NaOH.

Plates or cryst powder. d 1.16. mp 86-87°; sublimes at 100°. Soluble in 15 parts water; very sol in alcohol, chloroform, ether, sol in benzene. Gives a red color with H_2SO_4.

THERAP CAT: Anesthetic (local).

8462. Salicylaldehyde. [90-02-8] 2-Hydroxybenzaldehyde; salicylic aldehyde. $C_7H_6O_2$; mol wt 122.12. C 68.85%, H 4.95%, O 26.20%. Made by heating sodium phenolate and chloroform with NaOH. Toxicity data: Binet, *Rev. Med. Suisse Romande* **15**, 561 (1895).

Clear, colorless, oily liq; bitter almond-like odor; burning taste. d_4^{20} 1.167; mp −7°; bp 196-197°; n_D^{20} 1.5735. Slightly sol in water; sol in alc, ether. Gives an orange color with H_2SO_4. MLD in rats (mg/kg): 900-1000 s.c. (Binet).

USE: In perfumery.

8463. Salicylaldehyde Phenylhydrazone. [614-65-3] 2-Hydroxybenzaldehyde 2-phenylhydrazone; salicylic aldehyde phenylhydrazone. $C_{13}H_{12}N_2O$; mol wt 212.25. C 73.57%, H 5.70%, N 13.20%, O 7.54%. Indicator used for the titration of Grignard, organolithium, and hydridic reagents. Prepn from salicylaldehyde and phenylhydrazine: E. Fischer, *Ber.* **17**, 572 (1884); A. Rössing, *ibid.* 2988. Formation mechanism and kinetics: M. P. Bastos *et al.*, *J. Org. Chem.* **46**, 3342 (1981). Spectroscopy studies: M. H. M. Abou-El-Wafa, U. M. Rabie, *Rev. Roum. Chim.* **46**, 723 (2001); H. H. Hammud *et al.*, *Spectrochim. Acta A* **63**, 255 (2006). Polarography studies: V. Krishnaswamy, M. Anbu, *Acta Cienc. Indica Chem.* **17C**, 387 (1991). Tautomerism: G. N. Ledesma *et al.*, *J. Mol. Struct.* **415**, 115 (1997); and metal coordination properties: *idem et al.*, *Polyhedron* **17**, 1517 (1998); G. A. Ibañez *et al.*, *J. Mol. Struct.* **605**, 17 (2002). Metal chelate studies: P. Umpathy *et al.*, *Indian J. Chem.* **3**, 471 (1965); N. Thankarajan, K. P. Kumar, *ibid.* **30A**, 646 (1991); *eidem, J. Indian Chem. Soc.* **69**, 780 (1992). Organometallic reagent titrations: B. E. Love, E. G. Jones, *J. Org. Chem.* **64**, 3755 (1999).

Needles, mp 142-143°. bp$_{28}$ 234°. pKa (50% ethanol): 12.38. Absorption max in acetonitrile: 299, 343 nm (ε 10480, 29060); in chloroform: 303, 346 nm (ε 10720, 22240); in DMF: 303, 352 nm (ε 10720, 25360); in methanol: 303, 346 nm (ε 10840, 24470); in ethanol: 300, 348 nm (ε 7350, 15080); in 50% ethanol (neutral form): 348 nm; in 50% ethanol (ionic form): 370 nm. Emission: 435 nm.

USE: Titration indicator for organometallic reagents. Gravimetric reagent for palladium. Also for separation and determn of lanthanum.

8464. Salicylaldoxime. [94-67-7]; [21013-96-7] (*E*-form). 2-Hydroxybenzaldehyde oxime; Saldox. $C_7H_7NO_2$; mol wt 137.14. C 61.31%, H 5.15%, N 10.21%, O 23.33%.

(*E*)-form

Prisms, mp 57°. Slightly sol in cold water; more sol in hot water; freely sol in alcohol, ether, benzene, dil HCl. Insol in petr ether. On heating it dec into salicylaldehyde and hydroxylamine.

USE: As a reagent for copper and nickel.

8465. Salicylamide. [65-45-2] 2-Hydroxybenzamide; Salamid; Samid; Cidal; Salizell; Salymid; Urtosal. $C_7H_7NO_2$; mol wt 137.14. C 61.31%, H 5.15%, N 10.21%, O 23.33%. Prepn: A. Cahours, *Ann.* **48**, 60 (1843); D. S. Hoffenberg, C. R. Hauser, *J. Org. Chem.* **20**, 1496 (1955); J. B. Chattopadhyaya, A. V. Rama Rao, *Tetrahedron* **30**, 2899 (1974). Toxicity data: Hart, *J. Pharmacol. Exp. Ther.* **89**, 205 (1947). Clinical efficacy in headache: W. J. Murray, *J. Clin. Pharmacol. J. New Drugs* **7**, 150 (1967). Pharmacokinetics: A. G. de Boer *et al.*, *Biopharm. Drug Dispos.* **4**, 321 (1983).

White or slightly pink, crystalline powder, somewhat bitter taste. Sensation of warmth on tongue. mp 140°. Soly in water at 30° = 0.2%, at 47° = 0.8%; in glycerol at 5° = 2.0%, at 39° = 5.0%, at 60° = 10.0%; in propylene glycol at 5° = 10.0%. Freely sol in ether and in solns of alkalies; sol in hot water, alc, propylene glycol; slightly sol in water, chloroform. pH of satd aq soln at 28° about 5. Forms a water-soluble sodium salt at pH 9. LD$_{50}$ orally in mice: 1.4 g/kg (Hart).

THERAP CAT: Analgesic.

8466. Salicylamide *O*-Acetic Acid. [25395-22-6] 2-[2-(Aminocarbonyl)phenoxy]acetic acid; *o*-(carbamylphenoxy)acetic acid; α-(2-carbamoylphenoxy)acetic acid. $C_9H_9NO_4$; mol wt 195.17. C 55.39%, H 4.65%, N 7.18%, O 32.79%. Prepn: Merriman, *J. Chem. Soc.* **103**, 1838 (1913); Tzofin, *J. Gen. Chem. USSR* **3**, 17 (1933); Klosa, *Arch. Pharm.* **288**, 389 (1955); Izumi *et al.*, *JP 56 874* (1956 to Yoshitomi). Pharmacokinetics and efficacy: D. Loew *et al.*, *Int. J. Clin. Pharmacol. Ther. Toxicol.* **30**, 509 (1992).

Crystals, mp 221°. Sol in aq alkali.

Sodium salt. [3785-32-8] Salizell (amp.). $C_9H_8NNaO_4$; mol wt 217.16. Crystals, mp 212-215°.

Diethylamine salt. Akistin. $C_{13}H_{22}N_2O_4$; mol wt 270.33.

THERAP CAT: Analgesic; anti-inflammatory; antipyretic.

8467. Salicylanilide. [87-17-2] 2-Hydroxy-*N*-phenylbenzamide; *N*-phenylsalicylamide; Salinidol. $C_{13}H_{11}NO_2$; mol wt 213.24. C 73.22%, H 5.20%, N 6.57%, O 15.01%. Usually made by the reaction of salicylic acid with aniline in the presence of PCl$_3$ at an elevated temp. Better yields are obtained by using an inert organic solvent such as toluene or carbon tetrachloride as a reaction diluent. Novel process using ion-exchange resins: Majewski, Skelly, *US 3221051* (1965). Alternate process: Majewski *et al.*, *US 3231611* (1966 to Dow).

Odorless leaflets. mp 135.8-136.2°. Slightly sol in water; freely sol in alcohol, ether, chloroform, benzene.

Caution: In concd form may cause irritation of skin, mucous membranes. *See also* Salicylic Acid.

USE: Anti-mildew, fungicide.

THERAP CAT: Antifungal (topical).

THERAP CAT (VET): Antifungal (topical).

8468. Salicylhydroxamic Acid. [89-73-6] *N*,2-Dihydroxybenzamide; 2-hydroxybenzhydroxamic acid; 2-hydroxyphenylhydroxamic acid. $C_7H_7NO_3$; mol wt 153.14. C 54.90%, H 4.61%, N 9.15%, O 31.34%. Prepn: Jeanrenaud, *Ber.* **22**, 1272 (1889); Urbanski, *Nature* **166**, 267 (1950). Trypanocidal activity: F. Opperdoes *et al.*, *Exp. Parasitol.* **40**, 198 (1976); A. B. Clarkson, F. H. Brohn, *Science* **194**, 204 (1976); A. J. Barnicoat *et al.*, *Experientia* **37**, 1291 (1981).

Needles from acetic acid, mp 168° (slow heating), mp 176-178° (quick heating). Sublimes. pK (25°) 4.19.

Sodium salt. [2460-25-5] $C_7H_6NNaO_3$; mol wt 175.12. Plates. Freely sol in water. pH of 0.1*N* soln 7.7.

USE: Complexing agent.

8469. Salicylic Acid. [69-72-7] 2-Hydroxybenzoic acid; Acnisal; Duofilm; Duoplant; Keralyt; Occlusal; Verrugon. $C_7H_6O_3$; mol wt 138.12. C 60.87%, H 4.38%, O 34.75%. Occurs in the form of esters in several plants, notably in wintergreen leaves and the bark of sweet birch. Manuf by heating sodium phenolate with carbon dioxide under pressure: W. L. Faith *et al.*, *Industrial Chemicals* (Wiley, New York, 3rd ed., 1965) pp 652-655. Review of prepns, properties, and uses of salicylic acid and derivatives: S. H. Erickson in *Kirk-Othmer Encyclopedia of Chemical Technology* vol. **20** (John Wiley & Sons, New York, 3rd ed., 1982) pp 500-524. Toxicity data: K. Sota *et al.*, *J. Pharm. Soc. Jpn.* **89**, 1392 (1969). HPLC determn in foods: B. H. Chen, S. C. Fu, *Chromatographia* **41**, 43 (1995). Clinical evaluation to treat warts: B. J. Bart *et al.*, *J. Am. Acad. Dermatol.* **20**, 74 (1989); M. E. Bender *et al.*, *Cutis* **47**, 199 (1991). Review of use in acne: E. Zander, S. Weisman, *Clin. Ther.* **14**, 247 (1992). Comprehensive description: M. A. Abounassif *et al.*, *Anal. Profiles Drug Subs. Excip.* **23**, 421-470 (1994).

White crystals, fine needles, or fluffy white crystalline powder, mp 159°; sublimes at 76°. Gradually discolors in sunlight. d_4^{20} 1.443. bp$_{20}$ 211°. pKa 2.98. *n* 1.565. Flash point (closed cup): 157°C. uv max (4 mg% in ethanol): 210, 234, 303 nm (ε 8343, 5466, 3591). Soly (Wt %): water 0.20 (20°), 2.21 (80°); carbon tetrachloride 0.262 (25°); benzene 0.775 (25°); propanol 27.36 (21°); abs ethanol 34.87 (21°); acetone 396 (23°). pH of satd aq soln 2.4. Freely sol in ether; sol in boiling water; sparingly sol in chloroform. Log P (octanol/water) 2.188 ±0.0548. *Keep protected from light.* LD$_{50}$ i.v. in mice: 500 mg/kg (Sota).

Sodium salt. [54-21-7] Sodium salicylate; Enterosalicyl; Enterosalil; Idocyl; Saliglutin. $C_7H_5NaO_3$; mol wt 160.10. Clinical trial in rheumatoid arthritis: S. J. Preston *et al.*, *Br. J. Clin. Pharmacol.* **27**, 607 (1989). Toxicity data: E. I. Goldenthal, *Toxicol. Appl. Pharmacol.* **18**, 185 (1971). White crystalline powder, mp 440°. Soly (g/100 cc): water 125; methanol 17. *Protect from light.* LD$_{50}$ i.p. in rats: 780 mg/kg (Goldenthal).

Caution: Ingestion of large amounts can cause vomiting, abdominal pain, increased respiration, acidosis, mental disturbances. May cause skin rashes in sensitive individuals. *See: Clinical Toxicology of Commercial Products*, R. E. Gosselin *et al.*, Eds. (Williams & Wilkins, Baltimore, 5th ed., 1984) pp 368-375.

USE: In manuf aspirin, methyl salicylate, and other salicylates. Coupling agent for azo dyes. Has been used as food preservative. Buffer.

THERAP CAT: Keratolytic. Sodium salt as anti-inflammatory, analgesic, antipyretic.

THERAP CAT (VET): Keratolytic.

8470. 4-Salicyloylmorpholine. [3202-84-4] (2-Hydroxyphenyl)-4-morpholinyl methanone; 4-(2-hydroxybenzoyl)morpholine; salicyl morpholide; L-1102; Tardisal. $C_{11}H_{13}NO_3$; mol wt 207.23. C 63.76%, H 6.32%, N 6.76%, O 23.16%. Prepn: Carron, **FR 1096209** (1955 to Robert & Carrière); Schweitzer, **US 2906728** (1959 to Dow).

Crystals, mp 175°. Soly (g/100 ml): water 0.41; alcohol 3.3; ether 0.22. pH of satd aq soln 6.2.

THERAP CAT: Choleretic.

8471. Salicylsulfuric Acid. [89-45-2] 2-(Sulfooxy)benzoic acid; salicylic acid, acid sulfate; salicylic acid sulfuric acid ester. $C_7H_6O_6S$; mol wt 218.18. C 38.54%, H 2.77%, O 44.00%, S 14.69%. Prepd by treating salicylic acid with chlorosulfonic acid in pyridine: Loeper *et al.*, *C. R. Seances Soc. Biol. Ses Fil.* **135**, 917 (1941); *Chim. Ind. (Paris)* **49**, 99 (1943).

Monosodium salt. [6155-64-2] Sodium salicylsulfate; Salcyl; Salcylix. Fine needles. Sol in water. Insol in organic solvents.

THERAP CAT: Analgesic, anti-inflammatory.

8472. Salinazid. [495-84-1] 2-[(2-Hydroxyphenyl)methylene]hydrazide 4-pyridinecarboxylic acid; 1-isonicotinoyl-2-salicylidenehydrazine; isonicotinic acid salicylidenehydrazide; saliniazid; *N'*-*o*-hydroxybenzylidenepyridine-4-carboxyhydrazide; *o*-hydroxybenzal isonicotinylhydrazone; Nupa-Sal; Acozid. $C_{13}H_{11}N_3O_2$; mol wt 241.25. C 64.72%, H 4.60%, N 17.42%, O 13.26%. Prepd by the interaction of isoniazid and salicyl-aldehyde in water: Hart *et al.*, *Antibiot. Chemother.* **4**, 803 (1954); *cf.* Yale *et al.*, *J. Am. Chem. Soc.* **75**, 1933 (1953). Tuberculostatic activity and toxicity: Bavin *et al.*, *J. Pharm. Pharmacol.* **7**, 1032 (1955).

Crystals from ethanol, mp 232-233° (also given as 251°). Soly at 25° in 100 ml water: 0.005 g; in 100 ml abs ethanol: 0.18 g; in 100 ml propylene glycol: 0.212 g. Sol in dil aq acids and alkalies forming a yellow soln.

THERAP CAT: Antibacterial (tuberculostatic).

8473. Salinomycin. [53003-10-4] $C_{42}H_{70}O_{11}$; mol wt 751.01. C 67.17%, H 9.40%, O 23.43%. Polyether ionophoric antibiotic having a unique tricyclic spiroketal ring system and an unsaturated 6-membered ring in the molecule. Produced by a strain of *Streptomyces albus* (FERM-P No. 419 and ATCC 21838). Production: Y. Miyazaki *et al.*, **JP Kokai 72 25392** (1972 to Kaken Chem.), *C.A.* **78**, 41561 (1973). Structure: H. Kinashi *et al.*, *Tetrahedron Lett.* **1973**, 4955. Taxonomy, production, isolation and physicochemical and biological properties: Y. Miyazaki *et al.*, *J. Antibiot.* **27**, 814 (1974). Use as a coccidiostat: Y. Tanaka *et al.*, **DE 2253031**; *eidem*, **US 3857948** (1973, 1974 to Kaken Chem.). Anticoccidial efficacy: T. T. Migaki *et al.*, *Poult. Sci.* **58**, 1192 (1979). Total synthesis: Y. Kishi *et al.*, *Front. Chem., Plenary Keynote Lect. IUPAC Congr., 28th* **1981**, K. J. Laidler, Ed. (Pergamon, Oxford, 1982) pp 287-304; R. C. D. Brown, P. J. Kocienski, *Synlett* **1994**, 415, 417. HPLC determn in animal feeds: M. R. LaPointe, H. Cohen, *J. Assoc. Off. Anal. Chem.* **71**, 480 (1988).

mp 112.5-113.5°. pKa' 6.4 (DMF). $[\alpha]_D^{25}$ −63° (c = 1 in ethanol). uv max (ethanol-water, 2:1): 284 nm (ε 126). LD$_{50}$ in mice (mg/kg): 18 i.p.; 50 orally (Miyazaki).

Sodium salt. [55721-31-8] Bio-Cox; Sacox; Salocin. C$_{42}$H$_{69}$NaO$_{11}$; mol wt 772.99. mp 140-142°. $[\alpha]_D^{25}$ −37° (c = 1 in ethanol).

THERAP CAT (VET): Anticoccidial agent.

8474. Salinosporamides. Family of cytotoxic proteasome inhibitors produced by the marine actinomycete, *Salinispora tropica*. Characterized by a fused γ-lactam-β-lactone ring structure; salinosporamide A is the most abundant. Isoln, structure, and bioactivity of A: R. H. Feling *et al.*, *Angew. Chem. Int. Ed.* **42**, 355 (2003); W. Fenical *et al.*, **US 7176232** (2007 to Regents of Univ. Calif.). Isoln of B and C: P. G. Williams *et al.*, *J. Org. Chem.* **70**, 6196 (2005); of D-J: K. A. Reed *et al.*, *J. Nat. Prod.* **70**, 269 (2007). Total syntheses of A: L. R. Reddy *et al.*, *J. Am. Chem. Soc.* **126**, 6230 (2004); A. Endo, S. J. Danishefsky, *ibid.* **127**, 8298 (2005). Structure-activity study: V. R. Macherla *et al.*, *J. Med. Chem.* **48**, 3684 (2005). Pharmacology and antitumor activity of A: D. Chauhan *et al.*, *Cancer Cell* **8**, 407 (2005). Biosynthetic studies: G. Tsueng *et al.*, *Appl. Microbiol. Biotechnol.* **75**, 999 (2007); L. L. Beer, B. S. Moore, *Org. Lett.* **9**, 845 (2007).

Salinosporamide A

Salinosporamide A. [437742-34-2] (1R,4R,5S)-4-(2-Chloroethyl)-1-[(S)-(1S)-2-cyclohexen-1-ylhydroxymethyl]-5-methyl-6-oxa-2-azabicyclo[3.2.0]heptane-3,7-dione; NPI-0052. C$_{15}$H$_{20}$ClNO$_4$; mol wt 313.78. Soly and degradation study: N. Denora *et al.*, *J. Pharm. Sci.* **96**, 2037 (2007). Colorless needles from ethyl acetate/iso-octane, mp 169-171°. $[\alpha]_D^{25}$ −72.9° (c = 0.55 in CH$_3$OH). pKa 3.93. uv max (methanol): 225, 205 nm (log ε 3.3, 4.03). Sol in chloroform, methanol, DMSO, acetone, acetonitrile, benzene, pyridine, DMF. Soly in aq sodium acetate buffer (pH 5): 35.3±0.9 μg/ml.

Salinosporamide B. [863126-95-8] (1R,4R,5S)-1-[(S)-(1S)-2-Cyclohexen-1-ylhydroxymethyl]-4-ethyl-5-methyl-6-oxa-2-azabicyclo[3.2.0]heptane-3,7-dione; NPI-0047. C$_{15}$H$_{21}$NO$_4$; mol wt 279.34. Amorphous crystals from ethyl acetate, mp 143-145°. $[\alpha]_D$ −54.5° (c = 0.286 in CH$_3$OH). uv max (methanol): 256 nm (log ε 3.7).

Salinosporamide C. [863126-96-9] (4aR,8aS,9S,9aR)-2-(2-Chloroethyl)-4a,7,8,8a,9,9a-hexahydro-9-hydroxy-1-methyl-3H-pyrrolo[1,2-a]indole-3,6(5H)-dione. C$_{14}$H$_{18}$ClNO$_3$; mol wt 283.75. Colorless oil. $[\alpha]_D$ −33.6° (c = 0.268 in CH$_3$OH). uv max (methanol): 222 nm (log ε 3.9).

8475. Salmeterol. [89365-50-4] 4-Hydroxy-α^1-[[[6-(4-phenylbutoxy)hexyl]amino]methyl]-1,3-benzenedimethanol; ($\pm$)-4-hydroxy-α^1-[[[6-(4-phenylbutoxy)hexyl]amino]methyl]-m-xylene-α,α^1-diol; GR-33343X. C$_{25}$H$_{37}$NO$_4$; mol wt 415.57. C 72.26%, H 8.97%, N 3.37%, O 15.40%. β_2-Adrenergic agonist; structural ana-

log of albuterol, *q.v.* Prepn: I. F. Skidmore *et al.*, **DE 3414752**; *eidem*, **US 4992474** (1984, 1991 both to Glaxo). Pharmacology: M. Johnson, *Lung* **168**, Suppl., 115 (1990). Preliminary evaluations in asthma: A. Ullman *et al.*, *Am. Rev. Respir. Dis.* **142**, 571 (1990); O. P. Twentyman *et al.*, *Lancet* **336**, 1338 (1990). Review of pharmacology and therapeutic use: R. N. Brogden, D. Faulds, *Drugs* **42**, 895-912 (1991).

mp 75.5-76.5°.

1-Hydroxy-2-naphthoate. [94749-08-3] Salmeterol xinafoate; GR-33343G; Arial; Salmetedur; Serevent. C$_{25}$H$_{37}$NO$_4$.C$_{11}$H$_8$O$_3$; mol wt 603.76. mp 137-138°. Freely sol in methanol; slightly sol in ethanol, chloroform, isopropanol; sparingly sol in water.

THERAP CAT: Bronchodilator.

8476. Salmine Sulfate. [53597-25-4] Salmine sulfate (1:1) (salt). A protamine found in the sperm of salmon. Contains arginine, proline, serine, glycine, valine, leucine, alanine, threonine, isoleucine, lysine, histidine, aspartic and glutamic acids. Mol wt 6000 to 7000. N about 25%. Prepn and properties: Fisher, Scott, *J. Pharmacol. Exp. Ther.* **58**, 78 (1936); Felix, *Am. Sci.* **43**, 431 (1955); Callanan *et al.*, *J. Biol. Chem.* **229**, 279 (1957); Carroll *et al.*, *ibid.* **234**, 2314 (1959). Separation of components and amino acid sequence of major component: Ando, Watanabe, *Int. J. Protein Res.* **1**, 221 (1969). Toxicity study: Starbuck *et al.*, *Arch. Int. Pharmacodyn. Ther.* **165**, 374 (1967).

Salmine base is a powder sol in disodium phosphate buffer; the sulfate separates from water in the cold as a clear, immiscible liquid. LD$_{50}$ i.v. in rats: 75 mg/kg (Starbuck).

8477. Salsalate. [552-94-3] 2-Hydroxybenzoic acid 2-carboxyphenyl ester; disalicylic acid; salicylic acid bimolecular ester; salicyloxysalicylic acid; salicylsalicylic acid; NSC-49171; Disalcid; Disalgesic; Mono-Gesic; Salflex. C$_{14}$H$_{10}$O$_5$; mol wt 258.23. C 65.12%, H 3.90%, O 30.98%. Nonacetylated aspirin analog. Prepn: **DE 211403** and **DE 214044** (1909, both to Boehringer, Mann.), *Frdl.* **9**, 928; *C.A.* **4**, 368 (1910); W. Baker *et al.*, *J. Chem. Soc.* **1951**, 201. Metabolism: S. M. Dromgoole *et al.*, *J. Pharm. Sci.* **73**, 1657 (1984). Clinical evaluation in arthritis: T. C. McPherson, *Clin. Ther.* **6**, 388 (1984). Mechanism of action studies: C. A. Divincenzo, F. R. Venezio, *Curr. Ther. Res.* **42**, 720 (1987). HPLC determn in plasma and urine: L. I. Harrison *et al.*, *J. Pharm. Sci.* **69**, 1268 (1980). Review of chemistry, pharmacokinetics, safety and clinical efficacy: P. T. Singleton, *Clin. Ther.* **3**, 80-102 (1980).

Crystals from benzene. mp 148-149°. Practically insol in water but gradually hydrolyzed by it into 2 mols of salicylic acid. Sol in alc, ether; sparingly sol in benzene.

THERAP CAT: Analgesic, anti-inflammatory.

8478. Salsoline. [101467-40-7] (1R)-1,2,3,4-Tetrahydro-7-methoxy-1-methyl-6-isoquinolinol; 6-hydroxy-7-methoxy-1-methyl-1,2,3,4-tetrahydroisoquinoline; (+)-salsoline. C$_{11}$H$_{15}$NO$_2$; mol wt 193.25. C 68.37%, H 7.82%, N 7.25%, O 16.56%. In *Salsola richteri* Karel., *Chenopodiaceae*. Extraction procedure: A. Orechoff, N. Proskurnina, *Ber.* **66**, 841 (1933); N. Proskurnina, A. Orekhoff, *Bull. Soc. Chim. Fr.* **1937**, 1265. Structure and synthesis: E. Späth *et al.*, *Ber.* **67**, 1214 (1934); O. Kovács, G. Fodor, *Ber.* **84**, 795 (1951). Abs config: A. R. Battersby, T. P. Edwards, *J. Chem. Soc.* **1960**, 1214. Biosynthetic studies: I. McFarlane, M. Slaytor, *Phytochemistry* **11**, 235 (1972). Synthesis of enantiomers: S. Teitel *et al.*, *J. Med. Chem.* **17**, 134 (1974). HPLC study: J. Strömbom, J. G. Bruhn, *J. Chromatogr.* **147**, 513 (1978).

Crystals from alc, mp 221°. $[\alpha]_D^{20}$ +34.5° (c = 1 in 0.1N HCl). Sol in chloroform, hot alc, dil NaOH. Slightly sol in water, benzene. Almost insol in ether, petr ether.

Hydrochloride. [51424-33-0] mp 174-175°. $[\alpha]_D$ +31.0° (methanol). uv max (isopropanol): 204, 227, 284, 286 nm (ε 39400, 5900, 3540, 3530). Sol in water, hot alcohol. Very sparingly sol in acetone, chloroform. LD_{50} in mice (mg/kg): 140 i.v.; >1000 orally (Teitel).

(−)-Form. [89-31-6] (*S*)-1,2,3,4-Tetrahydro-7-methoxy-1-methyl-6-isoquinolinol. mp 214-215°.

8479. Salutaridine. [1936-18-1] 5,6,8,14-Tetradehydro-4-hydroxy-3,6-dimethoxy-17-methylmorphinan-7-one; floripavine. $C_{19}H_{21}NO_4$; mol wt 327.38. C 69.71%, H 6.47%, N 4.28%, O 19.55%. Intermediate in morphine synthesis; naturally occurring in (+)-form. Isolated from various species of *Croton* and *Papaver:* Barton *et al.*, *J. Chem. Soc.* **1965**, 2423; uv max (isopropanol): Structure and identity with floripavine: Mndzhoyan *et al.*, *C.A.* **68**, 114787w (1968). Synthesis: T. Kametani *et al.*, *J. Chem. Soc. C* **1969**, 2030; G. Horvath, S. Makleit, *Acta Chim. Acad. Sci. Hung.* **106**, 37 (1981); of racemate: T. Kametani *et al.*, *J. Chem. Soc. Perkin Trans. 1* **1972**, 1435; C. Szántay *et al.*, *J. Org. Chem.* **47**, 594 (1982); of racemate and (−)-form: W. Ludwig, H. J. Schafer, *Angew. Chem. Int. Ed.* **25**, 1025 (1986). *Review:* Stuart, *Chem. Rev.* **1971**, 47.

Rods from ethyl acetate, mp 197-198°. $[\alpha]_D^{12}$ +114° (c = 0.5 in methanol). uv max (methanol): 236, 279 nm (log ε 4.23, 3.76).

(−)-Form. Sinoaculine. $[\alpha]_D^{20}$ −98° (c = 0.55 g/100 ml CH_3OH). uv max (methanol): 276, 240 nm (log ε 3.76, 4.24).

8480. Salvia. Sage; Dalmatian sage; garden sage; common sage. Small, aromatic, perennial shrub, *Salvia officinalis* L., *Lamiaceae* (*Labiatae*). Widely used as a culinary flavoring and in traditional medicine as an antiseptic, anhidrotic, and carminative. Medicinal parts are the leaves, flowers, and essential oil. Other salvia species, such as *S. triloba* L. (Greek sage) and *S. lavandulaefolia* (L.) Vahl. (Spanish sage) are used similarly. *Habit.* Southern Europe; cultivated in North America. *Constit.* carnosol, rosmarinic acid, carnosic acid, rosmanol, 1-3.5% volatile oil primarily containing α-thujone, β-thujone, 1,8-cineole, borneol, camphor. Phenolic composition and antioxidant activity: M.-E. Cuvelier *et al.*, *J. Am. Oil Chem. Soc.* **73**, 645 (1996). Composition of essential oil: N. B. Perry *et al.*, *J. Agric. Food Chem.* **47**, 2048 (1999); A. Raal *et al.*, *Nat. Prod. Res.* **21**, 406 (2007). *Reviews:* J. Gruenwald *et al.*, *PDR for Herbal Medicines* (Thomson PDR, Montvale, 3rd Ed., 2004) pp 698-700; E. Small, *Culinary Herbs* (NRC Research Press, Ottawa, 2nd Ed., 2006) pp 794-802.

USE: Flavoring, condiment.

8481. Salvinorins. Divinorins. Family of neoclerodane diterpenes isolated from the hallucinogenic Mexican mint *Salvia divinorum.* Traditional medicine used by the Mazatec people of Oaxaca, Mexico. Isoln of salvinorin A: A. Ortega *et al.*, *J. Chem. Soc. Perkin Trans. 1* **1982**, 2505; of A and B and biologic activity: L. J. Valdés, III *et al.*, *J. Org. Chem.* **49**, 4716 (1984); of C: L. J. Valdés, III *et al.*, *Org. Lett.* **3**, 3935 (2001). Stereochemistry: M. Koreeda

et al., *Chem. Lett.* **1990**, 2015. Pharmacologic effects: D. J. Siebert, *J. Ethnopharmacol.* **43**, 53 (1994). *Review:* L. J. Valdés III, *J. Psychoactive Drugs* **26**, 277-283 (1994).

Salvinorin A

Salvinorin A. [83729-01-5] (2*S*,4a*R*,6a*R*,7*R*,9*S*,10a*S*,10b*R*)-9-(Acetyloxy)-2-(3-furanyl)dodecahydro-6a,10b-dimethyl-4,10-dioxo-2*H*-naphtho[2,1-*c*]-pyran-7-carboxylic acid methyl ester; divinorin A. $C_{23}H_{28}O_8$; mol wt 432.47. One of the most potent natural hallucinogens known. First naturally occurring non-nitrogenous κ opioid receptor (KOR) agonist identified. HPLC quantification in *S. divinorum* extracts: J. W. Gruber *et al.*, *Phytochem. Anal.* **10**, 22 (1999). LC determn in biological fluids: M. S. Schmidt *et al.*, *J. Chromatogr. B* **818**, 221 (2005). KOR binding profile: B. L. Roth *et al.*, *Proc. Natl. Acad. Sci. USA* **99**, 11934 (2002). Brief review: D. J. Sheffler, B. L. Roth, *Trends Pharmacol. Sci.* **24**, 107-109 (2003). Colorless crystals from methanol, mp 238-240° (Ortega); also reported as mp 242-244° (Valdes, 1984). $[\alpha]_D^{22}$ −45.3° (c = 8.530 in CHCl₃). $[\alpha]_D^{25}$ −41° (c = 1 in CHCl₃).

USE: In traditional medicine, the leaves are used for divination and for treatment of anemia and excretory functions.

8482. Samaderins. Antitumor agents isolated from the bark and seed of *Samadera indica* Gaertn, *Simaroubaceae:* van der Marck, *Arch. Pharm.* **239**, 96 (1901). Separable into three components, samaderins A, B and C: Polonsky *et al.*, *Bull. Soc. Chim. Fr.* **1962**, 1715. According to Zylber *et al.*, *ibid.* **1963**, 1322, samaderin C is identical with *samaderoside A* isolated by Mitra, Gregg, *Naturwissenschaften* **14**, 327 (1962). Structure of B and C: Zylber, Polonsky, *Bull. Soc. Chim. Fr.* **1964**, 2016. Structure of A: M. C. Wani *et al.*, *Chem. Commun.* **1977**, 295. Crystal structure of A: K. D. Onan, A. T. McPhail, *J. Chem. Res. Synop.* **1978**, 14.

Samaderin A   Samaderin B   R =O

                          Samaderin C   —OH

Samaderin A. $C_{18}H_{18}O_6$; mol wt 330.34. mp 253-255°. $[\alpha]_D^{25}$ −31.3° (c = 0.259 in pyridine). uv max (methanol): 288 nm (ε 13400).

Samaderin B. $C_{19}H_{22}O_7$; mol wt 362.38. Prisms from ethyl acetate, mp 235-240°.

Samaderin C. $C_{19}H_{24}O_7$; mol wt 364.39. Shiny plates from ethyl acetate+ alcohol, mp 265-268°. Sol in pyridine. Practically insol in ether. Very sparingly sol in other org solvents.

8483. Samandarine. [467-51-6] (2*S*,5*R*,5a*S*,5b*S*,7a*R*,9*S*,-10a*S*,10b*S*,12a*R*)-Octadecahydro-5a,7a-dimethyl-2,5-epoxycyclopenta[5,6]naphth[1,2-*d*]azepin-9-ol; (1*α*,4*α*,5*β*,16*β*)-1,4-epoxy-3-aza-A-homoandrostan-16-ol. $C_{19}H_{31}NO_2$; mol wt 305.46. C 74.71%, H 10.23%, N 4.59%, O 10.48%. Constitutes together with samandarone approx 75% of the alkaloid content in the skin glands of the European fire and Alpine salamanders: *Salamandra maculosa* resp. *atra* Laur., *Salamandridae*. Isoln: Schöpf, Braun, *Ann.* **514**, 69 (1934). Structure and configuration: Wölfel *et al.*, *Ber.* **94**, 2361 (1961).

Needles from abs methanol or 50% acetone, mp 187-188°. Solvated crystals from dil methanol. $[\alpha]_D^{17}$ +43.3° (acetone). Freely sol in most organic solvents. Practically insol in water and NaOH solns.

Hydrochloride. $C_{19}H_{31}NO_2$.HCl. Crystals from methanol, mp 312-313°; from water, mp 321-322° (monohydrate).

Samandarone. $C_{19}H_{29}NO_2$; mol wt 303.45. Ketone corresponding to the secondary alcohol in samandarine. Synthesis: Hara, Oka, *J. Am. Chem. Soc.* **89**, 1041 (1967). Needles from methyl ethyl ketone, mp 191-192°; from 35% acetone, mp 189-191°. $[\alpha]_D^{21}$ −115.7° (acetone).

8484. Samarium. [7440-19-9] Sm; at. wt 150.36; at. no. 62; valences 2, 3. A lanthanide belonging to the cerium group of rare earth metals; named for the mineral "samarskite" from which it was isolated. Naturally occurring isotopes (mass numbers): 144 (3.1%); 147 (15.0%), radioactive, $T_{1/2}$ 1.06×10^{11} years, α-emitter; 148 (11.3%), radioactive, $T_{1/2}$ 7×10^{15} years, α-emitter; 149 (13.8%); 150 (7.4%); 152 (26.7%); 154 (22.7%). Known artificial radioactive isotopes: 133-143; 145; 146; 151; 153; 155-158. Abundance in earth's crust: 6.47-7.0 ppm. Commercially important sources are the rare earth minerals monazite and bastnaesite; also occurs in samarskite, cerite, orthite, ytterbite, and fluorspar. Isoln: L. de Boisbaudran, *Compt. Rend.* **88**, 322 (1879); **89**, 212 (1880). Sepn by crystn of the nitrates: Demarcay, *ibid.* **122**, 728 (1896); Feit, Przibylla, *Z. Anorg. Chem.* **43**, 203 (1905). Sepn of metal: Schumacher, Harris, *J. Am. Chem. Soc.* **48**, 3108 (1926); by reduction of salts: Marsh, *J. Chem. Soc.* **1942**, 398, 523; **1943**, 8; Daane *et al.*, *J. Am. Chem. Soc.* **75**, 2272 (1953); Onstott, *ibid.* **75**, 5128 (1953); **77**, 812 (1955). Toxicity study: Haley, *J. Pharm. Sci.* **54**, 663 (1965). Reviews of prepn, properties and compds: Prandtl, *Z. Anorg. Allg. Chem.* **238**, 321-334 (1938); *The Rare Earths*, F. H. Spedding, A. H. Daane (Krieger, Huntington, N.Y., 1971, reprint of 1961 ed) 641 pp; Hulet, Bode, "Separation Chemistry of the Lanthanides and Transplutonium Actinides" in *MTP Int. Rev. Sci.: Inorg. Chem., Ser. One* **Vol. 7**, K. W. Bagnall, Ed. (University Park Press, Baltimore, 1972) pp 1-45; Moeller, "The Lanthanides" in *Comprehensive Inorganic Chemistry* **Vol. 4**, J. C. Bailar Jr. *et al.*, Eds. (Pergamon Press, Oxford, 1973) pp 1-101; F. H. Spedding in *Kirk-Othmer Encyclopedia of Chemical Technology* **vol. 19**, (John Wiley & Sons, New York, 3rd ed., 1982) pp 833-854; *Chemistry of the Elements*, N. N. Greenwood, A. Earnshaw, Eds. (Pergamon Press, New York, 1984) pp 1423-1449. Brief review of properties: G. T. Seaborg, *Radiochim. Acta* **61**, 115-122 (1993).

Yellow metal; tarnishes on exposure to air. Hardest metal of the cerium group. Crystalline forms: rhombohedral α-form, d 7.536, transforms to β-form at 917°; body-centered cubic β-form exists at >917°. mp 1074°. bp 1794°. Heat of fusion: 8.623 kJ/mol. Heat of sublimation (25°): 206.7 kJ/mol. E°(aq) Sm^{3+}/Sm −2.41 V (calc).

Oxide. Sm_2O_3. Yellowish-white powder. d 8.347.

Hydroxide. $Sm(OH)_3$. Gelatinous precipitate.

Trichloride. $SmCl_3$. White-yellowish powder. d 4.465, mp 686°. Forms addition compds with ammonia. Forms a hexahydrate, $SmCl_3.6H_2O$, d 2.382, yellow crystalline plates. By reducing the anhydr trichloride at high temps with hydrogen, ammonia or aluminum powder, samarium dichloride is obtained. LD_{50} in mice (mg/kg): 585 i.p.; >2000 orally (Haley).

Dichloride. $SmCl_2$. Dark brown crystalline mass. d^{22} 3.687. Practically insol in alcohol. Dec by water.

Sulfate. $Sm_2(SO_4)_3$. Octahydrate, light yellow crystals, d^{18} 2.930. Sparingly sol in water.

USE: Oxide in control rods of some commercial nuclear power reactors. Alloys with cobalt to produce extremely stong permanent magnets.

8485. Samarium Cobalt. [12017-68-4] Cobalt compd with samarium (5:1); samarium pentacobalt. Co_5Sm; mol wt 445.03. Co 66.21%, Sm 33.79%. $SmCo_5$. Prepd by sintering or mechanical alloying to yield a single phase, highly efficient permanent magnetic material. Prepn: K. J. Strnat, A. E. Ray, *Z. Metallkd.* **61**, 461 (1970). Synthesis by mechanical alloying: J. Wecker *et al.*, *J. Appl. Phys.* **69**, 6058 (1991). Prepn and magnetic properties of nanoparticles: E. M. Kirkpatrick *et al.*, *IEEE Trans. Magn.* **32**, 4502 (1996). Studies on use as a dental material: H. Tsutsui *et al.*, *J. Dent. Res.* **58**, 1597 (1979). Biocompatibility study: L. Bondemark *et al.*, *Am. J. Orthod. Dentofacial Orthop.* **105**, 568 (1994). Review of magnetic properties: R. A. McCurrie, *Cobalt* **1**, 23-24, 28 (1973); of properties, production and applications: H. J. Marik, K. Schlenk, *Powder Metallurgy Int.* **9**, 142-144 (1977).

mp ~1325°. Curie temp: 724° (Strnat); also reported as 680° (Tsutsui). d 8.60.

Samarium Cobalt. [12052-78-7] Cobalt compd with samarium (17:2). $Co_{17}Sm_2$; mol wt 1302.58. Sm_2Co_{17}. Prepn and crystal structure: W. Ostertag, K. J. Strnat, *Acta Crystallogr.* **21**, 560 (1966); and magnetic properties: K. Strnat *et al.*, *J. Appl. Phys.* **37**, 1252 (1966). ^{147}Sm and ^{149}Sm NMR study: H. Figiel *et al.*, *J. Magn. Magn. Mater.* **101**, 401 (1991). Orthodontic tooth movement study in animals: M. A. Darendeliler *et al.*, *Am. J. Orthod. Dentofacial Orthop.* **107**, 578 (1995).

USE: In permanent magnetic materials including: computer disk drives, sensors, traveling wave tubes, actuators, satellite systems, motors; in dental applications.

8486. Samarium Iodide. [32248-43-8] Diiodosamarium; samarium diiodide; samarium(II) iodide. I_2Sm; mol wt 404.17. I 62.80%, Sm 37.20%. SmI_2. One-electron reducing agent. Prepn: C. Matignon, E. Cazes, *Ann. Chim. Phys.* **8**, 417 (1906); and use in organic synthesis: J. L. Namy *et al.*, *Nouv. J. Chim.* **1**, 5 (1977); P. Girard *et al.*, *J. Am. Chem. Soc.* **102**, 2693 (1980). Coordination chemistry and crystal structure of adducts with THF: W. J. Evans *et al.*, *ibid.* **117**, 8999 (1995). Review of properties: S.-H. Wang, *Rev. Inorg. Chem.* **11**, 1-20 (1990); of uses in organic chemistry: G. A. Molander, C. R. Harris, *Chem. Rev.* **96**, 307-338 (1996); H. B. Kagan, *Tetrahedron* **59**, 10351-10372 (2003).

Black to light black solid, mp 526 ±3°. Deep blue-green soln in THF. Absorption max (THF): 565, 617 nm. Sol in ether, THF. Redox potential: E^0 −1.55 v. *Unstable in the presence of oxygen or moisture; store under an inert atmosphere.*

USE: In organic synthesis and electron transfer reactions; in discharge lamps.

8487. Samarium Sm 153 Lexidronam. [154427-83-5] *(OC-6-21)*-[[*P,P'*-[1,2-Ethanediylbis[(nitrilo-κN)bis(methylene)]]tetrakis[phosphonato-κO]](8−)]-samarate(5−)-^{153}Sm; samarium 153 ethylenediaminetetramethylenephosphonic acid; ^{153}Sm-EDTMP. $C_6H_{12}N_2O_{12}P_4$^{153}Sm. Tetraphosphonate chelating agent bound to a β-emitting radionuclide; targets areas of increased bone turnover. Prepn: J. Simon *et al.*, **EP 164843**; *eidem*, **US 4898724** (1985, 1990 both to Dow); W. F. Goeckeler *et al.*, *Nucl. Med. Biol.* **13**, 479 (1986). Biodistribution studies: *idem et al.*, *J. Nucl. Med.* **28**, 495 (1987). Pharmacokinetics and clinical evaluation of pain palliation in metastatic bone cancer: M. Farhanghi *et al.*, *ibid.* **33**, 1451 (1992). HPLC determn in urine: W. F. Goeckeler *et al.*, *Nucl. Med. Biol.* **20**, 657 (1993). Chromatographic determn of radiochemical purity: H. T. Gasiglia, H. Okada, *J. Radioanal. Nucl. Chem. Lett.* **199**, 295 (1995). Review of therapeutic potential: R. A. Holmes, *Semin. Nucl. Med.* **22**, 41-45 (1992). Clinical trial in palliation of bone pain in hormone-refractory prostate cancer: O. Sartor *et al.*, *Urology* **63**, 940 (2004).

Pentasodium salt. [160369-78-8] CYT-424; Quadramet. $C_6H_{12}N_2Na_5O_{12}P_4{}^{153}Sm$.

THERAP CAT: Antineoplastic.

8488. Sambucus. Elder; elderberry; sweet elder. Large, deciduous shrub, *Sambucus nigra* L., *Caprifoliaceae*, known as European elder; used in traditional medicine as a treatment for respiratory infection, as a diuretic, and as a topical anti-inflammatory. North American elder, *S. canadensis* L., is also used. Medicinal parts include the flowers, leaves, bark, roots, and berries. *Habit.* Europe, Asia, North Africa; naturalized in North America. *Constit.* Rutin, quercetin, isoquercitrin, hyperoside, α- and β-amyrin, oleanolic and ursolic acids, chlorogenic acid; also contains the cyanogenic glycoside, sambunigrin. Analysis of constituents: J. Lee, C. E. Finn, *J. Sci. Food Agric.* **87**, 2665 (2007); R. Veberic *et al., Food Chem.* **114**, 511 (2009); M. Scopel *et al., Planta Med.* **76**, 1026 (2010). Clinical evaluation of berry extracts in influenza: Z. Zakay-Rones *et al., J. Int. Med. Res.* **32**, 132 (2004). Review of use in traditional medicine: *Altern. Med. Rev.* **10**, 51-55 (2005); J. E. Vlachojannis *et al., Phyto-ther. Res.* **24**, 1-8 (2010).

USE: Dietary supplement, in traditional medicines, as dye for textiles, food colorant. Berries as food, esp in desserts and wine, and as flavoring agent.

8489. Sampatrilat. [129981-36-8] N^2-(Methylsulfonyl)-L-lysyl-1-[(2*S*)-3-amino-2-carboxypropyl]cyclopentanecarbonyl-L-tyrosine; UK-81252. $C_{26}H_{40}N_4O_9S$; mol wt 584.69. C 53.41%, H 6.90%, N 9.58%, O 24.63%, S 5.48%. Dual inhibitor of neutral endopeptidase (NEP) and angiotensin converting enzyme (ACE). Prepn: J. C. Danilewicz *et al.*, **EP 358398**; *idem*, **US 4975444** (both 1990 to Pfizer). Determn in plasma: R. F. Venn *et al., J. Pharm. Biomed. Anal.* **16**, 875 (1998); and urine: *eidem, ibid.* 883. Clinical evaluations in hypertension: E. J. Wallis *et al., Clin. Pharmacol. Ther.* **64**, 439 (1998); G. R. Norton *et al., Am. J. Hypertens.* **12**, 563 (1999).

THERAP CAT: Antihypertensive.

8490. Sandalwood. White saunders; yellow saunders. Heartwood of *Santalum album* L., *Santalaceae*. *Habit.* India; cultivated in Timor and Sunda Islands. *Constit.* Volatile oil (3-5%), resin, tannin. Botanical description and medicinal uses: J. Gruenwald *et al., PDR for Herbal Medicines* (Medical Economics, Montvale, 2nd Ed., 2000) pp 659-660.

Volatile oil. [8006-87-9] Oil of Santal; sandalwood oil. Obtained by steam distillation from the dried heartwood. *Constit.* Santalols, α-bergamotol, α-bergamotal. Fragrance monograph: *J. Food Cosmet. Toxicol.* **12**, 989 (1974). Pale yellow, somewhat viscid liquid; characteristic sandalwood odor and taste. d_{25}^{25} 0.965-0.980. n_D^{20} 1.500-1.510. Sol in most fixed oils, propylene glycol, mineral oil; in 5 vols 70% alcohol. Insol in glycerin. *Keep well closed, cool and protected from light.*

USE: Wood in boxes and carvings, as incense. Oil in perfumes and as fragrance in creams and lotions.

THERAP CAT: In treatment of urinary tract inflammations.

8491. Sandarac. Resin from *Callitris quadrivalvis* Vent., *Pinaceae. Habit.* Morocco. *Constit.* About 80% pimaric acid, about 10% callitrolic acid; sandaricinic acid.

Light yellow, brittle, elongated tears; translucent with vitreous fracture; crumbles to powder when masticated. Insol in water, benzene, petr ether. Sol in alcohol, ether, acetone, hot caustic alkalies; partially sol in chloroform, volatile oils, oil turpentine, carbon disulfide.

USE: Tooth cements, lacquers, varnishes; as an incense. Pharmaceutic aid (in ointments and plasters).

8492. Sanguinaria. Bloodroot; red puccoon; red root; puccoon root; tetterwort. Dried rhizome and roots of *Sanguinaria can-*

adensis L., *Papaveraceae. Habit.* North America. *Constit.* Sanguinarine, chelerythrine, protopine, homochelidonine, resin. An extract of *Sanguinaria canadensis* has been used in cough syrup; used experimentally as toothpaste base, in gingivitis and in periodontal disease: P. A. Ladanyi, **US 4145412** (1979 to Vipont Chem. Co.).

8493. Sanguinarine. [2447-54-3] 13-Methyl[1,3]benzodioxolo[5,6-*c*]-1,3-dioxolo[4,5-*i*]phenanthridinium; pseudochelerythrine; ψ-chelerythrine. $[C_{20}H_{14}NO_4]^+$. Principal alkaloid in sanguinaria extract. Isolated from the root of *Sanguinaria canadensis* L., and other *Papaveraceae:* Schmidt *et al., Arch. Pharm.* **231**, 145 (1893). Widely distributed in poppy-fumaria species; constituent of argemone oil: S. A. E. Hakim *et al., Nature* **189**, 198 (1961). Identity with pseudochelerythrine: Gadamer, Stichel, *ibid.* **262**, 488 (1924). Structure: Späth, Kuffner, *Ber.* **64**, 370, 2034 (1931); Beke, *Acta Chim. Acad. Sci. Hung.* **17**, 463 (1958). Biosynthesis: Leete, *J. Am. Chem. Soc.* **85**, 473 (1963). Synthesis: Dyke *et al., Tetrahedron Lett.* **1968**, 3933; Sainsbury *et al., J. Chem. Soc. C* **1970**, 1797; Onda *et al., Chem. Pharm. Bull.* **17**, 404 (1969); **19**, 31 (1971); M. Hanaoka *et al., Chem. Lett.* **22**, 739 (1986). Purification procedure: Stipanovic *et al., J. Heterocycl. Chem.* **9**, 1453 (1972). Studies of effect of sanguinarine and argemone oil on intraocular pressure: S. A. E. Hakim, *Br. J. Ophthalmol.* **38**, 193 (1954); G. C. Dobbie, M. E. Langham, *ibid.* **45**, 81 (1961). Anti-inflammatory activity in rats: J. Lenfeld *et al., Planta Med.* **43**, 161 (1981). Clinical study of sanguinarine oral rinse as antiplaque agent: G. L. Southard *et al., J. Am. Dent. Assoc.* **108**, 338 (1984). Acute toxicity study: P. J. Becci *et al., J. Toxicol. Environ. Health* **20**, 199 (1987). *Review:* M. Shamma, *The Isoquinoline Alkaloids* (Academic Press, New York, 1972) pp 315-343. Symposium on chemistry, antimicrobial effects, clinical studies and toxicology: *J. Can. Dent. Assoc.* **56**, Suppl 7, 5-47 (1990).

Monohydrate. $C_{20}H_{13}NO_4.H_2O$. Crystals from water, mp 278-280° (Stipanovic). Soluble in alcohol, chloroform, acetone, ethyl acetate.

Chloride. Viadent. $C_{20}H_{14}ClNO_4$; mol wt 367.79. LD_{50} in male mice (mg/kg): 15.9 i.v.; in female mice (mg/kg): 102.0 s.c. (Lenfeld). LD_{50} in rats (mg/kg): 29 i.v.; 1658 orally (Becci).

Chloride dihydrate. $C_{20}H_{14}ClNO_4.2H_2O$. Fine orange needles, mp 273-274° (dec). uv max (methanol): 234, 283, 325 nm (log ε 4.50, 4.52, 4.18).

USE: In mouthwash and toothpaste.

8494. α-Santalol. [115-71-9] (2*Z*)-5-[(1*R*,3*R*,6*S*)-2,3-Dimethyltricyclo[2.2.1.0²,⁶]hept-3-yl]-2-methyl-2-penten-1-ol. $C_{15}H_{24}O$; mol wt 220.36. C 81.76%, H 10.98%, O 7.26%. A sesquiterpene alcohol which together with β-santalol comprises about 90% of commercial sandalwood oil: F. W. Semmler, K. Bode, *Ber.* **40**, 1124 (1907); E. Gruenther, *The Essential Oils* vol. 2 (Van Nostrand, New York, 1949) pp 266-269. Structure: F. W. Semmler, *Ber.* **43**, 1893 (1910). Separation of α- and β-santalols of high purity: A. E. Bradfield *et al., J. Chem. Soc.* **1935**, 309. Synthesis of α-santalol and conversion to β-santalol: S. C. Bhattacharyya, *Sci. Cult.* **13**, 207 (1947); *C.A.* **43**, 5484gh (1949); *see also* V. M. Sathe *et al., Indian J. Chem.* **4**, 393 (1966); S. Y. Kamat *et al., Tetrahedron* **23**, 4487 (1967). Stereochemistry studies: G. Brieger, *Tetrahedron Lett.* **1963**, 2123. Natural α-santalol has a *cis*-(+)-configuration. Total synthesis: R. G. Lewis *et al., ibid.* **1967**, 401. Stereospecific synthesis: E. J. Corey *et al., J. Am. Chem. Soc.* **92**, 226, 6314 (1970); M. Julia, P. Ward, *Bull. Soc. Chim. Fr.* **1973**, 3065; M. Tamura, G. Suzukamo, *Tetrahedron Lett.* **22**, 577 (1981). Synthesis and separation of *cis*- and *trans*-isomers: K. Sato *et al., Bull. Chem. Soc. Jpn.* **49**, 3351 (1976). *Review:* J. Simonsen, D. H. R. Barton, *The Terpenes* vol. 3 (Univ. Press, Cambridge, 1961) pp 178-188; A. Bhati, *Flavour Ind.* **1**, 235-251 (1970).

Liquid. bp_{14} 166-167°. d_{25}^{25} 0.9770. n_D^{25} 1.5017. α_{5461} +10.3°. $[\alpha]_D^{20}$ +17.20° (c = 0.8 in chloroform). Sol in alcohol; slightly sol in propylene glycol, glycerine. Practically insol in water.

USE: In perfumes, soaps and detergents.

8495. β-Santalol. [77-42-9] (2Z)-2-Methyl-5-[(1S,2R,4R)-2-methyl-3-methylenebicyclo[2.2.1]hept-2-yl]-2-penten-1-ol; 2-methyl-5-(2-methyl-3-methylene-2-norbornyl)-2-penten-1-ol. $C_{15}H_{24}O$; mol wt 220.36. C 81.76%, H 10.98%, O 7.26%. A sesquiterpene alcohol from sandalwood oil. Isoln and purification and additional refs: see α-santalol. Structure: L. Ruzicka, G. Thomann, *Helv. Chim. Acta* **18**, 355 (1935). Synthesis: S. C. Bhattacharyya, *Sci. Cult.* **13**, 209 (1957); *C.A.* **43**, 5385a (1949); H. C. Kretschmar, W. F. Erman, US 3662008, US 3679756 (1972 to Procter & Gamble). Natural β-santalol has a *cis*-(−)-configuration. Total synthesis: *eidem*, *Tetrahedron Lett.* **1970**, 41. Synthesis of *dl*-form: P. A. Christenson, B. J. Willis, *J. Org. Chem.* **44**, 2012 (1979); M. Baumann, W. H. Hoffman, *Ann.* **1979**, 743; K. Sato *et al.*, *Chem. Lett.* **1981**, 1183.

Liquid. bp_{17} 177-178°. d_{25}^{25} 0.9717. n_D^{25} 1.5100. α_{5461} −87.1°. Sol in alcohol; practically insol in water.

dl-**Form.** $bp_{0.1}$ 101-103°.

USE: In perfumes, soaps and detergents.

8496. Santonica. Levant wormseed. Dried, unexpanded flower heads of *Artemisia maritima* L., sens. lat., *Compositae*. Habit. Iran, Turkestan. *Constit.* 2-4% santonin, 2-3% volatile oil; artemisin, resin.

THERAP CAT: Anthelmintic.

8497. Santonic Acid. [510-35-0] (αS,1R,3aS,4R,5S,7aS)-Octahydro-α,3a,5-trimethyl-6,8-dioxo-1,4-methano-1H-indene-1-acetic acid; hexahydro-α,3a,5-trimethyl-6,8-dioxo-1,4-methanoindan-1-acetic acid. $C_{15}H_{20}O_4$; mol wt 264.32. C 68.16%, H 7.63%, O 24.21%. Prepd from santonin by the action of hot concd aq bases, preferably potassium hydroxide: v. Oettingen, *Dissertation*, Göttingen, 1913; Woodward *et al.*, *J. Am. Chem. Soc.* **70**, 4216 (1948). Structure: Wedekind, Engel, *J. Prakt. Chem.* **139**, 115 (1934); Woodward *et al.*, *loc. cit.*; Woodward, Yates, *Chem. Ind. (London)* **1954**, 1391. Chemistry: Hortmann, Daniel, *J. Org. Chem.* **37**, 4446 (1972).

Crystals from water or alcohol, mp 170-172°; bp_{15} 285°; $[\alpha]_D^{20}$ −74.0° (chloroform). Soluble in 190 parts of water at 17°. Freely sol in alc, chloroform, ether, glacial acetic acid.

8498. α-Santonin. [481-06-1] (3S,3aS,5aS,9bS)-3a,5,5a,9b-Tetrahydro-3,5a,9-trimethylnaphtho[1,2-b]furan-2,8(3H,4H)-dione; 1,2,3,4,4a,7-hexahydro-1-hydroxy-a,4a,8-trimethyl-7-oxo-2-naphthaleneacetic acid γ-lactone; *l*-santonin. $C_{15}H_{18}O_3$; mol wt 246.31. C 73.15%, H 7.37%, O 19.49%. Anthelmintic isolated from the dried unexpanded flower heads of *Artemisia maritima* L., sens. lat., *Compositae* [Levant wormseed] and other species of *Artemisia*

found principally in Russian and Chinese Turkestan and the Southern Ural region. Structure: Clemo *et al.*, *J. Chem. Soc.* **1930**, 1110; Ruzicka, Steiner, *Helv. Chim. Acta* **17**, 614 (1934). Stereochemistry: Corey, *J. Am. Chem. Soc.* **77**, 1044 (1955); Cocker, McMurry, *Tetrahedron* **8**, 181 (1960); Asher, Sim, *Proc. Chem. Soc. London* **1962**, 111, 335; Nakazaki, Arakawa, *Bull. Chem. Soc. Jpn.* **37**, 464 (1964); Pregosin *et al.*, *J. Chem. Soc. Perkin Trans. 1* **1972**, 299. Total synthesis: Abe *et al.*, *Proc. Jpn. Acad.* **30**, 116 (1954); *J. Am. Chem. Soc.* **78**, 1422 (1956); US 2836604 (1958 to Takeda); J. A. Marshall, P. G. M. Wuts, *J. Org. Chem.* **43**, 1086 (1978). Mass spectral studies of the santonins and derivs: Woseda *et al.*, *Tetrahedron* **23**, 4623 (1967). *Review:* C. H. Heathcock in *The Total Synthesis of Natural Products* vol. **2**, J. W. ApSimon, Ed. (Wiley, New York, 1973) pp 315-324.

(−)-Form. Tabular crystals, orthorhombic sphenoidal, mp 170-173°. Almost tasteless with bitter aftertaste. $[\alpha]_D^{25}$ −170 to −175° (c = 2 in alc). Becomes yellow on exposure to light. Irritating to mucous membranes. d 1.187. One part dissolves in 5000 parts of cold water, in 250 parts of boiling water, in 280 parts of 50% alcohol at 17°, in 10 parts of boiling 50% alcohol, in 44 parts of cold 90% alcohol, in 3 parts of boiling 90% alcohol, in 125 parts of cold ether, in 72 parts of boiling ether, in 4.3 parts of cold chloroform.

(±)-Form. Colorless plates from methanol, mp 181°. uv max (ethanol): 241 nm (log ε 4.10).

(+)-Form. Colorless plates from methanol, mp 172°. $[\alpha]_D^{20}$ +165.9° (c = 1.92 in ethanol).

4-Hydroxysantonin see Artemisin.

THERAP CAT: Anthelmintic (Nematodes).

THERAP CAT (VET): Has been used as an anthelmintic.

8499. Saperconazole. [110588-57-3] 4-[4-[4-[4-[[2-(2,4-Difluorophenyl)-2-(1H-1,2,4-triazol-1-ylmethyl)-1,3-dioxolan-4-yl]methoxy]phenyl]-1-piperazinyl]phenyl]-2,4-dihydro-2-(1-methylpropyl)-3H-1,2,4-triazol-3-one; (±)-1-*sec*-butyl-4-[p-[4-[p-[[(2R*,4S*)-2-(2,4-difluorophenyl)-2-(1H-1,2,4-triazol-1-ylmethyl)-1,3-dioxolan-4-yl]methoxy]phenyl]-1-piperazinyl]phenyl]-Δ²-1,2,4-triazolin-5-one; R-66905. $C_{35}H_{38}F_2N_8O_4$; mol wt 672.74. C 62.49%, H 5.69%, F 5.65%, N 16.66%, O 9.51%. Fluorinated triazole antifungal; mixture of 4 *cis*-isomers. Prepn: J. Heeres *et al.*, EP 283992; *eidem*, US 4916134 (1988, 1990 both to Janssen). *In vitro* antifungal activity: F. C. Odds, *J. Antimicrob. Chemother.* **24**, 533 (1989); D. W. Denning *et al.*, *Eur. J. Clin. Microbiol. Infect. Dis.* **9**, 693 (1990). *In vivo* efficacy vs *Aspergillus* species: J. Van Cutsem *et al.*, *Antimicrob. Agents Chemother.* **33**, 2063 (1989).

Relative stereochemistry

Crystals from acetonitrile, mp 189.5°. Poorly sol in water.

THERAP CAT: Antifungal.

8500. Saponaria. Soapwort; soaproot; fuller's herb; bruisewort; bouncing bet. *Saponaria officinalis* L., *Caryophyllaceae*; herbaceous perennial plant having sharp-pointed leaves and large pale pink flowers. Habit. Europe to Middle Asia; naturalized in U.S. *Constit.* Saponins, saponarin, saporins, *q.q.v.* Produces soapy lather when bruised and agitated with water. Root has been used in traditional medicine as antitussive or as a wash for skin irritations. Brief review: M. Wichtl, N. G. Bisset, *Herbal Drugs and Phytopharmaceuticals*, English Ed. (CRC Press, Boca Raton, 1994) pp 453-454.

8501. Saponarin. [20310-89-8] 6-β-D-Glucopyranosyl-7-(β-D-glucopyranosyloxy)-5-hydroxy-2-(4-hydroxyphenyl)-4*H*-1-benzopyran-4-one. $C_{27}H_{30}O_{15}$; mol wt 594.52. C 54.55%, H 5.09%, O 40.37%. From leaves of *Saponaria officinalis* L., *Caryophyllaceae:* Barger, *J. Chem. Soc.* **89**, 1210 (1906); from *Hibiscus syriacus* L., *Malvaceae:* Nakoaki, *J. Pharm. Soc. Jpn.* **64**, 304 (1944), *C.A.* **46**, 108d (1952); from *Lemna* (*Spirodela*) *oligorrhiza* Kurz., *Lemnaceae:* Jurd *et al.*, *Arch. Biochem. Biophys.* **67**, 284 (1957). Structure: Seikel, Geissman, *ibid.* **71**, 17 (1957). Revised structure: Hörhammer *et al.*, *Tetrahedron Lett.* **1965**, 1707. On acidic hydrolysis yields the aglycone, *saponaretin*.

Monohydrate. Pale yellow granules, mp 228°. $[\alpha]_D$ −7.9° (aq pyridine). uv max (ethanol): 335, 272 nm. Practically insol in cold water. Sparingly sol in hot water, alcohol; sol in alkalies with yellow color and in concd H_2SO_4 with blue fluorescence; sol in pyridine.

8502. Saponins. Sapogenin glycosides. A type of glycoside widely distributed in plants. Each saponin consists of a sapogenin which constitutes the aglucon moiety of the molecule, and a sugar. The sapogenin may be a steroid or a triterpene and the sugar moiety may be glucose, galactose, a pentose, or a methylpentose. Poisonous towards the lower forms of life and used for killing fish by the aborigines of South America. *Review and bibliography:* R. J. McIlroy, *The Plant Glycosides* (Edward Arnold & Co., London, 1951) Chapter IX; Y. Birk, I. Peri in *Toxic Const. Plant Foodst.*, I. E. Liener, Ed. (Academic Press, New York, 2nd ed., 1980) pp 161-182.

Bitter taste. All saponins foam strongly when shaken with water. They form oil-in-water emulsions and act as protective colloids.

Caution: Although practically non-toxic to man upon oral ingestion, they act as powerful hemolytics when injected into the blood stream, dissolving the red corpuscles even at extreme dilutions.

8503. Saporins. Single chain ribosome inactivating proteins (type 1 RIPs); found in the seeds, leaves and roots of the soapwort plant, *Saponaria officinalis* L., *Caryophyllaceae*. Used to produce targeted immunotoxins by conjugating with a specific monoclonal antibody or other cell receptor ligand. At least 7 isoforms have been identified. Saporin-6 is the most abundant and is often referred to as "saporin". Isoln from seeds: F. Stirpe *et al.*, *Biochem. J.* **216**, 617 (1983). Purification and distribution in plant tissues: J. M. Ferreras *et al.*, *Biochim. Biophys. Acta* **1216**, 31 (1993). Production by plant cell culture: A. Di Cola *et al.*, *Plant Cell Rep.* **17**, 55 (1997). Review of use as neural lesioning tools: R. G. Wiley, *Trends Neurosci.* **15**, 285-290 (1992); in model for Alzheimer's disease: R. Schliebs *et al.*, *Prog. Brain Res.* **109**, 253-264 (1996). Discussion of therapeutic potential in cancer: D. J. Flavell, *Curr. Top. Microbiol. Immunol.* **234**, 57-61 (1998).

Saporin-6. [166944-16-7] SO6. Isoform constituting 7% of the total seed protein. Composed of 259 amino acid residues; mol wt 29 kDa. Review of isoln, properties and production by recombinant technology: M. R. Soria *et al.*, *Targeted Diagn. Ther.* **7**, 193-212 (1992). Crystal structure: C. Savino *et al.*, *Acta Crystallogr.* **D54**, 636 (1998). pI >9.5. LD$_{50}$ i.p. in mice: 4 mg/kg (Stirpe).

USE: Biological research tool.

8504. Sapphire. [1317-82-4] Al_2O_3; mol wt 101.96. Al 52.93%, O 47.07%. The corundum modification of aluminum oxide, Al_2O_3. Mined in Montana, Siam, Burma, Ceylon. The highest priced variety of the native gem has the blue color of cornflowers. Easily prepd synthetically. Hydrothermal synthesis: Laudise, Ballman, *J. Am. Chem. Soc.* **80**, 2655 (1958).

USE: In jewelry, as phonograph needles, industrial abrasive, watch and instrument bearings.

8505. Sapropterin. [62989-33-7] (6*R*)-2-Amino-6-[(1*R*,2*S*)-1,2-dihydroxypropyl]-5,6,7,8-tetrahydro-4(1*H*)-pteridinone; (6*R*)-L-*erythro*-tetrahydrobiopterin; dapropterin; *R*-THBP; 6*R*-BH$_4$. C_9H_{15}N$_5$O$_3$; mol wt 241.25. C 44.81%, H 6.27%, N 29.03%, O 19.90%. Natural cofactor of the aromatic amino acid hydroxylases required for catecholamine and serotonin biosynthesis. Identification of cofactor activity: S. Kaufman, *Proc. Natl. Acad. Sci. USA* **50**, 1085 (1963). Prepn of (6*R*,*S*)-BH$_4$: B. Schircks *et al.*, *Helv. Chim. Acta* **61**, 2731 (1978). Chromatographic separation of diastereoisomers: S. W. Bailey, J. E. Ayling, *J. Biol. Chem.* **253**, 1598 (1978). Absolute configuration of natural isomer: W. L. F. Armarego *et al.*, *Aust. J. Chem.* **35**, 785 (1982). Stereospecific synthesis: S. Matsuura *et al.*, *Heterocycles* **23**, 3115 (1985); H. Sakai, T. Kanai, **EP 191335**; *eidem*, **US 4713454** (1986, 1987 both to Shiratori; Suntory). Bioavailability: G. Kapatos, S. Kaufman, *Science* **212**, 955 (1981). Effect on neurotransmitter monoamine biosynthesis: S. Miwa *et al.*, *Arch. Biochem. Biophys.* **239**, 234 (1985). LC determn in biological samples: Y. Tani, T. Ishihara, *Life Sci.* **46**, 373 (1990). Therapeutic potential in hyperphenylalaninemia: S. Kaufman, *J. Nutr. Sci. Vitaminol.*, Suppl., 601 (1992). Clinical trial in phenylketonuria: H. L. Levy *et al.*, *Lancet* **370**, 504 (2007). Review of development and clinical experience: J. R. Burnett, *IDrugs* **10**, 805-813 (2007).

pK' 5.05. uv max (0.1*N* HCl): 265 nm (ε 14000).

Dihydrochloride. [69056-38-8] SUN-0588; Biopten; Kuvan. $C_9H_{15}N_5O_3$.2HCl; mol wt 314.17. Crystals from HCl, mp 245-246° (dec). $[\alpha]_D^{23}$ −6.81° (c = 0.665 in 0.1*M* HCl). uv max (2*M* HCl): 264 nm (ε 16770).

THERAP CAT: In treatment of hyperphenylalaninemia.

8506. Saquinavir. [127779-20-8] (2*S*)-*N*¹[(1*S*,2*R*)-3-[(3*S*,-4a*S*,8a*S*)-3-[[(1,1-Dimethylethyl)amino]carbonyl]octahydro-2(1*H*)-isoquinolinyl]-2-hydroxy-1-(phenylmethyl)propyl]-2-[(2-quinolinylcarbonyl)amino]butanediamide; (*S*)-*N*-[(α*S*)-α-[(1*R*)-2-[(3*S*,4a*S*,-8a*S*)-3-(*tert*-butylcarbamoyl)octahydro-2(1*H*)-isoquinolyl]-1-hydroxyethyl]phenethyl]-2-quinaldamido succinamide; *N*-*tert*-butyldecahydro-2-[2(*R*)-hydroxy-4-phenyl-3(*S*)-[[*N*-(2-quinolylcarbonyl)-L-asparaginyl]amino]butyl](4a*S*,8a*S*)-isoquinoline-3(*S*)-carboxamide; Ro-31-8959. $C_{38}H_{50}N_6O_5$; mol wt 670.86. C 68.03%, H 7.51%, N 12.53%, O 11.92%. Peptidomimetic HIV protease inhibitor. Prepn: J. A. Martin, S. Redshaw, **EP 432695**; *eidem*, **US 5196438** (1991, 1993 both to Hoffmann-LaRoche); K. E. B. Parkes *et al.*, *J. Org. Chem.* **59**, 3656 (1994). *In vitro* HIV proteinase inhibition: N. A. Roberts *et al.*, *Science* **248**, 358 (1990). Antiviral properties: J. C. Craig *et al.*, *Antiviral Res.* **16**, 295 (1991); S. Galpin *et al.*, *Antivir. Chem. Chemother.* **5**, 43-45 (1994). Clinical evaluation of tolerability and activity: V. S. Kitchen *et al.*, *Lancet* **345**, 952 (1995). Review of pharmacology and clinical experience: S. Kravcik, *Expert Opin. Pharmacother.* **2** 303-315 (2001).

White crystalline solid. $[\alpha]_D^{20}$ −55.9° (c = 0.5 in methanol). Soly (21°): 0.22 g/100 ml water.

Methanesulfonate salt. [149845-06-7] Saquinavir mesylate; Ro-31-8959/003; Fortovase; Invirase. $C_{38}H_{50}N_6O_5$.CH$_3$SO$_3$H; mol wt 766.96.

THERAP CAT: Antiretroviral.

8507. Sarafloxacin. [98105-99-8] 6-Fluoro-1-(4-fluorophenyl)-1,4-dihydro-4-oxo-7-(1-piperazinyl)-3-quinolinecarboxylic acid. $C_{20}H_{17}F_2N_3O_3$; mol wt 385.37. C 62.33%, H 4.45%, F 9.86%, N 10.90%, O 12.45%. Fluorinated quinolone antibacterial. Prepn: D. T. W. Chu, **EP 131839**; *idem*, **US 4730000** (1985, 1988 both to Abbott); D. T. W. Chu *et al.*, *J. Med. Chem.* **28**, 1558 (1985); H. Narita *et al.*, **JP Kokai 85 237069** (1985 to Toyama). Comparative *in vitro* antibacterial spectrum: A. Digranes, W. L. Dibb, *Chemotherapy* **34**, 298 (1988). *In vivo* antibacterial activity: P. B. Fernandes *et al.*, *Antimicrob. Agents Chemother.* **29**, 201 (1986). HPLC determn: J. F. Bauer *et al.*, *Pharm. Res.* **7**, 1177 (1990).

Monohydrochloride. [91296-87-6] Abbott 56620; A-56620. $C_{20}H_{17}F_2N_3O_3 \cdot HCl$; mol wt 421.83. mp as monohydrate >275°.

THERAP CAT (VET): Antibacterial.

8508. Sarafotoxins. SRTx. Family of vasoactive toxins isolated from the venom of the Israeli burrowing asp, *Atractaspis engaddensis*. Several isoforms exist, all are closely related in structure and biological activity to the mammalian endothelins, *q.v.*; both families show highly conserved sequences of 21 amino acids with 2 fixed disulfide bridges. Isoln: E. Kochva *et al.*, *Toxicon* **20**, 581 (1982). Purification of a-c: C. Takasaki *et al.*, *ibid.* **26**, 543 (1988); of d: A. Bdolah *et al.*, *FEBS Lett.* **256**, 1 (1989). Structure-activity study: A. M. Doherty *et al.*, *J. Cardiovasc. Pharmacol.* **17**, Suppl 7, S59 (1991). Conformational stability: A. R. Atkins *et al.*, *Int. J. Pept. Protein Res.* **44**, 372 (1994). Review of structure and evolution: E. Kochva *et al.*, *Toxicon* **31**, 541-568 (1993); of biological activity: M. Sokolovsky in *Toxins and Signal Transduction*, Y. Gutman, P. Lazarovici, Eds. (Harwood Academic Publishers, Netherlands, 1997) pp 53-67.

C—S—C—K—D—M—T—D—K—E—C—L—N—F—C—H—Q—D—V—I—W

Sarafotoxin A

8509. Saralasin. [34273-10-4] *N*-Methylglycyl-L-arginyl-L-valyl-L-tyrosyl-L-valyl-L-histidyl-L-prolyl-L-alanine; 1-(*N*-methylglycine)-5-L-valine-8-L-alanine angiotensin II; 1-sar-8-ala-angiotensin II. $C_{42}H_{65}N_{13}O_{10}$; mol wt 912.06. C 55.31%, H 7.18%, N 19.96%, O 17.54%. A specific antagonist of angiotensin II. Prepn: Sipos *et al.*, **DE 2127393** corresp to **US 3751404** (1972, 1973 to Norwich Pharmacal). Activity studies: Pals *et al.*, *Circ. Res.* **29**, 673 (1971); Pals, Fulton, *Arch. Int. Pharmacodyn. Ther.* **204**, 20 (1973); Solomon, Buckley, *J. Pharm. Sci.* **63**, 1109 (1974); Streeten *et al.*, *N. Engl. J. Med.* **292**, 657 (1975). Comprehensive review of the acetate: C. T. Huang *et al.*, in *Pharmacological and Biochemical Properties of Drug Substances* vol. 3, M. E. Goldberg, Ed. (Am. Pharm. Assoc., Washington, DC, 1981) pp 176-225.

Sar–Arg–Val–Tyr–Val–His–Pro–Ala

Hydrated acetate. [39698-78-7] Saralasin acetate; P-113; Sarenin. Described as $C_{42}H_{65}N_{13}O_{10} \cdot xC_2H_4O_2 \cdot xH_2O$. Fluffy white powder, mp 256°. Sol in water, 5% aq dextrose, 90-95% aq alcohol. LD_{50} i.v. in male mice: 1171 mg/kg (Huang).

THERAP CAT: Antihypertensive, diagnostic aid (renin-dependent hypertension).

8510. Sarcosine. [107-97-1] *N*-Methylglycine; *N*-methylaminoacetic acid; Sar; *N*-methylglycocoll; methylaminoethanoic acid. $C_3H_7NO_2$; mol wt 89.09. C 40.45%, H 7.92%, N 15.72%, O 35.92%. Has been found in starfish and sea urchins: A. Kossel, S. Edlbacher, *Z. Physiol. Chem.* **94**, 264 (1915); in rock lobsters: L. Novellie, H. M. Schwartz, *Nature* **173**, 450 (1954). Formation from caffeine by dec with barium hydroxide: Paulmann, *Arch. Pharm.* **232**, 601 (1894). Prepd on a large scale from formaldehyde, sodium cyanide, and methylamine: Eschweiler, *Ann.* **279**, 39 (1894); Baumann, *J. Biol. Chem.* **21**, 563 (1915); Caverly, **US 2720540** (1955 to du Pont); Leake, Brakebill, **US 3009954** (1961 to Allied Chem.). Synthesis: M. Ebata *et al.*, *Bull. Chem. Soc. Jpn.* **39**, 2535 (1966). Review of syntheses: T. Shirai, *Synthetic Production and Utilization of Amino Acids*, T. Kaneko *et al.*, Eds. (Wiley, New York, 1974) pp 184-186.

Orthorhombic, deliquescent crystals from dil methanol. Dec 212°. Sweetish taste. pK_1' 2.23; pK_2' 10.01. Sol in water: 5 ml of a satd aq soln contain 2.1412 g sarcosine. Slightly sol in alc.

Sodium salt. [4316-73-8] $C_3H_6NNaO_2$; mol wt 111.08. Usually sold as aq soln, pH 12.0.

Hydrochloride. [637-96-7] $C_3H_7NO_2 \cdot HCl$; mol wt 125.55. Needles from alcohol, dec 171°. Freely sol in water; slightly sol in alcohol, ether.

USE: Intermediate in the synthesis of antienzyme agents for toothpaste.

8511. Sarin. [107-44-8] *P*-Methylphosphonofluoridic acid 1-methylethyl ester; isopropoxymethylphosphoryl fluoride; isopropylmethylphosphonofluoridate; GB. $C_4H_{10}FO_2P$; mol wt 140.09. C 34.30%, H 7.20%, F 13.56%, O 22.84%, P 22.11%. Nerve gas; potent cholinesterase inhibitor similar in structure and activity to soman and tabun, *q.q.v.* Prepn: Holmstedt, *Acta Physiol. Scand.* **25**, suppl. 90, p 106 (1951); *Protar* **14**, 113 (1948); **16**, 131 (1950). Additional syntheses: Schrader, *B.I.O.S.* **714**, 41 (1947); B. C. Saunders, *Some Aspects of the Chemistry and Toxic Action of Organic Compounds Containing Phosphorus and Fluorine* (Cambridge, 1957) p 92 sqq; Bryant *et al.*, *J. Chem. Soc.* **1960**, 1553. Absolute configuration: Benschop *et al.*, *Rec. Trav. Chim.* **87**, 387 (1968). Toxicity study: B. Holmstedt, *Pharmacol. Rev.* **11**, 567 (1959). Epidemiological study of chronic effects: Y. Nishiwaki *et al.*, *Environ. Health Perspect.* **109**, 1169 (2001). Brief review: Schrader, *Die Entwicklung neuer insektizider Phosphorsäure-Ester* (Verlag Chemie, Weinheim, 1963) p 4; B. L. Harris in *Kirk-Othmer Encyclopedia of Chemical Technology* vol. 5 (Wiley-Interscience, New York, 4th ed., 1993) pp 795-816.

Liquid, d_4^{20} 1.10. mp −57°. bp_{760} 147°; bp_{16} 56°. Miscible with and hydrolyzed by water. Rapidly hydrolyzed by dil aq sodium hydroxide or sodium carbonate forming relatively non-toxic products. Water alone removes the fluorine atom producing the non-toxic acid $CH_3PO[OCH(CH_3)_2]OH$. LD_{50} i.p. in mice: 0.42 mg/kg (Holmstedt, 1959).

Caution: Potential symptoms of overexposure by inhalation are constriction of pupils of the eye, difficulty breathing followed by bronchial constriction, convulsions, and death. Also toxic due to absorption through skin and eyes. *See: Chem. Eng. News* **31**, 4676 (1953).

USE: Chemical warfare agent.

8512. Sarizotan. [351862-32-3] *N*-[[(2*R*)-3,4-Dihydro-2*H*-1-benzopyran-2-yl]methyl]-5-(4-fluorophenyl)-3-pyridinemethanamine; (*R*)-(−)-2-[5-(4-fluorophenyl)-3-pyridylmethylaminomethyl]chromane; (*R*)-(−)-1-*N*-[5-(4-fluorophenyl)-3-pyridylmethyl]-*N*-2-chromanylmethylamine; EMD-77697. $C_{22}H_{21}FN_2O$; mol wt 348.42. C 75.84%, H 6.08%, F 5.45%, N 8.04%, O 4.59%. Dual serotonin 5-HT$_{1A}$ receptor agonist and dopamine D$_2$ receptor antagonist. Prepn: H. Böttcher *et al.*, **EP 707007**; *eidem*, **US 5767132** (1996, 1998 both to Merck Patent GmbH). PET receptor binding study: E. A. Rabiner *et al.*, *J. Psychopharmacol.* **16**, 195 (2002). Neurochemical profile: G. D. Bartoszyk *et al.*, *J. Neural Transm.* **111**, 113 (2004). Clinical evaluation in dyskinesia in Parkinson's disease: C. W. Olanow *et al.*, *Clin. Neuropharmacol.* **27**, 58 (2004); W. Bara-Jimenez *et al.*, *Mov. Disord.* **20**, 932 (2005).

Dihydrochloride. [177976-12-4] EMD-128130. C$_{22}$H$_{21}$FN$_2$-O.2HCl; mol wt 421.34. mp 234-235°. [α]^{20} −65° (c = 1 in methanol).

THERAP CAT: Antidyskinetic.

8513. Sarkomycin. [11031-48-4] Sarcomycin. Antibiotic and antitumor substance produced by *Streptomyces erythrochromogenes*, strain W-115-C, from soil found at Kamakura, Japan: Umezawa *et al.*, *J. Antibiot.* **6A**, 101, 147, 153 (1953); *eidem, Antibiot. Chemother.* **4**, 514 (1954). Structure of sarkomycin A, the active principle: Hooper *et al.*, *ibid.* **5**, 585 (1955). Characterization of sarkomycins A, A′ and B: Maeda, Kondo, *J. Antibiot.* **11A**, 37 (1958). Sarkomycin reacts with H$_2$S to form sarkomycins S$_1$, S$_2$ and S$_3$: Tatsuoka *et al.*, *ibid.* **9B**, 107, 110 (1956). Structure of S$_1$ and S$_2$: *eidem, ibid.* **11B**, 275 (1958). Synthesis of *d*- and *l*-sarkomycin A: Toki, *ibid.* **10A**, 35, 226 (1957). Synthesis of *dl*-sarkomycin A: Toki, *Bull. Chem. Soc. Jpn.* **30**, 450 (1957); R. K. Boeckman, Jr. *et al.*, *J. Org. Chem.* **45**, 752 (1980); J. N. Marx, G. Minaskanian, *ibid.* **47**, 3306 (1982); B. A. Wexler *et al.*, *ibid.* 3333; A. P. Kozikowski, P. D. Stein, *J. Am. Chem. Soc.* **104**, 4023 (1982). Abs config of A: Sato *et al.*, *Chem. Pharm. Bull.* **11**, 829 (1963). Revised config of A: Hill *et al.*, *J. Org. Chem.* **32**, 2330 (1967). *Review:* S. A. Waksman, H. A. Lechevalier, *The Actinomycetes* vol. **III** (Williams & Wilkins, Baltimore, 1962) pp 362-364; Sung in *Antibiotics* vol. **I**, D. Gottlieb, P. Shaw, Eds. (Springer-Verlag, New York, 1967) pp 156-165.

Sarkomycin A

Oily liquid. [α]$^{15}_D$ −32.5° (c = 1 in methanol). uv max (water): 230 nm. Sol in water, methanol, ethanol, butanol, ethyl acetate. Sparingly sol in petr ether.

Sarkomycin A. [489-21-4] (1*R*)-2-Methylene-3-oxocyclopentanecarboxylic acid. C$_7$H$_8$O$_3$; mol wt 140.14.

8514. Sarmentose. [13484-14-5] 2,6-Dideoxy-3-*O*-methyl-*xylo*-hexose. C$_7$H$_{14}$O$_4$; mol wt 162.19. C 51.84%, H 8.70%, O 39.46%. A methyl ether of a 2-desoxyhexomethylose, isomeric with cymarose. Obtained on hydrolysis of sarmentocymarin, a glycoside isolated from seeds of *Strophanthus sarmentosus* DC., *Apocynaceae* by the enzymic method: Jacobs, Heidelberger, *J. Biol. Chem.* **81**, 765 (1929); Jacobs, Bigelow, *ibid.* **96**, 355 (1932). Synthesis: H. Hauenstein, T. Reichstein, *Helv. Chim. Acta* **33**, 446 (1950).

Prisms, plates from ether + petr ether, mp 78-79°. Shows mutarotation. [α]$^{20}_D$ +12° (20 min) → +15.8° (24 hrs c = 1.08): Elderfield, *Adv. Carbohydr. Chem.* **1**, 172 (1945).

8515. Sarpagine. [482-68-8] Sarpagan-10,17-diol; raupine. C$_{19}$H$_{22}$N$_2$O$_2$; mol wt 310.40. C 73.52%, H 7.14%, N 9.03%, O 10.31%. Isoln from *Rauwolfia serpentina* (L.) Benth., *Apocynaceae*: Stoll, Hofmann, *Helv. Chim. Acta* **36**, 1143 (1953). Structure: Stauffacher *et al.*, *ibid.* **40**, 508 (1957); Poisson *et al.*, *Bull. Soc. Chim. Fr.* **1957**, 610; Biemann, *J. Am. Chem. Soc.* **83**, 4801 (1961). Stereochemistry: Bartlett *et al.*, *ibid.* **84**, 622 (1962); Ohashi *et al.*, *Tetrahedron* **19**, 2241 (1963).

Needles from methanol or ethanol, long plates from acetone, dec >320°. [α]$^{20}_D$ +54° (c = 0.5 in pyridine). uv max (ethanol): 230, 278 nm (log ε 4.30, 3.92). One gram dissolves in 60 ml boiling ethanol, 250 ml boiling methanol, 400 ml boiling acetone. Practically insol in chloroform.

Hydrochloride. C$_{19}$H$_{22}$N$_2$O$_2$.HCl. Needles from alcohol, dec >220°.

8516. Sarsaparilla. Dried root of *Smilax aristolochiaefolia* Mill. (*S. medica* Cham. & Schlecht.) (Mexican Sarsap.), or *S. regelii* Killip & Morton (*S. ornata* Hook f.) (Jamaica or Central American Sarsap.), *Liliaceae*. *Habit.* Honduras, Jamaica, Mexico (Vera Cruz), Brazil, Guatemala. *Constit.* Sarsaponin, smilacin (parillin), paroaparic acid, resin, volatile oil.

USE: As flavor in beverages. Pharmaceutic aid (flavor).

8517. Sarsasapogenin. [126-19-2] (3β,5β,25*S*)-Spirostan-3-ol; parigenin. C$_{27}$H$_{44}$O$_3$; mol wt 416.65. C 77.83%, H 10.64%, O 11.52%. Steroid sapogenin from *Smilax ornata* Hooker, fil., *Liliaceae* (Sarsaparilla): Power, Salway, *J. Chem. Soc.* **105**, 201 (1914); Jacobs, Simpson, *J. Biol. Chem.* **105**, 501 (1934); **109**, 573 (1935). Partial structure: Hirschmann *et al.*, *J. Org. Chem.* **20**, 572 (1955). Stereochemistry: Taylor, *Chem. Ind. (London)* **1954**, 1066; Wall, Serota, *J. Am. Chem. Soc.* **76**, 2850 (1954); Callow, Massey-Beresford, *J. Chem. Soc.* **1957**, 4482; Rosen *et al.*, *J. Am. Chem. Soc.* **81**, 1687 (1959). *Reviews:* Turner, *The Steroidal Sapogenins* Pub. no. 361, (Univ. Microfilms, Ann Arbor, Mich., 1941) 249 pp; Shabica, *Sarsasapogenin and Related Compounds, Univ. Microfilms* (Ann Arbor, Mich.), Pub. no. 549, 89 pp (1943); L. F. Fieser, M. Fieser, *Steroids* (Reinhold, New York, 1959) p 810. *See also* Smilagenin.

Large prismatic needles from acetone, mp 199-199.5°. [α]$^{25}_D$ −75°, [α]$^{25}_{546}$ −89° (c = 0.5 in CHCl$_3$). uv spectrum: Smith, Eddy, *Anal. Chem.* **31**, 1539 (1959). Sol in alcohol, acetone, benzene, chloroform. Is precipitated by digitonin.

Acetate. C$_{29}$H$_{46}$O$_4$. Flat needles from methanol, mp 144-145°. [α]$^{25}_D$ −70.2°; [α]$^{25}_{546}$ −83.1° (c = 1.18 in CHCl$_3$).

USE: In the manuf of compds of the pregnane series.

8518. Sassafras. Saxifrax; ague tree; cinnamon wood; saloop. Deciduous tree, *Sassafras albidum* (Nutt.) Nees, *Lauraceae*; also known as *S. variifolium* (Salisb.) Kuntze, and *S. officinale* Nees & Eberm. Medicinal parts are the peeled and dried root bark, root wood, and essential oil of the root wood; traditionally used as carminative, diuretic, dermatologic and antirheumatic. *Habit.* North America. *Constit.* Volatile oil (5-9%), isoquinoline alkaloids (<0.1%), lignins such as sesamin, resin, sitosterol, starch, tannins, wax. Review of pharmacology: D. Hutson, M. J. Cupp in *Toxicology and Clinical Pharmacology of Herbal Products*, M. J. Cupp, Ed. (Humana Press, Totowa, 2000) pp 245-252; of constituents and traditional uses: J. Barnes *et al.*, *Herbal Medicines* (Pharmaceutical Press, London, 2nd Ed., 2002) pp 414-416.

Volatile oil. [8006-80-2] Oil of sassafras. Obtained from the root. *Constit.* 80-90% safrole, eugenol, α-pinene, α- and β-phellandrene, asarone, *d*-camphor, myristicin. Yellow to reddish-yellow liquid; characteristic odor and taste of sassafras. d_{25}^{25} 1.065-1.077. α_D^{25} +2 to +4°. n_D^{20} 1.5250-1.5350. Very slightly sol in water; sol in 2 vols 90% alcohol. *Keep well closed, cool and protected from light.*

Caution: Reported to be toxic and carcinogenic due to the content of safrole, *q.v.* Potential symptoms of overexposure include vomiting, stupor, spasm followed by paralysis (Barnes).

USE: Formerly as flavoring in beverages such as root beer. Oil as scent in perfumes and soaps.

8519. Sassy Bark. Saucy bark; Mancona bark; ordeal bark; red-water tree bark; casca bark; Saxon bark; doom bark; teli; bondou. Bark of *Erythrophleum guineense* G. Don, *Leguminosae*. *Habit.* Central and Western Africa. *Constit.* Erythrophleine, tannin, resin.

8520. Satraplatin. [129580-63-8] (*OC*-6-43)-Bis(acetato-κO)amminedichloro(cyclohexanamine)-platinum; bis-acetato-ammine-dichloro-cyclohexylamine-platinum(IV); BMS-182751; BMY-45594; JM-216; Orplatna. $C_{10}H_{22}Cl_2N_2O_4Pt$; mol wt 500.28. C 24.01%, H 4.43%, Cl 14.17%, N 5.60%, O 12.79%, Pt 38.99%. Orally active analog of cisplatin, *q.v.* Prepn: M. J. Abrams *et al.*, *EP 328274*; *eidem*, *US 5072011* (1989, 1991 both to Johnson Matthey). Clinical pharmacokinetics: S. Vouillamoz-Lorenz *et al.*, *Anticancer Res.* **23**, 2757 (2003). Clinical evaluation in small-cell lung cancer: E. Fokkema *et al.*, *J. Clin. Oncol.* **17**, 3822 (1999); in hormone-refractory prostate cancer: C. N. Sternberg *et al.*, *Oncology* **68**, 2 (2005). Review of pharmacology and clinical experience: A. Bhargava, U. N. Vaishampayan, *Expert Opin. Invest. Drugs* **18**, 1787-1797 (2009).

Soly (mg/ml): water ~0.3; saline 0.4; 1-octanol 0.7. Partition coefficient (octanol/water): 0.1. LD_{50} in mice (mg/kg): 30 i.p.; 330 orally (Kelland).

THERAP CAT: Antineoplastic.

8521. Satumomab. [144058-40-2] Anti-(human tumor-associated glycoprotein 72) immunoglobulin G1 (mouse monoclonal B72.3 γ₁-chain) disulfide with mouse monoclonal B72.3 light chain, dimer; B72.3; CYT-099. Murine monoclonal antibody that targets a high molecular weight tumor-associated glycoprotein, *TAG-72*, expressed on the surface of certain human colon, breast and ovarian carcinoma cell lines. Prepn and characterization of MAb: D. Colcher *et al.*, *Proc. Natl. Acad. Sci. USA* **78**, 3199 (1981); J. Schlom *et al.*, *US 4612282* (1986 to USA Secy. Dept. Health Human Services). Pharmacokinetics of ^{111}In labeled MAb: B. A. Brown *et al.*, *Cancer Res.* **47**, 1149 (1987). Clinical studies of immunoscintigraphy in colorectal cancer: B. D. Collier *et al.*, *Radiology* **185**, 179 (1992); in ovarian cancer: E. A. Surwit *et al.*, *Gynecol. Oncol.* **48**, 285 (1993). Review of prepn and diagnostic use: R. T. Maguire *et al.*, *Cancer*

72, 3453-3462 (1993). Book: *Diagnosis of Colorectal and Ovarian Carcinoma*, R. T. Maguire, D. Van Nostrand, Eds. (Marcel Dekker, Inc., New York, 1992) 237 pp.

Indium In 111 satumomab pendetide. [138955-27-8] Satumomab, N^6-[*N*-[2-[[2-[bis(carboxymethyl)amino]ethyl](carboxymethyl)amino]ethyl]-*N*-(carboxymethyl)glycyl]-N^2-(*N*-glycyl-L-tyrosyl)-L-lysine conjugate, indium-^{111}In chelate; ^{111}In-CYT-103; monoclonal antibody B72.3-GYK-DTPA-In-111; B72.3-glycyl-tyrosyl-*N*-ε-diethylenetriaminepentaacetic acid-In-111; OncoScint.

THERAP CAT: Indium In 111 satumomab pendetide as diagnostic aid (radioactive imaging agent).

8522. Saunders, Red. Red sandalwood; ruby wood. Heartwood of *Pterocarpus santalinus* L. f., *Leguminosae*. *Habit.* East Indies. *Constit.* Santalin (santalinic acid), santal pterocarpin, tannin.

USE: For coloring tinctures and similar medicinal prepns; formerly used as a dye.

8523. Savin. Young shoots of *Juniperus sabina* L., *Cupressaceae*; used in traditional medicine as emmenagogue and antirheumatic. *Habit.* Europe, Northern Asia, Northern U.S. *Constit.* 3-5% volatile oil, tannin, savinin, podophyllotoxin. Isoln and structure of *savinin*: Hartwell *et al.*, *J. Am. Chem. Soc.* **75**, 235 (1953); Schrecker, Hartwell, *ibid.* **76**, 4896 (1954). *Review*: J. Gruenwald *et al.*, *PDR for Herbal Medicines* (Thomson PDR, Montvale, 3rd Ed., 2004) pp 706-707.

Volatile oil. [8024-00-8] Oil of savin. *Constit.* Chiefly sabinyl acetate, sabinene; β-myrcene, terpin-4-ol, γ-terpinene, α-pinene, limonene. Colorless or pale yellow liquid. d_{15}^{15} 0.903-0.923. α_D +40 to +60°. n_D^{20} 1.470-1.478. Very slightly sol in water; very sol in alcohol. *Keep well closed, cool and protected from light.*

Caution: Potential symptoms of overexposure include queasiness, spasm, hematuria, central paralysis, unconsciousness. External application may cause skin irritation, blistering, necrosis (Gruenwald).

THERAP CAT: In treatment of warts.

8524. Saw Palmetto. Low shrubby palm, *Serenoa repens* (Bartr.) Small, *Palmae*; also known as *Sabal serrulata* Roem. et Schult. *Habit.* Southeastern U.S. Traditionally used in Native American medicine for genitourinary conditions. Medicinal formulations are prepared from the dried fruits. *Constit.* Phytosterols, fatty acids, polysaccharides, flavonoids. Description of botany, constituents and medicinal uses: E. Bombardelli, P. Morazzoni, *Fitoterapia* **68**, 99-113 (1997). Effect of climate on growth: H. L. Gholz *et al.*, *Can. J. For. Res.* **29**, 1248 (1999). Review of clinical trials in treatment of urinary tract symptoms in BPH: T. J. Wilt *et al.*, *J. Am. Med. Assoc.* **280**, 1604-1609 (1998). *Review:* A. Meadows, M. J. Cupp in *Toxicology and Clinical Pharmacology of Herbal Products*, M. J. Cupp, Ed. (Humana Press, Totowa, NJ, 2000) pp 133-139.

Lipidosterolic extract. Capistan; Libeprosta; Permixon; Sereprostat. *n*-Hexane extract of entire fruit (pulp and seed). Constituents of extract: G. Jommi *et al.*, *Gazz. Chim. Ital.* **118**, 823 (1988). Tissue distribution in rats: G. Chevalier *et al.*, *Eur. J. Drug Metab. Pharmacokinet.* **22**, 73 (1997). Pharmacology: F. Van Coppenolle *et al.*, *Prostate* **43**, 49 (2000). Review of clinical trials in BPH: P. Boyle *et al.*, *Urology* **55**, 533-539 (2000).

THERAP CAT: In treatment of micturitional disorders due to benign prostatic hypertrophy.

8525. Saxagliptin. [361442-04-8] (1*S*,3*S*,5*S*)-2-[(2*S*)-2-Amino-2-(3-hydroxytricyclo[3.3.1.1^{3,7}]dec-1-yl)acetyl]-2-azabicyclo[3.1.0]hexane-3-carbonitrile; (*S*)-3-hydroxyadamantylglycine-L-*cis*-4,5-methanoprolinenitrile; BMS-477118. $C_{18}H_{25}N_3O_2$; mol wt 315.42. C 68.54%, H 7.99%, N 13.32%, O 10.14%. Dipeptidyl peptidase-IV (DPP-IV) inhibitor. Prepn: J. A. Robl *et al.*, **WO 0168603**; *eidem*, **US 6395767** (2001, 2002 both to Bristol-Myers Squibb). Synthesis and pharmacology: D. J. Augeri *et al.*, *J. Med. Chem.* **48**, 5025 (2005). Mechanism of action: Y. B. Kim *et al.*, *Arch. Biochem. Biophys.* **445**, 9 (2006). Crystal structure of complex with DPP-IV variants: W. J. Metzler *et al.*, *Protein Sci.* **17**, 240 (2008). Clinical experience in treatment-naive type 2 diabetes: J. Rosenstock *et al.*, *Diabetes Obes. Metab.* **10**, 376 (2008). Review of discovery and development: B. Gallwitz, *IDrugs* **11**, 906-917 (2008).

Monohydrate. [945667-22-1] BMS-477118-11; Onglyza. C_{18}-$H_{25}N_3O_2 \cdot H_2O$; mol wt 333.43.

THERAP CAT: Antidiabetic.

8526. Saxitoxin. [35523-89-8] (3aS,4R,10aS)-2,6-Diamino-4-[[(aminocarbonyl)oxy]methyl]-3a,4,8,9-tetrahydro-1H-10H-pyrrolo[1,2-c]purine-10,10-diol; mussel poison; clam poison; paralytic shellfish poison; gonyaulax toxin; STX. $[C_{10}H_{17}N_7O_4]^{2+}$. Powerful neurotoxin produced by the dinoflagellates *Gonyaulax catenella*, or *G. tamarensis*, the consumption of which causes the California sea mussel *Mytilus californianus*, the Alaskan butterclam *Saxidomus giganteus* and the scallop to become poisonous: Sommer *et al.*, *Arch. Pathol.* **24**, 537, 560 (1937); Schantz *et al.*, *Can. J. Chem.* **39**, 2117 (1961); Ghazarossian *et al.*, *Biochem. Biophys. Res. Commun.* **59**, 1219 (1974). These poisonous shellfish have been connected to instances of toxic "red-tides" where the high concentrations of algae discoloring the water where the *Gonyaulax* genus. Isoln and partial characterization: Schantz *et al.*, *J. Am. Chem. Soc.* **79**, 5230 (1957); Mold *et al.*, *ibid.* 5235. Purifn and biochemical data: Schantz *et al.*, *Biochemistry* **5**, 1191 (1966); Schantz, *Ann. N.Y. Acad. Sci.* **90**, 843 (1960). pKa values: R. S. Rogers, H. Rapoport, *J. Am. Chem. Soc.* **102**, 7335 (1980). Toxicity: G. S. Wiberg, N. R. Stephenson, *Toxicol. Appl. Pharmacol.* **2**, 607 (1960). Structural studies: Schuett, Rapoport, *J. Am. Chem. Soc.* **84**, 2266 (1962); Wong *et al.*, *ibid.* **93**, 4633, 7344 (1971). Revised structure: Schantz *et al.*, *ibid.* **97**, 1238 (1975); Bordner *et al.*, *ibid.* **97**, 6008 (1975). Pharmacology including study of sodium transport inhibition: Cheymol *et al.*, *Arch. Int. Pharmacodyn. Ther.* **174**, 393 (1968); Evans, *Br. Med. Bull.* **25**, 263 (1969). Stereospecific total synthesis of *dl*-saxitoxin: H. Tanino *et al.*, *J. Am. Chem. Soc.* **99**, 2818 (1977); Y. Kishi, *Heterocycles* **14**, 1477 (1980). *Reviews:* Kao, *Pharmacol. Rev.* **18**, 997 (1966); Kao, *Fed. Proc.* **31**, 1117 (1972); Narahashi, *ibid.* 1124; Evans, *Int. Rev. Neurobiol.* **15**, 83 (1972); Y. Shimizu in *Progress in the Chemistry of Natural Products* vol. **45**, W. Herz *et al.*, Eds. (Springer-Verlag, New York, 1984) pp 239-246. *See also* Brevetoxins.

Dihydrochloride. [35554-08-6] $C_{10}H_{17}N_7O_4 \cdot 2HCl$; mol wt 372.21. White, hygroscopic solid. pKa in water: 8.24, 11.60. $[\alpha]_D^{25}$ +130°. Very sol in water, methanol; sparingly sol in ethanol, glacial acetic acid. Practically insol in lipid solvents. Stable in acid solns; decomposes rapidly in alkaline media. Boiling 3-4 hrs at pH 3 causes loss of activity. LD_{50} in mice (μg/kg): 10 i.p.; 263 orally; 3.4 i.v. (Wiberg, Stephenson).

USE: As a tool in neurochemical research.

8527. SBR Rubber. Styrene-butadiene rubber; GR-S; Government Rubber Styrene. General-purpose synthetic rubbers that were originally produced in government owned plants as GR-S. They are copolymers obtained by the polymerization of butadiene and styrene in a ratio of approx 3:1. The chains contain a random sequence of the two monomers. Very similar in composition to *Buna S*, *Kraton*, *Solprene*, *Stereon*. *Review:* P. Schneider *et al.*, in Ullmann, *Encyklopädie der technischen Chemie*, 3 Aufl., Bd. IX (München-Berlin, 1957) p 331; M. Morton, *Introduction to Rubber Technology* (Reinhold, 1959) pp 56 and 256-284; R. G. Bauer in

Kirk-Othmer Encyclopedia of Chemical Technology vol. **8** (Wiley-Interscience, New York, 3rd ed., 1979) pp 608-625.

USE: In tires, adhesives, chewing gum.

8528. Scammony Root. Root of *Convolvulus scammonia* L., *Convolvulaceae*, yielding not less than 8% resin. *Habit.* Eastern Mediterranean region, especially Syria. *Constit.* 8-13% resin; dihydroxycinnamic acid, β-methylesculetin, ipuranol, a reducing sugar, starch.

THERAP CAT: Cathartic.

8529. Scandium. [7440-20-2] Sc; at. wt 44.955912; at. no. 21; valence 3. Group IIIB (3). Rare earth metal. Naturally occurring isotope (mass number): 45; known artificial radioactive isotopes: 40-44, 46-51. Abundance in earth's crust: 5-25 ppm. Widely dispersed in nature. Occurs in the minerals thortveitite [(Sc,-$Y)_2Si_2O_7$] and in other rare earth minerals such as davidite, ytterbite, orthite and cerrite; frequently associated with tin or zirconium. Predicted and called "ekaboron" by Mendeleev. Discovered by Nilson: *Ber.* **12**, 551, 554 (1879); **13**, 1430, 1439 (1880). Sepn from wolframite: Lukens, *J. Am. Chem. Soc.* **35**, 1470 (1913); on basis of solubility: Fischer, Bock, *Z. Anorg. Chem.* **249**, 146 (1942). Toxicity of the chloride: Haley *et al.*, *J. Pharm. Sci.* **51**, 1043 (1962). Review of isolns including ion-exchange techniques: F. H. Spedding *et al.*, *J. Electrochem. Soc.* **105**, 683-686 (1958). Review of prepn, properties and compds: R. C. Vickery, *The Chemistry of Yttrium and Scandium* (Pergamon Press, New York, 1960) 123 pp; *idem*, "Scandium, Yttrium and Lanthanum" in *Comprehensive Inorganic Chemistry* vol. **3**, J. C. Bailar, Jr. *et al.*, Eds. (Pergamon Press, Oxford, 1973) pp 329-353; T. Moeller, "The Lanthanides", *ibid.* vol. **4**, pp 1-101; F. H. Spedding in *Kirk-Othmer Encyclopedia of Chemical Technology* vol. **19** (John Wiley & Sons, New York, 3rd ed., 1982) pp 833-854; *Chemistry of the Elements*, N. N. Greenwood, A. Earnshaw, Eds. (Pergamon Press, New York, 1984) pp 1102-1110, 1423-1449.

Metal. Reported to be dimorphic. Crystalline forms: hexagonal close-packed α-form, d 2.9890; evidence of existence of face-centered cubic β-form, d 3.19, is inconclusive. mp 1541°. bp 2836°. Heat of fusion: 14.10 kJ/mol. Heat of sublimation (25°): 377.8 kJ/mol. E°(aq) Sc^{3+}/Sc −2.08 V (calc). Salts are hydrolyzed in aq soln.

Oxide. Scandia. O_3Sc_2. Fine white powder, d 3.864. Obtained by igniting the metal or its compds. Readily sol in hot or concd acids.

Hydroxide. $Sc(OH)_3$. White gelatinous precipitate forming a hard, horny mass on exposure to air; obtained by the action of alkalies on solns of the salts; dissolves readily in dil acids.

Chloride. $ScCl_3$. White deliquescent solid. Prepd by the action of a mixture of sulfur chloride and chlorine on the heated oxide; mp 960°; crystallizes with 6 mols of water. Sol in water. Practically insol in alc. LD_{50} in mice: 755 mg/kg i.p.; 4 g/kg orally (Haley).

Sulfate. $Sc_2(SO_4)_3$. Pentahydrate, d 2.519. Most sol of the sulfates of the rare earths (54.6 g/100 ml at 25°). Converted to the dihydrate on heating above 100°.

Nitrate. $Sc(NO_3)_3$. Crystallizes as the tetrahydrate, prismatic deliquescent crystals, readily dissolves in water or alc.

8530. Scarlet Red. [85-83-6] 1-[2-[2-Methyl-4-[2-(2-methylphenyl)diazenyl]phenyl]diazenyl]-2-naphthalenol; C.I. Solvent Red 24; *o*-tolylazo-*o*-tolylazo-β-naphthol; 1-(4-*o*-tolylazo-*o*-tolylazo)-2-naphthol; C.I. 26105; Biebrich Scarlet Red; Sudan IV; Fat Ponceau R. $C_{24}H_{20}N_4O$; mol wt 380.45. C 75.77%, H 5.30%, N 14.73%, O 4.21%. Prepd by diazotizing *o*-aminoazotoluene and coupling with β-naphthol: *Colour Index* vol. **4** (3rd ed., 1971) p 4227. Review of carcinogenicity studies: *IARC Monographs* **8**, 217 (1975).

Dark brown powder. Softens at 175°; mp 181-188°; dec completely at 260°. Practically insol in water. One gram dissolves in 15 ml chloroform. Sol in oils, fats, warm petrolatum, paraffin, phenol. Slightly sol in acetone, alcohol, benzene.

USE: Fat stain.

THERAP CAT: Has been used to promote wound healing.

THERAP CAT (VET): Has been used to stimulate healing of wounds.

8531. Schiff Bases. $R-CH=N-C_6H_5$. Condensation products of aldehydes and ketones with primary amines. The compounds are stable if there is at least one aryl group on the nitrogen or carbon; H. Schiff, *Ann.* Suppl. **3**, 343 (1864-1865). *Reviews:* N. V. Sidgewick, *Organic Chemistry of Nitrogen* (Clarendon Press, Oxford, 3rd ed., 1966) p 164-166; R. L. Reeves, *Chemistry of the Carbonyl Group*, S. Patai, Ed. (Wiley, New York, 1966) pp 608-614; J. Fastrez, *Ind. Chim. Belge* **34**, 835 (1969); A. Bruylants, E. Feymantis-de Medicis, *Chemistry of the Carbon-Nitrogen Double Bond*, S. Patai, Ed. (Wiley, New York, 1970) pp 465-502. Metal complexes: S. Yamada, *Coord. Chem. Rev.* **1**, 415 (1966); E. Sinn, C. M. Harris, *ibid.* **4**, 391 (1969); L. F. Lindsy, *Q. Rev. Chem. Soc.* **25**, 379 (1971).

8532. Schradan. [152-16-9] $N,N,N',N',N'',N'',N''',N'''$-Octamethyldiphosphoramide; octamethyl pyrophosphoramide; bis[bisdimethylaminophosphonous] anhydride; bis-N,N,N',N'-tetramethylphosphorodiamidic anhydride; OMPA; Pestox III; Sytam. C_8H_{24}-$N_4O_3P_2$; mol wt 286.25. C 33.57%, H 8.45%, N 19.57%, O 16.77%, P 21.64%. Synthesis: Schrader, *BIOS Final Report* **No. 1808** (1948); Gardiner, Kilby, *J. Chem. Soc.* **1950**, 1769; *Biochem. J.* **51**, 79 (1952); Hartley *et al.*, *J. Sci. Food Agric.* **2**, 303, 310 (1951); **3**, 60 (1952); Toy, Costello, *US 2717249* (1955 to Victor Chem.); Toy, Walsh, *Inorg. Synth.* **7**, 73 (1963). Has been found to inhibit peripheral cholinesterase without pronounced effects on the central nervous system: DuBois *et al.*, *J. Pharmacol. Exp. Ther.* **99**, 376 (1950). Toxicity study: T. B. Gaines, *Toxicol. Appl. Pharmacol.* **14**, 515 (1969).

Viscous liquid, d_4^{25} 1.09. $bp_{0.5}$ 120-125°. $bp_{2.0}$ 154°. mp 14-20°. n_D^{25} 1.462. Vapor pressure at 25°: 1×10^{-3} mm Hg. Misc with water. Sol in most organic solvents, including ketones, nitriles, esters, aromatic hydrocarbons and alcohols. Practically insol in higher aliphatic hydrocarbons. Hydrolyzed by acids. LD_{50} in male, female rats (mg/kg): 9.1, 42 orally; 15, 44 dermally (Gaines).

USE: Insecticide.

8533. Schwartz's Reagent. [37342-97-5] Chlorobis(η^5-2,4-cyclopentadien-1-yl)hydrozirconium; $Cp_2Zr(H)Cl$; (η^5-C_5H_5)$_2$Zr-(H)Cl. $C_{10}H_{11}ClZr$; mol wt 257.87. C 46.58%, H 4.30%, Cl 13.75%, Zr 35.38%. Hydrozirconation reducing agent for a range of unsaturated compounds. Prepn: P. C. Wailes, H. Weigold, *J. Organomet. Chem.* **24**, 405 (1970); with high yield: S. L. Buchwald *et al.*, *Tetrahedron Lett.* **28**, 3895 (1987); *in situ* prepn: B. H. Lipshutz *et al.*, *ibid.* **31**, 7257 (1990). Stereoselectivity: E. Cesarotti *et al.*, *Inorg. Chim. Acta* **64**, L207 (1982). Use as general reducing agent: J. Schwartz, J. A. Labinger, *Angew. Chem. Int. Ed.* **15**, 333 (1976); reduction of halides in presence of triethylborane, *q.v.*: K. Fujita *et al.*, *J. Am. Chem. Soc.* **123**, 3137 (2001); in silylation of alkenes: J. Terao *et al.*, *Tetrahedron* **60**, 1301 (2004). Review of applications in phosphorous chemistry: J.-P. Majoral *et al.*, *Ber.* **129**, 879-886 (1996); of mechanism and reduction of amides: J. M. White *et al.*, *Chem. Innovation* **30**, 23-28 (2000).

White solid turning pink in light.

USE: Functionalization of unsaturated systems via an organozirconium intermediate.

8534. Schweizer's Reagent. [17500-49-1] Tetraamminecopper(2+) dihydroxide. $CuH_{14}N_4O_2$; mol wt 165.68. Cu 38.35%, H 8.52%, N 33.82%, O 19.31%. $[Cu(NH_3)_4]OH_2$. Usually applied in the form of an ammoniacal soln which is easily obtained by dissolving $Cu(OH)_2$ in 20% ammonia water. The soln is an intense azure color. Evaporation yields long, blue needles: Schweizer, *J. Prakt. Chem.* **72**, 109, 344 (1857).

USE: Dissolves cellulose; used in rayon production.

8535. Schwesinger P₄ Base. [111324-04-0] N'''-(1,1-Dimethylethyl)N,N',N''-tris[tris(dimethylamino)phosphoranylidene]-phosphorimidic triamide; t-Bu-P_4. $C_{22}H_{63}N_{13}P_4$; mol wt 633.73. C 41.70%, H 10.02%, N 28.73%, P 19.55%. Sterically hindered polyiminophosphazene base. Prepn: R. Schwesinger, H. Schlemper, *Angew. Chem. Int. Ed.* **26**, 1167 (1987); and properties: R. Schwesinger *et al.*, *Ann.* **1996**, 1055. Use as promoter for ethylene oxide polymerization: B. Esswein *et al.*, *Macromol. Symp.* **107**, 331 (1996); for copolymer formation: S. Förster, E. Krämer, *Macromolecules* **32**, 2783 (1999).

Colorless crystalline solid, mp 207°. Sublimes at 0.001 torr. Hydrophilic and lipophilic; deliquesces in moist air and hexane vapors at room temp.

USE: Promoter for polymerizations; nonnucleophilic base.

8536. Scillaren. [11003-70-6] A mixture of glycosides, scillaren A and B in the proportions in which they occur in fresh squill, *Urginea (Scilla) maritima* (L.) Baker, *Liliaceae*, about 2 parts of A to 1 part of B. Isolation of scillaren and separation of A and B: Stoll *et al.*, *Helv. Chim. Acta* **16**, 703 (1933).

Granular, very bitter powder. On drying in high vacuum at 78° for 15 hrs it loses not more than 6% of its wt. $[\alpha]_D^{20}$ -25 to $-35°$ (0.5 g in 25 ml of 75% w/w alcohol). One gram dissolves in 3000 ml water, in 5 ml abs alcohol, in 5 ml methanol. Practically insol in ether and chloroform. Aq solns are neutral to litmus.

Scillaren A. [124-99-2] (3β)-3-[(6-Deoxy-4-O-β-D-glucopyranosyl-α-L-mannopyranosyl)oxy]-14-hydroxybufa-4,20,22-trienolide; glucoproscillaridin A; transvaalin. $C_{36}H_{52}O_{13}$; mol wt 692.80. C 62.41%, H 7.57%, O 30.02%. Characterization and structure: Stoll *et al.*, *Z. Physiol. Chem.* **222**, 24 (1933); *Helv. Chim. Acta* **17**, 641, 1334 (1934); **18**, 82, 120, 401, 644, 1247 (1935); **24**, 1380 (1941); Zoller, Tamm, *ibid.* **36**, 1744 (1953). Pharmacology: K. K. Chen, F. G. Henderson, *J. Pharmacol. Exp. Ther.* **111**, 365 (1954); G.

Vogel, E. Kluge, *Arzneim.-Forsch.* **11**, 848 (1961). Very bitter taste. Two crystal modifications from methanol: prisms, mp 184-186°; leaflets, mp 208-211°. $[\alpha]_D^{23}$ −71.9° (c = 1.011 in methanol). Tendency to form solvated crystals. Six-sided plates, tablets contg 1 mol CH_3OH + 1 mol H_2O from dil methanol, mp 230-240°. Anhydr from 85% alcohol, solvent-free: mp 270°; $[\alpha]_D^{20}$ −72 to −78° (c = 1 to 3 in 75% alcohol), $[\alpha]_D^{20}$ −73.8°. Soluble in 350 parts alcohol, in 80 parts methanol, in 40 parts dil alcohol (4 vols ethanol + 1 vol water). Practically insol in chloroform, ether. Sparingly sol in water. LD_{50} i.v. in rats: 15.5 mg/kg (Vogel, Kluge).

Scillaren B. An undefined water-sol mixture of glucosides remaining after extraction of scillaren A. Amorphous, granular powder. Very bitter taste. Scillaren B dried in a high vacuum at 78°C for 15 hrs loses not more than 5% of its weight. Completely dried scillaren B contains approx 99.5% active glycosidal substance. $[\alpha]_D^{20}$ +35 to 41° (0.5 g in 25 ml of 75% w/w ethanol). Freely sol in water. Soluble in alcohol, methanol (about 1 in 5); very slightly sol in chloroform (about 1 in 10,000). Practically insol in ether. Aq solns are neutral to litmus.

THERAP CAT: Cardiotonic.

8537. Scillarenin. [465-22-5] (3β)-3,14-Dihydroxybufa-4,-20,22-trienolide. $C_{24}H_{32}O_4$; mol wt 384.52. C 74.97%, H 8.39%, O 16.64%. Prepn by adaptive enzymic decompn of proscillaridin A: Stoll *et al., Helv. Chim. Acta* **34**, 2301 (1951); from *Urginea burkei* Baker, *Liliaceae:* Zoller, Tamm, *ibid.* **36**, 1744 (1953). Structure: Stoll *et al., ibid.* **35**, 1934 (1952). Toxicity study: Chen, Henderson, *J. Pharmacol. Exp. Ther.* **111**, 365 (1954).

Prisms from methanol, mp 232-238°. $[\alpha]_D^{20}$ −16.8° (c = 0.357 in methanol); $[\alpha]_D^{20}$ +17.9° (c = 0.39 in chloroform). uv max: 300 nm (log ε 3.72). Mean LD i.v. in cats: 0.1567 mg/kg (Chen, Henderson).

3-Acetate. $C_{26}H_{34}O_5$. Crystals from methanol, dec 240-243°. $[\alpha]_D^{20}$ −23.4° (c = 1.365 in chloroform).

THERAP CAT: Cardiotonic.

8538. Scilliroside. [507-60-8] (3β,6β)-6-(Acetyloxy)-3-(β-D-glucopyranosyloxy)-8,14-dihydroxybufa-4,20,22-trienolide. $C_{32}H_{44}O_{12}$; mol wt 620.69. C 61.92%, H 7.15%, O 30.93%. Glycoside from red squill, the red variety of *Urginea maritima* (L.) Baker, *Liliaceae:* Stoll, Renz, *Helv. Chim. Acta* **25**, 43 (1942); Wichtl, Fuchs, *Arch. Pharm.* **295**, 361 (1962). Structure: Stoll *et al., Helv. Chim. Acta* **26**, 648 (1943); von Wartburg, Renz, *ibid.* **42**, 1620 (1959). Toxicity data: F. Dybing *et al., Acta Pharmacol. Toxicol.* **8**, 391 (1952).

Long prisms from dilute methanol. The crystals are solvated and lose about 8% of their wt in high vacuum, but retain ½ mol H_2O. For the hemihydrate, indistinct mp 168-170°, dec 200°. $[\alpha]_D^{20}$ −59 to −60° (methanol). Absorption max (96% ethanol): 300 nm (log ε 3.73); (98% H_2SO_4): 295, 505 nm ($E_{1cm}^{1\%}$ 260, 315). Freely sol in the lower alcohols, in ethylene glycol, dioxane, glacial acetic acid. Slightly sol in water, acetone, chloroform, ethyl acetate. Practically insol in ether, petr ether. LD_{50} in mice (mg/kg): 0.471 s.c.; 0.440 orally (Dybing).

Tetraacetyl scilliroside. $C_{40}H_{52}O_{16}$. Rosettes of long needles from methanol, mp 199°. Sol in chloroform. The four additional acetyl groups are on the sugar.

8539. Scoparin. [301-16-6] 8-β-D-Glucopyranosyl-5,7-dihydroxy-2-(4-hydroxy-3-methoxyphenyl)-4H-1-benzopyran-4-one; 8-glycosyl-4',5,7-trihydroxy-3'-methoxyflavone; scoparoside. $C_{22}H_{22}O_{11}$; mol wt 462.41. C 57.14%, H 4.80%, O 38.06%. From leaves and branches of *Spartium scoparium* L., *Leguminosae:* Stenhouse, *Ann.* **78**, 15 (1851); Mascré, Paris, *Compt. Rend.* **204**, 1270 (1937). Structural study: Hörhammer *et al., Naturwissenschaften* **49**, 393 (1962). Revised structure: Prox, *Tetrahedron* **24**, 3697 (1968).

Needles from 80% methanol, mp 253°. uv max (methanol): 345, 270 nm (log ε 4.27, 4.18). Practically insol in cold water, ether, chloroform, benzene. Sol in hot water, ethanol, methanol, acetic acid, ethyl acetate, acetone, pyridine.

Heptaacetate. $C_{36}H_{36}O_{18}$. Crystals, mp 240-241°.

8540. Scoparius. Broom; green broom; Scotch broom; Irish broom; hogweed; bannal. Dried tops of *Cytisus scoparius* (L.) Link, *Leguminosae.* The seeds and flowers are also used, though not officially. *Habit.* Western Asia, Southern and Western Europe; cultivated in U.S.A., particularly in Oklahoma. *Constit.* Sparteine, scoparin, genisteine, sarothamnine, volatile oil, tannin, fat, wax, sugar.

THERAP CAT: Diuretic, cathartic.

8541. Scoparone. [120-08-1] 6,7-Dimethoxy-2H-1-benzopyran-2-one; 6,7-dimethoxycoumarin; esculetin dimethyl ether. $C_{11}H_{10}O_4$; mol wt 206.20. C 64.07%, H 4.89%, O 31.04%. Isolation from *Zanthoxylum* and *Artemisia* spp: Araki, Miyashita, *J. Pharm. Soc. Jpn.* **48**, 437 (1928); Sera, Shibuye, *C.A.* **24**, 5742 (1930); Parihar, Dutt, *Proc. Indian Acad. Sci.* **25A**, 153 (1947); King *et al., J. Chem. Soc.* **1954**, 1392; Singh *et al., J. Sci. Ind. Res.* **15B**, 190 (1956); Ban'kovskaya, Ban'kovskaya, *C.A.* **55**, 17776c (1961). Structure and synthesis: Singh *et al., Chem. Ind. (London)* **1954**, 1294. Improved synthesis: R. S. Mali *et al., Indian J. Chem.* **21B**, 759 (1982).

Crystals, mp 145-146°. Absorption spectra: Cingolani, *Gazz. Chim. Ital.* **89**, 985, 999 (1959).

8542. Scopola. Dried rhizome of *Scopola carniolica* Jacq., *Solanaceae. Habit.* Germany (Bavaria), Hungary, Russia. *Constit.* Scopolamine (hyoscine), atropine, hyoscyamine; total about 0.7% alkaloids.

THERAP CAT: Anticholinergic.

8543. Scopolamine. [51-34-3] (αS)-α-(Hydroxymethyl)benzeneacetic acid (1α,2β,4β,5α,7β)-9-methyl-3-oxa-9-azatricyclo-

[3.3.1.0^{2,4}]non-7-yl ester; 6β,7β-epoxy-1αH,5αH-tropan-3α-ol (−)-tropate; 6β,7β-epoxy-3α-tropanyl S-(−)-tropate; 6,7-epoxytropine tropate; scopine tropate; tropic acid ester with scopine; hyoscine; l-scopolamine; Scopoderm TTS; Transcop; Transderm Scop. C$_{17}$H$_{21}$-NO$_4$; mol wt 303.36. C 67.31%, H 6.98%, N 4.62%, O 21.10%. Anticholinergic, tropane alkaloid isolated from *Datura metel* L., *Scopola carniolica* Jacq. and other *Solanaceae*. Constituent of impure duboisine from *Duboisia myoporoides* R. Br., pure duboisine is l-hyoscyamine, q.v. Isoln: A. Ladenburg, *Ann.* **206**, 274 (1881); E. Schmidt, *Arch. Pharm.* **230**, 207 (1892). Identity with hyoscine: O. Hesse, *Ann.* **271**, 100 (1892); idem, *J. Prakt. Chem.* **66**, 194 (1902). Absorption spectra: J. J. Dobbie, J. J. Fox, *J. Chem. Soc.* **103**, 1194 (1913). Resolution of isomers and review of early literature: H. King, *ibid.* **1919**, 476. Extraction procedure: F. Chemnitius, *J. Prakt. Chem.* **120**, 221 (1928). Structural studies: J. Gadamer, F. Hammer, *Arch. Pharm.* **259**, 122 (1921); K. Hess, O. Wahl, *Ber.* **55**, 1979 (1922); R. Willstätter, E. Berner, *ibid.* **56**, 1079 (1923); W. Steffens, *Arch. Pharm.* **262**, 205 (1924). Configuration: G. Fodor, *Nature* **170**, 278 (1952); J. Meinwald, *J. Chem. Soc.* **1953**, 712. Review of stereochemistry: G. Fodor, *Tetrahedron* **1**, 86 (1957). Absolute configuration of tropic acid moiety: G. Fodor, G. Csepreghy, *Tetrahedron Lett.* **1959**, 16; eidem, *J. Chem. Soc.* **1961**, 3222. Synthesis of DL-form: G. Fodor et al., *Chem. Ind. (London)* **1956**, 764; P. Dobo et al., *J. Chem. Soc.* **1959**, 3461. Clinical evaluation in motion sickness: J. J. Brand, P. Whittingham, *Lancet* **2**, 232 (1970); in peripheral vertigo: T. Rahko, P. Karma, *J. Laryngol. Otol.* **99**, 653 (1985). Acute toxicity of the hydrobromide: Stockhaus, Wick, *Arch. Int. Pharmacodyn. Ther.* **180**, 155 (1969). Review of CNS effects in humans: D. J. Safer, R. P. Allen, *Biol. Psychiatry* **3**, 347-355 (1971); of use in anesthesia: L. E. Shutt, J. B. Bowles, *Anaesthesia* **34**, 476-490 (1979); of pharmacology and clinical efficacy: S. P. Clissold, R. C. Heel, *Drugs* **29**, 189-207 (1985). Comprehensive description: F. J. Muhtadi, M. M. A. Hassan, *Anal. Profiles Drug Subs.* **19**, 477-551 (1990).

Viscous liquid. pKa 7.55-7.81. [α]$_D^{20}$ −28° (c = 2.7). Sol in 9.5 parts water at 15°. Forms a cryst monohydrate, mp 59°. Freely sol in hot water, in alcohol, ether, chloroform, acetone. Sparingly sol in benzene, petr ether. Easily hydrolyzed by acids or alkalies. Dec on standing.

Hydrobromide trihydrate. [114-49-8] Scopolammonium bromide. C$_{17}$H$_{21}$NO$_4$.HBr.3H$_2$O; mol wt 438.32. Orthorhombic sphenoidal crystals from water, slightly efflorescent in dry air. mp 195° (after drying at 105° for 3 hours). [α]$_D^{25}$ −24 to −26° (c = 5, calculated on anhydrous basis). uv max (methanol) 246, 252, 258, 264 nm (A$_{1cm}^{1\%}$ 3.5, 4.0, 4.5, 3.0). pH of 0.05*M* soln 5.85. One gram dissolves in 1.5 ml water, 20 ml alcohol. Slightly sol in chloroform. Practically insol in ether. LD$_{50}$ in rats (mg/kg): 3800 s.c. (Stockhaus, Wick).

Hydrochloride. [55-16-3] C$_{17}$H$_{21}$NO$_4$.HCl. Crystals from acetone, mp 200°. Dihydrate, prisms from water, melts in water of crystn at 80°. Very sol in water and alcohol. pH of 0.05*M* soln 5.85.

Methyl bromide *see* Methscopolamine Bromide.

Methyl nitrate. [6106-46-3] Methscopolamine nitrate. C$_{17}$H$_{21}$-NO$_4$.CH$_3$NO$_3$; mol wt 380.40. Crystals. Freely sol in water, dil alcohol; slightly sol in abs alcohol.

DL-Form. Atroscine. Dihydrate, chisel-shaped prisms from ethanol + water, mp 38-40°. Monohydrate, efflorescent crystals, mp 55-57°. Anhydrous, long prisms, mp 82-83°. Very slightly sol in water; sol in alc, chloroform, ether, oils.

THERAP CAT: In treatment of motion sickness; antiemetic; antispasmodic; mydriatic; preanesthetic medicant.

THERAP CAT (VET): Preanesthetic medicant.

8544. Scopolamine N-Oxide. [97-75-6] (αS)-α-(Hydroxymethyl)benzeneacetic acid (1α,2β,4β,5α,7β)-9-methyl-9-oxido-3-oxa-9-azatricyclo[3.3.1.0^{2,4}]non-7-yl ester; 6β,7β-epoxy-1αH,5αH-tropan-3α-ol (−)-tropate 8-oxide; scopolamine aminoxide. C$_{17}$H$_{21}$-NO$_5$; mol wt 319.36. C 63.94%, H 6.63%, N 4.39%, O 25.05%. Anticholinergic. Prepn: Polonovski, Polonovski, *Compt. Rend.* **180**, 1755 (1925). Absolute configuration: Huber et al., *Can. J. Chem.* **49**, 3258 (1971).

White powder, mp ~80°. [α]$_D^{20}$ −14° in water.

Hydrobromide. [6106-81-6] Genoscopolamine. C$_{17}$H$_{22}$Br-NO$_5$; mol wt 400.27. Prisms from water, mp 135-138°. [α]$_D^{25}$ −25° (c = 2). Soly in water about 10 g/100 ml. Slightly sol in alcohol, acetone. pH of 3% aq soln about 3.2.

THERAP CAT: Antiparkinsonian.

8545. Scopoletin. [92-61-5] 7-Hydroxy-6-methoxy-2*H*-1-benzopyran-2-one; 7-hydroxy-6-methoxycoumarin; 6-methoxyumbelliferone; β-methylesculetin; chrysatropic acid; gelseminic acid. C$_{10}$H$_8$O$_4$; mol wt 192.17. C 62.50%, H 4.20%, O 33.30%. The aglucone of scopolin. Occurs in root of *Scopolia japonica* Maxim., *Scopolia carniolica* Jacq., *Atropa belladonna* L., *Solanaceae*, *Convolvulus scammonia* L., *Convolvulaceae*. Isoln: Eykman, *Ber.* **17 III**, 442 (1884). Synthesis: Crosby, *J. Org. Chem.* **26**, 1215 (1961); Desai, Desai, *J. Indian Chem. Soc.* **40**, 456 (1963).

Needles or prisms from chloroform or acetic acid, mp 204°. uv max: 230, 254, 260, 298, 346 nm (log ε 4.11, 3.68, 3.63, 3.68, 4.07), Ballantyne et al., *Tetrahedron* **27**, 871 (1971). Slightly sol in water or cold alcohol; sol in hot alcohol or hot glacial acetic acid; moderately sol in chloroform. Practically insol in benzene. Its alcoholic soln has a blue fluorescence. Reduces Fehling's soln.

8546. Scopolin. [531-44-2] 7-(β-D-Glucopyranosyloxy)-6-methoxy-2*H*-1-benzopyran-2-one. C$_{16}$H$_{18}$O$_9$; mol wt 354.31. C 54.24%, H 5.12%, O 40.64%. Mono-β-glucopyranoside of scopoletin. From root of *Scopolia japonica* Maxim., *S. carniolica* Jacq., *Solanaceae* and *Nerium odorum* Sol., *Apocynaceae*. Isoln: Eykman, *Ber.* **17 III**, 442 (1884); Ritter et al., *Helv. Chim. Acta* **36**, 434 (1953). Synthesis: Merz, *Arch. Pharm.* **270**, 476 (1932). Possible identity with **murrayin**: Bose, Mookerjee, *J. Indian Chem. Soc.* **14**, 489 (1937). Chromatography, and spectrum of murrayin: Chakraborty, Bose, *ibid.* **33**, 905 (1956); Chakraborty, Chakraborty, *Trans. Bose Res. Inst. (Calcutta)* **24** (1), 15 (1961), *C.A.* **55**, 24228c (1961).
Needles. mp 218°. Soluble in water, alcohol. Practically insol in chloroform, ether.

Tetraacetate. C$_{24}$H$_{26}$O$_{13}$. Polymorphous crystals, prisms, mp 166° or flat plates, mp 184-185°.

8547. Scopoline. [487-27-4] rel-(2R,3aS,5R,6R,6aR)-Hexahydro-4-methyl-2,5-methano-2*H*-furo[3,2-*b*]pyrrol-6-ol; 3β,7β-epoxy-1βH,5βH-tropan-6α-ol; 3α,6α-epoxy-7β-hydroxytropane; Oscine. C$_8$H$_{13}$NO$_2$; mol wt 155.20. C 61.91%, H 8.44%, N 9.03%, O 20.62%. Decomposition product of scopolamine. Prepn: Kovács, *J. Chem. Soc.* **1953**, 2341. Prepn from teloidine: Zeile, Heusner, *Ber.* **90**, 2800, 2809 (1957); DE 1056617; DE 1062703 (1959 to Boehringer, Ing.). Structure: Fodor et al., *J. Chem. Soc.* **1955**, 3504. Stereochemistry: Fodor, *Tetrahedron* **1**, 86 (1957).

Relative stereochemistry

Crystals from petr ether, mp 107-109°. bp 248°. Sol in water, alcohol, acetone, ether.

Hydrochloride. $C_8H_{13}NO_2 \cdot HCl$. Crystals from ethanol, mp 255-257°.

Acetate. $C_{10}H_{15}NO_3$. Crystals from petr ether, mp 56-57°.

8548. Scotophobin. [33579-45-2] L-Seryl-L-aspartyl-L-asparaginyl-L-asparaginyl-L-glutaminyl-L-glutaminylglycyl-L-lysyl-L-seryl-L-alanyl-L-glutaminyl-L-glutaminylglycylglycyl-L-tyrosinamide. $C_{62}H_{97}N_{23}O_{26}$; mol wt 1580.59. C 47.11%, H 6.19%, N 20.38%, O 26.32%. Pentadecapeptide, isolated from the brains of rats with acquired fear of the dark, which induces dark avoidance in untrained mice. (From the Greek *skotos*, dark and *phobos*, fear.) Thought to be the first isolation, identification and synthesis of one of the "chemical code words" that control memory and learning. *See:* Ungar *et al., Nature* **238**, 198 (1972). Original training, testing and extraction procedures: Ungar *et al., ibid.* **217**, 1259 (1968); *eidem, Proc. West. Pharmacol. Soc.* **13**, 149 (1970). Structural elucidation: Desiderio *et al., Chem. Commun.* **1971**, 432. Synthesis of preliminary structures: Ali *et al., Experientia* **27**, 1138 (1971); *eidem, Int. J. Pept. Protein Res.* **4**, 395 (1972). Solid phase synthesis of revised structure: Parr, Holzer, *Z. Physiol. Chem.* **352**, 1043 (1971); *eidem, Experientia* **28**, 884 (1972). Passive transfer of dark avoidance to mice: Malin, Guttman, *Science* **178**, 1219 (1972); to goldfish: Guttman *et al., Nature New Biol.* **235**, 126 (1972); Bryant *et al., Science* **177**, 635 (1972). Chemistry and critical analysis of published data on scotophobin: Stewart, *Nature* **238**, 202-210 (1972).

Ser–Asp–Asn–Asn–Gln–Gln–Gly–Lys–Ser–Ala–Gln–Gln–Gly–Gly–TyrNH₂

8549. Scutellarein. [529-53-3] 5,6,7-Trihydroxy-2-(4-hydroxyphenyl)-4*H*-1-benzopyran-4-one; 4′,5,6,7-tetrahydroxyflavone; 6-hydroxypelargidenon 1465. $C_{15}H_{10}O_6$; mol wt 286.24. C 62.94%, H 3.52%, O 33.54%. By hydrolysis of scutellarin: Molisck, Goldschmidt, *Monatsh. Chem.* **22**, 679 (1901); Marsh, *Biochem. J.* **59**, 58 (1955). From *Digitalis lanata* Ehrh., *Scrophulariaceae:* Rangaswami, Rao, *Proc. Indian Acad. Sci.* **54A**, 51 (1961), *C.A.* **56**, 10076f (1962). Structure: Wessely, Moser, *Monatsh. Chem.* **56**, 97 (1930). Synthesis: Sastri, Seshadri, *Proc. Indian Acad. Sci.* **23A**, 262 (1946), *C.A.* **41**, 449a (1947); Zemplén *et al., Acta Chim. Acad. Sci. Hung.* **16**, 445 (1958); Jouanne, Mentzer, *C. R. Seances Acad. Sci. Ser. C* **263**, 1022 (1966).

Yellow leaflets from methanol, does not melt below 300°. uv max (ethanol): 286, 339 nm (ε 16600; 18300).

Tetraacetate. $C_{23}H_{18}O_{10}$. Prisms from ethyl acetate, mp 235-237°.

Glucuronide. Scutellarin. $C_{21}H_{18}O_{12}$. From leaves of *Scutellaria altissima* Linn., *Labiatae:* Goldschmiedt, Zerner, *Monatsh. Chem.* **31**, 439 (1910); from *Centaurea scabiosa* L., *Compositae:* Charaux, Rabaté, *J. Pharm. Chim.* **9**, 155 (1940); from *Scutellaria* spp.: Marsh, *Biochem. J.* **59**, 58 (1955); *Nature* **183**, 1824 (1959). Synthesis: Farkas *et al., Ber.* **107**, 3878 (1974). Needles from alc, darkens above 230°, does not melt below 300°. $[\alpha]_D^{18}$ −14° (water); $[\alpha]_D^{20}$ −139° (pyridine). uv max (ethanol): 285, 335 nm (ε 20000;

26100). Practically insol in water. Sol in alkali hydroxides, glacial acetic acid; slightly sol in organic solvents.

8550. Scutellaria. Skullcap; helmet flower. Dried overground portion of *Scutellaria lateriflora* L., *Labiatae. Habit.* North America. *Constit.* Scutellarin, volatile oil, tannin.

8551. Seaborgium. [54038-81-2] Element 106; unnilhexium. Sg, Unh; at. no. 106. Group VIB (6). No stable nuclides. Prepn of isotope ^{263}106 (α-emitter; $T_{1/2}$ 0.9 ±0.2 sec; longest-lived known isotope; rel. at. mass 263.1186) by ^{249}Cf (^{18}O,4n): A. Ghiorso *et al., Phys. Rev. Lett.* **33**, 1490 (1974). Four known isotopes: 259-261, 263. Confirmation of discovery: K. E. Gregorich *et al., ibid.* **72**, 1423 (1994). Review of history, prepn and properties: R. J. Silva in *The Chemistry of the Actinide Elements* vol. 2, J. J. Katz *et al.*, Eds. (Chapman and Hall, New York, 1986) pp 1108-1110; P. Armbruster, *Proc. Int. Sch. Phys. "Enrico Fermi"* 222-240 (1986); G. N. Flerov, G. M. Ter-Akopian, *Prog. Particle Nucl. Phys.* **19**, 197-239 (1987); G. T. Seaborg, W. D. Loveland, *The Elements Beyond Uranium* (John Wiley & Sons, Inc., New York, 1990) p 54-56; Transfermium Working Group, *Pure Appl. Chem.* **65**, 1757-1814 (1993).

8552. Sebacic Acid. [111-20-6] Decanedioic acid; 1,8-octanedicarboxylic acid. $C_{10}H_{18}O_4$; mol wt 202.25. C 59.39%, H 8.97%, O 31.64%. Obtained on a large scale by heating castor oil or ricinoleic acid with sodium hydroxide. Convenient lab procedures: Topchiev, Pavlov, *C.A.* **47**, 8002h (1953); Dominguez *et al., J. Chem. Educ.* **29**, 446 (1952). Also prepd by ozonization of undecylenic acid: Noorduyn, *Rec. Trav. Chim.* **38**, 326 (1919); Verkade *et al., ibid.* 385. Toxicity study: P. M. Jenner *et al., Food Cosmet. Toxicol.* **2**, 327 (1964). Review and bibliography: Jones, *Chimia* **5**, 169-173 (1951).

Monoclinic prismatic tablets, leaflets from acetone + petr ether (commercial product is a white, free-flowing powder having a mild fatty acid odor). d_4^{20} 1.207. mp 134.5°. bp$_{100}$ 294.5°; bp$_{50}$ 273°; bp$_{15}$ 243.5°; bp$_{10}$ 232°. Sp heat (solid) 0.48 cal/g; sp heat (liq) 0.53 cal/g. Sublimes slowly at 750 mm when heated to mp. n_D^{134} 1.422. pK$_1$ 4.59; pK$_2$ 5.59. Aq solns are neutral to methyl orange. Soly in water: 0.004% at 0°; 0.10% at 20°; 0.42% at 65°; 2.0% at 100°. Freely sol in alcohols, esters, ketones. Soly in ether: 0.1% at 17°. Sparingly sol in hydrocarbons and chlorinated hydrocarbons.

Diethyl ester. [110-40-7] Diethyl sebacate; ethyl sebacate. $C_{14}H_{26}O_4$; mol wt 258.36. Colorless to yellowish fluid. d_4^{20} 0.965. bp ~307° with some decompn. n_D^{20} 1.4369. Sol in about 700 parts cold, 50 parts boiling water; misc with alc, ether and other organic solvents. LD$_{50}$ orally in rats: 14470 mg/kg (Jenner).

USE: Raw material in the manuf of synthetic resins of the alkyd or polyester type, non-migrating plasticizers, polyester rubbers, synthetic fibers of the polyamide type.

8553. Sebacil. [96-01-5] 1,2-Cyclodecanedione. $C_{10}H_{16}O_2$; mol wt 168.24. C 71.39%, H 9.59%, O 19.02%. Prepd by oxidation of sebacoin with chromium trioxide in acetic acid: Blomquist *et al., J. Am. Chem. Soc.* **74**, 3640 (1952); *cf.* Prelog *et al., Helv. Chim. Acta* **35**, 1610 (1952); Blomquist, Goldstein, *Org. Synth.* **coll. vol. IV**, 838 (1963).

Crystals, mp 40-41°. bp$_{18}$ 120-125°; bp$_{10}$ 104-105°.

8554. Sebacoin. [96-00-4] 2-Hydroxycyclodecanone; 1-cyclodecanol-2-one. $C_{10}H_{18}O_2$; mol wt 170.25. C 70.55%, H 10.66%, O 18.79%. Prepd by cyclization of methyl or ethyl sebacate with sodium metal: Hansley, US 2228268 (1941 to du Pont); Prelog *et al., Helv. Chim. Acta* **35**, 1598 (1952); Allinger, *Org. Synth.* **coll. vol. IV**, 840 (1963).

Prisms from ether + petr ether, mp 42°. bp$_6$ 113-114°; bp$_{0.1}$ 77°.

8555. Secalonic Acids. C$_{32}$H$_{30}$O$_{14}$; mol wt 638.58. C 60.19%, H 4.74%, O 35.08%. Toxic mold metabolites which are diastereoisomeric dimeric hydroxanthone derivatives manifest as yellow pigments. First isolation studies: Stoll *et al.*, *Helv. Chim. Acta* **35**, 2022 (1952). Seven isomers have been isolated: *secalonic acid A*, *secalonic acid B*, and *secalonic acid C* from ergot: Aberhart *et al.*, *Tetrahedron* **21**, 1417 (1965); Franck *et al.*, *Ber.* **99**, 3842, 3875 (1966); *secalonic acid D* from *Penicillium oxalicum:* Steyn, *Tetrahedron* **26**, 51 (1970); *secalonic acid E* from *Phoma terrestris:* Hansen, Howard *et al.*, *Chem. Commun.* **1973**, 464; *secalonic acid F* from *Aspergillus aculeatus:* R. Andersen *et al.*, *J. Org. Chem.* **42**, 352 (1977); *secalonic acid G* from *Pyrenochaeta terrestris:* I. Kurobane, L. C. Vining, *Tetrahedron Lett.* **1978**, 4633. Secalonic acid A is enantiomeric with D; secalonic acid B is enantiomeric with E. Chromatographic separation of the ergot pigments: Franck, *Angew. Chem. Int. Ed.* **8**, 251 (1969); structures: ApSimon *et al.*, *J. Chem. Soc.* **1965**, 4144. Revised structures of secalonic acids A, B, C, D: Hooper *et al.*, *Chem. Commun.* **1971**, 111; *eidem*, *J. Chem. Soc. C* **1971**, 3580. Crystal and molecular structure of secalonic acid A: C. C. Howard *et al.*, *J. Chem. Soc. Perkin Trans. 1* **1976**, 1820. Identity of secalonic acid A with *entothein*: Yoshioka *et al.*, *Chem. Pharm. Bull.* **16**, 2090 (1968). Biosynthesis of secalonic acid A: I. Kurobane *et al.*, *Tetrahedron Lett.* **1978**, 1379. Review of secalonic acids and other ergochromes: Franck, Flasch in *Fortschr. Chem. Org. Naturst.* **30**, 151-206 (1973).

Secalonic acid A

Secalonic acid A. [35287-72-0] (3*R*,3'*R*,4*S*,4'*S*,4a*S*,4'a*S*)-2,2',-3,3',4,4',9,9'-Octahydro-1,1',4,4',8,8'-hexahydroxy-3,3'-dimethyl-9,9'-dioxo-[7,7'-bi-4a*H*-xanthene]-4a,4'a-dicarboxylic acid dimethyl ester; 6,6',7,7'-tetrahydro-1,1',5β,5'β,8,8'-hexahydroxy-6α,6'α-dimethyl-9,9'-dioxo-[2,2'-bixanthene]-10aβ,10'aβ(5*H*,-5'*H*)-dicarboxylic acid dimethyl ester; ergochrome AA(2,2'). Fine lemon-yellow needles from dioxane + petr ether + chloroform, mp 246-248° (dec); also reported as mp 260° (Yoshioka *et al.*, *loc. cit.*). [α]$_D$ −73° (CHCl$_3$). uv max (methanol): 247, 340 nm (ε 18140, 31200). pK$_{DMF}$ 8.7. Readily soluble in pyridine, dimethylformamide, dioxane; moderately sol in chloroform, methylene chloride, acetone. Gives a wine-red color reaction with 5% FeCl$_3$ in methanol.

8556. Secnidazole. [3366-95-8] α,2-Dimethyl-5-nitro-1*H*-imidazole-1-ethanol; 1-(2-hydroxypropyl)-2-methyl-5-nitroimidazole; 1-(2-methyl-5-nitroimidazol-1-yl)-2-propanol; PM-185184; RP-14539; Flagentyl. C$_7$H$_{11}$N$_3$O$_3$; mol wt 185.18. C 45.40%, H 5.99%, N 22.69%, O 25.92%. Analog of metronidazole, *q.v.* Prepn: **FR M3270** (1965 to Rhône-Poulenc), *C.A.* **63**, 11571d (1965); C. Cosar *et al.*, *Arzneim.-Forsch.* **16**, 23 (1966). Anti-amebic and trichomonacidal activities: F. Benazet, L. Guillaume, *Bull. Soc. Pathol. Exot. Ses Fil.* **69**, 309 (1976), *C.A.* **90**, 145922v (1979). Serum half-life: J. Symonds, *J. Antimicrob. Chemother.* **5**, 484 (1979). Therapeutic use: D. Videau *et al.*, *Br. J. Vener. Dis.* **54**, 77 (1978).

Cryst from toluene, mp 76° (Cosar).

THERAP CAT: Antiamebic. Antiprotozoal (Trichomonas).

8557. Secobarbital Sodium. [309-43-3] 5-(1-Methylbutyl)-5-(2-propen-1-yl)-2,4,6(1*H*,3*H*,5*H*)-pyrimidinetrione sodium salt (1:1); sodium 5-allyl-5-(1-methylbutyl)barbiturate; 5-allyl-5-(1-methylbutyl)barbituric acid sodium salt; 5-allyl-5-(1-methylbutyl)-malonylurea sodium salt; meballymal sodium; quinalbarbitone sodium; Barbosec; Immenoctal; Pramil; Quinalspan; Sedutain; Seconal Sodium; Seotalnatrium. C$_{12}$H$_{17}$N$_2$NaO$_3$; mol wt 260.27. C 55.38%, H 6.58%, N 10.76%, Na 8.83%, O 18.44%. Prepn: **US 1954429** (1934). Comprehensive description: I. Comer, *Anal. Profiles Drug Subs.* **1**, 343-365 (1972). Toxicity data: E. W. Schafer, *Toxicol. Appl. Pharmacol.* **21**, 315 (1972).

White, hygroscopic powder. Bitter taste. Very sol in water. Sol in alc. Practically insol in ether. Aq solns are alkaline to litmus. The soly in elixirs is increased by the presence of methenamine. LD$_{50}$ orally in rats: 125 mg/kg (Schafer).

Free acid. C$_{12}$H$_{18}$N$_2$O$_3$. mp 100°.

Note: This is a controlled substance (depressant): **21 CFR,** 1308.12 and 1308.13.

THERAP CAT: Sedative; hypnotic.

THERAP CAT (VET): Short-acting hypnotic; sedative. Pre-anesthetic.

8558. Secretin. [1393-25-5] A strongly basic polypeptide gastrointestinal hormone, produced primarily in the duodenum and jejunum. Released in response to gastric acid delivered into the duodenal lumen; stimulates exocrine pancreatic secretion of water and bicarbonate and inhibits gastric acid secretion and emptying. Discovered: W. M. Bayliss, E. H. Starling, *Ergeb. Physiol.* **5**, 664 (1905). Isoln of porcine secretin: J. E. Jorpes, V. Mutt, *Acta Chem. Scand.* **15**, 1790 (1961). Structure: J. E. Jorpes *et al.*, *Biochem. Biophys. Res. Commun.* **9**, 275 (1962); V. Mutt *et al.*, *Biochemistry* **4**, 2358 (1965). Synthesis: M. Bodanszky *et al.*, *J. Am. Chem. Soc.* **89**, 685, 6753 (1967); M. A. Ondetti *et al.*, *ibid.* **90**, 4711 (1968). Solid phase synthesis: B. Hemmasi, E. Bayer, *Int. J. Pept. Protein Res.* **9**, 63 (1977); D. H. Coy *et al.*, *Peptides* **3**, 137 (1982). Isoln of bovine secretin and amino acid sequence (identical to porcine): M. Carlquist *et al.*, *FEBS Lett.* **127**, 71 (1981); of human secretin: *idem et al.*, *IRCS Med. Sci.* **13**, 217 (1985). Comparison of natural porcine secretin with synthetic porcine and human forms in pancreatic function testing: L. Somogyi *et al.*, *Pancreas* **27**, 230 (2003). Review of radioimmunoassay methods: T. M. Chang, W. Y. Chey, *Dig. Dis. Sci.* **25**, 529-552 (1980). General reviews: W. H. Häcki, *Clin. Gastroenterol.* **9**, 609-632 (1980); W. Y. Chey, T.-M. Chang, *J. Gastroenterol.* **38**, 1025-1035 (2003).

Porcine Secretin

Porcine secretin. [17034-35-4] Secretin (swine); secretin (pig). C$_{130}$H$_{220}$N$_{44}$O$_{41}$; mol wt 3055.46.

Porcine secretin acetate salt. [17034-34-3] RG-1068; Secreflo. Synthetic porcine secretin; amino acid sequence identical to natural form. Efficacy in pancreatic function testing: L. Somogyi *et al.*, *Pancreas* **21**, 262 (2000); in diagnosis of Zollinger-Ellison syndrome: D. C. Metz *et al.*, *Aliment. Pharmacol. Ther.* **15**, 669 (2001). Use in pancreatic duct cannulation: B. M. Devereaux *et al.*, *Am. J. Gastroenterol.* **97**, 2279 (2002).

Human secretin. [108153-74-8] 15-L-Glutamic acid-16-glycine-secretin (swine). $C_{130}H_{220}N_{44}O_{40}$; mol wt 3039.46.

THERAP CAT: Diagnostic aid (pancreatic function).

8559.   Securinine. [5610-40-2] (6*S*,11a*R*,11b*S*)-9,10,11,11a-Tetrahydro-8*H*-6,11b-methanofuro[2,3-*c*]pyrido[1,2-*a*]azepin-2(6*H*)-one; securinan-11-one. $C_{13}H_{15}NO_2$; mol wt 217.27. C 71.87%, H 6.96%, N 6.45%, O 14.73%. From leaves and roots of *Securinega suffruticosa* Rehder, *Euphorbiaceae* found in the Ussuri region: Murav'eva, Ban'kovskii, *C.A.* **50**, 17335c (1956); *eidem*, *Dokl. Akad. Nauk SSSR* **110**, 998 (1956); *C.A.* **51**, 8121a (1957). Improved extraction process: Kogan *et al.*, **GB 1169471** (1959 to All-Union Sci. Res. Inst. of Medicinal Plants), *C.A.* **72**, 47349x (1970) corresp to **US 3538103** (1970). Structure: Satoda *et al.*, *Tetrahedron Lett.* **1962**, 1199; Mukherjee *et al.*, *Naturwissenschaften* **50**, 155 (1963); Saito *et al.*, *Tetrahedron* **19**, 2085 (1963). Stereochemistry: Nakano *et al.*, *Chem. Ind. (London)* **1963**, 1034; Parello *et al.*, *Bull. Soc. Chim. Fr.* **1963**, 898; Nakano *et al.*, *J. Org. Chem.* **28**, 2619 (1963); Horii *et al.*, *Tetrahedron* **19**, 2101 (1963); Imado *et al.*, *Chem. Ind. (London)* **1964**, 1691. Synthesis of the racemate: Saito *et al.*, *Chem. Pharm. Bull.* **14**, 313, 1059 (1966); Horii *et al.*, *Tetrahedron* **23**, 1165 (1967). Toxicology study: S. L. Friess *et al.*, *Toxicol. Appl. Pharmacol.* **3**, 347 (1961).

Yellow crystals from alcohol, mp 142-143°. $[\alpha]_D^{20}$ −1042° (c = 1 in alc). uv max (alc): 256, 330 nm (log ε 4.27, 3.30).

Hydrochloride. mp 230°. $[\alpha]_D^{20}$ −259.2° (alcohol).

Nitrate. mp 205°. $[\alpha]_D^{20}$ −312.12° (alcohol). LD_{50} i.v. in mice: 3.5 ± 0.9 mg/kg (Friess).

8560.   Seidlitz Mixture. [8014-63-9] (2*R*,3*R*)-2,3-Dihydroxybutanedioic acid monopotassium monosodium salt mixt with sodium hydrogen carbonate. $C_5H_5KNa_2O_9$; mol wt 294.16. C 20.42%, H 1.71%, K 13.29%, Na 15.63%, O 48.95%. Mixture of 3 parts Rochelle salt and 1 part sodium bicarbonate. Ten grams of the mixture are employed with 2.17 g tartaric acid for one Seidlitz powder.

THERAP CAT: Cathartic.

8561.   Selamectin. [220119-17-5] (5*Z*)-25-Cyclohexyl-4'-*O*-de(2,6-dideoxy-3-*O*-methyl-α-L-*arabino*-hexopyranosyl)-5-demethoxy-25-de(1-methylpropyl)-22,23-dihydro-5-(hydroxyimino)avermectin A$_{1a}$; 5-oxoimino-25-cyclohexyl avermectin B$_1$ monosaccharide; Revolution. $C_{43}H_{63}NO_{11}$; mol wt 769.97. C 67.08%, H 8.25%, N 1.82%, O 22.86%. Antiparasitic, semisynthetic avermectin, *q.v.* Prepn: B. F. Bishop *et al.*, **WO 9415944**; *eidem*, **US 5981500** (1994, 1999 both to Pfizer).

Waterfast, no unpleasant odor.

THERAP CAT (VET): Anthelmintic; ectoparasiticide.

8562.   Selectfluor®. [140681-55-6] 1-(Chloromethyl)-4-fluoro-1,4-diazoniabicyclo[2.2.2]octane tetrafluoroborate(1−) (1:2); F-TEDA-BF$_4$. $C_7H_{14}B_2ClF_9N_2$; mol wt 354.26. C 23.73%, H 3.98%, B 6.10%, Cl 10.01%, F 48.27%, N 7.91%. Electrophilic N—F fluorinating agent. Prepn: R. E. Banks, **US 5086178** (1992 to Air Prods. Chem.); *idem et al.*, *J. Chem. Soc. Chem. Commun.* **1992**, 595; and applications as an electrophilic fluorinating agent: *idem et al.*, *J. Chem. Soc. Perkin Trans. 1* **1996**, 2069. X-ray crystal structure: *idem et al.*, *Acta Crystallogr.* **C49**, 492 (1993). Review of properties and applications in organic synthesis: R. E. Banks, *J. Fluorine Chem.* **87**, 1-17 (1998); of commercial development: J. J. Hart, R. G. Syvret, *ibid.* **100**, 157-161 (1999); of applications and mechanism of fluorination: P. T. Nyffeler *et al.*, *Angew. Chem. Int. Ed.* **44**, 192-212 (2005).

White, free-flowing, non-hygroscopic crystalline solid, mp 225° (dec). Soly (g/l): water 176 (20°); acetonitrile 50. Slightly sol in lower alcohols, acetone. LD$_{50}$ in male rats: 640 mg/kg orally; >2 g/kg dermally (Banks, 1998).

Caution: Exothermic decompn can occur at temp >80° (Banks, 1996).

USE: Mild, versatile fluorinating reagent used in pharmaceutical and agricultural applications.

8563.   Selectins. Family of calcium-dependent cell adhesion molecules (CAMs) that initiate leukocyte adhesion to activated platelets or vascular endothelium in reponse to tissue injury or infection. Involved in lymphocyte homing to lymph nodes and in the extravasation of neutrophils into inflamed tissues. Binds to specific carbohydrate ligands on the surface of target cells. Three types have been identified and are designated according to the cell type from which they were originally discovered: L-selectin, from lymphocytes; E-selectin, from endothelial cells; and P-selectin, from platelets. Structures consist of an amino terminal lectin domain, an epidermal growth factor type repeat, and a discreet number of short consensus repeats similar to those found in complement-regulatory proteins (CR domains). The term "selectin" was proposed to highlight the presence of the lectin domain and to emphasize their selective expression and function. *Reviews:* D. Vestweber, *Semin. Cell Biol.* **3**, 211-220 (1992); M. P. Bevilacqua, R. M. Nelson, *J. Clin. Invest.* **91**, 379-387 (1993); R. P. McEver, *Curr. Opin. Immunol.* **6**, 75-84 (1994). Review of role in inflammation: M. A. Jutila, *Adv. Pharmacol.* **25**, 235-262 (1994); as potential therapeutic targets: J. K. Welply *et al.*, *Biochim. Biophys. Acta* **1197**, 215-226 (1994). Review of selectin ligands: A. Varki, *Proc. Natl. Acad. Sci. USA* **91**, 7390-7397 (1994).

L-Selectin. Lymphocyte adhesion molecule-1; LAM-1; leukocyte-endothelial cell adhesion molecule-1; LECAM-1; leu-8. Constitutively expressed on the surface of almost all leukocytes; rapidly shed upon cell activation. Binds to activated endothelial cells and to the high endothelial venule of lymph nodes. Human form contains 2 CR domains. Originally described as a lymphocyte homing receptor: W. M. Gallatin *et al.*, *Nature* **304**, 30 (1983).

E-Selectin. Endothelial-leukocyte adhesion molecule-1; ELAM-1. Expressed by endothelial cells following stimulation by endotoxin or inflammatory cytokines such as IL-1 or TNF. Binds to neutrophils, monocytes, eosinophils, certain lymphocytes and some tumor cells. Human form contains 6 CR domains. Identification: M. P. Bevilacqua *et al.*, *Proc. Natl. Acad. Sci. USA* **84**, 9238 (1987). Crystal structure of lectin domain: W. I. Weis, *Structure* **2**, 147 (1994).

P-Selectin. α-Granule membrane protein 140; GMP-140; platelet activation-dependent granule-external membrane protein; PAD-GEM. Synthesized constitutively by endothelial cells and platelets; stored in platelet α-granules or in Weibel-Palade bodies of endothelial cells. Rapidly mobilized to the cell surface following activation by histamine or thrombin. Binds to neutrophils, monocytes, and certain lymphocyte subsets. Human form contains 9 CR domains. Identification on activated platelets: S.-C. Hsu-Lin *et al., J. Biol. Chem.* **259**, 9121 (1984).

8564. Selectride®. Group of stereoselective trialkyl borohydride reducing agents. Reduction of cycloalkenones: J. M. Fortunato, B. Ganem, *J. Org. Chem.* **41**, 2194 (1976). Review of synthetic applications: S. Wittmann, B. Schönecker, *J. Prakt. Chem. Chem.-Ztg.* **338**, 759-762 (1996).

L-Selectride

L-Selectride. [38721-52-7] Lithium (*T*-4)-hydrotris(1-methylpropyl)borate(1−); lithium tri-*sec*-butylborohydride. $C_{12}H_{28}BLi$; mol wt 190.11. C 75.82%, H 14.85%, B 5.69%, Li 3.65%. Prepn and use in stereoselective reduction of cyclic ketones: H. C. Brown, S. Krishnamurthy, *J. Am. Chem. Soc.* **94**, 7159 (1972). Synthetic use in reduction of cyclic anhydrides: M. A. Makhlouf, B. Rickborn, *J. Org. Chem.* **46**, 4810 (1981); in cleavage of carbamates and *O*-demethylation of alkaloids: A. Coop *et al., ibid.* **63**, 4392 (1998).

K-Selectride. [54575-49-4] Potassium (*T*-4)-hydrotris(1-methylpropyl)borate(1−); potassium tri-*sec*-butylborohydride. $C_{12}H_{28}$-BK; mol wt 222.26. C 64.85%, H 12.70%, B 4.86%, K 17.59%. Prepn: C. A. Brown, *J. Am. Chem. Soc.* **95**, 4100 (1973); *idem, Inorg. Synth.* **17**, 26 (1977). Synthetic use in conjugate reduction and reductive alkylation of α,β-unsaturated cyclohexenones: B. Ganem, *J. Org. Chem.* **40**, 146 (1975); in reduction of diketo steroids: D. M. Tal *et al., Tetrahedron* **40**, 851 (1984). Prepd as a viscous, clear, water-white to yellow soln in THF. *Reacts vigorously with water.*

Caution: Solutions in THF may cause skin irritation and chemical burns (C. A. Brown, 1977).

USE: Mild reducing agents in organic synthesis.

8565. Selegiline. [14611-51-9] (α*R*)-*N*,α-Dimethyl-*N*-2-propyn-1-ylbenzeneethanamine; L-(−)-*N*,α-dimethyl-*N*-2-propynyl-phenethylamine; L-*N*-(2-phenylisopropyl)-*N*-methylprop-2-ynylamine; (−)-deprenil; L-deprenyl; Emsam. $C_{13}H_{17}N$; mol wt 187.29. C 83.37%, H 9.15%, N 7.48%. Monoamine oxidase-B inhibitor related structurally to pargyline, *q.v.* Prepn of racemate: **FR 1368136**; **GB 1031425** (1964, 1966 both to Chinoin); of (−)-form: **NL 6605956**; **GB 1153578** (1966, 1969 both to Chinoin); J. S. Fowler, *J. Org. Chem.* **42**, 2637 (1977). Pharmacology and toxicology of racemate: J. Knoll *et al., Arch. Int. Pharmacodyn.* **155**, 154 (1965); of isomers: K. Magyar *et al., Acta Physiol. Acad. Sci. Hung.* **32**, 377 (1967). Review of pharmacology: E. H. Heinonen, R. Lammintausta, *Acta Neurol. Scand. Suppl.* **136**, 44-59 (1991); of efficacy in Parkinson's disease: D. W. Robin, *Am. J. Med. Sci.* **302**, 392-395 (1991); K. L. Poston, C. Waters, *Expert Opin. Pharmacother.* **8**, 2615-2624 (2007). Review of veterinary use in canine cognitive dysfunction: W. W. Ruehl, B. L. Hart in *Psychopharmacology of Animal Behavior Disorders*, N. H. Dodman, L. Shuster, Eds. (Blackwell Sci., Malden, Mass., 1998) pp 283-304. Review of transdermal selegiline in major depressive disorder: D. S. Robinson, J. D. Amsterdam, *J. Affect. Disord.* **105**, 15-23 (2008).

Oil, $bp_{0.8}$ 92-93°. n_D^{20} 1.5180.

Hydrochloride. [14611-52-0] Anipryl; Antiparkin; Amindan; Carbex; Déprényl; Egibren; Eldepryl; Jumex; Movergan; Otrasel; Plurimen; Seledat; Selegam; Selepark; Selgimed; Xilopar; Zelapar. $C_{13}H_{17}N \cdot HCl$; mol wt 223.74. Crystals, mp 141-142°. $[\alpha]_D^{25}$ −10.8° (c = 6.48 in water). Freely sol in water, chloroform, methanol. LD_{50} in rats (mg/kg): 81 i.v., 280 s.c. (Magyar).

(±)-Form. [2323-36-6] Deprenyl; phenylisopropyl-*N*-methylpropinylamine. Oil, bp_5 104-110°. n_D^{20} 1.5229.

(±)-Form hydrochloride. [2079-54-1] E-250. Crystals from ethanol + ether, mp 131-131.5°. LD_{50} in rats (mg/kg): 63 i.v., 126 s.c., 385 orally (Knoll).

THERAP CAT: Antiparkinsonian; antidepressant.

THERAP CAT (VET): In treatment of canine cognitive dysfunction syndrome.

8566. Selenic Acid. [7783-08-6] H_2O_4Se; mol wt 144.97. H 1.39%, O 44.14%, Se 54.47%. H_2SeO_4. Prepd by treating lead selenate with hydrogen sulfide and concentrating the filtered soln by evaporation: Mitscherlich, *Pogg. Ann.* **9**, 623 (1827); **11**, 327 (1827); *Ann. Chim. Phys.* [2] **36**, 100 (1827); by treating calcium selenate with cadmium oxalate and hydrogen sulfide: von Hauer, *J. Prakt. Chem.* [1] **80**, 214, 317 (1860); by treating a soln of silver selenite with bromine and evaporating the filtered soln: Thomsen, *Ber.* **2**, 598 (1869); by oxidizing selenium oxide with 30% hydrogen peroxide and removing water: Gilbertson, King, *J. Am. Chem. Soc.* **58**, 180 (1936); *Inorg. Synth.* **3**, 137 (1950).

Hexagonal prisms; mp 58°; d_4^{15} 2.9508; bp 260°. *Corrosive.* Very sol in water; sol in sulfuric acid. Insol in ammonia. Dec in alcohol. Very deliquescent. Reduced by hydrobromic acid, hydriodic acid, hydrogen sulfide, hydroxylamine hydrochloride, phenylhydrazine, formic acid, oxalic acid, malonic acid, pyruvic acid, acetyl chloride, and several metals.

8567. Selenious Acid. [7783-00-8] Monohydrated selenium dioxide; selenous acid. H_2O_3Se; mol wt 128.97. H 1.56%, O 37.22%, Se 61.22%. H_2SeO_3. Prepd by dissolving selenium dioxide in hot water and cooling: Berzelius, *Acad. Handl. Stockholm* **39**, 13 (1818).

Deliquescent hexagonal prisms; d_4^{15} 3.004. Vapor press. (mm Hg): 2 at 15°; 4.5 at 35°; 7 at 40.3°. $K_1 = 0.0024$; $K_2 = 4.8 \times 10^{-9}$. Gives off water upon heating and selenium oxide sublimes. Oxidized to selenic acid by strong oxidizing agents such as ozone, hydrogen dioxide, chlorine. Reduced to selenium by most reducing agents including hydriodic acid, sulfurous acid, sodium hyposulfite, hydroxylamine salts, hydrazine salts, hypophosphorous acid, phosphorous acid. 90 parts dissolve in 100 parts water at 0°, 400 parts at 90°; very sol in alcohol; insol in ammonia.

USE: As a reagent for alkaloids; as oxidizing agent.

8568. Selenium. [7782-49-2] Se; at. wt 78.96; at. no. 34; valences 2, 4, 6. Group VIA (16). Six stable isotopes: 74 (0.87%); 76 (9.02%); 77 (7.58%); 78 (23.52%); 80 (49.82%); 82 (9.19%); artificial, radioactive isotopes: 70-73; 75; 79; 81; 83-85; 87. Discovered in 1817 by Berzelius. Constitutes about 0.09 ppm of the earth's crust. Occurs in nature usually in the sulfide ores of the heavy metals; found in small quantities in pyrite; in the minerals *clausthalite* (PbSe), *naumannite* [(Ag,Pb)Se], *tiemannite* (HgSe); in selenosulfur. Essential trace element in human and animal diets; constituent of selenoproteins. Prepn: Waitkins *et al., Ind. Eng. Chem.* **34**, 899 (1942). Purification: Nielsen, Heritage, *J. Electrochem. Soc.* **106**, 39 (1959); Oberbacher, Schlier, **US 2930678** (1960 to Norddeutsche Raffinerie). Symposium on organic selenium and tellurium compds: Y. Okamoto, W. H. H. Gunther, Eds., *Ann. N.Y. Acad. Sci.* **192**, 1-225 (1972). Reviews of selenium and its compds: K. W. Bagnall, *The Chemistry of Selenium, Tellurium and Polonium* (Elsevier, New York, 1966) 200 pp; Bagnall in *Comprehensive Inorganic Chemistry* vol. 2, J. C. Bailar, Jr. *et al.*, Eds. (Pergamon Press, Oxford, 1973) pp 935-1008; *Selenium*, R. A. Zingaro, W. C. Cooper, Eds. (Van Nostrand, Reinhold, New York, 1974) 835 pp; E. M. Elkin in *Kirk-Othmer Encyclopedia of Chemical Technology* vol. 20 (Wiley-Interscience, New York, 3rd ed., 1982) pp 575-601. Reviews of nutritional properties and toxicology: E. J. Underwood, *Trace Elements in Human and Animal Nutrition* (Academic Press, New York, 1971) pp 323-368; Frost, *Crit. Rev. Toxicol.* **1**, 467-514 (1972); Allaway, *Cornell Vet.* **63**, 151-170 (1973); Frost, Lish, *Annu. Rev. Pharmacol.* **15**, 259-284 (1975); of biochemistry: T. C. Stadtman, *Annu. Rev. Biochem.* **49**, 93-110 (1980); *idem, FASEB J.* **1**, 375-379 (1987). Review of role in human health: M. P. Rayman, *Lancet* **356**, 233-241 (2000); of toxicology and human exposure: *Toxicological Profile for Selenium* (PB2004-100005, 2003) 457 pp.

Exists in several allotropic forms: amorphous, cryst or red, and gray or metallic. Liquid is a brownish red, boils at 685° forming dark red vapors. Sol in dil aq caustic alkali solns; in aq potassium cyanide soln, in potassium sulfite soln. Burns in air with a bright blue flame forming the dioxide and emitting a characteristic odor resembling rotten horseradish. Combines directly with hydrogen, with the halogens (excluding iodine). Oxidized to selenious acid by nitric acid, to selenic acid by sulfuric acid. Reduces hot aqueous solutions of silver and gold salts with formation of silver selenide and metallic gold, respectively. Reacts with many metals.

Amorphous forms. Vitreous, black selenium; dark red-brown to bluish-black solid; formed when molten Se is cooled rapidly. d 4.28. Softens at 50-60° and becomes elastic at 70°. Red, amorphous form; formed by reduction of selenious acid in water; by condensation of Se vapor on a cold surface. d 4.26. Review of structural studies: Richter, Breitling, *Z. Naturforsch.* **26B**, 1699-1708, 2074-2075 (1971). When freshly precipitated, reacts with water at 50° forming selenious acid and hydrogen. Sol in carbon disulfide, methylene iodide, benzene, or quinoline.

Crystalline or red. Two monoclinic forms; dark red, transparent crystals. α-Form prepd by slow evaporation of CS_2 soln of Se; β-form by rapid evaporation of soln. mp below 200°. d (α-form) 4.46. Structure of α-form: Cherin, Unger, *Acta Crystallogr.* **28B**, 513 (1972); *see also* Bagnall, *loc. cit.* Both forms are metastable and change into gray form on heating.

Gray or metallic. Lustrous gray to black hexagonal crystals. The most stable form. d_4^{20} 4.81. mp 217°. Mohs' hardness: 2.0. Latent heat of fusion 16.4 cal/g. Latent heat of vaporization 20.6 kcal/mol. Linear coefficient of thermal expansion per degree C = 37 × 10^{-6}. Specific heat (28°): 0.084 cal/g/°C. Surface tension (217°): 92.5 dynes/cm. Thermal conductivity (25°): 0.0007-0.00183 cal/(cm)-(°C)(sec). Insol in water, alcohol; very slightly sol in carbon disulfide (2 mg/100 ml at ord temp). Sol in ether. Conducts electricity and rectifies alternating current; the conductivity increases up to a thousand times on exposure to light.

Caution: Potential symptoms of overexposure are irritation of eyes, skin, nose and throat; visual disturbance; headache; chills, fever; dyspnea, bronchitis; metallic taste, garlic breath, GI disturbance; dermatitis; skin and eye burns. *See NIOSH Pocket Guide to Chemical Hazards* (DHHS/NIOSH 97-140, 1997) p 276.

USE: As ingredient of toning baths in photography; as pigment in manuf ruby-, pink-, orange-, or red-colored glass; as metallic base in making electrodes for arc lights, electrical instruments and apparatus, as rectifier in radio and television sets; in selenium photocells, in semiconductor fusion mixtures, selenium cells, telephotographic apparatus; as vulcanizing agent in processing of rubber; as catalyst in determination of nitrogen by Kjeldahl method; for dehydrogenation of organic compds.

THERAP CAT (VET): Nutritional factor (interrelationship with vitamin E), chiefly to prevent muscle degenerative diseases.

8569. Selenium Chloride. [10025-68-0] Selenium monochloride. Cl_2Se_2; mol wt 228.82. Cl 30.99%, Se 69.01%. Se_2Cl_2. Prepd from the elements: Sacc *et al.*, cited in *Mellor's* **vol. X**, 894 (1930); by the action of PCl_5 on selenium, on the selenide of phosphorus or antimony, on SeO_2 or on $SeOCl_2$: Baudrimont, *Ann. Chim. Phys.* [4] **2**, 5 (1864); Michaelis, *Z. Chem.* [2] **6**, 460 (1870); by the action of HCl on a soln of selenium in fuming sulfuric acid: Divers, Shimose, *J. Chem. Soc.* **45**, 194, 198, 201 (1884); from Se, SeO_2 and HCl: Lenher, Kao, *J. Am. Chem. Soc.* **47**, 772 (1925); **48**, 1550 (1926).

Deep red, oily liquid. bp_{733} 127° (dec). d_4^{25} 2.7741. mp −85°. Sol in chloroform, benzene, carbon tetrachloride, carbon disulfide, fuming sulfuric acid; dec in water.

8570. Selenium Hexafluoride. [7783-79-1] (*OC*-6-11)-Selenium fluoride (SeF₆). F₆Se; mol wt 192.95. F 59.08%, Se 40.92%. SeF₆. Prepd by passing gaseous fluorine over finely divided selenium in a copper vessel: Klemm, Henkel, *Z. Anorg. Allg. Chem.* **207**, 74 (1932); Yost, Claussen, *J. Am. Chem. Soc.* **55**, 885 (1933); Yost, *Inorg. Synth.* **1**, 121 (1939). *Review:* Kemmitt, Sharp, *Adv. Fluorine Chem.* **4**, 235 (1965).

Gas. mp −50.8°; sublimes −63.8°. Vapor press. (mm Hg): 651.2 (−48.7°); 213.1 (−64.8°); 30.4 (−87.5°). *Poisonous; corrosive.* Insol in water. Reacts with ammonia gas at 200° to give selenium, nitrogen, and hydrogen fluoride. Covalently satd, does not attack glass.

Caution: Potential symptoms of overexposure in exptl animals are pulmonary irritation and edema. *See NIOSH Pocket Guide to Chemical Hazards* (DHHS/NIOSH 97-140, 1997) p 276.

USE: Gaseous electric insulator.

8571. Selenium Oxide. [7446-08-4] Selenium dioxide; selenious anhydride. O_2Se; mol wt 110.96. O 28.84%, Se 71.16%. SeO_2. Prepd by burning selenium in oxygen: Berzelius, *Acad. Handl. Stockholm* **39**, 13 (1818); *Ann. Chim. Phys.* [2] **9**, 160, 225, 337 (1818); by burning selenium in oxygen and nitrogen dioxide: Lenher, *J. Am. Chem. Soc.* **20**, 559 (1898); by oxidizing selenium with nitric acid: A. P. Julien, *ibid.* **47**, 1799 (1925). Use as oxidizing agent: L. F. Fieser, M. Fieser, *Reagents for Organic Chemistry* **vol. 1** (New York, Wiley, 1967) p 992. Toxicology: E. A. Cerwenka, W. C. Cooper, *Arch. Environ. Health* **3**, 189 (1961). *Review:* Waitkins, Clark, *Chem. Rev.* **36**, 235-289 (1945).

Lustrous, tetragonal needles. Acidic taste; leaves a burning sensation. Its yellowish green vapor has a pungent, sour smell. mp 340°; d_{15}^{15} 3.954. Vapor pressure (mm Hg): 12.5 at 70°; 20.2 at 94°; 39.0 at 181°; 760.0 at 315°; 848.0 at 320°. n_D^{20} <1.76. Soly (parts/100 parts solvent) in water: 38.4 (14°); in methanol: 10.16 (11.8°); in 93% ethanol: 6.67 (14°); in acetone: 4.35 (15.3°); in acetic acid: 1.11 (13.9°). Sol in concd H_2SO_4. Stable to light and heat. Rapidly absorbs dry hydrogen fluoride, hydrogen chloride, hydrogen bromide, and hydrogen iodide to form the corresponding selenium oxyhalide. Reacts with ammonia to form nitrogen and selenium; in alcohol soln forms ammonium ethyl selenite, $HN_4(C_2H_5)SeO_3$. Yields nitrogen and black amorphous selenium with hydrazine; nitrogen and reddish-brown amorphous selenium with hydroxylamine hydrochloride. Forms selenic acid with nitric acid. Reduced by carbon and other organic substances.

Caution: Direct contact may cause intense local irritation of skin and eyes (Cerwenka, Cooper).

USE: In the manuf of other selenium compounds; as a reagent for alkaloids; as catalyst and oxidizing agent.

8572. Selenium Oxychloride. [7791-23-3] Seleninyl chloride. Cl_2OSe; mol wt 165.86. Cl 42.75%, O 9.65%, Se 47.61%. $SeOCl_2$. Prepd by passing chlorine into SeO_2 suspended in carbon tetrachloride: Lenher, *J. Am. Chem. Soc.* **42**, 2498 (1920); Hönigschmid, Görnhardt, *Z. Naturforsch.* **1**, 661 (1946); by dehydration of dichloroselenious acid: Smith, Jackson, *Inorg. Synth.* **3**, 130 (1950); from $SOCl_2$ and SeO_2: Paetzold, Aurich, *Z. Anorg. Allg. Chem.* **315**, 72 (1962); from $AlCl_3$ and SeO_2: Drago, Whitten, *Inorg. Chem.* **5**, 677 (1966). Acute toxicity: C. G. Wilber, *Clin. Toxicol.* **17**, 171 (1980).

Nearly colorless or yellowish liquid; fumes in the air. d_4^{16} 2.44. Solidif ~5°. bp 180°. n_D^{20} 1.651. *Corrosive; poisonous.* Dec by water into HCl and selenious acid. Miscible with carbon tetrachloride, chloroform, carbon disulfide, benzene, toluene. It is an excellent solvent for many substances including metals. LD_{50} s.c. in rabbits: 7 mg/kg (Wilber).

Caution: Strong irritant, vesicant. *See Clinical Toxicology of Commercial Products*, R. E. Gosselin *et al.*, Eds. (Williams & Wilkins, Baltimore, 5th ed., 1984) Section II, p 129.

8573. Selenium Sulfides. Prepn from selenious acid and H_2S; product may be a mixture of the elements: *Gmelins, Selenium* (8th ed.) **10** (part B), 159-176 (1949). Prepn by fusion of the elements; solid solns of sulfur (S_8), selenium (Se_8) and selenium sulfides (S_n-Se_{8-n}; n = 1-7) formed: Fergusson *et al.*, *J. Inorg. Nucl. Chem.* **24**, 157 (1962); Hawes, *Nature* **198**, 1267 (1963); Cooper, Culka, *J. Inorg. Nucl. Chem.* **27**, 755 (1965); **29**, 1217 (1967).

Se₄S₄. Red tabular crystals from benzene, d 3.20. mp 113° (dec). Sol in CS_2. Solubility in benzene (20°): 0.4 g/l.

Se₂S₆. Light-orange needles from benzene, d 2.44. mp 121.5°. Sol in CS_2. Solubility in benzene (20°): 12 g/l.

Selenium disulfide. [7488-56-4] Exsel; Selsun. SeS_2. Prepn of shampoo: M. M. Baldwin, A. P. Young, Jr., US 2694669 (1954 to Abbott). Toxicity studies: L. M. Cummins, E. T. Kimura, *Toxicol. Appl. Pharmacol.* **20**, 89 (1971). Review of use as antidandruff agent: Matson, *J. Soc. Cosmet. Chem.* **7**, 459-466 (1956). Reddish-brown to bright orange powder. Practically insol in water and in organic solvents. Insol in 0.01N HCl. LD_{50} orally in rats: 138 mg/kg (Cummins, Kimura).

Caution: Selenium sulfide (as SeS) is reasonably anticipated to be a human carcinogen: *Report on Carcinogens, Twelfth Edition* (PB2011-111646, 2011) p 376. *See also Bioassay of Selenium Sulfide (Gavage) for Possible Carcinogenicity* (PB82-164955, 1980) 113p.

THERAP CAT: Disulfide as topical antiseborrheic.

THERAP CAT (VET): Topically in eczemas and dermatomycoses.

8574. Selenium Tetrabromide. [7789-65-3] (*T*-4)-Selenium bromide (SeBr₄). Br₄Se; mol wt 398.58. Br 80.19%, Se 19.81%. SeBr₄. Prepd by adding excessive bromine to selenium: Schneider, *Pogg. Ann.* **128**, 327 (1866); **129**, 450, 634 (1866); by dissolving selenium dioxide in hydrobromic acid: Muthmann, Schäfer, *Ber.* **26**, 1008 (1893).

Red-brown, cryst powder. Unpleasant odor. Dec 70-80°. Dec in moist air and water. Sol in carbon disulfide, chloroform, ethyl bromide.

8575. Selenium Tetrachloride. [10026-03-6] (*T*-4)-Selenium chloride (SeCl₄). Cl₄Se; mol wt 220.76. Cl 64.23%, Se 35.77%. SeCl₄. Prepd by the action of an excess of chlorine on selenium: Berzelius, cited in *Mellor's* **vol X**, 898 (1930); by the action of thionyl chloride or phosphorus trichloride on selenium oxychloride and by the action of phosphorus pentachloride on selenium oxide: Michaelis, *Z. Chem.* [2] **6**, 460 (1870); by the action of anhydr selenic acid on acetyl chloride at 0°: Lamb, *Am. Chem. J.* **30**, 209 (1903). Crystal structure: Shoemaker, Abrahams, *Acta Crystallogr.* **18**, 296 (1965).

White to pale yellow crystals. d 2.6. Sublimes when heated. Dec in water and moist air. Insol in liquid bromine. Dec by dry ammonia.

8576. Selenium Tetrafluoride. [13465-66-2] (*T*-4)-Selenium fluoride (SeF₄). F₄Se; mol wt 154.95. F 49.04%, Se 50.96%. SeF₄. Fluorinating agent. Prepn from SeCl₄ and AgF: Prideaux, Cox, *J. Chem. Soc.* **1927**, 928; **1928**, 1603; from Se and F₂: Aynsley *et al.*, *ibid.* **1952**, 1231; J. Carre *et al.*, *J. Fluorine Chem.* **14**, 139 (1979); from Se and AgF: Glemser *et al.*, *Naturwissenschaften* **52**, 130 (1965); from Se₂Cl₂ and F₂: Goggin, *J. Inorg. Nucl. Chem.* **28**, 661 (1966); from Se and ClF: Pitts, Jache, **US 3373000** (1968 to Olin Mathieson); from Se and ClF₃: Olah *et al.*, *J. Am. Chem. Soc.* **96**, 925 (1974).

Colorless liquid; fumes in air. mp −9.5°. bp 106°. d¹⁸ 2.75. Forms a white hygroscopic solid. Hydrolyzed violently by water. Attacks glass slowly when perfectly dry. Completely miscible with ether, ethanol, iodine pentafluoride, sulfuric acid; sol in chloroform, carbon tetrachloride.

USE: Fluorination of alcohols, carboxylic acids and carbonyl compds: Olah *et al.*, *loc. cit.*

8577. Selenocysteine. [10236-58-5] 3-Selenyl-L-alanine; selenium cysteine. C₃H₇NO₂Se; mol wt 168.05. C 21.44%, H 4.20%, N 8.34%, O 19.04%, Se 46.99%. Naturally occurring amino acid found in the active site of certain prokaryotic and eukaryotic enzymes. Highly oxidizable making synthesis difficult. Easily prepared from its oxidized form *selenocystine*. Prepn of *dl*-selenocystine: A. Fredga, *Sven. Kem. Tidskr.* **48**, 160 (1936); E. P. Painter, *J. Chem. Soc.* **69**, 229 (1947); of *l*-form: H. Tanaka, K. Soda, *Methods Enzymol.* **143**, 240 (1987). Reduction of selenocystine and separation of L-selenocysteine: J. N. Burnell *et al.*, *J. Inorg. Biochem.* **12**, 343 (1980). Chemical properties: R. E. Huber, R. S. Criddle, *Arch. Biochem. Biophys.* **122**, 164 (1967). Determn by HPLC/MS: H. E. Ganther *et al.*, *Methods Enzymol.* **107**, 582 (1984); colorimetric determn: N. Esaki, K. Soda, *ibid.* **143**, 148 (1987). Identification of Se as a component in protein: S.-H. Oh *et al.*, *Biochemistry* **13**, 1825 (1974). Characterization of selenocysteine as the Se containing amino acid moiety in various enzymes: J. E. Cone *et al.*, *Proc. Natl. Acad. Sci. USA* **73**, 2659 (1976); J. B. Jones *et al.*, *Arch. Biochem. Biophys.* **195**, 255 (1979); presence in the catalytic site of mammalian glutathione peroxidase: J. W. Forstrom *et al.*, *Biochemistry* **17**, 2639 (1978); J. J. Zabrowski *et al.*, *Biochem. Biophys. Res. Commun.* **84**, 248 (1978). Metabolism in animals: H. Tanaka *et al.*, *Curr. Top. Cell. Regul.* **27**, 487 (1985). Identification of selenocysteine-specific tRNA: W. C. Hawkes *et al.*, *Biochim. Biophys. Acta* **699**, 183 (1982); W. Leinfelder *et al.*, *Nature* **331**, 723 (1988). Review: T. C. Stadtman, *FASEB J.* **1**, 375-379 (1987).

pK 5.2. Unstable to acid hydrolysis.

8578. Selenomethionine. [3211-76-5] (2*S*)-2-Amino-4-(methylseleno)butanoic acid; α-amino-γ-(methylseleno)butyric acid; L-selenomethionine. C₅H₁₁NO₂Se; mol wt 196.11. C 30.62%, H 5.65%, N 7.14%, O 16.32%, Se 40.26%. Selenium analog of methionine. Prepn of DL-form: E. P. Painter, *J. Am. Chem. Soc.* **69**, 232 (1949); H. J. Klosterman, E. P. Painter, *ibid.* 2009; G. Zdansky, *Ark. Kemi* **19**, 559 (1962), *C.A.* **58**, 6919h (1963). Prepn of L-form: *idem, ibid.* **29**, 437 (1968); C. S. Pande *et al.*, *J. Org. Chem.* **35**, 1440 (1970). Biosynthesis of radioactive (⁷⁵Se) form: M. Blau, *Biochim. Biophys. Acta* **49**, 389 (1961). Metabolism: M. Blau, J. F. Holland in *Radioactive Pharmaceuticals*, G. A. Andrews *et al.*, Eds. (USAEC Symposium Series #6, 1976) pp 409-421. Review of use in pancreatic scanning: J. E. Agnew *et al.*, *Br. J. Radiol.* **49**, 979-995 (1976).

Crystals from aqueous acetone, mp 266-268° (dec). [α]D²² +21.6° (c = 0.5 in 2*N* HCl).

DL-Form. [1464-42-2] Transparent, hexagonal sheets or plates from methanol and water, metallic luster; mp 265° (dec). IR spectrum, *see* Zdansky (1962).

⁷⁵Se-Labeled form. [1187-56-0] Selenomethionine Se 75; Sethotope.

THERAP CAT: Selenomethionine Se 75 as diagnostic aid (radioactive imaging agent).

8579. Seliciclib. [186692-46-6] (2*R*)-2-[[9-(1-Methylethyl)-6-[(phenylmethyl)amino]-9*H*-purin-2-yl]amino]-1-butanol; 2-(*R*)-(1-ethyl-2-hydroxyethylamino)-6-benzylamino-9-isopropylpurine; roscovitine; CYC-202. C₁₉H₂₆N₆O; mol wt 354.46. C 64.38%, H 7.39%, N 23.71%, O 4.51%. Cyclin-dependent kinase (CDK) inhibitor. Prepn: L. Meijer *et al.*, **WO 9720842** (1997 to CNRS); *eidem*, **US 6316456** (2001 to CNRS; Inst. Exp. Bot. Acad. Sci. Czech Repub.); L. Havlicek *et al.*, *J. Med. Chem.* **40**, 408 (1997). Synthesis and crystal structure: S. Wang *et al.*, *Tetrahedron: Asymmetry* **12**, 2891 (2001). CDK inhibition: W. F. De Azevedo *et al.*, *Eur. J. Biochem.* **243**, 518 (1998); and cellular effects: L. Meijer *et al.*, *ibid.* 527. LC/MS/MS determn in plasma and urine: M. Vita *et al.*, *J. Chromatogr. B* **817**, 303 (2005). Protein binding and stability in plasma: *eidem, ibid.* **821**, 75 (2005). Metabolism and pharmacokinetics in mice: B. P. Nutley *et al.*, *Mol. Cancer Ther.* **4**, 125 (2005). Clinical evaluation in patients with solid tumors: C. Benson *et al.*, *Br. J. Cancer* **96**, 29 (2007). Review of pharmacology and clinical experience: I. T. Aldoss *et al.*, *Expert Opin. Invest. Drugs* **18**, 1957-1965 (2009).

Crystals from chloroform/ethyl ether, mp 102-104°; [α]D −34.6° (c = 0.43 in chloroform) (Havlicek). Also reported as colorless crystalline solid from ethyl acetate/hexane, mp 106-108°; [α]D²⁰ +56.3° (c = 0.56 in chloroform) (Wang). pKa 4.4.

(2S)-Form. [186692-45-5] Crystals from chloroform/ethyl ether, mp 118-120°; $[\alpha]_D$ +35.3° (c = 0.57 in chloroform) (Havlicek). Also reported as colorless crystalline solid from ethyl acetate/hexane, mp 108-109°. $[\alpha]_D^{20}$ −56.3° (c = 0.56 in chloroform) (Wang).

(±)-Form. [186692-44-8] Crystals from chloroform/ethyl ether, mp 137-139°.

THERAP CAT: Antineoplastic.

8580. Semagacestat. [425386-60-3] (2S)-2-Hydroxy-3-methyl-N-[(1S)-1-methyl-2-oxo-2-[[(1S)-2,3,4,5-tetrahydro-3-methyl-2-oxo-1H-3-benzazepin-1-yl]amino]ethyl]butanamide; N^2-[(2S)-2-hydroxy-3-methylbutanoyl]-N^1-[(1S)-3-methyl-2-oxo-2,3,4,5-tetrahydro-1H-3-benzazepin-1-yl]-L-alaninamide; N-(S)-2-hydroxy-3-methylbutyryl-1-L-alaninyl-(S)-1-amino-3-methyl-2,3,4,5-tetrahydro-1H-3-benazazepin-2-one; LY-450139. $C_{19}H_{27}N_3O_4$; mol wt 361.44. C 63.14%, H 7.53%, N 11.63%, O 17.71%. γ-Secretase inhibitor; blocks synthesis of amyloid β peptide. Prepn: T. M. Koenig et al., WO 0240451; J. E. Audia et al., US 7468365 (2002, 2008 both to Lilly). Synthesis: S. K. Boini et al., Synthesis 2009, 1983; Y. J. Pu et al., Org. Process Res. Dev. 13, 310 (2009). Clinical pharmacokinetics and metabolism: P. Yi et al., Drug Metab. Dispos. 38, 554 (2010). Clinical effect on amyloid β production in Alzheimer's patients: A. S. Fleisher et al., Arch. Neurol. 65, 1031 (2008). Review of development and clinical experience: D. B. Henley et al., Expert Opin. Pharmacother. 10, 1657-1664 (2009); B. P. Imbimbo, I. Peretto, Curr. Opin. Investig. Drugs 10, 721-730 (2009).

White solid, mp 208-212°.

THERAP CAT: Antiamyloidogenic agent; in treatment of Alzheimer's disease.

8581. Sematilide. [101526-83-4] N-[2-(Diethylamino)ethyl]-4-[(methylsulfonyl)amino]benzamide; CK-1752. $C_{14}H_{23}N_3O_3S$; mol wt 313.42. C 53.65%, H 7.40%, N 13.41%, O 15.31%, S 10.23%. Prepn: D. D. Davey et al., US 4544654 (1985 to Schering AG); W. C. Lumma, Jr. et al., J. Med. Chem. 30, 755 (1987). Pharmacology: S. S. Greenberg et al., Drug Dev. Res. 25, 283 (1992). Pharmacokinetics and HPLC determn in plasma: S. Dancik et al., J. Pharm. Sci. 80, 157 (1991). Clinical pharmacology: W. Wong et al., Am. J. Cardiol. 69, 206 (1992). Electrophysiologic profile: P. T. Sager et al., Circulation 88, 1072 (1993). Review: T. A. Argentieri, Cardiovasc. Drug Rev. 10, 182-198 (1992).

Monohydrochloride. [101526-62-9] CK-1752A. $C_{14}H_{23}N_3$-O_3S·HCl; mol wt 349.87. Crystals from acetone/methanol. mp 141-142° (Lumma). Also reported as having two polymorphic crystalline forms, mp 137° and mp 142° (Argentieri). Apparent pKa_1 7.60; pKa_2 9.51. Distribution coefficient (octanol/pH 7.4 buffer): 0.104 at 10 mM. uv max (0.1N NaOH): 289 nm (ε 1.91×10^4). uv max (0.1N HCl): 254 nm (ε 1.78×10^4). LD$_{50}$ i.p. in mice: 250-300 mg/kg (Lumma). LD$_{50}$ in mice, rats, dogs (mg/kg): 96, 92, 143-175 i.v.; 1,800, 3,200, 500-1,000 orally (Argentieri).

THERAP CAT: Antiarrhythmic (class III).

8582. Semduramicin. [113378-31-7] (2R,3S,4S,5R,6S)-Tetrahydro-2,4-dihydroxy-6-[(1R)-1-[(2S,5R,7S,8R,9S)-9-hydroxy-2,8-dimethyl-2-[(2S,2'R,3'S,5R,5'R)-octahydro-2-methyl-5'-[(2S,-3S,5R,6S)-tetrahydro-6-hydroxy-3,5,6-trimethyl-2H-pyran-2-yl]-3'-[[(2S,5S,6R)-tetrahydro-5-methoxy-6-methyl-2H-pyran-2-yl]oxy]-

[2,2'-bifuran]-5-yl]-1,6-dioxaspiro[4.5]dec-7-yl]ethyl]-5-methoxy-3-methyl-2H-pyran-2-acetic acid; UK-61689. $C_{45}H_{76}O_{16}$; mol wt 873.09. C 61.91%, H 8.77%, O 29.32%. Semi-synthetic polyether ionophore antibiotic. Prepn: A. C. Goudie, N. D. A. Walshe, EP 255335; eidem, US 4804680 (1988, 1989 both to Pfizer). LC determn in poultry liver: J. F. Ericson et al., J. AOAC Int. 77, 577 (1994). Metabolism and tissue residue studies in broiler chickens: M. J. Lynch et al., ACS Symp. Ser. 503, 49-69 (1992). Anticoccidial effects in chickens: M. E. McKenzie et al., Poult. Sci. 72, 2052 (1993); N. B. Logan et al., ibid. 2058.

Sodium salt. [119068-77-8] UK-61689-2; Aviax. $C_{45}H_{75}$-NaO_{16}; mol wt 895.07. Crystals from isopropyl ether. mp 175-176°. $[\alpha]_D^{25}$ +19.3° (c = 0.5 in methanol).

THERAP CAT (VET): Coccidiostat.

8583. Semicarbazide Hydrochloride. [563-41-7] Hydrazinecarboxamide hydrochloride (1:1); aminourea hydrochloride; carbamylhydrazine hydrochloride. CH_6ClN_3O; mol wt 111.53. C 10.77%, H 5.42%, Cl 31.79%, N 37.68%, O 14.35%. NH_2NH-$CONH_2$·HCl. Prepd by electrolytic reduction of nitrourea with cathodes of copper, nickel, lead, and mercury in hydrochloric acid solution: Ingersoll et al., Org. Synth. 5, 93 (1925). Commercial prepn from hydrazine hydrate, iron carbonyl, and carbon monoxide: Sampson, US 2589289 (1952 to du Pont).

Prisms from dil alc, dec 175-185°. Freely sol in water with acid reaction. Dissociation consts of free base: Kb at 20° = 29 × 10^{-11}; Ka = 1.6 × 10^{-11}. Very slightly sol in hot alcohol, insol in anhydr ether. Ammonolysis with liquid NH$_3$ gives a 95% yield of the free base; mp 96°: Audrieth, J. Am. Chem. Soc. 52, 1250 (1930).

Acetone semicarbazone. $C_4H_9N_3O$. Needles from water, mp 192° (heated stage); mp 188° (open capillary tube).

USE: As a reagent for ketones and aldehydes with which it affords cryst compds having characteristic melting points.

8584. Semioxamazide. [515-96-8] 2-Amino-2-oxoacetic acid hydrazide; aminooxamide; oxamic acid hydrazide. $C_2H_5N_3O_2$; mol wt 103.08. C 23.30%, H 4.89%, N 40.77%, O 31.04%. Made from hydrazine sulfate, ethyl oxamate and KOH. Ref: Wilson, Pickering, J. Chem. Soc. 123, 394 (1923); Leonard, Boyer, J. Org. Chem. 15, 42 (1950).

Lustrous leaflets, mp ~220° with decompn. Soluble in 400 parts water; more sol in hot water; easily sol in acids and alkalies with formation of salts; insol in alcohol, ether.

USE: As a reagent for aldehydes and ketones.

8585. Semotiadil. [116476-13-2] (2R)-2-[2-[3-[2-(1,3-Benzodioxol-5-yloxy)ethyl]methylamino]propoxy]-5-methoxyphenyl]-4-methyl-2H-1,4-benzothiazin-3(4H)-one; (+)-(R)-3,4-dihydro-2-[5-methoxy-2-[3-[N-methyl-[N-2-[3,4-(methylenedioxy)phenoxy]ethyl]amino]propoxy]phenyl]-4-methyl-3-oxo-2H-1,4-benzothiazine; sesamodil; DS-4823. $C_{29}H_{32}N_2O_6S$; mol wt 536.64. C 64.91%, H 6.01%, N 5.22%, O 17.89%, S 5.97%. Benzothiazine calcium antagonist. Prepn: J. Iwao et al., WO 8700838; eidem, US 4786635 (1987, 1988 both to Santen). Synthesis: M. Fujita et al., J. Med. Chem. 33, 1898 (1990). LC determn in plasma: K. Morishima et al., J. Chromatogr. 527, 381 (1990). Molecular conformation and crystal structure: A. Ota *et al.*, *Chem. Pharm. Bull.* **41**, 1681 (1993). Electrophysiology: N. Miyawaki *et al.*, *J. Cardiovasc. Pharmacol.* **16**, 769 (1990); and antiarrhythmic effects: *eidem*, *Drug Dev. Res.* **22**, 293 (1991). Antihypertensive effects in rats: A. Kanda, H. Hashimoto, *Jpn. J. Pharmacol.* **63**, 121 (1993). Mechanism of action study: K. Murakami *et al.*, *J. Cardiovasc. Pharmacol.* **25**, 262 (1995).

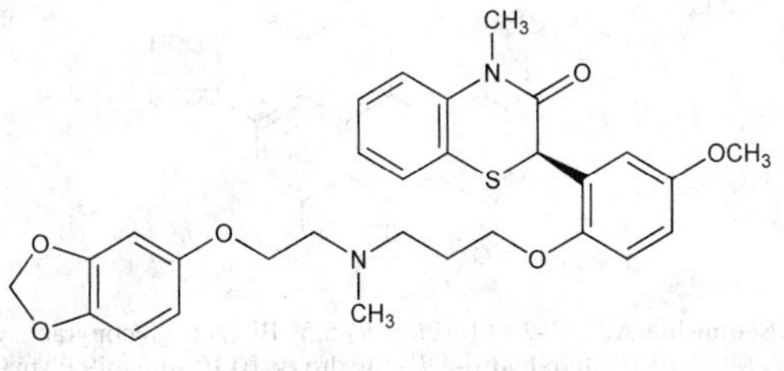

Fumarate. [116476-14-3] SD-3211. $C_{29}H_{32}N_2O_6S \cdot C_4H_4O_4$; mol wt 652.72. Crystals from EtOH, mp 134-135°. $[\alpha]_D^{25}$ +195° (in DMSO).

THERAP CAT: Antihypertensive; antianginal.

8586. Sempervirine. [6882-99-1] 2,3,4,13-Tetrahydro-1*H*-benz[*g*]indolo[2,3-*a*]quinolizin-6-ium; 3,4,5,6,14,15,20,21-octadehydroyohimbanium; sempervirene. $C_{19}H_{16}N_2$; mol wt 272.35. C 83.79%, H 5.92%, N 10.29%. Found in rhizome and roots of Carolina jessamine (yellow jessamine) *Gelsemium sempervirens* (L.) Ait. f., *Loganiaceae:* Stevenson, Sayre, *J. Am. Pharm. Assoc.* **4**, 60 (1915); **8**, 708 (1919); Hasenfratz, *Bull. Soc. Chim. Fr.* [4] **53**, 1084 (1933); Forsyth *et al.*, *J. Chem. Soc.* **1945**, 579; Prelog, *Helv. Chim. Acta* **31**, 588 (1948). Believed to exist primarily as inner salt depicted below: Woodward, Witkop, *J. Am. Chem. Soc.* **71**, 379 (1949). Synthesis: Woodward, McLamore, *ibid.* 379; Swan, *J. Chem. Soc.* **1958**, 2039; Ban, Seo, *Tetrahedron* **16**, 11 (1961); Potts, Mattingly, *J. Org. Chem.* **33**, 3985 (1968).

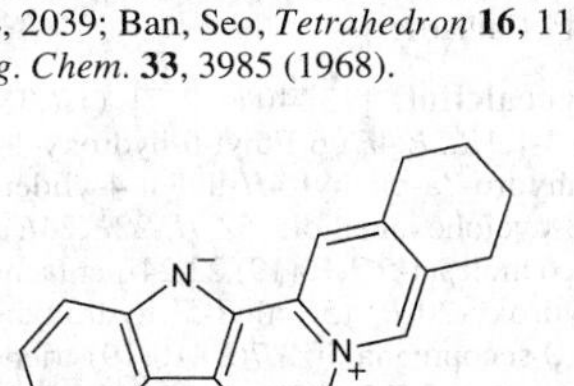

Monohydrate. Yellow needles from chloroform, mp 228°. Brownish-yellow leaflets from dil ethanol, mp 258°. uv max (ethanol): 243, 249, 297, 345, 387 nm (log ε 4.58, 4.57, 4.20, 4.26, 4.24). Dipole moment: 8.5 (dioxane); 7.5 (benzene). Soluble in alcohol, chloroform, pyridine. Slightly sol in acetone. Practically insol in ether, benzene.

Methochloride. $C_{20}H_{19}ClN_2$. Minute yellow needles from ethanol, mp 330-332°. Soluble in water giving yellow solns with purple fluorescence. uv max: 241, 292, 330, 395 nm (log ε 4.56, 4.20, 4.28, 4.22).

8587. Senecialdehyde. [107-86-8] 3-Methyl-2-butenal; 3,3-dimethylacrolein; 3-methylcrotonaldehyde; senecioaldehyde; β,β-dimethylacrolein. C_5H_8O; mol wt 84.12. C 71.39%, H 9.59%, O 19.02%. Prepn from isoamyl alcohol: Fischer *et al.*, *Ber.* **64**, 30 (1931); Fischer, Löwenberg, *Ann.* **494**, 272 (1932); *cf.* Burkhardt *et al.*, *GB 512465* (1939); Montavon, Saucy, *US 2902515* (1959 to Hoffmann-La Roche). Absorption spectrum: Forbes, Skilton, *J. Org. Chem.* **24**, 436 (1959).

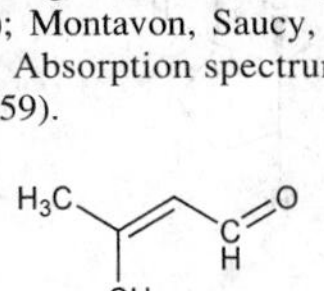

Liquid. Odor more pungent than that of isovaleraldehyde. Autooxidizes easily, but the pure substance can be stored under vacuum. d_4^{20} 0.8722; bp_{730} 133°; n_D^{20} 1.4526.

Caution: Irritant to mucous membranes. Narcotic in high concentrations.

8588. Senecic Acid. [13588-16-4] (2*R*,3*R*,5*Z*)-5-Ethylidene-2-hydroxy-2,3-dimethylhexanedioic acid. $C_{10}H_{16}O_5$; mol wt 216.23. C 55.55%, H 7.46%, O 37.00%. One of the most widely studied *necic acids* which are constituent acids of the hepatotoxic pyrrolizidine alkaloids and are isolated from these alkaloids by hydrolysis or by hydrogenolysis. Necic acids are C_5- to C_{10}-acids; many are γ- or δ-hydroxy acids and have structures that may be dissected into isoprene units. Synthesis of senecic acid: C. C. J. Culvenor, T. A. Geissman, *J. Am. Chem. Soc.* **83**, 1647 (1961). uv spectra of senecic acid and its isomers: V. Simanek *et al.*, *Collect. Czech. Chem. Commun.* **34**, 1832 (1969). Biosynthesis: D. H. G. Crout *et al.*, *J. Chem. Soc. Perkin Trans. 1* **1972**, 671. Comprehensive reviews on necic acids: F. L. Warren, *Fortschr. Chem. Org. Naturst.* **12**, 198-269 (1955); **24**, 329-406 (1966); D. J. Robins *ibid.* **41**, 115-203 (1982); F. L. Warren in *Alkaloids* Vol. **XII**, R. H. F. Manske, Ed. (Academic Press, New York, 1970) pp 245-331.

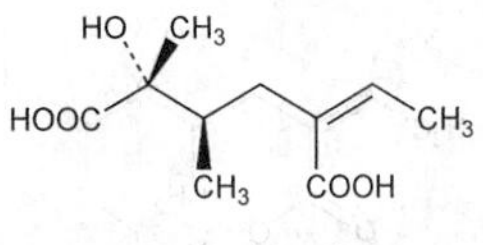

Needles from benzene or ethyl acetate, mp 145-146°. Sublimes readily at ~125°.

8589. Senecio. Golden ragwort; squaw weed; life root. Dried plant of *Senecio aureus* L., *Compositae. Habit.* Canada and Eastern U.S. *Constit.* The alkaloids, senecifoline, senecine, etc., resins.

THERAP CAT: Emmenagogue.

8590. Senecionine. [130-01-8] (3*Z*,5*R*,6*R*,14a*R*,14b*R*)-3-Ethylidene-3,4,5,6,9,11,13,14,14a,14b-decahydro-6-hydroxy-5,6-dimethyl[1,6]dioxacyclododecino[2,3,4-*gh*]pyrrolizine-2,7-dione; 12-hydroxysenecionan-11,16-dione; aureine. $C_{18}H_{25}NO_5$; mol wt 335.40. C 64.46%, H 7.51%, N 4.18%, O 23.85%. Hepatotoxic pyrrolizidine alkaloid from whole plant of *Senecio vulgaris* L., *Compositae* and several other *Senecio* spp. Isoln: Barger, Blackie, *J. Chem. Soc.* **1936**, 743. Structure: Adams, Govindachari, *J. Am. Chem. Soc.* **71**, 1953 (1949); Koekmoer, Warren, *J. Chem. Soc.* **1955**, 63. Conformation: Culvenor, *Tetrahedron Lett.* **1966**, 1091. Toxicity study: P. N. Harris *et al.*, *J. Pharmacol. Exp. Ther.* **75**, 69 (1942). Review of toxicology of senecionine and other pyrrolizidine alkaloids: E. K. McLean, *Pharmacol. Rev.* **22**, 429-483 (1970). Comprehensive reviews of pyrrolizidine alkaloids: L. Bull *et al.*, *The Pyrrolizidine Alkaloids* (North-Holland, Amsterdam, 1968) 293 pp; F. L. Warren in *The Alkaloids* vol. **12**, R. H. F. Manske, Ed. (Academic Press, New York, 1970) pp 245-331; D. J. Robins, *Fortschr. Chem. Org. Naturst.* **41**, 115-203 (1982).

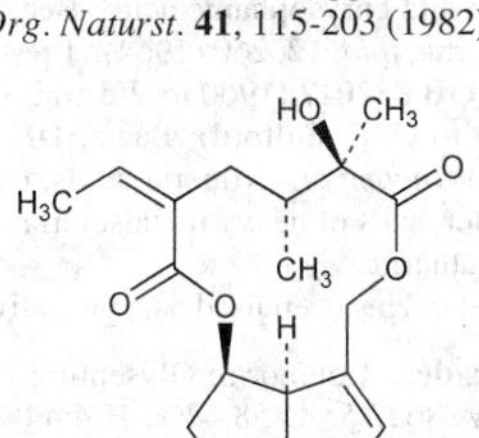

Bitter plates, mp 236°. $[\alpha]_D^{25}$ −55.1° (c = 0.034 in chloroform). Practically insol in water. Freely sol in chloroform; slightly sol in alcohol, ether. LD$_{50}$ i.v. in mice: 64.12 ±2.24 mg/kg (Harris).

C_{15}-*trans* **Isomer.** Integerrimine; squalidine; [15(20)*E*]-12-hydroxysenecionan-11,16,dione. Isolated from several *Senecio* and *Crotalaria* spp.: L. Bull *et al.*, *loc. cit.*; F. L. Warren, *loc. cit.* Identity of squalidine and integerrimine: A. G. Gonzalez, A. Calero, *Chem. Ind. (London)* **1958**, 126. mp 172-172.5°; $[\alpha]_D^{28}$ +4.3° (c = 0.8 in methanol).

8591. Seneciphylline. [480-81-9] (3*Z*,6*R*,14a*R*,14b*R*)-3-Ethylidene-3,4,5,6,9,11,13,14,14a,14b-decahydro-6-hydroxy-6-methyl-

5-methylene[1,6]dioxacyclododecino[2,3,4-*gh*]pyrrolizine-2,7-dione; 13,19-didehydro-12-hydroxysenecionan-11,16-dione; jacodine; α-longilobine. $C_{18}H_{23}NO_5$; mol wt 333.38. C 64.85%, H 6.95%, N 4.20%, O 24.00%. Hepatotoxic pyrrolizidine alkaloid, common constituent of *Senecio* species. Isoln from *Senecio platyphllus* DC. *Compositae:* A. Orechoff, W. Tiedebel, *Ber.* **68**, 650 (1935); from *S. jacobaea* L.: G. Barger, J. J. Blackie, *J. Chem. Soc.* **1937**, 584; from *Crotalaria juncea* L. *Leguminosae:* R. Adams, M. Gianturco, *J. Am. Chem. Soc.* **78**, 1919 (1956). Identity with α-longilobine: R. Adams, J. H. Looker, *ibid.* **73**, 134 (1951). Identity with jacodine: R. B. Bradbury, C. C. J. Culvenor, *Chem. Ind. (London)* **1954**, 1021. Structural study: R. Adams *et al.*, *J. Am. Chem. Soc.* **74**, 700 (1952). Revised structure: S. Masume, *Chem. Ind. (London)* **1959**, 21. Review and evaluation of toxicity and carcinogenicity studies: *IARC Monographs* **10**, 319-325, 333-342 (1976). Comprehensive reviews of seneciphylline and other pyrrolizidine alkaloids: L. Bull *et al.*, *The Pyrrolizidine Alkaloids* (North-Holland, Amsterdam, 1968) 293 pp; F. L.Warren in *The Alkaloids* **vol. 12**, R. H. F. Manske, Ed. (Academic Press, New York, 1970) pp 245-331.

Small rhombic platelets from hot alcohol or acetone, mp 217-218°. $[α]_D$ −128° (chloroform). Easily sol in chloroform, ethylene chloride; less sol in alc, acetone. Difficultly sol in ether, ligroin.

8592. Senega. Senega snakeroot; Seneca snakeroot; rattlesnake root. Dried root of *Polygala senega* L., *Polygalaceae. Habit.* North America. Used by native Americans as a treatment for snakebite. *Constit.* Triterpenoid saponins, 6-12%, with presenegenin as the main sapogenin; ~5% lipids; mono- and oligosaccharides; phenolic acid esters; traces of essential oil containing methyl salicylate which imparts a slight aromatic odor. Quantitative analysis of senega saponins: Kita *et al.*, *Yakugaku Zasshi* **89**, 1111 (1969). Brief review: M. Wichtl, N. G. Bisset, *Herbal Drugs and Phytopharmaceuticals*, English Ed. (CRC Press, Boca Raton, 1994) pp 384-385.

THERAP CAT: Expectorant.

8593. Senna. Dried leaflets of *Cassia senna* L. (*C. acutifolia* Delile), (Alexandria senna), or of *C. angustifolia* Vahl (India or Tinnevelly senna), *Leguminosae. Habit.* Egypt and neighboring districts, Southern India (Tinnevelly). *Constit.* Sennosides A and B; glucosides of rhein and chrysophanic acid: Becker, *Planta Med.* **7**, 390 (1959); Fairbairne, *ibid.* **12**, 260 (1964). Prepn of senna powder: Friedmann, Ryan, **GB 832017** (1960 to Westminster Labs.); Corran, **GB 852343** (1960 to C. E. Fulford); Baetz, **DE 1158211** (1963 to Ludwig Heumann). *Incompat:* Mineral acids, carbonates, infusion cinchona, lime water, salts of heavy metals, tartar emetic.

THERAP CAT: Cathartic.

THERAP CAT (VET): Has been used as a purgative.

8594. Sennosides. Colonorm; Glysennid; Nytilax; Pursennid. $C_{42}H_{38}O_{20}$; mol wt 862.75. C 58.47%, H 4.44%, O 37.09%. Anthraquinone glucosides found in senna in equal amts and in the rhubarbs where sennoside A predominates. Isoln from Tinnevelly senna (*Cassia angustifolia* Vahl, *Leguminosae):* Stoll *et al.*, *Helv. Chim. Acta* **32**, 1892 (1949); **GB 804232** (1958 to Byk-Gulden); Khorana, Sanghavi, *J. Pharm. Sci.* **53**, 110 (1964); Menssen *et al.*, **US 3517269** (1970 to Nattermann & Cie). Isoln from the rhizome of *Rheum palmatum* L.: Zwaving, *Planta Med.* **13**, 474 (1965). Isoln of sennoside A from rhubarb: Miyamoto *et al.*, *Yakugaku Zasshi* **87**, 1040 (1967). Sennosides B, C, D, E, and F have also been isolated from rhubarb. *See:* Oshio *et al.*, *Chem. Pharm. Bull.* **22**, 823 (1974). Structure: Stoll *et al.*, *Helv. Chim. Acta* **33**, 313 (1950). Sennoside A is built up from the dextrorotatory aglucon sennidin A and D-glucose. Sennoside B is built up from the intramolecularly compensated *meso*-sennidin B and D-glucose. *Review:* Stoll, Becker,

Fortschr. Chem. Org. Naturst. **7**, 248 (1950). Transport and mechanism of action studies: Dobbs *et al.*, *Farmaco Ed. Sci.* **30**, 147 (1975).

Sennoside A. [81-27-6] (9*R*,9′*R*)-5,5′-Bis(β-D-glucopyranosyloxy)-9,9′,10,10′-tetrahydro-4,4′-dihydroxy-10,10′-dioxo[9,9′-bianthracene]-2,2′-dicarboxylic acid. Rectangular yellow plates from dil acetone, dec 200-240°. $[α]_D^{20}$ −164° (c = 0.1 in 60% acetone); $[α]_D^{20}$ −147° (c = 0.1 in 70% acetone); $[α]_D^{20}$ −24° (c = 0.2 in 70% dioxane). Insol in water, benzene, ether, chloroform. Sparingly sol in methanol, Carbitol, acetone, dioxane. The soly increases tremendously in water-miscible organic solvents contg an optimum of 30% w/w of water. Sol in aq solns of sodium bicarbonate. Sennoside A can be slowly isomerized to B in NaHCO₃ soln at 80°.

Sennoside B. [128-57-4] (9*R*,9′*S*)-5,5′-Bis(β-D-glucopyranosyloxy)-9,9′,10,10′-tetrahydro-4,4′-dihydroxy-10,10′-dioxo[9,9′-bianthracene]-2,2′-dicarboxylic acid. Light yellow prisms from dil acetone; fine needles from water, dec 180-186°. $[α]_D^{20}$ −100° (c = 0.2 in 70% acetone); $[α]_D^{20}$ −67° (c = 0.4 in 70% dioxane). uv max (0.5% NaHCO₃): 270, 308, 355 nm. Has the same soly characteristics as sennoside A, but is more sol. Can be recrystallized from large amounts of hot water. *Keep protected from air and light.*

THERAP CAT: Cathartic.

8595. Seocalcitol. [134404-52-7] (1*R*,3*S*,5*Z*)-5-[(2*E*)-2-[(1*R*,3a*S*,7a*R*)-1-[(1*R*,2*E*,4*E*)-6-Ethyl-6-hydroxy-1-methyl-2,4-octadien-1-yl]octahydro-7a-methyl-4*H*-inden-4-ylidene]ethylidene]-4-methylene-1,3-cyclohexanediol. (5*Z*,7*E*,22*E*,24*E*)-24a,26a,27a-trihomo-9,10-secocholesta-5,7,10(19),22,24-pentaene-1α,3β,25-triol; 1(*S*),3(*R*)-dihydroxy-20(*R*)-(5′-ethyl-5′-hydroxyhepta-1′(*E*),3′(*E*)-dien-1′-yl)-9,10-secopregna-5(*Z*),7(*E*),10(19)-triene; EB-1089. $C_{30}H_{46}O_3$; mol wt 454.70. C 79.25%, H 10.20%, O 10.56%. Antiproliferative vitamin D analog with reduced effects on calcium metabolism. Prepn: E. Binderup, M. J. Calverley, **WO 9100855**; **US 5190935** (1991, 1993 both to Leo Pharm.). LC-MS-MS determn in serum: A.-M. Kissmeyer *et al.*, *J. Chromatogr. B* **740**, 117 (2000). Pharmacokinetics and metabolism: A.-M. Kissmeyer, J. T. Mortensen, *Xenobiotica* **30**, 815 (2000). Mechanism of action study: G. A. DeMasters *et al.*, *J. Steroid Biochem. Mol. Biol.* **92**, 365 (2004). Clinical evaluation in pancreatic cancer: T. R. J. Evans *et al.*, *Br. J. Cancer* **86**, 680 (2002); in hepatocellular carcinoma: K. Dalhoff *et al.*, *Br. J. Cancer* **89**, 252 (2003). Review of pharmacology and clinical experience: C. M. Hansen *et al.*, *Curr. Pharm. Des.* **6**, 803-828 (2000).

Needles from methyl formate-hexane, mp 123-125°. uv max: 232, 264 nm (ε 43745, 18060).

THERAP CAT: Antineoplastic.

8596. Sepia. Cuttle-fish bone. Calcareous substance found under the skin of the back of *Sepia officinalis* L., *Cephalopoda*. *Habit.* Mediterranean Sea, Atlantic and Pacific Oceans. *Constit.* Calcium carbonate and phosphate, gluten.

USE: As a polishing agent and in tooth powders.

8597. Sepiomelanin. The eumelanin isolated from ink sacs of the cuttlefish, *Sepia officinalis* L., *Cephalopoda*. Present in the raw ink as a suspension of small, dark granules in a colorless plasma. Structure is a macromolecule or probably a mixture of polyacid polymers in which the predominant chemical unit is of the indole type. Structure studies: Piattelli, Nicolaus, *Tetrahedron* **15**, 66 (1961); Piattelli *et al., ibid.* **19**, 2061 (1963). *Review:* R. A. Nicolaus, *Melanins* (Hermann, Paris, 1968) pp 68-91.

In the dry form, an amorphous, hygroscopic, black powder. Insol in all solvents.

USE: Pigment.

8598. Seratrodast. [112665-43-7] ζ-(2,4,5-Trimethyl-3,6-dioxo-1,4-cyclohexadien-1-yl)benzeneheptanoic acid; (±)-7-(3,5,6-trimethyl-1,4-benzoquinon-2-yl)-7-phenylheptanoic acid; AA-2414; Abbott 73001; Bronica. $C_{22}H_{26}O_4$; mol wt 354.45. C 74.55%, H 7.39%, O 18.06%. Thromboxane A_2-receptor antagonist. Prepn: S. Terao, Y. Maki, **EP 171251**; *eidem,* **US 5180742** (1986, 1993 both to Takeda); M. Shiraishi *et al., J. Med. Chem.* **32**, 2214 (1989). Effect on thromboxane A_2/prostaglandin endoperoxidase-receptor: Y. Imura *et al., Jpn. J. Pharmacol.* **52**, 35 (1990). Clinical effect in asthma: M. Fujimura *et al., J. Allergy Clin. Immunol.* **87**, 23 (1991). Clinical pharmacokinetics: Z. Hussein *et al., Clin. Pharmacol. Ther.* **55**, 441 (1994).

Crystals from ethanol, mp 128-129°.

THERAP CAT: Antiasthmatic.

8599. Serine. [56-45-1] L-Serine; Ser; S; 2-amino-3-hydroxypropanoic acid; β-hydroxyalanine; (S)-2-amino-3-hydroxypropanoic acid; α-amino-β-hydroxypropionic acid. $C_3H_7NO_3$; mol wt 105.09. C 34.29%, H 6.71%, N 13.33%, O 45.67%. A non-essential amino acid for human development. A major intracellular source of one carbon units for *de novo* purine synthesis. Found in the active site of serine proteases such as trypsin. Isoln from the silk protein, sericine: E. Cramer, *Prakt. Chem.* **96**, 76 (1865). Structure and synthesis of DL-form: E. Fischer, H. Leuchs, *Ber.* **35**, 3787 (1902). Early chemistry and biochemistry: *Amino Acids and Proteins,* D. M. Greenberg, Ed. (Charles C. Thomas, Springfield, IL, 1951) 950 pp., *passim;* J. P. Greenstein, M. Winitz, *Chemistry of the Amino Acids* **vols 1-3** (John Wiley and Sons, Inc., New York, 1961) pp. 2202-2237, *passim.* Enzymatic determn in biological samples: R. D. Hurst *et al., Anal. Biochem.* **117**, 339 (1981). Synthetic reviews: K. Toi in *Synthetic Production and Utilization of Amino Acids,* T. Kaneko *et al.,* Eds. (Halsted, New York, 1974) pp 187-195; of DL-forms: L. Bassignani *et al., Ber.* **112**, 148-160 (1979). Review of role in enzyme active sites: B. S. Hartley, *Ann. N.Y. Acad. Sci.* **227**, 438-445 (1974); J. Lamotte-Brasseur *et al., Biotechnol. Genet. Eng. Rev.* **12**, 189-230 (1994). Review of metabolism: K. Snell, D. A. Fell, *Adv. Enzyme Regul.* **30**, 13-32 (1990).

Hexagonal plates or prisms. Sweetish taste, insipid aftertaste. Dec 228°. Sublimes at 150° in high vac (10⁻⁴ mm Hg). $[\alpha]_D^{20}$ −6.83° (c = 10.41); $[\alpha]_D^{25}$ +14.95° (c = 9.34 in 1N HCl). Absorption spectrum: *Z. Physiol. Chem.* **176**, 257 (1928). Sol in water. Practically insol in abs alc, ether. Insol in most common neutral solvents.

DL-Form. [302-84-1] Monoclinic prismatic leaflets from water. d 1.537. Dec 246° (closed capillary, bath preheated to 225°). Sublimes at 150° in high vac (10⁻⁴ mm Hg). pK_1 2.21; pK_2 9.15. Soly in water (g/l) at 0° = 22.04; at 25° = 50.23; at 50° = 103; at 75° = 192; at 100° = 322. Insol in common neutral solvents.

D-Form. [312-84-5] Presence in rat brain: A. Hashimoto *et al., FEBS Lett.* **296**, 33 (1992). As synaptic modulator: M. J Schell *et al., Proc. Natl. Acad. Sci. USA* **92**, 3948 (1995).

8600. Sermorelin. [86168-78-7] 29-L-Argininamide-1-29-somatoliberin (human pancreatic islet); human growth hormone-releasing factor(1-29)amide; human pancreatic somatoliberin(1-29)-amide; GRF(1-29)NH_2; hpGRF(1-29)NH_2; SM-8144; Geref; Groliberin. $C_{149}H_{246}N_{44}O_{42}S$; mol wt 3357.93. C 53.30%, H 7.38%, N 18.35%, O 20.01%, S 0.95%. Amidated fragment of human somatoliberin, *q.v.* Characterization of *in vitro* activity: J. Rivier *et al., Nature* **300**, 276 (1982). Prepn: N. Ling *et al., Biochem. Biophys. Res. Commun.* **123**, 854 (1984); J. E. F. Rivier *et al.,* **US 4703035** (1987 to Salk Inst. Biol. Stud.). Clinical pharmacology: M. Losa *et al., Klin. Wochenschr.* **62**, 1140 (1984). Clinical evaluation in growth hormone deficiency: M. B. Ranke *et al., Eur. J. Pediatr.* **145**, 485 (1986); R. Hümmelink *et al., Acta Paediatr. Scand.* **Suppl. 331**, 48 (1987). Field trial in pigs and dairy heifers: D. Petitclerc *et al., J. Anim. Sci.* **65**, 996 (1987).

¹
Tyr–Ala–Asp–Ala–Ile–Phe–Thr–Asn–Ser–Tyr–Arg–Lys–Val–Leu–Gly

29
NH₂Arg–Ser–Met–Ile–Asp–Gln–Leu–Leu–Lys–Arg–Ala–Ser–Leu–Gln

$[\alpha]_D^{20}$ −63.1° (c = 1 in 30% acetic acid).

THERAP CAT: Growth hormone releasing factor.

8601. Serotonin. [50-67-9] 3-(2-Aminoethyl)-1H-indol-5-ol; 5-hydroxytryptamine; 3-(β-aminoethyl)-5-hydroxyindole; 5-hydroxy-3-(β-aminoethyl)indole; enteramine; thrombocytin; thrombotonin; 5-HT. $C_{10}H_{12}N_2O$; mol wt 176.22. C 68.16%, H 6.86%, N 15.90%, O 9.08%. Vasoactive amine found in tissues and fluids of vertebrates and invertebrates. Extraction of enteramine from rabbit tissue, pharmacology: V. Erspamer, *Arch. Exp. Pathol. Pharmakol.* **196**, 343 (1940). Isoln of 5-HT from beef serum: M. M. Rapport *et al., Science* **108**, 329 (1948); *eidem, J. Biol. Chem.* **176**, 1237, 1243 (1948); M. M. Rapport, *ibid.* **180**, 961 (1949). Identity with enteramine, the enterochromaffin cell hormone: V. Erspamer, B. Asero, *Nature* **169**, 800 (1952). Ultrastructural localization in enterochromaffin cells: R. D. Dey, J. Hoffpauir, *J. Histochem. Cytochem.* **32**, 661 (1984). Synthesis with 5-benzyloxyindole: M. E. Speeter *et al., J. Am. Chem. Soc.* **73**, 5514 (1951); K. E. Hamlin, **US 2715129** (1955 to Abbott). Alternate routes: R. Justoni, R. Pessina, **US 2947757** (1960 to Vismara). Pharmacology: G. Reid, M. Rand, *Nature* **169**, 801 (1952). Existence of multiple serotonin receptors: S. J. Peroutka, S. H. Snyder, *Mol. Pharmacol.* **16**, 687 (1979); G. Engel *et al., Arch. Pharmacol.* **324**, 116 (1983); and effect on smooth muscle: W. Feniuk *et al., Eur. J. Pharmacol.* **96**, 71 (1983). Extraneural synthesis in CNS: C. Maruki *et al., J. Neurochem.* **43**, 316 (1984). Activity as neurotransmitter: S. G. Griffith *et al., Brain Res.* **247**, 388 (1982). Role in hemostasis: F. De Clerck *et al., Agents Actions* **15**, 627 (1984). Series of articles on role in hypertension, Raynaud's phenomenon, and platelet activation: *J. Cardiovasc. Pharmacol.* **7**, Suppl. 7, 1-182 (1985). Review of role in migraine: E. T. MacKenzie *et al., Cephalalgia* **5**, 69-78 (1985). *Review:* V. Erspamer in E. Jucker, *Prog. Drug Res.* **3**, 151-367 (1961). Books: *Serotonin in Health and Disease* vols. I-V, W. B. Essman, Ed. (Spectrum, New York, 1978); *Adv. Exp. Med.* **vol. 133**, entitled "Serotonin: Current Aspects of Neurochemistry and Function," B. Haber, Ed. (1981) 824 pp; *Serotonin and the Cardiovascular System,* P. M. Vanhoutte, Ed. (Raven Press, New York, 1985) 280 pp. *See also* Bufotenine and Ketanserin.

Hydrochloride. $C_{10}H_{12}N_2O.HCl$. Hygroscopic crystals, sensitive to light, mp 167-168°. Water sol. Aq solns are stable at pH 2-6.4.

Complex with creatinine sulfate, monohydrate. Antemovis. $C_{14}H_{21}N_5O_6S.H_2O$; mol wt 405.43. Plates, 215° dec. uv max (water at pH 3.5): 275 nm (ε 15000). $pK_1' = 4.9$; $pK_2' = 9.8$. pH of 0.01 molar aq soln: 3.6. Sol in glacial acetic acid. Very sparingly sol in methanol, 95% ethanol. Insol in abs ethanol, acetone, pyridine, chloroform, ethyl acetate, ether, benzene.

8602. Serpentaria. Snakeroot; snakeweed; sangrel; birthwort. Dried rhizome and roots of *Aristolochia serpentaria* L. (Virginia snakeroot), and of *Aristolochia reticulata* Nutt. (Texas snakeroot), *Aristolochiaceae*. *Habit.* U.S. *Constit.* Aristolochic acid, volatile oil (~1%), resin, tannin.

THERAP CAT: Bitter tonic.

8603. Serpentine (Alkaloid). [18786-24-8] (19α)-3,4,5,6,-16,17-Hexadehydro-16-(methoxycarbonyl)-19-methyloxayohimbanium. $C_{21}H_{20}N_2O_3$; mol wt 348.40. C 72.40%, H 5.79%, N 8.04%, O 13.78%. From roots of *Rauwolfia serpentina* (L.) Benth., *Apocynaceae*: Schlittler, Schwarz, *Helv. Chim. Acta* **33**, 1463 (1950). Structure: Schlittler *et al., ibid.* **37**, 1912 (1954); Klohs *et al., J. Am. Chem. Soc.* **76**, 1332 (1954); Wenkert, Roychaudhuri, *ibid.* **80**, 1613 (1958). Stereoisomer of alstonine, *q.v.* Stereochemistry: Fritz, *Ann.* **655**, 148 (1962); Shamma, Richey, *J. Am. Chem. Soc.* **85**, 2507 (1963).

Solvated, yellow rods or leaflets from abs ethanol, mp 158° (air-dried), mp 175° (after drying at 120° in a high vacuum, but still not entirely solvent-free). Dec and turns red on drying at 150°. $[\alpha]_D^{25}$ +292° (c = 0.27 in methanol); $[\alpha]_D^{25}$ +267° (c = 0.21 in ethanol). uv max (ethanol): 252, 308, 370 nm (log ε 4.49, 4.30, 3.61). Freely sol in methanol. Sol in 10% acetic acid, in other organic solvents.

Nitrate. $C_{21}H_{20}N_2O_3.HNO_3$. Yellow crystals, mp 170-172°. Only slightly sol in water.

Hydrochloride monohydrate. $C_{21}H_{20}N_2O_3.HCl.H_2O$. Yellow crystals, mp 246-248°. $[\alpha]_D^{23}$ +178°. Sol in water.

Perchlorate. $C_{21}H_{20}N_2O_3.HClO_4$. Yellow crystals, mp 255-256°. $[\alpha]_D^{22}$ +185° (c = 0.5 in acetone). uv max (methanol): 252, 306-308, 365-370 nm (log ε 4.54, 4.35, 3.68).

8604. Sertaconazole. [99592-32-2] 1-[2-[(7-Chlorobenzo[*b*]-thien-3-yl)methoxy]-2-(2,4-dichlorophenyl)ethyl]-1*H*-imidazole; (±)-1-[2,4-dichloro-β-[(7-chlorobenzo[*b*]thien-3-yl)methoxy]phenethyl]imidazole; 7-chloro-3-[1-(2,4-dichlorophenyl)-2-(1*H*-imidazol-1-yl)ethoxy-methyl]benzo[*b*]thiophene; FI-7045. $C_{20}H_{15}Cl_3$-N_2OS; mol wt 437.76. C 54.87%, H 3.45%, Cl 24.29%, N 6.40%, O 3.65%, S 7.32%. Prepn: R. Foguet *et al.*, **EP 151477**; *idem*, **US 5135943** (1985, 1992 both to Ferrer). Synthesis, antifungal activity, pharmacology, toxicology and clinical evaluations: Arzneimittel-Forsch. **42**, 689-774 (1992). Physical props: C. Albet *et al., ibid.* 695. HPLC determn: *eidem, J. Pharm. Biomed. Anal.* **10**, 205 (1992).

mp 146-147°.

Nitrate. [99592-39-9] FI-7056; Dermofix; Zalain. $C_{20}H_{15}Cl_3$-$N_2OS.HNO_3$; mol wt 500.78. Odorless, white crystalline powder, mp 158-160°. Fairly sol in ethanol (1.7%); chloroform (1.5%); slightly sol in acetone (0.95%); very slightly sol in *n*-octanol (0.069%). Practically insol in water (<0.01%). pKb: 7.26. uv max (methanol): 302.3 ($A_{1cm}^{1\%}$ 79.8), 292.9, 260.3 nm.

THERAP CAT: Antifungal.

8605. Sertindole. [106516-24-9] 1-[2-[4-[5-Chloro-1-(4-fluorophenyl)-1*H*-indol-3-yl]-1-piperidinyl]ethyl]-2-imidazolidinone; 5-chloro-1-(4-fluorophenyl)-3-[1-[2-(2-imidazolidinon-1-yl)-ethyl]-4-piperidyl]-1*H*-indole; Lu-23-174; Serdolect. $C_{24}H_{26}$-ClFN$_4$O; mol wt 440.95. C 65.37%, H 5.94%, Cl 8.04%, F 4.31%, N 12.71%, O 3.63%. Combined dopamine (D$_2$) and serotonin (5-HT$_2$) receptor antagonist. Prepn: J. K. Perregaard, **EP 200322**; *idem*, **US 4710500** (1986, 1987 both to Lundbeck); *idem et al., J. Med. Chem.* **35**, 1092 (1992). Neuropharmacology and receptor binding study: C. Sanchez *et al., Drug Dev. Res.* **22**, 239 (1991). Reviews of clinical efficacy and safety: J. M. Kane, *Int. Clin. Psychopharmacol.* **13**, Suppl. 3, S59-S64 (1997); A. S. Hale, *ibid.* S65-S70.

Crystals from acetone, mp 154-155°. Second crystal modification from 2-propanol + ethyl acetate, mp 166°.

THERAP CAT: Antipsychotic.

8606. Sertraline. [79617-96-2] (1*S*,4*S*)-4-(3,4-Dichlorophenyl)-1,2,3,4-tetrahydro-*N*-methyl-1-naphthalenamine. $C_{17}H_{17}Cl_2N$; mol wt 306.23. C 66.68%, H 5.60%, Cl 23.15%, N 4.57%. Selective serotonin reuptake inhibitor (SSRI). Prepn: W. M. Welch *et al.*, **EP 30081**; *eidem*, **US 4536518** (1981, 1985 both to Pfizer); and stereospecific activity: W. M. Welch *et al., J. Med. Chem.* **27**, 1508 (1984). GC determn in plasma: H. G. Fouda *et al., J. Chromatogr.* **417**, 197 (1987). CE determn of stereoisomers: M. X. Zhou, J. P. Foley, *J. Chromatogr. A* **1052**, 13 (2004). Comprehensive description: B. M. Johnson, P.-T. L. Chang, *Anal. Profiles Drug Subs. Excip.* **24**, 443-486 (1996). Review of pharmacology and clinical efficacy in depression and obsessive-compulsive disorder: D. Murdoch, D. McTavish, *Drugs* **44**, 604-624 (1992); of clinical efficacy in post-traumatic stress disorder: A. C. Schwartz, B. O. Rothbaum, *Expert Opin. Pharmacother.* **3**, 1489-1499 (2002); of clinical pharmacokinetics: C. L. DeVane *et al., Clin. Pharmacokinet.* **41**, 1247-1266 (2002).

Hydrochloride. [79559-97-0] CP-51974-1; Lustral; Zoloft. $C_{17}H_{17}Cl_2N\cdot HCl$; mol wt 342.69. Crystals, mp 243-245°. $[\alpha]_D^{23}$ +37.9° (c = 2 in methanol). Soly room temp. (mg/ml): water 3.8; 0.1N HCl 0.51; 0.1N NaOH 0.002; ethanol 15.7; isopropyl alcohol 4.3; chloroform 110; acetone 1.1; N,N-dimethylformamide 88; dimethylsulfoxide 147; ethyl acetate 0.20; acetonitrile 0.85; methanolic 0.1N HCl 47; chloroform/methanol (1:1) 134. pKa (water): 9.48 ± 0.04. pKa (methanol:water, 40:60 v/v): 8.6. pKa (ethanol:water, 1:1 v/v): 8.5. d 1.37.

THERAP CAT: Antidepressant; antiobsessional.

THERAP CAT (VET): In treatment of canine and feline behavioral disorders.

8607. Serum Albumin. [9048-46-8] Blood serum albumin. Most abundant of the plasma proteins (comprising ~60%). Structure characterized as a single, carbohydrate-free polypeptide chain consisting of 580-585 amino acid residues; stabilized by seventeen S-S bridges. Main biological functions include maintenance of osmotic pressure and the transport of sparingly soluble physiological and pharmacological substances, esp long-chain fatty acids and bilirubin. Trace amounts present in normal urine, excreted as the major protein constit in the condition of proteinuria. Review of determn in serum and plasma: P. G. Hill, *Ann. Clin. Biochem.* **22**, 565-578 (1985); of biosynthesis and biological function: M. A. Rothschild *et al.*, *Hepatology* **8**, 385-401 (1988); of structural studies: D. C. Carter, J. X. Ho, *Adv. Protein Chem.* **45**, 153-203 (1994).

Human serum albumin. HSA; Albuminar; Buminate; Pro-Bumin. Review of biochemistry and clinical experience: *Proceedings of the Workshop on Albumin*, J. T. Sgouris, A. René, Eds. (DHEW Publication No. (NIH) 76-925, 1975) 355 pp; of therapeutic use: G. E. Hastings, P. G. Wolf, *Arch. Fam. Med.* **1**, 281-287 (1992). Mol wt 65,000. Net charge per molecule (pH 7.4) −19. High solubility at neutrality: 30% solns prepd in water, 40% solns prepd in methanol or ethanol. Below pH 3, sol in 80-100% methanol, ethanol or acetone. Low viscosity. Stable, even at high temperatures.

^{131}I-labelled form. Iodinated (^{131}I) human serum albumin; ^{131}I HSA; Albumotope-^{131}I. Normal human serum albumin mildly iodinated with radioactive iodine (^{131}I) which has a half-life of 8 days, and emits beta and gamma rays. Contains not more than one atom of iodine per molecule of albumin. Physical and chemical properties essentially the same as those of albumin. Stable at 10° for at least 3 weeks. Supplied for injection in an isotonic sodium chloride soln.

Sonicated human serum albumin. [150103-82-5] Albunex. Suspension of air filled albumin microspheres in a 5% (w/v) human albumin soln. Prepn: M. W. Keller *et al.*, *J. Am. Coll. Cardiol.* **12**, 1039 (1988); D. Cerny *et al.*, **EP 359246**; *eidem*, **US 4957656** (both 1990 to Molecular Biosystems). Pharmacology: M. W. Keller *et al.*, *J. Am. Soc. Echocardiogr.* **2**, 48 (1989). Diagnostic use in echocardiography: L. J. Crouse *et al.*, *J. Am. Coll. Cardiol.* **22**, 1494 (1993). Physical and biochemical properties: C. Christiansen *et al.*, *Biotechnol. Appl. Biochem.* **19**, 307 (1994). *Review:* N. Sponheim *et al.*, *IEEE Conf. Publ., Acoustic Sensing and Imaging* **369**, 103-108 (1993). pH 6.79-6.86. Osmolality 253-264 mmol/kg. Viscosity 1.44.

THERAP CAT: Plasma volume expander; treatment of hypoproteinemia. ^{131}I-labelled form as diagnostic aid (blood volume determination). Sonicated form as diagnostic aid (ultrasound contrast agent).

8608. Sesame Oil. Benne oil; teel oil; gingilli oil. From the seeds of cultivated varieties of *Sesamum indicum* L., *Pedaliaceae. Constit.* Olein, stearin, palmitin, myristin, linolein, sesamin, sesamolin. *Review:* Budowski, Markley, *Chem. Rev.* **48**, 125 (1951).

Pale yellow oil; almost odorless; bland taste. d 0.916-0.920. Solidif about −5°. $[\alpha]_D^{25}$ +1 to +9°. n_D^{40} 1.4650-1.4665. Sapon no. 188-193. Iodine no. 103-122. Miscible with ether, chloroform, solvent hexane, carbon disulfide. Soluble in petr ether; slightly sol in alc. Insol in water.

USE: Manuf oleomargarine, cosmetics, iodized oil, etc. Pharmaceutic aid (solvent).

THERAP CAT (VET): Formerly as emollient, pediculicide.

8609. Sesamin. [607-80-7] (1S,3aR,4S,6aR)-5,5'-(Tetrahydro-1H,3H-furo[3,4-c]furan-1,4-diyl)bis-1,3-benzodioxole; tetrahydro-1,4-bis[3,4-(methylenedioxy)phenyl]-1H,3H-furo[3,4-c]furan; 2,6-bis(3,4-methylenedioxyphenyl)-3,7-dioxabicyclo[3.3.0]octane. $C_{20}H_{18}O_6$; mol wt 354.36. C 67.79%, H 5.12%, O 27.09%. Isolated from the bark of various *Fagara* species; from sesame oil: Bertram *et al.*, *Biochem. Z.* **197**, 1 (1928); from fruit of *Piper lowong* Blume, *Piperaceae:* Peinemann, *Arch. Pharm.* **234**, 238 (1896). Structure: Cohen, *Rec. Trav. Chim.* **57**, 653 (1938). Synthesis of *dl*-form: Beroza, Schechter, *J. Am. Chem. Soc.* **78**, 1242 (1956); Freudenberg, *Naturwissenschaften* **43**, 16 (1956). Identity of *dl*-form with *fagarol*: Carnmalm, *Acta Chem. Scand.* **9**, 1111 (1955); identity of *d*-form with *pseudocubebin*: *idem, ibid.* **10**, 134 (1956). Stereochemistry of the *d*-isomer: Freudenberg, Sidhu, *Ber.* **94**, 851 (1961); Becker, Beroza, *Tetrahedron Lett.* **1962**, 157. *Review:* Budowski, *J. Am. Oil Chem. Soc.* **41**, 280 (1964).

(+)-Sesamin

d-Form. Needles from ethanol, mp 122-123°. $[\alpha]_D^{20}$ +64.5° (c = 1.75 in chloroform).

dl-Form. Crystals from ethanol, mp 125-126°. Freely sol in chloroform, benzene, acetic acid, acetone. Practically insol in water, alkaline solns and hydrochloric acid.

USE: Insecticide synergist.

8610. Sesamolin. [526-07-8] 5-[(1S,3aR,4R,6aR)-4-(1,3-Benzodioxolol-5-yloxy)tetrahydro-1H,3H-furo[3,4-c]furan-1-yl]-1,3-benzodioxole; tetrahydro-1-[3,4-(methylenedioxy)phenoxy]-4-[3,4-(methylenedioxy)phenyl]-1H,3H-furo[3,4-c]furan. $C_{20}H_{18}O_7$; mol wt 370.36. C 64.86%, H 4.90%, O 30.24%. Constituent of sesame seeds, the seeds of *Sesamum indicum* L., *Pedaliaceae.* Isoln: Canzoneri, Perciabosco, *Gazz. Chim. Ital.* **33**, II, 253 (1907). Structure and synthesis: J. Boseken, W. D. Cohen, *Rec. Trav. Chim.* **55**, 815 (1936); E. Haslam, R. D. Haworth, *J. Chem. Soc.* **1955**, 827. Stereochemistry: E. Haslam, *ibid.* **1970**, 2332.

White plates from ethanol, mp 93-94°. $[\alpha]_D^{20}$ +212° (chloroform).

USE: Synergist for pyrethrum insecticides.

8611. Setastine. [64294-95-7] 1-[2-[1-(4-Chlorophenyl)-1-phenylethoxy]ethyl]hexahydro-1H-azepine; 1-[2-[(p-chloro-α-methyl-α-phenylbenzyl)oxy]ethyl]hexahydro-1H-azepine; 1-[2-(4-chloro-α-methyl-α-phenylbenzyloxy)ethyl]azepam; N-[2-(α-methyl-p-chlorobenzhydryloxy)ethyl]hexamethyleneimine; N-[2-[(4-

chloro-α-methyl-α-phenylbenzyl)oxy]ethyl]hexamethylenimine. $C_{22}H_{28}ClNO$; mol wt 357.92. C 73.83%, H 7.89%, Cl 9.90%, N 3.91%, O 4.47%. Nonsedating-type histamine H_1-receptor antagonist; structurally similar to clemastine, *q.v.* Prepn: I. Beck *et al.*, **DE 2528194** (1976 to EGYT), *C.A.* **85**, 46439h (1976). *In vitro* antihistaminic activity: A. Németh, Z. Huszti, *Agents Actions* **23**, 194 (1988). Clinical pharmacology: Z. Vezekényi *et al.*, *Acta Med. Hung.* **44**, 55 (1987).

Hydrochloride. [59767-13-4] EGIS-2062; EGYT-2062; Loderix. $C_{22}H_{28}ClNO.HCl$; mol wt 394.38.
THERAP CAT: Antihistaminic.

8612. Sethoxydim. [74051-80-2] 2-[1-(Ethoxyimino)butyl]-5-[2-(ethylthio)propyl]-3-hydroxy-2-cyclohexen-1-one; BAS-9052; NP-55; Poast. $C_{17}H_{29}NO_3S$; mol wt 327.48. C 62.35%, H 8.93%, N 4.28%, O 14.66%, S 9.79%. Selective post-emergence herbicide. Prepn: **JP Kokai 81 75408** (1981 to Nippon Soda), *C.A.* **95**, 110175e (1981). Activity: C. M. Knott, *Proc. Br. Crop Prot. Conf. - Weeds* **1980**, 39; G. H. Ingram *et al.*, *ibid.* 91.

USE: Herbicide.

8613. Sevelamer. [52757-95-6] 2-Propen-1-amine polymer with 2-(chloromethyl)oxirane; allylamine polymer with 1-chloro-2,3-epoxypropane; allylamine-epichlorohydrin copolymer; poly(allylamine-co-*N*,*N*'-diallyl-1,3-diamino-2-hydroxypropane). Polymeric non-absorbed phosphate binder consisting of polyallylamine crosslinked with epichlorohydrin to form a hydrogel where 40% of the amines are protonated. Binds dietary phosphate leading to increased fecal excretion, decreased absorption and decreased serum phosphorous levels. Prepd not claimed: S. R. Holmes-Farley *et al.*, **WO 9505184**; *eidem*, **US 5496545** (1995, 1996 both to GelTex). Mechanism of action: S. R. Holmes-Farley *et al.*, *J. Macromol. Sci. Pure Appl. Chem.* **A36**, 1085 (1999). Phosphate binding capacity: J. R. Mazzeo *et al.*, *J. Pharm. Biomed. Anal.* **19**, 911 (1999). Clinical studies in end stage renal disease: E. A. Slatopolsky *et al.*, *Kidney Int.* **55**, 299 (1999). Review of clinical applications: J. Henderson, P. Altmann, *Nephron Clin. Pract.* **94**, 53-58 (2003).

Carbonate. [845273-93-0] Renvela. Insol in water. Hygroscopic.
Hydrochloride. [152751-57-0] PB-94; GT16-026A; Renagel. Insol in water. Hydrophilic.
THERAP CAT: Antihyperphosphatemic.

8614. Sevoflurane. [28523-86-6] 1,1,1,3,3,3-Hexafluoro-2-(fluoromethoxy)propane; fluoromethyl 2,2,2-trifluoro-1-(trifluoromethyl)ethyl ether; fluoromethyl 1,1,1,3,3,3-hexafluoro-2-propyl ether; SevoFlo; Sevofrane; Sevorane; Ultane. $C_4H_3F_7O$; mol wt 200.06. C 24.01%, H 1.51%, F 66.47%, O 8.00%. Rapid-acting

inhalational anesthetic. Prepn: B. M. Regan, J. C. Longstreet, **DE 1954268**; *eidem*, **US 3689571** (1970, 1972 both to Baxter Labs.). Pharmacology and physicochemical properties: R. F. Wallin *et al.*, *Anesth. Analg.* **54**, 758 (1975). Clinical pharmacology and metabolism: D. A. Holaday, F. R. Smith, *Anesthesiology* **54**, 100 (1981). Toxicity study: D. P. Strum *et al.*, *Anesth. Analg.* **66**, 769 (1987). Minimum alveolar concentration (MAC) in humans: T. Katoh, K. Ikeda, *Anesthesiology* **66**, 301 (1987). NMR studies: E. Maciega *et al.*, *J. Mol. Struct.* **785**, 139 (2006). Clinical trial in maintenance of anesthesia: C. Campbell *et al.*, *J. Clin. Anesth.* **8**, 557 (1996). Review of use in adult day-case anesthesia: S. Ghatge *et al.*, **47**, 917-931 (2003).

Clear, colorless, volatile liquid, bp_{751} 58.1-58.2°; bp_{760} 58.6°. d_4^{23} 1.505 d^{20} 1.520-1.525. Vapor pressure (mm Hg): 317 (36°); 197 (25°); 157 (20°). Nonflammable in air at ambient temps; flammability limit in O_2: 11 vol. percent; in N_2O: 10 vol. percent. Partition coefficient at 37° (blood/gas): 0.63-0.69; (water/gas): 0.36; (olive oil/gas): 47-54; (brain/gas): 1.15. Miscible with alc, chloroform, ether. Slightly sol in water.
THERAP CAT: Anesthetic (inhalation).
THERAP CAT (VET): Anesthetic (inhalation).

8615. Seyferth-Gilbert Reagent. [27491-70-9] *P*-(Diazomethyl)phosphonic acid dimethyl ester; dimethyl diazomethylphosphonate. $C_3H_7N_2O_3P$; mol wt 150.07. C 24.01%, H 4.70%, N 18.67%, O 31.98%, P 20.64%. Divalent carbon transfer reagent. Prepn: D. Seyferth, R. S. Marmor, *Tetrahedron Lett.* **11**, 2493 (1970); and applications in organophosphorus syntheses: D. Seyferth *et al.*, *J. Org. Chem.* **36**, 1379 (1971). Reactions with aldehydes and aryl ketones to form alkynes: J. C. Gilbert, U. Weerasooriya, *ibid.* **47**, 1837 (1982). Improved prepn: D. G. Brown *et al.*, *ibid.* **61**, 2540 (1996).

Distillable yellow liquid, $bp_{0.42}$ 59°. n_D^{25} 1.4585.
USE: Reagent for one-carbon homologation of aldehydes and ketones to alkynes.

8616. Shark Liver Oil. [68990-63-6] Robecote. Oil expressed from the livers of sharks and other *Elasmobranchii* species. *Constit:* Squalene, pristane, vitamins A and D, esters of fatty acids (esp. palmitic and oleic), glycerol ethers, triglycerides, cholesterol and fatty alcohols. Isoln: A. Ehrenreich, **GB 284657** (1927); C. E. Brown, S. A. Reed, **GB 1021542** (1966 to Marfleet Refining Co. Ltd.). Composition: N. A. Sörensen, J. Mehlum, *Acta Chem. Scand.* **2**, 140 (1948); E. Gelpi, J. Oro, *J. Am. Oil Chem. Soc.* **45**, 144 (1968); R. Lombardi *et al.*, *Ann. Pharm. Fr.* **29**, 429 (1971). Comparison with cod liver oil: A. O. Banjo, *J. Food Technol.* **14**, 107 (1979).

Clear, yellow oil. Phys props of liver oil from basking shark: n_D^{20} 1.4784; d^{20} 0.922; sapon no. 140-146; iodine no. 140-145 (Lombardi).

THERAP CAT: Topical protectant.

8617. Shellac. [9000-59-3] Lacca; lac. A resinous excretion of the insect *Laccifer (Tachardia) lacca* Kerr, order *Homoptera*, family *Coccidae*. Different resiniferous trees of India serve as host trees as the season progresses. The insects suck the juice of the tree and excrete "stick-lac" almost continuously. Whitest shellac is produced while the kusum tree (*Schleichera trijuga*) is the host. Most shellac is produced in the Central and in the United provinces of India and in the states of Bihar and Orissa. The major component of lac is a resin which upon mild hydrolysis gives a complex mixture of aliphatic and alicyclic hydroxy acids and their polyesters. The composition of the hydrolysate depends on the lac source and the time of collection. The major component of the aliphatic fraction is aleuritic acid, *q.v.*; the major component of the alicyclic fraction is shellolic acid, *q.v.*: Yates, Field, *Tetrahedron* **26**, 3135 (1970). Physical and chemical properties of shellac: Cockeran, Levine, *J. Soc. Cosmet. Chem.* **12**, 316 (1961). Trends in shellac chemistry: *eidem, Am. Ink Maker* **39** (7), 26 (1961); E. Hicks, *Shellac, Its Origin and Applications* (Chemical Publ. Co., New York, 1961). *Review:* J. Martin in *Kirk-Othmer Encyclopedia of Chemical Technology* **vol. 20** (Wiley-Interscience, New York, 3rd ed., 1982) pp 737-747.

Brittle, yellowish, transparent sheets or crushed pieces or powder. d 1.035-1.140. mp 115-120°. Sapon no. 185-210; iodine no. 10-18. Solubility in alcohol 85-95% w/w (very slowly sol); in ether 13-15%; in benzene 10-20%; in petr ether 2-6%. Sparingly sol in oil of turpentine. Insol in water; sol in aq solns of ethanolamines, of alkalies or borax with slightly purple color.

USE: Chiefly in lacquers and varnishes; also in manuf buttons, grinding wheels, sealing wax, cements, inks, phonograph records, paper; for stiffening hats; in electrical machines; coating confections and medicinal tablets; finishing leather.

8618. Shellolic Acid. [4448-95-7] (3*R*,3α*S*,4*S*,7*R*,8*S*,8a*S*)-2,-3,4,7,8,8a-Hexahydro-4-hydroxy-8-(hydroxymethyl)-8-methyl-1*H*-3a,7-methanoazulene-3,6-dicarboxylic acid; 10β,13-dihydroxycedr-8-ene-12,15-dioic acid. $C_{15}H_{20}O_6$; mol wt 296.32. C 60.80%, H 6.80%, O 32.40%. Major component of the alicyclic fraction of shellac hydrolysate of which it usually constitutes 5-8%. Isoln: Harries, Nagel, *Ber.* **55**, 3833 (1922); Nagel, Mertens, *ibid.* **70**, 2173 (1937); Kirk *et al., J. Am. Chem. Soc.* **63**, 1243 (1941). Structure studies: Carruthers *et al., J. Chem. Soc.* **1961**, 5251; Cookson *et al., Tetrahedron* **18**, 547, 1321 (1962); Yates, Field, *ibid.* **26**, 3135 (1970). Stereochemistry: Yates *et al., ibid.* **26**, 3159 (1970).

Crystals, mp 204-207°.
Dimethyl ester. Crysts from ethyl acetate, mp 153-154.5°. uv max (ethanol): 231 nm (ε 6200).

8619. Shikimic Acid. [138-59-0] (3*R*,4*S*,5*R*)-3,4,5-Trihydroxy-1-cyclohexene-1-carboxylic acid. $C_7H_{10}O_5$; mol wt 174.15. C 48.28%, H 5.79%, O 45.93%. Naturally occurring (−)-form is a major biosynthetic precursor of phenylalanine, tyrosine, and tryptophan and hence of the majority of plant alkaloids. It is also involved in the biosynthesis of lignin, *q.v.*, flavonoids and other important aromatic compounds. Isoln from the fruit of the oriental plant *Illicium religiosum* Sieb. et Zucc., *Magnoliaceae* (called in Japanese *shikimi-no-ki):* J. F. Eykman, *Rec. Trav. Chim.* **4**, 32 (1885); *idem, Ber.* **24**, 1278 (1891). Structural study: H. O. L. Fischer, G. Dangshat, *Helv. Chim. Acta* **17**, 1200 (1934). Configuration: *eidem, ibid.* **18**, 1206 (1935); **20**, 705 (1937). Conformation in soln: L. D. Hall, *J. Org. Chem.* **29**, 297 (1964). Enzymatic synthesis: P. R. Srinivasan *et al., J. Am. Chem. Soc.* **77**, 4943 (1955). Stereospecific synthesis: R. McCrindle *et al., J. Chem. Soc.* **1960**, 1560; E. E. Smissman *et al., J. Am. Chem. Soc.* **84**, 1040 (1962). Improved synthesis: J. L. Pawlak, G. A. Berchtold, *J. Org. Chem.* **52**, 1765 (1987). Synthesis of the (±)-form: R. Grewe, I. Hinrichs, *Ber.* **97**, 443 (1964); R. Grewe, S. Kersten, *Angew. Chem.* **77**, 859 (1965). Carcinogenicity study: I. A. Evans, M. A. Osman, *Nature* **250**, 348 (1974). *Reviews:* B. A. Bohm, *Chem. Rev.* **65**, 435-466 (1965); B. Ganem, *Tetrahedron* **34**, 3353-3383 (1978); J. B. Harborne, *Biosynthesis* **6**, 40-75 (1980). Books: E. Haslam, *The Shikimate Pathway* (Halsted Press, New York, 1974) 316 pp; *idem*, "Shikimic Acid Metabolites" in *Comprehensive Organic Chemistry* **vol. 5** (Pergamon Press, Oxford, 1979) pp 1167-1205; U. Weiss, J. M. Edwards, *The Biosynthesis of Organic Compounds* (Wiley, New York, 1980) 728 pp.

Needles from methanol/ethyl acetate, mp 183-184.5°. Sublimes. $[\alpha]_D$ −161° (c = 0.57 in methanol). $[\alpha]_D^{18}$ −183.8° (c = 4.03 in water). uv max (alcohol): 213 nm (ε 8900). pK (14.1°) 5.19. Soly in water about 18 g/100 ml. Soly at 23° (g/100 m): 2.25 in abs alcohol: 0.015 in anhydr ether. Practically insol in chloroform, benzene, petr ether. LD_5 i.p. in mice: 1000 mg/kg (Evans, Osman).

(±)-Form. Needles from methanol/ethyl acetate, mp 191-192°. uv max (ethanol): 212 nm (ε 8200).

8620. Shionone. [10376-48-4] (1*R*,4a*S*,4b*S*,6a*S*,8*R*,10a*R*,-10b*S*,12a*S*)-Hexadecahydro-1,4b,6a,8,10a,12a-hexamethyl-8-(4-methyl-3-penten-1-yl)-2(1*H*)-chrysenone; D:A-friedo-18,19-seco-lup-19-en-3-one; 3β,5α,8,17aβ-tetramethyl-3-(4-methyl-3-penten-yl)-D-homoandrostan-17-one. $C_{30}H_{50}O$; mol wt 426.73. C 84.44%, H 11.81%, O 3.75%. From *Aster tartaricus* L., *Compositae:* Nakaoki, *J. Pharm. Soc. Jpn.* **52**, 499 (1932); *C.A.* **26**, 4821⁴ (1932); Takahashi *et al., ibid.* **79**, 1281 (1959); *C.A.* **54**, 4667g (1960). Structure: Y. Tanahashi *et al., Bull. Soc. Chim. Fr.* **1966**, 1670. Conformation: Y. Moriyama *et al., ibid.* **1968**, 2890.

Crystals from alc, mp 161-162°. $[\alpha]_D^{26.5}$ −56.1° (c = 1.07 in chloroform). uv max (chloroform): 290 nm (log ε 1.47).

8621. Showdomycin. [16755-07-0] 3-β-D-Ribofuranosyl-1*H*-pyrrole-2,5-dione; 2-β-D-ribofuranosylmaleimide. $C_9H_{11}NO_6$; mol wt 229.19. C 47.17%, H 4.84%, N 6.11%, O 41.88%. Nucleoside antibiotic isolated from *Streptomyces showdoensis:* Nishimura *et al., J. Antibiot.* **17A**, 148 (1964); Nishimura, **FR M2751** corresp to **US 3316149** (1964, 1967, both to Shionogi). Structure: Nakagawa *et al., Tetrahedron Lett.* **1967**, 4105; Darnall *et al., Proc. Natl. Acad. Sci. USA* **57**, 548 (1967). Synthesis: Kalvoda *et al., Tetrahedron Lett.* **1970**, 2297; Trummlitz, Moffatt, *J. Org. Chem.* **38**, 1841 (1973); T. Sato *et al., Tetrahedron Lett.* **1978**, 1829; J. G. Buchanan *et al., J. Chem. Soc. Perkin Trans. 1* **1979**, 225; T. Inoue, I. Kuwajima, *Chem. Commun.* **1980**, 251; A. P. Kozikowski, A. Ames, *J. Am. Chem. Soc.* **103**, 3923 (1981); Y. Araki *et al., Tetrahedron Lett.* **29**, 351 (1988). Exhibits antitumor activity: Shinzo *et al., J. Antibiot.* **17A**, 234 (1964). *Review:* D. W. Visser, S. Roy-Burman in *Antibiotics* **vol. 5**(pt. 2), F. E. Hahn, Ed. (Springer-Verlag, New York, 1979) pp 363-371.

Leaflets from acetone + benzene, mp 153-154°. $[\alpha]_D^{22.5}$ +49.9° (H₂O). uv max: 220-221 nm ($E_{1cm}^{1\%}$ 422). Sol in water, alcohol, acetone, dioxane; insol in ethyl ether, benzene, petr ether. More stable in acid media than in neutral or alkaline. LD_{50} in mice (mg/kg): 25 i.p.; 18 s.c.; 110 i.v. (Nishimura).

8622. Shvo Catalyst. [104439-77-2] Tetracarbonyl-μ-hydro[(1,2,3,4,5-η)-1-hydroxylato-2,3,4,5-tetraphenyl-2,4-cyclopentadien-1-yl][(1,2,3,4,5-η)-1-hydroxy-2,3,4,5-tetraphenyl-2,4-cyclopentadien-1-yl]diruthenium; [(η⁴-tetracyclone)(CO)₂Ru]₂. $C_{62}H_{42}$-O_6Ru_2; mol wt 1085.15. C 68.62%, H 3.90%, O 8.85%, Ru 18.63%. Diruthenium catalyst especially for hydrogen-transfer reactions. Prepn: Y. Blum, Y. Shvo, *J. Organomet. Chem.* **282**, C7 (1985); Y. Shvo *et al., J. Am. Chem. Soc.* **108**, 7400 (1986). X-ray structure study: M. J. Mays *et al., Organometallics* **8**, 1162 (1989). Mechanism of hydrogen transfer: J. B. Johnson, J.-E. Bäckvall, *J. Org. Chem.* **68**, 7681 (2003). Catalytic disproportionation of aldehydes: N. Menashe, Y. Shvo, *Organometallics* **10**, 3885 (1991). Resolution of alcohols: B. A. Persson *et al.*, *J. Am. Chem. Soc.* **121**, 1645 (1999); oxidation of alcohols: J. H. Choi *et al.*, *Tetrahedron Lett.* **45**, 4607 (2004). Brief review: R. Prabhakaran, *Synlett* **2004**, 2048-2049.

Orange crystalline solid.

USE: Catalyst in organic syntheses for hydrogen transfers including disproportionation of aldehydes, reduction of ketones and aldehydes, oxidation and/or racemization of alcohols and amines.

8623. Sialic Acids. Nonulosaminic acids. Family of amino sugars having a 9 carbon backbone based on neuraminic acid, *q.v.*; both *N*- and *O*- substituents are known. Widely distributed throughout nature; important component of glycoproteins, gangliosides, glycosaminoglycans, and mucins. *N*-Acetylneuraminic acid is the most prevalent form in human glycoconjugates; other predominant forms have *N*-glycolyl- and *N*-acetyl-9-*O*-acetyl- substituents. *Reviews*: A. Gottschalk, *Rev. Pure Appl. Chem.* **12**, 46 (1962); R. Schauer, *Curr. Opin. Struct. Biol.* **19**, 507-514 (2009). Review of determn methods in biological samples: R. Lacomba *et al.*, *J. Pharm. Biomed. Anal.* **51**, 346-357 (2010). Review of role in brain development and cognition: B. Wang, *Annu. Rev. Nutr.* **29**, 177-122 (2009); of biology and chemistry: X. Chen, A. Varki, *ACS Chem. Biol.* **5**, 163-176 (2010).

N-Acetyl-β-neuraminic acid

N-Acetylneuraminic acid. [131-48-6]; [19342-33-7] (β-anomer); [884595-19-1] (α-anomer). 5-(Acetylamino)-3,5-dideoxy-D-*glycero*-D-*galacto*-2-nonulosonic acid; Neu5Ac; NANA; lactaminic acid. $C_{11}H_{19}NO_9$; mol wt 309.27. Isoln: Klenk, Faillard, *Z. Physiol. Chem.* **298**, 230 (1954). Synthesis: Cornforth *et al.*, *Biochem. J.* **68**, 57 (1958); R. Csuk *et al.*, *Helv. Chim. Acta* **71**, 609 (1988). First synthesis from non-carbohydrate sources: M. P. DeNinno, S. J. Danishevsky, *J. Org. Chem.* **51** 2615 (1986). Configuration: Kuhn, Brossmer, *Angew. Chem. Int. Ed.* **1**, 218 (1962). Crystals from water/acetic acid (0.8:12) at 5°, mp 180-182° (dec). $[\alpha]_D^{20}$ −33.1° (c = 0.9 in water).

8624. Sibutramine. [106650-56-0] 1-(4-Chlorophenyl)-*N*,*N*-dimethyl-α-(2-methylpropyl)cyclobutanemethanamine; *N*-1-[1-(4-chlorophenyl)cyclobutyl]-3-methylbutyl-*N*,*N*-dimethylamine. $C_{17}H_{26}ClN$; mol wt 279.85. C 72.96%, H 9.37%, Cl 12.67%, N 5.01%. Serotonin and noradrenaline reuptake inhibitor (SNRI); decreases calorie intake and increases energy expenditure. Prepn: J. E. Jeffery *et al.*, *DE 3212682*; *idem*, *US 4929629* (1982, 1990 both to Boots). HPLC determn in dietary supplements: Z. Huang *et al.*, *J. Chromatogr. Sci.* **46**, 707 (2008). LC-MS/MS determn in plasma: W. Kang *et al.*, *J. Pharm. Biomed. Anal.* **51**, 264 (2010). Clinical trial in obesity: C. Hanotin *et al.*, *Int. J. Obes.* **22**, 32 (1998). Review of pharmacology and mechanism of action: D. J. Heal *et al.*, *ibid.*

Suppl. 1, S18-S28; of cardiovascular risk: A. J. Scheen, *Diabetes Care* **34**, Suppl. 2, S114-S118 (2011).

Hydrochloride monohydrate. [125494-59-9] BTS-54524; Reductil; Meridia. $C_{17}H_{26}ClN \cdot HCl \cdot H_2O$; mol wt 334.33. Crystals from water, mp 193-195.5°. Partition coefficient (octanol/water): 30.9 (pH 5.0). Soly in water: 2.9 mg/ml at pH 5.2.

Note: This is a controlled substance (stimulant): **21 CFR, 1308.14.**

THERAP CAT: Antiobesity agent.

8625. Siccanin. [22733-60-4] (4a*S*,6a*S*,11b*R*,13a*S*,13b*S*)-2,-3,4,4a,5,6,6a,13b-Octahydro-4,4,6a,9-tetramethyl-1*H*,11b*H*,13*H*-benzo[*a*]furo[2,3,4-*mn*]xanthen-11-ol; Tackle. $C_{22}H_{30}O_3$; mol wt 342.48. C 77.16%, H 8.83%, O 14.01%. Antibiotic obtained from the cultured broth of *Helminthosporium siccans* Drechsler, a parasitic organism of rye grass (*Lolium multiflorum* Lam). Isoln: K. Ishibashi, **JP 62 3548** (1962 to Sankyo); *idem*, *J. Antibiot.* **15A**, 161 (1962). The *cis* formation of the decahydronaphthalene system may be the first example of the naturally occurring drimane skeleton. Structure: K. Hirai *et al.*, *Tetrahedron Lett.* **1967**, 2177; K. Hirai *et al.*, *Tetrahedron* **27**, 6057 (1971). Approach to synthesis: H. Oida *et al.*, *Chem. Pharm. Bull.* **20**, 2634 (1972); T. Yoshida *et al.*, *ibid.* 2642. Total synthesis of racemate: M. Kato *et al.*, *J. Am. Chem. Soc.* **103**, 2434 (1981); *idem*, *Tetrahedron* **43**, 711 (1987). CD spectral study: S. Nozoe *et al.*, *Tetrahedron* **30**, 2773 (1974). *In vitro* activity studies: M. Arai *et al.*, *Antimicrob. Agents Chemother.* **1969**, 247; M. G. Bellotti, L. Riviera, *Chemioterapia* **4**, 431 (1985); *in vivo* studies: S. Sugarwara, *Antimicrob. Agents Chemother.* **1969**, 253. Toxicology: Y. Suzuki *et al.*, *Sankyo Kenkyusho Nempo* **21**, 120 (1969); *C.A.* **72**, 130939k (1970). Mode of action: K. Nose, A. Endo, *J. Bacteriol.* **105**, 176 (1971). Biosynthesis: K. Suzuki *et al.*, *Bioorg. Chem.* **3**, 72 (1974). Clinical evaluation in surface mycosis: D. Crippa *et al.*, *G. Ital. Dermatol. Venereol.* **120**, 57 (1985).

White to pale yellow odorless crystals, mp 139-140°. $[\alpha]_D^{20}$ −136° (c = 2 in chloroform). uv max (ethanol): 210, 285 nm (ε 45690, 1717). pK 10.9. Very sol in chloroform, dimethylformamide, benzene; freely sol in acetone, ether, ethyl acetate; sol in alcohol. Practically insol in water. LD_{50} in mice, rats (mg/kg): >6000, >1000 orally; >3000, >600 s.c. or i.p. (Suzuki).

THERAP CAT: Antifungal.

8626. Siduron. [1982-49-6] *N*-(2-Methylcyclohexyl)-*N'*-phenylurea; Tupersan. $C_{14}H_{20}N_2O$; mol wt 232.33. C 72.38%, H 8.68%, N 12.06%, O 6.89%. Prepn: Luckenbaugh, **BE 631289** (1963 to du Pont), *C.A.* **60**, 15775h (1964), corresp to **GB 1028818**. Prepn of *trans*-form: Hayman *et al.*, *J. Pharm. Pharmacol.* **16**, 538 (1964).

Solid, mp 133-138°. Sol in water (18 ppm); sol to the extent of 10% or more in ethanol, dimethylacetamide, dimethylformamide, methylene chloride.

***trans*-Form.** Crystals, mp 157-159°.

USE: Herbicide.

8627. Silane. [7803-62-5] Silicon tetrahydride; silicane; monosilane. H$_4$Si; mol wt 32.12. H 12.55%, Si 87.44%. SiH$_4$. Prepd from aluminum silicide and hydrochloric acid or by electrolysis of an aq soln of sodium, ammonium, manganese, or ferrous chloride using an aluminum-silicon alloy as the positive pole: Buff, Wöhler, *Ann.* **102**, 128 (1857); from magnesium silicide and HCl: Wöhler, *ibid.* **106**, 56 (1858); Stock, Somiesky, *Ber.* **49**, 111 (1916); from zinc silicide and acid: Deville, Caron, *Compt. Rend.* **45**, 163 (1857); from lithium silicide and acid: Moissan, *ibid.* **135**, 1284 (1902); from magnesium silicide and ammonium bromide or chloride in liquid ammonia: Johnson, Isenberg, *J. Am. Chem. Soc.* **57**, 1349 (1935); Clasen, **DE 926069** (1952); *Angew. Chem.* **70**, 179 (1958); from lithium aluminum hydride in ether and silicon tetrachloride: Finholt *et al.*, *J. Am. Chem. Soc.* **69**, 2692 (1947); from silica gel, aluminum oxide and hydrogen: Jackson, **US 3068069** (1962 to du Pont); from silicon tetrachloride and sodium aluminum hydride: Mason, Kelly, **US 3043664** (1962 to Allied Chem.). Purification: Jacob, Trenner, **US 3019087** (1962 to Merck & Co.); Shoemaker, Sumrell, **US 3041141** (1962 to Baker Chem.).

Gas, repulsive odor. d^{-185} (liq) 0.68. *Flammable.* Solidifies at approx −200°. mp −185°; bp −112°. Stable at ordinary temps; completely dec ~400° into silicon and hydrogen; dec by an electric discharge; ignites in an atm of chlorine; ignites in air by raising the temp. Slowly dec in water. Practically insol in alc, ether, benzene, chloroform, silicochloroform and silicon tetrachloride. Dec in potassium hydroxide solns.

Caution: Potential symptoms of overexposure are irritation of eyes, skin, mucous membranes; nausea, headache. *See NIOSH Pocket Guide to Chemical Hazards* (DHHS/NIOSH 97-140, 1997) p 278.

USE: Source of hyperpure silicon for semiconductors.

8628. Sildenafil. [139755-83-2] 5-[2-Ethoxy-5-[(4-methyl-1-piperazinyl)sulfonyl]phenyl]-1,6-dihydro-1-methyl-3-propyl-7*H*-pyrazolo[4,3-*d*]pyrimidin-7-one; 1-[[3-(4,7-dihydro-1-methyl-7-oxo-3-propyl-1*H*-pyrazolo[4,3-*d*]pyrimidin-5-yl)-4-ethoxyphenyl]sulfonyl]-4-methylpiperazine; UK-92480. C$_{22}$H$_{30}$N$_6$O$_4$S; mol wt 474.58. C 55.68%, H 6.37%, N 17.71%, O 13.48%, S 6.76%. Orally active, selective type 5 cGMP phosphodiesterase inhibitor. Prepn: A. S. Bell *et al.*, **EP 463756**; *eidem*, **US 5250534** (1992, 1993 both to Pfizer). Structure-activity study: N. K. Terrett *et al.*, *Bioorg. Med. Chem. Lett.* **6**, 1819 (1996). Clinical trial in impotence: I. Goldstein *et al.*, *N. Engl. J. Med.* **338**, 1397 (1998); in pulmonary hypertension: B. K. S. Sastry *et al.*, *J. Am. Coll. Cardiol.* **43**, 1149 (2004). Review of pharmacology: M. Boolell *et al.*, *Int. J. Impot. Res.* **8**, 47-52 (1996); of clinical experience in pulmonary arterial hypertension: K. F. Croom, M. P. Curran, *Drugs* **68**, 383-397 (2008); in erectile dysfunction: A. Tsertsvadze *et al.*, *Urology* **74**, 831-836 (2009).

Crystals, mp 187-189°.

Citrate. [171599-83-0] Revatio; Viagra. C$_{22}$H$_{30}$N$_6$O$_4$S.C$_6$H$_8$O$_7$; mol wt 666.70.

THERAP CAT: In treatment of male erectile dysfunction. In treatment of pulmonary arterial hypertension.

8629. Silicic Acid. [1343-98-2] General term for hydrated forms of silicon dioxide, *q.v.* The most common and bioavailable form, orthosilicic acid, is found ubiquitously in water and is converted by marine algae (diatoms) into biogenic, amorphous silica. Silicic acids readily polymerize to form complex linear or cyclic structures which can occur as sols, gels, or particles. Colorimetric determn in seawater: J. B. Mullin, J. P. Riley, *Anal. Chim. Acta* **12**, 162 (1955); P. Rimmelin-Maury *et al.*, *ibid.* **587**, 281 (2007). Clinical pharmacokinetics and excretion of orthosilicic acid: D. M. Reffitt *et al.*, *J. Inorg. Biochem.* **76**, 141 (1999). Clinical effect on bone

formation: T. D. Spector *et al.*, *BMC Musculoskelet. Disord.* **9**, 85 (2008). Review of polymerization reactions: T. Tarutani, *Anal. Sci.* **5**, 245-252 (1989). Review of orthosilicic acid uptake by diatoms and role in the global silicon cycle: O. Ragueneau *et al.*, *Glob. Planet. Change* **26**, 317-365 (2000). Review of chemistry, condensation reactions, and characterization of polysilicic acid structures: D. J. Belton *et al.*, *FEBS J.* **279**, 1710-1720 (2012).

Orthosilicic acid. [10193-36-9] Silicic acid (H$_4$SiO$_4$); monosilicic acid; silicon tetrahydroxide; tetrahydroxysilane. H$_4$O$_4$Si; mol wt 96.11. Si(OH)$_4$. Sol in water. pKa 9.8.

Metasilicic acid. [7699-41-4] Silicic acid (H$_2$SiO$_3$). H$_2$O$_3$Si; mol wt 78.10. SiO(OH)$_2$.

Pyrosilicic acid. [20638-18-0] Silicic acid (H$_6$Si$_2$O$_7$); hexahydroxydisiloxane; disilicic acid. H$_6$O$_7$Si$_2$; mol wt 174.21. Si(OH)$_3$-OSi(OH)$_3$

8630. Silicon. [7440-21-3] Si; at. wt [28.084; 28.086]; conventional at. wt 28.085; at. no. 14; valence 4, also 2. Group IVA (14). Three naturally occurring isotopes: 28 (92.18%); 29 (4.71%); 30 (3.12%); artificial isotopes: 25-27; 31; 32. Does not occur free in nature; found as silica (quartz, sand, sandstone) or as silicate (feldspar or orthoclase, kaolinite, anorthite, etc.). Constitutes about 27.6% of the earth's crust; second most abundant element on earth, oxygen being first. Prepd industrially by carbon reduction of silica in an electric arc furnace. Purification by zone refining (*see* ref. *under* Germanium). Very pure silicon is obtained by decompn of silicon tetraiodide: Litton, Anderson, *J. Electrochem. Soc.* **101**, 287 (1954); *Chem. Eng. News* **34**, 5007 (1956). From silicon tetrachloride: Lyon *et al.*, *Trans. Electrochem. Soc.* **96**, 359 (1949); Klyuchnikov, *J. Appl. Chem. USSR* **29**, 139 (1956). By thermal decompn of a chlorosilane: Schering, **US 3041144** (1962 to Siemens-Schuckertwerke). Review of silicon and its compounds: Rochow, "Silicon" in *Comprehensive Inorganic Chemistry* vol. 1, J. C. Bailar, Jr. *et al.*, Eds. (Pergamon Press, Oxford, 1973) pp 1323-1467; W. Runyan in *Kirk-Othmer Encyclopedia of Chemical Technology* vol. 20 (Wiley-Interscience, New York, 3rd ed., 1982) pp 826-845. The uses of silicon compounds in organic chemistry: E. W. Colvin, *Chem. Soc. Rev.* **7**, 15 (1978); I. Fleming, *ibid.* **10**, 83 (1981); L. A. Paquette, *Science* **217**, 793 (1982).

Black to gray, lustrous, needle-like crystals or octahedral platelets (cubic system). The amorphous form is a dark brown powder. Poor conductor of electricity. d$_4^{25}$ 2.33. mp 1410°. Average heat capacity (16-100°): 0.1774 cal/g/°C. Lattice constant (25°): 5.41987 × 10^{-8} cm. Compressibility (V/V$_0$) at 25 × 10^3 kg/cm^2: 0.978; at 100 × 10^3 kg/cm^2: 0.940, *Gmelins, Silicon* (8th ed.) **15B** (1959) p 57. Dielectric const: 13. Covalent bond ionization energy at 0 K = 1.2 ev. Band gap: 1.106 ev. Impurity atom ionization energy: ~0.04 ev. Intrinsic resistivity at 300 K = 0.23 megohm. Electron mobility at 300 K: 1500 cm^2/volt/sec. Hole mobility at 300 K: 500 cm^2/volt/sec. Intrinsic charge density at 300 K: 1.5 × 10^{10}. Electron diffusion constant at 300 K: 38. Hole diffusion constant at 300 K: 13. Practically insol in water. Attacked by hydrofluoric or a mixture of hydrofluoric and nitric acids (depending upon cryst modifications). Soluble in molten alkali oxides. Burns in fluorine, chlorine.

Caution: Potential symptoms of overexposure are irritation of eyes, skin, upper respiratory system; cough. *See NIOSH Pocket Guide to Chemical Hazards* (DHHS/NIOSH 97-140, 1997) p 278.

USE: In making silanes and silicones, the Si—C bond being about as strong as a C—C bond. In the manuf of transistors, silicon diodes and similar semiconductors. For making alloys such as ferrosilicon, silicon bronze, silicon copper. As a reducing agent like aluminum in high temp reactions.

8631. Silicon Carbide. [409-21-2] CSi; mol wt 40.10. C 29.95%, Si 70.04%. SiC. Made by heating coke and sand (and salt as flux) in electric furnace.

Exceedingly hard, green to bluish-black, iridescent, sharp crystals; hardness = 9.5. d 3.23. Dielectric constant: 7.0. Electron mobility: >100 cm^2/volt-sec. Hole mobility: >20 cm^2/volt-sec. Band gap energy: 2.8 ev.

Caution: Potential symptoms of overexposure are irritation of eyes, skin, upper respiratory system; cough. *See NIOSH Pocket Guide to Chemical Hazards* (DHHS/NIOSH 97-140, 1997) p 278.

USE: Polishing glass, granite; smoothing bisque ware; in sharpening-stones; abrasive; manuf porcelain, "emery" paper and wheels, shoe soles, refractory brick, furnace linings; antiskid pavements; in semiconductor technology.

8632. Silicon Dioxide. [7631-86-9] Silica; silicic anhydride. O_2Si; mol wt 60.08. O 53.26%, Si 46.75%. Occurs in nature as **agate, amethyst, chalcedony, cristobalite, flint, quartz, sand, tridymite.** *Reviews: Kirk-Othmer Encyclopedia of Chemical Technology* vol. 18 (Interscience, New York, 2nd ed., 1969) pp 46-111; Rochow in *Comprehensive Inorganic Chemistry* **vol. 1,** J. C. Bailar, Jr. *et al.*, Eds. (Pergamon Press, Oxford, 1973) pp 1388-1402. Toxicology: L. T. Fairhall, *Industrial Toxicology* (Hafner, New York, 1969) pp 105-107.

Transparent, tasteless crystals, or amorphous powder. d (amorphous) 2.2. d^0 (quartz) 2.65. Melts to a glass. Silica has the lowest coefficient of expansion by heat of any known substance. Sol in hot solns of alkali hydroxides. Insol in water, alc, and in other organic solvents; practically insol in acids, except hydrofluoric acid in which it readily dissolves forming the gas silicon tetrafluoride; it is also slowly attacked by heating with concd phosphoric acid. The crystallized forms of silica are scarcely attacked by alkalies, while the amorphous is sol, especially when finely divided. *See also* Infusorial Earth.

Caution: Potential symptom of overexposure to amorphous silica are eye irritation and pneumoconiosis. Potential symptoms of overexposure to crystalline silica as respirable dust are cough, dyspnea, wheezing; decreased pulmonary function, progressive respiratory symptoms (silicosis). *See NIOSH Pocket Guide to Chemical Hazards* (DHHS/NIOSH 97-140, 1997) p 276-279. Crystalline silica (respirable size), primarily quartz dusts occurring in industrial and occupational settings, is listed as a known human carcinogen: *Report on Carcinogens, Twelfth Edition* (PB2011-111646, 2011) p 377.

USE: Manuf glass, water glass, refractories, abrasives, ceramics, enamels; decolorizing and purifying oils, petroleum products, etc.; in scouring- and grinding-compounds, ferrosilicon, molds for castings; as anticaking and defoaming agent. Pharmaceutic aid (dessicant, suspending and viscosity-increasing agent).

8633. Silicon Disulfide. [13759-10-9] Silicon sulfide (SiS_2). S_2Si; mol wt 92.21. S 69.54%, Si 30.46%. SiS_2. Prepd from SiO_2 and Al_2S_3: Tiede, *Ber.* **59,** 1703 (1926); Zintl, Loosen, *Z. Phys. Chem.* **A174,** 301 (1935).

White, fibrous mass or needles. Yields H_2S on contact with moist air. d 2.02. mp 1090° (sublimes). Burns when held in flame. Sol with decompn in water, alcohol, alkaline solns. Insol in benzene.

8634. Silicones. Broad family of synthetic polymers containing a repeating silicon-oxygen backbone with organic side groups attached via carbon-silicon bonds. Classified into three major types: fluids, resins and elastomers. The name was coined by F. S. Kipping under the supposition that the compounds were analogous to ketones; now used to designate a variety of complex polymeric siloxanes which may be linear, branched or cross-linked. Structural unit is represented by the formula, $[R_n SiO_{(4-n/2)}]_m$, where $n = 1$-3 and $m >$ 1. R groups are usually methyl, longer alkyl, fluoroalkyl, phenyl, vinyl, alkoxy or alkylamino. Silicones are characterized by thermal and oxidative stability, chemical inertness, resistance to weathering, good dielectric strength and low surface tension. Prepd commercially by the direct reaction of silicon with an organic halide, hydrolysis of the resulting organohalosilanes and condensation of the unstable diols. Review of chemistry, properties and uses: B. Hardman, A. Torkelson in *Encyclopedia of Polymer Science and Engineering* **vol. 15** (Wiley-Interscience, New York, 1989) pp 204-308. Review of silicone surfactants: B. Grüning, G. Koerner, *Tenside Surfactants Deterg.* **26,** 312-317 (1989); and their uses in cosmetics and toiletries: C. Gould, *Spec. Chem.* **11,** 354-359 (1991). Clinical use of silicone elastomers in reconstructive surgery: M. B. Habal in *Advances in Biomedical Polymers*, C. G. Gebelein, Ed. (Plenum Press, New York, 1987) pp 17-29. Review of properties and commercial uses of silicone fluids: D. H. Demby *et al.* in *Synthetic Lubricants and High-Performance Functional Liquids*, R. L. Shubkin, Ed. (Marcel Dekker, New York, 1993) pp 183-203.

Polydimethylsiloxane *see* Dimethicone.

Polyether copolymers. Silicone polyethers; dimethicone copolyols; silicone glycol copolymers; silicone glycols. Branched siloxanes modified at the chain ends with a polyether, *e.g.* polyoxyethylene and/or polyoxypropylene. Trade marks for such products include: **Abil, Stilwet L.**

USE: As lubricants; wetting agents; defoamers; surfactants; hydraulic oils; dielectric oils; protective coatings; adhesives and caulking compounds; release agents; in paints, enamels, and varnishes; water repellents for textiles, paper, masonry, and concrete; damping fluids; electrical insulation; dental molds; medical implants; contact lenses; and in cosmetic and pharmaceutical formulations. A specialty product known as "bouncing putty" or "silly putty" has unusual rheological properties which allow it to be stretched, shattered, molded, or to rebound like rubber when dropped on hard surface.

8635. Silicon Monoxide. [10097-28-6] Oxosilylene. OSi; mol wt 44.08. O 36.30%, Si 63.71%. SiO. Prepd by high vacuum sublimation of a mixture of silicon and quartz: Schenk in *Handbook of Preparative Inorganic Chemistry* **vol. 1,** G. Brauer, Ed. (Academic Press, New York, 2nd ed., 1963) pp 696-697. Existence below 1200° in doubt; may disproportionate to a mixture of Si and SiO_2: Benyon, *Vacuum* **20,** 293 (1970). *Review:* Rochow in *Comprehensive Inorganic Chemistry* **vol. 1,** J. C. Bailar, Jr. *et al.*, Eds. (Pergamon Press, Oxford, 1973) pp 1353-1355.

Brownish-black scales by sublimation. Has been crystallized. Cubic system. About as hard as silicon. Nonconductor of electricity even at red heat. d 2.18.

8636. Silicon Nitride. [12033-89-5] Trisilicon tetranitride. N_4Si_3; mol wt 140.28. N 39.94%, Si 60.06%. Si_3N_4. Non-oxide ceramic material of high thermal stability. Manuf by direct nitridation of pure silicon. Prepn by pyrolysis of silicon diimide (Si(N-$H_2)_2$): O. Glemser, P. Naumann, *Z. Anorg. Allg. Chem.* **298,** 134 (1954); and characterization: K. S. Mazdiyasni, C. M. Cooke, *J. Am. Ceram. Soc.* **56,** 628 (1973); *eidem,* US **3959446** (1976 to USA); by pyrolysis of silazanes: D. Seyferth *et al.*, *J. Am. Ceram. Soc.* **66,** C-13 (1983); *eidem,* US **4397828** (1983 to M.I.T.). Brief review: D. L. Segal, *Chem. Ind. (London)* **1985,** 544.

Exists in two crystalline forms, α-Si_3N_4 and β-Si_3N_4. α-Form occurs when powdered Si_3N_4 is heated to 1200° >4 hr; increasing the temperature to >1450°C for 2 hr produces the β-form. White when pure but often brown or black powder. mp 2173°K (sublimes). d 3.2 kg/dm³. Coefficient of thermal expansion K^{-1}: 2.5×10^{-6}.

USE: As high temp engineering material for use in gas turbines, diesel engines.

8637. Silicon Tetraacetate. [562-90-3] Acetic acid tetraanhydride with silicic acid (H_4SiO_4); tetraacetoxysilane; tetrakis(acetyloxy)silane. $C_8H_{12}O_8Si$; mol wt 264.26. C 36.36%, H 4.58%, O 48.43%, Si 10.63%. Prepd from $SiCl_4$ + $(CH_3CO)_2O$: Balthis, *Inorg. Synth.* **4,** 45 (1953).

Extremely hygroscopic cryst mass. Hydrolyzed rapidly by moist air. mp 110°. Dec at 160-170° with evolution of acetic anhydride. $bp_{6.0}$ 148°. Violent reaction on contact with water. Forms SiO_2 and ethyl acetate on contact with alcohol. Moderately sol in acetone, benzene.

8638. Silicon Tetrabromide. [7789-66-4] Tetrabromosilane; silicon bromide. Br_4Si; mol wt 347.70. Br 91.92%, Si 8.08%. $SiBr_4$. Prepd from the elements: Gattermann, *Ber.* **22,** 189 (1889).

Colorless, fuming liq; disagreeable odor; becomes yellow on exposure to air. d 2.8. Solidif -12°; mp 5°; bp 153°. Dec by water into silicic acid and HBr with great evolution of heat. Reacts violently with metallic potassium.

8639. Silicon Tetrachloride. [10026-04-7] Tetrachlorosilane; silicon chloride. Cl_4Si; mol wt 169.89. Cl 83.47%, Si 16.53%. $SiCl_4$. Prepd from the elements: Schenk in *Handbook of Preparative Inorganic Chemistry* **vol. 1,** G. Brauer, Ed. (Academic Press, New York, 2nd ed., 1963) pp 682-683.

Colorless, clear, mobile, fuming liq; suffocating odor. d_4^0 1.52. mp -70°; bp 59°. *Corrosive.* Dec by water with much heat into silicic acid and HCl. Miscible with benzene, ether, chloroform, petr ether.

Caution: May be irritating to eyes, respiratory tract.

USE: Producing smoke screens in warfare. In the prepn of pure silicon.

8640. Silicon Tetrafluoride. [7783-61-1] Tetrafluorosilane; silicon fluoride. F_4Si; mol wt 104.08. F 73.01%, Si 26.98%. SiF_4. Commercial prepn from waste gases of phosphate fertilizer production: Molstad, US 2833628 (1958 to W. R. Grace). Lab prepns: Booth, Swinehart, *J. Am. Chem. Soc.* **57**, 1337 (1935); Hoffman, Gutowski, *Inorg. Synth.* **4**, 145 (1953); Belf, *Chem. Ind. (London)* **1955**, 1296.

Colorless gas; very pungent odor similar to that of hydrogen chloride. *Poisonous; corrosive.* Forms heavy clouds with moist air. Dec by water into silicic acid and HF. Sublimes $-95.7°$; mp $-90.2°$ (under pressure): d (liq; $-80°$) 1.590. Critical temp $-1.5°$; critical pressure 50 atm.

8641. Silicotungstic Acid. [12027-38-2] [μ12-[Orthosilicato-$(4-)-\kappa O:\kappa O:\kappa O:\kappa O':\kappa O':\kappa O':\kappa O'':\kappa O'':\kappa O'':\kappa O''':\kappa O''':\kappa O''']$]tetracosa-$\mu$-oxododecaoxododecatungstate$(4-)$ hydrogen (1:4); tungstosilicic acid. $H_4O_{40}SiW_{12}$; mol wt 2878.16. H 0.14%, O 22.24%, Si 0.98%, W 76.65%. $H_4[SiO_4 \cdot (W_3O_9)_4]$. Prepn of hydrated compd: Grüttner, Jander in *Handbook of Preparative Inorganic Chemistry* vol. 2, G. Brauer, Ed. (Academic Press, New York, 2nd ed., 1965) pp 1717-1718.

Hydrate. White to slightly yellow, deliquesc crystals. Very sol in water, alcohol. *Keep tightly closed.*

USE: Preparing heavy solns for separating minerals; as a mordant for basic aniline dyes; as a reagent for pptn and determination of alkaloids.

8642. Silodosin. [160970-54-7] 2,3-Dihydro-1-(3-hydroxypropyl)-5-[(2R)-2-[[2-[2-(2,2,2-trifluoroethoxy)phenoxy]ethyl]amino]propyl]-1H-indole-7-carboxamide; KMD-3213; Rapaflo; Silodyx; Urief; Urorec. $C_{25}H_{32}F_3N_3O_4$; mol wt 495.54. C 60.60%, H 6.51%, F 11.50%, N 8.48%, O 12.91%. α_{1a}-Adrenoceptor antagonist. Prepn: M. Kitazawa *et al.*, EP 600675; *eidem*, US 5387603 (1994, 1995 both to Kissei). Adrenoceptor binding study: K. Shibata *et al.*, *Mol. Pharmacol.* **48**, 250 (1995); and tissue selectivity: S. Murata *et al.*, *J. Urol.* **164**, 578 (2000). LC/MS/MS determn in plasma: X. Zhao *et al.*, *J. Chromatogr.* B **877**, 3724 (2009). Clinical trial in treatment of benign prostatic hyperplasia (BPH): K. Kawabe *et al.*, *BJU Int.* **98**, 1019 (2006). Review of pharmacology and clinical experience in BPH: M. Yoshida *et al.*, *Expert Opin. Invest. Drugs* **16**, 1955-1965 (2007).

White to pale, yellowish white powder, mp 105-109°. $[\alpha]_D^{25}$ $-14.0°$ (c = 1.01 in methanol). Sol in acetic acid, alcohol; very slightly sol in water.

THERAP CAT: In treatment of benign prostatic hypertrophy.

8643. Silver. [7440-22-4] Ag; at. wt 107.8682; at. no. 47; valence 1, 2. Group IB (11). Occurrence in the earth's crust: 0.1 ppm; also present in seawater: 0.01 ppm. Natural isotopes: 107 (51.35%); 109 (48.65%); artificial isotopes (mass numbers): 100-106, 108, 110-117. One of the earliest known metals. Found native or associated with copper, gold and lead. Principle ores are *argentite*, *cerargyrite* or *horn silver* (mixture of halides), *proustite* ($3Ag_2\text{-}S.As_2S_3$), *pyrargyrite* ($Ag_2S.Sb_2S_3$). Extraction from ores: Percy *et al.*, cited by Mellor, *A Comprehensive Treatise on Inorganic and Theoretical Chemistry* **3**, 301 (1928); *Silver, Its Economics, Extraction, Use*, A. Butts, C. D. Coxe, Eds. (Van Nostrand, Princeton, 1967) 480 pp. *Reviews:* Thompson, "Silver" in *Comprehensive Inorganic Chemistry* vol. 3, J. C. Bailar, Jr. *et al.*, Eds. (Pergamon Press, Oxford, 1973) pp 79-128; G. H. Sistare and H. B. Lockhart in *Kirk-Othmer Encyclopedia of Chemical Technology* vol. 21 (Wiley-Interscience, New York, 3rd ed., 1983) pp 1-32. Review of toxicol-

ogy and human exposure: *Toxicological Profile for Silver* (PB91-180430, 1990) 157 pp.

White metal, face-centered cubic structure. More malleable and ductile than any other metal except gold; excellent conductor of heat and electricity. mp 960.5°. bp $\sim$2000°. d^{15} 10.49. Not attacked by water or atmospheric oxygen; blackened by ozone, by hydrogen sulfide, by sulfur. Inert to most acids; readily reacts with dil nitric acid, hot concd sulfuric acid; superficially attacked by hydrochloric acid. Sol in fused alkali hydroxides in presence of air, in fused alkali peroxides, in alkali cyanides in presence of air or oxygen. Most silver salts are light-sensitive.

Caution: Potential symptoms of overexposure to dust are blue-gray eyes, nasal septum, throat and skin; irritation and ulceration of skin; GI disturbance. *See NIOSH Pocket Guide to Chemical Hazards* (DHHS/NIOSH 97-140, 1997) p 280. Blue-gray discoloration, known as argyria or argyrosis, results from chronic exposure to silver or silver salts. *See Patty's Industrial Hygiene and Toxicology* vol. 2A, G. D. Clayton, F. E. Clayton, Eds. (Wiley-Interscience, New York, 3rd ed., 1981) pp 1881-1894.

USE: For coinage, most frequently alloyed with copper or gold; for manuf tableware, mirrors, jewelry, ornaments; for electroplating; for making vessels and apparatus used in manuf medicinal chemicals, in processing foods and beverages, in handling organic acids; as catalyst in hydrogenation and oxidation processes; as ingredient of dental alloys; in electrical and electronic products; in brazing alloys and solders. Has been used for purification of drinking water because of toxicity to bacteria and lower forms of life. Salts used in photographic materials.

8644. Silver Acetate. [563-63-3] Acetic acid silver(1+) salt (1:1). $C_2H_3AgO_2$; mol wt 166.91. C 14.39%, H 1.81%, Ag 64.63%, O 19.17%. CH_3COOAg.

White to slightly grayish, lustrous needles or cryst powder. d 3.26. Sol in 100 parts of cold water, 35 parts of boiling water; freely sol in dil nitric acid. *Protect from light.*

USE: Oxidizing agent for use in liquid ammonia: Kline, Kershner, *Inorg. Chem.* **5**, 932 (1966).

8645. Silver Bromide. [7785-23-1] AgBr; mol wt 187.77. Ag 57.45%, Br 42.55%.

Yellowish, odorless powder; darkens on exposure to light. d 6.47; mp 432°. Soly in water (25°): 0.135 mg/l. Insol in alcohol, or most acids. Slightly sol in dil ammonia, but moderately sol in concd ammonia. One liter of 10% ammonia dissolves 3.3 g at 12°; sol in solns of alkali cyanides, sparingly sol in solns of thiocyanates or thiosulfates; sol in 220 parts saturated NaCl, in 35 parts saturated KBr solns; slightly sol in ammonium carbonate soln. *Protect from light.*

USE: In photography.

THERAP CAT: Topical anti-infective; astringent.

8646. Silver Carbonate. [534-16-7] Carbonic acid silver(1+) salt (1:2). CAg_2O_3; mol wt 275.74. C 4.36%, Ag 78.24%, O 17.41%. Ag_2CO_3. Silver carbonate pptd on Celite, *q.v.*, is known as *Fétizon's reagent.* Thermal decompn studies: M. Centnerszwer, B. Bruzs, *J. Phys. Chem.* **29**, 733 (1925); P. Norby *et al.*, **41**, 3628 (2002). Prepn from silver nitrate and sol carbonate soln: Poyer *et al.*, *Inorg. Synth.* **5**, 19 (1957). Crystal structure: R. Masse *et al.*, *Acta Crystallogr.* **B35**, 1428 (1979). Mechanism of alcohol oxidation using silver carbonate pptd on celite: M. Fetizon, M. Golfier, *C. R. Acad. Sci. Ser. C* **267**, 900 (1968); M. Fetizon *et al.*, *Tetrahedron Lett.* **13**, 4445 (1972); F. J. Kakis *et al.*, *J. Org. Chem.* **39**, 523 (1974). Biological staining method: J. M. López-Cepero, *J. Histochem. Cytochem.* **52**, 211 (2004).

Light yellow powder when freshly pptd, but becomes darker on drying and on exposure to light. Dec at about 220° into Ag_2O and CO_2 and at higher temp into metal Ag. d 6.08. Sol in 30,000 parts cold water, 2000 parts boiling water; readily sol in dil nitric acid, ammonia, or alkali cyanides. *Protect from light.*

USE: Mild oxidizing agent for conversion of alcohols to aldehydes and ketones. Biological stain.

8647. Silver Chlorate. [7783-92-8] Chloric acid silver(1+) salt (1:1); argentous chlorate. $AgClO_3$; mol wt 191.32. Ag 56.38%, Cl 18.53%, O 25.09%. Prepd from $AgNO_3$ and $NaClO_3$: Nicholson, Holley, Jr., *Inorg. Synth.* **2**, 4 (1946).

White, tetragonal crystals. d_4^{20} 4.430. mp 230°, dec 270° forming silver chloride and oxygen. Solubility in water (g/100 ml): 10 (15°);

20 (27°); 50 (80°). Slightly sol in alc. Darkens upon exposure to light due to slow decompn. *Keep away from light, organic vapors and oxidizable substances.*

USE: In organic synthesis as an effective oxidizing agent for certain organic compds.

8648. Silver Chloride. [7783-90-6] AgCl; mol wt 143.32. Ag 75.26%, Cl 24.73%.

White powder; darkens on exposure to light. d 5.56; mp 455°; bp 1550°. Soly in water (25°): 1.93 mg/l; hydrochloric acid increases its soly. Sol in 250 parts of concd HCl, in 13 parts of 10% ammonia; more sol in stronger ammonia and also at higher temps. Sol in solns of alkali cyanides, thiosulfates, ammonium carbonates; appreciably sol in concd aq solns of ammonium chloride; mercuric nitrate, and silver nitrate. The freshly pptd chloride dissolves more readily than the dried precipitate. Insol in alcohol, or dil acids. *Protect from light.*

USE: In silver plating, in making antiseptic silver prepns.

8649. Silver Chromate(VI). [7784-01-2] Chromic acid (H₂-CrO₄) silver(1+) salt (1:2). Ag₂CrO₄; mol wt 331.73. Ag 65.03%, Cr 15.67%, O 19.29%.

Dark brownish-red, cryst powder. d^{25} 5.625. Soly in water (0°): 0.0014%. Sol in nitric acid and ammonia.

Caution: Chromium hexavalent (VI) compounds are listed as known human carcinogens: *Report on Carcinogens, Twelfth Edition* (PB2011-111646, 2011) p 106.

USE: Catalyst for formation of aldol from alcohol; formed at end point of Mohr titration of halides.

8650. Silver Citrate. [126-45-4] 2-Hydroxy-1,2,3-propane-tricarboxylic acid silver(1+) salt (1:3); Itrol. $C_6H_5Ag_3O_7$; mol wt 512.70. C 14.06%, H 0.98%, Ag 63.12%, O 21.84%.

White, odorless, heavy, crystalline powder; darkens in light. Sol in 3500 parts water, more sol in boiling water; readily sol in dil HNO₃, ammonia. *Protect from light.*

USE: Anti-infective dusting powder.

8651. Silver Cyanide. [506-64-9] CAgN; mol wt 133.89. C 8.97%, Ag 80.56%, N 10.46%. AgCN.

White or grayish, odorless powder; stable in dry air; darkens on exposure to light; dec at 320°. *Poisonous.* d 3.95. Insol in water, alcohol, or dil acids. Sol in alkali cyanides and in boiling concd nitric acid; converted by dil HCl into hydrocyanic acid and silver chloride; sparingly sol in dil, more in concd ammonia. *Protect from light.*

USE: For silver plating; formerly used for extemporaneous prepn of dil hydrocyanic acid by treatment with HCl.

8652. Silver Difluoride. [7783-95-1] Silver fluoride (AgF₂); argentic fluoride. AgF₂; mol wt 145.87. Ag 73.95%, F 26.05%. Prepd by the action of fluorine on silver: Ebert *et al., J. Am. Chem. Soc.* **55**, 3056 (1933); by the interaction of fluorine and silver halides: Ruff, Giese, *Z. Anorg. Allg. Chem.* **219**, 143 (1934); v. Wartenberg, *ibid.* **242**, 406 (1939); Struve *et al., Ind. Eng. Chem.* **39**, 353 (1947). Laboratory procedure according to the equation 2AgCl + 2F₂ → 2AgF₂ + Cl₂: Priest, Swinehart, *Inorg. Synth.* **3**, 176 (1950); Kwasnik in *Handbook of Preparative Inorganic Chemistry* **vol. 1**, G. Brauer, Ed. (Academic Press, New York, 2nd ed., 1963) pp 241-242. Prepd from silver chloride and chlorine trifluoride: Rochow, Kukin, *J. Am. Chem. Soc.* **74**, 1615 (1952).

White when pure. Usually obtained as a gray-black or brownish, amorphous solid showing a yellow bloom. Sensitive to light. d 4.7. mp 690°. Very hygroscopic, is converted to a greasy black mass on exposure to atmospheric moisture. Violent reaction with water (instant hydrolytic cleavage). Powerful oxidizing agent giving off some ozone when treated with dil acids. Liberates iodine from iodides. May be stored in quartz or iron ampuls.

Caution: Highly toxic. Symptoms due to fluoride.

USE: In the fluorination of hydrocarbons.

8653. Silver Fluoride. [7775-41-9] Argentous fluoride; silver monofluoride. AgF; mol wt 126.87. Ag 85.02%, F 14.97%. Prepared according to the equation Ag₂CO₃ + 2HF → 2AgF + CO₂ + H₂O; also prepd by reduction of AgF₂ with H₂: Ruff, *Die Chemie des Fluors* (Berlin, 1920) p 37; Emeléus in *J. H. Simons' Fluorine Chemistry* **vol. I** (Academic Press, New York, 1950) p 32; Kwasnik

in *Handbook of Preparative Inorganic Chemistry* **vol. 1**, G. Brauer, Ed. (Academic Press, New York, 2nd ed., 1963) pp 240-241.

Flexible leaflets (cubic, NaCl lattice). Very hygroscopic. Darkens on exposure to light. d 5.852. mp 435°. bp ~1150°. Soly in water when freshly prepd: 182 g/100 ml at 15.5°. In moist air it gradually becomes insol because of basic fluoride formation. Aq solns are neutral and are solvents for silver oxide which renders them alkaline. Also sol in HF, NH₃, CH₃CN. Forms several hydrates. The dihydrate is stable to 39.5° and the tetrahydrate is stable from −14° to +18.7°. In addition, an acid fluoride, AgF.3HF can be prepd by cooling a soln of AgF in HF. This loses HF at 0°, forming AgF.HF.

Caution: Prolonged absorption may cause mottling of teeth, skeletal changes.

USE: To convert organic Br and Cl compds to their fluoro analogs; as antiseptic.

THERAP CAT: Anti-infective.

8654. Silver Iodate. [7783-97-3] Iodic acid (HIO₃) silver(1+) salt (1:1). AgIO₃; mol wt 282.77. Ag 38.15%, I 44.88%, O 16.97%.

White, cryst powder. d 5.53. mp >200°. Sol in 1875 parts water at 25°, in about 1000 parts of 35% nitric acid, in 2.5 parts of 10% ammonia. *Protect from light.*

USE: As reagent for determining small quantities of chloride, e.g., in blood: Sendroy, *J. Biol. Chem.* **120**, 335-445 (1937).

8655. Silver Iodide. [7783-96-2] AgI; mol wt 234.77. Ag 45.95%, I 54.05%. Prepd according to the equation AgNO₃ + KI → AgI + KNO₃: Kolkmeijer, van Hengel, *Z. Kristallogr.* **A88**, 317 (1934).

Light yellow, odorless powder; slowly darkened by light. Crystals are hexagonal or cubic. d 5.67; mp 552°. Practically insol in water (0.03 mg/l); in acid (except concd HI in which it dissolves readily on heating); in ammonium carbonate. Freely sol in solns of alkali cyanides or iodides; 35 mg dissolve in a liter of 10% ammonia; appreciably sol in concd solns of alkali bromides, chlorides, thiocyanates, thiosulfates, mercuric and silver nitrates. It is slowly attacked by boiling concd acids, but not affected by hot solns of alkali hydroxides.

USE: In cloud precipitation (rain-making).

THERAP CAT: Local anti-infective.

THERAP CAT (VET): In colloidal suspensions as local antiseptic for mucous membranes.

8656. Silver Lactate. [15768-18-0] [2-(Hydroxy-κO)pro-panoato-κO]silver; 2-hydroxypropanoic acid silver salt. C_3H_5-AgO₃; mol wt 196.94. C 18.30%, H 2.56%, Ag 54.77%, O 24.37%. CH₃CH(OH)COOAg.

Monohydrate. White or slightly gray, cryst powder readily affected by light. Sol in about 15 parts water; slightly sol in alcohol. *Protect from light.*

THERAP CAT: Topical anti-infective; astringent.

8657. Silver Nitrate. [7761-88-8] Nitric acid silver(1+) salt (1:1). AgNO₃; mol wt 169.87. Ag 63.50%, N 8.25%, O 28.26%. Prepn: T. W. Richards, G. S. Forbes, *J. Am. Chem. Soc.* **29**, 808 (1907). Clinical prophylactic use in neonatal conjunctivitis: Credé, *Arch. Gynaekol.* **17**, 50 (1881). Raman and IR structural studies: Z. X. Shen, W. F. Sherman, *J. Mol. Struct.* **271**, 175 (1992). Clinical evaluation in molluscum contagiosum: K. Niizeki, K. Hashimoto, *Pediatr. Dermatol.* **16**, 395 (1999); of histological effects in epithelial cautery: J. Hanif *et al., Clin. Otolaryngol.* **28**, 368 (2003). Review of prophylactic use in neonatal gonococcal conjunctivitis: G. Schneider, *Can. Med. Assoc. J.* **131**, 193-196 (1984); of toxicology: S. D. M. Humphreys, P. A. Routledge, *Adverse Drug React. Toxicol. Rev.* **17**, 115-143 (1998); of chromatographic applications: C. M. Williams, L. N. Mander, *Tetrahedron* **57**, 425-447 (2001).

Colorless, rhombic crystals or white small crystals. *Oxidizer.* d 4.352. Dimorphic; transition temp: 159.8°. mp 212°, forming a yellowish liquid solidifying to a white, cryst mass on cooling. Dec at 440° into metallic silver, nitrogen and nitrogen oxides. Soly in water at 25° (g/100 g H₂O): 2.16×10^2. Soly (g/L): ethanol 20.8; methanol 35; benzene 2.2. Not photosensitive when pure; trace amts of organic material promote photoreduction.

Toughened silver nitrate. [8007-31-6] Nitric acid silver(1+) salt mixt with silver chloride (AgCl); lunar caustic; molded silver nitrate. Contains not less than 94.5% silver nitrate, remainder is silver chloride. White or grayish, hard rods or thin small cones. Darkens on exposure to light. *Protect from light.*

Caution: Potential symptoms of overexposure by direct contact are tissue corrosion and burns; absorption through mucous membranes or damaged areas of skin may be followed by systemic toxicity (Humphreys). *See also* Silver.

USE: In chromatography; photography; manuf of mirrors; other silver salts; silver plating; in sympathetic and indelible inks; dyeing hair; coloring porcelain; etching ivory; as a titrimetric reagent in analytical chemistry.

THERAP CAT: Astringent, antiseptic, caustic.

THERAP CAT (VET): Astringent, antiseptic, caustic.

8658. Silver Nitrite. [7783-99-5] Nitrous acid silver(1+) salt (1:1). $AgNO_2$; mol wt 153.87. Ag 70.10%, N 9.10%, O 20.80%.

Pale yellow, odorless needles; non-hygroscopic; becomes gray in light. d 4.45. Dec at 140°. Sol in 300 parts water; more sol in boiling water; less sol in aqueous solns of silver nitrate; partly dec on prolonged boiling with water; insol in alcohol; dec by dil acids. *Protect from light.*

USE: For prepn of standard $NaNO_2$ soln for water analysis; prepn of aliphatic nitro-compds; as reagent for primary, secondary, and tertiary alcohols.

8659. Silver Oxalate. [533-51-7] $[\mu\text{-}[Ethanedioato(2-)}\text{-}\kappa O^1,\kappa O^{2'}:\kappa O^{1'},\kappa O^2]]$disilver; ethanedioic acid disilver(1+) salt. C_2-Ag_2O_4; mol wt 303.75. C 7.91%, Ag 71.02%, O 21.07%. Ag-OOCCOOAg.

White, cryst powder; at 140° dec violently. d 5.03. Sol in 24,000 parts water; sol in moderately concd nitric acid, in ammonia.

USE: Exptl photographic emulsion (low sensitivity).

8660. Silver Oxide. [20667-12-3] Argentous oxide. Ag_2O; mol wt 231.74. Ag 93.09%, O 6.90%. Prepd by making sol silver salt solns alkaline: Madsen, *Z. Anorg. Allg. Chem.* **79**, 197 (1913). Toxicity study: H. F. Smyth *et al., Am. Ind. Hyg. Assoc. J.* **30**, 470 (1969).

Brownish-black, heavy, odorless powder. d_4^{25} 7.22. Begins to dec at about 200° and at 250-300° decompn is rapid; it also breaks up into its constituents in sunlight. Reduced by hydogen, carbon monoxide and most metals. Moist silver oxide absorbs carbon dioxide. Sol in 40,000 parts water; freely sol in dil nitric acid, in ammonia; somewhat sol in NaOH solns. Practically insol in alcohol. *Protect from light.* LD_{50} orally in rats: 2.82 g/kg (Smyth).

USE: As catalyst; in the purification of drinking water; in the glass industry (polishing, coloring glass yellow).

THERAP CAT (VET): Has been used as germicide and parasiticide.

8661. Silver(II) Oxide. [1301-96-8] Argentic oxide; silver peroxide; silver suboxide; Divasil. AgO; mol wt 123.87. Ag 87.08%, O 12.92%. Exists as a silver(I)-silver(III) oxide: McMillan, *Chem. Rev.* **62**, 65 (1962). Prepd by the oxidation of silver nitrate with potassium peroxydisulfate in an alk medium: Hammer, Kleinberg, *Inorg. Synth.* **4**, 12 (1953). Prepn of single crystals: *Chem. Eng. News* **47**, 32 (Aug 11, 1969).

Charcoal-gray powder. *Avoid contact with skin, organic matter, strong ammonia and alkalies.* Malleable. Cubic or orthorhombic system when cryst. d_4^{25} 7.483. The dry solid is stable at 100° for 18 hrs, dec above 100° into silver and oxygen. Possesses semiconductor properties and is diamagnetic. *Strong oxidizing agent.* Ka 7.9 × 10^{-13}. Practically insol in water: 27 mg/l at 25° (decompn). Sol in alkalies and NH_4OH (with decompn and evolution of N_2). In dil acids oxygen is evolved immediately, in concd acid intensely colored solns are formed (brown in nitric acid and olive green in sulfuric acid).

Caution: Highly irritating to skin, eyes, mucous membranes, respiratory tract.

USE: In the manuf of silver oxide-zinc alkali batteries.

8662. Silver Perchlorate. [7783-93-9] Perchloric acid silver-(1+) salt (1:1). $AgClO_4$; mol wt 207.31. Ag 52.03%, Cl 17.10%, O 30.87%. Prepn from $NOClO_4$ + AgBr: Markowitz *et al., J. Inorg. Nucl. Chem.* **16**, 159 (1960).

Deliquescent crystals, dec 486°. d^{25} 2.806. Freely sol in water: 557 g/100 ml; a satd aq soln contains 84.8% w/w at 25°: Smith, Ring, *J. Am. Chem. Soc.* **59**, 1889 (1937). Much less sol in 60% perchloric acid (5.63% w/w). Forms a monohydrate stable to 43°.

Sol in many organic solvents, e.g., aniline, pyridine, benzene, toluene, nitromethane, glycerol, nitrobenzene, and chlorobenzene. Solvated crystals have been obtained contg 6 mols aniline, 4 mols pyridine, 1 mol benzene, 1 mol toluene. These compds explode readily when struck: Brinkley, *J. Am. Chem. Soc.* **62**, 3524 (1940).

Caution: Irritating to skin, mucous membranes.

USE: In the explosives industry.

8663. Silver Permanganate. [7783-98-4] Permanganic acid $(HMnO_4)$ silver(1+) salt; silver manganate(VII); Koerbl's catalyst. $AgMnO_4$; mol wt 226.80. Ag 47.56%, Mn 24.22%, O 28.22%. Prepn: Büssem, Herrmann, *Z. Kristallogr.* **A74**, 459 (1930).

Violet, cryst powder; dec in light. d 4.49. Soly in water at room temps about 9 g/l. More sol in hot water; dec by alcohol. *Protect from light.*

USE: In gas masks. Combustion catalyst for microelementary analysis.

8664. Silver Phosphate. [7784-09-0] Silver orthophosphate. Ag_3O_4P; mol wt 418.57. Ag 77.31%, O 15.29%, P 7.40%. Ag_3PO_4.

Yellow, odorless powder; darkens in light. d 6.37. mp 849°. Reduced by hydrogen. Soluble in 15,500 parts water, the soly in water is decreased by presence of silver nitrate; slightly sol in dil acetic acid; freely sol in dil nitric acid, in ammonia, ammonium carbonate, also in alkali cyanides and thiosulfates. *Protect from light.*

USE: In photography for collodion emulsions.

8665. Silver Picrate. [146-84-9] 2,4,6-Trinitrophenol silver-(1+) salt (1:1); picric acid silver derivative; silver trinitrophenolate; Picragol; Picrotol. $C_6H_2N_3O_7$; mol wt 335.96. C 21.45%, H 0.60%, Ag 32.11%, N 12.51%, O 33.34%. $(O_2N)_3C_6H_2OAg$.

Monohydrate. Yellow crystals. *Flammable.* Sol in about 50 parts water; sparingly sol in alc; slightly sol in acetone or glycerol. Insol in chloroform, ether.

THERAP CAT: Antiprotozoal (Trichomonas).

THERAP CAT (VET): Has been used in bovine granular vaginitis.

8666. Silver Protein. [9015-51-4] Argentoproteinum; protargin; silver proteinate; silver nucleate; silver nucleinate. Group of compounds characterized as colloidal combinations of silver and protein. Variably prepd from silver oxide, silver nitrate or other silver salts with gelatin, serum albumin, casein or peptone. Examples of prepns: **DE 105866** (1897 to Bayer); **GB 18478** (1897); H. P. Kaufmann, *Arzneimittel-Synthese* (Springer, 1953) p 588. Generally classified as mild or strong based on their germicidal and irritant properties. Differentiated by the amount of ionizable silver present: Van Deripe, Konnerth, *Am. J. Pharm.* **111**, 65 (1939). Use to stain nerve fibers: D. Bodian, *Anat. Rec.* **65**, 89 (1936). Neuronal selectivity and use in light and electron microscopy: I. S. Zagon, J. H. Haring, *Acta Anat.* **114**, 193 (1982); in Alzheimer's disease: C. Duyckaerts *et al., Acta Neuropathol.* **73**, 167 (1987).

Mild silver protein. Argentum vitellinum; Argyrol; Silvol. Brown, dark-brown, or almost black, odorless, lustrous scales or granules. Contains ~20% silver. Somewhat hygroscopic. Freely sol in water. Almost insol in alcohol, chloroform, ether. *Keep well closed and protected from light.*

Strong silver protein. Albumose silver; Protargol. Pale yellowish-orange to brownish-black, odorless powder. Contains ~8% silver, most of which is present in ionic form. Somewhat hygroscopic. Freely sol in water. Almost insol in alcohol, ether, chloroform. *Keep well closed and protected from light.*

USE: Neuronal-selective histological stain; general contrast medium for electron microscopy.

THERAP CAT: Antiseptic.

THERAP CAT (VET): Antiseptic.

8667. Silver Selenide. [1302-09-6] Ag_2Se; mol wt 294.70. Ag 73.21%, Se 26.79%. Occurs in nature as the mineral naumannite. Prepd by melting silver and selenium together and by passing hydrogen selenide into a soln of silver nitrate: Berzelius cited by Mellor, *A Comprehensive Treatise on Inorganic and Theoretical Chemistry* **10**, 771 (1930); by combination of ions or elements in solns or melts: Kulifay, **US 3026175** (1962 to Monsanto).

Gray, hexagonal, microscopic needles. mp 880°. d^{15} 8.216. Two forms exist; transition temp at 133°. Forms metallic silver and selenium oxide when heated in oxygen; transformed by chlorine into silver and selenium chlorides; oxidized to silver selenite by fuming nitric acid. Sol in molten silver or bismuth without chemical change. Practically insol in water.

8668. Silver Sulfate. [10294-26-5] Sulfuric acid silver(1+) salt (1:2). Ag_2O_4S; mol wt 311.79. Ag 69.19%, O 20.53%, S 10.28%. Ag_2SO_4. Prepd from silver nitrate and sulfuric acid: Richards, Jones, *Z. Anorg. Allg. Chem.* **55**, 72 (1907).
Crystals or cryst powder; slowly darkens on exposure to light. mp 657°; dec 1085°. d 5.45. Slowly sol in 125 parts water, 71 parts boiling water; sol in nitric acid, ammonia, concd sulfuric acid. Soly in $AgNO_3$ solns: Lietzke, Stoughton, *J. Inorg. Nucl. Chem.* **28**, 1877 (1966). *Protect from light.*
USE: Titrimetric reagent in analytical chemistry.

8669. Silver Sulfide. [21548-73-2] Argentous sulfide. Ag_2S; mol wt 247.80. Ag 87.06%, S 12.94%. Occurs in nature as the mineral *argentite*. Usually prepd from the elements: Hönigschmid, Sachtleben, *Z. Anorg. Allg. Chem.* **195**, 207 (1931).
Grayish-black, heavy powder. The cryst structure is orthorhombic, changing to cubic when heated above 179°. d_4^{20} 7.234. mp 845°. Insol in water. Sol in nitric acid, in solns of alkali cyanides.
USE: In ceramics.

8670. Silver Tetraiodomercurate(II). [7784-03-4] (*T*-4)-Disilver(1+) tetraiodomercurate(2−); mercuric silver iodide. Ag_2-HgI_4; mol wt 923.94. Ag 23.35%, Hg 21.71%, I 54.94%.
Deep yellow thermochromic powder. Insol in water or dil acids. Sol in solns of alkali iodides or cyanides.
USE: Thermal sensor to detect overheating in journal bearings, etc., becomes blood-red at 40-50° and yellow again on cooling.

8671. Silybin. Silibin; silibinin; silybum substance E_6; silymarin I. Hepatoprotective flavonoid compounds isolated from the fruit or seeds of milk thistle, *Silybum marianum* L. Gaertn. (*Asteraceae*); major constituent of the therapeutic extract, silymarin. Exists as a mixture of diastereomers A and B. Isoln and chemistry: H. Wagner *et al.*, *Arzneim.-Forsch.* **18**, 688 (1968); **24**, 466 (1974). Structural studies: R. Hänsel, G. Schöpflin, *Tetrahedron Lett.* **8**, 3645 (1967); A. Pelter, R. Hänsel, *ibid.* **9**, 2911 (1968); R. Hänsel *et al.*, *J. Chem. Soc. Chem. Commun.* **1972**, 195. Biomimetic synthesis: L. Merlini *et al.*, *ibid.* **1979**, 695; *eidem*, *J. Chem. Soc. Perkin Trans. 1* **1980**, 775. Pharmacology and toxicology: G. Hahn *et al.*, *Arzneim.-Forsch.* **18**, 698 (1968). Structure and stereochemistry of silybin and isosilybin diastereomers: D. Y.-W. Lee, Y. Liu, *J. Nat. Prod.* **66**, 1171 (2003); N.-C. Kim *et al.*, *Org. Biomol. Chem.* **1**, 1684 (2003). HPLC determn in plasma: C. S. L. Hoh *et al.*, *J. Agric. Food Chem.* **55**, 2532 (2007); and clinical pharmacokinetics of complex with phosphatidylcholine: W. Li *et al.*, *J. Chromatogr. B* **862**, 51 (2008). Mechanism of antioxidant action: P. Trouillas *et al.*, *J. Phys. Chem. A* **112**, 1054 (2008). Review of mechanism of action and therapeutic potential: R. Gazak *et al.*, *Curr. Med. Chem.* **14**, 315-338 (2007).

Silybin A

Silybin A. [22888-70-6] (2*R*,3*R*)-2-[(2*R*,3*R*)-2,3-Dihydro-3-(4-hydroxy-3-methoxyphenyl)-2-(hydroxymethyl)-1,4-benzodioxin-6-yl]-2,3-dihydro-3,5,7-trihydroxy-4*H*-1-benzopyran-4-one. $C_{25}H_{22}$-O_{10}; mol wt 482.44. Yellowish flat crystals from $CH_3OH + H_2O$, mp 162-163°. $[\alpha]_D$ +20° (c = 0.21 in acetone) (Lee, Liu). Also reported as mp 158-160°. $[\alpha]_D$ +6.1° (c = 0.3 in methanol). uv max (methanol): 217, 230, 288 nm (log ε 5.2, 5.2, 5.1) (Kim). Sol in acetone, ethyl acetate, methanol, ethanol; sparingly sol in chloroform. Practically insol in water.
Silybin B. [142797-34-0] (2*R*,3*R*)-2-[(2*S*,3*S*)-2,3-Dihydro-3-(4-hydroxy-3-methoxyphenyl)-2-(hydroxymethyl)-1,4-benzodioxin-6-yl]-2,3-dihydro-3,5,7-trihydroxy-4*H*-1-benzopyran-4-one. $C_{25}H_{22}$-O_{10}. Yellow grain crystals from $CH_3OH + H_2O$, mp 158-160°. $[\alpha]_D$ −1.07° (c = 0.28 in acetone) (Lee, Liu). Also reported as mp 157-

159°. $[\alpha]_D$ +0.6° (c = 0.3 in methanol). uv max (methanol): 217, 230, 288 nm (log ε 5.2, 5.3, 5.2) (Kim).
Isosilybin A. [142796-21-2] (2*R*,3*R*)-2-[(2*R*,3*R*)-2,3-Dihydro-2-(4-hydroxy-3-methoxyphenyl)-3-(hydroxymethyl)-1,4-benzodioxin-6-yl]-2,3-dihydro-3,5,7-trihydroxy-4*H*-1-benzopyran-4-one. $C_{25}H_{22}O_{10}$. Colorless needles from $CH_3OH + H_2O$, mp 201-203°. $[\alpha]_D$ +48.15° (c = 0.27 in acetone) (Lee, Liu). Also reported as mp 158-160°. $[\alpha]_D$ +30.4° (c = 0.3 in methanol). uv max (methanol): 212, 230, 288 nm (log ε 5.4, 5.3, 5.1) (Kim).
Isosilybin B. [142796-22-3] (2*R*,3*R*)-2-[(2*S*,3*S*)-2,3-Dihydro-2-(4-hydroxy-3-methoxyphenyl)-3-(hydroxymethyl)-1,4-benzodioxin-6-yl]-2,3-dihydro-3,5,7-trihydroxy-4*H*-1-benzopyran-4-one. $C_{25}H_{22}O_{10}$. Colorless needles from $CH_3OH + H_2O$, mp 236-238°. $[\alpha]_D$ −23.55° (c = 0.31 in acetone) (Lee, Liu). Also reported as mp 194-196°. $[\alpha]_D$ −17.6° (c = 0.3 in methanol). uv max (methanol): 212, 230, 288 nm (log ε 5.4, 5.3, 5.1) (Kim).

8672. Simazine. [122-34-9] 6-Chloro-N^2,N^4-diethyl-1,3,5-triazine-2,4-diamine; 2-chloro-4,6-bis(ethylamino)-*s*-triazine; 2,4-bis(ethylamino)-6-chloro-*s*-triazine; G-27692; Gesatop; Princep; Simanex. $C_7H_{12}ClN_5$; mol wt 201.66. C 41.69%, H 6.00%, Cl 17.58%, N 34.73%. Selective pre-emergence herbicide. Prepn: Hofmann, *Ber.* **18**, 2776 (1885); Pearlman, Banks, *J. Am. Chem. Soc.* **70**, 3726 (1948); Thurston *et al.*, *ibid.* **73**, 2981 (1951). As herbicide: Gysin, Knüsli, US 2891855 (1959 to Geigy). Toxicity studies: G. W. Bailey, J. L. White, *Residue Rev.* **10**, 97 (1965).

Crystals from ethanol or methyl Cellosolve, mp 226-227°. Practically insol in water. Slightly sol in dioxane, ethyl Cellosolve. LD_{50} orally in rats: 5000 mg/kg (Bailey, White).
USE: Herbicide.

8673. Simeconazole. [149508-90-7] α-(4-Fluorophenyl)-α-[(trimethylsilyl)methyl]-1*H*-1,2,4-triazole-1-ethanol; 2-(4-fluorophenyl)-1-(1*H*-1,2,4-triazol-1-yl)-3-trimethylsilylpropan-2-ol; F-155; Mongarit; Sanlit. $C_{14}H_{20}FN_3OSi$; mol wt 293.42. C 57.31%, H 6.87%, F 6.47%, N 14.32%, O 5.45%, Si 9.57%. Broad spectrum triazole fungicide for use in food crops and turf; sterol demethylation inhibitor (DMI). Prepn: J. Tobitsuka *et al.*, EP 537957; *eidem*, US 5306712 (1993, 1994 both to Sankyo); H. Itoh *et al.*, *Chem. Pharm. Bull.* **48**, 1148 (2000). Comprehensive description: M. Tsuda *et al.*, *Brighton Crop Prot. Conf. - Pests Dis.* **2000**, 557. Systemic activity in plants: M. Tsuda *et al.*, *Pest Manage. Sci.* **60**, 881 (2004); and comparison with other DMI fungicides: *eidem*, *ibid*, 875. Pharmacokinetics in rodents: K. Wakabayashi *et al.*, *J. Pestic. Sci.* **30**, 90 (2005).

Colorless crystals from diisopropyl ether, mp 118-119°. Log P (*n*-octanol/water): 3.2. Vapor pressure at 25°: 5.4 × 10^{-5} Pa. Soly in water (20°): 57.5 mg/l. LD_{50} in male, female rats (mg/kg): 611, 682 orally (Tsuda).
USE: Agricultural fungicide.

8674. Simetryn. [1014-70-6] N^2,N^4-Diethyl-6-(methylthio)-1,3,5-triazine-2,4-diamine; 2,4-bis(ethylamino)-6-(methylthio)-*s*-triazine; 2,4-bis(ethylamino)-6-(methylthio)-1,3,5-triazine; 2-methylthio-4,6-bis(monoethylamino)-*s*-triazine; G-32911; Gy-bon. C_8H_{15}-N_5S; mol wt 213.30. C 45.05%, H 7.09%, N 32.83%, S 15.03%. Prepn: Gysin, Knüsli, CH 337019 (1959 to Geigy); *C.A.* **57**, 14226c (1962). Crystal structure: A. J. Graham *et al.*, *Cryst. Struct. Commun.* **7**, 227 (1978). Toxicity study: E. F. Edson *et al.*, *World Rev. Pest Control* **1**, 36 (1965).

Crystals, mp 82-83°. LD$_{50}$ orally in rats: 1830 mg/kg (Edson).
USE: Herbicide.

8675. Simfibrate. [14929-11-4] 2-(4-Chlorophenoxy)-2-methylpropanoic acid 1,1'-(1,3-propanediyl) ester; 2-(p-chlorophenoxy)-2-methylpropionic acid trimethylene ester; 1,3-propanediol bis[α-(p-chlorophenoxy)isobutyrate]; 1,3-propanediol bis[2-(4-chlorophenoxy)-2-methylpropionate]; sinfibrate; CLY-503; Cholesolvin; Liposolvin. C$_{23}$H$_{26}$Cl$_2$O$_6$; mol wt 469.36. C 58.86%, H 5.58%, Cl 15.11%, O 20.45%. Prepn: NL 6600044 corresp to Nakanishi *et al.*, US 3494957 (1966, 1970 to Yoshitomi). Activity and toxicology studies: Nakanishi *et al.*, *J. Pharm. Soc. Jpn.* **90**, 267, 921, 926 (1970). Activity on postheparin lipoprotein lipase: L. Cavallini *et al.*, *Clin. Ter.* **92**, 25 (1980).

Crystals, mp 51-53°. bp$_{0.03}$ 197-200°, bp$_{0.15}$ 220-230°. LD$_{50}$ in mice, rats (g/kg): 3.3-3.5; 7.3-8.0 orally (Nakanishi, *J. Pharm. Soc. Japan*).

THERAP CAT: Antilipemic.

8676. Simplesse®. Protein-based fat substitute. Prepn from microparticulated, denatured dairy whey protein: N. S. Singer *et al.*, EP 250623; *eidem*, US 4734287 (both 1988 to John LaBatt Ltd.). Brief description: D. D. Duxbury, *Food Process.* **49**, 21 (1988). *Review*: N. A. Bringe, D. R. Clark, *Front. Foods Food Ingredients* **1**, 51-68 (1993).

USE: Fat substitute for use in foods.

8677. Simvastatin. [79902-63-9] 2,2-Dimethylbutanoic acid (1S,3R,7S,8S,8aR)-1,2,3,7,8,8a-hexahydro-3,7-dimethyl-8-[2-[(2R,4R)-tetrahydro-4-hydroxy-6-oxo-2H-pyran-2-yl]ethyl]-1-naphthalenyl ester; 2,2-dimethylbutyric acid 8-ester with (4R,6R)-6-[2-[(1S,2S,6R,8S,8aR)-1,2,6,7,8,8a-hexahydro-8-hydroxy-2,6-dimethyl-1-naphthyl]ethyl]tetrahydro-4-hydroxy-2H-pyran-2-one; synvinolin; MK-733; Denan; Liponorm; Lodalès; Simovil; Sinvacor; Sivastin; Zocor; Zocord. C$_{25}$H$_{38}$O$_5$; mol wt 418.57. C 71.74%, H 9.15%, O 19.11%. HMG-CoA reductase inhibitor; synthetic analog of lovastatin, *q.v.* Prepn: A. K. Willard *et al.*, EP 33538; W. F. Hoffman *et al.*, US 4444784 (1981, 1984 both to Merck & Co.); W. F. Hoffman *et al.*, *J. Med. Chem.* **29**, 849 (1986). Comprehensive description: D. K. Ellison *et al.*, *Anal. Profiles Drug Subs. Excip.* **22**, 359-388 (1993). HPLC/MS determn in plasma and pharmacokinetics: B. Barrett *et al.*, *J. Pharm. Biomed. Anal.* **41**, 517 (2006). Effect on survival in CHD: Scandinavian Simvastatin Survival Study Group, *Lancet* **344**, 1383 (1994). Clinical trial in primary hypercholesterolemia: L. Ose *et al.*, *Clin. Cardiol.* **23**, 39 (2000); in mixed hyperlipidemia: E. Stein *et al.*, *Am. J. Cardiol.* **86**, 406 (2000). Review of development and clinical experience: J. G. Robinson, *Expert Opin. Pharmacother.* **8**, 2159-2172 (2007); of therapeutic potential in multiple sclerosis: O. Neuhaus, H.-P. Hartung, *Expert Rev. Neurother.* **7**, 547-556 (2007).

White to off-white, nonhygroscopic, crystalline powder from *n*-butyl chloride + hexane, mp 135-138°. Soly (mg/ml): chloroform 610; DMSO 540; methanol 200; ethanol 160; *n*-hexane 0.15; 0.1M HCl 0.06; polyethylene glycol-400 70; propylene glycol 30; 0.1M NaOH 70; water 0.03. uv max (acetonitrile): 231, 238, 247 nm (A$_{1 cm}^{1\%}$ 516, 604, 408). [α]$_D^{25}$ +292° (c = 0.5% in acetonitrile).

THERAP CAT: Antilipemic.

8678. Sinalbin. [20196-67-2] 2-[[3-(4-Hydroxy-3,5-dimethoxyphenyl)-1-oxo-2-propen-1-yl]oxy]-N,N,N-trimethylethanaminium 1-thio-β-D-glucopyranose 1-[4-hydroxy-N-(sulfooxy)benzeneethanimidate] ion(1−); sinapine glucosinalbate. C$_{30}$H$_{42}$N$_2$O$_{15}$S$_2$; mol wt 734.79. C 49.04%, H 5.76%, N 3.81%, O 32.66%, S 8.73%. Mustard oil glucoside from white or yellow mustard (*Sinapis alba* L., *Cruciferae*): Gadamer, *Arch. Pharm.* **235**, 38, 83, 570 (1897); Viehoever, Nelson, *J. Assoc. Off. Agric. Chem.* **21**, 488 (1938); *Am. J. Pharm.* **110**, 411 (1938). Structure: Schulz, Gmelin, *Z. Naturforsch.* **7b**, 500 (1952); Ettlinger, Lundeen, *J. Am. Chem. Soc.* **78**, 4172 (1956).

Needles from 95% ethanol, mp 100-102°. (Also reported as 83-84° for the hydrated, 138-140° for the anhydr.) [α]$_D^{20}$ −8.76° (c = 0.29). Soluble in hot alcohol, cold water.

8679. Sinapine. [18696-26-9] 2-[[3-(4-Hydroxy-3,5-dimethoxyphenyl)-1-oxo-2-propenyl]oxy]-N,N,N-trimethylethanaminium; sinapic acid choline ester. [C$_{16}$H$_{24}$NO$_5$]$^+$. From black mustard seeds, the seeds of *Brassica nigra* Koch, *Cruciferae*. Extraction procedure: Remsen, Coale, *Am. Chem. J.* **6**, 52 (1884). Synthesis: Späth, *Monatsh. Chem.* **41**, 271 (1920).

Chloride. (C$_{16}$H$_{24}$NO$_5$)Cl. Does not crystallize, very sol in water, alcohol.

Bromide trihydrate. (C$_{16}$H$_{24}$NO$_5$)Br.3H$_2$O. Needles, mp 92°; mp 107-115° when anhydr, very sol in water.

Iodide trihydrate. (C$_{16}$H$_{24}$NO$_5$)I.3H$_2$O. Crystals, becomes anhydr over H$_2$SO$_4$, mp 179° (anhydr). Sparingly sol in water.

Acid sulfate trihydrate. (C$_{16}$H$_{24}$NO$_5$)HSO$_4$.3H$_2$O. Tablets, dec 127°; dec 187° when anhydr. Freely sol in water, hot alcohol; very sparingly sol in ether.

Neutral sulfate pentahydrate. (C$_{16}$H$_{24}$NO$_5$)$_2$SO$_4$.5H$_2$O. Tablets from hot alcohol, mp 193° (anhydr). Not very stable; very sol in water, sparingly in alcohol.

8680. Sinapultide. [138531-07-4] KL4. C$_{126}$H$_{238}$N$_{26}$O$_{22}$; mol wt 2469.45. C 61.28%, H 9.71%, N 14.75%, O 14.25%. Synthetic pulmonary surfactant protein consisting of 21 amino acid residues in a repeating sequence of one lysine and four leucines. Mimics ***surfactant protein-B*** (SP-B), a naturally occurring, hydrophobic protein found in human lung surfactant. Prepn: C. G. Cochrane, S. D. Revak, WO 9222315; *eidem*, US 5407914 (1992, 1995 both to Scripps Res. Inst.). Surface behavior in phospholipid systems: J. Ma *et al.*, *Biophys. J.* **74**, 1899 (1998).

Lys–Leu–Leu–Leu–Leu–Lys–Leu–Leu–Leu–Leu–Lys–Leu–Leu–Leu–Leu
|
Lys–Leu–Gmeln–Leu–Leu–Lys

Lucinactant. KL$_4$-surfactant; Surfaxin. Mixture of sinapultide with dipalmitoyl phosphatidylcholine and palmitoyl oleoylphosphatidyl glycerol (3:1) and palmitic acid. Effect on pulmonary function in animal models of meconium aspiration syndrome: C. G. Cochrane *et al.*, *Pediatr. Res.* **44**, 705 (1998). Clinical evaluation in meconium aspiration syndrome: T. E. Wiswell *et al.*, *Pediatrics* **109**, 1081 (2002). Clinical trial in premature infant respiratory distress syndrome: S. K. Sinha *et al.*, *ibid.* **115**, 1030 (2005). Review of chemistry and clinical experience: S. M. Donn, *Expert Opin. Invest. Drugs* **14**, 329-334 (2005).

THERAP CAT: Pulmonary surfactant; in treatment of respiratory distress syndrome.

8681. Sincalide. [25126-32-3] 5-L-Methionine-3-10-caerulein; 1-de(5-oxo-L-proline)-2-de-L-glutamine-5-L-methioninecaerulein; cholecystokinin *C*-terminal octapeptide; CCK *C*-terminal octapeptide; pancreozymin *C*-terminal octapeptide; SQ-19844; Kinevac. $C_{49}H_{62}N_{10}O_{16}S_3$; mol wt 1143.27. C 51.48%, H 5.47%, N 12.25%, O 22.39%, S 8.41%. The *C*-terminal octapeptide of cholecystokininpancreozymin (*see* cholecystokinin). Structurally similar to caerulein, *q.v.*, it evokes a variety of biological responses similar to CCK-PZ, including smooth-muscle contraction of the gall bladder and small intestine, relaxation of the choledochoduodenal junction, protein secretion by the pancreas and acid secretion by the stomach. It is more active than cholecystokinin on a weight or molar basis. Initial isoln: V. Mutt, J. E. Jorpes, *Eur. J. Biochem.* **6**, 156 (1968). Prepn: M. A. Ondetti *et al.*, **DE 1922185** corresp to **US 3723406** (1969, 1970, both to Squibb); *eidem*, *J. Am. Chem. Soc.* **92**, 195 (1970). Mechanical and metabolic effects: K. E. Andersson *et al.*, *Acta Physiol. Scand.* **96**, 495 (1976). *In vitro* study: M. Deschodt-Lanckman *et al.*, *Gastroenterology* **68**, 819 (1975). Use in clinical radiology: W. M. Thompson, J. R. Amberg, *Gastrointest. Radiol.* **3**, 191 (1978). Effect on food intake: M. A. Della-Fera, C. A. Baile, *Science* **206**, 471 (1979). Review of synthesis: L. Balaspiri, *Acta Physiol. Acad. Sci. Hung.* **47**, 299 (1976). Review of pharmacology: S. L. Engel *et al.*, in *Pharmacological and Biochemical Properties of Drug Substances* vol. 2, M. E. Goldberg, Ed. (Am. Pharm. Assoc., Washington, DC, 1979) pp 516-526.

SO₃H
|
Asp–Tyr–Met–Gly–Trp–Met–Asp–PheNH₂

Solid. $[\alpha]_D^{23}$ −18.4° (c = 0.7 in 1*N* NH₃). uv max (0.1*N* NaOH): 280, 288 nm (ε 4850, 4230).

THERAP CAT: Choleretic.

8682. Sinefungin. [58944-73-3] 6,9-Diamino-1-(6-amino-9*H*-purin-9-yl)-1,5,6,7,8,9-hexadeoxy-D-glycero-α-L-*talo*-decofuranuronic acid; Compd 57926; Antibiotic 32232RP; A-9145. $C_{15}H_{23}N_7O_5$; mol wt 381.39. C 47.24%, H 6.08%, N 25.71%, O 20.97%. Adenine-containing antibiotic produced by *Streptomyces griseoleus* NRRL 3739. Isoln: R. L. Hamill, M. M. Hoehn, **US 3758681** (1973 to Lilly); *eidem*, *J. Antibiot.* **26**, 463 (1973). Antifungal activity: R. S. Gordee, T. F. Butler, *ibid.* 466. Antiprotozoal activity: D. K. Dube *et al.*, *Am. J. Trop. Med. Hyg.* **32**, 31 (1983); A. Ferrante *et al.*, *Trans. R. Soc. Trop. Med. Hyg.* **78**, 837 (1984); P. Paolantonacci *et al.*, *Antimicrob. Agents Chemother.* **28**, 528 (1985); E. Zweygarth *et al.*, *Trop. Med. Parasitol.* **37**, 255 (1986); *in vivo* antileishmanial activity: J. L. Avila *et al.*, *Am. J. Trop. Med. Hyg.* **43**, 139 (1990). Synthesis: M. Geze *et al.*, *J. Am. Chem. Soc.* **105**, 7638 (1983); stereoselective synthesis: M. P. Maguire *et al.*, *J. Org. Chem.* **55**, 948 (1990). Inhibition of methyltransferase: R. T. Borchardt *et al.*, *Biochem. Biophys. Res. Commun.* **89**, 919 (1979).

White powder. Sol in water, very slightly sol in methanol and ethanol. Insol in organic solvents. Stable at room temp in aq soln

from pH 1 to 11. pKa (66% DMF): 2.9, 3.9, 8.9, 10.2. uv max in neutral solns: 206 nm, 258 nm ($E_{1cm}^{1\%}$ 520, 325); in basic solns: 256 nm ($E_{1cm}^{1\%}$ 325). $[\alpha]_D^{25}$ −2.61° (c = 5 in water). Also reported as $[\alpha]_D^{23}$ +12 ±2° (c = 0.227 in water). LD₅₀ s.c. in mice: 185 mg/kg (Hamill, Hoehn, *J. Antibiot.*)

8683. Sinigrin. [3952-98-5] 1-Thio-β-D-glucopyranose 1-[*N*-(sulfooxy)-3-butenimidate] potassium salt (1:1); sinigroside; myronate potassium; potassium myronate; allyl glucosinolate. $C_{10}H_{16}KNO_9S_2$; mol wt 397.45. C 30.22%, H 4.06%, K 9.84%, N 3.52%, O 36.23%, S 16.13%. A β-glucopyranoside isolated from black mustard seeds and from the horseradish root, *Alliaria officinalis* Andrz., *Cruciferae*: A. Bussy, *J. Pharm. Chim.* **26**, 39 (1840); Gadamer, *Arch. Pharm.* **235**, 44 (1897); Stoll, Seebeck, *Helv. Chim. Acta* **31**, 1432 (1948). Structure: Ettlinger, Lundeen, *J. Am. Chem. Soc.* **78**, 4173 (1956); **79**, 1764 (1957). Hydrolysis: Ettlinger *et al.*, *Proc. Natl. Acad. Sci. USA* **47**, 1875 (1961). Crystal structure: Waser, Watson, *Nature* **198**, 1297 (1963). Synthesis: Benn, Ettlinger, *Chem. Commun.* **1965**, 445; A. Kjaer, *Acta Chem. Scand.* **22**, 3324 (1968).

Monohydrate. Crystals, mp 127-129°. When anhydr mp 179°. $[\alpha]_D^{18}$ −16.4°. Freely sol in water, hot alcohol; insol in benzene, chloroform, ether.

Tetraacetate. $C_{18}H_{24}KNO_{13}S_2$. Crystals, mp 193-195°. $[\alpha]_D^{26}$ −16° (c = 0.14).

8684. Sinomenine. [115-53-7] (9α,13α,14α)-7,8-Didehydro-4-hydroxy-3,7-dimethoxy-17-methylmorphinan-6-one; cucoline; coculine; kukoline. $C_{19}H_{23}NO_4$; mol wt 329.40. C 69.28%, H 7.04%, N 4.25%, O 19.43%. An optical isomer of methoxythebainone. Configuration at the asymmetric centers, C_9, C_{13}, and C_{14}, is the mirror image of those in morphine. From root of *Sinomenium acutum* (Thunb.) Rehd. & Wils. (*Cocculus diversifolius* DC.), *Menispermaceae*: Goto, *J. Chem. Soc. Jpn.* **44**, 795 (1923), *C.A.* **18**, 2710 (1924); Ohta, *Kitasato Arch. Exp. Med.* **6**, 259 (1925), *C.A.* **21**, 3233² (1927). Structure: Kondo, Ochiai, *Ann.* **470**, 224 (1929); Sasaki, *Yakugaku Zasshi* **80**, 270 (1960), *C.A.* **54**, 11702g (1960). Biogenesis: Cohen, *Chem. Ind. (London)* **1956**, 1391. Pharmacology: S. Fu *et al.*, *Yao Hsueh Hsueh Pao* **10**, 673 (1963), *C.A.* **60**, 8525h (1964). *Reviews:* Holmes in R. H. F. Manske, H. L. Holmes, *The Alkaloids* vol. **II** (Academic Press, New York, 1952) pp 219-260; K. W. Bentley, *The Chemistry of the Morphine Alkaloids* (Oxford, 1954).

Clusters of needles from benzene, mp 161°. After melting once, the mp is raised to 182°. $[\alpha]_D^{26}$ −71° (c = 2.1 in alc). Sol in alc, acetone, chloroform, dil alkali; slightly sol in water, ether, benzene. LD₅₀ orally in mice: 580 mg/kg (Fu).

Hydrochloride dihydrate. $C_{19}H_{24}ClNO_4.2H_2O$. Long prisms from water, dec 231° when anhydr. $[\alpha]_D^{17}$ −82° (c = 4.4). Sol in about 1.5 parts water.

8685. Sipuleucel-T. APC-8015; Provenge. Autologous dendritic cell product loaded *ex vivo* with the recombinant fusion protein

PA2024 consisting of human granulocyte macrophage-colony stimulating factor (GM-CSF) linked to human prostatic acid phosphatase (PAP). Therapeutic vaccine designed to stimulate immunity against PAP expressed on prostate cancer cells. Prepn of target antigen: R. Laus *et al.*, **WO 9724438**; *eidem*, **US 5976546** (1997, 1999 both to Dendreon); of antigen presenting cells and pharmacology: E. J. Small *et al.*, *J. Clin. Oncol.* **18**, 3894 (2000). Clinical evaluation in androgen-independent prostate cancer: P. A. Burch *et al.*, *Prostate* **60**, 197 (2004); P. F. Schellhammer, R. M. Hershberg, *World J. Urol.* **23**, 47 (2005). Review of development and therapeutic potential: F. H. Valone *et al.*, *Cancer J.* **7**, Suppl. 2, S53-S61 (2001); B. I. Rini, *Curr. Opin. Mol. Ther.* **4**, 76-79 (2002).

THERAP CAT: Antineoplastic, immunomodulator.

8686. Sirius Red. [2610-10-8] 7,7′-(Carbonyldiimino)bis[4-hydroxy-3-[2-[2-sulfo-4-[2-(4-sulfophenyl)diazenyl]phenyl]diazenyl]-2-naphthalenesulfonic acid] sodium salt (1:6); C.I. 35780; C.I. Direct Red 80; Sirius Red F3B. $C_{45}H_{26}N_{10}Na_6O_{21}S_6$; mol wt 1373.05. C 39.36%, H 1.91%, N 10.20%, Na 10.05%, O 24.47%, S 14.01%. Sulphonated tetrakisazo dye. Dye molecules bind parallel to helical native collagen. Water and light stability: J. J. Porter, *Text. Res. J.* **43**, 735 (1973). Spectral study: J. P. Hart, W. F. Smyth, *Spectrochim. Acta* **36A**, 279 (1980). Staining mechanisms: L. Feldskov Nielsen *et al.*, *Biotech. Histochem.* **73**, 71 (1998). Use as stain for collagen: F. Sweat *et al.*, *Arch. Pathol.* **78**, 69 (1964); D. A. Lee *et al.*, *J. Mater. Sci. Mater. Med.* **9**, 47 (1998).

uv max (buffered): 220, 235, 290 nm (ε 1.2×10^4, 1.3×10^4, 1.2×10^4). vis max (buffered): 420, 520, 545 nm (ε 0.8×10^4, 2.5×10^4, 2.6×10^4). $pK_{1-6} < 2$; $pK_{7,8} > 13$.

USE: Stain for connective tissue. Dye for cellulose fibers.

8687. Sisal. Henequem; yaxci; potosina. *Agave sisalana* Perrine, Amaryllidaceae; one of the plants native to the American desert. Process of treating juice and extracting steroidal material and waxes: Spray, **US 3019193** (1962). *Habit.* Principally the Mexican tableland. Abundant in the Mexican states Yucatan and Chiapas. *Constit.* Hemp, sapogenins such as hecogenin and tigogenin. *Review:* *Chem. Week* 41-48 (April 20, 1957).

Caution: The dust is irritating to respiratory tract.

USE: Manuf sisal ropes, steroid intermediates.

8688. Sisomicin. [32385-11-8] *O*-3-Deoxy-4-*C*-methyl-3-(methylamino)-β-L-arabinopyranosyl-(1 → 6)-*O*-[2,6-diamino-2,3,4,6-tetradeoxy-α-D-*glycero*-hex-4-enopyranosyl-(1 → 4)]-2-deoxy-D-streptamine; (2*S*-cis)-4-*O*-[3-amino-6-(aminomethyl)-3,4-dihydro-2*H*-pyran-2-yl]-2-deoxy-6-*O*-[3-deoxy-4-*C*-methyl-3-(methylamino)-β-L-arabinopyranosyl]-D-streptamine; rickamicin; antibiotic 6640; Sch-13475. $C_{19}H_{37}N_5O_7$; mol wt 447.53. C 50.99%, H 8.33%, N 15.65%, O 25.02%. Gentamicin-like aminoglycoside antibiotic produced by *Micromonospora inyoesis* (NRRL 3292): M. J. Weinstein *et al.* **DE 1932309**; *eidem*, **US 3832286** (1970, 1974 both to Schering); *eidem*, *J. Antibiot.* **23**, 551 (1970). Isoln: G. H. Wagman *et al.*, *ibid.* 555. Chemical and physical studies: Cooper *et al.*, *Chem. Commun.* **1971**, 285, 924; Kugelman *et al.*, *J. Antibiot.* **26**, 394 (1973). Structure consists of 2-deoxystreptamine, *q.v.*, linked to two saccharide units, **garosamine** and **sisosamine**: Keiman *et al.*, *J. Org. Chem.* **39**, 1451 (1974). Synthesis: D. H. Davies *et al.*, *J. Med. Chem.* **21**, 189 (1978). Synthesis of sisosamine derivs: Cleophax *et al.*, *Chem. Commun.* **1975**, 11. Has broad spectrum antibiotic activity: Waitz *et al.*, *J. Antibiot.* **23**, 559 (1970). Biotransformation of sisomicin to gentamicin: B. K. Lee *et al.*, *Antimicrob. Agents*

Chemother. **12**, 335 (1977). Clinical studies: R. Lorber *et al.*, *Clin. Ther.* **4**, 263 (1981); J. P. Sculier *et al.*, *J. Antimicrob. Chemother.* **9**, 63 (1982). *Review:* P. Noone, *Drugs* **27**, 548-578 (1984).

Monohydrate. Needles from ethanol, mp 198-201°. $[\alpha]_D^{26}$ +189° (c = 0.3).

Penta-*N*-acetyl derivative. $C_{29}H_{47}N_5O_{12}$. Amorphous white powder from methanol-ethyl ether, mp 188-198° (dec). $[\alpha]_D^{26}$ +200° (c = 0.3).

Sulfate. [53179-09-2] Baymicin; Extramycin; Mensiso; Siseptin; Sisobiotic; Sisolline; Sisomin. $(C_{19}H_{37}N_5O_7)_2.5H_2SO_4$; mol wt 1385.43. LD_{50} in mice (mg/kg): 34 i.v., 221 i.p., 288 s.c. (Weinstein).

THERAP CAT: Antibacterial.

8689. Sitafloxacin. [127254-12-0] 7-[(7*S*)-7-Amino-5-azaspiro[2.4]hept-5-yl]-8-chloro-6-fluoro-1-[(1*R*,2*S*)-2-fluorocyclopropyl]-1,4-dihydro-4-oxo-3-quinolinecarboxylic acid; DU-6859; Gracevit. $C_{19}H_{18}ClF_2N_3O_3$; mol wt 409.82. C 55.69%, H 4.43%, Cl 8.65%, F 9.27%, N 10.25%, O 11.71%. Fluorinated quinolone antibacterial. Prepn: I. Hayakawa, Y. Kimura, **EP 341493**; *eidem*, **US 5587386** (1989, 1996 both to Daiichi); Y. Kimura *et al.*, *J. Med. Chem.* **37**, 3344 (1994). Antibacterial spectrum: L. M. Deshpande, R. N. Jones, *Diagn. Microbiol. Infect. Dis.* **37**, 139 (2000). HPLC determn in serum and urine: H. Aoki *et al.*, *J. Chromatogr. B* **660**, 365 (1994). Clinical pharmacokinetics: M. Nakashima *et al.*, *Antimicrob. Agents Chemother.* **39**, 170 (1995). Clinical evaluation: N. Shetty, A. P. R. Wilson, *J. Antimicrob. Chemother.* **46**, 633 (2000).

Sesquihydrate. [163253-35-8] DU-6859a. Crystals from ethanol, water + ammonium hydroxide, mp 225° (dec). $[\alpha]_{589}$ −199.9° (c = 1 in 1*N* NaOH). Soly in water: 131 μg/ml.

THERAP CAT: Antibacterial.

8690. Sitagliptin. [486460-32-6] (3*R*)-3-Amino-1-[5,6-dihydro-3-(trifluoromethyl)-1,2,4-triazolo[4,3-*a*]pyrazin-7(8*H*)-yl]-4-(2,4,5-trifluorophenyl)-1-butanone; (2*R*)-4-oxo-4-[3-(trifluoromethyl)-5,6-dihydro[1,2,4]triazolo[4,3-*a*]pyrazin-7(8*H*)-yl]-1-(2,4,5-trifluorophenyl)butan-2-amine; 7-[(3*R*)-3-amino-1-oxo-4-(2,4,5-trifluorophenyl)butyl]-5,6,7,8-tetrahydro-3-(trifluoromethyl)-1,2,4-triazolo[4,3-*a*]pyrazine. $C_{16}H_{15}F_6N_5O$; mol wt 407.32. C 47.18%, H 3.71%, F 27.99%, N 17.19%, O 3.93%. Selective inhibitor of dipeptidyl peptidase IV (DPP-IV). Prepn: S. D. Edmondson *et al.*, **WO 03004498**; *eidem*, **US 6699871** (2003, 2004 both to Merck & Co.); and enzyme inhibitory profile: D. Kim *et al.*, *J. Med. Chem.* **48**, 141 (2005). Improved process: K. B. Hansen *et al.*, *Org. Process Res. Dev.* **9**, 634 (2005). Discovery and characterization: N. A. Thornberry, A. E. Weber, *Curr. Top. Med. Chem.* **7**, 557 (2007). HTLC determn in plasma: W. Zeng *et al.*, *Rapid Commun. Mass Spectrom.* **20**, 1169 (2006). Clinical pharmacokinetics and pharmacodynamics: G. A. Herman *et al.*, *Clin. Pharmacol. Ther.* **78**,

675 (2005). *Review*: C. F. Deacon, *Curr. Opin. Investig. Drugs* **6**, 419-426 (2005). Review of pharmacology and clinical experience: M. Choy, S. Lam, *Cardiology Rev.* **15**, 264-271 (2007).

Viscous oil.

Monophosphate monohydrate. [654671-77-9]; [654671-78-0] (anhydrous). Sitagliptin phosphate; MK-0431; Januvia. $C_{16}H_{15}F_6$-$N_5O.H_3PO_4.H_2O$; mol wt 523.33. White to off-white, crystalline, nonhygroscopic powder, mp 215-217°. $[\alpha]_D$ −74.4° (c = 1.0 in water). Sol in water, DMF; slightly sol in methanol; very slightly sol in ethanol, acetone, acetonitrile. Insol in isopropanol, isopropyl acetate.

THERAP CAT: Antidiabetic.

8691. Sitamaquine. [57695-04-2] N^1,N^1-Diethyl-N^6-(6-methoxy-4-methyl-8-quinolinyl)-1,6-hexanediamine; 8-(6'-diethylaminohexylamino)-6-methoxylepidine; 6-methoxy-8-(6-diethylaminohexylamine)-4-methylquinoline; kalazaquine; WR-6026. C_{21}-$H_{33}N_3O$; mol wt 343.52. C 73.43%, H 9.68%, N 12.23%, O 4.66%. 8-Aminoquinolone; analog of the antimalarial agent primaquine, *q.v.* Prepn: K. N. Campbell *et al., J. Am. Chem. Soc.* **68**, 1556 (1946); *idem,* US 2508937 (1950 to USA, as represented by the Secy. of War). Antileishmanial activity in hamsters: K. E. Kinnamon *et al., Am. J. Trop. Med. Hyg.* **27**, 751 (1978). HPLC determn in plasma: J. C. Anders *et al., J. Chromatogr.* **311**, 117 (1984). Clinical evaluation in AIDS-associated *Pneumocystis carinii* pneumonia: B. G. Petty *et al., J. Acquir. Immune Defic. Syndr.* **21**, 26 (1999); in visceral leishmaniasis: R. Dietze *et al., Am. J. Trop. Med. Hyg.* **65**, 685 (2001). *Ex vivo* use in prevention of transfusion-associated Chagas' disease: H. Moraes-Souza *et al., Rev. Soc. Bras. Med. Trop.* **35**, 563 (2002). Review of clinical development: C. Yeates, *Curr. Opin. Investig. Drugs* **3**, 1446-1452 (2002); H. Sangraula *et al., J. Assoc. Physicians India* **51**, 686-690 (2003).

bp$_{0.05}$ 190-200°.

Dihydrochloride. [5330-29-0] $C_{21}H_{33}N_3O.2HCl$; mol wt 416.43. Yellow crystals from propanol, mp 179-180°.

THERAP CAT: Antiprotozoal (Leishmania).

8692. Sitaxsentan. [184036-34-8] *N*-(4-Chloro-3-methyl-5-isoxazolyl)-2-[2-(6-methyl-1,3-benzodioxol-5-yl)acetyl]-3-thiophenesulfonamide; *N*-(4-chloro-3-methyl-5-isoxazolyl)-2-[3,4-(methylenedioxy)-6-methyl]phenylacetyl-3-thiophenesulfonamide; IPI-1040; TBC-11251. $C_{18}H_{15}ClN_2O_6S_2$; mol wt 454.90. C 47.53%, H 3.32%, Cl 7.79%, N 6.16%, O 21.10%, S 14.10%. Selective endothelin A (ET$_A$) receptor antagonist. Prepn: M. F. Chan *et al.,* **WO 9631492** ; *eidem,* **US 5594021** (1996, 1997 both to Texas Biotech.); C. Wu *et al., J. Med. Chem.* **40**, 1690 (1997). Pharmacology: R. G. Tilton *et al., Pulm. Pharmacol. Ther.* **13**, 87 (2000). Clinical evaluation in heart failure: M. M. Givertz *et al., Circulation* **101**, 2922 (2000); in pulmonary arterial hypertension: D. Langleben *et al., Chest* **126**, 1377 (2004). Review of development and clinical

experience in pulmonary arterial hypertension: R. J. Barst, *Expert Opin. Pharmacother.* **8**, 95-109 (2007).

Yellow powder, mp 42-45°.

Sodium salt. [210421-74-2] Thelin. $C_{18}H_{14}ClNaN_2O_6S_2$; mol wt 476.88. Yellow powder, mp >200° (dec). Sol in water, normal saline, 5% dextrose, and DMSO. Max soly 78 mg/ml at pH >5.5.

8693. **α_1-Sitosterol.** [474-40-8] (3β,4α,5α,24Z)-4-Methylstigmasta-7,24(28)-dien-3-ol; citrostadienol. $C_{30}H_{50}O$; mol wt 426.73. C 84.44%, H 11.81%, O 3.75%. Occurs in plants. Isoln from wheat germ oil: Anderson *et al., J. Am. Chem. Soc.* **48**, 2987 (1926); Wallis, Fernholz, *ibid.* **58**, 2446 (1936). Structure: Bernstein, Wallis, *ibid.* **61**, 2308 (1939); Mazur, Sondheimer, *ibid.* **80**, 6293, 6296 (1958). Identity with citrostadienol: Schreiber, Osske, *Experientia* **19**, 69 (1963); *eidem, Tetrahedron* **20**, 2575 (1964). Stereochemistry: C. Brooks *et al., Steroids* **20**, 487 (1972).

Needles from alcohol, mp 164-166°. $[\alpha]_D^{28}$ −1.7° (c = 2 in chloroform). Is pptd by digitonin.

8694. **β-Sitosterol.** [83-46-5] (3β)-Stigmast-5-en-3-ol; 22:23-dihydrostigmasterol; α-dihydrofucosterol; Δ^5-stigmasten-3β-ol; 24β-ethyl-Δ^5-cholesten-3β-ol; α-phytosterol; cinchol; cupreol; rhamnol; quebrachol; sitosterin; Harzol; Prostasal; Sito-Lande. C_{29}-$H_{50}O$; mol wt 414.72. C 83.99%, H 12.15%, O 3.86%. Common sterol in plants. Isoln from wheat germ oil, corn oil: Anderson *et al., J. Am. Chem. Soc.* **48**, 2987 (1926); from rye germ oil: Gloyer, Schuette, *ibid.* **61**, 1901 (1939); from cottonseed oil: Wallis, Chakravorty, *J. Org. Chem.* **2**, 335 (1937); from tall oil: Sandqvist, Bengtsson, *Ber.* **64**, 2167 (1931). Also occurs in soy and calabar beans, in rice embryos: Tanaka, *J. Biochem.* **17**, 483 (1933); in cascara, cinchona bark and cinchona wax: Dirscherl, *Z. Physiol. Chem.* **235**, 1 (1935). Prepn from tall oil: G. I. Fujimoto, A. E. Jacobson, *J. Org. Chem.* **29**, 3377 (1964). Identity with 22:23-dihydrostigmasterol: W. Dirscherl, H. Nahm, *Ann.* **555**, 57 (1944). Identity with cinchol: *eidem, Ann.* **558**, 231 (1947). Structure: Bernstein, Wallis, *J. Org. Chem.* **2**, 341 (1937); Bergmann, Low, *ibid.* **12**, 67 (1947); Shoppee, *J. Chem. Soc.* **1948**, 1032. Stereospecific synthesis: W. Sucrow, M. Slopianka, *Ber.* **108**, 3721 (1975). Clinical efficacy in treatment of type II hyperlipoproteinemia: R. S. Lees, A. M. Lees, in *Lipoprotein Metabolism*, H. Greten, Ed. (Springer-Verlag, New York, 1976) pp 119-124; P. Oster *et al., ibid.* pp 125-130. Studies on inhibition of cholesterol absorption: S. M. Grundy, H. Y. I. Mok, *ibid.* pp 112-118; I. Ikeda, M. Sugano, *Biochim. Biophys. Acta* **732**, 651 (1983). Inhibition of induced carcinogenesis: N. D. Nigro *et al., J. Natl. Cancer Inst.* **69**, 103 (1982). Clinical trial in treatment of prostatic adenoma: H.-P. Szutrely, *Med. Klin.* **77**, 520 (1982). Book: *Monographs on Atherosclerosis* **Vol. 10**, T. B. Clarkson *et al.,* Eds. entitled "Sitosterol" by O. J. Pollak, D. Kritchevsky (Karger, Basel, 1981) 219 pp.

mp 218-220°.

Sodium salt tetrahydrate. [201677-61-4]; [150374-95-1] (anhydrous). EI-546; ONO-5046; Elaspol. $C_{20}H_{21}N_2NaO_7S.4H_2O$; mol wt 528.50. LD_{50} in rats (mg/kg): 450 i.v. (Yanagi).

THERAP CAT: In treatment of lung injury resulting from systemic inflammatory response syndrome.

8697. Sizofiran. [9050-67-3] Schizophyllan; SPG; Sonifilan. Immunostimulant polysaccharide produced by the fungus *Schizophyllum commune* Fries. Polymer of repeating tetrasaccharide units composed of three β-(1 → 3)-linked D-glucopyranose residues, to one of which is attached a single β-(1 → 6)-linked D-glucopyranosyl side chain. Forms triple-stranded helix of mol wt ~450,000 in neutral aqueous solution. Prepn by fermentation: S. Kikumoto *et al.*, *JP 67 12000* (1967 to Taito); *C.A.* **67**, 107386r (1967). Synthesis of the 3 possible tetrasaccharide units: K. Takeo, S. Tei, *Carbohydr. Res.* **145**, 293 (1986). Host-mediated antitumor activity: N. Komatsu *et al.*, *Gann* **60**, 137 (1969). Pharmacology, toxicology and antitumor activity in mice: T. Matsuo *et al.*, *Arzneim.-Forsch.* **32**, 647 (1982). Macrophage stimulant activity *in vitro*: I. Sugawara *et al.*, *Cancer Immunol. Immunother.* **16**, 137 (1984). Clinical trial in cervical cancer: K. Okamura *et al.*, *Cancer* **58**, 865 (1986).

White, amorphous powder. Sol in water, forming highly viscous soln.

THERAP CAT: Antineoplastic.

8698. Skatole. [83-34-1] 3-Methyl-1*H*-indole. C_9H_9N; mol wt 131.18. C 82.41%, H 6.92%, N 10.68%. Constituent of feces, beetroot, nectandra wood, and coal tar. Obtained by fusing egg albumin with KOH. Toxicity data: Bin-Ichi, *Tohoku J. Exp. Med.* **25**, 407 (1935).

White to brownish scales; fecal odor; mp 95°; bp 265-266°. Sol in hot water, alc, benzene, chloroform, ether. With potassium ferrocyanide and sulfuric acid it gives a violet color. MLD in frogs (mg/kg): 1000 s.c. (Bin-Ichi).

8699. Skimmianine. [83-95-4] 4,7,8-Trimethoxyfuro[2,3-*b*]quinoline; β-fagarine; 7,8-dimethoxydictamnine. $C_{14}H_{13}NO_4$; mol wt 259.26. C 64.86%, H 5.05%, N 5.40%, O 24.68%. In *Skimmia japonica* Thunb., *Fagara* spp., *Glycosmis pentaphylla* Corr., *Ruta graveolens* L., *Rutaceae*: Honda, *Arch. Exp. Pathol. Pharmakol.* **52**, 83 (1904); Paris, Moyse-Mignon, *Ann. Pharm. Fr.* **5**, 410 (1947), *C.A.* **42**, 3909h (1948); Chatterjee, Majumdar, *J. Am. Chem. Soc.* **76**, 2459 (1954); Schneider, *Arzneim.-Forsch.* **14**, 435 (1964). Identity with β-fagarine: Deulofeu *et al.*, *J. Am. Chem. Soc.* **64**, 2326 (1942). Structure: Asahina, Inubuse, *Ber.* **63**, 2052 (1930).

Pyramids, octahedral rods from alcohol. mp 178°. uv max (ethanol): 212, 251, 321, 331 nm (log ε 4.10, 4.91, 3.89, 3.91). Neutral to litmus. Sol in alcohol, chloroform; slightly in ether, amyl alcohol and carbon disulfide; practically insol in water and petr ether.

8700. Skimmin. [93-39-0] 7-(β-D-Glucopyranosyloxy)-2*H*-1-benzopyran-2-one; 7-(glucosyloxy)coumarin; 7-hydroxycoumarin-7-glucoside; umbelliferone glucoside. $C_{15}H_{16}O_8$; mol wt 324.29. C 55.56%, H 4.97%, O 39.47%. In bark and wood of *Skimmia japonica* Thunb., *Rutaceae*. Isoln, structure and synthesis: Späth, Neufeld, *Rec. Trav. Chim.* **57**, 535 (1938).

Plates from alcohol, mp 140°. $[\alpha]_D^{25}$ −37° (c = 2 in chloroform).

Acetate. $C_{31}H_{52}O_2$. mp 127-128°. $[\alpha]_D^{25}$ −41° (c = 2 in chloroform).

THERAP CAT: Antilipemic. Treatment of prostatic adenoma.

8695. γ-Sitosterol. [83-47-6] (3β,24*S*)-Stigmast-5-en-3-ol; clionasterol. $C_{29}H_{50}O$; mol wt 414.72. C 83.99%, H 12.15%, O 3.86%. Principal sterol of soybean oil: Bonstedt, *Z. Physiol. Chem.* **176**, 269 (1928). Isoln from corn oil: Anderson, Shriner, *J. Am. Chem. Soc.* **48**, 2976 (1926); from venom of Chinese toad, *Bufo gargarizans:* Rukstuhl, Meyer, *Helv. Chim. Acta* **40**, 1270 (1957). Differs from β-sitosterol in spatial configuration at C-24: Bergmann, Low, *J. Org. Chem.* **12**, 67 (1947). Structure: Pavanaram, *Helv. Chim. Acta* **46**, 1377 (1963). Stereospecific synthesis: W. Sucrow, M. Slopianka, *Ber.* **108**, 3721 (1975).

Monohydrate. Plates from alcohol. When dry, mp 147-148°. $[\alpha]_D^{20}$ −43° (c = 1.9 in chloroform). Table of mp's and $[\alpha]_D$'s: Dirscherl, *Z. Physiol. Chem.* **257**, 242 (1939).

Acetate. $C_{31}H_{52}O_2$. mp 143-144°. $[\alpha]_D^{20}$ −45.3° (c = 1.9 in chloroform).

8696. Sivelestat. [127373-66-4] 4-[[[2-[[(Carboxymethyl)amino]carbonyl]phenyl]amino]sulfonyl]phenyl ester 2,2-dimethylpropanoic acid; *N*-[2-[[[4-(2,2-dimethyl-1-oxopropoxy)phenyl]sulfonyl]amino]benzoyl]glycine; *N*-[*o*-(*p*-pivaloyloxybenzene)sulfonylaminobenzoyl]glycine; *N*-[2-[4-(2,2-dimethylpropionyloxy)phenylsulfonylamino]benzoyl]aminoacetic acid. $C_{20}H_{22}N_2O_7S$; mol wt 434.46. C 55.29%, H 5.10%, N 6.45%, O 25.78%, S 7.38%. Neutrophil elastase inhibitor. Prepn: K. Imaki *et al.*, **EP 347168**; *eidem*, **US 5017610** (1989, 1991 both to Ono); and structure-activity study: K. Imaki *et al.*, *Bioorg. Med. Chem.* **4**, 2115 (1996). Pharmacology: K. Kawabata *et al.*, *Biochem. Biophys. Res. Commun.* **177**, 814 (1991). HPLC determn in plasma and urine: T. Shintani *et al.*, *J. Pharm. Biomed. Anal.* **12**, 397 (1994). Pharmacokinetics and metabolism in rats: F. Watanabe *et al.*, *Biol. Pharm. Bull.* **20**, 392 (1997). Enzyme inhibition in *ex vivo* human pleural effusion: T. Sakuma *et al.*, *Eur. J. Pharmacol.* **353**, 273 (1998). Acute toxicity: H. Yanagi *et al.*, *J. Toxicol. Sci.* **23**, Suppl. 3, 409 (1998); *C.A.* **130**, 20355 (1998). *Review*: L. Pradella, *Curr. Opin. Anti-Inflam. Immunomod. Invest. Drugs* **1**, 485-501 (1999).

β-D-glucose–O... [chemical structure]

Crystals with one H_2O from water. After drying at 100° and 10 mm: mp 219-221° (evac tube). $[\alpha]_D^{18}$ −80° (c = 10 in pyridine). uv max (ethanol): 320 nm. Moderately sol in hot water, alcohol; practically insol in ether, chloroform.

8701. Slendid®. Natural food ingredient made from pectin extracted from citrus peels. Analysis of interaction with flavor compounds: J. P. Schirle-Keller *et al.*, *J. Food Sci.* **59**, 813 (1994). Brief review: D. D. Duxbury, *Food Process.* **52**, 54-55 (1991).

Light cream to light tan, granular free-flowing powder. Tasteless and odorless. Completely sol in 25 parts of water at 60° with adequate agitation.

USE: Fat replacer in foods.

8702. SMEAH. [22722-98-1] Dihydrobis[2-(methoxy-κO)-ethanolato-κO]aluminate(1−) sodium (1:1); sodium bis(2-methoxyethoxy)aluminum hydride; Red-Al; Vitride. $C_6H_{16}AlNaO_4$; mol wt 202.16. C 35.65%, H 7.98%, Al 13.35%, Na 11.37%, O 31.66%. $NaAlH_2(OCH_2CH_2OCH_3)_2$. Prepn: B. Casensky *et al.*, **US 3507895** (1970). Properties: J. Vit, *Eastman Organic Chemical Bulletin* **42**(3), 1 (1970). Review of reductions by metal alkoxyaluminum hydrides: J. Málek, *Org. React.* **34**, 1-317 (1985); *idem, ibid.* **36**, 252-590 (1987).

Highly viscous liquid at room temp. Miscible in a great variety of inert solvents (e.g., aromatic hydrocarbons and ethers). Immiscible in paraffinic hydrocarbons. Thermally stable at 200°C; thermal decomposition starts at 205°C and is vigorous.

USE: Reducing agent.

8703. Smilagenin. [126-18-1] (3β,5β,25R)-Spirostan-3-ol; isosarsasapogenin. $C_{27}H_{44}O_3$; mol wt 416.65. C 77.83%, H 10.64%, O 11.52%. Steroid sapogenin from *Smilax ornata* Hooker, *Liliaceae* (Sarsaparilla): Ashew *et al.*, *J. Chem. Soc.* **1936**, 1399; from *Agave lecheguilla* Tou., *Amaryllidaceae:* Nord, **US 2897192** (1959). C_{25} isomer of sarsasapogenin: Scheer *et al.*, *J. Am. Chem. Soc.* **77**, 641 (1955). Obtained by acid isomerization of sarsasapogenin: Marker, Rohrmann, *ibid.* **61**, 846 (1939); Wall *et al.*, *ibid.* **77**, 3086 (1955). For structure and stereochemistry *see* Sarsasapogenin.

Silky needles from acetone, mp 185°. $[\alpha]_D^{25}$ −69° (c = 0.5 in chloroform). Absorption spectrum: Smith, Eddy, *Anal. Chem.* **31**, 1539 (1959).

Acetate. $C_{29}H_{46}O_4$. Needles from methanol, mp 152°.

USE: In making compds of the pregnane series.

8704. Soap. Any salt of a fatty acid, usually made by saponification of a vegetable oil with caustic soda. Hard soap consists largely of sodium oleate (low titer) or of sodium palmitate or stearate (high titer); soft soap, of potassium and sodium salts of coconut and/or palm oil fatty acids. Rosin acid salts have detergent properties and are often incorporated in laundry soaps. Shaving soaps contain glycerol and gum to prevent rapid drying. Medicated soaps contain antiseptics such as phenols or mercury salts. Soaps also may be prepd from the fatty acid and an amine, *e.g.,* triethanolamine. Non-alkali-metal salts of the fatty acids are water insol and as such have certain uses.

Hard soap (olive-oil castile) is a white or yellowish white powder or bar, sol in water, alc. The aq soln is alkaline due to hydrolysis; the alc soln is only slightly alkaline. Transparent hard soap is obtained by adding additional glycerol during saponification. Floating soaps are made by adding air after saponification during the so-called "crutching" process. Soft soap (green soap) is a yellowish-white to brownish-yellow (or green) soft mass, more sol than hard soap.

USE: Hard soap—detergent; in soln as a vehicle for liniments; as ingredient of pills contg resinous drugs like aloe, etc. Soft soap—detergent; extern. as vehicle for active medicaments applied in ointment form or liniment. Insoluble soaps—fungicides (Cu), disinfectants (Pb, Cu, Hg), face powders and ointments (Zn), waxes and polishes (Al), greases (Al, Ca) and surface coatings (Ca).

THERAP CAT: Topical anti-infective. Antidote (mineral acid and heavy metal poisoning).

THERAP CAT (VET): Topically as detergent (cleansing agent) and mild antiseptic. In liniments as counterirritant. Has been used inter-

nally as laxative, antacid and as an antidote for mineral acids and heavy metals.

8705. Sobrerol. [498-71-5] 5-Hydroxy-α,α,4-trimethyl-3-cyclohexene-1-methanol; *p*-menth-6-ene-2,8-diol; 6,8-carvomenthenediol; 1-*p*-menthene-6,8-diol; pinol hydrate. $C_{10}H_{18}O_2$; mol wt 170.25. C 70.55%, H 10.66%, O 18.79%. An oxidation product of oil of turpentine formed by the autooxidation of α-pinene, *q.v.,* in the presence of water and especially in sunlight. First obtained by Margueron in 1797. Confused with terpin hydrate, *q.v.,* until fully characterized by Sobrero in 1851. The naturally occurring form is the *trans*-form. For a review of early structure and prepn work, *see* J. L. Simonsen, *The Terpenes* **vol. II** (University Press, Cambridge, 2nd ed., 1949) pp 137-138. Summary of general methods of prepn: Lombard, Heywang, *Bull. Soc. Chim. Fr.* **1954**, 1210. Prepn of *dl-trans*-form: Ward, *J. Am. Chem. Soc.* **60**, 325 (1938); Schenck *et al., Ann.* **584**, 177 (1953); Moore *et al., J. Am. Chem. Soc.* **78**, 1173 (1956); Brook, Wright, *J. Org. Chem.* **22**, 1314 (1957); of *l-trans*-form: Klein, **US 2815378** (1958 to Glidden). Prepn of *cis*- and *trans*-forms and analysis of stereochemistry: Blumann, Wood, *J. Chem. Soc.* **1952**, 4420; H. Schmidt, *Ber.* **86**, 1437 (1953). Pharmacology and chemistry of *dl-trans*-form: **GB 1176817**; C. Corvi-Mora, **US 3592908** (1970, 1971 to Camillo Corvi); V. Dalla Valle, *Boll. Chim. Farm.* **109**, 761 (1970). Clinical evaluation: C. F. Marchioni, G. Monzali, *Clin. Ter.* **60**, 135 (1972). Activity on mucociliary transport: P. Braga, *Clin. Trials J.* **18**, 30 (1981). Pharmacokinetics: P. C. Braga *et al., Eur. J. Clin. Pharmacol.* **24**, 209 (1983). Metabolism: P. Ventura *et al., Xenobiotica* **13**, 139 (1983); *eidem, ibid.* **15**, 317 (1985). GLC determn in human plasma: U. A. Shulka, *J. Chromatogr.* **308**, 189 (1984).

[chemical structure]

(−)-*trans* form

***dl-trans*-Form.** Lysmucol; Sobrepin. Crystals, mp 130-131.5°. bp 270-271°. Soly in water: 3.3 g/100 ml (15°). LD_{50} in rats, mice (mg/kg): both 580 i.v. (Dalla Valle).
***d-trans*-Form.** mp 148-149°. $[\alpha]_D^{20}$ +150° (Schmidt).
***l-trans*-Form.** mp 148-149°. $[\alpha]_D^{20}$ −150° (Schmidt).
THERAP CAT: Mucolytic.

8706. Sobuzoxane. [98631-95-9] *C,C′*-[1,2-Ethanediylbis[(2,6-dioxo-4,1-piperazinediyl)methylene]]carbonic acid *C,C′*-bis(2-methylpropyl) ester; 1,2-ethanediylbis[(2,6-dioxo-4,1-piperazinediyl)methylene]carbonic acid bis(2-methylpropyl) ester; 4,4′-ethylenebis[1-(hydroxymethyl)-2,6-piperazinedione] bis(isobutyl carbonate) (ester); 4,4′-(1,2-ethanediyl)bis(1-isobutoxycarbonyloxymethyl)-2,6-piperazinedione; 1,2-bis(4-isobutoxycarbonyloxymethyl-3,5-dioxopiperazin-1-yl)ethane; MST-16; Perazolin. $C_{22}H_{34}N_4O_{10}$; mol wt 514.53. C 51.36%, H 6.66%, N 10.89%, O 31.09%. Noncleavable complex-stabilizing DNA topoisomerase II inhibitor. Prepn: J.-C. Cai, M. Takase, **EP 140327**; *eidem*, **US 4650799** (1985, 1987 both to Zenyaku Kogyo); *idem et al., Chem. Pharm. Bull.* **37**, 2976 (1989). Antitumor activity: T. Narita *et al., Cancer Chemother. Pharmacol.* **28**, 235 (1991). Clinical evaluation in non-Hodgkin's lymphoma: R. Ohno *et al., J. Natl. Cancer Inst.* **84**, 435 (1992); in adult T-cell leukemia-lymphoma: *idem et al., Cancer* **71**, 2217 (1993). Acute toxicity study: T. Tokiwa *et al., Iyakuhin Kenkyu* **22**, 435 (1991); *C.A.* **115**, 222916k (1991).

[chemical structure]

Crystals from ethylene glycol monomethylether, mp 128-130°. Also reported as crystals from ethanol, mp 132-133° (Cai, 1989). Practically insol in water. LD$_{50}$ in male, female mice, rats (mg/kg): 807, 960, 877, 567 i.p.; 400, 673, 3025, 2821 s.c.; >5 g/kg orally in all species (Tokiwa).

THERAP CAT: Antineoplastic.

8707. Soda Lime. [8006-28-8] Sodium hydroxide (Na(OH)) mixt with lime. A mixture of calcium oxide with 5-20% sodium hydroxide and contg 6-18% water. It absorbs 25-35% its wt of CO$_2$.

White or gray-white granules. Rapidly deteriorates on exposure to air through absorption of carbon dioxide. *Keep tightly closed.*

Caution: Strongly corrosive and irritating to skin, mucous membranes, eyes. Ingestion can cause severe damage to G.I. tract, death.

USE: To absorb carbon dioxide in basal metabolism tests, in rebreathing anesthesia systems, in submarines, and in carbon determinations.

8708. Sodium. [7440-23-5] Natrium. Na; at. no. 11; at. wt 22.98976928; valence 1. Group IA (1). Alkali metal. Occurrence in earth's crust: 2.83% by wt; principal cation in hydrosphere. Naturally occurring isotope: 23 (100%); radioactive isotopes (mass number): 19-22; 24-35. Prepd by Davy in 1807 by electrolysis of fused sodium hydroxide. Found in form of its compds, halides, silicates, carbonates; does not occur free. Industrial prepns primarily in Downs cells, also in Castner cells: Batsford, *Chem. Metall. Eng.* **26**, 888, 932 (1932); Regelsberger, *Chemische Technologie der Leichtmetalle* Leipzig, 1926; Hardie, *Ind. Chem.* **30**, 161 (1954). *Reviews: ACS Monograph Series* **no. 133**, entitled "Sodium," M. Sittig, Ed. (Reinhold, New York, 1956); Whaley, "Sodium, Potassium, Rubidium, Cesium and Francium" in *Comprehensive Inorganic Chemistry* Vol. **1**, J. C. Bailar, Jr. *et al.*, Eds. (Pergamon Press, Oxford, 1973) pp 369-529; *Chemistry of the Elements* N. N. Greenwood, A. Earnshaw, Eds. (Pergamon Press, New York, 1984) pp 75-116; C. H. Lemke, V. H. Markant in *Kirk-Othmer Encyclopedia of Chemical Technology* vol. **22** (Wiley-Interscience, New York, 4th ed., 1997) pp 327-354.

Light, silvery-white metal; body-centered cubic structure; lustrous when freshly cut; tarnishes on exposure to air, becoming dull and gray. Soft at ordinary temp, fairly hard at −20°. mp 97.82°; bp 881.4°; d^{20} 0.968. Heat capacity of solid: 0.292 cal/g deg; heat capacity of liquid at mp: 0.331 cal/g deg. Heat of fusion: 27.05 cal/g; thermal conductivity (cal/sec °C cm): 0.205 (97.82°) 0.170 (400°). E^0 (aqueous) Na/Na$^+$ 2.714 V. *Dangerous when wet.* Violently decomposes water, forming sodium hydroxide and hydrogen which may ignite spontaneously. Decomposes alc. Reacts vigorously with oxygen, burning with a yellow flame. Emits characteristic yellow color (589.2 nm) in flame. Combines directly with the halogens, with phosphorus. Reduces most oxides to the elemental form, reduces metallic chlorides. Dissolves in liquid ammonia to give a blue soln; when heated in ammonia gas yields sodamide. Dissolves in mercury, forming sodium amalgam. *Keep under liquids contg no oxygen, such as kerosene, naphtha.*

Caution: Direct contact with metal may be corrosive and cause skin and eye burns. *See: Fire Protection Guide to Hazardous Materials* (National Fire Protection Assoc., Quincy, MA, 12th ed., 1997) Section 49, p 117.

USE: Manuf of sodium compds, such as the cyanide, azide, peroxide, etc.; manuf of tetraethyllead; manuf of refractory metals; in org syntheses; for photoelectric cells; in sodium lamps; as catalyst for many polymerization reactions. Alloyed with potassium in heat transfer media. Prepn of alkoxide titrants.

8709. Sodium Acetate. [127-09-3] Acetic acid sodium salt (1:1). C$_2$H$_3$NaO$_2$; mol wt 82.03. C 29.28%, H 3.69%, Na 28.03%, O 39.01%. CH$_3$COONa. Crystal structure of the trihydrate: K.-T. Wei, D. L. Ward, *Acta Crystallogr.* **B33**, 522 (1977).

Hygroscopic powder. d 1.518. Very sol in water; sol in alc.

Trihydrate. [6131-90-4] C$_2$H$_3$NaO$_2$.3H$_2$O; mol wt 136.08. Transparent crystals or granules. Efflorescent in warm air. d 1.45. mp 58°; becomes anhyd at 120°; dec at higher temp. One gram dissolves in 0.8 ml water, 0.6 ml boiling water, 19 ml alc. pH of 0.1 molar aq soln at 25°: 8.9.

USE: Anhydrous form as auxiliary in acetylations. Trihydrate in photography; as reagent in anal. chemistry to eliminate effect of strong acids (buffer); for foot warmers and milk-bottle warmers.

Technical grades are used as a mordant in dyeing. As acidulant in food. Pharmaceutic aid (buffering agent in dialysis solutions).

THERAP CAT (VET): Has been used in bovine ketosis.

8710. Sodium Acid Pyrophosphate. [7758-16-9] Diphosphoric acid sodium salt (1:2); disodium dihydrogen pyrophosphate. H$_2$Na$_2$O$_7$P$_2$; mol wt 221.94. H 0.91%, Na 20.72%, O 50.46%, P 27.91%. Prepn: Bell, *Inorg. Synth.* **3**, 98 (1950).

White, fused masses or powder. Dec at 220°. d (hexahydrate) 1.86. Sol in water, the soln having an acid reaction.

USE: Chiefly in baking powders.

8711. Sodium Alizarinesulfonate. [130-22-3] 9,10-Dihydro-3,4-dihydroxy-9,10-dioxo-2-anthracenesulfonic acid sodium salt (1:1); alizarine S; alizarine carmine; C.I. Mordant Red 3; C.I. 58005. C$_{14}$H$_7$NaO$_7$S; mol wt 342.25. C 49.13%, H 2.06%, Na 6.72%, O 32.72%, S 9.37%. Prepn: *Colour Index* vol. **4** (3rd ed., 1971) p 4514.

Monohydrate. Orange-yellow powder. Freely sol in water with yellow color; sol in alcohol.

USE: As acid-base indicator and in the determination of fluorine, usually used as a 1% aq soln. pH: 3.7 yellow, 5.2 purple; as a reagent for aluminum and stain in microscopy.

8712. Sodium Aluminate. [1302-42-7] Aluminate (AlO$_2$$^{1-}$) sodium (1:1); aluminum sodium oxide. AlNaO$_2$; mol wt 81.97. Al 32.92%, Na 28.05%, O 39.04%.

White granular mass. mp 1650°. *Corrosive.* Very sol in water. Insol in alcohol. The aq soln is strongly alkaline.

USE: Printing on fabrics; manuf lake colors; sizing paper; manuf milk-glass, soap; hardening building stones; water softener.

8713. Sodium Amalgam. [11110-52-4] An alloy of sodium and mercury. Readily prepd by adding sodium metal in small portions at a time to mercury. The combination takes place with evolution of much heat. The sodium contents of the amalgam can be varied, as desired, up to about 20% sodium, corresponding to approx Na$_2$Hg. The consistency and mp vary with the sodium content; the amalgam corresponding to NaHg$_2$, 5.4% Na, has the highest mp, +360°.

Sodium amalgams are dec by water similarly to sodium metal, but more slowly. *Keep tightly closed.*

8714. Sodium Amide. [7782-92-5] Sodium amide (Na(NH$_2$)); sodamide. H$_2$NNa; mol wt 39.01. H 5.17%, N 35.91%, Na 58.93%. NaNH$_2$. Prepd from sodium metal and gaseous ammonia (or liq ammonia with ferric nitrate catalyst): Dennis, Browne, *Inorg. Synth.* **1**, 74 (1939); Greenlee, Henne, *ibid.* **2**, 128 (1946); Schenk in *Handbook of Preparative Inorganic Chemistry* vol. **1**, G. Brauer, Ed. (Academic Press, New York, 2nd ed., 1963) pp 465-468. Alternate method: Bergstrom, *Org. Synth.* **coll. vol. III**, 778 (1955).

Crystals, mp 210°. The commercial product may be a white to olive-green solid with sea-shell fracture. Begins to volatilize at 400° and dec into its elements between 500 and 600°. Heat of formation of solid (18°; 1 atm): −32.26 kcal/mole. Heat of soln (21°): −31.06 kcal/mole. Soly in liquid ammonia at 20° = about 0.1%. Reacts violently with water forming NaOH and NH$_3$. The reaction with alcohol is considerably slower. Should be stored in sealed containers which prevent all contact with air during storage: Bergstrom, Fernelius, *Chem. Rev.* **12**, 63, 75, 78 (1933). When exposed to the atmosphere, sodium amide rapidly absorbs H$_2$O and CO$_2$. When only limited absorption takes place, as in poorly sealed containers, products are formed which render the resulting mixture highly explosive. The formation of oxidation products is accompanied by the development of a yellow or brownish color. If such a change is noticed, the substance should be destroyed at once. This is conveniently accomplished by covering with much benzene, toluene, or kerosene and slowly adding dil ethanol with stirring.

Caution: Intensely irritating to skin, eyes, mucous membranes.

USE: Dehydrating agent. In the production of indigo and hydrazine. Intermediate in the prepn of sodium cyanide. In ammonolysis reactions, in Claisen condensations, alkylation of nitriles and ketones, synthesis of ethynyl compds, acetylenic carbinols. Fused $NaNH_2$ dissolves metallic Mg, Zn, Mo, W, quartz, glass, silicates and other substances.

8715. Sodium Arsenate, Dibasic. [7778-43-0] Arsenic acid (H_3AsO_4) sodium salt (1:2); disodium arsenate. $AsHNa_2O_4$; mol wt 185.91. As 40.30%, H 0.54%, Na 24.73%, O 34.42%. Na_2HAsO_4. Toxicity study: Franke, Moxon, *J. Pharmacol. Exp. Ther.* **58**, 454 (1936). Review of toxicology and human exposure: *Toxicological Profile for Arsenic* (PB2008-100002, 2007) 559 pp.

Powder. *Poisonous.* Very sol in water; slightly sol in alc.

Heptahydrate. [10048-95-0] $AsHNa_2O_4.7H_2O$; mol wt 312.01. Odorless crystals; effloresces in warm air. *Poisonous, but less so than the arsenite.* Loses 5 H_2O (29%) at about 50°; becomes anhydr at 100°. At 150° or higher it is converted into pyroarsenate. d 1.87. mp 57° when rapidly heated. Sol in 1.3 parts water, in glycerol; slightly sol in alc. The aq soln is alkaline to litmus. LD_{75} i.p. in rats: 14-18 mg/kg (Franke, Moxon).

Caution: Arsenic and inorganic arsenic compounds are listed as known human carcinogens: *Report on Carcinogens, Twelfth Edition* (PB2011-111646, 2011) p 50.

USE: The technical grade, about 98% pure, is used in dyeing with Turkey-red oil and in printing fabrics; manuf of other arsenates. Source of soluble arsenic.

8716. Sodium Arsenite. [7784-46-5] Arsenenous acid sodium salt (1:1); sodium meta-arsenite. $AsNaO_2$; mol wt 129.91. As 57.67%, Na 17.70%, O 24.63%. Product of commerce is 95-98% pure. Toxicity study: H. F. Smyth *et al.*, *Am. Ind. Hyg. Assoc. J.* **30**, 470 (1969). Review of toxicology and human exposure: *Toxicological Profile for Arsenic* (PB2008-100002, 2007) 559 pp.

White or grayish-white powder; somewhat hygroscopic. Absorbs CO_2 from air. *Poisonous.* Freely sol in water, slightly in alc. *Keep well closed.* LD_{50} orally in rats: 0.041 g/kg (Smyth).

Caution: Arsenic and inorganic arsenic compounds are listed as known human carcinogens: *Report on Carcinogens, Twelfth Edition* (PB2011-111646, 2011) p 50.

USE: Technical grade in manuf of arsenical soap for use on skins, for treating vines against certain scale diseases; as insecticide especially for termites.

THERAP CAT (VET): Topical acaricide.

8717. Sodium Azide. [26628-22-8] Sodium azide ($Na(N_3)$); hydrazoic acid sodium salt; Smite. N_3Na; mol wt 65.01. N 64.64%, Na 35.36%. NaN_3. Cytochrome oxidase inhibitor. Prepd from $NaNH_2 + N_2O$: Dennis, Browne, *Z. Anorg. Allg. Chem.* **40**, 95 (1904); Schenk in *Handbook of Preparative Inorganic Chemistry* **Vol. 1**, G. Brauer, Ed. (Academic Press, New York, 2nd ed., 1963) pp 474-475. Alternate procedures: *Inorg. Synth.* **1**, 79 (1939); **2**, 139 (1946). Large scale manuf processes: B. T. Fedoroff *et al.*, *Encyclopedia of Explosives and Related Items* **Vol. 1** (Picatinny Arsenal, Dover, N.J., 1960) pp A601-A619. *Review:* L. E. Audrieth, *Chem. Rev.* **15**, 169 (1934). Review of toxicology: K. A. Frederick, J. G. Babish, *Regul. Toxicol. Pharmacol.* **2**, 308-322 (1982); of human health effects: S. Chang, S.H. Lamm, *Int. J. Toxicol.* **22**, 175-186 (2003).

Colorless hexagonal crystals, d 1.846. On heating dec into sodium and nitrogen. Highly sol in water. Rapidly converted to hydrazoic acid, *q.v.* Soly in water: 40.16% at 10°, 41.7% at 17°. pK = 4.8, aq solns contains HN_3 which escapes readily at 37°. Slightly sol in alcohol. Insol in ether. Sol in liquid ammonia. *Poisonous. Heat and shock sensitive. Reacts with acids and heavy metals to form explosive compounds.* LD_{50} in rats (mg/kg): 45 orally (Frederick, Babish).

Caution: Potential symptoms of overexposure are irritation of eyes, skin; nausea, vomiting, restlessness, diarrhea; headache, dizziness, weakness, blurred vision; dyspnea; hypotension, tachycardia,

bradycardia, tachypnea; hypothermia; acidosis; convulsions; kidney changes. *See NIOSH Pocket Guide to Chemical Hazards* (DHHS/NIOSH 97-140, 1997) p 280; *Clinical Toxicology of Commercial Products*, R. E. Gosselin *et al.*, Eds. (Williams & Wilkins, Baltimore, 5th ed., 1984), Section II, p 114; *Prudent Practices for Handling Hazardous Chemicals in Laboratories* (National Academy Press, Washington, D.C., 1981) pp 145-147.

USE: In organic syntheses; in the preparation of hydrazoic acid, lead azide, pure sodium. In the differential selection of bacteria; in automatic blood counters; as preservative for laboratory reagents. Propellant for inflating automotive safety bags. Agricultural nematocide; herbicide; in fruit rot control.

8718. Sodium Benzoate. [532-32-1] Benzoic acid sodium salt (1:1). $C_7H_5NaO_2$; mol wt 144.10. C 58.35%, H 3.50%, Na 15.95%, O 22.21%. Toxicity: Smyth, Carpenter, *J. Ind. Hyg. Toxicol.* **30**, 63 (1948).

White, odorless granules or crystalline powder; sweetish, astringent taste. One gram dissolves in 1.8 ml water, 1.4 ml boiling water, about 75 ml alcohol, in 50 ml of a mixture of 47.5 ml alcohol and 3.7 ml water. The aq soln is slightly alkaline to litmus. pH about 8. *Incompat:* Acids, ferric salts. LD_{50} orally in rats: 4.07 g/kg (Smyth, Carpenter).

USE: Antimicrobial agent, flavoring agent and adjuvant in food; not to exceed a maximum level of 0.1% in food (**21 CFR,** 184.1733, 582.3733). Antifungal and bacteriostatic preservative in pharmaceuticals at concentrations of ~0.1%. Clinical reagent (bilirubin assay).

THERAP CAT: Diagnostic aid (hepatic function).

8719. Sodium Bicarbonate. [144-55-8] Carbonic acid sodium salt (1:1); sodium hydrogen carbonate; sodium acid carbonate; baking soda. $CHNaO_3$; mol wt 84.01. C 14.30%, H 1.20%, Na 27.37%, O 57.13%. $NaHCO_3$. The bicarbonate of commerce is about 99.8% pure. Prepd from sodium carbonate, water and carbon dioxide. Manuf: *Faith, Keyes & Clark's Industrial Chemicals*, F. A. Lowenheim, M. K. Moran, Eds. (Wiley-Interscience, New York, 4th ed., 1975) pp 702-705.

White cryst powder or granules. Begins to lose CO_2 at about 50° and at 100° it is converted into Na_2CO_3. Readily dec by weak acids. In aq soln it begins to break up into carbon dioxide and sodium carbonate at about 20° and completely on boiling. Sol in 10 parts water at 25°, in 12 parts water at about 18°. Insol in alcohol. Its aq soln prepd with cold water and without agitation is only slightly alkaline to litmus or phenolphthalein; on standing or rise in temp the alkalinity increases. pH of freshly prepd 0.1 molar aq soln at 25°: 8.3.

USE: Manuf many sodium salts; source of CO_2; ingredient of baking powder, effervescent salts and beverages; in fire extinguishers, cleaning compds. In buffers; in analytical chemistry for pH adjustment.

THERAP CAT: Antacid; systemic alkalinizing agent.

THERAP CAT (VET): Systemic and urinary alkalinizing agent.

8720. Sodium Bifluoride. [1333-83-1] Sodium fluoride (Na(HF_2)); sodium hydrogen fluoride. F_2HNa; mol wt 61.99. F 61.30%, H 1.63%, Na 37.09%. NaF.HF.

White, cryst powder. Sol in water. The aq soln corrodes glass.

USE: As a "sour" in laundering.

8721. Sodium Bismuthate(V). [12232-99-4] Bismuth sodium oxide ($BiNaO_3$). $BiNaO_3$; mol wt 279.97. Bi 74.64%, Na 8.21%, O 17.14%. $NaBiO_3$. The bismuthate of commerce contains about 85% $NaBiO_3$; the balance is chiefly water and Bi_2O_3.

Yellow to yellowish-brown, somewhat hygroscopic. Slowly dec on keeping; decompn accelerated by moisture and higher temp. Insol in cold, dec by hot water forming Bi_2O_3, NaOH, and liberating oxygen; dec by acids; with HCl chlorine is formed; with oxy-acids oxygen is liberated. LD_{100} orally in rats: 720 mg/kg, Hanzlik *et al.*, *J. Pharmacol. Exp. Ther.* **62**, 372 (1938).

USE: For the determination of manganese in iron and steel, etc., the manganese being oxidized by it in hot HNO_3 or H_2SO_4 soln to permanganate.

8722. Sodium Bisulfate. [7681-38-1] Sulfuric acid sodium salt (1:1); sodium acid sulfate; sodium hydrogen sulfate; sodium pyrosulfate. $HNaO_4S$; mol wt 120.05. H 0.84%, Na 19.15%, O 53.31%, S 26.71%. $NaHSO_4$.

Fused $NaHSO_4$, hygroscopic pieces. d 2.435. mp ~315°. Sol in 2 parts water, 1 part boiling water; dec by alcohol into sodium sulfate and free H_2SO_4. *Keep well closed.*

Monohydrate. [10034-88-5] $HNaO_4S.H_2O$; mol wt 138.07. Odorless crystals. When strongly heated it changes into pyrosulfate. Sol in about 0.8 part water; dec by alcohol into sodium sulfate and free H_2SO_4. The aq soln is strongly acid. pH of 0.1 molar soln: 1.4.

USE: Fusion of minerals to make them sol for analysis; for liberating CO_2 in carbonic acid baths. Technical grades are used for pickling metals, carbonizing wool, bleaching and swelling leather, manuf magnesia cements, etc.

8723. Sodium Bisulfide. [16721-80-5] Sodium sulfide (Na(SH)); sodium sulfhydrate; sodium hydrosulfide; sodium hydrogen sulfide. HNaS; mol wt 56.06. H 1.80%, Na 41.01%, S 57.19%. NaSH. Prepd from sodium ethylate and hydrogen sulfide: A. Rule, *J. Chem. Soc. Trans.* **99**, 558 (1911); Teichert, Klemm, *Z. Anorg. Allg. Chem.* **243**, 86 (1939); Eibeck, *Inorg. Synth.* **7**, 128 (1963). The technical grade may be obtained by reacting sodium bisulfide with calcium sulfide in the cold or by saturating NaOH solns with H_2S.

Colorless to yellow rhombohedric-cubic crystals. Odor of hydrogen sulfide. *Corrosive. Spontaneously combustible and highly flammable.* Very hygroscopic. Readily hydrolyzed in moist air to NaOH and Na_2S. d 1.79. Turns yellow upon heating in dry air, changing to orange at higher temps. mp 350° forming a black liquid. Sol in water, alcohol, ether. Gives a blue-green soln in DMF. Reacts with acids to release H_2S. *Store in cool, dry conditions in well sealed containers.*

Dihydrate. [12135-05-6] $HNaS.2H_2O$; mol wt 92.09. Needles or flakes, mp 55°. Completely and rapidly sol in water, alcohol, ether.

Trihydrate. [12135-06-7] $HNaS.3H_2O$; mol wt 110.10. Shiny rhombs, mp 22°.

USE: Dehairing hides; in paper-pulping; in the manuf of sulfur-contg dyes and other thio compds; in copper mining.

8724. Sodium Bisulfite. [7631-90-5] Sulfurous acid sodium salt (1:1); sodium acid sulfite; sodium hydrogen sulfite. $HNaO_3S$; mol wt 104.05. H 0.97%, Na 22.09%, O 46.13%, S 30.81%. $NaHSO_3$. The bisulfite of commerce is a mixture with sodium metabisulfite, $Na_2S_2O_5$, and for all practical purposes possesses the same properties as the true bisulfite. Toxicity: Hoppe, Goble, *J. Pharmacol. Exp. Ther.* **101**, 101 (1951).

White, crystalline powder; SO_2 odor; disagreeable taste; on exposure to air it loses some SO_2 and is gradually oxidized to sulfate. d 1.48. Sol in 3.5 parts cold water, 2 parts boiling water, in about 70 parts alcohol. Its aq soln is acid. *Keep well closed and in a cool place.* LD_{50} i.v. in rats: 115 mg/kg (Hoppe, Goble).

Caution: Potential symptoms of overexposure are irritation of eyes, skin, mucous membranes. *See NIOSH Pocket Guide to Chemical Hazards* (DHHS/NIOSH 97-140, 1997) p 282.

USE: As disinfectant and bleach, particularly for wool; in dyeing for preparing hot and cold indigo vats; in paper-making in place of sodium hyposulfite to remove Cl from bleached fibers; as stripper (reducer) in laundering; to remove permanganate stains from skin and clothing; to render certain dyes sol; manuf sodium hydrosulfite; coagulating rubber latex; as preservative for deteriorative liqs or solns used for technical purposes; as antiseptic in fermentation industries. As preservative and bleach in food. Pharmaceutic aid (antioxidant).

8725. Sodium Bitartrate. [526-94-3] (2*R*,3*R*)-2,3-Dihydroxybutanedioic acid sodium salt (1:1); sodium acid tartrate; sodium hydrogen tartrate. $C_4H_5NaO_6$; mol wt 172.07. C 27.92%, H 2.93%, Na 13.36%, O 55.79%. $NaHC_4H_4O_6$.

Monohydrate. [6131-98-2] $C_4H_5NaO_6.H_2O$; mol wt 190.08. White crystals. Sol in about 9 parts water, 2 parts boiling water. Almost insol in alcohol. The aq soln is acid.

USE: For detecting potassium; in nutrient media.

8726. Sodium Borate. [1330-43-4] Boron sodium oxide ($B_4Na_2O_7$); anhydrous borax; borax glass; fused borax; sodium bibo-

rate; sodium pyroborate; sodium tetraborate. $B_4Na_2O_7$; mol wt 201.21. B 21.49%, Na 22.85%, O 55.66%. $Na_2B_4O_7$. Toxicity: H. F. Smyth *et al.*, *Am. Ind. Hyg. Assoc. J.* **30**, 470 (1969). Review of toxicology and human exposure: *Toxicological Profile for Boron* (PB2010-100001, 2010) 249 pp.

Powder or glass-like plates becoming opaque on exposure to air. Freely sol in boiling water, glycerin; sol in water. Insol in alc.

Decahydrate. [1303-96-4] Borax; Jaikin. $B_4Na_2O_7.10H_2O$; mol wt 381.36. Hard odorless crystals, granules or cryst powder; efflorescent in dry air, the crystal often being coated with white powder. d 1.73. Melts when rapidly heated at 75°; at 100° loses $5H_2O$; at 150° loses $9H_2O$; becomes anhydr at 320°. One gram dissolves in 16 ml water, 0.6 ml boiling water, about 1 ml glycerol. Insol in alc. The aq soln is alkaline to litmus and phenolphthalein. pH about 9.5. Borax dissolves many metallic oxides when fused with them. LD_{50} orally in rats: 5.66 g/kg (Smyth).

Caution: Potential symptoms of overexposure are irritation of eyes, skin, upper respiratory system; dermatitis; epistaxis; cough, dyspnea. *See NIOSH Pocket Guide to Chemical Hazards* (DHHS/NIOSH 97-140, 1997) p 30.

USE: Soldering metals; manuf glazes and enamels; tanning; in cleaning compds; artificially aging wood; as preservative, either alone or with other antiseptics against wood fungus; fireproofing fabrics and wood; curing and preserving skins; in cockroach control. In buffers; as a complexing or masking agent in analytical chemistry. Pharmaceutic aid (alkalizer).

8727. Sodium Borohydride. [16940-66-2] Tetrahydroborate(1−) sodium (1:1). BH_4Na; mol wt 37.83. B 28.58%, H 10.66%, Na 60.77%. $NaBH_4$. Prepn: H. I. Schlesinger *et al.*, *J. Am. Chem. Soc.* **75**, 205 (1953); H. C. Brown *et al.*, *J. Am. Chem. Soc.* **79**, 5400 (1957). Reviews: R. M. Adams, A. R. Siedle, in *Boron, Metallo-Boron Compounds and Boranes* (Interscience, New York, 1964) 765 pp; B. D. James, M. G. H. Walbridge, *Prog. Inorg. Chem.* **11**, 99-231 (1970).

Hygroscopic, cubic crystals forming a dihydrate, mp 36-37°. *Dangerous when wet.* The anhydr material (d 1.074) is stable in dry air to 300°; dec slowly at 400° and rapidly at 500°. Supports combustion. Soly (g/100 g) in water at 25°: 55; at 60°: 88.5; liquid ammonia at 25°: 104; ethylenediamine at 75°: 22; morpholine at 25°: 1.4; pyridine at 25°: 3.1; methanol at 20°: 16.4 (reacts); ethanol at 20°: 4.0 (reacts slowly); tetrahydrofurfuryl alcohol at 20°: 14.0 (reacts slowly); tetrahydrofuran at 20°: 0.1; diglyme at 25°: 5.5; dimethylformamide at 20°: 18.0.

USE: Reducing agent for aldehydes, ketones and Schiff bases in nonaqueous solvents. Also reduces acids, esters, acid chlorides, disulfides, nitriles, inorganic anions. Further used to generate diborane, as foaming agent, as scavenger for traces of aldehyde, ketones and peroxides in organic chemicals.

8728. Sodium Bromate. [7789-38-0] Bromic acid sodium salt (1:1). $BrNaO_3$; mol wt 150.89. Br 52.96%, Na 15.24%, O 31.81%. $NaBrO_3$. The article of commerce contains about 99% $NaBrO_3$.

Colorless, odorless crystals, white granules or cryst powder. *Oxidizer.* d 3.34. mp 381° with dec and liberation of oxygen. Sol in 2.5 parts water, 1.1 parts boiling water. The aq soln is neutral. *Keep from contact with organic matter.*

Caution: See Potassium Bromate.

USE: As a mixture with sodium bromide for dissolving gold from its ores.

8729. Sodium Bromide. [7647-15-6] Sodium bromide (NaBr); Sedoneural. BrNa; mol wt 102.89. Br 77.66%, Na 22.34%. NaBr. Prepd commercially by adding some excess bromine to a sodium hydroxide soln forming a mixture of bromide and bromate. The reaction products are evaporated to dryness and treated with carbon to reduce the bromate to bromide. Prepn: J. H. van der Meulen, US 1775598 (1930). Toxicity study: Smith, Hambourger, *J. Pharmacol. Exp. Ther.* **55**, 200 (1935). Brief review of physical and chemical properties and pharmacology: T. O. Soine, C. O. Wilson, *Roger's Inorganic Pharmaceutical Chemistry* (Lea & Febiger, Philadelphia, 8th ed., 1967) pp 213-216.

White crystals, granules or powder; saline, feebly bitter taste. Absorbs moisture from air but is not deliquescent. d 3.21. mp 755°; volatilizes at somewhat higher temp. One gram dissolves in 1.1 ml water, about 16 ml alcohol, 6 ml methanol. The aq soln is practically neutral. pH 6.5-8.0. *Keep well closed.* From water of room temp, sodium bromide crystallizes with $2H_2O$ in the form of colorless crystals. *Incompat.* Acids, alkaloidal and heavy metal salts. LD_{50} orally in rats: 3.5 g/kg (Smith, Hambourger).

USE: In photography. Standard in ion chromatography.

THERAP CAT: Sedative; hypnotic; anticonvulsant.

THERAP CAT (VET): Sedative. Has been used to control convulsions, chorea, hysteria.

8730. Sodium Cacodylate. [124-65-2] *As,As*-Dimethylarsinic acid sodium salt (1:1); [(dimethylarsino)oxy]sodium *As*-oxide; sodium dimethylarsinate; cacodylic acid sodium salt; Arsecodile; Arsicodile; Arsycodile; Rad-e-cate; Silvisar. $C_2H_6AsNaO_2$; mol wt 159.98. C 15.02%, H 3.78%, As 46.83%, Na 14.37%, O 20.00%. An organic compd of arsenic yielding inorganic, trivalent arsenic in the body (is excreted partly unchanged, also yields dimethylarsine oxide). Prepd by the distillation of a mixture of arsenic trioxide and potassium acetate which yields Cadet's liquid, contg mostly dimethylarsine oxide. This is oxidized with mercuric oxide yielding crystals of cacodylic acid which is neutralized with Na_2CO_3 or NaOH: Cadet de Gassicourt, *Mem. Sav. Etrangers* **3**, 633 (1760); Valeur, Gaillot, *Compt. Rend.* **185**, 956 (1927). Use in treatment of psoriasis: S. I. Dawes, H. C. Jackson, *J. Am. Med. Assoc.* **48**, 2020 (1903). Metabolism and excretion: A. Heffter, *Arch. Exp. Pathol. Pharmakol.* **46**, 230 (1901). Use as herbicide: M. A. Sprague, US 3056663 (1962 to Ansul). Possible use in radiation dosimetry: H. S. Levinson, E. B. Garber, *Nature* **207**, 751 (1965). Review of toxicology and human exposure: *Toxicological Profile for Arsenic* (PB2008-100002, 2007) 559 pp.

$$H_3C - \overset{\overset{O}{\|}}{\underset{H_3C}{As}} - O^- \quad Na^+$$

Trihydrate. [6131-99-3] Crystals, granules. Slight odor. Liquefies in its water of hydration at about 60°. Becomes anhyd at 120°. Burns with a bluish flame, emitting a garlic-like odor. One gram dissolves in 0.5 ml water, 2.5 ml alc. pH about 8-9. *Keep well closed.*

USE: Herbicide.

THERAP CAT (VET): Has been used in chronic eczema, anemia, as a tonic.

8731. Sodium Carbonate. [497-19-8] Carbonic acid sodium salt (1:2). CNa_2O_3; mol wt 105.99. C 11.33%, Na 43.38%, O 45.28%. Na_2CO_3. Occurs in nature as the hydrate, *thermonatrite*, and the decahydrate, *natron* or *natrite*. Produced by the ammonia-soda or Solvay process, or from lake brines or sea water by electrolytic processes: *Faith, Keyes & Clark's Industrial Chemicals*, F. A. Lowenheim, M. K. Moran, Eds. (Wiley-Interscience, New York, 4th ed., 1975) pp 706-715. Toxicity: C. Norde *et al.*, *Compt. Rend.* **257**, 791 (1963). Reviews: Bailey in *Mellor's* **vol II**, suppl II, *The Alkali Metals* (part 1), 1058-1205 (1961).

Anhydrous. Solvay soda. The technical grade (about 99% pure) is known as *soda ash.* Odorless, hygroscopic powder; alkaline taste; d 2.53. mp 851° but begins to lose CO_2 even at 400°. On exposure to air it will gradually absorb one mol water, about 15%. Sol in glycerol; in 3.5 parts water at room temp, 2.2 parts water at 35°. Insol in alcohol. Dec by acids with effervescence. Combines with water with evolution of heat. Its aq soln is strongly alkaline. pH 11.6. *Keep well closed.* LD_{50} (30 day) i.p. in mice: 116.6 mg/kg (Norde).

Monohydrate. [5968-11-6] $CNa_2O_3 \cdot H_2O$; mol wt 124.00. Odorless, small crystals or cryst powder; alkaline taste. Stable at ordinary temps and atmospheric conditions; dries out somewhat in warm, dry air or above 50°; becomes anhyd at 100°. d 2.25; also reported as 1.55. Sol in 3 parts water, 1.8 parts boiling water, 7 parts glycerol. Insol in alcohol.

Decahydrate. [6132-02-1] Nevite; Soda. $CNa_2O_3 \cdot 10H_2O$; mol wt 286.14. The technical product is known as *sal soda* or *washing soda.* Transparent crystals; readily effloresces on exposure to air. d 1.46. mp 30°. Sol in 2 parts cold, 0.25 part boiling water, in glycerol. Insol in alcohol. The aq soln is strongly alkaline to litmus. *Keep well closed and in a cool place.*

Caution: Potential symptoms of overexposure to dusts or vapors of sodium carbonate are irritation of mucous membranes with subsequent coughing and shortness of breath. Chronic overexposure may lead to perforation of the nasal septum. Direct contact may cause skin irritation and redness, with concentrated solutions causing erythema. Chronic skin exposures may cause dermatitis and ulceration. *See Patty's Industrial Hygiene and Toxicology* **vol. 2A**, G. D. Clayton, F. E. Clayton, Eds. (Wiley-Interscience, New York, 4th ed., 1993) pp 769-771.

USE: In manuf of Na salts, glass, soap; for washing wool; textiles, etc.; in bleaching linen, cotton; general cleanser; in water-softening; in photography; as reagent in analytical chemistry for pH adjustment; buffer; standard for acid-base titrimetry. Pharmaceutic aid (alkalizer).

THERAP CAT (VET): Has been used as an emetic. In solution to cleanse skin, in eczema, to soften scabs of ringworm.

8732. Sodium Cellulose Phosphate. [9038-41-9] Cellulose phosphate sodium salt; SCP; Calcibind; Calcisorb. Non-absorbable ion exchange resin with high affinity for calcium ions. Prepn: G. P. Touey, US 2759924 (1956 to Eastman Kodak). Use in treatment of hypercalcemia: C. Y. C. Pak *et al.*, *J. Clin. Endocrinol. Metab.* **28**, 1828 (1968). Mechanism of action: C. Y. C. Pak, *J. Clin. Pharmacol.* **13**, 15 (1973). Clinical pharmacology: *idem, ibid.* **19**, 451 (1979). Use in treatment of calcium urolithiasis: U. Backman *et al.*, *J. Urol.* **123**, 9 (1980); C. Y. C. Pak, *Invest. Urol.* **19**, 187 (1981). Powder. Insol in water.

USE: As ion exchange resin.

THERAP CAT: Treatment of calcium urolithiasis.

8733. Sodium Chlorate. [7775-09-9] Chloric acid sodium salt (1:1); Atlacide; Defol; Dervan. $ClNaO_3$; mol wt 106.44. Cl 33.31%, Na 21.60%, O 45.09%. $NaClO_3$. Produced from sodium chloride by electrolysis: *Faith, Keyes & Clark's Industrial Chemicals*, F. A. Lowenheim, M. K. Moran, Eds. (Wiley-Interscience, New York, 4th ed., 1975) pp 716-721. Toxicity: Ulrich, *J. Pharmacol. Exp. Ther.* **35**, 1 (1929).

Colorless, odorless crystals or white granules. d 2.5. mp 248°; at about 300° liberates oxygen; entirely dec at higher temp. Sol in about 1 ml cold, 0.5 ml boiling water, about 130 ml alcohol, about 50 ml boiling alcohol, 4 ml glycerol. Sodium chloride diminishes its soly in water. The aq soln is neutral. *Keep out of contact with organic matter or other oxidizable substances.* LD orally in rats: 12000 mg/kg (Ulrich).

USE: An oxidizer, like potassium chlorate, in manuf of dyes; explosives and matches; dyeing and printing fabrics; tanning and finishing leather. Herbicide. Pharmaceutic aid (oxidizing agent). Oxidimetric reagent in analytical chemistry.

8734. Sodium Chloride. [7647-14-5] Sodium chloride (NaCl); salt; common salt. ClNa; mol wt 58.44. Cl 60.66%, Na 39.34%. NaCl. The article of commerce is also known as *table salt*, *rock salt* or *sea salt.* Occurs in nature as the mineral *halite.* Produced by mining (rock salt), by evaporation of brine from underground salt deposits and from sea water by solar evaporation: *Faith, Keyes & Clark's Industrial Chemicals*, F. A. Lowenheim, M. K. Moran, Eds. (Wiley-Interscience, New York, 4th ed., 1975) pp 722-730. Toxicity studies: E. M. Boyd, M. N. Shanas, *Arch. Int. Pharmacodyn.* **144**, 86 (1963). Comprehensive monograph: D. W. Kaufmann, "Sodium Chloride" in *ACS Monograph Series* no. **145** (Reinhold, New York, 1960) 743 pp.

Cubic, white crystals, granules, or powder; colorless and transparent or translucent when in large crystals. d 2.17. The salt of commerce usually contains some calcium and magnesium chlorides which absorb moisture and make it cake. mp 804°; begins to volatilize at a little above this temp. One gram dissolves in 2.8 ml water at 25°, in 2.6 ml boiling water, in 10 ml glycerol; very slightly sol in alcohol. Its soly in water is decreased by HCl. Almost insol in concd HCl. Its aq soln is neutral. pH: 6.7-7.3. d of satd aq soln at 25° is 1.202. A 23% aq soln of sodium chloride freezes at $-20.5°C$ (5°F). LD_{50} orally in rats: 3.75 ±0.43 g/kg (Boyd, Shanas).

Note: **Blusalt**, a brand of sodium chloride contg trace amounts of cobalt, iodine, iron, copper, manganese, zinc is used in farm animals.

USE: Natural salt is the source of chlorine and of sodium as well as of all, or practically all, their compds, e.g., hydrochloric acid, chlorates, sodium carbonate, hydroxide, etc.; for preserving foods;

manuf soap, to salt out dyes; in freezing mixtures; for dyeing and printing fabrics, glazing pottery, curing hides; metallurgy of tin and other metals; in analytical chemistry as electrolyte, buffer, and matrix modifier.

THERAP CAT: Electrolyte replenisher; emetic; topical anti-inflammatory.

THERAP CAT (VET): Essential nutrient factor. May be given orally as emetic, stomachic, laxative or to stimulate thirst (prevention of calculi). Intravenously as isotonic solution to raise blood volume, to combat dehydration. Locally as wound irrigant, rectal douche.

8735. Sodium Chlorite. [7758-19-2] Chlorous acid sodium salt (1:1). $ClNaO_2$; mol wt 90.44. Cl 39.20%, Na 25.42%, O 35.38%. $NaClO_2$. Prepd on a commercial scale by passing chlorine dioxide into a soln of sodium hydroxide contg carbonaceous matter and lime: Vincent, US 2092944; US 2092945 (both 1937 to Mathieson). Review of manuf and properties: Kesting, *Pulp Paper Mag. Can.* **53**, no. 8, 99-104 (1952); of toxicology and human exposure: *Toxicological Profile for Chlorine Dioxide and Chlorite* (PB2004-107332, 2004) 191 pp.

Slightly hygroscopic crystals or flakes, does not cake. Dec 180-200°. Powerful oxidizer, but will not explode on percussion unless in contact with oxidizable material. Soly in water (g/100 g soln): at 5°: 34; at 17°: 39; at 30°: 46; at 45°: 53; at 60°: 55.

Trihydrate. [49658-21-1] Triclinic leaflets, becomes anhydr at 38° or in desiccator over KOH at room temp.

USE: In the preparation of chlorine dioxide for immediate use; in water purification; as bleaching agent for textiles, paper pulp; disinfectant.

8736. Sodium Chromate. [7775-11-3] Chromic acid (H_2CrO_4) sodium salt (1:2); disodium chromate; sodium chromate(VI). $CrNa_2O_4$; mol wt 161.97. Cr 32.10%, Na 28.39%, O 39.51%. Na_2CrO_4. Hexavalent chromium compound produced by roasting chromite ore with soda ash. Physical properties: W. H. Hartford, *Ind. Eng. Chem.* **41**, 1993 (1949). Manuf: *Faith, Keyes & Clark's Industrial Chemicals*, F. A. Lowenheim, M. K. Moran, Eds. (Wiley-Interscience, New York, 4th ed., 1975) pp 731-736. Production from chromium-containing industrial wastes: A. Dettmer *et al.*, *Chem. Eng. J.* **160**, 8 (2010).

Yellow, orthorhombic crystals, mp 792°. *Poisonous; corrosive.* Hygroscopic, forms tetra-, hexa-, and decahydrates. Deliquescent in moist air. d_4^{25} 2.723. Sol in water.

Tetrahydrate. [10034-82-9] $CrNa_2O_4 \cdot 4H_2O$; mol wt 234.03. Yellow, somewhat deliquesc crystals. Sol in about 1 part water, slightly sol in alcohol. The aq soln is alkaline. *Keep well closed.*

Sodium chromate Cr 51. [10039-53-9] Sodium chromate-^{51}Cr; Chromitope Sodium. [51]$CrNa_2O_4$. ^{51}Cr-Labeled compound prepd from radioactive chromium which has a half-life of 27.7 days. The emission of gamma rays is applicable to biological tagging and tracing. Use for determn of blood volume: A. C. Ertl *et al.*, *Am. J. Med. Sci.* **334**, 32 (2007); H. J. Dworkin *et al.*, *ibid.*, 37.

Caution: Chromium hexavalent (VI) compounds are listed as known human carcinogens: *Report on Carcinogens, Twelfth Edition* (PB2011-111646, 2011) p 106.

USE: Corrosion inhibitor; wood preservative.

THERAP CAT: ^{51}Cr-labeled compound as diagnostic aid (blood volume determination; blood cell survival).

8737. Sodium Citrate. [68-04-2] 2-Hydroxy-1,2,3-propanetricarboxylic acid sodium salt (1:3); trisodium citrate; Citrosodine; Cystemme; Urisal. $C_6H_5Na_3O_7$; mol wt 258.07. C 27.92%, H 1.95%, Na 26.73%, O 43.40%. Crystal structure of dihydrate: A. Fischer, G. Palladino, *Acta Crystallogr.* **E59**, m1080 (2003). Anticoagulant use in blood transfusion: R. Weil, *J. Am. Med. Assoc.* **64**, 425 (1915); *cf. ibid.* **250**, 1901 (1983). Toxicity study: C. M. Gruber, Jr., W. A. Halbeisen, *J. Pharmacol. Exp. Ther.* **94**, 65 (1948). Clinical evaluation of antacid efficacy: C. P. Gibbs *et al.*, *Anesthesiology* **57**, 44 (1982); of effects on athletic performance: V. Oöpik *et al.*, *Br. J. Sports Med.* **37**, 485 (2003).

LD$_{50}$ in rats, mice (mmol/kg): 6.0, 5.5 i.p. (Gruber, Halbeisen).

Dihydrate. [6132-04-3] $C_6H_5Na_3O_7 \cdot 2H_2O$; mol wt 294.10. White, odorless crystals, granules or powder; cool, saline taste. Stable in air; becomes anhydrous at 150°. d 1.814. Sol in 1.3 parts water, 0.6 part boiling water. Insol in alcohol. The aq soln is slightly alkaline to litmus. pH about 8.

USE: Anticoagulant for collection of blood. In photography; as sequestering agent to remove trace metals; as emulsifier, acidulant and sequestrant in foods. Buffer.

THERAP CAT: Systemic alkalinizing agent.

8738. Sodium Cobaltinitrite. [13600-98-1] (*OC*-6-11)-Hexakis(nitrito-*κN*)cobaltate(3−) sodium (1:3); trisodium (*OC*-6-11)-hexakis(nitrito-*κN*)cobaltate(3−); sodium hexanitrocobaltate. $CoN_6Na_3O_{12}$; mol wt 403.93. Co 14.59%, N 20.81%, Na 17.07%, O 47.53%. $Na_3Co(NO_2)_6$.

Yellow to brownish-yellow, cryst powder. Very sol in water, slightly in alc. Dec by mineral acids, but unaffected by dil acetic or similar organic acids. The aq soln dec gradually but if a few drops of acetic acid are added it may be kept for about 3 months.

USE: For the detection of potassium with which it forms a slightly sol compd.

8739. Sodium Cocoyl Isethionate. [61789-32-0] Coco fatty acids 2-sulfoethyl esters sodium salts; SCI; Elfan AT 84; Geropon AS-200; Hostapon SCI; Jordapon CI; Tauranol I-78. Ionic surfactant prepd from a mixture of fatty acids derived from coconut oil, typically consisting of C_8 to C_{18} fatty acids. Prepn from sodium isethionate and coconut fatty acid chlorides: K. Daimler, K. Platz, US 1881172 (1932 to I. G. Farben); E. C. Britton, A. R. Sexton, US 2821535 (1958 to Dow); by direct esterification using zinc oxide catalyst: A. Cahn *et al.*, US 3320292 (1967 to Lever Brothers). Overview of preparations and quality specifications: R. B. Login, *Household Pers. Prod. Ind.* **21**, 56 (1984). Clinical tolerability in synthetic detergent (syndet) bars: H. C. Korting *et al.*, *Int. J. Cosmet. Sci.* **14**, 277 (1992). Review of safety assessment: M. M. Zondlo, *J. Am. Coll. Toxicol.* **12**, 459-479 (1993).

n = 1-11

White powder, coconut odor. *Eye irritant.* Soly in water (g/100 ml): 0.01 at 25°; >50 at 70°. Stable at pH 6-8. LD$_{50}$ orally in rats: 4.33 g/kg (Zondlo).

Sodium lauroyl isethionate. [7381-01-3] Dodecanoic acid 2-sulfoethyl ester sodium salt (1:1); 2-sulfoethyl laurate sodium salt; sodium 2-sulfonatoethyl laurate. $C_{14}H_{27}NaO_5S$; mol wt 330.41. Typically the predominant component in SCI formulations. Prepn and properties: R. G. Bistline, Jr. *et al.*, *J. Am. Oil Chem. Soc.* **48**, 657 (1971). White powder, mp 216-218°. Surface tension (0.1% aq soln): 46.6 dynes/cm at 25°. Critical micelle concentration: 6.4 mmol/l. Sol in ethanol. Insol in chloroform, toluene, hexane.

Ammonium cocoyl isethionate. [223705-57-5] Coco fatty acids 2-sulfoethyl esters ammonium salts; ACI; Jordapon ACI 30. Manuf process: R. G. Briody *et al.*, WO 9501331; *eidem*, US 5739365 (1995, 1998 both to PPG Ind.). Properties, uses and comparison with SCI: W. C. Allison, J. E. Neibaur, *Household Pers. Prod. Ind.* **32**, 92-100 (1995). Solid or stiff paste. Highly water soluble. Unstable in aq soln at acidic pH. Commercial formulation is a 30% aq soln, d 1.02; viscosity 40 cps; pH 6.8.

USE: In shampoos, cleansers, synthetic detergent bars.

8740. Sodium Cyanate. [917-61-3] Cyanic acid sodium salt (1:1); sodium isocyanate. $CNNaO$; mol wt 65.01. C 18.48%, N 21.55%, Na 35.36%, O 24.61%. $NaOCN$. Prepn and properties: *Gmelins, Sodium* (8th ed.) **21**, 799-801 (1928) and supplement, part 4, 1382-1386 (1967). Used experimentally in treatment of sickle cell anemia. Effect of cyanate on sickling: May *et al.*, *Lancet* **1**, 658 (1972); P. N. Gillette *et al.*, *N. Engl. J. Med.* **290**, 654 (1974). Pharmacology and toxicology: Cerami *et al.* *J. Pharmacol. Exp. Ther.* **185**, 653 (1973). Clinical studies: Peterson *et al.*, *ibid.* **189**, 577 (1974).

Colorless needles from alcohol. d_4^{20} 1.893. mp 550°. Sol in water; decomposes to form Na_2CO_3 and urea. Soly in alc (0°): 0.22 g/100 g solvent. Insol in ether. LD_{50} i.p. in mice: 260 mg/kg (Cerami).

8741. Sodium Cyanide. [143-33-9] Sodium cyanide (Na(CN)); cyanogran. CNNa; mol wt 49.01. C 24.51%, N 28.58%, Na 46.91%. NaCN. This cyanide of commerce is 95-98% pure. Mixtures of sodium cyanide with sodium chloride or carbonate for special uses are also marketed. Toxicity study: Smyth *et al.*, *Am. Ind. Hyg. Assoc. J.* **30**, 470 (1969). Review of toxicology and human exposure: *Toxicological Profile for Cyanide* (PB2007-100674, 2006) 341 pp.

White granules or fused pieces. *Poisonous.* Odorless when perfectly dry; somewhat deliquesc in damp air and emits slight odor of HCN. mp 563°. Freely sol in water; slightly sol in alc. The aq soln is strongly alkaline and rapidly decomposes; the soln readily dissolves gold and silver in presence of air. *Keep well closed.* LD_{50} orally in rats: 15 mg/kg (Smyth).

Caution: Potential symptoms of overexposure are irritation of eyes and skin; weakness, headache and confusion; nausea, vomiting; increased rate of respiration; slow gasping respiration; asphyxia; thyroid and blood changes. *See NIOSH Pocket Guide to Chemical Hazards* (DHHS/NIOSH 97-140, 1997) p 282.

USE: Extracting gold and silver from ores; electroplating baths; case hardening steel by liquid nitriding; manuf hydrocyanic acid and other cyanides. Complexing agent in analytical chemistry.

8742. Sodium Cyanoborohydride. [25895-60-7] $(T\text{-}4)\text{-}$(Cyano-κC)trihydroborate(1−) sodium (1:1); sodium (cyano-C)trihydroborate(1−); sodium borocyanohydride; sodium cyanohydridoborate. CH_3BNNa; mol wt 62.84. C 19.11%, H 4.81%, B 17.20%, N 22.29%, Na 36.58%. $NaBH_3CN$. Reducing agent prepared from $NaBH_4$ and HCN: R. C. Wade *et al.*, *Inorg. Chem.* **9**, 2146 (1970); R. C. Wade, **DE 2028569** corresp to **US 3667923** (1971, 1972 both to Ventron). Review of prepn, properties and use: C. F. Lane, *Synthesis* **1975**, 135-146.

White, hygroscopic powder, mp 240-242° (dec). d^{28} 1.199. Soly (g/100 g solvent) in water (29°): 212; in THF (28°): 37.2; in diglyme (25°): 17.6. Very sol in methanol; slightly sol in ethanol, isopropylamine; insol in ethyl ether, benzene, hexane. Stable in acid to pH 3; undergoes rapid hydrolysis in 12N HCl. Rate of hydrolysis 10^{-8} that of $NaBH_4$.

USE: Selective reducing agent for aldehydes, ketones, oximes, enamines; does not reduce amides, ethers, lactones, nitriles, nitro compds and epoxides. Also used for reductive amination of ketones and aldehydes, reductive alkylation of amines and hydrazines, reductive displacement of halides and tosylates, deoxygenation of aldehydes and ketones. *See* Lane, *loc. cit.*

8743. Sodium Diacetate. [126-96-5] Acetic acid sodium salt (2:1); sodium acid acetate; Dykon. $C_4H_7NaO_4$; mol wt 142.09. C 33.81%, H 4.97%, Na 16.18%, O 45.04%. $CH_3COONa.CH_3$-COOH. Described as a "bound" compd of sodium acetate and acetic acid.

White powder, dec above 150°. Sol in water, liberating 42.25% available acetic acid.

USE: Acetic acid in solid form; sequestrant; food preservative to inhibit molds and rope-forming bacteria in bread: Glabe, *Food Ind.* **14**, no. 2, 46 (1942).

8744. Sodium Dichromate(VI). [10588-01-9] Chromic acid $(H_2Cr_2O_7)$ sodium salt (1:2); sodium bichromate; bichromate of soda. $Cr_2Na_2O_7$; mol wt 261.96. Cr 39.70%, Na 17.55%, O 42.75%. $Na_2Cr_2O_7$. Usually prepd from Na_2CrO_4 and H_2SO_4. Description of industrial processes: Müller, Glissmann in *Ullmanns Encyklopädie der technischen Chemie* **vol. 5** (Munich, 3rd ed., 1954) p 575; *Faith, Keyes & Clark's Industrial Chemicals*, F. A. Lowenheim, M. K. Moran, Eds. (Wiley-Interscience, New York, 4th ed., 1975) pp 731-736.

Dihydrate. [7789-12-0] $Cr_2Na_2O_7.2H_2O$; mol wt 297.99. Reddish to bright orange, somewhat deliquescent crystals. Crystal system: monoclinic sphenoidal. Crystal habit: elongated prismatic. d_4^{25} 2.348. Bulk density: 96 lbs/cu ft. Becomes anhydr on prolonged heating at ~100°. The anhydr salt mp 356.7° and starts to dec at ~400°. Heat of soln −28.2 cal/g. Very sol in water. A satd

aq soln contains at 0°: 70.6% $Na_2Cr_2O_7.2H_2O$; at 20°: 73.18%; at 40°: 77.09%; at 60°: 82.04%; at 80°: 88.39%; at 100°: 91.43%. A 20% soln freezes at −3.5°, a 30% soln at −6°, a 60% soln at −26°, a 69% soln at −48°. Specific heat of 20% soln at 25°: 0.85 cal/g/°C. Solns are acidic: pH of 1% soln: 4.0; pH of 10% soln: 3.5.

Caution: Potential symptoms of overexposure by ingestion are violent gasteroenteritis, peripheral vascular collapse, vertigo, muscle cramps, coma, hemorrhagic diathesis, fever, liver damage and acute renal failure. Direct contact may be highly corrosive to skin and mucous membranes. *See Clinical Toxicology of Commercial Products*, R. E. Gosselin *et al.*, Eds. (Williams & Wilkins, 5th ed., 1984) Section II, p 108. Chromium hexavalent (VI) compounds are listed as known human carcinogens: *Report on Carcinogens, Twelfth Edition* (PB2011-111646, 2011) p 106.

USE: Oxidizing agent in manuf of dyes, many other synthetic organic chemicals, inks, etc.; in chrome-tanning of hides; in electric batteries; bleaching fats, oils, sponges, resins; refining petroleum; manuf chromic acid, other chromates and chrome pigments; in corrosion-inhibitors; corrosion-inhibiting paints; in many metal treatments; electroengraving of copper; mordant in dyeing; for hardening gelatin; for the defoliation of cotton plants and other plants and shrubs, La Lande, **US 2760854** (1956 to Pennsylvania Salt). Prepn of titrant in redox reactions.

THERAP CAT: Anti-infective (topical).

8745. Sodium Dicyanoaurate(I). [15280-09-8] Bis(cyano-κC)aurate(1−) sodium (1:1); gold sodium cyanide; sodium aurocyanide. C_2AuN_2Na; mol wt 271.99. C 8.83%, Au 72.42%, N 10.30%, Na 8.45%. $NaAu(CN)_2$.

White, cryst powder. Sol in water. *Poisonous.*

USE: Goldplating.

8746. Sodium Dithionate. [7631-94-9] Dithionic acid sodium salt (1:2); sodium sulfoxylate. $Na_2O_6S_2$; mol wt 206.09. Na 22.31%, O 46.58%, S 31.11%. Prepd according to the equations $MnO_2 + 2SO_2 \rightarrow MnS_2O_6$ and $MnS_2O_6 + Na_2CO_3 \rightarrow MnCO_3 + Na_2S_2O_6$: de Baat, *Rec. Trav. Chim.* **45**, 237 (1926); Pfanstiel, *Inorg. Synth.* **2**, 170 (1946).

Dihydrate. [10101-85-6] $Na_2O_6S_2.2H_2O$; mol wt 242.12. Colorless, water-clear, orthorhombic crystals. Very stable in air. d 2.189. Loses all of its water of crystn at 110°. When heated to 267° it is dissociated into Na_2SO_4 and SO_2. Soly in water at 0°: 6.05% (w/w); at 20°: 13.39%; at 30°: 17.32%. Insol in alc.

8747. Sodium Dithionite. [7775-14-6] Dithionous acid sodium salt (1:2); sodium sulfoxylate; sodium hydrosulfite; sodium hyposulfite. $Na_2O_4S_2$; mol wt 174.10. Na 26.41%, O 36.76%, S 36.83%. $Na_2S_2O_4$. Reducing agent for synthetic and industrial applications. Manuf: L. A. Pratt, *Ind. Eng. Chem.* **16**, 676 (1924). Crystal structure: J. B. Weinrach *et al.*, *J. Crystallogr. Spectrosc. Res.* **22**, 291 (1992). Proposed mechanism for the reduction of aldehydes and ketones: S.-K. Chung, *J. Org. Chem.* **46**, 5457 (1981). Use in the synthesis of benzimidazoles: D. Yang *et al.*, *Synthesis* **2005**, 47. Field evaluation for reduction of hexavalent chromium in groundwater: R. D. Ludwig *et al.*, *Environ. Sci. Technol.* **41**, 5299 (2007).

White or grayish-white, cryst powder; slight characteristic odor of sulfur dioxide. *Spontaneously combustible; irritant.* d 2.41. Soly in water (20°): 240 g/l. Oxidizes with the evolution of heat when mixed with water in the presence of air. *Store in a tightly closed container.*

USE: As reducing agent in organic synthesis, in textiles for reduction of vat dyes; in pulp and paper industry for bleaching.

8748. Sodium Dodecylbenzenesulfonate. [25155-30-0] Dodecylbenzenesulfonic acid sodium salt (1:1); dodecylbenzene sodium sulfonate; Conoco C-50; Conoco SD 40; Conoco C-60. $C_{18}H_{29}NaO_3S$; mol wt 348.48. C 62.04%, H 8.39%, Na 6.60%, O 13.77%, S 9.20%. $C_{12}H_{25}C_6H_4SO_3Na$. Manuf: *Chem. Eng.* **61**, no. 6, 372 (1954); Huber *et al., J. Am. Oil Chem. Soc.* **33**, 57 (1956); **GB 761095**; Seaton, **US 2782230** (1956, 1957 both to Monsanto); **GB 773423**; Gerhart, Karwacki, **US 2820056** (1957, 1958 both to Continental Oil). Toxicity study: Hopper *et al., J. Am. Pharm. Assoc. Sci. Ed.* **38**, 428 (1949).

LD$_{50}$ in mice: 2 g/kg orally; 105 mg/kg i.v. (Hopper).

Caution: May cause skin irritation. If swallowed will cause vomiting.

USE: Anionic detergent.

8749. Sodium Ethoxide. [141-52-6] Ethanol sodium salt (1:1); sodium ethylate; caustic alcohol. C_2H_5NaO; mol wt 68.05. C 35.30%, H 7.41%, Na 33.78%, O 23.51%. CH_3CH_2ONa.

White or yellowish, hygroscopic powder. Dec on exposure to air and becomes darker on keeping. Dec by water into NaOH and alcohol. Sol without decompn in abs alc. *Keep tightly closed, protected from light and in a cool place.*

8750. Sodium Ethyl Sulfate. [546-74-7] Sulfuric acid monoethyl ester sodium salt (1:1); ethyl sodium sulfate; sodium sulfovinate. $C_2H_5NaO_4S$; mol wt 148.11. C 16.22%, H 3.40%, Na 15.52%, O 43.21%, S 21.65%. $NaC_2H_5SO_4$.

Monohydrate. [6106-02-1] $C_2H_5NaO_4S.H_2O$; mol wt 166.12. White, very hygroscopic crystals. Sol in 0.7 part water, in alcohol. *Keep well closed.*

USE: In organic syntheses.

8751. Sodium Ferric Gluconate. [34089-81-1] D-Gluconic acid iron(3+) sodium salt (1:?:?). Bioavailable iron supplement containing a ferric hydroxide core stabilized and solubilized by simple carbohydrates. Prepn: W. Traube *et al., Ber.* **66B**, 1545 (1933). Intestinal absorption of clinical formulations: L. Samochowiec, *Clin. Ter.* **56**, 341 (1971).

Sodium ferric gluconate complex. Ferlixit; Ferrlecit; Ferrosprint; Sanifer. Aqueous iron formulation administered as an injectable, stable, macromolecular iron saccharate complex. Gluconate bridges the iron centers. Mol formula is approximately: $[NaFe_2O_3-(C_6H_{11}O_7)(C_{12}H_{22}O_{11})_5]_n$. Apparent mol wt 289-440 kDa. Safety profile: G. Faich, J. Strobos, *Am. J. Kidney Dis.* **33**, 464 (1999). Clinical evaluation in hemodialysis patients: A. R. Nissenson *et al., ibid.* 471; and comparison with iron dextran: B. Michael *et al., Kidney Int.* **61**, 1830 (2002). Clinical pharmacokinetics: P. A. Seligman *et al., Pharmacotherapy* **24**, 574 (2004). Clinical evaluation in renal transplant patients: R. S. Gillespie, J. M. Symons, *Pediatr. Transplant.* **9**, 43 (2005). Review of therapeutic applications: S. Fishbane, J. Wagner, *Am. J. Kidney Dis.* **37**, 879-883 (2001). Deep red color indicative of ferric oxide linkages.

THERAP CAT: Hematinic.

8752. Sodium Ferrocyanide. [13601-19-9] (*OC*-6-11)-Hexakis(cyano-κC)ferrate(4−) sodium (1:4); tetrasodium hexakis-(cyano-C)ferrate(4−); sodium hexacyanoferrate(II); yellow prussiate of soda; sodium prussiate yellow. $C_6FeN_6Na_4$; mol wt 303.91. C 23.71%, Fe 18.38%, N 27.65%, Na 30.26%. $Na_4Fe(CN)_6$. Review of properties, chemistry and syntheses: *The Chemistry of Ferrocyanides,* American Cyanamid Co. (Beacon Press, New York, 1953) 112 pp.

Decahydrate. [14434-22-1] $C_6FeN_6Na_4.10H_2O$; mol wt 484.06. Pale yellow, monoclinic, slightly efflorescent crystals. Steady dehydration occurs >50°. Becomes anhyd at 81.5°. Dec 435°, forming sodium cyanide, iron, carbon, and nitrogen. Soly in water at 1°: 10.2% (calcd as the anhydr salt); at 17°: 14.7%; at 25°: 17.6%; at 53°: 28.1%; at 85°: 39%; at 96.6°: 39.7%. Practically insol in most organic solvents. Addition of sodium ferrocyanide solns to slightly acidic solns of iron salts causes precipitation of insol Prussian blue (ferric ferrocyanide), $Fe_4[Fe(CN)_6]_3$. Alkaline solns yield sol Prussian blue, $NaFe[Fe(CN)_6]$. Sodium ferrocyanide forms gels with heavy metals in general.

Caution: Do not mix with hot or concd acids and do not expose solns to sunlight for any length of time to avoid generation of hydrogen cyanide. Waste ferrocyanides in streams and lakes should not exceed 2 ppm because irradiated solns become toxic to fish: G. E.

Burdick, M. Lipscheutz, *C.A.* **44**, 10939f (1950). Because of strong chemical bondage between the cyanide groups and the iron, ferrocyanides have a low order of toxicity.

USE: Anticaking agent in rock salt and food grade salt. Used in ore flotation. In manuf of Prussian blue pigment. In photography for bleaching, toning, and fixing. Additive to pickling baths. Peptizing agent in rubber. Arc stabilizer in welding rod coatings. Emulsion polymerization catalyst.

8753. Sodium Fluoborate. [13755-29-8] Tetrafluoroborate-(1−) sodium (1:1); sodium tetrafluoroborate; sodium borofluoride. BF_4Na; mol wt 109.79. B 9.85%, F 69.22%, Na 20.94%. $NaBF_4$. Prepd according to the equation $2H_3BO_3 + 8HF + Na_2CO_3 \rightarrow 2NaBF_4 + 7H_2O + CO_2$: Balz, Wilke-Dörfurt, *Z. Anorg. Allg. Chem.* **159**, 197 (1927); Kwasnik in *Handbook of Preparative Inorganic Chemistry* **vol. 1**, G. Brauer, Ed. (Academic Press, New York, 2nd ed., 1963) p 222.

Orthorhombic, stout rectangular prisms, d^{20} 2.47. mp 384° (slight decompn). Does not etch glass when absolutely dry. Soly in water (g/100 ml): 108 (26°); 210 (100°). Sparingly sol in alcohol. Aq solns have a bitter taste and are acid to litmus.

USE: Fluorinating agent. See: Lawton, Levy, *J. Am. Chem. Soc.* **77**, 6083 (1955).

8754. Sodium Fluoride. [7681-49-4] Sodium fluoride (NaF); Chemifluor; Duraphat; Florocid; Fluoros; Flura-Drops; Karidium; Lemoflur; Luride-SF; Ossalin; Ossin; Osteo-F; Osteofluor; Slow-Fluoride; Villiaumite; Zymafluor. FNa; mol wt 41.99. F 45.25%, Na 54.75%. NaF. Prepd by fusing cryolite with NaOH; by adding equiv amounts of NaOH or Na_2CO_3 to 40% HF (precipitation is instantaneous and crystal size depends on pH, but too much HF yields sodium bifluoride, $NaHF_2$): Müller, *Chem. Ztg.* **52**, 5 (1928); Kwasnik in *Handbook of Preparative Inorganic Chemistry* **vol. 1**, G. Brauer, Ed. (Academic Press, New York, 2nd ed., 1963) pp 235-236. Technical grades are 90% and 95% NaF, light (37 cu in/lb) and dense (23 cu in/lb), and 98%. The impurities are mainly sodium and aluminum fluosilicates. Pharmacology: Caruso *et al., Handb. Exp. Pharmakol.* **XX** (Part 2), F. Smith, Ed. (Springer, Berlin, 1970) pp 144-165. Toxicity: H. F. Smyth *et al., Am. Ind. Hyg. Assoc. J.* **30**, 470 (1969). Carcinogenicity studies: J. R. Bucher *et al., Int. J. Cancer* **48**, 733 (1991). Review of toxicology: D. W. Banting, *J. Am. Dent. Assoc.* **122**, 86-91 (1991); and human exposure: *Toxicological Profile for Fluorides, Hydrogen Fluoride, and Fluorine* (PB2004-100002, 2003) 404 pp. Review of clinical efficacy in prevention of dental caries: L. G. Petersson, *Caries Res.* **27**, Suppl 1, 35-42 (1993); of clinical experience in osteoporosis: M. Kleerekoper, D. B. Mendlovic, *Endocr. Rev.* **14**, 312-323 (1993).

Cubic or tetragonal crystals (NaCl lattice). d 2.78. mp 993°. bp 1704°. *Poisonous.* Soly in water (g/100 ml): 4.0 (15°); 4.3 (25°); 5.0 (100°). Insol in alc. Aq solns have an alkaline reaction caused by partial hydrolysis. pH of freshly prepd satd soln 7.4. Aq solns etch glass, but the dry crystals or powder may be kept in glass bottles. Sodium fluoride sold as household insecticide must be tinted Nile Blue. LD$_{50}$ orally in rats: 0.18 g/kg (Smyth).

Caution: Potential symptoms of overexposure by ingestion are salty or soapy taste; salivation, nausea, abdominal pain, vomiting, diarrhea; dehydration, thirst; sweating; stiff spine; muscle weakness, tremors; CNS depression; shock; arrhythmia. Direct contact may cause dermatitis; irritation of eyes, respiratory system. Potential symptoms of chronic ingestion are mottling of tooth enamel; osteosclerosis, calcification of ligaments of ribs, pelvis. *See NIOSH Pocket Guide to Chemical Hazards* (DHHS/NIOSH 97-140, 1997) p 282; *Clinical Toxicology of Commercial Products,* R. E. Gosselin *et al.,* Eds. (Williams & Wilkins, Baltimore, 5th ed., 1984) Section III, pp 185-193.

USE: As insecticide, particularly for roaches and ants; in other pesticide formulations; constituent of vitreous enamel and glass mixes; as a steel degassing agent; in electroplating; in fluxes; in heat-treating salt compositions; in the fluoridation of drinking water; for disinfecting fermentation apparatus in breweries and distilleries; preserving wood, pastes and mucilage; manuf of coated paper; frosting glass; in removal of HF from exhaust gases to reduce air pollution. Sequestering agent in analytical chemistry. Dental caries prophylactic.

THERAP CAT: In treatment of osteoporosis.

THERAP CAT (VET): Anthelmintic, pediculicide, acaricide.

8755. **Sodium Formaldehyde Sulfoxylate.** [149-44-0] 1-Hydroxymethanesulfinic acid sodium salt (1:1); formaldehyde sodium sulfoxylate; formaldehydesulfoxylic acid sodium salt; sodium hydroxymethanesulfinate; sodium methanalsulfoxylate; Bruggolite; Rongalite. CH_3NaO_3S; mol wt 118.08. C 10.17%, H 2.56%, Na 19.47%, O 40.65%, S 27.15%. $Na[HOCH_2SO_2]$. Prepn: Heyl, Greer, *Am. J. Pharm.* **94**, 80 (1922); Binns, **US 2013125** (1935 to Virginia Smelting); Postnikov, Kunin, *J. Appl. Chem. USSR* **13**, 185 (1940). Structure of dihydrate: Truter, *J. Chem. Soc.* **1955**, 3064; **1962**, 3400. Use in manuf of arsphenamines: Krumwiede, *J. Am. Pharm. Assoc.* **8**, 795 (1919); Heyl, Miller, *ibid.* **11**, 432 (1922). Of limited value in treatment of mercuric chloride poisoning: Modell *et al., J. Pharmacol. Exp. Ther.* **61**, 66 (1937). Use in vat color printing pastes: Borstelmann, Fordemwalt, **US 2597281** (1952 to Am. Cyanamid); in polymerization of ethylenic compds: **GB 816252**; **GB 852593** (1959 to Hercules Powder; 1960 to Air Reduction).

Dihydrate. [6035-47-8] $CH_3NaO_3S.2H_2O$; mol wt 154.11. Crystals, mp 63-64°, dec at higher temp. Odorless when freshly prepd, but quickly develops a characteristic (garlic) odor. Freely sol in water; slightly sol in alc, ether, chloroform, benzene. Readily dec by dil acids. Aq soln is practically neutral. *Keep well closed in a cool place.* LD s.c. in mice, 4.0 g/kg: Rosenthal, *Public Health Rep.* **49**, 908 (1934).

USE: Reducing agent in emulsion polymerization; industrial bleaching agent. Pharmaceutical aid (antioxidant).

THERAP CAT: Treatment of mercury poisoning.

8756. **Sodium Formate.** [141-53-7] Formic acid sodium salt (1:1). $CHNaO_2$; mol wt 68.01. C 17.66%, H 1.48%, Na 33.80%, O 47.05%. HCOONa.

White, deliquesc granules or cryst powder; slight odor of formic acid. d 1.92. mp 253°; at higher temp dec into sodium oxalate and hydrogen, then into sodium carbonate. Sol in ~1.3 parts water; sol in glycerol, slightly in alcohol. The aq soln is neutral. pH ~7. Has buffering action. *Keep well closed.*

USE: In dyeing and printing fabrics; also in anal. chemistry as a precipitant for the "noble" metals. Solubilizes trivalent metal ions in soln by forming complex ions. Buffering action adjusts the pH of strong mineral acids to higher values.

THERAP CAT: Caustic, astringent.

8757. **Sodium Glycerophosphate.** [1334-74-3] 1,2,3-Propanetriol mono(dihydrogen phosphate) sodium salt (1:2). $C_3H_7Na_2$-O_6P; mol wt 216.04. C 16.68%, H 3.27%, Na 21.28%, O 44.43%, P 14.34%. Three isomers exist: The β-glycerophosphoric acid disodium salt ($(HOCH_2)_2CHOPO_3Na_2$) and D(+)- and L(−)-α-glycerophosphoric acid disodium salt ($HOCH_2CH(OH)CH_2OPO_3Na_2$). Prepn: H. King, F. L. Pyman, *Pharm. J.* **92**, 511 (1914). Exptl use in diagnosis of prostatic carcinoma: M. K. Schwartz *et al., Ann. N.Y. Acad. Sci.* **166**, 775 (1969). Chronic toxicity study of β-form: K. L. Raheja *et al., Toxicology* **8**, 115 (1977). Efficacy of β-form as cariostatic agent: T. H. Grenby, J. M. Bull. *Arch. Oral Biol.* **20**, 717 (1975). GC determn: Y. Handa *et al., J. Chromatogr.* **206**, 387 (1981).

β-Form hemiundecahydrate. White, odorless, scale-like crystals; dec >130°. Sol in ~1.5 parts water; more sol in hot water. (pH of aq soln: ~9.5). Insol in alcohol.

THERAP CAT: Tonic.

THERAP CAT (VET): Has been used as a tonic.

8758. **Sodium Hexachloroplatinate(IV).** [16923-58-3] (*OC*-6-11)-Hexachloroplatinate(2−) sodium (1:2); disodium hexachloroplatinate(2−); sodium platinichloride; sodium chloroplatinate. Cl_6-Na_2Pt; mol wt 453.76. Cl 46.88%, Na 10.13%, Pt 42.99%. Na_2-$[PtCl_6]$. Prepn: Grube in *Handbook of Preparative Inorganic Chemistry* **vol. 2**, G. Brauer, Ed. (Academic Press, New York, 2nd ed., 1965) pp 1571-1572; Cox, Peters, *Inorg. Synth.* **13**, 173 (1971).

Yellow, hygroscopic crystals. Easily forms hexahydrate at 25° and relative humidity >50% (reconverted to anhydr salt by heat at 110° for one hour). uv max (1 formal HCl): 262 nm (ε 24500). Sol in water, alcohol.

USE: Catalyst.

8759. **Sodium Hexafluorosilicate.** [16893-85-9] Hexafluorosilicate(2−) sodium (1:2); sodium fluorosilicate; sodium fluosilicate; sodium silicofluoride; Salufer. F_6Na_2Si; mol wt 188.05. F

60.62%, Na 24.45%, Si 14.93%. Na_2SiF_6. Toxicity study: C. W. Muehlberger, *J. Pharmacol. Exp. Ther.* **39**, 246 (1930). Review of toxicology of fluoride compounds: G. L. Waldbott, *Acta Med. Scand. Suppl.* **400**, 1-44 (1963).

White, granular powder. *Poisonous.* d 2.68. Melts at red heat with decompn. Sol in 150 parts cold, 40 parts boiling water. Insol in alc. The soln in cold water is neutral. Readily hydrolyzes in water to fluoride ions and hydrated silica. LD in rabbits (mg F/kg): 76 intragastric; in rats (mg F/kg): 42 s.c. (Muehlberger). LD in guinea pigs (mg/kg): 250 orally, 500 s.c. (Waldbott).

USE: Fluoridating agent for drinking water. In enamels for china and porcelain; manuf opalescent glass; as insecticide, rodenticide; mothproofing of woolens. Intermediate in prodn of synthetic cryolite and other fluorosilicates.

8760. **Sodium Hydride.** [7646-69-7] Sodium hydride (NaH); sodium monohydride. HNa; mol wt 24.00. H 4.20%, Na 95.79%. NaH. Prepd by passing hydrogen into molten sodium dispersed in oil or mixed with a catalyst such as anthracene above 250°: Hansley, Carlisle, *Chem. Eng. News* **23**, 1332 (1945). Laboratory procedure by hydrogenating sodium dispersions: Mattson, Whaley, *Inorg. Synth.* **5**, 10 (1957). Book: J. Plesek, S. Hermanek, *Sodium Hydride, Its Use in the Laboratory and in Technology* (Iliffe Books, London, 1968) 185 pp.

Silvery needles; the commercial product is a gray-white powder. d 1.396. Dec 425°. *Dangerous when wet.* Reacts explosively with water, violently with lower alcohols, ignites spontaneously on standing in moist air. Sol in molten sodium hydroxide, insol in liquid ammonia but forms sodamide at moderate temps.

USE: At low temps where reducing properties of sodium are undesirable as in the condensation of ketones and aldehydes with acid esters; in soln with molten sodium hydroxide for the reduction of oxide scale on metals; at high temps as a reducing agent and reduction catalyst.

8761. **Sodium Hydroxide.** [1310-73-2] Sodium hydroxide (Na(OH)); caustic soda; soda lye; sodium hydrate. HNaO; mol wt 40.00. H 2.52%, Na 57.47%, O 40.00%. NaOH. By reacting calcium hydroxide with sodium carbonate; from sodium chloride by electrolysis; from sodium metal and water vapor at low temp. Description of industrial processes: *Faith, Keyes & Clark's Industrial Chemicals,* F. A. Lowenheim, M. K. Moran, Eds. (Wiley-Interscience, New York, 4th ed., 1975) pp 737-745. Toxicity: Fazekas, *Arch. Exp. Pathol. Pharmakol.* **184**, 587 (1937).

Fused solid with crystalline fracture. Rapidly absorbs carbon dioxide and water from the air. Very corrosive (caustic) to animal and vegetable tissue and to aluminum metal in the presence of moisture. Sold as lumps, sticks, pellets, chips, etc. When kept in tight containers, the usual grades contain 97-98% NaOH. mp 318°. d^{25} 2.13. One gram dissolves in 0.9 ml water, 0.3 ml boiling water, 7.2 ml abs alcohol, 4.2 ml methanol, also sol in glycerol. Generates considerable heat while dissolving, or when the soln is mixed with an acid. Volumetric NaOH solns used in the laboratory must be protected from air to avoid formation of carbonate. Concentrated NaOH solns dissolve practically no sodium carbonate. The pH of a 0.05% w/w soln ~12, of a 0.5% soln ~13, of a 5% soln ~14. Density, boiling and freezing pt data for (w/w) water solns. d_4^{15}: 5% 1.056, 10% 1.111, 20% 1.222, 30% 1.333, 40% 1.434, 50% 1.530. bp: 5% 102°, 10% 105°, 20% 110°, 30% 115°, 40% 125°, 50% 140°. fp: 5% −4°, 10% −10°, 20% −26°, 30% 1°, 40% 15°, 50% 12°. LD orally in rabbits: 500 mg/kg (10% soln) (Fazekas).

Caution: Potential symptoms of overexposure are irritation of eyes, skin and mucous membranes; pneumonitis; eye and skin burns; temporary loss of hair. *See NIOSH Pocket Guide to Chemical Hazards* (DHHS/NIOSH 97-1140, 1997) p 284.

USE: NaOH solutions are used to neutralize acids and make sodium salts, *e.g.*, in petroleum refining to remove sulfuric and organic acids; to treat cellulose in making viscose rayon and cellophane; in reclaiming rubber to dissolve out the fabric; in making plastics to dissolve casein. NaOH solns hydrolyze fats and form soaps; they precipitate alkaloids (bases) and most metals (as hydroxides) from water solns of their salts. Pharmaceutical aid (alkalizer). Analytical uses as alkalimetric titrant, buffer, and pH modifier.

THERAP CAT (VET): Caustic; dehorning of calves.

8762. **Sodium Hypochlorite.** [7681-52-9] Hypochlorous acid sodium salt (1:1). ClNaO; mol wt 74.44. Cl 47.62%, Na

30.88%, O 21.49%. NaClO. Strong oxidizing and hydrolyzing agent; used in aqueous solutions of various strengths for its bactericidal properties. Prepn as the pentahydrate from NaOH and Cl$_2$ in the presence of water: Sanfourche, Gardent, *Bull. Soc. Chim.* [4] **35**, 1089 (1924); Schmeisser in *Handbook of Preparative Inorganic Chemistry* vol. 1, G. Brauer, Ed. (Academic Press, New York, 2nd ed., 1963) pp 309-310. Review of prepn and uses of *Dakin's solution*, a diluted antiseptic formulation: H. Plagge, *Pharm. Ztg.* **138**, No. 14, 26-31 (1993). Review of toxicology and use as household bleach: F. Racioppi *et al.*, *Food Chem. Toxicol.* **32**, 845-861 (1994); of use in health care facilities: W. A. Rutala, D. J. Weber, *Clin. Microbiol. Rev.* **10**, 597-610 (1997). Review of use as endodontic irrigant: R. M. Clarkson, A. J. Moule, *Aust. Dent. J.* **43**, 250-256 (1998).

Prepd as the pentahydrate, crystals, mp 18°. Dec by CO$_2$ from air. Anhydr NaClO may be obtained by freeze-drying in a vacuum (over concd H$_2$SO$_4$). *Anhydr NaClO is very explosive.* Soly at 0°: 29.3 g/100 ml H$_2$O. Aqueous solutions for household bleach contain ~5.25%. Solutions for use as antiseptics contain ~0.5% sodium hypochlorite and are buffered or stabilized with various agents.

Caution: Potential symptoms of overexposure by ingestion are pain and inflammation of the mouth, pharynx, esophagus, stomach; vomiting; circulatory collapse, cold and clammy skin, cyanosis, shallow respirations; confusion, delirium, coma; edema of pharynx, larynx, glottis with stridor and obstruction; perforation of esophagus, stomach. Potential symptoms of overexposure by fume inhalation are severe respiratory tract irritation, pulmonary edema. Direct contact may cause vesicular eruptions on skin and eczematoid dermatitis. *See Clinical Toxicology of Commercial Products*, R. E. Gosselin *et al.*, Eds. (Williams & Wilkins, Baltimore, 5th ed., 1984) Section III, pp 202-205.

USE: Aq soln as bleach, disinfectant; chlorination of swimming pools; sanitation of drinking water. Analytical use as chlorine standard.

THERAP CAT: Antiseptic, disinfectant.

8763. Sodium Hypophosphite. [7681-53-0] Phosphinic acid sodium salt (1:1). H$_2$NaO$_2$P; mol wt 87.98. H 2.29%, Na 26.13%, O 36.37%, P 35.21%. NaH$_2$PO$_2$. Solubility data: Palit, *J. Am. Chem. Soc.* **69**, 3120 (1947).

Monohydrate. [10039-56-2] H$_2$NaO$_2$P.H$_2$O; mol wt 105.99. White, odorless, deliquesc granules; saline taste. When strongly heated, it dec with evolution of phosphine which ignites spontaneously in the air. *It explodes when triturated with chlorates or other oxidizing agents.* Sol in 1 part water, 0.15 part boiling water; freely sol in glycerol and in boiling alcohol; sol in cold alcohol, slightly in abs alcohol. Insol in ether. Soly of anhydr NaH$_2$PO$_2$ at 25° in ethylene glycol: 33.0 g/100 g; in propylene glycol: 9.7 g/100 g. The aq soln is neutral. *Keep well closed.*

USE: As reagent for arsenic and iodates; prepn of hypophosphites syrup.

8764. Sodium Iodate. [7681-55-2] Iodic acid (HIO$_3$) sodium salt (1:1). INaO$_3$; mol wt 197.89. I 64.13%, Na 11.62%, O 24.25%. NaIO$_3$. Crystal structure: I. Naray-Szabo, J. Neugebauer, *J. Am. Chem. Soc.* **69**, 1280 (1947). Use to induce experimental retinal degeneration: A. Grignolo *et al.*, *Exp. Eye Res.* **5**, 86 (1966); L. M. Franco *et al.*, *Invest. Ophthalmol. Visual Sci.* **50**, 4004 (2009). Acute toxicity study: S. H. Webster *et al.*, *J. Pharmacol. Exp. Ther.* **120**, 171 (1957). Review of safety assessment: *J. Am. Coll. Toxicol.* **14**, 231-239 (1995).

White, cryst powder. *Oxidizer.* d 4.28. Sol in acetone, acetic acid. Sol in ~11 parts water, 3 parts boiling water. Insol in alc. The aq soln is neutral. LD$_{50}$ in female mice (mg/kg): 119 ±4 i.p., 108 ±4 i.v., 505 ±26 orally (Webster).

USE: Retinotoxic agent for experimental vision studies.

8765. Sodium Iodide. [7681-82-5] Sodium Iodide (NaI); Iodopen. INa; mol wt 149.89. I 84.67%, Na 15.34%. NaI. Review of use as scintillator to detect ionizing radiation: T. D. Cradduck, *Semin. Nucl. Med.* **3**, 205-223 (1973).

White, odorless, deliquesc crystals or granules. mp 651°. d 3.67. Gradually absorbs up to about 5% (½ mol) moisture on exposure to air. Slowly becomes brown in the air due to liberation of iodine. Its aq soln is similarly affected. One gram dissolves in 0.5 ml water, ~2 ml alc, 1 ml glycerol; sol in acetone. *Keep well closed and*

protected from light. At ordinary room temp crystallizes from water with 2H$_2$O in the form of colorless, prismatic crystals. MLD i.v. in rats: 1.3 g/kg, Loeser, Konwiser, *J. Lab. Clin. Med.* **15**, 35 (1929).

Sodium iodide I 131. [7790-26-3] Hicon; Iodotope. Review of efficacy in treatment of thyroid carcinoma: K. L. Parthasarathy, E. S. Crawford, *J. Nucl. Med. Technol.* **30**, 165-171 (2002); J. Buscombe *et al.*, *Expert Rev. Anticancer Ther.* **8**, 1425-1431 (2008). Prepd with radioactive iodine (^{131}I) which has a half-life of 8 days and emits beta and gamma rays. Other properties identical with those of ordinary sodium iodide.

USE: In scintillation detectors.

THERAP CAT: Iodine supplement. Sodium iodide I 131 as diagnostic aid (thyroid function); antineoplastic (radiation source).

THERAP CAT (VET): In treatment of actinobacillosis and actinomycosis.

8766. Sodium Isopropyl Xanthate. [140-93-2] Carbonodithioic acid *O*-(1-methylethyl) ester sodium salt (1:1); isopropyl-xanthic acid sodium salt; SIPX; NAX 31. C$_4$H$_7$NaOS$_2$; mol wt 158.21. C 30.37%, H 4.46%, Na 14.53%, O 10.11%, S 40.53%. Prepd by the addition of CS$_2$ to a solution of NaOH in isopropanol. Use in the flotation recovery of metals: B. G. Cousins, R. S. MacPhail, *Miner. Eng.* **9**, 509 (1996). Interaction with silver: G. A. Hope *et al.*, *Colloids Surf. A* **178**, 157 (2001). Quantification of adsorbed collector under flotation conditions: D. Lascelles, J. A. Finch, *Miner. Eng.* **18**, 257 (2005).

Deliquescent, white to yellowish powder or lumps, dec 150°. *Irritant.* Slightly unpleasant odor. Soly in water: 30% at 4°; 46% at 24°; 54% at 35°. Sol in acetone.

USE: In mining as a collector reagent for the recovery of base and precious metals.

8767. Sodium Lactate. [72-17-3] 2-Hydroxypropanoic acid sodium salt (1:1); sodium α-hydroxypropionate; sodium DL-lactate; lacolin. C$_3$H$_5$NaO$_3$; mol wt 112.06. C 32.16%, H 4.50%, Na 20.52%, O 42.83%. Commercially available as a mixture with water containing 70-80% sodium lactate. *Ref:* Shaw, US 2856326 (1958 to Nat. Dairy Prod. Corp.).

Colorless or almost colorless, thick, odorless liquid. Miscible with water, alcohol. The soln is neutral.

USE: Instead of glycerol in calico printing; as a plasticizer for casein; as a corrosion inhibitor in alc antifreeze mixture.

THERAP CAT: Electrolyte replenisher; systemic and urinary alkalinizing agent.

THERAP CAT (VET): Has been used in bovine ketosis.

8768. Sodium Lauryl Sulfate. [151-21-3] Sulfuric acid monododecyl ester sodium salt (1:1); sodium dodecyl sulfate; SDS; Irium. C$_{12}$H$_{25}$NaO$_4$S; mol wt 288.38. C 49.98%, H 8.74%, Na 7.97%, O 22.19%, S 11.12%. CH$_3$(CH$_2$)$_{10}$CH$_2$OSO$_3$Na. Anionic detergent prepd by sulfation of lauryl alcohol, followed by neutralization with sodium carbonate: A. Lottermoser, F. Stoll, *Kolloid-Z.* **63**, 50 (1933). Surfactant properties: J. Powney, C. C. Addison, *Trans. Faraday Soc.* **33**, 1244 (1937); E. E. Dreger *et al.*, *Ind. Eng. Chem.* **36**, 610 (1944). Use in electrophoretic sepn and mol wt estimation of proteins: A. L. Shapiro *et al.*, *Biochem. Biophys. Res. Commun.* **28**, 815 (1967); K. Weber, M. Osborn, *J. Biol. Chem.* **244**, 4406 (1969); of glycopolypeptides: B. S. Leach *et al.*, *Biochemistry* **19**, 5734 (1980). Toxicity study: A. I. T. Walker *et al.*, *Food Cosmet. Toxicol.* **5**, 763 (1967). Review of toxicology: Ch. Gloxhuber, *Arch. Toxicol.* **32**, 245-270 (1974).

White or cream-colored crystals, flakes, or powder. mp 204-207°. Faint odor of fatty substances. Smooth feel. Neutral reaction. One gram dissolves in 10 ml water, giving an opalescent soln. Lowers the surface tension of aq solns. Emulsifies fats. LD_{50} orally in rats: 1288 mg/kg (Walker).

USE: Wetting agent, detergent, esp in the textile industry. Electrophoretic separation of proteins and lipids. Ingredient of toothpastes.

8769. Sodium-Lead Alloy. [12740-44-2] Sodium alloy, nonbase, Na, Pb; lead-sodium alloy; sodium-lead; Drynap; Dri-Na. Usually contains a minimum of 9.5% active sodium. For tetraethyllead manufacture the sodium-lead alloy is produced in large quantities by making a melt of 90 parts of lead with 10.5 parts of sodium (w/w). The reaction is strongly exothermic and starts at 225°. Prepn and use as drying agent: H. Soroos, *Ind. Eng. Chem. Anal. Ed.* **11**, 657 (1939). Use in quantitative analysis of multivalent cations: R. A. Edge, G. W. A. Fowles, *Anal. Chim. Acta* **32**, 191 (1965); for detection of heteroelements in organic compounds: R. C. Lance et al., *Microchem. J.* **20**, 103 (1975).

Obtained in brittle lumps which can be stored in an air-tight container. Corrosive. The alloy may be ground to a very fine powder under the surface of a non-polar solvent such as kerosine or ether. Finely ground powder, if not protected by a suitable liquid, may react with excess moisture from the air sufficiently to catch fire. *Protect from moisture; reacts violently with water.*

Caution: Lead and all lead compounds are listed as reasonably anticipated to be human carcinogens: *Report on Carcinogens, Twelfth Edition* (PB2011-111646, 2011) p 251.

USE: Analytical reagent. In manuf of tetraethyllead; for drying ether and for reductions: Tabei et al., *Bull. Chem. Soc. Jpn.* **40**, 1538 (1967).

8770. Sodium Metabisulfite. [7681-57-4] Disulfurous acid sodium salt (1:2). $Na_2O_5S_2$; mol wt 190.09. Na 24.19%, O 42.08%, S 33.73%. $Na_2S_2O_5$. Crystal structure: K. L. Carter et al., *Acta Crystallogr.* **B60**, 155 (2004).

White crystals or powder; odor of SO_2. d 2.34-2.36. Freely sol in water, glycerol; slightly sol in alcohol. The aq soln is acid.

Caution: Potential symptoms of overexposure are irritation of eyes, skin and mucous membranes. See *NIOSH Pocket Guide to Chemical Hazards* (DHHS/NIOSH 97-140, 1997) p 284.

USE: Pharmaceutic aid (antioxidant). Reducing agent.

8771. Sodium Metaborate. [7775-19-1] Boric acid (HBO_2) sodium salt (1:1). $BNaO_2$; mol wt 65.80. B 16.43%, Na 34.94%, O 48.63%. $NaBO_2$. Obtained by fusing equivalent mol wts of borax and sodium carbonate.

White pieces or powder. mp 966°. Sol in water, the soln being strongly alkaline.

8772. Sodium Metaperiodate. [7790-28-5] Periodic acid (HIO_4) sodium salt (1:1); sodium periodate. $INaO_4$; mol wt 213.89. I 59.33%, Na 10.75%, O 29.92%. $NaIO_4$. Synthesis starting with sodium iodide: *Inorg. Synth.* **2**, 212 (1946) and **1**, 170 (1939).

White, tetragonal crystals, d_4^{16} 3.865. Dec ~300°. Sol in cold water, sulfuric, nitric, acetic acids.

Trihydrate. [13472-31-6] $INaO_4.3H_2O$; mol wt 267.94. White, efflorescent, trigonal crystals, d_4^{18} 3.219. Dec 175°. One gram dissolves in 8 ml water at 20°.

USE: Oxidimetric standard. Determn of manganese.

8773. Sodium Metaphosphate. [10361-03-2] Metaphosphoric acid (HPO_3) sodium salt (1:1). Term used for a series of condensed inorganic phosphates prepd by dehydration of sodium orthophosphates; differing reaction conditions lead to various cyclic or linear polymeric structures. True metaphosphates, with the general formula, $(NaPO_3)_n$, are cyclic polymers. Of these, sodium trimetaphosphate, *q.v.*, is most commonly used. Linear sodium polyphosphates have the general formula, $Na_{(n+2)}P_nO_{(3n+1)}$, where n may range from 3 to 10^6. Short chain polyphosphates may be obtained in pure crystalline form, such as sodium tripolyphosphate, *q.v.* Long chain polyphosphates are usually obtained as mixtures with varying degrees of polymerization. The name, *sodium hexametaphosphate*, has been used for both the cyclic hexamer and for a mixture of soluble sodium phosphate polymers also known as sodium polymetaphosphate. Prepn: L. F. Audrieth, R. N. Bell, *Inorg. Synth.* **3**, 85 (1950). Separation of polymers by capillary gel electrophoresis: T. Wang, S. F. Y. Li, *J. Chromatogr. A* **802**, 159 (1998). *Reviews:* J. R. Van Waser, *Phosphorus and Its Compounds* vol. **1** (Interscience, New York, 1958) pp 601-800; E. Thilo in *Adv. Inorg. Chem.*

Radiochem. **4**, 1-75 (1962). Review of safety assessment for use in cosmetics: *Int. J. Toxicol.* **20**, Suppl. 3, 75-89 (2001). Review of structures, properties, and biochemistry: I. S. Kulaev et al., *The Biochemistry of Inorganic Polyphosphates*, 2nd Ed. (John Wiley & Sons, 2004) 294 pp.

Sodium polymetaphosphate. [68915-31-1] Polyphosphoric acids sodium salts; Graham's salt; glassy sodium metaphosphate; glassy sodium polyphosphate; sodium polyphosphate glass. Mixture of polymers of varying length prepd by heating NaH_2PO_4 above 600° followed by rapid chilling. Prepn: R. N. Bell, *Inorg. Synth.* **3**, 103 (1950); and mol wt distribution: H. N. Bhargava et al., *Colloid Polym. Sci.* **252**, 20 (1974). Review of use in dentifrices: C. Sensabaugh, M. E. Sagel, *J. Dent. Hyg.* **83**, 70-78 (2009). Clear, hygroscopic glass. mp 628°. Sol in water, but dissolves slowly. Depolymerizes in aqueous soln to form sodium trimetaphosphate and sodium orthophosphates.

Insoluble sodium metaphosphate. [50813-16-6] Maddrell's salt; insoluble sodium polyphosphate; IMP. High molecular weight linear polymer usually prepd by condensation of NaH_2PO_4 at 250-500°. Prepn by heating two parts $NaNO_3$ with one part H_3PO_4: R. Maddrell, *Ann.* **61**, 53 (1847). Mol wt distribution: U. P. Strauss, J. W. Day, *J. Am. Chem. Soc.* **81**, 79 (1959). Kinetics and mechanism of formation: C. Y. Shen, *Ind. Eng. Chem. Prod. Res. Dev.* **20**, 144 (1981). Manuf process: S. Kleeman et al., *US* **4780293** (1988 to Benckiser-Knapsack). White crystalline powder. Insol in water. Sol in mineral acids and in solutions of ammonium and potassium chlorides. Stable to about 500°, where it changes to sodium trimetaphosphate, and above 625° to glassy sodium polymetaphosphate.

USE: As water softeners and detergents. For leather tanning, dyeing, textile processing. In dentifrices, polishing agents, abrasive detergents. In foods as emulsifier, sequestrant, texturizer; in cosmetics as chelating agent.

8774. Sodium Metavanadate. [13718-26-8] Vanadate (VO_3^{1-}) sodium (1:1); sodium trioxovanadate; sodium vanadium trioxide; sodium vanadate(V). NaO_3V; mol wt 121.93. Na 18.85%, O 39.36%, V 41.78%. $NaVO_3$. Prepn: D. J. McAdam, Jr., *J. Am. Chem. Soc.* **32**, 1603 (1910). Occurs in nature as the minerals, munirite, which contains up to 2 molecules of H_2O, and metamunirite, which is anhydrous. Description of munirite: K. A. Butt, K. Mahmood, *Mineral. Mag.* **47**, 391 (1983); of metamunirite: H. T. Evans, Jr., *ibid.* **55**, 509 (1991). Use as catalyst: M. Jian et al., *J. Mol. Catal. A* **253**, 1 (2006); L. F. Zhu et al., *J. Catal.* **245**, 446 (2007).

Colorless to yellow crystals or cream colored solid. mp 610°. *Poisonous; irritant.* Sol in water.

Tetrahydrate. [10135-94-1] $NaO_3V.4H_2O$; mol wt 193.99. Yellowish-white, cryst powder. Sol in hot water. LD_{75} i.p. in rats: 4-5 mg V/kg, Franke, Moxon, *J. Pharmacol. Exp. Ther.* **58**, 454 (1936).

USE: In photography and manuf of inks; catalyst in organic synthesis.

8775. Sodium Methoxide. [124-41-4] Methanol sodium salt (1:1); sodium methylate. CH_3NaO; mol wt 54.02. C 22.23%, H 5.60%, Na 42.56%, O 29.62%. CH_3ONa. Prepn: Burness, *Org. Synth.* **39**, 51 (1959).

White, free-flowing powder. Sensitive to air and moisture. Dec by water. Sol in methanol, ethanol. Also exists in solvated form, $CH_3ONa.2CH_3OH$, white powder. Apparent density of solvent-free material ~4.6 lb/gal.

USE: In organic syntheses.

8776. Sodium Molybdate(VI). [7631-95-0] Sodium (T-4)-molybdate (MoO_4^{2-}) (2:1); sodium molybdenum oxide. $MoNa_2O_4$; mol wt 205.93. Mo 46.59%, Na 22.33%, O 31.08%. Na_2MoO_4. Thermal and X-ray diffraction studies: K. D. Singh Mudher et al., *J. Alloys Compd.* **396**, 275 (2005).

Dihydrate. [10102-40-6] Molyhibit 100. $MoNa_2O_4.2H_2O$; mol wt 241.96. Cryst powder. Loses its water of crystn at 100°. Sol in 1.7 parts cold water, ~0.9 part boiling water. pH of 5% aq soln at 25° = 9.0-10.0. Has been reported to be less toxic than the other corresponding compds of group 6B in the periodic table: Fairhall et al., *Public Health Bull.* no. 293 (1945).

USE: Manuf of inorganic and organic pigments, corrosion inhibitor, bath additive for metals finishing, reagent for alkaloids, micronutrient for plants and animals.

8777. Sodium β-Naphthoquinone-4-sulfonate. [521-24-4] 3,4-Dihydro-3,4-dioxo-1-naphthalenesulfonic acid sodium salt (1:1); sodium 1,2-naphthoquinone-4-sulfonate; 1,2-naphthoquinone-4-sulfonic acid sodium salt; β-naphthoquinone-4-sulfonic acid sodium salt. $C_{10}H_5NaO_5S$; mol wt 260.19. C 46.16%, H 1.94%, Na 8.84%, O 30.74%, S 12.32%. Prepn: Folin, *J. Biol. Chem.* **51**, 377 (1922).

Yellow crystals from dil alc. Readily sol in water; slightly sol in 95% alcohol; moderately sol in acetone; practically insol in ether, chloroform, carbon disulfide, benzene, petr ether. Aq solns fade slowly on exposure to light but may be stabilized by HCl.

USE: Colorimetric determn of amino acids and amines.

8778. Sodium Nitrate. [7631-99-4] Nitric acid sodium salt (1:1); chile saltpeter; cubic niter; soda niter. $NNaO_3$; mol wt 84.99. N 16.48%, Na 27.05%, O 56.47%. $NaNO_3$. The purified grade contains at least 99% $NaNO_3$. Occurs as a mineral in Chile. Toxicity data: Dollahite, Rowe, *Southwest. Vet.* **27**, 246 (1974).

Colorless, transparent crystals, white granules or powder; deliquesc in moist air. d 2.26. mp 308°. One gram dissolves in 1.1 ml water, 0.6 ml boiling water, 125 ml alcohol, 52 ml boiling alcohol, 3470 ml abs alcohol, 300 ml abs methanol. When dissolved in water the temp of the soln is lowered. The aq soln is neutral. *Oxidizer. Keep well closed.* LD_{50} orally in rabbits: 1.955 g anion/kg (Dollahite, Rowe).

USE: Manuf of nitric acid and as catalyst in the manuf of sulfuric acid. Manuf sodium nitrite, glass, enamels for pottery; in matches; for improving burning properties of tobacco; pickling meats; as color fixative in meats. Clinical reagent (parasites). The technical grade is used as fertilizer. Oxidizer.

8779. Sodium Nitrite. [7632-00-0] Nitrous acid sodium salt (1:1); erinitrit. $NNaO_2$; mol wt 68.99. N 20.30%, Na 33.32%, O 46.38%. Contains 96-98% $NaNO_2$. Thermodynamic properties: Plekhotkun, *Zh. Prikl. Khim.* **40**, 1843 (1967), *C.A.* **68**, 33797x (1968). Toxicity studies: Faccini *et al.*, *Ind. Aliment.* *(Pinerolo, Italy)* **8**, 77 (1969), *C.A.* **71**, 100583b (1969); H. F. Smyth *et al.*, *Am. Ind. Hyg. Assoc. J.* **30**, 470 (1969). Review of chemistry of $NaNO_2$ in a biological system as related to meat curing: Bard, Townsend, *Science of Meat & Meat Products*, J. F. Price, B. S. Schweigert, Eds. (W. H. Freeman, 2nd ed. 1971) pp 452-470.

White or slightly yellow, hygroscopic granules, rods, or powder. *Oxidizer; poisonous.* Very slowly oxidizes to nitrate in air. d 2.17. mp 271°; dec above 320°. Sol in 1.5 parts cold water, 0.6 part boiling water; sparingly sol in alc. Dec even by weak acids with evolution of brown fumes of N_2O_3. The aq soln is alkaline. pH ~9. *Keep well closed.* LD_{50} orally in rats: 180 mg/kg (Smyth). *Incompat:* Acetanilide, antipyrine, chlorates, hypophosphites, iodides, mercury salts, permanganate, sulfites, tannic acid, vegetable astringent decoctions, infusions or tinctures.

Caution: Consult latest government regulation to determine max amounts allowable in food.

USE: Manuf diazo dyes, nitroso compds, and in many other processes of manuf of organic chemicals; dyeing and printing textile fabrics; bleaching flax, silk, and linen; photography. In meat curing, coloring and preserving; in processing smoked chub. Reducing agent in analytical chemistry.

THERAP CAT: Vasodilator; antidote (cyanide poisoning).

THERAP CAT (VET): Antidote (cyanide poisoning). Has been used as a vasodilator, as a circulatory (blood pressure) depressant and to relieve smooth muscle spasm.

8780. Sodium Nitroprusside. [14402-89-2] Sodium (*OC*-6-22)-pentakis(cyano-κC)nitrosylferrate(2−) (2:1); sodium nitrosylpentacyanoferrate(III); sodium nitroferricyanide; sodium nitroprussiate. $C_5FeN_6Na_2O$; mol wt 261.92. C 22.93%, Fe 21.32%, N 32.09%, Na 17.55%, O 6.11%. $Na_2[Fe(CN)_5NO]$. Peripheral vasodilator. Prepn: L. Playfair, *Proc. R. Soc. London* **5**, 846 (1849).

Crystal structure: D. Schaniel *et al.*, *Phys. Rev. B* **68**, 104108 (2003); *idem et al.*, *ibid.* **71**, 174112 (2005). Pharmacology: Fernandez *et al.*, *Arch. Inst. Farmacol. Exp. Madrid* **23**, 1-51 (1971). *Review:* I. H. Tuzel, *J. Clin. Pharmacol.* **14**, 494-503 (1974). Review of pharmacology, toxicology and therapeutic uses: J. H. Tinker, J. D. Michenfelder, *Anesthesiology* **45**, 340 (1976). Comprehensive description: R. Rucki, *Anal. Profiles Drug Subs.* **6**, 487-513 (1977); A. Bult *et al.*, *ibid.* **15**, 781-792 (1986). Clinical trial in critically ill patients with heart failure: U. N. Khot *et al.*, *N. Engl. J. Med.* **348**, 1756 (2003). Review of therapeutic uses: J. A. Friederich, J. F. Butterworth, *Anesth. Analg.* **81**, 152-162 (1995).

Dihydrate. [13755-38-9] Nipride; Nipruss; Nitropress. $C_5FeN_6Na_2O.2H_2O$; mol wt 297.95. Ruby-red, practically odorless, transparent crystals. Sol in ~2.3 parts water; slightly sol in alcohol; very slightly sol in chloroform. Insol in benzene. Slowly dec in aq soln.

USE: Reagent for the detection of many organic compds, e.g., acetone, aldehydes, also of alkali sulfides, zinc, SO_2.

THERAP CAT: Antihypertensive.

THERAP CAT (VET): Antihypertensive.

8781. Sodium Oxalate. [62-76-0] Ethanedioic acid sodium salt (1:2); oxalic acid disodium salt. $C_2Na_2O_4$; mol wt 134.00. C 17.93%, Na 34.31%, O 47.76%. $Na_2C_2O_4$. Crystal structure and X-ray studies: E. A. Boldyreva *et al.*, *Z. Kristallogr.* **221**, 186 (2006). Thermal properties: B. Menczel *et al.*, *J. Chem. Thermodyn.* **36**, 41 (2004).

White, odorless, cryst powder. d^{30} 2.888. Soly in water (*m*): 0.210 at 0°, 0.270 at 25°. Insol in alcohol. The aq soln is practically neutral. Dielectric constant, ε 6.1.

Caution: Potential symptoms of overexposure by ingestion are burning pain in throat, esophagus, stomach; vomiting; weak, irregular pulse; hypotension, cardiovascular collapse; headache, muscle cramps, tetany, convulsions, stupor, coma; kidney damage. See *Clinical Toxicology of Commercial Products*, R. E. Gosselin *et al.*, Eds. (Williams & Wilkins, Baltimore, 5th ed., 1984) Section III, pp 326-328.

USE: Finishing textiles, tanning and finishing leather; for standardizing potassium permanganate soln. Titrimetry.

8782. Sodium Oxide. [1313-59-3] Sodium oxide (Na_2O); sodium monoxide. Na_2O; mol wt 61.98. Na 74.18%, O 25.81%.

White, amorphous pieces or powder. d 2.27. Melts at a dull red heat and begins to dec >400° into sodium peroxide and metal. It is very reactive and combines violently with water, forming sodium hydroxide. *Handle with tongs and not with bare hands, and keep tightly closed.*

USE: As a dehydrating agent; in certain chemical reactions as a polymerizing or condensing agent.

8783. Sodium Perborate. [7632-04-4] Perboric acid (HBO-(O_2)) sodium salt (1:1); dexol. $BNaO_3$; mol wt 81.80. B 13.22%, Na 28.10%, O 58.68%. $NaBO_3$. Contains, when reasonably fresh, ~95% of the perborate corresp to 9.9% available oxygen. Prepn from sodium metaborate and hydrogen peroxide: Leblon, Lambert, **US 3109706** (1963 to Solvay & Cie).

Tetrahydrate. White, odorless, cryst powder; saline taste. Stable when kept cool and dry, but is dec with liberation of oxygen in warm or moist air. Dec >60°. Sol in ~40 parts water, the soln being alkaline and dec with the liberation of H_2O_2, and then of oxygen. In the presence of acids, H_2O_2 is formed. *Keep well closed and in a cool place.*

Caution: Prevent swallowing of soln.

USE: Bleaching straw and other fibers, ivory, sponges, bristles, waxes, textiles; in laundering, dentifrices, soaps.

THERAP CAT: Antiseptic (topical).

THERAP CAT (VET): Mouthwash.

8784. Sodium Perchlorate. [7601-89-0] Perchloric acid sodium salt (1:1); Irenat. $ClNaO_4$; mol wt 122.44. Cl 28.95%, Na 18.78%, O 52.27%. $NaClO_4$.

Monohydrate. [7791-07-3] $ClNaO_4.H_2O$; mol wt 140.45. White, deliquesc crystals. Dec ~130°. d 2.02. Very sol in water. *Oxidizer; keep well closed.*

USE: In the explosives industry. Electrolyte. Oximetric standard.

THERAP CAT: Thyroid inhibitor.

Sodium Permanganate

8785. Sodium Permanganate. [10101-50-5] Permanganic acid ($HMnO_4$) sodium salt. $MnNaO_4$; mol wt 141.92. Mn 38.71%, Na 16.20%, O 45.09%. $NaMnO_4$.

Trihydrate. [10102-36-0] $MnNaO_4.3H_2O$; mol wt 195.97. Reddish-black, very hygroscopic granules. *Oxidizer.* Very sol in water; dec by alcohol.

8786. Sodium Peroxide. [1313-60-6] Sodium peroxide (Na_2-(O_2)); sodium dioxide; sodium superoxide; Solozone. Na_2O_2; mol wt 77.98. Na 58.96%, O 41.03%. The product of commerce contains 90-95% Na_2O_2. Prepd by heating sodium metal to 300° in aluminum vessels with a current of air from which carbon dioxide has been removed. Prepn of the octahydrate: Penneman, *Inorg. Synth.* **3**, 1 (1950).

Yellowish-white, granular powder. *Oxidizer.* Absorbs water and CO_2 from the air. Freely sol in water, forming sodium hydroxide and hydrogen peroxide, the latter quickly dec into oxygen and water. With dil acids H_2O_2 is formed which remains stable. In contact with organic matter or readily oxidizable substances ignition and explosion may take place. *Keep tightly closed and protected from contact with organic or oxidizable substances.*

Caution: Irritant and corrosive. *See* Sodium Hydroxide.

USE: Bleaching animal and vegetable fibers, feathers, bones, ivory, wood, wax, sponges, coral; rendering air charged with CO_2 respirable as in torpedo boats, submarines, diving bells, etc.; purifying air in sick rooms; dyeing and printing textiles; chemical analysis. General oxidizing agent.

8787. Sodium Persulfate. [7775-27-1] Peroxydisulfuric acid ([(HO)S(O)$_2$]$_2$O$_2$) sodium salt (1:2); sodium peroxydisulfate. Na_2-O_8S_2; mol wt 238.09. Na 19.31%, O 53.76%, S 26.93%. $Na_2S_2O_8$. Toxicity data: DaVal, *Arch. Ital. Sci. Farmacol.* **2**, 445 (1933).

White, cryst powder. *Oxidizer.* Gradually dec; decompn is promoted by moisture and higher temp. Initial soly in water at 20°: 549 g/l; dec by alcohol and silver ions. MLD in rabbits (mg/kg): 178 i.v. (DaVal).

Caution: Highly irritating to skin, mucous membranes.

USE: Bleaching and oxidizing agent; promoter for emulsion polymerization reactions.

8788. Sodium Pertechnetate Tc 99m. [23288-60-0] (*T*-4)-Technetate ($^{99}TcO_4^{1-}$) sodium; Pertscan; Ultra-Technekow. $NaO_4{}^{99m}Tc$. $Na^{99m}TcO_4$. Prepn: Keller, Kanellakopulos, *Radiochim. Acta* **1**, No. 2, 107 (1963); *C.A.* **59**, 1256a (1963); Kanellakopulos, **AEC Accession** No. **31424**, Rept. No. **KFK-197**, 73 pp (1964), *C.A.* **62**, 7350d (1965). Clinical application for labelling red blood cells: D. Ducassou *et al.*, *Br. J. Radiol.* **49**, 344 (1976). Diagnostic use in Meckel's diverticulum: D. R. Cooney *et al.*, *J. Pediatr. Surg.* **17**, 611 (1982); in thyroid neoplasm: M. Vorne, K. Jarve, *Eur. J. Nucl. Med.* **13**, 362 (1987). Review of diagnostic use in brain scanning: J. G. McAfee *et al.*, *J. Nucl. Med.* **5**, 811-827 (1964); in thyroid function: M. S. Sucupira *et al.*, *Int. J. Nucl. Med. Biol.* **10**, 29-33 (1983).

THERAP CAT: Diagnostic aid (radioactive imaging agent).

8789. Sodium Phosphate, Dibasic. [7558-79-4] Phosphoric acid sodium salt (1:2); dibasic sodium phosphate; disodium hydrogen phosphate; disodium orthophosphate; disodium phosphate; DSP; phosphate of soda; secondary sodium phosphate. HNa_2O_4P; mol wt 141.96. H 0.71%, Na 32.39%, O 45.08%, P 21.82%. Na_2HPO_4. Industrial production: *Faith, Keyes & Clark's Industrial Chemicals* (John Wiley, New York, 4th ed., 1975) pp 746-754. Toxicity of heptahydrate: H. F. Smyth *et al.*, *Am. Ind. Hyg. Assoc. J.* **30**, 470 (1969).

Anhydr, *exsiccated sodium phosphate.* Hygroscopic powder. On exposure to air will absorb from 2 to 7 mols H_2O, depending on the humidity and temp. Sol in ~8 parts water, much more sol in hot water. Soly per 100 gal water increases from ~14 lbs at slightly >0° to over 900 lbs at 95°. Insol in alc. pH of 1% aq soln at 25°: 9.1. *Keep well closed.*

Dihydrate. [10028-24-7] Sorensen's phosphate; Sorensen's sodium phosphate. $HNa_2O_4P.2H_2O$; mol wt 177.99.

Heptahydrate. [7782-85-6] $HNa_2O_4P.7H_2O$; mol wt 268.06. Crystals or granular powder. Stable in the air. d ~1.7. Sol in 4 parts water, more sol in boiling water; very slightly sol in alc. The aq soln is alkaline, pH ~9.5. LD_{50} orally in rats: 12.93 g/kg (Smyth).

Dodecahydrate. [10039-32-4] $HNa_2O_4P.12H_2O$; mol wt 358.14. Translucent crystals or granules; readily loses 5 mols of water on exposure to air at ordinary temp. mp 34-35° (when it contains the full 12 mols of H_2O). d ~1.5. Sol in 3 parts water; practically insol in alc. Aq soln is alkaline, pH ~9.5. *Keep well closed and in a cool place. Incompat:* Alkaloids, antipyrine, chloral hydrate, lead acetate, pyrogallol, resorcinol.

Caution: Anhydr form may cause mild irritation to skin, mucous membranes; intern. causes purging.

USE: As sequestrant, emulsifier and buffer in foods. As mordant in dyeing; for weighting silk; in tanning; in manuf of enamels, ceramics, detergents, boiler compds; as fireproofing agent; in soldering and brazing instead of borax; as reagent and buffer in analytical chemistry.

THERAP CAT: Cathartic.

THERAP CAT (VET): Laxative.

8790. Sodium Phosphate, Monobasic. [7558-80-7] Phosphoric acid sodium salt (1:1); sodium biphosphate; sodium dihydrogen phosphate; acid sodium phosphate; monosodium orthophosphate; primary sodium phosphate. H_2NaO_4P; mol wt 119.98. H 1.68%, Na 19.16%, O 53.34%, P 25.82%. NaH_2PO_4. It is about 99% pure.

Monohydrate. [10049-21-5] $H_2NaO_4P.H_2O$; mol wt 137.99. White, odorless, slightly deliquesc crystals or granules. At 100° loses all its water; when ignited it converts into metaphosphate. Freely sol in water; practically insol in alcohol. The aq soln is acid. pH of 0.1 molar aq soln at 25°: 4.5.

Dihydrate. [13472-35-0] $H_2NaO_4P.2H_2O$; mol wt 156.01. Orthorhombic bisphenoidal colorless crystals, mp 60°. d 1.915. At room temp crystallizes with 2H_2O. Directions for max yield: Beans, Kiehl, *J. Am. Chem. Soc.* **49**, 1878 (1927).

USE: In baking powders; in boiler water treatment; as dry acidulant and sequestrant for foods: Tidridge, Pals, US **3030213** (1962 to FMC). Buffering agent.

THERAP CAT: Urinary acidifier.

THERAP CAT (VET): Urinary acidifier.

8791. Sodium Phosphate P 32. [7635-46-3] Phosphoric-^{32}P acid disodium salt; disodium hydrogen phosphate ^{32}P; sodium phosphate (^{32}P); Phosphotope. $HNa_2O_4{}^{32}P$. Radiopharmaceutical that targets mitotically active cells in bone marrow. Review of use in myeloproliferative disorders: B. E. Roberts, A. H. Smith, *Blood Rev.* **11**, 146-153 (1997); in polycythemia vera: C. Parmentier, *Eur. J. Nucl. Med. Mol. Imaging* **30**, 1413-1417 (2003); in painful bone metastases: E. B. Silberstein, *Semin. Nucl. Med.* **35**, 152-158 (2005).

Prepd as an aq soln with a pH range of 5.0-6.0. Contains ^{32}P which is a pure beta emitter with a half-life of 14.3 days.

THERAP CAT: Antineoplastic; antipolycythemic.

8792. Sodium Phosphate, Tribasic. [7601-54-9] Phosphoric acid sodium salt (1:3); trisodium phosphate; trisodium phosphate; TSP; Oakite. Na_3O_4P; mol wt 163.94. Na 42.07%, O 39.04%, P 18.89%. Na_3PO_4. Crystallizes with 8 and 12 mols of H_2O.

Dodecahydrate. [10101-89-0] $Na_3O_4P.12H_2O$; mol wt 380.12. Colorless or white crystals. When rapidly heated melts at ~75°. Does not lose the last mol of water even on moderate ignition. d 1.6. Sol in 3.5 parts water, 1 part boiling water; insol in alcohol. The aq soln is strongly alkaline. pH of 0.1% soln: 11.5; of 0.5% soln: 11.7; of 1.0% soln: 11.9. Technical crystals are sometimes made with excess alkali to prevent caking and give more alkaline solutions. LD_{50} orally in rats: 7.40 g/kg, H. F. Smyth *et al.*, *Am. Ind. Hyg. Assoc. J.* **30**, 470 (1969).

USE: In photographic developers; clarifying sugar; removing boiler scale, softening water; manuf paper; laundering; tanning leather; in detergent mixture; buffer; sequestering agent.

8793. Sodium Phosphite. [13708-85-5] Phosphonic acid sodium salt (1:2); disodium hydrogen phosphite; disodium phosphite. HNa_2O_3P; mol wt 125.96. H 0.80%, Na 36.50%, O 38.10%, P 24.59%. Na_2HPO_3.

Pentahydrate. [13517-23-2] $HNa_2O_3P.5H_2O$; mol wt 216.03. White, hygroscopic cryst powder. Heat of formation (25°): −684.2 kcal/mole. Freely sol in water. *Keep well closed.*

8794. Sodium Phosphomolybdate. [1313-30-0] Tetracosa-μ-oxododecaoxo[μ_{12}-[phosphato(3−)-κO:κO:κO:$\kappa O'$:$\kappa O'$:$\kappa O'$:-

Consult the Name Index before using this section.

$\kappa O'':\kappa O'':\kappa O'':\kappa O''':\kappa O''':\kappa O''':\kappa O'''$]]dodecamolybdate(3−) sodium (1:3); sodium molybdophosphate. $Mo_{12}Na_3O_{40}P$; mol wt 1891.30. Mo 60.88%, Na 3.65%, O 33.84%, P 1.64%. $Na_3PO_4.12MoO_3$.

White crystals. Freely sol in water.

USE: As reagent in chemical analysis.

8795. Sodium Phosphotungstate. [51312-42-6] Sodium tungstophosphate. Approx $2Na_2O.P_2O_5.12WO_3.18H_2O$.

White, granular powder. Sol in water.

USE: As reagent for alkaloids, uric acid, potassium.

8796. Sodium Polyanethole Sulfonate. [91178-70-0] 2(or 5)-Methoxy-5(or 2)-(1-propenyl)benzenesulfonic acid homopolymer sodium salt; polyanetholesulfonic acid sodium salt; anetholesulfonic acid sodium salt polymer; sodium polyanetholsulfonate; SPS; Liquoid. Polymer of anetholesulfonic acid. Originally developed as an anticoagulant, possesses anticomplement activity and lowers the bactericidal action of blood. Description: V. Demole, M. Reinert, *Arch. Exp. Pathol. Pharmakol.* **158**, 211 (1930). Use in blood culture media: R. D. Stuart, *J. Clin. Pathol.* **1**, 311 (1948); J. Eng, *J. Clin. Microbiol.* **1**, 119 (1975). Use in identification of *Peptostreptococcus anaerobius*: M. H. Graves *et al.*, *Appl. Microbiol.* **27**, 1131 (1974); P. A. Wideman *et al.*, *J. Clin. Microbiol.* **4**, 330 (1976).

Light brown powder. *Irritant.* Insol in alcohol. Swells in water and slowly goes in soln with neutral reaction. Aq solns are stable to heat, dil alkalies and dil acids.

USE: Anticoagulant for blood culture specimens; reagent for differential identification of pathogenic microorganisms.

8797. Sodium Polystyrene Sulfonate. [9080-79-9] Ethenylbenzenesulfonic acid homopolymer sodium salt; poly(styrenesulfonic acid) sodium salt; Elutit-Natrium; Kayexalate; K-Exit; Kionex; Resonium A; SPS. Sulfonated cation exchange resin charged with sodium. Viscosity study: J. L. M. Ganter *et al.*, *Polymer* **33**, 113 (1992). *In vivo* study in treatment of lithium intoxication: J. G. Linakis *et al.*, *Pharmacol. Toxicol.* **65**, 387 (1989). Review of clinical use in hyperkalemia: I. Meyer, *ANNA J.* **20**, 93-95 (1993).

Fine cream to light brown anhydrous powder. Insol in water.

USE: Superplasticizer in concrete; ion exchange membrane in fuel cell applications.

THERAP CAT: Ion-exchange resin (potassium). In treatment of hyperkalemia.

THERAP CAT (VET): Ion-exchange resin (potassium). In treatment of hyperkalemia.

8798. Sodium Propionate. [137-40-6] Propanoic acid sodium salt (1:1); propionic acid sodium salt; sodium propanoate; Impedex. $C_3H_5NaO_2$; mol wt 96.06. C 37.51%, H 5.25%, Na 23.93%, O 33.31%. CH_3CH_2COONa.

Transparent crystals, granules. Deliquescent in moist air. Neutral or slightly alkaline reaction to litmus. One gram dissolves in ~1 ml water, in ~0.65 ml boiling water, in ~24 ml alcohol at 25°. Most

active at acid pH: Wolford, Andersen, *Food Ind.* **17**, 622 (1945); Olsen, Macy, *J. Dairy Sci.* **29**, 173 (1946).

USE: Food preservative, fungicide, mold inhibitor.

THERAP CAT: Antifungal (topical).

THERAP CAT (VET): In ketoses of ruminants (glucose precursor). Antifungal agent. Has been used in dermatoses, wound infections, conjunctivitis.

8799. Sodium Rhodizonate. [523-21-7] 5,6-Dihydroxy-5-cyclohexene-1,2,3,4-tetrone sodium salt (1:2); [(3,4,5,6-tetraoxo-1-cyclohexen-1,2-ylene)dioxy]disodium. $C_6Na_2O_6$; mol wt 214.04. C 33.67%, Na 21.48%, O 44.85%. Chromogenic reagent for the detection of univalent and divalent metal ions. Prepn from carbon monoxide and sodium dissolved in ammonia: A. F. Scott, *Science* **115**, 118 (1952); from diaminotetrahydroxybenzene: P. Möckel, G. Stärk, **DE 1095823** (1961 to VEB Berlin-Chemie). Analytical use for determn of metal ions: F. Feigl, H. A. Suter, *Ind. Eng. Chem.* **14**, 840 (1942). Use for determn of lead in air: Y. P. Grover, *Anal. Chim. Acta* **101**, 225 (1978); in drinking water: E. Jungreis, M. Nechama, *Microchem. J.* **34**, 219 (1986); in gunshot residue: M. R. Bartsch *et al.*, *J. Forensic Sci.* **41**, 1046 (1996); R. Zoja *et al.*, *Biotech. Histochem.* **81**, 151 (2006).

Violet crystals. Sol in water with an orange-yellow color; slightly sol in soda soln; insol in alc. Solns are unstable even in the refrigerator, and should be freshly prepd. *Keep tightly closed in cool, dry, ventilated place.*

USE: Analytical reagent for lead, strontium, and other metal ions. In forensics for the detection of gunshot residue.

8800. Sodium Selenate. [13410-01-0] Selenic acid sodium salt (1:2); disodium selenate. Na_2O_4Se; mol wt 188.94. Na 24.34%, O 33.87%, Se 41.79%. Na_2SeO_4. Naturally occuring form of selenium; found in alkaline soils and water. Acute toxicity study: C. Nofre *et al.*, *C. R. Hebd. Seances Acad. Sci.* **257**, 791 (1963). Review of role in ruminant nutrition: C. B. Ammerman, S. M. Miller, *J. Dairy Sci.* **58**, 1561-1577 (1975). Review of absorption, excretion, and toxicology: *NTP Technical Report on Toxicity Studies of Sodium Selenate and Sodium Selenite* (NIH 94-3387, 1994) 121 pp. Impact of soil supplementation on food chain and human health: H. Hartikainen, *J. Trace Elem. Med. Biol.* **18**, 309-318 (2005).

Decahydrate. [10102-23-5] White crystals; very sol in water. LD_{50} i.p. in mice: 18.45 mg/kg (Nofre).

USE: Insecticide in some horticultural applications; in manuf of glass.

THERAP CAT: Dietary supplement.

THERAP CAT (VET): Dietary supplement for poultry and livestock.

8801. Sodium Selenide. [1313-85-5] Sodium selenide (Na_2-Se); disodium monoselenide; disodium selenide. Na_2Se; mol wt 124.94. Na 36.80%, Se 63.20%. Prepd by adding selenium to a soln of sodium in liquid ammonia: Hugot, *Compt. Rend.* **129**, 299 (1899); *Ann. Chim. Phys.* [7] **21**, 34 (1900); Feher in *Handbook of Preparative Inorganic Chemistry* vol. 1, G. Brauer, Ed. (Academic Press, New York, 2nd ed., 1963) p 421.

Amorphous crystals. d^{10} 2.625. mp >875°. Turns red on exposure to air and deliquesces. Dec in water. Insol in ammonia.

Hemienneahydrate. [169558-84-3] $2Na_2Se.9H_2O$; mol wt 412.01.

Hexadecahydrate. [169558-83-2] $Na_2Se.16H_2O$; mol wt 413.18. Prisms. mp 40°. Dec in air to sodium carbonate, selenium and a small amount of sodium selenide.

8802. Sodium Selenite. [10102-18-8] Selenious acid sodium salt (1:2); Selenase. Na_2O_3Se; mol wt 172.94. Na 26.59%, O 27.75%, Se 45.66%. Na_2SeO_3. Prepd by evaporating an aqueous solution of sodium hydroxide and selenious acid between 60° and 100°: Krak, *J. Am. Ceram. Soc.* **12**, 530 (1929); by heating a mixture of sodium chloride and selenium oxide: Cameron, Macallan, *Proc. Roy. Soc.* **46**, 13 (1890). Metabolism: M. Sandholm, *Acta Pharmacol. Toxicol.* **33**, 6 (1973); H. W. Symonds *et al.*, *Br. J. Nutr.* **45**, 117 (1981). Mutagenicity study: M. Nodo *et al.*, *Mutat. Res.* **66**, 175 (1979). Toxicity study: Cummins, Kimura, *Toxicol. Appl. Pharmacol.* **20**, 89 (1971). Clinical effect of selenium supplementation on immune cell function: M. Roy *et al.*, *Biol. Trace Elem. Res.* **46**, 115 (1994); L. Kiremidjian-Schumacher *et al.*, *ibid.* 183. Clinical evaluation as immunostimulant in head and neck cancer: *eidem*, *ibid.* **73**, 97 (2000); in treatment of radiation-associated secondary lymphedema: O. Micke *et al.*, *Int. J. Radiat. Oncol. Biol. Phys.* **56**, 40 (2003).

Tetragonal prisms. Stable in air. Freely sol in water. Insol in alcohol. LD$_{50}$ orally in rats: 7 mg/kg (Cummins, Kimura).

USE: Removing green color from glass during its manuf; alkaloidal reagent; soil additive.

THERAP CAT: Selenium supplement.

THERAP CAT (VET): Selenium supplement for livestock.

8803. Sodium Sesquicarbonate. [533-96-0] Carbonic acid sodium salt (2:3); urao; trona. $C_2HNa_3O_6$; mol wt 189.99. C 12.64%, H 0.53%, Na 36.30%, O 50.53%. $Na_2CO_3 \cdot NaHCO_3$. Found in nature as the dihydrate, e.g., Owens Lake, Searles Lake (U.S.A.); Lake Magadi (Kenya). Produced on a large scale from sodium carbonate and a slight excess of sodium bicarbonate: Schenk in *Winnacker-Weingaertner, Chemische Technologie* vol. I (München, 1950) p 427.

Dihydrate. [6106-20-3] $C_2HNa_3O_6 \cdot 2H_2O$; mol wt 226.02. Monoclinic needles, d 2.112. Crystals are stable in air. Soly in water (g/100 ml) at 0°: 13; at 100°: 42. Aq solns are mildly alkaline. pH of 0.1M soln = 10.1.

Caution: Irritating to skin, mucous membranes.

USE: Chiefly in laundering in conjunction with soap.

8804. Sodium Silicate. [1344-09-8] Silicic acid sodium salt; soluble glass. General term refering to compounds containing varying amounts of sodium oxide (Na_2O) and silica (SiO_2); usually prepared by fusing quartz sand and sodium carbonate. Crystalline forms include sodium metasilicate (Na_2SiO_3), sodium disilicate ($Na_2Si_2O_5$), sodium orthosilicate (Na_4SiO_4) and sodium pyrosilicate ($Na_6Si_2O_7$). Prepn and properties: G. W. Morey, *J. Am. Chem. Soc.* **36**, 215 (1914). Manuf: *Faith, Keyes & Clark's Industrial Chemicals*, F. A. Lowenheim, M. K. Moran, Eds. (Wiley-Interscience, New York, 4th ed., 1975) pp 755-761. Thermodynamic properties: N. W. McCready, *J. Phys. Chem.* **52**, 1277 (1948). Review of industrial applications: R. C. Merrill, *Ind. Eng. Chem.* **41**, 337-345 (1949). Review of prepn and properties: *Gmelins, Sodium* (8th ed. supplement) **21**, pp 1474-1478 (1967); of structural chemistry: V. Kahlenberg, *Chimia* **64**, 716-722 (2010).

Sodium metasilicate. [6834-92-0] Silicic acid (H_2SiO_3) sodium salt (1:2). Na_2O_3Si; mol wt 122.06. Obtained as a glass or as orthorhombic crystals. d 2.614. mp 1089°. n_D^{25} (glass) 1.520. Specific heat (20°): 0.217. Heat of formation: −364.7 kcal/mol. Heat of soln (cryst) −7.45 kcal/mol. Heat of fusion: 10.3 kcal/mol. Sol in cold water, hydrolyzed by hot water. Insol in alcohol, acids, salt solns.

Sodium metasilicate nonahydrate. [13517-24-3] $Na_2O_3Si \cdot 9H_2O$; mol wt 284.20. Efflorescent, orthorhombic bipyramidal platelets, mp 48° in water of crystn. Heat of hydration: −24.15 kcal/mol.

Sodium silicate solution. Liquid glass; water glass. Aqueous solns of varying composition characterized by the weight ratio of SiO_2: Na_2O; values between 1.6 and 3.8 are most common. Prepn and properties: R. W. Harman, *J. Phys. Chem.* **32**, 44 (1928); I. Halasz *et al.*, *Catal. Lett.* **117**, 34 (2007). Determn of composition: J. S. Falcone, Jr. *et al.*, *Ind. Eng. Chem. Res.* **49**, 6287 (2010). Clear to hazy, colorless, thick liquids. Strongly alkaline, pH >11. *Corrosive.* Readily dec by acids with separation of silicic acid which when heated forms a hard, glassy silica gel.

USE: Starting material for synthesis of zeolites and silica catalysts. As adhesives and binders, corrosion inhibitor; penetrating sealant; in cements, drilling fluids. For fireproofing wood, paper or fabric; as a detergent booster; in waste water treatment. As shell coating to preserve eggs.

8805. Sodium Stannate(IV). [12058-66-1] Stannate (SnO_3^{2-}) sodium (1:2); disodium stannate; sodium tin oxide. Na_2O_3Sn; mol wt 212.69. Na 21.62%, O 22.57%, Sn 55.81%. Na_2SnO_3.

Trihydrate. [12209-98-2] $Na_2O_3Sn \cdot 3H_2O$; mol wt 266.73. White or colorless crystals; gradually dec in the air; dec by weak acids. Sol in ~1.7 parts water; insol in alcohol. The aq soln is alkaline. *Keep well closed.*

USE: As mordant in dyeing and printing calico; for fireproofing of curtains, etc.

8806. Sodium Stearate. [822-16-2] Octadecanoic acid sodium salt (1:1); stearic acid sodium salt; sodium octadecanoate. $C_{18}H_{35}NaO_2$; mol wt 306.47. C 70.54%, H 11.51%, Na 7.50%, O 10.44%. Anionic surfactant. Prepd from alcoholic solution of stearic acid with sodium hydroxide. Constituent of common soap prepared from tallow and lye. Commercial grades prepd from plant or animal fats may also contain other fatty acid sodium salts, especially the palmitate and oleate. Prepn and phase transition behavior: A. S. C. Lawrence, *Trans. Faraday Soc.* **1938**, 660. Prepn of anhydrous compd and phase rule study: J. W. McBain *et al.*, *J. Phys. Chem.* **44**, 1013 (1940). Safety assessment for use in cosmetics: *Int. J. Toxicol.* **1**, 143-177 (1982).

White powder; soapy feel; slight, tallow-like odor. Slowly soluble in cold water or cold alcohol; freely sol in the hot solvents. The aq soln is strongly alkaline, due to hydrolysis; the alcohol soln is practically neutral.

USE: Industrial and household soap. In cosmetics and personal care products as cleanser, emulsifier, thickener. In stick fragrances and deodorants as gelling agent. Pharmaceutic aid (emulsifying and stiffening agent).

8807. Sodium Stearyl Fumarate. [4070-80-8] (2*E*)-2-Butenedioic acid monooctadecyl ester sodium salt (1:1); fumaric acid monooctadecyl ester sodium salt; sodium monostearyl fumarate; sodium octadecyl fumarate; sodium octadecyl (*E*)-butenedioate; NaSF; Pruv. $C_{22}H_{39}NaO_4$; mol wt 390.54. C 67.66%, H 10.07%, Na 5.89%, O 16.39%. Lubricant in oral pharmaceutical formulations. Conditioning and stabilizing agent for food products. Prepn and uses in starch compositions: P. D. Thomas, US 3343964 (1967 to Pfizer). Effects as food additive: *idem et al.*, *Cereal Sci. Today* **11**, 46 (1966); B. A. Brachfeld *et al.*, *Baker's Dig.* **40**, 53 (1966). Evaluation of lubricant properties: A. W. Hölzer, J. Sjögren, *Int. J. Pharm.* **2**, 145 (1979); S. I. Saleh *et al.*, *Lab. Pharm. Prob. Tech.* **32**, 588 (1984); N. H. Shah *et al.*, *Drug Dev. Ind. Pharm.* **12**, 1329 (1986). Absorption and metabolism: S. K. Figdor, R. Pinson, *J. Agric. Food Chem.* **18**, 872 (1970). Determn in aqueous suspension by supercritical fluid chromatography: O. Gyllenhaal, *J. Pharm. Biomed. Anal.* **40**, 971 (2006).

Fine white powder, mp 238-242° with decomp. d 1.12-1.14. Soly in water (g/ml): 0.1 at 25°; 10 at 80°; 20 at 90°. Slightly sol in methanol. Practically insol in acetone, chloroform, ethanol. pH (5% w/v aq soln at 90°): 8.3.

USE: Pharmaceutic aid (capsule and tablet lubricant). Food additive for dough and other baked goods.

8808. Sodium Succinate. [150-90-3] Butanedioic acid sodium salt (1:2); succinic acid sodium salt; disodium succinate. $C_4H_4Na_2O_4$; mol wt 162.05. C 29.65%, H 2.49%, Na 28.37%, O

39.49%. Crystal structure of hexahydrate: I. Fonseca *et al.*, *Acta Crystallogr.* **C42**, 1123 (1986). Acute toxicity and clinical experience: M. Zuckerbrod, I. Graef, *Ann. Intern. Med.* **32**, 905 (1950). LD_{50} i.v. in mice: 4.5 g/kg (Zuckerbrod, Graef).

Hexahydrate. [6106-21-8] $C_4H_4Na_2O_4.6H_2O$; mol wt 270.14. Granules or crystalline powder. Loses all its water at 120°. Sol in ~5 parts water. Insol in alcohol. The aq soln is neutral or slightly alkaline.

USE: Acidulant and flavoring agent in foods. As protective coating for fruits and vegetables.

8809. Sodium Sulfate. [7757-82-6] Sodium sulfate sodium salt (1:2). Na_2O_4S; mol wt 142.04. Na 32.37%, O 45.05%, S 22.57%. Na_2SO_4. Occurs in nature as the minerals *mirabilite, thenardite*. Industrial production: *Faith, Keyes & Clark's Industrial Chemicals*, F. A. Lowenheim, M. K. Moran, Eds. (Wiley-Interscience, New York, 4th ed., 1975) p 762-768.

Anhydrous form, *salt cake* (technical grade). Powder or orthorhombic bipyramidal crystals, mp 884°. d 2.7. Sol in ~3.6 parts water. Max soly at 33°: 1 in 2. Above this temp the soly gradually decreases and at 100° requires 2.4 parts water. Sol in glycerin. Insol in alc.

Decahydrate. [7757-82-6] Glauber's salt. $Na_2O_4S.10H_2O$; mol wt 322.19. Odorless, efflorescent crystals or granules, mp 32.4°. d 1.46. Loses all its water at 100°. Sol in 1.5 parts water at 25°, in 3.3 parts water at 15°. Soly in water decreased by NaCl. Sol in glycerol; insol in alc. The aq soln is neutral. pH 6-7.5. *Keep well closed in a cool place.*

USE: For standardizing dyes; in freezing mixtures; in dyeing and printing textiles. The *anhydrous* form for drying organic liquids; in Kjeldahl nitrogen determination; in manuf of glass, ultramarine, paper pulp.

THERAP CAT: Cathartic.

THERAP CAT (VET): Purgative.

8810. Sodium Sulfide. [1313-82-2] Sodium sulfide (Na_2S); sodium monosulfide; sodium sulfuret. Na_2S; mol wt 78.04. Na 58.92%, S 41.08%. Best prepd from the elements in liq ammonia, also obtained by dehydration of the nonahydrate: Courtois, *Compt. Rend.* **207**, 1220 (1938); Klemm *et al.*, *Z. Anorg. Allg. Chem.* **241**, 281 (1939); Feher in *Handbook of Preparative Inorganic Chemistry* **vol 1**, G. Brauer, Ed. (Academic Press, New York, 2nd ed., 1963) pp 358-360. *Review:* C. Drum in *Kirk-Othmer Encyclopedia of Chemical Technology* **vol. 21** (Wiley-Interscience, New York, 3rd ed., 1983) pp 256-262.

Cubic crystals or granules. *Spontaneously combustible.* Extremely hygroscopic. Discolors upon exposure to air. d_4^{14} 1.856. mp 1180° *(in vacuo)*; also reported as mp 920°. Soly in water (g/100 g H_2O): 8.1 (−9.0°); 12.4 (0°); 18.6 (20°); 29.0 (40°); 35.7 (48°); 39.0 (50°). Slightly sol in alcohol. Insol in ether. Aq solns are strongly alkaline.

Pentahydrate. [1313-83-3] $Na_2S.5H_2O$; mol wt 168.11. Flat, shiny, four-sided, prismatic crystals. Loses 3 mols water at 100°. mp 120° (with loss of all water of crystn). Freely sol in water. Also sol in alcohol. Aq solns are strongly alkaline. Insol in ether. Dec by acids with evolution of H_2S.

Nonahydrate. [1313-84-4] $Na_2S.9H_2O$; mol wt 240.17. Tetragonal, deliquescent crystals. Odor of hydrogen sulfide. Discolors upon exposure to light and air (first yellow, then brownish-black). d_4^{16} 1.427. mp ~50°. One gram dissolves in 0.5 ml water, also reported as 18 g dissolve in 100 ml water at 25°. Slightly sol in alcohol, insol in ether. Dec by acids, even by carbonic acid. Aqueous solns are very alkaline and upon standing in contact with air are slowly converted according to the equation: $2Na_2S + 2O_2 + H_2O \rightarrow Na_2S_2O_3 + 2NaOH$. Exposure of the crystals to air produces hydrogen sulfide according to the equation $Na_2S + H_2O + CO_2 \rightarrow Na_2CO_3 + H_2S$. *Keep well closed and in a cool place. Do not handle with bare hands.*

USE: In dehairing hides and wool pulling; desulfurizing viscose rayon; in the manuf of rubber, sulfur dyes; in ore flotation, metal refining, engraving, cotton printing, as chemical intermediate, as laboratory reagent; in paper-pulping process. Source of hydrogen sulfide for heavy metal tests.

8811. Sodium Sulfite. [7757-83-7] Sulfurous acid sodium salt (1:2). Na_2O_3S; mol wt 126.04. Na 36.48%, O 38.08%, S 25.44%. Na_2SO_3.

Small crystals or powder. It is fairly stable and does not oxidize as readily as the hydrated sulfite. Sol in 3.2 parts water; sol in glycerol; very slightly sol in alc. pH about 9. *Keep well closed.* LD_{50} i.v. in mice: 175 mg/kg, Hoppe, Goble, *J. Pharmacol. Exp. Ther.* **101**, 101 (1951).

Heptahydrate. [10102-15-5] $Na_2O_3S.7H_2O$; mol wt 252.14. Efflorescent crystals. Unstable, oxidizing in the air to sulfate. Sol in 1.6 parts water, about 30 parts glycerol; sparingly sol in alc. The aq soln is alkaline and dissolves sulfur. pH ~9. *Keep well closed in a cool place. Note:* The commercial heptahydrate (about 90% pure, the remainder consisting chiefly of the sulfate) has been largely replaced by the commercial anhydr salt (about 96-99% pure) which is more stable.

USE: Chiefly in photographic developers and instead of "hypo" for fixing prints; bleaching wool, straw, silk; generating SO_2; as reducer in manuf dyes; silvering glass; removing traces of Cl in bleached textiles and paper; preserving meat, egg yolks, etc. Reducing agent.

8812. Sodium Tartrate. [868-18-8] (2*R*, 3*R*)-2,3-Dihydroxybutanedioic acid sodium salt (1:2). $C_4H_4Na_2O_6$; mol wt 194.05. C 24.76%, H 2.08%, Na 23.69%, O 49.47%. $Na_2C_4H_4O_6$.

Dihydrate. [6106-24-7] $C_4H_4Na_2O_6.2H_2O$; mol wt 230.08. White crystals or granules. Loses its water at about 120°. d 1.82. Sol in about 3 parts cold water, 1.5 parts boiling water; insol in alcohol. The aq soln is slightly alkaline to litmus, pH 7 to 9.

USE: Dihydrate as standard for standardizing Karl Fischer reagent (determination of water). Pharmaceutic aid (sequestering agent).

THERAP CAT: Cathartic.

8813. Sodium Tellurate. [10101-83-4] Telluric acid (H_2TeO_4) sodium salt (1:2); disodium tetraoxotellurate; sodium tellurate(VI). Na_2O_4Te; mol wt 237.58. Na 19.35%, O 26.94%, Te 53.71%. Na_2TeO_4. Toxicity data: Franke, *J. Pharmacol. Exp. Ther.* **58**, 454 (1936).

White powder. Sol in 130 parts cold water, 50 parts boiling water. *Poisonous; irritant.* MLD in rats (mg/kg): 37.2-55.8 i.p. (Franke).

8814. Sodium Tellurite. [10102-20-2] Telluric acid (H_2TeO_3) sodium salt (1:2); sodium tellurate(IV); disodium trioxotellurate; disodium tellurite. Na_2O_3Te; mol wt 221.58. Na 20.75%, O 21.66%, Te 57.59%. Na_2TeO_3. Review of toxicology: B. Venugopal, T. P. Luckey, *Environ. Qual. Saf. Suppl.* **1**, 4-73 (1975).

White powder. Soluble in water. *Poisonous.* LD_{50} in rats (mg/kg): 56.78 i.p., 55.85 i.v.; in rabbits (mg/kg): 104 orally (Venugopal, Luckey).

8815. Sodium Tetrachloroaluminate. [7784-16-9] (*T*-4)-Tetrachloroaluminate(1−) sodium (1:1); aluminum sodium chloride; sodium chloroaluminate. $AlCl_4Na$; mol wt 191.77. Al 14.07%, Cl 73.94%, Na 11.99%. $AlCl_3.NaCl$. Prepn: *Gmelins, Aluminum* (8th ed.) **35B**, p 376 (1934); Good, Batha, *US 2867499* (1959 to Olin Mathieson).

Yellowish powder. Sol in water.

USE: Catalyst for organic reactions.

8816. Sodium Tetrachloroaurate(III). [15189-51-2] (*SP*-4-1)-Tetrachloroaurate(1−) sodium (1:1); gold sodium chloride; sodium chloroaurate(III). $AuCl_4Na$; mol wt 361.76. Au 54.45%, Cl 39.20%, Na 6.35%. Prepd by evapn of a soln of $HAuCl_4$ and NaCl.

Dihydrate. [13874-02-7] $AuCl_4Na.2H_2O$; mol wt 397.79. Orange-yellow, rhombic, bipyramidal crystals. Stable to 100°. Dec before all H_2O of hydration lost. Freely sol in water; sol in alcohol, ether.

USE: For paper-toning in photography, in electroplating baths.

8817. Sodium Tetradecyl Sulfate. [139-88-8] 7-Ethyl-2-methyl-4-undecanol 4-(hydrogen sulfate) sodium salt (1:1); 7-ethyl-2-methyl-4-hendecanol sulfate sodium salt; sodium 2-methyl-7-ethyl-4-undecyl sulfate; sodium 7-ethyl-2-methylundecyl-4-sulfate; Niaproof 4; Sotradecol; Tergitol 4; Trombovar. $C_{14}H_{29}NaSO_4$; mol wt 316.43. C 53.14%, H 9.24%, Na 7.27%, S 10.13%, O 20.22%. Anionic surfactant. Prepn and detergent properties: J. N. Wickert, **US 2088020** (1937 to Union Carbide). Use in XLT4 agar for selective isolation of *Salmonella* spp: R. G. Miller *et al.*, *Poult. Sci.* **70**,

2429 (1991). Clinical trial in treatment of varicose and telangiectatic leg veins: M. P. Goldman, *Dermatol. Surg.* **28**, 52 (2002).

White, waxy solid. Sol in water, alcohol, ether. The pH of a 5% soln is from 6.5 to 9.0. Surface tension (dynes/cm) of aq soln at 25°: 56.5 dynes/cm (0.05% w/w); 52 (0.10%); 47 (0.20%); 40 (0.50%); 35 (1.0%). LD$_{50}$ orally in rats: 4.95 g/kg, H. F. Smyth, C. P. Carpenter, *J. Ind. Hyg. Toxicol.* **30**, 63 (1948).

USE: Wetting agent. In differential microbiological media.

THERAP CAT: Sclerosing agent.

8818. Sodium Tetraphenylborate. [143-66-8] Sodium tetraphenylborate(1−) (1:1); tetraphenylboron sodium; Kalignost. C$_{24}$H$_{20}$BNa; mol wt 342.22. C 84.23%, H 5.89%, B 3.16%, Na 6.72%. Prepn: Wittig, Raff, *Ann.* **573**, 195 (1950); US 2853525 (1958) Heyl, GB 705719 (1954); Kozlova, Pal'm, *Zh. Obshch. Khim.* **31**, 2922 (1961); Holtzapfel, Richter, *ibid.* **32**, 1358 (1962).

Snow-white crystals from chloroform. Freely sol in water, acetone. Less sol in ether, chloroform. Practically insol in petr ether. Aq solns should be adjusted to a pH of about 5 and can be stored at room temp or lower. Such solns have been stored at 45° for 5 days without deterioration. The soly in polar solvents increases as the temp decreases.

USE: As reagent for determination of potassium, ammonium, rubidium and cesium ions: Barnard, Buechl, *Chemist-Analyst* **48**, 44, 49 (1959); Montequi, Serrano, *An. R. Acad. Farm.* **26**, 107 (1960).

8819. Sodium Thioglycolate. [367-51-1] 2-Mercaptoacetic acid sodium salt (1:1); sodium mercaptoacetate. C$_2$H$_3$NaO$_2$S; mol wt 114.09. C 21.06%, H 2.65%, Na 20.15%, O 28.05%, S 28.10%. HSCH$_2$COONa.

Hygroscopic crystals. Slight characteristic odor. Discolors on exposure to air or iron. Freely sol in water; slightly sol in alc. LD$_{50}$ i.p. in rats: 148 mg/kg, Freeman, Rosenthal, *Fed. Proc.* **11**, 347 (1952).

USE: In cold-waving of hair; as depilatory; in bacteriology for the prepn of thioglycolate media; as analytical reagent, *see* Thioglycolic Acid.

8820. Sodium Thiophosphate. [10101-88-9] Phosphorothioic acid sodium salt (1:3); sodium phosphorothioate; sodium monothiophosphate. Na$_3$O$_3$PS; mol wt 180.00. Na 38.32%, O 26.67%, P 17.21%, S 17.81%. Na$_3$PO$_3$S. Conveniently prepd from NaPO$_3$ + Na$_2$S: Zintl, Bertram, *Z. Anorg. Allg. Chem.* **245**, 16 (1940); from PSCl$_3$ + NaOH: Tridot, Tudo, *Bull. Soc. Chim. Fr.* **1960**, 1231. Several alternate routes, e.g. from P$_2$S$_5$ and NaOH, *see* Klement in *Handbook of Preparative Inorganic Chemistry* **vol. 1**, G. Brauer, Ed. (Academic Press, New York, 2nd ed., 1963) pp 569-570.

Dodecahydrate. [51674-17-0] Na$_3$O$_3$PS.12H$_2$O; mol wt 396.18. Thin, six-sided leaflets from water. Effloresces in dry air. mp 60°. Freely sol in warm water. Aq solns are strongly alkaline.

8821. Sodium Thiosulfate. [7772-98-7] Thiosulfuric acid (H$_2$S$_2$O$_3$) sodium salt (1:2); sodium hyposulfite; "hypo"; antichlor; Sodothiol; Sulfothiorine; Ametox. Na$_2$O$_3$S$_2$; mol wt 158.10. Na 29.08%, O 30.36%, S 40.56%. Na$_2$S$_2$O$_3$. Review of mfg processes: *Faith, Keyes & Clark's Industrial Chemicals*, F. A. Lowenheim, M. K. Moran, Eds. (Wiley-Interscience, New York, 4th ed., 1975) pp

769-773. Crystal structure of the pentahydrate: A. A. Uraz, N. Armagan, *Acta Crystallogr. B* **33**, 1396 (1977). Toxicity study of pentahydrate: C. Voegtlin *et al.*, *J. Pharmacol. Exp. Ther.* **25**, 297 (1925).

Powder. Sol in water; practically insol in alc.

Pentahydrate. [10102-17-7] Na$_2$O$_3$S$_2$.5H$_2$O; mol wt 248.17. Odorless crystals or granules, mp 48° when rapidly heated. Effloresces in warm dry air; slightly deliquesces in moist air. d 1.69. Loses all its water at 100°; dec at higher temp. Very sol in water. Insol in alc. Slowly dec in aq soln at ordinary temp, more rapidly when heated. The aq soln is practically neutral. pH 6.5-8.0. It dissolves silver halides and many other salts of silver. *Incompat:* Iodine, acids; lead, mercury, and silver salts. LD i.v. in rats: >2.5 g/kg (Voegtlin).

USE: To remove chlorine from solns according to the eq: Na$_2$S$_2$O$_3$ + 4Cl$_2$ +5H$_2$O → 2NaHSO$_4$ + 8HCl and Na$_2$S$_2$O$_3$ + 2HCl → 2NaCl + H$_2$O + S + SO$_2$. As "antichlor" in bleaching of paper pulp; as fixer in photography; for extraction of silver from ores; as mordant in dyeing and printing textiles; reducer in chrome dyeing; manuf leather; bleaching bone, straw, ivory; also as titrimetry reagent in analytical chemistry. Pharmaceutic aid (antioxidant).

THERAP CAT: Antidote (cyanide).

THERAP CAT (VET): Antidote (cyanide). Has been used as a "general detoxifier", in bloat, and externally in ringworm, mange.

8822. Sodium Triacetoxyborohydride. [56553-60-7] (*T*-4)-Tris(acetato-κO)hydroborate(1−) sodium (1:1). C$_6$H$_{10}$BNaO$_6$; mol wt 211.94. C 34.00%, H 4.76%, B 5.10%, Na 10.85%, O 45.29%. Mild borohydride reagent. Prepn and use in reduction of aldehydes: G. W. Gribble, D. C. Ferguson, *J. Chem. Soc. Chem. Commun.* **1975**, 535. Reductive amination of aldehydes and ketones: A. F. Abdel-Magid *et al.*, *J. Org. Chem.* **61**, 3849 (1996); L.-X. Yang, K. G. Hofer, *Tetrahedron Lett.* **37**, 6081 (1996); of amino acid esters: J. M. Ramanjulu, M. M. Joullié, *Synth. Commun.* **26**, 1379 (1996). Reductive alkylation of amines: Y. Han, M. Chorev, *J. Org. Chem.* **64**, 1972 (1999). Reduction of conjugated aldehydes: J. Singh *et al.*, *Synth. Commun.* **30**, 1515 (2000). Stereoselective reduction of ketones: Y.-T. Liu *et al.*, *Tetrahedron Lett.* **45**, 6097 (2004).

USE: Selective reducing agent in organic synthesis.

8823. Sodium Trimetaphosphate. [7785-84-4] Metaphosphoric acid (H$_3$P$_3$O$_9$) sodium salt (1:3); trimetaphosphoric acid trisodium salt; trisodium trimetaphosphate; cyclic sodium trimetaphosphate; STMP. Na$_3$O$_9$P$_3$; mol wt 305.88. Na 22.55%, O 47.07%, P 30.38%. (NaPO$_3$)$_3$. One of the cyclic forms of sodium metaphosphate, *q.v.* Usually prepd by heating NaH$_2$PO$_4$ at high temp. Prepn: T. Fleitmann, W. Henneberg, *Ann.* **65**, 304 (1848); G. v. Knoore, *Z. Anorg. Allg. Chem.* **24**, 381 (1900); R. N. Bell, *Inorg. Synth.* **3**, 103 (1950); J. D. Lee, A. H. Bond, *J. Appl. Chem.* **18**, 345 (1968). Confirmation of ring structure: E. Thilo, R. Ratz, *Z. Anorg. Chem.* **258**, 33 (1949). Crystal structure: H. M. Ondik, *Acta Crystallogr.* **18**, 226 (1965). Toxicology: R. E. Gosselin *et al.*, *J. Pharmacol. Exp. Ther.* **108**, 117 (1953). Prepn, conversion to sodium tripolyphosphate, *q.v.*, and use in detergents: C. Y. Shen, *Ind. Eng. Chem. Prod. Res. Dev.* **5**, 272 (1966). NMR study of polysaccharide crosslinking mechanism: S. Lack *et al.*, *Carbohydr. Res.* **342**, 943 (2007).

White crystals or white, crystalline powder. d^{20} 2.49. Freely sol in water. pH of 1:100 aq soln is about 6. Hydrolyzes to sodium tripolyphosphate in dil alkaline soln. LD_{50} in rats (mg/kg): 3650 i.p. (Gosselin).

Hexahydrate. [29856-33-5] $Na_3O_9P_3 \cdot 6H_2O$; mol wt 413.97. Efflorescent, triclinic-rhombohedral prisms. d 1.786. mp 53°. Loses water rapidly at 50°. One gram dissolves in 4.5 ml water. Insol in alc.

USE: In detergent processing. Crosslinking agent for starch in foods and pharmaceuticals; buffering and chelating agent in cosmetics.

8824. Sodium Tripolyphosphate. [7758-29-4] Triphosphoric acid sodium salt (1:5); triphosphoric acid pentasodium salt; pentasodium triphosphate; sodium triphosphate; STPP. $Na_5O_{10}P_3$; mol wt 367.86. Na 31.25%, O 43.49%, P 25.26%. Prepd by molecular dehydration of mono- and disodium phosphates: L. F. Audrieth, R. N. Bell, *Inorg. Synth.* **3**, 85 (1950). Toxicity study: H. F. Smyth *et al.*, *Am. Ind. Hyg. Assoc. J.* **30**, 470 (1969). Determn in meat samples by capillary zone electrophoresis: A. Jastrzebska, *Talanta* **69**, 1018 (2006). Tooth whitening properties: R. P. Shellis *et al.*, *J. Dent.* **33**, 313 (2005). Use in preserving fish fillets: S. Kin *et al.*, *J. Food Sci.* **75**, S74 (2010). *See also:* Sodium Trimetaphosphate.

Slightly hygroscopic granules. *Irritant.* Soly in water (g/100 ml) at 25°: 20; at 100°: 86.5. pH of 1% soln at 25° = 9.7-9.8. Reverts to the orthophosphate with prolonged heating. LD_{50} orally in rats: 6.50 g/kg (Smyth).

USE: In water softeners and detergents; emulsifier and dispersing agent; as crosslinking agent for starch; as preservative, sequestrant and texturizer in foods; whitening agent in toothpaste.

8825. Sodium Trithiocarbonate. [534-18-9] Carbonotrithioic acid sodium salt (1:2); trithiocarbonic acid disodium salt; sodium sulfocarbonate; sodium thiocarbonate; Aquamet T; Tramfloc 904; Trimet. CNa_2S_3; mol wt 154.17. C 7.79%, Na 29.82%, S 62.39%. Na_2CS_3. Prepn from carbon disulfide with sodium sulfide: E. W. Yeoman, *J. Chem. Soc. Trans.* **119**, 38 (1921); with sodium sulfide or sodium hydroxide: G. Ingram, B. A. Toms, *J. Chem. Soc.* **1957**, 4328. Synthetic utility in the prepn of thiols from alkyl chlorides: D. J. Martin, C. C. Greco, *J. Org. Chem.* **33**, 1275 (1968); of thioureas: Y. Takikawa *et al.*, *Chem. Lett.* **1982**, 641; of alkenes from *vic*-dihaloalkanes: A. Sugawara *et al.*, *Bull. Chem. Soc. Jpn.* **62**, 2739 (1989). Use as heavy metal precipitant: G. S. Elfline, US **4612125** (1986 to CX/Oxytech); M. M. Matlock *et al.*, *J. Hazard. Mater. B* **92**, 129 (2002).

Amorphous pink powder. Extremely hygroscopic. *Corrosive.* Stable in dry air, but decomposes in the presence of moisture. Very sol in water; sol in ethanol. Insol in ether, benzene. Aq solns are alkaline with distinctive red color.

Dihydrate. [19086-11-4] $CNa_2S_3 \cdot 2H_2O$; mol wt 190.20. Yellow rectangular plates from ethanol-ether.

USE: Reagent in synthetic organic chemistry; precipitant for heavy metals in waste streams.

8826. Sodium Tungstate(VI). [13472-45-2] Sodium $(T\text{-}4)$-tungstate (WO_4^{2-}) (2:1). Na_2O_4W; mol wt 293.82. Na 15.65%, O 21.78%, W 62.57%. Na_2WO_4. Crystals have a spinel structure isomorphous with sodium molybdate(VI), *q.v.* Mfr: N. T. Gordon, A. F. Spring, *Ind. Eng. Chem.* **16**, 555 (1924). Properties of sodium tungstate derivatives: E. F. Smith, *J. Am. Chem. Soc.* **44**, 2027 (1922). Crystal structure: I. Lindqvist, *Acta Chem. Scand.* **4**, 1066 (1950). Polymerization study: M. L. Freedman, *J. Am. Chem. Soc.* **80**, 2072 (1958). Thermochemistry study: R. L. Graham, L. G. Hepler, *ibid.* 3538. HPLC and CZE analysis: K. Hettiarachchi *et al.*, *J. Pharm. Biomed. Anal.* **13**, 515 (1995). Pharmacokinetics in rats: S. Le Lamer-Déchamps *et al.*, *Int. J. Pharm.* **248**, 131 (2002). Antidiabetic effects in rats: A. Barberà *et al.*, *Diabetologia* **44**, 507 (2001). Review of toxicity: B. Venugopal, T. D. Luckey, *Environ. Qual. Safety,* Suppl. 1, 4-73 (1975); of toxicology and human expo-

sure: *Toxicological Profile for Tungsten* (PB2006-100007, 2005) 203 pp.

Dihydrate. [10213-10-2] $Na_2O_4W \cdot 2H_2O$; mol wt 329.85. Colorless crystals or white, cryst powder, mp 665°. Effloresces in dry air. Loses its water at 100°. Sol in about 1.1 parts water. Insol in alcohol. The aq soln is slightly alkaline, pH 8-9. LD_{50} (mg tungstate/kg) in rabbits: 105 i.m.; in rats: 240 s.c. (Venugopal, Luckey).

USE: Fireproofing and waterproofing fabrics; preparing complex compds such as phosphotungstate and silicotungstate; as a protective colloid in biological development methods; precipitant for alkaloids and proteins; catalyst in oxidation reactions; corrosion inhibitor for steel.

8827. Sodium Uranate(VI). [13721-34-1] Sodium uranium oxide $(Na_2U_2O_7)$; sodium diuranate; uranium oxide yellow; uranium yellow. $Na_2O_7U_2$; mol wt 634.03. Na 7.25%, O 17.66%, U 75.08%. $Na_2U_2O_7$.

Monohydrate. Yellow powder. Insol in water; sol in acids.

USE: Manuf yellowish-green fluorescent glass; painting on porcelain and enameling.

8828. Sofalcone. [64506-49-6] 2-[5-[[(3-Methyl-2-buten-1-yl)oxy]-2-[3-[4-[(3-methyl-2-buten-1-yl)oxy]phenyl]-1-oxo-2-propen-1-yl]phenoxy]acetic acid; 2'-carboxymethoxy-4,4'-bis(3-methyl-2-butenyloxy)chalcone; Su-88; Solon. $C_{27}H_{30}O_6$; mol wt 450.53. C 71.98%, H 6.71%, O 21.31%. Substituted chalcone with anti-ulcer activity. Prepn: K. Kyogoku *et al.*, **DE 2705603**; *eidem*, **US 4085135** (1977, 1978 both to Taisho); *eidem*, *Chem. Pharm. Bull.* **27**, 2943 (1979). HPLC determn in plasma: W. Aidong *et al.*, *J. Chromatogr. B* **856**, 348 (2007). Gastric cytoprotective effect in rats: T. Suwa *et al.*, *Jpn. J. Pharmacol.* **35**, 47 (1984); S. Kishimoto *et al.*, *Arzneim.-Forsch.* **31**, 944 (1987). Toxicity study: Y. Tarumoto *et al.*, *Yakuri to Chiryo* **10**, 61 (1982). Clinical trial in *Helicobacter pylori* infection: H. Isomoto *et al.*, *World J. Gastroenterol.* **11**, 1629 (2005); K. Higuchi *et al.*, *J. Gastroenterol. Hepatol.* **25**, Suppl 1, S155 (2010).

Light yellow needles from ethanol, mp 143-144°. LD_{50} in mice, rats: >10 g/kg orally (Tarumoto).

THERAP CAT: Antiulcerative.

8829. Solan. [2307-68-8] N-(3-Chloro-4-methylphenyl)-2-methylpentanamide; 3'-chloro-2-methyl-*p*-valerotoluidide; Niagara 4512. $C_{13}H_{18}ClNO$; mol wt 239.74. C 65.13%, H 7.57%, Cl 14.79%, N 5.84%, O 6.67%. Prepn: Dorschner *et al.*, **GB 869169**; **US 3020142** (1961, 1962 both to FMC). Toxicity study: G. W. Bailey, J. L. White, *Residue Rev.* **10**, 97 (1965).

Crystals, mp 79-80°. Sol in pine oil, diisobutylketone, isophorone, xylene. Practically insol in water. LD_{50} orally in rats: 10000 mg/kg (Bailey, White).

USE: Herbicide.

8830. Solanesol. [13190-97-1] (2E,6E,10E,14E,18E,22E,-26E,30E)-3,7,11,15,19,23,27,31,35-Nonamethyl-2,6,10,14,18,22,-26,30,34-hexatriacontanonaen-1-ol; nonaisoprenol. $C_{45}H_{74}O$; mol wt 631.09. C 85.64%, H 11.82%, O 2.54%. High molecular weight isoprenoid alcohol isolated from tobacco leaf. Associated with the particulate fraction of tobacco smoke. Isoln: R. L. Rowland *et al.*, *J. Am. Chem. Soc.* **78**, 4680 (1956). Synthesis: R. Rüegg *et al.*, *Helv. Chim. Acta* **43**, 1745 (1960). Stereoselective synthesis: K. Sato *et al.*, *J. Chem. Soc. Perkin Trans. 1* **1981**, 761. Structural analysis: N. J. Jensen, T. Sumpter, *Beitr. Tabakforsch. Int.* **16**, 85

(1995). Determn in tobacco smoke: M. W. Ogden, K. C. Maiolo, *LC-GC* **10**, 459 (1992); in tobacco extracts: K. Z. Liu *et al.*, *Anal. Lett.* **31**, 1947 (1998).

White waxy solid, mp 41.5-42.5°. Sol in organic solvents. Insol in water.

USE: Tracer for environmental exposure to tobacco smoke.

8831. Solanidine. [80-78-4] (3β)-Solanid-5-en-3-ol; solatubine. C$_{27}$H$_{43}$NO; mol wt 397.65. C 81.55%, H 10.90%, N 3.52%, O 4.02%. Steroidal alkaloid isolated from *Solanum* spp., *Solanaceae*. Aglycone of the toxic solanine glycoalkaloids found in potatoes. Extraction procedure: Soltys, Wallenfels, *Ber.* **69**, 811 (1936). Review on structure: Reichstein, Reich, *Annu. Rev. Biochem.* **15**, 155 (1946). Stereochemistry: Rosen, Rosen, *Chem. Ind. (London)* **1954**, 1581. Revised stereochemistry: Höhne *et al.*, *Tetrahedron* **22**, 673 (1966). Synthesis: Schreiber, Roensch, *Ber.* **97**, 2362 (1964); *Tetrahedron* **21**, 645 (1965); Kessar *et al.*, *ibid.* **27**, 2153 (1971).

Long needles from chloroform-methanol, mp 218-219°. Sublimes near mp with slight decompn. [α]$_D^{21}$ −29° (c = 0.5 in chloroform). Freely sol in benzene, chloroform; slightly in alcohol, methanol; almost insol in ether, water.

Hydrochloride. C$_{27}$H$_{43}$NO.HCl. Prisms from 80% alcohol, dec 345°.

Methyliodide. C$_{27}$H$_{43}$NO.CH$_3$I. Crystals from 50% alcohol, dec 286°.

Acetylsolanidine. C$_{29}$H$_{45}$NO$_2$. Crystals from alc, mp 208°.

8832. Solanine. [51938-42-2] Solatunine. Mixture of toxic glycoalkaloids isolated from *Solanum* species, especially from *S. tuberosum* L. (potato) and *S. nigrum* L. (woody nightshade), *Solanaceae*. Found throughout the potato plant, tubers contain 20-100 mg/kg with higher concentrations in green or damaged tubers. Originally thought to be a single substance, principle components are α-solanine and α-chaconine which are trisaccharide derivatives of the steroidal alkaloid, solanidine, *q.v.* β- and γ-forms are di- and monosaccharide metabolites. Isoln from berries of *S. nigrum*: Desfosses, *J. Pharm. Sci. Accessoires* **6**, 374 (1820). Identification as glycoside: C. Zwenger, *Ann.* **109**, 244 (1859). Extraction procedure from potatoes: L. C. Baker *et al.*, *J. Sci. Food Agric.* **6**, 197 (1955). Identification of components: R. Kuhn, I. Löw, *Angew. Chem.* **66**, 639 (1954); R. Kuhn *et al.*, *Ber.* **88**, 1492 (1955). Pharmacology, distribution and fate in animals: K. Nishie *et al.*, *Toxicol. Appl. Pharmacol.* **19**, 81 (1971). Toxicology: S. Chaube, C. A. Swinyard, *Toxicol. Appl. Pharmacol.* **36**, 227 (1976); P. Slanina, *Food Chem. Toxicol.* **28**, 759 (1990). *Reviews:* M. Friedman, G. M. McDonald, *Crit. Rev. Plant Sci.* **16**, 55-132 (1997); M. Friedman, *J. Agric. Food Chem.* **54**, 8655-8681 (2006).

α-Solanine

α-Solanine. [20562-02-1] (3β)-Solanid-5-en-3-yl *O*-6-deoxy-α-L-mannopyranosyl-(1 → 2)-*O*-[β-D-glucopyranosyl-(1 → 3)]-β-D-galactopyranoside. C$_{45}$H$_{73}$NO$_{15}$; mol wt 868.07. Trisaccharide consisting of D-galactose, D-glucose and L-rhamnose linked to solanidine. *Poisonous.* Colorless needles from methanol, mp 286°. [α]$_D^{20}$ −59° (pyridine). pKa 6.7. LD$_{50}$ i.p. in rats: 67.0 mg/kg (Chaube, Swinyard).

α-Chaconine. [20562-03-2] (3β)-Solanid-5-en-3-yl *O*-6-deoxy-α-L-mannopyranosyl-(1 → 2)-*O*-[6-deoxy-α-L-mannopyranosyl-(1 → 4)]-β-D-glucopyranoside. C$_{45}$H$_{73}$NO$_{14}$; mol wt 852.07. C 63.43%, H 8.64%, N 1.64%, O 26.29%. Trisaccharide consisting of D-glucose and 2 molecules of L-rhamnose linked to solanidine. *Poisonous.* Colorless needles from methanol, mp 243°. [α]$_D^{20}$ −85° (pyridine). LD$_{50}$ i.p. in rats: 84.0 mg/kg (Chaube, Swinyard).

8833. Solanocapsine. [639-86-1] (2*S*,4a*S*,4b*S*,6a*S*,6b*R*,7*S*,-7a*R*,10*R*,11a*S*,12a*R*,13a*S*,13b*R*,15a*S*)-2-Aminodocosahydro-4a,6a,-7,10-tetramethylnaphth[2″,1″:4′,5′]indeno[1′,2′:5,6]pyrano[3,2-*b*]-pyridin-11a(1*H*)-ol; (3β,5α,16α,22α,23β,25β)-3-amino-16,23-epoxy-16,28-secosolanidan-23-ol; solanocapsin; 3β-amino-22,26-imino-16β,23-oxido-5α,22α,23β,25α-cholestan-23-ol; 3β-amino-22,26-imino-16β,23-oxido-5α,22*R*,23*S*,25*R*-cholestan-23-ol. C$_{27}$-H$_{46}$N$_2$O$_2$; mol wt 430.68. C 75.30%, H 10.77%, N 6.50%, O 7.43%. Poisonous steroidal alkaloid isolated from Jerusalem cherry, *Solanum pseudocapsicum* L., *Solanaceae*. Isolation: G. Barger, H. L. Fraenkel-Conrat, *J. Chem. Soc.* **1936**, 1537; K. Schreiber, H. Ripperger, *Z. Naturforsch.* **17B**, 217 (1962). Structure: *eidem, Experientia* **16**, 536 (1960). Revised structure and synthesis: *eidem, Ann.* **723**, 159 (1969). Crystal structure: E. Höhne *et al.*, *Tetrahedron* **26**, 3569 (1970). Total synthesis: H. Ripperger *et al.*, *Tetrahedron Lett.* **11**, 5251 (1970); *eidem, Tetrahedron* **28**, 1629 (1972).

Monohydrate, needles from methanol, mp 213-215°. [α]$_D^{20}$ +26.3° (c = 1.80 in methanol); [α]$_D^{20}$ +24.9° (c = 2.13 in methanol).

N′-Acetylsolanocapsine. C$_{29}$H$_{48}$N$_2$O$_3$. Needles from ethanol, mp 233-234°. [α]$_D^{23}$ −33.7° (c = 1.59 in pyridine).

N,N′-Diacetylsolanocapsine. C$_{31}$H$_{50}$N$_2$O$_4$. Crystals from chloroform + acetone, mp 281-283°. [α]$_D^{22}$ −38.6° (c = 1.31 in pyridine).

N,N,N′-Trimethylsolanocapsine. C$_{30}$H$_{52}$N$_2$O$_2$. Needles from abs methanol, mp 222-224°. [α]$_D^{22}$ +53.2° (c = 1.09 in pyridine).

N-Benzylidenesolanocapsine. C$_{34}$H$_{50}$N$_2$O$_2$. Needles from pyridine + abs methanol, mp 249-251°. [α]$_D^{18}$ +31.4° (c = 0.78 in pyridine).

8834. Solanone. [1937-54-8] (5*S*,6*E*)-8-Methyl-5-(1-methylethyl)-6,8-nonadien-2-one; (−)-2-methyl-5-isopropyl-1,3-nonadien-8-one. C$_{13}$H$_{22}$O; mol wt 194.32. C 80.35%, H 11.41%, O 8.23%. Stereochemically unique terpene. Isoln from tobacco, structure and synthesis: Johnson, Nicholson *J. Org. Chem.* **30**, 2918 (1965). Stereoselective synthesis: A. Kohda, T. Sato, *Chem. Commun.* **1981**, 951.

Needles from methanol, sinters 296°, mp 301-303°. $[\alpha]_D^{23}$ $-88°$ (c = 1.01 in pyridine); $[\alpha]_D^{22}$ $-74.5°$ (c = 0.51 in methanol). Sol in hot alc, hot dioxane; slightly sol in hot water, dil acetic acid. Practically insol in chloroform, ether.

8838. Solifenacin. [242478-37-1] (1*S*)-3,4-Dihydro-1-phenyl-2(1*H*)-isoquinolinecarboxylic acid (3*R*)-1-azabicyclo[2.2.2]oct-3-yl ester; (1*S*,3′*R*)-3′-quinuclidinyl-1-phenyl-1,2,3,4-tetrahydro-2-isoquinolinecarboxylate. $C_{23}H_{26}N_2O_2$; mol wt 362.47. C 76.21%, H 7.23%, N 7.73%, O 8.83%. Muscarinic M_3 receptor antagoinst. Prepn: M. Takeuchi *et al.*, **WO 9620194**; *eidem*, **US 6017927** (1996, 2000 both to Yamanouchi). Receptor binding studies: K. Ikeda *et al.*, *Arch. Pharmacol.* **366**, 97 (2002); selectivity profile: A. Ohtake *et al.*, *Eur. J. Pharmacol.* **492**, 243 (2004). Clinical study in overactive bladder: C. R. Chapple *et al.*, *Br. J. Urol.* **93**, 303 (2004); F. Habb *et al.*, *Eur. Urol.* **47**, 376 (2005). Clinical comparison with tolterodine: C. R. Chapple *et al.*, *ibid.* **48**, 464 (2005). Review of efficacy and tolerability studies: S. Brunton, L. Kuritzky, *Curr. Med. Res. Opin.* **21**, 71-80 (2005); of clinical development: C. K. Payne, *Drugs* **66**, 175-190 (2006).

Yellow oil.
Succinate. [242478-38-2] YM-905; Vesicare. $C_{23}H_{26}N_2O_2 \cdot C_4H_6O_4$; mol wt 480.56. White to slightly yellowish crystals, mp ~145°. Freely sol in acetic acid, water, methanol, dimethylsulfoxide.
THERAP CAT: In treatment of urinary incontinence.

8839. Soman. [96-64-0] *P*-Methylphosphonofluoridic acid 1,2,2-trimethylpropyl ester; pinacoloxymethylphosphoryl fluoride; pinacolyl methylphosphonofluoridate; 1,2,2-trimethylpropoxyfluorophosphine oxide; 1,2,2-trimethylpropyl methylphosphonofluoridate; GD. $C_7H_{16}FO_2P$; mol wt 182.18. C 46.15%, H 8.85%, F 10.43%, O 17.56%, P 17.00%. Nerve gas; potent cholinesterase inhibitor similar in structure and activity to sarin and tabun, *q.q.v.* Prepn: Holmstedt, *Acta Physiol. Scand.* **25**, suppl. 90, p 106 (1951); *Protar* **16**, 131 (1950); G. Schrader, *Die Entwicklung neuer Insektizide auf Grundlage organischer Fluor- und Phosphor-Verbindungen* [Monographie Nr. 62 zu Angewandte Chemie und Chemie-Ingenieur-Technik (Verlag Chemie 1951)]; B. C. Saunders, *Some Aspects of the Chemistry and Toxic Action of Organic Compounds Containing Phosphorus and Fluorine* (Cambridge, 1957) p 94. Toxicity study: Loomis, Salafsky, *Toxicol. Appl. Pharmacol.* **5**, 685 (1963). Brief description: B. L. Harris in *Kirk-Othmer Encyclopedia of Chemical Technology* vol. 5 (Wiley-Interscience, New York, 4th ed., 1993) pp 795-816.

LD_{50} in mice (mg/100 g): 0.062 i.p.; 0.78 dermally (Loomis, Salafsky).
Caution: Potential symptoms of overexposure by inhalation are constriction of pupils of the eye, difficulty breathing followed by bronchial constriction, convulsions, death. Also toxic due to absorption through skin and eyes. *See: Chem. Eng. News* **31**, 4676 (1953).
USE: Chemical warfare agent.

8840. Somatoliberin. [83930-13-6] Somatoliberin (human pancreatic islet); GH-RF; GH-RH; GRF; growth hormone-releasing factor; growth hormone-releasing hormone; hGRF; hpGRF; somatocrinin. Stimulatory growth-hormone releasing factor of the hypothalamus that mediates, together with somatostatin, *q.v.*, the neuroregulation of somatotropin secretion. The concept of hypothalamic

---

Liquid. bp_1 60°. n_D^{20} 1.4755. d_4^{20} 0.870. $[\alpha]_D^{23}$ +13.6° (neat). uv max (ethanal and hexane): 230 nm (log ε 4.07).

8835. Solanum. Bull nettle; radical weed; sand-brier; horse nettle. Air-dried ripe fruit of *Solanum carolinense* L., Solanaceae. *Habit.* South America, Florida and other sections of U.S.A. *Constit.* Solanine, solanidine.
THERAP CAT: Sedative, anticonvulsant.

8836. Solasodine. [126-17-0] (3β,22α,25*R*)-Spirosol-5-en-3-ol; solasod-5-en-3β-ol; Δ^5-20β_F,22α_F,25α_F,27-azaspirosten-3β-ol; solancarpidine; solanidine-S; purapuridine. $C_{27}H_{43}NO_2$; mol wt 413.65. C 78.40%, H 10.48%, N 3.39%, O 7.74%. Steroidal alkaloid isolated from various *Solanum* species. By hydrolysis of solasonine: Rochelmeyer, *Arch. Pharm.* **277**, 329 (1939). *See also* ref *under* Solasonine and Solanidine. Structure: Briggs *et al.*, *J. Chem. Soc.* **1950**, 3013. Synthesis: Uhle, *J. Org. Chem.* **27**, 656 (1962); Schreiber, Rönsch, *Tetrahedron* **20**, 1939 (1964); Kessar *et al.*, *ibid.* **27**, 2869 (1971). Comprehensive description: G. Indrayanto *et al.*, *Anal. Profiles Drug Subs. Excip.* **24**, 487-522 (1996).

Hexagonal plates from methanol or by sublimation in high vacuum, mp 200-202°. $[\alpha]_D^{25}$ $-98°$ (c = 0.14 in methanol); $[\alpha]_D$ $-113°$ (CHCl$_3$). Alkaline reaction to litmus in alcoholic soln. pKb 6.30. uv max (methanol): 206 nm. Freely sol in benzene, pyridine, and chloroform. Practically insol in ether. Soly at 30° (mg/ml): methanol 9.5; 95% ethanol 5.0; acetone 3.5; *n*-hexane <1.0; water <1.0.
USE: Starting material for steroidal drugs.

8837. Solasonine. [19121-58-5] (3β,22α,25*R*)-Spirosol-5-en-3-yl *O*-6-deoxy-α-L-mannopyranosyl-(1 → 2)-*O*-[β-D-glucopyranosyl-(1 → 3)]-β-D-galactopyranoside; solanine-S; purapurine. $C_{45}H_{73}NO_{16}$; mol wt 884.07. C 61.14%, H 8.32%, N 1.58%, O 28.96%. From *Solanum aviculare* Forst F., *S. sodomeum* L., *S. xanthocarpum* Schrad & Wendl., Solanaceae: Bell, Briggs, *J. Chem. Soc.* **1942**, 1; Briggs *et al.*, *ibid.* **1942**, 3; Kuhn, Löw, *Ber.* **88**, 289 (1955). Isoln from other *S.* spp: Briggs, Cambie, *J. Chem. Soc.* **1958**, 1422; Briggs *et al.*, *ibid.* **1961**, 4645. Paper chromatography: Szendey, *Arch. Pharm.* **290**, 563 (1957). Structure: Briggs *et al.*, *J. Chem. Soc.* **1963**, 2848. Enzymic decompn: Guseva, Pasechnichenko, *Biokhimiya* **24**, 563 (1959).

regulation of growth hormone release was postulated on the basis of physiological and biochemical evidence, *cf.* A. V. Schally *et al.*, *Recent Prog. Horm. Res.* **24**, 497 (1968); J. B. Martin in *Frontiers in Neuroendocrinology*, L. Martini, W. F. Ganong, Eds. (Raven Press, New York, 1976) pp 129-168. Demonstration of the existence of a GH-RH: R. Deuben, J. Meites, *Endocrinology* **74**, 408 (1964); A. V. Schally *et al.*, *ibid.* **82**, 271 (1968); E. Dickermann *et al.*, *Neuroendocrinology* **4**, 75 (1969). Isoln of a decapeptide originally believed to be GRF: A. V. Schally *et al.*, *Endocrinology* **84**, 1493 (1969). Review of early literature: *eidem, Science* **179**, 341 (1973). Isoln and characterization of hpGRF, a fully bioactive, 44 amino acid peptide from human pancreatic tumor: R. Guillemin *et al.*, *Science* **218**, 585 (1982); J. Rivier *et al.*, *Nature* **300**, 276 (1982). Cloning and sequence analysis of cDNA for hpGRF precursor: U. Gubler *et al.*, *Proc. Natl. Acad. Sci. USA* **80**, 4311 (1983). Expression-cloning of cDNA for hpGRF: K. E. Mayo *et al.*, *Nature* **306**, 86 (1983). Isoln, primary structure and synthesis of native human hypothalamic GRF (hGRF), identity with hpGRF and comparison with GRF from other species: N. Ling *et al.*, *Proc. Natl. Acad. Sci. USA* **81**, 4302 (1984). Immunohistochemical detection of GRF in brain: B. Bloch *et al.*, *ibid.* **301**, 607 (1983). Potent interaction of glucocorticoids with GRF *in vivo:* W. B. Wehrenberg *et al.*, *Science* **221**, 556 (1983). Series of articles on clinical pharmacology: *Horm. Res.* **22**, 32-57 (1985). Clinical evaluation in treatment of hypopituitary dwarfism: M. O. Thorner *et al.*, *N. Engl. J. Med.* **312**, 4 (1985). *Reviews:* R. Guillemin *et al.*, *Recent Prog. Horm. Res.* **40**, 233-299 (1984); W. B. Wehrenberg *et al.*, *Annu. Rev. Pharmacol. Toxicol.* **25**, 463-483 (1985).

Porcine GRF is destroyed by trypsin, chymotrypsin, pepsin. Biological activity remains after incubation at pH 8.5 for 4 hours at 37°, but is destroyed after similar incubation at pH 2.

8841. Somatostatin. [38916-34-6] Somatostatin (sheep); growth hormone-release inhibiting factor; GH-RIF; somatotropin release inhibiting factor; SRIF; SRIF-14. $C_{76}H_{104}N_{18}O_{19}S_2$; mol wt 1637.90. C 55.73%, H 6.40%, N 15.39%, O 18.56%, S 3.91%. Widely occurring cyclic tetradecapeptide that mediates, together with somatoliberin, *q.v.*, the neuroregulation of somatotropin secretion. Inhibits release of growth hormone, insulin and glucagon. Also a potent inhibitor in a number of systems, including central and peripheral neural, gastrointestinal, and vascular smooth muscle. Isoln from ovine hypothalamic extracts: P. Brazeau *et al.*, *Science* **179**, 77 (1973). Structure: Burgus *et al.*, *Proc. Natl. Acad. Sci. USA* **70**, 684 (1973); Ling *et al.*, *Biochem. Biophys. Res. Commun.* **50**, 127 (1973). Synthesis: Rivier *et al.*, *C. R. Seances Acad. Sci. Ser. D* **276**, 2737 (1973); Sarantakis, McKinley, *Biochem. Biophys. Res. Commun.* **54**, 234 (1973); Yamashiro, Li, *ibid.* 882; Coy *et al.*, *ibid.* 1267; A. M. Felix *et al.*, *Int. J. Pept. Protein Res.* **15**, 342 (1980); B. Hartrodt *et al.*, *Pharmazie* **37**, 403 (1982). Synthesis and bacterial expression of somatostatin gene: K. Itakura *et al.*, *Science* **198**, 1056 (1977); K. Itakura, A. D. Riggs, BE **871782**; K. Itakura, US **4356270**; A. D. Riggs, US **4366246** (1979, 1982, 1982 all to Genentech). Isoln of a 28-amino acid somatostatin, termed *SRIF-28*, from porcine gastrointestinal tract: L. Pradayrol *et al.*, *Biochem. Biophys. Res. Commun.* **85**, 701 (1978). Primary structure: *eidem, FEBS Lett.* **109**, 55 (1980). Synthesis: E. Wunsch, *Z. Naturforsch.* **35B**, 911 (1980). Isoln of a 25-amino acid somatostatin, *SRIF-25*, and structure: P. Brazeau *et al.*, *C. R. Seances Acad. Sci. Ser. D* **290**, 1369 (1980). Both of these peptides consist of an *N*-terminal extension of SRIF-14. They are more potent than SRIF-14 in inhibition of insulin release but less potent in inhibition of glucagon release: L. Mandarino *et al.*, *Nature* **291**, 76 (1981). Synthetic linear SRIF-14 exhibits the same bioactivity as the natural cyclic form: P. Brazeau *et al.*, *Endocrinology* **94**, 184 (1974). Activity studies in humans: T. M. Siler *et al.*, *J. Clin. Endocrinol. Metab.* **37**, 632 (1973). Inhibitory effects on the secretion of thyrotropin: Vale *et al.*, *Fed. Proc.* **32**, 211 (1973); of insulin: K. G. Alberti *et al.*, *Lancet* **2**, 1299 (1973); Koeker *et al.*, *Science* **184**, 482 (1974). Role in diabetes: Maugh, *ibid.* **188**, 920 (1975). Clinical trial in controlling acute variceal hemorrhage: S. A. Jenkins *et al.*, *Br. Med. J.* **290**, 275 (1985). Review of distribution, secretion, physiology of gastrointestinal somatostatin: C. H. S. McIntosh, *Life Sci.* **37**, 2043-2058 (1985). Symposium of biosynthesis, bioactivity, and clinical applications: *Adv. Exp. Med. Biol.* **188**, 1-524 (1985). *Reviews:* Vale *et al.*, *Recent Prog. Horm. Res.* **31**, 365-397 (1975); Guillemin, Gerich, *Annu.*

Rev. Med. **27**, 379-388 (1976); R. L. Moss, *Annu. Rev. Physiol.* **41**, 617 (1979); A. Arimura, J. B. Fishback, *Neuroendocrinology* **33**, 246-256 (1981); several authors in *Gut Hormones*, S. R. Bloom, J. M. Polack, Eds. (Churchill, New York, 2nd ed., 1981) 605 pp; S. M. McCann, *Annu. Rev. Pharmacol. Toxicol.* **22**, 491-515 (1982); N. Bethge *et al.*, *J. Clin. Chem. Clin. Biochem.* **20**, 603-613 (1982). Book: *Somatostatin* Vol. 2, M. T. McQuillan, Ed. (Eden Press, Quebec, 1980) 238 pp.

Ala–Gly–Cys–Lys–Asn–Phe–Phe–Trp
| |
Cys–Ser–Thr–Phe–Thr–Lys

Acetate. SRIF-A; Aminopan; Modustatina; Somatofalk; Stilamin.

THERAP CAT: Treatment of severe, acute hemorrhage of gastroduodenal ulcers. Treatment of erosive or hemorrhagic gastritis. Experimental antidiabetic. Growth hormone inhibitor.

8842. Somatotropin. [9002-72-6] Adenohypophyseal growth hormone; GH; hypophyseal growth hormone; anterior pituitary growth hormone; growth hormone; phyone; pituitary growth hormone; somatotropic hormone; STH. Species-specific anabolic protein that promotes somatic growth, stimulates protein synthesis, and regulates carbohydrate and lipid metabolism. Increases serum levels of somatomedins, *q.v.* Secreted by the anterior pituitary under the regulation of the hypothalamic hormones, somatoliberin and somatostatin, *q.q.v.* Growth hormones from various species differ in amino acid sequence, antigenicity, isoelectric point, and in the range of animals in which they can produce biological responses. Isoln from bovine anterior pituitary substance: C. H. Li *et al.*, *J. Biol. Chem.* **159**, 353 (1945); A. E. Wilhelmi *et al.*, *ibid.* **176**, 735 (1948); C. H. Li, US **3118815** (1964 to Upjohn). Isoln from human pituitaries: Lewis, Brink, US **2974088** (1961 to Merck & Co.); Reisfeld *et al.*, *Endocrinology* **71**, 559 (1962). Amino acid sequence of human growth hormone (HGH): C. H. Li *et al.*, *J. Am. Chem. Soc.* **88**, 2050 (1966); *eidem, Arch. Biochem. Biophys.* **133**, 70 (1969). Total synthesis: C. H. Li, D. Yamashiro, *J. Am. Chem. Soc.* **92**, 7608 (1970). Revised sequence of HGH: H. D. Niall, *Nature New Biol.* **230**, 90 (1971). Review of structural studies and comparison with the lactogenic hormones: H. D. Niall *et al.*, *Recent Prog. Horm. Res.* **29**, 387-416 (1973). Review of protein chemistry: C. H. Li, *Mol. Cell. Biochem.* **46**, 31-41 (1982). Cloning and expression of cDNA for HGH in *E. coli:* D. V. Goeddel *et al.*, *Nature* **281**, 544 (1979); D. V. Goeddel, H. L. Heyneker, BE **884012**; *eidem*, US **4342832** (1980, 1982 both to Genentech); in mammalian cells: G. N. Pavakis *et al.*, *Proc. Natl. Acad. Sci. USA* **78**, 7398 (1981). Purification and bioactivity of recombinant HGH: K. C. Olson *et al.*, *Nature* **293**, 408 (1981). Clinical comparison of natural and recombinant HGH: R. L. Hintz *et al.*, *Lancet* **1**, 1276 (1982). Use of recombinant bovine GH to increase milk production in cows: D. E. Bauman *et al.*, *J. Dairy Sci.* **68**, 1352 (1985). Discussion of use of recombinant GH in applied animal agriculture: J. L. Burton, B. W. McBride, *J. Agric. Ethics* **2**, 129-159 (1989). Review of GH gene studies: D. D. Moore *et al.*, *Recent Prog. Horm. Res.* **38**, 197-225 (1982). Review of bioregulation of HGH secretion: D. G. Johnston *et al.*, *J. R. Soc. Med.* **78**, 319-327 (1985); of mechanism of action on target cells: O. G. P. Isaksson *et al.*, *Annu. Rev. Physiol.* **47**, 483-499 (1985). Symposium on pharmacology and clinical efficacy: *Horm. Res.* **33**, Suppl. 4, 1-107 (1990). Books: *Growth and Growth Hormone*, A. Pecile, E. Muller, Eds. (Excerpta Medica, Amsterdam, 1972); *Human Growth Hormone and Gonadotrophins in Health and Disease*, P. Franchimont, H. Burger, Eds. (Elsevier, New York, 1975) 494 pp; *Growth Hormone and Other Biologically Active Peptides*, A. Pecile, E. Muller, Eds. (Elsevier, New York, 1980); *Evaluation of Growth Hormone Secretion*, Z. Laron, O. Butenandt, Eds. (S. Karger, New York, 1983); *Use of Somatotropin in Livestock Production*, K. Sejrsen *et al.*, Eds. (Elsevier, New York, 1989).

Human growth hormone. [12629-01-5] HGH; somatropin; CB-311; Crescormon; Genotropin; Grorm; Humatrope; Nanormon; Norditropin; Nutropin; Nutropinaq; Saizen; Umatrope; Zomacton. $C_{990}H_{1529}N_{263}O_{299}S_7$; mol wt 22124.08. A single polypeptide chain of 191 amino acids having a molecular weight of 22,124. Isoelectric point 4.9. $[\alpha]^{25}$ $-38.7°$ (0.1*M* acetic acid).

Methionyl human growth hormone. [82030-87-3] Somatrem; met-HGH; Protropin; Somatonorm. $C_{995}H_{1537}N_{263}O_{301}S_8$; mol wt

22256.26. Produced in bacteria from recombinant DNA. Contains the complete amino acid sequence of the natural hormone plus an additional N-terminal methionine.

Bovine somatotropin. One of the four naturally occurring molecular variants is known as *somavubove*, $C_{976}H_{1533}N_{265}O_{286}S_8$. Several variants have been produced by recombinant DNA technology: *somagrebove*, $C_{987}H_{1550}N_{268}O_{291}S_9$, *CL-291894, Quest; sometribove*, $C_{978}H_{1537}N_{265}O_{286}S_9$; *somidobove*, $C_{1020}H_{1596}N_{274}O_{302}S_9$, *EL-349, LY-177837, Optiflex*.

Porcine somatotropin. Somacton. Several variants have been produced by recombinant DNA technology: *somalapor*, $C_{977}H_{1527}N_{265}O_{287}S_7$; *somenopor*, $C_{938}H_{1469}N_{255}O_{275}S_7$; *sometripor*, $C_{979}H_{1527}N_{265}O_{287}S_8$; *somfasepor*, $C_{938}H_{1465}N_{257}O_{278}S_6$, *Grolean, Leanstar*.

THERAP CAT: Growth stimulant.

THERAP CAT (VET): Growth stimulant. Bovine somatotropin as galactopoietic.

8843. Songorine. [509-24-0] (3R,6S,6aR,6bR,9R,11R,11aR,-12R,12aR,14R)-1-Ethyldodecahydro-6,11-dihydroxy-3-methyl-10-methylene-12,3,6a-ethanylylidene-9,11a-methanoazuleno[2,1-b]-azocin-8(9H)-one; (1α,15β)-21-ethyl-1,15-dihydroxy-4-methyl-16-methylene-7,20-cycloveatchan-12-one; napellonine; zongorine. $C_{22}H_{31}NO_3$; mol wt 357.49. C 73.92%, H 8.74%, N 3.92%, O 13.43%. From *Aconitum songoricum* Popov, *Ranunculaceae:* Yunusov, *J. Gen. Chem. USSR* **18**, 515 (1948); Kuzovkov, *ibid.* **23**, 504 (1953); *ibid.* **25**, 2006 (1955). Identity with napellonine: Kuzovkov, *Zh. Obshch. Khim.* **28**, 2283 (1958); *ibid.* **29**, 1728 (1959). Structure: Sugasawa, *Chem. Pharm. Bull.* **9**, 889, 897 (1961). Absolute configuration: Okamoto *et al.*, *ibid.* **13**, 1270 (1965). Synthesis of the aromatic intermediate: Wiesner *et al.*, *Can. J. Chem.* **51**, 3978 (1973). Pharmacology: Sadritdinov, *Farmakol. Alk.* No. 312 (1965), *C.A.* **66**, 93772d (1967). Mass spectra data: Yunusov *et al.*, *Khim. Prir. Soedin.* **6**, 101 (1970), *C.A.* **73**, 131178u (1970). Pharmacology and toxicity data: N. G. Bisset, *J. Ethnopharmacol.* **4**, 247-336 (1981).

Crystals, mp 201-202°. $[\alpha]_D^{20}$ −135.4°. uv max: 290 nm (log ε 2.6). LD_{50} in mice (mg/kg): 1575 orally, 630 s.c., 485 i.p., 142.5 i.v.; in rats (mg/kg): 407.5 i.p. (Bisset).

Hydrochloride. Crystals, mp 257-258°. $[\alpha]_D^{20}$ −114° (c = 2 in water).

8844. Sophorabioside. [2945-88-2] 3-[4-[[2-O-(6-Deoxy-α-L-mannopyranosyl)-β-D-glucopyranosyl]oxy]phenyl]-5,7-dihydroxy-4H-1-benzopyran-4-one; genistein-4'-glucosidorhamnoside. $C_{27}H_{30}O_{14}$; mol wt 578.52. C 56.06%, H 5.23%, O 38.72%. From fruits of *Sophora japonica* L., *Leguminosae.* Isoln and structure: Zemplén, Bognár, *Ber.* **75B**, 482 (1942). The biose is not identical with rutinose.

The anhydr substance mp 248° (slight decompn). $[\alpha]_D^{19}$ −73° (0.27 g in 10 ml pyridine). Freely sol in pyridine; sol in hot alcohol, hot acetone; slightly sol in boiling water. The alcoholic soln gives a purple color with ferric chloride.

Trihydrate. Needles from dil alcohol, mp 156-160°. The water of crystn can be removed by drying at 100° over P_2O_5 *in vacuo* for 12 hours.

8845. Sophorose. [534-46-3] 2-O-β-D-Glucopyranosyl-D-glucose. $C_{12}H_{22}O_{11}$; mol wt 342.30. C 42.11%, H 6.48%, O 51.41%. From pods of *Sophora japonica* L., *Leguminosae:* Rebaté, *Bull. Soc. Chim. Fr.* **7**, 565 (1940); Clancy, *J. Chem. Soc.* **1960**, 4213; Clancy in *Methods in Carbohydrate Chemistry* vol. I, R. L. Whistler, M. L. Wolfrom, Eds. (Academic Press, New York, 1962) pp 345-349. Structure and synthesis: Coxon, Fletcher, *J. Org. Chem.* **26**, 2892 (1961); Koeppen, *Carbohydr. Res.* **7**, 410 (1968). Crystal structure: J. Ohanessian *et al.*, *Acta Crystallogr.* **B34**, 3666 (1978).

Monohydrate. Needles from 80% aq methanol, mp 196-198°. $[\alpha]_D^{18}$ +19° (c = 1.2 in water).

Octa-O-acetyl-β-sophorose. $C_{28}H_{38}O_{19}$. Needles from ethanol, mp 193-194°. $[\alpha]_D^{18}$ −3.2° (c = 2.5 in chloroform).

8846. Sorafenib. [284461-73-0] 4-[4-[[[[4-Chloro-3-(trifluoromethyl)phenyl]amino]carbonyl]amino]phenoxy]-N-methyl-2-pyridinecarboxamide; N-4-(chloro-3-(trifluoromethyl)phenyl)-N'-(4-(2-(N-methylcarbamoyl)-4-pyridyloxy)phenyl)urea; Bay-43-9006. $C_{21}H_{16}ClF_3N_4O_3$; mol wt 464.83. C 54.26%, H 3.47%, Cl 7.63%, F 12.26%, N 12.05%, O 10.33%. Multiple kinase inhibitor targeting both RAF kinase and receptor tyrosine kinases that promote angiogensis. Prepn: B. Riedl *et al.*, WO 0041698 (2000 to Bayer); *eidem*, US 03139605 (2003); D. Bankston *et al.*, *Org. Process Res. Dev.* **6**, 777 (2002). Structure-activity study: U. R. Khire *et al.*, *Bioorg. Med. Chem. Lett.* **14**, 783 (2004). Characterization of kinase inhibition: S. M. Wilhelm *et al.*, *Cancer Res.* **64**, 7099 (2004). Mechanism of action study: D. J. Panka *et al.*, *Cancer Res.* **66**, 1611 (2006). LC/MS/MS determn in serum: M. Zhao *et al.*, *J. Chromatogr. B* **846**, 1 (2007). Clinical pharmacokinetics: H. Richly *et al.*, *Int. J. Clin. Pharmacol. Ther.* **41**, 620 (2003). Analysis of clinical safety and efficacy: D. Strumberg *et al.*, *Eur. J. Cancer* **42**, 548 (2006). Clinical evaluation and brief review: H. DeGrendele, *Clin. Colorectal Cancer* **3**, 16-18 (2003). Review of clinical development: B. I. Rini, *Expert Opin. Pharmacother.* **7**, 453-461 (2006).

White solid, mp 205.6°.

Tosylate. [475207-59-1] Nexavar. $C_{21}H_{16}ClF_3N_4O_3 \cdot C_7H_8O_3S$; mol wt 637.03.

THERAP CAT: Antineoplastic.

8847. Sorbic Acid. [110-44-1] (2E,4E)-2,4-Hexadienoic acid; (E,E)-1,3-pentadiene-1-carboxylic acid; 2-propenylacrylic acid. $C_6H_8O_2$; mol wt 112.13. C 64.27%, H 7.19%, O 28.54%. May be obtained from berries of the mountain ash, *Sorbus aucuparia* L., *Rosaceae* where it occurs as the lactone, called parasorbic acid: A. W. Hofmann, *Ann.* **110**, 129 (1859). Synthesis by condensing crotonaldehyde and malonic acid in pyridine soln: O. Doebner, *Ber.* **33**, 2140 (1900); C. F. H. Allen, J. Van Allan, *Org. Synth.* coll. vol. III, 783 (1955). Prepn from crotonaldehyde and ketene: H. J. Hagemeyer, Jr., *Ind. Eng. Chem.* **41**, 765 (1949); H. Fernholz, E. Mundlos, US 3021365 (1962 to Hoechst). Improved synthesis: T. Hattori *et al.*, *Chem. Commun.* **2000**, 73. Toxicity study: H. F. Smyth, C. P. Carpenter, *J. Ind. Hyg. Toxicol.* **30**, 63 (1948). GC-MS determn in urine: T. Renner *et al.*, *J. Chromatogr. A* **847**, 127 (1999). HPLC determn in cosmetics: E. Mikami *et al.*, *J. Pharm. Biomed. Anal.* **28**, 261 (2002); in foods: B. Saad *et al.*, *J. Chromatogr. A* **1073**, 393 (2005). Safety assessment for use in cosmetics: *J. Am. Coll. Toxicol.* **7**, 837 (1988). Review of food applications: E. Lück, *Food Addit. Contam.* **7**, 711-715 (1990). Review of antimicrobial action: J. N. Sofos *et al.*, *Int. J. Food Microbiol.* **3**, 1-17 (1986); of stability and degradation in foods: B. R. Thakur *et al.*, *Food Rev. Int.* **10**, 71-91 (1994).

Needles from water, mp 134.5°. Should be stored at temps below 40°. bp 228° (dec). Vapor pressure at 20° <0.01 mm, at 143° 50 mm. Flash pt 260°F (127°C). pK (25°) = 4.76. Sol in ether. Soly in water at 30° 0.25%, at 100° 3.8%. Soly at 20°: 5.5% in propylene glycol, 12.90% in abs ethanol or methanol, 0.29% in 20% ethanol, 11.5% in glacial acetic acid, 9.2% in acetone, 2.3% in benzene, 1.3% in carbon tetrachloride, 0.28% in cyclohexane, 11.0% in dioxane, 0.31% in glycerol, 8.4% in isopropanol, 2.7% in isopropyl ether, 6.1% in methyl acetate, 1.9% in toluene. LD$_{50}$ orally in rats: 7.36 g/kg (Smyth, Carpenter).

Calcium salt. [7492-55-9] Calcium sorbate. C$_{12}$H$_{14}$CaO$_4$; mol wt 262.32. Prepn: C. M. Gooding, US 3139378 (1964 to Corn Products Co.). Fine white crystalline powder. Sol in water. Practically insol in ethanol.

Sodium salt. [7757-81-5] Sodium sorbate. C$_6$H$_7$NaO$_2$; mol wt 134.11. Prepn: O. Horn, H. Fernholz, DE 1045390 (1958 to Hoechst).

Potassium salt. [24634-61-5] Potassium sorbate. C$_6$H$_7$KO$_2$; mol wt 150.22. Prepn: O. Probst, H. Oehme, US 3173948 (1965 to Hoechst). Crystals, d$_{20}^{25}$ 1.363. Dec above 270°. Soly in water at 20°: 58.2%; in alc: 6.5%.

USE: Preservative and antimicrobial agent for foods, cosmetics, and pharmaceuticals. To improve the characteristics of drying oils. In alkyd type coatings to improve gloss. To improve milling characteristics of cold rubber.

8848. Sorbic Alcohol. [17102-64-6]; [111-28-4] (unspecified stereo). (2*E*,4*E*)-2,4-Hexadien-1-ol; 1-hydroxy-2,4-hexadiene; hexadenol; sorbinol; sorbyl alcohol. C$_6$H$_{10}$O; mol wt 98.15. C 73.42%, H 10.27%, O 16.30%. Prepn from sorbic aldehyde and aluminum isopropylate: T. Reichstein *et al.*, *Helv. Chim. Acta* **15**, 264 (1932); from sorbic acid and lithium aluminum hydride: M. Jacobson, *J. Am. Chem. Soc.* **77**, 2461 (1955).

Long needles. Agreeable odor of new mown grass. mp 30.5-31.5°. bp$_{16}$ 78-79. bp$_{12}$ ~80°. n$_D^{25}$ 1.4971. *Irritant. Flammable.* Flash pt, closed cup: 161.6°F (72°C). Volatile with steam. Sol in alcohol, ether, oils. Insol in water. Light sensitive. Not stable to air; must be sealed in evacuated ampuls for storage.

USE: Reagent in organic chemistry.

8849. Sorbinil. [68367-52-2] (4*S*)-6-Fluoro-2,3-dihydrospiro[4*H*-1-benzopyran-4,4'-imidazolidine]-2',5'-dione; (+)-(4*S*)-6-fluorospiro[chroman-4,4'-imidazolidine]-2',5'-dione; CP-45634. C$_{11}$H$_9$FN$_2$O$_3$; mol wt 236.20. C 55.94%, H 3.84%, F 8.04%, N 11.86%, O 20.32%. Spirohydantoin aldose reductase inhibitor. Prepn of racemate and resolution of isomers: R. Sarges, DE 2821966; *idem*, US 4130714 (both 1978 to Pfizer). Synthesis: R. Sarges *et al.*, *J. Org. Chem.* **47**, 4081 (1982); N. L. Dirlam *et al.*, *J. Org. Chem.* **52**, 3587 (1987). Absolute configuration: R. Sarges *et al.*, *J. Med. Chem.* **28**, 1716 (1985). Effect on polyol accumulation in diabetic rats: M. J. Peterson *et al.*, *Metabolism* **28**, Suppl. 1, 456 (1979). Pharmacokinetics in humans: G. Foulds *et al.*, *Clin. Pharmacol. Ther.* **30**, 693 (1981). HPLC determn in human lens and plasma: P. Lloyd, M. J. C. Crabbe, *J. Chromatogr.* **343**, 402 (1985). Preliminary clinical trials in diabetic neuropathy: R. J. Young *et al.*, *Diabetes* **32**, 938 (1983); J. Jaspan *et al.*, *Lancet* **2**, 758 (1983). Symposium on pharmacology and clinical efficacy: *Metabolism* **35**, Suppl 1, 1-121 (1986).

Crystals from ethanol, mp 241-243°. [α]$_D^{25}$ +54.0° (c = 1 in methanol).

8850. Sorbitan Esters. Sorbitan fatty acid esters; SFAE. Nonionic surface active agents; partial esters of the common fatty acids (lauric, palmitic, stearic, and oleic) and hexitol anhydrides derived from sorbitol. Commercial products may be mixtures of fatty acid esters of 1,4- and 1,5 anhydrosorbitol and 1,4,3,6-dianhydrosorbitol, but generally conform to the structure depicted below. Prepn: K. R. Brown, US 2322820 (1943 to Atlas Powder Co.). Improved process: G. J. Stockburger, US 4297290 (1981 to ICI). Comprehensive description: P. Becher, "Polyol Surfactants" in *Nonionic Surfactants*, M. J. Schick, Ed. (Dekker, New York, 1967) pp 247-299. Description of prepn and uses: L. R. Chislett, J. Walford, *Int. Flavours Food Addit.* **7**, 61 (1976). GLC determn of sorbitan monolaurate in plasma: S. H. Giovanetto, *Anal. Lett.* **16**, 867 (1983). Analysis by HPLC: N. Garti *et al.*, *J. Am. Oil Chem. Soc.* **60**, 1151 (1983). Series of toxicity studies: *Food Cosmet. Toxicol.* **16**, 519-542 (1978). Review: Cosmetic, Toiletry and Fragrance Assoc., *J. Am. Coll. Toxicol.* **4**, 65-121 (1985).

Sorbitan Laurate	R = OOC(C$_{11}$H$_{23}$)
Sorbitan Stearate	R = OOC(C$_{17}$H$_{35}$)
Sorbitan Oleate	R = OOC(C$_{17}$H$_{33}$)

Polyoxyethylene derivatives *see* Polysorbates.

Sorbitan Laurate. [1338-39-2] Sorbitan monolaurate; Alkamuls SML; Arlacel 20; Emsorb 2515; Glycomul L; Span 20. C$_{18}$H$_{34}$O$_6$; mol wt 346.46. Yellow, oily liquid. Acid value: <8. Saponification value: 150-165. Hydroxyl value: 330-360. Sol in mineral oil, cottonseed oil, methanol, ethanol, isopropyl alc, ethylene glycol. Insol in water, propylene glycol.

Sorbitan Stearate. [1338-41-6] Sorbitan monostearate; Alkamuls SMS; Arlacel 60; Glycomul S; Span 60. C$_{24}$H$_{46}$O$_6$; mol wt 430.63. Cream-colored to tan waxy solid, mp 49-65°. Acid value: 5-11. Saponification value: 140-157. Hydroxyl value: 230-260. Dispersible in warm water. Sol in alcohols, carbon tetrachloride, toluene, mineral oil, ethyl acetate. Insol in cold water, acetone, mineral spirits.

Sorbitan Oleate. [1338-43-8] Sorbitan monooleate; Alkamuls SMO; Arlacel 80; Capmul O; Emsorb 2500; Glycomul O; Span 80. C$_{24}$H$_{44}$O$_6$; mol wt 428.61. Yellow to amber-colored oily liquid. Acid number: 5-8. Saponification value: 140-160. Hydroxyl value: 193-215. Miscible with mineral and vegetable oils. Sol in ethanol, isopropyl alc. Insol in water, propylene glycol.

USE: As emulsifiers, stabilizers, and thickeners in foods, cosmetics, and medicinal products. In the textile industry as fiber lubricants and softeners. Pharmaceutic aid (surfactant).

8851. Sorbitol. [50-70-4] D-Glucitol; D-sorbitol; L-gulitol; sorbit; Resulax; Sorbilax; Sorbo; Sorbostyl; Sorbilande. C$_6$H$_{14}$O$_6$; mol wt 182.17. C 39.56%, H 7.75%, O 52.69%. First found in the ripe berries of the mountain ash *Pyrus aucuparia* Ehrh. (L.) (*Sorbus aucuparia* L.), Rosaceae. Occurs also in many other berries (except grapes) and in cherries, plums, pears, apples, seaweed and algae. Has been detected in blackstrap molasses. Isoln from berries: Embden, Griesbach, *Z. Physiol. Chem.* **91**, 268 (1914). Prepd industrially from glucose by high pressure hydrogenation or by electrolytic reduction: Boye, *Chem. Ztg.* **82**, 657 (1958); Fedor *et al.*, *Ind. Eng.*

Chem. **52**, 282 (1960); from dextrose by catalytic hydrogenation: *Faith, Keyes & Clark's Industrial Chemicals*, F. A. Lowenheim, M. K. Moran, Eds. (Wiley-Interscience, New York, 4th ed., 1975) pp 774-778. Review of uses: Kempf, *Staerke* **6**, 269-274 and 303-306 (1954); **9**, 234-237 (1955).

$$CH_2OH$$
$$HC-OH$$
$$HO-CH$$
$$HC-OH$$
$$HC-OH$$
$$CH_2OH$$

Needles with ½ or $1H_2O$. Hygroscopic. Sweet taste, ~60% as sweet as sugar (w/w). In the healthy human organism 1.0 g of sorbitol yields 3.994 calories which is comparable to 3.940 calories from 1.0 g of cane sugar. Seventy percent of orally ingested sorbitol is converted to CO_2 without appearing as glucose in the blood: Adcock, Gray, *Biochem. J.* **65**, 554 (1957). The hydrated crystals melt somewhat below 100°. When completely anhyd mp 110-112°. $[\alpha]_D^{20}$ −2.0° (H_2O). In the presence of molybdate the rotation is reversed and increased to +56°. Very sol in water. High % sorbitol solns are much more viscous than corresp glycerol solns. Also sol in methanol, isopropanol, butanol, cyclohexanol, phenol, acetone, acetic acid, DMF, pyridine, acetamide solns. Sparingly sol in alc. Practically insol in ethyl ether and most organic solvents. Not attacked in the cold when mixed with dil acids, alkalies or mild oxidizing substances. pKa (17.5°): 13.6. pH about 7.0. A commercial 70% aq soln may have the following characteristics: d_{20}^{20} 1.2879; n_D^{25} 1.45831; $[\alpha]_D^{20}$ −2.10°; bp_{760} 105°; pH between 6 and 7; viscosity (25°): 110 cP. d_4^{20} for various % solns: 5% 1.014; 10% 1.038; 25% 1.099; 50% 1.198; 60% 1.249; 70% 1.299; 83% 1.391. Viscosity in cP at 20°: 5% soln 1.230; 10% 1.429; 25% 2.689; 50% 11.09; 60% 35.73; 70% 185; 83% >10,000.

USE: In manuf of sorbose, ascorbic acid, propylene glycol, synthetic plasticizers and resins; as humectant (moisture conditioner) on printing rolls, in leather, tobacco. In writing inks to insure a smooth flow and to prevent crusting on the point of the pen. In antifreeze mixtures with glycerol or glycols. In candy manuf to increase shelf life by retarding the solidification of sugar; as humectant and softener in shredded coconut and peanut butter; as texturizer in foods; as sequestrant in soft drinks and wines. Used to reduce the undesirable aftertaste of saccharin in foodstuffs; as sugar substitute for diabetics. Pharmaceutic aid (sweetening agent, tablet excipient, humectant, wetting and solubilizing agent); to increase absorption of vitamins and other nutrients in pharmaceutical preparations: *Chem. Eng. News* **36**, 59 (Feb. 24, 1958).

THERAP CAT: Laxative.

THERAP CAT (VET): In ruminant ketosis, osmotic diuretic, laxative.

8852. Sorbose. [87-79-6] L-Sorbose; sorbin; sorbinose. C_6-$H_{12}O_6$; mol wt 180.16. C 40.00%, H 6.71%, O 53.28%. From sorbitol by fermentation with *Acetobacter suboxydans*: Perlman, *J. Chem. Educ.* **36**, 60 (1959); Lockwood, *Methods Carbohyd. Chem.* **1**, 151 (Academic Press, New York, 1962).

$$CH_2OH$$
$$C=O$$
$$HO-CH$$
$$HC-OH$$
$$HO-CH$$
$$CH_2OH$$

Orthorhombic, bisphenoidal crystals. About as sweet as sucrose. d^{15} 1.65. mp 165°. $[\alpha]_D^{30}$ −42.7° (c = 5). pKa (17.5°): 11.55. Freely sol in water; almost insol in alcohol. Reduces Fehling's soln.

USE: In the manuf of vitamin C (accounts for nearly 1000 tons of ascorbic acid produced every year). For conversion of L-sorbose to

2-keto-L-gulonic acid. *See:* Reichstein, Grüssner, *Helv. Chim. Acta* **17**, 311 (1934).

8853. Sorivudine. [77181-69-2] 1-β-D-Arabinofuranosyl-5-[(1*E*)-2-bromoethenyl]-2,4(1*H*,3*H*)-pyrimidinedione; 1-β-D-arabinofuranosyl-(*E*)-5-(2-bromovinyl)uracil; 5-bromovinyl-araU; brovavir; BV-araU; BVAU; YN-72; SQ-32756; Usevir. $C_{11}H_{13}$-BrN_2O_6; mol wt 349.14. C 37.84%, H 3.75%, Br 22.89%, N 8.02%, O 27.49%. Orally active antiviral agent; inhibits viral DNA synthesis. Prepn: H. Machida, S. Sakata, *EP 31128*; *eidem*, *US 4386076* (1981, 1983 both to Yamassa Shoyu); S. Sakata *et al.*, *Nucleic Acids Symp. Ser.* **8**, s39 (1980); R. Busson *et al.*, *ibid.* **9**, 49 (1981). Antiviral activity and toxicity: H. Machida, S. Sakata, *Antiviral Res.* **4**, 135 (1984). Mechanism of action studies: T. Yokota *et al.*, *Mol. Pharmacol.* **36**, 312 (1989); H. Machida, *Adv. Exp. Med. Biol.* **278**, 255 (1990). Clinical trial in herpes zoster infection: M. Niimura, *ibid.* 267.

mp 182° (Sakata, 1980); also reported as white crystals from ethanol, mp 195-200° (dec) (Machida, Sakata, 1983). $[\alpha]_D^{25}$ +0.5° (1*N* NaOH). LD_{50} in mice (mg/kg): ~3300 i.p.; >5000 s.c.; >10000 orally (Machida, Sakata, 1984).

THERAP CAT: Antiviral.

8854. Sotalol. [3930-20-9] *N*-[4-[1-Hydroxy-2-[(1-methylethyl)amino]ethyl]phenyl]methanesulfonamide; 4′-[1-hydroxy-2-(isopropylamino)ethyl]methanesulfonanilide. $C_{12}H_{20}N_2O_3S$; mol wt 272.36. C 52.92%, H 7.40%, N 10.29%, O 17.62%, S 11.77%. β-Adrenergic blocker. Prepn: Uloth *et al.*, *J. Med. Chem.* **9**, 88 (1966). Pharmacology and toxicology: Larsen, Lish, *Nature* **203**, 1283 (1964); Lish *et al.*, *J. Pharmacol. Exp. Ther.* **149**, 161 (1965); Stanton *et al.*, *ibid.* 175; Kvam *et al.*, *ibid.* 183. Activity of the *d*- and *l*-forms: Somani, Watson, *ibid.* **164**, 317 (1968); Somani, Bachand, *Eur. J. Pharmacol.* **7**, 239 (1969). HPLC determn in body fluids: W. P. Gluth *et al.*, *Arzneim.-Forsch.* **38**, 408 (1988). Enantioselective LC/MS determn in plasma: E. Badaloni *et al.*, *J. Chromatogr. B* **796**, 53 (2003). Review of pharmacology and therapeutic use: B. N. Singh *et al.*, *Drugs* **34**, 311-349 (1987); S. H. Hohnloser, R. L. Woosley, *N. Engl. J. Med.* **331**, 31-38 (1994). Comprehensive description: R. T. Foster, R. A. Carr, *Anal. Profiles Drug Subs. Excip.* **21**, 501-533 (1992). Review of clinical experience: J. L. Anderson, E. N. Prystowsky, *Am. Heart J.* **137**, 388-409 (1999).

Crystals. uv max (chloroform): 242.2, 275.2 nm. pK_1 8.2, pK_2 9.8.

Hydrochloride. [959-24-0] MJ-1999; Beta-Cardone; Betapace; Darob; Rytmobeta; Sotacor; Sotalex. $C_{12}H_{20}N_2O_3S$.HCl; mol wt 308.82. White crystalline solid, mp 206.5-207° (dec), also reported as mp 218-219° (Singh). Freely sol in water; slightly sol in chloroform. Partition coefficient (water/*n*-octanol): 0.24; (octan-1-ol/pH 7.4 phosphate buffer) at 37°: 0.09. LD_{50} in male mice, rats (mg/kg): 2600, 3450 orally; 670, 680 i.p.; LD_{50} orally in rabbits: 1000 mg/kg; LD_{50} i.p. in dogs: 330 mg/kg (Lish).

THERAP CAT: Antianginal; antiarrhythmic (class III); antihypertensive.

THERAP CAT (VET): Antiarrhythmic (class III).

8855. Soterenol. [13642-52-9] *N*-[2-Hydroxy-5-[1-hydroxy-2-[(1-methylethyl)amino]ethyl]phenyl]methanesulfonamide; 2′-hydroxy-5′-[1-hydroxy-2-(isopropylamino)ethyl]methanesulfonanilide; MJ-1992. $C_{12}H_{20}N_2O_4S$; mol wt 288.36. C 49.98%, H 6.99%, N 9.71%, O 22.19%, S 11.12%. β-Adrenergic agonist. Prepn: Larsen *et al., J. Med. Chem.* **10**, 462 (1967). Pharmacological study: K. W. Dungan *et al., J. Pharmacol. Exp. Ther.* **164**, 290 (1968).

Monohydrochloride. [14816-67-2] $C_{12}H_{20}N_2O_4S$·HCl. Crystals from methanol-isopropyl ether, mp 195.5-196.5° (dec). LD_{50} (7 day) in mice (mg/kg): 41 i.v.; 315 i.p.; 660 orally (Dungan).

THERAP CAT: Bronchodilator.

8856. Soybean. Soya bean; soja bean; Lincoln bean; Manchurian bean; Chinese pea. Seed of *Glycine max* (L.) Merrill [*G. soja* Sieb. & Zucc., *G. hispida* (Moench) Maxim., *Soja hispida* Moench], *Leguminosae. Habit.* Eastern Asia, especially Manchuria. Cultivated in the midwestern U.S.A., Brazil, some European countries, such as Italy and Yugoslavia. *Constit.* Proteins 40%, carbohydrates 17%, oil 18%, ash 4.6%. The ash content equals K 1.67%, Na 0.34%, Ca 0.28%, Mg 0.22%, Cl 0.024%, I 0.000054%, Fe 0.0097%, Cu 0.0012%, Mn 0.0028%, Zn 0.0022%, P 0.66%, S 0.41%, Al 0.0007%. The chief proteins are glycinin, a globulin, phaseolin, another globulin, and 2 albumins: legumelin and soy legumelin. The carbohydrates are sucrose, raffinose, stachyose, and pentosans. Phosphorus compds (about 3%) are phospholipids, nucleic acid phosphates, phytin, and inorganic phosphates. The phospholipids contain lecithin, cephalin, and inositol. The vitamin content is moderate: Vitamin A 110 i.u./100g; thiamine 1.14 mg/100 g; riboflavin 0.31 mg/100 g; niacin 2.1 mg/100 g; ascorbic acid: trace. Food energy: 350 cal/100 g. Amino acid analysis of the protein fraction: Glutamic acid 18.4%, leucine 8.1%, arginine 7.5%, lysine 6.7%, valine 5.4%, isoleucine 5.3%, phenylalanine 5.2%, threonine 3.9%, histidine 2.3%, tryptophan 1.6%, methionine 1.4%. Ten kg of soybeans yield 15 g of the isoflavone glycoside genistin, the mother liquor contains daidzin. Soybeans contain a small amount of enzymes such as lipoxidase, urease, uricase, protease, antienzymes such as antitrypsin, and a growth-inhibiting factor. Bitter principles and saponins are also found. *Reviews:* K. S. Markley, *Soybeans and Soybean Products,* 2 vols. (Interscience, New York, 1951); Cowan in *Kirk-Othmer Encyclopedia of Chemical Technology* **vol. 18** (Interscience, New York, 2nd ed., 1969) pp 599-614.

Whitish or yellowish-green to brownish-black, ovoid beans, about 8 mm long, 7 mm wide, and 6 mm thick. Most varieties have a wt of 20 g/100 seeds.

USE: In the production of soybean oil. As food and feedstuff. Debittered soybean flour contains practically no starch and is widely used in dietetic foods. Soybean meal obtained after expressing the oil is a preferred source of protein for feedstuffs. Other products are soybean lecithin, genistin, monosodium glutamate, soybean milk; *tofu,* a soybean curd; *miso,* a fermented mixture with barley or rice; *natto,* which is soybean cheese. Soybean proteins are used also in the adhesive and plastics industries.

THERAP CAT: Nutrient.

8857. Soybean Oil. Obtained from soybeans by solvent extraction or by mechanical expression. *Constit.* Triglycerides of oleic acid 26%, of linoleic acid 49%, of linolenic acid 11%, of saturated acids 14%. Free fatty acids are usually less than 1%. Phospholipids (lecithin) 1.5-4%. Another 0.8% consists of stigmasterol, sitosterols, and tocopherols. Reviews and bibliographies: E. W. Eckey, *Vegetable Fats and Oils* (Reinhold, New York, 1954); W. J. Wolf in *Kirk-Othmer Encyclopedia of Chemical Technology* **vol. 21** (Wiley-Interscience, New York, 3rd ed., 1983) pp 417-442; E. G. Hammond *et al., Bailey's Industrial Oil and Fat Products* **vol. 2** (John Wiley & Sons, 6th Ed., 2005) pp 577-653.

Pale yellow to brownish-yellow oil. Slight characteristic odor and taste. d_{25}^{25} 0.916-0.922. Flash pt 540°F (282°C). Ignition temp 833°F (445°C). n_D^{25} 1.471-1.475. Titer 22-27. Solidifies at −10 to −16°.

Viscosity (cP): 172.9 (0°); 99.7 (10°); 50.09 (25°); 28.86 (40°). Acid value 0.3-3.0. Saponification value 189-195. Iodine value 127-138. Thiocyanogen value 77-85. Diene no. 0.7. Hydroxyl value 4-8. Reichert-Meissl value 0.2-0.7. Polenske value 0.2-1.0. Miscible with abs alcohol, ether, petr ether, chloroform, carbon disulfide.

USE: As cooking and salad oil. In the manuf of margarine, shortenings, mayonnaise, candy, soap. In paints, varnishes, resins, plastics, lubricants, biodiesel fuel.

8858. Soy Sauce. Soy sauce is a hydrolysis product of soybeans. A combination of mold fermentation and acid hydrolysis is used. The molds employed are *Aspergillus flavus, A. niger,* and *A. oryzae.* Soy sauce consists of a mixture of amino acids, peptides, polypeptides, peptones, simple proteins, purines, carbohydrates, and lesser organic compds suspended in an 18% sodium chloride soln. Adenine, arginine, choline, lysine, betaine, and glutamic acid have been isolated. In the manuf of soy sauce, during the fermentation process, 60-70% of the vitamins present in soybeans are destroyed. Review and bibliography: K. S. Markley, *Soybeans and Soybean Products,* 2 vols. (Interscience, N.Y., 1951).

8859. SPADNS. [23647-14-5] 4,5-Dihydroxy-3-[2-(4-sulfophenyl)diazenyl]-2,7-naphthalenedisulfonic acid sodium salt (1:3); 4,5-dihydroxy-3-[(4-sulfophenyl)azo]-2,7-naphthalenedisulfonic acid trisodium salt; 2-(4-sulfophenylazo)-1,8-dihydroxy-3,6-naphthalenedisulfonic acid trisodium salt; trisodium 2-(*p*-sulfophenylazo)-1,8-dihydroxynaphthalene-3,6-disulfonate. $C_{16}H_9N_2Na_3$-$O_{11}S_3$; mol wt 570.40. C 33.69%, H 1.59%, N 4.91%, Na 12.09%, O 30.85%, S 16.86%. Cation chelant; dye ligand that reacts with metal ions to form colored complexes. Prepn from chromotropic and sulfanilic acids, *q.q.v.*: G. Banerjee, *Fresenius Z. Anal. Chem.* **146**, 417 (1955). Metal chelate studies: *idem, ibid.* **147**, 105 (1955); *idem, ibid.* **148**, 349 (1955); *idem, Anal. Chim. Acta* **16**, 56, 62 (1957); A. K. Bhattacharjee, M. K. Mahanti, *Indian J. Chem.* **20A**, 427 (1981); M. S. Rizk *et al., Egypt. J. Chem.* **36**, 199 (1993); and antifungal activity: G. Pandey, K. K. Narang, *Bioinorg. Chem. Appl.* **3**, 217 (2005). Fluorine determn: G. Banerjee, *Anal. Chim. Acta* **13**, 409 (1955); R. P. Hollingworth, *Anal. Chem.* **29**, 1130 (1957); E. Bellack, P. J. Schouboe, *ibid.* **30**, 2032 (1958); L. C. Peck, V. C. Smith, *Talanta* **11**, 1343 (1964). Infrared and Raman spectroscopy studies: G. G. Siu, Z. L. Chen, *J. Raman Spectrosc.* **24**, 173 (1993).

Red powder. *Irritant.* Aq soln is scarlet, pH 6.9, pKa ~9. Absorption max (pH 5.3-6.9): 512 nm. Sol in water; slightly sol in ethanol.

USE: Chromogenic indicator for spectrophotometric determn of metal ions including beryllium, thorium, zirconium, cobalt, nickel, iron, aluminum, magnesium, calcium, and rare earth metals. Also in determn of fluoride ion, especially in biological samples.

8860. Spandex. Lycra; Vyrene. An elastic, segmented polyurethane fiber obtained by the interaction of a diisocyanate with a glycol. The term "segmented" indicates that it is made from a block copolymer in which reasonably long flexible chains are joined to shorter stiff chains through the urethane linkages. *Ref: Mod. Text. Mag.* **40**, 38 (Dec. 1959); Hicks Jr., *Am. Dyest. Rep.* **52**, no. 1, 33 (1963). Book: Moncrieff, *Man-Made Fibres* (Heywood, London, 4th ed., 1963) pp 396-403. *See:* **US 2692873** (1954 to du Pont) and **US 2751363** (1956 to U.S. Rubber). Compare also Smith, **US 3061574**; Frazer, Shivers **US 3071557** (1962, 1963 to du Pont). Historical review of industrial production with extensive patents list: L. Rose, *Rep. Prog. Appl. Chem.* **51**, 609-612 (1966). Monograph: M. McDonald, *Spandex Manufacture* (Noyes, Park Ridge, N.J., 1970) 190 pp.

All linkages are hydrolytically stable to acids, alkalies.

Caution: Can cause allergic contact dermatitis if any diisocyanate is present as impurity.

USE: Snap-back fiber, stronger and lighter than rubber. Used mainly for elastic garments such as belts, girdles, corsets, brassieres, garters, surgical stockings and sock tops. Can be used bare, giving lighter fabrics and eliminating the costly covering process needed for rubber.

8861. Sparassol. [520-43-4] 2-Hydroxy-4-methoxy-6-methylbenzoic acid methyl ester; 4-methoxy-2,6-cresotic acid methyl ester; everninic acid methyl ester; orsellinic acid methyl ester 4-methyl ether. $C_{10}H_{12}O_4$; mol wt 196.20. C 61.22%, H 6.17%, O 32.62%. Antibiotic substance produced by the fungus *Sparassis ramosa:* Falck, *Ber.* **56**, 2555 (1923). Also obtained in methanol extracts of the lichen *Evernia prunasti:* Stenhouse, *Ann.* **68**, 55 (1848); Späth, Jeschki, *Ber.* **57**, 471 (1924). Structure: Fischer, Hoesch, *Ann.* **391**, 347 (1912); Wedekind, Fleischer, *Ber.* **56**, 2556 (1923). Synthesis: G. Nicollier *et al., Helv. Chim. Acta* **61**, 2899 (1978).

Prisms from water, mp 67-68°. Slightly sol in hot water; freely sol in acetone, ether, chloroform; moderately sol in methanol, ethanol, petr ether. Alkaline hydrolysis yields free everninic acid, mp 170°.

8862. Sparfloxacin. [110871-86-8] *rel*-5-Amino-1-cyclopropyl-7-[(3R,5S)-3,5-dimethyl-1-piperazinyl]-6,8-difluoro-1,4-dihydro-4-oxo-3-quinolinecarboxylic acid; AT-4140; CI-978; PD-131501; Spara; Zagam. $C_{19}H_{22}F_2N_4O_3$; mol wt 392.41. C 58.16%, H 5.65%, F 9.68%, N 14.28%, O 12.23%. Fluorinated quinolone antibacterial. Prepn: J. Matsumoto *et al.,* **EP 221463**; *eidem,* **US 4795751** (1987, 1989 both to Dainippon). Synthesis and structure-activity relationship: T. Miyamoto *et al., J. Med. Chem.* **33**, 1645 (1990). Antibacterial activity: S. Nakamura *et al., Antimicrob. Agents Chemother.* **33**, 1167 (1989). Comparative *in vitro* antimicrobial spectrum: T. Kojima *et al., ibid.* 1988. Pharmacokinetics: S. Nakamura *et al., ibid.* **34**, 89 (1990). Subacute toxicity studies: M. Yasuba *et al., Chemotherapy* **39**, Suppl. 4, 180 (1991), *C.A.* **115**, 247561d (1991); M. Iida *et al., ibid.* 195, *C.A.* **115**, 247562e (1991).

Relative stereochemistry

Crystals from chloroform + ethanol, mp 266-269° (dec). pKa₁ 6.25, pKa₂ 9.30.

THERAP CAT: Antibacterial.

8863. Sparsomycin. [1404-64-4] (2E)-N-[(1S)-1-(Hydroxymethyl)-2-[(R)-[(methylthio)methyl]sulfinyl]ethyl]-3-(1,2,3,4-tetrahydro-6-methyl-2,4-dioxo-5-pyrimidinyl)-2-propenamide; (E)-(1S)-1,2,3,4-tetrahydro-N-[1-(hydroxymethyl)-2-[[(methylthio)methyl]-sulfinyl]ethyl]-6-methyl-2,4-dioxo-5-pyrimidineacrylamide; (+)-sparsomycin; NSC-59729; U-19183. $C_{13}H_{19}N_3O_5S_2$; mol wt 361.43. C 43.20%, H 5.30%, N 11.63% O 22.13%, S 17.74%. Protein synthesis inhibitor with antibiotic and antitumor activity. Isoln from fermentation broth of *Streptomyces sparsogenes* var *sparsogenes:* S. P. Owen *et al., Antimicrob. Agents Chemother.* **1962**, 772; **GB 974541** (1964 to Upjohn); from *S. cuspidosporus:* E. Higashide *et al., Takeda Kenkyusho Nempo* **25**, 1 (1966), *C.A.* **66**, 54238q (1967). Characterization and purification: A. D. Argoudelis, R. R. Herr, *Antimicrob. Agents Chemother.* **1962**, 780. Structural elucidation: P. F. Wiley, F. A. MacKellar, *J. Org. Chem.* **41**, 1858 (1976). Total synthesis: H. C. J. Ottenheijm *et al., ibid.* **46**, 3273 (1981); R. M. J. Liskamp *et al., ibid.* 5408. HPLC determn in plasma

and urine: B. Winograd *et al., J. Chromatogr.* **275**, 145 (1983). Pharmacokinetics and toxicology: Z. Zylicz *et al., Cancer Chemother. Pharmacol.* **20**, 115 (1987). Ribosomal binding studies: E. Lazaro *et al., Antimicrob. Agents Chemother.* **35**, 10 (1991). Biosynthetic studies: R. J. Parry *et al., J. Am. Chem. Soc.* **114**, 5946 (1992). Review of chemistry and biological activity: H. C. J. Ottenheijm *et al., Prog. Med. Chem.* **23**, 220-268 (1986).

Crystals, mp 208-209° (dec). $[\alpha]_D^{25}$ +69° (c = 0.5 in water). pKa' 8.67 in water, 9.05 in 40% ethanol. uv max: 302 nm (water or 0.1N aq sulfuric acid), 328 nm (0.1N aq potassium hydroxide). Slightly sol in water and lower alcohols. Insol in less polar organic solvents. LD₅₀ in dogs, rats, mice (mg/kg): 0.5-1.0, 2.25, 4.32 i.v. (Ottenheijm, 1986). LD₅₀ i.p. in mice: 2.4 mg/kg (Owen).

USE: Research tool for studying protein biosynthesis.

8864. Sparteine. [90-39-1] (7S,7aR,14S,14aS)-Dodecahydro-7,14-methano-2H,6H-dipyrido[1,2-a:1′,2′-e][1,5]diazocine; *l*-sparteine; lupinidine. $C_{15}H_{26}N_2$; mol wt 234.39. C 76.87%, H 11.18%, N 11.95%. In yellow and black lupin beans, *Lupinus luteus* L. and *L. niger* Hort.; also in *Cytisus scoparius* (L.) Link. and *Anagyris foetida* L., *Leguminosae.* Extraction procedure: Karrer *et al., Helv. Chim. Acta* **11**, 1062 (1928). Structure: Clemo *et al., J. Chem. Soc.* **1931**, 429; Ing, *ibid.* **1933**, 504; Schirm, Besendorf, *Arch. Pharm.* **280**, 64 (1942); Galinovsky, Stern, *Ber.* **77B**, 132 (1944); Clemo *et al., J. Chem. Soc.* **1949**, 663. Biosynthesis: Anet *et al., Nature* **165**, 35 (1950); Schöpf *et al., Angew. Chem.* **65**, 161 (1953); **69**, 69 (1957); van Tamelen, Foltz, *J. Am. Chem. Soc.* **82**, 2400 (1960). Absolute configuration: Okuda *et al., Chem. Ind. (London)* **1961**, 1116. Conformation: Bohlmann *et al., Tetrahedron Lett.* **1965**, 2705; Wiewiorowski *et al., Can. J. Chem.* **45**, 1447 (1967). Synthesis of racemate: Bohlmann *et al., Ber.* **106**, 3026 (1973); N. Takatsu *et al., Chem. Pharm. Bull.* **35** 4990 (1987). Chemistry: Binnig, *Arzneim.-Forsch.* **24**, 752 (1974). Studies of cardiovascular effects: Raschak, *ibid.* 753.

Viscous, oily liquid. bp₈ 173°. Volatile with steam. $[\alpha]_D^{21}$ −16.4° (c = 10 in abs alc). n_D^{20} 1.5312. d_4^{20} 1.020. pK₁ at 20°: 2.24; pK₂: 9.46; pH of 0.01 molar soln 11.6. One gram dissolves in 325 ml water. Freely sol in alcohol, chloroform or ether.

Sulfate pentahydrate. [6160-12-9] Depasan; Tocosamine. $C_{15}H_{26}N_2 \cdot H_2SO_4 \cdot 5H_2O$; mol wt 422.53. Columnar crystals, loses water of crystn at 100° turning brown, dec 136°. pH of 0.05 molar soln 3.3. One gram dissolves in 1.1 ml water, 3 ml alcohol. Practically insol in chloroform and ether.

l-α-**Isosparteine.** [446-95-7] Genisteine (alkaloid). 14aα-stereoisomer. Isoln from *Cytisus scoparius* (L.) Link, (*Spartium scoparium* L.), *Leguminosae:* Valeur, *Compt. Rend.* **167**, 23, 163 (1918); from *Lupinus caudatus* Kellog, *Leguminosae:* Marion *et al., Can. J. Chem.* **29**, 22 (1951). Identity as stereoisomer: Marion, Leonard, *ibid.* 297. Structure: Leonard *et al., J. Am. Chem. Soc.* **77**, 1552 (1955). Absolute configuration: Okuda, Tsuda, *Chem. Ind. (London)* **1961**, 1115. Monohydrate, needles from boiling acetone, mp 108-110°. $[\alpha]_D^{22}$ −51.6° (c = 0.7 in abs. ethanol).

THERAP CAT: In treatment of cardiac insufficiency.

8865. Spearmint. Green mint; curled mint. Hardy perennial herb, *Mentha spicata* L., *Labiatae.* Medicinal portions include the dried leaves and tops and essential oil. *Habit.* Mediterrean region; cultivated in U.S. and Europe. *Constit.* Volatile oil (0.8-2.5%), flavonoids incl. thymonin. Botanical description and medicinal uses: J. Gruenwald *et al.*, *PDR for Herbal Medicines* (Medical Economics, Montvale, 2nd Ed., 2000) pp 709-710. Extraction of volatile constituents and effect of drying method: M. C. Diaz-Maroto *et al.*, *J. Agric. Food Chem.* **51**, 1265 (2003).

Volatile oil. [8008-79-5] Oil of spearmint. Obtained by steam distillation of flowering tops. *Constit.* Primarily carvone (~50%), (−)-limonene, 1,8-cineole. Colorless, yellow or greenish-yellow liq; characteristic spearmint odor and taste. d_{25}^{25} 0.917-0.934. α_D^{20} −48 to −59°. n_D^{20} 1.484-1.491. Very slightly sol in water; sol in equal vol 80% alcohol. *Keep well closed, cool and protected from light.*

Spirit of spearmint. An alcoholic soln contg per liter 100 ml oil of spearmint and the alcohol-soluble principles from 10 g coarsely powdered spearmint leaves previously macerated with water.

USE: As flavor in foods, dental products, chewing gum. Pharmaceutic aid (flavor).

THERAP CAT: Carminative.

8866. Spectinomycin. [1695-77-8] (2*R*,4a*R*,5a*R*,6*S*,7*S*,8*R*,-9*S*,9a*R*,10a*S*)-Decahydro-4a,7,9-trihydroxy-2-methyl-6,8-bis(methylamino)-4*H*-pyrano[2,3-*b*][1,4]benzodioxin-4-one; actinospectacin; espectinomicina; CHX-3101; M-141. $C_{14}H_{24}N_2O_7$; mol wt 332.35. C 50.60%, H 7.28%, N 8.43%, O 33.70%. Antibiotic isolated from fermentation broth of *Streptomyces spectabilis*: D. J. Mason *et al.*, *Antibiot. Chemother.* **11**, 118 (1961); M. E. Bergy *et al.*, *ibid.* 661; M. E. Bergy, C. De Boer, *US 3234092* (1966 to Upjohn). Purification and crystallization: H. K. Jahnke, *US 3206360*; V. J. Peters, *US 3272706* (1965, 1966 both to Upjohn). Isoln and characterization: A. C. Sinclair, A. F. Winfield, *Antimicrob. Agents Chemother.* **1961**, 503. Structure: H. Hoeksema *et al.*, *J. Am. Chem. Soc.* **84**, 3212 (1962); P. F. Wiley *et al.*, *ibid.* **85**, 2652 (1963). Solubility data: J. R. Marsh, P. J. Weiss, *J. Assoc. Off. Anal. Chem.* **50**, 457 (1967). Stereochemistry and abs config: T. G. Cochran *et al.*, *Chem. Commun.* **1972**, 494. Biosynthesis: L. A. Mitscher *et al.*, *ibid.* **1971**, 1541; H. Otsuka *et al.*, *J. Am. Chem. Soc.* **102**, 6817 (1980). Enantioselective synthesis: D. R. White *et al.*, *Tetrahedron Lett.* **1979**, 2737; S. Hanessian, R. Roy, *Can. J. Chem.* **63**, 163 (1985). Mechanism of action study: M. F. Brink *et al.*, *Nucleic Acids Res.* **22**, 325 (1994). HPLC determn in animal plasma: N. Haagsma *et al.*, *J. Chromatogr.* **615**, 289 (1993). Clinical pharmacokinetics: J. G. Wagner *et al.*, *Int. Z. Klin. Pharmakol. Ther. Toxikol.* **1**, 261 (1968). Clinical evaluation in gonorrhea: Y. H. Kouri *et al.*, *Genitourin. Med.* **65**, 342 (1989); in chancroid: M. Guzmán *et al.*, *Sex. Transm. Dis.* **19**, 291 (1992). Co-treatment with lincomycin, *q.v.*, for footrot in sheep: C. M. Venning *et al.*, *Aust. Vet. J.* **67**, 258 (1990). *Reviews*: W. M. McCormack, M. Finland, *Ann. Intern. Med.* **84**, 712-716 (1976); W. J. Holloway, *Med. Clin. North Am.* **66**, 169-173 (1982).

Amorphous powder. $[\alpha]_D^{25}$ −20° (water). pKa$'_1$ 6.95, pKa$'_2$ 8.70. Sol in water, methanol, ethanol. Insol in acetone, hydrocarbon solvents.

Dihydrochloride pentahydrate. [22189-32-8] Spectam; Stanilo; Trobicin. $C_{14}H_{24}N_2O_7 \cdot 2HCl \cdot 5H_2O$; mol wt 495.34. Exists as a ketone hydrate and not in the carbonyl form (Cochran *et al.*). Colorless needles from aq acetone, mp 205-207° (dec). $[\alpha]_D$ +14.8° (c = 0.42 in water). Soly in water, methanol, propylene glycol, formamide, DMSO, 0.1*N* NaOH, 0.1*N* HCl: >20 mg/ml. Soly (mg/ml): chloroform 0.042; acetone 0.015; diethyl ether 0.010.

Sulfate tetrahydrate. [64058-48-6] $C_{14}H_{24}N_2O_7 \cdot H_2SO_4 \cdot 4H_2O$. White crystals from aq acetone, mp ~185° (dec). $[\alpha]_D^{25}$ +17.0° (c = 1 in water). pKa$'_1$ 7.00, pKa$'_2$ 8.75. Soly in water, DMSO: 225, 5-10 (mg/ml). Practically insol in pyridine, chloroform, ethanol, ethyl acetate, cyclohexane, benzene, acetone, DMF, dioxane, acetonitrile.

THERAP CAT: Antibacterial.

THERAP CAT (VET): Antibacterial.

8867. Spectrin. Tektin A. Major protein component of the membrane cytoskeleton of mammalian and avian erythroid cells. Responsible for maintaining the normal shape, strength and stability of the erythrocyte. A ubiquitous family of proteins, spectrin analogs have also been isolated from a variety of non-erythroid cells. Given the name spectrin because of its original isolation from hemoglobin-free red cell membranes known as erythrocyte "ghosts": V. T. Marchesi, E. Steers, *Science* **159**, 203 (1968). Comparison with actin, *q.v.*: E. Steers, V. Marchesi, *J. Gen. Physiol.* **54**, 65 S (1969). Isoln from human red cells and species comparison: T. W. Tillack *et al.*, *Biochim. Biophys. Acta* **200**, 125 (1970). Spectrin is a heterodimer composed of 2 high mol wt polypeptide subunits referred to as band 1 or α-spectrin (mol wt ~240 kDa) and band 2 or β-spectrin (mol wt ~220 kDa). Structural studies: M. Clarke, *Biochem. Biophys. Res. Commun.* **45**, 1063 (1971); T. L. Steck, *J. Cell Biol.* **62**, 1 (1974); J. M. Anderson, *J. Biol. Chem.* **254**, 939 (1979); D. W. Speicher *et al.*, *ibid.* **257**, 9093 (1982). The β-subunit of spectrin is bound to the red cell membrane by proteins known as **ankyrin**: V. Bennett, P. J. Stenbuck, *ibid.* **254**, 2533 (1979); or **syndeins**: J. Yu, S. R. Goodman, *Proc. Natl. Acad. Sci. USA* **76**, 2340 (1979). The spectrin heterodimers aggregate on the membrane surface and, together with actin and other proteins, form a filamentous network that covers the surface of the cytoplasmic membrane: S. E. Lux, *Nature* **281**, 426 (1979); *idem, Semin. Hematol.* **16**, 21 (1979). Identification of spectrin analogs from nonerythroid cells: S. R. Goodman *et al.*, *Proc. Natl. Acad. Sci. USA* **78**, 7580 (1981). Isoln of **fodrin**, also known as brain or neural cell spectrin: J. Levine, M. Willard, *J. Cell Biol.* **90**, 631 (1981); V. Bennett *et al.*, *Nature* **299**, 126 (1982). Isoln of *TW 260/240*, a spectrin analog from chicken intestinal epithelial cells: J. R. Glenney *et al.*, *Cell* **28**, 843 (1982). Comparison of erythroid and non-erythroid spectrins: J. R. Glenney *et al.*, *Proc. Natl. Acad. Sci. USA* **79**, 4002 (1982); J. R. Glenney, P. Glenney, *Eur. J. Biochem.* **144**, 529 (1984). Review of erythrocyte spectrin: S. R. Goodman, K. Shiffer, *Am. J. Physiol.* **244**, C121-C141 (1983); of neural cell spectrin: S. R. Goodman, I. S. Zagon, *ibid.* **250**, C347-C360 (1986). Comprehensive review: S. R. Goodman *et al.*, *Crit. Rev. Biochem.* **23**, 171-234 (1988).

8868. Spermaceti. Cetaceum; Spermwax. A waxy substance from the head of the sperm whale. *Constit.* Chiefly cetyl palmitate; free cetyl alcohol present in appreciable amounts; esters of lauric, stearic, and myristic acids; esters of higher alcohols also present.

White, somewhat translucent, slightly unctuous masses with crystalline fracture and pearly luster; almost odorless and tasteless but becomes yellow and rancid on long exposure to air. d 0.938-0.944; mp 42-50°; n_D^{80} ~1.4330. Sapon no. 120-136. Iodine no. 3-4.4. Insol in water or cold alcohol. Sol in chloroform, ether, carbon disulfide, oils, boiling alcohol; slightly sol in petr ether.

USE: As a base for ointments, cerates, etc., and as emulsion with egg yolk or expressed almond oil. In manuf of candles, soaps, cosmetics, laundry wax; finishing and lustering linens. Emollient.

8869. Spermidine. [124-20-9] *N*^{1}-(3-Aminopropyl)-1,4-butanediamine; *N*-(4-aminobutyl)-1,3-diaminopropane; *N*-(γ-aminopropyl)tetramethylenediamine; 1,8-diamino-4-azaoctane. $C_7H_{19}N_3$; mol wt 145.25. C 57.88%, H 13.19%, N 28.93%. Biogenic polyamine formed from putrescine; a precursor of spermine, *q.q.v.* First detected in human sperm, but occurs widely in nature. Essential for both normal and neoplastic tissue growth. Synthesis: H. W. Dudley *et al.*, *Biochem. J.* **20**, 1082 (1926); **21**, 97 (1927); J. V. Braun, W. Pinkernelle, *Ber.* **70**, 1234 (1937); M. Danzig, H. P. Schultz, *J. Am. Chem. Soc.* **74**, 1836 (1952). For reviews of the early literature, *see* M. Guggenheim, *Die biogenen Amine* (S. Karger, Basel, 4th ed., 1951) 619 pp; H. Tabor *et al.*, *Annu. Rev. Biochem.* **30**, 579-604 (1961). Role in cell growth processes: C. W. Tabor, H. Tabor, *ibid.* **45**, 285 (1976); J. Janne *et al.*, *Biochim. Biophys. Acta* **473**, 241 (1978); C. W. Porter, R. J. Bergeron, *Science* **219**, 1083 (1983). Formation and interconversion of spermidine and putrescine in mammalian cells: A. E. Pegg *et al.*, *Adv. Enzyme Regul.* **19**, 427 (1980). Effect on polypeptide chain elongation *in vitro*: A. K. Abraham, A. Pihl, *Eur. J. Biochem.* **106**, 257 (1980). Regulation of tRNA methyltransferase activity: M. Mach *et al.*, *Biochem. J.* **202**, 153 (1982). HPLC study: C. E. Prussak, D. H. Russell, *J. Chromatogr.* **229**, 47 (1982). Interaction with actin, *q.v.*: C. Oriol-Audit, *Biochem. Biophys. Res. Commun.* **105**, 1096 (1982). Studies on use as a biochemical tool in cancer research: A. Thyss *et al.*, *Eur. J.*

Cancer Clin. Oncol. **18**, 611 (1982); V. Quemener *et al., J. Nat. Prod.* **45**, 608 (1982); C. W. Porter *et al., Cancer Res.* **42**, 4072 (1982). Toxicity study: P. R. Langford *et al., J. Antibiot.* **35**, 1387 (1982). *Reviews:* T. C. Theoharides, *Life Sci.* **27**, 703-713 (1980); O. Heby, *Differentiation* **19**, 1-20 (1981); L. Stevens, *Med. Biol.* **59**, 308-313 (1981). Book: *Polyamines in Biology and Medicine,* D. R. Morris, L. J. Marton, Eds. (Dekker, New York, 1981) 512 pp.

Liq, bp_{14} 128-130°. *Corrosive.* Sol in water, ethanol, ether.
Trihydrochloride. [334-50-9] $C_7H_{19}N_3.3HCl$; mol wt 254.62. Cryst from ethanolic HCl (65:2). mp 256-258°.
USE: As a tool in biochemical research.

8870. Spermine. [71-44-3] N^1,N^4-Bis(3-aminopropyl)-1,4-butanediamine; N,N'-bis(3-aminopropyl)tetramethylenediamine; gerontine; musculamine; neuridine. $C_{10}H_{26}N_4$; mol wt 202.35. C 59.36%, H 12.95%, N 27.69%. Biogenic polyamine formed from spermidine, *q.v.,* and occurring in almost all tissues. Essential for both normal and neoplastic tissue growth. First observed in human semen and described as the cryst phosphate salt: A. von Leeuwenhoek, *Philos. Trans. R. Soc. London* **12**, 1040 (1678). Description of the history, occurrence, formation, and early prepns: *Beilstein* **vol. IV,** Suppl. 1, 704; M. Guggenheim, *Die biogenen Amine* (S. Karger, Basel, 4th ed., 1951) 619 pp; H. Tabor *et al., Annu. Rev. Biochem.* **30**, 579-604 (1961). Prepn of spermine and its tetrahydrochloride: Israel *et al., J. Med. Chem.* **7**, 710 (1964). Role in cell growth processes: C. W. Tabor, H. Tabor, *Annu. Rev. Biochem.* **45**, 285 (1976); J. Janne *et al., Biochim. Biophys. Acta* **473**, 241 (1978). Modulation of calcium-dependent immune processes: T. C. Theoharides, *Life Sci.* **27**, 703 (1980). Biosynthesis in fungi: L. Stevens, *Med. Biol.* **59**, 308 (1981). HPLC study: C. E. Prussak, D. H. Russell, *J. Chromatogr.* **229**, 47 (1982). Use as a biochemical marker for malignant tumors: Y. Horn *et al., Cancer Res.* **42**, 3248 (1982). Metabolic study: A. E. Pegg *et al., Biochemistry* **21**, 5082 (1982). Review of role in cell proliferation and differentiation: O. Heby, *Differentiation* **19**, 1-20 (1981). Book: *Polyamines in Biology and Medicine* D. R. Morris, L. J. Marton, Eds. (Dekker, New York, 1981) 512 pp.

Needles, mp 55-60°. Liq, $bp_{0.5}$ 141-142°. Strong base, absorbs carbon dioxide from air. *Corrosive. Keep well closed.* Sol in water, lower alcohols, chloroform. Practically insol in ether, benzene, petr ether.
Diphosphate hexahydrate. [58298-97-8] Spermine phosphate. $C_{10}H_{26}N_4.2H_3PO_4.6H_2O$; mol wt 506.42. Cryst from water, mp 230-234° (dec). Soly in water: 0.037% at 20°; 1% at 100°. Insol in alc, ether and other organic solvents. Sol in dilute acids and alkali. Known as ***Charcot-Neumann crystals,*** also found in spleen, blood, bone marrow in leukemia and secretions in asthma.
Tetrahydrochloride. [306-67-2] $C_{10}H_{26}N_4.4HCl$; mol wt 348.18. Cryst from ethanol, mp 312-314.5°.
USE: As a tool in biochemical research.

8871. Sperm Oil. From sperm whale. *Constit.* Esters of fatty acids; small quantities of spermaceti.
Yellow, thin liquid; slight fishy odor if not of good quality. d 0.875-0.884. Sapon no. 123-147. Iodine no. 80-84. Insoluble in water, cold alcohol, or petr ether; sol in chloroform, ether.
USE: As lubricant, in lamps, hardening steel, manuf soap.

8872. Sphingofungins. Family of six antifungal agents, sphingofungins A-F, isolated from *Aspergillus fumigatus* of which sphingofungin C is the most abundant. Specific inhibitors of serine palmitoyltransferase. Isoln and charaterization of sphingofungins A-D: F. VanMiddlesworth *et al., J. Antibiot.* **45**, 861 (1992); of sphingofungins E-F: W. S. Horn *et al., ibid.* 1692. Stereochemistry of A-D: F. VanMiddlesworth *et al., Tetrahedron Lett.* **33**, 297 (1992). Mechanism of action: M. M. Zweerink *et al., J. Biol. Chem.* **267**, 25032 (1992). Synthesis of D: K. Mori, K. Otaka, *Tetrahedron Lett.* **35**,

9207 (1994); and stereochemistry of E: B. Wang *et al., Synlett* **2001**, S1 904.

Sphingofungin C

Sphingofungin C. [121025-46-5] 5-(Acetyloxy)-2-amino-3,-4,14-trihydroxy-6-eicosenoic acid. $C_{22}H_{41}NO_7$; mol wt 431.57. Cream colored solid.

8873. Sphingomyelins. Important structural component of biological membranes and plasma lipoproteins. Occurs naturally as a mixture of N-acyl derivatives of a long chain base, predominantly sphingosine, coupled to phosphorylcholine, *q.q.v.* The acyl chain composition varies among tissues. Variants in the long chain base have also been identified. Discovery: J. L. W. Thudichum, *A Treatise on the Chemical Composition of the Brain* (London, 1884). Isoln from bovine brain and characterization: P. A. Levene, *J. Biol. Chem.* **18**, 453 (1914); S. J. Thannhauser, N. F. Boncoddo, *ibid.* **173**, 141 (1948). Structure: Y. Fujino, *J. Biochem. (Tokyo)* **39**, 45 (1952); G. Rouser *et al., J. Am. Chem. Soc.* **75**, 310 (1953); G. Marinetti *et al., ibid.* 313. Synthesis: D. Shapiro *et al., ibid.* **81**, 4360 (1959). Synthesis and configuration of naturally occuring D-*erythro*-form: D. Shapiro, H. M. Flowers, *ibid.* **84**, 1047 (1962). Review of stereospecific syntheses: A. L. Weis, *Chem. Phys. Lipids* **102**, 3-12 (1999). Analysis of molecular species in various tissues: A. A. Karlsson *et al., J. Mass Spectrom.* **33**, 1192 (1998); F.-F. Hsu, J. Turk, *J. Am. Soc. Mass Spectrom.* **11**, 437 (2000). Review of properties and behavior in biological membranes: Y. Barenholz, T. E. Thompson, *Biochim. Biophys. Acta* **604**, 129-158 (1980); of absorption and transport: A. Nilsson, R.-D. Duan, *J. Lipid Res.* **47**, 154-171 (2006); of metabolism at the plasma membrane: D. Milhas *et al., FEBS Lett.* **584**, 1887-1894 (2010).

N-Palmitoylsphingomyelin

N-Palmitoylsphingomyelin. [6254-89-3] (7S)-4-Hydroxy-7-[(1R,2E)-1-hydroxy-2-hexadecen-1-yl]-N,N,N-trimethyl-9-oxo-3,5-dioxa-8-aza-4-phosphatetracosan-1-aminium, inner salt, 4-oxide; N-palmitoyl-D-erythro-sphingosylphosphorylcholine; hexadecanoyl sphingomyelin; d-18:1/16:0-SM. $C_{39}H_{79}N_2O_6P$; mol wt 703.04. Slightly hygroscopic powder, mp 215-217°. $[\alpha]_D^{25}$ +6.1° (chloroform + methanol).
N-Stearoylsphingomyelin. [58909-84-5] (7S)-4-Hydroxy-7-[(1R,2E)-1-hydroxy-2-hexadecen-1-yl]-N,N,N-trimethyl-9-oxo-3,5-dioxa-8-aza-4-phosphahexacosan-1-aminium, inner salt, 4-oxide; N-stearoyl-D-erythro-sphingosylphosphorylcholine; octadecanoyl sphingomyelin; d-18:1/18:0-SM. $C_{41}H_{83}N_2O_6P$; mol wt 731.10. Slightly hygroscopic powder, mp 213-215°. $[\alpha]_D^{25}$ +6.1° (chloroform + methanol).

8874. Sphingosine. [123-78-4] (2S,3R,4E)-2-Amino-4-octadecene-1,3-diol; (E)-D-erythro-4-octadecene-1,3-diol; (−)-D-erythro-sphingosine; trans-4-sphingenine; $C_{18}H_{37}NO_2$; mol wt 299.50. C 72.19%, H 12.45%, N 4.68%, O 10.68%. Basic structural unit of ceramides, sphingomyelins, gangliosides, and other sphingolipids which are ubiquitous membrane components of eukaryotic cells. Participates in diverse cellular processes, inhibits protein kinase C and other signalling kinases, and induces apoptosis. First obtained by hydrolysis of cerebrosides from human brain: J. L. W. Thudichum, *Die Konstitution des Gehirns des Menschen und der Tiere* (Tübingen, 1901). Separation procedures: H. E. Carter *et al., J. Biol. Chem.* **170**, 269 (1947); J. B. Wittenberg, *ibid.* **216**, 379 (1955). Structure: H. E. Carter *et al., ibid.* **170**, 285 (1947). Configuration: H. E. Carter, D. Shapiro, *J. Am. Chem. Soc.* **75**, 5131 (1953); K.

Mislow, *ibid.* **74**, 5155 (1952); G. Marinetti, E. Stotz, *ibid.* **76**, 1347 (1954). Synthesis of DL-*erythro-trans*-form: D. Shapiro, K. Segal, *ibid.* **76**, 5894 (1954); C. A. Grob, F. Gadient, *Helv. Chim. Acta* **40**, 1145 (1957). Stereoselective synthesis: H. Newman, *J. Am. Chem. Soc.* **95**, 4098 (1973); B. Bernet, A. Vasella, *Tetrahedron Lett.* **24**, 5491 (1983). Scalable synthesis: H. Yang, L. S. Liebeskind, *Org. Lett.* **9**, 2993 (2007). Review of role in signal transduction: A. H. Merrill, Jr., V. L. Stevens, *Biochim. Biophys. Acta* **1010**, 131-139 (1989); of role in apoptosis: J. Woodcock, *IUBMB Life* **58**, 462-466 (2006). Review of syntheses: P. M. Koskinen, A. M. P. Koskinen, *Synthesis* **1998**, 1075-1091; of chemistry and biophysical properties: F. M. Goñi, A. Alonso, *Biochim. Biophys. Acta* **1758**, 1902-1921 (2006).

Crystals from ethyl acetate, mp 80-84°. $[\alpha]_D^{22}$ −3° (chloroform).

Triacetylsphingosine. [2482-37-3] *N*-[(1*S*,2*R*,3*E*)-2-(Acetyl-oxy)-1-[(acetyloxy)methyl]-3-heptadecen-1-yl]acetamide. $C_{24}H_{43}NO_5$; mol wt 425.61. Crystals, mp 101-102°. $[\alpha]_D^{25}$ −11.7° (chloroform).

DL-*erythro*-**Sphingosine.** [2733-29-1] Waxy crystals from ether + pentane, mp 67°.

DL-*erythro*-**Triacetylsphingosine.** [67113-24-0] $C_{24}H_{43}NO_5$. Crystals from pentane + ether, mp 91-92°.

8875. Spigelia. Pinkroot; Indian pink; Carolina pink; Maryland pink; wormgrass. Dried rhizome and roots of *Spigelia marilandica* L., *Loganiaceae*. *Habit.* North America (New Jersey to Florida and west to Wisconsin). *Constit.* Spigeline, resin, tannin, bitter principle, volatile oil.

THERAP CAT: Anthelmintic.

8876. α-Spinasterol. [481-18-5] (3β,5α,22*E*)-Stigmasta-7,22-dien-3-ol; α-spinasterin; hitodesterol. $C_{29}H_{48}O$; mol wt 412.70. C 84.40%, H 11.72%, O 3.88%. Stereoisomeric with chondrillasterol, *q.v.* Extracted from spinach leaves: Hart, Heyl, *J. Biol. Chem.* **95**, 311 (1932); from *Citrullus colocynthis* Schrad., *Cucurbitaceae*: Hamilton, Kermack, *J. Chem. Soc.* **1952**, 5051. Structure: Fieser *et al.*, *J. Am. Chem. Soc.* **71**, 2226 (1949). Prepn: Kircher, Rosenstein, *J. Org. Chem.* **38**, 2259 (1973); M. Anastasia *et al.*, *J. Chem. Soc. Perkin Trans. 1* **1981**, 2561.

Crystals from alcohol + light petr, mp 168-169°. $[\alpha]_D^{25}$ −3.6° (c = 2.8 in chloroform).

Benzoate. $C_{36}H_{52}O_2$. Plates from ethyl acetate, mp 200°. $[\alpha]_D^{25}$ +1.8° (c = 1.7 in chloroform).

8877. Spinosyns. [131929-60-7] Family of fermentation-derived, 12 membered macrocyclic lactones in a unique tetracyclic ring. At least 20 spinosyns have been isolated from the actinomycete, *Saccharopolyspora spinosa*; spinosyns A and D are the most abundant. Isolation and biological activity: L. D. Boeck *et al.*, **EP 375316** (1990 to Lilly); *eidem*, **US 5496931** (1996 to DowElanco); and structure determn: H. A. Kirst *et al.*, *Tetrahedron Lett.* **32**, 4839 (1991). Soil degradation: K. A. Hale, D. E. Portwood, *J. Environ. Sci. Health* **B31**, 477 (1996). Series of articles on determn methods in food and environmental matrices: *J. Agric. Food Chem.* **48**, 5131-

5153 (2000). Uptake and metabolism in larvae: T. C. Sparks *et al.*, *Proc. Beltwide Cotton Conf.* **2**, 1259 (1997). Mode of action study: V. L. Salgado *et al.*, *Pestic. Biochem. Physiol.* **60**, 103 (1998). Review of physical and biological properties: C. V. DeAmicis *et al.*, *ACS Symp. Ser.* **658**, 144-154 (1997). *Review:* G. D. Crouse, T. C. Sparks, *Rev. Toxicol.* **2**, 133-146 (1998).

Spinosyn A   R = H
Spinosyn D   R = CH_3

Spinosyn A. (2*R*,3a*S*,5a*R*,5b*S*,9*S*,13*S*,14*R*,16a*S*,16b*R*)-2-[(6-Deoxy-2,3,4-tri-*O*-methyl-α-L-mannopyranosyl)oxy]-13-[[(2*R*,5*S*,6*R*)-5-(dimethylamino)tetrahydro-6-methyl-2*H*-pyran-2-yl]oxy]-9-ethyl-2,3,3a,5a,5b,6,9,10,11,12,13,14,16a,16b-tetradecahydro-14-methyl-1*H-as*-indaceno[3,2-*d*]oxacyclododecin-7,15-dione; lepicidin A; A-83543A; LY-232105. $C_{41}H_{65}NO_{10}$; mol wt 731.97. Total synthesis: L. A. Paquette *et al.*, *J. Am. Chem. Soc.* **120**, 2553 (1998). White, odorless crystalline solid. mp 118°. pKa 8.1. uv max (methanol): 243 nm (ε 11000). $[\alpha]_D^{27}$ −262.7° (methanol). Vapor pressure: 2.4 × 10^{-10}. Soly in water (ppm): 290 (pH 5), 235 (pH 7), 16 (pH 9), distilled 20. Soly (w/v%): methanol 19, acetone 17, dichloromethane >50, hexane 0.45%. LD$_{50}$ in rats (mg/kg): 3783-5000 orally (Crouse).

Spinosyn D. [131929-63-0] A-83543D. $C_{42}H_{67}NO_{10}$; mol wt 746.00. Odorless, white crystalline solid. mp 169°. pKa 7.8. uv max (methanol): 243 nm (ε 11000). $[\alpha]_{436}^{27}$ −297.5° (methanol). Vapor pressure: 2.0 × 10^{-10}. Soly in water (ppm): 28 (pH 5), 0.329 (pH 7), 0.04 (pH 9), distilled 1.3. Soly (w/v%): methanol 0.25, acetone 1.0, dichloromethane 45, hexane 0.07.

Spinosad. [168316-95-8] XDE-105; DE-105; Comfortis; Conserve; Entrust; Justice; Natroba; Naturalyte; SpinTor; Success; Tracer. Mixture of spinosyns A and D. Effect on beneficial insects: D. Murray, R. Lloyd, *Australian Cottongrower* **18**, 62 (1997). Field trial as grain protectant: B. Subramanyam *et al.*, *Crop Prot.* **26**, 1021 (2007). Veterinary trial for treatment and control of fleas on dogs: D. E. Snyder *et al.*, *Vet. Parasitol.* **150**, 345 (2007). Clinical trial for treatment of head lice: D. Stough *et al.*, *Pediatrics* **124**, e389 (2009). Light grey to white crystals (tech). LD$_{50}$ in rats, mallard ducks, quail (mg/kg): >3600, >2000, >2000 orally (Crouse).

USE: Insecticide.

THERAP CAT (VET); Ectoparasiticide.

8878. Spiperone. [749-02-0] 8-[4-[4-(4-Fluorophenyl)-4-oxo-butyl]-1-phenyl-1,3,8-triazaspiro[4.5]decan-4-one; 8-[3-(*p*-fluoro-benzoyl)propyl]-1-phenyl-1,3,8-triazaspiro[4.5]decan-4-one; 4-phenyl-8-[3-(4-fluorobenzoyl)propyl]-1-oxo-2,4,8-triazaspiro[4.5]-decane; Spiropitan. $C_{23}H_{26}FN_3O_2$; mol wt 395.48. C 69.85%, H 6.63%, F 4.80%, N 10.63%, O 8.09%. Prepn: Janssen, **US 3155669**; **US 3155670** and **US 3161644** (all 1964 to Janssen).

Crystals, mp 190-193.6°.

THERAP CAT: Antipsychotic.

8879. **Spiramycin.** [8025-81-8] RP-5337; Selectomycin; Rovamicina; Rovamycin. Antibiotic substance classified in the erythromycin-carbomycin group and produced by *Streptomyces ambofaciens* from soil of northern France: Cosar *et al.*, *C. R. Seances Soc. Biol. Ses Fil.* **234**, 1498 (1952); Pinnert-Sindico *et al.*, *Antibiot. Annu.* **1954-1955**, 724; Ninet, Verrier, US 2943023 (1960 to Rhône-Poulenc), *see also* US 3000785 (1961 to Rhône-Poulenc). Antibacterial activity and toxicity: H. Sous *et al.*, *Arzneim.-Forsch.* **8**, 386 (1958). Separation into 3 components named spiramycin I, II and III: Preud'homme, Charpentier, US 2978380 and US 3011947 (1961 to Rhône-Poulenc). Structure: Kuehne, Benson, *J. Am. Chem. Soc.* **87**, 4660 (1965). Revised structure: Omura *et al.*, *ibid.* **91**, 3401 (1969); Mitscher *et al.*, *J. Antibiot.* **26**, 55 (1973). Revised configuration at C-9: Freiberg *et al.*, *J. Org. Chem.* **39**, 2474 (1974). Symposium on pharmacology, antibacterial spectrum, and clinical efficacy: *J. Antimicrob. Chemother.* **22**, Suppl. B, 1-213 (1988).

Spiramycin I   R = H
Spiramycin II   R = COCH$_3$
Spiramycin III R = COCH$_2$CH$_3$

Amorphous base, slightly sol in water. [α]$_D^{20}$ $-80°$ (methanol). uv max (ethanol): 231 nm. Sol in most organic solvents. Active on gram-positive bacteria and rickettsiae. Cross resistance between microorganisms resistant to erythromycin and carbomycin. LD$_{50}$ in rats (mg/kg): 9400 orally; 1000 s.c.; 170 i.v. (Sous).

Embonate. Spira 200.
Hexanedioate. Spiramycin adipate; Stomamycin; Suanovil.
Spiramycin I. [24916-50-5] Foromacidin A. C$_{43}$H$_{74}$N$_2$O$_{14}$; mol wt 843.07. Crystals, mp 134-137°. [α]$_D^{20}$ $-96°$.
Spiramycin I triacetate. Crystals, mp 140-142°. [α]$_D^{20}$ $-92.5°$.
Spiramycin II. [24916-51-6] Foromacidin B. C$_{45}$H$_{76}$N$_2$O$_{15}$; mol wt 885.10. Crystals, mp 130-133°. [α]$_D^{20}$ $-86°$.
Spiramycin II diacetate. Crystals from cyclohexane, mp 156-160°. [α]$_D^{20}$ $-98.4°$.
Spiramycin III. [24916-52-7] Foromacidin C. C$_{46}$H$_{78}$N$_2$O$_{15}$; mol wt 899.13. Crystals, mp 128-131°. [α]$_D^{20}$ $-83°$.
Spiramycin III diacetate. Crystals from cyclohexane, mp 140-142°. [α]$_D^{20}$ $-90.4°$.

THERAP CAT: Antibacterial.
THERAP CAT (VET): Antibacterial; growth promotant.

8880. **Spirapril.** [83647-97-6] (8S)-7-[(2S)-2-[[(1S)-1-(Ethoxycarbonyl)-3-phenylpropyl]amino]-1-oxopropyl]-1,4-dithia-7-azaspiro[4.4]nonane-8-carboxylic acid; (8S)-7[(S)-N-[(S)-1-carboxy-3-phenylpropyl]alanyl]-1,4-dithia-7-azaspiro[4.4]nonane-8-carboxylic acid 1-ethyl ester. C$_{22}$H$_{30}$N$_2$O$_5$S$_2$; mol wt 466.61. C 56.63%, H 6.48%, N 6.00%, O 17.14%, S 13.74%. Angiotensin-converting enzyme (ACE) inhibitor. Prepn: E. H. Gold *et al.*, US 4470972 (1984 to Schering); E. M. Smith *et al.*, *J. Med. Chem.* **32**, 1600 (1989). Pharmacology: E. J. Sybertz *et al.*, *Arch. Int. Pharmacodyn. Ther.* **286**, 216 (1987); T. Baum *et al.*, *ibid.* 230. Radioimmunoassay in plasma: M. Hossein-Nia *et al.*, *Ther. Drug Monit.* **14**, 234 (1992). Pharmacokinetics and evaluation in CHF: S. A. J. van den Broek *et al.*, *J. Cardiovasc. Pharmacol.* **18**, 614 (1991). Clinical evaluation in essential hypertension: G. P. Reams *et al.*, *J. Clin. Pharmacol.* **33**, 348 (1993). *Review:* P. Jerie, H.-J. Kremer, *Cor Vasa* **34**, 82-87 (1992).

Hemihydrate. White foam. [α]$_D^{26}$ $-29.5°$ (c = 0.2 in ethanol).
Hydrochloride. [94841-17-5] Sch-33844; TI-211-950; Renormax; Renpress; Sandopril. C$_{22}$H$_{30}$N$_2$O$_5$S$_2$.HCl; mol wt 503.07. White solid, mp 192-194° (dec). [α]$_D^{26}$ $-11.2°$ (c = 0.4 in ethanol).
Diacid. [83602-05-5] Spiraprilat; spiraprilic acid; Sch-33861. C$_{20}$H$_{26}$N$_2$O$_5$S$_2$; mol wt 438.56. Hemihydrate, mp 163-165°. [α]$_D^{26}$ $+4.1°$ (c = 0.4 in ethanol).

THERAP CAT: Antihypertensive.

8881. **Spirit of Ammonia, Aromatic.** [8013-59-0] Spirit of hartshorn, aromatic. A soln contg 34 g ammonium carbonate, 90 ml 10% ammonia water, 10 ml lemon oil, 1 ml oil of lavender, 1 ml oil of myristica, 700 ml alcohol and a sufficient quantity of water to make 1 liter. Absolute alc by vol, ~66%.
Almost colorless to slightly yellow liquid with an aromatic, pungent odor and the taste of ammonia. d 0.90. Forms an opalescent mixture with water; miscible with alcohol.

THERAP CAT: Reflex respiratory stimulant.
THERAP CAT (VET): By inhalation: respiratory and circulatory stimulant. Internally: expectorant, diaphoretic, antacid, carminative. Externally: counterirritant, and in diluted form to relieve the irritation of insect stings and bites.

8882. **Spirit of Ether.** [8013-43-2] Ethanol mixt. with 1,1'-oxybis[ethane]; Hoffmann's drops. A mixture of alcohol and ether. Made by mixing 325 ml ether U.S.P. with sufficient U.S.P. alcohol (~690 ml) to make 1000 ml. Absolute alc by vol, ~66%.
Colorless, volatile liquid. d 0.785.

THERAP CAT: Carminative.

8883. **Spirit of Ether Compound.** [8013-44-3] Hoffmann's anodyne. A mixture of 325 ml ether U.S.P., 25 ml ethereal oil and sufficient alcohol (~665 ml) to make 1000 ml. Absolute alc by vol, ~64%.
Colorless, clear liquid. d ~0.785.

THERAP CAT: Carminative.
THERAP CAT (VET): Has been used as a stomachic, carminative, in colic.

8884. **Spirodiclofen.** [148477-71-8] 2,2-Dimethylbutanoic acid 3-(2,4-dichlorophenyl)-2-oxo-1-oxaspiro[4.5]dec-3-en-4-yl ester; BAJ2740; Envidor. C$_{21}$H$_{24}$Cl$_2$O$_4$; mol wt 411.32. C 61.32%, H 5.88%, Cl 17.24%, O 15.56%. Tetronic acid acaricide for use in fruit, grapevines and nuts. Prepn: R. Fischer *et al.*, DE 4216814; *eidem*, US 5262383 (both 1993 to Bayer). Comprehensive description of properties and activity: U. Wachendorff *et al.*, *BCPC Conf. - Pests Dis.* **2000**, 53. Greenhouse trials vs spider mites: R. Nauen *et al.*, *ibid.* 453.

White powder, mp 101-108°. Log P (*n*-octanol/water): 5.8 (pH 4 at 20°). Vapor pressure (20°): 3 × 10^{-7} Pa. Soly in water (pH 4): 50 μg/l at 20°. LD$_{50}$ in rats (mg/kg): >2500 orally; >2000 dermally (24 hr). LC$_{50}$ (4 hr) in rats (mg/m^3): >5000 by inhalation; LC$_{50}$ (96 hr) in fish: >68 mg/l (Wachendorff).

USE: Acaricide.

8885. Spirogermanium. [41992-23-8] 8,8-Diethyl-*N*,*N*-dimethyl-2-aza-8-germaspiro[4.5]decane-2-propanamine; 2-[3-(dimethylamino)propyl]-8,8-diethyl-2-aza-8-germaspiro[4.5]decane. $C_{17}H_{36}GeN_2$; mol wt 341.12. C 59.86%, H 10.64%, Ge 21.29%, N 8.21%. Cytostatic germanium deriv. Prepn: L. M. Rice, **DE 2243550**; *idem*, **US 3825546** (1973, 1974 both to Geschicter Fund for Med. Res.); L. M. Rice *et al.*, *J. Heterocycl. Chem.* **11**, 1041 (1974). Toxicology study: M. C. Henry *et al.*, *Preclinical Toxicologic Evaluation* (PB-264117, 1977) 213 pp. Metabolism study: D. Garteiz *et al.*, *Drug Metab. Dispos.* **19**, 44 (1991). Clinical evaluation: P. S. Schein *et al.*, *Cancer Treat. Rep.* **64**, 1051 (1980); C. Tropie *et al.*, *ibid.* **65**, 119 (1981). Toxicity study: M. C. Henry *et al.*, *ibid.* **64**, 1207 (1980).

Oil, bp$_{0.03}$ 106-109°.
Dihydrochloride. [41992-22-7] NSC-192965; Spiro-32. $C_{17}H_{36}GeN_2 \cdot 2HCl$; mol wt 414.04. Crystals, mp 287-288°. LD_{50} in rats (mg/kg): 75 i.p.; in mice (mg/kg): 150 i.p., 324 orally (Rice *et al.*). LD_{50} in male, female mice (mg/kg): 44.5, 41.5 i.v.; 142.9, 119 i.m. (Henry, 1980).

THERAP CAT: Antineoplastic.

8886. Spiromesifen. [283594-90-1] 3,3-Dimethylbutanoic acid 2-oxo-3-(2,4,6-trimethylphenyl)-1-oxaspiro[4.4]non-3-en-4-yl ester; 3-(2,4,6-trimethylphenyl)-4-neopentylcarbonyloxy-5,5-tetramethylene-Δ^3-dihydrofuran-2-one; BSN-2060; Oberon. $C_{23}H_{30}O_4$; mol wt 370.49. C 74.56%, H 8.16%, O 17.27%. Tetronic acid acaricide and whitefly insecticide for use in cotton, vegetables and ornamentals. Prepn: R. Fischer *et al.*, **EP 528156**; *idem*, **US 5262383** (both 1993 to Bayer); *see also*: U. Wachendorff-Neumann, **US 6436988** (2002 to Bayer). Review of properties and activity: R. Nauen *et al.*, *BCPC Conf. - Pests Dis.* **2002**, 39-44. Series of articles on discovery, properties, ecological profile, and field trials: *Pflanzenschutz-Nachr. Bayer* **58**, 307-502 (2005).

White solid, mp 98°. d^{20} 1.13. Log P (*n*-octanol/water): 4.55 (pH 2 and 7.5). Vapor pressure (20°): 7×10^{-6} Pa. Soly in water: 0.13 mg/ml. LD_{50} in rats (mg/kg): >2500 orally; >2000 dermally (24 hr); LC_{50} (4 hr) in rats (mg/m³): >4873 by inhalation (Nauen).

USE: Insecticide.

8887. Spironolactone. [52-01-7] (7α,17α)-7-(Acetylthio)-17-hydroxy-3-oxopregn-4-ene-21-carboxylic acid γ-lactone; 17-hydroxy-7α-mercapto-3-oxo-17α-pregn-4-ene-21-carboxylic acid γ-lactone, acetate; 3-(3-oxo-7α-acetylthio-17β-hydroxy-4-androsten-17α-yl)propionic acid γ-lactone; SC-9420; Aldactone; Aquareduct; Practon; Osyrol; Sincomen; Spirobeta; Spiroctan; Spirolone; Spironone; Verospiron; Xenalon. $C_{24}H_{32}O_4S$; mol wt 416.58. C 69.20%, H 7.74%, O 15.36%, S 7.70%. Aldosterone antagonist. Prepn: Cella, Tweit, *J. Org. Chem.* **24**, 1109 (1959); **US 3013012** (1961 to Searle); Tweit *et al.*, *J. Org. Chem.* **27**, 3325 (1962). Activity and metabolic studies: Gerhards, Engelhardt, *Arzneim.-Forsch.* **13**, 972 (1963). Crystal and molecular structure: Dideberg, Dupont, *Acta Crystallogr.* **B28**, 3014 (1972). Comprehensive description: J. L. Sutter, E. P. K. Lau, *Anal. Profiles Drug Subs.* **4**, 431-451 (1975). Review of carcinogenetic risk: *IARC Monographs* **24**, 259-273 (1980). Review of antiandrogen effects and clinical use in hirsutism: R. R. Tremblay, *Clin. Endocrinol. Metab.* **15**, 363-371 (1986); of clinical efficacy in hypertension: A. N. Brest, *Clin. Ther.* **8**, 568-585 (1986). Review of pharmacology: H. A. Skluth, J. G. Gums, *DICP*

Ann. Pharmacother. **24**, 52-59 (1990). Clinical trial in congestive heart failure: B. Pitt *et al.*, *N. Engl. J. Med.* **341**, 709 (1999).

Crystals from methanol, mp 134-135° (resolidifies and dec 201-202°). $[\alpha]_D^{20}$ −33.5° (chloroform). uv max: 238 nm (ε 20200). Freely sol in benzene, chloroform; sol in alc, ethyl acetate; slightly sol in methanol, fixed oils. Practically insol in water. LD_{50} in rats, mice, rabbits (mg/kg): 790, 360, 870 i.p. (IARC, 1980).

THERAP CAT: Diuretic.
THERAP CAT (VET): Diuretic.

8888. Spirotetramat. [203313-25-1] Carbonic acid *cis*-3-(2,5-dimethylphenyl)-8-methoxy-2-oxo-1-azaspiro[4.5]dec-3-en-4-yl ethyl ester; *cis*-3-(2,5-dimethylphenyl)-8-methoxy-2-oxo-1-azaspiro[4.5]dec-3-en-4-yl ethyl carbonate; *cis*-4-(ethoxycarbonyloxy)-8-methoxy-3-(2,5-xylyl)-1-azaspiro[4.5]dec-3-en-2-one; BYI-8330; Movento. $C_{21}H_{27}NO_5$; mol wt 373.45. C 67.54%, H 7.29%, N 3.75%, O 21.42%. Insecticidal tetramic acid derivative for use in food crops. Transformed by the plant to the bioactive enol derivative which inhibits lipid biosynthesis. Prepn: F. Lieb *et al.*, **WO 9805638**; *idem*, **US 6114374** (1998, 2000 both to Bayer); R. Fischer, H.-C. Weiss, *Bayer CropSci. J.* **61**, 127 (2008). Properties and activity: X. van Waetermeulen *et al.*, *Proc. 16th Int. Plant Prot. Congr.* **Vol. 1**, 60 (2007). Determn of residues in water and photodegradation study: S. Sathiyanaranyanan *et al.*, *Open Catal. J.* **2**, 24 (2009). Insecticidal profile and field performance: E. Brück *et al.*, *Crop Prot.* **28**, 838 (2009).

Relative stereochemistry

Light beige powder, mp 142°. d$_4^{20}$ 1.22. pKa 10.7. Log P (octanol/water): 2.5. Vapor pressure (25°): 1.5×10^{-8} Pa. Soly at 20° (g/l): DCM >600; DMSO 200-300; acetone 100-120; ethyl acetate 67; toluene 60; ethanol 44; *n*-hexane 0.055. Soly in water at 20° (mg/l): 29.9 (pH 7); 33.5 (pH 4). LD_{50} in rats (mg/kg): >2000 orally; >2000 dermally (24 hr). LC_{50} in rats (mg/m³): >4381 by inhalation (van Waetermeulen).

USE: Insecticide.

8889. Spiroxamine. [118134-30-8] 8-(1,1-Dimethylethyl)-*N*-ethyl-*N*-propyl-1,4-dioxaspiro[4.5]decane-2-methamine; KWG-4168; Impulse. $C_{18}H_{35}NO_2$; mol wt 297.48. C 72.68%, H 11.86%, N 4.71%, O 10.76%. Spiroketalamine fungicide for use on cereal crops; inhibits ergosterol biosynthesis. Prepn: W. Kraemer *et al.*, **DE 3735555**; *idem*, **US 4851405** (1988, 1989 both to Bayer). Mode of action and field trials: S. Dutzmann *et al.*, *Brighton Crop Prot. Conf. - Pests Dis.* **1996**, 47. Series of articles on chemistry, activity, determn and ecobiology: *Pflanzenschutz-Nachr.* **50**, 5-98 (1997).

Colorless liquid, mixture of stereoisomers. bp 120° at 0.067 hPa. n_D^{20} 1.4662. Log P (*n*-octanol/water): <3. pKa: 6.9. Soly in acetone, acetonitrile, dichlormethane, *n*-hexane, 2-propanol, toluene, water (pH 3): all >200 g/l at 20°. LD_{50} in rats (mg/kg): ~595 orally; >1600 dermally. LC_{50} in rats (mg/m³): ~2772 by inhalation; LC_{50} (96 hr) in rainbow trout: 18.5 mg/l (Dutzmann).

USE: Agricultural fungicide.

8890. Splendipherin. [149152-95-4] Caerin 2.3. First reported aquatic sex pheromone from frogs or toads; specific for the magnificent tree frog, *Litoria splendida*. 25 L-amino acid peptide produced by males to attract females. Isoln from skin secretions of *Litoria caerula* as one of the hypotensive caerin peptides: D. J. M. Stone *et al.*, *J. Chem. Res. Synop.* **1993**, 138; from *L. splendida* and identification of pheromonal activity: P. A. Wabnitz *et al.*, *Nature* **401**, 444 (1999). Seasonal distribution and specificity of effect: *eidem, Eur. J. Biochem.* **267**, 269 (2000).

```
                          1
       Gly–Leu–Val–Ser–Ser–Ile–Gly–Lys–Ala–Leu–Gly
                                                  |
   25                                            Gly
   Ala–Pro–Gln–Gly–Lys–Ser–Lys–Val–Val–Asp–Ala–Leu–Leu
```

8891. Splenin. [1416-60-0] Thymopoietin III; SP. Polypeptide hormone of the spleen which induces differentiation and maturation of both T- and B-cell precursors. Originally detected by its cross reactivity with thymopoietin, *q.v.*, in radioimmunoassay. Human splenin (hSP) contains 48 amino acids; bovine splenin (bSP) contains 49. *Splenopentin*, a pentapeptide corresponding to residues 32-36, exhibits full bioactivity. The amino acid sequence of splenin is highly homologous with that of thymopoietin; the differences in bioactivity are attributed to a single amino acid substitution at position 34. Unlike the thymic hormone, splenin exhibits no neuromuscular effects. Isoln from bovine spleen and complete amino acid sequence: T. Audhya *et al.*, *Biochemistry* **20**, 6195 (1981). Synthesis: T. Abiko *et al.*, *Chem. Pharm. Bull.* **34**, 2133 (1986). Isoln and sequence of human splenin: T. Audhya *et al.*, *Proc. Natl. Acad. Sci. USA* **84**, 3545 (1987). Biological activities of bSP and comparison with thymopoietin: T. Audhya *et al.*, *ibid.* **81**, 2847 (1984). Effect on antibody production *in vivo*: W. Diezel *et al.*, *Biomed. Biochim. Acta* **45**, 1349 (1986). Effect of hSP on B-cell number *in vitro*: T. Abiko, H. Sekino, *Chem. Pharm. Bull.* **37**, 391 (1989).

```
               Arg–Lys–Ala–Val–Tyr
```

Human Splenopentin

Note: The term, splenin, formerly referred to a crude mixture of bioactive substances extracted from mammalian spleen: G. Ungar, *Endocrinology* **37**, 329 (1945).

8892. Sporidesmins. Toxic fungal mixture from *Pithomyces chartarum* Ellis (*Sporidesmium bakeri* Sydow), composed of sporidesmins A-H and J, as causative agent of "facial eczema" in sheep. First isoln of A (major metabolite): R. L. M. Synge, E. P. White, *Chem. Ind. (London)* **1959**, 1546. Isoln of A and B: J. W. Ronaldson *et al.*, *J. Chem. Soc. C* **1963**, 3172. Structure of A: J. Fridrichsons, A. M. Mathieson, *Tetrahedron Lett.* **1962**, 1265. Abs config of A: A. F. Beecham *et al.*, *ibid.* **1966**, 3131. Isoln and structure of D and F: W. D. Jamieson *et al.*, *J. Chem. Soc. C* **1969**, 1564; of E: R. Rahman *et al.*, *ibid.* 1665; of G: E. Francis *et al.*, *J. Chem. Soc. Perkin Trans. 1* **1972**, 470; of H and J: R. Rahman *et al.*, *ibid.* **1978**, 1476. Total synthesis of A: Y. Kishi *et al.*, *J. Am. Chem. Soc.* **95**, 6493 (1973).

Sporidesmin A

Sporidesmin A. [1456-55-9] $(3\alpha,5a\alpha,10b\alpha,11\beta,11a\alpha)$-9-Chloro-2,3,5a,6,10b,11-hexahydro-10b,11-dihydroxy-7,8-dimethoxy-2,-3,6-trimethyl-3,11a-epidithio-11a*H*-pyrazino[1',2':1,5]pyrrolo[2,3-*b*]indole-1,4-dione; sporidesmin. $C_{18}H_{20}ClN_3O_6S_2$; mol wt 473.94. Colorless needles with a faint green sheen, from aq methanol, mp 179° (dependent on rate of heating). $[\alpha]_D^{20}$ −45° (c = 0.98 in methanol). uv max: 219, 253, 305 nm ($E_{1cm}^{1\%}$ 700, 220, 40). Very slightly sol in water, light petroleum, CCl_4. Readily sol in most organic solvents.

8893. Sporidesmolides. Cyclodepsipeptides containing both D and L amino acids, isolated from the pasture fungus *Sporidesmium bakeri* Syd. (*Pithomyces chartarum* (Berk. & Curt.) Ellis): D. W. Russell, *Biochim. Biophys. Acta* **45**, 411 (1960); W. S. Bertaud *et al.*, *J. Gen. Microbiol.* **32**, 385 (1963); E. Bishop, D. W. Russell, *Biochem. J.* **92**, 19P (1964).

```
       ┌─────────────────────────────────────┐
       Hiv–Val–MeLeu–Hiv–D-Val–D-Leu
              Sporidesmolide I

         Hiv = α-hydroxyisovaline
```

Sporidesmolide I. [2900-38-1] $C_{33}H_{58}N_4O_8$; mol wt 638.85. Structure: D. W. Russell, *J. Chem. Soc.* **1962**, 753. Synthesis: M. M. Shemyakin *et al.*, *Tetrahedron* **19**, 995 (1963). Biosynthetic study: D. W. Russell, *Biochim. Biophys. Acta* **261**, 469 (1972). MS determn: B. C. Das *et al.*, *J. Antibiot.* **32**, 569 (1979). Needles from 70% acetic acid, mp 261-263°. $[\alpha]_D^{17}$ −217° (c = 1.5 in chloroform). Practically insol in water; very sol in chloroform; sparingly sol in other common organic solvents.

Sporidesmolide II. [3200-75-7] 2-D-Alloisoleucine sporidesmolide I. $C_{34}H_{60}N_4O_8$; mol wt 652.87. Structure and synthesis: M. M. Shemyakin *et al.*, *Tetrahedron Lett.* **1963**, 1927. Crystals, mp 228-230°. $[\alpha]_D^{20}$ −195° (c = 0.6 in chloroform).

Sporidesmolide III. [1803-67-4] 6-L-Leucine sporidesmolide I. $C_{32}H_{56}N_4O_8$; mol wt 624.82. Synthesis: Y. A. Ovchinnikov *et al.*, *ibid.* **1965**, 1111. Crystals, mp 277-278°. $[\alpha]_D$ −80° (c = 1.6 in acetic acid).

Sporidesmolide IV. [10252-34-3] 4-(4-Methyl-L-2-hydroxypentanoic acid) sporidesmolide I. $C_{34}H_{60}N_4O_8$; mol wt 652.87. Structure and synthesis: A. A. Kiryushkin *et al.*, *ibid.* **1965**, 143; E. Bishop, D. W. Russell, *J. Chem. Soc. C* **1967**, 634. Crystals, mp 227-228°. $[\alpha]_D$ −195° (c = 1 in chloroform). Practically insol in water, sol in chloroform, moderately sol in common organic solvents.

8894. Squalamine. [148717-90-2] $(3\beta,5\alpha,7\alpha,24R)$-3-[[3-[(4-Aminobutyl)amino]propyl]amino]-cholestane-7,24-diol 24-(hydrogen sulfate); 3β-(*N*-[3-aminopropyl]-1,4-butanediamine)-7α,24ζ-dihydroxy-5α-cholestane 24-sulfate; 3β-*N*-1-[*N*-[3-(4-aminobutyl)]-1,3-diaminopropane]-7α,24ζ-dihydroxy-5α-cholestane 24-sulfate. $C_{34}H_{67}N_3O_5S$; mol wt 629.99. C 64.82%, H 10.72%, N 6.67%, O 12.70%, S 5.09%. Broad-spectrum aminosterol antibiotic present in shark tissues; novel indication for steroid as vertebrate host-defense agent. Isoln from the stomach of the spiny dogfish shark, *Squalus acanthias*: M. Zasloff *et al.*, *US 5192756* (1993 to Children's Hospital of Penn.). Structural determn and antimicrobial activity: K. S. Moore *et al.*, *Proc. Natl. Acad. Sci. USA* **90**, 1354 (1993). NMR spectral studies: S. L. Wehrli *et al.*, *Steroids* **58**, 370 (1993). Synthesis of trihydrochloride: R. M. Moriarty *et al.*, **WO 9419366** (1994 to Magainin Pharm.).

Sol in water.

8895. Squalane. [111-01-3] 2,6,10,15,19,23-Hexamethyltetracosane; perhydrosqualene; dodecahydrosqualene; spinacane; Cosbiol; Robane; Wax-O-Sol. $C_{30}H_{62}$; mol wt 422.83. C 85.22%, H 14.78%. Prepn by complete hydrogenation of squalene, *q.v.*: Chapman, *J. Chem. Soc.* **123**, 770 (1923); Heilbron *et al.*, *ibid.* **1926**, 3135. Commercial grades are obtained by direct hydrogenation of shark liver oil and may contain some batyl alcohol: Tsujimoto, *Chem. Umschau Fette* **34**, 256 (1927), *C.A.* **21**, 4081 (1927). Synthesis: J. W. Scott, D. Valentine, *Org. Prep. Proced. Int.* **12**, 7 (1980); T. Mandai *et al.*, *Tetrahedron Lett.* **22**, 763 (1981).

Oil. Stable to air and oxygen. d_4^{15} 0.8115. mp ~−38°. bp_{760} ~350°; bp_{10} 263°; bp_5 248°. Flash pt 425°F (218°C). n_D^{15} 1.4530. Specific heat at 20° ~0.62. Viscosity (Engler) at 20° ~6.08. Miscible with ether, chloroform. Readily sol in gasoline, petr ether, benzene, oils; slightly sol in methanol, ethanol, acetone, glacial acetic acid; very slightly sol in absolute alc. Insol in water. Concd H_2SO_4 at 70° is discolored, but the squalane remains unchanged.

USE: Lubricant, transformer oil. Ingredient of watch and chronometer oils. Perfume fixative. In pharmacy and cosmetics as skin lubricant, ingredient of suppositories, carrier of lipid-soluble drugs.

THERAP CAT (VET): Cerumenolytic.

8896. Squalene. [111-02-4] (2*E*,6*E*,10*E*,14*E*,18*E*)-2,6,10,-15,19,23-Hexamethyl-2,6,10,14,18,22-tetracosahexaene; Spinacene; Supraene. $C_{30}H_{50}$; mol wt 410.73. C 87.73%, H 12.27%. All *trans* isoprenoid contg six isoprene units. Found in large quantities in shark liver oil. Occurs in smaller amounts (0.1 to 0.7%) in olive oil, wheat germ oil, rice bran oil, and yeast. Intermediate in biosynthesis of cholesterol. Isoln: Heilbron *et al.*, *J. Chem. Soc.* **1926**, 1630. Structure: *eidem, ibid.* **1929**, 873; Heilbron, Thompson, *ibid.* 883; Karrer *et al.*, *Helv. Chim. Acta* **13**, 1084 (1930); Karrer, Helfenstein, *ibid.* **14**, 78 (1931); Karrer *et al.*, *ibid.* 435. Crystal structure: J. Ernst *et al.*, *Angew. Chem. Int. Ed.* **15**, 778 (1976). Synthesis: Trippett, *Chem. Ind. (London)* **1956**, 80; Dicker, Whiting, *J. Chem. Soc.* **1958**, 1994; Cornforth *et al.*, *ibid.* **1959**, 2539; Johnson *et al.*, *J. Am. Chem. Soc.* **92**, 741 (1970); Hirai *et al.*, *Tetrahedron Lett.* **1971**, 4359; P. A. Grieco, Y. Masaki, *J. Org. Chem.* **39**, 2135 (1974). Synthesis of squalene and *trans-cis-trans-trans* isomer: Biellmann, Ducep, *Tetrahedron* **27**, 5861 (1971).

Oil. Faint, agreeable odor. Absorbs oxygen and becomes viscous like linseed oil. d_4^{20} 0.8584; bp_{25} 285°; bp_2 240°; $bp_{0.15}$ 203°. Crystals from ether/methanol, mp −5°. n_D^{20} 1.4965. Viscosity at 25°: 12 cP. Iodine no. 360-380. Practically insol in water. Freely sol in ether, petr ether, CCl_4, acetone, other fat solvents; sparingly sol in alc, glacial acetic acid.

USE: Bactericide; intermediate in manuf of pharmaceuticals, organic coloring materials, rubber chemicals, aromatics and surface active agents.

8897. Squaric Acid. [2892-51-5] 3,4-Dihydroxy-3-cyclobutene-1,2-dione; 1,2-dihydroxycyclobutenedione; diketocyclobutenediol. $C_4H_2O_4$; mol wt 114.06. C 42.12%, H 1.77%, O 56.11%. Dibasic acid; two-dimensional molecular antiferroelectric material. Prepn: S. Cohen *et al.*, *J. Am. Chem. Soc.* **81**, 3480 (1959); J. D. Park *et al.*, *ibid.* **84**, 2919 (1962). Aqueous dissociation study: L. M. Schwartz, L. O. Howard, *J. Phys. Chem.* **74**, 4374 (1970). Structure determn studies: D. Semmingsen, *Acta Chem. Scand.* **27**, 3961 (1973); Y. Wang *et al.*, *J. Chem. Soc. Perkin Trans. 2* **1974**, 35. Review of syntheses and reactivity: G. Maahs, P. Hegenberg, *Angew. Chem. Int. Ed.* **5**, 888-893 (1966); of coordination chemistry: L. A. Hall, D. J. Williams, *Adv. Inorg. Chem.* **52**, 249-291 (2001).

Crystals from water, dec 293°. d 1.90. uv max (H_2O): 269.5 nm (ε 37000). Soly: 7% boiling water; 2% room temp water. Insol in acetone, ether. pK_1 ~0.6 (25°). pK_2 3.480 ±0.023 (25°).

Dibutylester. [2892-62-8] 3,4-Dibutoxy-3-cyclobutene-1,2-dione; dibutyl squarate. $C_{12}H_{18}O_4$; mol wt 226.27. Prepn: G. Maahs, *Ann.* **686**, 55 (1965). Clinical evaluation in alopecia: G. Micali *et al.*, *Int. J. Dermatol.* **35**, 52 (1996); in treatment of warts: *idem et al.*, *Pediatr. Dermatol.* **17**, 315 (2000). $bp_{0.5}$ 138-139°. n_D^{20} 1.4943.

USE: Squaric acid and its derivatives are used in synthesis of pharmaceutical intermediates, squarylium dyes, and photoconducting squaraines.

THERAP CAT: Dibutylester as contact sensitizer for treatment of alopecia and warts.

8898. Squill. Sea onion; Bulbus Scillae; Meerzwiebel. The fleshy inner bulb scales of the white variety of *Urginea maritima* (L.) Baker (*Scilla maritima* L.), Liliaceae. *Habit.* Lands of the Mediterranean seacoast. *Constit.* Glucoscillaren A (scillarenin + rhamnose + glucose + glucose); scillaren A (scillarenin + rhamnose + glucose); proscillaridin A (scillarenin + rhamnose); scillaridin A; scilliglaucoside; scillipheoside; glucoscillipheoside; scillicyanoside; scillicoeloside; scilliazuroside; scillicryptoside. *Review:* G. Baumgarten, W. Förster, *Die Herzwirksamen Glykoside* (Thieme, Leipzig, 1963) pp 70-75 *et passim*.

USE: Rodenticide.

THERAP CAT: Diuretic, emetic, expectorant, cardiotonic.

THERAP CAT (VET): Has been used as expectorant, emetic.

8899. Stachydrine. [471-87-4] (2*S*)-2-Carboxy-1,1-dimethylpyrrolidinium inner salt; methyl hygrate betaine; hygric acid methylbetaine. $C_7H_{13}NO_2$; mol wt 143.19. C 58.72%, H 9.15%, N 9.78%, O 22.35%. Occurs widely in nature, especially in alfalfa, chrysanthemum, citrus and stachys species. May be prepd by methylation of proline. Isoln: Planta, Schulze, *Ber.* **26**, 939 (1893); Jahns, *ibid.* **29**, 2065 (1896); Yoshimura, *Z. Physiol. Chem.* **88**, 334 (1913); Vickery, *J. Biol. Chem.* **61**, 117 (1924). Structure: Schulze, Trier, *Z. Physiol. Chem.* **67**, 59 (1910). Synthesis: Karrer, Widmer, *Helv. Chim. Acta* **8**, 364 (1925). Biosynthesis: Robertson, Marion, *Can. J. Chem.* **38**, 396 (1960). *Review:* Marion in *The Alkaloids* vol. 1, R. H. F. Manske, H. L. Holmes, Eds. (Academic Press, New York, 1950) pp 101-103.

Monohydrate. Deliquescent crystals, sweetish taste, mp 235° when anhydr. Isomerizes at the mp to methyl hygrate. Sol in water, alcohol, dil acids. Practically insol in ether, chloroform.

Hydrochloride. $C_7H_{13}NO_2$.HCl. Large prisms from abs alcohol, dec 235°, very sol in water, sol in 13 parts alcohol.

Acid oxalate. $C_7H_{13}NO_2$.$C_2H_2O_4$. Needles, mp 106°. Practically insol in abs alcohol.

Aurichloride. $C_7H_{13}NO_2$.HAuCl$_4$. Yellow leaflets, mp 225° (rapid heating). Practically insol in cold water. Quite sol in hot water.

Platinichloride tetrahydrate. $(C_7H_{13}NO_2)_2$.H_2PtCl_6.$4H_2O$. Orange crystals, dec 210-220° (rapid heating), very sol in water and dil alc. Also obtained with 2 mols H_2O of crystn.

8900. Stannic Bromide. [7789-67-5] Tetrabromo stannane; tin tetrabromide; tin(IV) bromide. Br_4Sn; mol wt 438.33. Br 72.92%, Sn 27.08%. $SnBr_4$.

White, cryst mass; fumes strongly on exposure to air. d 3.34; mp 31°; bp 202°. Very sol in water with evolution of heat; also sol in alcohol. *Keep tightly closed.*

USE: In metallurgical separation of minerals.

8901. Stannic Chloride. [7646-78-8] Tetrachlorostannane; tin tetrachloride; tin(IV) chloride; fuming spirit of Libavius. Cl_4Sn; mol wt 260.51. Cl 54.43%, Sn 45.57%. $SnCl_4$. Improperly called *"tin bichloride"*.

Fuming, caustic liquid. *Corrosive.* d 2.26; mp −33°; bp 114°. Sol in water and evolution of much heat; sol in alcohol, carbon tetrachloride, benzene, toluene, acetone, kerosene, gasoline. *Keep tightly closed.*

Pentahydrate. White or slightly yellow crystals or fused small lumps; slight HCl odor. *Corrosive.* Very sol in H_2O; sol in alc.
Caution: May be highly irritating to eyes, mucous membranes.
USE: As mordant; reviving colors; stabilizer for colors and perfumes in soap; in dyeing of fabrics, weighting silk, tinning vessels; dehydrating agent in organic syntheses; in ceramics to produce abrasion-resistant or light-reflecting coatings.

8902. Stannic Chromate(VI). [38455-77-5] Chromic acid (H_2CrO_4) tin(2+) salt (2:1). Cr_2O_8Sn; mol wt 350.69. Cr 29.65%, O 36.50%, Sn 33.85%. $Sn(CrO_4)_2$.
Brownish-yellow, cryst powder; dec when heated. Sol in water.
Caution: Chromium hexavalent (VI) compounds are listed as known human carcinogens: *Report on Carcinogens, Twelfth Edition* (PB2011-111646, 2011) p 106.
USE: Decorating porcelain and china in rose and violet colors.

8903. Stannic Fluoride. [7783-62-2] Tetrafluorostannane; tin tetrafluoride; tin(IV) fluoride. F_4Sn; mol wt 194.70. F 39.03%, Sn 60.97%. SnF_4. Lewis acid. Prepd from $SnCl_4$ and HF: Ruff, Plato, *Ber.* **37**, 673 (1904); or from stannous fluoride and chlorine or bromine: Forbes, Anderson, *J. Am. Chem. Soc.* **67**, 1911 (1945); from stannous oxide or sulfide and fluorine: Haendler *et al., J. Am. Chem. Soc.* **76**, 2179 (1954). *Review:* Kemmitt, Sharp, *Adv. Fluorine Chem.* **4**, 185-186 (1965).
Snow-white, tetragonal crystals. Very hygroscopic. d_4^{19} 4.78. Sublimes at 705°. Hydrolyzes readily, but is more resistant to water than stannic chloride. Forms complexes with donor molecules.
USE: Friedel-Crafts catalyst.

8904. Stannic Iodide. [7790-47-8] Tetraiodostannane; tin tetraiodide; tin(IV) iodide. I_4Sn; mol wt 626.33. I 81.05%, Sn 18.95%. SnI_4. Toxicity data: Kolmer, *J. Pharmacol. Exp. Ther.* **43**, 515 (1931).
Yellow to reddish crystals. d 4.46. mp ~143°; sublimes at about 180°. bp 340°. Dec by water; sol in alcohol, benzene, chloroform, ether, carbon disulfide. MLD in rats (mg/kg): 200 i.v. (Kolmer).

8905. Stannic Oxide. [18282-10-5] Tin oxide (SnO_2); white tin oxide; tin dioxide; tin(IV) oxide; stannic anhydride; flowers of tin. O_2Sn; mol wt 150.71. O 21.23%, Sn 78.77%. SnO_2. Occurs in nature as the mineral *cassiterite*. The commercial grade is also known as *polishing powder, putty powder*, or *tin ash*. Review of toxicology and human exposure: *Toxicological Profile for Tin* (PB2006-100006, 2005) 426 pp.
White or slightly gray powder. d 6.95. Insol in water, alcohol, or cold acids. Slowly sol in hot concd potassium or sodium hydroxide soln.
Caution: Potential symptoms of overexposure are stannosis (benign pneumoconiosis); dyspnea, decreased pulmonary function. *See NIOSH Pocket Guide to Chemical Hazards* (DHHS/NIOSH 97-140, 1997) p 308.
USE: Polishing glass and metals; manuf milk-colored, ruby and alabaster glass, enamels, pottery, putty; mordant in printing and dyeing fabrics; in fingernail polishes.

8906. Stannic Selenide. [20770-09-6] Tin selenide ($SnSe_2$); tin diselenide. Se_2Sn; mol wt 276.63. Se 57.09%, Sn 42.91%. $SnSe_2$. Prepd by passing the vapor of selenium over heated tin: Little, *On Selenium and Some of the Metallic Selenides*, Göttingen (Thesis, 1859); by treating a soln of stannic chloride with hydrogen selenide: Berzelius, cited in *Mellor's* vol. **X**, 785 (1930); by treating a soln of an alkali selenostannate or sulfoselenostannate with hydrochloric acid: Ditte, *Compt. Rend.* **95**, 641 (1882).
Red-brown crystals. d 5.133 (Little); d 4.85. *See:* Schneider, *Pogg. Ann.* **127**, 624 (1866). mp 650°. Soluble in alkali, concd sulfuric acid, aqua regia, aq ammonia. Insol in water, dilute acids. Dec in nitric acid. Forms potassium selenostannate with potassium selenide; sodium selenostannate with sodium selenide.

8907. Stannic Sulfide. [1315-01-1] Tin sulfide (SnS_2); tin disulfide; tin(IV) sulfide; mosaic gold; tin bronze. S_2Sn; mol wt 182.83. S 35.07%, Sn 64.93%. SnS_2.
Golden leaflets with metallic luster; fatty feel to the touch. d 4.5. Insol in water or dil acids. Sol in aqua regia, in solns of alkali hydroxides or sulfides.

Note: The term "mosaic gold" is also used to designate an alloy consisting of 65.3% copper and 34.7% zinc.
USE: Gilding and bronzing metals, gypsum, wood and paper, usually by suspending in lacquer or varnish.

8908. Stannous Acetate. [638-39-1] Acetic acid tin(2+) salt (2:1); tin(II) acetate. $C_4H_6O_4Sn$; mol wt 236.80. C 20.29%, H 2.55%, O 27.03%, Sn 50.13%. $Sn(C_2H_3O_2)_2$. Prepd by refluxing granulated tin with 98% acetic acid: Colonna, *Gazz. Chim. Ital.* **35 II**, 224 (1905); by refluxing SnO with 50% (v/v) acetic acid under nitrogen: Donaldson *et al., J. Chem. Soc.* **1964**, 5942.
White, orthorhombic crystals; dec by water. mp 182.5-183°. d 2.31. Sol in dil HCl. *Keep well closed.*
USE: Reducing agent.

8909. Stannous Bromide. [10031-24-0] Tin bromide ($SnBr_2$); tin dibromide; tin(II) bromide. Br_2Sn; mol wt 278.52. Br 57.38%, Sn 42.62%. $SnBr_2$.
Yellowish powder; oxidizes in air. d 5.12; mp 215°; bp 623°. Sol in little water, gradually dec by much water; sol in alcohol, ether, acetone. *Keep tightly closed and protected from light.*

8910. Stannous Chloride. [7772-99-8] Tin chloride ($SnCl_2$); tin dichloride; tin protochloride; Stannochlor. Cl_2Sn; mol wt 189.61. Cl 37.39%, Sn 62.61%. $SnCl_2$. Prepn: Stephen, *J. Chem. Soc.* **1930**, 2786; Williams, *Org. Synth.* coll. vol. **III**, 627 (1955). Thermal properties: W. Fischer, R. Gewehr, *Z. Anorg. Allg. Chem.* **242**, 188 (1939). Metabolism and toxicity studies: M. Marciniak, *Acta Physiol. Pol.* **32**, 193 (1981); P. P. Singh, A. Y. Junnarkar, *Indian J. Pharmacol.* **23**, 153 (1991). Review of toxicology and human exposure: *Toxicological Profile for Tin* (PB2006-100006, 2005) 426 pp.
Orthorhombic cryst mass or flakes; fatty appearance. mp 247°; bp 652°; d 3.95. Dissolves in dilute hydrochloric acid. Freely sol in water, alc; sol in acetone, ether, methyl acetate, methyl ethyl ketone, isobutyl alcohol. Practically insol in mineral spirits, petr naphtha, xylene. LD_{50} in mice, rats (mg/kg): 1710.0, 2000.0 orally; 271.0, 316.0 i.p.; 34.8, 43.0 i.v. (Singh, Junnarkar).
Dihydrate. [10025-69-1] $Cl_2Sn \cdot 2H_2O$; mol wt 225.64. Crystals; absorbs oxygen from air and forms insol oxychloride. d 2.71. mp 37-38° when rapidly heated; dec on strong heating. Sol in less than its own wt of water; with much water it forms an insol basic salt; very sol in dil or concd hydrochloric acid; also sol in alc, ethyl acetate, glacial acetic acid, sodium hydroxide soln. *Keep tightly closed, in a cool place.*
USE: Powerful reducing agent, particularly in manuf of dyes and ^{99m}Tc radiopharmaceuticals; in tinning by galvanic methods; in liquor finishing of wire; in sensitizing of glass and plastics before metallizing; as soldering flux; as mordant in dyeing with cochineal; in manufacture of tin chemicals, color pigments, pharmaceuticals, sensitized paper, lubricating oil additives; as tanning agent; in removing ink stains; in yeast revivers; as reagent in analytical chemistry; as catalyst in organic reactions.

8911. Stannous Fluoride. [7783-47-3] Tin fluoride (SnF_2); tin difluoride; tin(II) fluoride; Fluoristan. F_2Sn; mol wt 156.71. F 24.25%, Sn 75.75%. SnF_2. Prepd by evaporating a soln of stannous oxide in hydrofluoric acid in the absence of oxygen: Gay-Lussac, Thénard, *Mém. Phys. Chim.* **2**, 317 (1809); Nebergall *et al., J. Am. Chem. Soc.* **74**, 1604 (1952); from tin and hydrogen fluoride: Muetterties, *Inorg. Chem.* **1**, 342 (1962). *Review:* Kemmitt, Sharp, *Adv. Fluorine Chem.* **4**, 186 (1965). Review of toxicology and human exposure: *Toxicological Profile for Tin* (PB2006-100006, 2005) 426 pp.
Monoclinic, lamellar plates. mp 213°. d^{25} 4.57. Freely sol in water. Practically insol in alc, ether, chloroform. Forms an oxyfluoride, $SnOF_2$, on exposure to air.
USE: Dental caries prophylactic.

8912. Stannous Iodide. [10294-70-9] Tin iodide (SnI_2); tin diiodide; tin(II) iodide. I_2Sn; mol wt 372.52. I 68.13%, Sn 31.87%. SnI_2. Preparation and crystal structure: Moser, Trevena, *Chem. Commun.* **1969**, 25.
Red, cryst powder or needles. d 5.28; mp 320°; bp 720° with decompn. Slightly sol in and dec by water; sol in solns of alkali chlorides or iodides, in benzene, chloroform, carbon disulfide. *Keep tightly closed.*

Consult the Name Index before using this section. **Page 1625**

8913. Stannous Oxalate. [814-94-8] Ethanedioic acid tin(2+) salt (1:1); oxalic acid tin(2+) salt (1:1); tin(II) oxalate. C_2O_4Sn; mol wt 206.73. C 11.62%, O 30.96%, Sn 57.42%. SnC_2O_4.

White, heavy powder. d 3.56. Insol in water. Sol in dil hydrochloric acid.

USE: Dyeing and printing textiles.

8914. Stannous Oxide. [21651-19-4] Tin oxide (SnO); tin monoxide; tin protoxide; tin(II) oxide. OSn; mol wt 134.71. O 11.88%, Sn 88.12%. SnO. Prepn of high purity SnO: Kwestroo, Vromans, *J. Inorg. Nucl. Chem.* **29**, 2187 (1967). Prepn of metastable, red, orthorhombic form: Donaldson *et al., J. Chem. Soc.* **1961**, 839.

Brownish-black powder; burns to SnO_2 on heating in air. d 6.45. Insol in water or alcohol. Sol in acids, in concd sodium or potassium hydroxide solns.

Caution: Potential symptoms of overexposure are stannosis (benign pneumoconiosis); dyspnea, decreased pulmonary function. *See NIOSH Pocket Guide to Chemical Hazards* (DHHS/NIOSH 97-140, 1997) p 308.

USE: Reducing agent; prepn of stannous salts.

8915. Stannous Pyrophosphate. [15578-26-4] Diphosphoric acid tin(2+) salt (1:2); ditin diphosphate; ditin pyrophosphate. $O_7-P_2Sn_2$; mol wt 411.36. O 27.23%, P 15.06%, Sn 57.72%. $Sn_2P_2O_7$. Prepd by thermal dehydration of stannous hydrogen phosphate: Jablczynski, Wieckowski, *Z. Anorg. Allg. Chem.* **152**, 207 (1926); from $Na_2H_2P_2O_7$ and $SnCl_2$: Klement, Haselbeck, *Ber.* **96**, 1022 (1958); from pyrophosphoric acid and stannous chloride: Nelson, US 3401012 (1968 to Monsanto). Clinical application for labelling red blood cells: D. Ducasson *et al., Br. J. Radiol.* **49**, 344 (1976). Chemical characterization of technetium complexes: J. Kroesbergen *et al., Nucl. Med. Biol.* **15**, 209 (1988). Diagnostic application in bone scanning: N. S. Anderton *et al., Am. J. Roentgenol. Radium Ther. Nucl. Med.* **124**, 625 (1975); in myocardial infarction: D. E. Jansen *et al., Circulation* **75**, 611 (1987).

Amorphous powder. $d^{16.4}$ 4.009. Insol in water. Sol in concd acid, excess alkali.

Technetium Tc 99m pyrophosphate. ^{99m}Tc-PPi; ^{99m}Tc-PYP; Phosphotec; TechneScan PYP. Sterile, aqueous solution of pyrophosphate labeled with ^{99m}Tc. Prepd from sodium pyrophosphate, stannous chloride or fluoride, and sodium pertechnetate Tc 99m prior to use.

USE: Ingredient in caries-preventing toothpaste: Norris, Schweizer, US 2946725 (1960 to Proctor & Gamble).

THERAP CAT: Technetium Tc 99m pyrophosphate as diagnostic aid (radioactive imaging agent).

8916. Stannous Selenide. [1315-06-6] Tin selenide (SnSe); tin monoselenide; tin(II) selenide. SeSn; mol wt 197.67. Se 39.95%, Sn 60.05%. SnSe. Prepd by direct fusing of the elements: Ditte, *Compt. Rend.* **95**, 641 (1882); **97**, 44 (1883); Berzelius cited in *Mellor's* **vol. X**, 784 (1930); by adding powdered selenium to molten, anhydr stannous chloride: Schneider, *Pogg. Ann.* **127**, 624 (1866).

Steel-gray prisms. d^0 6.18; mp 861°. Insol in water. Sol in aqua regia and alkali sulfides and selenides.

8917. Stannous Sulfate. [7488-55-3] Sulfuric acid tin(2+) salt (1:1); tin(II) sulfate. O_4SSn; mol wt 214.77. O 29.80%, S 14.93%, Sn 55.27%. $SnSO_4$. Prepn: Donaldson, Moser, *J. Chem. Soc.* **1960**, 4000.

Snow-white, orthorhombic crystals; dec at 378° to SnO_2 and SO_2. Sol in water, the soln soon decomposing with pptn of a basic sulfate; sol in dil H_2SO_4. *Keep well closed.*

USE: In tin plating; prepn of stannous salts.

8918. Stannous Sulfide. [1314-95-0] Tin sulfide (SnS); tin monosulfide; tin protosulfide; tin(II) sulfide. SSn; mol wt 150.77. S 21.26%, Sn 78.74%. SnS.

Dark-gray crystals or black, amorphous powder. d 5.08. Insol in water or alkali hydroxide or alkali sulfide soln. Sol in concd HCl, hot concd H_2SO_4.

USE: Polymerization catalyst.

8919. Stannous Tartrate. [815-85-0] (2R,3R)-2,3-Dihydroxybutanedioic acid tin(2+) salt (1:1); tartaric acid tin(2+) salt

(1:1); tin(2+) tartrate. $C_4H_4O_6Sn$; mol wt 266.78. C 18.01%, H 1.51%, O 35.98%, Sn 44.50%. $SnC_4H_4O_6$.

White powder. Sol in water, dil HCl. *Keep well closed.*

USE: Dyeing and printing textiles.

8920. Stanolone. [521-18-6] (5α,17β)-17-Hydroxyandrostan-3-one; 17β-hydroxy-3-androstanone; 3-oxo-17β-hydroxyandrostane; androstan-17β-ol-3-one; 4-dihydrotestosterone; androstanolone; Anabolex; Andractim; Androlone. $C_{19}H_{30}O_2$; mol wt 290.45. C 78.57%, H 10.41%, O 11.02%. Prepd by hydrogenation of testosterone: Butenandt *et al., Ber.* **68**, 2097 (1935); from dehydroepiandrosterone: Ruzicka, Kagi, *Helv. Chim. Acta* **20**, 1557 (1937); Ruzicka *et al., ibid.* **24**, 1151 (1941); from 3,17-androstandione: Oliveto, Hershburg, US 2927921 (1960 to Schering).

Crystals from ethyl acetate + hexane. Sublimes$_{0.01}$ 135°. mp 181°. $[\alpha]_D^{20}$ +32.4° (alcohol). Infrared absorption data: *J. Am. Chem. Soc.* **75**, 903 (1953). Sol in acetone, ether, alcohol, ethyl acetate. Practically insol in water.

Note: This is a controlled substance (anabolic steroid): **21 CFR,** 1308.13, as defined in 1300.01.

THERAP CAT: Androgen.

8921. Stanozolol. [10418-03-8] (2'H form); [302-96-5] (1'H form). (5α,17β)-17-Methyl-2'H-androst-2-eno[3,2-c]pyrazol-17-ol; 1,2,3,3a,3b,4,5,5a,6,8,10,10a,10b,11,12,12a-hexadecahydro-1,10a,-12a-trimethylcyclopenta[7,8]phenanthro[2,3-c]pyrazol-1-ol; 17β-hydroxy-17α-methylandrostano[3,2-c]pyrazole; androstanazole; stanazol; NSC-43193; Win-14833; Stromba; Strombaject; Winstrol. $C_{21}H_{32}N_2O$; mol wt 328.50. C 76.78%, H 9.82%, N 8.53%, O 4.87%. Prepn: Clinton *et al., J. Am. Chem. Soc.* **81**, 1513 (1959); **83**, 1478 (1961); Manson, US 3030358 (1962 to Sterling Drug).

Crystals from alc, mp 229.8-242.0°. $[\alpha]_D$ +35.7° (chloroform); $[\alpha]_D$ +48.6° (methanol). uv max: 223 nm (ε 4740). Sol in DMF; sparingly sol in alc, chloroform; slightly sol in ethyl acetate, acetone; very slightly sol in benzene. Insol in water.

Note: This is a controlled substance (anabolic steroid): **21 CFR,** 1308.13, as defined in 1300.01.

THERAP CAT: Androgen.

THERAP CAT (VET): Anabolic steroid.

8922. Staphisagria. Stavesacre. Ripe seed of *Delphinium staphisagria* L., *Ranunculaceae. Habit.* Mediterranean basin; cultivated in France, Italy. *Constit.* Delphinine, delphinoidine, delphisine, staphisagrine, staphisagroine, malic acid, fixed oil.

THERAP CAT: Parasiticide (external).

8923. Staphylokinase. [9040-61-3] Staphylokinase (enzyme-activating). Fibrin-specific plasminogen activator lacking enzymatic activity; produced by certain strains of *Staphylococcus aureus.* Consists of 136 amino acids in a single polypeptide chain without disulfide bridges; several molecular forms have been purified, mol wt 16,500-18,000. Forms a stoichiometric complex with the proenzyme plasminogen, *q.v.,* that becomes biologically active after the conversion of plasminogen to the proteolytic enzyme plasmin, *q.v.* Profibrinolytic properties: C. H. Lack, *Nature* **161**, 559 (1948); E.

B. Gerheim, *et al.*, *Proc. Soc. Exp. Biol. Med.* **68**, 246 (1948). Cloning and expression in *Escherichia coli:* T. Sako *et al.*, *Mol. Gen. Genet.* **190**, 271 (1983); in *Bacillus subtilis:* D. Behnke, D. Gerlach, *ibid.* **210**, 528 (1987). High yield production and purification of recombinant form: B. Schlott *et al.*, *Biotechnology* **12**, 185 (1994). Series of articles on isoln, characterization, structure, biochemistry and pharmacology of natural (*STAN*) and recombinant (*STAR*) forms: *Fibrinolysis* **6**, 203-242 (1992). Determn in plasma by bioimmunoassay: H. R. Lijnen *et al.*, *Thromb. Haemostasis* **70**, 491 (1993); by ELISA: A. Mike *et al.*, *Biol. Pharm. Bull.* **17**, 564 (1994). Clinical evaluation in peripheral arterial occlusion: S. Vanderschueren *et al.*, *Circulation* **92**, 2050 (1995). Comparative clinical trial vs alteplase, *q.v.: idem et al.*, *ibid.* 2044. Series of articles on recombinant variants with altered immunoreactivity: D. Collen *et al.*, *ibid.* **94**, 197-216 (1996); **95**, 455-472 (1997). Reviews: D. Collen, H. R. Lijnen, *Blood* **84**, 680-686 (1994); H. R. Lijnen, D. Collen, *Fibrinolysis* **10**, 119-126 (1996).

THERAP CAT: Thrombolytic.

8924. Star Anise. Chinese anise. Fruit of *Illicium verum* Hook. f., *Magnoliaceae*. *Habit.* Southeastern Asia and subtropical countries; commercial supply chiefly from China. *Constit.* About 5% volatile and fixed oils; anisic acid, tannin, resin, pectin. *Note:* Japanese star anise is *Illicium anisatum* L. (*I. religiosum* Sieb. & Zucc.; *I. japonicum* Sieb.) and contains a toxic lactone called anisatin. Chinese star anise does not contain this toxic principle. Shikimic acid has been found in both.

USE: Manufacture of liqueurs and the volatile oil. The fruit as source of oil of anise.

THERAP CAT: Hemostatic.

8925. Starch. Amylum. $(C_6H_{10}O_5)_n$. Carbohydrate polymer stored by plants; analogous to storage of fats by animals. Occurs as discrete granules in the mature grain of corn, *Zea mays* Linné, *Gramineae* or of wheat, *Triticum aestivum* Linné, *Gramineae* or tubers of potato, *Solanum tuberosum* Linné, *Solanaceae* or rice, *Oryza sativa* Linné, *Gramineae*. Starches are mixtures of two polymers: **amylose**, a linear $(1 \rightarrow 4)$-α-D-glucan and **amylopectin**, a branched D-glucan with mostly α-D-$(1 \rightarrow 4)$ and approx 4% α-D-$(1 \rightarrow 6)$ linkages. The starch in corn contains approx 27% amylose and 73% amylopectin, with these two polymers so associated in the crystal lattice that they are practically insol in cold water or alcohol. *Refs:* J. N. BeMiller, "Starch Amylose" in *Industrial Gums*, R. L. Whistler, Ed. (Academic Press, New York, 2nd ed., 1973) pp 545-566; E. L. Powell, "Starch Amylopectin", *ibid.* pp 567-576.

Irregular white masses or fine powder. Insol in cold water, alc. Although hydrolysis will not take place in cold water, and starch is comparatively resistant to naturally occurring enzymes, the reaction may be brought about by the use of acids or enzymes (α-amylase, β-amylase, amyloglucosidase). The hydrolysis reaction follows a different path depending on whether acids or enzymes are used. While acid hydrolysis produces a mixture of saccharides, the enzymes give more specific products. β-Amylase, for example, breaks off mostly maltose units, and amyloglucosidase yields mainly D-glucose. Chemistry and technology: R. L. Whistler, E. F. Paschall, Eds., *Starch: Chemistry and Technology* 2 vols. (Academic Press, New York, 1965); J. A. Radley, Ed., *Starch and Its Derivatives* (Chapman & Hall, London, 4th ed., 1968). Comprehensive description: A. W. Newman *et al.*, *Anal. Profiles Drug Subs. Excip.* **24**, 523-577 (1996).

Caution: Potential symptoms of overexposure are irritation of eyes, skin, mucous membranes; cough, chest pains; dermatitis; rhinorrhea. *See NIOSH Pocket Guide to Chemical Hazards* (DHHS/NIOSH 97-140, 1997) p 284.

USE: Starching and sizing fabrics, etc.; paste; as indicator in iodometric analyses. In the food industry. Pharmaceutic aid (tablet disintegrant, filler, binder); dusting powder. Dietetic grades of corn starch are marketed as **Maizena**; **Mondamin**.

THERAP CAT: Antidote (iodine poisoning).

THERAP CAT (VET): Internally: demulcent, mild astringent, in diarrhea, as an antidote for iodine poisoning. Externally: absorbent, emollient, in dusting powders and in ointments.

8926. Starch, Soluble. [9005-84-9] Amylodextrin; amylogen. Prepd by treating potato or corn starch with dilute hydrochloric acid.

White, odorless, tasteless powder. Readily soluble in hot water; forms transparent mobile liquid.

USE: For determination of diastatic power of malt, etc.; as indicator in iodometric analyses.

8927. Statine. [49642-07-1] (3*S*,4*S*)-4-Amino-3-hydroxy-6-methylheptanoic acid; AHMHA. $C_8H_{17}NO_3$; mol wt 175.23. C 54.84%, H 9.78%, N 7.99%, O 27.39%. Amino acid present in pepstatin, *q.v.* Synthesis: H. Morishima *et al.*, *J. Antibiot.* **26**, 115 (1973). Abs config and stereospecific synthesis of all four isomers: M. Kinoshita *et al.*, *ibid.* 249. Crystal structure: H. Nakamura *et al.*, *ibid.* 255. Biosynthesis: H. Morishima *et al.*, *ibid.* **27**, 267 (1974). Alternate syntheses: M. Kinoshita *et al.*, *Bull. Chem. Soc. Jpn.* **48**, 570 (1975); W.-S. Liu, G. I. Glover, *J. Org. Chem.* **43**, 754 (1978); D. H. Rich *et al.*, *ibid.* 3624; K. E. Rittle *et al.*, *ibid.* **47**, 3016 (1982). Distribution in rats: D. A. Grant *et al.*, *Biochem. Pharmacol.* **31**, 2302 (1982).

mp 201-203° (dec). $[\alpha]_D^{15} -20°$ (c = 0.64 in water).

8928. Staurosporine. [62996-74-1] (9*S*,10*R*,11*R*,13*R*)- 2,3,-10,11,12,13-Hexahydro-10-methoxy-9-methyl-11-(methylamino)-9,13-epoxy-1*H*,9*H*-diindolo[1,2,3-*gh*:3',2',1'-*lm*]pyrrolo[3,4-*j*][1,7]benzodiazonin-1-one; AM-2282; CGP-39360. $C_{28}H_{26}N_4O_3$; mol wt 466.54. C 72.09%, H 5.62%, N 12.01%, O 10.29%. Protein kinase C inhibitor; alkaloid isolated from *Streptomyces staurosporeus*. Isoln: S. Omura *et al.*, *J. Antibiot.* **30**, 275 (1977). Crystal and molecular structure: A. Furusaki *et al.*, *J. Chem. Soc. Chem. Commun.* **1978**, 800; *eidem*, *Bull. Chem. Soc. Jpn.* **55**, 3681 (1982). Corrected stereochemistry: N. Funato *et al.*, *Tetrahedron Lett.* **35**, 1251 (1994). Total synthesis: J. T. Link *et al.*, *J. Am. Chem. Soc.* **117**, 552 (1995); *idem et al.*, *ibid.* **118**, 2825 (1996). Biosynthetic studies: D. Meksuriyen, G. A. Cordell, *J. Nat. Prod.* **51**, 884, 893 (1988); S.-W. Yang *et al.*, *ibid.* **62** 1551 (1999). HPLC determn in blood and pharmacokinetics in rats: L. R. Gurley *et al.*, *J. Chromatogr. B* **712**, 211 (1998). Inhibition of protein kinase C: T. Tamaoki *et al.*, *Biochem. Biophys. Res. Commun.* **135**, 397 (1986); of other protein kinases: U. T. Rüegg, G. M. Burgess, *Trends Pharmacol. Sci.* **10**, 218 (1989). Induction of apoptosis: E. Falcieri *et al.*, *Biochem. Biophys. Res. Commun.* **193**, 19 (1993); R. Bertrand *et al.*, *Exp. Cell Res.* **211**, 314 (1994); of tyrosine phosphorylation: D. Rasouly, P. Lazarovici, *Eur. J. Pharmacol.* **269**, 255 (1994).

Pale yellow needles from chloroform-methanol as the methanol solvate, mp 270° (dec) (Omura). Also reported as yellow crystals from methanol, mp 288-291° (Meksuriyen, Cordell). $[\alpha]_D^{25} +35.0°$ (c = 1 in methanol); $[\alpha]_D^{22} +56.1°$ (c = 0.14 in methanol). uv max (methanol): 241.0, 266.0, 292.5, 321.5, 335.0, 355.0, 372.5 nm (log ε 4.25, 4.26, 4.53, 3.88, 3.96, 3.81, 3.85). Sol in DMSO, DMF. Slightly sol in chloroform, methanol.

Hydrochloride. $C_{28}H_{26}N_4O_3 \cdot HCl$; mol wt 503.00. LD_{50} in mice (mg/kg): 6.6 i.p. (Omura).

USE: Pharmacological tool to study signal transduction pathways, tyrosine phosphorylation and to induce apoptosis.

8929. Stavudine. [3056-17-5] 2',3'-Didehydro-3'-deoxythymidine; 1-(2,3-dideoxy-β-*glycero*-pent-2-enofuranosyl)thymine; 3'-

deoxy-2'-thymidinene; D4T; BMY-27857; Zerit. $C_{10}H_{12}N_2O_4$; mol wt 224.22. C 53.57%, H 5.39%, N 12.49%, O 28.54%. Analog of thymidine, *q.v.;* reverse transcriptase inhibitor. Prepn: J. P. Horwitz *et al., J. Org. Chem.* **31**, 205 (1966); J. W. Beach *et al., ibid.* **57**, 3887 (1992). Large scale production: J. E. Starrett *et al.,* **EP 334368**; *eidem,* **US 5130421** (1989, 1992 both to Bristol-Myers). *In vitro* activity against HIV and toxicology: M. M. Mansuri *et al., Antimicrob. Agents Chemother.* **34**, 637 (1990). Mechanism of action study: H.-T. Ho, M. J. M. Hitchcock, *ibid.* **33**, 844 (1989). Disposition and metabolism: E. M. Cretton *et al., ibid.* **37**, 1816 (1993). Clinical pharmacokinetics: M. N. Dudley *et al., J. Infect. Dis.* **166**, 480 (1992). HPLC determn in plasma and urine: J. S. Janiszewski *et al., J. Chromatogr.* **577**, 151 (1992). Clinical evaluation in AIDS and ARC: M. J. Browne *et al., J. Infect. Dis.* **167**, 21 (1993).

Colorless, granular solid from ethanol/benzene, mp 165-166° (Horwitz). Also reported as crystals from ethanol-ether, mp 174° (Beach). uv max (water): 266 nm (ε 10149). $[\alpha]_D^{25}$ −39.4° (c = 0.701 in water); $[\alpha]_D^{20}$ −46.1° (c = 0.7 in water). Sol in water, demethylacetamide, dimethyl sulfoxide; sparingly sol in methanol, alc, acetonitrile; slightly sol in dichloromethane. Insol in hexane.

THERAP CAT: Antiretroviral.

8930. Stearic Acid. [57-11-4] Octadecanoic acid; Emersol 132. $C_{18}H_{36}O_2$; mol wt 284.48. C 76.00%, H 12.76%, O 11.25%. Occurs as a glyceride in tallow and other animal fats and oils, as well as in some vegetable oils; also prepd synthetically by hydrogenation of cottonseed and other vegetable oils.

White or slightly yellow, crystal masses, or a white to slightly yellow powder; slight tallow-like odor. Does not congeal below 54°. d^{70} 0.847; mp 69-70°; bp 383°; n_D^{80} 1.4299. Slowly volatilizes at 90-100°. Stearic acid consists chiefly of a mixture of stearic and palmitic acids. Freely sol in ether. One gram dissolves in 21 ml alc, 5 ml benzene, 2 ml chloroform, 26 ml acetone, 6 ml carbon tetrachloride, 3.4 ml carbon disulfide; also sol in amyl acetate, toluene. Practically insol in water. LD_{50} i.v. in mice, rats: 23±0.7, 21.5±1.8 mg/kg; *see:* L. Orö, A. Wretlind, *Acta Pharmacol. Toxicol.* **18**, 141 (1961).

Ethyl ester. [111-61-5] Ethyl stearate. $C_{20}H_{40}O_2$; mol wt 312.54. White, cryst solid; odorless or practically so. mp 33-35°. bp 224°. Ethyl ester of commerce solidifies at 20-24°; bp_4 180°. Insol in water; sol in alc or ether.

Methyl ester. [112-61-8] Methyl stearate. $C_{19}H_{38}O_2$. White crystals. mp 38-39°. bp_{15} 215°. Sol in alc, ether. Insol in water.

USE: For suppositories, coating enteric pills, ointments, and for coating bitter remedies. Manuf stearates of aluminum, zinc, and other metals, stearin soap for opodeldoc, candles, phonograph records, insulators, modeling compds; impregnating plaster of Paris; in vanishing creams and other cosmetics.

8931. Stearidonic Acid. [20290-75-9] (6Z,9Z,12Z,15Z)-6,9,-12,15-Octadecatetraenoic acid; moroctic acid; 18:4(n-3). $C_{18}H_{28}O_2$; mol wt 276.42. C 78.21%, H 10.21%, O 11.58%. Naturally occurring, omega-3 polyunsaturated fatty acid (PUFA). Detected in small amounts in several species of algae, fungi, and in most fish oils. Also found in plant seed oils, especially black currants and various *Echium* species. Biosynthetic precursor of eicosapentaenoic acid (EPA); formed by desaturation of α-linolenic acid. Isoln from oil of the South African pilchard, *Sardina ocellata* J.: M. Matic, *Biochem. J.* **68**, 692 (1958). Quantification in *Echium* seeds: J. L. Guil-Guerrero *et al., Phytochemistry* **53**, 451 (2000). Production by genetically

remodeled canola: V. M. Ursin, *J. Nutr.* **133**, 4271 (2003); by transgenic soybeans: S. Sato *et al., Crop Sci.* **44**, 646 (2004); H. Eckert *et al., Planta* **224**, 1050 (2006). Metabolism in humans: M. J. James *et al., Am. J. Clin. Nutr.* **77**, 1140 (2003). Clinical effect of dietary supplementation on omega-3 index: W. S. Harris *et al., Lipids* **43**, 805 (2008). Review of natural sources, medical uses, and role in human nutrition: J. L. Guil-Guerrero, *Eur. J. Lipid Sci. Technol.* **109**, 1226-1236 (2007).

Pale-yellow oil, mp −57.4° to −56.6°. n_D^{16} 1.4888. Sol in ethanol. Soly (mg/ml): DMSO 100; DMF 100.

USE: Nutritional supplement.

8932. Stearyl Alcohol. [112-92-5] 1-Octadecanol; 1-hydroxyoctadecane; stenol. $C_{18}H_{38}O$; mol wt 270.50. C 79.93%, H 14.16%, O 5.91%. The official substance is a mixture of solid alcohols consisting chiefly of stearyl alcohol. Preparation from ethyl stearate: Brown, Rao, *J. Am. Chem. Soc.* **78**, 2582 (1956); Hesse, Schrödel, *Ann.* **607**, 24 (1954). Prepn of technical grade from sperm whale oil: Maiorov *et al., Zh. Prikl. Khim.* **37**, 1344 (1964).

Unctuous white flakes or granules, mp 56-60° (the pure substance, mp 59.4-59.8°, bp_{15} 210°). Sol in alc, ether, benzene, acetone. Insol in water.

USE: Substitute for cetyl alc in pharmaceutical dispensing, in cosmetic creams, for emulsions, textile oils and finishes, as antifoam agent, lubricant, and chemical raw material. Pharmaceutic aid (stiffening agent).

8933. Stellar®. [151031-37-7] Starch based fat replacer produced by controlled acid hydrolysis of corn starch. Shears to create a firm, deformable creme consisting of loosely aggregated submicron-sized starch particles. Consists of branched amylodextrins; average mol wt <20000. Prepn: D. W. Harris, K. D. Stanly, **EP 529891** (1992 to Staley Manuf.). Functional properties in relationship to structure: D. W. Harris, G. A. Day, *Starch/Staerke* **45**, 221 (1993). Brief description: *Prepared Foods* **161**, 77 (1992).

Bland taste, slick fat-like mouthfeel. Hydrophilic. pH of 10% slurry, 3.0-5.0. Caloric value: 16 kJ/g. Exhibits thixotropic rheological behavior similar to hydrogenated vegetable shortening.

USE: Fat substitute for reduced calorie foods.

8934. Stem Cell Factor. Steel factor; mast cell growth factor; kit ligand; SCF. Hematopoietic growth factor that interacts with other cytokines to stimulate primitive, undifferentiated blast cells and is particularly important in mast cell and erythroid lineages. The full length, membrane bound protein consists of 248 amino acids that is cleaved to produce a soluble form consisting of 165 residues. The soluble form exists as a glycosylated homodimer; mol wt ~45 kDa. Identification: D. E. Williams *et al., Cell* **63**, 167 (1990); K. M. Zsebo *et al., ibid.* 195. Review of discovery and bioactivity: D. E. Williams, S. D. Lyman, *Prog. Growth Factor Res.* **3**, 235-242 (1991). Crystal structure of human soluble dimer: Z. Zhang *et al., Proc. Natl. Acad. Sci. USA* **97**, 7732 (2000). Review of biology and clinical potential: I. K. McNiece, R. A. Briddell, *J. Leukocyte Biol.* **57**, 14-22 (1995); J. Glaspy *et al., Cancer Chemother. Pharmacol.* **38**, Suppl., S53-S57 (1996). Review of role in hematopoietic system: L. K. Ashman, *Int. J. Biochem. Cell Biol.* **31**, 1037-1051 (1999); and therapeutic potential: S. Nakagawa, T. Kitoh, *Curr. Opin. Hematol.* **7**, 133-142 (2000).

Ancestim. [163545-26-4] *N*-L-Methionyl-1-165-hematopoietic cell growth factor KL (human clone V19.8:hSCF162) dimer; rhSCF; Stemgen. Produced in *E. coli* by recombinant DNA technology. Prepn: K. M. Zsebo *et al.,* **EP 423980** (1991 to Amgen). Clinical trial in combination with filgrastim for hematopoietic reconstitution in chemotherapy patients: E. J. Shpall *et al., Blood* **9**, 2491 (1999).

THERAP CAT: Hematopoietic.

8935. Stenbolone. [5197-58-0] (5α,17β)-17-Hydroxy-2-methylandrost-1-en-3-one; 2-methyl-5α-androst-1-en-17β-ol-3-one; 2-methyl-17β-hydroxy-5α-androst-1-en-3-one; stenobolone (rescinded USAN). $C_{20}H_{30}O_2$; mol wt 302.46. C 79.42%, H 10.00%, O 10.58%. Preparation of free alcohol and acetate: R. Mauli *et al.*, *J. Am. Chem. Soc.* **82**, 5494 (1960); Kaspar *et al.*, **DE 1096356** (1961 to Schering AG), *C.A.* **55**, 27440b (1961); R. E. Counsell *et al.*, *J. Org. Chem.* **27**, 248 (1962); **GB 925849** (1963 to Syntex).

Crystals from acetone + hexane, mp 155-158°. $[\alpha]_D$ +52° (chloroform) (Mauli). $[\alpha]_D^{26}$ +47° (chloroform) (Counsell). uv max (95% ethanol): 241 nm (log ε 3.99).

Acetate. [1242-56-4] $C_{22}H_{32}O_3$; mol wt 344.50. Crystals from acetone + hexane, mp 146-149°. $[\alpha]_D$ +32° (chloroform) (Mauli). $[\alpha]_D^{26}$ +60° (chloroform) (Counsell). uv max (95% ethanol): 241 nm (log ε 4.03).

Note: This is a controlled substance (anabolic steroid): **21 CFR**, 1308.13, as defined in 1300.01.

8936. Stepronin. [72324-18-6] *N*-[1-Oxo-2-[(2-thienylcarbonyl)thio]propyl]glycine; 2-(α-thenoylthio)propionylglycine; *N*-(2-mercaptopropionyl)glycine 2-thiophenecarboxylate (ester); TTPG; prostenoglycine; Tiase. $C_{10}H_{11}NO_4S_2$; mol wt 273.32. C 43.94%, H 4.06%, N 5.12%, O 23.41%, S 23.46%. Prepn, properties and toxicology: F. Bolasco, **BE 875186**; *eidem*, **US 4242354** (1979, 1980 to Mediolanum Farmaceutici). Clinical trials in bronchitis: M. Lingetti *et al.*, *Int. J. Clin. Pharmacol. Res.* **1**, 273 (1981); F. Dotta *et al.*, *Clin. Ter.* **105**, 307 (1983); G. C. Morandini *et al.*, *Curr. Ther. Res.* **35**, 783 (1984). Effect on human lung mucus: M. C. Morale *et al.*, *Int. J. Tissue React.* **5**, 231 (1983).

Colorless crystals from acetonitrile, mp 168-170°. uv max: 292 nm. LD_{50} in mice (mg/kg): >2500 orally, >1250 i.v.; in rats (mg/kg): >2500 orally, 1801 i.m. (Bolasco).

Lysine salt. [113790-28-6] Masor; Mucodil.

Sodium salt. [78126-10-0] Broncoplus; Tioten. $C_{10}H_{10}NNaO_4S_2$; mol wt 295.30.

THERAP CAT: Mucolytic.

8937. Stevia. Candyleaf; kaà-hê-é; sweet herb; yerba dulce. Small, shrubby, flowering perennial, *Stevia rebaudiana* Bert., *Asteraceae.* Leaves are intensely sweet due to the presence of various glycosides of the tetracyclic diterpene, steviol. *Habit.* Paraguay, Brazil. *Constit.* Stevioside, rebaudiosides A-E, dulcoside A, steviolbioside; labdane diterpenes including jhanol, austroinulin; triterpenes such as β-amyrin acetate, lupeol; β-sitosterol, stigmasterol, flavonoids, sterebins, tannins, and volatile oil. Overview of agriculture and constituents: J. E. Brandle *et al.*, *Can. J. Plant Sci.* **78**, 527 (1998). Prepn of extracts and identification of constituents: S. K. Yoda *et al.*, *J. Food Eng.* **57**, 125 (2003). LC-MS method for characterization of extracts: J. Pól *et al.*, *J. Chromatogr. A* **1150**, 85 (2007). Biosynthesis of glycosides: J. E. Brandle, P. G. Telmer, *Phytochemistry* **68**, 1855 (2007). *Review:* A. Y. Leung, S. Foster, *Encyclopedia of Common Natural Ingredients* (Wiley-Interscience, Hoboken, 2nd Ed., 2003) pp 478-480. *Book:* A. D. Kinghorn, Ed., *Stevia: The Genus Stevia* (Taylor & Francis, New York, 2002) 224 pp.

USE: Source of steviol glycosides; sweetener for use in foods.

8938. Steviol. [471-80-7] (4α)-13-Hydroxykaur-16-en-18-oic acid; hydroxydehydrostevic acid. $C_{20}H_{30}O_3$; mol wt 318.46. C 75.43%, H 9.50%, O 15.07%. Aglycon of stevioside and the rebaudiosides, the sweet glycosides of the stevia plant. Isoln: M. Bridel,

R. Lavieille, *Compt. Rend.* **192**, 1123 (1931). Structure: E. Mosettig, W. R. Nes, *J. Org. Chem.* **20**, 884 (1955). Absolute configuration: E. Mosettig *et al.*, *J. Am. Chem. Soc.* **85**, 2305 (1963). Biosynthesis: M. Ruddat *et al.*, *Arch. Biochem. Biophys.* **110**, 496 (1965). Total synthesis of (±)-form: K. Mori *et al.*, *Tetrahedron Lett.* **11**, 2411 (1970); of (−)-form: I. F. Cook, J. R. Knox, *ibid.* 4091. Plant growth regulation activity: B. H. de Oliveira *et al.*, *Phytochemistry* **69**, 1528 (2008). Review of genetic toxicity studies: D. J. Brusick, *Food Chem. Toxicol.* **46**, Suppl. 1, S83-S91 (2008).

Needles from methanol, mp 212-213°. $[\alpha]_D$ −93.6° (ethanol).

8939. Stevioside. [57817-89-7] (4α)-13-[(2-*O*-β-D-Glucopyranosyl-β-D-glucopyranosyl)oxy]kaur-16-en-18-oic acid β-D-glucopyranosyl ester; steviosin. $C_{38}H_{60}O_{18}$; mol wt 804.88. C 56.71%, H 7.51%, O 35.78%. Intensely sweet, steviol glycoside isolated from the leaves of the stevia plant, *Stevia rebaudiana* Bert., *Asteraceae.* Structurally similar to the rebaudiosides, *q.v.* Isoln and characterization: M. Bridel, R. Lavieille, *Compt. Rend.* **192**, 1123 (1931). Structure of glucose moieties: H. B. Wood *et al.*, *J. Org. Chem.* **20**, 875 (1955). Total synthesis: T. Ogawa *et al.*, *Tetrahedron* **36**, 2641 (1980). Toxicology study: L. Xili *et al.*, *Food Chem. Toxicol.* **30**, 957 (1992). Stability and interaction with food ingredients: G. Th. Kroyer, *Lebensm. Wiss. Technol.* **32**, 509 (1999). Comparison of extraction methods: J. Pól *et al.*, *Anal. Bioanal. Chem.* **388**, 1847 (2007). Review of biosynthesis, metabolism and toxicology: J. M. C. Geuns, *Phytochemistry* **64**, 913-921 (2003); of pharmacology and therapeutic applications: V. Chatsudthipong, C. Muanprasat, *Pharmacol. Ther.* **121**, 41-54 (2009).

Hygroscopic crystals, mp 198°. $[\alpha]_D^{25}$ −39.3° (c = 5.7 in H_2O). 300 times as sweet as cane sugar. One gram dissolves in 800 ml water. Sol in dioxane. Slightly sol in alc.

USE: Non-nutritive sweetener.

8940. Stibine. [7803-52-3] Antimony hydride. H_3Sb; mol wt 124.78. H 2.42%, Sb 97.58%. SbH_3. Conveniently prepd by dissolving zinc-antimony or magnesium-antimony alloy in dil HCl: Hurd, *Chemistry of the Hydrides* (Wiley, New York, 1952) p 132. Detailed directions (including prepn of the alloy from powdered Sb and Mg): Schenk in *Handbook of Preparative Inorganic Chemistry* **vol. 1**, G. Brauer, Ed. (Academic Press, New York, 2nd ed., 1963) pp 606-608. Review of preparative methods: Jolly, Norman, "Hydrides of Groups IV and V" in *Preparative Inorganic Reactions* **vol. 4**, W. L. Jolly, Ed. (Interscience, New York, 1968) pp 1-58.

Colorless gas. *Poisonous; flammable.* Disagreeable odor. mp −88°. bp −18.4°. d 2.204 g/ml at bp. Heat of formation +34.68 kcal/mole: Gunn, Green, *J. Phys. Chem.* **65**, 779 (1961). Thermally less stable than arsine. Dec slowly on standing at room temp. Quickly destroyed at 200°. The decomposition products are hydrogen and metallic antimony, generally deposited in the form of a mirror. The gas is slightly sol in water. Freely sol in alcohol, carbon disulfide, other organic solvents. Lethal concn in air for mice: about 100 ppm.

Caution: Potential symptoms of overexposure are headache, weakness; nausea, abdominal pain; lumbar pain, hemoglobinuria, hematuria and hemolytic anemia; jaundice; pulmonary irritation. See *NIOSH Pocket Guide to Chemical Hazards* (DHHS/NIOSH 97-140, 1997) p 284.

USE: Has been used as fumigating agent.

8941. Stibocaptate. [3064-61-7] 2,2'-[(1,2-Dicarboxy-1,2-ethanediyl)bis(thio)]bis-1,3,2-dithiastibolane-4,5-dicarboxylic acid sodium salt (1:6); 2,3-dimercaptosuccinic acid cyclic ester with antimonic acid, diester with 2,3-dimercaptosuccinic acid hexasodium salt; 2,3-dimercaptosuccinic acid cyclic thioantimonate(III) *S,S*-diester with 2,3-dimercaptosuccinate hexasodium salt; sodium antimony-2,3-*meso*-dimercaptosuccinate; "antimony dimercaptosuccinate"; TWSb; Ro-4-1544/6; SB-58; Astiban. $C_{12}H_6Na_6O_{12}S_6Sb_2$; mol wt 915.99. C 15.74%, H 0.66%, Na 15.06%, O 20.96%, S 21.00%, Sb 26.59%. Prepn: Friedheim, **US 2880222** (1959). Prepn and antischistosomal activity: J.-K. Hsu, M.-K. Ch'en, *K'o Hsueh T'ung Pao* **1966**, 978, *C.A.* **65**, 17580e (1966). Toxicity data: Ercoli, *Proc. Soc. Exp. Biol. Med.* **129**, 284 (1968).

White or slightly yellowish-green powder. Hygroscopic, unstable when moistened. Sol in water. LD$_{50}$ s.c. in mice: 500 mg Sb/kg (Ercoli).

Tripotassium salt. [27279-76-1] Clinical experience in schistosomiasis: E. A. H. Friedheim *et al., Am. J. Trop. Med. Hyg.* **3**, 714 (1954).

THERAP CAT: Anthelmintic (Schistosoma).

8942. Stibophen. [15489-16-4] (*T*-4)-Bis[4,5-di(hydroxy-κ*O*)-1,3-benzenedisulfonato(4−)]antimonate(5−) sodium hydrate (1:5:7); sodium antimony bis(pyrocatechol-2,4-disulfonate) heptahydrate; antimony pyrocatechol sodium disulfonate heptahydrate; Sdt-91; Fuadin; Fouadin; Fantorin; Neoantimosan; Repodral. $C_{12}H_4Na_5O_{16}S_4Sb.7H_2O$; mol wt 895.20. C 16.10%, H 2.03%, Na 12.84%, O 41.11%, S 14.33%, Sb 13.60%. Prepn of potassium salt: **GB 213285** (1923 to Heyden). Prepn from sodium pyrocatechol-3,5-disulfonate and antimony trioxide in alkaline soln: H. Schmidt, **DE 448800** (1924 to I. G. Farbenind.), *Frdl.* **16**, 2546 (1931); *idem*, **US 1549154** (1925); *idem, Z. Angew. Chem.* **43**, 963 (1930); *idem*, **US 1873668** (1932 to Winthrop). Toxicity: H. Eagle *et al., J. Pharmacol. Exp. Ther.* **89**, 196 (1947). Clinical trial in schistosomiasis: J. Azar *et al., Am. J. Trop. Med.* **29**, 595 (1949); in leishmaniasis: H. S. Girgla *et al., Br. J. Dermatol.* **97**, 307 (1977). UV spectrophotometric analysis: A. Besada *et al., Anal. Lett.* **21**, 447 (1988).

Fine crystals. Almost insol in abs alcohol, ether, chloroform, acetone, petr ether. Readily sol in cold water. Unless the colorless neutral water soln is acidified it acquires a yellowish tint and finally reaches lemon-yellow color. LD$_{50}$ i.v. in rabbits: ~90 mg/kg (Eagle).

Potassium salt. Antimosan; Heyden 611.

THERAP CAT: Anthelmintic (Schistosoma).

THERAP CAT (VET): Has been used in schistosomiasis.

8943. Stigmastanol. [83-45-4] (3β,5α)-Stigmastan-3-ol; dihydro-β-sitosterol; β-sitostanol; 24α-ethylcholestanol; fucostanol. $C_{29}H_{52}O$; mol wt 416.73. C 83.58%, H 12.58%, O 3.84%. Occurs along with β-sitosterol and stigmasterol. Formation from β-sitosterol: Bernstein, Wallis, *J. Org. Chem.* **2**, 341 (1937); from β-sitostenone: Marker, Wittle, *J. Am. Chem. Soc.* **59**, 2704 (1937). Antihypercholesterolemic activity in rabbits: I. Ikeda *et al., J. Nutr. Sci. Vitaminol.* **27**, 243 (1981).

Monohydrate. Crystals, mp 138-139°. When dry mp 144-145°. [α]$_D^{20}$ +25° (c = 1.1 in chloroform).

Acetate. $C_{31}H_{54}O_2$. mp 137-138°. [α]$_D^{20}$ +14° (c = 1.8 in chloroform).

8944. Stigmasterol. [83-48-7] (3β,22E)-Stigmasta-5,22-dien-3-ol; 3β-hydroxy-24-ethyl-Δ^{5,22}-cholestadiene. $C_{29}H_{48}O$; mol wt 412.70. C 84.40%, H 11.72%, O 3.88%. Usually isolated from the phytosterol mixture from soy or calabar beans, *cf.* Thornton *et al., J. Am. Chem. Soc.* **62**, 2006 (1940); Byerrum, Ball, *Biochem. Prep.* **7**, 86 (1959). Structure: Fernholz, *Ann.* **507**, 128 (1933); **508**, 215 (1933); Fernholz, Chakravorty, *Ber.* **67**, 2021 (1934). Identity with *guinea-pig-anti-stiffness factor*: H. Rosenkrantz *et al., Proc. Soc. Exp. Biol. Med.* **76**, 408 (1951). Synthesis: W. Sucrow, M. Slopianka, *Ber.* **108**, 3721 (1975). HPLC separation from plant oils: H. Colin *et al., Anal. Chem.* **51**, 1661 (1979).

Monohydrate. Crystals from alcohol. When dry, mp 170°. [α]$_D^{22}$ −51° (c = 2 in chloroform). Insol in water. Sol in the usual organic solvents.

Acetate. $C_{31}H_{50}O_2$. mp 144°. [α]$_D^{20}$ −55.6° (c = 2 in chloroform).

p-**Nitrobenzoate.** $C_{36}H_{51}NO_4$. mp 203°. [α]$_D^{17}$ −13° (c = 2 in alcohol).

8945. Stilbamidine. [122-06-5] 4,4'-(1,2-Ethenediyl)bisbenzenecarboximidamide; 4,4'-stilbenedicarboxamidine; 4,4'-diamidinostilbene. $C_{16}H_{16}N_4$; mol wt 264.33. C 72.70%, H 6.10%, N 21.20%. Prepn: A. J. Ewins, J. N. Ashley, **GB 510097** (1939 to May & Baker); J. N. Ashley *et al., J. Chem. Soc.* **1942**, 103; of isethionate: G. Newbery, A. P. T. Easson, **US 2394003** (1946 to May & Baker). Trypanocidal activity: E. M. Lourie, W. Yorke, *Ann. Trop. Med. Parasitol.* **33**, 289 (1939). Preliminary pharmacology: R. Wien, *ibid.* **37**, 1 (1943). Mechanism of action studies: M. J. Pine, *Biochem. Pharmacol.* **17**, 75 (1968); G. Weissmann *et al., ibid.* **19**, 1251 (1970). Crystal structure: C. Courseille *et al., C. R. Seances Acad. Sci. Ser. C* **272**, 1115 (1971). DNA binding study: N. Gresh, B. Pullman, *Mol. Pharmacol.* **25**, 452 (1984). Early review of pharmacology, mode of action and clinical applications: E. B. Schoenbach, E. M. Greenspan, *Medicine* **27**, 327-377 (1948).

Dihydrochloride. $C_{16}H_{18}Cl_2N_4$. Needles from water. LD_{50} in mice (mg/g): 0.031 i.v.; 0.18 s.c. (Wien).

Isethionate. [140-59-0] M & B 744. $C_{20}H_{28}N_4O_8S_2$; mol wt 516.58. Crystals, discolored by light, dec 290°. uv max: 330 nm ($E_{1cm}^{1\%}$ 750). Soly in water: approx 1 g in 2.5 to 3.0 ml H_2O; in methanol: 1.5 g/l00 ml. pH of 0.5% aq soln 5.5-6.5.

THERAP CAT: Antiprotozoal (Leishmania, Trypanosoma).

8946. Stilbene. [588-59-0] 1,1'-(1,2-Ethenediyl)bisbenzene; α,β-diphenylethylene; bibenzal; bibenzylidene. $C_{14}H_{12}$; mol wt 180.25. C 93.29%, H 6.71%. *cis*-Form prepd by Clemmensen reduction of benzoin: Shriner, Berger, *Org. Synth.* **coll. vol. III**, 786 (1955); *cis*-form by copper-chromite decarboxylation of α-phenyl-cinnamic acid: Buckles, Wheeler, *ibid.* **coll. vol. IV**, 857 (1963). Synthesis of *cis*- and *trans*-forms by the Wittig reaction and by decarboxylation of phenylcinnamic acids: Wheeler, Batlle de Pabon, *J. Org. Chem.* **30**, 1473 (1965).

trans-Form   *cis*-Form

***trans*-Form.** [103-30-0] Crystals from 95% ethanol, mp 124°. bp_{760} 306-307°. uv max (95% ethanol): 296, 305 nm (ε 28100; 26700). Volatile with steam. Practically insol in water. Sol in 90 parts cold alc, 13 parts boiling alc, freely in benzene, ether.

***cis*-Form.** [645-49-8] Liquid, solidifies at $-5°$. bp_{10} 135°; $bp_{1.0}$ 96°. n_D^{25} 1.6188. uv max (95% ethanol): 278 nm (ε 10200). Completely sol in cold abs alcohol.

8947. Stillingia. Queen's root; yaw root; silver leaf. Dried roots of *Stillingia sylvatica* L., *Euphorbiaceae. Habit.* Southeastern U.S. *Constit.* Acrid resin (sylvacrol), fixed and volatile oils, a glucoside.

THERAP CAT: Emetic, cathartic.

8948. Stiripentol. [137767-55-6]; [49763-96-4] (unspecified stereo). (1*E*)-1-(1,3-Benzodioxol-5-yl)-4,4-dimethyl-1-penten-3-ol; (*E*)-(±)-4,4-dimethyl-1-(3,4-methylenedioxyphenyl)-1-penten-3-ol; (±)-1-(3,4-methylenedioxyphenyl)-4,4-dimethyl-1(*E*)-penten-3-ol; BCX-2600; D-306; Diacomit. $C_{14}H_{18}O_3$; mol wt 234.30. C 71.77%, H 7.74%, O 20.49%. Aromatic allylic alcohol with antiepileptic activity. Prepn: F. M. J. Vallet, **DE 2308494**; *idem,* **US 3910959** (1973, 1975 both to Unicler). Crystal structure: J. N. Lisgarten, R. A. Palmer, *Acta Crystallogr. C* **44**, 1992 (1988). Confirmation of (*E*)-configuration: R. Céolin *et al., Int. J. Pharm.* **74**, 77 (1991). Synthesis, configuration, and metabolism of enantiomers: K. Zhang *et al., Drug Metab. Dispos.* **22**, 544 (1994). Pharmacokinetics and activity of enantiomers: R. H. G. P. Arends *et al., Epilepsy Res.* **18**, 91 (1994). Activity on GABA$_A$-receptor channel: P. P. Quilichini *et al., Epilepsia* **47**, 704 (2006); J. L. Fisher, *Neuropharmacology* **56**, 190 (2009). Clinical evaluation in severe myoclonic epilepsy in infancy (Dravet syndrome): C. Chiron *et al., Lancet* **356**, 1638 (2000). Review of pharmacology: *idem, Neurotherapeutics* **4**, 123-125 (2007); of development and clinical experience: M. K. Trojnar *et al., Pharmacol. Rep.* **57**, 154-160 (2005); S. J. Czuczwar *et al., Expert Opin. Drug Discov.* **3**, 453-460 (2008).

(*R*)-form

Fine colorless needles from ethanol, mp 74°. Easily sol in acetone, alcohol; moderately sol in chloroform. Insol in water. LD_{50} in mice, rats (mg/kg): >3000, >3000 (orally); 1250, 1050 i.p. (Vallet).

(***R***)-**Form.** [144017-65-2] (+)-STP. $[\alpha]_D^{22} = +24.9°$ (c = 2.55 in CH_3OH).

(***S***)-**Form.** [144017-66-3] $(-)$-STP. Oil. $[\alpha]_D^{22} = -23.0°$ (c = 2.61 in CH_3OH).

THERAP CAT: Anticonvulsant.

8949. Storax. [8046-19-3] Storax (balsam); styrax; sweet oriental gum. Balsam obtained from the trunk of *Liquidambar orientalis* Mill., known as Levant Storax, or of *L. styraciflua* L., known as American Storax (family *Hamamelidaceae*). *Habit.* Levant storax is native to Asia Minor. American storax is produced chiefly in Honduras; found along the Atlantic coast from Connecticut to Central America. *Constit.* 33-50% α- and β-storesin and its cinnamic ester; 5-10% styracin; 10% phenylpropyl cinnamate; small amounts of ethyl cinnamate; benzyl cinnamate; 5-15% free cinnamic acid; styrene; 0.4% levorotatory oil; $C_{10}H_{16}O$, and traces of vanillin.

Semiliquid, grayish, sticky, opaque mass (Levant storax) or a semisolid, sometimes solid mass softened by warming (American storax). Storax is transparent in thin layers, has characteristic taste and odor, and is denser than water. Sol, usually incompletely, in equal wt of warm alc; sol in ether, acetone, carbon disulfide. Insol in water. Ingredient of Compound Benzoin Tincture, *see* Benzoin.

USE: In fumigating pastilles and powders; in perfumery; as imbedding material in microscopy.

THERAP CAT: Topical protectant. Expectorant.

THERAP CAT (VET): As a component of Compound Benzoin Tincture. Has been used as a parasiticide.

8950. Stramonium. Thorn apple; Jamestown weed; Jimpson weed; Jimson weed; stinkweed; devil's apple; apple of Peru. Dried leaves and flowering tops of *Datura stramonium* L., Solanaceae or its varieties. Used in traditional medicine in treatment of asthma. *Habit.* Europe, Asia, America. *Constit.* Leaves: 0.25-0.45% alkaloids consisting of atropine, hyoscyamine, and scopolamine; proteins, albumin. Seed: the same alkaloids but in lesser amounts; fixed oil, malic acid proteins. Comprehensive monograph: A. F. Blakeslee, *The Genus Datura* (Ronald Press, New York, 1959) 289 pp. *Review:* V. E. Tyler *et al.* in *Pharmacognosy,* 9th Ed., (Lea & Febiger, Philadelphia, 1988) pp 197-199.

8951. Strepogenin. [11000-03-6] A name for the biologically active principle capable of stimulating growth of certain microorganisms, and originally found present in products of natural origin, such as liver extract, flour, yeast, and tomato juice. Its peptide nature was indicated by its presence in hydrolyzates of purified proteins. To compare the activities of various protein hydrolyzates, a strepogenin unit was introduced and defined as: 1 mg of a standard liver extract (Wilson's liver fraction L) has the strepogenin activity of one unit (using *Lactobacillus casei* as test organism). The different structures of strepogenin-active peptides subsequently obtained from natural sources or by synthesis, indicate that strepogenin is not a distinct, chemically defined, biologically active peptide, but rather represents a biological phenomenon of a multitude of peptides. From the large number of strepogenin-active peptides known, no conclusion can be drawn with regard to structure-function relationships; these peptides have been suspected to be nothing more than easily accessible amino acids, particularly since they (peptides) are not essential for the growth of the microorganisms. Early isolation from liver extract and from tryptic digests of pure proteins: Sprince, Woolley, *J. Exp. Med.* **80**, 213 (1944); *idem, J. Am. Chem. Soc.* **67**, 1734 (1945). *Review:* E. Schröder, K. Lübke, *The Peptides* vol. **II** (Academic Press, New York, 1966) pp 267-270.

8952. Streptidine. [85-17-6] N^1,N^3-Bis(aminoiminomethyl)streptamine; N,N'-diamidinostreptamine; N,N'''-(2,4,5,6-tetrahydroxy-1,3-cyclohexanediyl)bisguanidine; 1,3-diguanido-2,4,5,6-cyclohexanetetrol. $C_8H_{18}N_6O_4$; mol wt 262.27. C 36.64%, H 6.92%, N 32.04%, O 24.40%. The aglycone component of the streptomycin molecule joined with streptobiosamine through a glycosidic linkage. Obtained from streptomycin by acid hydrolysis: Peck *et al., J. Am. Chem. Soc.* **68**, 29 (1946); Fried *et al., J. Biol. Chem.* **162**, 391 (1946). Structure: Carter *et al., Science* **103**, 53, 540 (1946). Synthesis: Wolfrom *et al., J. Am. Chem. Soc.* **72**, 1724 (1950). Absolute configuration: Dyer, Todd, *ibid.* **85**, 3896 (1963); Tatsuoka, Horii, *J. Antibiot.* **17A**, 88 (1964). Biosynthesis studies: Bruton *et al., J. Biol. Chem.* **242**, 813 (1967).

Optically inactive, diacidic base.

Dipicrate dihydrate. $C_{20}H_{24}N_{12}O_{18}.2H_2O$. Yellow needles from water, dec 283-284° (becomes anhydrous *in vacuo* at 56°, dec 284-285°).

Sulfate. $C_8H_{18}N_6O_4.H_2SO_4$. Crystals from dilute sulfuric acid + acetone, not melted at 300°. Forms solvated crystals from methanol.

Dihydrochloride. $C_8H_{18}N_6O_4.2HCl$. Hygroscopic, amorphous powder, dec 170-210°. Forms solvated crystals from methanol.

8953. Streptodornase. [37340-82-2] Streptococcal deoxyribonuclease. DNAase produced by various strains of β-hemolytic streptococci. Identification in culture medium: W. S. Tillett *et al., Proc. Soc. Exp. Biol. Med.* **68**, 184 (1948); M. McCarty, *J. Exp. Med.* **88**, 181 (1948). Purification: J. H. Mowat *et al.,* US 2753291 (1956 to Am. Cyanamid). Clinical use with streptokinase in wound debridement: J. Poulsen *et al., Acta Chir. Scand.* **149**, 245 (1983).

Mixture with streptokinase. [8048-16-6] Ernodasa; Varidase.

THERAP CAT: Debriding agent.

8954. Streptogramins. Family of related antibiotic complexes produced by variety of soil organisms of the genus *Streptomyces.* Consists of two types of antibacterial components: a polyunsaturated macrolactone (Group A, Type II) and a cyclic hexadepsipeptide (Group B, Type I). Both groups inhibit bacterial growth by preventing the translation of mRNA by binding to the bacterial ribosome and act synergistically. Isoln of complex from *Streptomyces graminofaciens:* J. Charney *et al., Antibiot. Chemother.* **3**, 1283 (1953). Each strain specific complex and its respective components were named to reflect the producing strain, however structural studies showed considerable overlap *see* Pristinamycin, Virginiamycin, *q.q.v.* Nomenclature: P. Crooy, R. De Neys, *J. Antibiot.* **25**, 371 (1972). Mechanism of action: B. T. Porse, R. A. Garrett, *J. Mol. Biol.* **286**, 375 (1999). Review of isolations, identification and activity: D. Vazquez, *Antibiotics* **vol. 1**, D. Gottlieb, P. D. Shaw, Eds. (Springer-Verlag, New York, 1967) pp 387-403; **vol. 3**, J. W. Corcoran, F. E. Hahn, Eds. (1975) pp 521-534. *Review:* J.-C. Pechère, *Drugs* **51**, Suppl. 1, 13-19 (1996); of chemistry and biological activity: J. C. Barrière *et al., Curr. Pharm. Des.* **4**, 155-180 (1998).

Streptogramin A *see* Virginiamycin M1.

Streptogramin B *see* Pristinamycin IA.

8955. Streptokinase. [9002-01-1] Streptokinase (enzyme-activating); streptococcal fibrinolysin; plasminokinase; Kabikinase; Streptase. Single-chain protein produced by β-hemolytic streptococci; mol wt ~47 kDa. Activates plasminogen to produce plasmin which dissolves fibrin. Discovery: W. S. Tillet, R. L. Garner, *J. Exp. Med.* **58**, 485 (1933). Isoln from streptococcal culture filtrates: L. R. Christensen, *J. Gen. Physiol.* **28**, 363 (1945). Purification: M. Siegel *et al.,* US 3042586 (1962 to Merck & Co.); P. K. Siiteri, R. D. Mills, US 3226304 (1965 to Am. Cyanamid); and characterization: E. C. De Renzo *et al., J. Biol. Chem.* **242**, 533 (1967). Review of isoln, assay and properties: F. B. Taylor, R. H. Tomar, *Methods Enzymol.* **19**, 807-821 (1970); F. Castellino, Jr. *et al., ibid.* **45B**, 244-257 (1976); of mechanism of action: K. N. N. Reddy in *Fibrinolysis,* D. L. Kline, K. N. N. Reddy, Eds. (CRC, Boca Raton, 1980) pp 71-94. Clinical trials in acute myocardial infarction: J. L. Anderson *et al., N. Engl. J. Med.* **308**, 1312 (1983); ISIS-2 Collaborative Group, *Lancet* **2**, 349 (1988). Clinical comparison with alteplase (tPA): GISSI-2 and Int. Study Group, *Eur. Heart J.* **13**, 1692 (1992). Pharmacokinetics: D. S. Grierson, T. D. Bjornsson, *Clin. Pharmacol. Ther.* **41**, 304 (1987). Review of clinical efficacy in deep venous thrombosis: L. Q. Rogers, C. L. Lutcher, *Am. J. Med.* **88**, 389-395 (1990); in myocardial infarction: K. L. Goa *et al., Drugs* **39**, 693-

719 (1990); P. E. Battershill *et al., Drugs Aging* **4**, 63-86 (1994). Review of mechanism of action and production methods: A. Banerjee *et al., Biotechnol. Adv.* **22**, 287-307 (2004).

THERAP CAT: Thrombolytic.

THERAP CAT (VET): Thrombolytic.

8956. Streptomycin. [57-92-1] *O*-2-Deoxy-2-(methylamino)-α-L-glucopyranosyl-(1 → 2)-*O*-5-deoxy-3-*C*-formyl-α-L-lyxofuranosyl-(1 → 4)-N^1,N^3-bis(aminoiminomethyl)-D-streptamine; streptomycin A. $C_{21}H_{39}N_7O_{12}$; mol wt 581.58. C 43.37%, H 6.76%, N 16.86%, O 33.01%. Antibiotic substance produced by the soil Actinomycete *Streptomyces griseus* (Krainsky) Waksman et Henrici (Fam. *Actinomycetaceae*). Isolation: Schatz *et al., Proc. Soc. Exp. Biol. Med.* **55**, 66 (1944). Production by aerobic fermentation and purification: Tishler in *Streptomycin,* Selman A. Waksman, Ed. (Williams & Wilkins, Baltimore, 1949) pp 32-54. Isoln and purification by ion exchange: Bartels *et al., Chem. Eng. Prog.* **54** (8), 49-51 (Aug. 1958); Bartels *et al.,* US 2868779 (1959 to Olin Mathieson). Structure: Brink, Folkers, *J. Am. Chem. Soc.* **69**, 1234 (1947); Wolfrom *et al., ibid.* **76**, 3675 (1950). Total synthesis: Umezawa *et al., J. Antibiot.* **27**, 997 (1974). Mechanism of action: B. J. Wallace *et al.,* in *Antibiotics* **vol. 5**(pt. 1), F. E. Hahn, Ed. (Springer-Verlag, New York, 1979) pp 272-303. *Review:* Lemieux, Wolfrom, "The Chemistry of Streptomycin" in W. W. Pigman, M. L. Wolfrom, *Adv. Carbohydr. Chem.* **3**, 337-384 (1948). Comprehensive description: J. Mossa *et al., Anal. Profiles Drug Subs.* **16**, 507-609 (1986).

Streptomycin is usually available as the trihydrochloride, trihydrochloride-calcium chloride double salt, phosphate, or sesquisulfate, which occur as granules or powder. Odorless or nearly so, with a slightly bitter taste. Most salts are hygroscopic and deliquesce on exposure to air, but are not affected by air or light. The salts are very sol in water; but almost insol in alc, chloroform, ether. Solns are levorotatory.

Trihydrochloride. Streptomycin hydrochloride. $C_{21}H_{39}N_7O_{12}.$ 3HCl. $[\alpha]_D^{25}$ −84°. Soly at about 28° (mg/ml): water >20; methanol >20; ethanol 0.90; isopropanol 0.12; isoamyl alcohol 0.117; petr ether 0.02; carbon tetrachloride 0.042; ether 0.01. *See:* Weiss *et al., Antibiot. Chemother.* **7**, 374 (1957).

Trihydrochloride-calcium chloride double salt. Streptomycin hydrochloride-calcium chloride complex. $(C_{21}H_{39}N_7O_{12}.3HCl)_2$·$CaCl_2$. Prepn from the trihydrochloride: Peck, US 2446102 (1948 to Merck & Co.). Very hygroscopic, dec about 200°. $[\alpha]_D^{25}$ −76°.

Pantothenate. Streptothenat. Prepn: GB 771338 (1957 to Grünenthal). The commercial prepn may contain the sulfate.

Sesquisulfate. [3810-74-0] Streptomycin sulfate; AgriStrep; Streptobrettin; Vetstrep. $(C_{21}H_{39}N_7O_{12})_2.3H_2SO_4$; mol wt 1457.38. White to light gray or pale buff powder with faint amine-like odor. Hygroscopic. Solubilities as determined by Weiss *et al., loc. cit.,* in mg/ml at about 28°: water >20; methanol 0.85; ethanol 0.30; isopropanol 0.01; petr ether 0.015; carbon tetrachloride 0.035; ether 0.035. Practically insol in chloroform.

THERAP CAT: Antibacterial (tuberculostatic).

THERAP CAT (VET): Antibacterial.

8957. Streptonigrin. [3930-19-6] (4*R*)-5-Amino-6-(7-amino-5,8-dihydro-6-methoxy-5,8-dioxo-2-quinolinyl)-4-(2-hydroxy-

3,4-dimethoxyphenyl)-3-methyl-2-pyridinecarboxylic acid; 5-amino-6-(7-amino-5,8-dihydro-6-methoxy-5,8-dioxo-2-quinolyl)-4-(2-hydroxy-3,4-dimethoxyphenyl)-3-methylpicolinic acid; bruneomycin; NSC-45383. $C_{25}H_{22}N_4O_8$; mol wt 506.47. C 59.29%, H 4.38%, N 11.06%, O 25.27%. Antitumor antibiotic produced by *Streptomyces flocculus:* Rao, Cullen, *Antibiot. Annu.* **1959-1960**, 950. Structure: Rao *et al., J. Am. Chem. Soc.* **85**, 2532 (1963). Crystal and molecular structure: Y. Y. H. Chiu, W. N. Lipscomb, *ibid.* **97**, 2525 (1975). Identity with bruneomycin: Brazhnikova *et al., Antibiotiki* **13**, 99 (1968), *C.A.* **68**, 89890q (1968). Synthetic studies: Kametani *et al., C.A.* **76**, 14272w (1972); **79**, 31817g (1973). Total synthesis: F. Z. Basha *et al., J. Am. Chem. Soc.* **102**, 3962 (1980); A. S. Kende *et al., ibid.* **103**, 1271 (1981); S. M. Weinreb *et al., ibid.* **104**, 536 (1982). Biosynthesis: S. J. Gould, D. E. Cane, *ibid.* 343. *Reviews:* N. S. Mizuno in *Antibiotics* **vol. 5**, pt. 2, F. E. Hahn, Ed. (Springer-Verlag, New York, 1979) pp 372-384; S. J. Gould, S. M. Weinreb, *Fortschr. Chem. Org. Naturst.* **41**, 77-111 (1982). Review of genotoxicity: A. D. Bolzán, M. S. Bianchi, *Mutat. Res.* **488**, 25-37 (2001).

Coffee-brown to almost black rectangular plates from acetone or dioxane, dec 275°; brown needles, mp 262-263° (Brazhnikova). uv max (methanol): 248, 375-380 nm (ε 38400, 17400). Weak acid. pKa 6.2-6.4 (1:1 aq dioxane). Sol in dioxane, pyridine, dimethylformamide, aq sodium bicarbonate with decompn; slightly sol in water, lower alcohols, ethyl acetate, chloroform.

Caution: Potential symptoms of overexposure include severe and prolonged bone marrow depression (Bolzán, Bianchi).

8958. L-Streptose. [13008-73-6] 5-Deoxy-3-*C*-formyl-L-lyxose; 3-*C*-formyl-5-deoxy-L-lyxofuranose. $C_6H_{10}O_5$; mol wt 162.14. C 44.45%, H 6.22%, O 49.34%. A sugar which is part of the streptomycin molecule: Kuehl *et al., J. Am. Chem. Soc.* **68**, 2096 (1946). Structure: *eidem, ibid.* **68**, 2679 (1946). Configuration: Wolfrom, DeWalt, *ibid.* **70**, 3148 (1948); Kuehl *et al., ibid.* **71**, 1445 (1949). Biosynthesis: Hough, Jones, *Nature* **167**, 180 (1951). Synthesis of α- and β-L-streptose: Dyer *et al., J. Am. Chem. Soc.* **87**, 654 (1965); Paulsen *et al., Ber.* **105**, 1978 (1972).

Colorless, strongly hygroscopic glass. $[\alpha]_D^{20}$ −18° (c = 0.65).

8959. Streptothricins. Racemomycins; yazumycins. A mixture of antibiotics, originally thought to be a single substance, differing only in the number of repeating residues in the peptide side chain. Streptothricins F, E, D, C, B, A, and X are known and have 1 to 7 β-lysine residues resp.: Khokhlov, Shutova, *J. Antibiot.* **25**, 501 (1972). Separation of isomers: Khokhlov, Reshetov, *J. Chromatogr.* **14**, 495 (1964). Identity with racemomycins: H. Taniyama *et al., Chem. Pharm. Bull.* **19**, 1627 (1971). Identity with yazumycins: *eidem, J. Antibiot.* **24**, 390 (1971). Activity studies: Germanova, Goncharskaya, *Antibiotiki* **14**, 48, 137 (1969), *C.A.* **70**, 66547, 85906t (1969). Biosynthetic studies: Voronina *et al., ibid.* 1063, *C.A.* **72**, 77387a (1970); Y. Sawada *et al., Chem. Pharm. Bull.* **26**, 885 (1978). Partial syntheses: M. Kinoshita, Y. Suzuki, *Bull. Chem. Soc. Jpn.* **50**, 2375 (1977); S. Kusumoto *et al., Chem. Lett.*

1981, 1317. Prepn of semi-synthetic racemomycins: Y. Sawada, H. Taniyama, *Chem. Pharm. Bull.* **25**, 1302 (1977). *Streptolin* was formerly used for the streptothricin components where n equals 2 or 3.

Streptothricin F. Racemomycin A; streptothricin VI; yazumycin A. $C_{19}H_{34}N_8O_8$; mol wt 502.53. n = 1. Originally known as streptothricin. Produced by variants of *Actinomyces lavendulae.* Isoln: Waksman, Woodruff, *Proc. Soc. Exp. Biol. Med.* **49**, 207 (1942). Purification: Peck *et al., J. Am. Chem. Soc.* **68**, 772 (1946); Kocholaty, Junowicz-Kocholaty, *Arch. Biochem.* **15**, 55 (1947). Identity with streptothricin VI: Hutchinson *et al., Arch. Biochem.* **22**, 16 (1949); Swart, *J. Am. Chem. Soc.* **71**, 2942 (1949). Prepn: Carter *et al., J. Am. Chem. Soc.* **76**, 566 (1954). Structure: van Tamelen *et al., ibid.* **83**, 4295 (1961); Johnson, Westley, *J. Chem. Soc.* **1962**, 1642. Revised structure: S. Kusumoto *et al., J. Antibiot.* **35**, 925 (1982). Biosynthesis: S. J. Gould *et al., J. Am. Chem. Soc.* **103**, 2871 (1981). Total synthesis: S. Kusumoto *et al., Tetrahedron Lett.* **23**, 2961 (1982). Amorphous, white powder. Thermostable. Water sol; slightly sol in most organic solvents. Stable between pH 1 and 8.5. LD_{50} i.v. in mice: 300 mg/kg (Taniyama).

Hydrochloride. White powder. $[\alpha]_D^{25}$ −51.3° (c = 1.4). Sol in water and dil mineral acids. Practically insol in ether, petr ether, chloroform.

8960. Streptovaricin. [1404-74-6] Streptovarycin. A complex of ansamycin antibiotics consisting of streptovaricins A, B, C, D, E, F, G, J and K of which streptovaricin C is the major component. Isolation from *Streptomyces spectabilis:* Siminoff *et al., Am. Rev. Tuberc. Pulm. Dis.* **75**, 576 (1957); Whitfield *et al., ibid.* 584; and resolution of components: Dietz *et al.,* US **3116202** (1963 to Upjohn). Characterization of complex: Rinehart *et al., Chem. Commun.* **1974**, 861. Preliminary structural studies: Rinehart *et al., J. Am. Chem. Soc.* **88**, 3149, 3150 (1966); **90**, 6241 (1968); Rinehart *et al., ibid.* **93**, 6273 (1971). Revised structure of streptovaricin C: Sasaki *et al., J. Antibiot.* **25** 68 (1972); Rinehart, Antosz, *ibid.* 71. Activity studies: Quintrell, McAuslan, *J. Virol.* **6**, 485 (1970); Tan, McAuslan, *Biochem. Biophys. Res. Commun.* **42**, 230 (1971); Carter *et al., Nature New Biol.* **232**, 212, 214 (1971). Biosynthetic study: P. V. Deshmukh *et al., J. Am. Chem. Soc.* **98**, 870 (1976). ^{13}C-NMR studies: K. B. Kakinuma *et al., J. Org. Chem.* **41**, 1358 (1976). Synthesis of a key intermediate for the total synthesis of streptovaricin A: P. A. McCarthy, *Tetrahedron Lett.* **23**, 4199 (1982). Synthesis of the naphthalene core of streptovaricin D: B. M. Trost, W. H. Pearson, *ibid.* **24**, 269 (1983). Review of the chemistry of streptovaricins: K. L. Rinehart *et al., Fortschr. Chem. Org. Naturst.* **33**, 231-307 (1976).

R = COOCH₃
R' = OCOCH₃

Streptovaricin C

The complex yields yellow crystals from ethyl acetate + hexane. Neutral or weakly acidic substance which hydrolyzes on treatment with dil alkali. Stable at room temp for 3-4 days at pH 2.0-6.0. Gradually loses its antibiotic activity while standing at room temp for 2-4 days at pH 7.8.

Streptovaricin A. [23344-16-3] $C_{42}H_{53}NO_{16}$; mol wt 827.88. mp 194-196°. $[\alpha]_D$ in chloroform +61°. uv max (95% ethanol): 245, 260, 320, 430 nm ($E_{1cm}^{1\%}$ 418.9, 352.1, 139.8, 136.2).

Streptovaricin B. [11031-82-6] $C_{42}H_{53}NO_{15}$; mol wt 811.88. mp 185-187°. $[\alpha]_D$ in chloroform +576° (Dietz, +454°). uv max (95% ethanol): 245, 266, 320, 432 nm ($E_{1cm}^{1\%}$ 408.3, 338.5, 140.1, 109.7).

Streptovaricin C. [23344-17-4] $C_{40}H_{51}NO_{14}$; mol wt 769.84. mp 189-191° (Dietz, 168-171°). $[\alpha]_D$ in chloroform +602° (Dietz, +317°). uv max (95% ethanol): 245.5, 260, 320, 430 nm ($E_{1cm}^{1\%}$ 443.5, 389.7, 153.1, 121.9). Solubility data: March, Weiss, *J. As-soc. Off. Anal. Chem.* **50**, 457 (1967).

Streptovaricin D. [32164-26-4] $C_{40}H_{51}NO_{13}$; mol wt 753.84. mp 167-170° (Dietz, 115-118°). $[\alpha]_D$ in chloroform +436° (Dietz, +102°). uv max (95% ethanol): 246, 264, 320, 433 nm ($E_{1cm}^{1\%}$ 431.4, 370.3, 147.7, 105.3).

Streptovaricin E. [35413-63-9] $C_{40}H_{49}NO_{14}$; mol wt 767.83. mp 102-105°. $[\alpha]_D^{24}$ in chloroform +164°, also reported as +613°. uv max (95% ethanol): 245, 273, 320, 437 nm ($E_{1cm}^{1\%}$ 382.2, 346.9, 190.4, 91.9).

Streptovaricin F. [35512-37-9] $C_{39}H_{47}NO_{14}$; mol wt 753.80.

Streptovaricin G. [11031-85-9] $C_{40}H_{51}NO_{15}$; mol wt 785.84. mp 190-192°. $[\alpha]_D$ in chloroform +473°.

Streptovaricin J. [52275-61-3] $C_{42}H_{53}NO_{15}$; mol wt 811.88. mp 177-180°. $[\alpha]_D$ +326° (c = 0.25 in ethanol); +436° (c = 0.094 in chloroform).

8961. Streptovirudin. [56833-74-0] Nucleoside antibiotic complex produced by *Streptomyces griseoflavus* subsp. *thuringiensis*, strain JA 10124, that has inhibitory activity vs gram-positive bacteria, mycobacteria, and several RNA and DNA viruses. Isoln: K. Eckardt *et al., Z. Allg. Mikrobiol.* **13**, 625 (1973); *eidem, J. Antibiot.* **28**, 274 (1975). Taxonomy, fermentation, production: H. Thrum *et al., ibid.* 514. Streptovirudin is related chemically to tunicamycin, *q.v.*, and contains at least 10 components, of which 8 have been isolated in pure form. The complex is divided into 2 series of compounds: series I (streptovirudins A$_1$ through D$_1$) contains dihydrouracil and series II (streptovirudins A$_2$ through D$_2$) contains uracil. Differentiation of streptovirudin and tunicamycin by gel chromatography, HPLC, and hydrolysis: K. Eckardt *et al., ibid.* **33**, 908 (1980). Structures of streptovirudins and identity of components B$_{2a}$ and C$_2$ with tunicamycin components: *eidem, ibid.* **34**, 1631 (1981); W. Ihn *et al., Tetrahedron* **38**, 1781 (1982). Biological activities of isolated streptovirudin and tunicamycin components, identity of streptovirudin C$_2$ with tunicamycin A: R. W. Keenan *et al., Biochemistry* **20**, 2968 (1981). Chemical and biological properties of streptovirudins: A. D. Elbein *et al., ibid.* 4210. Viral inhibition study: M. S. Kang *et al., Biochem. Biophys. Res. Commun.* **99**, 422 (1981).

Streptovirudin A$_1$

White amorphous solid. uv max (methanol): 212 nm. Sol in methanol, pyridine. Slightly sol in water, ethanol. Insol in chloroform, acetone, benzene, ethyl acetate. Stable up to 5 hours at 90°. No loss of activity was observed after standing at 37° for 4 weeks. LD$_{50}$ in mice (mg/kg) of the antibiotic complex from culture filtrate: 250 orally; 15 s.c.; 15-17.5 i.p.; 17.5 i.v.; of the complex from mycelium: 150 orally; 5-10 s.c.; 3.5-7.5 i.p. (Thrum).

8962. Streptozocin. [18883-66-4] 2-Deoxy-2-[[(methylnitrosoamino)carbonyl]amino]-D-glucose; 2-deoxy-2-(3-methyl-3-nitrosoureido)-D-glucopyranose; streptozotocin; NSC-85998; U-9889; Zanosar. $C_8H_{15}N_3O_7$; mol wt 265.22. C 36.23%, H 5.70%, N 15.84%, O 42.23%. Isoln from *Streptomyces achromogenes* fermentation broth: Herr *et al., Antibiot. Annu.* **1959-1960**, 236; Bergy *et al.*, FR 1434920 (1966 to Upjohn), *C.A.* **65**, 17661h (1966). Structure and synthesis: R. R. Herr *et al., J. Am. Chem. Soc.* **89**, 4808 (1967); E. Hardegger *et al., Helv. Chim. Acta* **52**, 2555 (1969); E. J. Hessler, H. K. Jahnke, *J. Org. Chem.* **35**, 245 (1970); P. F. Wiley *et al., ibid.* **44**, 9 (1979). Diabetogenic effect: Rakieten *et al., Cancer Chemother. Rep.* **29**, 91 (May 1963). Antileukemic activity: Bhuyan *et al., Cancer Chemother. Rep. Part 1* **56**, 709 (1972). Biosynthetic study: S. Singaram *et al., J. Antibiot.* **32**, 379 (1979). Review and evaluation of studies of carcinogenicity in laboratory animals: *IARC Monographs* **4**, 221-227 (1974). *Review:* B. Rudas, *Arzneim.-Forsch.* **22**, 830-861 (1972); P. F. Wiley, *Anticancer Agents Based on Natural Product Models*, J. M. Cassidy, J. D. Douros, Eds. (Academic Press, New York, 1980) pp 167-200. Book: *Streptozotocin: Fundamentals and Therapy*, M. K. Agarwal, Ed. (Elsevier/North Holland Biomedical Press, New York, 1981) 309 pp. Review of pharmacology and clinical efficacy: R. B. Weiss *et al., J. Org. Chem.* **44**, 9-16 (1978).

Pointed platelets or prisms, from 95% ethanol, mp 115° (dec). Sol in H$_2$O, lower alcohols and ketones. uv max (ethanol): 228 nm (ε 6360). A mixture of α and β anomers; aq solns rapidly undergo mutarotation to an equilibrium value of $[\alpha]_D^{25}$ +39°. LD$_{50}$ in female mice (mg/kg): 360 i.p.; 275 i.v.; in male dogs (mg/kg): 50 i.v. *See:* M. Iwasaki *et al., J. Med. Chem.* **19**, 918 (1976).

Tetraacetate. $C_8H_{11}N_3O_7(COCH_3)_4$. Crystals, mp 111-114° (dec). $[\alpha]_D^{25}$ +41° (c = 0.78 in 95% ethanol).

Caution: Streptozocin is reasonably anticipated to be a human carcinogen: *Report on Carcinogens, Twelfth Edition* (PB2011-111646, 2011) p 329.

USE: Production of experimental diabetes in laboratory animals.

THERAP CAT: Antineoplastic.

THERAP CAT (VET): Antineoplastic.

8963. Strigol. [11017-56-4] (3E,3aR,5S,8bS)-3-[[[(2R)-2,5-Dihydro-4-methyl-5-oxo-2-furanyl]oxy]methylene]-3,3a,4,5,6,7,8,-8b-octahydro-5-hydroxy-8,8-dimethyl-2H-indeno[1,2-b]furan-2-one. $C_{19}H_{22}O_6$; mol wt 346.38. C 65.88%, H 6.40%, O 27.71%. Potent seed germination stimulant for the root parasite, witchweed, *Striga lutea* Lour. isolated from root exudates of cotton, *Gossypium hirsutum* L.: C. E. Cook *et al., Science* **154**, 1189 (1966). Structure: *eidem, J. Am. Chem. Soc.* **94**, 6198 (1972). Crystal structure: P. Coggan *et al., J. Chem. Soc. Perkin Trans. 2* **1973**, 465. Synthesis: J. B. Heather *et al., J. Am. Chem. Soc.* **98**, 3661 (1976); of (±)-form: G. A. MacAlpine *et al., Chem. Commun.* **1974**, 834; *eidem, J. Chem. Soc. Perkin Trans. 1* **1976**, 410. Synthesis of analogs: A. W. Johnson *et al., ibid.* **1981**, 1734. Witchweed seed germination stimulating activity: A. L. Hsiao *et al., Weed Sci.* **29**, 101 (1981); A. B. Pepperman *et al., ibid.* **30**, 561 (1982).

White needles from benzene-hexane, mp 200-202° (dec). $[\alpha]_D^{25}$ +293° (c = 0.15 in CHCl$_3$). uv max: 234 nm (ε 17700). Sol in acetone, methylene chloride. Moderately sol in benzene; insol in hexane.

8964. Strobilurins. Class of antifungal antibiotics found in wood-rotting mushrooms. Strobilurins A-H have been isolated. Strobilurin A is the parent compound for a new class of agricultural antifungals which inhibit mitochondrial respiration at the cytochrome *bc$_1$* complex. *See* synthetic analogs: azoxystrobin, kresoxim methyl. Identification of activity in *Oudemansiella mucida*: V. Musilek *et al.*, *Folia Microbiol.* **14**, 377 (1969). Isoln of A and B from *Strobilurus tenacellus*: T. Anke *et al.*, *J. Antibiot.* **30**, 806 (1977); of F, G, H from *Bolinea lutea* Sacc.: A. Fredenhagen *et al.*, *ibid.* **43**, 655 (1990). Identity with mucidin and mechanism of action: G. Von Jagow *et al.*, *Biochemistry* **25**, 775 (1986). Revised stereostructure and synthesis of A: T. Anke *et al.*, *Ann.* **1984**, 1616. Stereosynthesis: K. Beautement, J. M. Clough, *Tetrahedron Lett.* **28**, 475 (1987). Brief review: T. Anke, *Can. J. Bot.* **73** Suppl. 1, S940-S945 (1995).

Strobilurin A

Strobilurin A. [52110-55-1] (*E,Z,E*)-2-(Methoxymethylene)-3-methyl-6-phenyl-3,5-hexadienoic acid methyl ester; mucidin. C$_{16}$H$_{18}$O$_3$; mol wt 258.32. Colorless, homogeneous oil. uv max (ethanol): 230, 237, 294 nm (ε 16900, 15330, 21850). Sol in methanol, ethanol. acetone, ethyl acetate, CHCl$_3$, CCl$_4$; very poorly sol in water.

8965. Strontium. [7440-24-6] Sr; at. wt 87.62; at. no. 38; valence 2. Group IIA (2). Alkaline earth metal. Naturally occurring, stable isotopes: 88 (82.58%); 86 (9.86%); 87 (7.00%); 84 (0.56%). Known radioactive isotopes: 76-83, 85, 89, 90 (longest lived isotope, T$_{1/2}$ 28.78 yr, β^- decay), 91-102. Occurs in the minerals, *celestine* (SrSO$_4$) and *strontianite* (SrCO$_3$); found in small quantities associated with calcium or barium minerals. Abundance in earth's crust 384 ppm. First prepared in 1807 by Davy. Commercial production by thermal reduction of strontium oxide with aluminum. Prepn: Glascock, *J. Am. Chem. Soc.* **32**, 1222 (1910); Matignon, *Compt. Rend.* **177**, 1116 (1923); *J. Chem. Soc.* **126**[ii], 44 (1924); Guntz *et al.*, cited in *Gmelins, Strontium* (8th ed.) **29**, 35 (1931). *Reviews*: Goodenough, Stenger, "Magnesium, Calcium, Strontium, Barium and Radium" in *Comprehensive Inorganic Chemistry* vol. 1, J. C. Bailar, Jr. *et al.*, Eds. (Pergamon Press, Oxford, 1973) pp 591-774; *Chemistry of the Elements* N. N. Greenwood, A. Earnshaw, Eds. (Pergamon Press, New York, 1984) pp 117-154; S. G. Hibbins in *Kirk-Othmer Encyclopedia of Chemical Technology* vol. **22** (Wiley-Interscience, New York, 4th ed., 1997) pp 947-955. Review of clinical use of ^{89}Sr in pain palliation for metastatic bone cancer: G. B. Altman, C. A. Lee, *Oncol. Nurs. Forum* **32**, 523-527 (1996). Review of toxicology and human exposure: *Toxicological Profile for Strontium* (PB2004-104400, 2004) 445 pp.

Silvery-white metal; face-centered cubic structure; rapidly becomes yellow on exposure to air and assumes an oxide film. The finely divided metal ignites spontaneously in air. d 2.6; mp 757 ±1°; bp 1366°. E^0 (aq) Sr^{2+}/Sr -2.89 V. Undergoes reactions characteristic of alkaline earth metals. Strong reducing agent. *Keep under liquid containing no oxygen.* The heated metal combines with hydrogen to form *strontium hydride* and with nitrogen to form *strontium nitride*. Metal and its salts emit characteristic brilliant red color in flame.

Note: The problems of internal radiation hazards from radiostrontium are discussed by A. Engström *et al.*, *Bone and Radiostrontium* (Wiley, New York, 1958).

USE: In fireworks, in red signal flares; on tracer bullets. Eutectic modifier in Al-Ag casting alloys to improve strength and ductility. Innoculant in ductile iron casting to control graphite formation.

THERAP CAT: ^{89}Sr as antineoplastic (radiation source).

8966. Strontium Acetate. [543-94-2] Acetic acid strontium salt (2:1). C$_4$H$_6$O$_4$Sr; mol wt 205.71. C 23.36%, H 2.94%, O 31.11%, Sr 42.59%. Sr(C$_2$H$_3$O$_2$)$_2$. Toxicity: V. V. Cole *et al.*, *J. Pharmacol. Exp. Ther.* **71**, 1 (1941).

Hemihydrate. White crystalline powder. Loses its water at 150°; on ignition is converted into SrCO$_3$. Sol in 2.5 parts water; slightly sol in alcohol. The aq soln is practically neutral to litmus. LD i.v. in rats: 1.16 mmol/kg (Cole).

8967. Strontium Bromide. [10476-81-0] Br$_2$Sr; mol wt 247.43. Br 64.59%, Sr 35.41%. SrBr$_2$. Acute toxicity: K. W. Cochran *et al.*, *Arch. Ind. Hyg. Occup. Med.* **1**, 637 (1950).

Hexahydrate. Colorless, deliquesc crystals or white granules; bitter saline taste. mp 88° when rapidly heated; mp 643° when anhydr. Loses all its water at 180°. Sol in 0.35 part water; sol in alcohol. Insol in ether. The aq soln is neutral. *Incompat:* Soluble sulfates. LD$_{50}$ i.p. in rats: 1000 mg/kg (Cochran).

THERAP CAT: Has been used as anticonvulsant.

THERAP CAT (VET): Has been used as sedative, anticonvulsant.

8968. Strontium Carbonate. [1633-05-2] Carbonic acid strontium salt (1:1). CO$_3$Sr; mol wt 147.63. C 8.14%, O 32.51%, Sr 59.35%. SrCO$_3$. Occurs in nature as the mineral *strontianite*.

White, odorless, tasteless powder. d 3.5. Dec at 1100° into SrO and CO$_2$. Sol in 100,000 parts water, in about 1000 parts of water saturated with CO$_2$; sol in dil acids.

USE: In pyrotechnics; manuf iridescent glass; refining sugar.

8969. Strontium Chlorate. [7791-10-8] Chloric acid strontium salt (2:1). Cl$_2$O$_6$Sr; mol wt 254.51. Cl 27.86%, O 37.72%, Sr 34.43%. Sr(ClO$_3$)$_2$.

Colorless or white crystals. d 3.15. mp 120° with decomposition and evolution of O$_2$. Sol in 0.6 part water, slightly in alcohol. *Oxidizer. Handle with caution like potassium chlorate.*

USE: In pyrotechnics to produce red fire.

8970. Strontium Chloride. [10476-85-4] Cl$_2$Sr; mol wt 158.52. Cl 44.73%, Sr 55.27%. SrCl$_2$. Acute toxicity: I. B. Syed, F. Hosain, *Toxicol. Appl. Pharmacol.* **22**, 150 (1972).

Hexahydrate. [10025-70-4] Cl$_2$Sr.6H$_2$O; mol wt 266.61. Colorless, odorless crystals or white granules. Effloresces in air; deliquesces in moist air. At 100° loses 5H$_2$O; at 150° all its H$_2$O; d 1.96; mp 61° when rapidly heated; the anhydr salt melts at 868°. Soluble in 0.8 part water, 0.5 part boiling water; sol in alcohol. The aq soln is neutral. *Keep well closed.* LD$_{50}$ i.v. in mice: 147.6 mg/kg (Syed, Hosain).

Strontium chloride Sr 89. [38270-90-5] Metastron. Clinical studies in pain palliation for metastatic bone cancer: R. G. Robinson *et al.*, *RadioGraphics* **9**, 271 (1989); K. M. Kälkner *et al.*, *Anticancer Res.* **20**, 1109 (2000). Contains ^{89}Sr which is a beta emitter with a half-life of 50.52 days.

USE: Manuf strontium salts; in pyrotechnics; pharmaceutic aid (dental desensitizer).

THERAP CAT: Strontium chloride Sr 89 as antineoplastic (radiation source).

8971. Strontium Chromate(VI). [7789-06-2] Chromic acid (H$_2$CrO$_4$) strontium salt (1:1). CrO$_4$Sr; mol wt 203.61. Cr 25.54%, O 31.43%, Sr 43.03%. SrCrO$_4$.

Yellow powder. d 3.89. Soluble in 840 parts cold water, about 5 parts boiling water; freely sol in dil hydrochloric, nitric or acetic acids.

Caution: Chromium hexavalent (VI) compounds are listed as known human carcinogens: *Report on Carcinogens, Twelfth Edition* (PB2011-111646, 2011) p 109.

USE: Corrosion inhibitor in pigments; in electrochemical processes to control sulfate concn of solns.

8972. Strontium Fluoride. [7783-48-4] F$_2$Sr; mol wt 125.62. F 30.25%, Sr 69.75%. SrF$_2$. Prepd by dissolving SrCO$_3$ in an excess of 40% HF and evaporating to dryness: Berzelius, *Pogg. Ann.* **1**, 20 (1824); final drying should be done at 150° *in vacuo*.

White powder or cubic crystals. d 4.24. mp ~1400°. bp 2460°. Stable in air up to 1000°. Above this temp it is oxidized to strontium oxide. Heat of evaporation 78.2. Trouton const 28.6. Soly in water (18°) 11.7 mg/100 ml. More sol in dil acids. Dec by strong acids. May be stored in glass bottles.

8973. Strontium Hydroxide. [18480-07-4] Strontium hydroxide (Sr(OH)$_2$); strontium hydrate. H$_2$O$_2$Sr; mol wt 121.63. H 1.66%, O 26.31%, Sr 72.04%. Sr(OH)$_2$.

Octahydrate. Colorless, deliquesc crystals or white powder; absorbs CO$_2$ from air forming carbonate. Loses part of its water at about 100°. Sol in 50 parts water, 2.1 parts boiling water. The soln is very alkaline. pH about 13.5. *Keep tightly closed.*

USE: Refining beet sugar; separating crystallizable sugar from molasses.

8974. Strontium Iodide. [10476-86-5] I$_2$Sr; mol wt 341.43. I 74.34%, Sr 25.66%. SrI$_2$. The article of commerce is hydrated. Toxicity data: K. W. Cochran *et al., Arch. Ind. Hyg. Occup. Med.* **1**, 637 (1950).

Hexahydrate. Colorless to yellowish, deliquesc, fused masses or granules with bitterish saline taste; becomes yellow on exposure to air and light due to liberation of iodine. Melts when rapidly heated, at about 120°. mp 402° (anhydrous). d 4.42. Sol in 0.2 part water; sol in alcohol. The aq soln is practically neutral. *Keep well closed and protected from light.* LD$_{50}$ i.p. in rats: 800 mg/kg (Cochran).

THERAP CAT: Iodine source.

THERAP CAT (VET): Has been used in iodide therapy.

8975. Strontium Nitrate. [10042-76-9] Nitric acid strontium salt (2:1). N$_2$O$_6$Sr; mol wt 211.63. N 13.24%, O 45.36%, Sr 41.40%. Sr(NO$_3$)$_2$.

White granules or powder. *Oxidizer.* d 2.99. mp 570°. n_D 1.588. Sol in 1.5 parts water; slightly sol in alcohol or acetone. The aq soln is neutral. At low temps strontium nitrate crystallizes with 4H$_2$O (25.4%). LD$_{50}$ i.p. in rats: 540 mg/kg, Cochran *et al., Arch. Ind. Hyg.* **1**, 637 (1950).

USE: In pyrotechnics (red fire), signal lights, marine signals, railroad flares, matches.

8976. Strontium Oxide. [1314-11-0] Strontia; strontium monoxide. OSr; mol wt 103.62. O 15.44%, Sr 84.56%. SrO.

White to grayish-white, porous, caustic mass. d 4.7. mp 2430°. When treated with water, forms the hydroxide with evolution of much heat. *Keep tightly closed.*

USE: Manuf of strontium salts.

8977. Strontium Peroxide. [1314-18-7] Strontium dioxide. O$_2$Sr; mol wt 119.62. O 26.75%, Sr 73.25%. SrO$_2$. The peroxide of commerce contains about 85% SrO$_2$.

White, odorless, tasteless powder; gradually dec on exposure to air. Almost insol in water, but is gradually dec by it with evolution of oxygen; forms hydrogen peroxide with dil acids. *Oxidizer. Keep well closed.*

THERAP CAT: Antiseptic.

8978. Strontium Sulfate. [7759-02-6] Sulfuric acid strontium salt (1:1). O$_4$SSr; mol wt 183.68. O 34.84%, S 17.45%, Sr 47.70%. SrSO$_4$. Occurs in nature as the mineral *celestine* or *celestite.*

White, odorless, crystalline powder. d 3.96. One gram dissolves in ~8800 ml water, 800 ml 2% HCl, 700 ml 3% HNO$_3$; appreciably sol in alkali chloride solns.

USE: In ceramics, pyrotechnics.

8979. Strontium Sulfide. [1314-96-1] SSr; mol wt 119.68. S 26.79%, Sr 73.21%. SrS. The article of commerce contains 65-75% SrS. Strontium sulfide made by reduction of the sulfate is not luminous; the luminous sulfide is prepared by heating strontium hydroxide with sulfur.

Gray powder; has the odor of H$_2$S in moist air. d 3.70. Slightly sol in water; sol in acids with decompn. *Keep dry.*

USE: In luminous paints; as a depilatory.

8980. Strontium Titanate. [12060-59-2] Strontium titanium oxide (SrTiO$_3$). O$_3$SrTi; mol wt 183.48. O 26.16%, Sr 47.75%, Ti 26.09%. SrTiO$_3$. Dielectric and photoelectric material. Occurs in nature as the mineral *tausonite.* Optical properties: H. A. Weakliem *et al., RCA Rev.* **36**, 149 (1975). Electrical conductivity: U. Balachandran, N. G. Eror, *J. Solid State Chem.* **39**, 351 (1981). Dielectric properties of glass-ceramics: S. L. Swartz, A. S. Bhalla, *Ferroelectrics* **87**, 141 (1988). Review of growth and properties of single crystals: K. Nassau, A. E. Miller, *J. Cryst. Growth* **91**, 373-381 (1988). Prepn by sol-gel method: J. Moreno *et al., J. Mater. Chem.*

5, 509 (1995); by solution-precipitation method: V. Kumar, *J. Am. Ceram. Soc.* **82**, 2580 (1999). Prepn of thin films by ECR plasma sputtering: S. Miyake *et al., Surf. Coat. Technol.* **169-170**, 27 (2003); by spray pyrolysis: G. Brankovic *et al., J. Eur. Ceram. Soc.* **24**, 989 (2004). Piezoelectric response at cryogenic temperatures: D. E. Grupp, A. M. Goldman, *Science* **276**, 392 (1997). IR characterization of thin films: B. G. Almeida *et al., Appl. Surf. Sci.* **238**, 395 (2004). Room-temperature ferroelectricity: J. H. Haeni *et al., Nature* **430**, 758 (2004).

d 5.12. n_D 2.409. Hardness 5½ on Moh's scale. Specific heat: ~10 J s^{-1} K^{-1} m^{-1}. Thermal conductivity: ~52 mW cm^{-1} K^{-1}.

USE: In electronics and electrochemical insulation; in photocatalysis; sputtering target for thin film capacitors; substrate for epitaxial growth of high temperature superconductor thin films. Diamond simulant gemstone.

8981. Strophanthidin. [66-28-4] (3β,5β)-3,5,14-Trihydroxy-19-oxocard-20(22)-enolide; apocynamarin; convallatoxigenin; corchorin; cymarigenin; cynotoxin; corchorgenin; corchsularin. C$_{23}$H$_{32}$O$_6$; mol wt 404.50. C 68.29%, H 7.97%, O 23.73%. By acid or enzymic hydrolysis of glycosides present in several species of *Strophanthus, Apocynaceae.* Isoln: Jacobs, Heidelberger, *J. Biol. Chem.* **54**, 253 (1922). Structure: Kon, *Chem. Ind. (London)* **53**, 593, 956 (1934); Jacobs, Elderfield, *Science* **80**, 533 (1934). Identity with corchorin, corchorgenin and corchsularin: Sen *et al., Helv. Chim. Acta* **40**, 588 (1957). Synthesis from pregnenolone acetate, *q.v.:* E. Yoshii *et al., J. Org. Chem.* **43**, 3946 (1978). Toxicity study: Graebner, Geisel, *Arzneim.-Forsch.* **22**, 1854 (1972). *Reviews:* Elderfield, *Chem. Rev.* **17**, 187 (1935); Tschesche, *Ergeb. Physiol.* **38**, 31 (1936); Stoll, *The Cardiac Glycosides* (Pharmaceutical Press, London, 1937); Strain in *Organic Chemistry* vol. **II**, Gilman, Ed. (New York, 2nd ed., 1943); Fieser, Fieser, *Steroids* (Reinhold, New York, 1959) pp 729-730, 736-750.

Orthorhombic tablets from 5 parts methanol and 10 parts water. *Poisonous.* Contains ½ mol H$_2$O which is given up at 110° over P$_2$O$_5$ at 20 mm Hg. mp 171-175° with effervescence. Occasionally a few isolated crystals are found which melt at ~230°. [α]$_D^{25}$ +43.1° (c = 2.8 in methanol). Sol in alcohol, acetone, chloroform, benzene, and glacial acetic acid. Practically insol in water, ether, petr ether. Color reaction with H$_2$SO$_4$ and Ac$_2$O like cholesterol. LD$_{100}$ i.v. in cats: 0.337 mg/kg (Graebner, Geisel).

3-Benzoylstrophanthidin. C$_{30}$H$_{36}$O$_7$. mp 230°. [α]$_D^{26}$ +47.8° (c = 1.07 in acetone).

3-(p-Bromobenzoyl)strophanthidin. C$_{30}$H$_{35}$BrO$_7$. mp 222-224°. [α]$_D^{20}$ +42.0° (c = 1.094 in acetone).

Dihydrostrophanthidin. C$_{23}$H$_{34}$O$_6$. mp 100-103°, resolidifies and again melts at 190-195°. [α]$_D^{23}$ +34.9° (methanol).

16-Hydroxystrophanthidin. Strophadogenin. C$_{23}$H$_{32}$O$_7$. Long prisms from ethanol, dec 238-241°. [α]$_D^{23}$ +52.9° (c = 1.24 in 96% alc). uv max: 219 nm (log ε 4.3).

8982. Strophanthin. [11005-63-3] K-Strophanthin; K-strophanthoside; Kombetin. A glycoside or a mixture of glycosides obtained from *Strophanthus kombé* Oliv., *Apocynaceae.* Properly speaking K-strophanthoside contains α-glucose 19%, β-glucose 19%, cymarose 15%, strophanthidin 47%. K-Strophanthin-β (Strophosid) contains β-glucose, cymarose, and strophanthidin: Geissberger, *Schweiz. Med. Wochenschr.* **91**, 241 (1961). Pharmacology and acute toxicity: H. Mehnert, *Arch. Pharmacol.* **184**, 181 (1937).

White or yellowish powder contg as much as 10% water, which it does not lose entirely without decompn. *Poisonous.* Stable in air, but affected by light. Soluble in water and in dil alcohol; less sol in abs alcohol. Nearly insol in chloroform, ether, benzene. Aq solns are neutral to litmus. When bioassayed, strophanthin shall possess a potency per mg equivalent to 0.5 mg of U.S.P. Ouabain Reference Standard. A deviation of 20% is permitted. MLD in cats, rats (mg/kg): 0.11, 9.4 i.v. (Mehnert).

G-strophanthin see Ouabain.

THERAP CAT: Cardiotonic.

THERAP CAT (VET): Has been used as a cardiotonic.

8983. Strophanthus. Dried ripe seeds of *Strophanthus kombé* Oliv., or of *S. hispidus* DC., *Apocynaceae*, deprived of the awns. *Habit.* East and Central Africa. *Constit.* 2-5% Strophanthin, kombic acid, choline, trigonelline, ~30% oil. *Poisonous!*

USE: As arrow poison by African natives.

THERAP CAT: Cardiotonic.

THERAP CAT (VET): Has been used as a cardiotonic.

8984. Struvite. [15490-91-2] Magnesium ammonium phosphate hexahydrate; guanite. $H_4MgNO_4P.6H_2O$; mol wt 245.40. H 6.57%, Mg 9.90%, N 5.71%, O 65.20%, P 12.62%. (NH_4)-$MgPO_4.6H_2O$. Phosphate mineral named in honor of H. C. G. von Sturve, a Russian naturalist, in 1845 by Ulex, a Swedish geologist. A component of urinary calculi and bat guano also found as a contaminant in canned foods, in waste management systems. Production by urease-splitting bacteria: H. Robinson, *Proc. Cambridge Philos. Soc.* **6**, 360 (1889); J. Beavon, N. G. Heatley, *J. Gen. Microbiol.* **31**, 167 (1962); M. T. Gonzalez-Munoz *et al.*, *Chemosphere* **26**, 1881 (1993). Identification in tinned seafood: M. T. Gilles, Seafood Processing in *Food Processing Review* **22**, (Noyes Data Corp., Park Ridge, NJ, 1971) pp 31-58; in home-canned beef: J. G. Sebranek *et al.*, *J. Muscle Foods* **4**, 81 (1993). Precipitation in sludge digesters: C. Maqueda *et al.*, *Water Res.* **28**, 411 (1994). Crystal growth: R. J. C. McLean *et al.*, *Urol. Res.* **18**, 39 (1990). Thermodynamics of formation: J. R. Buchanan *et al.*, *Trans. ASAE* **37**, 617 (1994). Chemical control in waste systems: J. R. Buchanan *et al.*, *ibid.* 1301. Review of struvite urolithiasis: formation, detection and dissolution: C. A. Osborne *et al.*, *Adv. Vet. Sci. Compar. Med.* **29**, 1-101 (1985). Review of struvite calculi: M. J. Gleeson, D. P. Griffith, *Br. J. Urol.* **71**, 503-511 (1993).

Glass like crystals.

8985. Strychnine. [57-24-9] Strychnidin-10-one. $C_{21}H_{22}$-N_2O_2; mol wt 334.42. C 75.42%, H 6.63%, N 8.38%, O 9.57%. Occurs most abundantly in seeds of *Strychnos nux-vomica* L., *Loganiaceae* and beans of *S. ignatti*, Berg. One of the first alkaloids isolated in pure form. Isoln: Pelletier, Caventou, *Ann. Chim. Phys.* **8**, 323 (1818); *eidem, ibid.* **10**, 142 (1819). Extraction procedure: C. Srinivasulu *et al.*, *Res. Ind.* **23**, 224 (1978). Structure elucidation: L. H. Briggs *et al.*, *J. Chem. Soc.* **1946**, 903; H. T. Openshaw, R. Robinson, *Nature* **157**, 438 (1946); R. B. Woodward *et al.*, *J. Am. Chem. Soc.* **69**, 2250 (1947). Total synthesis: R. B. Woodward *et al., ibid.* **76**, 4749 (1954); *eidem, Tetrahedron* **19**, 247 (1963). Abs config: A. F. Peerdeman, *Acta Crystallogr.* **9**, 824 (1956); K. Nagarajan *et al.*, *Helv. Chim. Acta* **46**, 1212 (1963). Stereoselective total synthesis: S. D. Knight *et al.*, *J. Am. Chem. Soc.* **115**, 9293 (1993). ^{1}H- and ^{13}C-NMR analysis: E. Wenkert *et al.*, *J. Org. Chem.* **43**, 1099 (1978). Pharmacology and pharmacokinetics: S. Weiss, R. A. Hatcher, *J. Pharmacol. Exp. Ther.* **19**, 419 (1921). Acute toxicity: I. Setnikar, M. J. Magistretti, *Arzneim.-Forsch.* **14**, 996 (1964). HPLC determn in urine and tissues: T. Egloff *et al.*, *J. Clin. Chem. Clin. Biochem.* **20**, 203 (1982); GC/EI-MS determn in whole blood: M. Barroso *et al.*, *J. Chromatogr. B* **816**, 37 (2005). *Review:* J. B. Hendrickson in *The Alkaloids* vol. **VI**, R. H. F. Manske, Ed. (Academic Press, New York, 1960) pp 179-195; G. F. Smith, *ibid.* vol. **VIII** (1965) pp 591-671. Comprehensive description: F. J. Muhtadi, M. S. Hifnawy, *Anal. Profiles Drug Subs.* **15**, 563-646 (1986). Brief review of synthesis: U. Beifuss, *Angew. Chem. Int. Ed.* **33**, 1144-1149 (1994).

Brilliant, colorless cubes from chloroform-ether, mp 275-285°. d^{18} 1.359. $[\alpha]_D^{18}$ −104.3° (c = 0.254 in alc); $[\alpha]_D^{25}$ −139° (c = 0.4 in chloroform). pKa (25°) 8.26: A. J. Everett *et al.*, *J. Chem. Soc.* **1957**, 1120. uv max (95% ethanol): 255, 280, 290 nm ($E_{1\ cm}^{1\%}$ 377, 130, 101): A. I. Biggs, *J. Pharm. Pharmacol.* **13**, 547 (1952). One gram dissolves in 182 ml ethanol, 6.5 ml chloroform, 150 ml benzene, 250 ml methanol, 83 ml pyridine; very slightly sol in ether, water. LD_{50} i.v. (slow infusion) in rats: 0.96 mg/kg (Setnikar, Magistretti).

Hydrochloride dihydrate. [6101-04-8] $C_{21}H_{22}N_2O_2.HCl.$-$2H_2O$; mol wt 406.91. Efflorescent, trimetric prisms. One gram dissolves in ~35 ml cold water, ~80 ml alc. Insol in ether. pH of $0.01M$ soln 5.4.

Nitrate. [66-32-0] $C_{21}H_{22}N_2O_2.HNO_3$; mol wt 397.43. Colorless, odorless needles or white, cryst powder. At 25°, one gram dissolves in 42 ml water, 120 ml alcohol, 156 ml chloroform. Insol in ether. pH ~5.7. *Protect from light.*

Sulfate. [60-41-3] $(C_{21}H_{22}N_2O_2)_2.H_2SO_4$; mol wt 766.91. Usually crystallizes as pentahydrate. Properties of pentahydrate: colorless, odorless, very bitter crystals, or white, cryst powder. Effloresces in dry air; loses all its water of crystn at 100°. mp when anhydr ~200° with decompn. At 25°, one gram dissolves in 31 ml water, 65 ml alcohol, 325 ml chloroform. Insol in ether. pH 5.5 in 1:100 soln. *Protect from light.* LD_{50} orally in rats: 5 mg/kg, E. W. Schafer, *Toxicol. Appl. Pharmacol.* **21**, 315 (1972).

Caution: Extremely poisonous, *see:* G. D. Osweiler, *Curr. Vet. Ther.* **6**, 115 (1977). Potential symptoms of overexposure to strychnine are stiff neck and facial muscles; restlessness, apprehension and increased acuity of perception; increased reflex excitability; cyanosis; tetanic convulsions with opisthotonos. *See NIOSH Pocket Guide to Chemical Hazards* (DHHS/NIOSH 97-140, 1997) p 286.

USE: Chiefly in poison baits for rodents.

THERAP CAT (VET): Has been used as a tonic and central stimulant.

8986. Strychnine N^6-Oxide. [7248-28-4] Strychnidin-10-one 19-oxide; Genostrychnine. $C_{21}H_{22}N_2O_3$; mol wt 350.42. C 71.98%, H 6.33%, N 7.99%, O 13.70%. Prepn: Pictet, Mattison, *Ber.* **38**, 2782 (1905); Bailey, Robinson, *J. Chem. Soc.* **1948**, 703; Zellner, US 2758113 (1956 to Donau-Pharmazie).

Monoclinic prisms from water, dec 207°. pK 5.17. Freely sol in alc, glacial acetic acid, chloroform; fairly sol in water; sparingly sol in benzene. Practically insol in ether, petr ether.

Hydriodide. Light yellow plates from hot water, dec 253°.

Nitrate. Small transparent prisms from water, dec 250°.

8987. Stryker's Reagent. [33636-93-0] Octahedrohexa-μ^3-hydrohexakis(triphenylphosphine)hexacopper; hexa-μ-hydrohexakis(triphenylphosphine) hexacopper; triphenylphosphine copper(I) hydride hexamer. $C_{108}H_{96}Cu_6P_6$; mol wt 1961.07. C 66.15%, H 4.93%, Cu 19.44%, P 9.48%. $[(C_6H_5)_3PCuH]_6$. Copper(I) hydride cluster; mild hydride donor in organic synthesis. Prepn and crystal characterization: S. A. Bezman *et al.*, *J. Am. Chem. Soc.* **93**, 2063 (1971); M. R. Churchill *et al.*, *Inorg. Chem.* **11**, 1818 (1972). Improved prepns: D. M. Brestensky *et al.*, *Tetrahedron Lett.* **29**, 3749 (1988); P. Chiu *et al.*, *ibid.* **44**, 455 (2003). Use in reduction of α,β-unsaturated carbonyl compds: W. S. Mahoney *et al.*, *J. Am. Chem. Soc.* **110**, 291 (1988); W. S. Mahoney, J. M. Stryker, *ibid.* **111**, 8818 (1989); in hydrosilylation of aldehydes and ketones: B. H. Lipshutz *et al.*, *J. Organomet. Chem.* **624**, 367 (2001). Brief review: A. de Fátima, *Synlett* **2005**, 1805-1806 (2005).

Isolated as $H_6Cu_6(PPh_3)_6.DMF$. Bright red crystals, mp 111° (dec). d 1.367. Sol in benzene; slightly sol in DMF. Insol in acetonitrile, water. Dec in alcohols, ethers, chlorinated hydrocarbons. *Air sensitive in soln. Store under nitrogen.*

USE: Stoichiometric reducing agent or catalyst in reduction reactions.

8988. Stylopine. [4312-32-7] 6,7,12b,13-Tetrahydro-4*H*-bis-[1,3]benzodioxolo[5,6-*a*:4',5'-*g*]quinolizine; 2,3:9,10-bis(methylenedioxy)-13a-berbine; tetrahydrocoptisine. $C_{19}H_{17}NO_4$; mol wt 323.35. C 70.58%, H 5.30%, N 4.33%, O 19.79%. From root of *Stylophorum diphyllum* (Michx.) Nutt, *Corydalis cava* Schwg. and *Chelidonium majus* L., *Papaveraceae*: Schlotterbeck, Watkins, *Ber.* **35**, 7 (1902); Späth, Julian, *Ber.* **64**, 1131 (1931); Manske, *Can. J. Res.* **B20**, 53 (1942); Slavik, *Collect. Czech. Chem. Commun.* **20**, 198 (1955); Bandelin, Malesh, *J. Am. Pharm. Assoc.* **45**, 702 (1956); Slavik, *Collect. Czech. Chem. Commun.* **26**, 2933 (1961).

***dl*-Form.** Crystals from chloroform + methanol, mp 222-224°.
Hydrochloride. $C_{19}H_{18}ClNO_4$. Needles from methanol + HCl, mp 266-269°.
***l*-Form.** Needles from chloroform + ethanol, mp 203°. $[\alpha]_D^{25}$ −300° (c = 0.53 in chloroform). Practically insol in water; sol in alc, glacial acetic acid; slightly sol in dil acids.
***d*-Form.** Crystals from alcohol, mp 203-204°. $[\alpha]_D^{25}$ +310° (c = 0.88 in chloroform). Sol in chloroform; slightly sol in ether, ethanol, methanol.

8989. Styphnic Acid. [82-71-3] 2,4,6-Trinitro-1,3-benzenediol; 2,4,6-trinitroresorcinol; 2,4-dihydroxy-1,3,5-trinitrobenzene. $C_6H_3N_3O_8$; mol wt 245.10. C 29.40%, H 1.23%, N 17.14%, O 52.22%. May be prepd from natural sources, such as Pernambuco wood extract or quebracho extract by the action of nitric acid: Einbeck, Jablonski, *Ber.* **54**, 1086 (1921). Prepd industrially by first sulfonating, then nitrating resorcinol: Merz, Zetter, *Ber.* **12**, 681, 2037 (1879); Datta, Varma, *J. Am. Chem. Soc.* **41**, 2043 (1919). Prepd by nitrating 2,4-dinitroresorcinol: Kametani, Ogasawara, *Chem. Pharm. Bull.* **15**, 893 (1967).

Hexagonal yellow crystals from dil alc. Astringent, but not bitter taste. Solvated crystals from acetic acid. Sublimes in high vacuum. Becomes almost colorless upon vacuum sublimation, but turns deep yellow on contact with air. When dry, mp 175.5°; also reported as mp 179-180°. One gram dissolves in 156 ml water at 14°, in 88 ml water at 62°. Freely sol in alcohol, ether. Acid to litmus. Dibasic acid. Deflagrates on rapid heating.
USE: Has been used in the manuf of explosives.

8990. Styrene. [100-42-5] Ethenylbenzene; styrol; styrolene; cinnamene; cinnamol; phenylethylene; vinylbenzene. C_8H_8; mol wt 104.15. C 92.26%, H 7.74%. Isolated from storax, *q.v.*, by Bonastre in 1831. Manuf from benzene and ethylene: *Faith, Keyes & Clark's Industrial Chemicals*, F. A. Lowenheim, M. K. Moran, Eds. (Wiley-Interscience, New York, 4th ed., 1975) pp 779-785. Synthesis starting with 1-phenylethanol and leading to polystyrene: Wilen *et al.*, *J. Chem. Educ.* **38**, 304 (1961). When heated to 200° it is converted into the polymer, polystyrene, which is a clear plastic having excellent insulating properties even at ultra-high radio frequencies. Toxicity data: H. J. Meyer, R. Kretzschmar, *Arzneim.-Forsch.* **19**, 617 (1969). Monograph: W. C. Teach, G. C. Kiessling, *Polystyrene* (Reinhold, New York, 1960). Reviews of styrene monomer and polymers: Boyer *et al.*, "Styrene Polymers" in *Encyclopedia of Polymer Science and Technology* **vol. 13** (Interscience, New York, 1970) pp 128-447; Coulter *et al.*, "Styrene and Related Monomers" in *Vinyl and Diene Monomers* (part 2), E. C. Leonard, Ed. (Wiley-Interscience, New York, 1971) pp 479-576. Review of neurotoxicity: R. Pahwa, J. Kalra, *Vet. Hum. Toxicol.* **35**, 516-520 (1993); of toxicology and carcinogenic risk: *IARC Sci. Publ.* **127**, 1-412 (1993); of toxicology and human exposure: *Toxicological Profile for Styrene* (PB2010-100007, 2010) 283 pp.

Colorless to yellowish, very refractive, oily liq; penetrating odor. On exposure to light and air it slowly undergoes polymerization and oxidation with formation of peroxides, etc. d^{20} 0.9059. mp −30.6°. bp 145-146°. n_D^{20} 1.5463. Flash pt, closed cup: 87°F (31°C). Sol in alc, ether, methanol, acetone, carbon disulfide; sparingly sol in water. LD_{50} in mice (mg/kg): 660 ± 44.3 i.p.; 90 ± 5.2 i.v. (Meyer, Kretzschmar).
Polystyrene. Dylene; Trycite. Physical properties of unmodified polystyrene: d_4^{20} 1.04-1.065; n_D^{25} 1.60; water-clear solid plastic, begins to soften at ~85°. Dielectric constant at 100 megacycles: 2.4-2.65.
Caution: Potential symptoms of overexposure are irritation of eyes, nose, respiratory system; headache, fatigue, dizziness, confusion, malaise, drowsiness, weakness and unsteady gait; narcosis; defatting dermatitis; liver injury, reproductive effects. *See NIOSH Pocket Guide to Chemical Hazards* (DHHS/NIOSH 97-140, 1997) p 286. Styrene is listed as reasonably anticipated to be a human carcinogen: *Report on Carcinogens, Twelfth Edition* (PB2011-111646, 2011) p 383.
USE: Manuf plastics; synthetic rubber; resins; insulator.

8991. Styrene Glycol. [93-56-1] 1-Phenyl-1,2-ethanediol; phenylglycol; phenylethyleneglycol; α,β-dihydroxyethylbenzene. $C_8H_{10}O_2$; mol wt 138.17. C 69.54%, H 7.30%, O 23.16%. Prepd by reduction of phenylglyoxylic acid with $LiAlH_4$: Nystrom, Brown, *J. Am. Chem. Soc.* **69**, 2548 (1947); by reduction of styrene peroxide: Russel, *J. Am. Chem. Soc.* **75**, 5011 (1953). Prepn of optically active styrene glycol: Wilhelm, Bright, *Helv. Chim. Acta* **37**, 221 (1954).

Needles from ligroin, mp 67-68°. bp_{755} 272-274°. Freely sol in water, alcohol, benzene, ether, chloroform, acetic acid; slightly sol in ligroin.
USE: Esters as plasticizers.

8992. Suberic Acid. [505-48-6] Octanedioic acid; 1,6-dicarboxyhexane; 1,6-hexanedicarboxylic acid. $C_8H_{14}O_4$; mol wt 174.20. C 55.16%, H 8.10%, O 36.74%. Made by heating castor oil or ricinoleic acid with nitric acid: Dominguez, Lopez, *Ciencia (Mexico City)* **20**, 73 (1959), *C.A.* **54**, 18990f (1960). Prepn from diethyl malonate + ω-bromocapronitrile: Cason *et al.*, *J. Org. Chem.* **14**, 37 (1949); by oxidation of oleic acid with nitric acid: Cavanaugh, Weir, **US 2560156** (1951 to du Pont).

Crystals, mp 140-144°. bp_{100} 279°. Sublimes at 300° without decompn. One gram dissolves in 625 ml water, 172 ml ether; sol in alcohol; almost insol in chloroform.
USE: In the plastics industry.

8993. Substance P. [33507-63-0] SP. $C_{63}H_{98}N_{18}O_{13}S$; mol wt 1347.65. C 56.15%, H 7.33%, N 18.71%, O 15.43%, S 2.38%. An undecapeptide belonging to a group of proteins named tachykinins characterized by contractile action on extravascular smooth muscle. Discovered by von Euler and Gaddum, *J. Physiol.* **72**, 74 (1931). Present in the brain of all vertebrates including man, in spinal ganglia, and in the intestines, esp the duodenum and jejunum. Isoln from cattle brain: Zuber, Jaques, *Angew. Chem. Int. Ed.* **1**, 160 (1962); from horse intestine: Studer *et al.*, *Helv. Chim. Acta* **56**, 860 (1973). Purification from bovine hypothalami and amino acid composition: N. N. Chang, S. E. Leeman, *J. Biol. Chem.* **245**, 4784 (1970). Amino acid sequence and solid-phase synthesis: Chang *et al.*, *Nature New Biol.* **232**, 86 (1971); Tregear *et al.*, *ibid.* 87; G. H. Fisher *et al.*, *J. Med. Chem.* **17**, 843 (1974). Additional syntheses: Yajima, Kitigawa, *Chem. Pharm. Bull.* **21**, 682 (1973); Bayer, Mutter, *Ber.* **107**, 1344 (1974); K. Neubert *et al.*, *Pharmazie* **36**, 10 (1981); A. Fournier *et al.*, *J. Med. Chem.* **25**, 64 (1982). Substance P acts as a vasodilator, a depressant, stimulates salivation and produces increased capillary permeability. It is also capable of producing both analgesia and hyperalgesia in animals, depending on dose and pain responsiveness of the animal, *see:* R. C. A. Frederickson *et al.*, *Science* **199**, 1359 (1978); P. Oehme *et al.*, *ibid.* **208**, 305 (1980); role in sensory transmission and pain perception: T. M. Jessell, *Adv. Biochem. Psychopharmacol.* **28**, 189 (1981). Mechanism of action studies: Stern *et al.*, *Arch. Pharmacol.* **281**, 233 (1974); B. G. Livett *et al.*, *Nature* **278**, 256 (1979). *Reviews:* Haefeli, Huelimann, *Experientia* **18**, 297 (1962); Lembeck, Zetler in *International Encyclopedia of Pharmacology and Therapeutics* **sect. 72, vol. 1**, J. M. Walker, Ed. (Pergamon Press, New York, 1971) pp 29-71; J. L. Barker, *Physiol. Rev.* **56**, 435 (1976); D. R. Brown, R. J. Miller, *Annu. Rep. Med. Chem.* **17**, 271-280 (1982). Books: U.S. Von Euler, B. Pernow, Eds., *Substance P* (Raven Press, New York, 1977) 360 pp; *Substance P* **vol. 2**, P. Skrabanek, D. Powell, Eds. (Eden Press, Quebec, 1980) 175 pp; *Substance P in The Nervous System: Ciba Foundation Symposium* **91**, R. Porter, M. O'Connor, Eds. (Pitman, London, 1982) 350 pp.

Arg–Pro–Lys–Pro–Gln–Gln–Phe–Phe–Gly–Leu–MetNH₂

Behaves during isoln as a basic polypeptide (R_f value = 0.5 in the thin layer chromatogram: silica gel *n*-butanol/pyridine/AcOH/H₂O = 30:20:6:24). Upon high voltage electrophoresis (pH 1.9; acetic acid/formic acid/H₂O = 15:3:100) it migrates approx as far as glutamic acid or serine, and at pH 9.5 (0.25*M* triethylammonium carbonate buffer) it behaves in a manner similar to arginine. Biological activity is destroyed by pepsin and chymotrypsin. Aqueous solns of highly purified material lose biological activity in a few minutes. This loss is prevented by storage at low pH, under nitrogen, or with various antioxidants. Addition of Tween 80, gelatin, human serum albumin, or bovine γ-globulin increases stability of aq SP solns. Crude SP solns are stable though rapidly destroyed above pH 8.

8994. Subtilin. [1393-38-0] L-Tryptophyl-L-lysyl-D-cysteinyl-L-α-glutamyl-2,3-didehydroalanyl-L-leucyl-L-cysteinyl-(2*R*)-2-amino-3-mercaptobutanoyl-L-prolylglycyl-L-cysteinyl-L-valyl-(2*R*)-2-amino-3-mercaptobutanoylglycyl-L-alanyl-L-leucyl-L-glutaminyl-2-amino-2-butenoyl-L-cysteinyl-L-phenylalanyl-L-leucyl-L-glutaminyl-(2*R*)-2-amino-3-mercaptobutanoyl-L-leucyl-(2*R*)-2-amino-3-mercaptobutanoyl-L-cysteinyl-L-asparaginyl-L-cysteinyl-L-lysyl-L-isoleucyl-2,3-didehydroalanyl-L-lysine cyclic (3 → 7), (8 → 11), (13 → 19), (23 → 26), (25 → 28)-pentakis(thioether). C₁₄₈H₂₂₇N₃₉O₃₈S₅; mol wt 3320.98. C 53.53%, H 6.89%, N 16.45%, O 18.31%, S 4.83%. Polypeptide antibiotic structurally similar to nisin, *q.v.*, but contains no methionine. Produced by *Bacillus subtilis* (NRRL no. B-543): E. F. Jansen, D. J. Hirshmann, *Arch. Biochem.* **4**, 297 (1944); Lewis *et al.*, *ibid.* **14**, 415 (1947); Dimick *et al.*, *ibid.* **15**, 1 (1947). Isoln by partition chromatography on silica gel: Alderton, Snell, *J. Am. Chem. Soc.* **81**, 701 (1959). Early structural studies: Stracher, Craig, *ibid.* 696. Structure is a heterodetic pentacyclic peptide consisting of 32 amino acid residues and containing unusual amino acids such as lanthionine, β-methyllanthionine, D-alanine, dehydroalanine and dehydrobutyrine: E. Gross *et al.*, *Z. Physiol. Chem.* **354**, 810 (1973). Biosynthetic studies: C. Nishio *et al.*, *Biochem. Biophys. Res. Commun.* **116**, 751 (1983); S. Banerjee, J. N. Hansen, *J. Biol. Chem.* **263**, 9508 (1988).

Amorphous powder. $[\alpha]_D^{23}$ −29 to −35° (c = 1 in 1% acetic acid). Diffuses quickly through cellophane. Readily sol in dilute acids; sparingly sol (less than 5 mg/ml) in water at pH 6-9. Sol in methanol, 80% alc. Soly in butanol satd with water: about 5 mg/ml. Easily salted out of soln by NaCl.

USE: Food preservative.

8995. Succimer. [304-55-2] *rel*-(2*R*,3*S*)-2,3-Dimercaptobutanedioic acid; *meso*-2,3-dimercaptosuccinic acid; DMS; DMSA; Ro-1-7977; Chemet. C₄H₆O₄S₂; mol wt 182.21. C 26.37%, H 3.32%, O 35.12%, S 35.19%. A water soluble chelating agent. Prepn: L. N. Owen, M. U. S. Sultanbawa, *J. Chem. Soc.* **1949**, 3109. Pharmacology, toxicity and activity in heavy metal poisoning in animals: E. Friedheim, C. Corvi, *J. Pharm. Pharmacol.* **27**, 624 (1975); J. H. Graziano *et al.*, *J. Pharmacol. Exp. Ther.* **207**, 1051 (1978). GC determn in urine: J. J. Knudsen, E. L. McGown, *J. Chromatogr.* **424**, 231 (1988). Clinical evaluations in heavy metal poisoning: E. Friedheim *et al.*, *Lancet* **2**, 1234 (1978); L. Fournier *et al.*, *Med. Toxicol.* **3**, 499 (1988). Diagnostic use of complex with technetium in urinary tract imaging: I. G. Verber *et al.*, *Arch. Dis. Child.* **63**, 1320 (1988); in tumor imaging: H. Ohta *et al.*, *Nucl. Med. Commun.* **9**, 105 (1988). *Review:* H. V. Aposhian, "Biological Chelation: 2,3-Dimercaptopropanesulfonic Acid and Meso-Dimercaptosuccinic Acid" in *Adv. Enzyme Regul.* **20**, G. Weber, Ed. (Pergamon Press, Oxford, 1982) pp 301-319. Review of pharmacology and clinical use in heavy metal poisoning: *idem*, *Annu. Rev. Pharmacol. Toxicol.* **23**, 193-215 (1983); J. H. Graziano, *Med. Toxicol.* **1**, 155-162 (1986).

Relative stereochemistry

White crystals from aqueous methanol, mp 192-194°. LD_{50} i.p. in mice: >3000 mg/kg (Friedheim, Corvi).

USE: Chelating agent.

THERAP CAT: Antidote (heavy metal poisoning). ⁹⁹ᵐTc complex as diagnostic aid (radioactive imaging agent).

8996. Succinamide. [110-14-5] Butanediamide; succinic acid diamide. C₄H₈N₂O₂; mol wt 116.12. C 41.37%, H 6.94%, N 24.13%, O 27.56%. Made by the action of ammonia water on the dimethyl or diethyl ester of succinic acid.

Needles. mp 260° with decompn; also stated as 242°. Sol in 220 parts cold, 9 parts boiling water; insol in alc, ether.

8997. Succinanil. [83-25-0] 1-Phenyl-2,5-pyrrolidinedione; *N*-phenylsuccinimide; *N*-phenylbutanimide. C₁₀H₉NO₂; mol wt 175.19. C 68.56%, H 5.18%, N 8.00%, O 18.26%. Prepd by treating succinanilic acid with acetyl chloride: Ruggli, *Ann.* **412**, 4 (1916); Warren, Briggs, *Ber.* **64**, 29 (1931); L. F. Fieser, *Experiments in Organic Chemistry* (Boston, 1955) p 106.

Monoclinic prisms from water or alcohol, mp 154.5°. d_4^{20} 1.356. bp₇₆₀ ~400°. Sol in ether, boiling alcohol. Slightly sol in boiling water.

8998. Succinanilic Acid. [102-14-7] 4-Oxo-4-(phenylamino)butanoic acid; succinic acid monoanilide; *N*-phenylsuccinamic acid; *N*-phenylbutanedioic acid monoamide. C₁₀H₁₁NO₃; mol wt 193.20. C 62.17%, H 5.74%, N 7.25%, O 24.84%. Prepd by boiling succinic anhydride with aniline in benzene: Menschutkin, *Ann.* **162**, 176 (1872); Warren, Briggs, *Ber.* **64**, 29 (1931); L. F. Fieser, *Experiments in Organic Chemistry* (Boston, 3rd ed., 1955) p 105.

Needles from benzene, mp 150°. pK (25°) 4.69. Soluble in alcohol, ether, boiling water (some dec).

8999. Succinic Acid. [110-15-6] Butanedioic acid; amber acid; ethylenesuccinic acid; Bernsteinsäure (German); Asuccin. $C_4H_6O_4$; mol wt 118.09. C 40.68%, H 5.12%, O 54.19%. Observed by Agricola in 1546, in the distillate from amber. Occurs in fossils, fungi, lichens, etc. Prepn from acetic acid: Coffman *et al.*, *J. Am. Chem. Soc.* **80**, 2864 (1958). Manuf: Crosby, Braunwarth, **US 2862028** (1958 to Pure Oil); Chafetz, Patterson, **US 2978473** (1961 to Texaco). Toxicity study: H. F. Smyth *et al.*, *Arch. Ind. Hyg. Occup. Med.* **4**, 119 (1951).

Odorless, monoclinic prisms; very acid taste; d 1.56. mp 185-187°. bp 235° with partial conversion into the anhydride. pKa_1 4.03±0.03; pKa_2 5.28±0.03. Log P (octanol/water): −0.59. One gram dissolves in 13 ml cold water, 1 ml boiling water, 18.5 ml alcohol, 6.3 ml methanol, 36 ml acetone, 20 ml glycerol, 113 ml ether. Practically insol in benzene, carbon disulfide, carbon tetrachloride, petr ether. pH of 0.1 molar aq soln 2.7.

Potassium salt trihydrate. [6100-18-1] Potassium succinate. $C_4H_4K_2O_4.3H_2O$; mol wt 248.31. Hygroscopic crystalline powder. d 1.564. Freely sol in water; the aq soln is practically neutral. *Keep tightly closed.*

Diethyl ester. [123-25-1] Ethyl succinate. $C_8H_{14}O_4$; mol wt 174.20. Prepn: A. Zurqiyah, C. E. Castro, *Org. Synth.* **coll. vol. V**, 993 (1973). Colorless, clear liq; faint, pleasant odor. d_4^{20} 1.040. mp −21°. bp_{760} 217-218°; bp_{44} 129°. n_D^{20} 1.4201. Insol in water. Miscible with alcohol, ether. LD_{50} orally in rats: 8.53 g/kg (Smyth).

Dimethyl ester. [106-65-0] Methyl succinate. $C_6H_{10}O_4$; mol wt 146.14. Colorless liquid. d_4^{18} 1.1202. mp 19.5°. bp_{25} 103.5°. n_D^{20} 1.41969. Sol in 120 parts water, 35 parts alcohol.

USE: Manuf lacquers, dyes, esters for perfumes, succinates; in photography. Internal standard in chromatography.

9000. Succinic Anhydride. [108-30-5] Dihydro-2,5-furandione; succinic acid anhydride; succinyl oxide; 2,5-diketotetrahydrofuran; butanedioic anhydride; Bernsteinsäureanhydrid (German). $C_4H_4O_3$; mol wt 100.07. C 48.01%, H 4.03%, O 47.96%. Usually prepd by warming succinic acid with acetic anhydride, acetyl chloride, or phosphorus oxychloride: Shriner, Struck, *Org. Synth.* **12**, 66 (1932); L. F. Fieser, *Experiments in Organic Chemistry* (Boston, 3rd ed., 1955) p 105.

Orthorhombic prisms from abs alc. d 1.503 (also given as 1.104 and 1.234). Rate of sublimation: Kempf, *J. Prakt. Chem.* [2] **78**, 257 (1908). mp 119.6°; bp_{760} 261°; bp_{400} 237°; bp_{200} 212°; bp_{100} 189°; bp_{60} 174°; bp_{40} 163°; bp_{20} 145°; bp_{10} 128°. Sublimes at 115° and 5 mm pressure, at 92° and 1.0 mm pressure. Soluble in chloroform, carbon tetrachloride, alcohol; very slightly sol in ether and water.

9001. Succinimide. [123-56-8] 2,5-Pyrrolidinedione; butanimide; 2,5-diketopyrrolidine; 3,4-dihydropyrrole-2,5-dione; dihydro-3-pyrroline-2,5-dione; 2,5-dioxopyrrolidine; Orotic. $C_4H_5NO_2$; mol wt 99.09. C 48.49%, H 5.09%, N 14.14%, O 32.29%.

Prepd by rapid distillation of ammonium succinate: Clarke, Behr, *Org. Synth.* **coll. vol. II** (1943) p 562; Fehling, *Ann.* **49**, 198 (1844); Bunge, *Ann. Suppl.* **7**, 118 (1870); Menschutkin, *Ann.* **162**, 166 (1872). Prepn by heating succinic acid with ammonia: Fehling, *loc. cit.;* Franchimont, Friedmann, *Rec. Trav. Chim.* **25**, 79 (1906); with urea: Ma, Sah, *C.A.* **28**, 6108 (1934); *see also* Tafel, Stern, *Ber.* **33**, 2232 (1900); Koller, *Ber.* **37**, 1598 (1904). Prevention of oxalic lithiasis in rats: J. M. Melon *et al.*, *Therapie* **26**, 991 (1971).

Orthorhombic bipyramidal crystals from acetone or alc. mp 125-127°. d 1.41. bp 287-289° (slight decompn). pKa 9.5. One gram dissolves in 3 ml water, 0.7 ml boiling water, 24 ml alc, 5 ml alc at 60°. Insol in ether, chloroform. LD_{50} orally in rats: 14 g/kg (Melon).

THERAP CAT: Antiurolithic.

9002. Succinobucol. [216167-82-7] Butanedioic acid 1-[4-[[1-[[3,5-bis(1,1-dimethylethyl)-4-hydroxyphenyl]thio]-1-methylethyl]thio]-2,6-bis(1,1-dimethylethyl)phenyl]ester; probucol monosuccinate; AGI-1067. $C_{35}H_{52}O_5S_2$; mol wt 616.92. C 68.14%, H 8.50%, O 12.97%, S 10.39%. Metabolically stable monosuccinic acid ester of probucol, *q.v.*, with antioxidant and vascular protectant activity. Inhibits expression of inducible vascular cell adhesion molecule-1 (VCAM-1). Prepn: P. K. Somers, **US 6147250** (2000 to AtheroGenics); C. Q. Meng *et al.*, *Bioorg. Med. Chem. Lett.* **12**, 2545 (2002). Clinical trial to reduce restenosis after angioplasty: J.-C. Tardif *et al.*, *Circulation* **107**, 552 (2003). Review of design and pharmacology: M. A. Wasserman *et al.*, *Am. J. Cardiol.* **91**, Suppl., 34A-40A (2003); of clinical experience: J.-C. Tardif, *ibid.* 41A-49A; V. L. Serebruany *et al.*, *Expert Rev. Cardiovasc. Ther.* **5**, 635-641 (2007).

White to off-white solid, mp 132-134°.
THERAP CAT: Antiatherosclerotic.

9003. Succinonitrile. [110-61-2] Butanedinitrile; ethylene dicyanide; succinic acid dinitrile; *sym*-dicyanoethane; ethylene cyanide; Dinile; Suxil. $C_4H_4N_2$; mol wt 80.09. C 59.99%, H 5.03%, N 34.98%. Prepd from ethylene bromide and KCN in alcohol: Fauconnier, *Bull. Soc. Chim. Fr.* **50**, 214 (1888).

Waxy isometric crystals. d_4^{45} 1.023. mp 57.15°. d_4^{60} (liq) 0.9868. n_D^{60} (liq) 1.41734. May be heated to 200° for 72 hrs without decompn. bp_{760} 265-267°; bp_{60} 185°; bp_{20} 158-160°. Sol in acetone, chloroform, dioxane. Slightly sol in water, ethanol, benzene, ether, carbon disulfide.

Caution: Potential symptoms of overexposure are irritation of eyes, skin, respiratory system; headache, dizziness, weakness, giddiness, confusion, convulsions; blurred vision; dyspnea; abdominal pain, nausea, vomiting. *See NIOSH Pocket Guide to Chemical Hazards* (DHHS/NIOSH 97-140, 1997) p 288.

9004. Succinyl Chloride. [543-20-4] Butanedioyl dichloride; succinic acid dichloride; succinyl dichloride. $C_4H_4Cl_2O_2$; mol wt 154.97. C 31.00%, H 2.60%, Cl 45.75%, O 20.65%.

Fuming, highly refractive liq; d_4^{15} 1.395; n_D^{15} 1.473; mp 17°; also stated as 20°; bp 192-193°. Solidif at 0° to cryst leaflets.

9005. Succinylcholine Bromide. [55-94-7] 2,2′-[(1,4-Dioxo-1,4-butanediyl)bis(oxy)]bis[N,N,N-trimethylethanaminium] bromide (1:2); bis[2-dimethylaminoethyl]succinate bis[methobromide]; 3,8-dioxadecane-4,7-dione-1,10-bis[trimethylammonium bromide]; suxamethonium bromide; IS-370; compd 48/268; LT-1; M & B 2207; Brevidil M. $C_{14}H_{30}Br_2N_2O_4$; mol wt 450.21. C 37.35%, H 6.72%, Br 35.50%, N 6.22%, O 14.21%. Prepd by reacting β-bromoethyl succinate with trimethylamine: Glick, *J. Biol. Chem.* **137**, 357 (1941); Fusco *et al.*, *Gazz. Chim. Ital.* **79**, 129, 837 (1949); Walker, *J. Chem. Soc.* **1950**, 193. Alternate prepns: Phillips, *J. Am. Chem. Soc.* **71**, 3264 (1949); Tammelin, *Acta Chem. Scand.* **7**, 185 (1953).

Slightly hygroscopic crystals, mp 225°. Freely sol in water or normal saline, giving solns which are very slightly acidic. Aq solns undergo progressive hydrolysis with corresponding loss of activity and increase in acidity. *Prepare solns just before using.* The rate of this decompn increases with temp and autoclaving or prolonged exposure to warmth should be avoided.

THERAP CAT: Neuromuscular blocking agent.

9006. Succinylcholine Chloride. [71-27-2] 2,2′-[(1,4-Dioxo-1,4-butanediyl)bis(oxy)]bis[N,N,N-trimethylethanaminium] chloride (1:2); bis[2-dimethylaminoethyl]succinate bis[methochloride]; 2-dimethylaminoethyl succinate dimethochloride; diacetylcholine dichloride; suxamethonium chloride; choline succinate dichloride; succinic acid bis[β-dimethylaminoethyl] ester dimethochloride; choline chloride succinate (2:1); listenon; Anectine; Lysthenon; Midarine; Quelicin; Scoline; Sucostrin. $C_{14}H_{30}Cl_2N_2O_4$; mol wt 361.30. C 46.54%, H 8.37%, Cl 19.62%, N 7.75%, O 17.71%. Prepn: R. Fusco *et al.*, *Gazz. Chim. Ital.* **79**, 129 (1949); Tammelin, *Acta Chem. Scand.* **7**, 185 (1953); O. Schmid, **AT 171411** (1952 to OSSW), *C.A.* **47**, 4902f (1953); C. H. Wang *et al.*, *Org. Prep. Proced. Int.* **11**, 93 (1979). Crystal structure: B. Jensen, *Acta Chem. Scand. B* **30**, 1002 (1976). Toxicity data: P. Anttila, P. Ertama, *Med. Biol.* **56**, 152 (1978). Comprehensive description: P. R. B. Foss, S. A. Benezra, *Anal. Profiles Drug Subs.* **10**, 691-704 (1981).

Exists as a dihydrate at room temp, white crystals, mp 156-163°. Anhydr form mp ~190°. Slightly bitter taste. Freely sol in water (about 1 g/1 ml water). Soly in 95% ethanol: 0.42 g/100 ml. Sparingly sol in benzene; slightly sol in chloroform. Practically insol in ether. The pH of a 2-5% aq soln may vary from 4.5 to 3.0. Solns without stabilizers (e.g., 0.1% methyl *p*-hydroxybenzoate) must be kept refrigerated, they are incompatible with alkaline agents, such as thiopental sodium. LD_{50} i.v. in mice: 0.45 mg/kg (Anttila, Ertama).

THERAP CAT: Neuromuscular blocking agent.

THERAP CAT (VET): Muscle relaxant (skeletal).

9007. Succinylcholine Iodide. [541-19-5] 2,2′-[(1,4-Dioxo-1,4-butanediyl)bis(oxy)]bis[N,N,N-trimethylethanaminium] iodide (1:2); *O,O*-succinyldicholine iodide; diacetylcholine iodide; diacetylcholine diiodide; suxamethonium iodide; bis(β-dimethylaminoethyl)succinate bis(methyl iodide); Celocurine. $C_{14}H_{30}I_2N_2O_4$; mol wt 544.21. C 30.90%, H 5.56%, I 46.64%, N 5.15%, O 11.76%. Prepd by reacting β-bromoethyl succinate with trimethylamine: Glick, *J. Biol. Chem.* **137**, 357 (1941); Fusco *et al.*, *Gazz. Chim. Ital.* **79**, 129, 837 (1949); Walker, *J. Chem. Soc.* **1950**, 193. Succinic acid chloride may be coupled with choline chloride directly or with dimethylaminoethanol, followed by quaternization with methyl iodide: Tammelin, *Acta Chem. Scand.* **7**, 185 (1953). A third method is to start with the diethyl ester of succinic acid, then bring about the exchange reaction with dimethylaminoethanol, and quaternize with methyl iodide: Phillips, *J. Am. Chem. Soc.* **71**, 3264 (1949).

Slightly hygroscopic crystals, mp 243-245°. Freely sol in water or normal saline, giving solns which are very slightly acidic. Most stable at pH 4-5. Aq solns undergo progressive hydrolysis with corresp loss of activity and increase in acidity. *Prepare solns just before using.* Incompatible with solns of alkaline salts.

THERAP CAT: Neuromuscular blocking agent.

THERAP CAT (VET): Muscle relaxant (skeletal).

9008. Succinylsulfathiazole. [116-43-8] 4-Oxo-4-[[4-[(2-thiazolylamino)sulfonyl]phenyl]amino]butanoic acid; 4′-(2-thiazolylsulfamoyl)succinanilic acid; *p*-2-thiazolylsulfamylsuccinanilic acid; 2-(N^4-succinylsulfanilamido)thiazole; Sulfasuxidine. $C_{13}H_{13}N_3O_5S_2$; mol wt 355.38. C 43.94%, H 3.69%, N 11.82%, O 22.51%, S 18.04%. Prepd by refluxing sulfathiazole with a slight excess of succinic anhydride in alcohol: Moore, Miller, *J. Am. Chem. Soc.* **64**, 1572 (1942); **US 2324013**; **US 2324014** (both 1943); by refluxing *p*-succinimidobenzenesulfonyl chloride with 2-aminothiazole in benzene: **GB 578004** (1946). Toxicology: A. D. Welch *et al.*, *J. Pharmacol. Exp. Ther.* **75**, 231 (1942).

Monohydrate. Crystals with unclear mp. Has been reported as mp 184-186°; also reported as mp 192-195°. Emits pungent fumes on heating to mp. One gram dissolves in about 4800 ml water. Soluble in solns of alkali hydroxides and in solns of sodium bicarbonate with the evolution of carbon dioxide; sparingly sol in alcohol and in acetone. Insol in chloroform and ether. LD_{50} i.p. in mice: ~5.7 g/kg (Welch).

THERAP CAT: Antibacterial.

THERAP CAT (VET): Antibacterial, esp in enteric infections.

9009. Succisulfone. [5934-14-5] 4-[[4-[(4-Aminophenyl)sulfonyl]phenyl]amino]-4-oxobutanoic acid; 4′-sulfanilylsuccinanilic acid; 4-amino-4′-(β-carboxypropionylamino)phenylsulfonylbenzene; 4-amino-4′-β-carboxypropionylaminodiphenylsulfone; 4-(β-carboxypropionylamino)-4′-aminodiphenyl sulfone; 4-succinylamido-4′-aminodiphenylsulfone; F-1500; Fourneau 1500; Exosulfonyl. $C_{16}H_{16}N_2O_5S$; mol wt 348.37. C 55.16%, H 4.63%, N 8.04%, O 22.96%, S 9.20%. Prepn: Fourneau, Tréfouel, **FR 866619** (1941 to Rhône-Poulenc); Kharasch, Reinmuth, **US 2268754** (1942 to Lilly); Bauer, *J. Am. Chem. Soc.* **70**, 2254 (1948); Rohls, Behnish, **DE 895600** (1953 to Bayer); Liberman, **FR 1020713** (1953 to Chimie et Atomistique).

Crystals, mp 157°. Soluble in ammonia.

2,2′-Iminodiethanol salt. [547-36-4] $C_{20}H_{27}N_3O_7S$. Prepn: **FR 992112** (1951 to Theraplix).

THERAP CAT: Antibacterial (leprostatic).

9010. Sucralfate. [54182-58-0] Hexadeca-μ-hydroxytetracosahydroxy[μ_8-[[1,3,4,6-tetra-*O*-sulfo-β-D-fructofuranosyl α-D-glucopyranoside tetrakis(sulfato-$\kappa O'$)](8−)]]hexadecaaluminum; β-D-fructofuranosyl-α-D-glucopyranoside octakis(hydrogen sulfate) aluminum complex; sucrose octakis(hydrogen sulfate) aluminum complex; Antepsin; Carafate; Citogel; Hexagastron; Keal; Succosa; Sucralfin; Sucrate; Sugast; Sulcrate; Ulcar; Ulcerlmin; Ulcogant. $C_{12}H_{54}Al_{16}O_{75}S_8$; mol wt 2086.67. C 6.91%, H 2.61%, Al 20.69%, O 57.50%, S 12.29%. A basic aluminum sucrose sulfate complex which inhibits peptic hydrolysis and stomach acidity. Prepn: M, Nametaka *et al.*, *Yakugaku Zasshi* **87**, 889 (1967), *C.A.* **68**, 20831d (1968); **FR 1500571**; N. Yoshihiro *et al.*, **US 3432489** (1967, 1969 both to Chugai). Structure and properties: R. Nagashima, N. Yoshida, *Arzneim.-Forsch.* **29**, 1668 (1979). *In vitro* and *in vivo* study of antipeptic and antiulcer activity: L. E. Borella *et al.*, *ibid.* 793. Series of articles on mode of action: *eidem*, *ibid.* **30**, 73-88 (1980). Pharmacokinetics, metabolism and selective binding studies: K. Steiner *et al.*, *ibid.* **32**, 512 (1982). Clinical studies: J. F. Mayberry *et al.*, *Br. J. Clin. Pract.* **32**, 291 (1978). Symposium on clinical studies: *J. Clin. Gastroenterol.* **3**, Suppl. 2, 103-184 (1981); *Scand. J. Gastroenterol.* **18**, Suppl. 83, 1-82 (1983). Review of pharmacology and clinical studies: R. N. Brogden *et al.*, *Drugs* **27**, 194 (1984). Series of articles on clinical efficacy in peptic ulcer disease and gastritis: *Am. J. Med.* **79**, Suppl. 2C, 1-64 (1985).

R = SO₃[Al₂(OH)₅]

R = SO$_3$[Al$_2$(OH)$_5$]

White amorphous powder. Sol in dil HCL and NaOH solns. Practically insol in water, ethanol, CHCl₃. pKa = 0.43 to 1.19. Dissolution of aluminum occurs at pH <3; of sucrose sulfate at pH >4.

THERAP CAT: Antiulcerative.

9011. Sucralose. [56038-13-2] 1,6-Dichloro-1,6-dideoxy-β-D-fructofuranosyl-4-chloro-4-deoxy-α-D-galactopyranoside; 4,1′,6′-trichloro-4,1′,6′-trideoxy-*galacto*-sucrose; 1′,4,6′-trichlorogalactosucrose; TGS; Splenda. $C_{12}H_{19}Cl_3O_8$; mol wt 397.63. C 36.25%, H 4.82%, Cl 26.75%, O 32.19%. Chlorinated sucrose derivative with enhanced sweetness. Prepn: P. H. Fairclough *et al.*, *Carbohydr. Res.* **40**, 285 (1975). Prepn of crystalline anhydrous and pentahydrate: M. R. Jenner, D. Waite, **EP 30804** (1981 to Tate & Lyle; Talres Dev.); *eidem*, **US 4343934** (1982 to Talres Dev.). Use as nonnutritive sweetener: **BE 850180**; L. Hough *et al.*, **US 4435440** (1977, 1984 both to Tate & Lyle). *In vitro* activity vs cariogenic bacteria: J. Verran, D. B. Drucker, *Arch. Oral Biol.* **27**, 693 (1982). Structure-sweetness relationship: M. Mathlouthi *et al.*, *Carbohydr. Res.* **152**, 47 (1986). *Review:* L. Hough, R. Khan, *Trends Biochem. Sci.* **3**, 61-63 (1978).

Syrup, $[\alpha]_D$ +68.2° (c = 1.1 in ethanol). Anhydrous crystalline form: orthorhombic, needle-like crystals, mp 130°. Intensely sweet taste.

Pentahydrate. mp 36.5°.

USE: Non-nutritive sweetener.

9012. Sucrose. [57-50-1] β-D-Fructofuranosyl-α-D-glucopyranoside; α-D-glucopyranosyl-β-D-fructofuranoside; sugar; saccharose; cane sugar; beet sugar. $C_{12}H_{22}O_{11}$; mol wt 342.30. C 42.11%, H 6.48%, O 51.41%. Obtained from sugar cane (*Saccharum officinarum* L., *Gramineae*) and sugar beet (*Beta valgaris* L., *Chenopodiaceae*). Sugar cane contains from 15-20% and sugar beet from 10-17% sucrose. Structure: Avery *et al.*, *J. Chem. Soc.* **1927**, 2308; Beevers, Cochrane, *Proc. Roy. Soc.* **190A**, 257 (1947). Synthesis: Pictet, Vogel, *Helv. Chim. Acta* **11**, 436 (1928); Lemieux, Huber, *J. Am. Chem. Soc.* **78**, 4117 (1956). Ref. with extensive bibliography: Bates, *Polarimetry, Saccharimetry, and the Sugars, National Bureau of Standards Circular* C440, Washington, 1942; W. Pigman, *The Carbohydrates* (Academic Press, New York, 1957) pp 501-506. *Reviews:* M. R. Jenner, *Dev. Food Carbohyd.* **2**, 91-143 (1980); R. Khan, *Pure Appl. Chem.* **56**, 833 (1984).

Monoclinic sphenoidal crystals, cryst masses, blocks, or powder. Sweet taste. Stable in air. Finely divided sugar is hygroscopic and absorbs up to 1% moisture which is given up on heating to 90°. d_4^{25} 1.587. Dec 160-186°. Chars and emits characteristic odor of caramel. $[\alpha]_D^{20}$ not less than +65.9° (c = 26); usual value $[\alpha]_D^{25}$ +66.47 to +66.49°. One gram dissolves in 0.5 ml water; in slightly more than 0.2 ml boiling water, in 170 ml alcohol; in about 100 ml methanol. Moderately sol in glycerol, pyridine. Practically insol in dehydrated alcohol. pKa 12.62. d_4^{20} of water solns (g/100 g): 2% 1.0060; 6% 1.0219; 10% 1.0381; 20% 1.0810; 30% 1.1270; 40% 1.1764; 50% 1.2296; 60% 1.2865; 70% 1.3471; 76% 1.3854. n_D^{20} of 10% soln 1.34783. Sucrose does not reduce Fehling's soln, form an osazone, or show mutarotation. It is hydrolyzed to glucose and fructose by dil acids and by invertase, a yeast enzyme. Upon hydrolysis the optical rotation falls and is negative when the hydrolysis is complete. The mixture of glucose and fructose is known as "invert sugar." Sucrose is fermentable, but resists bacterial decompn when in high concentrations.

Caution: Potential symptoms of overexposure are irritation of eyes, skin, upper respiratory system; cough. *See NIOSH Pocket Guide to Chemical Hazards* (DHHS/NIOSH 97-140, 1997) p 288.

USE: Sweetening agent and food. Starting material in the fermentative production of ethanol, butanol, glycerol, citric and levulinic acids. Pharmaceutic aid (flavor, preservative, antioxidant in the form of invert sugar, demulcent, substitute for glycerol, granulation agent and excipient for tablets, coating for tablets). In the plastics and cellulose industry, in rigid polyurethane foams, manuf of ink and of transparent soaps. Optical rotation standard.

9013. Sucrose Octaacetate. [126-14-7] 1,3,4,6-Tetra-*O*-acetyl-β-D-fructofuranosyl-α-D-glucopyranoside 2,3,4,6-tetraacetate. $C_{28}H_{38}O_{19}$; mol wt 678.59. C 49.56%, H 5.64%, O 44.80%. Prepn from sucrose: Linstead *et al.*, *J. Am. Chem. Soc.* **62**, 3260 (1940). Synthesis: Lemieux, Huber, *ibid.* **78**, 4117 (1956).

Hygroscopic, intensely bitter needles from alcohol, mp 89°; dec above 285°; bp₁ 260°. $[\alpha]_D^{25.4}$ +58.5° (c = 2.56 in abs alc). n_D 1.4660. Very sol in methanol, chloroform; sol in ether. Sol in 1100 parts water, 1.1 parts acetal, 0.7 part glacial acetic acid, 0.3 part acetone, 11 parts alcohol, 0.6 part benzene, 22 parts carbon tetrachloride, about 0.5 part methyl acetate, 7 parts paraldehyde, about 0.5 part toluene.

USE: Adhesive; impregnating and insulating papers; in lacquers and plastics; as a denaturant for alcohol.

9014. Sucrose Polyester. Olestra; SPE; Olean. Non-absorbable, non-digestible lipid; noncaloric substitute for fat in foods. Mixture of hexa-, hepta- and octa-fatty acid esters of sucrose. Analysis by gel permeation chromatography: C. G. Birch, F. E. Crowe, *J. Am. Oil Chem. Soc.* **53**, 581 (1976). Effect on absorption of dietary cholesterol in rats: F. H. Mattson *et al.*, *J. Nutr.* **106**, 747 (1976); L. Aust *et al.*, *Ann. Nutr. Metab.* **25**, 255 (1981). Effects as a dietary agent for lowering plasma cholesterol in man: R. W. Fallat *et al.*, *Am. J. Clin. Nutr.* **29**, 1024 (1976); C. J. Glueck *et al.*, *ibid.* **32**, 1636 (1979). Effects on cholesterol metabolism in man: J. R. Crouse, S. M. Grundy, *Metab. Clin. Exp.* **28**, 994 (1979); R. J. Jandacek *et al.*, *Am. J. Clin. Nutr.* **33**, 251 (1980). Caloric dilution study in obese patients: C. J. Glueck *et al.*, *ibid.* **35**, 1352 (1982). Review of properties and clinical safety studies: K. D. Lawson *et al.*, *Drug Metab. Rev.* **29**, 651-703 (1997); of clinical effects on GI functions: J. W. Freston *et al.*, *Regul. Toxicol. Pharmacol.* **26**, 210-218 (1997).

Smoke point 480°F. Flash point 550°F. Fire point 625°F.

USE: Dietary fat replacement in savory snack foods.

9015. Sudan III. [85-86-9] 1-[2-[4-(2-Phenyldiazenyl)phenyl]diazenyl]-2-naphthalenol; 1-(*p*-phenylazophenylazo)-2-naphthol; tetrazobenzene-β-naphthol; D & C Red No. 17; C.I. Solvent Red 23; C.I. 26100; Sudan Red BK; Tony Red. $C_{22}H_{16}N_4O$; mol wt 352.40. C 74.98%, H 4.58%, N 15.90%, O 4.54%. Weakly acidic azo dye; originally introduced as a fat stain in 1896. Prepn: R. Nietzki, *Ber.* **13**, 1838 (1880); *Colour Index* **vol. 4** (3rd ed., 1971) p 4227. Description: *H. J. Conn's Biological Stains*, R. D. Lillie, Ed. (Williams & Wilkins, Baltimore, 9th ed., 1977) pp 168-169, 576. HPLC determn in cosmetics: J. W. Wegener *et al.*, *Chromatographia* **24**, 865 (1987); LC/electrospray-MS determn as environmental contaminant: H.-Y. Lin, R. D. Voyksner, *Anal. Chem.* **65**, 451 (1993). Use as biological stain: W. Frisch-Niggemeyer, *J. Virol. Methods* **5**, 135 (1982); A. Kishida *et al.*, *J. Controlled Release* **13**, 83 (1990). Structure effects on dye adsorption: H. L. Needles *et al.*, *Colourage Annu.* **1995**, 115.

Brown leaves with green metallic shine, mp ~195°. Insol in water or lye. Slightly sol in hot glacial acetic acid. Sol in chloroform; moderately sol ether, acetone, petr ether, fixed and volatile oils, hot glycerol. Sol in alcohol 0.15%.

USE: Dye. Biological stain.

9016. Sudan Black B. [4197-25-5] 2,3-Dihydro-2,2-dimethyl-6-[2-[4-(2-phenyldiazenyl)-1-naphthalenyl]diazenyl]-1*H*-perimidine; C.I. Solvent Black 3; C.I. 26150; SBB; SSB; SBB-II; SSB-II. $C_{29}H_{24}N_6$; mol wt 456.55. C 76.29%, H 5.30%, N 18.41%. Lipid dye comprised of two major blue components SBB-I, SBB-II and at least 12 secondary components. SBB-II is defined by the chemical name and CAS registry number cited. Introduced: L. Lison, *C. R. Seances Soc. Biol. Ses Fil.* **115**, 202 (1934). TLC purification and histochemistry: A. G. W. Lansink, *Histochemie* **16**, 68 (1968); and solvent effects: W. M. Frederiks *et al.*, *Acta Histochem. Suppl.* **24**, 259 (1981). Structure elucidation: U. Pfüller *et al.*, *Histochemistry* **54**, 237 (1977). Staining technique: D. A. Nelson; F. R. Davey, "Sudan Black B Staining" in *Hematology*, W. J. Williams *et al.*, Eds. (McGraw-Hill Publishing Co., New York, 1990) pp 1747-1748. Use as lipid stain: H. N. Subramaniam, K. A. Chaubal, *J. Biochem. Biophys. Methods* **21**, 9 (1990).

Sudan Black B II

Basic compound completely decomposed at 180-186°.

Sudan Black B I. SBB-I; SSB-I; 2,3-dihydro-2,2-dimethyl-4[(4-phenylazo-1-naphthalenyl)-azo]-1*H*-perimidine. Neutral compound, mp 180-186° (dec). Sublimates 140-160° at 10^{-3} torr without decompositon.

USE: Biological stain for lipids.

9017. Suet, Prepared. Mutton suet. The internal fat of abdomen of sheep purified by melting and straining.

White, solid fat with slight odor and taste; bland if fresh, but rancid after long exposure to air. mp 45-50°; n_D^{60} 1.449-1.451. Sapon no. 192-195. Iodine no. 33-46. Insol in water or cold alcohol; one gram dissolves in 45 ml boiling alcohol, about 60 ml ether.

USE: Preparing ointments.

9018. Sufentanil. [56030-54-7] *N*-[4-(Methoxymethyl)-1-[2-(2-thienyl)ethyl]-4-piperidinyl]-*N*-phenylpropanamide; *N*-[4-(methoxymethyl)-1-[2-(2-thienyl)ethyl]-4-piperidyl]propionanilide; sufentanil; R-30730. $C_{22}H_{30}N_2O_2S$; mol wt 386.55. C 68.36%, H 7.82%, N 7.25%, O 8.28%, S 8.29%. Synthetic opioid analgesic; derivative of fentanyl, *q.v.* Prepn: P. A. J. Janssen, G. H. P. Van

Daele, **DE 2610228**; *eidem*, **US 3998834** (both 1976 to Janssen); G. H. P. Van Daele *et al.*, *Arzneim.-Forsch.* **26**, 1521 (1976). Analgesic activity and safety assessment: W. F. M. Van Bever *et al.*, *ibid.* 1548; C. J. E. Niemegeers *et al.*, *ibid.* 1551. *In vitro* binding properties to *mu*-opiate receptor: J. E. Leyson *et al.*, *Eur. J. Pharmacol.* **87**, 209 (1983). LC-MS/MS determn in plasma: R. Schmidt *et al.*, *J. Chromatogr. B* **836**, 98 (2006). Comparative review of clinical pharmacokinetics: J. Scholz *et al.*, *Clin. Pharmacokinet.* **31**, 275-292 (1996). Review of clinical efficacy in postoperative analgesia: J. A. Grass, *J. Pain Symptom Manage.* **7**, 271-286 (1992).

Crystals from petr ether, mp 96.6°. pKa 8.01. Partition coefficient (octanol/water): 1727. LD_{50} i.v. in mice: 18.7 mg/kg (Van Bever).

Citrate. [60561-17-3] R-33800; Fentatienil; Sufenta. $C_{22}H_{30}N_2O_2S.C_6H_8O_7$; mol wt 578.68. White powder. Sol in water; freely sol in methanol; sparingly sol in acetone, ethanol, chloroform.

Note: This is a controlled substance (opiate): **21 CFR**, 1308.12.

THERAP CAT: Analgesic; anesthesia adjunct.

9019. Sugammadex. [343306-71-8] 6A,6B,6C,6D,6E,6F,-6G,6H-Octakis-*S*-(2-carboxyethyl)-6A,6B,6C,6D,6E,6F,6G,6H-octathio-γ-cyclodextrin; 6-perdeoxy-6-per(2-carboxyethyl)thio-γ-cyclodextrin. $C_{72}H_{112}O_{48}S_8$; mol wt 2002.12. C 43.19%, H 5.64%, O 38.36%, S 12.81%. Selective relaxant binding agent (SRBA). Cyclodextrin-based host molecule that reverses the action of steroidal neuromuscular blocking drugs via selective binding and encapsulation. Prepn: M. Zhang *et al.*, **WO 0140316**; *eidem*, **US 6670340** (2001, 2003 both to Akzo Nobel). Synthesis, structure-activity, and x-ray crystallography: J. M. Adam *et al.*, *J. Med. Chem.* **45**, 1806 (2002). Crystal structure of complex with rocuronium: A. Bom *et al.*, *Angew. Chem. Int. Ed.* **41**, 266 (2002). LC/MS determn in plasma and urine: O. Epemolu *et al.*, *Rapid Commun. Mass Spectrom.* **16**, 1946 (2002). Clinical pharmacokinetics and safety assessment: F. Gijsenbergh *et al.*, *Anesthesiology* **103**, 695 (2005). Series of articles on pharmacology and clinical experience: *Anesth. Analg.* **104**, 555-586 (2007).

Octasodium salt. [343306-79-6] Sugammadex sodium; Org-25969; Bridion. $C_{72}H_{104}Na_8O_{48}S_8$; mol wt 2177.97. Colorless, orthorhombic crystals from DMF + water. d 1.446. Sol in water.

THERAP CAT: Reversal agent for neuromuscular blocking drugs.

9020. Suint. Portion of the sheep's fleece which is sol in cold water after the wax has been removed. Complex mixture of metallic ions, organic acids, peptides, weak bases, neutral substances, and inorganic cations. The following acids have been identified: acetic, propionic, butyric, valeric, oxalic, succinic, and glutaric. Paper chromatography indicates the presence of adipic and pimelic acids: Deane, Truter, *Biochim. Biophys. Acta* **18**, 435 (1955); *J. Chem. Soc.* **1959**, 2746. Determination of ion content: Mohsin, Shah, *Pak. J. Sci. Ind. Res.* **12**, 286 (1970), *C.A.* **73**, 4778q (1970).

9021. Sulbactam. [68373-14-8] (2S,5R)-3,3-Dimethyl-7-oxo-4-thia-1-azabicyclo[3.2.0]heptane-2-carboxylic acid 4,4-dioxide; penicillanic acid sulfone; penicillanic acid 1,1-dioxide; CP-45899. $C_8H_{11}NO_5S$; mol wt 233.24. C 41.20%, H 4.75%, N 6.01%, O 34.30%, S 13.75%. Semi-synthetic β-lactamase inhibitor. Prepn and use with β-lactam antibiotics: **BE 867859**; W. E. Barth, **US 4234579** (1978, 1980 both to Pfizer); R. A. Volkmann *et al.*, *J. Org. Chem.* **47**, 3344 (1982). β-Lactamase activity and antibacterial spectrum *in vitro*: A. R. English *et al.*, *Antimicrob. Agents Chemother.* **14**, 414 (1978); R. N. Jones *et al.*, *Diagn. Microbiol. Infect. Dis.* **3**, 489 (1985). HPLC determn in human plasma and urine: J. Haginaka *et al.*, *J. Chromatogr.* **341**, 115 (1985). Pharmacokinetics in humans: G. Foulds *et al.*, *Antimicrob. Agents Chemother.* **23**, 692 (1983). Clinical study of synergistic effect with ampicillin: S. Mehtar *et al.*, *J. Antimicrob. Chemother.* **17**, 389 (1986); B. V. Stromberg *et al.*, *Surg. Gynecol. Obstet.* **162**, 575 (1986). Review of activity and therapeutic use of sulbactam with ampicillin: D. M. Campoli-Richards, R. N. Brogden, *Drugs* **33**, 577-609 (1987).

White, crystalline solid, mp 148-151° (Barth); also reported as mp 154-155.5° (dec) (Barth); also reported as mp 170° (dec) (Volkmann). $[\alpha]_D^{20}$ +251° (c = 0.01 in pH 5.0 buffer). Sol in water.
Sodium salt. [69388-84-7] CP-45899-2; Betamaze. C_8H_{10}-NNaO$_5$S; mol wt 255.22. White to off-white, crystalline powder. Freely sol in water, dilute acid; sparingly sol in acetone, ethyl acetate, chloroform.
Mixture of sodium salt with ampicillin sodium. [117060-71-6] Loricin; Unacim; Unasyn (inj.).
Mixture of sodium salt with cefoperazone sodium. Sulperazone.
Compd with ampicillin *see* Sultamicillin.
THERAP CAT: In combination with β-lactam antibiotics as antibacterial.

9022. Sulbenicillin. [41744-40-5] (2S,5R,6R)-3,3-Dimethyl-7-oxo-6-[[(2R)-phenylsulfoacetyl]amino]-4-thia-1-azabicyclo-[3.2.0]heptane-2-carboxylic acid; α-sulfobenzylpenicillin; sulfocillin. $C_{16}H_{18}N_2O_7S_2$; mol wt 414.45. C 46.37%, H 4.38%, N 6.76%, O 27.02%, S 15.47%. Semi-synthetic antibiotic related to penicillin. Prepn: S. Morimoto *et al.*, **DE 1948943** corresp to **US 3660379** (1970, 1972 both to Takeda); *eidem, J. Med. Chem.* **15**, 1105, 1108 (1972). *In vitro* activity studies: Tsuchiya *et al.*, *J. Antibiot.* **24**, 607 (1971); *in vivo*: Yamazaki, Tsuchiya, *ibid.* 620. Metabolism: Tsuchiya *et al.*, *ibid.* **25**, 336 (1972). Toxicity: Y. Murata *et al.*, *Takeda Kenkyusho ho* **30**, 262 (1971), *C.A.* **76**, 147x (1972). Pharmacokinetics: A. Montanari *et al.*, *Clin. Ter.* **89**, 163 (1979); *eidem, Int. J. Clin. Pharmacol. Ther. Toxicol.* **18**, 225 (1980).

Disodium salt. Kedacillina; Sulpelin; Lilacillin. $C_{16}H_{16}N_2Na_2$-O_7S_2; mol wt 458.41. Prepd as a 3:1 mixture of D(−) and L(+)

isomers. Yellowish-white powder, mp 195-198° (dec). $[\alpha]_D^{22}$ +169-173°. uv max: 257, 262, 268 nm. Very sol in water; sol in methanol. Almost insol in *n*-propanol, acetone, chloroform, benzene, ethyl acetate. LD$_{50}$ in male, female mice, male, female rats (mg/kg): 7900, 8000, 6000, 6200 i.v.; 9600, 10000, 7200, 7500 i.p.; 11500, 13500, 11000, 11800 s.c.; 11000, 10500, 8300, 8600 i.m.; all >15000 orally (Murata).
THERAP CAT: Antibacterial.

9023. Sulbenox. [58095-31-1] *N*-(4,5,6,7-Tetrahydro-7-oxo-benzo[*b*]thien-4-yl)urea; CL-206576; Vigazoo. $C_9H_{10}N_2O_2S$; mol wt 210.25. C 51.41%, H 4.79%, N 13.32%, O 15.22%, S 15.25%. Prepn: **NL 7610270** (1977 to Am. Cyanamid), *C.A.* **88**, 22607x (1978). Activity: G. Asato, R. D. Wilbur, *Experientia* **35**, 1458 (1979).

Crystals, mp 245-246°. LD$_{50}$ orally in rats: >5000 mg/kg (Asato, Wilbur).
THERAP CAT (VET): Growth stimulant.

9024. Sulbentine. [350-12-9] Tetrahydro-3,5-bis(phenyl-methyl)-2*H*-1,3,5-thiadiazine-2-thione; 3,5-dibenzyltetrahydro-2*H*-1,3,5-thiadiazine-2-thione; 2-thioxo-3,5-dibenzyltetrahydro-1,3,5-thiadiazine; dibenzthione; Fungiplex; Refungine. $C_{17}H_{18}N_2S_2$; mol wt 314.47. C 64.93%, H 5.77%, N 8.91%, S 20.39%. Prepn: Rieche *et al.*, *Arch. Pharm.* **293**, 957 (1960); **DD 20634** (1961), *C.A.* **56**, 487e (1962).

Crystals from acetone or methanol, mp 101-102°.
THERAP CAT: Antifungal.

9025. Sulconazole. [61318-90-9] 1-[2-[[(4-Chlorophenyl)methyl]thio]-2-(2,4-dichlorophenyl)ethyl]-1*H*-imidazole; (±)-1-[2,4-dichloro-β-[(*p*-chlorobenzyl)thio]phenethyl]imidazole. C_{18}-$H_{15}Cl_3N_2S$; mol wt 397.74. C 54.36%, H 3.80%, Cl 26.74%, N 7.04%, S 8.06%. Prepn: K. A. M. Walker, **DE 2541833**; *idem*, **US 4055652** (1976, 1977 both to Syntex). HPLC determn in plasma: M. Fass *et al.*, *J. Pharm. Sci.* **70**, 1338 (1981). Mechanism of action study: W. H. Beggs, *Biochem. Arch.* **10**, 117 (1994). Clinical trial in tinea pedis: W. A. Akers *et al.*, *J. Am. Acad. Dermatol.* **21**, 686 (1989). Review of pharmacology and clinical efficacy: P. Benfield, S. P. Clissold, *Drugs* **35**, 143-153 (1988).

Nitrate. [61318-91-0] RS-44872; Exelderm; Myk; Sulcosyn. $C_{18}H_{15}Cl_3N_2S$.HNO_3; mol wt 460.75. Colorless crystals from acetone, mp 130.5-132°. Freely sol in pyridine; sparingly sol in methanol; slightly sol in alc, chloroform, acetone, methylene chloride; very slightly sol in water, toluene, dioxane.
THERAP CAT: Antifungal.

9026. Sulcotrione. [99105-77-8] 2-[2-Chloro-4-(methylsulfonyl)benzoyl]-1,3-cyclohexanedione; ICI-A-0051; SC-0051; Mikado. $C_{14}H_{13}ClO_5S$; mol wt 328.76. C 51.15%, H 3.99%, Cl 10.78%, O 24.33%, S 9.75%. Triketone bleaching herbicide for use in maize. Inhibits 4-hydroxyphenylpyruvate dioxygenase, an enzyme involved in the biosynthesis of plastoquinones. Prepn: W. J. Michaely, G. W. Kraatz, **EP 137963**; *eidem*, **US 4780127** (1983, 1988 both to Stauffer Chem.). Field trials in corn: J. S. Wilson, C. L. Foy, *Weed Technol.* **4**, 731 (1990). Soil adsorption: *eidem, ibid.* **6**, 583 (1992). Mode of action study: J. Secor, *Plant Physiol.* **106**, 1429 (1994). Comprehensive description: J. M. Beraud *et al., Brighton Crop Prot. Conf. - Weeds* **1991**, 51-56.

Light tan solid, mp 139°. Vapor pressure (25°): 4×10^{-8} mm Hg. Soly in water (25°): 165 mg/l. Sol in acetone, chlorobenzene.
USE: Herbicide.

9027. Sulesomab. [167747-19-5] Anti-(human NCA-90 granulocyte cell antigen) immunoglobulin G1 Fab' fragment (mouse monoclonal IMMU-MN3 γ_1-chain) disulfide with mouse monoclonal IMMU-MN3 light chain. Murine antigranulocyte monoclonal Fab' antibody fragment directed against nonspecific crossreactive antigen-90 (NCA-90), a surface glycoprotein on granulocytes, and cross reactive with carcinoembryonic antigen (CEA). [99m]Technetium radiolabeled immunoconjugate designed for imaging activated granulocytes. Prepn and characterization as an anti-CEA antibody: H. J. Hansen *et al., Cancer* **71**, 3478 (1993). Prepn of radiolabeled form: *idem et al.,* **US 5328679** (1994 to Immunomedics). Pharmacology and immunoscintigraphy: W. Becker *et al., Semin. Nucl. Med.* **24**, 142 (1994). Clinical study in detection of osteomyelitis: S. J. Harwood *et al., Clin. Infect. Dis.* **28**, 1200 (1999); of appendicitis: B. Baron *et al., Surgery* **125**, 288 (1999).
Complex with [99m]Tc. [157856-25-2] Technetium Tc 99m sulesomab; ImmuRAID-MN3-Tc-99m; LeukoScan.
THERAP CAT: ^{99m}Tc-complex as diagnostic aid (radioactive imaging agent).

9028. Sulfabenz. [127-77-5] 4-Amino-*N*-phenylbenzenesulfonamide; sulfanilanilide; *p*-aminobenzenesulfonanilide; N^1-phenylsulfanilamide. $C_{12}H_{12}N_2O_2S$; mol wt 248.30. C 58.05%, H 4.87%, N 11.28%, O 12.89%, S 12.91%. Prepn: Gelmo, *J. Prakt. Chem.* **77**, 369 (1908); Mastryukova *et al., Tetrahedron* **19**, 357 (1963).

Needles from dil alc, mp 200°. Practically insol in water; readily sol in warm alcohol, acetone. pKa: 10.94 (in 50% aq alc); 11.59 (in 80% aq alc).
THERAP CAT (VET): Antibacterial; coccidiostat (for poultry).

9029. Sulfabenzamide. [127-71-9] *N*-[(4-Aminophenyl)sulfonyl]benzamide; *N*-sulfanilylbenzamide; N^1-benzoylsulfanilamide; *N*-(*p*-aminobenzenesulfonyl)benzamide; Sulfabenzide. $C_{13}H_{12}N_2O_3S$; mol wt 276.31. C 56.51%, H 4.38%, N 10.14%, O 17.37%, S 11.60%. Prepn: Crossley *et al., J. Am. Chem. Soc.* **61**, 2950 (1939); Dvornikoff, **US 2240496** (1941 to Monsanto); **GB 541958** (1942 to Schering AG); Siebenmann, Schnitzer, *J. Am. Chem. Soc.* **65**, 2126 (1943).

Long, hexagonal prisms from 60% alcohol, mp 181.2-182.3°. pKa (25°): 4.57. One gram dissolves in 3225 ml of water at 30°, in 33 ml of 95% ethanol, in 9 ml of acetone. Soly in 1*N* NaOH or KOH: ~16 g/100 ml; pH just above 7.0. Aq solns of the water-soluble sodium salt have the same pH. Sol in alc, acetone, NaOH. Insol in water, ether.
THERAP CAT: Antibacterial.
THERAP CAT (VET): Antimicrobial.

9030. Sulfabromomethazine. [116-45-0] 4-Amino-*N*-(5-bromo-4,6-dimethyl-2-pyrimidinyl)benzenesulfonamide; N^1-(5-bromo-4,6-dimethyl-2-pyrimidinyl)sulfanilamide; 5-bromo-4,6-dimethyl-2-sulfanilamidopyrimidine; 2-sulfanilamide-5-bromo-4,6-dimethyl pyrimidine; 5-bromosulfamethazine; SN-3517. $C_{12}H_{13}BrN_4O_2S$; mol wt 357.23. C 40.35%, H 3.67%, Br 22.37%, N 15.68%, O 8.96%, S 8.97%. Prepn: English *et al., J. Am. Chem. Soc.* **68**, 453 (1946). Pharmacology: C. M. Stowe *et al., Am. J. Vet. Res.* **29**, 345 (1958).

Crystals, dec 250-252°. uv max (methanol): 238, 272 nm (A 428, 635); min: 250 nm (A 290). Sol in alkaline solns.
Sodium salt monohydrate. Sulfabrom. $C_{12}H_{12}BrN_4NaO_2S \cdot H_2O$; mol wt 397.22. Cream-colored powder. Sol in water.
THERAP CAT (VET): Antibacterial.

9031. Sulfacetamide. [144-80-9] *N*-[(4-Aminophenyl)sulfonyl]acetamide; *N*-sulfanilylacetamide; N^1-acetylsulfanilamide; *p*-aminobenzenesulfonoacetamide. $C_8H_{10}N_2O_3S$; mol wt 214.24. C 44.85%, H 4.71%, N 13.08%, O 22.40%, S 14.96%. Incorrectly called *sulfacetimide*. Prepn: M. L. Crossley *et al., J. Am. Chem. Soc.* **61**, 2950 (1939); M. Dohrn, P. Diedrich, *Muench. Med. Wochenschr.* **85**, 2017 (1938); *eidem*, **US 2411495** (1946 to Schering). Toxicity: R. S. Fisher, H. B. Haag, *J. Urol.* **47**, 183 (1942). Antimicrobial activity: R. D. Houlsby *et al., J. Pharm. Sci.* **72**, 1401 (1983). GC determn in animal tissues: A. E. Mooser, H. Koch, *J. AOAC Int.* **76**, 976 (1993). Clinical trial in conjunctivitis: J. A. Lohr *et al., Pediatr. Infect. Dis. J.* **7**, 626 (1988); in acne: D. L. Breneman, M. C. Ariano, *Int. J. Dermatol.* **32**, 365 (1993). Comprehensive description: I. Ahmad *et al., Anal. Profiles Drug Subs. Excip.* **23**, 471-509 (1994).

White or yellowish white prisms, mp 182-184°. Acidic, slightly saline taste. uv max (water, 0.1*M* HCl, 0.1*M* NaOH): 258, 217, 270, 256 (a_m 17700, 18900, 8900, 17300). Soluble in 150 parts water at 20°, in 15 parts alcohol, in 7 parts acetone. Freely sol in dil mineral acids, solns of potassium and sodium hydroxides; sol in solns of alkali hydroxide and carbonates; slightly sol in ether; very slightly sol in chloroform. Practically insol in benzene. An aq soln is acid to litmus. LD_{50} orally in dogs: 8000 mg/kg (Fisher).
Sodium salt. [127-56-0]; [6209-17-2] (monohydrate). Soluble sulfacetamide; Ak-Sulf; Albucid; Antébor; Beocid-Puroptal; Bleph-10; Cetamide; Prontamid; Sebizon; Sodium Sulamyd; Sulf-10; Sulten-10. $C_8H_9N_2NaO_3S$; mol wt 236.22. Minute prisms from dil alcohol, mp 257°. Occurs as the monohydrate; white, odorless, crystalline powder with slightly bitter taste. uv max (pH 7.0 phosphate buffer, 0.1*M* HCl, aq. acid, 0.1*M* NaOH, aq alkali): 255, 271, 271, 256, 256 ($A_{1cm}^{1\%}$ 660-720, 207, 260, 626, 750). Sol in 1.5 parts water. Slightly sol in 96% ethanol; sparingly sol in acetone. Practically insol in ether, chloroform. pKa$_1$ (amino group): 1.8; pKa$_2$ (sulfonamide group at 25°): 5.4.

THERAP CAT: Antibacterial.
THERAP CAT (VET): Antibacterial.

9032. Sulfachlorpyridazine. [80-32-0] 4-Amino-*N*-(6-chloro-3-pyridazinyl)benzenesulfonamide; *N*[1]-(6-chloro-3-pyridazinyl)-sulfanilamide; 3-chloro-6-sulfanilamidopyridazine; 3-sulfanilamido-6-chloropyridazine; 3-(*p*-aminophenylsulfonamido)-6-chloropyridazine; Ciba 10370; Ba-10370; Cosumix; Sonilyn. $C_{10}H_9ClN_4O_2S$; mol wt 284.72. C 42.19%, H 3.19%, Cl 12.45%, N 19.68%, O 11.24%, S 11.26%. Prepn: Lester, English, **US 2790798** (1957 to Am. Cyanamid). Comparative antibacterial spectrum: G. E. Burrows *et al.*, *J. Vet. Diagn. Invest.* **5**, 541 (1993). Pharmacokinetics and therapeutic potential in horses: E. van Duijkeren *et al.*, *Vet. Rec.* **137**, 483 (1995). ELISA determn in food of animal origin: C. A. Spinks *et al.*, *Food Addit. Contam.* **18**, 11 (2001).

Sodium salt. [23282-55-5] Vetisulid. $C_{10}H_8ClN_4NaO_2S$; mol wt 306.70.
THERAP CAT: Antibacterial.
THERAP CAT (VET): Antibacterial agent. In enteric infections.

9033. Sulfachrysoidine. [485-41-6] 3,5-Diamino-2-[[4-(aminosulfonyl)phenyl]azo]benzoic acid; 3,5-diamino-2-[*p*-(sulfamoylphenyl)azo]benzoic acid; carboxysulfamidochrysoidine; Azo Compd No. 4; Rubiazol. $C_{13}H_{13}N_5O_4S$; mol wt 335.34. C 46.56%, H 3.91%, N 20.88%, O 19.08%, S 9.56%. Prepn: Gley, Girard, *C. R. Seances Soc. Biol. Ses Fil.* **125**, 1027 (1937).

Deep red crystals. mp above 300°.
THERAP CAT: Antibacterial.

9034. Sulfacytine. [17784-12-2] 4-Amino-*N*-(1-ethyl-1,2-dihydro-2-oxo-4-pyrimidinyl)benzenesulfonamide; *N*′-(1-ethyl-1,2-dihydro-2-oxo-4-pyrimidinyl)sulfanilamide; 1-ethyl-*N*-sulfanilylcytosine; *N*-sulfanilyl-1-ethylcytosine; CL-636; Renoquid. $C_{12}H_{14}N_4O_3S$; mol wt 294.33. C 48.97%, H 4.79%, N 19.04%, O 16.31%, S 10.89%. Prepn: **NL 6610815** corresp to Doub, Krolls, **US 3375247** (1967, 1968 to Parke, Davis). Prepn and activity: Doub *et al.*, *J. Med. Chem.* **13**, 242 (1970).

Crystals from butyl alcohol, methanol, mp 166.5-168°, (monohydrate, mp 104°). uv max (methanol): 263, 297 nm ($E_{1cm}^{1\%}$ 584, 762). Equilibrium soly in pH 5 buffer at 37°: approx 175 mg/100 ml. Soly increases with increasing pH. pK′ 6.9.
THERAP CAT: Antibacterial.

9035. Sulfadiazine. [68-35-9] 4-Amino-*N*-2-pyrimidinylbenzenesulfonamide; 2-sulfanilamidopyrimidine; *N*[1]-2-pyrimidinylsulfanilamide; *N*[1]-2-pyrimidylsulfanilamide; 2-sulfanilylaminopyrimidine; sulfapyrimidine; Adiazine. $C_{10}H_{10}N_4O_2S$; mol wt 250.28. C 47.99%, H 4.03%, N 22.39%, O 12.78%, S 12.81%. Prepd by condensing 2-aminopyrimidine with acetylsulfanilyl chloride followed by hydrolysis of the acetyl group with NaOH: Roblin *et al.*, *J. Am. Chem. Soc.* **62**, 2002 (1940); Barber, **GB 557055** (1943); Sprague,

US 2407966 (1946 to Sharp & Dohme); Winnek, Roblin, **US 2410793** (1946 to Am. Cyanamid). Toxicological, chemotherapeutic and pharmacokinetic studies: Böhni *et al.*, *Chemotherapy* **14**, 195-226 (1969). *In vitro* activity of silver salt: Carr *et al.*, *Antimicrob. Agents Chemother.* **4**, 585 (1973). Use of silver salt in burns: C. L. Fox, *Arch. Surg.* **96**, 184 (1968). Comprehensive description: H. Stober, W. DeWitte, *Anal. Profiles Drug Subs.* **11**, 523-551 (1982).

White or slightly yellow powder. mp 252-256°. Sparingly sol in water at 37°: 13 mg/100 ml at pH 5.5; 200 mg/100 ml at pH 7.5. Sparingly sol in alc, acetone. One gram dissolves in about 620 ml of human serum at 37°. Freely sol in dil mineral acids and in solns of potassium and sodium hydroxides and in ammonia water.
Mixture with trimethoprim (usually 5:1). Co-trimazine; Norodine; Scorprin; Tribrissen; Triglobe; Vesuprim.
Sodium salt. Sulfadiazine sodium; soluble sulfadiazine. $C_{10}H_9$-N_4NaO_2S. White powder. On prolonged exposure to humid air, it absorbs carbon dioxide with the liberation of sulfadiazine and becomes incompletely sol in water. One gram dissolves in about 2 ml of water. Slightly sol in alc. Solns are alkaline to phenolphthalein (pH 9-11).
Silver salt. [22199-08-2] Flamazine; Flammazine; Silvadene. White, crystalline powder. Freely sol in 30% ammonium soln; slightly sol in acetone. Practically insol in alc, chloroform, ether. Decomposes in moderately strong mineral acids.
THERAP CAT: Antibacterial.
THERAP CAT (VET): Antibacterial.

9036. Sulfadicramide. [115-68-4] *N*-[(4-Aminophenyl)sulfonyl]-3-methyl-2-butenamide; 3-methyl-*N*-sulfanilylcrotonamide; *N*[1]-senecioylsulfanilamide; *N*-sulfanilylseneciamide; *N*[1]-dimethylacroylsulfanilamide; *N*-sulfanilyl-β,β-dimethylacrylamide; Irgamide. $C_{11}H_{14}N_2O_3S$; mol wt 254.30. C 51.95%, H 5.55%, N 11.02%, O 18.87%, S 12.61%. Prepd from β,β-dimethylacrylamide by treatment with sodamide and *p*-nitrobenzenesulfonyl chloride; the resulting 4-nitro-*N*-(β,β-dimethylacrylyl)benzenesulfonamide is reduced with Fe and dil AcOH: Martin, Häfliger, **US 2417005** (1947), *cf.* **US 2383874** (1945).

Crystals from alc. mp 184-185°. Slightly sol in water, ether. Freely sol in alcohol, acetone. For the prepn of solns the water-soluble sodium salt is marketed. A 20% aq soln of the sodium salt has a pH of 8.3.
THERAP CAT: Antibacterial.

9037. Sulfadimethoxine. [122-11-2] 4-Amino-*N*-(2,6-dimethoxy-4-pyrimidinyl)benzenesulfonamide; *N*[1]-(2,6-dimethoxy-4-pyrimidinyl)sulfanilamide; 2,6-dimethoxy-4-sulfanilamidopyrimidine; 2,4-dimethoxy-6-sulfanilamido-1,3-diazine; 6-sulfanilamido-2,4-dimethoxypyrimidine; 2,6-dimethoxy-4-(*p*-aminobenzenesulfonamido)pyrimidine; Albon; Sulforal; Ultrasulfon. $C_{12}H_{14}N_4O_4S$; mol wt 310.33. C 46.44%, H 4.55%, N 18.05%, O 20.62%, S 10.33%. Prepn: Bretschneider, Klötzer, **US 2703800** (1955 to Oesterr. Stickstoffwerke); *eidem, Monatsh. Chem.* **87**, 136 (1956); Langley, **GB 866843** (1961 to I.C.I.); Bretschneider *et al.*, *Monatsh. Chem.* **92**, 128 (1961); **US 3127398** (1964 to Hoffmann-La Roche); Shepherd *et al.*, *J. Org. Chem.* **26**, 2764 (1961). Toxicity data: Seki *et al.*, *Arzneim.-Forsch.* **15**, 1441 (1965). Toxicological, chemotherapeutic and pharmacokinetic studies: Böhni *et al.*, *Chemotherapy* **14**, 195-226 (1969); *see also* Aviado *et al.*, *ibid.* 37.

Crystals from dil alc, mp 201-203°. Soly in water at 37° (mg/100 ml): 4.6 at pH 4.10; 29.5 at pH 6.7; 58.0 at pH 7.06; 5170 at pH 8.71. Sol in dil HCl, in $2N$ sodium hydroxide, aq solns of sodium carbonate; sparingly sol in $2N$ hydrochloric acid; slightly sol in alc, ether, chloroform, hexane. Practically insol in water. LD_{50} orally in mice: >10 g/kg (Seki).

Sodium salt. [1037-50-9] Di-Methox. $C_{12}H_{13}N_4NaO_4S$; mol wt 332.31. Crystals from methanol + ether. Freely sol in water. pH of 5% solution: 8.1; of 10% soln: 8.6.

THERAP CAT (VET): Antibacterial.

9038. Sulfadoxine. [2447-57-6] 4-Amino-N-(5,6-dimethoxy-4-pyrimidinyl)benzenesulfonamide; N'-(5,6-dimethoxy-4-pyrimidinyl)sulfanilamide; 6-(4-aminobenzenesulfonamido)-4,5-dimethoxypyrimidine; 4-sulfanilamido-5,6-dimethoxypyrimidine; sulforthomidine; sulphormethoxine; Ro-4-4393; Fanasil. $C_{12}H_{14}N_4O_4S$; mol wt 310.33. C 46.44%, H 4.55%, N 18.05%, O 20.62%, S 10.33%. Dihydropteroate synthetase inhibitor. Prepn: **BE 618639**; H. Bretschneider *et al.*, **US 3132139** (1962, 1964 both to Hoffmann-La Roche). Toxicological, chemotherapeutic and pharmacokinetic studies: E. Böhni *et al.*, *Chemotherapy* **14**, 195-226 (1969). Comprehensive description: V. K. Kapoor, *Anal. Profiles Drug Subs.* **17**, 571-605 (1988). Enzyme inhibition *in vitro*: Y. Zhang, S. R. Meshnick, *Antimicrob. Agents Chemother.* **35**, 267 (1991). Determn in plasma using supercritical fluid chromatography: S. I. Bhoir *et al.*, *J. Chromatogr. B* **757**, 39 (2001). Clinical trial of combination with pyrimethamine, *q.v.*, for malaria: C. V. Plowe *et al.*, *Br. Med. J.* **328**, 545 (2004).

Crystals from 50% aq alc, mp 190-194°. Practically insol in ether. Very slightly sol in water, slightly sol in alcohol, methanol. Sol in dilute mineral acids, solutions of alkali hydroxides and carbonates. LD_{50} in mice (microcrystals, mg/kg): 5200 orally, 2900 s.c., 2900 i.p. (Böhni).

Mixture with trimethoprim. [39295-60-8] Animar; Bimotrim; Borgal.

THERAP CAT: Antibacterial; antimalarial.

THERAP CAT (VET): Antibacterial.

9039. Sulfaethidole. [94-19-9] 4-Amino-N-(5-ethyl-1,3,4-thiadiazol-2-yl)benzenesulfonamide; N'-(5-ethyl-1,3,4-thiadiazol-2-yl)sulfanilamide; 5-ethyl-2-sulfanilamido-1,3,4-thiadiazole; 2-(p-aminobenzenesulfonamido)-5-ethylthiadiazole; sulfaethylthiadiazole; VK-55; Sul-Spantab. $C_{10}H_{12}N_4O_2S_2$; mol wt 284.35. C 42.24%, H 4.25%, N 19.70%, O 11.25%, S 22.55%. The commercial prepn (Schering AG) is described in P.B. reports, no. 245, p 5, and no. 1361, p 73. Prepn: Wojahn, Wuckel, *Arch. Pharm.* **284**, 53 (1951).

Crystals, mp 185.5-186.0°. One gram dissolves in 4000 ml water, in 40 g methanol, in 30 g ethanol, in 10 g acetone, in 1350 g ether, in 2800 g chloroform, in 20,000 g benzene. Faintly acid to litmus.

Sodium salt. pH of 20% soln: 7.5.

THERAP CAT: Antibacterial.

THERAP CAT (VET): Antimicrobial.

9040. Sulfaguanidine. [57-67-0] 4-Amino-N-(aminoiminomethyl)benzenesulfonamide; 4-amino-N-(diaminomethylene)benzenesulfonamide; N'-amidinosulfanilamide; N'-guanylsulfanilamide; p-aminobenzenesulfonylguanidine; sulfanilylguanidine; RP-2275; Diacta; Ganidan; Guanicil; Resulfon; Shigatox. $C_7H_{10}N_4O_2$-S; mol wt 214.24. C 39.24%, H 4.71%, N 26.15%, O 14.94%, S 14.96%. Prepd by condensing acetylsulfanilyl chloride with guanidine nitrate in the presence of much NaOH in water-acetone: Winnek, **US 2218490** (1940); **US 2229784** (1941); **US 2233569** (1941); or by fusing N^4-acetylsulfanilamide and dicyanodiamide: Haworth, Rose, **GB 551524** (1943). Additional syntheses: *ACS Monograph Series* **no. 106**, entitled "Sulfonamides and Allied Compounds," E. H. Northey, Ed. (Reinhold, New York, 1948). Revised structure: G. R. Sullivan, J. D. Roberts, *J. Org. Chem.* **42**, 1095 (1977). Review of pharmacology: E. K. Marshall, Jr. *et al.*, *Bull. Johns Hopkins Hosp.* **67**, 163-188 (1940); E. Pick, *J. Mt. Sinai Hosp.* **10**, 343-354 (1943).

Monohydrate. Needles. When anhydr, mp 190-193°. One gram dissolves in about 1000 ml water at 25° and in about 10 ml at 100°. Sparingly sol in alcohol or acetone. Freely sol in dil mineral acids. Insol in NaOH solns at room temp.

THERAP CAT: Antibacterial.

THERAP CAT (VET): Antimicrobial. In enteric infections.

9041. Sulfaguanole. [27031-08-9] 4-Amino-N-[[(4,5-dimethyl-2-oxazolyl)amino]iminomethyl]benzenesulfonamide; N'-[(4,5-dimethyl-2-oxazolyl)amidino]sulfanilamide; 1-(4,5-dimethyloxazol-2-yl)-3-sulfanilylguanidine; 1-(p-aminophenylsulfonyl)-3-(4,5-dimethyl-2-oxazolyl)guanidine; sulfadimethyloxazolylguanidine; Enterocura. $C_{12}H_{15}N_5O_3S$; mol wt 309.34. C 46.59%, H 4.89%, N 22.64%, O 15.52% S 10.36%. Prepn: Loop *et al.*, **GB 1185139**, *C.A.* **73**, 3909w (1970); *eidem*, **US 3562258** (1970, 1971 both to Nordmark-Werke); Loop, Kohlmann, *Arzneim.-Forsch.* **23**, 171 (1973). Toxicology: J. Kuhne *et al.*, *ibid.* 178. Series of articles on pharmacology: *ibid.* 172-192.

Almost colorless crystals from 9:1 acetone-water, mp 233-236°; from water-methanol, mp 228-230°. pKa 7.76. Practically insol in water. Sol in dil NaOH solns. No deaths in mice or rats for oral doses of 5 g/kg (Kuhne).

THERAP CAT: Antibacterial.

9042. Sulfalene. [152-47-6] 4-Amino-N-(3-methoxypyrazinyl)benzenesulfonamide; N'-(3-methoxy-2-pyrazinyl)sulfanilamide; 2-(p-aminobenzenesulfonamido)-3-methoxypyrazine; 3-methoxy-2-sulfanilamidopyrazine; sulfamethopyrazine; 2-sulfanilamido-3-methoxypyrazine; sulfamethoxypyrazine; sulfapyrazinemethoxyine; Farmitalia 204/122; Dalysep; Kelfizina; Kelfizine W; Longum; Policydal; Vetkelfizina. $C_{11}H_{12}N_4O_3S$; mol wt 280.30. C 47.14%, H 4.32%, N 19.99%, O 17.12%, S 11.44%. Prepn: B. Camerino, G. Palamidessi, *Gazz. Chim. Ital.* **90**, 1815 (1960); **GB 928151**; **US 3098069** (both 1963 to Farmitalia). *Review*: Gasparini, *Veterinaria*

(Milan) **20**, 302 (1971). Also used in combination with trimetho-prim, *q.v.* Toxicology: V. De Pascale *et al.*, *Boll. Chim. Farm.* **116**, 155 (1977). Pharmacodynamics: D. S. Reeves *et al.*, *J. Antimicrob. Chemother.* **6**, 647 (1980). Clinical assessment: H. Harazim *et al.*, *J. Int. Med. Res.* **11**, 197 (1983); M. L. Colombo *et al.*, *Pharmather-apeutica* **3**, 556 (1984). Review of clinical efficacy: F. Celotti *et al.*, *J. Int. Med. Res.* **14**, 236-241 (1986).

Crystals from alcohol, mp 176°. LD_{50} in mice (g/kg): 2.164 oral-ly; 1.41 i.v. (**US 3098069**).

Mixture with trimethoprim. [50933-06-7] Kelfiprim. LD_{50} in mice, rats (mg/kg): 3500; 3550 orally (De Pascale).

THERAP CAT: Antibacterial.

THERAP CAT (VET): Antibacterial.

9043. Sulfallate. [95-06-7] *N,N*-Diethylcarbamodithioic acid 2-chloro-2-propen-1-yl ester; diethyldithiocarbamic acid 2-chloroal-lyl ester; 2-chloroallyl diethyldithiocarbamate; CDEC; CP-4742; Ve-gadex. $C_8H_{14}ClNS_2$; mol wt 223.78. C 42.94%, H 6.31%, Cl 15.84%, N 6.26%, S 28.65%. Selective pre-planting or pre-emer-gence herbicide. Prepn: Harman, D'Amico, *J. Am. Chem. Soc.* **75**, 4081 (1953); **GB 769222** (1957 to Monsanto). Toxicity data: G. W. Bailey, J. L. White, *Residue Rev.* **10**, 97 (1965). Review of carcino-genic risk: *IARC Monographs* **30**, 283-291 (1983).

Amber liquid, bp_1 128-130°. n_D^{25} 1.5822. d^{25} 1.088. Soly in water at 25°: 100 ppm. Sol in most organic solvents. LD_{50} orally in rats: 850 mg/kg (Bailey, White).

Caution: Prolonged contact with skin and eyes may cause mod-erate irritation. *See Clinical Toxicology of Commercial Products*, R. E. Gosselin *et al.*, Eds. (Williams & Wilkins, Baltimore, 5th ed., 1984) Section II, p 313. This substance is reasonably anticipated to be a human carcinogen: *Report on Carcinogens, Twelfth Edition* (PB2011-111646, 2011) p 392.

USE: Herbicide.

9044. Sulfaloxic Acid. [14376-16-0] 2-[[[4-[[[[(Hydroxy-methyl)amino]carbonyl]amino]sulfonyl]phenyl]amino]carbonyl]-benzoic acid; 4′-(carbamoylsulfamoyl)phthalanilic acid hydroxy-methyl derivative; 4′-(carbamoylsulfamoyl)-*N*-(hydroxymethyl)-phthalanilic acid; formophthaloylsulfanilyl urea; 2-[4-(hydroxy-methylureidosulfonyl)phenylcarbamoyl]benzoic acid; sulphaloxic acid. $C_{16}H_{15}N_3O_7S$; mol wt 393.37. C 48.85%, H 3.84%, N 10.68%, O 28.47%, S 8.15%. Prepn: Wiedemann, Strassburger, **DE 960190** (1957 to von Heyden), *C.A.* **53**, 6547i (1959).

Crystals, mp 160-165°. Soluble in dil bases.

Calcium salt. [59672-20-7] Enteromide; Intestin-Euvernil. $C_{32}H_{28}CaN_6O_{14}S_2$; mol wt 824.80.

THERAP CAT: Antibacterial.

9045. Sulfamerazine. [127-79-5] 4-Amino-*N*-(4-methyl-2-pyrimidinyl)benzenesulfonamide; N^1-(4-methyl-2-pyrimidyl)sulfa-nilamide; N^1-(4-methyl-2-pyrimidinyl)sulfanilamide; 2-sulfanilami-do-4-methylpyrimidine; sulfamethyldiazine; RP-2632; Mesulfa; Per-coccide. $C_{11}H_{12}N_4O_2S$; mol wt 264.30. C 49.99%, H 4.58%, N 21.20%, O 12.11%, S 12.13%. Prepd by condensing 2-amino-4-methylpyrimidine with acetylsulfanyl chloride followed by hydro-lysis of the acetyl group: Roblin *et al.*, *J. Am. Chem. Soc.* **62**, 2002 (1940); Sprague *et al.*, *ibid.* **63**, 3028 (1941); Sprague, **US 2407966** (1946 to Sharp & Dohme). For prepn of 2-amino-4-methylpyrimi-dine *see:* Benary, *Ber.* **63**, 2601 (1930); Backer, Grevenstuk, *Rec. Trav. Chim.* **61**, 291 (1942); *cf.* E. H. Northey, *Sulfonamides* (Rein-hold, New York, 1948). Antimicrobial activity: Gill *et al.*, *Indian J. Vet. Sci.* **32**, 240 (1962); Vaichulis, Vedros, *Chemotherapia* **11**, 315 (1966). Toxicity studies: Simunek *et al.*, *Vet. Med. (Prague)* **13**, 619 (1968). Kinetics of sulfamerazine decompn: Zajac, *Diss. Pharm. Pharmacol.* **22**, 455 (1970). Comprehensive description: R. D. G. Woolfenden, *Anal. Profiles Drug Subs.* **6**, 515-577 (1977).

Crystals, mp 234-238°. uv max (water): 243, 257 nm ($E_{1cm}^{1\%}$ 875, 822); (0.1M HCl): 243, 307 nm ($E_{1cm}^{1\%}$ 625, 200); (ethanol): 271 nm ($E_{1cm}^{1\%}$ 835). Slowly darkens on exposure to light. Soly in water at 37°: 35 mg/100 ml at pH 5.5; 170 mg/100 ml at pH 7.5. Readily sol in dil mineral acids and in solns of potassium, ammonium and sodium hydroxides. Sparingly sol in acetone, slightly sol in alcohol, very slightly sol in ether, chloroform.

Monosodium salt. [127-58-2] Soluble sulfamerazine; Solumé-dine. $C_{11}H_{11}N_4NaO_2S$; mol wt 286.28. Crystals. Bitter, caustic taste. Hygroscopic. On prolonged exposure to humid air, it absorbs CO_2 with the liberation of sulfamerazine and becomes incompletely sol in water. Its solns are alkaline to phenolphthalein (pH 10 or more). One gram dissolves in 3.6 ml water. Slightly sol in alc. Insol in ether, chloroform.

THERAP CAT: Antibacterial.

THERAP CAT (VET): Antibacterial.

9046. Sulfameter. [651-06-9] 4-Amino-*N*-(5-methoxy-2-py-rimidinyl)benzenesulfonamide; N^1-(5-methoxy-2-pyrimidinyl)sul-fanilamide; sulfa-5-methoxypyrimidine; 2-sulfanilamido-5-meth-oxypyrimidine; 2-(*p*-aminobenzenesulfonamido)-5-methoxypyrim-idine; 5-methoxysulfadiazine; sulfamethoxydiazine; sulfametin (re-scinded USAN); sulfametorine; methoxypyrimal; AHR-857; I-2586; Bayrena; Durenat; Kinecid; Kiron; Kirocid; Sulla; Ultrax. $C_{11}H_{12}$-N_4O_3S; mol wt 280.30. C 47.14%, H 4.32%, N 19.99%, O 17.12%, S 11.44%. Prepn: **DE 1101428**; P. Diedrich, **US 3214335** (1961, 1965 both to Schering AG); Horstmann *et al.*, *Arzneim.-Forsch.* **11**, 682 (1961); Budesinky *et al.*, *Experientia* **17**, 129 (1961); **CS 98818**; **BE 633513** (1963 to SPOFA); *Collect. Czech. Chem. Commun.* **29**, 2980 (1964). Toxicity study: G. Hecht *et al.*, *Arzneim.-Forsch.* **11**, 695 (1961). Toxicology, chemotherapeutics and pharmacokinetics: Böhne *et al.*, *Chemotherapy* **14**, 195-226 (1969).

Minute, bitter crystals, mp 214-216°. uv max: 230, 271 nm ($E_{1cm}^{1\%}$ 562, 726). Very sparingly sol in water, alcohol, ether. Sol in dil acids, alkalies.

Sodium salt. Water soluble. pK about 6.8. pH of 1% aq soln: about 8.9. LD_{50} in mice, rats (g/kg): 1.1 ±0.2, 1.2 i.v.; 1.5, 1.1 i.p.; 3.0, 1.0 orally (Hecht).

THERAP CAT: Antibacterial.

9047. Sulfamethazine. [57-68-1] 4-Amino-N-(4,6-dimethyl-2-pyrimidinyl)benzenesulfonamide; N^1-(4,6-dimethyl-2-pyrimidinyl)sulfanilamide; N^1-(4,6-dimethyl-2-pyrimidyl)sulfanilamide; 4,6-dimethyl-2-sulfanilamidopyrimidine; sulfamezathine; sulfadimerazine; sulfadimidine; sulfamidine; sulfadimethylpyrimidine; Diazil; Sulfadine; S-Dimidine; Dimidin-R; Neazina; Sulmet. $C_{12}H_{14}N_4O_2S$; mol wt 278.33. C 51.78%, H 5.07%, N 20.13%, O 11.50%, S 11.52%. Prepn: Caldwell et al., *J. Am. Chem. Soc.* **63**, 2188 (1941); Roblin et al., *ibid.* **64**, 567 (1942); Weisner, Katscher, **GB 546158** (1942 to Ward, Blenkinsop); Sprague, **US 2407966** (1946 to Sharp & Dohme). Alternate prepn: Haworth, Rose, **GB 552887** (1943 to I.C.I.); Garzia, **US 3119818** (1964 to Ist. Chemioterap. Ital.). Anticoccidiosis activity: Zarin, Feodorova, *C.A.* **66**, 114, 435p-436q (1967). Metabolic studies: Turco et al., *Clin. Pharmacol. Ther.* **7**, 603 (1966); Parker et al., *Hum. Hered.* **19**, 402 (1969). Solubility studies: Shkadova et al., *Farm. Zh. (Kiev)* **24**, 39 (1969). Acute toxicity: B. Bobranski et al., *Arch. Immunol. Ther. Exp.* **16**, 804 (1968). Comprehensive description: C. Papostephanou, M. Frantz, *Anal. Profiles Drug Subs.* **7**, 401-422 (1978).

Crystals from dioxane-water, mp 176° (also reported as mp 178-179°; mp 198-199°; mp 205-207°). uv max (water, pH 6.6): 241 nm ($E_{1cm}^{1\%}$ 670); (0.01N NaOH): 243, 257 nm ($E_{1cm}^{1\%}$ 765, 776); (0.01N HCl): 241, 297 nm ($E_{1cm}^{1\%}$ 561, 266). Soly in water at 29°: 150 mg/100 ml; at 37°: 192 mg/100 ml at pH 7.0. pK_1 7.4±0.2, pK_2 2.65±0.2. The soly increases rapidly with an increase in pH. Sol in acetone; slightly sol in alc; very slightly sol in ether. LD_{50} i.p. in mice: 1.06 g/kg (Bobranski).

Sodium salt. [1981-58-4] Intradine; Sulfoxine 33; Vesadin. $C_{12}H_{13}N_4NaO_2S$; mol wt 300.31.

THERAP CAT: Antibacterial.

THERAP CAT (VET): Antibacterial.

9048. Sulfamethizole. [144-82-1] 4-Amino-N-(5-methyl-1,3,4-thiadiazol-2-yl)benzenesulfonamide; N^1-(5-methyl-1,3,4-thiadiazol-2-yl)sulfanilamide; 2-sulfanilamido-5-methyl-1,3,4-thiadiazole; 5-methyl-2-sulfanilamido-1,3,4-thiadiazole; 2-(p-aminobenzenesulfonamido)-5-methylthiadiazole; sulfamethylthiadiazole; Lucosil; Rufol; Urolucosil. $C_9H_{10}N_4O_2S_2$; mol wt 270.33. C 39.99%, H 3.73%, N 20.73%, O 11.84%, S 23.72%. Prepn: Hübner, **US 2447702** (1948 to Lundbeck).

Crystals from water, mp 208°. pKa 5.45. One gram dissolves in 4000 ml water at pH 6.5, in 5 ml water at pH 7.5, in 40 g methanol, in 30 g ethanol, in 10 g acetone, in 1370 g ether, in 2800 g chloroform. Freely sol in solns of ammonium, potassium and sodium hydroxides; sol in dilute mineral acids. Practically insol in benzene.

THERAP CAT: Antibacterial.

THERAP CAT (VET): Antibacterial.

9049. Sulfamethomidine. [3772-76-7] 4-Amino-N-(6-methoxy-2-methyl-4-pyrimidinyl)benzenesulfonamide; N^1-(6-methoxy-2-methyl-4-pyrimidinyl)sulfanilamide; 2-methyl-4-methoxy-6-sulfanilamidopyrimidine; 4-sulfanilamido-2-methyl-6-methoxypyrimidine; 4-methoxy-6-sulfanilamido-1,3-diazine; 4-(p-aminobenzenesulfonyl)amino-2-methyl-6-methoxypyrimidine; Duroprocin; Methofadin; Télémid. $C_{12}H_{14}N_4O_3S$; mol wt 294.33. C

48.97%, H 4.79%, N 19.04%, O 16.31%, S 10.89%. Prepn: Loop, Lührs, *Ann.* **580**, 225 (1953); Loop, **DE 926131** (1955 to Nordmark).

Crystals, mp 146°. Also obtained as the monohydrate.

THERAP CAT: Antibacterial.

9050. Sulfamethoxazole. [723-46-6] 4-Amino-N-(5-methyl-3-isoxazolyl)benzenesulfonamide; N^1-(5-methyl-3-isoxazolyl)sulfanilamide; 5-methyl-3-sulfanilamidoisoxazole; 3-sulfanilamido-5-methylisoxazole; 3-(p-aminophenylsulfonamido)-5-methylisoxazole; sulfisomezole; sulfamethylisoxazole; sulfamethoxizole; Gantanol; Sinomin. $C_{10}H_{11}N_3O_3S$; mol wt 253.28. C 47.42%, H 4.38%, N 16.59%, O 18.95%, S 12.66%. Prepn starting with ethyl 5-methylisoxazole-3-carbamate: Kano et al., **US 2888455** (1959 to Shionogi); *Annu. Rep. Shionogi Res. Lab.* **7**, 1 (1957), *C.A.* **51**, 17889 (1957). Toxicity data: Yamamoto et al., *Chemotherapy (Tokyo)* **21**, 187 (1973), *C.A.* **79**, 73738n (1973). Clinical trial of mixture with trimethoprim in *Pneumocystis carinii* pneumonia: J. M. Wharton et al., *Ann. Intern. Med.* **105**, 37 (1986). Comprehensive description: B. C. Rudy, B. Z. Senkowski, *Anal. Profiles Drug Subs.* **2**, 467-486 (1973). Review of antibacterial activity and clinical efficacy of mixture with trimethoprim: G. P. Wormser et al., *Drugs* **24**, 459-518 (1982). Symposium on clinical intravenous therapy: *Rev. Infect. Dis.* **9**, Suppl. 2, S152-S229 (1987).

Bitter crystals from dil ethanol, mp 167°. Freely sol in acetone, dilute solns of sodium hydroxide; sparingly sol in alc. Practically insol in water, ether, chloroform. LD_{50} orally in mice: 3662 mg/kg (Yamamoto).

Mixture with trimethoprim. [8064-90-2] Co-trimoxazole; Abacin; Apo-Sulfatrim; Bactramin; Bactrim; Baktar; Chemotrim; Drylin; Eusaprim; Fectrim; Gantaprim; Gantrim; Imexim; Kepinol; Laratrim; Linaris; Microtrim; Nopil; Oraprim; Septra; Septrin; Sigaprim; Sulfotrim; Sulprim; Sumetrolim; Supracombin; Suprim; Teleprim; Thiocuran; Trigonyl; Trimesulf; Uroplus. LD_{50} orally in mice: 5513 mg/kg (Yamamoto).

N^4-**Acetylsulfamethoxazole.** [21312-10-7] Crystals from alcohol, mp 209-210°.

THERAP CAT: Antibacterial; antipneumocystic.

THERAP CAT (VET): Antibacterial.

9051. Sulfamethoxypyridazine. [80-35-3] 4-Amino-N-(6-methoxy-3-pyridazinyl)benzenesulfonamide; N^1-(6-methoxy-3-pyridazinyl)sulfanilamide; 6-methoxy-3-sulfanilamidopyridazine; 3-(p-aminobenzenesulfamido)-6-methoxypyridazine; RP-7522; Lederkyn; Midicel; Midikel. $C_{11}H_{12}N_4O_3S$; mol wt 280.30. C 47.14%, H 4.32%, N 19.99%, O 17.12%, S 11.44%. Prepn: Clark, **US 2712012** (1955 to Am. Cyanamid). Toxicity: Seki et al., *Arzneim.-Forsch.* **15**, 1441 (1965). Toxicological, chemotherapeutic and pharmacokinetic studies: Böhni et al., *Chemotherapy* **14**, 195-226 (1969).

Bitter crystals from water, mp 182-183°. pKa 6.7. Soly in water at 37° (mg/100 ml): 110 at pH 5; 120 at pH 6; 147 at pH 6.5. Slightly sol in methanol, ethanol (about 1:200); more sol in acetone (1:50), in dimethylformamide (1 g/1 ml). Freely sol in aq solns of alkali hydroxides. LD$_{50}$ orally in mice: 1750 mg/kg, (Seki).

Sodium salt. [2577-32-4] Davosin; Sulfoxine LA. C$_{11}$H$_{11}$N$_4$NaO$_3$S; mol wt 302.28.

THERAP CAT: Antibacterial.

THERAP CAT (VET): Antibacterial.

9052. Sulfametrole. [32909-92-5] 4-Amino-N-(4-methoxy-1,2,5-thiadiazol-3-yl)benzenesulfonamide; N^1-(4-methoxy-1,2,5-thiadiazol-3-yl)sulfanilamide; 3-methoxy-4-(4'-aminobenzenesulfonamido)-1,2,5-thiadiazole. C$_9$H$_{10}$N$_4$O$_3$S$_2$; mol wt 286.32. C 37.75%, H 3.52%, N 19.57%, O 16.76%, S 22.39%. Anti-infective sulfanilamide, related structurally to sulfadiazine, q.v. Prepn: **BE 629551** corresp to K. Menzl, **US 3247193** (1963, 1966 both to OSSW). Improved prepn: idem, **US 4151164** (1979 to Chemie Linz). Bacteriological study (of mixture with trimethoprim): G. Nabert-Bock, H. Grims, Arzneim.-Forsch. **27**, 1109 (1977). Tolerance, therapeutic effect, pharmacokinetics: S. Breyer et al., Wien. Med. Wochenschr. **130**, 448 (1980). Renal excretion mechanism, pharmacokinetics: T. B. Vree et al., Eur. J. Clin. Pharmacol. **20**, 283 (1981). Pharmacokinetics of sulfametrole and its metabolite in man: Y. A. Hester et al., J. Antimicrob. Chemother. **8**, 133 (1981). Crystal and molecular structures: C. H. Koo et al., Bull. Korean Chem. Soc. **3**, 9 (1982), C.A. **96**, 208771 (1982). Single-dose therapy of chancroid with trimethoprim-sulfametrole: F. A. Plummer et al., N. Engl. J. Med. **309**, 67 (1983).

Cryst from dilute acetic acid, mp 149-150°.

Mixture with trimethoprim. [63749-94-0] Lidaprim; Maderan. THERAP CAT: Antibacterial.

9053. Sulfamic Acid. [5329-14-6] Amidosulfonic acid. H$_3$NO$_3$S; mol wt 97.09. H 3.11%, N 14.43%, O 49.44%, S 33.02%. Obtained from chlorosulfonic acid and ammonia, or by heating urea with H$_2$SO$_4$. Purification: Sisler et al., Inorg. Synth. **2**, 178 (1946). Toxicity data: Ambrose, J. Ind. Hyg. Toxicol. **25**, 26 (1943). Reviews: Audrieth et al., Chem. Rev. **26**, 49 (1940); Burton, Nickless, "Amido- and Imido-Sulfonic Acids" in Inorganic Sulphur Chemistry, G. Nickless, Ed. (Elsevier, New York, 1968) pp 607-627, 661-667; E. B. Bell in Kirk-Othmer Encyclopedia of Chemical Technology vol. 21 (Wiley-Interscience, New York, 3rd ed., 1983) pp 949-960. Brief review of synthetic applications: B. Wang, Synlett **2005**, 1342-1343.

Orthorhombic crystals. d 2.15. mp ~205° (dec). Corrosive. Stable when dry but in soln slowly hydrolyzes forming ammonium bisulfate. Sol in 6.5 parts water at 0°, in about 2 parts water at 80°. Sulfuric acid decreases soly in water. Sparingly sol in alcohol, methanol; slightly sol in acetone. Insol in ether. Freely sol in nitrogenous bases, e.g., liquid ammonia, also in nitrogen contg organic solvents, e.g., pyridine, formamide, dimethylformamide. Immiscible with toluene, THF. A strong acid; pH of a 1% soln at 25° 1.18. Can be titrated with bases by means of indicators showing color change between pH 4.5 to 9. MLD orally in rats: 1.6 g/kg (Ambrose).

Caution: Moderately irritating to skin, mucous membranes.

USE: As standard in alkalimetry and acidimetry; in acid cleaning; in nitrite removal; in chlorine stabilization for use in swimming pools, cooling towers, paper mills. Solid-acid catalyst in organic synthesis. The acid or its ammonium salt has been recommended for flameproofing fabrics and wood. Metal salts are used in electroplating. Ammonium sulfamate, q.v., is also widely used as a weed killer.

9054. Sulfamide. [7803-58-9] H$_4$N$_2$O$_2$S; mol wt 96.10. H 4.20%, N 29.15%, O 33.30%, S 33.36%. H$_2$NSO$_2$NH$_2$. Prepd from sulfuryl chloride and gaseous ammonia: Traube, Ber. **25**, 2427 (1892); **26**, 610 (1893); Hantzsch, Holl, Ber. **34**, 3430 (1902); Schenk in Handbook of Preparative Inorganic Chemistry vol. 1, G. Brauer, Ed. (Academic Press, New York, 2nd ed., 1963) pp 482-483.

Orthorhombic, tasteless plates from abs alcohol, mp 93°; dec 250°. Dipole moment: 3.9. Freely sol in water, hot alcohol, acetone. Sparingly sol in cold alc. Solid and solns are stable.

Note: Not to be confused with ammonium sulfamate or with sulfanilamide.

USE: Similar to urea in many of its reactions, but is more acidic, and can act as a dibasic acid.

9055. Sulfamidochrysoidine. [103-12-8] 4-[2-(2,4-Diaminophenyl)diazenyl]benzenesulfonamide; 2,4-diaminoazobenzene-4'-sulfonamide; 4'-sulfamyl-2,4-diaminoazobenzene. C$_{12}$H$_{13}$N$_5$O$_2$S; mol wt 291.33. C 49.47%, H 4.50%, N 24.04%, O 10.98%, S 11.00%. Prototype compd leading to the development of sulfonamide antibacterials. Prepn: F. Mietzsch, J. Klarer, **DE 607537** (1935 to I. G. Farbenind.); eidem, **US 2085037** (1937 to Winthrop). Antibacterial activity in vivo: G. Domagk, Dtsch. Med. Wochenschr. **61**, 573 (1935); for English translation see: Rev. Infect. Dis. **8**, 163 (1986). Identification of sulfanilamide as active metabolite: J. Tréfouel et al., C. R. Seances Soc. Biol. Ses Fil. **120**, 756 (1935); L. Colebrook et al., Lancet **2**, 1323 (1936). Polymorphic forms in commercial prepns: L. Kofler, A. Kofler, Monatsh. Chem. **81**, 321 (1950). Use to visualize carbonic anhydrase during isoelectric focusing in polyacrylamide gels: W. Siffert et al., J. Biochem. Biophys. Methods **8**, 331 (1983). Historical review: M. H. Bickel, Gesnerus **45**, 67-86 (1988).

Hydrochloride. [33445-35-1] Prontosil; Prontosil flavum; Prontosil rubrum; Rubiazol I; Septosan; Streptozon. C$_{12}$H$_{13}$N$_5$O$_2$S.HCl; mol wt 327.79. Orange-red crystals, mp 248-250°. One gram dissolves in 400 ml water; much more sol in hot water. Sol in alcohol, acetone, fats, oils.

USE: Specific stain for carbonic anhydrase in polyacrylamide gels.

THERAP CAT: Antibacterial.

9056. Sulfamoxole. [729-99-7] 4-Amino-N-(4,5-dimethyl-2-oxazolyl)benzenesulfonamide; N^1-(4,5-dimethyl-2-oxazolyl)sulfanilamide; 2-(p-aminobenzenesulfonamido)-4,5-dimethyloxazole; 4,5-dimethyl-2-sulfanilamidooxazole; p-aminobenzenesulfonyl-2-amino-4,5-dimethyloxazole; sulfadimethyloxazole; Justamil; Sulfmidil; Sulfuno; Tardamide. C$_{11}$H$_{13}$N$_3$O$_3$S; mol wt 267.30. C 49.43%, H 4.90%, N 15.72%, O 17.96%, S 11.99%. Prepn: Loop et al., **US 2809966** (1957 to Nordmark-Werke). Series of articles on antimicrobial activity, pharmacology and clinical efficacy of mixture with trimethoprim: Arzneim.-Forsch. **26**, 595-684 (1976). Toxicity data: F. Lagler et al., ibid. 634.

Crystals, mp 193-194°. Soly in mg/100 ml at 20°: water 85; 0.01N HCl 163; 0.01N NaOH 196; methanol 2315; chloroform 240. uv max (4.91 μg/ml methanol): 210, 250, 270 nm (ε 740, 546, 857). LD$_{50}$ in mice, rats (g/kg): >10.0, >12.5 orally; 1.80, ~2.50 i.p. (Lagler).

Mixture with trimethoprim. [57197-43-0] Co-trifamole; CN-3123; Co-Fram; Dibactil; Nevin; Supristol. LD$_{50}$ in mice, rats (g/kg): >12.0, 14.0 orally; 1.87, 2.00 i.p. (Lagler).

THERAP CAT: Antibacterial.

9057. Sulfanilamide. [63-74-1] 4-Aminobenzenesulfonamide; *p*-anilinesulfonamide; *p*-sulfamidoaniline; 1162-F; Prontosil album; Prontylin; Streptocide. $C_6H_8N_2O_2S$; mol wt 172.20. C 41.85%, H 4.68%, N 16.27%, O 18.58%, S 18.62%. Active metabolite of the antibacterial dye, sulfamidochrysoidine, *q.v.* Prepn: P. Gelmo, *J. Prakt. Chem.* **77**, 369 (1908); A. Galat, *Ind. Eng. Chem.* **36**, 192 (1944); Hurdis, Yang, *J. Chem. Educ.* **46**, 697 (1969). Large-scale process: F. Mietzsch, J. Klarer, **US 2132178**; *eidem et al.*, **US 2276664** (1938, 1942 both to Winthrop). Antibacterial activity: J. Tréfouel *et al.*, *C. R. Seances Soc. Biol. Ses Fil.* **120**, 756 (1935); and clinical efficacy: L. Colebrook, A. W. Purdie, *Lancet* **233**, 1291 (1937). Toxicity study: E. K. Marshall, Jr. *et al.*, *J. Am. Med. Assoc.* **110**, 252 (1938). Historical review: M. H. Bickel, *Gesnerus* **45**, 67-86 (1988).

Crystals from boiling water, mp 164.5-166.5°. Neutral to litmus. pH (0.5% aq soln): 5.8-6.1. uv max: 255, 312 nm. One gram dissolves in about 37 ml alcohol; in about 5 ml acetone; in about 2 ml boiling water. Sol in glycerol, HCl, solns of K and Na hydroxides. Practically insol in chloroform, ether, benzene. LD_{50} orally in mice: 3.8 g/kg (Marshall).

N^4-**Acetylsulfanilamide.** [121-61-9] *N*-[4-(Aminosulfonyl)phenyl]acetamide. $C_8H_{10}N_2O_3S$; mol wt 214.24. Major metabolite of sulfanilamide: E. K. Marshall, Jr. *et al.*, *Science* **85**, 202 (1937). Needles from water, mp 219°.

THERAP CAT: Antibacterial.

THERAP CAT (VET): Antibacterial.

9058. Sulfanilic Acid. [121-57-3] 4-Aminobenzenesulfonic acid; *p*-anilinesulfonic acid. $C_6H_7NO_3S$; mol wt 173.19. C 41.61%, H 4.07%, N 8.09%, O 27.71%, S 18.51%. Readily prepd from aniline and sulfuric acid: I. G. Farben, *FIAT Final Rept.* **1313** (I), 255 (1948); Fierz-David, Blangey, *Fundamental Processes of Dye Chemistry* (Interscience, New York, 1949) pp 126-128; A. I. Vogel, *Practical Organic Chemistry* (Longmans, London, 3rd ed., 1959) p 586; R. Q. Brewster *et al.*, *Unitized Experiments in Organic Chemistry* (Van Nostrand, Princeton, 2nd ed., 1964) pp 162-163. Improved procedure: Jacobs *et al.*, *Ind. Eng. Chem.* **35**, 321-323 (1943).

Monohydrate. [6101-32-2] Orthorhombic plates from water (very slow crystn can yield the dihydrate), becomes anhydr at around 100°. Dec without melting at ~288° (also reported as >360°). pK_2 (25°): 3.23. Slowly sol in water: ~1% at 20°, 1.45% at 30°, 1.94% at 40° (w/w anhydr). Almost insol in ethanol, benzene, ether. Slightly sol in hot methanol.

Sodium salt dihydrate. [6106-22-5] $C_6H_6NNaO_3S.2H_2O$. Orthorhombic bipyramidal plates. Freely sol in water, sol in hot methanol.

Zinc salt tetrahydrate. [31884-76-1] Zinc sulfanilate tetrahydrate; Nizin. $C_{12}H_{12}N_2O_6S_2Zn.4H_2O$; mol wt 481.83.

Amide *see* Sulfanilamide.

USE: Manuf various dyes and organic chemicals; as reagent in anal. chemistry (Ehrlich's reagent, detn of nitrites).

THERAP CAT: Antibacterial.

9059. *p*-Sulfanilylbenzylamine. [4393-19-5] 4-[(4-Aminophenyl)sulfonyl]benzenemethanamine; *p*-sulfanilidobenzylamine; 4-aminomethyl-4′-aminodiphenyl sulfone; *p*-aminophenyl-*p*-aminomethylphenyl sulfone; Alphamide. $C_{13}H_{14}N_2O_2S$; mol wt 262.33.

C 59.52%, H 5.38%, N 10.68%, O 12.20%, S 12.22%. Prepn: Dewing, *J. Chem. Soc.* **1946**, 466; Wenner, *J. Org. Chem.* **22**, 1508 (1957).

Needles from water, mp 159°.

Monohydrochloride monohydrate. $C_{13}H_{14}N_2O_2S.HCl.H_2O$. Crystals, mp 195°. Very sol in water.

Dihydrochloride. $C_{13}H_{14}N_2O_2S.2HCl.$ mp 285°.

THERAP CAT: Antibacterial.

9060. Sulfanilyl Fluoride. [98-62-4] *p*-Aminobenzenesulfonyl fluoride. $C_6H_6FNO_2S$; mol wt 175.18. C 41.14%, H 3.45%, F 10.85%, N 8.00%, O 18.27%, S 18.30%. Prepd by heating potassium fluoride with 4-acetamidobenzenesulfonyl chloride: Parker, Hofmann, **US 2576037** (1951 to Am. Cyanamid).

Crystals, mp 68-69°.

USE: In the prepn of dyes which pick up light readily.

9061. N^4-Sulfanilylsulfanilamide. [547-52-4] 4-Amino-*N*-[4-(aminosulfonyl)phenyl]benzenesulfonamide; 4-aminobenzenesulfono-*p*-sulfamoylanilide; 4-(4′-aminobenzenesulfonamido)benzenesulfonamide; 4′-sulfamoylsulfanilide; DB-32; Diseptal C; Uliron C; Disulon; Neosanamid II; Albasil C; Disulfan. $C_{12}H_{13}N_3O_4S_2$; mol wt 327.37. C 44.03%, H 4.00%, N 12.84%, O 19.55%, S 19.59%. Prepn: **FR 817034** (to I. G. Farben.), *C.A.* **32**, 1714i (1938); Sakai, Yamamoto, *J. Pharm. Soc. Jpn.* **58**, 683 (1938); Golovchinskaya, *J. Appl. Chem. USSR* **18**, 647 (1945); Sasa, *J. Soc. Org. Syn. Chem. Jpn.* **12**, 211 (1954), *C.A.* **51**, 2780h (1957); Grigorovskii, Dykhanov, *J. Appl. Chem. USSR* **30**, 1284 (1957). Absorption max: Maschka *et al.*, *Monatsh. Chem.* **84**, 1071 (1953).

Needles from water, mp 133-134°. Slightly sol in cold water; considerably more sol in hot water. Sol in methanol, ethanol, ether, dil NH_3, dil HCl. Practically insol in petr ether, chloroform.

Note: The name Disulon is also used for disulfanilamide, $H_2NC_6H_4SO_2NHSO_2C_6H_4NH_2$, esp. in the French literature.

THERAP CAT: Antibacterial (topical).

9062. Sulfanilylurea. [547-44-4] 4-Amino-*N*-(aminocarbonyl)benzenesulfonamide; *N*-sulfanilylcarbamide; sulfacarbamide; sulfaurea; Euvernil; Uractyl; Uramid; Urenil; Urosulfan. $C_7H_9N_3O_3S$; mol wt 215.23. C 39.06%, H 4.22%, N 19.52%, O 22.30%, S 14.90%. Prepd by treating N^4-acetylsulfanilamide with potassium cyanate or with carbamyl chloride or with urea (or with nitrourea and sodium carbonate) in 80% alcohol. The *p*-AcNHC_6H_4SO_2NHCONH_2$ is saponified by slight warming with dil KOH and then acidified: Martin *et al.*, **US 2411661** (1946 to Geigy). By boiling sulfanilamide with urea and sodium carbonate in 75% alcohol: Haack, *Alien Prop. Custodian, Serial* **369**, 118 (1943). By warming calcium acetylsulfanilylcyanamide with dil HCl: Winnek *et al.*, *J. Am. Chem. Soc.* **64**, 1684 (1942); improved procedure: Leitch *et al.*, *Can. J. Res.* **23B**, 139 (1945).

Crystals from water. mp 146-148° (slight dec). Solubility in water at 37°: 811 mg/100 ml. Soluble in alkalies. Forms a very soluble sodium salt.

Monohydrate. mp 125-127°.

THERAP CAT: Antibacterial.

9063. N-Sulfanilyl-3,4-xylamide. [120-34-3] *N*-[(-4-Amino-phenyl)sulfonyl]-3,4-dimethylbenzamide; N^1-(3,4-dimethylben-zoyl)sulfanilamide; Geigy 867; Irgafen. $C_{15}H_{16}N_2O_3S$; mol wt 304.36. C 59.19%, H 5.30%, N 9.20%, O 15.77%, S 10.53%. Prepd from *p*-nitrobenzenesulfonamide and 3,4-dimethylbenzoyl chloride; the resulting *p*-nitro-*N*-(3,4-dimethylbenzoyl)benzenesulfonamide is reduced with Fe and dil AcOH: Martin *et al.*, **US 2383874** (1945).

Needles from alcohol, mp 222-223°. Sparingly sol in water. Marketed as the water-soluble sodium salt for making solns; a 5% aq soln having a pH of 8.2.

THERAP CAT: Antibacterial.

9064. Sulfanitran. [122-16-7] *N*-[4-[[(4-Nitrophenyl)amino]sulfonyl]phenyl]acetamide; 4′-[(*p*-nitrophenyl)sulfamoyl]acet-anilide; N^4-acetyl-N^1-(*p*-nitrophenyl)sulfanilamide; 4-acetamino-benzenesulfon-4′-nitroanilide; *N*-(*p*-acetylaminobenzene sulfonyl)-*p*-nitroaniline; APNPS. $C_{14}H_{13}N_3O_5S$; mol wt 335.33. C 50.15%, H 3.91%, N 12.53%, O 23.86%, S 9.56%. Prepn: Webster, Powers, *J. Am. Chem. Soc.* **60**, 1553 (1938); Kaufmann, Bückmann, *Arch. Pharm.* **279**, 194 (1941); Shepherd, *J. Org. Chem.* **12**, 275 (1947).

Crystals from dil alcohol, mp 239-240° (Kaufmann); mp 264° (Shepherd). Freely sol in acetone; sol in hot ethanol, methanol; sparingly sol in water, ether.

Note: Ingredient of **Unistat, Novastat W**.

THERAP CAT (VET): Antibacterial; coccidiostat (poultry).

9065. Sulfaperine. [599-88-2] 4-Amino-*N*-(5-methyl-2-pyrimidinyl)benzenesulfonamide; N^1-(5-methyl-2-pyrimidinyl)sulfanilamide; 5-methyl-2-sulfanilamidopyrimidine; 2-sulfanilamido-5-methylpyrimidine; isosulfamerazine; 5-methylsulfadiazine; Pallidin; Retardon; Rexulfa; Sintosulfa; Sulfatreis. $C_{11}H_{12}N_4O_2S$; mol wt 264.30. C 49.99%, H 4.58%, N 21.20%, O 12.11%, S 12.13%. Prepd by condensing 2-amino-5-methylpyrimidine with acetylsulfanilyl chloride followed by hydrolysis of the acetyl group with NaOH: Sprague, **US 2407966** (1946 to Sharp & Dohme).

Minute, cream-colored crystals, mp 262-263°. Very sparingly sol in water, ethanol; ~40 mg/100 ml H_2O at pH 5.5. Sol in aq solns of acids and alkalies. Forms a water-sol sodium salt.

THERAP CAT: Antibacterial.

9066. Sulfaphenazole. [526-08-9] 4-Amino-*N*-(1-phenyl-1*H*-pyrazol-5-yl)benzenesulfonamide; N'-(1-phenylpyrazol-5-yl)-sulfanilamide; 1-phenyl-5-sulfanilamidopyrazole; 3-(*p*-aminoben-zenesulfonamido)-2-phenylpyrazole. $C_{15}H_{14}N_4O_2S$; mol wt 314.36. C 57.31%, H 4.49%, N 17.82%, O 10.18%, S 10.20%. Synthesis: Schmidt, Druey, *Helv. Chim. Acta* **41**, 309 (1958); Druey, Schmidt, **US 2858309** (1958 to Ciba). Toxicity study: Seki *et al.*, *Arzneim.-Forsch.* **15**, 1441 (1965).

Crystals from alcohol, mp 179-183°. Very sparingly sol in water: 0.15 g/100 ml H_2O at pH 7.0 and 25°. More sol in methanol, ethanol, glacial acetic acid. LD_{50} orally in mice: 5800 mg/kg (Seki).

Sodium salt monohydrate. $C_{15}H_{13}N_4NaO_2S.H_2O$. White powder. Sol in water. A 10% soln (w/w) has a pH of ~9.

THERAP CAT: Antibacterial.

THERAP CAT (VET): Antibacterial.

9067. Sulfaproxyline. [116-42-7] *N*-[4-(Aminophenyl)sul-fonyl]-4-(1-methylethoxy)benzamide; N^1-(*p*-isopropoxybenzoyl)-sulfanilamide; *N*-(4-isopropoxybenzoyl)-*p*-aminobenzenesulfon-amide; sulphaproxyline. $C_{16}H_{18}N_2O_4S$; mol wt 334.39. C 57.47%, H 5.43%, N 8.38%, O 19.14%, S 9.59%. Prepn: Gysin, **US 2503820** (1950 to Geigy).

Crystals, mp 172-173°.

THERAP CAT: Antibacterial.

9068. Sulfapyrazine. [116-44-9] 4-Amino-*N*-(pyrazinyl)ben-zenesulfonamide; N^1-2-pyrazinylsulfanilamide; 2-sulfanilamidopyr-azine. $C_{10}H_{10}N_4O_2S$; mol wt 250.28. C 47.99%, H 4.03%, N 22.39%, O 12.78%, S 12.81%. Prepd by reacting acetylsulfanil chloride with 2-aminopyrazine in pyridine, followed by hydrolysis and neutralization: Ellingson, **US 2420703** (1947). Biological properties: Ghione *et al.*, *Chemotherapia* **6**, 344 (1962).

Crystals, dec 250-254°. Very slightly sol in alc; slightly sol in acetone. Sol in aq solns of sodium, potassium, and barium hydroxides, in ammonia water, and in dil and concd mineral acid solns. Practically insol in water (5 mg/100 ml at 25°, and 5.2 mg/100 ml at 37°).

Sodium salt monohydrate. $C_{10}H_9N_4NaO_2S.H_2O$. Bitter powder. Freely soluble in water (1 g/3.33 ml at 25°). Alkaline to litmus. pH of 10% aq soln 9.1. Very sol in acetone, slightly sol in alc. Insol in ether, chloroform. Aq solns of sulfapyrazine sodium may absorb carbon dioxide which causes precipitation of sulfapyrazine.

THERAP CAT: Antibacterial.

THERAP CAT (VET): Antibacterial.

9069. Sulfapyridine. [144-83-2] 4-Amino-*N*-2-pyridinylbenzenesulfonamide; dagenan; N^1-2-pyridylsulfanilamide; 2-sulfanilamidopyridine; 2-(*p*-aminobenzenesulfonamido)pyridine; M & B 693. $C_{11}H_{11}N_3O_2S$; mol wt 249.29. C 53.00%, H 4.45%, N 16.86%, O 12.84% S 12.86%. Prepn: A. J. Ewins, M. A. Phillips, **GB 512145** (1939); *eidem*, **US 2275354** (1942); R. Winterbottom, *J. Am. Chem. Soc.* **62**, 160 (1940). Metabolism: R. Wien, J. Hampton, *J. Pharmacol. Exp. Ther.* **84**, 211 (1945). Toxicology: R. Wien *et al.*, *ibid.* 203. GC-MS determn in edible animal tissues: A. E. Mooser, H. Koch, *J. AOAC Int.* **76**, 976 (1993). Efficacy and proposed mechanism of action in dermatological disease: O. J. Stone, *Med. Hypotheses* **31**, 99 (1990). Clinical evaluation in ocular cicatricial pemphigoid: M. J. Elder *et al.*, *Br. J. Ophthalmol.* **80**, 549 (1996).

Crystals from alc, mp 190-191°. One gram dissolves in ~3500 ml water, 440 ml alcohol, 65 ml acetone. Freely sol in dil mineral acids and in aq solns of KOH and NaOH. More sol in warm sugar solns than in water alone. The aq soln is neutral. LD_{50} orally in mice: 7.5 mg/g (Wien).

Sodium salt monohydrate. [127-57-1] Soluble sulfapyridine. $C_{11}H_{10}N_3NaO_2S.H_2O$; mol wt 289.28. On prolonged exposure to humid air it absorbs CO_2, liberates sulfapyridine, becomes incompletely sol in water. One gram dissolves in ~1.5 ml water, 10 ml alcohol. pH of 5% aq soln: 11.4. LD_{50} orally in mice: 2.7 mg/g (Wien).

THERAP CAT: In treatment of dermatitis herpetiformis; antibacterial.

THERAP CAT (VET): Antibacterial.

9070. Sulfaquinoxaline. [59-40-5] 4-Amino-*N*-2-quinoxalinylbenzenesulfonamide; N^1-(2-quinoxalinyl)sulfanilamide; 2-sulfanilamidoquinoxaline; N^1-(2-quinoxalyl)sulfanilamide; sulfabenzpyrazine; Compd 3-120; S.Q.; Sulquin. $C_{14}H_{12}N_4O_2S$; mol wt 300.34. C 55.99%, H 4.03%, N 18.65%, O 10.65% S 10.67%. Prepd by treating 2-aminoquinoxaline with acetylsulfanilyl chloride in the presence of pyridine and hydrolyzing the resulting acetyl derivative: J. Weijlard *et al.*, *J. Am. Chem. Soc.* **66**, 1957 (1944); J. Weijlard, M. Tishler, **US 2404199** (1946 to Merck & Co.). Anticoccidial spectrum: G. F. Mathis *et al.*, *Poult. Sci.* **63**, 1149 (1984). Evaluation of antibacterial and anticoccidial efficacy of mixture with trimethoprim: G. White, R. B. Williams, *Vet. Rec.* **113**, 608 (1983); D. W. T. Piercy *et al.*, *ibid.* **114**, 60 (1984). Drug residue study in poultry muscle and liver: M. Patthy, *J. Chromatogr.* **275**, 115 (1983). HPLC determn in rabbit plasma and urine: J. G. Eppel, J. J. Thiessen, *J. Pharm. Sci.* **73**, 1635 (1984).

Minute crystals, mp 247-248°. uv max (pH 6.6 in H_2O): 252, 360 nm ($E_{1cm}^{1\%}$ 1110, 275). Solubility in water at pH 7: 0.75 mg/100 ml; in 95% alcohol: 73 mg/100 ml; in acetone: 430 mg/100 ml. Sol in aq Na_2CO_3 and NaOH solns.

Sodium salt. Aviochina. $C_{14}H_{11}N_4NaO_2S$; mol wt 322.32. Very sol in water; pH of 1% soln ~10. The amorphous salt is deliquescent and absorbs CO_2 which liberates the practically insol sulfaquinoxaline.

THERAP CAT (VET): Coccidiostat.

9071. Sulfarside. [1134-98-1] [2-Amino-4-(aminosulfonyl)-phenyl]arsinous acid; 4-sulfamoyl-*o*-arsanilic acid; 2-amino-4-aminosulfonylphenylarsinic acid; RP-4482; Bemarside. $C_6H_9AsN_2O_5$-S; mol wt 296.13. C 24.34%, H 3.06%, As 25.30%, N 9.46%, O 27.01%, S 10.83%. Prepn: Trefouel, **US 2616913** (1953 to Rhône-Poulenc).

THERAP CAT: Sodium salt as antiamebic.

9072. Sulfarsphenamine. [618-82-6] 1,1'-[1,2-Diarsenediylbis[(6-hydroxy-3,1-phenylene)imino]]bismethanesulfonic acid sodium salt (1:2); disodium 3,3'-diamino-4,4'-dihydroxyarsenobenzene *N*-dimethylenesulfonate; sulfarsenobenzene; Karsulphan; Myosalvarsan; Myarsenol; Metarsenobillon; Thiosarmine. $C_{14}H_{14}$-$As_2N_2Na_2O_8S_2$; mol wt 598.21. C 28.11%, H 2.36%, As 25.05%, N 4.68%, Na 7.69%, O 21.40%, S 10.72%. Prepn: C. Voegtlin, J. M. Johnson, *J. Am. Chem. Soc.* **44**, 2573 (1922); W. J. C. Dyke, H. King, *J. Chem. Soc.* **1935**, 1745.

Yellow, odorless, or almost odorless powder. Very sol in water, slightly in alcohol. The aq soln is usually slightly acid. *Keep in evacuated or in inert gas-filled ampuls.*

THERAP CAT: Formerly as antisyphilitic.

THERAP CAT (VET): Has been used as antibacterial.

9073. Sulfasalazine. [599-79-1] 2-Hydroxy-5-[2-[4-[(2-pyridinylamino)sulfonyl]phenyl]diazenyl]salicylic acid; salazosulfapyridine; 5-[4-(2-pyridylsulfamoyl)phenylazo]salicylic acid; 5-[*p*-(2-pyridylsulfamoyl)phenylazo]-2-hydroxybenzoic acid; 4-(pyridyl-2-amidosulfonyl)-3'-carboxy-4'-hydroxyazobenzene; salicylazosulfapyridine; sulphasalazine; Azulfidine; Colo-Pleon; Salazopyrin. C_{18}-$H_{14}N_4O_5S$; mol wt 398.39. C 54.27%, H 3.54%, N 14.06%, O 20.08%, S 8.05%. Conjugate of 5-aminosalicylic acid and sulfapyridine, *q.q.v.* Prepn and bactericidal effect: E. E. A. Askelof *et al.*, **US 2396145** (1946 to Pharmacia); Doraswamy, Guha, *J. Indian Chem. Soc.* **23**, 278 (1946). Physicochemical properties: B. Nygard *et al.*, *Acta Pharm. Suec.* **3**, 313-342 (1966). Metabolism studies: H. Schroder, D. E. S. Campbell, *Clin. Pharmacol. Ther.* **13**, 539 (1972); M. A. Peppercorn, P. Goldman, *J. Pharmacol. Exp. Ther.* **181**, 555 (1972). HPLC determn in biological fluids: R. A. Hogezand *et al.*, *J. Chromatogr.* **305**, 470 (1984). Clinical trial in rheumatoid arthritis: D. E. Bax, R. S. Amos, *Ann. Rheum. Dis.* **44**, 194 (1985). Review of clinical pharmacokinetics: U. Klotz, *Clin. Pharmacokinet.* **10**, 285-302 (1985); of pharmacology and clinical use in inflammatory bowel disease: M. A. Peppercorn, *Ann. Intern. Med.* **3**, 377 (1984); G. Watkinson, *Drugs* **32**, Suppl. 1, 1-11 (1986). Comprehensive description: J. P. McDonnell, *Anal. Profiles Drug Subs.* **5**, 515-532 (1976).

Minute, brownish-yellow crystals, dec 240-245°. uv max: 237 ($E_{1cm}^{1\%}$ ~658) and 359 nm. Sol in aqueous solns of alkali hydroxides; very slightly sol in alc. Practically insol in water, benzene, chloroform, ether.

THERAP CAT: Anti-inflammatory (gastrointestinal).

THERAP CAT (VET): Has been used in granulomatous colitis.

9074. Sulfathiazole. [72-14-0] 4-Amino-*N*-2-thiazolylbenzenesulfonamide; *N*¹-2-thiazolylsulfanilamide; 2-sulfanilamidothiazole; 2-(sulfanilylamino)thiazole; 2-(*p*-aminobenzenesulfonamido)thiazole; norsulfazole; RP-2090; M & B 760; Thiazamide; Cibazol; Enterobiocine; Duatok; Sulfamul; Sulfavitina; Sulzol. $C_9H_9N_3O_2S_2$; mol wt 255.31. C 42.34%, H 3.55%, N 16.46%, O 12.53%, S 25.11%. Prepn: R. J. Fosbinder, L. A. Walter, *J. Am. Chem. Soc.* **61**, 2032 (1939); W. A. Lott, F. H. Bergeim, *ibid.* 3593; G. Newbery, P. Viaud, **GB 517272**; G. Newbery, **US 2362087** (1940, 1944 both to May & Baker). Antibacterial activity *in vivo*: O. W. Barlow, E. Homburger, *Proc. Soc. Exp. Biol. Med.* **42**, 792 (1939). Toxicology: H. B. van Dyke *et al.*, *ibid.* 410; P. H. Long *et al.*, *ibid.* **43**, 328 (1940). Review of early literature: *ACS Monograph Series* **no. 106**, entitled "Sulfonamides and Allied Compounds," E. H. Northey, Ed. (Reinhold, New York, 1948) 660 pp. HPLC determn in milk: V. K. Agarwal, *J. Liq. Chromatogr.* **16**, 3793 (1993). Oral bioavailability in ruminants: K. Weijkamp *et al.*, *Vet. Q.* **16**, 33 (1994). Comprehensive description: V. K. Kapoor, *Anal. Profiles Drug Subs. Excip.* **22**, 389-430 (1993).

Fine, white or faintly yellowish powder. mp 202-202.5°. pKa 7.2. Soly at 26° (mg/100 ml): water 60 (pH 6.03); alcohol 525. Sol in acetone, dil mineral acids, KOH and NaOH solns, and 6*N* ammonium hydroxide. Slightly sol in ethanol; very slightly sol in water. Practically insol in chloroform, ether.

Sodium salt sesquihydrate. Soluble sulfathiazole. $C_9H_8N_3NaO_2S_2.1\frac{1}{2}H_2O$. Crystals or white powder or granules. Also occurs as the monohydrate and pentahydrate. One gram dissolves in ~2.5 ml water, in ~15 ml alc. pH of 1% aq soln 9.35; of 10% soln 10.2. LD_{50} s.c. in mice: 1.45 g/kg (van Dyke); also reported as 1.95 g/kg (Long).

Polymer with formaldehyde. [12041-72-4] Formosulfathiazole; formaldehyde-sulfathiazole; Forbina. $(C_9H_9N_3O_2S_2.CH_2O)_x$. Contains ~11% formaldehyde. Prepn: J. Druey, A. Becker, *Helv. Chim. Acta* **31**, 2184 (1948). Intrauterine use in cows: D. P. Dobson, D. E. Noakes, *Vet. Rec.* **127**, 128 (1990). Amorphous powder. Practically insol in water.

THERAP CAT: Antibacterial.

THERAP CAT (VET): Antibacterial.

9075. Sulfathiourea. [515-49-1] 4-Amino-*N*-(aminothioxomethyl)benzenesulfonamide; 1-sulfanilyl-2-thiourea; sulfathiocarbamide; RP-2255; Badional; Fontamide. $C_7H_9N_3O_2S_2$; mol wt 231.29. C 36.35%, H 3.92%, N 18.17%, O 13.83%, S 27.72%. Prepd by reacting acetylsulfanilylcyanamide with hydrogen sulfide. Has also been prepd directly from sulfanilyl cyanamide and ammonium sulfide: Leitch *et al.*, *Can. J. Res.* **23B**, 139 (1945); alternate method: Földi *et al.*, **US 2332906** (1943).

Crystals, dec 171.5-172°. Soly in water at 37°: 1.1 g/100 ml.

Sodium salt. [6101-34-4] $C_7H_8N_3NaO_2S$. Dec 245-245.5°. Very sol in water. pH of 1% aq soln: 7.

Water sol form. Solufontamide. Marketed as a slightly viscous soln contg 33% of 1-sulfanilyl-2-thiourea. d_4^{15} 1.16. pH 6.8-7.2.

THERAP CAT: Antibacterial.

9076. Sulfazamet. [852-19-7] 4-Amino-*N*-(3-methyl-1-phenyl-1*H*-pyrazol-5-yl)benzenesulfonamide; *N*¹-(3-methyl-1-phenyl-pyrazol-5-yl)sulfanilamide; 5-(*p*-aminobenzenesulfonamido)-3-methyl-1-phenylpyrazole; 3-(*p*-aminobenzenesulfonamido)-2-phenyl-5-methylpyrazole; 3-methyl-1-phenyl-5-(sulfanilamido)pyrazole;

sulfamethylphenazole; sulfapyrazole; Vesulong. $C_{16}H_{16}N_4O_2S$; mol wt 328.39. C 58.52%, H 4.91%, N 17.06%, O 9.74%, S 9.76%. Prepn: Crippa, Guarneri, *Gazz. Chim. Ital.* **85**, 199 (1955); **GB 848627** (1960 to Ciba); Seydel, Krueger-Theimer, *Arzneim.-Forsch.* **14**, 1294 (1964). Amino-imino tautomerism: *eidem, ibid.*

Crystals from alc, mp 195° (Seydel, Krueger-Theimer); mp 181-182° (**GB 848627**). pKa' 5.69.

THERAP CAT (VET): Antibacterial.

9077. Sulfazecin. [77912-79-9] D-γ-Glutamyl-*N*-[(3*R*)-3-methoxy-2-oxo-1-sulfo-3-azetidinyl]-D-alaninamide; 3-γ-D-gluta-myl-D-alanylamino-3-methoxyazetidin-2-one-1-sulfonic acid; (3*R*)-3-(γ-D-glutamyl-D-alanylamino)-3-methoxy-2-oxoazetidine-1-sulfonic acid; antibiotic G-6302. $C_{12}H_{20}N_4O_9S$; mol wt 396.37. C 36.36%, H 5.09%, N 14.14%, O 36.33%, S 8.09%. Novel monocyclic β-lactam (monobactam) antibiotic, produced by *Pseudomonas acidophila* and *P. mesoacidophila*. Isoln, description of physicochemical and bacteriostatic properties: A. Imada *et al.*, **DE 2855949** corresp to **US 4229436** (1979, 1980 both to Takeda); *eidem, Nature* **289**, 590 (1981); M. Asai *et al.*, *J. Antibiot.* **34**, 621 (1981). X-ray crystallographic structure determn: K. Kamiya *et al.*, *Acta Crystallogr.* **B37**, 1626 (1981). Prepn of fluorinated sulfazecin analogs: K. Yoshioka *et al.*, *J. Org. Chem.* **49**, 1427 (1984); P. F. Bevilacqua *et al.*, *ibid.* 1430.

Crystallizes from methanol/water as the alcoholate-hemihydrate; colorless needles, mp 168-170°. $[\alpha]_D^{25}$ +82° (c = 1.0 in water). pKa: 3.4 (COO⁻); 9.2 (NH₃⁺). Sol in water, DMF, DMSO. Slightly sol in methanol, THF. Practically insol in ethanol, acetone, ethyl acetate, chloroform and other organic solvents. Relatively stable in neutral and weakly acidic solns; unstable in alkaline and strongly acidic solns.

Sodium salt monohydrate. $C_{12}H_{19}N_4NaO_9S.H_2O$. Colorless powder, browns at 170°, no sharp mp. $[\alpha]_D^{20}$ +85° (c = 0.37 in water).

9078. Sulfentrazone. [122836-35-5] *N*-[2,4-Dichloro-5-[4-(difluoromethyl)-4,5-dihydro-3-methyl-5-oxo-1*H*-1,2,4-triazol-1-yl]phenyl]methanesulfonamide; 2′,4′-dichloro-5′-(4-difluoromethyl-4,5-dihydro-3-methyl-5-oxo-1*H*-1,2,4-triazol-1-yl)methanesulfonanilide; F-6285; FMC-97285; Authority. $C_{11}H_{10}Cl_2F_2N_4O_3S$; mol wt 387.18. C 34.12%, H 2.60%, Cl 18.31%, F 9.81%, N 14.47%, O 12.40%, S 8.28%. Aryltriazolinone herbicide; inhibits protoporphyrinogen oxidase in the chlorophyll biosynthetic pathway. Prepn: G. Theodoridis, **US 4818275** (1989 to FMC); and structure-activity study: *idem et al.*, *ACS Symp. Ser.* **504**, 134 (1992). Metabolism and distribution in animals: L. Y. Leung *et al.*, *J. Agric. Food Chem.* **39**, 1509 (1991). Properties and field trials in soybeans: W. A. Van Saun *et al.*, *Brighton Crop Prot. Conf. - Weeds* **1991**, 77. Field trial in peanuts: W. C. Johnson, III, B. G. Mullinix, Jr., *Peanut Sci.* **21**, 65 (1994).

Tan solid, mp 75-78°. pKa 6.56. vapor pressure (25°): 1×10^{-9} mm Hg. Soly in water at 25° (mg/g): 0.11 (pH 6.0); 0.78 (pH 7.0); 16 (pH 7.5). LD_{50} orally in rats: 2855 mg/kg; dermally in rabbits: >2000 mg/kg; LC_{50} (96 hr) in bluegill sunfish, rainbow trout (ppm): 92.8, >130 (Van Saun).

USE: Herbicide.

9079. Sulfinpyrazone. [57-96-5] 1,2-Diphenyl-4-[2-(phenyl-sulfinyl)ethyl]-3,5-pyrazolidinedione; 1,2-diphenyl-3,5-dioxo-4-(2-phenylsulfinylethyl)pyrazolidine; 4-(phenylsulfoxyethyl)-1,2-diphenyl-3,5-pyrazolidinedione; 4-(2-benzenesulfinylethyl)-1,2-diphenylpyrazolidine-3,5-dione; sulfoxyphenylpyrazolidine; G-28315; Anturan; Anturane; Anturano; Enturen. $C_{23}H_{20}N_2O_3S$; mol wt 404.48. C 68.30%, H 4.98%, N 6.93%, O 11.87%, S 7.93%. Prepn: F. Häfliger, US 2700671 (1955 to Geigy AG); R. Pfister, F. Häfliger, *Helv. Chim. Acta* **44**, 232 (1961). Uricosuric effect: J. J. Burns *et al.*, *J. Pharmacol. Exp. Ther.* **119**, 418 (1957). Clinical trial in prevention of sudden cardiac death: Anturane Reinfarction Trial Research Group, *N. Engl. J. Med.* **298**, 289 (1978); **302**, 250 (1980); as antithrombotic: R. D. Sautter *et al.*, *J. Am. Med. Assoc.* **250**, 2649 (1983); D. T. Domoto *et al.*, *Thromb. Res.* **62**, 737 (1991). *Review:* E. H. Margulies, A. M. White, in *Pharmacological and Biochemical Properties of Drug Substances* vol. 2, M. E. Goldberg, Ed. (Am. Pharm. Assoc., Washington, DC, 1979) pp 255-278; E. H. Margulies *et al.*, *Drugs* **20**, 179-197 (1980).

Crystals from chloroform + heptane, mp 136-137°. Stable to light and air. uv max (1.0N NaOH): 255 nm. Sol in alc, acetone, ethyl acetate, chloroform; sparingly sol in dilute alkali; slightly sol in ether, mineral oils, fats. Practically insol in water, solvent hexane.

d-Form. Crystals from ethanol, mp 130-133°. $[\alpha]_D^{22}$ +67.1° (c = 2.04 in ethanol); $[\alpha]_D^{25}$ +109.3° (c = 0.5 in $CHCl_3$).

l-Form. Crystals, mp 130-133°. $[\alpha]_D^{23}$ −64.2° (c = 2.14 in ethanol); $[\alpha]_D^{26}$ −104.5° (c = 0.5 in $CHCl_3$).

THERAP CAT: Uricosuric; antithrombotic.

9080. Sulfiram. [95-05-6] N,N,N',N'-Tetraethylthiodicarbonic diamide ([$(H_2N)C(S)$]$_2$S); bis(diethylthiocarbamoyl)sulfide; tetraethylthiuram monosulfide; monosulfiram; TTMS; Kutkasin; Tetmosol. $C_{10}H_{20}N_2S_3$; mol wt 264.46. C 45.42%, H 7.62%, N 10.59%, S 36.37%. Prepn: Davies, Sexton, *Biochem. J.* **40**, 331 (1946); Ritter, US 2524081 (1950 to Sharples Chem.); Sahasrabudhey, Radhakrishnan, *J. Indian Chem. Soc.* **31**, 853 (1954); Zbirovsky, Ettel, *Chem. Listy* **51**, 2094 (1957); **52**, 95 (1958).

Crystals from benzene + petr ether, mp 30-32°.

USE: As vulcanization agent.

THERAP CAT: Ectoparasiticide.

THERAP CAT (VET): Ectoparasiticide.

9081. Sulfisomidine. [515-64-0] 4-Amino-N-(2,6-dimethyl-4-pyrimidinyl)benzenesulfonamide; N^1-(2,6-dimethyl-4-pyrimidinyl)sulfanilamide; 2,6-dimethyl-4-sulfanilamidopyrimidine; 6-sulfanilamido-2,4-dimethylpyrimidine; 6-(sulfanilamido)-2,4-dimethyl-1,3-diazine; 6-(p-aminophenylsulfonamido)-2,4-dimethylpyrimidine; 6-(p-aminobenzenesulfonyl)amino-2,4-dimethylpyrimidine; sulfadimetine; sulfaisodimidine; sulphasomidine; Elkosin; Elcosine; Elkosil; Domain; Aristamid. $C_{12}H_{14}N_4O_2S$; mol wt 278.33. C

51.78%, H 5.07%, N 20.13%, O 11.50%, S 11.52%. Prepn: **FR** 886009 (1943 to Nordmark); Gysin, US 2351333 (1944 to Geigy); Hartmann *et al.*, US 2386852; US 2429184 (1945, 1947, both to Ciba); Matsukawa *et al.*, *J. Pharm. Soc. Jpn.* **70**, 283 (1950); Loop, Lührs, *Ann.* **580**, 225 (1953).

Needles from ethanol, mp 243°. Soly in water at 15°: 0.12 g/100 ml, at 30°: 0.30 g/100 ml. More sol in hot water (about 1:60). Aq solns are neutral to litmus. Soly in urine at 37°: 360 mg/100 ml at pH 5.5 to 1100 mg/100 ml at pH 7.5. Slightly sol in alcohol, acetone. Practically insol in benzene, ether, chloroform. Freely sol in dil HCl and NaOH. Compared with sulfanilamide it is but slightly acetylated in the body.

Note: Do not confuse with sulfamethazine, q.v.

THERAP CAT: Antibacterial.

THERAP CAT (VET): Antimicrobial.

9082. Sulfisoxazole. [127-69-5] 4-Amino-N-(3,4-dimethyl-5-isoxazolyl)benzenesulfonamide; N^1-(3,4-dimethyl-5-isoxazolyl)sulfanilamide; 3,4-dimethyl-5-sulfanilamidoisoxazole; 5-(4-aminophenylsulfonamido)-3,4-dimethylisoxazole; 5-(p-aminobenzenesulfonamido)-3,4-dimethylisooxazole; sulfafurazole; sulphafurazole. $C_{11}H_{13}N_3O_3S$; mol wt 267.30. C 49.43%, H 4.90%, N 15.72%, O 17.96%, S 11.99%. Prepn: H. M. Wuest, M. Hoffer, US 2430094 (1947 to Hoffmann-La Roche). Toxicity study: Seki *et al.*, *Arzneim.-Forsch.* **15**, 1441 (1965). Comprehensive description: B. C. Rudy, B. Z. Senkowski, *Anal. Profiles Drug Subs.* **2**, 487-506 (1973). HPLC determn in biological fluids: D. Jung, S. Oie, *Clin. Chem.* **26**, 51 (1980). Clinical trial in otitis media: P. A. M. Bernard *et al.*, *Pediatrics* **88**, 215 (1991).

White to slightly yellowish crystalline powder. Bitter taste. mp 194°. pKa 5. Soly in water (25°): 0.13 mg/ml. Sol in boiling alc, 3N hydrochloric acid. LD_{50} orally in mice: 6800 mg/kg (Seki).

Diethanolamine salt. [4299-60-9] Sulfisoxazole diolamine. $C_{11}H_{13}N_3O_3S.C_4H_{11}NO_2$; mol wt 372.44. White to off-white, odorless, crystalline powder. Freely sol in water; sol in alc. A 4% soln is about isotonic with tears.

Acetyl sulfisoxazole. [80-74-0] N-[(4-Aminophenyl)sulfonyl]-N-(3,4-dimethyl-5-isoxazolyl)acetamide; N^1-monoacetyl sulfisoxazole; Gantrisin. $C_{13}H_{15}N_3O_4S$; mol wt 309.34. Prepn: M. Hoffer, US 2721200 (1955 to Hoffmann-La Roche). Should not be confused with N^4-acetyl sulfisoxazole which is a metabolite. Tasteless crystals, mp 193-194°. Soly (mg/ml): 0.07 in water; 4.93 in methanol; 5.7 in 95% ethanol; 0.94 in ether; 29.0 in chloroform.

THERAP CAT: Antibacterial.

THERAP CAT (VET): Antibacterial.

9083. 2-Sulfoacetic Acid. [123-43-3] Sulfoethanoic acid. $C_2H_4O_5S$; mol wt 140.11. C 17.15%, H 2.88%, O 57.09%, S 22.88%.

Monohydrate. [6155-83-5] $C_2H_4O_5S.H_2O$; mol wt 158.12. Hygroscopic crystals. mp 84-86° when anhydr. bp 245° with decompn. Sol in water, alcohol; insol in ether, chloroform. *Corrosive.*

Keep well closed. LD$_{50}$ orally in rats: 3.16 g/kg; *see:* Smyth *et al.*, *J. Ind. Hyg. Toxicol.* **31**, 60 (1949).

9084. Sulfobromophthalein Sodium. [71-67-0] 3,3′-(4,5,-6,7-Tetrabromo-3-oxo-1(3H)-isobenzofuranylidene)bis[6-hydroxy-benzenesulfonic acid] disodium salt; disodium phenoltetrabromo-phthalein sulfonate; bromosulfophthalein; BSP; Bromsulphalein; Bromthalein. C$_{20}$H$_8$Br$_4$Na$_2$O$_{10}$S$_2$; mol wt 837.99. C 28.67%, H 0.96%, Br 38.14%, Na 5.49%, O 19.09%, S 7.65%. Prepd by the condensation of phenol and tetrabromophthalic acid or its anhydride, followed by sulfonation and conversion to the disodium salt. Pharmacology and use as liver function test: S. M. Rosenthal, E. C. White, *J. Pharmacol.* **24**, 265 (1924); W. Häcki *et al.*, *J. Lab. Clin. Med.* **88**, 1019 (1976).

Hygroscopic crystals. Bitter taste. Sol in water. Insol in alcohol, acetone. Alkaline solns have an intense bluish-purple color.

THERAP CAT: Diagnostic aid (hepatic function).

THERAP CAT (VET): Diagnostic aid (hepatic function).

9085. Sulfolane. [126-33-0] Tetrahydrothiophene 1,1-dioxide; tetrahydrothiophene 1-dioxide; thiophan sulfone. C$_4$H$_8$O$_2$S; mol wt 120.17. C 39.98%, H 6.71%, O 26.63%, S 26.68%. Prepn by catalytic hydrogenation of sulfolene oxides: Mahan, Fauske, US **2578565** (1951 to Phillips Petroleum); from butadiene and sulfur dioxide: Staaterman *et al.*, *Chem. Eng. Prog.* **43**, 148 (1947); Evans, Morris, US **2360859** (1944 to Shell). Toxicity study: H. F. Smyth *et al.*, *Am. Ind. Hyg. Assoc. J.* **30**, 470 (1962).

Liquid, d$_4^{30}$ 1.2606. mp 27.4-27.8°. bp$_{760}$ 285°. n$_D^{30}$ 1.481. Viscosity at 30° = 10.34 cP. Flash pt, open cup: 350°F (176°C). At 30°, miscible with water, acetone, toluene; partially misc with octanes, olefins and naphthenes. LD$_{50}$ orally in rats: 1.54 ml/kg (Smyth).

USE: Selective solvent for liquid-vapor extractions.

9086. 3-Sulfolene. [77-79-2] 2,5-Dihydrothiophene 1,1-dioxide; 1-thia-3-cyclopentene 1,1-dioxide; butadiene sulfone. C$_4$H$_6$O$_2$S; mol wt 118.15. C 40.66%, H 5.12%, O 27.08%, S 27.13%. Preparation: Staudinger, Rizenthaler, *Ber.* **68**, 455 (1935); de Roy van Zuydewijn, *Rec. Trav. Chim.* **56**, 1047 (1937); Hooker *et al.*, US **2395050** (1946 to Dow); Morris, Finch, US **2420834** (1947 to Shell).

Crystals, mp 64-65.5°. Sol in water, organic solvents.

9087. Sulfometuron-methyl. [74222-97-2] 2-[[[[(4,6-Dimethyl-2-pyrimidinyl)amino]carbonyl]amino]sulfonyl]benzoic acid methyl ester; N-[(4,6-dimethylpyrimidin-2-yl)aminocarbonyl]-2-methoxycarbonylbenzenesulfonamide; 2-[3-(4,6-dimethylpyrimidin-2-yl)ureidosulfonyl]benzoic acid methyl ester; Aa-5648; DPX-5648; Oust. C$_{15}$H$_{16}$N$_4$O$_5$S; mol wt 364.38. C 49.44%, H 4.43%, N 15.38%, O 21.95%, S 8.80%. Nonselective pre- and post-emergence

sulfonylurea herbicide for noncropland use; inhibits branched chain amino acid biosynthesis. Prepn: G. Levitt, EP **7687**; *idem*, US **4394506** (1980, 1983 both to Du Pont); *idem*, *Pestic. Chem.: Hum. Welfare Environ., Proc. 5th Int. Congr. Pestic.* **1**, 243 (1983). Herbicidal activity and toxicity: R. H. Harding *et al.*, *Proc. West. Soc. Weed Sci.* **34**, 120 (1981). Mechanism of action: T. B. Ray, *Proc. Br. Crop Prot. Conf. - Weeds* **1985**, 131. HPLC determn in soil and water: E. W. Zahnow, *J. Agric. Food Chem.* **33**, 479 (1985). Degradation and bioaccumulation studies: J. Harvey *et al.*, *ibid.* 590; J. J. Anderson, J. J. Dulka, *ibid.* 596. Field trials: R. L. Atkins *et al.*, *Proc. South. Weed Sci. Soc.* **36**, 300 (1983); in afforested areas: A. Nir, Z. Arenstein, *Proc. Br. Crop Prot. Conf. - Weeds* **1987**, 787.

White solid, mp 198-202°. Soly in water at 25° (ppm): 10 at pH 5, 300 at pH 7. pKa 5.7. LD$_{50}$ in male, female rats (mg/kg): >5000, >5000 orally (Harding).

USE: Herbicide.

9088. Sulfonethylmethane. [76-20-0] 2,2-Bis(ethylsulfonyl)butane; diethylsulfonmethylethylmethane; methylsulfonal; Trional. C$_8$H$_{18}$O$_4$S$_2$; mol wt 242.35. C 39.65%, H 7.49%, O 26.41%, S 26.46%. Prepd by condensing ethylmercaptan with methyl ethyl ketone and oxidizing the resulting mercaptol with KMnO$_4$: Fromm, *Ann.* **253**, 135 (1889); *cf.* Baumann, *Ber.* **18**, 883 (1885); *Arch. Pharm.* **226**, 511 (1888); Bayer, *Pharm. Ztg.* **34**, 98 (1889); DE **46333**; Moragas, Uthoff, *Chim. Ind. (Paris)* **17**, 284 (1927); Ramberg, Samén, *C.A.* **28**, 6103 (1934); Suter, *Organic Chemistry of Sulfur* (New York, 1944) pp 735, 742.

Lustrous leaflets, scales from water. Bitter taste. mp 74-76°. Upon further heating dec with evolution of SO$_2$. One gram dissolves in 200 ml cold, about 30 ml boiling water, in 8 ml alc. Sol in ether. A satd aq soln is neutral to litmus.

Note: This is a controlled substance (depressant): **21 CFR**, 1308.13.

THERAP CAT: Sedative, hypnotic.

THERAP CAT (VET): Has been used as a hypnotic.

9089. Sulfoniazide. [3691-81-4] 4-Pyridinecarboxylic acid 2-[(3-sulfophenyl)methylene]hydrazide; isonicotinic acid m-sulfobenzylidene hydrazide; isonicotinyl hydrazonotoluene-m-sulfonic acid; G-605. C$_{13}$H$_{11}$N$_3$O$_4$S; mol wt 305.31. C 51.14%, H 3.63%, N 13.76%, O 20.96%, S 10.50%. Prepn: Girard, US **2727041** (1955 to Lab. Français Chimiother.); Sugimoto, JP **57 1117** (1957 to Tanake), *C.A.* **52**, 4696e (1958).

Needles, dec 250-253°. Slightly sol in water.

Sodium salt trihydrate. Sulfon-Niazone. C$_{13}$H$_{10}$N$_3$NaO$_4$S.-3H$_2$O; mol wt 381.33. Crysts from water. Soly in water (18°): 1 g/33 cc.

THERAP CAT: Antibacterial (tuberculostatic).

9090. Sulfonmethane. [115-24-2] 2,2-Bis(ethylsulfonyl)propane; diethylsulfondimethylmethane; propane-diethyl sulfone; sulfonal. $C_7H_{16}O_4S_2$; mol wt 228.32. C 36.82%, H 7.06%, O 28.03%, S 28.08%. Prepd by condensing ethyl mercaptan with acetone, then oxidizing with $KMnO_4$: Fromm, *Ann.* **253**, 140 (1889); *cf.* Baumann, *Ber.* **18**, 883 (1885); *Arch. Pharm.* **226**, 511 (1888); Bayer, *Pharm. Ztg.* **34**, 98 (1889); **DE 46333**; Moragas, Uthoff, *Chim. Ind. (Paris)* **17**, 284 (1927); Ramberg, Samén, *C.A.* **28**, 6103 (1934); Suter, *Organic Chemistry of Sulfur* (New York, 1944) pp 735, 742. Soly data: Falck, *J. Pharm. Chim.* [7] **21**, 279 (1920).

Crystals, almost tasteless, mp 124-126°. bp 300°. One gram dissolves in 365 ml water, 16 ml boiling water, 60 ml alcohol, 3 ml boiling alcohol, 64 ml ether, 11 ml chloroform. Soluble in benzene. Insol in glycerol.

Note: This is a controlled substance (depressant): **21 CFR**, 1308.13.

THERAP CAT: Hypnotic.

THERAP CAT (VET): Has been used as a hypnotic.

9091. Sulfonyldiacetic Acid. [123-45-5] 2,2′-Sulfonylbis-acetic acid; sulfonediacetic acid; dimethylenesulfone-α,α'-dicarboxylic acid. $C_4H_6O_6S$; mol wt 182.15. C 26.38%, H 3.32%, O 52.70%, S 17.60%. Prepd by oxidation of thiodiglycolic acid: Lovén, *Ber.* **17**, 2818 (1884).

Tabular crystals, mp 182°. Freely sol in water. Sol in alc. Slightly sol in ether.

USE: Detection of barium, lead, mercury and silver.

9092. Sulforaphane. [142825-10-3]; [4478-93-7] (unspecified stereo). 1-Isothiocyanato-4-[(*R*)-methylsulfinyl]butane; 4-methylsulfinylbutyl isothiocyanate. $C_6H_{11}NOS_2$; mol wt 177.28. C 40.65%, H 6.25%, N 7.90%, O 9.02%, S 36.17%. Potent, selective inducer of phase II detoxification enzymes with anticarcinogenic properties. Occurs naturally in broccoli. Prepn: H. Schmid, P. Karrer, *Helv. Chim. Acta* **31**, 1497 (1948). Absolute configuration: K. Mislow *et al.*, *J. Am. Chem. Soc.* **87**, 665 (1965). Isoln from SAGA broccoli and enzyme inducing activities: Y. Zhang *et al.*, *Proc. Natl. Acad. Sci. USA* **89**, 2399 (1992). Asymmetric synthesis: J. K. Whitesell, M.-S. Wong, *J. Org. Chem.* **59**, 597 (1994). *In vivo* antitumor activity: Y. Zhang *et al.*, *Proc. Natl. Acad. Sci. USA* **91**, 3147 (1994). Mechanism of action study: M. C. Myzak *et al.*, *Cancer Res.* **64**, 5767 (2004). Review of antioxidant activitiy: J. W. Fahey, P. Talalay, *Food Chem. Toxicol.* **37**, 973-979 (1999).

$[\alpha]_D^{22}$ −79.3 ±1° (c = 1.223 in chloroform). $[\alpha]_D^{25}$ −78.6° (c = 1.19 in chloroform). uv max (water): 238 nm (ε 910 $M^{-1}cm^{-1}$); upon addition of 0.1*M* NaOH uv max: 226 nm (ε 15300 $M^{-1}cm^{-1}$).

9093. Sulforidazine. [14759-06-9] 10-[2-(1-Methyl-2-piperidinyl)ethyl]-2-(methylsulfonyl)-10*H*-phenothiazine; 3-(methylsulfonyl)-10-[2-(1-methyl-2-piperidyl)ethyl]phenothiazine; thioridazine-2-sulfone; TPN-12; Imagotan; Inofal. $C_{21}H_{26}N_2O_2S_2$; mol wt 402.57. C 62.66%, H 6.51%, N 6.96%, O 7.95%, S 15.93%. Dopamine receptor blocker; metabolite of thioridazine and mesoridazine, *q.q.v.* Prepn: Renz *et al.*, **FR 1363683** and **FR 1459476** (1964, 1966, both to Sandoz). Pharmacology: Maruyama *et al.*, *C.A.* **68**, 20799z (1968). Effects on dopaminergic function in comparison with mesoridazine and thioridazine: D. M. Niedzwiecki *et al.*, *J. Pharmacol. Exp. Ther.* **228**, 636 (1984); C. D. Kilts *et al.*, *ibid.* **231**, 334 (1984). Clinical evaluation as antipsychotic: R. Axelsson,

Curr. Ther. Res. **21**, 587 (1977). HPLC determn in plasma: D. A. Ganes, K. K. Midha, *J. Chromatogr.* **423**, 227 (1987).

Crystals from acetone, mp 121-123°.
THERAP CAT: Antipsychotic.

9094. Sulfosalicylic Acid. [97-05-2] 2-Hydroxy-5-sulfobenzoic acid; 3-carboxy-4-hydroxybenzenesulfonic acid; 5-sulfosalicylic acid; 2-hydroxybenzoic-5-sulfonic acid; salicylsulfonic acid. $C_7H_6O_6S$; mol wt 218.18. C 38.54%, H 2.77%, O 44.00%, S 14.69%. Prepn: Pusca *et al.*, *Rev. Chim. (Bucharest)* **13**, no. 1, 49-50 (1962), *C.A.* **57**, 14994b (1962).

Dihydrate. [5965-83-3] $C_7H_6O_6S\cdot2H_2O$; mol wt 254.21. White crystals or cryst powder. Colored pink by traces of iron. mp ~120° when anhydr; at higher temp dec into phenol and salicylic acid. Very sol in water or alcohol; sol in ether. Generally sol in polar solvents. *Keep well closed and protected from light.*

Caution: Irritating to skin, mucous membranes.

USE: Clinical reagent for albumin in urine; colorimetric reagent for ferric ion with which it gives a violet color. It is a trifunctional aromatic compound, undergoing reactions typical of phenols, carboxylic and sulfonic acid groups. Suggested industrial uses: metal chelating agent, intermediate in the manuf of surface-active agents, organic catalysts and grease additives.

9095. Sulfosulfuron. [141776-32-1] *N*-[[(4,6-Dimethoxy-2-pyrimidinyl)amino]carbonyl]-2-(ethylsulfonyl)imidazo[1,2-*a*]pyridine-3-sulfonamide; 1-(2-ethylsulfonylimidazol[1,2-*a*]pyridin-3-ylsulfonyl)-3-(4,6-dimethoxypyrimidin-2-yl)urea; MON-37500; Maverick; Monitor; Sundance. $C_{16}H_{18}N_6O_7S_2$; mol wt 470.48. C 40.85%, H 3.86%, N 17.86%, O 23.80%, S 13.63%. Sulfonylurea herbicide to control grassy weeds in wheat. Prepn: Y. Ishida *et al.*, **EP 477808** (1992 to Takeda). Description of physical properties, toxicity and activity: S. K. Parrish *et al.*, *Brighton Crop Prot. Conf. - Weeds* **1995**, 57. Soil degradation: *idem et al.*, *ibid.* 667. Field trial in winter wheat: R. E. Blackshaw, W. M. Hamman, *Weed Technol.* **12**, 421 (1998). Absorption and translocation in grass species: B. L. S. Olson *et al.*, *Weed Sci.* **47**, 37 (1999).

White, odorless crystals, mp 201.1-201.7°. Soly in water (ppm): 18 (pH 5); 1627 (pH 7); 482 (pH 9). Vapor pressure: <10^{-6} Pa. Log P (octanol/water): <10. LD_{50} in rats (mg/kg): >5000 orally; >5000 dermally; LC_{50} (96 hr) in rainbow trout, carp (mg/l): >95, >91 (Parrish).

USE: Herbicide.

9096. Sulfotep. [3689-24-5] Thiodiphosphoric acid ([(HO)$_2$-P(S)]$_2$O) *OP,OP,OP',OP'*-tetraethyl ester; thiopyrophosphoric acid tetraethyl ester; sulfotepp; thiotepp; dithio; dithione; dithiophos; TEDP; ASP-47; Bayer E 393; ENT-16273; Bladafum. $C_8H_{20}O_5$-P_2S_2; mol wt 322.31. C 29.81%, H 6.25%, O 24.82%, P 19.22%, S 19.89%. Prepn: A. D. F. Toy, *J. Am. Chem. Soc.* **73**, 4670 (1951); G. Schrader, R. Mühlmann, **DE 848812** (1952 to Bayer), *C.A.* **47**, 5426a (1953); R. Appel, H. Einig, *Z. Anorg. Allg. Chem.* **414**, 236 (1975). Activity: F. F. Smith *et al.*, *J. Econ. Entomol.* **43**, 627 (1950). Toxicity data: B. Holmstedt, *Pharmacol. Rev.* **11**, 567 (1959).

Pale yellow liquid, bp$_2$ 136-139°. *Poisonous.* Vapor press. at 20°: 1.7×10^{-4} mm Hg. d_4^{25} 1.196. n_D^{25} 1.4753. Soly in water: 25 mg/l. Miscible with most organic solvents. Corrosive to iron. LD$_{50}$ s.c. in mice: 8 mg/kg (Holmstedt).

Caution: Potential symptoms of overexposure are eye pain, blurred vision, lacrimation and rhinorrhea; headache; cyanosis; anorexia, nausea, vomiting and diarrhea; local sweating, weakness, twitching, paralysis, Cheyne-Stokes respiration, convulsions, low blood pressure and cardiac irregularities; skin and eye irritation. *See NIOSH Pocket Guide to Chemical Hazards* (DHHS/NIOSH 97-140, 1997) p 294.

USE: Insecticide; miticide.

9097. Sulfoxide. [120-62-7] 5-[2-(Octylsulfinyl)propyl]-1,3-benzodioxole; 1,2-(methylenedioxy)-4-[2-(octylsulfinyl)propyl]benzene; *n*-octylsulfoxide of isosafrole; 1-methyl-2-(3,4-methylenedioxyphenyl)ethyl octyl sulfoxide; Sulfox-Cide. $C_{18}H_{28}O_3S$; mol wt 324.48. C 66.63%, H 8.70%, O 14.79%, S 9.88%. Prepn: M. E. Synerholm *et al.*, *Contrib. Boyce Thompson Inst.* **15**, 35 (1947); M. E. Synerholm, **US 2486579** (1949 to Boyce Thompson Inst. for Plant Res.). Synergistic use with pyrethrum, allethrin, rotenone, ryania, etc.: *idem*, **US 2486445** (1949 to Boyce Thompson Inst. for Plant Res.). Acute toxicity: T. B. Gaines, R. E. Linder, *Fundam. Appl. Toxicol.* **7**, 299 (1986).

Pale yellow, sweet-smelling, viscous oil. Practically insol in water. Sol in most organic solvents except petr ether. LD$_{50}$ in adult male, female rats (mg/kg): 3957, 3477 orally (Gaines, Linder).

USE: Insecticide.

9098. Sulfoxone Sodium. [144-75-2] 1,1'-[Sulfonylbis(4,1-phenyleneimino)]bismethanesulfinic acid sodium salt (1:2); disodium[sulfonylbis(*p*-phenylenimino)]dimethanesulfinate; 4,4'-diaminodiphenylsulfone disodium formaldehyde sulfoxylate; disodium formaldehydesulfoxylate-diaminodiphenylsulfone; aldesulfone sodium; Diazon; Novotrone; Diasone. $C_{14}H_{14}N_2Na_2O_6S_3$; mol wt 448.43. C 37.50%, H 3.15%, N 6.25%, Na 10.25%, O 21.41%, S 21.45%. Prepn: H. Bauer, *J. Am. Chem. Soc.* **61**, 617 (1939); S. M. Rosenthal, H. Bauer, **US 2234981** (1941); G. W. Raiziss *et al.*, *J. Am. Pharm. Assoc.* **33**, 43 (1944); **US 2256575** (1941). Therapeutic use in dermatitis herpetiformis: T. Cornbleet, *Arch. Dermatol. Syphilol.* **64**, 684 (1951); S. I. Katz, *Arch. Dermatol.* **118**, 809 (1982); in leprosy: J. H. Peters *et al.*, *Lepr. Rev.* **46**, 171 (1975). Comprehensive description: V. K. Kapoor, *Anal. Profiles Drug Subs.* **19**, 553-573 (1990).

Dihydrate. Needles or amorphous powder. After drying at 100-110° it dec 263-265°. Easily sol in water. Slightly sol in alcohol. Practically insol in the usual organic solvents. Must be stored in vacuum-sealed ampuls. The addition of 10% sodium bicarbonate helps to prevent oxidative changes.

THERAP CAT: Antibacterial (leprostatic).

9099. Sulfur. [7704-34-9] Brimstone; sulphur. S; at. wt [32.059; 32.076]; conventional at. wt 32.06; at. no. 16; valences 2, 4, 6. Group VIA (16). Four naturally occurring isotopes: 32 (95.0%); 33 (0.76%); 34 (4.22%); 36 (0.014%); artificial, radioactive isotopes: 29-31; 35; 37; 38. Has been known from very early times. Occurs both in the free state and in combination, mainly as sulfides and sulfates; constitutes 0.05% of the crust of the earth. Industrial prepn by Frasch process: *Faith, Keyes & Clark's Industrial Chemicals*, F. A. Lowenheim, M. K. Moran, Eds. (Wiley-Interscience, New York, 4th ed., 1975) pp 786-794. Exists in several allotropic modifications; at S.T.P. only the orthorhombic, cyclooctasulfur (S_α) is thermodynamically stable: Meyer, *Chem. Rev.* **64**, 429 (1964). Prepn of cyclohexasulfur: Bartlett *et al.*, *J. Am. Chem. Soc.* **83**, 10 (1961); of cyclohepta- and cyclodecasulfur: Schmidt *et al.*, *Angew. Chem. Int. Ed.* **7**, 632 (1968); of cyclononasulfur: Schmidt, Wilhelm, *Chem. Commun.* **1970**, 1111; of cyclododecasulfur: *eidem*, *Angew. Chem. Int. Ed.* **5**, 964 (1966). Review of structures of elemental sulfur: Meyer, *Adv. Inorg. Chem. Radiochem.* **18**, 287-317 (1976). Reviews of sulfur and its compds: W. N. Tuller, *The Sulphur Data Book* (McGraw-Hill, New York, 1954); B. Meyer, *Elemental Sulfur* (Interscience-Wiley, 1965); *Inorganic Sulphur Chemistry*, G. Nickless, Ed. (Elsevier, New York, 1968); Schmidt, Siebert in *Comprehensive Inorganic Chemistry* **vol. 2**, J. C. Bailar, Jr. *et al.*, Eds. (Pergamon, Oxford, 1973) pp 795-933. *See also* Sulfur, Pharmaceutical.

Yellow solid. Insol in water. Sparingly sol in alcohol, in ether; sol in carbon disulfide (one gram/2 ml); sol in benzene (about 2.4% at 30°, much more at higher temps.), in toluene. Liquid ammonia (anhydr) dissolves 38.5% S at −78°; acetone dissolves 2.65% at 25°; methylene iodide dissolves 9.1% at 10°; chloroform dissolves about 1.5% at 18°. For other solvents *see* Tuller (1954). *Flammable.* Ignites in air above 261°, in oxygen below 260°, burning to the dioxide; combines readily with hydrogen; combines in the cold with fluorine, chlorine, and bromine; combines with carbon at high temperatures; reacts with silicon, phosphorus, arsenic, antimony and bismuth at their melting points; combines with nearly all metals; with lithium, sodium, potassium, copper, mercury and silver in the cold on contact with the solid; with magnesium, zinc and cadmium very slightly in the cold, more readily on heating; with other metals at high temperatures. Does not react with iodine, nitrogen, tellurium, gold, platinum and iridium.

α-Sulfur. Orthorhombic cyclooctasulfur. Amber-colored crystals; the stable form at ordinary temperature. d 2.06; when heated to 94.5° becomes opaque owing to the formation of monoclinic sulfur.

β-Sulfur. Monoclinic cyclooctasulfur. Opaque, light-yellow, brittle, needle-like crystals; stable between 94.5-120°. Passes slowly into the rhombic form on standing. d 1.96; mp 115.21°: West, *J. Am. Chem. Soc.* **81**, 29 (1959). bp ~444.6°.

γ-Sulfur. Mother-of-pearl sulfur. A second monoclinic form, mp 106.8°.

Polymeric sulfur. Crystex. Amorphous form; mol wt approx 200,000. Metastable, gradually reverts to the α-form. Insol in solvents used for orthorhombic form. *Review:* A. V. Tobolsky, W. J. MacKnight, *Polymeric Sulfur and Related Polymers* (Interscience, New York, 1965) pp 87-97.

Caution: May cause irritation of skin, mucous membranes.

USE: In mfg sulfuric acid, carbon disulfide, sulfites, insecticides, plastics, enamels, metal-glass cements; in vulcanizing rubber; in syntheses of dyes; in making gunpowder, matches; for bleaching wood pulp, straw, wool, silk, felt, linen.

9100.   Sulfur Chloride. [10025-67-9] Disulfur dichloride; sulfur monochloride; sulfur subchloride. Cl_2S_2; mol wt 135.02. Cl 52.51%, S 47.49%. S_2Cl_2. Prepared by passing chlorine into molten sulfur, can be purified by fractional distillation: Fehér in *Handbook of Preparative Inorganic Chemistry* vol. 1, G. Brauer, Ed. (Academic Press, New York, 2nd ed., 1963) pp 371-372. Hydrolysis: H. L. Olin, *J. Am. Chem. Soc.* **48**, 167 (1926).

Non-flammable, light amber to yellowish red, fuming, oily liquid. Penetrating odor. $d_{15.5}^{15.5}$ 1.6885. n_D^{20} 1.670. mp $-77°$. bp_{760} 138°; bp_{100} 72.0°; bp_{10} 19.1°. Dielectric constant 4.9 at 22°. Dipole moment 1.60. Sol in alcohol, benzene, ether, carbon disulfide, carbon tetrachloride, oils. Readily dissolves sulfur (up to 67% at room temp). Dec by water yielding sulfur, hydrogen chloride, sulfur dioxide, hydrogen sulfide, sulfite, thiosulfate. In acid solns pentathionic and other polythionic acids are formed. *Corrosive. Keep tightly closed and out of contact with water.*

Caution: Potential symptoms of overexposure are irritation of eyes, skin and mucous membranes; lacrimation; coughing; eye and skin burns; pulmonary edema. *See NIOSH Pocket Guide to Chemical Hazards* (DHHS/NIOSH 97-140, 1997) p 290.

USE: Intermediate and chlorinating agent in the manuf of organic chemicals, sulfur dyes, insecticides, synthetic rubbers; in cold vulcanization of rubber; as polymerization catalyst for vegetable oils; for hardening soft woods.

9101.   Sulfur Dioxide. [7446-09-5] Sulfurous anhydride; sulfurous oxide. O_2S; mol wt 64.06. O 49.95%, S 50.05%. SO_2. Liquid SO_2 used as solvent: Waddington, *Non-aqueous Solvent Systems*, T. C. Waddington, Ed. (Academic Press, New York, 1965) pp 253-284. Review of toxicology and human exposure: *Toxicological Profile for Sulfur Dioxide* (PB99-122020, 1998) 223 pp. Brief review of synthetic applications: F. Fonquerne, *Synlett* **2005**, 1340-1341.

Colorless, nonflammable gas; strong suffocating odor. *Poisonous; corrosive.* Condenses at $-10°$ and ordinary pressure to a colorless liquid. d (liq) 1.5. mp $-72°$. bp $-10°$. Mixed with oxygen and passed over red-hot Pt, it is converted into SO_3. With water forms sulfurous acid (H_2SO_3). Bleaches vegetable colors. Soly: in water 17.7% at 0°, 11.9% at 15°, 8.5% at 25°, 6.4% at 35°; in alcohol 25%; in methanol 32%; also sol in chloroform or ether.

Caution: Potential symptoms of overexposure are irritation of eyes, nose and throat, rhinorrhea; choking, coughing; reflex bronchoconstriction; direct contact with liquid may cause frostbite. *See NIOSH Pocket Guide to Chemical Hazards* (DHHS/NIOSH 97-140, 1997) p 288.

USE: Preserving fruits, vegetables, etc.; disinfectant in breweries and food factories; bleaching textile fibers, straw, wicker ware, gelatin, glue, beet sugars. Solvent and reagent in organic synthesis.

9102.   Sulfuretin. [120-05-8] (2Z)-2-[(3,4-Dihydroxyphenyl)methylene]-6-hydroxy-3(2H)-benzofuranone; 2-(3,4-dihydroxybenzylidene)-6-hydroxy-3(2H)-benzofuranone; 3′,4′,6-trihydroxyaurone; 3′,4′,6-trihydroxybenzalcoumaranone. $C_{15}H_{10}O_5$; mol wt 270.24. C 66.67%, H 3.73%, O 29.60%. Flavonoid pigment responsible for the yellow color of certain species of *Compositae*. Isoln from *Cosmos sulphureus* Cav., *Compositae:* Shimokoriyama, Hattori, *J. Am. Chem. Soc.* **75**, 1900 (1953); from *Dahlia variabilis* Desf., *Compositae* and prepn by condensation of 6-hydroxy-3-coumaranone and protocatechuic alcohol: C. G. Nordström, T. Swain, *Arch. Biochem. Biophys.* **60**, 329 (1956). Prepn and structure studies: Geissman, Jurd, *J. Am. Chem. Soc.* **76**, 4475 (1954); Farkas *et al.*, *Ber.* **92**, 2847 (1959). Structure: Shimokoriyama, Geissman, *J. Org. Chem.* **25**, 1956 (1960).

Orange crystals from dil alc, mp 280-285° (Shimokoriyama), mp 315° (dec) (Nordström), mp 302-304° (Farkas). uv max: 398 nm.

Triacetate. $C_{21}H_{16}O_8$; mol wt 396.35. Pale yellow needles from methanol, mp 191-194° (Shimokoriyama), mp 167-168° (Nordström).

9103.   Sulfur Hexafluoride. [2551-62-4] (*OC*-6-11)-Sulfur fluoride (SF₆); hexafluorosulfur. F_6S; mol wt 146.05. F 78.05%, S 21.95%. SF_6. Prepd by direct fluorination of sulfur or sulfur dioxide: Moissan, Lebeau, *Compt. Rend.* **130**, 865, 984 (1900); *eidem, Ann. Chim. Phys.* [7] **26**, 147 (1902); Schumb, *Inorg. Synth.* **3**, 119 (1950); Kwasnik in *Handbook of Preparative Inorganic Chemistry* Vol. 1, G. Brauer, Ed. (Academic Press, New York, 2nd ed., 1963) pp 169-170. *Reviews:* Cady, "Fluorine-Containing Compounds of Sulfur" in *Adv. Inorg. Chem. Radiochem.* **2**, 105-157 (1960); Kemmitt, Sharp, *Adv. Fluorine Chem.* **4**, 218-219 (1965).

Colorless, odorless gas. *Non-flammable.* One of the heaviest known gases; density approx 5 times that of air. mp $-50.8°$. Sublimes at $-63.8°$. Crit temp 45.6°. d (liq; $-50.8°$) 1.88. Sparingly sol in water, somewhat more in alcohol. At 25° and 1 atm 0.297 ml SF_6 dissolves in 1.0 ml of transformer oil. Thermodynamically unstable but kinetically stable gas. This stability explained by symmetrical, octahedral structure of the molecule. Inert to nucleophilic attack. Does not attack glass. No fluorine exchange with HF. Stable to silent electrical discharge. Unchanged at 500°.

Caution: Potential symptoms of overexposure are asphyxia; increased breathing rate and pulse rate; slight muscle incoordination, emotional upset; fatigue, nausea, vomiting, convulsions. *See NIOSH Pocket Guide to Chemical Hazards* (DHHS/NIOSH 97-140, 1997) p 288.

USE: In electrical circuit interrupters. In electronic ultra-high frequency piping.

9104.   Sulfuric Acid. [7664-93-9] Oil of vitriol. H_2O_4S; mol wt 98.07. H 2.06%, O 65.26%, S 32.69%. H_2SO_4. Prepd by the Contact Process according to the reactions $2SO_2 + O_2 \rightarrow 2SO_3$, and $SO_3 + H_2O \rightarrow H_2SO_4$; by the Chamber Process according to the reactions $2NO + O_2 \rightarrow 2NO_2$, and $NO_2 + SO_2 + H_2O \rightarrow H_2SO_4 + NO$. Sulfuric acid of commerce contains 93-98% H_2SO_4; the remainder is water. Monograph: W. W. Duecker, J. R. West, *The Manufacture of Sulfuric Acid* (Reinhold, New York, 1959) 515 pp. Review of manuf: Pearce, "Sulphuric Acid: Physico-Chemical Aspects of Manufacture" in *Inorganic Sulphur Chemistry*, G. Nickless, Ed. (Elsevier, New York, 1968) pp 535-561; *Faith, Keyes & Clark's Industrial Chemicals*, F. A. Lowenheim, M. K. Moran, Eds. (Wiley-Interscience, New York, 4th ed., 1975) pp 795-806. Toxicity data: H. F. Smyth *et al.*, *Am. Ind. Hyg. Assoc. J.* **30**, 470 (1969). Review of toxicology and human exposure: *Toxicological Profile for Sulfur Trioxide and Sulfuric Acid* (PB99-122038, 1998) 224 pp.

Clear, colorless, odorless, oily liquid. *Corrosive; poisonous.* Has a very great affinity for water, abstracting it from the air and also from many organic substances; hence it chars sugar, wood, etc. d ~1.84. bp ~290°; dec 340° into sulfur trioxide and water. mp 10° (anhydrous acid). 98% H_2SO_4 freezes at +3°; 93% at $-32°$; 78% at $-38°$; 74% at $-44°$; 65% at $-64°$. Misc with water and alcohol with the generation of much heat and with contraction in vol. When diluting, the acid should be added to the diluent. *Handle with caution. Keep tightly closed.* LD_{50} orally in rats: 2.14 g/kg (Smyth).

Sulfuric acid, fuming. [8014-95-7] H_2SO_4 with free SO_3, designated in commerce as *oleum.* Available grades contain up to about 80% free SO_3. Colorless or slightly colored, viscous liquid, emitting choking fumes of sulfur trioxide. *Handle with caution. Corrosive; poisonous. Keep tightly closed in glass-stoppered bottles.*

Caution: Potential symptoms of overexposure are eye, skin, nose and throat irritation; pulmonary edema, bronchitis; emphysema; conjunctivitis; stomatis; dental erosion; tracheobronchitis; skin and eye burns; dermatitis. *See NIOSH Pocket Guide to Chemical Hazards* (DHHS/NIOSH 97-140, 1997) p 290. *See also Clinical Toxicology of Commercial Products,* R. E. Gosselin *et al.*, Eds. (Williams & Wilkins, Baltimore, 5th ed., 1984) section III, pp 8-12. Occupational exposure to strong inorganic acid mists containing sulfuric acid is listed as a known human carcinogen: *Report on Carcinogens, Twelfth Edition* (PB2011-111646, 2011) p 380.

USE: In manuf of fertilizers, explosives, dyestuffs, other acids, parchment paper, glue, purification of petroleum, pickling of metal. Dehydrating agent. Digestion of organic matter. pH modification. Titration of bases.

THERAP CAT: Dil acid formerly in treatment of gastric hypoacidity. Concd acid formerly as a topical caustic.

9105. Sulfur Iodide. [1312-15-8] Iodosulfane. Approx SI. Obtained by fusing together 4 parts iodine with 1 part sulfur. It probably contains free iodine as well as free sulfur.

Grayish-black masses of metallic luster and iodine odor. Insol in water; sol in carbon disulfide, alcohol, ether, in about 50 parts glycerol. KI soln dissolves a portion of the iodine leaving free sulfur. *Keep well closed.*

THERAP CAT: Dermatologic.

THERAP CAT (VET): Has been used in chronic eczema, ringworm, mange.

9106. Sulfurous Acid. [7782-99-2] Sulfur dioxide soln. A soln of ~6% sulfur dioxide in water.

Colorless, clear, acid liquid; suffocating odor of sulfur dioxide. d ~1.03. Gradually oxidizes in the air to sulfuric acid. *Corrosive. Keep in nearly full, tightly closed containers, in a cool place.*

USE: Dental bleach. Reducing agent.

THERAP CAT: Antiseptic.

9107. Sulfur, Pharmaceutical. [7704-34-9] Two forms of sulfur of 99.5% purity or better are recognized in pharmacy: *sublimed sulfur*, also known as *flowers of sulfur*, and *precipitated sulfur*, also known as *milk of sulfur*, made by boiling sublimed sulfur with lime and pptg with hydrochloric acid. Other sulfur preparations include: *washed sulfur*, made by treating sublimed sulfur with ammonia and washing with water to dissolve impurities, particularly arsenic; *sulfurated lime*, the active component of *Vleminckx's lotion*, made by boiling sublimed sulfur with lime, resulting in formation of calcium pentasulfide and calcium thiosulfate; and *sulfurated potash* ($K_2(S_x)$), also known as *liver of sulfur*, made by heating sublimed sulfur with potassium carbonate. Review of dermatological applications: C. W. McMurtry, *J. Cutan. Dis.* **1913**, 322-328, 399-408; of chemistry, biologic effects and clinical use: A. N. Lin *et al., J. Am. Acad. Dermatol.* **18**, 553-558 (1988).

Sublimed sulfur forms a fine, yellow crystalline powder. Insol in water, alcohol. Precipitated sulfur forms a very fine, pale yellow, amorphous or microcryst powder. Practically insol in water. Very slightly sol in alcohol. Soly in ether at 23°: 0.97%; in acetone at 25°: 2.084%; in carbon disulfide at 20°: 29.5%; in olive oil at 15°: 2.3%. Sulfurated potash forms irregular, liver-brown pieces; dec in air. Sol in water.

THERAP CAT: Keratolytic; scabicide.

THERAP CAT (VET): Antiseptic and parasiticidal; in lotions, ointments and dips.

9108. Sulfur Tetrafluoride. [7783-60-0] (*T*-4)-Sulfur fluoride (SF_4). F_4S; mol wt 108.05. F 70.33%, S 29.67%. SF_4. First prepd from cobalt trifluoride and elemental sulfur: Fischer, Jaenckner, *Angew. Chem.* **42**, 810 (1929); Kwasnik in *Handbook of Preparative Inorganic Chemistry* **vol. 1**, G. Brauer, Ed. (Academic Press, New York, 2nd ed., 1963) pp 168-169. Best prepared from SCl_2 and NaF suspended in acetonitrile; also from sulfur, chlorine and sodium fluoride: Tullock *et al., J. Am. Chem. Soc.* **82**, 539 (1960); Arth, Fried, US 3046094 (1962 to Merck & Co.); Fawcett, Tullock, *Inorg. Synth.* **7**, 119 (1963). Reviews of prepn and chemistry: Smith, *Angew. Chem. Int. Ed.* **1**, 467-475 (1962); Kemmitt, Sharp, *Adv. Fluorine Chem.* **4**, 220-222 (1965); Martin, *Ann. N.Y. Acad. Sci.* **145**, 161-168 (1967).

Colorless gas. *Poisonous; corrosive.* Thermostable to 600°. mp −121.0°. bp −38°. d (liq; −78°) 1.95. d (solid; −183°) 2.349. Freely sol in benzene. Violent reaction with water. Hydrofluoric acid released on exposure to moisture. Dec by concd H_2SO_4. Attacks glass but not quartz or mercury.

Caution: Potential symptoms of overexposure are irritation of eyes, mucous membranes; direct contact with liquid may cause frostbite; eye and skin burns from released hydrofluoric acid. *See NIOSH Pocket Guide to Chemical Hazards* (DHHS/NIOSH 97-140, 1997) p 290.

USE: Selective fluorinating agent.

9109. Sulfur Trioxide. [7446-11-9] Sulfuric anhydride. O_3S; mol wt 80.06. O 59.95%, S 40.04%. SO_3. Prepd by the contact process, *i.e.*, by the action of oxygen on sulfur dioxide in the presence of catalysts such as platinized asbestos, platinized magne-

sium sulfate, ferric oxide, or vanadium compds. Extensive review and bibliography: Fairlie, "Sulfuric Acid Manufacture" in *A.C.S. Monograph Series* no. **69** (New York, 1936). Exists in 3 modifications. May be prepd in the laboratory by heating fuming sulfuric acid and collecting the sublimate in a cooled receiver. If the vapor is condensed above 27°, the γ-form is obtained as a liquid. If the vapor is condensed below 27° and in the presence of a trace of moisture, a mixture of all 3 forms is obtained. The 3 forms can be separated by fractional distillation. The α-form is the stable modification, the β- and γ-forms are metastable. Review of toxicology and human exposure: *Toxicological Profile for Sulfur Trioxide and Sulfuric Acid* (PB99-122038, 1998) 224 pp.

Corrosive; poisonous. Melted SO_3 exists in the γ-form and on solidifying tends to the α-form. The difference in vapor pressures explains the so-called "alpha explosion" of sulfur trioxide. Heating of high-melting SO_3 in glass vessels should be avoided to prevent possible shattering of the container. Absolutely dry SO_3 is not corrosive to metals and shows no acid reaction. On exposure to air, it absorbs moisture rapidly, emitting dense white fumes. Combines with water with explosive violence (heat of dilution 504 cals/g) forming sulfuric acid. Due to this avidity for water, SO_3 chars many organic substances. On contact with wood shavings the heat produced by dehydration is sufficient to cause fire.

α-Form. Asbestos-like needles, mp 62.3°. Vapor pressure at 25° = 73 mm.

β-Form. Asbestos-like neeedles, mp 32.5°. Vapor pressure at 25° = 344 mm.

γ-Form. Ice-like mass, mp 16.8° or liquid, d 1.9224. bp_{760} 44.8°. Vapor pressure at 25° = 433 mm.

Caution: Potential symptoms of overexposure by inhalation are tooth erosion and respiratory tract irritation; by ingestion are mouth and throat burns, gastric erosion; direct contact may cause irritation or chemical burns of skin, eyes, respiratory and gastrointestinal tracts. *See* PB99-122038.

USE: Intermediate in sulfuric acid manuf; in sulfonations for formation of addition compds with amines; in the manuf of explosives.

9110. Sulfuryl Chloride. [7791-25-5] Sulfuryl dichloride; sulfonyl chloride; sulfonyl dichloride; sulfuric oxychloride. Cl_2O_2S; mol wt 134.96. Cl 52.53%, O 23.71%, S 23.76%. SO_2Cl_2. Prepd by passing a mixture of dry sulfur dioxide and chlorine through activated charcoal (other catalysts, such as camphor, can be used): Danneel, *Angew. Chem.* **39**, 1553 (1926); Durrans, *J. Soc. Chem. Ind.* **45**, 347 (1926); Meyer, *Angew. Chem.* **44**, 41 (1931); Danneel, Hesse, *Z. Anorg. Allg. Chem.* **212**, 214 (1933); Allen, Maxson, *Inorg. Synth.* **1**, 114 (1939).

Colorless, mobile liquid, very pungent odor. *Even the vapors are corrosive to human skin and mucous membranes.* Turns yellow upon prolonged standing because of slight dissociation into SO_2 and Cl_2. d_4^{20} 1.6674. d_4^0 1.7045. mp −54.1° (also given as −46°). bp 69.3°. n_D^{20} 1.4437. Dipole moment 1.86. Trouton constant 20.7. Slowly dec by water, forming H_2SO_4 and HCl. With ice-cold water it forms a hydrate, SO_2Cl_2.15H_2O, resembling camphor in appearance. Violent reaction on contact with alkalies. Miscible with benzene, toluene, ether, glacial acetic acid, other organic solvents.

USE: Chlorinating and sulfonating or chlorosulfonating agent in organic syntheses, e.g., in the manufacture of chlorophenol and chlorothymol. Has been used in war gas formulations.

9111. Sulfuryl Fluoride. [2699-79-8] Sulfuryl difluoride; sulfonyl fluoride; sulfonyl difluoride; sulfuric oxyfluoride; Profume; Vikane. F_2O_2S; mol wt 102.05. F 37.23%, O 31.36%, S 31.42%. SO_2F_2. Prepn: H. Moissan, P. Lebeau, *C. R. Hebd. Seances Acad. Sci.* **132**, 374 (1901); M. Trautz, K. Ehrmann, *J. Prakt. Chem.* [2] **142**, 79 (1935). Insecticidal properties and use as fumigant: E. E. Kenaga, US 2875127 (1959 to Dow). Review of prepn and properties: W. Kwasnik in *Handbook of Preparative Inorganic Chemistry* **vol. 1**, G. Brauer, Ed. (Academic Press, New York, 2nd ed., 1963) pp 173-174; R. D. W. Kemmitt, D. W. A. Sharp, *Adv. Fluorine Chem.* **4**, 225-227 (1965). Ion chromatographic determn in air: S. A. Bouyoucos *et al., Am. Ind. Hyg. Assoc. J.* **44**, 57 (1983). Efficacy vs termite species during structural fumigation: N.-Y. Su, R. H. Scheffrahn, *J. Econ. Entomol.* **79**, 903 (1986). Comparison with methyl bromide, *q.v.*, of residual potential in packaged foods: R. H. Scheffrahn *et al., Bull. Environ. Contam. Toxicol.* **48**, 821 (1992).

Odorless, colorless gas. Not very reactive. *Poisonous.* Stable to 400°. mp −135.82°. bp −55.38°. Vapor pressure (20°): 15.2 atm. Vapor density (20°): 4.3 g/l. Specific gravity (20°): 1.35. Soly at 16.5° (ml gaseous F_2SO_2/100 ml solvent): 4-5 (water); 24-27 (alcohol); 210-220 (toluene); 136-138 (CCl_4). Not hydrolyzed by water. Hydrolyzed by NaOH solns. May be stored in compressed form in steel cylinders or in a gasometer over H_2SO_4.

Caution: Potential symptoms of overexposure are conjunctivitis, rhinitis and pharyngitis; paresthesia; direct contact with liquid may cause frostbite. *See NIOSH Pocket Guide to Chemical Hazards* (DHHS/NIOSH 97-140, 1997) p 292.

USE: Fumigant insecticide; termiticide.

9112. Sulindac. [38194-50-2] (1*Z*)-5-Fluoro-2-methyl-1-[[4-(methylsulfinyl)phenyl]methylene]-1*H*-indene-3-acetic acid; *cis*-5-fluoro-2-methyl-1-1-[*p*-(methylsulfinyl)benzylidene]indene-3-acetic acid; MK-231; Arthrocine; Artribid; Clinoril. $C_{20}H_{17}FO_3S$; mol wt 356.41. C 67.40%, H 4.81%, F 5.33%, O 13.47%, S 9.00%. Nonsteroidal anti-inflammatory drug. Prepn: T.-Y. Shen *et al.*, **DE 2039426**; *eidem*, **US 3654349** (1971, 1972 both to Merck & Co.). Stereospecific synthesis: R. F. Shuman *et al.*, *J. Org. Chem.* **42**, 1914 (1977); enantioselective synthesis: A. R. Maguire *et al.*, *Synlett* **2001**, 41. ^{13}C-NMR study: A. W. Douglas, *Can. J. Chem.* **56**, 2129 (1978). Metabolism and disposition: H. B. Hucker *et al.*, *Drug Metab. Dispos.* **1**, 721 (1973). HPLC determn in biological fluids: D. G. Musson *et al.*, *J. Pharm. Sci.* **73**, 1270 (1984). Book: *Current Concepts on Anti-inflammatory Drugs*, K. Miehlke, Ed. (Biomedical Information Corp., New York, 1980) 240 pp. Review of pharmacology and efficacy in rheumatic disease: R. N. Brogden *et al.*, *Drugs* **16**, 97-114 (1978); in treatment of colorectal polyps: F. Tonelli *et al.*, *Dig. Dis.* **12**, 259-264 (1994). Review of clinical pharmacokinetics: N. M. Davies, M. S. Watson, *Clin. Pharmacokinet.* **32**, 437-459 (1997).

Yellow odorless crystals from ethyl acetate, mp 182-185° (dec). uv max (methanolic 0.1*N* HCl): 327, 285, 256, 226 nm ($E_{1cm}^{1\%}$ 375, 420, 410, 540). pKa (25°) 4.7. Slightly sol in methanol, alc, acetone, chloroform; very slightly sol in ethyl acetate, isopropanol. Practically insol in hexane, water at pH <4.5. Soly increases with rising pH to ∼3.0 mg/ml at pH 7. Stable in aq acid and base. Solid stable for at least three days in air at 100°.

THERAP CAT: Anti-inflammatory.

9113. Sulisobenzone. [4065-45-6] 5-Benzoyl-4-hydroxy-2-methoxybenzenesulfonic acid; 3-benzoyl-4-hydroxy-6-methoxybenzenesulfonic acid; 2-benzoyl-5-methoxy-1-phenol-4-sulfonic acid; 2-hydroxy-4-methoxybenzophenone-5-sulfonic acid; benzophenone-4; NSC-60584; Spectra-Sorb UV 284; Sungard; Uval; Uvinul MS-40. $C_{14}H_{12}O_6S$; mol wt 308.30. C 54.54%, H 3.92%, O 31,14%, S 10.40%. Prepn: A. J. Cofrancesco, **GB 1136525** (1968 to GAF). Activity: J. M. Knox *et al.*, *J. Invest. Dermatol.* **34**, 51 (1960).

Light-tan powder, mp 145°. 1 g dissolves in 2 ml methanol, 3.3 ml alc, 4 ml water, 100 ml ethyl acetate.

USE: Ultraviolet absorber for leather and textile fibers; in cosmetics and shampoos.

THERAP CAT: Ultraviolet screen.

9114. Sulmarin. [29334-07-4] 4-Methyl-6,7-bis(sulfooxy)-2*H*-1-benzopyran-2-one; 6,7-dihydroxy-4-methylcoumarin disulfate; 4-methylesculetin bis(hydrogen sulfate); 4-methyl-6,7-dihydroxycoumarindisulfate; 4-methylesculetindisulfonic acid; 4-methylesculetin-6,7-disulfuric ester; MG-143; Idro P_2. $C_{10}H_8O_{10}S_2$; mol wt 352.28. C 34.10%, H 2.29%, O 45.42%, S 18.20%. Prepn: Cavallini, Mazzucchi, *Farm. Sci. Tec.* **3**, 297 (1948), *C.A.* **42**, 8900d (1948). Description: Maggiorelli, *G. Med. Mil.* **101**, 365 (1951), *C.A.* **47**, 6412d (1951); Banchetti, *Farmaco Ed. Sci.* **10**, 970 (1955). Polarographic method for determn: A. M. Contri, *Farmaco Ed. Prat.* **25**, 231 (1970).

Disodium salt trihydrate. $C_{10}H_6Na_2O_{10}S_2$.$3H_2O$. Crystals from 60% alc, dec 252-253°. uv max (pH 11.85): 304 nm.

THERAP CAT: Hemostatic.

9115. Sulmazole. [73384-60-8] 2-[2-Methoxy-4-(methylsulfinyl)phenyl]-3*H*-imidazo[4,5-*b*]pyridine; AR-L 115BS; Vardax. $C_{14}H_{13}N_3O_2S$; mol wt 287.34. C 58.52%, H 4.56%, N 14.62%, O 11.14%, S 11.16%. Orally active non-glycoside, non-adrenergic inotropic agent. Prepn: **NL 7401254**; E. Kutler *et al.*, **US 3985891** (1974, 1976 both to Thomae). Toxicity study: W. Diederen, R. Kadatz, *Arzneim.-Forsch.* **31**, 141 (1981). Series of articles on pharmacology, pharmacokinetics, metabolism, clinical studies: *eidem, ibid.* 129-278. Brief review of pharmacology: H. Koch, *Pharm. Int.* **3**, 5 (1982).

Solid, mp 203-205°. LD_{50} in albino mice (mg/kg): 560 orally; 163 i.v. (Diederen, Kadatz).

THERAP CAT: Cardiotonic.

9116. Sulodexide. [57821-29-1] Glucuronylglucosamineglicane sulfate; KRX-101; Provenal; Ravenol; Vessel. Orally active heparinoid prepared from porcine intestinal mucosa. Mixture of 80% electrophoretically fast moving heparin and 20% dermatan sulfate; average mol wt <8000. Preparative method: P. Bianchini, **DE 2426586**; *idem*, **US 3936351** (1975, 1976 both to Opocrin). Review of pharmacology: F. A. Ofosu, *Semin. Thromb. Hemostasis* **24**, 127-138 (1998); J. Harenberg, *Med. Res. Rev.* **18**, 1-20 (1998). Effect on albuminuria in diabetic patients: J. Skrha *et al.*, *Diabetes Res. Clin. Pract.* **38**, 25 (1997). Clinical trial in venous leg ulcers: S. Coccheri *et al.*, *Thromb. Haemostasis* **87**, 947 (2002); in deep venous thrombosis: B. M. Errichi *et al.*, *Angiology* **55**, 243 (2004); in diabetic nephropathy: A. Achour *et al.*, *J. Nephrol.* **18**, 568 (2005).

THERAP CAT: Antithrombotic; in treatment of chronic leg ulcers and diabetic nephropathy.

9117. Sulphan Blue. [129-17-9] *N*-[4-[[4-(Diethylamino)phenyl](2,4-disulfophenyl)methylene]-2,5-cyclohexadien-1-ylidene]-*N*-ethylethanaminium inner salt sodium salt (1:1); C.I. Acid Blue 1; [4-[α-[*p*-(diethylamino)phenyl]-2,4-disulfobenzylidene]-2,5-cyclohexadien-1-ylidene]diethylammonium hydroxide inner salt sodium salt; anhydro-4,4′-bis(diethylamino)triphenylmethanol-2″,4″-disulfonic acid monosodium salt; C.I. Food Blue 3; C.I. 42045; Disulphine Blue. $C_{27}H_{31}N_2NaO_6S_2$; mol wt 566.66. C 57.23%, H 5.51%, N 4.94%, Na 4.06%, O 16.94%, S 11.32%. Prepn from 4-formylbenzene-1,3-disulfonic acid and diethylaniline: *Colour Index* vol. 4 (3rd ed., 1971) p 4382.

Violet powder. One gram dissolves in 20 ml water at 20°. Dilute aq solns are blue and turn yellow upon the addition of concd hydrochloric acid. In the absence of strong acid or caustic the blue color is stable over a wide pH range. Partly sol in alcohol.

2,5-Disulfophenyl isomer. [68238-36-8] Isosulfan blue; Lymphazurin.

USE: Coloring medicinal products; in dyeing and printing wool, silk.

THERAP CAT: Isosulfan blue as diagnostic aid (lymphangiography).

9118. Sulphenone. [80-00-2] 1-Chloro-4-(phenylsulfonyl)-benzene; *p*-chlorophenyl phenyl sulfone; 4-chlorodiphenyl sulfone; sulfenone; R-242. $C_{12}H_9ClO_2S$; mol wt 252.71. C 57.03%, H 3.59%, Cl 14.03%, O 12.66%, S 12.69%. May also contain 20% related diaryl sulfones. Prepn: Bender, Pitt, US 2593001 (1952 to Stauffer Chem.). Toxicology: L. W. Hazleton *et al., J. Agric. Food Chem.* **3**, 836 (1955).

Dimorphic crystals, mp 94°. Slight aromatic odor. Tasteless. Soly at 20° (g/100 ml): acetone 74.4; dioxane 65.6; isopropanol 21; hexane 0.4; benzene 44.4; toluene 29.4; xylene 18.2; carbon tetrachloride 4.9. Practically insol in water. Slightly sol in petr oils. Stable to oxidizing and reducing agents, acids and alkalies found in spray formulations. LD$_{50}$ orally in mice: 2.7 g/kg (Hazleton).

USE: Acaricide.

9119. Sulpiride. [15676-16-1] 5-(Aminosulfonyl)-*N*-[(1-ethyl-2-pyrrolidinyl)methyl]-2-methoxybenzamide; *N*-[(1-ethyl-2-pyrrolidinyl)methyl]-5-sulfamoyl-*o*-anisamide; *N*-[(1-ethyl-2-pyrrolidinyl)methyl]-2-methoxy-5-sulfamoylbenzamide; Abilit; Aiglonyl; Coolspan; Dobren; Dogmatil; Dogmatyl; Dolmatil; Guastil; Meresa; Miradol; Mirbanil; Misulvan; Neogama; Omperan; Pyrikappl; Sernevin; Splotin; Sulpitil; Sulpor; Sursumid; Synédil; Trilan. $C_{15}H_{23}$-N_3O_4S; mol wt 341.43. C 52.77%, H 6.79%, N 12.31%, O 18.74%, S 9.39%. Dopamine D$_2$ and D$_3$-receptor antagonist. Prepn: C. S. Miller *et al.,* US 3342826 (1964 to Soc. d'Etudes Sci. Ind. de l'Ile-de-France); of *l*-form: F. Mauri, DE 2903891 (1979 to Ravizza), *C.A.* **91**, 211259h (1979). Crystal structure: L. Y. Y. Ma *et al., Acta Crystallogr.* **B38**, 2861 (1982). Mechanism of action: P. Jenner *et al., J. Pharm. Pharmacol.* **32**, 39 (1980). Structure-activity study: P. Dostert *et al., Eur. J. Med. Chem. - Chim. Ther.* **17**, 437 (1982). Physical properties: D. Pitrè, E. Valoti, *Arch. Pharmacol.* **320**, 859 (1987). Comprehensive description: D. Pitrè *et al., Anal. Profiles Drug Subs.* **17**, 607-641 (1988). Review of clinical efficacy in psychiatry: A. J. Wagstaff *et al., CNS Drugs* **2**, 313-333 (1994); of levosulpiride in gastroenterology: M. Guslandi, *Curr. Ther. Res.* **53**, 484-501 (1993).

White, odorless crystalline powder, mp 178-180°. Sparingly sol in methanol. Practically insol in water, ether, chloroform, benzene. pKa$_1$ 9.00, pKa$_2$ 10.19. $[\alpha]_D^{25}$ −66.8° (c = 0.5 in DMF). LD$_{50}$ in mice (mg/kg): 170 i.p.; 2250 orally (Dostert).

*l***-Form.** [23672-07-3] Levosulpiride; *S*(−)-sulpiride; Levobren; Levopraid. Monomorphic crystalline solid, mp 185-187°.

THERAP CAT: Antipsychotic; antidepressant; antiemetic.

9120. Sulprostone. [60325-46-4] (5*Z*)-7-[(1*R*,2*R*,3*R*)-3-Hydroxy-2-[(1*E*,3*R*)-3-hydroxy-4-phenoxy-1-buten-1-yl]-5-oxocyclopentyl]-*N*-(methylsulfonyl)-5-heptenamide; CP-34089; SHB-286; ZK-57671; Nalador. $C_{23}H_{31}NO_7S$; mol wt 465.56. C 59.34%, H 6.71%, N 3.01%, O 24.06%, S 6.89%. Analog of prostaglandin E$_2$, *q.v.,* with uterine stimulant activity. Prepn: J. Bindra, M. R. Johnson, DE 2355540 (1974 to Pfizer), *C.A.* **81**, 49330u (1974); *eidem,* US 4024179 (1977 to Pfizer). Antifertility effects: H. Hess *et al., Experientia* **33**, 1076 (1977). Influence on platelet function: R. C. Briel, T. H. Lippert, *Adv. Prostaglandin Thromboxane Res.* **6**, 351 (1980). Biological action and half-life: *eidem, Prostaglandins Med.* **6**, 1 (1981). Induction of abortion: K. Schmidt-Gollwitzer, *Int. J. Fertil.* **26**, 86 (1981).

Colorless oil.

THERAP CAT: Abortifacient.

9121. Sultamicillin. [76497-13-7] (2*S*,5*R*,6*R*)-6-[[(2*R*)-2-Amino-2-phenylacetyl]amino]-3,3-dimethyl-7-oxo-4-thia-1-azabicyclo[3.2.0]heptane-2-carboxylic acid [[[(2*S*,5*R*)-3,3-dimethyl-4,4-dioxido-7-oxo-4-thia-1-azabicyclo[3.2.0]hept-2-yl]carbonyl]oxy]methyl ester; 1,1-dioxopenicillanoyloxymethyl 6-(D-α-amino-α-phenylacetamido)penicillanate; 6'-(2-amino-2-phenylacetamido)penicillanoyloxymethyl penicillanate 1,1-dioxide; CP-49952; VD-1827. $C_{25}H_{30}N_4O_9S_2$; mol wt 594.65. C 50.50%, H 5.09%, N 9.42%, O 24.21%, S 10.78%. Orally absorbed double ester of sulbactam and ampicillin, *q.q.v.* Prepn: E. C. Bigham, DE 3018590; *eidem,* US 4244951 (1980, 1981 both to Pfizer); and pharmacology: B. Baltzer *et al., J. Antibiot.* **33**, 1183 (1980). HPLC determn and pharmacokinetics: H. J. Rogers *et al., J. Antimicrob. Chemother.* **11**, 435 (1983). Clinical trial in pediatric otitis media: K. H. Chan *et al., Pediatr. Infect. Dis. J.* **12**, 24 (1993). Review of antibacterial activity, pharmacokinetics and therapeutic efficacy: H. A. Friedel *et al., Drugs* **37**, 491-522 (1989).

mp 190°.

Tosylate. [83105-70-8] Bacimex; Unacid PD oral; Unacim; Unasyn. $C_{25}H_{30}N_4O_9S_2.C_7H_8O_3S$; mol wt 766.85.

THERAP CAT: Antibacterial.

9122. Sulthiame. [61-56-3] 4-(Tetrahydro-1,1-dioxido-2*H*-1,2-thiazin-2-yl)benzenesulfonamide; 4-(tetrahydro-2*H*-1,2-thiazin-2-yl)benzenesulfonamide *S,S*-dioxide; *N*-(4'-sulfamylphenyl)-1,4-butanesultam; tetrahydro-2-(*p*-sulfamoylphenyl)-1,2-thiazine 1,1-dioxide; 2-(*p*-aminosulfonylphenyl)tetrahydro-1,2-thiazine dioxide; 1-(*p*-amidosulfonylphenyl)-2-thiapiperidine 2,2-dioxide; *N*-(*p*-aminosulfonylphenyl)-1,4-butanesultam; Elisal; Ospolot; Trolone. $C_{10}H_{14}N_2O_4S_2$; mol wt 290.35. C 41.37%, H 4.86%, N 9.65%, O 22.04%, S 22.08%. Prepn of this type of compd: Helferich, Behnisch, US 2916489 (1959 to Schenley).

Crystalline powder, mp 180-182°. Practically insol in cold water. Partly sol in boiling water; slightly sol in alcohol, acids; readily sol in alkalies.

THERAP CAT: Anticonvulsant.

9123. Sultopride. [53583-79-2] *N*-[(1-Ethyl-2-pyrrolidinyl)-methyl]-5-(ethylsulfonyl)-2-methoxybenzamide; *N*-[(1-ethyl-2-pyrrolidinyl)methyl]-5-(ethylsulfonyl)-*o*-anisamide. $C_{17}H_{26}N_2O_4S$; mol wt 354.47. C 57.60%, H 7.39%, N 7.90%, O 18.05%, S 9.04%. Dopamine D_2-receptor antagonist. Prepn: C. S. Miller *et al.*, **FR M 5916**; *see also:* G. Bulteau *et al.*, **US 3975434** (1966, 1976 both to Soc. Etudes Sci. Ind. l'Ile-de-France). Pharmacology: C. Lavelle, J. Margarit, *J. Pharmacol.* **5**, Suppl. 2, 58 (1974); B. Bruguerolle *et al.*, *ibid.* **12**, 27 (1981). Clinical trial in psychosis: D. Morel, *Sem. Hop.* **59**, 2337 (1983). GC determn in serum: A. Kamizono *et al.*, *J. Chromatogr.* **567**, 113 (1991). Pharmacokinetics of enantiomers: *idem et al.*, *Biol. Pharm. Bull.* **16**, 1121 (1993).

Hydrochloride. [23694-17-9] LIN-1418; Barnetil; Barnotil. $C_{17}H_{26}N_2O_4S$.HCl; mol wt 390.92. Crystals from methyl ethyl ketone, mp 181-182° (Miller); also reported as crystals from ethanol, mp 190° (Bulteau).

THERAP CAT: Antipsychotic.

9124. Sultosilic Acid. [57775-26-5] 2-Hydroxy-5-[[(4-methylphenyl)sulfonyl]oxy]benzenesulfonic acid; 2-hydroxy-5-tosyloxybenzenesulfonic acid; 2,5-dihydroxybenzenesulfonic acid 5-*p*-toluenesulfonate. $C_{13}H_{12}O_7S_2$; mol wt 344.35. C 45.34%, H 3.51%, O 32.52%, S 18.62%. Prepn: **NL 7305916** (1973 to Lab. del Esteve); J. Esteve *et al.*, *Eur. J. Med. Chem.* **11**, 43 (1976); of piperazine salt: A. Esteve-Subirana, **US 3954767** (1976). Toxicology: L. Rodriguez *et al.*, *Arch. Farmacol. Toxicol.* **5**, 281 (1979). Metabolism: S. G. Wood *et al.*, *Xenobiotica* **12**, 165 (1982). Clinical evaluation: H. Vinazzer, J. C. Farine, *Atherosclerosis* **49**, 109 (1983). HPLC determn and pharmacokinetics: R. Roser *et al.*, *Eur. J. Drug Metab. Pharmacokinet.* **11**, 1 (1986).

Piperazine salt. [57775-27-6] Diethylenediamine sultosylate; piperazine sultosylate; A-585; Mimedran. $C_{13}H_{12}O_7S_2.C_4H_{10}N_2$; mol wt 430.49. Crystals from ethanol, mp 174°. LD_{50} i.p. in beagles, male rats, female rats: 605, 833.6, 1272 mg/kg; orally in rats: >11 g/kg (Rodriguez).

THERAP CAT: Antilipemic.

9125. Sumach. Name used for several plants of the *Rhus* species. Leaves of *Rhus glabra* L. and *Rhus typhina* L., *Anacardiaceae* are known commercially as **North American Sumach**. They contain fisetin, dihydrofisetin, 27% tannin and gallic acid esters. Constituents of the *Rhus typhina* fruit: J. Tischer, *Pharmazie* **15**, 83 (1960).

USE: North American sumach as black dye; as scenting agent for tobacco.

9126. Sumatriptan. [103628-46-2] 3-[2-(Dimethylamino)-ethyl]-*N*-methyl-1*H*-indole-5-methanesulfonamide; GR-43175. $C_{14}H_{21}N_3O_2S$; mol wt 295.40. C 56.92%, H 7.17%, N 14.23%, O 10.83%, S 10.85%. Serotonin 5HT$_1$-receptor agonist. Prepn: M. D. Dowle, I. H. Coates, **DE 3320521**; *eidem*, **US 4816470**; A. W. Oxford, **GB 2162522** (1983, 1989, 1986 all to Glaxo). Receptor binding studies: P. P. A. Humphrey *et al.*, *Br. J. Pharmacol.* **94**, 1123 (1988); P. Schoeffter, D. Hoyer, *Arch. Pharmacol.* **340**, 135 (1989). LC-MS determn in plasma: J. Oxford, M. S. Lant, *J. Chromatogr.* **496**, 137 (1989). Clinical evaluations in migraine: A. Doenicke *et al.*, *Lancet* **1**, 1309 (1988); Subcutaneous Sumatriptan International Study Group, *N. Engl. J. Med.* **325**, 316 (1991); in acute cluster headache: Sumatriptan Cluster Headache Study Group, *ibid.* 322. Review of pharmacology and clinical experience: S. J. Peroutka, *Headache* **30** (Suppl. 2), 554-560 (1990).

mp 169-171°. Very slightly sol in water.

Succinate. [103628-48-4] GR-43175C; Imigran; Imitrex; Imiject. $C_{14}H_{21}N_3O_2S.C_4H_6O_4$; mol wt 413.49. mp 165-166°. Freely sol in water; sparingly sol in methanol. Practically insol in methylene chloride.

THERAP CAT: Antimigraine.

9127. Sumbul. Musk root. Dried rhizome and roots of *Ferula sumbul* (Kauffm.) Hook. f. (*Euryangium sumbul* Kauffm.), or of other closely related species of *Ferula* having a musk-like odor (*Umbelliferae*). *Habit.* Central Asia, East Indies. *Constit.* Angelic (sumbulic) acid, valeric acid, methylcrotonic acid, resin, 0.2-0.4% volatile oil.

THERAP CAT: Sedative.

9128. Sunflower Seed Oil. The oil obtained by milling the seeds of *Helianthus annuus* L., *Compositae*. Classified as a semidrying oil. Alternately classified as an oleic-linoleic acid oil. Produced on a large scale in U.S.S.R., the Baltic region, India, Egypt, Canada, Argentina. Composition: palmitic acid 6.4%, stearic acid 1.3%, arachidic acid 4.0%, behenic acid 0.8%, oleic acid 21.3%, linoleic acid 66.2% (the glycerides in sunflower oil consist mainly of mixed triglycerides, each contg one or two linoleic acid radicals), linolenic acid <0.1%. Vitamin E (mixed tocopherols) 75 mg/100 g. This approaches the tocopherol content of wheat germ oil which is 103 mg/100 g. *Reviews:* E. W. Eckey, *Vegetable Fats and Oils* (Reinhold, New York, 1954) pp 772-777; L. H. Bailey, *Standard Cyclopedia of Horticulture* **vol. II** (Macmillan, New York, 1953) pp 1445-1449.

Pale yellow oil. Bland, agreeable taste. d_{15}^{15} 0.922-0.926. d_{25}^{25} 0.915-0.919. mp −18°. Titer: 16-20°. Acid value: 0.6. Saponif value: 188-194. Iodine value: 125-136. Thiocyanogen value: 78. Hydroxyl value: 14-16. Reichert-Meissl value below 0.5. Polenske value below 0.5. Unsaponifiable below 1.5%. n_D^{25} 1.472-1.474. n_D^{40} 1.466-1.468. Slightly sol in alc. Miscible with benzene, chloroform, carbon tetrachloride. Forms a "skin" after exposure to air for 2-3 weeks.

USE: Food and salad oil, in candy manuf, in oleomargarine. Like wheat germ oil in dietary supplements. Industrially in oil-modified alkyd resins and soap manuf.

9129. Sunitinib. [557795-19-4] *N*-[2-(Diethylamino)ethyl]-5-[(Z)-(5-fluoro-1,2-dihydro-2-oxo-3*H*-indol-3-ylidene)methyl]-2,4-dimethyl-1*H*-pyrrole-3-carboxamide; 5-[5-fluoro-2-oxo-1,2-dihydroindol-(3Z)-ylidenemethyl]-2,4-dimethyl-1*H*-pyrrole-3-carboxylic acid (2-diethylaminoethyl)amide. $C_{22}H_{27}FN_4O_2$; mol wt 398.48. C 66.31%, H 6.83%, F 4.77%, N 14.06%, O 8.03%. Tyrosine kinase inhibitor targeting VEGF-R1, VEGF-R2, VEGF-R3, PDGF-Rα, PDGF-Rβ, KIT, FLT3, CSF-1R, and RET. Prepn: P. C. Tang *et al.*, **WO 0160814** (2001 to Sugen); *eidem*, **US 6573293** (2003 to Sugen; Pharmacia & Upjohn); and kinase inhibition activity: L. Sun *et al.*, *J. Med. Chem.* **46**, 1116 (2003). Pharmacology:

D. B. Mendel *et al.*, *Clin. Cancer Res.* **9**, 327 (2003). Clinical pharmacokinetics and evaluation in acute myeloid leukemia: A.-M. O'Farrell *et al.*, *ibid.* 5465. LC-MS-MS determn in biological samples: S. Barattè *et al.*, *J. Chromatogr. A* **1024**, 87 (2004). Formulation development: A. Sistla *et al.*, *Drug Dev. Ind. Pharm.* **30** 19 (2004). Review of development and therapeutic potential: K. M. Sakamoto, *Curr. Opin. Investig. Drugs* **5**, 1329-1339 (2004); of clinical evaluation in solid tumors: R. J. Motzer *et al.*, *Expert Opin. Invest. Drugs* **15**, 553-561 (2006).

Orange solid. pKa 8.5. Log P (octanol/water): 3.1. Soly at 22° (*μg*/ml): 2582 in 20mM KCl/HCl buffer (pH 2); 364 in 20 mM phosphate buffer (pH 6).

Malate. [341031-54-7] SU-11248; Sutent. $C_{22}H_{27}FN_4O_2.C_4$-H_6O_5; mol wt 532.57. Yellow to orange powder. pKa 8.95. Soly in aq. media (pH 1.2-6.8): >25 mg/ml. Log P (octanol/water): 5.2 (pH 7).

THERAP CAT: Antineoplastic.

9130. Sunset Yellow FCF. [2783-94-0] 6-Hydroxy-5-[2-(4-sulfophenyl)diazenyl]-2-naphthalenesulfonic acid sodium salt (1:2); 1-*p*-sulfophenylazo-2-naphthol-6-sulfonic acid disodium salt; FD & C Yellow No. 6; C.I. Food Yellow 3; C.I. 15985. $C_{16}H_{10}N_2Na_2$-O_7S_2; mol wt 452.36. C 42.48%, H 2.23%, N 6.19%, Na 10.16%, O 24.76%, S 14.17%. Prepn: P. Griess, *Ber.* **11**, 2191 (1878). Metabolism: J. L. Radomski, T. J. Mellinger, *J. Pharmacol. Exp. Ther.* **136**, 259 (1962). Toxicology: I. F. Gaunt *et al.*, *Food Cosmet. Toxicol.* **5**, 747 (1967); **7**, 9 (1969); **12**, 1 (1974). *Review: IARC Monographs* **8**, 257-266 (1967). *See also: Colour Index* vol. 4 (3rd ed., 1971) p 4087.

Orange-red crystals. Absorption max ($0.02N$ CH_3COONH_4): 480 nm. Sol in water; slightly sol in ethanol. Reddish-orange soln in conc H_2SO_4, changing to yellow on dilution. LD_{50} in rats, mice (g/kg): >10, >6 orally (Gaunt, 1967).

USE: Provisionally listed for use in food, drugs and cosmetics.

9131. Superoxide Dismutase. [9054-89-1] SOD; E.C.1.15.1.1. Family of naturally occurring metalloenzymes that act as free oxygen radical scavengers and protect against the deleterious effects of biologically generated superoxide oxygen radicals ($O_2^{\cdot-}$) by dismutation to hydrogen peroxide (H_2O_2) + molecular oxygen (O_2). Three metallotypes have been described; distinguished by the metal found at the active site: cupro-zinc containing (Cu/Zn SOD), manganese containing (Mn SOD), and iron containing (Fe SOD). Widely distributed, present in almost all aerobic organisms and some anaerobes. Isoln and purification of Cu/Zn SOD from bovine erythrocytes and identity with ***erythrocuprein*** (***hemocuprein***): J. M. Mc-Cord, I. Fridovich, *J. Biol. Chem.* **244**, 6049 (1969); of Mn SOD from *E. coli:* B. B. Keele *et al.*, *ibid.* **245**, 6176 (1970); of Fe SOD from *E. coli:* F. J. Yost, Jr., I. Fridovich, *ibid.* **248**, 4905 (1973). Enzyme immunoassay for determn of human Cu/Zn SOD: N. Kurobe *et al.*, *Clin. Chim. Acta* **187**, 11 (1990); of human Mn SOD: *idem et al.*, *ibid.* **192**, 171 (1990). Review of biological activity and therapeutic use: A. Petkau, *Cancer Treat. Rev.* **13**, 17-44 (1986); of protective effect vs radiation injury: *eidem*, *Br. J. Cancer* **55**, Suppl. 8, 87-95 (1987); of structure and function: J. V. Bannister *et al.*,

Crit. Rev. Biochem. **22**, 111-180 (1987); of chemical and biological characteristics: B. A. Omar *et al.*, *Adv. Pharmacol.* **23**, 109-161 (1992). Books: *Superoxide and Superoxide Dismutases*, A. M. Michelson *et al.*, Eds. (Academic Press, New York, 1977) 568 p.; *Superoxide Dismutase* **vols. I, II, III**, L. W. Oberley, Ed. (CRC Press, Inc., Boca Raton, 1982, 1982, 1985).

Orgotein. Artrolasi; Ormetein (rescinded); Ontosein; Oxinorm; Palosein; Peroxinorm. Water-soluble Cu/Zn SOD produced from beef liver. Mol wt ~33,000 with a compact conformation maintained by about 4 gram atoms of chelated divalent metal. Isoln: W. Huber, **ZA 6902983**; *idem*, **US 3579495** (1969, 1971 both to Diagnostic Data). Purification: *idem*, **DE 2101866**; *idem*, **US 3624251** (both 1971 to Diagnostic Data). Toxicology: S. Carson *et al.*, *Toxicol. Appl. Pharmacol.* **26**, 184 (1973). Clinical pharmacokinetics: G. Jadot *et al.*, *Clin. Pharmacokinet.* **28**, 17 (1995). Clinical studies in osteoarthritis: K. Lund-Olesen, K. B. Menander-Huber, *Arzneim.-Forsch.* **33**, 1199 (1983); in Peyronie's disease: G. Primus, *Int. Urol. Nephrol.* **25**, 169 (1993). Clinical use of liposomal form vs radiation-induced fibrosis: S. Delanian *et al.*, *Radiother. Oncol.* **32**, 12 (1994). Therapeutic use in animals: D. E. Breshears *et al.*, *Mod. Vet. Pract.* **55**, 85 (1974); S. Ahlengard *et al.*, *Equine Vet. J.* **10**, 122 (1978); J. R. Coffman *et al.*, *J. Am. Vet. Med. Assoc.* **174** 261 (1979). Review of clinical trials: L. Flohé, *Mol. Cell. Biochem.* **84**, 123-131 (1988). LD_{50} in mice (mg/kg): >5800 s.c., >60 i.p., >4000 i.v.; in rats (mg/kg): >400 s.c., >284 i.p. (Jadot).

Pegorgotein. [155773-57-2] PEG-SOD; Win-22118; Dismutec. Orgotein polyethylene glycol conjugate; mol wt 71-105 kDa. Prepn and anti-inflammatory activity: P. S. Pyatak *et al.*, *Res. Commun. Chem. Pathol. Pharmacol.* **29**, 113 (1980). Pharmacokinetics: E. Boccu *et al.*, *Pharmacol. Res. Commun.* **14**, 113 (1982). Toxicology: A. T. Viau *et al.*, *J. Free Radicals Biol. Med.* **2**, 283 (1986). Clinical trials in head injury: J. P. Muizelaar, *Adv. Exp. Med. Biol.* **366**, 389 (1994).

Recombinant human form. RhSOD. Pharmacokinetics: C. Tsao *et al.*, *Clin. Pharmacol. Ther.* **50**, 713 (1991). Clinical study in renal transplantation: W. Land *et al.*, *Transplantation* **57**, 211 (1994). Review of cloning and expression: D. Tuoati, *Free Radical Biol. Med.* **5**, 393-402 (1990).

USE: Research probe for $O_2^{\cdot-}$ in biological and chemical reactions.
THERAP CAT: Anti-inflammatory; radioprotective agent.
THERAP CAT (VET): Anti-inflammatory; antirheumatic.

9132. Suplatast Tosylate. [94055-76-2] [3-[[4-(3-Ethoxy-2-hydroxypropoxy)phenyl]amino]-3-oxopropyl]dimethylsulfonium 4-methylbenzenesulfonate (1:1); [3-[[4-(3-ethoxy-2-hydroxypropoxy)phenyl]amino]-3-oxopropyl]dimethylsulfonium salt with 4-methylbenzenesulfonic acid; (±)-[2-[[*p*-(3-ethoxy-2-hydroxypropoxy)phenyl]carbamoyl]ethyl]dimethylsulfonium *p*-toluenesulfonate; IPD-1151T; IPD. $C_{16}H_{26}NO_4S.C_7H_7O_3S$; mol wt 499.64. C 55.29%, H 6.66%, N 2.80%, O 22.41%, S 12.83%. Selective inhibitor of the IgE antibody response; inhibits interleukin-4 gene expression and mast cell differentiation. Prepn: A. Koda *et al.*, **DE 3408708**; *eidem*, **US 4556737** (1984, 1985 both to Taiho). Suppression of IL-4 production and the IgE response: Y. Yanagihara *et al.*, *Jpn. J. Pharmacol.* **61**, 23, 31 (1993). Effect on mast cell differentiation: S. Konno *et al.*, *Eur. J. Pharmacol.* **259**, 15 (1994). Series of articles on toxicology: *J. Toxicol. Sci.* **17**, Suppl. 2, 1-205, (1992). Acute toxicity: K. Yamishita *et al.*, *ibid.* 1. LC/ESIMS determn of active metabolite in plasma: L. Ding *et al.*, *Biomed. Chromatogr.* **21**, 1297 (2007). Clinical evaluation in asthma: J. Tamaoki *et al.*, *Lancet* **356**, 273 (2000).

Crystals, mp 70-73°. LD_{50} in male, female mice, male, female rats (mg/kg): 81, 96, 96, 93 i.v.; in mice, rats (g/kg): >12.5, >10 orally (Yamashita).

THERAP CAT: Antiallergic; antiasthmatic.

9133. Suprasterol II. [562-71-0] (2α,7α,8R,19α,22E)-7,-19:8,19-Dicyclo-9,10-secoergosta-5(10),22-dien-2-ol. $C_{28}H_{44}O$;

mol wt 396.66. C 84.78%, H 11.18%, O 4.03%. Ultraviolet irradiation product of vitamin D_2. Isoln: Windaus *et al.*, *Ann.* **483**, 17 (1930). Structure and stereochemistry: Dauben, Baumann, *Tetrahedron Lett.* **1961**, 565. Crystal structure: C. P. Saunderson *et al.*, *ibid.* 573. Improved isoln and spectral data: T. Kobayashi *et al.*, *J. Nutr. Sci. Vitaminol.* **23**, 291 (1977).

Rough prisms from acetone or methanol, mp 110°. bp$_{0.005}$ 190°. $[\alpha]_D^{19}$ +62.9° (methanol); $[\alpha]_D^{22}$ +47.8° (methanol). Sol in all organic solvents.

9134. Suprofen. [40828-46-4] α-Methyl-4-(2-thienylcarbonyl)benzeneacetic acid; *p*-(2-thenoyl)hydratropic acid; sutoprofen; R-25061; Masterfen; Srendam; Sulprotin; Supranol; Suprocil; Suprol; Topalgic. $C_{14}H_{12}O_3S$; mol wt 260.31. C 64.60%, H 4.65%, O 18.44%, S 12.32%. Prostaglandin biosynthesis inhibitor. Prepn: P. A. Janssen *et al.*, *DE 2353357*; *eidem*, *US 4035376* (1974, 1977 both to Janssen); P. G. H. Van Daele *et al.*, *Arzneim.-Forsch.* **25**, 1495 (1975). Pharmacology: C. J. E. Niemegeers *et al.*, *ibid.* 1537. HPLC determn in plasma and urine: H. Muller *et al.*, *ibid.* **32**, 257 (1982). Series of articles on pharmacology, toxicology and clinical trials: *ibid.* **25**, 1501-1542 (1975); *ibid.* **33**, 1322-1338 (1983); on pharmacology, efficacy and safety: *Pharmacology* **27**, Suppl. 1, 1-96 (1983); on pharmacokinetics and clinical efficacy: *Arzneim.-Forsch.* **35**, 738-759 (1985); *ibid.* **36**, 941-971 (1986).

White to slightly yellow microcryst powder, mp 124.3°. pKa 3.91. uv max (0.01*N* HCl-90% 2-propanol): 266, 292 nm (ε 15700, 15600). Freely sol in methanol, ethanol, chloroform, acetone, polyethylene glycol, 1.0*N* NaOH; sol in ether; sparingly sol in water; slightly sol in 0.1*N* NaOH. Practically insol in *n*-hexane. LD$_{50}$ (7 days post-drug) in mice, rats, guinea pigs, dogs (mg/kg): 590, 353, 280, 160 orally (Niemegeers).

THERAP CAT: Anti-inflammatory.

9135. Suramin Sodium. [129-46-4] 8,8'-[Carbonylbis[imino-3,1-phenylenecarbonylimino(4-methyl-3,1-phenylene)carbonylimino]]bis-1,3,5-naphthalenetrisulfonic acid sodium salt (1:6); hexasodium *sym*-bis(*m*-aminobenzoyl-*m*-amino-*p*-methylbenzoyl-1-naphthylamino-4,6,8-trisulfonate) carbamide; Bayer 205; Fourneau 309; Antrypol; Germanin; Naganol. $C_{51}H_{34}N_6Na_6O_{23}S_6$; mol wt 1429.15. C 42.86%, H 2.40%, N 5.88%, Na 9.65%, O 25.75%, S 13.46%. Trypanocide discovered in 1917 by O. Dressel and R. Kothe: J. Dressel, *J. Chem. Educ.* **38**, 620 (1961). Prepn: E. Fourneau *et al.*, *Compt. Rend.* **178**, 675 (1924); J. Trefouel, E. Fourneau, **GB 224849** (1923); B. Heymann, *Angew. Chem.* **37**, 585 (1924). Pharmacology, toxicology, and efficacy in onchocerciasis and trypanosomiasis: F. Hawking, *Adv. Pharmacol. Chemother.* **15**, 289-322 (1978). Inhibition of reverse transcriptase: E. De Clercq, *Cancer Lett.* **8**, 9 (1979); of infectivity of HIV: H. Mitsuya *et al.*, *Science* **226**, 172 (1984). HPLC determn in plasma: R. W. Klecker, J. M. Collins, *J. Liq. Chromatogr.* **8**, 1685 (1985). Review of biological activities and clinical experience in infection and carcinoma: T. E. Voogd *et al.*, *Pharmacol. Rev.* **45**, 177-203 (1993); C. A. Stein, *Cancer Res.* **53**, 2239-2248 (1993).

White or slightly pink or cream-colored powder. Slightly bitter taste. Hygroscopic. Freely sol in water, in physiological saline; sparingly sol in 95% alcohol. Insol in benzene, ether, petr ether, chloroform. Aq solns are neutral to litmus. LD$_{50}$ in mice (mg/kg): ~620 i.v. (Hawking).

THERAP CAT: Anthelmintic (Nematodes); antiprotozoal (Trypanosoma).

THERAP CAT (VET): Antiprotozoal (Trypanosoma).

9136. Surinamine. [537-49-5] *N*-Methyl-L-tyrosine; andirine; angeline; geoffroyine; ratanhine. $C_{10}H_{13}NO_3$; mol wt 195.22. C 61.53%, H 6.71%, N 7.17%, O 24.59%. Isoln from bark of *Andira retusa* Kunth. (*Geoffroya retusa* Lam.), *Leguminosae*: Hiller-Bombein, *Arch. Pharm.* **230**, 513 (1892). Prepn of inactive compd: Kanevskaya, *J. Prakt. Chem.* **124**, 48 (1929). Prepn of L-form: Kanao, *J. Pharm. Soc. Jpn.* **66**, 4 (1946); Huguenin, Boissonnas, *Helv. Chim. Acta* **44**, 213 (1961). Prepn of D-form: Izumiya, Nagamatsu, *Bull. Chem. Soc. Jpn.* **25**, 265 (1952).

Crystals, mp 292-295°. $[\alpha]_D^{21}$ +19.6° (c = 3.8 in 10% HCl). Almost insol in water. Slightly sol in alcohol; sol in dil acids.

D-Form. Crystals, mp 273-274°. $[\alpha]_D^{15}$ -18.9° (c = 1.74 in 3*N* HCl).

DL-Form. Crystals from water.

9137. Survivin. Mammalian apoptosis inhibitor; member of the IAP (inhibitor of apoptosis protein) family. Selectively overexpressed in common human cancers and correlates with unfavorable prognosis. Also present in embryonic cells but not detectable in normal differentiated tissues. Cytoplasmic, 142 residue protein; mol wt ~16.5 kDa. Expressed during mitosis in a cell cycle-dependent manner and localized to the mitotic spindle. Identification in human cancers: G. Ambrosini *et al.*, *Nat. Med.* **3**, 917 (1997). Developmentally regulated expression in fetal tissue: C. Adida *et al.*, *Am. J. Pathol.* **152**, 43 (1998). Mechanism of action study: F. Li *et al.*, *Nature* **396**, 580 (1998). Crystal structure of murine: S. W. Muchmore *et al.*, *Mol. Cell* **6**, 173 (2000); of human: L. Chantalat *et al.*, *ibid.* 183; M. A. Verdecia *et al.*, *Nat. Struct. Biol.* **7**, 602 (2000). Determn in urine and clinical use in diagnosis of bladder cancer: S. D. Smith *et al.*, *J. Am. Med. Assoc.* **285**, 324 (2001).

9138. Sutherlandia. Cancer bush; kankerbos; insiswa; gansies; unwele. Woody, flowering shrub, *Sutherlandia frutescens* (L.) R. Br., *Fabaceae*, also known as *Lessertia frutescens*. Widely used in African traditional medicine as anticancer, anti-infective, antidiabetic, immunostimulant, anxiolytic, and tonic. *Habit.* Southern Africa. *Constit.* Free amino acids: asparagine, proline, arginine, canavanine, γ-aminobutyric acid; pinitol; flavonoids; cycloartane glycosides known as **sutherlandiosides**. Chromatographic analysis of constituents and antiproliferative activity: J. Tai *et al.*, *J. Ethnopharmacol.* **93**, 9 (2004). Isoln of sutherlandiosides: X. Fu *et al.*, *J. Nat. Prod.* **71**, 1749 (2008). Anti-HIV activity: S. M. Harnett *et al.*, *J. Ethnopharmacol.* **96**, 113 (2005). Clinical pharmacology and safety: Q. Johnson *et al.*, *PLoS Clin. Trials* **2**, e16 (2007). Review of taxonomy, constituents, and medicinal uses: B.-E. van Wyk, C. Albrecht, *J. Ethnopharmacol.* **119**, 620-629 (2008).

9139. Suxibuzone. [27470-51-5] Butanedioic acid 1-[(4-butyl-3,5-dioxo-1,2-diphenyl-4-pyrazolidinyl)methyl] ester; succinic acid monoester with 4-butyl-4-(hydroxymethyl)-1,2-diphenyl-3,5-pyrazolidinedione; 4-butyl-4-(hydroxymethyl)-1,2-diphenyl-3,5-pyrazolidinedione hydrogen succinate (ester); 1,2-diphenyl-4-*n*-butyl-4-hydroxymethyl-3,5-dioxopyrazolidine hemisuccinate; 4-hydroxymethylbutazolidine hemisuccinate; AE-17; Calibène; Danilon; Flogos; Solurol. $C_{24}H_{26}N_2O_6$; mol wt 438.48. C 65.74%, H 5.98%, N 6.39%, O 21.89%. Prodrug of phenylbutazone, *q.v.* Prepd (not claimed) and use as anti-inflammatory: A. Esteve, **DE 1936747**; *idem*, **US 3752894** (1970, 1973 both to Lab. Esteve). Pharmacology: A. Esteve *et al.*, *Quim. Ind. (Madrid)* **17**, 107 (1971). HPLC determn in plasma and urine: T. Marunaka *et al.*, *J. Pharm. Sci.* **69**, 1258 (1980). Metabolism in humans: Y. Yasuda *et al.*, *ibid.* **71**, 565 (1982). Comparative clinical study with phenylbutazone in rheumatoid arthritis: Y. Mizushima *et al.*, *Int. J. Tissue React.* **5**, 35 (1983).

White, bitter crystalline powder from alcohol, mp 126-127°. Sol in most organic solvents. Insol in water. LD_{50} orally in mice: 5.683 micromol/kg (Esteve, 1973).

THERAP CAT: Anti-inflammatory.

THERAP CAT (VET): Analgesic, antipyretic, anti-inflammatory.

9140. Swertiamarin. [17388-39-5] (4a*R*,5*R*,6*S*)-5-Ethenyl-6-(β-D-glucopyranosyloxy)-4,4a,5,6-tetrahydro-4a-hydroxy-1*H*,3*H*-pyrano[3,4-*c*]pyran-1-one; 4,4a,5,6-tetrahydro-4aα-hydroxy-1-oxo-5β-vinyl-1*H*,3*H*-pyrano[3,4-*c*]pyran-6-yl β-D-glucopyranoside. $C_{16}H_{22}O_{10}$; mol wt 374.34. C 51.34%, H 5.92%, O 42.74%. Bitter principle of *Swertia japonica* (Maxim.) Makino, Gentianaceae. Yields erythrocentaurin on hydrolysis with emulsin. Isoln: Kubota, Tomita, *Chem. Ind. (London)* **1958**, 229; Inouye *et al.*, *Chem. Pharm. Bull.* **18**, 1856 (1970). Occurrence in gentianaceous plants: Inouye, Nakamura, *J. Pharm. Soc. Jpn.* **91**, 755 (1971), *C.A.* **75**, 95431b (1971). Structure: Kubota, Tomita, *Tetrahedron Lett.* **1961**, 176; *Bull. Chem. Soc. Jpn.* **34**, 1345 (1961); Koch *et al.*, *Bull. Soc. Chim. Fr.* **1964**, 405. Stereochemistry: Inouye *et al.*, *Tetrahedron Lett.* **1968**, 4429. Biosynthesis studies: Inouye *et al.*, *Chem. Pharm. Bull.* **18**, 2043 (1970).

Plates from ethanol + chloroform + ether, mp 113-114°. $[\alpha]_D^{20}$ −127° (c = 1 in 96% ethanol). uv max (methanol): 238 nm (log ε 3.93).

Tetraacetate. $C_{24}H_{30}O_{14}$; mol wt 542.49. Prisms, mp 190-191°. $[\alpha]_D$ −100.3° (CHCl$_3$). uv max: 206, 234 nm (log ε 3.20, 4.00).

9141. Sydnones. [50927-09-8] A term originally applied to the compd obtained by action of acetic anhydride on *N*-nitrosophenylglycine: Eade, Earl, *J. Chem. Soc.* **1946**, 591; Earl, *Chem. Ind. (London)* **1953**, 746. Sydnones are 5-membered heterocyclic mesoionic compounds that are best represented by polar resonance structures: W. Baker, W. D. Ollis, *Q. Rev. Chem. Soc.* **11**, 15 (1957); Y. Noel, *Bull. Soc. Chim. Fr.* **1964**, 173. Photochromic properties: S. Nespurek, M. Sorm, *Collect. Czech. Chem. Commun.* **42**, 811 (1977). ^{14}N-NMR study: L. Stefaniak, *Tetrahedron* **33**, 2571 (1977). ^{15}N-NMR study: L. Stefaniak *et al.*, *Org. Magn. Reson.* **13**, 274 (1980). Regioselectivity study: A. Padwa *et al.*, *J. Org. Chem.*

47, 786 (1982). Review of chemistry: F. H. C. Stewart, *Chem. Rev.* **64**, 129-147 (1964).

9142. Synephrine. [94-07-5] 4-Hydroxy-α-[(methylamino)methyl]benzenemethanol; *p*-hydroxy-α-[(methylamino)methyl]benzyl alcohol; 1-(4-hydroxyphenyl)-2-methylaminoethanol; *p*-methylaminoethanolphenol; β-methylamino-α-(4-hydroxyphenyl)ethyl alcohol; methylaminomethyl 4-hydroxyphenyl carbinol; Analeptin; Ethaphene; Oxedrine; Parasympatol; Simpalon; Synephrin; Synthenate. $C_9H_{13}NO_2$; mol wt 167.21. C 64.65%, H 7.84%, N 8.38%, O 19.14%. Prepd by hydrogenating ω-methylamino-4-hydroxyacetophenone in water in the presence of Pt or Pd: **DE 566578** (1931 to Boehringer, Ing.), *Frdl.* **18**, 3025.

Crystals, mp 184-185°. Stable to air and light.

Hydrochloride. [5985-28-4] $C_9H_{13}NO_2$.HCl. Crystals, mp 151-152°. Freely sol in water.

Tartrate. [16589-24-5] Corvasymton; Simpadren; Sympathol. $2C_9H_{13}NO_2.C_4H_6O_6$; mol wt 484.50. Crystals, mp 188-190° (some dec). Freely sol in water; sol in alcohol.

Tartaric acid monoester. [6414-49-9] *p*-Methylaminoethanolphenol tartrate; Neupentedrin; Pentedrin. $C_{13}H_{17}NO_7$; mol wt 299.28.

THERAP CAT: Adrenergic; vasopressor.

9143. Synhexyl. [117-51-1] 3-Hexyl-7,8,9,10-tetrahydro-6,-6,9-trimethyl-6*H*-dibenzo[*b,d*]pyran-1-ol; 1-hydroxy-3-hexyl-6,6,9-trimethyl-7,8,9,10-tetrahydro-6*H*-dibenzo[*b,d*]pyran; parahexyl; pyrahexyl. $C_{22}H_{32}O_2$; mol wt 328.50. C 80.44%, H 9.82%, O 9.74%. Psychotomimetic synthetic analog of the tetrahydrocannabinols, *q.v.* Prepn: Adams *et al.*, *J. Am. Chem. Soc.* **63**, 1971 (1941); Adams, **US 2419935** (1947); Hughes *et al.*, **US 3576887** (1971 to Am. Home Products). Comparison of physiological effects with those of tetrahydrocannabinol: Hollister *et al.*, *Clin. Pharmacol. Ther.* **9**, 783 (1968). Physical constants: Farmilo *et al.*, *Bull. Narc.* **6**, 7 (1954). Toxicity data: Loewe, *J. Pharmacol. Exp. Ther.* **88**, 154 (1946).

Liquid, bp$_{1.0}$ 190-192°; n_D^{20} 1.5504. uv and ir spectra: *Bull. Narc.* **6**, 27 (1954); *ibid.* **7**, 42 (1955). LD_{50} in mice, dogs, rabbits (mg/kg): 170, 223, 143 i.v. (Loewe).

Note: This is a controlled substance (hallucinogen): **21 CFR**, 1308.11.

9144. Synsorbs. [83382-98-3] Synthetic mimics of toxin receptor binding sites consisting of a trisaccharide covalently coupled via an 8-methoxycarbonyloctyl linker to **Chromosorb®** P, a commercially available SiO$_2$-based column matrix. Preparative methods for trisaccharides and linkers: R. U. Lemieux *et al.*, *J. Am. Chem. Soc.* **97**, 4076 (1975); D. D. Cox *et al.*, *Carbohydr. Res.* **63**, 139 (1978); P. J. Garegg, S. Oscarson, *ibid.* **136**, 207 (1985).

Synsorb Pk. [192230-81-2] Synthetic analog of the *E. coli* shigatoxin (verotoxin) receptor. Trisaccharide component is αGal(1 → 4)βGal(1 → 4)βGlc. Characterization of specificity: G. D. Armstrong *et al.*, *J. Infect. Dis.* **164**, 1160 (1991). Binding affinity: T. Takeda *et al.*, *Microbiol. Immunol.* **43**, 331 (1999). Clinical pharmacokinetics: G. D. Armstrong *et al. J. Infect. Dis.* **171**, 1042 (1995). Review of clinical experience: *idem et al.*, in *Escherichia coli O157:H7 and Other Shiga Toxin-Producing E. coli Strains*, J. B. Kaper, A. D. O'Brien, Eds. (American Society for Microbiology, Washington D.C., 1998) pp 374-384.

Synsorb CD. Synsorb 90. Synthetic analog of the *Clostridium difficile* toxin A receptor. Trisaccharide component is αGal(1 → 3)βGal(1 → 4)βGlc. Characterization of specificity: L. D. Heerze *et al.*, *J. Infect. Dis.* **169**, 1291 (1994); I. Castagliuolo *et al.*, *Gastroenterology* **111**, 433 (1996).

THERAP CAT: Synsorb Pk in the treatment of verotoxigenic *E. coli* gastroenteritis; Synsorb CD in the treatment of *Clostridium difficile* associated disease.

9145. Syringaldehyde. [134-96-3] 4-Hydroxy-3,5-dimethoxybenzaldehyde; syringic aldehyde; 3,5-dimethoxy-4-hydroxybenzene carbonal; gallaldehyde 3,5-dimethyl ether. $C_9H_{10}O_4$; mol wt 182.18. C 59.34%, H 5.53%, O 35.13%. Widely distributed in plants. Hydrolysis product of the naturally occurring glycosyringic aldehyde: Körner, *Gazz. Chim. Ital.* **18**, 215 (1888). Prepn from heat-treated beechwood and lignin: Kratzl, Silbernagel, *C.A.* **50**, 6040 (1956). Synthesis from pyrogallol 1,3-dimethyl ether: Pearl, *J. Am. Chem. Soc.* **70**, 1746 (1948), *cf.* Graebe, Martz, *Ber.* **36**, 1032 (1903); Allen, Leubner, *Org. Synth.* **31**, 92 (1951); **coll. vol. IV**, 866 (1963).

Very pale yellow needles from petr ether, mp 113°. bp$_{14}$ 192-193°. uv max (dioxane): 305 nm. Very sparingly sol in water, petr ether. Sol in alcohol, ether, chloroform, hot benzene, glacial acetic acid. Forms yellow sodium and potassium salts.

9146. Syringin. [118-34-3] 4-[(1*E*)-3-Hydroxy-1-propen-1-yl]-2,6-dimethoxyphenyl-β-D-glucopyranoside; 4-(3-hydroxypropenyl)-2,6-dimethoxyphenyl-D-glucoside; syringoside; ligustrin; lilacin; methoxyconiferine. $C_{17}H_{24}O_9$; mol wt 372.37. C 54.83%, H 6.50%, O 38.67%. First isolated by Meillet in 1841 from bark of *Syringa vulgaris* L. (lilac): *Ann.* **40**, 319 (1841). Prepn: Pauly, Strassberger, *Ber.* **62**, 2277 (1929); Freudenberg, Schraube, *Ber.* **88**, 16 (1955). Isoln from lilac bark: Freudenberg *et al.*, *Ber.* **84**, 472 (1951); from various plants: Plouvier, *Compt. Rend.* **254**, 4196 (1962); from cambial sap of spruce: Freudenberg, Harkin, *Phytochemistry* **2**, 189 (1963).

Monohydrate. Crystals from water, mp 192°. $[\alpha]_D^{20}$ −8.2° (c = 2.43 in chloroform), −17.25° (water). Slightly sol in cold water; sol in hot water, alc. Practically insol in ether.

9147. Syrosingopine. [84-36-6] (3β,16β,17α,18β,20α)-18-[[4-[(Ethoxycarbonyl)oxy]-3,5-dimethoxybenzoyl]oxy]-11,17-dimethoxyyohimban-16-carboxylic acid methyl ester; methyl carbethoxysyringoyl reserpate; carbethoxysyringoyl methyl reserpate; methyl *O*-(*O′*-carbethoxysyringoyl)reserpate; syringopine; Su-3118; Isotense; Londomin; Raunova; Seniramin; Siringina. $C_{35}H_{42}N_2O_{11}$; mol wt 666.72. C 63.05%, H 6.35%, N 4.20%, O 26.40%. Prepn: Lucas *et al.*, *J. Am. Chem. Soc.* **81**, 1928 (1959); US 2813871 (1957 to Ciba).

Crystals from acetone, mp 175-179°.
THERAP CAT: Antihypertensive.

9148. Systemin. [137181-56-7] L-Alanyl-L-valyl-L-glutaminyl-L-seryl-L-lysyl-L-prolyl-L-prolyl-L-seryl-L-lysyl-L-argininyl-L-α-aspartyl-L-prolyl-L-prolyl-L-lysyl-L-methionyl-L-glutaminyl-L-threonyl-L-aspartic acid. First known polypeptide signal in plants. Systemically induces plant defenses in response to wounding of leaves by insects or pathogens. Proline-rich palindromic structure. Identification of activity: C. A. Ryan, *Plant Physiol.* **54**, 328 (1974). Isolation from tomato leaves and characterization: G. Pearce *et al.*, *Science* **253**, 895 (1991). Purification by capillary electrophoresis: P. Mucha *et al.*, *J. Chromatogr. A* **734**, 410 (1996). Proton NMR assignments: D. J. Russell *et al.*, *J. Protein Chem.* **11**, 265 (1992). Structure-activity: T. Meindl *et al.*, *Plant Cell* **10**, 1561 (1998). Tertiary structure and mechanism of action: T. Specht *et al.*, *Plants Proteins Eur. Crops* **1998**, 41. Review: C. A. Ryan, G. Pearce, *Annu. Rev. Cell Dev. Biol.* **14**, 1-17 (1998).

Ala–Val–Gln–Ser–Lys–Pro–Pro–Ser–Lys–Arg–Asp–Pro–Pro–Lys–Met–Gln–Thr–Asp

Insol in lipid solvents.
USE: Wound hormone in plants.

T

9149. 2,4,5-T. [93-76-5] 2-(2,4,5-Trichlorophenoxy)acetic acid; Esteron 245; Trioxone. $C_8H_5Cl_3O_3$; mol wt 255.48. C 37.61%, H 1.97%, Cl 41.63%, O 18.79%. Post-emergence herbicide. Prepd from 2,4,5-trichlorophenol: Pokorny, *J. Am. Chem. Soc.* **63**, 1768 (1941); from benzenehexachloride: Galat, *ibid.* **74**, 3890 (1952). Activity: C. L. Hamner, T. B. Tukey, *Science* **100**, 154 (1944). Contains trace levels of TCDD, *q.v.*, as a contaminant: J. Smith, *Science* **203**, 1090 (1979); *Chem. Eng. News* **59**, 6 (Jan. 5, 1981). Toxicity: V. A. Rowe, T. A. Hymas, *Am. J. Vet. Res.* **15**, 622 (1954). *See also* 2,4-D.

Crystals from benzene, mp 153°. d_{20}^{20} 1.80. Soly in water at 30°: 238 mg/kg. Sol in alcohol. Forms water-soluble sodium and alkanolamine salts. Commercial products are usually in the form of amines or esters, often in mixture with 2,4-D. LD_{50} orally in mice, rats: 389, 500 mg/kg (Rowe, Hymas).
Caution: Potential symptoms of overexposure in exptl animals are ataxia; skin irritation, acne-like rash; liver damage. *See NIOSH Pocket Guide to Chemical Hazards* (DHHS/NIOSH 97-140, 1997) p 292.
Note: In March 1985 the E.P.A. terminated all registrations for the use of this herbicide on rice fields, orchards, sugarcane, rangeland and other noncrop sites. This follows the 1970 action of the Department of Agriculture halting the use of the pesticide on all food crops except rice: *Chem. Eng. News* **63**, 6 (Mar. 25, 1985).
USE: Formerly as herbicide.

9150. Tabernanthine. [83-94-3] 13-Methoxyibogamine. $C_{20}H_{26}N_2O$; mol wt 310.44. C 77.38%, H 8.44%, N 9.02%, O 5.15%. Indole alkaloid isolated from root of *Tabernanthe iboga* Baill., *Apocynaceae*: Delourme-Houdé, *Ann. Pharm. Fr.* **4**, 30 (1946); Dickel *et al., J. Am. Chem. Soc.* **80**, 123 (1958). Also in *Tabernaemontana* and *Stemmadenia* spp.; usually found in ibogaine mother liquors: Walls *et al., Tetrahedron* **2**, 173 (1958). Isoln from genus *Conopharingia*, *Apocynaceae*: Renner, Prins, **US 3008954** (1961 to Geigy). Structure: Bartlett *et al., J. Am. Chem. Soc.* **80**, 126 (1958). Mass spectrum: Biemann, Friedmann-Spiteller, *ibid.* **83**, 4805 (1961). Derivs: Taylor, **US 2877229** (1959 to Ciba). Interaction with benzodiazepine receptors: J.-H. Trouvin *et al., Eur. J. Pharmacol.* **140**, 303 (1987).

Needles or shiny leaflets from ethanol, mp 213.5-215°. Sublimes at 160° (0.005 mm pressure). $[\alpha]_D^{20}$ −40° (acetone). pKa 6.04 in 80% methylcellosolve. uv max (ethanol): 228, 271, 299 nm (log ε 4.53, 3.64, 3.77). Sol in alcohol, benzene, ether, chloroform. Practically insol in water.
Hydrochloride. $C_{20}H_{26}N_2O\cdot HCl$. Crystals from water, dec 275-277°. $[\alpha]_D^{25}$ −66° (methanol, Dickel, *loc. cit.*); mp 210°, $[\alpha]_D^{20}$ −76.5° (methanol, Delourme-Houdé). Sol in water. More sol in chloroform than ibogaine hydrochloride.

9151. Tabun. [77-81-6] *N,N*-Dimethylphosphoramidocyanidic acid ethyl ester; ethyl *N*-dimethylphosphoramidocyanidate; dimethylamidoethoxyphosphoryl cyanide; GA. $C_5H_{11}N_2O_2P$; mol wt 162.13. C 37.04%, H 6.84%, N 17.28%, O 19.74%, P 19.10%. Nerve gas; potent cholinesterase inhibitor similar in structure and activity to sarin and soman, *q.q.v.* Prepd from dimethylamidophosphoryl dichloride and sodium cyanide in the presence of ethanol: Holmstedt, *Acta Physiol. Scand.* **25**, Suppl. 90, 26 (1951). Synthesis

of dimethylamidophosphoryl dichloride: Michaelis, *Ann.* **326**, 129 (1903). Alternate synthetic route: B. C. Saunders, *Some Aspects of the Chemistry and Toxic Action of Organic Compounds Containing Phosphorus and Fluorine* (Cambridge, 1957) p 91. Toxicity study: B. Holmstedt, *Pharmacol. Rev.* **11**, 567 (1959). Brief review: Schrader, *Die Entwicklung neuer insektizider Phosphorsäure-Ester* (Verlag Chemie, Weinheim, 1963) p 3; B. L. Harris in *Kirk-Othmer Encyclopedia of Chemical Technology* vol. 5 (Wiley-Interscience, New York, 4th ed., 1993) pp 795-816.

Liquid. Fruity odor reminiscent of bitter almonds. d 1.077. mp −50°. bp_{760} 240°; bp_{10} 120°; bp_9 100-108°. n_D^{20} 1.4250. IR absorption: *Acta Chem. Scand.* **5**, 1179 (1951). Readily sol in organic solvents. Miscible with water, but quickly hydrolyzed. Destroyed by bleaching powder (chlorinated lime), but gives rise to cyanogen chloride. *Extremely poisonous.* LD_{50} i.p. in mice: 0.6 mg/kg (Holmstedt).
Caution: Potential symptoms of overexposure by inhalation are constriction of pupils of the eye, difficulty breathing followed by bronchial constriction, convulsions, death. Also toxic due to absorption through skin and eyes. *See: Chem. Eng. News* **31**, 4676 (1953).
USE: Chemical warfare agent.

9152. Tacalcitol. [57333-96-7] (1*R*,3*S*,5*Z*)-4-Methylene-5-[(2*E*)-2-[(1*R*,3a*S*,7a*R*)-octahydro-1-[(1*R*,4*R*)-4-hydroxy-1,5-dimethylhexyl]-7a-methyl-4*H*-inden-4-ylidene]ethylidene]-1,3-cyclohexanediol; (1α,3β,5*Z*,7*E*,24*R*)-9,10-secocholesta-5,7,10(19)-triene-1,-3,24-triol; 1α,24(*R*)-dihydroxycholecalciferol; 1α,24*R*-dihydroxyvitamin D_3; TV-02; Bonealfa; Curatoderm. $C_{27}H_{44}O_3$; mol wt 416.65. C 77.83%, H 10.64%, O 11.52%. Bioactive, synthetic vitamin D_3 analog; exhibits antiproliferative effect on keratinocytes. Prepn: T. Takeshita *et al.*, **DE 2526981**; *eidem*, **US 4022891** (1976, 1977 both to Teijin); M. Morisaki *et al., J. Chem. Soc. Perkin Trans. 1* **1975**, 1421; K. Ochi *et al., ibid.* **1979**, 165. Pharmacology: T. Matsunaga *et al., J. Dermatol.* **17**, 135 (1990). Clinical evaluation in psoriasis: M. J. P. Gerritsen *et al., Br. J. Dermatol.* **131**, 57 (1994). *Review:* M. Nishimura *et al., Eur. J. Dermatol.* **3**, 255-261 (1993).

White solid. uv max (ethanol): 265 nm.
THERAP CAT: Antipsoriatic.

9153. Tachysterol. [115-61-7] (1*S*)-3-[(1*E*)-2-[(1*R*,3a*R*,-7a*R*)-2,3,3a,6,7,7a-Hexahydro-7a-methyl-1-[(1*R*,2*E*,4*R*)-1,4,5-trimethyl-2-hexen-1-yl]-1*H*-inden-4-yl]ethenyl]-4-methyl-3-cyclohexen-1-ol; (3β,6*E*,22*E*)-9,10-secoergosta-5(10),6,8,22-tetraen-3-ol. $C_{28}H_{44}O$; mol wt 396.66. C 84.78%, H 11.18%, O 4.03%. From ergosterol or lumisterol by ultraviolet irradiation: Windaus *et al., Ann.* **492**, 226 (1932); *Ann.* **499**, 188 (1932); Dimroth, *Ber.* **70**, 1631 (1937). From calciferol by adsorption on acid clay: Thibaudet, *Compt. Rend.* **220**, 751 (1945). From precalciferol: Velluz, Goffinet, **US 2847426** (1958 to UCLAF). Structure: Grundmann, *Z. Physiol. Chem.* **252**, 151 (1938); Thibaudet, *loc. cit.* Stereochemistry of the tachysterol system: Inhoffen, *Ber.* **88**, 1424 (1955); Verloop,

Rec. Trav. Chim. **76**, 689 (1957); Delaroff et al., Bull. Soc. Chim. Fr. **1963**, 1739.

Oil. $[\alpha]_D^{18}$ −70° (24.6 mg in 2 ml petr ether); $[\alpha]_{546}^{18}$ −86.3° (petr ether). uv max: 280 nm. Not pptd by digitonin. Insol in water. Sol in organic solvents, but not in methanol. Very easily oxidized by air.

4-Methyl-3,5-dinitrobenzoate. $C_{36}H_{48}N_2O_6$. Pale yellow crystals, mp 155°.

Dihydro derivative see Dihydrotachysterol.

9154. Tacrine. [321-64-2] 1,2,3,4-Tetrahydro-9-acridinamine; 9-amino-1,2,3,4-tetrahydroacridine; 5-amino-1,2,3,4-tetra-hydroacridine; 1,2,3,4-tetrahydro-5-aminoacridine; $C_{13}H_{14}N_2$; mol wt 198.27. C 78.75%, H 7.12%, N 14.13%. Centrally active anticholinesterase; cognition enhancer. Prepn: Braun et al., Ber. **64**, 227 (1931); Albert, Gledhill, J. Soc. Chem. Ind. **64**, 169T (1945); Petrow, J. Chem. Soc. **1947**, 634; J. A. Moore, L. D. Kornreich, Tetrahedron Lett. **4**, 1277 (1963). Pharmacology and toxicity: S. Gershon, F. H. Shaw, J. Pharm. Pharmacol. **10**, 638 (1958). Enzyme inhibiting activity: P. N. Kaul, ibid. **14**, 243 (1962). Neuropharmacology: B. Drukarch et al., Life Sci. **42**, 1011 (1988). Spectrofluorimetric determn in serum: I. Aparico et al., Analyst **123**, 1575 (1998). Clinical trial in severe Alzheimer's disease: W. K. Summers et al., N. Engl. J. Med. **315**, 1241 (1986). See also: ibid. **316**, 1603-1606. Clinical evaluation of reduced mortality in dementia: B. R. Ott, K. L. Lapane, J. Am. Geriatr. Soc. **50**, 35 (2002). Review of pharmacology and clinical uses: W. K. Summers et al., Clin. Toxicol. **16**, 269-281 (1980); of cholinesterase inhibition in Alzheimer's disease: N. Qizilbash et al., J. Am. Med. Assoc. **280**, 1777-1782 (1998).

Octahedra from very dil alc, mp 183-184°.

Hydrochloride. [1684-40-8] Cognex; THA. $C_{13}H_{14}N_2 \cdot HCl$; mol wt 234.73. Yellow needles from concd hydrochloric acid, mp 283-284°. Bitter taste. Sol in water. pH of 1.5% soln: 4.5-6. uv max (acetate buffer, pH 1-7): 242 nm. Emission max: 362 nm.

THERAP CAT: Nootropic. Antidote to curare. Respiratory stimulant.

9155. Tacrolimus. [104987-11-3] (3S,4R,5S,8R,9E,12S,-14S,15R,16S,18R,19R,26aS)-5,6,8,11,12,13,14,15,16,17,18,19,24,-25,26,26a-Hexadecahydro-5,19-dihydroxy-3-[(1E)-2-[(1R,3R,4R)-4-hydroxy-3-methoxycyclohexyl]-1-methylethenyl]-14,16-dimeth-oxy-4,10,12,18-tetramethyl-8-(2-propen-1-yl)-15,19-epoxy-3H-pyr-ido[2,1-c][1,4]oxaazacyclotricosine-1,7,20,21(4H,23H)-tetrone; 17-allyl-1,14-dihydroxy-12-[2-(4-hydroxy-3-methoxycyclohexyl)-1-methylvinyl]-23,25-dimethoxy-13,19,21,27-tetramethyl-11,28-di-oxa-4-azatricyclo[22.3.1.0⁴,⁹]octacos-18-ene-2,3,10,16-tetraone. $C_{44}H_{69}NO_{12}$; mol wt 804.03. C 65.73%, H 8.65%, N 1.74%, O 23.88%. Macrolide calcineurin inhibitor. Isoln from Streptomyces tsukubaensis no. 9993: M. Okuhara et al., **EP 184162** (1986 to Fujisawa); and characterization: T. Kino et al., J. Antibiot. **40**, 1249

(1987). Structure determn: H. Tanaka et al., J. Am. Chem. Soc. **109**, 5031 (1987). Total synthesis of (−)-form: T. K. Jones et al., J. Am. Chem. Soc. **111**, 1157 (1989). In vitro immunosuppressant activity in comparison with cyclosporin, q.v.: T. Kino et al., J. Antibiot. **40**, 1256 (1987). Toxicology: K. Ohara et al., Transplant. Proc. **22**, 83 (1990). Symposium on pharmacology and clinical trials: ibid. **23**, 2709-3376 (1991). Review of mechanism of action: G. Wiederrecht et al., Ann. N.Y. Acad. Sci. **696**, 9-19 (1993); of clinical trials in comparison with cyclosporin in renal transplantation: G. A. Knoll, R. C. Bell, Br. Med. J. **318**, 1104-1107 (1999). Review of use in dermatoses: A. K. Gupta et al., J. Eur. Acad. Dermatol. Venereol. **16**, 100-114 (2002); treatment of inflammatory bowel disease: D. K. L. Chow, R. W. L. Leong, Expert Opin. Drug Saf. **6**, 479-485 (2007).

Monohydrate. [109581-93-3] FK-506; FR-900506; Prograf; Protopic. $C_{44}H_{69}NO_{12} \cdot H_2O$. Colorless prisms from acetonitrile, mp 127-129°. $[\alpha]_D^{23}$ −84.4° (c = 1.02 in chloroform). Sol in methanol, ethanol, acetone, ethyl acetate, chloroform, diethyl ether; sparingly sol in hexane, petroleum ether. Insol in water. LD_{50} i.p. in mice: >200 mg/kg (Kino). LD_{50} in male, female rats (mg/kg): 57.0, 23.6 i.v.; 134, 194 orally (Ohara).

THERAP CAT: Immunosuppressant; dermatological in treatment of atopic eczema.

9156. Tacryl®. Acrylic fiber of a specific multichain type. The structure could be described as a spider molecule with up to six long straight linear legs which can orient independently and build up a fibrous structure. The molecular structure is built up by a controlled cross-linking process, so that Tacryl is like wool as it has a specific chain molecular interlinking which contributes to its mechanical and elastic properties. Synthesis of multichain polymers: Schaefgen, Flory, J. Am. Chem. Soc. **70**, 2709 (1948). Tacryl has a higher shear modulus and a higher strength than linear acrylics with the same degrees of orientation. It undergoes very slow hydrolysis under hot acid aq conditions, and is very resistant to dry and wet heat (neutral conditions): Sundén, Tappi **41**, 173A (1958).

9157. Tadalafil. [171596-29-5] (6R,12aR)-6-(1,3-Benzo-dioxol-5-yl)-2,3,6,7,12,12a-hexahydro-2-methylpyrazino[1',2':1,6]-pyrido[3,4-b]indole-1,4-dione; (6R,12aR)-2,3,6,7,12,12a-hexahy-dro-2-methyl-6-[3,4-(methylenedioxy)phenyl]pyrazino[1',2':1,6]-pyrido[3,4-b]indole-1,4-dione; GF-196960; IC-351; Adcirca; Cialis. $C_{22}H_{19}N_3O_4$; mol wt 389.41. C 67.86%, H 4.92%, N 10.79%, O 16.43%. Selective type 5 cGMP phosphodiesterase (PDE-5) inhibitor. Prepn: A. C. Daugan, **WO 9519978** (1995 to Glaxo); eidem, **US 5859006** (1999 to ICOS); and PDE-5 inhibition: idem et al., J. Med. Chem. **46**, 4525, 4533 (2003). HPLC determn in plasma: C.-L. Cheng, C.-H. Chou, J. Chromatogr. B **822**, 278 (2005). Clinical pharmacokinetics: S. T. Forgue et al., Br. J. Clin. Pharmacol. **61**, 280 (2005). Clinical comparison with sildenafil for erectile disfunction: I. Eardley et al., BJU Int. **96**, 1323 (2005). Clinical trial in pulmonary arterial hypertension: N. Galié et al., Circulation **119**, 2894 (2009). Review of cardiovascular effects: T. Reffelmann et al., Expert Opin. Drug Saf. **7**, 43-52 (2008).

White crystals from 2-propanol, mp 302-303°. $[\alpha]_D^{20}$ +71.0° (c = 1.00 in CHCl$_3$). Very slightly sol in ethanol. Practically insol in water.

THERAP CAT: In treatment of male erectile dysfunction. In treatment of pulmonary arterial hypertension.

9158. TAED. [10543-57-4] N,N'-1,2-Ethanediylbis[N-acetylacetamide]; N,N,N',N'-tetraacetylethylenediamine; TAED 4303. $C_{10}H_{16}N_2O_4$; mol wt 228.25. C 52.62%, H 7.07%, N 12.27%, O 28.04%. Bleach activator for peroxide-based detergents. Prepn: A. P. N. Franchimont, J. V. Dubsky, *Rec. Trav. Chim.* **30**, 184 (1911); R. P. Mariella, K. H. Brown, *J. Org. Chem.* **36**, 735 (1971). Chemical degradation studies: D. M. Davies, M. E. Deary, *J. Chem. Soc. Perkin Trans. 2* **1991**, 1549; N. Brand *et al.*, *Chemosphere* **34**, 2637 (1997). Determn by iodometric titration in household detergents: L. Baini *et al.*, *Riv. Ital. Sostanze Grasse* **69**, 615 (1992). Biocidal efficacy: V. B. Cloud, I. M. George, *Household Pers. Prod. Ind.* **34**, 82 (1997). Enhancement of whiteness of cellulosic blends: A. J. Mathews, S. J. Scarborough, *Proc. Beltwide Cotton Conf.* **1998**, 732; of wool/cotton blends: P. A. Duffield *et al.*, *ibid.* 816; of paper pulp: C. Leduc *et al.*, *Appita J.* **51**, 306 (1998).

Solid, mp 149-150°. Sol in water up to 10^{-3} mol/l. uv max (water): 213 nm (ε 18000 mol^{-1}cm^{-1}).

USE: Peroxide bleach activator for household detergents, paper pulp.

9159. Tafenoquine. [106635-80-7] N^4-[2,6-Dimethoxy-4-methyl-5-[3-(trifluoromethyl)phenoxy]-8-quinolinyl]-1,4-pentanediamine; 8-[(4-amino-1-methylbutyl)amino]-2,6-dimethoxy-4-methyl-5-[3-(trifluoromethyl)phenoxy]quinoline; WR-238605. $C_{24}H_{28}F_3N_3O_3$; mol wt 463.50. C 62.19%, H 6.09%, F 12.30%, N 9.07%, O 10.36%. Analog of primaquine, *q.v.* Prepn: P. Blumbergs, M. P. LaMontagne, **US 4617394** (1986 to U.S. Sec. Army); M. P. LaMontagne *et al.*, *J. Med. Chem.* **32**, 1728 (1989). HPLC determn in blood and plasma: D. A. Kocisko *et al.*, *Ther. Drug Monit.* **22**, 184 (2000). Metabolism: O. R. Idowu *et al.*, *Drug Metab. Dispos.* **23**, 1 (1995). Clinical pharmacokinetics: M. D. Edstein *et al.*, *Br. J. Pharmacol.* **52**, 663 (2001). Clinical evaluation in prevention of malaria relapse: D. S. Walsh *et al.*, *J. Infect. Dis.* **180**, 1282 (1999); in malaria prophylaxis: B. Lell *et al.*, *Lancet* **355**, 2041 (2000); B. R. Hale *et al.*, *Clin. Infect. Dis.* **36**, 541 (2003).

Succinate. [106635-81-8] Etaquine. $C_{24}H_{28}F_3N_3O_3 \cdot C_4H_6O_4$. Crystals from acetonitrile, mp 146-149°. LD$_{50}$ in male, female rats (mg/kg): 102, 71 i.p.; 429, 416 orally (LaMontagne).

THERAP CAT: Antimalarial.

9160. Tafluprost. [209860-87-7] (5Z)-7-[(1R,2R,3R,5S)-2-[(1E)-3,3-Difluoro-4-phenoxy-1-buten-1-yl]-3,5-dihydroxycyclopentyl]-5-heptenoic acid 1-methylethyl ester; 16-phenoxy-15-deoxy-15,15-difluoro-17,18,19,20-tetranorprostaglandin F$_{2\alpha}$ isopropyl ester; AFP-168; Taflotan; Tapros. $C_{25}H_{34}F_2O_5$; mol wt 452.54. C 66.35%, H 7.57%, F 8.40%, O 17.68%. Fluorinated prostaglandin derivative; ester prodrug that reduces intraocular pressure. Prepn: E. Shirasawa *et al.*, **EP 850926**; *eidem*, **US 5886035** (1998, 1999 both to Asahi Glass and Santen); Y. Matsumura *et al.*, *Tetrahedron Lett.* **45**, 1527 (2004). Pharmacology: Y. Takagi *et al.*, *Exp. Eye Res.* **78**, 767 (2004). Clinical pharmacokinetics and tolerability: A. Sutton *et al.*, *Int. J. Clin. Pharmacol. Ther.* **46**, 400 (2008).

Sol in methyl acetate, ethanol, DMSO, DMF. Sparingly sol in aqueous buffers.

THERAP CAT: Antiglaucoma.

9161. D-Tagatose. [87-81-0] D-*lyxo*-Hexulose; Naturlose. $C_6H_{12}O_6$; mol wt 180.16. C 40.00%, H 6.71%, O 53.28%. Uncommon, but naturally occurring ketohexose. Isoln from *Sterculia setigera* gum: E. L. Hirst *et al.*, *J. Chem. Soc.* **1949**, 3145. Prepn from D-galactose: C. A. L. De Bruyn, W. A. Van Ekenstein, *Rec. Trav. Chim.* **16**, 262 (1897); T. Reichstein, W. Bosshard, *Helv. Chim. Acta* **17**, 753 (1934); by biochemical oxidation of D-talitol: E. L. Totton, H. A. Lardy, *J. Am. Chem. Soc.* **71**, 3076 (1949); from lactose in heated milk: S. Adachi, *Nature* **181**, 840 (1958). Synthesis: M. L. Wolfrom, R. B. Bennett, *J. Org. Chem.* **30**, 1284 (1965); A. A. H. Al-Jobore *et al.*, *Carbohydr. Res.* **16**, 474 (1971). Crystal structure: S. Takagi, R. D. Rosenstein, *ibid.* **11**, 156 (1969). Manuf process from whey: J. R. Beadle *et al.*, **US 5002612** (1991 to Biospherics). Stability and sweetening properties in toothpaste: Y. Lu, *Int. J. Cosmet. Sci.* **23**, 175 (2001). Review of use as bulk sugar substitute: G. V. Levin *et al.*, *Am. J. Clin. Nutr.* **62**, Suppl., 1161S-1168S (1995). Series of articles on toxicology and gastrointestinal tolerance: *Regul. Toxicol. Pharmacol.* **29**, S1-S93 (1999).

Crystals from aq ethanol, mp 131-133°. $[\alpha]_D^{25}$ $-5°$ (c = 1 in water). Sucrose-like taste, approx 92% as sweet as sucrose.

L-Tagatose. [17598-82-2] Prepn: C. Glatthaar, T. Reichstein, *Helv. Chim. Acta* **20**, 1537 (1937). Stereoselective synthesis: T. Mukaiyama *et al.*, *Chem. Lett.* **1982**, 1169. Crystals, mp 134-135°. $[\alpha]_D^{16}$ +1° (c = 2 in water).

USE: Non-nutritive sweetener. Sweetening agent for pharmaceuticals and personal aid products.

9162. Taka-Diastase. [9001-19-8] α-Amylase (*Aspergillus oryzae*); Koji; Aspergillus diastase; Sanzyme. A purified multienzyme produced by the microorganism *Aspergillus oryzae* (Ahl.) Cohn, *Aspergillaceae*, grown on sterilized wheat bran or on rice hulls. Represents more than 30 different enzymatic functions. It is not only amylolytic but digests proteins and fats also.

Whitish-yellow, very hygroscopic powder. Converts 450 times its wt of starch into maltose. *Keep tightly closed.*

USE: Preparing the Japanese national drink "sake"; in converting maizes into sugar in manuf of whiskey.

THERAP CAT: Amylolytic.

9163. Talactoferrin. [308240-58-6] Lactoferrin (recombinant human LF00); [11-L-threonine,29-L-arginine]lactoferrin (human); rhLF; talactoferrin alfa; LF00. Recombinant human lactoferrin produced in *Aspergillus niger* var. *awamori*. Antimicrobial, immunomodulatory glycoprotein of 692 amino acids; mol wt ~80 kDa. Activates natural killer (NK) and lymphokine-activated killer cells; enhances polymorphonuclear cells and macrophage cytotoxicity. Prepn: D. R. Headon *et al.*, **WO 9013642** (1990 to Granada Biosciences); O. M. Conneely *et al.*, **US 5849881** (1998 to Baylor College of Med.); and antibacterial activity: P. P. Ward *et al.*, *Biotechnology* **13**, 498 (1995). Overview of development: J. H. Andersen, *Curr. Opin. Mol. Ther.* **6**, 344 (2004). Clinical evaluation of wound healing properties: T. E. Lyons *et al.*, *Am. J. Surg.* **193**, 49 (2007). Clinical efficacy in metastatic cancers: T. G. Hayes *et al.*, *Invest. New Drugs* **24**, 233 (2006); E. Jonasch *et al.*, *Cancer* **113**, 72 (2008). Review of clinical experience in non-small cell lung cancer: R. J. Kelly, G. Giaccone, *Expert Opin. Biol. Ther.* **10**, 1379-1386 (2010).

THERAP CAT: Immunomodulator.

9164. Talampanel. [161832-65-1] (8*R*)-7-Acetyl-5-(4-aminophenyl)-8,9-dihydro-8-methyl-7*H*-1,3-dioxolo[4,5-*h*][2,3]-benzodiazepine; *R*-(−)-1-(4-aminophenyl)-3-acetyl-4-methyl-7,8-methylenedioxy-3,4-dihydro-5*H*-2,3-benzodiazepine; LY-300164. C$_{19}$H$_{19}$N$_3$O$_3$; mol wt 337.38. C 67.64%, H 5.68%, N 12.46%, O 14.23%. Noncompetitive AMPA receptor antagonist. Prepn: F. Andrasi *et al.*, **EP 492485**; *eidem*, **US 5536832** (1992, 1996 both to Gyogyszerkutato Intezet KFT). Enantioselective synthesis: I. Ling *et al.*, *J. Chem. Soc. Perkin Trans. 1* **1995**, 1423. HPLC determn in plasma: J. A. Eckstein, S. P. Swanson, *J. Chromatogr. B* **668**, 153 (1995). Stereoselective electropharmacology and neuroprotective effects: D. Lodge *et al.*, *Neuropharmacology* **35**, 1681 (1996). Review of manufacturing process: B. A. Anderson *et al.* in *Process Chemistry in the Pharmaceutical Industry*, K. G. Gadamasetti, Ed. (Marcel Dekker, New York, 1999) pp 263-282. Review of clinical development: C. Q. Meng, *Curr. Opin. Cent. Peripher. Nerv. Syst. Invest. Drugs* **1**, 637-643 (1999); J. F. Howes, C. Bell, *Neurotherapeutics* **4**, 126-129 (2007).

Crystals, mp 169-172°. [α]$_D$ −321.34° (c = 1 in methanol).

THERAP CAT: Anticonvulsant.

9165. Talampicillin. [47747-56-8] (2*S*,5*R*,6*R*)-6-[[(2*R*)-2-Amino-2-phenylacetyl]amino]-3,3-dimethyl-7-oxo-4-thia-1-azabicyclo[3.2.0]heptane-2-carboxylic acid 1,3-dihydro-3-oxo-1-isobenzofuranyl ester; (2*S*,5*R*,6*R*)-6-[(*R*)-2-amino-2-phenylacetamido]-3,3-dimethyl-7-oxo-4-thia-1-azabicyclo[3.2.0]heptane-2-carboxylic acid ester with 3-hydroxyphthalide; 6-[D-(−)-α-aminophenylacetamido]penicillanic acid phthalide ester; ampicillin 1-oxo-1,3-dihydroisobenzofuran-3-yl ester; phthalidyl D-α-aminobenzylpenicillanate. C$_{24}$H$_{23}$N$_3$O$_6$S; mol wt 481.52. C 59.87%, H 4.81%, N 8.73%, O 19.94%, S 6.66%. Semi-synthetic antibiotic related to penicillin. Prepn: H. Ferres, M. P. Clayton, **DE 2228012**; *eidem*, **US 3860579** (1972, 1975 both to Beecham). *See also:* M. Murakami *et al.*, **US 3951954** (1976 to Yamanouchi). Pharmacology: Clayton *et al.*, *Antimicrob. Agents Chemother.* **5**, 670 (1974). HPLC determn in urine and comparison of pharmacokinetics with other aminopenicillins: T. Uno *et al.*, *Chem. Pharm. Bull.* **29**, 1957 (1981). Clinical comparison with ampicillin, *q.v.*: H. N. Williams *et al.*, *Br. J. Clin. Pract.* **35**, 147 (1981).

Hydrochloride. [39878-70-1] BRL-8988; Talat; Talpen; Yamacillin. C$_{24}$H$_{23}$N$_3$O$_6$S.HCl; mol wt 517.98. White powder, mp 154-157° (dec).

THERAP CAT: Antibacterial.

9166. Talaporfin. [110230-98-3] *N*-[2-[(7*S*,8*S*)-3-Carboxy-7-(2-carboxyethyl)-13-ethenyl-18-ethyl-7,8-dihydro-2,8,12,17-tetramethyl-21*H*,23*H*-porphin-5-yl]acetyl]-L-aspartic acid; (2*S*,3*S*)-18-carboxy-20-[*N*-(*S*)-1,2-dicarboxyethyl]carbamoylmethyl-13-ethyl-3,7,12,17-tetramethyl-8-vinylchlorin-2-propanoic acid; mono-L-aspartyl chlorin e$_6$; NPe6. C$_{38}$H$_{41}$N$_5$O$_9$; mol wt 711.77. C 64.12%, H 5.81%, N 9.84%, O 20.23%. Semisynthetic derivative of chlorin e$_6$, *q.v.* Photosensitizer activated at 664 nm by laser or light-emitting diode-based light infusion device. Causes irreversible tumor blood vessel closure. Prepn: J. C. Bommer, B. F. Burnham, **EP 168831**; *eidem*, **US 4675338** (1986, 1987 both to Nippon Petrochemicals). Photophysical properties: J. D. Spikes, J. C. Bommer, *J. Photochem. Photobiol. B* **17**, 135 (1993); L. Li *et al.*, *ibid.* **67**, 51 (2002). Chemical and NMR structural studies: S. Gomi *et al.*, *Heterocycles* **48**, 2231 (1998). Safety assessment in treatment of refractory solid tumors: R. A. Lustig *et al.*, *Cancer* **98**, 1767 (2003). Clinical evaluation in lung cancer: H. Kato *et al.*, *Lung Cancer* **42**, 103 (2003).

Dark blue-green powder. Sol in water. Absorption max (phosphate buffer, pH 7.4): 400, 654 nm (ε 180000, 40000). Absorption max (*p*-dioxane): 401.7, 663.5 nm (E$_{mM}$ 111, 38). Hygroscopic and light sensitive. *Store under vacuum; protect from light.*

Tetrasodium salt. [220201-34-3] Talaporfin sodium; LS-11; ME-2906; Laserphyrin. C$_{38}$H$_{37}$N$_5$Na$_4$O$_9$; mol wt 799.70.

THERAP CAT: Antineoplastic (photosensitizer).

9167. Talbutal. [115-44-6] 5-(1-Methylpropyl)-5-(2-propen-1-yl)-2,4,6(1*H*,3*H*,5*H*)-pyrimidinetrione; 5-allyl-5-*sec*-butylbarbituric acid; 5-allyl-5-(1-methylpropyl)barbituric acid; Lotusate. C$_{11}$H$_{16}$N$_2$O$_3$; mol wt 224.26. C 58.91%, H 7.19%, N 12.49%, O 21.40%. Prepn: Volwiler, *J. Am. Chem. Soc.* **47**, 2236 (1925). Acute toxicity: E. W. Schafer, *Toxicol. Appl. Pharmacol.* **21**, 315 (1972).

Crystals from water or dil alcohol, mp 108-110°. Slightly bitter taste. Practically insol in water and petr ether. Sol in alcohol, chloroform, ether, acetone, glacial acetic acid, also in solns of fixed alkali hydroxides. A satd aq soln is acid to litmus. LD_{50} orally in rats: 57.5 mg/kg (Schafer).

Note: This is a controlled substance (depressant): **21 CFR,** 1308.13.

THERAP CAT: Sedative, hypnotic.

9168. Talc. [14807-96-6] Talcum; French chalk. $H_2Mg_3O_{12}Si_4$; mol wt 379.26. H 0.53%, Mg 19.23%, O 50.62%, Si 29.62%. The lumps are also known as *soapstone* or *steatite*. Finely powdered native hydrous magnesium silicate. Comprehensive description: A. W. Newman *et al., Anal. Profiles Drug Subs. Excip.* **23,** 511-542 (1994).

White to grayish-white, very fine odorless, crystalline powder; unctuous, and adheres readily to the skin. Insol in water, cold acids or in alkalies. Sol in hot concentrated phosphoric acid.

Caution: Potential symptoms of overexposure to talc containing no asbestos and less than 1% quartz are fibriotic pneumoconiosis; irritation of eyes. Potential symptoms of overexposure to soapstone containing less than 1% quartz are pneumoconiosis; cough, dyspnea; digital clubbing; cyanosis; basal crackles, cor pulmonale. *See NIOSH Pocket Guide to Chemical Hazards* (DHHS/NIOSH 97-140, 1997) p 292, 280.

USE: Dusting powder, either alone or with starch or boric acid, for medicinal and toilet prepns; excipient and filler for pills, tablets and for dusting tablet molds; clarifying liquids by filtration. As pigment in paints, varnishes, rubber; filler for paper, rubber, soap; in fireproof and cold-water paints for wood, metal and stone; lubricating molds and machinery; glove and shoe powder; electric and heat insulator.

9169. Taliglucerase Alfa. [1005808-93-4] Glucosylceramidase (human); L-glutamyl-L-phenylalanyl-[495(497)-L-histidine (R → H)]human glucosylceramidase (β-glucocerebrosidase) peptide with L-aspartyl-L-leucyl-L-leucyl-L-valyl-L-aspartyl-L-threonyl-L-methionine glycosylated peptide 1-506; prGCD; Uplyso. Mannose-terminated, recombinant human glucocerebrosidase produced in plant cells. Glycoprotein containing 506 amino acids. Sequence is homologous with the endogenous human enzyme with the addition of 2 amino acids at the *N*-terminus derived from the linker used for the signal peptide and an additional 7 amino acids at the *C*-terminus derived from the vacuolar targeting signal. Glycosylated by the transformed plant cell with mannose-rich terminal glycans which enhances uptake by tissue macrophages where glucosylceramide accumulates in Gaucher's disease. Prepn: Y. Shaaltiel *et al.,* **US 080038232** (2008 to Protalix). Expression in carrot cell suspension culture, characterization, and enzymatic activity: Y. Shaaltiel *et al., Plant Biotechnol. J.* **5,** 579 (2007). Clinical pharmacokinetics and safety: D. Aviezer *et al., PLoS One* **4,** e4792 (2009).

THERAP CAT: Enzyme replacement therapy in treatment of Gaucher's disease.

9170. Talinolol. [57460-41-0] *N*-Cyclohexyl-*N′*-[4-[3-[(1,1-dimethylethyl)amino]-2-hydroxypropoxy]phenyl]urea; 1-(4-cyclohexylureidophenoxy)-2-hydroxy-3-*tert*-butylaminopropane; (±)-1-[*p*-[3-(*tert*-butylamino)-2-hydroxypropoxy]phenyl]-3-cyclohexylurea; 02-115; Cordanum. $C_{20}H_{33}N_3O_3$; mol wt 363.50. C 66.09%, H 9.15%, N 11.56%, O 13.20%. Selective β_1-adrenergic blocker structurally related to practolol, *q.v.* Prepn: R. Eckardt *et al.,* **DE 2153024** (1972 to Arzneimittelwerke VEB), *C.A.* **77,** 61639b (1972); *idem et al., Pharmazie* **30,** 633 (1975). Series of articles on pharmacology, toxicology: *ibid.* 638-683. Acute toxicity: K. Femmer *et al., ibid.* 642. Clinical pharmacokinetics and bioavailability: B. Trausch *et al., Biopharm. Drug Dispos.* **16,** 403 (1995). Review of clinical experience: I. Assmann, *Curr. Med. Res. Opin.* **13,** 325-342 (1995).

Cryst from isopropanol, mp 142-144°. LD_{50} in rats, mice (mg/kg): 1180, 593 orally; 54.3, 74.7 i.p.; 29.7, 25.0 i.v. (Femmer).

THERAP CAT: Antihypertensive; antiarrhythmic.

9171. Talipexole. [101626-70-4] 5,6,7,8-Tetrahydro-6-(2-propen-1-yl)-4*H*-thiazolo[4,5-*d*]azepin-2-amine; 6-allyl-2-amino-5,-6,7,8-tetrahydro-4*H*-thiazolo[4,5-*d*]azepine. $C_{10}H_{15}N_3S$; mol wt 209.31. C 57.38%, H 7.22%, N 20.08%, S 15.32%. α_2-Adrenoceptor and dopamine D_2-receptor agonist. Prepn: G. Griss *et al.,* **DE 2040510** (1972 to Thomae); *eidem,* **US 3804849** (1974 to Boehringer, Ing.). Dopaminergic effects: P. A. Johansen *et al., Life Sci.* **43,** 515 (1988); S. K. Kulkarni, K. Chopra, *Methods Find. Exp. Clin. Pharmacol.* **12,** 99 (1990). Clinical evaluation in Parkinson's disease: Y. Mizuno *et al., Drug Invest.* **5,** 186 (1993).

Dihydrochloride. [36085-73-1] B-HT-920; Domin. $C_{10}H_{15}N_3S.2HCl$; mol wt 282.23. mp 245° (dec).

THERAP CAT: Antiparkinsonian.

9172. Tall Oil. Liquid rosin; Acintol C; tallol; talleol. "Tall" is Swedish for "pine". A by-product of the wood pulp industry. Usually recovered from pine wood "black liquor" of the sulfate or kraft paper process. Contains rosin acids, oleic and linoleic acids. Long chain alcohols and small amounts of sterols, especially phytosterol, have also been found. Comprehensive collection of 1660 abstracts: J. Weiner, *Tall Oil* (The Institute of Paper Chemistry, Appleton, Wisconsin, 3rd ed., 1959) 450 pp; J. Weiner, J. Byrne, 1st supplement (1965).

Dark brown liquid. Acrid odor similar to that of burnt rosin. d 0.95 to 1.0. n_D^{20} ~1.5. Acid no. 170-180. Sapon no. 172-185. Iodine no. 120-188. Fatty acids 50-60%. Rosin acids 34-40%. Unsaponifiable matter 5-10%.

USE: Mfg soap pastes, flotation agents, greases, paint, alkyd resins, linoleum, soaps, fungicides, asphalt emulsions, rubber formulations, cutting oils, sulfonated oils. Review of possible uses: Cannon, *Chem. Eng.* **61,** 142 (June 1954).

9173. Tallow. In North America designates the fat from the fatty tissue of bovine cattle and sheep only. It may be offered separately as beef tallow and as sheep or mutton tallow. The term horse tallow is generally no longer admitted. *Oleo stock* is the highest grade of beef tallow. Contains (as glycerides): Oleic acid (37-43%), palmitic (24-32%), stearic (20-25%), myristic (3-6%), linoleic (2-3%). Minor constituents are cholesterol, arachidonic, elaidic, and vaccenic acids. Perhaps the most observed characteristic of tallow is its titer (solidif pt) which ranges from 40° to 46°.

9174. Tallow Alcohol. A name for commercial mixtures of *n*-octadecanol and *n*-hexadecanol.

Fatty crystalline mass, mp 46-47°.

USE: Defoaming agent, emollient, intermediate for surface active agents.

9175. Tallysomycin. [67995-68-0] Talisomycin; BU-2231. Antitumor antibiotic complex and third generation analog of bleomycins, *q.v.,* produced by *Streptoalloteichus hindustanus* E 465-94. Prodn, isoln, properties of the major components, tallysomycins A and B: H. Kawaguchi *et al.,* **BE 845513**; *eidem,* **US 4051237** (both 1977 to Bristol-Myers); *eidem, J. Antibiot.* **30,** 779 (1977). Structure of A and B, based on originally proposed bleomycin structure: M. Konishi *et al., ibid.* 789; *cf.* bleomycin for revised structure. Antitumor activity: H. Imanishi *et al., ibid.* **31,** 667 (1978). Radioimmunoassay: A. Broughton *et al., Cancer Treat. Rep.* **63,** 1829 (1979). Pharmacokinetics: J. E. Strong *et al., ibid.* 1821. ^{13}C-NMR spectra of tallysomycin and its zinc complex: F. T. Greenaway *et al., Org. Magn. Reson.* **13,** 270 (1980). Biosynthetic derivs: T. Miyaki *et al., J. Antibiot.* **34,** 658, 665 (1981). Relative pulmonary toxicity: A. Broughton *et al., Cancer Treat. Rep.* **64,** 659 (1980). *Review:* S. T. Crooke *et al., Recent Results Cancer Res.* **76,** 83-90 (1981).

Tallysomycin A. R =

Tallysomycin B. R =

Tallysomycin A. [65057-90-1] N^1-[4-Amino-6-[[3-[(4-amino-butyl)amino]propyl]amino]-6-oxohexyl]-13-[(4-amino-4,6-dideoxy-α-L-talopyranosyl)oxy]-19-demethyl-12-hydroxybleomycinamide; BU-2231A. $C_{68}H_{110}N_{22}O_{27}S_2$; mol wt 1731.88. The term "talisomycin" has also been used to refer to tallysomycin A. White amorphous solid. No definite mp; gradually dec >210°. Sol in water, methanol, DMF; slightly sol in ethanol. Practically insol in other organic solvents. $[\alpha]_D^{23}$ −21° (c = 0.5 in water). uv max (water): 290 nm ($E_{1cm}^{1\%}$ 67). LD$_{50}$ s.c. in mice: 28 mg/kg (**US 4051237**).

Tallysomycin B. [65057-91-2] N^1-[3-[(4-Aminobutyl)amino]propyl]-13-[(4-amino-4,6-dideoxy-α-L-talopyranosyl)oxy]-19-demethyl-12-hydroxybleomycinamide; talisomycin B; BU-2231B. $C_{62}H_{98}N_{20}O_{26}S_2$; mol wt 1603.70. Mp, soly similar to tallysomycin A. $[\alpha]_D^{23}$ −19° (c = 0.5 in water). uv max (water): 289.5 ($E_{1cm}^{1\%}$ 77). LD$_{50}$ s.c. in mice: >50 mg/kg (**US 4051237**).

9176. Talniflumate. [66898-62-2] 2-[[3-(Trifluoromethyl)-phenyl]amino]-3-pyridinecarboxylic acid 1,3-dihydro-3-oxo-1-isobenzofuranyl ester; phthalidyl 2-(3-trifluoromethylanilino)nicotinate; phthalidyl 2-(α,α,α-trifluoro-*m*-toluidino)nicotinate; BA-7602-06; Somalgen. $C_{21}H_{13}F_3N_2O_4$; mol wt 414.34. C 60.88%, H 3.16%, F 13.76%, N 6.76%, O 15.45%. Prodrug of niflumic acid, *q.v.* Prepn: **BE 858864** (1978 to Bago), *C.A.* **89**, 109104 (1978); S. Bago, **US 4168313** (1979). Synthesis and pharmacologic study: M. Los *et al.*, *Farmaco Ed. Sci.* **36**, 372 (1981). HPLC determn in plasma: D.-J. Jang *et al.*, *Biomed. Chromatogr.* **19**, 32 (2005). Review of clinical development as a mucoregulator: D. Knight, *Curr. Opin. Investig. Drugs* **5**, 557-562 (2004).

White or pale yellow crystalline powder, mp 165-166°. uv max (chloroform): 287, 357 nm (ε 25600, 7800). LD$_{50}$ orally in rats: 12000 mg/kg (Los).

THERAP CAT: Anti-inflammatory.

9177. Taltirelin. [103300-74-9] (4*S*)-Hexahydro-1-methyl-2,6-dioxo-4-primidinecarbonyl-L-histidyl-L-prolinamide; (*S*)-*N*-(1-methyl-4,5-dihydroorotyl)-L-histidyl-L-prolinamide. $C_{17}H_{23}N_7O_5$; mol wt 405.42. C 50.36%, H 5.72%, N 24.18%, O 19.73%. Analog of thyrotropin releasing hormone TRH, *q.v.*, with pronounced CNS activity. Prepn: H. Sugano *et al.*, *EP 168042*; *eidem*, **US 4665056** (1986, 1987 both to Tanabe Seiyaku): M. Suzuki *et al.*, *J. Med. Chem.* **33**, 2130 (1990). Radioimmunoassay determn in plasma and urine: S. Morikawa *et al.*, *J. Pharm. Biomed. Anal.* **16**, 1267 (1998). CNS pharmacology: M. Yamamura *et al.*, *Jpn. J. Pharmacol.* **53**, 451 (1990). Effect on motor neuron survival: Y. Iwasaki *et al.*,

Neurol. Res. **19**, 613 (1997). Review of pharmacology, toxicology, and clinical trials: K. Kinoshita *et al.*, *CNS Drug Rev.* **4**, 25-41 (1998).

Crystals from water as the hemiheptahydrate, mp 72-75°. $[\alpha]_D^{25}$ −13.6° (c = 1 in water).

Tetrahydrate. [201677-75-0] TA-0910; Ceredist. $C_{17}H_{21}N_7$-O_5.4H$_2$O; mol wt 475.46. White, odorless crystals, bp 72-75°. Sol in water, acetic acid, ethanol. Slightly sol in methanol, acetonitrile. $[\alpha]_D^{23}$ −13.6°. LD$_{50}$ in mice, rats (mg/kg): >5000, >5000 orally; in mice, male, female rats (mg/kg): >2000, 799, 946 i.v. (Kinoshita).

THERAP CAT: In treatment of spinocerebellar degeneration.

9178. Tamarind. Partially dried ripe fruit of *Tamarindus indica* L., *Leguminosae*, preserved in sugar or syrup. *Habit*. East Indies, India, Africa; naturalized in West Indies. *Constit*. The pulp contains about 10% tartaric acid, also some citric and malic acids; 25-40% invert sugar, pectin. *Review*: Rao, Srivastava, in *Industrial Gums*, R. L. Whistler, Ed. (Academic Press, New York, 2nd ed., 1973) pp 369-411.

USE: The pulp as souring agent in Indian curries. The seed kernel powder with water as sizing agent.

9179. Tamibarotene. [94497-51-5] 4-[[(5,6,7,8-Tetrahydro-5,5,8,8-tetramethyl-2-naphthalenyl)amino]carbonyl]benzoic acid; *p*-(5,6,7,8-tetrahydro-5,5,8,8-tetramethyl-2-naphthylcarbamoyl)benzoic acid; AM-80; Tamibaro. $C_{22}H_{25}NO_3$; mol wt 351.45. C 75.19%, H 7.17%, N 3.99%, O 13.66%. Synthetic retinoic acid receptor-α/β-selective retinoid. Prepn: K. Shudo, **EP 170105** (1986 to Sumitomo; Yoshitomi); *idem*, **US 4703110** (1987); H. Kagechika *et al.*, *J. Med. Chem.* **31**, 2182 (1988). HPLC determn in plasma: M. Itoh, G. Kominami, *J. Immunoassay Immunochem.* **22**, 213 (2001). Receptor binding studies in human leukemia cells: Y. Hashimoto *et al.*, *Jpn. J. Cancer Res.* **79**, 473 (1988). Series of articles on pharmacokinetics and reproductive and developmental toxicity: K. Mizojiri *et al.*, *Arzneim.-Forsch.* **47**, 59, 195, 201, 259, 270 (1997). Inhibition of myeloma cell-induced angiogenesis: T. Sanda *et al.*, *Leukemia* **19**, 901 (2005). Clinical evaluation in acute promyelocytic leukemia: K. Shinjo *et al.*, *Int. J. Hematol.* **72**, 470 (2000). Review of development and clinical experience: I. Miwako, H. Kagechika, *Drugs of Today* **43**, 563-568 (2007).

Colorless prisms from ethyl acetate-*n*-hexane, mp 231-232°.

THERAP CAT: Antineoplastic.

9180. Tamoxifen. [10540-29-1] 2-[4-[(1Z)-1,2-Diphenyl-1-buten-1-yl]phenoxy]-*N*,*N*-dimethylethanamine; 1-*p*-β-dimethylaminoethoxyphenyl-*trans*-1,2-diphenylbut-1-ene. $C_{26}H_{29}NO$; mol wt 371.52. C 84.06%, H 7.87%, N 3.77%, O 4.31%. Nonsteroidal estrogen antagonist. Prepn: **BE 637389** (1964 to ICI). Identification and separation of isomers: G. R. Bedford, D. N. Richardson, *Nature* **212**, 733 (1966); **BE 678807**; M. J. K. Harper *et al.*, **US 4536516** (1966, 1985 both to ICI). Stereospecific synthesis: R. B. Miller, M. I. Al-Hassan, *J. Org. Chem.* **50**, 2121 (1985). Review of chemistry and pharmacology: B. J. A. Furr, V. C. Jordan, *Pharmacol. Ther.* **25**, 127-205 (1984). Reviews of clinical experience in treatment and prevention of breast cancer: I. A. Jaiyesimi *et al.*, *J.*

Clin. Oncol. **13**, 513-529 (1995); C. K. Osborne, *N. Engl. J. Med.* **339**, 1609-1618 (1998).

Crystals from petr ether, mp 96-98°.

Citrate. [54965-24-1] ICI-46474; Kessar; Nolvadex; Tamofène; Zemide; Zitazonium. $C_{26}H_{29}NO.C_6H_8O_7$; mol wt 563.65. Fine, white, odorless crystalline powder, mp 140-142°. Sol in methanol; very slightly sol in water, ethanol, acetone, chloroform. Hygroscopic at high relative humidities. Sensitive to uv light. LD_{50} in mice, rats (mg/kg): 200, 600 i.p.; 62.5, 62.5 i.v.; 3000-6000, 1200-2500 orally (Furr, Jordan).

(*E*)-Form. [13002-65-8] mp 72-74° from methanol.

(*E*)-Form citrate. ICI-47699. mp 126-128°.

Caution: Tamoxifen is listed as a known human carcinogen: *Report on Carcinogens, Twelfth Edition* (PB2011-111646, 2011) p 393.

THERAP CAT: Antineoplastic (hormonal).

9181. Tamsulosin. [106133-20-4] 5-[(2*R*)-2-[[2-(2-Ethoxyphenoxy)ethyl]amino]propyl]-2-methoxybenzenesulfonamide; amsulosin. $C_{20}H_{28}N_2O_5S$; mol wt 408.51. C 58.80%, H 6.91%, N 6.86%, O 19.58%, S 7.85%. Specific α_1-adrenoceptor antagonist. Prepn: K. Imai *et al.*, **EP 34432**; *eidem*, **US 4703063** (1981, 1987 both to Yamanouchi). Comparative pharmacology of enantiomers and racemate: K. Honda *et al.*, *Arch. Pharmacol.* **336**, 295 (1987). HPLC determn in plasma: Y. Soeishi *et al.*, *J. Chromatogr.* **533**, 291 (1990). Clinical trials in benign prostatic hypertrophy: K. Kawabe *et al.*, *J. Urol.* **144**, 908 (1990); P. Abrams *et al.*, *Br. J. Urol.* **76**, 325 (1995). Review of pharmacology and clinical experience in urological disorders: M. C. Michel, J. J. de la Rosette, *Expert Opin. Pharmacother.* **5**, 151-160 (2004).

Hydrochloride. [106463-17-6] LY-253351; YM-12617-1; YM-617; Flomax; Flomaxtra; Harnal; Omnic; Pradif. $C_{20}H_{28}N_2O_5S.HCl$; mol wt 444.97. White crystals. mp 228-230° (dec). $[\alpha]_D^{24}$ $-4.0°$ (c = 0.35 in methanol). Freely sol in formic acid; sparingly sol methanol; slightly sol in in water, glacial acetic acid, dehydrated alc. Practically insol in ether.

THERAP CAT: In treatment of benign prostatic hypertrophy.

9182. Tanacetin. [1401-54-3] (3a*S*,5a*S*,6*R*,9a*R*,9b*S*)-Decahydro-6,9a-dihydroxy-5a-methyl-3,9-bis(methylene)naphtho[1,2-*b*]furan-2(3*H*)-one; 1β,5α-dihydroxy-6β,7α*H*-selina-4(15),11(13)-dien-6,12-olide. $C_{15}H_{20}O_4$; mol wt 264.32. C 68.16%, H 7.63%, O 24.21%. Isoln from seed, herb, and flowers of *Tanacetum vulgare* L., *Compositae*: Homolle, *J. Pharm. Chim.* **7**, 57 (1845); Jaretzky, Kühne, *Arch. Pharm.* **271**, 353 (1933); Suchy, *Collect. Czech. Chem. Commun.* **27**, 1058 (1962). Structure and absolute config: Samek *et al.*, *ibid.* **38**, 1971 (1973).

Crystals, mp 205°. $[\alpha]_D^{22}$ +179.5° (c = 2.3 in ethanol).

9183. Tandospirone. [87760-53-0] *rel*-(3a*R*,4*S*,7*R*,7a*S*)-Hexahydro-2-[4-[4-[4-(2-pyrimidinyl)-1-piperazinyl]butyl]-4,7-methano-1*H*-isoindole-1,3(2*H*)-dione; (1*R**,2*S**,3*R**,4*S**)-*N*-[4-[4-(2-pyrimidinyl)-1-piperazinyl]butyl]-2,3-bicyclo[2.2.1]heptanedicarboximide. $C_{21}H_{29}N_5O_2$; mol wt 383.50. C 65.77%, H 7.62%, N 18.26%, O 8.34%. Serotonin (5-HT$_{1A}$) receptor agonist. Prepn: K. Ishizumi *et al.*, **EP 82402**; *eidem*, **US 4507303** (1983, 1985 both to Sumitomo); *idem et al.*, *Chem. Pharm. Bull.* **39**, 2288 (1991). Behavioral pharmacology: C. A. Sannerud *et al.*, *Drug Alcohol Depend.* **32**, 195 (1993). Clinical efficacy in treatment of bulimia: H. Tamai *et al.*, *Int. J. Obes.* **14**, 289 (1990). Clinical evaluation of potential adverse effects: M. Suzuki *et al.*, *Jpn. J. Psychopharmacol.* **13**, 213 (1993); of abuse liability: S. M. Evans *et al.*, *J. Pharmacol. Exp. Ther.* **271**, 683 (1994). Review of pharmacology: P. A. Seymour *et al.*, *Prog. Clin. Biol. Res.* **361**, 453-460 (1990).

Relative stereochemistry

Crystals from toluene/*n*-hexane, mp 112-113.5°.

Citrate. [112457-95-1] SM-3997; Sediel. $C_{21}H_{29}N_5O_2.C_6H_8O_7$; mol wt 575.62. mp 169.5-170°.

Hydrochloride. $C_{21}H_{29}N_5O_2.HCl$. Crystals from isopropanol, mp 227-229°.

THERAP CAT: Anxiolytic; antidepressant.

9184. Tannic Acid. [1401-55-4] Tannins; gallotannin; gallotannic acid. Incorrectly *"digallic acid"*. Tannic acid of commerce usually contains about 10% H_2O. Occurs in the bark and fruit of many plants, notably in the bark of the oak species, in sumac and myrobalan. It is produced from Turkish or Chinese nutgall, the former contg 50-60%, the latter about 70%. The chemistry of the tannins is most complex and non-uniform. Tannins may be divided into 2 groups: *(a)* derivatives of flavanols, so-called condensed tannins and *(b)* hydrolyzable tannins (the more important group) which are esters of a sugar, usually glucose, with one or more trihydroxybenzenecarboxylic acids. The structure given here is that of a tannin named *corilagin*: Schmidt *et al.*, *Ann.* **587**, 67 (1954). The empirical formula of corilagin is $C_{27}H_{22}O_{18}$. For the commercial tannic acid, whose specifications follow, the empirical formula is usually given as $C_{76}H_{52}O_{46}$. Toxicity study: Robinson, Graessle, *J. Pharmacol. Exp. Ther.* **77**, 63 (1943). Comprehensive reviews: M. Nierenstein, *The Natural Organic Tannins* (London, 1934); O. Th. Schmidt, "Gallotannine" in *Fortschr. Chem. Org. Naturst.* **13**, 70-136 (1956); *Symposium on the Chemistry of Vegetable Tannins* (Soc. Leather Trades Chemists, Croydon 1956).

Corilagin

Yellowish-white to light brown, amorphous, bulky powder or flakes, or spongy masses; faint characteristic odor; astringent taste. Gradually darkens on exposure to air and light; at 210-215° dec mostly into pyrogallol and CO_2. Gives insol ppts with albumin, starch, gelatin, most alkaloidal and metallic salts; produces a bluish-black color or precip with ferric salts. One gram dissolves in 0.35 ml water, 1 ml warm glycerol; very sol in alc, acetone; freely sol in diluted alc; slightly sol in dehydrated alc. Practically insol in benzene, chloroform, ether, petr ether, carbon disulfide, carbon tetrachloride, solvent hexane. *Keep well closed and protected from light.* LD_{100} orally in mice: 6.0 g/kg (Robinson, Graessle).

USE: Mordant in dyeing; manuf ink; sizing paper and silk; printing fabrics; with gelatin and albumin for manuf of imitation horn and tortoise shell; tanning; clarifying beer or wine; in photography; as coagulant in rubber manuf; manuf gallic acid and pyrogallol; as reagent in analytical chemistry for pH control.

THERAP CAT: Astringent.

THERAP CAT (VET): Astringent, hemostatic, in solutions for burns. Has been used internally as an astringent and as a heavy metal antidote.

9185. Tantalum. [7440-25-7] Ta; at. wt 180.94788; at. no. 73; valence 5, also 4, 3, 2. Group VB (5). Two naturally occurring isotopes: 181 (99.9877%); 180 (0.0123%), $T_{1/2} > 10^{12}$ years; artificial radioactive isotopes: 172-179; 182-186. Occurs almost invariably with niobium; less abundant than niobium. Found in the minerals *columbite* [(Fe,Mn)(Nb,Ta)$_2$O$_6$], *tantalite* [(Fe,Mn)(Ta, Nb)$_2$O$_6$] and *microlite* [(Na,Ca)$_2$Ta$_2$O$_6$(O,OH,F)]. Discovered by Ekeberg in 1802; first obtained pure by Bolton: *Z. Elektrochem.* **11**, 45 (1905). Prepn: Schoeller, Powell, *J. Chem. Soc.* **119**, 1927 (1921). Reviews of tantalum and its compounds: G. L. Miller, *Tantalum and Niobium* (Academic Press, New York, 1959) 770 pp; Brown, "The Chemistry of Niobium and Tantalum" in *Comprehensive Inorganic Chemistry* **vol. 3**, J. C. Bailar, Jr. *et al.*, Eds. (Pergamon Press, Oxford, 1973) pp 553-622.

Gray, very hard, malleable, ductile metal; can readily be drawn in fine wires. mp 2996°. bp 5429°. d 16.69. Spec heat (0°): 0.036 cal/g/°C. Electrical resistivity (18°): 12.4 μohm-cm. Insol in water. Very resistant to chemical attack; not attacked by acids other than hydrofluoric; not attacked by aq alkalies; slowly attacked by fused alkalies. Reacts with fluorine, chlorine, and oxygen only on heating. At high temps absorbs several hundred times its volume of hydrogen; combines with nitrogen, with carbon.

Caution: Potential symptoms of overexposure to metal and oxide dust are irritation of eyes and skin. *See NIOSH Pocket Guide to Chemical Hazards* (DHHS/NIOSH 97-140, 1997) p 294.

USE: In pen points; analytical weights; apparatus and instruments for chemical, surgical, and dental use instead of platinum, in tantalum capacitors (a type of electrolytic condenser, trademarked "Tantalytic").

9186. Tantalum Pentachloride. [7721-01-9] Tantalum chloride (TaCl$_5$). Cl$_5$Ta; mol wt 358.20. Cl 49.48%, Ta 50.52%. TaCl$_5$. Prepn: Rolsten, *J. Am. Chem. Soc.* **80**, 2952 (1958). Review of tantalum halides: Fairbrother in *Halogen Chemistry* **vol. 3**, V. Gutmann, Ed. (Academic Press, New York, 1967) pp 123-178. Acute toxicity: K. W. Cochran *et al.*, *Arch. Ind. Hyg. Occup. Med.* **1**, 637 (1950).

White or light yellow, cryst powder; monoclinic; dec in moist air. d 3.68; mp 216.5-220°. Begins to volatilize at 144°, bp 239.3°. Dec by water; sol in abs alcohol. LD_{50} in rats (mg/kg): 75 i.p.; 1900 orally (Cochran).

9187. Tantalum Pentafluoride. [7783-71-3] Tantalum fluoride (TaF$_5$). F$_5$Ta; mol wt 275.94. F 34.42%, Ta 65.58%. TaF$_5$. Prepd from tantalum pentachloride by the halide exchange method according to the equation TaCl$_5$ +5HF → TaF$_5$ + 5HCl: Ruff, Zedner, *Ber.* **42**, 492 (1909); Ruff, Schiller, *Z. Anorg. Allg. Chem.* **72**, 329 (1911); Kwasnik in *Handbook of Preparative Inorganic Chemistry* **vol. 1**, G. Brauer, Ed. (Academic Press, New York, 2nd ed, 1963) pp 255-256. Prepn from the elements: Fairbrother, Frith, *J. Chem. Soc.* **1951**, 3051. Review of transition metal pentafluorides: Peacock, *Adv. Fluorine Chem.* **7**, 113-145 (1973).

Deliquescent, strongly refractive prisms. d^{20} 4.74. mp 96.8°. Also reported as 95.1°: Fairbrother, Frith, *loc. cit.* bp 229.5°. Sol in water and ether with formation of oxyfluoro complexes. Also sol in concd nitric acid, more sol in fuming nitric acid. Sparingly sol in hot carbon disulfide and hot carbon tetrachloride. Etches glass slowly.

USE: Friedel-Crafts catalyst.

9188. Tantalum Pentoxide. [1314-61-0] Tantalum oxide (Ta$_2$O$_5$); tantalic acid anhydride. O$_5$Ta$_2$; mol wt 441.89. O 18.10%, Ta 81.90%. Ta$_2$O$_5$. Acute toxicity: K. W. Cochran *et al.*, *Arch. Ind. Hyg. Occup. Med.* **1**, 637 (1950).

White, microcrystalline, infusible powder. Insol in water, alcohol, mineral acids. Sol in HF. Dec by fusing with KHSO$_4$ or KOH,

forming potassium tantalate with the latter. LD_{50} orally in rats: 8000 mg/kg (Cochran).

9189. Tapentadol. [175591-23-8] 3-[(1R,2R)-3-(Dimethylamino)-1-ethyl-2-methylpropyl]phenol; CG-5503. C$_{14}$H$_{23}$NO; mol wt 221.34. C 75.97%, H 10.47%, N 6.33%, O 7.23%. Centrally acting analgesic; combined μ-opioid agonist and noradrenaline reuptake inhibitor. Prepn: H. Buschmann, **EP 693475**; *eidem*, **US 6248737**; **US RE39593** (1996, 2001, 2007 all to Grünenthal). Pharmacology: T. M. Tzschentke *et al.*, *J. Pharmacol. Exp. Ther.* **323**, 265 (2007). Absorption, metabolism and excretion study: R. Terlinden *et al*, *Eur. J. Drug Metab. Pharmacokinet.* **32**, 163 (2007). Clinical trial in post-surgical dental pain: R. Kleinert *et al.*, *Anesth. Analg.* **107**, 2048 (2008); in uncontrolled osteoarthritis: C. Hartrick *et al.*, *Clin. Ther.* **31**, 260 (2009).

Hydrochloride. [175591-09-0] Nucynta; Palexia. C$_{14}$H$_{23}$NO.-HCl; mol wt 257.80. mp 168-170°. $[\alpha]_D$ $-27.5°$ (c = 0.97 in methanol). Log P (*n*-octanol/water): 2.87. pKa 9.34; 10.45. Sol in acetone, acetonitrile, isopropanol.

Note: This is a controlled substance (opiate): **21 CFR, 1308.**12.

THERAP CAT: Analgesic.

9190. Taprostene. [108945-35-3] 3-[(Z)-[(3aR,4R,5R,6aS)-4-[(1E,3S)-3-Cyclohexyl-3-hydroxy-1-propen-1-yl]hexahydro-5-hydroxy-2H-cyclopenta[b]furan-2-ylidene]methyl]benzoic acid; α-[(2Z,3aR,4R,5R,6aS)-4-[(1E,3S)-3-cyclohexyl-3-hydroxypropenyl]-hexahydro-5-hydroxy-2H-cyclopenta[b]furan-2-ylidene]-m-toluic acid; [(5Z,13E,9α,11α,15S)-2,3,4-trinor-1,5-inter-m-phenylene-6,9-epoxy-11,15-dihydroxy-15-cyclohexyl-16,17,18,19,20-pentanor]-prosta-5,13-dienoic acid. C$_{24}$H$_{30}$O$_5$; mol wt 398.50. C 72.34%, H 7.59%, O 20.07%. Prostacyclin analog; platelet aggregation inhibitor. Prepn: U. Seipp *et al.*, **EP 45842**; *eidem*, **US 4372971** (1982, 1983 both to Grünenthal); L. Flohé *et al.*, *Arzneim.-Forsch.* **33**, 1240 (1983). As adjuvant to thrombolytic therapy in acute myocardial infarction: F. W. Bär *et al.*, *Eur. Heart J.* **14**, 1118 (1993). Review of pharmacology and clinical experience: J. Schneider *et al.*, *Cardiovasc. Drug Rev.* **11**, 479-500 (1993).

Sodium salt. [87440-45-7] CG-4203; Rheocyclan. C$_{24}$H$_{29}$-NaO$_5$; mol wt 420.48. $[\alpha]_D^{22}$ +249° (c = 0.68 in methanol). LD_{50} in mice, rats (mg/kg): 164, 20 i.v. (Schneider).

THERAP CAT: Antithrombotic.

9191. TAPS. [29915-38-6] 3-[[2-Hydroxy-1,1-bis(hydroxymethyl)ethyl]amino]-1-propanesulfonic acid; N-tris[(hydroxymethyl)methyl]-3-aminopropanesulfonic acid. C$_7$H$_{17}$NO$_6$S; mol wt 243.27. C 34.56%, H 7.04%, N 5.76%, O 39.46%, S 13.18%. Zwitterionic N-substituted sulfonic acid in the style of the "Good" buffers; active in the pH range 6-8.5. Prepn: I. Zeid, I. Ismail, *Ann.* **1974**, 667. Dissociation: A. M. El-Nady, H. A. Azab, *Acta Chim. Hung.* **130**, 665 (1993). Use as eluent: R. H. P. Reid, *J. Chromatogr. A* **684**, 221 (1994). Effect of buffer concentration on pH-activity relationship: C. G. Bevans, A. L. Harris, *J. Biol. Chem.* **274**, 3711 (1999).

Crystals, mp 194°. pKa: 8.55; pKa (37°): 8.1; pKa$_2$ (25°): 8.28.

TAPSO. [68399-81-5] 2-Hydroxy-3-[[2-hydroxy-1,1-bis(hydroxymethyl)ethyl]amino]-1-propanesulfonic acid. $C_7H_{17}NO_7S$; mol wt 259.27. Hydroxy analog of TAPS. Prepn: W. J. Ferguson *et al., Anal. Biochem.* **104**, 300 (1980). Crystal structure: J. Wouters, D. Stalke, *Acta Crystallogr.* **C52**, 1684 (1996). Thermodynamics of dissociation: R. N. Roy *et al., J. Chem. Eng. Data* **42**, 446 (1997). Triclinic colorless crystals, mp 226-228° (dec). pKa (20°): 7.7. ΔpKa/°C: −0.018. Soly in water (0°): 1M.

USE: Biological buffer.

9192. Taraxacum. Dandelion; lion's tooth. Dried rhizome and roots of *Taraxacum palustre* (Lyons) Lam. & DC. (*T. officinale* Weber, *Leontodon taraxacum* L.), *Compositae*. *Habit.* Europe; naturalized in North America. *Constit.* Taraxerol, choline, levulin, inulin, pectin.

9193. Taraxasterol. [1059-14-9] (3β,18α,19α)-Urs-20(30)-en-3-ol; taraxast-20(30)-en-3β-ol; anthesterin; α-lactucerol; taraxasterin. $C_{30}H_{50}O$; mol wt 426.73. C 84.44%, H 11.81%, O 3.75%. A monohydroxy triterpene. Isoln from *Taraxacum officinale*, Wiggers, *Compositae:* Power, Browning, *J. Chem. Soc.* **101**, 2411 (1912). Structure and configuration: Ames *et al., ibid.* **1954**, 1905. Identity with anthesterin: Power, Browning *ibid.* **105**, 1829 (1914); with α-lactucerol: Zellner, *Monatsh. Chem.* **47**, 681 (1926).

Needles from alcohol, mp 221-222°. $[\alpha]_D$ +96.3° (CHCl$_3$). Very sol in alcohol, ether, petr ether; slightly sol in chloroform, benzene, carbon disulfide, acetone.

Acetate. Lactucerin; lactucon. $C_{32}H_{52}O_2$. Hexagonal plates. mp 251-252° (from ethyl acetate + alcohol). $[\alpha]_D$ +100.5°.

9194. Taraxein. [9010-30-4] A protein complex isolated from the blood serum of schizophrenics; thought to be an antibrain antibody. Method of isolation: R. G. Heath *et al., Am. J. Psychiatry* **114**, 14 (1957). Studies of relationship to schizophrenia: *idem et al., Arch. Gen. Psychiatry* **16**, 1, 10, 24 (1967); J. R. Bergen *et al., Biol. Psychiatry* **15**, 369 (1980). Review: R. G. Heath in *Nutrients and Brain Function*, W. B. Essman, Ed. (Karger, Basel, 1987) pp 186-192.

9195. Taraxerol. [127-22-0] (3β,13α)-13-Methyl-27-norolean-14-en-3-ol; (3β)-D-friedoolean-14-en-3-ol; isoolean-14-en-3β-ol; skimmiol; alnulin; tiliadin. $C_{30}H_{50}O$; mol wt 426.73. C 84.44%, H 11.81%, O 3.75%. Found in *Tilia cordata* Mill., *Tiliaceae:* Bräutigam, *Arch. Pharm.* **238**, 555 (1900); in *Alnus glutinosa* (L.) Gaertn., *Betulaceae:* Zellner, Weiss, *Sitzungsber. Akad. Wiss. Wien* **132**, 258 (1923); in *Taraxacum officinale* Weber, *Compositae:* Burrows, Simpson, *J. Chem. Soc.* **1938**, 2042; in *Litsea dealbata* Nees, *Lauraceae:* Dunstan *et al., Aust. J. Chem.* **6**, 321 (1953); from *Befaria racemosa* (Vent.), *Ericaceae:* Euda *et al., J. Org. Chem.* **26**, 271 (1961). Structure: Beaton *et al., J. Chem. Soc.* **1955**, 2131. Partial synthesis from β-amyrin: *idem, Chem. Ind. (London)* **1955**, 35.

Plates from chloroform + methanol, needles from benzene, mp 282-285°. uv max (ethanol): 210, 215, 220, 223 nm (ε 3900, 2400, 700, 250). Sol in benzene, chloroform, ether, ethyl acetate, acetic anhydride, acetic acid, phenol, pyridine, xylene; less sol in alcohol.

Acetate. $C_{32}H_{52}O_2$. Plates from chloroform + methanol, mp 303-305°. $[\alpha]_D$ +10.5° (c = 1.8 in chloroform).

Benzoate. $C_{37}H_{54}O_2$. Needles from benzene or chloroform + alc, mp 292-293°. $[\alpha]_D^{23}$ +35.7° (c = 0.7 in chloroform).

9196. Tarenflurbil. [51543-40-9] (αR)-2-Fluoro-α-methyl-[1,1′-biphenyl]-4-acetic acid; (−)-2-(2-fluoro-4-biphenylyl)propionic acid; (R)-flurbiprofen; E-7869; MPC-7869; Flurizan. $C_{15}H_{13}FO_2$; mol wt 244.27. C 73.76%, H 5.36%, F 7.78%, O 13.10%. γ-Secretase inhibiting isomer of flurbiprofen, *q.v.* Lowers brain levels of the 42-residue isoform of amyloid β peptide (Aβ42) that is implicated in Alzheimer's disease pathology. Prepn of racemate: BE 658723; S. S. Adams *et al., US 3755427* (1965, 1973 both to Boots); of isomers: J. S. Nicholson, J. L. Turner, US 4188491 (1980 to Boots); R. Hardy *et al., US 5599969* (1997 to Boots). Determn in plasma using capillary electrophoresis: A. Rousseau *et al., Electrophoresis* **29**, 3641 (2008). Effect on γ-secretase and Aβ42 *in vivo*: J. L. Eriksen *et al., J. Clin. Invest.* **112**, 440 (2003). Clinical pharmacokinetics and safety: D. R. Galasko *et al., Alzheimer Dis. Assoc. Disord.* **21**, 292 (2007). Evaluation in Alzheimer's disease patients: G. K. Wilcock *et al., Lancet Neurol.* **7**, 483 (2008). Review of development and clinical experience: H. Geerts, *IDrugs* **10**, 121-133 (2007).

Crystals from *n*-heptane. $[\alpha]_D^{20}$ −29.5°. Sol in methanol.

THERAP CAT: Antiamyloidogenic agent; in treatment of Alzheimer's disease.

9197. Taribavirin. [119567-79-2] 1-β-D-Ribofuranosyl-1*H*-1,2,4-triazole-3-carboximidamide; 1-β-D-ribofuranosyl-1,2,4-triazole-3-carboxamidine; ribamidine. $C_8H_{13}N_5O_4$; mol wt 243.22. C 39.51%, H 5.39%, N 28.79%, O 26.31%. Liver-targeting antiviral nucleoside analog; 3-carboxamidine prodrug of ribavirin, *q.v.* Prepn of hydrochloride: J. T. Witkowski *et al., DE 2220246; eidem, US 3798209* (1972, 1974 both to ICN); *idem et al., J. Med. Chem.* **16**, 935 (1973). Improved synthesis: G. D. Kini *et al., ibid.* **32**, 1447 (1989); B. Gabrielsen *et al., ibid.* **35**, 3231 (1992). *In vitro* and *in vivo* efficacy vs *Phlebovirus*: R. W. Sidwell *et al., Antiviral Res.* **10**, 193 (1988); vs influenza: *idem et al., ibid.* **68**, 10 (2005). LC/MS/MS determn in plasma: Y. Liu *et al., J. Chromatogr. B* **832**, 17 (2006). Liver-targeting properties and safety profile in animals: C. Lin *et al., Antivir. Chem. Chemother.* **14**, 145 (2003). Mechanism of action study: J. Z. Wu *et al., Antimicrob. Agents Chemother.* **48**, 4006 (2004). Clinical pharmacokinetics in hepatitis C patients: S. Aora *et al., J. Clin. Pharmacol.* **45**, 275 (2005).

Hydrochloride. [40372-00-7] Viramidine. $C_8H_{13}N_5O_4.HCl$; mol wt 279.68. Crystals from acetonitrile + ethanol, mp 177-179° (dec).

THERAP CAT: Antiviral.

9198. Tariquidar. [206873-63-4] *N*-[2-[[[4-[2-(3,4-Dihydro-6,7-dimethoxy-2(1*H*)-isoquinolinyl)ethyl]phenyl]amino]carbonyl]-4,5-dimethoxyphenyl]-3-quinolinecarboxamide; XR-9576. $C_{38}H_{38}$-

N_4O_6; mol wt 646.74. C 70.57%, H 5.92%, N 8.66%, O 14.84%. Multidrug resistance modulator; selective inhibitor of P-glycoprotein (P-gp), *q.v.* Prepn: H. Ryder *et al.*, **WO 9817648**; *eidem*, **US 6218393** (1998, 2001 both to Xenova); M. Roe *et al.*, *Bioorg. Med. Chem. Lett.* **9**, 595 (1999). Molecular interaction with P-gp: C. Martin *et al.*, *Br. J. Pharmacol.* **128**, 403 (1999). Effect on multidrug resistant cell lines: P. Mistry *et al.*, *Cancer Res.* **61**, 749 (2001). Clinical pharmacology: A. Stewart *et al.*, *Clin. Cancer Res.* **6**, 4186 (2000). Effect on accumulation of ^{99m}Tc-sestamibi, *q.v.*, in patients with metastatic cancers: M. Agrawal *et al.*, *Clin. Cancer Res.* **9**, 650 (2003). Evaluation of ^{11}C-labeled compound as radiotracer for positron emission tomography: F. Bauer *et al.*, *Bioorg. Med. Chem.* **18**, 5489 (2010). Review of clinical experience: E. Fox, S. E. Bates, *Expert Rev. Anticancer Ther.* **7**, 447-459 (2007).

THERAP CAT: Antineoplastic adjunct (chemosensitizer).

9199. Tar Oil. Volatile oil distilled from wood tar. *Principal constit.* Phenolic substances and hydrocarbons.

Almost colorless liquid when fresh, but soon becomes dark brownish-red. d 0.860-0.900. Insol in water; sol in alcohol, ether.

9200. Tar Oil, Rectified. Pine tar oil. The volatile oil from pine tar rectified by steam distillation. Chief active constituents are phenolic substances.

Dark reddish-brown, thin liquid; strong empyreumatic odor and taste. d_{25}^{25} 0.960-0.990. Insol in water; miscible with alcohol.

THERAP CAT: Antiseptic (topical); dermatologic.

THERAP CAT (VET): Antiseptic, antipruritic. For chronic skin conditions and in hoof dressings. Has been used internally as an expectorant.

9201. Tarragon. Estragon. The dried leaves and flowering tops of the perennial herb, *Artemisia dracunculus* L., *Compositae. Habit.* Siberia, Caspian Sea region; cultivated in Western Europe. Yields up to 0.8% oil of tarragon. *Constit.* *p*-Allylanisole (estragole; methyl chavicol); ocimene; myrcene; phellandrene(?); *p*-methoxycinnamaldehyde.

USE: The herb for culinary purposes, the oil as flavoring agent in liqueurs, soups, sauces and salad dressings. In perfumery to improve the note of chypre type perfumes.

9202. *meso*-Tartaric Acid. [147-73-9] *rel*-(2*R*,3*S*)-2,3-Dihydroxybutanedioic acid; mesotartaric acid; internally compensated tartaric acid; unresolvable tartaric acid; Antiweinsäure (German). $C_4H_6O_6$; mol wt 150.09. C 32.01%, H 4.03%, O 63.96%. Prepd by boiling L-tartaric acid with alkali; as byproduct of racemization: Winther, *Z. Phys. Chem.* **56**, 507 (1906); Holleman, *Org. Synth.* **coll. vol. I** (2nd ed., 1941) p 497; Milas, **US 2414385** (1947). Microbial prepn: Martin, Foster, *J. Bacteriol.* **70**, 405 (1955); Foster, **US 2947665** (1960). Explanation of optical inactivity: Noller, *Science* **102**, 508 (1945); C. R. Noller, *Chemistry of Organic Compounds* (Philadelphia, 2nd ed., 1957) p 339.

Relative stereochemistry

Monohydrate. Rectangular plates, d_4^{20} 1.666 (also reported as 1.737). mp 140° (also reported as 159-160°). pKa$_1$ 3.11; pKa$_2$ 4.80. Maximum soly in water at 20°: 125 g/100 ml.

9203. D-Tartaric Acid. [147-71-7] (2*S*,3*S*)-2,3-Dihydroxybutanedioic acid; unusual tartaric acid; unnatural tartaric acid; *l*-tartaric acid; (−)-tartaric acid; levotartaric acid; D-*threo*-2,3-dihydroxysuccinic acid. $C_4H_6O_6$; mol wt 150.09. C 32.01%, H 4.03%, O 63.96%. Levorotatory tartaric acid having a dextro configuration. Although termed "unnatural," its occurrence in nature has been demonstrated. Obtained in small amounts from racemic tartaric acid through biochemical cleavage using *Penicillium notatum, Aspergillus griseus, A. niger* or other microorganisms: Pasteur, *Compt. Rend.* **51**, 298 (1860). Alternate route using salt formation with *d*-methylamphetamine: Walton, *J. Soc. Chem. Ind.* **64**, 219 (1945). Monograph: K. Freudenberg, *Stereochemie* **I**, (1933), reprinted by J. W. Edwards (Ann Arbor, 1945). Crystallographic data: A. N. Winchell, *The Optical Properties of Organic Compounds* (Academic Press, New York, 2nd ed., 1954) p 47.

Monoclinic sphenoidal prisms. d_4^{20} 1.7598. mp 168-170°. $[\alpha]_D^{20}$ −12.0° (c = 20 in H_2O). pKa$_1$ 2.93; pKa$_2$ 4.23. One gram dissolves in 0.75 ml water at room temp, in 0.5 ml boiling water, 1.7 ml methanol, 3 ml ethanol, 10.5 ml propanol, 250 ml ether. Also sol in glycerol. Insol in chloroform. Maximum soly in water at 20°: 139 g/100 ml.

9204. DL-Tartaric Acid. [133-37-9] *rel*-(2*R*,3*R*)-2,3-Dihydroxybutanedioic acid; racemic tartaric acid; racemic acid; *dl*-tartaric acid; resolvable tartaric acid; uvic acid; paratartaric acid; *dl*-Weinsäure (German); Vogessäure (German); Traubensäure (German). $C_4H_6O_6$; mol wt 150.09. C 32.01%, H 4.03%, O 63.96%. Probably never a natural product, although sometimes found in small amounts during wine-making. Prepn from L-tartaric acid by boiling with aq NaOH (*meso*-tartaric acid is obtained as a byproduct): Holleman, *Org. Synth.* **6**, 82 (1926); **coll. vol. I** (2nd ed., 1941) p 497. Synthesis by oxidation of fumaric acid: Milas, Terry, *J. Am. Chem. Soc.* **47**, 1412 (1925); Milas, Sussman, *ibid.* **58**, 1302 (1936); **US 2000213** (1935 to Standard Brands). From maleic acid: Church, Blumberg, *Ind. Eng. Chem.* **43**, 1780 (1951).

Relative stereochemistry

Anhyd acid, triclinic pinacoidal crystals from abs alc, from water above 73°, or by drying the monohydrate at 100°. mp 206°. pKa$_1$ 2.96; pKa$_2$ 4.24. Less soluble in water than L-tartaric acid. pH of 0.1*M* aq soln: 2.0. Soly in alcohol (g/100 g): 2.006 at 0°; 3.153 at 15°; 5.01 at 25°; 6.299 at 40°. Soly in ether about 1%.

Monohydrate. Triclinic pinacoidal crystals from water. d_4^{20} 1.697. One hundred parts (w/w) of water dissolve 14.00 parts at 10°; 20.60 at 20°; 29.10 at 30°; 43.32 at 40°; 99.88 at 70°; 184.91 at 100°.

9205. L-Tartaric Acid. [87-69-4] (2*R*,3*R*)-2,3-Dihydroxybutanedioic acid; ordinary tartaric acid; natural tartaric acid; *d*-tartaric acid; (+)-tartaric acid; dextrotartaric acid; L-2,3-dihydroxybutanedioic acid; *d*-α,β-dihydroxysuccinic acid; Weinsäure (German); Weinsteinsäure (German). $C_4H_6O_6$; mol wt 150.09. C 32.01%, H 4.03%, O 63.96%. Dextrorotatory tartaric acid having a levo configuration. Widely distributed in nature, classified as a fruit acid. Occurs in many fruits, free and combined with potassium, calcium and magnesium. Observed in antiquity as the acid potassium salt found deposited as a fine crystalline crust during fermentation of grape juice or tamarind juice and termed *faecula* (little yeast) by the Romans. The derivation from *Tartarus* is of medieval, alchemical origin. In modern processes the acid potassium tartrate obtained during wine-making is first converted to calcium tartrate which is then hydrolyzed to tartaric acid and calcium sulfate: Metzner, *Chem. Eng. Prog.* **43**, 160 (1947); several modifications, *e.g.*, **IT 490221** (1954 to Procedimenti Chimici), *C.A.* **50**, 11607c (1956). Extraction from tamarind pulp in about 10% yield: **IN 52167** (1955), *C.A.* **50**, 5249g (1956). Synthesis by hydroxylation of maleic acid: Church, Blumberg, *Ind. Eng. Chem.* **43**, 1780 (1951). Monograph: U. Roux, *La Grande Industrie des Acides Organiques* (Dounod, Paris, 1939). Example of a modern process: Dabul, **US 3114770** (1963 to Orandi & Massera).

Monoclinic sphenoidal prisms, mp 168-170°. Stable to air and light. Strong acid taste. Refreshing when in dil aq soln. d_4^{20} 1.7598. Odor of burnt sugar when heated to mp. $[\alpha]_D^{20}$ +12.0° (c = 20 in H_2O). Strong organic acid. At 25° pKa_1 2.98; pKa_2 4.34. pH of $0.1N$ soln: 2.2. Heat of combustion: -275.1 kcal/mol. Specific heat: 0.288 cal/g/°C at 21 to 51°; 0.296 at 0 to 99.6°. Dielectric constant 36.0 for 1200 cm waves. Freely sol in water. d_4^{15} of aq solns (w/w at 15°): 1% 1.0045; 10% 1.0469; 20% 1.0969; 30% 1.1505; 40% 1.2078; 50% 1.2696. Max soly in water in g/100 ml at various temps: 0° = 115; 10° = 126; 20° = 139; 30° = 156; 40° = 176; 50° = 195; 60° = 217; 70° = 244; 80° = 273; 90° = 307; 100° = 343. One gram dissolves in 0.75 ml water at room temp, in 0.5 ml boiling water, 1.7 ml methanol, 3 ml ethanol, 10.5 ml propanol, 250 ml ether. Also sol in glycerol. Insol in chloroform.

Caution: Can cause local irritation. *See Patty's Industrial Hygiene and Toxicology* vol. **2C**, G. D. Clayton, F. E. Clayton, Eds. (Wiley-Interscience, New York, 3rd ed., 1982) p 4937, 4943-4945.

USE: In the soft drink industry, confectionery products, bakery products, gelatin desserts, as an acidulant. In photography, tanning, ceramics, manuf tartrates. The common commercial esters are the diethyl and dibutyl derivs used for lacquers and in textile printing. Pharmaceutic aid (buffering agent). Complexing agent.

9206. Tartrazine. [1934-21-0] 4,5-Dihydro-5-oxo-1-(4-sulfophenyl)-4-[2-(4-sulfophenyl)diazenyl]-1*H*-pyrazole-3-carboxylic acid sodium salt (1:3); C.I. Acid Yellow 23; 3-carboxy-5-hydroxy-1-*p*-sulfophenyl-4-*p*-sulfophenylazopyrazole trisodium salt; 5-hydroxy-1-(*p*-sulfophenyl)-4-[(*p*-sulfophenyl)azo]pyrazole-3-carboxylic acid trisodium salt; hydrazine yellow; C.I. 19140; FD & C Yellow No. 5; C.I. Food Yellow 4. $C_{16}H_9N_4Na_3O_9S_2$; mol wt 534.36. C 35.96%, H 1.70%, N 10.49%, Na 12.91%, O 26.95%, S 12.00%. Prepn: **US 2457823** (1949 to Ilford); Freeman *et al.*, *J. Assoc. Off. Agric. Chem.* **33**, 937 (1950). *See also: Colour Index* vol. 4 (3rd ed., 1971) p 4132.

Bright orange-yellow powder. Freely sol in water. The aq soln is not changed by HCl but becomes redder with sodium hydroxide.

USE: As a dye for wool and silks; as colorant in food, drugs and cosmetics. In biochemistry as an adsorption-elution indicator for chloride estimations.

9207. Tartronic Acid. [80-69-3] 2-Hydroxypropanedioic acid; hydroxymalonic acid. $C_3H_4O_5$; mol wt 120.06. C 30.01%, H 3.36%, O 66.63%. Prepd by ozonization of malonic acid in aq soln: Dobinson, *Chem. Ind. (London)* **1959**, 853.

Colorless, odorless crystals. Crystallizes from water with ½ and $1H_2O$. Becomes anhyd at 60°. Sublimes 110-120°. mp 158-160° with decompn (CO_2 evolved). pK_1 2.42; pK_2 4.54, *see:* Grandjean, *Bull. Soc. R. Sci. Liege* **38**, 288 (1969), *C.A.* **72**, 42684t (1970). Very sol in water, alcohol; the anhydrous acid is sol in ether.

9208. Tasimelteon. [609799-22-6] *N*-[[(1*R*,2*R*)-2-(2,3-Dihydro-4-benzofuranyl)cyclopropyl]methyl]propanamide; (1*R-trans*)-*N*-[[2-(2,3-dihydrobenzofuran-4-yl)cycloprop-1-yl]methyl]propanamide; BMS-214778; VEC-162. $C_{15}H_{19}NO_2$; mol wt 245.32. C 73.44%, H 7.81%, N 5.71%, O 13.04%. Melatonin MT_1/MT_2 receptor agonist. Prepn: J. D. Catt *et al.*, **WO 9825606**; *eidem*, **US 5856529** (1998, 1999 both to Bristol-Myers Squibb). Large scale process: J. S. Prasad *et al.*, *Org. Process Res. Dev.* **7**, 821 (2003). Pharmacokinetics and metabolism in animals: N. N. Vachharajani *et al.*, *J. Pharm. Sci.* **92**, 760 (2003). Clinical trial for transient insomnia after sleep-time shift: S. M. W. Rajaratnam *et al.*, *Lancet* **373**, 482 (2009). Review of pharmacology and clinical experience: R. Hardeland, *Curr. Opin. Investig. Drugs* **10**, 691-701 (2009).

THERAP CAT: Sedative, hypnotic.

9209. Taspoglutide. [275371-94-3] 8-(2-Methylalanine)-35-(2-methylalanine)-36-L-argininamide-7-36-glucagon-like peptide I (human); [Aib^8,35^]hGLP-1(7-36)NH₂; BIM-51077; R-1583; RO-5073031. $C_{152}H_{232}N_{40}O_{45}$; mol wt 3339.76. C 54.66%, H 7.00%, N 16.78%, O 21.56%. Human glucagon-like peptide 1 (GLP-1) analogue with α-aminoisobutyric acid (Aib) substitutions at amino acids 8 and 35. Prepn: Z. X. Dong **WO 0034331**; *idem*, **US 6903186** (2000, 2005 both to Soc. Conseils Recher. Sci.). LC-MS/MS determn in plasma: K. Heinig, T. Wirz, *Anal. Chem.* **81**, 3705 (2009). Receptor binding and resistance to DPP-4 degradation: E. Sebokova *et al.*, *Endocrinology* **151**, 2474 (2010). Pharmacodynamics in patients with type 2 diabetes: C. Kapitza *et al.*, *Diabet. Med.* **26**, 1156 (2009). Clinical evaluation with metformin in patients with type 2 diabetes: M. A. Nauck *et al.*, *Diabetes Care* **32**, 1237 (2009). Review of development and clinical experience: K. Retterstol, *Expert Opin. Invest. Drugs* **18**, 1405-1411 (2009).

THERAP CAT: Antidiabetic.

9210. Tasquinimod. [254964-60-8] 1,2-Dihydro-4-hydroxy-5-methoxy-*N*,1-dimethyl-2-oxo-*N*-[4-(trifluoromethyl)phenyl]-3-quinolinecarboxamide; 4-hydroxy-5-methoxy-*N*,1-dimethyl-2-oxo-*N*-[4-(trifluoromethyl)phenyl]-1,2-dihydroquinoline-3-carboxamide; *N*-methyl-*N*-(4-trifluoromethylphenyl)-1,2-dihydro-4-hydroxy-5-methoxy-1-methyl-2-oxoquinoline-3-carboxamide; ABR-215050. $C_{20}H_{17}F_3N_2O_4$; mol wt 406.36. C 59.12%, H 4.22%, F 14.03%, N 6.89%, O 15.75%. Tumor angiogenesis inhibitor. Upregulates thrombospondin-1 (TSP1) expression in tumor cells. Prepn: A. Björk *et al.*, **WO 0003991**; *eidem*, **US 6133285** (both 2000 to Active Biotech); S. Jönsson *et al.*, *J. Med. Chem.* **47**, 2075 (2004). LC-MS/MS determn in plasma: G. P. Hansson *et al.*, *J. Chromatogr. B* **879**, 3401 (2011). Mechanism of action study: A. Olsson *et al.*, *Mol. Cancer* **9**, 107 (2010). Preclinical antiangiogenic activity: J. T. Isaacs *et al.*, *Prostate* **66**, 1768 (2006). Clinical evaluation in prostate cancer: R. Pili *et al.*, *J. Clin. Oncol.* **29**, 4022 (2011). Review of development and therapeutic potential: J. T. Isaacs, *Expert Opin. Invest. Drugs* **19**, 1235-1243 (2010).

Crystalline solid from heptane. Soly in water: 4 mg/ml at pH 7.5.
THERAP CAT: Antineoplastic.

9211. Taurine. [107-35-7] 2-Aminoethanesulfonic acid; Dibicor. $C_2H_7NO_3S$; mol wt 125.14. C 19.20%, H 5.64%, N 11.19%, O 38.35%, S 25.62%. Conditionally essential sulfonated amino acid synthesized in the liver from cysteine and methionine. Important during mammalian development, especially for cells of the cerebellum and retina. Physiological roles include: bile acid conjugation, osmoregulation, membrane stabilization, calcium homeostasis, and regulation of retinal and cardiac function. Isoln from ox bile: Hammarsten, *Z. Physiol. Chem.* **32**, 456 (1901); from abalone (*Haliotis*): C. L. A. Schmidt, T. Watson, *J. Biol. Chem.* **33**, 499 (1918). Prepn by treatment of salts of 2-chloroethane-1-sulfonic acid with NH_3: J. W. James, *J. Prakt. Chem. Chem.-Ztg.* **31**, 413 (1885); starting with 2-bromoethanesulfonate: C. S. Marvel, C. F. Bailey, *Org. Synth.* **coll. vol. II**, 563 (1943); by Na_2SO_3 sulfonation of ethylene chloride followed by ammonolysis with anhydr NH_3 or aq NH_3 and ammonium carbonate: J. W. Schick, F. Degering, *Ind. Eng. Chem.* **39**, 906 (1947); from aziridines: L. Hu *et al.*, *J. Org. Chem.* **72**, 4543 (2007). Crystal density study: D. E. Hibbs *et al.*, *Chem. Eur. J.* **9**, 1075 (2003). Review of separation and determn methods: S. Mou *et al.*, *J. Chromatogr. B* **781**, 251-267 (2002). Importance to retinal function in cats: K. C. Hayes *et al.*, *Invest. Ophthalmol.* **15**, 52 (1976). Review of role in nutrition: H. P. Redmond *et al.*, *Nutrition* **14**, 599-604 (1998); of role in retinal function: L. Lima, *Neurochem. Res.* **24**, 1333-1338 (1999); of therapeutic potential in diabetes: S. H. Hansen, *Diabetes Metab. Res. Rev.* **17**, 330-346 (2001); of immunomodulatory effects: G. B. Schuller-Levis, E. Park, *FEMS Microbiol. Lett.* **226**, 195-202 (2003).

Colorless crystals, mp 325° (dec). d 1.734. pKa 4.96. Sol in 15.5 parts of water at 12°. 100 parts of 95% alc dissolve 0.004 parts at 17°. Insol in abs alc.

USE: Nutritional supplement.

THERAP CAT (VET): In prevention of retinal degeneration and in prevention and treatment of taurine-deficiency cardiomyopathy in cats.

9212. Taurocholic Acid. [81-24-3] 2-[[(3α,5β,7α,12α)-3,-7,12-Trihydroxy-24-oxocholan-24-yl]amino]ethanesulfonic acid; *N*-choloyltaurine; cholaic acid; cholyltaurine. $C_{26}H_{45}NO_7S$; mol wt 515.71. C 60.55%, H 8.80%, N 2.72%, O 21.72%, S 6.22%. The product of conjugation of cholic acid with taurine. Its sodium salt is the chief ingredient of the bile of carnivorous animals. Prepn from dog bile: Hammarsten in Abderhalden, *Handbuch der Biol. Arbeitsmethoden*, Abt. **I**, Teil 6, p 219 (1925). Separation from bile acids: Ahrens, Craig, *J. Biol. Chem.* **195**, 763 (1952). Enzymic synthesis: Siperstein, Murray, *Science* **123**, 377 (1956). Prepn of the sodium salt: Cortese, *J. Am. Chem. Soc.* **59**, 2532 (1937); Norman, *Ark. Kemi* **8**, 331 (1955); of the barium salt: Kazuno, Yanazaki, *Z. Physiol. Chem.* **224** 160 (1934). Binding of taurocholic acid to serum proteins: Burke *et al.*, *Clin. Chim. Acta* **32**, 207 (1971). Toxicity data: Klaassen, *Toxicol. Appl. Pharmacol.* **24**, 37 (1973).

Clusters of slender, four-sided prisms from alcohol + ether, stable to air. (Commercial prepns are usually amorphous, very hygroscopic, and of yellow color.) Dec about 125°. $[\alpha]_D^{18}$ +38.8° (c = 2 in alcohol). pK 1.4. Freely sol in water, sol in alcohol. Almost insol in ether, ethyl acetate. Is hydrolyzed to cholic acid and taurine by acids and alkalies. LD_{50} in newborn rats: 380 mg/kg (Klaassen).

Sodium salt. Sodium taurocholate. $C_{26}H_{44}NNaO_7S$. Crystals with 1.5 and 2 mols H_2O. Sweet taste with bitter aftertaste. Dec

about 230°. $[\alpha]_D^{20}$ +24° (c = 3). Very freely sol in water or alcohol. Solvent action on cholesterol: Rosin, *Z. Physiol. Chem.* **124**, 282 (1923).

Barium salt. Barium taurocholate. $C_{52}H_{88}BaN_2O_{14}S_2$. Crystals with 5 mols H_2O, dec 225-227°. $[\alpha]_D^{20}$ +25.6°. Converted to taurocholic acid by treatment with sulfuric acid.

USE: Sodium salt is a lipase accelerator. *See also* Ox Bile Extract.

THERAP CAT: Choleretic.

9213. Taurolidine. [19388-87-5] 4,4′-Methylenebis[tetrahydro-2*H*-1,2,4-thiadiazine] 1,1,1′,1′-tetraoxide; 4,4′-methylenebis-(perhydro-1,2,4-thiadiazine 1,1-dioxide); bis(1,1-dioxoperhydro-1,-2,4-thiadiazin-4-yl)methane; Drainasept; Taurolin; Tauroflex. C_7-$H_{16}N_4O_4S_2$; mol wt 284.35. C 29.57%, H 5.67%, N 19.70%, O 22.51%, S 22.55%. Broad spectrum, synthetic formaldehyde carrier formed by the condensation of two molecules of taurine and three molecules of formaldehyde. Prepn: **FR 1458701**; R. W. Pfirrmann, **US 3423408** (1966, 1969 both to Ed. Geistlich Söhne). Antibacterial activity in mice: M. K. Browne *et al.*, *J. Appl. Bacteriol.* **41**, 363 (1976). Anti-endotoxin activity in lab animals: R. W. Pfirrmann, G. B. Leslie, *ibid.* **46**, 97 (1979). Mechanism of action: E. Myers *et al.*, *ibid.* **48**, 89 (1980). HPLC determn of metabolites in plasma: A. D. Woolfson *et al.*, *Int. J. Pharm.* **49**, 135 (1989). Pharmacokinetics: C. Steinbach-Lebbin *et al.*, *Arzneim.-Forsch.* **32**, 1542 (1982). Metabolism in humans: B. I. Knight *et al.*, *Br. J. Clin. Pharmacol.* **12**, 695 (1981). Clinical trials in peritonitis: M. K. Browne *et al.*, *Surg. Gynecol. Obstet.* **146**, 721 (1978); G. Wesch *et al.*, *Fortschr. Med.* **101**, 545 (1983); in wound sepsis: A. K. Halsall *et al.*, *Pharmatherapeutica* **2**, 673 (1981); in pleural infection: A. A. Conlan *et al.*, *S. Afr. Med. J.* **64**, 653 (1983).

White crystals, mp 154-158°. Sol in water.

THERAP CAT: Antibacterial.

9214. Taxicins. Acyl-free polyols from which the taxines are prepared by partial esterfication: Baxter *et al.*, *J. Chem. Soc.* **1962**, 2964. Structure of taxicin-I and -II: Eyre *et al*, *Proc. Chem. Soc. London* **1963**, 271. Structure and stereochemical studies of taxicin-I and -II: Dukes *et al.*, *Tetrahedron Lett.* **1965**, 4765; *J. Chem. Soc. C* **1967**, 448; Eyre *et al.*, *ibid.* 452. Structure of *O*-cinnamoyltaxicin-II triacetate (taxinine): Kurono *et al.*, *Tetrahedron Lett.* **1963**, 2153; Nakanishi, Kurono, *ibid.* **1963**, 2161; Ueda *et al.*, *ibid.* **1963**, 2167; Uyeo *et al.*, *J. Pharm. Soc. Jpn.* **84**, 762 (1964). Stereochemistry of taxinine: Kurono *et al.*, *Tetrahedron Lett.* **1965**, 1917; Shiro *et al.*, *Chem. Commun.* **1966**, 97.

	R	R′	R″
Taxicin-I	OH	H	H
O-cinnamoyltaxicin-I	OH	H	COCH=CHC₆H₅
O-cinnamoyltaxicin-I triacetate	OH	COCH₃	COCH=CHC₆H₅
Taxicin-II	H	H	H
O-cinnamoyltaxicin-II triacetate	H	COCH₃	COCH=CHC₆H₅

Taxicin-I. [5308-89-4] [3*S*-(3α,4aα,5α,6α,11β,12α,12aβ)]-1,-3,4,4a,5,6,7,11,12,12a-Decahydro-3,5,6,11,12-pentahydroxy-9,-12a,13,13-tetramethyl-4-methylene-6,10-methanobenzocyclodecen-8(2*H*)-one. $C_{20}H_{30}O_6$; mol wt 366.45.

***O*-Cinnamoyltaxicin-I.** [11034-45-0] $C_{29}H_{36}O_7$. mp 233-234°. $[\alpha]_D^{21}$ +285° (chloroform). uv max (alc): 282 nm (ε 28200).

***O*-Cinnamoyltaxicin-I triacetate.** [13452-36-3] $C_{35}H_{42}O_{10}$. Prisms from alc. mp 237-239°. $[\alpha]_D^{18}$ +218° (chloroform).

Taxicin-II. [5308-90-7] $C_{20}H_{30}O_5$; mol wt 350.46.

O-**Cinnamoyltaxicin-II triacetate.** [3835-52-7] Taxinine. C_{35}-$H_{42}O_9$. Prisms from alc, mp 265-267°. $[\alpha]_D^{18}$ +137° (chloroform). uv max: 217, 223, 278 nm (ε 20500; 16000; 28600).

9215. Taxine. [12607-93-1] A mixture of alkaloids from needles and berries of the yew tree, *Taxus baccata* L., *Taxaceae:* Lucas, *Arch. Pharm.* **85**, 145 (1856); Winterstein, Iatrides, *Z. Physiol. Chem.* **117**, 240 (1921); Winterstein, Guyer, *ibid.* **128**, 175 (1923); Callow *et al., J. Chem. Soc.* **1931**, 2138. Structural studies of *taxine-I*: Baxter *et al., ibid.* **1962**, 2964e; of *taxine-II*: Dukes *et al., ibid.* **1967**, 448; Eyre *et al., ibid.* 452. Sepn of cryst *taxine A* and amorph *taxine B* from the amorph mixture: E. Graf, H. Bertholdt, *Pharm. Zentralhalle* **96**, 385 (1957). Structure of taxine A: E. Graf *et al., Ann.* **1982**, 376. *Review:* B. Lythgoe, *The Alkaloids* vol. X, R. H. F. Manske, Ed. (Academic Press, New York, 1968) pp 597-626. Toxicity data: Y. Tekol, *Vet. Hum. Toxicol.* **33**, 337 (1991).

Taxine A

Granular amorphous powder, mp 121-124°. $[\alpha]_D^{17}$ +95.7° (c = 4.59 in ethanol). Sol in ether, chloroform, alcohol. Practically insol in water, petr ether. LD_{50} (as the sulfate salt) in mice (mg/kg): 19.72 orally, 21.88 i.p.; in rats (mg/kg): 20.18 s.c. (Tekol).

Taxine A. [1361-49-5] ($\alpha R,\beta S$)-β-(Dimethylamino)-α-hydroxy benzenepropanoic acid (1*S*,2*R*,3*E*,5*S*,7*S*,8*S*,10*R*,13*S*)-2,13-bis(acetyloxy)-7,10-dihydroxy-8,12,15,15-tetramethyl-9-oxotricyclo-[9.3.1.1$^{4.8}$]hexadeca-3,11-dien-5-yl ester. $C_{35}H_{47}NO_{10}$; mol wt 641.76. mp 204-206°. $[\alpha]_D$ −140° (CHCl₃).

Caution: Taxine is undoubtedly responsible for the poisonous properties of the yew. Fatalities among domestic animals due to yew poisoning are not uncommon. Human fatal symptoms are those of gastrointestinal irritation, cardiac and respiratory failure.

9216. Taxodione. [19026-31-4] (4b*S*,8a*S*)-4b,5,6,7,8,8a-Hexahydro-4-hydroxy-4b,8,8-trimethyl-2-(1-methylethyl)-3,9-phenanthrenedione; 11-hydroxy-13-isopropylpodocarpa-7,9(11),13-triene-6,12-dione. $C_{20}H_{26}O_3$; mol wt 314.43. C 76.40%, H 8.34%, O 15.26%. Diterpenoid quinone methide with tumor-inhibitory properties. Isoln of naturally occurring (+)-form from *Taxodium distichum* Rich, *Taxodiaceae:* S. M. Kupchan *et al., J. Am. Chem. Soc.* **90**, 5923 (1968). Structure: *eidem, J. Org. Chem.* **34**, 3912 (1969). Total synthesis of the racemate: K. Mori, M. Matsui, *Tetrahedron* **26**, 3467 (1970); T. Matsumoto *et al., Bull. Chem. Soc. Jpn.* **44**, 2766 (1971); **50**, 1575 (1977); D. L. Snitman *et al., Tetrahedron Lett.* **1979**, 2477; R. V. Stevens, G. S. Bisacchi, *J. Org. Chem.* **47**, 2396 (1982). Total synthesis of the (+)-form: T. Matsumoto *et al., Bull. Chem. Soc. Jpn.* **50**, 266 (1977); R. H. Burnell *et al., Can. J. Chem.* **65**, 775 (1987). Antitumor activity studies: Hanson *et al., Science* **168**, 378 (1970).

Golden plates from methanol, mp 115-116°. $[\alpha]_D^{28}$ +56° (c = 1 in CHCl₃). uv max (methanol): 320, 332, 400 nm (ε 25000, 26000, 2000).

9217. Tazanolast. [82989-25-1] 2-Oxo-2-[[3-(2*H*-tetrazol-5-yl)phenyl]amino]acetic acid butyl ester; butyl 3′-(1*H*-tetrazol-5-yl)-oxanilate; TO-188; WP-833; Tazalest; Tazanol. $C_{13}H_{15}N_5O_3$; mol wt 289.30. C 53.97%, H 5.23%, N 24.21%, O 16.59%. Mast-cell degranulation inhibitor. Prepn: **JP Kokai 82 11975** (1982 to Wakamoto), *C.A.* **96**, 217856a (1982). Immunopharmacology: M. Agata *et al., Jpn. J. Pharmacol.* **32**, 689 (1982). Series of articles on pharmacology and mechanism of action: *Arzneim.-Forsch.* **38**, 70-92 (1988).

THERAP CAT: Antiallergic.

9218. Tazarotene. [118292-40-3] 6-[2-(3,4-Dihydro-4,4-dimethyl-2*H*-1-benzothiopyran-6-yl)ethynyl]-3-pyridinecarboxylic acid ethyl ester; ethyl 6-[2-(4,4-dimethylthiochroman-6-yl)ethynyl] nicotinate; AGN-190168; Avage; Tazorac; Zorac. $C_{21}H_{21}NO_2S$; mol wt 351.46. C 71.77%, H 6.02%, N 3.99%, O 9.10%, S 9.12%. Acetylenic retinoid prodrug converted to the active metabolite, *tazarotenic acid*, with selective affinity for retinoic acid receptors RARβ and RARγ. Prepn: R. A. S. Chandrarratna, **EP 284288**; *idem*, **US 5089509** (1988, 1992 both to Allergan). Clinical evaluation in psoriasis: T. Esgleyes-Ribot *et al., J. Am. Acad. Dermatol.* **30**, 581 (1994). Clinical trial in treatment of photodamage: S. Kang *et al., ibid.* **52**, 268 (2005). Review of pharmacokinetics and metabolism: D. D.-S. Tang-Liu *et al., Clin. Pharmacokinet.* **37**, 273-287 (1999); of mechanism of action and treatment strategies: A. Roeder *et al., Skin Pharmacol. Physiol.* **17**, 111-118 (2004).

White solid.
THERAP CAT: Antiacne; antipsoriatic. In treatment of photodamaged skin.

9219. Tazettine. [507-79-9] (3*S*,4a*S*,6a*S*,13b*S*)-4,4a,5,6-Tetrahydro-3-methoxy-5-methyl-8*H*-[1,3]dioxolo[6,7][2]benzopyrano[3,4-*c*]indol-6a(3*H*)-ol; sekisanine; sekisanoline; ungernine. $C_{18}H_{21}NO_5$; mol wt 331.37. C 65.24%, H 6.39%, N 4.23%, O 24.14%. From *Narcissus tazetta* L., *Lycoris radiata* Herb., *Ungernia sewerzowi* (Rgl.) Fedtsch., and other *Amaryllidaceae:* Späth, Kahovec, *Ber.* **67**, 1501 (1934). Structure and stereochemistry: Ikeda *et al., J. Chem. Soc.* **1956**, 4749. Abs config: Highet, Highet, *Tetrahedron Lett.* **1966**, 4099. Synthesis: Hendrickson *et al., J. Am. Chem. Soc.* **92**, 5538 (1970); Tsuda *et al., Tetrahedron Lett.* **1972**, 3153. Biosynthesis: Fales, Wildman, *J. Am. Chem. Soc.* **86**, 294 (1964). Identity with sekisanine and sekisanoline: Ikeda *et al., loc. cit.* Stereospecific total synthesis: Hendrickson *et al., J. Am. Chem. Soc.* **96**, 7781 (1974); S. Danishefsky *et al., ibid.* **102**, 2838 (1980); **104**, 7591 (1982).

Crystals, mp 210-211° (evac tube); racemate reported as mp 237-238° (Tsuda) and mp 175-176° (Danishefsky). $[\alpha]_D^{25}$ +150.3° (82 mg in 2 ml chloroform). Sol in methanol, ethanol, choroform. Sparingly sol in ether.

Hydrochloride. Crystals, mp 206°, water soluble.

Methiodide. Crystals, dec 220° (evacuated tube).

9220. Tazobactam. [89786-04-9] (2*S*,3*S*,5*R*)-3-Methyl-7-oxo-3-(1*H*-1,2,3-triazol-1-ylmethyl)-4-thia-1-azabicyclo[3.2.0]heptane-2-carboxylic acid 4,4-dioxide; 2β-[(1,2,3-triazol-1-yl)methyl]-2α-methylpenam-3α-carboxylic acid 1,1-dioxide; YTR-830H; CL-298741. $C_{10}H_{12}N_4O_5S$; mol wt 300.29. C 40.00%, H 4.03%, N 18.66%, O 26.64%, S 10.68%. β-Lactamase inhibitor. Prepn: R. G. Micetich *et al.*, **EP 97446**; *eidem*, **US 4562073** (1984, 1985 both to Taiho); R. G. Micetich *et al.*, *J. Med. Chem.* **30**, 1469 (1987). Degradation in solution: T. Marunaka *et al.*, *Chem. Pharm. Bull.* **36**, 4478 (1988); in solid state: E. Matsushima *et al.*, *ibid.* 4593. β-Lactamase inhibiting activity in comparison with clavulanic acid and sulbactam, *q.q.v.*, vs aerobes: M. R. Jacobs *et al.*, *Antimicrob. Agents Chemother.* **29**, 980 (1986); vs anaerobes: P. C. Appelbaum *et al.*, *ibid.* **30**, 789. HPLC determn in biological materials: T. Marunaka *et al.*, *J. Chromatogr.* **431**, 87 (1988). Clinical trial in combination with piperacillin, *q.v.*: I. M. Gould *et al.*, *Drugs Exp. Clin. Res.* **17**, 187 (1991).

White to pale yellow crystalline powder. Sol in DMF; slightly sol in water, methanol, acetone, alc; very slightly sol in ethyl acetate, ethyl ether, chloroform. Insol in hexane.

Sodium salt. [89785-84-2] YTR-830; CL-307579. $C_{10}H_{11}N_4$-NaO_5S; mol wt 322.27. Amorphous solid, mp >170° (dec).

Combination of sodium salt with piperacillin sodium. Tazocilline; Tazocin; Zosyn.

THERAP CAT: In combination with β-lactam antibiotics as antibacterial.

9221. TCDD. [1746-01-6] 2,3,7,8-Tetrachlorodibenzo[*b*,*e*]-[1,4]dioxin; 2,3,7,8-tetrachlorodibenzo-*p*-dioxin; 2,3,6,7-tetrachlorodibenzodioxin; dioxin; TCDBD. $C_{12}H_4Cl_4O_2$; mol wt 321.96. C 44.77%, H 1.25%, Cl 44.04%, O 9.94%. Highly toxic contaminant; produced as a by-product during the manuf of chlorinated phenols (2,4,5-trichlorophenol, *q.v.*) and phenoxyherbicides (2,4-D and 2,-4,5-T, *q.q.v.*), chlorine bleaching of paper pulp and combustion of chlorine-containing waste. Prepn: W. Sandermann, *Ber.* **90**, 690 (1957); M. Tomita *et al.*, *Yakugaku Zasshi* **79**, 186 (1959), *C.A.* **53**, 13152d (1959). Crystal structure: F. P. Boer *et al.*, *Acta Crystallogr.* **28B**, 1023 (1972). Environmental degradation: D. G. Crosby, A. S. Wong, *Science* **195**, 1337 (1976). Comprehensive review of formation, chemistry, and toxic and environmental effects: *Adv. Chem. Ser.* **120**, entitled "Chlorodioxins: Origin and Fate," E. H. Blair, Ed. (ACS, Washington DC, 1973) 141 pp. Toxicity and metabolism: B. A. Schwetz *et al.*, *ibid.* 55-69; A. Poland, A. Kende, *Fed. Proc.* **35**, 2404 (1976). Special issue: *Chem. Eng. News* **61** (June 6, 1983). Review of toxicology and human exposure: *Toxicological Profile for Chlorinated Dibenzo-p-Dioxins* (PB99-121998, 1998) 721 pp; of epidemiological data: L. Tollefson, *Regul. Toxicol. Pharmacol.* **13**, 150-169 (1991); of receptor binding and mechanism of toxicity: J. P. Whitlock, Jr., *Annu. Rev. Pharmacol. Toxicol.* **30**, 251-277 (1990); of carcinogenicity: J. Huff *et al.*, *ibid.* **34**, 343-372 (1994). Review of GC high resolution mass spectrometric analysis: J.-F. Focant *et al.*, *J. Chromatogr. A* **1067**, 265-275 (2005).

Needles, mp 295° (Tomita); crystals from anisole, mp 320-325° (Sandermann). LD_{50} in male, female rats (mg/kg): 0.022, 0.045 orally (Schwetz).

Note: An industrial accident during the manufacture of 2,4,5-trichlorophenol in Seveso, Italy on July 10, 1976 caused the release

of an estimated two to ten pounds of TCDD into the environment. Concentrations as high as 51.3 ppm TCDD were found in some samples: R. Rawls, D. A. O'Sullivan, *Chem. Eng. News* **54**, 27 (Aug. 23, 1976); A. Hay, *Nature* **262**, 636 (1976). TCDD, as a contaminant created in the manufacture of *Agent Orange*, a widely used defoliant in Vietnam during the 1960's, has also been implicated as the causative agent of various symptoms described by veterans exposed to the defoliant; *see:* C. Holden, *Science* **205**, 770 (1979).

Caution: Potential symptoms of overexposure to TCDD are eye irritation; allergic dermatitis, chloracne; porphyria; GI disturbances; possible reproductive and teratogenic effects. *See NIOSH Pocket Guide to Chemical Hazards* (DHHS/NIOSH 97-140, 1997) p 298. Toxic effects in animals include the wasting syndrome, gastric ulcers, immunotoxicity, hepatotoxicity, hepatoporphyria, vascular lesions, chloracne, teratogenicity, fetotoxicity, impaired reproductive performance, endometriosis and delayed death. *See:* Poland, Kende, *loc. cit.*; C. D. Carter *et al.*, *Science* **188**, 738 (1975). This substance is listed as a known human carcinogen: *Report on Carcinogens, Twelfth Edition* (PB2011-111646, 2011) p 396.

Note: TCDD is listed as a persistent organic pollutant (POP) in Annex C of the *Stockholm Convention on Persistent Organic Pollutants* (United Nations, Stockholm, 2001) 43 pp; amended (Geneva, 2009) 63 pp.

9222. TCEP. [5961-85-3] 3,3′,3″-Phosphinidynetrispropanoic acid; tris(2-carboxyethyl)phosphine. $C_9H_{15}O_6P$; mol wt 250.19. C 43.21%, H 6.04%, O 38.37%, P 12.38%. Disulfide reducing agent. Prepn: M. M. Rauhut *et al.*, *J. Am. Chem. Soc.* **81**, 1103 (1959); and mechanism: J. A. Burns *et al.*, *J. Org. Chem.* **56**, 2648 (1991). Use as a reducing agent in measurement of redox activity: M. Neuburger *et al.*, *Eur. J. Biochem.* **267**, 2882 (2000); in determn of aminothiols: J. Krijt *et al.*, *Clin. Chem.* **47**, 1821 (2001); in protein sequencing: T.-Y. Yen *et al.*, *J. Mass Spectrom.* **37**, 15 (2002); B. P. English *et al.*, *J. Am. Chem. Soc.* **124**, 4995 (2002).

Hydrochloride. [51805-45-9] Odorless, air-stable, white crystalline solid. mp 176°. uv max: 192, 218 nm (ε 150, 180 L mol^{-1} cm^{-1}). Water soluble. pKa 7.66. When pH of dilute solutions is less than pKa, air oxidation occurs slowly.

USE: Biochemical tool for selective reduction of disulfide bridges at low pH; reductant for redox assays.

9223. TCMTB. [21564-17-0] Thiocyanic acid (2-benzothiazolylthio)methyl ester; 2-(thiocyanomethylthio)benzothiazole; Bulab 6009; Busan 72; Busan 1118. $C_9H_6N_2S_3$; mol wt 238.34. C 45.35%, H 2.54%, N 11.75%, S 40.35%. Antimicrobial agent used as a substitute for chlorophenols in industrial applications. Prepn: A. G. M. Willems *et al.*, *Rec. Trav. Chim.* **90**, 97 (1971). Antimicrobial spectrum and field study: W. G. Guthrie, R. Elsmore, *Spec. Chem.* **10**, 345 (1990). Environmental fate: T. Reemtsma *et al.*, *Environ. Sci. Technol.* **29**, 478 (1995). HPLC determn in wastewater: J. S. Warner *et al.*, *Determination of MBTS and TCMTB in Industrial and Municipal Wastewaters* (PB85-189025, 1985) 45 p; in seawater: K. Martínez *et al.*, *J. Chromatogr. A* **879**, 27 (2000). Use as wood preservative: K. Tsunoda, *Wood Res.* **77**, 58 (1990); as biocide vs zebra mussels: I. D. Martin *et al.*, *Arch. Environ. Contam. Toxicol.* **24**, 381 (1993); as fungicide on leather: L. Muthusubramanian *et al.*, *J. Soc. Leather Technol. Chem.* **82**, 22 (1998). Toxicity studies: R. E. Rush *et al.*, *Acute Toxic. Data* **1**, 206, 207 (1992).

Oil with pungent odor. d^{25} 1.05 (c = 0.30). Flash point (open cup): 66°C. Insol in water. Sol in most organic solvents. LD_{50} in male, female rats (mg/kg): 752, 679 orally (Rush).

USE: Wood preservative, marine biocide, fungicide.

9224. TDCPP. [13674-87-8] 1,3-Dichloro-2-propanol phosphate (3:1); phosphoric acid tris(1,3-dichloro-2-propyl)ester; tris-

(1,3-dichloroisopropyl)phosphate; tris[2-chloro-1-(chloromethyl)-ethyl]phosphate; Fyrol FR-2. $C_9H_{15}Cl_6O_4P$; mol wt 430.89. C 25.09%, H 3.51%, Cl 49.36%, O 14.85%, P 7.19%. Flame retardant formerly used in children's sleepwear; once considered as a potential replacement for Tris-BP, *q.v.* Prepn: W. J. Jones *et al., J. Chem. Soc.* **1946**, 824. Use in flameproofing: R. J. Polacek, **US 3041293** (1962 to Celanese). Thermal studies: K. Paciorek *et al., Am. Ind. Hyg. Assoc. J.* **39**, 633 (1978); N. Inagaki *et al., J. Appl. Polym. Sci.* **21**, 217 (1977); ibid. **24**, 1 (1979). Mutagenicity studies: M. D. Gold *et al., Science* **200**, 785 (1978); A. Nakamura *et al., Mutat. Res.* **66**, 373 (1979); D. Brusick *et al., J. Environ. Pathol. Toxicol.* **3**, 207 (1979). Toxicity study: J. K. Piotrowski *et al., Bromatol. Chem. Toksykol.* **9**, 141 (1976), *C.A.* **85**, 104825u (1976).

Viscous liquid, bp_5 236-237°. n_D^{20} 1.5022. Soly in water ~100 ppm. LD_{50} orally in rats: 1.85 g/kg (Piotrowski).

USE: Flame retardant.

9225. Tea. Evergreen shrub or tree, *Camellia sinensis* (L.) O. Kuntze, *Theaceae*, also known as *Thea sinensis*. Tender leaves and buds used commercially to produce the beverage; tea extracts are also used in traditional medicine, esp for their antineoplastic, stimulant, and diuretic properties. *Habit.* Discovered in China 4-6000 yrs ago, also indigenous to India, Japan, Sri Lanka, Indonesia and other tropical and subtropical countries. Approx. 2.5 million tons of dried leaves are used in tea production each yr. Leaves are quickly steamed to prevent enzymatic oxidation in production of **green tea**, partially oxidized to a green/brown color in **oolong tea**, and rolled, sifted and oxidized to a copper color in **black tea**. *Constit.* Caffeine, theobromine, theophylline, amino acids, proteins, flavonoids, lignin, organic acids, chlorophyll, polysaccharides, polyphenols. Beneficial health effects have been linked to the polyphenols known as catechins, esp epigallocatechin gallate (EGCG), *q.v.* Determn of catechins: J. J. Dalluge, B. C. Nelson, *J. Chromatogr. A* **881**, 411 (2000). Review of tea manufacture: M. A. Bokuchava, N. I. Skobeleva, *Crit. Rev. Food Sci. Nutr.* **12**, 303-370 (1980); and plant history, tea chemistry and consumption: H. N. Graham in *The Methylxanthine Beverages and Foods: Chemistry, Consumption, and Health Effects*, G. A. Spiller, Ed. (Alan R. Liss, New York, 1984) pp 29-74; of chromatographic determn: A. Finger *et al., J. Chromatogr.* **624**, 293-315 (1992); of potential therapeutic antioxidant properties: L. A. Mitscher *et al., Med. Res. Rev.* **17**, 327-365 (1997); of botany and horticulture: L. Manivel, *Hort. Rev.* **22**, 267-295 (1998); of potential therapeutic dermatologic applications: A. F. Alexis *et al., Int. J. Dermatol.* **38**, 735-743 (1999); of anti-cariogenic properties: J. M. T. Hamilton-Miller, *J. Med. Microbiol.* **50**, 299-302 (2001); of inhibition of carcinogenesis: C. S. Yang *et al., Annu. Rev. Pharmacol. Toxicol.* **42**, 25-54 (2002).

Astringent, bitter taste. Black tea is faintly aromatic; green tea is practically odorless.

USE: Stimulant beverage; flavoring in food products; ingredient in skincare products and shampoos.

9226. Tea Tree Oil. [68647-73-4] Melaleuca oil; tea tree (*Melaleuca alternifolia*) oil. Native Australian medicinal oil obtained from the leaves and terminal branchlets of the tea tree, *Melaleuca alternifolia* Myrtaceae. A complex mixture of hydrocarbons and terpenes of ~100 components; the two standardized components for commercial use are 1,6-cineole (max of 15%) and the 1-terpinen-4-ol (min of 30%). Description: A. R. Penfold, *J. Proc. R. Soc. N.S.W.* **59**, 306 (1925). Analysis and composition: J. J. Brophy *et al., J. Agric. Food Chem.* **37**, 1330 (1989). Toxicity study: D. Villar *et al., Vet. Hum. Toxicol.* **36**, 139 (1994). Antimicrobial activity: C. F. Carson, T. V. Riley, *J. Appl. Bacteriol.* **78**, 264 (1995); and mode of action: S. D. Cox *et al., ibid.* **88**, 170 (2000). Antifungal activity: K. A. Hammer *et al., Antimicrob. Agents Chemother.* **44**, 467 (2000).

Clinical evaluation in oral candidiasis in AIDS patients: A. Jandourek *et al., AIDS* **12**, 1033 (1998). Review of early work: A. R. Penfold, F. R. Morrison, "Tea Tree Oils" in *The Essential Oils* **4**, E. Guenther, Ed. (Van Nostrand Co., New York, 1950) pp 526-548. Review: J. A. Staton, *Proc. 3rd. Sci. Conf. Asian Soc. Cosmet. Sci. Taipei*, 18-24 (1997). Review of clinical trials: E. Ernst, A. Huntley, *Forsch. Komplementärmed.* **7**, 17-20 (2000); of antimicrobial properties: L. Halcón, K. Milkus, *Am. J. Infect. Control* **32**, 402-408 (2004); and medicinal properties: C. F. Carson *et al., Clin. Microbiol. Rev.* **19**, 50-62 (2006). Review of toxicity: K. A. Hammer *et al., Food Chem. Toxicol.* **44**, 616-625 (2006).

Colorless to pale-yellow, clear, mobile liquid with fresh terpene type odor with nutmeg associations and possibly with citrus or floral undertones. n_D^{20} 1.4760-1.4810. $[\alpha]_D$ +6°48′ to +9°48′. d_{15}^{15} 0.8950-0.9050. Soln in 80% alcohol (w/w): 0.6 to 0.8 volumes. Sparingly sol in water, misc in nonpolar solvents. LD_{50} orally in rats: 1.9-2.6 ml/kg; dermally in rabbits: 5.0 g/kg (Halcón).

USE: Flavoring and antiseptic agent in personal hygiene items such as toothpaste; as a cosmeceutical.

9227. Tebanicline. [198283-73-7] 5-[(2*R*)-2-Azetidinylmethoxy]-2-chloropyridine; ABT-594. $C_9H_{11}ClN_2O$; mol wt 198.65. C 54.42%, H 5.58%, Cl 17.85%, N 14.10%, O 8.05%. Nonopioid 3-pyridyl ether analgesic acting acting via a neuronal nicotinic acetylcholine receptor mediated mechanism. Prepn: M. W. Holladay *et al.,* **WO 9640682**; M. A. Abreo *et al.,* **US 5948793** (1996, 1999 both to Abbott); as hydrochloride: M. W. Holladay *et al., J. Med. Chem.* **41**, 407 (1998). Asymmetric synthesis: J. K. Lynch *et al., Tetrahedron: Asymmetry* **9**, 2791 (1998). Analgesic activity and receptor affinity: A. W. Bannon *et al., Science* **279**, 77 (1998). Pharmacology *in vitro*: D. L. Donnelly-Roberts *et al., J. Pharmacol. Exp. Ther.* **285**, 777 (1998); *in vivo*: A. W. Bannon *et al., ibid.* 787. Efficacy in neuropathic pain models: *idem et al., Brain Res.* **801**, 158 (1998).

Hydrochloride. [203564-54-9] mp 116-117°. $[\alpha]_D^{23}$ +8.6° (c = 0.52 in methanol).

Tosylate. [198283-74-8] $C_9H_{11}ClN_2O.C_7H_8O_3S$; mol wt 370.85.

USE: Analgesic agent in pain models.

9228. Tebbe Reagent. [67719-69-1] μ-Methylene-μ-chlorobis(η^5-2,4-cyclopentadien-1-yl)(dimethylaluminum)titanium; (μ-chloro)(μ-methylene)bis(cyclopentadienyl)(dimethylaluminum)-titanium. $C_{13}H_{18}AlClTi$; mol wt 284.59. C 54.87%, H 6.38%, Al 9.48%, Cl 12.46%, Ti 16.82%. Titanium-aluminum complex used in the Tebbe Olefination. Prepn: F. N. Tebbe *et al., J. Am. Chem. Soc.* **100**, 3611 (1978). Improved prepn: L. F. Cannizzo, R. H. Grubbs, *J. Org. Chem.* **50**, 2386 (1985). Ab initio structural model: M. M. Francl, W. J. Hehre, *Organometallics* **2**, 457 (1983). Carbonyl methylenation applications compared with Wittig reactions: S. H. Pine *et al., J. Org. Chem.* **50**, 1212 (1985); *idem et al., Synthesis* **1991**, 165. Review of methylenation chemistry: *Org. Synth. Highlights*, J. Mulzer *et al.,* Eds. (VCH, Weinheim, 1991) pp 192-196.

Reddish-orange crystals. Sol in toluene, benzene. *Sensitive to air and water.*

USE: Reagent for the methylenation of carbonyl groups.

9229. Tebuconazole. [107534-96-3] α-[2-(4-Chlorophenyl)-ethyl]-α-(1,1-dimethylethyl)-1*H*-1,2,4-triazole-1-ethanol; (*RS*)-1-(4-chlorophenyl)-4,4-dimethyl-3-(1*H*-1,2,4-triazol-1-ylmethyl)pentan-3-ol; ethyltrianol; fenetrazole; terbuconazole; terbutrazole; BAY

HWG 1608; HWG-1608; Corail; Elite; Folicur; Horizon; Lynx; Raxil. $C_{16}H_{22}ClN_3O$; mol wt 307.82. C 62.43%, H 7.20%, Cl 11.52%, N 13.65%, O 5.20%. Ergosterol biosynthesis inhibitor. Prepn: G. Holmwood et al., **EP 40345**; eidem, **US 4723984** (1981, 1988 both to Bayer). Synthesis of enantiomers: J. Kaulen, Angew. Chem. Int. Ed. **28**, 462 (1989). Photodegradation: H. Wamhoff et al., Z. Naturforsch. **49b**, 280 (1994). GC determn in plant material, soil and water: W. Maasfeld, Pflanzenschutz-Nachr. Bayer (Engl. Ed.) **40**, 29 (1987). Review of chemistry and biochemistry: D. Berg et al., ibid. 111-132. Series of articles on field trials: ibid. **42**, 91-222 (1989); BCPC Monograph **57**, 97-108 (1994).

Colorless crystals, mp 104.7°. Vapor pressure (20°) $<10^{-5}$ mbar. Soly (g/l at 20°): water 0.032, dichloromethane >200, n-hexane 2-5, 2-propanol, 100-200, toluene 50-100.

USE: Fungicide.

9230. Tebufenozide. [112410-23-8] 3,5-Dimethylbenzoic acid 1-(1,1-dimethylethyl)-2-(4-ethylbenzoyl)hydrazide; N-tert-butyl-N'-(4-ethylbenzoyl)-3,5-dimethylbenzoylhydrazide; RH-5992; Confirm; Mimic. $C_{22}H_{28}N_2O_2$; mol wt 352.48. C 74.97%, H 8.01%, N 7.95%, O 9.08%. Synthetic, nonsteroidal ecdysone agonist causing premature molting; insect growth regulator specific to lepidopteran species. Prepn: A. C. Hsu, H. E. Aller, **AU 64289**; eidem, **US 4985461** (1986, 1991 both to Rohm & Haas). Properties and field evaluation: J. J. Heller et al., Brighton Crop Prot. Conf. - Pests Dis. **1992**, 59. Determn by LC in environmental samples: K. M. S. Sundaram et al., J. AOAC Int. **76**, 668 (1993); by GC and LC in formulations: idem, et al., J. Chromatogr. A **687**, 323 (1994). Soil adsorption: idem, J. Environ. Sci. Health **B29**, 415 (1994). Insecticidal activity: G. Smagghe, D. Degheele, Pestic. Sci. **42**, 85 (1994). Mechanism of action: A. Retnakaran et al., Insect Biochem. Mol. Biol. **25**, 109 (1995). Aquatic toxicology: D. P. Kreutzweiser et al., Ecotoxicol. Environ. Saf. **28**, 14 (1994). Review of biological activity and use: H. Oberlander et al., Arch. Insect Biochem. Physiol. **28**, 209-223 (1995).

mp 191°. Also reported as mp 186-188° (Sundaram, 1081). Vapor pressure (25°): 3×10^{-8} mm Hg. Also reported as vapor pressure (25°): 1.7×10^{-6} mm Hg (Sundaram, 415). Soly (25°): 0.83 μg/ml water. Partition coefficient (octanol/water): 32000. LD_{50} orally in rats, mice: >5000 mg/kg; dermally in rats: >5000 mg/kg; LD_{50} in honey bees (96 hr, contact): >234 μg/bee; LC_{50} in mallard duck (8-day dietary): >5000 mg/kg; LC_{50} in rainbow trout (96 hr): 5.7 mg/l (Heller).

USE: Insecticide.

9231. Tebufenpyrad. [119168-77-3] 4-Chloro-N-[[4-(1,1-dimethylethyl)phenyl]methyl]-3-ethyl-1-methyl-1H-pyrazole-5-carboxamide; N-(4-t-butylbenzyl)-4-chloro-3-ethyl-1-methylpyrazole-5-carboxamide; MK-239; Pyranica. $C_{18}H_{24}ClN_3O$; mol wt 333.86. C 64.76%, H 7.25%, Cl 10.62%, N 12.59%, O 4.79%. Used to control plant-feeding mites on cotton, citrus, fruits, vegetables, and ornamentals; inhibits electron transport during mitochondrial respiration. Prepn: I. Okada et al., **EP 289879**; eidem, **US 4950668** (1988, 1990 both to Mitsubishi Kasei). GC/MS determn in water, soil, and urine samples: A. Navalón et al., Chromatographia **54**, 377 (2001). Field trial vs mites in apple orchards: W. G. Thwaite et

al., Exp. Appl. Acarol. **20**, 177 (1996); in honey bee colonies: M. A. Ali et al., J. Econ. Entomol. **96**, 259 (2003). Mite toxicity study: Y.-J. Kim et al., ibid. **92**, 187 (1999). Review: K. Inoue, T. Fukuchi, Agrochem. Jpn. **64**, 12-14 (1994).

Off-white solid, slight aromatic odor, mp 61-62°. Vapor pressure (25°): 1.0×10^{-5} Pa. LD_{50} in male, female rats (mg/kg): 595, 997 orally; >2000, >2000 dermally. LC_{50} in male rats (mg/m³): 2660 by inhalation (Inoue, Fukuchi).

USE: Acaricide.

9232. Tebuthiuron. [34014-18-1] N-[5-(1,1-Dimethylethyl)-1,3,4-thiadiazol-2-yl]-N,N'-dimethylurea; 1-(5-tert-butyl-1,3,4-thiadiazol-2-yl)-1,3-dimethylurea; EL-103; Graslan; Perflan; Spike. $C_9H_{16}N_4OS$; mol wt 228.31. C 47.35%, H 7.06%, N 24.54%, O 7.01%, S 14.04%. Broad spectrum pre- and post-emergent herbicide. Prepn: **GB 1266172** (1972 to Air Prod, Chem.); E. V. P. Tao, **BE 799575**; idem, **US 3803164** (1973, 1974 both to Lilly); idem, Synth. Commun. **4**, 249 (1974). Herbicidal activity: A. T. Perkins et al., Proc. South. Weed Sci. Soc. **30**, 326 (1977). Toxicity study: G. C. Todd et al., Food Cosmet. Toxicol. **12**, 461 (1974). Metabolism in animals: D. M. Morton, D. G. Hoffman, J. Toxicol. Environ. Health **1**, 757 (1976). Determn by HPLC: J. H. Kennedy, J. Chromatogr. Sci. **15**, 79 (1977); by GC: A. Loh et al., J. Agric. Food Chem. **26**, 410 (1978). Field trial against woody shrubs: C. H. Herbel et al., J. Range Manage. **38**, 391 (1985).

mp 160-163°. Moderately sol in water (2500 ppm); sol in methanol (170,000 ppm) (company literature). LD_{50} in mice, rats, rabbits (mg/kg): 579, 644, 286 orally (Todd).

USE: Herbicide.

9233. Tecadenoson. [204512-90-3] N-[(3R)-Tetrahydro-3-furanyl]adenosine; CVT-510. $C_{14}H_{19}N_5O_5$; mol wt 337.34. C 49.85%, H 5.68%, N 20.76%, O 23.71%. Selective adenosine A_1-receptor agonist. Prepn: R. T. Lum et al., **WO 9808855**; eidem, **US 5789416** (both 1998 to CV Therapeutics). Clinical effect on AV nodal conduction: B. B. Lerman et al., J. Cardiovasc. Pharmacol. Ther. **6**, 237 (2001). Clinical evaluation in paroxysmal supraventricular tachycardia: E. N. Prystowsky et al., J. Am. Coll. Cardiol. **42**, 1098 (2003); K. A. Ellenbogen et al., Circulation **111**, 3202 (2005). Review of pharmacology and clinical experience: A. Zaza, Curr. Opin. Investig. Drugs **3**, 96-100 (2002); J. W. Cheung, B. B. Lerman, Cardiovasc. Drug Rev. **21**, 277-292 (2003).

THERAP CAT: Antiarrhythmic.

9234. Technetium. [7440-26-8] Tc; at. no. 43. Usual valences 4 and 7; 3 less common. Group VIIB (7). Radioactive element; no stable nuclides. Discovery claimed by Noddack, Tacke, and Berg who called it "masurium"; the existence of masurium has never been confirmed by isoln of the element. First artificially produced element. Named from the Greek word for "artificial"; separated from a molybdenum plate that had been bombarded for a few months with a strong beam of deuterons in the Berkeley cyclotron: Perrier, Segré, *Nature* **140**, 193 (1937); *eidem, J. Chem. Phys.* **5**, 712 (1937); Cacciapuoti, Segré, *Phys. Rev.* **52**, 1252 (1937). Most commonly available isotope (mass number): 99 ($T_{1/2}$ 2.12×10^5 years, rel. at. mass 98.9063). Other long-lived isotopes: 97 ($T_{1/2}$ 2.6×10^6 years, rel. at. mass 96.9064); 98 ($T_{1/2}$ 4.2×10^6 years, longest-lived isotope, rel. at. mass 97.9072). Known artificial isotopes: 92-107. Prepn of metal: Cobble *et al., J. Am. Chem. Soc.* **74**, 1852 (1952). Comprehensive reviews: Boyd, *J. Chem. Educ.* **36**, 3-14 (1959); Schwochau, *Angew. Chem.* **76**, 9-19 (1964); Kotegov *et al.,* in *Adv. Inorg. Chem. Radiochem.* **11**, 1-90 (1968); Peacock in *Comprehensive Inorganic Chemistry* vol. 3, J. C. Bailar Jr. *et al.,* Eds. (Pergamon Press, Oxford, 1973) pp 877-903. Pharmacokinetics and organ distribution of radiopharmaceuticals: O. P. D. Noronha, K. S. Venkateswarlu, *Eur. J. Nucl. Med.* **6**, 121 (1980).

Close-packed hexagonal structure; isomorphous with rhodium, ruthenium, and osmium. The element obtained by hydrogen reduction of ammonium pertechnate is a silver-gray spongy mass which tarnishes slowly in moist air. mp 2250 ±50°. In its chemical behavior resembles rhenium. The element is precipitated by hydrogen sulfide from hydrochloric acid (diluted up to 5N); for higher acid concns the precipitation is not complete. Is oxidized by hydrogen peroxide in alkaline soln into sol anions: Perrier, Segré, *Atti R. Accad. Lincei* [6] **27**, 579 (1938); Segré, *Nature* **143**, 460 (1939). Reacts with dil or concd nitric acid, in aqua regia, in concd sulfuric acid; not with HCl. Burns in fluorine to form penta- and hexafluorides. Combines with sulfur at high temp to form disulfide; with carbon to form TcC.

USE: Minute quantities of TcO_4^- ion exert remarkable inhibition of the corrosion of soft iron in neutral aq soln: Cartledge, *J. Am. Chem. Soc.* **77**, 2658 (1955).

9235. Technetium Tc 99m Apcitide. [178959-14-3] [[*N*-(2-Mercaptoacetyl)-D-tyrosyl-*S*-(3-aminopropyl)-L-cysteinylglycyl-L-α-aspartyl-L-cysteinylglycylglycyl-*S*-[(acetylamino)methyl]-L-cysteinylglycyl-*S*-[(acetylamino)methyl]-L-cysteinylglycyl-κ*N*-glycyl-κ*N*-L-cysteinamide-κ*N*^{2},κ*S*] cyclic (1 → 5)-thioetherato(5−)oxotechnetate(2−)-^{99m}Tc sodium hydrogen (1:1:1); ^{99m}Tc-P246; AcuTect. Radiolabeled synthetic peptide that preferentially binds to glycoprotein GP IIb/IIIa receptors on the surface of activated platelets. Monomer designed as a scintigraphic imaging agent for detection and localization of acute deep vein thrombosis (DVT). Prepn: R. T. Dean, J. Lister-James, **WO 9325244**; *eidem,* **US 5508020** (1993, 1996 both to Diatech); D. A. Pearson *et al., J. Med. Chem.* **39**, 1372 (1996). Pharmacology and immunoscintigraphy: J. Lister-James *et al., J. Nucl. Med.* **37**, 755 (1996). Clinical study in DVT: R. Taillefer *et al., ibid.* **41**, 1214 (2000).

Dimer. [153507-46-1] Bibapcitide; P-280. $C_{112}H_{162}N_{36}O_{43}S_{10}$; mol wt 3021.34. C 44.52%, H 5.40%, N 16.69%, O 22.77%, S 10.61%. Synthetic peptide consisting of 26 amino acids; exhibits the same GP IIb/IIIa binding activity as the monomer. Splits to form monomer upon reaction with sodium pertechnetate ^{99m}Tc, *q.v.*

THERAP CAT: Diagnostic aid (radioactive imaging agent).

9236. Technetium Tc 99m Bicisate. [121281-41-2] (*SP*-5-35)-[[1-1′-Diethyl-*N*,*N*′-1,2-ethanediylbis[L-cysteinato-κ*N*,κ*S*]](3−)]oxotechnetium-^{99m}Tc; [*N*,*N*′-ethylenedi-L-cysteinato(3−)]oxo[^{99m}Tc]technetium(V) diethyl ester; technetium ^{99m}Tc ethyl cysteinate dimer; ^{99m}Tc-L,L-ECD; RP-217; Neurolite. $C_{12}H_{21}N_2O_5$-S_2^{99m}Tc. Neutral, lipophilic complex which rapidly crosses blood-brain barrier. Prepn: P. L. Bergstein *et al.,* **EP 279417** (1988 to DuPont); *eidem,* **US 5279811** (1994 to DuPont Merck). Pharmacology: R. C. Walovitch *et al., J. Nucl. Med.* **30**, 1892 (1989). Clinical pharmacokinetics: S. Vallabhajosula *et al., ibid.* 599. Metabolism in humans: R. C. Walovitch *et al., Neuropharmacology* **30**, 283 (1991). TLC determn of radiochemical purity: J. M. Green *et al., J. Nucl. Med. Technol.* **22**, 21 (1994). Series of articles on clinical pharmacology and diagnostic use in SPECT brain imaging: *J. Cereb. Blood Flow Metab.* **14**, Suppl. 1, S1-S120 (1994).

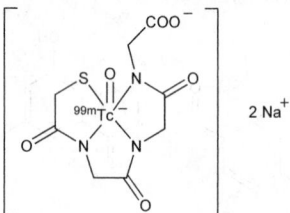

THERAP CAT: Diagnostic aid (radioactive imaging agent).

9237. Technetium Tc 99m Fanolesomab. [225239-31-6] Anti-(human CD15 (antigen)) immunoglobulin M (mouse monoclonal RB5 μ-chain) disulfide with mouse monoclonal RB5 light chain, pentamer, technetium-^{99m}Tc salt; ^{99m}Tc-anti-SSEA-1; LeuTech; NeutroSpec. Murine monoclonal antibody directed against stage-specific embryonic antigen-1 (SSEA-1); binds to cluster designation 15 (CD15) antigen expressed on human neutrophils. Antigranulocyte antibody designed for *in vivo* functional imaging and diagnosis of infection. Prepn of antibody: D. Solter, B. B. Knowles, *Proc. Natl. Acad. Sci. USA* **75**, 5565 (1978); of labeled form: M. L. Thakur *et al., J. Nucl. Med.* **29**, 1817 (1988). *See also* for diagnostic use: *idem,* **US 4917878** (1990 to Thomas Jefferson Univ.). *In vivo* labeling and imaging of neutrophils: *idem et al., J. Nucl. Med.* **37**, 1789 (1996). Clinical biodistribution and radiation dosimetry study: P. D. Mozley *et al., ibid.* **40**, 625 (1999). Clinical evaluation as imaging agent for osteomyelitis: C. J. Palestro *et al., J. Foot Ankle Surg.* **42**, 2 (2003); for appendicitis: E. B. Rypins *et al., Ann. Surg.* **235**, 232 (2002). Review of development and clinical use: C. Love, C. J. Palestro, *IDrugs* **6**, 1079-1085 (2003).

THERAP CAT: Diagnostic aid (radioactive imaging agent).

9238. Technetium Tc 99m Mertiatide. [125224-05-7]; [104348-91-6]. (*SP*-5-25)-[*N*-[*N*-[*N*-(Mercaptoacetyl)glycyl]glycyl]glycinato(5−)-*N*,*N*′,*N*″,*S*]oxotechnetate(2−)-^{99m}Tc disodium; ^{99m}Tc-MAG$_3$; ^{99m}Tc mercaptoacetyltriglycine; TechneScan MAG3. $C_8H_8N_3Na_2O_6S$^{99m}Tc. Nonisomeric mercaptide compound complexed with technetium; radioactive renal function imaging agent. Prepn and biodistribution: A. R. Fritzberg *et al., J. Nucl. Med.* **27**, 111 (1986); *idem et al.,* **EP 284071** (1988 to Neorx). Stability and HPLC determn of radiochemical purity: A. M. Millar *et al., Nucl. Med. Commun.* **11**, 405 (1990). Clinical pharmacokinetics: B. Bubeck *et al., J. Nucl. Med.* **31**, 1285 (1990). Diagnostic use in captopril renography: I. E. Datseris *et al., ibid.* **35**, 251 (1994); in renal transplant evaluation: J. P. O'Malley *et al., Clin. Nucl. Med.* **18**, 22 (1993). Review of synthesis and diagnostic use: D. Eshima, A. Taylor, Jr., *Semin. Nucl. Med.* **22**, 61-73 (1992).

THERAP CAT: Diagnostic aid (radioactive imaging agent).

9239. Technetium Tc 99m Sestamibi. [109581-73-9] (*OC*-6-11)-Hexakis(1-isocyano-2-methoxy-2-methylpropane)technetium(1+)-^{99m}Tc; hexakis(2-methoxy-2-methylpropyl isocyanide)-[^{99m}Tc] technetium(1+); ^{99M}Tc hexakis(2-methoxyisobutylisonitrile); ^{99m}Tc-HEXAMIBI; ^{99m}Tc-MIBI; ^{99m}Tc-RP-30A; Cardiolite. $C_{36}H_{66}N_6O_6$^{99m}Tc. Lipophilic, cationic isonitrile compound complexed with technetium; radioactive myocardial perfusion imaging agent. Prepn: P. L. Bergstein, N. Subramanyam, **EP 233368** (1987 to Du Pont); P. Angelberger, E. Zbiral, *Oesterr. Forschungszent. Seibersdorf* **4411**, I (1987). Preparative TLC purification: A. Proulx *et al., Appl. Radiat. Isot.* **40**, 95 (1989). Biodistribution and comparative clinical evaluation as myocardial perfusion imaging agent: F. J. Th. Wackers *et al., J. Nucl. Med.* **30**, 301 (1989). Clinical evaluation as parathyroid imaging agent: A. J. Coakley *et al., Nucl. Med. Commun.* **10**, 791 (1989). Symposium on pharmacology and clinical studies: *Am. J. Cardiol.* **66**, 1E-96E (1990). Review of cardiac imaging: J. A. Leppo *et al., J. Nucl. Med.* **32**, 2012-2022 (1991).

THERAP CAT: Diagnostic aid (radioactive imaging agent).

9240. Technetium Tc 99m Teboroxime. [104716-22-5] (*TPS*-7-1-232′4′54)-Chloro[bis[(1,2-cyclohexanedione 1-oxime 2-oximate-κO) (1−)][[1,2-cyclohexanedione 1-(oxime-κO) 2-oximato] (2−)]methylborato(2−)-κN,κN′,κN″,κN‴,κN⁗,κN⁗⁗]technetium-⁹⁹Tc; (*TPS*-7-1-232′4′54)-[bis[(1,2-cyclohexanedione dioximato)(1−)-O] [(1,2-cyclohexanedione dioximato)(2−)-O]methylborato(2−)-N,N′,N″,N‴,N⁗,N⁗⁗]chlorotechnetium-⁹⁹ᵐTc; CDO-MeB; SQ-30217; Cardiotec. $C_{19}H_{29}BClN_6O_6{}^{99m}Tc$. Neutral, lipophilic technetium complex classified as a boronic acid adduct of technetium dioxime (BATO); radioactive myocardial perfusion imaging agent. Prepn: A. D. Nunn *et al.*, **EP 199260**; *eidem*, **US 4705849** (1986, 1987 both to Squibb); and structural characterization: E. N. Treher *et al.*, *Inorg. Chem.* **28**, 3411 (1989). Pharmacokinetics: R. E. Stewart *et al.*, *J. Nucl. Med.* **31**, 1183 (1990). Reviews of physiologic properties and pharmacology: D. J. Meerdink, J. A. Leppo, *Am. J. Cardiol.* **66**, 9E-15E (1990); of clinical pharmacokinetics and comparative studies of myocardial imaging: L. L. Johnson, D. W. Seldin, *ibid.* 63E-67E; of diagnostic use in coronary artery disease: J. A. Leppo *et al.*, *J. Nucl. Med.* **32**, 2012-2022 (1991).

THERAP CAT: Diagnostic aid (radioactive imaging agent).

9241. Teclothiazide. [4267-05-4] 6-Chloro-3,4-dihydro-3-trichloromethyl-2*H*-1,2,4-benzothiadiazine-7-sulfonamide 1,1-dioxide; 6-chloro-3,4-dihydro-7-sulfamoyl-3-trichloromethyl-2*H*-1,2,4-benzothiadiazine 1,1-dioxide; 3-trichloromethylhydrochlorothiazide; tetrachlormethiazide. $C_8H_7Cl_4N_3O_4S_2$; mol wt 415.08. C 23.15%, H 1.70%, Cl 34.16%, N 10.12%, O 15.42%, S 15.45%. Prepn: Close *et al.*, *J. Am. Chem. Soc.* **82**, 1132 (1960); Novello *et al.*, *J. Org. Chem.* **25**, 970 (1960). Pharmacology and toxicology: M. Auclair *et al.*, *Therapie* **18**, 131, 137 (1963).

Crystals, mp 300-303° (Close); mp 287° (Novello).

Potassium salt. [5306-80-9] PS-207; K-33; Depleil. $C_8H_6Cl_4$-$KN_3O_4S_2$; mol wt 453.17. LD_{50} i.p. in mice: 4.75 g/kg (Auclair).

THERAP CAT: Diuretic.

9242. Teclozan. [5560-78-1] *N,N′*-[1,4-Phenylenebis(methylene)]-bis[2,2-dichloro-*N*-(2-ethoxyethyl)acetamide]; *N,N′*-bis(ethoxyethyl)-*N,N′*-bis(dichloroacetyl)-1,4-xylylenediamine; *N,N′*-bis(dichloroacetyl)-*N,N′*-bis(2-ethoxyethyl)-1,4-bis(aminomethyl)-benzene; teclosan; teclosine; teclozine; NSC-107433; Win-13146; Win-AM-13146; Falmonox. $C_{20}H_{28}Cl_4N_2O_4$; mol wt 502.25. C 47.83%, H 5.62%, Cl 28.23%, N 5.58%, O 12.74%. Prepn: Surrey, Mayer, *J. Med. Pharm. Chem.* **3**, 409 (1961). Amebicidal activity and toxicity: Berberian *et al.*, *Am. J. Trop. Med. Hyg.* **10**, 503 (1961).

Crystals, mp 137.6-143.9°. LD_{50} orally in mice: >8000 mg/kg (Berberian).

THERAP CAT: Antiamebic.

9243. Tecomanine. [6878-83-7] (4*R,7S,7aS*)-1,2,3,4,7,7a-Hexahydro-2,4,7-trimethyl-6*H*-cyclopenta[*c*]pyridin-6-one; [4*R*-(4α,7β,7aβ)]-1,2,3,4,7,7a-hexahydro-2,4,7-trimethyl-6*H*-2-pyridin-6-one; tecomine. $C_{11}H_{17}NO$; mol wt 179.26. C 73.70%, H 9.56%, N 7.81%, O 8.93%. Principal alkaloid isolated from *Tecoma stans* (L.) H.B.K. (*Bignonia stans* L.), Bignoniaceae: Y. Hammouda, M. M. Motawi, *Egypt. Pharm. Bull.* **41**, 73 (1959); *C.A.* **54**, 21646c (1960). Structure: G. Jones *et al.*, *Tetrahedron Lett.* **1963**, 397. Stereochemistry, abs config and crystal structure: G. Jones *et al.*, *Chem. Commun.* **1971**, 994; G. Ferguson, W. C. Marsh, *J. Chem. Soc. Perkin Trans. 2* **1975**, 1124. Hypoglycemic activity of salts: Y. Hammouda *et al.*, *J. Pharm. Pharmacol.* **16**, 833 (1964); Y. Hammouda, M. S. Amer, *J. Pharm. Sci.* **55**, 1452 (1966). Stereoselective total synthesis of (±)-form: T. Imanishi *et al.*, *Tetrahedron Lett.* **22**, 667 (1981). Facile synthesis of (+)-form: T. Kametani *et al.*, *Heterocycles* **26**, 1491 (1987).

Liquid. $bp_{0.1}$ 125°. $[\alpha]_D^{24}$ −175° (c = 1.17 in $CHCl_3$). uv max (alc): 226 nm (log ε 4.10).

Methiodide. $C_{12}H_{20}INO$. Crystals, dec 240-242°.

Methoperchlorate. $C_{12}H_{20}ClNO_5$. Crystals from methanol, mp 242°.

9244. Tecovirimat. [869572-92-9] *rel*-*N*-[(3a*R*,4*R*,4a*R*,5a*S*,-6*S*,6a*S*)-3,3a,4,4a,5,5a,6,6a-Octahydro-1,3-dioxo-4,6-ethenocycloprop[*f*]isoindol-2(1*H*)-yl]benzamide; 4-trifluoromethyl-*N*-(3,3a,4,4a,5,5a,6,6a-octahydro-1,3-dioxo-4,6-ethenocycloprop[*f*]isoindol-2(1*H*)-yl)benzamide; SIGA-246; ST-246. C_{19}-$H_{15}F_3N_2O_3$; mol wt 376.34. C 60.64%, H 4.02%, F 15.14%, N 7.44%, O 12.75%. Orthopoxvirus egress inhibitor; targets F13L protein in vaccinia virus and its homologs in other orthopoxvirus species. Prepn: R. Jordan *et al.*, **WO 04112718** (2004 to Viropharma); *eidem*, **US 7687641** (2010 to Siga Technologies); and structure-activity studies: T. R. Bailey *et al.*, *J. Med. Chem.* **50**, 1442 (2007). Synthesis and crystal structure: X.-B. Zhou *et al.*, *Chinese. J. Struct. Chem.* **29**, 1043 (2010). Preclinical antiviral activity: G. Yang *et al.*, *J. Virol.* **79**, 13139 (2005). Mode of action study in pathogenic orthopoxviruses: S. Duraffour *et al.*, *Antivir. Ther.* **13**, 977 (2008). Clinical safety and pharmacokinetics: R. Jordan *et al.*, *Antimicrob. Agents Chemother.* **54**, 2560 (2010). Review of development for biodefense and therapeutic potential vs smallpox: D. W. Grosenbach *et al.*, *Future Virol.* **6**, 653-671 (2011).

Relative stereochemistry

White solid from ethyl acetate and hexanes, mp 196°. Log P (octanol/water): 2.94. Nonhygroscopic. Soly (mg/ml): pH 6.8 phosphate buffer 0.00700; water 0.0261; 0.01N HCl 0.0375; 1% sodium laurel sulfate 0.0687; 1% Tween 80 0.102; methanol 60.8; ethanol 62.1; acetonitrile 64.0.

Monohydrate. [1162664-19-8] $C_{19}H_{15}F_3N_2O_3.H_2O$; mol wt 394.35. White solid , mp 196-197°.

THERAP CAT: Antiviral.

9245. Tectorigenin. [548-77-6] 5,7-Dihydroxy-3-(4-hydroxyphenyl)-6-methoxy-4H-1-benzopyran-4-one; 4′,5,7-trihydroxy-6-methoxyisoflavone. $C_{16}H_{12}O_6$; mol wt 300.27. C 64.00%, H 4.03%, O 31.97%. Phytoalexin aglycon isolated from rhizomes of *Iris tectorum* Maxim., Iridaceae, the extract of which is used as a traditional Chinese medicine. Isoln of the glycoside and aglycon: S. Shibata, *J. Pharm. Soc. Jpn*. **37**, 380 (1927), *C.A.* **21**, 3050[8] (1927); from *Iris germanica* Linnaeus: A. Kawase *et al.*, *Agric. Biol. Chem.* **37**, 145 (1973). Structure: Asahina *et al.*, *J. Pharm. Soc. Jpn*. **48**, 1087 (1928), *C.A.* **23**, 2718[1] (1929); Shriner *et al.*, *J. Am. Chem. Soc.* **61**, 2322 (1939). Synthesis: Farkas, Várady, *Ber.* **93**, 1269 (1960); W. Baker *et al.*, *J. Chem. Soc. C* **1970**, 1219. Inhibition of cyclooxygenase-2 induction: K. Ohuchi *et al.*, *Recent Adv. Nat. Prod. Res.*, *Proc. 3rd Int. Symp. 1999*, 12-24. Antioxidant effects: K.-T. Lee *et al.*, *Arch. Pharmacal Res.* **23**, 461 (2000).

Pale yellow needles from ethanol, mp 225-226°. uv max (ethanol): 267-269, 338 nm (log ε 4.41, 3.40).

Tectoridin. [611-40-5] Tectorigenin-7-glucoside; shekanin. $C_{22}H_{22}O_{11}$; mol wt 462.41. Identity with shekanin and structure: Mannich *et al.*, *Arch. Pharm.* **275**, 317 (1937). Synthesis: Várady, *Acta Chim. Acad. Sci. Hung.* **48**, 181 (1966). Colorless needles from ethanol, mp 256-258°. uv max (ethanol): 267-269, 336 nm (log ε 4.62, 3.65). $[\alpha]_D^{20}$ −29.4° (pyridine). Sol in pyridine, alcohol, water. Practically insol in organic solvents.

9246. Tedisamil. [90961-53-8] 3′,7′-Bis(cyclopropylmethyl)spiro[cyclopentane-1,9′-[3,7]diazabicyclo[3.3.1]nonane]. $C_{19}H_{32}N_2$; mol wt 288.48. C 79.11%, H 11.18%, N 9.71%. Potassium channel blocking bradycardic agent. Prepn: U. Schön *et al.*, **DE 3234457**; *eidem*, **US 4550112** (1984, 1985 both to Kali-Chemie); and pharmacology: *idem et al.*, *J. Med. Chem.* **41**, 318 (1998). Suppression of the outward transient K+ current in rat myocytes: I. D. Dukes, M. Morad, *Am. J. Physiol.* **275**, H1746 (1989). Toxicity in rats: E. Hayes *et al.*, *Pharmacol. Toxicol.* **73**, 257 (1993). Comparative clinical study in coronary artery disease: V. Mitrovic, *et al.*, *Clin. Cardiol.* **21**, 492 (1998). Clinical study in stable angina: K. M. Fox *et al.*, *Heart* **83**, 167 (2000). Review of pharmacology and clinical development: B. Freestone, G. Y. H. Lip, *Expert Opin. Invest. Drugs* **13**, 151-160 (2004).

Dihydrochloride. [132523-84-3] $C_{19}H_{32}N_2.2HCl$; mol wt 361.40. White crystals, mp 195-197°. bp 230° (0.1 Torr). Sol in water.

Sesquifumarate. [150501-62-5] KC-8857; Pulzium. $(C_{19}H_{32}N_2)_2.3C_4H_4O_4$; mol wt 925.17. Prepn: U. Schön *et al.*, **US 5324732** (1994 to Kali-Chemie). Crystals, mp 135.8-136.9°. Soly in water: 4.4g/100ml.

THERAP CAT: Antiarrhythmic (class III).

9247. Tedizolid. [856866-72-3] (5R)-3-[3-Fluoro-4-[6-(2-methyl-2H-tetrazol-5-yl)-3-pyridinyl]phenyl]-5-(hydroxymethyl)-2-oxazolidinone; (*R*)-3-[4-[2-(2-methyltetrazol-5-yl)pyridin-5-yl]-3-fluorophenyl]-5-hydroxymethyloxazolidin-2-one; torezolid; DA-7157; TR-700. $C_{17}H_{15}FN_6O_3$; mol wt 370.34. C 55.14%, H 4.08%, F 5.13%, N 22.69%, O 12.96%. Oxazolidinone antibacterial agent. Prepn: J. K. Rhee *et al.*, **WO 05058886**; *eidem*, **US 7816379** (2005, 2010 both to Dong-A); and pharmacology: W. B. Im *et al.*, *Eur. J. Med. Chem.* **46**, 1027 (2011). HPLC determn in plasma, stability, and pharmacokinetics in rats: S. K. Bae *et al.*, *J. Pharm. Pharmacol.* **59**, 955 (2007). Antibacterial spectrum: D. M. Livermore *et al.*, *J. Antimicrob. Chemother.* **63**, 713 (2009). Accumulation by human macrophages and intracellular antibacterial activity: S. Lemaire *et al.*, *ibid.* **64**, 1035 (2009). Clinical evaluation in skin and skin structure infections: P. Prokocimer *et al.*, *Antimicrob. Agents Chemother.* **55**, 583 (2011).

Solid, mp 201°. Soly in water: 0.00434 mg/ml. Log P (*n*-octanol/water): >1.3. MLD orally in male mice: >1000 mg/ml (Rhee).

Phosphate. [856867-55-5] (5R)-3-[3-Fluoro-4-[6-(2-methyl-2H-tetrazol-5-yl)-3-pyridinyl]phenyl]-5-[(phosphonooxy)methyl]-2-oxazolidinone; TR-701 FA. $C_{17}H_{16}FN_6O_6P$; mol wt 450.32. Prepn of purified crystalline form: D. Phillipson, **WO 10091131** (2010 to Trius Ther.). Crystals, mp 256.9° (dec).

Phosphate disodium salt. [856867-39-5] Tedizolid sodium phosphate; DA-7218; TR-701. $C_{17}H_{14}FN_6Na_2O_6P$; mol wt 494.29. Crystals from methanol, mp >200° (dec). Soly in water: >150 mg/ml.

THERAP CAT: Antibacterial.

9248. Teduglutide. [197922-42-2] L-Histidylglycyl-L-α-aspartylglycyl-L-seryl-L-phenylalanyl-L-seryl-L-α-aspartyl-L-α-glutamyl-L-methionyl-L-asparaginyl-L-threonyl-L-isoleucyl-L-leucyl-L-α-aspartyl-L-asparaginyl-L-leucyl-L-alanyl-L-alanyl-L-arginyl-L-α-aspartyl-L-phenylalanyl-L-isoleucyl-L-asparaginyl-L-tryptophyl-L-leucyl-L-isoleucyl-L-glutaminyl-L-threonyl-L-lysyl-L-isoleucyl-L-threonyl-L-aspartic acid; glucagon-like peptide II [2-glycine] (human); h[Gly[2]]GLP-2; ALX-0600. $C_{164}H_{252}N_{44}O_{55}S$; mol wt 3752.13. C 52.50%, H 6.77%, N 16.43%, O 23.45%, S 0.85%. Analog of glucagon-like peptide II, *q.v.*, that is resistant to enzymatic degradation by dipeptidylpeptidase (DPP) IV. Induces mucosal epithelial proliferation in the gastrointestinal tract. Prepn: D. J. Drucker *et al.*, **WO 97 39031**; *eidem*, **US 5789379** (1997, 1998 both to Allelix). Resistance to DPP-IV: P. L. Brubaker *et al.*, *Endocrinology* **138**, 4837 (1997). Intestinotrophic activity: C. Booth *et al.*, *Cell Prolif.* **37**, 385 (2004). Clinical trial in short bowel syndrome: P. B. Jeppesen *et al.*, *Gut* **54**, 1224 (2005). Review of pharmacology and clinical development: M. Ferrone, J. S. Scolapio, *Ann. Pharmacother.* **40**, 1105-1109 (2006).

```
1
His–Gly–Asp–Gly–Ser–Phe–Ser–Asp–Glu–Met–Asn–Thr–Ile–Leu–Asp–Asn–Leu
                                                                    |
33
Asp–Thr–Ile–Lys–Thr–Gln–Ile–Leu–Trp–Asn–Ile–Phe–Asp–Arg–Ala–Ala
```

THERAP CAT: In treatment of short bowel syndrome.

9249. Teflubenzuron. [83121-18-0] *N*-[[(3,5-Dichloro-2,4-difluorophenyl)amino]carbonyl]-2,6-difluorobenzamide; 1-(3,5-dichloro-2,4-difluorophenyl)-3-(2,6-difluorobenzoyl)urea; CME-134; Nomolt. $C_{14}H_6Cl_2F_4N_2O_2$; mol wt 381.11. C 44.12%, H 1.59%,

Cl 18.60%, F 19.94%, N 7.35%, O 8.40%. Insect growth regulator; inhibits chitin biosynthesis and moulting. Prepn: H.-M. Becher *et al.*, **EP 52833**; *eidem*, **US 4457943** (1982, 1984 both to Celamerck). Physical properties and bioactivity: *eidem et al.*, *Proc. 10th Int. Congr. Plant Prot.* **1**, 408 (1983). Field trials in soybeans: D. A. Herbert, J. D. Harper, *J. Econ. Entomol.* **78**, 333 (1985). Efficacy vs fire ants (*Solenopsis invicta* Buren): D. F. Williams *et al.*, *Fla. Entomol.* **80**, 84 (1997).

mp 221-224°. Vapor pressure (20°): 8×10^{-12} mbar. Soly at 20-23° (g/100 ml): hexane 0.005; toluene 0.085; ethanol 0.14; acetone 1.0; DMSO 6.6. Soly in water (20-23°): 20 ppb. LD_{50} orally in rats: >5000 mg/kg (Becher).

USE: USE: Insecticide.

9250. Teflurane. [124-72-1] 2-Bromo-1,1,1,2-tetrafluoroethane; 1,1,1,2-tetrafluoro-2-bromoethane; Abbott 16900; DA-708; Terflurane. C_2HBrF_4; mol wt 180.93. C 13.28%, H 0.56%, Br 44.16%, F 42.00%. FBrCHCF$_3$. Prepn: Larsen, **US 2971990** (1961 to Dow). Pharmacology: Black *et al.*, *Br. J. Anaesth.* **41**, 288 (1969).

Gas, bp 8°. Nonexplosive; nonflammable.

THERAP CAT: Anesthetic (inhalation).

9251. Tefluthrin. [79538-32-2] *rel*-(1*R*,3*R*)-3-[(1*Z*)-2-Chloro-3,3,3-trifluoro-1-propen-1-yl]-2,2-dimethylcyclopropanecarboxylic acid (2,3,5,6-tetrafluoro-4-methylphenyl)methyl ester; (±)-*cis*-4-methyltetrafluorobenzyl 3-(2-chloro-3,3,3-trifluoro-prop-1-en-1-yl)-2,2-dimethylcyclopropane carboxylate; 2,3,5,6-tetrafluoro-4-methylbenzyl-(Z)-(1*RS*,3*RS*)-3-(2-chloro-3,3,3-trifluoro-1-propenyl)-2,2-dimethylcyclopropanecarboxylate; JF-6064; PP-993; R-151993; Force; Forza; Komet. $C_{17}H_{14}ClF_7O_2$; mol wt 418.74. C 48.76%, H 3.37%, Cl 8.47%, F 31.76%, O 7.64%. Synthetic pyrethroid insecticide. Prepn: N. Punja, **EP 31199**; *idem*, **US 4405640** (1981, 1983 both to ICI). Evaluation as seed treatment: G. J. Marrs, R. F. S. Gordon, *Br. Crop Prot. Counc. Monogr.* **39**, 17 (1988); for sugar beet: D. Cooke *et al.*, *Br. Sugar Beet Rev.* **59**, 30 (1991). Effects on non-target organisms: J. M. Coulson *et al.*, *Brighton Crop Prot. Conf. - Pests Dis.* **1990**, 975; A. M. Dewar *et al.*, *ibid.* 987. Review of properties and insecticidal activity: M. Konradt, C. Hemmen, *Gesunde Pflanz.* **44**, 64-68 (1992); of toxicology and human exposure: *Toxicological Profile for Pyrethrins and Pyrethroids* (PB2004-100004, 2003) 332 pp.

Relative stereochemistry

Solid, mp 44°. d 1.48 g/cm³. bp$_{1.33hPa}$ 156°. Vapor pressure (20°): 8×10^{-6} mPa. Soly in water: 2×10^{-2} mg/l; in acetone, dichloromethane, hexane and toluene: >500 g/l. LD_{50} technical grade in rats (mg/kg): 35 orally; 200-1000 dermally (Marrs, Gordon).

USE: Insecticide.

9252. Tegafur. [17902-23-7] 5-Fluoro-1-(tetrahydro-2-furanyl)-2,4(1*H*,3*H*)-pyrimidinedione; 5-fluoro-1-(tetrahydro-2-furyl)uracil; *N$_1$*-(2'-furanidyl)-5-fluorouracil; FT-207; MJF-12264; NSC-148958; Citofur; Coparogin; Exonal; Fental; Franrose; Ftorafur; Fulaid; Fulfeel; Furafluor; Furofutran; Futraful; Lamar; Lifril; Neberk; Nitobanil; Riol; Sinoflurol; Sunfural; Tefsiel C. $C_8H_9FN_2O_3$; mol wt 200.17. C 48.00%, H 4.53%, F 9.49%, N 14.00%, O 23.98%. Prepn: S. A. Hillers *et al.*, *Dokl. Akad. Nauk SSSR* **176**, 332 (1967), *C.A.* **68**, 29664j (1968); *eidem*, **GB 1168391** (1969), *C.A.* **72**, 43715r (1970) and **FR 1574684** (1969), *C.A.* **73**, 77281g (1970). Alternate

synthesis: T. Kametani *et al.*, *J. Heterocycl. Chem.* **14**, 473 (1977). Synthesis, antitumor activity, toxicity study: M. Yasumoto *et al.*, *J. Med. Chem.* **21**, 738 (1978). Crystal structure: Y. Nakai, *Chem. Pharm. Bull.* **30**, 2629 (1982). Pharmacokinetics, metabolism in man: J. L. Au *et al.*, *Cancer Treat. Rep.* **63**, 343 (1979). Evaluation of efficacy and toxicity: C. R. Smart *et al.*, *Cancer* **36**, 103 (1975). Evaluation in colorectal cancer: T. Buroker *et al.*, *Cancer* **44**, 48 (1979). *In vivo* and *in vitro* studies on Walker carcinoma in rats: J. Mattern *et al.*, *Arzneim.-Forsch.* **30**, 981 (1980).

Crystals from ethanol, mp 164-165°. uv max: 270 nm [ε 8460 (pH 2); ε 8050 (pH 7); ε 6700 (pH 12)]. Easily sol in hot water, alcohol, DMF. Practically insol in ether. LD_{50} in mice (mg/kg): 900 orally (3 days) (Yasumoto); 750 i.p. (**FR 1574684**), also reported as 1150 i.p. (Smart).

THERAP CAT: Antineoplastic.

9253. Tegaserod. [145158-71-0] 2-[(5-Methoxy-1*H*-indol-3-yl)methylene]-*N*-pentylhydrazinecarboximidamide. $C_{16}H_{23}N_5O$; mol wt 301.39. C 63.76%, H 7.69%, N 23.24%, O 5.31%. Selective serotonin 5HT$_4$-receptor partial agonist. Prepn: R. K. A. Giger, H. Mattes, **EP 505322**; *eidem*, **US 5510353** (1992, 1996 both to Sandoz); K.-H. Buchheit *et al.*, *J. Med. Chem.* **38**, 2331 (1995). Clinical pharmacology: S. Appel *et al.*, *Clin. Pharmacol. Ther.* **62**, 546 (1997); and pharmacokinetics: *idem et al.*, *J. Clin. Pharmacol.* **37**, 229 (1997). Clinical trial in irritable bowel syndrome: S. A. Müller-Lissner *et al.*, *Aliment. Pharmacol. Ther.* **15**, 1655 (2001); in female patients: J. Novick *et al.*, *ibid.* **16**, 1877 (2002). Review of clinical efficacy: B. W. Jones *et al.*, *J. Clin. Pharm. Ther.* **27**, 343-352 (2002); of mechanism of action, efficacy and safety: M. Corsetti, J. Tack, *Expert Opin. Pharmacother.* **3**, 1211-1218 (2002).

mp 155°.

Maleate. [189188-57-6] SDZ-HTF-919; Zelmac; Zelnorm. $C_{16}H_{23}N_5O \cdot C_4H_4O_4$; mol wt 417.47. White to off-white crystalline powder. Slightly sol in ethanol; very slightly sol in water.

THERAP CAT: Gastroprokinetic; in treatment of irritable bowel syndrome.

9254. Teichoic Acids. Major components of walls and membranes of a number of bacteria, accounting for 20 to 60% of the dry weight of cell walls. Teichoic acids vary considerably in structure but are all rich in phosphodiester linkages. Depending on their location, they can be divided into two classes: membrane and cell-wall teichoic acids. *Reviews:* Baddiley, *Endeavour* **23**, 33 (1964); *idem*, *Proc. R. Soc. London* **170B**, 331 (1968); *idem*, *Acc. Chem. Res.* **3**, 98 (1970); M. Duckworth in *Surface Carbohydrates of the Procaryotic Cell*, I. W. Sutherland, Ed. (Academic Press, London, 1977) pp 177-208.

Membrane teichoic acids. Contain polyglycerol phosphate chains linking positions 1 and 3 on adjacent glycerol units through the phosphodiesters, with glycosyl substituents and alanine residues on some or all of the 2 positions. Structural studies: Kelemen, Baddiley, *Biochem. J.* **80**, 246 (1961). Biosynthetic studies: Burger, Glaser, *J. Biol. Chem.* **239**, 3168, 3187 (1964); **241**, 494 (1966). Synthesis of a membrane teichoic acid fragment of *Staphylococcus aureus*: J. Oltvoort *et al.*, *Rec. Trav. Chim.* **101**, 87 (1982).

Wall teichoic acids. Have greater structural diversity and include also polyribitol phosphate chains linking positions 1 and 5 on adjacent ribitol residues. Polymers may contain 6 to 20 repeating units. Structural studies: Armstrong *et al.*, *Biochem. J.* **76**, 610 (1960); Baddiley, *ibid.* **85**, 49 (1962). Biosynthetic studies: Glaser, *J. Biol. Chem.* **239**, 3178 (1964). Molecular arrangement in cell wall of *Staphylococcus lactis:* Archibald *et al.*, *Nature New Biol.* **241**, 29 (1973).

9255. Teicoplanin. [61036-64-8] Teicoplanin A₂; teichomycin A₂; MDL-507; Targocid; Targosid. Glycopeptide antibiotic complex produced by *Actinoplanes teichomyceticus* nov. sp.; structurally related to vancomycin, *q.v.* Comprised of 5 major components differentiated by a specific fatty acid moiety. Inhibits peptidoglycan synthesis in the cell wall of gram-positive bacteria. Isoln: **BE 839259**; C. Coronelli *et al.*, **US 4239751** (1976, 1980 both to Lepetit); F. Parenti *et al.*, *J. Antibiot.* **31**, 276 (1978); M. R. Bardone *et al.*, *ibid.* 170. Separation of components: A. Borghi *et al.*, **DE 3320342**; *eidem*, **US 4542018** (1983, 1985 both to Lepetit); *eidem*, *J. Antibiot.* **37**, 615 (1984). Structural studies: C. Coronelli *et al.*, *ibid.* 621; J. C. J. Barna *et al.*, *J. Am. Chem. Soc.* **106**, 4895 (1984). Comparative *in vitro* activity: M. H. Cynamon, P. A. Granato, *Antimicrob. Agents Chemother.* **21**, 504 (1982); H. C. Neu, P. Labthavikul, *idem*, **24**, 425 (1983). Human pharmacokinetics: L. Verbist *et al.*, *ibid.* **26**, 881 (1984); P. L. Carver *et al.*, *ibid.* **33**, 82 (1989). Mechanism of action: S. Somma *et al.*, *ibid.* **26**, 917 (1984). Synergism with rifampin, *q.v.*: C. Watanakunakorn, *J. Antimicrob. Chemother.* **19**, 439 (1987). HPLC determn in human serum: J. Levy *et al.*, *ibid.* 533. Clinical trial in gram-positive bacterial infections: Y. Glupczynski *et al.*, *Antimicrob. Agents Chemother.* **29**, 52 (1986). Brief review: A. H. Williams, R. N. Grüneberg, *J. Antimicrob. Chemother.* **14**, 441 (1984). Symposium on chemistry, pharmacology, activity, and clinical trials: *J. Hosp. Infect.* **7**, Suppl. A, 47-112 (1986).

Teicoplanin A₂-1	(Z)-4-decenoic acid
A₂-2	8-methylnonanoic acid
A₂-3	n-decanoic acid
A₂-4	8-methyldecanoic acid
A₂-5	9-methyldecanoic acid

Amorphous powder, mp 260° (dec). uv max in 0.1N HCl: 278 ($E_{1cm}^{1\%}$ 53); in 0.1N NaOH: 297 ($E_{1cm}^{1\%}$ 74). Sol in aq soln at pH 7.0; partially sol in methanol, ethanol. Insol in dil mineral acids, in nonpolar organic solvents.

Teicoplanin A₂-1. [91032-34-7] $C_{88}H_{95}Cl_2N_9O_{33}$; mol wt 1877.66. White amorphous powder, darkens at 220°, dec 255°.

Teicoplanin A₂-2. [91032-26-7] $C_{88}H_{97}Cl_2N_9O_{33}$; mol wt 1879.67. White amorphous powder, darkens at 210°, dec 250°.

Teicoplanin A₂-3. [91032-36-9] $C_{88}H_{97}Cl_2N_9O_{33}$; mol wt 1879.67. White amorphous powder, darkens at 210°, dec 250°.

Teicoplanin A₂-4. [91032-37-0] $C_{89}H_{99}Cl_2N_9O_{33}$; mol wt 1893.70. White amorphous powder, darkens at 210°, dec 250°.

Teicoplanin A₂-5. [91032-38-1] $C_{89}H_{99}Cl_2N_9O_{33}$; mol wt 1893.70. White amorphous powder, darkens at 210°, dec 250°.

THERAP CAT: Antibacterial.

9256. Telaprevir. [402957-28-2] (1*S*,3a*R*,6a*S*)-(2*S*)-2-Cyclohexyl-*N*-(2-pyrazinylcarbonyl)glycyl-3-methyl-L-valyl-*N*-[(1*S*)-1- [2-(cyclopropylamino)-2-oxoacetyl]butyl]octahydrocyclopenta[*c*]pyrrole-1-carboxamide; (1*S*,3a*R*,6a*S*)-2-[(2*S*)-2-[[(2*S*)-cyclohexyl[(pyrazinylcarbonyl)amino]acetyl]amino]-3,3-dimethylbutanoyl]-*N*-[(1*S*)-1-[2-(cyclopropylamino)oxoacetyl]butyl]octahydrocyclopenta[*c*]pyrrole-1-carboxamide; LY-570310; VX-950; Incivek; Incivo. $C_{36}H_{53}N_7O_6$; mol wt 679.86. C 63.60%, H 7.86%, N 14.42%, O 14.12%. Peptidomimetic inhibitor of the NS3-4A serine protease of hepatitis C virus (HCV). Prepn: R. E. Babine *et al.*, **WO 0218369**; *eidem*, **US 050197299** (2002, 2005 both to Lilly); and SAR: S.-H. Chen *et al.*, *Lett. Drug Des. Discovery* **2**, 118 (2005). Pharmacology and antiviral activity: R. B. Perni *et al.*, *Antimicrob. Agents Chemother.* **50**, 899 (2006). Clinical evaluation in hepatitis C patients: N. Forestier *et al.*, *Hepatology* **46**, 640 (2007); E. Lawtiz *et al.*, *J. Hepatol.* **49**, 163 (2008).

White to off-white powder. Soly in water: 0.0047 mg/ml.
THERAP CAT: Antiviral.

9257. Telavancin. [372151-71-8] $N^{3"}$-[2-(Decylamino)ethyl]-29-[[(phosphonomethyl)amino]methyl]vancomycin. $C_{80}H_{106}Cl_2N_{11}O_{27}P$; mol wt 1755.65. C 54.73%, H 6.09%, Cl 4.04%, N 8.78%, O 24.60%, P 1.76%. Broad spectrum, semi-synthetic glycopeptide antibiotic; derivative of vancomycin, *q.v.* Inhibits bacterial cell wall formation and disrupts cell membrane integrity. Prepn: M. R. Leadbetter, M. S. Linsell, **WO 0198328**; *eidem*, **US 6635618** (2001, 2003 both to Advanced Medicine). Manufacturing process: M. Leadbetter *et al.*, **WO 03029270** (2001 to Theravance). Chemistry and antibacterial properties: *idem et al.*, *J. Antibiot.* **57**, 326 (2004). Comparative *in vitro* antimicrobial activity: A. King *et al.*, *J. Antimicrob. Chemother.* **53**, 797 (2004). Mechanism of action study: D. L. Higgins *et al.*, *Antimicrob. Agents Chemother.* **49**, 1127 (2005). Clinical pharmacology: J. P. Shaw *et al.*, *Antimicrob. Agents Chemother.* **49**, 195 (2005). Clinical trial in complicated skin and soft-tissue infections: M. E. Stryjewski *et al.*, *Clin. Infect. Dis.* **40**, 1601 (2005). Review of clinical development: J. L. Pace, J. K. Judice, *Curr. Opin. Investig. Drugs* **6**, 216-225 (2005). Review of role in treatment of methicillin-resistant *S. aureus* (MRSA) infections: M. Bassetti *et al.*, *Expert Opin. Invest. Drugs* **18**, 521-529 (2009).

Hydrochloride. [380636-75-9] (unspecified composition); [560130-42-9] (monohydrochloride). TD-6424; Vibativ. $C_{80}H_{106}$-$Cl_2N_{11}O_{27}P.xHCl$ ($x = 1$-3). Off-white to slightly colored amorphous powder. Highly lipophilic. Slightly sol in water.

THERAP CAT: Antibacterial.

9258. Telbivudine. [3424-98-4] 1-(2-Deoxy-β-L-erythro-pentofuranosyl)-5-methyl-2,4(1H,3H)-pyrimidinedione; 1,2'-deoxy-β-L-ribofuranosylthymine; 2'-deoxy-L-thymidine; LdT; L-thymidine; Sebivo; Tyzeka. $C_{10}H_{14}N_2O_5$; mol wt 242.23. C 49.59%, H 5.83%, N 11.57%, O 33.02%. Nucleoside analog; specific inhibitor of hepatitis B virus (HBV) replication. Prepn: J. Smejkal, F. Sorm, *Collect. Czech. Chem. Commun.* **29**, 2809 (1964); A. Holy, *ibid.* **37**, 4072 (1972). Anti-HBV activity: M. L. Bryant *et al.*, *Nucleosides Nucleotides Nucleic Acids* **20**, 597 (2001). *In vitro* pharmacology: B. Hernandez-Santiago *et al.*, *Antimicrob. Agents Chemother.* **46**, 1728 (2002). Clinical evaluation in hepatitis B: C.-L. Lai *et al.*, *Hepatology* **40**, 719 (2004). Review of mechanism of action and pharmacology: D. N. Standring *et al.*, *Antivir. Chem. Chemother.* **12**, Suppl. 1, 119-129 (2001); of clinical development: S.-H. B. Han, *Expert Opin. Invest. Drugs* **14**, 511-519 (2005). *See also* Thymidine.

White to slightly yellowish powder. mp 189° (Holy). Also reported as crystals from ethanol, mp 186° (Smejkal, Sorm). $[\alpha]_D^{20}$ $-20.3°$ (c = 0.192 in water). Soly (mg/ml): water >20; ethanol 0.7; *n*-octanol 0.1. uv max (pH 2): 267 nm (ε 9800).

THERAP CAT: Antiviral.

9259. Telcagepant. [781649-09-0] N-[(3R,6S)-6-(2,3-Difluorophenyl)hexahydro-2-oxo-1-(2,2,2-trifluoroethyl)-1H-azepin-3-yl]-4-(2,3-dihydro-2-oxo-1H-imidazo[4,5-b]pyridin-1-yl)-1-piperidinecarboxamide; N-[(3R,6S)-6-(2,3-difluorophenyl)-2-oxo-1-(2,2,2-trifluoroethyl)azepan-3-yl]-4-(2-oxo-2,3-dihydro-1H-imidazo[4,5-b]pyridin-1-yl)piperidine-1-carboxamide; MK-0974. $C_{26}H_{27}$-$F_5N_6O_3$; mol wt 566.53. C 55.12%, H 4.80%, F 16.77%, N 14.83%, O 8.47%. Calcitonin gene-related peptide (CGRP) receptor antagonist. Prepn: C. S. Burgey *et al.*, **WO 04092168**; *eidem*, **US 6953790** (2004, 2005 both to Merck & Co.); D. V. Paone *et al.*, *J. Med. Chem.* **50**, 5564 (2007). Improved synthesis: C. S. Burgey *et al.*, *Org. Lett.* **10**, 3235 (2008). Large scale asymmetric synthesis: F. Xu *et al.*, *J. Org. Chem.* **75**, 7829 (2010). Determn by HTLC-MS/MS in plasma and urine: Y. Xu *et al.*, *J. Chromatogr. B* **863**, 64 (2008). Determn of isomers by LC-MS/MS in plasma: Y. Xu, D. G. Musson, *ibid.* **873**, 195 (2008). Pharmacology: C. A. Salvatore *et al.*, *J. Pharmacol. Exp. Ther.* **324**, 416 (2008). Clinical trial in migraine: T. W. Ho *et al.*, *Lancet* **372**, 2115 (2008).

$[\alpha]_D^{20}$ $-18.5°$ (c = 1.0 in CHCl$_3$). Sol in acetonitrile/water.

Potassium salt monoethanolate. [953077-35-5] Telcagepant potassium. $C_{26}H_{26}F_5KN_6O_3.CH_3OH$; mol wt 636.67. Prepn: K. Belyk, **WO 07120592** (2007 to Merck & Co.). White powder.

THERAP CAT: Antimigraine.

9260. Teleocidins. [78474-55-2] Indole alkaloid tumor promoters isolated from the mycelia of *Streptomyces*. Consists of *teleocidin A* (two isomers) and, the major component, *teleocidin B* (four isomers). The mixture is known as telocidin; the name is derived from its toxicity to fish (Teleostei). Isoln: M. Takashima, H. Sakai, *Bull. Agric. Chem. Soc. Jpn.* **24**, 647 (1960). Resolution into A and B and their isomers: H. Fujiki, T. Sugimura, *Cancer Surv.* **2**, 539 (1983). Structure: S. Sakai *et al.*, *Chem. Pharm. Bull.* **32**, 354 (1984). Abs config: Y. Endo *et al.*, *ibid.* 358. Structure determn of B isomers: Y. Hitotsuyanagi *et al.*, *ibid.* 4233. Structure-activity study: T. Kawai *et al.*, *J. Med. Chem.* **35**, 2248 (1992). Tumor promotion: H. Fujiki *et al.*, *Biochem. Biophys. Res. Commun.* **90**, 976 (1979); *eidem*, *Carcinogenesis* **3**, 895 (1982). *Review:* H. Fujiki, T. Sugimura, *Adv. Cancer Res.* **49**, 233-264 (1987).

Teleocidin B$_4$

White-crystal like powder, dec >61°. Easily sol in methanol, ethanol, ether, acetone, ethyl-acetate, benzene, chloroform, carbon tetrachloride; slightly sol in petroleum-ether, petroleum-benzine, ligroin. Almost insol in water, %5 HCl and 5% NaOH solution. LD$_{50}$ orally in mice: 2 mg/kg (Takashima).

Teleocidin B$_4$. [11032-05-6] (4S,7S,10R,13R)-13-Ethenyl-1,3,-4,5,7,8,10,11,12,13-decahydro-4-(hydroxymethyl)-8,10,13-trimethyl-7,10-bis(1-methylethyl)-6H-benzo[g][1,4]diazonino[7,6,5-cd]indol-6-one; olivoretin D. $C_{28}H_{41}N_3O_2$; mol wt 451.66.

Dihydroteleocidin B. Structure: H. Harada *et al.*, *Bull. Chem. Soc. Jpn.* **39**, 1773 (1966). Tumor promoting activity: H. Fujiki *et al.*, *Proc. Natl. Acad. Sci. USA* **78**, 3872 (1981).

USE: use: Tumor promoting agent.

9261. Telithromycin. [191114-48-4] (3aS,4R,7R,9R,10R,-11R,13R,15R,15aR)-4-Ethyloctahydro-11-methoxy-3a,7,9,11,13,15-hexamethyl-1-[4-[4-(3-pyridinyl)-1H-imidazol-1-yl]butyl]-10-[[3,-4,6-trideoxy-3-(dimethylamino)-β-D-xylo-hexopyranosyl]oxy]-2H-oxacyclotetradecino[4,3-d]oxazole-2,6,8,14(1H,7H,9H)-tetrone; 3-de[(2,6-dideoxy-3-C-methyl-3-O-methyl-α-L-ribohexopyranosyl)-oxy]-11,12-dideoxy-6-O-methyl-3-oxo-12,11-[oxycarbonyl[[4-[4-(3-pyridinyl)-1H-imidazol-1-yl]butyl]imino]]erythromycin; HMR-3647; RU-66647; Ketek. $C_{43}H_{65}N_5O_{10}$; mol wt 812.02. C 63.60%, H 8.07%, N 8.62%, O 19.70%. Semisynthetic macrolide antibiotic of the ketolide class, a group of erythromycin derivatives in which the L-cladinose residue has been replaced by a 3-keto group. Prepn: C. Agouridas *et al.*, **EP 680967**; *eidem*, **US 5635485** (1995, 1997 both to Roussel Uclaf); A. Denis *et al.*, *Bioorg. Med. Chem. Lett.* **9**, 3075 (1999). *In vitro* activity vs anaerobic bacteria: L. M. Ednie *et al.*, *Antimicrob. Agents Chemother.* **41**, 2019 (1997); vs gram-positive bacteria: K. Malathum *et al.*, *ibid.* **43**, 930 (1999). Time-kill kinetics: F. J. Boswell *et al.*, *J. Antimicrob. Chemother.* **41**, 149 (1998). HPLC analysis: B. Lingerfelt, W. S. Champney, *J. Pharm. Biomed. Anal.* **20**, 459 (1999). Review of pharmacology and clinical trials: G. Ackermann, A. C. Rodloff, *J. Antimicrob. Chemother.* **51**, 497-511 (2003).

Crystals from ether, mp 187-188°.
THERAP CAT: Antibacterial.

9262. Telluric(VI) Acid. [7803-68-1] Telluric acid (H_6-TeO_6); orthotelluric acid; tellurium hexahydroxide. H_6O_6Te; mol wt 229.64. H 2.63%, O 41.80%, Te 55.57%. $Te(OH)_6$. Most stable form. Prepd by oxidizing tellurium or its dioxide with chromic acid or potassium permanganate in nitric acid: Staudenmaier, Z. Anorg. Chem. **10**, 189 (1895); Mathers et al., Inorg. Synth. **3**, 145 (1950); from tellurium by oxidation with chloric acid: Meyer, Franke, Z. Anorg. Allg. Chem. **193**, 191 (1930); Meyer, Holowatyi, Ber. **81**, 119 (1948); Fehér in Handbook of Preparative Inorganic Chemistry vol. **1**, G. Brauer, Ed. (Academic Press, New York, 2nd ed., 1963) pp 451-453°. Forms polymetatelluric acid ($H_2TeO_4)_n$ when heated in air at 100-200°; n approx 11. Concd soln of polymerized form called allotelluric acid; formed when orthotelluric acid is heated in a sealed tube and dissolves in its water of constitution. Review: Datton, Cooper, Chem. Rev. **66**, 657-675 (1966).

White solid. Dimorphic: monoclinic, d 3.068; cubic, d 3.163: Avinens, Petit, C. R. Seances Acad. Sci. Ser. C **266**, 981 (1968). Crystallizes as tetrahydrate at temp below 10°. Very weak acid: $K_1 = 2 \times 10^{-8}$; $K_2 = 1 \times 10^{-11}$. Soly in water about 33% at 30°: Mylius, Ber. **34**, 2208 (1901). Strong tendency to polymerize (like stannic acid) as the mol wt increases the soly becomes less and less and aq solns become truly colloidal. Sparingly sol in concd nitric acid. Soly in dil nitric acid: Inorg. Syn. (loc. cit.).

9263. Tellurium. [13494-80-9] Aurum paradoxum; metallum problematum. Te; at. wt 127.60; at. no. 52; valence 2, 4, 6. Group VIA (16). Diatomic (Te_2) in the vapor state. Eight stable isotopes: 120 (0.089%); 122 (2.46%); 123 (0.87%); 124 (4.61%); 125 (6.99%); 126 (18.71%); 128 (31.79%); 130 (34.48%); artificial radioactive isotopes: 114-119; 121; 127; 129; 131-134. Present in the earth's crust to the extent of 0.002 ppm. Discovered by von Reichenstein in 1782; named by Klaproth in 1798. Occurs as tellurides in combination with metals in the minerals tetradymite, altaite, coloradolite; found as the dioxide, tellurite; found also native, associated with silver and gold. Prepn: Kracek, J. Am. Chem. Soc. **63**, 1989 (1941); Fehér in Handbook of Preparative Inorganic Chemistry vol. **1**, G. Brauer, Ed. (Academic Press, New York, 2nd ed., 1963) pp 437-438. Prepn of spectrally pure Te for semiconductor devices: Weidel, Z. Naturforsch. **9a**, 697 (1954). Symposium on organic selenium and tellurium compds: Y. Okamoto, W. H. H. Gunther, Eds., Ann. N.Y. Acad. Sci. **192**, 1-225 (1972). Reviews: Stone, Caron in Rare Metals Handbook, C. A. Hampel, Ed. (Reinhold, New York, 1954) pp 405-415; Bagnall in Comprehensive Inorganic Chemistry vol. **2**, J. C. Bailar, Jr. et al., Eds. (Pergamon Press, Oxford, 1973) pp 935-1008; E. M. Elkin in Kirk-Othmer Encyclopedia of Chemical Technology vol. **22** (Wiley-Interscience, New York, 3rd ed., 1983) pp 658-679.

Grayish-white, lustrous, brittle, crystalline solid, hexagonal, rhombohedral structure, or dark-gray to brown, amorphous powder with metal characteristics. d (cryst) 6.11-6.27. mp 449.8°. bp 989.9°. Electrical resistivity (19.6°): 200,000 μ-ohms-cm. Latent heat of fusion: 4.27 kcal/mole. Linear coefficient of thermal expansion: $16.8 \times 10^{-6}/°C$. Modulus of elasticity: 6,000,000 psi. Specific heat

(solid): 0.047 cal/g/°C. Magnetic susceptibility (18°): -0.31×10^{-6} cgs. Hardness (Mohs): 2.3. Thermal conductivity: 0.014 at 20°. Burns in air with a greenish-blue flame, forming the dioxide. Insol in water, in benzene, in carbon disulfide. Not attacked by hydrochloric acid; reacts with nitric acid; with concd or fuming sulfuric acids, forming a red soln; in presence of air dissolves in potassium hydroxide with formation of a deep-red soln. Combines with the halogens; does not react with sulfur or selenium.

Caution: Potential symptoms of overexposure are garlic odor on breath and sweat; dry mouth, metal taste; somnolence; anorexia, nausea and no sweating; dermatitis. See NIOSH Pocket Guide to Chemical Hazards (DHHS/NIOSH 97-140, 1997) p 294.

USE: As coloring agent in chinaware, porcelains, enamels, glass; reagent in producing black finish on silverware; in manuf special alloys of marked electrical resistance; in semiconductor research.

9264. Tellurium Dichloride. [10025-71-5] Tellurous chloride. Cl_2Te; mol wt 198.50. Cl 35.72%, Te 64.28%. $TeCl_2$. Prepn from CCl_2F_2 and Te: Aynsley, J. Chem. Soc. **1953**, 3016.

Black, amorphous solid; black liquid; purple vapor. mp 208°. bp 328°. Disproportionates in ether, dioxane, dibutyl ether. Insol in CCl_4.

9265. Tellurium Dioxide. [7446-07-3] Tellurium oxide (TeO_2). O_2Te; mol wt 159.60. O 20.05%, Te 79.95%. TeO_2. Prepd by oxidation of Te by HNO_3: Norris, J. Am. Chem. Soc. **28**, 1675 (1906); Marshall, Inorg. Synth. **3**, 143 (1950); Fehér in Handbook of Preparative Inorganic Chemistry vol. **1**, G. Brauer, Ed. (Academic Press, New York, 2nd ed., 1963) pp 447-449. The orthorhombic form occurs in nature as the mineral tellurite. Review: Dutton, Cooper, Chem. Rev. **66**, 657-675 (1966).

White crystals, dimorphic: tetragonal, d 5.75; orthorhombic, d 6.04. Turns yellow on heating, mp 733°, forming a deep yellow liquid. Soly in water about 1:150,000. Sol in sodium hydroxide solns and in hydrochloric acid. Practically insol in ammonia water.

9266. Tellurium Hexafluoride. [7783-80-4] (OC-6-11)-Tellurium fluoride (TeF_6). F_6Te; mol wt 241.59. F 47.18%, Te 52.82%. TeF_6. Prepd by direct fluorination of tellurium metal: Prideaux, J. Chem. Soc. **89**, 322 (1906); Klemm, Henkel, Z. Anorg. Allg. Chem. **207**, 74 (1932); Yost, Claussen, J. Am. Chem. Soc. **55**, 885 (1933); Yost, Inorg. Synth. **1**, 121 (1939).

Colorless gas. Repulsive odor. Poisonous; corrosive. mp $-37.6°$. Sublimes at $-38.9°$. Critical temperature 83°. d (solid; $-191°$) 4.006; d (liq; $-10°$) 2.499. Not as inert chemically as SeF_6 and SF_6 because the covalence maximum of tellurium is higher than 6. Slowly absorbed by water with hydrolysis to telluric acid, H_6TeO_6; more quickly hydrolyzed by aq KOH. Does not attack glass when pure. Corrodes mercury.

Caution: Potential symptoms of overexposure are headache; dyspnea; garlic odor on breath. See NIOSH Pocket Guide to Chemical Hazards (DHHS/NIOSH 97-140, 1997) p 294.

9267. Tellurium Tetrabromide. [10031-27-3] (T-4)-Tellurium bromide ($TeBr_4$); telluric bromide. Br_4Te; mol wt 447.22. Br 71.47%, Te 28.53%. $TeBr_4$. Prepd from the elements: Fehér in Handbook of Preparative Inorganic Chemistry vol. **1**, G. Brauer, Ed. (Academic Press, New York, 2nd ed., 1963) pp 445-446.

Orange crystals when cold, red when hot. d 4.3. mp ~380°. bp 414-420°, dec into dibromide and bromine; can be sublimed without dec at 300° in vacuum. Soluble in a little water, but hydrolyzed by much water. Sol in HBr, ether, glacial acetic acid. Keep well closed.

9268. Tellurium Tetrachloride. [10026-07-0] (T-4)-Tellurium chloride ($TeCl_4$); telluric chloride. Cl_4Te; mol wt 269.40. Cl 52.64%, Te 47.36%. $TeCl_4$. Prepd from the elements: Suttle, Smith, Inorg. Synth. **3**, 140 (1950).

White, very hygroscopic, cryst solid. d 3.01. mp 225°. Melts to a yellow liquid, becoming dark red at higher temp. bp 380° without decompn. Dec by water into TeO_2 and HCl. Sol in abs alcohol and toluene. Keep tightly closed.

9269. Tellurium Tetraiodide. [7790-48-9] (T-4)-Tellurium iodide (TeI_4). I_4Te; mol wt 635.22. I 79.91%, Te 20.09%. TeI_4. Prepd from $Te(OH)_6$ and HI, Gutbier, Flury, Z. Anorg. Allg. Chem. **32**, 108 (1902); Damiens, Ann. Chim. [9] **19**, 44 (1923); Fehér in

Handbook of Preparative Inorganic Chemistry **vol. 1**, G. Brauer, Ed. (Academic Press, New York, 2nd ed., 1963) p 447.

Gunmetal-gray crystals. Stable in moist air. d_4^{15} 5.05. mp 280°. Gives off I_2 when heated. Hydrolyzed slowly by cold, quickly by hot water, forming TeO_2 and HI. Sol in HI; somewhat sol in acetone.

9270. Tellurous Acid. [10049-23-7] Telluric(IV) acid. H_2O_3Te; mol wt 177.61. H 1.14%, O 27.02%, Te 71.84%. H_2TeO_3.

White crystals or cryst powder. d 3.0. Slightly sol in water; sol in dil acids or alkalies. Its potassium salt is reduced by many microorganisms, producing dark-colored solns.

9271. Telmisartan. [144701-48-4] 4'-[(1,4'-Dimethyl-2'-propyl[2,6'-bi-1H-benzimidazol]-1'-yl)methyl][1,1'-biphenyl]-2-carboxylic acid; 4'-[[4-methyl-6-(1-methyl-2-benzimidazolyl)-2-propyl-1-benzimidazolyl]methyl]-2-biphenylcarboxylic acid; BIBR 277; Kinzalmono; Micardis; Pritor. $C_{33}H_{30}N_4O_2$; mol wt 514.63. C 77.02%, H 5.88%, N 10.89%, O 6.22%. Angiotensin II receptor antagonist. Prepn: N. Hauel *et al.*, **EP 502314** (1992 to Thomae), *C.A.* **117**, 251352 (1992); U. J. Ries *et al.*, *J. Med. Chem.* **36**, 4040 (1993). Large-scale synthesis: K. S. Reddy *et al.*, *Org. Process Res. Dev.* **11**, 81 (2007). Pharmacology: W. Wienen *et al.*, *Br. J. Pharmacol.* **110**, 245 (1993). Binding study: W. Wienen, M. Entzeroth, *J. Hypertens.* **12**, 119 (1994). Clinical pharmacology, pharmacokinetics: J. M. Neutel, D. H. G. Smith, *Adv. Ther.* **15**, 206 (1998). Clinical trial in hypertension: D. H. G. Smith *et al.*, *ibid.* 229. Review of pharmacology and clinical efficacy: K. J. McClellan, A. Markham, *Drugs* **56**, 1039-1044 (1998).

White solid, mp 261-263°. Practically insol in water (pH 3-9). Sparingly sol in strong acid (except insol in hydrochloric acid); sol in strong base.

THERAP CAT: Antihypertensive.

9272. Telomerase. Telomere terminal transferase. RNA-containing enzyme found in unicellular organisms and germline cells. Also produced by tumor cells and by immortalized cell lines; not expressed in normal somatic cells. Specialized reverse transcriptase that adds short, tandemly repeated segments of DNA onto the ends of eukaryotic chromosomes (telomeres) using its RNA component as a template. Replenishes the small amount of telomeric DNA that is normally lost during replication. In the absence of telomerase, this incremental loss of DNA continues with each cell division until the chromosome can no longer replicate and ultimately results in cell death. Activation of telomerase is associated with the unlimited proliferation of tumor cells. Discovery in the ciliate, *Tetrahymena thermophila:* C. W. Greider, E. H. Blackburn, *Cell* **43**, 405 (1985). Overview and potential physiological significance: C. W. Greider, *BioEssays* **12**, 363-369 (1990). Proposed role in cell immortalization: C. M. Counter *et al.*, *EMBO J.* **11**, 1921 (1992). Detection in human ovarian carcinoma cells: *idem et al.*, *Proc. Natl. Acad. Sci. USA* **91**, 2900 (1994). Structural studies: A. Bhattacharyya, E. H. Blackburn, *EMBO J.* **13**, 5721 (1994); M. McCormick-Graham, D. P. Romero, *Nucleic Acids Res.* **23**, 1091 (1995). Clinical implications in cancer diagnosis and therapy: N. W. Kim, *Eur. J. Cancer* **33**, 781 (1997). Effect on life-span of normal human cells *in vitro:* A. G. Bodnar *et al.*, *Science* **279**, 349 (1998).

9273. Telomycin. [19246-24-3] (3R)-L-α-Aspartyl-L-seryl-L-threonyl-L-allothreonyl-L-alanylglycyl-(3S)-3-hydroxy-L-prolyl-α,β-didehydrotryptophyl-β-methyl-L-tryptophyl-(3S)-3-hydroxy-L-leucyl-3-hydroxy-L-proline (11 → 3)-lactone. $C_{59}H_{77}N_{13}O_{19}$; mol wt 1272.34. C 55.70%, H 6.10%, N 14.31%, O 23.89%. Polypep-

tide antibiotic produced by *Streptomyces* spp. from Florida soil: Misiek *et al.*, *Antibiot. Annu.* **1957-1958**, 852. Structure: Sheehan *et al.*, *J. Am. Chem. Soc.* **85**, 2867 (1963); **90**, 462 (1968). NMR spectrum and conformation: Kumar, Urry, *Biochemistry* **12**, 3811, 4392 (1973). Antibacterial spectrum: A. Gourevitch *et al.*, *Antibiot. Annu.* **1957-1958**, 856. Pharmacology: D. E. Tisch *et al.*, *ibid.* 863.

a-Thr = *allo*-threonine
t-3-Hyp = *trans*-3-hydroxyproline
c-3-Hyp = *cis*-3-hydroxyproline
Δ-Trp = α,β-didehydrotryptophan
e-3-Hyl = *erythro*-3-hydroxyleucine

Amorphous, gray solid. $[\alpha]_D^{28}$ −133° (c = 1 in 2:1 methanol-water). uv max (ethanol:water, 2:1): 222.5, 277, 290, 339 nm (ε 63732; 13746; 11890; 22058). Minimum soly in water at pH 3.0 to 3.3 = 4 mg/ml. Maximum soly is above pH 8.5 = >150 mg/ml. Sol in 10% sodium chloride soln less than 1 mg/ml. Moderately sol in methanol, ethanol. Very slightly sol in acetone, ethyl acetate. Insol in ether, chloroform, hydrocarbons. Aq solns are stable to heat. LD_{50} in mice (mg/kg): >1000 orally, i.v., i.p., i.m. (Tisch).

9274. Temafloxacin. [108319-06-8] 1-(2,4-Difluorophenyl)-6-fluoro-1,4-dihydro-7-(3-methyl-1-piperazinyl)-4-oxo-3-quinoline-carboxylic acid; T-1258; A-63004; Teflox; Temac; Omniflox. $C_{21}H_{18}F_3N_3O_3$; mol wt 417.39. C 60.43%, H 4.35%, F 13.66%, N 10.07%, O 11.50%. Trifluorinated quinolone antibacterial. Prepn: D. T. W. Chu, **EP 131839**; *idem*, **US 4730000** (1985, 1988 to Abbott). Antibacterial activities: D. J. Hardy *et al.*, *Antimicrob. Agents Chemother.* **31**, 1768 (1987); K. V. I. Rolston *et al.*, *Eur. J. Clin. Microbiol. Infect. Dis.* **7**, 684 (1988); in comparison with other fluoroquinolones: A. L. Barry, R. N. Jones, *J. Antimicrob. Chemother.* **23**, 527 (1989).

Hydrochloride. [105784-61-0] A-62254. $C_{21}H_{18}F_3N_3O_3.HCl$; mol wt 453.85.

9275. Temazepam. [846-50-4] 7-Chloro-1,3-dihydro-3-hydroxy-1-methyl-5-phenyl-2H-1,4-benzodiazepin-2-one; 3-hydroxy-diazepam; N-methyloxazepam; oxydiazepam; ER-115; Ro-5-5345; Wy-3917; Euhypnos; Euipnos; Gelthix; Levanxene; Levanxol; Normison; Perdorm; Planum; Remestan; Restoril. $C_{16}H_{13}$-ClN_2O_2; mol wt 300.74. C 63.90%, H 4.36%, Cl 11.79%, N 9.32%, O 10.64%. Pharmacologically active metabolite of diazepam, *q.v.* Prepn: S. C. Bell, S. J. Childress, *J. Org. Chem.* **27**, 1691 (1962); S. C. Bell, **US 3197467** (1965 to Am. Home Prod.). *See also:* E. Reeder *et al.*, **US 3340253** and **US 3374225** (1967, 1968, both to Hoffmann-La Roche). Metabolism: H. J. Schwandt *et al.*, *Xenobiotica* **4**, 733 (1974); S. H. Curry *et al.*, *Br. J. Pharmacol.* **57**, 427P (1976). Pharmacology: L. O. Randall *et al.*, *Arch. Int. Pharmacodyn.* **185**, 135 (1970); S. Garattini *et al.*, "Metabolic Studies on Benzodiazepines in Various Animal Species" in *Benzodiazepines*, S. Garattini, Ed. (Raven Press, New York, 1973) pp 73-97. Pharmacology and toxicity study: L. O. Randall *et al.*, *Curr. Ther. Res.* **7**, 590 (1965). Clinical study: P. Sarteschi *et al.*, *Arzneim.-Forsch.* **22**, 93 (1972). Review of pharmacology and therapeutic efficacy: R. C. Heel *et al.*, *Drugs* **21**, 321-340 (1981).

Crystals from cyclohexane, mp 119-121°. Sparingly sol in alc; very slightly sol in water.

Note: This is a controlled substance (depressant): **21 CFR,** 1308.14.

THERAP CAT: Sedative, hypnotic.

9276. Tembotrione. [335104-84-2] 2-[2-Chloro-4-(methylsulfonyl)-3-[(2,2,2-trifluoroethoxy)methyl]benzoyl]-1,3-cyclohexanedione; 2-[2-chloro-4-mesyl-3-[(2,2,2-trifluoroethoxy)methyl]-benzoyl]cyclohexane-1,3-dione; AE-0172747. $C_{17}H_{16}ClF_3O_6S$; mol wt 440.81. C 46.32%, H 3.66%, Cl 8.04%, F 12.93%, O 21.78%, S 7.27%. Triketone herbicide for the control of broadleaf and grassy weeds in corn crops; inhibits the enzyme, 4-hydroxyphenylpyruvate dioxygenase (HPPD): Prepn: A. van Almsick *et al.,* **WO 0021924** (2000 to Aventis CropSci); *eidem,* **US 6376429** (2002 to Hoechst Schering AgrEvo); A. van Almsick *et al., Bayer Crop Sci. J.* **62,** 5 (2009). Effect on corn quality: P. Bonis *et al., Cereal Res. Commun.* **36,** Suppl., 215 (2008). Review of use in combination with isoxadifen-ethyl, *q.v.:* M. Wegener, H. Roos, *J. Plant Dis. Prot.* **Sp. Iss. 21,** 629-634 (2008).

Beige powder. mp 123°; dec above 150°. d_4^{20} 1.56. pKa 3.18. Vapor pressure (25°): 2.9×10^{-10} hPa. Log P (octanol/water): 2.16 (pH 2); −1.09 (pH 7); −1.37 (pH 9). Soly in water at 20° (g/l): 0.22 (pH 4); 28.30 (pH 7); 29.69 (pH 9). Sol at 20° (g/l): ethanol 8.2; *n*-hexane 47.6; toluene 75.7; dichloromethane >600; acetone 300-600; ethyl acetate 180.2; DMSO >600. LD_{50} in rats (mg/kg): >5000 orally; 4000 dermally; LC_{50} in rats: >3.59 mg/l by inhalation (Wegener, Roos).

Mixture with isoxadifen-ethyl. [473278-62-5] Laudis; Soberan.
USE: Herbicide.

9277. TEMED. [110-18-9] N^1,N^1,N^2,N^2-Tetramethyl-1,2-ethanediamine; TMEDA. $C_6H_{16}N_2$; mol wt 116.21. C 62.01%, H 13.88%, N 24.11%. Prepn: M. Freund, H. Michaels, *Ber.* **30,** 1374 (1897); W. Hanhart, C. K. Ingold, *J. Chem. Soc.* **1927,** 997. Base strength: L. Spialter, R. W. Moshier, *J. Am. Chem. Soc.* **79,** 5955 (1957). Effect on kinetics of gel polymerization: Y. Pegon, C. Quincy, *J. Chromatogr.* **100,** 11 (1974); on gel pore size: H. Tamagawa *et al., Polymer* **41,** 7201 (2000). Use as initiator in electrophoresis: B. J. Bassam, S. Bentley, *BioTechniques* **19,** 568 (1995); as chelator in biological redox systems: H. C. Chang, J. A. Bumpus, *Proc. Natl. Sci. Counc. Repub. China Part B* **25,** 26 (2001); as solvent in RPLC: L. Palego *et al., Prog. Neuro-Psychopharmacol. Biol. Psychiatry* **25,** 519 (2001); in capillary isoelectric focusing: D. Mohan, C. S. Lee, *J. Chromatogr. A* **979,** 271 (2002).

Basic liquid. fp −55.1°. bp_{724} 119.4-119.5°. n_D^{18} 1.4196. pKa_1 5.85; pKa_2 8.97.

USE: Polymerization accelerator in gel electrophoresis, solvent and oxidizing reagent.

9278. Temephos. [3383-96-8] $O^P,O^{P'}$-(Thiodi-4,1-phenylene)phosphorothioic acid $O^P,O^P,O^{P'},O^{P'}$-tetramethyl ester; O,O'-(thiodi-4,1-phenylene)bis(O,O'-dimethylphosphorothioate); $O,O,-O',O'$-tetramethyl O,O'-thiodi-*p*-phenylene phosphorothioate; phosphorothioic acid O,O-dimethyl ester O,O-diester with 4,4′-thiodiphenol; ENT-27165; AC-52160; Abate. $C_{16}H_{20}O_6P_2S_3$; mol wt 466.46. C 41.20%, H 4.32%, O 20.58%, P 13.28%, S 20.62%. Organophosphate insecticide; cholinesterase inhibitor. Prepn: J. B. Lovell, R. W. Baer, **BE 648531**; *eidem,* **US 3317636** (1964, 1967 both to Am. Cyanamid). Metabolism: R. C. Blinn, *J. Agric. Food Chem.* **17,** 118 (1969). Toxicity study: T. B. Gaines, *Toxicol. Appl. Pharmacol.* **14,** 515 (1969). HPLC determn and degradation in water: S. Lacorte *et al., J. Chromatogr. A* **777,** 99 (1997).

Crystalline solid, mp 30.0-30.5°. Optimum stability at pH 5-7. Sol in acetonitrile, carbon tetrachloride, ether, dichloroethane, toluene. Almost insol in water, hexane. LD_{50} in male, female rats (mg/kg): 8600, 13000 orally (Gaines).

Caution: Potential symptoms of overexposure are eye irritation, blurred vision; dizziness, confusion; dyspnea; salivation; abdominal cramps, nausea, diarrhea, vomiting. *See NIOSH Pocket Guide to Chemical Hazards* (DHHS/NIOSH 97-140, 1997) p 296.

USE: Insecticide; used for mosquito control.

9279. Temocapril. [111902-57-9] (2*S*,6*R*)-6-[[(1*S*)-1-(Ethoxycarbonyl)-3-phenylpropyl]amino]tetrahydro-5-oxo-2-(2-thienyl)-1,4-thiazepine-4(5*H*)-acetic acid; (+)-(2*S*,6*R*)-6-[[(1*S*)-1-carboxy-3-phenylpropyl]amino]tetrahydro-5-oxo-2-(2-thienyl)-1,4-thiazepine-4(5*H*)-acetic acid, 6-ethyl ester. $C_{23}H_{28}N_2O_5S_2$; mol wt 476.61. C 57.96%, H 5.92%, N 5.88%, O 16.78%, S 13.45%. Angiotensin-converting enzyme (ACE) inhibitor. Hydrolyzed *in vivo* to the active diacid metabolite. Prepn: H. Yanagisawa *et al.,* **EP 161801**; *eidem,* **US 4699905** (1985, 1987 both to Sankyo); *idem et al., J. Med. Chem.* **30,** 1984 (1987). Pharmacology: K. Oizumi *et al., Jpn. J. Pharmacol.* **48,** 349 (1988). GC/MS determn in human plasma and urine: H. Shioya *et al., J. Chromatogr.* **496,** 129 (1989). Clinical evaluation: M. Arita *et al., Clin. Exp. Pharmacol. Physiol.* **21,** 195 (1994); and pharmacokinetics: S. Tokunaga *et al., Drug Invest.* **7,** 161 (1994). *Review:* H. Koike *et al., Annu. Rep. Sankyo Res. Lab.* **44,** 1-81 (1992).

Cryst powder, mp 168°. $[\alpha]^{23}$ +40° (c = 1.1 in DMF).

Hydrochloride. [110221-44-8] CS-622; Acecol. Solid from ethanol/ethyl acetate, mp 187° (dec). $[\alpha]_D^{25}$ +47.7° (c = 1 in DMF). LD_{50} in mice, rats, dogs (mg/kg): >5000, >5000, >800 orally (Koike).

Diacid. [110221-53-9] Temocaprilat; RS-5139. $C_{21}H_{24}N_2-O_5S_2$; mol wt 448.55. Solid from ethanol, mp 246° (dec). $[\alpha]_D^{25}$ +63.4° (c = 1 in DMF).

THERAP CAT: Antihypertensive.

9280. Temocillin. [66148-78-5] (2*S*,5*R*,6*S*)-6-[[2-Carboxy-2-(3-thienyl)acetyl]amino]-6-methoxy-3,3-dimethyl-7-oxo-4-thia-1-azabicyclo[3.2.0]heptane-2-carboxylic acid; (6*S*)-6-[2-carboxy-2-(3-thienyl)acetamido]-6-methoxypenicillanic acid. $C_{16}H_{18}N_2O_7S_2$; mol wt 414.45. C 46.37%, H 4.38%, N 6.76%, O 27.02%, S 15.47%. Semi-synthetic injectable penicillin deriv with high activity vs a large number of gram-negative bacteria, but with little activity vs gram-positive organisms. Prepn: J. P. Clayton, P. H. Bentley, **DE 2600866** (1976 to Beecham), *C.A.* **86,** 55420t (1977); P. H. Bentley *et al., J. Chem. Soc. Perkin Trans. 1* **1979,** 2455. *In vitro* antibacterial activity and β-lactamase susceptibility: I. Phillips *et al.*, *J. Antimicrob. Chemother.* **10**, 271 (1982). Comparison to other penicillins vs *H. influenzae* and intestinal gram-negative rods: H. Y. Chen, J. D. Williams, *ibid.* 279. Pharmacokinetics and tissue penetration in healthy volunteers: R. M. Brown *et al.*, *ibid.* 295. *In vivo* and *in vitro* comparison to ampicillin: R. Yogev *et al.*, *Antimicrob. Agents Chemother.* **23**, 182 (1983). Series of articles on microbiology, pharmacology and clinical studies: *Drugs* **29**, Suppl. 5, 1-243 (1985).

Disodium salt. [61545-06-0] BRL-17421; Temopen. $C_{16}H_{16}N_2Na_2O_7S_2$; mol wt 458.41. Amorphous solid.

THERAP CAT: Antibacterial.

9281. Temoporfin. [122341-38-2] 3,3′,3″,3‴-(7,8-Dihydro-21*H*,23*H*-porphine-5,10,15,20-tetrayl)tetrakisphenol; 2,3-dihydro-5,10,15,20-tetra(*m*-hydroxyphenyl)porphyrin; 5,10,15,20-tetra(*m*-hydroxyphenyl)chlorin; *m*-THPC; dihydro meta-HK7; EF-9; Foscan. $C_{44}H_{32}N_4O_4$; mol wt 680.76. C 77.63%, H 4.74%, N 8.23%, O 9.40%. Second generation systemic photosensitizing agent. Prepn: R. Bonnett, M. C. Berenbaum, **EP 337601**; *eidem*, **US 4992257** (1989, 1991 both to Efamol Holdings); R. Bonnett *et al.*, *Biochem. J.* **261**, 277 (1989). NMR study: *idem et al.*, *J. Chem. Soc. Perkin Trans. 2* **1994**, 1839. Toxicology study: R. B. Veenhuizen *et al.*, *Int. J. Cancer* **59**, 830 (1994). HPLC determn in plasma and tissue: M. Barberi-Heyob *et al.*, *J. Chromatogr. B* **688**, 331 (1997). Clinical pharmacokinetics: T. Glanzmann *et al.*, *Photochem. Photobiol.* **67**, 596 (1998). Clinical evaluation in photodynamic therapy of malignancies: J.-F. Savary *et al.*, *Arch. Otolaryngol. Head Neck Surg.* **123**, 162 (1997).

Purple solid from methanol/water. Absorption max (methanol): 284, 306, 415, 516, 543, 591, 650 nm (ε 16900, 15600, 146000, 11000, 7300, 4400, 22400 l/mol-cm). Sol in polar solvents.

THERAP CAT: Antineoplastic (photosensitizer).

9282. Temozolomide. [85622-93-1] 3,4-Dihydro-3-methyl-4-oxoimidazo[5,1-*d*]-1,2,3,5-tetrazine-8-carboxamide; 8-carbamoyl-3-methylimidazo[5,1-*d*]-1,2,3,5-tetrazin-4(3*H*)-one; methazolastone; M & B 39831; CCRG-81045; NSC-362856; Temodal; Temodar. $C_6H_6N_6O_2$; mol wt 194.15. C 37.12%, H 3.12%, N 43.29%, O 16.48%. Imidazotetrazine alkylating agent. Prepn: E. Lunt *et al.*, **DE 3231255** (1983 to May & Baker); *eidem*, **US 5260921** (1993 to Cancer Res. Campaign Technol.); M. F. G. Stevens *et al.*, *J. Med. Chem.* **27**, 196 (1984). Crystal structure and structure-activity study: P. R. Lowe *et al.*, *ibid.* **35**, 3377 (1992). Synthesis: Y. Wang *et al.*, *J. Org. Chem.* **62**, 7288 (1997). HPLC determn in plasma and urine: F. Shen *et al.*, *J. Chromatogr. B* **667**, 291 (1995). Review of pharmacology and clinical efficacy in melanoma and brain tumor: E. S. Newlands *et al.*, *Cancer Treat. Rev.* **23**,

35-61 (1997); S. J. Danson, M. R. Middleton, *Expert Rev. Anticancer Ther.* **1**, 13-19 (2001).

Crystals from dichloromethane, mp 212° (dec). uv max (95% ethanol): 327 nm.

THERAP CAT: Antineoplastic.

9283. TEMPO. [2564-83-2] 2,2,6,6-Tetramethyl-1-piperidinyloxy; 2,2,6,6-tetramethylpentamethylene nitroxide. $C_9H_{18}NO$; mol wt 156.25. C 69.18%, H 11.61%, N 8.96%, O 10.24%. Stable nitroxide free radical used as reversible capping agent in "living" free radical polymerizations and as catalyst for selective oxidation of alcohols. Prepn: E. G. Ruzantsev *et al.*, *Bull. Acad. Sci. USSR Div. Chem. Sci.* **1962**, 2152. Crystal structure: Z. Ciunik, *J. Mol. Struct.* **412**, 27 (1997). Thermal stability: M. V. Ciriano *et al.*, *J. Am. Chem. Soc.* **121**, 6375 (1999). Effect of acids on reaction rates: M. V. Baldovi *et al.*, *Macromolecules* **29**, 5497 (1996). Use in oxidation of alcohols: M. Zhao *et al.*, *J. Org. Chem.* **64**, 2564 (1999); in production of aminyl radicals: W. Huang *et al.*, *J. Am. Chem. Soc.* **121**, 3939 (1999); in polymerization: L. I. Gabaston *et al.*, *Polymer* **40**, 4505 (1999).

Monoclinic dark red crystals, mp 36-40°. Flash pt: 67°C (Acros Organics Data Sheet).

USE: In organic chemistry as a radical trap, a catalyst and in polymerization mediation.

9284. TEMPOL. [2226-96-2] 4-Hydroxy-2,2,6,6-tetramethyl-1-piperidinyloxy; 4-hydroxy-TEMPO; 4-hydroxy-2,2,6,6-tetramethyl piperidine *N*-oxide; 4-hydroxy-2,2,6,6-tetramethylpiperidinooxy. $C_9H_{18}NO_2$; mol wt 172.25. C 62.76%, H 10.53%, N 8.13%, O 18.58%. Stable nitroxyl radical; water-soluble analogue of TEMPO, *q.v.* Functions as a membrane-permeable radical scavenger. Prepn: E. G. Rozantsev, *Bull. Acad. Sci. USSR Div. Chem. Sci.* **12**, 2085 (1964). Energy transfer studies: N. N. Quan, A. V. Guzzo, *J. Phys. Chem.* **85**, 140 (1981). IR conformation study: W. A. Bueno, L. Degrève, *J. Mol. Struct.* **74**, 291 (1981). Solid state NMR spectra: C. J. Groombridge, M. J. Perkins, *J. Chem. Soc. Chem. Commun.* **1991**, 1164. LC/MS/MS determn: I. D. Podmore, *J. Chem. Res. Synop.* **2002**, 574. Use as a phase transfer catalyst: X.-Y. Wang *et al.*, *Synth. Commun.* **29**, 157 (1999). Review of effects in animal models for shock, ischemia-reperfusion injury, and inflammation: C. Thiemermann, *Crit. Care Med.* **31**, S76-S84 (2003).

Crystals from ether + hexane, mp 71.5°. uv max (hexane): 240, 450-500 (ε ∼1800, ∼5). uv max (ethanol): 242, 435-455 (ε ∼3800, ∼10). Sol in water.

USE: Spin label for EPR studies; phase transfer dehydration catalyst; antioxidant; inhibitor of olefin free radical polymerization.

9285. Temsirolimus. [162635-04-3] Rapamycin 42-[3-hydroxy-2-(hydroxymethyl)-2-methylpropanoate]; (3S,6R,7E,9R,10R,-12R,14S,15E,17E,19E,21S,23S,26R,27R,34aS)-9,10,12,13,14,21,-22,23,24,25,26,27,32,33,34,34a-hexadecahydro-9,27-dihydroxy-3-[(1R)-2-[(1S,3R,4R)-hydroxy-3-methoxycyclohexyl]-1-methylethyl]-10,21-dimethoxy-6,8,12,14,20,26-hexameth-3H-pyrido[2,1-c]-[1,4]oxaazacyclohentriacontine-1,5,11,28,29(4H,6H,31H)-pentone 4′-[2,2-bis(hydroxymethyl)propanoate]; rapamycin 42-ester with 2,2-bis-(hydroxymethyl)propionic acid; CCI-779; Torisel. $C_{56}H_{87}$-NO_{16}; mol wt 1030.30. C 65.28%, H 8.51%, N 1.36%, O 24.85%. Ester analog of rapamycin, *q.v.*; selectively inhibits mammalian target of rapamycin (mTOR). Prepn: J. S. Skotnicki *et al.*, **US 5362718** (1994 to Am. Home Prod.). Lipase-catalyzed synthesis from rapamycin: J. Gu *et al.*, *Org. Lett.* **7**, 3945 (2005). Clinical pharmacology: E. Raymond *et al.*, *J. Clin. Oncol.* **22**, 2336 (2004). Clinical study in advanced refractory renal cell carcinoma: M. B. Atkins *et al.*, *ibid.* 909. Clinical evaluation in glioblastoma multiforme: E. Galanis *et al.*, *J. Clin. Oncol.* **23**, 5294 (2005); in breast cancer: S. Chan *et al.*, *ibid.* 5314; in mantle cell lymphoma: T. E. Witzig *et al.*, *ibid.* 5347.

White to off-white powder. Lipophilic. Sol in alcohol. Practically insol in water. Soly is independent of pH.

THERAP CAT: Antineoplastic.

9286. Tenecteplase. [191588-94-0] 103-L-Asparagine-117-L-glutamine-296-L-alanine-297-L-alanine-298-L-alanine-299-L-alanineplasminogen activator (human tissue-type); TNK-tPA; Metalyse. Genetically engineered variant of human tissue plasminogen activator (t-PA), *q.v.*; expressed in Chinese hamster ovary cells. mol wt ~65 kDa. Constructed by oligonucleotide-directed mutagenesis at 3 specific sites. Prepn: W. F. Bennett *et al.*, **WO 9324635** (1993 to Genentech); B. A. Keyt *et al.*, *Proc. Natl. Acad. Sci. USA* **91**, 3670 (1994). Pharmacology: C. R. Benedict *et al.*, *Circulation* **92**, 3032 (1995). Clinical pharmacokinetics: N. B. Modi *et al.*, *Thromb. Haemostasis* **79**, 134 (1998). Clinical trial in acute myocardial infarction: ASSENT-2 Investigators, *Lancet* **354**, 716 (1999).

THERAP CAT: Thrombolytic.

9287. Tenidap. [120210-48-2] (3Z)-5-Chloro-2,3-dihydro-3-(hydroxy-2-thienylmethylene)-2-oxo-1H-indole-1-carboxamide; 5-chloro-2,3-dihydro-2-oxo-3-(2-thienylcarbonyl)-1H-indole-1-carboxamide; 5-chloro-3-(2-thenoyl)-2-oxindole-1-carboxamide; CP-66248. $C_{14}H_9ClN_2O_3S$; mol wt 320.75. C 52.43%, H 2.83%, Cl 11.05%, N 8.73%, O 14.96%, S 10.00%. Inhibitor of 5-lipoxygenase and interleukin-1 (IL-1) activity. Prepn: S. B. Kadin, **EP 156603**; *idem*, **US 4556672** (both 1985 to Pfizer). Effect on 5-lipoxygenase activity *in vitro*: K. Fogh *et al.*, *Arch. Dermatol. Res.* **280**, 430 (1988). Effect on IL-1 activity in patients with rheumatoid arthritis: B. McDonald *et al.*, *Arthritis Rheum.* **31**, Suppl., S52 (1988). Clinical evaluation: P. Katz *et al.*, *ibid.* S52.

Fluffy, yellow crystals from acetic acid, mp 230° (dec).

Sodium salt. [119784-94-0] CP-66248-2. $C_{14}H_8ClN_2NaO_3S$; mol wt 342.73. Crystals from methanol-isopropanol, mp 237-238°.

THERAP CAT: Anti-inflammatory.

9288. Teniposide. [29767-20-2] (5R,5aR,8aR,9S)-5,8,8a,9-Tetrahydro-5-(4-hydroxy-3,5-dimethoxyphenyl)-9-[[4,6-O-[(R)-2-thienylmethylene]-β-D-glucopyranosyl]oxy]furo[3′,4′:6,7]naphtho[2,3-d]-1,3-dioxol-6(5aH)-one; 4′-demethylepipodophyllotoxin 9-(4,6-O-2-thenylidene-β-D-glucopyranoside); 4′-demethylepipodophyllotoxin-β-D-thenylidine glucoside; ETP; NSC-122819; VM-26; Vehem-Sandoz; Vumon. $C_{32}H_{32}O_{13}S$; mol wt 656.66. C 58.53%, H 4.91%, O 31.67%, S 4.88%. Semi-synthetic derivative of podophyllotoxin, *q.v.* Prepn: A. Von Wartburg, **ZA 6607585**; C. Keeler-Juslen *et al.*, **US 3524844** (1968, 1970 both to Sandoz). Mechanism of action: H. Stählen, *Eur. J. Cancer* **6**, 303 (1970). Pharmacology: M. Hacker, D. Roberts, *Cancer Res.* **37**, 3287 (1977); S. M. Sieber *et al.*, *Teratology* **18**, 31 (1978); T. J. Vietti *et al.*, *Cancer Treat. Rep.* **62**, 1313 (1978). Metabolism: L. Allen, *Drug Metab. Rev.* **8**, 119 (1978); *Cancer Res.* **38**, 2549 (1978). Clinical studies: N. M. Gad-el-Mawla *et al.*, *Cancer Treat. Rep.* **62**, 993 (1978); R. E. Bellet *et al.*, *ibid.* 445. Studies on delayed toxicity in mice after i.p. injections: M. Hacker, D. Roberts, *Cancer Res.* **35**, 1756 (1975); H. Stählin, *Eur. J. Cancer* **12**, 925 (1976). Review of pharmacology, pharmacokinetics and assay methods: P. I. Clark, M. L. Slevin, *Clin. Pharmacokinet.* **12**, 223-252 (1987). Comprehensive description: J. J. Kettenes-van den Bosch *et al.*, *Anal. Profiles Drug Subs.* **19**, 575-600 (1990).

Crystals from abs ethanol, mp 242-246°. $[\alpha]_D^{20}$ −107° (9:1 chloroform/methanol). uv max (methanol): 283 nm ($E_{1cm}^{1\%}$ 64.1). pKa 10.13.

THERAP CAT: Antineoplastic.

9289. Tenofovir. [147127-20-6] P-[[(1R)-2-(6-Amino-9H-purin-9-yl)-1-methylethoxy]methyl]phosphonic acid; (R)-9-(2-phosphonomethoxypropyl)adenine; (R)-PMPA; GS-1278. $C_9H_{14}N_5O_4$-P; mol wt 287.22. C 37.64%, H 4.91%, N 24.38%, O 22.28%, P 10.78%. Acyclic phosphonate nucleotide analog; DNA polymerase and reverse transcriptase inhibitor. Prepn: I. Rosenberg *et al.*, *Collect. Czech. Chem. Commun.* **53**, 2753 (1988); A. Holy *et al.*, *ibid.* **60**, 1390 (1995); L. M. Schultze *et al.*, *Tetrahedron Lett.* **39**, 1853 (1998). Prepn of ester prodrug: M. N. Arimilli *et al.*, *Antivir. Chem. Chemother.* **8**, 557 (1997); *idem et al.*, **WO 9804569**; *eidem*, **US 5922695** (1998, 1999 both to Gilead Sci.). Antiretroviral activity *in vitro*: J. Balzarini *et al.*, *Biochem. Biophys. Res. Commun.* **219**, 337 (1996). Metabolism and pharmacokinetics of prodrugs: J.-P. Shaw *et al.*, *Pharm. Res.* **14**, 1824 (1997). Clinical trial in HIV-infected patients: J. E. Gallant *et al.*, *J. Am. Med. Assoc.* **292**, 191 (2004); in chronic hepatitis B: P. Marcellin *et al.*, *N. Engl. J. Med.* **359**, 2442 (2008). Review of pharmacology and clinical efficacy in HIV infection: P. A. Pham, J. E. Gallant, *Expert Opin. Drug Metab. Toxicol.* **2**, 459-469 (2006); in hepatitis B: A. M. Jenh *et al.*, *Pharmacotherapy* **29**, 1212-1227 (2009).

Crystals from boiling water + ethanol, mp 279°. $[\alpha]_D$ +21° (c = 1 in 0.1M HCl).

Bis(isopropyloxycarbonyloxymethyl) ester. [201341-05-1] 5-[[(1*R*)-2-(6-Amino-9*H*-purin-9-yl)-1-methylethoxy]methyl]-2,4,6,8-tetraoxa-5-phosphanonanedioic acid 1,9-bis(1-methylethyl) ester 5-oxide; tenofovir disoproxil; (*R*)-bis(POC)PMPA. $C_{19}H_{30}N_5O_{10}P$; mol wt 519.45. Log P (1-octanol/aqueous phosphate buffer): 1.3 (pH 6.5).

Disoproxil fumarate. [202138-50-9] Tenofovir DF; GS-4331-05; Viread. $C_{19}H_{30}N_5O_{10}P.C_4H_4O_4$; mol wt 635.52. White to off-white crystalline powder. Log P (octanol/phosphate buffer) at 25°: 1.25 (pH 6.5). Soly in distilled water (25°): 13.4 mg/ml.

THERAP CAT: Antiviral; antiretroviral.

9290. Tenonitrozole. [3810-35-3] *N*-(5-Nitro-2-thiazolyl)-2-thiophenecarboxamide; 2-(α-thenoylamino)-5-nitrothiazole; thenitrazole; TC-109; Atrican; Moniflagon. $C_8H_5N_3O_3S_2$; mol wt 255.27. C 37.64%, H 1.97%, N 16.46%, O 18.80%, S 25.12%. Prepd from 2-thenoyl chloride and 2-amino-5-nitrothiazole: **FR M715** (1961 to Chantereau), *C.A.* **59**, 7533g (1963).

Crystals from dioxane or DMF, mp 255-256°.

THERAP CAT: Antiprotozoal (Trichomonas); antifungal.

9291. Tenoxicam. [59804-37-4] 4-Hydroxy-2-methyl-*N*-2-pyridinyl-2*H*-thieno[2,3-*e*]-1,2-thiazine-3-carboxamide 1,1-dioxide; Ro-12-0068; Alganex; Dolmen; Liman; Mobiflex; Rexalgan; Tilatil; Tilcotil. $C_{13}H_{11}N_3O_4S_2$; mol wt 337.37. C 46.28%, H 3.29%, N 12.46%, O 18.97%, S 19.01%. Nonsteroidal anti-inflammatory agent. Prepn: O. Hromatka *et al.*, **DE 2537070** (1976 to Hoffmann-La Roche), *C.A.* **85**, 63077 (1976). Pharmacology: Y. Tanaka *et al.*, *Nippon Yakurigaku Zasshi* **77**, 531 (1981), *C.A.* **95**, 35473 (1981). HPLC determn in plasma: M. E. Pickup *et al.*, *J. Chromatogr.* **225**, 493 (1981). Comparative study vs aspirin in normal volunteers: H. A. Bird *et al.*, *Curr. Med. Res. Opin.* **8**, 9 (1982). Preliminary review of pharmacology, therapeutic efficacy and mechanism of action: J. P. Gonzalez, P. A. Todd, *Drugs* **34**, 289-310 (1987). Comprehensive description: A. M. Al-Obaid, M. S. Mian, *Anal. Profiles Drug Subs. Excip.* **22**, 431-459 (1993).

Crystals from xylene, mp 209-213° (dec). Soly (mg/ml): water 0.045; ethanol <1; methanol <1; acetone 2; dichloromethane 10; chloroform 8; DMSO 63. uv max (ethanol): 205, 265, 360 nm (ε

16422.12, 10127.42, 12544.21). pKa_1 5.3, pKa_2 1.1. Partition coefficient (octanol/water): 0.3 (pH 7.4); 3.5 (pH 1.2).

THERAP CAT: Anti-inflammatory.

9292. Tenuazonic Acid. [610-88-8] (5*S*)-3-Acetyl-1,5-dihydro-4-hydroxy-5-[(1*S*)-1-methylpropyl]-2*H*-pyrrol-2-one; L-3-acetyl-5-*sec*-butyl-4-hydroxy-3-pyrrolin-2-one; L-3-acetyl-5-*sec*-butyl-tetramic acid. $C_{10}H_{15}NO_3$; mol wt 197.23. C 60.90%, H 7.67%, N 7.10%, O 24.34%. Mycotoxin produced by *Alternaria alternata* and other fungal species; contaminant found in wheat and other grains, sunflower seeds, tomatoes and other fruits. Isoln from culture filtrates of *Alternaria tenuis* Auct.: T. Rosett *et al.*, *Biochem. J.* **67**, 390 (1957). Structure: C. E. Stickings, *Biochem. J.* **72**, 332 (1959). Biosynthesis: C. E. Stickings, R. J. Townsend, *ibid.* **78**, 412 (1961). Chemical synthesis: R. N. Lacey, *J. Chem. Soc.* **1954**, 850; S. A. Harris *et al.*, *J. Med. Chem.* **8**, 478 (1965); R. Schobert *et al.*, *Org. Biomol. Chem.* **2**, 3524 (2004). Tumor inhibitory properties: E. A. Kaczka *et al.*, *Biochem. Biophys. Res. Commun.* **14**, 54 (1964); C. O. Gitterman *et al.*, *Cancer Res.* **24**, 440 (1964). Toxicity of isolates from agricultural samples: R. A. Meronuck *et al.*, *Appl. Microbiol.* **23**, 613 (1972). Review of analytical methods and occurrence in foods: P. M. Scott, *J. AOAC Int.* **84**, 1809-1817 (2001).

Pale brown, viscous, gummy substance, $[\alpha]_D^{20}$ −128° (c = 1.0 in methanol); −132 ± 2° (c = 0.5 in $CHCl_3$). $bp_{0.035}$ 117°. Readily sol in organic solvents incl petr ether; sparingly sol in water. On long standing, changes into the crystalline *iso*-form.

Sodium salt. [1013-59-8] Sodium L-tenuazonate. $C_{10}H_{14}NNaO_3$. $[\alpha]_{546}$ −96.7° (c = 2.0 in methanol). uv max (pH 7): 280, 241 nm ($E_{1\ cm}^{1\%}$ 573, 400).

9293. Tephrosin. [76-80-2] (7a*R*,13a*R*)-13,13a-Dihydro-7a-hydroxy-9,10-dimethoxy-3,3-dimethyl-3*H*-[1]benzopyrano[3,4-*b*]-pyrano[2,3-*h*][1]benzopyran-7(7a*H*)-one; hydroxydeguelin; toxicarol. $C_{23}H_{22}O_7$; mol wt 410.42. C 67.31%, H 5.40%, O 27.29%. Occurs in leaves of *Tephrosia vogelii* Hook. f., *Leguminosae*, in derris root, cubé root: Hanriot, *Compt. Rend.* **144**, 150 (1907); Clark, *J. Am. Chem. Soc.* **53**, 729 (1931). Structure: Butenandt, Hilgetag, *Ann.* **495**, 172 (1932).

Prisms. mp 198° (218-220°). Practically insol in water. Sol in chloroform, ether, acetone; sparingly sol in methanol.

9294. Tepoxalin. [103475-41-8] 5-(4-Chlorophenyl)-*N*-hydroxy-1-(4-methoxyphenyl)-*N*-methyl-1*H*-pyrazole-3-propanamide; 3-[5-(4-chlorophenyl)-1-(4-methoxyphenyl)-3-pyrazolyl]-*N*-hydroxy-*N*-methylpropanamide; ORF-20485; RWJ-20485; Zubrin. $C_{20}H_{20}ClN_3O_3$; mol wt 385.85. C 62.26%, H 5.22%, Cl 9.19%, N 10.89%, O 12.44%. Dual inhibitor of cyclooxygenase and 5-lipoxygenase. Prepn: M. P. Wachter, M. P. Ferro, **EP 248594**; *eidem*, **US 4826868** (1987, 1989, both to Ortho); W. V. Murray, S. K. Hadden, *J. Org. Chem.* **57**, 6662 (1992). LC/MS characterization of impurities and degradation products: D. J. Burinsky *et al.*, *J. Pharm. Sci.* **85**, 159 (1996). Mechanism of action: S. S. C. Tam *et al.*, *J. Biol. Chem.* **270**, 13948 (1995). Toxicity study in rats, dogs: E. V.

Knight *et al.*, *Fundam. Appl. Toxicol.* **33**, 38 (1996). Pharmacological profile in animal models: D. C. Argentieri *et al.*, *J. Pharmacol. Exp. Ther.* **271**, 1399 (1994); in man: M. Depré *et al.*, *Int. J. Clin. Pharmacol. Res.* **16**, 1 (1996). Bioavailability in dogs: L. M. Homer *et al.*, *J. Vet. Pharmacol. Ther.* **28**, 287 (2005).

Crystals from ethyl acetate + hexane, mp 124-126°.

THERAP CAT (VET): Anti-inflammatory.

9295. Tepraloxydim. [149979-41-9] 2-[1-[[[(2*E*)-3-Chloro-2-propen-1-yl]oxy]imino]propyl]-3-hydroxy-5-(tetrahydro-2*H*-pyran-4-yl)-2-cyclohexen-1-one; 2-[1-[(2*E*)-3-chloroallyloxyimino]-propyl]-3-hydroxy-5-perhydropyran-4-ylcyclohex-2-en-1-one; caloxydim; BAS-620H; Aramo; Equinox. $C_{17}H_{24}ClNO_4$; mol wt 341.83. C 59.73%, H 7.08%, Cl 10.37%, N 4.10%, O 18.72%. Graminicide for use in broad-leaf crops. Inhibits acetyl CoA carboxylase which catalyzes the first step in fatty acid formation. General prepn: R. Becker *et al.*, **EP 071707**; *eidem* **US 4422864** (both 1983 to BASF). Comprehensive description: E. Kibler *et al.*, *Brighton Crop Prot. Conf. - Weeds* **1999**, 59-64. Efficacy vs black-grass: R. E. Ruske, S. R. Moss, *ibid.* **191**. HPLC-UV determn in drinking water: P. Sandin-España *et al.*, *Chromatographia* **55**, 681 (2002). Mechanism of action study: A. Takahashi *et al.*, *Weed Biol. Manage.* **2**, 84 (2002).

Beige solid, slight characteristic odor. mp 71.5°. Vapor pressure (20°): 1.1×10^{-5} Pa. Soly in water: 0.14 mg/l. LD_{50} in rats (mg/kg): > 2200 orally; > 2000 dermally; LC_{50} in rats (mg/l): 5.1 by inhalation. LC_{50} (96 hr) in trout: > 100 mg/l (Kibler).

USE: Herbicide.

9296. Teprenone. [6809-52-5] 6,10,14,18-Tetramethyl-5,9,-13,17-nonadecatetraen-2-one; geranylgeranylacetone; GGA; E-0671; E36U31; Selbex. $C_{23}H_{38}O$; mol wt 330.56. C 83.57%, H 11.59%, O 4.84%. Acyclic polyisoprenoid 2:3 mixture of (5*Z*,9*E*,-13*E*) and all *trans* (5*E*,9*E*,13*E*) isomers. Prepn from geranylgeranyl bromide: L. Ruzicka, L. Castro, *Helv. Chim. Acta* **28**, 590 (1945); from substituted acetylenes: K. Sato *et al.*, *J. Org. Chem.* **35**, 565 (1970). Anti-ulcer and toxicity studies: M. Murakami *et al.*, *Arzneim.-Forsch.* **31**, 799 (1981). *In vivo* effects on aspirin induced gastric ulcer: *eidem*, *Jpn. J. Pharmacol.* **32**, 299 (1982); on stress induced gastric ulcer: *eidem*, *ibid.* **33**, 549 (1983). Mechanism of action studies: Y. Nishizawa *et al.*, *Biochem. Biophys. Res. Commun.* **103**, 706 (1981); K. Oketani *et al.*, *Jpn. J. Pharmacol.* **33**, 593 (1983). Metabolism in rats: Y. Nishizawa *et al.*, *Xenobiotica* **17**, 575 (1987). Determn in serum by GC-MS: M. Tanaka *et al.*, *J. Chromatogr.* **231**, 301 (1982); by HPLC: T. Seki *et al.*, *ibid.* **424**, 410 (1988).

all *trans*-form

Yellow oil, $bp_{0.01}$ 155-160°. $d_4^{20.5}$ 0.9081. n_D^{20} 1.4947.

THERAP CAT: Antiulcerative.

9297. Terazosin. [63590-64-7] [4-(4-Amino-6,7-dimethoxy-2-quinazolinyl)-1-piperazinyl](tetrahydro-2-furanyl)methanone; 2-[(4-tetrahydro-2-furoyl)-1-piperazinyl]-4-amino-6,7-dimethoxy-quinazoline. $C_{19}H_{25}N_5O_4$; mol wt 387.44. C 58.90%, H 6.50%, N 18.08%, O 16.52%. α_1-Adrenergic blocker related to prazosin, *q.v.* Prepn: M. Winn *et al.*, **DE 2646186**; *eidem*, **US 4026894** (both 1977 to Abbott); of the hydrochloride dihydrate: R. Roteman, **DE 2831112**; *idem*, **US 4251532** (1979, 1981 both to Abbott). HPLC determn in biological fluids: S. E. Patterson, *J. Chromatogr.* **311**, 206 (1984). Toxicity in rats: F. L. Fort *et al.*, *Drug Chem. Toxicol.* **7**, 435 (1984). Clinical study in essential hypertension: P. A. Abraham *et al.*, *Pharmacotherapy* **5**, 285 (1985); in treatment of benign prostatic hyperplasia: H. Lepor *et al.*, *J. Urol.* **148**, 1467 (1992). Symposium on pharmacology, pharmacokinetics, and clinical efficacy in hypertension: *Am. J. Med.* **80**, Suppl. 5B, 1-105 (1986). Comprehensive description: Z. L. Chang, J. F. Bauer, *Anal. Profiles Drug Subs.* **20**, 693-727 (1991). Review of pharmacology and therapeutic efficacy: R. Achari, A. Laddu, *J. Clin. Pharmacol.* **32**, 520-523 (1992).

Fine, white, odorless crystals, mp 272.6-274°. Soly at 25°C (mg/ml): methanol 33.7; water 29.7; 95% ethanol 4.1; 0.1*N* HCl 3.8; chloroform 1.2; acetone 0.01. Practically insol in hexane. uv max (0.005% in water): 212, 245, 330 nm (a 65.7, 127.5, 24.0). pKa (0.1*N* NaOH): 7.1.

Hydrochloride. $C_{19}H_{25}N_5O_4 \cdot HCl$. Hygroscopic crystals from isopropyl alc, mp 278-279°. Soly in water: 761.2 mg/ml. Freely sol in isotonic saline soln; sol in methanol; slightly sol in alc, 0.1*N* hydrochloric acid; very slightly sol in chloroform. Practically insol in acetone, hexanes. LD_{50} in mice (mg/kg): 259.3 i.v. (Winn).

Hydrochloride dihydrate. [70024-40-7] Abbott 45975; Heitrin; Hytracin; Hytrin; Hytrinex; Itrin; Urodie; Vasocard; Vasomet; Vicard. $C_{19}H_{25}N_5O_4 \cdot HCl \cdot 2H_2O$; mol wt 459.93. mp 271-274°. Soly in water: 24.2 mg/ml. LD_{50} in male, female rats (mg/kg): 277, 293 i.v. (Fort).

THERAP CAT: Antihypertensive. In treatment of benign prostatic hyperplasia.

9298. Terbacil. [5902-51-2] 5-Chloro-3-(1,1-dimethylethyl)-6-methyl-2,4(1*H*,3*H*)-pyrimidinedione; 3-*tert*-butyl-5-chloro-6-methyluracil; Du Pont Herbicide 732; Sinbar. $C_9H_{13}ClN_2O_2$; mol wt 216.67. C 49.89%, H 6.05%, Cl 16.36%, N 12.93%, O 14.77%. Prepn: Loux, **US 3235357** (1966 to du Pont). Herbicidal activity and physical properties: G. D. Hill, Jr. *et al.*, *2nd Symp. New Herbic.* **1965**, 313, *C.A.* **66**, 18234b (1967). Metabolism and mode of action: Herboldt, *Diss. Abstr. Int. B* **30**, 1978 (1969).

Crystals, mp 175-177°. Soly in water at 25°: 710 ppm. Sol in DMF, dimethylacetamide, cyclohexanone; moderately sol in methyl isobutyl ketone, butyl acetate, xylene. Approx. LD orally in rats: >5000 mg/kg (Hill).

USE: Herbicide.

9299. Terbinafine. [91161-71-6] *N*-[(2*E*)-6,6-Dimethyl-2-hepten-4-yn-1-yl]-*N*-methyl-1-naphthalenemethanamine; *trans-N*-methyl-*N*-(1-naphthylmethyl)-6,6-dimethylhept-2-en-4-ynyl-1-amine. $C_{21}H_{25}N$; mol wt 291.44. C 86.55%, H 8.65%, N 4.81%. Orally active, antimycotic allylamine related to naftifine, *q.v.* Specific inhibitor of squalene epoxidase, a key enzyme in fungal ergosterol biosynthesis. Prepn: A. Stütz, **EP 24587**; *idem*, **US 4755534** (1981, 1988 both to Sandoz); A. Stütz, G. Petranyi, *J. Med. Chem.* **27**, 1539 (1984). Mode of action: G. Petranyi *et al.*, *Science* **224**, 1239 (1984); N. S. Ryder, *Antimicrob. Agents Chemother.* **27**, 252 (1985). *In vitro* antifungal activity: S. Shadomy *et al.*, *Sabouraudia* **23**, 125 (1985). CE determn in pharmaceutical formulations: P. Mikus *et al.*, *Talanta* **65**, 1031 (2005). Symposium on pharmacology and clinical trials: *J. Am. Acad. Dermatol.* **23**, Suppl., 775-812 (1990). Clinical trial as systemic treatment of toenail onychomycosis: E. G. V. Evans *et al.*, *Br. Med. J.* **318**, 1031 (1999); A. Tavkkol *et al.*, *Am. J. Geriatr. Pharmacother.* **4**, 1 (2006). Toxicology: U. Ganzinger *et al.*, *Proc. 13th Int. Congr. Chemother.* **6**, 116/52 (1983).

Hydrochloride. [78628-80-5] SF-86-327; Lamisil. $C_{21}H_{25}N$·HCl; mol wt 327.90. Crystals from 2-propanol + diethyl ether, mp 195-198° (change in crystal structure begins ~150°). Freely sol in dehydrated alc, methanol, methylene chloride; sol in ethanol; slightly sol in water, acetone. LD_{50} in mice, rats (mg/kg): 4000, 4000 orally; 393, 213 i.v. (Ganzinger).

THERAP CAT: Antifungal.

THERAP CAT (VET): Antifungal.

9300. Terbium. [7440-27-9] Tb; at. wt 158.92535; at. no. 65; valences 3, 4. A lanthanide; belongs to the yttrium group of rare earth metals. Naturally occurring isotope (mass number): 159; known artificial radioactive isotopes: 144-158; 160-164. Abundance in earth's crust: 0.91-1.2 ppm; occurs in small quantities in monazite, cerite, gadolinite and other rare earth minerals. Discovered by Mosander, *Skand. Naturför. Förh.* **3**, 387 (1842); *Philos. Mag.* [3] **23**, 241 (1843). Sepn by fractional crystn and precipitation: Urbain, *Compt. Rend.* **139**, 736 (1904); **141**, 521 (1905); James, Bissel, *J. Am. Chem. Soc.* **36**, 2060 (1914); by ion exchange: Spedding *et al.*, *ibid.* **76**, 2557 (1954). Prepn of metal by electrodeposition: *eidem*, *J. Electrochem. Soc.* **100**, 442 (1953). Absorption spectrum: Urbain, *loc. cit.* Toxicity study: Haley, *J. Pharm. Sci.* **54**, 663 (1965). Reviews of prepn, properties and compds: *The Rare Earths*, F. H. Spedding, A. H. Daane, Eds. (Krieger, Huntington, N.Y., 1971, reprint of 1961 ed.) 641 pp; Hulet, Bode, "Separation Chemistry of the Lanthanides and Transplutonium Actinides" in *MTP Int. Rev. Sci.: Inorg. Chem., Ser. One* vol. 7, K. W. Bagnall, Ed. (University Park Press, Baltimore, 1972) pp 1-45; Moeller, "The Lanthanides" in *Comprehensive Inorganic Chemistry* vol. 4, J. C. Bailar, Jr. *et al.*, Eds. (Pergamon Press, Oxford, 1973) pp 1-101; F. H. Spedding in *Kirk-Othmer Encyclopedia of Chemical Technology* vol. 19 (John Wiley & Sons, New York, 3rd ed., 1982) pp 833-854; *Chemistry of the Elements*, N. N. Greenwood, A. Earnshaw, Eds. (Pergamon Press, New York, 1984) pp 1423-1449. Brief review of properties: G. T. Seaborg, *Radiochim. Acta* **61**, 115-122 (1993).

Silver-gray metal, easily oxidized in air. Hexagonal close-packed crystals at room temp; d 8.27 (Spedding, Daane, *loc. cit.* p. 183). mp 1356°. bp 3230°; also reported as bp 2480° (Moeller).

Oxide. Terbia. O_3Tb_2. A white solid.

Oxide. Non-stoichiometric, approx composition, Tb_4O_7, for this formula Tb^{3+} and Tb^{4+} present in equal amounts. Dark brown or black solid; obtained by igniting the oxalate or the sulfate. Dissolves in hot concd acids with formation of salts; loses oxygen on heating; forms the oxide when heated in hydrogen.

Nitrate. $Tb(NO_3)_3$. Occurs as the hexahydrate, monoclinic crystals, mp 89.3°. *See:* Urbain, cited by Mellor, *A Comprehensive Treatise on Inorganic and Theoretical Chemistry* **5**, 695 (1929). LD_{50} in rats (mg/kg): 260 i.p.; >5000 orally (Haley).

Chloride hexahydrate. $TbCl_3·6H_2O$. Prismatic deliquesc crystals. Very sol in water; forms supersaturated solns. Dehydrated on

heating in hydrogen chloride at 180-200°. The anhydr chloride, crystals, d^0 4.35; mp 588°; dissolves in water without hydrolysis. LD_{50} in mice (mg/kg): 550 i.p.; 5100 orally (Haley).

9301. Terbufos. [13071-79-5] Phosphorodithioic acid S-[[(1,1-dimethylethyl)thio]methyl] O,O-diethyl ester; phosphorodithioic acid S-[(tert-butylthio)methyl] O,O-diethyl ester; S-tert-butylthiomethyl O,O-diethyl phosphorodithioate; AC-92100; Counter. $C_9H_{21}O_2PS_3$; mol wt 288.42. C 37.48%, H 7.34%, O 11.09%, P 10.74%, S 33.35%. Organophosphate insecticide, cholinesterase inhibitor. Prepn: S. Takahashi *et al.*, **JP 66 11859** (1966 to Hokko), *C.A.* **65**, 15231a (1966). As insecticidal soil treatment: F. M. Gordon, **DE 2258528**; *idem*, **US 4065558** (1973, 1977 both to Am. Cyanamid). Multiresidue determn in fruits and vegetables: C. Jansson *et al.*, *J. Chromatogr. A* **1023**, 93 (2004). Photodegradation in water: R.-J. Wu *et al.*, *J. Hazard. Mater.* **162**, 945 (2009). Toxicity to birds: E. F. Hill, M. B. Camardese, *Ecotoxicol. Environ. Saf.* **8**, 551 (1984).

Technical product (85 to 88% purity): clear, colorless to pale yellow liq, d^{24} 1.105. $bp_{0.01}$ 69°. mp −29.2°. Flash pt 88°C (tag open cup). Sol in acetone, alcs, aromatic and chlorinated hydrocarbons. Soly in water: 10-15 ppm. *Poisonous.* LD_{50} orally in quail: 15 mg/kg (Hill, Camardese).

USE: Soil insecticide.

9302. Terbutaline, [23031-25-6] 5-[2-[(1,1-Dimethylethyl)-amino]-1-hydroxyethyl]-1,3-benzenediol; α-[(tert-butylamino)-methyl]-3,5-dihydroxybenzyl alcohol; 1-(3,5-dihydroxyphenyl)-2-(tert-butylamino)ethanol; mol wt 225.29. C 63.98%, H 8.50%, N 6.22%, O 21.30%. β-Adrenergic agonist. Prepn: K. Wetterlin, L. A. Svensson, **BE 704932**; *eidem*, **US 3937838** (1968, 1976 both to Draco). Pharmacology: Bergman *et al.*, *Experientia* **25**, 899 (1969). Resolution of isomers and activity studies: K. Wetterlin, *J. Med. Chem.* **15**, 1182 (1972). HPLC determn of enantiomers in urine: K. H. Kim *et al.*, *J. Chromatogr. B* **751**, 69 (2001). Clinical study in treatment of preterm labor: S. N. Caritis *et al.*, *Am. J. Obstet. Gynecol.* **150**, 7 (1984). *Review:* J. J. McPhillips in *Pharmacological and Biochemical Properties of Drug Substances* vol. 1, M. E. Goldberg, Ed. (Am. Pharm. Assoc., Washington, DC, 1977) pp 311-328. Review of clinical toxicology: J. D. Truwit, *Crit. Care Clin.* **7**, 639-657 (1991). Comprehensive description: S. Ahuja, J. Ashman, *Anal. Profiles Drug Subs.* **19**, 601-625 (1990).

Crystals from abs ether, mp 119-122°.

Sulfate. [23031-32-5] Brethaire; Brethine; Bricanyl; Butaliret; Monovent; Terbasmin; Terbasmin. $(C_{12}H_{19}NO_3)_2·H_2SO_4$; mol wt 548.65. mp 246-248°. uv max (0.1N HCl): 276 nm ($A_{1cm}^{1\%}$ 67.6). pKa_1 8.8, pKa_2 10.1, pKa_3 11.2. Soly at 25° (mg/ml): water >20; 0.1N HCl >20; 0.1N NaOH >20; ethanol 1.2; 10% ethanol >20; methanol 2.7. Insol in chloroform.

THERAP CAT: Bronchodilator; tocolytic.

THERAP CAT (VET): Bronchodilator.

9303. Terconazole. [67915-31-5] rel-1-[4-[[(2R,4S)-2-(2,4-Dichlorophenyl)-2-(1H-1,2,4-triazol-1-ylmethyl)-1,3-dioxolan-4-yl]methoxy]phenyl]-4-(1-methylethyl)piperazine; triaconazole; R-42470; Fungistat; Gyno-Terazol; Terazol; Tercospor. $C_{26}H_{31}Cl_2$-N_5O_3; mol wt 532.47. C 58.65%, H 5.87%, Cl 13.32%, N 13.15%, O 9.01%. Topical triazole antifungal. Prepn: J. Heeres *et al.*, **DE 2804096**; *eidem*, **US 4144346**; **US 4223036** (1978, 1979, 1980 all to Janssen); *eidem*, *J. Med. Chem.* **26**, 611 (1983). Pharmacology: J.

Van Cutsem *et al.*, *Chemotherapy* **29**, 322 (1983). Clinical comparison with clotrimazole in vaginal candidiasis: A. Kjaeldgaard, *Pharmatherapeutica* **4**, 525 (1986).

Relative stereochemistry

Crystals from isopropyl ether, mp 126.3°.
THERAP CAT: Antifungal.

9304. Terebic Acid. [79-91-4] Tetrahydro-2,2-dimethyl-5-oxo-3-furancarboxylic acid; tetrahydro-2,2-dimethyl-5-oxo-3-furoic acid; terebinic acid; (1-hydroxy-1-methylethyl)succinic acid γ-lactone. $C_7H_{10}O_4$; mol wt 158.15. C 53.16%, H 6.37%, O 40.47%. Prepared from fumaric or maleic acid: Schenck, Steinmetz, *Tetrahedron Lett.* **21**, 1 (1960); Lipp *et al.*, *Ann.* **644**, 37 (1961). Prepn of optical isomers: Fredga, *C.A.* **42**, 123g (1948); Delépine, Badoche, *Compt. Rend.* **235**, 1069 (1952).

Crystals, mp 174-175°, but begins to volatilize at 100°. d 0.815. Slightly sol in cold water, freely in boiling water or warm alcohol.
(+)-Form. $[\alpha]_D^{25}$ +13.2° (c = 0.03 in acetone).
(−)-Form. mp 201-205° (dec). $[\alpha]_D^{25}$ −13.2° (c = 0.03 in acetone).

9305. Terephthalic Acid. [100-21-0] 1,4-Benzenedicarboxylic acid; *p*-phthalic acid; TPA; Tephthol. $C_8H_6O_4$; mol wt 166.13. C 57.84%, H 3.64%, O 38.52%. Prepd by oxidation of *p*-methylacetophenone: Koelsch, *Org. Synth.* **coll. vol. III**, 791 (1955). Manuf processes: **US 3014961** (1959 to VEB Chemie Werke Buna); Sherwood, *Chem. Ind. (London)* **1960**, 1096. *Review: Faith, Keyes & Clark's Industrial Chemicals*, F. A. Lowenheim, M. K. Moran, Eds. (Wiley-Interscience, New York, 4th ed., 1975) pp 807-813; A. G. Bemis *et al.*, "Phthalic Acids" in *Kirk-Othmer Encyclopedia of Chemical Technology* **vol. 17** (Wiley-Interscience, New York, 3rd ed., 1982) pp 732-777. Review of toxicology and risk assessment: G. L. Ball *et al.*, *Crit. Rev. Toxicol.* **42**, 28-67 (2012).

Crystals. Sublimes at 402°. Log P (octanol/water): 2. Soly in water (20°): 15 mg/l. Slightly sol in cold alcohol, more in hot alcohol; sol in alkalies. Practically insol in chloroform, ether, acetic acid.
Dimethyl ester. [120-61-6] Dimethyl terephthalate; DMT. $C_{10}H_{10}O_4$. White crystals, mp 140.6°. bp 288°. Log P (octanol/water): 2.25. Soly in water (25°): 19 mg/l.
USE: Forms polyesters with glycols which are made into plastic films and sheets used in photography and packaging for foods and beverages; in manuf of polyester fibers for carpet yarns, clothing, fiber-fill, industrial filaments; in analytical chemistry.

9306. Terfenadine. [50679-08-8] α-[4-(1,1-Dimethylethyl)-phenyl]-4-(hydroxydiphenylmethyl)-1-piperidinebutanol; 1-(*p-tert*-butylphenyl)-4-[4′-(α-hydroxydiphenylmethyl)-1′-piperidyl]butanol; α-(*p-tert*-butylphenyl)-4-(α-hydroxy-α-phenylbenzyl)-1-piperidinebutanol; MDL-9918; Allerplus; Cyater; Seldane; Teldane; Teldanex; Terfex; Ternadin; Triludan. $C_{32}H_{41}NO_2$; mol wt 471.69. C 81.48%, H 8.76%, N 2.97%, O 6.78%. Nonsedating-type histamine H_1-receptor antagonist. Prepn: A. A. Carr, C. R. Kinsolving, **DE**

2303306; *eidem*, **US 3878217** (1973, 1975 both to Richardson-Merrell); A. A. Carr, D. R. Meyer, *Arzneim.-Forsch.* **32**, 1157 (1982). Series of articles on chemistry, pharmacology, toxicology and clinical studies: *ibid.* **32**, 1153-1218 (1982). Metabolism in human liver: M. Jurima-Romet *et al.*, *Drug Metab. Dispos.* **22**, 849 (1994). Comprehensive description: A. A. Badwan *et al.*, *Anal. Profiles Drug Subs.* **19**, 627-662 (1990). *Review:* H. C. Masheter, *Clin. Rev. Allergy* **11**, 5-34 (1993).

Crystals from acetone, mp 146.5-148.5°; may exist in three polymorphic forms, mp 149-152°, 146-148°, 142-144°. Soly at 30° (g/100ml): water 0.001; ethanol 3.780; methanol 3.750; hexane 0.034; 0.1M HCl 0.012; 0.1M citric acid 0.110; 0.1M tartaric acid 0.045. uv max (methanol): 260 nm (A 660.4); (ethanol): 260 nm (A 671.4); (dichloromethane): 260 nm (A 762.2). LD_{50} orally in mice: >2000 mg/kg (Carr, Meyer).
THERAP CAT: Antihistaminic.

9307. Terguride. [37686-84-3] *N,N*-Diethyl-*N′*-[(8α)-6-methylergolin-8-yl]urea; *N*-(D-6-methyl-8-isoergolin-1-yl)-*N′,N′*-diethylurea; 6-methyl-8α-(diethylcarbamoylamino)ergoline; 9,10α-dihydrolisuride; transdihydrolisuride; TDHL. $C_{20}H_{28}N_4O$; mol wt 340.47. C 70.56%, H 8.29%, N 16.46%, O 4.70%. Ergot derivative; dihydrogenated analog of lisuride, *q.v.* Exhibits dopamine agonist and antagonist activity. Prepn: V. Zikán *et al.*, *Collect. Czech. Chem. Commun.* **37**, 2600 (1972); *eidem*, **DE 2238540**; *eidem*, **US 3953454** (1973, 1976 both to Spofa). Physical properties: A. Cerny *et al.*, *Collect. Czech. Chem. Commun.* **52**, 1331 (1987). Receptor binding studies in rat brain: M. W. Valchár *et al.*, *Eur. J. Pharmacol.* **136**, 97 (1987). Radioreceptor assay in biological fluids: R. Lapka *et al.*, *J. Pharmacol. Methods* **11**, 263 (1984). HPLC determn in tablets and plasma: J. Sochor *et al.*, *J. Chromatogr. B* **663**, 309 (1995). Pharmacokinetics in humans: W. Krause *et al.*, *Eur. J. Clin. Pharmacol.* **27**, 335 (1984). Clinical evaluation in Parkinson's disease: T. Brücke *et al.*, *Adv. Neurol.* **45**, 573 (1986); I. Suchy *et al.*, *ibid.* 577; in hyperprolactinemia and acromegaly: D. Dallabonzana *et al.*, *J. Clin. Endocrinol. Metab.* **63**, 1002 (1986).

Crystals from ethanol, mp 203-204° (dec). $[\alpha]_D^{20}$ +30° (c = 1 in pyridine). Also reported as crystals from ethanol, mp 205-207° (dec) (Cerny). $[\alpha]_D^{20}$ +29.0° (c = 0.2 in pyridine). uv max: 292, 281, 224 nm (log ε 3.72, 3.81, 4.42). Practically insol in water.
Hydrogen maleate. [37686-85-4] SH-406; VUFB-6638; ZK-31224; Mysalfon. $C_{20}H_{28}N_4O.C_4H_4O_4$; mol wt 456.54. Crystals from ethanol, mp 190-191°.
Hydrogen maleate hydrate. Crystals from ethanol, mp 150-153°. $[\alpha]_D^{20}$ −15.0° (c = 0.1 in H_2O). Soly in water: 1.26 mg/ml.
THERAP CAT: Antiparkinsonian; antihyperprolactinemic.

9308. Teriflunomide. [163451-81-8]; [108605-62-5] (unspecified stereo). (2Z)-2-Cyano-3-hydroxy-*N*-[4-(trifluoromethyl)phenyl]-2-butenamide; A-771726; Aubagio. $C_{12}H_9F_3N_2O_2$; mol wt 270.21. C 53.34%, H 3.36%, F 21.09%, N 10.37%, O 11.84%. Disease modifying antirheumatic drug (DMARD); active metabolite of leflunomide, *q.v.* Prepn: R. R. Bartlett, F.-J. Kämmerer, **WO 9117748**; *eidem*, **US 5494911** (1991, 1996 both to Hoechst). Identification as active metabolite: J. W. Patterson *et al.*, *J. Med. Chem.* **35**, 507 (1992). Phamacokinetics: J. Lucien *et al.*, *Ther. Drug Monit.* **17**, 454 (1995). Inhibits dihydroorotate dehydrogenase: J. P. Davis *et al.*, *Biochemistry* **35**, 1270 (1996). Configurational study: G. Bertolini *et al.*, *J. Med. Chem.* **40**, 2011 (1997); and crystal structure: C. Papageorgiou *et al.*, *Bioorg. Chem.* **25**, 233 (1997). LC-MS/MS determn in plasma: H. Rakhila *et al.*, *J. Pharm. Biomed. Anal.* **55**, 325 (2011). Clinical trial in relapsing multiple sclerosis: P. O'Connor *et al.*, *N. Engl. J. Med.* **365**, 1293 (2011).

THERAP CAT: Immunomodulator; in treatment of multiple sclerosis.

9309. Teriparatide Acetate. [99294-94-7] L-Seryl-L-valyl-L-seryl-L-α-glutamyl-L-isoleucyl-L-glutaminyl-L-leucyl-L-methionyl-L-histidyl-L-asparaginyl-L-leucylglycyl-L-lysyl-L-histidyl-L-leucyl-L-asparaginyl-L-seryl-L-methionyl-L-α-glutamyl-L-arginyl-L-valyl-L-α-glutamyl-L-tryptophyl-L-leucyl-L-arginyl-L-lysyl-L-lysyl-L-leucyl-L-glutaminyl-L-α-aspartyl-L-valyl-L-histidyl-L-asparaginyl-L-phenylalanine acetate hydrate (1:?:?); hPTH 1-34 acetate; MN-10T; Parathar; Forteo. $C_{181}H_{291}N_{55}O_{51}S_2 \cdot yC_2H_4O_2 \cdot xH_2O$. Synthetic, biologically active polypeptide consisting of the 1-34 amino-terminal fragment of human parathyroid hormone (hPTH), *q.v.* Amino acid sequence: H. B. Brewer *et al.*, *Proc. Natl. Acad. Sci. USA* **69**, 3585 (1972). Prepn: R. H. Andreatta *et al.*, *Helv. Chim. Acta* **56**, 470 (1973). Revised structure: H. D. Niall *et al.*, *Proc. Natl. Acad. Sci. USA* **71**, 384 (1974). Prepn: R. L. Colescott, **DE 2649727**; *idem*, **DE 2649848**; G. W. Tregear, **US 4086196** (1977, 1977, 1978 all to Armour Pharm.). Solid phase synthesis and biological activity: G. W. Tregear *et al.*, *Z. Physiol. Chem.* **355**, 415 (1974). Solution synthesis and biological activity: M. Takai *et al.*, *Pept. Chem.* **17**, 187 (1980). Clinical pharmacology: R. M. Neer *et al.*, *J. Clin. Endocrinol. Metab.* **38**, 420 (1977). RIA determn in plasma: J. M. Zanelli *et al.*, *J. Immunoassay* **1**, 289 (1980). Clinical pharmacokinetics: G. N. Kent *et al.*, *Clin. Sci.* **68**, 171 (1985). Clinical evaluation in osteoporosis: J. Reeve *et al.*, *Br. Med. J.* **280**, 1340 (1980); D. M. Slovik *et al.*, *J. Clin. Invest.* **68**, 1261 (1981). Diagnostic use: T. Igarashi *et al.*, *Pharmatherapeutica* **3**, 79 (1982); L. E. Mallette *et al.*, *J. Clin. Endocrinol. Metab.* **67**, 964 (1988). Review of diagnostic use in modified Ellsworth-Howard test: L. E. Mallette, *Ann. Intern. Med.* **109**, 800-804 (1988); of pharmacology and clinical efficacy: H. Dobnig, *Expert Opin. Pharmacother.* **5**, 1153-1162 (2004). THERAP CAT: Treatment of osteoporosis. Diagnostic aid (hypocalcemia).

9310. Terlipressin. [14636-12-5] *N*-(Glycylglycylglycyl)-8-L-lysinevasopressin; N^α-glycylglycylglycyl-8-lysine-vasopressin; triglycyl-lysine-vasopressin; Glypressin; Lucassin; Remestyp. $C_{52}H_{74}N_{16}O_{15}S_2$; mol wt 1227.38. C 50.89%, H 6.08%, N 18.26%, O 19.55%, S 5.22%. Synthetic analog of the antidiuretic hormone, vasopressin, *q.v.*, with increased selectivity for V_1-receptors. Prodrug converted *in vivo* to lypressin, *q.v.* Prepn: E. Kasafirek *et al.*, *Collect. Czech. Chem. Commun.* **31**, 4581 (1966); Z. Procházka *et al.*, *ibid.* **43**, 1285 (1978). Pharmacological studies: J. Kyncl *et al.*, *Eur. J. Pharmacol.* **28**, 294 (1974). Hemostatic effects in cirrhosis patients: C. V. Prowse *et al.*, *Eur. J. Clin. Invest.* **10**, 49 (1980). Clinical trial in hepatorenal syndrome: A. J. Sanyal *et al.*, *Gastroenterology* **134**, 1360 (2008). Review of clinical trials in acute esophageal variceal hemorrhage: G. N. Ioannou *et al.*, *Aliment. Pharmacol. Ther.* **17**, 53-64 (2003); of pharmacology and clinical experience in septic shock: A. Morelli *et al.*, *Expert Opin. Pharmacother.* **10**, 2569-2575 (2009).

Gly–Gly–Gly–Cys–Tyr–Phe–Gln–Asn–Cys–Pro–Lys–GlyNH$_2$

Diacetate pentahydrate. Glycylpressin. $C_{56}H_{82}N_{16}O_{19}S_2 \cdot 5H_2O$; mol wt 1437.56. $[\alpha]_D^{25}$ −82° (c = 0.2 in 1*M* acetic acid). THERAP CAT: Vasoconstrictor.

9311. Terpin. [80-53-5] 4-Hydroxy-α,α,4-trimethylcyclohexanemethanol; *p*-menthane-1,8-diol; dipenteneglycol. $C_{10}H_{20}O_2$; mol wt 172.27. C 69.72%, H 11.70%, O 18.57%. Both *cis*-and *trans*-modifications are known. The *cis*-compd is obtained most readily in the hydrated form, *cis*-terpin hydrate. Prepn of *cis*-form from oil of turpentine: Hempel, *Ann.* **180**, 71 (1876); Wallach, *Ann.* **230**, 225 (1885); Schmitt, *Manuf. Chem.* **26**, 350 (1955). From *d*-limonene: Sword, *J. Chem. Soc.* **127**, 1632 (1925). Prepn of *trans*-form from 1,8-cineole, α-terpineol or *cis*-terpin hydrate: Matsuura *et al.*, *Bull. Chem. Soc. Jpn.* **31**, 990 (1958); Lombard, Ambroise, *Bull. Soc. Chim. Fr.* **1961**, 230. Structure of *cis*- and *trans*-forms: Baeyer, *Ber.* **26**, 2861 (1893); Wagner, *ibid.* **27**, 1636 (1894).

cis-form

*cis***-Form hydrate.** [2451-01-6] Terpin hydrate; terpinol. Rhombic pyramids from water, mp 116-117°; sublimes at ~100° when heated slowly; slight characteristic odor and slightly bitter taste; efflorescent in dry air. Anhydr *cis*-form: mp 104-105°; bp 258°; rapidly re-forms hydrate on exposure to air. One gram dissolves in 34 ml boiling water, 13 ml alcohol, 3 ml boiling alcohol, 135 ml chloroform, 140 ml ether, ~1 ml boiling glacial acetic acid. At 20°, one gram dissolves in 13 ml methanol, 13 ml ethyl acetate, 250 ml water, 77 ml benzene, 290 ml carbon tetrachloride, 250 ml carbon disulfide. Practically insol in petr ether.

*trans***-Form.** [565-50-4] Monoclinic prisms, mp 158-159°. One gram dissolves at 20° in 11 ml methanol, 20 ml ethyl acetate, 100 ml water, 250 ml benzene, 250 ml carbon tetrachloride, 500 ml carbon disulfide.

THERAP CAT: *cis*-Form hydrate as expectorant.
THERAP CAT (VET): Expectorant.

9312. Terpinene. $C_{10}H_{16}$; mol wt 136.24. C 88.16%, H 11.84%. Mixture of three isomeric hydrocarbons: α-terpinene and γ-terpinene which occur naturally and β-terpinene which has been prepd synthetically. Isoln of α-terpinene from cardamom and marjoram oils: Weber, *Ann.* **238**, 101 (1887); Beltz, *Ber.* **32**, 996 (1899). Isoln of α-terpinene from oils of *Mosla japonica, M. grosserrata* and *Cupressus macrocarpa*: Richter, Wolff, *Ber.* **60**, 477 (1927); Briggs, Sutherland, *J. Org. Chem.* **7**, 397 (1942). Prepn of β-terpinene from sabinine: Wallach, *Ann.* **357**, 64 (1907); **362**, 285 (1908). Structure: Wallach, *ibid.* **362**, 293 (1908). *Review:* J. L. Simonsen, *The Terpenes* vol. I (University Press, Cambridge, 2nd ed., 1947) pp 172-193.

α   β   γ

α-Terpinene. Oil, pleasant odor of lemons. bp 173.5-174.8°; $bp_{13.5}$ 65.4-66°, $d_4^{19.6}$ 0.8375. n_D^{20} 1.4784. Practically insol in water. Miscible with alcohol, ether.
β-Terpinene. Oil, bp 173-174°. d^{22} 0.838. n_D^{22} 1.4754.
γ-Terpinene. Oil, bp 183°. d_4^{15} 0.853. $n_D^{15.6}$ 1.4754.
Dihydrochloride. $C_{10}H_{18}Cl_2$. Crystals, mp 51-52°.

9313. α-Terpineol. [98-55-5] α,α,4-Trimethyl-3-cyclohexene-1-methanol; *p*-menth-1-en-8-ol. $C_{10}H_{18}O$; mol wt 154.25. C 77.87%, H 11.76%, O 10.37%. Terpineol exists as three isomers, α-, β-, and γ-terpineol: J. L. Simonsen, *The Terpenes* vol. I (University Press, Cambridge, 2nd ed., 1947) pp 256-274. Isoln of *d*-α-terpineol from petitgrain oil: Walbaum, Hüthig, *J. Prakt. Chem.* **67**, 322 (1903). Isoln from *l*-α-terpineol from long leaf pine oil: Teeple,

J. Am. Chem. Soc. **30**, 412 (1908). Isoln of *dl*-α-terpineol from cajeput oil: Voiry, *Compt. Rend.* **106**, 1540 (1888). Synthesis of *d*-α-terpineol: Cologne, Crabalona, *Bull. Soc. Chim. Fr.* **1960**, 102. Synthesis of *l*-α-terpineol: *eidem, ibid.* **1959**, 1505. Stereochemistry: Henbest, McElkinney, *J. Chem. Soc.* **1959**, 1834. *Review:* Wagner, *Manuf. Chem.* **22**, 98, 153 (1951).

d-Form. Liquid. bp$_{4.5}$ 81-82°; bp$_{731}$ 206-207°. d$_4^{20}$ 0.9338. n$_D^{20}$ 1.4818. [α]$_D^{20}$ +92.45°. Solidifies at 31°.

Phenylurethan. C$_{17}$H$_{23}$NO$_2$. Crystals from petr ether, mp 111°. [α]$_D^{20}$ +30.50° (benzene).

l-Form. Liquid. bp$_5$ 80-81.5°. d$_4^{20}$ 0.935. n$_D^{20}$ 1.4820. [α]$_D^{20}$ −100° (c = 20 in alc). Solidifies at 36.4°.

Dinitrobenzoate. C$_{17}$H$_{20}$N$_2$O$_6$. Crystals from alcohol, mp 101.5°. [α]$_D^{20}$ −31° (c = 9.5 in carbon tetrachloride).

dl-Form. Liquid. bp$_3$ 85°; bp$_{752}$ 218.8-219.4°, d^{15} 0.9386. n$_D^{20}$ 1.4831.

Phenylurethan. C$_{17}$H$_{23}$NO$_2$. Crystals, mp 113°.

USE: Perfumes; denaturing fats for soap manufacture.

THERAP CAT: Antiseptic.

9314. Terrecyclic Acid. [83058-94-0] (3a*S*,4*R*,7*R*,7a*R*)-Octahydro-8,8-dimethyl-3-methylene-2-oxo-3a,7-ethano-3a*H*-indene-4-carboxylic acid; terrecyclic acid A. C$_{15}$H$_{20}$O$_3$; mol wt 248.32. C 72.55%, H 8.12%, O 19.33%. Cytotoxic, tricyclic sesquiterpene produced by *Aspergillus terreus*; biosynthetic precursor of quadrone, *q.v.* Modulates multiple cellular stress response pathways and induces the heat shock response. Isoln: M. Nakagawa *et al., J. Antibiot.* **35**, 778 (1982). Structure elucidation: A. Hirota *et al., ibid.*, 783. Synthesis and abs config of (−)-form: K. Kon *et al., Tetrahedron Lett.* **25**, 3739 (1984). Abs config of naturally occuring (+)-form: A. Hirota *et al., J. Antibiot.* **39**, 149 (1986). Biosynthesis: A. Hirota *et al., Agric. Biol. Chem.* **48**, 835 (1984); D. E. Cane *et al., Bioorg. Chem.* **14**, 417 (1986). Mechanisms of anticancer activity: T. J. Turbyville *et al., Mol. Cancer Ther.* **4**, 1569 (2005).

Crystals from hexane and ethyl ether, mp 122°. [α]$_D^{20}$ +29.1° (c = 4 in ethanol). uv max: 236 nm (ε 6325). Easily sol in methanol, acetone, ethyl acetate, ethyl ether, chloroform, benzene. Difficultly sol in water, *n*-hexane. LD$_{50}$ in mice (mg/kg): 63-125 i.p. (Nakagawa).

(−)-Form. [93219-11-5] [α]$_D^{25}$ −28.0° (c = 0.175 in chloroform).

USE: Biochemical probe to study cellular stress response.

9315. Terreic Acid. [121-40-4] (1*R*,6*S*)-3-Hydroxy-4-methyl-7-oxabicyclo[4.1.0]hept-3-ene-2,5-dione; 2-hydroxy-3-methyl-1,4-benzoquinone 5,6-epoxide; 5,6-epoxy-3-hydroxy-*p*-toluquinone. C$_7$H$_6$O$_4$; mol wt 154.12. C 54.55%, H 3.92%, O 41.52%. Antibiotic metabolite produced by the mold *Aspergillus terreus*; naturally occurring as the (−)-form: Wilkins, Harris, *Br. J. Exp. Pathol.* **23**, 166 (1942); Abraham, Florey, in H. W. Florey *et al., Antibiotics* vol. I (Oxford Univ. Press, New York, 1949) p 337; Kaplan *et al., Antibiot. Chemother.* **4**, 746 (1954). Structure: Sheehan *et al., J. Am. Chem. Soc.* **80**, 5536 (1958). Synthesis of the racemate: Rashid, Read, *J. Chem. Soc. C* **1967**, 1323. Alternate synthesis and resolution of isomers: Sheehan, Lo, *J. Med. Chem.*

17, 371 (1974). Improved synthesis: A. Enhsen *et al., J. Org. Chem.* **55**, 1177 (1990).

Pale yellow plates from benzene or hexane. Easily sublimed *in vacuo.* mp 127-127.5°. Rotation varies considerably with the solvent: [α]$_D^{22}$ −16.6° (chloroform); [α]$_D^{22}$ −28.6° (methanol-benzene 1:1); [α]$_D^{22}$ +74.3° (pH 7 phosphate buffer). uv max (ethanol): 214, 316 nm (log ε 4.03, 3.88). Enol-type acid, pKa 4.5. Slightly sol in water. Soluble in ether, lower alcohols, acetone, hot cyclohexane. Moderately stable to mineral acid, but dec rapidly in alkaline soln.

9316. Tertatolol. [83688-84-0] 1-[(3,4-Dihydro-2*H*-1-benzothiopyran-8-yl)oxy]-3-[(1,1-dimethylethyl)amino]-2-propanol; (±)-1-*tert*-butylamino-3-(1-thiachroman-8-yloxy)-2-propanol; *dl*-8-[2-hydroxy-3-[(*tert*-butylamino)propyl]oxy]thiochromane. C$_{16}$H$_{25}$-NO$_2$S; mol wt 295.44. C 65.05%, H 8.53%, N 4.74%, O 10.83%, S 10.85%. Nonselective β-adrenergic blocker. Prepn: C. Malen, M. Laubie, DE **2115201**; *eidem*, US **3960891** (1971, 1976 both to Sci. Union et Cie). Pharmacology in animals: M. Laubie *et al., Arch. Int. Pharmacodyn. Ther.* **201**, 323, 334 (1973); B. R. Walker *et al., J. Cardiovasc. Pharmacol.* **7**, 1193 (1985). Synergistic effect with indapamide, *q.v.*: E. Marmo *et al., Drugs Exp. Clin. Res.* **11**, 709 (1985). GC-MS determn in plasma and urine: S. Staveris *et al., J. Chromatogr.* **339**, 97 (1985). Pharmacology in humans and mechanism of action study: A. De Blasi *et al., Clin. Pharmacol. Ther.* **39**, 245 (1986). Preliminary evaluation in hypertension: J. P. Degaute *et al., Am. J. Hypertens.* **1**, 263S (1988).

Crystals from hexane, mp 70-72°.

Hydrochloride. [33580-30-2] S-2395; SE-2395; Artex; Artexal; Prenalex. C$_{16}$H$_{25}$NO$_2$S.HCl; mol wt 331.90. Crystals from acetonitrile, mp 180-183°. pKa 9.8. LD$_{50}$ in rats, mice (mg/kg): 40, 37 i.v.; 90, 120 i.p. (Laubie).

THERAP CAT: Antihypertensive.

9317. α-Terthienyl. [1081-34-1] 2,2′:5′2″-Terthiophene; 5-(2-thienyl)-2,2′-bithiophene. C$_{12}$H$_8$S$_3$; mol wt 248.38. C 58.03%, H 3.25%, S 38.72%. Biocidal constituent of various species of marigolds. Isoln from *Tagetes erecta* L., *Compositae:* L. Zechmeister, J. W. Sease, *J. Am. Chem. Soc.* **69**, 273 (1947). Isoln and distribution in *Tageteae*: K. R. Downum, G. H. N. Towers, *J. Nat. Prod.* **46**, 98 (1983). Synthesis: W. Steinkopf *et al., Ann.* **546**, 180 (1941); H. J. Kooreman, H. Wynberg, *Rec. Trav. Chim.* **86**, 37 (1967); J.-P. Beny *et al., J. Org. Chem.* **47**, 2201 (1982). Biological activity as nematocide: J. H. Uhlenbroek, J. D. Bijloo, *Rec. Trav. Chim.* **77**, 1004 (1958); J. D. Bijloo *et al.*, DE **1075891**; *eidem*, US **3050442** (1960, 1962, both to North American Philips); J. Bakker *et al., J. Biol. Chem.* **254**, 1841 (1979); as herbicide: J. Harvey, Jr., US **3086854** (1963 to Du Pont); G. Campbell *et al., J. Chem. Ecol.* **8**, 961 (1982); as antimicrobial: J. R. Kagan *et al., Photochem. Photobiol.* **31**, 465 (1980); T. Arnason *et al., ibid.* **33**, 821 (1981); F. DiCosmo *et al., Pestic. Sci.* **13**, 589 (1982).

Yellow-orange plates from methanol, mp 93-94°. uv max (methanol): 254, 350 nm (ε 7100, 21300). Sol in carbon bisulfide, ether, benzene, acetone, petr ether; slightly sol in methanol, ethanol. Insol in water.

9318. Tertiapin. [58694-52-3] L-Alanyl-L-leucyl-L-cysteinyl-L-asparaginyl-L-cysteinyl-L-asparaginyl-L-arginyl-L-isoleucyl-L-isoleucyl-L-isoleucyl-L-prolyl-L-histidyl-L-methionyl-L-cysteinyl-L-tryptophyl-L-lysyl-L-lysyl-L-cysteinylglycyl-L-lysyl-L-lysinamide; TPN. $C_{106}H_{180}N_{34}O_{23}S_5$; mol wt 2459.12. C 51.77%, H 7.38%, N 19.37%, O 14.96%, S 6.52%. Neurotoxic 21 residue peptide isolated from the venom of the European honey bee, *Apis mellifera*. Characterized by 2 disulfide linkages and the complete absence of negatively charged residues, it selectively inhibits G-protein inwardly rectifying potassium channels (GIRK) by binding to the external vestibule of the channel. Isoln: J. Gauldie *et al.*, *Eur. J. Biochem.* **61**, 369 (1976); B. E. C. Banks *et al.*, *Bull. Inst. Pasteur* **74**, 137 (1976). Primary structure: Y. A. Ovchinnikov *et al.*, *Soviet J. Bioorg. Chem.* **6**, 187 (1980); 3D-structure: X. Xu, J. W. Nelson, *Proteins Struct. Funct. Genet.* **17**, 124 (1993). Inhibition of GIRK-channels: W. Jin, Z. Lu, *Biochemistry* **37**, 13291 (1998); of muscarinic inwardly rectifying potassium channels ($I_{K(ACh)}$): M.-D. Drici *et al.*, *Br. J. Pharmacol.* **131**, 569 (2000); D. E. Benavides-Haro *et al.*, *Arch. Pharm.* **368**, 309 (2003).

Ala-Lys-Cys-Asn-Cys-Asn-Arg-Ile-Ile-Ile-Phe-His-Met-Cys-Trp-Lys-Lys-Cys-Gly-Lys-LysNH₂

Tertiapin Q. [252198-49-5] TPN_Q. Non-air-oxidizable derivative of TPN in which Met-13 is replaced with Gln. Synthesis: W. Jin, Z. Lu, *Biochemistry* **38**, 14286 (1999). Mechanism of inhibition: W. Jin *et al.*, *ibid.* 14294. Proton titration of TPN_Q binding: Y. Ramu *et al.*, *ibid.* **40**, 3601 (2001).
USE: Biochemical probe for potassium channels.

9319. TES. [7365-44-8] 2-[[2-Hydroxy-1,1-bis(hydroxymethyl)ethyl]amino]ethanesulfonic acid; *N*-[2-hydroxy-1,1-bis(hydroxymethyl)ethyl]taurine; *N*-tris(hydroxymethyl)methyl-2-aminoethanesulfonic acid. $C_6H_{15}NO_6S$; mol wt 229.25. C 31.44%, H 6.60%, N 6.11%, O 41.87%, S 13.98%. One of the zwitterionic amino acids known as "Good" buffers; active in the pH range 6.5 to 8.0. Prepn: N. E. Good *et al.*, *Biochemistry* **5**, 467 (1966). Thermodynamic constants: C. A. Vega, R. G. Bates, *Anal. Chem.* **48**, 1293 (1976). Effect of temperature and solvent systems on pK₂: H. A. Azab *et al.*, *Monatsh. Chem.* **125**, 233 (1994). Use as buffer: F. C. Molinia *et al.*, *Reprod. Nutr. Dev.* **34**, 491 (1994); K. Isobe, H. Nishise, *J. Biotechnol.* **75**, 265 (1999).

Crystals from alcohol/water, mp 226-228°, dec 231°. pKa of 0.1M soln: (0°) 7.92; (20°) 7.5; (37°) 7.14. Δ pKa/°C −0.020.
USE: Biological buffer.

9320. Tesamorelin. [218949-48-5] *N*-[(3*E*)-1-Oxo-3-hexen-1-yl]-somatoliberin (human pancreatic islet); (3*E*)-hex-3-enoylsomatoliberin (human); TH-9507; Egrifta. $C_{221}H_{366}N_{72}O_{67}S$; mol wt 5135.86. C 51.68%, H 7.18%, N 19.64%, O 20.87%, S 0.62%. Synthetic analog of somatoliberin, *q.v.*, modified with a *trans*-3-hexenoyl adduct on the terminal nitrogen of tyrosine. Modification yields resistance to degradation by dipeptidyl peptidase-IV (DPP-IV) while maintaining characteristics of human growth hormone-releasing factor (hGRF). Reduces visceral adipose tissue and increases muscle mass. Prepn: M. Ibea *et al.*, **WO 9637514**; *eidem*, **US 5861379** (1996, 1999 both to Theratechnologies). Pharmacology and LC-MS determn in plasma: E. S. Ferdinandi *et al.*, *Basic Clin. Pharmacol. Toxicol.* **100**, 49 (2007). Clinical experience in HIV-infected patients with abdominal fat accumulation: J. Falutz *et al.*, *AIDS* **22**, 1719 (2008); *eidem*, *J. Acquir. Immune Defic. Syndr.* **53**, 311 (2010). Review of discovery: P. Dubreuil, C. Désévaux, *Comb. Chem. High Throughput Screen.* **9**, 171-174 (2006); of development and clinical experience: Y. Wang, B. Tomlinson, *Expert Opin. Invest. Drugs* **10**, 303-310 (2009); J. Falutz *et al.*, *J. Clin. Endocrinol. Metab.* **95**, 4291 (2010).
Acetate. [901758-09-6] $C_{221}H_{366}N_{72}O_{67}S.xC_2H_4O_2$
THERAP CAT: In treatment of HIV-associated lipodystrophy.

9321. Tesmilifene. [98774-23-3] *N,N*-Diethyl-2-[4-(phenylmethyl)phenoxy]ethanamine; DPPE. $C_{19}H_{25}NO$; mol wt 283.42. C 80.52%, H 8.89%, N 4.94%, O 5.64%. Intracellular histamine antagonist with chemopotentiating and cytoprotective activity. Structurally similar to tamoxifen, *q.v.*, although binds anti-estrogen binding site (AEBS) with no affinity for the estrogen receptor. Prepn: L. J. Brandes, M. W. Hermonat, *Biochem. Biophys. Res. Commun.* **123**, 724 (1984); and use as antineoplastic: *eidem*, **US 4803227** (1989 to Univ. Manitoba); and study of binding affinity: M. Poirot *et al.*, *Bioorg. Med. Chem.* **8**, 2007 (2000). Spectral analysis of interaction with P450 isozymes: L. J. Brandes *et al.*, *Cancer Chemother. Pharmacol.* **45**, 298 (2000). Clinical evaluation in combination with cyclophosphamide in prostate cancer: L. J. Brandes *et al.*, *J. Clin. Oncol.* **13**, 1398 (1995); in combination with doxorubicin in breast cancer: L. Reyno *et al.*, *J. Clin. Oncol.* **22**, 269 (2004).

Hydrochloride. [92981-78-7] BMS-217380-01; BMY-33419. $C_{19}H_{25}NO.HCl$; mol wt 319.87. White crystals from isopropanol + acetone (3:1), mp 156-158°. pKa 10.9.
THERAP CAT: Antineoplastic adjunct (chemosensitizer).

9322. Tesofensine. [195875-84-4] (1*R*,2*R*,3*S*,5*S*)-3-(3,4-Dichlorophenyl)-2-(ethoxymethyl)-8-methyl-8-azabicyclo[3.2.1]octane; (1*R*,2*R*,3*S*)-2-ethoxymethyl-3-(3,4-dichlorophenyl)tropane; NS-2330. $C_{17}H_{23}Cl_2NO$; mol wt 328.28. C 62.20%, H 7.06%, Cl 21.60%, N 4.27%, O 4.87%. Presynaptic reuptake inhibitor of noradrenaline, serotonin and dopamine. Prepn: J. Scheel-Krüger *et al.*, **WO 9730997**; *eidem*, **US 6288079** (1997, 2001 both to NeuroSearch). Pharmacology: H. H. Hansen *et al.*, *Eur. J. Pharmacol.* **636**, 88 (2010). Mechanism of action study: A. M. Axel *et al.*, *Neuropsychopharmacology* **35**, 1464 (2010). Clinical pharmacokinetics: T. Lehr *et al.*, *Br. J. Clin. Pharmacol.* **64**, 36 (2007). Clinical trial in obesity: A. Sjödin *et al.*, *Int. J. Obesity* **34**, 1634 (2010). Review of development and therapeutic potential: N. T. Bello, M. R. Zahner, *Curr. Opin. Investig. Drugs* **10**, 1105-1116 (2009).

Citrate. [861205-83-6] $C_{17}H_{23}Cl_2NO.C_6H_8O_7$; mol wt 520.40. Crystals from ethanol, mp 153-155.5°.
THERAP CAT: Antiobesity agent.

9323. Testolactone. [968-93-4] (4a*S*,4b*R*,10a*R*,10b*S*,12a*S*)-3,4,4a,5,6,10a,10b,11,12,12a-Decahydro-10a,12a-dimethyl-2*H*-phenanthro[2,1-*b*]pyran-2,8(4b*H*)-dione; D-homo-17a-oxaandrosta-1,4-diene-3,17-dione; 13-hydroxy-3-oxo-13,17-secoandrosta-1,4-dien-17-oic acid δ-lactone; 1,2,3,4,4a,4b,7,9,10,10a-decahydro-2-hydroxy-2,4b-dimethyl-7-oxo-1-phenanthrenepropionic acid δ-lactone; delta-1-testololactone; 1-dehydrotestololactone; 17α-oxo-D-homo-1,4-androstadiene-3,17-dione; Δ¹-testolactone; SQ-9538; Fludestrin; Teslac. $C_{19}H_{24}O_3$; mol wt 300.40. C 75.97%, H 8.05%, O 15.98%. Obtained by microbial transformation of progesterone, Reichstein's substance S, or testosterone: Fried *et al.*, *J. Am. Chem. Soc.* **75**, 5764 (1953); *eidem*, **US 2744120** (1956 to Olin Mathieson); Brannon *et al.*, *J. Org. Chem.* **30**, 760 (1965). Comprehensive description: K. Florey, *Anal. Profiles Drug Subs.* **5**, 533-553 (1976).

Crystals from acetone, mp 218-219°. $[\alpha]_D^{23}$ −45.6° (c = 1.24 in chloroform). uv max (ethanol): 242 nm (ε 15800). Sol in alc, chloroform; slightly sol in water, benzyl alc. Insol in ether, solvent hexane.

Note: This is a controlled substance (anabolic steroid): **21 CFR,** 1308.13, as defined in 1300.01.

THERAP CAT: Antineoplastic.

9324. Testosterone. [58-22-0] (17β)-17-Hydroxyandrost-4-en-3-one; Δ^4-androsten-17β-ol-3-one; *trans*-testosterone; Andro; Androderm; Andropatch; Mertestate; Oreton; Testoderm; Testogel; Testolin; Testro AQ; Virosterone. $C_{19}H_{28}O_2$; mol wt 288.43. C 79.12%, H 9.79%, O 11.09%. Principal hormone of the testes, produced by the interstitial cells. Major circulating androgen; converted by 5α-reductase in androgen-dependent target tissues to *5α-dihydrotestosterone* which is required for normal male sexual differentiation. Also converted by aromatization to estradiol, *q.v.* Isoln from bull testes: David *et al.*, *Z. Physiol. Chem.* **233**, 281 (1935). Prepn from cholesterol and confirmation of structure: A. Butenandt, G. Hanisch, *Ber.* **68**, 1859 (1935); *eidem, Z. Physiol. Chem.* **237**, 89 (1935); from dehydroandrosterone: L. Ruzicka, A. Wettstein, *Helv. Chim. Acta* **18**, 1264 (1935); from mixed esters: L. Ruzicka *et al.*, *ibid.* 1478. Crystal structure: P. J. Roberts *et al.*, *J. Chem. Soc. Perkin Trans. 2* **1973**, 1978. Historical review: J. M. Hoberman, C. E. Yesalis, *Sci. Am.* **272**, 76-81 (Feb. 1995). Review of role in aging males: F. E. Kaiser, J. E. Morley, *Neurobiol. Aging* **15**, 559-563 (1994); of clinical relevance in females: R. S. Rittmaster, *Am. J. Med.* **98**, Suppl. 1A, 17S-21S (1995).

Needles from dil acetone, mp 155°. $[\alpha]_D^{24}$ +109° (c = 4 in alc). uv max: 238 nm. Freely sol in dehydrated alc, chloroform; sol in vegetable oils, alc, dioxane and other organic solvents; slightly sol in ether. Practically insol in water.

Acetate. [1045-69-8] $C_{21}H_{30}O_3$. mp 140-141°.

17β-Cyclopentanepropionate. [58-20-8] Testosterone cypionate; depAndro; Depotest; Depo-Testosterone; Depovirin; Pertestis; Virilon. $C_{27}H_{40}O_3$; mol wt 412.61. Pharmacology: A. C. Ott *et al.*, *J. Clin. Endocrinol. Metab.* **12**, 15 (1952). Crystals, mp 101-102°. $[\alpha]_D^{25}$ +87° (CHCl$_3$). Freely sol in alc, chloroform, dioxane, ether; sol in vegetable oils. Insol in water.

Enanthate. [315-37-7] Andro LA; Androtardyl; Delatestryl; Everone; Primoteston; Testinon; Testo-Enant. $C_{26}H_{40}O_3$; mol wt 400.60. Prepn: Junkmann *et al.*, **US 2840508** (1958 to Schering AG). Comprehensive description: K. Florey, *Anal. Profiles Drug Subs.* **4**, 452-465 (1975). White crystals, mp 36-37.5°. Very sol in ether; sol in vegetable oils. Insol in water.

Propionate. [57-85-2] Anertan; Enarmon; Neo-Hombreol (amp.); Orchisterone; Synandrol; Testex; Testoviron; Virormone. $C_{22}H_{32}O_3$; mol wt 344.50. Prepn: K. Miescher *et al.*, **US 2109400** (1938 to Soc. Chem. Ind. Basel). Stout prisms from alcohol + water, mp 118-122°. $[\alpha]_D^{25}$ +83 to +90° (100 mg in 10 ml dioxane). Freely sol in alc, ether, pyridine, dioxane and other organic solvents; sol in vegetable oils. Insol in water.

Undecanoate. [5949-44-0] Testosterone 17β-undecylate; Andriol; Pantestone; Restandol. $C_{30}H_{48}O_3$; mol wt 456.71.

Note: This is a controlled substance (anabolic steroid): **21 CFR,** 1308.13, as defined in 1300.01.

THERAP CAT: Androgen.

THERAP CAT (VET): Androgen.

9325. Tetraamminecopper Sulfate. [14283-05-7] Tetraamminecopper(2+) sulfate (1:1); cuprammonium sulfate; ammonium cupric sulfate; cupric sulfate, ammoniated; Eau Celeste. $CuH_{12}N_4$-O_4S; mol wt 227.73. Cu 27.90%, H 5.31%, N 24.60%, O 28.10%, S 14.08%. [Cu(NH$_3$)$_4$]SO$_4$. Prepd by dissolving copper sulfate in ammonia water and pptg with alcohol: Mazzi, *Acta Crystallogr.* **8**, 137 (1955). Crystal structure of monohydrate: Morosin, *ibid.* **25B**, 19 (1969).

Monohydrate. Large, dark blue crystals. d_4^{20} 1.81. Ammonia odor, dec in air. Loses H$_2$O and 2NH$_3$ on heating to 120°, remaining 2NH$_3$ at 160°. Soly in water at 21.5°: 18.5 g/100 ml. Practically insol in the lower alcohols.

USE: In textile printing, especially in calico finishing. As fungicide.

9326. Tetrabenazine. [58-46-8] *rel*-(3R,11bR)-1,3,4,6,7,-11b-Hexahydro-9,10-dimethoxy-3-(2-methylpropyl)-2H-benzo[a]-quinolizin-2-one; 2-oxo-3-isobutyl-9,10-dimethoxy-1,2,3,4,6,7-hexahydro-11bH-benzo[a]quinolizine; Ro-1-9569; Nitoman; Xenazine. $C_{19}H_{27}NO_3$; mol wt 317.43. C 71.89%, H 8.57%, N 4.41%, O 15.12%. Dopamine depleting agent with antipsychotic activity. Prepn: A. Brossi *et al.*, *Helv. Chim. Acta* **41**, 119 (1958); **US 2830993** (1958 to Hoffmann-La Roche); J. M. Osbond, *J. Chem. Soc.* **1961**, 4711. Pharmacology: I. Leusen *et al.*, *Arch. Int. Pharmacodyn.* **119**, 225 (1959). Metabolism: D. E. Schwartz *et al.*, *Biochem. Pharmacol.* **15**, 645 (1966). Clinical trial in hyperkinetic movement disorders: C. Kenney *et al.*, *Mov. Disord.* **22**, 193 (2007); in Huntington's disease: J. P. Marshall *et al.*, *Neurology* **66**, 366 (2006). Review of pharmacology and therapeutic efficacy in Tourette's syndrome: M. Porta *et al.*, *Clin. Drug Invest.* **28**, 443-459 (2008).

(R,R)-Form

Mixture of (R,R)- and (S,S)- isomers. Prisms from methanol, mp 126-129°. pKa 6.51. Sparingly sol in water; sol in ethanol.

Hydrochloride. [2105-47-7] $C_{19}H_{27}NO_3$·HCl. Crystals, mp 208-210°. Soluble in hot water. Practically insol in acetone. uv max (alcohol): 230, 284 nm (ε 7780, 3820).

Methanesulfonate. [804-53-5] $C_{19}H_{27}NO_3$·CH$_3$SO$_3$H; mol wt 413.53. Bitter crystals, sensitive to light, mp 126-130°. Sparingly sol in water, sol in alcohol. Practically insol in acetone.

R,R-Form. [1026016-83-0] Prepn: M. J. Rishel *et al.*, *J. Org. Chem.* **74**, 4001 (2009). Colorless crystals from DME-hexanes, mp 126.0°. $[\alpha]_D^{26}$ +37.2° (c = 0.41 in CH$_2$Cl$_2$).

THERAP CAT: Antidyskinetic.

9327. Tetraborane(10). [18283-93-7] Tetraboron decahydride; borobutane. B_4H_{10}; mol wt 53.32. B 81.10%, H 18.90%. Prepd by the reaction of magnesium boride with hydrochloric or phosphoric acid: Stock, Kuss, *Ber.* **56B**, 789 (1923); from dihydropentaborane: Burg, Schlesinger, *J. Am. Chem. Soc.* **55**, 4009 (1933); from diborane: Stock, Mathing, *Ber.* **69B**, 1456 (1936); Klein *et al.*, *J. Am. Chem. Soc.* **80**, 4149 (1958). *Review:* Greenwood in *Comprehensive Inorganic Chemistry* vol. **1**, J. C. Bailar, Jr. *et al.*, Eds. (Pergamon Press, Oxford, 1973) pp 785-791.

Gas; disagreeable odor; mp −120°; bp 18°. Vapor pressure 580 mm Hg at 6°; 388 mm at 0°. Dec at room temp in a few hours. Dec rapidly at 100°. Ignites spontaneously in air or oxygen. Hydrolyzes in water to boric acid and hydrogen. Reacts with ammonia to form a tetraammoniate.

9328. Tetrabromobisphenol A. [79-94-7] 4,4′-(1-Methylethylidene)bis[2,6-dibromophenol]; 4,4′-(1-isopropylidene)bis[2,6-dibromophenol]; 2,2-bis(3,5-dibromo-4-hydroxyphenyl)propane; TBBA; TBBPA; BA-59P; Fire Guard 2000; Saytex CP-2000. C_{15}-$H_{12}Br_4O_2$; mol wt 543.88. C 33.13%, H 2.22%, Br 58.77%, O 5.88%. Brominated flame retardant used in the formulation of thermoplastic resins for electronics. Prepn: T. Zincke, M. Grüters, *Ann.* **343**, 75 (1905). Manufacturing process: A. J. Dietzler, **US 3029291** (1962 to Dow); H. E. Hennis, *Ind. Eng. Chem. Prod. Res. Dev.* **2**, 140 (1963). Fire retardant properties: J. A. Schneider *et al.*, *ibid.* **9**, 559 (1970). Crystal structure: J. Eriksson, L. Eriksson, *Acta Crystallogr.* **C57**, 1308 (2001). Polymerization reaction kinetics: F. Barontini *et al.*, *Ind. Eng. Chem. Res.* **39**, 855 (2000). Thermal degradation: *idem et al.*, *ibid.* **43**, 1952 (2004). Photodegradation: J. Eriksson *et al.*, *Chemosphere* **54**, 117 (2004). Toxicokinetics: U. M. D. Schauer *et al.*, *Toxicol. Sci.* **91**, 49 (2006). Review of toxicology: *Tetrabromobisphenol A* (National Toxicology Program, USDHHS, 2002) 33pp. Review of environmental analysis methods: A. Covaci *et al.*, *J. Chromatogr. A* **1216**, 346-363 (2009).

Crystals from isopropyl alcohol, mp 181-182°. bp ~316°. pKa$_1$ 7.5; pKa$_2$ 8.5. uv max (water): 310 nm (ε 9170). Crystal density: 2.158. Log P (octanol/water): 4.5. Soly in water (mg/l): 0.72 (15°); 4.16 (25°); 1.77 (35°). Sol in methanol, ether, chloroform, acetone, benzene, methylene chloride, methyl ethyl ketone. LD$_{50}$ in rats, mice (mg/kg): >2000, >2000 orally; ≥3200, ≥3200 i.p. (USDHHS).

USE: Reactive flame retardant in epoxy and polycarbonate resins; additive flame retardant in acrylonitrile-butadiene-styrene (ABS) resins.

9329. 3,4,5,6-Tetrabromo-*o*-cresol. [576-55-6] 2,3,4,5-Tetrabromo-6-methylphenol. C$_7$H$_4$Br$_4$O; mol wt 423.72. C 19.84%, H 0.95%, Br 75.43%, O 3.78%. Prepd by bromination of *o*-cresol or its methyl or ethyl ether: Zincke, Hedenström, *Ann.* **350**, 269 (1906); Bonneaud, *Bull. Soc. Chim. Fr.* **7**, 776 (1910); Treacy, **US 2319960** (1943 to Merck & Co.).

White to buff, cryst powder, mp 205-208° with dec. Practically insol in water. Sol in alcohol, ether, alkali hydroxides.

Acetate. C$_9$H$_6$Br$_4$O$_2$. Needles from dil alcohol, mp 154°. Slightly sol in alcohol and glacial acetic acid.

Caution: Irritating to skin, mucous membranes.

USE: Fungicide.

9330. *sym*-Tetrabromoethane. [79-27-6] 1,1,2,2-Tetrabromoethane; acetylene tetrabromide; Muthmann's liquid. C$_2$H$_2$Br$_4$; mol wt 345.65. C 6.95%, H 0.58%, Br 92.47%. Br$_2$CHCHBr$_2$. Manuf by bromination of acetylene: **GB 889649** (1962 to Associated Ethyl Co.). Toxicity studies: Gray, *Arch. Ind. Hyg. Occup. Med.* **2**, 407 (1950); D. L. Wolff, *Acta Biol. Med. Ger.* **41**, 945 (1982).

Yellowish, heavy, very refractive liquid; odor of camphor and iodoform. d 2.964; bp$_{54}$ 151°; mp 0°; n$_D^{20}$ 1.638. Insol in water. Miscible with alc, chloroform, ether, aniline, glacial acetic acid. LD$_{50}$ i.p. in mice: 443.3 mg/kg (Wolff).

Caution: Potential symptoms of overexposure are irritation of eyes and nose; anorexia, nausea; severe headaches; abdominal pain; jaundice; monocytosis; CNS depression. *See NIOSH Pocket Guide to Chemical Hazards* (DHHS/NIOSH 97-140, 1997) p 6.

USE: In microscopy, as solvent, separating minerals by density.

9331. Tetrabromophenolphthalein. [76-62-0] 3,3-Bis(3,5-dibromo-4-hydroxyphenyl)-1(3*H*)-isobenzofuranone; 3′,3″,5′,5″-tetrabromophenolphthalein; TBP. C$_{20}$H$_{10}$Br$_4$O$_4$; mol wt 633.91. C 37.89%, H 1.59%, Br 50.42%, O 10.10%. Prepn: A. Baeyer, *Ann.* **202**, 68 (1880); F. F. Blicke *et al.*, *J. Am. Chem. Soc.* **54**, 1465 (1932). Use as contrast agent in cholecystography: E. A. Graham, W. H. Cole, *J. Am. Med. Assoc.* **82**, 613 (1924). Use in drug screening tests in urine: T. Sakai, N. Ohno, *Anal. Sci.* **2**, 275 (1986).

Crystals from alcohol, glacial acetic acid, or ether, mp 295-297°. *Irritant.* Practically insol in water. Slightly sol in alc, glacial acetic acid; sol in ether, in alkali with a violet color.

Ethyl ester. [1176-74-5] 2-[(3,5-Dibromo-4-hydroxyphenyl)-(3,5-dibromo-4-oxo-2,5-cyclohexadien-1-ylidene)methyl]benzoic acid ethyl ester; bromophthalein magenta E; ethyl tetrabromophenolphthalein; TBPE. C$_{22}$H$_{14}$Br$_4$O$_4$; mol wt 661.97. Prepn: R. Nietzki, E. Burckhardt, *Ber.* **30**, 175 (1897). Red prisms from alcohol, mp 210-215°. *Irritant.*

Ethyl ester potassium salt. [62637-91-6] C$_{22}$H$_{13}$Br$_4$KO$_4$; mol wt 700.06. Deep-violet powder. Sol in ethanol. Yellow at pH 3.0; light green at pH 3.5; blue-green at pH 4.2; blue-purple at pH 4.9-5.2.

USE: Indicator; analytical reagent.

9332. Tetrabutylammonium Fluoride. [429-41-4] *N,N,N*-Tributyl-1-butanaminium fluoride (1:1); TBAF. C$_{16}$H$_{36}$FN; mol wt 261.47. C 73.50%, H 13.88%, F 7.27%, N 5.36%. Catalytic reagent which is both a potent base and a source of nucleophilic fluoride. Prepn as hydrated form: D. L. Fowler *et al.*, *J. Am. Chem. Soc.* **62**, 1140 (1940). Instability of anhydrous form: R. K. Sharma, J. L. Fry, *J. Org. Chem.* **48**, 2112 (1983). Use as fluoride source: D. P. Cox *et al.*, *ibid.* **49**, 3216 (1984); T. Kobayashi *et al.*, *ibid.* **67**, 3156 (2002). Synthesis and/or use as catalyst in cyclization reactions: J. Pless, *ibid.* **39**, 2644 (1974); A. R. Gangloff *et al.*, *Tetrahedron Lett.* **42**, 1441 (2001); in addition reactions: P. Molina *et al.*, *Synlett* **2003**, 714. Deprotecting agent: M. Namikoshi *et al.*, *J. Org. Chem.* **56**, 5464 (1991); J. J. Parlow *et al.*, *Bioorg. Med. Chem. Lett.* **8**, 2391 (1998). Brief review in cleavage of silyl protecting groups: M. B. Kumar, *Synlett* **2002**, 2125-2126.

Colorless oil which crystallizes on exposure to humidity.

USE: Reagent in organic syntheses in addition, condensation, base-catalyzed cyclization reactions, fluorination and desulfonylation reactions and as a deprotecting agent. Typically dried prior to use.

9333. Tetracaine Hydrochloride. [136-47-0] 4-(Butylamino)benzoic acid 2-(dimethylamino)ethyl ester hydrochloride (1:1); *p*-butylaminobenzoyl-2-dimethylaminoethanol hydrochloride; 2-dimethylaminoethyl 4-*n*-butylaminobenzoate hydrochloride; dicain; amethocaine hydrochloride; Anethaine; Decicain; Pantocaine; Pontocaine Hydrochloride; Tonexol. C$_{15}$H$_{25}$ClN$_2$O$_2$; mol wt 300.83. C 59.89%, H 8.38%, Cl 11.78%, N 9.31%, O 10.64%. Prepn: **US 1889645** (1932); **GB 815144** (1959 to Abbott). Prepn of pharmaceutical dosage forms: Shupe, **US 3272700** (1966 to Sterling Drug). Mechanism of action studies: Y.-W. Leung *et al.*, *J. Infect. Dis.* **136**, 679 (1977). Toxicity study: Dawes, *Br. J. Pharmacol. Chemother.* **1**, 90 (1946). Acute toxicity: B. A. Bopp *et al.*, *J. Pharm. Sci.* **67**, 882 (1978). Comprehensive description: M. Riaz, *Anal. Profiles Drug Subs.* **18**, 379-411 (1989).

Faintly bitter, hygroscopic crystals producing transient numbness of the tongue. mp 147-150°. pKa 8.39. uv max (water): 225, 310 nm (ε 14108, 26352); (0.1*N* H$_2$SO$_4$): 229, 281, 312 nm (E$_{1cm}^{1\%}$ 509, 55, 76); (methanol): 226, 310 nm (ε 7586, 29512); (chloroform): 308 nm (ε 27542). Sol at 20° in 7.5 parts water, 40 parts alcohol; 30 parts chloroform. Practically insol in acetone. Insol in ether, benzene. The aq soln is neutral to litmus. Aq solns are stable and may be sterilized by brief boiling. LD$_{50}$ i.p. in mice: 70 mg/kg (Dawes); also reported as LD$_{50}$ in female mice (mg/kg): 13 i.v.; 35 s.c. (Bopp).

THERAP CAT: Anesthetic (local).

THERAP CAT (VET): Anesthetic (topical).

9334. Tetrachloroethane. [79-34-5] 1,1,2,2-Tetrachloroethane; *sym*-tetrachloroethane; acetylene tetrachloride; Cellon; Bonoform. $C_2H_2Cl_4$; mol wt 167.84. C 14.31%, H 1.20%, Cl 84.49%. $Cl_2CHCHCl_2$. Manuf by catalytic addition of chlorine to acetylene: Peters, Neumann, *Angew. Chem.* **45**, 261 (1932); by chlorination of ethylene: Pye, **US 2752402** (1956 to Dow); by catalytic chlorination of ethane: Joseph, **US 2752401** (1956 to Dow); by chlorination of 1,2-dichloroethane: Conrad, **US 2725412** (1955 to Ethyl Corp.); Fox, **US 2846484** (1958 to Monsanto). Toxicity study: H. F. Smyth *et al., Am. Ind. Hyg. Assoc. J.* **30**, 470 (1969). Review of toxicology and human exposure: *Toxicological Profile for 1,1,2,2-Tetrachloroethane* (PB2009-100008, 2008) 258 pp.

Nonflammable, heavy, mobile liquid. Sweetish, suffocating, chloroform-like odor. d_4^{25} 1.58658. mp $-44°$. bp_{760} 146.5°. n_D^{20} 1.49419. Very sparingly sol in water. At 25° one gram dissolves in 350 ml H_2O. Misc with methanol, ethanol, benzene, ether, petr ether, carbon tetrachloride, chloroform, carbon disulfide, dimethylformamide, oils. Has the highest solvent power of the chlorinated hydrocarbons. LD_{50} orally in rats: 0.20 ml/kg (Smyth).

Caution: Potential symptoms of overexposure are nausea, vomiting and abdominal pain; finger tremors; jaundice, hepatitis, tender liver; dermatitis; monocytosis; kidney damage. Potential occupational carcinogen. *See NIOSH Pocket Guide to Chemical Hazards* (DHHS/NIOSH 97-140, 1997) p 300.

USE: Nonflammable solvent for fats, oils, waxes, resins, cellulose acetate, rubber, copal, phosphorus, sulfur. As solvent in certain types of Friedel-Crafts reactions or phthalic anhydride condensations. In the manuf of paint, varnish, and rust removers. In soil sterilization and weed killer and insecticide formulations. In the determination of theobromine in cacao. As immersion fluid in crystallography. In the biological laboratory to produce pathological changes in gastrointestinal tract, liver, and kidneys. Intermediate in the manuf of trichloroethylene and other chlorinated hydrocarbons having two carbon atoms.

9335. Tetrachloroethylene. [127-18-4] 1,1,2,2-Tetrachloroethene; tetrachloroethene; perchloroethylene; ethylene tetrachloride; tetrachlorethylene; Nema; Tetracap; Tetropil; Perclene; Ankilostin; Didakene. C_2Cl_4; mol wt 165.82. C 14.49%, Cl 85.51%. $Cl_2C=CCl_2$. Prepd by Faraday in 1821. Manuf by catalytic oxidation of 1,1,2,2-tetrachloroethane: Ellsworth, Vancamp, **US 2951103** (1960 to Columbia-Southern Chem.); Feathers, Rogerson, **US 3040109** (1962 to Pittsburgh Plate Glass); by catalytic chlorination of acetylene: Thermet, Parvi, **US 2938931** (1960 to Société d'électrochimie, d'électrométallurgie et des aciéries électriques d'Ugine). Review of mfg processes: *Faith, Keyes & Clark's Industrial Chemicals*, F. A. Lowenheim, M. K. Moran, Eds. (Wiley-Interscience, New York, 4th ed., 1975) pp 604-611. Physical properties: Mumford, Phillips, *J. Chem. Soc.* **1950**, 75. Toxicity data: Dybing, *Acta Pharmacol. Toxicol.* **2**, 223 (1946); Lazarew, *Arch. Exp. Pathol. Pharmakol.* **141**, 19 (1929). Review of toxicology and human exposure: *Toxicological Profile for Tetrachloroethylene* (PB98-101181, 1997) 318 pp.

Colorless, nonflammable liq; ethereal odor. *Poisonous.* d_4^{15} 1.6311; d_4^{20} 1.6230. bp 121°. mp $\sim-22°$. n_D^{20} 1.5055. Sol in about 10,000 vol water; misc with alcohol, ether, chloroform, benzene. LD_{50} orally in mice: 8.85 g/kg (Dybing). LC for mice in air: 5925 ppm (Lazarew).

Caution: Potential symptoms of overexposure are irritation of eyes, nose and throat; nausea; flushing of face and neck; vertigo, dizziness, incoordination; headache, somnolence; skin erythema; liver damage. *See NIOSH Pocket Guide to Chemical Hazards* (DHHS/NIOSH 97-140, 1997) p 300. This substance is reasonably anticipated to be a human carcinogen: *Report on Carcinogens, Twelfth Edition* (PB2011-111646, 2011) p 398.

USE: Dry cleaning; textile processing; degreasing metals; solvent; chemical intermediate in production of fluorocarbons. Insulating fluid and cooling gas in electrical transformers.

THERAP CAT: Anthelmintic (Nematodes, Trematodes).

THERAP CAT (VET): Anthelmintic.

9336. 3,3',4',5-Tetrachlorosalicylanilide. [1154-59-2] 3,5-Dichloro-*N*-(3,4-dichlorophenyl)-2-hydroxybenzamide; Irgasan BS200. $C_{13}H_7Cl_4NO_2$; mol wt 351.00. C 44.49%, H 2.01%, Cl 40.40%, N 3.99%, O 9.12%. Prepn of polyhalosalicylanilides: Bin-

dler, Model, **US 2703332** (1955 to Geigy). Bacteriostat in the manuf of thermoplastic articles: Teller, **US 3005720** (1961 to Weco Products).

Crystals, mp 161°. Fluoresces under ultraviolet light. Practically insol in water. Sol in alkaline aq solns and in solns of wetting agents. Sol in many organic solvents.

USE: Bacteriostat in formulations of surgical soaps, laundry soaps, rinses, polishes, shampoos, deodorants. Also as preservative in textile finishes, certain petroleum products, cellulose esters, cutting oils, coolants.

9337. Tetrachlorvinphos. [22248-79-9] Phosphoric acid (*Z*)-2-chloro-1-(2,4,5-trichlorophenyl)ethenyl dimethyl ester; (*Z*)-2-chloro-1-(2,4,5-trichlorophenyl)vinyl dimethyl phosphate; stirofos; ENT-25841; SD-8447; Equitrol; Gardona; Rabon; Rabond. $C_{10}H_9$-Cl_4O_4P; mol wt 365.95. C 32.82%, H 2.48%, Cl 38.75%, O 17.49%, P 8.46%. Organophosphate insecticide; cholinesterase inhibitor. Prepn: D. D. Phillips, L. F. Ward, Jr., **US 3102842**; D. E. Ramey, **US 3553297** (1963, 1971 both to Shell). Properties: R. R. Whetstone *et al., J. Agric. Food Chem.* **14**, 352 (1966). Metabolism in cows: M. H. Akhtar, T. S. Foster, *ibid.* **28**, 698 (1980). Pharmacology in horses: J. Berger *et al., Vet. Res. Commun.* **32**, 75 (2008). Determn of residues in foods: V. Leoni *et al., J. AOAC Int.* **75**, 511 (1992). Toxicology studies: T. B. Gaines, *Toxicol. Appl. Pharmacol.* **14**, 515 (1969); A. I. T. Walker *et al., Pestic. Sci.* **3**, 517 (1972). Review of agricultural use: A. G. Jennings, *Int. Pest Control* **12**, 28-33 (1970); of carcinogenic risk: *IARC Monographs* **30**, 197-206 (1983).

mp 97-98°. Vapor pressure at 20° = 4.2×10^{-8}. Soly: 11 ppm in water; <15% in xylene; 40-50% in chloroform at room temp. LD_{50} in male, female rats (mg/kg): 1100, 1125 orally (Gaines).

USE: Insecticide.

THERAP CAT (VET): Ectoparasiticide.

9338. Tetraconazole. [112281-77-3] 1-[2-(2,4-Dichlorophenyl)-3-(1,1,2,2-tetrafluoroethoxy)propyl]-1*H*-1,2,4-triazole; (*RS*)-2-(2,4-dichlorophenyl)-3-(1*H*-1,2,4-triazol-1-yl)propyl 1,1,2,2-tetrafluoroethyl ether; M-14360; Domark; Eminent; Lospel. $C_{13}H_{11}$-$Cl_2F_4N_3O$; mol wt 372.14. C 41.96%, H 2.98%, Cl 19.05%, F 20.42%, N 11.29%, O 4.30%. Systemic triazole antifungal; inhibitor of sterol biosynthesis. Prepn: **JP Kokai 87 169773**; R. Colle *et al.*, **US 5081141** (1987, 1992 both to Montedison). Chemical and biological properties: C. Garavaglia *et al., Brighton Crop Prot. Conf. - Pests Dis.* **1988**, 49. Chemoenzymatic stereospecific synthesis and biological activity of racemate and enantiomers: D. Bianchi *et al., J. Agric. Food Chem.* **39**, 197 (1991). Stereoselective inhibition of ergosterol biosynthesis: F. Gozzo *et al., Pestic. Biochem. Physiol.* **53**, 10 (1995). ELISA of commercial prepns: F. Forlani *et al., J. Agric. Food Chem.* **40**, 328 (1992). Field trial on wheat: D. Pancaldi, I. Alberti, *Informat. Fitopatolog.* **44**, 48 (1994).

Viscous colorless oil. Vapor pressure (20°): 1.2×10^{-5} mmHg. Log P (octan-1-ol/water): 3.1 (pH 7). Stable in dil. aq. soln at pH 5-9. Stable in sunlight. Soly (20°) mg/l: water 150. Readily sol in dichloromethane, acetone and methanol. LD_{50} orally in rats: 1150 mg/kg (Garavaglia).

USE: Agricultural fungicide.

9339. Tetracosamethylhendecasiloxane. [107-53-9] 1,1,1,- 3,3,5,5,7,7,9,9,11,11,13,13,15,15,17,17,19,19,21,21,21-Tetracosa-methylundecasiloxane. $C_{24}H_{72}O_{10}Si_{11}$; mol wt 829.77. C 34.74%, H 8.75%, O 19.28%, Si 37.23%. Prepd by reaction of hexamethyl-disiloxane with octamethylcyclotetrasiloxane and sulfuric acid: Pat-node, Wilcock, *J. Am. Chem. Soc.* **68**, 362 (1946); by cohydrolysis of ethoxytrimethylsilane and diethoxydimethylsilane: Hyde, **US 2457677** (1948 to Corning Glass).

$$(H_3C)_3Si-\left[O-\underset{\underset{CH_3}{|}}{\overset{\overset{CH_3}{|}}{Si}}\right]_9 O-Si(CH_3)_3$$

Liquid; $bp_{4.7}$ 201°; $bp_{0.5}$ 152°; d_4^{25} 0.9247; n_D^{20} 1.3994. Stable. Inert to most chemical reagents and rubber. Maintains about the same viscosity over a wide temperature range. Sol in benzene and the lighter hydrocarbons; slightly sol in alcohol and the heavy hydro-carbons.

USE: As a basis for silicone oils or fluids designed to withstand extremes of temperature; as a foam suppressant in petroleum lubri-cating oil.

9340. Tetracyanoethylene. [670-54-2] 1,1,2,2-Ethenetetra-carbonitrile; 1,1,2,2-tetracyanoethene; $\Delta^{2,2'}$-bimalononitrile; TCNE. C_6N_4; mol wt 128.09. C 56.26%, N 43.74%. Prepd by debromina-tion of dibromomalononitrile with copper powder in boiling ben-zene: Cairns *et al.*, *J. Am. Chem. Soc.* **80**, 2775 (1958).

$$\underset{NC}{\overset{NC}{>}}C=C\underset{CN}{\overset{CN}{<}}$$

Crystals, mp 200°. Begins to sublime at 120°.

USE: In the synthesis of spiro compds, in modified Diels-Alder reactions, as aromatizing agent: Longone, Smith, *Tetrahedron Lett.* **1962**, 205.

9341. Tetracycline. [60-54-8] (4S,4aS,5aS,6S,12aS)-4-(Di-methylamino)-1,4,4a,5,5a,6,11,12a-octahydro-3,6,10,12,12a-pen-tahydroxy-6-methyl-1,11-dioxo-2-naphthacenecarboxamide; des-chlorobiomycin; tsiklomitsin; Liquamycin. $C_{22}H_{24}N_2O_8$; mol wt 444.44. C 59.46%, H 5.44%, N 6.30%, O 28.80%. Antibiotic sub-stance produced by *Streptomyces spp.* Prepn: J. H. Boothe *et al.*, *J. Am. Chem. Soc.* **75**, 4621 (1953); L. H. Conover *et al.*, *ibid.* 4622; Conover, **US 2699054** (1955). Production by *Streptomyces viridi-faciens:* Gourevitch, Lein, Heinemann *et al.*, **US 2712517; US 2886595** (1955, 1959 both to Bristol Labs.); by *S. aureofaciens:* Miller, Arishima, Sekizwa, **US 3005023; US 3019173** (1961, 1962 both to Am. Cyanamid). Purification: Kaplan, Granatek, **US 3301899** (1967 to Bristol-Myers). Total synthesis of tetracyclines: J. H. Boothe *et al.*, *J. Am. Chem. Soc.* **81**, 1006 (1959); L. H. Con-over *et al.*, *ibid.* **84**, 3222 (1962). Graphic outline of Woodward synthesis: *Chem. Eng. News* **40**, 36 (Oct. 8, 1962). Abs config: V. N. Dobrynin *et al.*, *Tetrahedron Lett.* **1962**, 901. Solubility studies: Weiss *et al.*, *Antibiot. Chemother.* **7**, 374 (1957). Toxicity: E. I. Goldenthal, *Toxicol. Appl. Pharmacol.* **18**, 185 (1971). Review: "Tetracycline" in *The Technology of the Tetracyclines* vol. **I**, R. C. Evans, Ed. (Quadrangle Press, New York, 1968) pp 209-426. Mech-anism of action: A. Kaji, M. Ryoji in *Antibiotics* vol. 5 (pt. 1), F. E. Hahn, Ed. (Springer-Verlag, New York, 1979) pp 304-328. Review of biosynthesis: C. R. Hutchinson in *Antibiotics* vol. 4, J. W. Cor-coran, Ed. (Springer-Verlag, New York, 1981) pp 1-12. Review of anticollagenase activity of tetracyclines: L. M. Golub *et al.*, *Crit. Rev. Oral Biol. Med.* **2**, 297-322 (1991).

Trihydrate, crystals. Swells at 165°. Dec 170-175°. Becomes an-hydr by drying *in vacuo* at 60° for 8 hrs. $[\alpha]_D^{25}$ $-257.9°$ (0.1N HCl); $[\alpha]_D^{25}$ $-239°$ (methanol). uv max (0.1N HCl): 220, 268, 355 nm (ε 13000, 18040, 13320). pKa (50% aq DMF): 8.3, 10.2. Stable in neutral and in alkaline soln. Soly at about 28°: 1.7 mg/ml water; >20 mg/ml methanol. Freely sol in dilute acid, alkali hydroxide solns; sparingly sol in alc. Practically insol in chloroform, ether. LD_{50} in rats, mice (mg/kg): 807, 808 orally (Goldenthal).

Hydrochloride. [64-75-5] Achromycin; Ambramicina; Diocy-clin; Helvecyclin; Hexacyclin; Hostacyclin; Imex; Panmycin; Rob-itet; Steclin; Sumycin; Supramycin; Sustamycin; Tefilin; Tetracyn; Tetralution; Tetsol; Topicycline. $C_{22}H_{24}N_2O_8 \cdot HCl$; mol wt 480.90. Crystals from butanol + HCl, dec 214°. Moderately hygroscopic. $[\alpha]_D^{25}$ $-257.9°$ (c = 0.5 in 0.1N HCl). Sol in water, methanol, solns of alkali hydroxides and carbonates; slightly sol in alc. Practically insol in chloroform, ether. pH (2% aq soln): 2.1-2.3. LD_{50} orally in rats: 6443 mg/kg (Goldenthal).

Phosphate complex. [1336-20-5] Tetrex. Prepn: Seiger, Wei-denheimer, **US 3053892** (1962 to Am. Cyanamid). Yellow, odorless powder. Sparingly sol in water; slightly sol in ethanol.

THERAP CAT: Antiamebic; antibacterial; antirickettsial.

THERAP CAT (VET): Antibacterial.

9342. Tetradecamethylhexasiloxane. [107-52-8] 1,1,1,3,3,-5,5,7,7,9,9,11,11,11,-Tetradecamethylhexasiloxane. $C_{14}H_{42}O_5Si_6$; mol wt 459.00. C 36.63%, H 9.22%, O 17.43%, Si 36.71%. Prepd by hydrolysis of dimethyldichlorosilane and trimethylchlorosilane: Patnode, Wilcock, *J. Am. Chem. Soc.* **68**, 358 (1946); by reaction of hexamethyldisiloxane with octamethylcyclotetrasiloxane and sulfu-ric acid: *eidem, ibid.* 362; Patnode, **US 2469888** (1949 to General Electric); by cohydrolysis of ethoxytrimethylsilane and diethoxydi-methylsilane: Hunter *et al.*, *J. Am. Chem. Soc.* **68**, 2284 (1946); Hyde, **US 2457677** (1948 to Corning Glass).

$$(H_3C)_3Si-O-\underset{\underset{CH_3}{|}}{\overset{\overset{CH_3}{|}}{Si}}-O-\underset{\underset{CH_3}{|}}{\overset{\overset{CH_3}{|}}{Si}}-O-\underset{\underset{CH_3}{|}}{\overset{\overset{CH_3}{|}}{Si}}-O-\underset{\underset{CH_3}{|}}{\overset{\overset{CH_3}{|}}{Si}}-O-Si(CH_3)_3$$

Liquid; bp_{20} 142°; d_4^{20} 0.8910; n_D^{20} 1.3948; flash pt 118.33°C. Freezes below $-100°$. Stable. Inert to most chemical reagents and rubber. Maintains about the same viscosity over a wide temp range. Sol in benzene and the lighter hydrocarbons; slightly sol in alcohol and the heavy hydrocarbons.

USE: As a basis for silicone oils or fluids designed to withstand extremes of temp; as a foam suppressant in petroleum lubricating oil.

9343. Tetradifon. [116-29-0] 1,2,4-Trichloro-5-[(4-chloro-phenyl)sulfonyl]benzene; p-chlorophenyl 2,4,5-trichlorophenyl sul-fone; 2,4,5,4'-tetrachlorodiphenyl sulfone; tedion; Tedion-V18. $C_{12}H_6Cl_4O_2S$; mol wt 356.04. C 40.48%, H 1.70%, Cl 39.83%, O 8.99%, S 9.00%. Prepn: Meltzer, Huisman, **US 2812281** (1957 to Phillips); and soly data: Huisman *et al.*, *Rec. Trav. Chim.* **77**, 103 (1958). Toxicity study: Ben-dyke *et al.*, *World Rev. Pest Control* **9**, 119 (1970).

Tetraethylammonium Bromide

Crystals from benzene; mp 146.5-147.5°. Stable to concd and dil alkalies, mineral acids, high temp, and uv light. LD$_{50}$ orally in rats: 556 mg/kg (Ben-dyke).

USE: Acaricide. Ovicide on deciduous fruits, citrus, cotton and other crops.

9344. Tetraethylammonium Bromide. [71-91-0] *N,N,N*-Triethylethanaminium bromide (1:1); TEAB; TMD-10; Etylon; Etambro; Sympatektoman; Tetranium. C$_8$H$_{20}$BrN; mol wt 210.16. C 45.72%, H 9.59%, Br 38.02%, N 6.66%. Ganglion blocking agent. Prepd from triethylamine and ethyl bromide: Hofmann, *Ann.* **78**, 263 (1851). Review of the pharmacology of the tetraethylammonium ion: Moe, Freyburger, *Pharmacol. Rev.* **2**, 61-95 (1950).

Deliquesc crystals. Freely sol in water, alc, chloroform, acetone. Slightly sol in benzene. pH of a 10% aq soln 6.5. The pH is not changed by heating for 28 hrs at 95°.

9345. Tetraethylammonium Chloride. [56-34-8] *N,N,N*-Triethylethanaminium chloride (1:1); etamon chloride; TEA chloride. C$_8$H$_{20}$ClN; mol wt 165.71. C 57.99%, H 12.17%, Cl 21.39%, N 8.45%. Potassium-selective ion channel blocker. Prepn: H. Schiff, U. Monsacchi, *Z. Phys. Chem.* **24**, 513 (1897). Synthetic applications: G. Coppens *et al., J. Org. Chem.* **27**, 3299 (1962); S. Gervat *et al., Tetrahedron Lett.* **34**, 2115 (1993). Hydrogen bonding structural studies: K. M. Harmon, J. M. Gabriele, *Inorg. Chem.* **20**, 4013 (1981); *idem et al., J. Mol. Struct.* **216**, 53 (1990). Toxicity studies: J. F. Keith, Jr. *et al., Circulation* **6**, 902 (1952). Mechanistic study of selective potassium channel binding: W. Jarolimek *et al., Mol. Pharmacol.* **49**, 165 (1996).

Deliquescent crystals. d$_4^{21}$ 1.0801. Freely sol in water, alcohol, chloroform, acetone; slightly sol in benzene. pH of 10% aq soln 6.48. LD$_{50}$ i.v. in dogs: 55.7-72.4 mg/kg (Keith, Jr.).

Tetrahydrate. [6024-76-6] C$_8$H$_{20}$ClN.4H$_2$O; mol wt 237.77. Monoclinic prismatic crystals. mp 37.5°. d 1.084.

USE: Dehydrogenation catalyst; phase transfer catalyst in polymerization reactions. Electrolyte in membrane transport studies.

9346. Tetraethylammonium Hydroxide. [77-98-5] *N,N,N*-Triethylethanaminium hydroxide (1:1). C$_8$H$_{21}$NO; mol wt 147.26. C 65.25%, H 14.37%, N 9.51%, O 10.86%. Made from the corresp halide by treating with silver oxide or with a soln of potassium hydroxide in methanol.

Marketed as an aq soln. A 10% soln has a d$_4^{25}$ ~1.01. The free base is known only in soln or as hydrates. Dec on boiling. It is a very strong base readily absorbing CO$_2$ from the air. The aq soln is colorless, odorless, bitter, caustic, strongly alkaline, and imparts a soapy feel to the skin. *Keep well closed.*

Tetrahydrate. [5990-65-8] C$_8$H$_{21}$NO.4H$_2$O; mol wt 219.32. mp 49-50°.

Hexahydrate. [6024-77-7] C$_8$H$_{21}$NO.6H$_2$O; mol wt 255.35. mp 55°.

9347. Tetraethyllead. [78-00-2] Tetraethylplumbane; lead tetraethyl; TEL. C$_8$H$_{20}$Pb; mol wt 323.45. C 29.71%, H 6.23%, Pb 64.06%. Prepd by the action of PbCl$_2$ on zinc ethyl or on a Grignard reagent; by heating C$_2$H$_5$Cl and sodium-lead alloy in an autoclave. Production from lead, ethylene, and hydrogen using triethylaluminum as intermediate was first described by K. Ziegler at the 14th International Congress of Pure and Applied Chemistry (July 1955): *Chem. Eng. News* **33**, 3486 (1955). Alternate synthesis using nonhalide compds: Pearson *et al.* in *Adv. Chem. Ser.* **23**, entitled "Metal-Organic Compounds," M. Sittig, Ed. (ACS, Washington DC, 1959) pp 299-305. *See also* Milde, Beatty, *ibid.*, 306-318. Toxicity study: Schroeder *et al., Experientia* **28**, 923 (1972). Review of toxicology and human exposure: *Toxicological Profile for Lead* (PB2008-100007, 2007) 582 pp.

Colorless liq; burns with an orange-colored flame with green margin. *Extremely poisonous.* d^{20} 1.653. bp ~200° also stated as 227.7° with decompn. n_D^{20} 1.5198. Sol in benzene, petr ether, gasoline; slightly sol in alc. Practically insol in water. LD$_{50}$ orally in rats: 12.3 mg/kg (Schroeder).

Caution: Potential symptoms of overexposure are insomnia, lassitude and anxiety; tremor, hyper-reflexia and spasticity; bradycardia, hypotension, hypothermia, pallor, nausea, anorexia and weight loss; confusion, disorientation, hallucinations, psychosis, mania, convulsions and coma; eye irritation. *See NIOSH Pocket Guide to Chemical Hazards* (DHHS/NIOSH 97-140, 1997) p 302. *See also Patty's Industrial Hygiene and Toxicology* vol. 2A, G. D. Clayton, R. E. Clayton, Eds. (Wiley-Interscience, New York, 3rd ed., 1981) pp 1687-1728; *Clinical Toxicology of Commercial Products*, R. E. Gosselin *et al.*, Eds. (Williams & Wilkins, Baltimore, 5th ed., 1984) Section II, p 139, Section III, pp 226-239. Lead and all lead compounds are listed as reasonably anticipated to be human carcinogens: *Report on Carcinogens, Twelfth Edition* (PB2011-111646, 2011) p 251.

USE: As a gasoline additive to prevent "knocking" in motors.

9348. *N,N,N',N'*-Tetraethylphthalamide. [83-81-8] *N*^{1},-*N*^{1},*N*^{2},*N*^{2}-Tetraethyl-1,2-benzenedicarboxamide; orthophthalic acid didiethylamide; *o*-phthalic acid bis[diethylamide]; tetraethylbis-(phthalamide); Analetil; Neo-Cardiamine; Neospiran; Unispiran. C$_{16}$H$_{24}$N$_2$O$_2$; mol wt 276.38. C 69.53%, H 8.75%, N 10.14%, O 11.58%. Prepd by treating phthalyl chloride with diethylamine: **FR 785428**; **GB 443396**; **US 2057145** (1936 to Chem. Fabrik Grünau); by heating sodium phthalate with diethylamine phosphate: **FR 866229** (1941 to Corbière).

Crystals, mp 39°. bp 175-180°. Soluble in water, physiol saline.
THERAP CAT: Analeptic.

9349. Tetraethyl Pyrophosphate. [107-49-3] Diphosphoric acid *P,P,P',P'*-tetraethyl ester; pyrophosphoric acid tetraethyl ester; bis-*O,O*-diethylphosphoric anhydride; TEPP; Bladan; Nifos T; Kilmite 40; Vapotone; Tetron; Killax; Mortopal. C$_8$H$_{20}$O$_7$P$_2$; mol wt 290.19. C 33.11%, H 6.95%, O 38.59%, P 21.35%. Cholinesterase inhibitor. Prepd commercially by controlled hydrolysis of *O,O*-diethylphosphoric acid chloride: Kosolapoff, **US 2479939** (1947 to Monsanto); Toy, *J. Am. Chem. Soc.* **70**, 3882 (1948). Chemical history and comparison of various syntheses: G. Schrader, *Die Entwicklung neuer insektizider Phosphorsäure-Ester* (Verlag Chemie, Weinheim, 3rd ed., 1963) pp 68-79. Toxicity study: T. B. Gaines, *Toxicol. Appl. Pharmacol.* **14**, 515 (1969).

Mobile liquid. Agreeable odor. Hygroscopic. d_4^{20} 1.185. Thermal decompn range 170-213° with copious formn of ethylene. $bp_{0.05}$ 82°; $bp_{1.0}$ 124°; $bp_{2.3}$ 138°. Vapor pressure at 30° = 4.7 × 10^{-4} mm Hg. n_D^{20} 1.4196. Misc with water, but quickly hydrolyzed by it (half life at 25° about 7 hrs in a 50 v/v mixt.). Also misc with acetone, methanol, ethanol, benzene, chloroform, carbon tetrachloride, glycerol, ethylene glycol, propylene glycol, toluene, xylene. Not misc with petr ether, kerosene, other petr oils. LD_{50} orally in male rats: 1.1 mg/kg (Gaines).

Caution: Potential symptoms of overexposure are eye pain, blurred vision, lacrimation, rhinorrhea, headache, tight chest and cyanosis; anorexia, nausea, vomiting and diarrhea; weakness, twitching, paralysis, Cheyne-Stokes respiration and convulsions; low blood pressure, cardiac irregularities; sweating. *See NIOSH Pocket Guide to Chemical Hazards* (DHHS/NIOSH 97-140, 1997) p 296.

USE: Insecticide.

9350. Tetrafluoroethylene. [116-14-3] 1,1,2,2-Tetrafluoroethene; perfluoroethene; TFE. C_2F_4; mol wt 100.02. C 24.02%, F 75.98%. $CF_2{=}CF_2$. Monomer of the plastic, polytetrafluoroethylene, *q.v.* Prepn: O. Ruff, O. Bretschneider, *Z. Anorg. Allg. Chem.* **210**, 173 (1933). Structural study: T. T. Broun, R. L. Livingston, *J. Am. Chem. Soc.* **74**, 6084 (1952). Physical properties: M. M. Renfrew, E. E. Lewis, *Ind. Eng. Chem.* **38**, 870 (1946). Addition reactions: D. D. Coffman *et al., J. Org. Chem.* **14**, 747 (1949). Thermodynamics and kinetics of mfr by pyrolysis of chlorodifluoromethane: P. B. Chinoy, P. D. Sunavala, *Ind. Eng. Chem. Res.* **26**, 1340 (1987). Review of mfr: Y. Venkateswarlu, P. S. Murti, *Chem. Process Eng.* **4**, 25-36 (1970); of chemistry: M. Saijwani, Dinkar, *Labdev J. Sci. Tech.* **9-A**, 1-12 (1971); of toxicology: G. L. Kennedy, Jr., *Crit. Rev. Toxicol.* **21**, 149-170 (1990).

Colorless, odorless gas, bp_{760} −76.3°. mp −142.5°. d 1.519. Critical temp: 33.3°. Critical press.: 572 lb/sq in. Critical d 0.58. Insol in water. LC_{50} (4 hr) in rats, mice, hamsters, guinea pigs (ppm): 37500-45000, 35000, 28500, 28000 (Kennedy). *Highly flammable when exposed to heat or flame. Reacts violently with oxygen.*

Caution: This substance is reasonably anticipated to be a human carcinogen: *Report on Carcinogens, Twelfth Edition* (PB2011-111646, 2011) p 401.

USE: In mfr of polymers and synthesis of fluorinated refrigerants, dielectric media and solvents. In vinyl polymerization, cycloalkylation and addition reactions.

9351. 2,2,3,3-Tetrafluoro-1-propanol. [76-37-9] C_3-Fluoroalcohol; 1,1,2,2-tetrafluoro-3-hydroxypropane. $C_3H_4F_4O$; mol wt 132.06. C 27.29%, H 3.05%, F 57.54%, O 12.11%. Prepn: Bestian, Rehn, **DE 1007771** (1957 to Hoechst).

Liquid. d_4^{20} 1.4853. mp −15°. bp_{760} 109-110°. n_D^{20} 1.3197. Surface tension at 20° = 27.6 dyn/cm.

p-Nitrobenzoate. [790-43-2] $C_{10}H_7F_4NO_4$; mol wt 281.16. mp 47°.

USE: To introduce fluoroalkyl groups into an organic molecule. Proposed intermediate for plastics, surface active agents, lubricants, elastomers.

9352. Tetraglycine Hydroperiodide. [7097-60-1] Glycine hydriodide compd with iodine (16:4:5); Globaline. $C_{16}H_{42}I_7N_8O_{16}$; mol wt 1490.88. C 12.89%, H 2.84%, I 59.58%, N 7.52%, O 17.17%. $2[(NH_2CH_2COOH)_4HI].2\frac{1}{2}I_2$. Prepn: Frost, Eddy, *J. Am. Chem. Soc.* **74**, 1346 (1952); Morris *et al., Ind. Eng. Chem.* **45**, 1013 (1953).

Flat needles with brassy-bronze metallic luster in reflected light, dec 162-167°. Soly in water at 25° = 380 g/l.

USE: Decontamination of drinking water in emergencies. Used in amounts sufficient to yield 8 ppm of active iodine. A tablet contg 20 mg plus 96 mg $Na_2H_2PO_7$ plus 4 mg talc will decontaminate one quart of water. Such tablets after 7 days' storage at 60° retained 60% of their original active iodine. Less stable than aluminum hexaurea sulfate triiodide.

9353. Tetraglyme. [143-24-8] 2,5,8,11,14-Pentaoxapentadecane; bis[2-(2-methoxyethoxy)ethyl] ether; tetraethylene glycol dimethyl ether; dimethoxytetraethylene glycol. $C_{10}H_{22}O_5$; mol wt 222.28. C 54.04%, H 9.98%, O 35.99%. Prepd from ethylene glycol methyl ether and 2,2′-dichlorodiethyl ether: Zellhoefer, **US 2111234** (1935); *Ind. Eng. Chem.* **29**, 550 (1937). Purification: Vogel, *J. Chem. Soc.* **1948**, 618. Toxicity study: H. F. Smyth *et al., J. Ind. Hyg. Toxicol.* **23**, 259 (1941).

Liquid. d_4^{20} 1.0087; d_4^{86} 0.9514. mp −27°. bp_{760} 275.3°; bp_2 118°. n_D^{20} 1.4325. Sol in water. Misc with hydrocarbon solvents. LD_{50} orally in rats: 5.14 g/kg (Smyth).

USE: Solvent.

9354. Tetrahydrocannabinols. Psychoactive constituents of cannabis, *q.v.* (*Cannabis sativa* L. *Cannabinaceae*). The (−)-*trans*-Δ^9 isomer (Δ^9-THC) is the principal active constituent; the (−)-*trans*-Δ^8 isomer (Δ^8-THC), although physiologically active, is present to a lesser extent. Δ^9 analogs with shorter alkyl side chains have also been identified. Isoln from cannabis resin: H. J. Wollner *et al., J. Am. Chem. Soc.* **64**, 26 (1942). Isoln and structure of Δ^9-THC: Y. Gaoni, R. Mechoulam, *ibid.* **86**, 1646 (1964). Isoln of Δ^8-THC: R. L. Hively *et al., ibid.* **88**, 1832 (1966). Abs config of (−)-*trans*-Δ^9 isomer: R. Mechoulam, Y. Gaoni, *Tetrahedron Lett.* **1967**, 1109. Stereospecific synthesis of Δ^8- and Δ^9-THC: R. Mechoulam *et al., J. Am. Chem. Soc.* **89**, 4552 (1967); of Δ^9-THC: R. K. Razden *et al., ibid.* **96**, 5860 (1974); *eidem, Experientia* **31**, 16 (1975). IR, NMR, MS data: T. Petrzilka, C. Sikemeier, *Helv. Chim. Acta* **50**, 1416, 2111 (1967). Toxicity studies: H. Rosenkranz *et al., Toxicol. Appl. Pharmacol.* **28**, 18 (1974); H. Yoshimura *et al., J. Med. Chem.* **21**, 1079 (1978). Clinical experience with Δ^9-THC in chemotherapy induced nausea and vomiting: D. S. Poster *et al., J. Am. Med. Assoc.* **245**, 2047 (1981); as appetite stimulant in AIDS-associated anorexia: J. E. Beal *et al., J. Pain Symptom Manage.* **14**, 7 (1997). Chromatographic determn in cannabis products: L. Vollner *et al., Regul. Toxicol. Pharmacol.* **6**, 348-358 (1986). GC/MS determn in hair: M. J. Baptista *et al., Forensic Sci. Int.* **128**, 66 (2002). Review of pharmacology: W. L. Dewey, *Pharmacol. Rev.* **38**, 151-178 (1986); of pharmacokinetics and metabolism: F. Grotenhermen, *Clin. Pharmacokinet.* **42**, 327-360 (2003).

Δ^9-THC

(−)-*trans*-Δ^9-Form. [1972-08-3] (6aR,10aR)-6a,7,8,10a-Tetrahydro-6,6,9-trimethyl-3-pentyl-6H-dibenzo[*b,d*]pyran-1-ol; (−)-Δ^1-3,4-*trans*-tetrahydrocannabinol; Δ^9-THC; Δ^1-THC; dronabinol; QCD-84924; Marinol. $C_{21}H_{30}O_2$; mol wt 314.47. C 80.21%, H 9.62%, O 10.18%. Light yellow resinous oil; hardens upon refrigeration. $bp_{0.02}$ 200°. $[\alpha]_D^{20}$ −150.5° (c = 0.53 in $CHCl_3$). uv max (ethanol): 283, 276 nm (log ε 3.21, 3.20). pKa 10.6. Essentially insol in water. LD_{50} (sesame oil emulsion) in male rats (mg/kg): 800 orally; 35.5. i.v.; 672 i.p. (Rosenkranz).

(−)-*trans*-Δ^8-Form. [5957-75-5] (−)-Δ^6-3,4-*trans*-Tetrahydrocannabinol; Δ^8-THC; Δ^6-THC. $bp_{0.001}$ 200°. $[\alpha]_D$ −264° (c = 0.11 in ethanol). uv max (ethanol): 282, 275 nm (log ε 3.22, 3.22); shoulder at 230 nm (log ε 4.07). LD_{50} i.v. in mice: 27.5 mg/kg (Yoshimura).

Note: This is a controlled substance (hallucinogen): **21 CFR,** 1308.11.

THERAP CAT: Antiemetic; appetite stimulant.

9355. Tetrahydrocortisone. [53-05-4] $(3\alpha,5\beta)$-3,17,21-Trihydroxypregnane-11,20-dione; $3\alpha,17\alpha,21$-pregnanetriol-11,20-dione; 11,20-pregnanedione-$3\alpha,17\alpha,21$-triol; THE. $C_{21}H_{32}O_5$; mol wt 364.48. C 69.20%, H 8.85%, O 21.95%. A normal mammalian metabolite of cortisone: Schneider, *J. Biol. Chem.* **183**, 365 (1950). Prepn by microbial reduction using a *Streptomyces* sp.: Barkemeyer *et al., Appl. Microbiol.* **8**, 237 (1960). Prepn of the triacetate: **GB 737291** (1955 to Merck & Co.). Prepn of the 21-acetate: Julian *et al.*, **US 2752339** (1956 to Glidden).

Crystals from ethyl acetate, mp 190°. $[\alpha]_D^{25}$ +85.5° (abs ethanol).
21-Acetate. $C_{23}H_{34}O_6$. Crystals from acetone, mp 227°.
Triacetate. $C_{27}H_{38}O_8$. Crystals from methanol, mp 150-152°; solvated crystals from ethyl acetate, mp 112-118°.

9356. Tetrahydrofuran. [109-99-9] Diethylene oxide; tetramethylene oxide. C_4H_8O; mol wt 72.11. C 66.63%, H 11.18%, O 22.19%. Prepn from 1,4-butanediol: Schmoyer, Case, *Nature* **187**, 592 (1960). Manuf by catalytic hydrogenation of maleic anhydride: Gilbert, Howk, **US 2772293** (1956 to du Pont); of furan: Banford, Manes, **US 2846449** (1958 to du Pont); Manly, **US 3021342** (1962 to Quaker Oats). Stabilization to prevent excessive peroxide formation on storage with 0.05-1.0% *p*-cresol, 0.05-0.1% hydroquinone, or less than 0.01-0.1% 4,4'-thiobis(6-*tert*-butyl-*m*-cresol): Bordner, Hinegardner, **US 2489260**; **US 2525410**; Campbell, **US 3029257** (1949, 1950, 1962 all to du Pont). Review of toxicology and biological effects: D. E. Moody, *Drug Chem. Toxicol.* **14**, 319-342 (1991).

Liquid. Ether-like odor. *Flammable.* mp −108.5°. d_4^{20} 0.8892. bp_{760} 66°; bp_{176} 25°. Flash pt 1°F. n_D^{20} 1.4070. Dipole moment: 1.70. uv cut-off for spectro grade: 220 nm. Miscible with water, alcohols, ketones, esters, ethers, and hydrocarbons. *Caution:* Distil only in presence of a reducing agent, such as ferrous sulfate; peroxide explosions have occurred: *Angew. Chem.* **68**, 182 (1956).
Caution: Potential symptoms of overexposure are irritation of eyes and upper respiratory system; nausea, dizziness and headache; CNS depression. *See NIOSH Pocket Guide to Chemical Hazards* (DHHS/NIOSH 97-140, 1997) p 302.
USE: Solvent for high polymers, esp polyvinyl chloride. As reaction medium for Grignard and metal hydride reactions. In the synthesis of butyrolactone, succinic acid, 1,4-butanediol diacetate. Solvent in histological techniques.

9357. 2,5-Tetrahydrofurandimethanol. [104-80-3] 2,5-Anhydro-3,4-dideoxyhexitol; 2,5-bis(hydroxymethyl)tetrahydrofuran. $C_6H_{12}O_3$; mol wt 132.16. C 54.53%, H 9.15%, O 36.32%. Prepd from diallyl by oxidation with perbenzoic acid in chloroform, boiling with dil sulfuric acid and hydrolyzing the reaction product with KOH soln: Böeseken, *Rec. Trav. Chim.* **45**, 838 (1926); by Raney nickel reduction of 5-hydroxymethylfurfural or of dimethyl furan-2,5-dicarboxylate: Cope, Baxter, *J. Am. Chem. Soc.* **77**, 393 (1955). The usual form obtained is the *cis* form, described here.

Hygroscopic liquid. Faint odor. *Avoid contact with eyes.* d_4^0 1.1719; d_4^{25} 1.1542; d_4^{50} 1.1359. mp below −50°. bp_{760} 265°; bp_{96} 200°; bp_{11} 155°; $bp_{0.25}$ 105°. n_D^{25} 1.4766. Viscosity (cP): 1926 at 0°; 225 at 25°; 51.9 at 50°. Coefficient of expansion: 0.00063 per °C at

25°. Specific heat 0.5 cal/g/°C. Heat of vaporization 115 cal/g. Miscible with water, methanol, ethanol, acetone, benzene, methyl acetate, methyl ethyl ketone, chloroform. Moderately sol in ether, toluene. Almost insol in heptane, methylcyclohexane.
Caution: Highly irritating to eyes, skin, mucous membranes.
USE: Solvent, softener, humectant. In the synthesis of plasticizers, resins, surfactants, agricultural chemicals.

9358. Tetrahydrofurfuryl Alcohol. [97-99-4] Tetrahydro-2-furanmethanol; tetrahydro-2-furancarbinol; tetrahydro-2-furylmethanol; THFA. $C_5H_{10}O_2$; mol wt 102.13. C 58.80%, H 9.87%, O 31.33%. Prepn by catalytic hydrogenation of furfuryl alcohol: Lukes, Nelson, *J. Org. Chem.* **21**, 1096 (1956). Manuf by catalytic hydrogenation of furfural or furfuryl alcohol: Dunlop, Schegulla, **US 2838523** (1958 to Quaker Oats). Occurs in two isomeric forms: D-isomer (levorotatory), L-isomer (dextrorotatory). Abs config: Gagnaire, Butt, *Bull. Soc. Chim. Fr.* **1961**, 312; Hartman, Barker, *J. Org. Chem.* **29**, 873 (1964). *Review:* A. P. Dunlop, F. N. Peters, *The Furans* (Reinhold, New York, 1953).

Liquid. Hygroscopic. d_{20}^{20} 1.0543; d_{24}^{24} 1.0511; d_{31}^{31} 1.0450. Melts below −80°. bp_{760} 178°. n_D^{20} 1.4520; n_D^{25} 1.4499. Flash pt, open cup: 183°F (84°C). Flammability in air: Upper limit 9.7% by vol, lower limit 1.5% by vol. Heat capacity at 30-37°: 0.432 cal/g/°C. Heat of combustion at constant vol: 708.6 cal/g mole. Viscosity at 20°: 6.24 cP. Surface tension at 25°: 37 dyn/cm. Octane no. 82.5. Evaporation rate: 7 (*n*-butyl acetate = 100). Kauributanol value 71.5. Dilution ratio (lacquer ingredients): 4.5. Dielectric constant at 23°: 13.6. Miscible with water, alcohol, ether, acetone, chloroform, benzene.
L-Isomer. Prepn *see* Hartman, Barker, *loc. cit.* $[\alpha]_D^{24}$ +14.9° (c = 5.0 in nitromethane).
Caution: Moderately irritating to skin, mucous membranes.
USE: Solvent for fats, waxes, resins. In organic synthesis: Undergoes the reactions of a primary alcohol, while the ring exhibits characteristics of a saturated cyclic ether.

9359. Tetrahydrogestrinone. [618903-56-3] (17α)-13-Ethyl-17-hydroxy-18,19-dinorpregna-4,9,11-trien-3-one; 18a-homopregna-4,9,11-trien-17β-ol-3-one; THG. $C_{21}H_{28}O_2$; mol wt 312.45. C 80.73%, H 9.03%, O 10.24%. Androgen and progestin; hydrogenation of gestrinone, *q.v.* Discovery, synthesis, LC/MS/MS and GC/MS determn in urine: D. H. Catlin *et al., Rapid Commun. Mass Spectrom.* **18**, 1245 (2004). Hormonal properties: A. K. Death *et al., J. Clin. Endocrinol. Metab.* **89**, 2498 (2004). *In vitro* metabolism study: J.-F. Lévesque *et al., Anal. Chem.* **77**, 3164 (2005).

Absorption max: 345 nm.
Note: This is a controlled substance (anabolic steroid): **21 CFR,** 1308.13, as defined in 1300.01.

9360. Tetrahydropalmatine. [2934-97-6] 5,8,13,13a-Tetrahydro-2,3,9,10-tetramethoxy-6*H*-dibenzo[*a,g*]quinolizine; 2,3,9,10-tetramethoxyberbine; 2,3,9,10-tetramethoxydibenzo[*a,g*]quinolizidine; hyndarin. $C_{21}H_{25}NO_4$; mol wt 355.43. C 70.97%, H 7.09%, N 3.94%, O 18.01%. Synthesis of *dl*-form: Haworth *et al., J. Chem. Soc.* **1927**, 548; Bradsher, Dutta, *J. Org. Chem.* **26**, 2231 (1961); T. Kametani, M. Ihara, *J. Chem. Soc. C* **1967**, 530; G. D. Pandey, K. P.Tiwari, *Indian J. Chem.* **18B**, 545 (1979); Z. Kiparissides *et al., Can. J. Chem.* **58**, 2770 (1980); N. S. Narasimhan *et al., Tetrahedron Lett.* **22**, 2797 (1981). Biosynthesis: D. S. Bhakuni *et al., Tetrahedron* **36**, 2491 (1980). *See also* Palmatine. Both optically active forms are found in plants.

l-**Form.** Caseanine; gindarine; rotundine. Crystals from dil methanol, mp 147°. $[\alpha]_D^{20}$ −291° (c = 0.8 in 95% alcohol). Also a hydrate, mp 115° (effervescence).

Hydrochloride. $C_{21}H_{25}NO_4 \cdot HCl$. Crystals.

d-**Form.** Crystals from methanol, mp 141-142° (evac tube). $[\alpha]_D^{14}$ +292° (c = 0.8 in 95% alcohol).

Hydrochloride. Crystals, mp 266°.

dl-**Form.** Crystals from methanol upon addition of water. mp 148-149°.

Hydrochloride. Needles from methanol, mp 215-216°.

9361. Tetrahydropapaveroline. [4747-99-3] 1-[(3,4-Dihydroxyphenyl)methyl]-1,2,3,4-tetrahydro-6,7-isoquinolinediol; 1-(3,4-dihydroxybenzyl)-1,2,3,4-tetrahydro-6,7-isoquinolinediol; 6,7-dihydroxy-1-(3,4-dihydroxybenzyl)-1,2,3,4-tetrahydroisoquinoline; norlaudanosoline; THP. $C_{16}H_{17}NO_4$; mol wt 287.32. C 66.89%, H 5.96%, N 4.88%, O 22.27%. Alkaloid deriv of dopamine and a biosynthetic precursor of morphine, *q.v.* Prepn of the racemic hydrochloride: F. L. Pyman, *J. Chem. Soc.* **95**, 1610 (1909); of the (+)- and (−)-form hydrochlorides: S. Teitel *et al.*, *J. Med. Chem.* **15**, 845 (1972). Chromatographic study: K. D. McMurtrey *et al.*, *J. Liq. Chromatogr.* **3**, 663 (1980). Possible role of THP in the biochemical mediation of alcohol addiction: V. E. Davis, M. J. Walsh, *Science* **167**, 1005 (1970); G. Cohen, M. Collins, *ibid.* 1749; R. D. Meyers, C. L. Melchior, *ibid.* **196**, 554 (1977); M. Sandler *et al.*, *Prog. Clin. Biol. Res.* **90**, 215 (1982). Biosynthetic study: D. K. Choudhary, B. L. Kaul, *Indian Drugs* **19**, 229 (1982). *In vivo* and *in vitro* effects on rat pituitary function: D. R. Britton *et al.*, *Biochem. Pharmacol.* **31**, 1205 (1982). Effect on dopaminergic neurons: I. S. Hoffman, L. X. Cubeddu, *J. Pharmacol. Exp. Ther.* **220**, 16 (1982). Comparison to opioid effects in brain regions: G. R. Siggins *et al.*, *Prog. Clin. Biol. Res.* **90**, 275 (1982).

(±)-Form hydrochloride. $C_{16}H_{18}ClNO_4$. Colorless microscopic prisms, mp 291-293° (dec). Practically insol in water, alc.

(−)-Form hydrochloride. Cryst from 6*N* HCl, mp 285-286°. $[\alpha]_D^{25}$ −32.4° (c = 1 in water). uv max (ethanol): 230, 286 nm (ε 11100, 6700).

(+)-Form hydrochloride. Cryst, mp 285-286°. $[\alpha]_D^{25}$ +32.1° (c = 1 in water).

USE: As a research tool in neurological biochemistry.

9362. Tetrahydropyran. [142-68-7] Tetrahydro-2*H*-pyran; pentamethylene oxide. $C_5H_{10}O$; mol wt 86.13. C 69.73%, H 11.70%, O 18.58%. Prepd by hydrogenation of dihydropyran (from furfural): Paul, *Bull. Soc. Chim. Fr.* [4] **53**, 1489 (1933); Andrus, Johnson, *Org. Synth.* **23**, 90 (1943); Cass, *Ind. Eng. Chem.* **40**, 219 (1948).

Mobile liquid. *Flammable.* Pungent, sweetish odor. d_4^{20} 0.8814. mp −49.2°. bp 88°. n_D^{20} 1.4211. Dipole moment 1.87. Flash pt

−4.0°F. Azeotrope with water, bp 71°, contains 8.5% H_2O. Sol in water [relative solubilities: Bennett, Philip, *J. Chem. Soc.* **1928**, 1937]. Miscible with alcohol, ether, many other organic solvents. Forms peroxides on exposure to air. All technical tetrahydropyran is stabilized against peroxide formation.

9363. Tetrahydrothiophene. [110-01-0] Tetramethylene sulfide; thiacyclopentane; THT; THTP. C_4H_8S; mol wt 88.17. C 54.49%, H 9.15%, S 36.36%. Prepn: J. v. Braun, A. Trümpler, *Ber.* **43**, 545 (1910); J. K. Lawson *et al.*, *Org. Synth.* **coll. vol. IV**, 892 (1963). Method for odorizing liquified natural gas: D. K. Mulliner, *Proc., Opec. Sect., Am. Gas Assoc.* **D**, 27 (1974). GC/MS determn in water: G. Carlucci *et al.*, *J. Chromatogr.* **287**, 425 (1984). On-site GC determn in natural gas: R. C. M. de Nijs *et al.*, *J. High Resolut. Chromatogr.* **16**, 379 (1993).

Colorless liquid. *Flammable.* bp 119-121°. n_D^{25} 1.5000-1.5014. LD₅₀ by inhalation in mice: 26.7 mg/l (Carlucci).

USE: Natural gas odorant.

9364. Tetrahydrozoline. [84-22-0] 4,5-Dihydro-2-(1,2,3,4-tetrahydro-1-naphthalenyl)-1*H*-imidazole; 2-(1,2,3,4-tetrahydro-1-naphthyl)-2-imidazoline; tetryzoline. $C_{13}H_{16}N_2$; mol wt 200.29. C 77.96%, H 8.05%, N 13.99%. Prepn from ethylenediamine, ethylenediamine hydrochloride, and methyl 1,2,3,4-tetrahydro-1-naphthoate: Synerholm *et al.*, **US 2731471** (1956 to Sahyun Labs.). Use as potentiator for veterinary depressants: Gardocki *et al.*, **US 2842478** (1958 to Pfizer).

Hydrochloride. [522-48-5] Rhinopront; Tyzine; Visine; Yxin. $C_{13}H_{16}N_2 \cdot HCl$; mol wt 236.74. Crystals from alcohol, dec 256-257°. uv max: 264.5, 271.5 nm (A$_{1cm}^{1\%}$ 17.5, 15.5). Freely sol in water, alcohol. Very slightly sol in chloroform. Practically insol in ether. pH of a 1% aq soln 5.0 to 6.5.

THERAP CAT: Adrenergic (vasoconstrictor); nasal decongestant.

9365. Tetraiodoethylene. [513-92-8] 1,1,2,2-Tetraiodoethene; diiodoform; ethylene periodide; ethylene tetraiodide. C_2I_4; mol wt 531.64. C 4.52%, I 95.48%. Prepd by the action of iodine on diodoacetylene obtained from calcium carbide and iodine. Crystal and molecular structure: B. C. Haywood, R. Shirley, *Acta Crystallogr.* **B33**, 1765 (1977).

Light yellow, heavy, small, practically odorless crystals; characteristic odor. On exposure to light turns brown. d 2.98. mp 187°. Insol in water; sol in benzene, chloroform, toluene, CS_2, slightly in ether. *Protect from light.*

9366. Tetrakis(dimethylamino)ethylene. [996-70-3] N^1,-N^1,N'^1,N'^1,N^2,N^2,N'^2,N'^2-Octamethyl-1,1,2,2-ethenetetramine; TDAE; tetrakis(dimethylamino)ethene. $C_{10}H_{24}N_4$; mol wt 200.33. C 59.96%, H 12.08%, N 27.97%. Organic reductant; alternative to zinc as an electron donor. Prepn: R. L. Pruett *et al.*, *J. Am. Chem. Soc.* **72**, 3646 (1950). Structural study: H. Bock *et al.*, *Angew. Chem. Int. Ed.* **28**, 1684 (1989). Thermodynamic properties: W. V. Steele *et al.*, *J. Chem. Eng. Data* **42**, 1037 (1997). Cyclic voltammetry and applications in organofluorine synthesis: C. Burkholder *et al.*, *J. Org. Chem.* **63**, 5385 (1998). Synthetic applications: G. Pawelke, *J. Fluorine Chem.* **42**, 429 (1989); M. Kuroboshi *et al.*, *J. Org. Chem.* **68**, 3938 (2003); M. Médebielle *et al.*, *Tetrahedron Lett.* **44**, 7871 (2003); Y. Nishiyama, A. Kobayashi, *ibid.* **47**, 5565 (2006).

Review of reactivity: N. Wiberg, *Angew. Chem. Int. Ed.* **7**, 766-779 (1968).

Pale green liquid, bp$_{0.9}$ 59°. n_D^{25} 1.4785. d$_4^{25}$ 0.8612. Flash pt, closed cup: 127.4°F (53°C). *Corrosive, flammable.* Ionization potential: 6.13 eV.

USE: In reduction of organofluorine compounds; as a radical generator in organic synthesis; in detection of scintillation light from gas scintillation proportional counters.

9367. Tetrakis(triphenylphosphine)palladium(0). [14221-01-3] (*T*-4)-Tetrakis(triphenylphosphine)palladium; palladium-tetrakis(triphenylphosphine); Pd(PPh$_3$)$_4$. C$_{72}$H$_{60}$P$_4$Pd; mol wt 1155.59. C 74.84%, H 5.23%, P 10.72%, Pd 9.21%. Pd[P(C$_6$-H$_5$)$_3$]$_4$. Palladium-based catalyst for a variety of transformations, including carbon-carbon bond formation reactions. Prepn: L. Malatesta, M. Angoletta, *J. Chem. Soc.* **1957**, 1186; D. R. Coulson, *Inorg. Synth.* **13**, 121 (1972). Synthetic utility in carbon-carbon bond formation: M. Kosugi *et al.*, *Chem. Lett.* **1977**, 301; N. Miyaura *et al.*, *Tetrahedron Lett.* **20**, 3437 (1979); E. Negishi *et al.*, *ibid.* **24**, 3823 (1983); M. Alami *et al.*, *ibid.* **34**, 6403 (1993); in removal of the allyloxycarbonyl protecting group: F. Guibe, Y. S. M'Leux, *ibid.* **22**, 3591 (1981); in carbonylations: V. P. Baillargeon, J. K. Stille, *J. Am. Chem. Soc.* **105**, 7175 (1983). *Review*: R. W. Friesen in *Encyclopedia of Reagents for Organic Synthesis* **7**, L. A. Paquette, Ed. (Wiley, New York, 1995) pp 4788-4796.

Yellow crystalline solid from ethanol, dec 100-105° (Malatesta); also reported as dec 116° (Coulson). Moderately sol in chloroform, dichloromethane, DMF, toluene, dimethyl ether; less sol in acetone, THF, acetonitrile. Insol in saturated hydrocarbons. *Protect from light.* Air stable for only short periods of time; turns orange upon prolonged exposure to air. Solns in benzene absorb molecular oxygen to form an insoluble green complex. Heat and moisture sensitive. Store under inert gas at 2-8°C.

USE: Catalyst in synthetic organic chemistry.

9368. Tetralin®. [119-64-2] 1,2,3,4-Tetrahydronaphthalene; Tetranap. C$_{10}$H$_{12}$; mol wt 132.21. C 90.85%, H 9.15%. Prepd by catalytic hydrogenation of purified naphthalene. Toxicity study: Smyth *et al.*, *Arch. Ind. Hyg. Occup. Med.* **4**, 119 (1951). *See* references under Decalin.

Liquid. Odor resembling that of a mixture of benzene and menthol. d$_4^{20}$ 0.9702; d$_4^{25}$ 0.9662. Volatile with steam; mp −31.0°; bp$_{760}$ 207.2°; bp$_{400}$ 181.8°; bp$_{200}$ 157.2°; bp$_{100}$ 135.3°; bp$_{60}$ 121.3°; bp$_{40}$ 110.4°; bp$_{20}$ 93.8°; bp$_{10}$ 79.0°; bp$_5$ 65.3°; bp$_{1.0}$ 38.0°. n_D^{20} 1.54135; n_D^{25} 1.53919. Flash pt, open cup 171°F (77°C), closed cup 180°F (82°C). Insol in water; miscible with ethanol, butanol, acetone, benzene, ether, chloroform, petr ether, Decalin; soluble in methanol: 50.6% w/w. Prolonged, intimate contact with air may cause the formn of tetralin peroxide which may cause explosion of tetralin distn residues. Peroxide formn is prevented by the addn of an antioxidant, such as hydroquinone. LD$_{50}$ orally in rats: 2.86 g/kg (Smyth).

Caution: Irritating to skin, eyes, mucous membranes, and, in high concns, narcotic. In exptl animals has produced cataracts: E. Browning, *Toxicity and Metabolism of Industrial Solvents* (Elsevier, New York, 1965) pp 119-124.

USE: Degreasing agent. Solvent for naphthalene, fats, resins, oils, waxes, used instead of turpentine in lacquers, shoe polishes, floor waxes.

9369. Tetralol. [530-91-6] 1,2,3,4-Tetrahydro-2-naphthalenol; *ac*-tetrahydro-β-naphthol; *ac*-β-tetralol. C$_{10}$H$_{12}$O; mol wt

148.21. C 81.04%, H 8.16%, O 10.79%. Prepd by hydrogenation of 2-naphthol in the presence of a palladium catalyst: Foreman, Stork, **US 2526859** (1950 to Lakeside Labs.); by reduction of 2-naphthol with sodium: Hueckel *et al.*, *Ann.* **645**, 162 (1961). Toxicity study: Draize *et al.*, *J. Pharmacol. Exp. Ther.* **93**, 26 (1948).

Liquid, bp$_{12}$ 140°; mp 15.5°. Crystallizes on prolonged storage at −50° and then at −15°. LD$_{50}$ orally in rats: 1.0 ml/kg (Draize).

p-**Tosylate.** C$_{10}$H$_{11}$O$_3$S.C$_6$H$_4$CH$_3$. Crystals from petr ether + benzene, mp 86°.

Methyl ether. C$_{10}$H$_{11}$OCH$_3$. bp$_{11}$ 114.5°, d$_4^{20}$ 1.0239, n_D^{20} 1.5326.

9370. Tetramethrin. [7696-12-0] 2,2-Dimethyl-3-(2-methyl-1-propen-1-yl)cyclopropanecarboxylic acid (1,3,4,5,6,7-hexahydro-1,3-dioxo-2*H*-isoindol-2-yl)methyl ester; 2,2-dimethyl-3-(2-methyl-propenyl)cyclopropanecarboxylic acid ester with *N*-(hydroxymethyl)-1-cyclohexene-1,2-dicarboximide; *N*-(3,4,5,6-tetrahydrophthalimide)methyl-*cis,trans*-chrysanthemate; *N*-(chrysanthemoxymethyl)-1-cyclohexene-1,2-dicarboximide; phthalthrin; FMC-9260; SP-1103; Neo-Pynamin. C$_{19}$H$_{25}$NO$_4$; mol wt 331.41. C 68.86%, H 7.60%, N 4.23%, O 19.31%. Synthetic pyrethroid insecticide. Prepn of racemic mixture: T. Kato *et al.*, *JP 65 8535*; *eidem*, **US 3268398** (1965, 1966 both to Sumitomo). Activity: *eidem*, *Agric. Biol. Chem.* **28**, 914 (1965). Comparative activity of isomers: Y. Okuno *et al.*, **DE 2348930**; *eidem*, **US 3934023** (1973, 1976 both to Sumitomo). Metabolism: J. Miyamoto *et al.*, *Agric. Biol. Chem.* **32**, 628 (1968). Photodecompn: Y.-L. Chen, J. E. Casida, *J. Agric. Food Chem.* **17**, 208 (1969). Toxicity study: T. Kato *et al.*, *Agric. Biol. Chem.* **28**, 914 (1965). Review of toxicology and human exposure: *Toxicological Profile for Pyrethrins and Pyrethroids* (PB2004-100004, 2003) 332 pp.

(*1R-trans*)-form

The commercial product is a mixture of isomers. White crystalline solid, mp 65-80°. d$_{20}^{20}$ 1.108; n$_D^{21.5}$ 1.5175. LD$_{50}$ orally in mice: 1000 mg/kg (Kato).

USE: Insecticide.

9371. Tetramethylammonium Hydroxide. [75-59-2] *N,N,-N*-Trimethylmethanaminium hydroxide (1:1); TMAH. C$_4$H$_{13}$NO; mol wt 91.15. C 52.71%, H 14.38%, N 15.37%, O 17.55%. Strong base, usually marketed as a soln in water or methanol. Prepn: J. Walker, J. Johnston, *J. Chem. Soc. Trans.* **87**, 955 (1905); A. Zwierzak, M. Kluba, *Tetrahedron* **27**, 3163 (1971); O. Yagi, S. Shimizu, *Chem. Lett.* **1993**, 2041. Silicon etching studies: J. T. L. Thong *et al.*, *Sens. Actuators A* **63**, 243 (1997). Use in sample prepn for trace element determn: P. Martins *et al.*, *Anal. Chim. Acta* **470**, 195 (2002).

The free base is known only in soln or as a solid pentahydrate, forming colorless, deliquesc needles, mp 63°. Strong ammonia-like odor. Soln d$_4^{25}$ ~ 1.00. Dec to trimethylamine and methanol upon distillation. Readily absorbs CO$_2$ from the air. *Corrosive; highly toxic by ingestion and skin absorption; keep container well closed.*

USE: Buffer; titrant; ion pair reagent; anisotropic etchant for silicon; in digestion of biological samples.

9372. Tetramethylammonium Iodide. [75-58-1] *N,N,N*-Trimethylmethanaminium iodide (1:1). C$_4$H$_{12}$IN; mol wt 201.05. C 23.90%, H 6.02%, I 63.12%, N 6.97%.

Pale yellow crystals. Begins to dec at about 230°. d 1.84. Sparingly sol in water, freely in abs alcohol; insol in chloroform, ether.

USE: Emergency disinfection of drinking water. Required dosage: 8 ppm of iodine.

9373. Tetramethyldiaminobutane. [111-51-3] $N^1,N^1,N^4,$-N^4-Tetramethyl-1,4-butanediamine; 1,4-bis(dimethylamino)butane; N,N,N',N'-tetramethylputrescine. $C_8H_{20}N_2$; mol wt 144.26. C 66.61%, H 13.97%, N 19.42%. From root and herb of *Hyoscyamus reticulatus* L. and *H. muticus* L., *Solanaceae:* Willstätter, Heuber, *Ber.* **40**, 3869 (1907); Konowalowa, Magidson, *Arch. Pharm.* **266**, 449 (1928). Synthesis: Lunsford *et al., J. Org. Chem.* **22**, 1225 (1957); Solov'ev, Skoldinov, *Zh. Obshch. Khim.* **33**, 1821 (1963), *C.A.* **59**, 7360f (1963).

White crystals; penetrating odor and sharp, scratching taste. bp 169°; bp$_{28}$ 78-80°; bp$_7$ 43°. n_D^{20} 1.4280. d^{20} 0.7861. Volatile with steam. Sol in water, alc, ether.

Dihydrochloride. [78204-83-8] $C_8H_{20}N_2$·2HCl; mol wt 217.18. Crystals from alc, mp 273°.

Dipicrate. [102454-36-4] $C_8H_{20}N_2$·2$C_6H_3N_3O_7$; mol wt 602.47. Needles from hot water, mp 203-205°.

9374. Tetramethylenedisulfotetramine. [80-12-6] 2,6-Dithia-1,3,5,7-tetraazatricyclo[3.3.1.1^{3,7}]decane 2,2,6,6-tetraoxide; 2,6-dithia-1,3,5,7-tetraazaadamantane 2,2,6,6-tetraoxide. $C_4H_8N_4$-O_4S_2; mol wt 240.25. C 20.00%, H 3.36%, N 23.32%, O 26.64%, S 26.69%. Prepd from sulfamide, $H_2NSO_2NH_2$ and formaldehyde in 60% H_2SO_4: Hecht, Henecka, *Angew. Chem.* **61**, 365 (1949). Toxicity: Hagen, *Dtsch. Med. Wochenschr.* **75**, 183 (1950).

Cubic crystals from acetone, dec 255-260°. *Violent convulsive poison.* Stable to acids and alkalies in dilutions up to 0.1N. Dec upon prolonged boiling of aq solns. Soly in water about 0.25 mg/ml. Slightly sol in acetone. Insol in methanol, ethanol. LD in mice (mg/kg): 0.20 orally or s.c. (Hagen).

9375. 1,1,3,3-Tetramethylguanidine. [80-70-6] N,N,N',N'-Tetramethylguanidine; TMG. $C_5H_{13}N_3$; mol wt 115.18. C 52.14%, H 11.38%, N 36.48%. Base used to catalyze a variety of organic transformations. Prepn: M. Schenck, *Z. Phys. Chem.* **77**, 328 (1912). Improved prepn: M. Schenck, F. von Graevenitz, *ibid.* **141**, 132 (1924). Properties as nonaqueous solvent: M. L. Anderson, R. N. Hammer, *J. Chem. Eng. Data* **12**, 442 (1967). Synthetic applications as a base catalyst in Michael reactions: G. P. Pollini *et al., Synthesis* **1972**, 44; in Henry reactions: D. Simoni *et al., Tetrahedron Lett.* **38**, 2749 (1997); in Wittig and Horner-Wadsworth-Emmons reactions: *idem et al., Org. Lett.* **2**, 3765 (2000); in Baylis-Hillman reactions: N. E. Leadbeater, C. van der Pol, *J. Chem. Soc. Perkin Trans. 1* **2001**, 2831. *Review*: T. Ishikawa, T. Kumamoto, *Synthesis* **2006**, 737-752.

Colorless liquid. *Flammable, corrosive.* bp 165°; bp$_{745}$ 159.5°; bp$_{11}$ 52-54°. fp −78° (glass). d^{25} 0.9136. n_D^{25} 1.4659; n_D^{20} 1.4690.

Flash pt, closed cup: 140°F (60°C). pKa (50% aq ethanol): 13.3; pKa (25°) as iodide salt: 13.6. Freely sol in most organic solvents; sol in water. Dielectric constant at 25°: 11.5. Viscosity (25°C): 1.40 cP.

USE: Reagent in synthetic organic chemistry.

9376. Tetramethyl-*p*-phenylenediamine. [100-22-1] $N^1,$-N^1,N^4,N^4-Tetramethyl-1,4-benzenediamine; Wurster's reagent; Wurster's blue. $C_{10}H_{16}N_2$; mol wt 164.25. C 73.13%, H 9.82%, N 17.06%. Prepn: Meyer, *Ber.* **36**, 2979 (1903); Cox, Smith, *J. Org. Chem.* **29**, 488 (1964).

Crystals from petr ether, mp 51-52°. bp 260°. Slightly sol in cold water; more sol in hot water; freely sol in alc, chloroform, ether, petr ether.

USE: In the form of the hydrochloride as a reagent in analytical chemistry.

9377. Tetramethylsilane. [75-76-3] Silicon tetramethyl; TMS. C_4H_{12}Si; mol wt 88.23. C 54.45%, H 13.71%, Si 31.83%. Two crystalline forms exist, α and β; β form is more stable. Prepn and properties: E. Krause, A. von Grosse in *Die Chemie der metallorganischen Verbindungen*, (Gebrüder Borntraeger, Berlin, 1937) pp 258-262; J. G. Aston *et al., J. Am. Chem. Soc.* **63**, 2343 (1941); S. Tannenbaum *et al., ibid.* **75**, 3753 (1953). Dielectric properties: A. P. Altshuller, L. Rosenblum, *ibid.* **77**, 272 (1955). Temperature dependence of magnetic susceptibility, ^{1}H, and ^{13}C chemical shifts: F. G. Morin *et al., J. Magn. Reson.* **48**, 138 (1982). Thermodynamic properties: W. V. Steele, *J. Chem. Thermodyn.* **15**, 595 (1983). Use in chemical vapor deposition of films of silicon carbide, *q.v.*: N. Herlin *et al., J. Phys. Chem.* **96**, 7063 (1992); A. Grill, V. Patel, *J. Appl. Phys.* **85**, 3314 (1999). Definition of ^{1}H resonance of TMS as the NMR primary standard: R. K. Harris *et al., Pure Appl. Chem.* **73**, 1795 (2001).

Liquid. *Flammable.* mp −102.12° (α-form), −99.04° (β-form). bp 26.6°. d 0.636 at bp. d$_4^0$ 0.6688. d$_4^{20}$ 0.6464. n_D^{20} 1.3588. Heat of vaporization at normal bp: 5785.1 ± 16.0 cal/mole. Dielectric constant at 20°: 1.921.

USE: NMR reference standard. In semiconductor applications (chemical vapor deposition). Silylation of samples for GC analysis.

9378. Tetramethylurea. [632-22-4] N,N,N',N'-Tetramethylurea; tetramethylcarbamide; TMU; Temur. $C_5H_{12}N_2O$; mol wt 116.16. C 51.70%, H 10.41%, N 24.12%, O 13.77%. Prepn: Lawson, Croom, *J. Org. Chem.* **28**, 232 (1963). Review of prepn, manuf, properties and use as solvent and as reagent: Lüttringhaus, Dirksen, *Angew. Chem. Int. Ed.* **3**, 260 (1964).

Liquid with faint, pleasant odor, bp 176.5°, bp$_{740}$ 174.5°, bp$_{12}$ 63-64°. mp −1.2°. Flash pt ~75°C. d$_4^{20}$ 0.9687. n_D^{25} 1.4493. Dipole moment: 3.47 D in benzene. pKb 2. uv max: 217.5 nm (ε 1940). Miscible with water and with all common organic solvents including petr ether. LD$_{50}$ i.v. in rats: 1.1 g/kg (Lüttringhaus, Dirksen).

USE: As solvent and reagent.

9379. Tetrandrine. [518-34-3] (4aS,16aS)-3,4,4a,5,16a,17,-18,19-Octahydro-12,21,22,26-tetramethoxy-4,17-dimethyl-16H-1,-24:6,9-dietheno-11,15-metheno-2H-pyrido[2',3':17,18][1,11]dioxacycloeicosino[2,3,4-*ij*]isoquinoline. $C_{38}H_{42}N_2O_6$; mol wt 622.76. C 73.29%, H 6.80%, N 4.50%, O 15.41%. Analgesic and antipyretic alkaloid from root of *Stephania tetrandra* S. Moore, *Menisperma-*

ceae. Component of the Chinese drug, han-fang-chi. Isoln: Kondo, Yano, *Ann.* **497**, 90 (1932). Structure: Fujita, Murai, *J. Pharm. Soc. Jpn.* **71**, 1039 (1951). Synthesis: Kataoka, *C.A.* **51**, 16501i (1957). Total synthesis: Inubushi *et al., Tetrahedron Lett.* **1968**, 3399; *eidem, J. Chem. Soc. C* **1969**, 1547.

Needles. mp 217-218°. $[\alpha]_D^{26}$ +252.4° (chloroform). Practically insol in water, petr ether; sol in ether and some other organic solvents.

l-**Form.** [1263-79-2] Phaenthine. mp 210°. $[\alpha]_D^{20}$ −278° (chloroform).

9380. Tetranectin. [109489-77-2] Tetrameric protein isolated from human plasma. Enhances plasminogen activation catalyzed by tissue plasminogen activator, *q.v.* Composed of four identical, non-covalently bound, polypeptide chains each containing 181 amino acids, with mol wt 20,100 daltons. Isoln from human serum: I. Duhl Clemmensen, C. Kluft, **EP 206400** (1986 to Ned. Cent. Org. Toegepast-Natuurwetenschappelijk Onderzoek). Purification, characterization and plasminogen binding activity: I. Clemmensen *et al., Eur. J. Biochem.* **156**, 327 (1986). Primary structure: J. Fuhlendorff *et al., Biochemistry* **26**, 6757 (1987). Enzyme-linked immunoassay in human plasma: B. A. Jensen *et al., J. Lab. Clin. Med.* **110**, 612 (1987). Distribution in human endocrine tissue: L. Christensen *et al., Histochemistry* **87**, 195 (1987). Possible role in cancer metastasis: B. A. Jensen, I. Clemmensen, *Cancer* **62**, 869 (1988).

Isoelectric point 5.8. $A_{280nm}^{1\%}$ 12.5.

9381. Tetranitromethane. [509-14-8] CN_4O_8; mol wt 196.03. C 6.13%, N 28.58%, O 65.29%. Prepd by nitration of acetic anhydride with anhydrous nitric acid: Liang, *Org. Synth.* **coll. vol. III**, p 803 (1955). Energy of decompn: Tschinkel, *Ind. Eng. Chem.* **48**, 732 (1956).

Pale yellow liquid. *Oxidizer; poisonous.* d_4^{25} 1.6229; d_4^{25} 1.638 (tech). mp 13.8°. bp_{760} 126°; $bp_{25.8}$ 40°; $bp_{14.9}$ 30°; $bp_{8.4}$ 20°; $bp_{5.7}$ 13.8°; $bp_{1.9}$ 0°. n_D^{20} 1.4384; n_D^{25} 1.4358. Viscosity at 20° = 1.76 cP. Insol in water. Freely sol in alcohol, ether, alcoholic KOH. Attacks iron, copper, brass, zinc, rubber. Highly explosive in the presence of impurities.

Caution: Potential symptoms of overexposure are irritation of eyes, skin, nose and throat; dizziness, headache; chest pain, dyspnea; methemoglobinuria, cyanosis; skin burns. *See NIOSH Pocket Guide to Chemical Hazards* (DHHS/NIOSH 97-140, 1997) p 304. *See also Patty's Industrial Hygiene and Toxicology* **vol. 2A**, G. D. Clayton, F. E. Clayton, Eds. (John Wiley & Sons, New York, 4th ed., 1994) 614-617. This substance is reasonably anticipated to be a human carcinogen: *Report on Carcinogens, Twelfth Edition* (PB2011-111646, 2011) p 402.

USE: Oxidizer in rocket propellants. As explosive in admixture with toluene. To increase cetane number of diesel fuels. Reagent for detecting the presence of double bonds in organic compds and for mild nitrations.

9382. Tetrantoin. [52094-70-9] 3',4'-Dihydrospiro[imidazolidine-4,2'(1'*H*)-naphthalene]-2,5-dione; 7,8-benzo-1,3-diazaspiro-

[4.5]decane-2,4-dione; S-2-676; Spirodon. $C_{12}H_{12}N_2O_2$; mol wt 216.24. C 66.65%, H 5.59%, N 12.96%, O 14.80%. Prepn: Novelli, *An. Farm. Bioquim.* **21**, 81 (1954), *C.A.* **50**, 4922 (1956); Faust *et al., J. Am. Pharm. Assoc.* **46**, 118 (1957); Jules *et al.,* **US 2716648** (1955 to Cutter Labs.).

Crystals from ethanol or glacial acetic acid, mp 267-268°.
THERAP CAT: Anticonvulsant.

9383. Tetraphenylarsonium Chloride. [507-28-8] Tetraphenylarsonium chloride (1:1); phenylarsonium chloride (Ph_4AsCl). $C_{24}H_{20}AsCl$; mol wt 418.80. C 68.83%, H 4.81%, As 17.89%, Cl 8.46%. Prepn: F. F. Blicke, E. Monroe, *J. Am. Chem. Soc.* **57**, 720 (1935). Use as analytical reagent: H. H. Willard, G. M. Smith, *Ind. Eng. Chem. Anal. Ed.* **11**, 186 (1939).

Crystals from absolute ether, mp 256-257°. *Poisonous.* Soly in water (30°): 32.50 g/100 ml. Sol in alcohol or methanol, sparingly sol in acetone.

Tetraphenylarsonium bromide. [507-27-7] Tetraphenylarsonium bromide. $C_{24}H_{20}AsBr$; mol wt 463.25. Long, colorless, glistening needles from water, mp 273-275°. *Poisonous.* Soly in water (25°): 1.29 g/100 ml. Sol in alcohol or methanol, sparingly sol in acetone.

USE: As a reagent for Cd, Hg, Zn, perchlorate, periodate, and other ions.

9384. Tetraphenylphosphonium. [18198-39-5] TPP^+. $[C_{24}H_{20}P]^+$. Membrane-permeable, lipophilic cation. Prepd as bromide: J. Dodonov, H. Medox, *Ber.* **61B**, 907 (1928); D. B. Denney, F. J. Gross, *J. Org. Chem.* **32**, 3710 (1967). As analytical reagent for identification of metals and their salts: H. H. Willard, L. R. Perkins, *Anal. Chem.* **25**, 1634 (1953); R. Neeb, *Z. Anal. Chem.* **154**, 17 (1957). Enhanced metal -TPP^+ complex determn by extraction-visible spectrophotometry: M. Siroki, L. Maric, *ibid.* **276**, 371 (1975). Solvent interactions: J. F. Coetzee, W. R. Sharpe, *J. Phys. Chem.* **75**, 3141 (1971). Use in determn of membrane potential in liposomes: K. Nakazato *et al., Biochim. Biophys. Acta* **946**, 143 (1988); in hepatocytes: S. Saito *et al., ibid.* **1111**, 221 (1992). NMR determn of TPP^+ driven potential: C. M. Franzin, P. M. Macdonald, *Biochemistry* **35**, 851 (1996).

USE: As analytical reagent for metals and their salts. As a probe for measuring membrane potential.

9385. Tetraphosphorus Trisulfide. [1314-85-8] 3,5,7-Trithia-1,2,4,6-tetraphosphatricyclo[2.2.1.0^2,^{6}]heptane; phosphorus sesquisulfide; trisulfurated phosphorus. P_4S_3; mol wt 220.08. P 56.30%, S 43.70%. Prepn by fusing red phosphorus with sulfur:

Stock, *Ber.* **43**, 150 (1910); also from white phosphorus and sulfur in a high-boiling solvent such as α-chloronaphthalene: Frary, **DE 309618** (1918); *Chem. Zentralbl.* **1919**, II, 55.

Yellowish-green, long, rhombic needles from benzene. Stable to air. d_4^{20} 2.03. mp 172.5°. bp 407.5°. Insol in cold water. Dec by hot water, yielding H_2S. Soly in carbon disulfide (20°): about 60% (w/w). Sol in benzene, similar hydrocarbons, phosphorus trichloride.

USE: In match tips.

9386. Tetrapropylammonium Perruthenate. [114615-82-6] *N,N,N*-Tripropyl-1-propanaminium (*T*-4)-tetraoxoruthenate(1−) (1:1); tetrapropylammonium tetraoxoruthenate; TPAP; (*n*-Pr₄N)-(RuO₄). $C_{12}H_{28}NO_4Ru$; mol wt 351.43. C 41.01%, H 8.03%, N 3.99%, O 18.21%, Ru 28.76%. Reagent for the mild oxidation of primary and secondary alcohols. Prepn and use in oxidation reactions: W. P. Griffith *et al.*, *J. Chem. Soc. Chem. Commun.* **1987**, 1625. Improved prepn: A. J. Bailey *et al.*, *Inorg. Chem.* **32**, 268 (1993). Kinetic studies: D. G. Lee *et al.*, *J. Org. Chem.* **57**, 3276 (1992). Use of oxygen as co-oxidant: R. Lenz, S. V. Ley, *J. Chem. Soc. Perkin Trans. 1* **1997**, 3291. Reviews: W. P. Griffith, S. V. Ley, *Aldrichim. Acta* **23**, 13-19 (1990); S. V. Ley *et al.*, *Synthesis* **1994**, 639-666; P. Langer, *J. Prakt. Chem.* **342**, 728-730 (2000).

Dark green solid. *Oxidizer. Irritant.* Sol in dichloromethane, acetonitrile; partially sol in benzene. Stable in air at room temp; small quantities dec when heated to 150-160°C. Hygroscopic.

USE: Catalyst and reagent in synthetic organic chemistry.

9387. Tetrasilane. [7783-29-1] Tetrasilicon decahydride; silicobutane; tetrasilicane; tetrasilicobutane. $H_{10}Si_4$; mol wt 122.42. H 8.23%, Si 91.77%. Si_4H_{10}. Prepn by the action of hydrochloric acid on magnesium silicide: Stock, Somiesky, *Ber.* **49**, 111 (1916); **54B**, 524 (1921); **56B**, 247 (1923); Stock *et al.*, *Ber.* **56**, 1695 (1923); Emeleus, Maddock, *J. Chem. Soc.* **1946**, 1131.

Liquid. mp ∼−90°; bp 109°; vapor pressure 7.8 mm Hg at 0° (Stock *et al.*, *loc. cit.*). mp −84.3°; bp (calc) 107.4°; vapor pressure 9.1 mm Hg at 0° (Emeleus, Maddock, *loc. cit.*). d^0 0.825. Dec at room temp; explodes in air. Reacts vigorously with CCl_4 and $CHCl_3$. Dec in water.

9388. Tetrasodium Pyrophosphate. [7722-88-5] Diphosphoric acid sodium salt (1:4); TSPP; pyro; sodium pyrophosphate. $Na_4O_7P_2$; mol wt 265.90. Na 34.58%, O 42.12%, P 23.30%. Na_4-P_2O_7. Available alkalinity as Na_2O 4.4%, total alkalinity 22.7%. Produced by molecular dehydration of dibasic sodium phosphate at 500°: Bell, *Inorg. Synth.* **3**, 98 (1950).

Crystals, d 2.534. mp 988°. Soly in water (g/100 ml) at 0°: 2.61; at 25°: 6.70; at 100°: 42.2. pH of a 1% soln = 10.2. Hydrolyzes to orthophosphate in aq soln, but the rate of hydrolysis is much slower than for the more acid pyrophosphate. No noticeable hydrolysis within 60 hrs at 70°: Bell, *Ind. Eng. Chem.* **39**, 136 (1947).

Decahydrate. Crystals, d 1.82. mp 79.5°. Slight efflorescence in dry air. Soly in water (g/100 ml) at 0°: 3.16; at 20°: 6.23; at 25°: 8.14; at 60°: 21.83; at 80°: 30.04. pH of 1% soln at 25° = 10.2. Insol in alc.

Caution: Potential symptoms of overexposure are irritation of eyes, skin, nose, throat; dermatitis. *See NIOSH Pocket Guide to Chemical Hazards* (DHHS/NIOSH 97-140, 1997) p 304.

USE: In cleansing compds, oil-well drilling, water treatment, cheese emulsification, as general sequestering agent, to remove rust stains, as ingredient of one-fluid ink eradicators, in electrodeposition of metals.

9389. Tetrasulfur Tetranitride. [28950-34-7] $1\lambda^4\delta^2, 5\lambda^4\delta^2$-1,3,5,7,2,4,6,8-Tetrathiatetrazocine; $1\lambda^4,3,5\lambda^4,7$-tetrathia-2,4,6,8-tetrazacycloocta-1,4,5,8-tetraene; tetranitrogen tetrasulfide; schwefelstickstoff. N_4S_4; mol wt 184.27. N 30.41%, S 69.59%. S_4N_4. Prepd by the interaction of disulfur dichloride and ammonia: Becke-Goehring, *Inorg. Synth.* **6**, 124 (1960).

Orange-red, monoclinic needles from benzene, mp 178°. Additional purification by sublimation in high vacuum, (bath temp 100°), mp 180°. bp_{760} ∼185°. Further heating results in deflagration and explosion. Practically insol in cold water, hydrolyzed by boiling water. Slightly sol in benzene, abs ethanol, carbon disulfide. *Handle with caution.* May dec explosively on striking or at temps much above 100°.

9390. Tetrathiafulvalene. [31366-25-3] 2-(1,3-Dithiol-2-ylidene)-1,3-dithiole; Δ 2,2′-bi-1,3-dithiole; bis-1,3-dithiole; 1,4,5,8-tetrathiafulvalene; TTF. $C_6H_4S_4$; mol wt 204.34. C 35.27%, H 1.97%, S 62.76%. Of interest in solid state chemistry as a conductor, catalyst or sensor due to its unusual electronic and magnetic properties. Primarily used as the parent compound for supramolecular assemblies. Prepn: F. Wudl *et al.*, *Chem. Commun.* **1970**, 1453. In combination with TCNQ (*tetracyano-p-quinodimethane*) to form first "organic metal": J. Ferraris *et al.*, *J. Am. Chem. Soc.* **95**, 948 (1973); electronic structure: R. Gleiter *et al.*, *J. Electron. Spectrosc. Relat. Phenom.* **2**, 207 (1973). Use as a catalyst in radical reactions: R. J. Fletcher *et al.*, *J. Chem. Soc. Perkin Trans. 1* **1995**, 623; J. A. Murphy, S. J. Roome, *ibid.* 1349. TTF-mediated biosensors for glucose: T. Yu *et al.*, *J. Appl. Polym. Sci.* **58**, 973 (1995); for NADH: X. Zhang *et al.*, *Anal. Commun.* **33**, 111 (1996); for gln and glu: A. Mulchandani, A. S. Bassi, *Biosens. Bioelectron.* **11**, 271 (1996). *Review:* T. Jorgensen *et al.*, *Chem. Soc. Rev.* **23**, 41-51 (1994). Review of electrical properties and structural data of unsymmetric TTF cmpds: J. M. Fabre *et al.*, *Synth. Met.* **35**, 57-64 (1990); of role in superconductors: M. R. Bryce, *J. Mater. Chem.* **5**, 1481-1496 (1995).

Orange solid in its neutral state; loses electrons to form purple cation radical and subsequent yellow dication. Also reported as yellow solid, mp 118.5-119°. Sublimes 100°, 0.3 mm. uv max (CH_2-Cl_2): 290, 310 nm (ε 4 × 10^4, 4 × 10^4). Insol in water. Readily photo-oxidized in air to violet water-soluble radical cation. The cation is a deep purple crystalline solid, mp 155-165° (dec). uv max (H_2O): 250 nm (ε 4.7 × 10^4); absorption max (H_2O): 340, 405, 435, 575 nm (ε 4.4 × 10^4; 3.6 × 10^4; 5.6 × 10^4; 1.6 × 10^4). A counter ion is necessary to prevent further oxidation to the dication. Ethanolic solutions of the cation are stable for prolonged periods of time (>48hr).

USE: Molecular sensors; radical catalyst.

9391. Tetrazepam. [10379-14-3] 7-Chloro-5-(1-cyclohexen-1-yl)-1,3-dihydro-1-methyl-2*H*-1,4-benzodiazepin-2-one; 7-chloro-5-(1-cyclohexenyl)-1-methyl-2-oxo-2,3-dihydro-1*H*-[1,4]benzo[*f*]-diazepine; CB-4261; Musaril; Muskelat; Myolastan. $C_{16}H_{17}ClN_2$-O; mol wt 288.78. C 66.55%, H 5.93%, Cl 12.28%, N 9.70%, O 5.54%. Prepn: J. Schmitt, **NL 6600095**; *idem*, **US 3426014**; *idem*, **US 3551412** (1966, 1969, 1970 to Clin-Byla); J. Schmitt *et al.*, *Chim. Ther.* **2**, 254 (1967). Spectroscopic and chromatographic studies: Lafargue *et al.*, *Ann. Pharm. Fr.* **28**, 343, 477 (1970). GC-MS determn in urine: H. Maurer, K. Pfleger, *J. Chromatogr.* **422**, 85 (1987). Pharmacokinetics: H. Bun *et al.*, *Arzneim.-Forsch.* **37**, 199 (1987). Pharmacology and benzodiazepine receptor study: P. E. Keane *et al.*, *J. Pharmacol. Exp. Ther.* **245**, 692, 699 (1988). Clinical evaluation in post-stroke spasticity: I. Milanov, *Acta Neurol. Belg.* **92**, 5 (1992).

Yellow-brown crystals from ethyl acetate, mp 144°. uv max (ethanol): 227 nm (ε 28500). LD_{50} in mice (mg/kg): 415 i.p.; 2000 orally (Schmitt).

Note: This is a controlled substance (depressant): **21 CFR,** 1308.14.

THERAP CAT: Muscle relaxant (skeletal).

9392. Tetrazolium Blue. [1871-22-3] 2,2'-(3,3'-Dimethoxy[1,1'-biphenyl]-4,4'-diyl)bis[3,5-diphenyl-2*H*-tetrazolium] chloride (1:2); 3,3'-dianisolebis[4,4'-(3,5-diphenyl)tetrazolium chloride]; blue tetrazolium; dimethoxy neotetrazolium; ditetrazolium chloride; BT. $C_{40}H_{32}Cl_2N_8O_2$; mol wt 727.65. C 66.03%, H 4.43%, Cl 9.74%, N 15.40%, O 4.40%. Prepn: A. M. Rutenburg *et al., Cancer Res.* **10**, 113 (1950); L. J. Pannone, J. B. Rust, US **2713581** (1955 to Montclair Res. Corp. and Ellis-Foster Co.).

Lemon-yellow crystals, dec 242-245°. Freely sol in methanol, ethanol, chloroform. Slightly sol in water. Insol in ethyl acetate, acetone, ether. Reduction potential about −0.08 volt. Yields a dark blue diformazan pigment in the presence of a reducing agent.

USE: For research in seed germination, as stain for bacteria and molds, in histochemical studies, to demonstrate oxidation-reduction enzymes in normal and cancerous tissues. *See also* Triphenyltetrazolium Chloride.

9393. Tetrin. Polyene antifungal antibiotic produced by Streptomyces *Illinois* #155-2: Pote, *Diss. Abstr.* **19**, 2778 (1959); Gottlieb, Pote, *Phytopathology* **50**, 817 (1960). Isoln of the two tetraenes, tetrin A and B: Rinehart *et al., Ann.* **668**, 77 (1963); German, *Diss. Abstr.* **25**, 97 (1964). Structure of tetrin A: Pandey *et al., J. Am. Chem. Soc.* **93**, 3738 (1971); of tetrin B: Rinehart *et al., ibid.* 3747. Revised structure: R. C. Pandey, K. L. Rinehart, *J. Antibiot.* **29**, 1035 (1976). Mode of action of tetrin A: van Etten, Gottlieb, *J. Gen. Microbiol.* **46**, 377 (1967).

Tetrin A   R = H
Tetrin B   R = OH

Tetrin A. $C_{34}H_{51}NO_{13}$; mol wt 681.78. Fine, colorless needles from methanol or aqueous *n*-butanol, mp >350° (dec). $[\alpha]_D^{28}$ +8.3° (c = 0.72 in pyridine). $[\alpha]_D^{28}$ +27.5° (c = 1.0 in pyridine). uv max: 214, 278, 290, 303, 318 nm ($E_{1cm}^{0.1\%}$ 19.4, 44.2, 81.2, 115.0, 110.9). Monobasic, pKa' 8.30 in 60% ethanol. Sol in pyridine, dil alkalies, dil mineral acids; moderately sol in lower alcohols; practically insol in acetone, ether, water.

Tetrin B. $C_{34}H_{51}NO_{14}$; mol wt 697.78. Brown, amorphous powder, mp >360° (darkens at 160-165°, blackens at 250-295°). $[\alpha]_D^{24}$ +43.5° (c = 0.14 in methanol); $[\alpha]_D^{28}$ +45° (c = 0.3 in pyridine). uv max: 214, 278, 290, 303, 318 nm ($E_{1cm}^{0.1\%}$ 18.6, 51.4, 80.1, 112.8, 108.9). Readily sol in ethanol + water, dioxane + water; fairly sol in water, lower alcohols, dioxane, pyridine, dimethyl sulfoxide; slightly sol in acetone. Practically insol in ethyl acetate, chloroform, ether, ethylene dichloride.

9394. Tetrodotoxin. [4368-28-9] (4*R*,4a*R*,5*R*,7*S*,9*S*,10*S*,-10a*R*,11*S*,12*S*)-2-Amino-1,4,4a,5,9,10-hexahydro-12-(hydroxymethyl)-5,9:7,10a-dimethano-10a*H*-[1,3]dioxocino[6,5-*d*]pyrimidine-4,7,10,11,12-pentol; maculotoxin; spheroidine; tarichatoxin; tetrodontoxin; fugu poison; TTX. $C_{11}H_{17}N_3O_8$; mol wt 319.27. C

41.38%, H 5.37%, N 13.16%, O 40.09%. Toxin from the ovaries and liver of many species of *Tetraodontidae*, esp the globe fish (*Spheroides rubripes*): Yokoo, *J. Chem. Soc. Jpn.* **71**, 590 (1950), *C.A.* **45**, 6759c (1951). Identity with tarichatoxin: Buchwald *et al., Science* **143**, 474 (1963); with maculotoxin: D. D. Sheumack *et al., ibid.* **199**, 188 (1978). Structure studies: Goto *et al., Tetrahedron Lett.* **1963**, 2105, 2115; **1964**, 779, 1831. Structure: Woodward, *Pure Appl. Chem.* **9**, 49 (1964); Tsuda *et al., Chem. Pharm. Bull.* **12**, 1357 (1964); Goto *et al., Tetrahedron* **21**, 2059 (1965). Synthetic studies: Kishi *et al., Tetrahedron Lett.* **1970**, 5127, 5129. Total synthesis: Kishi *et al., J. Am. Chem. Soc.* **94**, 9219 (1972). Pharmacology: Evans, *Br. Med. Bull.* **25**, 263 (1969); Kao, *Fed. Proc.* **31**, 1117 (1972). Mechanism of action: Narahashi, *ibid.* 1124. Toxicity study: C. Y. Kao, F. A. Fuhrman, *J. Pharmacol. Exp. Ther.* **140**, 31 (1963). *Review:* Scheuer, *Fortschr. Chem. Org. Naturst.* **22**, 265 (1964); Mosher *et al., Science* **144**, 1100 (1964); Kao, *Pharmacol. Rev.* **18**, 997 (1966); Evans, *Int. Rev. Neurobiol.* **15**, 83 (1972); of total systheses: U. Koert, *Angew. Chem. Int. Ed.* **43**, 5572-5576 (2004).

Darkens above 220° without dec. $[\alpha]_D^{25}$ −8.64° (c = 8.55 in dil acetic acid). pKa: 8.76 (water); 9.4 (50% alc). Sol in dil acetic acid; slightly sol in water, dry alc, ether. Practically insol in other organic solvents. Toxin destroyed in strong acids and in alkaline solns. LD$_{50}$ i.p. in mice: 10 μg/kg (Kao, Fuhrman).

9395. Tetrofosmin. [127502-06-1] 6,9-Bis(2-ethoxyethyl)-3,12-dioxa-6,9-diphosphatetradecane; ethylenebis[bis(2-ethoxyethyl)phosphine]; P53. $C_{18}H_{40}O_4P_2$; mol wt 382.46. C 56.53%, H 10.54%, O 16.73%, P 16.20%. Prepn: J. D. Kelly *et al.,* **EP 337654**; *eidem,* **US 5045302** (1989, 1991 both to Amersham). Pharmacology and determn of radiochemical purity: *idem et al., J. Nucl. Med.* **34**, 222 (1993). Clinical biodistribution: B. Higley *et al., ibid.* 30. Clinical trial as a myocardial perfusion imaging agent: B. L. Zaret *et al., Circulation* **91**, 313 (1995).

Complex with ^{99m}Tc. [127455-27-0] Technetium Tc 99m tetrofosmin; technetium-99m tetrofosmin; [^{99m}Tc(tetrofosmin)$_2$O$_2$]$^+$; PPN1011; Myoview. $C_{36}H_{80}O_{10}P_4$^{99m}Tc.

THERAP CAT: ^{99m}Tc complex as diagnostic aid (radioactive imaging agent).

9396. Tetronasin. [75139-06-9] 4-Hydroxy-3-[(2*S*)-2-[(1*S*,-2*S*,6*R*)-2-[(1*E*)-3-hydroxy-2-[(2*R*,3*R*,6*S*)-tetrahydro-3-methyl-6-[(1*E*,3*S*)-3-[(2*R*,3*S*,5*R*)-tetrahydro-5-[(1*S*)-1-methoxyethyl]-3-methyl-2-furanyl]-1-buten-1-yl]-2*H*-pyran-2-yl]-1-propen-1-yl]-6-methylcyclohexyl]-1-oxopropyl]-2(5*H*)-furanone; antibiotic M139603; ICI-139603; M-139603. $C_{35}H_{54}O_8$; mol wt 602.81. C 69.74%, H 9.03%, O 21.23%. Polyether antibiotic produced by *Streptomyces longisporoflavus* NCIB 11426. Possesses a biosynthetically rare acid grouping in the form of an acyl tetronic acid moiety. Isoln and use in ruminants: D. H. Davies, G. L. F. Norris, **GB 2027013**; *eidem,* **US 4279894** (1980, 1981 both to ICI). Physical data and crystal structure: D. H. Davies *et al., Chem. Commun.* **1981**, 1073. Solution conformation and cation-binding properties: J. Grandjean, P. Laszlo, *Tetrahedron Lett.* **24**, 3319 (1983). Synthetic studies: A. M. Doherty, S. V. Ley, *ibid.* **27**, 105 (1986). Biosynthetic studies: J. M. Bulsing *et al., Chem. Commun.* **1984**, 1301; D. M. Doddrell *et al., ibid.* 1302; A. K. Demetriadou *et al., ibid.* **1985**, 408. Antimicrobial activity: C. J. Newbold *et al., Appl. Environ. Microbiol.* **54**, 544

(1988). Effect on gain efficiency in cattle: S. J. Bartle *et al.*, *J. Anim. Sci.* **66**, 1502 (1988).

Sodium salt. $C_{35}H_{53}NaO_8$. mp 176-178°. $[\alpha]_D^{23}$ −82° (c = 0.2 in methanol). uv max (ethanol): 234, 270 nm (ε 13000, 11000). pKa 1.8 ±0.3 (methanol/H$_2$O 1:9). Sol in most organic solvents. Insol in water.

THERAP CAT (VET): Ruminant performance enhancer.

9397. Tetroquinone. [319-89-1] 2,3,5,6-Tetrahydroxy-2,5-cyclohexadiene-1,4-dione; tetrahydroxy-*p*-benzoquinone; tetrahydroxyquinone; THQ; HPEK-1; NSC-112931. $C_6H_4O_6$; mol wt 172.09. C 41.88%, H 2.34%, O 55.78%. Prepn from glyoxal: A. J. Fatiadi, W. F. Sager, *Org. Synth.* **coll. vol. V**, 1011 (1973). Clinical evaluation for ophthalmic fibrosis: R. Ching, *Eye Ear Nose Throat Mon.* **50**, 101, 144 (1971).

Dark, odorless, tasteless crystals with blue-black metallic luster in reflected light. mp 280°. Sol (1:200) in cold water; sol (1:100) in hot water or in alcohol; slightly sol in ether. It acts like a strong dibasic acid.

Disodium salt. $Na_2C_6H_2O_6$. Almost black crystals with a green metallic luster, sparingly sol in water.

USE: Disodium salt is used as an indicator in the volumetric determination of sulfate by means of barium chloride solution.

9398. Tetroxoprim. [53808-87-0] 5-[[3,5-Dimethoxy-4-(2-methoxyethoxy)phenyl]methyl]-2,4-pyrimidinediamine; 2,4-diamino-5-[3,5-dimethoxy-4-(2-methoxyethoxy)benzyl]pyrimidine; HE-781. $C_{16}H_{22}N_4O_4$; mol wt 334.38. C 57.47%, H 6.63%, N 16.76%, O 19.14%. Analog of trimethoprim, *q.v.* Prepn: W. Liebenow, J. Prikryl, **FR 2221147**; *eidem*, **US 3992379** (1974, 1976 both to Heumann). Series of articles on tetroxoprim and other antibacterial folate inhibitors: *J. Antimicrob. Chemother.* **5**, Suppl. B, 1-239 (1979); on synthesis of radioactive tetroxoprim, tissue distribution, kinetics, HPLC determn: *Arzneim.-Forsch.* **30**, 307-319 (1980). Effect on bacterial growth kinetics: J. K. Seydel, E. Wempe, *Chemotherapy* **26**, 361 (1980). *In vitro* activity of the tetroxoprim-sulfadiazine combination: H. Hahn, A. Kirov, *Arzneim.-Forsch.* **30**, 1047 (1980).

Crystals from water, mp 153-156° (Liebenow, Prikryl); also reported as mp 160.1° (*eidem*, *J. Antimicrob. Chemother.* **5**, Suppl B, 15 (1979)). Soly at 30° (mg/ml): water 2.65; chloroform 69; *n*-octanol 1.61. pKb 8.25. LD$_{50}$ orally in rats: 1357 mg/kg (Liebenow, Prikryl).

Mixture with sulfadiazine. [73173-12-3] Co-tetroxazine; Biroxin; Sterinor; Tibirox; Troximin.

THERAP CAT: Antibacterial.

9399. Texanol Isobutyrate. [6846-50-0] 2-Methylpropanoic acid 1,1′-[2,2-dimethyl-1-(1-methylethyl)-1,3-propanediyl] ester; 2,-2,4-trimethyl-1,3-pentanediol diisobutyrate; TXIB. $C_{16}H_{30}O_4$; mol wt 286.41. C 67.10%, H 10.56%, O 22.34%. Plasticizer for PVC and other resins. Prepn: U. Schwenk, A. Becker, *Ann.* **756**, 162 (1972). GC determn in PVC plastisol formulations: W. P. Hayes *et al.*, *J. Chromatogr.* **139**, 395 (1977). Toxicology: B. D. Astill *et al.*, *Toxicol. Appl. Pharmacol.* **22**, 387 (1972). Technical data sheet: "Eastman TXIB Plasticizer" (Eastman Chemical Co., publication L-151K, April 1998) 18 pp.

Colorless liquid, slightly fruity odor. bp$_{760}$ 281.5°. fp −70°. Flash pt, closed cup: 262°F (128°C). d$_{20}^{20}$ 0.942-0.948. n_D^{25} 1.4300. Soly in water (20°): 0.42 g/l. Viscosity (25°): 9 cP.

USE: In mfr of vinyl flooring, toys and other vinyl products.

9400. Tezosentan. [180384-57-0] *N*-[6-(2-Hydroxyethoxy)-5-(2-methoxyphenoxy)-2-[2-(1*H*-tetrazol-5-yl)-4-pyridinyl]-4-pyrimidinyl]-5-(1-methylethyl)-2-pyridinesulfonamide; 5-isopropyl-pyridine-2-sulfonic acid 6-(2-hydroxyethoxy)-5-(2-methoxyphenoxy)-2-(2-1*H*-tetrazol-5-ylpyridin-4-yl)pyrimidin-4-ylamide. $C_{27}H_{27}N_9O_6S$; mol wt 605.63. C 53.55%, H 4.49%, N 20.82%, O 15.85%, S 5.29%. Dual endothelin (ET$_A$/ET$_B$) receptor antagonist. Prepn: V. Breu *et al.*, **WO 9619459**; *eidem*, **US 6004965** (1996, 1999 both to Hoffmann-La Roche). Pharmacology and receptor binding: M. Clozel *et al.*, *J. Pharmacol. Exp. Ther.* **290**, 840 (1999). Clinical pharmacokinetics: J. Dingemanse *et al.*, *Br. J. Clin. Pharmacol.* **53**, 355 (2002). Clinical evaluation in advanced heart failure: C. Schalcher *et al.*, *Am. Heart J.* **142**, 340 (2001). LC-MS determn in plasma: P. L. M. van Giersbergen *et al.*, *J. Chromatogr. B* **792**, 369 (2003).

White solid from acetonitrile, mp 198-200°.

Sodium salt. [180384-58-1] Ro-61-0612; Veletri. $C_{27}H_{25}N_9Na_2O_6S$; mol wt 649.59. White powder from sodium methylate. pKa$_1$ 4.4; pKa$_2$ 4.1. Sol in water.

THERAP CAT: In treatment of congestive heart failure.

9401. TFM. [88-30-2] 4-Nitro-3-(trifluoromethyl)phenol; α,α,α-trifluoro-4-nitro-*m*-cresol; 3-trifluoromethyl-4-nitrophenol; HOE-2770; Lamprecid. $C_7H_4F_3NO_3$; mol wt 207.11. C 40.60%, H 1.95%, F 27.52%, N 6.76%, O 23.17%. A substance toxic to the parasitic sea lamprey, *Petromyzon marinus*, which preys upon commercial fish species of the Great Lakes: Scherer *et al.*, **DE 1068505** *C.A.* **55**, 9774d (1961) corresp to **US 3157571** (1959, 1964, both to Hoechst). Toxicological studies in fish, and bibliography: Kawatski, McDonald, *Comp. Gen. Pharmacol.* **5**, 67 (1974). Physical properties: Smith, *J. Chem. Eng. Data* **6**, 607 (1961). *Review*: Schnick, *Investigations in Fish Control*, no. 44 (Sport Fisheries and Wildlife Bureau) 31 pp.

mp 76°. pK 6.07. uv max (acid): 280 nm (ε 1930); in 1% NaOH: 300, 395 nm (ε 4650, 13,130); in 95% ethanol: 290 nm (ε 14,700).

USE: Lamprey killer.

9402. Thalicarpine. [5373-42-2] (6aS)-9-[4,5-Dimethoxy-2-[[(1S)-1,2,3,4-tetrahydro-6,7-dimethoxy-2-methyl-1-isoquinolinyl]-methyl]phenoxy]-5,6,6a,7-tetrahydro-1,2,10-trimethoxy-6-methyl-4H-dibenzo[de,g]quinoline. $C_{41}H_{48}N_2O_8$; mol wt 696.84. C 70.67%, H 6.94%, N 4.02%, O 18.37%. Tumor-inhibitory alkaloid from *Thalictrum dasycarpum* Fisch. & Lall., *Ranunculaceae:* Kupchan *et al., J. Pharm. Sci.* **52**, 985 (1963). Structure: Kupchan, Yokoyama, *J. Am. Chem. Soc.* **86**, 2177 (1964); Tomita *et al., Tetrahedron Lett.* **1965**, 4309. Total synthesis: Kupchan, Liepa, *Chem. Commun.* **1971**, 599; Kupchan *et al., J. Am. Chem. Soc.* **95**, 2995 (1973). Pharmacology: Herman, Chadwick, *Toxicol. Appl. Pharmacol.* **26**, 137 (1973); S. M. Sieber *et al., Cancer Treat. Rep.* **60**, 1127 (1976). Antimicrobial and hypotensive activity: W. N. Wu *et al., Lloydia* **40**, 508 (1977). Biosynthesis: D. S. Bhakuni, S. Jain, *Tetrahedron* **38**, 729 (1982).

Needles from ethyl acetate, mp 160-161°. $[\alpha]_D^{25}$ +133° (c = 0.83 in methanol); $[\alpha]_D^{25}$ +89° (c = 0.88 in chloroform). uv max (methanol): 282, 302 nm (ε 17000, 13000).

9403. Thalidomide. [50-35-1] 2-(2,6-Dioxo-3-piperidinyl)-1H-isoindole-1,3(2H)-dione; N-(2,6-dioxo-3-piperidyl)phthalimide; α-phthalimidoglutarimide; 3-phthalimidoglutarimide; 2,6-dioxo-3-phthalimidopiperidine; N-phthalylglutamic acid imide; N-phthaloyl-glutamimide; K-17; Contergan; Neurosedyn; Softenon; Thalomid. $C_{13}H_{10}N_2O_4$; mol wt 258.23. C 60.47%, H 3.90%, N 10.85%, O 24.78%. Selective inhibitor of tumor necrosis factor α (TNF-α). Formerly used as sedative, hypnotic. Prepn: *GB 768821* (1957 to Chemie Grünenthal). Teratogenicity studies: I. D. Fratta *et al., Toxicol. Appl. Pharmacol.* **7**, 268 (1965). Review of proposed mechanisms of embryopathy: T. D. Stephens, *Teratology* **38**, 229-239 (1988). HPLC determn in plasma: A. Delon *et al., J. Liq. Chromatogr.* **18**, 297 (1995). Stereospecific determn and pharmacokinetics of enantiomers: T. Eriksson *et al., Chirality* **7**, 44 (1995). Clinical trial in HIV wasting syndrome: G. Reyes-Terán *et al., AIDS* **10**, 1501 (1996); in aphthous ulcers related to HIV infection: J. M. Jacobson *et al., N. Engl. J. Med.* **336**, 1487 (1997). Review of chemistry, pharmacokinetics and clinical safety: V. Günzler, *Drug Saf.* **7**, 116-134 (1992); of effect on TNF-α and clinical use in leprosy and tuberculosis: J. D. Klausner *et al., Clin. Immunol. Immunopathol.* **81**, 219-223 (1996); of pharmacology, history and potential clinical uses: D. Stirling *et al., J. Am. Pharm. Assoc.* **NS37**, 307-313 (1997); of use in multiple myeloma: S. V. Rajkumar, *Expert Rev. Anticancer Ther.* **1**, 20-28 (2001); of pharmacology and toxicology: C. Meierhofer, C. J. Wiedermann, *Curr. Opin. Drug Discov. Devel.* **6**, 92-99 (2003); of clinical experience: S. J. Matthews, C. McCoy, *Clin. Ther.* **25**, 342-395 (2003).

Needles, mp 269-271°. uv max (neutral soln): 220, 300 nm. Soly in water: ~2×10^{-4} mol/L; 45-60 mg/L. Sparingly sol in water, methanol, ethanol, acetone, ethyl acetate, butyl acetate, glacial acetic

acid. Very sol in dioxane, DMF, pyridine. Practically insol in ether, chloroform, benzene.

THERAP CAT: Immunomodulator.

9404. Thallium. [7440-28-0] Tl; at. wt [204.382; 204.385]; conventional at. wt 204.38; at. no. 81; valence 1, 3. Group IIIA(13). Naturally occurring isotopes: 203 (29.50%), 205 (70.50%); artificial, radioactive isotopes: 191-202; 204; 206-210. Occurs in crookesite, $(Cu,Tl,Ag)_2Se$, found in Sweden; in lorandite, $TlAsS_2$, found in Greece; in hutchinsonite, $(Tl,Cu,Ag)_2S.PbS.2As_2S_3$, found in Switzerland. Occurrence in the earth's crust: 0.7 ppm. Discovered by Crookes in 1861. Prepn: Sanderson, *Can. Mining J.* **65**, 624 (1944). Use in organic syntheses: McKillop *et al., Tetrahedron Lett.* **1970**, 5281; Taylor *et al., ibid.* 5285. *Review:* Wade, Banister in *Comprehensive Inorganic Chemistry* vol. 1, J. C. Bailar, Jr. *et al.,* Eds. (Pergamon Press, Oxford, 1973) pp 997-1000, 1119-1172. Review of toxicity: E. Browning, *Toxicity of Industrial Metals* (Appleton-Century-Crofts, New York, 2nd ed., 1969) pp 317-322; of toxicology and human exposure: *Toxicological Profile for Thallium* (PB93-110856, 1992) 114 pp.

Bluish-white, very soft, inelastic, easily fusible, heavy metal; leaves a streak on paper. Oxidizes superficially in air forming a coating of Tl_2O. Forms alloys with other metals and readily amalgamates with mercury. d 11.85. Begins to volatilize at 174°. mp 303.5°. bp 1457°. Specific heat at 20° 0.031 cal/g/°C. Latent heat of fusion 5.04 cal/g. Brinell hardness: 2. May be distilled in a stream of hydrogen. Insol in water; reacts with nitric or sulfuric acid; difficultly with hydrochloric acid.

Caution: Potential symptoms of overexposure are nausea, diarrhea, abdominal pain and vomiting; ptosis, strabismus; peripheral neuritis, tremor; retrosternal tightness, chest pain and pulmonary edema; seizure, chorea and psychosis; liver and kidney damage; alopecia; paresthesia of legs. *See NIOSH Pocket Guide to Chemical Hazards* (DHHS/NIOSH 97-140, 1997) p 304. *See also Patty's Industrial Hygiene and Toxicology* vol. 2A, G. D. Clayton, F. E. Clayton, Eds. (Wiley-Interscience, New York, 3rd ed., 1981) 1914-1931.

USE: In semi-conductor industry; alloyed with mercury for switches and closures which operate at subzero temps. In manuf of highly refractive optical glass. Has been used in admixture with 97-98% of inert substances as poison for rats and other rodents.

9405. Thallium Acetate. [563-68-8] Acetic acid thallium(1+) salt (1:1); thallous acetate. $C_2H_3O_2Tl$; mol wt 263.42. C 9.12%, H 1.15%, O 12.15%, Tl 77.59%. Toxicity study: W. L. Downs *et al., Am. Ind. Hyg. Assoc. J.* **21**, 399 (1960). Brief review: *Dangerous Prop. Ind. Mater. Rep.* **7**, 92-94 (1987).

White, deliquesc crystals, mp 131°. Specific gravity 3.68. *Poisonous.* Sol in water, alcohol. *Keep well closed.* LD$_{50}$ in female rats (mg Tl/kg): 23 i.p., 32 orally (Downs).

USE: Rodenticide, insecticide; fireworks dye; in optical glass; depilatory.

9406. Thallium(III) Acetate. [2570-63-0] Acetic acid thallium(3+) salt (3:1); thallic acetate; thallium triacetate; tris(acetato)-thallium; $Tl(OAc)_3$. $C_6H_9O_6Tl$; mol wt 381.51. C 18.89%, H 2.38%, O 25.16%, Tl 53.57%. $(CH_3COO)_3Tl$. Oxidizing reagent for numerous functional groups. Prepn: R. J. Meyer, E. Goldschmidt, *Ber.* **36**, 238 (1903). Improved prepn: J. K. Kochi, T. W. Bethea, III, *J. Org. Chem.* **33**, 75 (1968). Crystal structure: R. Faggiani, I. D. Brown, *Acta Crystallogr. B* **34**, 2845 (1978). Utility in oxidation chemistry: H.-J. Kabbe, *Ann.* **656**, 204 (1962); M. E. Kuehne, T. J. Giacobbe, *J. Org. Chem.* **33**, 3359 (1968); S. Uemura *et al., Tetrahedron* **37**, 291 (1981). Acute toxicity studies: C.-H. Lan, T.-S. Lin, *Ecotoxicol. Environ. Saf.* **61**, 432 (2005). *Review:* A. K. Banerjee in *Encyclopedia of Reagents for Organic Synthesis* **7**, L. A. Paquette, Ed. (Wiley, New York, 1995) pp 4833-4837.

White crystalline solid. *Poisonous.* Begins slow decompn ca. 75-80°. d 2.57. Sol in water, alcohols, hot acetic acid. Insol in benzene. Moisture sensitive. Protect from light. LC$_{50}$ (48 hr) in *Daphnia magna*: 203 µg/l (Lan, Lin).

USE: Reagent in synthetic organic chemistry.

9407. Thallium Bromide. [7789-40-4] Thallium bromide (TlBr); thallous bromide. BrTl; mol wt 284.28. Br 28.11%, Tl 71.89%. TlBr.

Pale yellow, cryst powder. *Poisonous*. d 7.5. mp ~460°. Sol in 2360 parts water.

9408. Thallium Carbonate. [6533-73-9] Carbonic acid thallium(1+) salt (1:2); thallous carbonate. CO_3Tl_2; mol wt 468.77. C 2.56%, O 10.24%, Tl 87.20%. Tl_2CO_3.

White crystals. *Poisonous*. d 7.1. mp 272°. Soluble in 24 parts water, 3.7 parts boiling water. Insol in alc.

USE: Manuf imitation diamonds.

9409. Thallium Chloride. [7791-12-0] Thallium chloride (TlCl); thallous chloride. ClTl; mol wt 239.83. Cl 14.78%, Tl 85.22%. TlCl.

White, cryst powder. *Poisonous*. d 7.0. mp 430°. Sol in 260 parts cold water, 70 parts boiling water. Insol in alcohol. HCl decreases its soly in water.

USE: As catalyst in chlorinations.

9410. Thallium Cyanide. [13453-34-4] Thallium cyanide (Tl(CN)); thallous cyanide. CNTl; mol wt 230.40. C 5.21%, N 6.08%, Tl 88.71%. TlCN. Prepn: *Gmelins, Thallium* (8th ed.) **38**, 390 (1940); E. C. Taylor *et al.*, *J. Org. Chem.* **43**, 2280 (1978).

White, hexagonal platelets. d 6.523. Soly in water: 16.8 g/100 ml. Sol in alcohol, acid. Aqueous soln is alkaline.

USE: In organic synthesis.

9411. Thallium Fluoride. [7789-27-7] Thallium fluoride (TlF); thallous fluoride. FTl; mol wt 223.38. F 8.50%, Tl 91.49%. TlF. Prepared from Tl_2CO_3 and HF: Ketelaar, *Z. Kristallogr.* **92**, 30 (1935); Hayek, *Z. Anorg. Allg. Chem.* **225**, 47 (1935); Barrow *et al.*, *Trans. Faraday Soc.* **51**, 1650 (1955); from Tl and HF: Keneshea, Cubicciotti, *J. Phys. Chem.* **69**, 3910 (1965); Tranquard, *Bull. Soc. Chim. Fr.* **1967**, 2578.

Hard, shiny, crystals which deliquesce when breathed upon, but which resolidify immediately in dry air. Not hygroscopic in the usual sense. Orthorhombic (deformed NaCl lattice). d_4^{25} 8.36. mp 322°. Begins to sublime at 300°. Very freely sol in water. Concd solns show strong alkalinity.

USE: In the prepn of fluoro esters.

9412. Thallium Hydroxide. [12026-06-1] Thallium hydroxide (Tl(OH)); thallous hydroxide. HOTl; mol wt 221.39. H 0.46%, O 7.23%, Tl 92.32%. TlOH.

Yellow needles. *Poisonous*. Very sol in water; the soln is strongly alkaline and turns turmeric paper brown.

9413. Thallium Iodide. [7790-30-9] Thallium iodide (TlI); thallous iodide. ITl; mol wt 331.28. I 38.31%, Tl 61.69%. TlI.

Yellow, cryst powder. *Poisonous*. d 7.1; mp 440°; bp 824°. Almost insol in water. Insol in alcohol. Sol in KI soln.

9414. Thallium Mononitrate. [10102-45-1] Nitric acid thallium(1+) salt (1:1); thallium(I) nitrate; thallous nitrate. NO_3Tl; mol wt 266.38. N 5.26%, O 18.02%, Tl 76.72%. $TlNO_3$. Toxicity study: M. W. Williams *et al.*, *Toxicol. Appl. Pharmacol.* **63**, 461 (1982).

White crystals. *Poisonous*. d 5.55. mp 206°; dec at 450°. Sol in 10 parts cold water, 0.3 part boiling water. Insol in alcohol. LD_{50} i.p. in mice: 0.14 mmol/kg (Williams).

USE: As a reagent in analytical chemistry, esp for the determination of iodine in presence of Br and Cl; also with $KClO_3$, $HgCl$ and resin for green fire for signalling at sea.

9415. Thallium Oxide. [1314-12-1] Thallium oxide (Tl_2O); thallous oxide. OTl_2; mol wt 424.76. O 3.77%, Tl 96.23%. Tl_2O.

Black powder. *Poisonous*. mp about 300°. Sol in water, forming the hydroxide; also sol in alcohol. On exposure to air it gradually oxidizes to thallic oxide and becomes insol.

USE: In manuf of glass of a high coefficient of refraction for optical purposes (thallium flint glass) and for artificial gems.

9416. Thallium Selenide. [15572-25-5] Thallium selenide (Tl_2Se); thallous selenide. $SeTl_2$; mol wt 487.72. Se 16.19%, Tl 83.81%. Tl_2Se. Prepared by the action of hydrogen selenide on a soln of thallous carbonate: Kuhlmann, *Bull. Soc. Chim.* [2] **1**, 330 (1864).

Dark gray plates with a metallic luster. mp 340° (Kuhlmann); mp 338° [Palabon, *Compt. Rend.* **145**, 118 (1907); **173**, 142 (1921)]. Insol in water and acids.

9417. Thallium Sesquioxide. [1314-32-5] Thallium oxide (Tl_2O_3); thallic oxide; thallium peroxide. O_3Tl_2; mol wt 456.76. O 10.51%, Tl 89.49%. Tl_2O_3.

Brown powder. d 9.65. mp 717°. Insol in water. Dec by HCl with evolution of chlorine and by H_2SO_4 with evolution of oxygen.

9418. Thallium Sulfate. [7446-18-6] Sulfuric acid thallium-(1+) salt (1:2); thallous sulfate; Eccothal. O_4STl_2; mol wt 504.82. O 12.68%, S 6.35%, Tl 80.97%. Tl_2SO_4.

White, rhomboid prisms. *Poisonous*. d 6.77; mp 632°. Soly in 100 ml water at 0°: 2.70 g; at 20°: 4.87 g; at 100°: 18.45 g. LD_{50} orally in rats: 25 mg/kg, E. W. Schafer, *Toxicol. Appl. Pharmacol.* **21**, 315 (1972).

Caution: Potential symptoms of overexposure are nausea, diarrhea, abdominal pain, vomiting; ptosis, strabismus; peripheral neuritis, tremor; retrosternal tightness, chest pain, pulmonary edema; seizures, chorea, psychosis; liver, kidney damage; alopecia; leg paresthesia. *See NIOSH Pocket Guide to Chemical Hazards* (DHHS/ NIOSH 97-140, 1997) p 304; *Clinical Toxicology of Commercial Products*, R. E. Gosselin *et al.*, Eds. (Williams & Wilkins, Baltimore, ·5th ed., 1984) section III, pp 379-383. *See also* Thallium.

USE: Rodenticide, insecticide. Reagent in analytical chemistry; in the semi-conductor industry.

9419. Thallium Sulfide. [1314-97-2] Thallium sulfide (Tl_2S); thallous sulfide. STl_2; mol wt 440.82. S 7.27%, Tl 92.73%. Tl_2S.

Bluish-black, cryst powder. d 8.39. mp 448.5°. Almost insol in water, alkali hydoxides, sulfides or cyanides. Sol in mineral acids.

9420. Thallium Trinitrate. [13746-98-0] Nitric acid thallium(3+) salt (3:1); thallium(III) nitrate. TlN_3O_9; mol wt 390.39. Tl 52.35%, N 10.76%, O 36.88%. $Tl(NO_3)_3$. Oxidizing reagent. Prepn and use for olefin rearrangement: A. McKillop *et al.*, *Tetrahedron Lett.* **1970**, 5275. Review of early uses: *idem*, E. C. Taylor, *Endeavour* **35**, 88-93 (1976). Phenolic oxidation for isodityrosine functionality: K. Nakamura *et al.*, *Tetrahedron Lett.* **42**, 5799 (2001); promotion of ring contractions for indans: H. M. C. Ferraz *et al.*, *Tetrahedron* **57**, 1709 (2001). *Review*: *idem et al.*, *Synthesis* **1999**, 2001-2023; T. O. Vieira, *Synlett* **2002**, 1017-1018.

Colorless crystals. Readily sol in methanol, dilute mineral acids and mixed sovent systems such as aqueous glyme.

USE: In organic syntheses for a wide variety of oxidations including rearrangement of ketones, olefins, electrophilic cyclization transformations, phenolic oxidative couplings and hydrolysis of dithianes.

9421. Thapsigargin. [67526-95-8] Octanoic acid (3S,3aR,-4S,6S,6aR,7S,8S,9bS)-6-(acetyloxy)-2,3,3a,4,5,6,6a,7,8,9b-decahydro-3,3a-dihydroxy-3,6,9-trimethyl-8-[[(2Z)-2-methyl-1-oxo-2-buten-1-yl]oxy]-2-oxo-4-(1-oxobutoxy)azuleno[4,5-b]furan-7-yl ester; thapsigargine; Tg. $C_{34}H_{50}O_{12}$; mol wt 650.76. C 62.75%, H 7.74%, O 29.50%. Sesquiterpene lactone which inhibits intracellular Ca^{2+} transport ATPases. Isolated from the root of *Thapsia garganica* L., *Apiaceae*, traditionally used as a counter-irritant: U. Rasmussen *et al.*, *Acta Pharm. Suec.* **15**, 133 (1978). Structure elucidation: S. B. Christensen *et al.*, *Tetrahedron Lett.* **21**, 3829 (1980); stereochemistry from x-ray analysis: *idem et al.*, *J. Org. Chem.* **47**, 649 (1982). Absolute configuration: S. B. Christensen, E. Norup, *Tetrahedron Lett.* **26**, 107 (1985). Inhibition of ATPases: J. Lytton *et al.*, *J. Biol. Chem.* **266**, 17067 (1991). Effect on gene expression: K. D. Rodland *et al.*, *Mol. Endocrinol.* **11**, 281 (1997). Review of mechanism of action: G. Inesi, Y. Sagara, *Arch. Biochem. Biophys.* **298**, 313-317 (1992). Review of use as molecular probe: O. Thastrup *et al.*, *Agents Actions* **27**, 17-23 (1989); T. B. Rogers *et al.*, *Biosci. Rep.* **15**, 341-349 (1995).

Colorless amorphous powder.
USE: Reagent for defining and manipulating intracellular calcium pools.

9422. Thaumatin. [53850-34-3] Thaumatins proteins; Talin. Sweet-tasting basic protein extracted from the fruit of the tropical plant, *Thaumatococcus danielli* Benth., *Marantaceae*, found in Western Africa from Sierra Leone to Zaire, in Sudan and Uganda. Composed of five different forms, thaumatins I, II, III, b and c; thaumatins I and II predominate. All are nearly 100,000 times sweeter than sucrose and have molecular weights of about 22,000. Isoln and characterization: H. van der Wel, K. Loeve, *Eur. J. Biochem.* **31**, 221 (1972). Extraction process: J. Higgenbotham, **US 4011206** (1976 to Tate and Lyle Ltd.). Electrophysical study of effects on taste receptors: Brouwer *et al.*, *Acta Physiol. Scand.* **89**, 550 (1973). Spectrometric investigation: O. Korver *et al.*, *Eur. J. Biochem.* **35**, 554 (1973). Studies on the primary structure of thaumatin I: R. B. Iyengar *et al.*, *ibid.* **96**, 193 (1979). Thaumatins I and II consist of almost identical sequences of 207 amino acids. Thaumatin I crystallizes in two different forms: H. van der Wel *et al.*, *FEBS Lett.* **56**, 316 (1975). The crystal structure shows two distinct regions with the amino acids either in sheets or in complex loops. Crystal structure: A. M. de Vos *et al.*, *Proc. Natl. Acad. Sci. USA* **82**, 1406 (1985). Cloning and expression in *E. coli* of the structural gene of thaumatin II: L. Edens *et al.*, *Gene* **18**, 1 (1982); C. T. Verrips, **EP 54330**; **EP 54331** (both 1982 to Unilever). Cloning of the natural gene: A. M. Ledeboer *et al.*, *Gene* **30**, 23 (1984). *Reviews:* R. Cagan, *Science* **181**, 32 (1973); H. van der Wel, *Chem. Ind. (London)* **1983**, 19.

Intensely sweet taste, licorice aftertaste. Strongly cationic, isoelectric pt greater than or equal to 11.7. uv max: 278 nm (pH 5.6); 283, 290 nm (pH 13.0). About 750-1600 times sweeter than sucrose on a wt basis; 30,000-100,000 times on a molar basis. Threshold values are near 10^{-4}%. The proteins lose sweetness on heating, on splitting of disulfide bridges and also at pH < 2.5 which points to the importance of the tertiary structure for the sweetness.
USE: Potential low-calorie sweetener.

9423. Theaflavine. [4670-05-7] 1,8-Bis[(2R,3R)-3,4-dihydro-3,5,7-trihydroxy-2H-1-benzopyran-2-yl]-3,4,6-trihydroxy-5H-benzocyclohepten-5-one. $C_{29}H_{24}O_{12}$; mol wt 564.50. C 61.70%, H 4.29%, O 34.01%. From black tea extracts: Roberts, Myers, *J. Sci. Food Agric.* **10**, 176 (1959). Structure: Takino *et al.*, *Tetrahedron Lett.* **1965**, 4019. Configuration: Brown *et al.*, *ibid.* **1966**, 1193.

Crystals from water, dec 237-240°. Absorption max (ethanol): 216, 271, 384, 470 nm (ε 35500, 19500, 8700, 3600).
Nonaacetate. $C_{47}H_{42}O_{21}$. Crystals, mp 167-168°. uv max (alc): 211, 250, 314, 353 nm (ε 25100, 13200, 7100, 4700).

9424. Theanine. [3081-61-6] N-Ethyl-L-glutamine; Suntheanine. $C_7H_{14}N_2O_3$; mol wt 174.20. C 48.26%, H 8.10%, N 16.08%, O 27.55%. Predominant amino acid found in leaves of the tea plant, *Camellia sinensis* (L.) O. Kuntze, *Theaceae*; umami tasteenhancing component of green tea. Used in nutraceutical formulations for its relaxation effect. Prepn: N. Lichtenstein, *J. Am. Chem. Soc.* **64**, 1021 (1942). Isoln from Japanese green tea leaves: Y. Sakato, *Nippon Nogei Kagaku Kaishi* **23**, 262 (1950), *C.A.* **45**, 3528*f* (1951). Synthesis: H. Kawagishi, K. Sugiyama, *Biosci. Biotechnol. Biochem.* **56**, 689 (1992); H. Gu *et al.*, *Org. Prep. Proced. Int.* **36**, 182 (2004). HPLC determn in tea leaves: Y. Ying *et al,. J. Liq.*

Chromatogr. Relat. Technol. **28**, 727 (2005). LC/MS determn in plasma and urine and pharmacokinetics of enantiomers: M. J. Desai *et al.*, *Chirality* **17**, 154 (2005). Impact on taste of green tea: S. Kaneko *et al.*, *J. Agric. Food Chem.* **54**, 2688 (2006). Review of physiological effects: M. Ozeki *et al.*, in *Nutraceutical Proteins and Peptides in Health and Disease* (CRC Press, Boca Raton, 2006) pp 377-390. Review of psychological effects on anxiety and cognitive performance: J. Bryan, *Nutr. Rev.* **66**, 82-90 (2008).

White crystals from ethanol + water, mp 214-216°. $[\alpha]_D^{20}$ +8.6°(c = 1.0 in water). Sol in 2.6 parts water at 0°; in 1.8 parts at 100°. Insol in ethanol, ether.
USE: Dietary supplement; food additive (flavor).

9425. Thearubigins. [12698-96-3] Thearubigin. Weakly acidic class of orange-brown phenolic pigments produced during the fermentation step of tea manufacture. Accounts for as much as 20% of dry weight of black tea; in combination with theaflavine, *q.v.*, determines the color, brightness and body of a tea infusion. Isolation: E. A. H. Roberts *et al.*, *J. Sci. Food Agric.* **8**, 72 (1957); E. A. H. Roberts, *ibid.* **9**, 212 (1958). Effects on tea characteristics: R. F. Smith, *ibid.* **19**, 530 (1968); M. Hazarika *et al.*, *ibid.* **35**, 1208 (1984). Structure studies: A. G. Brown *et al.*, *Phytochemistry* **8**, 2333 (1969); J. E. Berkowitz *et al.*, *ibid.* **10**, 2271 (1971); D. J. Cattell, H. E. Nursten, *ibid.* **15**, 1967 (1976). HPLC analysis: B. L. Wedzicha, T. J. Donovan, *J. Chromatogr.* **478**, 217 (1989); R. G. Bailey *et al.*, *ibid.* **542**, 115 (1991).

9426. Thebacon. [466-90-0] (5α)-6,7-Didehydro-4,5-epoxy-3-methoxy-17-methylmorphinan-6-ol 6-acetate; demethyldihydrothebaine acetate (ester); dihydrocodeinone enol acetate; acetyldemethyldihydrothebaine; acetyldihydrocodeinone. $C_{20}H_{23}NO_4$; mol wt 341.41. C 70.36%, H 6.79%, N 4.10%, O 18.74%. Semisynthetic opioid agonist. Prepn by refluxing dihydrocodeinone with acetic anhydride and anhydr sodium acetate: Small *et al.*, *J. Org. Chem.* **3**, 204 (1938).

Needles from methanol, mp 154°. Practically insol in water. Soluble in most organic solvents.
Hydrochloride. [20236-82-2] Acedicone. $C_{20}H_{23}NO_4 \cdot HCl$; mol wt 377.87. Crystals, mp 132-135° (dec). Very sol in water. Stable in boiling water.
Note: This is a controlled substance (opium derivative): **21 CFR**, 1308.11.
THERAP CAT: Antitussive.

9427. Thebaine. [115-37-7] (5α)-6,7,8,14-Tetrahydro-4,5-epoxy-3,6-dimethoxy-17-methylmorphinan; paramorphine. $C_{19}H_{21}NO_3$; mol wt 311.38. C 73.29%, H 6.80%, N 4.50%, O 15.41%. From opium, which contains from 0.3-1.5% depending on its origin. Discussion of structure and bibliography: Small, Lutz, "Chemistry of the Opium Alkaloids" in *U.S. Public Health Reports* **Suppl. No. 103** (Washington, 1932); Small, Browning, *J. Org. Chem.* **3**, 618 (1938); Cherbuliez, Araqui, *Helv. Chim. Acta* **26**, 2251 (1943); Ghosh, Robinson, *J. Chem. Soc.* **1944**, 506; K. W. Bentley, *The Chemistry of the Morphine Alkaloids* (Oxford, 1954) p 184 sqq. Config: Kalvoda *et al.*, *Helv. Chim. Acta* **38**, 1847 (1955). Syntheses: Rapoport *et al.*, *J. Am. Chem. Soc.* **89**, 1942 (1967); Schwartz, Mami, *ibid.* **97**, 1239 (1975); Barber, Rapoport, *J. Med. Chem.* **18**, 1074 (1975). Absorption spectrum: Csokán, *Z. Anal. Chem.* **124**,

344 (1942). Toxicity data: Eddy, *J. Pharmacol. Exp. Ther.* **66**, 182 (1939).

Orthorhombic, rectangular plates by sublimation at 170-180° under atmospheric pressure and a 1 mm distance. mp 193° (rapid heating). $[\alpha]_D^{15}$ −219° (p = 2 in alc); $[\alpha]_D^{23}$ −230° (p = 5 in chloroform). pK at 15°: 6.05. pH of satd water soln 7.6. One gram dissolves in 1460 ml water at 15°; Kolthoff, *Biochem. Z.* **162**, 336 (1925); in about 15 ml hot alcohol or 13 ml chloroform, about 200 ml ether, 25 ml benzene, 12 ml pyridine; not very sol in petr ether.

Hydrochloride monohydrate. $C_{19}H_{21}NO_3 \cdot HCl \cdot H_2O$; mol wt 365.85. Orthorhombic prisms from alcohol, $[\alpha]_D^{23}$ −164° (p = 2). Sol in about 12 parts water, in alcohol; pH of 0.05 molar soln 4.95. LD_{50} s.c. in rabbits: 14 mg/kg (Eddy).

Oxalate hexahydrate. $2C_{19}H_{21}NO_3 \cdot C_2H_2O_4 \cdot 6H_2O$; mol wt 820.89. Prisms, sol in about 10 parts water, in alcohol. Almost insol in ether.

Binoxalate monohydrate. $C_{19}H_{21}NO_3 \cdot C_2H_2O_4 \cdot H_2O$; mol wt 419.43. Prisms, sol in 45 parts water.

Bitartrate monohydrate. $C_{19}H_{21}NO_3 \cdot C_4H_6O_6 \cdot H_2O$; mol wt 479.48. Prisms, sol in 130 parts water, quite sol in hot water, hot alcohol.

Salicylate. $C_{19}H_{21}NO_3 \cdot C_7H_6O_3$; mol wt 449.50. Crystals, sol in 750 parts water.

Note: This is a controlled substance (opiate): **21 CFR**, 1308.12.

9428. Thebainone. [467-98-1] 7,8-Didehydro-4-hydroxy-3-methoxy-17-methylmorphinan-6-one; thebainone-A. $C_{18}H_{21}NO_3$; mol wt 299.37. C 72.22%, H 7.07%, N 4.68%, O 16.03%. Prepn from thebaine, codeinone or β-ethylthiocodide: Morris, Small, *J. Am. Chem. Soc.* **56**, 2159 (1934). Earlier references and discussion of structure: Small, Lutz, "Chemistry of the Opium Alkaloids" in *U.S. Public Health Reports* **Suppl. No. 103** (Washington, 1932). About anomalies in nomenclature and difference from metathebainone *see* Henry, *Plant Alkaloids* (London, 1939) p 249. Description of all thebainones: K. W. Bentley, *The Chemistry of the Morphine Alkaloids* (Oxford, 1954) p 219.

Crystals from ethyl acetate, mp 146°. $[\alpha]_D^{28}$ −47° (c = 1.16 in 95% alc). One gram dissolves in 250 ml water, about 120 ml boiling water. Sol in chloroform, benzene, acetone; sparingly sol in ether, alcohol, methanol.

Sesquihydrate. Crystals from water, mp 90°.

Methanolate. Crystals from methanol, mp 118°.

Hydrochloride. $C_{18}H_{21}NO_3 \cdot HCl$. Crystals from alcohol, mp 256° (turns red at mp). $[\alpha]_D^{30}$ −25° (c = 1.63).

Hydriodide. $C_{18}H_{21}NO_3 \cdot HI$. Crystals, mp 165°, solidifies and remelts 260°.

Methiodide. mp 251°.

9429. Thenaldine. [86-12-4] 1-Methyl-*N*-phenyl-*N*-(2-thienylmethyl)-4-piperidinamine; 1-methyl-4-*N*-2-thenylanilinopiperidine; 1-methyl-4-amino-*N*-phenyl-*N*-(2-thenyl)piperidine; thenophenopiperidine; 1-methyl-4-[phenyl-(2-thenyl)amino]piperidine; thenalidine; Sandostene. $C_{17}H_{22}N_2S$; mol wt 286.44. C 71.28%, H 7.74%, N 9.78%, S 11.19%. Prepn: Stoll, Bourquin, **US 2717251** and **US 2757175** (1955 and 1956 to Sandoz).

bp$_{0.02}$ 158-160°, mp 95-97°.

Tartrate. Crystals, mp 170-172°.

THERAP CAT: Tartrate as antihistaminic; antipruritic.

9430. 3-Thenoic Acid. [88-13-1] 3-Thiophenecarboxylic acid; β-thiophenic acid. $C_5H_4O_2S$; mol wt 128.15. C 46.86%, H 3.15%, O 24.97%, S 25.02%. Prepd by the oxidation of 3-thenaldehyde with silver oxide: Campaigne, LeSuer, *Org. Synth.* **33**, 94 (1953).

Crystals from water, mp 137-138°. Soly in water at 25°: 0.43 g/100 g. pKa 6.23. Volatile with steam.

9431. Thenyldiamine. [91-79-2] N^1,N^1-Dimethyl-N^2-2-pyridinyl-N^2-(3-thienylmethyl)-1,2-ethanediamine; 2-[(2-dimethylaminoethyl)-3-thenylamino]pyridine; *N,N*-dimethyl-*N'*-(α-pyridyl)-*N'*-(3-methylthienyl)ethylenediamine; *N*-(α-pyridyl)-*N*-(β-thenyl)-*N',N'*-dimethylethylenediamine; *N*-(2-dimethylaminoethyl)-*N*-2-pyridyl-3-thenylamine; dethylandiamine; Win-2848; Thenfadil. $C_{14}H_{19}N_3S$; mol wt 261.39. C 64.33%, H 7.33%, N 16.08%, S 12.27%. Prepn: Campaigne, LeSuer, *J. Am. Chem. Soc.* **71**, 333 (1949). Toxicity study: Hoppe, Lands, *J. Pharmacol. Exp. Ther.* **97**, 371 (1949).

Free base, liquid. bp$_{1.0}$ 169-172°. n_D^{20} 1.5915.

Hydrochloride. $C_{14}H_{19}N_3S \cdot HCl$; mol wt 297.85. Crystals from methanol, mp 169.5-170°. Bitter taste. Sol in water up to 20%. Slightly sol in alc. pH of 1% aq soln 6.5. LD_{50} orally in rats: 525 mg/kg (Hoppe, Lands).

THERAP CAT: Hydrochloride as antihistaminic.

9432. Theobroma Oil. Cacao butter; cocoa butter. From roasted seeds of *Theobroma cacao* L., *Sterculiaceae. Constit.* Chiefly glycerides of stearic, palmitic, oleic, arachidic, and linoleic acids.

Yellowish-white solid; brittle below 25°; chocolate odor and taste; d_{25}^{100} 0.858-0.864; mp 30-35°; n_D^{40} 1.4537-1.4578. Sapon. no. 188-195. Iodine no. 35-40. Insoluble in water. Slightly sol in alcohol; sol in boiling abs alc; very sol in chloroform, ether, benzene, petr ether.

USE: Lubricant in massage; base for suppositories and ointments. Manuf chocolate, toilet soaps, creams, etc.

9433. Theobromine. [83-67-0] 3,7-Dihydro-3,7-dimethyl-1*H*-purine-2,6-dione; 3,7-dimethylxanthine. $C_7H_8N_4O_2$; mol wt 180.17. C 46.67%, H 4.48%, N 31.10%, O 17.76%. The principal alkaloid of the cacao bean which contains 1.5-3% of the base. Also present in cola nuts and in tea. Usually extracted from the hull of cacao beans which contains 0.7-1.2%. Extraction process: Schwyzer, *Die Fabrikation pharmazeutischer und chemisch-technischer Produkte* (Berlin, 1931). Synthesis starting with 3-methyluric acid: Fischer, Ach, *Ber.* **31**, 1980 (1898); *cf.* Gebner, Krebs, *J. Gen. Chem. USSR* **16**, 179-186 (1946). Comparison with theophylline, *q.v.*, of metabolism in humans: D. J. Birkett *et al., Drug Metab.*

Dispos. **13**, 725 (1985). Pharmacokinetics: A. Lelo *et al.*, *Br. J. Clin. Pharmacol.* **22**, 177 (1986). Bronchodilator effect in asthma: F. E. R. Simons *et al.*, *J. Allergy Clin. Immunol.* **76**, 703 (1985).

Monoclinic needles (lamellar twinning on 001) from water, mp 357°. Sublimes 290-295°. Kb at 18°: 1.3×10^{-14}; Ka 0.9×10^{-10}. Absorption spectrum: Hartley, *J. Chem. Soc.* **87**, 1803, 1810 (1905). One gram dissolves in about 2000 ml water, 150 ml boiling water, 2220 ml 95% alcohol; sol in the fixed alkali hydroxides, concd acids, in about 22 parts of 20% aq tribasic sodium phosphate soln; moderately sol in ammonia. Almost insol in benzene, ether, chloroform, carbon tetrachloride. Forms salts which are dec by water, and compds with bases which are more stable.

Calcium salicylate. [8065-51-8] Theocalcin; Calcium Diuretin. Double salt or mixture of calcium theobromine and calcium salicylate. Contains no less than 44% theobromine. White, amorphous powder. Slightly saline taste. Partly sol in water. Aq solns are alkaline to phenolphthalein.

Sodium acetate. [8002-88-8] Thesodate. Equimolar mixture of sodium theobromine and sodium acetate, containing $1H_2O$. Theobromine 59.6%, anhydr sodium acetate 27.1%. White, odorless or almost odorless, hygroscopic powder. Absorbs CO_2 from the air becoming incompletely sol. Very sol in water, sparingly sol in cold alcohol; the solns are strongly alkaline. pH about 10. *Keep tightly closed.*

Sodium salicylate. [8048-31-5] Diuretin. Equimolar mixture of sodium theobromine and sodium salicylate, containing $1H_2O$. Theobromine 47.3%, sodium salicylate 42.1%. White, odorless or almost odorless, hygroscopic powder. Absorbs CO_2 from the air and becomes incompletely sol. Soluble in 1 part water; slightly sol in alcohol; the solns are strongly alkaline. pH about 10.

THERAP CAT: Diuretic, bronchodilator, cardiotonic.

THERAP CAT (VET): Diuretic, myocardial stimulant, vasodilator.

9434. 1-Theobromineacetic Acid. [5614-56-2] 2,3,6,7-Tetrahydro-3,7-dimethyl-2,6-dioxo-1*H*-purine-1-acetic acid; 3,6-dihydro-3,7-dimethyl-2,6-dioxo-1(2*H*)-purineacetic acid; 3,6-dihydro-2,6-diketo-3,7-dimethyl-1(2)-purineacetic acid. $C_9H_{10}N_4O_4$; mol wt 238.20. C 45.38%, H 4.23%, N 23.52%, O 26.87%. Prepd from sodium theobromine + chloroacetic acid: E. Merck *et al.*, **DE 352980**, *C.A.* **17**, 1307[1] (1923).

Crystals, mp 260°.

Sodium salt. [32245-40-2] Sodium theobromine acetate; Técarine. $C_9H_9N_4NaO_4$; mol wt 260.18. Crystalline powder, freely sol in water, forming a neutral soln.

THERAP CAT: Bronchodilator.

9435. Theofibrate. [54504-70-0] 2-(4-Chlorophenoxy)-2-methylpropanoic acid 2-(1,2,3,6-tetrahydro-1,3-dimethyl-2,6-dioxo-7*H*-purin-7-yl)ethyl ester; 2-(*p*-chlorophenoxy)-2-methylpropionic acid ester with 7-(2-hydroxyethyl)theophylline; 1-(theophyllin-7-yl)ethyl 2-(*p*-chlorophenoxy)isobutyrate; etofylline clofibrate; ML-1024; Duolip. $C_{19}H_{21}ClN_4O_5$; mol wt 420.85. C 54.23%, H 5.03%, Cl 8.42%, N 13.31%, O 19.01%. Deriv of clofibric acid, *q.v.*, with antilipemic, antithrombotic, and platelet-aggregation inhibitory activity: Prepn: G. Metz, M. Specker, **DE 2308826**; *eidem*, **US 3984413** (1974, 1976 both to Merckle); *eidem*, *Arzneim.-Forsch.* **25**, 1686 (1975). Hypolipemic activity: G. Metz *et al.*, *ibid.* **27**, 1173

(1977). Series of articles on chemistry, biopharmaceutic evaluation, pharmacology, toxicology, clinical studies: *ibid.* **30**, 2013-2074 (1980).

Colorless crystals from ethanol, mp 133-135°. Practically insol in water at pH 2-7.4 and in cold alcohols. Sol in acetone, chloroform, hot alcohols. LD_{50} in mice, rats, dogs (g/kg): 11.7, 17.0, >10.0 orally (Metz).

THERAP CAT: Antilipemic.

9436. Theophylline. [58-55-9] 3,9-Dihydro-1,3-dimethyl-1*H*-purine-2,6-dione; 1,3-dimethylxanthine; Aerobin; Aerolate; Bronchoretard; Diffumal; Duraphyllin; Elixophyllin; Etheophyl; Euphylline; Euphylong; Lasma; Nuelin; PulmiDur; Pulmo-Timelets; Respbid; Slo-Phyllin; Solosin; Teosona; Theobid; Theochron; Theolair; Theon; Theograd; Theostat; Unifyl; Uniphyl; Uniphyllin; Xanthium. $C_7H_8N_4O_2$; mol wt 180.17. C 46.67%, H 4.48%, N 31.10%, O 17.76%. Xanthine derivative with diuretic, cardiac stimulant and smooth muscle relaxant activities; isomeric with theobromine, *q.v.* Small amounts occur in tea. Synthesis starting with dimethylurea and ethyl cyanoacetate: Traube, *Ber.* **33**, 3035 (1900); Grinberg, *J. Appl. Chem. USSR* **13**, 1461 (1940); Gebner, Krebs, *J. Gen. Chem. USSR* **16**, 179 (1946). Bioavailability: K. Svedmyr *et al.*, *Allergy* **37**, 111 (1982). Comprehensive description: J. L. Cohen, *Anal. Profiles Drug Subs.* **4**, 466-493 (1975). Symposium on clinical experience in asthma and allergy: *Am. J. Med.* **79** (6A), 1-78 (1985); on pharmacology, bioavailability, pharmacokinetics and efficacy in obstructive pulmonary disease: *J. Allergy Clin. Immunol.* **78** (4 Part 2), 669-824 (1986). Review of clinical toxicology: J. D. Truwit, *Crit. Care Clin.* **7**, 639-657 (1991).

Monohydrate, thin monoclinic tablets from water, mp 270-274°. Bitter taste. pKa (25°): 8.77; pKb: 13.5, 11.5. One gram dissolves in 120 ml water, 80 ml alc, about 110 ml chloroform; more sol in hot water. Freely sol in solns of alkali hydroxides, ammonia; sol in dil HCl or HNO_3; sparingly sol in ether. uv max (0.1*N* NaOH): 274 nm.

Ethanolamine. [573-41-1] $C_9H_{15}N_5O_3$; mol wt 241.25. White, crystalline, nonhygroscopic powder contg 75% anhydr theophylline and 25% ethanolamine. Sol in water.

Isopropanolamine. [5600-19-1] $C_{10}H_{17}N_5O_3$; mol wt 255.28. Prepn: Greenbaum, *Am. J. Pharm.* **109**, 550 (1937). Prisms from alcohol. Very sol in water. pH of aq solns 9.20.

Lysine salt. Paidomal.

Sodium acetate. [8002-89-9] Equimolar mixture of theophylline sodium and sodium acetate, containing $1H_2O$. White, odorless, crystalline powder; bitter taste. Sol in about 25 parts water. Insol in alcohol, chloroform or ether.

Sodium glycinate. [8000-10-0] Afonilum. Prepn: Krantz *et al.*, *J. Am. Pharm. Assoc.* **36**, 248 (1947). Equimolar mixture of theophylline sodium and glycine. Darkens at 180°. Soly in water (w/v): 18%. Very slightly sol in alc. Practically insol in chloroform. Saturated soln has a pH of 8.7-9.1, d^{20} 1.05, and is stable to carbon dioxide.

THERAP CAT: Bronchodilator.

THERAP CAT (VET): Bronchodilator.

9437. Thermolysin. [9073-78-3] Bacillus thermoproteolyticus neutral proteinase; EC 3.4.24.27. Proteolytic enzyme of mol wt 37,500 that hydrolyzes protein bonds on the *N*-terminal side of hydrophobic amino acid residues. Contains a zinc atom essential for activity and four Ca^{2+} ions essential for thermal and conformational stability. Isoln from *Bacillus thermoproteolyticus:* S. Endo, *J. Ferment. Technol.* **40**, 346 (1962). Properties and amino acid composition: Y. Ohta *et al., J. Biol. Chem.* **241**, 5919 (1966). Site of enzymatic hydrolysis: Y. Ohta, Y. Ogura, *J. Biochem.* **58**, 607 (1965). Substrate specificity studies: H. Matsubara *et al., Biochem. Biophys. Res. Commun.* **21**, 242 (1965); **24**, 427 (1966); K. Morihara, H. Tsuzuki, *Biochim. Biophys. Acta* **118**, 215 (1966). Stability studies: Y. Ohta, *J. Biol. Chem.* **242**, 509 (1967). Inhibition studies: H. Matsubara *et al., Biochem. Biophys. Res. Commun.* **34**, 719 (1969); J. Murphy *et al., Arch. Biochem. Biophys.* **202**, 405 (1980). Purification: H. Matsubara, *Methods Enzymol.* **19**, 642 (1970). Structure studies: K. Titani *et al., Nature* **238**, 35 (1972); B. W. Matthews *et al., ibid.* **37**, 41; P. M. Colman *et al., J. Mol. Biol.* **70**, 701 (1972). Function of the metal ions: J. Feder *et al., Biochemistry* **10**, 4552 (1971); G. Voordouw, R. S. Roche, *ibid.* **14**, 4667 (1975); R. S. Roche, G. Voordouw, *Crit. Rev. Biochem.* **5**, 1 (1978). Effect of the histidyl residue on the mechanism of action: S. Blumberg *et al., Isr. J. Chem.* **12**, 643 (1974); M. K. Pangburn, K. A. Walsch, *Biochemistry* **14**, 4050 (1975). *Review:* *Experientia* **26**, Suppl., 31-59 (1976).

Crystals. uv max: 280 nm (ε 66,300). Optimum pH 7.0-8.5; stable at pH 6.0-9.0. The refrigerated lyophilized enzyme is stable for months; frozen enzyme soln can be kept for weeks without significant loss of activity. Not deactivated at 65°, but loses half of its activity upon heating at 80° for 1 hr.

USE: In studies of protein sequences.

9438. Thermorubin. [11006-83-0] Antibiotic complex isolated from *Thermoactinomyces antibioticus:* Craveri *et al., Clin. Med.* **71**, 511 (1964); *eidem,* DE 1180891; *eidem,* US 3300379 (1964, 1967 both to Lepetit). Structure of thermorubin A, the major component: Moppett *et al., J. Am. Chem. Soc.* **94**, 3269 (1972); revised structure: F. Johnson *et al., ibid.* **102**, 5580 (1980). Inhibition of bacterial protein synthesis: G. Pirali *et al., Biochim. Biophys. Acta* **366**, 310 (1974). Studies on mechanism of action: F. Lin, A. Wishnia, *Biochemistry* **21**, 477, 484 (1982).

Thermorubin A

Orange-red rosettes and needles from chloroform, darkens 190°, chars 300° without melting. $[\alpha]_D^{25}$ −14° (c = 0.4 in dioxane). Absorption max [dioxane-cyclohexane (1:1)]: 300, 328, 414, 430 nm ($E_{1cm}^{1\%}$ 1041, 1066, 313, 332). Sol in dioxane, pyridine, tetrahydrofuran, DMF, DMSO, concd alkalies, concd H_2SO_4, glacial acetic acid; slightly sol in methanol, ethanol, butanol, ethyl acetate, acetone, chloroform, benzene, cyclohexane. Practically insol in water, ether, petr ether, hexane. LD_{50} i.p. in mice: 300 mg/kg (Craveri, 1967).

Diacetate. Crystals from ethyl acetate. Absorption max [0.1*N* HCl-dioxane (1:1)]: 337, 436, 462 nm ($E_{1cm}^{1\%}$ 1500, 113, 162).

Thermorubin A. $C_{32}H_{24}O_{12}$; mol wt 600.53. Dihydrate, orange-red rosettes from ethyl acetate. Absorption max (ethanol): 250, 300, 328, 435 nm (ε 33700, 54640, 49980, 16300). pKa values: 4.7, 7.0, 9.0. Easily tautomerizes in soln. Thermally labile, decomposes rapidly in soln at temps >60°.

9439. Thevetin A. [37933-66-7] (3β,5β)-3-[(*O*-β-D-Glucopyranosyl-(1 → 6)-*O*-D-glucopyranosyl-(1 → 4)-6-deoxy-3-*O*-methyl-α-L-glucopyranosyl)oxy]-14-hydroxy-19-oxocard-20(22)-enolide. $C_{42}H_{64}O_{19}$; mol wt 872.96. C 57.79%, H 7.39%, O 34.82%. Glycoside isolated from *Thevetia neriifolia* Juss., *Apocynaceae:* Bloch *et al., Helv. Chim. Acta* **43**, 652 (1960). Separation

from thevetin B: Delalande, Baisse, US 3030355; US 3043829 (1962). Structure: K. Tori *et al., Tetrahedron Lett.* **1977**, 717.

Crystals from water. mp 208-210°. $[\alpha]_D^{24}$ −72.0 ±1.5° (c = 1.48 in methanol).

Acetyl derivative. $C_{58}H_{80}O_{27}$. Crystals from methanol + ether, mp 143-149°. $[\alpha]_D^{26}$ −54.2 ±1° (c = 1.86 in chloroform).

9440. Thiabendazole. [148-79-8] 2-(4-Thiazolyl)-1*H*-benzimidazole; 4-(2-benzimidazolyl)thiazole; MK-360; Equizole; Mertect; Mintezol; Tecto. $C_{10}H_7N_3S$; mol wt 201.25. C 59.68%, H 3.51%, N 20.88%, S 15.93%. Prepd by the reaction of 4-thiazolecarboxamide with *o*-phenylenediamine in polyphosphoric acid: H. D. Brown *et al., J. Am. Chem. Soc.* **83**, 1764 (1961); L. H. Sarett, H. D. Brown, US 3017415 (1962 to Merck & Co.). Synthesis of labeled thiabendazole: D. J. Tocco *et al., J. Med. Chem.* **7**, 399 (1964). Alternate route of synthesis: V. J. Grenda *et al., J. Org. Chem.* **30**, 259 (1965). Anthelmintic props: H. D. Brown *et al., loc. cit.;* K. C. Kates *et al., J. Parasitol.* **57**, 356 (1971). Fungicidal props: H. J. Robinson *et al., J. Invest. Dermatol.* **42**, 479 (1966). Systemic props in plants: D. C. Erwin *et al., Phytopathology* **58**, 860 (1968). Toxicity: H. J. Robinson *et al., Toxicol. Appl. Pharmacol.* **7**, 53 (1965). Residue analysis: IUPAC Appl. Chem. Div., *Pure Appl. Chem.* **52**, 2567 (1980). Comprehensive description: V. K. Kapoor, *Anal. Profiles Drug Subs.* **16**, 611-639 (1986).

White to off-white, odorless powder, mp 304-305°. uv max (methanol): 298 nm (ε 23330). Fluorescence max in acid soln: 370 nm (310 nm excitation). Soluble in DMF, DMSO; slightly soluble in acetone, alcohols, esters, chlorinated hydrocarbons; very slightly sol in chloroform, ether. Max soly in water at pH 2.2: 3.84%. LD_{50} in mice, rats, rabbits (g/kg): 3.6, 3.1, >3.8 orally (Robinson).

Monophosphinate. [28558-32-9] Thiabendazole hypophosphite; Arbotect. $C_{10}H_7N_3S.H_3PO_2$; mol wt 267.24. Amber liquid. d^{25} 1.103.

USE: Fungicide for spoilage control of citrus fruit; for treatment and prevention of Dutch elm disease in trees; for control of fungal diseases of seed potatoes.

THERAP CAT: Anthelmintic (Nematodes).

THERAP CAT (VET): Anthelmintic, fungicide.

9441. Thiacetarsamide. [531-72-6] 2,2′-[[[4-(Aminocarbonyl)phenyl]arsinidene]bis(thio)]bisacetic acid; [[(*p*-carbamoylphenyl)arsylene]dithio]diacetic acid; bis[carboxymethylmercapto](*p*-carbamylphenyl)-arsine; *p*-[bis(carboxymethylmercapto)arsino]benzamide; dithioglycolyl *p*-arsenobenzamide; 4-carbamylphenyl bis[carboxymethylthio]arsenite; arsenamide. $C_{11}H_{12}AsNO_5S_2$; mol wt 377.26. C 35.02%, H 3.21%, As 19.86%, N 3.71%, O 21.20%, S 17.00%. Prepd by the condensation of *p*-arsenosobenzamide and thioglycolic acid: G. A. C. Gough, H. King *J. Chem. Soc.* **1930**, 669; T. H. Maren *et al., Science* **68**, 1864 (1946). Veterinary trial for heartworm infection in dogs: Z. S. Polizopoulou *et al., Vet. Rec.* **146**, 466 (2000).

Fine needles from boiling water, mp 168-169° (Gough, King); also reported as mp 158-162° (Maren). pKa 4. Sol in warm methanol, warm dehydrated alc; sparingly sol in cold dehydrated alc, cold methanol, cold water; more sol in water above 90°. Insol in warm isopropyl ether.

Disodium salt. [14433-82-0] Sodium thiacetarsamide; Caparsolate. $C_{11}H_{10}AsNNa_2O_5S_2$; mol wt 421.22. Stoichiometrically formed in aq soln with NaOH at pH 7-8.

THERAP CAT (VET): Anthelmintic for heartworm infection in dogs, especially the adult worm.

9442. Thiacetazone. [104-06-3] N-[4-[[(2-Aminothioxomethyl)hydrazinylidene]methyl]phenyl]acetamide; 2-[[4-(acetylamino)phenyl]methylene]hydrazinecarbothioamide; 4′-formylacetanilide thiosemicarbazone; p-acetamidobenzaldehyde thiosemicarbazone; p-acetylaminobenzaldehyde thiosemicarbazone; p-acetaminobenzylidenethiosemicarbazone; amithiozone; thibone; thioacetazone; Tb I-698; Conteben; Livazone; Myrizone; Neustab; Panrone; Seroden; Tebethion; Thiocarbazil; Thioparamizone; Tibione; Tiobicina. $C_{10}H_{12}N_4OS$; mol wt 236.29. C 50.83%, H 5.12%, N 23.71%, O 6.77%, S 13.57%. Prepn: G. Domagk et al., Naturwissenschaften **33**, 315 (1946); Behnisch et al., Angew. Chem. **60A**, 113 (1948); Chabrier, Cattelain, Bull. Soc. Chim. Fr. **1950M**, 52; Das, Mukherjee, J. Am. Chem. Soc. **75**, 1241 (1953). Toxicity data: Bavin et al., J. Pharm. Pharmacol. **2**, 764 (1950). HPLC determn in body fluids: P. J. Jenner, J. Chromatogr. **276**, 463 (1983). Review of clinical experience: H. C. Hinshaw, W. McDermott, Am. Rev. Tuberc. **61**, 145-157 (1950); of clinical safety and efficacy: P. Nunn et al., Trans. R. Soc. Trop. Med. Hyg. **87**, 578 (1993).

Minute, pale yellow crystals from abs alc. Bitter taste. Darkens on exposure to light. Dec 225-230°. uv max (ethanol): 328 nm, extinction 0.580. Sol in hot alc; very sparingly sol in cold alc. Insol in water, acetone, benzene, carbon tetrachloride, chloroform, carbon disulfide, petr ether. Practically insol in the other common organic solvents except glycols. Soly in propylene glycol ~1%. LD50 s.c. in mice: 1-2 g/kg (Bavin).

THERAP CAT: Antibacterial (tuberculostatic).

9443. Thiacloprid. [111988-49-9] [N(Z)]-[3-[(6-Chloro-3-pyridinyl)methyl]-2-thiazolidinylidene]cyanamide; (2Z)-3-[(6-chloro-3-pyridinyl)methyl]-1,3-thiazolidin-2-ylidenecyanamide; BAY YRC 2894; Alanto; Bariard; Biscaya; Calypso. $C_{10}H_9ClN_4S$; mol wt 252.72. C 47.53%, H 3.59%, Cl 14.03%, N 22.17%, S 12.69%. Chloronicotinyl insecticide for use in fruits and vegetables. Prepn: K. Shiokawa et al., EP 235725; eidem, US 4849432 (1987, 1989 both to Nihon Tokushu Noyaku Seizo KK). LC-MS determn in water samples: S. Seccia et al., Anal. Chim. Acta **553**, 21 (2005); in fruit and vegetables: A. Di Muccio et al., J. Chromatogr. A **1108**, 1 (2006). Review of properties and bioactivity: A. Elbert et al., Brighton Crop Prot. Conf. - Pests Dis. **2000**, 21-26; of synthesis and chemical structure: P. Jeschke et al., Pflanzenschutz-Nachr. Bayer (Engl. Ed.) **54**, 147-160 (2001).

Crystals from ether, mp 128-129°. Vapor pressure (20°): 3 × 10^{-10} Pa. Soly in water (20°): 185 mg/l. Log P (octanol/water): 1.26. LD50 in male, female rats (mg/kg): 836, 444 orally; >2000,

>2000 dermally. LC50 (4 hr) in male, female rats (mg/m³): >2535, 1223. LC50 (96 hr) in rainbow trout (mg/l): 30.5 (Elbert).

USE: Insecticide.

9444. Thialbarbital. [467-36-7] 5-(2-Cyclohexen-1-yl)dihydro-5-(2-propen-1-yl)-2-thioxo-4,6(1H,5H)-pyrimidinedione; 5-allyl-5-(2-cyclohexen-1-yl)-2-thiobarbituric acid; 5-(2-cyclohexen-1-yl)-5-allyl-2-thiobarbituric acid; thialbarbitone; Kemithal. $C_{13}H_{16}N_2O_2S$; mol wt 264.34. C 59.07%, H 6.10%, N 10.60%, O 12.10%, S 12.13%. Prepn: Volwiler, Tabern, US 2153730 (1939 to Abbott).

Crystals, mp 148-150°.

Sodium salt. $C_{13}H_{15}N_2NaO_2S$; mol wt 286.32. Pale yellow crystals from methanol, mp 130-132°. Very sol in water, alc. Practically insol in benzene, ether. pH of 2.5% soln in water: 10.5.

Note: This is a controlled substance (depressant): **21 CFR**, 1308.13.

THERAP CAT: Anesthetic (intravenous).

THERAP CAT (VET): Anesthetic.

9445. Thiambutene. [86-14-6] N,N-Diethyl-4,4-di-2-thienyl-3-butene-2-amine; 3-diethylamino-1,1-di(2′-thienyl)but-1-ene; diethylthiambutene; 191C49; NIH-4185. $C_{16}H_{21}NS_2$; mol wt 291.47. C 65.93%, H 7.26%, N 4.81%, S 22.00%. Prepn: Adamson, J. Chem. Soc. **1950**, 885; US 2561899 (1951 to Burroughs Wellcome). Toxicity data: J. S. McKenzie, N. R. Beechey, Arch. Int. Pharmacodyn. Ther. **135**, 376 (1962). Use: Hayes, Vet. Rec. **83**, 528 (1968). Evaluation as analgesic and preanesthetic agent: W. D. Harbison et al., Aust. Vet. J. **50**, 543 (1974). HPLC determn: I. Jane et al., J. Chromatogr. **323**, 191 (1985).

bp0.03 122-128°. LD50 i.p. in mice: 90 mg/kg (McKenzie, Beechey).

Hydrochloride. Themalon. $C_{16}H_{21}NS_2 \cdot HCl$; mol wt 327.93. Crystals, mp 152-153°.

Note: This is a controlled substance (opiate): **21 CFR**, 1308.11.

THERAP CAT (VET): Analgesic (narcotic).

9446. Thiamethoxam. [153719-23-4] 3-[(2-Chloro-5-thiazolyl)methyl]tetrahydro-5-methyl-N-nitro-4H-1,3,5-oxadiazin-4-imine; CGA-293343; Actara; Adage; Cruiser. $C_8H_{10}ClN_5O_3S$; mol wt 291.71. C 32.94%, H 3.46%, Cl 12.15%, N 24.01%, O 16.45%, S 10.99%. Broad-spectrum chloronicotinyl insecticide. Prepn: P. Maienfisch, L. Gsell, CA 2100924 (1994 to Ciba-Geigy); eidem, US 5852012 (1998 to Novartis); T. Göbel et al., Pestic. Sci. **55**, 343 (1999). Effect on feeding behavior of aphids: P. Harrewijn et al., Brighton Crop Prot. Conf. - Pests Dis. **1998**, 813. LC-MS determn in fruits and vegetables: D. Zywitz et al., Dtsch. Lebensm. Rundsch. **99**, 188 (2003). Review of physical properties and field trials: R. Senn et al., ibid. 27-36; of biological activity, metabolism, and physical properties: D. S. Lawson et al., Proc. Beltwide Cotton Conf. **2**, 1106-1109 (1999).

Crystalline powder, mp 139.1°. Soly in water at 25°: 4100 mg/l. Vapor pressure at 25°: 6.6×10^{-9} Pa. Log P (octanol/water) at 25°: -0.13. LD_{50} in rats (mg/kg): 1563 orally, >2000 dermally; LD_{50} in bobwhite quail, mallard duck (mg/kg): 1552, 576 orally. LC_{50} (96hr) in rainbow trout, bluegill (mg/l): >100, >114 (Senn).

USE: Insecticide.

9447. Thiamine. [59-43-8] 3-[(4-Amino-2-methyl-5-pyrimidinyl)methyl]-5-(2-hydroxyethyl)-4-methylthiazolium chloride (1:1); vitamin B_1; aneurin; thiamine monochloride; thiaminium chloride. $C_{12}H_{17}ClN_4OS$; mol wt 300.81. C 47.91%, H 5.70%, Cl 11.78%, N 18.63%, O 5.32%, S 10.66%. Essential nutrient required for carbohydrate metabolism; also involved in nerve function. Biosynthesized by microorganisms and plants. Dietary sources include whole grains, meat products, vegetables, milk, legumes and fruit. Also present in rice husks and yeast. Converted *in vivo* to thiamine diphosphate, a coenzyme in the decarboxylation of α-keto acids. Chronic deficiency may lead to neurological impairment, beriberi, Wernicke-Korsakoff syndrome. Isoln from rice bran: B. C. P. Jansen, W. F. Donath, *Chem. Weekbl.* **23**, 201 (1926). Structure: R. R. Williams, *J. Am. Chem. Soc.* **58**, 1063 (1936); R. R. Williams, J. K. Cline, *ibid.* 1504; R. R. Williams *et al., ibid.* **59**, 526 (1937). Review of syntheses: Knobloch in H. Vogel, *Chemie und Technik der Vitamine* **vol. II** (Stuttgart, 1953) pp 1-128. Toxicity data: D. Winter *et al., Int. Z. Vitaminforsch.* **37**, 82 (1967). HPLC determn in foods, pharmaceuticals, body tissues: T. Kawaski, *Methods Enzymol.* **122**, 15 (1986); in plasma and pharmacokinetics: H. Mascher, C. Kikuta, *J. Pharm. Sci.* **82**, 56 (1993). Review of bioavailability, absorption, and role in nutrition: F. L. Iber *et al., Am. J. Clin. Nutr.* **36**, 1067-1082 (1982). *Reviews:* "Thiamin: Twenty Years of Progress", *Ann. N.Y. Acad. Sci.* **378**, H. Z. Sable, C. J. Grubier, Eds. (1982) 470 pp; "Thiamin, Vitamin B_1, Aneurin" in *Vitamins*, W. Friedrich, Ed. (de Gruyter, Berlin, 1988) pp 339-401.

Hydrochloride. [67-03-8] Thiamine chloride hydrochloride; thiamine dichloride; Benerva; Betabion; Betalin S; Betaxin; Bewon; Metabolin; Vitaneurin. $C_{12}H_{17}ClN_4OS \cdot HCl$; mol wt 337.26. Comprehensive description: K. A. M. Al-Rashood *et al., Anal. Profiles Drug Subs.* **18**, 413-458 (1989). Monoclinic plates in rosette-like clusters. Slight thiazole odor. Bitter taste. dec 248°. One gram dissolves in ~1 ml water, 18 ml glycerol, 100 ml 95% alcohol, 315 ml abs alcohol; more sol in methanol. Sol in propylene glycol. Practically insol in hexane, chloroform; insol in ether, benzene. pH of a 1% w/v soln in water 3.13; pH of a 0.1% w/v soln in water 3.58. On exposure to air of average humidity, the vitamin absorbs an amount of water corresponding to nearly one mol, forming a hydrate. LD_{50} in mice (mg/kg): 89.2 i.v.; 8224 orally (Winter).

Mononitrate. [532-43-4] $C_{12}H_{17}N_5O_4S$. Prepn: R. J. Turner, G. J. Stammer, **US 2844579** (1958 to Am. Cyanamid). White crystals, mp 196-200° (dec). Practically nonhygroscopic. pKa 4.8. Soly in water (g/100 ml): 2.7 (25°); ~30 (100°). pH of 2% aq soln 6.5 to 7.1. Slightly sol in alc; very slightly sol in chloroform. More stable than the hydrochloride; suitable for enrichment of flours and feeds, multivitamin prepns.

THERAP CAT: Vitamin (enzyme cofactor).

THERAP CAT (VET): Vitamin (enzyme cofactor).

9448. Thiamine Diphosphate. [154-87-0] 3-[(4-Amino-2-methyl-5-pyrimidinyl)methyl]-4-methyl-5-(4,6,6-trihydroxy-4,6-dioxido-3,5-dioxa-4,6-diphosphahex-1-yl)thiazolium chloride (1:1); thiamine trihydrogen pyrophosphate (ester); cocarboxylase; thiamine pyrophosphate; thiamine diphosphate ester chloride; TDP; TPP; Berolase; Bivitasi; Cocalose; Cocarvit; Nutrase; Pyrolase. $C_{12}H_{19}ClN_4O_7P_2S$; mol wt 460.76. C 31.28%, H 4.16%, Cl 7.69%, N 12.16%, O 24.31%, P 13.44%, S 6.96%. Coenzyme required for the decarboxylation of α-keto acids. Most abundant form of thiamine in animal tissues. Identification as cofactor and enzymatic synthesis: K. Lohmann, P. Schuster, *Biochem. Z.* **294**, 183 (1937). Chemical synthesis: Weijlard, Tauber, *J. Am. Chem. Soc.* **60**, 2263 (1938);

Weil-Malherbe, *Biochem. J.* **34**, 980 (1940); Weijlard, *J. Am. Chem. Soc.* **63**, 1160 (1941); Karrer, Viscontini, *Helv. Chim. Acta* **29**, 711 (1946). LC determn in blood: J. W. I. Brunnekreeft *et al., J. Chromatogr.* **491**, 89 (1989). Role in pathogenesis of Wernicke-Korsakoff Syndrome: R. F. Butterworth *et al., Alcohol. Clin. Exp. Res.* **17**, 1084 (1993). Review of enzyme activity: J. Ullrich *et al., Vitam. Horm.* **28**, 365 (1970); of mechanism of action: R. Kluger, *Chem. Rev.* **87**, 863-876 (1987); M. Louloudi, N. Hadjiliadis, *Coord. Chem. Rev.* **135**, 429-468 (1994).

Monohydrate. Crystals from alc contg some HCl, dec 240-244°. mp 238-240° from abs ethanol. uv max: 242 nm. Soluble in water. pH of 0.3% soln 2.23. The dry substance is very stable.

Free ester tetrahydrate. $C_{12}H_{18}N_4O_7P_2S \cdot 4H_2O$. Prepn: Wenz *et al.*, **US 2991284** (1961 to E. Merck). Dec 220-225°.

Thiamine monophosphate. [532-40-1] 3-[(4-Amino-2-methyl-5-pyrimidinyl)methyl]-4-methyl-5-[2-(phosphonooxy)ethyl]thiazolium chloride; thiamine monophosphate chloride. $C_{12}H_{18}ClN_4O_4PS$; mol wt 380.78. Prepd by sulfuric acid hydrolysis of thiamine diphosphate: Lohmann, Schuster, *loc. cit.* Monohydrate; prisms from dil acetone or water, mp ~200°. Freely sol in water.

THERAP CAT: Vitamin (enzyme cofactor).

9449. Thiamine Disulfide. [67-16-3] N,N'-[Dithiobis[2-(2-hydroxyethyl)-1-methyl-2,1-ethenediyl]]bis[N-[(4-amino-2-methyl-5-pyrimidinyl)methyl]formamide]; aneurin disulfide; vitamin B_1 disulfide; Aktivin; Neolamin. $C_{24}H_{34}N_8O_4S_2$; mol wt 562.71. C 51.23%, H 6.09%, N 19.91%, O 11.37%, S 11.39%. Isoln: Zima, Williams, *Ber.* **73**, 941 (1940). Prepn: Warnat, **US 2458453** (1949 to Hoffmann-La Roche); Matsukawa, Hirano, *J. Pharm. Soc. Jpn.* **73**, 379 (1953); Hirano, Iwatsu, *ibid.* **73**, 1115 (1953); Kawasaki, *ibid.* **76**, 706 (1956). Bioavailability: N. Aoyagi *et al., Chem. Pharm. Bull.* **34**, 281 (1986).

(Z,Z)-isomer

When anhydr, mp 177° with intense yellow coloration. Very sparingly sol in benzene, acetone, ether, ethanol; freely sol if the crystals contain solvated acetone or water. Soly in water at 37° (mg/ml): 37.7 (pH 1.2); 3.0 (pH 3); 1.33 (pH 5); 0.568 (pH 7.2).

Hydrochloride. [18642-10-9] Crystals from methanol + water, dec 231°.

Phosphate. [992-46-1] Thiamine monophosphate disulfide; Biotinin; Vitamogen. $C_{24}H_{36}N_8O_{10}P_2S_2$; mol wt 722.67.

***O,O*-Diisobutyrate.** [3286-46-2] *O*-Isobutyrylthiamine disulfide; bisibutiamine; sulbutiamine; Arcalion; Neodaian; Vitaberin. $C_{32}H_{46}N_8O_6S_2$; mol wt 702.89.

THERAP CAT: Vitamin (enzyme cofactor).

9450. Thiamine Triphosphate. [3475-65-8] 3-[(4-Amino-2-methyl-5-pyrimidinyl)methyl]-4-methyl-5-(4,6,8,8-tetrahydroxy-6,8-trioxido-3,5,7-trioxa-4,6,8-triphosphaoct-1-yl)thiazolium inner salt; thiamine triphosphoric acid ester; TTP. $C_{12}H_{19}N_4O_{10}P_3S$; mol wt 504.28. C 28.58%, H 3.80%, N 11.11%, O 31.73%, P 18.43%, S 6.36%. One of the bioactive forms of thiamine; concentrated in neurons and in excitable tissues such as skeletal muscle. Believed to have a noncofactor role in nerve function. Prepn: L. Velluz *et al., Bull. Soc. Chim. Fr.* **15**, 871 (1948); Roux *et al., Bull. Soc. Chim. Biol.* **30**, 592 (1948); M. Viscontini *et al., Helv. Chim. Acta* **32**, 1478

(1949). Relation to thiamine diphosphate: L. Velluz *et al.*, *J. Biol. Chem.* **180**, 1137 (1949). Identification in rat liver: A. Rossi-Fanelli *et al.*, *Science* **116**, 711 (1952). Biosynthesis by baker's yeast: K.-H. Kiessling, *Nature* **172**, 1187 (1953). Enzymatic synthesis: J. R. Cooper, K. Nishino, *Methods Enzymol.* **122**, 24 (1986). Chromatographic determn in brain extracts: J. R. Cooper, T. Matsuda, *ibid.* 20. Review of metabolism, tissue distribution and potential neurophysiological role: J. R. Cooper, J. H. Pincus, *Neurochem. Res.* **4**, 223-239 (1979); L. Bettendorff, *Metab. Brain Dis.* **9**, 183-209 (1994).

Hemihydrate. Very hygroscopic, rhomb-shaped microcrystals from water-alcohol-acetone, dec 228-232°. Freely sol in water. Insol in the usual organic solvents.

9451. Thiamiprine. [5581-52-2] 6-[(1-Methyl-4-nitro-1*H*-imidazol-5-yl)thio]-9*H*-purin-2-amine; 2-amino-6-[(1-methyl-4-nitroimidazol-5-yl)thio]purine; 2-amino-6-(1'-methyl-4'-nitro-5'-imidazolyl)mercaptopurine; Guaneran. $C_9H_8N_8O_2S$; mol wt 292.28. C 36.98%, H 2.76%, N 38.34%, O 10.95%, S 10.97%. Prepd from thioguanine and 1-methyl-4-nitro-5-chloroimidazole: Hitchings, Elion, **US 3056785** (1962 to Burroughs Wellcome).

Crystals, dec slowly >200°. uv max: 320 nm at pH 1; 315 nm at pH 11.

THERAP CAT: Antineoplastic.

9452. Thiamorpholine. [123-90-0] Thiomorpholine; tetrahydro-1,4-thiazine; 1,4-thiazan; parathiazan. C_4H_9NS; mol wt 103.18. C 46.56%, H 8.79%, N 13.58%, S 31.07%. Prepd by heating β,β'-dichlorodiethyl sulfide with alcoholic ammonia under pressure: Davies, *J. Chem. Soc.* **117**, 298, 306 (1920).

Mobile liquid. Strong odor resembling that of piperidine. Absorbs CO_2 from air. bp$_{758}$ 169°; bp$_{743}$ 166-167°. Volatile with steam. Miscible with water and many organic liquids.

Hydrochloride. $C_4H_9NS.HCl$. Hygroscopic needles from alcohol + ether, dec 163°. Freely sol in water.

Picrolonate. $C_4H_9NS.C_{10}H_8N_4O_5$. Deep orange prisms from alcohol, dec 242° (darkens at 210°).

9453. Thiamphenicol. [15318-45-3] 2,2-Dichloro-*N*-[(1*R*,2*R*)-2-hydroxy-1-(hydroxymethyl)-2-[4-(methylsulfonyl)phenyl]-ethyl]acetamide; D-*d*-*threo*-2-dichloroacetamido-1-(4-methylsulfonyl)phenyl-1,3-propanediol; dextrosulphenidol; Win-5063-2; 8053CB; Hyrazin; Igralin; Neomyson; Rigelon; Thiamcol; Thionicol; Thiophenicol; Urfamycine; Urophenil. $C_{12}H_{15}Cl_2NO_5S$; mol wt 356.21. C 40.46%, H 4.24%, Cl 19.90%, N 3.93%, O 22.46%, S 9.00%. Synthesis: Cutler *et al.*, *J. Am. Chem. Soc.* **74**, 5475 (1952); Suter *et al.*, *ibid.* **75**, 4330 (1953); Suter, **US 2759927**; **US 2759970**; **US 2759971**; **US 2759972**; **US 2759976** (all 1956 to Sterling Drugs); **GB 770277** (1957 to Parke, Davis).

Crystals, mp 164.3-166.3°. $[\alpha]_D^{25}$ +12.9° (ethanol). uv max (95% ethanol): 224, 266, 274 nm (ε 13,700, 800, 700). Appreciably sol in water. Soluble in alcohol.

DL-Form. [847-25-6] Raceophenidol; racephenicol; Dexawin; Thiocymetin.

THERAP CAT: Antibacterial.

THERAP CAT (VET): D-Form as antimicrobial; DL-form in control of fowl cholera.

9454. Thiamylal. [77-27-0] Dihydro-5-(1-methylbutyl)-5-(2-propen-1-yl)-2-thioxo-4,6(1*H*,5*H*)-pyrimidinedione; 5-allyl-5-(1-methylbutyl)-2-thiobarbituric acid; thioseconal. $C_{12}H_{18}N_2O_2S$; mol wt 254.35. C 56.67%, H 7.13%, N 11.01%, O 12.58%, S 12.60%. Prepn: **GB 613704** (1948 to Lilly); Abe *et al.*, *J. Pharm. Soc. Jpn.* **75**, 891 (1955); Izumi, Nakanishi, **JP 56 8785** (1956 to Yoshitomi); Donnison, **US 2876225** (1959 to Abbott).

Crystals from dil ethanol, mp 132-133°.

Sodium salt. [337-47-3] Surital. $C_{12}H_{17}N_2NaO_2S$; mol wt 276.33.

Note: This is a controlled substance (depressant): **21 CFR**, 1308.13.

THERAP CAT: Anesthetic (intravenous).

THERAP CAT (VET): Anesthetic (intravenous).

9455. Thianaphthene. [95-15-8] Benzo[*b*]thiophene; benzothiofuran. C_8H_6S; mol wt 134.20. C 71.60%, H 4.51%, S 23.89%. Occurs in lignite tar. Catalytic synthesis from styrene and hydrogen sulfide: Moore, Greensfelder, *J. Am. Chem. Soc.* **69**, 2008 (1947); from ethyl benzene and hydrogen sulfide: Hansch, Hawthorne, *ibid.* **70**, 2495 (1948).

Leaflets. Odor similar to that of naphthalene. mp 32°; bp$_{760}$ 221°; bp$_{20}$ 103-105°. Volatile with steam. Sol in the usual organic solvents.

USE: Manuf of pharmaceuticals, thioindigo.

9456. Thiarubrines. Family of dark red, bioactive, dithiacyclohexadiene polyacetylenes isolated from various plant sources. Isoln from *Eriophyllum caespitosum* Dougl., *Compositae*: F. Bohlmann, K.-M. Kleine, *Ber.* **98**, 3081 (1965); and separation by HPLC: R. A. Norton *et al.*, *Phytochemistry* **24**, 356 (1985); from *Ambrosia chamissonis* (Less.) Greene: S. Ellis *et al.*, *ibid.* **33**, 224 (1993). Isoln of thiarubrine A from leaves of African medicinal plants of the genus *Aspilia*: E. Rodriguez *et al.*, *Experientia* **41**, 419 (1985). NMR spectra and biosynthetic study of A: M. L. Gomez-Barrios *et al.*, *Phytochemistry* **31**, 2703 (1992). Synthesis of A: M. Koreeda, W. Yang, *J. Am. Chem. Soc.* **116**, 10793 (1994); of B: E. Block *et al.*, *ibid.* **9403**; of C: Y. Wang *et al.*, *J. Org. Chem.* **63**, 8644 (1998). Antibiotic properties of A: G. H. N. Towers *et al.*, *Planta Med.* **51**, 225 (1985); photo-enhancement of activity by conversion to thiophenes: C. P. Constabel, G. H. N Towers, *ibid.* **55**, 35 (1989). Antiviral properties of A: J. B. Hudson *et al.*, *ibid.* **52**, 51 (1986); photo-enhancement of activity against HIV: *idem et al.*, *Photochem. Photobiol.* **57**, 675 (1993).

Thiarubrine A

Thiarubrine A. [63543-09-9] 3-(5-Hexene-1,3-diynyl)-6-(1-propynyl)-1,2-dithiin. $C_{13}H_8S_2$; mol wt 228.33. C 68.38%, H 3.53%, S 28.08%. Red liquid. Absorption max (ethanol): 490, 345, 233 nm.

Thiarubrine B. [71539-72-5] 3-(3-Buten-1-ynyl)-6-(1,3-pentadiynyl)-1,2-dithiin. $C_{13}H_8S_2$; mol wt 228.33. Absorption max (ethanol): 490, 345, 243 nm.

Thiarubrine C. [65710-89-6] 3-(3,5-Hexadien-1-ynyl)-6-(1-propynyl)-1,2-dithiin. $C_{13}H_{10}S_2$; mol wt 230.34. C 67.79%, H 4.38%, S 27.84%. Red oil. Absorption max (ethanol): 486.0, 345.0, 276.0 nm (ε 1645, 10660, 1534).

9457. Thiazesim. [5845-26-1] 5-[2-(Dimethylamino)ethyl]-2,3-dihydro-2-phenyl-1,5-benzothiazepin-4(5H)-one; thiazenone; tiazesim. $C_{19}H_{22}N_2OS$; mol wt 326.46. C 69.90%, H 6.79%, N 8.58%, O 4.90%, S 9.82%. Prepn of the hydrochloride: Krapcho *et al.*, *J. Med. Chem.* **6** (5), 544 (1963); Krapcho, *US 3075967* (1963 to Olin Mathieson). Metabolism: J. Dreyfuss *et al.*, *J. Pharm. Sci.* **57**, 1497, 1505 (1968).

Hydrochloride. [3122-01-8] SQ-10496; Altinil. $C_{19}H_{22}N_2OS$.HCl; mol wt 362.92. Crystals from acetonitrile, mp 222-224°.
THERAP CAT: Antidepressant.

9458. Thiazinamium Methylsulfate. [58-34-4] N,N,N,α-Tetramethyl-10H-phenothiazine-10-ethanaminium methyl sulfate (1:1); trimethyl(1-methyl-2-phenothiazin-10-ylethyl)ammonium methyl sulfate; trimethyl[1-methyl-2-(10-phenothiazinyl)ethyl]ammonium methyl sulfate; N-[β-(10-phenothiazinyl)propyl]trimethylammonium methyl sulfate; RP-3554; Multergan; Padisal. $C_{19}H_{26}N_2O_4S_2$; mol wt 410.55. C 55.59%, H 6.38%, N 6.82%, O 15.59%, S 15.62%. Prepd by treating promethazine with dimethyl sulfate: GB 641452 (1950 to Rhône-Poulenc). Crystal structure: P. Marsau, Y. Cam, *Acta Crystallogr.* **B29**, 980 (1973). Determn in body fluids: J. H. G. Jonkman *et al.*, *J. Pharm. Pharmacol.* **27**, 849 (1975). Bioavailability: *idem et al.*, *Clin. Pharmacol. Ther.* **21**, 457 (1977). Pharmacokinetics: *idem et al.*, *Arzneim.-Forsch.* **23**, 223 (1983); *idem et al.*, *Int. J. Clin. Pharmacol. Ther. Toxicol.* **21**, 454 (1983).

Crystals, mp 206-210° (some dec). Discolors on exposure to light. Soly in water (25°): ~10%. Freely sol in abs ethanol; sparingly sol in acetone. Practically insol in ether, benzene.
THERAP CAT: Antihistaminic.

9459. Thiazole. [288-47-1] C_3H_3NS; mol wt 85.12. C 42.33%, H 3.55%, N 16.46%, S 37.66%. First described as the pyridine of the thiophene series: A. Hantzsch, J. Weber, *Ber.* **20**, 3118

(1887); A. Hantzsch, *Ann.* **249**, 1 (1888). Book: *The Chemistry of Heterocyclic Compounds* **vol. 34**, A. Weissberger, E. C. Taylor, Eds. (New York, Wiley, 1979).

Colorless or pale yellow liquid. Characteristic foul odor; d^{17} ~1.20; bp 115-118°. Slightly sol in water; sol in many organic solvents. Forms compds with auric, mercuric, and platinic chlorides.

9460. Thiazole Orange. 1-Methyl-4-[(3-methyl-2(3H)-benzothiazolylidene)methyl]quinolinium; TO. $[C_{19}H_{17}N_2S]^+$. Asymmetric cyanine dye; fluoresces when intercalated with DNA or RNA. Prepn of tosylate and use in reticulocyte analysis: L. G. Lee *et al.*, *Cytometry* **7**, 508 (1986). Interactions with DNA: J. Nygren *et al.*, *Biopolymers* **46**, 39 (1998). Use in detection of reticulated platelets: T. Fujii *et al.*, *Thromb. Res.* **97**, 431 (2000); L. Joutsi-Korhonen *et al.*, *Eur. J. Haematol.* **65**, 66 (2000); in fluorescent intercalator displacement assays: D. L. Boger, W. C. Tse, *Bioorg. Med. Chem.* **9**, 2511 (2001).

Dimer. [143413-84-7] 1,1'-[1,3-Propanediylbis[(dimethylimino)-3,1-propanediyl]]bis[4-(3-methyl-2(3H)-benzothiazolylidene)methyl]quinolinium tetraiodide; TOTO. $C_{49}H_{58}I_4N_6S_2$; mol wt 1302.78. Synthesis and characterization: H. S. Rye *et al.*, *Nucleic Acids Res.* **20**, 2803 (1992). ^{1}H NMR soln structure of complex with DNA: H. P. Spielmann *et al.*, *Biochemistry* **34**, 8542 (1995). Interactions with DNA: J. Bunkenborg *et al.*, *Bioconjugate Chem.* **11**, 861 (2000). Absorption max (4 mM TAE, pH 8.2): 481 nm. ε_{507} 131.7×10^3 in methanol.

Iodide. [24147-36-2] $C_{19}H_{17}IN_2S$; mol wt 432.32. Absorption max (4 mM TAE, pH 8.2): 501 nm. ε_{502} 77.0×10^3 in methanol.

Methyl Sulfate. [172668-52-9] $C_{19}H_{17}N_2S.CH_3O_4S$; mol wt 416.51. Prepn: T. G. Deligeorgiev *et al.*, *Dyes Pigm.* **29**, 315 (1995). Crystals from acetone, mp 225-227°. Absorption max (methanol): 500 nm (ε 80700).

Tosylate. [107091-89-4] $C_{19}H_{17}N_2S.C_7H_7O_3S$; mol wt 476.61. Sol in methanol. Absorption max (methanol): 512 nm.

Note: The term "thiazole orange" is used to describe various salt forms in the chemical literature.

USE: Stains DNA or RNA in flow cytometry reticulocyte analysis, in agarose gels, in capillary electrophoresis.

9461. Thiazolsulfone. [473-30-3] 5-[(4-Aminophenyl)sulfonyl]-2-thiazolamine; 2-amino-5-sulfanilylthiazole; 4-aminophenyl-2'-aminothiazolyl-5'-sulfone; thiazosulfone; thiazolesulfone; Promizole. $C_9H_9N_3O_2S_2$; mol wt 255.31. C 42.34%, H 3.55%, N 16.46%, O 12.53%, S 25.11%. Prepd from the reaction products of *p*-nitrobenzenesulfonyl chloride and 2 mols of 2-aminothiazole: Bambas, *J. Am. Chem. Soc.* **67**, 671 (1945); Bambas, *US 2389126* (1945 to Parke, Davis).

Fine needles from alc, mp 219-221° (some decompn). Sol in water at pH 6.5 to the extent of 30-40 mg per 100 ml at 28-30°. Freely sol in acetone, dioxane, 70% alc, dil acids. Moderately sol in abs alcohol, ethyl acetate and ether. It is a very weak acid, dissolves in 10% alkali (decompn) and forms an alkali metal salt at pH 10.
THERAP CAT: Antibacterial.

9462. Thiazol Yellow G. [1829-00-1] 2,2'-(1-Triazene-1,3-diyldi-4,1-phenylene)bis[6-methyl-7-benzothiazolesulfonic acid] sodium salt (1:2); C.I. Direct Yellow 9; 2,2'-[(diazoamino)di-*p*-phenylene]bis[6-methyl-7-benzothiazolesulfonic acid] disodium salt; C.I. 19540; Chlorazol Yellow 2G; Clayton yellow; Diazamine Golden Yellow T; Titan yellow. $C_{28}H_{19}N_5Na_2O_6S_4$; mol wt 695.71. C 48.34%, H 2.75%, N 10.07%, Na 6.61%, O 13.80%, S 18.43%. Prepn from 2-(*p*-aminophenyl)-6-methyl-7-benzothiazolesulfonic acid ("dehydrothio-*p*-toluidinesulfonic acid"): *Beilstein* **vol. 27**, 2nd suppl, 509; F. J. Welcher, *Organic Analytical Reagents* **vol. 4** (Van Nostrand, New York, 1948) p 391; H. King *et al.*, *Analyst* **92**, 695 (1967). *See also: Colour Index* **vol. 2** (3rd ed., 1971) p 2010; *H. J. Conn's Biological Stains*, R. D. Lillie, Ed. (Williams & Wilkins, Baltimore, 9th ed., 1977) p 371.

Yellowish-brown powder. Sol in water, or alc (yellow solns), NaOH (reddish-yellow soln), H_2SO_4 (brownish-yellow soln). *Protect from light.*

USE: Dyeing cotton, viscose rayon, natural silk; as biological stain; as an analytical reagent, most often for the determination of Mg; as pH indicator, yellow 11.0 to red 13.0; as fluorescent dye for microscopy.

9463. Thiazopyr. [117718-60-2] 2-(Difluoromethyl)-5-(4,5-dihydro-2-thiazolyl)-4-(2-methylpropyl)-6-(trifluoromethyl)-3-pyridinecarboxylic acid methyl ester; RH-123652; MON-13200; Mandate; Visor. $C_{16}H_{17}F_5N_2O_2S$; mol wt 396.38. C 48.48%, H 4.32%, F 23.96%, N 7.07%, O 8.07%, S 8.09%. Pre-emergent herbicide for use in crops. Prepn: Y.-L. L. Sing, L. F. Lee, **EP 278944**; *eidem*, **US 4988384** (1988, 1991 both to Monsanto). GC determn in soil and plant samples: R. A. Pérez *et al.*, *J. Chromatogr. A* **778**, 193 (1997). Metabolism in plants: P. C. C. Feng *et al.*, *Pestic. Sci.* **45**, 203 (1995). Field trials in perennial crops: H. L. Warner, J. A. Holmdal, *Brighton Crop Prot. Conf. - Weeds* **1995**, 942.

Pale orange crystals from hexanes, mp 79-81°.
USE: Herbicide.

9464. Thibenzazoline. [6028-35-9] 1,3-Dihydro-1,3-bis(hydroxymethyl)-2*H*-benzimidazole-2-thione; 1,3-bis(hydroxymethyl)-2-benzimidazolinethione; 2-mercaptobenzimidazole-1,3-dimethylol; Thyreocordon. $C_9H_{10}N_2O_2S$; mol wt 210.25. C 51.41%, H 4.79%, N 13.32%, O 15.22%, S 15.25%. Prepn: Monti, Venturi, *Gazz. Chim. Ital.* **76**, 365 (1946).

Very bitter. mp 160-162°. Soluble dil in alkalies.
THERAP CAT: Antihyperthyroid.

9465. Thidiazuron. [51707-55-2] *N*-Phenyl-*N'*-1,2,3-thiadiazol-5-ylurea; 5-(*N*-phenylcarbamoylamino)-1,2,3-thiadiazole; thiadiazuron; TDZ; SN-49537; Dropp; Ginstar. $C_9H_8N_4OS$; mol wt 220.25. C 49.08%, H 3.66%, N 25.44%, O 7.26%, S 14.56%. Non-purine containing urea derivative with cytokinin activity. Prepn: H. Schulz, F. Arndt, **DE 2214632**; *eidem*, **US 3883547** (1973, 1975 both to Schering AG). Field trial in cotton: A. R. Hopkins, R. F.

Moore, *J. Econ. Entomol.* **73**, 768 (1980). Effect of temperature and adjuvants on activity in cotton: C. E. Snipes, G. D. Wills, *Weed Sci.* **42**, 13 (1994). Review of effect on shoot formation and axillary shoot proliferation: C.-Y. Lu, *In Vitro Cell. Dev. Biol.* **29P**, 92-96 (1993); on woody plant tissue culture: C. A. Huetteman, J. E. Preece, *Plant Cell Tissue Organ Cult.* **33**, 105-119 (1993).

Crystals from isopropanol, mp 217° (dec). Very sol in DMSO, DMF, acetone, cyclohexanone, isophorone. Slightly sol in aliphatic and aromatic hydrocarbons and water.
USE: Cotton defoliant; plant growth regulator in tissue culture.

9466. Thienamycin. [59995-64-1] (5*R*,6*S*)-3-[(2-Aminoethyl)thio]-6-[(1*R*)-1-hydroxyethyl]-7-oxo-1-azabicyclo[3.2.0]hept-2-ene-2-carboxylic acid. $C_{11}H_{16}N_2O_4S$; mol wt 272.32. C 48.52%, H 5.92%, N 10.29%, O 23.50%, S 11.77%. The first member of a family of des-thia-carbapenem nucleus antibiotics having a thioethylamine side-chain on the enamine portion of the fused 5-membered ring. Produced by *Streptomyces cattleya:* J. S. Kahan *et al.*, **US 3950357** (1976 to Merck & Co.). Discovery, isoln, taxonomy, physical properties: *eidem*, *J. Antibiot.* **32**, 1 (1979). Structure, absolute configuration: G. Albers-Schönberg *et al.*, *J. Am. Chem. Soc.* **100**, 6491 (1978). Synthesis of the (±)-form: D. B. R. Johnston *et al.*, *ibid.* 313; F. A. Bouffard *et al.*, *J. Org. Chem.* **45**, 1130 (1980); S. M. Schmitt *et al.*, *ibid.* 1135, 1142; D. G. Melillo *et al.*, *Tetrahedron Lett.* **21**, 2783 (1980); *eidem, ibid.* **22**, 913 (1981); M. Shiozaki, T. Hiraoka, *Tetrahedron* **38**, 3457 (1982). Alternate prepn of key intermediate: J. D. Buynak *et al.*, *Chem. Commun.* **1986**, 941. Stereocontrolled synthesis of the naturally occurring (+)-form: T. Salzmann *et al.*, *J. Am. Chem. Soc.* **102**, 6161 (1980); S. Karady *et al.*, *ibid.* **103**, 6765 (1981). Alternate synthetic routes: S. T. Hodgson *et al.*, *Tetrahedron* **41**, 5871 (1985); T. Iimori, M. Shibasaki, *Tetrahedron Lett.* **26**, 1523 (1985); T. Chiba, T. Nakai, *ibid.* 4647; D. J. Hart, D. C. Ha, *ibid.* 5493; T. Kametani *et al.*, *J. Org. Chem.* **50**, 2327 (1985); H. Maruyama, T. Hiraoka, *ibid.* **51**, 399 (1986); D. G. Melillo *et al.*, *ibid.* 1498. Continuous production in immobilized cell systems: E. J. Arcuri *et al.*, *Biotechnol. Bioeng.* **28**, 842 (1986). Biosynthesis: J. M. Williamson *et al.*, *J. Biol. Chem.* **260**, 4637 (1985). *In vitro* antibacterial activity: F. P. Tally *et al.*, *Antimicrob. Agents Chemother.* **14**, 436 (1978); S. S. Weaver *et al.*, *ibid.* **15**, 518 (1979). Evaluation of *in vitro* and *in vivo* activities: H. Kropp *et al.*, *Antimicrob. Agents Chemother.* **17**, 993 (1980). Comparative study vs gram-positive and gram-negative aerobic and anaerobic species and β-lactamase stability: H. C. Neu, P. Labthavikul, *ibid.* **21**, 180 (1982). *N*-Acyl derivatives as anti-inflammatory agents: J. B. Doherty *et al.*, **US 4465687** (1984 to Merck & Co.). Pharmacokinetics, bacteriological efficacy: P. Patamasucon, G. H. McCracken, *Antimicrob. Agents Chemother.* **21**, 390 (1982). Review of early syntheses of the carbapenems: T. Kametani, *Heterocycles* **17**, 463-506 (1982).

White hygroscopic solid. $[\alpha]_D^{27}$ +82.7° (c = 1.0 in water). uv max (water, pH 4-8): 296.5 nm (ε 7900); (pH 2): 309 nm; (pH 12): 300.5 nm. Freely sol in water, sparingly sol in methanol. In dilute soln, stability is optimal between pH 6-7, declining with unusual rapidity above that range. Susceptible to inactivation by dilute solns of hydroxylamine and cysteine.

N-Formimidoylthienamycin monohydrate see Imipenem.

9467. Thiethylperazine. [1420-55-9] 2-(Ethylthio)-10-[3-(4-methyl-1-piperazinyl)propyl]-10*H*-phenothiazine; 3-ethylmercapto-10-(1'-methylpiperazinyl-4'-propyl)phenothiazine. $C_{22}H_{29}N_3S_2$; mol wt 399.62. C 66.12%, H 7.31%, N 10.52%, S 16.05%. Prepn: Bourquin *et al.*, *Helv. Chim. Acta* **41**, 1072 (1958).

Crystals from acetone, mp 62-64°. bp$_{0.01}$ 227°.

Dimaleate. [1179-69-7] GS-95; NSC-130044; Torecan; Toresten. $C_{22}H_{29}N_3S_2.2C_4H_4O_4$; mol wt 631.76. Crystals from methanol, dec 188-190°. Slightly sol in methanol. Practically insol in water, chloroform.

Dihydrochloride. $C_{22}H_{29}N_3S_2.2HCl$. Crystals from ethanol, mp 214-216°.

Dimalate. [52239-63-1] $C_{22}H_{29}N_3S_2.2C_4H_6O_5$. Crystals from ethanol, mp 139°.

THERAP CAT: Antiemetic.

9468. Thifensulfuron-methyl. [79277-27-3] 3-[[[[(4-Methoxy-6-methyl-1,3,5-triazin-2-yl)amino]carbonyl]amino]sulfonyl]-2-thiophenecarboxylic acid methyl ester; methyl 3-(3-(4-methoxy-6-methyl-1,3,5-triazin-2-yl)ureidosulfonyl)thiophene-2-carboxylate; thiameturon-methyl; DPX-M6316; Pinnacle; Harmony. $C_{12}H_{13}N_5$-O_6S_2; mol wt 387.39. C 37.21%, H 3.38%, N 18.08%, O 24.78%, S 16.55%. Sulfonylurea herbicide for postemergence broadleaf weed control in food crops. Prepn: G. Levitt, **EP 30142**; *idem*, **US 4481029** (1981, 1984 both to DuPont). Degradation in soil: H. M. Brown *et al.*, *J. Agric. Food Chem.* **45**, 955 (1997). HPLC determn in cotton seed: J. J. Stry *et al.*, *J. AOAC Int.* **83**, 651 (2000). Metabolism in soybeans: H. M. Brown *et al.*, *Pestic. Biochem. Physiol.* **37**, 303 (1990). Review of physical properties, mode of action and field trials: S. D. Sionis *et al.*, *Proc. Br. Crop Prot. Conf. - Weeds* **1985**, 49-54.

White solid, mp 186°. pKa at 25°: 4.0. Soly in water at 25° (mg/l): 24 (pH 4); 260 (pH 5); 2400 (pH 6). Vapor pressure at 25°: 2.7 × 10^{-6} mmHg. Partition coefficient (octanol/water): 0.027.

USE: Herbicide.

9469. Thifluzamide. [130000-40-7] *N*-[2,6-Dibromo-4-(trifluoromethoxy)phenyl]-2-methyl-4-(trifluoromethyl)-5-thiazolecarboxamide; 2′,6′-dibromo-2-methyl-4′-trifluoromethoxy-4′-trifluoromethyl-1,3-thiazole-5-carboxanilide; MON-24000; RH-130753; Pulsor. $C_{13}H_6Br_2F_6N_2O_2S$; mol wt 528.06. C 29.57%, H 1.15%, Br 30.26%, F 21.59%, N 5.31%, O 6.06%, S 6.07%. Succinic dehydrogenase inhibitor (SDH) in fungal tricarboxylic acid cycle; especially active against basidiomycete fungi. Prepn: G. H. Alt *et al.*, **EP 371950**; *eidem*, **US 5045554** (1989, 1991 both to Monsanto). Field studies: P. O'Reilly *et al.*, *Brighton Crop Prot. Conf. - Pests Dis.* **1992**, 427; R. K. Moudgal *et al.*, *Pestology* **33**, 34 (2009). Soil persistence and leaching studies: S. Gupta, V. T. Gajbhiye, *Chemosphere* **57**, 471 (2004). GLC determn in rice: *eidem*, *Pestic. Res. J.* **16**, 43 (2004). GC determn in agricultural samples: Y. Hirahara *et al.*, *J. Health Sci.* **51**, 617 (2005).

Crystals from ethyl acetate + cyclohexane, mp 172-173°. Also reported as white to light brown powder, mp 177.9-178.6°. Log P

(*n*-octanol/water): 4.1. Soly in water (20°): 1.6 mg/l. LD$_{50}$ orally in rats: >5000 mg/kg. LC$_{50}$ (96 hr) in blue sunfish, rainbow trout, carp (mg/l): 1.2, 1.3, 2.9; LC$_{50}$ (48 hr) in *Daphnia magna* (mg/l): 1.6 (O'Reilly).

USE: Agricultural fungicide.

9470. Thimerosal. [54-64-8] Ethyl[2-(mercapto-κS)benzoato(2−)-κO]mercurate(1−) sodium (1:1); [(*o*-carboxyphenyl)thio]ethylmercury sodium salt; sodium ethylmercurithiosalicylate; thiomersalate; mercurothiolate; Merthiolate; Merzonin; Vitaseptol. $C_9H_9HgNaO_2S$; mol wt 404.81. C 26.70%, H 2.24%, Hg 49.55%, Na 5.68%, O 7.90%, S 7.92%. Prepd by reacting ethylmercuric chloride (or ethylmercuric hydroxide) with thiosalicyclic acid: Kharasch, **US 1672615** (1928); Trikojus, *Nature* **158**, 472 (1946); Swirska *et al.*, *Przem. Chem.* **39**, 371 (1960), *C.A.* **55**, 3507a (1961). Toxicity: Mason *et al.*, *Clin. Toxicol.* **4**, 185 (1971). Review of health effects from vaccine exposure: M. Bigham, R. Copes, *Drug Saf.* **28**, 89-101 (2005).

Cream-colored, crystalline powder. Stable in air, but not in sunlight. One gram dissolves in about 1 ml water, in about 8 ml alcohol. Practically insol in ether and benzene. Stabilization of solns with EDTA: Davisson, **US 2864844** (1958 to Lilly), pH of 1% aq soln: 6.7. LD$_{50}$ s.c. in rats: 98 mg/kg (Mason).

USE: Pharmaceutic aid (preservative).

THERAP CAT: Anti-infective.

THERAP CAT (VET): Antibacterial, antifungal (topical).

9471. Thioacetaldehyde. [2765-04-0] 2,4,6-Trimethyl-1,3,5-trithiane; trithioacetaldehyde. $C_6H_{12}S_3$; mol wt 180.34. C 39.96%, H 6.71%, S 53.33%. Occurs in α- (*trans*) and β- (*cis*) forms. Prepn: Baumann, Fromm, *Ber.* **22**, 2600 (1889); Fromm, Engler, *Ber.* **58**, 1916 (1925). Molecular structure: Hassel, Viervoll, *Acta Chem. Scand.* **1**, 164 (1947). Stereochemistry: Schönberg, Barakat, *J. Chem. Soc.* **1947**, 693.

α-form

α-Form. [23769-39-3] (2α,4α,6β)-2,4,6-Trimethyl-1,3,5-trithiane; *trans*-thioacetaldehyde. Monoclinic plates, mp 101°.

β-Form. [23769-40-6] (2α,4α,6α)-2,4,6-Trimethyl-1,3,5-trithiane; *cis*-thioacetaldehyde. Needles from acetone, mp 126°.

9472. Thioacetamide. [62-55-5] Ethanethioamide; acetothioamide. C_2H_5NS; mol wt 75.13. C 31.97%, H 6.71%, N 18.64%, S 42.67%. Prepd by heating ammonium acetate and aluminum sulfide: Kindler, Finndorf, *Ber.* **54**, 1080 (1921); from acetonitrile + H_2S: Kindler, *Ann.* **431**, 203 (1923); from acetamide + K_3PS_4: Schultz, Ranke, *Arch. Pharm.* **294**, 82 (1961). Toxicity data: Ambrose *et al.*, *J. Ind. Hyg. Toxicol.* **31**, 158 (1949). Review of uses: H. F. Walton, *Thioacetamide as Analytical Reagent*, Arapahoe Chemicals, Inc., Boulder, Colorado.

Crystals from benzene, mp 113-114°. Slight odor of mercaptans. uv max (water): 210, 261, 318 nm (log ε 3.66, 4.08, 1.8). Soly in water at 25° 16.3 g/100 ml; in ethanol 26.4 g/100 g. Sparingly sol in ether. MLD orally in rats: 200 mg/kg (Ambrose).

Caution: This substance is reasonably anticipated to be a human carcinogen: *Report on Carcinogens, Twelfth Edition* (PB2011-111646, 2011) p 403.

USE: Substitute for H_2S in laboratory qualitative analyses.

9473. Thioacetic Acid. [507-09-5] Ethanethioic acid; thiolacetic acid; thiacetic acid. C_2H_4OS; mol wt 76.11. C 31.56%, H 5.30%, O 21.02%, S 42.12%. CH_3COSH. Prepn by distilling glacial acetic acid with phosphorus pentasulfide: Kekulé, *Ann.* **90**, 309 (1854); from acetic anhydride and hydrogen sulfide: Ellingboe, *Org. Synth.* **31**, 105 (1951).

Yellow liquid; pungent odor. *Flammable.* d_4^{10} 1.075; bp 93°. Not solidified at −17°. Sol in water, particularly hot; very sol in alcohol.

9474. Thiobarbital. [77-32-7] 5,5-Diethyldihydro-2-thioxo-4,6(1*H*,5*H*)-pyrimidinedione; 5,5-diethyl-2-thiobarbituric acid; Ibition. $C_8H_{12}N_2O_2S$; mol wt 200.26. C 47.98%, H 6.04%, N 13.99%, O 15.98%, S 16.01%. Prepn: Fischer, Dilthey, *Ann.* **335**, 350 (1904); Carrington, *J. Chem. Soc.* **1944**, 124. Metabolism in man: Bush *et al.*, *J. Pharmacol. Exp. Ther.* **134**, 110 (1961).

Pale yellow needles from water, mp 180°. Sol in about 88 parts hot water, in ethanol, chloroform, ether, acetone, ammonia, alkalies. Sparingly sol in toluene. Practically insol in benzene.

THERAP CAT: Antihyperthyroid.

9475. Thiobenzyl Alcohol. [100-53-8] Benzenemethanethiol; benzyl mercaptan; phenylmethanethiol; α-toluenethiol. C_7H_8S; mol wt 124.20. C 67.69%, H 6.49%, S 25.81%.

Colorless liquid; odor of leek; d^{20} 1.058; bp 194-195°. Oxidizes in air to dibenzyl disulfide.

Caution: Can cause mild irritation to mucous membranes.

9476. Thiobutabarbital. [2095-57-0] 5-Ethyldihydro-5-(1-methylpropyl)-2-thioxo-4,6(1*H*,5*H*)-pyrimidinedione; 5-*sec*-butyl-5-ethyl-2-thiobarbituric acid; 5-ethyl-5-(1-methylpropyl)-2-thiobarbituric acid; thibutabarbital. $C_{10}H_{16}N_2O_2S$; mol wt 228.31. C 52.61%, H 7.06%, N 12.27%, O 14.02%, S 14.04%. Prepn: Volwiler, Tabern, US 2153729; US 2153731 (both 1939 to Abbott).

Crystals, mp 163-165°.

Sodium salt. [947-08-0] Inactin; Brevinarcon. $C_{10}H_{15}N_2NaO_2$-S; mol wt 250.29. Crystalline, slightly hygroscopic solid. Readily sol in water.

Note: This is a controlled substance (depressant): **21 CFR, 1308.13.**

THERAP CAT: Anesthetic (intravenous).

9477. Thiocolchicine. [2730-71-4] *N*-[(7*S*)-5,6,7,9-Tetrahydro-1,2,3-trimethoxy-10-(methylthio)-9-oxobenzo[*a*]heptalen-7-yl]-acetamide. $C_{22}H_{25}NO_5S$; mol wt 415.50. C 63.60%, H 6.06%, N 3.37%, O 19.25%, S 7.72%. Prepn: Velluz, Muller, *Bull. Soc. Chim. Fr.* **1954**, 755; Muller, Velluz, US 2820029 (1958 to UCLAF). Prepn of 2-glucoside analog: Velluz, Muller, *Bull. Soc. Chim. Fr.* **1955**, 194; Muller, Velluz, *loc. cit.*

Yellow cubic crystals from ethyl acetate, mp 192-194°. $[\alpha]_D^{20}$ −221°. Sol in ethanol, acetone, chloroform. Practically insol in water, ether.

2-Glucoside analog. [602-41-5] Thiocolchicoside; 2-demethoxy-2-glucosidoxythiocolchicine; Coltramyl; Coltrax; Miorel; Musco-Ril. $C_{27}H_{33}NO_{10}S$; mol wt 563.62. Crystals from ethanol + 1*N* NaOH, dec 220°. $[\alpha]_D$ −609° (water), −240° (ethanol).

THERAP CAT: Thiocolchicoside as muscle relaxant (skeletal).

9478. Thiocresol. [26445-03-4] Methylbenzenethiol; methylthiophenol; tolylmercaptan; toluenethiol. C_7H_8S; mol wt 124.20. C 67.69%, H 6.49%, S 25.81%. $CH_3C_6H_4SH$. Prepn of *m*-isomer: D. S. Tarbell, D. K. Fukushima, *Org. Synth.* **Coll. Vol. 3**, 809 (1955).

All three isomers boil at ~195°. Insol in water. Sol in alcohol or ether.

o-**Thiocresol.** [137-06-4] Colorless to pale yellow liquid, mp 15°; volatile with steam. Flash point, closed cup: 147°F (64°C).

m-**Thiocresol.** [108-40-7] Liquid, mp below −20°. bp_{25} 90-93°. d_4^{20} 1.044. n_D^{25} 1.568-1.571. Flash point, closed cup: 163°F (73°C).

p-**Thiocresol.** [106-45-6] Leaflets, mp 43-44°. Flash point, closed cup: 154°F (68°C).

Caution: May be irritating to skin, eyes, and respiratory tract.

9479. Thioctic Acid. [62-46-4] 1,2-Dithiolane-3-pentanoic acid; 1,2-dithiolane-3-valeric acid; 6,8-thioctic acid; α-lipoic acid; 5-(1,2-dithiolan-3-yl)valeric acid; 5-[3-(1,2-dithiolanyl)]pentanoic acid; δ-[3-(1,2-dithiacyclopentyl)]pentanoic acid; protogen A; acetate replacing factor; pyruvate oxidation factor; Biletan; Thioctacid; Thioctan; Tioctan. $C_8H_{14}O_2S_2$; mol wt 206.32. C 46.57%, H 6.84%, O 15.51%, S 31.08%. Growth factor for many bacteria and protozoa; prosthetic group, coenzyme, or substrate in plants, microorganisms, and animal tissues. Isoln of naturally occurring *d*-form: L. J. Reed *et al.*, *Science* **114**, 93 (1951); *eidem*, *J. Am. Chem. Soc.* **75**, 1267 (1953); Patterson *et al.*, *ibid.* **76**, 1823 (1954). Syntheses of *dl*-form: Bullock *et al.*, *ibid.* **74**, 1868, 3455 (1952); Hornberger *et al.*, *ibid.* 2382; Reed, US 2980716 and US 3049549 (1961, 1962 to Res. Corp.); Lewis, Raphael, *J. Chem. Soc.* **1962**, 4263; Ose *et al.*, US 3223712 (1965 to Yamanouchi); J. Tsuji *et al.*, *J. Org. Chem.* **43**, 3606 (1978). Biosynthesis via linoleic acid: J. P. Carreau *Methods Enzymol.* **62**, 152-158 (1974). Enantioselective synthesis of *d*-form: P. C. Bulmanpage *et al.*, *Chem. Commun.* **1986**, 1408. Clinical study in treatment of Wilson's disease: S. F. Gomes da Costa, *Arzneim.-Forsch.* **20**, 1210 (1970). Use in treatment of mushroom poisoning: R. Plotzker *et al.*, *Am. J. Med. Sci.* **283**, 79 (1982); J. P. Hanrahan, M. A. Gordon, *J. Am. Med. Assoc.* **251**, 1057 (1984). *Reviews:* Wagner, Folkers, *Vitamins and Coenzymes* (Interscience, New York, 1964) pp 244-263; Schmidt *et al.*, *Angew. Chem. Int. Ed.* **4**, 846 (1965); Schmidt *et al.*, *Adv. Enzymol. Relat. Areas Mol. Biol.* **32**, 423 (1969).

d-Form

Sodium salt. [2319-84-8] $C_8H_{13}NaO_2S_2$. White powder, sol in water. pH of aq solns about 7.4.

d-**Form.** [1200-22-2] Crystals by vacuum sublimation (at 85-90° and 25 microns). mp 46-48° (microblock). $[\alpha]_D^{23}$ +104° (c = 0.88 in benzene). uv max (methanol): 333 nm (ε 150). pKa 5.4. Practically insol in water. Sol in fat solvents.

dl-**Form.** [1077-28-7] Yellow needles from cyclohexane, mp 60-61°. bp 160-165°. uv spectrum: Calvin, *Fed. Proc.* **13**, 703 (1954). Practically insol in water. Sol in fat solvents. Forms a water-soluble sodium salt.

l-**Form.** [1077-27-6] Crystals from cyclohexane, mp 45-47.5° (microblock). $[\alpha]_D^{23}$ −113° (c = 1.88 in benzene). uv max (methanol): 330 nm (ε 140).

Ethylenediamine. Tioctidasi.

THERAP CAT: Treatment of liver disease; antidote to poisonous mushrooms (*Amanita* species).

9480. Thiocyanate Sodium. [540-72-7] Thiocyanic acid sodium salt (1:1); sodium sulfocyanate; sodium rhodanide; sodium thiocyanate. CNNaS; mol wt 81.07. C 14.82%, N 17.28%, Na 28.36%, S 39.55%. NaSCN. Contains at least 98% NaSCN. Toxicity study: Anderson, Chen, *J. Am. Pharm. Assoc.* **29**, 152 (1940).

Colorless or white, deliquesc crystals. mp about 300°. Sol in about 0.6 part water; freely sol in alcohol, acetone. When dissolved in water the temp is considerably lowered. The solns are neutral. *Keep well closed.* LD$_{50}$ orally in rats: 764 mg/kg; i.v. in mice: 484 mg/kg (Anderson, Chen).

USE: Manuf other thiocyanates, especially organic. In titrimetry.

9481. Thiocyanic Acid. [463-56-9] Hydrogen thiocyanate; Rhodanwasserstoffsäure (German). CHNS; mol wt 59.09. C 20.33%, H 1.71%, N 23.70%, S 54.26%. HSCN. Thiocyanic acid is believed to be a tautomeric mixture of HSCN and HNCS (*isothiocyanic acid*): Beard, Dailey, *J. Chem. Phys.* **18**, 1437 (1950). Conveniently prepd in the laboratory from KNCS + KHSO$_4$: Birckenbach, Bucher, *Ber.* **73**, 1153 (1940). Dil aq solns may be prepd from ammonium thiocyanate by the action of ion exchange resins: Klement, *Z. Anorg. Allg. Chem.* **260**, 268 (1949). Toxicology of organic thiocyanates: Y. Yokoi, *Jpn. J. Pharmacol.* **3**, 99 (1954); *C.A.* **48**, 13965f (1954).

Colorless gas or white solid depending upon degree of polymerization. Freely sol in water. Very strong acid. A 5% aq soln may be kept refrigerated for several weeks. Also sol in some organic solvents.

Ethyl ester. Ethyl thiocyanate; ethyl sulfocyanate. C$_3$H$_5$NS. Liquid. d$_4^{23}$ 1.007. bp 146°. n$_D^{15}$ 1.4684. Insol in water. Misc with alcohol, ether. MLD in mice (mg/kg): 52 orally; 39.1 s.c.; 18.3 i.p.; 6 i.v. (Yokoi).

9482. Thiodicarb. [59669-26-0] 2,4,8-Trimethyl-5-oxo-6-oxa-3,9-dithia-2,4,7-triazadec-7-enoic acid [1-(methylthio)ethylidene]azanyl ester; bis-[*O*-(1-methylthioethylimino)-*N*-methylcarbamic acid]-*N*,*N*′-sulfide; *N*,*N*′-bis[1-methylthioacetaldehyde *O*-(*N*-methylcarbamoyl)oxime]sulfide; *O*-[[*N*-[*N*′-(1-methylthioethylideneiminooxycarbonyl)-*N*′-methylaminosulfenyl]-*N*-methylcarbamoyl]]-*S*-methylacetohydroximate; *N*,*N*′-[thiobis[(methylimino)carbonyloxy]]bisethanimidothioic acid dimethyl ester; dicarbosulf; bismethomyl thioether; UC-51762; CGA-45156; Larvin. C$_{10}$H$_{18}$-N$_4$O$_4$S$_3$; mol wt 354.46. C 33.89%, H 5.12%, N 15.81%, O 18.05%, S 27.13%. Prepn: **NL 7508197** (1976 to Ciba-Geigy); T. D. J. D'Silva, **DE 2654331**; *eidem*, **US 4382957** (1977, 1983 both to Union Carbide). Activity: A. A. Sousa *et al.*, *J. Econ. Entomol.* **70**, 803 (1977). Field trials: E. P. Pieters, D. L. Pitts, *J. Ga. Entomol. Soc.* **15**, 207 (1980); J. R. Bradley, Jr., A. M. Agnello, *J. Econ. Entomol.* **81**, 706 (1988).

Crystals, mp 173-174°. LD$_{50}$ in rats (mg/kg): 160 orally, >1600 dermally (Sousa).

USE: Insecticide.

9483. 2,2′-Thiodiethanol. [111-48-8] 2,2′-Thiobisethanol; thiodiglycol; thiodiethylene glycol; bis(hydroxyethyl)sulfide. C$_4$-H$_{10}$O$_2$S; mol wt 122.18. C 39.32%, H 8.25%, O 26.19%, S 26.24%. Prepd from ethylene oxide and hydrogen sulfide: Chichibabin, Bestuzher, *Compt. Rend.* **200**, 242 (1935); Nenitzescu, Scarlatescu, *Ber.* **68**, 587 (1935); Headlee, cited in G. O. Curme, F. Johnston, *Glycols* (Reinhold, New York, 1952) p 103.

Liquid. d$_4^{20}$ 1.1824. mp −16°. bp$_{14}$ 168°. n$_D^{20}$ 1.519. Flash pt, open cup: 320°F (160°C). Misc with water, alc. Slightly sol in ether.

9484. Thiodiglycolic Acid. [123-93-3] 2,2′-Thiobis[acetic acid]; dimethylsulfide-α,α′-dicarboxylic acid; mercaptodiacetic acid. C$_4$H$_6$O$_4$S; mol wt 150.15. C 32.00%, H 4.03%, O 42.62%, S 21.35%. Prepn from sodium chloroacetate and hydrogen sulfide: Loven, *Ber.* **27**, 3059 (1894).

Crystals from water. mp 129°. Sol in water, alcohol.

USE: Detection of copper, lead, mercury, silver: Dubsky *et al.*, *C.A.* **34**, 6185 (1940).

9485. 3,3′-Thiodipropionic Acid. [111-17-1] 3,3′-Thiobis-[propanoic acid]; β,β-thiodipropionic acid; thiodihydracrylic acid; diethyl sulfide 2,2′-dicarboxylic acid. C$_6$H$_{10}$O$_4$S; mol wt 178.20. C 40.44%, H 5.66%, O 35.91%, S 17.99%. Prepn: **GB 571628** (1945 to Am. Cyanamid); Gresham, Shaver, **US 2449992** (1948 to B. F. Goodrich).

Nacreous leaflets from hot water, mp 134°. pK (25°) 4.11. One gram dissolves in 26.9 ml water at 26°. Freely sol in hot water, alcohol, acetone.

USE: Antioxidant for soap products and polymers of ethylene. In plasticizers and lubricants. Proposed for edible fats, oils, other foods.

9486. Thioformamide. [115-08-2] Methanethioamide. CH$_3$-NS; mol wt 61.10. C 19.66%, H 4.95%, N 22.92%, S 52.47%. Prepd by the action of phosphorus pentasulfide on formamide: Willstätter, Wirth, *Ber.* **42**, 1911 (1909). Synthesis: Erlenmeyer, Menzi, *Helv. Chim. Acta* **31**, 2071 (1948); Schmitz, **US 2682558** (1954 to du Pont); R. Tull, L. M. Weinstock, *Angew. Chem. Int. Ed.* **8**, 278 (1969). Spectroscopic studies: R. Sugisaki *et al.*, *J. Mol. Spectrosc.* **49**, 241 (1974); R. H. Judge *et al.*, *Can. J. Chem.* **65**, 2100 (1987). Use in synthesis of thiazoles: T. J. Cousineau, J. A. Secrist III, *J. Org. Chem.* **44**, 4351 (1979); H. Fukatsu *et al.*, *Heterocycles* **29**, 1517 (1989).

Prisms from ethyl acetate or from ether + petr ether, mp 26-27°. uv max (CH$_3$OH): 263 nm (ε 12500). Turns yellow and dec to a dark-colored mass. May be stored as soln in abs ether, preferably over P$_2$O$_5$. Sol in water (with considerable cooling). Freely sol in tetrahydrofuran, alcohol, ether, acetone, ethyl acetate; sparingly sol in cold chloroform, benzene, petr ether, carbon disulfide. The decompn products are hydrogen cyanide, hydrogen sulfide, ammonia, and some solid, amorphous, sulfur-contg compounds.

Monohydrate. [5967-97-5] CH$_3$NS.H$_2$O; mol wt 79.12. Yellow oil.

USE: In the synthesis of the thiazole containing compds.

9487. 5-Thio-D-glucose. [20408-97-3] α-D-Glucothiopyranose. C$_6$H$_{12}$O$_5$S; mol wt 196.22. C 36.73%, H 6.16%, O 40.77%, S 16.34%. Glucose analog thought to interfere with cellular transport systems utilizing D-glucose due to this structural similarity. The first chemical known, other than a hormone or alkylating agent, which can interfere reversibly with spermatogenesis and which also has been effective against malignant cultured cells. Prepn: M. S. Feather, R. L. Whistler, *Tetrahedron Lett.* **1962**, 667; Rowell, Whistler, *J. Org. Chem.* **31**, 1514 (1966); U. G. Nayak, R. L. Whistler, *ibid.* **34**, 97 (1969); Abd El-Rahman, Whistler, *Org. Prep. Proced. Int.* **5**, 245 (1973); H. Driguez, B. Henrissat, *Tetrahedron Lett.* **22**, 5061 (1981). Pharmacology: D. J. Hoffman, R. L. Whistler, *Biochemistry* **7**, 4479 (1968); R. L. Whistler, W. C. Lake, *Biochem. J.* **130**, 919 (1972). Radioprotective effect on tissue *in vivo*: V. L. Schuman *et al.*, *Int. J. Radiat. Oncol. Biol. Phys.* **8**, 589 (1982). Brief review: *Science* **186**, 431 (1974).

Crystals from methanol, mp 135-136°. $[\alpha]_D^{20}$ +188° (c = 1.56 in water). LD_{50} i.p. in mice: 5.5 g/kg (Schuman).

USE: Tool for examination of D-glucose biochemistry.

9488. Thioglycerol. [96-27-5] 3-Mercapto-1,2-propanediol; α-monothioglycerol; thioglycerin. $C_3H_8O_2S$; mol wt 108.16. C 33.31%, H 7.46%, O 29.58%, S 29.64%. Prepn: L. Carius, Ann. **124**, 221 (1862); B. Sjöberg, Ber. **75**, 13 (1942). Use in fast atom bombardment MS: C. Dass, J. Mass Spectrom. **31**, 77 (1996); as antioxidant in cell culture media: M. Brielmeier et al., Nucleic Acids Res. **26**, 2082 (1998).

Yellowish, very viscous, hygroscopic liquid; slight sulfidic odor. bp_1 100-101°. d 1.246. n_D^{20} 1.527. Flash point, closed cup: 235°F (113°C). Miscible with alc. Freely sol in water. Insol in ether.

USE: Antioxidant preservative; reagent in analytical chemistry, cell culture research. Matrix substrate in fast atom bombardment mass spectrometry.

9489. Thioglycolic Acid. [68-11-1] 2-Mercaptoacetic acid; thioglycollic acid. $C_2H_4O_2S$; mol wt 92.11. C 26.08%, H 4.38%, O 34.74%, S 34.81%. Prepd by the action of sodium sulfhydrate on sodium chloroacetate; by electrolysis of dithioglycollic acid (from sodium sulfide and sodium chloroacetate). Toxicity study: Deichmann, Mergard, J. Ind. Hyg. Toxicol. **30**, 373 (1948).

Liquid, strong, unpleasant odor. *Corrosive.* Readily oxidized by air. d 1.325. mp −16.5°. bp_{29} 123°. bp_{15} 108°. bp_5 96°. K_1 at 25° = 2.1×10^{-4}; $K_2 = 2.1 \times 10^{-11}$. Miscible with water, alcohol, ether, chloroform, benzene, and many other organic solvents. LD_{50} orally in rats: 0.15 ml/kg (Deichmann, Mergard).

Caution: Potential symptoms of overexposure are irritation of eyes, skin, nose, throat; lacrimation, corneal damage; skin burns, blisters. See NIOSH Pocket Guide to Chemical Hazards (DHHS/ NIOSH 97-140, 1997) p 306.

USE: Sensitive reagent for iron, molybdenum, silver, tin. With ferric iron a blue color appears, and when an alkali hydroxide is added to a soln contg ferrous salts and thioglycolic acid, a yellow precipitate forms. Used in the manuf of thioglycolates. The ammonium and sodium salts are commonly used for cold waving and the calcium salt is a depilatory. The sodium salt also is used in bacteriology in the prepn of thioglycolate media.

9490. Thioguanine. [154-42-7] 2-Amino-1,9-dihydro-6H-purine-6-thione; 2-aminopurine-6-thiol; Lanvis; Tabloid. $C_5H_5N_5S$; mol wt 167.19. C 35.92%, H 3.01%, N 41.89%, S 19.18%. Purine antimetabolite. Prepn: Elion, Hitchings, J. Am. Chem. Soc. **77**, 1676 (1955); Hitchings, Elion, US 2697709; US 2800473; US 2884667; Hitchings et al., US 3019224; US 3132144 (1954, 1957, 1959, 1962, and 1964, all to Burroughs Welcome). HPLC determn in DNA: D. J. Warren, L. Slordal, Anal. Biochem. **215**, 278 (1993). Clinical evaluation in refractory psoriasis: C. Mason, G. G. Krueger, J. Am. Acad. Dermatol. **44**, 67 (2001); in inflammatory bowel disease: A. Qasim et al., Scand. J. Gastroenterol. **42**, 194 (2007).

Needles from water, mp >360°. Freely sol in dilute solns of alkali hydroxides. Insol in water, alc, chloroform.

THERAP CAT: Antineoplastic; immunosuppressant.

THERAP CAT (VET): Antineoplastic.

9491. Thioguanosine. [85-31-4] 6-Thioguanosine; 2-amino-6-mercapto-9-β-D-ribofuranosylpurine. $C_{10}H_{13}N_5O_4S$; mol wt

299.31. C 40.13%, H 4.38%, N 23.40%, O 21.38%, S 10.71%. Prepn: Fox et al., J. Am. Chem. Soc. **80**, 1669 (1948).

Hemihydrate. Tiny tapered prisms from water, dec 224-227°. $[\alpha]_D^{22}$ −64° (c = 1.3 in 0.1N NaOH). uv max (pH 4-6): 257, 342 nm (ε 8820, 24800). pKa 8.33.

USE: In cancer research, cf. 6-Mercaptopurine and Nebularine.

9492. Thiokol®. Polysulfide rubber; thiorubber. Polysulfide polymers prepd from dihaloalkanes and sodium polysulfide. Reviews of prepn, chemistry and applications: Berenbaum in High Polymers, H. Mark et al., Eds., vol. **13** entitled Polyethers part 3, N. G. Gaylord, Ed. (Interscience, New York, 1962) pp 43-114; Panek, ibid. pp 115-224; Berenbaum in Encyclopedia of Polymer Science and Technology vol. **11** (Interscience, New York, 1969) pp 425-447.

Thiokol A

Thiokol A. Ethanite; Perduren. The first commercial polysulfide polymer, prepd from ethylene dichloride and sodium polysulfide. Sulfur content 84%; d about 1.6. Mixes with natural rubber. Cured polymer retains unpleasant odor; irritating fumes evolve during manuf. Stable to the usual organic solvents and dil mineral acids. Unstable to alkalies and oxidizing substances. Of low tensile strength and abrasion resistance. Not recommended where tropic or arctic climates prevail.

Thiokol FA. Prepd from ethylene dichloride, dichlorodiethyl formal and sodium polysulfide. Sulfur content 47%; d 1.34. No odor. Excellent solvent resistant characteristics but not as good as Thiokol A. Low temperature flexibility to −50°F.

USE: In rubber and resin manuf. As lining for flexible oil pipes and self-sealing gasoline tanks.

9493. Thiolactic Acid. [79-42-5] 2-Mercaptopropanoic acid; 2-mercaptopropionic acid; 2-thiolpropionic acid; α-mercaptopropanoic acid. $C_3H_6O_2S$; mol wt 106.14. C 33.95%, H 5.70%, O 30.15%, S 30.21%. First prepd from α-chloropropionic acid and potassium hydrosulfide: Schacht, Ann. **129**, 3 (1864); prepn from α-chloropropionic acid and sodium thiosulfate: Martin, US 2413361 (1946 to Martin Labs). From sodium sulfide, sulfur and α-bromopropionic acid: Dumesnil, US 2985557 (1961 to Frank E. Jonas). Review: Reid, Organic Chemistry of Bivalent Sulfur vol. **I** (Chemical Publ. Co., New York, 1958) p 449.

Oil, disagreeable odor. *Poisonous.* Solidif on cooling, crystals mp about 10°. d_4^{15} 1.220. bp_{16} 117°. n_D^{16} 1.4823. Miscible with water, alcohol, ether, acetone.

Copper salt. [6416-10-0] $C_3H_5CuO_2S$; mol wt 168.68. Yellow precipitate. Practically insol in water and in dil HNO_3.

Barium salt. [6028-21-3] $C_6H_{10}BaO_4S_2$; mol wt 347.59. Rubbery compd, very sol in water. Practically insol in alcohol.

Mercury salt. $C_6H_{10}HgO_4S_2$; mol wt 410.85. Small shiny platelets, sparingly sol in water, very sol in alcohol.

Sodium salt. [22535-46-2] $C_3H_5NaO_2S$; mol wt 128.12. Crystals as pentahydrate. Dec about 250°. Sol in water, methanol, ethanol. Less sol in propanol.

Calcium salt. [35440-78-9] $C_6H_{10}CaO_4S_2$; mol wt 250.34. Crystals as decahydrate. Dec 270°. Sol in water, methanol, ethanol. Less sol in propanol.

Platinum salt. [6416-12-2] $C_6H_{10}O_4PtS_2$; mol wt 405.35. Greenish-yellow precipitate. Practically insol in water and dil acids. Sol in sodium hydroxide and sodium carbonate.

Silver salt. $C_3H_5AgO_2S$; mol wt 213.00. Yellow precipitate. Practically insol in water and dil HNO_3.

USE: In depilatory and hair waving prepns.

9494. 2-Thiolhistidine. [13552-61-9] α-Amino-2,3-dihydro-2-thioxo-1*H*-imidazole-4-propanoic acid; 2-mercaptohistidine; α-amino-2-mercapto-4-imidazolepropionic acid. $C_6H_9N_3O_2S$; mol wt 187.22. C 38.49%, H 4.85%, N 22.44%, O 17.09%, S 17.12%. Prepn from histidine: Ashley, Harington, *J. Chem. Soc.* **1930**, 2586; from aspartic acid: Harington, Overhoff, *Biochem. J.* **27**, 338 (1933). Alternate procedures: Hegedüs, *Helv. Chim. Acta* **38**, 22 (1955); Marei, Raphael, *J. Chem. Soc.* **1958**, 2624. *Review:* J. P. Greenstein, M. Winitz, *Chemistry of the Amino Acids* **vol. 3** (John Wiley, New York, 1961) pp 2671-2675.

L-Form. Plates from water, darkens at 290°, not melted at 310°. $[\alpha]_D^{20}$ $-10°$ (c = 2 in *N* HCl). pK_1^1 1.84; pK_2^1 8.47; pK_3^1 11.4. Readily sol in hot water, sparingly sol in water at room temp. Tends to form supersaturated solns. Practically insol in alcohol, other organic solvents.

Dihydrochloride. $C_6H_9N_3O_2.2HCl$. Large prisms, dec 197-199°. Freely sol in water, sparingly sol in alcohol.

9495. Thiolutin. [87-11-6] 6-(Acetamido)-4-methyl-1,2-dithiolo[4,3-*b*]pyrrol-5(4*H*)-one; *N*-(4,5-dihydro-4-methyl-5-oxo-1,2-dithiolo[4,3-*b*]pyrrol-6-yl)acetamide; 3-acetamido-5-methylpyrrolin-4-one[4,3-*d*]-1,2-dithiole; acetopyrrothine. $C_8H_8N_2O_2S_2$; mol wt 228.28. C 42.09%, H 3.53%, N 12.27%, O 14.02%, S 28.09%. Antibiotic isolated from several strains of *Streptomyces albus:* F. W. Tanner, Jr. *et al.*, 118th Am. Chem. Soc. Meet. (Chicago, Sept. 1950), *Abstracts of Papers*, p 18A; F. W. Tanner, Jr. *et al.*, **US 2689854** (1954 to Pfizer). Characterization, similarity to aureothricin, *q.v.*: W. D. Celmer *et al.*, *J. Am. Chem. Soc.* **74**, 6304 (1952). Structure: W. D. Celmer, I. A. Solomons, *ibid.* **77**, 2861 (1955). Total synthesis: U. Schmidt, F. Geiger, *Ann.* **664**, 168 (1963); K. Hagio, N. Yoneda, *Bull. Chem. Soc. Jpn.* **47**, 1484 (1974). Bactericidal, protozoicidal, fungicidal properties: H. Seneca *et al.*, *Antibiot. Chemother.* **2**, 357 (1952). Mode of action: A. Jimenez *et al.*, *Antimicrob. Agents Chemother.* **3**, 729 (1973). Inhibition of microbiological growth in beer: J. B. Bockelmann, F. B. Strandskov, **US 2798811** (1957 to Schaefer Brewing Co.). Activity against soil borne pathogens: P. R. Deb, B. K. Dutta, *Curr. Sci.* **53**, 659 (1984). As allergy inhibitor: P. Stahl *et al.*, **DE 3434562** (1986 to Boehringer, Mannheim), *C.A.* **105**, 54609k (1986).

Brilliant yellow needles from *n*-butanol, dec 273-276°. Sublimes 200°/0.1 mm. uv max (methanol): 250, 311, 388 nm (ε 6300, 5700, 11,000) (Celmer, Solomons). Sparingly sol in water (210 mg/l); more sol in methanol, ethanol, chloroform, acetone (1% solns in acetone have been prepd), glacial acetic acid, methyl isobutyl ketone. Less sol in ether, benzene, hexane. Stable in acid and neutral solns, decomp in alkaline soln. Effective vs gram-positive and gram-neg-

ative bacteria, fungi and ameboid parasites. LD_{50} in mice (mg/kg): 25 s.c.; 25 orally (Seneca).

9496. Thiomalic Acid. [70-49-5] Mercaptobutanedioic acid; mercaptosuccinic acid. $C_4H_6O_4S$; mol wt 150.15. C 32.00%, H 4.03%, O 42.62%, S 21.35%.

White crystals; sulfidic odor. mp 149-150°. Soly in water at 40° about 50 g/100 ml; in ethanol at 25° about 50 g/100 ml. Also sol in acetone. Moderately sol in ether. Practically insol in benzene. The aq soln gives a transitory blue color with ferric chloride.

9497. Thionalide. [93-42-5] 2-Mercapto-*N*-2-naphthalenyl-acetamide; thioglycollic-β-aminonaphthalide. $C_{12}H_{11}NOS$; mol wt 217.29. C 66.33%, H 5.10%, N 6.45%, O 7.36%, S 14.75%.

White to ivory-colored needles. mp 111-112°. Insol in water. Freely sol in most organic solvents.

USE: As a reagent for copper, mercury, silver, thallium, and bismuth.

9498. Thionaphthene-2-carboxylic Acid. [6314-28-9] Benzo[*b*]thiophene-2-carboxylic acid; 2-benzothiophenecarboxylic acid; TNCA; BL-5583. $C_9H_6O_2S$; mol wt 178.21. C 60.66%, H 3.39%, O 17.96%, S 17.99%. Prepn: R. Weissgerber, O. Kruber, *Ber.* **53**, 1551 (1920); *eidem*, **DE 341837** (1921), *Frdl.* **13**, 279 (1921); D. A. Shirley, M. D. Cameron, *J. Am. Chem. Soc.* **72**, 2788 (1950); T. Higa, A. J. Krubsack, *J. Org. Chem.* **41**, 3399 (1976). Use as anti-osteoporotic agent: C. M. Samour, J. A. Vida, **US 4101668** (1978 to Bristol-Myers). Effect on total skeletal calcium in mice: J. C. Robin *et al.*, *J. Med.* **11**, 15 (1980). Mechanism of action study: J. C. Robin *et al.*, *Calcif. Tissue Int.* **36**, 194 (1984). Hypocalcemic effect: A. J. Johannesson *et al.*, *Endocrinology* **117**, 1508 (1985).

Prisms from alcohol or needles from water, mp 236° (Weissgerber). Also reported as crystals from benzene-ethyl acetate, mp 236-238° (Higa).

9499. Thionine. [581-64-6] 3,7-Diaminophenothiazin-5-ium chloride (1:1); Lauth's violet; C.I. 52000. $C_{12}H_{10}ClN_3S$; mol wt 263.74. C 54.65%, H 3.82%, Cl 13.44%, N 15.93%, S 12.16%. Prepn from *p*-phenylenediamine: Lauth, *Compt. Rend.* **82**, 1441 (1876); *idem, Bull. Soc. Chim. Fr.* [2] **26**, 422 (1876); Bernthsen, **DE 25150** (1883 to BASF), *Frdl.* **1**, 253; Loiseleur, *Ann. Inst. Pasteur* **86**, 262 (1954); Loiseleur, Petit, *Compt. Rend.* **250**, 2573 (1960); Balestic, Magat, *C.A.* **55**, 23542g (1961). *See also: Colour Index* **vol. 4** (3rd ed., 1971) p 4469.

Blackish-green glistening needles. Absorption max (water): 602.5 nm. Difficultly sol in cold, easily sol in hot water, giving first a blue, then a violet soln. The soln becomes somewhat bluer when HCl is added. Brownish-red precipitate with NaOH. Yellowish-green soln with concd H_2SO_4 which upon dilution turns blue, then violet on further dilution.

USE: For general nuclear staining; for counting bacteria in milk; as antioxidant for linseed oil. Has been used as catalyst in condensation of 1,3-butadienyl ethyl ether with maleic anhydride.

9500. Thionyl Bromide. [507-16-4] Br_2OS; mol wt 207.87. Br 76.88%, O 7.70%, S 15.42%. $SOBr_2$. Prepd from thionyl chloride and hydrogen bromide: Besson, *Compt. Rend.* **122**, 320 (1896); **123**, 884 (1896); Hibbert, Pullman, *Inorg. Synth.* **1**, 113 (1939).

Orange-yellow liquid. d_4^{20} 2.688. mp $-52°$. bp_{773} 138°; bp_{20} 48°. Somewhat less stable than thionyl chloride. Dec slowly on standing in glass-stoppered bottles. Hydrolyzed by water. Miscible with benzene, chloroform, carbon tetrachloride.

Caution: Highly irritating to skin, eyes, mucous membranes, respiratory tract.

9501. Thionyl Chloride. [7719-09-7] Sulfurous oxychloride. Cl_2OS; mol wt 118.96. Cl 59.60%, O 13.45%, S 26.95%. $SOCl_2$. Prepn by the oxidation of sulfur dichloride with sulfur trioxide: Michaelis, *Ann.* **274**, 173 (1893); Edwards, *US 2362057* (1944 to Hooker Electrochemical); Fehér in *Handbook of Preparative Inorganic Chemistry* **vol. 1**, G. Brauer, Ed. (Academic Press, New York, 2nd ed., 1963) pp 382-383; Macaluso, "Sulfur Compounds" in *Kirk-Othmer Encyclopedia of Chemical Technology* **vol. 19** (Interscience, New York, 2nd ed., 1969) pp 398-401. Purification by distilling from quinoline and boiled linseed oil: Martin, Fieser, *Org. Synth.* **coll. vol. II**, 570 (1943). Alternate methods of purification: Cottle, *J. Am. Chem. Soc.* **68**, 1380 (1946); Kunkel; Rosenberg, Flaxman, *US 3155457*; *US 3156529* (both 1964 to Hooker Chem.); Friedman, Wetter, *J. Chem. Soc. A* **1967**, 36.

Colorless to pale yellow or reddish, fuming, refractive liquid. *Corrosive. Suffocating odor.* d_4^0 1.676; d_4^{10} 1.655; d_4^{20} 1.638. mp $-104.5°$. bp_{760} 76°; $bp_{96.6}$ 20°. n_D^{20} 1.517. Dec when heated above 140° forming Cl_2, SO_2, and S_2Cl_2. Hydrolyzed by water, forming SO_2 and HCl. Miscible with benzene, chloroform, carbon tetrachloride.

Caution: Potential symptoms of overexposure are irritation of eyes, skin, mucous membranes; eye, skin burns. *See NIOSH Pocket Guide to Chemical Hazards* (DHHS/NIOSH 97-140, 1997) p 306.

USE: For making acyl chlorides, to replace OH or SH groups with chlorine atoms; reacts with Grignard reagents to form the corresp sulfoxides. Review of use in organic synthesis: J. S. Pizey, *Synthetic Reagents* **vol. 1** (John Wiley, New York, 1974) pp 321-357.

9502. Thionyl Fluoride. [7783-42-8] Thionyl difluoride. F_2OS; mol wt 86.06. F 44.15%, O 18.59%, S 37.25%. SOF_2. Usually prepd by the action of antimony trifluoride on thionyl chloride in the presence of antimony pentafluoride: Booth, Mericola, *J. Am. Chem. Soc.* **62**, 640 (1940); Smith, Muetterties, *Inorg. Synth.* **6**, 162 (1960). May also be prepared by the action of other fluorides on thionyl chloride: Kemmitt, Sharp, *Adv. Fluorine Chem.* **4**, 228-229 (1965).

Colorless gas. Suffocating odor. Does not attack glass. Stored in compressed form in steel cylinders. mp $-129.5°$. bp $-43.8°$. Crit temp $+88°$. d (liq; $-100°$) 1.780. d (solid; $-183°$) 2.095. Trouton constant 22.6. Very slowly hydrolyzed by cold water. Sol in ether, benzene.

Caution: Highly irritating to eyes, respiratory tract.

9503. Thiopental Sodium. [71-73-8] 5-Ethyldihydro-5-(1-methylbutyl)-2-thioxo-4,6(1*H*,5*H*)-pyrimidinedione sodium salt (1:1); 5-ethyl-5-(1-methylbutyl)-2-thiobarbituric acid sodium salt; thiomebumal sodium; penthiobarbital sodium; thiopentone sodium; thionembutal; Intraval Sodium; Nesdonal Sodium; Pentothal Sodium; Trapanal. $C_{11}H_{17}N_2NaO_2S$; mol wt 264.32. C 49.99%, H 6.48%, N 10.60%, Na 8.70%, O 12.11%, S 12.13%. Prepn: *US 2153729* (1939); *US 2876225* (1959). Prepn of nonhygroscopic crystals: Hartop, *US 3109001* (1963 to Abbott). Acute toxicity: Christensen, Lee, *Toxicol. Appl. Pharmacol.* **26**, 495 (1973). Comprehensive description: M. J. McLeish, *Anal. Profiles Drug Subs. Excip.* **21**, 535-572 (1992).

Yellowish-white, hygroscopic powder. Alliaceous, garlic-like odor. Sol in water, alc. Insol in absolute ether, benzene, petr ether,

solvent hexane. Aq solns are alkaline to litmus. Solns dec on standing; on boiling precipitation occurs. LD_{50} in mice (mg/kg): 149 i.p.; 78 i.v. (Christensen, Lee).

Note: This is a controlled substance (depressant): **21 CFR**, 1308.13.

THERAP CAT: Anesthetic (intravenous).

THERAP CAT (VET): Short-acting anesthetic.

9504. Thiopeptin. [12609-84-6] Thiofeed. Sulfur containing peptide antibiotic complex produced by *Streptomyces tateyamensis* no. 7906: Miyairi *et al.*, **DE 1929355** (1969 to Fujisawa), *C.A.* **72**, 88921w (1970). Consists of the a and b series of thiopeptins A_1, A_2, A_3, A_4, and B (major component). Structurally similar to thiostrepton, *q.v.* Biological and chemical studies: Miyairi *et al.*, *Antimicrob. Agents Chemother.* **1**, 192 (1972). Characterization of thiopeptins B: Miyairi *et al.*, *J. Antibiot.* **23**, 113 (1970); structural studies: Muramatsu *et al.*, *ibid.* **25**, 537 (1972); **30**, 383 (1977). Total structures of components: O. D. Hensens, G. Albers-Schönberg, *Tetrahedron Lett.* **1978**, 3649.

Thiopeptins B. Faint yellow crystals, mp 219-222° (dec). $[\alpha]_D^{23}$ $-80°$ (c = 1 in chloroform). uv spectrum (methanol): shoulders at 230-250, 295, 305 nm. Sol in dioxane, DMSO, DMF, pyridine, chloroform. Insol in ether, benzene, *n*-hexane, petr ether, water. Fairly sol in methanol, acetone, ethyl acetate.

THERAP CAT (VET): Antibiotic feed additive.

9505. Thiophanate. [23564-06-9] *N,N'*-[1,2-Phenylenebis-(iminocarbonothioyl)]biscarbamic acid *C,C'*-diethyl ester; 4,4'-*o*-phenylenebis[3-thioallophanic acid]diethyl ester; 1,2-bis(3-ethoxy-carbonyl-2-thioureido)benzene; Cercobin; Topsin; Nemafax. $C_{14}H_{18}N_4O_4S_2$; mol wt 370.44. C 45.39%, H 4.90%, N 15.12%, O 17.28%, S 17.31%. Prepn: T. Noguchi *et al.*, **DE 1806123**; *eidem*, *US 4020095* (1969, 1977 both to Nippon Soda Co.). Photodegradation: H. Buchenauer *et al.*, *Pestic. Sci.* **5**, 343 (1973). Activity vs nematode infection in ruminants: Eichler, *Br. Vet. J.* **129**, 533 (1973). Toxicology in animals: *idem, ibid.* **130**, 570 (1974).

Colorless plates from acetone, dec 194°. LD_{50} orally in mice and rats: >15 g/kg (Eichler).

*O,O-***Dimethyl analog.** [23564-05-8] Thiophanate-methyl; Cercobin-M; Topsin-M. $C_{12}H_{14}N_4O_4S_2$; mol wt 342.39. Prepn: T. Noguchi *et al., loc. cit.* Persistence in soil: J. R. Fleeker *et al.*, *J. Agric. Food Chem.* **22**, 592 (1974). Pharmacology, toxicity: Y. Hashimoto *et al.*, *Toxicol. Appl. Pharmacol.* **23**, 606, 616 (1972). Colorless prisms, mp 181.5-182.5°. Sol in acetone, methanol, chloroform, acetonitrile; slightly sol in other organics. Practically insol in water. LD_{50} in rats, mice, guinea pigs, rabbits (g/kg): 3.40, 6.64, 3.64, 2.27 orally (Hashimoto).

USE: Systemic fungicide.

THERAP CAT (VET): Anthelmintic (Nematodes).

9506. Thiophene. [110-02-1] Thiofuran; thiofurfuran; thiole; thiotetrole; divinylene sulfide. C_4H_4S; mol wt 84.14. C 57.10%, H 4.79%, S 38.10%. Found in coal tar, in coal gas, and in technical benzene: V. Meyer, *Ber.* **16**, 1471 (1883); **17**, 2642 (1884). Made available in commercial quantities by a process utilizing the dehydrogenation of butane with sulfur as the dehydrogenating agent, followed by cyclization with sulfur to form the thiophene ring: Rasmussen, Ray, *Chem. Inds.* **60**, 593, 620 (1947). Laboratory prepn by heating sodium succinate with phosphorus trisulfide: R. Phillips, *Org. Synth.* **coll. vol. II**, 578 (1943). Also prepd by passing ethylene or acetylene into boiling sulfur; or by passing acetylene and hydrogen sulfide over hot bauxite or nickel hydroxide: *US 1421743 C.A.* **16**, 3093 (1922). *Review:* B. Buchholz in *Kirk-Othmer Encyclopedia of Chemical Technology* **vol. 22** (Wiley-Interscience, New York, 3rd ed., 1983) pp 965-973.

Liquid. Slight aromatic odor resembling that of benzene. *Flammable*. d_4^0 1.0873; d_4^{25} 1.0573; d_4^{50} 1.0285. mp $-38.3°$. bp$_{760}$ 84.4°; bp$_{400}$ 64.7°; bp$_{200}$ 46.5°; bp$_{100}$ 30.5°; bp$_{60}$ 20.1°; bp$_{40}$ 12.5°; bp$_{20}$ 0.0°; bp$_{10}$ $-10.9°$; bp$_5$ $-20.8°$; n_D^{25} 1.52684. Absorption spectrum: Purvis, *J. Chem. Soc.* **97**, 1653, 1656 (1910). Insol in water; miscible with most organic solvents. May be heated to 850° without decompn.

USE: Solvent similar to benzene, but suitable for lower and higher temps; manuf of resins from thiophene-phenol mixtures and formaldehyde; manuf of dyes and pharmaceuticals.

9507. 2-Thiophenecarboxylic Acid. [527-72-0] 2-Thenoic acid. $C_5H_4O_2S$; mol wt 128.15. C 46.86%, H 3.15%, O 24.97%, S 25.02%. Prepn: Voerman, *Rec. Trav. Chim.* **26**, 293 (1907); Sy, de Malleray, *Bull. Soc. Chim. Fr.* **1963**, 1276; Gross *et al.*, *Ber.* **96**, 1382 (1963).

Needles from water, mp 128.5°. Very sol in ether, alc, hot water; moderately sol in CHCl$_3$; slightly sol in petr ether.

Sodium salt. Sodium 2-thiophenecarboxylate; sodium 2-thenoate; Trophires. $C_5H_3NaO_2S$; mol wt 150.13.

USE: Sodium salt as lubricating-grease thickener, Morway, Kolfenbach, *US 2576031* (1951 to Standard Oil).

9508. Thiophenol. [108-98-5] Benzenethiol; phenylmercaptan; phenylthiol. C_6H_6S; mol wt 110.17. C 65.41%, H 5.49%, S 29.10%. Prepd by the reduction of benzenesulfonyl chloride with zinc dust in sulfuric acid: Adams, Marvel, *Org. Synth.* **1**, 71 (1921).

Liquid. Repulsive, penetrating, garlic-like odor, esp when impure. d_4^{25} 1.0728; bp$_{760}$ 168.3°; bp$_{100}$ 103.6°; bp$_{50}$ 86.2°; bp$_{20}$ 69.7°; bp$_{1.0}$ 18.6°; n_D^{25} 1.58603. Heat of fusion 24.90 cal/g; spec heat at 25°: 0.3829; entropy at 25°: 52.6. Insol in water. Very sol in alc; miscible with ether, benzene, CS$_2$. Feebly acidic. Oxidizes in air, esp when dissolved in alcoholic ammonia, forming diphenyl disulfide, $C_6H_5SSC_6H_5$. The hydrogen of the SH group is easily replaced by metals.

Caution: Potential symptoms of overexposure are irritation of eyes, skin and respiratory system; dermatitis; cyanosis; coughing, wheezing, dyspnea, pulmonary edema, pneumonitis; headache, dizziness, CNS depression; nausea, vomiting; kidney, liver and spleen damage. *See NIOSH Pocket Guide to Chemical Hazards* (DHHS/ NIOSH 97-140, 1997) p 26.

9509. Thiopropazate. [84-06-0] 4-[3-(2-Chloro-10*H*-phenothiazin-10-yl)propyl]-1-piperazineethanol 1-acetate; 2-chloro-10-[3-[1-(2-acetoxyethyl)-4-piperazinyl]propyl]phenothiazine; 10-[3-[1-(2-acetoxyethyl)-4-piperazinyl]propyl]-2-chlorophenothiazine; *N*-(β-acetoxyethyl)-*N'*-[γ-(2'-chloro-10'-phenothiazinyl)propyl]piperazine; 1-(2-acetoxyethyl)-4-[3-(2-chloro-10-phenothiazinyl)propyl]-piperazine. $C_{23}H_{28}ClN_3O_2S$; mol wt 446.01. C 61.94%, H 6.33%, Cl 7.95%, N 9.42%, O 7.17%, S 7.19%. Prepd from 2-chloro-10-(γ-chloropropyl)phenothiazine and piperazine in butanone followed by treatment with β-bromoethyl acetate in toluene: Cusic, *US 2766235* (1956); Anderson *et al.*, *Arzneim.-Forsch.* **12**, 937 (1962).

Free base, bp$_{0.1}$ 214-218°. Sol in ether.

Dihydrochloride. [146-28-1] Dartal; Dartalan. $C_{23}H_{28}ClN_3$- $O_2S.2HCl$; mol wt 518.92. Crystals from 95% ethanol, dec 223-229°. Freely sol in water; much less sol in alc, chloroform; almost insol in ether.

Dimaleate. [104999-18-0] $C_{23}H_{28}ClN_3O_2S.2C_4H_4O_4$. Crystals from methanol + ethanol, mp 167-169°.

THERAP CAT: Antipsychotic.

9510. Thioproperazine. [316-81-4] *N,N*-Dimethyl-10-[3-(4-methyl-1-piperazinyl)propyl]-10*H*-phenothiazine-2-sulfonamide; 3-dimethylsulfamoyl-10-[3-(4-methylpiperazino)propyl]phenothiazine; 2-dimethylsulfamoyl-10-[3'-(4''-piperazino)propyl]phenothiazine; thioperazine; RP-7843; SKF-5883. $C_{22}H_{30}N_4O_2S_2$; mol wt 446.63. C 59.16%, H 6.77%, N 12.54%, O 7.16%, S 14.36%. Prepn: GB 814512 (1959 to Rhône-Poulenc).

Crystals, mp 140°.

Fumarate. $C_{22}H_{30}N_4O_2S_2.C_4H_4O_4$. Crystals, mp 182°.

Dimethanesulfonate. [2347-80-0] Thioproperazine mesylate; Majeptil; Vontil. $C_{22}H_{30}N_4O_2S_2.2CH_3SO_3H$; mol wt 638.83.

THERAP CAT: Antipsychotic; antiemetic.

9511. Thioredoxin. [52500-60-4] Ubiquitous, small, hydrogen carrier protein that participates in a wide variety of biochemical reactions, e.g., ribonucleotide reduction, methionine sulfoxide, sulfate and disulfide reduction, phosphate transfer reactions. Thioredoxin is oxidized from a dithiol to a disulfide during ribonucleotide reduction, and regenerated by reduced triphosphopyridine nucleotide ('TPNH) and the flavoprotein thioredoxin reductase: Laurent *et al.*, *J. Biol. Chem.* **239**, 3436 (1964). Prepn from *Escherichia coli*: Williams *et al.*, *ibid.* **242**, 5226 (1967). Complete amino acid sequence: A. Holmgren, *Eur. J. Biochem.* **6**, 475 (1968). Purification and characterization: A. Holmgren *et al.*, *J. Biol. Chem.* **256**, 3118 (1981). Prepn of bacteriophage T4-induced thioredoxin: Berglund, Sjoberg, *ibid.* **245**, 6030 (1970). Amino acid sequence: Sjoberg, Holmgren, *ibid.* **247**, 8063 (1972). The T4 thioredoxin shows no structural homology with thioredoxin from *E. coli*. Purification, characterization and amino acid sequence of thioredoxin from *Corynebacterium nephridii*: M. Meng, H. P. C. Hogenkamp, *ibid.* **256**, 9174 (1981).

9512. Thioridazine. [50-52-2] 10-[2-(1-Methyl-2-piperidinyl)ethyl]-2-(methylthio)-10*H*-phenothiazine; 2-methylmercapto-10-[2-(*N*-methyl-2-piperidyl)ethyl]phenothiazine; 3-methylmercapto-*N*-[2'-(*N'*-methyl-2-piperidyl)ethyl]phenothiazine. $C_{21}H_{26}N_2S_2$; mol wt 370.57. C 68.07%, H 7.07%, N 7.56%, S 17.30%. Dopamine receptor blocker; parent compound of sulforidazine and mesoridazine, *q.q.v.* Prepn: Bourquin *et al.*, *Helv. Chim. Acta* **41**, 1072 (1958). Toxicity: E. I. Goldenthal, *Toxicol. Appl. Pharmacol.* **18**, 185 (1971). Effect on dopaminergic function: D. M. Niedzwiecki *et al.*, *J. Pharmacol. Exp. Ther.* **228**, 636 (1984); C. D. Kilts *et al.*, *ibid.* **231**, 334 (1984). GLC determn in plasma: E. C. Dinovo *et al.*, *J. Pharm. Sci.* **65**, 667 (1976). HPLC sepn of isomers: R. Whelpton *et al.*, *J. Chromatogr.* **426**, 223 (1988). Clinical evaluation: R. Axelsson *et al.*, *Curr. Ther. Res.* **21**, 587 (1977). Comprehensive description: E. M. Abdel-Moety, K. A. Al-Rashood, *Anal. Profiles Drug Subs.* **18**, 459-525 (1989).

Crystals from acetone, mp 72-74°. bp$_{0.02}$ 230°. uv max (95% ethanol): 263, 314 nm (ε 38172, 4595); (0.1N HCl): 230, 263 nm (ε 20939, 45954); (0.1 N NaOH): 313 nm (ε 5226). pKa 9.5. Very sol in chloroform (1 in 0.81); freely sol in dehydrated alc; sol in alcohol (1 in 6), ether (1 in 3). Practically insol in water. LD$_{50}$ orally in rats: 995 ±39 mg/kg (Goldenthal).

Hydrochloride. [130-61-0] TP-21; Aldazine; Mellaril; Melleretten; Melleril; Mallorol; Novoridazine; Orsanil; Ridazin; Stalleril. C$_{21}$H$_{26}$N$_2$S.HCl; mol wt 407.03. Crystals from acetone, mp 158-160°. uv max (water): 262, 310 nm (ε 41842, 3215); (95% ethanol): 264, 310 nm (ε 41598, 3256); (0.1N HCl): 264, 305 nm (ε 42371, 5495); (0.1N NaOH): 263 nm (ε 18392). Freely sol in water (1 in 9), ethanol (1 in 10), methanol, chloroform (1 in 5). Insol in ether.

THERAP CAT: Antipsychotic.

9513.   Thiosalicylic Acid. [147-93-3] 2-Mercaptobenzoic acid; *o*-sulfhydrylbenzoic acid. C$_7$H$_6$O$_2$S; mol wt 154.18. C 54.53%, H 3.92%, O 20.75%, S 20.79%. Prepd by heating *o*-halogenated benzoic acids with an alkaline hydrosulfide in the presence of copper: **DE 189200** (1906); by reduction of dithiosalicylic acid: Claasz, *Ber.* **45**, 2427 (1912); **DE 205450**; C. F. H. Allen, D. D. MacKay, *Org. Synth.* **coll. vol. II**, 580 (1943).

Sulfur-yellow flakes, plates, needles from glacial acetic acid or alcohol. Softens at 158°. mp 164-165°. Sublimes. Slightly sol in hot water; freely in glacial acetic acid, alc. Exposure to air yields dithiosalicylic acid, which is produced also when an alcoholic soln of thiosalicylic acid comes in contact with FeCl$_3$ (resulting in a transitory blue color).

Sodium salt. Jecto-Sal; Thiocyl. C$_7$H$_5$NaO$_2$S; mol wt 176.16. Crystals, sol in water.

USE: Manuf thioindigo dyes, reagent for determn of iron.

9514.   Thiosemicarbazide. [79-19-6] Hydrazinecarbothioamide; 1-aminothiourea. CH$_5$N$_3$S; mol wt 91.13. C 13.18%, H 5.53%, N 46.11%, S 35.18%.

White, cryst powder. mp 182-184°. Sol in water or alc. LD$_{50}$ orally in adult Norway rats: 13 mg/kg; *see:* Dieke, *Proc. Soc. Exp. Biol. Med.* **70**, 688 (1949).

USE: As a reagent for detection of metals.

9515.   Thiosinamine. [109-57-9] *N*-2-Propen-1-ylthiourea; allyl thiourea; allylthiocarbamide; Aminosin; Rhodalline. C$_4$H$_8$N$_2$-S; mol wt 116.18. C 41.35%, H 6.94%, N 24.11%, S 27.60%. Made by warming a mixture of equal parts of allyl mustard oil and abs alcohol with an equal amount of 30% ammonia. Growth regulating activity: Karanov, Vasilev, *Izv. Inst. Fiziol. Rast. Bulg. Akad. Nauk* **16**, 167 (1970), *C.A.* **73**, 119454y (1970). Mechanism of action studies as an egg suppressive agent in schistosomiasis, A. B. Machado *et al.*, *J. Parasitol.* **56**, 392 (1970). *In vivo* and *in vitro* effects: I. Popiel, D. A. Erasmus, *Trans. R. Soc. Trop. Med. Hyg.* **75**, 287 (1981). Toxicity study: S. H. Dieke *et al.*, *J. Pharmacol. Exp. Ther.* **90**, 260 (1947).

White crystals; bitter taste; slight garlic odor; d 1.22; mp 78°. Sol in about 30 parts water; sol in alcohol, slightly sol in ether. Insol in benzene. LD$_{50}$ orally in rats: 200 mg/kg (Dieke).

THERAP CAT (VET): Has been used to minimize scar tissue.

9516.   1-Thiosorbitol. [24531-57-5] 1-Thio-D-glucitol; 1-de-oxy-1-mercapto-D-sorbitol. C$_6$H$_{14}$O$_5$S; mol wt 198.23. C 36.35%,

H 7.12%, O 40.35%, S 16.17%. Prepd by heating an aq soln of glucose with a catalyst, sulfur, and hydrogen at 150° and 1500 psi for 3 hrs: Farlow *et al.*, *J. Am. Chem. Soc.* **70**, 1392 (1948).

Crystals from abs alcohol, mp 92-93°. Nonhygroscopic. [α]$_D^{27}$ −1.9° (c = 2). Readily sol in water, pyridine, ethylene glycol, formamide; practically insol in benzene, petr ether, carbon tetrachloride; carbon disulfide. Solubilities at 20° in g/100 ml solvent: 1.7 in abs alcohol; 1.2 in dioxane; 0.016 in ethyl ether; 0.016 in trichloroethylene; 0.010 in acetone. Strong reducing agent.

USE: Suggested as a polymer stabilizer; anticorrosion agent for pickling baths; in plating baths.

9517.   Thiostrepton. [1393-48-2] *N*-[(7*R*,8*S*)-2-Carboxy-7,8-dihydro-8-hydroxy-4-[(1*S*)-1-hydroxyethyl]-7-quinolinyl]-L-isoleucyl-L-alanyl-2,3-didehydroalanyl-L-alanyl-2-[(4a*R*,11*S*,14*Z*,18*S*,-21*S*,28*S*,32a*S*)-4a-amino-21-[(1*S*,2*R*)-1,2-dihydroxy-1-methylpropyl]-14-ethylidene-3,4,4a,9,10,11,12,13,14,18,19,20,21,27,28,32a-hexadecahydro-11,28-bis[(1*R*)-1-hydroxyethyl]-9,12,19,26-tetraoxo-17*H*,26*H*-8,5:18,15:25,22:32,29-tetranitrilo-5*H*,15*H*-pyrido-[3,2-*m*][1,11,17,24,4,7,20,27]tetrathiatetraazacyclotriacontin-2-yl]-4-thiazolecarbonyl-2,3-didehydroalanyl-2,3-didehydroalaninamide (1 → 528) lactone; Bryamycin; Thiactin. C$_{72}$H$_{85}$N$_{19}$O$_{18}$S$_5$; mol wt 1664.89. C 51.94%, H 5.15%, N 15.99%, O 17.30%, S 9.63%. Polypeptide antibiotic contg sulfur. Produced by *Streptomyces azureus* isolated from New Mexican soil: Pagano *et al.*, *Antibiot. Annu.* **1955-56**, 554; Vandeputte, Dutcher, *ibid.* 560; **GB 795570** (1958); Donovick *et al.*; Platt, **US 2982689**; **US 2982698** (1961 all to Olin Mathieson). Structure studies: Bodanszky *et al.*, *J. Am. Chem. Soc.* **84**, 2003 (1962); **86**, 2478 (1964); Anderson *et al.*, *Nature* **225**, 233 (1970). Total structure determn: K. Tori *et al.*, *Tetrahedron Lett.* **1976**, 185; *eidem, J. Antibiot.* **32**, 1072 (1979). ^{13}C-NMR study: *eidem, ibid.* **34**, 124 (1981). Total synthesis: K. C. Nicolaou *et al.*, *Angew. Chem. Int. Ed.* **43**, 5087, 5092 (2004). Mode of action as an inhibitor of protein synthesis: Cannon, Burns, *FEBS Lett.* **18**, 1 (1971); Cundliffe, *Biochem. Biophys. Res. Commun.* **44**, 912 (1971). Identity with thiactin: Bodanszky *et al.*, *J. Antibiot.* **16A**, 76 (1963). Comprehensive description: K. Florey, *Anal. Profiles Drug Subs.* **7**, 423-444 (1978). *Review:* Pestka, Bodley, in *Antibiotics* **vol. 3**, J. W. Corcoran, F. E. Hahn, Eds. (Springer-Verlag, New York, 1975) pp 551-573; E. Cundliffe, *ibid.* **vol. 5** (pt. 1), F. E. Hahn, Ed. (1979) pp 329-343.

Crystals from chloroform + methanol, dec 246-256°. [α]$_D^{23}$ −98.5° (glacial acetic acid); −61° (dioxane); −20° (pyridine). Sol in chloroform, dioxane, pyridine, glacial acetic acid, DMF, dimethyl sulfoxide. Practically insol in water, the lower alcohols, nonpolar organic solvents, dil aq acids or alkali. Dissolved by methanolic acid or base with decompn. Stable in the presence of gastric and intestinal juices and urine. No uv maxima but shows characteristic shoulders at 225, 250, 280 nm ($E_{1cm}^{1\%}$ 520, 380, 255).

Hemisuccinate. Prepn: Bodanszky, Fried, **US 3181995** (1965 to Olin Mathieson). mp 200-220°, forms a water-soluble potassium salt.

THERAP CAT: Antibacterial.

THERAP CAT (VET): Antibacterial.

9518. Thiotepa. [52-24-8] 1,1′,1″-Phosphinothioylidynetrisaziridine; triethylenethiophosphoramide; tris(1-aziridinyl)phosphine sulfide; Tepadina; Tespamin. $C_6H_{12}N_3PS$; mol wt 189.22. C 38.09%, H 6.39%, N 22.21%, P 16.37%, S 16.94%. Preparation: Kuh, Seeger, **US 2670347** (1954 to Am. Cyanamid); Saijo, Endo, **JP 55 218** (1955 to Semimoto Chem.). Toxicity data: Scherf *et al.*, *Arzneim.-Forsch.* **20**, 1467 (1970). GC determn in urine and plasma: R. J. van Maanen *et al.*, *J. Chromatogr. B* **719**, 103 (1998). Review of carcinogenicity studies: *IARC Monographs* **9**, 85-94 (1975); of pharmacology, bioanalytic methods and pharmacokinetics: M. J. van Maanen *et al.*, *Cancer Treat. Rev.* **26**, 257-268 (2000). Clinical trial as conditioning regimen for stem cell transplant in patients with lymphoma: F. Waheed *et al.*, *Leuk. Lymphoma* **45**, 2253 (2004); and other hematological malignancies: E. P. Alessandrino *et al.*, *Bone Marrow Transplant.* **34**, 1039 (2004).

Crystals from pentane or ether, mp 51.5°. Soly in water at 25°: 19 g/100 ml. Freely sol in alc, ether, chloroform; sol in benzene. LD_{50} i.v. in rats: 15 mg/kg (Scherf).

Caution: This substance is listed as a known human carcinogen: *Report on Carcinogens, Twelfth Edition* (PB2011-111646, 2011) p 406.

USE: Insect sterilant.

THERAP CAT: Antineoplastic.

9519. Thiothixene. [5591-45-7] *N,N*-Dimethyl-9-[3-(4-methyl-1-piperazinyl)propylidene]-9*H*-thioxanthene-2-sulfonamide; *cis*-9-[3-(4-methyl-1-piperazinyl)propylidene]-2-(dimethylsulfonamido)thioxanthene; tiotixene; Navane; Orbinamon. $C_{23}H_{29}N_3O_2S_2$; mol wt 443.62. C 62.27%, H 6.59%, N 9.47%, O 7.21%, S 14.45%. Prepn of *cis/trans*-isomer mixture and sepn of isomers: Bloom, Muren, **BE 647066** corresp to **US 3310553** (1964, 1967 both to Pfizer); Muren, Bloom, *J. Med. Chem.* **13**, 17 (1970). Only *cis*-isomer exhibits therapeutic activity. Structure studies: J. P. Schaefer, *Chem. Commun.* **1967**, 743. Pharmacological studies: A. Weissman, *Psychopharmacologia* **12**, 142 (1968). HPLC determn in serum: C. Dilger *et al.*, *Arzneim.-Forsch.* **38**, 1522 (1988). Book: T. A. Ban, *Psychopharmacology of Thiothixene* (Raven Press, New York, 1978) 485 pp. Comprehensive description: D. K. Wyatt, L. T. Grady, *Anal. Profiles Drug Subs.* **18**, 527-565 (1989).

cis-form

***cis*-Isomer.** Crystals, mp 147.5-149°. uv max (methanol): 228, 260, 310 nm (log ε 4.6, 4.3, 3.9). Has greater pharmacologic activity than *trans*-isomer. LD_{50} in mice, rats (mg/kg): 100, 55 i.p. (Weissman).

***trans*-Isomer.** mp 123-124.5°. uv max (methanol): 229, 252, 301 nm (log ε 4.5, 4.2, 3.9). LD_{50} i.p. in mice: 235 mg/kg (Weissman).

***cis-trans*-Isomer mixture.** Crystals, mp 114-118°. White to tan crystals. Very sol in chloroform; slightly sol in methanol, acetone. Practically insol in water.

Dimaleate. mp 158-160.5°.

Dioxalate. mp 229°.

THERAP CAT: Antipsychotic.

9520. 2-Thiouracil. [141-90-2] 2,3-Dihydro-2-thioxo-4(1*H*)-pyrimidinone; 2-mercapto-4-hydroxypyrimidine; 4-hydroxy-2(1*H*)-pyrimidinethione; 2-mercapto-4(1*H*)-pyrimidinone; 6-hydroxy-2-mercaptopyrimidine; 2-mercapto-4-pyrimidone; Deracil. $C_4H_4N_2OS$; mol wt 128.15. C 37.49%, H 3.15%, N 21.86%, O 12.48%, S 25.02%. Occurs in seeds of *Brassica, Cruciferae*: Purves, *Br. J. Exp. Pathol.* **22**, 241 (1941). Prepd by condensing ethyl formylacetate with thiourea: Wheeler, Liddle, *Am. Chem. J.* **40**, 550 (1908). Antithyroid activity results from its interference with the iodination of thyroxine precursors: *see* Maloof, Soodak, *Pharmacol. Rev.* **15**, 72-79 (1963). Inhibition of nucleic acid metabolism: Cardeilhac, *Proc. Soc. Exp. Biol. Med.* **125**, 692 (1967). Toxicology: K. K. Carroll, R. L. Noble, *J. Pharmacol. Exp. Ther.* **97**, 478 (1949).

Minute, bitter crystals, no definite mp. Very slightly soluble in water (1:2000); practically insol in alcohol, ether, acids. Readily sol in alkaline solns. LD_{100} i.p. in rats: 1500 mg/kg (Carroll, Noble).

THERAP CAT: Treatment of hyperthyroidism; angina pectoris; congestive heart failure.

THERAP CAT (VET): Thyroid depressant. In hyperthyroidism and to promote fattening.

9521. Thiourea. [62-56-6] Thiocarbamide. CH_4N_2S; mol wt 76.12. C 15.78%, H 5.30%, N 36.80%, S 42.12%. Made by fusing ammonium thiocyanate: R. E. Powers, **US 2552584**; R. E. Powers, J. Mitchell, **US 2560596** (both 1951 to Koppers); by treating cyanamide with hydrogen sulfide: R. O. Roblin, Jr., **US 2173067** (1940 to Am. Cyanamid); W. F. Lewis, **US 2393917** (1946 to Monsanto); J. van de Kamp, **US 2357149** (1944 to Merck & Co.). Toxicity data: Dieke *et al.*, *J. Pharmacol. Exp. Ther.* **90**, 262 (1947).

Crystals, mp 176-178°. d 1.405. Soluble in 11 parts water, in alcohol; sparingly sol in ether. Neutral reaction. Forms addition compds with metallic salts. LD_{50} orally in wild Norway rats: 1830 mg/kg (Dieke).

Caution: Chronic administration in rats has resulted in hepatic tumors, bone marrow depression and goiters: *Clinical Toxicology of Commercial Products*, R. E. Gosselin *et al.*, Eds. (Williams & Wilkins, Baltimore, 5th ed., 1984) Section II, p 350. This substance is reasonably anticipated to be a human carcinogen: *Report on Carcinogens, Twelfth Edition* (PB2011-111646, 2011) p 407.

USE: In animal glue liquifiers and silver tarnish removers. Photographic fixing agent and to remove stains from negatives; manuf resins; vulcanization accelerator; a reagent for determn of bismuth, selenite ions.

9522. Thioxanthene. [261-31-4] 9*H*-Thioxanthene; thiaxanthene; dibenzothiopyran; dibenzopenthiophene; diphenylenemethane sulfide. $C_{13}H_{10}S$; mol wt 198.28. C 78.75%, H 5.08%, S 16.17%. Conveniently obtained from thioxanthone in 74% yield by reduction wth lithium aluminum hydride: Mustafa, Hilmy, *J. Chem. Soc.* **1952**, 1345.

Consult the Name Index before using this section.

Needles or rods from alcohol + chloroform. Sublimes easily. mp 128°. bp$_{730}$ 340°. Dipole moment: 1.44 D. Freely sol in chloroform. Moderately sol in alcohol, ether.

9523. Thioxanthone. [492-22-8] 9*H*-Thioxanthen-9-one; thiaxanthone; 9-oxothioxanthene. C$_{13}$H$_8$OS; mol wt 212.27. C 73.56%, H 3.80%, O 7.54%, S 15.10%. Prepd by the interaction of diphenyl sulfide and phosgene with aluminum chloride as catalyst: Szmant *et al.*, *J. Org. Chem.* **18**, 745 (1953).

Yellow needles from chloroform, mp 211°. bp$_{715}$ 273°. Freely sol in benzene, chloroform, carbon disulfide, hot glacial acetic acid. Slightly sol in alc. Practically insol in water, alkaline solns. Sol in concd sulfuric acid, giving a yellow soln with strong green fluorescence in visible light.

9524. Thiphenamil. [82-99-5] α-Phenylbenzeneethanethioic acid *S*-[2-(diethylamino)ethyl]ester; diphenylthioacetic acid *S*-(2-diethylaminoethyl) ester; diphenylthiolacetic acid 2-diethylaminoethyl ester; *S*-[2-(diethylamino)ethyl]diphenylthioacetate; β-diethylaminoethyl diphenylthioacetate. C$_{20}$H$_{25}$NOS; mol wt 327.49. C 73.35%, H 7.69%, O 4.89%, S 9.79%. Smooth muscle relaxant. Prepn: Richardson, US 2390555 (1945 to Wm. P. Poythress); Clinton, US 2510773 (1950 to Sterling Drug); Richardson, *J. Pharm. Sci.* **55**, 1316 (1966). Mechanism of action study: Y. Kimoto *et al.*, *J. Urol.* **144**, 1497 (1990). Clinical evaluation in detrusor incontinence: C. E. Constantinou, *Urol. Int.* **48**, 42 (1992).

Hydrochloride. [548-68-5] Trocinate. C$_{20}$H$_{25}$NOS.HCl; mol wt 363.94. Rosettes of tiny needles from benzene + petr ether, large prisms from abs ethanol + ethyl acetate, mp 129-130°. Sol in water. Aq solns are about neutral to litmus.

THERAP CAT: Antispasmodic.

9525. Thiram. [137-26-8] *N*,*N*,*N'*,*N'*-Tetramethylthioperoxydicarbonic diamide ([(H$_2$N)C(S)]$_2$S$_2$); bis(dimethylthiocarbamoyl) disulfide; bis(dimethylthiocarbamyl) disulfide; tetramethylthiuram disulfide; TMTD; ENT-987; SQ-1489; NSC-1771; Pomarsol; Spotrete F; Tuads. C$_6$H$_{12}$N$_2$S$_4$; mol wt 240.42. C 29.98%, H 5.03%, N 11.65%, S 53.34%. Prepn: v. Braun, Stechele, *Ber.* **35**, 820 (1902); **36**, 2280 (1903); Romani, *C.A.* **16**, 854 (1922); Cummings, Simmons, *Ind. Eng. Chem.* **20**, 1173 (1928). Acute toxicity: Gaines, *Toxicol. Appl. Pharmacol.* **14**, 515 (1969). Review of carcinogenic risk: *IARC Monographs* **12**, 225 (1976). Review of analytical methods, degradation, and agricultural applications: V. K. Sharma *et al.*, *J. Environ. Monit.* **5**, 717-723 (2003).

Crystals from chloroform + alcohol, mp 155-156°. d 1.29. Insol in water, dil caustic, gasoline. Soly in alcohol and in ether less than 0.2%; soly in acetone 1.2%; in benzene 2.5%. Sol in chloroform, acetone. LD$_{50}$ orally in rats: 640 mg/kg (Gaines).

Caution: Potential symptoms of overexposure are irritation of eyes, skin, mucous membranes; dermatitis; disulfiram-like effects. *See NIOSH Pocket Guide to Chemical Hazards* (DHHS/NIOSH 97-140, 1997) p 306. A potent skin sensitizer. Toxicity greater in presence of fats, oils, fat solvents. *See Clinical Toxicology of Commercial Products*, R. E. Gosselin *et al.*, Eds. (Williams & Wilkins, Baltimore, 5th ed., 1984) section III, pp 383-386.

USE: Rubber accelerator; vulcanizer; seed disinfectant; fungicide; bacteriostat in soap; animal repellent.

9526. Thomas Phosphate. [1306-01-0] Calcium oxide phosphate (Ca$_4$O(PO$_4$)$_2$); Thomas flour; tetracalcium oxide diphosphate; calcium phosphate, tetrabasic. Ca$_3$(PO$_4$)$_2$.CaO also represented by 4CaO.P$_2$O$_5$. A byproduct of the Thomas steel process used in Europe: Iron ore of the "Minette" type from Alsace and the Rhineland contains about 2% phosphates which are combined with calcium and separated during the refining process.

USE: Agricultural fertilizer.

9527. Thonzylamine Hydrochloride. [63-56-9] *N*^{1}-[(4-Methoxyphenyl)methyl]-*N*^{2},*N*^{2}-dimethyl-*N*^{1}-2-pyrimidinyl-1,2-ethanediamine hydrochloride (1:1); 2-[[2-(dimethylamino)ethyl](*p*-methoxybenzyl)amino]pyrimidine hydrochloride; *N*,*N*-dimethyl-*N'*-(*p*-methoxybenzyl)-*N'*-(2-pyrimidyl)ethylenediamine hydrochloride; Anahist; Tonamil. C$_{16}$H$_{23}$ClN$_4$O; mol wt 322.84. C 59.53%, H 7.18%, Cl 10.98%, N 17.35%, O 4.96%. Histamine H$_1$-receptor antagonist. Prepd by treating the sodium salt of 2-(*p*-methoxybenzyl)aminopyrimidine with *N*,*N*-dimethyl-2-chloroethylamine: Friedman, Tolstoouhov, US 2465865 (1949). Toxicity study: Reinhard, Seudi, *Proc. Soc. Exp. Biol. Med.* **66**, 512 (1947). Spectroscopic determn in nasal drops: S. M. Sabry *et al.*, *J. Pharm. Biomed. Anal.* **22**, 257 (2000).

Crystals, mp 173-176° (free base, oily liq, bp$_{2.2}$ 185-187°). Freely sol in water; sol in alcohol, chloroform. Practically insol in ether. pH 5.1-5.7 (2% aq solution). LD$_{50}$ orally in guinea pigs: 493 mg (base)/kg (Reinhard, Seudi).

THERAP CAT: Antihistaminic.

9528. Thorium. [7440-29-1] Th; at. wt 232.03806 (characteristic terrestrial isotopic composition); at. no. 90; valence 4. No stable nuclides. Naturally occurring isotopes (mass numbers): 232 (~100%, T$_{1/2}$ 1.40 × 10^{10} years, longest-lived known isotope, α-decay, rel. at. mass 232.0380); 228 (natural decay product of ^{232}Th); 227, 231 (natural decay products of ^{235}U); 230 (T$_{1/2}$ 7.54×10^4 years, rel. at. mass 230.0331), 234 (natural decay products of ^{238}U); other isotopes: 213-226, 229, 233, 235, 236. ^{232}Th natural decay by the emission of α-, β-, and γ-rays, eventually forming ^{208}Pb. Occurs in the minerals thorite, thorianite, orangite, yttrocrasite; in monazite sand (principal ore of commercial significance); present in the earth's crust 8-15 ppm. Isoln from thorite (ThSiO$_4$): J. J. Berzelius, *K. Sven. Ventenskapsakad. Handl.* **9**, 1 (1829); *Pogg. Ann.* **16**, 385 (1829). Prepn of metal: *idem, Ann. Phys.* **16**, 385 (1829); Marden, *Trans. Electrochem. Soc.* **66**, 39 (1934). Monographs: F. L. Cuthbert, *Thorium Production Technology* (Addison-Wesley, Reading, Mass., 1958); L. Grainger, *Uranium and Thorium* (Pitman, London, 1958). *Reviews: Mellor's* vol. 7, 174-253 (1930); "The Actinides," in *Comprehensive Inorganic Chemistry* vol. 5, J. C. Bailar, Jr., *et al.*, Eds. (Pergamon Press, Oxford, 1973) pp 1-715; L. I. Katzin in *Kirk-Othmer Encyclopedia of Chemical Technology* vol. 22 (Wiley-Interscience, New York, 3rd ed., 1983) pp 989-1002; L. I. Katzin, D. C. Sonnenberger in *The Chemistry of the Actinide Elements* vol. 1, J. J. Katz *et al.*, Eds. (Chapman and Hall, New York, 1986) pp 41-101. Review of prepn and purification: J. C Spirlet *et al.*, *Adv. Inorg. Chem.* **31**, 1-40 (1987). Review of toxicology and health effects: *Toxicological Profile for Thorium* (PB91-180448, 1990) 174 pp; *Patty's Industrial Hygiene and Toxicology* vol. 2C, G. D. Clayton, F. E. Clayton, Eds. (Wiley-Interscience, New York, 4th ed., 1994) pp 2249-2258.

Grayish-white, lustrous, radioactive metal; somewhat ductile and malleable. Two allotropic phases: face-centered cubic α-form, d^{25} 11.724, transforms to β-form at 1360°; body-centered cubic β-form transforms to liquid at mp 1750° (Katzin, Sonnenberger). Also reported as mp 1690° (Cuthbert). bp ~3800°. Heat of fusion <19.2 kJ/mol. Heat of vaporization ~586 kJ/mol. Heat capacity (25°) 27.32 J/mol·K. Darkens on prolonged exposure to air. Finely divided metal is pyrophoric in air. HCl attacks metal vigorously, leaving up to 25% as an undissolved residue. Nitric acid passivates metal. Dilute HF and H_2SO_4 and concentrated H_3PO_4 and $HClO_4$ attack thorium slowly, with evolution of H. Metal not attacked by alkali hydroxides.

USE: As fuel in nuclear reactors, as source of fissionable ^{233}U. In manuf incandescent gas-light mantles, welding electrodes, ceramics. As hardener in Mg alloys; for filament coatings in incandescent lamps and vacuum tubes; as chemical catalyst.

9529. Thorium Chloride. [10026-08-1] Thorium tetrachloride; tetrachlorothorium. Cl_4Th; mol wt 373.84. Cl 37.93%, Th 62.07%. $ThCl_4$. It may contain from 7 to 9 mols H_2O. Prepn of the octahydrate: Kremer, *J. Am. Chem. Soc.* **64**, 1009 (1942).

White, odorless crystals; d 4.59; mp 770°. bp 921°. Sol in water, alcohol. Sol in ethylenediamine.

9530. Thorium Nitrate. [13823-29-5] Nitric acid thorium-(4+) salt (4:1). $N_4O_{12}Th$; mol wt 480.05. N 11.67%, O 39.99%, Th 48.34%. $Th(NO_3)_4$. Commercial form obtained as tetrahydrate; also crystallizes with 6 and 12 mols H_2O. Obtained from monazite: Pearce *et al.*, *Inorg. Synth.* **II**, 38 (1946).

Tetrahydrate. [13470-07-0] $N_4O_{12}Th.4H_2O$; mol wt 552.11. White, slightly deliquesc crystals. Very sol in water, alcohol, the soln having an acid reaction. *Keep well closed.*

USE: Thorium nitrate with 1% cerium nitrate constitutes the usual impregnating liq for incandescent mantles. As a titrant for determination of fluorine.

9531. Thorium Oxide. [1314-20-1] Thorium dioxide; thoria. O_2Th; mol wt 264.04. O 12.12%, Th 87.88%. ThO_2. *Review:* Keller in *Comprehensive Inorganic Chemistry* **vol. 5**, J. C. Bailar, Jr. *et al.*, Eds. (Pergamon Press, Oxford, 1973) pp 221-223.

White, heavy, infusible, cryst powder; when heated is incandescent. d 10.0. mp 3390°. Insol in water or alkalies. Sol in acids with difficulty.

Suspension contg 20-25% ThO_2. Thorotrast. Reviews of tumor induction by Thorotrast: Looney, *Am. J. Roentgenol. Radium Ther. Nucl. Med.* **83**, 163-185; Smoron, Battifora, *Cancer* **30**, 1252-1259 (1972).

Caution: Thorium oxide is listed as a known human carcinogen: *Report on Carcinogens, Twelfth Edition* (PB2011-111646, 2011) p 243.

THERAP CAT: Has been used as diagnostic aid (radiopaque medium).

9532. Thorium Tetracyanoplatinate(II). [14481-33-5] Thorium(4+) tetrakis(cyano-*C*)platinate(2−) (1:2); platinous thorium cyanide; thorium platinocyanide. $C_8N_8Pt_2Th$; mol wt 830.35. C 11.57%, N 13.50%, Pt 46.99%, Th 27.94%. $Th[Pt(CN)_4]_2$.

Hexadecahydrate. Yellow crystals. Sparingly sol in water.

USE: For fluorescent screens, *see* Barium Platinous Cyanide.

9533. Thozalinone. [655-05-0] 2-(Dimethylamino)-5-phenyl-4(5*H*)-oxazolone; 2-(dimethylamino)-5-phenyl-2-oxazolin-4-one; 5-phenyl-2-(dimethylamino)-2-oxazolin-4-one; tozalinone; Stimsen. $C_{11}H_{12}N_2O_2$; mol wt 204.23. C 64.69%, H 5.92%, N 13.72%, O 15.67%. Prepn: Hardy, Jr. *et al.*, US 3037990 (1962 to Am. Cyanamid). Pharmacological studies: B. M. Bernstein, C. N. Latimer, *Psychopharmacologia* **12**, 338 (1968); U. H. Lindberg, J. Pedersen, *Acta Pharm. Suec.* **5**, 15 (1968). In treatment of Parkinsonism: W. D. Gray, C. E. Edward, US 3665075 (1972 to Am. Cyanamid).

Crystals from water, mp 133-136°.
THERAP CAT: Antidepressant.

9534. Threonine. [72-19-5] L-Threonine; Thr; T; (2*S*,3*R*)-2-amino-3-hydroxybutyric acid; α-amino-β-hydroxybutyric acid. $C_4H_9NO_3$; mol wt 119.12. C 40.33%, H 7.62%, N 11.76%, O 40.29%. An essential amino acid for human development; last of the common amino acids to be discovered. Identified in oat protein: S. B. Schryver, H. W. Buston, *Proc. R. Soc. London* **99B**, 476 (1925/6); isolated from blood fibrin: W. C. Rose, *J. Biol. Chem.* **109**, 77 (1935). Structure: C. E. Meyer, W. C. Rose, *ibid.* **115**, 721 (1936). Early chemistry and biochemistry: *Amino Acids and Proteins*, D. M. Greenberg, Ed. (Charles C. Thomas, Springfield, IL, 1951) 950 pp., *passim*; J. P. Greenstein, M. Winitz, *Chemistry of the Amino Acids* **vols 1-3** (John Wiley and Sons, Inc., New York, 1961) pp. 2238-2258, *passim*. GC/MS metabolism study: O. Ballevre *et al.*, *Anal. Biochem.* **193**, 212 (1991). Clinical evaluation in amyotrophic lateral sclerosis: J. B. Roufs, *Med. Hypotheses* **34**, 20 (1991); O. Blin *et al.*, *J. Neurol.* **239**, 79 (1992). Review of microbial production: Daoust, *Dev. Ind. Microbiol.* **7**, 41 (1966); R. Bhattacharyya *et al.*, *Hind. Antibiot. Bull.* **30**, 54-65 (1988). Review of biosynthesis: G. N. Cohen in *Amino Acids, Biotechnol. Ser.* **3**, 147-171 (1983); of regulation of biosynthesis: G. Galili, *Plant Cell* **7**, 899-906 (1995).

Crystals, dec 255-257°. $[\alpha]_D^{26}$ −28.3° (c = 1.09). pK_1' 2.63; pK_2' 10.43. Freely sol in water. Insol in absolute alc, ether, chloroform, common neutral solvents.

DL-Form hemihydrate. Orthorhombic crystals, dec 229-230°.

9535. D-Threose. [95-43-2] (2*S*,3*R*)-2,3,4-Trihydroxybutanal. $C_4H_8O_4$; mol wt 120.10. C 40.00%, H 6.71%, O 53.29%. From calcium D-xylonate by oxidation with H_2O_2: Ruff, *Ber.* **34**, 1370 (1901). Improved procedure using strontium D-xylonate and ferric acetate catalyst: Hockett, *J. Am. Chem. Soc.* **57**, 2260, 2265 (1935). From tetraacetyl-D-xylononitrile: Maquenne, *Compt. Rend.* **130**, 1403 (1900); *Ann. Chim.* [7] **24**, 404 (1901); Bonner, Roth, *J. Am. Chem. Soc.* **81**, 5454 (1959); from monobenzylidene-D-arabitol: Steiger, Reichstein, *Helv. Chim. Acta* **19**, 1016 (1939); from 1,1-diethylsulfonyl-D-*threo*-3,4,5-trihydroxypent-1-ene: Hough, Taylor, *J. Chem. Soc.* **1955**, 1212; from D-galactose: Perlin, Brice, *Can. J. Chem.* **34**, 541 (1956). Synthesis of DL-threose: Lake, Glattfeld, *J. Am. Chem. Soc.* **66**, 1091 (1944); Schmid, Grob, *Helv. Chim. Acta* **32**, 77 (1949); Sonogashira, Nakagawa, *Bull. Chem. Soc. Jpn.* **45**, 2616 (1972).

Syrup. Shows mutarotation. Final $[\alpha]_D^{20}$ −12.3° (20 min, c = 4). Very sol in water; slightly in alcohol. Practically insol in ether, petr ether.

Phenylosazone. $C_{16}H_{18}N_4O_2$. Dec 164-165°. Identical with D-erythrose phenylosazone.

Triacetate. $C_{10}H_{14}O_7$. Prisms from abs ethanol, mp 117-118°. $[\alpha]_D^{25}$ +34.4° (c = 2 in chloroform). Soluble in hot water, chloroform, acetone, ethyl acetate; sparingly sol in abs alcohol, methanol, ether.

9536. L-Threose. [95-44-3] (2*R*,3*S*)-2,3,4-Trihydroxybutanal. $C_4H_8O_4$; mol wt 120.10. C 40.00%, H 6.71%, O 53.29%. By degradation of L-xylose through the oxime and tetraacetyl-L-xylononitrile to L-threose diacetamide: Hockett *et al.*, *J. Am. Chem. Soc.* **60**, 278 (1938); from D-glucitol: Hutson, Weigel, *J. Chem. Soc.* **1961**, 1546.

Shows mutarotation; final $[\alpha]_D^{20}$ +13.2° (c = 4.5). Sol in water.
Phenylosazone. $C_{16}H_{18}N_4O_2$. Dec 162°.

9537. Thrombin. [9002-04-4] Blood-coagulation factor IIa; factor IIa; fibrinogenase; EC 3.4.21.5. Multifunctional serine protease generated at the site of vascular injury from prothrombin, *q.v.* Key enzyme in the coagulation cascade; converts fibrinogen into fibrin and activates factor XIII, *q.v.*, which cross-links and stabilizes the fibrin polymer. Enhances inflammation and tissue repair via receptor-mediated cellular responses; activates platelets and endothelial cells and stimulates proliferation of fibroblasts and smooth muscle cells. The mature enzyme is glycosylated and consists of two polypeptide chains, designated A and B, covalently linked by a disulfide bridge. Exists in several forms, the largest and most active being the 39 kDa α-thrombin. The human α-thrombin A chain contains 36 amino acids; the B chain contains 259 residues. Purification by conversion from bovine prothrombin: W. H. Seegers *et al., J. Biol. Chem.* **126**, 91 (1938); W. H. Seegers, D. A. McGinty *ibid.* **146**, 511 (1942); H. P. Smith, **US 2398077** (1946 to Parke, Davis). Sequence of bovine A chain: S. Magnussen, *Biochem. J.* **110**, 25P (1968). Primary structure of human α-thrombin: R. J. Butkowski *et al., J. Biol. Chem.* **252**, 4942 (1977). Review of structure: M. T. Stubbs, W. Bode, *Thromb. Res.* **69**, 1-58 (1993). Review of cloning, structure and function of thrombin receptors: S. R. Coughlin *et al., Cold Spring Harbor Symp. Quant. Biol.* **57**, 149-154 (1992); C. Tapparelli *et al., Trends Pharmacol. Sci.* **14**, 426-428 (1993). Review of role in hemostasis: J. W. Fenton *et al., Hematol. Oncol. Clin. North Am.* **7**, 1107-1119 (1993); in cell adhesion and proliferation: R. Bar-Shavit *et al., Am. J. Respir. Cell Mol. Biol.* **6**, 123-130 (1992); in inflammation and healing and effects in RA: R. Morris *et al., Ann. Rheum. Dis.* **53**, 72-79 (1994).

Topical thrombin. Thrombinar; Thrombogen; Thrombostat. A standardized prepn of bovine thrombin. Review of clinical experience: R. T. Tidrick *et al., Surgery* **14**, 191-196 (1943); in combination with fibrin: M. Brennan, *Blood Rev.* **5**, 240-244 (1991). White powder. Sol in water, isotonic saline.

THERAP CAT: Hemostatic (local).

9538. Thrombomodulin. Thrombin-binding protein present on the surface of vascular endothelial cells that modulates the provs anti-coagulant activities of thrombin, *q.v.* Alters the macromolecular specificity of thrombin, decreasing its ability to catalyze clot formation and converting it to a potent activator of protein C, *q.v.* Also expressed by platelets, synovial cells, and certain tumor cells. Identification: C. T. Esmon, W. G. Owen, *Proc. Natl. Acad. Sci. USA* **78**, 2249 (1981). Isoln from rabbit lung: N. L. Esmon *et al., J. Biol. Chem.* **257**, 859 (1982); from human placenta: H. H. Salem *et al., ibid.* **259** 12246 (1984). Human thrombomodulin is a single-chain glycoprotein containing 559 amino acids; mol wt ~75 kDa. A soluble, truncated form has been isolated from human plasma and urine: H. Ishii, P. W. Majerus, *J. Clin. Invest.* **76**, 2178 (1985). Review: N. L. Esmon, *Prog. Hemostasis Thromb.* **9**, 29-55 (1989); of structure and function: W. A. Dittman, P. W. Majerus, *Blood* **75**, 329-336 (1990); of tissue distribution and expression by tumor cells: L. M. Fink *et al., Int. J. Dev. Biol.* **37**, 221-226 (1993).

9539. Thromboplastin. [9035-58-9] Blood-coagulation factor III; Factor III; cytozyme; thrombokinase; thrombokinin; tissue factor; tissue thromboplastin; zymoplastic substance; thrombostop; Tachostyptan. An integral membrane glycoprotein which, in the presence of Ca²⁺ ions, initiates coagulation by augmenting the proteolytic attack of factor VII on factors IX and X, *q.q.v.*: Nemerson, *Biochemistry* **5**, 601 (1966). Prepn from cattle brains: Hess, *J. Am. Med. Assoc.* **66**, 558 (1916). From rabbit brains or lung tissue: Singher, Swart, **US 2842480** (1958 to Ortho); *see also* **US 2847347** and **US 2847350** (1958 to Ortho). Purification of bovine tissue factor: R. Bach *et al., J. Biol. Chem.* **256**, 8324 (1981). The tissue thromboplastin activity seems to be brought about by a lipoprotein comprising a combination of phospholipids: phosphatidyl ethanolamine (PE), phosphatidyl serine (PS), and phosphatidyl choline (PC): R. B. Hunter, *et al., Fibrinogen and Fibrin Turnover of Clotting Factors* (Schattauer, Stuttgart, 1963) pp 429-454. Removal of phospholipids from tissue factor results in loss of coagulant activity as well as solubilizing the protein component. Purification and characterization of the protein component: Nemerson, Pitlick, *Biochemistry* **9**, 5100 (1970); of tissue factor apoprotein: F. A. Pitlick, Y. Nemerson, *Methods Enzymol.* **45B**, 37-48 (1976). Located on the plasma membrane of endothelial cells readily available for complex-

ing with clotting factors: Zeldis *et al., Science* **175**, 766 (1972). Synthesis by cultured cells: Zacharski, McIntyre, *J. Med.* **4**, 118 (1973). Possibly a vitamin K dependent clotting factor: L. R. Zacharski *et al., Ann. N.Y. Acad. Sci.* **370**, 311 (1981). Use as clinical reagent: F. D. Ziegler *et al., Ann. Clin. Lab. Sci.* **11**, 202 (1981); A. L. Suchman, P. F. Griner, *Ann. Intern. Med.* **104**, 810 (1986).

THERAP CAT: Hemostatic.

9540. Thrombopoietin. Thrombocytopoiesis-stimulating factor; TSF; TPO; c-Mpl ligand; megakaryocyte potentiating factor. Glycopeptide hormone which is the major physiological regulator of platelet production. Promotes the maturation of megakaryocytes and increases platelet size and number. Produced in the kidney; detectable in plasma and urine of thrombocytopenic animals and humans. Identification of humoral factor that increases platelet levels: E. Kelemen *et al., Acta Haematol.* **20**, 350 (1958). Partial purification from rabbit plasma: B. L. Evatt *et al., J. Lab. Clin. Med.* **83**, 364 (1974); from human plasma: A. Vannucchi *et al., Leukemia* **2**, 236 (1988). Production by cultured human embryonic kidney cells: T. P. McDonald *et al., J. Lab. Clin. Med.* **85**, 59 (1975). Purification: *idem et al., Exp. Hematol.* **17**, 865 (1989). Review of biological activity and properties: *idem, Am. J. Pediatr. Hematol. Oncol.* **14**, 8-21 (1992). Recombinant TPO acts both as a proliferative and a maturation factor for megakaryocytes. Cloning and characterization of human TPO: F. J. de Sauvage *et al., Nature* **369**, 533 (1994); of murine: S. Lok *et al., ibid.* 565; K. Kaushansky *et al., ibid.* 568; F. Wendling *et al., ibid.* 571.

9541. Thrombospondin. Glycoprotein G; thrombin-sensitive protein; TSP. A major glycoprotein constituent of human platelet α-granules that is released in response to platelet activation by α-thrombin, and plays an important role in mediating platelet aggregation. Discovery and initial isoln and properties: N. L. Baenziger *et al., Proc. Natl. Acad. Sci. USA* **68**, 240 (1971); *eidem, J. Biol. Chem.* **247**, 2723 (1972). Thrombospondin was initially thought to be a single polypeptide of mol wt about 190,000, but has been shown to be a disulfide-linked trimer of mol wt 450,000: I. Hagen, *Biochim. Biophys. Acta* **392**, 242 (1975); D. R. Phillips, P. P. Agin, *J. Biol. Chem.* **252**, 2121 (1977); J. W. Lawler *et al., Thromb. Haemostasis* **37**, 355 (1977); *eidem, J. Biol. Chem.* **253**, 8609 (1978); S. S. Margossian *et al., ibid.* **256**, 7495 (1981). Synthesis and secretion by cells in culture: G. J. Raugi *et al., J. Cell Biol.* **95**, 351 (1982). Quantitative analysis: J. N. George *et al., J. Lab. Clin. Med.* **92**, 430 (1978). The binding of secreted thrombospondin to platelet membranes is dependent on Ca²⁺: D. R. Phillips *et al., J. Biol. Chem.* **255**, 11629 (1980). Evidence for calcium-sensitive structure: J. Lawler *et al., ibid.* **257**, 12257 (1982). Radioimmunoassay: S. D. Saglio, H. S. Slayter, *Blood* **59**, 162 (1982). Identity of thrombospondin with the endogenous lectin secreted by activated platelets: E. A. Jaffee *et al., Nature* **295**, 246 (1982). *See also* Fibrinogen.

Partial specific volume: 0.714 ml/g. Intrinsic viscosity: 40 ml/g in buffered saline at pH 7.6, 20°.

9542. Thromboxanes. Compounds derived from prostaglandin endoperoxides that cause platelet aggregation, contraction of arteries and other biological effects. Found in platelets, leucocytes, lung tissue, spleen, kidney, and umbilical artery. They are important mediators of the actions of polyunsaturated fatty acids transformed by cyclooxygenase. Discovery and structure of thromboxane B₂ (originally referred to as PHD): M. Hamberg, B. Samuelsson, *Proc. Natl. Acad. Sci. USA* **71**, 3400 (1974). Discovery and structure of thromboxane A₂ and identity with the unstable component of RCS (rabbit aorta contracting substance) *see* P. J. Piper, J. R. Vane, *Nature* **223**, 29 (1969); M. Hamberg *et al., Proc. Natl. Acad. Sci. USA* **72**, 2994 (1975). Biosynthesis and biological properties: P. Needleman *et al., Science* **193**, 163 (1976). Physiological review: J. R. Vane *et al., Int. Rev. Exp. Pathol.* **23**, 161-207 (1982). *Reviews:* B. Samuelsson in *Organic Chemistry*, A. T. Blomquist, H. H. Wasserman, Eds. **vol. 36**, entitled "Prostaglandin Research", P. Crabbe, Ed. (Academic Press, New York, 1977) pp 17-46; E. Granström *et al., Adv. Prostaglandin Thromboxane Leukotriene Res.* **10**, 15-58 (1982); L. J. Roberts *et al., ibid.* 211-225. Books: *Advances in Prostaglandin and Thromboxane Research* **vols. 1-8**, B. Samuelsson, R. Paoletti, Eds. (Raven Press, New York, 1976-1980); *New Synthetic Routes to Prostaglandins and Thromboxanes* S. M. Roberts, F. Scheinmann, Eds. (Academic Press, New York, 1982).

Thromboxane A₂

Thromboxane B₂

Thromboxane A₂. [57576-52-0] (5Z,9α,11α,13E,15S)-9,11-Ep-oxy-15-hydroxythromboxa-5,13-dien-1-oic acid; [1S-[1α,3α(1E,-3R*)4β(Z),5α]]-7-[3-(3-hydroxy-1-octenyl)-2,6-dioxabicyclo-[3.1.1]hept-1-yl]-5-heptenoic acid; TXA₂. C₂₀H₃₂O₅; mol wt 352.47. Highly unstable, biologically active bicyclic oxitane-oxane compound derived from the endoperoxide ***PGG*₂** and rapidly converted to thromboxane B₂ by addition of water. Formed by incubation of arachidonic acid or PGG₂ with washed platelets. It induces irreversible platelet aggregation and causes contraction of the isolated rabbit aorta and release of serotonin and ADP from platelets in platelet-rich plasma. Synthesis of stable analogs: E. J. Corey *et al.*, *Tetrahedron Lett.* **1980**, 137; K. M. Massey, G. M. Bundy, *ibid.* 445; S. Ohuchida *et al.*, *J. Am. Chem. Soc.* **103**, 4597 (1981); V. N. Kale, D. L. J. Clive, *J. Org. Chem.* **49**, 1554 (1984). Synthesis of biologically active unstable analogs: S. S. Bhagwat *et al.*, *Tetrahedron Lett.* **1985**, 1955. Total synthesis and structure of thromboxane A2: *eidem*, *Nature* **315**, 511 (1985). Formation and effects in human platelets: J. Svensson *et al.*, *Acta Physiol. Scand.* **98**, 285 (1970). Biological half-life: 32 ±2 sec at 37°.

Thromboxane B₂. [54397-85-2] (5Z,9α,13E,15S)-9,11,15-Tri-hydroxythromboxa-5,13-dien-1-oic acid; [2R-[2α(1E,3S*),3β-(Z),4β,6α]]-7-[tetrahydro-4,6-dihydroxy-2-(3-hydroxy-1-octenyl)-2H-pyran-3-yl]-5-heptenoic acid; TXB₂; PHD. C₂₀H₃₄O₆; mol wt 370.49. A stable metabolite of thromboxane A₂ in platelets, initially considered biologically inactive. It is released during anaphylaxis in isolated guinea pig lungs and has been isolated from guinea pig brain homogenates and carrageenin-induced granuloma. TXB₂ has also been reported as possessing chemotactic properties. Total synthesis: N. A. Nelson, R. W. Jackson, *Tetrahedron Lett.* **1976**, 3275; R. C. Kelly *et al.*, *ibid.* 3279; from a prostaglandin F₂$_\alpha$ derivative: W. P. Schneider, R. A. Morge, *ibid.* 3283; stereospecific synthesis from D-glucose: S. Hanessian, P. Lavallee, *Can. J. Chem.* **55**, 562 (1977); E. J. Corey *et al.*, *Tetrahedron Lett.* **1977**, 1625; S. Hanessian, P. Lavallee, *Can. J. Chem.* **59**, 870 (1981). Metabolism: L. J. Roberts *et al.*, *J. Biol. Chem.* **252**, 7415 (1966). Biological properties: J. R. Boot *et al.*, *J. Physiol.* **257**, 47P (1976); L. S. Wolfe *et al.*, *Biochem. Biophys. Res. Commun.* **70**, 907 (1976); W.-C. Chang *et al.*, *Prostaglandins* **13**, 3 (1977). Plates from ethyl acetate/ether/petr ether, mp 95-96°. [α]$_D^{25}$ +57.4° (c = 0.26 in ethyl acetate).

9543. Thuja. Arbor vitae; white cedar; tree of life. Coniferous, pyramidal evergreen tree, *Thuja occidentalis* L., *Cupressaceae*, used as an ornamental plant and in traditional medicine to treat respiratory infection, warts, scurvy and rheumatism. *Habit.* North America, Europe. Medicinal portions include the dried, leafy young branches and the oil extracted from the leaves and branch tips. *Constit.* Volatile oil (1.4-4%); *p*-coumaric acid; umbelliferone; flavonoids incl. quercetin, mearusitrin; tannic acid (~1.3%); polysaccharides and proteins (~4%). Botanical description and medicinal uses: J. Gruenwald *et al.*, *PDR for Herbal Medicines* (Medical Economics, Montvale, 2nd Ed., 2000) pp 759-760. Review of pharmacology and clinical experience: B. Naser *et al.*, *Evid. Based Complement. Alternat. Med.* **2**, 69-78 (2005).

Volatile oil. [8007-20-3] Oil of thuja; cedar leaf oil; oil of white cedar. *Constit.* α-thujone (69%), β-thujone (7-10%), fenchone (10-15%). Fragrance monograph: *Food Cosmet. Toxicol.* **12**, 843-844 (1974). Colorless to yellow liquid. d$_{25}^{25}$ 0.906-0.916. n_D^{20} 1.4560-

1.4590. Sol in most fixed oils, mineral oil, propylene glycol; 1 ml dissolves in 3 ml 70% alcohol. Practically insol in glycerin.

Caution: Symptoms of overexposure may include queasiness, vomiting, painful diarrhea, mucous membrane hemorrhaging, death (Gruenwald).

USE: As fragrance in soaps, detergents, perfumes.

THERAP CAT: Immunostimulant.

9544. β-Thujaplicin. [499-44-5] 2-Hydroxy-4-(1-methyleth-yl)-2,4,6-cycloheptatrien-1-one; hinokitiol; 4-isopropyltropolone. C₁₀H₁₂O₂; mol wt 164.20. C 73.15%, H 7.37%, O 19.49%. Tropolone derivative found in the heartwood of cupressaceous plants including western red cedar, eastern white cedar, and hiba; β-thujaplicin antimicrobial activity contributes to the decay-resistance of these trees. Isoln from western red cedar, *Thuja plicata* D. Don: H. Erdtman, J. Gripenberg, *Nature* **161**, 719 (1948); and characterization: A. B. Anderson, J. Gripenberg, *Acta Chem. Scand.* **2**, 644 (1948). Synthesis: W. von E. Doering, L. H. Knox, *J. Am. Chem. Soc.* **75**, 297 (1953). Biosynthesis in *Cupressus lusitanica*: J. Zhao, K. Sakai, *J. Exp. Bot.* **54**, 647 (2003). HPLC determn in cosmetics: M. Endo *et al.*, *J. Chromatogr.* **455**, 430 (1988). CZE determn in aqueous solns: L. Dyrskov *et al.*, *J. Agric. Food Chem.* **52**, 1452 (2004). Antibacterial activity in atopic dermatitis: Y. Arima *et al.*, *J. Antimicrob. Chemother.* **51**, 113 (2003).

mp 52-52.5° (Erdtman); also reported as mp 50-51° (Doering). uv max (isooctane): 236, 322, 353 nm (log ε 4.37, 3.72, 3.66).

USE: Antibacterial additive in foods, cosmetics, eye drops and toothpaste.

9545. Thujic Acid. [499-89-8] 5,5-Dimethyl-1,3,6-cyclohep-tatriene-1-carboxylic acid; 4,4-dimethylcyclohepta-2,5,7-triene-carboxylic acid; dehydroperillic acid. C₁₀H₁₂O₂; mol wt 164.20. C 73.15%, H 7.37%, O 19.49%. Antibiotic substance from heartwood of *Thuja plicata* D. Don, *Cupressaceae* (Western red cedar). Isoln: Anderson, Sherrard, *J. Am. Chem. Soc.* **55**, 3813 (1933); Erdtman, Gripenberg, *Acta Chem. Scand.* **2**, 625 (1948). Structure: Gripenberg, *ibid.* **3**, 1137 (1949); **5**, 995 (1951); **10**, 487 (1956); Davis, Tulinsky, *Tetrahedron Lett.* **1962**, 839.

Crystals from petr ether, mp 88-89°. uv max: 220, 280 nm (log ε 4.3, 3.7); min: 240 nm (log ε 3.0).

Methyl ester. C₁₁H₁₄O₂. Crystals, mp 34.5-35° or liquid with pleasant odor, bp₁₄ 112-113°. d$_4^{22}$ 1.0225. n_D^{22} 1.5130.

Hexahydrothujic acid. C₁₀H₁₈O₂. Liq, bp₁₄ 150-152°. n_D^{20} 1.4671.

9546. Thujone. 4-Methyl-1-(1-methylethyl)bicyclo[3.1.0]-hexan-3-one; 3-thujanone. C₁₀H₁₆O; mol wt 152.24. C 78.90%, H 10.59%, O 10.51%. A constituent of many essential oils; present in thuja, etc. Equilibrium mixture contains 33% α-thujone and 67% β-thujone: Eastman, Winn, *J. Am. Chem. Soc.* **82**, 5908 (1960). α- and β-Thujones differ only in the stereochemistry of the 4-methyl group. Conformation: Hach *et al.*, *Tetrahedron Lett.* **1970**, 3175. Chemistry: J. P. Kutney *et al.*, *Bioorg. Chem.* **7**, 289 (1978); *eidem*, *Can. J. Chem.* **57**, 3145 (1979); **58**, 2641 (1980). Toxicity study: K. C. Rice, R. S. Wilson, *J. Med. Chem.* **19**, 1054 (1976). Review: J. L. Simonsen, *The Terpenes* vol. **II** (University Press, Cambridge, 1949) pp 32-52.

α-Thujone　　　　　　　β-Thujone

Colorless or almost colorless liquid. uv max (isooctane): 300 nm (ε 23). Practically insol in water. Sol in alc and many other organic solvents. LD_{50} s.c. in mice: 134.2 mg/kg (Rice, Wilson).

α-Thujone. [546-80-5] ($1S,4R,5R$)-(−)-3-Thujanone. bp_{17} 83.8-84.1°. d_4^{25} 0.9109. n_D^{15} 1.4490. $[\alpha]_D^{20}$ −19.2°. LD_{50} s.c. in mice: 87.5 mg/kg (Rice, Wilson).

β-Thujone. [471-15-8] d-Isothujone; ($1S,4S,5R$)-(+)-3-thujanone. bp_{17} 85.7-86.2°. d_4^{25} 0.9135. n_D^{25} 1.4500. $[\alpha]_D^{15}$ +72.5°. LD_{50} s.c. in mice: 442.2 mg/kg (Rice, Wilson).

Caution: Ingestion may cause convulsions.

9547. Thujopsene. [470-40-6] ($1aS,4aS,8aS$)-1,1a,4,4a,5,6,-7,8-Octahydro-2,4a,8,8-tetramethylcyclopropa[d]naphthalene; widdrene. $C_{15}H_{24}$; mol wt 204.36. C 88.16%, H 11.84%. From wood oil of the Japanese Hiba tree, *Thujopsis dolobrata* Sieb. and Zucc., *Cupressaceae:* Yano, *J. Soc. Chem. Ind. Jpn.* **16**, 443 (1913); Uchida, *ibid.* **31**, 501 (1928). Identity with widdrene: Erdtman, Thomas, *Acta Chem. Scand.* **12**, 267 (1958). Structure: Norin, *ibid.* **15**, 1676 (1961). Stereochemistry: Sisido *et al., J. Org. Chem.* **26**, 1964 (1961); Norin *Acta Chem. Scand.* **17**, 738 (1963). Synthesis: Dauben, Ashcraft, *J. Am. Chem. Soc.* **85**, 3673 (1963); Büchi, White, *ibid.* **86**, 2884 (1964).

Liquid. bp_{10} 120°. $[\alpha]_D$ −110° (c = 2 in chloroform). n_D^{25} 1.5031. d^{24} 0.932. uv max (alc): 212 nm (ε 4680).

9548. Thulium. [7440-30-4] Tm; at. wt 168.93421; at. no. 69; valences 2, 3. A rare-earth metal of the yttrium group; member of the lanthanide series. Naturally occurring isotope (mass number): 169; known artificial radioactive isotopes: 147; 148; 150-168; 170-176. Estimated abundance in earth's crust: 0.2-0.5 ppm. Found in small quantities in euxenite, ytterspar, sipylite, gadolinite, and other rare earth minerals. Discovered in crude erbium oxide: Cleve, *Compt. Rend.* **89**, 478, 521, 708 (1879). Obtained in a state of high purity by fractional crystn of its bromide: James, *J. Am. Chem. Soc.* **32**, 517 (1910); **33**, 1332 (1911). Sepn from other rare earths by ion exchange: Spedding *et al., ibid.* **76**, 2557 (1954). Toxicity study: Haley, *J. Pharm. Sci.* **54**, 663 (1965). Review of prepn, properties and compds: *The Rare Earths,* F. H. Spedding, A. H. Daane, Eds. (Krieger, Huntington, N.Y., 1971, reprint of 1961 ed.) 641 pp; Hulet, Bode, "Separation Chemistry of the Lanthanides and Transplutonium Actinides" in *MTP Int. Rev. Sci.: Inorg. Chem., Ser. One* vol. 7, K. W. Bagnall, Ed. (University Park Press, Baltimore, 1972) pp 1-45; Moeller, "The Lanthanides" in *Comprehensive Inorganic Chemistry* vol. 4, J. C. Bailar, Jr. *et al.*, Eds. (Pergamon Press, Oxford, 1973) pp 1-101; F. H. Spedding in *Kirk-Othmer Encyclopedia of Chemical Technology* vol. 19 (John Wiley & Sons, New York, 3rd ed., 1982) pp 833-854; *Chemistry of the Elements,* N. N. Greenwood, A. Earnshaw, Eds. (Pergamon Press, New York, 1984) pp 1423-1449. Brief review of properties: G. T. Seaborg, *Radiochim. Acta* **61**, 115-122 (1993).

Silvery-white, easily worked metal. Hexagonal close-packed crystals, d 9.3208. mp 1545°. bp 1725°; also reported as bp 1950° (Spedding, 1982). Heat of fusion: 16.84 kJ/mol. Heat of sublimation (25°): 232.2 kJ/mol. Solns of thulium salts show a characteristic absorption spectrum: Exner, Haschek, cited in *Mellor's* **vol. V,** 698 (1929).

Oxide. [12036-44-1] Thulia. O_3Tm_2; mol wt 385.87. Dense powder of greenish-white color. Prepd by igniting the oxalate; dissolves slowly in strong acids; exhibits a reddish glow on gentle heating.

Hydroxide. [1311-33-7] $Tm(OH)_3$; mol wt 219.96. White precipitate.

Chloride heptahydrate. [10025-92-0] $TmCl_3.7H_2O$; mol wt 401.39. Deliquesc crystals. Sol in water, in alcohol. LD_{50} in mice: 485 mg/kg i.p.; 6.25 g/kg orally (Haley).

Sulfate octahydrate. [13778-40-0] $Tm_2(SO_4)_3.8H_2O$; mol wt 770.16. Obtained by ppting an aq soln of thulium chloride and sulfuric acid with alc.

Oxalate hexahydrate. [26677-68-9] $Tm_2(C_2O_4)_3.6H_2O$; mol wt 710.01. Greenish-white precipitate. Sol in aq alkali oxalates with formation of double oxalates.

9549. Thunder God Vine. Lei gong teng; TGV; TWHF. Perennial vine, *Tripterygium wilfordii* Hook. f., *Celastraceae*; used in traditional Chinese medicine to treat inflammatory and autoimmune diseases. Medicinal formulations are prepared from the skinned root; other plant parts such as the leaves, stem, flowers and root skin are poisonous. Has also been used as an insecticide. *Habit.* China, Japan, Korea. *Constit.* >350 terpenoids, especially triptolide, celastrol, *q.q.v.*, pristimerin, friedelanes, oleananes, β-sitosterol. Comprehensive description of chemistry and pharmacology of constituents: A. M. Brinker *et al., Phytochemistry* **68**, 732-766 (2007). Antiangiogenic activity: M.-F. He *et al., J. Ethnopharmacol.* **121**, 61 (2009). Clinical trials in rheumatoid arthritis: P. E. Lipsky, X.-L. Tao, *Semin. Arthritis Rheum.* **26**, 713 (1997); X. Tao *et al., Arthritis Rheum.* **46**, 1735 (2002); P. H. Canter *et al., Phytomedicine* **13**, 371 (2006).

9550. Thurfyl Nicotinate. [70-19-9] 3-Pyridinecarboxylic acid (tetrahydro-2-furanyl)methyl ester; nicotinic acid tetrahydrofurfuryl ester; tetrahydrofurfuryl nicotinate; nicotafuryl; Trafuril. $C_{11}H_{13}NO_3$; mol wt 207.23. C 63.76%, H 6.32%, N 6.76%, O 23.16%. Prepn: Hartmann, Merz, *US 2485152* (1949 to Ciba). Brief review: A. A. Fisher, *Cutis* **51**, 225-227 (1993).

Oil, $bp_{0.25}$ 114-116°. Sol in water, oil.

THERAP CAT: Rubefacient.

9551. Thyme. Dried leaves and flowering tops of *Thymus vulgaris* L., *Labiatae.* *Habit.* Southern Europe; cultivated in gardens. *Constit.* Volatile oil; tannin, gum.

USE: As a seasoning in foods.

9552. Thymidine. [50-89-5] 1-(2-Deoxy-β-D-ribofuranosyl)-5-methyluracil; thymine-2-desoxyriboside. $C_{10}H_{14}N_2O_5$; mol wt 242.23. C 49.59%, H 5.83%, N 11.57%, O 33.02%. Constituent of deoxyribonucleic acid, *q.v.* Isoln from thymonucleic acid: Levene, London, *J. Biol. Chem.* **83**, 793 (1929). Structure: Levene, Tipson, *ibid.* **109**, 623 (1935). Conformation: Lemieux, *Can. J. Chem.* **39**, 116 (1961); Tollin *et al., Nature* **217**, 1148 (1968). Prepn of thymidine-3'-phosphate and of thymidine-5'-phosphate: Tener, *J. Am. Chem. Soc.* **83**, 165 (1961). *Review: Basic Principles in Nucleic Acid Chemistry* vol. 1, P. O. P. Ts'o, Ed. (Academic Press, New York, 1974) *passim.*

Rosettes of needles from ethyl acetate. mp 185°. Yields a sublimate of thymine when heated. $[\alpha]_D^{25}$ +30.6° (c = 1.029). uv max (pH 7.2): 206.5, 267 nm ($\varepsilon \times 10^3$ 9.8, 9.7). Absorption spectra: D. Voet *et al.*, *Biopolymers* **1**, 193 (1963). Sol in water, methanol, hot alcohol, hot acetone, hot ethyl acetate, pyridine, glacial acetic acid; sparingly sol in hot chloroform.

Monotrityl thymidine. $C_{29}H_{28}N_2O_5$; mol wt 484.55. Prepd by the action of triphenylmethyl chloride on thymidine in pyridine. mp 125°. $[\alpha]_D^{24}$ +11.4° (c = 1.01 in acetone).

9553. Thymine. [65-71-4] 5-Methyl-2,4(1*H*,3*H*)-pyrimidinedione; 5-methyluracil; 2,4-dihydroxy-5-methylpyrimidine. C_5H_6-N_2O_2; mol wt 126.12. C 47.62%, H 4.80%, N 22.21%, O 25.37%. A pyrimidine derivative; constituent of nucleic acids. Originally isolated from thymus nucleic acid: Levene, *Z. Physiol. Chem.* **39**, 4 (1903). Prepn by heating 2-ethylmercapto-4-hydroxy-5-methylpyrimidine: Wheeler, Merriam, *Am. Chem. J.* **29**, 478 (1903); **43**, 29 (1910). From methylcyanacetylurea by catalytic reduction: Bergmann, Johnson, *J. Am. Chem. Soc.* **55**, 1733 (1933). From β-methylmalic acid: Scherp, *J. Am. Chem. Soc.* **68**, 912 (1946). Crystal structure of monohydrate: Gerdil, *Acta Crystallogr.* **14**, 333 (1961). *Review:* Ts'o, "Bases, Nucleosides and Nucleotides" in *Basic Principles in Nucleic Acid Chemistry* **vol. 1**, P. O. P. Ts'o, Ed. (Academic Press, New York, 1974) pp 453-584.

Dendritic or star-shaped plates from water, sometimes short needles. Sublimes in platelets. Dec 335-337° (Kofler stage). Weak acid, pK at 25° = 9.94. uv max (pH 7.0): 205, 264.5 nm ($\varepsilon \times 10^3$ 9.5, 7.9). Absorption spectra: D. Voet *et al.*, *Biopolymers* **1**, 193 (1963). Sol in hot water; slightly sol in cold water (4 g/l at 25°). Somewhat sol in alc; sparingly sol in ether; readily sol in alkalies with formation of salts. Oxidation yields urea, ethanal, pyruvic acid, formic acid. Hydrazine reacts with thymine forming urea and 4-methylpyrazolone. Thymine forms a silver salt which is sol in excess ammonia. Its mercuric and lead salts are insol.

Thymine-2-desoxyriboside *see* Thymidine.

USE: In biochemical research.

9554. Thymol. [89-83-8] 5-Methyl-2-(1-methylethyl)phenol; 5-methyl-2-isopropyl-1-phenol; 1-methyl-3-hydroxy-4-isopropylbenzene; 3-*p*-cymenol; 3-hydroxy-*p*-cymene; thyme camphor; *m*-thymol. $C_{10}H_{14}O$; mol wt 150.22. C 79.96%, H 9.39%, O 10.65%. Isolated by Neumann in 1719. Obtained from the essential oil of *Thymus vulgaris* L. and *Monarda punctata* L., *Labiatae*: Arppe, *Ann.* **58**, 41 (1846); Meyer, *Pharm. Ztg.* **81**, 192, 205 (1936). Also occurs in other volatile oils. Produced synthetically from *p*-cymene, piperitone, or *m*-cresol: Austerweil, **GB 221227** (1923); Jennen, Verdroncken, *Compt. Rend.* **245**, 183 (1957); Bottoms, **US 2840616** (1958 to Natl. Cylinder Gas). Bactericidal activity: J. M. Schaffer, F. W. Tilley, *J. Bacteriol.* **14**, 259 (1927). Mold elimination on surfaces: O. W. Richards, K. J. Hawley, *J. Chem. Educ.* **16**, 6 (1939). In vitro antifungal activity: H. B. Myers, *J. Am. Med. Assoc.* **89**, 1834 (1927). Effectiveness as antifungal preservative: M. Dersarkissian, M. Goodberry: *Stud. Conserv.* **25**, 28 (1980). Use as clinical preservative: T. Z. Liu, *Clin. Chem.* **25**, 336 (1979); T. Z. Liu *et al.*, *ibid.* **27**, 1144 (1981). Toxicity: P. M. Jenner *et al.*, *Food Cosmet. Toxicol.* **2**, 327 (1964).

Crystals, mp 51.5°. bp ~233°. Appreciably volatile at 100°; volatilizes in water vapors. Characteristic odor; pungent, somewhat

caustic taste. d_4^{25} 0.9699. n_D^{20} 1.5227; n_D^{25} 1.5204. One gram dissolves in ~1000 ml water, 1 ml alcohol, 0.7 ml chloroform, 1.5 ml ether, 1.7 ml olive oil at 25°. Sol in glacial acetic acid, oils, fixed alkali hydroxides. LD_{50} orally in rats: 980 mg/kg (Jenner). *Incompat:* Acetanilide, antipyrine, camphor, monobromated camphor, chloral hydrate, menthol, quinine sulfate, salol, urethane, spirit nitrous ether; in triturations because of liquefaction.

Acetate. [528-79-0] Acetylthymol; thymyl acetate. $C_{12}H_{16}O_2$. Yellowish, oily liq; thymol odor. d^0 1.009. bp 243.5-245.5°. Practically insol in water. Miscible with alcohol, benzene chloroform, ether.

Carbonate. [552-93-2] $C_{21}H_{26}O_3$. White crystals; thymol odor; volatilizes with steam. mp 49°. Insol in water, acids, alkalies. Sol in hot alcohol, chloroform, ether, carbon tetrachloride.

Caution: Mild irritant.

USE: For destroying mold; preserving documents, art objects and urine. Stabilizer (antioxidant) for trichloroethylene, halothane.

THERAP CAT: Antiseptic (topical); anthelmintic (Nematodes).

THERAP CAT (VET): Has been used as anthelmintic, and as an antiseptic, external and internal.

9555. Thymol Blue. [76-61-9] 4,4′-(1,1-Dioxido-3*H*-2,1-benzoxathiol-3-ylidene)bis[5-methyl-2-(1-methylethyl)phenol]; 4,4′-(3*H*-2,1-benzoxathiol-3-ylidene)bis[5-methyl-2-(1-methylethyl)phenol] *S,S*-dioxide; α-hydroxy-α,α-bis(5-hydroxycarvacryl)-*o*-toluenesulfonic acid γ-sultone; thymolsulfonephthalein. $C_{27}H_{30}O_5$-S; mol wt 466.59. C 69.50%, H 6.48%, O 17.14%, S 6.87%.

Brownish-green, cryst powder; characteristic odor. Insol in water. Sol in alcohol, dil alkali solns.

USE: As as acid-base indicator; pH: red 1.2 to yellow 2.8; also yellow 8.0 to blue 9.6.

9556. Thymolphthalein. [125-20-2] 3,3-Bis[4-hydroxy-2-methyl-5-(1-methylethyl)phenyl]-1(3*H*)-isobenzofuranone; 5′,5″-diisopropyl-2′,2″-dimethylphenolphthalein. $C_{28}H_{30}O_4$; mol wt 430.54. C 78.11%, H 7.02%, O 14.86%. Obtained by heating phthalic anhydride with thymol at 110° in the presence of stannic chloride.

Needles, mp ~253°. Insol in water; sol in alcohol, acetone; also sol in dil alkalies with a blue color, in H_2SO_4 with a carmine-red color.

USE: As pH indicator: colorless 9.3 to blue 10.5. Also as reagent for blood after decolorizing the alkaline soln by boiling with zinc dust.

9557. Thymomodulin. [90803-92-2] Leucotrofina. Cell-free thymic hormone preparation extracted from calf thymus by acid hydrolysis. Composed of a mixture of biologically active acidic peptides of mol wt <10,000. Modulates the maturation of T-cells. Prepn of crude extract from calf thymus: B. Brunetti, E. Pini, **US 3657417** (1972 to Ellem). Electrophoretic characterization: C. Secchi *et al.*, *Riv. Eur. Sci. Med. Farmacol.* **4**, 499 (1982), *C.A.* **101**, 21878m

(1984). Biological activity and comparison with other thymic hormones: J. J. Twomey, N. M. Kouttab, *Cell. Immunol.* **72**, 186 (1982). Purification and pharmacodynamics: *eidem, Drugs Exp. Clin. Res.* **10**, 921 (1984). Antileukopenic activity in myelodepressed cancer patients: Z. Uray *et al., Agressologie* **21**, 215 (1980); C. Gallo Curcio *et al., Int. J. Immunother.* **2**, 189 (1986). Use in prophylaxis of infantile asthma: R. Genova, A. Guerra, *Pediatr. Med. Chir.* **5**, 395 (1983). Preliminary study in AIDS: G. Valesini *et al., Eur. J. Cancer Clin. Oncol.* **22**, 531 (1986).

THERAP CAT: Immunoregulator.

9558. Thymopentin. [69558-55-0] L-Arginyl-L-lysyl-L-α-aspartyl-L-valyl-L-tyrosine; thymopoietin pentapeptide; TP-5; ORF-15244; Immunox; Sintomodulina; Timunox. $C_{30}H_{49}N_9O_9$; mol wt 679.78. C 53.01%, H 7.27%, N 18.54%, O 21.18%. Thymic hormone analog corresponding to residues 32-36 of thymopoietin, *q.v.*, which exhibits the full biological activity of the natural hormone. Synthesis: G. Goldstein *et al., Science* **204**, 1309 (1979); *eidem*, US 4190646 (1980 to Sloan-Kettering). Bioavailability: T. Audhya, G. Goldstein, *Int. J. Pept. Protein Res.* **22**, 187 (1983). Pharmacology: K. Bolla *et al., Int. J. Clin. Pharmacol. Res.* **4**, 431 (1984). Comparison of biological activity with thymopoietin and splenin: T. Audhya *et al., Proc. Natl. Acad. Sci. USA* **81**, 2847 (1984). Clinical study in treatment of primary immunodeficiencies: F. Aitui *et al., Lancet* **1**, 551 (1983); in AIDS: N. Clumeck *et al., Int. J. Clin. Pharmacol. Res.* **4**, 459 (1984); in rheumatoid arthritis: M. G. Malaise *et al., Lancet* **1**, 832 (1985). *Review:* E. A. Boyse, *Surv. Immunol. Res.* **4**, 6-10 (1985).

Arg–Lys–Asp–Val–Tyr

THERAP CAT: Immunoregulator.

9559. Thymopoietin. Thymin (formerly); TP. A polypeptide thymic hormone which selectively induces T-cell maturation from prothymocytes and inhibits B-cell differentiation. mol wt ~5500 daltons. First detected as a factor causing impaired neuromuscular transmission in myasthenia gravis, a neuromuscular disease associated with thymic abnormalities: G. Goldstein, S. Wittingham, *Lancet* **2**, 315 (1966); G. Goldstein, *ibid.* **2**, 119 (1968). Isoln from bovine thymus as two closely homologous forms, thymopoietins I and II: *idem, Nature* **247**, 11 (1974); *idem, Ann. N.Y. Acad. Sci.* **249**, 177 (1975); *idem*, US 4077949 (1978 to Sloan-Kettering). Bovine TPs contain 49 amino acids; I differs from II only in residues 1, 2, and 43. The active site corresponds to residues 32-36. Human TP contains 48 amino acids. Human and bovine forms are highly homologous and are identical at the active site. Proposed sequence of bTP-II: D. H. Schlesinger, G. Goldstein, *Cell* **5**, 361 (1975). Revised sequence of II, sequence of I, isoln and sequence of splenin, *q.v.*, a thymopoietin-like peptide found in spleen: T. Audhya *et al., Biochemistry* **20**, 6195 (1981). Isoln and sequence of human thymopoietin: T. Audhya *et al., Proc. Natl. Acad. Sci. USA* **84**, 3545 (1987). First synthesis of a bioactive fragment: D. H. Schlesinger, *Cell* **5**, 367 (1975). Identification of active site and synthesis of thymopentin, *q.v.*, the pentapeptide corresponding to the active site: G. Goldstein *et al., Science* **204**, 1309 (1979). Total synthesis of originally proposed amino acid sequence of II: M. Fujino *et al., Chem. Pharm. Bull.* **25**, 1486 (1977); of revised sequence: T. Abiko, H. Sekino, *ibid.* **35**, 2016 (1987). Effect on T-cell differentiation: R. S. Basch, G. Goldstein, *Proc. Natl. Acad. Sci. USA* **71**, 1474 (1974); M. P. Scheid *et al., Ann. N.Y. Acad. Sci.* **249**, 531 (1975). Comparison of bioactivities of thymopoietin and splenin: T. Audhya *et al., Proc. Natl. Acad. Sci. USA* **81**, 2847 (1984). Review of thymopoietin and other thymic hormones: J. F. Bach, *J. Immunopharmacol.* **1**, 277-310 (1979); of thymopoietin and immunoregulation: G. Goldstein, C. Lau in *Polypeptide Hormones*, R. F. Beers, E. G. Bassett, Eds. (Raven Press, New York, 1980) pp 459-467; E. A. Boyse, *Surv. Immunol. Res.* **4**, 6-10 (1985).

9560. Thymosins. Family of thymic peptide hormones that regulate immune and endocrine events. Extraction from calf thymus and partial purification: A. L. Goldstein *et al., Proc. Natl. Acad. Sci. USA* **56**, 1010 (1966). Purification and characterization of thymosins and isoln of one of the active components, **thymosin fraction 5**: *eidem, ibid.* **69**, 1800 (1972). Improved isoln and properties: J. A. Hooper *et al., Ann. N.Y. Acad. Sci.* **249**, 125 (1975). Thymosin

fraction 5 is a mixture of polypeptides ranging in mol wt from ~1000 to 15,000; it has a very low content of non-protein material; it is heat stable up to 80°. It is an immunopotentiating agent that can act in place of the thymus gland in immuno-deprived or thymus-deprived patients to restore some immune functions. Nomenclature of the polypeptide fragments of fraction 5 is based on isoelectric points: α-region below 5.0, β-region 5.0-7.0, and γ-region above 7.0. **Thymosin α₁** is an acidic (pI 4.2), 28 amino acid polypeptide (mol wt 3108) found in the highly acidic region of fraction 5. It is a biologically active proteolytic fragment of prothymosin α_1. Isoln and preliminary structural determination of thymosin α_1: A. L. Goldstein *et al., Proc. Natl. Acad. Sci. USA* **74**, 725 (1977). Isoln, characterization, biological activities, and amino acid sequence analysis of α_1 and **polypeptide β₁**, a 74 amino acid peptide identical with ubiquitin, *q.v.*: T. L. K. Low *et al., J. Biol. Chem.* **254**, 981 (1979); T. L. K. Low, A. L. Goldstein, *ibid.* 987. Synthesis of α_1 by solution methods: S. S. Wang *et al., J. Am. Chem. Soc.* **101**, 253 (1979). **Thymosin β₄** is a biologically active, G-actin sequestering peptide consisting of 43 amino acids; mol wt 4.9 kDa; pI 5.1. Amino acid sequence of β_4: T. L. K. Low *et al., Proc. Natl. Acad. Sci. USA* **78**, 1162 (1981). Automated solid phase synthesis of β_4: S. S. Wang *et al., Int. J. Pept. Protein Res.* **18**, 413 (1981). Role of β_4 in wound healing: K. M. Malinda *et al., J. Invest. Dermatol.* **113**, 364 (1999). Series of articles on prepn, purification and characterization of thymic hormones and peptides: *Methods Enzymol.* **116**, 213-291 (1985). Review of immune and endocrine modulation: P. H. Naylor, A. L. Goldstein, *Adv. Pigment Cell Res.* **256**, 489-502 (1988); of immunological activities and therapeutic uses of thymic hormones and factors: S. Ben-Efraim *et al., Crit. Rev. Immunol.* **19**, 261-284 (1999); of β-thymosins: V. T. Nachmias, *Curr. Opin. Cell Biol.* **5**, 56-62 (1993); T. Huff *et al., Int. J. Biochem. Cell Biol.* **33**, 205-220 (2001).

AcSer-Asp-Ala-Ala-Val-Asp-Thr-Ser-Ser-Glu-Ile-Thr-Thr
|
Lys
|
Asp
|
HO-Asn-Glu-Ala-Glu-Glu-Glu-Val-Val-Glu-Lys-Lys-Glu-Lys-Leu

Human Thymosin α1

Thymalfasin. [62304-98-7] Human thymosin α1; bovine thymosin α1; Zadaxin; Timosina. Synthetic peptide; mol wt 3108. Synthesis: S. S. Wang *et al., Int. J. Pept. Protein Res.* **15**, 1 (1980). Clinical trial in influenza vaccine enhancement: S. Gravenstein *et al., J. Am. Geriatr. Soc.* **37**, 1 (1989). Clinical pharmacokinetics: K. L. Rost *et al., Int. J. Clin. Pharmacol. Ther.* **37**, 51 (1999). Review of use in combined chemo-immunotherapy in treatment of cancer: E. Garaci *et al., Int. J. Immunopharmacol.* **22**, 1067-1076 (2000); of clinical efficacy in hepatitis: P. H. Naylor, *Expert Opin. Invest. Drugs* **8**, 281-287 (1999); A. Billich, *Curr. Opin. Investig. Drugs* **3**, 698-707 (2002). pI 3.8.

THERAP CAT: Thymalfasin as immunomodulator.

9561. Thymostatin. Calf thymic extract which markedly inhibits *in vitro* and *in vivo* the incorporation of labelled nucleosides into the DNA and RNA of lymphocytes. Also inhibited incorporation of labelled nucleosides *in vitro* into other cell types both lymphoid and non-lymphoid, but did not affect incorporation *in vitro* of labelled amino acids into cellular protein. A carbohydrate containing peptide with a particle size <2000. Prepn and properties: Goldstein *et al., Proc. Natl. Acad. Sci. USA* **57**, 821 (1967).

Relatively heat stable; insol in chloroform but sol in chloroform-methanol (2:1).

9562. Thymostimulin. [117149-90-3] TP-1. Biological response modifier; partially purified extract of calf thymus composed of a mixture of bovine thymic peptides. Extraction and purification: G. Bergesi, R. Falchetti, *Folia Allergol. Immunol. Clin.* **24**, 204 (1977). Pharmacology and biological properties: R. Falchetti *et al., Drugs Exp. Clin. Res.* **3**, 39 (1977). Clinical studies in combination with antimicrobial therapy: P. Periti *et al., J. Chemother.* **5**, 37 (1993); as an adjunct to cancer chemotherapy: M. Federico *et al.,*

Am. J. Clin. Oncol. **18**, 8 (1995); with zidovudine vs HIV infection: G. Barbaro *et al. Curr. Ther. Res.* **56**, 369 (1995). Review of pharmacology and therapeutic use: K. L. Dechant, H. M. Bryson, *Clin. Immunother.* **1**, 378-398 (1994).

THERAP CAT: Immunomodulator.

9563. *o*-Thymotic Acid. [548-51-6] 2-Hydroxy-6-methyl-3-(1-methylethyl)benzoic acid; 3-hydroxy-2-*p*-cymenecarboxylic acid; *o*-thymotinic acid; 6-methyl-3-isopropylsalicylic acid. $C_{11}H_{14}O_3$; mol wt 194.23. C 68.02%, H 7.27%, O 24.71%. Prepn from thymol: Kolbe, Lautemann, *Ann.* **115**, 205 (1861); Spallino, Provenzal, *Gazz. Chim. Ital.* **39** (II), 326 (1909); Royer *et al., Bull. Soc. Chim. Fr.* **1955**, 1421.

Monoclinic prismatic needles from water, mp 127°. Volatile with steam. One gram dissolves in 10 liters of water at 20°. Sol in alc, ether, chloroform, benzene, petr ether.

Sodium salt. $C_{11}H_{13}NaO_3$. White cryst mass, freely sol in water.

Amide. $C_{11}H_{15}NO_2$. Needles from alc, mp 137°, dec 205°. Sol in org solvents and in aq solns of sodium carbonate.

Acetonyl ester. $C_{14}H_{18}O_4$. Needles from alcohol, mp 75°. Sol in alcohol.

USE: As salicylic acid and salicylates.

9564. Thymyl *N*-Isoamylcarbamate. [578-20-1] Isoamylcarbamic acid thymyl ester; isopropyl-*m*-cresyl ester of isoamylcarbamic acid; Egressin. $C_{16}H_{25}NO_2$; mol wt 263.38. C 72.97%, H 9.57%, N 5.32%, O 12.15%. Prepd by interaction of isoamylamine and thymyl chloroformate: Zima, v. Werder, US 2524185 (1949 to E. Merck).

Needles from petr ether, mp 57°. Practically insol in water (<1:50,000). Upon alkaline saponification, it is split into isoamylamine and thymol (along with CO_2).

THERAP CAT: Anthelmintic (Nematodes).

9565. Thyroid. Tiroidina; Thyradin; Thyrocrine. Thyroid gland of domesticated animals that are used as food by man, freed from connective tissue and fat, dried and powdered. Contains not less than 0.17% and not over 0.23% iodine in thyroid combination. 1 part ≈ 5 parts fresh gland. Chemistry and physiology: Rawson *et al., The Hormones* vol. III (Academic Press, New York, 1955) pp 433-519.

Yellowish powder; slight meat-like odor; saline taste.

THERAP CAT: Thyroid hormone.

THERAP CAT (VET): In myxedema. Has been used in obesity, renal insufficiency, chronic skin conditions and to increase spermatogenesis, libido, lactation.

9566. Thyroidin. Iodothyrin. A dried extract of thyroid diluted with milk sugar or other suitable diluent, equal in potency to official thyroid.

THERAP CAT: Thyroid hormone.

9567. Thyropropic Acid. [51-26-3] 4-(4-Hydroxy-3-iodophenoxy)-3,5-diiodobenzenepropanoic acid; 4-(4-hydroxy-3-iodophenoxy)-3,5-diiodohydrocinnamic acid; 3,3',5-triiodothyropropionic acid; β-[4-(3'-iodo-4'-hydroxyphenoxy)-3,5-diiodophenyl]-propionic acid; Birodan. $C_{15}H_{11}I_3O_4$; mol wt 635.96. C 28.33%, H

1.74%, I 59.86%, O 10.06%. Thyroid hormone analog. Prepd by iodination of 3,5-diiodothyropropionic acid: Tomita, Lardy, *J. Biol. Chem.* **219**, 595 (1956).

Crystals from abs ethanol, mp 200°.

THERAP CAT: Antilipemic.

9568. Thyroprotein. [9005-97-4] Thyroactive protein; Protamone-D. Iodinated casein. Use in feed supplements: Kohler, "The Use of Thyroactive Protein in Dairy Cattle Feeding" in *The Texas Nutrition Conference*, Oct. 1953, pp 5-7. Prepn: Reineke *et al., J. Biol. Chem.* **143**, 285 (1942); **147**, 115 (1943); Turner, Reineke, US 2329445, US 2379842, US 2385117, US 2478065 (1944, 1945, 1945, 1949 to Am. Dairies & Quaker Oats); Whitmoyer, Moore, US 2382193 (1945 to Whitmoyer Labs.); West, Van Bruggen, US 2642426, US 2709671 (1953, 1955 to Feed Prods).

USE: Feed supplement, *Fed. Regist.* **45**, 41360 (1980).

9569. Thyrotropin. [9002-71-5] TSH; thyroid-stimulating hormone; thyrotropic hormone. Glycoprotein hormone secreted by the anterior pituitary under the regulation of TRH, *q.v.* Stimulates the uptake of iodine, synthesis of thyroglobulin and release of thyroxine, *q.v.*, by the thyroid. Structure is a heterodimer of α- and β-subunits; analogous with chorionic gonadotropin, luteinizing hormone, and follicle-stimulating hormone, *q.q.v.* Reviews of early literature: W. T. Salter, *Hormones* **2**, 301-349 (1950); M. Sonnenberg, *Vitam. Horm.* **16**, 205-261 (1958). Purification from bovine pituitaries: T.-H. Liao *et al., J. Biol. Chem.* **244**, 6458 (1969). Amino acid sequence of bovine TSH: B. Shome *et al., ibid.* **246**, 833 (1971); T.-H. Liao, J. G. Pierce, *ibid.* 850; of human α and β subunits: J. G. Pierce *et al., Recent Prog. Horm. Res.* **27**, 165 (1971). Use in diagnosis of thyroid function: C. W. H. Havard, M. Boss, *Br. Med. J.* **3**, 678 (1974). Review of structure and bioactivity: J. A. Magner, *Endocr. Rev.* **11**, 354-385 (1990); and therapeutic potential: M. Grossmann *et al., ibid.* **18**, 476-501 (1997).

Biological activity is destroyed by heating and by proteolysis with pepsin, trypsin, and chymotrypsin. Also inactivated by oxidizing agents, such as potassium permanganate and elemental iodine.

Thyrotropin alfa. [194100-83-9] Thyrotropin (human β-subunit) complex with chorionic gonadotropin (human α-subunit); Thyrogen. Recombinant human TSH produced in Chinese hamster ovary cells. Prepn and pharmacology: E. S. Cole *et al., Biotechnology* **11**, 1014 (1993). *See also:* C. A. Kelton *et al.*, US 5240832 (1993 to Genzyme). Use in diagnosis of differentiated thyroid carcinoma (DTC): P. W. Ladenson *et al., N. Engl. J. Med.* **337**, 888 (1997). Clinical evaluation as adjunct to radioablative treatment of DTC: M. Luster *et al., J. Clin. Endocrinol. Metab.* **85**, 3640 (2000). Review of clinical studies: R. J. Robbins, A. K. Robbins, *ibid.* **88**, 1933-1938 (2003).

THERAP CAT: Diagnostic aid (thyroid tumor imaging).

THERAP CAT (VET): Diagnostic aid (thyroid function).

9570. Thyroxine. [51-48-9] *O*-(4-Hydroxy-3,5-diiodophenyl)-3,5-diiodo-L-tyrosine; (−)-3-[4-(4-hydroxy-3,5-diiodophenoxy)-3,5-diiodophenyl]alanine; 3,5,3',5'-tetraiodo-L-thyronine; levothyroxine; T_4. $C_{15}H_{11}I_4NO_4$; mol wt 776.87. C 23.19%, H 1.43%, I 65.34%, N 1.80%, O 8.24%. One of the thyroid hormones involved in the maintenance of metabolic homeostasis. Synthesized and stored as amino acid residues of *thyroglobulin*, the major protein component of the thyroid follicular colloid. Synthesis and secretion are regulated by the pituitary hormone, thyrotropin, *q.v.* Circulates in the serum bound to specific proteins, primarily *thyroxine-binding globulin, transthyretin*, and albumin. Deiodinated in peripheral tissues to the active metabolite, liothyronine, *q.v.* The D-form has very little activity as a thyroid hormone, but has been used to treat hyperlipidemia. Isoln from thyroid: E. C. Kendall, *J. Am. Med. Assoc.* **64**, 2042 (1915); *idem, J. Biol. Chem.* **39**, 125 (1919). Structure: C. R. Harington, *Biochem. J.* **20**, 293, 300 (1926). Synthesis: C. R. Harington, G. Barger, *ibid.* **21**, 169 (1927); of naturally occurring L-form: J. R. Chalmers *et al., J. Chem. Soc.* **1949**, 3424; of isomers and racemate: L. G. Ginger, P. Z. Anthony, **US 2889363**; **US 2889364** (both 1959 to Baxter Labs.). Direct determn in serum by RIA: J. C. Nelson, R. T. Tomel, *Clin. Chem.* **34**, 1737 (1988). Comprehensive description of levothyroxine sodium: A. Post, R. J. Warren, *Anal. Profiles Drug Subs.* **5**, 225-281 (1976). Review of pharmacology: E. Sypniewski, *Ann. Thorac. Surg.* **56**, S2-S8 (1993); of therapeutic use: A. D. Toft, *N. Engl. J. Med.* **331**, 174-180 (1994). Review of thyroid physiology: D. A. Fisher, *Clin. Chem.* **42**, 135-139 (1996); R. R. Cavalieri, *Thyroid* **7**, 177-181 (1997).

Crystals, dec 235-236°. $[\alpha]_{546}^{25}$ $-3.2°$ (0.66 g in 6.07 g of 0.5N NaOH and 13.03 g alc); $[\alpha]_D^{20}$ $-4.4°$ (3% in 0.13N NaOH in 70% EtOH).

Sodium salt. [55-03-8] Levothyroxine sodium; Eferox; Eltroxin; Euthyrox; Levaxin; Levothroid; Levothyrox; Levoxyl; Oroxine; Soloxine; Synthroid; Thyro-Tabs. $C_{15}H_{10}I_4NNaO_4$; mol wt 798.86. Occurs as the hydrate. Triclinic crystals or cream-colored powder. Odorless, tasteless, somewhat hygroscopic. d_4^{20} 2.381. $[\alpha]_D^{20}$ $-4.4°$ (c = 3 in 70% ethanol). Soly in water (25°): about 15 mg/100 ml. Sol in mineral acids and in solns of alkali hydroxides and hot solns of alkali carbonates; slightly sol in alcohol. Insol in acetone, chloroform, ether. pH of a satd water soln: 8.35 to 9.35.

DL-Thyroxine. [300-30-1] Needle-like crystals. Dec 231-233°. Insol in water, in alcohol, and in the other usual organic solvents, but in the presence of mineral acids or alkalies it dissolves in alcohol; sol in solns of the alkali hydroxides and in hot solns of the alkali carbonates.

D-Thyroxine. [51-49-0] Dextrothyroxine. Crystals, dec 237°. $[\alpha]_{546}^{21}$ +2.97° (0.74 g in 6 g of 0.5N NaOH and 14 g of alcohol).

D-Thyroxine sodium salt. [137-53-1] Dextrothyroxine sodium; Choloxin; Debetrol; Dethyrona; Dextroid; Dynothel; Eulipos.

THERAP CAT: Thyroid hormone. Dextrothyroxine as antilipemic.
THERAP CAT (VET): Thyroid hormone.

9571. Tiadenol. [6964-20-1] 2,2'-[1,10-Decanediylbis(thio)]bisethanol; 2,2'-(decamethylenedithio)diethanol; 1,10-bis(2-hydroxyethylthio)decane; LL-1558; Delipid; Eulip; Fonlipol; Tiaden; Tiaterol. $C_{14}H_{30}O_2S_2$; mol wt 294.51. C 57.10%, H 10.27%, O 10.86%, S 21.77%. Prepn: Williams, Cossar, **US 3021215** (1962 to Eastman Kodak); Lafon, **DE 2038836** (1971 to Orsymonde), *C.A.* **75**, 35528d (1971). Series of articles on pharmacology, metabolism, and toxicology: *Therapie* **27**, 395-444 (1972).

Crystals, mp 69.5°. uv max (ethanol): 212 nm. Soluble in ethanol, chloroform. Practically insol in water.

THERAP CAT: Antilipemic.

9572. Tiagabine. [115103-54-3] (3R)-1-[4,4-Bis(3-methyl-2-thienyl)-3-buten-1-yl]-3-piperidinecarboxylic acid; (−)-(R)-1-[4,4-bis(3-methyl-2-thienyl)-3-butenyl]nipecotic acid; TGB; NO-328; NO-05-0328; NNC-05-0328; A-70569. $C_{20}H_{25}NO_2S_2$; mol wt 375.55. C 63.96%, H 6.71%, N 3.73%, O 8.52%, S 17.07%. Selective GABA reuptake inhibitor. Prepn: F. C. Groenvald, C. Braestrup, **WO 8700171** (1987 to Novo); *eidem*, **US 5010090** (1991 to Novo Nordisk); K. E. Andersen *et al.*, *J. Med. Chem.* **36**, 1716 (1993). HPLC determn in human plasma: L. E. Gustavson, S. Chu, *J. Chromatogr.* **574**, 313 (1992). Review of pharmacology: W. J. Giardina, *ibid.* **7**, 161-166 (1994); and chemistry: H. Mengel, *Epilepsia* **35**, Suppl. 5, S81-S84 (1994). Review of clinical experience in epilepsy: J. P. Leach, M. J. Brodie, *Lancet* **351**, 203-207 (1998); of pharmacology and therapeutic potential in anxiety disorders: T. L. Schwartz, N. Nihalani, *Expert Opin. Pharmacother.* **7**, 1977-1987 (2006).

Hydrochloride. [145821-59-6] Gabitril. $C_{20}H_{25}NO_2S_2$·HCl; mol wt 412.00. White to off-white, odorless crystalline powder, mp 192° (dec) (Mengel). $[\alpha]_D^{20}$ $-11°$. pKa_1 3.3; pKa_2 9.4. Partition coefficient (octanol/water): 39.3 (pH 7.4). Soly in water: 3%. Freely sol in methanol, alc; sol in isopropanol, aqueous base; very slightly sol in chloroform. Practically insol in n-heptane; insol in hexane.

THERAP CAT: Anticonvulsant.

9573. Tiamenidine. [31428-61-2] N-(2-Chloro-4-methyl-3-thienyl)-4,5-dihydro-1H-imidazol-2-amine; 2-[(2-chloro-4-methyl-3-thienyl)amino]-2-imidazoline; 2-chloro-4-methyl-3-(2'-imidazolin-2'-ylamino)thiophene; thiamenidine. $C_8H_{10}ClN_3S$; mol wt 215.70. C 44.55%, H 4.67%, Cl 16.43%, N 19.48%, S 14.86%. A thiophene analog of clonidine, *q.v.* Prepn: R. Rippel *et al.*, **DE 1941761**; *eidem*, **US 3758476** (1971, 1973, both to Hoechst). Structural studies: J. M. Leger, *C. R. Seances Acad. Sci. Ser. C* **289**, 93 (1979); A. Carpy *et al.*, *Mol. Pharmacol.* **21**, 400 (1981). GC mass spectrometry determn in human plasma: T. A. Bryce, J. L. Burrows, *Biomed. Mass Spectrom.* **6**, 27 (1979). Pharmacology: E. Lindner, J. Kaiser, *Arch. Int. Pharmacodyn. Ther.* **211**, 305 (1974). Activity as α-adrenoceptor agonist: A. G. Roach *et al.*, *J. Pharmacol. Exp. Ther.* **227**, 421 (1983).

Crystals from isopropanol + petr ether, mp 152°.

Hydrochloride. [51274-83-0] HOE-440; Sundralen. $C_8H_{10}ClN_3S$·HCl; mol wt 252.16. Crystals from isopropanol + petr ether, mp 228-229°. LD_{50} in rats, mice (mg/kg): 40, 45 i.v.; in mice: 170 s.c., 400 orally (Lindner, Kaiser).

THERAP CAT: Antihypertensive.

9574. Tiamulin. [55297-95-5] [[2-(Diethylamino)ethyl]thio]acetic acid (3aS,4R,5S,6S,8R,9R,9aR,10R)-6-ethenyldecahydro-5-hydroxy-4,6,9,10-tetramethyl-1-oxo-3aH-cyclopentacycloocten-8-yl ester; 14-desoxy-14-[(2-diethylaminoethyl)-mercaptoacetoxy]mutilin; thiamutilin; SQ-14055. $C_{28}H_{47}NO_4S$; mol wt 493.75. C 68.11%, H 9.60%, N 2.84%, O 12.96%, S 6.49%. Derivative of pleuromutilin, *q.v.* Prepn: H. Egger, **DE 2248237** corresp to **US 3919290** (1973, 1975 both to Sandoz); H. Egger, H. Reinshagen, *J. Antibiot.* **29**, 915 (1976). Biosynthesis: F. Knauseder, E. Brandl, *ibid.* **31**, 756 (1978). *In vitro* activity: H. Werner *et al.*, *ibid.* **31**, 756 (1978). Metabolism: J. Dreyfuss *et al.*, *ibid.* **32**, 496 (1979). Effect vs *Mycoplasma*: C. O. Baughn *et al.*, *Avian Dis.* **22**, 620 (1978); R. F. Goodwin, *Vet. Rec.* **104**, 194 (1979). Mechanism of action: G. Högenauer in *Antibiotics* **vol. 5** (pt. 1), F. E. Hahn, Ed. (Springer-Verlag, New York, 1979) pp 344-360. Treatment of swine dysentery: M. D. Anderson, *Vet. Med. Small Anim. Clin.* **78**, 98 (1983).

Sticky, translucent yellowish mass; slightly hygroscopic. Very sol in dichloromethane; freely sol in dehydrated alc. Practically insol in water.

Fumarate. [55297-96-6] 81723 hfu; SQ-22947; Denagard; Dynamutilin; Tiamutin. $C_{32}H_{51}NO_8S$; mol wt 609.82. Crystals from acetone, mp 147-148° (after stirring in ethyl acetate and drying at 60° and 80° overnight).

THERAP CAT (VET): Antibacterial.

9575. Tianeptine. [72797-41-2] 7-[(3-Chloro-6,11-dihydro-6-methyl-5,5-dioxidodibenzo[c,f][1,2]thiazepin-11-yl)amino]heptanoic acid; 7-[(3-chloro-6,11-dihydro-6-methyldibenzo[c,f][1,2]-thiazepin-11-yl)amino]heptanoic acid S,S-dioxide. $C_{21}H_{25}ClN_2O_4$-S; mol wt 436.95. C 57.73%, H 5.77%, Cl 8.11%, N 6.41%, O 14.65%, S 7.34%. Tricyclic compound with psychostimulant, anti-ulcer and anti-emetic properties. Prepn: C. Malen *et al.,* **DE 2011806** corresp to **US 3758528** (1970, 1973 both to Sci. Union et Cie-Soc. Franc. Rech. Med.). Neuropharmacology study: C. Malen, J. Poignant, *Experientia* **28**, 811 (1972).

Sodium salt. [30123-17-2] S-1574; Stablon. $C_{21}H_{24}ClN_2$-NaO_4S; mol wt 458.93. Solid, mp 180°.

THERAP CAT: Antidepressant.

9576. Tiapride. [51012-32-9] *N*-[2-(Diethylamino)ethyl]-2-methoxy-5-(methylsulfonyl)benzamide; *N*-[2-(diethylamino)ethyl]-5-(methylsulfonyl)-*o*-anisamide; thiapride; FLC-1374. $C_{15}H_{24}N_2$-O_4S; mol wt 328.43. C 54.86%, H 7.37%, N 8.53%, O 19.49%, S 9.76%. Dopamine receptor antagonist structurally related to sulpiride, *q.v.* Prepn: G. Bulteau *et al.,* **DE 2327192; GB 1394563** (1973, 1975 both to Soc. d'Etudes Sci. Ind. de L'Ile-de-France). Crystal structure: C. Houttemane *et al., Acta Crystallogr.* **C39**, 585 (1983). Dopamine receptor binding studies: T. Arima *et al., Jpn. J. Pharmacol.* **41**, 419 (1986). Pharmacokinetics: E. Rey *et al., Int. J. Clin. Pharmacol. Ther. Toxicol.* **20**, 62 (1982). Clinical evaluation in neuroleptic-induced dyskinesia: W. Greil *et al., Neuropsychobiology* **14**, 17 (1985); in alcohol withdrawal syndrome: R. Agricola *et al., J. Int. Med. Res.* **10**, 160 (1982); G. K. Shaw *et al., Br. J. Psychiatry* **150**, 164 (1987).

Crystals, mp 123-125°.

Hydrochloride. [51012-33-0] Gramalil; Italprid; Luxoben; Sereprile; Tiapridal; Tiapridex. $C_{15}H_{24}N_2O_4S$.HCl; mol wt 364.89.

THERAP CAT: Antidyskinetic.

9577. Tiaprofenic Acid. [33005-95-7] 5-Benzoyl-α-methyl-2-thiopheneacetic acid; α-methyl-5-benzoyl-2-thienylacetic acid; FC-3001; RU-15060; Suralgan; Surgam. $C_{14}H_{12}O_3S$; mol wt 260.31. C 64.60%, H 4.65%, O 18.44%, S 12.32%. Non-steroidal anti-inflammatory drug (NSAID). Prepn: F. Clémence, O. Le Martret, **DE 2055264** and **FR 2112111** (1971 and 1972 to Roussel-UCLAF), *C.A.* **75**, 63597u (1971) and **78**, 97473c (1973); and pharmacology: F. Clémence *et al., Eur. J. Med. Chem. - Chim. Ther.* **9**, 390 (1974). Review of pharmacology and efficacy in rheumatic diseases and pain control: E. M. Sorkin, R. N. Brogden, *Drugs* **29**, 208-235 (1985).

mp 96° (isopropyl ether).

THERAP CAT: Anti-inflammatory.

9578. Tiaprost. [71116-82-0] 7-[3,5-Dihydroxy-2-[3-hydroxy-4-(3-thienyloxy)-1-butenyl]cyclopentyl]-5-heptenoic acid; (15-*R,S*)-16-(3-thienyloxy)-ω-tetranor-PGF$_{2α}$. $C_{20}H_{28}O_6S$; mol wt 396.50. C 60.59%, H 7.12%, O 24.21%, S 8.09%. Analog of prostaglandin F$_{2α}$, *q.v.* Prepn: W. Bartmann *et al.,* **DE 2524955** corresp to **US 4258053** (1977, 1981 both to Hoechst); *eidem, Prostaglandins* **17**, 301 (1979). Efficacy in bovine endometritis: W. Bentele *et al., Tieraerztl. Umsch.* **35**, 676, 678, 683 (1980). Pharmacological effects: W. v. Rechenberg *et al., Blue Book for the Vet. Profession* (Hoechst AG) **30**, 417 (1981). Administration in cattle: R. Humble, *ibid.* 425.

Tromethamine salt. Tiaprost trometamol; Iliren. $C_{24}H_{39}NO_9S$; mol wt 517.63.

THERAP CAT (VET): Luteolytic.

9579. Tiaramide. [32527-55-2] 5-Chloro-3-[2-[4-(2-hydroxyethyl)-1-piperazinyl]-2-oxoethyl]-2(3*H*)-benzothiazolone; 4-[(5-chloro-2-oxo-3(2*H*)-benzothiazolyl)acetyl]-1-piperazineethanol; 5-chloro-3-[4-(2-hydroxyethyl)-1-piperazinyl]carbonylmethyl-2-benzothiazolinone; tialamide. $C_{15}H_{18}ClN_3O_3S$; mol wt 355.84. C 50.63%, H 5.10%, Cl 9.96%, N 11.81%, O 13.49%, S 9.01%. Prepn: Umio *et al.,* **JP 71 15302**, *C.A.* **75**, 36127j (1971); *eidem,* **JP 71 18752**, *C.A.* **75**, 63824r (1971); Umio, **US 3661921** (1971, 1971, 1972 all to Fujisawa). Pharmacology: Takashima *et al., Arzneim.-Forsch.* **22**, 711 (1972); Tsurumi *et al., ibid.* 716, 724. Metabolism: Noda *et al., ibid.* 732. Toxicity studies: Watanabe *et al., ibid.* **23**, 65 (1973). Mode of antiasthmatic action: G. C. Folco *et al., Pharmacol. Res. Commun.* **11**, 703 (1979).

pKa 6.2.

Hydrochloride. [35941-71-0] NTA-194; FK-1160; Solantal. $C_{15}H_{18}ClN_3O_3S$.HCl; mol wt 392.30. White, odorless, bitter tasting crystalline powder, mp 159-161°. pH (10% aqueous soln): 3.4-3.7. Very soluble in water; slightly sol in organic solvents. LD_{50} in male mice, rats (mg/kg): 178, 203 i.v.; 298, 540 i.p.; 564, 3600 orally (Watanabe).

THERAP CAT: Antiasthmatic; anti-inflammatory.

9580. Tiazofurin. [60084-10-8] 2-β-D-Ribofuranosyl-4-thiazolecarboxamide; riboxamide; TCAR; CI-909; NSC-286193; Tiazole. $C_9H_{12}N_2O_5S$; mol wt 260.26. C 41.54%, H 4.65%, N 10.76%, O 30.74%, S 12.32%. Nucleoside analog that inhibits inosine monophosphate dehydrogenase (IMPDH). Prepn: M. Fuertes *et al., J. Org. Chem.* **41**, 4074 (1976); and antiviral activity: P. C. Srivastava *et al., J. Med. Chem.* **20**, 256 (1977). Structure-activity study: G. Gebeyehu *et al., ibid.* **28**, 99 (1985). HPLC determn in plasma: R. W. Klecker, Jr., J. M. Collins, *J. Chromatogr.* **307**, 361 (1984). Clinical pharmacokinetics: D. Raghavan *et al., Cancer Chemother. Pharmacol.* **16**, 160 (1986). Clinical evaluation in leukemia: G. Tricot *et al, Int. J. Cell Cloning* **8**, 161 (1990). Series of articles on pharmacology and clinical experience: *Anticancer Res.* **16**, 3307-3354 (1996).

Crystals from ethanol-ethyl acetate, mp 145-146°. $[\alpha]_D^{25}$ $-9°$ (c = 0.5 in ethanol). uv max in ethanol: 215, 237 nm (ε 9450, 7625).

THERAP CAT: Antineoplastic.

9581. Tibezonium Iodide. [54663-47-7] *N,N*-Diethyl-*N*-methyl-2-[[4-[4-(phenylthio)phenyl]-3*H*-1,5-benzodiazepin-2-yl]-thio]ethanaminium iodide (1:1); diethylmethyl[2-[[4-[*p*-(phenylthio)phenyl]-3*H*-1,5-benzodiazepin-2-yl]thio]ethyl]ammonium iodide; 2-[β-(*N*-diethylamino)ethylthio]-4-(*p*-phenylthio)phenyl-3*H*-1,5-benzodiazepine methiodide; thiabenzazonium iodide; Rec-15-0691; Antoral. $C_{28}H_{32}IN_3S_2$; mol wt 601.61. C 55.90%, H 5.36%, I 21.09%, N 6.98%, S 10.66%. 1,5-Benzodiazepine deriv with bactericidal activity. Prepn: D. Nardi *et al.,* **CH 555347** (1974 to Recordati), *C.A.* **82**, 43480 (1975). Synthesis, physical characteristics, antibacterial activity: *eidem, Experientia* **31**, 440 (1975); *eidem, Farmaco Ed. Sci.* **30**, 248 (1975). Structure activity study: C. Greico *et al., ibid.* **32**, 909 (1977). Antimicrobial activity: M. Veronese *et al., Chemotherapy* **23**, 90 (1977). Toxicity study: D. Nardi *et al., Experientia* **31**, 440 (1975).

Crystals from isopropanol, mp 162°. LD$_{50}$ in mice, rats (mg/kg): 9000, >10000 orally; 42, 35 i.p. (Nardi).

THERAP CAT: Antibacterial.

9582. Tibolone. [5630-53-5] (7α,17α)-17-Hydroxy-7-methyl-19-norpregn-5(10)-en-20-yn-3-one; 7α-methyl-17α-ethynyl-17β-hydroxy-19-norandrost-5(10)-en-3-one; 7α-methyl-17α-ethynyl-17β-hydroxyestr-5(10)-en-3-one; Org-OD-14; Livial. $C_{21}H_{28}O_2$; mol wt 312.45. C 80.73%, H 9.03%, O 10.24%. Synthetic steroid with weak estrogenic, androgenic and progestogenic activity. Prepn: **NL 6406797**; H. P. de Jongh, N. P. van Vliet, **US 3340279** (1965, 1967 both to Organon). Improved process: M. S. de Winter, E. A. Harryvan, **US 3475465** (1969 to Organon). Endocrinological profile: J. de Visser *et al., Arzneim.-Forsch.* **34**, 1010 (1984). Series of articles on pharmacology and clinical efficacy in post-menopausal women: *Maturitas* Suppl 1, 1-72 (1987). Clinical effect in osteoporosis: P. Geusens *et al., ibid.* **13**, 155 (1991).

Crystals, mp 165-169°.

THERAP CAT: In treatment of menopausal syndrome.

9583. Ticagrelor. [274693-27-5] (1*S*,2*S*,3*R*,5*S*)-3-[7-[[(1*R*,-2*S*)-2-(3,4-Difluorophenyl)cyclopropyl]amino]-5-(propylthio)-3*H*-1,2,3-triazolo[4,5-*d*]pyrimidin-3-yl]-5-(2-hydroxyethoxy)-1,2-cyclopentanediol; AZD-6140; Brilinta; Brilique. $C_{23}H_{28}F_2N_6O_4S$; mol wt 522.57. C 52.86%, H 5.40%, F 7.27%, N 16.08%, O 12.25%, S 6.14%. Specific purinoceptor P2Y$_{12}$ antagonist; inhibits ADP-induced platelet aggregation. Prepn: S. Guile *et al.,* **WO 0034283**; D. Hardern *et al.,* **US 6525060** (2000, 2003 both to AstraZeneca); and SAR: B. Springthorpe *et al., Bioorg. Med. Chem. Lett.* **17**, 6013 (2007). Clinical pharmacokinetics, pharmacodynamics: S. Husted *et al., Eur. Heart J.* **27**, 1038 (2006). Comparative clinical trial with clopidogrel in patients with acute coronary syndromes: L. Wallentin *et al., N. Engl. J. Med.* **361**, 1045 (2009). *Review*: S. A. Doggrell, *IDrugs* **12**, 309-317 (2009).

Crystalline powder. Soly in water: ~10 µg/ml. Sol in ethyl acetate.

THERAP CAT: Antithrombotic.

9584. Ticarcillin. [34787-01-4] (2*S*,5*R*,6*R*)-6-[[(2*R*)-2-Carboxy-2-(3-thienyl)acetyl]amino]-3,3-dimethyl-7-oxo-4-thia-1-azabicyclo[3.2.0]heptane-2-carboxylic acid; *N*-(2-carboxy-3,3-dimethyl-7-oxo-4-thia-1-azabicyclo[3.2.0]hept-6-yl)-3-thiophenemalonamic acid; 6-[D(−)-α-carboxy-3-thienylacetamido]penicillanic acid; α-carboxy-3-thienylmethylpenicillin. $C_{15}H_{16}N_2O_6S_2$; mol wt 384.42. C 46.87%, H 4.20%, N 7.29%, O 24.97%, S 16.68%. Broad spectrum semi-synthetic antibiotic related to penicillin. Prepn: **BE 646991**; E. G. Brain, J. H. Nayler, **US 3282926** (1964, 1966 to Beecham Group Ltd.). *In vitro* studies: H. C. Neu, E. B. Winshell, *Antimicrob. Agents Chemother.* **1970**, 385; R. Sutherland *et al., ibid.* 390; N. J. Legakis, J. Papavassiliou, *J. Antibiot.* **28**, 912 (1975). *In vivo* studies: P. Acred *et al., Antimicrob. Agents Chemother.* **1970**, 396. Absorption and excretion: R. Sutherland, P. J. Wise, *ibid.* 402. Clinical pharmacology: V. Rodriguez *et al., Antimicrob. Agents Chemother.* **4**, 31 (1973); R. D. Libke *et al., Clin. Pharmacol. Ther.* **17**, 441 (1975). Review of pharmacology and therapeutic efficacy: R. N. Brogden *et al., Drugs* **20**, 325-352 (1980).

Disodium salt. [4697-14-7] BRL-2288; Monapen; Ticar; Ticarpen; Ticillin. $C_{15}H_{14}N_2Na_2O_6S_2$; mol wt 428.38. Creamy-white hygroscopic non-crystalline powder. Readily sol in water (>100 g/100 ml water) giving a clear soln with pH between 6.0 and 8.0. Aq solns are relatively stable; acid solns relatively unstable (Sutherland).

THERAP CAT: Antibacterial.

THERAP CAT (VET): Antibacterial.

9585. Ticlopidine. [55142-85-3] 5-[(2-Chlorophenyl)methyl]-4,5,6,7-tetrahydrothieno[3,2-*c*]pyridine; 5-(*o*-chlorobenzyl)-4,5,6,7-tetrahydrothieno[3,2-*c*]pyridine. $C_{14}H_{14}ClNS$; mol wt 263.78. C 63.75%, H 5.35%, Cl 13.44%, N 5.31%, S 12.15%. Platelet aggregation inhibitor. Prepn: **DE 2404308**; A. R. J. Castaigne, **US 4051141** (1974, 1977 both to Cent. Etudes Ind. Pharm.); E. Braye, **US 4127580** (1978 to Parcor). Metabolism: P. Godard *et al.,*

Eur. J. Drug Metab. Pharmacokinet. **3**, 67 (1978); *eidem, ibid.* **4**, 133 (1979); A. Tuong *et al., ibid.* **6**, 91 (1981). Mode of action: G. Leblondel, P. Allain, *Biochem. Pharmacol.* **27**, 2099 (1978); J. R. O'Brien *et al., Thromb. Res.* **13**, 245 (1978); J. J. Bruno, *ibid.* **1983**, Suppl. 4, 59. Pharmacology: A. Akashi *et al., Arzneim.-Forsch.* **30**, 409, 415 (1980). Clinical studies: J. J. Thebault *et al., J. Int. Med. Res.* **5**, 405 (1977); C. Lecrubier *et al., Therapie* **32**, 189 (1977); T. Katsumura *et al., Angiology* **33**, 357 (1982). Review of pharmacodynamics, pharmacokinetics and therapeutic use: E. Saltiel, A. Ward, *Drugs* **34**, 222-262 (1987). Comprehensive description: F. J. Al-Shammary, N. A. A. Mian, *Anal. Profiles Drug Subs. Excip.* **21**, 573-609 (1992).

Hydrochloride. [53885-35-1] 4-C-32; 53-32 C; Anagregal; Caudaline; Panaldine; Ticlid; Ticlodix; Ticlodone; Ticlosin; Tiklid. C_{14}-$H_{14}ClNS.HCl$; mol wt 300.24. Crystals from ethanol, mp 190°. uv max (water): 214, 268, 295 nm ($A_{1cm}^{1\%}$ 303.8, 13.14, 2). pKa 7.64. Almost sol in water; sol in 95% alcohol, methanol, chloroform. Insol in ether. LD_{50} in mice (mg/kg/24 hrs): 55 i.v.; >300 orally (Castaigne).

THERAP CAT: Antithrombotic.

9586. Ticrynafen. [40180-04-9] 2-[2,3-Dichloro-4-(2-thienylcarbonyl)phenoxy]acetic acid; [2,3-dichloro-4-(2-thenoyl)phenoxy]acetic acid; [2,3-dichloro-4-(2-thiophenecarbonyl)phenoxy]acetic acid; tienilic acid; thienylic acid; ANP-3624; CE-3624; SKF-62698; Diflurex; Selacryn. $C_{13}H_8Cl_2O_4S$; mol wt 331.16. C 47.15%, H 2.44%, Cl 21.41%, O 19.32%, S 9.68%. A heterocyclic derivative of phenoxyacetic acid. Prepn: J. Godfroid, J. Thuillier, **DE 2048372**; **US 3758506** (1971, 1973 both to C.E.R.P.H.A.) and **FR 2115042** (1972 to C.E.R.P.H.A.). Synthesis and pharmacology: G. Thuillier *et al., Eur. J. Med. Chem.* **9**, 625 (1974). Pharmacokinetics in healthy volunteers: A. L. Kerremans *et al., Eur. J. Clin. Pharmacol.* **22**, 515 (1982). Comparative study in hypertensive patients: B. T. Emmerson *et al., ibid.* 203. Hepatotoxicity study: J. W. Manier *et al., Am. J. Gastroenterol.* **77**, 401 (1982).

Crystals from 50% ethanol, mp 148-149°; also reported as mp 157°. LD_{50} in mice (mg/kg): 225 i.v., 1275 orally (**US 3758506**).

THERAP CAT: Diuretic; uricosuric; antihypertensive.

9587. Tiemonium Iodide. [144-12-7] 4-[3-Hydroxy-3-phenyl-3-(2-thienyl)propyl]-4-methylmorpholinium iodide (1:1); *N*-methyl-*N*-[3-hydroxy-3-phenyl-3-(α-thienyl)propyl]morpholinium iodide; Visceralgina. $C_{18}H_{24}INO_2S$; mol wt 445.36. C 48.54%, H 5.43%, I 28.49%, N 3.15%, O 7.18%, S 7.20%. Anticholinergic. Prepn: **GB 953386** (1964 to C.E.R.M.).

Solid, mp 189-191°.
THERAP CAT: Antispasmodic.

9588. Tigecycline. [220620-09-7] (4*S*,4a*S*,5a*R*,12a*S*)-4,7-Bis(dimethylamino)-9-[[2-[(1,1-dimethylethyl)amino]acetyl]amino]-1,4,4a,5,5a,6,11,12a-octahydro-3,10,12,12a-tetrahydroxy-1,11-dioxo-2-naphthacenecarboxamide; 9-*t*-butylglycylamidominocy-

cline; TBG-MINO; GAR-936; Tygacil. $C_{29}H_{39}N_5O_8$; mol wt 585.66. C 59.47%, H 6.71%, N 11.96%, O 21.85%. Broad spectrum glycylcycline antibiotic; semisynthetic tetracycline analogue. Prepn: J. J. Hlavka *et al.,* **EP 536515**; *eidem,* **US 5494903** (1993, 1996 both to Am. Cyanamid); P.-E. Sum, P. Petersen, *Bioorg. Med. Chem. Lett.* **9**, 1459 (1999). *In vivo* pharmacodynamics: M. L. van Ogtrop *et al., Antimicrob. Agents Chemother.* **44**, 943 (2000). *In vitro* activity vs clinical bacterial isolates: A. C. Gales, R. N. Jones, *Diagn. Microbiol. Infect. Dis.* **36**, 19 (2000); D. Milatovic *et al., Antimicrob. Agents Chemother.* **47**, 400 (2003). Mode of action study: G. Bauer *et al., J. Antimicrob. Chemother.* **53**, 592 (2004). Clinical pharmacology: A. K. Meagher *et al., Diagn. Microbiol. Infect. Dis.* **52**, 165 (2005). Clinical evaluation in complicated skin and skin-structure infections: R. G. Postier *et al., Clin. Ther.* **26**, 704 (2004); E. J. Ellis-Grosse *et al., Clin. Infect. Dis.* **41**, S341 (2005); in complicated intra-abdominal infections: T. Babinchak *et al., ibid.* S354. Review of chemistry, pharmacology and clinical development: E. Rubinstein, D. Vaughan, *Drugs* **65**, 1317-1336 (2005); R. A. Squires, R. G. Postier, *Expert Opin. Invest. Drugs* **15**, 155-162 (2006).

Orange powder or cake.
THERAP CAT: Antibacterial.

9589. Tiglic Acid. [80-59-1] (2*E*)-2-Methyl-2-butenoic acid; (*E*)-2-methylcrotonic acid; *trans*-2,3-dimethylacrylic acid. $C_5H_8O_2$; mol wt 100.12. C 59.98%, H 8.05%, O 31.96%. The stable isomer of angelic acid. Found as glyceride in croton oil, as butyl ester in the oil of the Roman camomile, *Anthemis nobilis* L., *Compositae*, and as geranyl tiglate in oil of geranium. Is formed during the charcoaling of maple wood. Formation by the intestinal roundworm, *Ascaris lumbricoides:* Bueding, *J. Biol. Chem.* **202**, 505 (1953). Has been found in crude sodium penicillin: Cram, Tishler, *J. Am. Chem. Soc.* **70**, 4238 (1948). Sepn by partition chromatography: Bueding, *loc. cit.* Synthesis from 2-hydroxy-2-methylbutyronitrile: Crawford, *J. Soc. Chem. Ind. London* **64**, 231 (1945). Review and bibliography: Buckles *et al., Chem. Rev.* **55**, 659-677 (1955).

Triclinic plates, rods from water. Spicy odor. *Vesicant.* d 0.972. mp 63.5-64°. bp_{760} 198.5°; $bp_{11.5}$ 95.0-96°. Volatile with steam. n_D^{81} 1.4342. pK (25°) 5.02. uv max (H_2O): 216-217 nm (ε 10700). Molar heat of combustion 635.1 kcal. Sparingly sol in cold water; freely sol in hot water. Sol in alcohol, ether.

Calcium salt trihydrate. $Ca(C_5H_7O_2)_2.3H_2O$; mol wt 292.34. Leaflets. Much less sol in water than calcium angelate: 100 parts of aq soln satd at 17° contains 6.05 parts of anhydr calcium tiglate.

Amide. C_5H_9NO; mol wt 99.13. Crystals, mp 75-76°.

Methyl ester. $C_6H_{10}O_2$; mol wt 114.14. Liquid; d_4^{20} 0.9498; bp_{766} 139.6°; n_D^{20} 1.4370.

Ethyl ester. $C_7H_{12}O_2$; mol wt 128.17. Liquid; $d_4^{19.5}$ 0.9247; bp_{752} 156°; bp_{11} 55.5°. n_D^{20} 1.4350. Heat of formn at constant vol: 953.2 kcal, at constant pressure: 954.4 kcal.

Geranyl ester. $C_{15}H_{24}O_2$; mol wt 236.36. Liquid; pleasant odor; d_{15}^{15} 0.9279; bp_7 149-151°.

USE: The esters in perfumes and flavoring agents. The free acid as a breaker of emulsions.

9590. Tigloidine. [495-83-0] (2*E*)-2-Methyl-2-butenoic acid (3-*exo*)-8-methyl-8-azabicyclo[3.2.1]oct-3-yl ester; (*E*)-1α*H*,5α*H*-tropan-3β-ol 2-methylcrotonate; tiglylpseudotropeine; 3β-tigloyl-oxytropane; tiglic acid ester with pseudotropine. $C_{13}H_{21}NO_2$; mol wt 223.32. C 69.92%, H 9.48%, N 6.27%, O 14.33%. Isoln from *Duboisia myoporoides* R. Br., *Solanaceae* and prepn from tropine and tigloyl chloride: Barger *et al.*, *J. Chem. Soc.* **1937**, 1820; isoln from *Datura innoxia* Miller, *Solanaceae:* Evans, Wellendorf, *ibid.* **1959**, 1406. Pharmacology: Sanghvi *et al.*, *Eur. J. Pharmacol.* **4**, 246 (1968).

Hydrobromide. Tiglyssin. $C_{13}H_{21}NO_2 \cdot HBr$; mol wt 304.23. mp 234-235°. Soluble in chloroform.

THERAP CAT: Antispasmodic.

9591. Tigogenin. [77-60-1] $(3\beta,5\alpha,25R)$-Spirostan-3-ol. $C_{27}H_{44}O_3$; mol wt 416.65. C 77.83%, H 10.64%, O 11.52%. The aglycon of tigonin; C-5 epimer of smilagenin, *q.v.* Obtained from leaves of *Digitalis lanata* Ehrh., *Scrophulariaceae:* Windhaus, *Z. Physiol. Chem.* **150**, 205 (1925); Jacobs, Fleck, *J. Biol. Chem.* **88**, 545 (1930). From the sisal plant *Agave sisalana* L., *Amaryllidaceae:* Rubin, *US 2991282* (1961); *US 3303187* (1967). Structure: Marker, Rohrmann, *J. Am. Chem. Soc.* **62**, 898 (1940); and synthesis: Mazur, Sondheimer, *ibid.* **81**, 3161 (1959); Caglioti, Magi, *Tetrahedron* **19**, 1127 (1963). In synthesis of steroids: T. Ohta *et al.*, *Org. Process Res. Dev.* **1**, 420 (1997); L. Amiranshvili *et al.*, *Bull. Georgian Acad. Sci.* **161**, 252 (2000).

Crystals from dil methanol, mp 203°. $[\alpha]_D^{20}$ −62°. More sol in acetone, in ether, and esp in petr ether than gitogenin. Pptd by digitonin.

Tigonin. [1329-83-5] $C_{56}H_{92}O_{27}$; mol wt 1197.33. Isoln from *Digitalis* sp: Tschesche, *Ber.* **69**, 1665 (1936); Liang, Noller, *J. Am. Chem. Soc.* **57**, 525 (1935). Built from 2 glucose, 2 galactose, 1 xylose, and 1 tigogenin unit. Hygroscopic, amorphous flakes from 95% alc. After drying at 118° over P_2O_5 *in vacuo* sinters at 220°. mp ~260° *(in vacuo)*. Sol in water. Not pptd by ether from water (difference from digitonin); amorphous ppt with amyl alcohol.

9592. Tiletamine. [14176-49-9] 2-(Ethylamino)-2-(2-thienyl)cyclohexanone. $C_{12}H_{17}NOS$; mol wt 223.33. C 64.54%, H 7.67%, N 6.27%, O 7.16%, S 14.36%. Dissociative anesthetic similiar to ketamine, *q.v.* NMDA receptor antagonist used in veterinary medicine in combination with zolazepam, *q.v.* Prepn: *NL 6603587*; R. F. Parcell, *US 3522273* (1966, 1970 both to Parke, Davis). Pharmacology: G. Chen *et al.*, *J. Pharmacol. Exp. Ther.* **168**, 171 (1969). Effect on NMDA receptors in rat brain: J. M. H. ffrench-Mullen *et al.*, *ibid.* **243**, 915 (1987). GC/MS determn in plasma: A. Kumar *et al.*, *J. Chromatogr. B* **842**, 131 (2006). Review of pharmacology and veterinary use in combination with zolazepam: H. C. Lin *et al.*, *J. Vet. Pharmacol. Ther.* **16**, 383-418 (1992).

Hydrochloride. [14176-50-2] CI-634. $C_{12}H_{17}NOS \cdot HCl$; mol wt 259.79. Crystals from isopropanol-ether, mp 196-197°. Soly in

water: 30%. Freely sol in 0.1*N* hydrochloric acid; sol in methanol; slightly sol in chloroform. Practically insol in ether. LD_{50} in mice (mg/kg): 116.0 ±2.0 i.p. (Chen).

Combination of hydrochloride with zolazepam hydrochloride. [75418-09-6] Telazol.

Note: This is a controlled substance (depressant): **21CFR**, 1308.13.

THERAP CAT (VET): Anesthetic.

9593. Tiliacorine. [27073-72-9] (4a*S*,16a*R*)-3,4,4a,5,16a,-17,18,19-Octahydro-12,21-dimethoxy-4,17-dimethyl-2*H*,16*H*-22,26-epoxy-1,24-etheno-6,10:11,15-dimethenopyrido[2',3':17,18]-oxacycloeicosino[2,3,4-*ij*]isoquinolin-9-ol; $(1\alpha,1'\alpha)$-6',7-didemethoxy-6',7-epoxyrodiasine. $C_{36}H_{36}N_2O_5$; mol wt 576.69. C 74.98%, H 6.29%, N 4.86%, O 13.87%. From bark of *Tiliacora acuminata* Miers, and *T. racemosa* Colebr., *Menispermaceae.* Isoln: Van Itallie, Steenhauer, *Pharm. Weekbl.* **59**, 1381 (1922). Structure: Anjaneyulu *et al.*, *Chem. Ind. (London)* **1959**, 1119; Rao, Row, *J. Org. Chem.* **25**, 981 (1960). Revised structure: Anjaneyulu *et al.*, *J. Sci. Ind. Res.* **21B**, 602 (1962); *eidem*, *Tetrahedron* **25**, 3091 (1969); M. Shamma, J. E. Foy, *J. Org. Chem.* **41**, 1293 (1976). Abs config and biosynthesis: D. S. Bhakuni *et al.*, *Chem. Commun.* **1978**, 226.

Bitter crystals, mp 271-272°. $[\alpha]_D$ +105.3°. uv max: 295, 265 nm (log ε 3.91, 3.48). Sol in alcohol, benzene, chloroform, ether.

9594. Tilidine. [51931-66-9] *rel*-(1*R*,2*S*)-2-(Dimethylamino)-1-phenyl-3-cyclohexene-1-carboxylic acid ethyl ester; ethyl 2-(dimethylamino)-1-phenyl-3-cyclohexene-1-carboxylate; 3-*trans*-dimethylamino-4-phenyl-4-*trans*-carbethoxy-Δ^1-cyclohexene. $C_{17}H_{23}NO_2$; mol wt 273.38. C 74.69%, H 8.48%, N 5.12%, O 11.70%. Opioid analgesic. Prepn: *ZA 6606476*; *GB 1120186*; G. Satzinger, *US 3557127* (1967, 1968, 1971, all to Warner-Lambert). Synthesis of ^{14}C-labelled hydrochloride: K.-O. Vollmer, F. W. Koss, *Arzneim.-Forsch.* **20**, 990 (1970). Pharmacology and toxicity data: M. Hermann *et al.*, *ibid.* **977**, 983. Metabolism: K.-O. Vollmer, H. Achenbach, *ibid.* **24**, 1237 (1974). HPLC determn in pharmaceutics: D. Zivanov-Stakic *et al.*, *Farmaco* **44**, 759 (1989). Clinical trial in comparison with nefopam, *q.v.*: J. Abeloos *et al.*, *Acta Anaesthesiol. Belg.* **34**, 283 (1983); in combination with naloxone, *q.v.*: H. Van Cauwenberge *et al.*, *Int. J. Clin. Pharmacol. Res.* **12**, 1 (1992).

Relative stereochemistry

$bp_{0.01}$ 95.5-96°.

Hydrochloride hemihydrate. Gö-1261C; W-5759A; Lucayan; Valoron. $C_{17}H_{23}NO_2 \cdot HCl \cdot \frac{1}{2}H_2O$; mol wt 318.84. Crystals, mp 125°. Easily sol in water. LD_{50} (7 day) in mice, rats (mg/kg): 437.0, 417.7 i.g.; 490.0, 400.0 s.c.; 52.0, 74.1 i.v. (Herrmann).

Hydrochloride. [27107-79-5] Crystals from ethyl acetate + methyl ethyl ketone, mp 159°.

cis-(±)-**Form.** [112244-09-4] $bp_{0.01}$ 97.5-98°.

cis-(±)-**Hydrochloride.** $C_{17}H_{23}NO_2 \cdot HCl \cdot 1\frac{1}{2}H_2O$. Crystals from ethyl acetate-methyl ethyl ketone, mp 84°.

Note: This is a controlled substance (opiate): **21 CFR**, 1308.11.
THERAP CAT: Analgesic.

9595. Tilisolol. [85136-71-6] 4-[3-[(1,1-Dimethylethyl)amino]-2-hydroxypropoxy]-2-methyl-1*(2H)*-isoquinolinone; 4-[3-(*tert*-butylamino)-2-hydroxypropoxy]-*N*-methylisocarbostyril. $C_{17}H_{24}$-N_2O_3; mol wt 304.39. C 67.08%, H 7.95%, N 9.20%, O 15.77%. β-Adrenergic blocker. Prepn: H. Fukushima, Y. Suzuki, **DE 2631080**; *eidem*, **US 4129565** (1977, 1978 both to Nisshin). Antiarrhythmic activity: K. Hashimoto *et al., J. Pharmacol. Exp. Ther.* **223**, 801 (1982); and pharmacological profile: Y. Nakagawa *et al., Arzneim.-Forsch.* **34**, 194 (1984). Receptor binding study: T. Nagatomo *et al., Jpn. J. Pharmacol.* **34**, 249 (1984). Vasodilating activity in humans: T. Imaizumi *et al., Arzneim.-Forsch.* **38**, 1342 (1988). HPLC determn in plasma: K. Yonezawa *et al., J. Chromatogr.* **339**, 219 (1985). Toxicity data: S. Kawase *et al., Oyo Yakuri* **34**, 499 (1987), *C.A.* **108**, 161293m (1988).

Hydrochloride. [62774-96-3] N-696; Selecal. $C_{17}H_{24}N_2O_3$·HCl; mol wt 340.85. White crystals from ethyl acetate, mp 203-205°. LD$_{50}$ in male, female mice and rats (mg/kg): 1393, 1290, 145, 188 orally; 1219, 1245, 176, 169 s.c.; 578, 557, 39.5, 29.2 i.p.; 74.3, 104.7, 75.8, 38.1 i.v. (Kawase).
THERAP CAT: Antiarrhythmic; antihypertensive.

9596. Tilmicosin. [108050-54-0] 4^A-*O*-De(2,6-dideoxy-3-*C*-methyl-α-L-*ribo*-hexopyranosyl)-20-deoxo-20-[(3*R*,5*S*)-3,5-dimethyl-1-piperidinyl]tylosin; 20-deoxo-20-(3,5-dimethylpiperidin-1-yl)-desmycosin; EL-870; LY-177370; Micotil; Pulmotil. $C_{46}H_{80}N_2$-O_{13}; mol wt 869.15. C 63.57%, H 9.28%, N 3.22%, O 23.93%. Macrolide antibiotic; structurally related to tylosin, *q.v.* Prepd as 85:15 mixture of *cis/trans* isomers: M. Debono, H. A. Kirst, **EP 103465**; *eidem*, **US 4820695** (1984, 1989 both to Lilly); M. Debono *et al., J. Antibiot.* **42**, 1253 (1989). *In vitro* antibacterial activity: E. E. Ose, *ibid.* **40**, 190 (1987). Metabolism and tissue residue studies in cattle: A. L. Donoho *et al., ACS Symp. Ser.* **503**, 158-167 (1992). Trial in treatment of calf pneumonia: E. E. Ose, L. V. Tonkinson, *Vet. Rec.* **123**, 367 (1988). Prophylactic efficacy in bovine respiratory tract disease: D. W. Morck *et al., J. Am. Vet. Med. Assoc.* **202**, 273 (1993).

White to off-white, amorphous solid. uv max: 283 nm (ε 22643). pKa′ (66% DMF): 7.4, 8.5. $[α]_D^{23}$ +12.75° (c = 0.010004 in CHCl$_3$, 5 cm). Slightly sol in water, *n*-hexane.
THERAP CAT (VET): Antibacterial.

9597. Tilorone. [27591-97-5] 2,7-Bis[2-(diethylamino)ethoxy]-9*H*-fluoren-9-one; bis-DEAE-fluorenone. $C_{25}H_{34}N_2O_3$; mol wt 410.56. C 73.14%, H 8.35%, N 6.82%, O 11.69%. The first recognized synthetic, small mol wt compound that is an orally active interferon inducer: Krueger, Mayer, *Science* **169**, 1213 sqq. (1970). Prepn: Fleming *et al.,* **DE 1964761** corresp to **US 3592819** (1970, 1971 to Richardson Merrell); E. R. Andrews *et al., J. Med. Chem.* **17**, 882 (1974); H. M. Burke, M. M. Joullie, *ibid.* **21**, 1084 (1978). Mechanism of action: De Clercq, Merigan, *J. Infect. Dis.* **123**, 190

(1971). Toxicity studies: Kaufman *et al., Proc. Soc. Exp. Biol. Med.* **137**, 357 (1971). *Review:* P. Chandra *et al.,* "Tilorone Hydrochloride" in *Antibiotics* **vol. 5**(pt. 2), F. E. Hahn, Ed. (Springer-Verlag, New York, 1979) pp 385-413.

Dihydrochloride. [27591-69-1] $C_{25}H_{34}N_2O_3$·2HCl. Crystals from butanone-methanol, mp 235-237°. uv max (water): 269 nm ($E_{1cm}^{1\%}$ 1600). LD$_{50}$ in mice, rats (single dose): 959, 852 mg/kg orally; 145, 244 mg/kg i.p. (Krueger, Mayer).

9598. Tiludronic Acid. [89987-06-4] *P,P′*-[[(4-Chlorophenyl)thio]methylene]bisphosphonic acid; ACPMD; Cl-TMBP; ME-3737. $C_7H_9ClO_6P_2S$; mol wt 318.60. C 26.39%, H 2.85%, Cl 11.13%, O 30.13%, P 19.44%, S 10.06%. Bisphosphonate antiresorptive agent. Prepn: J. C. Breliere *et al.,* **EP 100718**; *eidem*, **US 4876248** (1984, 1989 both to Sanofi). HPLC determn in plasma and urine: J.-P. Fels *et al., J. Chromatogr.* **430**, 73 (1988). Determn of dissociation constants: M. Bonnery *et al., Bull. Soc. Chim. Fr.* **1988**, 49. Symposium on pharmacology and clinical experience: *Bone* **17**, Suppl., 471S-519S (1995). Clinical trial in Paget's disease: W. D. Fraser *et al., Postgrad. Med. J.* **73**, 496 (1997).

pK$_1$ 10.85; pK$_2$ 6.90; pK$_3$ 2.95; pK$_4$ 1.30.
Disodium salt. [149845-07-8]; [155453-10-4] (hemihydrate). Tiludronate disodium; SR-41319B; Skelid. $C_7H_7ClNa_2O_6P_2S$; mol wt 362.56. Occurs as the hemihydrate.
Di-*tert*-butylamine salt. mp 253° (dec).
THERAP CAT: Bone resorption inhibitor.

9599. Timepidium Bromide. [35035-05-3] 3-(Di-2-thienylmethylene)-5-methoxy-1,1-dimethylpiperidinium bromide (1:1); SA-504; Mepidium; Sesden. $C_{17}H_{22}BrNOS_2$; mol wt 400.39. C 51.00%, H 5.54%, Br 19.96%, N 3.50%, O 4.00%, S 16.01%. Anticholinergic. Prepn: M. Kawazu *et al.,* **DE 2128808** corresp to **US 3764607** (1971, 1973 both to Tanabe). Pharmacological properties: H. Tamaki *et al., Jpn. J. Pharmacol.* **22**, 685 (1972), *C.A.* **78**, 52812 (1973). Metabolism: J. Sugihara, N. Taga, *Radioisotopes* **26**, 238 (1977). Absorption, distribution, excretion: M. Yoshikawa, *Oyo Yakuri* **14**, 179 (1977), *C.A.* **88**, 163772 (1978). Teratological study: Y. Fujisawa *et al., ibid.* **7**, 1293 (1973), *C.A.* **81**, 99413 (1974). Chronic toxicity study: K. Doi *et al., ibid.* **13**, 851 (1977), *C.A.* **88**, 32014 (1978).

Colorless cryst from acetone/ether, mp 198-200°.
THERAP CAT: Antispasmodic.

9600. Timiperone. [57648-21-2] 4-[4-(2,3-Dihydro-2-thioxo-1*H*-benzimidazol-1-yl)-1-piperidinyl]-1-(4-fluorophenyl)-1-butanone; 1-[1-[3-(4-fluorobenzoyl)propyl]-4-piperidyl]-2-mercaptobenzimidazole; 4-fluoro-4-[4-(2-thioxo-1-benzimidazolinyl)piperidino]butyrophenone; 1-[1-[3-(4-fluorobenzoyl)propyl]-4-piperidyl]-

2,3-dihydrobenzimidazole-2-thione; DD-3480; Tolopelon. $C_{22}H_{24}$-FN_3OS; mol wt 397.51. C 66.47%, H 6.09%, F 4.78%, N 10.57%, O 4.02%, S 8.07%. Butyrophenone derivative with neuroleptic activity. Prepn: M. Sato *et al.*, **JP Kokai 75 84578**; K. Ueno *et al.*, **US 3963727** (1975, 1976 both to Daiichi); M. Sato *et al.*, *J. Med. Chem.* **21**, 1116 (1978). Improved synthesis: M. Sato, M. Arimoto, *Chem. Pharm. Bull.* **30**, 719 (1982). Pharmacology: T. Yamasaki *et al.*, *Jpn. J. Pharmacol.* **27** (Suppl.), 124P (1977); *eidem*, *Arzneim.-Forsch.* **31**, 701, 707 (1981); T. Shibuya *et al.*, *Int. J. Clin. Pharmacol. Ther. Toxicol.* **20**, 251 (1982). Pharmacokinetics and metabolism: H. Tachizawa *et al.*, *Drug Metab. Dispos.* **9**, 442 (1981); K. Sudo *et al.*, *Xenobiotica* **11**, 685 (1981). Multicenter controlled clinical trials in schizophrenia: R. Takahashi *et al.*, *J. Int. Med. Res.* **10**, 257 (1982); T. Kariya *et al.*, *ibid.* **11**, 66 (1983).

Crystals from acetone, mp 201-203°. uv max (ethanol): 226.5, 246, 309 nm. Slightly sol in water. LD_{50} in male rats, mice (mg/kg): 232, 478 orally (Yamasaki).

THERAP CAT: Antipsychotic.

9601. Timolol. [26839-75-8] (2*S*)-1-[(1,1-Dimethylethyl)-amino]-3-[[4-(4-morpholinyl)-1,2,5-thiadiazol-3-yl]oxy]-2-propanol; *S*-(−)-3-(3-*tert*-butylamino-2-hydroxypropoxy)-4-morpholino-1,2,5-thiadiazole; (−)-3-morpholino-4-(3-*tert*-butylamino-2-hydroxypropoxy)-1,2,5-thiadiazole. $C_{13}H_{24}N_4O_3S$; mol wt 316.42. C 49.35%, H 7.65%, N 17.71%, O 15.17%, S 10.13%. β-Adrenergic blocker. Prepn: B. K. Wasson, **DE 1925956**; *idem*, **US 3655663** (1969, 1972 both to Frosst). Manufacturing process: L. M. Weinstock *et al.*, **DE 1925955**; *eidem*, **US 3657237** (1970, 1972 both to Frosst). Synthesis and activity data: Wasson *et al.*, *J. Med. Chem.* **15**, 651 (1972). HPLC determn in plasma: H. He *et al.*, *J. Chromatogr. B* **661**, 351 (1994). Pharmacology: Franciosa *et al.*, *Clin. Pharmacol. Ther.* **13**, 138 (1972); Ulrych *et al.*, *ibid.* 232. Review of efficacy in glaucoma: R. C. Heel *et al.*, *Drugs* **17**, 38-55 (1979). Clinical evaluation in hypertension: B. A. Rofman *et al.*, *Hypertension* **2**, 643 (1980). Multicenter study of effect in myocardial infarction: *N. Engl. J. Med.* **304**, 801 (1981); of efficacy in limiting infarct size: M. Sederholm *et al.*, *ibid.* **310**, 9 (1984). Comprehensive description: D. J. Mazzo, A. E. Loper, *Anal. Profiles Drug Subs.* **16**, 641-692 (1987). Review of clinical experience with dorzolamide, *q.v.* in glaucoma and ocular hypertension: J. E. Frampton, C. M. Perry, *Drugs Aging* **23**, 977 (2006).

Hemihydrate. White, odorless, crystalline powder. Slightly sol in water; freely sol in ethanol.

Hydrogen maleate salt. [26921-17-5] MK-950; Betim; Betimol; Blocadren; Nyogel; Temserin; Timabak; Timacar; Timacor; Timoptic; Timoptol. $C_{13}H_{24}N_4O_3S.C_4H_4O_4$; mol wt 432.49. White crystals from ethanol, mp 201.5-202.5°. $[\alpha]_{405}^{24}$ −12.0° (c = 5 in 1*N* HCl), $[\alpha]_D^{25}$ −4.2°. uv max (0.1*N* HCl): 294 nm ($A_{1cm}^{1\%}$ 200). Sol in water, ethanol, methanol; sparingly sol in chloroform, propylene glycol; very slightly sol in cyclohexane. Practically insol in isooctane; insol in ether, cyclohexane. Stable in soln up to pH 12.

(±)-Free base. Crystalline solid from isopropyl ether, mp 71.5-72.5°.

THERAP CAT: Antihypertensive; antiarrhythmic (class II); antianginal; antiglaucoma agent.

9602. Timonacic. [444-27-9] 4-Thiazolidinecarboxylic acid; ATC; norgamen; thioproline; NSC-25855; Detoxepa; Hepalidine;

Heparegen; Tiazolidin. $C_4H_7NO_2S$; mol wt 133.17. C 36.08%, H 5.30%, N 10.52%, O 24.03%, S 24.07%. Prepd from DL- or L-cysteine and formaldehyde. Prepn of *l*-form: Ratner, Clarke, *J. Am. Chem. Soc.* **59**, 200 (1937); of *dl*- and *l*-forms: Werner *et al.*, *Helv. Chim. Acta* **30**, 432 (1947); **FR M3184** (1965 to Sogespar S.A.), *C.A.* **63**, 12980h (1965). Proton magnetic resonance and conformation: Martin, Mathur, *J. Am. Chem. Soc.* **87**, 1065 (1965). Series of articles on pharmacology and toxicology: *Gazz. Med. Ital.* **131**, 251-286 (1972). Toxicity data: Bertrand, Piton, *ibid.* 265. Induces reverse transformation of tumor cells to normal. Clinical evaluation in cancer: A. Brugarolas, M. Gosalvez, *Lancet* **1**, 68 (1980).

dl-**Form.** Crystals, mp 195°. LD_{50} orally in mice: 400 mg/kg (Bertrand, Piton).

l-**Form.** Crystals from water, dec 196-197°. Readily sol in hot water, acid, alkali; sparingly sol in cold water. Practically insol in alcohol.

THERAP CAT: Hepatoprotectant.

9603. Tin. [7440-31-5] Sn; at. wt 118.710; at. no. 50; valence 2, 4. Group IVA (14). Naturally occurring isotopes: 112 (0.95%); 114 (0.65%); 115 (0.34%); 116 (14.24%); 117 (7.57%); 118 (24.01%); 119 (8.59%); 120 (32.97%); 122 (4.71%); 124 (5.98%); artificial, radioactive isotopes: 108-111; 113; 121; 123; 125-132. Found in cassiterite, stannite, and tealite. Occurrence in earth's crust: $6 \times 10^{-4}\%$. The metal of commerce is about 99.8% pure. Prepn of high purity tin: Baralis, Marone, *Met. Ital.* **59**, 494 (1967), *C.A.* **67**, 119613a (1967). Physical properties: Kirshenbaum, Cahill, *J. Inorg. Nucl. Chem.* **25**, 232 (1963). *Monograph:* C. L. Mantell, *Tin: Its Mining, Production, Technology and Applications* (Reinhold, New York, 1949). *Reviews:* Abel in *Comprehensive Inorganic Chemistry* **vol. 2**, J. C. Bailar, Jr. *et al.*, Eds. (Pergamon Press, Oxford, 1973) pp 43-104; W. Germain *et al.*, in *Kirk-Othmer Encyclopedia of Chemical Technology* **vol. 23** (Wiley-Interscience, New York, 3rd ed., 1983) pp 18-42. Review of toxicology and human exposure: *Toxicological Profile for Tin* (PB-2006-100006, 2005) 426 pp.

Silver-white, lustrous, soft, very malleable and ductile metal; only slightly tenacious; easily powdered. When being bent, emits the crackling "tin cry". Brittle at 200°. At −40° crumbles to gray amorphous powder ("gray tin"), slowly changing back above 20° to white tin. Available in the form of bars, foil, powder, shot, etc. Stable in air, but when in powder form it oxidizes, esp in presence of moisture. d 7.31. mp 231.9°. bp 2507° (2780 K). Specific heat (25°) 0.053 cal/g/°C. Brinell hardness 2.9. Insol in water. Reacts slowly with cold dil HCl or dil HNO_3, hot dil H_2SO_4; readily with concd HCl, aqua regia; very slowly attacked by acetic acid; slowly attacked by cold, more readily by hot caustic alkali; concd HNO_3 converts it into insol metastannic acid.

Caution: Potential symptoms of overexposure to metallic tin are irritation of eyes, skin, respiratory system. Potential symptoms of overexposure to organic tin compounds are irritation of eyes, skin, respiratory system; headache, vertigo; psychoneurologic disturbances; sore throat, cough; abdominal pain, vomiting; urine retention; paresis, focal anesthesia; skin burns; pruritis. *See NIOSH Pocket Guide to Chemical Hazards* (DHHS/NIOSH 97-140, 1997) p 308. *See also Patty's Industrial Hygiene and Toxicology* **vol. 2A**, G. D. Clayton, F. E. Clayton, Eds. (Wiley-Interscience, New York, 3rd ed., 1981) pp 1940-1968.

USE: Chiefly for tin-plating and manuf of food, beverage and aerosol containers, soldering alloys, babbitt and type metals, manuf tin salts, collapsible tubes, coating for copper wire. Principle component in pewter. Alloys as dental materials (silver-tin-mercury), nuclear reactor components (tin-zirconium), aircraft components (tin-titanium), bronze (copper-tin), brass. Reducing agent.

9604. Tinidazole. [19387-91-8] 1-[2-(Ethylsulfonyl)ethyl]-2-methyl-5-nitro-1*H*-imidazole; ethyl[2-(2-methyl-5-nitro-1-imidazolyl)ethyl]sulfone; CP-12574; Fasigin; Fasigyn; Simplotan; Tindamax; Tricolam; Trimonase. $C_8H_{13}N_3O_4S$; mol wt 247.27. C 38.86%, H 5.30%, N 16.99%, O 25.88%, S 12.97%. Prepn: K.

Butler, **US 3376311** (1968 to Pfizer); M. W. Miller *et al., J. Med. Chem.* **13**, 849 (1970). Series of articles on prepn, antiprotozoal activity and pharmacokinetics: *Antimicrob. Agents Chemother.* **1969**, 257-270. Review of antiprotozoal activity: P. R. Sawyer *et al., Drugs* **11**, 423-440 (1976); of antibacterial activity, pharmacology and therapeutic efficacy vs anaerobes: A. A. Carmine *et al., ibid.* **24**, 85-117 (1982). Symposium on clinical experience in anaerobic infection: *J. Antimicrob. Chemother.* **10**, Suppl. A, 1-184 (1982). Clinical trial in *Helicobacter pylori* positive gastritis: G. Oderda *et al., Gut* **33**, 1328 (1992).

Colorless crystals from benzene, mp 127-128°. Sol in acetone, methylene chloride; sparingly sol in methanol. Practically insol in water. LD_{50} in mice (mg/kg): >3600 orally; >2000 i.p. (Miller).

THERAP CAT: Antiprotozoal (Trichomonas, Giardia); antiamebic; antibacterial.

9605. Tin Phosphides. Several stoichiometries reported: *Mellor's* vol. **VIII**, 847-849 (1931). Crystal structure of Sn_4P_3 and discussion of other phases: Olofsson, *Acta Chem. Scand.* **21**, 1659 (1967); **24**, 1153 (1970).

USE: Tin phosphide with a phosphorus content >10% used in manuf of phosphor bronze.

9606. Tinzaparin. Innohep; Logiparin. Low molecular weight heparin prepared by enzymatic depolymerization of porcine mucosal heparin with heparinase from *Flavobacterium heparinum.* Approx mol wt 4900 Da. Manufacturing process: J. I. Nielsen, P. B. Ostergaard, **EP 244236**; J. I. Nielsen, **US 5106734** (1987, 1992 both to Novo). Pharmacokinetics: P. C. Pedersen *et al., Thromb. Res.* **61**, 477 (1991). Multicenter clinical trial: A. Liezorovicz *et al., Br. J. Surg.* **78**, 412 (1991).

THERAP CAT: Antithrombotic.

9607. Tioclomarol. [22619-35-8] 3-[3-(4-Chlorophenyl)-1-(5-chloro-2-thienyl)-3-hydroxypropyl]-4-hydroxy-2*H*-1-benzopyran-2-one; 3-[5-chloro-α-(*p*-chloro-β-hydroxyphenethyl)-2-thienyl]-4-hydroxycoumarin; Apegmone; $C_{22}H_{16}Cl_2O_4S$; mol wt 447.33. C 59.07%, H 3.61%, Cl 15.85%, O 14.31%, S 7.17%. Slow-acting anticoagulant, related structurally to warfarin, *q.v.* Prepn: E. Boschetti *et al., ZA 6707267* corresp to **US 3574234** (1968, 1971 both to Lipha). Synthesis and anticoagulant activity: *eidem, Chim. Ther.* **7**, 20 (1972).

White cryst from methanol, mp 104°.
THERAP CAT: Anticoagulant.

9608. Tioconazole. [65899-73-2] 1-[2-[(2-Chloro-3-thienyl)-methoxy]-2-(2,4-dichlorophenyl)ethyl]-1*H*-imidazole; 1-[2,4-dichloro-β-[(2-chloro-3-thienyl)oxy]phenethyl]imidazole; UK-20349; Fungibacid; Gyno-Trosyd; Trosyd; Trosyl; Vagistat; Zoniden. $C_{16}H_{13}Cl_3N_2OS$; mol wt 387.70. C 49.57%, H 3.38%, Cl 27.43%, N 7.23%, O 4.13%, S 8.27%. Antimycotic imidazole derivative. Prepn: G. E. Gymer, **BE 841309**; *idem, US 4062966* (1976, 1977 both to Pfizer). Antifungal spectrum: S. Jevons, *Antimicrob. Agents Chemother.* **15**, 597 (1979); F. C. Odds, *J. Antimicrob. Chemother.* **6**, 749 (1980). Pharmacology: M. S. Marriott *et al., Dermatologica* **166**, Suppl. 1, 1 (1983). Clinical trial in dermatomycosis: Y. M. Clayton *et al., Clin. Exp. Dermatol.* **7**, 543 (1982). Series of articles

on pharmacology and clinical efficacy in gynecological use: *Gynak. Rundsch.* **23**, Suppl. 1, 1-60 (1983).

Hydrochloride. $C_{16}H_{13}Cl_3N_2OS.HCl$. Crystals, mp 168-170°.
THERAP CAT: Antifungal (topical).

9609. Tiopronin. [1953-02-2] *N*-(2-Mercapto-1-oxopropyl)-glycine; *N*-(2-mercaptopropionyl)glycine; α-mercaptopropionylglycine; Acadione; Captimer; Thiola; Thiosol. $C_5H_9NO_3S$; mol wt 163.19. C 36.80%, H 5.56%, N 8.58%, O 29.41%, S 19.65%. Synthetic thiol; undergoes thiol-disulfide exchange to form a water-soluble complex with biological cystine. Prepn: Mita *et al.*, **US 3246025** (1966 to Santen). Antidotal effects in mice: H. Fujimura *et al., Nippon Yakurigaku Zasshi* **60**, 278 (1964), *C.A.* **62**, 972h (1965). Anti-inflammatory pharmacology in animals: F. Capasso, *Agents Actions* **11**, 741 (1981). Chemiluminescence determn in pharmaceutical formulations: J. Lu *et al., J. Pharm. Biomed. Anal.* **33**, 1033 (2003). LC/MS determn in plasma: J. Ma *et al., J. Chromatogr. A* **1113**, 55 (2006). Clinical trial in chronic hepatitis: F. Ichida *et al., J. Int. Med. Res.* **10**, 325 (1982). Clinical comparison with D-penicillamine in rheumatoid arthritis: B. Amor *et al., Arthritis Rheum.* **25**, 698 (1982); G. Pasero *et al., ibid.* 923.

Crystals from ethyl acetate, mp 95-97°. Freely sol in water. LD_{50} i.v. in mice: 2.1 g/kg (Fujimura).

THERAP CAT: Antidote (heavy metal poisoning); hepatoprotectant; mucolytic.

THERAP CAT (VET): In treatment of cystine urolithiasis.

9610. Tiotropium Bromide. [136310-93-5] (1α,2β,4β,5α,-7β)-7-[(2-Hydroxy-2,2-di-2-thienylacetyl)oxy]-9,9-dimethyl-3-oxa-9-azoniatricyclo[3.3.1.0²,⁴]nonane bromide (1:1); 6β,7β-epoxy-3β-hydroxy-8-methyl-1αH,5αH-tropanium bromide di-2-thienylglycolate; Ba-679 BR; Ba-679. $C_{19}H_{22}BrNO_4S_2$; mol wt 472.41. C 48.31%, H 4.69%, Br 16.91%, N 2.97%, O 13.55%, S 13.57%. Muscarinic receptor antagonist. Prepn: R. Banholzer *et al., EP 418716*; *eidem, US 5610163* (1991, 1997 both to Boehringer Ingelheim). Prepn of the crystalline monohydrate: R. Banholzer *et al., US 6908928* (2005 to BI Pharma). Pharmacology: T. Takahashi *et al., Am. J. Respir. Crit. Care Med.* **150**, 1640 (1994). Binding study: E.-B. Haddad *et al., Mol. Pharmacol.* **45**, 899 (1994). HPLC determn in plasma: L. Ding *et al., J. Chromatogr. Sci.* **46**, 445 (2008). Comparative clinical trial with salmeterol in chronic obstructive pulmonary disease (COPD): J. F. Donohue *et al., Chest* **122**, 47 (2002). Clinical trial in uncontrolled asthma: S. P. Peters *et al., N. Engl. J. Med.* **363**, 1715 (2010). Review of pharmacology and clinical experience in COPD: P. Santus, F. Di Marco, *Expert Opin. Drug Saf.* **8**, 387-395 (2009).

mp 218-220°.

Monohydrate. [411207-31-3] Spiriva. $C_{19}H_{22}BrNO_4S_2 \cdot H_2O$; mol wt 490.43. White to yellowish-white powder. Sol in methanol; sparingly sol in water.

THERAP CAT: Bronchodilator.

9611. Tioxidazole. [61570-90-9] N-(6-Propoxy-2-benzothiazolyl)carbamic acid methyl ester; methyl 6-propoxy-2-benzothiazolylcarbamate; Sch-21480; Tiox. $C_{12}H_{14}N_2O_3S$; mol wt 266.32. C 54.12%, H 5.30%, N 10.52%, O 18.02%, S 12.04%. Structurally related to benzimidazole anthelmintics. Prepn and anthelmintic activity: M. M. Nafissi-Varchei, **BE 840945**; **US 4006242** (1976, 1977 both to Schering). Physical data, activity against roundworms: E. Panitz et al., Experientia **34**, 733 (1978). Anthelmintic effect against gastrointestinal parasites in naturally infected horses: J. H. Drudge et al., Am. J. Vet. Res. **41**, 1383 (1980); E. T. Lyons et al., ibid. **42**, 1048 (1981).

White odorless solid, mp 178-180°. Insoluble in water. Slightly sol in organic solvents.

THERAP CAT (VET): Anthelmintic for horses.

9612. Tioxolone. [4991-65-5] 6-Hydroxy-1,3-benzoxathiol-2-one; 6-hydroxy-2-oxo-1,3-benzoxathiole; thioxolone; Camyna; Stepin. $C_7H_4O_3S$; mol wt 168.17. C 50.00%, H 2.40%, O 28.54%, S 19.06%. Prepd from resorcinol and ammonium thiocyanate in presence of copper sulfate: Werner, **US 2332418** (1943 to Winthrop); Urushibara, Koga, Bull. Chem. Soc. Jpn. **29**, 419 (1956); Berg, Fiedler, **US 2886488** (1959 to Thomae). Formerly believed to be the 4-hydroxy isomer: Fiedler, Ber. **95**, 1771 (1962).

Crystals from water, mp 160°. Practically insol in water. Sol in ethanol, isopropanol, propylene glycol, ether, benzene, toluene. Hydrolyzed by alkali.

Combination with hydrocortisone. [8064-73-1] Psoil.

THERAP CAT: Antiseborrheic; antiacne.

9613. Tipepidine. [5169-78-8] 3-(Di-2-thienylmethylene)-1-methylpiperidine; 1-methyl-3-piperidylidenedi(2-thienyl)methane; tipedine; AT-327; CR-662. $C_{15}H_{17}NS_2$; mol wt 275.43. C 65.41%, H 6.22%, N 5.09%, S 23.28%. Prepn: Okumura et al., Tanabe Seiyaku Kenkyu Nempo **3**, 30 (1958), C.A. **53**, 10214 (1959); Ponomarev, Martem'yanova, **SU 176903** (1965), C.A. **64**, 12648b (1966). Synthesis: eidem, Khim. Geterotsikl. Soedin. **1967**, 174, C.A. **67**, 73501g (1967). Prepn of hibenzate: M. Yamamoto, H. Yoshikawa, **JP 62 17988**, C.A. **59**, 11446a (1963); **GB 924544** (1962, 1963 both to Tanabe Seiyaku). Pharmacology and clinical efficacy: Higaki et al., Tanabe Seiyaku Kenkyu Nempo **4**, 35 (1959), C.A. **54**, 3725h (1960); Kase et al., Chem. Pharm. Bull. **7**, 372 (1959). Metabolism: Watanabe et al., Yakugaku Zasshi **89**, 29 (1969); Sasaki et al., ibid. 345.

Yellow crystals from petr ether, mp 64-65°. bp$_{4.5}$ 178-184°. LD$_{50}$ in mice (mg/kg): 294 i.p.; 308 i.m.; 867 orally (Higaki).

Citrate monohydrate. Bithiodine. $C_{15}H_{17}NS_2 \cdot C_6H_8O_7 \cdot H_2O$; mol wt 485.57. Yellow crystals, mp 138-139°. Sol in water, ethanol, propylene glycol.

Hibenzate. [31139-87-4] Asverin; Sotal. $C_{29}H_{27}NO_4S_2$; mol wt 517.66. Crystals from acetone, mp 187-190°. Sparingly sol in water.

THERAP CAT: Antitussive.

9614. Tipifarnib. [192185-72-1]; [192185-68-5] (unspecified stereo). 6-[(R)-Amino(4-chlorophenyl)(1-methyl-1H-imidazol-5-yl)methyl]-4-(3-chlorophenyl)-1-methyl-2(1H)-quinolinone; R-115777; Zarnestra. $C_{27}H_{22}Cl_2N_4O$; mol wt 489.40. C 66.26%, H 4.53%, Cl 14.49%, N 11.45%, O 3.27%. Farnesyl transferase inhibitor. Prepn: M. G. Venet et al., **WO 9721701**; eidem, **US 6037350** (1997, 2000 both to Janssen). Review of syntheses: P. R. Angibaud et al., Eur. J. Org. Chem. **2004**, 479-486. Inhibition of farnesyl protein transferase and antitumor effects in vivo: D. W. End et al., Cancer Res. **61**, 131 (2001). Clinical pharmacology and pharmacokinetics: J. Zujewski et al., J. Clin. Oncol. **18**, 927 (2000). Accelerator mass spec determn in biological samples: R. C. Garner et al., Drug Metab. Dispos. **30**, 823 (2002). Clinical evaluation in hematologic malignancies: J. Cortes et al., Blood **101**, 1692 (2003). Review of clinical experience: P. K. Epling-Burnette, T. P. Loughran, Jr., Expert Opin. Invest. Drugs **19**, 689-698 (2010).

Crystals from 2-propanol, mp 234°. $[\alpha]_D^{20}$ +22.86° (c = 0.98 in methanol).

THERAP CAT: Antineoplastic.

9615. Tipranavir. [174484-41-4] N-[3-[(1R)-1-[(6R)-5,6-Dihydro-4-hydroxy-2-oxo-6-(2-phenylethyl)-6-propyl-2H-pyran-3-yl]propyl]phenyl]-5-(trifluoromethyl)-2-pyridinesulfonamide; 5-trifluoromethyl-N-[3(R)-[1-[5,6-dihydro-4-hydroxy-2-oxo-6(R)-(2-phenethyl)-6(R)-n-propyl-2H-pyran-3-yl]propyl]phenyl]-2-pyridinesulfonamide; PNU-140690; Aptivus. $C_{31}H_{33}F_3N_2O_5S$; mol wt 602.67. C 61.78%, H 5.52%, F 9.46%, N 4.65%, O 13.27%, S 5.32%. Nonpeptidic HIV protease inhibitor (NPPI). Prepn: K. R. Romines et al., **WO 9530670** (1995 to Upjohn); eidem, **US 5852195** (1998 to Pharmacia & Upjohn); S. R. Turner et al., J. Med. Chem. **41**, 3467 (1998); K. S. Fors et al., J. Org. Chem. **63**, 7348 (1998). Antiviral activity: S. M. Poppe et al., Antimicrob. Agents Chemother. **41**, 1058 (1997); vs multidrug resistant clinical isolates: B. A. Larder et al., AIDS **14**, 1943 (2000). HPLC determn in plasma: E. Dailly et al., J. Chromatogr. B **832**, 317 (2006). Clinical pharmacokinetics in combination with ritonavir: T. R. MacGregor et al., HIV Clin. Trials **5**, 371 (2004). Clinical evaluation in HIV infection: S. McCallister et al., J. Acquir. Immune Defic. Syndr. **35**, 376 (2004). Review of pharmacologic interactions with other HIV medications: M. Boffito et al., J. Clin. Pharmacol. **46**, 130-139 (2006); of clinical development: B. Best, R. Haubrich, Expert Opin. Invest. Drugs **15**, 59-70 (2006).

White solid from ethyl acetate + heptane, mp 86-89°. $[\alpha]_D$ +20° (ethanol). Freely sol in dehydrated alcohol, propylene glycol. Insol in aq. buffer, pH 7.5.

THERAP CAT: Antiretroviral.

9616. Tiquizium Bromide. [71731-58-3] rel-(5R,9aR)-3-(Di-2-thienylmethylene)octahydro-5-methyl-2H-quinolizinium bromide (1:1); 3-(di-2-thienylmethylene)-5-methyl-trans-quinolizidinium bromide; HSR-902; HS-902; Thiaton. $C_{19}H_{24}BrNS_2$; mol wt 410.43. C 55.60%, H 5.89%, Br 19.47%, N 3.41%, S 15.62%. Quaternary ammonium salt with anticholinergic activity. Prepn: H. Kato *et al.*, **BE 866988** (1978 to Hokuriku), C.A. **90,** 151996p (1979); H. Kato, E. Koshinaka, **US 4205074** (1980 to Hokuriku). Synthesis and anticholinergic activity: E. Koshinaka *et al.*, *Chem. Pharm. Bull.* **27,** 1454 (1979). Pharmacological comparison with other antispasmodics: S. Kubo *et al.*, *Jpn. J. Pharmacol.* **30,** 103P (1980). Pharmacology: M. Yamazaki *et al.*, *Oyo Yakuri* **23,** 417, 423 (1981), *C.A.* **97,** 66016u, 66017v (1982). HPLC determn in biological fluids: T. Yamada *et al.*, *Yakugaku Zasshi* **103,** 1319 (1983), *C.A.* **100,** 131957y (1984).

Relative stereochemistry

Needles from methanol-acetone, mp 278-281° (dec).
THERAP CAT: Antispasmodic.

9617. Tirapazamine. [27314-97-2] 1,2,4-Benzotriazin-3-amine 1,4-dioxide; 3-amino-1,2,4-benzotriazine 1,4-dioxide; NSC-130181; SR-4233; Win-59075. $C_7H_6N_4O_2$; mol wt 178.15. C 47.19%, H 3.39%, N 31.45%, O 17.96%. Bioreductive anticancer agent; selectively toxic to cells under hypoxic conditions. Prepn: R. F. Robbins, K. Schofield, *J. Chem. Soc.* **1957,** 3186; J. C. Mason, G. Tennant, *J. Chem. Soc. B* **1970,** 911. *See also:* K. Ley *et al.*, **DE 2204574**; *eidem*, **US 3868371** (1973, 1975 both to Bayer). HPLC determn in plasma: H. Robin, Jr. *et al.*, *Cancer Chemother. Pharmacol.* **36,** 266 (1995). Review of pharmacology and mechanism of action studies: J. M. Brown, *Br. J. Cancer* **67,** 1163-1170 (1993); L. Marcu, I. Oliver, *Curr. Clin. Pharmacol.* **1,** 71-79 (2006). Review of clinical experience: S. B. Reddy, S. K. Williamson, *Expert Opin. Invest. Drugs* **18,** 77-87 (2009).

Orange needles from methanol, mp 229-230° (dec) (Robbins, Scholfield). Also reported as reddish-golden crystals from water + acetic acid, mp 220° (dec) (Ley). uv max: 272, 474 nm (log ε 4.24, 3.52).
THERAP CAT: Antineoplastic.

9618. Tiratricol. [51-24-1] 4-(4-Hydroxy-3-iodophenoxy)-3,5-diiodobenzeneacetic acid; [4-(4-hydroxy-3-iodophenoxy)-3,5-diiodophenyl]acetic acid; 3,3′,5-triiodo-4-(4-hydroxyphenoxy)phenylacetic acid; 3,5-diiodo-4-(3-iodo-4-hydroxyphenoxy)phenylacetic acid; 3,3′,5-triiodothyroacetic acid; Triac; Triacana. $C_{14}H_9I_3O_4$; mol wt 621.94. C 27.04%, H 1.46%, I 61.21%, O 10.29%. Thyroid hormone analog. Prepd by iodination of 4-(4′-hydroxyphenoxy)-3,5-diiodophenylacetic acid: Wilkinson, *Biochem. J.* **63,** 601 (1956); Hems, **GB 803149** (1958 to Glaxo); Wilkinson, **GB 805761** (1958 to Nat. Res. Dev. Corp.); Meltzer *et al.*, *J. Org. Chem.* **26,** 1418 (1961). Clinical evaluation in thyroidectomy: S. I. Sherman, P. W. Ladenson, *J. Clin. Endocrinol. Metab.* **75,** 901 (1992).

Needles from methanol+ water, mp 65°, resolidifies 110°, mp 180-183°.
THERAP CAT: Antihypothyroid.

9619. Tirilazad. [110101-66-1] (16α)-21-[4-(2,6-Di-1-pyrrolidinyl-4-pyrimidinyl)-1-piperazinyl]-16-methylpregna-1,4,-9(11)-triene-3,20-dione. $C_{38}H_{52}N_6O_2$; mol wt 624.87. C 73.04%, H 8.39%, N 13.45%, O 5.12%. Member of a novel class of nonglucocorticoid, 21-aminosteroid antioxidants known as *lazaroids* which inhibit lipid peroxidation and have cytoprotectant activity. Prepn: J. M. McCall *et al.*, **WO 8701706**; *eidem*, **US 5175281** (1987, 1992 both to Upjohn). Inhibition of iron-dependent lipid peroxidation *in vitro*: J. M. Braughler *et al.*, *J. Biol. Chem.* **262,** 10438 (1987). HPLC determn in plasma: J. W. Cox, R. H. Pullen, *J. Chromatogr.* **424,** 293 (1988). Clinical pharmacokinetics: J. C. Fleishaker *et al.*, *J. Clin. Pharmacol.* **33,** 175, 182 (1993). Veterinary evaluation in endotoxemia of neonatal calves: M. L. Rose, S. D. Semrad, *Am. J. Vet. Res.* **53,** 2305 (1992). *Review:* E. D. Hall, *Ann. Neurol.* **32,** S137-S142 (1992). Review of efficacy and mechanism of action in experimental models of subarachnoid hemorrhage: *idem*, *Eur. J. Anaesthesiol.* **13,** 279-289 (1996); of clinical trials in acute ischemic stroke: Tirilazad International Steering Committee, *Stroke* **32,** 2257-2265 (2000).

Methanesulfonate monohydrate. [111793-42-1] Tirilazad mesylate; U-74006F; Freedox. $C_{38}H_{52}N_6O_2.CH_3SO_3H.H_2O$; mol wt 738.99. mp 181-185° (dec). uv max: 234, 285 nm (ε 52000, 17000). Log P (octanol/water): 8.

9620. Tirofiban. [144494-65-5] *N*-(Butylsulfonyl)-*O*-[4-(4-piperidinyl)butyl]-L-tyrosine; *N*-(butylsulfonyl)-4-[4-(4-piperidyl)-butoxy]-L-phenylalanine; 2-*S*-(*n*-butylsulfonylamino)-3-[4-(piperidin-4-yl)butyloxyphenyl]propionic acid. $C_{22}H_{36}N_2O_5S$; mol wt 440.60. C 59.97%, H 8.24%, N 6.36%, O 18.16%, S 7.28%. Specific nonpeptide platelet fibrinogen receptor (GPIIb/IIIa) antagonist. Prepn: M. S. Egbertson *et al.*, **EP 478363**; M. E. Duggan *et al.*, **US 5292756** (1992, 1994 both to Merck & Co.); J. Y. L. Chung *et al.*, *Tetrahedron* **49,** 5767 (1993). Physicochemical properties: J. A. McCauley *et al.*, *J. Phys. D: Appl. Phys.* **26,** B85 (1993). LC-MS/MS determn in plasma: J. D. Ellis *et al.*, *J. Pharm. Biomed. Anal.* **15,** 561 (1997). Pharmacology: J. J. Lynch *et al.*, *J. Pharmacol. Exp. Ther.* **272,** 20 (1995). Clinical pharmacokinetics and pharmacodynamics: J. S. Barrett *et al.*, *Clin. Pharmacol. Ther.* **56,** 377 (1994). Clinical trials in angina: P. Théroux *et al.*, *N. Engl. J. Med.* **338,** 1488 (1998); H. D. White *et al.*, *ibid.* 1498. Clinical comparison with abciximab, *q.v.*, in coronary revascularization: D. J. Moliterno *et al.*, *Lancet* **360,** 355 (2002). Review of clinical experience in acute coronary syndrome: M. Valgimigli, M. Tebaldi, *Expert Opin. Drug Saf.* **9,** 801-819 (2010).

White solid, mp 223-225°.
Hydrochloride monohydrate. [150915-40-5]; [142373-60-2] (hydrochloride). MK-383; Aggrastat. $C_{22}H_{36}N_2O_5S.HCl.H_2O$; mol wt 495.07. Solid, mp 131-132°. $[\alpha]_D^{25}$ −14.4° (c = 0.92 in methanol). Very slightly sol in water.
THERAP CAT: Antithrombotic; in treatment of unstable angina.

9621. Tiron. [149-45-1] 4,5-Dihydroxy-1,3-benzenedisulfonic acid sodium salt (1:2); 1,2-dihydroxybenzene-3,5-disulfonic acid disodium salt; disodium-1,2-dihydroxybenzene-3,5-disulfonate; sodium catechol disulfonate; sodium pyrocatechol-2,4-disulfonate; disodium pyrocatechol-3,5-disulfonate. $C_6H_4Na_2O_8S_2$; mol wt 314.19. C 22.94%, H 1.28%, Na 14.63%, O 40.74%, S 20.41%. Chelating agent. Prepn: Fukayama *et al.*, **JP 52 4327** (1952 to Sanwa Pure Chem.), *C.A.* **48**, 5215c (1954). Use in HPLC determn of iron, titanium, osmium and aluminum: E. Wang, A. Liu, *Microchem. J.* **43**, 191 (1991); in spectrophotometric determn of titanium: S. Honshi *et al.*, *Anal. Sci.* **13**, 863 (1997).

HO—[benzene ring]—SO_3^- Na^+, with HO and SO_3^- Na^+ substituents

Crystals, nonhygroscopic. Very freely sol in water. Produces water-sol, colored compds with metal salts. With ferric chloride the resulting complex yields a deep blue soln at pH below 5. Titanium salts give an orange color, copper salts produce a green-yellow, and hexavalent molybdenum a canary-yellow color.

USE: In determination of trace metals.

9622. Tiropramide. [55837-29-1] α-(Benzoylamino)-4-[2-(diethylamino)ethoxy]-*N*,*N*-dipropylbenzenepropanamide; DL-α-benzamido-*p*-[2-(diethylamino)ethoxy]-*N*,*N*-dipropylhydrocinnamamide; *O*-(2-diethylaminoethyl)-*N*-benzoyl-DL-tyrosyl-di-*n*-propylamide; CR-605. $C_{28}H_{41}N_3O_3$; mol wt 467.65. C 71.91%, H 8.84%, N 8.99%, O 10.26%. Smooth muscle relaxant; deriv of tyrosine. Prepn: F. Makovec *et al.*, **DE 2503992**; *eidem*, **US 4004008** (1975, 1977 both to Rotta). Activity on gastrointestinal motility: P. Senin *et al.*, *Gastrointestinal Motility in Health and Disease*, H. L. Duthie, Ed. (University Park Press, Baltimore, 1978) pp 417-427. Mechanism of action: R. R. Vidal y Plana *et al.*, *J. Pharm. Pharmacol.* **33**, 19 (1981).

[Chemical structure diagram]

Cryst from petr ether, mp 65-67°. LD_{50} in rats: 33.9 mg/kg i.v. (Makovec).

Hydrochloride. [53567-47-8] Alfospas; Maiorad. $C_{28}H_{41}N_3$-O_3·HCl; mol wt 504.11.

THERAP CAT: Antispasmodic.

9623. Tissue Factor Pathway Inhibitor. TFPI; lipoprotein-associated coagulation inhibitor; LACI. Endogenous inhibitor of the tissue factor mediated pathway of blood coagulation. Glycoprotein containing 276 amino acid residues; mol wt ~42 kDa. Structure consists of an acidic amino terminal region followed by three tandem Kunitz-type protease inhibitory domains and a basic carboxy terminal region. Forms a quaternary inhibitory complex with tissue factor, factor VIIa, and factor Xa to block the formation of thrombin from prothrombin. Isoln of factor produced by hepatoma cells: G. J. Broze, Jr., J. P. Miletich, *Proc. Natl. Acad. Sci. USA* **84**, 1886 (1987); from human plasma: W. F. Novotny *et al.*, *J. Biol. Chem.* **264**, 18832 (1989). Review of structure and physiology: G. J. Broze, Jr., *Annu. Rev. Med.* **46**, 103-112 (1995); of pharmacology and therapeutic potential: A. K. Lindahl, *Cardiovasc. Res.* **33**, 286-291 (1997); B. Kaiser *et al.*, *Emerg. Drugs* **5**, 73-87 (2000).

Tifacogin. [148883-56-1] *N*-L-Alanyl-blood coagulation factor LACI (human clone λP9 protein moiety reduced); SC-59735. Recombinant human form; differs from the native protein by an additional alanine residue at the amino terminus and is not glycosylated. Prepn by expression in *E. coli*: G. J. Broze, Jr. *et al.*, **EP 318451**; *eidem*, **US 4966852** (1989, 1990 both to Monsanto Co.; Washington Univ.); M. E. Gustafson *et al.*, *Protein Expression Purif.* **5**, 233

(1994). Production in *Saccharomyces cerevisiae*: M. A. Innis, A. A. Creasey, **WO 9604377**; *eidem*, **US 6103500** (1996, 2000 both to Chiron). Clinical pharmacokinetics: M. J. B. Kemme *et al.*, *Clin. Pharmacol. Ther.* **67**, 504 (2000). Clinical evaluation in sepsis: E. Abraham, *Crit. Care Med.* **28**, Suppl., S31 (2000). *Review*: N. Sne *et al.*, *IDrugs* **5**, 91-97 (2002).

THERAP CAT: Antithrombotic.

9624. Tissue Plasminogen Activator. Fibrinokinase; extrinsic plasminogen activator; t-PA; TPA. Serine protease catalyzing the enzymatic conversion of plasminogen to plasmin, *q.q.v.* Component of the mammalian fibrinolytic system; major physiological function is to provide plasmin that will dissolve fibrin clots. Shows high affinity and specificity for fibrin-bound plasminogen and weak affinity for circulating plasminogen. Plasminogen activation and fibrinolysis are therefore localized to the clot site, with limited systemic enzymatic activity. t-PA exists as 2 bioactive structural variants and can be isolated in either form depending on method of purification. Synthesized in vascular endothelial cells and released into circulation as the native single chain form, 527 amino acids, mol wt ~70,000. Converted to the two chain form by cleavage of Arg275-Ile276 bond to form two polypeptides connected by a disulfide bond. Discovery of *in vitro* plasminogen-activating ability in tissue fragments: T. Astrup, P. M. Permin, *Nature* **159**, 681 (1947); H. J. Tagnon, M. L. Petermann, *Proc. Soc. Exp. Biol. Med.* **70**, 359 (1949). Review of early literature: T. Astrup, *Fed. Proc.* **25**, 42-51 (1966). Purifn from human tissue: D. C. Rijken *et al.*, *Biochim. Biophys. Acta* **580**, 140 (1979). Identification and purifn of 2 forms: P. Wallen *et al.*, *Biochim. Biophys. Acta* **719**, 318 (1982); and comparison of biological activity: D. C. Rijken *et al.*, *J. Biol. Chem.* **257**, 2920 (1982). Review of role in fibrinolytic system: D. Collen, *Thromb. Haemostasis* **43**, 77-89 (1980); of pharmacology and clinical efficacy: S. D. Rogers *et al.*, *Pharmacotherapy* **7**, 111-121 (1987); D. Collen *et al.*, *Drugs* **38**, 346-388 (1989); of production by cell cultures: J. B. Griffiths, A. Electricwala, *Adv. Biochem. Eng. Biotechnol.* **34**, 147-166 (1987).

Practically insol in solns of low ionic strength at neutral pH. Sol in 1.6*M* KSCN at neutral pH; 0.3*M* K-acetate at pH 4.2.

Alteplase. [105857-23-6] rt-PA; Actilyse; Actiplas; Activase. Recombinant form of human t-PA; purified glycoprotein of 527 amino acids with a sequence identical to the naturally occurring form. Cloning and expression of cDNA: D. Pennica *et al.*, *Nature* **301**, 214 (1983). Review of pharmacokinetics: P. Tanswell *et al.*, *Arzneim.-Forsch.* **41**, 1310-1319 (1991). Comparative clinical trial in myocardial infarction: C. P. Cannon *et al.*, *J. Am. Coll. Cardiol.* **24**, 1602 (1994); in acute ischemic stroke: The Natl. Inst. Neurological Disorders and Stroke rt-PA Stroke Study Group, *N. Engl. J. Med.* **333**, 1581 (1994).

Duteplase. [120608-46-0] 245-L-Methionine-plasminogen activator (human tissue-type 2-chain form protein moiety); Solclot. Two-chain form of recombinant human t-PA. Clinical pharmacokinetics: R. W. Koster *et al.*, *Clin. Pharmacol. Ther.* **50**, 267 (1991). Clinical efficacy in myocardial infarction: Z. G. Turi *et al.*, *Am. J. Cardiol.* **71**, 1009 (1993).

THERAP CAT: Thrombolytic.

9625. Titanic(IV) Acid. [20338-08-3] (*T*-4)-Titanium hydroxide (Ti(OH)₄); orthotitanic acid; tetrahydroxytitanium; titanic hydroxide; titanium tetrahydroxide. $Ti(OH)_4 \cdot xH_2O$. The H_2O content depends upon the method and time of drying. White, amorphous powder. Insol in water; when freshly pptd it is sol in dil HCl or H_2SO_4, and in boiling caustic alkali solns; after drying it is almost insol.

9626. Titanium. [7440-32-6] Ti; at. wt 47.867; at. no. 22; valence 2, 3, 4 (mostly tetravalent). Group IVB (4). Five naturally occurring isotopes (mass numbers): 48 (73.94%); 46 (7.93%); 47 (7.28%); 49 (5.51%); 50 (5.34%). Artificial isotopes: 43-45; 51. Ninth most abundant element in earth's crust; 0.63% by wt. Occurs as the oxide in the minerals rutile, ilmenite, perovskite, anatase, or octahedrite and brookite; other minerals include *sphene* or *titanite* (CaTiSiO₅) and *benitoite* (BaTiSi₃O₉). Discovered by Gregor in 1789; investigated and named by Klaproth in 1795; isolated by Berzelius in 1825. Prepn: de Boer, *Ind. Eng. Chem.* **19**, 1256 (1927); Fast, *Z. Anorg. Chem.* **241**, 42 (1939); Ehrlich in *Handbook of Preparative Inorganic Chemistry* **vol. 2**, G. Brauer, Ed. (Academic

Press, New York, 2nd ed., 1965) pp 1161-1172. *Reviews: Gmelins, Titanium* (8th ed.) **41** (1951); Everhart, *Titanium and Titanium Alloys* (Reinhold, New York, 1954); Brophy *et al., Titanium Bibliography* 1900-1951 + suppl (Washington, 1954); McQuillan & McQuillan, *Titanium* (Butterworth's, London, 1956); Barksdale, *Titanium, Its Occurrence, Chemistry and Technology* (Ronald Press, New York, 2nd ed), Clark, "Titanium" in *Comprehensive Inorganic Chemistry* vol. 3, J. C. Bailar, Jr. *et al.*, Eds. (Pergamon Press, Oxford, 1973) pp 355-417.

Dark gray, lustrous metal. Dimorphic; α-form: hexagonal structure below 882.5°; β-form: body-centered cubic crystals above 882.5°. Brittle when cold; malleable when hot. Ductile only when free of oxygen; traces of oxygen or nitrogen increase strength. mp 1677°. bp 3277°. Calculated d (α- form): 4.506 (25°); (β-form): 4.400 (900°). Specific heat (25°): 5.98 cal/g-atom/°C. Decomposes steam at 700-800°. Combines with oxygen at a red heat. Can burn in an atm of oxygen under certain conditions: *Chem. Eng. News* **36**, 36 (Aug. 4, 1958). Attacked by acids only on heating; oxidized by nitric acid to the dioxide. Reacts with fluorine at 150°, with chlorine at 300°, with bromine at 360°, with iodine above 360°. Combines with nitrogen at 800°. Forms alloys with aluminum, chromium, cobalt, copper, iron, lead, nickel, tin.

USE: As alloy with copper and iron in titanium bronze; as addition to steel to impart great tensile strength; to aluminum to impart resistance to attack by salt solns and by organic acids; to remove traces of oxygen and nitrogen from incandescent lamps. Surgical aid (fracture fixation).

9627. Titanium Dichloride. [10049-06-6] Titanium chloride ($TiCl_2$); unitane; dichlorotitanium. Cl_2Ti; mol wt 118.77. Cl 59.70%, Ti 40.30%. $TiCl_2$. Prepn: Clifton, McWood, *J. Phys. Chem.* **60**, 311 (1956); Ehrlich *et al., Z. Anorg. Allg. Chem.* **292**, 139 (1957).

Black crystals. mp 1035°. d 3.13. Burns like tinder when heated in air; dec by water. Sol in alcohol. Practically insol in chloroform, ether, carbon disulfide. *Keep dry and well closed.*

9628. Titanium Dioxide. [13463-67-7] Titanium oxide (TiO_2); unitane; C.I. Pigment White 6; C.I. 77891. O_2Ti; mol wt 79.87. O 40.06%, Ti 59.93%. TiO_2. Found in nature as the minerals *rutile* (tetragonal), *anatase* or *octahedrite* (tetragonal), *brookite* (orthorhombic), *ilmenite* ($FeTiO_3$), and *perovskite* ($CaTiO_3$). May be prepd by direct combination of titanium and oxygen; by treatment of titanium salts in aq soln; by the reaction of volatile, inorganic titanium compds with oxygen; by oxidation or hydrolysis of organic compds of titanium. Industrial prepn from ilmenite or rutile: *Faith, Keyes & Clark's Industrial Chemicals*, F. A. Lowenheim, M. K. Moran, Eds. (Wiley-Interscience, New York, 4th ed., 1975) pp 814-821. Prepn of synthetic rutile: Merker, US 2760874 (1956 to National Lead). Prepn of spectroscopically pure material by dissolving titanium in an ammoniacal soln of 90% H_2O_2: Czanderna *et al., J. Am. Chem. Soc.* **79**, 5407 (1957). Comprehensive description: H. G. Brittain *et al., Anal. Profiles Drug Subs. Excip.* **21**, 659-691 (1992).

White powder, mp 1855°. d (rutile): 4.23; (anatase): 3.90; (brookite): 4.13. Sol in hot concd H_2SO_4, HF. Insol in water, HCl, HNO_3, $2N$ H_2SO_4. The reactivity depends on a previous heat treatment; prolonged heating produces a less sol material. Also made sol by fusion with potassium bisulfate or with alkali hydroxides or carbonates to form alkali titanates. Possesses perhaps the greatest hiding power of all inorganic white pigments. *Titania* is a name applied to large TiO_2 crystals (translucent water-white or with yellowish cast) suitable for use in jewelry. These crystals have a refractive index (2.7) higher than diamonds (2.4), but lack the hardness of diamonds. When substantially pure, a massive single crystal (boule) of rutile has the properties of a precious gem with a very light straw color and with reflectance, refraction and brilliance measuring greater than those of a diamond.

Note: The rutile structure is common among metal fluorides and oxides of the type MF_2 and MO_2.

Caution: Potential symptom of overexposure is lung fibrosis. Potential occupational carcinogen. *See NIOSH Pocket Guide to Chemical Hazards* (DHHS/NIOSH 97-140, 1997) p 310.

USE: Airfloated ilmenite is used for titanium pigment manuf. Rutile sand is suitable for welding-rod-coating materials, as ceramic colorant, as source of titanium metal. As color in the food industry. Anatase titanium dioxide is used for welding-rod-coatings, acid re-sistant vitreous enamels, in specification paints, exterior white house paints, acetate rayon, white interior air-dry and baked enamels and lacquers, inks and plastics, for paper filling and coating, in water paints, tanners' leather finishes, shoe whiteners, and ceramics. High opacity and tinting values are claimed for rutile-like pigments. Pharmaceutic aid (coating agent).

THERAP CAT: Protectant (topical); ultraviolet screen.

9629. Titanium Hydride. [7704-98-5] Titanium dihydride. H_2Ti; mol wt 49.88. H 4.04%, Ti 95.96%. TiH_2. Prepd by the reduction of titanium oxide with calcium hydride in the presence of hydrogen above 600°: Alexander, US 2427338 (1947).

Gray-black metallic powder. *Flammable.* Stable in air. Dissociation appreciable at 450°.

USE: Additive in powder metallurgy; getter for oxygen and nitrogen in electronic tubes; wetting agent for ceramic to metal seals; source of pure hydrogen.

9630. Titanium Nitride. [25583-20-4] Titanium mononitride. NTi; mol wt 61.87. N 22.64%, Ti 77.37%. TiN. Refractory transition metal nitride with high chemical and thermal stability, good wear resistance, and high electrical conductivity. Prepn by plasma-enhanced chemical vapor deposition: N. J. Ianno *et al., J. Electrochem. Soc.* **136**, 276 (1989). Hydrazide sol-gel synthesis: I. Kim, P. N. Kumta, *J. Mater. Chem.* **13**, 2028 (2003). Low temp induced synthesis: X. Feng *et al., Inorg. Chem.* **43**, 3558 (2004). Prepn and catalytic properties: S. Kaskel *et al., J. Mol. Catal. A* **208**, 291 (2004). Physical-mechanical properties of TiN nanostructures: R. A. Andrievski, *NanoStruct. Mater.* **9**, 607 (1997). Bioactivity of TiN-coated hip prostheses: S. Piscanec *et al., Acta Mater.* **52**, 1237 (2004). Review of chemistry, properties and prepn methods: L. E. Toth, *Transition Metal Carbides and Nitrides* (Academic Press, New York, 1971) 279 pp; of structure and properties of TiN coatings: J.-E. Sundgren, *Thin Solid Films* **128**, 21-44 (1985); S. Chatterjee *et al., J. Mater. Sci.* **27**, 1989-2006 (1992).

Golden-yellow solid, mp 2950°. d 5.39. Microhardness: 2000 kg/mm².

USE: As protective coating for cutting tools; in diffusion barriers for microelectronic devices; as crucibles in metal smelting; as a gold-colored surface for jewelry. In high-performance ceramic materials. As a coating for orthopedic and dental implants.

9631. Titanium Oxysulfate. [13825-74-6] Oxo[sulfato(2−)-$κO,κO'$]titanium; titanyl sulfate; titanium sulfate. O_5STi; mol wt 159.92. O 50.02%, S 20.05%, Ti 29.93%. $TiO(SO_4)$. Prepn: E. Hayek, A. Engelbrecht, *Monatsh. Chem.* **80**, 640 (1949). Crystal structure: B. M. Gatehouse *et al., Acta Crystallogr.* **B49**, 428 (1993). Catalytic hydrolysis of chlorofluorocarbon: X. Deng *et al., J. Catal.* **204**, 200 (2001). Prepn of titanium dioxide, *q.v.,* via thermal hydrolysis: A. V. Tolchev *et al., Russ. J. Appl. Chem.* **74**, 1631 (2001); L. G. Gerasimova *et al., ibid.* **75**, 875 (2002). Sol-gel synthesis of nanosized *anatase*, a mineral form of titanium dioxide: S. Sivakumar *et al., Mater. Lett.* **57**, 330 (2002). Brief review as whitening agent in tanning: S. Bangaruswamy *et al., J. Soc. Leather Technol. Chem.* **68**, 13-14 (1984).

Clear, colorless elongated prisms.

Note: The name titanium sulfate is used for both this compound and titanium(IV) sulfate ($Ti(SO_4)$).

USE: In tanning; in production of pigment grade and of anatase-structured titanium dioxide.

9632. Titanium Sesquisulfate. [10343-61-0] Sulfuric acid titanium(3+) salt (3:2); dititanium trisulfate; titanous sulfate. $O_{12}S_3Ti_2$; mol wt 383.90. O 50.01%, S 25.05%, Ti 24.94%. $Ti_2(SO_4)_3$.

Green, cryst powder. Insol in water, alc, concd H_2SO_4. Sol in dil HCl or dil H_2SO_4, giving violet-colored solns. Also crystallizes as a hydrate with $8H_2O$.

USE: See titanium trichloride.

9633. Titanium Tetrabromide. [7789-68-6] Br_4Ti; mol wt 367.48. Br 86.98%, Ti 13.03%. $TiBr_4$. Prepn: Olsen, Ryan, *J. Am. Chem. Soc.* **54**, 2215 (1932).

Amber-yellow or orange, very hygroscopic crystals. mp 28.25°. bp 233.45°. d²⁰ 3.25. Also reported as 2.6: Duppa, *Proc. R. Soc. London* **8**, 42 (1857). Dec by water; sol in abs ether, abs alcohol. *Keep tightly closed.*

9634. Titanium Tetrachloride. [7550-45-0] Cl₄Ti; mol wt 189.67. Cl 74.76%, Ti 25.24%. TiCl₄. Purification: Baxter *et al., J. Am. Chem. Soc.* **45**, 1228 (1923); **48**, 3117 (1926). Survey of preparative methods: Ehrlich in *Handbook of Preparative Inorganic Chemistry* Vol. 2, G. Brauer, Ed. (Academic Press, New York, 2nd ed., 1965) pp 195-199. Review of toxicology and human exposure: *Toxicological Profile for Titanium Tetrachloride* (PB98-101082, 1997) 145 pp.

Colorless liquid; penetrating acid odor. Absorbs moisture from the air and evolves dense white fumes. Reacts with water to produce titanium dioxide and hydrochloric acid. d 1.726. mp −24.1°. bp 136.4°. Vapor pressure (20°): 10.0 mm; (22°) 9.6 mm. Sol in cold water, alcohol; dec by hot water. *Corrosive; poisonous. Keep tightly closed.*

Caution: Potential symptoms of overexposure by inhalation are coughing, tightness in chest, chemical bronchitis or pneumonia, congestion of mucous membranes of upper resp tract. Direct contact may cause irritation or burns of eyes, skin, mucous membranes and lungs (PB98-101082).

USE: Manuf titanium compounds, iridescent glass and artificial pearls. Detection of peroxides. Prepn of standards. Formerly used with potassium bitartrate as a mordant in textile industry, and with dyewoods in dyeing leather; also as smoke-producing screen with ammonia.

9635. Titanium Tetrafluoride. [7783-63-3] F₄Ti; mol wt 123.86. F 61.35%, Ti 38.65%. TiF₄. Prepd by the reaction of hydrogen fluoride with titanium tetrachloride: Ruff, Ipsen, *Ber.* **36**, 1777 (1903); Ruff, Plato, *ibid.* **37**, 673 (1904); by the action of fluorine on the metal or the dioxide: Haendler *et al., J. Am. Chem. Soc.* **76**, 2177 (1954).

Powdery white mass. d₄²⁰ 2.798. Sublimes at 284°. mp >400°. Hisses on contact with water. Very hygroscopic; the dihydrate may be crystallized from aqueous soln (also obtained from solns of TiO₂ in HF). Aq solns hydrolyze slowly, and the existence of an oxyfluoride, TiOF₂, has been established. Also sol in alcohol and pyridine from which the compds TiF₄.2EtOH and TiF₄.C₅H₅N have been isolated. Insol in ether. Dry ammonia is absorbed at room temp to form TiF₄.4NH₃, but at 120° TiF₄.2NH₃ is the stable phase. This is sol in water and is sufficiently stable to sublime. TiF₄ is also sol in phosphorus oxychloride, but at 30° a reaction occurs and POF₃ is evolved. *See also* Eméleus in *Fluorine Chemistry* vol. I, J. H. Simons, Ed. (Academic Press, New York, 1950) p 47.

9636. Titanium Tetraisopropoxide. [546-68-9] 2-Propanol titanium(4+) salt (4:1); isopropyl titanate(IV); titanium isopropylate; titanium isopropoxide. C₁₂H₂₈O₄Ti; mol wt 284.22. C 50.71%, H 9.93%, O 22.52%, Ti 16.84%. Ti[OCH(CH₃)₂]₄. Mild Lewis acid catalyst. Prepn: F. Bischoff, H. Adkins, *J. Am. Chem. Soc.* **46**, 256 (1924). Vibrational spectra: P. D. Moran *et al., Inorg. Chem.* **37**, 2741 (1998). Kinetic behavior in asymmetric epoxidation: S. S. Woodard *et al., J. Am. Chem. Soc.* **113**, 106 (1991). Kinetics in prepn of titanium dioxide, *q.v.*, thin films: V. Gourinchas-Courtecuisse *et al., High Press. Chem. Eng.* **1996**, 133; G. A. Battiston *et al., Metalurgija* **8**, 183 (2002); of sol-gel process: A. Soloviev *et al., J. Mater. Sci.* **38**, 3315 (2003). Use as complexing agent in sol-gel process: B. Wang *et al., Polym. Commun.* **32**, 400 (1991); H. Arce *et al., Rev. Soc. Quim. Mex.* **47**, 73 (2003). Review as reagent in stereoselective synthesis: R. Mahrwald, *J. Prakt. Chem.* **341**, 191-194 (1999); as catalyst: F. Lake, C. Moberg, *Russ. J. Org. Chem.* **39**, 436-452 (2003). Brief review in a variety of organic syntheses: O. López, *Synlett* **2003**, 2261-2262.

Moisture-sensitive liquid. *Flammable.* mp 18-20°. bp₇₄₀ 230°. Sol in a wide range of solvents including ethers, organohalides, alcohols and benzenes.

USE: Catalyst especially for asymmetric induction in organic syntheses; in preparation of nanosized TiO₂. Complexing agent in sol-gel process.

9637. Titanium Trichloride. [7705-07-9] Titanium chloride (TiCl₃); titanous chloride; trichlorotitanium. Cl₃Ti; mol wt 154.22. Cl 68.96%, Ti 31.04%. TiCl₃. Prepn by reduction of the tetrachloride: Ingraham *et al., Inorg. Synth.* **6**, 52 (1960); Sherfey, *ibid.* 57.

Dark red-violet, unstable, deliquesc crystals; dec on heating above 500°. The dry powder is pyrophoric in air. Very reactive and readily

dissociated by moisture in air. Sol in water, exothermic process; sol in alcohol. Practically insol in ether. *Keep tightly closed.*

USE: A powerful reducing agent. Reduces nitrate to ammonia; when boiled with aq SO₂, sulfur is separated; hence is used as an aq soln for estimation of nitro groups, ferric ions, per-salts, etc. Removes stains, etc. (stripper) in laundering.

9638. Titanocene Dichloride. [1271-19-8] Dichlorobis(η⁵-2,4-cyclopentadien-1-yl)titanium; dichlorodi-π-cyclopentadienyltitanium; biscyclopentadienyltitanium(IV) dichloride. C₁₀H₁₀Cl₂Ti; mol wt 248.96. C 48.24%, H 4.05%, Cl 28.48%, Ti 19.23%. Cp₂-TiCl₂. Prepn: Wilkinson, Birmingham, *J. Am. Chem. Soc.* **76**, 4281 (1954). Structure: Clearfield *et al., Can. J. Chem.* **53**, 1622 (1975).

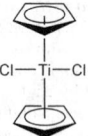

Bright red acicular crystals from toluene. mp 289 ± 2°. d 1.60. Moderately sol in toluene, chloroform, alcohol, other hydroxylic solvents; sparingly sol in water, petr ether, benzene, ether, carbon disulfide, carbon tetrachloride.

USE: Catalyst; with aluminum alkyls as Ziegler-Natta polymerization catalyst.

9639. Titin. Connectin. Giant, elastic protein found in vertebrate skeletal and cardiac muscles; comprises ~10% of the myofibrillar mass. One of the largest polypeptides known; mol wt ~3 × 10⁶ Da. Structure consists of repeats of 2 motifs, each with approximately 100 amino acid residues, organized into a super-repeating pattern. Enriched in proline (8-9 mole %) which contributes to its flexibility. Integral component of the cytoskeletal lattice of muscle cells (*see also* Nebulin). Thought to act as a "molecular ruler" that regulates the assembly of myosin and other thick filament proteins; connects the ends of the thick filament with the Z-line. Also responsible for the generation of resting tension in stretched muscle fibers. Identification in rabbit muscle: K. Maruyama *et al., Nature* **262**, 58 (1976). Characterization and localization in myofibrils: K. Wang *et al., Proc. Natl. Acad. Sci. USA* **76**, 3698 (1979). Purification methods: K. Wang, *Methods Enzymol.* **85B**, 264 (1982). Distribution and organization in the cytoskeletal matrix of striated muscle: *idem, Adv. Exp. Med. Biol.* **170**, 285-305 (1984). Quantitative determn in muscle fibers by gel electrophoresis: H. L. Granzier, K. Wang, *Electrophoresis* **14**, 56 (1993). *Reviews:* J. Trinick, *Trends Biochem. Sci.* **19**, 405-409 (1994); K. Maruyama, *Biophys. Chem.* **50**, 73-85 (1994).

9640. Tivantinib. [905854-02-6] (3R,4R)-3-(5,6-Dihydro-4H-pyrrolo[3,2,1-ij]quinolin-1-yl)-4-(1H-indol-3-yl)-2,5-pyrrolidinedione; (−)-trans-3-(5,6-dihydro-4H-pyrrolo[3,2,1-ij]quinolin-1-yl)-4-(1H-indol-3-yl)-pyrrolidine-2,5-dione; ARQ-197. C₂₃H₁₉-N₃O₂; mol wt 369.42. C 74.78%, H 5.18%, N 11.37%, O 8.66%. Human mesenchymal-epithelial transition factor (c-Met) receptor tyrosine kinase inhibitor. Prepn: C. J. Li *et al.,* WO 0686484; *eidem,* US 7713969 (2006, 2010 both to ArQule). Binding studies and mechanism of inhibition: S. Eathiraj *et al., J. Biol. Chem.* **286**, 20666 (2011). Preclinical antineoplastic activity: N. Munchi *et al., Mol. Cancer Ther.* **9**, 1544 (2010). Clinical pharmacology and pharmacokinetics: T. A. Yap *et al., J. Clin. Oncol.* **29**, 1271 (2011). Clinical evaluation in non-small cell lung cancer: L. V. Sequist *et al., ibid.,* 3307. Review of development and therapeutic potential: A. A. Adjei *et al., Oncologist* **16**, 788-799 (2011).

Crystals from ethanol or trifluoroethanol.

THERAP CAT: Antineoplastic.

9641. Tixocortol. [61951-99-3] (11β)-11,17-Dihydroxy-21-mercaptopregn-4-ene-3,20-dione; 11β,17α-dihydroxy-21-thio-3,20-dioxo-4-pregnene. $C_{21}H_{30}O_4S$; mol wt 378.53. C 66.63%, H 7.99%, O 16.91%, S 8.47%. Synthesis and biological activity: S. S. Simons *et al.*, *J. Steroid Biochem.* **13**, 311 (1980). Use in fluorescent chemoaffinity labeling: *eidem*, *Biochemistry* **18**, 4915 (1979). Prepn of the 21-pivalate: D. R. Torossian *et al.*, **DE 2357778**; *eidem*, **US 4014909** (1974, 1977 both to Jouveinal). Series of articles on pharmacology, *in vitro* and *in vivo* activity of the pivalate: *Arzneim.-Forsch.* **31**, 453-469 (1981).

Fine white solid, dec 220-221°. uv max (95% ethanol): 241 nm (ε 1.65×10⁴).

21-Pivalate. [55560-96-8] (11β)-21-[(2,2-Dimethyl-1-oxopropyl)thio]-11,17-dihydroxypregn-4-ene-3,20-dione; JO-1016; Pivalone; Rectovalone; Tiovalon. $C_{26}H_{38}O_5S$; mol wt 462.65. Crystals from ethanol, mp 195-200°. [α]$_D^{20}$ +145° (c = 1 in dioxane). uv max (methanol): 229 nm (log ε 4.259).

USE: Free thiol in chemoaffinity labeling.

THERAP CAT: Anti-inflammatory.

9642. Tizanidine. [51322-75-9] 5-Chloro-N-(4,5-dihydro-1H-imidazol-2-yl)-2,1,3-benzothiadiazol-4-amine; 5-chloro-4-(2-imidazolin-2-ylamino)-2,1,3-benzothiadiazole. $C_9H_8ClN_5S$; mol wt 253.71. C 42.61%, H 3.18%, Cl 13.97%, N 27.60%, S 12.64%. α₂-Adrenergic agonist; centrally active myotonolytic. Prepn and antitremor activity: P. Neumann, **NL 7306228**; *eidem*, **US 3843668** (1973, 1974 both to Wander-Sandoz). Pharmacology: A. C. Sayers *et al.*, *Arzneim.-Forsch.* **30**, 793 (1980); and mechanism of action studies: L. Turski *et al.*, *Brain Res.* **379**, 367 (1986); H. Ono *et al.*, *Gen. Pharmacol.* **17**, 137 (1986). Pharmacokinetics: V. Heazlewood *et al.*, *Eur. J. Clin. Pharmacol.* **25**, 65 (1983). Clinical evaluation in spasticity of multiple sclerosis: P. M. Newman *et al.*, *ibid.* **23**, 31 (1982); M. C. Hoogstraten *et al.*, *Acta Neurol. Scand.* **77**, 224 (1988).

Crystals from methanol, mp 221-223°. LD₅₀ orally in mice: 235 mg/kg (Sayers).

Hydrochloride. [64461-82-1] AB-021; DS-103-282; Sirdalud; Ternelin; Zanaflex. $C_9H_8ClN_5S$·HCl; mol wt 290.17. Almost white to slightly yellow, crystalline powder. Slightly sol in water, methanol.

THERAP CAT: Muscle relaxant (skeletal).

9643. TMD. [60761-10-6] *rel*-(2R,4aR,8aS)-Decahydro-1,1,4a-trimethyl-2-naphthalenol; 1,1,10-trimethyl-*trans*-2-decalol; 4,4,-10β-trimethyl-*trans*-decal-3β-ol. $C_{13}H_{24}O$; mol wt 196.33. C 79.53%, H 12.32%, O 8.15%. Prepn: B. Gaspert *et al.*, *J. Chem. Soc.* **1958**, 624; C. Djerassi, D. Marshall, *J. Am. Chem. Soc.* **80**, 3986 (1958); T. G. Halsall *et al.*, *J. Chem. Soc.* **1959**, 2798; S. L. Mukherjee, P. C. Dutta, *ibid.* **1960**, 67. Inhibitor of cholesterol biosynthesis: J. A. Nelson *et al.*, *J. Am. Chem. Soc.* **100**, 4900 (1978); T.-Y. Chang *et al.*, *J. Biol. Chem.* **254**, 11258 (1979).

Relative stereochemistry

Crystals, mp 60-65° (subl). Sol in alcohol; slightly sol in water.

d-Form. Crystals, mp 87-88° (subl). [α]$_D^{26}$ +15.9° (CHCl₃).

l-Form. Crystals, mp 87-88° (subl). [α]$_D^{26}$ -16.9° (CHCl₃).

9644. TNAZ. [97645-24-4] 1,3,3-Trinitroazetidine. $C_3H_4N_4O_6$; mol wt 192.09. C 18.76%, H 2.10%, N 29.17%, O 49.97%. Melt-castable nitramine explosive. Synthesis and crystal structure: T. G. Archibald *et al.*, *J. Org. Chem.* **55**, 2920 (1990). Alternate synthesis: A. R. Katritzky *et al.*, *J. Heterocycl. Chem.* **31**, 271 (1994); T. Axenrod *et al.*, *J. Org. Chem.* **60**, 1959 (1995). Decomposition studies: P. Politzer, J. M. Seminario, *Chem. Phys. Lett.* **207**, 27 (1993). Thermolysis: J. C. Oxley *et al.*, *J. Phys. Chem.* **98**, 7004 (1994). *Review:* S. Borman, *Chem. Eng. News* **72**, 18-22 (Jan. 17, 1994).

Explosive. White crystals from CCl₄, mp 100-101°. d 1.84 g/cm³.

USE: Munitions.

9645. TOAC. [15871-57-5] 4-Amino-4-carboxy-2,2,6,6-tetramethyl-1-piperidinyloxy; 2,2,6,6-tetramethylpiperidine-N-oxyl-4-amino-4-carboxylic acid. $C_{10}H_{19}N_2O_3$; mol wt 215.27. C 55.80%, H 8.90%, N 13.01%, O 22.30%. Conformationally-constrained, nitroxide, achiral, Cᵅ-tetrasubstituted α-amino acid used as a spin-label for protein structure determination. Prepn: A. Rassat, P. Rey, **FR 1501917** (1967 to Commissariat à l'Energie Atomique); *eidem*, *Bull. Soc. Chim. Fr.* **1967**, 815. Use as probe for structure conformation studies: T. A. Pertinhez *et al.*, *Biopolymers* **42**, 821 (1997); T. T. T. Bui *et al.*, *J. Chem. Soc. Perkin Trans.* **2000**, 1043; of interchain interactions: A. Polese *et al.*, *J. Am. Chem. Soc.* **121**, 11071 (1999).

Pale yellow crystals from water-ethanol (20%/80%), mp 228-230° (dec). uv max (water): 239, 416 nm (ε 1480, 8.4).

USE: Electron paramagnetic resonance (EPR) probe.

9646. Tobacco Mosaic Virus. TMV. Single stranded RNA virus that infects tobacco and other *Solanaceae*; causes a characteristic discoloration of the leaves. First virus to be studied. TMV is a rod-shaped particle, ~300 nm long; mol wt 39 × 10⁶. The capsid protein consists of 2130 polypeptide chains arranged to form a tubular molecule containing a core of RNA. Each peptide chain consists of 158 amino acid residues; their mol wt is 17,533. Isoln from diseased tobacco plants: Stanley, *Science* **81**, 644 (1935). Prepn and characterization of essentially uniform TMV particles: Boedtker, Simmons, *J. Am. Chem. Soc.* **80**, 2550 (1958); Knight, *Biochem. Prep.* **9**, 132 (1962). Sequential arrangement of the 158 amino acid residues of the protein subunit: Tsugita *et al.*, *Proc. Natl. Acad. Sci. USA* **46**, 1463 (1960). Synthesis of TMV-RNA by cell free extracts from infected tobacco leaves: Kim, Wildman, *Biochem. Biophys. Res. Commun.* **8**, 394 (1962); Cochran, *Chem. Eng. News* **40**, 64 (Sept. 17, 1962). *Reviews:* Anderer, *Adv. Protein Chem.* **18**, 1 (1963); Caspar, *ibid.* 37; Klug, Caspar, *Adv. Virus Res.* **7**, 233-277 (1960); Lauffer, Stevens, *ibid.* **13**, 1 (1968); Reddi, *ibid.* **17**, 51 (1972); L. Hirth, K. E. Richards, *ibid.* **26**, 145-199 (1981).

USE: Popular tool for studying the correlation between chemical structure and biological function.

9647. Tobramycin. [32986-56-4] O-3-Amino-3-deoxy-α-D-glucopyranosyl-(1 → 6)-O-[2,6-diamino-2,3,6-trideoxy-α-D-*ribo*-hexopyranosyl-(1 → 4)]-2-deoxy-D-streptamine; 4-[2,6-diamino-2,-3,6-trideoxy-α-D-glycopyranosyl]-6-[3-amino-3-deoxy-α-D-glycopyranosyl]-2-deoxystreptamine; nebramycin factor 6; NF 6; Tobracin; Tobralex; Tobramaxin; Tobrex. $C_{18}H_{37}N_5O_9$; mol wt 467.52. C 46.24%, H 7.98%, N 14.98%, O 30.80%. Aminoglycoside antibiotic; component of the **nebramycin** complex produced by *Streptomyces tenebrarius. See also* apramycin. Series of articles on isolation, separation and evaluation of the nebramycin complex: *Antimicrob. Agents Chemother.* **1967**, 314-348. Elucidation of structure: K. F. Koch, J. A. Rhoades, *Antimicrob. Agents Chemother.* **1970**, 309. Synthesis: Y. Takagi *et al., Bull. Chem. Soc. Jpn.* **49**, 3649 (1976); M. Tanabe *et al.,Tetrahedron Lett.* **1977**, 3607. Toxicology: J. S. Welles *et al.,Toxicol. Appl. Pharmacol.* **22**, 332 (1972). Comprehensive description: A. K. Dash, *Anal. Profiles Drug Subs. Excip.* **24**, 579-613 (1996). Clinical trial in cystic fibrosis patients: B. W. Ramsey *et al., N. Engl. J. Med.* **340**, 23 (1999). Review of clinical experience: H. Lode, *Curr. Ther. Res.* **59**, 420-453 (1998).

Basic substance. White to off-white, hygroscopic powder. Freely sol in water (1 in 1.5 parts). Very slightly sol in ethanol (1 in 2000 parts). Practically insol in chloroform, ether. $[\alpha]_D^{20}$ +129° (c = 1 in water). LD_{50} in mice, rats (mg/kg): 441, 969 s.c. (Welles).

Sulfate. [79645-27-5] Gernebcin; Nebcin; Nebicina; Obracin; Tobra; Tobradistin.

THERAP CAT: Antibacterial.

9648. Tocainide. [41708-72-9] 2-Amino-N-(2,6-dimethylphenyl)propanamide; 2-aminopropiono-2',6'-xylidide. $C_{11}H_{16}N_2$-O; mol wt 192.26. C 68.72%, H 8.39%, N 14.57%, O 8.32%. Antiarrhythmic agent related to lidocaine, *q.v.* Prepn: R. N. Boys *et al.*, **DE 2235745** (1973 to Astra), *C.A.* **78**, 140411 (1973); and resolution of isomers: E. W. Byrnes *et al., J. Med. Chem.* **22**, 1171 (1979). *In vitro* effects on muscle contractions: R. Dengler, R. Rüdel, *Arzneim.-Forsch.* **29**, 270 (1979). Inhibition of leucocyte locomotion in dogs: G. J. Stewart *et al., Lab. Invest.* **42**, 302 (1980). Biotransformation in humans: A. T. Elvin *et al., J. Pharm. Sci.* **69**, 47 (1980). Kinetics: C. Groffner, *Clin. Pharmacol. Ther.* **27**, 64 (1980). Use in refractory ventricular arrhythmias: D. M. Roden *et al.,Am. Heart J.* **100**, 15 (1980). HPLC resolution of enantiomers: K. M. McErlane, G. K. Pillai, *J. Chromatogr.* **274**, 129 (1983); determn in plasma: A. J. Sedman, J. Gal, *ibid.* **306**, 155 (1984). Pharmacokinetics of enantiomers: A. H. Thomson *et al., Br. J. Pharmacol.* **21**, 149 (1986). *In vitro* and *in vivo* pharmacodynamics of enantiomers: A. J. Block *et al., J. Cardiovasc. Pharmacol.* **11**, 216 (1988). Review of pharmacology and therapeutic efficacy: B. Holmes *et al., Drugs* **26**, 93-123 (1983); D. M. Roden, R. L. Woosley, *N. Engl. J. Med.* **315**, 41-45 (1986).

(±)-Form hydrochloride. [71395-14-7] W-36095; Tonocard; Xylotocan. $C_{11}H_{16}N_2O.HCl$; mol wt 228.72. Crystals from ethanol/ether, mp 246-247°. Freely sol in water, alc. Practically insol in chloroform, ether.

R-(−)-Form hydrochloride. [53984-74-0] Crystals from ethanol-diethyl ether, mp 265-266°. $[\alpha]_D$ −42.16° (c = 2.63 in methanol). Freely sol in water, alc. Practically insol in chloroform, ether.

S-(+)-Form hydrochloride. [53984-76-2] Crystals from ethanol-diethyl ether, mp 264.5°. $[\alpha]_D$ +42.35° (c = 2.63 in methanol). Freely sol in water, alc. Practically insol in chloroform, ether.

THERAP CAT: Antiarrhythmic (class IB).

THERAP CAT (VET): Antiarrhythmic (class IB).

9649. Tocamphyl. [5634-42-4] 1,2,2-Trimethyl-1,3-cyclopentanedicarboxylic acid 1-[1-(4-methylphenyl)ethyl] ester, compd with 2,2′-iminobis[ethanol] (1:1); camphoric acid 1-(p,α-dimethylbenzyl) ester, compd with 2,2′-imidodiethanol (1:1); p,α-dimethylbenzyl camphorate diethanolamine salt; p-toluoylmethylcarbinol-mono-d-camphoric acid ester diethanolamine salt; methyl p-tolylcarbinol camphorate diethanolamine salt; p-tolylmethylcarbinol camphoric acid ester diethanolamine salt; diethanolamine p-tolylmethylcarbinol camphorate; diethanolamine d-methyltoluylcarbinol camphorate; Biliphorine; Hepatoxane; Syncuma. $C_{23}H_{37}NO_6$; mol wt 423.55. C 65.22%, H 8.81%, N 3.31%, O 22.66%. Prepn: **CH 211203** (1940 to Chemiewerk Homburg); *Chem. Zentralbl.* **1941**, I, 2972.

Crystals, sol in water.

THERAP CAT: Choleretic.

9650. Toceranib. [356068-94-5] 5-[(Z)-(5-Fluoro-1,2-dihydro-2-oxo-3H-indol-3-ylidene)methyl]-2,4-dimethyl-N-[2-(1-pyrrolidinyl)ethyl]-1H-pyrrole-3-carboxamide; PHA-291639; SU-11654. $C_{22}H_{25}FN_4O_2$; mol wt 396.47. C 66.65%, H 6.36%, F 4.79%, N 14.13%, O 8.07%. Tyrosine kinase inhibitor with antiangiogenesis activity. Prepn: P. C. Tang *et al.*, **WO 0160814** (2001 to Sugen); *eidem*, **US 6573293** (2003 to Sugen, Pharmacia & Upjohn). Pharmacokinetics and efficacy in spontaneous canine tumors: C. A. London *et al., Clin. Cancer Res.* **9**, 2755 (2003). Mechanism of action study: N. K. Pryer *et al., ibid.* 5729. Clinical trial in canine mast cell tumors: C. A. London *et al., ibid.* **15**, 3856 (2009).

Phosphate. [874819-74-6] PHA-291639E; Palladia. $C_{22}H_{25}$-$FN_4O_2.H_3O_4P$; mol wt 494.46.

THERAP CAT (VET): Antineoplastic.

9651. Tocilizumab. [375823-41-9] Anti-(human interleukin 6 receptor) immunoglobulin G1 (human-mouse monoclonal MRA heavy chain) disulfide with human-mouse monoclonal MRA κ-chain, dimer; atlizumab; MRA; rhPM-1; R-1569; Actemra; Ro-Actemra. Recombinant humanized monoclonal antibody directed against human interleukin 6 receptor (IL-6R); designed to inhibit biological activities of IL-6. Prepn: M. Tsuchiya *et al.*, **WO 9219759**; *eidem*, **US 5795965** (1992, 1998 both to Chugai); K. Sato *et al., Cancer Res.* **53**, 851 (1993). Pharmacology: H. Shinkura *et al., Anticancer Res.* **18**, 1217 (1998). Clinical pharmacokinetics: N. Nishimoto *et al., J. Rheumatol.* **30**, 1426 (2003). Clinical trial in rheumatoid arthritis: J. S. Smolen *et al., Lancet* **371**, 987 (2008); in juvenile arthritis: S. Yokota *et al., ibid.*, 998. Review of pharmacology and clinical experience: V. Oldfield *et al., Drugs* **69**, 609-632 (2009).

THERAP CAT: Anti-inflammatory.

9652. Tocol. [119-98-2] 3,4-Dihydro-2-methyl-2-(4,8,12-trimethyltridecyl)-2H-1-benzopyran-6-ol; 2-methyl-2-(4,8,12-trimethyltridecyl)-6-chromanol; 2-methyl-2-phytyl-6-chromanol; 6-hydroxy-2-methyl-2-phytylchroman; 2-methyl-2-phytyl-6-hydroxychroman. $C_{26}H_{44}O_2$; mol wt 388.64. C 80.35%, H 11.41%, O

Consult the Name Index before using this section.

8.23%. Synthesis by the condensation of hydroquinone and phytol in the presence of anhydr formic acid: Pendse, Karrer, *Helv. Chim. Acta* **40**, 1837 (1957). Antioxidant activity of tocol and its methyl derivs: Olcott, van der Veen, *Lipids* **3**, 331 (1968).

Colorless, viscous oil. bp$_{0.001}$ 165-175°.
Acetate. C$_{28}$H$_{46}$O$_3$. Viscous oil. bp$_{0.001}$ 180-185°.
USE: Antioxidant.

9653. α-Tocopherol. [59-02-9] (2*R*)-3,4-Dihydro-2,5,7,8-tetramethyl-2-[(4*R*,8*R*)-4,8,12-trimethyltridecyl]-2*H*-1-benzopyran-6-ol; (+)-2,5,7,8-tetramethyl-2-(4′,8′,12′-trimethyltridecyl)-6-chromanol; *R*,*R*,*R*-α-tocopherol; *d*-α-tocopherol; 5,7,8-trimethyltocol; Optovit; Tocovital. C$_{29}$H$_{50}$O$_2$; mol wt 430.72. C 80.87%, H 11.70%, O 7.43%. Most bioactive of the naturally occurring forms of vitamin E, *q.v.* Richest sources are green vegetables, grains, and oils, particularly palm, safflower and sunflower oils. Isoln from wheat germ: H. M. Evans *et al.*, *J. Biol. Chem.* **113**, 319 (1936). Structure: E. Fernholz, *J. Am. Chem. Soc.* **59**, 1154 (1937); **60**, 700 (1938). Synthesis of *dl*-form: P. Karrer *et al.*, *Helv. Chim. Acta* **21**, 520, 820 (1938); F. Bergel *et al.*, *J. Chem. Soc.* **1938**, 1382. Distillation from vegetable oils and prepn of esters: J. G. Baxter *et al.*, *J. Am. Chem. Soc.* **65**, 918 (1943). Prepn of crystalline natural form: C. D. Robeson, *ibid.* 1660; of crystalline acetate: *idem, ibid.* **64**, 1487 (1942). Abs config of natural α-tocopherol: H. Mayer *et al.*, *Helv. Chim. Acta* **46**, 963 (1963). Stereoselective synthesis: K.-K. Chan *et al.*, *J. Org. Chem.* **43**, 3435 (1978). Total synthesis of all 8 stereoisomers: N. Cohen *et al.*, *Helv. Chim. Acta* **64**, 1158 (1981). Clinical trial in Alzheimer's disease: M. Sano *et al.*, *N. Engl. J. Med.* **336**, 1216 (1997); to improve immune function in healthy elderly: S. N. Meydani *et al.*, *J. Am. Med. Assoc.* **277**, 1380 (1997). Review of bioavailability from vitamin E supplements: M. G. Traber, *Bio-Factors* **10**, 115-120 (1999). Review of clinical trials in heart disease: W. A. Pryor, *Free Radical Biol. Med.* **28**, 141-164 (2000).

Transparent needles, mp 2.5-3.5°. [α]$_{546.1}^{25}$ −3.0° (benzene); [α]$_{546.1}^{25}$ +0.32° (ethanol).
Acetate. [58-95-7] Spondyvit. C$_{31}$H$_{52}$O$_3$; mol wt 472.75. Light yellow oil. Crystallized at −30° as needle-like crystals, mp 26.5-27.5°. [α]$_D^{25}$ +0.25° (c = 10 in chloroform); [α]$_D^{25}$ +3.2° (in ethanol).
Succinate. [4345-03-3] *d*-α-Tocopheryl acid succinate; Tocovite. Needles from petr ether, mp 76-77°. uv max (ethanol): 286 nm (E$_{1cm}^{1\%}$ 38.5). Practically insol in water.
***dl*-α-Tocopherol.** [10191-41-0] *all-rac*-α-Tocopherol. Equimolar mixture of all four racemates. Slightly viscous, pale yellow oil. d$_4^{25}$ 0.950; bp$_{0.1}$ 200-220°; n$_D^{25}$ 1.5045. uv max: 294 nm (E$_{1cm}^{1\%}$ 71). Practically insol in water. Freely sol in oils, fats, acetone, alcohol, chloroform, ether, other fat solvents. Stable to heat and alkalies in the absence of oxygen. Not affected by acids up to 100°. Slowly oxidized by atm oxygen, rapidly by ferric and silver salts. Gradually darkens on exposure to light.
***dl*-α-Tocopherol acetate.** [52225-20-4] *dl*-α-Tocopheryl acetate; Detulin; Ephynal; Eusovit; Evion. Comprehensive description: B. C. Rudy, B. Z. Senkowski, *Anal. Profiles Drug Subs.* **3**, 111-126 (1974). Pale yellow, viscous liquid. mp −27.5°. d$_4^{21.3}$ 0.9533. bp$_{0.01}$ 184°; bp$_{0.025}$ 194°; bp$_{0.3}$ 224°. n$_D^{20}$ 1.4950-1.4972. uv max (cyclohexane): 285.5 nm. Practically insol in water. Freely sol in acetone, chloroform, ether. Less readily sol in alc.
USE: As an antioxidant in vegetable oils and shortening.
THERAP CAT: Vitamin E supplement.
THERAP CAT (VET): Vitamin E supplement.

9654. β-Tocopherol. [16698-35-4]; [148-03-8] (*dl*-form). (2*R*)-3,4-Dihydro-2,5,8-trimethyl-2-[(4*R*,8*R*)-4,8,12-trimethyltri-

decyl]-2*H*-1-benzopyran-6-ol; (+)-2,5,8-trimethyl-2-(4,8,12-trimethyltridecyl)-6-chromanol; 5,8-dimethyltocol; cumotocopherol; neotocopherol; *p*-xylotocopherol. C$_{28}$H$_{48}$O$_2$; mol wt 416.69. C 80.71%, H 11.61%, O 7.68%. One of the naturally occurring forms of vitamin E, *q.v.* Is biologically less active than α-tocopherol. May be separated by fractional crystn: Emerson *et al.*, *Science* **83**, 421 (1936); *J. Biol. Chem.* **113**, 319 (1936); Baxter *et al.*, *J. Am. Chem. Soc.* **65**, 918 (1943).

Pale yellow, viscous oil. bp$_{0.1}$ 200-210°. [α]$_{546.1}^{25}$ +2.9° (c = 7.15 in ethanol). uv max: 297 nm (E$_{1cm}^{1\%}$ 87.6). Insol in water. Freely sol in oils, fats, acetone, alcohol, chloroform, ether, other fat solvents. Very stable to heat and alkalies. Slowly oxidized by atmospheric oxygen, rapidly by ferric and silver salts. Gradually darkens on exposure to light.

9655. γ-Tocopherol. [54-28-4]; [7616-22-0] (*dl*-form). (2*R*)-3,4-Dihydro-2,7,8-trimethyl-2-[(4*R*,8*R*)-4,8,12-trimethyltridecyl]-2*H*-1-benzopyran-6-ol; (+)-2,7,8-trimethyl-2-(4,8,12-trimethyltridecyl)-6-chromanol; (*R*,*R*,*R*)-γ-tocopherol; 7,8-dimethyltocol; *o*-xylotocopherol. C$_{28}$H$_{48}$O$_2$; mol wt 416.69. C 80.71%, H 11.61%, O 7.68%. One of the naturally occurring forms of vitamin E, *q.v.* Most abundant tocopherol in soybean and corn oils. Isoln by fractional crystn: Emerson *et al.*, *Science* **83**, 421 (1936); *J. Biol. Chem.* **113**, 319 (1936); J. G. Baxter *et al.*, *J. Am. Chem. Soc.* **65**, 918 (1943). Prepn of crystalline natural form: C. D. Robeson, *J. Am. Chem. Soc.* **65**, 1660 (1943). Comparison of bioactivity with α-tocopherol, *q.v.*: J. G. Bieri, R. P. Evarts, *J. Nutr.* **104**, 850 (1974). Protective effects vs reactive nitrogen oxide species: R. V. Cooney *et al.*, *Proc. Natl. Acad. Sci. USA* **90**, 1771 (1993); S. Christen *et al.*, *ibid.* **94**, 3217 (1997). HPLC determn in serum: A. Sobczak *et al.*, *J. Chromatogr. B* **730**, 265 (1999). Review of bioavailability, metabolism, and activity: Q. Jiang *et al.*, *Am. J. Clin. Nutr.* **74**, 714-722 (2001).

Pale yellow, viscous oil. Has been crystallized as transparent needles, mp −3 to −2°. bp$_{0.1}$ 200-210°. [α]$_{546.1}^{25}$ −2.4° (c = 8.59 in benzene); [α]$_{546.1}^{25}$ +2.2° (c = 9.32 in ethanol). uv max: 298 nm (E$_{1cm}^{1\%}$ 92.8). Insol in water. Freely sol in oils, fats, acetone, alcohol, chloroform, ether, other fat solvents. Very stable to heat and alkalies. Slowly oxidized by atmospheric oxygen, rapidly by ferric and silver salts. Gradually darkens on exposure to light.

9656. δ-Tocopherol. [119-13-1] (2*R*)-3,4-Dihydro-2,8-dimethyl-2-[(4*R*,8*R*)-4,8,12-trimethyltridecyl]-2*H*-1-benzopyran-6-ol; 8-methyltocol. C$_{27}$H$_{46}$O$_2$; mol wt 402.66. C 80.54%, H 11.52%, O 7.95%. One of the naturally occurring forms of vitamin E, *q.v.* Isoln from soybean oil: Stern *et al.*, *J. Am. Chem. Soc.* **69**, 869 (1947). Synthesis: Green *et al.*, *J. Chem. Soc.* **1959**, 3374; GB 900085 (1961 to Hoffmann-La Roche).

Pale yellow, viscous oil. [α]$_{546}^{25}$ +3.4° (c = 15.5 in alc); [α]$_{546}^{25}$ +1.1° (c = 10.9 in benzene). uv max: 298 nm (E$_{1cm}^{1\%}$ 91.2).

9657. Tocoretinate. [40516-48-1] Retinoic acid (±)-(2*R*)-3,4-dihydro-2,5,7,8-tetramethyl-2-[(4*R*,8*R*)-4,8,12-trimethyltridecyl]-2*H*-1-benzopyran-6-yl ester; (±)-(2*R**)-2,5,7,8-tetramethyl-2-[(4*R**,8*R**)-4,8,12-trimethyltridecyl]-6-chromanyl retinoate; tretinoin tocoferil; DL-α-tocopheryl retinoate; L-300; N-021; Olcenon. $C_{49}H_{76}O_3$; mol wt 713.14. C 82.53%, H 10.74%, O 6.73%. Wound healing agent that stimulates proliferation of normal skin fibroblasts. Prepn: H. Fukawa, K. Tanaka, **JP Kokai 73 00469**; *eidem*, **US 3878202** (1973, 1975 both to Nisshin Flour Milling Co.). Pharmacology: K. Sakyo *et al.*, *Oyo Yakuri* **43**, 111 (1992), *C.A.* **116**, 228223z (1992). Pharmacokinetics: T. Nakazawa *et al.*, *ibid.* 205, *C.A.* **117**, 19823 (1993). Mechanism of action study: N. Kawamura *et al.*, *Dig. Dis. Sci.* **39**, 2191 (1994). Clinical evaluation in skin ulcers: T. Doi, *Skin Res.* **36**, 384 (1994); K. Nakagawa *et al.*, *ibid.* 209. Acute toxicity study: Y. Harada *et al.*, *Oyo Yakuri* **43**, 227 (1992), *C.A.* **117**, 20721 (1992).

Relative stereochemistry

Light yellow oil. uv max (ethanol): 365 nm ($E_{1cm}^{1\%}$ 642). LD_{50} in mice (mg/kg): >1000 i.v.; >2000 orally (Fukawa, 1975).

THERAP CAT: Vulnerary.

9658. α-Tocotrienol. [58864-81-6]; [1721-51-3] (unspecified stereo). (2*R*)-3,4-Dihydro-2,5,7,8-tetramethyl-2-[(3*E*,7*E*)-4,8,12-trimethyl-3,7,11-tridecatrien-1-yl]-2*H*-1-benzopyran-6-ol; 2,5,7,8-tetramethyl-2-(4,8,12-trimethyl-3,7,11-tridecatrienyl)-6-chromanol; ζ_1-tocopherol. $C_{29}H_{44}O_2$; mol wt 424.67. C 82.02%, H 10.44%, O 7.53%. One of the naturally occurring forms of vitamin E, *q.v.* Isoln from wheat bran: Green *et al.*, *J. Sci. Food Agric.* **6**, 274 (1955); Green *et al.*, *Chem. Ind. (London)* **1960**, 73. Structure: Green *et al.*, *J. Chem. Soc.* **1959**, 3362. Attempt at synthesis: McHale *et al.*, *ibid.* **1963**, 784. *Review:* M. Kofler *et al.*, "Physicochemical Properties and Assay of the Tocopherols" in R. S. Harris, I. G. Wood, *Vitam. Horm.* **20**, 407-439 (1962). Synthesis of all-*trans*-form: Schudel *et al.*, *Helv. Chim. Acta* **46**, 2517 (1963).

uv max (ethanol): 292.5 nm ($E_{1cm}^{1\%}$ 91).

9659. β-Tocotrienol. [490-23-3] (2*R*)-3,4-Dihydro-2,5,8-trimethyl-2-[(3*E*,7*E*)-4,8,12-trimethyl-3,7,11-tridecatrien-1-yl]-2*H*-1-benzopyran-6-ol; 2,5,8-trimethyl-2-(4,8,12-trimethyltrideca-3,7,11-trienyl)chroman-6-ol; 5-methyltocol; ε-tocopherol. $C_{28}H_{42}O_2$; mol wt 410.64. C 81.90%, H 10.31%, O 7.79%. One of the naturally occurring forms of vitamin E, *q.v.* Isoln from wheat germ oil and from bran: Eggitt, Ward, *J. Sci. Food Agric.* **4**, 569 (1953); Eggitt, Norris, *ibid.* **6**, 689 (1955); **7**, 496 (1956). Structure: Green *et al.*, *J. Chem. Soc.* **1959**, 3362; *Chem. Ind. (London)* **1960**, 73; McHale *et al.*, *J. Chem. Soc.* **1963**, 784. Synthesis: Schudel *et al.*, *Helv. Chim. Acta* **46**, 2517 (1963).

Pale yellow oil. uv max (ethanol): 296 nm ($E_{1cm}^{1\%}$ 87).

9660. Todralazine. [14679-73-3] 2-(1-Phthalazinyl)hydrazinecarboxylic acid ethyl ester; 3-(1-phthalazinyl)carbazic acid ethyl ester; N^1-carbethoxy-N^2-phthalazinehydrazine; carboethoxyphthal-

azinohydrazine; ecarazine. $C_{11}H_{12}N_4O_2$; mol wt 232.24. C 56.89%, H 5.21%, N 24.13%, O 13.78%. Prepn: S. Biniecki *et al.*, *Bull. Acad. Pol. Sci. Ser. Sci.* **6**, 227 (1958), *C.A.* **52**, 18424g (1958); *eidem*, **BE 647722**; *eidem*, **US 3591588** (1964, 1971 both to Polfa). Spectrofluorometric determn in plasma: A. Ishii, T. Deguchi, *Chem. Pharm. Bull.* **26**, 2241 (1978). Pharmacological study: M. Filczewski, E. Boguka, *Pol. J. Pharmacol. Pharm.* **31**, 127 (1979), *C.A.* **91**, 204508 (1979). Absorption, distribution, excretion in rats and humans: A. Ishii *et al.*, *Oyo Yakuri* **18**, 61 (1979), *C.A.* **92**, 104012 (1980). Clinical study: W. Reiterer, H. Czitober, *Arzneim.-Forsch.* **27**, 2163 (1977). Toxicity study: F. Parravincini *et al.*, *Farmaco Ed. Sci.* **34**, 299 (1979).

Hydrochloride. [3778-76-5] CEPH; BT-621; Apiracohl; Aperdor; Apride; Atapren; Binazin; Illcut; Propat. $C_{11}H_{12}N_4O_2\cdot HCl$; mol wt 268.70. LD_{50} i.p. in mice: 500 mg/kg (Parravincini).

THERAP CAT: Antihypertensive.

9661. Tofacitinib. [477600-75-2] (3*R*,4*R*)-4-Methyl-3-(methyl-7*H*-pyrrolo[2,3-*d*]pyrimidin-4-ylamino)-β-oxo-1-piperidinepropanenitrile; 3-[(3*R*,4*R*)-4-methyl-3-[methyl-(7*H*-pyrrolo[2,3-*d*]pyrimidin-4-yl)amino]piperidin-1-yl]-3-oxopropanenitrile; tasocitinib; CP-690550. $C_{16}H_{20}N_6O$; mol wt 312.38. C 61.52%, H 6.45%, N 26.90%, O 5.12%. Janus kinase 3 (JAK3) inhibitor. Prepn (unspecified stereo): T. A. Blumenkopf *et al.*, **WO 0142246**; *eidem*, **US 7265221** (2001, 2007 both to Pfizer); and resolution of isomers: M. E. Flanagan, M. J. Munchhof, **US 7301023** (2007 to Pfizer). Large scale process: K. E. Price *et al.*, *Org. Lett.* **11**, 2003 (2009). Prepn and kinase selectivity: J. Jiang *et al.*, *J. Med. Chem.* **51**, 8012 (2008). LC/LC-MS determn in blood: R. Paniagua *et al.*, *Ther. Drug Monit.* **27**, 608 (2005). Clinical pharmacokinetics: E. van Gurp *et al.*, *Am. J. Transplant.* **8**, 1711 (2008). Clinical trial in rheumatoid arthritis: J. H. Coombs *et al.*, *Ann. Rheum. Dis.* **69**, 413 (2010). Review of pharmacology and clinical experience in immune-mediated disorders: K. West, *Curr. Opin. Investig. Drugs* **10**, 491-504 (2009); L. Vijayakrishnan *et al.*, *Trends Pharmacol. Sci.* **32**, 25-34 (2011).

Yellow foam. $[\alpha]_D^{25}$ +10.4° (c = 0.64 in methanol).

Monocitrate. [540737-29-9] Tofacitinib citrate; CP-690550-10. $C_{16}H_{20}N_6O\cdot C_6H_8O_7$; mol wt 504.50. Prepn: M. E. Flannagan, Z. J. Li, **US 6965027** (2005 to Pfizer). White crystalline solid, mp 199-206°.

THERAP CAT: Immunosuppressant.

9662. Tofisopam. [22345-47-7] 1-(3,4-Dimethoxyphenyl)-5-ethyl-7,8-dimethoxy-4-methyl-5*H*-2,3-benzodiazepine; EGYT-341; Grandaxin; Seriel. $C_{22}H_{26}N_2O_4$; mol wt 382.46. C 69.09%, H 6.85%, N 7.32%, O 16.73%. The first 5*H*-2,3-benzodiazepine. Prepn: J. Korosi, **HU 155572** (1969 to Pharm. Res. Inst.), *C.A.* **70**, 115026a (1969); **GB 1202579** corresp to **US 3736315** (1970, 1973 both to EGYT); J. Korosi, T. Lang, *Ber.* **107**, 3883 (1974). Synthesis and conformation: *eidem*, *Ther. Hung.* **23**, 132 (1975). FT ^{13}C NMR study: A. Neszmelyi *et al.*, *Ber.* **107**, 3894 (1974). Pharmacology: L. Petocz, I. Kosoczky, *Ther. Hung.* **23**, 134 (1975). Human pharmacokinetics: S. Ronai *et al.*, *ibid.* 139. Comparative efficacy: H. L. Goldberg, R. J. Finnerty, *Am. J. Psychiatry* **136**, 196 (1979).

Colorless to light cream cryst powder from isopropyl alcohol, mp 156-157°. uv max (methanol): 310, 272, 239 nm (ε 16100, 11200, 26300).

THERAP CAT: Anxiolytic.

9663. Togni's Reagents. Electrophilic hypervalent iodine reagents used in mild and selective trifluoromethylation reactions. Prepn and preliminary reactivity studies: P. Eisenberger *et al.*, *Chem. Eur. J.* **12**, 2579 (2006). Improved prepn of the dimethyl compd and utility in the trifluoromethylation of esters and thiols: I. Kieltsch *et al.*, *Angew. Chem. Int. Ed.* **46**, 754 (2007). Improved prepn of the carbonyl compd and utility in the trifluoromethylation of phenols: K. Stanek *et al.*, *J. Org. Chem.* **73**, 7678 (2008). Trifluoromethylation synthetic applications with phosphines: P. Eisenberger *et al.*, *Chem. Commun.* **2008**, 1575; with alcohols: R. Koller *et al.*, *Angew. Chem. Int. Ed.* **48**, 4332 (2009); with arenes and heteroarenes: R. Shimizu *et al.*, *Tetrahedron Lett.* **51**, 5947 (2010); M. S. Wiehn *et al.*, *J. Fluorine Chem.* **131**, 951 (2010). Demonstration of enantioselective reactions with aldehydes: A. E. Allen, D. W. C. MacMillan, *J. Am. Chem. Soc.* **132**, 4986 (2010). *Reviews:* I. Kieltsch *et al.*, *Chimia* **62**, 260-263 (2008); D. K. Yadav, *Synlett* **2010**, 2523-2524.

Dimethyl compound: X = C(CH₃)₂

Carbonyl compound: X = C=O

Dimethyl compound. [887144-97-0] 3,3-Dimethyl-1-(trifluoromethyl)-1,2-benziodoxole; 1,3-dihydro-3,3-dimethyl-1-(trifluoromethyl)-1,2-benziodoxole; 1-trifluoromethyl-1,3-dihydro-3,3-dimethyl-1,2-benziodoxole. C₁₀H₁₀F₃IO; mol wt 330.09. White solid from sublimation, mp 76-79°. *Irritant*. Heat and moisture sensitive. Store under inert gas at 2-8°C. Can be exposed to moist air for short periods of time without any apparent alteration.
   Carbonyl compound. [887144-94-7] 1-(Trifluoromethyl)-1,2-benziodoxol-3(1H)-one; C₈H₄F₃IO₂; mol wt 316.02. White crystals, mp 122° (dec). d 2.365. Can be purified by sublimation. Air and moisture stable over several months at room temperature.

USE: Reagents in synthetic organic chemistry.

9664. Tolan. [501-65-5] 1,1'-(1,2-Ethynediyl)bisbenzene; diphenylacetylene; diphenylethyne. C₁₄H₁₀; mol wt 178.23. C 94.35%, H 5.66%. Prepd by the oxidation of benzil dihydrazone with mercuric oxide: Schlenk, Bergmann, *Ann.* **463**, 76 (1928); by dehydrohalogenation of stilbene dibromide: Söderbäck, *Ann.* **443**, 161 (1925); Smith, Hoehn, *J. Am. Chem. Soc.* **63**, 1180 (1941); Smith, Falkof, *Org. Synth.* **coll. vol. III**, 350 (1955). Improved procedure: L. F. Fieser, *Experiments in Organic Chemistry* (Boston, 3rd ed., 1955) p 181.

Monoclinic, pseudo-rhombic rods or large spears from 95% ethanol. mp 60-61° (also reported as 62.5°). bp₇₆₀ 300°; bp₁₉ 170°. Dipole moment 0.3. Specific heat at 20°: 0.297. uv max: 216, 221,

269, 272, 279, 288, 297 nm (ε 20600, 20300, 23450, 25200, 33000, 23250, 29400). Insol in water. Freely sol in ether, hot alcohol.

9665. Tolazamide. [1156-19-0] *N*-[[(Hexahydro-1*H*-azepin-1-yl)amino]carbonyl]-4-methylbenzenesulfonamide; 1-(hexahydro-1*H*-azepin-1-yl)-3-(*p*-tolylsulfonyl)urea; *N*-(*p*-toluenesulfonyl)-*N'*-hexamethyleniminourea; tolazolamide; U-17835; Tolanase. C₁₄H₂₁N₃O₃S; mol wt 311.40. C 54.00%, H 6.80%, N 13.49%, O 15.41%, S 10.30%. Prepn: J. B. Wright, **GB 887886** (1962 to Upjohn); J. B. Wright, R. E. Willette, *J. Med. Pharm. Chem.* **5**, 815 (1962). Pharmacology: W. E. Dulin *et al.*, *Proc. Soc. Exp. Biol. Med.* **107**, 245 (1961). Mode of action study: Marshall *et al.*, *Metab. Clin. Exp.* **19**, 1046 (1970). Clinical experience and review of early literature: Balodimos, Marble, *Curr. Ther. Res.* **13**, 6-12 (1971). Clinical bioavailability and pharmacokinetics: P. G. Welling *et al.*, *J. Pharm. Sci.* **71**, 1259 (1982). Structural studies: C. H. Koo *et al.*, *Arch. Pharmacal Res.* **11**, 74 (1988). HPLC determn in serum: B. J. Starkey *et al.*, *J. Liq. Chromatogr.* **12**, 1889 (1989). Comprehensive description: J. K. Lee *et al.*, *Anal. Profiles Drug Subs. Excip.* **22**, 489-516 (1993).

Crystals, mp 170-173°. Very slightly sol in water; slightly sol in alcohol; sol in acetone; freely sol in chloroform. pKa (25°): 3.6, pKa (37.5°): 5.68. LD₅₀ in rats, mice (mg/kg): >5000 orally, 2239 i.p. (Dulin).

THERAP CAT: Antidiabetic.

9666. Tolazoline. [59-98-3] 4,5-Dihydro-2-(phenylmethyl)-1*H*-imidazole; 2-benzyl-2-imidazoline; 2-benzyl-2-iminazoline; benzazoline; 2-benzyl-4,5-imidazoline; phenylmethylimidazoline. C₁₀H₁₂N₂; mol wt 160.22. C 74.97%, H 7.55%, N 17.48%. α-Adrenergic blocker. Prepn: Sonn, **US 2161938** (1939 to Ciba). HPLC determn in serum: L. M. L. Todesco *et al.*, *Ther. Drug Monit.* **9**, 78 (1987). Review of pharmacology and clinical use: R. M. Ward, *Clin. Perinatol.* **11**, 703-713 (1984).

Hydrochloride. [59-97-2] Priscol; Priscoline; Tolazine; Vaso-Dilatan. C₁₀H₁₂N₂.HCl; mol wt 196.68. Bitter crystals, mp 174°. Freely sol in water, alcohol; sol in chloroform; very slightly sol in ether, ethyl acetate. pH of 2.5% soln 4.9-5.3.

THERAP CAT: Vasodilator (peripheral).

THERAP CAT (VET): Reversal agent for xylazine.

9667. Tolbutamide. [64-77-7] *N*-[(Butylamino)carbonyl]-4-methylbenzenesulfonamide; 1-butyl-3-(*p*-tolylsulfonyl)urea; tolylsulfonylbutylurea; 3-(*p*-tolyl-4-sulfonyl)-1-butylurea; *N*-*n*-butyl-*N'*-tosylurea; *N'*-4-methylbenzenesulfonyl-*N''*-butylurea; *N*-(sulfonyl-*p*-methylbenzene)-*N'*-*n*-butylurea; D-860; U-2043; Artosin; Diaben; Dolipol; Mobenol; Orabet; Orinase; Oterben; Pramidex; Rastinon. C₁₂H₁₈N₂O₃S; mol wt 270.35. C 53.31%, H 6.71%, N 10.36%, O 17.75%, S 11.86%. Description: Ehrhart, *Naturwissenschaften* **43**, 93 (1956). Prepn: **GB 808071**; Aumüller, Herr, **DE 1066575** (both 1959 to Hoechst); Ruschig *et al.*, **US 2968158** (1961 to Upjohn). Comprehensive description: W. F. Beyer, E. H. Jensen, *Anal. Profiles Drug Subs.* **3**, 513-543 (1974).

White, or practically white crystals or powder, mp 128.5-129.5°. Sol in alc, chloroform. Practically insol in water.

Sodium salt. [473-41-6] $C_{12}H_{17}N_2NaO_3S$. White to off-white, crystalline powder. mp 130-133°. Tetrahydrate, mp 41-43°. Freely sol in alc; sol in alc, chloroform; very slightly sol in ether.

THERAP CAT: Antidiabetic.

THERAP CAT (VET): Hypoglycemic agent.

9668. Tolcapone. [134308-13-7] (3,4-Dihydroxy-5-nitrophenyl)(4-methylphenyl)methanone; 3,4-dihydroxy-4'-methyl-5-nitrobenzophenone; Ro-40-7592; Tasmar. $C_{14}H_{11}NO_5$; mol wt 273.24. C 61.54%, H 4.06%, N 5.13%, O 29.28%. Orally active inhibitor of central and peripheral catechol-O-methyltransferase (COMT). Prepn: K. Bernauer *et al.*, **EP 237929**; *eidem*, **US 5236952** (1987, 1993 both to Hoffmann-La Roche). Pharmacology: G. Zürcher *et al.*, *Adv. Neurol.* **53**, 497 (1990). HPLC determn in plasma: U. Timm, R. Erdin, *J. Chromatogr.* **593**, 63 (1992). Clinical pharmacokinetics: J. Dingemanse *et al.*, *Clin. Pharmacol. Ther.* **57**, 508 (1995). Clinical evaluation as adjunct to levodopa: P. Limousin *et al.*, *Clin. Neuropharmacol.* **18**, 258 (1995). Review of pharmacology and clinical experience: G. M. Keating, K. A. Lyseng-Williamson, *CNS Drugs* **19**, 165-184 (2005); of safety and efficacy: N. Borges, *Expert Opin. Drug Saf.* **4**, 69-73 (2005).

Crystals from methylene chloride, mp 146-148°. Freely sol in acetone, tetrahydrofuran; sol in methanol, ethyl acetate; sparingly sol in chloroform, dichloromethane. Insol in water, *n*-hexane.

THERAP CAT: Antiparkinsonian.

9669. Tolciclate. [50838-36-3] *N*-Methyl-*N*-(3-methylphenyl)carbamothioic acid *O*-(1,2,3,4-tetrahydro-1,4-methanonaphthalen-6-yl) ester; *O*-(1,4-methano-1,2,3,4-tetrahydro-6-naphthyl)-*N*-methyl-*N*-(*m*-tolyl)thiocarbamate; KC-9147; Fungifos; Kilmicen; Tolmicen. $C_{20}H_{21}NOS$; mol wt 323.45. C 74.27%, H 6.54%, N 4.33%, O 4.95%, S 9.91%. Topical antimycotic agent with high liposolubility. Prepn: P. Melloni *et al.*, **DE 2313845** corresp to **US 3855263** (1973, 1974 both to Carlo Erba). *In vitro* and *in vivo* study: I. deCarneri *et al.*, *Arzneim.-Forsch.* **26**, 769 (1976). Antimycotic studies: A. Bianchi *et al.*, *Antimicrob. Agents Chemother.* **12**, 429 (1977). Clinical studies in dermatomycosis: L. C. Cucé *et al.*, *J. Int. Med. Res.* **8**, 144 (1980); C. Intini *et al.*, *Pharmatherapeutica* **2**, 439 (1980).

White cryst powder from isopropanol, mp 92-94°. Practically insol in water. Soly (mg/ml): 14.9 in *n*-hexane; 23.9 in *n*-octanol. LD_{50} in mice, rats, dogs (mg/kg): 4000, 6000, 5000 orally (deCarneri).

THERAP CAT: Antifungal.

9670. Tolclofos-methyl. [57018-04-9] *O*-(2,6-Dichloro-4-methylphenyl)phosphorothioic acid *O*,*O*-dimethyl ester; *O*-(2,6-dichloro-*p*-tolyl) *O*,*O*-dimethyl phosphorothioate; tolclophos-methyl; S-3349; Rizolex. $C_9H_{11}Cl_2O_3PS$; mol wt 301.12. C 35.90%, H 3.68%, Cl 23.55%, O 15.94%, P 10.29%, S 10.65%. Organophosphorus fungicide; inhibits phospholipid biosynthesis. Prepn: T. Kato *et al.*, **DE 2501040**; **GB 1467561** (1975, 1977 both to Sumitomo). Properties and biological activity: S. Ohtsuki, A. Fujinami, *Jpn. Pestic. Inf.* **41**, 21 (1982). Synthesis: M. Sasaki *et al.*, *J. Pestic. Sci.* **9**, 737 (1984). Mode of action: S. Nakamura, T. Kato, *ibid.* 725; P. Leroux *et al.*, *Pestic. Sci.* **36**, 255 (1992). GC determn in soil, lettuce: M. Gennari *et al.*, *J. AOAC Int.* **80**, 1298 (1997). Control of bottom rot on lettuce: J. R. Coley-Smith *et al.*, *Plant Pathol.* **40**, 359 (1991).

White crystals from methanol, mp 79-79.5°. Vapor pressure at 20°: 4.27×10^{-4} mm Hg. Soly at 23°: water 0.3-0.4 ppm. Easily sol in xylene, acetone, cyclohexanone, chloroform. LD_{50} in male, female rats, male, female mice (mg/kg): ~5000, ~5000, 3500, 3600 orally; all >5000 dermally; ~5000, 4900, 1070, 1260 i.p.; all >5000 s.c. (Ohtsuki, Fujinami).

USE: Agricultural fungicide.

9671. Tolcyclamide. [664-95-9] *N*-[(Cyclohexylamino)carbonyl]-4-methylbenzenesulfonamide; 1-cyclohexyl-3-*p*-tolylsulfonylurea; tolhexamide; glycyclamide; cyclamide; K-386; Diaboral. $C_{14}H_{20}N_2O_3S$; mol wt 296.39. C 56.73%, H 6.80%, N 9.45%, O 16.19%, S 10.82%. Prepn: Logemann, Artini, *Ber.* **90**, 2527 (1957).

Crystals from trichloroethylene, mp 174-176°.

THERAP CAT: Antidiabetic.

9672. Toldimfos Sodium. [575-75-7] *P*-[4-(Dimethylamino)-2-methylphenyl]phosphinic acid sodium salt (1:1); (4-dimethylamino-*o*-tolyl)phosphonous acid sodium salt; sodium (4-dimethylamino-*o*-tolyl)phosphonate; *p*-dimethylamino-*o*-toluenephosphonous acid sodium salt; Foston; Tonofosfan. $C_9H_{13}NNaO_2P$; mol wt 221.17. C 48.88%, H 5.92%, N 6.33%, Na 10.39%, O 14.47%, P 14.00%. Prepd from *N*,*N*-dimethyl-*m*-toluidine and phosphorus trichloride: Benda, Schmidt, **DE 397813** (1924 to Cassella), *Frdl.* **14**, 1409.

Trihydrate. [5787-63-3] Scales, needles, or prisms from alc. Freely sol in cold water, hot alcohol.

THERAP CAT (VET): Phosphorus source.

9673. Tolfenamic Acid. [13710-19-5] 2-[(3-Chloro-2-methylphenyl)amino]benzoic acid; *N*-(3-chloro-*o*-tolyl)anthranilic acid; *N*-(2-methyl-3-chlorophenyl)anthranilic acid; GEA-6414; Clotam; Tolfedine; Tolfine. $C_{14}H_{12}ClNO_2$; mol wt 261.71. C 64.25%, H 4.62%, Cl 13.55%, N 5.35%, O 12.23%. Deriv of anthranilic acid, related structurally to mefenamic and flufenamic acids, *q.q.v.* Prepn: **NL 6600251** (1966 to Gea A/S), *C.A.* **66**, 2377 (1967); R. A. Scherrer, F. W. Short, **US 3313848** (1967 to Parke, Davis). Inhibition of prostaglandin biosynthesis: I. B. Linden *et al.*, *Scand. J. Rheumatol.* **5**, 129 (1976). HPLC determn: F. Nielsen-Kudsk, *Acta Pharmacol. Toxicol.* **47**, 267 (1980). Metabolism: T. Kuninaka *et al.*, *Yakugaku Zasshi* **101**, 232 (1981); *C.A.* **95**, 168 (1981). Human pharmacokinetics: P. Pentikaeinen *et al.*, *Eur. J. Clin. Pharmacol.* **19**, 359 (1981). Pharmacological studies: S. Yamashita *et al.*, *Toho Igakkai Zasshi* **28**, 76-105 (1981), *C.A.* **95**, 16183, 180846 (1981). Clinical study: V. Rejholec *et al.*, *Scand. J. Rheumatol.* **Suppl. 33**, 50 (1980); *ibid.* **Suppl. 36**, 1 (1980).

Crystals from abs ethanol, mp 207-207.5°.
THERAP CAT: Anti-inflammatory; analgesic.
THERAP CAT (VET): Anti-inflammatory.

9674. *o*-**Tolidine.** [119-93-7] 3,3'-Dimethyl-[1,1'-biphenyl]-4,4'-diamine; 3,3'-dimethylbenzidine; 4,4'-diamino-3,3'-dimethyl-biphenyl. $C_{14}H_{16}N_2$; mol wt 212.30. C 79.21%, H 7.60%, N 13.20%. Made by alkaline reduction of *o*-nitrotoluene with zinc, and subsequent rearrangement of the *o*-hydrazotoluene formed, by boiling with HCl: Van Loon, *Chem. Weekbl.* **5**, 689 (1907). *See also* Schultz *et al.*, *Ann.* **352**, 111 (1907). Crystal and molecular structure: Chawdhury *et al.*, *Acta Crystallogr.* **B24**, 1222 (1968). Metabolism: Dieteren, *Arch. Environ. Health* **12**, 30 (1966). Carcinogenic activity: Pliss, Zebenzhinskii, *J. Natl. Cancer Inst.* **45**, 283 (1970).

White to reddish crystals or cryst powder. mp 129-131°. Slightly sol in water; sol in alcohol, ether, dil acids. *Keep well closed and protected from light.*
Sulfate. [531-20-4] $C_{14}H_{16}N_2 \cdot H_2SO_4$; mol wt 310.37. White to gray mass. Slightly sol in water, alcohol; sol in dil acids.
Dihydrochloride. [612-82-8] $C_{14}H_{16}N_2 \cdot 2HCl$; mol wt 285.21.
Caution: Potential symptoms of overexposure to *o*-tolidine are irritation of eyes and nose. *See NIOSH Pocket Guide to Chemical Hazards* (DHHS/NIOSH 97-140, 1997) p 310. *o*-Tolidine is reasonably anticipated to be a human carcinogen: *Report on Carcinogens, Twelfth Edition* (PB2011-111646, 2011) p 168.

USE: Manuf dyes; also as very sensitive reagent for gold (1:10 million detectable), and for free chlorine in water.

9675. **Tolindate.** [27877-51-6] *N*-Methyl-*N*-(3-methylphenyl)carbamothioic acid *O*-(2,3-dihydro-1*H*-inden-5-yl) ester; *m*,*N*-dimethylthiocarbanilic acid *O*-5-indanyl ester; *O*-(5-indanyl) *m*,*N*-dimethylthiocarbanilate; Dalnate. $C_{18}H_{19}NOS$; mol wt 297.42. C 72.69%, H 6.44%, N 4.71%, O 5.38%, S 10.78%. Prepared by treating 5-indanyl thionochloroformate with *N*-methyl-*m*-toluidine: Elpern, Youlus, *US 3509200* (1970 to USV).

Crystals, mp 94-95°.
THERAP CAT: Antifungal.

9676. **Tollens Reagent.** A solution prepd from equal amounts of 10% silver nitrate and 10% sodium hydroxide solutions to which enough dilute ammonia solution has been added to dissolve the precipitated silver oxide. Tollens reagent oxidizes aldehydes to the corresponding acids; during the reaction the silver, bound in form of a complex, is reduced to metallic silver and forms a characteristic silver mirror. *Refs:* B. Tollens, *Ber.* **15**, 1635 (1882); W. Ponndorf, *ibid.* **64B**, 1913 (1937); S. Siggia, E. Segel, *Anal. Chem.* **25**, 640 (1953); J. M. Kolthoff, P. J. Elving, *Treatise on Analytical Chemistry* vol. **13** (New York, 1966) p 183.
Caution: Tollens reagent should always be prepared freshly; old, opaque or "dried out" solutions are explosive: H. Waldmann, *Chimia* **13**, 297 (1959).
USE: Reagent in characterization of sugars, aldehydes, hydrazides. As oxidizing agent.

9677. **Tolmetin.** [26171-23-3] 1-Methyl-5-(4-methylbenzoyl)-1*H*-pyrrole-2-acetic acid; 1-methyl-5-*p*-toluoylpyrrole-2-acetic acid; 5-(*p*-toluoyl)-1-methylpyrrole-2-acetic acid; McN-2559. $C_{15}H_{15}NO_3$; mol wt 257.29. C 70.02%, H 5.88%, N 5.44%, O 18.65%. Prepn: Carson, *FR 1574570* (1969 to McNeil Labs.), *C.A.* **72**, 100498y (1969). Pharmacology: Carson *et al.*, *J. Med. Chem.* **14**, 646 (1971); S. Wong *et al.*, *J. Pharmacol. Exp. Ther.* **185**, 127 (1973). *Review:* S. Wong in *Pharmacological and Biochemical*

Properties of Drug Substances vol. **1**, M. E. Goldberg, Ed. (Am. Pharm. Assoc., Washington, DC, 1977) pp 233-255. Review of pharmacology and therapeutic efficacy: R. N. Brogden *et al.*, *Drugs* **15**, 429-450 (1978).

Crystals from acetonitrile, mp 155-157° (dec).
Sodium salt dihydrate. [64490-92-2] McN-2559-21-98; Reutol; Tolectin; Tolmene. $C_{15}H_{14}NNaO_3 \cdot 2H_2O$; mol wt 315.30. Light yellow to light orange, crystalline powder. Freely sol in water, methanol; slightly sol in alc; very slightly sol in chloroform.
THERAP CAT: Anti-inflammatory.

9678. **Tolnaftate.** [2398-96-1] *N*-Methyl-*N*-(3-methylphenyl)carbamothioic acid *O*-2-naphthalenyl ester; *m*,*N*-dimethylthiocarbanilic acid *O*-2-naphthyl ester; *O*-2-naphthyl *m*,*N*-dimethylthiocarbanilate; 2-naphthyl *N*-methyl-*N*-(3-tolyl)thionocarbamate; Sch-10144; Aftate; Chinofungin; Fungistop; Hi-Alarzin; Sporiline; Timoped; Tinactin; Tinaderm; Tonoftal. $C_{19}H_{17}NOS$; mol wt 307.41. C 74.24%, H 5.57%, N 4.56%, O 5.20%, S 10.43%. Prepn: **FR 1337797**; Miyazaki *et al.*, *US 3334126* (1963, 1967 to Japan Soda); Noguchi *et al.*, *J. Pharm. Soc. Jpn.* **88**, 335 (1968). Pharmacology and toxicology: Noguchi *et al.*, *Antimicrob. Agents Chemother.* **1962**, 259; Hashimoto *et al.*, *Toxicol. Appl. Pharmacol.* **8**, 380 (1966); Noguchi *et al.*, *ibid.* 368. Clinical comparison with undecylenic acid: J. F. Fuerst *et al.*, *Cutis* **25**, 544 (1980); F. Battistini *et al.*, *Int. J. Dermatol.* **22**, 388 (1983). Mode of action study: M. P. Gupta *et al.*, *J. Vet. Med. Mycol.* **29**, 45 (1991). LC determn in pharmaceuticals: A. K. Dash, *J. Pharm. Biomed. Anal.* **11**, 847 (1993). Comprehensive description: *idem*, *Anal. Profiles Drug Subs. Excip.* **23**, 543-570 (1994).

Crystals from alcohol, mp 110.5-111.5°. Freely sol in acetone, chloroform; sol in CCl₄ (1:9); sparingly sol in methanol, ether; slightly sol in alc. Practically insol in water. uv max (methanol): 258, 222 nm. LD_{50} in mice, rats (g/kg): >10, >6 orally; >6, >4 s.c. (Hashimoto).
THERAP CAT: Antifungal.
THERAP CAT (VET): Antifungal.

9679. **Tolonium Chloride.** [92-31-9] 3-Amino-7-(dimethylamino)-2-methylphenothiazin-5-ium chloride (1:1); 3-amino-7-dimethylamino-2-methylphenazathionium chloride; toluidine blue O; dimethyltoluthionine chloride; C.I. Basic Blue 17; C.I. 52040; Blutene; Klot; Tolazul. $C_{15}H_{16}ClN_3S$; mol wt 305.82. C 58.91%, H 5.27%, Cl 11.59%, N 13.74%, S 10.48%. Prepd from dimethyl-*p*-phenylenediamine, sodium thiosulfate, and *o*-toluidine: Dändliker, Bernthsen, **US 416055** (1888 to BASF). Prepn of hemostatic compositions contg tolonium chloride: D. A. Hoff, **US 2809913** (1957 to Warren-Teed). Prepn of clear, colorless, stable, isotonic solns of purified leucotoluidine blue O by reducing toluidine blue O with sodium hydrosulfite at pH 2.5-3.5: B. March, E. E. Moore, **US 2571593** (1951 to Abbott). Clinical studies in bleeding disorders: J. Allen *et al.*, *Surg. Gynecol. Obstet.* **89**, 692 (1949). Clinical use for parathyroid identification during thyroidectomy: R. M. Yeager, E. T. Krementz, *Ann. Surg.* **169**, 829 (1969). Acute and chronic toxicity study: T. J. Haley, F. Stolarsky, *Stanford Med. Bull.* **9**, 96 (1951). Review of therapeutic and diagnostic use: A. Mashberg, *J. Am. Dent. Assoc.* **106**, 319-323 (1983). *See also Colour Index* vol. **4** (3rd ed., 1971) p 4471; *H. J. Conn's Biological Stains*, R. D. Lillie, Ed. (Williams & Wilkins, Baltimore, 9th ed., 1977) p 428.

Dark green powder. Sol in water (3.82 g/100 ml), giving a blue to violet soln; sol in alc (0.57 g/100 ml), giving a blue soln. Absorption max (water): 640.4 nm. LD_{50} in mice, rats, rabbits (mg/kg): 27.56, 28.93, 13.44 i.v. (Haley, Storlarsky).

USE: Direct dyeing, printing of wool, silk. Biological stain.

THERAP CAT: Hemostatic. Diagnostic aid (oral carcinoma).

9680. Toloxatone. [29218-27-7] 5-(Hydroxymethyl)-3-(3-methylphenyl)-2-oxazolidinone; 5-(hydroxymethyl)-3-*m*-tolyl-2-oxazolidinone; MD-69276; Humoryl; Perenum. $C_{11}H_{13}NO_3$; mol wt 207.23. C 63.76%, H 6.32%, N 6.76%, O 23.16%. Reversible monoamine oxidase type A inhibitor. Prepn: C. Fauran et al., **DE 2012120**; eidem, **US 3655687** (1970, 1972 both to Delalande); eidem, Chim. Ther. **8**, 324 (1973). Pharmacology: G. Raynaud et al., ibid. 328. Psychopharmacological profile: J.-P. Kan et al., Eur. J. Med. Chem. **12**, 13 (1977); H. Giono-Barber et al., Arzneim.-Forsch. **27**, 1188 (1977). Pharmacokinetics: M. S. Benedetti et al., ibid. **32**, 276 (1982). Metabolism: A. Malnoe, M. S. Benedetti, Xenobiotica **9**, 281 (1979). GLC determn in plasma: S. Vajta et al., J. Chromatogr. **274**, 139 (1983). Clinical evaluation in panic disorder: G. Perna et al., J. Clin. Psychopharmacol. **14**, 414 (1994).

Crystals from isopropyl alcohol, mp 76°. LD_{50} orally in mice (mg/kg): 1850 (Fauran); also reported as 1500 (Raynaud).

THERAP CAT: Antidepressant.

9681. Tolperisone. [728-88-1] 2-Methyl-1-(4-methylphenyl)-3-(1-piperidinyl)-1-propanone; 2,4′-dimethyl-3-piperidinopropiophenone; 1-piperidino-2-methyl-3-(*p*-tolyl)-3-propanone; 2-methyl-3-piperidino-1-*p*-tolylpropan-1-one; mydetone. $C_{16}H_{23}NO$; mol wt 245.37. C 78.32%, H 9.45%, N 5.71%, O 6.52%. Prepn: Nádor et al., **HU 144997** (1956); Yokoyama et al., **JP 65 20390** (1965 to Eisai). Pharmacology and toxicity: J. Porszasz et al., Acta Physiol. Acad. Sci. Hung. **18**, 149 (1960); eidem, Arzneim.-Forsch. **11**, 257 (1961).

Hydrochloride. [3644-61-9] N-553; Abbsa; Atmosgen; Arantoick; Besnoline; Isocalm; Kineorl; Menopatol; Metosomin; Minacalm; Muscalm; Mydocalm; Naismeritin; Tolisartine. $C_{16}H_{23}NO\cdot HCl$; mol wt 281.82. Crystals from methyl ethyl ketone, mp 176-177°. LD_{50} s.c. in mice: 620 mg/kg (Porszasz, 1961).

THERAP CAT: Muscle relaxant (skeletal).

9682. Tolpropamine. [5632-44-0] N,N,4-Trimethyl-γ-phenylbenzenepropanamine; N,N-dimethyl-3-phenyl-3-*p*-tolylpropylamine; 3-dimethylamino-1-phenyl-1-*p*-tolylpropane. $C_{18}H_{23}N$; mol wt 253.39. C 85.32%, H 9.15%, N 5.53%. Prepn: Bockmühl, Stein, **DE 925468** (1955 to Hoechst); Klosa, J. Prakt. Chem. **34**, 312 (1966). Pharmacology: Sendrail, Gleizes, Therapie **15**, 119 (1960).

Hydrochloride. Pragman Gelee. $C_{18}H_{23}N\cdot HCl$; mol wt 289.85. mp 182-184°.

THERAP CAT: Topical antihistaminic, antipruritic.

9683. Tolrestat. [82964-04-3] N-[[6-Methoxy-5-(trifluoromethyl)-1-naphthalenyl]thioxomethyl]-N-methylglycine; tolrestatin; AY-27773; Alredase; Lorestat. $C_{16}H_{14}F_3NO_3S$; mol wt 357.35. C 53.78%, H 3.95%, F 15.95%, N 3.92%, O 13.43%, S 8.97%. Orally active aldose reductase inhibitor. Prepn and pharmacology: K. Sestanj et al., **EP 59596** (1982 to Ayerst); eidem, J. Med. Chem. **27**, 255 (1984). Prevention of cataracts in galactosemic rats: N. Simard-Duquesne et al., Proc. Soc. Exp. Biol. Med. **178**, 599 (1985). Clinical pharmacokinetics and metabolism: D. R. Hicks et al., Clin. Pharmacol. Ther. **36**, 493 (1984). Ultraviolet and HPLC determn in serum: D. R. Hicks, M. Kraml, Ther. Drug Monit. **6**, 328 (1984). Effect on erythrocyte sorbitol levels in diabetic patients: P. Raskin et al., Clin. Pharmacol. Ther. **38**, 625 (1985).

mp 164-165°.

Methyl ester. $C_{17}H_{16}F_3NO_3S$. mp 109-110°.

THERAP CAT: Treatment of diabetic neuropathy.

9684. Tolterodine. [124937-51-5] 2-[(1*R*)-3-[Bis(1-methylethyl)amino]-1-phenylpropyl]-4-methylphenol; (+)-(*R*)-2-[α-[2-(diisopropylamino)ethyl]benzyl]-*p*-cresol; (+)-N,N-diisopropyl-3-(2-hydroxy-5-methylphenyl)-3-phenylpropylamine; Kabi 2234. $C_{22}H_{31}NO$; mol wt 325.50. C 81.18%, H 9.60%, N 4.30%, O 4.92%. Muscarinic receptor antagonist. Prepn: N. A. Jönsson et al., **EP 325571** (1989 to KabiVitrum); eidem, **US 5382600** (1995 to Pharmacia). Asymmetric total synthesis: C. Hedberg, P. G. Andersson, Adv. Synth. Catal. **347**, 662 (2005). Pharmacology: L. Nilvebrant et al., Life Sci. **60**, 1129 (1997). Receptor binding study: idem et al., Eur. J. Pharmacol. **327**, 195 (1997). GC-MS determn in biological fluids: L. Palmér et al., J. Pharm. Biomed. Anal. **16**, 155 (1997). Clinical pharmacokinetics: N. Brynne et al., Int. J. Clin. Pharmacol. Ther. **35**, 287 (1997). Review of clinical trials: R. A. Appell, Urology **50**, Suppl. 6A, 90-96 (1997); of use in overactive bladder: E. S. Rovner, Expert Opin. Pharmacother. **6**, 653-666 (2005).

$[\alpha]_D^{25}$ +72° (c = 1.0 in CH_2Cl_2).

Tartrate. [124937-52-6] PNU-200583E; Detrol; Detrusitol. $C_{22}H_{31}NO\cdot C_4H_6O_6$; mol wt 475.58. Crystals from ethanol. $[\alpha]_{546}^{25}$ +36.0°. pKa 9.87. Soly in water: 12 mg/ml. Sol in methanol; slightly sol in ethanol. Practically insol in toluene. Partition coefficient (*n*-octanol/water): 1.83 (pH 7.3). LD_{50} i.v. in male mice: 10-20 mg/kg (Jönsson).

THERAP CAT: In treatment of urinary incontinence.

9685. Toltrazuril. [69004-03-1] 1-Methyl-3-[3-methyl-4-[4-[(trifluoromethyl)thio]phenoxy]phenyl]-1,3,5-triazine-2,4,6(1*H*,3*H*,5*H*)-trione; 1-methyl-3-[4-[*p*-[(trifluoromethyl)thio]phenoxy]-*m*-tolyl]-*s*-triazine-2,4,6(1*H*,3*H*,5*H*)-trione; Bay Vi 9142; Baycox. $C_{18}H_{14}F_3N_3O_4S$; mol wt 425.38. C 50.82%, H 3.32%, F 13.40%, N 9.88%, O 15.04%, S 7.54%. Triazinetrione anticoccidial. General prepn: **BE 826900**; J. H. Reisdorff et al., **US 3966725** (1975, 1976 both to Bayer). Prepn and use as animal growth promotant: **BE**

866389; A. Haberkorn *et al.*, **US 4219552** (1978, 1980 both to Bayer). Series of articles on efficacy vs coccidia in chickens: E. Kutzer *et al.*, *Wien. Tieraerztl. Monatssch.* **72**, 321-340 (1985). Field trial in sheep: B. Gjerde, O. Helle, *Acta Vet. Scand.* **27**, 124 (1986); in chickens: H. D. Chapman, *J. Comp. Pathol.* **97**, 21 (1987).

mp 194°.

THERAP CAT (VET): Coccidiostat.

9686. *o*-Tolualdehyde. [529-20-4] 2-Methylbenzaldehyde; *o*-toluylaldehyde. C_8H_8O; mol wt 120.15. C 79.97%, H 6.71%, O 13.32%. Prepd by reacting nitropropane with *o*-xylyl bromide in the presence of sodium ethanoate: Hass, Bender, *Org. Synth.* **30**, 99 (1950).

Liquid. d_4^{19} 1.0386. bp$_{760}$ 200-202°; bp$_{15}$ 94-96°; bp$_6$ 68-72°. n_D^{25} 1.5430; n_D^{19} 1.549; n_α^{19} 1.5423; n_β^{19} 1.5650; n_γ^{19} 1.5798.

9687. *o*-Toluamide. [527-85-5] 2-Methylbenzamide. C_8H_9-NO; mol wt 135.17. C 71.09%, H 6.71%, N 10.36%, O 11.84%. Prepd by reacting *o*-toluinitrile, hydrogen peroxide, 95% alcohol, and sodium hydroxide at 40-50°: Noller, *Org. Synth.* **coll. vol. II**, 586 (1943). Also prepared by reacting the nitrile with boron fluoride in dil acetic acid: Hauser, Hoffenberg, *J. Org. Chem.* **20**, 1448 (1955).

Crystals from water, mp 144-145°. Very sol in alcohol, hot water, concd HCl, less sol in ether. Sparingly sol in benzene. Practically insol in cold water. *Explosive. Keep away from open flame.*

9688. **Toluene.** [108-88-3] Methylbenzene; toluol; phenylmethane; Methacide. C_7H_8; mol wt 92.14. C 91.25%, H 8.75%. Obtained mainly from tar oil. Review of mfg processes: *Faith, Keyes & Clark's Industrial Chemicals*, F. A. Lowenheim, M. K. Moran, Eds. (Wiley-Interscience, New York, 4th ed., 1975) pp 822-830. Solubility: F. P. Schwarz, *Anal. Chem.* **52**, 10 (1980). Myelotoxic potential: L. Greenburg *et al.*, *J. Am. Med. Assoc.* **118**, 573 (1942). Comparison with benzene of effects on hematopoiesis and bone marrow metabolism: H. W. Gerarde, *AMA Arch. Ind. Health* **13**, 468 (1956). Acute toxicity: H. F. Smyth *et al.*, *Am. Ind. Hyg. Assoc. J.* **30**, 470 (1969). Evaluation of chronic occupational exposure: H. Tahti *et al.*, *Int. Arch. Occup. Environ. Health* **48**, 61 (1981). *Review:* M. C. Hoff in *Kirk-Othmer Encyclopedia of Chemical Technology* vol. 23 (Wiley-Interscience, New York, 3rd ed., 1983) pp 246-273. Review of reproductive toxicity: J. M. Donald *et al.*, *Environ. Health Perspect.* **94**, 237-244 (1991); of toxicology and human exposure: *Toxicological Profile for Toluene* (PB2000-108028, 2000) 357 pp.

Refractive liq; benzene-like odor. *Flammable.* d_4^{20} 0.866. mp −95°. bp 110.6°. n_D^{20} 1.4967. Flash pt, closed cup: 40°F (4.4°C). Soly in water at 23.5°C (w/w): 0.067%. Very slightly sol in water; misc with alc, chloroform, ether, acetone, glacial acetic acid, carbon disulfide. LD_{50} orally in rats: 7.53 g/kg (Smyth).

Caution: Readily absorbed by inhalation, ingestion and somewhat by skin contact. Direct contact may cause severe dermatitis due to drying and defatting action. May present lung aspiration hazard if ingested. Potential symptoms of acute overexposure by inhalation may include local irritation; CNS excitation and depression. Low concentrations may result in transitory mild upper respiratory tract irritation, mild eye irritation, lacrimation, metallic taste, slight nausea, hilarity, lassitude, drowsiness and impaired balance. High concentrations may cause paresthesia, vision disturbances, dizziness, nausea, headache, narcosis and collapse; death from respiratory failure or sudden ventricular fibrillation. Chronic overexposure by inhalation has been associated with hepatotoxicity and nephrotoxicity. Syndromes following chronic inhalation involve severe muscle weakness, cardiac arrhythmias, gastrointestinal and respiratory complaints. *See Patty's Industrial Hygiene and Toxicology* **vol. 2B**, G. D. Clayton, F. E. Clayton, Eds. (Wiley-Interscience, New York, 4th ed., 1994) pp 1326-1332; *Clinical Toxicology of Commercial Products*, R. E. Gosselin *et al.*, Eds. (Williams & Wilkins, Baltimore, 5th ed., 1984) Section II, p 153, Section III, p 397-404.

USE: In manuf benzoic acid, benzaldehyde, explosives, dyes, and many other organic compds; as a solvent for paints, lacquers, gums, resins; thinner for inks, perfumes, dyes; in the extraction of various principles from plants; as gasoline additive.

9689. **Toluene 2,4-Diisocyanate.** [584-84-9] 2,4-Diisocyanato-1-methylbenzene; 2,4-diisocyanatotoluene; 2,4-tolylene diisocyanate; TDI; Nacconate 100. $C_9H_6N_2O_2$; mol wt 174.16. C 62.07%, H 3.47%, N 16.09%, O 18.37%. Usually prepd from toluene-2,4-diamine and phosgene. *Review:* Astle, *Industrial Organic Nitrogen Compounds* (New York, 1961) pp 284-313; *Faith, Keyes & Clark's Industrial Chemicals*, F. A. Lowenheim, M. K. Moran, Eds. (Wiley-Interscience, New York, 4th ed., 1975) pp 831-835.

Liquid at room temperature. Sharp, pungent odor. mp 19.5-21.5°. d_4^{20} liq 1.2244. bp$_{760}$ 251°; bp$_{11}$ 126°. Darkens on exposure to sunlight. Reacts with water with evolution of carbon dioxide. Flash pt, open cup: 132°C (270°F). Misc with alcohol (decompn), diglycol monomethyl ether, ether, acetone, carbon tetrachloride, benzene, chlorobenzene, kerosene, olive oil. Concd alkaline compds such as NaOH or *tert*-amines may cause run-away polymerization.

Caution: Potential symptoms of overexposure are irritation of eyes, skin, nose and throat; choking, paroxysmal cough; chest pain, retrosternal soreness; nausea, vomiting and abdominal pain; bronchitis, bronchospasm, pulmonary edema; dyspnea, asthma; conjunctivitis, lacrimation; dermatitis, skin sensitization. *See NIOSH Pocket Guide to Chemical Hazards* (DHHS/NIOSH 97-140, 1997) p 312. *See also Clinical Toxicology of Commercial Products*, R. E. Gosselin *et al.*, Eds. (Williams & Wilkins, Baltimore, 5th ed., 1984) Section II, p 414. This substance is reasonably anticipated to be a human carcinogen: *Report on Carcinogens, Twelfth Edition* (PB2011-111646, 2011) p 414.

USE: In the manuf of polyurethane foams and other elastomers.

9690. **Toluene-3,4-dithiol.** [496-74-2] 1,2-Dimercapto-4-methylbenzene; "dithiol". $C_7H_8S_2$; mol wt 156.26. C 53.81%, H 5.16%, S 41.03%. Prepd from toluene-3,4-disulfonyl chloride with tin and hydrochloric acid: Mills, Clark, *J. Chem. Soc.* **1936**, 178.

Crystals, mp 31°. bp$_{84}$ 185-187°. Sol in benzene, in aq alkali hydroxide solns.

USE: For the detection of bismuth, molybdenum, rhenium, tin, tungsten, *see*: Bickford *et al., J. Am. Pharm. Assoc. Sci. Ed.* **37**, 255 (1948).

9691. *p*-Toluenesulfinic Acid. [536-57-2] 4-Methylbenzenesulfinic acid. C$_7$H$_8$O$_2$S; mol wt 156.20. C 53.83%, H 5.16%, O 20.49%, S 20.52%. Prepd by reduction of *p*-toluenesulfonyl chloride with zinc dust: Whitmore, Hamilton, *Org. Synth.* **2**, 89 (1922). Because the sulfinic acid is difficult to dry without partial conversion to the sulfonic acid, the sodium salt, CH$_3$C$_6$H$_4$SO$_2$Na.2H$_2$O, is usually prepd. The free sulfinic acid is then obtained as needed by dissolving the sodium salt in cold water and carefully acidifying the soln with the exact amt of HCl needed.

Long, rhombic plates or needles from water. mp 85°. Freely sol in alc, ether; sparingly sol in water, hot benzene.

9692.   *p*-Toluenesulfonic Acid. [104-15-4] 4-Methylbenzenesulfonic acid; tosic acid. C$_7$H$_8$O$_3$S; mol wt 172.20. C 48.83%, H 4.68%, O 27.87%, S 18.62%. Prepd by sulfonation of toluene with H$_2$SO$_4$. Convenient lab prepn: L. F. Fieser, *Experiments in Organic Chemistry* (Boston, 3rd ed., 1955) p 144. The separation of toluene from petroleum fractions can be accomplished by sulfonation with H$_2$SO$_4$ at 60°.

Monoclinic leaflets or prisms. Also reported as crystallizing with 1H$_2$O or 4H$_2$O. When anhydrous, mp 106-107°. Metastable form, mp 38°. bp$_{20}$ 140°. bp$_{0.1}$ 185-187°. Freely sol in water, about 67 g/ 100 ml. Sol in alc and ether.
Sodium salt. [6263-41-8] C$_7$H$_7$NaO$_3$S; mol wt 194.18. Orthorhombic plates, very sol in water.
Monohydrate. [6192-52-5] C$_7$H$_8$O$_3$S.H$_2$O; mol wt 190.21. Freely sol in water.
Caution: Highly irritating to skin, mucous membranes.
USE: In dye chemistry; in manuf of oral antidiabetic drugs. Derivatizing agent.

9693.   *p*-Toluenesulfonyl Chloride. [98-59-9] 4-Methylbenzenesulfonyl chloride; tosyl chloride. C$_7$H$_7$ClO$_2$S; mol wt 190.64. C 44.10%, H 3.70%, Cl 18.60%, O 16.78%, S 16.82%. Made by treating toluene with chlorosulfonic acid.

Crystals, mp 69-71°. bp$_{15}$ 146°. Insol in water; freely sol in alcohol, benzene, ether.

9694.   *p*-Toluenesulfonylhydrazide. [1576-35-8] 4-Methylbenzenesulfonic acid hydrazide; *p*-toluenesulfonic acid hydrazide; *p*-toluenesulfonic hydrazide; 4-toluenesulfonylhydrazine; tosylhydrazide; tosylhydrazine; TsNHNH$_2$. C$_7$H$_{10}$N$_2$O$_2$S; mol wt 186.23. C 45.15%, H 5.41%, N 15.04%, O 17.18%, S 17.22%. Reagent utilized in the prepn of tosylhydrazone intermediates, useful synthetic precursors to other functional groups. Prepn: K. Freudenberg, F. Blümmel, *Ann.* **440**, 45 (1924); A. Albert, R. Royer, *J. Chem. Soc.* **1949**, 1148; L. Friedman *et al., Org. Synth.* **coll. vol. V**, 1055 (1973). Crystal structure: P. Lightfoot *et al., J. Chem. Soc. Perkin Trans. 2*

1993, 1625. Synthetic applications in the prepn of tosylhydrazones from aldehydes and ketones: S. H. Bertz, G. Dabbagh, *J. Org. Chem.* **48**, 116 (1983); from keto esters: P. Vinczer *et al., Synth. Commun.* **14**, 281 (1984); from nitriles: M. Tóth, L. Somsák, *Tetrahedron Lett.* **42**, 2723 (2001).

Fluffy white crystalline needles, mp 109-110°. *Flammable, poisonous.* Flash pt, closed cup: 140°F (60°C). Sol in most organic solvents. Insol in water, hydrocarbons.
USE: Reagent in synthetic organic chemistry; scavenger of carbonyl compds when bonded to silica.

9695.   Toluic Acid. Methylbenzoic acid. C$_8$H$_8$O$_2$; mol wt 136.15. C 70.58%, H 5.92%, O 23.50%. Prepn of *m*- and *o*-forms by oxidation of corresponding xylene: Toland, **US 2903480** (1959 to California Res. Corp.); Hay *et al., J. Org. Chem.* **25**, 616 (1960). Prepn of *p*-form by reaction of *p*-tolyldiazonium tetrafluoroborate with nickel carbonyl and acetic acid: Clark, Cookson, *J. Chem. Soc.* **1962**, 686; by oxidation of *p*-xylene: Taves, **US 3030413** (1962 to Hercules Powder). Manuf of *p*-form from toluene: Braunworth, **US 3046305** (1962 to Pure Oil).

***m*-Toluic acid.** Prisms from water, mp 111-113°, bp 263°. Sublimes. Sol in 1170 parts water at 15°, 60 parts boiling water; very sol in alcohol, ether.
***o*-Toluic acid.** Crystals, mp 107-108°; bp 258-260°; volatile with steam. Slightly sol in cold water; sol in 35 parts boiling water; very sol in alcohol.
***p*-Toluic acid.** Crystals, mp 180-181°; bp 274-275°. Sparingly sol in hot water; very sol in alcohol, ether, methanol.

9696.   Toluidine. C$_7$H$_9$N; mol wt 107.16. C 78.46%, H 8.47%, N 13.07%. Prepn: J. S. Muspratt, A. W. Hofmann, *Ann.* **54**, 1 (1845); of each isomer: F. Beilstein, A. Kuhlberg, *ibid.* **156**, 66 (1870); P. Kovacic, J. L. Foote, *J. Am. Chem. Soc.* **83**, 743 (1961); P. Kovacic *et al., ibid.* **84**, 759 (1962). Toxicity data: H. F. Smyth, *Am. Ind. Hyg. Assoc. J.* **23**, 95 (1962). GC determn in urine: K. El-Bayoumy *et al., Cancer Res.* **46**, 6064 (1986).

***m*-Toluidine.** [108-44-1] 3-Methylbenzamine; 3-aminotoluene; 3-methylaniline. Liquid, mp ~ −50°. bp 203-204°. d$_{25}^{25}$ 0.990. n$_D^{22}$ 1.5711. Slightly sol in water; sol in alcohol, ether, dil acids.
***o*-Toluidine.** [95-53-4] 2-Methylbenzamine; 2-aminotoluene; 2-methylaniline. Light yellow liquid becoming reddish brown on exposure to air and light. bp 200-202°. d$_{20}^{20}$ 1.008. n$_D^{20}$ 1.5688. Flash pt, closed cup: 185°F (85°C). Slightly sol in water; sol in alcohol, ether, dil acids. Keep well closed and protected from light. LD$_{50}$ orally in rats: 0.94 g/kg (Smyth).
***p*-Toluidine.** [106-49-0] 4-Methylbenzamine; 4-aminotoluene; 4-methylaniline. Lustrous plates or leaflets, mp 44-45°. bp 200-201°. d$_{20}^4$ 1.046. n$_D^{59}$ 1.5532. Flash pt, closed cup: 188°F (86°C). Sol in about 135 parts water; freely sol in alcohol, ether, acetone, methanol, carbon disulfide, oils, dil acids.
Caution: Potential symptoms of overexposure to *o*-toluidine are eye irritation; anoxia, headache, cyanosis; weakness, dizziness, drowsiness; microhematuria; eye burns; dermatitis. Potential symptoms of overexposure to *m*- or *p*-toluidine are eye, skin irritation;

hematuria, methemoglobinemia; cyanosis, nausea, vomiting, low blood pressure, convulsions; anemia, weakness. *p*-Toluidine is a potential occupational carcinogen. *See NIOSH Pocket Guide to Chemical Hazards* (DHHS/NIOSH 97-140, 1997) p 312. *o*-Toluidine and its hydrochloride are reasonably anticipated to be human carcinogens: *Report on Carcinogens, Twelfth Edition* (PB2011-111646, 2011) p 416.

USE: Manufacture of various dyes and other organic chemicals. *o*-Isomer also in printing textiles blue black; making colors fast to acids. *p*-Isomer also as a reagent for lignin, nitrite, phloroglucinol.

9697. *o*-Tolunitrile. [529-19-1] 2-Methylbenzonitrile; 2-methylbenzenecarbonitrile; *o*-cyanotoluene; *o*-methylbenzonitrile. C_8H_7N; mol wt 117.15. C 82.02%, H 6.02%, N 11.96%. Prepd from *o*-toluidine by diazotization in HCl soln and treatment of the diazonium chloride with potassium cuprocyanide: Herb, *Ann.* **258**, 9 (1890); Clarke, Read, *Org. Synth.* coll. vol. I (2nd ed., 1941) p 514. Absorption spectrum: Baly, Ewbank, *J. Chem. Soc.* **87**, 1357 (1905); Purvis, *ibid.* **107**, 503 (1915).

Liquid; d_4^{20} 0.9955; d_4^{45} 0.9737; d_4^{75} 0.9481; mp −13°; bp_{760} 205.2°; bp_{100} 135°; bp_{40} 110°; bp_{20} 93°; bp_{10} 77.9°; bp_5 64°; $bp_{1.0}$ 36.7°; n_D^{23} 1.52720. Insol in water. Miscible with alc, ether.

9698. *p*-Tolunitrile. [104-85-8] 4-Methylbenzonitrile. C_8-H_7N; mol wt 117.15. C 82.02%, H 6.02%, N 11.96%. Prepd from *p*-toluidine in the manner described for *o*-tolunitrile.

Needles from alc. d_4^{30} 0.9785; d_4^{45} 0.9640; d_4^{60} 0.9512; d_4^{75} 0.9390; mp 29.5°; bp_{760} 217.6°; bp_{100} 145.2°; bp_{60} 130°; bp_{40} 109.5°; bp_{20} 101.7°; bp_{10} 85.8°; bp_5 71.3°; $bp_{1.0}$ 42.5°. Absorption spectrum: Baly, Ewbank, *J. Chem. Soc.* **87**, 1357 (1905); Purvis, *ibid.* **107**, 503 (1915). Insol in water. Very sol in alcohol, ether.

9699. Tolvaptan. [150683-30-0] *N*-[4-[(7-Chloro-2,3,4,5-tetrahydro-5-hydroxy-1*H*-1-benzazepin-1-yl)carbonyl]-3-methylphenyl]-2-methylbenzamide; 7-chloro-5-hydroxy-1-[2-methyl-4-(2-methylbenzoylamino)benzoyl]-2,3,4,5-tetrahydro-1*H*-1-benzazepine; OPC-41061; Samsca. $C_{26}H_{25}ClN_2O_3$; mol wt 448.95. C 69.56%, H 5.61%, Cl 7.90%, N 6.24%, O 10.69%. Nonpeptide arginine vasopressin V_2-receptor antagonist. Prepn: H. Ogawa *et al.*, **WO 9105549**; *eidem*, **US 5258510** (1991, 1993 both to Otsuka); K. Kondo *et al.*, *Bioorg. Med. Chem.* **7**, 1743 (1999). Pharmacology: Y. Yamamura *et al.*, *J. Pharmacol. Exp. Ther.* **287**, 860 (1998). Clinical trial in heart failure: M. Gheorghiade *et al.*, *J. Am. Med. Assoc.* **291**, 1963 (2004); in hyponatremia: R. W. Schrier *et al.*, *N. Engl. J. Med.* **355**, 2099 (2006). Review of pharmacology and clinical experience in congestive heart failure: A. Ambrosy *et al.*, *Expert Opin. Pharmacother.* **12**, 961-976 (2011).

Colorless prisms, mp 225.9°.
THERAP CAT: In treatment of congestive heart failure.

9700. Tolycaine. [3686-58-6] 2-[[2-(Diethylamino)acetyl]-amino]-3-methylbenzoic acid methyl ester; 2-(2-diethylaminoacetamido)-*m*-toluic acid methyl ester; 2-methyl-6-carbomethoxy-*N*-diethylaminoacetanilide; methyl 2-diethylaminoacetamido-*m*-toluate; 3-methyl-2-diethylaminoacetylaminobenzoic acid methyl ester. $C_{15}H_{22}N_2O_3$; mol wt 278.35. C 64.73%, H 7.97%, N 10.06%, O 17.24%. Prepn: Hiltmann *et al.*, **DE 1018070** (1957 to Bayer); **US 2921077** (1960 to Schenley).

Oil, bp_5 190-192°.
Hydrochloride. [7210-92-6] Baycain. $C_{15}H_{22}N_2O_3 \cdot HCl$; mol wt 314.81. Crystals, mp 139-140.5°.
THERAP CAT: Anesthetic (local).

9701. Tolylhydrazine. $C_7H_{10}N_2$; mol wt 122.17. C 68.82%, H 8.25%, N 22.93%. Prepd by stannous chloride reduction of the diazonium salt of the corresponding toluidine: Hunsberger *et al.*, *J. Org. Chem.* **21**, 394 (1956).

m-Tolylhydrazine. Oily liquid. bp_{760} 243°, bp_{16} 132-134°. d_{15}^{15} 1.061-1.062, d_4^{20} 1.057-1.058. Insol in water; sol in alc, chloroform, ether.
Nitrate. $C_7H_{10}N_2 \cdot HNO_3$. Needles, mp 145-147°.
o-Tolylhydrazine. Needles, mp 56-59°. Sparingly sol in water; sol in alcohol, chloroform, ether, slightly in cold petr ether.
Nitrate. Leaflets, mp 98-100°. Very sol in water, alcohol; insol in ether.
p-Tolylhydrazine. Rhombic bipyramids, mp 61° or 65-66°. bp 240-244° with slight decompn. Sparingly sol in water; sol in alcohol, benzene, ether.
Nitrate. Leaflets, mp 152-153°.
USE: *o*-Tolylhydrazine as reagent for galactose.

9702. *p*-Tolylsulfonylmethylnitrosamide. [80-11-5] *N*,4-Dimethyl-*N*-nitrosobenzenesulfonamide; *N*-methyl-*N*-nitroso-*p*-toluenesulfonamide; Diazald. $C_8H_{10}N_2O_3S$; mol wt 214.24. C 44.85%, H 4.71%, N 13.08%, O 22.40%, S 14.96%. Prepd by the action of nitrous acid on *p*-tolylsulfonylmethylamide: **DE 224388** (1910 to Bayer); *Frdl.* **10**, 1216; *Chem. Zentralbl.* **1910**, II, 609; Takizawa, *J. Pharm. Soc. Jpn.* **70**, 490 (1950); de Boer, Backer, *Rec. Trav. Chim.* **73**, 229 (1954); *Org. Synth.* **34**, 96 (1954).

Yellow crystals from benzene + petr ether, mp 62°. Stable in an ordinary brown bottle for several years. A white coating (formed of *p*-tolylsulfonylmethylamide) does no harm. Insol in water. Sol in ether, petr ether, benzene, chloroform, carbon tetrachloride. Yields diazomethane on treatment with alkali.

USE: In the laboratory prepn of diazomethane. Directions for use: de Boer, Backer, *Rec. Trav. Chim.* **73**, 232 (1954).

9703. Tomatidine. [77-59-8] (3β,5α,22β,25S)-Spirosolan-3-ol; 5α-tomatidan-3β-ol; 5α,20β_F,22α_F,25β_F,27-azaspirostan-3β-ol. $C_{27}H_{45}NO_2$; mol wt 415.66. C 78.02%, H 10.91%, N 3.37%, O 7.70%. By hydrolysis of tomatine: Kuhn *et al.*, *Ber.* **83**, 448 (1950). Isoln from the roots of Rutgers tomato plant [*Lycopersicon esculentum* Mill., cultivar. "Rutgers"]: Brink, Folkers, *J. Am. Chem. Soc.*

73, 4018 (1951); Fontaine *et al.*, *ibid.* 878; Sato *et al.*, *ibid.* 880; Kuhn, Low, **US 2770618** (1956 to Amer. Home Prod.). Structure: Sato *et al.*, *J. Org. Chem.* **25**, 783 (1960); Schreiber, Adams, *Experientia* **17**, 13 (1961). Synthesis: Uhle, Moore, *J. Am. Chem. Soc.* **76**, 6412 (1954); Uhle, *ibid.* **83**, 1460 (1961); Kessar *et al.*, *Tetrahedron* **27**, 2869 (1971).

Plates from ethyl acetate, mp 202-206°. $[\alpha]_D^{25}$ +8° (chloroform).
Hydrochloride. $C_{27}H_{45}NO_2$.HCl. Crystals from abs ethanol, mp 265-270°. $[\alpha]_D^{25}$ −5° (methanol).

9704. Tomatine. [17406-45-0] (3β,5α,22β,25S)-Spirosolan-3-yl *O*-β-D-glucopyranosyl-(1 → 2)-*O*-[β-D-xylopyranosyl-(1 → 3)]-*O*-β-D-glucopyranosyl-(1 → 4)-β-D-galactopyranoside; lycopersicin. $C_{50}H_{83}NO_{21}$; mol wt 1034.20. C 58.07%, H 8.09%, N 1.35%, O 32.49%. Occurs in the extract of leaves of wild tomato plants: Fontaine *et al.*, *Arch. Biochem.* **18**, 467 (1948); Kuhn, Low, *Ber.* **81**, 552 (1948); Kuhn *et al.*, *ibid.* **83**, 448 (1950); Bognar, Makleit, *Pharmazie* **11**, 376 (1956). Yields on partial hydrolysis, besides α-tomatine, the main constituent, $β_1$-, $β_2$-, γ- and δ-tomatine: Kuhn *et al.*, *Ber.* **90**, 203 (1957). α-Tomatine consists of one mol tomatidine linked to a tetrasaccharide composed of 2 mols D-glucose, 1 mol D-xylose and 1 mol D-galactose: Kuhn *et al.*, *Angew. Chem.* **68**, 212 (1956). Proposed as an alternate precipitant to digitonin: Schultz, Sander, *Z. Physiol. Chem.* **308**, 122 (1957). Structure: Reichstein, *ibid.* **74**, 887 (1962). Toxicity study: Wilson *et al.*, *Toxicol. Appl. Pharmacol.* **3**, 39 (1961).

Needles from methanol, mp 263-268°. $[\alpha]_D^{20}$ −18° (c = 0.55 in pyridine). Sol in ethanol, methanol, dioxane, propylene glycol. Practically insol in water, ether, petr ether. Stable to strong alkali but hydrolyzed by acids to produce cryst tomatidine and a soln rich in reducing sugars. Has been found to inhibit the growth of various fungi and bacteria. LD orally in rats: 900-1000 mg/kg (Wilson).
USE: Precipitating agent for steroids.

9705. Tonin. [53414-68-9] β-Angiotensin I converting enzyme (formerly); kallikrein rK2. A converting enzyme that differs from renin, *q.v.*, in its ability to form angiotensin II directly from angiotensinogen by cleaving the Phe-His bond; it can also convert angiotensin I to angiotensin II. Its presence was discovered in rat submaxillary glands: R. Boucher *et al.*, *Hypertension 72*, J. Genest, E. Koiw, Eds. (Springer-Verlag, New York, 1972) p 512; *eidem*, *Circ. Res.* **Suppl. I**, 203 (1974). Purification and characterization: S. Demassieux *et al.*, *Can. J. Biochem.* **54**, 788 (1976). Crystal data: K. Hayakawa *et al.*, *J. Mol. Biol.* **123**, 107 (1978). Radioimmunoassay: J. Gutkowska *et al.*, *Can. J. Biochem.* **56**, 769 (1978). Purification by affinity chromatography: M. Ikeda *et al.*, *Hypertension* **3**, 81 (1981); by gel permeation and HPLC: C. Lazure *et al.*, *Anal. Biochem.* **125**, 406 (1982). Isoln using chromatofocusing: E. S. P. Cheng, B. J. Morris, *ibid.* **126**, 295 (1982). *N*-Terminal amino acid sequence of rat tonin: N. G. Seidah *et al.*, *Can. J. Biochem.* **56**, 920

(1978). Substrate specificity studies: *eidem*, *Proc. Am. Peptide Symp. 6th*, E. Gross, J. Meienhofer, Eds. (Pierce Chem. Co., Rockford, Ill., 1979) p 921; M. Chretien *et al.*, *FEBS Lett.* **113**, 173 (1980). Formation of angiotensin II by tonin from partially purified human angiotensinogen: C. Grise *et al.*, *Can. J. Biochem.* **59**, 250 (1981). Pressor effect in anephric animals: E. L. Schiffrin *et al.*, *Can. J. Physiol. Pharmacol.* **59**, 864 (1981). Sequence homologies between tonin and other peptides: C. Lazure *et al.*, *Nature* **292**, 383 (1981). Immunohistochemical study: T. B. Oerstavik *et al.*, *J. Histochem. Cytochem.* **30**, 1123 (1982). Role as renin activator: J. Gutkowska *et al.*, *Can. J. Biochem.* **60**, 843 (1982). Role in exptl hypertension: R. Garcia *et al.*, *Hypertension, Int. Symp.*, *3rd*, H. Villarreal, Ed. (Wiley, New York, 1981) p 79. *Reviews:* R. Boucher *et al.*, *Circ. Res.* **Suppl. II**, 26-29 (1977); R. Boucher, J. Genest, *Endocrine Functions of the Brain*, M. Motta, Ed. (Raven Press, New York, 1980) pp 373-384; J. Genest, *Heterogeneity of Renin and Renin Substrate*, M. P. Sambhi, Ed. (Elsevier, New York, 1981) pp 11-24.

Mol wt determn is 31,400 by gel filtration and 28,700 by sedimentation equilibrium. Activity is not affected by pepstatin. Can be incubated at 20° for 150 min, between pH 3.4-8 without significant loss of enzymatic activity; loses 15% of its original activity at pH 2.8. After 5 min of incubation at 100°, 60-65% activity remains.

9706. Topiramate. [97240-79-4] 2,3:4,5-Bis-*O*-(1-methylethylidene)-β-D-fructopyranose 1-sulfamate; 2,3:4,5-di-*O*-isopropylidene-β-D-fructopyranose sulfamate; McN-4853; RWJ-17021-000; Topamax. $C_{12}H_{21}NO_8S$; mol wt 339.36. C 42.47%, H 6.24%, N 4.13%, O 37.72%, S 9.45%. Sulfamate substituted monosaccharide; structurally distinct antiepileptic agent. Prepn: B. E. Maryanoff, J. F. Gardocki, **US 4513006** (1985 to McNeil); and anticonvulsant activity: B. E. Maryanoff *et al.*, *J. Med. Chem.* **30**, 880 (1987). GC determn in plasma: M. L. Holland *et al.*, *J. Chromatogr.* **433**, 276 (1988). Comparative pharmacokinetics: M. Bialer, *Clin. Pharmacokinet.* **24**, 441 (1993). Series of articles on pharmacology and clinical experience in epilepsy: *Epilepsia* **38**, Suppl. 1, 1-62 (1997). Review of clinical trials in migraine prevention: G. Bussone *et al.*, *Int. J. Clin. Pract.* **59**, 961-968 (2005); of pharmacology and clinical experience: S. D. Silberstein *et al.*, *Clin. Ther.* **27**, 154-165 (2005). Clinical trial in binge eating disorder: S. L. McElroy *et al.*, *Biol. Psychiatry* **61**, 1039 (2007); in chronic migraine: S. D. Silberstein *et al.*, *Headache* **47**, 170 (2007).

White crystalline powder; bitter taste. Crystals from ethyl acetate + hexane, mp 125-126°. $[\alpha]_D^{23}$ −34.0° (c = 0.4 in methanol). Most sol in alkaline solns containing NaOH or sodium phosphate, pH 9-10. Freely sol in acetone, chloroform, DMSO, ethanol, dichloromethane. Soly in water: 9.8mg/ml.
THERAP CAT: Anticonvulsant; antimigraine.
THERAP CAT (VET): Anticonvulsant.

9707. Topotecan. [123948-87-8] (4S)-10-[(Dimethylamino)-methyl]-4-ethyl-4,9-dihydroxy-1*H*-pyrano[3′,4′:6,7]indolizino[1,2-*b*]quinoline-3,14(4*H*,12*H*)-dione; 9-[(dimethylamino)methyl]-10-hydroxy-(20S)-camptothecin; hycamptamine; SKF-104864. $C_{23}H_{23}N_3O_5$; mol wt 421.45. C 65.55%, H 5.50%, N 9.97%, O 18.98%. DNA topoisomerase I inhibitor; semisynthetic analog of camptothecin, *q.v.* Prepn: J. C. Boehm *et al.*, **EP 321122**; *eidem*, **US 5004758** (1989, 1991 both to SmithKline Beecham); W. D. Kingsbury *et al.*, *J. Med. Chem.* **34**, 98 (1991). HPLC determn in plasma: J. H. Beijnen *et al.*, *J. Pharm. Biomed. Anal.* **8**, 789 (1990). Clinical pharmacology: E. K. Rowinsky *et al.*, *J. Clin. Oncol.* **10**, 647 (1992); and pharmacokinetics: L. J. C. van Warmerdam *et al.*, *Cancer Chemother. Pharmacol.* **38**, 254 (1996). Clinical evaluation in ovarian cancer: A. P. Kudelka *et al.*, *J. Clin. Oncol.* **14**, 1552 (1996); in small cell lung cancer: J. H. Schiller *et al.*, *ibid.* 2345. Review of clinical toxicity: K. Seiter, *Expert Opin. Drug Saf.* **4**, 45-53 (2005); of clinical evaluation in cervical cancer as single and combination therapy: L. M. Randall-Whitis, B. J. Monk, *Expert Opin. Pharmacother.* **8**, 227-236 (2007).

Hydrochloride. [119413-54-6] NSC-609669; SKF-104864A; Hycamtin. $C_{23}H_{23}N_3O_5 \cdot HCl$; mol wt 457.91. Light yellow to greenish powder, mp 213-218° (dec). Soluble in water up to 1 mg/ml. Hygroscopic. Heat and light sensitive.

THERAP CAT: Antineoplastic.

9708. Torcetrapib. [262352-17-0] (2R,4S)-4-[[[3,5-Bis-(trifluoromethyl)phenyl]methyl](methoxycarbonyl)amino]-2-ethyl-3,4-dihydro-6-(trifluoromethyl)-1(2H)-quinolinecarboxylic acid ethyl ester; (2R,4S)-4-[(3,5-bis-trifluoromethylbenzyl)methoxycarbonylamino]-2-ethyl-6-trifluoromethyl-3,4-dihydro-2H-quinoline-1-carboxylic acid ethyl ester; CP-529414. $C_{26}H_{25}F_9N_2O_4$; mol wt 600.48. C 52.01%, H 4.20%, F 28.47%, N 4.67%, O 10.66%. Cholesteryl ester transfer protein (CETP) inhibitor. Prepn: M. P. DeNinno *et al.*, **WO 0017164**; *eidem*, **US 6197786** (2000, 2001 both to Pfizer); of crystalline forms: D. J. M. Allen *et al.*, **WO 0140190** (2001 to Pfizer). Mechanism of action study: R. W. Clark *et al.*, *J. Lipid Res.* **47**, 537 (2006). Clinical evaluation of effects on HDL cholesterol levels: R. W. Clark *et al.*, *Arterioscler. Thromb. Vasc. Biol.* **24**, 490 (2004); M. E. Brousseau *et al.*, *N. Engl. J. Med.* **350**, 1505 (2004). Review of clinical development in combination with atorvastatin: J. R. Burnett, *Curr. Opin. Investig. Drugs* **6**, 944-950 (2005).

Anhydrous, non-hygroscopic crystals, mp 89-90°. d 1.406.
Ethanolate. [343798-00-5] $C_{26}H_{25}F_9N_2O_4 \cdot C_2H_6O$; mol wt 646.55. White crystalline powder, mp 54-58°. $[\alpha]_D$ −93.3° (c = 1.08 in methanol). d 1.402. Non-hygroscopic. Higher aqueous soly than anhydrous form.

THERAP CAT: Antilipemic; antiatherosclerotic.

9709. Toremifene. [89778-26-7] 2-[4-[(1Z)-4-Chloro-1,2-diphenyl-1-buten-1-yl]phenoxy]-N,N-dimethylethanamine; (Z)-4-chloro-1,2-diphenyl-1-[4-[2-(N,N-dimethylamino)ethoxy]phenyl]-1-butene. $C_{26}H_{28}ClNO$; mol wt 405.97. C 76.92%, H 6.95%, Cl 8.73%, N 3.45%, O 3.94%. Nonsteroidal antiestrogen structurally similar to tamoxifen, *q.v.* Prepn: R. J. Toivola *et al.*, **EP 95875**; *eidem*, **US 4696949** (1983, 1987 both to Farmos). Pharmacology: S. Kallio *et al.*, *Cancer Chemother. Pharmacol.* **17**, 103 (1986). Antitumor effects *in vitro* and *in vivo*: L. Kangas *et al.*, *ibid.* 109. HPLC determn in plasma: W. M. Holleran *et al.*, *Anal. Lett.* **20**, 871 (1987). Clinical evaluation in high-grade prostatic intraepithelial neoplasia: M. S. Steiner, C. R. Pound, *Clin. Prostate Cancer* **2**, 24 (2003). Review of pharmacology and clinical efficacy in advanced breast cancer: L. R. Wiseman, K. L. Goa, *Drugs* **54**, 141-160 (1997); in postmenopausal breast cancer: J. U. Mäenpää, S.-L. Ala-Fossi, *Drugs Aging* **11**, 261-270 (1997).

mp 108-110°.
Citrate. [89778-27-8] FC-1157a; Acapodene; Fareston. $C_{26}H_{28}ClNO \cdot C_6H_8O_7$; mol wt 598.09. Off-white powder. mp 160-162°. pKa 8.0. Soly at 37° (mg/ml): water 0.63; 0.02N HCl 0.38.

THERAP CAT: Antineoplastic.

9710. Torilin. [13018-10-5] (2Z)-2-Methyl-2-butenoic acid (5S,6R,8S,8aR)-5-[1-(acetyloxy)-1-methylethyl]-1,2,4,5,6,7,8,8a-octahydro-3,8-dimethyl-2-oxo-6-azulenyl ester; (Z)-8β,11-dihydroxy-1β-guai-4-en-3-one 11-acetate 2-methylcrotonate. $C_{22}H_{32}O_5$; mol wt 376.49. C 70.19%, H 8.57%, O 21.25%. Guaiane sesquiterpene and major constituent of Japanese hedge parsley, *Torilis japonica* (Houtt.) DC, also known as *Torilis anthriscus* (L.) Gmel., *Apiaceae*, which is used in Asian traditional medicine as an anthelmintic, anti-inflammatory, and antimicrobial. Isoln from seeds of *T. japonica* and structure: M. Nakazaki *et al.*; *Tetrahedron Lett.* **7**, 4499 (1966); H. Chikamatsu *et al.*, *Tetrahedron* **25**, 4751 (1969). Isoln from Japanese elm, *Ulmus davidiana* var. *japonica*, *Ulmaceae*, and inhibitory effect on nitric oxide production: Y. C. Kim *et al.*, *Fitoterapia* **78**, 196 (2007). Reversal of multi-drug resistance in cancer cells: S. E. Kim *et al.*, *Planta Med.* **64**, 332 (1998). Antimicrobial activity: W.-I. Cho *et al.*, *J. Food Sci.* **73**, M37 (2008).

Crystals from petr ether, mp 77-78°. $[\alpha]_D^{14}$ −45.3° (c = 0.848 in ethanol). uv max (ethanol): 236 nm (log ε 4.29).

9711. Torsemide. [56211-40-6] N-[[(1-Methylethyl)amino]carbonyl]-4-[(3-methylphenyl)amino]-3-pyridinesulfonamide; 1-isopropyl-3-[(4-m-toluidino-3-pyridyl)sulfonyl]urea; 3-isopropylcarbamylsulfonamido-4-(3'-methylphenyl)aminopyridine; torasemide; AC-4464; BM-02015; JDL-464; Demadex; Toradiur; Torem; Unat. $C_{16}H_{20}N_4O_3S$; mol wt 348.42. C 55.16%, H 5.79%, N 16.08%, O 13.78%, S 9.20%. Sulfonylurea loop diuretic. Prepn: J. E. DeLarge *et al.*, **DE 2516025**; *eidem*, **US 4018929** (1975, 1977 both to A. Christiaens, S.A.); J. DeLarge, C. L. Lapiere, *Ann. Pharm. Fr.* **36**, 369 (1978). Pharmacokinetics in humans: M. Lesne *et al.*, *Int. J. Clin. Pharmacol. Ther. Toxicol.* **20**, 382 (1982). Preliminary evaluation in acute heart failure: R. Stroobandt *et al.*, *Arch. Int. Pharmacodyn.* **260**, 151 (1982). Clinical pharmacology: D. C. Brater *et al.*, *Clin. Pharmacol. Ther.* **42**, 187 (1987). Series of articles on pharmacology, mode of action and renal effects in animals: *Arzneim.-Forsch.* **35**, 1520-1541 (1985); on pharmacology, pharmacokinetics and clinical studies: *Eur. J. Clin. Pharmacol.* **31**, Suppl., 1-55 (1986); *Arzneim.-Forsch.* **38**, 143-214 (1988). Clinical comparison with furosemide, *q.v.*, in congestive heart failure: J. Cosin *et al.*, *Eur. J. Heart Fail.* **4**, 507 (2002).

White to off-white, crystalline powder. mp 163-164°. pKa 6.44. Slightly sol in 0.1*N* sodium hydroxide, in 0.1*N* hydrochloric acid, alc, methanol; very slightly sol in acetone, chloroform. Practically insol in water, ether.

THERAP CAT: Diuretic.

9712. Torularhodin. [514-92-1] 3′,4′-Didehydro-β,ψ-caroten-16′-oic acid. $C_{40}H_{52}O_2$; mol wt 564.85. C 85.06%, H 9.28%, O 5.66%. Carotenoid pigment found in *Torula rubra* and *Rhodotorula mucilaginosa* yeasts. Isoln: Karrer, Rutschmann, *Helv. Chim. Acta* **26**, 2109 (1943). Structure and synthesis: Isler *et al.*, *ibid.* **42**, 864 (1959).

Fine dark purple needles from methanol + ether or toluene, mp 210-212° (vac, some decompn). Absorption max in CS_2: 582, 541, 502 nm; in methanol: 529, 493, 460 nm. Freely sol in carbon disulfide, chloroform, pyridine; less sol in ether, benzene, hot ethanol; sparingly sol in methanol. Practically insol in petr ether.

Methyl ester. $C_{41}H_{54}O_2$. Dark red needles from benzene + methanol, mp 172-173°.

9713. Tositumomab. [208921-02-2] Anti-(human CD20 antigen) immunoglobulin G2a (mouse monoclonal clone B1R1 γ_{2a}-chain) disulfide with mouse monoclonal clone B1R1 λ_x-chain, dimer; anti-B1 antibody. Murine monoclonal antibody targeted against CD-20 antigen located on mature B lymphocytes but not on normal stem cells or progenitor cells. [131]Iodine radioimmunoconjugate designed for tumor-targeted treatment of non-Hodgkin's lymphoma (NHL). Prepn: M. S. Kaminski *et al.*, US 5595721 (1997 to Coulter); and clinical evaluation: *idem et al.*, *N. Engl. J. Med.* **329**, 459 (1993). Clinical study of radioimmunotherapy in NHL: *eidem*, *Blood* **96**, 1259 (2000). Review of mechanism of action and clinical application: A. K. Gopal, O. W. Press, *J. Lab. Clin. Med.* **134**, 445 (1999); of use in follicular lymphoma: B. M. William, P. J. Bierman, *Expert Opin. Biol. Ther.* **10**, 1271-1278 (2010).

Conjugate with [131]I. [192391-48-3] Iodine I 131 tositumomab; SB-393229; Bexxar.

THERAP CAT: Antineoplastic. Conjugate with ^{131}I in radioimmunotherapy.

9714. Tosufloxacin. [100490-36-6] 7-(3-Amino-1-pyrrolidinyl)-1-(2,4-difluorophenyl)-6-fluoro-1,4-dihydro-4-oxo-1,8-naphthyridine-3-carboxylic acid; A-61827. $C_{19}H_{15}F_3N_4O_3$; mol wt 404.35. C 56.44%, H 3.74%, F 14.10%, N 13.86%, O 11.87%. Trifluorinated quinolone antibacterial. Prepn: H. Narita *et al.*, DE 3514076 (1985 to Toyama), *C.A.* **104**, 129888r (1986); D. T. W. Chu, EP 153580; *idem*, US 4616019 (1985, 1986 both to Abbott); and activity: *idem et al.*, *J. Med. Chem.* **29**, 2363 (1986); H. Narita *et al.*, *Yakugaku Zasshi* **106**, 802 (1986), *C.A.* **106**, 196291v (1987). *In vitro* activity studies of the base: P. B. Fernandes *et al.*, *Antimicrob. Agents Chemother.* **32**, 27 (1988); and *in vivo* animal studies of the toluenesulfonate: M. Takahata *et al.*, *J. Antimicrob. Chemother.* **22**, 143 (1988). Series of articles on antibacterial activity and clinical evaluation: *Chemotherapy (Tokyo)* **36**, Suppl. 9, 1-1538 (1988).

Hydrochloride. [104051-69-6] A-60969. $C_{19}H_{15}F_3N_4O_3 \cdot HCl$; mol wt 440.81. Crystals from conc HCl-ethanol (1:3), mp 247-250° (dec).

Toluenesulfonate. [115964-29-9] Tosufloxacin tosylate; A-64730; T-3262; Ozex; Tosuxacin. $C_{19}H_{15}F_3N_4O_3 \cdot C_7H_8O_3S$; mol wt 576.55. Prepd as the monohydrate, mp 258-260°.

THERAP CAT: Antibacterial.

9715. Tosylmethyl Isocyanide. [36635-61-7] 1-[(Isocyanomethyl)sulfonyl]-4-methylbenzene; (4-methylphenylsulfonyl)-methyl isocyanide; *p*-toluenesulfonylmethyl isocyanide; TosMIC. $C_9H_9NO_2S$; mol wt 195.24. C 55.37%, H 4.65%, N 7.17%, O 16.39%, S 16.42%. Isonitrile reagent with reversed polarity; serves as a carbonyl anion equivalent in organic synthesis. Prepn: U. Schöllkopf *et al.*, *Ann.* **766**, 130 (1972); A. M. van Leusen *et al.*, *Tetrahedron Lett.* **13**, 2367 (1972); B. E. Hoogenboom *et al.*, *Org. Synth.* **57**, 102 (1977). Review of applications in organic synthesis: C. Lamberth, *J. Prakt. Chem. Chem.-Ztg.* **340**, 483-485 (1998); V. K. Tandon, S. Rai, *Sulfur Rep.* **24**, 307-385 (2003).

Crystals from methanol, mp 116-117° (dec).

USE: Versatile synthon in organic chemistry, esp in the synthesis of heterocyclic compds.

9716. Toxaphene. [8001-35-2] Chlorinated camphene; camphechlor; polychlorocamphene; Hercules 3956; Alltox; Geniphene; Motox; Phenacide; Phenatox; Strobane-T; Toxakil. Organochlorine pesticide. A very complex, but reproducible mixture of at least 177 C_{10} polychloro derivs, having an approx overall empirical formula of $C_{10}H_{10}Cl_8$. Produced by the chlorination of camphene to 67-69% chlorine by weight and made up of compds of $C_{10}H_8Cl_{10}$, $C_{10}H_{18-n}Cl_n$ (mostly polychlorobornanes) and $C_{10}H_{16-n}Cl_n$ (polychlorobornenes and/or polychlorotricyclenes) with n = 6 to 9. Prepn: Buntin, US 2565471 (1951 to Hercules Powder). Isoln of components in crystalline form: Casida *et al.*, *Science* **183**, 520 (1974); *eidem*, *J. Agric. Food Chem.* **22**, 939 (1974). Synthesis and photostability of major components, *B7-1001* and *B8-1412*: M. Coelhan, M. Maurer, *ibid.* **53**, 10105 (2005). Livestock toxicity and tissue residues: L. Penumarthy *et al.*, *Vet. Toxicol.* **18**, 60 (1976). Acute toxicity data: T. B. Gaines, *Toxicol. Appl. Pharmacol.* **14**, 515 (1969). Mutagenicity studies: N. K. Hooper *et al.*, *Science* **205**, 591 (1979). *Reviews:* Liebmann *et al.*, *Arch. Pflanzenschutz* **7**, 131-150 (1971); F. Korte *et al.*, *Pure Appl. Chem.* **51**, 1583-1601 (1979); M. A. Saleh, *Rev. Environ. Contam. Toxicol.* **118**, 1-85 (1990). Review of toxicology and human exposure: *Toxicological Profile for Toxaphene* (PB97-121057, 1996) 252 pp; of environmental distribution, toxicology and analytical methods: H.-J. De Geus *et al.*, *Environ. Health Perspect.* **107**, 115-144 (1999).

Yellow waxy solid, mp 65-90°. Pleasant piney odor. Vapor pressure at 20°: 3×10^{-7} mm Hg. d^{25} 1.630. Log P (octanol/water): 6.44. Dehydrochlorinates in the presence of alkali, prolonged exposure to sunlight, and at temps about 155°. Soly in water: 3 mg/l. Freely sol in aromatic hydrocarbons. Corrosive to iron. LD_{50} in male, female rats (mg/kg): 90, 80 orally; 1075, 780 dermally (Gaines).

Caution: Potential symptoms of overexposure are nausea, confusion, agitation, tremors, convulsions and unconsciousness; dry, red skin. *See NIOSH Pocket Guide to Chemical Hazards* (DHHS/NIOSH 97-140, 1997) p 58. *See also Clinical Toxicology of Commercial Products*, R. E. Gosselin *et al.*, Eds. (Williams & Wilkins, Baltimore, 5th ed., 1984) Section III, pp 386-387. This substance is reasonably anticipated to be a human carcinogen: *Report on Carcinogens, Twelfth Edition* (PB2011-111646, 2011) p 418.

Note: Toxaphene is listed as a persistent organic pollutant (POP) in Annex A of the *Stockholm Convention on Persistent Organic Pollutants* (United Nations, Stockholm, 2001) 43 pp; amended (Geneva, 2009) 63 pp.

USE: Insecticide.

9717. Toxiferine I. [6888-23-9] (1*S*,3a*S*,10*S*,11a*S*,12*S*,14a*S*,-19a*S*,20b*S*,21*S*,22a*S*,23*E*,26*E*)-2,3,11,11a,13,14,22,22a-Octahydro-23,26-bis(2-hydroxyethylidene)-1,12-dimethyl-10*H*,21*H*-1,21:10-diethano-19a*H*,20b*H*-dipyrrolo[3,2-*f*:3′,2′-*f*′][1,5]-diazocino[3,2,1-*jk*:7,6,5-*j*′*k*′]dicarabazolium; *C*-Toxiferine I. $[C_{40}H_{46}N_4O_2]^{2+}$.

Naturally occurring neuromuscular blocker. From calabash curare: Schmid, Karrer, *Helv. Chim. Acta* **30**, 1162 (1947); from *Strychnos toxifera* Schomb., *Loganiaceae:* Wieland *et al.*, *Ann.* **547**, 156 (1941); King, *J. Chem. Soc.* **1949**, 3263. Identity with *toxiferine V* and *toxiferine XI:* Battersby *et al.*, *ibid.* **1960**, 1848. Structure: Arnold *et al.*, *Helv. Chim. Acta* **42**, 394 (1959); Grdinic *et al.*, *J. Am. Chem. Soc.* **86**, 3357 (1964). Pharmacokinetics: P. G. Waser, J. Reller, *Agents Actions* **2**, 170 (1972). ^{13}C-NMR study: E. Wenkert *et al.*, *J. Org. Chem.* **43**, 1099 (1978).

Dichloride. $C_{40}H_{46}Cl_2N_4O_2$. Crystals. $[\alpha]_D^{22}$ −546° (c = 0.30). uv max (ethanol): 292 nm (log ε 4.62). Sol in water.
C-Toxiferine II. [7257-29-6] *C*-Calebassine; *C*-strychnotoxine. $[C_{40}H_{48}N_4O_2]^{2+}$. From calabash-curare: Karrer, Schmidt, *Helv. Chim. Acta* **29**, 1853 (1946); Zürcher *et al.*, *J. Am. Chem. Soc.* **80**, 1500 (1958). Identity of *C*-toxiferine II and *C*-strychnotoxine: Wieland, Merz, *Ber.* **85**, 731 (1952). Structure: Hesse *et al.*, *Helv. Chim. Acta* **44**, 2211 (1961); Fehlmann *et al.*, *ibid.* **48**, 303 (1965).

9718. Toxoflavin. [84-82-2] 1,6-Dimethylpyrimido[5,4-*e*]-1,2,4-triazine-5,7(1*H*,6*H*)-dione; 1,6-dimethyl-5,7-dioxo-1,5,6,7-tetrahydropyrimido[5,4-*e*]-*as*-triazine; xanthothricin. $C_7H_7N_5O_2$; mol wt 193.17. C 43.52%, H 3.65%, N 36.26%, O 16.56%. Highly toxic antibiotic from cultures of *Pseudomonas cocovenenans* which also produces bongkrekic acid, *q.v.* Isoln: van Veen, Mertens, *Rec. Trav. Chim.* **53**, 257, 398 (1934); R. A. Machlowitz *et al.*, *Antibiot. Chemother.* **4**, 259 (1954). Structure: Van Damm *et al.*, *Rec. Trav. Chim.* **79**, 255 (1960). Synthesis: Daves *et al.*, *J. Am. Chem. Soc.* **83**, 3904 (1961); **84**, 1724 (1962); Yoneda *et al.*, *Tetrahedron Lett.* **1971**, 851. Mode of action: Latuasan, Berends, *Biochim. Biophys. Acta* **52**, 502 (1961). Biosynthesis: Levenberg, Linton, *J. Biol. Chem.* **241**, 846 (1966). Production by *Burkholderia glumae* and pathogenicity to rice seedlings: K. Yoneyama *et al.*, *Ann. Phytopathol. Soc. Jpn.* **64**, 91 (1998).

Bright yellow platelets from propanol, dec 172-173°. uv max: 257.5, 394 nm (ε 16400, 2500). Acts as a pH indicator with a sharp loss of color at pH 10.5, but is destroyed by alkali. Sol in water, chloroform, ethyl acetate, ethanol. LD_{50} in mice (mg/kg): 1.7 i.v.; 8.4 orally (Machlowitz).

9719. Toxohormone. [9014-44-2] Name given in 1948 by Nakahara and Fukuoka to a factor produced by living cancer cells and released into the circulation to produce decreases in liver catalase activity, tryptophan pyrrolase activity, liver ferritin and plasma iron. *See:* Nakahara, Fukuoka, *Jpn. J. Cancer Res.* **40**, 45 (1949). This tumor-specific concept has been questioned: Greenfield, Meister, *J. Natl. Cancer Inst.* **11**, 997 (1951); Olivares *et al.*, *Science* **157**, 327 (1967); Kampschmidt, Upchurch, *Proc. Soc. Exp. Biol. Med.* **127**, 632 (1968). Described as a polypeptide of low mol wt having 30-40 amino acid residues of which 12-13 are different, and with a high

content of glycine, glutamic acid, aspartic acid, alanine and leucine. Amino acid and lipid composition of a highly purified toxohormone prepd from human malignant tissue: Yunoki, Griffin, *Cancer Res.* **21**, 537 (1961); from cell-free fluid of ascites sarcoma 180: H. Masuno *et al.*, *ibid.* **41**, 284 (1981). *Reviews:* W. Nakahara, F. Fukuoka, *Chemistry of Cancer Toxin Toxohormone* (C. C. Thomas, Springfield, Illinois, 1961) 75 pp; Nakahara, *Methods Cancer Res.* **2**, 203-237 (1967); Olivares, Kampschmidt in *Oncology, Proc. Int. Cancer Congr., 10th, Houston 1970* **3**, 158-170 (1971); S. Fujii, *Gann Monogr. Cancer Res.* **24**, 215-222 (1979).
Thermostable, non-heat-coagulable, water-sol and alcohol precipitable.

9720. Toxopyrimidine. [73-67-6] 4-Amino-2-methyl-5-pyrimidinemethanol; 6-amino-5-hydroxymethyl-2-methylpyrimidine; 4-amino-5-hydroxymethyl-2-methylpyrimidine; pyramin; pyramine. $C_6H_9N_3O$; mol wt 139.16. C 51.79%, H 6.52%, N 30.20%, O 11.50%. A metabolite of thiamine. Prepd from 4-amino-5-aminomethyl-2-methylpyrimidine dihydrochloride or ethyl 4-amino-2-methyl-5-pyrimidinecarboxylate: Dornow, Petsch, *Ber.* **86**, 1404 (1953); DiBella, Hennessy, *J. Org. Chem.* **26**, 2017 (1961).

Needles from water, mp 193-198°; crystals from methanol + ether, mp 198-200°. Sublimes at 0.01 mm between 155-170°.

9721. Toyocamycin. [606-58-6] 4-Amino-7-β-D-ribofuranosyl-7*H*-pyrrolo[2,3-*d*]pyrimidine-5-carbonitrile; 4-amino-5-cyano-7-(D-ribofuranosyl)-7*H*-pyrrolo[2,3-*d*]pyrimidine; uramycin B; vengicide; antibiotic 1037; E-212. $C_{12}H_{13}N_5O_4$; mol wt 291.27. C 49.48%, H 4.50%, N 24.04%, O 21.97%. Antibiotic substance extracted from the culture filtrate and mycelium of *Streptomyces toyocaensis*. Isoln: Nishimura *et al.*, *J. Antibiot.* **9A**, 60 (1956). Structure: Ohkuma, *ibid.* **14A**, 343 (1961). Synthesis of aglycone: Taylor, Hendess, *J. Am. Chem. Soc.* **86**, 951 (1964). Total synthesis: Tolman *et al.*, *ibid.* **90**, 524 (1968); **91**, 2102 (1969). Biosynthesis: Uematsu, Suhadolnik, *Biochemistry* **9**, 1260 (1970); *eidem*, *J. Biol. Chem.* **245**, 4365 (1970). Crystal and molecular structure: P. Prusiner, M. Sundaralingam, *Acta Crystallogr.* **B34**, 517 (1978).

Fine needles from methanol or acetone, mp 243°. Recrystallization from water yields the hydrate, $C_{12}H_{13}N_5O_4·H_2O$, mp 239-243°. $[\alpha]_D^{16}$ −45.7° (c = 1.05 in 0.1*N* HCl). uv max (H_2O): 230, 277 nm ($E_{1cm}^{1\%}$ 400, 548). Soluble in acetic acid, acidic solns. Moderately sol in methanol, ethanol, acetone, dioxane, butanol, water, ether. Practically insol in chloroform, ethyl acetate, petr ether. LD_{100} s.c. in mice: 10-20 mg/kg (Nishimura).

9722. TPMPA. [182485-36-5] *P*-Methyl-*P*-(1,2,3,6-tetrahydro-4-pyridinyl)phosphinic acid; (1,2,5,6-tetrahydropyridine-4-yl)-methylphosphinic acid. $C_6H_{12}NO_2P$; mol wt 161.14. C 44.72%, H 7.51%, N 8.69%, O 19.86%, P 19.22%. Selective antagonist for $GABA_c$ receptors. Prepn: Y. Murata *et al.*, *Bioorg. Med. Chem. Lett.* **6**, 2073 (1996); R. Miledi *et al.*, *US 5627169* (1997 to Univ. California). Design and *in vitro* pharmacology: D. Ragozzino *et al.*, *Mol. Pharmacol.* **50**, 1024 (1996). Specificity vs human receptors and alternate synthesis: M. Chebib *et al.*, *Eur. J. Pharmacol.* **357**, 227 (1998). Use as antagonist: A. Rozzo *et al.*, *Neuroscience* **90**, 1085 (1999).

Off-white solid from ethanol, mp 252-254°. Soly in water 16 mg/ml. Insol in DMSO.

USE: Neurochemical tool.

9723. Trabedersen. [925681-61-4] d(P-Thio)(C-G-G-C-A-T-G-T-C-T-A-T-T-T-T-G-T-A)DNA; AP-12009. Synthetic, 18 base phosphorothioate antisense oligonucleotide (ASO). Designed to inhibit expression of transforming growth factor β_2 (TGF-β2), q.v., which is overexpressed in certain cancers, such as malignant glioma and pancreatic cancer. Prepn: G.-F. Schlingensiepen et al., **WO 94 25588**; eidem, **US 6455689** (1994, 2002 both to Biognostik GmbH). Pharmacology and toxicology: R. Schlingensiepen et al., Oligonucleotides **15**, 94 (2005). Clinical evaluation of intratumor delivery in patients with malignant glioma: P. Hau et al., ibid. **17**, 201 (2007). Review of development and clinical experience: L. Vallières, IDrugs **12**, 445-453 (2009).

Sol in normal saline. LD_{50} in mice, rats (mg/kg): 706, 1175 i.v. (R. Schlingensiepen).

THERAP CAT: Antineoplastic.

9724. Tralkoxydim. [87820-88-0] 2-[1-(Ethoxyimino)propyl]-3-hydroxy-5-(2,4,6-trimethylphenyl)-2-cyclohexen-1-one; 2-[1-(ethoxyimino)propyl]-3-hydroxy-5-mesitylcyclohex-2-en-1-one; PP-604; Achieve; Grasp. $C_{20}H_{27}NO_3$; mol wt 329.44. C 72.92%, H 8.26%, N 4.25%, O 14.57%. Cereal selective post-emergent herbicide. Prepn: R. B. Warner et al., **EP 80301**; eidem, **US 4717418** (1983, 1988 both to ICI). Physical properties and herbicidal activity: R. B. Warner et al., Proc. Br. Crop Prot. Conf. - Weeds **1987**, 19. Field trials on grass weeds: J. Rola, ibid. 363; P. B. Sutton et al., ibid. 389. Mechanism of action study: J. Secor, C. Cseke, Plant Physiol. **86**, 10 (1988).

White crystalline solid, mp 106°. Vapor pressure at 20°: 4×10^{-10} kPa. Soly at 20° (mg/l): water 6 at pH 6.5, 5 at pH 5.0; at 24° (g/l): hexane 18; toluene 213; dichloromethane >500; methanol 25; acetone 89; ethyl acetate 110. LD_{50} orally in male and female rats, male and female mice, male rabbit (mg/kg): 1324, 934, 1231, 1100, 519; dermally in male and female rats: >2000, >2000 mg/kg (Warner 1987).

USE: Post-emergent herbicide.

9725. Tralomethrin. [66841-25-6] 2,2-Dimethyl-3-(1,2,2,2-tetrabromoethyl)cyclopropanecarboxylic acid cyano(3-phenoxyphenyl)methyl ester; RU-25474; OMS-3048; Saga; Scout; Tracker; Tralate; Tralox. $C_{22}H_{19}Br_4NO_3$; mol wt 665.01. C 39.74%, H 2.88%, Br 48.06%, N 2.11%, O 7.22%. Synthetic pyrethroid consisting of two active diastereomers whose absolute configurations differ at the monobrominated carbon atom. Prepn: **BE 873201**; J. Martel et al., **US 4279835** (1979, 1981 both to Roussel-Uclaf). Insecticidal activity: M. Benoit et al., Pestic. Biochem. Physiol. **26**, 284 (1986). Mechanism of action study: M. Roche et al., ibid. **24**, 306 (1985). HPLC determn in water, sediment and fish tissue: J. Mao et al., J. Agric. Food Chem. **41**, 596 (1993). Field trials: D. D. Amalraj et al., Indian J. Malariol. **28**, 141 (1991); W. R. Halliday et al., J. Agric. Entomol. **9**, 145 (1992). Review of toxicology and human exposure: Toxicological Profile for Pyrethrins and Pyrethroids (PB2004-100004, 2003) 332 pp.

USE: Insecticide.

9726. Tramadol. [27203-92-5] rel-(1R,2R)-2-[(Dimethylamino)methyl]-1-(3-methoxyphenyl)cyclohexanol; E-265; CG-315E; U-26225A. $C_{16}H_{25}NO_2$; mol wt 263.38. C 72.97%, H 9.57%, N 5.32%, O 12.15%. Synthetic opioid analgesic which also blocks norepinephrine and serotonin reuptake; racemate of (1RS,-2RS) forms. Prepn: **GB 997399**; K. Flick, E. Frankus, **US 3652589** (1965, 1972 both to Grünenthal); K. Flick et al., Arzneim.-Forsch. **28**, 107 (1978). Series of articles on pharmacology and clinical studies: ibid. 114-219. Toxicology: F. Lagler et al., ibid. 164. Mechanism of action study: R. B. Raffa et al., J. Pharmacol. Exp. Ther. **267**, 331 (1993). HPLC determn of enantiomers in plasma: M. A. Campanero et al., J. Chromatogr. A **1031**, 219 (2004). HPLC determn in pharmaceutical formulations: M. Zecevic et al., ibid. **1119**, 251 (2006). Use as probe for cytochrome P450 2D6 phenotyping: R. S. Pedersen et al., Clin. Pharmacol. Ther. **77**, 458 (2005). Review of pharmacology and clinical efficacy: S. Grond, A. Sablotzki, Clin. Pharmacokinet. **43**, 879-923 (2004); of abuse potential: D. H. Epstein et al., Biol. Psychol. **73**, 90-99 (2006); of therapeutic potential in osteoarthritis: M. S. Cepeda et al., J. Rheumatol. **34**, 543-555 (2007).

Relative stereochemistry

Hydrochloride. [36282-47-0] Amadol; Contramal; Tradonal; Tramal; Ultram; Zamudol; Zydol. $C_{16}H_{25}NO_2.HCl$; mol wt 299.84. White, odorless crystalline powder. Bitter taste. mp 180-181°. Readily sol in water, ethanol. pKa 9.41. LogP (n-octanol/water): 1.35 (pH 7). LD_{50} in mice, rats (mg/kg): 350, 228 orally; 200, 286 s.c. (Lagler).

THERAP CAT: Analgesic.

THERAP CAT (VET): Analgesic.

9727. Tramazoline. [1082-57-1] 4,5-Dihydro-N-(5,6,7,8-tetrahydro-1-naphthalenyl)-1H-imidazol-2-amine; 2-[(5,6,7,8-tetrahydro-1-naphthyl)amino]-2-imidazoline. $C_{13}H_{17}N_3$; mol wt 215.30. C 72.52%, H 7.96%, N 19.52%. α-Adrenergic agonist. Prepn: Berg, **DE 1191381**; **DE 1195323** (both 1965 to Thomae), C.A. **63**, 8373c; 13274d (1965). Pharmacology and toxicity data: R. Engelhorn, H. Klupp, Arzneim.-Forsch. **12**, 971 (1962). Activity studies: Sachsenröder et al., ibid. **22**, 392 (1972).

Crystals from isopropanol, mp 142-143°.
Hydrochloride monohydrate. [3715-90-0] KB-227; Biciron; Ellatun; Rhinaspray; Rhinogutt; Rhinospray; Rinogutt; Towk. $C_{13}H_{17}N_3.HCl.H_2O$; mol wt 269.77. Crystals from alc + ether or acetone + ether, mp 172-174°. Sol in water. LD_{50} orally in mice: 195 mg/kg (Engelhorn, Klupp).

THERAP CAT: Decongestant.

9728. Tramiprosate. [3687-18-1] 3-Amino-1-propanesulfonic acid; γ-aminopropanesulfonic acid; 3-sulfopropylamine; homotaurine; NC-531; NC-758; Alzhemed; Cerebril. $C_3H_9NO_3S$; mol wt 139.17. C 25.89%, H 6.52%, N 10.06%, O 34.49%, S 23.04%. Sulfonic acid analog of γ-aminobutyric acid (GABA), q.v. Binds to soluble amyloid β peptide (Aβ), q.v., and inhibits fibrillization and plaque formation. Prepn: S. Gabriel, W. E. Lauer, Ber. **23**, 87 (1890); P. Rumpf, Compt. Rend. **204**, 592 (1937). Large scale synthesis from 1,3-propane sultone: X. Kong et al., **WO 04113391**; eidem, **US 050143462** (2004, 2005 both to Neurochem). Crystal structure: K. Tomita, Tetrahedron Lett. **12**, 2587 (1971). GABA mimetic activity: R. G. Fariello, G. T. Golden, *Brain Res. Bull.* **5**, Suppl. 2, 691 (1980). Pharmacology and binding to Aβ: F. Gervais *et al.*, *Neurobiol. Aging* **28**, 537 (2007). Pharmacokinetics in patients with cerebral amyloid angiopathy: S. M. Greenberg *et al.*, *Alzheimer Dis. Assoc. Disord.* **20**, 269 (2006). Clinical effect on Aβ$_{42}$ levels in CSF of Alzheimer's patients: P. S. Aisen *et al.*, *Neurology* **67**, 1757 (2006).

White crystalline powder, mp 292°. *Irritant.* Crystal density: 1.610 (α-form); 1.599 (β-form). pKa$_2$ 10.05. Sol in water; slightly sol in ethanol.

Sodium salt. [14650-46-5] C$_3$H$_8$NNaO$_3$S; mol wt 161.15. White crystalline powder from ethanol, mp 198-199°.

THERAP CAT: Antiamyloidogenic agent; in treatment of Alzheimer's disease.

9729. Trandolapril. [87679-37-6] (2*S*,3a*R*,7a*S*)-1-[(2*S*)-2-[[(1*S*)-1-(Ethoxycarbonyl)-3-phenylpropyl]amino]-1-oxopropyl]-octahydro-1*H*-indole-2-carboxylic acid; (3a*R*,7a*S*)-1-[*N*-[1(*S*)-(ethoxycarbonyl)-3-phenylpropyl]-(*S*)-alanyl]octahydroindole-2(*S*)-carboxylic acid; (2*S*,3a*R*,7a*S*)-1-[(*S*)-*N*-[(*S*)-1-carboxy-3-phenylpropyl]alanyl]hexahydro-2-indolinecarboxylic acid 1-ethyl ester; RU-44570; Mavik; Odrik; Gopten. C$_{24}$H$_{34}$N$_2$O$_5$; mol wt 430.55. C 66.95%, H 7.96%, N 6.51%, O 18.58%. Angiotensin converting enzyme (ACE) inhibitor. Prepn: H. Urbach *et al.*, **EP 84164**; *eidem*, **US 4933361** (1983, 1990 both to Hoechst). Enzyme inhibition and pharmacology: N. L. Brown *et al.*, *Eur. J. Pharmacol.* **148**, 79 (1988). Clinical pharmacology: F. De Ponti *et al.*, *Eur. J. Clin. Pharmacol.* **40**, 149 (1991). Clinical trial in prevention of death after myocardial infarction: L. Kober *et al.*, *N. Engl. J. Med.* **333**, 1670 (1995). Series of articles on pharmacology and clinical trials: *Am. J. Hypertens.* **8**, 63S-74S (1995). Clinical trial in diabetic neuropathy: R. A Malik *et al.*, *Lancet* **352**, 1978 (1998).

Colorless, crystalline solid, mp 125°. Sol in chloroform, dichloromethane, methanol.

Diacid. [87679-71-8] Trandolaprilat; RU-44403. C$_{22}$H$_{30}$N$_2$O$_5$; mol wt 402.49.

THERAP CAT: Antihypertensive.

9730. Tranexamic Acid. [1197-18-8] *trans*-4-(Aminomethyl)cyclohexanecarboxylic acid; AMCHA; RP-18429; Anvitoff; Cyklokapron; Exacyl; Lysteda; Spotof; Tranex; Transamin; Ugurol. C$_8$H$_{15}$NO$_2$; mol wt 157.21. C 61.12%, H 9.62%, N 8.91%, O 20.35%. Antifibrinolytic agent; blocks lysine binding sites of plasminogen. Prepn: A. Einhorn, C. Ladisch, *Ann.* **310**, 194 (1900); M. Levine, R. Sedlecky, *J. Org. Chem.* **24**, 115 (1959); NL 6503605; T. Naito *et al.*, US 3499925 (1965, 1970 to both Daiichi Seiyaku and Mitsubishi Chem.). Pharmacology: Andersson *et al.*, *Scand. J. Haematol.* **2**, 230 (1965). Resoln of isomers and antiplasmin activity: M. Shimizu *et al.*, *Chem. Pharm. Bull.* **16**, 357 (1968); T. Naito *et al.*, *ibid.*, 728. Toxicity data: B. Melander *et al.*, *Acta Pharmacol. Toxicol.* **22**, 340 (1965). LC-MS/MS determn in serum: S. G. Delyle *et al.*, *Clin. Chim. Acta* **411** 438 (2010). Clinical study in treatment of acute upper gastrointestinal tract bleeding: D. Barer *et al.*, *N. Engl. J. Med.* **308**, 1571 (1983); in menorrhagia: J. Bonnar, B. L. Sheppard, *Br. Med. J.* **313**, 579 (1996). Review of pharmacology and therapeutic use: C. J. Dunn, K. L. Goa, *Drugs* **57**, 1005-1032 (1999); of clinical experience in menorrhagia: M. A. Lumsden, L. Wedisinghe, *Expert Opin. Pharmacother.* **12**, 2089-2095 (2011).

Relative stereochemistry

mp 386-392° (dec). Soly in water: about 1 g/6 ml. Very slightly sol in alcohol, ether. Practically insol in most other organic solvents. Chemically stable; not hygroscopic. LD$_{50}$ in mice, rats (mg/kg): 1500, 1200 i.v. (Melander).

cis-form. [1197-17-7] mp 236-238° (dec).

THERAP CAT: Hemostatic.

9731. Tranilast. [53902-12-8] 2-[[3-(3,4-Dimethoxyphenyl)-1-oxo-2-propen-1-yl]amino]benzoic acid; *N*-(3',4'-dimethoxycinnamoyl)anthranilic acid; N-5'; Rizaben. C$_{18}$H$_{17}$NO$_5$; mol wt 327.34. C 66.05%, H 5.23%, N 4.28%, O 24.44%. Orally active anti-allergic agent. Prepn: K. Harita *et al.*, **DE 2402398**; *idem*, **US 3940422** (1974, 1976 both to Kissei). Pharmacological properties: H. Azuma *et al.*, *Br. J. Pharmacol.* **58**, 483 (1976). Mechanism of action study: Y. Iijima *et al.*, *Biochem. Biophys. Res. Commun.* **93**, 912 (1980). Clinical study in pediatric bronchial asthma: H. Shioda, *Allergy* **34**, 213 (1979). Toxicity studies: M. Nakazawa *et al.*, *Oyo Yakuri* **12**, 385, 407 (1976), *C.A.* **88**, 115327-8 (1978). Series of articles on teratogenicity tests: *Iyakuhin Kenkyu* **9**, 148-193 (1978), *C.A.* **88**, 130930f, 146300m-303q (1978).

Crystals from chloroform, mp 211-213°. LD$_{50}$ in male, female mice, male, female rats (mg/kg): 780, 680, 1600, 1100 orally; 410, 385, 405, 395 i.p.; 2630, 2820, 3630, 3060 s.c. (Nakazawa, p 385).

THERAP CAT: Antiallergic.

9732. Transferrins. A group of homologous non-heme, iron-binding glycoproteins of approx mol wts of 76,000-81,000. They are widely distributed in a variety of physiological fluids and cells, esp in the sera of most vertebrates, in egg whites and in mammalian milk, tears and leukocytes. They are involved in iron transport to developing red cells for hemoglobin synthesis. Each protein molecule specifically binds with two Fe^{3+} ions to form salmon-pink complexes; bicarbonate or carbonate ions are involved in the formation of these colored complexes. Reviews of isoln, properties and biological functions: Feeney, Komatsu, *Struct. Bonding* **1**, 149-206 (1966); Aisen, "The Transferrins" in *Inorganic Biochemistry* vol. 1, G. L. Eichhorn, Ed. (Elsevier, New York, 1973) pp 280-305; Bezkorovainy, Zschocke, *Arzneim.-Forsch.* **24**, 476-485, 726-737 (1974); P. Aisen, A. Leibman, *Bioinorg. Chem.* **II**, K. N. Raymond, Ed. (ACS, Washington, 1977) pp 104-126; P. Aisen, I. Listowsky, *Annu. Rev. Biochem.* **49**, 357-393 (1980). Review of transferrin receptors: R. Newman *et al.*, *Trends Biochem. Sci.* **7**, 397-399 (1982).

Serum transferrin. β$_1$-Metal-combining protein; siderophilin. Commonly called transferrin. Structure of carbohydrate moiety: Jamieson *et al.*, *J. Biol. Chem.* **246**, 3686 (1971). Composed of two homologous domains, each containing a binding site for metal ions. The sites are similar, but not identical, in their metal-binding properties. X-ray studies: L. J. DeLucas *et al.*, *J. Mol. Biol.* **123**, 285 (1978). Resolution of the two sites by Eu(III) excitation spectroscopy: P. B. O'Hara, R. Bersohn, *Biochemistry* **21**, 5269 (1982). N-Terminal amino acid sequence of human serum transferrin: M.-H. Metz-Boutigue *et al.*, *Biochim. Biophys. Acta* **670**, 243 (1981). Absorption max of human serum Fe^{3+}-transferrin: about 465 nm (E$_{1cm}^{1\%}$ 0.57); uv max: 280 nm (E$_{1cm}^{1\%}$ 14.3).

Conalbumin. Ovotransferrin; Diarconal. Isolated from egg white; distinguished from ovalbumin by its lower thermal coagulation point: Osborne, Campbell, *J. Am. Chem. Soc.* **22**, 422 (1900). Sepn from other egg-white proteins: Longworth *et al.*, *ibid.* **62**, 2580 (1940). Primary structure of hen ovotransferrin: J. Williams *et al.*, *Eur. J. Biochem.* **122**, 297 (1982). Purification, characterization and function of the iron-binding fragments: W.-M. Keung *et al.*, *J. Biol.*

Chem. **257**, 1177, 1184 (1982). Antibacterial activity: P. Valenti *et al.*, *Antimicrob. Agents Chemother.* **21**, 840 (1982). Absorption max of Fe^{3+}-complex: 470 nm ($E_{1cm}^{1\%}$ 0.62).

Lactoferrin. Lactotransferrin. Important component of the human milk bacteriostatic system; also found in human and bovine tear proteins. Isoln from human whey by a single chromatographic step: L. Bläckberg, O. Hernell, *FEBS Lett.* **109**, 180 (1980). Sequential purification of lactoferrin, lysozyme, and secretory IgA from human milk: M. Boesman-Finkelstein, R. A. Finkelstein, *ibid.* **144**, 1 (1982). Partial C-terminal amino acid sequence of human lactoferrin: M.-H. Metz-Boutigue *et al.*, *ibid.* **142**, 107 (1982).

9733. Transforming Growth Factor-β. TGF-β. Family of multifunctional cytokines that regulate cellular differentiation, motility and growth. Also regulate the synthesis and deposition of the extracellular matrix. Involved in various physiological processes including embryogenesis, immunoregulation, bone remodeling and wound healing. Secreted by virtually all cell types as a biologically inactive (latent) form which is stored at the cell surface and in the extracellular matrix. The mature, active cytokine is a homodimer; mol wt 25 kDa. At least 5 isoforms have been identified. The three mammalian isoforms (TGF-β_1, β_2, and β_3) exhibit 70-80% sequence homology, bind to the same receptors and exert similar biological effects. TGF-β_1 is the most abundant. Biological effects are mediated by binding to membrane receptors that exist on virtually all cells. Released from degranulating platelets at the site of a wound, TGF-β initiates angiogenesis and collagen synthesis. Acts as a chemoattractant and activator of macrophages and fibroblasts. Dysregulation is implicated in the pathogenesis of fibrotic diseases. Inhibits the growth of most epithelial and lymphoid cells. Escape from this normal control mechanism is implicated in carcinogenic transformation. Initial description: J. E. DeLarco, G. J. Todaro, *Proc. Natl. Acad. Sci. USA* **75**, 4001 (1978). Crystal structure of TGF-β_2: S. Daopin *et al.*, *Science* **257**, 369 (1992). Review of physiological actions: A. B. Roberts, M. B. Sporn, *Growth Factors* **8**, 1-9 (1993); of latent forms, binding proteins and receptors: K. Miyazono *et al.*, *ibid.* 11-22. Proposed mechanisms for tumor cell transformation: M. J. Newman, *Cancer Metastasis Rev.* **12**, 239-254 (1993). Role in embryogenesis: N. L. McCartney-Francis, S. M. Wahl, *J. Leukocyte Biol.* **55**, 401-409 (1994); in tissue fibrosis: W. A. Border, N. A. Noble, *N. Engl. J. Med.* **331**, 1286-1292 (1994); in kidney disease: K. Sharma, F. N. Ziyadeh, *Am. J. Physiol.* **266**, F829-F842 (1994). Review of potential clinical applications in oncology: J. Kekow, G. J. Wiedemann, *Int. J. Oncol.* **7**, 177-182 (1995).

9734. Tranylcypromine. [155-09-9] *rel*-(1*R*,2*S*)-2-Phenylcyclopropanamine; SKF-385. $C_9H_{11}N$; mol wt 133.19. C 81.16%, H 8.32%, N 10.52%. Monoamine oxidase inhibitor. Prepn: Burger, Yost, *J. Am. Chem. Soc.* **70**, 2198 (1948); R. E. Tedeschi, US **2997422** (1961 to SK & F).

Relative stereochemistry

Liquid, bp$_{1.5-1.6}$ 79-80°.
Hydrochloride. $C_9H_{11}N.HCl$. Crystals from ethyl acetate + ether, mp 164-166°.
Sulfate. [13492-01-8] Parnate; Tylciprine. $(C_9H_{11}N)_2.H_2SO_4$; mol wt 364.46. Crystals, sol in water; very slightly sol in alcohol, ether. Practically insol in chloroform.
THERAP CAT: Antidepressant.

9735. Trapidil. [15421-84-8] *N*,*N*-Diethyl-5-methyl-[1,2,4]-triazolo[1,5-*a*]pyrimidin-7-amine; 7-diethylamino-5-methyl-*s*-triazolo[1,5-*a*]pyrimidine; trapymin; AR-12008; Avantrin; Rocornal. $C_{10}H_{15}N_5$; mol wt 205.27. C 58.51%, H 7.37%, N 34.12%. First triazolopyrimidine registered as a drug. Prepn: E. Tenor *et al.*, **DD 55956** (1967), *C.A.* **67**, 90830f (1967); E. Tenor, R. Ludwig, *Pharmazie* **26**, 534 (1971). Physical and chemical properties: S. Pfeifer *et al.*, *ibid.* 539. Pharmacology and toxicology: H. Fuller *et al.*, *ibid.* 554. Metabolic studies: S. Pfeifer *et al.*, *ibid.* 549; *ibid.* **27**, 752 (1972); I. Bornschein *et al.*, *ibid.* **33**, 51 (1978). Mechanism of ac-

tion: K. Satoh *et al.*, *Arzneim.-Forsch.* **30**, 1264 (1980). Antiarrhythmic activities in rabbits: M. Sakanashi *et al.*, *ibid.* **33**, 215 (1983). Clinical hemodynamic effects: M. Di Donato *et al.*, *ibid.* **35**, 1295 (1985). GC determn in biological fluids: A. Marzo *et al.*, *ibid.* **37**, 947 (1987).

White to yellowish, odorless and bitter crystalline powder, mp 98-99.4° (Pfeifer); 102-104° from heptane (Tenor). Eutectic temp of mixture with azobenzene: 48°. Very sol in water, 1*N* sulfuric acid, 10% ammonium hydroxide; easily sol in methanol, isopropanol, *n*-butanol, chloroform and benzene; sol in ether. Practically insol in hexane, heptane. pK$_s$ = 2.79. uv max (methanol): 222, 270, 307 nm (log ε 4.28, 3.83, 4.28). Very stable except under extremely alk conditions. LD$_{50}$ in mice, rats (mg/kg): 115, 76 i.v.; 380, 235 orally; 155, 100 i.p.; 132, 100 s.c. (Fuller).
Hydrochloride. $C_{10}H_{15}N_5.HCl$. mp 212°.
THERAP CAT: Vasodilator (coronary).

9736. Trastuzumab. [180288-69-1] Anti-(human p185neu receptor) immunoglobulin G1 (human-mouse monoclonal rhuMab HER2 γ_1-chain) disulfide with human-mouse monoclonal rhuMab HER2 light chain, dimer; rhuMab HER2; Herceptin. Humanized monoclonal antibody directed against the protein product of the HER2/*neu* oncogene which is homologous to the human epidermal growth factor receptor and is overexpressed by certain tumor cells. Constructed by inserting the antigen binding regions of murine monoclonal antibody 4D5 into the framework of a consensus human immunoglobulin G$_1$. Prepn: R. M. Hudziak *et al.*, **WO 8906692**; *eidem*, **US 5677171** (1989, 1997 both to Genentech); P. Carter *et al.*, *Proc. Natl. Acad. Sci. USA* **89**, 4285 (1992). Analysis by capillary electrophoresis: G. Hunt *et al.*, *J. Chromatogr. A* **744**, 295 (1996). Clinical evaluation in combination with cisplatin in breast cancer: M. D. Pegram *et al.*, *J. Clin. Oncol.* **16**, 2659 (1998). Review of clinical safety and tolerability in breast cancer: S. Rueckert *et al.*, *Expert Opin. Biol. Ther.* **5**, 853-866 (2005). Clinical effect on disease-free survival: L. Gianni *et al.*, *Lancet Oncol.* **12**, 236 (2011).

Trastuzumab emtansine. [1018448-65-1] Anti-(human p185neu receptor) immunoglobulin G1 (human-mouse monoclonal rhuMab HER2 γ_1-chain) disulfide with human-mouse monoclonal rhuMab HER2 light chain, dimer tetramide with $N^{2'}$-[3-[[1-[(4-carboxycyclohexyl)methyl]-2,5-dioxo-3-pyrrolidinyl]thio]-1-oxopropyl]-$N^{2'}$-deacetylmaytansine; trastuzumab-MCC-DM1; T-DM1; PRO-132365. Tumor activated prodrug of maytansinoid DM1, *q.v.*, conjugated to trastuzumab via a nonreducible thioether linker. Delivers the cytotoxic drug to HER2-expressing tumor cells where the immunoconjugate is converted to the active moiety. Prepn: A. A. Wakankar *et al.*, *Bioconjugate Chem.* **21**, 1588 (2010). Clinical evaluation in HER2-positive breast cancer: I. E. Krop *et al.*, *J. Clin. Oncol.* **28**, 2698 (2010).
THERAP CAT: Antineoplastic.

9737. Traumatic Acid. [6402-36-4] (2*E*)-2-Dodecenedioic acid; *trans*-2-dodecenedioic acid; 1-decene-1,10-dicarboxylic acid. $C_{12}H_{20}O_4$; mol wt 228.29. C 63.14%, H 8.83%, O 28.03%. A wound hormone of plants. The naturally occurring form is *trans*. Isoln from pods of green beans: English *et al.*, *Proc. Natl. Acad. Sci. USA* **25**, 323 (1939). Synthesis: *eidem*, *J. Am. Chem. Soc.* **61**, 3434 (1939); **US 2339259** (1944); **US 2391824** (1945); Truscheit, Eiter, *Ann.* **658**, 86 (1962); Dolezal, *Collect. Czech. Chem. Commun.* **35**, 1932 (1970); Schreurs *et al.*, *Recl. Trav. Chim. Pays-Bas Belg.* **90**, 1331 (1971); Prakasa Rao, Nayak, *Synthesis* **1975**, 608; J. H. Babler, R. K. Moy, *Synth. Commun.* **9**, 669 (1979).

Crystals from alc, acetone, or 1,2-dimethoxyethane, mp 166-167°. bp$_{0.001}$ 150-160°. Very sparingly sol in water; sol in alc, ether, benzene, chloroform.

cis-**Form.** [6556-35-0] (2*Z*)-2-Dodecenedioic acid. Synthesis: Lauer, Gensler, *J. Am. Chem. Soc.* **67**, 1171 (1945). Crystals from ethyl acetate and petr ether, mp 67-68°.

9738. Travoprost. [157283-68-6] (5*Z*)-7-[(1*R*,2*R*,3*R*,5*S*)-3,5-Dihydroxy-2-[(1*E*,3*R*)-3-hydroxy-4-[3-(trifluoromethyl)phenoxy]-1-buten-1-yl]cyclopentyl]-5-heptenoic acid 1-methylethyl ester; (+)-16-[3-(trifluoromethyl)phenoxy]-17,18,19,20-tetranorprostaglandin F$_{2\alpha}$ isopropyl ester; (+)-9α,11α,15-trihydroxy-16-(3-trifluoromethylphenoxy)-17,18,19,20-tetranor-5-*cis*-13-*trans*-prostadienoic acid isopropyl ester; AL-6221; Travatan. C$_{26}$H$_{35}$F$_3$O$_6$; mol wt 500.56. C 62.39%, H 7.05%, F 11.39%, O 19.18%. Selective FP prostaglandin receptor agonist. Isopropyl ester of (+)-fluprostenol, *q.v.* General prepn (not claimed): J. W. Stjernschantz, **EP 364417** (1989 to Pharmacia). Large scale synthesis: L. T. Boulton *et al.*, *Org. Process Res. Dev.* **6**, 138 (2002). Pharmacology: M. R. Hellberg *et al.*, *J. Ocul. Pharmacol. Ther.* **17**, 421 (2001). LC/MS/MS determn in plasma: B. A. McCue *et al.*, *J. Pharm. Biomed. Anal.* **28**, 199 (2002). Ocular hypotensive effects in dogs: A. B. Carvalho *et al.*, *Vet. Ophthalmol.* **9**, 121 (2006). Clinical trial in glaucoma or ocular hypertension: R. L. Fellman *et al.*, *Ophthalmology* **109**, 998 (2002); in combination with timolol: J. S. Schuman *et al.*, *Am. J. Ophthalmol.* **140**, 242-250 (2005).

Colorless oil. $[\alpha]_D^{20}$ +14.6° (c = 1.0 in methylene chloride). Very sol in acetonitrile, methanol, octanol, chloroform. Practically insol in water.

THERAP CAT: Antiglaucoma.

THERAP CAT (VET): Antiglaucoma.

9739. Traxanox. [58712-69-9] 9-Chloro-7-(1*H*-tetrazol-5-yl)-5*H*-[1]benzopyrano[2,3-*b*]pyridin-5-one; 9-chloro-7-(5-1*H*-tetrazolyl)-5-oxo-5*H*-[1]benzopyrano[2,3-*b*]pyridine. C$_{13}$H$_6$ClN$_5$O$_2$; mol wt 299.67. C 52.10%, H 2.02%, Cl 11.83%, N 23.37%, O 10.68%. Antiallergic agent which selectively inhibits release of allergic mediators from mast cells. Prepn: T. Oe, M. Tsuruda, **DE 2521980**; *eidem*, **US 4085111** (1975, 1978 both to Yoshitomi). Antianaphylactic activity in animals: K. Goto *et al.*, *Jpn. J. Pharmacol.* **30**, 537 (1980). Mechanism of action study: K. Goto *et al.*, *Int. Arch. Allergy Appl. Immunol.* **68**, 332 (1982). Clinical pharmacology and pharmacokinetics: A. Ebihara *et al.*, *Arzneim.-Forsch.* **37**, 1388 (1987). Metabolism in humans: M. Tateno *et al.*, *Yakuri to Chiryo* **16**, 3251 (1988), *C.A.* **110**, 18040n (1989).

mp >300°.

Sodium salt pentahydrate. Y-12141; Clearnal. C$_{13}$H$_5$ClN$_5$NaO$_2$.5H$_2$O; mol wt 411.73.

THERAP CAT: Antiallergic; antiasthmatic.

9740. Trazodone. [19794-93-5] 2-[3-[4-(3-Chlorophenyl)-1-piperazinyl]propyl]-1,2,4-triazolo[4,3-*a*]pyridin-3(2*H*)-one. C$_{19}$-H$_{22}$ClN$_5$O; mol wt 371.87. C 61.37%, H 5.96%, N 9.53%, N 18.83%, O 4.30%. Serotonin-2 antagonist/reuptake inhibitor (SARI). Prepn: Palazzo, Silvestrini, **US 3381009** (1968 to Angelini Francesco). Pharmacology: Catanese, Lisciani, *Boll. Chim. Farm.* **109**, 369 (1970); B. Silvestrini, E. Quadri, *Eur. J. Pharmacol.* **12**, 231 (1970). Analytical data: Baiocchi *et al.*, *Arzneim.-Forsch.* **23**,

400 (1973). Crystal structure: J. P. Fillers, S. W. Hawkinson, *Acta Crystallogr.* **B35**, 498 (1979). *Pharmacological and Biochemical Properties of Drug Substances* **vol. 3**, M. E. Goldberg, Ed. (Am. Pharm. Assoc., Washington, DC, 1981) pp 94-119. Comprehensive description: D. Gorecki, R. Verbeeck, *Anal. Profiles Drug Subs.* **16**, 693-729 (1986). Comparative clinical trial in major depressive disorder: C. Munizza *et al.*, *Curr. Med. Res. Opin.* **22**, 1703 (2006). Review of therapeutic use in depression: M. Haria *et al.*, *Drugs Aging* **4**, 331-355 (1994).

Crystals, mp 86-87°. Also reported as mp 96° (Baiocchi). pKa (50% ethanol): 6.14. Highly lipophilic.

Hydrochloride. [25332-39-2] AF-1161; Desyrel; Molipaxin; Oleptro; Thombran; Trazolan; Trittico. C$_{19}$H$_{22}$ClN$_5$O.HCl; mol wt 408.33. White, odorless plates from ethanol, mp 223°. Sparingly sol in water, ethanol, methanol, chloroform. Practically insol in common organic solvents. uv max (water): 211, 246, 274, 312 nm (ε 50100, 11730, 3840, 3840). LD$_{50}$ i.v. in mice: 96 mg/kg (Silvestrini, Quadri).

THERAP CAT: Antidepressant.

9741. Trebananib. [894356-79-7] Immunoglobulin Gl (synthetic human Fc domain fragment) fusion protein with angiopoietin 1/angiopoietin 2-binding peptide (synthetic); anti-(human angiopoietin-2/angiopoietin-1) peptibody (synthetic clone 2xCon4 (C)); tovasanib; AMG-386. Recombinant peptide-Fc fusion protein (peptibody) that binds to and neutralizes angiopoetin 1 and 2, supressing angiogenensis. Prepn: J. Oliner, H. Min, **WO 03057134**, *eidem*, **US 7138370** (2003, 2006 both to Amgen); and angiopoietin neutralization: J. Oliner *et al.*, *Cancer Cell* **6**, 507 (2004). Preclinical antiangiogenic activity and tumor supression: A. Coxon *et al.*, *Mol. Cancer Ther.* **9**, 2641 (2010). Clinical pharmacokinetics and evaluation in solid tumors: R. S. Herbst *et al.*, *J. Clin. Oncol.* **27**, 3557 (2009); in combination with chemotherapy: A. C. Mita *et al.*, *Clin. Cancer Res.* **16**, 3044 (2010). Review of development and therapeutic potential: J. Neal, H. Wakelee, *Curr. Opin. Mol. Ther.* **12**, 487-495 (2010).

THERAP CAT: Antineoplastic.

9742. Trehalose. [99-20-7] α-D-Glucopyranosyl-α-D-glucopyranoside; mushroom sugar; mycose; α,α-trehalose. C$_{12}$H$_{22}$O$_{11}$; mol wt 342.30. C 42.11%, H 6.48%, O 51.41%. Non-reducing disaccharide found in fungi, bacteria, yeasts, and insects; 45% as sweet as sucrose. Provides the energy source for flight in many insects. Incorporated into mycobacterial structural glycolipids such as cord factors, *q.v.* Isoln from the ergot of rye: H. A. L. Wiggers, *Ann.* **1** 129 (1832). Prepn and review of early history: T. S. Harding, *Sugar* **25**, 476-478 (1923). Isoln from yeast: E. M. Koch, F. C. Koch, *Science* **61**, 570 (1925); L. C. Stewart *et al.*, *J. Am. Chem. Soc.* **72**, 2059 (1950). Synthesis: Lemieux, Bauer, *Can. J. Chem.* **32**, 340 (1954). Crystal structure: G. A. Jeffrey, R. Nanni, *Carbohydr. Res.* **137**, 21 (1985). Review of metabolism: A. D. Elbein, *Adv. Carbohydr. Chem. Biochem.* **30**, 227-256 (1974). *In vitro* evaluation in cryopreservation of human oocytes: A. Eroglu *et al.*, *Fertil. Steril.* **77**, 152 (2002). Use in freeze-drying human platelets: W. F. Wolkers *et al.*, *Cryobiology* **42**, 79 (2001); *eidem*, *Cell Preservation Technol.* **1**, 175 (2003). Review of properties, toxicity and safety studies: A. B. Richards *et al.*, *Food Chem. Toxicol.* **40**, 871-898 (2002); of stabilizing functions and applications: T. Higashiyama, *Pure Appl. Chem.* **74**, 1263-1269 (2002).

Dihydrate. [6138-23-4] Orthorhombic, bisphenoidal crystals from dil alcohol. Sweet taste. mp 96.5-97.5°. The water of crystn escapes around 130°. Anhydrous trehalose melts at 203°. $[\alpha]_D^{20}$ +178° (c = 7 of the dihydrate). Sol in water, hot alcohol. Insol in ether. Does not reduce Fehling's soln. Is fermented by yeast. Is not split by α-glucosidase. Acid hydrolysis gives 2 mols D-glucose.

USE: Stabilizes cells during freezing, freeze-drying and air-drying. Sweetener and stabilizer in foods; cryoprotectant for freeze-dried foods. Additive in cosmetics and personal care products.

9743. Tremetone. [4976-25-4] 1-[(2R)-2,3-Dihydro-2-(1-methylethenyl)-5-benzofuranyl]ethanone; 2,3-dihydro-2-isopropen-yl-5-benzofuranyl methyl ketone; 2-isopropenyl-2,3-dihydro-5-ace-tylbenzofuran. $C_{13}H_{14}O_2$; mol wt 202.25. C 77.20%, H 6.98%, O 15.82%. Principal ketone suspected of being the active toxin of *Eupatorium urticaefolium* Reichard, *Compositae* (white snakeroot). Isoln and structure: Bonner, DeGraw, *Tetrahedron* **18**, 1295 (1962). Synthesis of dihydrotremetone: DeGraw, Bonner, *ibid.* 1311. Synthesis of racemic tremetone: DeGraw *et al., ibid.* **19**, 19 (1963); Bohlmann, Buehmann, *Ber.* **105**, 863 (1972). Abs config: Bonner *et al., Tetrahedron* **20**, 1419 (1964). *Review:* Christensen, *Econ. Bot.* **19**, 293-300 (1965).

Liquid. $[\alpha]_D^{28}$ −59.6° (c = 5.52 in absolute ethanol). n_D^{25} 1.5658. d_4^{28} 1.080. uv max (ethanol): 227, 280, 285 nm (ε 11,950, 12,600, 12,300).

Dihydrotremetone. (R)-1-[2,3-Dihydro-2-(1-methylethyl)-5-benzofuranyl]ethanone. $C_{13}H_{16}O_2$. Liquid. bp 216-221°. $[\alpha]_D^{25}$ −47.0° (c = 1.78 in abs ethanol). uv max (ethanol): 231, 279 nm (ε 39,500; 18,800).

9744. Tremorine. [51-73-0] 1,1'-(2-Butyne-1,4-diyl)bispyr-rolidine; 1,1'-(2-butynylene)dipyrrolidine; 1,4-dipyrrolidino-2-bu-tyne. $C_{12}H_{20}N_2$; mol wt 192.31. C 74.95%, H 10.48%, N 14.57%. Prepn from pyrrolidine + 1,4-dichloro-2-butyne: Maier, **DE 896810** (1953 to BASF); Reppe *et al., Ann.* **596**, 79 (1955); Biel, DiPierro, *J. Am. Chem. Soc.* **80**, 4609 (1958). Review of chemical and biological studies: Karlen, *Acta Pharm. Suec.* **1970**, 169-200; *see also* Kolla, Obvintseva, *Farmakol. Toksikol.* **36**, 736-745 (1973).

Liquid. $bp_{2.5}$ 116-116.5°; bp $_{0.1}$ 93-95°.
Methiodide. $C_{12}H_{20}N_2.2CH_3I$. Crystals, mp 239-240°.
USE: To produce exptl parkinsonism: Everett *et al., Science* **124**, 79 (1956).

9745. Trenbolone. [10161-33-8] (17β)-17-Hydroxyestra-4,-9,11-trien-3-one; 4,9,11-estratrien-17β-ol-3-one; 17β-hydroxy-19-norandrosta-4,9,11-trien-3-one; 19-norandrosta-4,9,11-trien-17β-ol-3-one; trienbolone; trienolone. $C_{18}H_{22}O_2$; mol wt 270.37. C 79.96%, H 8.20%, O 11.83%. Prepn of base: Velluz *et al., C. R. Hebd. Seances Acad. Sci.* **257**, 569 (1963); Heller *et al., Steroids* **10**, 211 (1967). Base and 17-acetate: **FR M1958** and **GB 1035683** (1963 and 1966 to Roussel-UCLAF), *C.A.* **60**, 3039h (1964); **65**, 17027c (1966). 17-Cyclohexylmethylcarbonate: Nedelec, Coster-ousse, **FR M5979** (1968 to Roussel-UCLAF), *C.A.* **71**, 50356g (1969). Pharmacology of the acetate: Krüskemper *et al., Arzneim.-Forsch.* **17**, 449 (1967). Animal studies: Beranger, Malterre, *C. R. Seances Soc. Biol. Ses Fil.* **162**, 1157 (1968); Best, *Vet. Rec.* **91**, 624 (1972). Environmental degradation study: B. Schiffer *et al., Environ. Health Perspect.* **109**, 1145 (2001).

Crystals, mp 186°. $[\alpha]_D^{20}$ +19° (c = 0.45 in ethanol). Also reported as mp 183-186° from acetone-water. uv max: 239, 340.5 nm (ε 5260, 28000) (Heller).

Acetate. [10161-34-9] (17β)-17-Acetyloxyestra-4,9,11-trien-3-one; 17β-acetoxy-3-oxoestra-4,9,11-triene; 17β-acetoxyestra-4,-9,11-trien-3-one; Finaplix; Parabolan. $C_{20}H_{24}O_3$; mol wt 312.41. Crystals, mp 96-97°. $[\alpha]_D^{20}$ +36.8° (c = 0.37 in methanol).

Cyclohexylmethylcarbonate. [23454-33-3] (17β)-17-[(Cyclo-hexylmethoxy)carbonyl]oxyestra-4,9,11-trien-3-one; trenbolone hexahydrobenzylcarbonate. $C_{26}H_{34}O_4$; mol wt 410.55. Crystals from cyclohexane-petr ether, mp 90-95°. $[\alpha]_D^{20}$ +41.6° (c = 0.5 in ethanol).

Note: This is a controlled substance (anabolic steroid): **21 CFR**, 1308.13, as defined in 1300.01.

THERAP CAT (VET): Anabolic.

9746. Trengestone. [5192-84-7] (9β,10α)-6-Chloropregna-1,4,6-triene-3,20-dione; 6-chloro-1,6-didehydroretroprogesterone; 6-chloro-1,6-bisdehydroretroprogesterone; Ro-4-8347; Retroid. $C_{21}H_{25}ClO_2$; mol wt 344.88. C 73.14%, H 7.31%, Cl 10.28%, O 9.28%. A member of a class of hormonally active steroids termed "retrosteroids", which are characterized by a (9β, 10α)-configuration in contrast to the usual (9α,10β)-configuration of steroids. Prepn: Threadgold, **BE 652597** corresp to Reerink *et al.*, **US 3422122** (1965, 1969 to Phillips). Activity studies: Sadovsky *et al., Gynecol. Invest.* **1**, 319 (1970); Kalra *et al., J. Endocrinol.* **51**, 675 (1971). Metabolism: Breuer *et al., Acta Endocrinol.* **74**, 127 (1973); Dixon *et al., Steroids* **22**, 35 (1973).

Crystals from acetone, mp 208-209° (dec). uv max: 229, 253, 302 nm (ε 11500, 10520, 10650).

THERAP CAT: Progestogen.

9747. Trepibutone. [41826-92-0] 2,4,5-Triethoxy-γ-oxoben-zenebutanoic acid; 3-(2,4,5-triethoxybenzoyl)propionic acid; AA-149; Supacal. $C_{16}H_{22}O_6$; mol wt 310.35. C 61.92%, H 7.15%, O 30.93%. Prepn: T. Murata *et al.*, **DE 2244324** corresp to **US 3943169** (1973, 1976 both to Takeda). Properties and stabilities: M. Mitani *et al., Takeda Kenkyushoho* **36**, 206 (1977), *C.A.* **88**, 126284f (1978). Pharmacokinetics: *eidem, ibid.* 215, *C.A.* **88**, 115067m (1978). Biotransformation: T. Kobayashi *et al., Xenobiotica* **8**, 535 (1978). Metabolism studies: S. Tanayama *et al., ibid.* 365, 377. Spasmolytic action in dogs: H. Satoh *et al., Eur. J. Pharmacol.* **48**, 309 (1978). Mechanism of choleretic action: *eidem, ibid.* 125. Toxicity studies: S. Sato *et al., Takeda Kenkyushoho* **36**, 263 (1977), *C.A.* **88**, 11535x (1978).

Colorless needles from aq ethanol or plates from aq acetone, mp 150-151°. Stable to heat, humidity, indoor diffused sunlight. Aq solns heated to 100° for 10 hr showed no degradation.

THERAP CAT: Choleretic; antispasmodic.

9748. Treprostinil. [81846-19-7] 2-[[(1R,2R,3aS,9aS)-2,3,-3a,4,9,9a-Hexahydro-2-hydroxy-1-[(3S)-3-hydroxyoctyl]-1H-benz-[f]inden-5-yl]oxy]acetic acid; 9-deoxy-2',9α-methano-3-oxa-4,5,6-trinor-3,7-(1',3'-interphenylene)-13,14-dihydroprostaglandin F_1. $C_{23}H_{34}O_5$; mol wt 390.52. C 70.74%, H 8.78%, O 20.48%. Synthetic analog of prostacyclin, *q.v.* Prepn: P. A. Aristoff *et al.*, **GB 2070596**; *idem*, **US 4306075** (both 1981 to Upjohn). Pharmacology:

R. P. Steffen, M. de la Mata, *Prostaglandins Leukotrienes Essent. Fatty Acids* **43**, 277 (1991). RIA determn in plasma: J. W. A. Findlay *et al.*, *ibid.* **48**, 167 (1993). Clinical pharmacokinetics: K. Laliberte *et al.*, *J. Cardiovasc. Pharmacol.* **44**, 209 (2004). Clinical trial in pulmonary arterial hypertension: G. Simonneau *et al.*, *Am. J. Respir. Crit. Care Med.* **165**, 800 (2002); in combination with bosentan: R. L. Benza *et al.*, *Chest* **134**, 139 (2008). Review of clinical experience: N. Skoro-Sajer *et al.*, *Vasc. Health Risk Manag.* **4**, 507-513 (2008).

Crystals from ethyl acetate in hexane, mp 121-123°.

Sodium salt. [289480-64-4] 15AU81; BW15AU; U-62840; UT-15; Remodulin; Tyvaso. $C_{23}H_{33}NaO_5$; mol wt 412.50.

THERAP CAT: In treatment of pulmonary hypertension.

9749. Tretoquinol. [21650-42-0] 1,2,3,4-Tetrahydro-1-[(3,-4,5-trimethoxyphenyl)methyl]-6,7-isoquinolinediol; 1-(3′,4′,5′-trimethoxybenzyl)-6,7-dihydroxy-1,2,3,4-tetrahydroisoquinoline; trimethoquinol. $C_{19}H_{23}NO_5$; mol wt 345.40. C 66.07%, H 6.71%, N 4.06%, O 23.16%. Synthesis of *dl*-form: Yamato *et al.*, *Tetrahedron Suppl.* **8**, 129 (1966). Pharmacology: Fogelman, Grundy, *Br. J. Pharmacol.* **38**, 416 (1970). Metabolic studies: Meshi *et al.*, *Biochem. Pharmacol.* **19**, 2937 (1970); Satoh *et al.*, *Chem. Pharm. Bull.* **19**, 667 (1971).

***dl*-Form hydrochloride.** [18559-63-2] Pale yellow crystals from methanol + ether, decomp 224.5-226°.

***l*-Form hydrochloride.** [18559-59-6] AQ-110; Inolin; Vems. $C_{19}H_{23}NO_5 \cdot HCl$; mol wt 381.85. Pale yellow crystals, freely sol in water; sol in alcohol.

THERAP CAT: Bronchodilator.

9750. TRH. [24305-27-9] 5-Oxo-L-prolyl-L-histidyl-L-prolinamide; thyrotropin-releasing factor; thyrotropin releasing hormone; TRF; protirelin; lopremone (rescinded USAN); TSH-releasing factor; pyroglutamylhistidylprolinamide; thyroliberin; Antepan; Stimu-TSH; Thypinone; Thyrefact. $C_{16}H_{22}N_6O_4$; mol wt 362.39. C 53.03%, H 6.12%, N 23.19%, O 17.66%. A hypothalamic neurohormone which stimulates the release and synthesis of TSH, *q.v.*, from the anterior pituitary via the hypophyseal portal system; the first of the hypothalamic regulatory hormones to be isolated, characterized, and synthesized. Isoln from bovine hypothalami: Schreiber *et al.*, *Experientia* **18**, 338 (1962); Schally *et al.*, *Endocrinology* **78**, 726 (1966). Purification from ovine hypothalami: Guillemin *et al.*, *C. R. Seances Acad. Sci. Ser. D* **262**, 2278 (1966). Isoln from porcine hypothalami and properties: Schally *et al.*, *Biochem. Biophys. Res. Commun.* **25**, 165 (1966); Schally *et al.*, *J. Biol. Chem.* **244**, 4077 (1969). Structural studies: Folkers *et al.*, *Biochem. Biophys. Res. Commun.* **37**, 123 (1969); Burgus *et al.*, *C. R. Seances Acad. Sci. Ser. D* **269**, 226 (1969). Identity of isolated TRH with synthetic tripeptide: Burgus *et al.*, *ibid.* 1870; Bowers *et al.*, *Endocrinology* **86**, 573, 1143 (1970). Solubility: Burgus *et al.*, *Experientia* **23**, 417 (1967). Synthesis: Flouret, *J. Med. Chem.* **13**, 843 (1970); Chang *et al.*, *ibid.* **14**, 481 (1971); E. Gross *et al.*, *Angew. Chem. Int. Ed.* **12**, 664 (1972); P. G. Pietta *et al.*, *J. Org. Chem.* **39**, 44 (1974).

Review of synthetic methods: Rivier, *Methods Enzymol.* **37**, 408 (1975). TRH is also believed to induce the secretion of the pituitary lactogenic hormone, prolactin, *q.v.*: Tasjian *et al.*, *Biochem. Biophys. Res. Commun.* **43**, 516 (1971); Bowers *et al.*, *ibid.* **51**, 512 (1973). It has been shown to block and reverse leukotriene-induced hypotension in the unanesthetized guinea pig: W. E. Lux *et al.*, *Nature* **302**, 822 (1983). Clinical studies: Hershman, *N. Engl. J. Med.* **290**, 886 (1974). Reviews of TRH and other hypothalamic releasing hormones: Schally *et al.*, *Recent Prog. Horm. Res.* **24**, 497 (1968); Burgus, Guillemin, *Annu. Rev. Biochem.* **39**, 499 (1970); *Polypeptide Hormones*, R. F. Beers, E. G. Bassett, Eds. (Raven Press, New York, 1980) pp 165-278.

5-oxo-Pro–His–ProNH$_2$

Purified TRH is partially sol in chloroform, highly sol in absolute methanol. Completely insol in pyridine. Inactivated by diazotized sulfanilic acid (Pauly reagent) and by plasma, serum, or whole blood *in vitro*. Resists inactivation by proteolytic enzymes.

Tartrate. [54974-58-4] Irtonin; Xantium. $C_{16}H_{22}N_6O_4 \cdot xC_4H_6O_6$

THERAP CAT: Prohormone.

9751. Triacetin. [102-76-1] 1,2,3-Propanetriol 1,2,3-triacetate; glyceryl triacetate; 1,2,3-triacetoxypropane; triacetyl glycerine; Enzactin; Fungacetin. $C_9H_{14}O_6$; mol wt 218.21. C 49.54%, H 6.47%, O 43.99%. Prepd by acetylation of glycerol: Dunbar, Bolstad, *J. Org. Chem.* **21**, 1041 (1956); by reaction of oxygen with a liquid phase mixture of allyl acetate and acetic acid using a bromide as catalyst: Keith, US 2911437 (1959 to Sinclair Refining). As an antifungal: GB 845029 (1960 to Wisc. Alumni Res. Found.). Acute toxicity: A. Wretlind, *Acta Physiol. Scand.* **40**, 338 (1957).

Colorless, somewhat oily liquid having a slight, fatty odor and a bitter taste. d_4^{25} 1.1562, d_4^{20} 1.1596, d_{20}^{20} 1.163. mp $-78°$. bp 258-260°, bp$_{40}$ 172°. n_D^{20} 1.4307. Sol in 14 parts water. Miscible with alcohol, ether, chloroform. Slightly sol in carbon disulfide. LD$_{50}$ i.v. in mice: 1600 ±81 mg/kg (Wretlind).

USE: As fixative in perfumery; solvent in manuf celluloid, photographic films. Technical triacetin (a mixture of mono-, di-, and small quantities of triacetin) as a solvent for basic dyes, particularly indulines, and tannin in dyeing.

THERAP CAT: Antifungal (topical).

9752. Triacetone Triperoxide. [17088-37-8] 3,3,6,6,9,9-Hexamethyl-1,2,4,5,7,8-hexoxonane; 3,3,6,6,9,9-hexamethyl-1,4,7-cyclononatriperoxane; TATP; trimeric acetone peroxide. $C_9H_{18}O_6$; mol wt 222.24. C 48.64%, H 8.16%, O 43.19%. Explosive solid; decomposes via an entropically favorable mechanism to form 4 gas phase molecules. Prepn: R. Wolffenstein, *Ber.* **28**, 2265 (1895); A. Baeyer, V. Villiger, *ibid.* **33**, 858 (1900); F. Acree, Jr., H. L. Haller, *J. Am. Chem. Soc.* **65**, 1652 (1943). Crystal structure determn: P. Groth, *Acta Chem. Scand.* **23**, 1311 (1969). Spectroscopic characterization and detection by ion mobility spectrometry: G. A. Buttigieg *et al.*, *Forensic Sci. Int.* **135**, 53 (2003). Decompn study: F. Dubnikova *et al.*, *J. Am. Chem. Soc.* **127**, 1146 (2005).

Crystals from methyl alcohol, mp 98°. d 1.22. *Explosive! Sensitive to mechanical stress and open flame. Highly volatile.*

USE: In explosives.

9753. 1-Triacontanol. [593-50-0] Melissyl alcohol; myricyl alcohol; 1-hydroxytriacontane. $C_{30}H_{62}O$; mol wt 438.83. C 82.11%, H 14.24%, O 3.65%. Present in plant cuticle waxes and in beeswax as the palmitate. Prepn: Robinson, *J. Chem. Soc.* **1934**, 1545; K. Maruyama *et al.*, *J. Org. Chem.* **45**, 737 (1980). Use as a plant growth regulator: S. K. Ries *et al.*, *Science* **195**, 1339 (1977); S. K. Ries, **BE 854587** corresp to **US 4150970** (1977, 1979 to Michigan State Univ.). Use of colloidally dispersed triacontanol in plant growth enhancement studies: R. G. Laughlin *et al.*, *Science* **219**, 1219 (1983).

Crystals. d_{95} 0.777. mp 87°. Practically insol in water; freely sol in benzene, ether; very slightly sol in cold, more sol in hot alcohol.

USE: Plant growth regulator.

9754. Triadimefon. [43121-43-3] 1-(4-Chlorophenoxy)-3,3-dimethyl-1-(1*H*-1,2,4-triazol-1-yl)-2-butanone; BAY MEB 6447; Bayleton. $C_{14}H_{16}ClN_3O_2$; mol wt 293.75. C 57.24%, H 5.49%, Cl 12.07%, N 14.31%, O 10.89%. Systemic triazole fungicide active against mildews and rusts of grains, fruits, vegetables and ornamentals. Prepn and activity: W. Meiser *et al.*, **DE 2201063**; *eidem*, **US 3912752** (1973, 1975 both to Bayer). Fungal metabolism: M. Gasztonyi, *Pestic. Sci.* **12**, 433 (1981). Photodegradation study: J. P. DaSilva *et al.*, *J. Photochem. Photobiol. A* **154**, 293 (2003). GC determn and persistence in soils: N. Singh, *J. Agric. Food Chem.* **53**, 70 (2005). Effect on monoamine uptake and release: Q. D. Walker, R. B. Mailman, *Toxicol. Appl. Pharmacol.* **139**, 227 (1996). *Reviews:* F. Michel, P. Pourcharesse, *Def. Veg.* **31**, 97-109 (1977); T. J. Martin, D. B. Morris, *Pflanzenschutz-Nachr.* **32**, 31-79 (1979).

Crystals, mp 82°. Soly in water at 20°: 260 mg/l. Moderately sol in most organic solvents except aliphatics. LD_{50} in male, female rats (mg/kg): 568, 363 orally (Michel, Pourcharesse).

USE: Systemic agricultural fungicide.

9755. Triadimenol. [55219-65-3] β-(4-Chlorophenoxy)-α-(1,1-dimethylethyl)-1*H*-1,2,4-triazole-1-ethanol; 1-(4-chlorophenoxy)-3,3-dimethyl-1-(1*H*-1,2,4-triazol-1-yl)butan-2-ol; BAY KWG 0519; Bayfidan; Baytan; Spinnaker; Summit. $C_{14}H_{18}ClN_3O_2$; mol wt 295.77. C 56.85%, H 6.13%, Cl 11.99%, N 14.21%, O 10.82%. Agricultural fungicide systemically active against powdery mildews and rusts of grains. Commercial product is mixture of 2 diastereoisomers. Prepn: **BE 814831**; W. Kramer *et al.*, **US 3952002** (1974, 1976 both to Bayer AG). Comprehensive description: P. E. Frohberger, *Pflanzenschutz-Nachr.* **31**, 11 (1978). Field trials in cereal diseases: J. Trägner-Born, T. van den Boom, *ibid.* 25. Inhibits fungal ergosterol biosynthesis: H. Buchenauer, *Pestic. Sci.* **9**, 507 (1978). Active fungal metabolite of triadimefon, *q.v.*: M. Gasztonyi, *ibid.* **12**, 433 (1981). GC separation of enantiomers: T. Clark, A. H. B. Deas, *J. Chromatogr.* **329**, 181 (1985).

Crystals, mp 112-117°. Slight, non-specific odor. Soly in water at 20°: 0.012 g/100 g. Soluble in alcohol, ketones. LD_{50} in male, female rats (mg/kg): 1161, 1105 orally; >5000 dermally, 24-hr; LD_{50} in quail: >10000 mg/kg (Frohberger).

USE: Systemic agricultural fungicide; cereal seed protectant.

9756. Triallate. [2303-17-5] *N,N*-Bis(1-methylethyl)carbamothioic acid *S*-(2,3,3-trichloro-2-propen-1-yl) ester; diisopropylthiocarbamic acid *S*-(2,3,3-trichloroallyl) ester; 2,3,3-trichloro-2-propene-1-thiol diisopropylcarbamate; *S*-2,3,3-trichloroallyl diisopropylthiocarbamate; *S*-(2,3,3-trichloro-2-propenyl) bis(1-methylethyl)carbamothioate; CP-23426; Avadex BW; Far-Go. $C_{10}H_{16}Cl_3$-NOS; mol wt 304.65. C 39.43%, H 5.29%, Cl 34.91%, N 4.60%, O 5.25%, S 10.52%. Selective herbicide for control of wild oats in wheat. Prepn: M. W. Harman, J. J. D'Amico, **US 3330821** (1967 to Monsanto). Herbicidal activity: R. Grover *et al.*, *Weed Res.* **19**, 363 (1979). Mutagenic evaluation *in vitro* (Ames Test): F. De Lorenzo *et al.*, *Cancer Res.* **38**, 13 (1978); G. R. Douglas *et al.*, *Mutat. Res.* **85**, 45 (1981). Soil persistence: A. E. Smith, B. J. Hayden, *Bull. Environ. Contam. Toxicol.* **29**, 240 (1982). Field studies for pre-emergent use: T. G. Reeves, C. L. Touhey, *Aust. J. Exp. Agric. Anim. Husb.* **12**, 55 (1972); E. M. Randall, R. H. Jarvis, *Exp. Husb.* **38**, 32 (1982); post-emergent use: R. P. Garnett, *Aspects Appl. Biol.* **13**, 73 (1980). Determn in soils by GC: A. E. Smith, *J. Chromatogr.* **97**, 103 (1974); by HPLC: A. Pena Heras, F. Sanchez-Rasero, *ibid.* **358**, 302 (1986). Review of genotoxicity: C. E. Healy *et al.*, *Int. J. Toxicol.* **22**, 233-251 (2003).

USE: Herbicide.

9757. Triamcinolone. [124-94-7] (11β, 16α)-9-Fluoro-11,-16,17,21-tetrahydroxypregna-1,4-diene-3,20-dione; Δ¹-9α-fluoro-16α-hydroxyhydrocortisone; 9α-fluoro-16α-hydroxyprednisolone; Δ¹-16α-hydroxy-9α-fluorohydrocortisone; 16α-hydroxy-9α-fluoroprednisolone; CL-19823; Aristocort; Kenacort; Ledercort (tabl.); Omcilon; Tricortale; Volon. $C_{21}H_{27}FO_6$; mol wt 394.44. C 63.95%, H 6.90%, F 4.82%, O 24.34%. Prepn: Bernstein *et al.*, *J. Am. Chem. Soc.* **78**, 5693 (1956); **81**, 1689 (1959); Thoma *et al.*, *ibid.* **79**, 4818 (1957); Bernstein *et al.*, Allen *et al.*, **US 2789118**; **US 3021347** (1957, 1962, both to Am. Cyanamid). Comprehensive description: K. Florey, *Anal. Profiles Drug Subs.* **1**, 367-396, 423-442 (1972); D. H. Sieh, *ibid.* **11**, 593-614, 651-661 (1982).

White or practically white crystals, mp 269-271°. mp also reported as 260-262.5°. $[\alpha]_D^{25}$ +75° (acetone). uv max: 238 nm (ε 15800). Slightly sol in alc, methanol; very slightly sol in water, chloroform, ether.

16,21-Diacetate. [67-78-7] (11β,16α)-16,21-Bis(acetyloxy)-9-fluoro-11,17-dihydroxypregna-1,4-diene-3,20-dione; 16α,21-diacetoxy-9α-fluoro-11β,17α-dihydroxy-1,4-pregnadiene-3,20-dione; Cenocort; CINO-40; Tracilon. $C_{25}H_{31}FO_8$; mol wt 478.51. Solvated crystals, mp 186-188° (with effervescence, mp 235° after drying). $[\alpha]_D^{25}$ +22° (chloroform). uv max: 239 nm (ε 15200). Sol in chloroform; sparingly sol in alc, methanol; slightly sol in ether. Practically insol in water.

THERAP CAT: Glucocorticoid.

THERAP CAT (VET): Glucocorticoid.

9758. Triamcinolone Acetonide. [76-25-5] (11β,16α)-9-Fluoro-11,21-dihydroxy-16,17-[(1-methylethylidene)bis(oxy)]pregna-1,4-diene-3,20-dione; 9α-fluoro-11β,16α,17,21-tetrahydroxypregna-1,4-diene-3,20-dione cyclic 16,17-acetal with acetone; 9α-fluoro-16α-hydroxyprednisolone acetonide; triamcinolone 16α,17-acetonide; 9α-fluoro-11β,21-dihydroxy-16α,17α-isopropylidenedioxy-1,4-pregnadiene-3,20-dione; 9α-fluoro-16α,17-isopropylidenedioxyprednisolone; Adcortyl; Azmacort; Delphicort; Extracort; Ftorocort; Kenacort-A; Kenalog; Ledercort Cream; Nasacort; Respicort; Rineton; Solodelf; Tramacin; Triam; Tricinolon; Vetalog; Volon A; Volonimat. $C_{24}H_{31}FO_6$; mol wt 434.50. C 66.34%, H 7.19%, F 4.37%, O 22.09%. Prepd by stirring a suspension of triamcinolone in acetone in the presence of a trace of perchloric acid: Fried *et al.*, *J. Am. Chem. Soc.* **80**, 2338 (1958); Bernstein *et al.*, *ibid.* **81**, 1689 (1959); Bernstein, Allen, **US 2990401** (1961 to Am. Cyanamid). Alternate synthesis using 2,3-dibromo-5,6-dicyanoquinone: Hydorn, **US 3035050** (1962 to Olin Mathieson). Clinical trial in chronic asthma: I. L. Bernstein *et al.*, *Chest* **81**, 20 (1982). Comprehensive description: K. Florey, *Anal. Profiles Drug Subs.* **1**, 397-421 (1972); D. H. Sieh, *ibid.* **11**, 615-649 (1982).

Crystals, mp 292-294°. $[\alpha]_D^{23}$ +109° (c = 0.75 in chloroform). uv max (abs alc.): 238 nm (ε 14600). Sparingly sol in methanol, acetone, ethyl acetate, chloroform, dehydrated alc. Practically insol in water.

21-Acetate. Crystals, mp 268-270°. $[\alpha]_D^{23}$ +92° (c = 0.59 in chloroform).

21-Disodium phosphate. [1997-15-5] Aristosol. $C_{24}H_{30}FNa_2$-O_9P; mol wt 558.45.

21-Hemisuccinate. Solutedarol. $C_{28}H_{35}FO_9$; mol wt 534.58.

THERAP CAT: Glucocorticoid; antiasthmatic (inhalant); antiallergic (nasal).

THERAP CAT (VET): Glucocorticoid.

9759. Triamcinolone Hexacetonide. [5611-51-8] (11β, 16α)-21-(3,3-Dimethyl-1-oxobutoxy)-9-fluoro-11-hydroxy-16,17-[(1-methylethylidene)bis(oxy)]pregna-1,4-diene-3,20-dione; 9-fluoro-11β,16α,17,21-tetrahydroxypregna-1,4-diene-3,20-dione cyclic 16,17-acetal with acetone, 21-(3,3-dimethylbutyrate); 21-*tert*-butylacetate-9α-fluoro-11β-hydroxy-16α,17α-(isopropylidenedioxy)-pregna-1,4-diene-3,20-dione; 21-(3,3-dimethylbutyryloxy)-9α-fluoro-11β-hydroxy-16α,17α-(isopropylidenedioxy)pregna-1,4-diene-3,20-dione; triamcinolone acetonide *tert*-butyl acetate; TATBA; CL-34433; Aristospan; Hexatrione; Lederlon; Lederspan. $C_{30}H_{41}FO_7$; mol wt 532.65. C 67.65%, H 7.76%, F 3.57%, O 21.03%. The hexacetonide ester of the potent glucocorticoid, triamcinolone, *q.v.* Prepn of syringeable suspension: Nash, Naeger, **US 3457348** (1969 to Am. Cyanamid). Anti-inflammatory activity in rabbits: I. M. Hunneyball, *Agents Actions* **11**, 490 (1981). Early clinical studies: Bilka, *Minn. Med.* **50**, 483 (1967); Layman, Peterson, *ibid.* 669. Clinical studies of intra-articular therapy in arthritis: R. C. Allen *et al.*, *Arthritis Rheum.* **29**, 997 (1986); M. Talke, *Fortschr. Med.* **104**, 742 (1986). Toxicity study: Tonelli, *Steroids* **8**, 857 (1966). Comprehensive description: V. Zbinovsky, G. P. Chrekian, *Anal. Profiles Drug Subs.* **6**, 579-595 (1977).

Fine, white, needle-like crystals, mp 295-296° (dec), also reported as mp 271-272° (dec). uv max (ethanol): 238 nm (ε 15500). $[\alpha]_D^{25}$ +90±2° (c = 1.13% in chloroform). Soly in g/100 ml at 25°: chloroform and dimethylacetamide >5; ethyl acetate 0.77, methanol 0.59,

diethyl carbonate 0.50, glycerin 0.42, propylene glycol 0.13; absolute alcohol 0.03; water 0.0004.

THERAP CAT: Anti-inflammatory.

9760. Triamterene. [396-01-0] 6-Phenyl-2,4,7-pteridinetriamine; 2,4,7-triamino-6-phenylpteridine; 6-phenyl-2,4,7-triaminopteridine; ademin(e); pterofen; pterophene; NSC-77625; SKF-8542; Dyren; Dyrenium; Dytac; Jatropur; Teriam; Triteren; Urocaudal. $C_{12}H_{11}N_7$; mol wt 253.27. C 56.91%, H 4.38%, N 38.71%. Prepn: Spickett, Timmis, *J. Chem. Soc.* **1954**, 2887; Pachter, *J. Org. Chem.* **28**, 1191 (1963); Weinstock, Wiebelhaus, **US 3081230** (1963 to SK & F); Osdene *et al.*, *J. Med. Chem.* **10**, 431 (1967). *Review: Ther. Triamteren Wien. Symp.* **1966**, K. Fellinger, Ed. (Georg Thieme, Vienna, 1967). Comprehensive description: V. K. Kapoor, *Anal. Profiles Drug Subs. Excip.* **23**, 571-605 (1994).

Yellow plates from butanol, mp 316° (Spickett, Timmis); also reported as crystals from DMF, mp 327° (Osdene *et al.*). pKa 6.2. Sol in formic acid; sparingly sol in methoxyethanol; slightly sol in water (1 in 1000), ethanol (1 in 3000), chloroform (1 in 4000); very slightly sol in acetic acid, dilute mineral acids. Practically insol in ether, benzene, dilute alkali hydroxides. uv max (4.5% formic acid): 356 nm (ε 21000).

THERAP CAT: Diuretic.

9761. Triasulfuron. [82097-50-5] 2-(2-Chloroethoxy)-*N*-[[(4-methoxy-6-methyl-1,3,5-triazin-2-yl)amino]carbonyl]benzenesulfonamide; 1-[2-(2-chloroethoxy)phenylsulfonyl]-3-(4-methoxy-6-methyl-1,3,5-triazin-2-yl)urea; CGA-131036; Amber; Logran. $C_{14}H_{16}ClN_5O_5S$; mol wt 401.82. C 41.85%, H 4.01%, Cl 8.82%, N 17.43%, O 19.91%, S 7.98%. Sulfonylurea herbicide for broadleaf weeds. Prepn: W. Meyer, W. Föry, **EP 44808**; *eidem*, **US 4514212** (1982, 1985 both to Ciba-Geigy). Comprehensive description: J. Amrein, H. R. Gerber, *Proc. Br. Crop Prot. Conf. - Weeds* **1985**, 55-62. Uptake and metabolism in wheat: A. M. Meyer, F. Müller, *Brighton Crop Prot. Conf. - Weeds* **1989**, 441. Behavior in soil: F. K. Oppong, G. R. Sagar, *Weed Res.* **32**, 157, 167 (1992). Immunoassay in soil: J. A. Schlaeppi *et al.*, *J. Agric. Food Chem.* **42**, 1914 (1994).

White crystals, mp 186°. Vapor pressure (20°): 7.5×10^{-13} mm Hg. Soly in water (20°): 1.5 g/l (pH 7). LD_{50} in rats (mg/kg): >5000 orally; >2000 dermally; LC_{50} (4 hr) in rats: >5185 mg/m³ by inhalation (Amrein, Gerber).

USE: Herbicide.

9762. Triazamate. [112143-82-5] 2-[[1-[(Dimethylamino)-carbonyl]-3-(1,1-dimethylethyl)-1*H*-1,2,4-triazol-5-yl]thio]acetic acid ethyl ester; ethyl (3-*tert*-butyl-1-dimethylcarbamoyl-1*H*-1,2,4-triazol-5-ylthio)acetate; 1-dimethylcarbamoyl-3-*tert*-butyl-5-carboethoxymethylthio-1,2,4-triazole; RH-7988; WL-145158; Aphistar; Aztec. $C_{13}H_{22}N_4O_3S$; mol wt 314.40. C 49.66%, H 7.05%, N 17.82%, O 15.27%, S 10.20%. Cholinesterase inhibitor. Prepn: R. M. Jacobson *et al.*, **EP 213718**; R. M. Jacobson, M. Thirugnanam, **US 4742072** (1987, 1988 both to Rohm and Haas). Properties and biological activity: A. Murray *et al.*, *Brighton Crop Prot. Conf. - Pests Dis.* **1988**, 73. Field studies on sugar beets: A. M. Dewar *et al.*, *ibid.* **1994**, 407. HPLC determn in crops: R. Weitzel, U. Zimmermann, *ibid.* **1996**, 983. Review of metabolism and aphicidal spectrum: R. M. Jacobson, M. Thriugnanam, *ACS Symp. Ser.* **443**, 322-339 (1991).

Light tan solid, mp 60°. Vapor pressure 4.8 ×10⁻⁶ torr. Soly of technical grade in water <1%; sol in methylene chloride, ethyl acetate. LD₅₀ (14 d) in mice, rats (mg/kg): 61, 50-200 orally; in rats (mg/kg): >5000 dermally (Murray). LC₅₀ (48 h) in daphnia: 0.048 mg/l; (96 h) in bluegill, trout: 1.0, 0.43 mg/l (Murray).

USE: Pesticide.

9763. *s*-**Triazine.** [290-87-9] 1,3,5-Triazine. C₃H₃N₃; mol wt 81.08. C 44.44%, H 3.73%, N 51.83%. Originally prepd in 1895 and incorrectly identified as "dimeric hydrocyanic acid," C₂H₂N₂. Prepn: C. Grundmann, A. Kreutzberger, *J. Am. Chem. Soc.* **76**, 632, 5646 (1954). Reactivity and use in synthesis of heterocycles: *eidem, ibid.* **77**, 44, 6559 (1955). IR and Raman spectroscopy: J. Goubeau *et al., J. Phys. Chem.* **58**, 1078 (1954). X-ray crystal structure: P. J. Wheatley, *Acta Crystallogr.* **8**, 224 (1955). Determn of enthalpy of combustion and sublimation: K. Byström, *J. Chem. Thermodyn.* **14**, 865 (1982). Molecular structure: C. A. Morrison *et al., J. Phys. Chem. A* **101**, 10029 (1997). Review of applications in organic chemistry: T. Güthner, *Chim. Oggi* **17**, 12-14 (1999); G. Giacomelli *et al., Curr. Org. Chem.* **8**, 1497-1519 (2004).

Rhombohedral crystals, mp 86°. bp₇₆₀ 114°. *Extremely volatile. Hydrolyzes to formamide upon contact with water. Store in a tightly sealed container.*

USE: Reagent in oxidation reactions, in synthesis of heterocycles. Triazine derivatives are used as herbicides, pharmaceuticals, complexation agents, peptidomimetic building blocks, and dyes.

9764. **Triaziquone.** [68-76-8] 2,3,5-Tri-1-aziridinyl-2,5-cyclohexadiene-1,4-dione; 2,3,5-tris(1-aziridinyl)-*p*-benzoquinone; 2,3,5-tris(aziridino)-1,4-benzoquinone; 2,3,5-tris(ethyleneimino)-benzoquinone; trenimon; Bayer 3231. C₁₂H₁₃N₃O₂; mol wt 231.26. C 62.32%, H 5.67%, N 18.17%, O 13.84%. Polyfunctional alkylating agent. Prepn: Gauss, Domagk, US 2976279 (1961 to Schenley). Review of biochemical, physiological and genetic effects: G. Obe, B. Beek, *Mutat. Res.* **65**, 21-70 (1979); of structure and chemical reactivity: P. Rademacher, G. Obe, *ibid.* **340**, 37-49 (1995).

Purple acicular crystals from ethyl acetate, mp 162.5-163°. Sparingly sol in cold water; sol in acetone, benzene, chloroform, ethyl acetate, methanol and warm acetic acid.

USE: Alkylating reagent in mutation research.

9765. **Triazolam.** [28911-01-5] 8-Chloro-6-(2-chlorophenyl)-1-methyl-4*H*-[1,2,4]triazolo[4,3-*a*][1,4]benzodiazepine; 8-chloro-6-(*o*-chlorophenyl)-1-methyl-4*H*-*s*-triazolo[4,3-*a*][1,4]benzodiazepine; clorazolam; U-33030; Halcion; Songar. C₁₇H₁₂Cl₂N₄; mol wt 343.21. C 59.49%, H 3.52%, Cl 20.66%, N 16.32%. Prepn: J. B. Hester, DE 2012190; *eidem*, US 3701782 (1970, 1972 both to Upjohn); J. B. Hester *et al., J. Med. Chem.* **14**, 1078 (1971). Pharmacology: T. Furukawa *et al., Igaku Kenkyu* **45**, 285 (1975), *C.A.*

84, 130354p. Metabolism: F. S. Eberts, *Drug Metab. Dispos.* **5**, 547 (1977). HPLC determn in urine: T. Inoue, S. Suzuki, *J. Chromatogr.* **422**, 197 (1987). Clinical studies: K. K. Okawa, G. S. Allens, *J. Int. Med. Res.* **6**, 343 (1978); A. J. Bowen, *ibid.* 337; A. J. Puech *et al., Therapie* **33**, 287 (1978). Toxicity: *Pharm. Weekbl.* **113**, 725 (1978). Review of pharmacology and therapeutic efficacy: G. E. Pakes *et al., Drugs* **22**, 81-110 (1981).

Tan crystals from 2-propanol, mp 233-235°. Sol in chloroform; slightly sol in alc. Practically insol in water, ether. LD₅₀ in mice, rats (mg/kg): >100, >5000 orally *(Pharm. Weekblad.).*

Note: This is a controlled substance (depressant): **21 CFR**, 1308.14.

THERAP CAT: Sedative, hypnotic.

9766. **1*H*-1,2,4-Triazole.** [288-88-0] Pyrrodiazole; 3,4-diazapyrrole. C₂H₃N₃; mol wt 69.07. C 34.78%, H 4.38%, N 60.84%. Prepn starting with thiosemicarbazide and formic acid: Ainsworth, *Org. Synth.* **40**, 99 (1960).

Needles from ethanol + benzene, mp 120-121°. bp₇₆₀ 260° (decompn). Appreciably sol in water, alcohol.

Hydrochloride. C₂H₃N₃.HCl. Platelets, dec 169°.

9767. **Triazophos.** [24017-47-8] Phosphorothioic acid *O,O*-diethyl *O*-(1-phenyl-1*H*-1,2,4-triazol-3-yl) ester; 1-phenyl-3-(*O,O*-diethylthionophosphoryl)-1,2,4-triazole; *O,O*-diethyl *O*-1-phenyl-1*H*-1,2,4-triazol-3-yl phosphorothioate; HOE-2960; Hostathion. C₁₂H₁₆N₃O₃PS; mol wt 313.31. C 46.00%, H 5.15%, N 13.41%, O 15.32%, P 9.89%, S 10.23%. Organophosphate insecticide; cholinesterase inhibitor. Prepn: O. Scherer, H Mildenberger, ZA 6803471; *eidem*, US 3686200 (1968, 1972 both to Hoechst). Properties and analytical methods: W. G. Thier *et al. Anal. Methods Pestic. Plant Growth Regul.* **10**, 127 (1978). Immunochromatographic determn in soil and water: W.-J. Gui *et al., Anal. Biochem.* **377**, 202 (2008). Field studies as insecticide and acaricide: S. J. B. Hay, *Proc. 6th Brit. Crop Prot. Council - Insect. Fungic.* 597 (1972). Nematocidal activity: A. Sirohi, Siyanand, *Indian J. Nematol.* **22**, 134 (1992). Residues and dissipation rates in soil and wheat crops: W. Li *et al., Ecotoxicol. Environ. Saf.* **69**, 312 (2008).

Light brown oil, mp 0-5°. Vapor pressure 2.9 × 10⁻⁶ mbar (303 K). d 1.247. Soly at 296 K: 39 ppm in water. Soly at 298 K (g/100 ml): *n*-hexane 0.7, toluene 30, ethyl acetate 30, acetone 30, ethanol 30. LD₅₀ in rats (mg/kg): 82 orally, 1100 dermally (Thier). LD₅₀ in rats (mg/kg): 107 i.p.; in rabbits (mg/kg): 280 dermally (Hay).

USE: Insecticide; acaricide; nematocide.

9768. **Triazoxide.** [72459-58-6] 7-Chloro-3-(1*H*-imidazol-1-yl)-1,2,4-benzotriazine 1-oxide; 3-(imidazol-1-yl)-7-chlorobenzo-1,-2,4-triazine-1-oxide; BAY SAS 9244. C₁₀H₆ClN₅O; mol wt 247.64. C 48.50%, H 2.44%, Cl 14.32%, N 28.28%, O 6.46%. Prepn: K. Sasse *et al.*, DE 2802488; *eidem*, US 4239760 (1979,

1980 both to Bayer). GC determn in water: R. Brennecke, K. Vogeler, *Pflanzenschutz-Nachr.* **37**, 46 (1984). HPLC determn of adsorption coeff on soil: W. Kördel *et al.*, *Sci. Total Environ.* **162**, 119 (1995). Physicochemical properties, toxicology, and fungicidal activity: S. Dutzmann, *Br. Crop Prot. Counc. Monogr.* **57**, 85 (1994). Field trials in combination with tebuconazole, *q.v.*: A. Wainwright *et al.*, *ibid.* 103.

Light yellow crystals, mp 182° (Dutzmann); also reported as mp 172° (Sasse). Vapor pressure at 25°: 5.2×10^{-12} Pa. Soly at 20° (g/l): water 0.03; *n*-hexane <1; dichloromethane 50-100; 2-propanol 2-5; toluene 20-50. LD_{50} in rats (mg/kg): >5000 dermally; 100-200 orally. LC_{50} (4hr) in rats (mg/l): 0.8-3.2 by inhalation (Dutzmann). USE: Fungicide.

9769. Tribenoside. [10310-32-4] Ethyl 3,5,6-tris-*O*-(phenylmethyl)-D-glucofuranoside; Ba-21401; Alven; Flebosan; Glyvenol; Hemocuron; Venex. $C_{29}H_{34}O_6$; mol wt 478.59. C 72.78%, H 7.16%, O 20.06%. Prepn: Druey, Huber, US 3157634 (1964 to Ciba). Pharmacology: Lecomte, *C. R. Seances Soc. Biol. Ses Fil.* **163**, 1469 (1969); Helfer, Jaques, *Pharmacology* **5**, 23 (1971). Increase in capillary resistance in patients with rheumatoid arthritis: W. C. Dick *et al.*, *Ann. Rheum. Dis.* **28**, 187 (1969). GLC determn in plasma: A. Sioufi, F. Pommier, *J. Pharm. Sci.* **69**, 167 (1980). Excretion as hippuric acid: *eidem, Eur. J. Drug Metab. Pharmacokinet.* **7**, 223 (1982). Review of pharmacology: R. Jaques, *ibid.* **15**, 445-460 (1977).

bp$_{1.2}$ 270-280°. $[\alpha]_D^{26}$ +8° (chloroform).
THERAP CAT: Sclerosing agent.

9770. Tribromoacetic Acid. [75-96-7] 2,2,2-Tribromoacetic acid. $C_2HBr_3O_2$; mol wt 296.74. C 8.10%, H 0.34%, Br 80.78%, O 10.78%. CBr$_3$COOH. Prepn: Müller, US 2057964 (1936 to Winthrop Chem.); A. M. Kovalevskaya, S. A. Shkylar, *Zh. Org. Khim.* **1**, 1540 (1965). Study of far IR spectrum: G. Statz, E. Lippert, *Ber. Bunsen-Ges. Phys. Chem.* **71**, 673 (1967).

Monoclinic prisms, mp 129-135°. bp 245° (dec). Sol in water, alcohol, ether; slightly sol in petr ether. Dec in boiling water to bromoform.
USE: Catalyst for polymerization; as brominating agent: W. J. Szczepek, *Pol. J. Chem.* **55**, 709 (1981).

9771. 2,4,6-Tribromoaniline. [147-82-0] 2,4,6-Tribromobenzenamine; aniline tribromide. $C_6H_4Br_3N$; mol wt 329.82. C 21.85%, H 1.22%, Br 72.68%, N 4.25%. Prepd by controlled bromination of aniline: Suthers *et al.*, *J. Org. Chem.* **27**, 447 (1962).

Needles; d 2.35; mp 120-122°; bp 300°. Insol in water. Sol in hot alc, chloroform, ether; slightly sol in cold alcohol.

9772. 2,4,6-Tribromoanisole. [607-99-8] 1,3,5-Tribromo-2-methoxybenzene; 1-methoxy-2,4,6-tribromobenzene; TBA. $C_7H_5Br_3O$; mol wt 344.83. C 24.38%, H 1.46%, Br 69.52%, O 4.64%. Halogenated natural product produced by the microbiological *O*-methylation of 2,4,6-tribromophenol, *q.v.* One of the causative compounds of cork taint in wine; imparts an unpleasant moldy or musty taste and odor. Prepn by bromination of 4-methoxybenzoic acid: A. Reinecke, *Z. Chem.* **1866**, 366; by methylation of tribromophenol: M. Kohn, A. Fink, *Monatsh. Chem.* **44**, 183 (1923); S. M. Gerber, D. Y. Curtin, *J. Am. Chem. Soc.* **71**, 1499 (1949); D. Maes *et al.*, *Org. Prep. Proced. Int.* **39**, 395 (2007). Identification in shellfish: T. Miyazaki *et al.*, *Bull. Environ. Contam. Toxicol.* **26**, 577 (1981); in river and marine sediment: I. Watanabe *et al.*, *ibid.* **35**, 272 (1985). Production by *Rhodococcus* and *Acinetobacter* spp.: A.-S. Allard *et al.*, *Appl. Environ. Microbiol.* **53**, 839 (1987); by the red alga, *Polysiphonia sphaerocarpa*: C. Flodin, F. B. Whitfield, *Phytochemistry* **53**, 77 (2000). Properties and bioactivity study: W. Vetter *et al.*, *Arch. Environ. Contam. Toxicol.* **48**, 1 (2004). Identification as musty contaminant in packaging materials and food: F. B. Whitfield *et al.*, *J. Agric. Food Chem.* **45**, 889 (1997); in wine: P. Chatonnet *et al.*, *ibid.* **52**, 1255 (2004). GC-MS/MS determn in water: L. Zhang *et al.*, *J. Chromatogr. A* **1098**, 7 (2005); in wine: C. Pizarro *et al.*, *ibid.* **1218**, 1576 (2011).

Crystals from ethanol, mp 87°. bp 297-299°. Odor threshold concentration: 2×10^{-5} µg/l in water. Soly in water (25°): 12.2 mg/l. Log P (octanol/water): 4.4. Subcooled liquid vapor pressure: 0.06562 Pa.

9773. Tribromo-*tert*-butyl Alcohol. [76-08-4] 1,1,1-Tribromo-2-methyl-2-propanol; acetone-bromoform; brometone. $C_4H_7Br_3O$; mol wt 310.81. C 15.46%, H 2.27%, Br 77.12%, O 5.15%. Prepd from acetone and bromoform in the presence of sodium or potassium hydroxide: Aldrich, *J. Am. Chem. Soc.* **33**, 386 (1911); Viehe *et al.*, *Ber.* **96**, 426 (1963).

Crystals from dil alcohol, mp 167-176° (Aldrich), from ether, mp 169° (Viehe). Camphor-like odor and taste. Volatilizes in air. Slightly sol in water; sol in alcohol, ether.
USE: As modifier in polymerization of vinyl chloride, Seymour, US 2716112 (1955 to U.S. Rubber).

9774. 2,4,6-Tribromo-*m*-cresol. [4619-74-3] 2,4,6-Tribromo-3-methylphenol; 2,4,6-tribromo-3-hydroxytoluene; Micatex. $C_7H_5Br_3O$; mol wt 344.83. C 24.38%, H 1.46%, Br 69.52%, O 4.64%. Prepn by bromination of *m*-cresol: Biilmann, Rimbert, *Bull. Soc. Chim. Fr.* [4] **33**, 1473 (1923).

Crystals from 50% aq alcohol, mp 84°.
THERAP CAT: Antifungal (topical).

9775. 2,4,6-Tribromophenol. [118-79-6] 2,4,6-TBP; FR-613; PH-73FF; Bromol. $C_6H_3Br_3O$; mol wt 330.80. C 21.79%, H 0.91%, Br 72.46%, O 4.84%. Microbicidal natural product produced by marine algae, enteropneust worms, and other marine organisms; dietary component contributing to the flavor of saltwater fish and crustaceans. Prepn: H. E. Armstrong, *J. Chem. Soc.* **25**, 865 (1872); R. L. Datta, J. C. Bhoumik, *J. Am. Chem. Soc.* **43**, 303 (1921); by controlled bromination of phenol: J. O. Konecny, *ibid.* **76**, 4993 (1954). Isoln from marine tube worms, *Phoronopsis viridis:* Y. M. Sheikh, C. Djerassi, *Experientia* **31**, 265 (1975); from *Enteropneusta* species: T. Higa *et al., Comp. Biochem. Physiol. B* **65**, 525 (1980). Distribution in marine organisms and role as flavor component: F. B. Whitfield *et al., J. Agric. Food Chem.* **46**, 3750 (1998); *idem et al., ibid.* **47**, 4756 (1999). GC-MS determn in water samples: M. Polo *et al., J. Chromatogr. A* **1124**, 11 (2006). ELISA determn in treated wood: M. Nichkova *et al., J. Agric. Food Chem.* **56**, 29 (2008). Degradation in soil: J. R. Nyholm *et al., Environ. Pollut.* **158**, 2235 (2010). Toxicity study: E. F. Stohlman, *Public Health Rep.* **66**, 1303 (1951); G. L. Phipps *et al., Bull. Environ. Contam. Toxicol.* **26**, 585 (1981). Review of properties, uses, and toxicology: P. D. Howe *et al., Concise International Chemical Assessment Document* **No. 66** (WHO, Geneva, 2005) 47 pp.

Crystals from glacial acetic acid, mp 96°. d 2.55; bp 282-290°. Log P (octanol/water): 4.13. Iodoform-like taste. Flavor threshold in water: 0.6 ng/g. Soly in water (25°): 0.007 g/100 ml. Sol in ethanol, methanol, chloroform, ether, methylene chloride. LD_{50} orally in rats: <2000 mg/kg (Stohlman). LC_{50} (96 hr) in fathead minnow: 6.6 mg/l (Phipps). *Irritant; avoid inhalation of dust.*

Sodium salt. [2666-53-7] Sodium 2,4,6-tribromophenolate. $C_6H_2Br_3NaO$; mol wt 352.78.

USE: Wood preservative; reactive flame retardant and intermediate in the production of brominated flame retardants.

9776. 1,2,3-Tribromopropane. [96-11-7] *sym*-Tribromopropane; glycerol tribromohydrin; tribromohydrin; allyl tribromide. $C_3H_5Br_3$; mol wt 280.79. C 12.83%, H 1.79%, Br 85.37%. Prepd by the addn of bromine to allyl bromide in carbon tetrachloride: Perkin, Simonsen, *J. Chem. Soc.* **87**, 859 (1905); Johnson, McEwen, *Org. Synth.* **5**, 99 (1925). By gentle heating of propylene bromide with bromine in the presence of iron wire: Kronstein, *Ber.* **54**, 7 (1921); Tapley, Giesy, *J. Am. Pharm. Assoc.* **15**, 173 (1926); by reaction of bromotrichloromethane with allyl bromide initiated by γ-rays: Heiba, Anderson, *J. Am. Chem. Soc.* **79**, 4940 (1957).

Liquid; d^{23} 2.436; mp 16.5°; bp_{760} 220°; bp_{200} 170°; bp_{100} 148°; bp_{60} 134°; bp_{40} 123°; bp_{20} 106°; bp_{10} 90°; bp_5 76°; $bp_{1.0}$ 47.5°; n_D^{18} 1.58436. Insol in water; sol in alcohol, ether, chloroform.

USE: As a nematocide: Youngson, Goring, **US 3003914** (1959 to Dow).

9777. Tribromosilane. [7789-57-3] Tribromomonosilane; silicobromoform. Br_3HSi; mol wt 268.81. Br 89.18%, H 0.37%, Si 10.45%. $SiHBr_3$. Prepd by passing hydrogen bromide over heated silicon or a silicide such as copper silicide: Schumb, Young, *Inorg. Synth.* **1**, 38 (1939); Schenk in *Handbook of Preparative Inorganic Chemistry* vol. **1**, G. Brauer, Ed. (Academic Press, New York, 2nd ed., 1963) pp 692-694.

Mobile liquid. Spontaneously flammable when the contact surface with air is large. d_4^{17} 2.7. mp −73.5°. bp_{760} 111.8°. The vapor pressure at 0° is given as 8.8 mm Hg. Dipole moment 0.79. Hydrolyzed by water with the formation of silicoformic anhydride, $H_2Si_2O_3$ and HBr. Sol in chlorinated hydrocarbons.

9778. Tribromsalan. [87-10-5] 3,5-Dibromo-*N*-(4-bromophenyl)-2-hydroxybenzamide; 3,4′,5-tribromosalicylanilide; TBS; Temasept IV. $C_{13}H_8Br_3NO_2$; mol wt 449.92. C 34.70%, H 1.79%, Br 53.28%, N 3.11%, O 7.11%. Prepn: **GB 840366** (1960 to Unilever); Lemaire *et al., J. Pharm. Sci.* **50**, 831 (1961); Lamberti, **US**

2967885; Schramm, Lemaire, **US 3064048**; Schramm, **US 3057920** (1961, 1962, 1962 all to Lever Bros.); Majewski, **US 3254121** (1966 to Dow).

Crystals, mp 227-228°. Practically insol in water. Sol in hot acetone; very sol in DMF.

USE: Bacteriostat in detergents.

9779. Tribufos. [78-48-8] Phosphorotrithioic acid *S,S,S*-tributyl ester; *S,S,S*-tributyl phosphorotrithioate; B-1776; Def. $C_{12}H_{27}OPS_3$; mol wt 314.50. C 45.83%, H 8.65%, O 5.09%, P 9.85%, S 30.58%. Organophosphate defoliant used in preparation of cotton for harvesting. Preparative method: K. H. Rattenbury, J. R. Costello, **US 2943107** (1960 to Chemagro). Prepn: D. Voigt *et al., C. R. Hebd. Seances Acad. Sci.* **260**, 2210 (1965). Comprehensive description: D. MacDougall, *Anal. Methods Pestic. Plant Growth Regul. Food Addit.* **4**, 89-93 (1964). Oxidative chemistry and toxicology: J. H. Hur *et al., J. Agric. Food Chem.* **40**, 1703 (1992). GC determn in water and fish tissue: C. Habig *et al., J. AOAC Int.* **70**, 103 (1987). Acute toxicity: T. B. Gaines, *Toxicol. Appl. Pharmacol.* **14**, 515 (1969). Human exposure studies: W. Kilgore *et al., Residue Rev.* **91**, 71 (1984). Soil dissipation study: T. L. Potter *et al., J. Agric. Food Chem.* **50**, 3795 (2002).

Colorless to pale yellow liquid, mp <−25°. $bp_{0.3}$ 150°. bp_4 166.5°. d_4^{20} 1.0552. n_D^{20} 1.534. Sol in aliphatic, aromatic, and chlorinated hydrocarbon solvents; alcohols. LD_{50} in male, female rats (mg/kg): 233, 150 orally; 360, 168 dermally (Gaines).

USE: Defoliant.

9780. Tributylamine. [102-82-9] *N,N*-Dibutyl-1-butanamine. $C_{12}H_{27}N$; mol wt 185.36. C 77.76%, H 14.68%, N 7.56%. Prepd by vapor phase alkylation of ammonia with butanol: Lemon, Myerly, **US 3022349** (1962 to Union Carbide Corp.).

Hygroscopic liquid; characteristic odor. d_{20}^{20} 0.7782; bp 216-217°. Sparingly sol in water; very sol in alcohol, ether. *Poisonous. Keep well closed.*

Caution: Causes CNS stimulation, skin irritation, sensitization.

9781. Tributyl Phosphate. [126-73-8] Phosphoric acid tributyl ester; tributoxyphosphine oxide; tributyl phosphate. $C_{12}H_{27}O_4P$; mol wt 266.32. C 54.12%, H 10.22%, O 24.03%, P 11.63%. Prepd by reaction of $POCl_3$ with butyl alcohol: Pianfetti, Janey, **US 3020303** (1962 to FMC). Toxicity study: Smyth, Carpenter, *J. Ind. Hyg. Toxicol.* **26**, 269 (1944).

Colorless, odorless liq; d_{25}^{25} 0.976; bp 289° with decompn; bp_{27} 177-178°; mp below −80°. Flash pt 146°C. n_D^{25} 1.4215. One ml dissolves in about 165 ml water; miscible with usual organic solvents. LD$_{50}$ orally in rats: 3.0 g/kg (Smyth, Carpenter).

Caution: Potential symptoms of overexposure are irritation of eyes, respiratory system and skin; headache; nausea. *See NIOSH Pocket Guide to Chemical Hazards* (DHHS/NIOSH 97-140, 1997) p 314.

USE: Plasticizer for cellulose esters, lacquers, plastics, and vinyl resins.

9782. Tributyltin Chloride. [1461-22-9] Tributylchlorostannane; chlorotributylstannane; chlorotributyltin; tributylchlorotin; tributylstannyl chloride; TBTC; Bu$_3$SnCl. C$_{12}$H$_{27}$ClSn; mol wt 325.51. C 44.28%, H 8.36%, Cl 10.89%, Sn 36.47%. Organotin compd used as a precursor to other tributyltin reagents. Prepn not claimed: E. W. Rugeley, W. M. Quattlebaum, US 2344002 (1944 to Carbide and Carbon Chem). Prepn: W. J. Jones *et al., J. Chem. Soc.* **1947**, 1446. Utility in prepn of tributyltin reagents: H. G. Kuivila, O. F. Beumel, Jr., *J. Am. Chem. Soc.* **83**, 1246 (1961); H. G. Kuivila, M. A. Weiner, *J. Org. Chem.* **26**, 4797 (1961); G. E. Keck *et al., Tetrahedron Lett.* **28**, 139 (1987). Synthetic applications in heterocycle formation: A. Srikrishna *et al., Tetrahedron* **53**, 10479 (1997); K. V. V. P. Rao *et al., Synlett* **2007**, 1289. Aquatic toxicity studies on *Daphnia magna:* M. L. Bao *et al., Bull. Environ. Contam. Toxicol.* **59**, 671 (1997); on the freshwater mollusk, *Lamellidens marginalis:* J. T. Jagtap *et al., World J. Fish Mar. Sci.* **3**, 100 (2011).

Colorless liquid. *Poisonous, irritant.* bp$_{25}$ 172°; bp$_3$ 128-130°; bp$_{0.4}$ 97°. d$_4^{20}$ 1.2072. n$_D^{20}$ 1.4909. Flash pt, closed cup: 235°F (113°C). Sol in diethyl ether, THF, hexane, dichloromethane, methanol. Air stable; slowly decomposes upon exposure to moisture. LC$_{50}$ in *Lamellidens marginalis* (ppm): 5.3327 (24 hr); 4.0258 (48 hr); 3.0532 (72 hr); 2.1244 (96 hr) (Jagtap).

USE: Reagent in synthetic organic chemistry. Biocide in antifouling paints and in wood protection.

9783. Tributyltin Hydride. [688-73-3] Tributylstannane; tri-*n*-butyltin hydride; TBTH; Bu$_3$SnH. C$_{12}$H$_{28}$Sn; mol wt 291.07. C 49.52%, H 9.70%, Sn 40.78%. Organotin compd that is the source of the tributyltin radical; utilized in a wide variety of free radical chain reactions. Prepn: G. J. M. van der Kerk *et al., J. Appl. Chem.* **7**, 366 (1957). Alkyl halide reduction and free radical mechanistic studies: H. G. Kuivila *et al., J. Am. Chem. Soc.* **84**, 3584 (1962); L. W. Menapace, H. G. Kuivila, *ibid.* **86**, 3047 (1964). *Reviews:* W. P. Neumann, *Synthesis* **1987**, 665-683; D. P. Curran, *ibid.* **1988**, 417-439; B. K. Banik, *Curr. Org. Chem.* **3**, 469-496 (1999).

Colorless liquid. *Poisonous. Flammable. Irritant.* bp$_{1.5}$ 123°; bp$_{0.5}$ 80-83°; bp$_{0.3}$ 68-74°. d$_4^{20}$ 1.104. n$_D^{20}$ 1.4726; n$_D^{30}$ 1.4682. Flash point, closed cup : 104°F (40°C). Freely sol in organic solvents.

USE: Reagent in synthetic organic chemistry.

9784. Tributyltin Methoxide. [1067-52-3] Tributylmethoxystannane; methoxytributylstannane; methoxytributyltin; tributylmethoxytin; Bu$_3$SnOMe. C$_{13}$H$_{30}$OSn; mol wt 321.09. C 48.63%, H 9.42%, O 4.98%, Sn 36.97%. Organotin reagent used in a variety of substitution, addition, and oxidation reactions. Prepn: D. Cleverdon *et al., US 2623892* (1952 to Distillers); A. N. Nesmeyanov *et al., Dokl. Akad. Nauk SSSR* **124**, 1073 (1959); D. L. Alleston, A. G. Davies, *J. Chem. Soc.* **1962**, 2050. Synthetic utility in addition reactions: A. J. Bloodworth *et al., J. Chem. Soc. C* **1967**, 1309; in glycoside chemistry: T. Ogawa, M. Matsui, *Carbohyd. Res.* **51**, C13 (1976); in deprotection of acetylated sugars: J. Herzig *et al., ibid.* **177**, 21 (1988); in cross-coupling reactions: V. Nair *et al., J. Org. Chem.* **53**, 3051 (1988). Application as catalyst for the prepn of large ring ketones: I. Minami *et al., Tetrahedron* **42**, 2971 (1986). Use as initiator of polymerizations: H. R. Kricheldorf *et al., Macromolecules* **24**, 1944 (1991).

Colorless liquid. *Poisonous.* bp$_{15}$ 148°; bp$_2$ 101-102°; bp$_1$ 98-100°; bp$_{0.06}$ 97-97.5°. d$_4^{20}$ 1.1690. n$_D^{25}$ 1.4710; n$_D^{20}$ 1.4745. Flash pt, closed cup: 210°F (99°C). Sol in common organic solvents. Moisture sensitive.

USE: Reagent and catalyst in synthetic organic chemistry.

9785. Tributyrin. [60-01-5] Butanoic acid 1,2,3-propanetriyl ester; butyryl triglyceride; glycerol tributanoate; glyceryl tributyrate; tributyrylglycerol. C$_{15}$H$_{26}$O$_6$; mol wt 302.37. C 59.58%, H 8.67%, O 31.75%. Prepd by esterification of glycerol with excess butyric acid: Weatherby *et al., J. Am. Chem. Soc.* **47**, 2249 (1925).

Oily liq; bitter taste; d$_4^{20}$ 1.032; mp −75°; bp$_{760}$ 305-310°. bp$_{15}$ about 190°; n$_D^{20}$ 1.4358. Insol in water. Very sol in alc, ether.

9786. Tricaine. [886-86-2] 3-Aminobenzoic acid ethyl ester methanesulfonate (1:1); tricaine methanesulfonate; ethyl *m*-aminobenzoate methanesulfonate; MS-222; Finquel. C$_{10}$H$_{15}$NO$_5$S; mol wt 261.29. C 45.97%, H 5.79%, N 5.36%, O 30.62%, S 12.27%. Prepn: DE 454698; O. Billeter *et al., US 1678317* (both 1928 to Sandoz). Pharmacology in fish: D. W. Jolly *et al., Vet. Rec.* **91**, 424 (1972). Blood concentrations during anesthetization and recovery: A. H. Houston, R. J. Woods, *J. Fish Res. Board Can.* **29**, 1344 (1972). pH effects on anesthetic potency: E. A. Ohr, *Comp. Biochem. Physiol.* **54C**, 13 (1976). Toxicity study: J. M. McKim *et al., Environ. Toxicol. Chem.* **6**, 295 (1987). Use as anesthetic in fish: P. K. Bourne, *Aquaculture* **36**, 313 (1984); on fish larvae: K. C. Massee *et al., ibid.* **134**, 351 (1995); in amphibians: J. Letcher, *Zoo Biol.* **11**, 243 (1992).

Fine white needles from alcohol + ethyl acetate, mp 149-150°. Soly in water to 11%; it forms clear colorless acid solns. In ocean water solns are almost neutral. LC$_{50}$ (96 hr) in rainbow trout: 50.50 mg/l (McKim).

THERAP CAT (VET): Anesthetic for fish.

9787. Tricarballylic Acid. [99-14-9] 1,2,3-Propanetricarboxylic acid; β-carboxyglutaric acid. C$_6$H$_8$O$_6$; mol wt 176.12. C 40.92%, H 4.58%, O 54.50%. Natural product produced by rumen bacteria; structural component of fumonisin mycotoxins. Prepn from aconitic acid by reduction with sodium amalgam: Fittig, *Ann.* **314**, 15 (1901); by hydrolysis of ethyl propane-1,1,2,2-tetracarboxylate: Clarke, Murray, *Org. Synth.* coll. vol. **I** (2nd ed., 1941) p 523. Absorption spectrum: Bielecki, Henri, *Ber.* **46**, 2596 (1913). Crystal structure: J. C. Barnes, J. D. Paton, *Acta Crystallogr.* **C44**, 758 (1988). Production by rumen microorganisms: J. B. Russell, N. Forsberg, *Br. J. Nutr.* **56**, 153 (1986). Review of chemistry: A. S. H. Elgazwy, *Curr. Org. Chem.* **8**, 1405-1423 (2004).

Large, orthorhombic prisms from water or ether, mp 166°. pK$_1$ (30°): 3.49; pK$_2$: 4.58; pK$_3$: 5.83. At 18° 50 g dissolve in 100 ml water and 0.9 g dissolve in 100 ml ether. Quite sol in alc.

Trimethyl ester. [6138-26-7] $C_9H_{14}O_6$. d_4^{20} 1.1822; bp_{13} 150°; n_D^{20} 1.4398.

9788. Trichlorfon. [52-68-6] P-(2,2,2-Trichloro-1-hydroxyethyl)phosphonic acid dimethyl ester; O,O-dimethyl-1-hydroxy-2,-2,2-trichloroethylphosphonate; O,O-dimethyl 2,2,2-trichloro-1-hydroxyethylphosphonate; chlorofos; metrifonate; trichlorphene; Bayer L 1359; Cekufon; Dipterex; Dylox. $C_4H_8Cl_3O_4P$; mol wt 257.43. C 18.66%, H 3.13%, Cl 41.31%, O 24.86%, P 12.03%. Organophosphorus insecticide; transformed nonenzymatically to the active cholinesterase inhibitor, dichlorvos, *q.v.* Prepn: W. Lorenz, **US 2701225** (1955 to Bayer); W. F. Barthel *et al., J. Am. Chem. Soc.* **76**, 4186 (1954); W. Lorenz *et al., ibid.* **77**, 2554 (1955). Insecticidal activity: R. L. Metcalf *et al., J. Econ. Entomol.* **52**, 44 (1959). Transformation and mode of action: E. Reiner *et al., Biochem. Pharmacol.* **24**, 717 (1975); I. Nordgren *et al., Arch. Toxicol.* **41**, 31 (1978). HPLC determn in plasma: L. K. Unni *et al., J. Chromatogr.* **573**, 99 (1992). Determn by ^{31}P-NMR in insecticidal formulations and crop residues: Z. Talebpour *et al., Anal. Chim. Acta* **576**, 290 (2006). Clinical trial in Alzheimer's disease: R. E. Becker *et al., Alzheimer Dis. Assoc. Disord.* **10**, 124 (1996). Evaluation of carcinogenic risk: *IARC Monographs* **30**, 207-231 (1983). Review of clinical experience in schistosomiasis: H. Feldmeier, E. Doehring, *Acta Trop.* **44**, 357 (1987).

White crystals, mp 83-84°. d_4^{20} 1.73. n_D^{20} 1.3439. Vapor pressure at 20°: 7.8 × 10^{-6} mm Hg. Soly in water (25°): 15.4 g/100 ml. Very sol in methylene chloride; freely sol in acetone, alc, chloroform, ether; sol in benzene; very slightly sol in hexane, pentane, CCl_4, diethyl ether. Insol in petroleum oils. Dec by alkali. LD_{50} in rats (mg/kg): 450 orally; 255 i.p.; in mice (ng/kg): 400 s.c.; 500 i.p. (Lorenz).

Butanoic acid ester. [126-22-7] 2,2,2-Trichloro-1-(dimethoxyphosphinoyl)ethyl butyrate; butonate; T-113. $C_8H_{14}Cl_3O_5P$; mol wt 327.52. Prepn: Arthur, Casida, *J. Agric. Food Chem.* **6**, 360 (1958). Liquid. $bp_{0.5}$ 129°.

USE: Insecticide.

THERAP CAT: Anthelmintic (schistosoma).

THERAP CAT (VET): Anthelmintic (Nematodes); ectoparasiticide.

9789. Trichlormethiazide. [133-67-5] 6-Chloro-3-(dichloromethyl)-3,4-dihydro-2*H*-1,2,4-benzothiadiazine-7-sulfonamide 1,1-dioxide; 6-chloro-3-dichloromethyl-7-sulfamyl-3,4-dihydro-1,-2,4-benzothiadiazine 1,1-dioxide; 3-dichloromethyl-6-chloro-7-sulfamyl-3,4-dihydro-1,2,4-benzothiadiazine 1,1-dioxide; 3-dichloromethylhydrochlorothiazide; hydrotrichlorothiazide; trichloromethiazide; Achletin; Anatran; Anistadin; Aponorin; Carvacron; Diurese; Esmarin; Fluitran; Flutra; Intromene; Kubacron; Metahydrin; Naqua; Salurin; Tachionin; Tolcasone; Triflumen. $C_8H_8Cl_3N_3O_4S_2$; mol wt 380.64. C 25.24%, H 2.12%, Cl 27.94%, N 11.04%, O 16.81%, S 16.85%. Prepn: deStevens *et al., Experientia* **16**, 113 (1960); Sherlock *et al., ibid.* **184**. Toxicity: E. I. Goldenthal, *Toxicol. Appl. Pharmacol.* **18**, 185 (1971).

Crystals from methanol + acetone + water, dec 266-273°, also reported as 248-250°. Soly (mg/ml) at 25°: water 0.8; ethanol 21; methanol 60. Freely sol in acetone; very slightly sol in ether, chloroform. LD_{50} orally in rats: >20000 mg/kg (Goldenthal).

THERAP CAT: Diuretic; antihypertensive.

THERAP CAT (VET): Diuretic.

9790. Trichlormethine. [555-77-1] 2-Chloro-*N,N*-bis(2-chloroethyl)ethanamine; 2,2′,2″-trichlorotriethylamine; tris(β-chloroethyl)amine; trimustine; HN-3. $C_6H_{12}Cl_3N$; mol wt 204.52. C 35.24%, H 5.91%, Cl 52.00%, N 6.85%. Nitrogen mustard formerly used as a chemical warfare agent. Prepd by the action of thionyl chloride on triethanolamine: Ward, *J. Am. Chem. Soc.* **57**, 914 (1935); **US 2072348** (1937 to Hercules Powder). Improved procedure: Wilson, Tishler, *J. Am. Chem. Soc.* **73**, 3635 (1951); Witten in *Kirk-Othmer Encyclopedia of Chemical Technology* **vol. 7** (Interscience, New York, 1951) p 130. Clinical use in treatment of Hodgkin's disease and leukemias: L. S. Goodman *et al., J. Am. Med. Assoc.* **132**, 126 (1946). Evaluation of carcinogenic risk: *IARC Monographs* **50**, 143 (1990). GC determn in air: J. R. Stuff *et al., J. Chromatogr. A* **849**, 529 (1999).

Mobile liquid. Faint odor of fish + soap. *Vesicant, necrotizing irritant. Never use without appropriate gas mask.* Volatility at 25° = 0.120 mg/l. d_4^{25} 1.2347, mp −4°. bp_{15} 144°. n_D^{25} 1.4925. Very slightly sol in water. Miscible with dimethylformamide, carbon disulfide, carbon tetrachloride, many other organic solvents and oils. The undiluted liquid dec on standing and forms polymeric quaternary ammonium salts which are insol in the free base.

Hydrochloride. [817-09-4] Trillekamin. $C_6H_{12}Cl_3N.HCl$; mol wt 240.98. Crystals, mp 130-131°. Freely sol in water, sol in alcohol.

THERAP CAT: Antineoplastic.

9791. Trichloroacetaldehyde. [75-87-6] 2,2,2-Trichloroacetaldehyde; trichloroethanal; chloral; anhydr chloral. C_2HCl_3O; mol wt 147.38. C 16.30%, H 0.68%, Cl 72.16%, O 10.86%. Cl_3-CCHO. Prepn by chlorinating alcohol, treating with H_2SO_4 and then distilling: Liebig, *Ann.* **1**, 189 (1832); Personne, *J. Pharm. Chim.* [4] **51**, 350 (1869); *idem, ibid.* **11**, 205 (1870); Brochet, *Bull. Soc. Chim. Fr.* [3] **17**, 228 (1897); *idem, Ann. Chim.* [7] **10**, 332 (1897); Trillat, *Bull. Soc. Chim. Fr.* [3] **17**, 230 (1897); Besson, **DE 133021** (1902); *idem, Chem. Zentralbl.* **1902**, II 553; Ohse, **DE 734723** (1943), *C.A.* **38**, 3671 (1944). Prepn by chlorination of a mixture of alcohol and acetaldehyde: Société d'Electrochimie, **FR 612396**; *Chim. Ind. (Paris)* **21**, 567 (1929); from chloral hydrate by azeotropic distn: Mahoney, Pierson, **US 2584036** (1952 to Merck & Co.); from hypochlorous acid and trichloroethylene: Stevens *et al.,* **US 2759978** (1956 to Columbia-Southern). In the laboratory anhydr chloral may be quickly obtained by shaking pharmaceutical grade chloral hydrate with concd H_2SO_4, separating the two layers, and distilling: *cf.* Gattermann, Wieland, *Praxis des Organischen Chemikers* (de Gruyter, Berlin, 40th ed., 1961) p 334. *See also* Chloral Hydrate.

Oily liquid. Pungent, irritating odor; d_4^{20} 1.510; d_4^{25} 1.5050. mp −57.5°. bp_{760} 97.8°. n_D^{20} 1.45572; $n_{He}^{21.4}$ 1.45412. Freely sol in water forming chloral hydrate. Sol in alcohol forming chloral alcoholate; sol in ether. Polymerizes under the influence of light and in presence of sulfuric acid forming a white solid trimer called *metachloral*.

USE: Manuf chloral hydrate, DDT.

9792. Trichloroacetic Acid. [76-03-9] 2,2,2-Trichloroacetic acid; TCA. $C_2HCl_3O_2$; mol wt 163.38. C 14.70%, H 0.62%, Cl 65.09%, O 19.59%. CCl_3COOH. Prepd by oxidation of chloral hydrate with nitric acid: Parkes, Hollingshead, *Chem. Ind. (London)* **1954**, 222. Manuf by chlorination of acetic acid: Eaker, **US 2832803** (1958 to Monsanto). Toxicity study: G. W. Bailey, J. L. White, *Residue Rev.* **10**, 97 (1965).

Very deliquesc crystals; slight characteristic odor. d_4^{61} 1.629; mp 57-58°; bp 196-197°. Very sol in water, alcohol, ether; dec by heating with caustic alkalies into chloroform and alkali carbonate. pKa ~0.7. pH of 0.1 molar aq soln 1.2. *Corrosive. Keep tightly closed in a cool place.* Storage of trichloroacetic acid solns in water of less than 30% strength is not recommended. Decompn products are chloroform, hydrochloric acid, carbon dioxide and carbon monoxide. LD_{50} orally in rats: 5000 mg/kg (Bailey, White).

Ethyl ester. [515-84-4] Ethyl trichloroacetate. $C_4H_5Cl_3O_2$. Clear liq; odor resembling menthol. d_4^{20} 1.383. bp 168°. n_D^{20} 1.4507. Insol in water. Misc with alcohol, ether.

Sodium salt. [650-51-1] Sodium trichloroacetate; Konesta; Varitox. $C_2Cl_3NaO_2$; mol wt 185.36. Yellow deliquesc powder, mp >300°. Soly in water at 25°: 1.2 kg/l. Sol in ethanol.
Caution: Potential symptoms of overexposure are irritation of eyes, skin, nose, throat, respiratory system; cough, dyspnea, delayed pulmonary edema; eye and skin burns; dermatitis; salivation, vomiting, diarrhea. *See NIOSH Pocket Guide to Chemical Hazards* (DHHS/NIOSH 97-140, 1997) p 314.
USE: As a decalcifier and fixative in microscopy; also as a precipitant of protein, esp in albumin detection. As herbicide.
THERAP CAT: Caustic.
THERAP CAT (VET): Caustic, vesicant.

9793. Trichloroacetonitrile. [545-06-2] 2,2,2-Trichloroacetonitrile; trichloromethylnitrile; cyanotrichloromethane. C_2Cl_3N; mol wt 144.38. C 16.64%, Cl 73.66%, N 9.70%. CCl_3CN. Prepd from ethyl trichloroacetate and aq ammonia, and physical and thermodynamic properties: Davies, Jenkin, *J. Chem. Soc.* **1954**, 2374; by action of phosphorus pentoxide on trichloroacetamide: Carpenter, *J. Org. Chem.* **27**, 2085 (1962). Manuf by reaction of methylnitrile, HCl and chlorine gas: Käbisch, US 2745868 (1956 to Degussa). Toxicity study: H. F. Smyth *et al.*, *Am. Ind. Hyg. Assoc. J.* **23**, 95 (1962). Toxicology and metabolism: E. L. C. Lin *et al.*, *Environ. Health Perspect.* **69**, 67 (1986). Review of use in organic syntheses: S. M. Sherif, A. W. Erian, *Heterocycles* **43**, 1083-1118 (1996).
Liquid, bp_{760} 85.7°; d_4^{25} 1.4403, d_4^{35} 1.4223; $n_D^{20.0}$ 1.4409, $n_D^{27.0}$ 1.4375. Flash pt, closed cup: 195°C (383°F). *Poisonous; flammable; corrosive.* LD_{50} orally in rats: 0.25 g/kg (Smyth).
USE: Reagent in organic synthesis.

9794. 2,4,6-Trichloroanisole. [87-40-1] 1,3,5-Trichloro-2-methoxybenzene; 1-methoxy-2,4,6-trichlorobenzene; Tyrene. $C_7H_5Cl_3O$; mol wt 211.47. C 39.76%, H 2.38%, Cl 50.29%, O 7.57%. Prepd by reaction of 2,4,6-trichlorophenol with dimethyl sulfate: Kohn, Heller, *Monatsh. Chem.* **46**, 91 (1925).

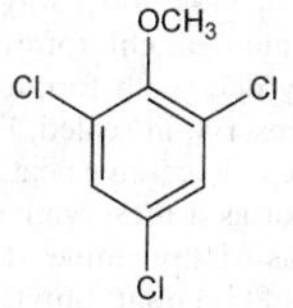

Monoclinic needles from alc, mp 60°, $bp_{738.2}$ 240°, bp_{28} 132°. Faint odor similar to that of acetophenone. Sublimes slowly at room temp. Volatile with steam. Practically insol in water. Sol in methanol, dioxane, benzene, cyclohexanone.
USE: Formerly as a dye assistant for polyester fibers.

9795. 1,2,3-Trichlorobenzene. [87-61-6] *vic*-Trichlorobenzene. $C_6H_3Cl_3$; mol wt 181.44. C 39.72%, H 1.67%, Cl 58.61%. Prepared from 3,4,5-trichloroaniline by diazotization: Cohen, Hartley, *J. Chem. Soc.* **87**, 1365 (1905); Holleman, *Rec. Trav. Chim.* **37**, 196 (1918); from 2,3,4-trichloroaniline by diazotization: Dadien *et al.*, *Monatsh. Chem.* **61**, 431 (1932); from 2,3,4-trichloroaniline by treatment with ethyl nitrite: Beilstein, Kurbatow, *Ann.* **192**, 234 (1878).

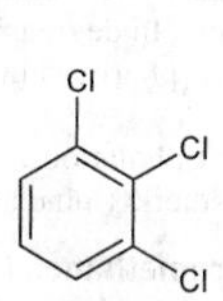

Platelets from alc, d 1.69. mp 52.6°. bp 221°. n_D^{19} 1.5776. Flash pt 113°C (235.4°F). Volatile with steam. Insol in water. Sparingly sol in alc. Freely sol in benzene, carbon disulfide.
Caution: Irritating to eyes, mucous membranes.
USE: A commercial grade (mixture of isomeric trichlorobenzenes) is used to combat termites.

9796. 1,2,4-Trichlorobenzene. [120-82-1] *unsym*-Trichlorobenzene. $C_6H_3Cl_3$; mol wt 181.44. C 39.72%, H 1.67%, Cl 58.61%. Prepd from 2,4-dichloroaniline or 2,5-dichloroaniline or 3,4-dichloroaniline by diazotization and treatment with Cu_2Cl_2:

Beilstein, Kurbatow, *Ann.* **192**, 230 (1878); van der Lande, *Rec. Trav. Chim.* **51**, 104, 110 (1932); from 1,3-diaminobenzene by tetrazotization and treatment with Cu_2Cl_2: Cohn, Fischer, *Monatsh. Chem.* **21**, 278 (1900).

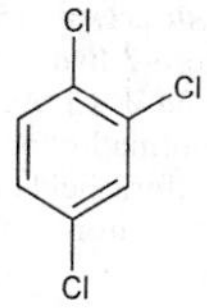

Liquid; mp 17°; d_{25}^{25} 1.4634; bp 213°; n_D^{25} 1.5524. Flash pt 110°C (230°F). Volatile with steam. Insol in water. Sparingly sol in alc. Miscible with ether, benzene, petr ether, carbon disulfide.
Caution: Potential symptoms of overexposure are irritation of eyes, skin, mucous membranes. *See NIOSH Pocket Guide to Chemical Hazards* (DHHS/NIOSH 97-140, 1997) p 314.
USE: Solvent. Prepn of standards.

9797. 1,3,5-Trichlorobenzene. [108-70-3] *sym*-Trichlorobenzene. $C_6H_3Cl_3$; mol wt 181.44. C 39.72%, H 1.67%, Cl 58.61%. Prepd from 2,4,6-trichloroaniline by diazotization and treatment with alcohol: Jackson, Lamar, *Am. Chem. J.* **18**, 667 (1896); Backer, van der Baan, *Rec. Trav. Chim.* **56**, 1177 (1937).

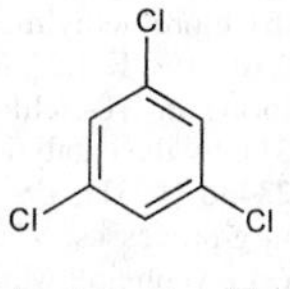

Crystals; mp 63.4°; bp 208.4°. n_D^{19} 1.5662. Flash pt 107°C (224.6°F). Volatile with steam. Insol in water. Sparingly sol in alc. Freely sol in ether, benzene, petr ether, carbon disulfide, glacial acetic acid.

9798. 2,4,6-Trichlorobenzoyl Chloride. [4136-95-2] Yamaguchi reagent. $C_7H_2Cl_4O$; mol wt 243.89. C 34.47%, H 0.83%, Cl 58.14%, O 6.56%. Acid chloride used for the regioselective prepn of functionalized esters, including large ring lactones. Prepn: J. J. Sudborough, *J. Chem. Soc., Trans.* **65**, 1028 (1894); and use in esterification reactions: J. Inanaga *et al.*, *Bull. Chem. Soc. Jpn.* **52**, 1989 (1979); M. A. Sutter, D. Seebach, *Ann.* **1983**, 939. Applications in macrolactonization reactions: M. Hikota *et al.*, *Tetrahedron Lett.* **31**, 6367 (1990); J. P. Marino *et al.*, *J. Am. Chem. Soc.* **124**, 1664 (2002); P. A. Wender *et al.*, *ibid.* 13648.

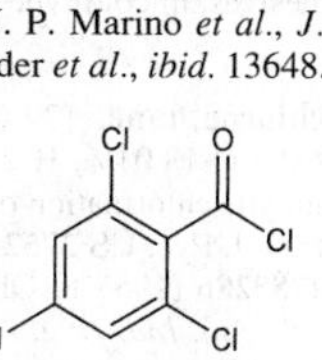

Colorless oil with pungent odor. *Corrosive, lachrymator.* bp 275°; bp_9 110-114°; bp_6 107-107.5°. Flash pt, closed cup: 235°F (113°C). Sol in most organic solvents. Moisture sensitive.
USE: Reagent in synthetic organic chemistry.

9799. α,α,β-Trichloro-*n*-butyraldehyde. [76-36-8] 2,2,3-Trichlorobutanal; butylchloral; anhydrous butylchloral; butyrchloral; crotonchloral. $C_4H_5Cl_3O$; mol wt 175.43. C 27.39%, H 2.87%, Cl 60.62%, O 9.12%. Prepd by the action of chlorine on acetaldehyde: Krämer, Pinner, *Ber.* **3**, 883 (1870); or on paraldehyde: Pinner, *Ann.* **179**, 24 (1875); on crotonaldehyde after saturation with gaseous HCl: High, US 2280290 (1942); Brown, Plump, US 2351000 (1944). From crotonaldehyde and chlorine: Ropp *et al.*, *Org. Synth. coll. vol. IV*, 130 (1963).

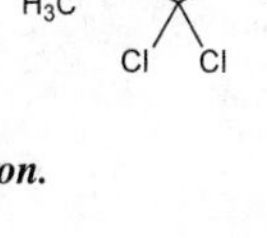

Oily liquid. Pungent, disagreeable odor; d_4^{20} 1.3956; bp_{760} 164.5-165.5°; bp_{25} 57-60°. n_D^{20} 1.47554. Freely sol in water forming butylchloral hydrate; sol in alcohol forming an alcoholate; sol in ether. Polymerizes.

9800. 2,4,6-Trichloro-*m*-cresol. [551-76-8] 2,4,6-Trichloro-3-methylphenol; C₇H₅Cl₃O; mol wt 211.47. C 39.76%, H 2.38%, Cl 50.29%, O 7.57%. Prepd by chlorination of *m*-cresol, chlorination of thymol in the presence of iron, or by action of concd sulfuric acid on 2,4,4-trichloro-3-methyl-6-isopropyl-$\Delta^{2,5}$-cyclohexadienone: Crowther, McCombie, *J. Chem. Soc.* **103**, 536 (1913).

Needles or plates from water or petr ether. mp 45-47°; bp 265°, bp_{14} 142-144°. Volatile with steam. Slightly sol in water, benzene, petr ether, glacial acetic acid. Freely sol in methanol, alc, chloroform, xylene, ether and in alkali solns.

9801. 1,1,1-Trichloroethane. [71-55-6] Methylchloroform; methyltrichloromethane; trichloromethylmethane; Chlorothene. C₂H₃Cl₃; mol wt 133.40. C 18.01%, H 2.27%, Cl 79.72%. CH₃CCl₃. Prepd by the action of chlorine on 1,1-dichloroethane: Sutton, *Proc. Roy. Soc.* **A133**, 673 (1931); by the catalytic addition of HCl to 1,1-dichloroethylene: **DE 523436** (1931); **US 2209000** (Nutting, Huscher, 1940). Review of mfg processes: *Faith, Keyes & Clark's Industrial Chemicals*, F. A. Lowenheim, M. K. Moran, Eds. (Wiley-Interscience, New York, 4th ed., 1975) pp 836-843; of toxicology and human exposure: *Toxicological Profile for 1,1,1-Trichloroethane* (PB95-264396, 1995) 307 pp.

Liquid; sweet sharp odor. Nonflammable. mp −32.5°. d_4^{20} 1.3376. bp_{760} 74.1°. n_D^{20} 1.43838. Insol in water. Absorbs some water. Sol in acetone, benzene, carbon tetrachloride, methanol, ether.

Caution: Potential symptoms of overexposure are irritation of eyes and skin; headache, lassitude, CNS depression, poor equilibrium; dermatitis; cardiac arrhythmias; liver damage. *See NIOSH Pocket Guide to Chemical Hazards* (DHHS/NIOSH 97-140, 1997) p 202.

USE: Solvent for adhesives, in cold type metal cleaning, in cleaning plastic molds.

9802. 1,1,2-Trichloroethane. [79-00-5] Vinyl trichloride. C₂H₃Cl₃; mol wt 133.40. C 18.01%, H 2.27%, Cl 79.72%. CH₂ClCHCl₂. Prepd by catalytic chlorination of ethane or ethylene: W. J. Joseph, **US 2752401**; D. J. Pye, **US 2752402** (both 1956 to Dow); M. B. Reynolds, **US 2783286** (1957 to Olin Mathieson). Toxicity data: H. F. Smyth *et al.*, *Am. Ind. Hyg. Assoc. J.* **30**, 470 (1969). Review of toxicology and human exposure: *Toxicological Profile for 1,1,2-Trichloroethane* (PB90-196411, 1989) 120 pp.

Nonflammable liquid; pleasant odor; d_4^{20} 1.4416. mp −35°. bp 113-114°. n_D^{20} 1.4711. Insol in water. Misc with alcohol, ether, and many other organic liquids. LD₅₀ orally in rats: 0.58 ml/kg (Smyth).

Caution: Potential symptoms of overexposure are irritation of nose and eyes; CNS depression; liver and kidney damage; dermatitis. Potential occupational carcinogen. *See NIOSH Pocket Guide to Chemical Hazards* (DHHS/NIOSH 97-140, 1997) p 314.

USE: Solvent for fats, waxes, natural resins, alkaloids.

9803. 2,2,2-Trichloroethanol. [115-20-8] Trichloroethyl alcohol; (hydroxymethyl)trichloromethane. C₂H₃Cl₃O; mol wt 149.40. C 16.08%, H 2.02%, Cl 71.18%, O 10.71%. CCl₃CH₂OH. Prepd by reduction of the corresponding ester, acid chloride, or acid with lithium aluminum hydride: Sroog *et al.*, *J. Am. Chem. Soc.* **71**, 1710 (1949). Manufacture by reduction of chloral hydrate with an amine borane: Chamberlain, Schechter, **US 2898379** (1959 to Callery Chem.). Pharmacology: H. Molitor, H. Robinson, *Curr. Res. Anesth. Analg.* **17**, 258 (1938).

Hygroscopic liquid, ethereal odor. At low temps it crystallizes in rhombic tablets. mp at 18°; bp 151-153°; d_{20}^{20} 1.55. Sol in about 12 parts water; miscible with alcohol or ether. pH of aq soln is 5-6, but on prolonged contact with water some free acid is formed. *Keep well closed and protected from light.*

THERAP CAT: Sedative; hypnotic.

9804. Trichloroethylene. [79-01-6] 1,1,2-Trichloroethene; ethinyl trichloride; ethylene trichloride; Tri-Clene; Trilene; Trichloren; Algylen; Trimar; Triline; Trethylene; Westrosol; Chlorylen; Gemalgene; Germalgene. C₂HCl₃; mol wt 131.38. C 18.28%, H 0.77%, Cl 80.95%. Cl₂C=CHCl. Prepn from *sym*-tetrachloroethane by elimination of HCl (by boiling with lime): **DE 171900**; by passing tetrachloroethane vapor over CaCl₂ catalyst at 300°: **DE 263457**; without catalyst at 450-470°: **GB 575530** (1946 to du Pont). Physical properties: E. W. McGovern, *Ind. Eng. Chem.* **35**, 1230 (1943); S. A. Mumford, J. W. C. Phillips, *J. Chem. Soc.* **1950**, 75. Review of mfg processes: S. A. Miller, *Chem. Process Eng.* **47**, 268 (1966); *Faith, Keyes & Clark's Industrial Chemicals*, F. A. Lowenheim, M. K. Moran, Eds. (Wiley-Interscience, New York, 4th ed., 1975) pp 844-848. Toxicity data: H. S. Smyth *et al.*, *Am. Ind. Hyg. Assoc. J.* **30**, 470 (1969). Review of carcinogenic risk: *IARC Monographs* **20**, 545-572 (1979); B. L. Van Duuren, *Environ. Res.* **49**, 333 (1989). Review of use as anesthetic: J. V. Farman, *Br. J. Anaesth.* **53**, Suppl. 3, 3S-9S (1981). Review of metabolism, toxicity and carcinogenicity: J. V. Bruckner *et al.*, *Crit. Rev. Toxicol.* **20**, 31-50 (1989); of toxicology and human exposure: *Toxicological Profile for Trichloroethylene* (PB98-101165, 1997) 335 pp. Series of articles on toxicology and risk assessment: *Environ. Health Perspect.* **108**, Suppl. 2, 159-366 (2000).

Nonflammable, mobile liquid. Characteristic odor resembling that of chloroform. *Poisonous. Use with adequate ventilation.* d_4^4 1.4904; d_4^{15} 1.4695; d_4^{20} 1.4642; d_4^{25} 1.4559. Vapor density: 4.53 (air = 1.00). mp −84.8°. bp_{760} 86.9°; bp_{400} 67.0°; bp_{200} 48.0°; bp_{100} 31.4°; bp_{60} 20.0°; bp_{20} −1.0°; bp_{10} −12.4°; bp_5 −22.8°; $bp_{1.0}$ −43.8°; n_D^{17} 1.47914; n_D^{20} 1.4775; n_D^{25} 1.45560. Soly in water (25°): 0.11 g/100 g. Misc with ether, alcohol, chloroform. Dissolves most fixed and volatile oils. Slowly dec (with formn of HCl) by light in the presence of moisture. Preserve in sealed, light-resistant containers; avoid prolonged exposure to excessive heat. Medicinal trichloroethylene may contain thymol as a preservative; industrial grades may contain stabilizers such as triethylamine. LD₅₀ orally in rats: 4.92 ml/kg; LC (4 hrs) in rats: 8000 ppm (Smyth).

Caution: Potential symptoms of overexposure are headache, vertigo; visual disturbance, fatigue, giddiness, tremors, somnolence, nausea and vomiting; irritation of eyes, skin; dermatitis; cardiac arrhythmias, paresthesia; liver injury. *See NIOSH Pocket Guide to Chemical Hazards* (DHHS/NIOSH 97-140, 1997) p 316. *See also Patty's Industrial Hygiene and Toxicology* vol. **2B**, G. D. Clayton, F. E. Clayton, Eds. (Wiley-Interscience, 3rd ed., 1982) pp 3553-3560; *Clinical Toxicology of Commercial Products*, R. E. Gosselin *et al.*, Eds. (Williams & Wilkins, Baltimore, 5th ed., 1984) Section II, pp 165-166. This substance is reasonably anticipated to be a human carcinogen: *Report on Carcinogens, Twelfth Edition* (PB2011-111646, 2011) p 420.

USE: Solvent for fats, waxes, resins, oils, rubber, paints, and varnishes. Solvent for cellulose esters and ethers. Used for solvent extraction in many industries. In degreasing, in dry cleaning. In the manuf of organic chemicals, pharmaceuticals, such as chloroacetic acid.

THERAP CAT: Anesthetic (inhalation).
THERAP CAT (VET): Anesthetic (inhalation).

9805. Trichlorofluoromethane. [75-69-4] Trichloromonofluoromethane; fluorotrichloromethane; fluorocarbon 11; trichloromethyl fluoride; FC 11; Arcton 11; Freon 11; Frigen 11. CCl₃F; mol wt 137.36. C 8.74%, Cl 77.42%, F 13.83%. Prepn: Henne, *Org. React.* **2**, 64 (1944). Manuf: *Faith, Keyes & Clark's Industrial Chemicals*, F. A. Lowenheim, M. K. Moran, Eds. (Wiley-Interscience, New York, 4th ed., 1975) pp 325-330.

Liquid at temps below 23.7°. Faint ethereal odor. Nonflammable. $d_4^{17.2}$ 1.494; d_{gas}^{25} 5.04 (air = 1). mp −111°. bp_{760} 23.7°; bp_{400} 6.8°; bp_{200} −9.1°; bp_{100} −23.0°; bp_{60} −32.3°; bp_{40} −39.0°; bp_{20} −49.7°; bp_{10} −59.0°; bp_5 −67.6°; $bp_{1.0}$ −84.3°. Crit temp 198°; crit press. 43.2 atm (635 lb/sq inch, abs). $n_D^{18.5}$ 1.3865. Dipole moment 0.45.

Practically insol in water. Sol in alcohol, ether, other, organic solvents.

Caution: Potential symptoms of overexposure are incoordination, tremors; dermatitis; cardiac arrhythmias, cardiac arrest; asphyxia; direct contact with liquid may cause frostbite. *See NIOSH Pocket Guide to Chemical Hazards* (DHHS/NIOSH 97-140, 1997) p 146. *See also Patty's Industrial Hygiene and Toxicology* **vol. 2B,** G. D. Clayton, F. E. Clayton, Eds. (Wiley-Interscience, New York, 3rd ed., 1981) p 3073-3091. *Note:* Consult latest Government regulations on use as aerosol propellant.

USE: In refrigeration machinery requiring a refrigerant effective at negative pressures. As aerosol propellant.

9806. Trichloroisocyanuric Acid. [87-90-1] 1,3,5-Trichloro-1,3,5-triazine-2,4,6(1H,3H,5H)-trione; trichloroiminocyanuric acid; TCCA; ACL-85; Chloreal; Symclosene. $C_3Cl_3N_3O_3$; mol wt 232.40. C 15.50%, Cl 45.76%, N 18.08%, O 20.65%. Industrially significant bleaching agent and biocide. Prepn: F. D. Chattaway, J. M. Wadmore, *J. Chem. Soc.* **81**, 191 (1902); Hands, Whitt, *J. Soc. Chem. Ind.* **67**, 66 (1948). Structure determn: R. C. Petterson *et al., J. Org. Chem.* **25**, 1595 (1960). Review of synthetic applications as a chlorinating agent or oxidant: U. Tilstam, H. Weinmann, *Org. Process Res. Dev.* **6**, 384-393 (2002); J. C. Barros, *Synlett* **2005,** 2115-2116.

Needles from ethylene chloride, mp 246-247° (dec). *Oxidizer.* May be stored in the dry state for at least a year. Releases hypochlorous acid on contact with water. pH of aq solns about 4.4. Soly in water at 25° about 0.2%. Sol in chlorinated and highly polar solvents.

Caution: Moderately irritating to eyes, skin, mucous membranes. USE: Chlorinating agent, disinfectant, industrial deodorant. In household cleansers. In removing oil and protein in stainless steel. In nonshrink treatment for wool. Oxidant and chlorinating reagent in organic synthesis.

THERAP CAT: Anti-infective (topical).

9807. 3,4,6-Trichloro-2-nitrophenol. [82-62-2] 2-Nitro-3,-4,6-trichlorophenol; 2,4,5-trichloro-6-nitrophenol; Dowlap. C_6H_2-Cl_3NO_3; mol wt 242.44. C 29.73%, H 0.83%, Cl 43.87%, N 5.78%, O 19.80%. Prepd by dissolving 2,4,5-trichlorophenol in glacial acetic acid and treating with concd nitric acid: Kohn, Fink, *Monatsh. Chem.* **58**, 73 (1931); Harrison *et al., J. Chem. Soc.* **1943**, 235. Larvicidal activity vs sea lampreys: V. C. Applegate *et al., Science* **127**, 336 (1958).

Pale yellow crystals from petr ether, mp 92-93°.
USE: Larvicide to combat the sea lamprey, an eel-like fish which attacks trout, especially in the Great Lakes region.

9808. 2,4,5-Trichlorophenol. [95-95-4] Collunosol; Dowicide 2. $C_6H_3Cl_3O$; mol wt 197.44. C 36.50%, H 1.53%, Cl 53.86%, O 8.10%. Prepd by treating 1,2,4,5-tetrachlorobenzene with methanolic NaOH in autoclave at 160° for several hrs: Harrison *et al., J. Chem. Soc.* **1943**, 235; Agfa, **DE 411052** (1925); *Chem. Zentralbl.* **1925**, I, 2411. Toxicity data: Deichmann, *Fed. Proc.* **2**, 76 (1943). Review of toxicology and human exposure: *Toxicological Profile for Chlorophenols* (PB99-166639, 1999) 260 pp.

Needles from alcohol or ligroin. Strong phenolic odor. mp 67°. Sublimes. bp_{746} 248°. bp_{760} 253°. Weak monobasic acid. pK (25°) 7.37. Soly (g/100 g of solvent at 25°): acetone 615; benzene 163; carbon tetrachloride 51; ether 525; denatured alcohol formula 30, 525; methanol 615; liquid petrolatum (at 50°) 56; soybean oil 79; toluene 122; water <0.2. LD_{50} orally in rats: 0.82 g/kg (Deichmann).

Sodium salt sesquihydrate. Dowicide B. Prepn: **US 1991329** (1935 to Dow). Flakes. Solubility (g/100 g solvent at 25°): acetone 163; denatured alcohol formula 30, 186; ethylene glycol 33; methanol 241; water 113. pH of satd aq soln 11.0-13.0.

USE: Fungicide, bactericide, biocide; intermediate in production of herbicides.

9809. 2,4,6-Trichlorophenol. [88-06-2] 1,3,5-Trichloro-2-hydroxybenzene; Dowicide 2S; Omal. $C_6H_3Cl_3O$; mol wt 197.44. C 36.50%, H 1.53%, Cl 53.86%, O 8.10%. Prepd by direct chlorination of phenol: Tiessens, *Rec. Trav. Chim.* **50**, 115 (1931); Chulkov *et al., Org. Chem. Ind. USSR* **3**, 97 (1937); *eidem, Chem. Zentralbl.* **1938,** I, 1419; *C.A.* **31**, 4967 (1937). Prepn of sodium salt monohydrate: Hunter, Seyfried, *J. Am. Chem. Soc.* **43**, 154 (1921). Review of toxicology and human exposure: *Toxicological Profile for Chlorophenols* (PB99-166639, 1999) 260 pp.

Crystals from ligroin. Strong phenolic odor. Volatile with steam, but not from alkaline soln. d 1.4901. mp 69°. bp_{760} 246°. Soly (g/100 g of solvent): Acetone 525; benzene 113; carbon tetrachloride 37; diacetone alcohol 335; ether 354; denatured alcohol formula 30, 400; methanol 525; pine oil 163; Stoddard solvent 16; toluene 100; turpentine 37; water <0.1.

Sodium salt monohydrate. Flaky crystals. Freely sol in water, alcohol, ether, acetone. pH of satd aq soln 11.0-13.0.

Caution: 2,4,6-Trichlorophenol is reasonably anticipated to be a human carcinogen: *Report on Carcinogens, Twelfth Edition* (PB2011-111646, 2011) p 424.

USE: Fungicide, bactericide, wood preservative, biocide; intermediate in production of higher chlorinated phenols.

9810. 1,1,1-Trichloro-2-propanol. [76-00-6] 1,1,1-Trichloroisopropyl alcohol; trichloroisopropanol; Isopral. $C_3H_5Cl_3O$; mol wt 163.42. C 22.05%, H 3.08%, Cl 65.08%, O 9.79%. Prepd by reaction of chloral and methylmagnesium bromide: Kharasch *et al., J. Am. Chem. Soc.* **63**, 2305 (1941). Toxicity study: Burtner, Lehmann, *J. Pharmacol. Exp. Ther.* **63**, 183 (1938).

Monoclinic crystals; camphor-like odor; pungent taste; mp 50°; bp 161-162°. Sublimes at ordinary temp. Sol in about 35 parts water; freely sol in alcohol or ether. *Keep well closed in a cool place.* LD_{50} orally in rats: 1 g/kg (Burtner, Lehmann).

9811. Trichlorosilane. [10025-78-2] Trichloromonosilane; silicochloroform; hydrotrichlorosilane. Cl_3HSi; mol wt 135.44. Cl 78.52%, H 0.74%, Si 20.74%. $SiHCl_3$. Prepd from Si and HCl or from SiH_4 and HCl in presence of $AlCl_3$: Gattermann, *Ber.* **22**, 190 (1889); Ruff, Albert, *ibid.* **38**, 2226 (1905); Stock, Zeidler, *ibid.* **56**,

986 (1923); Schenk in *Handbook of Preparative Inorganic Chemistry* vol. 1, G. Brauer, Ed. (Academic Press, New York, 2nd ed., 1963) pp 691-692. Toxicity study: H. F. Smyth *et al.*, *J. Ind. Hyg. Toxicol.* **31**, 60 (1949).

Volatile, mobile liquid. Fumes in air. Supports combustion. *Dangerous when wet; flammable; corrosive.* d_4^0 1.3830; d_4^{20} 1.3417; d_4^{25} 1.3313. mp $-126.5°$. bp_{760} 31.8°, also reported as 36.5°. Viscosity in cP: 0.397 at 0°; 0.332 at 20°; 0.316 at 25°. n_D^{20} 1.4020; n_D^{25} 1.3983. Dec by water. Sol in benzene, carbon disulfide, chloroform, carbon tetrachloride. Dipole moment 0.97. LD_{50} orally in rats: 1.03 g/kg (Smyth).

USE: In organic synthesis.

9812. Trichodermin. [4682-50-2] (4β)-12,13-Epoxytrichothec-9-en-4-ol acetate; WG-696. $C_{17}H_{24}O_4$; mol wt 292.38. C 69.84%, H 8.27%, O 21.89%. Antifungal metabolite from *Trichoderma viride* ND8: **NL 302527** (1964 to Loevens); *C.A.* **62**, 1050f (1965); also isolated from *Myrothecium roridum*. Structure: Godtfredsen, Vangedal, *Proc. Chem. Soc. London* **1964**, 188; Gutzwiller *et al.*, *Helv. Chim. Acta* **47**, 2234 (1964); Godtfredsen, Vangedal, *Acta Chem. Scand.* **19**, 1088 (1965); Abrahamsson, Nilsson, *ibid.* **20**, 1044 (1966). Total synthesis: Colvin *et al.*, *Chem. Commun.* **1971**, 858; *eidem*, *J. Chem. Soc. Perkin Trans. 1* **1973**, 1989. Toxicity studies: Yamamoto *et al.*, *Takeda Kenkyusho Nempo* **28**, 69 (1969), *C.A.* **72**, 76058g (1970). Inhibition of protein synthesis: F. Hernandez, M. Cannon, *J. Antibiot.* **35**, 875 (1982).

Crystals from pentane at −70°. mp 46°, 58-60° (Colvin). $bp_{0.05}$ 110-112°. $[\alpha]_D^{20}$ −11° (c = 1 in chloroform). uv max (ethanol): 205 nm (ε 2400). Sparingly soluble in water; soluble in all common organic solvents. LD_{50} in mice (mg/kg): 500-1000 s.c.; >1000 orally (**NL 302527**).

Trichodermol. [2198-93-8] Roridan C. $C_{15}H_{22}O_3$; mol wt 250.34. Crystals, mp 124-125°. *See also* verrucarins.

9813. Trichosanthin. [60318-52-7] GLQ223. Active component of the traditional Chinese medicine, Tian Hua Fen (Radix Trichosanthis; known since 300 A.D. and used in gynecological disorders and for termination of pregnancy. Basic polypeptide of 234 amino acids, M_r 26,000 Da, isolated from the roots of *Trichosanthes kirilowii* Maxim, *Cucurbitaceae*. Purification, characterization, and amino acid sequence: W. Yu *et al.*, *Pure Appl. Chem.* **58**, 789 (1986); J. M. Maraganore *et al.*, *J. Biol. Chem.* **262**, 11628 (1987). Conformation: S. Kubota *et al.*, *Biochim. Biophys. Acta* **871**, 101 (1986); and homology with ricin, *q.v.*: *eidem*, *Int. J. Pept. Protein Res.* **30**, 646 (1987). Inhibition of protein synthesis: H. W. Yeung *et al.*, *ibid.* **31**, 265 (1988). Inhibition of HIV replication: M. S. McGrath *et al.*, *Proc. Natl. Acad. Sci. USA* **86**, 2844 (1989). Abortifacient activity in animals: M. C. Chang *et al.*, *Contraception* **19**, 175 (1979); S. K. Saksena *et al.*, *ibid.* **20**, 367 (1979); I. F. Lau *et al.*, *ibid.* **21**, 77 (1980). Clinical trial as abortifacient: K. F. Cheng, *Obstet. Gynecol.* **59**, 494 (1982).

Crystals. pI 9.4.

THERAP CAT: Abortifacient.

9814. Trichostatin. Antibiotic A-300; A-300-I. Antifungal antibiotic isolated from metabolites of *Streptomyces hygroscopicus*. It is composed of trichostatins A (major), B, the ferric chelate of trichostatin A, and C, the first glycosyl hydroxamate from a natural source. Isoln of A and B: N. Tsuji *et al.*, **JP Kokai 74 14691**, *C.A.* **81**, 48547h (1974). Structural elucidation of A and B: *eidem*, *J. Antibiot.* **29**, 1 (1976). Isoln and structure of C: N. Tsuji, M. Kobayashi, *ibid.* **31**, 939 (1978).

Trichostatin A

Trichostatin A. [58880-19-6] [R-(E,E)]-7-[4-(Dimethylamino)-phenyl]-N-hydroxy-4,6-dimethyl-7-oxo-2,4-heptadienamide. C_{17}-$H_{22}N_2O_3$; mol wt 302.37. Crystals from ethyl acetate, mp 150-151°. $[\alpha]_D^{20.5}$ +62.8° ±1.1° (c = 1.007 in ethanol). uv max (ethanol): 252, 265, 341 nm ($E_{1cm}^{1\%}$ 531, 582, 648). Sol in lower alcohols, sparingly sol in chloroform, ethyl acetate, acetone, benzene.

Trichostatin B. [58895-00-4] Tris[7-[4-(dimethylamino)phenyl]-N-hydroxy-4,6-dimethyl-7-oxo-2,4-heptadienamidato-O^NO^1]-iron. $C_{51}H_{63}FeN_6O_9$; mol wt 959.94. Trihydrate, dark reddish purple prisms from methanol, mp 192° (dec). uv max (ethanol): 253, 277, 341, 450 nm ($E_{1cm}^{1\%}$ 624, 651, 918, 50).

Trichostatin C. [68676-88-0] 7-[4-(Dimethylamino)phenyl]-N-(β-D-glucopyranosyloxy)-4,6-dimethyl-7-oxo-2,4-heptadienamide. $C_{23}H_{32}N_2O_8$; mol wt 464.52. Colorless prisms from methanol, mp 171-173°. $[\alpha]_D^{24}$ +50.5 ±0.9° (c = 0.987 in methanol). uv max (methanol): 268, 344 nm (ε 14600; 14300).

9815. Trichothecin. [6379-69-7] (4β)-12,13-Epoxy-4-[[(2Z)-1-oxo-2-butenyl]oxy]trichothec-9-en-8-one; 12,13-epoxy-4-hydroxytrichothec-9-en-8-one crotonate. $C_{19}H_{24}O_5$; mol wt 332.40. C 68.65%, H 7.28%, O 24.07%. Mycotoxin with antibiotic activity. Produced by *Trichothecium roseum*: Freeman, Morrison, *Nature* **162**, 30 (1948). Activity: G. G. Freeman, *J. Gen. Microbiol.* **12**, 213 (1955). Structure: Godtfredsen, Vangedal, *Proc. Chem. Soc. London* **1964**, 188; Gutzwiller *et al.*, *Helv. Chim. Acta* **47**, 2234 (1964). Biogenesis: Jones, Lowe, *J. Chem. Soc.* **1960**, 3959. Biosynthetic studies: Achilladelis *et al.*, *J. Chem. Soc. Perkin Trans. 1* **1972**, 1425; Machida, Nozoe, *Tetrahedron Lett.* **1972**, 1969; *eidem*, *Tetrahedron* **28**, 5113 (1972).

Slender needles from petrol ether, mp 118°. $[\alpha]_D^{18}$ +44° (c = 1 in $CHCl_3$). uv max (hexane): 217 nm (ε 18000); (methanol): 215 nm (ε 19000). Slightly sol in water (400 mg/l at 25°). Freely sol in most organic solvents. Aq solns are stable at pH 1-10 for at least 48 hrs at 20°. At pH 12 the antifungal activity is destroyed rapidly. Aq solns at pH 7 can be maintained at 100° for at least one hour without loss of activity. LD_{50} i.v. in mice: ~300 mg/kg (Freeman).

9816. Tricine. [5704-04-1] N-[2-Hydroxy-1,1-bis(hydroxymethyl)ethyl]glycine; N-[tris(hydroxymethyl)methyl]glycine. C_6-$H_{13}NO_5$; mol wt 179.17. C 40.22%, H 7.31%, N 7.82%, O 44.65%. One of the zwitterionic amino acids known as "Good" buffers; active in the pH range 6-8.5. Prepn: N. E. Good, *Arch. Biochem. Biophys.* **96**, 653 (1962); and characterization: N. E. Good *et al.*, *Biochemistry* **5**, 467 (1966). Temperature effects on pK: R. N. Roy *et al.*, *J. Am. Chem. Soc.* **95**, 8231 (1973); M. L. Soni, R. C. Kapoor, *Int. J. Quantum Chem.* **20**, 385 (1981); and on pH: R. N. Roy *et al.*, *Cryobiology* **22**, 589 (1985). Scavenger of hydroxyl radicals: M. Hicks, J. M. Gebicki, *FEBS Lett.* **199**, 92 (1986); B. Halliwell *et al.*, *Anal. Biochem.* **165**, 215 (1987). Use as buffer: R. S. Gardner, *J. Cell Biol.* **42**, 320 (1969); R. S. Spendlove *et al.*, *Proc. Soc. Exp. Biol. Med.* **137**, 258 (1971); H. Schaegger, G. Von Jagow, *Anal. Biochem.* **166**, 368 (1987).

Crystals from alcohol/water, mp 187°. pKa_1 ~2.3; pKa_2 (0.1M): 0°, 8.6; 20°, 8.15; 37°, 7.8. pKa_2 (20°): 8.15 (0.2M); 8.15 (0.01M). ΔpKa/°C −0.021. Saturated aqueous soln is 0.8M at 0°.

USE: Biological buffer.

9817. Triclabendazole. [68786-66-3] 6-Chloro-5-(2,3-dichlorophenoxy)-2-(methylthio)-1H-benzimidazole; CGA-89317; Fasinex. $C_{14}H_9Cl_3N_2OS$; mol wt 359.65. C 46.75%, H 2.52%, Cl 29.57%, N 7.79%, O 4.45%, S 8.91%. Benzimidazole derivative effective against liver fluke. Prepn: J.-J. Gallay *et al.*, **BE 865870** corresp to **US 4197307** (1978, 1980 both to Ciba-Geigy). Efficacy against immature and mature *Fasciola hepatica* in sheep, goats: K. Wolff *et al.*, *Vet. Parasitol.* **13**, 145 (1983); in sheep: J. C. Boray *et al.*, *Vet. Rec.* **113**, 315 (1983). Effective also against *F. gigantica* in sheep: N. Güralp, R. Tinar, *J. Helminthol.* **58**, 113 (1984). Brief review: J. Eckert *et al.*, *Biol. Muench. Tieraerztl. Wochenschr.* **97**, 349 (1984).

Crystals, mp 175-176°.

THERAP CAT (VET): Anthelmintic (fasciola).

9818. Triclobisonium Chloride. [79-90-3] N^1,N^1,N^6,N^6-Tetramethyl-N^1,N^6-bis[1-methyl-3-(2,2,6-trimethylcyclohexyl)propyl]-1,6-hexanediaminium chloride (1:2); hexamethylenebis[dimethyl[1-methyl-3-(2,2,6-trimethylcyclohexyl)propyl]ammonium chloride]; N,N'-bis[1-methyl-3-(2,2,6-trimethylcyclohexyl)propyl]-N,N'-dimethyl-1,6-hexanediamine bis(methochloride); Ro-5-0810/1; Triburon. $C_{36}H_{74}Cl_2N_2$; mol wt 605.90. C 71.36%, H 12.31%, Cl 11.70%, N 4.62%. Obtained as the hemihydrate. Prepn: Goldberg, Teitel, **US 3064052** (1962 to Hoffmann-La Roche). Clinical application: Edelson *et al.*, *Antibiot. Annu.* **1958-1959**, 110; Robinson, Harmon, *ibid.* 113. Comprehensive description: B. C. Rudy, B. Z. Senkowski, *Anal. Profiles Drug Subs.* **2**, 507-521 (1973).

White cryst powder, mp 243-253° (dec). Sol in water, chloroform, alcohol.

THERAP CAT: Antiseptic.

9819. Triclocarban. [101-20-2] N-(4-Chlorophenyl)-N'-(3,4-dichlorophenyl)urea; 3,4,4'-trichlorocarbanilide; 1-(3',4'-dichlorophenyl)-3-(4'-chlorophenyl)urea; TCC; Cutisan; Nobacter; Solubacter. $C_{13}H_9Cl_3N_2O$; mol wt 315.58. C 49.48%, H 2.87%, Cl 33.70%, N 8.88%, O 5.07%. Prepn from 3,4-dichloroaniline and 4-chlorophenyl isocyanate: Beaver *et al.*, *J. Am. Chem. Soc.* **79**, 1236 (1957); Beaver, Stoffel, **US 2818390** (1957 to Monsanto); **GB 769273** (1957).

Fine white plates, mp 255.2-256°.

USE: Bacteriostat and antiseptic in soaps and other cleansing compositions.

THERAP CAT: Antiseptic, disinfectant.

9820. Triclofos. [306-52-5] 2,2,2-Trichloroethanol dihydrogen phosphate; trichloroethyl phosphate. $C_2H_4Cl_3O_4P$; mol wt 229.37. C 10.47%, H 1.76%, Cl 46.37%, O 27.90%, P 13.50%. Prepn: Hems *et al.*, *Br. Med. J.* **1**, 1834 (1962).

Monosodium salt. [7246-20-0] Sch-10159. $C_2H_3Cl_3NaO_4P$; mol wt 251.36. Sol in water. LD_{50} orally in mice: 1.4 g/kg (Hems).

THERAP CAT: Sedative, hypnotic.

9821. Triclopyr. [55335-06-3] 2-[(3,5,6-Trichloro-2-pyridinyl)oxy]acetic acid; Dowco 233. $C_7H_4Cl_3NO_3$; mol wt 256.46. C

32.78%, H 1.57%, Cl 41.47%, N 5.46%, O 18.72%. Selective postemergence herbicide. Prepn: L. D. Markley, **US 3862952** (1975 to Dow). Activity: B. C. Byrd *et al.*, *Proc. South. Weed Sci. Soc.* **28**, 251 (1975). Acute toxicity: E. E. Kenaga, *Down Earth* **35**, 25 (1979).

Fluffy solid, mp 148-150°. Vapor pressure at 25°: 1.26×10^{-6} mm Hg. pKa 2.68. Subject to photolysis. Soly in water at 25°: 440 mg/l. Soly at 25° (g/kg): acetone 989; 1-octanol 307. LD_{50} orally in rats: 713 mg/kg (Kenaga).

Butoxyethyl Ester. [64700-56-7] Garlon; Pathfinder; Remedy; Turflon. $C_{13}H_{16}Cl_3NO_4$; mol wt 356.62.

Triethylamine Salt. [57213-69-1] Grandstand. $C_7H_4Cl_3NO_3 \cdot C_6H_{15}N$; mol wt 357.66.

USE: Herbicide.

9822. Triclosan. [3380-34-5] 5-Chloro-2-(2,4-dichlorophenoxy)phenol; 2,4,4'-trichloro-2'-hydroxydiphenyl ether; CH-3635; Aquasept; Gamophen; Irgacare MP; Irgasan DP 300; Sapoderm; Ster-Zac. $C_{12}H_7Cl_3O_2$; mol wt 289.54. C 49.78%, H 2.44%, Cl 36.73%, O 11.05%. Broad spectrum antimicrobial; inhibits enoyl-acyl carrier protein reductase (ENR), an enzyme critical for bacterial cell-wall synthesis. Prepn: **NL 6401526**; E. Model, J. Bindler, **US 3506720** (1964, 1970 to Geigy). Physical and bacteriostatic properties: C. A. Savage, *Drug Cosmet. Ind.* **109** (3), 36 (1971); *eidem*, *ibid.* 161 . Metabolism: J. G. Black *et al.*, *Toxicology* **3**, 33 (1975). Use as disinfectant and textile preservative: E. Model, J. Bindler, **US 3629477** (1971 to Geigy). GC-MS determn in water samples: C.-Y. Cheng *et al.*, *Anal. Sci.* **27**, 197 (2011). Review of safety assessment: J. V. Rodricks *et al.*, *Crit. Rev. Toxicol.* **40**, 422-484 (2010); of environmental exposure and toxicity: A. B. Dann, A. Hontela, *J. Appl. Toxicol.* **31**, 285-311 (2011).

White to off-white crystalline powder, slight, faintly aromatic odor. mp 54-57.3°. Vapor pressure (20°C) 4×10^{-6} mm Hg. pKa 7.9. Readily sol in alkaline solns and many organic solvents; sol in methanol, alc, acetone; slightly sol in hexane. Practically insol in water.

USE: Bacteriostat in personal care products, cosmetics, and detergents; material preservative for industrial and household plastics and textiles.

THERAP CAT: Antiseptic, disinfectant.

THERAP CAT (VET): Antiseptic.

9823. Tricromyl. [85-90-5] 3-Methyl-4H-1-benzopyran-4-one; 3-methylchromone; 3-methyl-γ-benzopyrone; Cromonalgina. $C_{10}H_8O_2$; mol wt 160.17. C 74.99%, H 5.03%, O 19.98%. Based on an ancient Egyptian drug now termed *bezr el khelda*. Prepn from *o*-hydroxypropiophenone: Clerc-Bory *et al.*, *Bull. Soc. Chim. Fr.* **1955**, 1083; Mentzer, Meunier, **FR 980785** (1951 to Lab. Franc. Chimiother.); Mentzer, **US 2769015** (1956 one-half to Laroche-Navarron).

Crystals from ethanol, mp 68°. uv max (alc): 304 nm.

THERAP CAT: Antispasmodic; vasodilator (coronary).

9824. Tridecylbenzene. [123-02-4] 1-Phenyltridecane; Detergent Alkylate #5; Tridane. $C_{19}H_{32}$; mol wt 260.47. C 87.61%, H

12.38%. Prepn: Ziegler *et al.*, *Ann.* **511**, 13 (1934). Manufacture: Williamson, Bieneman, **US 3207800** (1965 to Continental Oil).

Liquid, bp 346°, bp$_{10}$ 188-189.5°. mp 10°. d$_4^{20}$ 0.8550, d$_4^{25}$ 1.8515. n_D^{20} 1.4821, n_D^{25} 1.4800. Forms stable foams in the presence of fat.
USE: In manuf of detergents and surface-active agents. Can be sulfonated.

9825. Tridemorph. [24602-86-6] 2,6-Dimethyl-4-tridecyl-morpholine; *N*-tridecyl-2,6-dimethylmorpholine; 2,6-dimethyl-4-tridecyltetrahydro-1,4-oxazine. C$_{19}$H$_{39}$NO; mol wt 297.53. C 76.70%, H 13.21%, N 4.71%, O 5.38%. Prepn: W. Sanne *et al.*, **BE 614214**; *eidem*, **US 3468885** (1962, 1969 both to BASF); K.-H. König *et al.*, *Angew. Chem. Int. Ed.* **4**, 336 (1965). Structure determn of two diastereomers: D. Kost, E. Gurfinkel, *J. Chromatogr.* **108**, 207 (1975). Fungicidal activity: J. Kradel *et al.*, *Proc. 5th Brit. Insectic. Fungic. Conf.*, 16 (1969); *in vitro* activity comparison of tridemorph and commercial formulation Calixin: A. Kerkenaar, A. A. Sijpesteijn, *Pestic. Biochem. Physiol.* **12**, 124 (1979). Uptake and distribution in barley: E.-H. Pommer *et al.*, *Proc. 5th Brit. Insectic. Fungic. Conf.*, 347 (1969); R. H. Waring, M. S. Wolfe, *Pestic. Sci.* **6**, 169 (1975). Behavior in soil: S. Otto, N. Drescher, *Proc. 7th Brit. Insectic. Fungic. Conf.*, 57 (1973). Metabolic fate in rats: D. R. Hawkins *et al.*, *Pestic. Sci.* **5**, 535 (1974); of Calixin: R. H. Waring, *Xenobiotica* **4**, 717 (1974). Mode of action: A. Kerkenaar *et al.*, *Pestic. Biochem. Physiol.* **12**, 195 (1979). Comparison of toxicity of tridemorph and Calixin: J. Merkle *et al.*, *Teratology* **29**, 259 (1984). Field trials: P. Lakshmanan, S. Mohan, *Pesticides* **22**, 27 (1988).

Oil, bp$_{0.7}$ 130-133°; bp$_{1.3}$ 139-142°. n_D^{25} 1.4568.
Commercial product. Calixin. Reaction mixture of C$_{11}$-C$_{14}$ 4-alkyl-2,6-dimethylmorpholine homologs containing 60-70% of tridemorph.
USE: Systemic agricultural fungicide.

9826. Tridihexethyl Iodide. [125-99-5] γ-Cyclohexyl-*N,N,*-triethyl-γ-hydroxybenzenepropanaminium iodide (1:1); (3-cyclohexyl-3-hydroxy-3-phenylpropyl)triethylammonium iodide; 3-diethylamino-1-cyclohexyl-1-phenyl-1-propanol ethiodide; 3-diethylamino-1-phenyl-1-cyclohexyl-1-propanol ethiodide; α-(2-diethylaminoethyl)-α-phenylcyclohexanemethanol ethiodide; propethonum iodide; tridihexethide; 921 C; Claviton. C$_{21}$H$_{36}$INO; mol wt 445.43. C 56.63%, H 8.15%, I 28.49%, N 3.14%, O 3.59%. Anticholinergic. Prepn: Lobby, **US 2913494** (1959 to Am. Cyanamid).

Bitter crystals, mp 179-184°. Solubility in water at 25°: 1.1 g/100 ml. Freely sol in alc, chloroform. Very slightly sol in ether. pH of a 1% aq soln 5.5-7.
Tridihexethyl chloride. [4310-35-4] Pathilon. C$_{21}$H$_{36}$ClNO; mol wt 353.98. White, odorless crystalline powder. Freely sol in water, chloroform, mthanol. Practically insol in acetone, ether.
THERAP CAT: Antispasmodic.

9827. Tridiphane. [58138-08-2] 2-(3,5-Dichlorophenyl)-2-(2,2,2-trichloroethyl)oxirane; Dowco 356. C$_{10}$H$_7$Cl$_5$O; mol wt 320.42. C 37.49%, H 2.20%, Cl 55.32%, O 4.99%. Prepn: L. D. Markley, E. J. Norton, **DE 2519073**; *eidem*, **US 4211549** (1975,

1980 both to Dow). Herbicidal activity: J. A. Jagschitz, *Proc. Annu. Meet. Northeast. Weed Sci. Soc.* **39**, 274 (1985); D. B. Vitolo *et al.*, *ibid.* **40**, 272 (1986). Field trial in combination with atrazine, *q.v.:* P. J. Dryden, M. J. Watson, *Proc. 38th N. Z. Weed Pest Control Conf.* **1985**, 191. Mechanism of action: G. L. Lamoureux, D. G. Rusness, *Pestic. Biochem. Physiol.* **26**, 323 (1986). Metabolism in mouse: J. Magdalou, B. D. Hammock, *Toxicol. Appl. Pharmacol.* **91**, 439 (1987). Toxicological studies: T. R. Hanley *et al.*, *Fundam. Appl. Toxicol.* **8**, 179 (1987); J. A. John-Greene *et al.*, *Toxicology* **43**, 325 (1987).

Yellow oil, n_D^{25} 1.5720. LD$_{50}$ for technical grade (89-91% pure) in male, female mice, male, female rats (mg/kg): ~1200, ~740, 1700-2300, 1500-1900 orally (Hanley).
USE: Herbicide.

9828. Trientine. [112-24-3] N^1,N^2-Bis(2-aminoethyl)-1,2-ethanediamine; triethylenetetramine; 1,8-diamino-3,6-diazaoctane; 3,6-diazaoctane-1,8-diamine; 1,4,7,10-tetraazadecane; trien; TETA; TECZA. C$_6$H$_{18}$N$_4$; mol wt 146.24. C 49.28%, H 12.41%, N 38.31%. Oral copper chelator. Prepn: A. W. von Hoffmann, *Ber.* **23**, 3711 (1890); J. van Alphen, *Rec. Trav. Chim.* **55**, 412 (1936); G. D. Jones *et al.*, *J. Org. Chem.* **9**, 125 (1944); of the dihydrochloride: H. B. Dixon *et al.*, *Lancet* **1**, 853 (1972). Effects on the metabolism of copper in rats: F. W. Sunderman *et al.*, *Toxicol. Appl. Pharmacol.* **38**, 177 (1976); H. Harders *et al.*, *Arzneim.-Forsch.* **30**, 254 (1980). Use of the dihydrochloride in the treatment of Wilson's disease: J. M. Walshe, *Prog. Clin. Biol. Res.* **34**, 271 (1979); R. H. Haslam *et al.*, *Dev. Pharmacol. Ther.* **1**, 318 (1980); J. M. Walshe, *Lancet* **1**, 643 (1982). *Review:* J. M. Walshe, *Orphan Drugs*, F. E. Karch, Ed. (Marcel Dekker, New York, 1982) pp 57-71.

Oily liquid, mp 12°. bp 266-267°. bp$_{31}$ 174°. bp$_1$ 112-113°. n_D^{25} 1.4951; d^{15} 0.9817. Flash pt 290°F (143°C). pH 14. Sol in water, alc. *Corrosive!*
Dihydrochloride. [38260-01-4] Syprine. C$_6$H$_{18}$N$_4$·2HCl; mol wt 219.15. White to pale yellow hygroscopic crystals, mp 115-118°. Freely sol in water; sol in methanol; slightly sol in ethanol. Insol in chloroform, ether.
USE: Epoxy curing agent; lubricating oil and fuel additive; in production of polyamides; analytical reagent for Cu, Ni.
THERAP CAT: Chelating agent (copper); Wilson's disease treatment.
THERAP CAT (VET): Chelating agent (copper).

9829. Trietazine. [1912-26-1] 6-Chloro-N^2,N^2,N^4-triethyl-1,3,5-triazine-2,4-diamine; 2-chloro-4-diethylamino-6-ethylamino-*s*-triazine; 2-ethylamino-4-diethylamino-6-chloro-*s*-triazine; G-27901; NC-1667; Aventox. C$_9$H$_{16}$ClN$_5$; mol wt 229.71. C 47.06%, H 7.02%, Cl 15.43%, N 30.49%. Prepn: Pearlman, Banks, *J. Am. Chem. Soc.* **70**, 3726 (1948). Toxicity study: G. W. Bailey, J. L. White, *Residue Rev.* **10**, 97 (1965).

Crystals from propanol, mp 100-102°. Soly at 25°: 20 ppm in water; 17% in acetone; 20% in benzene; >50% in chloroform; 10%

in dioxane; 3% in ethanol. LD$_{50}$ orally in rats: 1750 mg/kg (Bailey, White).

USE: Herbicide.

9830. Triethanolamine. [102-71-6] 2,2′,2″-Nitrilotrisethanol; trihydroxytriethylamine; tris(hydroxyethyl)amine; triethylolamine; trolamine. C$_6$H$_{15}$NO$_3$; mol wt 149.19. C 48.30%, H 10.13%, N 9.39%, O 32.17%. Produced along with mono- and diethanolamine by ammonolysis of ethylene oxide. See references under Ethanolamine. Monograph: E. J. Fischer, *Triäthanolamin und andere Alkanolamine* (Heidelberg, 4th ed., 1954).

Very hygroscopic, viscous liq. Slight ammoniacal odor. Turns brown on exposure to air and light. d$_4^{20}$ 1.1242; d$_4^{60}$ 1.0985. One gallon weighs 9.37 lbs. mp 21.57°. bp$_{760}$ 335.4°. *See:* McDonald *et al., J. Chem. Eng. Data* **4**, 311 (1959). Viscosity (cP): 590.5 (25°); 65.7 (60°). Strong base. pK (25°) 9.50. pH of 0.1N aq soln 10.5. n$_D^{20}$ 1.4852. Flash pt 365°F. Miscible with water, methanol, acetone. Soly at 25° in benzene, 4.2%; in ether, 1.6%; in carbon tetrachloride, 0.4%; in *n*-heptane, <0.1%.

Hydrochloride. [637-39-8] C$_6$H$_{15}$NO$_3$.HCl; mol wt 185.65. Crystals from ethanol, mp 177°.

Salicylate. [2174-16-5] Trolamine salicylate; Mobisyl; Myoflex. C$_6$H$_{15}$NO$_3$.C$_7$H$_6$O$_3$; mol wt 287.31.

USE: Intermediate in the manuf of surface active agents, textile specialties, waxes, polishes, herbicides, petroleum demulsifiers, toilet goods, cement additives, cutting oils. In making emulsions with mineral and vegetable oils, paraffin and waxes. Solvent for casein, shellac, dyes; manuf synthetic resins; increasing the penetration of organic liquids into wood and paper. In the production of lubricants for the textile industry. Pharmaceutic aid (alkalizer).

THERAP CAT: Analgesic.

9831. Triethylamine. [121-44-8] *N,N*-Diethylethanamine; (diethylamino)ethane. C$_6$H$_{15}$N; mol wt 101.19. C 71.22%, H 14.94%, N 13.84%. Prepd by reaction of *N,N*-diethylacetamide with lithium aluminum hydride: Uffer, Schlittler, *Helv. Chim. Acta* **31**, 1397 (1948). Manuf by vapor phase alkylation of ammonia with ethanol: Lemon, Myerly, **US 3022349** (1962 to Union Carbide). Toxicity study: H. F. Smyth *et al., Arch. Ind. Hyg. Occup. Med.* **4**, 119 (1951).

Liquid; strong ammoniacal odor; d$_4^{25}$ 0.7255; mp −115°; bp 89-90°; n$_D^{20}$ 1.4003. Flash pt, closed cup: 20°F (−6°C). Slightly sol in water above 18.7°; misc with alcohol, ether, also with water below 18.7°. *Flammable; corrosive. Keep well closed.* LD$_{50}$ orally in rats: 0.46 g/kg (Smyth).

Hydrochloride. C$_6$H$_{15}$N.HCl; mol wt 137.65. Crystals from alcohol, mp 253-254°, sublimes at 245°; d 1.069. Sol in 0.7 parts water; sol in alcohol, chloroform; very slightly sol in benzene. Practically insol in ether.

Caution: Potential symptoms of overexposure to triethylamine are irritation of eyes, respiratory system and skin. *See NIOSH Pocket Guide to Chemical Hazards* (DHHS/NIOSH 97-140, 1997) p 318.

USE: In the prepn of quaternary ammonium compds.

9832. Triethylammonium Formate. [585-29-5] Formic acid compd with *N,N*-diethylethanamine (1:1). C$_7$H$_{17}$NO$_2$; mol wt 147.22. C 57.11%, H 11.64%, N 9.51%, O 21.73%. (C$_2$H$_5$)$_3$N.HOCHO. Prepd by neutralizing 50% formic acid with triethylamine and evaporating the resulting soln on a steambath for twelve hours at 20 mm: Alexander, Wildman, *J. Am. Chem. Soc.* **70**, 1187 (1948).

Light brown syrup. Very sol in water. Sol in alc.

USE: In syntheses employing the Leuckart reaction.

9833. Triethylborane. [97-94-9] Triethylboron. C$_6$H$_{15}$B; mol wt 98.00. C 73.54%, H 15.43%, B 11.03%. Primarily as an initiator in radical reactions. Prepn: A. Stock, F. Zeidler, *Ber.* **78**,

531 (1921); E. C. Ashby, *J. Am. Chem. Soc.* **81**, 4791 (1959). Use as protecting group: C. Lutz *et al., Tetrahedron* **54**, 10317 (1998); as transfer reagent in polymerizations: V. Beraud *et al., Macromol. Rapid Commun.* **21**, 901 (2000); as stereoselective intiator for hydrometalation: K. Takami *et al., J. Org. Chem.* **68**, 6627 (2003). Review of radical reactivity in generation of carbon-centered radicals: C. Ollivier, P. Renaud, *Chem. Rev.* **101**, 3415-3434 (2001). Brief review of application in radical initiation and in non-radical processes: G. O'Mahony, *Synlett* **2004**, 572-573.

Colorless liquid. *Spontaneously flammable! Burns in air with green flame.*

USE: Reagent in organic synthesis as blocking agent and radical initiator.

9834. Triethylenediamine. [280-57-9] 1,4-Diazabicyclo-[2.2.2]octane; Dabco. C$_6$H$_{12}$N$_2$; mol wt 112.18. C 64.24%, H 10.78%, N 24.97%. Prepn: Krause *et al.,* **GB 871754** (1958 to Houdry Process).

Crystals, extremely hygroscopic. Sublimes readily at room temp. mp 158°. bp 174°. pKa$_1$ 3.0; pKa$_2$ 8.7. Soly at 25° (g/100g): water 45; acetone 13; benzene 51; ethanol 77; methyl ethyl ketone 26.1.

USE: Catalyst in making urethane foams.

9835. Triethylene Glycol. [112-27-6] 2,2′-[1,2-Ethanediyl-bis(oxy)]bisethanol; 2,2′-ethylenedioxybis(ethanol). C$_6$H$_{14}$O$_4$; mol wt 150.17. C 47.99%, H 9.40%, O 42.62%. Prepd from ethylene oxide and ethylene glycol in the presence of sulfuric acid: Matignon *et al., Bull. Soc. Chim.* [5] **1**, 1308 (1934). Manuf by forming etherester of HOCH$_2$COOH with glycol and then hydrogenating: Gresham, **US 2654786** (1953 to du Pont). Toxicity data: Stenger *et al., Arzneim.-Forsch.* **18**, 1536 (1968).

Colorless, hygroscopic, practically odorless liquid. d$_4^{15}$ 1.1274; bp 285°; bp$_{14}$ 165°. mp −7.2°. Viscosity at 20°: 47.8 cP; n$_D^{15}$ 1.4578. Misc with water, alcohol, benzene, toluene; sparingly sol in ether. Practically insol in petr ether. LD$_{50}$ in mice, rats (g/kg): 21, 15-22 orally; 7.3-9.5, 11.7 i.v. (Stenger).

USE: In various plastics to increase pliability; in air disinfection.

9836. Triethylenemelamine. [51-18-3] 2,4,6-Tris(1-aziridinyl)-1,3,5-triazine; 2,4,6-tris(ethylenimino)-*s*-triazine; 2,4,6-triethylenimino-1,3,5-triazine; triethanomelamine; tretamine; triamelin; TEM; NSC-9706; Persistol. C$_9$H$_{12}$N$_6$; mol wt 204.24. C 52.93%, H 5.92%, N 41.15%. Prepd from ethylenimine and cyanuric chloride: Bestian, *Ann.* **566**, 210 (1950); Wystrach, Kaiser, **US 2520619** (1950 to Am. Cyanamid); Kaiser, Schaefer, **US 2653934** (1953); Wystrach *et al., J. Am. Chem. Soc.* **77**, 5915 (1955). Toxicology: F. S. Philips, J. B. Thiersch, *J. Pharmacol. Exp. Ther.* **100**, 398 (1950). Review of carcinogenic risk: *IARC Monographs* **9**, 95-105 (1975); of comparative mutagenicity: C. Ramel, *Environ. Sci. Res.* **24**, 943-976 (1981).

Minute crystals from chloroform, dec 139°. Soly (w/w) at 26°: in water 40%, chloroform 28.1%, methylene chloride 19.7%, methanol 12.5%, acetone 10.6%, dioxane 9.6%, ethanol 7.7%, benzene 5.6%, dimethyl Cellosolve 4.8%, methyl ethyl ketone 4.7%, ethyl acetate 4.5%, carbon tetrachloride 3.6%. Stable at reduced temps; ampuled aq solns stored at 4° are stable for about 3 months. At room temp, aq solns polymerize. LD_{50} in mice, rats (mg/kg): 2.8, 1.0 i.p.; 15, 13 orally (Philips).

USE: Manuf of resinous products, textile finishing agents. Insect sterilant. Research tool used as positive control for mutagenicity assays.

THERAP CAT: Antineoplastic.

9837. Triethylenephosphoramide. [545-55-1] 1,1′,1″-Phosphinylidynetrisaziridine; tris(1-aziridinyl)phosphine oxide; phosphoric acid triethyleneimide; aphoxide; APO; TEPA. $C_6H_{12}N_3OP$; mol wt 173.16. C 41.62%, H 6.99%, N 24.27%, O 9.24%, P 17.89%. Prepn: Bestian, *Ann.* **566**, 231 (1950). Toxicity study: Gaines, *Toxicol. Appl. Pharmacol.* **14**, 515 (1969).

Crystals, mp 41°. bp_{23} 90-91°. Extremely sol in water. Very sol in alcohol, ether, acetone. LD_{50} orally in male rats: 37 mg/kg (Gaines).

USE: Insect chemosterilant; in dyeing, creaseproofing and flameproofing textiles; stabilizer for polymers; in photographic emulsion hardening.

THERAP CAT: Antineoplastic.

9838. Triethyl Phosphate. [78-40-0] Phosphoric acid triethyl ester; ethyl phosphate; triethoxyphosphine oxide. $C_6H_{15}O_4P$; mol wt 182.16. C 39.56%, H 8.30%, O 35.13%, P 17.00%. Prepd from tetraethyl hypophosphate with ethanol in the presence of aluminum ethoxide: Mukaiyama *et al.*, *J. Org. Chem.* **27**, 1815 (1962). Manuf by treating triethyl phosphite with diethyl hydrogen phosphate: McCall, Coover, **US 2960529** (1960 to Kodak).

Liquid; d^{19} 1.0725; bp 215-216°; bp_{10} 90-95°; n_D^{17} 1.4067. Sol in water with some decompn; sol in alcohol, ether.

USE: As ethylating agent; formation of polyesters which are used as insecticides.

9839. Triethylphosphine. [554-70-1] Triethyphosphorus. $C_6H_{15}P$; mol wt 118.16. C 60.99%, H 12.80%, P 26.21%. Prepd from ethyllithium and phosphorus trichloride: Screttas, Isbell, *J. Org. Chem.* **27**, 2573 (1962). Manuf from white phosphorus, ethylene and hydrogen at elevated pressures: Oppegard, **US 2687437** (1954 to du Pont).

Colorless liquid; odor of hyacinths; d_4^{15} 0.800; bp_{744} 127-128°. Practically insol in water. Miscible with alcohol, ether.

USE: In organic syntheses.

9840. Triethyl Phosphite. [122-52-1] Phosphorous acid triethyl ester; ethyl phosphite; triethoxyphosphine; $P(OEt)_3$. $C_6H_{15}O_3P$; mol wt 166.16. C 43.37%, H 9.10%, O 28.89%, P 18.64%. Organophosphorus reagent used to synthesize phosphonate and phosphate derivatives. Prepn: C. Zimmermann, *Ann.* **175**, 1 (1875); A. H. Ford-Moore, B. J. Perry, *Org. Synth.* **coll. vol. IV**, 955 (1963). Use in the synthesis of phosphonates: V. K. Yadav, *Synth. Commun.* **20**, 239 (1990); G. W. Kabalka, S. K. Guchhait, *Org. Lett.* **5**, 729

(2003); and subsequent phosphonate utility in Horner-Wadsworth-Emmons reactions: A. Ianni, S. R. Waldvogel, *Synthesis* **2006**, 2103. Phosphorylation application to make phosphate triesters: J. K. Stowell, T. S. Widlanski, *Tetrahedron Lett.* **36**, 1825 (1995). Use in the desulfurization of thiiranes: R. D. Schuetz, R. L. Jacobs, *J. Org. Chem.* **26**, 3467 (1961); in the deoxygenation of diaryl peroxides: A. J. Burn *et al.*, *J. Chem. Soc.* **1963**, 1527.

Colorless liquid with pungent odor. *Flammable, irritant.* bp_{760} 154°; bp_{16} 57-58°; bp_{13} 51-52°; bp_{10} 43-44°; $bp_{0.1}$ 20-22°. d_4^{20} 0.963. n_D^{25} 1.4104-1.4106; n_D^{20} 1.4126. Flash pt, closed cup: 129°F (54°C). Sol in most organic solvents.

USE: Reagent in synthetic organic chemistry.

9841. Triethylsilane. [617-86-7] Triethylhydrosilane; triethylsilicon hydride; TES. $C_6H_{16}Si$; mol wt 116.28. C 61.98%, H 13.87%, Si 24.15%. Versatile organosilicon hydride reducing agent; used in silylations, radical chain reactions, and transfer hydrogenations. Prepn: A. Ladenburg, *Ann.* **164**, 300 (1872); C. A. Kraus, W. K. Nelson, *J. Am. Chem. Soc.* **56**, 195 (1934); and physical properties: S. Tannenbaum *et al.*, *ibid.* **75**, 3753 (1953). Synthetic applications: M. P. Doyle *et al.*, *J. Org. Chem.* **55**, 6082 (1990); S. J. Cole *et al.*, *J. Chem. Soc. Perkin Trans. 1* **1991**, 103; V. Gevorgyan *et al.*, *J. Org. Chem.* **66**, 1672 (2001); P. K. Mandal, J. S. McMurray, *ibid.* **72**, 6599 (2007). *Reviews*: J. L. Fry in *Encyclopedia of Reagents for Organic Synthesis* 7, L. A. Paquette, Ed. (Wiley, New York, 1995) pp 5118-5122; J. de Almeida Rodrigues Jr., *Synlett* **2008**, 3249-3250.

Colorless liquid with faint garlic-like odor. *Flammable. Irritant.* fp −156.90. bp 108.77°. d^{20} 0.7318. n_D^{20} 1.4119. Flash pt, closed cup: 26.6°F (−3.0°C). Sol in most organic solvents. Insol in water.

USE: Reagent in synthetic organic chemistry.

9842. Triflic Acid. [1493-13-6] 1,1,1-Trifluoromethanesulfonic acid; TFMSA. CHF_3O_3S; mol wt 150.07. C 8.00%, H 0.67%, F 37.98%, O 31.98%, S 21.36%. F_3CSO_2OH. One of the strongest acids known. Prepn: R. N. Haszeldine, J. M. Kidd, *J. Chem. Soc.* **1954**, 4228. Acid-base equilibria study: Y. Y. Fialkov, V. I. Ligus, *J. Gen. Chem. USSR* **42**, 256 (1972). Deprotecting agent in peptide chemistry: H. Yajima *et al.*, *J. Chem. Soc. Chem. Commun.* **1974**, 107. Hammett acidity function: S. Saito *et al.*, *Chem. Pharm. Bull.* **39**, 2718 (1991). Electrochemical determ of ionic form: F. Favier, J. L. Pascal, *Analyst* **116**, 479 (1991). Ion chromatographic determn of triflic and trifluoroacetic acids: N. Simonzadeh, *J. Chromatogr.* **634**, 125 (1993). *Reviews*: R. D. Howells, J. D. McCown, *Chem. Rev.* **77**, 69-92 (1977); P. J. Stang, M. R. White, *Aldrichim. Acta* **16**, 15-22 (1983).

Clear, colorless, hygroscopic liquid; bp_{760} 162°; $bp_{57.5}$ 91°; $bp_{37.5}$ 81°; bp_1 42°. d^{25} 1.6980. n_D^{25} 1.325. Miscible with water; sol in alcohols, ketones, ethers, esters and many other polar organic solvents. Viscosity at 25°: 2.87 cP. Electrical conductivity at 25°: 2 × 10^{-4} Ω^{-1} cm^{-1}. Thermally stable; nonoxidizing.

Caution: Direct contact may cause skin burns (Haszeldine).

USE: As a catalyst in Friedel-Crafts type acylation, alkylation and polymerization reactions; as a solvent for ESR; as a nonaqueous strong acid titrant; with trifluoroacetic acid in solid-phase peptide synthesis.

9843. Trifloxystrobin. [141517-21-7] (αE)-α-(Methoxyimino)-2-[[[(E)-1-[3-(trifluoromethyl)phenyl]ethylidene]amino]oxy]methyl]benzeneacetic acid methyl ester; (E,E)-methoxyimino-[2-[1-(3-trifluoromethylphenyl)ethylideneaminooxymethyl]phenyl]acetic acid methyl ester; CGA-279202; Flint; Compass; Gem; Twist. $C_{20}H_{19}F_3N_2O_4$; mol wt 408.38. C 58.82%, H 4.69%, F 13.96%, N 6.86%, O 15.67%. Strobilurin fungicide. Prepn: J. M. Clough *et al.*, **EP 472300**; *eidem*, **US 5238956** (1992, 1993 both to ICI). Field trial in fruit trees: M. Reuveni *et al.*, *Crop Prot.* **19**, 335 (2000).

Review of physical properties, biological properties and field trials: P. Margot *et al.*, *Brighton Crop Prot. Conf. - Pests Dis.* **1998**, 375-382.

Odorless white powder, mp 72.9°. bp~312°; decompn starting at 285°. Soly in water at 25°: 610 μg/l. Vapor pressure at 25°: 3.4 × 10⁻⁶ Pa. Log P at 25° (*n*-octanol/water): 4.5. LD₅₀ in rats (mg/kg): >5000 orally; >2000 dermally. LC₅₀ by inhalation in rats (mg/m³): >4646. LC₅₀ in bobwhite quail (mg/kg): >2000. LC₅₀ in rainbow trout (mg/l): 0.015 (Margot).

USE: Agricultural fungicide.

9844. Triflumuron. [64628-44-0] 2-Chloro-*N*-[[[4-(trifluoromethoxy)phenyl]amino]carbonyl]benzamide; *N*-(2-chlorobenzoyl)-*N'*-[4-(trifluoromethoxy)phenyl]urea; trifluron; SIR-8514; BAY SIR 8514; Alsystin; Baycidal; Starycide. C₁₅H₁₀ClF₃N₂O₃; mol wt 358.70. C 50.23%, H 2.81%, Cl 9.88%, F 15.89%, N 7.81%, O 13.38%. Arthropod insecticide which inhibits chitin biosynthesis: Prepn: W. Sirrenberg *et al.*, **DE 2601780**; *eidem*, **US 4139636** (1977, 1979 both to Bayer AG); **BE 867046**; R. H. Rigterink, **US 4170657** (1978, 1979 both to Dow Chemical). Insecticidal activity: C. H. Schaefer *et al.*, *J. Econ. Entomol.* **71**, 427 (1978); S. C. Chang, *ibid.* **72**, 479 (1979). Stability under field conditions: C. H. Schaefer, E. F. Dupras, *J. Agric. Food Chem.* **27**, 1031 (1979). Comprehensive description: G. Zoebelein *et al.*, *Z. Angew. Entomol.* **89**, 289-297 (1980).

Crystals, mp 198° (Sirrenberg). Also reported as mp 188-190° (Rigterink). Slightly toxic to fish. Safe to bees. LD₅₀ in rats, mice (mg/kg): >5000, >5000 i.p., s.c., orally (Zoebelein).

USE: Insecticide (larvicide).

9845. Trifluomeprazine. [2622-37-9] *N,N,β*-Trimethyl-2-(trifluoromethyl)-10*H*-phenothiazine-10-propanamine; 10-[3-(dimethylamino)-2-methylpropyl]-2-(trifluoromethyl)phenothiazine; RP-7746. C₁₉H₂₁F₃N₂S; mol wt 366.45. C 62.28%, H 5.78%, F 15.55%, N 7.64%, S 8.75%. Prepn: **GB 813861** corresp to G. E. Ullyot, **US 2921069** (1959, 1962 both to SK&F). Metabolism study: T. L. Flanagan *et al.*, *J. Pharm. Sci.* **51**, 996 (1962). Chromatographic studies: M. W. Anders, G. J. Mannering, *J. Chromatogr.* **7**, 258 (1962); R. J. Warren *et al.*, *J. Pharm. Sci.* **55**, 144 (1966); B. B. Wheals, *J. Chromatogr.* **187**, 65 (1980). Pharmacological studies: R. C. Kelsey, G. J. Frishmuth, *Arch. Int. Pharmacodyn. Ther.* **173**, 44 (1968); E. W. Baur, *J. Pharmacol. Exp. Ther.* **177**, 219 (1971).

uv max (95% ethanol): 308, 258 nm (log ε 3.60, 4.55).
Maleate. Nortran. C₁₉H₂₁F₃N₂S.C₄H₄O₄; mol wt 482.52.
THERAP CAT (VET): Tranquilizer.

9846. Trifluoperazine. [117-89-5] 10-[3-(4-Methyl-1-piperazinyl)propyl]-2-(trifluoromethyl)-10*H*-phenothiazine; 2-trifluoromethyl-10-[3'-(1-methyl-4-piperazinyl)propyl]phenothiazine. C₂₁-

H₂₄F₃N₃S; mol wt 407.50. C 61.90%, H 5.94%, F 13.99%, N 10.31%, S 7.87%. Prepn: Craig *et al.*, *J. Org. Chem.* **22**, 709 (1957); **GB 813861**; G. E. Ullyot, **US 2921069** (1959 and 1960 to SK & F). Metabolism: T. L. Flanagan *et al.*, *J. Pharm. Sci.* **51**, 996 (1962); C. L. Huang, K. G. Bhansali, *ibid.* **57**, 1511 (1968). Toxicity: P. J. Fowler *et al.*, *Arzneim.-Forsch.* **27**, 866 (1977). Comprehensive description: A. Post *et al.*, *Anal. Profiles Drug Subs.* **9**, 543-581 (1980).

bp₀.₆ 202-210°. uv max (ethanol): 258, 307.5 nm (log ε 4.50, 3.50). LD₅₀ orally in rats, mice: 542.7, 424.0 mg/ml (Fowler).

Dihydrochloride. [440-17-5] Triftazin; triphthasine; Eskazinyl; Eskazine; Jatroneural; Modalina; Stelazine; Terfluzine. C₂₁H₂₄F₃-N₃S.2HCl; mol wt 480.42. Cream colored fine powder from abs alc, mp 242-243°. Hygroscopic. Freely soln in water; sol in alc; sparingly sol in chloroform. Insol in dil base, ether, benzene. pK₁ 3.9, pK₂ 8.1. pH of 5% aq soln 2.2.

THERAP CAT: Antipsychotic.

9847. Trifluoroacetic Acid. [76-05-1] 2,2,2-Trifluoroacetic acid; perfluoroacetic acid. C₂HF₃O₂; mol wt 114.02. C 21.07%, H 0.88%, F 49.99%, O 28.06%. CF₃COOH. Prepn: F. Swarts, *Bull. Sci. Acad. R. Belg.* **8**, 343 (1922); by oxidation of fluorine olefins: A. L. Henne, **US 2371757** (1945 to du Pont); A. L. Henne *et al.*, *J. Am. Chem. Soc.* **67**, 918 (1945). Improved prepn: A. L. Henne, P. Trott, *ibid.* **69**, 1820 (1947). Photochemical prepn: R. N. Hazeldine, F. Nyman, *J. Chem. Soc.* **1959**, 387; *eidem, ibid.* 420. Toxicity and metabolism: M. M. Airaksinen, T. Tammisto, *Ann. Med. Exp. Biol. Fenn.* **46**, 242 (1968). *Review:* J. B. Milne in *Chem. Non-Aqueous Solvents* **vol. 5B**, J. J. Lagowski, Ed. (Academic Press, New York, 1978) pp 1-52.

Liquid, sharp biting odor. bp 72.4°. mp −15.4°. d²⁰ 1.5351. Miscible with ether, acetone, ethanol, benzene, CCl₄, hexane. *Corrosive.* Strong, non-oxidizing acid. pka 0.3. LD₅₀ i.v. in mice: 1200 mg/kg (Airaksinen, Tammisto).

USE: In organic synthesis; dissolves protein when mixed with liquid SO₂.

9848. Trifluoroethanol. [75-89-8] 2,2,2-Trifluoroethanol; TFE; 2,2,2-trifluoroethyl alcohol. C₂H₃F₃O; mol wt 100.04. C 24.01%, H 3.02%, F 56.97%, O 15.99%. CF₃CH₂OH. Weakly acidic alcohol. Prepn: F. Swarts, *Bull. Soc. Chim. Belg.* **43**, 471 (1934). Structural study: L. A. Curtiss *et al.*, *J. Am. Chem. Soc.* **100**, 1979 (1978). Interaction with peptides and proteins: R. Rajan, P. Balaram, *Int. J. Pept. Protein Res.* **48**, 328 (1996). Review of production and applications: S. Arai, *Spec. Chem.* **18**, 294-297 (1998); of effects on peptide and protein structures: M. Buck, *Q. Rev. Biophys.* **31**, 297-355 (1998).

Transparent, colorless liquid, odor similar to ethanol; bp 74.05°; mp −43.5°. d⁰ 1.4106. d²² 1.3739. n_D²² 1.2907. pK 12.4. Dielectric constant: 26.67. Dipole moment: 2.52 D. Miscible with water, alcohols, ethers, ketones, chloroform.

USE: In synthesis of medical anaesthetics, pharmaceuticals, and agrochemicals; in polymerizations. Protein denaturant; stabilizes peptide structures. Cleaning solvent; eluent in HPLC separations; working fluid in Rankine heat cycle systems. Environmentally friendly alternative to CFCs.

9849. Trifluoroiodomethane. [2314-97-8] Iodotrifluoromethane; trifluoromethyl iodide. CF₃I; mol wt 195.91. C 6.13%, F 29.09%, I 64.78%. Non-ozone depleting gas with very low global warming potential. Prepn from IF₅ and CCl₄: A. A. Banks *et al.*, *J. Chem. Soc.* **1948**, 2188; from silver trifluoroacetate and iodine: A. L. Henne, W. G. Finnegan, *J. Am. Chem. Soc.* **72**, 3806 (1950); from ZnBr(CF₃).2DMF and ICl: D. Naumann *et al.*, *J. Fluorine Chem.* **67**, 91 (1994); from CHF₃ and I₂: N. Nagasaki *et al.*, *Catal. Today* **88**, 121 (2004). Physical properties: E. A. Nodiff *et al.*, *J. Org.*

Chem. **18**, 235 (1953). Microwave spectrum: A. P. Cox *et al.*, *J. Chem. Soc. Faraday Trans. 2* **76**, 339 (1980). Powder neutron diffraction structure: S. J. Clarke *et al.*, *Z. Kristallogr.* **206**, 87 (1993). Raman spectra: A. J. Beardsall *et al.*, *J. Raman Spectrosc.* **25**, 761 (1994). Spectroscopic evaluation of ozone depletion and global warming potentials: S. Solomon *et al.*, *J. Geophys. Res.* **99**, 20929 (1994). Thermal decompn: S. S. Kumaran *et al.*, *Chem. Phys. Lett.* **243**, 59 (1995). ^{19}F- and ^{13}C-NMR spectra: W. Tyrra *et al.*, *Z. Anorg. Allg. Chem.* **623**, 1857 (1997). Thermodynamic properties: Y. Y. Duan *et al.*, *Int. J. Thermophys.* **21**, 393 (2000). FTIR spectroscopy: S. M. Webb *et al.*, *J. Quant. Spectrosc. Radiat. Transfer* **94**, 425 (2005). Use as a dielectric etchant: R. A. Levy *et al.*, *J. Mater. Res.* **13**, 2643 (1998). Synthetic use with tetrakis(dimethylamino)ethylene as a trifluoromethylation reagent: W. Xu, W. R. Dolbier, Jr., *J. Org. Chem.* **70**, 4741 (2005).

bp_{760} −22.5°. mp ~ −130°. $d_4^{78.5}$ 2.5485. $d_4^{-32.5}$ 2.3608. $n_D^{-29.8}$ 1.3710. $n_D^{-42.2}$ 1.3790. Misc with mineral oil; compat with materials of refrigeration systems. *Light sensitive.*

USE: Trifluoromethylating agent in organic synthesis; etchant gas; halon alternative in fire extinguishers and refrigerant applications.

9850. Trifluoromethanesulfonic Anhydride. [358-23-6] 1,-1,1-Trifluoromethanesulfonic acid 1,1′-anhydride; triflate anhydride; triflic anhydride; Tf_2O. $C_2F_6O_5S_2$; mol wt 282.13. C 8.51%, F 40.40%, O 28.35%, S 22.73%. Electrophilic reagent used to introduce the trifluoromethanesulfonate functional group. Prepn: T. J. Brice, P. W. Trott, US 2732398 (1956 to Minnesota Mining & Manuf.); J. Burdon *et al.*, *J. Chem. Soc.* **1957**, 2574; and use in the catalysis of esterification reactions: T. Gramstad, R. N. Haszeldine, *ibid.* 4069; and use in the formation of vinyl triflates: P. J. Stang, T. E. Dueber, *Org. Synth.* **coll. vol. VI**, 757 (1988). Utility in the synthesis of triflates from alcohols: C. D, Beard *et al.*, *J. Org. Chem.* **38**, 3673 (1973); of triflates from phenols: A. M. Echavarren, J. K. Stille, *J. Am. Chem. Soc.* **109**, 5478 (1987); of triflamides from amines: J. B. Hendrickson, R. Bergeron, *Tetrahedron Lett.* **14**, 4607 (1973). Application as an oxidizing agent: G. Maas, P. J. Stang, *J. Org. Chem.* **46**, 1606 (1981). Use in glycosylation reactions: D. Kahne *et al.*, *J. Am. Chem. Soc.* **111**, 6881 (1989). *Review*: I. L. Baraznenok *et al.*, *Tetrahedron* **56**, 3077-3119 (2000).

Colorless liquid. *Corrosive.* bp 81-84°. d^{25} 1.677. Sol in dichloromethane. Insol in hydrocarbons. Hygroscopic. Reacts violently with water. Store under inert gas. Does not fume in air.

USE: Reagent in synthetic organic chemistry

9851. Trifluoromethyltrimethylsilane. [81290-20-2] Trimethyl(trifluoromethyl)silane; Ruppert's reagent; TMS-CF_3. C_4H_9-F_3Si; mol wt 142.20. C 33.79%, H 6.38%, F 40.08%, Si 19.75%. Reagent used for fluoride ion catalyzed trifluoromethylation of organic substrates. Prepn: I. Ruppert *et al.*, *Tetrahedron Lett.* **25**, 2195 (1984); and use in trifluoromethylation of aldehydes and ketones: G. K. S. Prakash *et al.*, *J. Am. Chem. Soc.* **111**, 393 (1989). Review of chemistry: G. K. S. Prakash, A. K. Yudin, *Chem. Rev.* **97**, 757-786 (1997); R. P. Singh, J. M. Shreeve, *Tetrahedron* **56**, 7613-7632 (2000); R. S. Bastos, *Synlett* **2008**, 1425-1426.

Colorless, volatile liquid, bp 45° (Ruppert); also reported as bp 54-55° (Prakash, Yudin).

USE: Nucleophilic trifluoromethylating reagent.

9852. Trifluperidol. [749-13-3] 1-(4-Fluorophenyl)-4-[4-hydroxy-4-[3-(trifluoromethyl)phenyl]-1-piperidinyl]-1-butanone; 4′-fluoro-4-[4-hydroxy-4-(α,α,α-trifluoro-*m*-tolyl)piperidino]butyrophenone; *p*-fluoro-4-[4′-hydroxy-4′-(3″-trifluoromethyl)phenyl]-piperidinobutyrophenone; 1-(3′-*p*-fluorobenzoylpropyl)-4-hydroxy-4-(3″-trifluoromethylphenyl)piperidine; ω-[4-hydroxy-4-(*m*-trifluoromethylphenyl)piperidino]-*p*-fluorobutyrophenone; flumoperone. $C_{22}H_{23}F_4NO_2$; mol wt 409.42. C 64.54%, H 5.66%, F 18.56%, N

3.42%, O 7.82%. Prepn: P. A. J. Janssen, **GB 895309**; *idem*, **US 3438991** (1962, 1969 both to Janssen). Pharmacology: *idem*, *Arzneim.-Forsch.* **11**, 932 (1961).

Hydrochloride. [2062-77-3] R-2498; Psicoperidol; Psychoperidol; Triperidol. $C_{22}H_{23}F_4NO_2 \cdot HCl$; mol wt 445.88. Crystals from acetone, mp 200.5-201.3°. Sol in water. LD_{50} in rats (mg/kg): 14 i.v.; 70 s.c. (Janssen, 1961).

THERAP CAT: Antipsychotic.

9853. Triflupromazine. [146-54-3] *N,N*-Dimethyl-2-(trifluoromethyl)-10*H*-phenothiazine-10-propanamine; 10-[3-(dimethylamino)propyl]-2-trifluoromethylphenothiazine; 2-trifluoromethyl-10-(γ-dimethylaminopropyl)phenothiazine; fluopromazine. C_{18}-$H_{19}F_3N_2S$; mol wt 352.42. C 61.35%, H 5.43%, F 16.17%, N 7.95%, S 9.10%. Prepn: Yale *et al.*, *J. Am. Chem. Soc.* **79**, 4375 (1957); **GB 813861** (1959 to SK & F). Comprehensive description of the hydrochloride: K. Florey, *Anal. Profiles Drug Subs.* **2**, 523-550 (1973).

Viscous, light amber-colored, oily liquid, $bp_{0.7}$ 176°; $bp_{0.4}$ 162-164°. n_D^{23} 1.5780. Practically insol in water.

Hydrochloride. [1098-60-8] Psyquil; Vesprin. $C_{18}H_{19}F_3N_2S.$-HCl; mol wt 388.88. Crystals from xylene, dec 173-174°. uv max: 255, 305 nm ($E_{1cm}^{1\%}$ 700, 90). Sol in water, ethanol, acetone. Insol in ether. pH of 2% aq soln 4.1, if the pH is raised to 6.4, pptn of the free base results.

THERAP CAT: Antipsychotic.

THERAP CAT (VET): Tranquilizer.

9854. Trifluralin. [1582-09-8] 2,6-Dinitro-*N,N*-dipropyl-4-(trifluoromethyl)benzenamine; α,α,α-trifluoro-2,6-dinitro-*N,N*-dipropyl-*p*-toluidine; 2,6-dinitro-*N,N*-dipropyl-α,α,α-trifluoro-*p*-toluidine; 2,6-dinitro-*N,N*-dipropyl-4-trifluoromethylaniline; *N,N*-dipropyl-2,6-dinitro-4-trifluoromethylaniline; L-36352; Lilly 36352; Treflan; Tri-4; Triflurex. $C_{13}H_{16}F_3N_3O_4$; mol wt 335.28. C 46.57%, H 4.81%, F 17.00%, N 12.53%, O 19.09%. Pre-emergence herbicide used for grass control in crops. Prepn: **GB 917253**; Soper, **US 3403180** (1963, 1968 both to Lilly). Photodecomposition: E. Leitis, D. G. Crosby, *J. Agric. Food Chem.* **22**, 842 (1974). Soil degradation: T. Golab *et al.*, *ibid.* **27**, 163 (1979). Toxicity data: Goldenthal, *Toxicol. Appl. Pharmacol.* **18**, 185 (1971). Carcinogenicity study: J. F. Robens, *Vet. Hum. Toxicol.* **22**, 328 (1980). Review of toxicology: E. Ebert *et al.*, *Food Chem. Toxicol.* **30**, 1031-1044 (1992).

Yellow crystals, mp 46-47°. $bp_{4.2}$ 139-140°. Slightly sol in water (0.0024 g/100 ml); freely sol in acetone, Stoddard solvent, xylene. LD_{50} orally in rats: 500 mg/kg (Goldenthal).

USE: Herbicide.

9855. Trifluridine. [70-00-8] α,α,α-Trifluorothymidine; 2′-deoxy-5-(trifluoromethyl)uridine; 5-(trifluoromethyl)-2′-deoxyuridine; F3TDR; NSC-75520; TFT Thilo; Virophtha; Viroptic. $C_{10}H_{11}$-$F_3N_2O_5$; mol wt 296.20. C 40.55%, H 3.74%, F 19.24%, N 9.46%, O 27.01%. Prepn: C. Heidelberger *et al., J. Am. Chem. Soc.* **84**, 3597 (1962); *eidem, J. Med. Chem.* **7**, 1 (1964); C. Heidelberger, **US 3201387** (1965 to U.S. Dept. HEW). Crystal structure: A. H. Tench, *Diss. Abstr. Int. B* **33**, 3587 (1973). NMR study: R. J. Cushley *et al., J. Am. Chem. Soc.* **90**, 709 (1968). Metabolism: D. L. Dexter *et al., Cancer Res.* **32**, 247 (1972); W. J. O'Brien, H. F. Edelhauser, *Invest. Ophthalmol. Visual Sci.* **16**, 1093 (1977). Pharmacodynamics: B. L. Wigdahl, J. R. Parkhurst, *Antimicrob. Agents Chemother.* **14**, 470 (1978); G. J. Smith *et al., Biochem. Biophys. Res. Commun.* **83**, 1538 (1978). Teratogenicity study: M. Itoi *et al., Arch. Ophthalmol.* **93**, 46 (1975). Cytotoxicity and mutagenicity study: E. Huberman, C. Heidelberger, *Mutat. Res.* **14**, 130 (1972). Clinical studies: H. E. Kaufman, *Invest. Ophthalmol. Visual Sci.* **17**, 941 (1978); R. A. Hyndiuk *et al., Arch. Ophthalmol.* **96**, 1839 (1978). Review of mechanism of antiviral activity: C. Heidelberger, *Ann. N.Y. Acad. Sci.* **255**, 317 (1975). Review of pharmacology and therapeutic use: A. A. Carmine *et al., Drugs* **23**, 329-353 (1982).

Cryst from ethyl acetate, mp 186-189°. uv max (0.1*N* HCl): 260 nm (ε 9960); (0.1*N* NaOH): 260 nm (ε 6590).

THERAP CAT: Antiviral (ophthalmic).

9856. Triflusal. [322-79-2] 2-(Acetyloxy)-4-(trifluoromethyl)benzoic acid; α,α,α-trifluoro-2,4-cresotic acid acetate; acetyl-4-trifluoromethylsalicylic acid; UR-1501; Disgren. $C_{10}H_7F_3O_4$; mol wt 248.16. C 48.40%, H 2.84%, F 22.97%, O 25.79%. Analog of aspirin, *q.v.*; inhibits platelet aggregation. Prepn of free acid: M. Hauptschein, **US 3019253** (1962 to Pennsalt). Inhibition of platelet aggregation in man and rat: J. Garcia-Rafanell, J. Morell, *Therapie* **32**, 337 (1977). Effect on blood coagulation: M. Rutllant *et al., Curr. Ther. Res.* **22**, 510 (1977). Pharmacokinetics: V. Rimbau *et al., Arch. Farmacol. Toxicol.* **7**, 11 (1981). Clinical studies: R. M. Masso *et al., Curr. Ther. Res.* **25**, 791 (1979); E. Sala-Planell *et al., Angiologia* **33**, 71 (1981).

White cryst solid from petr ether/ether, mp 120-122° (upon slow heating); 110-112° (upon quick heating). Misc with ethanol. Practically insol in water.

THERAP CAT: Antithrombotic.

9857. Triflusulfuron-methyl. [126535-15-7] 2-[[[[[4-(Dimethylamino)-6-(2,2,2-trifluoroethoxy)-1,3,5-triazin-2-yl]amino]carbonyl]amino]sulfonyl]-3-methyl benzoic acid methyl ester; DPX-66037; Debut; Safari; Upbeet. $C_{17}H_{19}F_3N_6O_6S$; mol wt 492.43. C 41.47%, H 3.89%, F 11.57%, N 17.07%, O 19.49%, S 6.51%. Sulfonylurea herbicide for use in beets; acetolactate synthase inhibitor. Prepn: M. P. Moon, **WO 8909214**; *idem,* **US 5090993** (1989, 1992 both to DuPont). Comprehensive description: K. A. Peeples *et al., Brighton Crop Prot. Conf. - Weeds* **1991**, 25-30. Selectivity in sugar beets: V. A. Wittenbach *et al., Pestic. Biochem. Physiol.* **49**, 72 (1994).

White solid, mp 160-163°. Log P (octanol/water pH 7) at 25°: 9.2. pKa 4.4. Soly in water at 25° (ppm): 1 (pH 3); 3 (pH 5); 110 (pH 7); 11000 (pH 9). LD_{50} orally in rats: >5000 mg/kg; dermally in rabbits: >2000 mg/kg (Peeples).

USE: Herbicide.

9858. Trifolium. Meadow clover; red clover; purple clover; cow clover. Perennial herb, *Trifolium pratense* L., *Leguminosae.* Important in agriculture as a forage legume; usually as a companion crop for cereal grains or pasture grass. Used in traditional medicine for coughs and bronchitis and as a dermatological agent. Medicinal formulations are prepared from the dried and fresh flowerheads (inflorescence). *Habit.* Europe, Asia, Northern Africa; naturalized in U.S. *Constit.* Diphenolic isoflavones, primarily biochanin A, formononetin, genistein, daidzein, *q.q.v.*; coumarin derivatives; volatile oil containing furfural, methyl salicylate; tannin, resins, polysaccharides. Description of botony and uses in agriculture: N. L. Taylor, K. H. Quesenberry, *Red Clover Science* (Kluwer, Dordrecht, 1996) 226 pp. HPLC-ES-MS determn of flavonoids in botanical extracts: X. He *et al., J. Chromatogr. A* **755**, 127 (1996). Review of medicinal uses: J. Barnes *et al., Herbal Medicines* (Pharmaceutical Press, London, 2nd ed., 2002) pp 399-400.

Ethanolic extract. Menoflavin; Promensil; Rimostil. Estrogenic activity: E. Dornstauder *et al., J. Steroid Biochem. Mol. Biol.* **78**, 67 (2001). Clinical effect on lipid and bone metabolism: P. B. Clifton-Bligh *et al., Menopause* **8**, 259 (2001); on menopausal symptoms: P. H. M. van der Weijer, R. Barentsen, *Maturitas* **42**, 187 (2002).

USE: Dietary supplement; source of phytoestrogens.

THERAP CAT: In treatment of menopausal symptoms.

9859. Triforine. [26644-46-2] *N,N'*-[1,4-Piperazinediyl-bis(2,2,2-trichloroethylidene)]bisformamide; 1,4-di(2,2,2-trichloro-1-formamidoethyl)piperazine; 1,4-bis(1-formamido-2,2,2-trichloroethyl)piperazine; Cela W-524; CME-74770; Basforin; Funginex; Saprol. $C_{10}H_{14}Cl_6N_4O_2$; mol wt 434.95. C 27.61%, H 3.24%, Cl 48.90%, N 12.88%, O 7.36%. Prepn: W. Ost *et al.,* **DE 1901421** corresp to **US 3595916** (1969, 1971, both to Boehringer, Ing.).

White crystals, mp 155°. Vapor pressure at 25° = 2×10^{-7} mm Hg. Soly in water at 20°: 27-29 ppm. Sol in CMF, DMSO, N-methylpyrrolidone; moderately sol in tetrahydrofuran. Insol in acetone, benzene, carbon tetrachloride, chloroform, methylene chloride, petroleum ether. Dec to chloral and piperazine salts by conc H_2SO_4 and HCl; slowly dec to chloroform and piperazine by conc alk. Half life in the soil ~3 weeks. Low toxicity to fish and bees.

USE: Fungicide.

9860. Triglycidyl Isocyanurate. [2451-62-9] 1,3,5-Tris(2-oxiranylmethyl)-1,3,5-triazine-2,4,6(1*H*,3*H*,5*H*)-trione; tris(2,3-epoxypropyl) isocyanurate; TGIC; Tepic; Araldite PT810. $C_{12}H_{15}$-N_3O_6; mol wt 297.27. C 48.49%, H 5.09%, N 14.14%, O 32.29%. Trifunctional polyepoxide that serves as a crosslinker in the synthesis of polymers. Exists as a mixture of two pairs of diastereomer racemates, designated α and β, that are separable on the basis of their solubility properites. Prepn: M. Budnowski, *Kunststoffe* **55**, 641 (1965); A. Wende, H. Priebe, **GB 1036279** (1966 to VVB Elektrochemie & Plaste). Improved process for the prepn and sepn of the α and β-forms by crystallization: M. Budnowski, US 3300490 (1967 to Henkel & Cie.). Prepn of the (R,R,R) and (S,S,S) isomers: H. Ikeda *et al.*, US 6605717 (2003 to Nissan Chem. Ind.). Evaluation of α and β-form physical properties: D. Joel, H. Becker, *Plaste Kautsch* **23**, 237 (1976). Determn of morphology and crystallinity: H. Beyer *et al.*, *Eur. Polym. J.* **42**, 2350 (2007). Binary solid-liquid phase diagrams: V. Vargha, *ibid.* 4762. MS and GC/MS analysis: G. Audisio *et al.*, *Biomed. Environ. Mass Spectrom.* **13**, 519 (1986). Collagen crosslinking studies: Y. Di., R. J. Heath, *Polym. Degrad. Stab.* **94**, 1684 (2009).

White crystalline solid, mp ~96°. bp$_{0.2}$ 210°. *Poisonous, irritant, skin sensitizer.*

α-Form. [59653-73-5] Consists of a mixture of the (R,R,S) and (S,S,R) stereoisomers. Crystals from methanol, mp 103-104.5° (Joel). Also reported as crystals from fractional crystallization from methanol, mp 95-103° (Beyer). d 1.446. n_D^{21} 1.532. Enthalpy of fusion: 77.0 J/g.

β-Form. [59653-74-6] Consists of a mixture of the (R,R,R) and (S,S,S) stereoisomers. Crystals from chloroform, mp 156-157.5° (Joel). Also reported as methanol-insoluble crystals, mp 149-154° (Beyer). d 1.523. n_D^{21} 1.591. Enthalpy of fusion: 142.5 J/g.

USE: Crosslinking agent in polymer synthesis; additive in plastic, rubber, and adhesives; hardener in polyester powder coatings; in protective coatings of electronic devices; in top-coated and solder-resistant inks.

9861. Triglyme. [112-49-2] 2,5,8,11-Tetraoxadodecane; 1,2-bis(2-methoxyethoxy)ethane; triethylene glycol dimethyl ether. C$_8$H$_{18}$O$_4$; mol wt 178.23. C 53.91%, H 10.18%, O 35.91%. Prepd by high pressure hydrogenation of 1,1,2,2-tetrakis(2-methoxyethoxy)-ethane: McNamee, MacDowell, US 2425042 (1947 to Carbide & Carbon Chem.).

Liquid. d$_4^{20}$ 0.990. Flash pt 111°C. mp −45°. bp$_{760}$ 216°; bp$_{10}$ 103.5°; bp$_{0.9}$ 20°. n_D^{20} 1.4233. Miscible with water, hydrocarbon solvents.

USE: Solvent.

9862. Trigonellamide Chloride. [1005-24-9] 3-(Aminocarbonyl)-1-methylpyridinium chloride (1:1); N^1-methylnicotinamide chloride; 1-methylpyridine-3-carboxylic acid amide chloride; nicotinamide chloromethylate; nicotinamide methyl chloride. C$_7$H$_9$-ClN$_2$O; mol wt 172.61. C 48.71%, H 5.26%, Cl 20.54%, N 16.23%, O 9.27%. One of the principal excretion products of the metabolism of nicotinic acid in man, dog, and rat. Coenzyme action: Warburg, Christian, *Biochem. Z.* **287**, 291 (1936). Synthesis by refluxing nicotinamide with methyl iodide in methanol, then shaking the nicotinamide methiodide with AgCl: Karrer *et al.*, *Helv. Chim. Acta* **19**, 826 (1936); and isolation from urine: Huff, Perizweig, *J. Biol. Chem.* **150**, 395 (1943). Differentiation from NAD: Carpenter, Kodicek, *Biochem. J.* **46**, 421 (1950). *In vitro* metabolism: G. S. Johnson, *Eur. J. Biochem.* **112**, 635 (1980); H. Hoshino *et al.*, *Biochim. Biophys. Acta* **801**, 250 (1984). HPLC determn in urine: M. A. Kutnink *et al.*, *J. Liq. Chromatogr.* **7**, 969 (1984).

Crystals from methanol. Dec 240°. Moderately sol in water. More sol in alcohol, butanol, isobutanol. Insol in amyl alcohol, octyl alcohol, benzene, chlorobenzene, chloroform. Destroyed upon boiling the aq soln, more rapidly in the presence of alkali. At room temp it is destroyed in alkaline soln. Reacts with ketones in aq alkaline soln to produce a greenish-blue fluorescence; on acidification the fluorescence changes to blue, and is intensified by heating.

9863. Trigonelline. [535-83-1] 3-Carboxy-1-methylpyridinium inner salt; nicotinic acid N-methylbetaine; coffearine; caffearine; gynesine; trigenolline. C$_7$H$_7$NO$_2$; mol wt 137.14. C 61.31%, H 5.15%, N 10.21%, O 23.33%. In seeds of *Trigonella foenumgraecum* L., *Leguminosae*, in coffee beans, in seeds of *Strophanthus* spp, *Apocynaceae* and of *Cannabis sativa* L., *Moraceae*, in seeds of many other plants; also in sea urchin, *Arabacia pustulosa*, and in jellyfish, *Velella spirans*. Excreted in urine after taking nicotinic acid: Ackermann, *Z. Biol.* **59**, 17 (1912). Isoln from normal urine: Linnewah, Renwein, *Z. Physiol. Chem.* **207**, 48 (1932); **209**, 110 (1932). Syntheses: Turnau, *Monatsh. Chem.* **26**, 551 (1905); Sarett *et al.*, *J. Biol. Chem.* **135**, 483 (1940); Green, Tong, *J. Am. Chem. Soc.* **78**, 4896 (1956); Kosower, Patton, *J. Org. Chem.* **26**, 1318 (1961). Toxicity study: Brazda, Coulson, *Proc. Soc. Exp. Biol. Med.* **62**, 19 (1946).

Monohydrate. Crystals from ethanol, mp 230-233°. Salty taste. Very sol in water; sol in alcohol. Practically insol in ether, chloroform. LD$_{50}$ s.c. in rats: 5.0 g/kg (Brazda, Coulson).

Hydrochloride. C$_7$H$_7$NO$_2$.HCl. Crystals from 90% alcohol, mp 258-259°. Very sol in water; slightly in alcohol. Practically insol in ether, benzene.

9864. Trihexyphenidyl Hydrochloride. [52-49-3] α-Cyclohexyl-α-phenyl-1-piperidinepropanol hydrochloride (1:1); 3-(1-piperidyl)-1-cyclohexyl-1-phenyl-1-propanol hydrochloride; 1-phenyl-1-cyclohexyl-3-piperidyl-1-propanol hydrochloride; benzhexol chloride; Aparkane; Artane; Broflex; Cyclodol; Pacitane; Paralest; Pargitan; Parkinane; Parkopan; Peragit; Pipanol; Sedrena; Tremin; Triphedinon; Triphenidyl; Tsiklodol. C$_{20}$H$_{32}$ClNO; mol wt 337.93. C 71.09%, H 9.55%, Cl 10.49%, N 4.14%, O 4.73%. Anticholinergic. Synthesis: Denton *et al.*, *J. Am. Chem. Soc.* **71**, 2053 (1949); Adamson, Wilkinson, US 2682543 (1954 to Burroughs Wellcome); Denton, US 2716121 (1955 to Am. Cyanamid). Resolution into isomers: Adamson, Duffin, GB 750156 (1956 to Wellcome Found.).

Crystals, dec 258.5°. Free base, mp 114.3-115.0°. Soly (g/100 ml): water at 25°, 1.0; alcohol 6; chloroform 5. More sol in methanol; very slightly sol in ether, benzene. pH of a 1% aq soln 5.5-6.0.

***l*-Form.** Crystals from isopropyl alc, mp 264°. [α]$_D^{20}$ −30° (c = 0.4 in chloroform). Free base, mp 112-113°. [α]$_D^{20}$ −25° (c = 0.4 in ethanol).

THERAP CAT: Antiparkinsonian.

9865. Trillium. Beth root; Indian balm; ground lily; birthroot. Dried rhizome of *Trillium erectum* L. and other spp of *Trillium*, *Liliaceae*. *Habit.* Canada, south to Tennessee and Missouri; also Japan. *Constit.* Trilline, fixed oil, tannin.

THERAP CAT: Formerly in metrorrhagia, menorrhagia, various types of hemorrhage; astringent in diarrhea.

9866. Trilobine. [6138-73-4] (4aS,16aS)-3,4,4a,5,16a,17,-18,19-Octahydro-9,21-dimethoxy-4-methyl-2H-12,26-epoxy-1,24:22,15-dietheno-6,10-metheno-16H-pyrido[2′:3′:17,18][1,10]-dioxacycloeicosino[2,3,4-*ij*]isoquinoline; (1′α)-6′,7-epoxy-6,12′-dimethoxy-2′-methyloxyacanthan. $C_{35}H_{34}N_2O_5$; mol wt 562.67. C 74.71%, H 6.09%, N 4.98%, O 14.22%. From root of *Cocculus trilobus* DC. and *C. sarmentosus* Diels, *Menispermaceae:* Tomita *et al., J. Pharm. Soc. Jpn.* **48**, 83 (1928); **50**, 127 (1930); **62**, 468, 481 (1942). Structure: Kondo, Tomita, *Ann.* **497**, 104 (1932); Inubushi, Nomura, *Tetrahedron Lett.* **1962**, 1133. Total synthesis of trilobine and isotrilobine: Y. Inubushi *et al., Chem. Pharm. Bull.* **25**, 1636 (1977).

Trilobine R = H
Isotrilobine R = CH₃

Crystals from benzene, mp 237°. $[\alpha]_D$ +307° (chloroform). Practically insol in water. Sparingly sol in alcohol, acetone, ether. Sol in chloroform.

Isotrilobine. [26195-62-0] Homotrilobine; *N*-methyltrilobine. $C_{36}H_{36}N_2O_5$. Crystals from acetone, mp 215°. $[\alpha]_D$ +317° (chloroform).

9867. Trilostane. [13647-35-3] $(4\alpha,5\alpha,17\beta)$-4,5-Epoxy-3,17-dihydroxyandrost-2-ene-2-carbonitrile; 4,5-epoxy-17-hydroxy-3-oxoandrostane-2-carbonitrile; 2α-cyano-$4\alpha,5\alpha$-epoxyandrostan-17β-ol-3-one; Win-24540; Desopan; Modrenal; Vetoryl. $C_{20}H_{27}$-NO_3; mol wt 329.44. C 72.92%, H 8.26%, N 4.25%, O 14.57%. Synthetic steroid analog; competitive inhibitor of 3β-hydroxysteroid dehydrogenase. Prepn: R. O. Clinton, A. J. Manson, **US 3296255** (1967 to Sterling); H. C. Neumann *et al., J. Med. Chem.* **13**, 948 (1970). Inhibition of steroid biosynthesis: G. O. Potts *et al., Steroids* **32**, 257 (1978). Disposition in animals: J. F. Baker *et al., Arch. Int. Pharmacodyn. Ther.* **243**, 4 (1980). Metabolism in rats: Y. Mori *et al., Chem. Pharm. Bull.* **29**, 2646 (1981); in humans: D. T. Robinson *et al., J. Steroid Biochem.* **21**, 601 (1984). HPLC determn in plasma: P. Powles *et al., J. Chromatogr.* **311**, 434 (1984); and of major metabolite, *17-ketotrilostane*: R. R. Brown *et al., ibid.* **339**, 440 (1985). Experience in treatment of canine Alopecia X: R. Cerundolo *et al., Vet. Dermatol.* **15**, 285 (2004). Clinical study in Cushing's syndrome: P. Komanicky *et al., J. Clin. Endocrinol. Metab.* **47**, 1042 (1978). Review of clinical experience in advanced breast cancer: J. R. Puddefoot *et al., Expert Opin. Pharmacother.* **7**, 2413-2419 (2006).

Tan crystals from pyridine/dioxane, mp 257.8-270° (dec). $[\alpha]_D^{25}$ +137.4° (c = 1 in pyridine). uv max (ethanol): 252 nm (ε 8300).

THERAP CAT: Adrenocortical suppressant; in treatment of breast cancer.

THERAP CAT (VET): Adrenocortical suppressant.

9868. Trimebutine. [39133-31-8] 3,4,5-Trimethoxybenzoic acid 2-(dimethylamino)-2-phenylbutyl ester. $C_{22}H_{29}NO_5$; mol wt 387.48. C 68.20%, H 7.54%, N 3.61%, O 20.64%. Opioid receptor agonist. Prepn by esterification: C. P. J. Roux, D. R. Torossian, **FR 1344455** (1963 to Jouveinal), *C.A.* **60**, 15777g (1964); *see also:* D. R. Torossian, G. G. Aubard, **GB 1342547** (1974 to Jouveinal). HPLC determn in plasma: M. Lavit *et al., Arzneim.-Forsch.* **50**, 640

(2000). Review of pharmacology and clinical experience: M. Delvaux, D. Wingate, *J. Int. Med. Res.* **25**, 225-246 (1997).

Crystals from ethanol, mp 78-80°C. Soluble in methylene chloride.

Maleate. [34140-59-5] TM-906; Cerekinon; Debridat; Digerent; Modulon; Polibutin; Spabucol; Transacalm. $C_{22}H_{29}NO_5.C_4H_4O_4$; mol wt 503.55. Crystals from water, mp 105-106°.

THERAP CAT: Antispasmodic.

9869. Trimecaine. [616-68-2] 2-(Diethylamino)-*N*-(2,4,6-trimethylphenyl)acetamide; 2-diethylamino-2′,4′,6′-trimethylacetanilide; *N-sym*-trimethylphenyldiethylaminoacetamide; 2-diethylaminoacetyl-2′,4′,6′-trimethylanilide; S-203; Mesocaine; Mesidicaine; Mesokain. $C_{15}H_{24}N_2O$; mol wt 248.37. C 72.54%, H 9.74%, N 11.28%, O 6.44%. Prepn: Löfgren, Lundqvist, **US 2441498** (1948 to Astra); A. Borovansky *et al., J. Am. Pharm. Assoc.* **48**, 402 (1959).

Crystals, mp 44°. bp₆ 187°; bp₀.₆ 154-155°. LD_{50} s.c. in mice: 295 mg/kg (Borovansky).

Hydrochloride. $C_{15}H_{24}N_2O.HCl$. Crystals from acetone, mp 140°.

THERAP CAT: Anesthetic.

9870. Trimedlure. [12002-53-8] 4(or 5)-Chloro-2-methyl-cyclohexanecarboxylic acid 1,1-dimethylethyl ester; *tert*-butyl 4(or 5)-chloro-2-methylcyclohexanecarboxylate. $C_{12}H_{21}ClO_2$; mol wt 232.75. C 61.93%, H 9.09%, Cl 15.23%, O 13.75%. Synthetic sex pheromone developed as attractant for the Mediterranean fruit fly, or medfly, *Ceratitis capitata* (Weidemann). Prepn and physical props: M. Beroza *et al., J. Agric. Food Chem.* **9**, 361 (1961). Commercial product consists mainly of the four isomers having the methyl and ester substituents on the ring in a *trans* configuration. Separation and identification of isomers: T. P. McGovern, M. Beroza, *J. Org. Chem.* **31**, 1472 (1966). Extension of activity by fixatives: *eidem, J. Econ. Entomol.* **60**, 379 (1967). Controlled release: S. Nakagawa *et al., ibid.* **72**, 625 (1979). Repellent effect of high concns: *eidem, ibid.* **64**, 762 (1971). Rate of loss from wicks: J. R. King, P. J. Landolt, *ibid.* **77**, 221 (1984). Acute toxicity studies: M. Beroza *et al., Toxicol. Appl. Pharmacol.* **31**, 421 (1975).

trans-form

Oil, bp 107-113°. n_D^{20} 1.460. LD_{50} in rats (mg/kg): 4556 ±1136 orally; in rabbits (mg/kg): >2025 dermally; LC_{50} (24 hr) in rainbow trout, bluegill sunfish (ppm): 11.5, 14.7 (Beroza).

USE: As attractant in medfly traps.

9871. Trimellitic Acid. [528-44-9] 1,2,4-Benzenetricarboxylic acid; 1,2,4-tricarboxybenzene. $C_9H_6O_6$; mol wt 210.14. C 51.44%, H 2.88%, O 45.68%. Obtained by oxidation of coal with nitric acid: Grosskinsky, *Glueckauf* **88**, 376 (1952), *C.A.* **46**, 7731 (1952); by oxidation of β-indancarboxylic acid with nitric acid: Braun *et al., Ber.* **53**, 1160 (1920). Manuf by oxidation of pseudocumene: Backlund, **US 3009953** (1961 to Union Oil of Calif.).

Crystals from acetic acid or dilute alcohol. mp 218-220° (also reported 229-234° dec). Solubilities at 25° in g/100 g solvent: carbon tetrachloride 0.004; ligroin 0.03; mixed xylenes 0.006; dimethylformamide 31.3; ethyl acetate 1.7; acetone 7.9; water 2.1; ethanol 25.3. Practically insol in chloroform, benzene, carbon disulfide.

USE: Intermediate in the prepn of resins, plasticizers, dyes, inks, adhesives.

9872. Trimellitic Anhydride. [552-30-7] 1,3-Dihydro-1,3-dioxo-5-isobenzofurancarboxylic acid; trimellitic acid 1,2-anhydride; anhydrotrimellitic acid; 1,3-dioxo-5-phthalancarboxylic acid. $C_9H_4O_5$; mol wt 192.13. C 56.26%, H 2.10%, O 41.64%. Prepd by subliming trimellitic acid above its mp: Alder, Dortmann, *Ber.* **85**, 556 (1952); by heating crude trimellitic acid with V_2O_5: McKinnis, US 2998431 (1959 to Union Oil of Calif.).

Crystals, mp 161-163.5°. bp_{14} 240-245°. Solubilities at 25° in g/100 g solvent: carbon tetrachloride 0.002; ligroin 0.06; mixed xylenes 0.4; dimethylformamide 15.5; acetone 49.6; ethyl acetate 21.6.

Caution: Potential symptoms of overexposure are irritation of eyes, skin, nose, respiratory system; pulmonary edema, respiratory sensitization; rhinitis, asthma, cough, wheezing, dyspnea, malaise, fever, muscle aches, sneezing. *See NIOSH Pocket Guide to Chemical Hazards* (DHHS/NIOSH 97-140, 1997) p 318.

USE: In the preparation of resins, adhesives, polymers, dyes, printing inks.

9873. Trimeprazine. [84-96-8] N,N,β-Trimethyl-10H-phenothiazine-10-propanamine; 10-[3-(dimethylamino)-2-methylpropyl]phenothiazine; 10-(2-methyl-3-dimethylaminopropyl)phenothiazine; alimemazine; methylpromazine; Bayer 1219; RP-6549. $C_{18}H_{22}N_2S$; mol wt 298.45. C 72.44%, H 7.43%, N 9.39%, S 10.74%. Prepn: Jacob, Robert, US 2837518 (1958 to Rhône-Poulenc). Metabolism studies: Robinson, *J. Pharm. Pharmacol.* **18**, 19 (1966).

Crystals, mp 68°. $bp_{0.3}$ 150-175°.
Tartrate. [4330-99-8] Panectyl; Repeltin; Temaril; Theralene; Vallergan. $(C_{18}H_{22}N_2S)_2 \cdot C_4H_6O_6$; mol wt 746.98. White to off-white crystals. Freely sol in water, chloroform; sol in alc; very slightly sol in ether, benzene.

THERAP CAT: Antipruritic.
THERAP CAT (VET): Antipruritic.

9874. Trimetazidine. [5011-34-7] 1-[(2,3,4-Trimethoxyphenyl)methyl]piperazine. $C_{14}H_{22}N_2O_3$; mol wt 266.34. C 63.14%, H 8.33%, N 10.52%, O 18.02%. Coronary vasodilator. Prepn: Regnier, Canevari, FR 1302958 (1962 to Sci. Union et Cie, Soc. Franç. Rech. Med.); FR M805; J. Servier, US 3262852 (1961, 1966 to Biofarma). Pharmacology: Fujita, *Jpn. J. Pharmacol.* **17**, 19 (1967); Nagata *et al., ibid.* **19**, 628 (1969); **21**, 337 (1971). GC-MS determn in biological fluids: L. Fay *et al., J. Chromatogr.* **490**, 198 (1989). Review of pharmacology, toxicology and clinical studies: C. Harpey *et al., Cardiovasc. Drug Rev.* **6**, 292-312 (1989).

bp_2 200-205°.
Dihydrochloride. [13171-25-0] Kyurinett; Vastarel F; Yoshimilon. $C_{14}H_{22}N_2O_3 \cdot 2HCl$; mol wt 339.26. Crystals, mp 225-228°. LD_{50} in male, female mice, rats (mg/kg): 91, 107, 124, 124 i.v.; 264, 245, 327, 288 i.p.; 528, 608, 1147, 987 orally (Harpey).

THERAP CAT: Antianginal.

9875. Trimethadione. [127-48-0] 3,5,5-Trimethyl-2,4-oxazolidinedione; 3,5,5-trimethyl-2,4-dioxooxazolidine; troxidone; Absentol; Epidione; Petidon; Ptimal; Tridione. $C_6H_9NO_3$; mol wt 143.14. C 50.35%, H 6.34%, N 9.79%, O 33.53%. Synthesis: Spielman, *J. Am. Chem. Soc.* **66**, 1244 (1944); US 2575692 (1951 to Abbott); Davies, Hook, US 2559011 (1951 to Brit. Schering). Metabolism to dimethadione, *q.v.:* T. C. Butler, *J. Pharmacol. Exp. Ther.* **108**, 11 (1953); T. C. Butler, W. J. Waddell, *ibid.* **110**, 241 (1954). Anticonvulsant activity: H. Ferngren, *Acta Pharmacol. Toxicol.* **26**, 177 (1968); H. H. Frey, *ibid.* **27**, 295 (1969). Pharmacodynamics and metabolism in rats: D. O. Thueson *et al., Epilepsia* **15**, 563 (1974). Evaluation in teratogenesis: J. German *et al., Teratology* **3**, 349 (1970); A. B. Rifkind, *Toxicol. Appl. Pharmacol.* **30**, 452 (1974). GC determn in serum: E. Tanaka, S. Misawa, *J. Chromatogr.* **413**, 376 (1987); LC determn in serum: M. Okamoto *et al., Chromatographia* **23**, 325 (1987). Use in treatment of petit mal epilepsy: S. Livingston *et al., J. Am. Med. Assoc.* **194**, 227 (1965). Use in dissolution of pancreatic stones in humans and dogs: A. Noda *et al., Lancet* **2**, 351 (1984); *eidem, Gastroenterology* **93**, 1002 (1987).

Granules, crystals, mp 46-46.5°. bp_5 78-80°. Slight camphor-like odor. Burning, faintly bitter taste. Soly in water about 5%; increased by the addition of urethan. Freely sol in alcohol, benzene, chloroform, ether. Practically insol in petr ether. The pH of a 5% soln is ~6.0.

THERAP CAT: Anticonvulsant.
THERAP CAT (VET): Anticonvulsant.

9876. Trimethaphan Camsylate. [68-91-7] Octahydro-2-oxo-1,3-bis(phenylmethyl)thieno[1′,2′:1,2]thieno[3,4-d]imidazol-5-ium (1S,4R)-7,7-dimethyl-2-oxobicyclo[2.2.1]heptane-1-methanesulfonate (1:1); 4,6-dibenzyl-5-oxo-1-thia-4,6-diazatricyclo-[6.3.0.0³·⁷]undecanium (+)-β-camphorsulfonate; d-3,4-(1′,3′-dibenzyl-2′-ketoimidazolido)-1,2-trimethylenethiophanium d-camphorsulfonate; trimethaphan camphorsulfonate; trimetaphan camphorsulfonate; methioplegium; Nu-2222; Arfonad. $C_{32}H_{40}N_2O_5S_2$; mol wt 596.80. C 64.40%, H 6.76%, N 4.69%, O 13.40%, S 10.74%. May be prepd by treating (+)-3,4-(1,3-dibenzylureylene)tetrahydro-2-oxothiophene with 3-ethoxypropylmagnesium bromide, followed by reduction, ring closure, and conversion into the (+)-β-camphorsulfonate. Pharmacology: L. O. Randall *et al., J. Pharmacol. Exp. Ther.* **97**, 48 (1949). Clinical use in hypertension: C. F. Scurr, J. B. Wyman, *Lancet* **263**, 338 (1954). Comprehensive description: K. W. Blessel *et al., Anal. Profiles Drug Subs.* **3**, 545-564 (1974).

Bitter crystals, dec ~245°. $[\alpha]_D^{20}$ +22.0° (c = 4 in water). One gram dissolves in less than 5 ml water and in less than 2 ml alcohol. Slightly sol in acetone, ether; pH of a 1% aq soln is 5.0-6.0.

THERAP CAT: Antihypertensive.

9877. Trimethobenzamide. [138-56-7] *N*-[[4-[2-(Dimethylamino)ethoxy]phenyl]methyl]-3,4,5-trimethoxybenzamide; *N*-[(2-dimethylaminoethoxy)benzyl]-3,4,5-trimethoxybenzamide; 4-(2-dimethylaminoethoxy)-*N*-(3,4,5-trimethoxybenzoyl)benzylamine. $C_{21}H_{28}N_2O_5$; mol wt 388.46. C 64.93%, H 7.27%, N 7.21%, O 20.59%. Prepn: M. W. Goldberg, S. Teitel, US 2879293 (1959 to Hoffmann-La Roche). Pharmacology: K. W. Blessel *et al.*, *Anal. Profiles Drug Subs.* **2**, 551-570 (1973).

Hydrochloride. [554-92-7] Ro-2-9578; Anaus; Tigan; Xametina. $C_{21}H_{28}N_2O_5 \cdot HCl$; mol wt 424.92. White crystals, mp 187.5-190°. Sol in water (approx soly at 25° >50%), warm alc. Insol in ether, benzene. A 5% aq soln is stable to autoclaving at 120° for 20 min at pH 3-7.

THERAP CAT: Antiemetic.

9878. Trimethoprim. [738-70-5] 5-[(3,4,5-Trimethoxyphenyl)methyl]-2,4-pyrimidinediamine; 2,4-diamino-5-(3,4,5-trimethoxybenzyl)pyrimidine; Instalac; Monotrim; Proloprim; Syraprim; Tiempe; Trimanyl; Trimogal; Trimopan; Trimpex; Uretrim; Wellcoprim. $C_{14}H_{18}N_4O_3$; mol wt 290.32. C 57.92%, H 6.25%, N 19.30%, O 16.53%. Prepn from guanidine and β-ethoxy-3,4,5-trimethoxybenzylbenzalnitrile: Stenbuck, Hood, US 3049544 (1962 to Burroughs Wellcome); Hoffer, US 3341541 (1967 to Hoffmann-La Roche). Improved synthesis: B. Roth *et al.*, *J. Med. Chem.* **23**, 379, 535 (1980). Toxicity data: Yamamoto *et al.*, *Chemotherapy (Tokyo)* **21**, 187 (1973). *Review:* Burchall in *Antibiotics* **vol. 3**, J. W. Corcoran, F. E. Hahn, Eds. (Springer-Verlag, New York, 1975) pp 304-320. Comprehensive description: G. J. Manius, *Anal. Profiles Drug Subs.* **7**, 445-475 (1978). Review of antibacterial activity, pharmacokinetics and therapeutic use: R. N. Brogden *et al.*, *Drugs* **23**, 405-430 (1982).

White to cream, bitter crystalline powder, mp 199-203°. Soly in g/100 ml at 25°: DMAC 13.86; benzyl alcohol 7.29; propylene glycol 2.57; chloroform 1.82; methanol 1.21; water 0.04; ether 0.003; benzene 0.002. Slightly sol in alc, acetone. Practically insol in carbon tetrachloride. pKa 6.6. LD_{50} orally in mice: 7000 mg/kg (Yamamoto).

Note: See Sulfamethoxazole, Sulfadiazine, Sulfametrole, Sulfamoxole, and Sulfalene for lists of trade names of mixtures with Trimethoprim.

THERAP CAT: Antibacterial.

THERAP CAT (VET): Antibacterial.

9879. Trimethylaluminum. [75-24-1] Aluminum trimethyl; aluminum trimethanide; TMA; Me₃Al. C_3H_9Al; mol wt 72.09. C 49.98%, H 12.58%, Al 37.43%. Organoaluminum Lewis acid used as a methylating agent. Prepn: G. B. Buckton, W. Odling, *Ann. Suppl.* **4**, 109 (1865); K. S. Pitzer, H. S. Gutowsky, *J. Am. Chem. Soc.* **68**, 2204 (1946). X-ray structural studies of the dimer: P. H. Lewis, R. E. Rundle, *J. Chem. Phys.* **21**, 986 (1953). Methylation applications in epoxide ring openings: W. R. Roush *et al.*, *Tetrahedron Lett.* **24**, 1377 (1983); in cyclopropanations: K. Maruoka *et al.*,

J. Org. Chem. **50**, 4412 (1985); in palladium-catalyzed cross-coupling reactions: K. Hirota, *J. Chem. Soc. Perkin Trans. 1* **1989**, 2513; in conjugate addition reactions: J. Kabbara *et al.*, *Tetrahedron* **51**, 743 (1995); in ketone synthesis: E.-A. Chung *et al.*, *J. Org. Chem.* **63**, 7590 (1998). Use in the prepn of Tebbe Reagent, *q.v.*: F. N. Tebbe *et al.*, *J. Am. Chem. Soc.* **100**, 3611 (1978). *Review:* F. Bracher, *J. Prakt. Chem.* **341**, 88-91 (1999).

Clear colorless liquid, bp₇₅₅ 125-126°; bp₂₀ 48-52°. Crystallizes when chilled in ice water bath, mp 15.0°. d_4^{20} 0.752. *Pyrophoric, corrosive, reacts violently with water and protic solvents.* Freely miscible with aromatic and saturated aliphatic hydrocarbons. Air sensitive. Store under inert gas in cool place.

USE: Reagent and catalyst in synthetic organic chemistry. Source of aluminum in semiconductors. Precursor to methylaluminoxane, a polymerization catalyst. Released from rockets to study the near-space environment.

9880. Trimethylamine. [75-50-3] *N,N*-Dimethylmethanamine. C_3H_9N; mol wt 59.11. C 60.96%, H 15.35%, N 23.70%. Together with other amines it is a degradation product of nitrogenous plant and animal substances. Formed during the distillation of sugar beet residues which contain betaine. Widely distributed as conjugated form in animal tissue and especially in fish. Converted to the free tertiary amine during putrefaction. Detected in menstrual blood, and in urine which was stored at room temp. Prepn from paraformaldehyde and ammonium chloride: Adams, Brown, *Org. Synth.* **1**, 75 (1921); **coll. vol. I**, 2nd ed., p 528; by the action of formaldehyde and formic acid on ammonia: Sommelet, Ferrand, *Bull. Soc. Chim.* [4] **35**, 446 (1924). Physical properties: J. G. Aston *et al.*, *J. Am. Chem. Soc.* **66**, 1171 (1944). Review as indicator of spoilage in stored fish: J. Oehlenschläger, *Dev. Food Sci.* **37**, 571-586 (1997).

Gas. Pungent, fishy, ammoniacal odor, saline taste. *Flammable.* mp −117.08°. bp₇₆₀ 2.87°; bp₇₄₇ 3.2-3.8°. d_0^0 0.6709. Strong base. pK_b (25°): 4.13. Readily absorbed by water, alcohol with which it is miscible; also sol in ether, benzene, toluene, xylene, ethylbenzene, chloroform. Liquefiable by pressure at ordinary temp or by condensation. *Aq solns corrosive.*

Hydrochloride. [593-81-7] Trimethylammonium chloride. $C_3H_9N \cdot HCl$; mol wt 95.57. Prepn: *Org. Synth.* **coll. vol. I**, 2nd ed., p 531. Monoclinic deliquesc crystals from alc. Odor less intense than that of base. Dec 277-278°. Sinters and sublimes at 200°. Sol in water, alcohol; moderately sol in chloroform. Insol in ether. *Keep well closed.*

Caution: Potential symptoms of overexposure to trimethylamine are irritation of eyes, skin, nose, throat, respiratory system; cough, dyspnea, delayed pulmonary edema; blurred vision, corneal necrosis; skin burns; direct contact with liquid may cause frostbite. *See NIOSH Pocket Guide to Chemical Hazards* (DHHS/NIOSH 97-140, 1997) p 318.

USE: In the manuf of quaternary ammonium compds; as insect attractant; as warning agent for natural gas.

9881. Trimethylamine *N*-Oxide. [1184-78-7] *N,N*-Dimethylmethanamine *N*-oxide; TMAO. C_3H_9NO; mol wt 75.11. C 47.97%, H 12.08%, N 18.65%, O 21.30%. Naturally occurring osmolyte found in marine organisms; counteracts protein-destabilizing factors such as urea, temperature, salt, and hydrostatic pressure. The enzymatic reduction to trimethylamine, *q.v.*, is used as an indication of spoilage in stored fish. Prepn of the dihydrate: W. R. Dunstan, E. Goulding, *J. Chem. Soc.* **75**, 792 (1899); of the anhydrous compd: J. A. Soderquist, C. L. Anderson, *Tetrahedron Lett.* **27**, 3961 (1986). Determ in natural waters and biological media: A. D. Hatton, S. W. Gibb, *Anal. Chem.* **71**, 4886 (1999). Review of use as reagent in organometallic chemistry: T.-Y. Luh, *Coord. Chem. Rev.* **60**, 255-276 (1984); in organic synthesis: H.-J. Knölker, *J. Prakt. Chem.*

338, 190-192 (1996). Review of distribution in marine animals of various habitats: R. H. Kelly, P. H. Yancey, *Biol. Bull.* **196**, 18-25 (1999); of biosynthesis, distribution and biological role: B. A. Seibel, P. J. Walsh, *J. Exp. Biol.* **205**, 297-306 (2002); P. H. Yancey *et al.*, *Comp. Biochem. Physiol. A* **133**, 667-676 (2002).

Free-flowing large colorless crystals, mp 225-227°. Very hygroscopic.

Dihydrate. [62637-93-8] $C_3H_9NO.2H_2O$; mol wt 111.14. Radiating needles, mp 96°. Sol in water, methanol. Insol in ether.

Caution: May be irritating to eyes, skin, respiratory system.
USE: Oxidizing reagent in organic synthesis.

9882. Trimethyl Borate. [121-43-7] Boric acid trimethyl ester (H_3BO_3); boron trimethoxide; trimethoxyboron; methyl borate. $C_3H_9BO_3$; mol wt 103.91. C 34.68%, H 8.73%, B 10.40%, O 46.19%. $B(OCH_3)_3$. Prepn from pyridine-boron trichloride complex: ⸱ard, Lappert, *Chem. Ind. (London)* **1952**, 53; from methanol and boric oxide, borax, or boric acid: Schlesinger, *J. Am. Chem. Soc.* **75**, 213 (1953); from methyl orthosilicates and boron halide: Wiberg, Krüerke, *Z. Naturforsch.* **8b**, 608 (1953); from boric acid and methanol: Steinberg, Hunter, *Ind. Eng. Chem.* **49**, 174 (1957). Several mfg processes: **US 2689259; US 2884439; US 2937195** (to Callery Chem.); **US 2880227** and **US 2884440** (to Olin Mathieson); **US 2855427** (to Am. Potash & Chem.); **US 2739979** (to USAEC). Acute toxicity: H. F. Smyth *et al.*, *Am. Ind. Hyg. Assoc. J.* **23**, 95 (1962).

Liquid, d 0.915. bp 67-68°. Flash pt 29°C (84.2°F). *Flammable.* Miscible with tetrahydrofuran, ether, isopropylamine, hexane, methanol, Nujol and other organic liquids. Stable in the absence of moisture, but hydrolyzes in the presence of water to methanol and boric acid. Forms an azeotrope with methanol: 70% $B(OCH_3)_3$ +30% methanol, d 0.87; bp 52-54°; flash pt 34°C (93.2°F). LD_{50} orally in rats: 6.14 ml/kg (Smyth).

USE: As solvent for waxes, resins, oils; catalyst in the manuf of ketones; analysis of paint and varnish ingredients; as neutron detector gas in the presence of a scintillation counter; as a promoter of diborane reactions.

9883. 2,2,3-Trimethylcyclopentanebutanoic Acid. [957136-80-0] 4-(2,2,3-trimethylcyclopentyl)butanoic acid; GIV-3727. $C_{12}H_{22}O_2$; mol wt 198.31. C 72.68%, H 11.18%, O 16.14%. Human bitter taste receptor antagonist. Inhibits activation of bitterness receptor hTAS2R31 by the artifical sweeteners, saccharin and acesulfame K. Prepn and use as food additive: I. M. Ungureanu *et al.*, **WO 08119196**; *eidem*, **WO 08119197** (both 2008 to Givaudan). Modulation of bitter taste perception and receptor binding study: J. P. Slack *et al.*, *Curr. Biol.* **20**, 1104 (2010).

Colorless viscous liquid.
USE: Food additive to block bitter taste.

9884. Trimethylene Bromide. [109-64-8] 1,3-Dibromopropane; α,γ-dibromopropane; ω,ω'-dibromopropane; trimethylene dibromide. $C_3H_6Br_2$; mol wt 201.89. C 17.85%, H 3.00%, Br 79.16%. Prepd by the action of hydrobromic acid on trimethylene glycol in the presence of sulfuric acid: Kamm, Marvel, *Org. Synth.* **coll. vol. I**, p 30 (1941); *cf.* Derick, Hess, *J. Am. Chem. Soc.* **40**, 545 (1918); Norris, Mulliken, *ibid.* **42**, 2096 (1920); Kamm, Newcomb, *ibid.* **43**, 2229 (1921). From trimethylene glycol and PBr_3: Bogert, Slocum, *ibid.* **46**, 765 (1924).

Colorless liquid; d_4^{25} 1.9712; bp_{760} 167° (mp −36°); n_D^{15} 1.5249. Slightly sol in water (1.68 g/l at 30°); sol in alc, ether. Upon pro-

longed heating trimethylene bromide dec and part of it is converted to propylene bromide (1,2-dibromopropane). Boiling with water yields trimethylene glycol.

9885. Trimethylene Glycol. [504-63-2] 1,3-Propanediol; 1,3-dihydroxypropane; 1,3-propylenediol; 1,3-propylene glycol. $C_3H_8O_2$; mol wt 76.10. C 47.35%, H 10.60%, O 42.05%. Prepd by reduction of ethyl glycidate with lithium aluminum hydride: Walborsky, Colombini, *J. Org. Chem.* **27**, 2387 (1962).

Colorless to pale yellow, very viscid, sweet liquid. d_4^{20} 1.0597; bp 210-212°; n_D^{20} 1.4398. Miscible with water, alc.

9886. Trimethylene Oxide. [503-30-0] Oxetane; 1,3-epoxypropane. C_3H_6O; mol wt 58.08. C 62.04%, H 10.41%, O 27.55%. Prepd by dropwise addition of 3-chloropropyl acetate to hot potassium hydroxide soln: Noller, *Org. Synth.* **29**, 92 (1949); modified procedure: Searles, *J. Am. Chem. Soc.* **73**, 124 (1951).

Oil, agreeable aromatic odor. d_0^0 0.8975; d_4^{25} 0.8930. bp_{750} 48°; bp_{736} 45-46°. n_D^{25} 1.3895; n_D^{23} 1.3905. Reacts with Grignard reagents and organolithium compds to give, after hydrolysis, 3-substituted propanols.

9887. Trimethyl Isopropyl Butanamide. [51115-67-4] N,-2,3-Trimethyl-2-(1-methylethyl)butanamide; 2-isopropyl-N,2,3-dimethylbutanamide; WS23. $C_{10}H_{21}NO$; mol wt 171.28. C 70.12%, H 12.36%, N 8.18%, O 9.34%. Prepn: D. G. Rowsell *et al.*, **DE 2317538**; *eidem*, **US 4296255** (1973, 1981 both to Wilkinson Sword). Structure-activity study: H. R. Watson *et al.*, *J. Soc. Cosmet. Chem.* **29**, 185 (1978). Brief review: M. A. Parrish, *Manuf. Chem.* **58**, 31-32 (February, 1987).

Colorless solid, mp 58-61° (Rowsell); also reported as white, crystalline solid, melting range 63° (Parrish). $bp_{0.35mm}$ 83-85°.
USE: Physiological coolant in foods, beverages, tobacco products, toiletries, cosmetics and pharmaceuticals.

9888. Trimethyl Orthovalerate. [13820-09-2] 1,1,1-Trimethoxypentane; methyl orthovalerate. $C_8H_{18}O_3$; mol wt 162.23. C 59.23%, H 11.18%, O 29.59%. Prepn: S. M. McElvain *et al.*, *J. Am. Chem. Soc.* **68**, 1922 (1946); J. W. Scheeren *et al.*, *Synthesis* **1978**, 283.

Liquid. d_4^{27} 0.9413. bp_{760} 164-166°; bp_{14} 60°. n_D^{24} 1.4090. n_D^{20} 1.4105.

9889. Trimethyloxonium Tetrafluoroborate. [420-37-1] Trimethyloxonium tetrafluoroborate(1−) (1:1); trimethyloxonium fluoborate. $C_3H_9BF_4O$; mol wt 147.91. C 24.36%, H 6.13%, B 7.31%, F 51.38%, O 10.82%. Oxonium salt that serves as a methylating agent for a variety of functional groups. Prepn: H. Meerwein *et al.*, *J. Prakt. Chem.* **147**, 257 (1937). Large scale prepn: H. Meerwein, *Org. Synth.* **coll. vol. V**, 1096 (1973); T. J. Curphey, *ibid.* **coll. vol. VI**, 1019 (1988). Synthetic utility in the formation of esters from carboxylic acids: D. J. Raber *et al.*, *J. Org. Chem.* **44**, 1149 (1979); from aryl amides: G. E. Keck *et al.*, *Tetrahedron* **56**, 9875 (2000). Application in the deprotection of thioacetals: I. Stahl *et al.*, *Tetrahedron Lett.* **12**, 4077 (1971); in regioselective indazole alkylations: M. Cheung *et al.*, *J. Org. Chem.* **68**, 4093 (2003).

H₃C, + ,CH₃   BF₄⁻

Solid, mp 124.5° (Meerwein); also reported as white crystalline solid, mp 179.6-180.0° (dec, sealed tube) (Curphey). *Corrosive, reacts violently with water.* Sol in nitrobenzene, nitromethane, chloroform, hot acetone, liquid sulfur dioxide; slightly sol in dichloromethane. Insol in common organic solvents. Nonhygroscopic. Store under inert gas at 2-8°C. May be handled in the air for short periods of time. Batches stored in a desiccator at −20°C for over one year have been successfully used in reactions.

USE: Reagent in synthetic organic chemistry.

9890. 2,4,6-Trimethylpyridine. [108-75-8] γ-Collidine; *sym*-collidine; 2,4,6-collidine; α,γ,α'-collidine. C₈H₁₁N; mol wt 121.18. C 79.29%, H 9.15%, N 11.56%. Found in small amts in coal tar, in shale oil. Produced commercially from coal tar to some extent; also produced by the Hantzsch pyridine synthesis: Hantzsch, *Ann.* **215**, 1 (1882); Mosher in *Heterocyclic Compounds* **vol. 1**, R. C. Elderfield, Ed., (John Wiley, New York, 1950) pp 462-472; from acetone and ammonia: Mosher in *Kirk-Othmer Encyclopedia of Chemical Technology* **vol. 11** (New York, 1953) p 287, Dürkopf, *Ber.* **21**, 2713 (1888); from ammonium acetate, and ammonia water: **DE 349267**; *Frdl.* **14**, 539; from 3,5-dimethyl-2-cyclohexen-1-one, ammonium acetate, and ammonia water: Frank, Meikle, *J. Am. Chem. Soc.* **72**, 4184 (1950).

Liquid. Aromatic odor. d₄¹⁶·⁴ 0.9191; d₄²²·¹ 0.9166 (commercial grade: d₁₅.₅¹⁵·⁵ 0.920-0.935, approx 7.74 lbs/gal). mp −46°. bp₇₆₂ 170.5°; bp₇₆₀ 171°; bp₃₁ 65°; bp₂.₇ 10°. n_D²⁵ 1.4959; n_D²²·¹ 1.49770. pKa (25°): 6.69. Flash pt 136°F. Dielectric constant at 22° = 6.6 (λ = 70 cm). More sol in cold water than in hot water: 20.8 g/100 ml at 6°; 3.5 g/100 ml at 20°; 1.8 g/100 ml at 100°. Miscible with ether. Sol in methanol, ethanol, chloroform, benzene, toluene, dil acids.

9891. Trimethylsilyl Triflate. [27607-77-8] 1,1,1-Trifluoromethanesulfonic acid trimethylsilyl ester; trimethylsilanol trifluoromethanesulfonate; trimethylsilyltrifluoromethane sulfonate. C₄H₉-F₃O₃SSi; mol wt 222.25. C 21.62%, H 4.08%, F 25.64%, O 21.60%, S 14.43%, Si 12.64%. Prepn: M. Schmeisser *et al.*, *Ber.* **103**, 868 (1970); H. W. Roesky, H. H. Giere, *Z. Naturforsch.* **25B**, 773 (1970); D. Haebich, F. Effenberger, *Synthesis* **1978**, 755; T. Morita *et al.*, *ibid.* **1981**, 745. As catalytic reagent: H. Vorbrueggen, K. Krolikiewicz, *Angew. Chem. Int. Ed.* **14**, 421 (1975); as silylating agent: G. A. Olah *et al.*, *J. Org. Chem.* **46**, 5212 (1981).

Liquid, bp₇₆₀ 140°, bp₁₀ 36.5°. n_D²⁰ 1.3630. Fumes in air; sensitive to atm moisture.

USE: Catalyst and silylating agent for organic syntheses.

9892. Trimetozine. [635-41-6] 4-Morpholinyl(3,4,5-trimethoxyphenyl)methanone; 4-(3,4,5-trimethoxybenzoyl)morpholine; *N*-(3,4,5-trimethoxybenzoyl)tetrahydro-1,4-oxazine; V-7; Opalène; Trioxazine. C₁₄H₁₉NO₅; mol wt 281.31. C 59.78%, H 6.81%, N 4.98%, O 28.44%. Prepd from morpholine and 3,4,5-trimethoxybenzoyl chloride: **GB 872350** (1961 to Egyesült Gyogyszer és Tápszergyár); Pettit *et al.*, *J. Med. Pharm. Chem.* **5**, 800 (1962).

Crystals, mp 120-122°. Slightly sol in water, alcohol.

THERAP CAT: Anxiolytic.

9893. Trimetrexate. [52128-35-5] 5-Methyl-6-[[(3,4,5-trimethoxyphenyl)amino]methyl]-2,4-quinazolinediamine; 2,4-diamino-5-methyl-6-[(3,4,5-trimethoxyanilino)methyl]quinazoline; TMQ; NSC-249008; JB-11; CI-898. C₁₉H₂₃N₅O₃; mol wt 369.43. C 61.77%, H 6.28%, N 18.96%, O 12.99%. Lipophilic dihydrofolate reductase inhibitor structurally related to methotrexate, *q.v.*, with antimicrobial and antitumor activity. Prepn: E. F. Elslager, L. M. Werbel, **GB 1345502** (1974 to Parke, Davis); E. F. Elslager *et al.*, *J. Med. Chem.* **26**, 1753 (1983); of water soluble salts: N. L. Colbry, **EP 51415**; *idem*, **US 4376858** (1982, 1983 both to Warner-Lambert). *In vitro* antifolate activity and *in vivo* antitumor effect: J. R. Bertino *et al.*, *Biochem. Pharmacol.* **28**, 1983 (1979). Pharmacology: E. C. Weir *et al.*, *Cancer Res.* **42**, 1696 (1982); R. C. Jackson *et al.*, *Adv. Enzyme Regul.* **22**, 187 (1984). GC-MS determn in human plasma: P. L. Stetson, W. D. Ensminger, *J. Chromatogr.* **383**, 69 (1986). Clinical evaluation with leucovorin vs *Pneumocystis carinii* in patients with AIDS: C. J. Allegra *et al.*, *N. Engl. J. Med.* **317**, 978 (1987). Review of pharmacology and clinical efficacy: J. T. Lin, J. R. Bertino, *J. Clin. Oncol.* **5**, 2032-2040 (1987); P. J. O'Dwyer *et al.*, *NCI Monogr.* **5**, 105-109 (1987).

Monoacetate monohydrate. [117381-09-6] C₁₉H₂₃N₅O₃.C₂-H₄O₂.H₂O. Crystals from aqueous acetic acid, mp 215-217°. Poorly sol in water. LD₅₀ i.p. in mice: 175 mg/kg (Jackson).

D-Glucuronate. [82952-64-5] NSC-352122; Neutrexin. C₁₉-H₂₃N₅O₃.C₆H₁₀O₇; mol wt 563.56. Tan colored solid. Soly in water: >50 mg/ml.

THERAP CAT: Antineoplastic; antipneumocystis.

9894. Trimipramine. [739-71-9] 10,11-Dihydro-*N,N,β*-trimethyl-5*H*-dibenz[*b,f*]azepine-5-propanamine; 5-[3-(dimethylamino)-2-methylpropyl]-10,11-dihydro-5*H*-dibenz[*b,f*]azepine; 5-(3-dimethylamino-2-methylpropyl)iminodibenzyl; trimeprimine; trimeproprimine; RP-7162; Sapilent. C₂₀H₂₆N₂; mol wt 294.44. C 81.59%, H 8.90%, N 9.51%. Prepn: Jacob, Messer, *Compt. Rend.* **252**, 2117 (1961). Toxicity studies: Okamoto, *C.A.* **72**, 77299y (1970). Chemistry: Bever, Bredenstein, *Dtsch. Apoth. Ztg.* **113**, 1562 (1973). Comprehensive description of the maleate: A. A. Al-Badr, *Anal. Profiles Drug Subs.* **12**, 683-712 (1983).

Crystals, mp 45°.

Maleate. [521-78-8] Stangyl (tabl); Surmontil (tabl). C₂₀H₂₆-N₂.C₄H₄O₄; mol wt 410.51. Crystals, mp 142°. Sol in chloroform; slightly sol in water, ethanol. Practically insol in ether.

Methanesulfonate. Stangyl (amp); Surmontil (amp).

THERAP CAT: Antidepressant.

9895. Trimoprostil. [69900-72-7] (5*Z*,11α,13*E*,15*R*)-15-Hydroxy-11,16,16-trimethyl-9-oxoprosta-5,13-dien-1-oic acid; (*Z*)-7-[(1*R*,2*R*,3*R*)-2-[(*E*)-(3*R*)-3-hydroxy-4,4-dimethyl-1-octenyl]-3-methyl-5-oxocyclopentyl]-5-heptenoic acid; *nat*-11*R*,16,16-trimethyl-15*R*-hydroxy-9-oxoprosta-*cis*-5-*trans*-13-dienoic acid; 11*R*,16,16-trimethyl-(11-desoxyprostaglandin E₂); 11-deoxy-11α,16,16-trimethyl-PGE₂; TM-PGE₂; Ro-21-6937; Ulstar. C₂₃H₃₈O₄; mol wt 378.55. C 72.98%, H 10.12%, O 16.91%. Synthetic prostaglandin

E_2 analog with antisecretory activity. Prepn: G. W. Holland *et al.*, **DE 2437622**; *eidem*, **US 4052446**; **US 4190587** (1975, 1977, 1980 all to Hoffmann-La Roche). Pharmacology: D. E. Wilson, S. L. Winter, *Prostaglandins* **16**, 127 (1978); S. P. Lee *et al.*, *Eur. J. Clin. Invest.* **17**, 1 (1987). Effect on bicarbonate secretion: M. Feldman, *J. Clin. Invest.* **72**, 295 (1983); on inhibition of gastric acid secretion: R. J. Wills *et al.*, *Clin. Pharmacol. Ther.* **37**, 113 (1985). Metabolism: S. J. Kolis *et al.*, *Drug Metab. Dispos.* **14**, 465 (1986). Clinical pharmacokinetics: R. J. Wills *et al.*, *J. Clin. Pharmacol.* **26**, 48 (1986). Clinical evaluation: H. G. Dammann *et al.*, *Arzneim.-Forsch.* **36**, 500 (1986). Multicenter clinical comparison with cimetidine, *q.v.*: K. D. Bardhan *et al.*, *Scand. J. Gastroenterol.* **23**, 134 (1988); with aldioxa, *q.v.*: A. Ishimori *et al.*, *Acta Ther.* **15**, 27 (1989). Toxicity study: M. Shimizu *et al.*, *Shin'yaku to Rinsho* **35**, 2199 (1986), *C.A.* **106**, 150180e (1987).

Colorless oil. $[\alpha]_D$ −51.54° (c = 1 in $CHCl_3$). LD_{50} in mice, rats (mg/kg): 41, 23 orally; 70, 21 i.p.; 68, 29 s.c. (Shimizu).

THERAP CAT: Antiulcerative.

9896. Trimyristin. [555-45-3] Myristin; glyceryl trimyristate. $C_{45}H_{86}O_6$; mol wt 723.18. C 74.74%, H 11.99%, O 13.27%. Occurs in many vegetable fats and oils, notably in coconut oil and nutmeg butter.

White to yellowish-gray solid; d_4^{60} 0.885; mp 56-57°; n_D^{60} 1.4429. Insol in H_2O; sol in alc, benzene, chloroform, ether.

9897. Trinexapac-ethyl. [95266-40-3] 4-(Cyclopropylhydroxymethylene)-3,5-dioxocyclohexanecarboxylic acid ethyl ester; cimectacarb; CGA-163935; Palisade. $C_{13}H_{16}O_5$; mol wt 252.27. C 61.90%, H 6.39%, O 31.71%. Gibberellin biosynthesis inhibitor. Prepn: H.-G. Brunner, **EP 126713**; *idem*, **US 4693745** (1984, 1987 both to Ciba-Geigy). Comprehensive description: E. Kerber *et al.*, *Brighton Crop Prot. Conf. - Weeds* **1989**, 83-88. Field trials in cereals and oilseed rape: J. Amrein *et al.*, *ibid.* 89; in turfgrass: D. B. Vitolo *et al.*, *Proc. Plant Growth Regul. Soc. Am.* **18**, 136 (1991). Mechanism of action: R. Adams *et al.*, *Curr. Plant Sci. Biotechnol. Agric.* **13**, 818 (1992).

mp 36°. n_D^{30} 1.5350. pKa 4.7. Vapor pressure (20°): 1.6×10^{-3} Pa. Soly in water at 20° (g/l): 27 (pH 7); 5 (pH 5). Soly at 20° (g/ml): methanol >1; acetonitrile >1; cyclohexanone >1; isopropanol 0.9; *n*-octanol 0.18; hexane 0.035. Log P (*n*-octanol/water) pH 3: 2.1; pH 7: −0.4. LD_{50} in rats (mg/kg): 4460 orally; >4000 dermally (Kerber).

USE: Plant growth regulator.

9898. sym-Trinitrobenzene. [99-35-4] 1,3,5-Trinitrobenzene; benzite. $C_6H_3N_3O_6$; mol wt 213.11. C 33.82%, H 1.42%, N 19.72%, O 45.04%. Prepared by decarboxylation of trinitrobenzoic acid, obtained by oxidation of TNT: Clarke, Hartman, *Org. Synth.* **2**, 93 (1922); by the action of alkali on 2,4,6-trinitrobenzaldehyde: Secareanu, *Bull. Soc. Chim.* **51**, 591 (1932). Use as explosive, less sensitive to impact than TNT but more powerful and brisant: Robertson, *J. Chem. Soc.* **119**, 8 (1921); van Duin, *Rec. Trav. Chim.* **39**, 687 (1920). Review of toxicology and human exposure: *Toxicological Profile for 1,3-Dinitrobenzene and 1,3,5-Trinitrobenzene* (PB95-264289, 1995) 169 p.

Orthorhombic bipyramidal plates from glacial acetic acid. mp 122.5°. bp 315°. d_4^{20} 1.76; d_4^{152} 1.4775. Log P (octanol:water) 1.18. Dimorphic, the other (rare) form melts at 61°. Can be sublimed by careful heating, explodes when heated rapidly. Absorption spectrum: Hatzsch, Picton, *Ber.* **42**, 2121 (1909). Soly (g/100 g solvent): water 0.035; benzene 6.2; methanol 4.9; alcohol 1.9; ether 1.5; carbon disulfide 0.25; petr ether 0.05. Freely sol in dil Na_2SO_3 soln. LD_{50} orally in rats: 275 mg/kg (*Toxicological Profile*).

USE: Explosive. Vulcanizing agent for natural rubber. Indicator for pH 12.0-14.0.

9899. 2,4,6-Trinitrobenzoic Acid. [129-66-8] 1-Carboxy-2,-4,6-trinitrobenzene; *sym*-trinitrobenzoic acid. $C_7H_3N_3O_8$; mol wt 257.11. C 32.70%, H 1.18%, N 16.34%, O 49.78%. Prepd by chromic acid oxidation of 2,4,6-trinitrotoluene: Clarke, Hartman, *Org. Synth.* **2**, 95 (1922); Kastens, Kaplan, *Ind. Eng. Chem.* **42**, 402 (1950).

Orthorhombic crystals from water, mp 228.7°. Sublimes with decompn forming CO_2 and trinitrobenzene. Soly at 25°: 2.05% (w/w) in water, 26.6% in alcohol, 14.7% in ether. Also sol in acetone, methanol; slightly sol in benzene.

9900. 2,4,7-Trinitrofluorenone. [129-79-3] 2,4,7-Trinitro-9*H*-fluoren-9-one; TNF. $C_{13}H_5N_3O_7$; mol wt 315.20. C 49.54%, H 1.60%, N 13.33%, O 35.53%. Prepn by nitration of fluorenone: Schmidt, Bauer, *Ber.* **38**, 3758 (1905); Orchin *et al.*, *J. Am. Chem. Soc.* **69**, 1225 (1947); by nitration of 2,5-dinitrofluorenone: Ray, Francis, *J. Org. Chem.* **8**, 58 (1943). Structure: Bell, *J. Chem. Soc.* **1928**, 1990. Crystal structure: D. L. Dorset *et al.*, *Acta Crystallogr.* **B28**, 3122 (1972); H. L. Ammon, *ibid.* **B29**, 2314 (1973). Carcinogenicity study: C. Huggins, N. C. Yang, *Science* **137**, 257 (1962). Evaluation of employee exposure to TNF in workplace using HPLC: M. J. Seymour, *J. Chromatogr.* **236**, 530 (1982); R. E. McCullen, A. N. Sanghvi, *ASTM Spec. Tech. Publ.* **786**, 26 (1982).

Yellow needles from acetic acid, mp 175.2-176°.

USE: In photocopiers; forms charge-transfer complexes with aromatic hydrocarbons and amines.

9901. Trinitromethane. [517-25-9] Nitroform. CHN_3O_6; mol wt 151.03. C 7.95%, H 0.67%, N 27.82%, O 63.56%. Prepn from tetranitromethane and $K_4[Fe(CN)_6]$ in aq soln: Chattaway, Harrison, *J. Chem. Soc.* **109**, 171 (1916); by nitration of acetylene with nitric acid: Hager, *Ind. Eng. Chem.* **41**, 2168 (1949).

Crystals, mp 15° (the unstable *aci*-form, mp 50°). d_4^{25} (liq) 1.469. Heat of combustion 746 cal/g. Dipole moment 2.61 (benzene). Dec above 25°. Explodes when heated rapidly. Sol in water, giving an intensely yellow soln, although the dry crystals are pure white.

Potassium salt. [14268-23-6] CKN_3O_6; mol wt 189.12. Moderately stable crystals. Soly in water at 0°: 16.7 g/100 ml; at 60°: 193.8 g/100 ml. Soly in ethanol: 5.29 g/l.

Caution: Slightly irritating to eyes, mucous membranes.

USE: In the manuf of explosives and propellants.

9902. 2,4,6-Trinitrotoluene. [118-96-7] 2-Methyl-1,3,5-trinitrobenzene; TNT; *α*-trinitrotoluol; *sym*-trinitrotoluene; 1-methyl-2,4,6-trinitrobenzene; trotyl; tolit; trilit. $C_7H_5N_3O_6$; mol wt 227.13. C 37.02%, H 2.22%, N 18.50%, O 42.26%. Prepn by nitration of toluene with mixed acid ($HNO_3 + H_2SO_4$) in three steps or by continuous flow according to the Schmid-Meissner and Biazi processes: Swift, Tittensor, *J. Soc. Chem. Ind.* **59**, 92 (1940); Johnston in *Mc-Graw-Hill Encyclopedia of Science and Technology* **9**, 104 (1960). Physical constants and applications: Lothrop, Handrick, *Chem. Rev.* **44**, 419-445 (1949); *ACS Monograph Series* **no. 139**, entitled "The Science of High Explosives," M. A. Cook, Ed. (Reinhold, New York, 1958). Review of toxicology and human exposure: *Toxicological Profile for 2,4,6-Trinitrotoluene* (PB95-264297, 1995) 208 pp.

Monoclinic rhombohedra from alcohol. The commercial crystals (needles) are yellow. mp 80.1°. d_4^{20} 1.654. Burns at 295° when not confined. Can be distilled under reduced pressure. Vapors are toxic. Dipole moment 1.37. Very sparingly sol in water: About 0.01% at 25°, one gram dissolves in 700 ml of boiling water. Sol in acetone, benzene. Less sol than 2,4,6-trinitrophenol in alcohol, ether, carbon disulfide. Reacts vigorously with reducing agents.

Caution: Potential symptoms of overexposure are irritation of skin, mucous membranes; liver damage; jaundice; cyanosis; sneezing; coughing, sore throat; peripheral neuropathy; muscle pain; kidney damage; cataract; sensitization dermatitis; leukocytosis; anemia; cardiac irregularities. *See NIOSH Pocket Guide to Chemical Hazards* (DHHS/NIOSH 97-140, 1997) p 322.

USE: High explosive in military munitions. Must be detonated by a high velocity initiator such as nitramine or by efficient concussion.

9903. Trinitrotriazidobenzene. [29306-57-8] 1,3,5-Triazido-2,4,6-trinitrobenzene; TNTA. $C_6N_{12}O_6$; mol wt 336.14. C 21.44%, N 50.00%, O 28.56%. Nontoxic, metal-free, high energy density material. Explodes to produce gaseous nitrogen and carbon dioxide. Prepn: I. O. Turek, *Chim. Ind. (Paris)* **26**, 785 (1931). Use as a fixative in immunofluorescence microscopy: E. McBeath, K. Fujiwara, *J. Cell Biol.* **99**, 2061 (1984). Thermal decompn studies: A. D. Yoffe, *Proc. R. Soc. London Ser. A* **208**, 188 (1951); D. Adam *et al.*, *Heteroat. Chem.* **10**, 548 (1999). Synthesis, crystal structure, and detonation properties: *idem et al.*, *Propellants Explos. Pyrotech.* **27**, 7 (2002).

Bright yellow plates from acetic acid, mp 128-130° (dec). d 1.84. *Explosive.*

Caution: Sensitive to electrostatic discharge, friction and impact; rapid heating above 168° causes explosion (Adam, 2002).

USE: Initiator or solid rocket propellant in military explosives.

9904. Triolein. [122-32-7] 9-Octadecenoic acid 1,2,3-propanetriyl ester; olein; glyceryl trioleate. $C_{57}H_{104}O_6$; mol wt 885.45.

C 77.32%, H 11.84%, O 10.84%. One of the chief constituents of nondrying oils and fats. From Palestine olive oil: Hilditch, Madison, *J. Soc. Chem. Ind.* **60**, 258 (1941); from cacao butter: Meara, *J. Chem. Soc.* **1949**, 2154. Prepn by esterification of oleic acid: Wheeler *et al.*, *J. Biol. Chem.* **132**, 687 (1940); Swicklik *et al.*, *J. Am. Oil Chem. Soc.* **32**, 69 (1955). Synthesis: Serebrennikova *et al.*, *Dokl. Akad. Nauk SSSR* **140**, 1083 (1961).

Colorless to yellowish, oily liquid; tasteless, odorless. d_4^{15} 0.915. mp −5 to −4°. bp_{15} 235-240°. n_D^{20} 1.4676; n_D^{60} 1.4561. Practically insol in water; sol in chloroform, ether, carbon tetrachloride; slightly sol in alcohol.

9905. Triostins. Quinoxaline antibiotic complex similar to echinomycin, *q.v.* Powerful, selective inhibitor of nucleic acid synthesis *in vitro*. Isoln of triostin C from *Streptomyces* S-2-210 resembling *S. aureus*: J. Shoji, K. Katagiri, *J. Antibiot.* **14A**, 335 (1961). Isoln of triostins A and B: H. Otsuka, J. Shoji, *ibid.* **19A**, 128 (1966). Structure of C: *eidem*, *Tetrahedron* **21**, 2931 (1965); of minor components: *eidem*, *ibid.* **23**, 1535 (1967); of A: H. Otsuka *et al.*, *J. Antibiot.* **29**, 107 (1976). Biosynthesis: T. Yoshida, K. Katagiri, *Biochemistry* **8**, 2645 (1969). Synthesis of A: P. K. Chakravarty, R. K. Olsen, *Tetrahedron Lett.* **1978**, 1613; M. Shin *et al.*, *Pept. Chem.* **18**, 207 (1980). Conformation of A in soln: J. R. Kalman *et al.*, *J. Chem. Soc. Perkin Trans. 1* **1979**, 1313. Review of chemistry and biochemistry: M. J. Waring in *Antibiotics* vol. 5 (pt. 2), F. E. Hahn, Ed. (Springer-Verlag, New York, 1979) pp 173-194.

Triostin A	R = CH_3
Triostin C	R = $CH(CH_3)_2$

Needles from chloroform + methanol.

Triostin A. $C_{50}H_{62}N_{12}O_{12}S_2$; mol wt 1087.24. mp 245-248° (dec). $[\alpha]_D^{25}$ −157° (c = 0.97 in chloroform). uv max (methanol): 243, 320 nm (log ε 4.75, 4.11).

Triostin C. $C_{54}H_{70}N_{12}O_{12}S_2$; mol wt 1143.35. mp >260° (dec). $[\alpha]_D^{24}$ −143.9° (c = 1.2 in chloroform). uv max (methanol): 243, 320 nm (log ε 4.87, 4.13).

9906. *s*-Trioxane. [110-88-3] 1,3,5-Trioxane; 1,3,5-trioxacyclohexane; metaformaldehyde; trioxymethylene. $C_3H_6O_3$; mol wt 90.08. C 40.00%, H 6.71%, O 53.28%. *Review:* Walker, Carlisle, *Chem. Eng. News* **21**, 1250 (1943).

Stable, cyclic trimer of formaldehyde possessing characteristic chloroform-like odor. Crystalline solid, mp 64°, bp$_{759}$ 114.5° without dec. Sublimes readily. Easily sol in water (17.2 g/100 ml at 18°, 21.1 g/100 ml at 25°), alcohols, ketones, ether, acetone, chlorinated and aromatic hydrocarbons, and other organic solvents. Slightly sol in pentane, petroleum ether, and lower paraffins. d^{65} 1.17. Flash pt 45°C. On distillation with water, forms an azeotrope boiling at 91.4° containing approx 70% trioxane by wt. Slowly depolymerized by strong acids in aq soln, but inert to alkalies. In non-aqueous systems, readily converted to monomeric formaldehyde by small concentrations of strong acids at a rate determined by the acid concentration.

9907. Trioxsalen. [3902-71-4] 2,5,9-Trimethyl-7*H*-furo[3,2-*g*][1]benzopyran-7-one; 6-hydroxy-*β*,2,7-trimethyl-5-benzofuranacrylic acid δ-lactone; 4,5',8-trimethylpsoralen; NSC-71047; Trisoralen. C$_{14}$H$_{12}$O$_3$; mol wt 228.25. C 73.67%, H 5.30%, O 21.03%. Synthetic trimethyl psoralen deriv. Prepn: K. D. Kaufman, *J. Org. Chem.* **26**, 117 (1961); US 3201421 (1965). Protective effect vs UV-B erythema: P. G. Agache, L. Coupez, *Arch. Dermatol. Res.* **268**, 85 (1980). Clinical study: N. Vaatainen *et al., Clin. Exp. Dermatol.* **6**, 133 (1981). Comprehensive description: M. M. A. Hassan, M. A. Loutfy, *Anal. Profiles Drug Subs.* **10**, 705-727 (1981). *See also* Psoralen, Methoxsalen.

Prisms from chloroform, mp 234.5-235°. uv max (methanol): 250, 295, 335 nm (log ε 4.35, 3.99, 3.80). Sol in methylene chloride; sparingly sol in chloroform; slightly sol in alc. Practically insol in water.

THERAP CAT: Pigmentation agent (photosensitizer).

9908. Tripalmitin. [555-44-2] Hexadecanoic acid 1,1',1''-(1,2,3-propanetriyl) ester; glyceryl trihexadecanoate; glyceryl tripalmitate; palmitic triglyceride; palmitin. C$_{51}$H$_{98}$O$_6$; mol wt 807.34. C 75.87%, H 12.24%, O 11.89%. Occurs in fats. Prepd from glycerol and palmitic acid in the presence of Twitchell reagent: Ozaki, *Biochem. Z.* **177**, 159 (1926); or in the presence of trifluoroacetic anhydride: Bourne *et al., J. Chem. Soc.* **1949**, 2976.

Needles from ether, mp 66° (occurs also in lower-melting, unstable forms). bp 310-320°; d$_4^{70}$ 0.8730; d$_4^{80}$ 0.8663. n$_D^{80}$ 1.43807. Saponification value 208.5. Insol in water. Practically insol in alcohol (0.0043 parts/100 parts of abs alcohol at 21°). Freely sol in ether, benzene, chloroform.

9909. Tripamide. [73803-48-2] *rel*-3-(Aminosulfonyl)-4-chloro-*N*-[(3a*R*,4*R*,7*S*,7a*S*)-octahydro-4,7-methano-2*H*-isoindol-2-yl]benzamide; (3aα,4α,7α,7aα)-3-(aminosulfonyl)-4-chloro-*N*-(octahydro-4,7-methano-2*H*-isoindol-2-yl)benzamide; 4-chloro-*N*-(*endo*-hexahydro-4,7-methanoisoindolin-2-yl)-3-sulfamoylbenzamide; toripamide; ADR-033; E-614; Normonal. C$_{16}$H$_{20}$ClN$_3$O$_3$S; mol wt 369.86. C 51.96%, H 5.45%, Cl 9.58%, N 11.36%, O 12.98%, S 8.67%. Sulfonamide with diuretic and peripheral vasodilator activity. Prepn: H. Hamano *et al.,* JP 73 5585 (1973 to Eisai), *C.A.* **78**, 136070r (1973). Synthesis of ^{14}C-tripamide: T. Nakamura *et al., J. Labelled Compd. Radiopharm.* **14**, 191 (1978). Pharmacological study: T. Satoh *et al., Oyo Yakuri* **21**, 607 (1981), *C.A.* **96**, 28414u (1982). Metabolism: T. Horie *et al., Xenobiotica* **11**, 197, 693 (1981).

Relative stereochemistry

Colorless needles.
THERAP CAT: Antihypertensive; diuretic.

9910. Tripelennamine. [91-81-6] *N*^{1},*N*^{1}-Dimethyl-*N*^{2}-(phenylmethyl)-*N*^{2}-2-pyridinyl-1,2-ethanediamine; 2-[benzyl(2-dimethylaminoethyl)amino]pyridine; *N*-benzyl-*N'*,*N'*-dimethyl-*N*-(2-pyridyl)ethylenediamine; *N*,*N*-dimethyl-*N'*-benzyl-*N'*-(α-pyridyl)-ethylenediamine; *β*-dimethylaminoethyl-2-pyridylbenzylamine; *β*-dimethylaminoethyl-2-pyridylaminotoluene. C$_{16}$H$_{21}$N$_3$; mol wt 255.37. C 75.25%, H 8.29%, N 16.45%. Prepn: C. P. Huttrer *et al., J. Am. Chem. Soc.* **68**, 1999 (1946); C. Djerassi *et al.,* US 2406594 (1946 to Ciba); R. J. Horclois, US 2502151 (1950 to Rhône-Poulenc). Crystal structure: M. Parvez, *Acta Crystallogr.* **C43**, 1408 (1987). Toxicity studies: D. P. Waller *et al., Clin. Toxicol.* **16**, 17 (1980). Comprehensive description: H. G. Piskorik, *Anal. Profiles Drug Subs.* **14**, 108-133 (1985).

Yellow oil. Amine odor. bp$_{0.1}$ 138-142°, bp$_{1.7}$ 185-190°, bp$_{20}$ 193-205°. n$_D^{25}$ 1.5759-1.5765. Miscible with water.

Hydrochloride. [154-69-8] Dehistin; Azaron; Pyribenzamine; PBZ; Vetibenzamina. C$_{16}$H$_{21}$N$_3$·HCl; mol wt 291.82. Crystals from ethyl acetate + methanol, mp 192-193°. Bitter taste, produces temporary numbness of the tongue. uv max (water): 244, 305 (ε 14470, 4780). One gram dissolves in 0.77 ml water, in 6 ml alcohol, in 6 ml chloroform, in about 350 ml acetone. Practically insol in benzene, ethyl acetate, ether. About neutral to litmus. pH of aq soln contg 25 mg/ml: 6.71; 50 mg/ml: 6.67; 100 mg/ml: 5.56. LD$_{50}$ in mice (mg/kg): 47 i.p. (Walker).

Citrate. [6138-56-3] C$_{16}$H$_{21}$N$_3$·C$_6$H$_8$O$_7$. Crystals, mp 106-110°. Less bitter than the hydrochloride. Freely sol in water, alc. Very slightly sol in ether. Practically insol in benzene, chloroform. 1% aq soln has a pH of 4.25.

THERAP CAT: Antihistaminic.
THERAP CAT (VET): Antihistaminic.

9911. Triphenylbismuth. [603-33-8] Triphenylbismuthine. C$_{18}$H$_{15}$Bi; mol wt 440.30. C 49.10%, H 3.43%, Bi 47.46%. Homoleptic Bi(III) reagent. Prepn from phenylmagnesium bromide and bismuth halides: P. Pfeiffer *et al., Ber.* **37**, 4620 (1904); A. Classen, O. Ney, *Z. Anorg. Allg. Chem.* **15**, 253 (1921). Thermal properties: W. V. Steele, *J. Chem. Thermodyn.* **11**, 187 (1979). Crystal structure: D. M. Hawley, G. Ferguson, *J. Chem. Soc. A* **1968**, 2059; P. G. Jones *et al., Z. Kristallogr.* **210**, 377 (1995). Synthetic applications in prepn of bismuth thiolates and carboxylates: P. C. Andrews *et al., J. Chem. Soc. Dalton Trans.* **2002**, 4634. Effects on radiopacity and flexural properties of polymeric dental resins: L. A. Lang *et al., J. Prosthodont.* **9**, 23 (2000).

Crystals from abs alcohol, mp 77.6°. bp_{14} 242°. d^{25} 1.585. Specific heat capacity: 0.75 J/gK. Sol in methylene chloride, chloroform, carbon tetrachloride, benzene, toluene, ether, THF, ethyl acetate. Insol in water, alcohols. Dec in acidic solvents. Also reported as fine white needles from acetonitrile, crystal d 1.95 (Hawley, Ferguson). LD_{50} orally in dogs: 180 g/kg (Andrews). *Protect from light, moisture and acidic vapors.*

Caution: Irritating to eyes, respiratory system.

USE: Radiopacifying agent in dental resins and medical devices. Polymerization catalyst; source of phenyl radicals under photolytic conditions. Catalyst for glycol cleavage; phenylating agent in organic synthesis.

9912. Triphenylcarbinol. [76-84-6] α,α-Diphenylbenzenemethanol; hydroxytriphenylmethane; triphenylmethanol; tritanol. $C_{19}H_{16}O$; mol wt 260.34. C 87.66%, H 6.19%, O 6.15%. Prepd by the action of phenylmagnesium bromide on benzophenone: Acree, *Ber.* **37**, 2755 (1904); Peters *et al., J. Am. Chem. Soc.* **47**, 452 (1925); Dubsky, Jacot-Guillarmod, *Helv. Chim. Acta* **53**, 1965 (1970); by the action of potassium permanganate on triphenylfluoro- or triphenylchloromethane: Blicke, *J. Am. Chem. Soc.* **46**, 1518 (1924). Convenient lab prepn from methyl benzoate and phenylmagnesium bromide: L. F. Fieser, *Experiments in Organic Chemistry* (Boston, 3rd ed., 1955) p 77. Absorption spectrum in alc: Orndorff *et al., J. Am. Chem. Soc.* **49**, 1543 (1927).

Trigonal crystals from benzene; d_4^0 1.199; mp 164.2°. Distills between 360 and 380° without decompn. Insol in water, petr ether. Easily sol in alc, ether, benzene. Sol in concd H_2SO_4 with an intensely yellow color, in glacial acetic acid without color.

9913. Triphenylene. [217-59-4] 9,10-Benzphenanthrene; 1,2,3,4-dibenznaphthalene; isochrysene. $C_{18}H_{12}$; mol wt 228.29. C 94.70%, H 5.30%. Occurs in coal tar: Kaffer, *Ber.* **68**, 1812 (1935). Synthesis from cyclohexanone: Mannich, *Ber.* **40**, 163 (1907); Nenitzescu, Curcaneanu, *Ber.* **70**, 346 (1937); from 9-phenanthrylmagnesium bromide and succinic anhydride: Bergmann, *J. Am. Chem. Soc.* **59**, 1441 (1937); from 1,2,3,4-tetrahydrophenanthrene: Bachmann, Struve, *J. Org. Chem.* **4**, 472 (1937); by reaction of 3 mols *o*-bromoiodobenzene in the presence of lithium: Heaney, Millar, *Org. Synth.* **40**, 105 (1960).

Long needles from alc or chloroform. Sublimes; d 1.302; mp 199°; bp 425°. Absorption spectrum: Clar, Lombardi, *Ber.* **65**, 1414 (1932). Solns have blue fluorescence. Does not react with maleic anhydride.

9914. Triphenylmethane. [519-73-3] 1,1',1''-Methylidenetrisbenzene; Tritan. $C_{19}H_{16}$; mol wt 244.34. C 93.40%, H 6.60%. Prepd by ether reduction of triphenylchloromethane obtained by reaction of carbon tetrachloride and benzene in the presence of aluminum chloride: Norris, *Org. Synth.* **4**, 81 (1925); from diphenylchloromethane and phenylmagnesium bromide: Sayles, Kharasch, *J. Org. Chem.* **26**, 4210 (1961). Absorption spectrum: Orndorff, *J. Am. Chem. Soc.* **49**, 1543 (1927); Anderson, *ibid.* **50**, 209 (1928); **51**, 1890 (1929)

Orthorhombic pyramidal, solvated crystals containing one mol benzene, mp 78.2°, dries on exposure to air. When dry, mp 93.4° (stable form; there are 2 metastable forms). d_4^{100} 1.0134; bp_{760} 360°; bp_{200} 239.7°; bp_{100} 228°; bp_{60} 221°; bp_{40} 215.5°; bp_{20} 206.8°; bp_{10} 197°; bp_5 188°; $bp_{1.0}$ 170°. n_D^{100} 1.59546. Very sol in ether, hot alcohol, chloroform. Sol in petr ether, benzene, CS_2. Slightly sol in glacial acetic acid. Soly (parts/100 parts satd soln (w/w) at 30°) 48.6 in chloroform; 12.5 in hexane; 53 in CS_2; 7.24 parts in benzene (19°).

9915. Triphenyl Phosphate. [115-86-6] Phosphoric acid triphenyl ester; triphenoxyphosphine oxide. $C_{18}H_{15}O_4P$; mol wt 326.29. C 66.26%, H 4.63%, O 19.61%, P 9.49%. Prepd from P_2O_5 and phenol: Prahl, **US 2805240** (1957); by reaction of triethyl phosphite with chloramine-T: Cadogan, Moulden, *J. Chem. Soc.* **1961**, 3079. Toxicology: Sutton *et al., Arch. Environ. Health* **1**, 33 (1960).

Nonflammable needles; mp 49-50°; bp_{11} 245°. Insol in water. Sol in benzene, chloroform, ether, acetone, moderately sol in alcohol.

Caution: Potential symptoms of overexposure are minor changes in blood enzymes. *See NIOSH Pocket Guide to Chemical Hazards* (DHHS/NIOSH 97-140, 1997) p 322.

USE: Noncombustible substitute for camphor in celluloid; rendering acetylcellulose, nitrocellulose, airplane "dope", etc., stable and fireproof; impregnating roofing paper; plasticizer in lacquers and varnishes.

9916. Triphenylphosphine. [603-35-0] $C_{18}H_{15}P$; mol wt 262.29. C 82.43%, H 5.76%, P 11.81%. Prepd from phenylmagnesium bromide and phosphorus trichloride: Pfeiffer, Pietsch, *Ber.* **37**, 4621 (1904); Sauvage, *Compt. Rend.* **139**, 675 (1904); Dodonon, Medox, *Ber.* **61**, 910 (1928); Denney *et al., J. Am. Chem. Soc.* **83**, 1729 (1961). Physical constants: Forward *et al., J. Chem. Soc.* **1949**, 5121.

Odorless monoclinic platelets or prisms from ether, mp 80.5°. bp >360° (in inert gas). d_4^{25} 1.194; d_4^{80}(liq) 1.075. Is triboluminescent. Freely sol in ether; sol in benzene, chloroform, glacial acetic acid; less sol in alcohol. Practically insol in water.

USE: In organic synthesis; polymerization initiator.

9917. Triphenylphosphine Dibromide. [1034-39-5] Dibromotriphenylphosphorane; bromotriphenylphosphonium bromide; triphenylphosphorus-dibromine; triphenyldibromophosphorane; TPPDB; PPh_3Br_2. $C_{18}H_{15}Br_2P$; mol wt 422.10. C 51.22%, H 3.58%, Br 37.86%, P 7.34%. Versatile brominating reagent. The molecular charge transfer species exists in the solid state as the four-coordinate spoke compd, Ph_3P—Br—Br; the adduct ionizes in dichloromethane to form $[Ph_3PBr]^+Br^-$. Prepn and use in formation of alkyl and acyl bromides: L. Horner *et al., Ann.* **626**, 26 (1959).

Prepn and use in ether cleavage: A. G. Anderson, Jr., F. J. Freenor, *J. Org. Chem.* **37**, 626 (1972). Prepn in situ and use in synthesis of phosphoranes: I. Mathieu-Pelta, S. A. Evans, Jr., *ibid.* **59**, 2234 (1994). Synthetic utility in aziridine synthesis: I. Okada *et al.*, *Bull. Chem. Soc. Jpn.* **43**, 1185 (1970); in alcohol oxidations: A. Bisai *et al.*, *Tetrahedron Lett.* **43**, 8355 (2002); in aziridine ring openings: M. Kumar *et al.*, *ibid.* **50**, 363 (2009); in esterification reactions: C. Salomé, H. Kohn, *Tetrahedron* **65**, 456 (2009). Solid state X-ray crystallography studies of the molecular four-coordinate spoke structure: N. Bricklebank *et al.*, *J. Chem. Soc. Chem. Commun.* **1992**, 355. *Review*: Garima, *Synlett* **2010**, 1426-1427.

Colorless crystalline solid, mp 235° (dec). *Corrosive.* d 1.650. Sol in dichloromethane, acetonitrile, benzonitrile, DMF; slightly sol in benzene, chlorobenzene. Forms light yellow suspension in diethyl ether. Typically prepared as soln or suspension immediately before use. Hygroscopic; may decompose upon exposure to moisture.

USE: Reagent in synthetic organic chemistry.

9918. Triphenyltetrazolium Chloride. [298-96-4] 2,3,5-Triphenyl-2*H*-tetrazolium chloride (1:1); red tetrazolium; TPTZ; TTC; RT; VitaStain. $C_{19}H_{15}ClN_4$; mol wt 334.81. C 68.16%, H 4.52%, Cl 10.59%, N 16.73%. Prepn: H. v. Pechmann, P. Runge, *Ber.* **27**, 2920 (1894); Atkinson *et al.*, *Science* **111**, 385 (1950); R. Kuhn, D. Jerchel, *Ber.* **74**, 945 (1941); R. Price, *J. Chem. Soc. A* **1971**, 3379. Review of prepn and use of tetrazolium salts as indicators in biochemical and biological oxidation-reduction processes: W. Ried, *Angew. Chem.* **14**, 391 (1952).

Solvated nearly colorless needles from alcohol or chloroform, dry at 105°. Turns yellow on exposure to light. Dec 243°. Sol in water, alcohol, acetone. Insol in ether. Oxidizes aldoses and ketoses, as well as other α-ketols, and is thereby reduced to a water-insoluble, deep red pigment, a triphenylformazan.

USE: In analytical chemistry as a sensitive reagent for reducing sugars, and to distinguish between α-ketols and simple aldehydes. For staining plant and animal tissue. Germination indicator in testing the ability of seeds to germinate: G. Lakon, *Ber. Dtsch. Bot. Ges.* **60**, 299, 434 (1942). In histochemical studies. In determination of antibiotics; of dehydrogenases.

9919. Triphenyltin Hydride. [892-20-6] Triphenylstannane; triphenylstannyl hydride; Ph_3SnH. $C_{18}H_{16}Sn$; mol wt 351.04. C 61.59%, H 4.59%, Sn 33.82%. Organotin reagent utilized in a variety of reactions involving free radical chemistry. Prepn: R. F. Chambers, P. C. Scherer, *J. Am. Chem. Soc.* **48**, 1054 (1926); G. J. M. van der Kerk *et al.*, *J. Appl. Chem.* **7**, 366 (1957); H. G. Kuivila, O. F. Beumel, Jr., *J. Am. Chem. Soc.* **83**, 1246 (1961). Synthetic applications in reduction reactions: E. J. Kupchik, R. J. Kiesel, *J. Org. Chem.* **31**, 456 (1966); K. C. Nicolaou *et al.*, *J. Am. Chem. Soc.* **109**, 2504 (1987); K. Nozaki *et al.*, *Bull. Chem. Soc. Jpn.* **64**, 2585 (1991); in hydrostannylations: *eidem*, *Tetrahedron* **45**, 923 (1989). Review of reductions by organotin hydrides: H. G. Kuivila, *Synthesis* **1970**, 499-509.

Slightly viscous, colorless liquid. *Poisonous.* $bp_{0.5}$ 168-172°; $bp_{0.1}$ 145-149°. Crystallizes when stored at 0°C. White needles from 90% methanol, mp 26-28°. d_4^{25} 1.3771. n_D^{25} 1.6342. Flash pt, closed cup: 235°F (113°C). Sol in many organic solvents. Moisture sensitive. *Protect from light.* Store at 2-8°C.

USE: Reagent in synthetic organic chemistry.

9920. Triphenyltin Hydroxide. [76-87-9] Hydroxytriphenylstannane; hydroxytriphenyltin; fentin hydroxide; fenolovo; Du-Ter. $C_{18}H_{16}OSn$; mol wt 367.04. C 58.90%, H 4.39%, O 4.36%, Sn 32.34%. Prepn by alkaline hydrolysis of $(C_6H_5)_3SnCl$: Kushlefsky *et al.*, *Inorg. Chem.* **2**, 187 (1963); of $(C_6H_5)_3SnI$: Poller, *J. Inorg. Nucl. Chem.* **24**, 593 (1963). GC determn in vegetable crops: H. H. Van den Broek *et al.*, *Analyst* **113**, 1237 (1988). Review of analytical methods: A. Van Rossum *et al.*, *Anal. Methods Pestic. Plant Growth Regul.* **11**, 227-246 (1980). *Reviews*: Ingham, *Chem. Rev.* **60**, 459-539 (1960); R. Bock, *Residue Rev.* **79**, 1-270 (1981). Review of toxicology and human exposure: *Toxicological Profile for Tin* (PB2006-100006, 2005) 426 pp.

Crystals. mp 122-123.5°. Thermally dec to $(C_6H_5)_4Sn$, $(C_6H_5)_2$-SnO and H_2O. Slightly sol in alcohol, toluene. Practically insol in water.

Acetate. [900-95-8] Triphenyltin acetate; acetoxytriphenyltin; acetatotriphenylstannane; fentin acetate; Brestan. $C_{20}H_{18}O_2Sn$; mol wt 409.07. Prepn from acetic acid and $(C_6H_5)_3SnOH$: van der Kirk, Luijten, *J. Appl. Chem.* **6**, 49 (1956); and $(C_6H_5)_3SnH$: Weber, Becker, *J. Org. Chem.* **27**, 1258 (1962). Small needles, mp 122-124°. Sol in ether, slightly sol in alcohol, benzene.

USE: Antifeeding compounds for insect pest control; non-systemic fungicide.

9921. Triphosgene. [32315-10-9] 1,1,1-Trichloromethanol 1,1′-carbonate; bis(trichloromethyl) carbonate; hexachlorodimethyl carbonate; trichloromethyl carbonate; BTC. $C_3Cl_6O_3$; mol wt 296.73. C 12.14%, Cl 71.68%, O 16.18%. Versatile substitute for phosgene, *q.v.*; reagent used for oxidations, chlorinations, and nucleophilic substitutions. Prepn: C. Councler, *Ber.* **13**, 1696 (1880); and synthetic applications: H. Eckert, B. Forster, *Angew. Chem. Int. Ed. Engl.* **26**, 894 (1987). Additional synthetic applications: M. J. Coghlan, B. A. Caley, *Tetrahedron Lett.* **30**, 2033 (1989); C. Palomo *et al.*, *J. Org. Chem.* **56**, 5948 (1991); R. Wilder, S. Mobashery, *ibid.* **57**, 2755 (1992). Crystal structure: A. M. Sorenson, *Acta Chem. Scand.* **25**, 169 (1971). *Reviews*: L. Cotarca *et al.*, *Synthesis* **1996**, 553-576; W. Su *et al.*, *Org. Prep. Proced. Int.* **36**, 499-547 (2004).

White crystalline solid, mp 81-83°. bp_{760} 203°; bp_{50} 124°; bp_{36} 117°; bp_{22} 105°. *Poisonous. Corrosive. Lachrymator.* d 1.78; d (calc) 2.01; d^{80} 1.629. Sol in methanol, ethanol, benzene, diethyl ether, hexane, tetrahydrofuran, ethyl acetate, chloroform. Dec slowly in cold water.

USE: Reagent in synthetic organic chemistry.

9922. Triprolidine. [486-12-4] 2-[(1*E*)-1-(4-Methylphenyl)-3-(1-pyrrolidinyl)-1-propen-1-yl]pyridine; *trans*-2-[3-(1-pyrrolidin-yl)-1-*p*-tolylpropenyl]pyridine; *trans*-1-(2-pyridyl)-3-pyrrolidino-1-*p*-tolylprop-1-ene; *trans*-1-(4-methylphenyl)-1-(2-pyridyl)-3-pyrrol-idinoprop-1-ene. $C_{19}H_{22}N_2$; mol wt 278.40. C 81.97%, H 7.97%, N 10.06%. Histamine H_1-receptor antagonist. Prepn: Adamson, **US 2712020**; **US 2712023** (both 1955 to Burroughs Wellcome); Adamson *et al.*, *J. Chem. Soc.* **1958**, 312. Structure-activity studies: Ison, Casy, *J. Pharm. Pharmacol.* **23**, 848 (1971). Crystal and molecular structure: James, Williams, *Can. J. Chem.* **52**, 1880 (1974). Pharmacokinetics and antihistaminic effects in humans: K. J. Simons *et al.*, *J. Allergy Clin. Immunol.* **77**, 326 (1986). Comprehensive description: S. A. Benezra, C.-H. Yang, *Anal. Profiles Drug Subs.* **8**, 509-528 (1979).

Crystals from light petr, mp 59-61°. uv max (ethanol): 236, 285 nm (ε 15300, 6800).

Hydrochloride monohydrate. [6138-79-0] 295C51; Actidil; Actidilon; Pro-Actidil; Pro-Entra; Venen. $C_{19}H_{22}N_2 \cdot HCl \cdot H_2O$; mol wt 332.87. Crystals from water, mp 116-118°. uv max (ethanol): 235, 283 nm (ε 15000, 7400). Moderately sol in water, ethanol, methanol.

Oxalate. $C_{19}H_{22}N_2 \cdot C_2H_2O_4$. Crystals from methanol, dec 173-174°. uv max (ethanol): 233, 283 nm (ε 16200, 8200).

THERAP CAT: Antihistaminic.

9923. Triptolide. [38748-32-2] (3b*S*,4a*S*,5a*S*,6*R*,6a*R*,7a*S*,-7b*S*,8a*S*,8b*S*)-3b,4,4a,6,6a,7a,7b,8b,9,10-Decahydro-6-hydroxy-8b-methyl-6a-(1-methylethyl)trisoxireno[4b,5:6,7:8a,9]phenanthro[1,2-*c*]furan-1(3*H*)-one; PG490. $C_{20}H_{24}O_6$; mol wt 360.41. C 66.65%, H 6.71%, O 26.63%. Diterpenoid triepoxide with immunosuppressant and antitumor properties; isolated from the Chinese medicinal herb, *Tripterygium wilfordii* Hook. f., *Celastraceae*, commonly known as thunder god vine, *q.v.* Isoln and structure determn: S. M. Kupchan *et al.*, *J. Am. Chem. Soc.* **94**, 7194 (1972). Synthesis of racemate: R. S. Buckanin *et al.*, *ibid.* **102**, 1200 (1980); of naturally occurring (−)-form: D. Yang *et al.*, *ibid.* **121**, 5579 (1999). HPLC and SPE determn in *Tripterygium* root extracts: A. M. Brinker, I. Raskin, *J. Chromatogr. A* **1070**, 65 (2005). Animal studies of male antifertility effects: A. P. S. Hikim *et al.*, *J. Androl.* **21**, 431 (2000); P. N. Huynh *et al.*, *ibid.* 689. Mechanistic studies of apoptotic and antitumor activities: W.-T. Chang *et al.*, *J. Biol. Chem.* **276**, 2221 (2001); T. M. Kiviharju *et al.*, *Clin. Cancer Res.* **8**, 2666 (2002); S. J. Leuenroth, C. M. Crews, *Chem. Biol.* **12**, 1259 (2005). Review of anti-inflammatory and immunosuppressive properties: B. J. Chen, *Leuk. Lymphoma* **42**, 253-265 (2001).

mp 226-227°. $[\alpha]_D^{25}$ −154° (c = 0.369 in CH_2Cl_2). uv max (ethanol): 218 nm (ε 14000). Sol in DMSO, ethanol. LD_{50} i.p. in mice: 1.93 mg/kg (Chen).

14-Succinyl triptolide sodium salt. [195883-09-1] PG490-88. $C_{24}H_{27}NaO_9$; mol wt 482.46. Semi-synthetic water soluble prodrug. Prepn: Y. M. Qi, J. H. Musser, **US 5663335** (1997 to Phar-

magenesis). Animal model studies of prevention of graft-vs-host disease: B. J. Chen *et al.*, *Transplantation* **70**, 1442 (2000); of anti-tumor effects: J. M. Fidler *et al.*, *Mol. Cancer Ther.* **2**, 855 (2003); of allograft survival in organ transplantation: F. Pan *et al.*, *Transplant. Proc.* **37**, 134 (2005).

9924. Triptorelin. [57773-63-4] 6-D-Tryptophanluteinizing hormone-releasing factor (swine); 6-D-tryptophan-LH-RH; D-trp⁶-LHRH; D-Trp⁶LRH; D-trp⁶-gonadorelin; détryptoréline; AY-25650; Wy-42462; Wy-42422. $C_{64}H_{82}N_{18}O_{13}$; mol wt 1311.47. C 58.61%, H 6.30%, N 19.22%, O 15.86%. Synthetic decapeptide agonist analog of gonadotropin-releasing hormone, *q.v.* Prepn: A. V. Schally, D. H. Coy, **DE 2625843**; *eidem*, **US 4010125** (1976, 1977); D. H. Coy *et al.*, *J. Med. Chem.* **19**, 423 (1976). Comparison with LH-RH of *in vitro* activity: D. H. Coy *et al.*, *Biochem. Biophys. Res. Commun.* **67**, 576 (1975). Pharmacokinetics and metabolism in humans: J. L. Barron *et al.*, *J. Clin. Endocrinol. Metab.* **54**, 1169 (1982). HPLC analysis: D. C. Serti *et al.*, *J. Liq. Chromatogr.* **4**, 1135 (1981). RIA in human serum: M. Mason-Garcia *et al.*, *Proc. Natl. Acad. Sci. USA* **82**, 1547 (1985). Clinical trial for *in vitro* fertilization: A. Hazout *et al.*, *Fertil. Steril.* **59**, 596 (1993). Clinical trial in prostate cancer: C. F. Heyns *et al.*, *BJU Int.* **92**, 226 (2003); in severe endometriosis: A. Y. K. Wong, L. Tang, *Fertil. Steril.* **81**, 1522 (2004).

5-oxoPro–His–Trp–Ser–Tyr–D-Trp–Leu–Arg–Pro–GlyNH₂

Fluffy, white solid. $[\alpha]_D^{23}$ −58.8° (c = 0.33 in acetic acid).

Acetate. [140194-24-7] Gonapeptil. $C_{64}H_{82}N_{18}O_{13} \cdot C_2H_4O_2$; mol wt 1371.53.

Pamoate. [124508-66-3] Decapeptyl; Salvacyl; Trelstar. $C_{64}H_{82}N_{18}O_{13} \cdot C_{23}H_{16}O_6$; mol wt 1699.85.

THERAP CAT: Antineoplastic (hormonal); in treatment of endometriosis and infertility.

9925. Triptycene. [477-75-8] 9,10-Dihydro-9,10-[1′,2′]-benzenoanthracene; tribenzobicyclo[2.2.2]octatriene. $C_{20}H_{14}$; mol wt 254.33. C 94.45%, H 5.55%. Synthesis by three different methods: Bartlett *et al.*, *J. Am. Chem. Soc.* **64**, 2649 (1942); Wittig, *Org. Synth.* **39**, 75 (1959); Friedman, Logullo, *J. Am. Chem. Soc.* **85**, 1549 (1963).

Crystals from cyclohexane or methylcyclohexane, mp 253-254°.

9926. 2,4,6-Tripyridyl-*s*-triazine. [3682-35-7] 2,4,6-Tri-2-pyridinyl-1,3,5-triazine; 2,4,6-tri-2-pyridinyl-*s*-triazine; tripyridyl-triazine; TPTZ. $C_{18}H_{12}N_6$; mol wt 312.34. C 69.22%, H 3.87%, N 26.91%. Synthesis: Case, Kroft, *J. Am. Chem. Soc.* **81**, 905 (1959). Preparation: Schaefer, **US 3294798** (1966 to Am. Cyanamid). Thermal stability data: Johns *et al.*, *J. Chem. Eng. Data* **7**, 227 (1962). *Review:* "2,4,6-Tripyridyl-*s*-triazine" in Diehl *et al.*, *The Iron Reagents* (The G. Frederick Smith Chem. Co., Columbus, Ohio, 1965) pp 41-56.

Crystals, mp 210-220° (Schaefer); trihydrate from aqueous ethanol, mp 244-245° (Case, Kroft). Reacts with ferrous ions to yield intense violet color over pH range 3.4-5.8. Absorption max $Fe(TPTZ)_2^{2+}$ (water): 593 nm (ε 22600), Collins *et al.*, *Anal. Chem.* **31**, 1862 (1959).

USE: Reagent for the spectrophotometric determn of iron.

9927. Tris-BP. [126-72-7] 2,3-Dibromo-1-propanol 1,1',1''-phosphate; phosphoric acid tris(2,3-dibromopropyl) ester; tris(2,3-dibromopropyl) phosphate; Apex 462-5; Flammex AP; Flammex T 23P; Firemaster LV-T 23P; Firemaster T 23P; T 23P; Fyrol HB 32. $C_9H_{15}Br_6O_4P$; mol wt 697.61. C 15.50%, H 2.17%, Br 68.72%, O 9.17%, P 4.44%. Prepn: G. E. Walter, I. Hornstein, US 2574515 (1951 to Glenn. L. Martin Co.); D. E. Overbeek, R. C. Nametz, US 3046297 (1962 to Michigan Chem. Co.); R. W. Rimmer, US 3223755 (1965 to duPont). Use in flameproofing: W. D. Paist, N. Van Gorder, US 2662834 (1953 to Celanese). Mutagenicity studies: M. J. Prival *et al.*, *Science* **195**, 76 (1977); A. Nakamura *et al.*, *Mutat. Res.* **66**, 373 (1979). Carcinogenicity studies: B. L. Van Duuren *et al.*, *Cancer Res.* **38**, 3236 (1978); G. Reznik *et al.*, *J. Natl. Cancer Inst.* **63**, 205 (1979). Review of toxicology: F. A. Daniher, *Proc. Symp. Text. Flammability* **4**, 126-143 (1976). *Review:* A. Blum, B. N. Ames, *Science* **195**, 17 (1977).

Viscous liquid. LD_{50} orally in rats: >5.0 g/kg (Daniher).
Caution: This substance is reasonably anticipated to be a human carcinogen: *Report on Carcinogens, Twelfth Edition* (PB2011-111646, 2011) p 428.
USE: Flame retardant. Formerly used in children's sleepwear.

9928. Tris(ethylenediamine)cadmium Dihydroxide. [14874-24-9] (OC-6-11)-Tris(1,2-ethanediamine-$\kappa N^1,\kappa N^2$)cadmium(2+) hydroxide (1:2); tris(ethylenediamine)cadmium hydroxide; tri(en)cadmium hydroxide; Cadoxen. $C_6H_{26}CdN_6O_2$; mol wt 326.73. C 22.06%, H 8.02%, Cd 34.40%, N 25.72%, O 9.79%. $[Cd(H_2NCH_2CH_2NH_2)_3](OH)_2$. Prepared by shaking a given amount of cadmium oxide in 10 times its wt of 30% aq ethylenediamine soln for 15 minutes and centrifuging; the supernatant liquor is the product: Jayme, Neuschäfer, *Naturwissenschaften* **44**, 62 (1957). The soln contains about 4.5% Cd (w/w) and dissolves about 3% (w/w) cellulose, giving a clear, highly viscous soln.
USE: Solvent for cellulose: Jayme, DE 1079318 (1960 to E. Merck), *C.A.* **55**, 18107d (1961); solvent for sulfite pulps.

9929. Tris(hydroxymethyl)nitromethane. [126-11-4] 2-(Hydroxymethyl)-2-nitro-1,3-propanediol; 2-nitro-2-(hydroxymethyl)-1,3-propanediol; trimethylolnitromethane. $C_4H_9NO_5$; mol wt 151.12. C 31.79%, H 6.00%, N 9.27%, O 52.93%. Prepn from trioxymethylene and nitromethane: Boileau, *Mem. Poudres* **35**, Annexe 7-76 (1953).

Crystals from ethyl acetate + benzene, mp 214° (pure); mp 180° (usual laboratory product); mp 175-176° (tech). Soly in water at 20°: 220 g/100 ml. Freely sol in alcohols. Sparingly sol in benzene, other hydrocarbons. pH of 0.1M aq soln 4.5.
Caution: Irritating to skin, mucous membranes.
USE: Bactericide for inanimate objects. To inhibit bacterial growth in circulating industrial water systems, cutting oils, nonprotein glues and sizings.

9930. Trisilane. [7783-26-8] Trisilicopropane; trisilicon octahydride; silicopropane; trisilicane. H_8Si_3; mol wt 92.32. H 8.73%, Si 91.26%. Si_3H_8. Obtained by separation of mixed silanes prepared from magnesium silicide and hydrochloric acid: Stock, Somiesky, *Ber.* **49**, 111 (1916); **54B**, 524 (1921); **56B**, 247 (1923); Culbertson, US 2551571 (1951 to Union Carbide); prepared by conversion of silane to higher silanes in an ozonizer type of electric discharge: Spanier, MacDiarmid, *Inorg. Chem.* **1**, 432 (1962).
Liquid. mp −117.4°; bp 52.9°; d^0 0.743; vapor pressure 95.5 mm Hg at 0°. Much less stable than silane or disilane. Detonates in air. Dec in water. Reacts vigorously with CCl_4 and $CHCl_3$.

9931. Tris(pentafluorophenyl)boron. [1109-15-5] Tris(2,-3,4,5,6-pentafluorophenyl)borane. $C_{18}BF_{15}$; mol wt 511.98. C 42.23%, B 2.11%, F 55.66%. Lewis acid catalyst. Prepn and characterization: A. G. Massey, A. J. Park, *J. Organomet. Chem.* **2**, 245 (1964); and NMR study of adducts: *eidem, ibid.* **5**, 218 (1966). Catalytic role in olefin polymerization: X. Yang *et al.*, *J. Am. Chem. Soc.* **116**, 10015 (1994). Lewis acid properties: H. Jacobsen *et al.*, *Organometallics* **18**, 1724 (1999). Prepn and properties of $B(C_6F_5)_3$ radical anion: R. J. Kwaan *et al.*, *ibid.* **20**, 3818 (2001). Review of chemistry and uses: W. E. Piers, T. Chivers, *Chem. Soc. Rev.* **26**, 345-354 (1997).

White solid, mp 126-128°. Sol in many organic solvents. Monomeric in benzene at 30°. Hygroscopic. Thermally stable.
USE: Catalyst in organic transformations; cocatalyst with group IV metallocene alkyls in olefin polymerization.

9932. Tristearin. [555-43-1] Octadecanoic acid 1,1',1''-(1,-2,3-propanetriyl) ester; stearin; glyceryl tristearate. $C_{57}H_{110}O_6$; mol wt 891.50. C 76.79%, H 12.44%, O 10.77%. Present in many animal and vegetable fats, especially the hard ones like cacao butter and tallow. Prepd from stearic acid and glycerol in the presence of Al_2O_3: Ingram, GB 663566 (1951 to I.C.I.); by catalytic hydrogenation of many oils: *Bailey's Industrial Oil and Fat Products* (Wiley, New York, 3rd ed., 1964) pp 881-882.
White powder; d_4^{80} 0.862; mp ~55°; on further heating solidifies and melts again at 72°. n_D^{80} 1.4385. Insol in water. Sol in benzene, chloroform, hot alcohol; almost insol in cold alcohol, ether, petr ether.
USE: In textile sizes. Formerly in making candles.

9933. Tris(tribromophenoxy)triazine. [25713-60-4] 2,4,6-Tris(2,4,6-tribromophenoxy)-1,3,5-triazine; tris(tribromophenyl)cyanurate; TTA; FR-245. $C_{21}H_6Br_9N_3O_3$; mol wt 1067.43. C 23.63%, H 0.57%, Br 67.37%, N 3.94%, O 4.50%. Brominated flame retardant for polystyrene plastics used in electronics. Prepn: J.-P. Zimmermann, FR 1566675 (1968 to Soc. d'Etudes Chim. Indust. Agric.). Manufacturing process: H. Klinkenberg *et al.*, US 4039538 (1977 to Dynamit Nobel). Crystal structure: F. Li *et al.*, *Acta Crystallogr.* **E62**, o3303 (2006). Flame retardancy and compatibility with poly(butylene terephthalate) composites: J. Chen *et al.*, *J. Appl. Polym. Sci.* **102**, 1291 (2006). Thermal properties: B. A. Howell, *J. Therm. Anal. Calorim.* **89**, 373 (2007).

Crystals from boiling methyl cellosolve, mp 229-231°. Dec 398°. d 2.44. Soly at 25° (g/100g): toluene 28; methyl ethyl ketone 4;

methyl acetate 3; carbon tetrachloride 17; methanol 0.2. Insol in water.

USE: Additive flame retardant for styrenic copolymers.

9934. Tris(trimethylsilyl)silane. [1873-77-4] 1,1,1,3,3,3-Hexamethyl-2-(trimethylsilyl)trisilane. $C_9H_{28}Si_4$; mol wt 248.66. C 43.47%, H 11.35%, Si 45.18%. Radical based mediator. Prepn: H. Gilman *et al., Chem. Ind. (London)* **1965**, 848; H. Bürger, W. Kilian, *J. Organomet. Chem.* **18**, 299 (1969). Si-H bond dissociation energies: J. M. Kanabus-Kaminska *et al., J. Am. Chem. Soc.* **109**, 5268 (1987). Identification as reducing agent: C. Chatgilialoglu *et al., J. Org. Chem.* **53**, 3641 (1988). Use as a reducing agent: M. Ballestri *et al., ibid.* **56**, 678 (1991); D. H. R. Barton *et al., Tetrahedron Lett.* **33**, 6629 (1992); in carbonylation: I. Ryu *et al., Synlett* **1993**, 143.

Liquid, bp_{8mm} 80-83°.
USE: Reducing and hydrosilylating agent.

9935. Tristriphenylphosphine Rhodium Carbonyl Hydride. [17185-29-4] Carbonylhydrotris(triphenylphosphine)rhodium; hydridocarbonyltris(triphenylphosphine)rhodium(I). $C_{55}H_{46}OP_3Rh$; mol wt 918.80. C 71.90%, H 5.05%, O 1.74%, P 10.11%, Rh 11.20%. HRh(CO)[P(C_6H_5)$_3$]$_3$. Prepn: Bath, Vaska, *J. Am. Chem. Soc.* **85**, 3500 (1963); Levison, Robinson, *J. Chem. Soc. A* **1970**, 2947; Ahmad *et al., J. Chem. Soc. Dalton Trans.* **1972**, 843; Ahmad *et al., Inorg. Synth.* **15**, 45 (1974). Structure: LaPlaca, Ibers, *J. Am. Chem. Soc.* **85**, 3501 (1963); *eidem, Acta Crystallogr.* **18**, 511 (1965). Chemistry: Evans *et al., J. Chem. Soc. A* **1968**, 2660; O'Connor, Wilkinson, *ibid.* 2665.

Yellow microcrystals, d 1.33. mp 120-122° in air; 172-174° under nitrogen. Moderately sol in benzene, chloroform, dichloromethane; sparingly sol in cyclohexane; insol in light petroleum. Dissociates extensively in organic solvents.
USE: Catalyst.

9936. Trithiocarbonic Acid. [594-08-1] Carbonotrithioic acid; dihydrogen thiocarbonate. CH_2S_3; mol wt 110.21. C 10.90%, H 1.83%, S 87.27%. Prepared by the treatment of $BaCS_3$ with ice-cold 10% HCl: v. Halban *et al., Z. Elektrochem.* **29**, 445 (1923); Mills, Robinson, *J. Chem. Soc.* **1928**, 2326; Gattow, Krebs, *Angew. Chem.* **74**, 29 (1962).

Highly refractive red oil. mp −26.9°. bp 57.8°. d_4^{20} 1.483; d_4^{25} 1.476. n_D^{20} 1.8225. Dec by water, alcohol. Addition of sulfur produces tetrathiocarbonic acid CH_2S_4.

9937. Trithiozine. [35619-65-9] 4-Morpholinyl(3,4,5-trimethoxyphenyl)methanethione; 4-(3,4,5-trimethoxythiobenzoyl)-morpholine; sulmetozine (rescinded INN); tritiozine; ISF-2001; Tresanil. $C_{14}H_{19}NO_4S$; mol wt 297.37. C 56.55%, H 6.44%, N 4.71%, O 21.52%, S 10.78%. Non-anticholinergic gastric secretion inhibitor. Prepn: G. Pifferi, *DE 2102246; idem, US 3862138* (1972, 1975 both to ISF). Synthesis and pharmacology: G. Pifferi *et al., Chim. Ther.* **8**, 462 (1973). HPLC determn in plasma and urine: T. Crolla *et al., J. Chromatogr.* **222**, 257 (1981). Clinical studies: M. Elakovic *et al., J. Int. Med. Res.* **8**, 347 (1980); U. Marini *et al., Clin. Ter.* **92**, 399 (1980). Evaluation of gastric acid suppression: K. Gibinski, *Curr. Med. Res. Opin.* **7**, 516 (1981).

Pale yellow solid from ethanol, mp 141-143°. LD_{50} i.p. in mice: 2000 mg/kg (Pifferi, 1973).
THERAP CAT: Antisecretory (gastric).

9938. Triticum. Agropyrum; couch grass; dog grass; graminis; quick grass. Dried rhizome and roots of *Agropyron repens* L., Beauv., *Gramineae*, gathered in spring. *Habit.* Europe, Northern Asia; naturalized in the U.S. *Constit.* Triticin, glucose, mannite, inosite.

9939. Tritium. [10028-17-8] Triterium. T or $^3_1$H. at. wt 3.016. Exists in the diatomic state, T_2; mol wt 6.032. Naturally occurring radioactive isotope of hydrogen, *q.v.*, ($T_{1/2}$ 12.32 yr, low energy β^- emitter). Under normal conditions the total atmospheric content of molecular T_2 gas is only 11 g. First prepd by the bombardment of deuterophosphoric acid with fast deuterons: M. L. E. Oliphant *et al., Proc. Roy. Soc.* **A144**, 692 (1934); T. W. Bonner, *Phys. Rev.* **53**, 711 (1938). Produced commercially from ^{6}Li by slow neutron bombardment: $^6_3$Li + ^1_0n → $^3_1$H + $^4_2$He. Prepn by neutron irradiation of lithium fluoride: Jenks *et al., US 3079317* (1963). *Reviews:* E. A. Evans, *Tritium and Its Compounds* (Butterworth, London, 1966) 441 pp; Mackay, Dove, "Deuterium and Tritium" in *Comprehensive Inorganic Chemistry* **vol. 1**, J. C. Bailar Jr. *et al.,* Eds. (Pergamon Press, Oxford, 1973) pp 77-116; *Chemistry of the Elements* N. N. Greenwood, A. Earnshaw, Eds. (Pergamon Press, New York, 1984) pp 38-74; J. J. Katz in *Kirk-Othmer Encyclopedia of Chemical Technology* **vol. 8** (Wiley-Interscience, New York, 4th ed., 1993) pp 17-30.

Gas, having properties similar to hydrogen. mp −254.54° (20.62 K) at 162 mm (triple point). bp −248.12° (25.04 K). Crit temp −232.56°. Crit press. 18.317 atm. Molar density of liquid: 45.35 moles/l (20.62 K).
USE: In fusion-based thermonuclear weapons (hydrogen bombs). Energy is released by deuteron bombardment according to the reaction: $^3_1$H + $^2_1$H → $^4_2$He + ^1_0n + 18 meV. Widely used as a radioactive tracer in chemical, biochemical, biological and hydrological research.

9940. Tritolyl Phosphate. [1330-78-5] Phosphoric acid tris(methylphenyl) ester; tricresyl phosphate; TCP; PX-917; Celluflex 179; Kronitex TCP; Lindol. $C_{21}H_{21}O_4P$; mol wt 368.37. C 68.47%, H 5.75%, O 17.37%, P 8.41%. A mixture of isomeric tritolyl phosphates, usually excluding the very toxic *ortho*-isomer as much as possible. Prepd from cresol and phosphoric oxychloride, phosphoric acid or pentachloride: Prahl, *US 2805240* (1957); Bondy, Gumb, *GB 890642* (1962 to Coalite and Chem. Prod.); *Faith, Keyes & Clark's Industrial Chemicals*, F. A. Lowenheim, M. K. Moran, Eds. (Wiley-Interscience, New York, 4th ed., 1975) pp 849-853.

Oily, flame resistant liquid. d_{25}^{25} 1.16; bp_{10} ~265°. Pour point −28°. n_D^{25} 1.55. Insol in water (<0.002% at 85°). Sp heat 0.38. Misc with all the common organic solvents and thinners, linseed oil, china wood oil, castor oil.
USE: As plasticizer in vinyl plastics manuf, as flame-retardant, solvent for nitrocellulose, in cellulosic molding compositions, as additive to extreme pressure lubricants, as a nonflammable fluid in hydraulic systems, as lead scavenger in gasoline: Yust, Bame, *US 2889212* (1959 to Shell); to sterilize certain surgical instruments.

9941. Tri-*o*-tolyl Phosphate. [78-30-8] Phosphoric acid tris(2-methylphenyl) ester; tri-*o*-cresyl phosphate. $C_{21}H_{21}O_4P$; mol wt 368.37. C 68.47%, H 5.75%, O 17.37%, P 8.41%.

Colorless or pale yellow liquid. *Poisonous.* mp 25.6°. bp ~410° with slight decompn. Flash pt 225°C. d 1.1955. n 1.5575. Sparingly sol in water, slightly sol in alcohol, sol in ether.

Caution: Potential symptoms of overexposure are GI disturbance; peripheral neuropathy; cramps in calves, paresthesia in feet or hands; weak feet, wrist drop and paralysis. *See NIOSH Pocket Guide to Chemical Hazards* (DHHS/NIOSH 97-140, 1997) p 322. Ingestion may cause acute nausea, vomiting, diarrhea. Following a latent period, polyneuritis progressing to paralysis of extremities has been seen in severe poisoning. *See Clinical Toxicology of Commercial Products,* R. E. Gosselin *et al.,* Eds. (Williams & Wilkins, Baltimore, 5th ed.) Section III, pp 388-393; *Patty's Industrial Hygiene and Toxicology* **vol. 2D,** G. D. Clayton, F. E. Clayton, Eds. (John Wiley & Sons, New York, 4th ed., 1994) pp 3063-3085.

USE: As plasticizer in lacquers and varnishes.

9942. Tritoqualine. [14504-73-5] 7-Amino-4,5,6-triethoxy-3-(5,6,7,8-tetrahydro-4-methoxy-6-methyl-1,3-dioxolo[4,5-*g*]isoquinolin-5-yl)-1(3*H*)-isobenzofuranone; 7-amino-4,5,6-triethoxy-3-(5,6,7,8-tetrahydro-4-methoxy-6-methyl-1,3-dioxolo[4,5-*g*]isoquinolin-5-yl)phthalide; tritocaline; L-554; Hypostamine; Inhibostamin; Livalfa. $C_{26}H_{32}N_2O_8$; mol wt 500.55. C 62.39%, H 6.44%, N 5.60%, O 25.57%. Prepn: **FR 1295309** (1962 to Lab. de Recherches Biol. Laborec). Activity: Hahn *et al., Arzneim.-Forsch.* **20,** 1490 (1970).

Crystals, mp 183°.
THERAP CAT: Antihistaminic.

9943. Tritosulfuron. [142469-14-5] *N*-[[[4-Methoxy-6-(trifluoromethyl)-1,3,5-triazin-2-yl]amino]carbonyl]-2-(trifluoromethyl)benzenesulfonamide; 1-(4-methoxy-6-trifluoromethyl-1,3,5-triazin-2-yl)-3-(2-trifluoromethylbenzenesulfonyl)urea; BAS-635; Biathlon; Tooler. $C_{13}H_9F_6N_5O_4S$; mol wt 445.30. C 35.06%, H 2.04%, F 25.60%, N 15.73%, O 14.37%, S 7.20%. Post-emergence sulfonylurea herbicide for use in food crops; acetolactate synthase inhibitor. Prepn: H. Mayer *et al.,* **DE 4038430**; *eidem,* **US 5478798** (1992, 1995 both to BASF). Skin penetration study: B. van Ravenzwaay, E. Leibold, *Hum. Exp. Toxicol.* **23,** 421 (2004). Toxicity, metabolism, and environmental impact: D. Barcelo Culleres *et al., EFSA J.* **621,** 2 (2007).

Crystals, mp 164-169°. Log P (octanol/water): 2.9. Soly in water (20°): 44 mg/l. LD50 in rats (mg/kg): 4700 orally; >2000 dermally. LC50 (4 hr) in rats: >5.9 mg/l (Barcelo Culleres).

USE: Herbicide.

9944. Trityl Chloride. [76-83-5] 1,1′,1″-(Chloromethylidyne)trisbenzene; chlorotriphenylmethane; triphenylchloromethane; triphenylmethyl chloride; TPCM. $C_{19}H_{15}Cl$; mol wt 278.78. C 81.86%, H 5.42%, Cl 12.72%. Reagent for the introduction of the triphenylmethyl (trityl) protecting group, commonly used to protect alcohols as trityl ethers. Prepn: W. Hemilian, *Ber.* **7,** 1203 (1874); W. E. Bachmann, *Org. Synth.* **coll. vol. III,** 841 (1955). Crystal structure analysis: R. Gerdil, A. Dunand, *Acta Crystallogr.* **B31,**

936 (1975); A. Dunand, R. Gerdil, *ibid.* **B38,** 570 (1982). Polymorphism studies: A. H. Brunetti, *Solid State Nuc. Magn. Reson.* **25,** 167 (2004). Hydroxyl group protection: S. K. Chaudhary, O. Hernandez, *Tetrahedron Lett.* **20,** 95 (1979); S. Colin-Messager *et al., ibid.* **33,** 2689 (1992). Additional synthetic applications: M. S. Newman, C. H. Chen, *J. Am. Chem. Soc.* **95,** 278 (1973); A. Khalafi-Nezhad *et al., Synthesis* **2008,** 617.

Colorless crystals, mp 111-112°. bp20 230-235°. d 1.26. *Corrosive. Lachrymator.* Sol in most organic solvents. Moisture sensitive.

USE: Reagent and catalyst in synthetic organic chemistry.

9945. Triuret. [556-99-0] Diimidotricarbonic diamide; *N,N′*-bis(aminocarbonyl)urea; 1,3-dicarbamylurea; carbonyldiurea. $C_3H_6N_4O_3$; mol wt 146.11. C 24.66%, H 4.14%, N 38.35%, O 32.85%. Prepared by the action of phosgene on urea: Schiff, *Ann.* **291,** 374 (1896); Blair, *J. Am. Chem. Soc.* **48,** 101 (1926); Haworth, Mann, *J. Chem. Soc.* **1943,** 603; Werner, Gray, *Sci. Proc. R. Dublin Soc.* **24,** 111 (1946).

Crystals from ammonia water, dec 233°. Freely sol in liquid ammonia. Forms a mono- and dipotassium salt, *see* Blair, *loc. cit.*

9946. Troclosene Potassium. [2244-21-5] 1,3-Dichloro-1,-3,5-triazine-2,4,6(1*H*,3*H*,5*H*)-trione potassium salt (1:1); 3,5-dichlorotetrahydro-2,4,6-trioxo-*s*-triazin-1(2*H*)-yl potassium; potassium troclosene; potassium dichloroisocyanurate; ACL-59. C_3Cl_2-KN_3O_3; mol wt 236.05. C 15.26%, Cl 30.04%, K 16.56%, N 17.80%, O 20.33%. Structure studies and prepn: Petterson *et al., J. Org. Chem.* **25,** 1595 (1960). Prepn from trisodium isocyanurate and gaseous Cl: Symes *et al.,* **US 3035056; US 3035057** (both 1962 to Monsanto).

USE: Useful source of available Cl in solid bleach and detergent formulations.
THERAP CAT: Anti-infective (topical).

9947. Trofosfamide. [22089-22-1] *N,N,3*-Tris(2-chloroethyl)tetrahydro-2*H*-1,3,2-oxazaphosphorin-2-amine 2-oxide; 2-[bis(2-chloroethyl)amino]-3-(2-chloroethyl)tetrahydro-2*H*-1,3,2-oxaphosphorine 2-oxide; *N,N,N′*-tris(2-chloroethyl)-*N′,O*-propylene phosphoric acid ester diamide; trilophosphamide; trophosphamide; NSC-109723; Z-4828; Ixoten. $C_9H_{18}Cl_3N_2O_2P$; mol wt 323.58. C 33.41%, H 5.61%, Cl 32.87%, N 8.66%, O 9.89%, P 9.57%. The 3-(2-chloroethyl) deriv of cyclophosphamide, *q.v.* Prepn: **GB 1188159**; Arnold *et al.,* **DE 2107936,** (1970, 1972 both to Asta-Werke), *see C.A.* **73,** 44892d (1970) and *C.A.* **77,** 152238m (1972); K. Pankiewicz *et al., J. Am. Chem. Soc.* **101,** 7712 (1979). Pharmacology: N. Brock, *Int. Congr. Chemother., Proc. 5th,* K. H. Spitzy, H. Haschek, Eds. (Verlag Wiener Med. Akad., Vienna, 1967) **II** (1), pp 155-161; J. Potel, N. Brock, *Arzneim.-Forsch.* **21,** 1250 (1971); N. Brock, J. Potel, *ibid.* **24,** 1149 (1974); Harrison, Fuquay, *Proc. Soc. Exp. Biol. Med.* **139,** 957 (1972). Crystal and molecular structure: A. Perales, S. Garcia-Blanco, *Acta Crystallogr.* **33B**, 1939 (1977).

Crystals from ether, mp 50-51°. $[\alpha]_D^{25}$ −28.6° (c = 2 in CH$_3$OH). LD$_{50}$ i.p. in mice: 212 mg/kg (Brock, Potel).

THERAP CAT: Antineoplastic.

9948. Troglitazone. [97322-87-7] 5-[[4-[(3,4-Dihydro-6-hydroxy-2,5,7,8-tetramethyl-2*H*-1-benzopyran-2-yl)methoxy]phenyl]-methyl]-2,4-thiazolidinedione; (±)-5-[4-[(6-hydroxy-2,5,7,8-tetramethylchroman-2-yl)methoxy]benzyl]-2,4-thiazolidinedione; romglizone; CS-045; CI-991; Noscal; Prelay; Rezulin. C$_{24}$H$_{27}$NO$_5$S; mol wt 441.54. C 65.29%, H 6.16%, N 3.17%, O 18.12%, S 7.26%. Oral hypoglycemic agent which improves insulin sensitivity and decreases hepatic glucose production. Prepn: **JP Kokai 85 51189**; T. Yoshioka *et al.*, **US 4572912** (1985, 1986 both to Sankyo); T. Yoshioka *et al.*, *J. Med. Chem.* **32**, 421 (1989). Mechanism of action studies: T. P. Ciaraldi *et al.*, *Metabolism* **39**, 1056 (1990); M. Kellerer *et al.*, *Diabetes* **43**, 447 (1994). Clinical evaluation: T. Kuzuya *et al.*, *Diabetes Res. Clin. Pract.* **11**, 147 (1991). Clinical metabolic effects: S. L. Suter *et al.*, *Diabetes Care* **15**, 193 (1992).

Crystals from benzene-acetone, mp 184-186°.

THERAP CAT: Antidiabetic.

9949. Troleandomycin. [2751-09-9] Oleandomycin 2″,4′,11-triacetate; triacetyloleandomycin; NSC-108166; Cyclamycin; Wytrion; Evramycin; TAO; Triocetin. C$_{41}$H$_{67}$NO$_{15}$; mol wt 813.98. C 60.50%, H 8.30%, N 1.72%, O 29.48%. Semi-synthetic macrolide antibiotic. Prepd from oleandomycin: Celmer *et al.*, *Antibiot. Annu.* **1957-1958**, p 476; **GB 877730** (1958 to Pfizer); M. Khristov, N. Petkov, *Farmatsiya (Sofia)* **27**, 1 (1977), *C.A.* **90**, 23458c (1979). *Review:* S. Ross, *Antimicrobial Therapy*, B. M. Kagan, Ed. (Saunders, Philadelphia, 1970) pp 134-144.

Crystals from isopropanol. Practically tasteless. Dec 176°. $[\alpha]_D^{25}$ −23° (methanol). pKa 6.6. Soly in water: <0.1 g/100 ml. Freely sol in alc; sol in chloroform; slightly sol in ether.

THERAP CAT: Antibacterial.

9950. Tromantadine. [53783-83-8] 2-[2-(Dimethylamino)-ethoxy]-*N*-tricyclo[3.3.1.1^{3,7}]dec-1-ylacetamide; *N*-1-adamantyl-*N*-[2-(dimethylamino)ethoxy]acetamide; 1-(dimethylaminoethoxyacet-amido)adamantane. C$_{16}$H$_{28}$N$_2$O$_2$; mol wt 280.41. C 68.53%, H 10.07%, N 9.99%, O 11.41%. Prepn: Scherm, Peteri, **DE 1941218** (1971 to Merz), *C.A.* **74**, 99516k (1971); May, Peteri, *Arzneim.-Forsch.* **23**, 718 (1973). Chemistry and toxicology: Peteri, Sterner, *ibid.* 577.

Hydrochloride. [41544-24-5] D-41; Viru-Merz; Viruserol. C$_{16}$-H$_{28}$N$_2$O$_2$.HCl; mol wt 316.87. Crystals, mp 157-158°. LD$_{50}$ orally in rats: 630 mg/kg; i.v. in mice: 71.0 mg/kg (Peteri, Sterner).

THERAP CAT: Antiviral.

9951. Tromethamine. [77-86-1] 2-Amino-2-(hydroxymethyl)-1,3-propanediol; trimethylol aminomethane; tris(hydroxymethyl)aminomethane; trisamine; tris buffer; trometamol; tromethane; TRIS; THAM; Trizma. C$_4$H$_{11}$NO$_3$; mol wt 121.14. C 39.66%, H 9.15%, N 11.56%, O 39.62%. May be prepd by reduction or catalytic hydrogenation of the corresp nitro compd. Prepn of similar compds: Hass, Vanderbilt, **US 2174242** (1940); Johnson, Degering, *J. Org. Chem.* **8**, 7 (1943); Boileau, *Mem. Poudres* **35**, Annexe 7-76 (1953). Prepn by electrolytic reduction: McMillan, **US 2485982** (1949 to Comm. Solvents Corp.). Titrimetric standard: Whitehead, *J. Chem. Educ.* **36**, 297 (1959). Monograph: *Ann. N.Y. Acad. Sci.* **92**, Art. 2, pp 333-812 (June 17, 1961). Crystal structure: R. Rudman *et al.*, *Science* **200**, 531 (1978). GC determn in plasma: H. Hulshoff, H. B. Kostenbauder, *J. Chromatogr.* **145**, 155 (1978). Pharmacokinetics in rabbits: H. Brasch, H. Iven, *Arch. Int. Pharmacodyn.* **254**, 4 (1981). Use as a biological buffer: R. A. Durst, B. R. Staples, *Clin. Chem.* **18** 206 (1972); S. P. Fling, D. S. Gregerson, *Anal. Biochem.* **155**, 83 (1986); T. Higa, D. M. Desiderio, *ibid.* **173**, 463 (1988). Interaction with hydroxyl radicals: M. Hicks, J. M. Gebicki, *FEBS Lett.* **199**, 92 (1986). Review of clinical experience in treatment of acidemia: G. G. Nahas *et al.*, *Drugs* **55**, 191-224 (1998).

Crystalline powder or mass, mp 171-172°. bp$_{10}$ 219-220°. Weak, monoacidic base: pKb (25°): 5.91. pKa (20°): 8.3. pKa (37°): 7.82. pH of 0.1 molar aq soln 10.4. Aq solns do not absorb CO$_2$ from the air. Soly at 25° (mg/ml): water 550; ethylene glycol 79.1; methanol 26; anhydr ethanol 14.6; 95% ethanol 22.0; DMF 14; acetone 2.0; ethyl acetate 0.5; olive oil 0.4; cyclohexane 0.1; chloroform 0.05; carbon tetrachloride <0.05. Freely sol in low molecular weight aliphatic alcohols. Practically insol in benzene.

USE: In the synthesis of surface-active agents, vulcanization accelerators, pharmaceuticals. As emulsifying agent for cosmetic creams and lotions, mineral oil and paraffin wax emulsions, leather dressings, textile specialties, polishes, cleaning compds, so-called soluble oils. Absorbent for acidic gases. Biological buffer. Acidimetric standard.

THERAP CAT: Alkalinizing agent.

9952. Tropacine. [6878-98-4] α-Phenylbenzeneacetic acid (3-*endo*)-8-methyl-8-azabicyclo[3.2.1]oct-3-yl ester; 1α*H*,5α*H*-tropan-3α-ol diphenylacetate; diphenylacetic acid 3α-tropanyl ester; 3α-tropanyl diphenylacetate; tropine diphenylacetate. C$_{22}$H$_{25}$NO$_2$; mol wt 335.45. C 78.77%, H 7.51%, N 4.18%, O 9.54%. Anticholinergic. Prepd from tropine and diphenylacetyl chloride: **CH 202181** (1939 to Ciba), *C.A.* **33**, 8922^8 (1939); Friess *et al.*, *Toxicol. Appl. Pharmacol.* **2**, 574 (1960).

Hydrochloride. Crystals from chloroform + ether, mp 217-218°. THERAP CAT: Antiparkinsonian

9953. Tropacocaine. [537-26-8] (3-*exo*)-8-Methyl-8-azabicyclo[3.2.1]octan-3-ol 3-benzoate; 1α*H*,5α*H*-tropan-3β-ol benzoate; benzoylpseudotropeine; benzoyl-ψ-tropeine; pseudotropine benzoate; ψ-tropine benzoate; tropacaine. $C_{15}H_{19}NO_2$; mol wt 245.32. C 73.44%, H 7.81%, N 5.71%, O 13.04%. From Javanese coca leaves. Prepared by heating pseudotropine with water and benzoic anhydride: Wilstätter, *Ber.* **29**, 943 (1896). Stereochemistry: Beyerman *et al.*, *Rec. Trav. Chim.* **75**, 1445 (1956).

Plates, tablets, mp 49°. Distills *in vacuo* without dec. pK (15°) 4.72. pH of 0.06*M* soln 8.4. Freely sol in alc, ether, chloroform, benzene, petr ether, dil acids; slightly sol in water. MLD i.v. in rats: 15-20 mg/kg, Hirschfelder, *Physiol. Rev.* **12**, 262 (1932).

Hydrochloride. $C_{15}H_{19}NO_2$·HCl. Strongly refractive prisms from alcohol, dec 283°. Sol in water, slightly in abs alc, practically insol in ether. Aq solns are stable to boiling water for ~20 minutes. pH of 0.1*M* soln 5.8.

9954. Tropaeolin O. [547-57-9] 4-[2-(2,4-Dihydroxyphenyl)diazenyl]benzenesulfonic acid sodium salt (1:1); C.I. Acid Orange 6; sodium *p*-(2,4-dihydroxyphenylazo)benzenesulfonate; sodium azoresorcinolsulfanilate; C.I. Food Yellow 8; C.I. 14270; Resorcinol Yellow; Yellow T; Tropeolin O; Tropaeolin R; Chrysoine; Gold Yellow. $C_{12}H_9N_2NaO_5S$; mol wt 316.26. C 45.57%, H 2.87%, N 8.86%, Na 7.27%, O 25.29%, S 10.14%. Prepd from resorcinol and a *p*-sulfobenzenediazonium salt, followed by reaction with an inorganic sodium salt: Sisley, *Bull. Soc. Chim. Fr.* [3] **25**, 869 (1901). *See also: Colour Index* vol. 4 (3rd ed., 1971) p 4064. Brief review: H. J. Conn's *Biological Stains*, R. D. Lillie, Ed. (Williams & Wilkins, Baltimore, 9th ed., 1977) p 106.

Brown powder. Sol in water with reddish-yellow color; sol in alcohol.

USE: As indicator. pH: yellow 11, orange-brown 12.7. Occasionally used as plasma stain.

9955. Tropaeolin OO. [554-73-4] 4-[2-[4-(Phenylamino)phenyl]diazenyl]benzenesulfonic acid sodium salt (1:1); C.I. Acid Orange 5; *p*-[(*p*-anilinophenyl)azo]benzenesulfonic acid sodium salt; sodium *p*-[(*p*-anilinophenyl)azo]benzenesulfonate; Diphenylamine Orange; Orange GS; Orange N; Orange IV; Fast Yellow; Acid Yellow D; C.I. 13080. $C_{18}H_{14}N_3NaO_3S$; mol wt 375.38. C 57.59%, H 3.76%, N 11.19%, Na 6.12%, O 12.79%, S 8.54%. Prepd from diphenylamine and *p*-sulfobenzenediazonium chloride, followed by reaction with NaOH: Witt, *Ber.* **12**, 258 (1879); *Colour Index* vol. 4 (3rd ed., 1971) p 4045.

Orange-yellow scales or yellow powder. Sol in water.
USE: As indicator. pH: red 1.4 to yellow 2.6.

9956. Tropane. [529-17-9] 8-Methyl-8-azabicyclo[3.2.1]octane; 1α*H*,5α*H*-tropane; 2,3-dihydro-8-methylnortropidine. C_8H_{15}-N; mol wt 125.22. C 76.74%, H 12.07%, N 11.19%. Structure and synthesis: Ladenburg, *Ber.* **16**, 1408 (1883); Merling, *Ber.* **25**, 3124 (1892); Willstätter, Iglauer, *Ber.* **30**, 721 (1897); **33**, 1170 (1900); *eidem*, *Ann.* **317**, 315, 350 (1901); Robinson, *J. Chem. Soc.* **111**, 762, 876 (1917); Hess, *Ber.* **51**, 1007 (1918); Schöpf, *Angew. Chem.* **50**, 779, 797 (1937); Ruggli, Maeder, *Helv. Chim. Acta* **27**, 436 (1944); Keagle, Hartung, *J. Am. Chem. Soc.* **68**, 1608 (1946).

Liquid; bp 163-169°; d_{15}^{15} 0.9259. Sparingly sol in water, the soly decreasing with an increase in temp. Tropane and water will mix provided the volume of tropane is in excess.

9957. Tropesin. [65189-78-8] 1-(4-Chlorobenzoyl)-5-methoxy-2-methyl-1*H*-indole-3-acetic acid 2-carboxy-2-phenylethyl ester; *d,l*-2-phenyl-3-[1-(4-chlorobenzoyl)-5-methoxy-2-methylindole-3-acetoxy]propionic acid; 1-*p*-chlorobenzoyl-5-methoxy-2-methyl-3-indolylacetic acid 2-phenyl-2-carboxyethyl ester; Repanidal. $C_{28}H_{24}ClNO_6$; mol wt 505.95. C 66.47%, H 4.78%, Cl 7.01%, N 2.77%, O 18.97%. Tropic acid ester of indomethacin, *q.v.* Prepn: L. Fisnerova *et al.*, *DE 2727629*; *eidem*, **US 4136194** (1978, 1979 both to Spofa). Clinical pharmacokinetics: I. Janku *et al.*, *Drugs Exp. Clin. Res.* **IX**, 407 (1983). Review of pharmacology and clinical experience: J. Grimová *et al.*, *Drugs Today* **27**, 391-400 (1991).

Crystals from nitromethane, mp 130-132°. Also reported as mp 127-129°; 128-130°. LD$_{50}$ in female mice, rats (mg/kg): 190, 140 orally (Fisnerova, 1979).
THERAP CAT: Anti-inflammatory.

9958. Tropic Acid. [529-64-6] α-(Hydroxymethyl)benzeneacetic acid; 2-phenylhydracrylic acid; α-phenyl-β-hydroxypropionic acid; tropaic acid; tropeic acid. $C_9H_{10}O_3$; mol wt 166.18. C 65.05%, H 6.07%, O 28.88%. Degradation product of tropane alkaloids, esp atropine: Lossen, *Ann.* **133**, 351, 370 (1865). Resolution of isomers: McKenzie, Wood, *J. Chem. Soc.* **115**, 828 (1919). Absolute config of isomers: Fodor, Csepreghy, *ibid.* **1961**, 3222; Watson, Youngson, *J. Chem. Soc. Perkin Trans. 1* **1972**, 1597. Prepn: Sletzinger, Paulsen, **US 2390278** (1945 to Merck & Co.); Blicke, **US 2716650** (1955 to U. of Michigan); **DE 923426** (1955 to Sterling Drug). Biosynthetic studies: Louden, Leete, *J. Am. Chem. Soc.* **84**, 1510, 4507 (1962).

(±)-Form. Needles or plates from water or benzene, mp 118°. K at 25° = 7.5×10^{-5}. Absorption spectrum: Dobbie, Fox, *J. Chem. Soc.* **103**, 1194 (1913). 1 gram dissolves in 50 ml water; freely sol in boiling water; sol in alcohol, ether, slightly in benzene; practically insol in petr ether.

(+)-Form. mp 107°. $[\alpha]_D^{20}$ +72° (c = 0.5 in water).

(−)-Form. mp 126-128°. $[\alpha]_D^{20}$ −72° (c = 0.5 in water).

9959. Tropicamide. [1508-75-4] N-Ethyl-α-(hydroxymethyl)-N-(4-pyridinylmethyl)benzeneacetamide; N-ethyl-2-phenyl-N-(4-pyridylmethyl)hydracrylamide; N-ethyl-N-(γ-picolyl)tropamide; Mydriacyl; Mydriaticum. $C_{17}H_{20}N_2O_2$; mol wt 284.36. C 71.81%, H 7.09%, N 9.85%, O 11.25%. Ophthalmic anticholinergic. Prepn: Rey-Bellet, Spiegelberg, **US 2726245** (1955 to Hoffmann-La Roche). Comprehensive description: K. W. Blessel *et al.*, *Anal. Profiles Drug Subs.* **3**, 565-580 (1974).

White or practically white crystals, mp 96-97°. uv max (0.025 mg/ml in 0.1N HCl): 254 nm (ε 5.1×10^3). Freely sol in chloroform and in solns of strong acids; slightly sol in water.

THERAP CAT: Mydriatic.
THERAP CAT (VET): Mydriatic.

9960. Tropine. [120-29-6] (3-endo)-8-Methyl-8-azabicyclo-[3.2.1]octan-3-ol; 1αH,5αH-tropan-3α-ol; 2,3-dihydro-3α-hydroxy-8-methylnortropidine; 2,3-dihydro-3α-hydroxytropidine. C_8H_{15}-NO; mol wt 141.21. C 68.05%, H 10.71%, N 9.92%, O 11.33%. Prepd by reaction of tropidine and HBr, followed by hydrolysis and separation of isomers, tropine and pseudotropine: Ladenburg, *Ber.* **35**, 1159 (1902); by hydrogenation of tropinone in the presence of Raney nickel: Van de Kamp, Sletzinger, **US 2366760** (1945 to Merck & Co.); Stoll *et al.*, **US 2746976** (1956 to Sandoz). Yield of tropine and pseudotropine under varying conditions of reduction: Beckett *et al.*, *Tetrahedron* **6**, 319 (1959). Separation of tropine and pseudotropine by gold salt formation: Ladenburg, *loc. cit.*; by fractional distillation under reduced pressure: Friess *et al.*, *Toxicol. Appl. Pharmacol.* **2**, 574 (1960). Stereochemistry: Beyerman *et al.*, *Rec. Trav. Chim.* **75**, 1445 (1956).

Hygroscopic plates from ether, mp 63°. bp 233°. pK at 15° = 3.80. pH of 0.05 molar soln 11.5. Freely sol in water and alcohol; sol in ether and chloroform.

Note: Esters of tropine are known as *tropeines*.

9961. Tropine Benzylate. [3736-36-5] α-Hydroxy-α-phenylbenzeneacetic acid (3-endo)-8-methyl-8-azabicyclo[3.2.1]oct-3-yl ester; 1αH,5αH-tropan-3α-ol benzilate; benzilic acid 3α-tropanyl ester; glykin; BAT; BTE; BETE. $C_{22}H_{25}NO_3$; mol wt 351.45. C 75.19%, H 7.17%, N 3.99%, O 13.66%. Prepn: Hromatka *et al.*, *Monatsh. Chem.* **83**, 1321 (1952).

Prisms from ether or benzene, mp 152-153°.
Hydrochloride. $C_{22}H_{25}NO_3$·HCl. mp 239-240°.

9962. Tropisetron. [89565-68-4] 1H-Indole-3-carboxylic acid (3-endo)-8-methyl-8-azabicyclo[3.2.1]oct-3-yl ester; 3α-tropanyl-1H-indole-3-carboxylic acid ester; 1αH,5αH-tropan-3α-yl indole-3-carboxylate; ICS-205-930. $C_{17}H_{20}N_2O_2$; mol wt 284.36. C 71.81%, H 7.09%, N 9.85%, O 11.25%. Specific serotonin (5-HT$_3$) receptor antagonist. Prepn: P. Donatsch *et al.*, **DE 3322574** (1983 to Sandoz); *idem et al.*, **US 4789673** (1988). Pharmacology: B. P. Richardson *et al.*, *Nature* **316**, 126 (1985); F. M. Williams *et al.*, *J. Cardiovasc. Pharmacol.* **7**, 550 (1985). Receptor binding studies: C. Waeber *et al.*, *Neuroscience* **31**, 393 (1989). Clinical pharmacokinetics: V. Fischer *et al.*, *Drug Metab. Dispos.* **20**, 603 (1992). Series of articles on antiemetic efficacy in chemotherapy and radiotherapy: *Drugs* **43**, Suppl. 3, 6-39 (1992). Review of pharmacology: C. Seynaeve *et al.*, *Anti-Cancer Drugs* **2**, 343-355 (1991); and toxicology: K. Kutz, *Ann. Oncol.* **4**, Suppl. 3, S15-S18 (1993).

mp 201-202° (from methylene chloride/ethyl acetate).
Monohydrochloride. [105826-92-4] Navoban; Novaban. $C_{17}H_{20}N_2O_2$·HCl; mol wt 320.82. mp 283-285° (dec).
THERAP CAT: Antiemetic.

9963. Tropolone. [533-75-5] 2-Hydroxy-2,4,6-cycloheptatrien-1-one; 2-hydroxytropone; purpurocatechol. $C_7H_6O_2$; mol wt 122.12. C 68.85%, H 4.95%, O 26.20%. Naturally occuring antibiotic with a characteristic seven-membered aromatic ring system. Prototype structure for a family of bioactive natural products known as tropolonoids. Proposed structure: M. J. S. Dewar, *Nature* **155**, 50 (1945). Synthesis: W. von E. Doering, L. H. Knox, *J. Am. Chem. Soc.* **72**, 2305 (1950). Isoln from *Pseudomonas* sp.: G. D. Lindberg *et al.*, *J. Nat. Prod.* **43**, 592 (1980). Use as ligand for platelet labeling: M. K. Dewanjee *et al.*, *J. Nucl. Med.* **22**, 981 (1981). Antifungal and insecticidal activity: Y. Morita *et al.*, *Biol. Pharm. Bull.* **26**, 1487 (2003). Review of chemistry, bioactivities, and biosynthesis of tropolonoids: J. Zhao, *Curr. Med. Chem.* **14**, 2597-2621 (2007); R. Bentley, *Nat. Prod. Rep.* **25**, 118-138 (2008).

Colorless needles from hexane, mp 48°. Sublimes easily. Flash point, closed cup: 234°F (112°C). Sol in water and in most organic solvents. pKa 6.7. uv max: 235, 327, 368, 392 nm (log ε 4.26, 3.78, 3.59, 3.11).

USE: Ligand for cell radiolabeling.

9964. Tropomyosins. Fibrous proteins involved in the regulation of muscle relaxation. They are present in all forms of striated and smooth muscles and probably in nonmuscle cells as well. Native tropomyosin consisting of two proteins, tropomyosin and *troponin*, is the Ca^{2+}-sensitive regulatory protein that controls the interaction between actin and myosin necessary for the production of force in muscle. In all skeletal tissues, there are two forms of tropomyosin chains, designated α and β. Their ratio depends on the muscle source. Troponin consists of three subunits, troponin T (the tropomyosin binding subunit), troponin I (the actomyosin ATPase inhibitory subunit) and troponin C (the Ca^{2+}-binding subunit). All three subunits, in addition to tropomyosin, are responsible for the native tropomyosin activity. Isoln of tropomyosin from skeletal muscle and cardiac muscle: K. Bailey, *Nature* **157**, 368 (1946); *idem, Biochem.*

J. **43**, 271 (1948). Structure studies: R. S. Hodges, L. B. Smillie, *Biochem. Biophys. Res. Commun.* **41**, 987 (1970). Amino acid sequence studies: J. Sodek *et al., Proc. Natl. Acad. Sci. USA* **69**, 3800 (1972). X-ray crystal structure of troponin C: M. Sundaralingam *et al., Science* **227**, 945 (1985). *Reviews:* C. E. Bodwell, K. Laki, in *Contractile Proteins and Muscle*, K. Laki, Ed. (Dekker, New York, 1971); W. F. Harrington in *The Proteins* **vol. 4**, H. Neurath, R. L. Hill, Eds. (Academic Press, New York, 1979) pp 317-327. Comprehensive review of isolation, preparation, identification and role in the contractile process: "Structural and Contractile Proteins" in *Methods Enzymol.* **85**, Part B, 1-774 (1982).

9965. Tropylium Bromide. [5376-03-4] Cycloheptatrienylium bromide (1:1); cycloheptatrienocarbonium bromide. C₇H₇Br; mol wt 171.04. C 49.16%, H 4.13%, Br 46.72%. Prepd by bromination of 1,3,5-cycloheptatriene in carbon tetrachloride, followed by removal of the CCl₄ and heating the residue *in vacuo* for several days: von Doering, Knox, *J. Am. Chem. Soc.* **79**, 352 (1957); King, Stone, *Inorg. Synth.* **7**, 99 (1963).

Yellow prisms from ethanol, mp 203°. Freely sol in water. Practically insol in ether.

9966. Trospium Chloride. [10405-02-4] (1α,3β,5α)-3-[(2-Hydroxy-2,2-diphenylacetyl)oxy]spiro[8-azoniabicyclo[3.2.1]octane-8,1′-pyrrolidinium] chloride (1:1); 3α-hydroxyspiro[1αH,5αH-nortropane-8,1′-pyrrolidinium] chloride benzilate; azoniaspiro(3α-benziloyloxynortropane-8,1′-pyrrolidine) chloride; azoniaspiro(3α-diphenylglycoloyloxynortropan-8,1′-pyrrolidine) chloride; 3α-benziloyloxyspiro(nortropane-8,1′-pyrrolidinium) chloride; Regurin; Relaspium; Sanctura; Spasmex; Spasmolyt. C₂₅H₃₀ClNO₃; mol wt 427.97. C 70.16%, H 7.07%, Cl 8.28%, N 3.27%, O 11.22%. Tropine derivative with anticholinergic activity. Prepn: **NL 6402155**; R. Pfleger *et al.*, **US 3480626** (1964, 1969 both to Pfleger); H. Bertholdt *et al., Arzneim.-Forsch.* **17**, 719 (1967). Pharmacology and toxicology in animals: H. Antweiler *et al., ibid.* **16**, 1581 (1966). Inhibition of gastric motility and acid secretion in humans: G. Lux, P. Frühmorgen, *Fortschr. Med.* **96**, 2113 (1978). Fluorimetric determn in plasma and urine: G. Schladitz-Keil *et al., J. Chromatogr.* **345**, 99 (1985). Bioavailability: *eidem, Arzneim.-Forsch.* **36**, 984 (1986). Clinical trial in bladder hyper-reflexia: H. Madersbacher *et al., Br. J. Urol.* **75**, 452 (1995). Review in overactive bladder: N. R. Zinner, *Expert Opin. Pharmacother.* **6**, 1409-1420 (2005).

Crystals from ethanol-ether, mp 255-257° (dec). Soly in water: ~1 g/2 ml. LD₅₀ in mice (mg/kg): 12.3 i.v. (Antweiler).

THERAP CAT: Antispasmodic; in treatment of urinary incontinence.

9967. Trovafloxacin. [147059-72-1] 7-[(1α,5α,6α)-6-Amino-3-azabicyclo[3.1.0]hex-3-yl]-1-(2,4-difluorophenyl)-6-fluoro-1,4-dihydro-4-oxo-1,8-naphthyridine-3-carboxylic acid; CP-99219. C₂₀H₁₅F₃N₄O₃; mol wt 416.36. C 57.70%, H 3.63%, F 13.69%, N 13.46%, O 11.53%. Fluorinated quinolone antibacterial. Prepn: K. E. Brighty, **US 5164402** (1992 to Pfizer). Voltammetric determn in urine and serum: J. L. Vílchez *et al., J. Pharm. Biomed. Anal.* **31**, 465 (2003). Antibacterial spectrum *in vitro:* H. C. Neu, N.-X. Chin, *Antimicrob. Agents Chemother.* **38**, 2615 (1994). Clinical phar-

macokinetics: R. Teng *et al., J. Antimicrob. Chemother.* **36**, 385 (1995). Evaluation in gonorrhea: E. W. Hook *et al., Antimicrob. Agents Chemother.* **40**, 1720 (1996). Mechanism of hepatotoxicity: M. J. Liguori *et al., Hepatology* **41**, 177 (2005).

Hydrochloride. [146961-34-4] C₂₀H₁₅F₃N₄O₃.HCl. Pale yellow crystals from acetonitrile/methanol, mp 246° (dec).

Methanesulfonate. [147059-75-4] Trovafloxacin mesylate; CP-99219-27; Trovan. C₂₀H₁₅F₃N₄O₃.CH₃SO₃H; mol wt 512.46.

THERAP CAT: Antibacterial.

9968. Troxacitabine. [145918-75-8] 4-Amino-1-[(2S,4S)-2-(hydroxymethyl)-1,3-dioxolan-4-yl]-2(1H)-pyrimidinone; (−)-2′-deoxy-3′-oxacytidine; β-L-(−) dioxolane cytidine; (−)-(2S,4S)-1-[2-(hydroxymethyl)-1,3-dioxolan-4-yl]cytosine; L-OddC; (−)-BCH-204; BCH-4556; Troxatyl. C₈H₁₁N₃O₄; mol wt 213.19. C 45.07%, H 5.20%, N 19.71%, O 30.02%. Synthetic nucleoside analog with stereochemically unnatural β-L configuration. Prepn: Y.-C. Cheng *et al.*, **WO 9218517** (1992 to Yale Univ. and Univ. Georgia Res. Found.); B. R. Belleau *et al., Tetrahedron Lett.* **33**, 6949 (1992); H. O. Kim *et al., J. Med. Chem.* **36**, 519 (1993). Resolution of enantiomers: M. P. DiMarco *et al., J. Chromatogr.* **645**, 107 (1993). Structure activity study: T. S. Mansour *et al., Nucleosides Nucleotides* **14**, 627 (1995). Clinical pharmacokinetics: J. S. deBono *et al. J. Clin. Oncol.* **20**, 96 (2001). Population pharmacokinetics: C. K. K. Lee *et al., Clin. Cancer Res.* **12**, 2158 (2006). Clinical evaluation in refractory leukemia: F. J. Giles *et al., ibid.* **20**, 656 (2002); in imatinib-resistant chronic myelogenous leukemia: *idem et al., Leuk. Res.* **27**, 1091 (2003); in pancreatic cancer: R. Lapointe *et al., Ann. Oncol.* **16**, 289 (2005). Review of pharmacology and clinical development: *idem, Expert Rev. Anticancer Ther.* **2**, 261-266 (2002); G. Ecker, *Curr. Opin. Investig. Drugs* **3**, 1533-1538 (2002).

mp 176-177°. [α]D²⁵ −38.33° (c = 0.43 in MeOH). uv max (water): 270.0 (pH 7), 278.0 (pH 2), 269.0 (pH 11) (ε 7770, 11970, 8380).

THERAP CAT: Antineoplastic.

9969. Troxerutin. [7085-55-4] 2-[3,4-Bis(2-hydroxyethoxy)phenyl]-3-[[6-O-(6-deoxy-α-L-mannopyranosyl)-β-D-glucopyranosyl]oxy]-5-hydroxy-7-(2-hydroxyethoxy)-4H-1-benzopyran-4-one; 7,3′,4′-tris[O-(2-hydroxyethyl)]rutin; trioxyethylrutin; tri(hydroxyethyl)rutoside; Posorutin; Ruven; Vastribil; Veinamitol; Veniten. C₃₃H₄₂O₁₉; mol wt 742.68. C 53.37%, H 5.70%, O 40.93%. The principal component of a mixture, the *O-(β-hydroxyethyl)rutosides*, which also contains mono-, di-, tetra- and other trihydroxyethyl derivs of rutin, *q.v.* The mixture is prepd by the hydroxyethylation of the phenolic groups of rutin with glycochlorohydrin in alk medium: J. Favre, **CH 349614** (1957); *see also* **GB 833174** (1960 to Zyma), *C.A.* **54**, 21135i (1960). Isolation and identification of major components of the mixture: P. Courbat *et al., Helv. Chim. Acta* **49**, 1203, 1420 (1966). Prepn of troxerutin: P. J. Courbat, **US 3420815** (1969 to Zyma). Metabolism in man: A. M. Hackett *et al., Arzneim.-Forsch.* **26**, 925 (1976).

Yellow powder, mp 181°. Sol in water, glycerol, propylene glycol. Practically insol in cold ethanol, methanol (forms alcoholate), ether, benzene, chloroform.

***O*-(β-Hydroxyethyl)rutinosides (mixture).** [55965-63-4] HR; Paroven; Relvene; Varemoid; Venoruton. Yellow powder, mp 156°. Sol in water, methanol, glycerol, propylene glycol. Practically insol in cold ethanol (forms alcoholate), ether, benzene, chloroform.

THERAP CAT: Treatment of venous disorders.

9970. Troxipide. [30751-05-4] 3,4,5-Trimethoxy-*N*-3-piperidinylbenzamide; 3-(3,4,5-trimethoxybenzamido)piperidine; KU-54; Aplace. $C_{15}H_{22}N_2O_4$; mol wt 294.35. C 61.21%, H 7.53%, N 9.52%, O 21.74%. Prepn: **BE 736840**; T. Irikura *et al.*, **US 3647805** (1969, 1972 both to Kyorin). Anti-ulcer activity: T. Irikura, K. Kasuga, *J. Med. Chem.* **14**, 357 (1971). Pharmacokinetics in man: T. Irikura *et al.*, *Iyakuhin Kenkyu* **12**, 971 (1981), *C.A.* **96**, 79377s (1982). Effect on gastric mucosa in rats: Y. Abe *et al.*, *Nippon Yakurigaku Zasshi* **83**, 317 (1984), *C.A.* **101**, 456g (1984); on glucosamine synthetase: Y. Abe *et al.*, *Oyo Yakuri* **27**, 521 (1984), *C.A.* **101**, 17001c (1984). Metabolism in rats: K. Tagaki *et al.*, *ibid.* 1151, *C.A.* **101**, 103555t (1984); K. Tagaki, K. Endo, *ibid.* 1167, *C.A.* **101**, 103556u (1984). Toxicity: T. Irikura *et al.*, *Kiso to Rinsho* **12**, 3422 (1978).

Needles from acetonitrile, mp 179-181.5°. Sol in ethanol. LD_{50} in male, female rats, male, female mice (mg/kg): 500, 2100, 2200, 2000 orally; >4150, >4150, 1600, 1550 s.c.; 340, 340, 300, 305 i.p. (Irikura, 1978).

Hydrochloride hemihydrate. $C_{15}H_{22}N_2O_4 \cdot HCl \cdot \frac{1}{2}H_2O$. Needles from acetonitrile, mp 206-209°. Sol in ethanol.

THERAP CAT: Antiulcerative.

9971. Truxillic Acid. [4462-95-7] 2,4-Diphenyl-1,3-cyclobutanedicarboxylic acid. $C_{18}H_{16}O_4$; mol wt 296.32. C 72.96%, H 5.44%, O 21.60%. Cinnamic acid polymers obtained from the minor alkaloids of cocaine: Liebermann, *Ber.* **21**, 2342 (1888). Five stereoisomers have been obtained: α-, γ-, ε-, *peri*- and *epi*-isomers. Stereochemical configurations: Stoermer, Bacher, *Ber.* **57B**, 15-23 (1924).

	a	b	c	d	e	f
α-Isomer	COOH	H	H	C₆H₅	H	COOH
γ-Isomer	COOH	H	H	C₆H₅	COOH	H
ε-Isomer	H	COOH	C₆H₅	H	H	COOH
peri-Isomer	COOH	H	C₆H₅	H	COOH	H
epi-Isomer	COOH	H	C₆H₅	H	H	COOH

α-Isomer. γ-Isatropaic acid; cocaic acid. Prepd by irradiation of cinnamic acid in water: White, Dunathan, *J. Am. Chem. Soc.* **78**, 6055 (1956). Crystals from acetic acid, mp 284-285°. Sparingly sol in boiling water, in ether, benzene, carbon disulfide; sol in hot glacial acetic acid, hot alc; sparingly sol in acetone.

γ-Isomer. ε-Isatropaic acid. γ-Truxillic anhydride, obtained by heating α-truxillic acid with acetic anhydride and sodium acetate, was heated with alkali and the free acid pptd with HCl: Liebermann, *Ber.* **22**, 124 (1889). Needles from dil alc, mp 228°. Very slightly sol in hot water; sol in ether.

ε-Isomer. β-Cocaic acid. Prepd by fusion of α-truxillic acid with KOH: Hesse, *Ann.* **271**, 180 (1892). Needles from ether, mp 192°. Freely sol in glacial acetic acid, abs alcohol, chloroform; less sol in benzene. Practically insol in ligroin.

***peri*-Isomer.** η-Truxillic acid. Prepn: γ-Truxillic anhydride heated at low pressure forms the *peri*-anhydride, converted to the acid by warming with alcoholic KOH: Stoermer, Bacher, *loc. cit.* Crystals from benzene + ligroin, mp 266° (effervescence). Soluble in alcohol. Practically insol in ether, benzene.

***epi*-Isomer.** Prepd by boiling *peri*-truxillic acid with excess 10% NaOH: Stoermer, Bacher, *loc. cit.* Crystals from dil alc, mp 285-287°. By melting or heating with acetic anhydride in a tube, ε-isomer forms. Practically insol in ether, benzene.

9972. Trypan Blue. [72-57-1] 3,3′-[(3,3′-Dimethyl[1,1′-biphenyl]-4,4′-diyl)bis(2,1-diazenediyl)]bis[5-amino-4-hydroxy-2,7-naphthalenedisulfonic acid] sodium salt (1:4); C.I. Direct Blue 14; C.I. 23850; 3,3′-[(3,3′-dimethyl-4,4′-biphenylene)bis(azo)]bis(5-amino-4-hydroxy-2,7-naphthalenedisulfonic acid) tetrasodium salt; tetrasodium 3,3′-[(3,3′-dimethyl-4,4′-biphenylene)bis(azo)]bis(5-amino-4-hydroxy-2,7-naphthalenedisulfonate); sodium ditolyl-diazobis-8-amino-1-naphthol-3,6-disulfonate; Benzamine Blue; Diamine Blue; Benzo Blue; Congo Blue; Dianil Blue; Naphthylamine Blue; Niagara Blue. $C_{34}H_{24}N_6Na_4O_{14}S_4$; mol wt 960.79. C 42.50%, H 2.52%, N 8.75%, Na 9.57%, O 23.31%, S 13.35%. Prepd by coupling diazotized *o*-tolidine with 5-amino-4-hydroxy-2,7-naphthalenedisulfonic acid in sodium carbonate soln: Lewers, Lowy, *Ind. Eng. Chem.* **17**, 1289-1290 (1925); *Colour Index* **vol. 4** (3rd ed., 1971) p 4198. Toxicity study: Anderson *et al.*, *Proc. Soc. Exp. Biol. Med.* **31**, 825 (1934). Review of teratogenicity studies: R. L. Cahen, *Clin. Pharmacol. Ther.* **5**, 480 (1964).

Bluish-gray powder. Sol in water forming a deep blue soln with violet tinge. Almost insol in alcohol. LD_{100} i.v. in rats: 300 mg/kg (Anderson).

USE: Biological stain.

9973. Trypan Red. [574-64-1] 4,4′-[(3-Sulfo[1,1′-biphenyl]-4,4′-diyl)bis(2,1-diazenediyl)]bis[3-amino-2,7-naphthalenedisulfonic acid] sodium salt (1:5); 4,4′-[(3-sulfo-4,4′-biphenylene)bis(azo)]bis(3-amino-2,7-naphthalenedisulfonic acid) pentasodium salt; pentasodium 4,4′-[(3-sulfo-4,4′-biphenylene)bis(azo)]bis(3-amino-2,7-naphthalenedisulfonate); C.I. 22850. $C_{32}H_{19}N_6Na_5O_{15}S_5$; mol wt 1002.78. C 38.33%, H 1.91%, N 8.38%, Na 11.46%, O 23.93%, S 15.99%. Prepd by coupling diazotized 4,4′-diaminobiphenyl-3-sulfonic acid with sodium 2-amino-3,6-naphthalenedisulfonate: Kuss, *J. Am. Chem. Soc.* **36**, 961 (1914); *Colour Index* **vol. 4** (3rd ed., 1971) p 4181.

Reddish-brown powder. Sol in water. Practically insol in alcohol.
USE: Biological stain.

THERAP CAT: Antiprotozoal (Trypanosoma).

THERAP CAT (VET): Has been used as a trypanocide.

9974. Tryparsamide. [554-72-3] *As*-[4-[(2-Amino-2-oxoethyl)amino]phenyl]arsonic acid sodium salt (1:1); *N*-(carbamoylmethyl)arsanilic acid monosodium salt; monosodium *N*-phenylglycinamide-*p*-arsonate; Glyphenarsine; Tryparsone; Tryponarsyl; Trypothane. $C_8H_{10}AsN_2NaO_4$; mol wt 296.09. C 32.45%, H 3.40%, As 25.30%, N 9.46%, Na 7.76%, O 21.61%. Prepd by heating a soln of arsanilic acid in aq NaOH, Na_2CO_3 and chloroacetamide: W. A. Jacobs *et al.*, **US 1280119**; *eidem*, **US 1280124** (both 1918 to Rockefeller Inst.); W. A. Jacobs, M. Heidelberger, *J. Am. Chem. Soc.* **41**, 1587 (1919); *Org. Synth.* **8**, 100 (1928).

Hemihydrate. [6159-29-1] Platelets, slowly affected by light, stable to air. One gram dissolves in about 2 ml water. Slightly sol in alcohol. Insol in ether, chloroform. pH of 1:20 aq soln 6.5. *Keep in tight containers, preferably at a temp not above 20° and protected from light.*

THERAP CAT: Antiprotozoal (Trypanosoma).

9975. Trypsin. [9002-07-7] Parenzymol; Trypure. Mol wt 24,000. Proteolytic enzyme formed in the small intestine by the action of a peptidase, enterokinase, on the pancreatic cell product, trypsinogen. Acts on lysyl and arginyl bonds of peptide chains and hydrolyzes even esters and amides. *Reviews:* Desnuelle, "Trypsin" in *The Enzymes* vol. 4, P. D. Boyer *et al.*, Eds. (Academic Press, New York, 2nd ed., 1960) pp 119-132; Keil, *ibid.* vol. 3 (3rd ed., 1971) pp 250-275; Inagami, "Trypsin" in *Proteins, Structure and Function* vol. 1, M. Funatsu *et al.*, Eds. (Kodansha, Tokyo, Wiley, New York, 1972) pp 1-83.

Yellow to grayish-yellow powder or crystals. Stable indefinitely in dry form at room temp. Sol in water. Practically insol in alcohol or glycerol. Readily sol in Sorensen's sodium phosphate buffer soln. Acts optimally at pH values between 7 and 9. Solns lose 75% of their potency within 3 hrs at room temp. Prepn of stabilized trypsin compositions contg partially hydrolyzed gelatin: Sullivan, Martin, **US 2930736** (1960 to National Drug).

THERAP CAT: Enzyme (proteolytic).

THERAP CAT (VET): Enzyme (proteolytic).

9976. Tryptamine. [61-54-1] 1*H*-Indole-3-ethanamine; 3-(2-aminoethyl)indole; 2-(3-indolyl)ethylamine. $C_{10}H_{12}N_2$; mol wt 160.22. C 74.97%, H 7.55%, N 17.48%. Occurs in plants. Synthesis starting with nitroethylene and indole: Noland, Hartman, *J. Am. Chem. Soc.* **76**, 3227 (1954). Alternate routes: Thesing, Schulde, *Ber.* **85**, 324 (1952); Jackson, Smith, *J. Chem. Soc.* **1965**, 3498; Tacconi, *Farmaco Ed. Sci.* **20**, 902 (1965); S. Takano *et al.*, *Heterocycles* **6**, 1167 (1977); I. Fleming, M. Woolias, *J. Chem. Soc. Perkin Trans. I* **1979**, 829. X-ray structure determn: Wakahara *et al.*, *Tetrahedron Lett.* **1970**, 4999. *Review* of tryptamine syntheses: J. E. Saxton in R. H. F. Manske, *The Alkaloids* vol. VIII (1965) pp 8-10.

Needles from petr ether, mp 118°. uv max (ethanol): 222, 282, 290 nm (log ε 4.56, 3.78, 3.71). Sol in ethanol, acetone. Practically insol in water, ether, benzene, chloroform.

Hydrochloride. $C_{10}H_{12}N_2$.HCl. Needles from ethanol + ethyl acetate, mp 248°. uv max (95% ethanol): 221, 275, 281, 290 nm (log ε 4.52, 3.73, 3.75, 3.69).

9977. Tryptophan. [73-22-3] L-Tryptophan; Trp; W; (*S*)-α-amino-1*H*-indole-3-propanoic acid; *l*-α-aminoindole-3-propionic acid; *l*-α-amino-3-indolepropionic acid; 2-amino-3-indolylpropanoic

acid; *l*-β-3-indolylalanine; Ardeytropin; Kalma; Optimax; Pacitron; Sedanoct; Trofan; Tryptan. $C_{11}H_{12}N_2O_2$; mol wt 204.23. C 64.69%, H 5.92%, N 13.72%, O 15.67%. An essential amino acid for human development; precursor of serotonin, *q.v.* Isoln from casein: F. G. Hopkins, S. W. Cole, *J. Physiol.* **27**, 418 (1902). Structure: A. Ellinger, *Ber.* **39**, 2515 (1906); A. Ellinger, A. C. Flamand, *Ber.* **40**, 3029 (1907). Early chemistry and biochemistry: *Amino Acids and Proteins*, D. M. Greenberg, Ed. (Charles C. Thomas, Springfield, IL, 1951) 950 pp., *passim*; J. P. Greenstein, M. Winitz, *Chemistry of the Amino Acids* **vols 1-3** (John Wiley and Sons, Inc., New York, 1961) pp. 2316-2347, *passim*. Intrinsic fluorescent/phosphorescent moiety in proteins; used in characterizing structure and conformational changes: E. A. Burstein *et al.*, *Photochem. Photobiol.* **18**, 263 (1973); C. Pokalsky *et al.*, *J. Biol. Chem.* **270**, 3809 (1995); review of phosphorescence: S. Papp, J. M. Vanderkooi, *Photochem. Photobiol.* **49**, 775-784 (1989). Review of microbial production: T. K. Maiti, S. P. Chatterjee, *Hind. Antibiot. Bull.* **33**, 26-61 (1991). Review of biosynthesis: I. P. Crawford, G. V. Stauffer, *Annu. Rev. Biochem.* **49**, 163-195 (1980); of nutrition and metabolism: J. C. Peters. *Adv. Exp. Med. Biol.* **294**, 345-358 (1991). Review of use in depression: S. N. Young, *J. Psychiatry Neurosci.* **16**, 241-246 (1991); in neuropsychiatric disorders: R. Sandyk, *Int. J. Neurosci.* **67**, 127-144 (1992). Review as toxic agent in eosinophilia-myalgia syndrome (EMS): D. S. Milburn, C. W. Myers, *DICP Ann. Pharmacother.* **25**, 1259 (1991); L. D. Kaufman, R. M. Philen, *Drug Saf.* **8**, 89-98 (1993).

Leaflets or plates from dil alc, dec 289° (rapid heating). $[\alpha]_D^{23}$ −31.5° (c = 1); $[\alpha]_D^{20}$ +2.4° (0.5*N* HCl); $[\alpha]_D^{20}$ +0.15° (c = 2.43 in 0.5*N* NaOH). pK_1 2.38; pK_2 9.39. Soly in water (g/l): 8.23 at 0°; 10.57 at 20°; 11.36 at 25°; 17.06 at 50°; 27.95 at 75°; 49.87 at 100°. Sol in hot alc, alkali hydroxides, dilute hydrochloric acid. Insol in chloroform.

Hydrochloride. $C_{11}H_{12}N_2O_2$.HCl. Needles from methanol, dec 251°.

USE: Probe for studying protein structure and dynamics.

THERAP CAT: In treatment of depression, schizophrenia and other neuropsychiatric disorders.

9978. Tryptophol. [526-55-6] 1*H*-3-Indole-3-ethanol; 3-ω-hydroxyethylindole; 2-(3-indolyl)ethyl alcohol; 2-indolyl(3)-ethanol; β-indolylethyl alcohol; 3-β-hydroxyethylindole. $C_{10}H_{11}NO$; mol wt 161.20. C 74.51%, H 6.88%, N 8.69%, O 9.92%. Prepd by treatment of indolemagnesium bromide with ethylene oxide: Oddo, Cambieri, *Gazz. Chim. Ital.* **60**, 19 (1939); Snyder, Pilgrim, *J. Am. Chem. Soc.* **70**, 1962 (1948); by Bouveault-Blanc reduction of 3-indoleacetic ester: Jackson, *J. Biol. Chem.* **88**, 659 (1930); by treatment of indolemagnesium iodide with ethylene chlorohydrin: Majima, Hoshino, *Ber.* **58**, 2042 (1925); from gramine: Snyder, Pilgrim, *J. Am. Chem. Soc.* **70**, 3770 (1948). By lithium aluminum hydride reduction of methyl 3-indoleglycolate: Speeter, Anthony, **US 3076814** (1963 to Upjohn).

Platelets from ether + petr ether, mp 59°. bp$_{2.0}$ 174°. Slightly sol in water. Sol in methanol, ethanol, ether, acetone, chloroform, ethyl acetate, glacial acetic acid. Moderately sol in benzene, amyl alcohol, hot carbon disulfide. Sparingly sol in petr ether.

9979. TSQ. [109628-27-5] *N*-(6-Methoxy-8-quinolinyl)-4-methylbenzenesulfonamide; 6-methoxy-8-(*p*-toluenesulfonyl)aminoquinoline. $C_{17}H_{16}N_2O_3S$; mol wt 328.39. C 62.18%, H 4.91%, N 8.53%, O 14.62%, S 9.76%. Fluorescent membrane permeant probe selective for Zn^{2+} in the presence of Ca^{2+} and Mg^{2+}. Prepn:

G. B. Bachman *et al.*, *J. Org. Chem.* **15**, 1278 (1950). Atomic absoportion determn of Zn^{2+} to picomole levels in biological systems: J. G. Reyes *et al.*, *Biol. Res.* **27**, 49 (1994). Structure/mechanism analysis: M. S. Nasir *et al.*, *J. Biol. Inorg. Chem.* **4**, 775 (1999). Visualization of reactive zinc in the brain: C. J. Frederickson *et al.*, *J. Neurosci. Methods* **20**, 91 (1987); in extracellular areas of CNS: A. A. Larson *et al.*, *Pain* **86**, 177 (2000); in apoptosis: G. R. Sauer *et al.*, *J. Cell. Biochem.* **88**, 954 (2003).

White crystalline solid from ethanol, mp 133-134°.

USE: Determn of extracellular and intracellular Zn^{2+} levels in biological systems.

9980. Tsuduranine. [517-97-5] (6a*R*)-5,6,6a,7-Tetrahydro-1,2-dimethoxy-4*H*-dibenzo[*de,g*]quinolin-10-ol; 1,2-dimethoxy-6aβ-noraporphin-10-ol; tuduranine. $C_{18}H_{19}NO_3$; mol wt 297.35. C 72.71%, H 6.44%, N 4.71%, O 16.14%. In sinomenine mother liquors. Isoln: Gotô, *Ann.* **521**, 175 (1935). Structure: Gotô, Shishido, *Ann.* **539**, 262 (1939). Synthesis of *dl*-form: Narayanaswami *et al.*, *Indian J. Chem.* **7**, 945 (1969).

Difficult to crystallize. Minute needles from slowly evaporating ether, mp about 125° (softens at 105°), or 204° depending on cryst form. $[\alpha]_D^{20} -127.5°$ (c = 0.855 in ethanol). Freely sol in the usual organic solvents.

Hydrochloride. Shiny scales from water, dec 286°. $[\alpha]_D^{15} -148°$ (c = 0.88 in water + methanol). Sparingly sol in water.

9981. T-2 Toxin. [21259-20-1] (3α,4β,8α)-12,13-Epoxytrichothec-9-ene-3,4,8,15-tetrol 4,15-diacetate 8-(3-methylbutanoate); 3α-hydroxy-4β,15-diacetoxy-8α-(3-methylbutyryloxy)-12,13-epoxy-Δ^9-tricothecene; 8α-(3-methylbutyryloxy)-4β,15-diacetoxyscirp-9-en-3α-ol; fusariotoxin T-2; insariotoxin; mycotoxin T-2; NSC-138780. $C_{24}H_{34}O_9$; mol wt 466.53. C 61.79%, H 7.35%, O 30.86%. *Tricothecene mycotoxin* isolated from *Fusarium tricinctum*: J. R. Bamburg *et al.*, *Tetrahedron* **24**, 3329 (1968). Physicochemical data: A. E. Pohland *et al.*, *Pure Appl. Chem.* **54**, 2119 (1982). Synthesis: M. C. Wani *et al.*, *J. Org. Chem.* **52**, 3468 (1987). Biosynthesis study: F. Van Middlesworth *et al.*, *J. Org. Chem.* **55**, 1237 (1990). Toxicology studies: W. F. O. Marasas *et al.*, *Toxicol. Appl. Pharmacol.* **15**, 471 (1969); H. B. Schiefer, D. S. Hancock, *ibid.* **76**, 464 (1984); D. A. Creasia *et al.*, *Fundam. Appl. Toxicol.* **14**, 54 (1990). Implicated as a chemical warfare agent in Southeast Asia with nivalenol, *q.v.*: N. Wade, *Science* **214**, 34 (1981); R. T. Rosen, J. D. Rosen, *Biomed. Mass Spectrom.* **9**, 443 (1982). *Review: Developments in Food Science* vol. **4**, Y. Ueno, Ed., entitled "Trichothecenes: Chemical, Biological and Toxicological Aspects" (Kodansha Ltd. and Elsevier, New York, 1983) 310 pp. Review of pharmacokinetics and metabolism: B. Yagen, M. Bialer, *Drug Metab. Rev.* **25**, 281-323 (1993).

Crystals, mp 151-152°. $[\alpha]_D^{26} +15°$ (c = 2.58 in ethanol). Freely sol in ethyl alcohol, ethyl acetate, chloroform, DMSO and other organic solvents; slightly sol in petroleum ether; very slightly sol in water. LD_{50} orally in female rats: 4.0 mg/kg (Marasas). LD_{50} (mg/kg) in mice: 5.2 i.p., 4.2 i.v.; in rats: 7.0 intragastric, 0.9-1.3 i.p., 0.9 i.v., 2.0 s.c.; in guinea pigs: 3.0-4.0 orally, 5.3 intragastric, 1.0 i.m., 1.0-2.0 i.v., 1.0-2.0 s.c.; in pigs: 5.0 orally, 3.0 i.v. (Yagen, Bailer).

Caution: May be highly irritating to skin and mucous membranes. Direct contact may cause extensive inflammation and tissue necrosis (Marasas). Topical exposure has lead to systemic toxicity and death in experimental animals (Schiefer, Hancock).

9982. Tuaminoheptane. [123-82-0] 2-Heptanamine; 1-methylhexylamine; 2-aminoheptane; Tuamine. $C_7H_{17}N$; mol wt 115.22. C 72.97%, H 14.87%, N 12.16%. α-Adrenergic agonist; topical vasoconstrictor. Prepd by heating 2-bromoheptane on steam bath with alcoholic ammonia in pressure tube: Clarke, *J. Am. Chem. Soc.* **21**, 1027 (1899); by hydrogenation of methyl amyl ketone and ammonia in the presence of Raney nickel: Norton *et al.*, *J. Org. Chem.* **19**, 1054 (1954).

Volatile liq; d_4^{25} 0.7600-0.7660; bp_{760} 142-144°; n_D^{25} 1.4150-1.4200. Slightly soluble in water. pH of 1% aq soln 11.45. Freely sol in alcohol, ether, petr ether, chloroform, benzene.

Hydrochloride. [6159-35-9] $C_7H_{17}N \cdot HCl$; mol wt 151.68. Needles, mp 133°. Soluble in water. pH about 5.4.

Sulfate. [6411-75-2] Heptedrine. $C_{14}H_{34}N_2 \cdot H_2SO_4$; mol wt 328.51. Crystals, readily sol in water. The pH of a 1% soln is about 5.4.

THERAP CAT: Nasal decongestant.

9983. Tuberactinomycin. [11075-36-8] Polypeptide antibiotic mixture produced by *Streptomyces griseoverticillatus* var *tuberacticus*: Nagata *et al.*, *J. Antibiot.* **21**, 681 (1968); *eidem*, **US 3639580** (1972). Composed of tuberactinomycins A, B, N, and O. Structures: Yoshioka *et al.*, *Tetrahedron Lett.* **1971**, 2043; Wakamiya *et al.*, *Bull. Chem. Soc. Jpn.* **46**, 949 (1973); Wakamiya, Shiba, *J. Antibiot.* **27**, 900 (1974); *eidem*, **28**, 292 (1975). Total synthesis of tuberactinomycin O: T. Teshima *et al.*, *Tetrahedron Lett.* **1976**, 2343; *eidem*, *J. Antibiot.* **30**, 1073 (1977). Chemical studies on tuberactinomycin: T. Wakamiya *et al.*, *Heterocycles* **15**, 999 (1981).

Tuberactinomycins

	A	B	N	O
R_1	OH	H	OH	H
R_2	OH	OH	H	H

Hydrochloride. White solid, mp 244-264° (dec). $[\alpha]_D^{25} -31.5°$ (c = 1 in water). uv maxima in water: 268 nm ($E_{1cm}^{1\%}$ 330); in 1*N* HCl: 268.5 nm ($E_{1cm}^{1\%}$ 313); in 0.1*N* NaOH: 285 nm ($E_{1cm}^{1\%}$ 206.5). Sol in water; weakly sol in methanol. Practically insol in ethanol, pyridine, ether, chloroform, dioxane; insol in acetone, benzene. pKa_1 7.2; pKa_2 10.3.

Tuberactinomycin A. [33103-21-8] $C_{25}H_{43}N_{13}O_{11}$; mol wt 701.70.

Tuberactinomycin B *see* Viomycin.

Tuberactinomycin N *see* Enviomycin.

Tuberactinomycin O. [33137-73-4] (2*S*)-3-Amino-*N*-[(3*S*)-3,6-diamino-1-oxohexyl]-L-alanyl-L-seryl-L-seryl-(2*Z*)-3-[(aminocarbonyl)amino]-2,3-didehydroalanyl-2-[(4*R*)-2-amino-3,4,5,6-tetrahydro-4-pyrimidinyl]glycine (5 → 13)-lactam. $C_{25}H_{43}N_{13}O_9$; mol wt 669.70.

THERAP CAT: Antibacterial (tuberculostatic).

9984. Tubercidin. [69-33-0] 7-β-D-Ribofuranosyl-7H-pyrrolo[2,3-d]pyrimidin-4-amine; 4-amino-7-β-D-ribofuranosyl-7H-pyrrolo[2,3-d]pyrimidine; 7-deazaadenosine; sparsamycin A; U-10071. $C_{11}H_{14}N_4O_4$; mol wt 266.26. C 49.62%, H 5.30%, N 21.04%, O 24.04%. Antibiotic substance produced in the culture broth of *Streptomyces tubericidus*. Isoln: Anzai *et al., J. Antibiot.* **10A**, 201 (1957). Structure: Susuki, Marumo, *ibid.* **14A**, 34 (1961). Total synthesis: Tolman *et al., J. Am. Chem. Soc.* **91**, 2102 (1969). Crystal structure: Stroud, *Acta Crystallogr.* **29B**, 690 (1973); Abola, Sundaralingham, *ibid.* 697.

Needles from water, dec 247-248°. $[\alpha]_D^{17}$ −67° (50% acetic acid). uv max (0.01N NaOH): 270 nm (ε 12,100). Sol in acidic and alkaline soln. One gram dissolves in 330 ml water, 200 ml methanol, 2000 ml ethanol. Practically insol in acetone, ethyl acetate, chloroform, benzene, petr ether. LD_{50} i.v. in mice: 45 mg/kg (Anzai).

9985. Tuberculin. A filtrate from triturated *Mycobacteria tuberculosis*. **Old Tuberculin** is the culture filtrate prepd by boiling and then filtering the *M. tuberculosis* culture. Purified protein derivative is a more refined tuberculin prepd by precipitating the culture filtrate with ammonium sulfate. Purification and dermal reactivity: F. B. Seibert, J. T. Glenn, *Am. Rev. Tuberc.* **44**, 9 (1941). Review of tuberculin skin test: G. W. Comstock *et al., Am. Rev. Respir. Dis.* **124**, 356 (1981).

Clear, brownish liquid. Readily misc with water.
Purified Protein Derivative. [92129-86-7] PPD; Aplisol; Tubersol.

THERAP CAT: Diagnostic aid (tuberculosis).

9986. Tuberin. [2501-37-3]; [53643-53-1] (unspecified stereo). N-[(1E)-2-(4-Methoxyphenyl)ethenyl]formamide; N-trans-(p-methoxystyryl)formamide; N-formyl trans-p-methoxystyrylamine. $C_{10}H_{11}NO_2$; mol wt 177.20. C 67.78%, H 6.26%, N 7.90%, O 18.06%. Antitubercular antibiotic isolated from the broth filtrate of *Streptomyces amakusaensis*: K. Ohkuma *et al., J. Antibiot. A* **15**, 115 (1962); Sumiki *et al.,* **JP 64 7399** (1964 to Inst. Phys. & Chem. Res.), *C.A.* **62**, 8355g (1965). Structure and synthesis: K. Anzai *et al., J. Antibiot. A* **15**, 110, 117, 123 (1962). Alternate synthesis: I. J. Massey, I. T. Harrison, *Chem. Ind. (London)* **1977**, 920. Biosynthetic studies: K. M. Cable *et al., J. Chem. Soc. Perkin Trans. 1* **1987**, 1593.

Prisms from benzene, mp 132-133°. Stable in weakly acidic or weakly alkaline solns. uv max (methanol): 219, 285 nm ($E_{1cm}^{1\%}$ 870, 1710). Sol in the lower alcohols, ethyl acetate, acetone; moderately sol in carbon tetrachloride, chloroform; sparingly sol in water, benzene. Practically insol in petr ether.
Dihydrotuberin. $C_{10}H_{13}NO_2$. Liquid. n_D^{19} 1.5349.

9987. Tubocurarine Chloride. [57-94-3] (13aR,25aS)-2,3,-13a,14,15,16,25,25a-Octahydro-9,19-dihydroxy-18,29-dimethoxy-1,14,14-trimethyl-13H-4,6:21,24-dietheno-8,12-metheno-1H-pyrido[3′,2′:14,15][1,11]dioxacycloeicosino[2,3,4-ij]isoquinolinium chloride hydrochloride (1:1:1); dextrotubocurarine chloride; Jexin; Tubarine. $C_{37}H_{42}Cl_2N_2O_6$; mol wt 681.65. C 65.20%, H 6.21%, Cl 10.40%, N 4.11%, O 14.08%. Arrow tip poison of South American Indians. Induces neuromuscular block through interactions with acetylcholine receptor. Identification as active principle from museum specimen: H. King, *J. Chem. Soc.* **1935**, 1381. Isoln from *Chondodendron tomentosum* R. & P. *Menispermaceae:* J. D.

Dutcher, *J. Am. Chem. Soc.* **68**, 419 (1946). Purification and structure determn: *idem, ibid.* **74**, 2221 (1952). Revised structure: A. J. Everett *et al., Chem. Commun.* **1970**, 1020. Conformational analysis: B. S. Zhorov, N. B. Brovtsyna, *J. Membr. Biol.* **135**, 19 (1993); by NMR: Y. Fraenkel *et al., Biochemistry* **33**, 644 (1994). Binding characterization: N. Shaker *et al., J. Pharmacol. Exp. Ther.* **220**, 172 (1982); M. E. O'Leary *et al., Am. J. Physiol.* **266**, C648 (1994). Clinical pharmacokinetics and dynamics: D. M. Fisher, *Anesthesiology* **57**, 203 (1982). Toxicity data: H. Rosen *et al., Proc. Soc. Exp. Biol. Med.* **120**, 511 (1965). Clinical trial in prevention of muscle fasciculations: S. C. Harvey *et al., Anesth. Analg.* **87**, 719 (1998). Review of clinial use: G. S. Perotti, *J. Am. Assoc. Nurse Anesth.* **45**, 182-186 (1977). Comprehensive description: C. Papastephanou, *Anal. Profiles Drug Subs.* **7**, 477-500 (1978).

Hexagonal and pentagonal microplatelets from water; can exist in the form of various hydrates. The anhydrous material (dec 274-275°) takes up water in moist atm until it reaches the pentahydrate stage, dec ~270°. uv max (H_2O): 280 nm ($E_{1cm}^{1\%}$ 118). $[\alpha]_D^{20-25}$ +215° (c = 0.25-0.3 g/100 ml). Soly (25°): ~50 mg/ml water; but supersatd solns are formed readily. Presence of 1.0N HCl diminishes soly by about one-third. Sparingly sol in ethanol, methanol. Insol in pyridine, chloroform, benzene, acetone, ether. pK: 7.4. LD_{50} in mice, rats (mg/kg): 33.2, 27.8 orally in DMSO; 59.5, 36.9 orally in water (Rosen).

l-**Form.** Isoln of *l*-form from *Ch. tomentosum:* H. King, *Nature* **158**, 515 (1946); *idem, J. Chem. Soc.* **1947**, 936. Needles from water as pentahydrate, mp 268° (effervescence). $[\alpha]_D^{20}$ −258° (c = 0.38) for the anhydr salt.

THERAP CAT: Neuromuscular blocking agent.
THERAP CAT (VET): Neuromuscular blocking agent.

9988. Tubulin. Colchicine-binding protein. The subunit protein of *microtubules*, which are large protein assemblies that play an important role in eukaryotic cell form determination and dynamics. Microtubules have the general structure of long hollow cylinders within which 13 protofilaments of tubulin are arranged in a parallel manner to the cylinder axis. The axial arrangement of the protofilaments with respect to each other results in the appearance of a helical structure, and the *in vitro* microtubule assembly process generally follows the laws of helical protein polymerization, *cf. Thermodynamics of the Polymerization of Proteins*, F. Oosawa, S. Asakura, Eds. (Academic Press, New York, 1975). Tubulin is an asymmetric dimer consisting of two nearly identical molecules, α-*tubulin* and β-*tubulin*, each having mol wts of about 55,000. The two molecules can be separated due to differences in electrophoretic mobilities. Isoln from mammalian brain using colchicine binding: R. C. Weisenberg *et al., Biochemistry* **7**, 4466 (1968). Discovery of conditions for microtubule assembly *in vitro:* R. C. Weisenberg, *Science* **177**, 1104 (1972); G. G. Borisy, J. B. Olmstead, *ibid.* 1196. Prepn of large quantities of brain tubulin through successive assembly-disassembly cycles: M. L. Shelanski *et al., Proc. Natl. Acad. Sci. USA* **70**, 765 (1973). Purification of tubulin from rat pancreas: J. F. Launay *et al., Biochem. Biophys. Res. Commun.* **111**, 253 (1983). Structure of two human α-tubulin genes: C. Wilde *et al., Proc. Natl. Acad. Sci. USA* **79**, 96 (1982). Structure and arrangement of protofilaments in microtubules and tubulin sheets: B. F. McEwen, *Diss. Abstr. B* **43**, 942 (1982). Series of articles on prepn, isoln, and purification of tubulin from various sources: *Methods Enzymol.* **85**, Pt. B, 376-417 (1982). Reviews: J. A. Snyder, J. R. McIntosh, *Annu. Rev. Biochem.* **45**, 699-720 (1976); S. N. Timasheff, L. M. Grisham, *ibid.* **49**, 565-591 (1980); M. F. Carlier, *Mol. Cell. Biochem.* **47**, 97-113 (1982).

Purified calf brain tubulin retains many of its *in vivo* biochemical characteristics, such as the ability to self-assemble into microtubules and the response of the assembly reaction to inhibitory effects of cold temperature, Ca^{2+}, and anti-microtubule agents, *e.g.* vinblastine and colchicine, *q.q.v.* uv max (PG buffer): 278 nm (ε 1.33 ml mg^{-1} cm^{-1}).

9989. Tuftsin. [9063-57-4] L-Threonyl-L-lysyl-L-prolyl-L-arginine; N^2-[1-(N^2-L-threonyl-L-lysyl)-L-prolyl]-L-arginine. $C_{21}H_{40}$-N_8O_6; mol wt 500.60. C 50.39%, H 8.05%, N 22.38%, O 19.18%. A naturally occurring tetrapeptide having a variety of immunopotentiating properties, especially stimulation and enhancement of phagocytosis. It also exhibits antitumor and antibacterial activity and has been shown to possess chemotactic, migration-enhancing, and mitogenic properties for leukocytes. Discovered during research on the physiological role of cytophilic gamma-globulin: V. A. Najjar, K. Nishioka, *Nature* **228**, 672 (1970). Produced in the spleen; present in mammalian blood in the gamma globulin fraction as part of the larger molecule **leucokinin**. Isoln and characterization: K. Nishioka *et al.*, *Biochim. Biophys. Acta* **310**, 217 (1973); V. A. Najjar, US 3778426 (1973 to Research Corp.). Solid phase synthesis: K. Nishioka *et al.*, *Biochem. Biophys. Res. Commun.* **47**, 172 (1972); *eidem, Biochim. Biophys. Acta* **310**, 230 (1973). Synthesis by fragment condensation: J. Vicar *et al.*, *Collect. Czech. Chem. Commun.* **41**, 3467 (1976); by liquid phase method: S. Nozaki *et al.*, *Bull. Chem. Soc. Jpn.* **50**, 422 (1977). ^{13}C-NMR and circular dichroism studies: I. Z. Siemion *et al.*, *Eur. J. Biochem.* **112**, 339 (1980). Conformational studies have provided conflicting evidence on the structure: M. Blumenstein *et al.*, *Biochemistry* **18**, 4247 (1979). Specific receptors on macrophages, monocytes, and granulocytes are thought to mediate the biological activity of tuftsin: A. Constantopoulos, V. A. Najjar, *J. Biol. Chem.* **248**, 3819 (1973); R. M. G. Nair *et al.*, *Immunochemistry* **15**, 901 (1978); Z. Bar-Shavit *et al.*, *Biochem. Biophys. Res. Commun.* **94**, 1445 (1980). Its physiological significance has been shown in patients in whom tuftsin deficiency has resulted in a human syndrome with increased incidence of severe infections: *Macrophages and Lymphocytes, Part A*, M. R. Escobar, H. Friedman, Eds. (Plenum Press, New York, 1980) pp 131-147; *Lymphokine Reports*, E. Pick, Ed. (Academic Press, New York, 1980) pp 157-159; V. A. Najjar, *Med. Biol.* **59**, 134 (1981). General biological properties: V. A. Najjar, *Mol. Cell. Biochem.* **41**, 1 (1981). Antitumor activity: K. Nishioka *et al.*, *ibid.* 13. Bactericidal activity: J. Martinez, F. Winternitz, *ibid.* 123. Analogs: F. Z. Siemion, *ibid.* 99. *Reviews:* V. A. Najjar, *Exp. Cell Biol.* **46**, 114-126 (1978); *eidem, Adv. Exp. Med. Biol.* **121A**, 131-147 (1980); K. Nishioka *et al.*, *Life Sci.* **28**, 1081-1090 (1981); V. A. Najjar, *Mol. Cell. Biochem.* **41**, 73-98 (1981). Conference proceedings: *Ann. N.Y. Acad. Sci.* **419**, entitled "Antineoplastic, Immunogenic and Other Effects of the Tetrapeptide Tuftsin: a Natural Macrophage Activator", V. A. Najjar, M. Fridkin, Eds. (1983) pp 1-273.

Thr–Lys–Pro–Arg

9990. Tulathromycin. Draxxin. $C_{41}H_{79}N_3O_{12}$; mol wt 806.09. C 61.09%, H 9.88%, N 5.21%, O 23.82%. Triamilide antibiotic for treatment of bovine and porcine respiratory disease. Exists as an equilibrium mixture of two isomeric forms, tulathromycin A (~90%) and B (~10%). HPLC determn in livestock plasma and lung homogenates: D. Gáler *et al.*, *J. Agric. Food Chem.* **52**, 2179 (2004). Pharmacokinetics in cattle: M. A. Nowakowski *et al.*, *Vet. Ther.* **5**, 60 (2004); in swine: H. A. Benchaoui *et al.*, *J. Vet. Pharmacol. Ther.* **27**, 203 (2004).

Tulathromycin A

Tulathromycin A. [217500-96-4] ($2R,3S,4R,5R,8R,10R,11R$,-$12S,13S,14R$)-13-[[2,6-Dideoxy-3-C-methyl-3-O-methyl-4-C-[(propylamino)methyl]-α-L-*ribo*-hexopyranosyl]oxy]-2-ethyl-3,4,10-trihydroxy-3,5,8,10,12,14-hexamethyl-11-[[3,4,6-trideoxy-3-(dimethylamino)-β-D-*xylo*-hexopyranosyl]oxy]-1-oxa-6-azacyclopentadecan-15-one; CP-472295. Prepn: B. S. Bronk *et al.*, **WO 9856802**; *eidem*, **US 6420536** (1998, 2002 both to Pfizer); and crystal structure: M. A. Letavic *et al.*, *Bioorg. Med. Chem. Lett.* **12**, 2771 (2002).

Tulathromycin B. [280755-12-6] ($2R,3R,6R,8R,9R,10S,11S$,-$12R$)-11-[[2,6-Dideoxy-3-C-methyl-3-O-methyl-4-C-[(propylamino)methyl]-α-L-*ribo*-hexopyranosyl]oxy]-2-[($1R,2R$)-1,2-dihydroxy-1-methylbutyl]-8-hydroxy-3,6,8,10,12-pentamethyl-9-[[3,4,6-trideoxy-3-(dimethylamino)-β-D-*xylo*-hexopyranosyl]oxy]-1-oxa-4-azacyclotridecan-13-one; CP-547272. Prepn: R. J. Rafka *et al.*, **WO 0031097**; *eidem*, **US 6329345** (2000, 2001 both to Pfizer).

THERAP CAT (VET): Antibacterial.

9991. Tulobuterol. [41570-61-0] 2-Chloro-α-[[(1,1-dimethylethyl)amino]methyl]benzenemethanol; α-[(*tert*-butylamino)methyl]-o-chlorobenzyl alcohol. $C_{12}H_{18}ClNO$; mol wt 227.73. C 63.29%, H 7.97%, Cl 15.57%, N 6.15%, O 7.03%. A β-adrenergic receptor agonist, related structurally to terbutaline, *q.v.* Prepn: H. Kato, S. Kurata, DE 2244737 (1973 to Hokuriku), *C.A.* **78**, 147538a (1973). Pharmacology: S. Kubo *et al.*, *Arzneim.-Forsch.* **25**, 1028 (1975); *eidem, ibid.* **27**, 1433 (1977); I. Uesaka *et al.*, *ibid.* 1439. Metabolism: T. Fujiihashi *et al.*, *Oyo Yakuri* **18**, 347 (1979), *C.A.* **92**, 121781 (1980); K. Matsumura *et al.*, *Yakugaku Zasshi* **101**, 198 (1981), *C.A.* **94**, 167404 (1981). Determn of tulobuterol and its metabolites in human urine: *eidem, J. Chromatogr.* **222**, 53 (1981). Toxicological studies: S. Kubo *et al.*, *Oyo Yakuri* **13**, 197, 317 (1977), *C.A.* **88**, 83591-2 (1978).

Crystals, mp 89-91°. LD$_{50}$ in male mice, rats, rabbits (mg/kg): 305, 850, 563 orally; 170, 417, 164 s.c. (Kubo, 1975).

Hydrochloride. [56776-01-3] C-78; Atenos; Berachin; Brelomax; Bremax; Hokunalin; Respacal. $C_{12}H_{18}ClNO.HCl$; mol wt 264.19. White crystalline powder, mp 161-163°.

THERAP CAT: Bronchodilator.

9992. Tumor Necrosis Factor. TNF. Cytokine produced by activated macrophages as part of the cellular immune response. Originally characterized by its selective hemorrhagic necrosis of tumor cells. Identification in the sera of endotoxin treated mice previously sensitized with bacillus Calmette-Guerin (BCG): E. A. Carswell *et al.*, *Proc. Natl. Acad. Sci. USA* **72**, 3666 (1975). Partial purification: S. Green *et al.*, *ibid.* **73**, 381 (1976). Preliminary characterization of murine TNF: T. Haranaka, N. Satomi, *Jpn. J. Exp. Med.* **51**, 191 (1981); F. C. Kull, P. Cuatrecasas, *J. Immunol.* **126**, 1279 (1981); of rabbit TNF: M. R. Ruff, G. E. Gifford, *ibid.* **125**, 1671 (1980); N. Matthews *et al.*, *Br. J. Cancer* **42**, 416 (1980). Identification of macrophages as cellular source of TNF: D. N. Männel *et al.*, *Infect. Immun.* **30**, 523 (1980); N. Santomi *et al.*, *Jpn. J. Exp. Med.* **51**, 317 (1981). Differentiation from interferon: N. Bloksma *et al.*, *Cancer Immunol. Immunother.* **14**, 41 (1982). Cytotoxic effect on the malaria parasite *Plasmodium falciparum*: C. G. Haidaris *et al.*, *Infect. Immun.* **42**, 385 (1983); A. O. Wozencraft *et al.*, *ibid.* **43**, 664 (1984). Activity against transplanted human and murine tumors in mice: T. Haranaka *et al.*, *Int. J. Cancer* **34**, 263 (1984). Human TNF is a trimer of 3 identical subunits with 157 amino acid residues and mol wt 17,350 Da. Cloning and expression of cDNA for human TNF in *E. coli*: D. Pennica *et al.*, *Nature* **312**, 724 (1984); T. Shirai *et al.*, *ibid.* **313**, 803 (1985); A. M. Wang *et al.*, *Science* **228**, 149 (1985). Identity with **cachectin**: B. Buetler, A. Cerami, *Nature* **320**, 584 (1986). Structure: E. Y. Jones *et al.*, *ibid.* **338**, 225 (1989). Preliminary evaluation with interferon-γ in metastatic melanoma: S. Retsas *et al.*, *Br. Med. J.* **298**, 1290 (1989). *Reviews:* M. R. Ruff, G. E. Gifford, *Lymphokines* **2**, 235 (1981); L. J. Old, *Sci. Am.* **258** (5), 59-60, 70-75 (1988).

Note: **Lymphotoxin**, a TNF-like factor produced by lymphocytes has been referred to as **TNF-β**.

9993. Tung Oil. China wood oil. A drying oil from seeds of *Aleurites cordata* Steud., *Euphorbiaceae*, indigenous to China and Japan, but now grown also in Florida. Chief fatty acid component is eleostearic acid. Unlike linseed and soybean oils, it need not be refined.

Pale yellow liquid; characteristic disagreeable odor. On long keeping, or on heating for a short time at 300°, polymerizes to a stiff jelly. d 0.936-0.943. Iodine no. 163-171. Sapon no. 190-197. Sol in chloroform, ether, carbon disulfide, oils; the polymerized product is practically insol in the usual organic solvents.

USE: Manuf quick-drying wood varnishes, linoleum, and floor cloth; in India rubber substitutes, insulating masses; for water-proofing paper and other tissues.

9994. Tungsten. [7440-33-7] Wolfram. W; at. wt 183.84; at. no. 74; valences 6, 5, 4, 3, 2. Group VIB (6). Naturally occurring isotopes: 180 (0.135%); 182 (26.4%); 183 (14.4%); 184 (30.6%); 186 (28.4%); artificial radioactive isotopes: 173-179; 181; 185; 187-189. Discovered by C. W. Scheele in 1781, isolated in 1783 by J. J. and F. de Elhuyar. One of the rarer metals, comprises about 1.5 ppm of the earth's crust. Chief ores are **wolframite** [(Fe,Mn)WO$_4$] and scheelite (CaWO$_4$). Found chiefly in China, Malaya, Mexico, Alaska, South America and Portugal. Scheelite ores mined in the U.S. carry from 0.4-1.0% WO$_3$. Description of isoln processes: K. C. Li, C. Y. Wang, "Tungsten" in *A.C.S. Monograph Series* no. **94** (Reinhold, New York, 3rd ed., 1955) pp 113-269; G. D. Rieck, *Tungsten and Its Compounds* (Pergamon Press, New York, 1967) 154 pp. Reviews: Parish, *Adv. Inorg. Chem. Radiochem.* **9**, 315-354 (1966); Rollinson, "Chromium, Molybdenum and Tungsten" in *Comprehensive Inorganic Chemistry* vol. 3, J. C. Bailar, Jr. *et al.*, Eds. (Pergamon Press, Oxford, 1973) pp 623-624, 742-769. Review of toxicology and human exposure: *Toxicological Profile for Tungsten* (PB2006-100007, 2005) 203 pp.

Steel-gray to tin-white metal; body centered cubic structure. d$_4^{20}$ 18.7-19.3; depends on extent of working. Hardness 6.5-7.5. mp 3410°. bp$_{760}$ 5900°. Spec heat (20°): 0.032 cal/g/°C. Heat of fusion 44 cal/g. Heat of vaporization 1150 cal/g. Electrical resistivity (20°): 5.5 μohm-cm. Stable in dry air at ordinary temps, forms the trioxide at red heat. Not attacked by water, but oxidized to the dioxide by steam. Very stable to acids, attacked only superficially by concd nitric acid or aqua regia. Powdered tungsten can be pyrophoric under the right conditions. Slowly sol in fused potassium hydroxide or sodium carbonate in presence of air; sol in a fused mixture of NaOH and nitrate. Attacked by fluorine at room temp; by chlorine at 250-300° giving the hexachloride in absence of air, and the trioxide and oxychloride in the presence of air.

Caution: Potential symptoms of overexposure to tungsten are irritation of eyes, skin, respiratory system; diffuse pulmonary fibrosis; loss of appetite, nausea, cough; blood changes. Potential symptoms of overexposure to tungsten carbide (cemented) are irritation of eyes, skin, respiratory system; possible skin sensitization to cobalt, nickel; diffuse pulmonary fibrosis; loss of appetite, nausea, cough; blood changes. *See NIOSH Pocket Guide to Chemical Hazards* (DHHS/NIOSH 97-140, 1997) p 324.

USE: To increase hardness, toughness, elasticity, and tensile strength of steel; manuf alloys; manuf filaments for incandescent lamps and in electron tubes; in contact points for automotive, telegraph, radio and television apparatus; in phonograph needles.

9995. Tungsten Carbide. [12070-12-1] CW; mol wt 195.85. C 6.13%, W 93.87%. WC. Stable, wear-resistant material with hardness close to diamond. Prepn: S. Hilpert, M. Ornstein, *Ber.* **46**, 1669 (1913); M. R. Andrews, *J. Phys. Chem.* **27**, 270 (1923); A. E. Newkirk, I. Aliferis, *J. Am. Chem. Soc.* **79**, 4629 (1957). Heat of combustion: L. D. McGraw *et al.*, *ibid.* **69**, 329 (1947). Development of WC dental burs: J. Osborne *et al.*, *Br. Dent. J.* **90**, 229 (1951). Review of industrial-scale production and applications: A. S. Hester *et al.*, *Ind. Eng. Chem.* **52**, 94-100 (1960). Structural and electronic properties: A. Y. Liu *et al.*, *Phys. Rev. B* **38**, 9483 (1988). Catalytic properties: E. Iglesia *et al.*, *Catal. Today* **15**, 307 (1992). Physical properties and calculated electronic structure: H. W. Hugosson, H. Engqvist, *Int. J. Refract. Met. Hard Mater.* **21**, 55 (2003). FTIR characterization of thin films and bulk samples: P. Hoffmann *et al.*, *Mater. Charact.* **50**, 255 (2003). Combustion synthesis: G. Jiang *et al.*, *Ceram. Int.* **30**, 185, 191 (2004). Review of toxicology

and human exposure: *Toxicological Profile for Tungsten* (PB2006-100007, 2005) 203 pp.

mp 2600-2850°. Hardness: 16-22 GPa. Youngs modulus: 696 GN m^{-2}. Fracture toughness: 28 MPa m$^{1/2}$. Compressive strength (20°): 5 GPa. Heat of combustion ($\Delta H_{298.16}$): −285.80 ±0.07 kcal/mol.

Caution: Cobalt-tungsten carbide powders and hard metals are listed as reasonably anticipated to be human carcinogens: *Report on Carcinogens, Twelfth Edition* (PB2011-111646, 2011) p 115.

USE: In drill tips, mining, metal cutting tools, dental burs, surgical instruments, wear-resistant coatings, wire-drawing dies. In mfr of cemented carbide. Substitute for noble metals in catalysis.

9996. Tungsten Hexafluoride. [7783-82-6] (*OC*-6-11)-Tungsten fluoride (WF$_6$). F$_6$W; mol wt 297.83. F 38.27%, W 61.73%. WF$_6$. Prepd by direct fluorination of powdered tungsten. Can be purified by distillation, preferably under pressure: Ruff, Ascher, *Z. Anorg. Allg. Chem.* **196**, 413 (1931); Henkel, Klemm, *ibid.* **222**, 68 (1935); Marchi, *Inorg. Synth.* **3**, 181 (1950); from ClF and tungsten: Pitts, Jache, *Inorg. Chem.* **7**, 1661 (1968). Review of toxicology and human exposure: *Toxicological Profile for Tungsten* (PB2006-100007, 2005) 203 pp.

Colorless gas or pale yellow liquid. Orthorhombic, deliquescent crystals when solid. *Poisonous; corrosive.* d$_{liq}^{15}$ 3.441. mp 2.3°. bp 17.5°. Soly in anhyd HF: 3.14 moles/1000 g HF, Frlec, Hyman, *Inorg. Chem.* **6**, 1596 (1967). May be stored in glass pressure ampuls.

USE: In the electronics industry as a source of tungsten metal that connects the aluminum layers within semiconductor devices.

9997. Tungsten Trioxide. [1314-35-8] Tungsten oxide (WO$_3$); tungstic anhydride. O$_3$W; mol wt 231.84. O 20.70%, W 79.30%. WO$_3$. Prepn from sodium tungstate: Hein, Herzog, in *Handbook of Preparative Inorganic Chemistry* vol. 2, G. Brauer, Ed. (Academic Press, New York, 2nd ed., 1965) pp 1423-1424. Review of toxicology and human exposure: *Toxicological Profile for Tungsten* (PB2006-100007, 2005) 203 pp.

Canary yellow, heavy powder; dark orange when heated, regaining the original color on cooling. Insol in water. Sol in caustic alkalies; very slightly sol in acids.

USE: Manuf tungstates which are used for x-ray screens and for fireproofing fabrics.

9998. Tungstic(VI) Acid. [7783-03-1] (*T*-4)-Tungstate (WO$_4^{2-}$) hydrogen (1:2). H$_2$O$_4$W; mol wt 249.85. H 0.81%, O 25.61%, W 73.58%. H$_2$WO$_4$. Prepn: Morley, *J. Chem. Soc.* **1930**, 1990.

Yellow or greenish-yellow powder. Insol in water and acid except hydrofluoric acid. Slowly sol in solns of caustic alkalies. When freshly pptd from a soluble tungstate it contains 1 mol H$_2$O, and is appreciably sol in water.

9999. Tunicamycin. [11089-65-9] A family of nucleoside antibiotics produced by *Streptomyces lysosuperificus*. Isoln, characterization: A. Takatsuki *et al.*, *J. Antibiot.* **24**, 215 (1971). Biological properties: A. Takatsuki, G. Tamura, *ibid.* **224**. Effect on microorganisms: A. Takatsuki *et al.*, *ibid.* **25**, 75 (1972). Tunicamycin is produced as a mixture of at least 10 homologous antibiotics, the main components being tunicamycins V, VII, II and X, also referred to respectively as A, B, C, D. They contain uracil, *N*-acetylglucosamine, an 11-carbon aminodialdose called **tunicamine**, and a fatty acid linked to the amino group of tunicamine. The homologs differ in their fatty acid components, which vary in degree of saturation and chain length and branching. Structural elucidation: A. Takatsuki *et al.*, *Agric. Biol. Chem.* **41**, 2307 (1977); T. Ito *et al.*, *ibid.* **44**, 695 (1980). HPLC sepn of components: W. C. Mahoney, D. Duskin, *J. Biol. Chem.* **254**, 6572 (1979). Approaches to synthesis: Y. Fukuda *et al.*, *Bull. Chem. Soc. Jpn.* **55**, 880 (1982). Tunicamycin has been shown to interfere with glycoprotein synthesis in yeast and mammalian systems: A. Takatsuki *et al.*, *Agric. Biol. Chem.* **39**, 2089 (1975); S. Kuo, J. O. Lampen, *Arch. Biochem. Biophys.* **172**, 574 (1976). Effect on epidermal glycoprotein and glycosaminoglycan synthesis *in vitro*: I. A. King, A. Tabiowo, *Biochem. J.* **198**, 331 (1981). Enhancement of antiviral and anticellular activity of interferon, *q.v.*: R. K. Maheshwari *et al.*, *Science* **219**, 1339 (1983). *Review:* A. D. Elbein, *Trends Biochem. Sci.* **6**, 219-221 (1981).

Book: *Tunicamycin*, G. Tamura, Ed. (Japan Sci. Soc., Tokyo, 1982) 220 pp. *See also* Streptovirudin.

Tunicamycins II, V, VII, X
(n = 8, 9, 10, 11)

White cryst powder, mp 234-235° (dec). $[\alpha]_D^{20}$ +52° (c = 0.5 in pyridine). uv max (methanol): 205, 260 nm ($E_{1cm}^{1\%}$ 230, 110). Sol in alk water, pyridine, hot methanol. Slightly sol in ethanol, butanol. Practically insol in acetone, ethyl acetate, chloroform, benzene, acidic water. When dissolved in water at 100° and held for 30 min, stable at neutral and alk pH, unstable at acidic pH.

USE: As a tool in studying glycoproteins in a wide variety of biological systems.

10000. Tunichrome B-1. [97689-87-7] (αE)-3,5-Dihydroxy-L-tyrosyl-α,β-didehydro-3,5-dihydroxy-N-[(1Z)-2-(3,4,5-trihy-droxyphenyl)ethenyl]tyrosinamide; TB-1. $C_{26}H_{25}N_3O_{11}$; mol wt 555.50. C 56.22%, H 4.54%, N 7.56%, O 31.68%. Major component of a group of polyphenolic yellow blood pigments isolated from the sea squirt (tunicate) *Ascidia nigra* (Linnaeus). The *tunichromes* are strong biological reducing agents involved in the selective accumulation of vanadium by *A. nigra*. Isoln of crude tunichrome mixture from the blood cells of *A. nigra*, action in converting vanadium-(V) to vanadium(III): I. G. Macara *et al.*, *Biochem. J.* **181**, 457 (1979). Isoln of tunichrome B-1 by HPLC under anaerobic conditions, characterization and structure determn by mass spectra, NMR: R. C. Bruening *et al.*, *J. Am. Chem. Soc.* **107**, 5298 (1985). Brief account of the role of tunichromes in studies into the biochemical nature of vanadium and other trace metals: R. J. Seltzer, *Chem. Eng. News* **63**, 67 (Sept. 16, 1985). *Review:* R. C. Bruening *et al.*, *J. Nat. Prod.* **49**, 193 (1986).

Yellow solid, sol in methanol. uv max (methanol): 210, 340 nm (ε 68000, 19600).

10001. Turanose. [547-25-1] 3-O-α-D-Glucopyranosyl-D-fructose; 3-(α-D-glucosido)-D-fructose. $C_{12}H_{22}O_{11}$; mol wt 342.30. C 42.11%, H 6.48%, O 51.41%. Prepd from melezitose: Tanret, *Compt. Rend.* **142**, 1424 (1906); Bridel, Aagard, *Bull. Soc. Chim. Biol.* **9**, 884 (1927); Hudson, Pacsu, *J. Am. Chem. Soc.* **52**, 2522 (1930); Pacsu in *Methods in Carbohydrate Chemistry* vol. I (Academic Press, New York, 1962) p 353. Structure: Isbell, Pigman, *J. Res. Natl. Bur. Stand.* **20**, 787 (1938); Isbell, *ibid.* **26**, 35 (1941); Hudson, *J. Org. Chem.* **9**, 117, 470 (1944); Hassid, Ballou in *The*

Carbohydrates, W. Pigman, Ed. (Academic Press, New York, 1957) p 508. Crystal structure: A. Neuman *et al.*, *Acta Crystallogr.* **B34**, 242 (1978).

Nonhygroscopic prisms from water + alc. Very sweet taste. Dec 157°. Shows mutarotation. $[\alpha]_D^{20}$ +27.3° → +75.8° (c = 4 in water). Freely sol in water, methanol. One gram dissolves in 19 ml of 95% alc. Acid hydrolysis yields 1 mol D-glucose and 1 mol D-fructose. Reducing power about ½ of D-glucose.

10002. 2-(*p*-Toluyl)benzoic Acid. [85-55-2] 2-(4-Methylben-zoyl)benzoic acid; *o*-*p*-toluylbenzoic acid; 4'-methylbenzophenone-2-carboxylic acid. $C_{15}H_{12}O_3$; mol wt 240.26. C 74.99%, H 5.03%, O 19.98%. Prepd from phthalic anhydride and toluene in the presence of aluminum chloride: Friedel, Crafts, *Ann. Chim. Phys.* [6] **14**, 447 (1888); Fieser, *Org. Synth.* **4**, 73 (1925).

Monohydrate. Triclinic pinacoidal crystals from alcohol. Sweet taste. Becomes anhydr at 100°, then melts at 146°. Slightly sol in boiling water; freely sol in alcohol, benzene, ether, acetone, boiling toluene.

Methyl ester. $C_{16}H_{14}O_3$. Plates from methanol, mp 53°. Soluble in alcohol, benzene.

10003. Toluylene Blue. [97-26-7] N-[4-[(2,4-Diamino-5-methylphenyl)imino]-2,5-cyclohexadien-1-ylidene]-N-methylmeth-anaminium chloride (1:1); [4-(4,6-diamino-*m*-tolyl)imino-2,5-cyclo-hexadien-1-ylidene]dimethylammonium chloride; C.I. 49410. $C_{15}H_{19}ClN_4$; mol wt 290.80. C 61.95%, H 6.59%, Cl 12.19%, N 19.27%. Prepd by condensation of *p*-nitrosodimethylaniline with 2,4-diaminotoluylene: Witt, *Ber.* **12**, 933 (1879); *DE 15272*; *Frdl.* **1**, 274. Also formed by irradiating a mixture of dimethyl-*p*-phenyl-enediamine and *m*-toluylenediamine: Loiseleur, *Compt. Rend.* **237**, 461 (1953). *See also Colour Index* vol. **4** (3rd ed., 1971) p 4443.

Monohydrate. Prismatic, copper-brown, shiny crystals. Gives blue soln with cold water, alc, acetic acid.

USE: Biological stain.

10004. Turkey-Red Oil. Sulfated castor oil; red oil. Prepd by treating castor oil with 15-30% H_2SO_4 at 25-30° for several hrs, followed by washing and neutralizing with NaOH soln. The product is an anion-active wetting agent. The commercial mfg process actually esterifies the 12-hydroxyl group in ricinoleic acid (the main constit of castor oil) thus producing a true sulfate ester, and not a sulfonate.

USE: In finishing cotton and linen; to obtain bright, clear colors in dyeing fabrics.

10005. Turmeric. Curcuma longa; tumeric; saffron Indian. From rhizome of *Curcuma longa* Linn. (*C. domestica* Valeton), *Zin-*

ar-Turmerone

giberaceae. Habit. India, China, East Indies. *Constit.* Yellow coloring matter (curcumin), *p,p*-dihydroxydicinnamoylmethane, *p*-hydroxycinnamoylferuloylmethane, *p*,α-dimethylbenzyl alcohol, 1-methyl-4-acetyl-1-cyclohexene, turmerone, α-phellandrene, sabinene, zingiberene, cineol, borneol, caprylic acid. Isoln of curcumin from turmeric: Janaki, Bose, *J. Indian Chem. Soc.* **44**, 985 (1967).

Turmeric has an aromatic pepper-like but somewhat bitter taste and gives curry dishes their characteristic yellowish color.

USE: Condiment (as curry powder), color for ointments.

10006. *ar*-Turmerone. [532-65-0] (6*S*)-2-Methyl-6-(4-methylphenyl)-2-hepten-4-one; 2-methyl-6-*p*-tolyl-2-hepten-4-one. C_{15}-$H_{20}O$; mol wt 216.32. C 83.29%, H 9.32%, O 7.40%. Isoln of naturally occurring (+)-form from Curcuma oil *(Curcuma longa* Linn., *Zingiberaceae):* H. Rupe *et al., Helv. Chim. Acta* **17**, 372 (1934). Structure: H. Rupe, A. Gassmann, *ibid.* **19**, 569 (1936). Abs config: V. K. Honwad, A. S. Rao, *Tetrahedron* **20**, 2921 (1964). Total synthesis: A. I. Meyers, R. K. Smith, *Tetrahedron Lett.* **1979**, 2749; T. Sato *et al., ibid.* **1980**, 3377.

Oil. bp_{10} 159-160°. $[\alpha]_D^{20}$ +82.21° (Rupe, Gassman); also reported as $[\alpha]_D^{22}$ +59.9° (c = 4.5 in hexane) (Sato). d_4^{20} 0.9634.

(±)-form. [38142-58-4] Synthesis: J. Colonge, J. Chambion, *C. R. Hebd. Seances Acad. Sci.* **222**, 557 (1946); T.-L. Ho, *Synth. Commun.* **11**, 579 (1981). n_D^{21} 1.5218.

10007. Turpentine. [9005-90-7] Gum thus; pine resin. Oleoresin from *Pinus palustris* Mill. and from other species of *Pinus*, *Pinaceae*. "Turpentine" also is used to designate the volatile oil. Review of production and properties: J. J. W. Coppen, G. A. Hone, *Gum Naval Stores: Turpentine and Rosin from Pine Resin* (FAO, Rome, 1995) 71 pp; of production and uses: S. K. Srivastava, M. C. Nigam, *Curr. Res. Med. Aromat. Plants* **3**, 49-70 (1981); T. Plocek, *Perfum. Flavor.* **23**, 1-6 (1998).

Yellowish, opaque, sticky masses; characteristic odor and taste. *Flammable.* Insol in water. Sol in alc, chloroform, ether, glacial acetic acid.

Volatile oil. [8006-64-2] Oil of turpentine; spirit of turpentine; turpentine oil. Distilled from the oleoresin yielding only terpene oils. *Constit.* α- and β-pinenes, limonene, 3-carene. May be rectified to remove unpleasant odor and taste by treatment with NaOH and distillation. A mixture of 3 parts oil with 1 part sulfurated linseed oil is known as **Haarlem oil.** Colorless liq; characteristic odor and taste, both becoming more pronounced and less agreeable on aging or exposure to air. d_{25}^{25} 0.854-0.868. Greater part distills between 154-170°. n_D^{20} 1.4680-1.4780. Rotation is variable. Insol in water; sol in 5 vols alcohol; miscible with benzene, chloroform, ether, carbon disulfide, petr ether and oils. *Flammable.*

Caution: Absorbed through skin, lungs, intestine. Potential symptoms of overexposure are irritation of eyes, skin, nose, throat; headache, vertigo, convulsions; skin sensitization; hematuria, albuminuria; kidney damage; abdominal pain, nausea, vomiting, diarrhea; cough, choking, dyspnea, cyanosis; excitement, ataxia, confusion, stupor. Aspiration of liquid may cause chemical pneumonia. *See NIOSH Pocket Guide to Chemical Hazards* (DHHS/NIOSH 97-140, 1997) p 324; *Clinical Toxicology of Commercial Products*, R. E. Gosselin *et al.*, Eds. (Williams & Wilkins, Baltimore, 5th ed., 1984) section III, pp 393-395.

USE: Solvent and thinner for paints, varnishes, polishes. In manufacture of aroma chemicals such as camphor, myrcene, linalool; source of pine oil.

10008. Tutin. [2571-22-4] (1a*S*,1b*R*,2*S*,2'*R*,5*R*,6*S*,6a*R*,7a*R*,8*R*)-Hexahydro-1b,6-dihydroxy-6a-methyl-8-(1-methylethenyl)-spiro[2,5-methano-7*H*-oxireno[3,4]cyclopent[1,2-*d*]oxepin-7,2'-oxiran]-3(2*H*)-one. $C_{15}H_{18}O_6$; mol wt 294.30. C 61.22%, H 6.17%, O 32.62%. Poisonous constituent of *Coriaria ruscifolia* L. or *C. japonica* A. Gray, (tutu) *Coriariaceae*. Naturally occurring form is (+)-tutin. Isoln: Easterfield, Aston, *J. Chem. Soc.* **79**, 125 (1901). Structure: Johns, Markham, *ibid.* **1961**, 3006; Craven, *Nature* **197**,

1193 (1963); Mackay, Mathieson, *Tetrahedron Lett.* **1963**, 1399. Absolute configuration: Okuda, Yoshida, *ibid.* **1965**, 2137. Biosynthesis: A. Corbella *et al., Chem. Commun.* **1969**, 634. NMR analysis: J. W. Blunt *et al., Aust. J. Chem.* **32**, 1339 (1979). Stereospecific synthesis: K. Watamatsu *et al., Tetrahedron* **42**, 5551 (1986). Determn in toxic honey: J. L. Love *et al., N.Z. J. Technol.* **2**, 179 (1986). Effect on central nervous system: D. R. Curtis *et al., Brain Res.* **63**, 419 (1973); A. Nistri *et al., Lancet* **1**, 996 (1974). Toxicity data: C. H. Jarboe *et al., J. Med. Chem.* **11**, 729 (1968).

White, odorless crystals. mp 209-212°; also reported as mp 204-205° (Wakamatsu). $[\alpha]_D^{20}$ +9.25° (alc); $[\alpha]_D^{17}$ +13.9° (c = 0.75, methanol). Sol in water, alcohol, ether. LD_{50} i.p. in female mice: 3.0 mg/kg (Jarboe).

Diacetate. $C_{19}H_{22}O_8$. Crystals from ethanol-water, mp 197-201°.

Caution: Poisoning from ingestion of *C. ruscifolia* has been reported and is due to presence of tutin. Symptoms include giddiness, stupor, coma, delirium, convulsions.

10009. Twistane. [253-14-5] Tricyclo[4.4.0.0^{3,8}]decane. C_{10}-H_{16}; mol wt 136.24. C 88.16%, H 11.84%. Synthesis: Whitlock, *J. Am. Chem. Soc.* **84**, 3412 (1962); Gautier, Deslongchamps, *Can. J. Chem.* **45**, 297 (1967); Whitlock, Siefken, *J. Am. Chem. Soc.* **90**, 4929 (1968). Absolute configuration: Keiichi *et al., Tetrahedron Lett.* **1968**, 5467.

Crystals, mp 163-164.8°.

10010. Tybamate. [4268-36-4] *N*-Butylcarbamic acid 2-[[(aminocarbonyl)oxy]methyl]-2-methylpentyl ester; carbamate of 2-(hydroxymethyl)-2-methylpentyl ester of butylcarbamic acid; *N*-butyl-2-methyl-2-propyl-1,3-propanediol dicarbamate; 2-methyl-2-propyltrimethylene butylcarbamate carbamate; Nospan; Solacen; Tybatran. $C_{13}H_{26}N_2O_4$; mol wt 274.36. C 56.91%, H 9.55%, N 10.21%, O 23.33%. Prepd from 2-methyl-2-propyl-3-hydroxypropyl carbamate + butyl isocyanate: Berger, Ludwig, US 2937119 (1960 to Carter Prod.). Comprehensive description: P. Reisberg *et al., Anal. Profiles Drug Subs.* **4**, 494-515 (1975).

Crystals from 1,1,2-trichloroethane + hexane (1:2), mp 49-51°, $bp_{0.06}$ 150-152°.

THERAP CAT: Anxiolytic.

10011. Tylocrebrine. [6879-02-3] 9,11,12,13,13a,14-Hexahydro-2,3,5,6-tetramethoxydibenzo[*f,h*]pyrrolo[1,2-*b*]isoquinoline; 2,3,5,6-tetramethoxyphenanthro[9,10:6',7']indolizidine. $C_{24}H_{27}$-NO_4; mol wt 393.48. C 73.26%, H 6.92%, N 3.56%, O 16.26%. Phenanthroindolizidine alkaloid; isomeric with tylophorine, *q.v.* Isoln of *l*-isomer from *Tylophora crebriflora* S. T. Blake, *Asclepiadaceae:* E. Gellert *et al., J. Chem. Soc.* **1962**, 1008; K. V. Rao *et al., J. Pharm. Sci.* **59**, 1501 (1970). Antileukemic activity: E. Gellert, R. Rudzats, *J. Med. Chem.* **7**, 361 (1964). *d*-Isomer isolated from

Consult the Name Index before using this section.

Ficus septica, Moraceae. Synthesis of the *dl*-form and structure: E. Gellert *et al., loc. cit.;* B. Chauncy, E. Gellert, *Aust. J. Chem.* **23**, 2503 (1970). Inhibits protein synthesis: G. R. Donaldson *et al., Biochem. Biophys. Res. Commun.* **31**, 104 (1968); E. Battaner, D. Vasquez, *Biochim. Biophys. Acta* **254**, 316 (1971). Mode of action studies: M. T. Huang, A. P. Grollman, *Mol. Pharmacol.* **8**, 538 (1972). *Review:* R. S. Gupta, *Antibiotics* **6**, 47 (1983).

dl-**Form.** Needles from chloroform + methanol, mp 219-221°.

l-**Form.** Crystals from methanol, dec 218-220°. uv max: 263, 342, 360 nm (log ε 4.81, 3.25, 3.09). [α]$_D^{24}$ −45° (c = 0.74 in chloroform). pKa (50% aq ethanol): 6.7.

d-**Form.** Crystals from methanol, mp 220-222°. [α]$_D^{22}$ +20.5°.

10012. Tylophorine. [482-20-2] (13a*S*)-9,11,12,13,13a,14-Hexahydro-2,3,6,7-tetramethoxydibenzo[*f,h*]pyrrolo[1,2-*b*]isoquinoline; 2,3,6,7-tetramethoxyphenanthro[9,10:6',7']indolizidine. C$_{24}$H$_{27}$NO$_4$; mol wt 393.48. C 73.26%, H 6.92%, N 3.56%, O 16.26%. Major alkaloid from *Tylophora asthmatica* Wight et Arn., *Asclepiadaceae.* Also found in other *Asclepiadaceae, Moraceae, Urticaceae* and *Lauraceae.* Naturally occurring form originally isolated and reported to be levorotatory; later corrected to dextrorotatory. Isoln: A. N. Ratnagiriswaran, K. Venkatachalam, *Indian J. Med. Res.* **22**, 433 (1935); R. N. Chopra *et al., Arch. Pharm.* **275**, 236 (1937); T. R. Govindachari *et al., J. Chem. Soc.* **1954**, 2801. Structure: *eidem, Tetrahedron* **9**, 53 (1960). Absolute configuration: *eidem, J. Chem. Soc. Perkin Trans. 1* **1974**, 1161. Synthesis of the *dl*-form: *eidem, Chem. Ind. (London)* **1960**, 664; N. A. Khatri *et al., J. Am. Chem. Soc.* **103**, 6387 (1981). Synthesis and verification of dextrorotation of naturally occurring form: T. F. Buckley 3rd, H. Rapoport, *J. Org. Chem.* **48**, 4222 (1983); J. E. Nordlander, F. G. Njoroge, *ibid.* **52**, 1627 (1987). Stereoselective synthesis of *l*-form: M. Ihara *et al., Tetrahedron Lett.* **29**, 4135 (1988). Biosynthesis: Mulchandani *et al., Phytochemistry* **8**, 1931 (1969); *ibid.* **10**, 1047 (1971); D. S. Bhakuni, V. K. Mangla, *Tetrahedron* **37**, 401 (1981). Pharmacology: C. Gopalakrishnan *et al., Indian J. Med. Res.* **69**, 513 (1979); *ibid.* **71**, 940 (1980).

Crystals, dec 282-284°. [α]$_D^{23}$ +15° (c = 0.7 in chloroform); [α]$_D^{21}$ +73° (c = 0.7 in chloroform). uv max in ethanol: 257, 286, 339, 356 nm (log ε 4.7, 4.42, 3.28, 3.19). Unstable in solutions, decomposition with yellowing sets in promptly accompanied by decreasing rotatory strength.

l-**Form.** Colorless crystals, dec 286-287°. [α]$_D^{27}$ −11.6° (c = 1.07 in chloroform). uv max 255, 290, 340, 352 nm (log ε 4.74, 4.49, 3.30, 2.93). Sol in chloroform; slightly sol in abs alc, ether, cold benzene. Practically insol in water.

dl-**Form.** Crystals from chloroform + ethanol, mp 292°.

10013. Tylosin. [1401-69-0] Tylan. C$_{46}$H$_{77}$NO$_{17}$; mol wt 916.11. C 60.31%, H 8.47%, N 1.53%, O 29.69%. Macrolide antibiotic isolated from a strain of *Streptomycetes fradiae* found in soil

from Thailand: Hamill *et al., Antibiot. Chemother.* **11**, 328 (1961); *eidem,* US 3178341 (1965 to Lilly). Prodn in batch and chemostat cultures: P. P. Gray, S. Bhuwapathanapun, *Biotechnol. Bioeng.* **22**, 1785 (1980). Partial structure: Morin, Gorman, *Tetrahedron Lett.* **1964**, 2339. Structure: Morin *et al., ibid.* **1970**, 4737; Achenbach *et al., Ber.* **108**, 2481 (1975). Configurational study: S. Omura *et al., Tetrahedron Lett.* **1977**, 1045. Abs config: *eidem, J. Antibiot.* **33**, 915 (1980); N. D. Jones *et al., ibid.* **35**, 420 (1982). Synthesis of *tylonolide,* the aglycone: S. Masamune *et al., J. Am. Chem. Soc.* **98**, 7874 (1976); K. Tatsuta *et al., Tetrahedron Lett.* **22**, 3997 (1981). Relationship of ribosomal binding and antibacterial properties: J. W. Corcoran *et al., J. Antibiot.* **30**, 1012 (1977). Biosynthesis studies: E. T. Seno *et al., Antimicrob. Agents Chemother.* **11**, 455 (1977); S. Omura *et al., J. Antibiot.* **31**, 254 (1978).

Crystals from water, mp 128-132°. [α]$_D^{25}$ −46° (c = 2 in methanol). uv max: 282 nm (E$_{1cm}^{1\%}$ 245). Soly in water at 25°: 5 mg/ml. Freely sol in methanol; sol in lower alcohols, esters and ketones; in chloroform, dilute mineral acids, amyl acetate, chlorinated hydrocarbons, benzene, ether. Solns are stable at pH 4-9; at pH <4 another active compd, *desmycosin,* is formed.

Hydrochloride. C$_{46}$H$_{77}$NO$_{17}$.HCl. Crystals from ethanol + ether, mp 141-145°.

THERAP CAT (VET): Antibacterial.

10014. Tyloxapol. [25301-02-4] Formaldehyde polymer with oxirane and 4-(1,1,3,3-tetramethylbutyl)phenol; oxyethylated tertiary octylphenol formaldehyde polymer; *p*-isooctylpolyoxyethylenephenol formaldehyde polymer; tyloxypal; Alevaire; Superinone; Triton A-20; Triton WR-1339. Nonionic detergent with surface-tension-reducing properties. Prepn: Bock, Rainey, US 2454541 (1948 to Rohm & Haas); J. W. Cornforth *et al., Nature* **168**, 150 (1951). Use to induce exptl hyperlipidemia: A. Kellner *et al., J. Exp. Med.* **93**, 373 (1951); P. E. Schurr *et al., Lipids* **7**, 68 (1972). Use as ophthalmic excipient: D. E. Guttman *et al., J. Pharm. Sci.* **50**, 305 (1961). Phase behavior of mixtures with water: K. Westesen, *Int. J. Pharm.* **102**, 91 (1994); K. Westesen, M. H. J. Koch, *ibid.* **103**, 225 (1994). Tissue distribution and excretion: R. L. DeAngelis *et al., Xenobiotica* **25**, 521 (1995).

R = (CH$_2$CH$_2$O)$_x$H
x = 8 - 10
m < 6

d^{20} 1.0963. Cloud point: 92-97°. Slowly but freely misc with water. Sol in benzene, toluene, chloroform, carbon tetrachloride, carbon disulfide, glacial acetic acid. Alkaline pH. Oxidized by metals.

Combination with colfosceril palmitate and hexadecanol *see* Exosurf®.

USE: Pharmaceutic aid (dispersing, solubilizing, wetting agent; excipient). To induce exptl hyperlipidemia in animal models.

THERAP CAT: Mucolytic.

10015. Tymazoline. [24243-97-8] 4,5-Dihydro-2-[[5-methyl-2-(1-methylethyl)phenoxy]methyl]-1*H*-imidazole; 2-[(thymyloxy)methyl]-2-imidazoline; 2-(thymyloxymethyl)glyoxalidine; 2-[(*p*-mentha-1,3,5-trien-2-yloxy)methyl]-2-imidazoline. $C_{14}H_{20}N_2O$; mol wt 232.33. C 72.38%, H 8.68%, N 12.06%, O 6.89%. Prepd from an alkyl thymyloxyacetimidate and ethylenediamine: Sonn, **US 2149473** (1939 to Ciba); Djerassi, Scholz, *J. Am. Chem. Soc.* **69**, 1688 (1947). Pharmacology: Pham-Huu-Chanh *et al.*, *Therapie* **24**, 797 (1969).

Hydrochloride. Pernazene. $C_{14}H_{20}N_2O \cdot HCl$; mol wt 268.79. Crystals, mp 215-217° (Sonn); from water or methyl ethyl ketone + ethanol, mp 223.5-225° (Djerassi, Scholz).

THERAP CAT: Nasal decongestant.

10016. Tyramine. [51-67-2] 4-(2-Aminoethyl)phenol; 4-hydroxyphenethylamine; tyrosamine; 2-*p*-hydroxyphenylethylamine; *p*-β-aminoethylphenol; α-(4-hydroxyphenyl)-β-aminoethane. $C_8H_{11}NO$; mol wt 137.18. C 70.05%, H 8.08%, N 10.21%, O 11.66%. Decarboxylation product of tyrosine. Found in mistletoes, putrefied animal tissue, ripe cheese, ergot. Synthesis: Barger, *J. Chem. Soc.* **95**, 1127 (1909); Waser, *Helv. Chim. Acta* **8**, 766 (1925); Buck, *J. Am. Chem. Soc.* **55**, 3389 (1933). Crystal and molecular structure: A. Podder *et al.*, *Acta Crystallogr.* **B35**, 649 (1979).

Crystals from benzene or alcohol, mp 164-165°. bp$_{25}$ 205-207°; bp$_2$ 166°. Alkaline reaction. One gram dissolves in 95 ml water at 15°; in 10 ml boiling alcohol. Sparingly sol in benzene, xylene.

Hydrochloride. [60-19-5] Mydrial. $C_8H_{11}NO \cdot HCl$; mol wt 173.64. Crystals from alcohol + ether, mp 269°. Sol in water with neutral reaction.

THERAP CAT: Adrenergic.

10017. Tyrocidine. [8011-61-8] Peptide antibiotic mixture produced by *Bacillus brevis:* major constituent of tyrothricin, *q.v.* Isoln: Hotchkiss *et al.*, *J. Biol. Chem.* **141**, 155, 163 (1941); Moses, **US 3265572** (1966 to Penick). Separated into the three components, tyrocidines A, B, and C. Review of chemistry and biosynthesis: E. Katz, A. L. Demain, *Bacteriol. Rev.* **41**, 449-474 (1977). Regulatory role in bacterial sporogenesis: H. Ristow *et al.*, *Nature* **280**, 165 (1979); *eidem, Eur. J. Biochem.* **129**, 395 (1982); W. Pschorn *et al.*, *ibid.* 403. Structure-activity relationship: W. Danders *et al.*, *Antimicrob. Agents Chemother.* **22**, 785 (1982).

Tyrocidine A

Hydrochloride mixture. Brevicidin; Rapicidin. Rods or needles from methanol, dec 240°. $[\alpha]_D^{20} -101°$ (c = 1.2 in 95% alc). Soluble in 95% alc, acetic acid, pyridine; slightly sol in water, acetone, abs alcohol. Practically insol in ether, chloroform, hydrocarbons.

Tyrocidine A. [1481-70-5] $C_{66}H_{87}N_{13}O_{13}$; mol wt 1270.50. Separation from the tyrocidine mixture: Battersby, Craig, *J. Am. Chem. Soc.* **74**, 4019, 4023 (1952). Structure: Paladine, Craig, *ibid.* **76**, 688 (1954). Synthesis: Ohno *et al.*, *Bull. Chem. Soc. Jpn.* **39**, 1738 (1966); K. Okamoto *et al.*, *ibid.* **50**, 231 (1977).

Tyrocidine A hydrochloride. Crystals from methanol + water, mp 240-242°. $[\alpha]_D^{25} -111°$ (c = 1.37 in 50% alc). Freely sol in aq methanol or alc; slightly sol in methanol, ethanol. Practically insol in chloroform, acetone, ether.

Tyrocidine B. [865-28-1] $C_{68}H_{88}N_{14}O_{13}$; mol wt 1309.54. Purification and amino acid sequence determination: King, Craig, *J. Am. Chem. Soc.* **77**, 6624, 6627 (1955). Synthesis: Kuromizu, Izumiya, *Experientia* **26**, 587 (1970). Possesses the same structure as tyrocidine A except that L-tryptophan replaces the L-phenylalanine. Crystals from methanol + isopropyl ether.

Tyrocidine B hydrochloride pentahydrate. Crystals, mp 236-237°. $[\alpha]_D -93.0°$ (c = 0.5 in methanol).

Tyrocidine C. [3252-29-7] $C_{70}H_{89}N_{15}O_{13}$; mol wt 1348.57. Separation from the tyrocidine mixture: Ruttenberg *et al.*, *Biochemistry* **4**, 11 (1965). Possesses same structure as tyrocidine B except that D-tryptophan replaces D-phenylalanine attached to L-asparagine. Synthesis: Kuromizu, Izumiya, *Tetrahedron Lett.* **1970**, 1471.

THERAP CAT: Antibacterial.

10018. Tyropanoate Sodium. [7246-21-1] α-Ethyl-2,4,6-triiodo-3-[(1-oxobutyl)amino]benzenepropanoic acid sodium salt (1:1); 3-butyramido-α-ethyl-2,4,6-triiodohydrocinnamic acid sodium salt; α-ethyl-β-(2,4,6-triiodo-3-butyramidophenyl)propionic acid sodium salt; sodium tyropanoate; Win-8851-2; Bilopaque; Lumopaque; Tyropaque. $C_{15}H_{17}I_3NNaO_3$; mol wt 663.01. C 27.17%, H 2.58%, I 57.42%, N 2.11%, Na 3.47%, O 7.24%. Prepn: S. Archer, J. O. Hoppe, **US 2895988** (1959 to Sterling Drug); and pharmacology: J. O. Hoppe *et al.*, *J. Med. Chem.* **13**, 997 (1970). Clinical trials: S. Y. Han, D. M. Witten, *Radiology* **112**, 529 (1974); R. F. Thoeni, A. A. Moss, *ibid.* **144**, 271 (1982).

Colorless solid from ethyl ether, mp 208-210°. Soluble in water. LD$_{50}$ i.v. in mice: 720 mg/kg (Hoppe).

Free acid. $C_{15}H_{18}I_3NO_3$. Crystals from ethyl oxyacetate, mp 182-184°.

THERAP CAT: Diagnostic aid (radiopaque medium—cholecystographic).

10019. Tyrosinase. [9002-10-2] Monophenol monooxygenase. A copper-containing enzyme widely distributed in plants, animals, and man. Catalyzes the hydroxylation of tyrosine in the liver and in melanin-forming cells to 3,4-dihydroxyphenylalanine (dopa). Causes the cut surface of many fruits and plants to darken. The enzyme, as isolated from the common edible mushroom or potato, is characterized by its ability to catalyze aerobic oxidation of both monohydric and *o*-dihydric phenols. These activities are commonly referred to as *cresolase* (*monophenolase, monophenoloxidase*) and *catecholase* (*o-dihydric phenolase*) activities. Since monophenolase activity is frequently lost during purification of the enzyme, the name *polyphenolase* (*polyphenoloxidase*) is preferred by some workers for prepns which have mainly *o*-dihydric phenolase activity. *Ref:* C. R. Dawson, W. B. Tarpley "Copper Oxidases" in J. B. Sumner, K. Myrbäck, *The Enzymes* vol. **II**, (Academic Press, New York, 1951) pp 456-483. Prepn from potato peels: Kubowitz, *Biochem. Z.* **299**, 32 (1938). Extraction from mushrooms: Cohen, Lerner, **US 2956929** (1960 to Gillette). Separation of α-, β-, γ-, and δ-tyrosinases of mushroom tyrosinase: Bouchilloux *et al.*, *J. Biol. Chem.* **238**, 1699 (1963). Isoln and properties of crystalline tyrosinase from *Neurospora:* Fling *et al.*, *ibid.* p 2045. Isoln and properties of β-tyrosinase: Kumagai *et al.*, *J. Biol. Chem.* **245**, 1767 (1970).

THERAP CAT: Antihypertensive.

10020. Tyrosine. [60-18-4] L-Tyrosine; Tyr; Y; β-(*p*-hydroxyphenyl)alanine; α-amino-*p*-hydroxyhydrocinnamic acid; (*S*)-α-amino-4-hydroxybenzenepropanoic acid. $C_9H_{11}NO_3$; mol wt 181.19. C 59.66%, H 6.12%, N 7.73%, O 26.49%. Non-essential amino acid for human development; precursor for the synthesis of thyroid hormones and select neurotransmitters, such as dopamine and norepinephrine. May be considered essential by the brain. Name derived from the Greek "tyros" for "cheese" from which it was first identified and isolated: J. Liebig, *Ann.* **57**, 127 (1846); *idem, ibid.* **62**, 257 (1847). Early chemistry and biochemistry: *Amino Acids and Proteins*, D. M. Greenberg, Ed. (Charles C. Thomas, Springfield, IL, 1951) 950 pp., *passim*; J. P. Greenstein, M. Winitz, *Chemistry of the Amino Acids* **vol 1-3** (John Wiley and Sons, Inc., New York, 1961) pp. 2348-2367, *passim*. Synthesis of labeled form: Y. Watanabe *et al.*, *Acta Radiol.* **376** (Suppl.), 110 (1991). Intrinsic fluorophore in proteins; used in characterizing structure and conformational changes: V. Giancotti *et al.*, *Biochim. Biophys. Acta* **624**, 60 (1980); S. T. Ferreira *et al.*, *Biophys. J.* **66**, 1185 (1994); B. Kierdaszuk *et al.*, *Photochem. Photobiol.* **61**, 319 (1995). Review of determn in blood: E. Robins, *Methods Biochem. Anal.* **17**, 287-309 (1969). Review of neurotransmitter synthesis: C. J. Gibson, *Retina* **2**, 332-340 (1982). Review of toxicity: C. Laberge *et al.*, *Adv. Exp. Med. Biol.* **206**, 209-221 (1986). Review of post-translational phosphorylation: S. Atherton-Fessler *et al.*, *Semin. Cell Biol.* **4**, 433-442 (1993); sulfation: C. Niehrs *et al.*, *Chem. Biol. Interact.* **92**, 257-271 (1994).

Fine silky needles, dec 342-344° (closed capillary, bath preheated to 280°, rapid heating). d 1.456. pK_1 2.20; pK_2 9.11; pK_3 10.07. $[\alpha]_D^{22}$ −10.6° (c = 4 in HCl); $[\alpha]_D^{18}$ −13.2° (c = 4 in 3N NaOH). Soly in water (g/100 g): 0.02 at 0°; 0.045 at 25°; 0.105 at 50°; 0.244 at 75°; 0.565 at 100°. Insol in common neutral solvents, such as abs alcohol, ether, acetone, except water. Sol in alkaline solns.

DL-Form. Synthetic product. Stout needles. Dec 316°. Soly in water (g/100 g): 0.0147 at 0°; 0.351 at 25°; 0.0836 at 50°.

D-Form. Crystals. Dec 310-314°. $[\alpha]_D^{25}$ +10.3° (c = 4 in HCl). Soly in water (g/100 g): 0.196 at 0°; 0.1052 at 50°.

USE: Probe for studying protein structure and dynamics.

10021. *m*-Tyrosine. [587-33-7] 3-Hydroxy-L-phenylalanine; α-amino-3-hydroxyhydrocinnamic acid; metatyrosine. $C_9H_{11}NO_3$; mol wt 181.19. C 59.66%, H 6.12%, N 7.73%, O 26.49%. A possible precursor of catecholamines: Sourkes *et al.*, *Nature* **189**, 577 (1961). An intermediate in an alternate pathway for the biosynthesis of catecholamines, where with the existing hydroxylating enzymes *m*-hydroxylation of phenylalanine to *m*-tyrosine occurs before *p*-hydroxylation (forming dopa) and is followed by subsequent decarboxylation to dopamine. Formation *in vitro* of dopa from L-*m*-tyrosine: Tong *et al.*, *Biochem. Biophys. Res. Commun.* **43**, 819 (1971); *in vivo:* Hollunger, Persson, *Acta Pharmacol. Toxicol.* **34**, 391 (1974). Biosynthesis and metabolism studies: D'Iorio *et al.*, *Adv. Neurol.* **5**, 265 (1974). Has also been isolated from a plant source, *Euphorbia myrsinites* L. *Euphorbiaceae:* Mothes *et al.*, *Z. Naturforsch.* **19b**, 1161 (1964). *m*-Tyrosine has the ability to cross the blood-brain barrier and is decarboxylated to *m*-tyramine which stimulates dopamine receptors, presumably accounting for the demonstrated pharmacological effects of *m*-tyrosine. Pharmacological studies: Carlsson, Lindqvist, *Eur. J. Pharmacol.* **2**, 187 (1967); Rubenson, *J. Pharm. Pharmacol.* **23**, 228, 412 (1971); Sandler *et al.*, *Nature* **229**, 414 (1971); Ungerstedt *et al.*, *Eur. J. Pharmacol.* **21**, 230 (1973). Crystal and molecular structure: Byrkjedal *et al.*, *Acta Chem. Scand.* **28B**, 750 (1974).

mp 267-270° (dec). $[\alpha]_D^{22}$ −14.5° (70% ethanol); $[\alpha]_D^{22}$ +8.9° (70% ethanol, 2N HCl).

10022. Tyrosol. [501-94-0] 4-Hydroxybenzeneethanol; 4-(2-hydroxyethyl)phenol. $C_8H_{10}O_2$; mol wt 138.17. C 69.54%, H 7.30%, O 23.16%. Principal phenolic antioxidant in virgin olive oil; also found in wine and other foods. Prepn: F. Ehrlich, *Ber.* **40**, 1027 (1907); F. Ehrlich, P. Pistschimuka, *ibid.* **45**, 2428 (1912); S. Yamada *et al.*, *Chem. Pharm. Bull.* **11**, 258 (1963). GC/MS determn in urine: E. Miró-Casas *et al.*, *Anal. Biochem.* **294**, 63 (2001). CZE determn in extra-virgin olive oil: M. Bonoli *et al.*, *J. Chromatogr. A* **1011**, 163 (2003). HPLC determn in virgin olive oil and antioxidant capacity: A. Carrasco-Pancorbo *et al.*, *J. Agric. Food Chem.* **53**, 8918 (2005). Bioavailability: M. I. Covas *et al.*, *Drugs Exp. Clin. Res.* **29**, 203 (2003). *In vitro* protective effects vs oxidized LDL: C. Giovannini *et al.*, *J. Nutr.* **129**, 1269 (1999); R. Di Benedetto *et al.*, *Nutr. Metab. Cardiovasc. Dis.* **17**, 535 (2007).

Colorless oil, bp_4 158°. bp_{18} 195°. Colorless needles from $CHCl_3$, mp 91-92°. Absorption max (water + acetonitrile): 230, 276 nm.

10023. Tyrothricin. [1404-88-2] "Dubos crude crystals"; Hydrotricine; Dermotricine; Tyri 10; Tyrosur. Polypeptide antibiotic mixture extracted from cultures of soil bacilli belonging to the *Tyrothrix* group of bacteria. Usually obtained by extracting acidified cultures of *Bacillus brevis* with alc and precipitating with NaCl soln. Surface, submerged, and aerated cultures are used. Tyrothricin contains from 10 to 20% gramicidin and from 40 to 60% tyrocidine. Solubility data: Weiss *et al.*, *Antibiot. Chemother.* **7**, 374 (1957). Function in the sporulation process: Sarkar, Paulus, *Nature New Biol.* **239**, 228 (1972). Toxicity data: H. Sous *et al.*, *Arzneim.-Forsch.* **7**, 98 (1957).

Gray to brown powder, decomp 215-220°. Has 2-3% absorbed or combined moisture. Stable to light, air and temps up to 50°. Practically insol in water. Sol in alcohol (about 28 mg/ml), in methanol and in propylene glycol. One part of a 2% alcoholic soln contg 10% formaldehyde soln U.S.P. dissolves in 49 parts of water without cloudiness. Tyrothricin should not be dissolved in ether or acetone as only the gramicidin component is soluble. Soly in mg/ml at about 28°: water 2.1; isopropanol 5.6; benzene 0.30; isooctane 0.042; carbon tetrachloride 0.455; ethyl acetate 2.65; acetone 6.8; ether 3.25; dioxane 11.1; chloroform 1.6. Aqueous suspensions may be attacked by microorganisms. Solutions and emulsions should not be sterilized by heat. Incompatible with alkalies and strong acids. Tyrothricin gives a positive biuret and Hopkins-Cole test. LD_{50} in mice (mg/kg): >1500 s.c.; 100 i.p.; >3000 orally (Sous).

THERAP CAT: Topical antibacterial.

THERAP CAT (VET): Topical antibacterial.

U

10024. Ubenimex. [58970-76-6] N-[(2S,3R)-3-Amino-2-hydroxy-1-oxo-4-phenylbutyl]-L-leucine; NK-421; Bestatin. $C_{16}H_{24}$-N_2O_4; mol wt 308.38. C 62.32%, H 7.84%, N 9.08%, O 20.75%. Dipeptide antitumor antibiotic produced by *Streptomyces olivoreticuli* with immunostimulant activity; inhibits leucine aminopeptidase and aminopeptidases B and N. Prepn from fermentation broth: H. Umezawa *et al.*, **DE 2528984**; *eidem*, **US 4052449** (1976, 1977 both to Microbiochem. Res. Found., Japan); *eidem*, *J. Antibiot.* **29**, 97 (1976). Structure: H. Suda *et al.*, *ibid.* 100. Synthesis of stereoisomers and structure-activity study: R. Nishizawa *et al.*, *J. Med. Chem.* **20**, 510 (1977). Stereocontrolled synthesis: S. Kobayashi *et al.*, *Tetrahedron Lett.* **25**, 5079 (1984). Crystal structure: J. S. Ricci, Jr. *et al.*, *J. Org. Chem.* **47**, 3063 (1982). Cell surface binding studies: H. Umezawa *et al.*, *J. Antibiot.* **29**, 857 (1976); W. E. Müller *et al.*, *Int. J. Immunopharmacol.* **4**, 393 (1982). Acute toxicity: T. Sakakibara *et al.*, *Jpn. J. Antibiot.* **36**, 2971 (1983). Review of pharmacology: G. Mathé, *Biomed. Pharmacother.* **45**, 49-54 (1991); of clinical studies: K. Ota, *ibid.*. 55-60. Clinical trial in squamous-cell lung carcinoma: Y. Ichinose *et al.*, *J. Natl. Cancer Inst.* **95**, 605 (2003).

Colorless needles, mp 233-236°. $[\alpha]_D^{20}$ -15.5° (c = 1.0 in 1N HCl). pKa 8.1, 3.1. uv max: 241.5, 248, 253, 258, 264.5, 268 nm ($E_{1cm}^{1\%}$ 3.8, 4.0, 5.0, 6.0, 4.6, 2.7). Sol in acetic acid, DMSO, methanol. Less sol in water. Insol in ethyl acetate, benzene, hexane, chloroform. LD_{50} in male, female mice, male, female rats (g/kg): 1.3, 1.9, 1.9, 2.1 s.c.; 0.19, 0.19, 0.90, 0.78 i.p.; >4.0, >4.0, >2.0, >2.0 orally (Sakakibara).

THERAP CAT: Immunomodulator; antineoplastic.

10025. Ubiquinones. Coenzymes Q; mitoquinones; SA; Q-275. A group of lipid-soluble benzoquinones involved in electron transport in mitochondria, *i.e.*, in the oxidation of succinate or reduced nicotine adenine dinucleotide (NADH) via the cytochrome system. Occurs in the majority of aerobic organisms, from bacteria to higher plants and animals. Ubiquinone structures, analogous to the menaquinones, *q.v.*, are based on the 2,3-dimethoxy-5-methylbenzoquinone nucleus with a variable terpenoid side chain contg one to twelve mono-unsaturated *trans*-isoprenoid units with 10 units being the most common in animals. According to the existing dual system of nomenclature the compds can be described as: coenzyme Q_n in which n = 1-12, or ubiquinone(x) in which x designates the total number of carbon atoms in the side chain and can be any multiple of 5. Differences in properties are due to the difference in length of the side chain. Naturally occurring members are the coenzymes Q_6-Q_{10}. The entire series has been prepd synthetically. Recent syntheses of ubiquinone 50: S. Terao *et al.*, *J. Org. Chem.* **44**, 868 (1979); Y. Naruta, *ibid.* **45**, 4097 (1980); K. Sato *et al.*, *Chem. Commun.* **1982**, 153. *Reviews: Ciba Foundation Symposium on Quinones in Electron Transport*, Wolstenholme, O'Connor, Eds. (Churchill, London, 1961) 453 pp; Wagner, Folkers, *Vitamins and Coenzymes* (Interscience, New York, 1964) pp 435-468; Crane in *Progr. Chem. Fats Lipids* **vol. 7**(2), Holman, Ed. (Pergamon Press, 1964) pp 267-289; *Methods Enzymol.* **18C**, 135-237 (1971); R. A. Morton in *The Vitamins* **vol. V**, W. H. Sebrell, R. S. Harris, Eds. (Academic Press, New York, 2nd ed., 1972) pp 355-391. Series of books: *Biomedical and Clinical Aspects of Coenzyme Q* **vols. 1-3**, K. Folkers, Y. Yamamura, Eds. (Elsevier, New York, 1977, 1980, 1981).

Ubiquinone 50

Ubiquinone 50. [303-98-0] Coenzyme Q_{10}; CoQ_{10}; coenzyme Q-199; ubidecarenone; ubiquinone 10; NSC-140865; Adelir; Caomet; Decafar; Decorenone; Dymion; Heartcin; Inokiten; Iuvacor; Mitocor; Neuquinon; Taidecanone; Ubicardio; Ubicor; Ubidenone; Ubifactor; Ubimaior; Ubisan; Ubiten-50; Ubivis; Udekinon. C_{59}-$H_{90}O_4$; mol wt 863.37. Clinical evaluation in early Parkinson disease: C. W. Shults *et al.*, *Arch. Neurol.* **59**, 1541 (2002).

Ubichromenol 50. [2382-48-1] Cyclic isomer of ubiquinone 50.

THERAP CAT: Cardiotonic; coenzyme Q_{10} as antiparkinsonian.

10026. Ubiquitin. ATP-dependent proteolytic factor; APF-1; Ub; ubiquitous immunopoietic polypeptide; UBIP. Ubiquitous polypeptide, mol wt 8500 daltons, so named due to its presence in all eukaryotes including plants; widely distributed in the organism. Highly conserved sequence of 76 amino acids is identical in a wide variety of sources including humans, fish, and insects. Participates in diverse cellular functions, such as protein degradation, chromatin structure, and heat shock, by conjugation to other proteins through its carboxyl terminus. Isoln: G. Goldstein *et al.*, *Proc. Natl. Acad. Sci. USA* **72**, 11 (1975); as APF-1 and recognition of role as marker in protein degradation: A. Ciechanover *et al.*, *Biochem. Biophys. Res. Commun.* **81**, 1100 (1978); A. Ciechanover *et al.*, *Proc. Natl. Acad. Sci. USA* **77**, 1365 (1980); A. Hershko *et al.*, *ibid.* 1783. Amino acid sequence: D. H. Schlesinger *et al.*, *Biochemistry* **14**, 2214 (1975). Identity of APF-1 with ubiquitin and revised amino acid sequence: A. Ciechanover *et al.*, *J. Biol. Chem.* **255**, 7525 (1980); K. D. Wilkinson *et al.*, *ibid.* 7529; K. D. Wilkinson, T. K. Audhya, *ibid.* **256**, 9235 (1981). Structural studies: W. J. Cook *et al.*, *J. Mol. Biol.* **130**, 353 (1979); S. Vijay-Kumar *et al.*, *J. Biol. Chem.* **262**, 6396 (1987); P. L. Weber *et al.*, *Biochemistry* **26**, 7282 (1987). Role of chromatin structure: D. C. Watson *et al.*, *Nature* **276**, 196 (1978); R. D. Mueller *et al.*, *J. Biol. Chem.* **260**, 5147 (1985); in receptor structure: M. Siegelman *et al.*, *Science* **231**, 823 (1986). Identity as heat shock protein: U. Bond, M. J. Schlesinger, *Mol. Cell. Biol.* **5**, 949 (1985). Role as marker for protein degradation in polymeric form: A. Hershko, H. Heller, *Biochem. Biophys. Res. Commun.* **128**, 1079 (1985); V. Chau *et al.*, *Science* **243**, 1576 (1989). Enzymic determn: I. A. Rose, J. V. B. Warms, *Proc. Natl. Acad. Sci. USA* **84**, 1477 (1987). Brief review of functions and mechanisms: D. Finley, A. Varshavsky, *Trends Biochem. Sci.* **10**, 343 (1985). Review of protein degradation: M. Rechsteiner, *Annu. Rev. Cell Biol.* **3**, 1-30 (1987); A. Hershko, *J. Biol. Chem.* **263**, 15327-15240 (1988). *Review:* K. D. Wilkinson, *Anti-Cancer Drug Des.* **2**, 211-229 (1987). Book: *Ubiquitin*, M. Rechsteiner, Ed. (Plenum Press, New York, 1988).

10027. Udenafil. [268203-93-6] 3-(6,7-Dihydro-1-methyl-7-oxo-3-propyl-1H-pyrazolo[4,3-d]pyrimidin-5-yl)-N-[2-(1-methyl-2-pyrrolidinyl)ethyl]-4-propoxybenzenesulfonamide; 5-[2-propyloxy-5-(1-methyl-2-pyrrolidinylethylamidosulfonyl)phenyl]-1-methyl-3-propyl-1,6-dihydro-7H-pyrazolo[4,3-d]pyrimidin-7-one; DA-8159; Zydena. $C_{25}H_{36}N_6O_4S$; mol wt 516.66. C 58.12%, H 7.02%, N 16.27%, O 12.39%, S 6.21%. Selective phosphodiesterase type 5 (PDE5) inhibitor. Prepn: M. Yoo *et al.*, **WO 0027848**; *eidem*, **US 6583147** (2000, 2003 both to Dong A). Erectogenic potential: T. Y. Oh *et al.*, *Arch. Pharmacal Res.* **24**, 471 (2000). Inhibition of PDE5: H. Doh *et al.*, *ibid.* **25**, 873 (2002). HPLC determn in plasma and urine: H. J. Shim *et al.*, *J. Pharm. Biomed. Anal.* **30**, 527 (2002). Review of pharmacology and clinical experience: E. A. Salem *et al.*, *Curr. Opin. Investig. Drugs* **7**, 661-669 (2006).

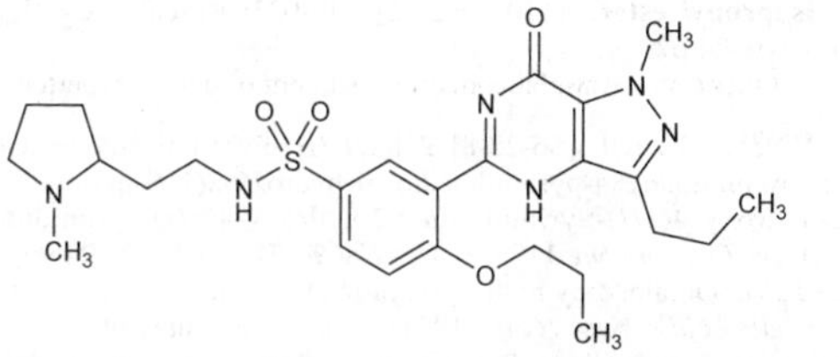

mp 162-164°. pKa$_1$ 6.5; pKa$_2$ 12.5. Log P (octanol/buffer): 0.76 (pH 1); 0.75 (pH 3); 0.81 (pH 5); 1.85 (pH 7). Soly in water (mg/ml): 27.4 (pH 2); 12.5 (pH 5); 0.82 (pH 7); 0.033 (pH 10); 0.019 (distilled H$_2$O). LD$_{50}$ in rats (g/kg): >1 orally (Yoo).

THERAP CAT: In treatment of erectile dysfunction.

10028. Uliginosins. Antibiotics isolated from *Hypericum uliginosum* HBK, a woody herb found in Mexico and Central America. Isoln: Taylor, Brooker, *Lloydia* **32**, 217 (1969). Structure: Parker, Johnson, *J. Am. Chem. Soc.* **90**, 4716 (1968); Parker *et al.*, *ibid.* 4723. Synthesis of uliginosin A and dihydrouliginosin B: Meikle, Stevens, *Chem. Commun.* **1972**, 123; *eidem, J. Chem. Soc. Perkin Trans. 1* **1978**, 1303. Synthesis of uliginosin B: *eidem, Tetrahedron Lett.* **1972**, 4787; *eidem, J. Chem. Soc. Perkin Trans. 1* **1979**, 2563.

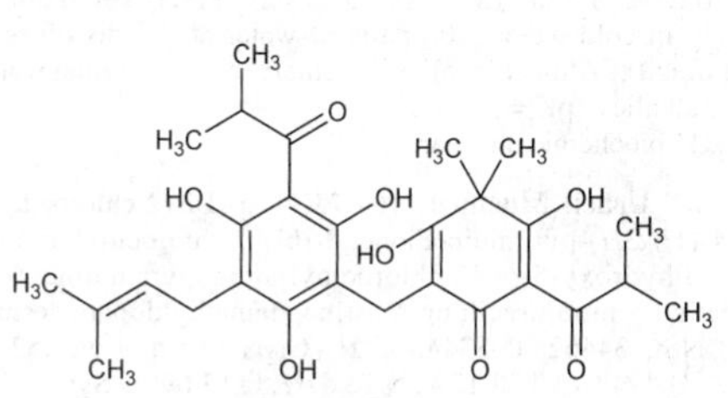

Uliginosin A

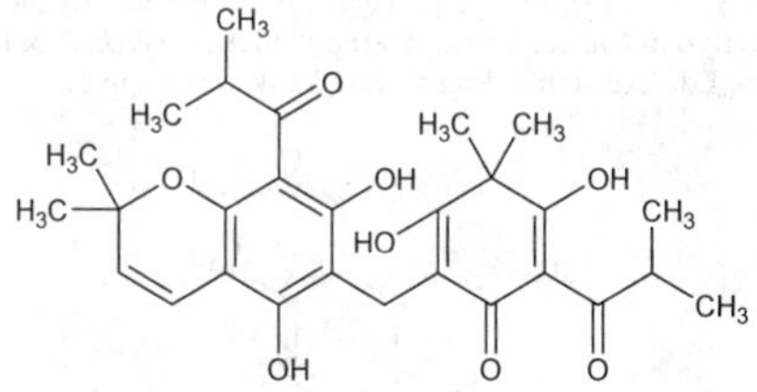

Uliginosin B

Uliginosin A. [19809-78-0] 3,5-Dihydroxy-4,4-dimethyl-2-(2-methyl-1-oxopropyl)-6-[[2,4,6-trihydroxy-3-(3-methyl-2-butenyl)-5-(2-methyl-1-oxopropyl)phenyl]methyl]-2,5-cyclohexadien-1-one. C$_{28}$H$_{36}$O$_8$; mol wt 500.59. Pale yellow crystals, mp 160.5-161.5° from acetonitrile-chloroform (4:1). uv max (cyclohexane): 229, 293 nm (ε 31500, 25000).
Uliginosin B. [19809-79-1] 2-[[5,7-Dihydroxy-2,2-dimethyl-8-(2-methyl-1-oxopropyl)-2*H*-1-benzopyran-6-yl]methyl]-3,5-dihydroxy-4,4-dimethyl-6-(2-methyl-1-oxopropyl)-2,5-cyclohexadien-1-one. C$_{28}$H$_{34}$O$_8$; mol wt 498.57. Pale yellow crystals, mp 139.5-142.0° from nitromethane. uv max (cyclohexane): 230, 270 nm (ε 34000, 37000).

10029. Ulimorelin. [842131-33-3] (2*S*)-*N*-[(2*R*)-2-[2-(3-Aminopropyl)phenoxy]propyl]-2-cyclopropylglycyl-*N*-methyl-D-alanyl-4-fluoro-D-phenylalanine (3 → 1)-lactam. C$_{30}$H$_{39}$FN$_4$O$_4$; mol wt 538.66. C 66.89%, H 7.30%, F 3.53%, N 10.40%, O 11.88%. Ghrelin receptor agonist that promotes gastric emptying. Prepn: P. Deslongchamps *et al.*, **WO 05012331**; *eidem*, **US 7452862** (2005, 2008 both to Tranzyme); of crystalline salts: H. R. Hoveyda *et al.*, **WO 11041369** (2011 to Tranzyme). Pharmacology and receptor binding study: G. L. Fraser *et al.*, *Endocrinology* **149**, 6280 (2008). Clinical pharmacokinetics, pharmacodynamics, and safety profile:

K. C. Lasseter *et al.*, *J. Clin. Pharmacol.* **48**, 193 (2008). Clinical evaluation in diabetic gastroparesis: N. Ejskjaer *et al.*, *Neurogastroenterol. Motil.* **22**, 1069 (2010); J. M. Wo *et al.*, *Aliment. Pharmacol. Ther.* **33**, 679 (2011).

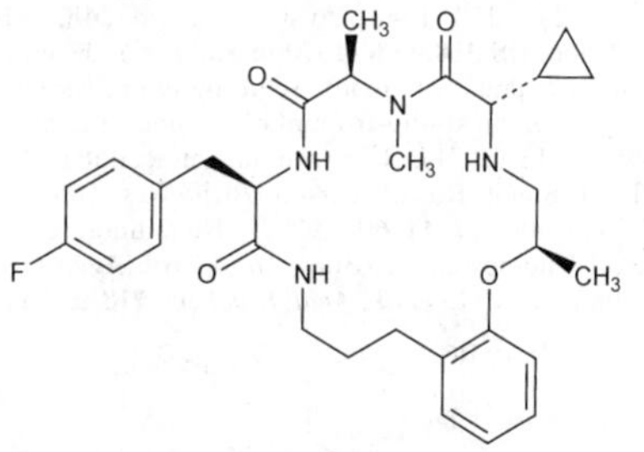

Hydrochloride monohydrate. [951326-02-6] TZP-101. C$_{30}$H$_{39}$FN$_4$O$_4$·HCl.H$_2$O; mol wt 593.14. White crystalline solid. d 1.252. Soly in water (mg/ml): 7 (pH 4.0); 2 (pH 6.0); 0.2 (pH 7). Sol in methanol; slightly sol in ethanol, acetonitrile, THF, 2-butanol; sparingly sol in DMSO; very slightly sol in isopropanol, ethyl acetate. uv max: 217, 266, 272, 278 nm (c = 0.1 in methanol).

THERAP CAT: Gastroprokinetic.

10030. Uliprist Acetate. [126784-99-4] (11β)-17-(Acetyloxy)-11-[4-(dimethylamino)phenyl]-19-norpregna-4,9-diene-3,20-dione; 17α-acetoxy-11β-(4-dimethylaminophenyl)-19-norpregna-4,9-dien-3,20-dione; uliprisnil acetate; CDB-2914; PGL-4001; RTI-3021-012; VA-2914; Ella. C$_{30}$H$_{37}$NO$_4$; mol wt 475.63. C 75.76%, H 7.84%, N 2.94%, O 13.45%. Selective progesterone receptor modulator (SPRM); synthetic 19-norprogesterone derivative. Prepn: C. E. Cook *et al.*, **WO 8912448**; *eidem*, **US 4954490** (1989, 1990 both to Research Triangle Inst.). Large scale synthesis: P. N. Rao *et al.*, *Steroids* **65**, 395 (2000). Pharmacology: E. E. Gainer, A. Ulmann, *ibid.* **68**, 1005 (2003); D. L. Blithe *et al.*, *ibid.*, 1013. Clinical evaluation as ovulation inhibitor: N. Chabbert-Buffet *et al.*, *J. Clin. Endocrinol. Metab.* **92**, 3582 (2007); in treatment of uterine fibroids: E. D. Levens *et al.*, *Obstet. Gynecol.* **111**, 1129 (2008); as emergency comtraceptive: P. Fine *et al.*, *ibid.* **115** 257 (2010). *Review*: P. A. Orihuela, *Curr. Opin. Investig. Drugs* **8**, 859-866 (2007).

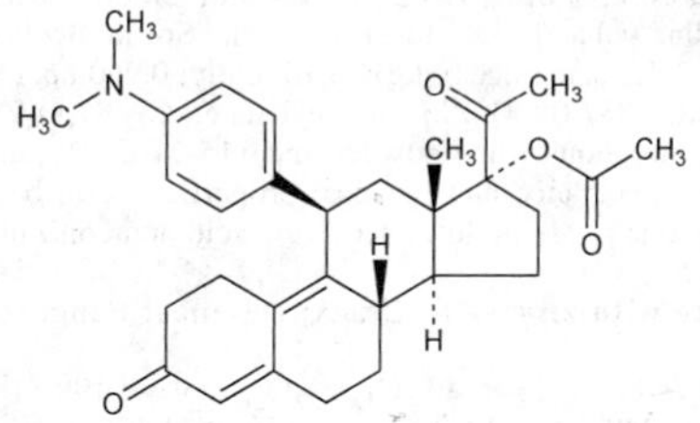

Crystals from aq ethanol, mp 183-185° (Rao); also reported as crystals from methanol + water, mp 118-121° (Cook). uv max (methanol): 261 nm.
Uliprist. [159811-51-5] (11β)-11-[4-(Dimethylamino)phenyl]-17-hydroxy-19-norpregna-4,9-diene-3,20-dione. C$_{28}$H$_{35}$NO$_3$; mol wt 433.59. Crystals from ether, mp 125-128°.

THERAP CAT: In treatment of uterine fibroids; oral contraceptive.

10031. Ultramarine. [57455-37-5] C.I. Pigment Blue 29; C.I. 77007. A blue pigment occurring naturally as the mineral lapis lazuli. Made by igniting a mixture of kaolin, Na$_2$CO$_3$ (or Na$_2$SO$_4$), S and carbon. The resulting aluminosulfosilicates resemble zeolites structurally and have the approx formula Na$_7$Al$_6$Si$_6$O$_{24}$S$_3$. *See: Colour Index* vol. 4 (3rd ed., 1971) p 4653.

Blue lumps or powder. Insol in water; readily dec by acids, even carbonic acid, with liberation of H$_2$S.

USE: As a pigment in calico printing, wall paper, mottled soap; bluing in laundry use; for coloring tiles, cements, rubber, but is now largely replaced by coal tar dyes; as a color in food.

10032. Umbelliferone. [93-35-6] 7-Hydroxy-2*H*-1-benzopyran-2-one; 7-hydroxycoumarin; hydrangin; skimmetin. $C_9H_6O_3$; mol wt 162.14. C 66.67%, H 3.73%, O 29.60%. The aglucon of skimmin. Present in many plants. Obtained by distillation of resins from umbelliferae: Zwenger, *Ann.* **115**, 1, 15 (1860). Prepn: Bert, *Compt. Rend.* **214**, 230 (1942); Austerweil, *ibid.* **248**, 1810 (1959); Dressler, Reabe, **US 3503996** (1970 to Koppers). Main product of metabolism of coumarin in man: Schilling *et al.*, *Nature* **221**, 664 (1969). Metabolism studies of umbelliferone: Indahl, Scheline, *Xenobiotica* **1**, 13 (1971). Use in brain intracellular pH measurements: T. M. Sundt, R.E. Anderson, *Brain Res.* **186**, 355 (1980); *eidem*, *J. Neurophysiol.* **44**, 60 (1980). Use in fluorescent immunoassays: S. G. Thompson, J. F. Burd, *Antimicrob. Agents Chemother.* **18**, 264 (1980); T. M. Li *et al.*, *Anal. Biochem.* **118**, 102 (1981).

Needles from water, mp 225-228°. Develops odor of coumarin on heating. Sublimes. Absorption spectrum: Sen, Bagchi, *J. Org. Chem.* **24**, 316 (1959). One gram dissolves in ~100 ml boiling water; freely sol in alcohol, chloroform, acetic acid; sol in dil alkalies; sparingly sol in ether. Solns show blue fluorescence.

USE: In sunscreen lotions and creams; as intracellular and pH sensitive fluorescent indicator and blood-brain barrier probe.

10033. Undecylenic Acid. [112-38-9] 10-Undecenoic acid; 10-hendecenoic acid; 9-undecylenic acid; Fungoid Solution; Mycodécyl Solution. $C_{11}H_{20}O_2$; mol wt 184.28. C 71.70%, H 10.94%, O 17.36%. Prepn from castor oil: Krafft, *Ber.* **10**, 2035 (1877); Perkins, Cruz, *J. Am. Chem. Soc.* **49**, 1073 (1927); G. Das *et al.*, *J. Am. Oil Chem. Soc.* **66**, 938 (1989). Toxicology: G. W. Newell *et al.*, *J. Invest. Dermatol.* **13**, 145 (1949). Clinical trials: A. L. Shapiro, S. Rothman, *Arch. Dermatol. Syphilol.* **52**, 166 (1945), republ. in *Arch. Dermatol.* **119**, 345 (1983); J. H. Chretien *et al.*, *Int. J. Dermatol.* **19**, 51 (1980). Historical review: J. W. Landau, *Arch. Dermatol.* **119**, 351-353 (1983).

Liquid or crystals. Odor suggestive of perspiration. d_4^{24} (vac) 0.9072; d_{25}^{25} 0.9102; d_{45}^{45} 0.8993; $d_4^{79.9}$ (vac) 0.8653. mp 24.5°; bp_{760} 275° (dec); bp_{182} 232-235°; bp_{130} 230-235°; bp_{100} 213.5°; bp_{90} 198-200°; bp_{15} 168.3°; $bp_{1.0}$ 131°. n_D^{25} 1.4486. Neutralization value 304.5; iodine value 137.8. Insol in water. Sol in alcohol, chloroform, ether. LD_{50} in mice (g/kg): 8.15 orally; 0.960 i.p. (Newell).

Zinc salt. [557-08-4] Zinc undecylenate. $C_{22}H_{38}O_4Zn$; mol wt 431.95. Amorphous white powder, mp 115-116°. Resembles zinc stearate in appearance and physical properties. Can be prepd by dissolving zinc oxide in dil undecylenic acid and concentrating the solution.

Mixture with zinc salt. Cruex; Desenex; Fungex; Turexan Crème.

Methyl ester. [111-81-9] $C_{12}H_{22}O_2$; mol wt 198.31. Liq; d_4^{15} 0.889; bp_{760} 248°; bp_{10} 124°; mp −27.5°; n_D^{20} 1.43928; n_D^{25} 1.43737. Sol in alcohol, chloroform, ether, petr ether, oils.

THERAP CAT: Antifungal (topical).

THERAP CAT (VET): Antifungal (topical).

10034. Unoprostone. [120373-36-6] (5*Z*)-7-[(1*R*,2*R*,3*R*,5*S*)-3,5-Dihydroxy-2-(3-oxodecyl)cyclopentyl]-5-heptenoic acid; 13,14-dihydro-15-keto-20-ethyl-PGF$_{2\alpha}$. $C_{22}H_{38}O_5$; mol wt 382.54. C 69.08%, H 10.01%, O 20.91%. Prepn: R. Ueno *et al.*, **EP 289349**; *eidem*, **US 5221763** (1988, 1993 both to Ueno). Pharmacological characterization: Y. Goh, J. Kishino, *Jpn. J. Ophthalmol.* **38**, 236 (1994). Mechanism of action study: M. Sakurai *et al.*, *ibid.* **37**, 252 (1993). Comparative clinical trial in glaucoma and ocular hypertension: J.-P. Nordmann *et al.*, *Am. J. Ophthalmol.* **133**, 1 (2002).

Isopropyl ester. [120373-24-2] UF-021; Rescula. $C_{25}H_{44}O_5$; mol wt 424.62.

THERAP CAT: Antiglaucoma; in treatment of ocular hypertension.

10035. Uracil. [66-22-8] 2,4(1*H*,3*H*)-Pyrimidinedione; 2,4-dioxopyrimidine; 2,4-pyrimidinediol; 4-hydroxy-2(1*H*)-pyrimidinone; 2-hydroxy-4(1*H*)-pyrimidinone; 2-hydroxy-4(3*H*)-pyrimidinone. $C_4H_4N_2O_2$; mol wt 112.09. C 42.86%, H 3.60%, N 24.99%, O 28.55%. Obtained by hydrolysis of nucleic acids, *cf.* Levene, Bass, *Nucleic Acids* (New York, 1931). Alternate routes of formation: Johnson, *J. Am. Chem. Soc.* **63**, 263 (1941); Fox, Harada, *Science* **133**, 1923 (1961); Takemoto, Yamamoto, *Synthesis* **1971**, 154. Several desmotropic forms. Crystal structure: Stewart, *Acta Crystallogr.* **23**, 1102 (1967). *Review:* Ts'o "Bases, Nucleosides and Nucleotides" in *Basic Principles in Nucleic Acid Chemistry* **vol. 1**, P. O. P Ts'o, Ed. (Academic Press, New York, 1974) pp 453-584.

Needles from water, mp 335° with effervescence. uv max (pH 7.0): 202.5, 259.5 nm ($\varepsilon \times 10^{-3}$ 9.2, 8.2). Freely sol in hot water; sparingly in cold water (100 parts of water at 25° dissolves 0.358 part of uracil). Almost insol in alc, ether; sol in ammonia water and in other alkalies. pK = 9.45.

USE: In biochemical research.

10036. Uracil Mustard. [66-75-1] 5-[Bis(2-chloroethyl)amino]-2,4-(1*H*,3*H*)-pyrimidinedione; 5-[bis(2-chloroethyl)amino]uracil; 2,6-dihydroxy-5-bis[2-chloroethyl]aminopyramidine; 5-[di(β-chloroethyl)amino]uracil; uramustine; demethyldopan; desmethyldopan; NSC-34462; U-8344. $C_8H_{11}Cl_2N_3O_2$; mol wt 252.10. C 38.12%, H 4.40%, Cl 28.12%, N 16.67%, O 12.69%. Synthesis from 5-aminouracil: Lyttle, Petering, *J. Am. Chem. Soc.* **80**, 6459 (1958); Lyttle, **US 2969364** (1961 to Upjohn). Teratology study: S. Chaube, M. L. Murphy, "The Teratogenic Effects of the Recent Drugs Active in Cancer Chemotherapy" in *Adv. Teratol.* **3**, D. H. M. Woollam, Ed. (Academic Press, New York, 1968) pp 181-237.

Crystals from methanol + water, dec 206°. Sparingly sol in water. uv max (0.01*N* H$_2$SO$_4$ in 95% ethanol): 257 nm (ε 5675). LD_{50} i.p. in rats: ~1.25-2.5 mg/kg (Chaube, Murphy).

THERAP CAT: Antineoplastic.

10037. Uramil. [118-78-5] 5-Amino-2,4,6(1*H*,3*H*,5*H*)-pyrimidinetrione; 5-aminobarbituric acid; dialuramide; 5-amino-2,4,6-pyrimidinetriol. $C_4H_5N_3O_3$; mol wt 143.10. C 33.57%, H 3.52%, N 29.36%, O 33.54%. Prepd by boiling alloxantin with ammonium chloride: Wöhler, Liebig, *Ann.* **26**, 241 (1838); by reduction of 5-nitrobarbituric acid with tin and HCl: Hartman, Sheppard, *Org. Synth.* **coll. vol. II**, 617 (1943); from barbituric acid, sodium nitrite and sodium hydrosulfite: Koppel, Robins, *J. Am. Chem. Soc.* **80**, 2751 (1958).

Crystals. Discolors in the air. Does not melt below 400°. Insol in cold water, in ether, chloroform; sparingly sol in hot water; sol in solns of alkali hydroxides, in ammonia and in sulfuric acid. Forms a slightly sol lead salt.

Consult the Name Index before using this section.

10038. Uranediol. [516-51-8] (1*S*,2*R*,4a*S*,4b*R*,6a*S*,8*S*,10a*S*,-10b*S*,12a*S*)-Octadecahydro-2,10a,12a-trimethyl-1,8-chrysenediol; (3β,5α,17α,17aβ)-17-methyl-*D*-homoandrostane-3,17a-diol. C$_{21}$-H$_{36}$O$_2$; mol wt 320.52. C 78.69%, H 11.32%, O 9.98%. From pregnant mares' urine: Marker *et al.*, *J. Am. Chem. Soc.* **60**, 210, 1061, 1561, 2719 (1938); Klyne, *Biochem. J.* **43**, 611 (1948); Brooks *et al.*, *ibid.* **51**, 694 (1952). Structure: Klyne, *Nature* **166**, 559 (1950). Formation by acid hydrolysis of 5α-pregnane-3β,20β-diol: Hirschmann, Williams, *J. Biol. Chem.* **238**, 2305 (1963). Configuration: Hirschmann *et al.*, *J. Org. Chem.* **31**, 375 (1966).

Needles from aq ethanol, mp 216-219°. Sublimes at 180° and 0.06-0.1 mm. [α]$_D^{15}$ +3.7° (c = 1.8 in chloroform).

Diacetate. C$_{25}$H$_{40}$O$_4$. Plates from aq methanol, mp 159.5-160.5°. Sublimes at 190° and 0.1 mm. [α]$_D^{20}$ −30.4° (c = 1.4 in chloroform); [α]$_D^{22}$ −29.6° (ethanol).

Dibenzoate. C$_{35}$H$_{44}$O$_4$. Leaflets from chloroform + methanol, mp 209-210°. [α]$_D^{29}$ +18.6° (c = 1.05 in chloroform). uv max: 230, 272 nm (log ε 4.43, 3.26).

10039. Uranium. [7440-61-1] U; at. wt 238.0289 (characteristic naturally occurring isotopic mixture); at. no. 92; valence 6, 5, 4, 3. No stable nuclides; three naturally occurring isotopes (mass numbers): 238, T$_{½}$ 4.47 × 10^9 years, rel. at. mass 238.0508 (99.275%); 235, T$_{½}$ 7.04 × 10^8 years, rel. at. mass 235.0439 (0.718%); 234, T$_{½}$ 2.46 × 10^5 years, rel. at. mass 234.0409 (0.005%); twelve artificial isotopes: 226-233; 236; 237; 239; 240. Occurrence in the earth's crust 2.1 ppm. Mined as uranium ore; main ores of commercial interest are carnotite, pitchblende, coffinite, uraninite, tobernite and autunite. Commercially important mines located in Elliot Lake−Blind River area in Canada, Rand gold fields in South Africa, Colorado and Utah in U.S., in Australia and France. Discovery from pitchblende: M. H. Klaproth, *Chem. Ann.* **II**, 387 (1789). Prepn of metal: E. Péligot, *C. R. Hebd. Seances Acad. Sci.* **12**, 735 (1841); *idem, Ann. Chim. Phys.* **5**, 5 (1842). Flowsheet and details of prepn of pure uranium metal: *Chem. Eng.* **62**, no. 10, 113 (1955); Spedding *et al.*, *US 2852364* (1958 to U.S.A. E.C.). *Reviews: Mellor's* **vol. XII**, 1-138 (1932); C. D. Harrington, A. R. Ruehle, *Uranium Production Technology* (Van Nostrand, Princeton, 1959); E. H. P. Cordfunke, *The Chemistry of Uranium* (Elsevier, New York, 1969) 250 pp; several authors in *Handb. Exp. Pharmakol.* **36**, 3-306 (1973); "The Actinides," in *Comprehensive Inorganic Chemistry* vol. 5, J. C. Bailar, Jr., *et al.*, Eds. (Pergamon Press, Oxford, 1973) *passim*; F. Weigel in *Kirk-Othmer Encyclopedia of Chemical Technology* vol. 23 (Wiley-Interscience, New York, 3rd ed., 1983) pp 502-547; *idem* in *The Chemistry of the Actinide Elements* vol. 1, J. J. Katz *et al.*, Eds. (Chapman and Hall, New York, 1986) pp 169-442; J. C. Spirlet *et al.*, *Adv. Inorg. Chem.* **31**, 1-40 (1987). Review of toxicology and human exposure: *Toxicological Profile for Uranium* (PB99-163362, 1999) 462 pp.

Silver-white, lustrous, radioactive metal; malleable and ductile. Tarnishes rapidly in air, forming a layer of dark-colored oxide. Three allotropic modifications: orthorhombic α-form, d 19.07, transforms to β-form at 667.8° ±1.3°; tetragonal, β-form d 18.11, transforms to γ-form at 774.9° ±1.6°; body-centered cubic γ-form, d 18.06, transforms to liquid at mp. mp 1132.8 ±0.8°. Heat of vaporization 446.7 kJ/mol; heat of fusion 19.7 kJ/mol; heat of sublimation 487.9 kJ/mol. Finely divided U metal and some U compounds may ignite spontaneously in air or oxygen. Rapidly soluble in aqueous HCl. Non-oxidizing acids, such as sulfuric, phosphoric and hydrofluoric, react only very slowly with U; nitric acid dissolves massive U at a moderate rate. Dissolution of finely divided U in nitric acid may approach explosive violence. Uranium metal is inert to alkalies.

Caution: Uranium is a chemical hazard as well as a radiological hazard. Potential symptoms of overexposure to U metal or insoluble U compds are dermatitis; kidney damage; blood changes. Potential symptoms of overexposure to soluble U compds are lacrimation, conjunctivitis; shortness of breath, cough, chest rales; nausea, vomiting, skin burns; red blood cells and casts in urine; albuminuria; high blood urea nitrogen. Radiation hazard is caused by the direct emission of α-particle radiation and by α-particles emitted from radon gas and its particulate daughters formed during the natural decay of U. U and U compds are potential occupational carcinogens. *See NIOSH Pocket Guide to Chemical Hazards* (DHHS/NIOSH 97-140, 1997) p 326; *Patty's Industrial Hygiene and Toxicology* vol. 2C, G. D. Clayton, F. E. Clayton, Eds. (Wiley-Interscience, New York, 4th ed., 1994) pp 2297-2317.

USE: ^{235}U in nuclear power reactors and nuclear weapons. Uranium depleted of ^{235}U to manuf armor-piercing ammunition, in inertial guidance devices and gyro compasses, as a counterweight for missile reentry vehicles, as radiation shielding material, and x-ray targets.

10040. Uranium Dioxide. [1344-57-6] Uranium oxide (UO$_2$); uranous oxide; black uranium oxide. O$_2$U; mol wt 270.03. O 11.85%, U 88.15%. UO$_2$. Occurs in nature as the minerals *uraninite* or *pitchblende*. Review of uranium oxides: Keller in *Comprehensive Inorganic Chemistry* vol. 5, J. C. Bailar, Jr. *et al.*, Eds. (Pergamon Press, Oxford, 1973) pp 224-233.

Brown to black powder, or cubic crystals. d 10.97. mp 2865°. When obtained by heating the urano-uranic oxide or the oxalate in hydrogen, it is brown or copper-red in color and is pyrophoric. Sol in concd acids. Insol in water, dil acids.

USE: Nuclear fuel; in filaments of incandescent lamps.

10041. Uranium Hexafluoride. [7783-81-5] (*OC*-6-11)-Uranium fluoride (UF$_6$); hexafluorouranium. F$_6$U; mol wt 352.02. F 32.38%, U 67.62%. UF$_6$. Prepd by the action of fluorine on uranium metal or carbide; on uranium pentachloride; on uranium tetrafluoride; on triuranium octaoxide in the presence of carbon. Uranium compds when heated with fluorine to a sufficiently high temp give UF$_6$: National Nuclear Energy Series **VIII-5**, J. J. Katz, E. Rabinowitch, *The Chemistry of Uranium*, Part I (New York, 1951) pp 396-449. Review of toxicology and human exposure: *Toxicological Profile for Uranium* (PB99-163362, 1999) 462 pp.

Volatile, white monoclinic crystal solid. d$_4^{20.7}$ 5.09. d^{70} (liq) 3.595. mp 64.8°. Sublimes at 56.5°. Critical temp 230.2°, crit pressure 45.5 atm: Oliver *et al.*, *J. Am. Chem. Soc.* **75**, 2827 (1953). Reacts vigorously with water, forming mainly UO$_2$F$_2$ and HF. Sol in liquid chlorine and bromine. Dissolves in nitrobenzene to give a dark red soln fuming in air. Sol in carbon tetrachloride, chloroform, and *sym*-tetrachloroethane, Cl$_2$CHCHCl$_2$, which forms the most stable soln, extensive reaction occurring only after several days at room temp. Sol in fluorocarbons (C$_6$F$_{12}$ or C$_7$F$_{16}$) without reaction. Best handled in copper apparatus.

USE: Production of depleted uranium metal.

10042. Uranium Tetrachloride. [10026-10-5] Uranium chloride (UCl$_4$); uranium(IV) chloride. Cl$_4$U; mol wt 379.83. Cl 37.33%, U 62.67%. UCl$_4$. Prepd from uranium(VI) oxide dihydrate and hexachloropropene: Hermann, Suttle, *Inorg. Synth.* **5**, 143 (1957).

Dark green octahedral crystals (tetragonal symmetry). Oxidizes in air and dec on contact with water. Should be stored in sealed ampuls. d$_4^{25}$ 4.725. mp 590°. bp$_{760}$ 791°. Heat of formation (solid) 250.9 kcal/mol at 0°. Freely sol in water (dec); sol in polar organic solvents. Insol in nonpolar solvents such as hydrocarbons and ethyl ether.

10043. Uranium Tetrafluoride. [10049-14-6] (*T*-4)-Uranium fluoride (UF$_4$); tetrafluorouranium. F$_4$U; mol wt 314.02. F 24.20%, U 75.80%. UF$_4$. Prepd from uranyl oxide and dichlorodifluoromethane: Booth *et al.*, *J. Am. Chem. Soc.* **68**, 1969 (1946); also prepd by the action of hydrogen fluoride on uranium dioxide or by addition of hydrofluoric acid to a U(IV) soln: Eméleus in *Fluorine Chemistry* **vol. I**, J. H. Simons, Ed. (Academic Press, New York, 1950) p 59.

Monoclinic green crystals, mp >1100°. Changes to U$_3$O$_8$ when heated in air. Insol in water. Sol in concd acids and alkalies with decompn.

Hemipentahydrate. Orthorhombic green crystals, insol in water.

10044. Uranium Trichloride. [10025-93-1] Cl₃U; mol wt 344.38. Cl 30.88%, U 69.12%. UCl₃. Prepd by hydrogen reduction of uranium(IV) chloride: Suttle, *Inorg. Synth.* **5**, 145 (1957).

Dark purple crystals. Less hygroscopic than uranium(IV) chloride. d 5.51. mp 842°. Heat of formation at 25° 212.0 kcal/mol. Freely sol in water, giving a purple soln, which evolves hydrogen and turns green because of oxidation to uranium(IV). Less sol in polar organic solvents than uranium(IV) chloride and insol in nonpolar solvents.

10045. Uranium Trioxide. [1344-58-7] Uranium oxide (UO₃); uranic oxide; red uranium oxide. O₃U; mol wt 286.03. O 16.78%, U 83.22%. UO₃. Review of uranium oxides: Keller in *Comprehensive Inorganic Chemistry* vol. 5, J. C. Bailar, Jr. *et al.*, Eds. (Pergamon Press, New York, 1973) pp 224-233.

Red or brownish-yellow powder. d 7.29. Insol in water; sol in acids.

10046. Uranyl Acetate. [541-09-3] (*T*-4)-Bis(acetato-κO)dioxouranium. C₄H₆O₆U; mol wt 388.11. C 12.38%, H 1.56%, O 24.73%, U 61.33%. UO₂(C₂H₃O₂)₂.

Dihydrate. [6159-44-0] C₄H₆O₆U.2H₂O; mol wt 424.14. Yellow, cryst powder; slight acetic odor. d 2.89. Sol in 10 parts of water, usually incompletely, due to the presence of basic salt. Freely sol in water acidulated with acetic acid, slightly in alcohol.

USE: As a reagent for precipitation of sodium; in dry copying inks, and as activator in bacterial oxidation processes.

10047. Uranyl Chloride. [7791-26-6] (*T*-4)-Dichlorodioxouranium; uranium dioxydichloride. Cl₂O₂U; mol wt 340.93. Cl 20.80%, O 9.39%, U 69.82%. UO₂Cl₂. Prepd by the reaction of O₂ with uranium(IV) chloride at 300-350°: Leary, Suttle, *Inorg. Synth.* **5**, 148 (1957).

Bright yellow crystals. Orthorhombic system. Very hygroscopic. Appreciably volatile above 775°. Dec *in vacuo* above 450° yielding chlorine and UO₂ plus U₃O₈. Very freely sol in water. Forms cryst hydrates. Sol in polar organic solvents such as acetone and alcohol, but insol in less polar solvents such as benzene. Aq solns are unstable.

10048. Uranyl Nitrate. [36478-76-9] (*OC*-6-11)-Bis(nitrato-κO,κO')dioxouranium. N₂O₈U; mol wt 394.03. N 7.11%, O 32.48%, U 60.41%. UO₂(NO₃)₂.

Hexahydrate. [13520-83-7] N₂O₈U.6H₂O; mol wt 502.12. Yellow crystals; greenish luster by reflected light. When shaken, rubbed, or crushed, the crystals show remarkable triboluminescence with occasional detonations. d 2.807; mp 60°. Soly in water: 122 g/100 g. Freely sol in alcohol, ether. The aq soln is acid. Solns of uranium nitrate in ether should not be allowed to stand in sunlight as explosion may occur.

USE: As intensifier in photography; manuf uranium glaze, decorating porcelain; also as reagent in analytical chemistry for pptn of sodium and magnesium.

10049. Uranyl Sulfate. [1314-64-3] Dioxo[sulfato(2−)-κO]uranium. O₆SU; mol wt 366.08. O 26.22%, S 8.76%, U 65.02%. UO₂SO₄.

Trihydrate. Lemon-yellow, cryst mass. d 3.28. Sol in ~5 parts water, 25 parts alcohol.

10050. Urapidil. [34661-75-1] 6-[[3-[4-(2-Methoxyphenyl)-1-piperazinyl]propyl]amino]-1,3-dimethyl-2,4(1*H*,3*H*)-pyrimidinedione; 6-[[3-[4-(*o*-methoxyphenyl)-1-piperazinyl]propyl]amino]-1,3-dimethyluracil; B-66256; Ebrantil; Eupressyl; Uraprene. C₂₀H₂₉N₅O₃; mol wt 387.48. C 62.00%, H 7.54%, N 18.07%, O 12.39%. α₁-Adrenergic antagonist; deriv of uracil, *q.v.* Prepn: W. Pruesse *et al.*, DE 1942405; *eidem*, US 3957786 (1971, 1976 both to Byk Gulden); K. Klemm *et al.*, *Arzneim.-Forsch.* **27**, 1895 (1977). Series of articles on pharmacology, pharmacokinetics, metabolism: *ibid.* 1898-1932. Toxicity data: J. Koenig *et al.*, *ibid.* 1919. Mode of action: M. Eltze, *Eur. J. Pharmacol.* **59**, 1 (1979); H. R. Kaplan, R. D. Smith, *Fed. Proc.* **40**, 2268 (1981). Hemodynamic responses in man: G. G. Belz *et al.*, *Clin. Pharmacol. Ther.* **37**, 48 (1985). Clinical trials: A. Barankay *et al.*, *Arzneim.-Forsch.* **31**, 849 (1981); H. Liebau *et al.*, *J. Hypertens.* **4**, Suppl. 6, S141 (1986). Series of articles on clinical efficacy in hypertension: *Drugs* **35**, Suppl. 6, 147-192 (1988).

Crystals from water, mp 156-158°. uv max (methanol): 237, 268 nm (ε 1.10×10⁴, 2.67×10⁴). pKa 7.10. LD₅₀ in male mice, rats (mg/kg): 750, 550 orally; 260, 145 i.v. (Koenig).

THERAP CAT: Antihypertensive.

10051. Urazole. [3232-84-6] 1,2,4-Triazolidine-3,5-dione; bicarbamimide; 1*H*-1,2,4-triazole-3,5(2*H*,4*H*)-dione; hydrazodicarbonimide; 3,5-diketotriazolidine. C₂H₃N₃O₂; mol wt 101.07. C 23.77%, H 2.99%, N 41.58%, O 31.66%. Prepd by heating biuret with hydrazine hydrate at 108°: Stolle, Krauch, *J. Prakt. Chem.* [2] **88**, 313 (1913). From methyl or ethyl allophanate: Gordon, Audrieth, *Inorg. Synth.* **5**, 52 (1957).

Leaflets from water, dec 249-250°. Weak acid, Ka: 1.6×10⁻⁶. pH of satd aqueous soln at 25° = 3.15. Soly in water (g/100 ml): 2.83 (0°); 23.7 (65°). Sparingly sol in alc, practically insol in ether. Sol in concd HCl from which it crystallizes without change. Forms ammonium and sodium salts.

10052. Urea. [57-13-6] Carbamide; carbonyldiamide; Aquacare; Aquadrate; Basodexan; Hyanit; Keratinamin; Nutraplus; Onychomal; Pastaron; Ureaphil; Urepearl. CH₄N₂O; mol wt 60.06. C 20.00%, H 6.71%, N 46.64%, O 26.64%. Physiological regulator of nitrogen excretion in mammals; synthesized in the liver as an end-product of protein catabolism and excreted in urine. Also occurs normally in skin. First organic compd to be synthesized from inorganic reagents: Wöhler, *Ann. Phys.* **12**, 253 (1828). Review of mechanism of ammonium cyanate-urea conversion: J. Shorter, *Chem. Soc. Rev.* **7**, 1-14 (1978). Prepn from ammonia, carbon monoxide and sulfur in methanol: Applegath, US 2857430 (1958 to Monsanto); Franz, Applegath, *J. Org. Chem.* **26**, 3304, 3306, 3309 (1961). Review of commercial processes: *Faith, Keyes & Clark's Industrial Chemicals*, F. A. Lowenheim, M. K. Moran, Eds. (Wiley-Interscience, New York, 4th ed., 1975) pp 854-861; of uses and manufacture: I. Mavrovic, A. R. Shirley, Jr. in *Kirk-Othmer Encyclopedia of Chemical Technology* vol. 23 (Wiley-Interscience, New York, 3rd ed., 1983) pp 548-575. Review of physiological synthesis: A. J. Meijer in *Nitrogen Metabolism and Excretion*, P. J. Walsh, P. Wright, Eds. (CRC Press, Boca Raton, 1995) pp 193-204; of analytical methods: A. J. Taylor, P. Vadgama, *Ann. Clin. Biochem.* **29**, 245-264 (1992). Clinical use as emollient: G. Swanbeck, *Acta Derm. Venereol. Suppl.* **177**, 7 (1992). Diagnostic use in *Helicobacter pylori* infection: A. K. Hamlet *et al.*, *Scand. J. Gastroenterol.* **30**, 1058 (1995).

Tetragonal prisms. Develops odor of NH₃. Cooling, saline taste. mp 132.7°. On further heating it decomposes to biuret, NH₃, and cyanuric acid. d₄¹⁸ 1.32; d₄¹⁸ of water solns (w/w): 10% 1.027; 20% 1.054; 50% 1.145. pH of a 10% water soln: 7.2. One gram dissolves in 1 ml water, 10 ml 95% alc, 1 ml boiling 95% alc, 20 ml abs alc, 6 ml methanol, 2 ml glycerol. Practically insol in chloroform, ether. Sol in concd HCl. Water solns hydrolyze slowly to ammonium carbonate with eventual decomp to ammonia and carbon dioxide.

Hydrochloride. [506-89-8] $CH_4N_2O.HCl$; mol wt 96.51. White to faintly yellow, deliquesc crystals. Dec at 145°. Sol in water.

USE: Nitrogen-release fertilizer. Starting material for resins and plastics. Condensed with malonic ester to form barbituric acid. Used in the manuf of paper. Analytical use in buffer solns.

THERAP CAT: Emollient; diuretic.

THERAP CAT (VET): Nutritional factor (partial source of dietary nitrogen in ruminants); debriding agent; diuretic.

10053. Ureaform. [9011-05-6] Urea polymer with formaldehyde; Uramite. Condensation product of urea and formaldehyde in the approx ratio of 1.2-1.5 to 1. Consists of a mixture of methyleneurea polymers of varying length. Total nitrogen content at least 35%. Prepn: E. T. Darden, **US 2766283** (1956 to Du Pont). Long-term trial in turf grass: D. V. Waddington *et al.*, *Soil Sci. Soc. Am. J.* **40**, 593 (1976). Availability and mineralization in soil: P. Sasson, *Soil Sci.* **128**, 285 (1979). *Review:* A. Alexander, H.-U. Helm, *Z. Pflanzenernaehr. Bodenkd.* **153**, 249-255 (1990).

n = 1-10

Yellow, granular material of small particle size. Cold water sol fraction is composed primarily of methylenediurea and dimethylenetriurea.

USE: Slow-release nitrogen fertilizer for turf grass.

10054. Urea Nitrate. [124-47-0] Urea nitrate (1:1); acidogen nitrate. $CH_5N_3O_4$; mol wt 123.07. C 9.76%, H 4.10%, N 34.14%, O 52.00%. $CO(NH_2)_2.HNO_3$. Prepn: **DE 285259** (1914 to Osterr. Ver. Chem. Metallurg. Prod.), *Frdl.* **12**, 97; and physicochemical props: T. R. Narayanan Kutty, A. R. Vasudeva Murthy, *Indian J. Technol.* **10**, 305 (1972). Crystal structure: S. Harkema, D. Feil, *Acta Crystallogr.* **B25**, 589 (1969). Use to solubilize mineral phosphates in soils: D. N. Pathak, *Fert. Technol.* **15**, 154 (1978). Use for nitration of aromatic amines: T. P. Sura *et al.*, *Synth. Commun.* **18**, 2161 (1988).

White, odorless leaflets. mp 152° with decompn. *Flammable, explosive.* Soly (g/100g) at 0°: water 9.30, alcohol 1.35; at 65.3°: water 39.84, alcohol 8.84. The aq soln is acid.

USE: As sensitizer for explosives; fertilizer; organic reagent.

10055. Urease. [9002-13-5] Urea amidohydrolase. Nickel-dependent metalloenzyme which hydrolyzes urea to ammonium carbonate. First enzyme crytallized; mol wt approx 489,000 Da. Isolation from jack bean, *Canavalia ensiformis* and crystallization: J. B. Sumner, *J. Biol. Chem.* **69**, 435 (1926). Early structure-activity studies: G. Gorin *et al.*, *Biochemistry* **1**, 911 (1962); C. J. Bailey, D. Boulter, *Biochem. J.* **113**, 669 (1969). Identification as nickel containing: N. E. Dixon *et al.*, *J. Am. Chem. Soc.* **97**, 4131 (1975). Crystal structure: E. Jabri *et al.*, *Science* **268**, 998 (1995). *Review:* J. E. Varner, "Urease" in *The Enzymes* vol. **4**, P. D. Boyer *et al.*, Eds. (Academic Press, New York, 2nd ed., 1960) pp 247-256; F. J. Reithel, *ibid.* vol. **4** (3rd ed., 1971) pp 1-21.

Colorless, octahedral crystals. Soluble in water, in dil alkali and dil ammonia. uv max: 278.5 nm. Isoelectric point pH 5.0-5.1; also reported as pH 4.8 (Reithel).

USE: Clinical reagent in determination of urea in body fluid.

10056. Urea Stibamine. [1340-35-8] Carbostibamide. Pentavalent antimonial. Therapeutic efficacy in kala-azar: U. N. Brahmachari, *Indian J. Med. Res.* **10**, 492 (1922). Prepn and proposed structure as $NH_2CONHC_6H_4SbO(OH)ONH_4$: U. N. Brahmachari, J. Das, *ibid.* **12**, 423 (1924). Shown to be a mixture of products with *sym*-diphenylcarbamido-4,4'-distibonic acid as the active principal: W. H. Gray *et al.*, *Proc. Roy. Soc.* **B108**, 54 (1931). Description of development: R. C. Guha *et al.*, *Nature* **151**, 108 (1943); W. Peters, *Indian J. Med. Res.* **73** Suppl., 1 (1981). Improved prepn, characterization and activity: S. B. Mahato *et al.*, *Biochem. Med. Metab. Biol.* **38**, 47 (1987).

Sol in water. Partially sol in alc, ether.

THERAP CAT: Antiprotozoal (Leishmania).

10057. Uredepa. [302-49-8] N-[Bis(1-aziridinyl)phosphinyl]-carbamic acid ethyl ester; ethyl [bis(1-aziridinyl)phosphinyl]carbamate; ethyl N-[bis(ethyleneimido)phosphoro]carbamate; bis(ethylenimido)phosphorylurethane; urethimine; AB-100; NSC-37095; Avinar. $C_7H_{14}N_3O_3P$; mol wt 219.18. C 38.36%, H 6.44%, N 19.17%, O 21.90%, P 14.13%. Outline of prepn: Bardos *et al.*, *Nature* **183**, 399 (1959); Bardos, Papanastassiou, **US 3201313** (1965 to Armour Pharm.).

Crystals from benzene + cyclohexane, mp 88-90°. Readily sol in water but dec in aq soln.

USE: Has been used experimentally as an insect chemosterilant.

THERAP CAT: Antineoplastic.

10058. Urena. Cadillo; caesarweed. A bast fiber from the plant *Urena lobata* Dill. ex L., *Malvaceae*, a common, perennial grass growing in tropical and subtropical areas. *Habit.* Brazil, India, Madagascar, and Africa. Harvested when in full bloom (pink flowers). The fiber is creamy white, lustrous, soft and flexible, comparable to jute. *Ref:* E. E. Stout, *Introduction to Textiles* (John Wiley, New York, 1960) pp 56-57; A. S. Roy, *Indian Textile J.* **81**, 73 (1981). Evaluation for use in pulp and paper: R. K. Dubey *et al.*, *IPPTA* **18**, 52 (1981).

USE: In cordage and course textiles.

10059. Urethan. [51-79-6] Carbamic acid ethyl ester; ethyl aminoformate; ethyl carbamate; urethane; ethyl urethan. $C_3H_7NO_2$; mol wt 89.09. C 40.45%, H 7.92%, N 15.72%, O 35.92%. Naturally occurring contaminant in fermented foods, particularly wine, stone-fruit brandies, and bread. Prepd by heating urea with alcohol under pressure; by warming urea nitrate with alcohol and sodium nitrite. Toxicity data: K. J. Franklin, *J. Pharmacol. Exp. Ther.* **42**, 1 (1931). Review of carcinogenic action and metabolism: S. S. Mirvish in *Adv. Cancer Res.* **11**, 1-42 (1968); of mutagenicity, metabolism and interactions with DNA: R. E. Sotomayor, T. F. X. Collins, *Toxicol. Ind. Health* **6**, 71-108 (1990); of physiological effects in animals: K. J. Field, C. M. Lang, *Lab. Anim.* **22**, 255-262 (1988). Review of analysis, occurrence and formation in foodstuffs: R. Battaglia *et al.*, *Food Addit. Contam.* **7**, 477-496 (1990); B. Zimmerli, J. Schlatter, *Mutat. Res.* **259**, 325-350 (1991).

Crystals, mp 48-50°. Cooling saline taste, d 1.1. bp 182-184°. Sublimes readily at 103° and 54 mm pressure. One gram dissolves in 0.5 ml water, 0.8 ml alcohol, 0.9 ml chloroform, 1.5 ml ether, 2.5 ml glycerol, 32 ml olive oil. The aq soln is neutral. MLD i.p. in mice: 2.1-2.2 g/kg (Franklin).

Caution: This substance is reasonably anticipated to be a human carcinogen: *Report on Carcinogens, Twelfth Edition* (PB2011-111646, 2011) p 434.

USE: Intermediate in organic synthesis. In the prepn and modification of amino resins. As solvent, solubilizer and cosolvent for various organic materials. Animal anesthetic in laboratory procedures.

10060. Uric Acid. [69-93-2] 7,9-Dihydro-1H-purine-2,6,8-(3H)-trione; 8-hydroxyxanthine; purine-2,6,8-triol; purine-2,6,8-(1H,3H,9H)-trione; 2,6,8-trioxypurine. $C_5H_4N_4O_3$; mol wt 168.11. C 35.72%, H 2.40%, N 33.33%, O 28.55%. Discovered by Scheele and independently by Bergman in 1776. It forms the chief end-product of the nitrogenous metabolism of birds and of scaly reptiles and is found in their excrement; present in the urine of all carnivorous animals. Prepn from urea: Bills *et al.*, *J. Org. Chem.* **27**, 4633

(1962). Role in biological processes: Bishop, Talbott, *Pharmacol. Rev.* **5**, 231 (1953).

White, odorless, tasteless crystals; dec by heat without melting and with evolution of HCN. d 1.89. One gram dissolves in about 15,000 parts cold water, about 2000 parts boiling water; sol in glycerol, in solns of alkali hydroxides, their carbonates, sodium acetate and sodium phosphate. Insol in alcohol, ether. Gives murexide reaction.

10061. Uricase. [9002-12-4] Urate oxidase; uric acid oxidase; urate:oxygen oxidoreductase; EC 1.7.3.3. Enzyme that catalyzes the conversion of uric acid to 5-hydroxyisourate which spontaneously decomposes to form allantoin as the end product of purine catabolism. The enzyme is endogenous in most mammals but is absent in higher primates, including humans. Exists as a tetramer with identical subunits; primarily found in the liver. Also found in microorganisms and plants; key enzyme in the nitrogen fixation pathway in legumes. Isoln from porcine liver: Holmberg, *Biochem. J.* **33**, 1901 (1939); Miller *et al.*, *J. Biol. Chem.* **216**, 625 (1955); Robbins, Grant, **US 2878161** (1959 to Armour). Purification from soybean root nodules: K. Lucas *et al.*, *Arch. Biochem. Biophys.* **226**, 190 (1983). Spectroscopic characterization of reaction intermediates in soybeans: K. Kahn, P. A. Tipton, *Biochemistry* **37**, 11651 (1998). *Review:* Mahler, "Uricase" in *The Enzymes* vol. 8, P. D. Boyer *et al.*, Eds. (Academic Press, New York, 1963) pp 285-296. Review of use in determn of uric acid in serum: Y. Zhao *et al.*, *Microchim. Acta* **164**, 1-6 (2009); of therapeutic potential in treatment of gout: J. S. Sundy, M. S. Hershfield, *Curr. Rheumatol. Rep.* **9**, 258-264 (2007).

Pale, brownish-green crystals or shiny, transparent, striated plates. Practically insol in water. Slightly sol in buffered alkali solns. Solns at pH 7.5-10.5 are relatively stable. Shows unusually high absorption in the region of 330-350 nm (for highly purified uricase: $A_{276}^{1\%}$ = 11.3, $A_{330}^{1\%}$ = 2.0, both in 1% Na_2CO_3; A_{280}/A_{330} = 5.6). Isoelec. pt. pH 6.3.

USE: In the determination of serum and urine uric acid.

10062. Uridine. [58-96-8] 1-β-D-Ribofuranosyluracil; uracil riboside. $C_9H_{12}N_2O_6$; mol wt 244.20. C 44.27%, H 4.95%, N 11.47%, O 39.31%. Nucleoside; widely distributed in nature. Prepd by hydrolysis of yeast nucleic acid with weak alkali, *cf.* Levene, Bass, *Nucleic Acids* (New York, 1931). Improved isolns: Harris, Thomas, *J. Chem. Soc.* **1948**, 1936; Elmore, *ibid.* **1950**, 2084; Lorine, Ploeser, *J. Biol. Chem.* **178**, 439 (1949). Crystal structure: Green *et al.*, *Chem. Commun.* **1971**, 53. *Review: Basic Principles in Nucleic Acid Chemistry* vol. 1, P. O. P. Ts'o, Ed. (Academic Press, New York, 1974) *passim*.

Needles from dil alc, mp 165°. $[\alpha]_D^{20}$ +4° (c = 2). uv max (pH 7.3): 261, 205 nm (ε × 10^{-3} 10.1, 9.8), Voet *et al.*, *Biopolymers* **1**, 193 (1963). Sol in water. Upon prolonged refluxing with HCl, furfurol is formed.

10063. Uridine 5'-Diphosphate. [58-98-0] Uridine 5'-(trihydrogen diphosphate); UDP; uridine 5'-pyrophosphate; uridine-5-pyrophosphoric acid. $C_9H_{14}N_2O_{12}P_2$; mol wt 404.16. C 26.75%, H 3.49%, N 6.93%, O 47.50%, P 15.33%. Can be isolated from calf's liver, thymus, and yeast. The commercial product is derived from yeast. Pentose nucleic acids (isolated from yeast) are digested with rattlesnake venom (freed of 5'-monoesterase) and the nucleotides are separated by chromatography: Cohn, Volkin, *Arch. Biochem. Biophys.* **35**, 465 (1952); *J. Biol. Chem.* **203**, 319 (1953). For alternate procedures *see* the refs under Uridine Diphosphate Glucose. Syntheses: Chambers, *J. Am. Chem. Soc.* **81**, 3032 (1959); Moffatt, Khorana, *ibid.* **83**, 649 (1961).

Lithium salt monohydrate. $C_9H_{12}N_2O_{12}P_2Li_2.H_2O$. Crystals, sol in water.

Sodium salt trihydrate. $C_9H_{12}N_2O_{12}P_2Na_2.3H_3O$. Crystals. pKa'_1 6.5; pKa'_2 9.4. uv max (pH 7): 262 nm; (pH 11): 261 nm. Freely sol in water.

10064. Uridine Diphosphate Glucose. [133-89-1] Uridine 5'-(trihydrogen diphosphate) P'-α-D-glucopyranosyl ester; UDPG; UDP-Glucose; uridine 5'-pyrophosphate glucose ester; uridine-5'-diphosphoglucose; co-waldenase; co-galactoisomerase. $C_{15}H_{24}N_2O_{17}P_2$; mol wt 566.30. C 31.81%, H 4.27%, N 4.95%, O 48.03%, P 10.94%. The coenzyme of the galactowaldenase system which catalyzes the conversion of galactose-1-phosphate into glucose-1-phosphate. Isoln from baker's yeast: Caputto *et al.*, *J. Biol. Chem.* **184**, 333 (1950). Also present in animal tissue. Synthesis: Michelson, Todd, *J. Chem. Soc.* **1956**, 3459; Moffatt, Khorana, *J. Am. Chem. Soc.* **80**, 3756 (1958). *Reviews:* Leloir, Cardini in *The Enzymes* vol. 2A, P. D. Boyer *et al.*, Eds. (Academic Press, New York, 2nd ed., 1960) pp 39-61; A. M. Michelson, *The Chemistry of Nucleosides and Nucleotides* (Academic Press, New York, 1963) pp 153-250; D. W. Hutchison, *Nucleotides and Coenzymes* (John Wiley, New York, 1964) pp 36-82.

Isolated as the barium or calcium salt, white powder, sol in water.

Sodium salt dihydrate. $C_{15}H_{23}N_2O_{17}P_2Na.2H_2O$. White powder. uv max (pH 2.0): 262 nm. Freely sol in water.

10065. Uridine 5'-Triphosphate. [63-39-8] Uridine 5'-(tetrahydrogen triphosphate); UTP. $C_9H_{15}N_2O_{15}P_3$; mol wt 484.14. C 22.33%, H 3.12%, N 5.79%, O 49.57%, P 19.19%. Pyrimidine analog of ATP. Activates chloride channels in epithelial cells; increases ciliary beat frequency and induces degranulation of goblet cells in airway epithelia. Also effects inflammatory cell function and vascular reactivity. Isoln from rabbit muscle: Lipton *et al.*, *J. Am. Chem. Soc.* **75**, 5450 (1953). Synthesis: Kenner *et al.*, *J. Chem. Soc.* **1954**, 2288; Hall, Khorana, *J. Am. Chem. Soc.* **76**, 5056 (1954). Review of bioactivity mediated by nucleotide receptors: S. E. O'Connor, *Life Sci.* **50**, 1657-1664 (1992). Clinical effect on chloride secretion in cystic fibrosis: M. R. Knowles *et al.*, *Am. J. Respir. Crit. Care Med.* **151**, S65 (1995); on mucociliary clearance in combination with amiloride, *q.v.*: W. D. Bennett *et al.*, *ibid.* **153**, 1796 (1996).

Trisodium salt dihydrate. Uteplex. $C_9H_{12}N_2Na_3O_{15}P_3.2H_2O$; mol wt 586.11. White powder, sol in water, very sparingly sol in alcohol. pKa_1 6.6; pKa_2 9.5. uv max (pH 7): 262 nm (α_M 10000); (pH 11): 261 nm (α_M 8100).

THERAP CAT: In treatment of cystic fibrosis.

10066. 5'-Uridylic Acid. [58-97-9] Uridine 5'-phosphoric acid; uridine 5'-monophosphate; UMP. $C_9H_{13}N_2O_9P$; mol wt 324.18. C 33.35%, H 4.04%, N 8.64%, O 44.42%, P 9.55%. Nucleotide; widely distributed in nature. Synthesis by phosphorylation of 2',3'-O-benzylidene uridine with diphenyl phosphorochloridate: Brown et al., J. Chem. Soc. **1950**, 408; Smith, Biochem. Prep. **8**, 130 (1961). Monograph on the synthesis of nucleotides: G. R. Pettit, Synthetic Nucleotides vol. 1 (Van Nostrand Reinhold, New York, 1972) 252 pp. Crystal structure of hydrated barium salt: Shefter, Trueblood, Acta Crystallogr. **18**, 1067 (1965). Reviews: see Uridine.

Free acid: pKa_1' 6.4, pKa_2' 9.5. uv max (pH 7.0): 262 nm (α_M 10000).

Disodium salt dihydrate. $C_9H_{11}N_2Na_2O_9P.2H_2O$. White powder, characteristic meaty taste. uv max (0.1M HCl): 262 nm (ε 10000). Soly in water at 20°: about 41 g/100 ml H_2O.

10067. Urinastatin. [80449-31-6] Bikunin trypsin inhibitor; mingin; urinary trypsin inhibitor; UTI; ulinastatin; Miraclid. Acid-stable glycoprotein (pI 2.8), mol wt ~67,000 by gel filtration. Consists of a single polypeptide chain of 147 amino acid residues containing two Kunitz-type protease inhibitor domains, and two carbohydrate side chains. Normally present in serum and urine. Isoln from human urine: N. R. Shulman, J. Biol. Chem. **213**, 655 (1955); and purification: G. J. Proksch, J. I. Routh, J. Lab. Clin. Med. **79**, 491 (1972). Characterization: M. Balduyck et al., Eur. J. Biochem. **158**, 417 (1986). Protease inhibition: H. Sumi, N. Toki, Proc. Soc. Exp. Biol. Med. **167** 530 (1981). Determn by enzyme immunoassay: N. Nishino et al., Haemostasis **19**, 112 (1989). Biodistribution and clinical kinetics: B.-M. Jönsson-Berling, K. Ohlsson, Scand. J. Clin. Lab. Invest. **51**, 549 (1991). Protective effect in exptl pancreatitis: T. Hirano, T. Manabe, Arch. Surg. **128**, 1322 (1993). Clinical evaluation vs ischemic myocardial injury: S. Shimai et al., Jpn. Circ. J. **53**, 1144 (1989); vs postoperative hyperamylasemia: D. Korenaga et al., Eur. Surg. Res. **23**, 214 (1991). Receptor binding studies: H. Kobayashi et al., J. Biol. Chem. **269**, 20642 (1994). Review of early literature: H. J. Faarvang, Scand. J. Clin. Lab. Invest. **17**, Suppl. 83, 11-77 (1965). Review of characterization as subunit of inter-α-trypsin inhibitor (inter-α-inhibitor): W. Gebhard, K. Hochstrasser in Proteinase Inhibitors, A. J. Barrett, G. Salvesen, Eds. (Elsevier, Amsterdam, 1986) pp 389-401; W. Gebhard et al., Biol. Chem. Hoppe-Seyler **371**, Suppl., 13-22 (1990).

THERAP CAT: Protease inhibitor.

10068. Urobilins. Tetrapyrrole bile pigments produced by the microbial catabolism of bilirubin, q.v., in the gastrointestinal tract. Sequential reduction of bilirubin produces **urobilinogen** which can be oxidized to the yellow pigment, urobilin, for excretion in urine, or further reduced to **stercobilinogen** which is converted to the dark brown pigment, stercobilin, for excretion in the feces. Isoln: C. J. Watson, Z. Physiol. Chem. **208**, 101 (1932); C. J. Watson et al., J. Biol. Chem. **200**, 697 (1953). Reviews of occurrence, proposed structures, stereoisomerism, and interrelationships: A. H. Jackson et al., Nature **209**, 581 (1966); C. H. Gray, D. C. Nicholson, Medicine **46**, 83 (1967); W. Rüdiger, Fortschr. Chem. Org. Naturst. **29**, 60-139 (1971). Structure of stercobilin and d-urobilin: C. H. Gray, D. C. Nicholson, J. Chem. Soc. **1958**, 3085. Revised structure of d-urobilin: S. D. Killilea, P. O'Carra, Biochem. J. **129**, 1179 (1972). Stereochemistry: W. J. Cole et al., J. Chem. Soc. **1965**, 4085. Spectral study: W. J. Cole et al., J. Chem. Soc. C **1966**, 1321. Synthesis of optically active stercobilin: H. Plieninger, J. Ruppert, Ann. **736**, 43 (1970); of optically active urobilins: H. Plieninger et al., ibid., 62. Abs config of stercobilin: H. Brockmann, Jr. et al., Proc. Natl. Acad. Sci. USA **68**, 2141 (1971). HPLC determn in feces: R. V. A. Bull et al., J. Chromatogr. **218**, 647 (1981). Identification of bilirubin reduction products: L. Vitek et al., J. Chromatogr. B **833**, 149 (2006). Determn in environmental water samples by HPLC-ES-MS: T. L. Jones-Lepp, J. Environ. Monit. **8**, 472 (2006).

Stercobilin

Stercobilin. [34217-90-8] (2R,3R,4S,16S,17R,18R)-3,18-Diethyl-1,2,3,4,5,15,16,17,18,19,22,24-dodecahydro-2,7,13,17-tetramethyl-1,19-dioxo-21H-biline-8,12-dipropanoic acid; (−)-stercobilin. $C_{33}H_{46}N_4O_6$; mol wt 594.75.

Urobilin. [1856-98-0] 3,18-Diethyl-1,4,5,15,16,19,22,24-octahydro-2,7,13,17-tetramethyl-1,19-dioxo-21H-biline-8,12-dipropanoic acid; i-urobilin; urobilin IXα. $C_{33}H_{42}N_4O_6$; mol wt 590.72.

USE: Marker for environmental fecal waste contamination.

10069. Urochloralic Acid. [97-25-6] 2,2,2-Trichloroethyl β-D-glucopyranosiduronic acid; 2,2,2-trichloroethyl β-D-glucosiduronic acid; β,β,β-trichloroethyl-D-glucuronide. $C_8H_{11}Cl_3O_7$; mol wt 325.52. C 29.52%, H 3.41%, Cl 32.67%, O 34.40%. Isolated from urine of man and dog after ingestion of chloral hydrate: von Mering, Musculus, Ber. **8**, 663 (1875); from urine of calves fed trichloroethylene: Seto, Schultze, J. Am. Chem. Soc. **78**, 1616 (1956).

Needles, mp 142°. $[\alpha]_D^{27}$ −50.0° (c = 0.7). Freely soluble in water, alcohol. One gram dissolves in 234 ml of anhydr ether at 20°.

10070. Urocortin. [176591-49-4] (human); [171543-83-2] (rat). Mammalian neuropeptide related to corticotropin releasing factor (CRF), q.v. Composed of 40 amino acid residues. Evokes secretion of ACTH; binds to CRF receptors and to CRF binding protein. Identification in rat brain: J. Vaughan et al., Nature **378**, 287 (1995). Cloning and characterization of human urocortin: C. J. Donaldson et al., Endocrinology **137**, 2167 (1996). Interaction with CRF binding protein: D. P. Behan et al., Brain Res. **725**, 263 (1996).

Appetite suppressing effects: M. Spina *et al.*, *Science* **273**, 1561 (1996).

10071. Urodilatin. [115966-23-9] Renal natriuretic peptide. Natriuretic hormone produced by the kidney and secreted into the urine; regulates sodium and water reabsorption in the medullary collecting duct. Derived from the same prohormone as atrial natriuretic peptide, *q.v.*, with similar pharmacological effects. Isoln from human urine: P. Schulz-Knappe *et al.*, *Klin. Wochenschr.* **66**, 752 (1988). RIA determn in biological samples: C. Drummer *et al.*, *Pfluegers Arch.* **423**, 372 (1993). Role in body fluid regulation: C. Drummer, *Semin. Nephrol.* **21**, 239 (2001). Review of pharmacology and clinical implications: W.-G. Forssmann *et al.*, *Cardiovasc. Res.* **51**, 450-462 (2001). Comparison of biological actions of natriuretic peptides in experimental heart failure: H. H. Chen *et al.*, *Am. J. Physiol. Regul. Integr. Comp. Physiol.* **288**, 1093 (2005).

```
1
Thr–Ala–Pro–Arg–Ser–Leu–Arg–Arg–Ser–Ser–Cys–Phe–Gly–Gly–Arg–Met–Asp–Arg
                                    |
                                    |                                Ile
                                  32                                 |
             Tyr–Arg–Phe–Ser–Asn–Cys–Gly–Leu–Gly–Ser–Gln–Ala–Gly
```

human Urodilatin

Ularitide. [118812-69-4] Human urodilatin; ANP 95-126; CDD 95-126. $C_{145}H_{234}N_{52}O_{44}S_3$; mol wt 3505.97. Nonglycosylated peptide of 32 amino acid residues. Solid phase synthesis: W.-G. Forssmann *et al.*, **WO 8806596**; *eidem*, **US 5449751** (1988, 1995 both to Bissendorf Peptide).

10072. Urokinase. [9039-53-6] Urokinase (enzyme-activating); E.C. 3.4.21.73; Win-22005; Abbokinase; Actosolv; Breokinase; Persolv; Purochin; Ukidan; Uronase; Win-Kinase. Serine protease which activates plasminogen to plasmin; present in mammalian blood and urine. Produced as prourokinase, *q.v.*, and converted to active form by plasmin or kallikrein, *q.q.v.* Description of fibrinolytic activity: J. R. B. Williams, *Br. J. Exp. Pathol.* **32**, 530 (1951). Isolation from human urine and activity: T. Astrup, I. Sterndorff, *Proc. Soc. Exp. Biol. Med.* **81**, 675 (1952); G. W. Sobel *et al.*, *Am. J. Physiol.* **171**, 768 (1952). Species specificity and distribution in mammals: S. R. Mohler *et al.*, *Am. J. Physiol.* **192**, 186 (1958), Isoln from human male urine: H. O. Singher, L. Zuckerman, **US 2961382** and **US 2989440** (1960, 1961 to Ortho); N. O. Kjeldgaard, J. Ploug, **US 2983647** (1961 to Lövens Kemiske Fabrik); J. Doczi, **US 3081236** (1963 to Warner-Lambert). Prepn of crystalline form: A. Lesuk *et al.*, *Science* **147**, 880 (1965). Two variants of bioactive urokinase, high molecular weight (HMW-UK, ~50 KDa) and low molecular weight (LMW-UK, ~30 KDa) have been identified: W. F. White *et al.*, *Biochemistry* **5**, 2160 (1966). Both are disulfide-linked dimers consisting of a heavy chain (B) and a light chain (A). HMW-UK is converted to LMW-UK by proteolytic cleavage of the A-chain to form the A_1-chain. Series of articles on plasminogen activation, fibrinolysis and clinical efficacy: *Proc. Serono Symp., Thrombosis and Urokinase* **9**, 1-257 (1977). *Review:* F. Duckert, *Handb. Exp. Pharmacol.* **46**, 209-237 (1978). Structural characterization: M. Nobuhara *et al.*, *J. Biochem.* **90**, 225 (1981). Structure and amino acid sequences: W. A. Günzler *et al.*, *Z. Physiol. Chem.* **363**, 133 (1982); G. J. Steffens *et al.*, *ibid.* 1043 (1982); W. A. Günzler *et al.*, *ibid.* 1155. Expression of gene coding for human urokinase in *E. coli:* B. Ratzkin *et al.*, *Proc. Natl. Acad. Sci. USA* **78**, 3313 (1981). Clinical evaluation in pulmonary embolism: P. Petitpretz *et al.*, *Circulation* **70**, 861 (1984). Clinical trial following myocardial infarction: H. Kambara *et al.*, *Jpn. Circ. J.* **51**, 1072 (1987). Clinical applications of urokinase-treated tubing: T. Ohshiro *et al.*, *Methods Enzymol.* **137**, 529 (1988). Literature review of thrombolytic therapy: J. A. Kaufman, M. A. Bettmann, *Semin. Intervent. Radiol.* **9**, 159-165 (1992).

THERAP CAT: Thrombolytic.

10073. Urothion. [19295-31-9] 2-Amino-7-(1,2-dihydroxyethyl)-6-(methylthio)thieno[3,2-*g*]pteridin-4(3*H*)-one. $C_{11}H_{11}N_5$-O_3S_2; mol wt 325.36. C 40.61%, H 3.41%, N 21.53%, O 14.75%, S 19.71%. Constituent of normal human urine: Koschara, *Z. Physiol.*

Chem. **277**, 284 (1943). Structure: Tschesche *et al.*, *Ber.* **88**, 1251 (1955). Revised structure: Goto *et al.*, *Tetrahedron Lett.* **1967**, 4507; *eidem*, *J. Biochem. (Tokyo)* **65**, 611 (1969). Synthesis: Sakurai, Goto, *Tetrahedron Lett.* **1968**, 2941; *eidem*, *J. Biochem. (Tokyo)* **65**, 755 (1969).

Orange-colored clusters of crystals, not melted at 360°. Shows mutarotation: $[\alpha]_D^{20} -20°$ (after 15 hrs in 0.05*N* NaOH). Soly in water at pH 6.6 about 1:10,000. More sol in alkaline soln. Also sol in acids. Practically insol in the usual organic solvents.

10074. Ursodiol. [128-13-2] (3α,5β,7β)-3,7-Dihydroxycholan-24-oic acid; 17β-(1-methyl-3-carboxypropyl)etiocholane-3α,-7β-diol; 3α,7β-dioxycholanic acid; ursodeoxycholic acid; UDCA; Actigall; Arsacol; Cholit Ursan; Delursan; Desol; Destolit; Deursil; Litursol; Peptarom; Urdes; Ursacol; Urso; Ursochol; Ursofalk; Ursolvan. $C_{24}H_{40}O_4$; mol wt 392.58. C 73.43%, H 10.27%, O 16.30%. Epimeric with chenodiol, *q.v.*, with respect to the hydroxyl group at C_7. Found in bear bile (combined with taurine). Isoln: Shoda, *J. Biochem. (Tokyo)* **7**, 505 (1927). Structure: Kaziro, *Z. Physiol. Chem.* **185**, 151 (1929); **197**, 206 (1931); Iwasaki, *ibid.* **244**, 181 (1936). Toxicity data: M. Ardenne, P. G. Reitnauer, *Arzneim.-Forsch.* **20**, 323 (1970). Effect on cholesterol and bile acid metabolism: G. S. Tint *et al.*, *Gastroenterology* **91**, 1007 (1986). Clinical trial in primary biliary cirrhosis: R. E. Poupon *et al.*, *N. Engl. J. Med.* **324**, 1548 (1991); R. Jorgensen *et al.*, *Am. J. Gastroenterol.* **97**, 2647 (2002); in prevention of gallstones following gastric bypass: H. J. Sugerman *et al.*, *Am. J. Surg.* **169**, 91 (1995). Review of pharmacology, toxicology, efficacy: A. Ward *et al.*, *Drugs* **27**, 95-131 (1984). Brief review of clinical effects and comparison with chenodiol: H. Fromm, *Gastroenterology* **87**, 229-233 (1984).

Bitter plates from alc. mp 203°. $[\alpha]_D^{20} +57°$ (c = 2 in abs ethanol). Freely sol in ethanol, glacial acetic acid; sparingly sol in chloroform; slightly sol in ether. Practically insol in water. LD_{50} in mice (g/kg): 0.1 i.v. (Ardenne, Reitnauer); in rats, mice (mg/kg): 2000, 6000 s.c.; 1000, 1200 i.p.; 310, 260 i.v. (Ward).

Diformate. $C_{26}H_{40}O_6$. Crystals, mp 170°.

Diacetate. $C_{28}H_{44}O_6$. Crystals, mp 98-102°.

THERAP CAT: Anticholelithogenic.

THERAP CAT (VET): Anticholelithogenic.

10075. Ursolic Acid. [77-52-1] (3β)-3-Hydroxyurs-12-en-28-oic acid; urson; prunol; micromerol; malol. $C_{30}H_{48}O_3$; mol wt 456.71. C 78.90%, H 10.59%, O 10.51%. In leaves and berries of *Arctostaphylos uva-ursi* (L.) Spreng (bearberry), of *Vaccinium macrocarpon* Ait. (cranberry), *Rhododendron hymenanthes* Makino, *Ericaceae*. In the protective wax-like coating of apples, pears, prunes, and other fruits. Isoln from apple peelings: Sando, *J. Biol. Chem.* **56**, 457 (1923). Structure: Ruzicka *et al.*, *Helv. Chim. Acta* **28**, 199 (1945); Zurcher *et al.*, *ibid.* **37**, 2145 (1954). Conversion from α-amyrin: Boar *et al.*, *J. Chem. Soc. C* **1970**, 678. Chemistry: Mezzetti *et al.*, *Planta Med.* **20**, 244 (1971).

Large, lustrous prisms from abs alcohol, fine hair-like needles from dil alcohol, mp 285-288°. $[\alpha]_D^{21}$ +67.5° (c = 1 in N alc KOH). Soly at 15°: One part dissolves in 88 parts methanol, 178 alcohol (35 boiling alcohol), 140 ether, 388 chloroform, 1675 carbon disulfide. Moderately sol in acetone. Sol in hot glacial acetic acid and in 2% alcoholic NaOH. Insol in water and petr ether.

Acetate. $C_{32}H_{50}O_4$. mp 289-290°. $[\alpha]_D$ +62.3° (c = 1.15 in chloroform).

Methyl ester. $C_{31}H_{50}O_3$. mp 171°. $[\alpha]_D^{20}$ +58° (c = 1.2 in pyridine).

Methyl ester acetate. $C_{33}H_{52}O_4$. mp 246-247°.

USE: As emulsifying agent in pharmaceuticals, foods.

10076. Urushiol. [53237-59-5] Allergenic principle of the irritant oil of poison ivy, *Toxicodendron radicans* (L.) Kuntze, poison oak, *T. diversilobum*, the Japanese lacquer tree, *T. verniciferum* D.C., and other plants of the family, *Anacardiaceae*. Consists of a mixture of 3-n-alk(en)ylcatechols, typically with 0-3 double bonds having the *cis*- configuration. The distribution of the congeners varies by species. Poison ivy urushiol contains primarily C_{15} substituents, while urushiol from poison oak contains primarily the C_{17} analogs. Allergenic potency increases with increasing unsaturation. Identification in urishi, the milky exudate from *T. verniciferum*: H. Yoshida, *J. Chem. Soc.* **43**, 472 (1883). Isoln of hydrourushiol and identification of basic structures: R. Majima *et al., Ber.* **55**, 172 (1922). Identification as toxic principle of poison ivy: G. A. Hill *et al., J. Am. Chem. Soc.* **56**, 2736 (1934). Confirmation of structures in urushiol from poison ivy: W. F. Symes, C. R. Dawson, *J. Am. Chem. Soc.* **76**, 2959 (1954); from Japanese lac: S. V. Sunthankar, C. R. Dawson, *ibid.* **76**, 5070 (1954). Separation of components in poison ivy: K. H. Markiewitz, C. R. Dawson, *J. Org. Chem.* **30**, 1610 (1965); in poison oak: M. D. Corbett, S. Billets, *J. Pharm. Sci.* **64**, 1715 (1975). Isoln of urushiol from poison ivy or oak extracts and separation of congeners by reversed phase chromatography: M. A. El Sohly *et al., J. Nat. Prod.* **45**, 532 (1982). LC-MS-MS determn of congeners from poison oak: W. M. Draper *et al., J. Agric. Food Chem.* **50**, 1852 (2002). Allergenic properties of congeners: R. A. Johnson *et al., J. Allergy Clin. Immunol.* **49**, 27 (1972); E. S. Watson *et al., J. Pharm. Sci.* **70**, 785 (1981). Clinical use in hyposensitization therapy: W. L. Epstein *et al., J. Allergy Clin. Immunol.* **68**, 20 (1981).

I	R = $(CH_2)_7CH_3$
II	R = CH=CH($CH_2)_5CH_3$
III	R = CH=CHCH$_2$CH=CH($CH_2)_2CH_3$
IV	R = CH=CHCH$_2$CH=CHCH$_2$CH=CH$_2$

Poison ivy urushiol

Pale yellow liq; $d_4^{21.5}$ 0.9687; bp 200-210°. Soluble in alcohol, ether, benzene. Moderately sol in petr ether. Sensitive to air oxidation and polymerization.

(15:0)-Urushiol. [492-89-7] 3-Pentadecyl-1,2-benzenediol; 3-n-pentadecylcatechol; tetrahydrourushiol; hydrourushiol. $C_{21}H_{36}O_2$;

mol wt 320.52. Synthesis: H. S. Mason, *J. Am. Chem. Soc.* **67**, 1538 (1945). Pale yellow crystals from petr ether, mp 57.5-58.5°. uv max (methanol): 278, 230 nm (log ε 3.23, 3.19).

(15:1)-Urushiol. [35237-02-6]; [2764-91-2] (unspecified stereochem.). 3-(8Z)-8-Pentadecen-1-yl-1,2-benzenediol; (Z)-3-(8-pentadecenyl)catechol; urushenol; urushiol monoene. $C_{21}H_{34}O_2$; mol wt 318.50. Synthesis: B. Loev, C. R. Dawson, *J. Org. Chem.* **24**, 980 (1959); J. H. P. Tyman *et al., Chem. Phys. Lipids* **120**, 101 (2002). Yellow oil. n_D^{25} 1.5083. uv max (methanol): 275, 218 nm (log ε 3.42, 3.99).

(17:0)-Urushiol. [5862-27-1] 3-Heptadecyl-1,2-benzenediol; 3-heptadecylcatechol. $C_{23}H_{40}O_2$; mol wt 348.57. White amorphous solid, mp 63-65°. uv max (methanol): 275, 230 nm (log ε 3.21, 3.19).

Caution: An extremely active allergen causing irritation, inflammation, and blistering of the skin of sensitive individuals.

THERAP CAT: Antiallergic (hyposensitization therapy).

10077. Uscharidin. [24321-47-9] $(2\alpha,3\beta,5\alpha)$-14-Hydroxy-19-oxo-2,3-[[(2S,3R,6R)-tetrahydro-3-hydroxy-6-methyl-4-oxo-2H-pyran-3,2-diyl]bis(oxy)]card-20(22)-enolide. $C_{29}H_{38}O_9$; mol wt 530.61. C 65.65%, H 7.22%, O 27.14%. African arrow poison and cardiac glycoside produced by *Calotropis procera* R. Br., *Asclepiadaceae*: Hesse *et al., Ann.* **566**, 130 (1950). Prepn by hydrolysis of uscharin with 2N H_2SO_4 in methanol: Hesse *et al., Ann.* **537**, 67 (1938); by treatment of voruscharin with mercuric chloride: Hesse, Ludwig, *Ann.* **632**, 158 (1960). Structure: Crout *et al., J. Chem. Soc.* **1964**, 2187. Revised structure: Brüschweiler *et al., Helv. Chim. Acta* **52**, 2276 (1969). Stereochemistry: H. T. A. Cheung, T. R. Watson, *J. Chem. Soc. Perkin Trans. 1* **1980**, 2162.

Rhombic plates, decomp 290°. $[\alpha]_D^{20}$ +38° (c = 0.9 in methanol).

10078. Usnic Acid. [125-46-2] 2,6-Diacetyl-7,9-dihydroxy-8,9b-dimethyl-1,3(2H,9bH)-dibenzofurandione; usninic acid; usnein. $C_{18}H_{16}O_7$; mol wt 344.32. C 62.79%, H 4.68%, O 32.53%. Antibacterial substance found in lichens. Isoln from varieties of *Usnea barbata* (L.) Wigg., *Usneaceae*: Rochleder, Heldt, *Ann.* **48**, 11 (1843); Widman, *Ann.* **310**, 230 (1900); **324**, 139 (1902). Isoln from *Ramalina reticulata*: Marshak, *Public Health Rep.* **62**, 3 (1947); Stark *et al., J. Am. Chem. Soc.* **72**, 1819 (1950). Occurs in nature in both the *d*- and *l*-forms as well as a racemic mixture. Structure: Curd, Robertson, *J. Chem. Soc.* **1937**, 894; Schöpf, Ross, *Ann.* **546**, (1941); Barton, Brunn, *J. Chem. Soc.* **1953**, 603. Resolution of (±)-usnic acid: Dean *et al., ibid.* 1250. Synthesis: Barton *et al., ibid.* **1956**, 530. Biosynthesis *in vitro*: Penttila, Fales, *Chem. Commun.* **1966**, 656. Absolute configuration of (+)-form: S. Huneck *et al., Tetrahedron Lett.* **22**, 351 (1981). Brief review: *Antibiotics* I, D. Gottlieb, P. Shaw, Eds. (Springer Verlag, New York, 1967) p 611.

d-form

d-**Form.** [7562-61-0] Yellow orthorhombic prisms from acetone, mp 204°. $[\alpha]_D^{16}$ +509.4° (c = 0.697 in chloroform). Soly at 25° (g/

100 ml): water <0.01; acetone 0.77; ethyl acetate 0.88; ethanol 0.02; methyl Cellosolve 0.22; ethyl Cellosolve 0.32; furfural 7.32; furfuryl alcohol 1.21. LD_{50} i.v. in mice: 25 mg/kg (Gottlieb, Shaw).

10079. Ustekinumab. [815610-63-0] Anti-(human interleukin 12 p40 subunit) immunoglobulin G1 (human monoclonal CNTO 1275 γ_1-chain) disulfide with human monoclonal CNTO 1275 κ-chain, dimer; CNTO-1275; Stelara. Human monoclonal antibody directed against the shared p40 subunit of interleukins IL-12 and IL-23. Prepn: J. Giles-Komar *et al.*, **WO 0212500**; *eidem*, **US 6902734** (2002, 2005 both to Centocor). Pharmacology: M. Reddy *et al.*, *Cell. Immunol.* **247**, 1 (2007). Clinical pharmacokinetics: C. L. Kauffman *et al.*, *J. Invest. Dermatol.* **123**, 1037 (2004). Clinical trials in psoriasis: C. L. Leonardi *et al.*, *Lancet* **371**, 1665 (2008); K. A. Papp *et al.*, *ibid.* 1675.

THERAP CAT: Antipsoriatic; immunomodulator.

10080. Ustilagic Acid. [8002-36-6] Ustizeain B. $C_{37}H_{62\text{-}66}O_{17}$; mol wt about 780. A mixture of partially acylated derivatives of a di-D-glucosyldihydroxyhexadecanoic acid. Antibiotic substance produced by the corn smut fungus *Ustilago zeae:* Haskins, *Can. J. Res.* **28C**, 213 (1950); Lemieux *et al.*, *Can. J. Chem.* **29**, 409, 415 (1951); Reed, Holder, *Can. J. Med. Sci.* **31**, 505 (1953); Haskins, **US 2698843** (1955 to National Research Council, Canada). Biosynthesis: Boothroyd *et al.*, *Can. J. Biochem. Physiol.* **33**, 289 (1955). Production by submerged culture: Lemieux, **CA 600121** (1960 to National Research Council, Canada).

Long crystals from ether, mp 146-147°. $[\alpha]_D^{23}$ +7° (pyridine). Freely sol in methanol, pyridine, 2,3-butanediol, 1,2-propanediol. Sparingly sol in ethanol, butanol, acetone. Insol in water, glycerol, ethyl acetate, ether, benzene, petr ether. Shows *in vitro* activity against *Cryptococcus neoformans*, *Candida albicans*, and some saprophytic fungi. Practically ineffective in rabbits and mice suffering from fungus diseases.

10081. Utrophin. Dystrophin-related protein; DRP; DMDL protein. Autosomal homolog of dystrophin, *q.v.*, the protein affected by mutation in Duchenne muscular dystrophy (DMD). Cytoplasmic protein; mol wt 395 kDa. Localized in the neuromuscular junctions of normal adult muscle. Widely distributed in a variety of other tissues including brain, stomach, kidney, spleen, liver, and lung. The name is derived from its ubiquitous pattern of expression. Like dystrophin, utrophin binds to actin and to a transmembrane glycoprotein complex, known as dystrophin-associated proteins (DAPs), to link the cytoskeleton with the extracellular matrix. Thought to partially compensate for the lack of dystrophin in DMD patients. Identification in mouse skeletal muscle: D. R. Love *et al.*, *Nature* **339**, 55 (1989). Primary structure: J. M. Tinsley *et al.*, *ibid.* **360**, 591 (1992). Association with DAPs: K. Matsumura *et al.*, *ibid.*, 588; J. M. Tinsley *et al.*, *Proc. Natl. Acad. Sci. USA* **91**, 8307 (1994). Analysis of actin-binding: S. J. Winder *et al.*, *J. Cell Sci.* **108**, 63 (1995). Expression in DMD patients: Y. Mizuno *et al.*, *J. Neurol. Sci.* **119**, 43 (1993). Review and comparison with dystrophin: J. M. Tinsley *et al.*, *Curr. Opin. Genet. Dev.* **3**, 484-490 (1993); D. R. Love *et al.*, *Neuromuscul. Disord.* **3**, 5-21 (1993); J. M. Tinsley, K. E. Davies, *ibid.* 537-539.

10082. Uva Ursi. Bearberry. Dried leaves of *Arctostaphylos uva-ursi* (L.), Spreng., *Ericaceae*, a small evergreen shrub. *Habit.* Temperate regions of Europe, North America, Asia. *Constit.* Hydroquinone derivatives, primarily arbutin, *q.v.*, and methylarbutin; tannins of the gallic and ellagic acid types; flavonoids (quercetin, myricetin); piceoside; monotropein; ursolic, gallic and quinic acids; volatile oil. Discussion of active components: D. Frohne, *Planta Med.* **18**, 1 (1970). Bioavailability: D. H. Paper *et al.*, *Pharm. Pharmacol. Lett.* **3**, 63 (1993). Brief review of use in phytomedicine: A. Y. Leung, S. Foster, *Encyclopedia of Common Natural Ingredients* (John Wiley & Sons, New York, 2nd ed., 1996) pp 505-506.

Leaf extract. Arctuvan; Uvalysat.

THERAP CAT: Antiseptic (urinary).

10083. Uzarin. [20231-81-6] $(3\beta,5\alpha)$-3-[(2-*O*-β-D-Glucopyranosyl-β-D-glucopyranosyl)oxy]-14-hydroxycard-20(22)-enolide. $C_{35}H_{54}O_{14}$; mol wt 698.80. C 60.16%, H 7.79%, O 32.05%. Isoln from the dried root of a *Gomphocarpus* sp, *Asclepiadaceae:* Windaus, Haack, *Ber.* **63**, 1377 (1930); Tschesche, Brathge, *ibid.* **85**, 1042 (1952); Schmid *et al.*, *Helv. Chim. Acta* **42**, 72 (1959). Yields uzarigenin by enzymic cleavage. Structure: Tschesche, Bohle, *Ber.* **68**, 2252 (1935). Structure of uzarigenin: Rangaswami, Reichstein, *Helv. Chim. Acta* **32**, 939 (1949); Russel *et al.*, *ibid.* **44**, 1320 (1961). Unlike all other known cardiac glycosides, has the A/B-*trans* configuration: L. F. Fieser, M. Fieser, *Steroids* (Reinhold, New York, 1959) pp 762-763. *Review:* Heusser, *Fortschr. Chem. Org. Naturst.* **7**, 101 (1950).

Uzarigenin

Prisms from pyridine + water, mp 266-270°; stout needles from methanol + ether, mp 206-208°. $[\alpha]_D^{20}$ −27° (c = 1.075 in pyridine); $[\alpha]_D^{19}$ −1.4° (c = 0.85 in methanol). uv max: 217 nm (log ε 4.23). Sol in pyridine, hot methyl Cellosolve; sparingly sol in water. Practically insol in ether, chloroform, acetone.

Uzarigenin. [466-09-1] $(3\beta,5\alpha)$-3,14-Dihydroxycard-20(22)-enolide; odorigenin. $C_{23}H_{34}O_4$; mol wt 374.52. Synthesis: Stache *et al.*, *Ann.* **726**, 136 (1969); Kamano *et al.*, *J. Org. Chem.* **39**, 2319 (1974); *eidem*, *J. Chem. Soc. Perkin Trans. 1* **1975**, 1972. Crystals from methanol, mp 240-256°. $[\alpha]_D^{20}$ +10.5° (c = 1.056 in alc).

3-*O*-Acetyluzarigenin. $C_{25}H_{36}O_5$. Hexagonal plates from methanol, mp 262-266°. $[\alpha]_D^{22}$ +4.6° (c = 1.09 in chloroform).

THERAP CAT: Antidiarrheal.

V

10084. Vaccenic Acid. [693-72-1] (11*E*)-11-Octadecenoic acid; *trans*-Δ^{11}-octadecenoic acid. $C_{18}H_{34}O_2$; mol wt 282.47. C 76.54%, H 12.13%, O 11.33%. Found in butterfat and in other animal fats. Growth-promoting factor for rats. Isoln: Bertram, *Biochem. Z.* **197**, 433 (1928). Synthesis: Böeseken, Hoagland, *Rec. Trav. Chim.* **46**, 632 (1927); Ahmad *et al., J. Am. Chem. Soc.* **70**, 3391 (1948). Configuration and ir spectrum: Rao, Daubert, *ibid.* 1102.

H_3C ～～～～～～COOH

Platelets from acetone, mp 43-44°; n_D^{60} 1.4439; n_D^{70} 1.4402. Neutralization equivalent 282.5; iodine no. 89.9.

Methyl ester. [6198-58-9] Methyl vaccenate. $C_{19}H_{36}O_2$; mol wt 296.50. bp₃ 172-173°.

10085. Valacyclovir. [124832-26-4] L-Valine 2-[(2-amino-1,6-dihydro-6-oxo-9*H*-purin-9-yl)methoxy]ethyl ester; L-valine ester with 9-[(2-hydroxyethoxy)methyl]guanine; valaciclovir; Val-ACV. $C_{13}H_{20}N_6O_4$; mol wt 324.34. C 48.14%, H 6.22%, N 25.91%, O 19.73%. L-Valine ester prodrug of acyclovir, *q.v.* Prepn: T. A. Krenitsky *et al.,* **EP 308065;** L. M. Beauchamp, **US 4957924** (1989, 1990 both to Wellcome). Evaluation as prodrug: L. M. Beauchamp *et al., Antivir. Chem. Chemother.* **3**, 157 (1992). Clinical pharmacokinetics: S. Weller *et al., Clin. Pharmacol. Ther.* **54**, 595 (1993). Review of pharmacology and clinical efficacy in herpes virus infections: C. M. Perry, D. Faulds, *Drugs* **52**, 754-772 (1996). Clinical trial to prevent cytomegalovirus disease in renal transplantation: D. Lowance *et al., N. Engl. J. Med.* **340**, 1462 (1999); to prevent transmission of genital herpes: L. Corey *et al., ibid.* **350**, 11 (2004).

Hydrochloride. [124832-27-5] 256U; BW-256U87; BW-256; Valtrex. Crystalline solid, occurs as hydrate. uv max (water): 252.8 nm (ε 8530). Soly in water: 174 mg/ml.

THERAP CAT: Antiviral.

10086. Valdecoxib. [181695-72-7] 4-(5-Methyl-3-phenyl-4-isoxazolyl)benzenesulfonamide; SC-65872; Bextra. $C_{16}H_{14}N_2O_3S$; mol wt 314.36. C 61.13%, H 4.49%, N 8.91%, O 15.27%, S 10.20%. Selective cyclooxygenase-2 (COX-2) inhibitor. Active metabolite of parecoxib, *q.v.* Prepn: J. J. Talley *et al.,* **WO 9625405** (1996 to Searle); *eidem,* **US 5633272** (1997); and activity: *eidem, J. Med. Chem.* **43**, 775 (2000). Chromatographic determn of purity: D. A. Roston *et al., J. Pharm. Biomed. Anal.* **26**, 339 (2001). Gastrointestinal tolerability study: G. M. Eisen *et al., Aliment. Pharmacol. Ther.* **21**, 591 (2005). Clinical trial in hip arthroplasty: F. Camu *et al., Am. J. Ther.* **9**, 43 (2002). Clinical comparison with oxycodone/acetaminophen in dental pain: S. E. Daniels *et al., J. Am. Dent. Assoc.* **133**, 611 (2002). Clinical trial in migraine: D. Kudrow *et al., Headache* **45**, 1151 (2005). Review of clinical experience: M. Goldman, S. Schutzer, *Formulary* **37**, 68-77 (2002); of clinical efficacy and safety: G. P. Joshi, *Expert Rev. Neurother.* **5**, 11-24 (2005).

Crystals, mp 155-157°. Soly at 25°(μg/ml): water 10 (pH 7.0). Sol in methanol, ethanol; freely sol in organic solvents and alkaline (pH = 12) aqueous solns.

THERAP CAT: Anti-inflammatory.

10087. Valdetamide. [512-48-1] 2,2-Diethyl-4-pentenamide; diethylallylacetamide; Novonal. $C_9H_{17}NO$; mol wt 155.24. C 69.63%, H 11.04%, N 9.02%, O 10.31%. Description: Bockmühl, Schaumann, *Dtsch. Med. Wochenschr.* **54**, 270 (1928). Pharmacokinetics and metabolism: H. Uehleke, M. Brinkschulte-Freitas, *Arch. Pharmacol.* **302**, 11 (1978). TLC determn in urine: E. Klug, P. Toffel, *Arzneim.-Forsch.* **29**, 1651 (1979).

White powder, mp 75-76°. Sol in 120 parts water; freely sol in alcohol, ether.

THERAP CAT: Sedative, hypnotic.

10088. *n*-Valeraldehyde. [110-62-3] Pentanal; valeral; valeric aldehyde. $C_5H_{10}O$; mol wt 86.13. C 69.73%, H 11.70%, O 18.58%. Prepn: Lieben, Rossi, *Ann.* **159**, 70 (1871); Olsen, **US 2548171** (1951 to GAF); Sisti *et al., J. Org. Chem.* **27**, 279 (1962). Toxicity study: H. F. Smyth *et al., Am. Ind. Hyg. Assoc. J.* **30**, 470 (1969).

Liquid, bp 102-103°. d_4^{20} 0.8095. n_D^{20} 1.3944. Very slightly sol in water; miscible with many organic solvents. LD₅₀ orally in rats: 5.66 ml/kg (Smyth).

Caution: Potential symptoms of overexposure are irritation of eyes, skin, nose, throat. *See NIOSH Pocket Guide to Chemical Hazards* (DHHS/NIOSH 97-140, 1997) p 326.

USE: In flavoring compds, resin chemistry, rubber accelerators.

10089. Valerian. Perennial herb, *Valeriana officinalis* L., *Valerianaceae*. Medicinal portions are the dried rhizomes and roots; traditionally used to treat insomnia and anxiety. *Habit.* Europe, Northern Asia; naturalized in eastern U.S. Numerous other species exist throughout the world which are also used medicinally. *Constit.* Volatile oil (0.5-2%); iridoids (valepotriates) such as valtrate, didrovaltrate, isovaltrate; alkaloids incl. valerianine and chatinine; tannin, resin. Review of pharmacology: P. J. Houghton, *J. Ethnopharmacol.* **22**, 121-142 (1988); of constituents and medicinal uses: J. Barnes *et al., Herbal Medicines* (Pharmaceutical Press, London, 2nd Ed., 2002) pp 468-476; J. Gruenwald *et al., PDR for Herbal Medicines* (Thomson PDR, Montvale, 3rd Ed., 2004) pp 852-856.

Volatile oil. [8008-88-6] Oil of valerian. *Constit.* Monoterpenes, incl. α- and β-pinene, camphene, borneol, eugenol; sesquiterpenes incl. β-bisabolene, caryophyllene, valeratone. Yellowish-green to brownish liquid. d_{15}^{15} 0.93-0.96. α_D −8 to −13°. Sapon no. 100-150. Acid no. 20-50. Slightly sol in water; very sol in alcohol, chloroform, ether. *Keep well closed, cool and protected from light.*

THERAP CAT: Sedative.

10090. *n*-Valeric Acid. [109-52-4] Pentanoic acid; valerianic acid; propylacetic acid. $C_5H_{10}O_2$; mol wt 102.13. C 58.80%, H 9.87%, O 31.33%. Obtained by decompn of *n*-propylmalonic acid: Fürth, *Monatsh. Chem.* **9**, 308 (1888); from *n*-butyl chloride: Gilman, Kirby, *Org. Synth.* **coll. vol. I**, 363 (2nd ed., 1941). Industrial synthesis by oxidation of amyl alcohol or by fermentation processes. Toxicity study: L. Orö, A. Wretlind, *Acta Pharmacol. Toxicol.* **18**, 141 (1961).

$$H_3C \quad\quad\quad COOH$$

Colorless liquid; unpleasant odor; d_4^{20} 0.939; mp $-34.5°$; bp 186-187°; bp_{23} 96°; n_D^{20} 1.4086. Sol in 30 parts water; freely sol in alcohol, ether. LD_{50} i.v. in mice: 1290 $\pm$53 mg/kg (Orö, Wretlind).

Ethyl ester. [539-82-2] Ethyl *n*-valerate. $C_7H_{14}O_2$. Liquid. d_4^{20} 0.877. bp 145-146°. n_D^{20} 1.3732. Insol in water. Misc with alcohol, ether.

USE: Intermediate in perfumery.

10091. Valeronitrile. [110-59-8] Pentanenitrile; 1-butyl cyanide; 1-cyanobutane. C_5H_9N; mol wt 83.13. C 72.24%, H 10.91%, N 16.85%. Prepn: A. Lieben, A. Rossi, *Ann.* **158**, 137 (1871); W. Kantlehner *et al., ibid.* **1980**, 389; H. G. Thomas, H. D. Greyn, *Synthesis* **1990**, 129. Acute toxicity and metabolism: H. Tanii, K. Hashimoto, *Arch. Toxicol.* **55**, 47 (1984). Comparative toxicity of aliphatic nitriles: M. A. Wallig *et al., Food Chem. Toxicol.* **26**, 149 (1988).

$$H_3C \quad\quad\quad CN$$

bp_{15} 45-47°. $bp_{739.3}$ 140.4°. bp 141°. d^0 0.8164. n_D^{20} 1.3962. Log P (*n*-octanol/water): 0.94. LD_{50} orally in male mice: 2.297 mmol/kg (Tanii).

USE: Solvent.

10092. Valethamate Bromide. [90-22-2] *N,N*-Diethyl-*N*-methyl-2-[(3-methyl-1-oxo-2-phenylpentyl)oxy]ethanaminium bromide (1:1); 3-methyl-2-phenylvaleric acid diethyl(3-hydroxyethyl)methylammonium bromide ester; 2-phenyl-3-methylvaleric acid β-(diethylamino)ethyl ester bromomethylate; 3-methyl-2-phenylvaleric acid 2-diethylaminoethyl ester methyl bromide; 2-diethylaminoethyl 2-phenyl-3-methylvalerate methyl bromide; diethyl(2-hydroxyethyl)methylammonium 3-methyl-2-phenylvalerate bromide; Resitan; Epidosin. $C_{19}H_{32}BrNO_2$; mol wt 386.37. C 59.06%, H 8.35%, Br 20.68%, N 3.63%, O 8.28%. Anticholinergic. Prepn: Stühmer, Funke, **DE 969245**; **DE 971136**; **DE 1112989** (all 1958 to Kali-Chemie); Martin, Habicht, **DE 1091124**; **US 2987517** (1960, 1961 both to Cilag-Chemie).

Crystals from ethanol + ether or acetone, mp 100-101°. Freely sol in water, very sol in alcohol. Practically insol in ether. Aq solns are stable to storage; 0.6% ampuled solns showed no loss after one year at room temp.

THERAP CAT: Antispasmodic.

10093. Valganciclovir. [175865-60-8] L-Valine 2-[(2-amino-1,6-dihydro-6-oxo-9*H*-purin-9-yl)methoxy]-3-hydroxypropyl ester; 2-[(2-amino-1,6-dihydro-6-oxo-9*H*-purin-9-yl)methoxy]-3-hydroxypropanyl-L-valinate. $C_{14}H_{22}N_6O_5$; mol wt 354.37. C 47.45%, H 6.26%, N 23.72%, O 22.57%. Valine ester prodrug of ganciclovir, *q.v.* Prepn: J. J. Nestor *et al.*, **EP 694547**; *eidem*, **US 6083953** (1996, 2000 both to Hoffmann-La Roche). HPLC determn in plasma: R. Chan *et al., J. Pharm. Biomed. Anal.* **21**, 647 (1999). Clinical pharmacokinetics in HIV- and CMV-positive patients: F.

Brown *et al., Clin. Pharmacokinet.* **37**, 167 (1999); in liver transplant patients: M. D. Pescovitz *et al., Antimicrob. Agents Chemother.* **44**, 2811 (2000). Clinical trial in cytomegalovirus retinitis: D. F. Martin *et al., N. Engl. J. Med.* **346**, 1119 (2002); in human herpesvirus-8 infection: C. Casper *et al., J. Infect. Dis.* **198**, 23 (2008). Review of use in prevention and treatment of CMV in transplant recipients: A. Asberg *et al., Expert Opin. Pharmacother.* **11**, 1159-1166 (2010).

Hydrochloride. [175865-59-5] Ro-107-9070/194; RS-79070-194; Cymeval; Darilin; Rovalcyte; Valcyte. $C_{14}H_{22}N_6O_5 \cdot HCl$; mol wt 390.83. White to off-white crystalline powder. Crystals from water + isopropanol, undergoes phase change at 142°; dec 175°. pKa 7.6. Soly in water (25°): 70 mg/ml (pH7). Very sol in 2-propanol; freely sol in alc; slightly sol in hexane. Practically insol in acetone, ethyl acetate. Partition coefficient (octanol/water): 0.0095 (pH7).

THERAP CAT: Antiviral.

10094. Validamycins. Antibiotic complex produced by *Streptomyces hygroscopicus* var *limoneus*. Consists of validamycins A (major component), B, C, D, E, and F. Isoln, characterization, and biological properties of validamycins A and B: Iwasa *et al., J. Antibiot.* **24**, 107, 119 (1971). Isoln and characterization of validamycins C, D, E, F: Horii *et al., ibid.* **25**, 48 (1972). Manuf: Horii *et al.*, **JP 72 39607** (1972 to Takeda), *C.A.* **78**, 122657a (1973). Structure of validamycin A: Horii, Kameda, *Chem. Commun.* **1972**, 747; revised structure: T. Suami *et al., J. Antibiot.* **33**, 98 (1980). Bioassay methods: Iwasa *et al., ibid.* **24**, 114 (1971). Total synthesis of (±)-validoxylamines A and B, constituents of validamycins: S. Ogawa *et al., J. Org. Chem.* **49**, 2594 (1984). Total synthesis of validamycin B: S. Ogawa, Y. Miyamoto, *Chem. Commun.* **1987**, 1843; of validamycin A: *eidem, Chem. Lett.* **1988**, 889.

Validamycin A

Validamycin A. [37248-47-8] [1*S*-(1α,4α,5β,6α)]-1,5,6-Tri-deoxy-4-*O*-β-D-glucopyranosyl-5-(hydroxymethyl)-1-[[4,5,6-trihydroxy-3-(hydroxymethyl)-2-cyclohexen-1-yl]amino]-D-*chiro*-inositol; Validacin; Valimon. $C_{20}H_{35}NO_{13}$; mol wt 497.49. Colorless hydrophilic powder. Does not show sharp mp; softens at 100°, dec at 135°. $[\alpha]_D^{24}$ +110° (c = 1 in water or pyridine), +92° (c = 1 in DMF). pKa 6.0. Sol in water, methanol, DMF, DMSO; sparingly sol in ethanol, acetone. Insol in ethyl acetate, diethyl ether.

Hydrochloride. $C_{20}H_{35}NO_{13} \cdot HCl$. Crystalline powder, mp 95° (dec). $[\alpha]_D^{22}$ +49° (c = 1 in water). Sol in water, alcohols. Insol in acetone, diethyl ether.

USE: Fungicide.

10095. Valine. [72-18-4] L-Valine; Val; V; 2-aminoisovaleric acid; 2-amino-3-methylbutyric acid; α-aminoisovaleric acid; (*S*)-2-amino-3-methylbutanoic acid. $C_5H_{11}NO_2$; mol wt 117.15. C 51.26%, H 9.46%, N 11.96%, O 27.31%. An essential amino acid for human development. Identified from organ extracts in 1856 by von Group-Besanez; isolated from proteins in 1879 by Schützenberger who proposed that it was aminovaleric acid. Structure confirmation: E. Fischer, *Ber.* **39**, 2320 (1906). Early chemistry and biochemistry: *Amino Acids and Proteins*, D. M. Greenberg, Ed. (Charles C. Thomas, Springfield, IL, 1951) 950 pp., *passim*; J. P. Greenstein, M. Winitz, *Chemistry of the Amino Acids* **vols 1-3** (John Wiley and Sons, Inc., New York, 1961) pp. 2368-2380, *passim*. Conformation study: R. H. Yun, J. Hermans, *Protein Eng.* **4**, 761 (1991). Nutritional assessment study: V. R. Young *et al.*, *J. Nutr.* **102**, 1159 (1972); in parenteral nutrition: P. Reiderer *et al.*, *Nutr. Metab.* **24**, 209 (1980); in hemodialysis patients: G. A. Young *et al.*, *Kidney Int.* **21**, 492 (1982). Brief review of metabolism: P. Kamoun, *Trends Biochem. Sci.* **17**, 175-176 (1992).

Leaflets from water + alcohol, mp 315° (closed capillary). d 1.230. Sublimes. $[M]_D$ +33.1° (5N HCl); +72.6° (glacial acetic acid); $[\alpha]_D^{23}$ +22.9° (c = 0.8 in 20% HCl). pK$_1$ 2.32; pK$_2$ 9.62. Soly at 25°: 5.74±0.05 g/100 g water. Practically insol in ether, alc, acetone; insol in common neutral solvents.

DL-Form. Sublimes without melting at ordinary speed of heating. Dec 298° (closed capillary, very rapid heating). One part dissolves in 11.7 parts of water at 15°, in 14.1 parts of water at 25°. Insol in common neutral solvents.

10096. Valinomycin. [2001-95-8] $C_{54}H_{90}N_6O_{18}$; mol wt 1111.34. C 58.36%, H 8.16%, N 7.56%, O 25.91%. Cyclododecadepsipeptide ionophore antibiotic produced by *Streptomyces fulvissimus* and related to the enniatins, *q.v.* Composed of 3 moles each of L-valine, D-α-hydroxyisovaleric acid, D-valine, and L-lactic acid linked alternately to form a 36-membered ring: Brockmann *et al.*, *Ber.* **88**, 57 (1955); *Ann.* **603**, 216 (1957). Structural studies: Shemyakin *et al.*, *Tetrahedron Lett.* **1963**, 351; *Tetrahedron* **19**, 995 (1963). Proposed structure: Brockmann *et al.*, *Naturwissenschaften* **50**, 689 (1963). Structure and synthesis: Shemyakin *et al.*, *Tetrahedron Lett.* **1963**, 1921. Solid phase synthesis: Gisin *et al.*, *J. Am. Chem. Soc.* **91**, 2691 (1969); Losse, Klengel, *Tetrahedron* **27**, 1423 (1971). Biosynthesis: Smirnova *et al.*, *C.A.* **73**, 97347m (1970); Ristow *et al.*, *FEBS Lett.* **42**, 127 (1974). Conformation: Ivanov *et al.*, *Biochem. Biophys. Res. Commun.* **34**, 803 (1969); Onishi, Urry, *ibid.* **36**, 194 (1969); Duax *et al.*, *Science* **176**, 911 (1972). *Review:* Y. A. Ovchinnokov, V. T. Ivanov, "The Cyclic Peptides: Structure, Conformation, and Function", in *The Proteins* **vol. V**, H. Neurath, R. L. Hill, Eds. (Academic Press, New York, 3rd ed., 1982) pp 563-573.

Shiny rectangular platelets from dibutyl ether, mp 190° (hot stage). $[\alpha]_D^{20}$ +31.0° (c = 1.6 in benzene). Neutral reaction. Practically insol in water. Freely sol in petr ether, ether, benzene, chloroform, glacial acetic, butyl acetate, acetone. Active *in vitro* against *Mycobacterium tuberculosis.*

USE: Insecticide, nematocide, Patterson, Wright, US 3520973 (1970 to Am. Cyanamid).

10097. Valnemulin. [101312-92-9] [[2-[[(2R)-2-Amino-3-methyl-1-oxobutyl]amino]-1,1-dimethylethyl]thio]acetic acid (3aS,4R,5S,6S,8R,9R,9aR,10R)-6-ethenyldecahydro-5-hydroxy-4,6,9,10-tetramethyl-1-oxo-3a,9-propano-3aH-cyclopentacycloocten-8-yl ester; 14-O-[1-(D-2-amino-3-methylbutyrylamino)-2-methylpropan-2-ylthioacetyl]mutilin. $C_{31}H_{52}N_2O_5S$; mol wt 564.83. C 65.92%, H 9.28%, N 4.96%, O 14.16%, S 5.68%. Semisynthetic antibiotic; derivative of pleuromutilin, *q.v.* Prepn: H. Berner, H. Vyplel, **EP 153277**; *eidem*, **US 4675330** (1985, 1987 both to Sandoz). Comparative *in vitro* activity vs porcine bacterial pathogens: I. A. Aitken *et al.*, *Vet. Rec.* **144**, 128 (1999). Mechanism of action study: S. M. Poulsen *et al.*, *Mol. Microbiol.* **41**, 1091 (2001). HPLC determn in animal feeds: H. Guo *et al.*, *J. Chromatogr. B* **879**, 181 (2011). Pharmacokinetics in pigs: S. Horkovics-Kovats, F. Schatz, *J. Pharm. Med.* **6**, 149 (1996). Efficacy vs *Mycoplasma gallisepticum* in chickens: F. T. W. Jordan *et al.*, *Avian Dis.* **42**, 738 (1998); vs *Mycoplasma bovis* in calves: L. Stipkovits *et al.*, *Res. Vet. Sci.* **78**, 207 (2005).

Hydrochloride. [133868-46-9] Econor. $C_{31}H_{52}N_2O_5S.HCl$; mol wt 601.28.

THERAP CAT (VET): Antibacterial.

10098. Valnoctamide. [4171-13-5] 2-Ethyl-3-methylpentanamide; 2-ethyl-3-methylvaleramide; α-ethyl-β-methylvaleramide; valmethamide; McN-X-181; Axiquel; Nirvanil. $C_8H_{17}NO$; mol wt 143.23. C 67.09%, H 11.96%, N 9.78%, O 11.17%. Isomer of valpromide, *q.v.* Prepn: Freifelder *et al.*, *J. Org. Chem.* **26**, 203 (1961). GLC determn in plasma: M. Bialer, B. Hoch, *J. Chromatogr.* **337**, 408 (1985). Clinical pharmacokinetics: M. Bialer *et al.*, *Eur. J. Clin. Pharmacol.* **38**, 289 (1990).

Crystals, mp 113.5-114°. Sol in water.
THERAP CAT: Anxiolytic.

10099. Valproic Acid. [99-66-1] 2-Propylpentanoic acid; 2-propylvaleric acid; di-n-propylacetic acid; Convulex; Depakene. $C_8H_{16}O_2$; mol wt 144.21. C 66.63%, H 11.18%, O 22.19%. Antiepileptic; increases levels of γ-aminobutyric acid (GABA) in the brain. Prepn: B. S. Burton, *Am. Chem. J.* **3**, 385 (1882); E. Oberreit, *Ber.* **29**, 1998 (1896); M. Tiffeneau, Y. Deux, *Compt. Rend.* **212**, 105 (1941). LC-MS/MS determn in plasma: D. S. Jain *et al.*, *Talanta* **72**, 80 (2007). Anticonvulsant activity: H. Meunier *et al.*, *Therapie* **18**, 435 (1963). Toxicity study: Jenner *et al.*, *Food Cosmet. Toxicol.* **2**, 327 (1964). Comprehensive description: Z. L. Chang, *Anal. Profiles Drug Subs.* **8**, 529-556 (1979). Review of teratogenicity studies: H. Nau *et al.*, *Pharmacol. Toxicol.* **69**, 310-321 (1991); R. Alsdorf, D. F. Wyszynski, *Expert Opin. Drug Saf.* **4**, 345-353 (2005). Review of pharmacology and clinical experience in epilepsy: E. M. Rimmer, A. Richens, *Pharmacotherapy* **5**, 171-184 (1985); in psychiatric disease: D. R. P. Guay, *ibid.* **15**, 631-647 (1995); in migraine prophylaxis: C. E. Shelton, J. F. Connelly, *Ann. Pharmacother.* **30**, 865-866 (1996). Review of pharmacodynamics and mechanisms of action: W. Löscher, *Prog. Neurobiol.* **58**, 31-59 (1999).

Colorless liquid with characteristic odor. bp 219.5°. $n_D^{24.5}$ 1.425. d_4^0 0.9215. pKa 4.6. Very sol in organic solvents. Soly in water: 1.3 mg/ml. Freely sol in 1*N* sodium hydroxide, methanol, alc, acetone, chloroform, benzene, ether, *n*-heptane; slightly sol in 0.1*N* hydrochloric acid. LD_{50} orally in rats: 670 mg/kg (Jenner).

Sodium salt (1:1). [1069-66-5] Sodium valproate; Depacon; Depakin; Dépakine; Epilim; Ergenyl; Leptilan; Orfiril. $C_8H_{15}NaO_2$; mol wt 166.20. White, odorless, crystalline, deliquescent powder. pKa 4.8. Hygroscopic. One gram is sol in 0.4 ml water; 1.5 ml ethanol; 5 ml methanol. Practically insol in common organic solvents. LD_{50} orally in mice: 1700 mg/kg (Meunier).

Sodium salt (2:1). [76584-70-8] Sodium hydrogen bis(2-propylpentanoate); divalproex sodium; valproate semisodium; Abbott 50711; Depakote; Valcote. $C_{16}H_{31}NaO_4$; mol wt 310.41.

Magnesium salt. Depamag. $C_{16}H_{30}MgO_4$; mol wt 310.72.

THERAP CAT: Anticonvulsant; antimanic; antimigraine.

THERAP CAT (VET): Anticonvulsant.

10100. Valpromide. [2430-27-5] 2-Propylpentanamide; 2-propylvaleramide; 2-propylpentamide; dipropylacetamide; Depamide. $C_8H_{17}NO$; mol wt 143.23. C 67.09%, H 11.96%, N 9.78%, O 11.17%. Prodrug of valproic acid, *q.v.* Prepn: E. Fischer, A. Dilthey, *Ber.* **35**, 844 (1902); M. Tiffeneau, Y. Deux, *Compt. Rend.* **212**, 105 (1941). GC determn in plasma: M. Pokrajac *et al.*, *Pharm. Acta Helv.* **67**, 237 (1992). Review of pharmacology and clinical activity: M. Bialer, *Clin. Pharmacokinet.* **20**, 114-122 (1991).

White, odorless and bitter crystalline powder, mp 125-126°. Practically insol in water.

THERAP CAT: Anticonvulsant.

10101. Valrubicin. [56124-62-0] Pentanoic acid 2-[(2*S*,4*S*)-1,-2,3,4,6,11-hexahydro-2,5,12-trihydroxy-7-methoxy-6,11-dioxo-4-[[2,3,6-trideoxy-3-[(2,2,2-trifluoroacetyl)amino]-α-L-*lyxo*-hexopyranosyl]oxy]-2-naphthacenyl]-2-oxoethyl ester; *N*-trifluoroacetyladriamycin-14-valerate; AD-32; NSC-246131; Valstar. $C_{34}H_{36}F_3$-NO_{13}; mol wt 723.65. C 56.43%, H 5.01%, F 7.88%, N 1.94%, O 28.74%. Semisynthetic doxorubicin, *q.v.*, analog. Prepn: M. Israel *et al.*, *Cancer Res.* **35**, 1365 (1975); M. Israel, E. J. Modest, US 4035566 (1977 to S. Farber Cancer Inst.). TLC determn in biological samples: B. Barbieri *et al.*, *J. Chromatogr.* **163**, 195 (1979). Clinical pharmacokinetics: R. E. Greenberg *et al.*, *Urology* **49**, 471 (1997). Clinical trial in refractory bladder cancer: G. Steinberg *et al.*, *J. Urol.* **163**, 761 (2000). Review of pharmacology and clinical development: C. Doehn, *Curr. Opin. Oncol. Endocr. Metab. Invest. Drugs* **1**, 407-415 (1999); S. V. Onrust, H. M. Lamb, *Drugs Aging* **15**, 69-75 (1999).

Orange or orange-red powder, mp 135-136°. Highly lipophilic. Sol in methylene chloride, ethanol, methanol, acetone. Very slightly sol in water, hexane, petroleum ether.

THERAP CAT: Antineoplastic.

10102. Valsartan. [137862-53-4] *N*-(1-Oxopentyl)-*N*-[[2'-(2*H*-tetrazol-5-yl)[1,1'-biphenyl]-4-yl]methyl]-L-valine; *N*-[*p*-(*o*-1*H*-tetrazol-5-ylphenyl)benzyl]-*N*-valeryl-L-valine; (*S*)-*N*-(1-carboxy-2-methylprop-1-yl)-*N*-pentanoyl-*N*-[2'-(1*H*-tetrazol-5-yl)-biphenyl-4-ylmethyl]amine; CGP-48933; Diovan; Tareg. $C_{24}H_{29}$-N_5O_3; mol wt 435.53. C 66.19%, H 6.71%, N 16.08%, O 11.02%. Nonpeptide angiotensin II AT_1-receptor antagonist. Prepn: P. Bühlmayer *et al.*, EP 443983; *eidem*, US 5399578 (1991, 1995 both to Ciba Geigy); *idem et al.*, *Bioorg. Med. Chem. Lett.* **4**, 29 (1994). Pharmacological profile: L. Criscione *et al.*, *Br. J. Pharmacol.* **110**, 761 (1993). HPLC determn in human plasma: A. Sioufi *et al.*, *J. Liq. Chromatogr.* **17**, 2179 (1994). Clinical pharmacology: P. Müller *et al.*, *Eur. J. Clin. Pharmacol.* **47**, 231 (1994). Clinical comparison with captopril, *q.v.*, in high risk patients following myocardial infarction: M. A. Pfeffer *et al.*, *N. Engl. J. Med.* **349**, 1893 (2003). Review of pharmacology and clinical experience in heart failure: R. Latini *et al.*, *Expert Opin. Pharmacother.* **5**, 181-193 (2004).

Crystals from diisopropyl ether, mp 116-117°. Partition coefficient (*n*-octanol/aq phosphate buffer): 0.033. Sol in water at 25°.

THERAP CAT: Antihypertensive.

10103. Valspodar. [121584-18-7] 6-[(2*S*,4*R*,6*E*)-4-Methyl-2-(methylamino)-3-oxo-6-octenoic acid]-7-L-valine-cyclosporin A; 6-[(2*S*,4*R*,6*E*)-4-methyl-2-(methylamino)-3-oxo-6-octenoic acid]cyclosporin D; [3'-desoxy-3'-oxo-MeBmt]¹-[Val]²-cyclosporin; [3'-keto-Bmt1]-[Val2]-cyclosporin; PSC-833; SDZ-PSC-833; Amdray. $C_{63}H_{111}N_{11}O_{12}$; mol wt 1214.65. C 62.30%, H 9.21%, N 12.68%, O 15.81%. Nonimmunosuppressive cyclosporin analog that reverses multidrug resistance (MDR) by inhibiting cellular drug efflux mediated by P-glycoprotein, *q.v.* Prepn: P. Bollinger *et al.*, EP 296122; *eidem*, US 5525590 (1988, 1996 both to Sandoz). Pharmacology: D. Boesch *et al.*, *Exp. Cell Res.* **196**, 26 (1991). Clinical pharmacokinetics: R. M. Lush *et al.*, *J. Clin. Pharmacol.* **37**, 123 (1997). HPLC determn in blood: M. G. Scott *et al.*, *Clin. Chem.* **43**, 505 (1997). Clinical evaluation in multiple myeloma: P. Sonneveld *et al.*, *Leukemia* **10**, 1741 (1996); in acute myelogenous leukemia: R. Advani *et al.*, *Blood* **93**, 787 (1999). Review of pharmacology and clinical efficacy: F. Loor, *Expert Opin. Invest. Drugs* **8**, 807-835 (1999).

Insol in water. Sol in ethanol, DMSO. $[\alpha]_D^{20}$ −255.1° (c = 0.5 in $CHCl_3$).

THERAP CAT: Antineoplastic adjunct (chemosensitizer).

10104. Vanadium. [7440-62-2] V; at. wt 50.9415; at. no. 23; valencies 2, 3, 4, 5. Group VB (5). Two naturally occurring isotopes: ⁵¹V (99.75%); ⁵⁰V (0.25%); the latter is radioactive: $T_{1/2}$ 6 × 10¹⁵ years. Artificial isotopes: 46-49; 52-54. Abundance in earth's crust: 0.01% by wt. Widespread in nature; over 65 minerals known including *patronite* (polysulfide), *vanadinite* ($9PbO.3V_2O_5.PbCl_2$), *roscoelite* [$2K_2O.2Al_2O_3.(Mg,Fe)O.3V_2O_5.10SiO_2.4H_2O$] and *carnotite* ($K_2O.2U_2O_3.V_2O_5.3H_2O$) which is also an important source of uranium. Discovered by Selström in 1830; prepd by Roscoe in 1869. Prepn: Prandtl, Manz, *Z. Anorg. Allg. Chem.* **79**, 209 (1912); Marden, Rich, *J. Ind. Eng. Chem.* **19**, 786 (1927); McKechnie, Seybolt, *J. Electrochem. Soc.* **97**, 311 (1950); Gregory, Lilliendahl, *ibid.* **98**, 395 (1951); *Handbook of Preparative Inorganic Chemistry* vol. 2, G. Brauer, Ed. (Academic Press, New York, 2nd ed., 1965) pp 1252-1255. *Review:* Clark, "Vanadium" in *Comprehensive Inorganic Chemistry* vol. 3, J. C. Bailar, Jr. *et al.*, Eds. (Pergamon Press, Oxford, 1973) pp 491-551. Review of toxicology and human exposure: *Toxicological Profile for Vanadium and Compounds* (PB93-110880, 1992) 130 pp.

Light gray or white lustrous powder, fused hard lumps or body-centered cubic crystals; not tarnished in air and not appreciably affected by moisture at ordinary temp. mp 1917°. $d^{18.7}$ 6.11. Sp heat (20-100°) 0.12 cal/g/°C. Electrical resistivity 24.8 microhms/cm. Insol in water. Not attacked by hot or cold hydrochloric acid, by cold sulfuric acid. Reacts with hot sulfuric acid, hydrofluoric acid, nitric acid, aqua regia. Not attacked by bromine water, or by aq alkalies. The metal precipitates gold, silver and platinum from their salts; reduces mercuric salts to mercurous, ferric salts to ferrous. *Caution:* Potential symptoms of overexposure to dust or fumes are irritation of eyes, skin, throat; green tongue; metallic taste, eczema; cough; fine rales, wheezing, bronchitis; dyspnea. *See NIOSH Pocket Guide to Chemical Hazards* (DHHS/NIOSH 97-140, 1997) p 328.

USE: Alloying agent in manuf of rust-resistant vanadium steel.

10105. Vanadium Carbonyl. [20644-87-5] (*OC*-6-11)-Hexacarbonylvanadate(1−); vanadium hexacarbonyl. C_6O_6V; mol wt 219.00. C 32.91%, O 43.83%, V 23.26%. $V(CO)_6$. Prepn: Natta *et al.*, *C.A.* **54**, 16252 (1960); Ercoli *et al.*, *J. Am. Chem. Soc.* **82**, 2966 (1960); Hileman in *Prep. Inorg. React.* **1**, 107 (1964).

Blue-green pyrophoric crystals. Sensitive to air. Should be stored under nitrogen. Dec 60-70° in a nitrogen-filled tube.

10106. Vanadium Pentafluoride. [7783-72-4] Vanadium fluoride (VF_5). F_5V; mol wt 145.93. F 65.09%, V 34.91%. VF_5. Prepd from vanadium tetrafluoride by thermal disproportionation at 650° in N_2 current according to the eq $2VF_4 \rightarrow VF_5 + VF_3$: Ruff, Lickfett, *Ber.* **44**, 2548 (1911); Kwasnik in *Handbook of Preparative Inorganic Chemistry* vol. 1, G. Brauer, Ed. (Academic Press, New York, 2nd ed., 1963) p 253; from the elements: Trevorrow *et al.*, *J. Am. Chem. Soc.* **79**, 5167 (1957); Clark, Emeleus, *J. Chem. Soc.* **1957**, 2119. Review of transition metal pentafluorides: Peacock, *Adv. Fluorine Chem.* **7**, 113-145 (1973).

Liquid. mp 19.5°; bp 47.9°. $d_4^{19.5}$ 2.502. Heat of vaporization 10.60 kcal/mole. Appreciable vapor pressure at room temp. Hydrolyzed by water, dil alkali. Soly in anhydrous HF: 3.3 moles/l. Freely sol in alcohol, chloroform, acetone, ligroin. Insol in carbon disulfide. Dec toluene and ether. Etches glass slowly at room temp; more rapidly in presence of moisture with formation of yellow color. May be stored in sealed vessels made from iron, nickel, copper, platinum.

10107. Vanadium Pentoxide. [1314-62-1] Vanadium oxide (V_2O_5); vanadic anhydride. O_5V_2; mol wt 181.88. O 43.98%, V 56.02%. V_2O_5. Prepd by heating ammonium metavanadate, NH_4VO_3.

Yellow to rust-brown orthorhombic crystals; d 3.35; mp 690°. *Poisonous.* Loses oxygen reversibly in the region 700-1125°. One gram dissolves in ~125 ml water; sol in concd acids, forming red to yellow solns; sol in alkalies, forming vanadates; insol in alcohol. Its acid solutions are reduced by SO_2, Zn + HCl, and by evaporation with HCl.

Caution: Reported to be a respiratory irritant and to cause skin pallor, greenish-black tongue, chest pain, cough, dyspnea, palpitation, lung changes. When ingested, causes G.I. disturbances. May also cause a papular skin rash: Zenz *et al.*, *Arch. Environ. Health* **5**, 542 (1962); E. Browning, *Toxicity of Industrial Metals* (Appleton-Century-Crofts, New York, 2nd ed., 1969) pp 340-347.

USE: As catalyst in the oxidation of SO_2 to SO_3, alcohol to acetaldehyde, etc.; for the manuf of yellow glass; inhibiting ultraviolet light transmission in glass; depolarizer; as developer in photography; in form of ammonium vanadate as mordant in dyeing and printing fabrics and in manuf of aniline black.

10108. Vanadium Trifluoride. [10049-12-4] F_3V; mol wt 107.94. F 52.80%, V 47.19%. VF_3. Prepd from VCl_3 and HF: Ruff, Lickfett, *Ber.* **44**, 2539 (1911); from VCl_2 and HF: Emeleus, Gutmann, *J. Chem. Soc.* **1949**, 2979; by thermal decompn in an inert atm of $(NH_4)_3VF_6$: Sturm, Sheridan, *Inorg. Synth.* **7**, 87 (1963).

Greenish-yellow powder. d 3.363. mp approx 1406°. Sublimes at bright red heat. Almost insol in water, alcohol, acetone, ethyl acetate, acetic anhydride, glacial acetic acid, toluene, carbon tetrachloride, chloroform, carbon disulfide. Black color with NaOH soln. Aq solns have strong reducing properties.

Trihydrate. Dark green rhombohedral crystals, obtained by crystallizing vanadium trioxide dissolved in hydrofluoric acid, or by electrolytic reduction of a soln of vanadium pentoxide in aq HF. Loses one molecule of water at 100°. Dissolves in water to some extent, forming autocomplexes.

10109. Vanadium Trioxide. [1314-34-7] Vanadium oxide (V_2O_3); vanadium sesquioxide; vanadic oxide. O_3V_2; mol wt 149.88. O 32.02%, V 67.98%. V_2O_3. Prepd by reduction of V_2O_5 with hydrogen or carbon monoxide.

Black powder. On exposure to air it is gradually converted into indigo-blue crystals of V_2O_4. d 4.87; mp 1940°. Insol in water; difficultly sol in acids.

USE: Catalyst, e.g., when making ethanol from ethylene.

10110. Vanadium Trisulfide. [1315-03-3] Vanadium sulfide (V_2S_3); vanadium sesquisulfide. S_3V_2; mol wt 198.06. S 48.56%, V 51.44%. V_2S_3.

Greenish-black powder; d 4.7. Dec when heated. Insol in water, cold HCl, dil H_2SO_4; sol in hot HCl, hot dil H_2SO_4, HNO_3.

10111. Vanadocene. [1277-47-0] Bis(cyclopentadienyl)vanadium; dicyclopentadienylvanadium; Cp_2V; η^5-$(C_5H_5)_2V$. $C_{10}H_{10}V$; mol wt 181.13. C 66.31%, H 5.57%, V 28.12%. Electronically and coordinatively unsaturated metallocene of the early transition metals. Prepn: E. O. Fischer, W. Hafner, *Z. Naturforsch.* **9b**, 503 (1954). Improved prepn: C. Floriani, V. Mange, *Inorg. Synth.* **28**, 263 (1990). Crystal structure: R. D. Rogers *et al.*, *J. Cryst. Mol. Struct.* **11**, 183 (1981); M. Y. Antipin, R. Boese, *Acta Crystallogr.* **B52**, 314 (1996). Magnetism properties: E. König *et al.*, *J. Organomet. Chem.* **187**, 61 (1980). Thermal behavior: L. Poirier *et al.*, *J. Anal. Appl. Pyrolysis* **36**, 121 (1996). NMR studies: H. Eicher *et al.*, *J. Chem. Phys.* **86**, 1829 (1987). Review of chemistry: R. Choukroun, C. Lorber, *Eur. J. Inorg. Chem.* **2005**, 4683-4692.

Violet crystals, mp 167-168°. d 1.37. *Flammable. Irritant.* Reacts exothermically with oxygen and water. Sublimes at 120-140° under vacuum (~10^{-1} torr).

USE: Organometallic reagent.

10112. Vanadyl Dichloride. [10213-09-9] Dichlorooxovanadium; vanadium oxydichloride. Cl_2OV; mol wt 137.84. Cl 51.44%, O 11.61%, V 36.96%. $VOCl_2$. Usually contains some water. Prepd according to the eq $V_2O_5 + 3VCl_3 + VOCl_3 \rightarrow 6VOCl_2$: Funk, Weiss, *Z. Anorg. Allg. Chem.* **295**, 327 (1958); Oppermann, *ibid.* **351**, 113 (1967).

Green, very deliquesc tabular crystals. d 2.88. Disproportionates at 384° to VOCl and $VOCl_3$. Slowly dec by water. Sol in abs alcohol, glacial acetic acid. *Keep tightly closed.*

USE: Has been used as mordant in printing fabrics.

10113. Vanadyl Sulfate. [27774-13-6] Oxo[sulfato(2−)-κO]-vanadium; vanadium oxysulfate. O_5SV; mol wt 163.00. O 49.08%, S 19.67%, V 31.25%. $VOSO_4$.

Dihydrate. Blue, cryst powder. Sol in water.

USE: Dihydrate as mordant in dyeing and printing textiles; manuf colored glass; for blue and green glazes on pottery.

10114. Vanadyl Trichloride. [7727-18-6] (*T*-4)-Trichloro-oxovanadium; vanadium(V) trichloride oxide; vanadium oxytrichloride. Cl$_3$OV; mol wt 173.29. Cl 61.37%, O 9.23%, V 29.40%. VOCl$_3$. Prepn by the action of dry Cl$_2$ on V$_2$O$_3$ or V$_2$O$_5$: Brown, Griffitts, *Inorg. Synth.* **1**, 106 (1939); **4**, 80 (1953); Oppermann, *Z. Anorg. Allg. Chem.* **351**, 113 (1967). Toxicity data: Smyth *et al.*, *Am. Ind. Hyg. Assoc. J.* **30**, 470 (1969).

Yellow liquid, emitting red fumes on exposure to moist air. d 1.84. bp 126-127°. mp −77°. Dec in presence of moisture into vanadic acid and HCl. When a small quantity of water is added, it becomes thick and almost blood-red because of formation of vanadic acid. *Keep tightly closed.* LD$_{50}$ orally in rats: 0.14 g/kg (Smyth).

10115. Vanaspati. Vegetable ghee. Partially hydrogenated vegetable shortening used in traditional Indian and Middle Eastern cooking. Typical formulations contain elaidic, palmitic, oleic, stearic, and linoleic acids. Manuf by hydrogenation of peanut, cottonseed, and sesame oils in the presence of activated-Ni catalyst: S. S. P. Gupta, *Indian Chem. Eng.* **3**, 163 (1961), *C.A.* **56**, 15895h (1962). Fatty acid composition and physicochemical properties of commercial preparations: T. Jeyarani, S. Yella Reddy, *J. Food Lipids* **12**, 232 (2005). Determn of fatty acid and sterol content by GC-MS: A. A. Kandhro *et al.*, *Pak. J. Sci. Ind. Res.* **53**, 316 (2010). Prepn of *trans*-fatty acid free formulations and evaluation in traditional cooking: S. Sampurna, S. R. Yella Reddy, *Food Sci. Technol. Res.* **17**, 219 (2011).

Slip melting point, 39-40°. Hardness index (g/mm) : 8.2-17.0 (3°); 1.85-5.4 (20°); 1.1-2.7 (25°).

USE: Shortening used in cooking and baking.

10116. Vancomycin. [1404-90-6] C$_{66}$H$_{75}$Cl$_2$N$_9$O$_{24}$; mol wt 1449.27. C 54.70%, H 5.22%, Cl 4.89%, N 8.70%, O 26.49%. Amphoteric glycopeptide antibiotic produced by *Streptomyces orientalis* discovered in soil from Borneo. Inhibits bacterial cell wall synthesis by binding to peptidoglycan. Isoln: M. H. McCormick *et al.*, *Antibiot. Annu.* **1955-56**, 606; *eidem*, US 3067099 (1962 to Lilly). Pharmacology and toxicology: R. C. Anderson *et al.*, *Antibiot. Annu.* **1956-57**, 75. Purif: H. M. Higgins *et al.*, *ibid.* **1957-58**, 906. Structure studies: F. J. Marshall, *J. Med. Chem.* **8**, 18 (1965); W. D. Weringa *et al.*, *J. Chem. Soc. Perkin Trans. 1* **1972**, 443; K. A. Smith *et al.*, *ibid.* **1974**, 2369. Total structure determination by x-ray analysis: G. M. Sheldrick *et al.*, *Nature* **271**, 223 (1978). Revised configuration: M. P. Williamson, D. H. Williams, *J. Am. Chem. Soc.* **103**, 6580 (1981). Further revision of structure: C. M. Harris *et al.*, *ibid.* **105**, 6915 (1983). Total synthesis: K. C. Nicolaou *et al.*, *Angew. Chem. Int. Ed.* **38**, 240 (1999). Review of mechanism of action studies: H. P. Perkins, H. Nieto, *Ann. N.Y. Acad. Sci.* **235**, 348-363 (1974). Series of articles on antimicrobial activity, pharmacokinetics and clinical usage: *J. Antimicrob. Chemother.* **14**, Suppl. D, 1-109 (1984). HPLC determn in serum: L. Li *et al.*, *Ther. Drug Monit.* **17**, 366 (1995). CE determn in formulations: A. Musenga *et al.*, *J. Pharm. Biomed. Anal.* **42**, 32 (2006). Review of clinical experience: B. A. Cunha, *Med. Clin. North Am.* **79**, 817-831 (1995); of pharmacokinetics and pharmacodynamics: M. J. Ryback, *Clin. Infect. Dis.* **42**, S35-S39 (2006).

Monohydrochloride. [1404-93-9] Lyphocin; Vanco; Vancocin. C$_{66}$H$_{75}$Cl$_2$N$_9$O$_{24}$.HCl; mol wt 1485.72. White solid; uv max (H$_2$O): 282 nm (E$_{1cm}^{1\%}$ 40). Soly in water: >100 mg/ml. Moderately sol in dil methanol. Insol in the higher alcohols, acetone, ether, chloroform. LD$_{50}$ in mice (mg/kg): 489 i.v.; 1734 i.p.; 5000 s.c.; 5000 orally (Anderson).

THERAP CAT: Antibacterial.

THERAP CAT (VET): Antibacterial.

10117. Vandetanib. [443913-73-3] *N*-(4-Bromo-2-fluorophenyl)-6-methoxy-7-[(1-methyl-4-piperidinyl)methoxy]-4-quinazolinamine; 4-(4-bromo-2-fluoroanilino)-6-methoxy-7-(1-methylpiperidin-4-ylmethoxy)quinazoline; *N*-(4-bromo-2-fluorophenyl)-6-methoxy-7-[(1-methylpiperidin-4-yl)methoxy]quinazolin-4-amine; ZD-6474; Caprelsa; Zactima. C$_{22}$H$_{24}$BrFN$_4$O$_2$; mol wt 475.36. C 55.59%, H 5.09%, Br 16.81%, F 4.00%, N 11.79%, O 6.73%. Inhibitor of VEGFR (vascular endothelial growth factor receptor) and EGFR (epidermal growth factor receptor) tyrosine kinases; inhibits angiogenesis. Prepn: L. F. A. Hennequin *et al.*, WO 01032651; *eidem*, US 7173038 (2001, 2007 both to AstraZeneca); L. F. Hennequin *et al.*, *J. Med. Chem.* **45** 1300 (2002). Pharmacokinetics in cancer patients: S. N. Holden *et al.*, *Ann. Oncol.* **16**, 1391 (2005); K. D. Miller *et al.*, *Clin. Cancer Res.* **11**, 3369 (2005). LC-MS/MS determn in plasma and CSF: F. Bai *et al.*, *J. Chromatogr. B* **879**, 2561 (2011). Review of pharmacology: A. J. Ryan, S. R. Wedge, *Br. J. Cancer* **92**, Suppl. 1, S6-S13 (2005). Clinical trial in medullary thyroid cancer: S. A. Wells *et al.*, *J. Clin. Oncol.* **28**, 767 (2010). Review of clinical experience in non-small cell lung cancer: J. Flanigan *et al.*, *Biol. Targets Ther.* **4**, 237-243 (2010).

Soly in water: 330 μM (pH 7.4); ~650 μM (pH ~6.5). Log P (octanol/water): 5.0.

THERAP CAT: Antineoplastic.

10118. Vanilla. Cured, full-grown, unripe fruit of *Vanilla planifolia* Andr., *Orchidaceae*. *Habit.* Mexico, West Indies, Reunion, Mauritius, Seychelles. *Constit.* 2-2.75% vanillin; ~4% resin; vanillic acid, ~10% sugar.

USE: In manuf of confectionery and in various bakery products; perfumery; flavor for beverages; pharmaceutic aid (flavor).

10119. Vanillic Acid. [121-34-6] 4-Hydroxy-3-methoxybenzoic acid. C$_8$H$_8$O$_4$; mol wt 168.15. C 57.14%, H 4.80%, O 38.06%. Obtained from vanillin by oxidation with silver oxide or by controlled caustic fusion: Pearl, *Org. Synth.* **30**, 101 (1950).

White, odorless needles. mp 210°. Sublimes undecomposed. Sol in 860 parts water; very sol in alc; sol in ether. Not colored by FeCl$_3$. Its salts are freely sol in water.

10120. Vanillin. [121-33-5] 4-Hydroxy-3-methoxybenzaldehyde; methylprotocatechuic aldehyde; vanillic aldehyde; Rhonavil Extra Pure. C$_8$H$_8$O$_3$; mol wt 152.15. C 63.15%, H 5.30%, O 31.55%. Occurs naturally in a wide variety of foods and plants such as orchids; major commercial source of natural vanillin is from vanilla bean extract. Synthetically produced in-bulk from lignin-based byproducts of paper processes or from guaicol. Known prior to literature identification, first reported in the literature by Bucholtz in 1816 but not isolated until 1858. Isoln from vanilla beans: M. Gobley, *Jahresber. Fortschr. Chem. Verw. Theile Anderer Wiss.* **1858**, 534. LC determn: S. Kahan, D. A. Krueger, *J. AOAC Int.* **80**, 564 (1997). Acute toxicity: P. M. Jenner *et al.*, *Food Cosmet. Toxicol.* **2**, 327 (1964). Brief review of toxicology: D. L. J. Opdyke, *ibid.* **15**, 633-638 (1977). Review of NMR studies: G. J. Martin, *Ind. Chem. Libr.* **8**, 506-527 (1996). Review of synthetic prepn: M. B. Hocking, *J. Chem. Educ.* **74**, 1055-1059 (1997). Reviews: G. S. Clark, *Perfum. Flavor.* **15**, 45-54 (1990); L. J. Esposito *et al.* in *Kirk-Othmer Encyclopedia of Chemical Technology* **vol. 24** (Wiley-Interscience, New York, 4th ed., 1997) pp 812-825.

White or off-white nonhygroscopic crystalline powder. Pleasant aromatic vanilla odor and taste. Affected by light. d 1.056; bulk density ~0.6. mp 81-83°. bp$_{101.3 \text{ kPa}}$ 285°; bp$_{10 \text{ mm Hg}}$ 154°. Solubility in water >2%; in ethanol is 1:2 vanillin:ethanol. Freely sol in chloroform, ether, in solns of fixed alkali hydroxides; sol in glycerin and hot water. Solns are acid to litmus. LD$_{50}$ orally in rats, guinea pigs: 1580, 1400 mg/kg (Jenner).

USE: Pharmaceutic aid (flavor). As a flavoring agent in confectionery, beverages, foods and animal feeds. Fragance and flavor in cosmetics. Reagent for synthesis.

10121. Vanilmandelic Acid. [55-10-7] α,4-Dihydroxy-3-methoxybenzeneacetic acid; 3-methoxy-4-hydroxymandelic acid; 4-hydroxy-3-methoxymandelic acid; VMA. C$_9$H$_{10}$O$_5$; mol wt 198.17. C 54.55%, H 5.09%, O 40.37%. Catecholamine metabolite. Urine levels elevated in various pathologies. Misnamed *vanillinemandelic acid* and *vanillylmandelic acid*. Prepn from vanillin cyanohydrin: Gardner, Hibbert, *J. Am. Chem. Soc.* **66**, 608 (1944). Improved procedure: Shaw *et al.*, *J. Org. Chem.* **23**, 30 (1958); E. F. Recondo, H. Rinderknecht, *ibid.* **25**, 2248 (1960); I. Goodman *et al.*, *Biochem. Prep.* **13**, 75 (1971). Resolution: Armstrong *et al.*, *Biochim. Biophys. Acta* **25**, 422 (1957). Determination in urine: T. C. Stewart, J. A. Freeman, *Vanilmandelic Acid & Catecholamine Determinations* (Am. Soc. Clin. Pathol., Chicago, 1976) pp 1-81.

Scales from ether + benzene, dec 131-133°; also reported as mp 134-135° (Goodman). uv max (0.1*N* HCl): 230, 279 nm (ε 6320, 2810); (0.1*N* NaOH): 247, 285, 345 nm (ε 6860, 3960, 630). Readily resinifies on heating or on prolonged exposure to air. Freely sol in water, acetone. Mod sol in ether, acetonitrile. Sparingly sol in benzene. Mass spectral data: T. R. Sharp, *Org. Mass Spectrom.* **15**, 381 (1980).

L-Form. [313952-20-4] Crystals, dec 152°. $[\alpha]_D^{22}$ +128° (c = 0.7).

D-Form. [313952-19-1] Crystals, dec 152°. $[\alpha]_D^{23}$ −131°.

10122. Vaniprevir. [923590-37-8] (1*R*,2*R*)-*N*-[[[6-(2-Carboxy-2,3-dihydro-1*H*-isoindol-4-yl)-2,2-dimethylhexyl]oxy]carbonyl]-3-methyl-L-valyl-(4*R*)-4-hydroxy-L-prolyl-1-amino-*N*-(cyclopropylsulfonyl)-2-ethyl-cyclopropanecarboxamide (1 → 2)-lactone; (5*R*,7*S*,10*S*)-*N*-[(1*R*,2*R*)-1-[(cyclopropylsulfonyl)carbamoyl]-2-ethylcyclopropyl]-10-(1,1-dimethylethyl)-15,15-dimethyl-3,9,12-trioxo-6,7,9,10,11,12,14,15,16,17,18,19-dodecahydro-1*H*,3*H*,5*H*-2,-23:5,8-dimethano-4,13,2,8,11-benzodioxatriazacyclohenicosine-7-carboxamide; MK-7009. C$_{38}$H$_{55}$N$_5$O$_9$S; mol wt 757.94. C 60.22%, H 7.31%, N 9.24%, O 19.00%, S 4.23%. Hepatitis C virus NS3/4a protease inhibitor. Prepn: M. K. Holloway *et al.*, **WO 07015787**; *eidem*, **US 7470664** (2007, 2008 both to Merck & Co.). Practical

synthesis: Z. J. Song *et al.*, *J. Org. Chem.* **76**, 7804 (2011). HPLC-MS/MS determn in biological fluids: M. D. G. Anderson *et al.*, *J. Chromatogr. B* **877**, 1047 (2009). Preclinical pharmacology and pharmacokinetics: N. J. Liverton *et al.*, *Antimicrob. Agents Chemother.* **54**, 305 (2010). Review of discovery and antiviral activity: J. A. McCauley *et al.*, *J. Med. Chem.* **53**, 2443-2463 (2010).

White powder, mp 175-177°. $[\alpha]_{589}^{25}$ −40.84° (1.4 in methanol).
Potassium salt. [1269195-08-5] Vaniprevir potassium. C$_{38}$H$_{54}$-KN$_5$O$_9$S; mol wt 796.03. Off-white crystalline solid. $[\alpha]_{365}^{25}$ −199.9° (1.1 in 95% methanol).
THERAP CAT: Antiviral.

10123. Vapreotide. [103222-11-3] D-Phenylalanyl-L-cysteinyl-L-tyrosyl-D-tryptophyl-L-lysyl-L-valyl-L-cysteinyl-L-tryptophanamide cyclic (2 → 7)-disulfide; BMY-41606; RC-160; Octastatin. C$_{57}$H$_{70}$N$_{12}$O$_9$S$_2$; mol wt 1131.38. C 60.51%, H 6.24%, N 14.86%, O 12.73%, S 5.67%. Octapeptide somatostatin analog. Prepn: A. V. Schally, R. Z. Cai, **EP 203031** (1986 to Tulane); *eidem*, **US 4650787** (1987). Inhibition of growth hormone, insulin and glucagon release: T. Karashima *et al.*, *Life Sci.* **41**, 1011 (1987); on pancreatic secretion: S. J. Konturek *et al.*, *Proc. Soc. Exp. Biol. Med.* **187**, 241 (1988). RIA determn in serum: M. Mason-Garcia *et al.*, *Proc. Natl. Acad. Sci. USA* **85**, 5688 (1988). Clinical pharmacology: M. A. Ritz *et al.*, *Br. J. Clin. Pharmacol.* **47**, 195 (1999). Clinical trial for variceal bleeding in cirrhosis: P. Calès *et al.*, *N. Engl. J. Med.* **344**, 23 (2001).

Acetate. [849479-74-9] Sanvar.
THERAP CAT: Hemostatic.

10124. Vardenafil. [224785-90-4] 2-[2-Ethoxy-5-[(4-ethyl-1-piperazinyl)sulfonyl]phenyl]-5-methyl-7-propylimidazo[5,1-*f*][1,2,-4]triazin-4(1*H*)-one; 1-[[3-(1,4-dihydro-5-methyl-4-oxo-7-propyl-imidazo[5,1-*f*][1,2,4]triazin-2-yl)-4-ethoxyphenyl]sulfonyl]-4-ethyl-piperazine. C$_{23}$H$_{32}$N$_6$O$_4$S; mol wt 488.61. C 56.54%, H 6.60%, N 17.20%, O 13.10%, S 6.56%. Orally active, selective phosphodiesterase type 5 (PDE5) inhibitor. Prepn: U. Niewöhner *et al.*, **WO 9924433**; *eidem*, **US 6362178** (1999, 2002 both to Bayer AG). Effect on cGMP in human corpus cavernosum muscle cells: N. N. Kim *et al.*, *Life Sci.* **69**, 2249 (2001). Pharmacology: E. Bischoff *et al.*, *J. Urol.* **165**, 1316 (2001). Clinical trial in impotence: H. Porst *et al.*, *Int. J. Impot. Res.* **13**, 192 (2001); in erectile dysfunction: C. Stief *et al.*, *Int. J. Clin. Pract.* **58**, 230 (2004); F. Giuliano *et al.*, *BJU Int.* **95**, 110 (2004). Review of clinical experience in erectile dysfunction: S. Markou *et al.*, *Int. J. Impot. Res.* **16**, 470-478 (2004); M. Kendirci *et al.*, *Expert Opin. Pharmacother.* **5**, 923-932 (2004).

Hydrochloride trihydrate. [330808-88-3]; [224785-91-5]. Bay-38-9456; Levitra; Nuviva. $C_{23}H_{32}N_6O_4S.HCl.3H_2O$; mol wt 579.11. Nearly colorless solid, mp 218°. Soly in water: 0.11 mg/ml.

THERAP CAT: In treatment of male erectile dysfunction.

10125. Varenicline. [249296-44-4] 7,8,9,10-Tetrahydro-6,10-methano-6*H*-pyrazino[2,3-*h*][3]benzazepine; 5,8,14-triazatetracyclo[10.3.1.0^{2,11}.0^{4,9}]hexadeca-2(11)-3,5,7,9-pentaene; CP-526555. $C_{13}H_{13}N_3$; mol wt 211.27. C 73.91%, H 6.20%, N 19.89%. Nicotinic $\alpha 4\beta 2$ acetylcholine receptor partial agonist. Prepn: P. R. P. Brooks, J. W. Coe, **WO 0162736** (2001 to Pfizer). Synthesis, receptor binding studies, and *in vivo* dopaminergic activity: J. W. Coe *et al.*, *J. Med. Chem.* **48**, 3474 (2005). Metabolism: R. S. Obach *et al.*, *Drug Metab. Dispos.* **34**, 121 (2006). Series of clinical trials in smoking cessation: *J. Am. Med. Assoc.* **296**, 47-71 (2006). Review as aid for smoking cessation: E. D. Glover, J. M. Rath, *Expert Opin. Pharmacother.* **8**, 1757-1767 (2007).

Tartrate. [375815-87-5] Champix; Chantix. $C_{13}H_{13}N_3.C_4H_6O_6$; mol wt 361.35. White to off-white to slightly yellow solid. Highly sol in water.

THERAP CAT: Aid in smoking cessation.

10126. Varespladib. [172732-68-2] 2-[[3-(2-Amino-2-oxoacetyl)-2-ethyl-1-(phenylmethyl)-1*H*-indol-4-yl]oxy]acetic acid; 2-[[3-(2-amino-1,2-dioxoethyl)-2-ethyl-1-(phenylmethyl)-1*H*-indol-4-yl]oxy]acetic acid; [[3-(aminooxoacetyl)-1-benzyl-2-ethyl-1*H*-indol-4-yl]oxy]acetic acid; LY-315920. $C_{21}H_{20}N_2O_5$; mol wt 380.40. C 66.31%, H 5.30%, N 7.36%, O 21.03%. Inhibitor of secretory phospholipase A_2 (sPLA$_2$), an enzyme that hydrolyzes the *sn*-2 ester bond in phospholipids to release fatty acids. Prepn: N. J. Bach *et al.*, **EP 675110**; *eidem*, **US 5654326** (1995, 1997 both to Lilly); and SAR: S. E. Draheim *et al.*, *J. Med. Chem.* **39**, 5159 (1996). Pharmacology and effect on sPLA$_2$: D. W. Snyder *et al.*, *J. Pharmacol. Exp. Ther.* **288**, 1117 (1999). Clinical effect on plasma lipoproteins and sPLA$_2$ levels in patients with coronary heart disease: R. S. Rosenson *et al.*, *Lancet* **373**, 649 (2009). Review of pharmacology and clinical experience: *idem et al.*, *Expert Opin. Invest. Drugs* **19**, 1245-1255 (2010).

Crystals from ethyl acetate, mp 230-234°.
Methyl ester. [172733-08-3] Varespladib methyl; methyl [[3-(aminooxoacetyl)-1-benzyl-2-ethyl-1*H*-indol-4-yl]acetate; A-002; LY-333013; S-3013. $C_{22}H_{22}N_2O_5$; mol wt 394.43. Crystals from ethyl acetate, mp 172-187°.

THERAP CAT: Antiatherosclerotic.

10127. Vasicine. [6159-55-3] (3*S*)-1,2,3,9-Tetrahydropyrrolo-[2,1-*b*]quinazolin-3-ol; peganine. $C_{11}H_{12}N_2O$; mol wt 188.23. C 70.19%, H 6.43%, N 14.88%, O 8.50%. Bioactive alkaloid isolated from the Aruyvedic medicinal plant, *Adhatoda vasica* Nees, *Acanthaceae*, also known as Malabar nut. Isoln: Hooper, *Pharm. J.* **18**, 841 (1888); Sen, Ghose, *J. Indian Chem. Soc.* **1**, 315 (1924); Mehta *et al.*, *J. Org. Chem.* **28**, 445 (1963). Isoln from *Peganum harmala* L., *Zygophyllaceae*: Späth, Nikawitz, *Ber.* **67**, 45 (1934); Späth, Kuffner, *ibid.* **67**, 868 (1934). Structure and synthesis: Späth *et al.*, *ibid.* **68**, 699 (1935); Späth, Platzer, *ibid.* **69**, 255 (1936). Synthesis of *dl*-vasicine: Southwick, Casanova, *J. Am. Chem. Soc.* **80**, 1168 (1958). ^{1}H- and ^{13}C-NMR: B. S. Joshi *et al.*, *J. Nat. Prod.* **57**, 953

(1994). Abs config: *idem et al.*, *Tetrahedron: Asymmetry* **7**, 25 (1996). Determn in *A. vasica* leaves by RP-HPLC: A. Suthar *et al.*, *Acta Chromatogr.* **22**, 599 (2010). Bioavailability and stability in medicinal formulations: T. Vyas *et al.*, *Fitoterapia* **82**, 446 (2011).

Needles from alc, mp 212°. $[\alpha]_D^{14}$ −254° (c = 2.4 in CHCl$_3$); $[\alpha]_D^{14}$ −62° (c = 2.4 in alc).
Hydrochloride. [7174-27-8] $C_{11}H_{12}N_2O.HCl$; mol wt 224.69. Needles, mp 208°.
Hydriodide. [4966-84-1] $C_{11}H_{12}N_2O.HI$; mol wt 316.14. Needles, mp 195°.
(±)-**Form.** [6159-56-4] Needles from alc. mp 210°. Sublimes in high vacuum. Sol in acetone, alcohol, chloroform; slightly sol in water, ether, benzene.

10128. Vaska's Compound. [15318-31-7]; [14871-41-1] (unspecified stereo). (*SP*-4-3)-Carbonylchlorobis(triphenylphosphine)-iridium; *trans*-bis(triphenylphosphine)iridium(I) carbonyl chloride; *trans*-chlorocarbonylbis(triphenylphosphine)iridium(I); *trans*-iridium(I)bis(triphenylphosphine)carbonyl chloride; Vaska's complex; *trans*-IrCl(CO)(PPh$_3$)$_2$. $C_{37}H_{30}ClIrOP_2$; mol wt 780.26. C 56.96%, H 3.88%, Cl 4.54%, Ir 24.63%, O 2.05%, P 7.94%. Square-planar iridium complex; homogenous catalyst for a variety of synthetic applications. Model compd for reaction mechanism studies that led to the concepts of oxidative addition and reductive elimination in organotransition metal chemistry. Prepn: L. Vaska, J. W. DiLuzio, *J. Am. Chem. Soc.* **83**, 2784 (1961). Improved prepn: M. Rahim, K. J. Ahmed, *Inorg. Chem.* **33**, 3003 (1994). Equilibrium constant determn studies: R. G. Pearson, C. T. Kresge, *ibid.* **20**, 1878 (1981). Reactivity studies: G. G. Eberhardt, L. Vaska, *J. Catal.* **8**, 183 (1967); L. Vaska, *Acc. Chem. Res.* **1**, 335 (1968); C. Douvris, C. A. Reed, *Organometallics* **27**, 807 (2008). Catalysis applications: Z. Aizenshtat *et al.*, *J. Org. Chem.* **42**, 2386 (1977); H. A. Zahalka, H. Alper, *Organometallics* **5**, 2497 (1986); H. Lebel, C. Ladjel, *ibid.* **27**, 2676 (2008). Brief description and review of synthetic uses: S. A. Westcott, "*trans*-Carbonyl(chloro)bis(triphenylphosphine)iridium(I)" in *Encyclopedia of Reagents for Organic Synthesis* **2**, L. A. Paquette, Ed. (John Wiley & Sons, New York, 1995) pp 1000-1001.

Yellow solid, mp 323-325° (dec). Sol in benzene, toluene, chloroform. Insol in ether, alcohols. Stable toward air and heat. Readily takes up oxygen in soln.

USE: Catalyst.

10129. Vasopressin. [11000-17-2] Antidiuretic hormone; ADH; β-hypophamine. Nonapeptide pituitary hormone with antidiuretic and vasopressor activity; also acts as a neuromodulator with effects on social behavior and memory. Structure is similar to oxytocin, *q.v.*, and is highly conserved throughout the animal kingdom. Most mammals, including humans, produce arginine vasopressin; pigs produce lysine vasopressin where lysine is substituted for the arginine residue. The non-mammalian hormone is known as *vasotocin* and consists of the peptide ring of oxytocin with the side chain of vasopressin. Most invertebrates produce only one peptide with both oxytocin-like and vasopressin-like activities. Synthetic analogs have been prepared. Isoln from pituitary glands and differentiation from oxytocin: O. Kamm *et al.*, *J. Am. Chem. Soc.* **50**, 573 (1928). Purification: R. A. Turner *et al.*, *J. Biol. Chem.* **191**, 21 (1951). Structure of vasopressins: V. du Vigneaud *et al.*, *J. Am. Chem. Soc.*

75, 4880 (1953). Review of phyletic distribution and function: W. H. Sawyer, *Am. J. Med.* **42**, 678-686 (1967); of bioactivities: E. Frank, R. Landgraf, *Eur. J. Pharmacol.* **583**, 226-242 (2008); of role in social behavior and cognition: Z. R. Donaldson, L. J. Young, *Science* **322**, 900-904 (2008); T. R. Insel, *Neuron* **65**, 768-779 (2010).

Cys–Tyr–Phe–Gln–Asn–Cys–Pro–Arg–GlyNH₂

Arginine Vasopressin

Arginine vasopressin. [113-79-1] 8-L-Argininevasopressin; 3-(L-phenylalanine)-8-L-arginineoxytocin; argipressin; AVP; Pitressin; Pressyn. $C_{46}H_{65}N_{15}O_{12}S_2$; mol wt 1084.24. Synthesis: V. du Vigneaud *et al.*, *J. Am. Chem. Soc.* **80**, 3355 (1958); D. A. Jones, Jr. *et al.*, *J. Org. Chem.* **38**, 2865 (1973). Review of therapeutic applications: N. F. Holt, K. L. Haspel, *J. Cardiothorac. Vasc. Anesth.* **24**, 330-347 (2010); of use in treatment of septic shock: S. R. Bauer, S. W. Lam, *Pharmacotherapy* **30**, 1057-1071 (2010).

Lysine vasopressin *see* Lypressin.

Ornithine vasopressin *see* Ornipressin.

Arginine vasotocin. [113-80-4] 8-L-Arginineoxytocin; 8-L-argininevasotocin; 3-isoleucine-8-arginine vasopressin; AVT. $C_{43}H_{67}N_{15}O_{12}S_2$; mol wt 1050.22. Predominant form of vasotocin found in nonmammalian vertebrates such as fish, birds, and reptiles. Synthesis: P. G. Katsoyannis, V. du Vigneaud, *J. Biol. Chem.* **233**, 1352 (1958). Identification as a natural hormone in chickens: R. A. Munsick *et al.*, *Endocrinology* **66**, 860 (1960).

THERAP CAT: Antidiuretic and vasopressor hormone; hemostatic.

THERAP CAT (VET): Antidiuretic hormone.

10130. Vatalanib. [212141-54-3] *N*-(4-Chlorophenyl)-4-(4-pyridinylmethyl)-1-phthalazinamine; 1-(4-chloroanilino)-4-(4-pyridylmethyl)phthalazine; CGP-79787. $C_{20}H_{15}ClN_4$; mol wt 346.82. C 69.26%, H 4.36%, Cl 10.22%, N 16.15%. Vascular endothelial growth factor (VEGF) receptor tyrosine kinase inhibitor; vascular endothelial angiogenesis inhibitor. Prepn: G. Bold *et al.*, **WO 9835958**; *eidem*, **US 6258812** (1998, 2001 both to Novartis); *eidem*, *J. Med. Chem.* **43**, 2310 (2000). Enzyme inhibition study and pharmacology: J. M. Wood *et al.*, *Cancer Res.* **60**, 2178 (2000). Antiangiogenic activity: J. Drevs *et al.*, *ibid.* **62**, 4015 (2002). Pharmacokinetics: B. Morgan *et al.*, *J. Clin. Oncol.* **21**, 3955 (2003). Review of mechanism of action: P. Furet, P. W. Manley *ACS Symp. Ser.* **796**, 282-298 (2001); and therapeutic potential: A. L. Thomas *et al.*, *Semin. Oncol.* **30**, Suppl. 6, 32-38 (2003).

Crystals from water + diethyl ether, mp 209-212°.

Succinate. [212142-18-2] CGP-79787D; PTK-787; ZK-222584. $C_{20}H_{15}ClN_4 \cdot C_4H_6O_4$; mol wt 464.91. Crystals from ethanol, mp 195°.

THERAP CAT: Antineoplastic.

10131. Veatchine. [76-53-9] $C_{22}H_{33}NO_2$; mol wt 343.51. C 76.92%, H 9.68%, N 4.08%, O 9.32%. From bark of *Garrya veatchii* Kellogg, *Garryaceae*, where it occurs together with garryine and other alkaloids: Oneto, *J. Am. Pharm. Assoc.* **35**, 204 (1946); Wiesner *et al.*, *Can. J. Chem.* **30**, 608 (1952). Structure: Wiesner *et al.*, *J. Am. Chem. Soc.* **76**, 6068 (1954); Pelletier, *ibid.* **82**, 2398 (1960). Stereochemistry: Vorbrüggen, Djerassi, *Tetrahedron Lett.* **1961**, 119; *J. Am. Chem. Soc.* **84**, 2990 (1962). Coexistence of epimers

found in crystal and molecular structure: S. W. Pelletier *et al.*, *ibid.* **100**, 7976 (1978). ¹³C-NMR study of epimers: N. V. Mody, S. W. Pelletier, *Tetrahedron* **34**, 2421 (1978). Racemic syntheses and resolution: Nagata *et al.*, *ibid.* **86**, 929 (1964), **89**, 1499 (1967); Guthrie *et al.*, *Collect. Czech. Chem. Commun.* **31**, 602 (1966). Total synthesis of optically active form: Wiesner *et al.*, *Experientia* **26**, 471 (1970).

Crystals from dil acetone, mp 119-120°. Bitter taste. pH 11.5. $[\alpha]_D^{27.5}$ −69.01° (c = 1.06 in ethanol). Readily sol in water, ethanol.

Hydrochloride. $C_{22}H_{33}NO_2 \cdot HCl$. Crystals from abs ethanol + ether, dec 267-271°. Sol in water.

10132. Vecuronium Bromide. [50700-72-6] 1-[(2β,3α,5α,-16β,17β)-3,17-Bis(acetyloxy)-2-(1-piperidinyl)androstan-16-yl]-1-methylpiperidinium bromide (1:1); NC-45; Org-NC-45; Musculax; Norcuron. $C_{34}H_{57}BrN_2O_4$; mol wt 637.74. C 64.03%, H 9.01%, Br 12.53%, N 4.39%, O 10.03%. Aminosteroid, competitive neuromuscular blocker. Prepn: W. R. Buckett *et al.*, *J. Med. Chem.* **16**, 1116 (1973); I. C. Carlyle *et al.*, **EP 8824** (1980 to Akzo), *C.A.* **93**, 155850 (1980). Series of articles on pharmacology, pharmacokinetics, pharmacodynamics, clinical studies: *Br. J. Anaesth.* **52**, Suppl. 1, 1S-72S (1980). Clinical pharmacology: M. R. Fahey *et al.*, *Anesthesiology* **55**, 6 (1981). Comparison with pancuronium bromide, *q.v.*: S. L. Son *et al.*, *ibid.* 12.

Crystals, mp 227-229°. Sparingly sol in alc; slightly sol in water, acetone. LD₅₀ in mice: 0.061 mg/kg i.v. (Buckett).

THERAP CAT: Neuromuscular blocking agent.

THERAP CAT (VET): Neuromuscular blocking agent.

10133. Vedaprofen. [71109-09-6] 4-Cyclohexyl-α-methyl-1-naphthaleneacetic acid; 2-(4-cyclohexyl-1-naphthyl)propionic acid; CERM-10202; PM-150; Quadrisol. $C_{19}H_{22}O_2$; mol wt 282.38. C 80.82%, H 7.85%, O 11.33%. Non-steroidal anti-inflammatory drug (NSAID); propionic acid derivative. Prepn: J. J. Godfroid, E. Steiner, **BE 870553**; *eidem*, **US 4218473** (1979, 1980 both to CERM). Summary of field trials in dogs: J. G. H. Bergman, P. van Laar, *Vet. Q.* **18**, S20 (1996). Pharmacodynamics and pharmacokinetics in horses: P. Lees *et al.*, *J. Vet. Pharmacol. Ther.* **22**, 96 (1999).

Crystals from ethyl alcohol/water, mp 150°. LD₅₀ orally in rat: 400 mg/kg (Godfroid).

THERAP CAT (VET): Anti-inflammatory.

10134. Vedejs Reagent. [23319-63-3] (*PB-7-34-1222′2′*)-(*N*,-*N*,*N′*,*N′*,*N″*,*N″*-Hexamethylphosphoric triamide-*κO*)oxodiperoxy-(pyridine)molybdenum; oxodiperoxymolybdenum(pyridine)(hexamethylphosphoric triamide); MoOPH. $C_{11}H_{23}MoN_4O_6P$; mol wt 434.25. C 30.43%, H 5.34%, Mo 22.10%, N 12.90%, O 22.11%, P 7.13%. Oxidizing agent. Prepn and structural characterization: H. Mimoun *et al.*, *Bull. Soc. Chim. Fr.* **1969**, 1481. Prepn and synthetic use in hydroxylation of enolates: E. Vedejs, *J. Am. Chem. Soc.* **96**, 5944 (1974); *idem et al.*, *J. Org. Chem.* **43**, 188 (1978); E. Vedejs, S. Larsen, *Org. Synth.* **64**, 127 (1986). Improved prepn: A. R. Daniewski, W. Wojciechowska, *Synth. Commun.* **16**, 535 (1986). Synthetic applications: K. Krohn *et al.*, *Ber.* **123**, 1729 (1990); K. Makino *et al.*, *Tetrahedron* **58**, 9737 (2002). Brief review: D. B. Weibel, *Synlett* **2000**, 1076.

Yellow crystalline powder, mp 103-105° (evolution of gas). *Protect from light and moisture; keep refrigerated.*

USE: Reagent for the hydroxylation of enolates.

10135. Velaglucerase Alfa. [884604-91-5] Glucosylceramidase (human HT-1080 cell); GA-GCB; Vpriv. High mannose, recombinant form of human β-glucocerebrosidase, an enzyme involved in glycolipid metabolism that is deficient in patients with Gaucher's disease. Monomeric glycoprotein composed of 497 amino acids with 5 potential glycosylation sites; mol wt ~63 kDa. Sequence is identical to that of the natural enzyme. Produced via targeted recombination with a promoter that activates the endogenous glucocerebrosidase gene in a continuous human cell line. Glycosylation is controlled by the mannosidase I inhibitor, kifunensine, *q.v.*, yielding high mannose type glycans which enhances uptake by tissue macrophages where the glycolipid accumulates. Manufacturing process: C. A. Kinoshita *et al.*, **WO 0215927** (2002 to Transkaryotic Ther.); P. F. Daniel, **US 7138262** (2006 to Shire Human Genetic Ther.). Characterization, crystal structure, and uptake by macrophages: B. Brumshtein *et al.*, *Glycobiology* **20**, 24 (2010). Clinical pharmacokinetics: A. Zimran *et al.*, *Blood Cells Mol. Dis.* **39**, 115 (2007). Clinical trial in adults with type I Gaucher's disease: *idem et al.*, *Blood* **115**, 4651 (2010). Review of development and clinical experience: G. M. Pastores, *Curr. Opin. Investig. Drugs* **11**, 472-478 (2010).

THERAP CAT: Enzyme replacement therapy in treatment of Gaucher's disease.

10136. Velimogene Aliplasmid. [296251-72-4] Allovectin; Allovectin-7. DNA plasmid-based gene transfer product consisting of a DNA eukaryotic expression vector (plasmid VCL-1005) complexed with DMRIE/DOPE, a cationic lipid mixture containing *1,2-dimyristyloxypropyl-3-dimethylhydroxyethyl ammonium bromide* and *dioleoylphosphatidylethanolamine*. The 4853-base-pair DNA plasmid encodes the human leukocyte antigen B7 (HLA-B7) and β2-microglobulin and is designed to facilitate expression of the major histocompatability complex (MHC) class I antigen on the cell surface. The lipid mixture facilitates transfection of tumor cells and is optimized to increase the immune response. Prepn: G. J. Nabel et al., **WO 9429469**; *eidem*, **US 5910488** (1994, 1999 both to Vical). Clinical evaluation in metastatic melanoma by direct intralesional injection: A. T. Stopeck *et al.*, *Clin. Cancer Res.* **7**, 2285 (2001); A. Y. Bedikian *et al.*, *Melanoma Res.* **20**, 218 (2010). Review of development and clinical experience: H. P. Soares, J. Lutzky, *Expert Opin. Biol. Ther.* **10**, 841-851 (2010).

THERAP CAT: Antineoplastic.

10137. Vellosimine. [6874-98-2] Sarpagan-17-al; velosimine. $C_{19}H_{20}N_2O$; mol wt 292.38. C 78.05%, H 6.90%, N 9.58%, O

5.47%. From bark of *Geissosperum vellosii* Allem., *Apocynaceae*: Rapoport *et al.*, *J. Am. Chem. Soc.* **80**, 1601 (1958). Structure: Rapoport, Moore, *J. Org. Chem.* **27**, 2981 (1962); Ohashi *et al.*, *Tetrahedron* **19**, 2241 (1963).

Crystals from methanol, mp 305-306°. Can be sublimed at 180-100° (0.01 mm). $[\alpha]_D^{26}$ +48°. uv max (ethanol): 280, 289 nm (ε 8000, 6430).

10138. Vemurafenib. [918504-65-1] *N*-[3-[[5-(4-Chlorophenyl)-1*H*-pyrrolo[2,3-*b*]pyridin-3-yl]carbonyl]-2,4-difluorophenyl]-1-propanesulfonamide; PLX-4032; RG-7204; RO-5185426; Zelboraf. $C_{23}H_{18}ClF_2N_3O_3S$; mol wt 489.92. C 56.39%, H 3.70%, Cl 7.24%, F 7.76%, N 8.58%, O 9.80%, S 6.54%. Inhibitor of oncogenic B-Raf kinase; selectively targets the protein product of the *BRAF* V600E mutant oncogene which is associated with metastatic melanomas and certain other cancers. Prepn: P. N. Ibrahim *et al.*, **WO 07002325**; *eidem*, **US 7863288** (2007, 2011 both to Plexxikon). Selective effect on BRAF mutant melanoma cells: W. D. Tap *et al.*, *Neoplasia* **12**, 637 (2010); E. W. Joseph *et al.*, *Proc. Natl. Acad. Sci. USA* **107**, 14903 (2010). Overview of discovery: G. Bollag *et al.*, *Nature* **467**, 596 (2010). Clinical trial in metastatic melanoma: P. B. Chapman *et al.*, *N. Engl. J. Med.* **364**, 2507 (2011).

White solid, mp 264°.

THERAP CAT: Antineoplastic.

10139. Venice Turpentine. Larch turpentine. Oleoresin from *Larix decidua* Mill. (*L. europaea* Lam. & DC.), *Pinaceae*. Habit. Middle and Southern Europe. Constit. Volatile oil, resin. Use in light microscopy: D. A. Johansen, *Plant Microtechnique* (McGraw Hill, New York, 1940) pp 115-116.

Yellow, sometimes greenish, limpid, tenacious, thick liquid; pleasant, aromatic odor; hot, pungent, somewhat bitter taste. Becomes hard and brittle on prolonged exposure to air. Insol in water; sol in glacial acetic acid, amyl alcohol, acetone, caustic alkalies; slowly but freely sol in alc.

USE: As clearing agent and mounting medium for light microscopy.

10140. Venlafaxine. [93413-69-5] 1-[2-(Dimethylamino)-1-(4-methoxyphenyl)ethyl]cyclohexanol; (±)-1-[α-[(dimethylamino)methyl]-*p*-methoxybenzyl]cyclohexanol; *N*,*N*-dimethyl-2-(1-hydroxycyclohexyl)-2-(4-methoxyphenyl)ethylamine; venlafaxine. $C_{17}H_{27}NO_2$; mol wt 277.41. C 73.60%, H 9.81%, N 5.05%, O 11.53%. Serotonin noradrenaline reuptake inhibitor (SNRI). Prepn: G. E. M. Husbands *et al.*, **EP 112669**; **US 4535186** (1984, 1985 both to Am. Home Prods.); and resolution of isomers: J. P. Yardley *et al.*, *J. Med. Chem.* **33**, 2899 (1990). Receptor binding studies: E. A. Muth *et al.*, *Biochem. Pharmacol.* **35**, 4493 (1986). HPLC-MS/ESI determn in plasma: W. Liu *et al.*, *J. Chromatogr. B* **850**, 405 (2007). Clinical pharmacokinetics: K. J. Klamerus *et al.*, *J. Clin. Pharmacol.* **32**, 716 (1992). Evaluation of suicide risk: A. Rubino *et al.*, *Br. Med. J.* **334**, 242 (2007). Review of pharmacology and clinical efficacy in depression: S. A. Montgomery, *J. Clin. Psychiatry* **54**, 119-126 (1993); in anxiety disorders: M. E. Thase, *Expert Rev. Neurother.* **6**, 269-282 (2006).

Hydrochloride. [99300-78-4] Wy-45030; Effexor. $C_{17}H_{27}$-$NO_2 \cdot HCl$; mol wt 313.87. White to off-white crystalline solid from methanol/ethyl acetate, mp 215-217°. Soly (mg/ml): 572 water. Partition coefficient (octanol/water): 0.43.

(+)-Form. Crystals from ethyl acetate, mp 102-104°. $[\alpha]_D^{25}$ +27.6° (c = 1.07 in 95% ethanol).

(+)-Form hydrochloride. Wy-45655. Crystals from methanol/ether, mp 240-240.5°. $[\alpha]_D^{25}$ −4.7° (c = 0.945 in ethanol).

(−)-Form. Crystals from ethyl acetate, mp 102-104°. $[\alpha]_D^{25}$ −27.1° (c = 1.04 in 95% ethanol).

(−)-Form hydrochloride. Wy-45651. Crystals from methanol/ether, mp 240-240.5°. $[\alpha]_D^{25}$ +4.6° (c = 1.0 in ethanol).

THERAP CAT: Antidepressant; anxiolytic.

10141. Venturicidins. Antifungal antibiotics isolated from *Streptomyces aureofaciens* strains. Preliminary isolation work: Rhodes *et al.*, *Nature* **192**, 952 (1962). Isoln of venturicidins A and B and activity studies: Brufani *et al.*, *Helv. Chim. Acta* **51**, 1293 (1968); *see also* Langcake *et al.*, *Biochem. Soc. Trans.* **2**, 202 (1974). Final structures: Brufani *et al.*, *Experientia* **27**, 604 (1971); *eidem*, *Helv. Chim. Acta* **55**, 2329 (1972).

Venturicidin A   R = NH_2CO
Venturicidin B   R = H

Venturicidin A. [33538-71-5] Venturicidin B 3′-carbamate. $C_{41}H_{67}NO_{11}$; mol wt 749.98. Needles from chloroform-petr ether, mp 145-147°. Also reported as mp 140-142°. $[\alpha]_D$ +119° (c = 0.5 in chloroform). uv max (alcohol): 206, 247 (shoulder), ~300 nm (shoulder).

Venturicidin B. [33538-72-6] (3-Decarbamoyloxy)-3-hydroxy-venturicidin A. $C_{40}H_{66}O_{10}$; mol wt 706.96. Amorphous white powder from chloroform-ether, mp 168-170°. Also reported as mp 145-149° (ethyl acetate-petr ether). $[\alpha]_D$ +100° (c = 0.847 in chloroform).

10142. Veralipride. [66644-81-3] 5-(Aminosulfonyl)-2,3-dimethoxy-*N*-[[1-(2-propen-1-yl)-2-pyrrolidinyl]methyl]benzamide; *N*-[(1-allyl-2-pyrrolidinyl)methyl]-5-sulfamoyl-*o*-veratramide; LIR-1660; Agréal; Agradil. $C_{17}H_{25}N_3O_5S$; mol wt 383.46. C 53.25%, H 6.57%, N 10.96%, O 20.86%, S 8.36%. Dopamine D_2-receptor antagonist. Prepn: **NL 7707982** (1978 to Soc. d'Etudes Sci. Ind. de l'Ile-de-France), *C.A.* **89**, 30768 (1978). Pharmacological studies: P. Bouyard *et al.*, *Sem. Hop.* **56**, 1475 (1980); J. C. Czyba, *ibid.* 1483. Clinical studies: R. Renaud, J. Macler, *ibid.* **57**, 353 (1981);

S. Angeli, P. Fougère, *ibid.* **58**, 111 (1982). Review of safety and tolerability: V. De Leo *et al.*, *Expert Opin. Drug Saf.* **5**, 695-701 (2006).

THERAP CAT: Treatment of menopausal disorders.

10143. Veralkamine. [17155-31-6] (3β,16β,17α,22S)-17-Methyl-20-[(2S,5S)-5-methyl-2-piperidinyl]-18-norpregna-5,12-diene-3,16-diol; (3β,16β,17α,22α)-17-methyl-18-nor-16,28-seco-solanida-5,12-diene-3,16-diol; 17-methyl-20α-((2S,5S)-5-methyl-2-piperidyl)-18-nor-17α-pregna-5,12-diene-3β,16β-diol; (22S:25S)-12,26-epimino-17β-methyl-18-norcholesta-5,12-diene-3β,16β-diol; (17S:12S:25S)-22,26-epimino-18(13 → 17)-*abeo*-cholesta-5,12-diene-3β,16β-diol; veralcamine. $C_{27}H_{43}NO_2$; mol wt 413.65. C 78.40%, H 10.48%, N 3.39%, O 7.74%. Steroidal alkaloid isolated from *Veratrum album* sp. *lobelianium* (Bernh.) Suessenguth, *Liliaceae*: Tomko *et al.*, *Pharm. Zentralhalle* **99**, 373 (1960), *C.A.* **55**, 2013e. Structure studies: Tomko *et al.*, *Collect. Czech. Chem. Commun.* **27**, 1404 (1962). Complete structure: Tomko *et al.*, *Tetrahedron Lett.* **1967**, 3907; *eidem*, *Tetrahedron* **24**, 4865 (1968); Hoehne *et al.*, *ibid.* 4875.

Crystals from ethanol, mp 119-123° and 165-169°; $[\alpha]_D^{24}$ −84.1 ±3° (c = 0.533 in $CHCl_3$).

N,O,O-**Triacetate.** mp 152-154°. $[\alpha]_D^{27}$ −8.0° ($CHCl_3$).

N-**Monoacetate.** mp 191-193°. $[\alpha]_D^{23}$ −79.1° ($CHCl_3$).

10144. Verapamil. [52-53-9] α-[3-[[2-(3,4-Dimethoxyphenyl)ethyl]methylamino]propyl]-3,4-dimethoxy-α-(1-methylethyl)-benzeneacetonitrile; 5-[(3,4-dimethoxyphenethyl)methylamino]-2-(3,4-dimethoxyphenyl)-2-isopropylvaleronitrile; α-isopropyl-α-[(*N*-methyl-*N*-homoveratryl)-γ-aminopropyl]-3,4-dimethoxyphenylacetonitrile; iproveratril; D-365. $C_{27}H_{38}N_2O_4$; mol wt 454.61. C 71.34%, H 8.43%, N 6.16%, O 14.08%. Prototype calcium antagonist; vasodilating activity resides primarily in the (S)-isomer. Both isomers inhibit the P-glycoprotein efflux pump in multidrug resistant tumor cells. Prepn: **BE 615861**; Dengel, **US 3261859** (1962, 1966 both to Knoll). Pharmacology: H. Haas, G. Härtfelder, *Arzneim.-Forsch.* **12**, 549 (1962). Physical and chemical data: W. Appel, *ibid.* 562. Synthesis and absolute configuration of enantiomers: H. Ramuz, *Helv. Chim. Acta* **58**, 2050 (1975). Stereospecific synthesis: L. J. Theodore, W. L. Nelson, *J. Org. Chem.* **52**, 1309 (1987). LC-MS/MS determn of enantiomers and metabolites in plasma: M. Hedeland *et al.*, *J. Chromatogr. B* **804**, 303 (2004). CE analysis of metabolism: P. T. T. Ha *et al.*, *J. Chromatogr. A* **1120**, 94 (2006). Comprehensive description: Z. L. Chang, *Anal. Profiles Drug Subs.* **17**, 643-674 (1988). Review of pharmacology and therapeutic use in arrhythmias: B. N. Singh *et al.*, *Drugs* **25**, 125-153 (1983); in hypertension: D. McTavish, E. M. Sorkin, *ibid.* **38**, 19-76 (1989). Clinical trial of dexverapamil as adjunct to cancer chemotherapy: G. Kornek *et al.*, *Cancer* **76**, 1356 (1995); R. J. Motzer *et al.*, *J. Clin. Oncol.* **13**, 1958 (1995).

Viscous, pale yellow oil, $bp_{0.01}$ 243-246°. n_D^{25} 1.5448. Freely sol in the lower alcohols, acetone, ethyl acetate, chloroform. Sparingly sol in hexane; sol in benzene, ether. Practically insol in water.

Hydrochloride. [152-11-4] Berkatens; Calan; Cardibeltin; Covera-HS; Dignover; Falicard; Flamon; Geangin; Isoptin; Quasar; Securon; Tarka; Univer; Veracim; Veramex; Veraptin; Verelan; Vertab; Zolvera. $C_{27}H_{38}N_2O_4 \cdot HCl$; mol wt 491.07. Crystals, dec 138.5-140.5° (corr). pH of 0.1% aq soln: 5.25. uv max: 232, 278 nm. Soly (mg/ml): water 83, ethanol (200 proof) 26, propylene glycol 93, ethanol (190 proof) >100, methanol >100, 2-propanol 4.6, ethyl acetate 1.0, DMF >100, methylene chloride >100, hexane 0.001. Freely sol in chloroform. Practically insol in ether. pKa 8.6. LD_{50} in mice, rats (mg/kg): 7.6, 16 i.v.; 68, 107 s.c.; 68, 67 i.p.; 163, 114 orally (Haas, Härtfelder).

(R)-Form. [38321-02-7] Dexverapamil. Colorless oil. $[\alpha]_D^{25}$ +27.2° (c = 1.00 in benzene).

(R)-Form hydrochloride. [38176-02-2] Dexverapamil hydrochloride. Crystals, mp 131-133°. $[\alpha]_D^{25}$ +8.9° (c = 5.01 in ethanol).

(S)-Form. [36622-29-4] Very pale yellow oil. $[\alpha]_D^{25}$ −26.4° (c = 1.00 in benzene).

(S)-Form hydrochloride. [36622-28-3] White crystalline solid from 2-propanol/cyclohexane, mp 131-133°. $[\alpha]_D^{24}$ −8.9° (c = 5.03 in ethanol).

THERAP CAT: Antihypertensive; antianginal; antiarrhythmic (class IV).

10145. Veratraldehyde. [120-14-9] 3,4-Dimethoxybenzaldehyde; 3,4-dimethoxybenzenecarbonal; veratric aldehyde; protocatechualdehyde dimethyl ether. $C_9H_{10}O_3$; mol wt 166.18. C 65.05%, H 6.07%, O 28.88%. Prepd by methylation of vanillin: Kostanecki, Tambor, *Ber.* **39**, 4022 (1906); Buck, *Org. Synth.* **13**, 102 (1933); Alt, *US 3007968* (1957 to Monsanto); by oxidation of veratryl alcohol with chromium(VI) oxide-pyridine complex: Holum, *J. Org. Chem.* **26**, 4814 (1961).

Needles from ether, petr ether, toluene, or carbon tetrachloride. Odor of vanilla beans; mp 42-43°; bp_{760} 281°; bp_{53} 201°; bp_{10} 155°. Slightly sol in hot water; freely sol in alcohol and ether. Solns are oxidized to veratric acid under the influence of light.

10146. Veratramine. [60-70-8] (2S,3R,5S)-5-Methyl-2-[(1S)-1-[(3S,6aR,11aS,11bR)-2,3,4,6,6a,11,11a,11b-octahydro-3-hydroxy-10,11b-dimethyl-1H-benzo[a]fluoren-9-yl]ethyl]-3-piperidinol; (3β,23β)-14,15,16,17-tetradehydroveratraman-3,23-diol. $C_{27}H_{39}$-NO_2; mol wt 409.61. C 79.17%, H 9.60%, N 3.42%, O 7.81%. Secondary base from *Veratrum grandiflorum* (Maxim.) Loes. f., and from *V. viride* Ait., *Liliaceae*. Isoln and structure: Saito, *Bull. Chem. Soc. Jpn.* **15**, 22 (1940); Jacobs, Craig, *J. Biol. Chem.* **160**, 555 (1945); Jacobs, Sato, *ibid.* **181**, 55 (1949); **191**, 71 (1951); Tamm, Wintersteiner, *J. Am. Chem. Soc.* **74**, 3842 (1952); Wintersteiner, *Festschrift Arthur Stoll* (Birkhäuser-Verlag, Basel) pp 166-176. Total synthesis: Masamune *et al., J. Am. Chem. Soc.* **89**, 4521 (1967); Johnson *et al., ibid.* 4523; Masamune *et al., Tetrahedron* **27**, 3369 (1971); Kutney *et al., Can. J. Chem.* **53**, 1796 (1975). Stereochemistry: Sicher, Tichy, *Tetrahedron Lett.* **1959** (12), 6 (1959); Kataoka, *Chem. Ind. (London)* **1961**, 512; Bailey *et al., Tetrahedron Lett.* **1963**, 555. Revised stereochemistry: Scott *et al., ibid.* **1967**, 2381; Kupchan, Suffness, *J. Am. Chem. Soc.* **90**, 2730 (1968);

Sprague *et al., Tetrahedron* **27**, 4857 (1971). *See also:* Veratrum Viride.

Crystals, mp 206-207°. Slightly sol in water. Sol in methanol, alcohol. Precipitated by digitonin. $[\alpha]_D^{25}$ −71.8° (c = 1.21); $[\alpha]_D^{25}$ −70° (c = 1.56 in methanol). uv max: 268 nm.

Dihydroveratramine. Crystals, mp 192.5-194°. $[\alpha]_D^{25}$ +26° (c = 1.26 in acetic acid).

10147. Veratric Acid. [93-07-2] 3,4-Dimethoxybenzoic acid; dimethylprotocatechuic acid. $C_9H_{10}O_4$; mol wt 182.18. C 59.34%, H 5.53%, O 35.13%. Isolated from seed of *Schoenocaulon officinale* (Schlecht. & Cham.) A. Gray (*Sabadilla officinarum* Brandt). Prepn: Arthur, Ng, *J. Chem. Soc.* **1959**, 3094.

Monohydrate. Odorless crystals. At 100° becomes anhydr; mp 180-181° when anhydr. Sublimes in rhombic crystals. Sol in 2150 parts cold, 165 parts boiling water; very sol in alcohol or ether. Its barium salt is but slightly sol in water.

10148. Veratridine. [71-62-5] (3β,4α,16β)-4,9-Epoxycevane-3,4,12,14,16,17,20-heptol 3-(3,4-dimethoxybenzoate); 3-veratroyl-veracevine. $C_{36}H_{51}NO_{11}$; mol wt 673.80. C 64.17%, H 7.63%, N 2.08%, O 26.12%. From seed of *Schoenocaulon officinale* (Schlecht. & Cham.) A. Gray and also from the rhizome of *Veratrum album* L. *Liliaceae*. Isoln: Blount, *J. Chem. Soc.* **1935**, 122; Vejdelek *et al., Chem. Listy* **50**, 603 (1956); *Collect. Czech. Chem. Commun.* **22**, 98 (1957). Purification and properties: L. C. McKinney *et al., Anal. Biochem.* **153**, 33 (1986). Toxicity studies: O. Krayer *et al., J. Pharmacol. Exp. Ther.* **82**, 167 (1944); K. Tanaka, *ibid.* **113**, 89 (1955); Swiss, Bauer, *Proc. Soc. Exp. Biol. Med.* **76**, 847 (1951). Review of chemistry and structure of veratridine and other *Veratrum* alkaloids: Kupchan, By, in *The Alkaloids* vol. 10, R. H. F. Manske, Ed. (Academic Press, New York, 1968) pp 193-285.

Yellowish-white, amorphous powder. Tenaciously retains water. mp 180° (after drying at 130°). $[\alpha]_D^{20}$ +8.0° (ethanol). pKa 9.54 ±0.02. Insol in water. Slightly sol in ether. LD_{50} in mice (mg/kg): 1.35 i.p. (Swiss, Bauer); 0.42 i.v. (Krayer); 6.3 s.c. (Tanaka).

Nitrate. Amorphous powder, sparingly sol in water.

Sulfate. Slender needles, very hygroscopic.

Perchlorate. Long thin needles from water, mp 259-260° (after drying at 120° *in vacuo*).

10149. Veratrine (Mixture). [8051-02-3] Mixture of alkaloids from seeds of *Schoenocaulon officinale* (Schlecht & Cham.) A. Gray, *Liliaceae*; used as a topical counterirritant. *Constit.* Cevadine, veratridine, cevadilline, sabadine, cevine. Veratrine is also used as a synonym for cevadine, *q.v.*

White or grayish-white powder. mp 145-155°. One gram dissolves in about 1800 ml water, 1000 ml boiling water, 2.8 ml alcohol, 0.7 ml chloroform, 4.2 ml ether, 80 ml olive oil; freely sol in dil acids, benzene, amyl alcohol, slightly in glycerol; insol in petr ether. *Caution: Poisonous!* Exceedingly irritating to mucous membranes, causing violent sneezing when inhaled.

10150. Veratrole. [91-16-7] 1,2-Dimethoxybenzene; *o*-dimethoxybenzene; pyrocatechol dimethyl ether. $C_8H_{10}O_2$; mol wt 138.17. C 69.54%, H 7.30%, O 23.16%. Prepd by methylation of pyrocatechol: Ullmann, *Ann.* **327**, 104 (1903); Drahowzal, Klamann, *Monatsh. Chem.* **82**, 588 (1951). Toxicity study: P. M. Jenner *et al.*, *Food Cosmet. Toxicol.* **2**, 327 (1964).

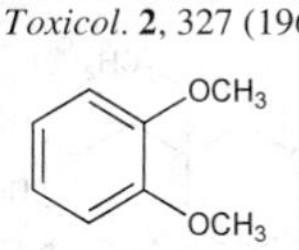

Crystals or liquid; d_{25}^{25} 1.084; mp 22-23°; bp 206-207°. Slightly sol in water; sol in alcohol, ether, fatty oils. LD_{50} in rats, mice (mg/kg): 1360, 2020 orally (Jenner).

10151. Veratrum viride. American hellebore; green hellebore; American veratrum; Indian poke. Dried rhizome and roots of *Veratrum viride* Ait., *Liliaceae*. *Habit.* North America. *Constit.* The alkaloids jervine, pseudojervine, rubijervine, cevadine, germitrine, germidine, veratralbine, veratroidine. A standardized extract of the alkaloids, known as *alkavervir*, has been used as an antihypertensive. Determn of constituents: Seiferle *et al.*, *J. Econ. Entomol.* **35**, 35 (1942); and hypotensive effects: Fried *et al.*, *J. Am. Chem. Soc.* **72**, 4621 (1950). Review of structures and activity of hypotensive principles: Kupchan, *J. Pharm. Sci.* **50**, 273-287 (1961); of medicinal uses: J. Gruenwald, *PDR for Herbal Medicines* (Thomson PDR, Montvale, 3rd Ed., 2004) pp 27-28. Review of toxicology and symptoms and treatment of poisoning: L. J. Schep *et al.*, *Toxicol. Rev.* **25**, 73-78 (2006). *Caution: Poisonous*; severely irritating to mucous membranes.

10152. Verbascose. [546-62-3] β-D-Fructofuranosyl *O*-α-D-galactopyranosyl-(1 → 6)-*O*-α-D-galactopyranosyl-(1 → 6)-*O*-α-D-galactopyranosyl-(1 → 6)-α-D-glucopyranoside; *O*-α-D-galactopyranosyl-(1 → 6)-[*O*-α-D-galactopyranosyl-(1 → 6)-]$_2$-*O*-α-D-glucopyranosyl-(1 → 2)-β-D-fructofuranoside. $C_{30}H_{52}O_{26}$; mol wt 828.72. C 43.48%, H 6.32%, O 50.19%. Oligosaccharide isolated from roots of mullein *Verbascum thapsus* L., *Scrophulariaceae*: Bourquelot, Bridel, *Compt. Rend.* **151**, 760 (1910); from birch wood, *Betula verrucosa* Ehrh., *Betulaceae*: Lindberg, Selleby, *Acta Chem. Scand.* **12**, 1512 (1958). Structure: Wickström *et al.*, *Bull. Soc. Chim. Fr.* **1956**, 827. *Review:* D. French "The Raffinose Family of Oligosaccharides" in *Adv. Carbohydr. Chem.* **9**, 180-181 (1954).

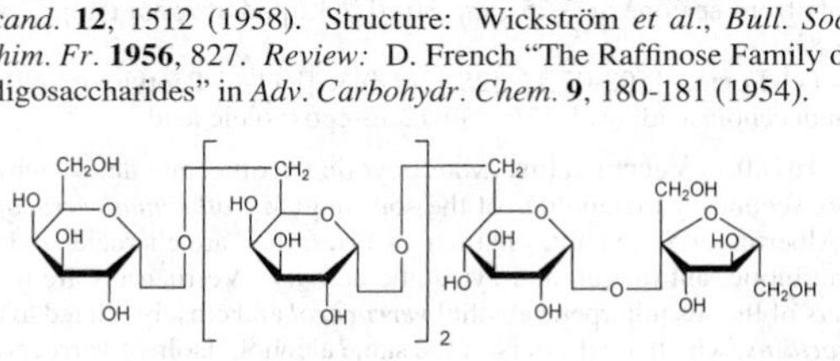

Needles, mp 219-220°, mp 253°. $[\alpha]_D^{20}$ +170° (water); $[\alpha]_D^{25}$ +146° (c = 2.1 in water).

10153. Verbenalin. [548-37-8] (1*S*,4a*S*,7*S*,7a*R*)-1-(β-D-Glucopyranosyloxy)-1,4a,5,6,7,7a-hexahydro-7-methyl-5-oxocyclopenta[*c*]pyran-4-carboxylic acid methyl ester; cornin. $C_{17}H_{24}O_{10}$; mol wt 388.37. C 52.58%, H 6.23%, O 41.20%. A glycoside isolated from *Verbena officinalis* L., *Verbenaceae, Cornus florida* L., *Cornaceae*. *Refs*: Bourdier, *J. Pharm. Chim.* **27** (6), 49, 101 (1908); Holste, *Arch. Exp. Pathol. Pharmakol.* **101**, 46 (1924); Miller, *J. Am. Pharm. Assoc.* **17**, 744 (1928). Identity with cornin:

Reichert, *Arch. Pharm.* **273**, 357 (1935). Structure: Büchi, Manning, *Tetrahedron Lett.* **1960**, 5 (1960); *eidem*, *Tetrahedron* **18**, 1049 (1962). Synthesis of *verbenalol*, the aglycone: P. Callant *et al.*, *ibid.* **37**, 2085 (1981).

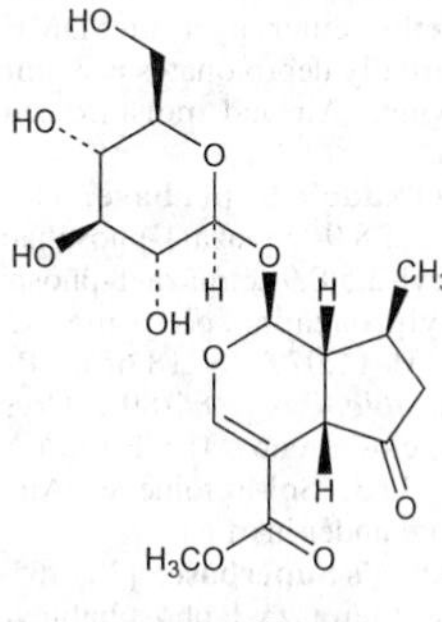

Bitter needles, mp 182-183°. $[\alpha]_D^{25}$ −173° (water). uv max: 238, 290 nm (ε 9600, 105) in ethanol. Freely sol in water, slightly in alcohol, ethyl acetate, acetone. Practically insol in chloroform, ether.

10154. *d*-Verbenone. [18309-32-5] (1*R*,5*R*)-4,6,6-Trimethyl-bicyclo[3.1.1]hept-3-en-2-one; (1*R*,5*R*)-(+)-2-pinen-4-one. $C_{10}H_{14}O$; mol wt 150.22. C 79.96%, H 9.39%, O 10.65%. A constituent of Spanish verbena oil (from *Verbena triphylla* L., *Verbenaceae*): Kerschbaum, *Ber.* **33**, 885 (1900). Prepn from α-pinene: Bain, Gary, **US 2911442** (1959 to Glidden). Structure: Blumann, Zeitschel, *Ber.* **46**, 1178 (1913); Weinhaus, Schumm, *Ann.* **439**, 20 (1924). Absolute configuration: Hurst, Whitham, *J. Chem. Soc.* **1960**, 2864. *Review:* J. L. Simonsen, *The Terpenes* vol. II (University Press, Cambridge, 2nd ed., 1949) p 232-239.

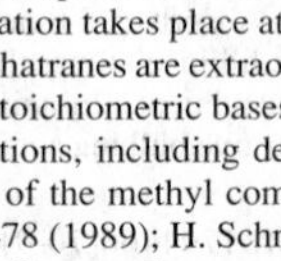

Oil. Characteristic odor. mp 6.5°. bp_{12} 102-105°; bp_{760} 227-228°. d^{20} 0.9780. n_D^{18} 1.4957. $[\alpha]_D^{18}$ +249.62°. uv max (ethanol): 253 nm (ε 6730). Practically insol in water. Miscible in all proportions with the usual organic solvents.

10155. Verkade's Superbases. Bicyclic, nonionic bases in which electron pair donation takes place at the phosphorus atom; the protonated proazaphosphatranes are extraordinarily stable. The class of compds is useful as stoichiometric bases and catalysts in a variety of synthetic transformations, including deprotonations and nucleophilic reactions. Prepn of the methyl compd: C. Lensink *et al.*, *J. Am. Chem. Soc.* **111**, 3478 (1989); H. Schmidt *et al.*, *Z. Anorg. Allg. Chem.* **578**, 75 (1989). Improved prepn: J. Tang, J. G. Verkade, *Tetrahedron Lett.* **34**, 2903 (1993). Prepn and reactivity of the isopropyl analog: A. E. Wróblewski *et al.*, *Main Group Chem.* **1**, 69 (1995); of the isobutyl analog: P. B. Kisanga, J. G. Verkade, *Tetrahedron* **57**, 467 (2001). Lewis basicity studies: M. A. H. Laramay, J. G. Verkade, *Z. Anorg. Allg. Chem.* **605**, 163 (1991). Photoelectron spectra: L. Nyulászi *et al.*, *Inorg. Chem.* **35**, 6102 (1996). Synthetic applications: B. A. D'Sa *et al.*, *J. Org. Chem.* **63**, 3961 (1998); P. Ilankumaran, J. G. Verkade, *ibid.* **64**, 3086 (1999); P. B. Kisanga, J. G. Verkade, *ibid.* 4298; P. B. Kisanga *et al.*, *ibid.* **67**, 3555 (2002); S. Urgaonkar, J. G. Verkade, *ibid.* **69**, 9135 (2004). *Review:* J. G. Verkade, P. B. Kisanga, *Tetrahedron* **59**, 7819-7858 (2003).

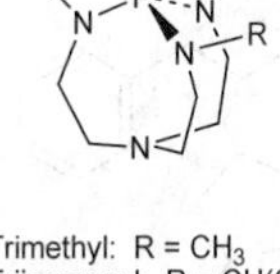

Trimethyl: R = CH$_3$
Triisopropyl: R = CH(CH$_3$)$_2$
Triisobutyl: R = CH$_2$CH(CH$_3$)$_2$

Trimethyl-Verkade's Superbase. [120666-13-9] 2,8,9-Tri-methyl-2,5,8,9-tetraaza-1-phosphabicyclo[3.3.3]undecane; proazaphosphatrane. $C_9H_{21}N_4P$; mol wt 216.27. C 49.98%, H 9.79%, N 25.91%, P 14.32%. White waxy solid from sublimation of thick oil. mp 110-115°. pKa of conjugate acid (acetonitrile): 32.90. Sol in tetrahydrofuran, diethyl ether, pyridine, DMF, benzene, toluene, pentane, hexane. Partially deprotonates acetonitrile, DMSO, and alcohols upon dissolving. Air and moisture sensitive. Handle and store under inert gas.

Triisopropyl-Verkade's Superbase. [175845-21-3] 2,8,9-Tris(1-methylethyl)-2,5,8,9-tetraaza-1-phosphabicyclo[3.3.3]undecane; 2,8,9-triisopropyl-2,5,8,9-tetraaza-1-phosphabicyclo[3.3.3]undecane; triisopropylproazaphosphatrane. $C_{15}H_{33}N_4P$; mol wt 300.43. C 59.97%, H 11.07%, N 18.65%, P 10.31%. Colorless viscous oil. *Combustible.* bp$_{0.08}$ 84-89°. Crystals upon standing, mp 45.5°. Flash pt, closed cup: 185°F (85°C). pKa of conjugate acid (acetonitrile): 33.63. Sol in toluene. Air and moisture sensitive. Handle and store under inert gas.

Triisobutyl-Verkade's Superbase. [331465-71-5] 2,8,9-Tris(2-methylpropyl)-2,5,8,9-tetraaza-1-phosphabicyclo[3.3.3]undecane; 2,8,9-triisobutyl-2,5,8,9-tetraaza-1-phosphabicyclo[3.3.3]undecane; triisobutylproazaphosphatrane. $C_{18}H_{39}N_4P$; mol wt 342.51. C 63.12%, H 11.48%, N 16.36%, P 9.04%. Colorless oil. *Combustible.* bp$_{0.210}$ 132°. Flash pt, closed cup: 154.9°F (68.3°C). pKa of conjugate acid (acetonitrile): 33.53. Sol in diethyl ether. Solidifies upon storage at −4°C for 24 h to form white solid. Air and moisture sensitive. Handle and store under inert gas.
USE: Reagent and catalyst in synthetic organic chemistry.

10156. Vermiculite. [1318-00-9] Hydrated magnesium-aluminum-iron silicate. Naturally occurring sheet silicate mineral; usually found in association with numerous other minerals. Typical analysis of Montana vermiculite ore: SiO_2 38.64%, MgO 22.68%, Al_2O_3 14.94%, Fe_2O_3 9.29%, K_2O 7.84%, CaO 1.23%, Cr_2O_3 0.29%, Mn_3O_4 0.11%, Cl 0.28%. Vermiculites should be classed as montmorillonoids: Rustum Roy, *Chem. Eng. News* **32**, 4842 (1954). Review of mineralogy and health effects: J. Addison, *Regul. Toxicol. Pharmacol.* **21**, 397-405 (1995). *Review:* R. V. Gaines *et al.*, *Dana's New Mineralogy* (Wiley, New York, 8th Ed., 1997) pp 1474-1476.
Monoclinic crystals, pseudo-hexagonal characteristics. Soft, resilient, glabrous feel. d 2.2-2.6. Hardness: ~1.5. Expands on heating to vermiform masses. Thermally expanded forms can absorb large quantities of liquids and remain free flowing. Dissolves in hot concd H_2SO_4. The ore is not sol in weak acids or bases, but expanded grades may be attacked.
Verxite. Highly purified, food grade form of vermiculite. Typical preparations are light brown, soft granules. Water content < 1%. Bulk density: 5-9 lbs/ft^3. Absorbs 2-5 times its weight in liquids.
USE: As catalyst; loose fill insulation, filler and packing material. In horticulture as soil amendment. In animal feeds as anti-caking agent, dispersant, non-nutritive bulking agent; roughage replacement for ruminants.

10157. Vernakalant. [794466-70-9] (3R)-1-[(1R,2R)-2-[2-(3,4-Dimethoxyphenyl)ethoxy]cyclohexyl]-3-pyrrolidinol; (1R,2R)-2-[(3R)-hydroxypyrrolidinyl]-1-(3,4-dimethoxyphenethoxy)cyclohexane. $C_{20}H_{31}NO_4$; mol wt 349.47. C 68.74%, H 8.94%, N 4.01%, O 18.31%. Atrial selective antifibrillary agent. Mixed ion channel antagonist; blocks frequency- and voltage-dependent sodium channels and Kv1.5 potassium channels. Prepn: G. N. Beatch *et al.*, **WO 04099137**; *eidem*, **US 7057053** (2002, 2005 to Cardiome). Comparative binding with inner pore of Kv1.5 channel: J. Eldstrom *et al.*, *Mol. Pharmacol.* **72**, 1522 (2007). Mechanism of action: D. Fedida *et al.*, *J. Cardiovasc. Electrophysiol.* **16**, 1227 (2005). Clinical trial for rapid conversion of atrial fibrillation: D. Roy *et al.*, *Circulation* **117**, 1518 (2008). Review of pharmacology and clinical experience: J. W. M. Cheng, *Ann. Pharmacother.* **42**, 533-542 (2008); S. T. Duggan, L. J. Scott, *Drugs* **71**, 237-252 (2011).

Hydrochloride. [748810-28-8] RSD-1235; Brinavess; Kynapid. $C_{20}H_{31}NO_4$.HCl; mol wt 385.93. Crystals from ethyl acetate, mp 144-150°. $[\alpha]_D$ −46.7° (c = 1.52 in CH_3OH); $[\alpha]_D$ −39.6° (c = 1.00 in $CHCl_3$).
THERAP CAT: Antiarrhythmic.

10158. Vernolepin. [18542-37-5] (3aR,4S,5aR,9aR,9bR)-5a-Ethenyloctahydro-4-hydroxy-3,9-bis(methylene)-2H-furo[2,3-f][2]-benzopyran-2,8(3H)-dione. $C_{15}H_{16}O_5$; mol wt 276.29. C 65.21%, H 5.84%, O 28.95%. The first recognized naturally occurring elemanolide dilactone; a sesquiterpene lactone with tumor inhibiting properties. Isoln from *Vernonia hymenolepis* A. Rich, *Compositae* and structure: S. M. Kupchan *et al.*, *J. Am. Chem. Soc.* **90**, 3596 (1968); structural elucidation: *eidem*, *J. Org. Chem.* **34**, 3903 (1969). Biological properties: S. M. Kupchan *et al.*, *J. Med. Chem.* **14**, 1147 (1971). Approach to stereosynthesis of (+)-form: H. Iio *et al.*, *Tetrahedron* **35**, 941 (1979); K. Kondo *et al. Tetrahedron Lett.* **34**, 4219 (1993). Total synthesis of (±)-form: P. A. Grieco *et al.*, *J. Am. Chem. Soc.* **98**, 1612 (1976); F. Zutterman *et al.*, *Tetrahedron* **35**, 2389 (1979).

Crystals from chloroform-petr ether, mp 181-182°. $[\alpha]_D^{28}$ +72° (c = 1.04 in acetone). uv max (methanol): 208 nm (ε 20300).
(±)-**Form.** [59598-29-7] Crystals from chloroform, mp 210-211°.

10159. Vernolic Acid. 11-(3-Pentyloxiranyl)-9-undecenoic acid; 12,13-epoxyoleic acid; cis-12,13-epoxyoctadec-cis-9-enoic acid; octadec-cis-12,13-epoxy-cis-9-enoic acid. $C_{18}H_{32}O_3$; mol wt 296.45. C 72.93%, H 10.88%, O 16.19%. Naturally occurring as the (+) and (−) forms. Principal seed-oil acid from *Vernonia anthelmintica* Willd. (*Serratula anthelmintica* Roxb.), *Conyza anthelmintica* Linn.) *Compositae*. Isoln and structure: Gunstone, *J. Chem. Soc.* **1954**, 1611; from *Plectranthus spp.*: C. M. J. Daulatabad, A. M. Mirajkur, *J. Chem. Technol. Biotechnol.* **45**, 143 (1989). Synthesis and absorption spectrum: Osbond, *ibid.* **1961**, 5270; *idem*, **GB 909354** (1962 to Roche). Enantiospecific synthesis of (−)-form: C. A. Moustakis *et al.*, *Tetrahedron Lett.* **27**, 303 (1986).

(+)-form

(−)-**Form.** [32381-42-3] [2R-[2α(Z),3α]]-11-(3-Pentyloxiranyl)-9-undecenoic acid; (12R,13S)-(−)-12,13-epoxyoleic acid. Crystals from acetone at −25°, mp 30-31°. Liquid at room temp. $[\alpha]_D$ −8°.
(+)-**Form.** [503-07-1] [2S-[2α(Z),3α]]-11-(3-Pentyloxiranyl)-9-undecenoic acid; (12S,13R)-(+)-12,13-epoxyoleic acid.

10160. Verrucarins. Macrocyclic tricothecane derivs which are secondary metabolites of the soil fungi *Myrothecium verrucaria* (Albertini et Schweinitz) Ditmar ex Fries; they are characterized by antibiotic, antifungal, and cytostatic activity. Verrucarins are triesters of the sesquiterpene alcohol **verrucarol** and closely related to the **roridins**, which are diesters of the same alcohol. Isoln of verrucarins A (major), B, C, D, E, F, G: E. Haerri *et al.*, *Helv. Chim. Acta* **45**, 840 (1962). Isoln of A: Symth, Kraskin, **US 3087859** (1963 to Rohm & Haas); Vittimberga, *J. Org. Chem.* **28**, 1786 (1963). Identity with muconomycin A: Vittimberga, Vittimberga, *ibid.* **30**, 746 (1965). Structure of A: J. Gutzwiller, C. Tamm, *Helv. Chim. Acta* **48**, 157 (1965). Stereochemistry: W. Zürcher *et al.*, *ibid.* 840; A. T. McPhail, G. T. Sim, *J. Chem. Soc.* **1966**, 1394. Synthesis of A: W. C. Still, H. Ohmizu, *J. Org. Chem.* **46**, 5242 (1981). Isoln of B: W. Loeffler *et al.*, **BE 627002** (1963 to Sandoz), *C.A.* **60**, 10485b (1964). Structure of B: J. Gutzwiller, C. Tamm, *Helv. Chim. Acta* **48**, 177 (1965). Toxicological studies of A and B: Guarino *et al.*,

Biotechnol. Bioeng. **10**, 457 (1968); of A: Mortimer *et al.*, *Res. Vet. Sci.* **12**, 508 (1971). Biosynthetic studies of A and B: B. Müller, C. Tamm, *Helv. Chim. Acta* **58**, 483 (1975); G. A. Cordell, *Chem. Rev.* **76**, 425 (1976). Isoln of J (muconomycin B): Vittimberga, Vittimberga, *loc. cit.*; of H and J: B. Boehner *et al.*, *Helv. Chim. Acta* **48**, 1079 (1965). Structure of J: E. Fetz *et al.*, *ibid.* 1669. Partial synthesis of tetrahydroverrucarin J and revised structure of J: W. Breitenstein, C. Tamm, *ibid.* **61**, 1975 (1978). Isoln and structure of K: *eidem*, *ibid.* **60**, 1522 (1977). Isoln of L: B. B. Jarvis *et al.*, *J. Antibiot.* **34**, 121 (1981). Total synthesis of verrucarol: R. H. Schlessinger, R. A. Nugent, *J. Am. Chem. Soc.* **104**, 1116 (1982). Toxicity study: E. M. Rüsch, H. Stählin, *Arzneim.-Forsch.* **15**, 893 (1965). Comprehensive review of the verrucarins and roridins: C. Tamm, *Fortschr. Chem. Org. Naturst.* **31**, 64-117 (1974). Description of the stereochemistry of the roridins: B. R. Jarvis *et al.*, *J. Nat. Prod.* **45**, 440 (1982).

Verrucarin A

Verrucarin A. [3148-09-2] Muconomycin A. $C_{27}H_{34}O_9$; mol wt 502.56. Colorless rectangular plates from ether/acetone, mp >360° (dec). $[\alpha]_D^{23}$ +206° (c = 1.012 in chloroform); uv max (ethanol): 260 nm (log ε 4.25). LD_{50} in mice, rats, rabbits (mg/kg): 1.5, 0.87, 0.54 i.v. (Rüsch, Stählin).

Verrucarin B. [2290-11-1] (2'S,3'R)-2'-Deoxy-2',3'-epoxyverrucarin A. $C_{27}H_{32}O_9$; mol wt 500.54. Colorless needles from acetone/ether, decomp >330°. $[\alpha]_D^{23}$ +147° (c = 1.066 in benzene). $[\alpha]_D^{22}$ +101 ±1.5° (c = 1.416 in dioxane). uv max (methanol): 258.5 nm (log ε 4.37).

Verrucarin J. [4643-58-7] (2'E)-2',3'-Didehydro-2'-deoxyverrucarin A; muconomycin B. $C_{27}H_{32}O_8$; mol wt 484.55. Crystals from ether, dec >235°. $[\alpha]_D^{19}$ +54° (benzene). $[\alpha]_D^{20}$ +20 ± 2° (c = 1.011 in chloroform). uv max: 261, 220.5 nm (ε 22000, 21600).

Verrucarin K. [63739-93-5] 12,13-Deepoxy-12,13-didehydroverrucarin A. $C_{27}H_{34}O_8$; mol wt 486.56. Crystals from methylene chloride/ether, dec >320°. $[\alpha]_D^{23}$ +218 ±2° (c = 0.58 in chloroform). uv max (ethanol): 259 nm (log ε 4.19).

Caution: Verrucarins are extremely toxic; can cause severe local irritation and inflammation of the skin.

10161. Versalide®. [88-29-9] 1-(3-Ethyl-5,6,7,8-tetrahydro-5,5,8,8-tetramethyl-2-naphthalenyl)ethanone; 3'-ethyl-5',6',7',8'-tetrahydro-5',5',8',8'-tetramethyl-2'-acetonaphthone. $C_{18}H_{26}O$; mol wt 258.41. C 83.66%, H 10.14%, O 6.19%. Prepn: Davidson, Lusskin, US 3045047 (1962 to Trubek Lab.). GC determn in fragrances: H. H. Wisneski *et al.*, *J. Assoc. Off. Anal. Chem.* **65**, 598 (1982). Toxicity: K. R. Butterworth, P. L. Mason, *Food Cosmet. Toxicol.* **19**, 753 (1981).

Crystals, mp 46.5°. bp_2 130°. Soluble in alc.
USE: As musk for perfumes, cosmetics, soaps.

10162. Versenol®. [139-89-9] N-[2-[Bis(carboxymethyl)amino]ethyl]-N-(2-hydroxyethyl)glycine sodium salt (1:3); N-(carboxy-

methyl)-N'-(2-hydroxyethyl)-N,N'-ethylenediglycine trisodium salt; N-hydroxyethylethylenediaminetriacetic acid trisodium salt; trisodium N-hydroxyethylethylenediaminetriacetate; HEDTA. $C_{10}H_{15}N_2Na_3O_7$; mol wt 344.21. C 34.89%, H 4.39%, N 8.14%, Na 20.04%, O 32.54%. Prepn: Young, US 2811550 (1957 to Refined Prod.); Kroll, Dexter, US 2845457 (1958 to Geigy).

USE: As chelating agent for trivalent iron from pH 7 to pH 10. Forms strong 1:1 ferric chelates. The iron chelates formed will then chelate other heavy metal or alkaline earth ions.

10163. Verteporfin. [129497-78-5] rel-(4R,4aS)-18-Ethenyl-4,4a-dihydro-3,4-bis(methoxycarbonyl)-4a,8,14,19-tetramethyl-$24H_D$,26H-benzo[b]porphine-9,13-dipropanoic acid monomethyl ester; BPD-MA; CL-318952; Visudyne. $C_{41}H_{42}N_4O_8$; mol wt 718.81. C 68.51%, H 5.89%, N 7.79%, O 17.81%. Benzoporphyrin photosensitizer. Prepd as a mixture of the monomethyl esters. Preparative method: A. M. Richter *et al.*, *J. Natl. Cancer Inst.* **79**, 1327 (1987). Photosensitizing efficiency: A. M. Richter *et al.*, *Biochem. Pharmacol.* **43**, 2349 (1992). Spectral characterization: R. Gillies *et al.*, *J. Photochem. Photobiol. B* **33**, 87 (1996). Tissue uptake parameters: R. K. Chowdhary *et al.*, *Biopharm. Drug Dispos.* **19**, 395 (1998). Clinical trial in macular degeneration: TAP Study Group, *Arch. Ophthalmol.* **117**, 1329 (1999). Review of pharmacology and clinical experience: K. J. Mellish, S. B. Brown, *Expert Opin. Pharmacother.* **2**, 351-361 (2001); S. J. Bakri, P. K. Kaiser, *ibid.* **5**, 195-203 (2004).

Relative stereochemistry

Ester A   R = H R^1= CH_3
Ester B   R = CH_3 R^1= H

Virtually insol in water. Absorption max: ~690 nm.
THERAP CAT: In treatment of age-related macular degeneration.

10164. Verticillins. Three antibiotics, verticillins A, B and C, produced by a species of *Verticillium* (strain TM-759), an imperfect fungus isolated from a basidiocarp of *Coltricia cinnamomea* (*Polystictus cinnamomeus*). Isoln, IR, NMR spectra of verticillin A: Katagiri *et al.*, *J. Antibiot.* **23**, 420 (1970); *see also* Chepenko *et al.*, *C.A.* **78**, 156654n (1973). Structure: Minato *et al.*, *Chem. Commun.* **1971**, 44. Structure of verticillins A, B and C and absolute configuration of verticillins A and B: *eidem*, *J. Chem. Soc. Perkin Trans. 1* **1973**, 1819. Synthetic studies: Häusler, Schmidt, *Ber.* **107**, 2804 (1974); Schmidt *et al.*, *ibid.* 2816.

Verticillin A

Verticillin A. [32164-16-2] $C_{30}H_{28}N_6O_6S_4$; mol wt 696.83. Pale yellow amorphous powder from tetrahydrofuran, mp 203-214° (dec). $[\alpha]_D$ +703.7° (c = 0.422 in dioxane). uv max (dioxane): 306 nm (ε 5960). LD_{50} i.p. in mice: 7.6 mg/kg (Katagiri).

Verticillin B. [52212-86-9] $C_{30}H_{28}N_6O_7S_4$; mol wt 712.83. The mono-3-hydroxymethyl analog of verticillin A. Pale yellow prisms from chloroform, mp 230-233° (dec). $[\alpha]_D^{21}$ +704.7° (c = 0.493 in dioxane). uv max (dioxane): 306 nm (ε 5600).

Verticillin C. $C_{30}H_{28}N_6O_7S_5$; mol wt 744.89. Differs from verticillin B by having a trisulfide rather than disulfide bridge in one of the two dioxopiperazine rings. Pale yellow amorphous powder from methanol-water, mp 230-235° (dec). $[\alpha]_D^{21}$ +765.0° (c = 0.506 in dioxane). uv max (dioxane): 303 nm (ε 5500).

10165.   Verticine. [23496-41-5] $(3\beta,5\alpha,6\alpha)$-Cevane-3,6,20-triol; peimine. $C_{27}H_{45}NO_3$; mol wt 431.66. C 75.13%, H 10.51%, N 3.24%, O 11.12%. A member of the hexacyclic *Ceveratrum* group of alkaloids. Isoln from *Fritillaria verticillata* Willd. var. *Thunbergii* Baker, *Liliaceae:* Fukuda: *Sci. Rep. Res. Inst. Tohoku Univ. Ser. A* **18**, 323 (1929); Morimoto, Kimata, *Chem. Pharm. Bull.* **8**, 302 (1960). Also isolated from *F. roylei* Hook: Chou, Chen, *Chin. J. Physiol.* **6**, 265 (1932), *C.A.* **26**, 5703 (1932). Structure: Ito *et al., ibid.* **11**, 1337 (1963). Abs config: S. Ito *et al., Tetrahedron Lett.* **1968**, 5373. Total synthesis: J. P. Kutney *et al., J. Am. Chem. Soc.* **99**, 964 (1977).

Needles from ethanol, mp 223-224°. $[\alpha]_D^{16}$ −19.4° (ethanol); $[\alpha]_D^{17}$ −20° (CHCl₃). pK'a 9.5. uv max (ethanol + HCl): 215 nm (ε 10).

Hydrochloride. $C_{27}H_{45}NO_3 \cdot HCl$; mol wt 468.12. Prisms from water, dec 291-294°. $[\alpha]_D^{18}$ −18.5°.

Perchlorate. $C_{27}H_{45}NO_3 \cdot HClO_4$; mol wt 532.12. Prisms from water, dec 273°. $[\alpha]_D^{23}$ −15° (c = 0.05).

Methiodide. $C_{27}H_{45}NO_3 \cdot CH_3I$; mol wt 573.60. Crystals from butanol + ether, dec 205-210°. $[\alpha]_D^{18}$ −14.0° (water). Sol in water, methanol, ethanol; practically insol in chloroform, benzene, acetone.

3-Glucoside. Peiminoside. $C_{33}H_{55}NO_8$; mol wt 593.80. Pale brown powder. Sol in water, methanol, alc; sparingly sol in benzene.

10166.   Vesnarinone. [81840-15-5] 6-[4-(3,4-Dimethoxybenzoyl)-1-piperazinyl]-3,4-dihydro-2(1*H*)-quinolinone; 1-(3,4-dimethoxybenzoyl)-4-(1,2,3,4-tetrahydro-2-oxo-6-quinolinyl)piperazine; 3,4-dihydro-6-[4-(3,4-dimethoxybenzoyl)-1-piperazinyl]-2(1*H*)-quinolinone; 6-[4-(3,4-dimethoxybenzoyl)-1-piperazinyl]-3,4-dihydrocarbostyril; 1-(1,2,3,4-tetrahydro-2-oxo-6-quinolyl)-4-veratroylpiperazine; piteranometozine; OPC-8212; Arkin. $C_{22}H_{25}N_3O_4$; mol wt 395.46. C 66.82%, H 6.37%, N 10.63%, O 16.18%. Positive inotropic agent. Prepn: Y. H. Yang *et al.*, **BE 890942**; M. Tominaga *et al.*, **US 4415572** (1982, 1983 both to Otsuka); M. Tominaga *et al., Chem. Pharm. Bull.* **32**, 2100 (1984). Physicochemical properties: T. Shimizu *et al., Arzneim.-Forsch.* **34**, 334 (1984). Series of articles on mechanism of action and pharmacology: *ibid.* 342-402. Effects on canine arrhythmias: K. Hashimoto, H. Mitsuhashi, *Br. J. Pharmacol.* **88**, 915 (1986). Metabolism: G. Miyamoto *et al., Xenobiotica* **18**, 1143 (1988). Clinical pharmacokinetics: A. Ohnishi, T. Ishizaki, *J. Clin. Pharmacol.* **28**, 719 (1988). HPLC determn in plasma and urine: G. Miyamoto *et al., J. Chromatogr.* **338**, 450 (1985). Clinical evaluations in congestive heart failure: H. Asanoi *et al., J. Am. Coll. Cardiol.* **9**, 865 (1987); A. M. Feldman *et al., Am. Heart J.* **116**, 771 (1988).

Colorless granules from ethanol-chloroform, mp 238.1-239.5° (Tominaga); also reported as odorless, tasteless, pale yellow crystalline powder, mp 238.1-239.8° (Shimizu). uv max (methanol, ethanol, chloroform): 271 nm (ε = 2.51 × 10⁴, 2.52 × 10⁴, 2.30 × 10⁴, c = 10 μg/ml). % solubility (25°): glacial acetic acid 18.68; chloroform 14.19; benzyl alcohol 5.913; *N*-methyl-2-pyrrolidone 3.407; DMSO 2.509; 60% sulfolane 1.229; DMF 1.179; dioxane 0.1653; methanol 0.1151; acetone 0.06389; ethanol 0.04005; water 0.002086. pKa 2.86.

THERAP CAT: Cardiotonic.

10167.   Vetivones. $C_{15}H_{22}O$; mol wt 218.34. C 82.52%, H 10.16%, O 7.33%. Ketonic components of vetiver oil; from roots of *Vetiveria zizanioides (Andropogon muricatus* Retz., *Gramineae).* Two isomers, the α-form having the stronger odor. Isoln and structure: Pfau, Plattner, *Helv. Chim. Acta* **22**, 640 (1939); Naves, Perrottet, *ibid.* **24**, 3 (1941).

α-Vetivone   β-Vetivone

α-**Vetivone.** [15764-04-2] (4*R-cis*)-4,4a,5,6,7,8-Hexahydro-4,4a-dimethyl-6-(1-methylethylidene)-2(3*H*)-naphthalenone; 4β*H*,-5α-eremophila-1(10),7(11)-dien-2-one; isonootkatone. Structure: Endo, DeMayo, *Chem. Commun.* **1967**, 89; Marshall, Andersen, *Tetrahedron Lett.* **1967**, 1611. Total syntheses of the racemate: Marshall *et al., Chem. Commun.* **1967**, 753; Marshall, Warne, *J. Org. Chem.* **36**, 178 (1971); Vandergen *et al., Rec. Trav. Chim.* **90**, 1034, 1045 (1971); Dastur, *J. Am. Chem. Soc.* **96**, 2605 (1974). Crystals from pentane, strong agreeable odor. mp 51.5° (Naves, Perrottet); mp of racemate 30-35° (Marshall, Andersen). n_D^{20} 1.5384. $[\alpha]_D^{20}$ +248° (chloroform). d_4^{20} 1.003 (liquid, supercooled). uv max (ethanol): 233 nm (ε 13200).

β-**Vetivone.** [18444-79-6] (5*R-cis*)-6,10-Dimethyl-2-(1-methylethylidene)spiro[4.5]dec-6-en-8-one; 2-isopropylidene-6,10-dimethylspiro[4.5]dec-6-en-8-one. Structure: Marshall *et al.*, *J. Am. Chem. Soc.* **89**, 2748, 2750 (1967). Syntheses of racemate: Marshall, Johnson, *Chem. Commun.* **1968**, 391; *eidem, J. Org. Chem.* **35**, 192 (1970); K. Uneyama *et al.*, *Chem. Lett.* **1977**, 493; S. Torii *et al.*, *Bull. Chem. Soc. Jpn.* **51**, 3590 (1978); E. Wenkert *et al.*, *J. Am. Chem. Soc.* **100**, 1267 (1978). Stereospecific synthesis of (−)-β-vetivone: Deighton *et al.*, *Chem. Commun.* **1975**, 662; M. Asaoka *et al.*, *Chem. Lett.* **1988**, 1225. Crystals from pentane, mp 44-46°. d_4^{20} 1.000 (liquid, supercooled). d_4^{45} 0.9804. $[\alpha]_D^{20}$ −38.9° (c = 10 in alcohol). n_D^{20} 1.5309 (liquid, supercooled), n_D^{45} 1.5216.

USE: Of potential interest in the perfume industry.

10168. Vetrabutine. [3735-45-3] α-(3,4-Dimethoxyphenyl)-*N*,*N*-dimethylbenzenebutanamine; *N*,*N*-dimethyl-α-(3-phenylpropyl)veratrylamine; 1-(3,4-dimethoxyphenyl)-1-dimethylamino-4-phenylbutane; 3,4-dimethoxy-*N*,*N*-dimethyl-α-(3-phenylpropyl)-benzylamine; dimophebumine. $C_{20}H_{27}NO_2$; mol wt 313.44. C 76.64%, H 8.68%, N 4.47%, O 10.21%. Prepn: **GB 802723** (1958 to Thomae).

bp$_{0.1}$ 166-168°.

Hydrochloride. [5974-09-4] SP-281; Monzal; Monzaldon. C_{20}-$H_{27}NO_2 \cdot HCl$; mol wt 349.90. Solid, mp 146-148°.

THERAP CAT (VET): Uterine relaxant.

10169. Viburnum opulus. European cranberrybush; cramp bark; high bush cranberry; cranberry tree; Guelder rose; snowball bush. Deciduous flowering shrub with bitter, edible red berries, *Viburnum opulus* L., *Caprifoliaceae*. The berries have been used in traditional medicine to treat respiratory infections, as a digestive aid, and as a source of antioxidants; the bark has been used as an antispasmodic. *Habit.* Europe, Northern Asia, naturalized in northern North America as *V. opulus* var. *americanum* Ait. *Constit.* Salicin, amentoflavone, chlorogenic acid; cyanidin glycosides, catechin, epicatechin, malic, oxalic, and ascorbic acids. Analysis of constituents: M. L. Altun, B. S. Yilmaz, *Chem. Nat. Compd.* **43**, 205 (2007); M. Cam *et al., ibid.* 460; L. Cesoniene *et al.*, *Cent. Eur. J. Biol.* **5**, 864 (2010). Antioxidant activity: O. Rop *et al.*, *Molecules* **15**, 4467 (2010); and anticholinesterase properties: I. Erdogan-Orhan *et al.*, *J. Med. Food* **14**, 434 (2011). Characterization of aroma profile: V. Kraujalyte *et al.*, *Food Chem.* **132**, 717 (2012).

USE: In landscaping as ornamental shrub. Berries for juice, sauces, jams and jellies.

10170. Viburnum prunifolium. Black haw; sweet haw; stag bush; sloe-leaved viburnum. Deciduous, flowering shrub with shiny, black berries, *Viburnum prunifolium* L., *Caprifoliaceae*; used in traditional medicine to treat dysmenorrhea and to prevent miscarriage. Medicinal parts are the bark and root. *Habit.* Eastern U.S. *Constit.* Scopoletin, aesculetin, scoplin, salicylic acid, salicin, chlorogenic acid, iridoid glycosides. Uterine relaxant properties: C. H. Jarboe *et al.*, *Nature* **212**, 837 (1966). Isoln of scopeletin, the principle antispasmodic component: *idem et al.*, *J. Med. Chem.* **10**, 488 (1967). Evaluation of spasmolytic activity of root extracts and isoln of iridoids: M. F. Cometa *et al.*, *J. Ethnopharmacol.* **123**, 201 (2009). Description, constituents, and medicinal uses: J. Gruenwald, *PDR for Herbal Medicines* (Thomson PDR, Montvale, 3rd Ed., 2004) pp 99-100.

10171. Vicine. [152-93-2] 2,6-Diamino-5-(β-D-glucopyranosyloxy)-4(1*H*)-pyrimidinone; 2,4-diamino-6-oxypyrimidine-5-(β-D-glucopyranoside); divicine 5-glucoside; vicioside; divicine-β-glucoside. $C_{10}H_{16}N_4O_7$; mol wt 304.26. C 39.48%, H 5.30%, N 18.41%, O 36.81%. Isoln from vetch seeds (*Vicia sativa* L., *Leguminosae*): Ritthausen, *Ber.* **9**, 301 (1876); Levene, *J. Biol. Chem.* **18**, 306 (1914); Levene, Senior, *ibid.* **25**, 611 (1916); Herissey, Cheymol, *Bull. Soc. Chim. Biol.* **13**, 29 (1931); Gmelin, Hasenmaier, *Arzneim.-Forsch.* **7**, 755 (1957). Structure: Bendich, *Trans. N.Y. Acad. Sci.*

15, 58 (1952); Bendich, Clements, *Biochim. Biophys. Acta* **12**, 462 (1953).

Needles from water, dec 243-244°. $[\alpha]_D^{26}$ −11.7° (c = 3.9 in 0.2*N* NaOH). uv max (0.1*N* HCl): 274 nm (ε 16400); (0.1*N* NaOH): 269, 248, 235, 230 nm (ε 9520, 4360, 5180, 5090). One gram dissolves in about 100 ml water; slightly sol in alcohol; readily sol in dil acids and alkalies.

Note: Another *vicin*, a rhamnosidoglucoside from vetch flowers, is described by Karrer, Widmer, *Helv. Chim. Acta* **10**, 67 (1927). It has the formula: $C_{21}H_{21}O_{12} \cdot 2H_2O$.

10172. Vicriviroc. [306296-47-9] (4,6-Dimethyl-5-pyrimidinyl)[4-[(3*S*)-4-[(1*R*)-2-methoxy-1-[4-(trifluoromethyl)phenyl]ethyl]-3-methyl-1-piperazinyl]-4-methyl-1-piperidinyl]methanone; 1-[(4,6-dimethyl-5-pyrimidinyl)carbonyl]-4-[(3*S*)-4-[(1*R*)-2-methoxy-1-[4-(trifluoromethyl)phenyl]ethyl]-3-methyl-1-piperazinyl]-4-methylpiperidine; SCH-D. $C_{28}H_{38}F_3N_5O_2$; mol wt 533.64. C 63.02%, H 7.18%, F 10.68%, N 13.12%, O 6.00%. CCR5 chemokine receptor antagonist; HIV viral entry inhibitor. Prepn: B. M. Baroudy *et al.*, **WO 0066558**; *eidem*, **US 6391865** (2000, 2002 both to Schering Corp.); J. R. Tagat *et al.*, *J. Med. Chem.* **47**, 2405 (2004); D.-Z. Feng *et al.*, *Org. Biomol. Chem.* **5**, 2690 (2007). Pharmacology: J. M. Strizki *et al.*, *Antimicrob. Agents Chemother.* **49**, 4911 (2005). Pharmacokinetics in HIV-infected adults: D. Schürmann *et al.*, *AIDS* **21**, 1293 (2007). Clinical evaluation in HIV infection: R. M. Gulick *et al.*, *J. Infect. Dis.* **196**, 304 (2007); in treatment-naive patients: R. J. Landovitz *et al.*, *ibid.* **198**, 1113 (2008).

White foam. $[\alpha]_D^{22}$ +6.6° (c = 0.5 in methanol). Sol in ethyl acetate, isopropyl acetate, isopropanol, methanol.

Hydrochloride. [541503-48-4] $C_{28}H_{38}F_3N_5O_2 \cdot HCl$; mol wt 570.10. Off-white solid from ethyl acetate, mp 175°. $[\alpha]_D^{25}$ +13.3° (c = 2.44 in methanol).

Maleate. [599179-03-0] Sch-417690. $C_{28}H_{38}F_3N_5O_2 \cdot C_4H_4O_4$; mol wt 649.71. White crystals.

THERAP CAT: Antiretroviral.

10173. Vidarabine. [5536-17-4]; [24356-66-9] (monohydrate). 9-β-D-Arabinofuranosyl-9*H*-purine-6-amine; 9-β-D-arabinofuranosyladenine; arabinosyladenine; adenine arabinoside; spongoadenosine; ara-A; CI-673; Arasena-A; Vira-A. $C_{10}H_{13}N_5O_4$; mol wt 267.25. C 44.94%, H 4.90%, N 26.21%, O 23.95%. Purine nucleoside first synthesized as a potential anticancer agent: Lee *et al.*, *J. Am. Chem. Soc.* **82**, 2648 (1960); Reist *et al.*, *J. Org. Chem.* **27**, 3274 (1962); Glaudemans, Fletcher, *ibid.* **28**, 3004 (1963); Reist *et al.*, *ibid.* **29**, 3725 (1964). Fermentation process using a strain of *Streptomyces antibioticus*: **GB 1159290**; J. D. Howells, A. Ryder, **US 3616208** (1969, 1971 both to Parke, Davis). Crystal and molecular structure: Bunick, Voet, *Acta Crystallogr.* **30B**, 1641 (1974). Series of articles on antiviral activity: *Antimicrob. Agents Chemother.* **1968**, 136-179. Toxicity study: S. M. Kurtz *et al.*, *ibid.* **180**. HPLC determn in plasma and urine: W. P. McCann *et al.*, *ibid.* **28**, 265

(1985). Clinical trial in immunocompromised patients: R. J. Whitley *et al., J. Infect. Dis.* **165**, 450 (1992). Book: *Adenine Arabinoside: An Antiviral Agent,* D. Paven-Langston *et al.,* Eds. (Raven Press, New York, 1975) xviii + 425 pp. Review of pharmacology and clinical experience: R. A. Buchanan, F. Hess, *Pharmacol. Ther.* **8**, 143-171 (1980). Comprehensive description: W. Hong *et al., Anal. Profiles Drug Subs.* **15**, 647-672 (1986).

Crystals from water, mp 257.0-257.5° (0.4 H$_2$O). $[\alpha]_D^{27}$ $-5°$ (c = 0.25). uv max (pH 1): 257.5 nm (ε 12700); pH 7: 259 nm (ε 13400); pH 13: 259 nm (ε 14000). Slightly sol in DMF; very slightly sol in water. LD$_{50}$ in mice (mg/kg): 4677 i.p.; >7950 orally (Kurtz).

THERAP CAT: Antiviral.

10174. Viehe's Salt. [33842-02-3] Dichloromethylmethanaminium chloride (1:1); *N*-(dichloromethylene)-*N*-methylmethanaminium chloride; *N*-(dichloromethylene)-*N*,*N*-dimethyliminium chloride; *N*,*N*-dimethylphosgeniminium chloride; phosgeniminium chloride. C$_3$H$_6$Cl$_3$N; mol wt 162.44. C 22.18%, H 3.72%, Cl 65.47%, N 8.62%. Electrophilic chlorinating reagent. Prepn: H. G. Viehe, Z. Janousek, *Angew. Chem.* **83**, 614 (1971); and chemistry: *eidem, Angew. Chem. Int. Ed.* **12**, 806 (1973). ^{35}Cl-NMR study: G. Jugie *et al., J. Chem. Soc. Perkin Trans. 2* **1975**, 925. Review of chemistry: Z. Janousek, H. G. Viehe, *J. Prakt. Chem. Chem.-Ztg.* **336**, 561-562 (1994); of applications in synthesis of heterocyclic compounds: J. M. Quintela *et al., Recent Res. Dev. Org. Chem.* **2**, 409-418 (1998).

Colorless, hygroscopic solid, fp ~180°.

USE: Chlorinating agent; introduces amide chloride groups into activated substrates. Building block in organic synthesis.

10175. Vigabatrin. [68506-86-5] 4-Amino-5-hexenoic acid; γ-vinyl-γ-aminobutyric acid; gamma-vinyl GABA; γ-vinyl GABA; GVG; MDL-71754; RMI-71754; Sabril. C$_6$H$_{11}$NO$_2$; mol wt 129.16. C 55.80%, H 8.58%, N 10.84%, O 24.77%. Irreversible inhibitor of γ-aminobutyric acid transaminase, the enzyme responsible for the degradation of the neurotransmitter, γ-aminobutyric acid (GABA). Prepn: B. W. Metcalf, M. Jung, US 3960927 (1976 to Richardson-Merrell); and *in vitro* enzyme inactivation: B. Lippert *et al., Eur. J. Biochem.* **74**, 441 (1977). Mechanism of action study: P. J. Schechter *et al., Eur. J. Pharmacol.* **45**, 319 (1977). Anticonvulsant activity and toxicity studies: W. Löscher, *Neuropharmacology* **21**, 803 (1982). HPLC determn in plasma and urine: J. A. Smithers *et al., J. Chromatogr.* **341**, 232 (1985). The S(+)-enantiomer is the pharmacologically active form. Pharmacokinetics of enantiomers in humans: K. D. Haegele, P. J. Schechter, *Clin. Pharmacol. Ther.* **40**, 581 (1986). Clinical studies in treatment resistant epilepsy: C. A. Tassinari *et al., Arch. Neurol.* **44**, 907 (1987); T. R. Browne *et al., Neurology* **37**, 184 (1987). Series of articles on clinical use in adult and childhood epilepsy: *J. Child Neurol.* **6**, Suppl. 2, S3-S69 (1991). Reviews of early literature and mechanism of action: M. J. Iadarola, K. Gale, *Mol. Cell. Biochem.* **39**, 305-330 (1981); of pharmacology and toxicology: E. J. Hammond, B. J. Wilder, *Clin. Neuropharmacol.* **8**, 1-12 (1985). *Review:* S. M. Grant, R. C. Heel, *Drugs* **41**, 889-926 (1991).

Crystals from acetone/water, mp 209°. Freely sol in water. LD$_{50}$ i.p. in mice: >2500 mg/kg (Löscher).

THERAP CAT: Anticonvulsant.

10176. Vilazodone. [163521-12-8] 5-[4-[4-(5-Cyano-1*H*-indol-3-yl)butyl]-1-piperazinyl]-2-benzofurancarboxamide; 1-[4-(5-cyanoindol-3-yl)butyl]-4-(2-carbamoylbenzofuran-5-yl)piperazine. C$_{26}$H$_{27}$N$_5$O$_2$; mol wt 441.54. C 70.73%, H 6.16%, N 15.86%, O 7.25%. Combined selective serotonin reuptake inhibitor (SSRI) and serotonin (5-HT$_{1A}$) receptor partial agonist. Prepn: H. Böttcher *et al.,* DE 4333254; *eidem,* US 5532241 (1995, 1996 both to Merck Patent GmbH); T. Heinrich *et al., J. Med. Chem.* **47**, 4684 (2004). Pharmacology: G. D. Bartoszyk *et al., Eur. J. Pharmacol.* **322**, 147 (1997); Z. A. Hughes *et al., ibid.* **510**, 49 (2005). Clinical trial in major depressive disorder (MDD): K. Rickels *et al., J. Clin. Psychiatry* **70**, 326 (2009). Review of pharmacology and clinical experience in MDD: A. Khan, *Expert Opin. Invest. Drugs* **18**, 1753-1764 (2009).

Solid from ethyl acetate.

Hydrochloride. [163521-08-2] EMD-68843; SB-659746A; Viibryd. C$_{26}$H$_{27}$N$_5$O$_2$·HCl; mol wt 477.99. From hydrochloride-saturated 2-propanol, mp 277-279°.

THERAP CAT: Antidepressant.

10177. Vildagliptin. [274901-16-5] (2*S*)-1-[2-[(3-Hydroxytricyclo[3.3.1.1^{3,7}]dec-1-yl)amino]acetyl]-2-pyrrolidinecarbonitrile; 1-[[(3-hydroxy-1-adamantyl)amino]acetyl]-2-cyano-(*S*)-pyrrolidine; LAF-237; NVP-LAF237; Equa; Galvus. C$_{17}$H$_{25}$N$_3$O$_2$; mol wt 303.41. C 67.30%, H 8.31%, N 13.85%, O 10.55%. Dipeptidyl peptidase-IV (DPP-IV) inhibitor. Prepn: E. Villhauer, WO 0034241; *idem,* US 6166063 (both 2000 to Novartis); and pharmacology: *idem et al., J. Med. Chem.* **46**, 2774 (2003). Clinical trial in type 2 diabetes: J. E. Foley, S. Sreenan, *Horm. Metab. Res.* **41**, 905 (2009); in combination with metformin: M. Goodman *et al., ibid.* **368**. Review of pharmacology and clinical experience: M. Banerjee *et al., Expert Opin. Pharmacother.* **10**, 2745-2757 (2009).

White solid from ethyl acetate, mp 138-140°; from 2-propanol, mp 148-150°. $[\alpha]$ $-78.3°$ (c = 9.73 in methanol).

THERAP CAT: Antidiabetic.

10178. Viloxazine. [46817-91-8] 2-[(2-Ethoxyphenoxy)methyl]morpholine; 2-(2-ethoxyphenoxymethyl)tetrahydro-1,4-oxazine; ICI-58834. C$_{13}$H$_{19}$NO$_3$; mol wt 237.30. C 65.80%, H 8.07%, N 5.90%, O 20.23%. Prepn: K. B. Mallion *et al.,* GB 1138405; *eidem,* US 3714161 (1969, 1973 both to I.C.I.). Manufacturing process: S. A. Lee, GB 1260886; *idem,* US 3712890 (1972, 1973 both to I.C.I.). Pharmacology: Mallion *et al., Nature* **238**, 157 (1972); Bereen, *Lancet* **1**, 379 (1973). Toxicity: R. D. Brosnan *et al., J. Int. Med. Res.* **4**, 83 (1976). Clinical evaluation in narcolepsy: C. Guilleminault *et* al., *Sleep* **9**, 275 (1986). Review of pharmacology and therapeutic efficacy: R. M. Pinder *et al.*, *Drugs* **13**, 401 (1977).

Hydrochloride. [35604-67-2] Vivalan; Vicilan; Vivarint. $C_{13}H_{19}NO_3 \cdot HCl$; mol wt 273.76. mp 185-186°. LD_{50} in mice (mg/kg): 1000 orally, 60 i.v. (Brosnan).

THERAP CAT: Antidepressant.

10179. Vinblastine. [865-21-4] Vincaleukoblastine; VLB. $C_{46}H_{58}N_4O_9$; mol wt 810.99. C 68.13%, H 7.21%, N 6.91%, O 17.75%. Antitumor alkaloid isolated from periwinkle, *Vinca rosea* Linn., *Apocynaceae*; inhibits microtubule assembly. Identification: R. L. Noble *et al.*, *Ann. N.Y. Acad. Sci.* **76**, 882-894 (1958). Isolation and characterization: M. Gorman *et al.*, *J. Am. Chem. Soc.* **81**, 4745, 4754 (1959); C. T. Beer *et al.*, **US 3097137** (1963 to CPD Ltd.). Structure: N. Neuss *et al.*, *J. Am. Chem. Soc.* **86**, 1440 (1964). ^{13}C-NMR spectral analysis: E. Wenkert *et al.*, *Helv. Chim. Acta* **58**, 1560 (1975). Synthesis from catharanthine and vindoline, *q.q.v.*: P. Mangeney *et al.*, *J. Am. Chem. Soc.* **101**, 2243 (1979); J. P. Kutney *et al.*, *Heterocycles* **27**, 1845 (1988). Enantioselective synthesis: M. E. Kuehne *et al.*, *J. Org. Chem.* **56**, 513 (1991). Toxicology: C. Lu, M. Meistrich, *Cancer Res.* **39**, 3575 (1979). Clinical trial in combination with bleomycin, methotrexate in Hodgkin's disease: P. G. Gobbi *et al.*, *J. Clin. Oncol.* **14**, 527 (1996). Review of pharmacology, toxicology and pharmacokinetics: W. P. Brade, *Beitr. Onkol.* **6**, 95-123 (1981). Comprehensive description: F. J. Muhtadi, A. F. A. Afify, *Anal. Profiles Drug Subs. Excip.* **21**, 611-658 (1992).

Solvated needles from methanol, mp 211-216°. $[\alpha]_D^{23}$ −32° (c = 0.88 in methanol). uv max (ethanol): 214, 259 nm (log ε 4.74, 4.22). pKa$_1$ 5.4; pKa$_2$ 7.4. Practically insol in water, petr ether. Sol in alcohols, acetone, ethyl acetate, CHCl$_3$.

Sulfate. [143-67-9] 29060-LE; Exal; Velban; Velbe. $C_{46}H_{58}N_4O_9 \cdot H_2SO_4$; mol wt 909.06. Crystallizes as hydrate, mp 284-285°. $[\alpha]_D^{26}$ −28° (c = 1.01 in methanol). uv max (methanol): 212, 262, 284, 292 nm (log ε 4.75, 4.28, 4.22, 4.18). Very slightly sol in ethanol. Practically insol in ether. One part is sol in 10 parts of water, 50 parts of chloroform. LD_{50} i.v. in mice: 9.5 mg/kg (Lu, Meistrich).

THERAP CAT: Antineoplastic.

THERAP CAT (VET): Antineoplastic.

10180. Vincamine. [1617-90-9] (3α,14β,16α)-14,15-Dihydro-14-hydroxyeburnamenine-14-carboxylic acid methyl ester; 13a-ethyl-2,3,5,6,12,13,13a,13b-octahydro-12-hydroxy-1*H*-indolo[3,2,1-*de*]pyrido[3,2,1-*ij*][1,5]naphthyridine-12-carboxylic acid methyl ester; Angiopac; Arteriovinca; Cetal; Oxygeron; Pervincamine; Vincafor; Vincagil; Vincimax; Vraap. $C_{21}H_{26}N_2O_3$; mol wt 354.45. C 71.16%, H 7.39%, N 7.90% O 13.54%. Major indole alkaloid of *Vinca minor* L., *Apocynaceae*, occurring naturally in the *d*-form: E. Schlittler, A. Furlenmeier, *Helv. Chim. Acta* **36**, 2017 (1953); Pailer,

Belohlav, *Monatsh. Chem.* **85**, 1055 (1954); King *et al.*, *J. Chem. Soc.* **1955**, 4206; J. Trojanek *et al.*, *Collect. Czech. Chem. Commun.* **26**, 867 (1961). Isoln from *Tabernaemontana rigida* Miers, *Apocynaceae*: M. P. Cava *et al.*, *J. Org. Chem.* **33**, 1055 (1968). Structure: J. Trojanek *et al.*, *Tetrahedron Lett.* **1961**, 702; J. Mokry *et al.*, *ibid.* **1962**, 433; O. Clauder, *ibid.* 1147; Plat *et al.*, *Bull. Soc. Chim. Fr.* **1962**, 1082. Abs config: J. Trojanek *et al.*, *Chem. Ind. (London)* **1965**, 1261; Blaha *et al.*, *Collect. Czech. Chem. Commun.* **33**, 3833 (1968). X-ray determn of molecular structure: H. P. Weber, T. J. Petcher, *J. Chem. Soc. Perkin Trans. 2* **1973**, 2001. Total synthesis: C. Szantay *et al.*, *Tetrahedron Lett.* **1973**, 191; *eidem*, *Tetrahedron* **33**, 1803 (1977); P. Pfäffli *et al.*, *Helv. Chim. Acta* **58**, 1131 (1975); W. Oppolzer *et al.*, *ibid.* **60**, 1801 (1977); of *dl*-form: M. E. Kuehne, *J. Am. Chem. Soc.* **86**, 2946 (1964); J. L. Herrmann *et al.*, *ibid.* **96**, 3702 (1974); K. H. Gibson, J. E. Saxton, *Tetrahedron* **33**, 833 (1977); K. Irie, Y. Ban, *Heterocycles* **18**, 255 (1982); T. R. Govindachari, S. Rajeswari, *Indian J. Chem.* **22B**, 531 (1983). Synthesis of stereoisomers: J. Warnant *et al.*, **DE 2115718**; *eidem*, **US 3770724** (1971, 1973 both to Roussel-UCLAF); G. Rossey, A. Wick, *J. Org. Chem.* **47**, 4745 (1982). HPLC determn: A. Amato *et al.*, *J. Chromatogr.* **270**, 387 (1983). Toxicity data: L. Szporny, K. Szász, *Arch. Exp. Pathol. Pharmakol.* **236**, 296 (1959). Hypotensive activity in dogs and man: Z. Szabo, Z. Nagy, *Arzneim.-Forsch.* **10**, 811 (1960). Clinical pharmacology: C. C. Lim *et al.*, *Br. J. Clin. Pharmacol.* **9**, 100 (1980). Clinical pharmacokinetics: H. Millart, D. Lamiable *et al.*, *Int. J. Clin. Pharmacol. Ther. Toxicol.* **21**, 581 (1983). Brief review: P. Cook, I. James, *N. Engl. J. Med.* **305**, 1562 (1981).

Yellow crystals from acetone or methanol, mp 232-233°. $[\alpha]_D^{23}$ +41° (in pyridine). uv max: 225, 278 nm (log ε 4.14, 3.61). LD_{50} in mice (mg/kg): 75 i.v.; >1000 s.c. (Szporny, Szász); 1000 orally (Szabo, Nagy).

Hydrochloride. [10592-03-7] Esberidin. $C_{21}H_{26}N_2O_3 \cdot HCl$; mol wt 390.91.

(±)-Form. Crystals from methanol, mp 228-229° (Gibson, Saxton).

Note: Other alkaloids found in the "vincamine fraction" of *Vinca minor* L. are: *vincine*, *vincaminine* and *vincinine*: Holubek *et al.*, *Tetrahedron Lett.* **1963**, 897.

THERAP CAT: Vasodilator.

10181. Vinclozolin. [50471-44-8] 3-(3,5-Dichlorophenyl)-5-ethenyl-5-methyl-2,4-oxazolidinedione; 3-(3,5-dichlorophenyl)-5-methyl-5-vinyloxazolidine-2,4-dione; BAS-352F; Ronilan; Vorlan. $C_{12}H_9Cl_2NO_3$; mol wt 286.11. C 50.38%, H 3.17%, Cl 24.78%, N 4.90%, O 16.78%. Prepn: D. Mangold *et al.*, **DE 2207576** (1973 to BASF), *C.A.* **79**, 137120q (1973). Activity: C. Hess, F. Locher, *Proc. 8th Br. Insectic. Fungic. Conf.* **2**, 693 (1975). Comparative mechanism of action: A. C. Pappas, D. J. Fisher, *Pestic. Sci.* **10**, 239 (1979).

Crystalline solid, mp 108°. *Irritant.* Soly in water at 20°: 1 g/l. Soly (g/kg): acetone 435; benzene 146; chloroform 319; ethyl acetate 253. Slowly hydrolyzed in alkaline soln. LD_{50} orally in rats: 10 g/kg (Hess, Locher).

USE: Agricultural fungicide.

10182. Vinconate. [70704-03-9] 3-Ethyl-2,3,3a,4-tetrahydro-1*H*-indolo[3,2,1-*de*][1,5]naphthyridine-6-carboxylic acid methyl ester; chanodesethylapovincamine. $C_{18}H_{20}N_2O_2$; mol wt 296.37. C 72.95%, H 6.80%, N 9.45%, O 10.80%. Synthetic hexahydrocanthane alkaloid; vincamine analog. Prepn: **BE 870887**; J. A. A. Hannart, **US 4200638** (1979, 1980 both to Omnium Chim.). Pharmacology: K. Ninomiya *et al.*, *Iyakuhin Kenkyu* **23**, 620 (1992), *C.A.* **118**, 32823 (1992). Pharmacokinetics: M. Kinbara, Y. Satou, *ibid.* 275, *C.A.* **117**, 123938 (1992). Nootropic activity: H. Kinoshita *et al.*, *Res. Commun. Chem. Pathol. Pharmacol.* **84**, 175 (1994). Clinical pharmacology: B. Saletu *et al.*, *Arch. Gerontol. Geriatr.* **3**, 127 (1984). Acute toxicity study: N. Nishimura *et al.*, *Yakuri to Chiryo* **20**, Suppl. 8, S1995 (1992), *C.A.* **118**, 52241 (1992).

mp 120-121°.
Monohydrochloride. [119600-43-0] OC-340; OM-853. C_{18}-$H_{20}N_2O_2$·HCl; mol wt 332.83. mp 194-195°. LD_{50} in male, female mice, rats (mg/kg): 947, 699, 2582, 2348 orally; 127, 131, 112, 117 i.v.; 4073, 3446, >6000, >6000 s.c. (Nishimura).

THERAP CAT: Nootropic.

10183.   **Vincristine.** [57-22-7] 22-Oxovincaleukoblastine; leurocristine; VCR; LCR. $C_{46}H_{56}N_4O_{10}$; mol wt 824.97. C 66.97%, H 6.84%, N 6.79%, O 19.39%. Antitumor alkaloid isolated from *Vinca rosea* Linn. (*Catharanthus roseus* G. Don), *Apocynaceae:* Svoboda, *Lloydia* **24**, 173 (1961). Structure: Neuss *et al.*, *J. Am. Chem. Soc.* **86**, 1440 (1964); Moncrief, Lipscomb, *ibid.* **87**, 4963 (1965). Pharmacology: R. H. Adamson *et al.*, *Arch. Int. Pharmacodyn. Ther.* **157**, 299 (1965); S. M. Sieber *et al.*, *Cancer Treat. Rep.* **60**, 127 (1976). Prepn and pharmacology of [³H]vincristine: Owellen, Donigian, *J. Med. Chem.* **15**, 894 (1972). LC-MS determn in plasma: M. S. Schmidt *et al.*, *J. Pharm. Biomed. Anal.* **41**, 540 (2006). Symposium on vincristine: *Cancer Chemother. Rep.* **52**, 453-535 (1968). Biosynthesis from catharanthine and vindoline, *q.q.v.:* A. Rahman *et al.*, *Tetrahedron Lett.* **1976**, 2351; P. Mangeney *et al.*, *J. Am. Chem. Soc.* **101**, 2243 (1979). Comprehensive description of the sulfate: J. H. Burns, *Anal. Profiles Drug Subs.* **1**, 463-480 (1972); F. J. Muhtadi, A. F. A. Afify, *Anal. Profiles Drug Subs. Excip.* **22**, 517-533 (1993). Review of clinical experience: C. E. M. Gidding *et al.*, *Crit. Rev. Oncol. Hematol.* **29**, 267-287 (1999).

Blades from methanol, mp 218-220°. $[\alpha]_D^{25}$ +17°; $[\alpha]_D^{25}$ +26.2° (ethylene chloride). pKa: 5.0, 7.4 in 33% DMF. uv max (ethanol): 220, 255, 296 nm (log a_m 4.65, 4.21, 4.18). LD_{50} i.p. in mice: 5.2 mg/kg (Adamson).
Sulfate. [2068-78-2] Oncovin. $C_{46}H_{56}N_4O_{10}$·H_2SO_4; mol wt 923.04. Crystals from ethanol. Hygroscopic. mp 273-281°. One part sol in 2 parts water, in 30 parts chloroform. Sol in methanol; slightly sol in ethanol. Practically insol in ether. $[\alpha]_D^{26}$ +8.5° (c = 0.8). uv max (methanol): 218, 252, 285, 293 nm (log ε 4.72, 4.24, 4.18, 4.23).

THERAP CAT: Antineoplastic.
THERAP CAT (VET): Antineoplastic.

10184.   **Vindesine.** [53643-48-4] 3-(Aminocarbonyl)-O^4-deacetyl-3-de(methoxycarbonyl)vincaleukoblastine; desacetylvinblastine amide; VDS; Compound 112531; NSC-245467. $C_{43}H_{55}N_5O_7$; mol wt 753.94. C 68.50%, H 7.35%, N 9.29%, O 14.85%. Synthetic deriv of the dimeric vinca alkaloid, vinblastine, *q.v.* Prepn: G. J. Cullinan, K. Gerzon, **DE 2415980** (1974 to Lilly), *C.A.* **82**, 72191r (1975); C. J. Burnett *et al.*, *J. Med. Chem.* **21**, 88 (1978). ¹³C-NMR spectroscopy: D. E. Dorman, J. W. Paschal, *Org. Magn. Reson.* **8**, 413 (1976). Antitumor activity in rodents: M. J. Sweeney *et al.*, *Cancer Res.* **38**, 2886 (1978). Pharmacokinetics: R. J. Owellen *et al.*, *ibid.* **37**, 2603 (1977). Use in acute lymphocytic leukemia of childhood: W. Krivit *et al.*, *Cancer Chemother. Pharmacol.* **2**, 267 (1979). Brief review of preclinical and early clinical data: R. W. Dyke *et al.*, *ibid.* 229. Toxicology: G. C. Todd *et al.*, *Toxicol. Environ. Health* **1**, 843 (1976); R. J. Owellen *et al.*, *Biochem. Pharmacol.* **26**, 1213 (1977).

Crystals from ethanol-methanol, mp 230-232°. $[\alpha]_D^{25}$ +39.4° (c = 1.0 in methanol). uv max (methanol): 214, 266, 288, 296 nm (ε 53400, 17450, 13950, 12500). pKa′ (DMF 66%) 5.39, 7.36; (H_2O) 6.04, 7.67.
Sulfate salt. [59917-39-4] LY-099094; Eldisine; Fildesin. C_{43}-$H_{55}N_5O_7$·H_2SO_4; mol wt 852.01. Amorphous solid from ethanol-isopropyl alcohol, mp >250° (solvated form). LD_{50} in mice, rats (mg/kg): 6.3 ±0.6, 2.0 ±0.2 i.v. (Todd); in mice: 8.8 ±2.5 i.p. (Owellen).

THERAP CAT: Antineoplastic.

10185.   **Vindoline.** [2182-14-1] (2β,3β,4β,5α,12β,19α)-4-(Acetyloxy)-6,7-didehydro-3-hydroxy-16-methoxy-1-methylaspidospermidine-3-carboxylic acid methyl ester. $C_{25}H_{32}N_2O_6$; mol wt 456.54. C 65.77%, H 7.07%, N 6.14%, O 21.03%. Major alkaloid from the leaves of *Vinca rosea* Linn. (*Catharanthus roseus* G. Don.), *Apocynaceae;* occurs naturally as the (−)-form: Gorman *et al.*, *J. Am. Pharm. Assoc.* **48**, 256 (1959); Svoboda *et al.*, *ibid.* 659; Moza, Trojánek, *Collect. Czech. Chem. Commun.* **28**, 1419 (1963). Structure: Gorman *et al.*, *J. Am. Chem. Soc.* **84**, 1058 (1962); Neuss, *Bull. Soc. Chim. Fr.* **1963**, 1509. Stereochemistry: J. W. Moncrief, W. N. Lipscomb, *J. Am. Chem. Soc.* **87**, 4963 (1965). Review of chemistry: Neuss *et al.*, *Adv. Chemother.* **1**, 133 (1964). Total synthesis of (±)-vindoline: M. Ando *et al.*, *J. Am. Chem. Soc.* **97**, 6880 (1975); Y. Ban *et al.*, *Tetrahedron Lett.* **1978**, 151; J. P. Kutney *et al.*, *J. Am. Chem. Soc.* **100**, 4220 (1978). Lacks physiological activity alone but is contained as the pentacyclic moiety in the antineoplastic agents vinblastine and vincristine, *q.q.v.*

Vindoline   R = OCH_3
Demethoxyvindoline   R = H

Needles from acetone + petr ether, mp 164-165°; prisms, mp 174-175°; $[\alpha]_D^{20}$ −18° (chloroform); pKa 5.5 in 66% DMF (Moza, Trojánek). Also reported as crystals, mp 154-155°; $[\alpha]_D^{27}$ +42° (chloroform) (Gorman). uv max (ethanol): 212, 250, 304 nm (log ε 4.49, 3.74, 3.57).

Hydrochloride. $C_{25}H_{32}N_2O_6$.HCl; mol wt 493.00. Crystals from acetone, mp 161-164°.

Demethoxyvindoline. [5231-60-7] Vindorosine; vindolidine. $C_{24}H_{30}N_2O_5$; mol wt 426.51. Structure: Moza, Trojánek, *Collect. Czech. Chem. Commun.* **28**, 1427 (1963). Needles from benzene + petr ether, mp 167°. $[\alpha]_D^{16}$ −31° (chloroform). uv max (methanol): 250, 302 nm (log ε 3.98, 3.52).

10186. Vinflunine. [162652-95-1] (2β,3β,4β,5α,12R,19α)-4-(Acetyloxy)-6,7-didehydro-15-[(2R,4R,6S,8S)-4-(1,1-difluoroethyl)-1,3,4,5,6,7,8,9-octahydro-8-(methoxycarbonyl)-2,6-methano-2H-azecino[4,3-b]indol-8-yl]-3-hydroxy-16-methoxy-1-methylaspidospermidine-3-carboxylic acid methyl ester; 4'-deoxy-20',20'-difluoro-8'-norvincaleukoblastine; 20',20'-difluoro-3',4'-dihydrovinorelbine. $C_{45}H_{54}F_2N_4O_8$; mol wt 816.94. C 66.16%, H 6.66%, F 4.65%, N 6.86%, O 15.67%. Semisynthetic vinca alkaloid with microtubule destabilizing and antiangiogenic activity; derivative of vinorelbine, *q.v.* Prepn: J.-C. Jacquesy *et al.*, **FR 2707988**; *eidem*, **US 5620985** (1995, 1997 both to Pierre Fabre); J. Fahy *et al.*, *J. Am. Chem. Soc.* **119**, 8576 (1997). NMR study of tubulin binding: C. Fabre *et al.*, *Biochem. Pharmacol.* **64**, 733 (2002). Clinical pharmacokinetics and evaluation in solid tumors: J. Bennouna *et al.*, *Ann. Oncol.* **14**, 630 (2003). Clinical evaluation in bladder cancer: S. Culine *et al.*, *Br. J. Cancer* **2006**, 1. Review of clinical development and mechanism of action: A. Kruczynski, B. T. Hill, *Crit. Rev. Oncol. Hematol.* **40**, 159-173 (2001); J. Bennouna *et al. Expert Opin. Invest. Drugs* **14**, 1259-1267 (2005).

Ditartrate. [194468-36-5] Javlor. $C_{45}H_{54}F_2N_4O_8$.2$C_4H_6O_6$; mol wt 1117.12.

THERAP CAT: Antineoplastic.

10187. Vinorelbine. [71486-22-1] (2β,3β,4β,5α,12R,19α)-4-(Acetyloxy)-6,7-didehydro-15-[(2R,6R,8S)-4-ethyl-1,3,6,7,8,9-hexahydro-8-(methoxycarbonyl)-2,6-methano-2H-azecino[4,3-b]indol-8-yl]-3-hydroxy-16-methoxy-1-methylaspidospermidine-3-carboxylic acid methyl ester; 3',4'-didehydro-4'-deoxy-C'-norvincaleukoblastine; nor-5'-anhydrovinblastine; NVB; KW-2307. $C_{45}H_{54}N_4O_8$; mol wt 778.95. C 69.39%, H 6.99%, N 7.19%, O 16.43%. Semisynthetic vinca alkaloid; structurally related to vinblastine, *q.v.* Prepn: **JP Kokai 80 31096**; N. Langlois *et al.*, **US 4307100** (1980, 1981 both to Agence Nat. Valorisation Recherche); P. Mangeney *et al.*, *Tetrahedron* **35**, 2175 (1979). Pharmacology: G. Mathé, P. Reizenstein, *Cancer Lett.* **27**, 285 (1985). HPLC determn in biological fluids: F. Jehl *et al.*, *J. Chromatogr.* **525**, 225 (1990). Veterinary trial in dogs with spontaneous neoplasia: V. J. Poirier *et al.*, *J. Vet. Intern. Med.* **18**, 536 (2004). Review of clinical pharmacokinetics and analytical methodology: D. Levêque, F. Jehl, *Clin. Pharmacokinet.* **31**, 184-197 (1996). Review of clinical efficacy in metastatic breast cancer: M. S. Aapro *et al.*, *Drugs* **67**, 657-667 (2007); in non-small cell lung cancer: M. C. Piccirillo *et al.*, *Expert Opin. Drug Saf.* **9**, 493-510 (2010).

Residue. $[\alpha]_D^{20}$ +52.4° (c = 0.3 in CHCl₃). uv max (ethanol): 215, 268, 282, 293, 310 nm (ε 3700, 11000, 9500, 7600, 4400). Partition coefficient (octanol/buffer pH 7.2): 16. LD₅₀ in mice (mg/m²): 72 i.v.; 78 orally (Mathé, Reizenstein).

Ditartrate. [125317-39-7] Eunades; Navelbine. $C_{45}H_{54}N_4O_8$.2$C_4H_6O_6$; mol wt 1079.12. Yellow-white amorphous powder. Hygroscopic. Freely sol in water; sol in ethanol, methanol. Practically insol in hexane. *Protect from light.*

THERAP CAT: Antineoplastic.

THERAP CAT (VET): Antineoplastic.

10188. Vinpocetine. [42971-09-5] (3α,16α)-Eburnamenine-14-carboxylic acid ethyl ester; 3α,16α-apovincaminic acid ethyl ester; ethyl apovincamin-22-oate; RGH-4405; Cavinton. $C_{22}H_{26}N_2O_2$; mol wt 350.46. C 75.40%, H 7.48%, N 7.99%, O 9.13%. Deriv of vincamine, *q.v.*, with vasodilating activity. Prepn: C. Lörincz *et al.*, **DE 2253750**; *eidem*, **US 4035370** (1973, 1977 both to Gedeon Richter); *eidem*, *Arzneim.-Forsch.* **26**, 1907 (1976). Series of articles on pharmacology, biochemistry, metabolism, pharmacokinetics, clinical studies: *ibid.* 1908-1989. Toxicity studies: E. Pálosi, L. Szporny, *ibid.* 1926; E. Cholnoky, L. I. Dömök, *ibid.* 1939. HPLC studies: G. Szepesi, M. Gazdag, *J. Chromatogr.* **205**, 57, 341 (1981). Evaluation of effectiveness as antimotion drug: E. I. Matsnev, D. Bodo, *Aviat. Space Environ. Med.* **55**, 281 (1984). One-pot synthesis from vincamine: Y. Kuge *et al.*, *Synth. Commun.* **24**, 759 (1994).

Crystals from benzene, mp 147-153° (dec). $[\alpha]_D^{20}$ +114° (c = 1 in pyridine). uv max (96% ethanol): 229, 275, 315 nm (log ε 4.45, 4.08, 3.85). LD₅₀ in mice, rats (mg/kg): 534, 503 orally; 240, 133.8 i.p.; 58.7, 42.6 i.v. (Pálosi, Szporny), also reported as 161.2 mg/kg i.p. in mice (Cholnoky, Dömök).

THERAP CAT: Vasodilator (cerebral).

10189. Vinyl Acetate. [108-05-4] Acetic acid ethenyl ester; acetic acid vinyl ester; acetoxyethene; acetoxyethylene; ethenyl acetate. $C_4H_6O_2$; mol wt 86.09. C 55.81%, H 7.03%, O 37.17%. Prepn: Schnizer, **US 2859241** (1958 to Celanese); Sharp, Steitz, **US 2860159** (1958 to Pan Am. Petroleum); Foster, Tobler, *J. Am. Chem. Soc.* **83**, 851 (1961). Toxicity data: Smyth, Carpenter, *J. Ind. Hyg. Toxicol.* **30**, 63 (1948). Study of chronic human exposure: Deese, Joyner, *Am. Ind. Hyg. Assoc. J.* **30**, 449 (1969). *Reviews:* Leonard "Vinyl Acetate" in *Vinyl and Diene Monomers* (part 1), E. C. Leonard, Ed. (Wiley-Interscience, New York, 1970) pp 263-328; *Faith, Keyes & Clark's Industrial Chemicals*, F. A. Lowenheim, M. K. Moran, Eds (Wiley-Interscience, New York, 4th ed., 1975) pp 862-867; W. Daniels, "Poly(Vinyl Acetate)" in *Kirk-Othmer Encyclopedia of Chemical Technology* vol. 23 (Wiley-Interscience, New York, 3rd ed., 1983) pp 817-847. Review of toxicology and human exposure: *Toxicological Profile for Vinyl Acetate* (PB93-110898, 1992) 166 pp.

Liquid; sweet, fruity odor. Polymerizes in light to a colorless, transparent mass. bp 72.7°. mp −100°; also reported as mp −93° (Daniels). d_4^{20} 0.932. *Flammable*. Flash pt, closed cup: 18°F (−8°C). Soly in water (20°): 1 g/50 ml. Misc with alc, ether. Sol in acetone, benzene, chloroform. LD_{50} orally in rats: 2.92 g/kg (Smyth, Carpenter).

Caution: Potential symptoms of overexposure are irritation of eyes, skin, nose, throat; hoarseness, cough; loss of smell; eye burns, skin blisters. *See NIOSH Pocket Guide to Chemical Hazards* (DHHS/NIOSH 97-140, 1997) p 328.

USE: In polymerized form for plastic masses, films and lacquers; in plastic film for food packaging. As modifier for food starch.

10190. Vinylbital. [2430-49-1] 5-Ethenyl-5-(1-methylbutyl)-2,4,6(1*H*,3*H*,5*H*)-pyrimidinetrione; 5-(1-methylbutyl)-5-vinylbarbituric acid; 5-vinyl-5-(1-methylbutyl)barbituric acid; butyvinal; Speda; Optanox. $C_{11}H_{16}N_2O_3$; mol wt 224.26. C 58.91%, H 7.19%, N 12.49%, O 21.40%. Prepn: Seefelder, *Festschr. Carl Wurster* **1960**, 71, *C.A.* **56**, 9928d (1962). Mechanism of action study: U. J. Jovanovic, *Arzneim.-Forsch.* **17**, 1369 (1967). Pharmacokinetics: D. D. Breimer, A. G. de Boer, *ibid.* **26**, 448 (1976). GC-MS screening method in urine: H. H. Maurer, *J. Chromatogr.* **530**, 307 (1990).

Crystals, mp 90-91.5°.

Note: This is a controlled substance (depressant): **21 CFR,** 1308.13.

THERAP CAT: Sedative, hypnotic.

10191. Vinyl Chloride. [75-01-4] Chloroethylene. C_2H_3Cl; mol wt 62.50. C 38.44%, H 4.84%, Cl 56.72%. CH_2=CHCl. Prepd from ethylene dichloride and alcoholic potassium: Regnault, *Ann.* **14**, 22 (1835); by halogenation of ethylene: Miller, Jenks, US 2896000 (1959 to National Distillers and Chemical Corp.). Review of mfg processes: *Faith, Keyes & Clark's Industrial Chemicals*, F. A. Lowenheim, M. K. Moran, Eds. (Wiley-Interscience, New York, 4th ed., 1975) pp 868-873. Acute toxicity study: L. Prodan *et al.*, *Ann. N.Y. Acad. Sci.* **246**, 154 (1975). Series of articles on toxicology and "vinyl chloride disease": *ibid.* 1-337. *Review:* in *Kirk-Othmer Encyclopedia of Chemical Technology* **vol. 12** (John Wiley & Sons, 3rd ed., 1983) pp 865-885. Review of carcinogenic risk: *IARC Monographs* **19**, 377-437 (1979); of toxicology and human exposure: *Toxicological Profile for Vinyl Chloride* (PB2007-100671, 2006) 329 pp.

Colorless gas; mild, sweet odor. Liquefies in a freezing mixture. Polymerizes in light or in presence of catalyst. mp −153.8°. bp −13.37°. d_4^{20} 0.9106. n_D^{20} 1.3700. Vapor pressure at 20°: 2530 mm Hg. *Flammable*. Flash pt, closed cup: −78°C (−112°F). Sol in alc, ether, carbon tetrachloride, benzene. Slightly sol in water. LD_{50} in mice, rats, guinea pigs, rabbits (mg/l): 293.75, 390, 595, 295 by inhalation (Prodan).

Caution: Potential symptoms of overexposure are weakness; abdominal pain, GI bleeding; hepatomegaly; pallor or cyanosis of extremities; direct contact with liquid may cause frostbite. *See NIOSH Pocket Guide to Chemical Hazards* (DHHS/NIOSH 97-140, 1997) p 330. *See also Patty's Industrial Hygiene and Toxicology* **vol. 2E**, G. D. Clayton, F. E. Clayton, Eds. (John Wiley & Sons, Inc., New York, 4th ed., 1994) pp 4169-4177. This substance is listed as a known human carcinogen: *Report on Carcinogens, Twelfth Edition* (PB2011-111646, 2011) p 438.

USE: In the plastics industry to manuf polyvinyl chloride; in organic syntheses. Has been used as refrigerant, spray can propellant.

10192. Vinylidene Chloride. [75-35-4] 1,1-Dichloroethene; 1,1-dichloroethylene; *asym*-dichloroethylene. $C_2H_2Cl_2$; mol wt 96.94. C 24.78%, H 2.08%, Cl 73.14%. CH_2=CCl_2. Prepn from

ethylene chloride: Reilly, US 2140548 (1938 to Dow); by dehydrochlorination of 1,1,2-trichloroethane: Conrad, Gould, US 2989570 (1961 to Ethyl Corp.). Review of production, properties and uses: D. S. Gibbs *et al.* in *Kirk-Othmer Encyclopedia of Chemical Technology* **vol. 23** (Wiley-Interscience, New York, 3rd ed., 1982) pp 764-798; of toxicology and human exposure: *Toxicological Profile for 1,1-Dichloroethene* (PB 95-100152, 1994) 200 pp.

Liquid. Mild, sweet odor resembling that of chloroform. *Flammable*. d_4^{20} 1.2129. mp −122.5°. bp_{760} 31.7°. n_D^{20} 1.4249. Practically insol in water. Sol in organic solvents. At temps >0° and especially in the presence of oxygen or other suitable catalysts polymerizes to a plastic. Several inhibitors to preserve the monomer have been invented. Uncontrolled polymerization may lead to explosive reaction products with oxygen or ozone: Reinhardt, *Chem. Eng. News* **25**, 2136 (1947).

Caution: Potential symptoms of overexposure are irritation of eyes, skin, throat; dizziness, headache, nausea; dyspnea; liver and kidney dysfunction; pneumonitis. Potential occupational carcinogen. *See NIOSH Pocket Guide to Chemical Hazards* (DHHS/NIOSH 97-140, 1997) p 332; *Patty's Industrial Hygiene and Toxicology* **vol. 2E**, G. D. Clayton, F. E. Clayton, Eds. (John Wiley & Sons, Inc., New York, 4th ed., 1994) pp 4181-4189.

USE: Intermediate in the production of vinylidene polymer plastics such as **Saran**.

10193. Violacein. [548-54-9] (3*E*)-3-[1,2-Dihydro-5-(5-hydroxy-1*H*-indol-3-yl)-2-oxo-3*H*-pyrrol-3-ylidene]-1,3-dihydro-2*H*-indol-2-one; 3-[2-(5-hydroxyindol-3-yl)-5-oxo-2-pyrrolin-4-ylidene]-2-indolinone; 3-[2-(5-hydroxyindol-3-yl)-5-oxopyrrolin-4-ylidene]oxindole; 5-(5-hydroxy-3-indolyl)-3-(3-oxindolylidene)-2-oxopyrroline. $C_{20}H_{13}N_3O_3$; mol wt 343.34. C 69.97%, H 3.82%, N 12.24%, O 13.98%. Pigment isolated from *Chromobacterium violaceum (Bacillus violaceus):* Tobie, *J. Bacteriol.* **29**, 223 (1935); Strong, *Science* **100**, 287 (1944); Ballantine *et al.*, *J. Chem. Soc.* **1958**, 755. Structure and synthesis: Ballantine *et al.*, *ibid.* **1960**, 2292. *Review:* DeMoss in *Antibiotics* **vol. 2**, D. Gottlieb, P. D. Shaw, Eds. (Springer, New York, 1967).

Purplish-black needles, prisms. Practically insol in water; slightly sol in alcohol; moderately sol in dioxane.

10194. Violaxanthin. [126-29-4] (3*S*,3′*S*,5*R*,5′*R*,6*S*,6′*S*)-5,-6:5′,6′-Diepoxy-5,5′,6,6′-tetrahydro-β,β-carotene-3,3′-diol; zeaxanthin diepoxide. $C_{40}H_{56}O_4$; mol wt 600.88. C 79.96%, H 9.39%, O 10.65%. Widely distributed carotenoid pigment. Formed in plants from zeaxanthin. Isoln from yellow pansies *(Viola tricolor):* Kuhn, Winterstein, *Ber.* **64**, 326 (1931). Structure: Karrer *et al.*, *Helv. Chim. Acta* **14**, 1044 (1931); **16**, 977 (1933); **19**, 1024 (1936); **27**, 1684 (1944). Partial synthesis: Karrer, Jucker, *ibid.* **28**, 300 (1945). Abs config: Bartlett *et al.*, *J. Chem. Soc. C* **1969**, 2527. Isoln from *Viola tricolor* and configuration of the 15-*cis*-isomer: P. Molnar, J. Szabolcs, *Phytochemistry* **19**, 623 (1980).

Orange prisms from methanol, reddish-brown acicular crystals from CS_2. mp 200°. $[\alpha]_{Cd}^{20}$ +35° (c = 0.08 in chloroform). Absorption max (in alc): 471.5, 442.5, 417.5 nm. Sol in alcohol, methanol, carbon disulfide, ether; almost insol in petr ether.

15-*cis*-Isomer. Irregular yellow plates from benzene/petrol, mp 109°. uv max (benzene): 479, 448, 423, 337 nm (log ε 4.91, 4.98, 4.83, 4.77). Friction causes crystals to form orange prisms.

10195. Viologen. A term coined by Michaelis to designate the chlorides of certain quaternary bases derived from γ,γ'-dipyridyl. Presently used as dichlorides, dibromides and diiodides. Prepn: Michaelis, *Biochem. Z.* **250**, 564 (1932); Michaelis, Hill, *J. Am. Chem. Soc.* **55**, 1481 (1933); *J. Gen. Physiol.* **16**, 859-873 (1933). Viologens are useful as oxidation-reduction indicators because their potential range is very negative. In contrast to other redox indicators, the color is exhibited by the reduced form, whereas usually the oxidized form is the colored one and secondly, the redox potential of these substances is independent of pH. Review of electrochemistry: C. L. Bird, A. T. Kuhn, *Chem. Soc. Rev.* **10**, 49-82 (1981).

Ethyl viologen. 1,1'-Diethyl-4,4'-bipyridinium; *N,N'*-diethyl-γ,γ'-dipyridylium. Normal potential at 30°: −0.449 volts.

Benzyl viologen. 1,1'-Bis(phenylmethyl)-4,4'-bipyridinium; *N,N'*-dibenzyl-γ,γ'-dipyridylium. Normal potential at 30°: −0.359 volts.

Betaine viologen. *N,N'*-Dibetaine-γ,γ'-dipyridylium. Normal potential at 30°: −0.444 volts.

Dimethyl analog *see* Paraquat.

USE: As biological oxidation-reduction indicators.

10196. Violuric Acid. [87-39-8] 2,4,5,6(1*H*,3*H*)-Pyrimidinetetrone 5-oxime; alloxan 5-oxime; 5-isonitrosobarbituric acid; 5-(hydroxyimino)barbituric acid. $C_4H_3N_3O_4$; mol wt 157.09. C 30.58%, H 1.93%, N 26.75%, O 40.74%. Known to exist as a mixture of keto-enol tautomers. Prepd from alloxan and hydroxylamine: Cresole, *Ber.* **16**, 1133 (1883); Guinchard, *Ber.* **32**, 1723 (1899).

Orthorhombic crystals, dec 240-241°. pK 4.7. Sparingly sol in water to a violet-colored soln; sol in alcohol. With $FeCl_3$ produces blue color.

USE: Analytical reagent for chromatographic separation of cations. Forms chelates: Leermakers, Hoffman, *J. Am. Chem. Soc.* **80**, 5663 (1958).

10197. Viomycin. [32988-50-4] (2*S*)-3-Amino-*N*-[(3*S*)-3,6-diamino-1-oxohexyl]-L-alanyl-L-seryl-L-seryl-(2*Z*)-3-[(aminocarbonyl)amino]-2,3-didehydroalanyl-2-[(4*R*,6*S*)-2-amino-1,4,5,6-tetrahydro-6-hydroxy-4-pyrimidinyl]glycine (5 → 13)-lactam; celiomycin; florimycin; tuberactinomycin B. $C_{25}H_{43}N_{13}O_{10}$; mol wt 685.70. C 43.79%, H 6.32%, N 26.56%, O 23.33%. Polypeptide antibiotic produced by various *Streptomyces* species including *S. puniceus, S. floridae, S. vinaceus.* Isoln: A. C. Finlay *et al., Am. Rev. Tuberc.* **63**, 1 (1951); Bartz *et al., ibid.* 4. Production: Marsh *et al.,* US **2633445** (1953 to Ciba); Freaney, US **2828245** (1958 to C.S.C.). Purification and chemical characterization: T. Kitagawa *et al., Chem. Pharm. Bull.* **20**, 2176 (1972). Proposed structure: Yoshioka *et al., Tetrahedron Lett.* **1971**, 2043. Alternate structure: B. W. Bycroft *et al., ibid.* **1968**, 5901; *eidem, Experientia* **27**, 501 (1971); *eidem, J. Chem. Soc. Perkin Trans. I* **1972**, 820, 827. Confirmed structure: Noda *et al., J. Antibiot.* **25**, 427 (1972); T. Kitagawa *et al., Chem. Pharm. Bull.* **20**, 2215 (1972). X-ray crystallography: B. W. Bycroft, *Chem. Commun.* **1972**, 660. Biosynthesis: J. H. Carter II *et al., Biochemistry* **13**, 1221, 1227 (1974). Toxicological study: H. Keller *et al., Arzneim.-Forsch.* **6**, 61 (1956). Enzyme immunoassay for determn in body fluids: T. Kitagawa *et al., Chem. Pharm. Bull.* **30**, 2487 (1982).

Sulfate. [37883-00-4] Viocin. Hygroscopic plates, mp 266° (dec). $[\alpha]_D^{18}$ −29.5° (c = 1 in H_2O). uv max (water or 0.1*N* HCl; 0.1*N* NaOH): 268 nm (log ε 4.4); 285 nm (log ε 4.2). LD_{50} in mice (mg/kg): 240 i.v. (Finlay); 1750 s.c. (Keller). Sol in water. Relatively insol in most organic solvents. Solns adjusted to pH 5-6 are quite stable. Soly data: Weiss *et al., Antibiot. Chemother.* **7**, 374 (1957).

Pantothenate sulfate. [1401-79-2] $C_{25}H_{43}N_{13}O_{10}$·$C_9H_{17}NO_5$·H_2SO_4; mol wt 1003.01. Prepn: H. Keller, H. Mückter, DE **1011800** (1957 to Grünenthal), *C.A.* **53**, 13519c (1959). Crystals, mp 242° (dec).

Hydrochloride. Hygroscopic plates, mp 270° (dec). $[\alpha]_D^{18}$ −16.6° (c = 1 in H_2O). uv max (water; 0.1*N* HCl; 0.1*N* NaOH): 268 nm (log ε 4.5); 268 nm (log ε 4.4); 285 nm (log ε 4.3).

THERAP CAT: Antibacterial (tuberculostatic).

10198. VIP. [37221-79-7] Vasoactive intestinal polypeptide; vasoactive intestinal peptide. Neuropeptide and gastrointestinal hormone related to glucagon and secretin, *q.q.v.* Structure consists of a highly conserved, 28 amino acid sequence which is identical in most mammals, including humans. Initially isolated from porcine intestinal wall and characterized by its activity as a gastrointestinal and vascular smooth muscle relaxant. Subsequently identified in the central and peripheral nervous systems and recognized as a neuromodulator. Widely distributed throughout the body including brain, intestines, pancreas, heart, lung, lymphoid tissue, and immune cells. Exhibits a wide variety of biological activities, including vasodilation, bronchodilation, stimulation of pancreatic secretion and gastrointestinal motility, enhancement of cardiac contractility and coronary blood flow, and modulation of the inflammatory response. Isoln: S. I. Said, V. Mutt, *Science* **169**, 1217 (1970); *eidem, Eur. J. Biochem.* **28**, 199 (1972). Amino acid sequence of porcine VIP: V. Mutt, S. I. Said, *Eur. J. Biochem.* **42**, 581 (1974). Synthesis: M. Bodanszky *et al., J. Am. Chem. Soc.* **96**, 4973 (1974). Solid-phase synthesis: D. H. Coy, J. Gardner, *Int. J. Pept. Protein Res.* **15**, 73 (1980); R. Colombo, *Experientia* **38**, 773 (1982). Isoln of chicken VIP: A. Nilsson, *FEBS Lett.* **47**, 284 (1974); and amino acid sequence: *idem, ibid.* **60**, 322 (1975). Isoln of human VIP: M. Carlquist *et al., Horm. Metab. Res.* **14**, 28 (1982); sequence and identity with porcine: N.W. Bunnett *et al., J. Clin. Endocrinol. Metab.* **59**, 1133 (1984). Distribution in brain and other tissues: M. G. Bryant *et al., Lancet* **307**, 991 (1976). Review of pharmacology and role as immunomodulator: M. Delgado *et al., Pharmacol. Rev.* **56**, 249-290 (2004). Review of cardiovascular effects: R. J. Henning, D. R. Sawmiller, *Cardiovasc. Res.* **49**, 27-37 (2001); of role in the mammalian circadian system: A. M. Vosko *et al., Gen. Comp. Endocrinol.* **152**, 165-175 (2007). Review of therapeutic potential in inflammatory disease: S. G. R. Smalley *et al., Clin. Exp. Immunol.* **157**, 225-234 (2009); in neurological disorders: C. M. White *et al., CNS Neurol. Disord. Drug Targets* **9**, 661-666 (2010); in pulmonary hypertension and asthma: D. Wu *et al., Respir. Res.* **12**, 45 (2011).

[1]
His–Ser–Asp–Ala–Val–Phe–Thr–Asp–Asn–Tyr–Thr–Arg–Leu–Arg

[28]
Asn–Leu–Ile–Ser–Asn–Leu–Tyr–Lys–Lys–Val–Ala–Met–Gln–Lys

Human VIP

Human VIP. [40077-57-4] Vasoactive intestinal octacosapeptide (swine). $C_{147}H_{238}N_{44}O_{42}S$; mol wt 3325.85. Peptide sequence common to humans, pigs, and most other mammals.

10199. Viquidil. [84-55-9] 3-[(3*R*,4*R*)-3-Ethenyl-4-piperidinyl]-1-(6-methoxy-4-quinolinyl)-1-propanone; quinicine; 1-(6-methoxy-4-quinolyl)-3-(3-vinyl-4-piperidyl)-1-propanone; chinicine; mequiverine; quinotoxine; quinotoxol; LM-192. $C_{20}H_{24}N_2O_2$; mol wt 324.42. C 74.05%, H 7.46%, N 8.64%, O 9.86%. An isomer of quinine; occurs naturally as the *d*-form. Present in small quantities in cinchona barks; formed by heating quinine with glycerol at 180°: Howard, *J. Chem. Soc.* **24**, 61 (1871); **25**, 101 (1872); Miller, Rohde,

Ber. **33**, 3214 (1900); Howard, Chick, *Pharm. J.* **99**, 143 (1917). Conversion to quinine: Rabe, Kindler, *Ber.* **51**, 466 (1918). Partial synthesis: Prostenik, Prelog, *Helv. Chim. Acta* **26**, 1965 (1943). Total synthesis: Woodward, Doering, *J. Am. Chem. Soc.* **66**, 849 (1944); **67**, 860 (1945); US 2500444 (1950 to Polaroid); Grethe *et al.*, *Helv. Chim. Acta* **56**, 1485 (1973). Review of synthesis, chemistry and pharmacology: Quevauviller *et al.*, *Ann. Pharm. Fr.* **24**, 39 (1966). Series of articles on pharmacology and metabolism: *Arzneim.-Forsch.* **22**, 1334-1346 (1972).

Yellow viscous oil, $[\alpha]_D$ +43°. Slightly sol in water; freely sol in alcohol, chloroform, ether.

Hydrochloride. [52211-63-9] Desclidium; Permiran. $C_{20}H_{24}$-N_2O_2·HCl; mol wt 360.88. Yellow, odorless and bitter tasting powder, mp 184 ±4°. uv max (chloroform): 246, 355 nm. Sol in alc; sparingly sol in water. Practically insol in acetone.

THERAP CAT: Vasodilator (cerebral).

10200. Virginiamycin. [11006-76-1] Staphylomycin; virgimycin; SKF-7988; Eskalin; Stafac; Staphylomycine. One of the streptogramins, *q.v.*, produced by a *Streptomyces virginiae*: P. De Somer, P. Van Dijck, *Antibiot. Chemother.* **5**, 632 (1955). Two principal components S_1 and M_1. Separation of factors M and S: H. Vanderhaeghe *et al.*, *ibid.* **7**, 606 (1957); of the complex into six components: Gosselinckx, Parmentier, *Chromatogr. Symp., 2nd* **1962**, 1817. Nomenclature and identification with other streptogramins: P. Crooy, R. De Neys, *J. Antibiot.* **25**, 371 (1972). HPLC determn in animal feeds: F. Gossele *et al.*, *Analyst* **116**, 1373 (1991). Quantitative determn of components from fermentation broth: B. V. Tyaglov *et al.*, *J. Planar Chromatogr. Mod. TLC* **8**, 374 (1995). Mode of action: B. T. Porse, R. A. Garrett, *J. Mol. Biol.* **286**, 375 (1999). *Review:* A. M. Biot, *Drugs Pharm. Sci.* **22**, 695-720 (1984). Review of use in feed: W. Witte *et al.*, *Acta Vet. Scand.* Suppl. **93**, 37-45 (2000).

Virginiamycin S_1

Virginiamycin M_1

Amorphous, reddish-yellow powder, mp 115-120°. Sol in methanol, ethanol, chloroform. Poorly sol in water. Practically insol in hexane and petroleum ether.

Virginiamycin S_1. [23152-29-6] Staphylomycin S. $C_{43}H_{49}N_7$-O_{10}; mol wt 823.90. C 62.69%, H 5.99%, N 11.90%, O 19.42%. Hexacyclicdepsipeptide antibiotic. Structure: H. Vanderhaeghe, G. Parmentier, *J. Am. Chem. Soc.* **82**, 4414 (1960). Crystals from methanol, mp 240-242°. $[\alpha]_D^{20}$ −28° (c = 1 in ethanol). uv max (ethanol): 305 nm (log ε 3.85).

Virginiamycin M_1. [21411-53-0] Mikamycin A; ostreogrycin A; pristinamycin II_A; staphylomycin M_1; streptogramin A; vernamycin A. $C_{28}H_{35}N_3O_7$; mol wt 525.60. C 63.99%, H 6.71%, N 7.99%, O 21.31%. Macrolactone antibiotic, *see also* Pristinamycin. Structure: G. R. Delpierre *et al.*, *Tetrahedron Lett.* **1966**, 369; *eidem*, *J. Chem. Soc. C* **1966**, 1653; D. G. I. Kingston *et al.*, *ibid.* 1669. Biosynthesis: *idem et al.*, *Rev. Latinoam. Quim.* **20**, 128 (1989). Molecular dynamics: E. Surcouf *et al.*, *Stud. Phys. Theor. Chem.* **71**, 719 (1990). Colorless laths from ethyl acetate, mp 203-205°. $[\alpha]_D^{20}$ −218° (c = 0.34 in ethanol). uv max (ethanol): 228 nm (log ε 4.51).

THERAP CAT: Antibacterial.

THERAP CAT (VET): Antibacterial; growth promotant.

10201. Viridicatin. [129-24-8] 3-Hydroxy-4-phenyl-2(1*H*)-quinolinone; 3-hydroxy-4-phenylcarbostyril; 2,3-dihydroxy-4-phenylquinoline. $C_{15}H_{11}NO_2$; mol wt 237.26. C 75.94%, H 4.67%, N 5.90%, O 13.49%. Antibiotic substance from the mycelium of *Penicillium viridicatum* Westling, the chief mold on stored corn: Cunningham, Freeman, *Biochem. J.* **53**, 328 (1953); from various strains of *P. cyclopium* Westling: Bracken *et al.*, *ibid.* **57**, 587 (1954); from *P. puberulum* Bainier: Austin, Meyers, *J. Chem. Soc.* **1964**, 1197. Biosynthesis: Luckner, *Tetrahedron Lett.* **1962**, 1035; Luckner, Mothes, *Arch. Pharm.* **296**, 18 (1963); Framm *et al.*, *Eur. J. Biochem.* **37**, 78 (1973). Synthesis: Eistert, Selzer, *Z. Naturforsch.* **17b**, 202 (1962). Mass spectra: Luckner *et al.*, *Tetrahedron* **25**, 2575 (1969). Antibacterial activity and inhibitory effect on plant growth: Taniguchi, Satomura, *Agric. Biol. Chem.* **34**, 506 (1970).

Lustrous needles from methanol or ethanol, mp 268°. Sublimes unchanged in high vacuum at 160-170°. Very weak acid. uv and ir curves: Cunningham, Freeman, *loc. cit.* Practically insol in water, aq NaHCO₃. Sparingly sol in cold organic solvents, cold concd HCl, dil mineral acids; sol in cold, aq 2*N* KOH, glacial acetic acid.

Sodium salt. $C_{15}H_{10}NNaO_2$; mol wt 259.24. Prismatic needles from aq NaOH, dec 260-265°. Sol in water.

Monoacetylviridicatin. $C_{17}H_{13}NO_3$; mol wt 279.30. Crystals from aq ethanol, mp 200-201°. Dec slowly on standing.

O,O-Dimethylviridicatin. $C_{17}H_{15}NO_2$; mol wt 265.31. Plates from ethanol, mp 86-87°.

O,N-Dimethylviridicatin. $C_{17}H_{15}NO_2$; mol wt 265.31. Plates from ethanol, mp 197-198°.

10202. Viridin. [3306-52-3] (1β,2β)-1-Hydroxy-2-methoxy-18-norandrosta-5,8,11,13-tetraeno[6,5,4-*bc*]furan-3,7,17-trione. $C_{20}H_{16}O_6$; mol wt 352.34. C 68.18%, H 4.58%, O 27.24%. Antibiotic substance from *Gliocladium virens*: Grove *et al.*, *J. Chem. Soc.* **1965**, 3803. Previously reported isolation from *Trichoderma viride*: Brian, McGowan, *Nature* **156**, 144 (1945); Brian *et al.*, *Ann. Appl. Biol.* **33**, 190 (1946). Epimerizes to β-viridin where methoxyl group is equatorial. Separation of the two isomers: Vischer *et al.*, *Nature* **165**, 528 (1950). Structure: Grove *et al.*, *Chem. Commun.* **1965**, 343. Configuration: *eidem*, *J. Chem. Soc. C* **1966**, 743; Neidle, Hursthouse, *J. Chem. Soc. Perkin Trans. 2* **1972**, 760. Biosynthesis: Blight *et al.*, *Chem. Commun.* **1968**, 1117; Grove, *J. Chem. Soc. C* **1969**, 549.

Prisms from benzene, mp 245° (dec). $[\alpha]_D^{19}$ −224°. uv max: 242, 300 nm (log ε 4.49, 4.22). Needles from acetone, mp 222-224° (dec). Prisms from glacial acetic acid, mp 200-205° (dec). Sol in water, chloroform; sparingly sol in carbon disulfide, carbon tetrachloride. Practically insol in ether, camphor. Aq solns lose activity rapidly unless acidified to pH 3.

β-Isomer. Prisms from benzene, mp 240-245° (dec). $[\alpha]_D^{16}$ −23°. uv max: 243, 300 nm (log ε 4.45, 4.25).

THERAP CAT: Antifungal.

10203. Virulizin®. [216586-46-8] Virulizin-2γ. Biological response modifier extracted from bovine bile. Prepn: R. Rang, **WO 9507089** (1995 to Imutec); *idem,* **US 6551623** (2003 to Lorus). *In vivo* antitumor activity: N. Feng *et al., Cancer Chemother. Pharmacol.* **51**, 247 (2003). Clinical evaluation in pancreatic cancer: E. Warner *et al., Clin. Invest. Med.* **17**, 37 (1994). Mechanism of action studies: M. Y. Cao *et al., Cancer Immunol. Immunother.* **54**, 229 (2005); H. Li *et al., ibid.* 1115. Review of pharmacology and therapeutic potential: E. S. Ferdinandi *et al., Expert Opin. Invest. Drugs* **8**, 1721-1735 (1999).

Aqueous soln containing 5% (w/v) solid material comprised of a mixture of inorganic salts (95-99% dry wt) and organic compds of mol wts <3000 Da (1-5% dry wt).

THERAP CAT: Immunomodulator.

10204. Viscose. A viscous orange-red aq soln of sodium cellulose xanthogenate obtained by dissolving wood pulp cellulose in sodium hydroxide soln and treating with carbon disulfide. Manuf: Bachlott, **US 2855321** (1958 to du Pont); Von Kohorn, **US 2985647** (1961).

USE: Intermediate in the manuf of rayon and cellophane.

10205. Viscosin. [27127-62-4] *N*-[(2*R*)-3-Hydroxy-1-oxodecyl]-L-leucyl-D-α-glutamyl-D-allothreonyl-D-valyl-L-leucyl-D-seryl-L-leucyl-D-seryl-L-isoleucine (9 → 3)-lactone. $C_{54}H_{95}N_9O_{16}$; mol wt 1126.40. C 57.58%, H 8.50%, N 11.19%, O 22.73%. Cyclic lipopeptide biosurfactant produced by various species of *Pseudomonas.* Isoln from *Pseudomonas viscosa:* M. Kochi *et al., Bact. Proc.* **1951**, 29; T. Ohno *et al., J. Agric. Chem. Soc. Jpn.* **27**, 665 (1953), *C.A.* **49**, 3012 (1955); and proposed structure: K. Toki, T. Ohno, *Nippon Nogei Kagaku Kaishi* **29**, 370 (1955), *C.A.* **53**, 243b (1959). Revised structure: M. Hiramoto *et al., Tetrahedron Lett.* **11**, 1087 (1970); *eidem, Chem. Pharm. Bull.* **19**, 1308, 1315 (1971). Solid phase synthesis: T. R. Burke, Jr. *et al., Tetrahedron Lett.* **30**, 519 (1989). Isoln from *P. fluorescens:* M. V. Laycock *et al., J. Agric. Food Chem.* **39**, 483 (1991). Isoln from *P. libanensis* and surfactant properties: H. S. Saini *et al., J. Nat. Prod.* **71**, 1011 (2008).

Powder, dec 270-273°. $[\alpha]_D^{29}$ −168.3° (ethanol). Critical micelle concentration: 54 mg/l. Stable to heat. Practically insol in water. Sol in methanol, ethanol, ether, acetone, alkaline phosphate buffer solns.

10206. Viscotoxin. [76822-96-3] Mixture of amphipathic, highly basic, cysteine-rich, toxic polypeptides isolated from European mistletoe, *Viscum album* L., *Viscaceae.* Structurally classified with the thionin family of plant proteins. Several isoforms are known, containing 46 amino acid residues and 3 disulfide bridges. Viscotoxin A₃ is the most abundant and cytotoxic. Viscotoxins interact with cell membranes, leading to permeabilization and cell death. Similar peptides, ***phoratoxins,*** have been isolated from American mistletoe, *Phoradendron leucarpum, Viscaceae.* Isoln of crude viscotoxin: K. Winterfeld, L. H. Bijl, *Ann.* **561**, 107 (1948). Improved isoln method: G. Samuelsson, *Sven. Farm. Tidskr.* **65**, 481 (1961). Separation of components: *idem, Acta Chem. Scand.* **20**, 1546 (1966). Purification and comparison of cytotoxicity in Yoshida sarcoma cells: G. Schaller *et al., Phytother. Res.* **10**, 473 (1996). Comparison of sequences and 3-D structure of A₃: S. Romagnoli *et al., Biochem. J.* **350**, 569 (2000). Mechanism of action studies: A. Coulon *et al., Biochim. Biophys. Acta* **1559**, 145 (2002); M. Giudici *et al., FEBS J.* **273**, 72 (2006).

¹Lys–Ser–Cys–Cys–Pro–Asn–Thr–Thr–Gly–Arg–Asn–Ile–Tyr–Asn–Ala–Cys–Arg–Leu–Thr
Gly
Ala
Cys–Thr–Ser–Gly–Ser–Ile–Lys–Cys–Gly–Ser–Leu–Lys–Ala–Cys–Thr–Pro–Arg–Pro
⁴⁶
Pro–Ser–Tyr–Pro–Asp–Lys   Viscotoxin A₃

LD₅₀ of partially purified viscotoxin A (mg/kg): 0.78 ±0.07 i.p. in mice (Samuelsson, 1961).

Viscotoxin A₁. [198228-20-5] $C_{200}H_{330}N_{62}O_{68}S_6$; mol wt 4883.57. Amino acid sequence: S. Orru *et al., Biol. Chem.* **378**, 989 (1997).

Viscotoxin A₂. [54651-39-7] $C_{199}H_{318}N_{62}O_{66}S_6$; mol wt 4827.46. Amino acid sequence: T. Olson, G. Samuelsson, *Acta Chem. Scand.* **26**, 585 (1972).

Viscotoxin A₃. [55465-79-7] $C_{201}H_{328}N_{62}O_{64}S_6$; mol wt 4829.57. Amino acid sequence: G. Samuelsson *et al., Acta Chem. Scand.* **22**, 2624 (1968).

Viscotoxin B. [11088-20-3] $C_{197}H_{328}N_{64}O_{67}S_6$; mol wt 4857.53. Amino acid sequence: G. Samuelsson, B. M. Pettersson, *Eur. J. Biochem.* **21**, 86 (1971).

10207. Vismodegib. [879085-55-9] 2-Chloro-*N*-[4-chloro-3-(2-pyridinyl)phenyl]-4-(methylsulfonyl)benzamide; GDC-0449. $C_{19}H_{14}Cl_2N_2O_3S$; mol wt 421.29. C 54.17%, H 3.35%, Cl 16.83%, N 6.65%, O 11.39%, S 7.61%. Smoothened (Smo) receptor antagonist; inhibits the hedgehog signalling pathway involved in tumorigenesis. Prepn: J. Gunzer *et al.,* **WO 06028958** (2006 to Genentech; Curis); *eidem,* **US 060063779** (2006). Structure activity studies: K. D. Robarge *et al., Bioorg. Med. Chem. Lett.* **19**, 5576 (2009). LC-MS/MS determn in plasma: X. Ding *et al., J. Chromatogr. B* **878**, 785 (2010). Smo receptor binding study: C. M. Rominger *et al., J. Pharmacol. Exp. Ther.* **329**, 995 (2009). Pharmacokinetics and metabolism: H. Wong *et al., Xenobiotica* **39**, 850 (2009). Clinical evaluation in basal cell carcinoma: D. D. Von Hoff *et al., N. Engl. J. Med.* **361**, 1164 (2009). Review of development and clinical experience: E. De Smaele *et al., Curr. Opin. Investig. Drugs* **11**, 707-718 (2010).

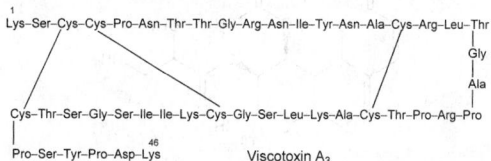

Solid. pKa 3.5. Sol in DMF.

Hydrochloride. $C_{19}H_{14}Cl_2N_2O_3S\cdot HCl$; mol wt 457.75. Crystals, mp 264°. Sol in water.

THERAP CAT: Antineoplastic.

10208. Visnadine. [477-32-7] (2R)-2-Methylbutanoic acid (9R,10R)-10-(acetyloxy)-9,10-dihydro-8,8-dimethyl-2-oxo-2H,8H-benzo[1,2-b:3,4-b']dipyran-9-yl ester; 2-methylbutyric acid 9-ester with 9,10-dihydro-9,10-dihydroxy-8,8-dimethyl-2H,8H-benzo[1,2-b:3,4-b']dipyran-2-one acetate; 8,8-dimethyl-2-oxo-9,10-dihydro-2H,8H-benzo[1,2-b:3,4-b']dipyran-9,10-diyl-10-acetate-9-(α-methylbutyrate); 3,4,5-trihydroxy-2,2-dimethyl-6-chromanacrylic acid δ-lactone 4-acetate 3-(2-methylbutyrate); 4'-acetoxy-3'-(-α-methylbutyryloxy)-2',2'-dimethyldihydropyrano(7,8:6',5')coumarin; 3-(α-methylbutyryloxy)-4-acetoxy-3,4-dihydroseseline; Cardine; Carduben; Vibeline; Visnamine. $C_{21}H_{24}O_7$; mol wt 388.42. C 64.94%, H 6.23%, O 28.83%. Isoln from seeds of *Ammi visnaga* L., *Umbelliferae* (Bishop's weed): Smith *et al.*, *Science* **115**, 520 (1952); *eidem*, *J. Am. Chem. Soc.* **79**, 3534 (1957); Smith, Haber, **US 2816118**; Smith, Hosansky, **US 2980699** (1957, 1961 both to Penick); Le Men, **US 2995574** (1961 to Lab. Roger Bellon); Baddar *et al.*, *J. Chem. Soc.* **1963**, 4522. Synthesis of (±)-form: Shanbhag *et al.*, *Tetrahedron* **21**, 3591 (1965). Pharmacology and toxicology: Nkondi *et al.*, *Therapie* **21**, 1267 (1966); Erbring *et al.*, *Arzneim.-Forsch.* **17**, 283 (1967); Eyraud, Aurousseau, *ibid.* **23**, 201 (1973).

Needles from light petr, or ether + hexane, mp 85-88°. $[\alpha]_D^{20}$ +9.2° (alc), $[\alpha]_D^{30}$ +42.5° (c = 2 in dioxane). Slightly sol in water; quite sol in ethanol, methanol. Very sol in chloroform, acetone, ether, benzene, DMF. LD_{50} in mice (mg/kg): 2240 orally; >370 s.c. (Erbring).

(±)-**Form.** Crystals from petr ether, mp 150-152° (Shanbhag).

THERAP CAT: Vasodilator (coronary).

10209. Visnagin. [82-57-5] 4-Methoxy-7-methyl-5H-furo-[3,2-g][1]benzopyran-5-one; 5-methoxy-2-methylfuranochromone. $C_{13}H_{10}O_4$; mol wt 230.22. C 67.82%, H 4.38%, O 27.80%. Spasmolytic constituent of *Ammi visnaga* Lam., *Umbelliferae*. One of the active principles in khella, a traditional Egyptian medicine used as a diuretic and vasodilator. *See also:* visnadine, khellin. Isoln and structure: Späth, Gruber, *Ber.* **74**, 1492 (1941). Synthesis: Aneja *et al.*, *Tetrahedron* **3**, 230 (1958); Badawi, Fayez, *Tetrahedron Lett.* **1967**, 1029. HPLC determn in *A. visnaga* fruit: M. M. El-Domiaty, *J. Pharm. Sci.* **81**, 475 (1992).

Thread-like needles from water. mp 142-145°. Very slightly sol in water; sparingly sol in alc; freely sol in chloroform.

10210. Vital Red. [574-65-2] 3-Amino-4-[2-[4'-[2-(7-amino-3-sulfo-1-naphthalenyl)diazenyl]-3,3'-dimethyl[1,1'-biphenyl]-4-yl]diazenyl]-2,7-naphthalenedisulfonic acid sodium salt (1:3); 3-amino-4-[[4'-[(2-amino-6-sulfo-1-naphthalenyl)azo]-3,3'-dimethyl-[1,1'-biphenyl]-4-yl]azo]-2,7-naphthalenedisulfonic acid trisodium salt; ditolyldiazo-3,6-disulfo-β-naphthylamine-β-naphthylamine-6-sulfonic acid sodium salt; C.I. Direct Red 34; C.I. 23570; brilliant vital red; Brilliant Congo R; Vital Red Evans. $C_{34}H_{25}N_6Na_3O_9S_3$; mol wt 826.76. C 49.39%, H 3.05%, N 10.17%, Na 8.34%, O 17.42%, S 11.63%. Prepd by diazotization of *o*-tolidine and coupling the resulting tetrazotoluidine first with 2-naphthylamine-3,6-disulfonic acid (amino R salt) and next with 2-naphthylamine-6-sulfonic acid (Brönner acid): Palkin, Evans, *J. Am. Chem. Soc.* **47**, 429 (1925); **DE 41095**; *Frdl.* **1**, 476; *Colour Index* vol. 4 (3rd ed., 1971) p 4191.

Cherry-red crystals from water. Absorption max: 500 nm (aq soln). Sol in water giving a brownish-red soln. Gives blue solution with concd HCl. Slightly sol in alc.

USE: For blood volume determinations.

10211. Vitamin A. [68-26-8] Retinol; *all-trans*-retinol; (all-E)-3,7-dimethyl-9-(2,6,6-trimethyl-1-cyclohexen-1-yl)-2,4,6,8-nonatetraen-1-ol; anti-infective vitamin; antixerophthalmic vitamin; axerophthol; lard-factor; oleovitamin A; ophthalamin (obsolete); vitamin A₁; vitamin A alcohol; Aquasol A; Avibon; Avitol; Vogan. $C_{20}H_{30}O$; mol wt 286.46. C 83.86%, H 10.56%, O 5.59%. Fat soluble, essential micronutrient. Occurs preformed only in animals; metabolized from carotenoids, such as β-carotene, *q.v.*, in the intestinal mucosa. Dietary sources include liver, milk, butter, cheese, eggs and fish liver oils or as carotenoids from fruits and vegetables. Stored primarily in the liver in esterified form; transported in the blood by **retinol binding protein** (RBP). Oxidized to retinal, *q.v.*, a key component of the visual system, and to retinoic acid, *q.v.*, which effects gene expression via specific nuclear receptors. Identification as essential nutrient in butter and egg: E. V. McCollum, M. Davis, *J. Biol. Chem.* **15**, 167 (1913). Isoln from fish oil: P. Karrer *et al.*, *Helv. Chim. Acta* **14**, 1036 (1931). Structure: *eidem*, *ibid.* **1431**; I. M. Heilbron *et al.*, *Biochem. J.* **26**, 1194 (1932). Synthesis: O. Isler *et al.*, *Helv. Chim. Acta* **30**, 1911 (1947); N. L. Wendler *et al.*, *J. Am. Chem. Soc.* **72**, 234 (1950); T. Mukaiyama, A. Ishida, *Chem. Lett.* **1975**, 1201. Stereospecific synthesis: P. S. Manchand *et al.*, *Helv. Chim. Acta* **59**, 567 (1976); G. Cardillo *et al.*, *J. Chem. Soc. Perkin Trans. 1* **1979**, 1729. Toxicology: J. J. Kamm, *J. Am. Acad. Dermatol.* **6**, 652 (1982). Review: W. Friedrich in *Vitamins* (Walter de Gruyter, New York, 1988) pp 63-140. Book: *The Retinoids*, M. B. Sporn *et al.*, Eds. (Raven Press, New York, 1994) 679pp. Review of role in vision: R. K. Crouch *et al.*, *Photochem. Photobiol.* **64**, 613-621 (1996); of physiological functions and dietary requirements in humans: H. Gerster, *Int. J. Vitam. Nutr. Res.* **67**, 71-90 (1997); of metabolism and transport: E. H. Harrison, *Annu. Rev. Nutr.* **18**, 259-276 (1998).

Solvated crystals from polar solvents, such as methanol or ethyl formate. mp 62-64°. Distills at 120-125° at 5×10⁻³ mm pressure. n_D^{22} 1.6410 (calculated from refractive indexes of 20-70% solns in mineral oil). uv max (ethanol): 324-325 nm ($E_{1cm}^{1\%}$ 1835). Practically insol in water or glycerol. Sol in abs alcohol, methanol, chloroform, ether, fats and oils. Exhibits a yellow-green fluorescence when irradiated with Xuv light. LD_{50} (10 day) in mice (mg/kg): 1510 i.p.; 2570 orally (Kamm).

Acetate. [127-47-9] $C_{22}H_{32}O_2$. Pale yellow prismatic crystals from methanol, mp 57-58°. uv max (ethanol): 326 nm ($E_{1cm}^{1\%}$ 1550). LD_{50} (10 day) orally in mice: 4100 mg/kg (Kamm).

Palmitate. [79-81-2] Arovit; Optovit-A. $C_{36}H_{60}O_2$; mol wt 524.87. Predominant storage form of vitamin A in the liver. Amorphous or cryst. mp 28-29°. uv max (ethanol): 325-328 nm ($E_{1cm}^{1\%}$ 975). LD_{50} (10 day) in mice, rats (mg/kg): 6060, 7910 orally (Kamm).

13-cis Form see Neovitamin A.

THERAP CAT: Vitamin (antixerophthalmic).

THERAP CAT (VET): Nutritional factor.

10212. Vitamin B₁₂. [68-19-9] Cyanocobalamin; 5,6-dimethylbenzimidazolyl cyanocobamide; cobinamide cyanide phosphate

3'-ester with 5,6-dimethyl-1-α-D-ribofuranosylbenzimidazole inner salt; LLD factor; *Lactobacillus lactis* Dorner factor; extrinsic factor; antipernicious anemia principle; Anacobin; Bedoz; Behepan; Berubi; Betalin-12; Betolvex; Cobalin; Crystamine; Cytacon; Cytamen; Cytobion; Docémine; Docigram; Fresmin; Millevit; Redisol; Rubesol; Rubramin PC; Vitarubin. $C_{63}H_{88}CoN_{14}O_{14}P$; mol wt 1355.39. C 55.83%, H 6.54%, Co 4.35%, N 14.47%, O 16.53%, P 2.29%. Prototype of the family of naturally occurring, cobalt coordination compds known as *corrinoids*. Analogs of vitamin B₁₂ which differ only in the β-ligand of the cobalt are termed *cobalamins*. Synthesized almost exclusively by bacteria. Dietary sources include fish, meat, liver, and dairy products; plants have little or no cobalamins. Also found in soil and water, the richest sources being activated sewage sludge (*see* Milorganite) or manure. Converted by the body into its bioactive forms, methylcobalamin and cobamamide, *q.q.v.*, which serve as enzyme cofactors. Severe deficiency may result in megaloblastic anemia and/or neurological impairment. Isoln from mammalian liver: E. L. Rickes *et al.*, *Science* **107**, 396 (1948); from cultures of *Streptomyces griseus*: *eidem, ibid.* **108**, 634 (1948); E. L. Rickes, T. R. Wood, US 2563794 (1951 Merck & Co.). Fermentation process using *Pseudomonas denitrificans*: R. A. Long, US 3018225 (1962 to Merck & Co.). Purification from sewage sludge: P. J. Van Melle, US 3057851 (1962 to Armour). Structure: D. C. Hodgkin *et al.*, *Nature* **176**, 325 (1955); R. Bonnett *et al.*, *ibid.* 328. X-ray structure analysis: D. C. Hodgkin, *Fortschr. Chem. Org. Naturst.* **15**, 167-229 (1958). Stereochemistry: Stora, *Bull. Soc. Chim. Fr.* **1959**, 1421. Total synthesis: R. B. Woodward, *Pure Appl. Chem.* **33**, 145 (1973). Nomenclature: IUPAC rules, *ibid.* **48**, 497 (1976). Gastrointestinal absorption is dependent on Castle's intrinsic factor, *q.v.*, and **haptocorrin** (R-protein); transport of the absorbed vitamin is mediated by **transcobalamins**: B. Seetharam, D. H. Alpers, *Annu. Rev. Nutr.* **2**, 343-369 (1982); R. Gräsbeck, *Clin. Biochem.* **17**, 99-107 (1984). Role of Co—C bond in B₁₂-dependent reactions: J. M. Pratt, *Pure Appl. Chem.* **65**, 1513 (1993). Review of biosynthesis: A. I. Scott, *Angew. Chem. Int. Ed.* **32**, 1223-1243 (1993). Comprehensive description: J. Kirschbaum, *Anal. Profiles Drug Subs.* **10**, 183-288 (1981). Book: *B₁₂* **vols. 1 and 2**, D. Dolphin, Ed. (Wiley-Interscience, New York, 1982) 672 and 506 pp. *Reviews:* B. T. Golding in *Comprehensive Organic Chemistry* **vol. 5**, E. Haslam, Ed. (Pergamon, New York, 1979) pp 549-584; "Vitamin B₁₂" in *Vitamins*, W. Friedrich, Ed. (de Gruyter, Berlin, 1988) pp 837-928. Review of chemistry and enzymology: K. L. Brown, *Chem. Rev.* **105**, 2075-2149 (2005).

Hygroscopic, dark red crystals. When exposed to air, may absorb ~12% water. The hydrated crystals are stable to air. Darkens at

210-220°. Not melted at 300°. $[\alpha]_{656}^{23}$ −59 ± 9° (dil aq soln). Absorption max (water): 278, 361, 550 nm (A$_{1cm}^{1\%}$ 115, 204, 64). Odorless and tasteless. One gram dissolves in ~80 ml water. Aq solns are neutral, maximum stability in the pH range 4.5-5. Solns in this pH range can be autoclaved for 20 min at 120°. Soluble in alc. Insol in acetone, CHCl₃, ether.

THERAP CAT: Vitamin (hematopoietic).

THERAP CAT (VET): Vitamin (hematopoietic).

10213. Vitamin B₁₂, Radioactive. Radioactive cyanocobalamin; radiocyanocobalamin. Prepn of ⁶⁰Co labelled compound: Chaiet *et al.*, *Science* **111**, 601 (1950). Properties identical with unlabelled vitamin B₁₂, except for the presence of radioactive cobalt. Evaluation of dual-isotope Schilling test for pernicious anemia: L. S. Zuckier, L. R. Chervu, *J. Nucl. Med.* **25**, 1032 (1984).

⁶⁰Co Vitamin B₁₂. [13422-53-2] Racobalamin-60; Rubratope-60. T₁/₂ 5.27 years, emitting beta and gamma rays.

⁵⁸Co Vitamin B₁₂. [18195-32-9] T₁/₂ 70.8 days, emitting beta, gamma and x-rays.

⁵⁷Co Vitamin B₁₂. [13115-03-2] Rubratope-57. T₁/₂ 271 days, emitting gamma and x-rays.

Combination of ⁵⁷Co and ⁵⁸Co labelled Vitamin B₁₂. Dicopac.

THERAP CAT: Diagnostic aid (pernicious anemia).

10214. Vitamin D₁. [520-91-2] (3β,9β,10α,22E)-Ergosta-5,-7,22-trien-3-ol compd with (1S,3Z)-4-methylene-3-[(2E)-2-[(1R,-3aS,7aR)-octahydro-7a-methyl-1-[(1R,2E,4R)-1,4,5-trimethyl-2-hexen-1-yl]-4H-inden-4-ylidene]ethylidene]cyclohexanol (1:1). $C_{56}H_{88}O_2$; mol wt 793.32. C 84.78%, H 11.18%, O 4.03%. C$_{28}$-H₄₄O.C₂₈H₄₄O. A 1:1 molecular compd of lumisterol and vitamin D₂, *q.q.v.* Prepn: Windaus *et al.*, *Ann.* **489**, 252 (1931); **493**, 259 (1932).

Crystals from acetone, mp 124-125°. Sublimes in high vacuum at 135°. Dec 180°. Soly at 16° (g/ml): 0.037 of acetone; 0.024 of petr ether (bp 35-40°); 0.020 of methanol. uv max (0.02% sol in petr ether): 265 nm (ε 1.56). $[\alpha]_D$ +140.5° (8.9 mg in 2 ml acetone), +140.5° (ethanol), +127° (chloroform).

10215. Vitamin D₂. [50-14-6] (1S,3Z)-4-Methylene-3-[(2E)-2-[(1R,3aS,7aR)-octahydro-7a-methyl-1-[(1R,2E,4R)-1,4,5-trimethyl-2-hexen-1-yl]-4H-inden-4-ylidene]ethylidene]cyclohexanol; (3β,5Z,7E,22E)-9,10-secoergosta-5,7,10(19),22-tetraen-3-ol; calciferol; ergocalciferol; oleovitamin D₂; viosterol; Deltalin; Devitol; Drisdol; Ostelin; Osto-D2; Sterogyl; Uvesterol-D; Vita D. C₂₈H₄₄-O; mol wt 396.66. C 84.78%, H 11.18%, O 4.03%. Antirachitic principle formed by UVB irradiation of ergosterol, *q.v.*, which is present in yeasts, fungi, and certain other plants; first form of vitamin D to be identified. Production in irradiated food: H. Steenbock, M. T. Nelson, *J. Biol. Chem.* **62**, 209 (1924); H. Steenbock, US 1680818 (1928 to Wisc. Alumni Res. Found.). Identification of ergosterol as precursor substance: O. Rosenheim, T. A. Webster, *Biochem. J.* **21**, 389 (1927); A. F. Hess, A. Windaus, *Proc. Soc. Exp. Biol. Med.* **24**, 461 (1927). Prepn by distillation of irradiated ergosterol: T. C. Angus *et al.*, *Proc. R. Soc. London B* **108**, 340 (1931). Stability study: W. Huber, O. W. Barlow, *J. Biol. Chem.* **149**, 125 (1943). Stereochemistry: D. Crowfoot, J. D. Dunitz, *Nature* **162**, 608 (1948); *Chem. Ind. (London)* **1957**, 1149. Direct total synthesis: B. Lythgoe *et al.*, *Tetrahedron Lett.* **18**, 3685 (1977). Identification of *25-hydroxyergocalciferol* as a bioactive metabolite: T. Suda *et al.*, *Biochem. Biophys. Res. Commun.* **35**, 182 (1969); *eidem, Biochemistry* **8**, 3515 (1969). LC-MS/MS determn in biological samples: J. Adamec *et al.*, *J. Sep. Sci.* **34**, 11 (2011). Rodenticidal properties: J. H. Greaves *et al.*, *J. Hyg. (Lond.)* **73**, 341 (1974). Review of occurence in food and effect of supplementation: L. O'Mahony *et al.*, *Nutrients* **3**, 1023-1041 (2011); of bioactivity in comparison with vitamin D₃: L. Tripkovic *et al.*, *Am. J. Clin. Nutr.* **95**, 1357-1364 (2012).

Prisms from acetone, mp 115-118°. Sublimes in very high vacuum (0.0006 mm) without dec. $[\alpha]_D^{25}$ +82.6° (c = 3 in acetone); $[\alpha]_D^{20}$ +102.5° (alcohol); $[\alpha]_D^{20}$ +52° (CHCl₃). uv max (hexane): 264.5 nm (E$_{1cm}^{1\%}$ 458.9 ±7.5). Not precipitated by digitonin (diff from ergosterol). Sol in alc, chloroform, ether, fatty oils and the usual organic solvents. 1 ml acetone dissolves 0.0695 g at 7°. Slightly sol in vegetable oils. Insol in water.

USE: Rodenticide.
THERAP CAT: Vitamin (antirachitic).
THERAP CAT (VET): Nutritional factor (antirachitic).

10216. Vitamin D₃. [67-97-0] (1*S*,3*Z*)-3-[(2*E*)-2-[(1*R*,3a*S*,-7a*R*)-1-[(1*R*)-1,5-Dimethylhexyl]octahydro-7a-methyl-4*H*-inden-4-ylidene]ethylidene]-4-methylenecyclohexanol; (3β,5*Z*,7*E*)-9,10-secocholesta-5,7,10(19)-trien-3-ol; cholecalciferol; colecalciferol; oleovitamin D₃; Neo Dohyfral D₃; D₃-Vicotrat; Vi-De-3; Vigantol. $C_{27}H_{44}O$; mol wt 384.65. C 84.31%, H 11.53%, O 4.16%. One of the naturally occurring forms of vitamin D; produced by the solar UVB irradiation of 7-dehydrocholesterol, *q.v.*, in sun-exposed skin. May also be obtained in the diet from fish and egg yolk. Hydroxylated first in the liver to calcifediol, *q.v.*, and then in the kidney to calcitriol, *q.v.*, which is the physiologically active form. Prepn by irradiation of 7-dehydrocholesterol: A. Windaus *et al.*, *Z. Physiol. Chem.* **241**, 100 (1936); and reaction mechanism: N. Akhtar, C. J. Gibbons, *Tetrahedron Lett.* **6**, 509 (1965). Absorption spectra: W. Huber *et al.*, *J. Am. Chem. Soc.* **67**, 609 (1945). Stability study: W. Huber, O. W. Barlow, *J. Biol. Chem.* **149**, 125 (1943). Direct total synthesis: B. Lythgoe *et al.*, *Tetrahedron Lett.* **18**, 3685 (1977). Prepn by laser photolysis: V. Malatesta *et al.*, *J. Am. Chem. Soc.* **103**, 6781 (1981). Review of metabolism to active form: M. R. Haussler, H. Rasmussen, *J. Biol. Chem.* **247**, 2328-2335 (1972); T. C. B. Stamp, *Nature* **245**, 180-182 (1973). Review of physiology: P. Lips, *Prog. Biophys. Mol. Biol.* **92**, 4-8 (2006); of dietary sources and supplementation: C. Lamberg-Allardt, *ibid.* 33-38; of role in human nutrition: M. F. Holick, *Nutr. Rev.* **66**, Suppl. 2, S182-S194 (2008).

Fine needles from dilute acetone, mp 84-85°. $[\alpha]_D^{20}$ +84.8° (c = 1.6 in acetone); $[\alpha]_D^{20}$ +51.9° (c = 1.6 in chloroform). uv max (alcohol or hexane): 264.5 nm ($E_{1cm}^{1\%}$ 450-490). Not precipitated by digitonin (diff from 7-dehydrocholesterol). Practically insol in water. Sol in the usual organic solvents; slightly sol in vegetable oils. Oxidized and inactivated by moist air within a few days.

THERAP CAT: Vitamin (antirachitic).
THERAP CAT (VET): Nutritional factor (antirachitic).

10217. Vitamin D₄. [511-28-4] (1*S*,3*Z*)-4-Methylene-3-[(2*E*)-2-[(1*R*,3a*S*,7a*R*)-octahydro-7a-methyl-1-[(1*R*,4*S*)-1,4,5-trimethylhexyl]-4*H*-inden-4-ylidene]ethylidene]cyclohexanol; (3β,5*Z*,7*E*)-9,10-secoergosta-5,7,10(19)-trien-3-ol; 22:23-dihydrovitamin D₂; 22,23-dihydroergocalciferol. $C_{28}H_{46}O$; mol wt 398.68. C 84.36%, H 11.63%, O 4.01%. Prepd from 22:23-dihydroergosterol by irradiation with light of the magnesium arc: Windaus, Trautmann, *Z. Physiol. Chem.* **247**, 185 (1937). Synthesis: P. J. Kocienski *et al.*, *J. Chem. Soc. Perkin Trans. 1* **1979**, 1290.

Platelets from dil acetone, mp 96-98°. Originally given as mp 107-108°, *see* Windaus, Guntzel, *Ann.* **538**, 122 (1939). $[\alpha]_D^{18}$ +89.3° (c = 0.47 in acetone). uv max: 265 nm. Not precipitated by digitonin. Practically insol in water. Sol in the usual organic solvents except petr ether; slightly sol in vegetable oils.

10218. Vitamin E. [1406-18-4] General term referring to 2 groups of closely related 6-chromanol derivatives known as tocopherols and tocotrienols, each comprised of four homologs (α, β, γ, and δ) that differ in the position and number of methyl substituents on the chromanol ring. All eight compounds are found in nature, occurring as the *R,R,R-* or *R,Z,Z-* forms. Dietary sources include vegetable oils and green leafy vegetables. α-Tocopherol, *q.v.*, is the most abundant and has the greatest bioactivity. Vitamin E acts as an antioxidant to prevent free radical damage to tissues and unsaturated lipids; also thought to stabilize membrane structures. Deficiency in healthy humans is rare and is characterized by peripheral neuropathy and spinocerebellar ataxia. Discovery as nutritional factor required for normal reproduction in animals: H. M. Evans, K. S. Bishop, *Science* **56**, 650 (1922). Comparison of bioactivities of tocopherols and tocotrienols: T. Leth, H. Sondergaard, *J. Nutr.* **107**, 2236 (1977). HPLC determn in infant formula: B. R. Mendoza *et al.*, *J. Chromatogr. A* **1018**, 197 (2003). Review of chemistry, nomenclature and activity: W. Friedrich, "Vitamin E" in *Vitamins* (Walter de Gruyter, Berlin, 1988) pp 217-283. Review of biological occurrence and functions: R. S. Parker, *Adv. Food Nutr. Res.* **33**, 157-232 (1989); of nutritional requirements: E. E. J. Valk, G. Hornstra, *Int. J. Vitam. Nutr. Res.* **70**, 31-42 (2000); of metabolism: F. Brigelius-Flohé, M. G. Traber, *FASEB J.* **13**, 1145-1155 (1999). Review of antioxidant functions and role in membranes: X. Wang, P. J. Quinn, *Prog. Lipid Res.* **38**, 309-336 (1999); of non-antioxidant roles: A. Azzi, A. Stocker, *ibid.* **39**, 231-255 (2000); of use in dairy cows to prevent free-radical mediated tissue damage: A. Baldi, *Livestock Prod. Sci.* **98**, 117-122 (2005); of synthesis in plants: D. DellaPenna, *J. Plant Physiol.* **162**, 729-737 (2005).

Note: The international unit of vitamin E is equal to one mg of standard *dl*-α-tocopheryl acetate. Activity may also be expressed as α-tocopherol equivalents (α-TE) where one α-TE is equal to the activity of 1 mg of α-tocopherol.

10219. Vitamin K. General term referring to a group of naphthoquinone derivatives required for the bioactivation of proteins involved in hemostasis. The designation "K" was derived from the German "Koagulationsvitamin." Vitamin K compds are classified into 3 groups: phylloquinone (K₁), *q.v.*, found in green plants; menaquinones (K₂), *q.v.*, primarily produced by intestinal bacteria; and menadione (K₃), *q.v.*, and derivatives which are synthetic, lipid soluble compounds. Reduced *in vivo* to dihydrovitamin K (KH₂) which serves as a coenzyme in the conversion of glutamic acid residues to γ-carboxyglutamic acid (Gla), *q.v.*, in the post-translational modification of blood coagulation factors II, VII, IX and X, and the anticoagulant proteins C and S. Other Gla-containing proteins, such as the bone matrix protein, osteocalcin, *q.v.*, have been identified in a wide variety of tissues. This γ-carboxylation is accompanied by the oxidation of KH₂ to vitamin K epoxide which is then recycled back to vitamin K. Discovery: H. Dam, *Biochem. Z.* **215**, 475 (1929); **220**, 158 (1930); *Nature* **135**, 652 (1935). Historical survey: H. Dam, *Vitam. Horm.* **24**, 295-306 (1966). Menadione and phylloquinone are metabolized by animals to menaquinone-4: C. Martius, H. O. Esser, *Biochem. Z.* **331**, 1 (1958); H. H. W. Thijssen, M. J. Drittij-

Reijnders, *Br. J. Nutr.* **72**, 415 (1994). HPLC determn in plasma: M. Kamao *et al.*, *J. Chromatogr. B* **816**, 41 (2005). Comprehensive review: W. Friedrich in *Vitamins* (de Gruyter, New York, 1988) pp 285-338. Review of metabolism and role in human nutrition: M. J. Shearer, *Blood Rev.* **6**, 92-104 (1992); J. W. Suttie, *J. Am. Diet. Assoc.* **92**, 585-590 (1992); of mechanism of action: P. Dowd *et al.*, *Nat. Prod. Rep.* **11**, 251-264 (1994); of pharmacology and therapeutic use: J. A. Thorp *et al.*, *Drugs* **49**, 376-387 (1995). Physiological review: M. J. Shearer, *Lancet* **345**, 229-234 (1995).

10220. Vitamin K₅. [83-70-5] 4-Amino-2-methyl-1-naphthalenol; 4-amino-2-methyl-1-naphthol; 2-methyl-4-amino-1-hydroxynaphthalene; 3-methyl-4-hydroxy-1-naphthylamine; Synkamin. $C_{11}H_{11}NO$; mol wt 173.22. C 76.27%, H 6.40%, N 8.09%, O 9.24%. Prepn: Sah, *Rec. Trav. Chim.* **59**, 458 (1940); **60**, 373 (1941); Veldstra, Wiardi, *ibid.* **61**, 547 (1942); **62**, 75 (1943). Insulin-mimicking effect on glucose transport: J. N. Livingston *et al.*, *Am. J. Physiol.* **234**, E484 (1978). Polargraphic determn in aq soln: K. Takamura, F. Watanabe, *Methods Enzymol.* **67**, 134 (1980).

Hydrochloride. [130-24-5] $C_{11}H_{11}NO.HCl$. Needles from dil HCl, darkens at 262°, dec 280-282°. Turns pink to dark violet on exposure to air and light. Freely sol in water; slightly sol in alc. Insol in ether.
USE: Chemical probe to study insulin-activated hexose transport.
THERAP CAT: Vitamin (prothrombogenic).

10221. Vitamin T. [1407-73-4] Tegotin; termitin; torutilin; factor T; vitamin T Goetsch; Goetsch's vitamin; Temina. A complex of growth-promoting substances, originally obtained from termites: Goetsch, *Oesterr. Zool. Z.* **1**, 58-85 (1946). Also obtainable from roaches, yeasts and fungi. Isoln of "vitamin T complexes" from yeast: Koch *et al.*, *DE 1000962* (1957 to Aschaffenburger Zellstoffwerke), *C.A.* **54**, 18898d (1960). Vitamin T may be a mixture of known vitamins and growth-promoting factors: Koch *et al.*, *Naturwissenschaften* **38**, 339 (1951). Polemic: Barkow, Goetsch, *ibid.* **42**, 346 (1955).

10222. Vitronectin. Serum spreading factor; S-protein. Multifunctional adhesive glycoprotein found in serum and in various tissues. Has binding sites for integrins, collagen, heparin, complement components, and perforin, *q.q.v.* Thought to play a regulatory role in hemostasis, wound healing, tissue remodeling and cancer. Synthesized by hepatocytes as a single chain polypeptide, mol wt ~78 kDa. Binds readily to proteoglycans of the extracellular matrix and is deposited in normal loose connective tissues, in dermal elastin fibers, and in the vascular wall. Also synthesized by platelets, macrophages and megakaryocytes. Released by activated platelets in different multimeric forms. Promotes cellular attachment, spreading and migration of a wide variety of cell types and serves as a matrix-associated regulator of blood coagulation. Identification: R. Holmes, *J. Cell Biol.* **32**, 297 (1967). Isoln and comparison with fibronectin, *q.v.*: D. Barnes *et al.*, *J. Supramol. Struct.* **14**, 47 (1980). Purification and tissue distribution: E. G. Hayman *et al.*, *Proc. Natl. Acad. Sci. USA* **80**, 4003 (1983). Amino acid sequence: S. Suzuki *et al.*, *EMBO J.* **4**, 2519 (1985). Identity with S-protein: K. T. Preissner *et al.*, *Biochem. Biophys. Res. Commun.* **134**, 951 (1986); B. R. Tomasini, D. F. Mosher, *Blood* **68**, 737 (1986). Review of structure and biological function in hemostasis: K. T. Preissner, *Annu. Rev. Cell Biol.* **7**, 275-310 (1991); K. T. Preissner, D. Jenne, *Thromb. Haemostasis* **66**, 123-132; 189-194 (1991). Review of vitronectin receptors and role in cell growth and differentiation: B. Felding-Habermann, D. A. Cheresh, *Curr. Opin. Cell Biol.* **5**, 864-868 (1993).

10223. Voacamine. [3371-85-5] 12-Methoxy-13-[(3α)-17-methoxy-17-oxovobasan-3-yl]ibogamine-18-carboxylic acid methyl ester; voacanginine. $C_{43}H_{52}N_4O_5$; mol wt 704.91. C 73.27%, H 7.44%, N 7.95%, O 11.35%. Bisindole alkaloid found in *Voacanga*

africana Stapf., *V. thouarsii* R. & Sch., var *obtusa* (K. Sch.) Pichon and *V. schweinfurthii* Staph., *Apocynaceae:* Janot, Goutarel, *Compt. Rend.* **240**, 1719 (1955); Rao, *J. Org. Chem.* **23**, 1455 (1958); Janot, Goutarel, *US 2823204* (1958 to Lab. Gobey). Identity with voacanginine: LeBarre, Gillo, *Bull. Acad. R. Med. Belg.* **20**, 194 (1956). Cleaved by acid catalysis to voacangine: Winkler, *Naturwissenschaften* **48**, 694 (1961). Structure: Goutarel *et al.*, *Compt. Rend.* **243**, 1670 (1956); Percheron, *Ann. Chim. (Paris)* **4**, 303 (1959); Büchi *et al.*, *J. Am. Chem. Soc.* **85**, 1893 (1963). *Review:* Gorman *et al.* in *The Alkaloids* **1**, J. E. Saxton, Ed. (The Chem. Soc., London, 1971) pp 242-249.

voacangine

Prisms from acetone + methanol, dec 223°. $[\alpha]_D^{20}$ −52° (chloroform). uv max: 225, 295 nm (log ε 4.72, 4.28). Sol in chloroform, acetone; slightly sol in methanol, ethanol.

Voacangine. Carbomethoxyibogaine; 12-methoxyibogamine-18-carboxylic acid methyl ester. $C_{22}H_{28}N_2O_3$; mol wt 368.48. Structure: Bartlett *et al.*, *J. Am. Chem. Soc.* **80**, 126 (1958). Prismatic needles from ethanol, mp 136-137°. Sublimes₀.₀₁ 135°. $[\alpha]_D^{20}$ −42° (c = 1.26 in chloroform). pKa 7.4 (40% aq methanol); pKa 5.73 (33% DMF). uv max (methanol): 225, 287 300 nm (log ε 4.43, 3.97, 3.93). Freely sol in acetone, chloroform; slightly sol in methanol, ethanol.

10224. Voclosporin. [515814-01-4] 6-[(2S,3R,4R,6E)-3-Hydroxy-4-methyl-2-(methylamino)-6,8-nonadienoic acid]cyclosporin A; cyclo[L-alanyl-D-alanyl-N-methyl-L-leucyl-N-methyl-L-leucyl-N-methyl-L-valyl-[(2S,3R,4R,6E)-3-hydroxy-4-methyl-2-(methylamino)nona-6,8-dienoyl]-(2S)-2-aminobutanoyl-N-methylglycyl-N-methyl-L-leucyl-L-valyl-N-methyl-L-leucyl]; ISA-247; ISAtx-247. $C_{63}H_{111}N_{11}O_{12}$; mol wt 1214.65. C 62.30%, H 9.21%, N 12.68%, O 15.81%. Cyclic peptide calcineurin inhibitor; semi-synthetic analog of cyclosporin A, *q.v.*, containing a modified MeBmt side chain. Prepn: S. Naicker *et al.*, *WO 9918120; eidem*, *US 6605593* (1999, 2003 both to Isotecknika). Stereoselective syntheses: S. Naiker *et al.* *WO 03033526* (2003 to Isotechnika; Roche); B. Molino *et al.*, *WO 06014872* (2006 to AMR Technology). Pharmacology: W. P. Maksymowych *et al.*, *J. Rheumatol.* **29**, 1646 (2002). Pharmacokinetics: M. Stalder *et al.*, *J. Heart Lung Transplant.* **22**, 1343 (2003). Efficacy in primate renal transplants: C. R. Gregory *et al.*, *Transplantation* **78**, 681 (2004). Review of preclinical studies and therapeutic potential: L. Aspeslet *et al.*, *Transplant. Proc.* **33**, 1048-1051 (2001); F. J. Dumont, *Curr. Opin. Investig. Drugs* **5**, 542-550 (2004). Clinical trial in plaque psoriasis: R. Bissonnette *et al.*, *J. Am. Acad. Dermatol.* **54**, 472 (2006).

White solid.
THERAP CAT: Immunosuppressant; antipsoriatic.

10225. Vodka. Little water. Defined as ethyl alcohol obtained from grain or potatoes, filtered through charcoal, and diluted with distilled water. Manuf by "spirit flow" process: Jacobs, **US 2946687** (1960 to Heublein).

10226. Voglibose. [83480-29-9] 3,4-Dideoxy-4-[[2-hydroxy-1-(hydroxymethyl)ethyl]amino]-2-C-(hydroxymethyl)-D-epiinositol; N-(1,3-dihydroxy-2-propyl)valiolamine; AO-128; Basen. $C_{10}H_{21}NO_7$; mol wt 267.28. C 44.94%, H 7.92%, N 5.24%, O 41.90%. α-Glucosidase inhibitor. Prepn: S. Horii et al., **EP 56194**; eidem, **US 4701559** (1982, 1987 both to Takeda); and structure-activity study: eidem et al., J. Med. Chem. **29**, 1038 (1986). Synthesis: H. Fukase, S. Horii, J. Org. Chem. **57**, 3651 (1992). Enzyme inhibition and pharmacology: T. Matsuo et al., Am. J. Clin. Nutr. **55**, 314S (1992). Clinical pharmacology: B. Göke et al., Digestion **56**, 493 (1995). Clinical evaluation in hyperinsulinemia: K. Shinozaki et al., Metabolism **45**, 731 (1996).

Colorless crystals, mp 162-163°. $[\alpha]_D^{25}$ +26.2° (c = 1 in water).
THERAP CAT: Antidiabetic.

10227. Volicitin. [191670-18-5] N^2-[(9Z,12Z,15Z,17S)-17-Hydroxy-1-oxo-9,12,15-octadecatrien-1-yl]-L-glutamine; N-(17-hydroxylinolenoyl)-L-glutamine. $C_{23}H_{38}N_2O_5$; mol wt 422.57. C 65.37%, H 9.06%, N 6.63%, O 18.93%. Key component in plant recognition of damage from insect herbivory. Isolated from the oral secretions of the beet armyworm caterpillar, it acts as a chemical signal to corn seedlings to produce volatile compounds to attract the caterpillar predator, parasitic wasps. Isolation and structure elucidation: H. T. Alborn et al., Science **276**, 945 (1997). Biosynthesis: P. W. Paré et al., Proc. Natl. Acad. Sci. USA **95**, 13971 (1998). Stereoselective synthesis: G. Phonert et al., Chem. Commun. **1999**, 1087.

10228. Vomicine. [125-15-5] (4aR,6aS,13aS,13bR,13cS)-4a,-5,13,13a,13b,13c-Hexahydro-10-hydroxy-16-methyl-12H-6a,4-(ethaniminomethano)indolo[3,2,1-ij]oxepino[2,3,4-de]quinoline-6,-12(2H)-dione; 4-hydroxy-19-methyl-16,19-secostrychnidine-10,16-dione; 12-hydroxy-N-methylpseudostrychnine. $C_{22}H_{24}N_2O_4$; mol wt 380.44. C 69.46%, H 6.36%, N 7.36%, O 16.82%. From seed of Strychnos nux vomica L., Loganiaceae: Wieland, Oertel, Ann. **469**, 193 (1929). Structure: Huisgen et al., ibid. **573**, 121 (1951). Synthesis: Rosenmund, Angew. Chem. **75**, 1127 (1963). Reviews: R. Robinson in Progress in Organic Chemistry vol. I (Butterworths, London, 1952) pp 2-21; J. B. Hendrickson in The Alkaloids vol. VI, R. H. F. Manske, H. L. Holmes, Eds. (Academic Press, New York, 1960) pp 195-204.

Hexagonal prisms from acetone, mp 284°. $[\alpha]_D^{22}$ +80° (c = 0.5 in alc). Weak, mono-acidic base forming salts with an acid reaction. Freely sol in chloroform, sol in hot alcohol, acetone, slightly in ether, ethyl acetate.
Hydrochloride. $C_{22}H_{24}N_2O_4$·HCl. Prisms from water, dec 245°. uv max (ethanol): 222, 263, 297 nm (log ε 4.27, 3.77, 3.52). Sparingly sol in water.
Methyl ether. $C_{23}H_{26}N_2O_4$. Needles from alcohol, dec 290°. $[\alpha]_D^{20}$ +16° (c = 0.5 in alcohol).

10229. Vomitoxin. [51481-10-8] (3α,7α)-12,13-Epoxy-3,-7,15-trihydroxytrichothec-9-en-8-one; deoxynivalenol; dehydronivalenol. $C_{15}H_{20}O_6$; mol wt 296.32. C 60.80%, H 6.80%, O 32.40%. Isoln of the trichothecene mycotoxin from Fusarium roseum and structure: N. Morooka et al., J. Food Hyg. Soc. Jpn. **13**, 368 (1972); T. Yoshizawa, N. Morooka, Agric. Biol. Chem. **37**, 2933 (1973). Isoln from F. graminearum: R. F. Vesonder et al., Appl. Microbiol. **26**, 1008 (1973); eidem, Appl. Environ. Microbiol. **31**, 280 (1976). Emetic and refusal activity in swine: D. M. Forsyth et al., ibid. **34**, 547 (1977). HPLC analysis: G. A. Bennett et al., J. Am. Oil Chem. Soc. **58**, 1002A (1981). Implicated as a chemical warfare agent with nivalenol, q.v. in Southeast Asia: N. Wade, Science **214**, 34 (1981).

Fine needles from ethyl acetate + petr ether, mp 151-153°. $[a]_D^{25}$+6.35° (c = 0.07 in ethanol). uv max (ethanol): 218 nm (ε 4500). LD_{50} i.p. in male, female mice (mg/kg): 70.0, 76.7 (Yoshizawa, Morooka).

10230. Vorapaxar. [618385-01-6] N-[(1R,3aR,4aR,6R,8aR,-9S,9aS)-9-[(1E)-2-[5-(3-Fluorophenyl)-2-pyridinyl]ethenyl]dodecahydro-1-methyl-3-oxonaphtho[2,3-c]furan-6-yl]carbamic acid ethyl ester; ethyl [(3aR,4aR,8aR,9aS)-9(S)-[(E)-2-[5-(3-fluorophenyl)-2-pyridinyl]ethenyl]dodecahydro-1(R)-methyl-3-oxonaphtho[2,3-c]-furan-6(R)-yl]carbamate. $C_{29}H_{33}FN_2O_4$; mol wt 492.59. C 70.71%, H 6.75%, F 3.86%, N 5.69%, O 12.99%. Thrombin receptor antagonist, selective for the platelet protease-activated receptor (PAR)-1. Structurally related to the natural product, himbacine, q.v. Prepn: S. Chackalamannil et al., **WO 03089428**; eidem, **US 7304078** (2003, 2007 both to Schering Corp.); and antiplatelet activity: eidem, J. Med. Chem. **51**, 3061 (2008). Improved process: T. K. Thiruvengadam et al., **US 7541471** (2009 to Schering Corp.). Metabolism and pharmacokinetic studies: Y. Hsieh et al., Curr. Pharm. Des. **15**, 2262 (2009). Safety and tolerability study in patients undergoing percutaneous coronary intervention: R. C. Becker et al., Lancet **373**, 919 (2009). Review of pharmacology and clinical studies: T. E. Macaulay et al., Expert Opin. Pharmacother. **11**, 1015-1022 (2010).

Solid, mp 125°. $[\alpha]_D^{20}$ −6.6° (c = 0.5 in methanol).
Sulfate. [705260-08-8] SCH-530348. $C_{29}H_{33}FN_2O_4$·H_2SO_4; mol wt 590.66. White solid from acetonitrile, mp 217°.
THERAP CAT: Antithrombotic.

10231. Voriconazole. [137234-62-9] (αR,βS)-α-(2,4-Difluorophenyl)-5-fluoro-β-methyl-α-(1H-1,2,4-triazol-1-ylmethyl)-4-pyrimideethanol; 2R,3S-2-(2,4-difluorophenyl)-3-(5-fluoropyrimi-

Consult the Name Index before using this section.

din-4-yl)-1-(1*H*-1,2,4-triazol-1-yl)butan-2-ol; UK-109496; Vfend. $C_{16}H_{14}F_3N_5O$; mol wt 349.32. C 55.01%, H 4.04%, F 16.32%, N 20.05%, O 4.58%. Ergosterol biosynthesis inhibitor. Prepn: S. J. Ray, K. Richardson, **EP 440372**; *eidem*, **US 5278175** (1991, 1994 both to Pfizer); R. P. Dickinson *et al.*, *Bioorg. Med. Chem. Lett.* **6**, 2031 (1996). Mechanism of action: H. Sanati *et al.*, *Antimicrob. Agents Chemother.* **41**, 2492 (1997). *In vitro* antifungal spectrum: F. Marco *et al.*, *ibid.* **42**, 161 (1998). HPLC determn in plasma: R. Gage, D. A. Stopher, *J. Pharm. Biomed. Anal.* **17**, 1449 (1998). Review of pharmacology and clinical development: P. E. Verweij *et al.*, *Curr. Opin. Anti-Infect. Invest. Drugs* **1**, 361-372 (1999); J. A. Sabo, S. M. Abdel-Rahman, *Ann. Pharmacother.* **34**, 1032-1043 (2000). Clinical pharmacokinetics: L. Purkins *et al.*, *Antimicrob. Agents Chemother.* **46**, 2546 (2002). Clinical comparison with amphotericin B: T. J. Walsh *et al.*, *N. Engl. J. Med.* **346**, 225 (2002).

mp 127°. $[\alpha]_D^{25}$ −62° (c = 1 in methanol).

THERAP CAT: Antifungal (systemic)

10232. Vorinostat. [149647-78-9] N^1-Hydroxy-N^8-phenyloctanediamide; suberoylanilide hydroxamic acid; SAHA; Zolinza. $C_{14}H_{20}N_2O_3$; mol wt 264.33. C 63.62%, H 7.63%, N 10.60%, O 18.16%. Second generation hybrid polar compound. Histone deacetylase (HDAC) inhibitor that induces cell cycle arrest, differentiation and apoptosis in tumor cells. Prepn: R. Breslow *et al.*, **WO 9307148**; *eidem*, **US 5369108** (1993, 1994 both to Sloan-Kettering Inst.; Columbia Univ.); J. C. Stowell *et al.*, *J. Med. Chem.* **38**, 1411 (1995). Improved synthesis: A. Mai *et al.*, *Org. Prep. Proced. Int.* **33**, 391 (2001). HTLC determn in serum: L. Du *et al.*, *Rapid Commun. Mass Spectrom.* **19**, 1779 (2005). Clinical pharmacokinetics and activity in cancer patients: W. K. Kelly *et al.*, *J. Clin. Oncol.* **23**, 3923 (2005). Clinical trial in cutaneous T-cell lymphoma: E. A. Olsen *et al.*, *J. Clin. Oncol.* **25**, 3109 (2007). Review of mechanism of action: V. M. Richon *et al.*, *Blood Cells Mol. Dis.* **27**, 260-264 (2001); of discovery and development: P. A. Marks, *Oncogene* **26**, 1351-1356 (2007). Review of clinical experience in acute myeloid leukemia and myelodysplastic syndromes: T. Prebet, N. Vey, *Expert Opin. Invest. Drugs* **20**, 287-295 (2011).

White to light orange powder. Crystals from acetonitrile, mp 159-160.5°. pKa 9.2. Freely sol in DMSO; slightly sol in ethanol, iso-

propanol, acetone; very slightly sol in water. Insol in methylene chloride. pH of saturated soln in water: 6.6.

THERAP CAT: Antineoplastic.

10233. Vorozole. [129731-10-8] 6-[(*S*)-(4-Chlorophenyl)-1*H*-1,2,4-triazol-1-ylmethyl]-1-methyl-1*H*-benzotriazole; (+)-(*S*)-6-[4-chloro-α-(1*H*-1,2,4-triazol-1-yl)benzyl]-1-methyl-1*H*-benzotriazole; R-83842; Rivizor. $C_{16}H_{13}ClN_6$; mol wt 324.77. C 59.17%, H 4.03%, Cl 10.92%, N 25.88%. Nonsteroidal aromatase inhibitor. Prepn: A. H. M. Raeymaekers *et al.*, **EP 293978** (1988 to Janssen); resolution of enantiomers: A. L. A. Willemsens *et al.*, **WO 9411364** (1994 to Janssen). Aromatase inhibition by racemate: W. Wouters *et al.*, *J. Steroid Biochem.* **32**, 781 (1989); and enantiomers: *eidem*, *J. Steroid Biochem. Mol. Biol.* **37**, 1049 (1990). Abs config: O. M. Peeters *et al.*, *Acta Crystallogr.* **C49**, 1958 (1993). Clinical studies in breast cancer: S. R. D. Johnston *et al.*, *Cancer Res.* **54**, 5875 (1994); P. E. Goss *et al.*, *Clin. Cancer Res.* **1**, 287 (1995). Review of clinical pharmacology and pharmacokinetics: W. Wouters *et al.*, *Breast Cancer Res. Treat.* **30**, 89-94 (1994).

Crystals from 2-propanol, mp 130-135°. $[\alpha]_D^{20}$ +8.0° (c = 10 in CH_3OH).

(±)-Form. [118949-22-7] R-76713. Crystals from 2-propanone/1,1-oxybisethane, mp 178.9°.

THERAP CAT: Antineoplastic.

10234. VX. [50782-69-9] *P*-Methylphosphonothioic acid *S*-[2-[bis(1-methylethyl)amino]ethyl] *O*-ethyl ester; *O*-ethyl *S*-[2-(diisopropylamino)ethyl]methylphosphonothioate; Tx 60. $C_{11}H_{26}NO_2$-PS; mol wt 267.37. C 49.42%, H 9.80%, N 5.24%, O 11.97%, P 11.58%, S 11.99%. Nerve gas. Prepn: R. V. Ley, G. L. Sainsbury, **GB 1346409**; A. W. Wardrop, C. Stratford, **GB 1346410** (both 1974 to U.K. Sec. of State for Defence), *C.A.* **81**, 4068y, 4069z (1974); S. R. Eckhaus *et al.*, **US 3911059** (1975 to U.S.A. Sec. for Army). Activity studies in humans: F. R. Sidell, W. A. Groff, *Toxicol. Appl. Pharmacol.* **27**, 241 (1974); F. N. Craig *et al.*, *J. Invest. Dermatol.* **68**, 357 (1977). Toxicity study: J. J. Gordon, L. Leadbeater, *Toxicol. Appl. Pharmacol.* **40**, 109 (1977).

Nonvolatile, odorless liquid. pKa' 7.9. LD_{50} s.c. in rabbits: 15.4 μg/kg (Gordon, Leadbeater).

Caution: Cholinesterase inhibitor; more potent than sarin, *q.v.*

USE: Chemical warfare agent.

W

10235. Warburganal. [62994-47-2] (1S,4aS,8aS)-1,4,4a,5,6,-7,8,8a-Octahydro-1-hydroxy-5,5,8a-trimethyl-1,2-naphthalenedicarboxaldehyde. $C_{15}H_{22}O_3$; mol wt 250.34. C 71.97%, H 8.86%, O 19.17%. Drimane sesquiterpene with antifeedant activity against the African army worm. Biological activity includes plant growth regulation, cytotoxic, antimicrobial and molluscicidal properties. Isoln from *Warburgia ugandensis, Canellaceae* and structure: I. Kubo *et al., Chem. Commun.* **1976**, 1013. Relationship between structure and antifeedant activity: K. Nakanishi, I. Kubo, *Isr. J. Chem.* **16**, 28 (1977). Synthesis of (±)-warburganal: S. P. Tanis, K. Nakanishi, *J. Am. Chem. Soc.* **101**, 4398 (1979); T. Nakata *et al., ibid.* 4400; A. S. Kende, T. J. Blacklock, *Tetrahedron Lett.* **1980**, 3119; P. A. Wender, S. L. Eck, *ibid.* **1982**, 1871; D. M. Hollinshead *et al., J. Chem. Soc. Perkin Trans. 1* **1983**, 1579. *See also:* **JP Kokai 80 136238**, and **JP 80 136240** (both 1980 to Inst. Phys. Chem. Res.); **JP Kokai 81 43236** (1981 to Suntory Ltd.); **JP Kokai 83 38232** (1983 to Teikoku Zoki). Synthesis of (−)-warburganal: H. Okawara *et al., Tetrahedron Lett.* **1982**, 1087.

mp 98-99°. uv max (methanol): 224 nm (ε 6300). $[\alpha]_D^{21}$ −260° (c = 0.350 in CHCl₃).

10236. Warfarin. [81-81-2] 4-Hydroxy-3-(3-oxo-1-phenylbutyl)-2H-1-benzopyran-2-one; 3-(α-acetonylbenzyl)-4-hydroxycoumarin; 1-(4′-hydroxy-3′-coumarinyl)-1-phenyl-3-butanone; 3-α-phenyl-β-acetylethyl-4-hydroxycoumarin; WARF compound 42; Rodex; Sakarat X; Warfotox. $C_{19}H_{16}O_4$; mol wt 308.33. C 74.01%, H 5.23%, O 20.76%. Coumarin anticoagulant. Marketed as the racemate; the S(−)-form is the more potent isomer. Prepn: M. A. Stahmann *et al.,* **US 2427578** (1947 to Wisconsin Alumni Res. Found.). Resolution and abs configuration: B. D. West *et al., J. Am. Chem. Soc.* **83**, 2676 (1961). Effect on vitamin K metabolism: R. G. Bell *et al., Biochemistry* **11**, 1959 (1972). Conformation in soln: E. J. Valente *et al., J. Med. Chem.* **20**, 1489 (1977). Toxicity studies: E. C. Hagan, J. L. Radomski, *J. Am. Pharm. Assoc. Sci. Ed.* **42**, 379 (1953); N. Back *et al., Pharmacol. Res. Commun.* **10**, 445 (1978). Comprehensive description: S. A. Babhair *et al., Anal. Profiles Drug Subs.* **14**, 423-452 (1985). Stereospecific HPLC determn in plasma: P. R. Ring, J. M. Bostick, *J. Pharm. Biomed. Anal.* **22**, 573 (2000). Review of rodenticide mechanisms of action and resistance: H. H. W. Thijssen, *Pestic. Sci.* **43**, 73-78 (1995). Review of pharmacology and therapeutic efficacy: J. Hirsh *et al., Chest* **102**, Suppl., 312S-326S (1992); of pharmacokinetics and effect of genetic variants: H. Takahashi, H. Echizen, *Clin. Pharmacokinet.* **40**, 587-603 (2001). Clinical comparison with aspirin following myocardial infarction: M. Hurlen *et al., N. Engl. J. Med.* **347**, 969 (2002).

Crystals from alc, mp 161°. uv max (water, pH 10): 308 nm (ε 13610). Soluble in acetone, dioxane. Moderately sol in methanol,

ethanol, isopropanol, some oils. Freely sol in alkaline aq solns (forms a water-soluble sodium salt). Practically insol in water, benzene, cyclohexane, Skellysolves A and B. Warfarin has an acidic enol which forms metallic salts and an acetate, mp 117-118°, and a ketone which forms an oxime, mp 182-183° and a 2,4-dinitrophenylhydrazone, mp 215-216°.

Sodium salt. [129-06-6] Coumadin; Marevan; Warfilone. $C_{19}H_{15}NaO_4$; mol wt 330.31. Slightly bitter, crystalline powder. Discolored by light. Very sol in water; freely sol in alcohol; very slightly sol in chloroform, ether. LD₅₀ in male rats, female rats, mice, rabbits (mg/kg): 323, 58, 374, ~800 orally (Hagen); also reported as LD₅₀ in male, female rats (mg/kg): 100.3, 8.7 orally (Back).

Caution: Potential symptoms of overexposure to warfarin are hematuria, back pain; hematoma of arms and legs; epistaxis, bleeding lips and mucous membrane hemorrhage; abdominal pain, vomiting and fecal blood; petechial rash; abnormal hematologic indices. *See NIOSH Pocket Guide to Chemical Hazards* (DHHS/NIOSH 97-140, 1997) p 334. *See also Clinical Toxicology of Commercial Products,* R. E. Gosselin *et al.,* Eds. (Williams & Wilkins, Baltimore, 5th ed., 1984) Section III, pp 395-397.

USE: Rodenticide.

THERAP CAT: Anticoagulant.

THERAP CAT (VET): Anticoagulant.

10237. Water. [7732-18-5] Hydrogen oxide. H_2O; mol wt 18.02. H 11.19%, O 88.78%. The most universal solvent known. *Reviews:* N. E. Dorsey, "Properties of Ordinary Water-Substance" in *A.C.S. Monograph Series* no. **81**, (Reinhold, New York, 1940) 673 pp; D. Eisenberg, W. Kauzmann, *The Structure and Properties of Water* (Oxford University Press, New York, 1969) 296 pp; Ebsworth *et al.,* in *Comprehensive Inorganic Chemistry* vol. **2**, J. C. Bailar, Jr. *et al.,* Eds. (Pergamon Press, Oxford, 1973) pp 741-747. *Pyrogen-free water* (water for injection) is distilled water rendered free of fever-producing proteins (bacteria and their metabolic products). Method of prepn: Ishizuka *et al., C.A.* **49**, 15177 (1955).

Liquid. mp 0°. bp 100°. Expands on freezing. Temp of max density 3.98°. d³·⁹⁸ 1.000000 g/ml (0.999972 g/cc). d₄²⁵ 0.997. d⁰ (ice) 0.917 g/cc; d₄⁰ (liq) 0.999868. Density tables: Bigg, *Br. J. Appl. Phys.* **18**, 521 (1967); Kell, *J. Chem. Eng. Data* **12**, 66 (1967). One liter satd vapor weighs 0.5974 g at 100° and 760 mm. Crit temp 374.2°; crit pressure 218 atm. Sp. heat (liq; 14°) 1.000 cal/g/°C. Latent heat of fusion: 1.436 kcal/mole. Latent heat of vaporization: 9.717 kcal/mole. n_D^{20} 1.3330. Dielectric const (0°) 87.740. Dipole moment (25°) in benzene 1.76; in dioxane 1.86. K for pure water only (25°): 1.008×10^{-14}; at moderate concn of solutes (e.g. 1.0M KOH): K (25°) 0.971×10^{-14}.

10238. Watermelon. Arbuse. *Citrullus vulgaris* Schrad., *Cucurbitaceae,* cultivated in hot and temperate zones the world over. Contains diuretic principles: Bliss *et al., Am. J. Pharm.* **105**, 53 (1933); Roby *et al., ibid.* **111**, 68 (1939).

10239. Wheat Germ Oil. Cav-Ecol; Myopone; Denamone. Obtained by hydraulic expression or solvent extraction of wheat germ which constitutes ~2% of a wheat grain, the seed of *Triticum aestivum* L. (*T. sativum* Lam., *T. vulgare* Vill.), *Gramineae.* Constit. (of the oil): Linoleic acid 44.1%, oleic acid 30.0%, satd acids 15.1%, linolenic acid 10.8%, unsaponifiable matter 4.7%. The unsaponifiable matter contains vitamin E-active tocopherols (reported as 0.5% of the oil and as 2 international vitamin E units per gram of the oil), sitosterols, dihydrositosterol and other cryst alcohols and phospholipids. *Review:* E. W. Eckey, *Vegetable Fats and Oils* (Reinhold, New York, 1954) pp 291-293.

Bland yellow oil resembling corn oil. d₂₅²⁵ 0.925-0.933. n_D^{40} 1.469-1.478. Acid value 6-20. Saponif value 179-190. Iodine value 115-129. Thiocyanogen value 80-85. Hydroxyl value 10-48. Reichert-Meissl value 0.3-1.4. Polenske value 0.4-2.1. Hehner value 76-95. Miscible with chloroform, ether, benzene, petr ether. Slightly sol in alc.

USE: Nutritional supplement. Source of natural vitamin E and unsatd fatty acids (vitamin F).

THERAP CAT (VET): Dietary source of vitamin E.

10240. Whisky. Whiskey. A liquid produced by distillation of the fermented mash of malted cereal grains, which has been stored in wood containers for not less than 4 years. Straight whisky contains 47-53% abs alcohol by vol, 0.05-0.16% acid calculated as acetic acid, and 0.038-0.15% esters as ethyl acetate. Whisky marked 100 proof contains 50% ethanol (v/v). It also contains small quantities of other natural constituents (congeners), which vary according to the grain used, method of fermentation, etc., and which are largely responsible for the characteristic aroma and flavor. Blended whisky contains at least 40% ethanol (v/v) and is made from at least 20% of 100 proof straight whisky mixed with neutral spirits, *q.v.* The most popular blends are made from 65 vols neutral spirits and 35 vols straight whisky.

Light to deep amber liquid; characteristic odor and taste. d 0.923-0.935 at 25°.

10241. White Pine. Deal pine; Northern pine; Weymouth pine. The dried inner bark of *Pinus strobus* L., *Pinaceae. Constit.* Coniferin glycoside, coniferyl alcohol, tannin, oleoresin, volatile oils.

Weak yellowish-orange to light yellowish-brown when powdered, with a slightly terebinthinate odor and a slightly mucilaginous, sweet, then bitter and astringent taste.

THERAP CAT: The bark as expectorant.

10242. Wieland-Gumlich Aldehyde. [466-85-3] (4*S*,17*R*)-19,20-Didehydro-17,18-epoxycuran-17-ol; 1-deacetyldiaboline; caracurine VII. $C_{19}H_{22}N_2O_2$; mol wt 310.40. C 73.52%, H 7.14%, N 9.03%, O 10.31%. Decompn product of isonitrosostrychnine: Wieland, Kaziro, *Ann.* **506**, 60 (1933). Isoln from *Strychnos toxifera* Benth., and *S. subcordata* Spruce, *Loganiaceae:* Asmis *et al., Helv. Chim. Acta* **37**, 1983 (1954); Penna *et al., Gazz. Chim. Ital.* **87**, 1163 (1957). Identity with caracurine VII: Bernauer *et al., Helv. Chim. Acta* **41**, 1405 (1958). Structure: Deyrup *et al., ibid.* **45**, 2266 (1962). Chemistry of the degradation with strychnine, *q.v.:* Hymon *et al., ibid.* **52**, 1564-1602 (1969). Synthesis: L. Szabo, O. Clauder, *Acta Chim. Acad. Sci. Hung.* **95**, 85 (1977). ^{13}C-NMR study: E. Wenkert *et al., J. Org. Chem.* **43**, 1099 (1978).

Crystals from acetone + methanol, decomp 213-214°. $[\alpha]_D^{22}$ −133.8° (c = 0.52 in methanol). Sol in methanol, ethanol, chloroform; slightly sol in acetone, ethyl acetate.

Hydrochloride hemihydrate. Crystals from ethanol, mp >300°. uv max (water): 240, 290 nm (log ε 3.80, 3.40). Sol in water, ethanol, methanol, warm chloroform; slightly sol in acetone, chloroform.

10243. Wild Cherry. Wild black cherry bark. Dried stem bark of *Prunus serotina* Ehrh., *Rosaceae,* collected in autumn. *Habit.* North America. *Constit.* Prunasin; the enzyme emulsin capable of hydrolyzing prunasin to benzaldehyde, glucose, and hydrocyanic acid; benzoic, trimethylgallic and *p*-coumaric acids; tannin, volatile oil.

USE: Flavoring in foods. Pharmaceutic aid (flavor).

10244. Wildfire Toxin. [32190-57-1] *N*-[2-Amino-4-(3-hydroxy-2-oxo-3-azetidinyl)-1-oxobutyl]-L-threonine. $C_{11}H_{19}N_3O_6$; mol wt 289.29. C 45.67%, H 6.62%, N 14.53%, O 33.18%. Isoln from *Pseudomonas tabaci,* the bacterium responsible for wildfire disease of tobacco: Woolley *et al., J. Biol. Chem.* **197**, 409 (1952). Highly toxic to a variety of organisms including bacteria, algae, higher plants and animals: Sinden *et al., Toxicol. Appl. Pharmacol.* **14**, 82 (1969). Mode of action: Lovrekovich *et al., Nature* **197**, 917; **198**, 710 (1963). Structural studies: Woolley *et al., J. Biol. Chem.* **215**, 485 (1955); Stewart, *J. Am. Chem. Soc.* **83**, 435 (1961). Revised structure: *idem, Nature* **229**, 174 (1971).

Colorless, very hygroscopic substance. Very soluble in water; readily sol in methanol. Moderately sol in ethanol; sparingly sol in *n*-butanol and higher alcohols. Practically insol in acetone, ethyl acetate, and less polar organic liquids. Shows marked instability. Slowly loses potency in aq soln at pH 6. Complete loss of activity at these conditions and 0° in one month. Inactivation more rapid in methanol or when pH is shifted to either side of 6, esp in alkaline range.

10245. Wilkinson's Catalyst. [14694-95-2] (*SP*-4-2)-Chlorotris(triphenylphosphine)rhodium. $C_{54}H_{45}ClP_3Rh$; mol wt 925.23. C 70.10%, H 4.90%, Cl 3.83%, P 10.04%, Rh 11.12%. Dimorphic compd exists in two forms, red and orange, both possess the same chemical properties. Prepn: J. F. Young *et al., Chem. Commun.* **1965**, 131; J. A. Osborn, G. Wilkinson, *Inorg. Synth.* **10**, 67 (1967); *eidem, ibid.* **28**, 77 (1990). Crystal structure: M. J. Bennett, P. B. Donaldson, *Inorg. Chem.* **16**, 655 (1977). Structure and properties: Y. Nishihara *et al., Organometallics* **21**, 825 (2002). Mechanistic study: S. B. Duckett *et al., J. Am. Chem. Soc.* **116**, 10548 (1994). Reduction of nitrobenzenes: H. R. Brinkman *et al., Synth. Commun.* **26**, 973 (1996); selective hydrogenation of olefins: A. Jourdant *et al., J. Org. Chem.* **67**, 3163 (2002). Comprehensive review of properties and reactions: F. H. Jardine, *Prog. Inorg. Chem.* **28**, 63-202 (1981); of hydrogenations: F. Joo, *Acc. Chem. Res.* **35**, 738-745 (2002).

Dark, burgundy-red crystals from hot ethanol. Orange crystals from < 200 ml ethanol soln refluxed for five minutes; converted to red form on further refluxing. mp 157°. Soln in chloroform, dichloromethane (25°): ~20 g/l; in benzene, toluene (25°): ~2 g/l; much less sol in acetic acid, acetone and other ketones, methanol, lower aliphatic alcohols. Virtually insol in alkanes and cyclohexane. Donor solvents such as pyridine, dimethyl sulfoxide, acetonitrile dissolve the complex with reaction. Solns are air sensitive. Orange form d 1.363; red form d 1.379.

USE: Homogeneous hydrogenation catalyst.

10246. Withaferin A. [5119-48-2] (4β,5β,6β,22R)-5,6-Epoxy-4,22,27-trihydroxy-1-oxoergosta-2,24-dien-26-oic acid δ-lactone; 4β,27-dihydroxy-1-oxo-5β,6β-epoxywitha-2,24-dienolide. $C_{28}H_{38}O_6$; mol wt 470.61. C 71.46%, H 8.14%, O 20.40%. Isolated from the leaves of *Withania somnifera* Dun., *Solanaceae:* Lavie, Yarden, *J. Chem. Soc.* **1962**, 2925; Kirson *et al., Tetrahedron* **26**, 2209 (1970); Subramanian, Sethi, *Indian J. Pharm.* **32**, 16 (1970); from roots of *Withania coagulans: eidem, Curr. Sci.* **38**, 267 (1969); from *Acnistus arborescens* (L.): Kupchan *et al., J. Org. Chem.* **34**, 3858 (1969). Structure: Lavie *et al., J. Chem. Soc.* **1965**, 7517; Kupchan, *loc. cit.* Crystal structure: McPhail, Sim, *J. Chem. Soc. C*

1968, 962. Antitumor activity studies: B. Shohat *et al.*, *Cancer Chemother. Rep.* **51**, 271 (1967); B. Shohat, H. Joshua, *Int. J. Cancer* **8**, 487 (1971). Synthetic studies: M. Hirayama *et al.*, *J. Chem. Soc. Perkin Trans. 1* **1981**, 88. Stereoselective synthesis: *eidem*, *Tetrahedron Lett.* **23**, 4725 (1982).

White prisms from acetone-petroleum ether, mp 252-253° (Kupchan). Also reported as mp 243-245° (ethyl acetate). $[\alpha]_D^{28}$ +125° (c = 1.30 in CHCl$_3$). uv max (ethanol): 214, 335 nm (ε 17300, 165).

10247. Woodruff. Woodward herb; petit muguet; Waldmeister (German). Leaves of *Asperula odorata* L., *Rubiaceae*. *Habit.* Europe, Siberia, North Africa, Australia. *Constit.* Coumarin, tannin, asperuloside, fatty oil, essential oil, bitter principle. Discussion of its fragrance, prepn of *asperule absolute*, use of *Asperula* in various fragrant formulations: *Perfum. Essent. Oil Rec.* **1963**, (June) p 382.

USE: Flavoring agent; the fresh leaves in flavoring May wine, the dried leaves in sachets.

10248. Woodward's Reagent K. [4156-16-5] 2-Ethyl-5-(3-sulfophenyl)isoxazolium inner salt; *N*-ethyl-5-phenylisoxazolium-3'-sulfonate. C$_{11}$H$_{11}$NO$_4$S; mol wt 253.27. C 52.17%, H 4.38%, N 5.53%, O 25.27%, S 12.66%. Prepd by reacting 5-phenylisoxazole with chlorosulfonic acid, followed by alkylation with triethyloxonium fluoroborate, and acid hydrolysis: Woodward *et al.*, *J. Am. Chem. Soc.* **83**, 1010 (1961).

Crystals, dec 207-208°. Stable. Non-hygroscopic.
USE: Peptide-bond former.

10249. Woollins' Reagent. [122039-27-4] 2,4-Diphenyl-1,3,-2,4-diselenadiphosphetane 2,4-diselenide. C$_{12}$H$_{10}$P$_2$Se$_4$; mol wt 532.00. C 27.09%, H 1.89%, P 11.64%, Se 59.37%. [PhP(Se)(μ-Se)]$_2$. Selenium analog of Lawesson's reagent, *q.v.* Prepn: J. C. Fitzmaurice *et al.*, *J. Chem. Soc. Chem. Commun.* **1988**, 741; P. T. Wood, J. D. Woollins, *ibid.* 1190; M. J. Pilkington *et al.*, *Heteroat. Chem.* **1**, 351 (1990). Synthesis and characterization by ^{31}P and ^{77}Se NMR: G. Grossmann *et al.*, *Z. Anorg. Allg. Chem.* **627**, 1269 (2001). Improved synthesis: I. P. Gray *et al.*, *Chem. Eur. J.* **11**, 6221 (2005). X-ray crystal structure: P. Bhattacharyya *et al.*, *J. Chem. Soc. Dalton Trans.* **2001**, 300. Synthetic applications: I. Baxter *et al.*, *Chem. Commun.* **1997**, 2049; P. Bhattacharyya, J. D. Woollins, *Tetrahedron Lett.* **42**, 5949 (2001); *idem et al.*, *Chem. Eur. J.* **8**, 2705 (2002); J. Bethke *et al.*, *Tetrahedron Lett.* **44**, 6911 (2003).

Red crystals from toluene, mp 192-204°.
USE: Selenium transfer reagent.

10250. Wormwood. Absinthium; green ginger. Shrubby, aromatic perennial plant, *Artemisia absinthium* L., *Compositae*. Medicinal portions include the essential oil and the dried leaves and flowering tops. *Habit.* Europe, northern Africa, Asia, North and South America. *Constit.* Absinthin, anabsinthin, artabsine, matricine, volatile oil. Description: E. Guenther, *The Essential Oils* V, 487 (Van Nostrand, New York 1952). Isolation of various constituents: Cekan, Herout, *Collect. Czech. Chem. Commun.* **21**, 79 (1956); Herout *et al.*, *ibid.* 1485. Characterization of oil: T. Sacco, F. Chialva, *Planta Med.* **54**, 93 (1988). Extraction of antioxidant components: J. M. Canadanovic-Brunet *et al.*, *J. Sci. Food Agric.* **85**, 265 (2005). Botanical description and medicinal uses: J. Gruenwald *et al.*, *PDR for Herbal Medicines* (Medical Economics, Montvale, 2nd Ed., 2000) pp 829-831; A. J. Skyles, B. V. Sweet, *Am. J. Health Syst. Pharm.* **61**, 239-241 (2004).

Very strong odor, acrid taste.

Volatile oil. [8008-93-3] Oil of wormwood; artemisia oil. Obtained by steam distillation of leaves and flowering tops. *Constit.* Highly variable among different strains, chiefly thujone, *cis*-epoxyocimene, *trans*-sabinyl acetate, chrysanthenyl acetate. Brownishgreen liq. d$_{15}^{15}$ 0.925-0.955. n_D^{20} 1.460-1.4741. Sol in ether, in 2 vols 80% alcohol; very slightly sol in water. *Keep well closed, cool and protected from light.*

Caution: Ingestion of large doses may cause vomiting, stomach and intestinal cramps, headache, dizziness and CNS disturbances (Gruenwald).

USE: Oil as flavoring in alcoholic beverages, *e.g.* **vermouth**, which is a blend of white wines, contg traces of absinthium and other flavors; formerly in absinthe.

THERAP CAT: Aromatic bitter; anthelmintic; externally for insect bites and wound healing.

10251. Wortmannin. [19545-26-7] (1*S*,6b*R*,9a*S*,11*R*,11b*R*)-11-(Acetyloxy)-1,6b,7,8,9a,10,11,11b-octahydro-1-(methoxymethyl)-9a,11b-dimethyl-3*H*-furo[4,3,2-*de*]indeno[4,5-*h*]-2-benzopyran-3,6,9-trione. C$_{23}$H$_{24}$O$_8$; mol wt 428.44. C 64.48%, H 5.65%, O 29.87%. Antifungal antibiotic from *Penicillium wortmanni* Klocker: Brian *et al.*, *Trans. Br. Mycol. Soc.* **40**, 365 (1957). Similar to viridin, *q.v.* Structure: MacMillan *et al.*, *Chem. Commun.* **1968**, 613; *eidem, J. Chem. Soc. Perkin Trans. 1* **1972**, 2898. Absolute stereochemistry: Petcher *et al.*, *Chem. Commun.* **1972**, 1061; MacMillan *et al.*, *ibid.* 1063.

Neutral solid, mp 240°. Unstable in aq solns pH 3-8.

X

10252. Xaliproden. [135354-02-8] 1,2,3,6-Tetrahydro-1-[2-(2-naphthalenyl)ethyl]-4-[3-(trifluoromethyl)phenyl]pyridine. $C_{24}H_{22}F_3N$; mol wt 381.44. C 75.57%, H 5.81%, F 14.94%, N 3.67%. Serotonin 5HT$_{1A}$-receptor agonist; enhances production of neurotrophic factors and extends motor neuron survival. Prepn: D. Nisato *et al.*, **EP 101381**; *eidem*, **US 4521428** (1984, 1985 both to Sanofi); of radiolabelled compound: N. Robic, J.-P. Noël, *J. Labelled Compd. Radiopharm.* **42**, 109 (1999). Pharmacology and receptor binding: L. Cervo *et al.*, *Eur. J. Pharmacol.* **253**, 139 (1994). Effect on neuron survival: Y. Iwasaki *et al.*, *J. Neurol. Sci.* **160**, Suppl. 1, S92 (1998). Neuroprotective effect in animal model of multiple sclerosis: B. Bourrié *et al.*, *Proc. Natl. Acad. Sci. USA* **96**, 12855 (1999). Review of pharmacology and clinical potential in treatment of amyotrophic lateral sclerosis: J. Fournier *et al.*, *CNS Drug Rev.* **3**, 148-167 (1997).

Hydrochloride. [90494-79-4] SR-57746A. $C_{24}H_{22}F_3N$·HCl; mol wt 417.90. mp 255-260°.

THERAP CAT: Neuroprotective.

10253. Xamoterol. [81801-12-9] *N*-[2-[[2-Hydroxy-3-(4-hydroxyphenoxy)propyl]amino]ethyl]-4-morpholinecarboxamide; 1-(4-hydroxyphenoxy)-3-[2-(4-morpholinocarboxamido)ethylamino]-2-propanol. $C_{16}H_{25}N_3O_5$; mol wt 339.39. C 56.62%, H 7.43%, N 12.38%, O 23.57%. β_1-Adrenoceptor partial agonist. Prepn: **BE 867376**; B. G. Main, J. J. Barlow, **US 4143140** (1978, 1979 both to ICI); prepn and adrenoceptor stimulant activity: J. J. Barlow *et al.*, *J. Med. Chem.* **24**, 315 (1981). Cardiovascular activity: A. Nuttall, H. M. Snow, *Br. J. Pharmacol.* **77**, 381 (1982). Hemodynamics, cardioselectivity and kinetics: G. Jennings *et al.*, *Clin. Pharmacol. Ther.* **35**, 594 (1984). Hemodynamic effects: J.-M. R. Detry *et al.*, *Eur. Heart J.* **4**, 584 (1983); H. Sato *et al.*, *Circulation* **75**, 213 (1987). Metabolism: T. R. Marten *et al.*, *Drug Metab. Dispos.* **12**, 652 (1984). HPLC determn in biological fluids: C. J. Oddie *et al.*, *J. Chromatogr.* **308**, 370 (1984). Clinical evaluations: F. L. Tseu *et al.*, *Br. Heart J.* **56**, 469 (1986); L. Barrios *et al.*, *Eur. J. Clin. Pharmacol.* **29**, 667 (1986). Clinical comparison with atenolol, *q.v.*, in asthmatic patients: J. W. J. Lammers *et al.*, *Br. J. Clin. Pharmacol.* **22**, 595 (1986).

Hemifumarate. [73210-73-8] ICI-118587; Corwin. $(C_{16}H_{25}N_3O_5)_2$·$C_4H_4O_4$; mol wt 794.86. Crystals from ethanol, mp 168-169° (dec).

THERAP CAT: Cardiotonic.

10254. Xanomeline. [131986-45-3] 3-[4-(Hexyloxy)-1,2,5-thiadiazol-3-yl]-1,2,5,6-tetrahydro-1-methylpyridine. $C_{14}H_{23}N_3OS$; mol wt 281.42. C 59.75%, H 8.24%, N 14.93%, O 5.69%, S 11.39%. Selective muscarinic M$_1$-receptor agonist. Prepn: P. Sauerberg, P. H. Olesen, **EP 384288** (1990 to Ferrosan); *eidem*, **US 5043345** (1991 to Novo Nordisk); *eidem et al.*, *J. Med. Chem.* **35**, 2274 (1992). Prepn of crystalline tartrate: L. M. Osborne *et al.*, **WO 9429303** (1994 to Novo Nordisk). Muscarinic receptor binding study: H. E. Shannon *et al.*, *J. Pharmacol. Exp. Ther.* **269**, 271 (1994). Pharmacology: F. P. Bymaster *et al.*, *ibid.* **282**. HPLC determn in plasma: C. L. Hamilton *et al.*, *J. Chromatogr.* **613**, 365 (1993). Review of clinical pharmacology and clinical experience: N. R. Mirza *et al.*, *CNS Drug Rev.* **9**, 159-186 (2003).

Oxalate. [141064-23-5] $C_{14}H_{23}N_3OS$·$C_2H_2O_4$. Crystals from acetone, mp 148°.

(+)-L-Hydrogen tartrate. [152854-19-8] Xanomeline tartrate; LY-246708; NNC-11-0232; Lomeron; Memcor. $C_{14}H_{23}N_3OS$·$C_4H_6O_6$; mol wt 431.50. Crystals from 2-propanol, mp 95.5°.

THERAP CAT: Cholinergic; nootropic.

10255. Xanthan Gum. [11138-66-2] Polysaccharide B-1459; Kelgum; Keltrol; Kelzan; Xantural. Mol wt >10⁶. Polysaccharide gum produced by the bacterium *Xathomonas campestris:* A. Jeanes *et al.*, *J. Appl. Polym. Sci.* **5**, 519 (1961). Composed of D-glucosyl, D-mannosyl and D-glucosyluronic acid residues and differing proportions of *O*-acetyl and pyruvic acid acetal. The primary structure consists of a cellulose backbone with trisaccharide side chains, the repeating unit being a pentasaccharide. Structural studies: J. H. Sloneker *et al.*, *Can. J. Chem.* **40**, 2066, 2188 (1962); **42**, 1261 (1964); P. E. Jansson *et al.*, *Carbohydr. Res.* **45**, 275 (1975); L. D. Melton *et al.*, *ibid.* **46**, 245 (1976). Review of structure and use in chemically enhanced oil recovery: G. Holzwarth, *Dev. Ind. Microbiol.* **26**, 271-280 (1985); of macromolecular properties: B. T. Stokke *et al.* in *Polysaccharides*, S. Dumitriu, Ed. (Dekker, New York, 1998) pp 433-472. Review of production and industrial uses: F. Garcia-Ochoa *et al.*, *Biotechnol. Adv.* **18**, 549-579 (2000).

Cream-colored, odorless, free-flowing powder. Dissolves readily in water with stirring to give highly viscous solns at low concns. Forms strong films on evaporation of aq solns. Resistant to heat degradation. Aq solns are highly pseudoplastic.

USE: In foods, pharmaceuticals, and cosmetics as stabilizer and thickening agent. For rheology control in water-based systems. In oil and gas drilling and completion fluids.

10256. Xanthatin. [26791-73-1] (3a*R*,7*S*,8a*S*)-3,3a,4,7,8,8a-Hexahydro-7-methyl-3-methylene-6-[(1*E*)-3-oxo-1-buten-1-yl]-2*H*-cyclohepta[*b*]furan-2-one. $C_{15}H_{18}O_3$; mol wt 246.31. C 73.15%, H 7.37%, O 19.49%. Antimicrobial agent from *Xanthium strumarium* L., sens. lat.; *X. pennsylvanicum* Wallr., *Compositae; X. riparium.* Isoln: J. E. Little *et al.*, *Arch. Biochem.* **27**, 247 (1950); T. A. Geissman *et al.*, *J. Am. Chem. Soc.* **76**, 685 (1954). Structure: P. G. Deuel, T. A. Geissman, *ibid.* **79**, 3778 (1957). Evaluation as repellent of blue mussel: A. Harada *et al.*, *Agric. Biol. Chem.* **49**, 1887 (1985).

Flat needles from methanol or ethanol, mp 114.5-115°. uv max: 275, 213 nm (ε 22800, 7300). $[\alpha]_D^{30}$ −20° (ethanol). Sol in ether, acetone, alcohol; slightly sol in hot water (neutral pH). Practically insol in petr ether, 5% NaOH, 5% HCl.

10257. Xanthine. [69-89-6] 3,7-Dihydro-1*H*-purine-2,6-dione; 2,6(1*H*,3*H*)-purinedione; 2,6-dioxopurine. $C_5H_4N_4O_2$; mol wt 152.11. C 39.48%, H 2.65%, N 36.83%, O 21.04%. Occurs in animal organs, yeast, potatoes, coffee beans, tea. First isolated from urinary bladder stones: *Beilstein* **26**, 447 (1937). Prepd by treating a sulfuric acid soln of guanine with sodium nitrite: Fischer, *Ann.* **215**, 309 (1882). Several other syntheses, *cf.* Levene, Bass, *Nucleic Acids* (New York, 1931).

Scales, plates from water. Dec on heating without melting and with partial sublimation. Kb at 40° = 6.09×10^{-14}; Ka at 40° = 1.19×10^{-10}. Absorption spectrum: Kalckar, *J. Biol. Chem.* **167**, 429 (1947). One gram dissolves in 14.5 liters of water at 16°, in 1.4 liters boiling water; less sol in alcohol. Sol in mineral acids, freely sol in NH_4OH and in NaOH solns.

1,3-Dimethylxanthine *see* Theophylline.

3,7-Dimethylxanthine *see* Theobromine.

1,3,7-Trimethylxanthine *see* Caffeine.

10258. Xanthinol Niacinate. [437-74-1] 3-Pyridinecarboxylic acid compd with 3,7-dihydro-7-[2-hydroxy-3-[(2-hydroxyethyl)-methylamino]propyl]-1,3-dimethyl-1*H*-purine-2,6-dione (1:1); 7-[2-hydroxy-3-[(2-hydroxyethyl)methylamino]propyl]theophylline compd with nicotinic acid; 7-[3-[*N*-(2-hydroxyethyl)amino]-2-hydroxypropyl]theophylline nicotinate; xanthinol nicotinate; SK-331-A; Complamin; Xavin. $C_{19}H_{26}N_6O_6$; mol wt 434.45. C 52.53%, H 6.03%, N 19.34%, O 22.10%. Prepn: Bestian, **DE 1102750** (1961 to Wuelfing), *C.A.* **56**, 11602i (1962). Alternate synthesis: Korbonits *et al.*, *Acta Pharm. Hung.* **38**, 98 (1968). Pharmacology: Hemmer, Diezemann, *Arzneim.-Forsch.* **12**, 672 (1962); Lennartz, *ibid.* 675; Stamm, *ibid.* 679.

Crystals, mp 180°. Freely sol in water.

THERAP CAT: Vasodilator (peripheral).

10259. Xanthocillin. [11042-38-9] Xantocillin; Brevicid. Antibiotic complex produced by *Penicillium notatum* Westling: Rothe, *Pharmazie* **5**, 190 (1950); Barwald, **GB 898498** (1962 to Arzneimittelwerk VEB), *C.A.* **57**, 9013a (1962). Consists of at least three antibiotics, xanthocillins X, Y$_1$ and Y$_2$; the first being the major component (about 70%).

Xanthocillin X   R = H   R' = H
Xanthocillin Y$_1$ R = OH   R' = H
Xanthocillin Y$_2$ R = OH   R' = OH

Xanthocillin X. [580-74-5] 4,4'-[(1Z,3Z)-2,3-Diisocyano-1,3-butadiene-1,4-diyl]bisphenol; bis(*p*-hydroxybenzylidene)ethylene isocyanide; 1,4-bis(*p*-hydroxyphenyl)-2,3-diisonitrilo-1,3-butadiene. $C_{18}H_{12}N_2O_2$; mol wt 288.31. Structure: Hagedorn, Tönjes, *Pharmazie* **11**, 409 (1956); *eidem,*, *ibid.* **12**, 567 (1957); Hagedorn *et al.*, *Ber.* **93**, 1584 (1960). Synthesis: Hagedorn, Eholzer, *Angew. Chem.* **74**, 215 (1962). Biosynthetic studies: Achenbach, Grisebach, *Z. Naturforsch.* **20b**, 137 (1965); Achenbach, König, *Ber.* **105**, 784 (1972). Crystal structure: D. Britton *et al.*, *Cryst. Struct. Commun.* **10**, 1497 (1981). Studies on antiviral activity: Takatsuki *et al.*, *J. Antibiot.* **21**, 671 (1968); *eidem, ibid.* **22**, 151 (1969); Kitahara, *ibid.* **34**, 1556 (1981). Clusters of yellow needles from alc; yellow rhombs from ethyl acetate. Chars at about 210°. Sol (up to 1%) in alc, ether, acetone, dioxane. Freely sol in aqueous alkaline solns. Practically insol in water, petr ether, benzene, chloroform. Forms a water-sol dipotassium salt, the concd solns of which have a pH of 10.5.

Dimethyl ether. $C_{20}H_{16}N_2O_2$. Yellow needles from dioxane + alcohol, dec 181°.

Xanthocillin Y$_1$. [38965-69-4] 4-[(1Z,3Z)-4-(4-Hydroxyphenyl)-2,3-diisocyano-1,3-butadien-1-yl]-1,2-benzenediol. $C_{18}H_{12}$-N_2O_3; mol wt 304.31.

Xanthocillin Y$_2$. [38965-70-7] 4,4'-[(1Z,3Z)-2,3-Diisocyano-1,3-butadiene-1,4-diyl]bis-1,2-benzenediol. $C_{18}H_{12}N_2O_4$; mol wt 320.30. Structure of xanthocillins Y$_1$ and Y$_2$: Achenbach *et al.*, *Ber.* **105**, 3061 (1972).

USE: In feed supplements.

THERAP CAT: Antibacterial.

10260. Xanthone. [90-47-1] 9*H*-Xanthen-9-one; 9-oxoxanthene; diphenylene ketone oxide; dibenzo-γ-pyrone; Genicide. C_{13}-H_8O_2; mol wt 196.21. C 79.58%, H 4.11%, O 16.31%. Prepd by heating phenyl salicylate: Graebe, *Ann.* **254**, 279 (1889); Holleman, *Org. Synth.* **coll. vol. I** (2nd ed., 1941) p 552.

Polymorphic needles from alcohol, mp 174°. bp$_{730}$ 351°. Sparingly volatile with steam. Dipole moment 3.0. Solubility: 0.55 g/100 ml of cold alcohol; 6.71 g/100 ml of boiling alcohol. Freely sol in chloroform. Slightly sol in hot water, ether, petr ether, benzene, toluene, xylene. Dissolves in concd H_2SO_4 to a yellow soln with pale blue fluorescence.

Oxime. Xanthoxime. mp 161°.

USE: In the prepn of xanthydrol. Ovicide for codling moth eggs; less effective as larvicide.

10261. Xanthophyll. [127-40-2] (3*R*,3'*R*,6'*R*)-β,ε-Carotene-3,3'-diol; lutein; vegetable lutein; vegetable luteol; Bo-Xan. C_{40}-$H_{56}O_2$; mol wt 568.89. C 84.45%, H 9.92%, O 5.62%. One of the most widespread carotenoid alcohols in nature. Originally isolated from egg yolk, also isolated by chromatography from nettles, algae, and petals of many yellow flowers. Occurs also in colored feathers of birds: Volker, *Z. Physiol. Chem.* **288**, 20 (1951). Extraction from petals of *Tagetes patula* L., *Compositae:* Karrer *et al.*, *Helv. Chim. Acta* **30**, 531 (1947). Occurs together with zeaxanthin, *q.v.* Dipalmitate occurs in *Helenium autumnale* L., *Compositae* and other flowers: Kuhn, Winterstein, *Naturwissenschaften* **18**, 754 (1930). Conversion to zeaxanthin with sodium alcoholate: Karrer, Jucker, *ibid.* 266. Does not possess vitamin A potency: Schumacher *et al.*, *Poult. Sci.* **23**, 529 (1944). Stereochemistry: Zechmeister, *Chem. Rev.* **34**, 267 (1944). Structure: Karrer, *Helv. Chim. Acta* **34**, 2160 (1951). Abs config: Goodfellow *et al.*, *Chem. Commun.* **1970**, 1578; R. Buchecker *et al.*, *Chimia* **25**, 192 (1971); *eidem, Helv. Chim. Acta* **57**, 631 (1974). Synthesis: H. Mayer, A. Rüttimann, *ibid.* **63**, 1451 (1980). Sepn and determn of configurational isomers: A. Rüttiman *et al.*, *J. High Resolut. Chromatogr. Chromatogr. Commun.* **6**, 612 (1983). Reviews: Zechmeister, *Carotinoide* (Berlin, 1934); Mayer, *The Chemistry of Natural Coloring Matters* (New York, 1943); Karrer, Jucker, *Carotenoids* (New York, 1950).

Yellow prisms with metallic luster from ether + methanol, mp 190° (corr), (a higher mp indicates impure material). Also reported as mp 183° [Buchecker (1974)]. [α]$_{Cd}^{18}$ +165° (c = 0.7 in benzene). Absorption max (dioxane): 481, 453, 429, 333, 268 nm (ε 142000, 152000, 100000, 15500, 35000). Insol in water, sol in fats and in fat solvents. More sol in boiling methanol (1:700) than zeaxanthin.

Dipalmitate. Helenien; Adaptinol. $C_{72}H_{116}O_4$; mol wt 1045.72. Red needles from alcohol, mp 92°.

10262. Xanthopterin. [119-44-8] 2-Amino-3,5-dihydro-4,6-pteridinedione; 2-amino-4,6-pteridinediol; 2-amino-4,6-dihydroxypteridine; 2-amino-4,6-dihydroxypyrimido[4,5-*b*]pyrazine. C_6H_5-N_5O_2; mol wt 179.14. C 40.23%, H 2.81%, N 39.10%, O 17.86%. Pigment first found in wings of butterflies. Widely distributed in insects and in animals. Has been separated from urine and from the crab. Isoln and structure: Schöpf, Becker, *Ann.* **507**, 266 (1933); **524**, 55, 126 (1936); Schöpf, Kottler, *Ann.* **539**, 128 (1939); Wieland, Purrmann, *Ann.* **544**, 163 (1940). *See also* Fukushima, Shiota, *J. Biol. Chem.* **247**, 4549 (1972). Synthesis: Purrmann, *Ann.* **546**, 98 (1940); **548**, 284 (1941); Koschara, *Z. Physiol. Chem.* **277**, 159 (1943); Totter, *J. Biol. Chem.* **154**, 105 (1944); Elion *et al.*, *J. Am. Chem. Soc.* **71**, 741 (1949); Koschara, **DE 859471** (1952 to Bayer), *C.A.* **52**, 10222f (1958); Stuart, Wood, *J. Chem. Soc.* **1963**, 4186. Facile synthesis: Taylor, Jacobi, *J. Am. Chem. Soc.* **95**, 4455 (1973); Taylor *et al.*, *J. Org. Chem.* **40**, 2341 (1975). Exhibits tumor inhibitory properties.

Monohydrate. Orange-yellow crystals, sinters around 360°, decomp above 410°. Practically insol in water. Freely sol in dil NH$_4$OH and NaOH giving yellow solns and in 2N HCl giving colorless solns. uv max at pH 11: 255, 390 nm (E$_{1cm}^{1\%}$ 0.92, 0.355). Can be converted by yeast into folic acid.

10263. Xanthosine. [146-80-5] Xanthine riboside; 9-β-D-ribofuranosyl xanthine; 9-β-D-ribofuranosyl-9H-purine-2,6-diol; 9-β-D-ribofuranosyl-9H-purine-2,6-(1H,3H)-dione. C$_{10}$H$_{12}$N$_4$O$_6$; mol wt 284.23. C 42.26%, H 4.26%, N 19.71%, O 33.77%. Prepd from guanosine by treatment with sodium nitrite and acetic acid: Levene, Jacobs, *Ber.* **43**, 3163 (1910); Gulland, Macrae, *J. Chem. Soc.* **1933**, 662. Synthesis: Howard *et al.*, *ibid.* **1949**, 232. Biosynthesis: Magasanik, Brooke, *J. Biol. Chem.* **206**, 83 (1954); Korn, Buchanan, *ibid.* **217**, 183 (1955); Bolis *et al.*, *ibid.* **219**, 917 (1956).

Dihydrate. Long prisms from water. Anhydr felted clusters (warts) from alc. Dec on heating, no distinct melting range. [α]$_D^{30}$ −51.2° (p = 8 in 0.3N NaOH). uv max: 253 nm (ε 8.790). α_M (molar absorbancy): 11.4×10^3 at 248.5 nm in water at pH 8.0. Sparingly sol in cold water; freely sol in hot water; sol in hot dil alcohol. Easily hydrolyzed by mineral acids.

10264. Xanthoxyletin. [84-99-1] 5-Methoxy-8,8-dimethyl-2H,8H-benzo[1,2-b:5,4-b']dipyran-2-one; 7-hydroxy-5-methoxy-2,2-dimethyl-2H-1-benzopyran-6-acrylic acid δ-lactone; xanthoxylin N; xanthoxyloin. C$_{15}$H$_{14}$O$_4$; mol wt 258.27. C 69.76%, H 5.46%, O 24.78%. Found in *Xanthoxylum americanum* Mill., *Melicope ternata* Forst., *M. mantelli* Buch., *Halifordia scleroxyla* F. Muell., *Rutaceae; Chloroxylon swietenia* DC., *Meliaceae.* Isoln: Staples, *Am. J. Pharm.* **1829**, 123; Witte, *Arch. Pharm.* **212**, 283 (1878); Lloyd, *Am. J. Pharm.* **1890**, 229; Gordin, *J. Am. Chem. Soc.* **28**, 1649 (1906); Bell *et al.*, *J. Chem. Soc.* **1936**, 627; Robertson, Subramaniam, *ibid.* **1937**, 286; Dieterle, Kruta, *Arch. Pharm.* **275**, 45 (1937); Briggs, Locker, *J. Chem. Soc.* **1951**, 3131; King *et al.*, *ibid.* **1954**, 1392; Hegarty, Lahey, *Aust. J. Chem.* **9**, 120 (1956); Cambie, *J. Chem. Soc.* **1960**, 2376. Synthesis: Joshi, Kamat, *Tetrahedron Lett.* **1966**, 5767.

Elongated prisms from methanol, mp 132-133°. uv max: 277, 269, 322, 347 nm (log ε 4.28, 4.32, 3.99, 4.05 in ethanol). Very easily sol in benzene, chloroform, hot alcohol; sol in acetone; sparingly sol in ether. Sol in 49 parts cold acetone and 25,000 parts cold water.

10265. Xanthoxylin. [90-24-4] 1-(2-Hydroxy-4,6-dimethoxy-phenyl)ethanone; 2'-hydroxy-4',6'-dimethoxyacetophenone; phloracetophenone 4,6-dimethyl ether; 2,4-dimethoxy-6-hydroxyacetophenone; brevifolin. C$_{10}$H$_{12}$O$_4$; mol wt 196.20. C 61.22%, H 6.17%, O 32.62%. Found in *Xanthoxylum piperitum* DC., *X. alatum* Roxb., *Rutaceae, Artemisia brevifolia* Wallich, *Compositae, Hippomane mancinella* L., *Sapium sebiferum* Roxb., *Euphorbiaceae.* Isoln: Stenhouse, *Ann.* **89**, 251 (1854); **104**, 236 (1857); Semmler, Schossberger, *Ber.* **44**, 2885 (1911); Smith, Smith, *Pharm. J.* **119**, 688 (1927); **123**, 604, 611 (1929); Schaeffer *et al.*, *J. Am. Pharm. Assoc.* **43**, 43 (1954); Chu *et al.*, *C.A.* **53**, 10532g (1959). Synthesis: Kostanecki, Lambor, *Ber.* **32**, 2260 (1899); Canter *et al.*, *J. Chem. Soc.* **1931**, 1245; Belton *et al.*, *Sci. Proc. R. Dublin Soc.* **25**, 19 (1949); Schmid, Bolleter, *Helv. Chim. Acta* **33**, 917 (1950); MacKenzie *et al.*, *J. Chem. Soc.* **1950**, 2965; Dean, Robertson, *ibid.* **1953**, 1244; Kawano, *Chem. Ind. (London)* **1959**, 368.

Crystals from ethanol, mp 81-83°. Practically insol in water. Sol in alcohol, ether.

10266. Xanthurenic Acid. [59-00-7] 4,8-Dihydroxy-2-quinolinecarboxylic acid; 4,8-dihydroxyquinaldic acid. C$_{10}$H$_7$NO$_4$; mol wt 205.17. C 58.54%, H 3.44%, N 6.83%, O 31.19%. Excreted by pyridoxine-deficient animals after the ingestion of tryptophan. Isoln from the urine of albino rats fed almost exclusively on fibrin: Musajo, *Gazz. Chim. Ital.* **67**, 165 (1937). Synthesis from ethyl oxalacetate and *o*-anisidine: Musajo, Minchilli, *Ber.* **74**, 1842 (1941); Mebane, Oroshnik, *J. Am. Chem. Soc.* **73**, 3520 (1951); Furst, Olsen, *J. Org. Chem.* **16**, 412 (1951).

Sulfur-yellow crystals, mp 286°. uv max (water): 243, 342 nm (ε 30000, 6500). Insol in water. Sol in aq alkali hydroxides and carbonates (yellow solns) and in hot dil HCl. It gives a red color with Millon reagent, intense ruby-red with alkali diazobenzenesulfonates, and intense green with FeSO$_4$. When the acid is dissolved in very dil NaHCO$_3$, the green color with FeSO$_4$ is still visible at a diln of 1:200,000.

Methyl ester. C$_{11}$H$_9$NO$_4$. Yellow crystals, dec 262°. Sol in NH$_4$OH, alkali hydroxides and carbonates.

10267. Xanthyletin. [553-19-5] 8,8-Dimethyl-2H,8H-benzo-[1,2-b:5,4-b']dipyran-2-one; 7-hydroxy-2,2-dimethyl-2H-1-benzopyran-6-acrylic acid δ-lactone; 2,2-dimethylchromenocoumarin. C$_{14}$H$_{12}$O$_3$; mol wt 228.25. C 73.67%, H 5.30%, O 21.03%. Linear pyranocoumarin found in *Xanthoxylum americanum* Mill., *Luvanga scandens* Ham., *Citrus acida* Roxb., *Rutaceae; Chloroxylon swietenia* DC., *Meliaceae.* Isoln: Bell, Robertson, *J. Chem. Soc.* **1936**, 1828; Bell *et al.*, *ibid.* **1937**, 1542; Späth *et al.*, *Ber.* **72**, 1450 (1939); Bose, Mookerjee, *J. Indian Chem. Soc.* **21**, 181 (1944); Mookerjee, *ibid.* **23**, 41 (1946); King *et al.*, *J. Chem. Soc.* **1954**, 1392. Synthesis: Späth, Hillel, *Ber.* **72**, 2093 (1939); Steck, *Can. J. Chem.* **49**, 2297 (1971); P. Waykole *et al.*, *Indian J. Chem.* **19B**, 238 (1980); V. K. Ahluwalia *et al.*, *Monatsh. Chem.* **112**, 119 (1981); J. Banerji *et al.*, *Indian J. Chem.* **21B**, 496 (1982).

Elongated flat prisms from methanol, mp 130-131°. bp$_{0.1}$ 140-145°. uv max: 266, 348 nm (log ε 4.34, 4.15).

10268. Xantphos. [161265-03-8] 1,1'-(9,9-Dimethyl-9*H*-xanthene-4,5-diyl)bis[1,1-diphenylphosphine]; 4,5-bis(diphenylphosphino)-9,9-dimethylxanthene; 9,9-dimethyl-4,5-bis(diphenylphosphino)xanthene. C$_{39}$H$_{32}$OP$_2$; mol wt 578.63. C 80.95%, H 5.57%, O 2.76%, P 10.71%. Diphosphine ligand with a wide bite angle; chelates with transition metals to form complexes that are used as catalysts. Prepn and X-ray crystal structure analysis: S. Hillebrand *et al.*, *Tetrahedron Lett.* **36**, 75 (1995); and use of rhodium complexes as catalysts in hydroformylations: M. Kranenburg *et al.*, *Organometallics* **14**, 3081 (1995). Synthetic utility of complexes in hydroformylations: G. Petöcz *et al.*, *J. Organomet. Chem.* **689**, 1188 (2004); in amidations: J. Yin, S. L. Buchwald, *J. Am. Chem. Soc.* **124**, 6043 (2002); in cross-coupling reactions: D. Seomoon, P. H. Lee, *J. Org. Chem.* **73**, 1165 (2008); in carbonylations: J. R. Martinelli *et al.*, *ibid.* 7102; in carbonyl transfer hydrogenations: A. N. Kharat *et al.*, *Inorg. Chem. Commun.* **14**, 1161 (2011). Review of xantphos ligands: P. C. J. Kamer *et al.*, *Acc. Chem. Res.* **34**, 895-904 (2001).

Colorless crystals from chloroform + 2-propanol, mp 230-232° (Hillebrand). Also reported as yellow-white powder, mp 221-222° (Kranenburg). d 1.21.

USE: Ligand in synthetic organic chemistry.

10269. Xemilofiban. [149820-74-6] (3*S*)-3-[[4-[[4-(Aminoiminomethyl)phenyl]amino]-1,4-dioxobutyl]amino]-4-pentynoic acid ethyl ester; ethyl (3*S*)-3-[3-[(*p*-amidinophenyl)carbamoyl]propionamido]-4-pentynoate. C$_{18}$H$_{22}$N$_4$O$_4$; mol wt 358.40. C 60.32%, H 6.19%, N 15.63%, O 17.86%. Platelet fibrinogen receptor (GPIIb/IIIa) antagonist. Prodrug converted *in vivo* to the active acid form. Prepn: P. R. Bovy *et al.*, **WO 9307867**; *eidem*, **US 5344957** (1993, 1994 both to Monsanto; Searle); J. Cossy *et al.*, *Bioorg. Med. Chem. Lett.* **7**, 1699 (1997). Structure-activity study: J. A. Zablocki *et al.*, *J. Med. Chem.* **38**, 2378 (1995). Pharmacology: N. S. Nicholson *et al.*, *Circulation* **91**, 403 (1995). Clinical pharmacokinetics: D. J. Kereiakes *et al.*, *ibid.* **94**, 906 (1996). Clinical evaluation in unstable angina: C. Simpfendorfer *et al.*, *ibid.* **96**, 76 (1997).

Hydrochloride. [156586-91-3] SC-54684A. C$_{18}$H$_{22}$N$_4$O$_4$.·HCl; mol wt 394.86.

Acid. [149193-61-3] SC-54701A. C$_{16}$H$_{18}$N$_4$O$_4$; mol wt 330.34. [α]$_D$ −33.7° (c = 1.45 in methanol).

THERAP CAT: Antithrombotic.

10270. Xenon. [7440-63-3] Xe; at. wt 131.29; at. no. 54. Valences 2, 4, 6, 8. Group VIIIA (18), also known as Group 0. A noble gas characterized by an electronic structure in which the outer *p* subshell is entirely filled. Naturally occurring stable isotopes (mass numbers): 124 (0.10%); 126 (0.09%); 128 (1.91%); 129 (26.4%); 130 (4.1%); 131 (21.2%); 132 (26.9%); 134 (10.4%); 136 (8.9%);

known artificial, radioactive isotopes: 110-123; 125; 127; 133; 135; 137-145. Discovered in the final residues obtained after evaporating liq air: Ramsay, Travers, *Proc. Roy. Soc.* **63** [A], 405 (1898). Occurs frequently in gases evolved from thermal springs; concentration in air: 0.087 ppm by vol. Obtained commercially from the atmosphere by distillation-liquefaction process. Extraction from liq air residues: Allen, Moore, *J. Am. Chem. Soc.* **53**, 2512 (1931). *Xenon platinum hexafluoride* was the first reported xenon compound: N. Bartlett, *Proc. Chem. Soc. London* **1962**, 218. Teratogenicity study: G. A. Lane *et al.*, *Science* **210**, 899 (1980). Review of biology, chemistry and anesthetic properties: R. M. Featherstone, C. A. Muelbaecher, *Pharmacol. Rev.* **15**, 97 (1963). Review of diagnostic use of radioactive compds for pulmonary studies: F. Fazio, P. Wollman, *Clin. Physiol.* **1**, 323 (1981); for cerebral blood flow: H. Yonas *et al.*, *Adv. Tech. Stand. Neurosurg.* **15**, 3 (1987). Reviews of chemistry and compds: *Noble-Gas Compounds*, H. H. Hyman, Ed. (Univ. Chicago Press, Chicago, 1963) 404 pp; J. H. Holloway, *Noble-Gas Chemistry* (Methuen, London, 1968) 213 pp; Sladky, "Noble Gases" in *MTP Int. Rev. Sci.: Inorg. Chem., Ser. One* vol. 3, V. Gutman, Ed. (Butterworths, London, 1972) pp 1-52; Cockett, Smith, "The Monatomic Gases" in *Comprehensive Inorganic Chemistry* vol. 1, J. C. Bailar, Jr. *et al.*, Eds. (Pergamon Press, Oxford, 1973) pp 139-211; Bartlett, Sladky, *ibid.* pp. 213-330; S.-C. Hwang, W. R. Weltmer, Jr. in *Kirk-Othmer Encyclopedia of Chemical Technology* vol. **13** (John Wiley & Sons, 4th ed., 1995) pp 1-38; G. J. Schrobilgen, J. M. Whalen, *ibid.* pp 38-53; *Chemistry of the Elements* N. N. Greenwood, A. Earnshaw, Eds. (Pergamon Press, New York, 1984) pp 1042-1059.

Colorless, odorless, tasteless, relatively inert, monatomic gas; will form compds with highly electronegative elements such as O, F, Cl. *Non-flammable.* Soly of gas in water (20°): 108.1 cm^3/kg water. Triple temp 161.35 K, press 81.66 kPa. Critical temp 289.74 K, critical press 5840 kPa, critical d 1100 kg/m^3. Gas: d^0 (101.3 kPa) 5.8971 kg/m^3, d (normal bp) 11 kg/m^3. Liquid: normal bp −108.13°, d (normal bp) 3057 kg/m^3, d (triple pt) 3084 kg/m^3, heat of vaporization (normal bp) 12.640 kJ/mol. Solid: d (triple pt) 3540 kg/m^3, heat of vaporization (triple pt) 15.1 kJ/mol, heat of fusion (triple pt) 2.3 kJ/mol. Solid form exists as face-centered cubic crystals at normal pressure. Spectrum: Collie, *Proc. Roy. Soc.* **97** [A], 349 (1920). Emission spectra: T. Jacksier, R. M. Barnes, *Appl. Spectrosc.* **48**, 65 (1994).

Xenon Xe 133. [14932-42-4] Xeneisol Xe 133.

Hydrate. [60212-94-4] Xe.*x*H$_2$O, mp 24°, and a deuterate, Xe.6D$_2$O, have been prepd: R. de Forcrand, *Compt. Rend.* **176**, 355 (1923); **181**, 15 (1925); Tamman, Krige, *Z. Anorg. Chem.* **146**, 179 (1925).

Xenon difluoride. [13709-36-9] F$_2$Xe; mol wt 169.29. Prepd from the elements: Weeks *et al.*, *J. Am. Chem. Soc.* **84**, 4612 (1962); Hoppe *et al.*, *Z. Anorg. Allg. Chem.* **324**, 214 (1963). Colorless crystals; d 4.32. Triple pt. 129.03°: Schreiner *et al.*, *J. Phys. Chem.* **72**, 1162 (1968). Sublimes without decompn. Soly in water at 0°: 25 g/l.

Xenon tetrafluoride. [13709-61-0] F$_4$Xe; mol wt 207.29. First prepd by direct combination of the elements at 6 atm and 400°: Claassen *et al.*, *J. Am. Chem. Soc.* **84**, 3593 (1962). Colorless crystals; d 4.04. Triple pt 117.10°: Schreiner *et al.*, *loc. cit.* Sublimes without decompn. Hydrolyzes to form Xe, O$_2$, HF and XeO$_3$.

Xenon hexafluoride. [13693-09-9] F$_6$Xe; mol wt 245.28. Laboratory prepn: Chernick *et al.*, *Inorg. Synth.* **8**, 258 (1966). Colorless solid; greenish-yellow vapor; vapor press. about 30 mm at 25°. mp 49.48°; bp 75.57°; d$^{24.4}$(solid) 3.411; d$^{55.2}$(liq) 3.173: Schreiner *et al.*, *J. Chem. Phys.* **51**, 4838 (1969). Hydrolyzed by water to form XeOF$_4$ and XeO$_3$. More powerful oxidizing and fluorinating agent than XeF$_2$ and XeF$_4$. Cannot be stored in glass or quartz containers.

Xenon trioxide. [13776-58-4] O$_3$Xe; mol wt 179.29. Prepn: Williamson, Koch, *Science* **139**, 1046 (1963); Jaselskis *et al.*, *J. Am. Chem. Soc.* **88**, 2149 (1966). *Powerful explosive*, formed when XeF$_4$ and XeF$_6$ are hydrolyzed. Colorless, hygroscopic solid; d 4.55. Aqueous solns, "*xenic acid*", may be prepd in concns >2*M*.

Caution: Can act as a simple asphyxiant by displacing air. *See: Matheson Gas Data Book* (Matheson Co., Inc., 4th ed., East Rutherford, NJ, 1966) pp 499-500.

USE: Gas in lamps designed to resemble natural daylight; in lamps of extremely high brilliance. Isotopes in leak detection systems for nuclear reactors.

THERAP CAT: Anesthetic (inhalation). Xenon Xe 133 as diagnostic aid (radioactive imaging agent).

10271. p-Xenylcarbimide. [92-95-5] 4-Isocyanato-1,1'-biphenyl; isocyanic acid 4-biphenylyl ester; p-diphenyl isocyanate. $C_{13}H_9NO$; mol wt 195.22. C 79.98%, H 4.65%, N 7.17%, O 8.20%. Prepd from the corresponding amine and phosgene: Kaplan, *J. Chem. Eng. Data* **6**, 272 (1961).

Needles, mp 56°; bp 283° (decomp), bp$_{2-3}$ 137-140°. Very sol in ether.

USE: As a reagent for the identification of hydroxy compounds, with which it gives crystalline derivatives of definite melting point.

10272. X-gal. [7240-90-6] 5-Bromo-4-chloro-1H-indol-3-yl β-D-galactopyranoside. $C_{14}H_{15}BrClNO_6$; mol wt 408.63. C 41.15%, H 3.70%, Br 19.55%, Cl 8.68%, N 3.43%, O 23.49%. Chromogenic substrate of the *lacZ* gene product, β-galactosidase; produces a rich blue color when cleaved by the enzyme. Prepn: J. P. Horwitz *et al.*, *J. Med. Chem.* **7**, 574 (1964). Colorimetric reaction in tissue to localize β-galactosidase: B. Pearson *et al.*, *Lab. Invest.* **12**, 1249 (1963); Z. Lojda, *Histochemie* **23**, 266 (1970). Protocol for blue-white α-complementation screening: J. Sambrook *et al.*, *Molecular Cloning: A Laboratory Manual* (Cold Spring Harbor, New York, 2nd ed., 1989) pp 1.7-1.9, 1.85-1.87. Optimization of detection of exogenous bacterial *lacZ* expression: D. J. Weiss *et al.*, *Hum. Gene Ther.* **8**, 1545 (1997); of fixative conditions in X-gal staining: W. Ma *et al.*, *J. Histochem. Cytochem.* **50**, 1421 (2002); of blue-white screening procedure: A. L. Sherwood, *BioTechniques* **34**, 644 (2003).

Amorphous solid from methanol, mp 237-239° (dec). $[\alpha]_D^{24}$ −69° (c = 1.0 in 50% DMF). Sol in DMF, DMSO.

USE: In histochemical staining, in blue-white selection of bacterial colonies with *lacZ* gene activity during cloning experiments.

10273. Xibornol. [13741-18-9] rel-4,5-Dimethyl-2-[(1R,2S,-4S)-1,7,7-trimethylbicyclo[2.2.1]hept-2-yl]phenol; 6-isobornyl-3,4-dimethylphenol; 6-(2-isobornyl)-3,4-xylen-1-ol; 3,4-dimethyl-6-isobornylphenol; Nanbacine. $C_{18}H_{26}O$; mol wt 258.41. C 83.66%, H 10.14%, O 6.19%. Syntheses: Moldovanskaya *et al.*, *C.A.* **66**, 76163p (1967); Starkov, Glushkova, *J. Appl. Chem. USSR* **40**, 209 (1967), *C.A.* **67**, 72960u (1967), *C.A.* **71**, 70225s (1969). Prepn of cryst substance: **GB 1206774** and Mardiguian, Fournier, **DE 2032170** (1970 and 1971 to MARPHA Soc. d'Etudes et d'Exploitation de Marques), *C.A.* **74**, 13299g, 88171t (1971). Activity study: Capponi, *Bull. Soc. Pathol. Exot.* **62**, 658 (1969), *C.A.* **73**, 118883a (1970).

Relative stereochemistry

Crystals from petr ether, mp 94-96°, bp$_3$ 165-168°. Also reported as very viscous, pale yellow liquid; bp$_9$ 185-189°. d$_4^{20}$ 1.0240. n_D^{20} 1.5382 (Starkov, Glushkova).

USE: Rubber antioxidant.

THERAP CAT: Antibacterial.

10274. Ximelagatran. [192939-46-1] N-[(1R)-1-Cyclohexyl-2-[(2S)-2-[[[[4-[(hydroxyamino)iminomethyl]phenyl]methyl]-amino]carbonyl]-1-azetidinyl]-2-oxoethyl]glycine ethyl ester; H-376/95; Exanta. $C_{24}H_{35}N_5O_5$; mol wt 473.57. C 60.87%, H 7.45%, N 14.79%, O 16.89%. Orally active, direct thrombin inhibitor; prodrug of melagatran, *q.v.* Prepn: T. Antonsson *et al.*, **WO 9723499**; D. Gustafsson *et al.*, **US 5965692** (1997, 1999 both to Astra). CE determn in drug formulations: P. K. Owens *et al.*, *J. Pharm. Biomed. Anal.* **27**, 587 (2002). LC-MS determn in biological samples: M. Larsson *et al.*, *J. Chromatogr. B* **783**, 335 (2003). Clinical pharmacokinetics: K. Wahlander *et al.*, *Thromb. Res.* **107**, 93 (2002); in deep vein thrombosis: M. Cullberg *et al.*, *Clin. Pharmacol. Ther.* **77**, 279 (2005). Clinical study vs warfarin as thromboprophylactic: C. W. Francis *et al.*, *Ann. Intern. Med.* **137**, 648 (2002). Review of clinical experience in acute coronary syndromes: J. S. Kalus, M. F. Caron, *Expert Opin. Invest. Drugs* **13**, 465-477 (2004); of clinical development: C. J. Boos *et al.*, *Eur. J. Int. Med.* **16**, 267-278 (2005).

THERAP CAT: Antithrombotic.

10275. Ximoprofen. [56187-89-4] 4-[3-(Hydroxyimino)cyclohexyl]-α-methylbenzeneacetic acid; p-(3-oxocyclohexyl)hydratropic acid oxime; XIFAM; 2-[4-(3-oximinocyclohexyl)phenyl]propionic acid; 13832-JL. $C_{15}H_{19}NO_3$; mol wt 261.32. C 68.94%, H 7.33%, N 5.36%, O 18.37%. Nonsteroidal anti-inflammatory. Prepn: J. G. Maillard, **DE 2442910**; *idem*, **US 3935255** (1975, 1976 both to Lab. Jacques Logeais). Metabolism: B. C. Mayo *et al.*, *Xenobiotica* **20**, 233 (1990). Clinical pharmacokinetics: I. W. Taylor *et al.*, *Br. J. Clin. Pharmacol.* **32**, 242 (1991). GC determn in plasma: I. W. Taylor, L. F. Chasseaud, *J. Chromatogr.* **495**, 275 (1989). HPLC determn in urine: G. R. Morris *et al.*, *ibid.* **530**, 377 (1990). Clinical evaluation: M. Dougados *et al.*, *J. Rheumatol.* **16**, 1167 (1989).

Crystals from water/methanol, mp 178°.
THERAP CAT: Anti-inflammatory.

10276. Xipamide. [14293-44-8] 5-(Aminosulfonyl)-4-chloro-N-(2,6-dimethylphenyl)-2-hydroxybenzamide; 4-chloro-2',6'-dimethyl-5-sulfamoylsalicylanilide; 4-chloro-5-sulfamylsalicyloyl-2',6'-dimethylanilide; Bei-1293; Aquaphor; Chronexan; Diurexan; Lumitens. $C_{15}H_{15}ClN_2O_4S$; mol wt 354.81. C 50.78%, H 4.26%, Cl 9.99%, N 7.90%, O 18.04%, S 9.04%. Salicylic acid derivative. Prepn: **NL 6607680**; Liebenow, **US 3567777** (1966, 1970 both to P. Beiersdorf). Clinical studies: O. Hammer, U. Dembowski, *Med. Klin.* **64**, 1862 (1969); R. Fischer, A. Lenhartz, *Med. Welt* **1970**, 270. Review of pharmacology and therapeutic efficacy: B. N. C. Prichard, R. N. Brogden, *Drugs* **30**, 313-332 (1985).

Crystals from methanol-water, mp 256°.
THERAP CAT: Diuretic; antihypertensive.

10277. Xylazine. [7361-61-7] N-(2,6-Dimethylphenyl)-5,6-dihydro-4H-1,3-thiazin-2-amine; 5,6-dihydro-2-(2,6-xylidino)-4H-

1,3-thiazine; 2-(2,6-dimethylphenylamino)-4H-5,6-dihydro-1,3-thi-azine; Bay 1470; Bay Va 1470; Wh-7286. $C_{12}H_{16}N_2S$; mol wt 220.33. C 65.42%, H 7.32%, N 12.71%, S 14.55%. Prepn: O. Behner et al., **BE 634552**; eidem, **DE 1173475**; eidem, **US 3235550** (1964, 1964, 1966 all to Bayer). Pharmacology: G. Sagner et al., Dtsch. Tieraerztl. Wochenschr. **75**, 565 (1968); Kroneberg, Schlossman, Arch. Pharmakol. Exp. Pathol. **268**, 348 (1971). Metabolic studies: Duhm et al., Berl. Muench. Tieraerztl. Wochenschr. **82**, 104 (1969). Veterinary clinical studies: Clarke, Hall, Vet. Rec. **85**, 512 (1969); Burns, McMullan, Vet. Med. **67**, 77 (1972).

Colorless, almost tasteless crystals from benzene-ligroin, mp 140-142°. Also reported as mp 136-139°. Soluble in benzene; sparingly sol in dilute acids, acetone, chloroform, petr ether. Practically insol in water; insol in dilute alkali. LD_{50} in mice (mg/kg): 43 i.v.; 121 s.c.; 240 orally; LD_{50} in rats (mg/kg): 130 orally (Sagner).

Hydrochloride. [23076-35-9] Narcoxyl; Rompun; Solvazine; Xylapan; Xylasol. $C_{12}H_{16}N_2S \cdot HCl$; mol wt 256.79. Colorless to white crystals. Sparingly sol in dilute acid, acetone, methanol. Insol in dilute alkali.

THERAP CAT (VET): Sedative, analgesic, muscle relaxant.

10278. Xylene. [1330-20-7] Dimethylbenzene; xylol. C_8H_{10}; mol wt 106.17. C 90.50%, H 9.49%. First isolated from a crude wood distillate: Cahours, Compt. Rend. **30**, 319 (1850). Obtained from coal tar: Fittig, Ann. **153**, 265 (1870). The xylene of commerce is a mixture of the three isomers o-, m- and p-xylene, the m-isomer predominating. Manuf from pseudocumene: Seubold, **US 2960545** (1960 to Union Oil); by catalytic isomerization of a hydrocarbon fraction: Berger, **US 3078318** (1963 to Universal Oil Prod.). Separation of isomers by clathration: Schaeffer, **US 3029300** (1962 to Union Oil). Use as clearing agent: K. Kubota, J. Polym. Sci. **5**, 1179 (1967); J. B. Matthews, J. Clin. Pathol. **34**, 103 (1981). Toxicity data: H. F. Smyth et al., Am. Ind. Hyg. Assoc. J. **23**, 95 (1962). Series of articles on toxicity: Environ. Health Perspect. **101**, Suppl. 6, 115-149 (1993). Review of toxicology and human exposure: Toxicological Profile for Xylene (PB2008-100008, 2007) 385 pp; of mfg processes and properties: W. J. Cannella in Kirk-Othmer Encyclopedia of Chemical Technology **Suppl.** (Wiley-Interscience, New York, 4th ed., 1998) pp 831-863.

Mobile liquid. *Flammable.* d ~0.86; bp 137-140°. Flash pt 29°C. Practically insol in water. Miscible with abs alcohol, ether, and many other organic liquids.

m-**Xylene.** [108-38-3] Colorless liquid; d_4^{15} 0.8684; mp −47.4°; bp 139.3°; n_D^{20} 1.4973. Flash pt, closed cup: 81°F (27°C). Insol in water. Miscible with alc, ether, and many other organic solvents. LD_{50} orally in rats: 7.71 ml/kg (Smyth).

o-**Xylene.** [95-47-6] Colorless liquid; d_4^{20} 0.8801; mp −25°; bp 144°; n_D^{20} 1.5058. Flash pt, closed cup: 90°F (32°C). Miscible with alc, ether.

p-**Xylene.** [106-42-3] Colorless plates or prisms at low temp; d_4^{20} 0.86104; mp 13-14°; bp 137-138°; n_D^{20} 1.49575: Thorne et al., Ind. Eng. Chem. Anal. Ed. **17**, 481 (1945). Flash pt, closed cup: 81°F (27°C). Sol in alc, ether, organic solvents. Insol in water.

Caution: Potential symptoms of acute overexposure by inhalation to m-, o-, or p-isomers are flushing and reddening of the face, a feeling of increased heat due to dilation of superficial blood vessels, disturbed vision, dizziness, tremors, salivation, cardiac stress, drowsiness, incoordination and staggering gait, CNS depression, confusion, coma. Symptoms of chronic inhalation exposure may include respiratory irritation, CNS excitation followed by CNS depression,

paresthesia, tremors, apprehension, impaired memory, weakness, nervous irritation, vertigo, headache, anorexia, nausea, flatulence, anemia, mucosal hemorrhage. Overexposure by ingestion may cause severe GI distress. Aspiration into lungs may cause chemical pneumonitis, pulmonary edema, hemorrhage. Direct contact may cause eye irritation, conjunctivitis, corneal burns; skin irritation and dermatitis due to defatting action. See Patty's Industrial Hygiene and Toxicology **vol. 2B**, G. D. Clayton, F. E. Clayton, Eds. (Wiley-Interscience, New York, 4th ed., 1994) p 1332-1339; NIOSH Pocket Guide to Chemical Hazards (DHHS/NIOSH 97-140, 1997) pp 334-337.

USE: As solvent; raw material for production of benzoic acid, phthalic anhydride, isophthalic and terephthalic acids as well as their dimethyl esters used in the manufacture of polyester fibers; manuf dyes and other organics; sterilizing catgut; with Canada balsam as oil-immersion in microscopy; clearing agent in microscope technique.

10279. Xylenol. [1300-71-6] Dimethylphenol. $C_8H_{10}O$; mol wt 122.17. C 78.65%, H 8.25%, O 13.10%. Constituent of "cresylic acid." There are 6 isomers of xylenol. They are only slightly sol in water but freely sol in alcohol, chloroform, ether, benzene, etc.; also sol in NaOH soln. Prepn and physical properties: R. J. L. Andon et al., J. Chem. Soc. **1960**, 5246.

2,3-Dimethylphenol. [526-75-0] vic-o-Xylenol. Needles from water or dil alc, mp 75°, bp 218°: Thöl, Ber. **18**, 2561 (1885); also reported as mp 72.57 ±0.02°, bp_{760} 216.87 ±0.001° (Andon).

2,4-Dimethylphenol. [105-67-9] as-m-Xylenol. Crystals, bp_{766} 211.5°, mp 25.4-26°: Jacobsen, Ber. **11**, 17 (1878); **18**, 3463 (1885); also reported as mp 24.54 ±0.01°, bp_{760} 210.931 ±0.001° (Andon).

2,5-Dimethylphenol. [95-87-4] p-Xylenol. Crystals from alcohol + ether, mp 74.5°, bp_{762} 211.5° (Jacobsen); bp 213.5°: Würtz, Ann. **147**, 372 (1868); also reported as mp 74.85 ±0.02°, bp_{760} 211.132 ±0.002° (Andon).

2,6-Dimethylphenol. [576-26-1] vic-m-Xylenol. Needles, mp 49°, bp 203°: Gattermann, Ann. **357**, 313 (1907); also reported as mp 45.62 ±0.01°, bp_{760} 201.030 ±0.001° (Andon).

3,4-Dimethylphenol. [95-65-8] as-o-Xylenol. Needles from water, mp 62.5°, bp 225°: Jacobsen, Ber. **17**, 159 (1884); also reported as mp 65.11 ±0.01°, bp_{760} 226.947 ±0.001° (Andon).

3,5-Dimethylphenol. [108-68-9] sym-m-Xylenol. Needles from water, mp 64°, bp 219.5°: Thöl, Ber. **18**, 359 (1885); also reported as mp 63.27 ±0.02°, bp_{760} 221.962 ±0.003° (Andon). LD_{50} orally in mice: 620 mg/kg: W. A. McOmie et al., J. Am. Pharm. Assoc. **38**, 366 (1949).

USE: For the prepn of coal tar disinfectants; manuf of artificial resins.

10280. Xylenol Blue. [125-31-5] 4,4′-(1,1-Dioxido-3H-2,1-benzoxathiol-3-ylidene)bis[2,5-dimethylphenol]; p-xylenolsulfonephthalein; 1,4-dimethyl-5-hydroxybenzenesulfonephthalein; α,4,4′-trihydroxy-2,5,2′,5′-tetramethyltriphenylmethane-2″-sulfonic acid γ-sultone. $C_{23}H_{22}O_5S$; mol wt 410.48. C 67.30%, H 5.40%, O 19.49%, S 7.81%. Prepd from o-sulfobenzoic acid dichloride or o-sulfobenzoic acid anhydride and p-xylenol in the presence of zinc chloride: Cohen, Biochem. J. **16**, 31 (1922).

Brown crystals from alcohol.

USE: Indicator, used in 0.02% soln: pH 1.2 red, 2.8 yellow, 9.6 blue.

10281. Xylenol Orange. [1611-35-4] N,N'-[(1,1-Dioxido-3H-2,1-benzoxathiol-3-ylidene)bis[(6-hydroxy-5-methyl-3,1-phenylene)methylene]]bis[N-(carboxymethyl)glycine]; 3-3′-bis[N,N-bis-(carboxymethyl)aminomethyl]-o-cresolsulfonephthalein. $C_{31}H_{32}N_2O_{13}S$; mol wt 672.66. C 55.35%, H 4.80%, N 4.16%, O 30.92%, S 4.77%. Cation chelant; sulfonephthalein titration indicator that forms soluble, colored complexes with metal ions. Exists in multiple sodium salt forms. Prepn from cresol red, $q.v.$: J. Körbl, R. Pribil, *Chem. Ind. (London)* **1957**, 233; and chromatography: M. Murakami *et al.*, *Talanta* **14**, 1293 (1967); H. Nakayama *et al.*, *Anal. Sci.* **5**, 619 (1989). Polarography studies: M. Véber, L. J. Csányi, *J. Electroanal. Chem.* **101**, 419 (1979). Ionization constant determn: M. B. Gholivand *et al.*, *Talanta* **46**, 875 (1998). Spectroscopy studies: S. Kiciak, H. Gontarz, *ibid.* **33**, 341 (1986). Metal chelate studies: K. Ogura *et al.*, *J. Inorg. Nucl. Chem.* **43**, 1243 (1981). Hydroperoxide assay studies: C. Gay *et al.*, *Anal. Biochem.* **273**, 149 (1999); C. A. Gay, J. M. Gebicki, *ibid.* **315**, 29 (2003); K. Arab, J.-P. Steghens, *ibid.* **325**, 158 (2004). Use in fluorometric detection of calcified tissue: B. A. Rahn, S. M. Perren, *Stain Technol.* **46**, 125 (1971); Y.-H. Wang *et al.*, *Biotechnol. Prog.* **22**, 1697 (2006); in fluorine determn: J. A. Ruzicka, *et al.*, *Talanta* **13**, 1341 (1966); in DNA determn: X. Chen *et al.*, *Spectrochim. Acta A* **61**, 2215 (2005). Reviews of analytical uses: B. Budesinsky in *Chelates in Analytical Chemistry*, H. A. Flaschka, A. J. Barnard, Jr., Eds. (Marcel Dekker, Inc., 1967) pp 15-47; S. Rani *et al.*, *Chem. Era* **13**, 372-388 (1978).

Dark solid. pKa$_1$ −1.74; pKa$_2$ −1.09; pKa$_3$ 0.76; pKa$_4$ 1.15; pKa$_5$ 2.58; pKa$_6$ 3.23; pKa$_7$ 6.40; pKa$_8$ 10.46; pKa$_9$ 12.58. Color of aq soln transitions from yellow to red between pH 6.4 and 10.4; color is dependent upon protonation state.

Tetrasodium salt. [3618-43-7] $C_{31}H_{28}N_2Na_4O_{13}S$; mol wt 760.59. Dark solid. Absorption max (0.1 N sodium hydroxide): 580 nm. Excitation max: 440 nm; 570 nm. Emission max: 610 nm. Freely sol in water; very slightly sol in ethanol.

USE: Chromogenic indicator for complexometric and spectrophotometric determn of numerous metal ions (restricted to pH range 0.0 to 6.0). Stain to demonstrate calcification in biological tissue. In the detection of fluorine, hydroperoxides, and DNA.

10282. Xylidine. [1300-73-8] ar,ar-Dimethylbenzenamine; dimethylaniline; aminodimethylbenzene. $C_8H_{11}N$; mol wt 121.18. C 79.29%, H 9.15%, N 11.56%. Class of aromatic amines; six isomeric xylidines exist depending on the position of the methyl groups. Prepd by reduction of corresponding nitro-compounds: Allchin, **US 1867962** (to I.C.I.); physical properties of all six isomers also given: S. F. Birch *et al.*, *J. Am. Chem. Soc.* **71**, 1362 (1949); van Loon *et al.*, *Rec. Trav. Chim.* **79**, 977 (1960); E. D. Bergmann, S. Berkovic, *J. Org. Chem.* **26**, 919 (1961).

All except 3,4-dimethylbenzenamine are liquids above 20°. d 0.97-0.99, and bp 213-226°. *Poisonous.* They are sparingly sol in water, sol in alcohol and form more or less sol salts with the strong mineral acids.

2,3-Dimethylbenzenamine. [87-59-2] 3-Amino-o-xylene; 2,3-dimethylaniline; 2,3-xylidine. bp$_{25}$ 118-119°; bp$_{11}$ 98-100°. Flash pt, closed cup: 205°F (96°C).

2,4-Dimethylbenzenamine. [95-68-1] 4-Amino-1,3-xylene; 2,4-dimethylaniline; 2,4-xylidine. mp 16°. bp 214.3°; bp$_3$ 76-78°. d^{25} 0.9723. Flash pt, closed cup: 208°F (98°C).

2,5-Dimethylbenzenamine. [95-78-3] 2-Amino-1,4-xylene; 2,5-dimethylaniline; 2,5-xylidine. bp$_{760}$ 217°. Flash pt, closed cup: 205°F (96°C).

2,6-Dimethylbenzenamine. [87-62-7] 2-Amino-1,3-xylene; 2,6-dimethylaniline; 2,6-xylidine. bp$_{735}$ 216°. d^{15} 0.980. Flash pt, closed cup: 196°F (91°C).

3,4-Dimethylbenzenamine. [95-64-7] 4-Amino-o-xylene; 3,4-dimethylaniline; 3,4-xylidine. mp 50-51°. bp$_{45}$ 132-134°. Flash pt, closed cup: 225°F (107°C).

3,5-Dimethylbenzenamine. [108-69-0] 5-Amino-1,3-xylene; 3,5-dimethylaniline; 3,5-xylidine. bp 222°. Flash pt, closed cup: 199°F (93°C).

Caution: Potential symptoms of overexposure are anoxia, cyanosis, methemoglobinemia; lung, liver and kidney damage. *See NIOSH Pocket Guide to Chemical Hazards* (DHHS/NIOSH 97-140, 1997) p 336.

USE: Reagents in synthetic organic chemistry, chiefly in the manuf of dyes.

10283. Xylitol. [87-99-0] *xylo*-Pentane-1,2,3,4,5-pentol; xylite; Eutrit; Kannit; Klinit; Kylit; Newtol; Torch; Xyliton. $C_5H_{12}O_5$; mol wt 152.15. C 39.47%, H 7.95%, O 52.58%. Intermediate in metabolism of D-glucose through glucuronate cycle in livers. Prepd by reduction of xylose: G. Bertrand, *Bull. Soc. Chim. Fr.* [3] **5**, 555 (1891); E. Fischer, R. Stahel, *Ber.* **24**, 538 (1891). Prepn of metastable crystals: M. L. Wolfrom, E. J. Kohn, *J. Am. Chem. Soc.* **64**, 1739 (1942); of stable form: J. F. Carson *et al.*, *ibid.* **65**, 1777 (1943). Crystal structure: H. S. Kim, G. A. Jeffrey, *Acta Crystallogr.* **25B**, 2607 (1969). Use in prevention of dental caries: E. Grunberg *et al.*, *Int. J. Vitam. Nutr. Res.* **43**, 227 (1973); A. Scheinin, K. K. Makinen, **DE 2606533** (1976 to Hoffmann-La Roche), *C.A.* **85**, 149140h (1976). Acute toxicity: S. Salminen *et al.*, *Toxicol. Lett.* **18**, Suppl. 1, 37 (1983). Reviews of toxicity, metabolism and use as dietary additive: *International Symposium on Metabolism, Physiology and Clinical Uses of Pentoses and Pentitols*, B. L. Horecker *et al.*, Eds. (Springer-Verlag, New York, 1969) 408 pp; *Sugars in Nutrition*, H. L. Sipple, K. W. McNutt, Eds. (Academic Press, New York, 1974) *passim;* G. E. Demetrakopoulos, H. Amos, *World Rev. Nutr. Diet* **32**, 96-122 (1978); R. Ylikahri, *Adv. Food Res.* **25**, 159-180 (1979). Book: *Xylitol*, J. N. Counsel, Ed. (Applied Science, London, 1978) 191 pp.

Stable form: orthorhombic needles from THF, prisms from ethanol; mp 93-94.5°; d 1.52. Metastable form: colorless, monoclinic, lath-shaped crystals from anhydrous methanol; hygroscopic; mp 61-61.5°. Soly of stable form (g/100 g soln): abs methanol 6.0; abs ethanol 1.2; water 64.2. Relative sweetness equal to sucrose. LD$_{50}$ orally in mice: approx 22 g/kg (Salminen).

USE: As oral and intravenous nutrient; in anticaries preparations.

10284. Xylometazoline. [526-36-3] 2-[[4-(1,1-Dimethylethyl)-2,6-dimethylphenyl]methyl]-4,5-dihydro-1H-imidazole; 2-(4-*tert*-butyl-2,6-dimethylbenzyl)-2-imidazoline. $C_{16}H_{24}N_2$; mol wt 244.38. C 78.64%, H 9.90%, N 11.46%. α-Adrenergic agonist; topical vasoconstrictor. Prepn: Hüni, **US 2868802** (1959 to Ciba). Pharmacology: S. Morimoto, H. Tanaka, *Osaka-shiritsu Daigaku Igaku Zasshi* **18**, 211 (1969), *C.A.* **72**, 20437n (1970). Comprehensive description: Y. Golander, W. J. DeWitte, *Anal. Profiles Drug Subs.* **14**, 135-156 (1985). GC determn in plasma and urine: A. Sioufi *et al.*, *J. Chromatogr.* **487**, 81 (1989). Clinical trial in allergic rhinitis: M. Fradis *et al.*, *J. Laryngol. Otol.* **101**, 666 (1987); in nasal surgery: J. P. Campbell *et al.*, *Otolaryngol. Head Neck Surg.* **107**, 697 (1992).

mp 131-133°.

Hydrochloride. [1218-35-5] Novorin; Olynth; Otrivine; Xymelin. $C_{16}H_{24}N_2$·HCl; mol wt 280.84. Soly in water: up to 3%. Freely sol in ethanol; sol in methanol; sparingly sol in chloroform. Practically insol in ether, benzene.

THERAP CAT: Decongestant.

10285. Xylose. [58-86-6] D-Xylose; wood sugar; Xylomed; Xylo-Pfan. $C_5H_{10}O_5$; mol wt 150.13. C 40.00%, H 6.71%, O 53.28%. Widely distributed in plant materials, especially in wood (maple, cherry), in straw, in hulls. Not found in free state, but in form of xylan, a polysaccharide built from D-xylose units and occurring in association with cellulose. Xylose occurs also as part of glycosides. Isoln from corn cobs by boiling with 8% H_2SO_4: Monroe, *J. Am. Chem. Soc.* **41**, 1002 (1919). Peanut shells and cottonseed hulls also are practical sources of xylose: Ling, Nanji, *J. Chem. Soc.* **1923**, 620. Configuration: Hudson, Yanovsky, *J. Am. Chem. Soc.* **39**, 1029 (1917); Haworth, *Nature* **116**, 430 (1925). Review on history, constitution and prepn: Harding, *Sugar* **24**, 14 (1922).

Monoclinic needles or prisms. Very sweet taste. mp 144-145° (Wheeler, Tollens, *Ann.* **254**, 304); mp 153-154° (Hébert, *Compt. Rend.* **110**, 970). d_4^{20} 1.525. Shows mutarotation. $[\alpha]_D^{20}$ +92° → +18.6° (16 hrs c = 10). One gram dissolves in 0.8 ml water. Sol in pyridine, hot alc; slightly sol in alc. pKa (18°): 12.14. Reduces warm Fehling's soln. Upon heating with water in closed tube to 140° or by boiling with dil H_2SO_4, furfurol is formed.

USE: In tanning, dyeing, and as a diabetic food.

THERAP CAT: Diagnostic aid (intestinal function).

10286. Xylulose. [5962-29-8] *threo*-2-Pentulose. $C_5H_{10}O_5$; mol wt 150.13. C 40.00%, H 6.71%, O 53.28%. L-Form has been found in the urine of humans with pentosuria. Prepn of DL-form: Gascoigne, *Chem. Ind. (London)* **1959**, 402; of D-form: Mendicino, *J. Am. Chem. Soc.* **82**, 4975 (1960); of L-form: Wolfrom, Bennett, *J. Org. Chem.* **30**, 458 (1965). Isoln of DL-form from the acid hydrolysate of bagasse hemicellulose: Banerjee *et al.*, *Sci. Cult.* **27**, 498 (1961), *C.A.* **56**, 11682d (1962). Enzymic prepn of L-form: Hough, Jones, *Chem. Ind. (London)* **1952**, 907; *eidem, J. Chem. Soc.* **1952**, 4047. Formation of L-form in normal humans and guinea pigs, and its utilization by guinea-pig liver prepns: Touster *et al.*, *J. Am. Chem. Soc.* **76**, 5005 (1954). *Reviews: The Carbohydrates*, W. Pigman, Ed. (Academic Press, New York, 1957) pp 80, 86-87, 759, 795; *Methods in Carbohydrate Chemistry* vol. **1**, R. L. Whistler, M. L. Wolfrom, Eds. (Academic Press, New York, 1962) pp 94-101.

L-isomer

D-Isomer. Syrup. $[\alpha]_D^{18}$ −33° (c = 2.5).

D-Isomer p-bromophenylhydrazone. $C_{11}H_{15}BrN_2O_4$. Pale yellow crystals from abs ethanol + water, mp 128-129°. $[\alpha]_D^{20}$ +24° (15 min) → −31° (7 days, in pyridine).

L-Isomer. Syrup. $[\alpha]_D^{21}$ +31°.

L-Isomer p-bromophenylhydrazone. Yellow plates from dil alc, mp 128°. $[\alpha]_D^{20}$ −20° (10 min) → +22° (5 hrs, c = 0.5 in ethanol).

10287. 1-Xylylazo-2-naphthol. [85-82-5] 1-[2-(2,5-Dimethylphenyl)diazenyl]-2-naphthalenol; C.I. Solvent Orange 7; 1-(2,5-xylylazo)-2-naphthol; FD & C Red no. 32; Oil Red XO; Ext. D & C Red no. 14; C.I. 12140. $C_{18}H_{16}N_2O$; mol wt 276.34. C 78.24%, H 5.84%, N 10.14%, O 5.79%. Once reported as 2,5-xylylazo deriv. Prepd by coupling diazotized *m*-xylidene with 2-naphthol: J. M. Tedder, *J. Chem. Soc.* **1957**, 4003; R. B. Smyth, G. G. McKeown, *J. Chromatogr.* **5**, 395 (1961). Metabolism: J. L. Radomski, *J. Pharmacol. Exp. Ther.* **136**, 378 (1962).

Red needles, mp 166°. Insol in water; sol in ethanol, acetone, benzene.

10288. Xylyl Bromide. C_8H_9Br; mol wt 185.06. C 51.92%, H 4.90%, Br 43.18%. Prepn of *m*-isomer: Wenner, *J. Org. Chem.* **17**, 523 (1952); of *o*-isomer: Dev, *J. Indian Chem. Soc.* **32**, 403 (1955); of *p*-isomer: Cockburn *et al.*, *J. Chem. Soc.* **1960**, 3340.

***m*-Xylyl bromide.** [620-13-3] 1-(Bromomethyl)-3-methylbenzene; α-bromo-*m*-xylene; *m*-methylbenzyl bromide; ω-bromo-*m*-xylene. Liquid, bp 212-215° with slight dec. d^{23} 1.371. Practically insol in water; sol in alcohol, ether.

***o*-Xylyl bromide.** [89-92-9] Prisms, mp 21°. bp 223-234°, bp$_{742}$ 216-217°, bp$_{15}$ 102°. n_D^{27} 1.5730. d^{23} 1.381. Practically insol in water. Sol in alcohol, ether.

***p*-Xylyl bromide.** [104-81-4] Needles from alcohol, mp 38°. bp$_{740}$ 218-220°, bp$_{15}$ 120°. d 1.324. Practically insol in water. Very sol in chloroform, hot ether.

Caution: Powerful lacrimator.

USE: In organic syntheses; in war-gas formulations.

10289. Xylyl Chloride. C_8H_9Cl; mol wt 140.61. C 68.34%, H 6.45%, Cl 25.21%. Prepn of *o*-isomer: Rabjohn, *J. Am. Chem. Soc.* **76**, 5479 (1954); of *m*-isomer: van Zanten, Nauta, *Rec. Trav. Chim.* **79**, 1211 (1960); of *p*-isomer: Newman, George, *J. Org. Chem.* **26**, 4306 (1961).

***m*-Xylyl chloride.** [620-19-9] 1-(Chloromethyl)-3-methylbenzene; α-chloro-*m*-xylene; *m*-methylbenzyl chloride; ω-chloro-*m*-xylene. Liquid, bp 195-196°. bp$_{10}$90-92°. d^{20} 1.064. Practically insol in water; miscible with abs alcohol, ether.

***o*-Xylyl chloride.** [552-45-4] Liquid, bp 195-203°, bp$_{25}$ 95-96°. n_D^{20} 1.5391. Practically insol in water; miscible with abs alcohol, ether.

***p*-Xylyl chloride.** [104-82-5] Fuming liquid, characteristic odor. bp 200-202°, bp$_1$ 48-50°. n_D^{17} 1.5360. Practically insol in water; miscible with abs alcohol, ether.

Caution: Powerful lacrimator.

Y

10290. Yam, Mexican. Giant yam; cabeza de negro. *Dioscorea macrostachya* Benth., *Dioscoreaceae* or one of ~600 other *Dioscoreae*, native to subtropical countries. The underground tubers weigh as much as 90 lbs. and contain steroidal sapogenins. Account of nature, origins, cultivation and utilization of the useful members of the *Dioscoreaceae:* D. G. Coursey, *Yams* (Longmans, London, 1967) 230 pp. *See also* Dioscorea and Barbasco.

USE: In the partial synthesis of hormones having a steroid structure.

10291. Yangonin. [500-62-9] 4-Methoxy-6-[(1*E*)-2-(4-methoxyphenyl)ethenyl]-2*H*-pyran-2-one; 4-methoxy-6-(*p*-methoxystyryl)-2*H*-pyran-2-one; 5-hydroxy-3-methoxy-7-(*p*-methoxyphenyl)-2,4,6-heptatrienoic acid δ-lactone; 4-methoxy-6-[β-(*p*-anisyl)vinyl]-α-pyrone; 6-(*p*-methoxystyryl)-4-methoxy-α-pyrone. $C_{15}H_{14}O_4$; mol wt 258.27. C 69.76%, H 5.46%, O 24.78%. From root of *Piper methysticum* Forst., *Piperaceae* (kava): Winzhermer, *Arch. Pharm.* **246**, 338 (1908); Borsche, Gerhardt, *Ber.* **47**, 2902 (1914). From *Ranunculus quelpaertensis* Nakai, *Ranunculaceae:* Shibata *et al., Bull. Chem. Soc. Jpn.* **45**, 930 (1972). Structure: Chmielewska *et al., Tetrahedron* **4**, 36 (1958); Bu'Lock, Smith, *J. Chem. Soc.* **1960**, 502. Synthesis: Harris, Combs, *J. Org. Chem.* **33**, 2399 (1968); R. Bacardit, *J. Heterocycl. Chem.* **19**, 157 (1982). Molecular and crystal structure: Engel, Nowacki, *Z. Kristallogr. Kristallgeom. Kristallphys. Kristallchem.* **134**, 180 (1971). Activity studies: Kretzschmar *et al., Arch. Int. Pharmacodyn. Ther.* **180**, 471 (1969).

Crystals from methanol, mp 155-157°. uv max (ethanol): 360 nm (log ε 4.33). Practically insol in water; sol in hot alcohol, glacial acetic acid, ethyl acetate, acetone; slightly sol in benzene, ether.
Dihydroyangonin. $C_{15}H_{16}O_4$. Needles from alc, mp 106-107°. uv max (methanol): 274, 228 nm (log ε 3.82, 3.71).

10292. Yeast. The moist, living cells of a fungus or fungi whose usual and dominant growth form is unicellular. The term yeast is not one with an exact botanical meaning, *see Henrici's Molds, Yeasts and Actinomycetes* (New York, 1947) p 264 sqq. Normally produces alcoholic fermentation in fluids contg sugar. Moist yeast is usually combined with a starchy or absorbent base and comes on the market in the form of white or yellowish-white, soft, easily broken masses of a characteristic, slightly sour odor. Description of a modern process of manuf: Schultz, Swift, **US 2717837** (1955 to Standard Brands).

The dried yeast of pharmacopoeias consists of the dry cells of any suitable strain of *Saccharomyces cerevisiae* Meyen, *Saccharomytaceae* or *Candida utilis* (Hanneberg) Lodder and Kreger-Van Rij, *Cryptococcaceae* (torula yeast), usually obtained as a by-product from the brewing of beer made from an extract of cereal grains and hops. The yeast cells are washed free of beer and dried, and may be debittered. These yeasts are commonly known as "Brewer's Dried Yeast" and "Debittered Brewer's Dried Yeast." Dried yeast may be obtained also by growing suitable strains of yeast, using media other than those required for the production of beer, and under appropriate environmental conditions. The yeast thus obtained is commonly known as "Primary Dried Yeast." Dried yeast that fulfills pharmacopeal requirements contains not less than 40% protein, and, in each gram, the equivalent of not less than 0.12 mg of thiamine hydrochloride, 0.04 mg of riboflavin, and 0.25 mg of nicotinic acid. It occurs as yellowish-white to weak yellowish-orange flakes, granules or powder, having an odor indicative of the type. It is inactive in fermenting power.

USE: Moist yeast in baking bread, brewing; producing alcohol by fermentation of sugar, molasses, and cereals; as a source of vitamins. Dried yeast as a source of vitamins; in baking.

THERAP CAT: Source of protein and vitamin B complex.
THERAP CAT (VET): Dietary source of B vitamins.

10293. Yellow AB. [85-84-7] 1-(2-Phenyldiazenyl)-2-naphthalenamine; C.I. Solvent Yellow 5; FD & C Yellow no. 3; Ext. D & C Yellow no. 9; C.I. 11380. $C_{16}H_{13}N_3$; mol wt 247.30. C 77.71%, H 5.30%, N 16.99%. Prepd from diazotized aniline and β-naphthylamine: Lawson, *Ber.* **18**, 796 (1885); *Colour Index* **vol. 4** (3rd ed., 1971) p 4021. Toxicity studies: Allmark *et al., J. Pharm. Pharmacol.* **7**, 591 (1955); Hansen *et al., Toxicol. Appl. Pharmacol.* **5**, 16 (1963). Formerly used in the U.S. for coloring oleomargarine.

Red platelets from abs alc, mp 102-104°. Practically insol in water. Sol in alcohol, carbon tetrachloride, acetic acid, vegetable oils. An alcoholic soln becomes redder with HCl addition, and is unaltered by NaOH addition.

10294. Yellow OB. [131-79-3] 1-[2-(2-Methylphenyl)diazenyl]-2-naphthalenamine; C.I. Solvent Yellow 6; 1-*o*-tolylazo-2-naphthylamine; FD & C Yellow no. 4; Ext. D & C Yellow no. 10; C.I. 11390. $C_{17}H_{15}N_3$; mol wt 261.33. C 78.13%, H 5.79%, N 16.08%. Prepd from diazotized *o*-toluidine and β-naphthylamine: Krüss, *Z. Phys. Chem.* **51**, 270 (1905); Norman, *J. Chem. Soc.* **101**, 1913 (1912); Fischer, *J. Prakt. Chem.* **104**, 102 (1922); Hodgson, Foster, *J. Chem. Soc.* **1942**, 30. Toxicity studies: Allmark *et al., J. Pharm. Pharmacol.* **7**, 591 (1955); Hansen *et al., Toxicol. Appl. Pharmacol.* **5**, 16 (1963). Formerly used in the U.S. for coloring oleomargarine.

Deep red crystals from alc. Also reported as bright yellow needles (Krüss). mp 125-126° (Fischer). Practically insol in water. Sol in alc, ether, benzene, carbon tetrachloride, vegetable oils, glacial acetic acid. An alcoholic soln becomes redder with HCl addition, but is unaltered by NaOH addition.

10295. Yig. [12063-56-8] Iron yttrium oxide ($Fe_5Y_3O_{12}$); yttrium iron garnet. $Fe_5O_{12}Y_3$; mol wt 737.93. Fe 37.84%, O 26.02%, Y 36.14%. $Y_3Fe_5O_{12}$. Man-made mineral. Prepn by molten salt technique: Nielsen, *Electronics*, Jan. 31, **1964**, pp 44-45. Advances in manufacture and applications: Gundlach, *ibid.*, Oct. 14, **1968**, pp 104-118. Electronic structure: Y.-N. Xu *et al., J. Appl. Phys.* **87**, 4867 (2000).

Hard, brittle, garnet-type crystals. Hardness about equal to that of quartz. mp >1040°. The melt is easily amenable to doping and crystal nucleation by epitaxy. Crystals may be oriented along the simple axes (100), (110) and (111) using the natural (110) face of the crystal.

USE: In microwave tunable filters and limiters.

10296. Yingzhaosu. Family of sesquiterpenes isolated from the roots of *Yingzhao*, (*Artabotrys uncinatus* (L.) Merr.) a traditional Chinese herbal medicine for treatment of malaria. Yingzhaosu A & C contain a peroxy group which is needed for antimalarial activity. Isoln of A: X.-T. Liang *et al., Acta Chim. Sin.* **37**, 215 (1979), *C.A.* **92**, 146954 (1980); of C: L. Zhang *et al., J. Chem. Soc. Chem. Commun.* **1988**, 523. Structure of C: *eidem, Sci. China Ser. B* **32**, 800 (1989). Total synthesis of A: X.-X. Xu *et al., Tetrahedron Lett.* **32**, 5785 (1991); stereocontrolled synthesis: W.-S. Zhou in *Newer Trends Essent. Oils. Flavours*, Sel. Pap. Int. Symp., K. L. Dhar *et al.*, Eds. (Tata McGraw-Hill, New Delhi, India, 1993) pp 43-50. Enantioselective synthesis of C: X.-X. Xu, H.-Q. Dong, *J. Org. Chem.* **60**, 3039 (1995).

Yingzhaosu A

Yingzhaosu A. [73301-54-9] (3*S*,4*E*)-5-[(1*S*,4*S*,5*S*,8*R*)-4,8-Dimethyl-2,3-dioxabicyclo[3.3.1]non-4-yl]-2-methyl-4-pentene-2,3-diol; (+)-yingzhosu A. $C_{15}H_{26}O_4$; mol wt 270.37. Contains a unique dioxabicyclo[3.3.1]nonane ring system. Isolated as solid, mp 95-96°. $[\alpha]_D^{25}$ +226° (in CHCl$_3$).

Yingzhaosu C. [121067-52-5] *rel*-(3*R*,6*S*)-α,α,6-Trimethyl-6-(4-methylphenyl)-1,2-dioxane-3-methanol. $C_{15}H_{22}O_3$; mol wt 250.34. Occurs naturally as an enantiomeric mixture of the (3*R*,6*S*), (3*S*,6*R*) with a predominance of the former. Oily substance. $[\alpha]_D^{14}$ +2.89° (c = 2.15 in methanol); (6*S*,3*R*): $[\alpha]_D$ +189.3° (c = 0.93 in CHCl$_3$); (3*S*,6*R*): $[\alpha]_D$ −185.4° (c = 0.78 in CHCl$_3$).

THERAP CAT: Antimalarial.

10297. Ylangene. [14912-44-8] 1,3-Dimethyl-8-(1-methylethyl)tricyclo[4.4.0.0^{2,7}]dec-3-ene; (1*S*,2*R*,6*R*,7*R*,8*S*)-(+)-8-isopropyl-1,3-dimethyltricyclo[4.4.0.0^{2,7}]dec-3-ene; α-ylangene. $C_{15}H_{24}$; mol wt 204.36. C 88.16%, H 11.84%. From ylang-ylang oil: Heraut, Dimitrov, *Chem. Listy* **46**, 432 (1952); from oil of birch buds: Holub *et al.*, *Collect. Czech. Chem. Commun.* **24**, 3730 (1959). From oil of *Juniperus oxycedrus* Linn., *Cupressaceae:* Motl *et al.*, *ibid.* **25**, 1656 (1960). Structure: Motl *et al.*, *Chem. Ind. (London)* **1963**, 1759; Hunter, Brogden, *J. Org. Chem.* **29**, 982 (1964); Motl *et al.*, *Tetrahedron Lett.* **1965**, 451. Stereoisomer of copaene, *q.v.* Total synthesis of (±)-form: Heathcock *et al.*, *J. Am. Chem. Soc.* **89**, 4133 (1967); Corey, Watt, *ibid.* **95**, 2303 (1973).

Oil. d_4^{20} 0.9091. n_D^{20} 1.4934. $[\alpha]_D^{20}$ +15.4°.

10298. Ylang-Ylang Oil. Cananga oil. Volatile oil distilled in Madagascar, Reunion Island, Comoro Islands and the Philippine Islands from freshly picked flowers of *Cananga odorata* Hook & Thoms., *Anonaceae.* The first distillate yields the so-called "ylang-ylang extra" and is the finest oil. The first, second and third fractions follow, the first two having insignificant commercial value. *Constit.* Geraniol and linalool esters of acetic and benzoic acids; *p*-cresol methyl ether, cadinene, a sesquiterpene, a phenol.

Light yellow, very fragrant liquid. d_{20}^{20} 0.930-0.950. Rotation: −27° to −50° in 100-mm tube.

USE: In delicate perfumes.

10299. Yogurt. Yoghurt; joghurt; kisselo-mleko; mazun; leben raib; dahi. Fermented whole milk. Produced by evaporating milk to ~50% of its volume and maintaining at 50°C for 12 hrs after adding "maya" a mixture of *Lactobacillus bulgaricus*, *L. acidophilus*, and *Streptococcus lactis*. Used as food. Contains antibacterial principles. Differentiation of yogurt and *clabber* as produced in Kentucky: *J. Am. Med. Assoc.* **209**, 778 (1969).

10300. Yohimbine. [146-48-5] (16α,17α)-17-Hydroxyyohimban-16-carboxylic acid methyl ester; quebrachine; corynine; aphrodine. $C_{21}H_{26}N_2O_3$; mol wt 354.45. C 71.16%, H 7.39%, N 7.90%, O 13.54%. Indole alkaloid with α$_2$-adrenergic blocking activity. Found in *Corynanthe johimbe* K. Schum., *Rubiaceae* and related trees, also in *Rauwolfia serpentina* (L.) Benth., *Apocynaceae:* Raymond-Hamet, *J. Pharm. Chim.* **19**, 209 (1934); Hofmann, *Helv. Chim. Acta* **37**, 849 (1954); Stoll, Jucker, *Ullmanns Encyklopädie der technischen Chemie* **vol. 3** (Munich, 3rd ed.), 1953) p 266; Bader *et al.*, *J. Am. Chem. Soc.* **76**, 1695 (1954). Structure: Witkop, *Ann.* **554**, 83 (1943); Clemo, Swan, *J. Chem. Soc.* **1946**, 617. Stereo-

chemistry: Janot *et al.*, *Bull. Soc. Chim. Fr.* **1952**, 1085; Godfredsen, Vandegal, *Acta Chem. Scand.* **10**, 1414 (1956); Van Tamelen *et al.*, *J. Am. Chem. Soc.* **78**, 4628 (1956); Ban, Yonemitsu, *Tetrahedron* **20**, 2877 (1964). Synthesis: Van Tamelen *et al.*, *J. Am. Chem. Soc.* **80**, 5006 (1958); Liljegren, Potts, *J. Org. Chem.* **27**, 377 (1962). Total synthesis: Van Tamelen *et al.*, *J. Am. Chem. Soc.* **91**, 7315 (1969); Stork, Guthikonda, *ibid.* **94**, 5109 (1972); T. Kametani *et al.*, *Chem. Pharm. Bull.* **24**, 2500 (1976); E. Wenkert *et al.*, *J. Am. Chem. Soc.* **100**, 4894 (1978); **101**, 5370 (1979); **104**, 2244 (1982); I. Ninomiya *et al.*, *Heterocycles* **14**, 631 (1980). Pharmacokinetics in humans: J. A. Owen *et al.*, *Eur. J. Clin. Pharmacol.* **32**, 577 (1987). Clinical studies in impotence: K. Reid *et al.*, *Lancet* **2**, 421 (1987); A. Morales *et al.*, *J. Urol.* **137**, 1168 (1987). Review of pharmacology and use in molecular studies of α$_2$-adrenoreceptor: M. R. Goldberg, D. Robertson, *Pharmacol. Rev.* **35**, 143-180 (1987). Comprehensive description: A. G. Mekkawi, A. A. Al-Badr, *Anal. Profiles Drug Subs.* **16**, 731-768 (1986).

Orthorhombic needles from dil alc, mp 234°. Also mp 235-237°. $[\alpha]_D^{20}$ +50.9 to +62.2° (ethanol); $[\alpha]_D^{20}$ +108° (pyridine); $[\alpha]_{546}^{20}$ +129° (c = 0.5 in pyridine). uv max (methanol): 226, 280, 291 nm (log ε 4.56, 3.88, 3.80). Sparingly sol in water. Sol in alcohol, chloroform, hot benzene; moderately sol in ether.

Hydrochloride. [65-19-0] Antagonil; Aphrodyne; Erex; Yobine; Yocon; Yohimex; Yohydrol; Yovital. $C_{21}H_{26}N_2O_3 \cdot$HCl; mol wt 390.91. Orthorhombic plates, prisms from alc; dec 302°. $[\alpha]_D^{22}$ +105° (H$_2$O). Sol in ~120 ml water, 400 ml alc. The aq soln is about neutral.

USE: Pharmacological probe for the study of α$_2$-adrenoceptor.

THERAP CAT: Mydriatic. In treatment of impotence.

THERAP CAT (VET): Xylazine reversing agent.

10301. α-Yohimbine. [131-03-3] (16β,17α,20α)-17-Hydroxyyohimban-16-carboxylic acid methyl ester; corynanthidine; isoyohimbine; mesoyohimbine; rauwolscine. $C_{21}H_{26}N_2O_3$; mol wt 354.45. C 71.16%, H 7.39%, N 7.90%, O 13.54%. From bark of *Corynanthe johimbe* K. Schum., *Rubiaceae:* Hahn, Brandenburg, *Ber.* **60**, 669 (1927); Wilbaut, van Gastel, *Rec. Trav. Chim.* **54**, 88 (1935); from *Rauwolfia canescens* L., *Apocynaceae:* Mookerjee, *J. Indian Chem. Soc.* **18**, 33 (1941), *C.A.* **35**, 7967 (1941); Stoll *et al.*, *Helv. Chim. Acta* **38**, 270 (1955). Identity with corynanthidine: Janot, Goutarel, *Bull. Soc. Chim. Fr.* **1946**, 535. Identity with isoyohimbine: Heinemann, *Ber.* **60**, 15 (1934). Identity with rauwolscine: Chatterjee *et al.*, *Chem. Ind. (London)* **1954**, 491. Structure and stereochemistry: Le Hir *et al.*, *Bull. Soc. Chim. Fr.* **1953**, 1027; Wenkert, Liu, *Experientia* **11**, 302 (1955); Aldrich *et al.*, *J. Am. Chem. Soc.* **81**, 2481 (1959); Janot *et al.*, *Bull. Soc. Chim. Fr.* **1961**, 637. Total synthesis: Töke *et al.*, *J. Org. Chem.* **38**, 2496 (1973).

Crystals, mp 243-244°. $[\alpha]_D^{19}$ −18° (pyridine); $[\alpha]_D^{19}$ −27° (abs alcohol). pKa 6.34. uv max (methanol): 227, 281 nm (log ε 4.50, 3.93). Sol in warm methanol, ethanol; slightly sol in ether, benzene; practically insol in petr ether, water.

Hydrochloride. $C_{21}H_{26}N_2O_3.HCl$. Crystals, mp 288°. $[\alpha]_D^{20}$ +55.5° (water).

10302. allo-Yohimbine. [522-94-1] (16α,17α,20α)-17-Hydroxyyohimban-16-carboxylic acid methyl ester; alloyohimbine; dihydroyohimbine. $C_{21}H_{26}N_2O_3$; mol wt 354.45. C 71.16%, H 7.39%, N 7.90%, O 13.54%. From bark of *Corynanthe johimbe* K. Schum., *Rubiaceae:* Warnat, *Ber.* **59**, 2388 (1926); Hahn, Brandenberg, *ibid.* **60**, 699 (1927). Identity with dihydroyohimbine: Heinemann, *ibid.* **67**, 15 (1934). Structure and stereochemistry: LeHir *et al.*, *Bull. Soc. Chim. Fr.* **1953**, 1027; Janot *et al.*, *ibid.* **1961**, 637. Revised structure and total synthesis: Töke *et al.*, *J. Org. Chem.* **38**, 2496 (1973). Total synthesis: E. Wenkert *et al.*, *J. Am. Chem. Soc.* **101**, 5370 (1979). *Review:* A. Chatterjee *et al.*, "Rauwolfia Alkaloids" in Zechmeister, *Progress in the Chemistry of Organic Natural Products* vol. **XIII** (Springer Verlag, Vienna, 1956) pp 354-356.

Trihydrate, needles from 50% alcohol, mp 98-99°. After drying at 100° and 0.01 mm Hg for 24 hrs, the anhydrous form melts at 135-140°. $[\alpha]_D^{19}$ +84° (c = 0.40 in pyridine). uv max (methanol): 225, 280, 290 nm (log ε 4.52, 3.91, 3.74). Sol in ethanol, methanol, pyridine, dil acids; slightly sol in benzene. Practically insol in water.

10303. Ytterbium. [7440-64-4] Yb; at. wt 173.054; at. no. 70; valences 2, 3. A rare earth metal of the yttrium group; member of the lanthanide series. Naturally occurring isotopes (mass numbers): 168 (0.13%); 170 (3.05%); 171 (14.3%); 172 (21.9%); 173 (16.12%); 174 (31.8%); 176 (12.7%); known artificial radioactive isotopes: 152-167; 169; 175; 177; 178. Estimated abundance in earth's crust: 2.66-3.1 ppm. Occurs in the rare earth minerals xenotime, ytterbite (gadolinite), monazite. Discovered independently: Urbain, *Compt. Rend.* **145**, 759 (1907); and called *aldebaranium*: von Welsbach, *Monatsh. Chem.* **29**, 181 (1908); **34**, 1713 (1913). Sepn and purification: Urbain, *Congress of Applied Chemistry* [X] **94** (1909); Prandtl, *Z. Anorg. Chem.* **238**, 321 (1938); Spedding *et al.*, *J. Am. Chem. Soc.* **74**, 2783 (1952); **76**, 2557 (1954). Spectrum: Exner, Haschek *et al.*, cited by Mellor, *A Comprehensive Treatise on Inorganic and Theoretical Chemistry* **5**, 706 (1929). Toxicity study: Haley, *J. Pharm. Sci.* **54**, 663 (1965). Reviews of prepn, properties and compds: *The Rare Earths*, F. H. Spedding, A. H. Daane, Eds. (Krieger, Huntington, N.Y., 1971, reprint of 1961 ed.) 641 pp; Hulet, Bode, "Separation Chemistry of the Lanthanides and Transplutonium Actinides", in *MTP Int. Rev. Sci.: Inorg. Chem., Ser. One* vol. **7**, K. W. Bagnall, Ed. (University Park Press, Baltimore, 1972) pp 1-45; Moeller, "The Lanthanides", in *Comprehensive Inorganic Chemistry* vol. **4**, J. C. Bailar Jr. *et al.*, Eds. (Pergamon Press, Oxford, 1973) pp 1-101; F. H. Spedding in *Kirk-Othmer Encyclopedia of Chemical Technology* vol. **19** (John Wiley & Sons, New York, 3rd ed., 1982) pp 833-854; *Chemistry of the Elements*, N. N. Greenwood, A. Earnshaw, Eds. (Pergamon Press, New York, 1984) pp 1423-1449. Brief review of properties: G. T. Seaborg, *Radiochim. Acta* **61**, 115-122 (1993).

Silvery, ductile metal. Crystalline forms: face-centered cubic α-form, d 6.977, transforms to β-form at 798°; body-centered cubic β-form, d 6.54, exists at >798°. mp 819°. bp 1196°. Heat of fusion: 7.657 kJ/mol. Heat of sublimation (25°): 152.1 kJ/mol. Forms both di- and trivalent salts.

Oxide. Ytterbia. O_3Yb_2. Colorless mass, sol in dil acids.
Hydroxide. $Yb(OH)_3$. Colorless gelatinous precipitate.
Chloride. $YbCl_3$. Hexahydrate, deliquescent crystals, d 2.575; mp 150-155°. LD_{50} in mice: 395 mg/kg i.p.; 6.7 g/kg orally (Haley).
Nitrate. $Yb(NO_3)_3$. Tetrahydrate, transparent hygroscopic prisms from concd nitric acid. LD_{50} (hexahydrate) in rats: 255 mg/kg i.p.; 3.1 g/kg orally (Haley).
Sulfate. $Yb_2(SO_4)_3$. Octahydrate, lustrous colorless crystals, soly decreases with rise in temp.

10304. Ytterbium Triflate. [54761-04-5] 1,1,1-Trifluoromethanesulfonic acid ytterbium(3+) salt (3:1); ytterbium(III) trifluoromethanesulfonate. $C_3F_9O_9S_3Yb$; mol wt 620.23. C 5.81%, F 27.57%, O 23.22%, S 15.51%, Yb 27.90%. $Yb(SO_3CF_3)_3$. Recyclable, water-stable Lewis acid catalyst. Prepn: J. Massaux, G. Duyckaerts, *Anal. Chim. Acta* **73**, 416 (1974). Thermal decompn of nonahydrate: J. E. Roberts, J. S. Bykowski, *Thermochim. Acta* **25**, 233 (1978). Catalytic applications in solvent-free carbamate synthesis: M. Curini *et al.*, *Tetrahedron Lett.* **43**, 4895 (2002); in Aldol-Grob synthesis of (E)-alkenes: idem *et al.*, *Eur. J. Org. Chem.* **2003**, 1631; in imino ene reactions: M. Yamanaka *et al.*, *J. Org. Chem.* **68**, 3112 (2003); in Fries-type rearrangements of acylanilides: W. Su, C. Jin, *J. Chem. Res.* **2004**, 611; in Friedel-Crafts reactions: eidem, *Synth. Commun.* **34**, 4199, 4249 (2004); in intramolecular Cannizzaro reactions of aryl methyl ketones: M. Curini *et al.*, *Org. Lett.* **7**, 1331 (2005); in aqueous Michael addition reactions: E. Keller, B. L. Feringa, *Tetrahedron Lett.* **37**, 1879 (1996); R. Ding *et al.*, *J. Org. Chem.* **71**, 352 (2006). Review of synthetic applications of lanthanide triflates: S. Kobayashi *et al.*, *Chem. Rev.* **102**, 2227-2302 (2002).

USE: Versatile rare earth metal catalyst in organic synthesis.

10305. Yttrium. [7440-65-5] Y; at. wt 88.90585; at. no. 39; valence 3. Group IIIB (3). A rare earth metal. Naturally occurring isotope (mass number): 89; known artificial radioactive isotopes: 80-88; 90-100. Estimated abundance in earth's crust: 28.1-31 ppm. Natural sources: xenotime, fergusonite, samarskite, gadolinite, and other rare earth minerals. Discovered in 1794 by Gadolin; separated as yttria in 1843 by Mosander; named by Ekeberg. Sepn by fractional precipitation: Bonardi, James, *J. Am. Chem. Soc.* **37**, 2642 (1915); Willand, James, *ibid.* **38**, 1198 (1916); Wichers *et al.*, *ibid.* **40**, 1615 (1918); by ion exchange: Spedding *et al.*, *ibid.* **69**, 2812 (1947); Mayer, Freiling, *ibid.* **75**, 5647 (1953). Toxicity study: Cochran *et al.*, *Arch. Ind. Hyg. Occup. Med.* **1**, 637 (1950). Reviews of prepn, properties and compds: *The Rare Earths*, F. H. Spedding, A. H. Daane, Eds. (Krieger, Huntington, N.Y., 1971, reprint of 1961 ed.) 641 pp; Vickery, "Scandium, Yttrium and Lanthanum", in *Comprehensive Inorganic Chemistry* vol. **3**, J. C. Bailar Jr. *et al.*, Eds. (Pergamon Press, Oxford, 1973) pp 329-353; Moeller, "The Lanthanides", *ibid.* vol. **4**, pp 1-101; F. H. Spedding in *Kirk-Othmer Encyclopedia of Chemical Technology* vol. **19** (John Wiley & Sons, New York, 3rd ed., 1982) pp 833-854; *Chemistry of the Elements* N. N. Greenwood, A. Earnshaw, Eds. (Pergamon Press, New York, 1984) pp 1102-1110, 1423-1449.

Iron-gray, lustrous powder; darkens on exposure to light. Forms hexagonal close-packed crystals. d 4.4689. mp 1522°. bp 3338°. Heat of fusion: 11.43 kJ/mol. Heat of sublimation (25°): 424.7 kJ/mol. E°(aq) Y^{3+}/Y −2.37 V (calc). Oxidizes on heating in air or oxygen; dec cold water slowly, boiling water rapidly.
Oxide. Yttria. O_3Y_2. White powder, body-centered cubic structure, d 5.03. Obtained by igniting yttrium or its salts. Sol in dil acids; readily absorbs ammonia from the air; displaces ammonia from ammonium salts. LD_{50} i.p. in rats: 500 mg/kg (Cochran).
Hydroxide. $Y(OH)_3$. White gelatinous ppt, dries to a white powder which absorbs CO_2 from the air, obtained by the action of ammonium or alkali hydroxides on a soln of an yttrium salt.
Chloride. YCl_3. Hexahydrate, colorless, deliquesc crystals. Sol in water, alc. Anhydr chloride obtained by heating in a stream of HCl.
Carbonate. $Y_2(CO_3)_3$. Trihydrate, white to reddish-white powder, prepd by hydrolysis of yttrium trichloroacetate. Insol in water. Sol in dil mineral acids.
Sulfate. $Y_2(SO_4)_3$. Octahydrate, monoclinic crystals. Sol in concd sulfuric acid with formation of $Y(HSO_4)_3$. Soly in water decreases with increase in temp. Forms double salts with alkali sulfates.
Nitrate. $Y(NO_3)_3$. Hexahydrate, deliquesc crystals, sol in water, produces basic nitrates on partial decompn. LD_{50} i.p. in rats: 350 mg/kg (Cochran).
Caution: Potential symptom of overexposure to metal is eye irritation. *See NIOSH Pocket Guide to Chemical Hazards* (DHHS/NIOSH 97-140, 1997) p 338.

USE: Yttrium doped with rare earths as phosphors for color television receivers. Oxide for mantles in gas and acetylene lights. Chloride in prepn of pure metal. Yttrium aluminum garnets (YAGS) in lasers and for making artificial diamonds.

Z

10306. **Zafirlukast.** [107753-78-6] *N*-[3-[[2-Methoxy-4-[[[(2-methylphenyl)sulfonyl]amino]carbonyl]phenyl]methyl]-1-methyl-1*H*-indol-5-yl]carbamic acid cyclopentyl ester; cyclopentyl 3-[2-methoxy-4-[(*o*-tolylsulfonyl)carbamoyl]benzyl]-1-methylindole-5-carbamate; *N*-[4-[5-(cyclopentyloxycarbonyl)amino-1-methylindol-3-ylmethyl]-3-methoxybenzoyl]-2-methylbenzenesulfonamide; ICI-204219; Accolate. C₃₁H₃₃N₃O₆S; mol wt 575.68. C 64.68%, H 5.78%, N 7.30%, O 16.67%, S 5.57%. Cysteinyl leukotriene type 1 receptor antagonist. Prepn: F. J. Brown *et al.*, **EP 199543**; P. R. Bernstein *et al.*, **US 4859692** (1986, 1989 both to ICI); V. G. Matassa *et al.*, *J. Med. Chem.* **33**, 1781 (1990). Improved prepn: K. Srinivas *et al.*, *Org. Process Res. Dev.* **8**, 952 (2004). Electrochemical characterization and voltammetric determn in pharmaceutical formulations: I. Süslü, S. Altinöz, *J. Pharm. Biomed. Anal.* **39**, 535 (2005). Preclinical pharmacology: R. D. Krell *et al.*, *Am. Rev. Respir. Dis.* **141**, 978 (1990). Effect on allergen-induced bronchoconstriction: S. R. Findlay *et al.*, *J. Allergy Clin. Immunol.* **89**, 1040 (1992); K. M. O'Shaughnessy *et al.*, *Am. Rev. Respir. Dis.* **147**, 1431 (1993). Clinical evaluation in bronchial asthma and allergic rhinitis: G. Piatti *et al.*, *Pharmacol. Res.* **47**, 541 (2003); in acute asthma: R. A. Silverman *et al.*, *Chest* **126**, 1480 (2004). Review of pharmacology and clinical efficacy in asthma: C. J. Dunn, K. L. Goa, *Drugs* **61**, 285-315 (2001); of clinical pharmacokinetics: P. N. R. Dekhuijzen, P. P. Koopmans, *Clin. Pharmacokinet.* **41**, 105-114 (2002).

White solid from methanol, mp 138-140°. Slightly sol in methanol; freely sol in tetrahydrofuran, acetone, DMSO. Practically insol in water.

THERAP CAT: Antiasthmatic.

10307. **Zalcitabine.** [7481-89-2] 2′,3′-Dideoxycytidine; dideoxycytidine; ddC; Hivid. C₉H₁₃N₃O₃; mol wt 211.22. C 51.18%, H 6.20%, N 19.89%, O 22.72%. Pyrimidine nucleoside reverse transcriptase inhibitor. Prepn: J. P. Horwitz *et al.*, *J. Org. Chem.* **32**, 817 (1967); R. Marumoto, M. Honjo, *Chem. Pharm. Bull.* **22**, 128 (1974). Prepn and structure-activity study: T.-S. Lin *et al.*, *J. Med. Chem.* **30**, 440 (1987). *In vitro* inhibition of HIV-1 virus replication: H. Mitsuya, S. Broder, *Proc. Natl. Acad. Sci. USA* **83**, 1911 (1986). Review of pharmacology and clinical experience in HIV infection: G. Skowron, *Adv. Exp. Med. Biol.* **394**, 257-269 (1996); J. C. Adkins *et al.*, *Drugs* **53**, 1054-1080 (1997).

Crystals from ethanol + benzene, mp 215-217° (Horwitz); also reported as mp 209-210° (Lin). [α]²⁵_D +81° (c = 0.635 in water). uv max in 0.1*N* HCl: 280 nm (ε 17720); in 0.1*N* NaOH: 270 nm (ε 8410). Soly in water (25°): 76.4 mg/ml. Sol in methanol; sparingly sol in alc, acetonitrile, chloroform, methylene chloride; slightly sol in cyclohexane.

THERAP CAT: Antiretroviral.

10308. **Zaldaride.** [109826-26-8] 1,3-Dihydro-1-[1-[(4-methyl-4*H*,6*H*-pyrrolo[1,2-*a*][4,1]benzoxazepin-4-yl)methyl]-4-piperidinyl]-2H-benzimidazol-2-one. C₂₆H₂₈N₄O₂; mol wt 428.54. C 72.87%, H 6.59%, N 13.07%, O 7.47%. Non-specific ion channel and calmodulin blocker. Prepn: J. W. F. Wasley, J. Norman, **EP 233483**; *eidem*, **US 4758559** (1987, 1988 both to Ciba-Geigy). Synthesis: S. Boyer *et al.*, *J. Heterocycl. Chem.* **25**, 1003 (1988). Calmodulin inhibition: J. A. Norman *et al.*, *Mol. Pharmacol.* **31**, 535 (1987); and ion channel inhibition: R. Neuhaus, B. F. X. Reber, *Eur. J. Pharmacol.* **226**, 183 (1992). Inhibition of gastric acid secretion: E. W. Black *et al.*, *J. Pharmacol. Exp. Ther.* **248**, 208 (1988); of secretory diarrhea in mice and rats: J. E. Shook *et al.*, *ibid.* **251**, 247 (1989). LC determn in human plasma: D. Chollet, M. Salanon, *J. Chromatogr.* **593**, 73 (1992). Clinical trial in travelers' diarrhea: H. L. DuPont *et al.*, *Gastroenterology* **104**, 709 (1993).

Crystals, mp 173-175° as hydrate.

    Maleate. [109826-27-9] CGS-9343B; ZY-17617B. C₂₆H₂₈N₄O₂·C₄H₄O₄; mol wt 544.61. Crystals, mp 189-190°.

THERAP CAT: Antidiarrheal.

10309. **Zaleplon.** [151319-34-5] *N*-[3-(3-Cyanopyrazolo[1,5-*a*]pyrimidin-7-yl)phenyl]-*N*-ethylacetamide; CL-284846; Sonata. C₁₇H₁₅N₅O; mol wt 305.34. C 66.87%, H 4.95%, N 22.94%, O 5.24%. Selective non-benzodiazepine GABAₐ receptor agonist. Prepn: J. P. Dusza *et al.*, **US 4626538** (1986 to Am. Cyanamid). Clinical pharmacology: D. Allen *et al.*, *Eur. J. Clin. Pharmacol.* **45**, 313 (1993); B. Beer *et al.*, *J. Clin. Pharmacol.* **34**, 335 (1994). Clinical pharmacokinetics: A. S. Rosen *et al.*, *Biopharm. Drug Dispos.* **20**, 171 (1999). Clinical comparison with zolpidem, *q.v.*: R. Elie *et al.*, *J. Clin. Psychiatry* **60**, 536 (1999). CE-LIF determn of metabolites in urine: C. Horstkötter *et al.*, *J. Chromatogr. A* **1014**, 71 (2003). Review of pharmacology and clinical experience: W. E. Heydorn, *Expert Opin. Invest. Drugs* **9**, 841-858 (2000).

White to off-white powder. mp 186-187°. Log P (octanol/water): 1.23 (pH 1-7). Sparingly sol in alcohol, propylene glycol. Practically insol in water.

Note: This is a controlled substance (depressant): **21 CFR**, 1308.14.

THERAP CAT: Sedative, hypnotic.

10310. **Zaltoprofen.** [74711-43-6] 10,11-Dihydro-α-methyl-10-oxodibenzo[*b*,*f*]thiepin-2-acetic acid; 2-(10,11-dihydro-10-oxodibenzo[*b*,*f*]thiepin-2-yl)propionic acid; CN-100; Soreton; Peon. C₁₇H₁₄O₃S; mol wt 298.36. C 68.44%, H 4.73%, O 16.09%, S 10.75%. Nonsteroidal anti-inflammatory drug; activity resides in (*S*)-enantiomer. Prepn: Y. Fujimoto, S. Yamabe, **DE 2941869**; *eidem*, **US 4247706** (1980, 1981 both to Nippon Chemiphar). Series of articles on pharmacology: *Arzneim.-Forsch.* **36**, 1796-1822 (1986). Resolution of racemate: M. Yamamoto *et al.*, *Chirality* **2**, 280 (1990). HPLC determn in human plasma and urine; pharmacokinetics: J. Oshima *et al.*, *J. Chromatogr.* **414**, 381 (1987). Determn in plasma by HPLC-MS/MS: H. W. Lee *et al.*, *Rapid Commun. Mass Spectrom.* **20**, 2675 (2006). Clinical trial in rheumatoid arthritis: M. Hatori, S. Kokubun, *Curr. Med. Res. Opin.* **14**, 79 (1998).

Antipyretic and analgesic effects in upper respiratory infection: A. Azuma *et al.*, *Pharmacology* **87**, 204 (2011).

Pale yellow crystals from benzene/*n*-hexane, mp 130.5-131.5°; also reported as mp 131-133°. Tasteless, odorless. Freely sol in acetone, chloroform; sol in methanol; slightly sol in ethanol, benzene. Practically insol in water, cyclohexane.

(*S*)-**form.** [89482-01-9] Crystals from ethylenedichloride, mp 129.5-131°. $[\alpha]_D^{25}$ +32.4° (c = 1 in CHCl$_3$).

THERAP CAT: Anti-inflammatory; analgesic

10311. Zanamivir. [139110-80-8] 5-(Acetylamino)-4-[(aminoiminomethyl)amino]-2,6-anhydro-3,4,5-trideoxy-D-*glycero*-D-*galacto*-non-2-enonic acid; 4-guanidino-2,4-dideoxy-2,3-dehydro-*N*-acetylneuraminic acid; 4-guanidino-Neu5Ac2en; 5-acetamido-4-guanidino-2,3,4,5-tetradeoxy-D-*glycero*-D-*galacto*-non-2-enopyranosonic acid; GG-167; GR-121167X; Relenza. C$_{12}$H$_{20}$N$_4$O$_7$; mol wt 332.31. C 43.37%, H 6.07%, N 16.86%, O 33.70%. Influenza viral neuraminidase inhibitor; structural analog of the sialic acids, *q.v.* Prepn: L. M. von Itzstein *et al.*, **WO 9116320**; *eidem*, **US 5360817** (1991, 1994 both to Biota); M. von Itzstein *et al.*, *Carbohydr. Res.* **259**, 301 (1994). Structure-activity study: *eidem et al.*, *Nature* **363**, 418 (1993). Large-scale synthesis: M. Chandler *et al.*, *J. Chem. Soc. Perkin Trans. I* **1995**, 1173. Inhibition of sialidase (neuraminidase) from influenza virus: M. S. Pegg, M. von Itzstein, *Biochem. Mol. Biol. Int.* **32**, 851 (1994). *In vitro* antiviral activity: J. M. Woods *et al.*, *Antimicrob. Agents Chemother.* **37**, 1473 (1993). LC-MS/MS determn in serum: G. D. Allen *et al.*, *J. Chromatogr. B* **732**, 383 (1999). Clinical trial in influenza A and B infection: F. G. Hayden *et al.*, *N. Engl. J. Med.* **337**, 874 (1997); in prevention of influenza in healthy adults: A. S. Monto *et al.*, *J. Am. Med. Assoc.* **282**, 31 (1999). Review of clinical experience: D. M. Fleming, *Expert Opin. Pharmacother.* **4**, 799-805 (2003).

Colorless crystals as the sesquihydrate, mp 256° (dec). $[\alpha]_D^{20}$ +40.9° (c = 0.9 in water). Zwitterionic at physiological pH. Soly in water (20°): ~18 mg/ml.

THERAP CAT: Antiviral.

10312. Zaragozic Acids. Squalestatins. Family of fungal metabolites with a core 4,6,7-trihydroxy-2,8-dioxobicyclo[3.2.1]-octane-3,4,5-tricarboxylic acid ring system differing among C6-O-acyl and C1-alkyl side chains. Inhibitors of squalene synthase, which catalyzes the first committed step of the sterol biosynthetic pathway. Originally isolated as zaragozic acid from a fungus found in Jalon River in Zaragosa, Spain and as squalestatins from *Phoma* sp. C2932 found in a soil sample from Armacao de Pera, Portugal. Isolation and properties of squalestatins 1, 2, and 3: M. J. Dawson *et al.*, *J. Antibiot.* **45**, 639 (1992); structure elucidation: P. J. Sidebottom *et al.*, *ibid* 648. Isolation, structure, and properties of zaragozic acids A, B, and C: J. D. Bergstrom *et al.*, *Proc. Natl. Acad. Sci. USA* **90**, 80 (1993). Review of discovery and mechanism of action: J. D. Bergstrom *et al.*, *Annu. Rev. Microbiol.* **49**, 607-639 (1995); and antihypercholesterolemia and antifungal properties: N. S. Watson, P. A. Procopiou, *Prog. Med. Chem.* **33**, 331-378 (1996); of chemistry and biology: A. Nadin, K.C. Nicolaou, *Angew. Chem. Int. Ed.* **35**, 1623-1656 (1996); of synthetic studies: A. Armstrong, T. J. Blench, *Tetrahedron* **58**, 9321-9349 (2002).

Zaragozic acid A

Zaragozic Acid A. [142561-96-4] (7*S*)-2,7-Anhydro-3,4-di-*C*-carboxy-8,9,10,12,13-pentadeoxy-10-methylene-12-(phenylmethyl)-L-*erythro*-L-*glycero*-D-*altro*-7-triduculo-7,4-furanosonic acid 11-acetate 5-[(2*E*,4*S*,6*S*)-4,6-dimethyl-2-octenoate]; squalestatin 1; squalestatin S1. C$_{35}$H$_{46}$O$_{14}$; mol wt 690.74. Absolute stereochemistry: K. E. Wilson *et al.*, *J. Org. Chem.* **57**, 7151 (1992). Biosynthesis: K. M. Byrne *et al.*, *ibid* **58**, 1019 (1993). Synthesis of bicyclic core: H. Abdel-Rahman *et al.*, *J. Chem. Soc. Chem. Commun.* **24**, 1841 (1993). Total synthesis: K. C. Nicolaou *et al.*, *Chem. Eur. J.* **1**, 467 (1995). Accumulation of farnesol-derived dicarboxylic acids in liver: S. Vaidya *et al.*, *Arch. Biochem. Biophys.* **355**, 84 (1998). Induction of CYP2B enzymes: T. A. Kocarek *et al.*, *Mol. Pharmacol.* **54**, 474 (1998). White foam. $[\alpha]_D^{25}$ +18.3° (c = 0.60 in CHCl$_3$); $[\alpha]_D^{25}$ +37° (c = 1.29 in methanol). uv max (methanol): 214 nm (log ε 4.29); uv max (MeCN): 210 nm (E$_{1cm}^{1\%}$ 314).

Zaragozic Acid B. [146389-61-9] (7*S*,10ξ,11ξ,12ξ)-2,7-Anhydro-3,4-di-*C*-carboxy-8,9,10,12,13,14,15-heptadeoxy-10,12-dimethyl-15-phenyl-L-*glycero*-D-*altro*-pentadec-14-en-7-ulo-7,4-furanosonic acid 5-[(6*E*,12*E*)-6,12-tetradecadienoate]. C$_{39}$H$_{54}$O$_{13}$; mol wt 730.85. Accumulation of farnesol contributing to antifungal activity: J. M. Hornby *et al.*, *Antimicrob. Agents Chemother.* **47**, 2366 (2003). $[\alpha]_D^{25}$ +5.9° (c = 0.27 in methanol). uv max (methanol): 205, 251 nm (log ε 4.44, 4.32).

Zaragozic Acid C. [146389-62-0] (7*S*)-2,7-Anhydro-3,4-di-*C*-carboxy-8,9,10,12,13-pentadeoxy-12-(phenylmethyl)-L-*erythro*-L-*glycero*-D-*altro*-7-trideculo-7,4-furanosonic acid 11-acetate 5-[(4*E*,6*R*)-6-methyl-9-phenyl-4-nonenoate]. C$_{40}$H$_{50}$O$_{14}$; mol wt 754.83. Isolation and structure elucidation: C. Dufresne *et al.*, *Tetrahedron* **48**, 10221 (1992). Synthesis of bicyclic core: I. Paterson *et al.*, *Tetrahedron Lett.* **38**, 4301 (1997). Total synthesis based on aldol approach: S. Nakamura *et al.*, *Tetrahedron* **61**, 11078 (2005). Review of syntheses: S. Nakamura, *Chem. Pharm. Bull.* **53**, 1-10 (2005). White powder. $[\alpha]_D^{20}$ +9.6° (c = 0.29 in ethanol). uv max (ethanol): 209, 259 nm (log ε 4.16, 2.81).

10313. Zatebradine. [85175-67-3] 3-[3-[[2-(3,4-Dimethoxyphenyl)ethyl]methylamino]propyl]-1,3,4,5-tetrahydro-7,8-dimethoxy-2*H*-3-benzazepin-2-one; 1-(7,8-dimethoxy-1,3,4,5-tetrahydro-2*H*-3-benzazepin-2-on-3-yl)-3-[*N*-methyl-*N*-[2-(3,4-dimethoxyphenyl)ethyl]amino]propane. C$_{26}$H$_{36}$N$_2$O$_5$; mol wt 456.58. C 68.40%, H 7.95%, N 6.14%, O 17.52%. Specific bradycardic agent; sinus node inhibitor. Prepn: M. Reiffen *et al.*, **EP 65229**; *eidem*, **US 4490369** (1982, 1984 both to Thomae); and structure-activity study: M. Reiffen *et al.*, *J. Med. Chem.* **33**, 1496 (1990). Pharmacology: W. Kobinger, C. Lillie, *Eur. J. Pharmacol.* **104**, 9 (1984). Mechanism of action study: M. Goethals *et al.*, *Circulation* **88**, 2389 (1993). Pharmacokinetics: W. Roth *et al.*, *J. Pharm. Sci.* **82**, 99 (1993). GC-MS determn in plasma: J. Schmid *et al.*, *J. Chromatogr.* **658**, 93 (1994). Clinical trial in stable angina: S. P. Glasser *et al.*, *Am. J. Cardiol.* **79**, 1401 (1997).

Hydrochloride. [91940-87-3] UL-FS-49. C$_{26}$H$_{36}$N$_2$O$_5$.HCl; mol wt 493.04. Two crystalline modifications, mp 188° and mp 168°. Sol in water.

THERAP CAT: Antianginal.

10314. Zearalenone. [17924-92-4] (3*S*,11*E*)-3,4,5,6,9,10-Hexahydro-14,16-dihydroxy-3-methyl-1*H*-2-benzoxacyclotetradecin-1,7(8*H*)-dione; 6-(10-hydroxy-6-oxo-*trans*-1-undecenyl)-β-resorcylic acid lactone; Compd F-2; FES. $C_{18}H_{22}O_5$; mol wt 318.37. C 67.91%, H 6.97%, O 25.13%. Estrogenic mycotoxin produced by *Fusarium* fungi commonly found in grains. One of a group of compounds known as **resorcylic acid lactones**. Isoln from the mycelia of *Gibberella zeae (Fusarium graminearum):* M. Stob *et al., Nature* **196**, 1318 (1962); F. N. Andrews, M. Stob, **US 3196019** (1965 to Purdue Res. Found.). Structure: W. H. Urry *et al., Tetrahedron Lett.* **1966**, 3109. Synthesis of *dl*-form: D. Taub *et al., Chem. Commun.* **1967**, 225; **NL 6812148**; N. N. Girotra, N. L. Wendler, **US 3551455** (1968, 1970 both to Merck & Co.); Vlattas *et al., J. Org. Chem.* **33**, 4176 (1968); T. Takahashi *et al., Tetrahedron Lett.* **22**, 1363 (1981). Total synthesis and abs config: D. Taub *et al., Tetrahedron* **24**, 2443 (1968). Physicochemical data: A. E. Pohland *et al., Pure Appl. Chem.* **54**, 2219 (1982). Natural occurrence in agricultural products: T. Tanaka *et al., J. Agric. Food Chem.* **36**, 979 (1988). GC determn in cereals: K. Schwadorf, H.-M. Müller, *J. Chromatogr.* **595**, 259 (1992).

Crystals, mp 164-165°. $[\alpha]_{546}^{25}$ −170.5° (c = 1.0 in CH_3OH). uv max (CH_3OH): 236, 274, 316 nm (ε 29700; 13909; 6020). Sol in aq alkali, ether, benzene, alcohols. Practically insol in water.

dl-**Form.** Crystals, mp 187-189°.

10315. Zeatin. [1637-39-4] (2*E*)-2-Methyl-4-(9*H*-purin-6-yl-amino)-2-buten-1-ol; *trans*-zeatin. $C_{10}H_{13}N_5O$; mol wt 219.25. C 54.78%, H 5.98%, N 31.94%, O 7.30%. Naturally occurring plant growth hormone; cytokinin originally isolated from sweet corn kernels, *Zea mays* L. *Gramineae*. Isoln and structure determn: D. S. Letham *et al., Proc. Chem. Soc. London* **1964**, 230. Synthesis: G. Shaw, D. V. Wilson, *ibid.* 231; G. Shaw *et al., J. Chem. Soc. C* **1966**, 921; J. Corse, J. Kuhnle, *Synthesis* **1972**, 618; G. M. Gray, *ibid.* **1983**, 488; *idem*, **EP 86454** (1983 to J. T. Baker). Inhibition of mitochondrial function: C. O. Miller, *Plant Physiol.* **69**, 1274 (1982); translocation in soybean explants: L. Noodén, D. S. Letham, *J. Plant Growth Regul.* **2**, 265 (1984). Reviews: D. S. Letham, *Annu. Rev. Plant Physiol.* **18**, 349-363 (1967); D. S. Letham, L. M. S. Palni, *ibid.* **34**, 163-197 (1983).

Crystals from water, mp 207-208°. uv max in 0.1*M* HCl: 207, 275 nm (ε 14500, 14650); at pH 7.2: 212, 270 nm (ε 17050, 16150); in 0.1*M* NaOH: 220, 276 nm (ε 15900, 14650).

10316. Zeaxanthin. [144-68-3] (3*R*,3′*R*)-β,β-Carotene-3,3′-diol; *all-trans*-β-carotene-3,3′-diol; (3*R*,3′*R*)-dihydroxy-β-carotene; zeaxanthol; anchovyxanthin. $C_{40}H_{56}O_2$; mol wt 568.89. C 84.45%, H 9.92%, O 5.62%. One of the most widespread carotenoid alcohols in nature. Occurs together and is isomeric with xanthophyll, *q.v.* It is the pigment of yellow corn *Zea mays* L., *Gramineae*. Isoln from yellow corn grits: Karrer *et al., Helv. Chim. Acta* **13**, 268 (1930). Isoln by chromatography: Kuhn, Grundmann, *Ber.* **67**, 596 (1934). Isoln from algae, bacteria: A. J. Aasen *et al., Acta Chem. Scand.* **26**, 404 (1972). Isoln of dipalmitate from *Physalis* petal: Kuhn *et al., Ber.* **63**, 1489 (1930). Does not possess vitamin A potency: Schumacher *et al., Poult. Sci.* **23**, 529 (1944). Stereochemistry: Zechmeister, *Chem. Rev.* **34**, 267 (1944). Abs config of natural zeaxanthin: T. E. de Ville *et al., Chem. Commun.* **1969**, 1311; J. R. Hlubucek *et al., J. Chem. Soc. Perkin Trans. 1* **1974**, 848. Total synthesis: Isler *et al., Helv. Chim. Acta* **39**, 2041 (1956); *ibid.* **40**, 456 (1957); Loeber *et al., J. Chem. Soc. C* **1971**, 404; A. Rüttimann, H. Mayer, *Helv. Chim. Acta* **63**, 1456 (1980); P. R. Ellis *et al., ibid.*

64, 1092 (1980); E. Widmer *et al., ibid.* **65**, 944, 958 (1982). Sepn of configurational isomers by HPLC: A. Rüttimann *et al., J. High Resolut. Chromatogr. Chromatogr. Commun.* **6**, 612 (1983). First isoln of enantiomeric and *meso*-zeaxanthin in nature: T. Maoka *et al., Comp. Biochem. Physiol.* **83B**, 121 (1986). Mfg procedure: Isler *et al.,* **US 2917539** (1959 to Hoffmann-La Roche). *Reviews:* Zechmeister, *Carotinoide* (Berlin, 1934); Mayer, *The Chemistry of Natural Coloring Matters* (New York, 1943); Karrer, Jucker, *Carotenoids* (New York, 1950).

Yellow rhombic plates with steel-blue metallic luster from ethanol, mp 207° (Zechmeister); mp 215.5° (Kuhn). Optically inactive. Absorption bands (ethanol) 483, 451.5 nm. Practically insol in water. Slightly sol in petr ether, methanol; sol in carbon disulfide, benzene, chloroform, carbon tetrachloride, pyridine, ethyl acetate; sol in glacial acetic acid upon addition of hexane. Less sol in boiling methanol (1:1550) than xanthophyll.

Dipalmitate. Physalien; physalin. $C_{72}H_{116}O_4$. Fine yellow or red needles from benzene, mp 97°.

10317. Zein. A prolamine; an alcohol-soluble protein present in amounts of 2.5-10% (dry basis) in corn (*Zea mays* L., *Gramineae*). The greater part of zein has a mol wt of 38,000. Does not contain lysine or tryptophan. Extracted from gluten meal with dil isopropanol: Swallen, *Ind. Eng. Chem.* **33**, 394 (1941). Improved method: Carter, Reck, **DE 2002337** (1971 to Nutrilite Prod.), *C.A.* **75**, 117341b (1971). *Review:* Mossé, *Ann. Physiol. Végétale* **3**, 105 (1961).

White to slightly yellow powder. d 1.226. When completely dry may be heated to 200° without visible signs of decompn. Readily sol in acetone-water mixtures between the limits of 60% and 80% of acetone by volume. Soluble in aq alcohols, the glycols, the ethyl ether of ethylene glycol, furfuryl alcohol, tetrahydrofurfuryl alcohol, aq alkaline solns of pH 11.5 (or greater). Tends to become denatured when in soln and becomes insol. Insol in water, acetone and in anhydr alcohols except methanol.

USE: Manuf plastics, paper coatings, adhesives, substitutes for shellac, laminated board, in solid color printing, films, edible coatings for foodstuffs. Pharmaceutic aid (coating agent).

10318. Zeolites. Crystalline, hydrated alkali-aluminum silicates of the general formula $M_{2/n}O.Al_2O_3.ySiO_2.wH_2O$ where M represents a group IA or IIA element, n is the cation valence, y is 2 or greater and w is the number of water molecules contained in the channels or interconnected voids within the zeolite. The cations are mobile and capable of undergoing ion exchange. Zeolites occur naturally in sedimentary and volcanic rocks, altered basalts, ores, clay deposits. Some 40 known zeolite minerals and a great number of synthetic zeolites are available commercially. Ref: Milton, **US 2882243** (1959 to Union Carbide). Environmentally friendly synthesis: H. Lee *et al., Nature* **425**, 385 (2003). *Reviews:* D. W. Breck, R. A. Anderson, "Molecular Sieves" in *Kirk-Othmer Encyclopedia of Chemical Technology* vol. 15 (Wiley-Interscience, New York, 3rd ed., 1981) pp 638-669; P. L. Layman, *Chem. Eng. News* **60**, 10 (Sept. 27, 1982); J. Haggin, *ibid.* **60**, 9 (Dec. 13, 1982). *Books: Adv. Chem. Ser.* **101 102**, entitled "Molecular Sieve Zeolites - I, II," E. M. Flanigan, Ed. (ACS, Washington DC, 1971) 526 pp, 459 pp; *Adv. Chem. Ser.* **121**, entitled "Molecular Sieves," W. M. Meier, Ed. (ACS, Washington DC, 1973) 634 pp; D. W. Breck, *Zeolite Molecular Sieves* (John Wiley, New York, 1974) 771 pp; *Natural Zeolites: Occurrence, Properties, Use*, L. B. Sand, F. A. Mumpton, Eds. (Pergamon, Oxford, 1978). Studies on fibrous zeolites as possible environmental hazards: A. N. Rohl *et al., Science* **216**, 518 (1982); Y. I. Baris, *Arch. Environ. Health* **37**, 177 (1982). Use as microscopic chemical reactor vessels: R. Pool, *Science* **263**, 1698 (1994).

USE: As molecular sieves, filters, adsorbents, catalysts, drying agents, cation exchangers, dispersing agents, detergent builders.

10319. Zeranol. [26538-44-3] (3*S*,7*R*)-3,4,5,6,7,8,9,10,11,12-Decahydro-7,14,16-trihydroxy-3-methyl-1*H*-2-benzoxacyclotetradecin-1-one; 6-(6,10-dihydroxyundecyl)-β-resorcylic acid μ-lactone; α-zearalanol; MK-188; P-1496; Ralgro; Ralabol; Ralone. C_{18}-$H_{26}O_5$; mol wt 322.40. C 67.06%, H 8.13%, O 24.81%. Animal growth promotant prepd from zearalenone, *q.v.*: E. B. Hodge *et al.*, **US 3239345** (1966 to Commercial Solvents). Stereoselective prepn and abs config: G. Snatzke *et al.*, *Helv. Chim. Acta* **69**, 734 (1986). TLC determn in plasma and tissues: M. B. Medina, N. Nagdy, *J. Chromatogr.* **614**, 315 (1993). Growth regulating and repartitioning effects in cattle: P. G. Lemieux *et al.*, *J. Anim. Sci.* **68**, 1702 (1990). *Review:* R. S. Baldwin *et al.*, *Regul. Toxicol. Pharmacol.* **3**, 9-25 (1983).

Crystals from isopropanol/water, mp 182-183°; $[\alpha]_D$ +46.3° (c = 1.0 in methanol). LD_{50} in mice (mg/kg): 4400 i.p.; >40000 orally (Baldwin).

THERAP CAT (VET): Anabolic.

10320. Zibotentan. [186497-07-4] *N*-(3-Methoxy-5-methyl-2-pyrazinyl)-2-[4-(1,3,4-oxadiazol-2-yl)phenyl]-3-pyridinesulfon-amide; ZD-4054. $C_{19}H_{16}N_6O_4S$; mol wt 424.44. C 53.77%, H 3.80%, N 19.80%, O 15.08%, S 7.55%. Specific endothelin A (ET_A) receptor antagonist. Prepn: R. H. Bradbury *et al.*, **WO 9640681**; *eidem*, **US 6060475** (1996, 2000 both to Zeneca). Crystal structure: B. Stensland, B. J. Roberts, *Acta Crystallogr. E* **60**, o1817 (2004). Clinical pharmacokinetics and tolerability: M. R. Ranson *et al.*, *Int. J. Clin. Pharmacol. Ther.* **48**, 708 (2010). Clinical trial in hormone-resistant prostate cancer and bone metastases: N. D. James *et al.*, *Eur. Urol.* **55**, 1112 (2009). Review of ET_A binding specificity: C. D. Morris *et al.*, *Br. J. Cancer* **92**, 2148-2152 (2005); of pharmacology: J. W. Growcott, *Anti-Cancer Drugs* **20**, 83-88 (2009); of discovery and clinical experience in castrate-resistant prostate cancer: D. R. Shepard, R. Dreicer, *Expert Opin. Invest. Drugs* **19**, 899-908 (2010).

Colorless, rhombohedral crystals from *N*-methylpyrrolidin-2-one. d (cryst): 1.455.

THERAP CAT: Antineoplastic.

10321. Ziconotide. [107452-89-1] L-Cysteinyl-L-lysylglycyl-L-lysylglycyl-L-alanyl-L-lysyl-L-cysteinyl-L-seryl-L-arginyl-L-leu-cyl-L-methionyl-L-tyrosyl-L-α-aspartyl-L-cysteinyl-L-cysteinyl-L-threonylglycyl-L-seryl-L-cysteinyl-L-arginyl-L-serylglycyl-L-lysyl-L-cysteinamide cyclic (1 → 16), (8 → 20), (15 → 25)-tris(disul-fide); ω-conopeptide MVIIA; SNX-111; CI-1009; Prialt. $C_{102}H_{172}$-$N_{36}O_{32}S_7$; mol wt 2639.14. C 46.42%, H 6.57%, N 19.11%, O 19.40%, S 8.50%. Selective antagonist of N-type voltage sensitive calcium channels (VSCC). Blocks neurotransmitter release by preventing depolarization-induced calcium influx. Synthetic, 25-residue peptide corresponding to one of the conotoxins, *q.v.*, derived from the venom of the marine cone snail, *Conus magus* L. Synthesis: B. M. Olivera *et al.*, *Biochemistry* **26**, 2086 (1987). Structure-activity study: L. Nadasdi *et al.*, *ibid.* **34**, 8076 (1995). Binding study and use as biochemical tool: R. Kristipati *et al.*, *Mol. Cell. Neurosci.* **5**, 219 (1994). NMR studies: L. K. MacLachlan *et al.*, *Methods Mol. Biol.* **60**, 337 (1997). Radioimmunoassay: C. M. Barksdale *et al.*, *J. Clin. Ligand Assay* **19**, 229 (1996). Clinical pharmacology: D. McGuire *et al.*, *J. Cardiovasc. Pharmacol.* **30**, 400 (1997). Clinical trial in pain relief from cancer or AIDS: P. S. Staats *et al.*, *J. Am. Med. Assoc.* **291**, 63 (2004).

USE: Ligand for binding studies of voltage sensitive calcium channels.

THERAP CAT: Analgesic; neuroprotective.

10322. Zidovudine. [30516-87-1] 3'-Azido-3'-deoxythymi-dine; azidothymidine; AZT; BW-A509U; Retrovir. $C_{10}H_{13}N_5O_4$; mol wt 267.25. C 44.94%, H 4.90%, N 26.21%, O 23.95%. Pyrim-idine nucleoside analog; reverse transcriptase inhibitor. Prepn: J. P. Horwitz *et al.*, *J. Org. Chem.* **29**, 2076 (1964); R. P. Glinski *et al.*, *ibid.* **38**, 4299 (1973). *See also:* J. L. Rideout *et al.*, **US 4724232** (1988 to Burroughs Wellcome). Total synthesis: C. K. Chu *et al.*, *Tetrahedron Lett.* **29**, 5349 (1988). HPLC determn and stability assay: A. Dunge *et al.*, *J. Pharm. Biomed. Anal.* **37**, 1109 (2005). *In vitro* antiviral and antitumor activity: E. De Clercq *et al.*, *Biochem. Pharmacol.* **29**, 1849 (1980); vs HIV-1 virus: H. Mitsuya *et al.*, *Proc. Natl. Acad. Sci. USA* **82**, 7096 (1985). Clinical pharmaco-kinetics: R. W. Klecker *et al.*, *Clin. Pharmacol. Ther.* **41**, 407 (1987). Toxicology studies: K. M. Ayers, *Am. J. Med.* **85**, Suppl. 2A, 186 (1988). Comprehensive description: M. L. Sethi, *Anal. Profiles Drug Subs.* **20**, 729-765 (1991). Review of clinical experi-ence: G. X. McLeod, S. M. Hammer, *Ann. Intern. Med.* **117**, 487-501 (1992); in prevention of perinatal HIV transmission: R. Sper-ling, *Infect. Dis. Obstet. Gynecol.* **6**, 197-203 (1998).

Needles from petr ether, mp 106-112° (Horwitz). Also reported as crystals from water, mp 120-122° (Glinski). $[\alpha]_D^{25}$ +99° (c = 0.5 in water). Sol in alc. Soly in water (25°C): 25 mg/ml. uv max (water): 266.5 nm (ε 11650). LD_{50} in male, female mice, male, female rats (mg/kg): 3568, 3062, 3084, 3683 orally; >750 i.v. (all species) (Ayers).

THERAP CAT: Antiretroviral.

THERAP CAT (VET): Antiviral in treatment of feline immuno-deficiency virus and feline leukemia virus.

10323. Zileuton. [111406-87-2] *N*-(1-Benzo[*b*]thien-2-yleth-yl)-*N*-hydroxyurea; (±)-*N*-hydroxy-*N*-(1-benzo[*b*]thien-2-ylethyl)-urea; A-64077; Abbott 64077; Leutrol; Zyflo. $C_{11}H_{12}N_2O_2S$; mol wt 236.29. C 55.91%, H 5.12%, N 11.86%, O 13.54%, S 13.57%. Inhibitor of 5-lipoxygenase, the initial enzyme in the biosynthesis of leukotrienes from arachidonic acid. Prepn: J. B. Summers, Jr. *et al.*, **EP 279263**; *eidem*, **US 4873259** (1988, 1989 both to Abbott). Chro-matographic resolution of enantiomers: S. B. Thomas *et al.*, *J. Chromatogr.* **623**, 390 (1992). Inhibition of 5-lipoxygenase: G. W. Carter *et al.*, *J. Pharmacol. Exp. Ther.* **256**, 929 (1991). Pharmaco-kinetics: P. Rubin *et al.*, *Agents Actions Suppl.* **35**, 103 (1991). Clinical evaluation in rheumatoid arthritis: M. E. Weinblatt *et al.*, *J. Rheumatol.* **19**, 1537 (1992). Clinical trial in asthma: E. Israel *et al.*, *Ann. Intern. Med.* **119**, 1059 (1993). Reviews: R. L. Bell *et al.*, *Int. J. Immunopharmacol.* **14**, 505-510 (1992); K. McGill, W. W. Busse, *Lancet* **348**, 519-524 (1996).

Crystals, mp 157-158°.

THERAP CAT: Antiasthmatic.

10324. Zilpaterol. [119520-05-7] *rel*-(6*R*,7*R*)-4,5,6,7-Tetra-hydro-7-hydroxy-6-[(1-methylethyl)amino]imidazo[4,5,1-*jk*][1]-benzazepin-2(1*H*)-one; (±)-*trans*-4,5,6,7-tetrahydro-7-hydroxy-6-(isopropylamino)imidazo[4,5,1-*jk*][1]benzazepin-2(1*H*)-one; RU-42173. $C_{14}H_{19}N_3O_2$; mol wt 261.33. C 64.35%, H 7.33%, N 16.08%, O 12.24%. β-Adrenergic agonist. Prepn: D. Fréchet *et al.*, **FR 2534257**; *eidem*, **US 4585770** (1984, 1986 both to Roussel Uclaf). Effects on growth performance and carcass characteristics of steers: A. Plascencia *et al.*, *Proc. West. Sect. Am. Soc. Anim. Sci.* **50**, 331-334 (1999). GC/MS determn in commercial feeds: B. Bocca *et* al., *J. Chromatogr. B* **783**, 141 (2003); in calf eye: *idem et al.*, *J. AOAC Int.* **86**, 8 (2003). Residue analysis in cattle and pigs: C. S. Stachel *et al.*, *Anal. Chim. Acta* **493**, 63 (2003). ELISA screening method: W. L. Shelver, D. J. Smith, *J. Agric. Food Chem.* **52**, 2159 (2004).

Relative stereochemistry

mp ≈166°.

Hydrochloride. [119520-06-8]; [186046-04-8] (monohydrate); [186046-05-9] (trihydrate). Zilmax. $C_{14}H_{19}N_3O_2$.HCl; mol wt 297.78. mp ≈270°.

THERAP CAT (VET): Growth promotant.

10325. Zimeldine. [56775-88-3] (2Z)-3-(4-Bromophenyl)-*N,N*-dimethyl-3-(3-pyridinyl)-2-propen-1-amine; (Z)-3-(4'-bromophenyl)-3-(3''-pyridyl)dimethylallylamine; zimelidine; *cis*-H-102/09. $C_{16}H_{17}BrN_2$; mol wt 317.23. C 60.58%, H 5.40%, Br 25.19%, N 8.83%. Serotonin uptake inhibitor. Prepn: P. B. Berntsson *et al.*, **ZA 7201503**; *eidem*, **US 3928369** (1972, 1975 both to AB Hassle). X-ray crystallographic study: S. Abrahamsson *et al.*, *Acta Chem. Scand.* **A30**, 609 (1976). *In vivo* study in rats and humans: S. B. Ross *et al.*, *Life Sci.* **19**, 205 (1976). Pharmacokinetics: D. Brown, *Eur. J. Clin. Pharmacol.* **17**, 111 (1980). Mode of action: K. Fuxe *et al.*, *Neurosci. Lett.* **13**, 307 (1979). Metabolism: J. Lundström *et al.*, *Arzneim.-Forsch.* **31**, 486 (1981). Clinical studies: A. Georgotas *et al.*, *Am. J. Psychiatry* **139**, 1057 (1982); E. Syvalahti *et al.*, *J. Int. Med. Res.* **10**, 250 (1982). Profile of antidepressant action: S. A. Montgomery *et al.*, *Adv. Biochem. Psychopharmacol.* **32**, 35 (1982). Reversal of ethanol-induced memory impairment in human subjects: H. Weingartner *et al.*, *Science* **221**, 472 (1983). Review of pharmacology and therapeutic efficacy: R. C. Heel *et al.*, *Drugs* **24**, 169-206 (1982). Symposium on pharmacology, pharmacokinetics, metabolism, clinical studies: *Br. J. Clin. Pract.* **1982**, Suppl. 19, 1-122.

Dihydrochloride monohydrate. [61129-30-4] Normud; Zelmid. $C_{16}H_{17}BrN_2$.2HCl.H$_2$O; mol wt 408.16. Crystals, mp 193°.

THERAP CAT: Antidepressant.

10326. Zinc. [7440-66-6] Zn; at. wt 65.38; at. no. 30; valence 2. Group IIB (12). Essential nutritional trace element necessary for function of many metallogenzymes. Abundance in earth's crust: 0.02% by wt. Natural isotopes: 64 (48.89%); 66 (27.81%); 68 (18.57%); 67 (4.11%); 70 (0.62%); eight radioactive isotopes and two isomers. Occurs in smithsonite or zinc spar, sphalerite or zinc blende, zincite, willemite, *franklinite*, [(Zn,Mn,Fe)O.(Fe.Mn$_2$)O$_3$] or *gahnite* (ZnAl$_2$O$_4$). Has been known since very early times. Commercial forms: ingots; lumps; sheets; wire; shot; strips; sticks; granules; mossy; powder (dust). Prepn: Gowland, Bannister, *Metallurgy of Non-Ferrous Metals* (Griffin, London, 1930); *Zinc Production, Properties and Uses* (Zinc Development Association, London, 1968). *Reviews: Zinc*, C. H. Mathewson, Ed., A.C.S. Monograph Series no. **142** (Reinhold, New York, 1959) 721 pp; Schlechter, Thompson, "Zinc and Zinc Alloys" in *Kirk-Othmer Encyclopedia of Chemical Technology* **vol. 22** (Interscience, New York, 2nd ed., 1970) pp 555-603; Aylett, "Group IIB" in *Comprehensive Inorganic Chemistry* **vol. 3**, J. C. Bailar, Jr. *et al.*, Eds. (Pergamon Press,

Oxford, 1973) pp 187-328. Review of toxicology and human exposure: *Toxicological Profile for Zinc* (PB2006-100008, 2005) 352 pp.

Bluish-white, lustrous metal; distorted hexagonal close-packed structure; stable in dry air; becomes covered with a white coating of basic carbonate on exposure to moist air. mp 419.5°. bp 908°. d^{25} 7.14. Heat capacity at constant pressure (25°): 6.07 cal/mole deg. Mohs' hardness 2.5. When heated to 100-150° becomes malleable, at 210° becomes brittle and pulverizable. Burns in air with a bluish-green flame. Loses electrons in aqueous systems to form Zn^{2+} E° (aq) Zn/Zn^{2+} 0.763 V. Slowly attacked by H_2SO_4 or HCl; oxidizing agents or metal ions, e.g. Cu^{2+}, Ni^{2+}, Co^{2+}, accelerate the process. Reacts slowly with ammonia water and acetic acid; rapidly with HNO_3. Reacts with alkali hydroxides to form *"zincates"*, ZnO_2^{2-}, which are actually hydroxo complexes such as $Zn(OH)_3^-$; $Zn(OH)_4^{2-}$, $[Zn(OH)_4(H_2O)_2]^{2-}$. Insol in water.

Caution: Potential symptoms of overexposure by fume inhalation are metal fume fever; by acute oral overexposure are stomach cramps, nausea, vomiting; by chronic oral overexposure are anemia, pancreatic damage, decreased HDL cholesterol levels (PB2006-100008).

USE: Galvanizing sheet iron; as ingredient of alloys such as bronze, brass, Babbitt metal, German silver, and special alloys for die-casting; as a protective coating for other metals to prevent corrosion; in paint coatings; for electrical apparatus, especially dry cell batteries, household utensils, castings, printing plates, building materials, railroad car linings, automotive equipment; as reducing agent and precipitating agent in organic chemistry; for deoxidizing bronze; extracting gold by the cyanide process, purifying fats for soaps; bleaching bone glue; manuf sodium hydrosulfite; insulin zinc salts; as reagent in analytical chemistry, e.g., in the Marsh and Gutzeit test for arsenic; as a reducer in the determination of iron. It is a nutritional trace element.

10327. Zinc Acetate. [557-34-6] Acetic acid zinc salt (2:1); Galzin. $C_4H_6O_4Zn$; mol wt 183.50. C 26.18%, H 3.30%, O 34.88%, Zn 35.65%. $Zn(C_2H_3O_2)_2$. Prepn of anhydr salt from zinc nitrate and acetic anhydride: Späth, *Monatsh. Chem.* **33**, 235 (1912). Clinical evaluations in Wilson's disease: G. M. Hill *et al.*, *Hepatology* **7**, 522 (1987); G. J. Brewer *et al.*, *J. Lab. Clin. Med.* **109**, 526 (1987). Toxicity: H. F. Smyth *et al.*, *Am. Ind. Hyg. Assoc. J.* **30**, 470 (1969).

Dihydrate. [5970-45-6] Wilzin. $C_4H_6O_4Zn.2H_2O$; mol wt 219.53. Crystallizes from dil acetic acid; faint, acetous odor; astringent taste; slightly efflorescent. d 1.735; mp 237°. One gram dissolves in 2.3 ml water, 1.6 ml boiling alcohol, 30 ml alcohol, about 1 ml boiling alcohol. The aq soln is neutral or slightly acid to litmus; pH about 5-6. *Keep in well-closed containers.* LD$_{50}$ orally in rats: 2.46 g/kg (Smyth).

USE: Preserving wood; as mordant in dyeing; manuf glazes for painting on porcelain. As a reagent in testing for albumin, tannin, urobilin, phosphate, blood.

THERAP CAT: Styptic, astringent. In treatment of Wilson's disease.

THERAP CAT (VET): Antiseptic, astringent, protective (topical). Has been used as an emetic.

10328. Zinc Bromide. [7699-45-8] Br_2Zn; mol wt 225.22. Br 70.96%, Zn 29.04%. $ZnBr_2$. Usually contains at least 97% $ZnBr_2$, the remainder being chiefly water.

Very hygroscopic, granular powder; sharp, metallic taste. d 4.22; mp 394°; bp 697° with partial decompn. One gram dissolves in 0.25 ml water, 0.5 ml 90% alcohol; sol in ether, solns of alkali hydroxides. The aq soln is acid to litmus; pH about 4. *Keep tightly closed.*

USE: Making silver bromide collodion emulsions for photography; in the shielding of viewing windows for nuclear reactions.

10329. Zinc Caprylate. [557-09-5] Octanoic acid zinc salt (2:1). $C_{16}H_{30}O_4Zn$; mol wt 351.82. C 54.62%, H 8.60%, O 18.19%, Zn 18.59%. $Zn(C_8H_{15}O_2)_2$. Prepd from ammonium caprylate and zinc sulfate: van Renesse, *Ann.* **171**, 380 (1874).

Lustrous scales from alc, mp 136°. Sparingly sol in boiling water; moderately sol in boiling alcohol. *Keep well closed.* Dec in moist atm giving off caprylic acid.

USE: As fungicide like zinc propionate.

10330. Zinc Carbonate. [3486-35-9] Carbonic acid zinc salt (1:1). CO_3Zn; mol wt 125.42. C 9.58%, O 38.27%, Zn 52.15%.

ZnCO₃. Occurs in nature as the minerals **smithsonite**, **zincspar**. Prepn: Hüttig *et al.*, *Monatsh. Chem.* **72**, 31 (1939).

Rhombohedral structure. Solubility in water at 15° 0.001 g/100 g; sol in dil acids, alkalies, solns of NH₄⁺ salts.

Basic carbonate. Zinc carbonate hydroxide; zinc subcarbonate. Variable composition, usually characterized as 3Zn(OH)₂.2ZnCO₃. Occurs as the mineral **hydrozincite**. Reagent specification: 70% ZnO minimum.

USE: As pigment; manuf of porcelains, pottery, rubber.

THERAP CAT: Astringent, topical antiseptic.

THERAP CAT (VET): Astringent, antiseptic, protective (topical). Also used in rations to prevent Zn deficiency diseases.

10331. Zinc Chloride. [7646-85-7] Butter of zinc. Cl₂Zn; mol wt 136.31. Cl 52.01%, Zn 47.99%. ZnCl₂. Usually contains at least 95% ZnCl₂; remainder is chiefly water and oxychloride. Toxicity: Bruner, *Fed. Proc.* **9**, 260 (1950).

White, odorless, very deliquesc granules, or fused pieces or rods. d²⁵ 2.907; mp 327.9°; bp 732°. Solubility in H₂O: 432 g/100 g (25°); 614 g/100 g (100°). One gram dissolves in 0.25 ml of 2% HCl, in 1.3 ml alcohol, 2 ml glycerol; freely sol in acetone. With much water some zinc oxychloride is formed. The aq soln is acid to litmus; pH about 4. *Corrosive. Keep tightly closed.* LD i.v. in rats: 60-90 mg/kg (Bruner).

Caution: Potential symptoms of overexposure to fumes are conjunctivitis; irritation of skin, eyes, nose and throat; coughing, copious sputum; dyspnea, chest pain, pulmonary edema and bronchopneumonia; pulmonary fibrosis, cor pulmonale; fever; cyanosis; tachypnea; skin burns. *See NIOSH Pocket Guide to Chemical Hazards* (DHHS/NIOSH 97-140, 1997) p 338.

USE: Deodorant, disinfecting and embalming material; wood preservative; alone or with phenol and other antiseptics for preserving railway ties; fireproofing lumber; with ammonium chloride as flux for soldering; etching metals; manuf parchment paper, artificial silk, dyes, activated carbon, cold-water glues, vulcanized fiber; browning steel, galvanizing iron, copper-plating iron; in magnesia cements; petroleum oil refining; cement for metals and for facing stone; mordant in printing and dyeing textiles; carbonizing woolen goods; producing crepe and crimping fabrics; mercerizing cotton; sizing and weighting fabrics; vulcanizing rubber; solvent for cellulose; preserving anatomical specimens; in microscopy for separating silk, wool, and plant fibers; as dehydrating agent in chemical syntheses. Dentin desensitizer. Component in smoke bombs.

THERAP CAT: Astringent.

THERAP CAT (VET): Antiseptic, astringent. Has been used in ulcers, fistulas, pododermatitis.

10332. Zinc Chromate(VI) Hydroxide. [13530-65-9] Chromic acid (H₂CrO₄) zinc salt (1:1); zinc yellow; buttercup yellow; C.I. Pigment Yellow 36. A basic salt of somewhat variable composition. Approx Zn₂CrO₄(OH)₂.

Hydrate. Yellow, odorless, fine powder. Slightly sol in water; sol in dil acids, including acetic acid.

Caution: Chromium hexavalent (VI) compounds are listed as known human carcinogens: *Report on Carcinogens, Twelfth Edition* (PB2011-111646, 2011) p 109.

USE: As pigment in paints, varnishes, oil colors, linoleum, rubber, etc.

10333. Zinc Citrate. [546-46-3] 2-Hydroxy-1,2,3-propanetricarboxylic acid zinc salt (2:3). C₁₂H₁₀O₁₄Zn₃; mol wt 574.43. C 25.09%, H 1.75%, O 38.99%, Zn 34.16%. Zn₃(C₆H₅O₇)₂. Prepd from zinc carbonate and citric acid: Heldt, *Ann.* **47**, 157 (1843).

Dihydrate. Odorless powder. Slightly sol in water; sol in dil mineral acids, in alkali hydroxides.

USE: In toothpaste and mouthwash.

10334. Zinc Cyanide. [557-21-1] C₂N₂Zn; mol wt 117.45. C 20.45%, N 23.85%, Zn 55.69%. Zn(CN)₂. Usually contains about 85% zinc cyanide, some water and oxide.

White powder. *Poisonous.* Insol in water. Sol in solns of alkali cyanides or hydroxides; not appreciably attacked by organic acids, but readily dec by dil mineral acid with evolution of hydrogen cyanide.

USE: Electroplating; removing NH₃ from producer gas.

10335. Zinc Fluoride. [7783-49-5] F₂Zn; mol wt 103.41. F 36.74%, Zn 63.25%. ZnF₂. Prepd from ZnCO₃ and HF: Ruff, *Die*

Chemie des Fluors (Berlin, 1920) p 36; Emeleus in *Fluorine Chemistry* vol. 1, J. H. Simons, Ed. (Academic Press, New York, 1950) p 38; Kwasnik in *Handbook of Preparative Inorganic Chemistry* **vol. 1**, G. Brauer, Ed. (Academic Press, New York, 2nd ed., 1963) p 242.

Tetragonal needles (rutile lattice) or white cryst mass. d²⁵ 5.00: Haendler *et al.*, *J. Am. Chem. Soc.* **74**, 3167 (1952). mp 872°. bp 1500°. Soly in water: 5 × 10⁻⁵ moles/l, Kwasnik, *loc. cit.* Slightly sol in aq HF, more sol in HCl, HNO₃, NH₄OH. Zinc fluoride used for fluorinations should be slightly hydrated.

Tetrahydrate. Rhombohedral crystals, becomes anhydr at 100°. Soly in water: 1.516 g/100 ml. May be stored in glass bottles.

USE: In the fluorination of organic compds, manuf phosphors for fluorescent electric lights, glazes and enamels for porcelain, preserving wood, in electroplating baths.

10336. Zinc Formate. [557-41-5] Formic acid zinc salt (2:1). C₂H₂O₄Zn; mol wt 155.44. C 15.45%, H 1.30%, O 41.17%, Zn 42.08%. Zn(HCOO)₂. Forms dihydrate readily. Prepd from zinc carbonate and formic acid: Kendall, Adler, *J. Am. Chem. Soc.* **43**, 1470 (1921).

Dihydrate. Crystals. d 2.207. Solubility in H₂O (20°): 5.2 g/100 g. Practically insol in alc.

10337. Zinc Hexafluorosilicate. [16871-71-9] Hexafluorosilicate(2−) zinc (1:1); zinc fluosilicate; zinc silicofluoride. F₆SiZn; mol wt 207.48. F 54.94%, Si 13.54%, Zn 31.53%. ZnSiF₆.

Hexahydrate. White crystals. Soluble in water. pH of 1% aqueous solution 3.2.

USE: Mothproofing agent; laundry sour; hardener for concrete.

10338. Zinc Iodate. [7790-37-6] Iodic acid (HIO₃) zinc salt. I₂O₆Zn; mol wt 415.21. I 61.13%, O 23.12%, Zn 15.75%. Zn(IO₃)₂. Prepn of dihydrate: F. Mylius, R. Funk, *Ber.* **30**, 1716 (1897). Soly studies: J. E. Ricci, G. J. Nesse, *J. Am. Chem. Soc.* **64**, 2305 (1942); J. N. Spencer *et al.*, *J. Chem. Eng. Data* **19**, 140 (1974). X-ray, IR and Raman spectroscopic study: S. Peter *et al.*, *Z. Anorg. Allg. Chem.* **626**, 208 (2000).

White, crystalline powder. Soly in water at 25°: 0.01542 M.

Dihydrate. [13986-19-1] I₂O₆Zn.2H₂O; mol wt 451.24.

10339. Zinc Iodide. [10139-47-6] I₂Zn; mol wt 319.22. I 79.51%, Zn 20.49%. ZnI₂. Usually contains at least 98% ZnI₂.

White or almost white, odorless, hygroscopic, granular powder; sharp, saline taste; becomes brown on exposure to air and light, due to liberation of iodine. d²⁵ 4.74; mp ~446°; bp ~625° with decompn. One gram dissolves in 0.3 ml water, 0.2 ml boiling water, 2 ml glycerol; freely sol in alcohol, ether. The aq soln is acid to litmus; pH about 5. *Keep well closed and protected from light.*

THERAP CAT: Topical antiseptic, astringent.

10340. Zinc Iodide-Starch. [9010-71-3] Trommsdorff's starch. A soln is prepd by heating 4 parts starch, 20 ZnCl₂, and 2 ZnI₂ with 1 liter water. Deteriorates with age and acquires a blue color.

USE: For detecting nitrites, free Cl, and other oxidizing agents.

10341. Zinc Lactate. [16039-53-5] (*T*-4)-Bis[2-(hydroxy-*κO*)propanoato-*κO*]zinc. C₆H₁₀O₆Zn; mol wt 243.55. C 29.59%, H 4.14%, O 39.41%, Zn 26.86%. Zn(C₃H₅O₃)₂. Prepd from zinc carbonate and lactic acid: Pederson *et al.*, *J. Biol. Chem.* **68**, 151 (1926).

Trihydrate. Crystals. Soluble in 60 parts cold, 6 parts boiling water.

10342. Zinc Meta-arsenite. [10326-24-6] Arsenenous acid zinc salt; ZMA. As₂O₄Zn; mol wt 279.25. As 53.66%, O 22.92%, Zn 23.42%. Zn(AsO₂)₂.

White powder. Sol in acids.

USE: Wood preservative, insecticide.

10343. Zinc Nitrate. [7779-88-6] Nitric acid zinc salt (2:1). N₂O₆Zn; mol wt 189.42. N 14.79%, O 50.68%, Zn 34.53%.

Hexahydrate. Colorless, odorless crystals; d 2.065; mp ~36°. Sol in about 0.5 part water, freely in alcohol. The aq soln is acid to litmus; pH of 5% aq soln 5.1. *Keep well closed and in a cool place.* Also available in the form of fused pieces. In this form it contains only about 20% water. Technical flake usually contains 25.6% H₂O.

USE: As a mordant in dyeing.

10344. Zinc Oleate. [557-07-3] (9Z)-9-Octadecenoic acid zinc salt (2:1). $C_{36}H_{66}O_4Zn$; mol wt 628.33. C 68.82%, H 10.59%, O 10.19%, Zn 10.41%. $(C_{17}H_{33}CO_2)_2Zn$. Usually contains small amounts of the zinc salts of other fatty acids. Prepn: Grabner, *Monatsh. Chem.* **42**, 287 (1921).

White, dry, greasy powder. Insol in water; sol in alcohol, ether, carbon disulfide, benzene, petr ether.

10345. Zincon. [135-52-4] 2-[2-[[2-(2-Hydroxy-5-sulfophenyl)diazenyl]phenylmethylene]hydrazinyl]benzoic acid; 2-carboxy-2'-hydroxy-5'-sulfoformazylbenzene; 2-[[α-(2-hydroxy-5-sulfophenylazo)-benzylidene]-hydrazino]-benzoic acid; o-[1-(2-hydroxy-5-sulfophenyl)-3-phenyl-5-formazane]benzoic acid; 2-[5-(2-hydroxy-5-sulfophenyl)-3-phenyl-1-formazyl]benzoic acid. $C_{20}H_{16}N_4O_6S$; mol wt 440.43. C 54.54%, H 3.66%, N 12.72%, O 21.80%, S 7.28%. Cation chelant dye; forms deep blue-colored complexes with metal ions. Preparative method: R. Wizinger, V. Biro, *Helv. Chim. Acta* **32**, 901 (1949). Metal chelate studies: J. H. Yoe, R. M. Rush, *Anal. Chim. Acta* **6**, 526 (1952); R. M. Rush, J. H. Yoe, *Anal. Chem.* **26**, 1345 (1954); A. Ringbom *et al.*, *Anal. Chim. Acta* **19**, 525 (1958); G. K. Singhal, K. N. Tandon, *Talanta* **14**, 1351 (1967); E. Hilario *et al.*, *J. Biochem. Biophys. Methods* **21**, 197 (1990); A. V. Rossi, M. Tubino, *J. Braz. Chem. Soc.* **7**, 161 (1996). Spectrophotometric metal determn studies: M. A. Koupparis, P. I. Anagnostopoulou, *Analyst* **111**, 1311 (1986); L. K. Shpigun *et al.*, *Anal. Chim. Acta* **573-574**, 360 (2006). Determn of calcium in biological samples: C. M. Corns, *Ann. Clin. Biochem.* **24**, 591 (1987). Protein staining applications: G. B. Smejkal, H. F. Hoff, *BioTechniques* **34**, 486 (2003); J.-K. Choi *et al.*, *Electrophoresis* **25**, 1136 (2004). Review of use as a biological stain: *Conn's Biological Stains*, R. W. Horobin, J. A. Kiernan, Eds. (BIOS Scientific Publishers Ltd, Oxford, UK, 10th ed., 2002) 167-168.

Purple powder, mp 203° (dec). pKa_1 4.0; pKa_2 7.8; pKa_3 15.0. Absorption max (0.1*N* sodium hydroxide): 490 nm. Moderately sol in acetone, alkali hydroxide solns; sparingly sol in water; slightly sol in ethanol.

Sodium salt. [62625-22-3] $C_{20}H_{15}N_4NaO_6S$; mol wt 462.41. Purple powder. *Irritant.* Absorption max (methanol): 490, 523(sh) nm. Sol in alkali hydroxide solns; slightly sol in water, ethanol. Insol in most organic solvents. Alkaline soln is orange-red.

USE: Chromogenic indicator for complexometric and spectrophotometric determn of zinc, copper, nickel, cobalt, and mercury. Also in chelometry with EDTA. Determn of calcium in biological samples. Dye for staining proteins separated by polyacrylamide gel electrophoresis.

10346. Zinc Oxalate. [547-68-2] [Ethanedioato(2−)-$κO^1$, $κO^2$]zinc. C_2O_4Zn; mol wt 153.43. C 15.66%, O 41.71%, Zn 42.63%. ZnC_2O_4. Prepn: Chatterjee, Dhar, *J. Phys. Chem.* **28**, 1020 (1924).

Dihydrate, powder. Very slightly sol in water; sol in dil mineral acids, ammonia.

10347. Zinc Oxide. [1314-13-2] Flowers of zinc; philosopher's wool; zinc white; C.I. Pigment White 4; C.I. 77947. OZn; mol wt 81.41. O 19.65%, Zn 80.35%. ZnO. Occurs as the mineral *zincite*. Prepd by vaporization of metallic zinc and oxidation of the vapors with preheated air (French process); also from franklinite, (American process) or from zinc sulfide: *Faith, Keyes & Clark's Industrial Chemicals*, F. A. Lowenheim, M. K. Moran, Eds. (Wiley-Interscience, New York, 4th ed., 1975) pp 882-888. Purification: Depew, **US 2372367** (1945 to American Zinc, Lead & Smelting). The medicinal grade contains 99.5% or more ZnO; technical grades contain 90-99% ZnO and a few tenths of 1% of lead. *See also:*

Colour Index vol. **4** (3rd ed., 1971) p 4687. Efficacy as sunblock: S. R. Pinnell *et al.*, *Dermatol. Surg.* **26**, 309 (2000).

White or yellowish-white, odorless powder. Hexagonal crystals: d 5.67. Also reported as d_4^{20} 5.607. Sublimes at normal pressure. n_D 2.0041, 2.0203. American process zinc oxide pH 6.95. French process zinc oxide pH 7.37. Sol in dil acetic or mineral acids, ammonia, ammonium carbonate, fixed alkali hydroxide solns. Insol in water, alc.

Calamine. Eczederm. Zinc oxide with a small proportion of ferric oxide as a coloring agent. Pink powder. Almost completely sol in mineral acids. Insol in water.

Caution: Potential symptoms of overexposure are metal fume fever (chills, muscle aches, nausea, fever, dry throat, cough, weakness, lassitude); metallic taste; headache; blurred vision; low back pain; vomiting; fatigue; malaise; tight chest, dyspnea, rales, decreased pulmonary function. *See NIOSH Pocket Guide to Chemical Hazards* (DHHS/NIOSH 97-140, 1997) p 338.

USE: As pigment in white paints instead of lead carbonate; in rubber industry as vulcanization activator and accelerator; in cosmetics, driers, quick-setting cements; with syrupy phosphoric acid or $ZnCl_2$ in dental cements; manuf opaque glass and certain types of transparent glass; manuf enamels, automobile tires, white glue, matches, white printing inks, porcelains, zinc green; as a reagent in analytical chemistry; in electrostatic copying paper; as flame retardant; in electronics as semiconductor.

THERAP CAT: Astringent; topical protectant; ultraviolet screen.

THERAP CAT (VET): Antiseptic; astringent; topical protectant.

10348. Zinc Perchlorate. [13637-61-1] Perchloric acid zinc salt (2:1). Cl_2O_8Zn; mol wt 264.30. Cl 26.83%, O 48.43%, Zn 24.75%. $Zn(ClO_4)_2$.

Hexahydrate, deliquesc crystals, mp 106°. Freely sol in water. Sol in alcohol. Also forms a tetrahydrate.

10349. Zinc Peroxide. [1314-22-3] ZPO; zinc superoxide. O_2Zn; mol wt 97.41. O 32.85%, Zn 67.15%. ZnO_2. The peroxide of commerce contains 50-60% ZnO_2, the remainder is ZnO.

White to yellowish-white, odorless powder. *Oxidizer.* Dec above 150°. Insol in, but gradually dec by, water. Sol in dil acids, liberating hydrogen peroxide.

USE: Accelerator in rubber compounding; curing agent for synthetic elastomers. Deodorant for wounds and skin diseases.

THERAP CAT: Antiseptic (topical), astringent.

10350. Zinc p-Phenolsulfonate. [127-82-2] 4-Hydroxybenzenesulfonic acid zinc salt (2:1); 1-phenol-4-sulfonic acid zinc salt; zinc p-hydroxybenzenesulfonate; zinc sulfocarbolate; zinc sulfophenate; Phenozin. $C_{12}H_{10}O_8S_2Zn$; mol wt 411.73. C 35.01%, H 2.45%, O 31.09%, S 15.57%, Zn 15.89%. $Zn(HOC_6H_4SO_3)_2$. It is at least 99.5% pure. Prepn: Rojahn, *Dtsch. Apoth. Ztg.* **50**, 1095 (1935), *C.A.* **31**, 6816[8] (1937).

Octahydrate, crystals or cryst powder; odorless. Effloresces in dry air; loses all its H_2O at about 120°. One gram dissolves in 1.6 ml water, 0.4 ml boiling water, 1.8 ml alcohol. The aq soln is acid to litmus; pH about 4.

USE: In insecticide formulations.

THERAP CAT: Astringent.

THERAP CAT (VET): Has been used as an intestinal antiseptic, and externally to promote healing of ulcers, slowly granulating wounds.

10351. Zinc Phosphate. [7779-90-0] Phosphoric acid zinc salt (2:3). $O_8P_2Zn_3$; mol wt 386.17. O 33.14%, P 16.04%, Zn 50.81%. $Zn_3(PO_4)_2$. The tetrahydrate occurs in nature as the mineral *hopeite*; the article of commerce is about 98% pure.

Tetrahydrate, white, odorless powder. Insol in water or alcohol. Sol in dil mineral acids, acetic acid, ammonia, and in alkali hydroxide solns.

USE: In dental cements.

10352. Zinc Phosphide. [1314-84-7] Trizinc diphosphide; Eraze; Rattoff. P_2Zn_3; mol wt 258.17. P 23.99%, Zn 76.01%. Zn_3P_2. Inorganic rodenticide. Prepn: M. von Stackelberg, R. Paulus, *Z. Phys. Chem. B* **28**, 30 (1935); K. R. Murali, B. S. V. Gopalam, *J. Mater. Sci. Lett.* **5**, 989 (1986). Sublimation thermodynamics: J. H. Greenberg *et al.*, *J. Chem. Thermodyn.* **6**, 1005 (1974). Crystal structure: I. E. Zanin *et al.*, *J. Struct. Chem.* **45**, 844 (2004). Practical syntheses and crystal growth: J. Misiewicz, F. Królicki, *Mater.*

Sci. **11**, 39 (1985). Synthesis of nanostructures and use in optoelectronic devices: R. Yang *et al.*, *Nano Lett.* **7**, 269 (2007). GC determn in environmental and biological samples: M. P. Heenan *et al.*, *Bull. Environ. Contam. Toxicol.* **71**, 1019 (2003). Toxicity to rats: L. Y. Ming, *Malays. Agric. J.* **52**, 166 (1979). Field trial for rodent control in corn fields: S. E. Hygnstrom *et al.*, *Int. Biodeterior. Biodegrad.* **45**, 215 (2000); in teak plantations: D. F. Rivera *et al.*, *Crop Prot.* **27**, 877 (2008).

Dark gray tetragonal crystals; lustrous or dull powder; faint phosphorus odor. mp 1193°. Undergoes phase transition ($\beta \rightarrow \alpha$) at 880°. d 4.54±0.04 (X-ray); d 4.21-4.76 (experimental). Sol in benzene, carbon disulfide. Insol in water, alcohol. Reacts violently with water and acids to liberate phosphine, a flammable and toxic gas. Stable when dry. *Poisonous.* LD_{50} orally in male, female rats: 12.00, 15.72 mg/kg (Ming).

USE: Rodenticide; light absorber in solar cells.

10353. Zinc Propionate. [557-28-8] Propanoic acid zinc salt (2:1). $C_6H_{10}O_4Zn$; mol wt 211.55. C 34.07%, H 4.76%, O 30.25%, Zn 30.92%. $Zn(C_3H_5O_2)_2$. Prepd by dissolving zinc oxide in dil propionic acid and concentrating the soln: Gaze, *Arch. Pharm.* **229**, 488 (1891).

Clusters of plates, tablets; crystallizes also in needles as the monohydrate. The soly of the anhydr salt is 32% (w/w) in water at 15°; 2.8% in alc at 15°; 17.2% in boiling alc. *Keep well closed.* Dec in moist atm, giving off propionic acid.

USE: As fungicide on adhesive tape to reduce plaster irritation caused by molds, fungi, and bacterial action.

THERAP CAT: Antifungal (topical).

10354. Zinc Pyrophosphate. [7446-26-6] Diphosphoric acid, zinc salt (1:2). $O_7P_2Zn_2$; mol wt 304.76. O 36.75%, P 20.33%, Zn 42.92%. $Zn_2P_2O_7$.

White, crystalline powder. d^{23} 3.75. Insoluble in water. Soluble in dilute mineral acids.

10355. Zinc Salicylate. [16283-36-6] (*T*-4)-Bis[2-(hydroxy-*κO*)benzoato-*κO*]zinc. $C_{14}H_{10}O_6Zn$; mol wt 339.64. C 49.51%, H 2.97%, O 28.26%, Zn 19.26%. Prepd from sodium salicylate and zinc sulfate: Clark, Kao, *J. Am. Chem. Soc.* **70**, 2151 (1948).

Trihydrate, needles or cryst powder. Sol in water, alc. The aq soln is practically neutral to litmus.

THERAP CAT: Antiseptic; astringent.

10356. Zinc Selenide. [1315-09-9] SeZn; mol wt 144.37. Se 54.69%, Zn 45.31%. ZnSe. Prepd by mixing a soln of a zinc salt and potassium selenide: Berzelius, cited in *Mellor's* **10**, 776 (1930); by passing selenium vapor over zinc heated *in vacuo:* Moser, Doctor, *Z. Anorg. Chem.* **118**, 284 (1921); by passing H_2Se into a methanol soln of zinc acetate: Nitsche, US 2805917 (1957 to du Pont); from $ZnSe.N_2H_4$ (zinc selenide hydrazinate): Conn *et al.*, US 3014779 (1961 to Merck & Co.).

Yellow, cubic crystals. d_4^{15} 5.42. mp >1100°. Dec in air. Insol in water. Dec in dil nitric acid.

USE: Commercial phosphor.

10357. Zinc Silicate. [13597-65-4] Silicic acid (H_4SiO_4) zinc salt (1:2); zinc orthosilicate. O_4SiZn_2; mol wt 222.90. O 28.71%, Si 12.60%, Zn 58.69%. Zn_2SiO_4. Occurs in nature as the mineral **willemite**. Prepd by heating the proper amounts of ZnO and SiO_2 at about 1200°.

White powder, insol in water or dil acids.

USE: In television screens.

10358. Zinc Stearate. [557-05-1] Octadecanoic acid zinc salt (2:1). $C_{36}H_{70}O_4Zn$; mol wt 632.36. C 68.38%, H 11.16%, O

10.12%, Zn 10.34%. $Zn(C_{18}H_{35}O_2)_2$. Usually occurs as a mixture of the zinc salts of stearic and palmitic acids, and usually with some excess of the zinc oxide. Contains 13.5-15% ZnO. Prepd from stearic acid and zinc chloride: Vold, Hattiangdi, *Ind. Eng. Chem.* **41**, 2311 (1949).

Fine, soft, bulky powder; slight characteristic odor; neutral reaction. Repels water. mp about 120°. Sol in benzene; dec by dil acids. Insol in water, alc, ether.

Caution: Potential symptoms of overexposure are irritation of eyes, skin, upper respiratory system; cough. *See NIOSH Pocket Guide to Chemical Hazards* (DHHS/NIOSH 97-140, 1997) p 338.

USE: In tablet manuf as tablet and capsule lubricant; in cosmetic and pharmaceutical powders and ointments; as a flatting and sanding agent in lacquers; as a drying lubricant and dusting agent for rubber; as a plastic mold releasing agent; as a waterproofing agent for concrete, rock wool, paper, textiles.

THERAP CAT (VET): Antiseptic, astringent, protective (topical).

10359. Zinc Sulfate. [7733-02-0] Sulfuric acid zinc salt (1:1); white vitriol; zinc vitriol. O_4SZn; mol wt 161.47. O 39.63%, S 19.86%, Zn 40.51%. $ZnSO_4$. Prepn and physical properties: *Gmelins, Zink* (8th ed.) **32**, 936-960 (1956). Effects of oral zinc sulfate in acne treatment: G. Michaelsson *et al.*, *Arch. Dermatol.* **113**, 31 (1977). Clinical evaluations in Wilson's Disease: T. U. Hoogenraad, C. J. A. Van den Hamer, *Acta Neurol. Scand.* **67**, 356 (1983); T. U. Hoogenraad *et al.*, *J. Neurol. Sci.* **77**, 137 (1987).

Monohydrate. [7446-19-7] Dried zinc sulfate; Zincaps; Z Span. Powder or granules. Loses H_2O above 238°. Freely sol in water. Practically insol in alc. The monohydrate does not cake as the heptahydrate does, and hence is more convenient for use during the warm season and in warm climates.

Heptahydrate. [7446-20-0] Collazin; Keratol; Op-Thal-Zin; Redeema; Solvazinc; Solvezink; Virudermin; Zincate; Zincomed. Odorless crystals or granules or powder; astringent taste. Efflorescent in dry air. d 1.97. mp 100°. At 280° loses all H_2O; dec above 500°. One gram dissolves in 0.6 ml water, 2.5 ml glycerol. Insol in alc. The aq soln is acid to litmus; pH about 4.5. *Keep well closed.*

Caution: Irritating to skin, mucous membranes: H. E. Stokinger, *Patty's Industrial Hygiene and Toxicology* **2A**, G. D. Clayton, F. E. Clayton, Eds. (John Wiley & Sons, New York, 1981) pp 2033-2049.

USE: As mordant in calico-printing; preserving wood and skins; as crenilating agent in manuf of rayon; with hypochlorite for bleaching paper; manuf lithopone and other zinc salts; clarifying glue; electrodepositing Zn; also as reagent in analytical chemistry.

THERAP CAT: Ophthalmic astringent. Zinc supplement.

THERAP CAT (VET): Astringent. Has been used as an emetic.

10360. Zinc Sulfide. [1314-98-3] Zinc blende. SZn; mol wt 97.47. S 32.89%, Zn 67.11%. ZnS.

Occurs in nature as the minerals **wurtzite** (hexagonal, d 4.087) and **sphalerite** (cubic, d 4.102). White to grayish-white or yellowish powder. When contg water, it slowly oxidizes in air to sulfate. Insol in water, alkalies. Sol in dil mineral acids. Precipitated zinc sulfide of commerce usually contains 15-20% water of hydration. The dried precipitate may have been heated to 725° in the absence of air to obtain substantial conversion to wurtzite, the form preferred by the pigment industry.

USE: Pigment for paints, oilcloths, linoleum, leather, dental rubber, etc., especially in the form of lithopone; mixed with ZnO as "mineral white." Anhyd zinc sulfide is used in x-ray screens and with a trace of a radium or mesothorium salt in luminous dials of watches; as phosphor in watches and television screens.

10361. Zinc Tartrate. [551-64-4] (2*R*,3*R*)-2,3-Dihydroxybutanedioic acid zinc salt (1:1). $C_4H_4O_6Zn$; mol wt 213.48. C 22.51%, H 1.89%, O 44.97%, Zn 30.64%. $ZnC_4H_4O_6$. Prepd from potassium tartrate and zinc chloride. Prepn and solubility in water: Cantoni, Zachoder, *Bull. Soc. Chim.* [3] **33**, 747 (1905).

Dihydrate, cryst powder.

10362. Zinc Telluride. [1315-11-3] TeZn; mol wt 193.01. Te 66.11%, Zn 33.89%. ZnTe. Prepd by fusing Zn and Te: Braithwaite, *Proc. Phys. Soc. London* **64B**, 274 (1951); Bube, *Proc. IRE* **43**, 1836 (1955); *see also* Dennis, Anderson, *J. Am. Chem. Soc.* **36**, 882 (1914). From zinc oxide and powdered tellurium in aq alkaline medium: Montignie, *Bull. Soc. Chim. Fr.* **1947**, 750.

Gray or brownish-red powder. Ruby-red crystals (cubic system) by sublimation. Stable in dry air. d_4^{15} 6.34. mp 1239°. Forms a monohydrate. Prolonged contact with water or dil HCl yields H_2 and H_2Te.

USE: In semiconductor research, as photoconductor.

10363. Zinc Thiocyanate. [557-42-6] Thiocyanic acid zinc salt (2:1); zinc sulfocyanate. $C_2N_2S_2Zn$; mol wt 181.57. C 13.23%, N 15.43%, S 35.31%, Zn 36.02%. $Zn(SCN)_2$.

White, deliquesc crystals. Sol in water, alcohol. The aq soln is only slightly acid to litmus. *Keep well closed and protected from light.*

USE: To assist in textile dyeing.

10364. Zinc Valerate. [556-38-7] Pentanoic acid zinc salt (2:1). $C_{10}H_{18}O_4Zn$; mol wt 267.66. C 44.87%, H 6.78%, O 23.91%, Zn 24.44%. $Zn(C_5H_9O_2)_2$. The article of commerce is usually basic. Prepd from valeric acid and zinc hydrate: Lieben, Rossi, *Ann.* **159**, 58 (1871).

Dihydrate, lustrous scales or powder; valerian odor; sweetish taste. Gradually dec on exposure to air. One gram dissolves in 70 ml water or in 22 ml alcohol when free from basic salt; dec by acid. *Keep tightly closed and in cool place.*

10365. Zineb. [12122-67-7] [N-2-[(Dithiocarboxy)amino]-ethyl]carbamodithioato(2−)-κS, κS′]zinc; [ethylenebis(dithiocarbamato)]zinc; zinc ethylenebis(dithiocarbamate); ethylenebis(dithiocarbamic acid) zinc salt; ENT-14874; Tritoftorol. $(C_4H_6N_2S_4Zn)_n$. Polymeric salt of ethylenebisdithiocarbamic acid; related to maneb and mancozeb, *q.q.v.* Prepd from a zinc salt and nabam, *q.v.*: C. B. Luginbuhl, **US 2690448** (1954 to du Pont). Fungicidal activity is due to degradation products, principally **ethylenethiuram monosulfide**: R. A. Ludwig *et al.*, *Can. J. Bot.* **33**, 42 (1955). Activity against potato blight: B. K. De, S. B. Chattopadhyay, *Pesticides* **18**, 52 (1984). Decomposition of zineb to ethylenebisdicarbamate and ethylene thiourea, *q.v.*, in tomatoes: B. D. Ripley, D. F. Cox, *J. Agric. Food Chem.* **26**, 1137 (1978). HPLC determn: K. Gustafsson *et al.*, *ibid.* **29**, 729 (1981). Toxicity data: T. B. Gaines, *Toxicol. Appl. Pharmacol.* **14**, 515 (1969). Aquatic toxicology: C. J. Van Leeuwen *et al.*, *Aquat. Toxicol.* **7**, 145 (1985).

Powder or crystals from chloroform + alcohol. Practically insol in water. The powder spreads easily on water, also forms aq suspensions. Sol in carbon disulfide, pyridine. LD_{50} orally in rats: >5000 mg/kg (Gaines).

Caution: Irritation of skin and mucous membranes has been reported. *See: Clinical Toxicology of Commercial Products*, R. E. Gosselin *et al.*, Eds. (Williams & Wilkins, Baltimore, 5th Ed., 1984) Section II, p 313.

USE: Agricultural fungicide.

10366. Zingerone. [122-48-5] 4-(4-Hydroxy-3-methoxyphenyl)-2-butanone; (4-hydroxy-3-methoxyphenyl)ethyl methyl ketone; vanillylacetone; zingherone; zingiberone. $C_{11}H_{14}O_3$; mol wt 194.23. C 68.02%, H 7.27%, O 24.71%. Isolated from ginger root or prepd from vanillin and acetone followed by catalytic hydrogenation: Nomura, *J. Chem. Soc.* **111**, 769 (1917); *idem*, **US 1263796** (1918); **US 1306710** (1919); Cotton, **US 2381210** (1945 to Penn. Coal Prod.); K. Banno, T. Mukaiyama, *Bull. Chem. Soc. Jpn.* **49**, 1453 (1976).

Crystals from acetone, petr ether, ether + petr ether, mp 40-41°. bp_{14} 187-188°. Spicy, pungent odor, characteristic of ginger. Flash point: 208°F (98°C). Sparingly sol in water, petr ether; sol in ether, dil alkalies.

USE: In fragrances, flavors and cosmetics; in artificial spice oils.

10367. Zinostatin. [9014-02-2] Neocarzinostatin; neocarcinostatin; NCS; NSC-69856. An antitumor acidic antibiotic consisting of 2 components: a protein of mol wt 10,700 containing 109 amino acid residues of 18 kinds and a labile non-protein chromophore which possesses the full biological activity of NCS. Isoln from the culture filtrate of *Streptomyces carcinostaticus* var. F-41: N. Ishida *et al.*, *J. Antibiot.* **18A**, 68 (1965). Characterization: H. Maeda *et al.*, *ibid.* **19A**, 253 (1966). Prodn: N. Ishida *et al.*, **US 3334022** (1967 to Empire Corp.). Amino acid sequence: J. Meienhofer *et al.*, *Science* **178**, 875 (1972). Review of protein structural studies: *eidem, Progress in Peptide Research* **vol. II**, S. Lande, Ed. (Gordon and Breach, New York, 1972) pp 295-306. Spectral characterization and separation of the chromophore: M. A. Napier *et al.*, *Biochem. Biophys. Res. Commun.* **89**, 635 (1979). Partial structure of the chromophore: G. Albers-Schönberg *et al.*, *ibid.* **95**, 1351 (1980). Roles of chromophore and apo-protein in biological activity: L. S. Kappen *et al.*, *Proc. Natl. Acad. Sci. USA* **77**, 1970 (1980). Molecular basis of action: I. H. Goldberg *et al.*, *Mol. Biol. Biochem. Biophys.* **32**, 308 (1980). Absorption, distribution, excretion: K. Toriyama *et al.*, *Jpn. J. Antibiot.* **28**, 64 (1975). Review of pharmacology: T. A. Beerman *et al.*, in *Advances in Enzyme Regulation* **vol. 14**, G. Weber, Ed. (Pergamon Press, New York, 1976) pp 207-228.

White powder, mp 260° (dec). uv max (water): 278 nm ($E_{1cm}^{1\%}$ 15). Isoelectric pt: pH 3.26. *See: Jpn. Med. Gaz.* **14** (3), 11 (1977). Sol in water. Insol in most common org solvents. Stability of aq solns: M. Kohno *et al.*, *Jpn. J. Antibiot.* **27**, 707 (1974). LD_{50} in mice (mg/kg): 1050 orally, 0.96 i.v. (Toriyama).

THERAP CAT: Antineoplastic.

10368. Zinpyr. Family of fluorescent sensors for Zn^{2+} in biological applications. Prepn of ZP-1: G. K. Walkup *et al.*, *J. Am. Chem. Soc.* **122**, 5644 (2000); and characterization of ZP-1 and ZP-2: S. C. Burdette *et al.*, *ibid.* **123**, 7831 (2001); of ZP-4: *idem et al.*, *ibid.* **125**, 1778 (2003).

Zinpyr-1 R = Cl
Zinpyr-2 R = H

Zinpyr-1. [288574-78-7] 4′,5′-Bis[[bis(2-pyridinylmethyl)amino]methyl]-2′,7′-dichloro-3′,6′-dihydroxy-spiro[isobenzofuran-1(3H),9′-[9H]xanthen]-3-one; 9-(o-carboxyphenyl)-2,7-dichloro-4,5-bis[bis(2-pyridylmethyl)aminomethyl]-6-hydroxy-3-xanthanone; ZP-1. $C_{46}H_{36}Cl_2N_6O_5$; mol wt 823.73. C 67.07%, H 4.41%, Cl 8.61%, N 10.20%, O 9.71%. Salmon-pink solid, mp 185-187° (dec). pKa 2.8. Quantum yield 0.38.

Zinpyr-2. [357916-12-2] 9-(o-Carboxyphenyl)-4,5-bis[bis(2-pyridylmethyl)aminomethyl]-6-hydroxy-3-xanthanone; ZP-2. $C_{46}H_{38}N_6O_5$; mol wt 754.85. C 73.19%, H 5.07%, N 11.13%, O 10.60%. Orange solid, mp 195-197° (dec). pKa 3.9. Quantum yield 0.25.

USE: Probe for measurement of cellular Zn^{2+}.

10369. Zinquin. [151606-29-0] 2-[[2-Methyl-8-[[(4-methyl-phenyl)sulfonyl]amino]-6-quinolinyl]oxy]acetic acid; Zinquin A. $C_{19}H_{18}N_2O_5S$; mol wt 386.42. C 59.06%, H 4.70%, N 7.25%, O 20.70%, S 8.30%. Membrane permeant, fluorescent probe derived from TSQ, *q.v.*; selective for Zn^{2+} in the presence of Ca^{2+} and Mg^{2+}. Synthesis: I. B. Mahadevan *et al.*, *Aust. J. Chem.* **49**, 561 (1996). Coordination chemistry: C. J. Fahrni, T. V. O'Halloran, *J. Am. Chem. Soc.* **121**, 11448 (1999); potentiometric coordination of fluorophore: K. M. Hendrickson *et al.*, *ibid.* **125**, 3889 (2003). Visualization of intracellular Zn^{2+} pools in pancreatic islets: P. D. Zalewski *et al.*, *J. Histochem. Cytochem.* **42**, 877 (1994); in apoptosis: P. J. Smith *et al.*, *Am. J. Physiol. Cell Physiol.* **283**, C609 (2002); G. R. Sauer *et al.*, *J. Cell. Biochem.* **88**, 954 (2003).

Crystals from ethyl acetate or ethyl acetate/light petroleum, mp 198-200°. pK_1 (quinoline nitrogen): <2; pK (carboxylic acid proton): 5.77 ±0.03; pK (sulfonamide proton): 9.91±0.02.

Ethyl Ester. [181530-09-6] Zinquin E. $C_{21}H_{22}N_2O_5S$; mol wt 414.48. Colorless crystals from dichloromethane/light petroleum, mp 111-113°. pK_1 (quinoline nitrogen): <2; pK (sulfonamide proton): 9.32 ±0.02.

USE: Determn of intracellular Zn^{2+} and labile Zn^{2+} ions.

10370. Zipeprol. [34758-83-3] 4-(2-Methoxy-2-phenylethyl)-α-(methoxyphenylmethyl)-1-piperazineethanol; α-(α-methoxybenzyl)-4-(β-methoxyphenethyl)-1-piperazineethanol; 1-(2-hydroxy-3-methoxy-3-phenylpropyl)-4-(2-methoxy-2-phenylethyl)piperazine; 1-(2-methoxy-2-phenylethyl)-4-(2-hydroxy-3-methoxy-3-phenylpropyl)piperazine. $C_{23}H_{32}N_2O_3$; mol wt 384.52. C 71.84%, H 8.39%, N 7.29%, O 12.48%. Prepn: R. Y. Mauvernay *et al.*, **DE 2109366**; *eidem*, **US 3718650** (1971, 1973 both to Mauvernay). Pharmacology: G. Rispat *et al.*, *Arzneim.-Forsch.* **26**, 523 (1976); D. Cosnier *et al.*, *ibid.* 848. Clinical abuse potential: L. Janiri *et al.*, *Drug Alcohol Depend.* **27**, 121 (1991). GC-MS determn in plasma and urine: P. Kintz *et al.*, *J. Pharm. Biomed. Anal.* **11**, 335 (1993).

Crystals from abs ethanol, mp 83°.

Dihydrochloride. [34758-84-4] CERM-3024; Antituxil-Z; Citizeta; Mirsol; Respirase; Robnin; Zitoxil. $C_{23}H_{32}N_2O_3 \cdot 2HCl$; mol wt 457.44. Crystals from abs ethanol, mp 231°. Very stable in aq soln. LD_{50} orally in mice: 301 mg/kg (Rispat).

THERAP CAT: Antitussive.

10371. Ziprasidone. [146939-27-7] 5-[2-[4-(1,2-Benzisothiazol-3-yl)-1-piperazinyl]ethyl]-6-chloro-1,3-dihydro-2*H*-indol-2-one; 5-(2-(4-(1,2-benzisothiazol-3-yl)piperazinyl)ethyl)-6-chlorooxindole; CP-88059. $C_{21}H_{21}ClN_4OS$; mol wt 412.94. C 61.08%, H 5.13%, Cl 8.58%, N 13.57%, O 3.87%, S 7.76%. Combined serotonin ($5HT_2$) and dopamine (D_2) receptor antagonist. Prepn: J. A. Lowe III, A. A. Nagel, **EP 281309**; *eidem*, **US 4831031** (1988, 1989 both to Pfizer). Clinical pharmacology: C. J. Bench *et al.*, *Psychopharmacology* **112**, 308 (1993). HPLC determn in serum: J. S. Janiszewski *et al.*, *J. Chromatogr. B* **668**, 133 (1995). Receptor binding profile: A. W. Schmidt *et al.*, *Eur. J. Pharmacol.* **425**, 197 (2001). Review of pharmacology and clinical experience: G. L. Stimmel *et al.*, *Clin. Ther.* **24**, 21-37 (2002); P. D. Harvey, C. R. Bowie, *Expert Opin. Pharmacother.* **6**, 337-346 (2005).

Hydrochloride monohydrate. [138982-67-9]; [122883-93-6] (anhydrous). CP-88059-1; Geodon; Zeldox. $C_{21}H_{21}ClN_4OS \cdot HCl \cdot H_2O$; mol wt 467.41. Prepn: D. J. M. Allen *et al.*, **EP 586191**; *eidem*, **US 5312925** (1993, 1994 both to Pfizer). White to slightly pink powder. Also prepd as the hemihydrate, mp >300°.

THERAP CAT: Antipsychotic.

10372. Ziram. [137-30-4] (*T*-4)-Bis(*N,N*-dimethylcarbamodithioato-$\kappa S,\kappa S'$)zinc; bis(dimethyldithiocarbamato)zinc; zinc dimethyldithiocarbamate; dimethyldithiocarbamic acid zinc salt; zinc bis(dimethylthiocarbamoyl) disulfide; methyl cymate; Crittam; Mezene; Pomarsol Z; Thionic; Triscabol. $C_6H_{12}N_2S_4Zn$; mol wt 305.83. C 23.56%, H 3.96%, N 9.16%, S 41.93%, Zn 21.39%. Prepd from zinc oxide, dimethylamine, and carbon disulfide: Olin, Deger, **US 2492314** (1949 to Sharples Chemicals). Crystal structure: Klug, *Acta Crystallogr.* **21**, 536 (1966). Toxicity study: Hodge *et al.*, *J. Pharmacol. Exp. Ther.* **118**, 174 (1956). Field analysis of residues in air: J. E. Woodrow *et al.*, *J. Agric. Food Chem.* **43**, 1524 (1995). Review and evaluation of toxicity studies: *IARC Monographs* **53**, 423-438 (1991).

Crystals from hot chloroform + alcohol, mp 250°. *Can form a flammable dust.* d_4^{25} 1.66. Practically insol in water. Soly per 100 ml of solvent at 25°: <0.2 g, alcohol; <0.5 g, acetone; <0.5 g, benzene; <0.2 g, carbon tetrachloride, more sol in chloroform; <0.2 g, ether; 0.5 g, naphtha. Sol in dil caustic solns. LD_{50} orally in rats: 1.4 g/kg (Hodge).

Caution: May be irritating to skin and mucous membranes. *See: Clinical Toxicology of Commercial Products*, R. E. Gosselin *et al.*, Eds. (Williams & Wilkins, Baltimore, 5th ed., 1984) Section II, p. 314.

USE: Rubber vulcanization accelerator; agricultural fungicide.

10373. Zirconium. [7440-67-7] Zr; at. wt 91.224; at. no. 40; valence 4; also 3. Group IVB (4). Five naturally occurring isotopes: 90 (51.46%); 91 (11.23%); 92 (17.11%); 94 (17.40%); 96 (2.80%); artificial radioactive isotopes: 81-89, 93, 95, 97-99. Occurrence in earth's crust: 0.023%. Occurs in the minerals zircon, malacon, baddeleyite, zirkelite, eudialyte; frequently found in the rare-earth minerals; in monazite sand. Discovered by Klaproth in 1789; prepd by Berzelius in 1824. Prepn: Fast, *Z. Anorg. Chem.* **239**, 145 (1938); purification of zirconium by ion exchange columns: Ayres, *J. Am. Chem. Soc.* **69**, 1879 (1947). Sepn of zirconium and hafnium: Fischer *et al.*, *Angew. Chem. Int. Ed.* **5**, 15 (1966). Reviews of zirconium and its compds: W. B. Blumenthal, *The Chemical Behavior of Zirconium* (Van Nostrand, Princeton, 1958); *Gmelins, Zirconium* (8th ed.) **42**, (1958) 448 pp; Larsen, "Zirconium and Hafnium Chemistry" in *Adv. Inorg. Chem. Radiochem.* **13**, 1-333 (1970); Bradley, Thornton, "Zirconium and Hafnium" in *Comprehensive Inorganic Chemistry* vol. 3, J. C. Bailar, Jr. *et al.*, Eds. (Pergamon Press, Oxford, 1973) pp 419-490.

Bluish-black, amorphous powder or grayish-white lustrous metal (platelets or flakes) of hexagonal lattice below 865°, body-centered cubic above 865°, mp 1857°; bp 3577°. d 6.5. Brinnell hardness: 85. Can absorb up to 10 atoms per cent of oxygen or nitrogen. *Flammable; spontaneously combustible.* Reacts with hydrofluoric acid, aqua regia, hot phosphoric acid. Not attacked by cold, very slightly attacked by hot, concd sulfuric or hydrochloric acid; not attacked by nitric acid. Attacked by fused potassium hydroxide or nitrate. On prolonged heating the compact form combines with oxygen, nitrogen, carbon, and the halogens. The powder form has a very low ignition temp and is very explosive when mixed with oxidizing agents.

Caution: Zirconium and its salts generally have low systemic toxicity. A granulomatous disease of the skin, particularly in the axilla, has been reported in users of a deodorant contg sodium zirconium lactate: *see* E. Browning, *Toxicity of Industrial Metals* (Appleton-Century-Crofts, New York, 2nd ed., 1969) pp 356-360. Consult latest Government regulations on use in aerosol antiperspirants.

USE: As an ingredient in priming or explosive mixtures; flashlight powders; as deoxidizer in metallurgy; as "getter" in vacuum tubes; in constructing rayon spinnerets in lamp filaments, flash bulbs. Pure zirconium (hafnium-free) is a valuable structural material for atomic reactors because of its low nuclear cross-section and high corrosion and heat resistance. Because of hafnium's high neutron absorption characteristics, it must be removed from zirconium which is to be used in nuclear reactors.

10374. Zirconium Chloride. [10026-11-6] (*T*-4)-Zirconium chloride ($ZrCl_4$); zirconium tetrachloride. Cl_4Zr; mol wt 233.02. Cl 60.85%, Zr 39.15%. $ZrCl_4$. In large-scale prepns zirconium oxide is converted to the carbide, which is chlorinated to yield the tetrachloride: Kroll *et al.*, *J. Electrochem. Soc.* **94**, 1 (1948). Lab prepn based on the equation $ZrO_2 + 2CCl_4 \rightarrow ZrCl_4 + 2COCl_2$: Hummers *et al.*, *Inorg. Synth.* **4**, 121 (1953). Toxicology study: N. A. Zhilova, A. A. Kasparov *Hyg. Sanit.* **31**, 328 (1966). *Review:* Blumenthal, *J. Chem. Educ.* **39**, 604-610 (1962).

Lustrous monoclinic crystals: Krebs, *Angew. Chem. Int. Ed.* **8**, 146 (1969); *idem, Z. Anorg. Allg. Chem.* **378**, 263 (1970). Tetrahedral symmetry in gas phase with the Zr-Cl distance of 2.33 Å. Lewis acid. Extremely hygroscopic, forms HCl vapor and gives off fumes in moist air. d 2.803. Sublimes at 331°. mp 437° under its own pressure which is about 25 atm at this temp. Decomposed by water to form $ZrOCl_2$ and HCl; sol in alcohol, ether. LD_{50} in mice, rats (mg/kg): 665, 1688 orally (Zhilova, Kasparov).

USE: Friedel-Crafts catalyst. Component of Ziegler-type catalysts in the condensation of ethylene. Starting material in the synthesis of a number of organic derivs of zirconium, such as alkoxides and zircocene. The alkoxides have been shown to be of value in the curing of silicone plastic films. The alkoxyzirconium carboxylates are said to be useful in the water-repellent treatment of textiles and other fibrous materials.

10375. Zirconium Fluoride. [7783-64-4] (*T*-4)-Zirconium fluoride (ZrF_4); zirconium tetrafluoride. F_4Zr; mol wt 167.22. F 45.45%, Zr 54.55%. ZrF_4. Prepd by thermal decompn of $(NH_4)_2$-ZrF_6: v. Hevesy, Dullenkamp, *Z. Anorg. Allg. Chem.* **221**, 161 (1934); according to the equation $ZrCl_4 + 4HF \rightarrow ZrF_4 + 4HCl$: Wolter, *Chem. Ztg.* **51**, 607 (1908); from zirconium oxide and fluorine: Haendler *et al.*, *J. Am. Chem. Soc.* **76**, 2177 (1954).

Strongly refractive, crystalline mass (monoclinic system). d^{16} 4.6. Sublimes above 600°. Solubility in water (20°): 1.32 g/100 ml. Does not react with water; forms stable trihydrate. Freely soluble in hydrofluoric acid.

10376. Zirconium Hydride. [7704-99-6] H_2Zr; mol wt 93.24. H 2.16%, Zr 97.84%. Ideal composition: ZrH_2. Prepd by the reduction of zirconium oxide with calcium hydride in the presence of hydrogen above 600°: Alexander, **US 2427339** (1947 to Metal Hydride).

Gray-black metallic powder. Stable, no reaction with water. *Flammable.*

USE: Powerful reducing agent in acid solution or at high temps; hydrogenation catalyst; in the vacuum tube industry and powder metallurgy.

10377. Zirconium Hydroxide. [14475-63-9] (*T*-4)-Zirconium hydroxide ($Zr(OH)_4$). H_4O_4Zr; mol wt 159.25. H 2.53%, O 40.19%, Zr 57.28%. $Zr(OH)_4$. Proposed structures of freshly pptd and aged compds, cyclic tetramers: Zaitsev, *Russ. J. Inorg. Chem.* **11**, 900 (1966).

White, bulky, amorphous powder; d 3.25. Insol in water; sol in mineral acids when freshly pptd; less sol when aged. It colors turmeric paper brown.

USE: In the pigment, dye, and glass industries.

10378. Zirconium Iodide. [13986-26-0] Zirconium tetraiodide. I_4Zr; mol wt 598.84. I 84.77%, Zr 15.23%. ZrI_4. Prepd from

the elements at a furnace temp of 450°: Eberly, *Inorg. Synth.* **7**, 52 (1963).

Orange-colored, crystalline solid. Sublimes at 431°. mp 499° (elevated pressure). Fumes heavily in air. Dissolves in water with the liberation of steam. Heat of formation: 90 kcal/mol at 25°. Magnetic susceptibility -0.238×10^{-6} c.g.s. electromagnetic units.

10379. Zirconium Nitrate. [13746-89-9] Nitric acid zirconium(4+) salt (4:1). $N_4O_{12}Zr$; mol wt 339.24. N 16.52%, O 56.59%, Zr 26.89%. $Zr(NO_3)_4$. Prepn according to the equation $ZrCl_4 + 4N_2O_5 \rightarrow Zr(NO_3)_4 + 4NO_2Cl$: Field, Hardy, *Proc. Chem. Soc. London* **1962**, 76. Pentahydrate obtained from strong nitric acid. The zirconium nitrate of commerce is usually somewhat basic.

Pentahydrate. White, very hygroscopic crystals or white pieces or scales. Very sol in water; sol in alcohol. The aq soln is acid to litmus. *Keep tightly closed.*

10380. Zirconium Oxide. [1314-23-4] Zirconia; zirconium dioxide; zirconic anhydride. O_2Zr; mol wt 123.22. O 25.97%, Zr 74.03%. ZrO_2. Occurs in nature as the mineral **baddeleyite**. Prepn: Clark, Reynolds, *Ind. Eng. Chem.* **29**, 711 (1937); Henderson, Higbie, *J. Am. Chem. Soc.* **76**, 5878 (1954).

White, heavy, amorphous, odorless, tasteless powder or monoclinic crystals. Also forms tetragonal crystals above ~1100°; cubic above ~1900°. d 5.85. mp 2680°. bp 4300°. Practically insol in water; slightly sol in HCl, HNO_3; slowly sol in HF; dissolves on heating with a mixt of 2 parts H_2SO_4 and 1 part H_2O. Fusion with sodium carbonate results in the formation of sodium zirconate which dec hydrolytically with water to form sodium hydroxide and practically insol zirconium hydroxide.

USE: Instead of lime for the oxyhydrogen light; with earths of the yttrium group in incandescent lighting (Nernst lamps); as pigment, abrasive; manuf enamels, white glass, refractory crucibles, and furnace linings.

THERAP CAT: Dermatologic.

10381. Zirconium Silicate. [10101-52-7] Silicic acid (H_4-SiO_4) zirconium(4+) salt (1:1); zirconium orthosilicate. O_4SiZr; mol wt 183.31. O 34.91%, Si 15.32%, Zr 49.76%. $ZrSiO_4$. Occurs in nature as the mineral **zircon**. Widely scattered in all kinds of rocks and in beach sand (South Carolina, Northeast Florida). Usually separated from sand by electrostatic and electromagnetic elutriation, yielding 99% pure $ZrSiO_4$. Prepn from SiO_2 and ZrO_2 in an arc furnace: Curtis, Sowman, *J. Am. Ceram. Soc.* **36**, 195 (1953); can also be made from solns of zirconium salts and sodium silicate.

Tetragonal, bipyramidal crystals. Should be colorless. Colors are from impurities and radioactive bombardment, often removable by calcination. Dissociates to ZrO_2 and SiO_2 when heated above 1540°. Recombines when cooled slowly, but forms a mixture of monoclinic ZrO_2 and vitreous SiO_2 when rapidly quenched. Undissociated (α-form) $ZrSiO_4$ has a high sp gr, 4.7, and a high birefringence; the dissociated, γ-form has a lower sp gr, 3.9-4.0, and a low birefringence. Hardness of α-form about the same as that of quartz (6-7.5). Very inert chemically. Unaffected by aq reagents.

USE: In refractories, ceramics, glazes, cements, coatings for casting molds, polishing materials, gem stones, catalyst in alkyl and alkenyl hydrocarbon manuf, in fritted glass filters, as stabilizer in silicone rubbers. In Europe in cosmetic creams and powders.

10382. Zirconium Sulfate. [14644-61-2] Sulfuric acid zirconium(4+) salt (2:1); disulfatozirconic acid. O_8S_2Zr; mol wt 283.34. O 45.17%, S 22.63%, Zr 32.20%. $Zr(SO_4)_2$. Tetrahydrate usually prepd by treating zirconium oxychloride with hot concd H_2SO_4: Blumenthal, *J. Chem. Educ.* **39**, 604 (1962). Toxicity study: Cochran *et al.*, *Arch. Ind. Hyg. Occup. Med.* **1**, 637 (1950).

Tetrahydrate. Crystalline solid. Converted to monohydrate at 100°; to anhyd form at 380°. Soly in water (18°): 52.5 g/100 g of soln. A soln at room temp deposits a solid on standing. The more dilute the soln, the more rapid the deposition. The composition of the solid is $4ZrO_2 \cdot 3SO_3 \cdot 15H_2O$, known as **Hauser's salt**: Hauser, Herzfeld, *Z. Anorg. Allg. Chem.* **67**, 369 (1910). This pptn does not take place if the soln has been heated above 64°. LD_{50} in rats: 3.5 g/kg orally; 175 mg/kg i.p. (Cochran).

USE: Catalyst support; precipitation of amino acids and proteins; in the tanning industry.

10383. Zirconyl Acetate. [5153-24-2] Bis(acetato-κO)oxozirconium; diacetatozirconic acid. Approx formula: $Zr(OH)_2(C_2H_3$-

$O_2)_2$. Obtained as aq soln by adding acetic acid to an aq slurry of carbonated hydrous zirconia: Blumenthal, *J. Chem. Educ.* **39**, 604 (1962). Toxicity data: Cochran *et al., Arch. Ind. Hyg. Occup. Med.* **1**, 637 (1950).

Available as 22% ZrO_2 soln. d 1.46. mp $-7°$. pH 3.8-4.2. Stable at room temp. Also available as 13% ZrO_2 soln. d 1.20. pH 3.3-4.0. Undergoes exchange with anion exchange resins, but not with cation exchangers. Evaporation of the solns under reduced pressure yields the solid compd. It is resinous or glue-like in appearance and amorphous under x-rays. Its properties indicate it to be a highly polymerized product. The powdered solid readily dissolves when added slowly to rapidly swirling water. On heating, the solns yield a solid hydrolyzate, which is a more basic acetate with the compn $ZrOOHC_2H_3O_2$. When the mixt is allowed to cool and stand at room temp, this ppt slowly redissolves. LD_{50} in rats: 4.1 g/kg orally; 300 mg/kg i.p. (Cochran).

USE: Precipitating agent for gelatin and starch on textiles and paper; water-repellent for textiles (especially in combination with silicones).

10384. Zirconyl Chloride. [7699-43-6] Dichlorooxozirconium; zirconium oxychloride; basic zirconium chloride. Cl_2OZr; mol wt 178.12. Cl 39.80%, O 8.98%, Zr 51.21%. $ZrOCl_2$. Prepn from $ZrCl_4$ and Cl_2O: Dehnicke, Meyer, *Z. Anorg. Allg. Chem.* **331**, 121 (1964). Octahydrate conveniently prepd by crystn of an aq soln of zirconium chloride: v. Siemens, Zander, *Wiss. Veroeff. Siemens-Werken* **2**, 484 (1922); Blumenthal, *J. Chem. Educ.* **39**, 607 (1962). Toxicity study: Cochran *et al., Arch. Ind. Hyg. Occup. Med.* **1**, 637 (1950).

Octahydrate. [13520-92-8] Tetragonal crystals from water. Crystals contain tetramers of the form $[Zr_4(OH)_8(H_2O)_{16}]^{8+}$; d 1.91: Mak, *Can. J. Chem.* **46**, 3491 (1968). Freely sol in water, alcohol. The pH is about equal to that of HCl of the same molarity. LD_{50} in rats: 400 mg/kg i.p.; 3.5 g/kg orally (Cochran).

USE: To make other zirconium compds; to precipitate acid dyes; to prepare high quality pigment toners; to improve the properties of color lakes. Aq $ZrOCl_2$ solns have considerable solvent action on sparingly sol sulfates, such as calcium sulfate. The free acid in $ZrOCl_2$ solns may be neutralized, and a sol compd $ZrOOHCl·xH_2O$ can be recovered. It is highly polymerized in soln and amorphous in the solid state; has been used in the prepn of body deodorants and antiperspirant preparations. Octahydrate as reagent for fluoride determn in water analysis and reduction of nitro compds to primary amines.

10385. Zoapatanol. [71117-51-6] 9-[(2S,3R,6E)-3-Hydroxy-6-(2-hydroxyethylidene)-2-methyl-2-oxepanyl]-2,6-dimethyl-2-nonen-5-one. $C_{20}H_{34}O_4$; mol wt 338.49. C 70.97%, H 10.12%, O 18.91%. Oxepane diterpenoid isolated from the leaves of the zoapatle plant, *Montanoa tomentosa, Compositeae,* which has been used in Mexican traditional medicine to induce menses and labor. Isoln and structure: M. P. Wachter, R. M. Kanojia, *US 4086358* (1978 to Ortho); S. D. Levine *et al., J. Am. Chem. Soc.* **101**, 3404 (1979); R. M. Kanojia *et al., J. Org. Chem.* **47**, 1310 (1982). Spasmogenic effects: J. B. Smith *et al., Life Sci.* **28**, 2743 (1981). Total syntheses of (±)-form: R. Chen, D. A. Rowand, *J. Am. Chem. Soc.* **102**, 6609 (1980); K. C. Nicolau *et al., ibid.* 6611; V. V. Kane, D. L. Doyle, *Tetrahedron Lett.* **22**, 3027, 3031 (1981). Mass spec: C. J. Shaw, *Org. Mass Spectrom.* **16**, 281 (1981). ^{13}C-NMR study: M. L. Cotter, *Org. Magn. Reson.* **17**, 14 (1981).

Pale yellow oil.

10386. Zofenopril. [81872-10-8] (4S)-1-[(2S)-3-(Benzoylthio)-2-methyl-1-oxopropyl]-4-(phenylthio)-L-proline; [1(S),4(S)]-1-(3-mercapto-2-methyl-1-oxopropyl)-4-phenyl-thio-L-proline-S-benzoylester. $C_{22}H_{23}NO_4S_2$; mol wt 429.55. C 61.52%, H 5.40%, N 3.26%, O 14.90%, S 14.93%. Angiotensin-converting enzyme (ACE) inhibitor; de-esterified *in vivo* to its active sulfhydryl-containing metabolite, *zofenoprilat.* Prepn: M. A. Ondetti, J. Krapcho, *GB 2028327; eidem, US 4316906* (1980, 1982 both to E. R. Squibb &

Sons). LC/MS/MS determn in plasma: L. Dal Bo *et al., J. Chromatogr. B* **749**, 287 (2000). Role of sulfhydryl group in improvement of endothelial dysfunction: H. Buikema *et al., Br. J. Pharmacol.* **130**, 1999 (2000); L. Cominacini *et al., Am. J. Hypertens.* **15**, 891 (2002). Clinical comparison with lisinopril of effect on post-infarction survival: C. Borghi, E. Ambrosioni, *Am. Heart J.* **145**, 80 (2003). Review of properties and pharmacokinetics: A. Subissi *et al., Cardiovasc. Drug Rev.* **17**, 115-133 (1999); of clinical efficacy in cardiovascular diseases: C. Borghi *et al., Expert Opin. Pharmacother.* **5**, 1965-1977 (2004).

Foamy solid, mp 42-44°. $[\alpha]_D^{25}$ $-36.5°$ (c = 1 in methanol).

Calcium salt. [81938-43-4] MEN-8029; SQ-26991; Bifril; Zofenil. $C_{44}H_{44}CaN_2O_8S_4$; mol wt 897.16. White crystalline powder, mp >250°. $[\alpha]_D^{23}$ $-67.6°$ (c = 1 in methanol/HCl). Partition coefficient (octanol/water): 3.5.

THERAP CAT: Antihypertensive.

10387. Zolazepam. [31352-82-6] 4-(2-Fluorophenyl)-6,8-dihydro-1,3,8-trimethylpyrazolo[3,4-e][1,4]diazepin-7(1H)-one; flupyrazapon. $C_{15}H_{15}FN_4O$; mol wt 286.31. C 62.93%, H 5.28%, F 6.64%, N 19.57%, O 5.59%. Pyrazolodiazepinone used in combination with tiletamine, *q.v.,* for veterinary anesthesia or to immobilize large animals. Prepn: H. A. DeWald, D. E. Butler, **DE 2023453**; *eidem,* **US 3558605** (1970, 1971 both to Parke, Davis); H. A. DeWald *et al., J. Med. Chem.* **20**, 1562 (1977). Metabolism in animals: J. Baukema *et al., Res. Commun. Chem. Pathol. Pharmacol.* **10**, 227 (1975). GC/MS determn in plasma: A. Kumar *et al., J. Chromatogr. B* **842**, 131 (2006). Review of pharmacology and veterinary use in combination with tiletamine: H. C. Lin *et al., J. Vet. Pharmacol. Ther.* **16**, 383-418 (1992).

Crystals from methanol-ether, mp 183-185°.

Hydrochloride. [33754-49-3] $C_{15}H_{15}FN_4O·HCl$; mol wt 322.77. Crystals from 2-propanol, mp 248° (dec). Freely sol in water, 0.1N hydrochloric acid; sol in methanol; slightly sol in chloroform. Practically insol in ether.

Note: This is a controlled substance (depressant): **21CFR**, 1308.13.

THERAP CAT (VET): Tranquillizer.

10388. Zoledronic Acid. [118072-93-8] P,P'-[1-Hydroxy-2-(1H-imidazol-1-yl)ethylidene]bisphosphonic acid; 2-(imidazol-1-yl)-1-hydroxyethane-1,1-diphosphonic acid. $C_5H_{10}N_2O_7P_2$; mol wt 272.09. C 22.07%, H 3.70%, N 10.30%, O 41.16%, P 22.77%. Bisphosphonate antiresorptive agent. Prepn: **JP Kokai 88 150291**; K. A. Jaeggi, L. Wilder, **US 4939130** (1988, 1990 both to Ciba-Geigy). Effect on bone metabolism: J. R. Green *et al., J. Bone Miner. Res.* **9**, 745 (1994). Determn in plasma by enzyme inhibition assay: F. Risser *et al., J. Pharm. Biomed. Anal.* **15**, 1877 (1997). Stability-indicating HPLC method: B. Mallikarjuna Rao *et al., J. Pharm. Biomed. Anal.* **39**, 781 (2005). Series of articles on pharmacology and clinical development: *Br. J. Clin. Pract. Suppl.* **87**, 15-22 (1996). Review of clinical experience in metastatic bone disease: D. Santini *et al., Expert Opin. Biol. Ther.* **6**, 1333-1348 (2006); in Paget's disease: G. M. Keating, L. J. Scott, *Drugs* **67**, 793-804 (2007). Clinical trial of once-yearly treatment of osteoporosis: D.

M. Black *et al.*, *N. Engl. J. Med.* **356**, 1809 (2007). Clinical trial in multiple myeloma: G. J. Morgan *et al.*, *Lancet* **376**, 1989 (2010).

Monohydrate. [165800-06-6] CGP-42446; Aclasta; Reclast; Zometa. $C_5H_{10}N_2O_7P_2.H_2O$; mol wt 290.10. White crystalline powder from water, mp 239° (dec). Highly sol in $0.1N$ NaOH; sparingly sol in water, $0.1N$ HCl. Practically insol in organic solvents. pH of 0.7% aq soln is approx 2.0.

Disodium salt tetrahydrate. [165800-07-7] Zoledronate disodium; CGP-42446A. $C_5H_8N_2Na_2O_7P_2.4H_2O$; mol wt 388.11.

Trisodium salt hydrate. [165800-08-8] Zoledronate trisodium; CGP-42446B. $(C_5H_7N_2Na_3O_7P_2)_5.2H_2O$; mol wt 1726.20.

THERAP CAT: Bone resorption inhibitor; in treatment of hypercalcemia of malignancy and bone metastases.

10389. Zolimidine. [1222-57-7] 2-[4-(Methylsulfonyl)phenyl]imidazo[1,2-*a*]pyridine; zoliridine; Solimidin. $C_{14}H_{12}N_2O_2S$; mol wt 272.32. C 61.75%, H 4.44%, N 10.29%, O 11.75%, S 11.77%. Gastroprotective agent. Prepn: **GB 991589**; L. Almirante *et al.*, **US 3318880** (1965, 1967 both to Selvi); *eidem, J. Med. Chem.* **8**, 305 (1965). Metabolism: *eidem, Farmaco Ed. Sci.* **29**, 941 (1974). Series of articles on pharmacology: *Panminerva Med.* **16**, 301-359 (1974). Pharmacokinetics: E. Schraven, D. Trottnow, *Arzneim.-Forsch.* **26**, 213 (1976). Clinical mucopoietic activity: S. Abate *et al.*, *Int. J. Tissue React.* **4**, 319 (1982); M. C. Parodi *et al.*, *Scand. J. Gastroenterol.* **19**, Suppl. 92, 163 (1984). Clinical study: A. Materia *et al.*, *Clin. Ter.* **97**, 183 (1981).

Crystals, mp 242-244°. LD_{50} orally in rats: 3710 mg/kg (Almirante, 1967).

THERAP CAT: Antiulcerative.

10390. Zolmitriptan. [139264-17-8] (4*S*)-4-[[3-[2-(Dimethylamino)ethyl]-1*H*-indol-5-yl]methyl]-2-oxazolidinone; (*S*)-*N,N*-dimethyl-2-[5-(2-oxo-1,3-oxazolidin-4-ylmethyl)-1*H*-indol-3-yl]ethylamine; 311C90; BW-311C90; Zomig. $C_{16}H_{21}N_3O_2$; mol wt 287.36. C 66.88%, H 7.37%, N 14.62%, O 11.14%. Serotonin $5HT_{1D}$-receptor agonist. Prepn: A. D. Robertson *et al.*, **WO 9118897** (1991 to Wellcome Foundation); *eidem*, **US 5466699** (1995 to Burroughs Wellcome). Structure-activity and receptor binding study: R. C. Glen *et al.*, *J. Med. Chem.* **38**, 3566 (1995). Pharmacology: P. J. Goadsby, L. Edvinsson, *Headache* **34**, 394 (1994). Clinical pharmacokinetics: E. Seaber *et al.*, *Br. J. Clin. Pharmacol.* **41**, 141 (1996). Clinical trial in migraine: S. J. Tepper *et al.*, *Curr. Med. Res. Opin.* **15**, 254 (1999). Review of clinical experience and pharmacokinetics: M. Gawel, I. Worthington, *Expert Opin. Pharmacother.* **6**, 1019-1024 (2005).

White crystals from isopropanol as the 0.9 isopropanolate hemihydrate, mp 139-141°. $[\alpha]_D^{22}$ −5.79° (c = 0.5 in methanol). pKa 9.64. Stable; nonhygroscopic. Soly in aq soln at neutral pH: >20 mg/ml.

THERAP CAT: Antimigraine.

10391. Zolpidem. [82626-48-0] *N,N,*6-Trimethyl-2-(4-methylphenyl)imidazo[1,2-*a*]pyridine-3-acetamide; *N,N,*6-trimethyl-2-*p*-tolylimidazo[1,2-*a*]pyridine-3-acetamide; SL-80.0750. $C_{19}H_{21}N_3$-O; mol wt 307.40. C 74.24%, H 6.89%, N 13.67%, O 5.20%. Selective non-benzodiazepine $GABA_A$ receptor agonist. Prepn: J. P. Kaplan, P. George, **EP 50563**; *eidem*, **US 4382938** (1982, 1983 both to Synthelabo). Neuropharmacology: H. Depoortere *et al.*, *J. Pharmacol. Exp. Ther.* **237**, 649 (1986). Neurochemical profile: B. Scatton *et al., ibid.* 659. Binding study in rat brain: S. Arbilla *et al., Eur. J. Pharmacol.* **130**, 257 (1986). HPLC determn in plasma: P. Guinebault *et al., J. Chromatogr.* **383**, 206 (1986); P. R. Ring, J. M. Bostick, *J. Pharm. Biomed. Anal.* **22**, 495 (2000). Review of pharmacology and pharmacokinetics: H. D. Langtry, P. Benfield, *Drugs* **40**, 291-313 (1990); of abuse and dependence potential: G. Hajak *et al., Addiction* **98**, 1371-1378 (2003). Review of clinical experience in insomnia: T. S. Harrison, G. M. Keating, *CNS Drugs* **19**, 65-89 (2005); of extended release formulation: R. L. Barkin, *Am. J. Ther.* **14**, 299-305 (2007).

mp 196°. pKa 6.2. Log P (octanol/water): 2.43.

L-(+)-Hemitartrate. [99294-93-6] Zolpidem tartrate; SL-80.0750-23N; Ambien; Intermezzo; Ivadal; Myslee; Niotal; Stilnoct; Stilnox; Tovalt. $(C_{19}H_{21}N_3O)_2.C_4H_6O_6$; mol wt 764.88. White to off-white crystalline powder. Soly in water (20°): 23 mg/ml. Sparingly sol in alcohol, propylene glycol.

Note: This is a controlled substance (depressant): **21 CFR,** 1308.14.

THERAP CAT: Sedative, hypnotic.

10392. Zonisamide. [68291-97-4] 1,2-Benzisoxazole-3-methanesulfonamide; 3-(sulfamoylmethyl)-1,2-benzisoxazole; AD-810; CI-912; Excegran; Trerief; Zonegran. $C_8H_8N_2O_3S$; mol wt 212.22. C 45.28%, H 3.80%, N 13.20%, O 22.62%, S 15.11%. Sulfonamide antiseizure agent; blocks repetitive firing of voltage-sensitive sodium channels and reduces voltage-sensitive T-type calcium currents. Prepn: H. Uno *et al.*, **JP Kokai 78 77057**; *eidem*, **US 4172896** (1978, 1979 both to Dainippon); *eidem, J. Med. Chem.* **22**, 180 (1979). Pharmacology and toxicity: Y. Masuda *et al., Arzneim.-Forsch.* **30**, 477 (1980). Mode of action study: T. Ito *et al., ibid.* 603. HPLC determn in serum: U. Juergens, *J. Chromatogr.* **385** 233 (1987). Clinical trial in epilepsy: D. Schmidt *et al., Epilepsy Res.* **15**, 67 (1993); in Parkinson's Disease: M. Murata *et al., Neurology* **68**, 45 (2007). Review of neuropharmacology: E. J. Hammond *et al., Gen. Pharmacol.* **18**, 303 (1987); of pharmacokinetics and mechanism of action: I. E. Leppik, *Seizure* **13S**, S5-S9 (2004); of pharmacology and clinical experience: S. V. Kothare, J. Kaleyias, *Expert Opin. Drug Metab. Toxicol.* **4**, 493-506 (2008).

White needles from ethyl acetate, mp 160-163° (Uno); also reported as mp 162-166° (Masuda, 1980). pKa 10.2. Soly (mg/ml): water 0.80; $0.1N$ HCl 0.50. Sparingly sol in chloroform, *n*-hexane. Sol in methanol, ethanol, ethyl acetate, and acetic acid. LD_{50} in mice, rats (mg/kg): 1892, 2001 orally; 1273, 2569 s.c.; 699, 733 i.p.; 604, 748 i.v. (Masuda, 1980).

THERAP CAT: Anticonvulsant; antiparkinsonian.

THERAP CAT (VET): Anticonvulsant.

10393. Zopiclone. [43200-80-2] 4-Methyl-1-piperazinecarboxylic acid 6-(5-chloro-2-pyridinyl)-6,7-dihydro-7-oxo-5*H*-pyrrolo[3,4-*b*]pyrazin-5-yl ester; 6-(5-chloropyrid-2-yl)-5-(4-methylpiperazin-1-yl)carbonyloxy-7-oxo-6,7-dihydro-5*H*-pyrrolo[3,4-*b*]pyrazine; RP-27267; Amoban; Imovane; Limovan; Sopivan; Ximovan; Zimovane. $C_{17}H_{17}ClN_6O_3$; mol wt 388.81. C 52.52%, H 4.41%,

Cl 9.12%, N 21.62%, O 12.34%. Cyclopyrrolone member of a family of non-benzodiazepine GABA$_A$ receptor agonists. Prepn: C. Cotrel *et al.*, **DE 2300491**; *eidem*, **US 3862149** (1973, 1975 both to Rhone-Poulenc); C. Jeanmart, C. Cotrel, *C. R. Seances Acad. Sci. Ser. C* **287**, 377 (1978). *In vitro* and *in vivo* inhibition of benzodiazepine binding: J. C. Blanchard *et al.*, *Life Sci.* **24**, 2417 (1979); *see also:* P. H. Wu *et al.*, *ibid.* **28**, 1023 (1981). Comparative study with benzodiazepines: E. Wickström, K. E. Giercksky, *Eur. J. Clin. Pharmacol.* **17**, 93 (1980). HPLC determn of enantiomers in plasma: C. Fernandez *et al.*, *J. Chromatogr.* **572**, 195 (1991); in urine: *eidem, ibid.* **617**, 271 (1993). Pharmacokinetics of zopiclone and enantiomers: *eidem, Drug Metab. Dispos.* **21**, 1125 (1993). Mechanism of action: A. Doble *et al.*, *Eur. Psychiatry* **10**, Suppl 3, 117s (1995). Sedative and anxiolytic effects of enantiomers and metabolite in rats: J. N. Carlson *et al.*, *Eur. J. Pharmacol.* **415**, 181 (2001). Series of articles on pharmacokinetics, pharmacology and efficacy in insomnia: *Sleep* **10**, Suppl 1, 1-87 (1987). Clinical overview: B. Musch, F. Maillard, *Int. J. Clin. Psychopharmacol.* **5**, 147-158 (1990). Review of pharmacology, tolerability and therapeutic efficacy: S. Noble *et al.*, *Drugs* **55**, 277-302 (1998).

Crystals from acetonitrile/diisopropyl ether (1:1), mp 178°.

S-Form. [138729-47-2] Eszopiclone; Estorra; Lunesta. Prepn: C. Cotrel, G. Roussel, **WO 9212980** (1992 to Rhone-Poulenc Rorer); *see also: eidem*, **US 6319926** (2001 to Sepracor). Crystals from acetonitrile, mp 206.5°. $[\alpha]_D^{20}$ +135 ±3° (c = 1.0 in acetone).

Note: This is a controlled substance (depressant): **21 CFR**, 1308.14.

THERAP CAT: Sedative, hypnotic.

10394. Zopolrestat. [110703-94-1] 3,4-Dihydro-4-oxo-3-[[5-(trifluoromethyl)-2-benzothiazolyl]methyl]-1-phthalazineacetic acid; 2-[4-oxo-3-[5-(trifluoromethyl)benzothiazol-2-ylmethyl]-3,4-dihydrophthalazin-1-yl]acetic acid; CP-73850. $C_{19}H_{12}F_3N_3O_3S$; mol wt 419.38. C 54.42%, H 2.88%, F 13.59%, N 10.02%, O 11.44%, S 7.64%. Aldose reductase inhibitor. Prepn: B. L. Mylari *et al.*, **EP 222576**; E. R. Larson, B. L. Mylari, **US 4939140** (1987, 1990 both to Pfizer); B. L. Mylari *et al. J. Med. Chem.* **34**, 108 (1991). Pharmacology: B. Tesfamariam *et al.*, *J. Cardiovasc. Pharmacol.* **21**, 205 (1993); *idem et al.*, *Am. J. Physiol.* **265**, H1189 (1993). Clinical pharmacokinetics: P. B. Inskeep *et al.*, *J. Clin. Pharmacol.* **34**, 760 (1994).

Crystals, mp 197-198°. pKa (dioxane/water): 5.46 (1:1); 6.38 (2:1). Log P (*n*-octanol/water): 3.43.

THERAP CAT: Treatment of diabetic complications.

10395. Zorubicin. [54083-22-6] Benzoic acid 2-[1-[(2S,4S)-4-[(3-amino-2,3,6-trideoxy-α-L-*lyxo*-hexopyranosyl)oxy]-1,2,3,4,-6,11-hexahydro-2,5,12-trihydroxy-7-methoxy-6,11-dioxo-2-napthacenyl]ethylidene]hydrazide; benzoic acid hydrazide 3-hydrazone with daunorubicin; RP-22050. $C_{34}H_{35}N_3O_{10}$; mol wt 645.67. C 63.25%, H 5.46%, N 6.51%, O 24.78%. Semi-synthetic antibiotic related to daunorubicin, *q.v.* Prepn: G. Jolles, **DE 2327211** (1974 to Rhone-Poulenc), *C.A.* **82**, 171381x (1975). Biological activity: R. Maral *et al.*, *C. R. Seances Acad. Sci. Ser. D* **275**, 301 (1972); R.

Maral, *Cancer Chemother. Pharmacol.* **2**, 31 (1979). Distribution and metabolism in mice: R. Baurain *et al.*, *ibid.* 37. Mechanism of action: G. P. Sartiano *et al.*, *J. Antibiot.* **32**, 1038 (1979). Acute cardiovascular effects in dogs: E. H. Herman, R. S. Young, *Cancer Treat. Rep.* **63**, 1771 (1979). Clinical study in breast cancer: J. N. Ingle, *ibid.* 1701.

Hydrochloride. [36508-71-1] NSC-164011; Rubidazone. $C_{34}H_{35}N_3O_{10}$·HCl; mol wt 682.12. Red-orange crystalline powder from ethanol. $[\alpha]_D^{20}$ −50° (c = 0.2 in water). uv max (methanol): 232.5, 253, 480, 495 nm (ε 40225, 35300, 10480, 10300). LD$_{50}$ in mice (mg/kg): 13.66 s.c., 4.42 i.p., 8.50 i.v. (Maral, 1972).

THERAP CAT: Antineoplastic.

10396. Zosuquidar. [167354-41-8] (αR)-4-[(1aα,6α,10bα)-1,1,-Difluoro-1,1a,6,10b-tetrahydrodibenzo[*a,e*]cyclopropa[*c*]cyclohepten-6-yl]-α-[(5-quinolinyloxy)methyl]-1-piperazineethanol; (2R)-*anti*-5-[3-[4-(10,11-difluoromethanodibenzosuber-5-yl)piperazin-1-yl]-2-hydroxypropoxy]quinoline. $C_{32}H_{31}F_2N_3O_2$; mol wt 527.62. C 72.85%, H 5.92%, F 7.20%, N 7.96%, O 6.06%. Multidrug resistance (MDR) modulator; selective inhibitor of P-glycoprotein (P-gp), *q.v.* Prepn: J. R. Pfister, D. L. Slate, **WO 9424107**; *eidem*, **US 5654304** (1994, 1997 both to Syntex); J. R. Pfister *et al.*, *Bioorg. Med. Chem. Lett.* **5**, 2473 (1995). Improved synthesis: C. J. Barnett *et al.*, *J. Org. Chem.* **69**, 7653 (2004). Effect on multidrug resistant cell lines: L. J. Green *et al.*, *Biochem. Pharmacol.* **61**, 1393 (2001). Specificity for P-gp: R. L. Shepard *et al.*, *Int. J. Cancer* **103**, 121 (2003). Clinical pharmacology and pharmacokinetics in patients with advanced malignancies: E. H. Rubin *et al.*, *Clin. Cancer Res.* **8**, 3710 (2002). Effect on pharmacokinetics of doxorubicin: S. Callies *et al.*, *Cancer Chemother. Pharmacol.* **51**, 107 (2003).

Trihydrochloride. [167465-36-3] LY-335979; RS-33295-198. $C_{32}H_{31}F_2N_3O_2$·3HCl; mol wt 636.99. mp 190°.

THERAP CAT: Antineoplastic adjunct (chemosensitizer).

10397. Zotepine. [26615-21-4] 2-[(8-Chlorodibenzo[*b,f*]-thiepin-10-yl)oxy]-*N,N*-dimethylethanamine; 2-chloro-11-(2-dimethylaminoethoxy)dibenzo[*b,f*]thiepine; Lodopin; Nipolept; Setous; Zoleptil. $C_{18}H_{18}ClNOS$; mol wt 331.86. C 65.15%, H 5.47%, Cl 10.68%, N 4.22%, O 4.82%, S 9.66%. Combined dopamine (D$_2$) and serotonin (5-HT$_2$) receptor antagonist. Prepn: S. Umio *et al.*, **DE 1907670**; *eidem*, **US 3704245** (1969, 1972 both to Fujisawa); I. Ueda *et al.*, *Chem. Pharm. Bull.* **26**, 3058 (1978). Pharmacology:

S. Uchida *et al., Arzneim.-Forsch.* **29**, 1588 (1979). Pharmacokinetics and metabolism: K. Noda *et al., ibid.* 1595. Toxicology: K. Fukuhara *et al., ibid.* 1600. Effects on serotonin receptors: A. Czyrak *et al., Pharmacopsychiatry* **26**, 53 (1993). GC-MS determn in plasma: O. Tanaka *et al., Ther. Drug Monit.* **18**, 294 (1996). Clinical trials in schizophrenia: C. Barnas *et al., Int. Clin. Psychopharmacol.* **7**, 23 (1992); A. Meyer-Lindenberg *et al., Pharmacopsychiatry* **30**, 35 (1997).

Crystals from cyclohexane, mp 90-91°. uv max (95% ethanol): 266 nm. LD$_{50}$ in male mice, rats (mg/kg): 108, 458 orally; 43.3, 39.7 i. +0.0, 97.0 i.p.; 84.9, 2080 s.c. (Fukuhara).

THERAP CAT: Antipsychotic.

10398. Zoxamide. [156052-68-5] 3,5-Dichloro-*N*-(3-chloro-1-ethyl-1-methyl-2-oxopropyl)-4-methylbenzamide; *N*-[3′-(1′-chloro-3′-methyl-2′-oxopentan)]-3,5-dichloro-4-methylbenzamide; RH-7281; Zoxium. C$_{14}$H$_{16}$Cl$_3$NO$_2$; mol wt 336.64. C 49.95%, H 4.79%, Cl 31.59%, N 4.16%, O 9.51%. Fungicide for foliar use on potatoes, vines and vegetables, targeting Oomycete fungi. Prepn: E. L. Michelotti, D. H. Young, **US 5304572** (1994 to Rohm and Haas); H. L. Rayle, L. Fellmeth, *Org. Process Res. Dev.* **3**, 172 (1999). Comprehensive description: A. R. Egan *et al., Brighton Crop Prot. Conf. - Pests Dis.* **1998**, 335-342. Mode of action: D. H. Young, R. A. Slawecki, *Pestic. Biochem. Physiol.* **69**, 100 (2001). Laboratory resistance risk study: D. H. Young *et al., Pest Manage. Sci.* **57**, 1081 (2001).

White solid. mp 158-160° (Rayle); also reported as mp 159.5-160.5° (Egan). Log P (octanol/water): 3.76 (20°). Vapor pressure at 25°, 35°, 45°: <1x10^{-7} torr. Soly in water at 20°: 0.681 mg/l. LD$_{50}$ in rats (mg/kg): >5000 orally; >2000 dermally. LC$_{50}$ in rats (mg/l): >5.3 by inhalation; in mallard ducks, bobwhite quail (mg/kg): >5250, >5250 dietary; in trout (μg/l): 160 (Egan).

USE: Agricultural fungicide.

10399. Zoxazolamine. [61-80-3] 5-Chloro-2-benzoxazolamine; 2-amino-5-chlorobenzoxazole; McN-485; Flexin. C$_7$H$_5$-ClN$_2$O; mol wt 168.58. C 49.87%, H 2.99%, Cl 21.03%, N 16.62%, O 9.49%. Centrally acting myorelaxant; formerly used as an antispasmodic and uricosuric. Prepn: T. Nagana *et al., J. Am. Chem. Soc.* **75**, 2770 (1953); J. Sam, **US 2780633** (1957 to McNeil). Pharmacology and structure-activity studies: C. K. Cain, A. P. Roszkowski, "Benzoxazoles, Benzothiazoles and Benzimidazoles" in *Medicinal Chemistry: A Series of Monographs* **4**, G. DeStevens, Ed. (Academic Press, New York, 1964) pp 325-357. Use in assay of microsomal oxidase activity: J. E. Tomazeski *et al., Arch. Biochem. Biophys.* **176**, 788 (1976). GLC determn and pharmacokinetics: M van der Graaff *et al., Drug Metab. Dispos.* **14**, 331 (1986).

Crystals from benzene, mp 185-185.5°. uv max (methanol): 244, 285 nm. Slightly sol in water; sol in alcohol. LD$_{50}$ in mice, rats (mg/kg): 376, 102 i.p.; 678, 730 orally (Cain, Roszkowski).

Hydrochloride. C$_7$H$_5$ClN$_2$O.HCl. Needles, mp 229° (dec).

USE: Tool for assessing hepatic cytochrome P-450 activity in rodents.

10400. Zymosan. [9010-72-4] Zymosans. Anticomplementary factor. Crude yeast cell wall prepns consisting chiefly of protein-carbohydrate complexes. Prepn from whole yeast cells: Pillemer, Ecker, *J. Biol. Chem.* **137**, 139 (1941); Pillemer *et al., J. Exp. Med.* **103**, 1 (1956); from yeast cell walls: Z. Holan, *Folia Microbiol.* **25**, 501 (1980). Composition of zymosan: F. J. DiCarlo, J. V. Fiore, *Science* **127**, 756 (1958). Enhances the specific defenses of an organism through the activation of the properdin system, *q.v.* Activity studies: Brade *et al., J. Immunol.* **111**, 1389 (1973); **112**, 1115 (1974). Induces aggregation and a release reaction in human platelet-rich plasma: Zucker, Grant, *ibid.* 1219. Review: Fitzpatrick, DiCarlo, *Ann. N.Y. Acad. Sci.* **118**, art. 4, 233 (1964).

Light gray powder. Practically insol in water, but readily disperses to give a homogeneous suspension.

USE: In the assay of properdin.

SUPPLEMENTAL TABLES

ALPHABETICAL LIST OF SUPPLEMENTAL TABLES

TABLE-1

Glossary

Listed below are brief definitions for selected terms used in the monographs
and/or frequently encountered within the references cited.

abzyme Catalytic antibody; antibody that can act like an enzyme in a specific chemical reaction.

adjuvant Substance administered with an antigen to augment its immunogenicity.

adrenergic System of autonomic innervation activated by epinephrine, norepinephrine or dopamine; particularly referring to substances that mimic the actions of the sympathetic nervous system.

agonist Receptor-specific activator; also a substance that mimics the effects of an endogenous regulatory compound.

aldose reductase Aldehyde reductase, EC 1.1.1.21; an oxidoreductase that reversibly catalyzes NADPH-dependent reduction of aldehydes, ketones, trioses, and triose phosphates.

alkaloid A nitrogenous, basic, usually bitter, substance found in plants. Term also applied to synthetic structural analogs.

allomone Chemical messenger used for communication between different species. Causes a reaction in the other species which is favorable to the organism secreting the substance.

allotrope One of several possible forms of an element; each of which differs in the atom bonding pattern.

aminoglycoside Carbohydrate typically consisting of 3 or 4 rings which has primary or secondary amines as ring substituents. Classification of antibiotics derived from various species of *Streptomyces,* which interfere with the function of bacterial ribosomes, e.g. streptomycin, gentamicin.

amyloid Complex proteinaceous fibrillar material deposited in blood vessels, heart, liver, brain, and other tissues in various disease states; gives characteristic starch-like reaction with iodine.

angiogenesis The development and formation of new blood vessels; postulated that tumor growth is preceded by an increase in new capillaries which converge upon the tumor.

antagonist Receptor-specific blocker; also a compound devoid of intrinsic regulatory activity which blocks or diminishes the action of an agonist.

anti Stereodescriptor restricted to bicyclo[X.Y.Z]anes in which the Z bridge reference group is oriented toward the Y bridge. Analogous to *trans* in double-bonded nitrogen systems such as oximes.

antibiotic Substance produced and secreted by various microorganisms which in low concentrations is toxic to other species.

antibody Immunoglobulin molecule produced in vertebrates by B-cells as a response to antigenic challenge; exist in different forms, each with a different binding site specifically recognizing the antigen that induced its formation.

antigen Immunogen; an agent that elicits an immune response including the production of an antibody; also considered to be any substance to which an antibody can bind.

antisense The DNA strand which is not transcribed; therefore, it has the same oligonucleotide sequence as the mRNA which codes for the protein.

antipyretic An agent that reduces or relieves fever.

apoprotein The protein portion of a conjugated protein without its prosthetic group.

apoptosis A regulated series of events leading to the death and removal of a cell in response to specific developmental or physiological signals; also known as "programmed cell death".

aromatase EC 1.14.14.1; enzyme that catalyzes the conversion of androgens to estrogens.

atriopeptidase Neutral endopeptidase, EC 3.4.24.11; an endopeptidase that catalyzes the degradation of atrial natriuretic factor.

bacteriocin A protein produced by bacteria that is toxic to another closely related strain.

balsam Resinous plant exudate containing the aromatic substances benzoic acid or cinnamic acid or their esters resulting from either a physiological or pathological response.

biological response modifiers Agents that alter the immune response such as interleukins, interferons, colony-stimulating factors, and tumor necrosis factor.

bitter A liquid, often alcoholic, that is prepared by steeping or distilling of a bitter herb or herb part that is used medicinally.

carbene A neutral organic radical containing a divalent carbon atom having only six electrons in its outer shell ($R_2C:$).

carbonic anhydrase Carbonate dehydratase, EC 4.2.1.1; catalyzes interconversion (reversible hydration) of carbonic acid to carbon dioxide and water.

carcinoma A malignant tumor arising from epithelial cells.

catechol *O*-methyl transferase EC 2.1.1.6; a methyltransferase that catalyzes the transfer of a methyl group from S-adenosyl-L-methionine to catechol or catecholamine.

chelate The cyclic structure formed by unshared electrons on neighboring atoms complexing with a metal ion; used as a sequestrant in metal poisoning.

chemoselectivity Pertains to the preferential reaction of a reagent with one functional group of a multifunctional molecule.

chemotaxis Migration of cells or organisms within a chemical concentration gradient.

chimeric Individual organism with cell populations from genetically different sources.

chiral The arrangements of atoms in space such that the resulting molecule is nonsuperimposable on its mirror image.

cholinesterase EC 3.1.1.8; catalyzes hydrolysis of acetylcholine and other choline esters to form choline and a carboxylic acid.

chromophore The group of atoms in a molecule that absorbs light and is responsible for the color of the molecule.

cis Relative stereodescriptor limited to cyclic systems with two stereocenters. Configuration where the two groups of higher priority are on the same side of the reference plane. See also *zusammen.*

TABLE-2

Glossary (Continued)

clinical trial, Phase I Safety testing and pharmacological profiling in humans; usually involves 20-80 healthy volunteers and takes about one year.

clinical trial, Phase II Effectiveness testing in volunteer patients; usually involves 100-300 volunteers and takes about two years.

clinical trial, Phase III Extensive clinical testing in humans to determine efficacy and to identify adverse reactions; usually involves 1000-3000 patients and lasts about three years.

clinical trial, Phase IV Additional post-marketing studies to evaluate long term effects.

cloning Proliferation of a single cell to a large colony; replication from a single gene introduced into a host cell.

coenzyme Nonprotein molecule that is essential for the activity of an enzyme; functions as a cofactor of an enzyme.

conformers Different three-dimensional arrangements of atoms within a molecule that can be interconverted by rotation around single bonds; not usually separable. May also be referred to as conformation isomers or rotamers. (e.g. the chair, boat and twist forms of cyclohexane).

corticosteroid A 21-carbon steroid hormone produced by the adrenal cortex in response to the release of corticotropin (ACTH) or angiotensin.

cyclooxygenase An activity of prostaglandin endoperoxide synthase (EC 1.14.99.1) that catalyzes the oxidation of arachidonic acid to produce prostaglandins and thromboxanes.

cytochromes Family of colored proteins, related by the presence of a porphyrin metal (usually Fe), which undergo rapid oxidation and reduction in organisms.

cytokines Diverse group of soluble, nonantibody proteins secreted by a variety of cell types, which modulate the functional activities of individual cells by interaction with specific cell surface receptors, e.g. interferon, interleukin.

cytokinin Plant hormone that stimulates cell division and metabolism.

dative bond Semipolar linkage between two atoms where one atom contributes both electrons; also referred to as a coordinate covalent bond.

depsipeptide Polypeptide that contains ester bonds as well as peptide bonds, commonly but not necessarily alternating. Naturally occurring depsipeptides are often cyclic. Common metabolic products of microorganisms; often have antibiotic activity.

eicosanoids Collective term applied to substances such as prostaglandins, thromboxanes, leukotrienes and lipoxins, which are all derived from arachidonic acid; play an important role in inflammation.

enantioselectivity Pertains to a stereoselective reaction in which the potential products are mirror images.

endo Stereodescriptor used only for ring positions on [X.Y.Z.]anes in which X≥Y>Z>0 and the group of higher priority is on the opposite side of the ring reference plane to the Z-bridgehead.

endopeptidase Endoproteinase, proteinase; catalyzes the cleavage of non-terminal peptide bonds in a polypeptide or protein.

endorphins Naturally occurring neuropeptides that bind to opioid receptors with pain relieving action; the name is derived from endogenous morphine.

endotoxins Heat-stable lipopolysaccharide-protein complexes contained in cell walls of gram-negative bacteria including non-infectious gram-negatives; released only when the cells are disrupted.

entactogen Non-hallucinogenic psychoactive compound which is purported to facilitate communication and introspective states. Derived from the Greek roots of *en* for within or inside, *gen* to produce or originate and the Latin root *tactus* for touch.

enterotoxin Exotoxin produced by enterobacteria, which, when ingested or produced within the intestine, affects its function, inducing nausea, diarrhea, vomiting, etc.

entgegen (E) German for opposite. Relative stereodescriptor for configuration about double bonds; used when groups of higher priority are on different sides of the vertical reference plane.

epitope Structural entity on the antigen that forms the recognition site for binding.

ergot The hardened web-like mass of threads of *Claviceps purpurea* developed on rye attacked by the fungus. Ergot alkaloids are used as oxytocics and in treatment of migraine.

essential fatty acids A group of polyunsaturated fatty acids, such as linoleic acid, required for growth in mammals; they cannot be synthesized and must be obtained in the diet.

essential oil Volatile, aromatic oil physically extracted from plants.

excipient Any more or less inert substance added to a drug to give suitable consistency or form to the drug.

excitotoxin Neurotoxic analogs of glutamic acid which mimic its excitatory effects on neurons of the CNS.

exo Stereodescriptor used only for ring positions on [X.Y.Z.]anes in which X≥Y>Z>0 and the group of higher priority is on the same side of the ring reference plane to the Z-bridgehead.

exotoxins Disease-producing proteins secreted by species of certain bacteria into surrounding medium.

flavoprotein Flavin dependent enzymes which normally contain FAD or FMN as tightly bound coenzymes.

galenic Prepared from a plant rather than chemically synthesized.

geminal (*gem*) Indicates that the two substituents in a disubstituted compound are on the same carbon atom.

glioma A tumor arising from the glial tissue of the brain or spinal cord.

glucocorticoid A 21-carbon steroid hormone secreted by adrenal cortex. Acts on carbohydrate, lipid and protein metabolism and inhibits the release of corticotropin (ACTH).

α-glucosidase Maltase, glucoinvertase, EC 3.2.1.20; a glucohydrolase that catalyzes hydrolysis of terminal non-reducing 1,4-linked α-D-glucose residues to release α-D-glucose.

green chemistry Development of chemical products, procedures, and technologies with the goal of minimizing or eliminating the generation of hazardous substances.

TABLE-3

Glossary (Continued)

hapten Small nonantigenic molecule which becomes an antigen when combined with a larger carrier molecule.

herbaceous Plants or parts of plants that do not have woody stems and typically die down annually.

HMG CoA reductase β-Hydroxyl-β-methyl glutarate CoA reductase, EC 1.1.1.88; oxidoreductase that catalyzes 3-hydroxy-3-methylglutaryl CoA reduction to mevalonate with NADPH as the electron donor.

homeopathy The theory that "like cures like"; wherein small amounts of compounds which in large amounts would produce undesirable symptoms are beneficial.

hybridoma Cell line produced by fusing an antibody-secreting cell with a myeloma cell; used to produce specific monoclonal antibodies in great quantity.

hyperplasia Abnormal increase in the number of normal cells in a tissue or organ.

hypertrophy Abnormal enlargement of an organ or tissue, not due to tumor formation; caused by the increase in size of its cellular components.

idiotype The segment of an immunoglobulin molecule that determines its antigenic specificity.

immunoglobulins Proteins produced by plasma cells derived from B cells of the immune system; function as antibodies to aid in fighting infection.

immunotoxins Molecules that comprise a cytotoxic agent linked to an antibody specific for target cell antigens.

inosine monophosphate dehydrogenase IMPH; EC 1.1.1.205; NAD$^+$-dependent enzyme that catalyzes oxidation of inosine 5′-monophosphate (IMP) to xanthosine 5′-monophosphate (XMP); rate limiting step in *de novo* guanine nucleotide synthesis.

isoform Protein that has the same function as another protein but which is encoded by a different gene or alternative splicing of variant alleles and may have small differences in its primary structure.

kairomones Chemical messengers used for communication between different species which are either nonadaptive or actually detrimental to the organisms producing them.

ketolide Semi-synthetic derivative of the macrolide erythromycin A.

knockout mutation A mutation in which a single gene of choice is inactivated without affecting other genes.

lake Organic pigment made of an oil-soluble organic dye, a precipitant, and an absorptive inorganic substrate.

lantobiotics Polycyclic bacteriocins consisting of the thioether amino acids lanthionine and beta-methyllanthionine.

liposome Closed vesicles of lipid bilayer formed by emulsification of cell membranes; the interior may be used to encapsulate drugs for targeted cellular delivery.

5-lipoxygenase EC 1.13.11.34; actin-binding protein that initiates the synthesis of leukotrienes via the oxidation of arachidonic acid.

lymphoma A malignant tumor of the lymphoid tissue.

lymphokine Nonimmunoglobulin protein synthesized primarily by T-lymphocytes that modulates an immune response, e.g. interleukin-2, γ-interferon.

major histocompatibility complex Large, multigene cluster that encodes for the cell surface glycoproteins which are essential in the cellular immune response to distinguish "self" and "non-self".

matrix metalloproteinase Zinc-dependent endopeptidases that catalyze the hydrolysis of extracellular proteins, particularly collagen and elastin.

melanoma A malignant tumor derived from melanocytes.

meso compound Pertaining to a compound which contains equal numbers of enantiomeric groups which are identically linked. Compound having at least two chiral centers, yet no optical rotation because it contains a plane of symmetry.

mesomerism The existence of chemical structures having the same molecular arrangements that differ only in the location of electrons; the actual electronic distribution is not represented by any one structure, but is a hybrid of contributing structures.

monoamine oxidase MAO, amine oxidase, EC 1.4.3.4; flavoprotein enzyme that catalyzes oxidative deamination of monoamines to form aldehydes or ketones with the release of hydrogen peroxide and ammonia.

monoclonal antibody Antibody produced by a cloned hybridoma cell, active against a single target antigen.

mordant Substance used to fix a dyestuff on a material by combining with the dye to form an insoluble compound.

morphogens Soluble substances that control the embryonic differentiation of cells and tissues.

multidrug resistance The resistance of tumor cells to varied chemotherapeutic agents mediated by the expression of a transmembrane pump, P-glycoprotein, which actively pumps the drugs out of the cells, lowering the intracellular concentration and cytotoxic effectiveness.

mutagen Agent that causes a permanent inheritable change in the DNA of the organism.

mutein Name applied to a protein arising as a result of a mutation.

mycotoxins Fungal toxins that are harmful to other organisms.

myeloma A malignant tumor derived from hemopoietic tissue.

neuraminidase Acetyl-neuraminyl hydrolase; sialidase; EC 3.2.1.18; enzyme that catalyzes the hydrolysis of terminal sialic acid groups from glycoproteins and oligosaccharides; found on the surface of the influenza virus.

neuroleptic Agent that pharmacologically manages the manifestations of a psychotic disorder, affecting cognition and behavior; also called major tranquilizer or anti-psychotic.

nitrene Carbene analog consisting of an electron-deficient, neutral monovalent nitrogen species.

neurotrophic factors Polypeptides that support the growth and survival of specific populations of neurons.

nootropic Agent having positive effects on organically impaired cognition or nervous system function; from the Greek *noos* or *nous* for mind and *-tropic* for nourishing, stimulating.

TABLE-4

Glossary (Continued)

nor Originally the abbreviation for *Nitrogen ohne Radikal.* Currently a subtractive prefix for stereoparents indicating ring contraction or the loss of a methyl group from the carbon skeleton.

nucleic acids Biologically occurring polynucleotides in which the nucleotide residues are linked in a specific sequence by phosphodiester bonds; DNA and RNA.

nucleoside A purine or pyrimidine base in glycosidic linkage with a pentose sugar (e.g. adenosine, cytosine, guanosine, thymidine).

nucleotide The phosphate ester of a nucleoside.

oleoresin Plant exudate containing essential oil and resin.

orexigenic Appetite stimulating.

peptidyl dipeptidase A Angiotensin converting enzyme; EC 3.4.15.1; a zinc-containing hydrolase that catalyzes cleavage of a C-terminal dipeptide; catalyzes conversion of angiotensin I to yield the active angiotensin II.

peroxidases Oxidoreductase enzymes that catalyze oxidation of substrates by hydrogen peroxide.

pharmacodynamics Study of the biochemical and physiological effects of drugs on living systems and their mechanism of action.

pharmacognosy Study of biological and chemical properties of naturally occurring medicinal products.

pharmacokinetics Study of the action of drugs in the body over time, including absorption, biotransformation, tissue distribution, and elimination of the drugs and their metabolites.

pharmacophore An abstract concept that defines the set molecular characteristics (steric, and electronic) and chemical functionalities necessary to ensure the optimal interaction with the target site.

pheromones Chemical substances used for communication between individual organisms of the same species. Perceived primarily by the olfactory sense and to a lesser extent the gustatory sense. From the Greek *pherein* to transfer, and *hormon* to excite or stimulate.

phytosterols Sterols obtained from higher plants; differ from sterols occurring in animals by having C_1 or C_2 residues at C-24 and/or a double bond at C-22.

prebiotic Food ingredients that function by stimulating growth of beneficial bacteria of the digestive system.

primary structure (protein) Amino acid sequence.

prion A small proteinaceous infectious particle which resists inactivation by procedures that modify nucleic acids.

probiotic Beneficial bacteria of the digestive system.

prochiral Achiral compound or group that would become chiral if one of two identical substituents is replaced by a new ligand.

prodrug A drug precursor; a substance which is inactive by itself and which must undergo chemical conversion by metabolic processes to become an active pharmacological agent.

prostaglandin endoperoxide synthase Prostaglandin synthase; EC 1.14.99.1; heme-containing enzyme that with dioxygenase and peroxidase activity; catalyzes the biosynthesis of prostaglandins G and H from arachidonic acid.

prosthetic group Tightly bound non-protein portion of a conjugated protein essential for the activity of a protein or enzyme.

pyrogens Fever-producing substances; agents which produce a febrile response when administered parenterally to man and certain animals.

quaternary structure (protein) Three-dimensional structure of a protein, in particular how subunits are integrated; typically classified as fibrous, globular, or conjugated.

receptor Protein or protein complex that binds a specific ligand such as a neurotransmitter, hormone, lymphokine, etc., producing a secondary biological signal within the cell.

rectus (R)—(right) Opposite of (*S*). Absolute term describing the spatial arrangement about an asymmetric carbon when the substituent of lowest priority lies below the reference plane and the priority of the remaining substituents decreases in a clockwise direction.

5α-reductase EC 1.3.99.5; membrane-bound enzyme that catalyzes the irreversible reduction of testosterone to the more potent androgen, dihydrotestosterone.

regioselectivity Pertains to the preferential formation of one structural isomer when two or more are possible.

reverse transcriptase EC 2.7.7.49; RNA-directed DNA polymerase; required for the replication of retroviruses.

RNA interference (RNAi) Process in which short interfering RNA (siRNA) binds to specific mRNA and silences gene expression.

saponins Foaming glycoside of sterols or triterpenes that are widely distributed in plants. Causes cell lysis.

sarcoma Malignant tumor arising from connective tissue or muscle cells.

secondary structure (protein) Residue by residue conformation of the backbone of the polypeptide chain. Typical conformations include the α helices, β sheets and random coils; given rise to by the regular hydrogen-bond interactions within contiguous stretches of polypeptide chain.

semiconductor Material with electric conductivity range between metals and insulators.

sinister (S)—(left) Opposite of (*R*). Absolute term describing the spatial arrangement about an asymmetric carbon when the substituent of lowest priority lies below the reference plane and the priority of the remaining substituents decreases in a counterclockwise direction.

sol-gel process Solution process for the transformation of colloidal liquids into ultrahomogeneous ceramic and glass gels used in material science.

stereoselectivity Pertains to the preferential formation of one stereoisomer when two or more are possible.

stereospecific reaction A reaction where starting materials, which differ only in their configuration, are converted to stereoisomerically distinct products. Applies when the reaction processes are constrained to proceed in a stereochemically defined manner.

superconductors Materials that below a certain critical temperature show zero resistance to electrical current flow. Ideally if resistance is truly zero, then electrons flowing in a loop of superconducting wire would never lose any energy.

TABLE-5

Glossary (Continued)

syn Stereodescriptor restricted to bicyclo[X.Y.Z]anes in which the Z bridge reference group is oriented toward the X bridge; analogous to *cis* in double-bonded nitrogen systems such as oximes.

teratogen Agent that causes abnormal fetal development.

terpene Hydrocarbon derived from essential oils, resins, etc., containing multiples of the C_5 isoprene unit as its carbon skeleton. Monoterpene is $C_{10}H_{16}$.

tertiary structure (protein) Three dimensional conformation of the polypeptide chain after folding.

tocolytic Compound to reduce uterine contractions.

topoisomerase Enzyme that can remove (or create) super-coiling in double stranded DNA by creating transitory breaks in one (type I) or both (type II) strands of the sugar-phosphate backbone.

toxoid A toxin which has been treated by a chemical agent or by heat to destroy its toxic properties while retaining its ability to stimulate antibody production.

trans Relative stereodescriptor limited to cyclic systems with two stereocenters. Configuration where the two groups of highest priority are on opposite sides of the reference plane.

transduction Transfer of genetic material from one cell to another by a viral vector.

transfection Introduction of exogenous DNA (usually viral or bacterial) into mammalian cells or bacterial cells.

transgenic Containing artificially introduced DNA as a result of experimental splicing of a segment of DNA from one genome into the germ line of another.

vasopeptidase inhibitor Cardiovascular drug that simultaneously inhibits both neutral endopeptidase and angiotensin-converting enzyme.

vicinal (*vic*) Indicates that two substituents in a disubstituted compound are on adjacent carbon atoms.

ylide Compounds in which a carbon atom with an unshared electron pair is connected to cation from Group 15 (Va) or 16 (VIa) of the periodic table.

zusammen (Z) German for together. Relative stereodescriptor for configuration about double bonds; used when groups of higher priority are on the same side of the vertical reference plane.

zwitterion Dipolar ion in which positive and negative charges are physically separated.

zymogen Inactive precursor molecule to an enzyme which is activated as a result of catalysis. Also known as a proenzyme.

TABLE-6

Acronyms

Following are the expanded meanings of acronyms or acronym-like abbreviations for chemical, agricultural, biomedical, or pharmaceutical terms, agencies, or organizations related to information found in *The Merck Index*.

AAAS	American Association for the Advancement of Science
AALAS	American Association for Laboratory Animal Science
AAPS	American Association of Pharmaceutical Scientists
AAS	Atomic absorption spectroscopy
AAV	Adeno-associated virus
AB	Aktiebolag
ABC	ATP binding cassette
ABPI	Association of the British Pharmaceutical Industry
ACAT	Acyl-CoA cholesterol transferase
ACE	Angiotensin converting enzyme
ACP	Acyl carrier protein
	American College of Physicians
ACRP	Association of Clinical Research Professionals
ACS	American Cancer Society
	American Chemical Society
	American College of Surgeons
ACTH	Adrenocorticotropic hormone
AD	Alzheimer's disease
ADA	American Dental Association
ADCC	Antibody-dependent cellular cytotoxicity
ADD	Attention deficit disorder
ADE	Adverse drug event
ADH	Alcohol dehydrogenase
ADHD	Attention deficit hyperactivity disorder
ADI	Acceptable daily intake
ADL	Activities of daily living
ADME	Absorption, distribution, metabolism, excretion
ADR	Adverse drug reaction
AE	Adverse event
AFC	Antibody forming cells
AFLP	Amplified fragment length polymorphism
AFM	Atomic force microscopy
AG	Aktiengesellschaft
AGE	Advanced glycation endproduct
AHA	American Heart Association
AHFS	American Hospital Formulary Service
AIChE	American Institute of Chemical Engineers
AIDS	Acquired immune deficiency syndrome
AIF	Apoptosis-inducing factor
AII	Angiotensin II
ALA	Aminolevulinic acid
ALG	Anti-lymphocyte globulin
ALL	Acute lymphoblastic leukemia
ALS	Amyotrophic lateral sclerosis
	Anti-lymphocyte serum
ALT	Alanine aminotransferase
AMA	American Medical Association
AMD	Age-related macular degeneration
AMI	Acute myocardial infarction
AML	Acute myeloid leukemia
ANDA	Abbreviated New Drug Application
ANOVA	Analysis of variance
ANP	Atrial natriuretic peptide
ANSI	American National Standards Institute
AOAC	Association of Official Analytical Chemists
APC	Antigen presenting cells
APhA	American Pharmaceutical Association
API	Active pharmaceutical ingredient
	American Petroleum Institute
APOE	Apolipoprotein E
APP	Amyloid precursor protein
APPI	Academy of Pharmaceutical Physicians and Investigators
APT	Alum precipitated toxoid

APTT	Activated partial thromboplastin time
ARC	AIDS-related complex
	American Red Cross
ARDS	Acute respiratory distress syndrome
	Adult respiratory distress syndrome
ARMS	Alveolar rhabdomyosarcoma
ASBMB	American Society for Biochemistry and Molecular Biology
ASHG	American Society for Human Genetics
ASHP	American Society of Hospital-System Pharmacists
ASIST	American Society for Information Science and Technology
ASM	American Society for Microbiology
ASPET	American Society for Pharmacology and Experimental Therapeutics
AST	Asparate amino transferase
ASTM	American Society for Testing and Materials
ATCC	American Type Culture Collection
AUC	Area under the curve
AV	Atrioventricular
AVMA	American Veterinary Medical Association
BAC	Bacterial artificial chromosome
BACE	Beta-site amyloid precursor protein cleaving enzyme
BACR	British Association for Cancer Research
BAL	Blood alcohol level
BAN	British Approved Name
BBB	Blood-brain barrier
BBSRC	Biotechnology and Biological Sciences Research Center (UK)
BCG	Bacillus Calmette-Guerin
BCPC	British Crop Protection Council
BCRP	Breast cancer resistance protein
BES	Balanced electrolyte solution
BGG	Bovine gamma globulin
BIBRA	British Industrial Biological Research Association
BLA	Biologics License Application (US)
BMA	British Medical Association
BMD	Bone mineral density
BMI	Body mass index
BMR	Basal metabolic rate
BNF	British National Formulary
BOD	Biochemical oxygen demand
BP	British Pharmacopeia
BPH	Benign prostatic hyperplasia
	Benign prostatic hypertrophy
BRM	Biological response modifier
BSA	Bovine serum albumin
BSE	Bovine spongiform encephalopathy
BSI	British Standard Institutes
BUN	Blood urea nitrogen
BV	Besloten Vennootschap
BVA	British Veterinary Association
CA	Chemical Abstracts
CABG	Coronary artery bypass grafts
CAD	Coronary artery disease
CAM	Cell adhesion molecules
	Complementary and alternative medicine
cAMP	Cyclic adenosine monophosphate
CAS	Chemical Abstracts Service
CAT	Computerized axial tomography
CB	Cannabinoid
CBC	Complete blood count
CBER	Center for Biologics Evaluation and Research (US)

TABLE-7

CBF	Cerebral blood flow	CVM	Center for Veterinary Medicine (US)
CCD	Charge-coupled device	DC	Dendritic cells
CCK	Cholecystokinin	DCL	Diffuse cutaneous leishmaniasis
CD	Circular dichroism	DEA	Drug Enforcement Administration (US)
	Clusters of differentiation	DEFRA	Department for Environment, Food & Rural Affairs (UK)
CDC	Centers for Disease Control and Prevention (US)	DXA	Dual-energy X-ray absorptiometry
CDER	Center for Drug Evaluation and Research (US)	DFA	Direct fluorescent antibody
CDK	Cyclin dependent kinase	DHA	Docosahexaenoic acid
cDNA	Complementary deoxyribonucleic acid	DHFR	Dihydrofolate reductase
CE	Capillary electrophoresis	DHHS	Department of Health and Human Services (US)
CEA	Carcinoembryonic antigen		
CETP	Cholesteryl ester transfer protein	DIC	Disseminated intravascular coagulation
CF	Cystic fibrosis	DKA	Diabetic ketoacidosis
CFA	Complete Freund's adjuvant	DM	Diabetes mellitus
CFC	Chlorofluorocarbon	DMARD	Disease modifying antirheumatic drug
CFR	Code of Federal Regulations (US)	DNA	Deoxyribonucleic acid
CFSAN	Center for Food Safety and Applied Nutrition (US)	DOD	Department of Defense (US)
		DPP	Dipeptidyl peptidase
CFT	Complement fixation test	DPT	Diphtheria, pertussis, tetanus
CFTR	Cystic fibrosis transmembrane regulator	DRG	Diagnosis related grouping
CFU	Colony forming unit	DRSP	Drug-resistant *Streptococcus pneumonia*
cGMP	Cyclic guanosine monophosphate	ds	Double stranded
CHD	Coronary heart disease	DSC	Differential scanning calorimetry
CHF	Chronic heart failure	DTA	Differential thermal analysis
	Congestive heart failure	DTaP	Diphtheria-Tetanus-acellular-Pertussis
CHO	Chinese hamster ovary (cells)	DTC	Direct to consumer
CI	Colour Index	DTH	Delayed type hypersensitivity
CIMS	Chemical ionization mass spectroscopy	DVT	Deep venous thrombosis
CIP	Cahn-Ingol-Prelog	EAA	Excitatory amino acid
CIR	Cosmetic Ingredient Review (US)	EAC	Erythrocyte coated by antibody and complement
CJD	Creutzfeldt-Jakob disease	EAE	Experimental allergic encephalomyelitis
CKD	Chronic kidney disease		Experimental autoimmune encephalomyelitis
CLA	Cutaneous lymphocyte antigen		
CLL	Chronic lymphocytic leukemia	EBI	European Bioinformatics Institute
CMC	Critical micelle concentration	EBT	Electron beam tomography
CMI	Cell mediated immunity	EBV	Epstein Barr virus
CML	Cell-mediated lymphocytotoxicity	EC	Effective concentration
	Chronic myelogenous leukemia		Electron capture
CMV	Cytomegalovirus		Enzyme Commission
CNS	Central nervous system		European Commission
CoA	Coenzyme A	ECB	European Chemicals Bureau
COD	Chemical oxygen demand		European Congress on Biotechnology
COMP	Committee for Orphan Medicinal Products (EU)	ECETOC	European Center for Ecotoxicology and Toxicology of Chemicals
COMT	Catechol-*o*-methyltransferase	ECF	Extracellular fluid
COPD	Chronic obstructive pulmonary disease	ECF-A	Eosinophil chemotactic factor of anaphylaxis
COSY	Correlation spectroscopy	ECG	Electrocardiogram
COX	Cyclo-oxygenase	ECM	Extracellular matrix
CP	Canadian Pharmacopeia	ED	Effective dose
	Chemically pure		Erectile dysfunction
CPAP	Continuous positive airway pressure	EDG	Electron donating group
CPR	Cardiopulmonary resuscitation	EDHF	Endothelial-derived hyperpolarizing factor
CRADA	Cooperative Research and Development Agreement (US)	EDRF	Endothelium-derived relaxing factor
		EDTA	Ethylenediaminetetraacetic acid
CRF	Chronic renal failure	EEG	Electroencephalogram
CRH	Corticotropin-releasing hormone	EFA	Essential fatty acids
CRO	Contract research organization	EFB	European Federation of Biotechnology
CSF	Cerebrospinal fluid	EFPIA	European Federation of Pharmaceutical Industries and Associations
	Colony stimulating factors		
CSIRO	Commonwealth Scientific and Industrial Research Organization (Australia)	EGA	Evolved gas analysis
		EGF	Epidermal growth factor
CSM	Committee on Safety of Medicines (UK)	EGFR	Epidermal growth factor receptor
CSP	Cold shock protein	EHR	Electronic Health Records
CSPS	Canadian Society for Pharmaceutical Sciences	EINECS	European Inventory of Existing Commercial Chemical Substances
CT	Charge transfer		
	Computed tomography	EKG	Elektrokardiographie
CTFA	Cosmetic, Toiletry, and Fragrance Association	ELINCS	EHR-Lab Interoperability and Connectivity Standards
CTL	Cytotoxic T lymphocyte		
CV	Ceiling value		European List of Notified Chemical Substances
CVA	Cerebral vascular accidents	ELISA	Enzyme-linked immunosorbent assay
CVD	Cardiovascular disease	EMBO	European Molecular Biology Organization
	Chemical vapor deposition		

TABLE-8

EMEA	European Medicines Evaluation Agency	GBS	Guillain-Barré syndrome
EMG	Electromyography	GC	Gas chromatography
EMSA	Electrophoretic mobility shift assay	GCPF	Global Crop Protection Federation
ENDOR	Electron-nuclear double resonance	GEM	Genetically engineered microorganism
EORTC	European Organization for Research and Treatment of Cancer		Genetically engineered mouse
		GERD	Gastroesophageal reflux disease
EPA	Eicosapentaenoic acid	GF	Growth factor
	Environmental Protection Agency	GFP	Green fluorescent protein
EPO	Erythropoietin	GFR	Glomerular filtration rate
	European Patent Office	GGT	Gamma glutamyl transferase
	European Pharmacopoeia Office	GH	Growth hormone
EPPO	European and Mediterranean Plant Protection Organization	GHRH	Growth hormone releasing hormone
		GI	Gastrointestinal
EPR	Electron paramagnetic resonance	GLA	Gamma-linolenic acid
EPS	Extracellular polysaccharide	GLC	Gas-liquid chromatography
ER	Endoplasmic reticulum	GLP	Glucagon like peptide
	Estrogen receptor	GLP	Good laboratory practices
ERCP	Endoscopic retrograde cholangiopancreatography	GmbH	Gesellschaft mit Beschrankter Haftung
		GMO	Genetically modified organism
ES	Embryonic stem (cells)	GMP	Good manufacturing practices
ESA	Ecological Society of America	GnRH	Gonadotropin-releasing hormone
	Entomological Society of America	GPC	Gel permeation chromatography
	European Space Agency	GPCR	G protein-coupled receptor
ESF	European Science Foundation	GRAS	Generally recognized as safe
ESHG	European Society of Human Genetics	GTP	Guanosine triphosphate
ESP	End sequence profiling	GVH	Graft-versus-host
ESR	Electron spin resonance	HAART	Highly active anti-retroviral therapy
	Erythrocyte sedimentation rate	HAT	Histone acetyl transferase
ESRD	End-stage renal disease	HBV	Hepatitis B virus
EST	Expressed sequence tag	HC	Heavy chain
EU	European Union		Hematopoietic cell
EUP	Experimental Use Permit (US)	HCFC	Hydrochlorofluorocarbon
EVM	European Vaccine Manufacturers	HCL	Hairy cell leukemia
EWG	Electron withdrawing group	HCV	Hepatitis C virus
FA	Fanconi anemia	HD	Huntington's Disease
FAB	Fast atom bombardment	HDAC	Histone deacetylase
FABP	Fatty acid binding protein	HDL	High density lipoprotein
FACS	Fluorescence activated cell sorting	HEK	Human embryonic kidney (cells)
FAO	Food and Agriculture Organization of the United Nations	hESC	Human embryonic stem cell
		HETE	Hydroxyeicosatetraenoic acid
FASEB	Federation of American Societies for Experimental Biology	HFC	Hydrofluorocarbon
		HGF	Hepatocyte growth factor
FCA	Freund's complete adjuvant	HGG	Human gamma globulin
FCC	Food Chemicals Codex (US)	HGH	Human growth hormone
FCS	Fetal calf serum	HHS	Health and Human Services (US)
FDA	Food and Drug Administration (US)	HiB	Hemophilus influenza B
FEBS	Federation of European Biochemical Societies	HIPAA	Health Insurance Portability and Accountability Act (US)
FEMS	Federation of European Microbiological Societies		
		HIV	Human immunodeficiency virus
FEV	Forced expiratory volume	HLA	Human leukocyte antigen
FFDCA	Federal Food, Drug, and Cosmetic Act (US)	HLB	Hydrophilic/Lipophilic balance
FGF	Fibroblast growth factor	HMO	Health maintenance organization
FIA	Freund's incomplete adjuvant	HOMO	Highest occupied molecular orbital
FIP	Fédération Internationale Pharmaceutique	HPA	Hypothalamic-pituitary-adrenal
FISH	Fluorescence *in situ* hybridization	HPB	Health Protection Branch (Canada)
FOI	Freedom of information	HPBL	Human peripheral blood lymphocyte
FP	Fusion protein	HPLC	High performance liquid chromatography
FPLC	Fast performance liquid chromatography		High pressure liquid chromatography
	Fast protein liquid chromatography	HPV	Human papillomavirus
FRET	Fluorescent resonance energy transfer	HRT	Hormone replacement therapy
FSH	Follicle-stimulating hormone	HSA	Human serum albumin
FT	Fourier transform	HSC	Hematopoietic stem cell
FTC	Federal trade commission (US)	HSP	Heat shock protein
FTIR	Fourier transform infrared (spectroscopy)	HSV	Herpes simplex virus
FWHM	Full width at half maximum	HTLV	Human T lymphotropic virus
FXR	Farnesoid X receptor	HUGO	Human Genome Organization
FXS	Fragile X syndrome	HUVEC	Human umbilical vein endothelial cell
GABA	Gamma amino butyric acid	IACR	International Association of Cancer Registries
GALT	Gut associated lymphoid tissue	IACUC	Institutional Animal Care and Use Committee (US)
GAP	Good agricultural practices		
	GTPase activating protein		

TABLE-9

IAEA	International Atomic Energy Agency	LC	Lethal concentration
IARC	International Agency for Research on Cancer		Liquid chromatography
IBD	Inflammatory bowel disease	LD	Lethal dose
IBS	Irritable bowel syndrome	LDH	Lactic dehydrogenase
IC	Inhibitory concentration	LDL	Low density lipoprotein
ICAM	Intercellular adhesion molecules	LED	Light-emitting diode
ICC	Interstate Commerce Commission (US)		Lowest effective dose
ICF	Intracellular fluid	LES	Lower esophageal sphincter
ICFA	Incomplete Freund's adjuvant	LFA	Leukocyte functional age
ICGEB	International Center for Genetic Engineering	LH	Luteinizing hormone
	and Biotechnology	LHRH	Luteinizing hormone releasing hormone
ICP	Intracranial pressure	LLC	Limited Liability Company
ICSI	Intracytoplasmic sperm injection	LM	Light microscopy
ID	Inhibitory dose	LOAEL	Lowest observed adverse effect level
	Intradermal	LOEL	Lowest observed effect level
IDDM	Insulin dependent diabetes mellitus	LOH	Loss of heterozygosity
IFN	Interferon	LPS	Lipopolysaccharide
Ig	Immunoglobulin	LT	Leukotriene
IGF	Insulin-like growth factor	LTR	Long terminal repeat
IGR	Insect growth regulator	LUMO	Lowest unoccupied molecular orbital
IHD	Ischemic heart disease	LVEDD	Left ventricular end-diastolic diameter
IL	Interleukin	LVEF	Left ventricular ejection fraction
	Ionic liquids	LXR	Liver X receptor
ILD	Interstitial lung diseases	mAb	Monoclonal antibody
IM	Intramuscular	MAC	Maximum allowable concentration
IMB	Irish Medicine Board	MALDI	Matrix-assisted laser desorption/ionization
IN	Intranasal	MALT	Mucosal-associated lymphoid tissue
INCI	International Nomenclature Cosmetic Ingredient	MAOI	Monoamine oxidase inhibitor
IND	Investigational New Drug (application) (US)	MAPK	Mitogen-activated protein kinase
INN	International Nonproprietary Name	MBP	Myelin basic protein
iNOS	Inducible nitric oxide synthase	MCHC	Mean corpuscular hemoglobin concentration
INRIA	Institut National de Recherche en Informatique	MCV	Mean corpuscular volume
	et en Automatique (Fr)	MD	Muscular dystrophy
IOBC	International Organization for Biological	MDD	Major depressive disorder
	Control	MDR	Multidrug resistant
IOTF	International Obesity Task Force	MEFV	Maximal expiratory flow volume
IP	Intraperitoneal	MEL	Maximum exposure limit
IPPB	Intermittent positive pressure breathing	MF	Molecular formula
IPT	Individual perception threshold	MGD	Mouse genome database
IPV	Inactivated polio vaccine	MHC	Major histocompatibility complex
IRB	Institutional review board	MHRA	Medicines and Healthcare products Regulatory
ISO	International Organization for Standardization		Agency (UK)
ISS	Ion scattering spectroscopy	MHW	Ministry of Health and Welfare (Japan)
ITP	Idiopathic thrombocytopenic purpura	MI	Myocardial infarction
IU	International Units	MLCT	Metal to ligand charge transfer
IUPAC	International Union of Pure and Applied	MLD	Minimum lethal dose
	Chemistry	MLR	Mixed lymphocyte reaction
IV	Intravenous	MLV	Modified live virus
IVF	*in vitro* fertilization	MMP	Matrix metalloproteinase
IVP	Intravenous pyelogram	MMR	Measles-mumps-rubella (vaccine)
JAN	Japanese Accepted Name	MPL	Maximum permissible level
JCIC	Japanese Cosmetic Ingredients Codex	MPS	Mucopolysaccharidosis
JCPA	Japan Crop Protection Association	MRI	Magnetic resonance imaging
JNK	c-Jun NH2-terminal kinase	MRL	Maximum residue level
JP	Japanese Pharmacopoeia	mRNA	Messenger ribonucleic acid
JPMA	Japan Pharmaceutical Manufacturers Association	MRP	Multidrug resistance protein
JSCI	Japanese Standards of Cosmetic Ingredients	MRSA	Methicillin-resistant *Staphylococcus aureus*
JSFA	Japanese Standards of Food Additives	MS	Mass spectrometry
KG	Kommanditgesellschaft		Multiple sclerosis
KGaA	Kommanditgesellschaft auf Aktien	MSDS	Material Safety Data Sheet
KK	Kabushiki Kaishi	MTC	Maximum tolerable concentration
KLH	Keyhole limpet hemocyanin	MTD	Maximum tolerated dose
KPA	Kidney plasminogen activator	MTEL	Maximum tolerated exposure level
LAB	Lactic acid bacteria	MW	Molecular weight
LAD	Lactic acid dehydrogenase	NABR	National Association for Biomedical Research
	Left anterior descending		(US)
	Lymphocyte-activating determinant	NACHGR	National Advisory Council for Human Genome
LAIV	Live attenuated influenza vaccine		Research (US)
LAK	Lymphokine activated killer (cell)	NAD	Nicotinamide adenine dinucleotide
LAMMA	Laser microprobe mass analysis	NADH	Nicotinamide adenine dinucleotide (reduced
LATS	Long acting thyroid stimulator		form)

TABLE-10

NADP	Nicotinamide adenine dinucleotide phosphate
NADPH	Nicotinamide adenine dinucleotide phosphate (reduced form)
NAL	National Agricultural Library (US)
NAS	National Academy of Sciences (US)
NASA	National Aeronautics and Space Administration (US)
NBS	National Bureau of Standards (US)
NCBI	National Center for Biotechnology Information (US)
NCE	New Chemical Entity
NCGR	National Center for Genome Resources (US)
NCHGR	National Center for Human Genome Research (US)
NCI	National Cancer Institute (US)
NCIC	National Cancer Institute of Canada
NCID	National Center for Infectious Diseases (US)
NCTC	National Collection of Type Cultures (UK)
NDA	New Drug Application (US)
NDO	Non-digestible oligosaccharides
NEP	Neutral endopeptidase
NER	Nucleotide excision repair
NF	National Formulary (US)
NFkB	Nuclear factor kappa B
NGF	Nerve growth factor
NHGRI	National Human Genome Research Institute (US)
NICE	National Institute for Health and Clinical Excellence (UK)
NIDDM	Non-insulin dependent diabetes mellitus
NIH	National Institutes of Health (US)
NIMH	National Institute of Mental Health (US)
NIOSH	National Institute for Occupational Safety and Health (US)
NIST	National Institute for Standards and Technology (US)
NK	Natural killer (cell)
NK	Neurokinin
NLM	National Library of Medicine (US)
NMDA	N-methyl-D-aspartate
NMP	Nuclear matrix protein
NMR	Nuclear magnetic resonance
NNRTI	Non-nucleoside reverse transcriptase inhibitor
NOAEL	No observed adverse effect level
NOE	Nuclear Overhauser effect
NOEL	No observed effect level
NOS	Nitric oxide synthase
NPA	National Pharmaceutical Association (UK)
	National Pharmacy Association (UK)
NPY	Neuropeptide Y
NRC	National Research Council (US)
	Nuclear Regulatory Commission (US)
NRTI	Nucleoside reverse transcriptase inhibitor
NSAID	Nonsteroidal anti-inflammatory drug
NSCLC	Non-small-cell lung cancer
NSF	National Science Foundation (US)
NTP	National Toxicology Program (US)
OA	Osteoarthritis
OC	Oral contraceptive
OCD	Obsessive compulsive disorder
OD	Optical density
OES	Occupational exposure standard
OGTT	Oral glucose tolerance test
OPPTS	Office of Prevention, Pesticides and Toxic Substances (US)
ORD	Optical rotary dispersion
ORDA	Office of Recombinant DNA Activities (US)
ORF	Open reading frame
OSD	Oral solid dosage
OSHA	Occupational Safety and Health Administration
OTC	Over the counter

PAF	Platelet activating factor
PAGE	Polyacrylamide gel electrophoresis
PAH	Poly-aromatic hydrocarbons
PAW	Pulmonary artery wedge
PBC	Peripheral blood cell
PBS	Phosphate buffered saline
PBSC	Peripheral blood stem cell
PCA	Passive cutaneous anaphylaxis
PCB	Polychlorinated biphenyls
PCP	Pneumocystis carinii pneumonia
	Primary care physician
	Primary care provider
PCR	Polymerase chain reaction
PCT	Patent Co-operation Treaty
PCV	Packed cell volume
PD	Pharmacodynamics
PDE	Phosphodiesterase
PDGF	Platelet-derived growth factor
PDR	Physicians Desk Reference
PDT	Photodynamic therapy
PE	Phosphatidylethanolamine
PEEP	Peak end expiratory pressure
PEF	Peak expiratory flow
PEG	Polyethylene glycol
PES	Photoelectron spectroscopy
PET	Positron emission tomography
PFC	Plaque-forming cell
PFGE	Pulsed-field gel electrophoresis
PG	Prostaglandin
PGES	Prostaglandin E synthase
PHA	Phytohemagglutinin
PhRMA	Pharmaceutical Research and Manufacturers of America (US)
PI	Protease inhibitor
PID	Pelvic inflammatory disease
PK	Pharmacokinetics
PKC	Protein kinase C
PKD	Polycystic kidney disease
PKU	Phenylketonuria
PMA	Premarket Approval (US)
PMN	Polymorphonuclear leukocyte
PMS	Post-marketing surveillance
	Premenstrual syndrome
PNS	Peripheral nervous system
PPAR	Peroxisome proliferator-activated receptor
PPV	Positive predictive value
PSA	Prostate specific antigen
PSD	Pesticides Safety Directorate (UK)
PSVT	Paroxysmal supraventricular tachycardia
PTC	Percutaneous transhepatic cholangiography
PTCA	Percutaneous transluminal coronary angioplasty
PTFE	Polytetrafluoroethylene
PTH	Parathyroid hormone
PTK	Protein tyrosine kinase
PTO	Patent and Trademark Office (US)
PTT	Partial thromboplastin time
PUFA	Polyunsaturated fatty acids
PVC	Polyvinyl chloride
PVDF	Polyvinylidene difluoride
PVR	Peripheral vascular resistance
PVT	Pressure-volume-temperature
QED	Quod erat demonstrandum (which was to be proved)
QSAR	Quantitative structure-activity relationship
RA	Rheumatoid arthritis
RAGE	Receptor of advanced glycation endproduct
RANTES	Regulated on activation, normal T expressed and secreted
RAR	Retinoic acid receptor
RAS	Renin-angiotensin system
RBC	Red blood cell

TABLE-11

RCC	Renal cell carcinoma	SRBC	Sheep red blood cell
RCT	Randomized clinical trial	Src	Sociedad Regular Colectiva
RCVS	Royal College of Veterinary Surgeons (UK)	Srl	Società a Responsabilità Limitata
RDA	Recommended daily allowance		Societate cu Raspoudere Limitata
RDI	Recommended dietary intake	SRM	Standard reference material
REM	Rapid eye movement	SSEA	Stage specific embryonic antigen
RFLP	Restriction fragment length polymorphism	SSRI	Selective serotonin reuptake inhibitor
RIA	Radioimmunoassay	STD	Sexually transmitted disease
RIND	Reversible ischemic neurological deficit	STEM	Scanning transmission electron microscopy
rINN	Recommended International Non-proprietary Name	STM	Scanning tunneling microscopy
		STP	Standard temperature and pressure
RMR	Resting metabolic rate	STS	Sequence-tagged site
RNAi	Ribonucleic acid interference	SV	Stroke volume
ROS	Reactive oxygen species	TB	Tuberculosis
RRS	Resonance raman spectroscopy	TCA	Tricarboxylic acid
RS	Raman spectroscopy		Tricyclic antidepressants
RSC	Royal Society of Chemistry (UK)	TCTP	Translationally controlled tumor protein
RSM	Royal Society of Medicine (UK)	TDL	Thoracic duct lymphocyte
RSV	Respiratory syncytical virus	TDN	Total digestible nutrients
	Rous sarcoma virus	TF	Transcription factor
RT	Room temperature	TGF	Transforming growth factor
RTECS	Registry of Toxic Effects of Chemical Substances (US)	TIA	Transient ischemic attack
		TK	Toxicokinetics
RTI	Reverse transcriptase inhibitor	TLC	Thin-layer chromatography
RXR	Retinoid X receptor	TLR	Toll-like receptor
SA	Sociedad Anónima	TLV	Threshold limit value
	Sociedade por Ações	TNF	Tumor necrosis factor
	Societate pe Actiuni	TPN	Total parenteral nutrition
	Société Anonyme	TRAIL	TNF related apoptosis inducing ligand
	Spolka Akcyjna	tRNA	Transfer ribonucleic acid
SAE	Society of Automotive Engineers	TSA	Tumor specific antigen
SAM	S-adenosyl methionine	TSG	Tumor suppressor gene
SApA	Societá in accomandita per azioni	TSH	Thyroid stimulating hormone
SAR	Structure activity relationship	TSTA	Tumor-specific transplantation antigens
SARS	Severe acute respiratory syndrome	TTP	Thrombotic thrombocytopenic purpura
SC	Subcutaneous	TUNEL	Terminal deoxynucleotidyl transferase (TdT)-mediated dUTP nick-end labeling
SCF	Supercritical fluid		
SCFA	Short chain fatty acids	TZD	Thiazolidinedione
SCID	Severe combined immune deficiency	UES	Upper esophageal sphincter
SCP	Single cell protein	UNEP	United Nations Environment Program
SD	Standard deviation	URI	Upper respiratory infection
SDH	Sorbitol dehydrogenase	USAN	United States Adopted Name
SE	Standard error	USDA	United States Department of Agriculture
SEC	Size-exclusion chromatography	USP	United States Pharmacopeia
SEM	Scanning electron microscopy	USPHS	United States Public Health Service
SERM	Selective estrogen-receptor modulator	USPTO	United States Patent and Trademark Office
SGGT	Serum gamma glutamyl transferase	UTI	Urinary tract infection
SGOT	Serum glutamic oxaloacetic transaminase	UTR	Untranslated region
SGPT	Serum glutamic-pyruvic transaminase	UV	Ultraviolet
SHIV	Simian–human immunodeficiency virus	VCAM	Vascular cell adhesion molecule
SIADH	Syndrome of inappropriate antidiuretic hormone (secretion)	VEGF	Vascular endothelial growth factor
		VLA	Very late antigens
SIB	Society for Industrial Biology	VLDL	Very low density lipoproteins
SIDS	Sudden infant death syndrome	VNTR	Variable number of tandem repeats
SIM	Society for Industrial Microbiology	VOC	Volatile organic compound
siRNA	Small interfering RNA	VPD	Ventricular premature depolarization
SIRS	Systemic inflammatory response syndrome	VSEPR	Valence shell electron pair repulsion
SIV	Simian immunodeficiency virus	VUV	Vacuum ultraviolet
SLE	Systemic lupus erythematosus	VZV	Varicella-zoster virus
SMART	Somatic mutation and recombination test	WBC	White blood cell
SNP	Single nucleotide polymorphism	WHO	World Health Organization
SNRI	Serotonin-noradrenaline reuptake inhibitor	WIPO	World Intellectual Property Organization
SOCMA	Synthetic Organic Chemical Manufacturers Association	WNL	Within normal limits
		WNV	West Nile virus
SOD	Superoxide dismutase	WSSA	Weed Science Society of America
SP	Substance P	WT	Wild type
SpA	Società per Azioni	WVA	World Veterinary Association
SPECT	Single photon emission computed tomography	XAFS	X-ray absorption fine structure spectroscopy
SPF	Sun protection factor	XPS	X-ray photoelectron spectroscopy
SPPARM	Selective PPAR modulator	YAC	Yeast artificial chromosome
SPRM	Selective progesterone receptor modulator	ZFP	Zinc finger protein

TABLE-12

Vaccines

The following table lists selected preparations for use in the prevention of infectious diseases by immunization.

Single vaccines

Trade Name	Product Name	Company	Description
ACAM2000®	Smallpox (Vaccinia) Vaccine, Live	Sanofi Pasteur	Live vaccinia virus derived from plaque purification cloning from Dryvax® vaccinia virus vaccine and grown in VERO cells
ActHIB®	Haemophilus b Conjugate Vaccine (Tetanus Toxoid Conjugate)	Sanofi Pasteur	Purified polyribose-ribitol-phosphate capsular polysaccharide of *Haemophilus influenzae* type b (strain 1482), covalently bound to tetanus toxoid
ACWY Vax®	Meningococcal Polysaccharide Vaccine (Groups A, C, Y, W_{135})	GlaxoSmithKline	Purified polysaccharide antigens from *Neisseria meningitidis* groups A, C, Y, and W_{135}
Afluria®	Influenza Virus Vaccine	CSL	Influenza virus types A and B grown in allantoic fluid of embryonated chicken eggs, inactivated by beta-propiolactone (split viron, trivalent)
Agrippal®	Influenza Vaccine (Surface Antigen, Inactivated)	Novartis	Purified surface antigen from influenza virus types A and B propagated in embryonated chicken eggs (trivalent, inactivated)
Agriflu®	Influenza Virus Vaccine	Novartis	Influenza virus types A and B grown in embryonated chicken eggs, inactivated by formaldehyde (trivalent, inactivated)
Arepanrix®	AS03-Adjuvanted H1N1 Pandemic Influenza Vaccine	GlaxoSmithKline	Immunizing antigen from influenza virus type A (H1N1) grown in embryonated chicken eggs, inactivated by ultraviolet light and formaldehyde (split virion); mixed with AS03 adjuvant
Attenuvax®	Measles Virus Vaccine, Live	Merck & Co.	Live, attenuated measles virus derived from Enders' attenuated Edmonston strain grown in chick embryo cell culture
Avaxim®	Hepatitis A Vaccine Inactivated	Sanofi Pasteur	Hepatitis A virus inactivated by formaldehyde, adsorbed onto aluminum hydroxide and propagated in human diploid MRC-5 cells
BCG Vaccine	BCG Vaccine U.S.P.	Organon	Live, attenuated TICE® BCG (Bacillus of Calmette and Guérin) strain of attenuated *Mycobacterium bovis*
BioThrax®	Anthrax Vaccine Adsorbed	Emergent BioDefense	Cell-free culture filtrate containing the 83 kDa protective antigen produced by an avirulent, nonencapsulated strain of *Bacillus anthracis*, adsorbed onto aluminum hydroxide
Cervarix®	Human Papillomavirus Bivalent (Types 16 and 18) Vaccine, Recombinant	GlaxoSmithKline	Purified virus-like particles of the major capsid protein (L1) of human papillomavirus types 16 and 18 produced by recombinant DNA technology in *Trichoplusia ni*, adsorbed onto aluminum hydroxide and adjuvanted with AS04 containing 3-*O*-desacyl-4′-monophosphoryl lipid A (MPL)
Dukoral®	Oral, Inactivated Travellers' Diarrhea and Cholera Vaccine	Sanofi Pasteur	Heat or formalin inactivated *Vibrio cholerae* serogroup O1 (Inaba and Ogawa serotypes; classical and El Tor biotypes) and recombinant cholera toxin B subunit
Encepur®	Tick-borne Encephalitis Virus Vaccine, Inactivated, Adsorbed	Novartis	Tick-borne encephalitis virus (strain K23) grown in chicken fibroblast cell culture, inactivated with formaldehyde, adsorbed onto aluminum hydroxide (adjuvant)
Engerix-B®	Hepatitis B Vaccine (Recombinant)	GlaxoSmithKline	Purified hepatitis B surface antigen produced by recombinant DNA technology in *Saccharomyces cerevisiae*, adsorbed onto aluminum hydroxide

TABLE-13

Single vaccines (continued)

Trade Name	Product Name	Company	Description
Epaxal®	Hepatitis A Vaccine (Inactivated, Virosome)	Crucell	Hepatitis A virus (RG-SB strain) propagated in human diploid MRC-5 cells, inactivated by formaldehyde, and bound to lipid/protein virosomes
Ervevax®	Live Attenuated Rubella Vaccine (RA27/3 strain)	GlaxoSmithKline	Live, attenuated RA 27/3 strain of rubella virus propagated in human diploid MRC-5 cells
Euvax B®	Hepatitis B Vaccine, Recombinant	Sanofi Pasteur	Purified hepatitis B surface antigen produced by recombinant DNA technology in *Saccharomyces cerevisiae*, adsorbed onto aluminum hydroxide
Fendrix®	Hepatitis B (rDNA) Vaccine (Adjuvanted, Adsorbed)	GlaxoSmithKline	Purified hepatitis B surface antigen produced by recombinant DNA technology in *Saccharomyces cerevisiae,* adsorbed on aluminum phosphate, and adjuvanted by AS04C containing 3-*O*-desacyl-4′-monophosphoryl lipid A (MPL)
Fluad®	Influenza Vaccine, Surface Antigen, Inactivated, Adjuvanted with MF59C.1	Novartis	Subunit antigens (hemagglutinin and neuraminidase) from three strains of influenza virus types A and B grown in embryonic chicken eggs, combined with MF59C.1 adjuvant emulsion
Fluarix®	Influenza Virus Vaccine	GlaxoSmithKline	Influenza virus types A and B propagated in embryonated chicken eggs, inactivated by sodium deoxycholate and formaldehyde (split virion, trivalent)
FluLaval®	Influenza Virus Vaccine	ID Biomed. Corp. of Quebec	Influenza virus types A and B propagated in embryonated chicken eggs, inactivated by ultraviolet light and formaldehyde, and disrupted with sodium deoxycholate (split virion, trivalent)
FluMist®	Influenza Vaccine, Live, Intranasal	MedImmune	Live, attenuated influenza virus reassortants (types A and B) grown in specific pathogen-free eggs (trivalent)
Fluvirin®	Influenza Virus Vaccine	Novartis	Influenza virus types A and B grown in embryonated chicken eggs, inactivated by beta-propiolactone and nonylphenol ethoxylate (split virus, trivalent)
Fluzone®	Influenza Virus Vaccine	Sanofi Pasteur	Influenza virus types A and B grown in embryonated chicken eggs, inactivated by formaldehyde, and disrupted with octylphenol ethoxylate (split virus, trivalent)
FSME-Immun®	Tick-borne Encephalitis Virus Vaccine (Inactivated)	Baxter	Inactivated tick-borne encephalitis virus (Neudörfl strain) grown in chick embryo fibroblast cells, adsorbed on aluminum hydroxide
Gardasil®	Human Papillomavirus Quadrivalent (Types 6, 11, 16, 18) Vaccine, Recombinant	Merck & Co.	Purified virus-like particles of the major capsid (L1) protein of human papillomavirus types 6, 11, 16 and 18 produced by recombinant DNA technology in *Saccharomyces cerevisiae*, adsorbed onto aluminum-containing adjuvant
GenHevac B®	Recombinant Hepatitis B Vaccine	Sanofi Pasteur	Recombinant hepatitis B virus surface antigen (S and pre-S_2 proteins) produced on CHO cell line
Havrix®	Hepatitis A Vaccine	GlaxoSmithKline	Purified lysate of hepatitis A virus strain HM175 grown in human diploid MRC-5 cells, inactivated by formalin, adsorbed on aluminum hydroxide
Hiberix®	Haemophilus b Conjugate Vaccine (Tetanus Toxoid Conjugate)	GlaxoSmithKline	Purified polyribosyl-ribitol-phosphate capsular polysaccharide of *Haemophilus influenzae* b (Hib) strain 20,752, covalently bound to tetanus toxoid

TABLE-14

Single vaccines (continued)

Trade Name	Product Name	Company	Description
HibTiter®	Haemophilus b Conjugate Vaccine (Diphtheria CRM$_{197}$ Protein Conjugate)	Nuron	Conjugate of oligosaccharides of the capsular antigen of *Haemophilus influenzae* type b and diphtheria CRM$_{197}$ protein
Imovax® Rabies	Rabies Vaccine	Sanofi Pasteur	Wistar rabies virus strain PM-1503-3M grown in human diploid MRC-5 cells, purified and inactivated by beta-propiolactone
Inflexal® V	Influenza Vaccine (Surface Antigen, Inactivated, Virosome)	Crucell	Purified surface antigens from 3 strains of influenza virus types A and B, propagated in embryonated chicken eggs, inactivated with beta-propiolactone, and bound to lipid virosomes
Influvac®	Influenza Vaccine (Surface Antigen, Inactivated)	Abbott	Purifed surface antigens from 3 strains of influenza virus types A and B, grown in embryonated chicken eggs
IPOL®	Poliovirus Vaccine Inactivated	Sanofi Pasteur	Purified, formalin-inactivated poliovirus type 1 (Mahoney), type 2 (MEF-1) and type 3 (Saukett) grown in VERO cell culture
Ixiaro®	Japanese Encephalitis Vaccine, Inactivated, Adsorbed	Intercell	Japanese encephalitis virus (strain SA$_{14}$-14-2) propagated in VERO cell culture, inactivated by formaldehyde, adsorbed on aluminum hydroxide
JE-Vax®	Japanese Encephalitis Virus Vaccine Inactivated	Sanofi Pasteur	Partially purified, formaldehyde-inactivated Japanese encephalitis virus (Nakayama-NIH strain), grown in mice
Menactra®	Meningococcal (Groups A, C, Y, W-135) Polysaccharide Diphtheria Toxoid Conjugate Vaccine	Sanofi Pasteur	*Neisseria meningitidis* serogroup A, C, Y, and W-135 capsular polysaccharide antigens individually conjugated to diphtheria toxoid protein
Meningitec®	Meningococcal Serogroup C Oligosaccharide Conjugate Vaccine (Adsorbed)	Pfizer	Capsular polysaccharide antigen of *Neisseria meningitidis* group C (strain C11) conjugated to *Corynebacterium diphtheria* protein CRM$_{197}$, adsorbed onto aluminum phosphate
Menjugate®	Meningococcal Group C–CRM197 Conjugate Vaccine	Novartis	Meningococcal group C oligosaccharides conjugated to *Corynebacterium diphtheriae* CRM$_{197}$ protein, adsorbed onto aluminum hydroxide
Menomune®	Meningococcal Polysaccharide Vaccine, Groups A, C, Y, W-135 Combined	Sanofi Pasteur	Purified polysaccharide antigens from *Neisseria meningitidis* groups A, C, Y, and W-135
Meruvax® II	Rubella Virus Vaccine Live	Merck & Co.	Wistar RA 27/3 strain of live, attenuated rubella virus propagated in WI-38 human diploid lung fibroblasts
Menveo®	Meningococcal (Groups A, C, Y, and W-135) Oligosaccharide Diphtheria CRM$_{197}$ Conjugate Vaccine	Novartis	*Neisseria meningitides* groups A, C, Y, and W-135 oligosaccharides conjugated individually to *Corynebacterium diphtheriae* CRM$_{197}$ protein
Mumpsvax®	Mumps Virus Vaccine Live	Merck & Co.	Jeryl Lynn™ (B level) strain of mumps virus propagated in chick embryo cell culture
NeisVac-C®	Meningococcal Serogroup C Polysaccharide Conjugate Vaccine	Baxter	*Neisseria meningitidis* serogroup C (strain C11) polysaccharide conjugated to tetanus toxoid, adsorbed onto aluminum hydroxide
Okavax®	Live Attenuated Varicella Virus Vaccine	Sanofi Pasteur	Live, attenuated varicella zoster virus (Oka strain) grown in human diploid cells
Pandemrix®	Influenza Vaccine (H1N1)v (Split Virion, Inactivated, Adjuvanted)	GlaxoSmithKline	Influenza virus type A (H1N1) grown in embryonated chicken eggs, inactivated (split virion); contains AS03 adjuvant
PedvaxHIB®	Haemophilus b Conjugate Vaccine (Meningococcal Protein Conjugate)	Merck & Co.	Purified capsular polysaccharide (polyribosylribitol phosphate) of *Haemophilus influenzae* type b (Ross strain), covalently bound to an outer membrane protein complex of *Neisseria meningitidis* serogroup B (B11 strain)

TABLE-15

Vaccines (Continued)

Trade Name	Product Name	Company	Description
Pneumo 23®	Pneumococcal Polysaccharide Vaccine	Sanofi Pasteur	Purified capsular polysaccharides of *Streptococcus pneumoniae* serotypes 1, 2, 3, 4, 5, 6B, 7F, 8, 9N, 9V, 10A, 11A, 12F, 14, 15B, 17F, 18C, 19A, 19F, 20, 22F, 23F, 33F
Pneumovax® 23	Pneumococcal Vaccine Polyvalent	Merck & Co.	Mixture of highly purified capsular polysaccharides from *Streptococcus pneumoniae* serotypes 1, 2, 3, 4, 5, 6B, 7F, 8, 9N, 9V, 10A, 11A, 12F, 14, 15B, 17F, 18C, 19A, 19F, 20, 22F, 23F, 33F
Polio Sabin™	Live Attenuated Poliomyelitis (Sabin) Vaccine (Oral)	GlaxoSmithKline	Mixture of poliovirus type 1 (LS-c, 2ab), type 2 (P712, Ch, 2ab), and type 3 (Leon 12a1b) live attenuated Sabin strains of poliomyelitis virus grown in VERO cell culture
Prevnar® 13 Prevenar® 13	Pneumococcal 13-valent Conjugate Vaccine (Diphtheria CRM$_{197}$ Protein)	Wyeth	Saccharides of the capsular antigens of *Streptococcus pneumoniae* serotypes 1, 3, 4, 5, 6A, 6B, 7F, 9V, 14, 18C, 19A, 19F, and 23F individually conjugated to diphtheria CRM$_{197}$ protein
Pumarix®	Pandemic Influenza Vaccine (H5N1) (Split Virion, Inactivated, Adjuvanted)	GlaxoSmithKline	Influenza virus type A (H5N1) grown in chicken eggs, inactivated by formaldehyde and sodium deoxycholate (split virion); contains AS03 adjuvant
RabAvert®	Rabies Vaccine	Novartis	Rabies virus (Flurry LEP strain) grown in chicken fibroblast cultures, inactivated with beta-propiolactone, purified
RabiPur®	Rabies Vaccine	Novartis	Rabies virus (Flurry LEP strain) grown in chicken embryo cells, inactivated, purified
Recombivax HB®	Hepatitis B Vaccine (Recombinant)	Merck & Co.	Purified hepatitis B surface antigen produced by recombinant DNA technology in *Saccharomyces cerevisiae*
Rotarix®	Rotavirus Vaccine, Live, Oral	GlaxoSmithKline	Live, attenuated rotavirus derived from human 89-12 strain (G1P[8] type) propagated in VERO cell culture
RotaTeq®	Rotavirus Vaccine, Live, Oral, Pentavalent	Merck & Co.	Mixture of 5 live reassortant rotoviruses: four expressing attachment protein P7 and one of capsid proteins G1, G2, G3 or G4 and one expressing attachment protein P1A and capsid protein G6; propagated in VERO cell culture
Rouvax®	Live Hyperattenuated Virus Vaccine Against Measles	Sanofi Pasteur	Live, attenuated measles virus (Schwarz strain) grown in chick embryo cell culture
Rudivax®	Attenuated Rubella Vaccine	Sanofi Pasteur	Live, attenuated rubella virus (Wistar RA 27/3M strain) grown in human diploid cells
Stamaril®	Yellow Fever Vaccine (Live)	Sanofi Pasteur	Live, attenuated yellow fever virus (17 D-204 strain) grown in specific pathogen free chick embryos
Synflorix®	Pneumococcal Polysaccharide Conjugate Vaccine (Adsorbed)	GlaxoSmithKline	Pneumococcal polysaccharide serotypes 1, 4, 5, 6B, 7F, 9V, 14, 23F conjugated to *Hemophilus influenzae* protein D, 18C conjugated to tetanus toxoid carrier protein, and 19F conjugated to diphtheria toxoid carrier protein; adsorbed onto aluminum phosphate
Tetavax®	Adsorbed Tetanus Toxoid Vaccine	Sanofi Pasteur	Tetanus toxoid adsorbed on aluminum hydroxide dihydrate
Typherix®	Typhoid Polysaccharide Vaccine	GlaxoSmithKline	Purified Vi polysaccharide of *Salmonella enterica* serovar Typhi (Ty2 strain)
Typhim Vi®	Typhoid Vi Polysaccharide Vaccine	Sanofi Pasteur	Purified Vi polysaccharide of *Salmonella enterica* serovar Typhi (Ty2 strain)

TABLE-16

Single vaccines (continued)

Trade Name	Product Name	Company	Description
Vaqta®	Hepatitis A Vaccine, Inactivated	Merck & Co.	Purified Hepatitis A virus grown in human diploid MRC-5 cells, inactivated by formalin and adsorbed onto aluminum hydroxyphosphate sulfate
Varilrix®	Varicella Vaccine, Live Attenuated	GlaxoSmithKline	Live, attenuated varicella-zoster (Oka strain) virus propagated in human diploid MRC-5 cells
Varivax®	Varicella Virus Vaccine Live	Merck & Co.	Live, attenuated varicella virus (Oka/Merck strain) propagated in human diploid MRC-5 cells
Vaxigrip®	Inactivated Influenza Vaccine Trivalent Types A and B (Split Virion)	Sanofi Pasteur	Inactivated influenza virus types A and B, propagated in embryonated eggs, split by Triton® X-100, inactivated by formaldehyde (trivalent, split virion)
Verorab®	Purified Inactivated Rabies Vaccine	Sanofi Pasteur	Purified rabies virus (Wistar strain PM/WI 38-1503-3M) produced in VERO cell culture, inactivated with beta-propiolactone
Vivotif®	Typhoid Vaccine Live Oral Ty21a	Crucell	Live, attenuated *Salmonella enterica* serovar Typhi (Ty21a strain)
YF-Vax®	Yellow Fever Vaccine	Sanofi Pasteur	Live, attenuated yellow fever virus (17D-204 strain) grown in ALV-free chick embryos
Zostavax®	Zoster Vaccine Live	Merck & Co.	Live, attenuated varicella-zoster virus (Oka/Merck strain) propagated in human diploid MRC-5 cells

Combination vaccines

Trade Name	Product Name	Company	Description
Adacel®	Tetanus Toxoid, Reduced Diphtheria Toxoid and Acellular Pertussis Vaccine Adsorbed	Sanofi Pasteur	Tetanus toxoid, diphtheria toxoid, and 5 pertussis antigens (detoxified pertussis toxin, filamentous hemagglutinin, pertactin, fimbriae types 2 and 3) individually adsorbed onto aluminum phosphate
Ambirix®	Hepatitis A (inactivated) and Hepatitis B (rDNA) Vaccine (Adsorbed)	GlaxoSmithKline	Hepatitis A virus grown in human diploid MRC-5 cells, inactivated, and adsorbed onto aluminum hydroxide; plus hepatitis B surface antigen produced by recombinant DNA technology in *Saccharomyces cerevisiae*, adsorbed on aluminum phosphate
Boostrix®	Tetanus Toxoid, Reduced Diphtheria Toxoid and Acellular Pertussis Vaccine, Adsorbed	GlaxoSmithKline	Diphtheria toxoid, tetanus toxoid, and 3 pertussis antigens (inactivated pertussis toxin, filamentous hemagglutinin, and pertactin) adsorbed onto aluminum hydroxide
Comvax®	Haemophilus b Conjugate (Meningococcal Protein Conjugate) and Hepatitis B (Recombinant) Vaccine	Merck & Co.	*Haemophilus* influenzae type b capsular polysaccharide (polyribosylribitol phosphate) covalently bound to an outer membrane protein complex of *Neisseria meningitidis* serogroup B; plus hepatitis B surface antigen produced by recombinant DNA technology in *Saccharomyces cerevisiae*; adjuvanted with aluminum hydroxyphosphate sulfate
Daptacel®	Diphtheria and Tetanus Toxoids and Acellular Pertussis Vaccine Adsorbed	Sanofi Pasteur	Tetanus toxoid, diphtheria toxoid, and 5 pertussis antigens (detoxified pertussis toxin, filamentous hemagglutinin, pertactin, fimbriae types 2 and 3) individually adsorbed onto aluminum phosphate
Decavac®	Tetanus and Diphtheria Toxoids Adsorbed	Sanofi Pasteur	Alum-precipitated tetanus and diphtheria toxoids (tetanus and diphtheria toxins detoxified by formaldehyde)
Dultavax®	Diphtheria, Tetanus and Poliomyelitis (Inactivated) Vaccine (Adsorbed)	Sanofi Pasteur	Diphtheria toxoid, tetanus toxoid, and inactivated poliovirus types 1, 2, and 3; adsorbed on aluminum hydroxide

TABLE-17

Vaccines (Continued)

Combination vaccines (continued)

Trade Name	Product Name	Company	Description
Hepatyrix®	Hepatitis A (Inactivated) and Ty-phoid Polysaccharide Vaccine (Adsorbed)	GlaxoSmithKline	Hepatitis A virus (HM175 strain) grown in MRC-5 cells, inactivated; plus Vi polysac-charide of *Salmonella typhi* (Ty2 strain); ad-sorbed on aluminum hydroxide
Imovax® dT	Diphtheria and Tetanus Vaccine (Adsorbed)	Sanofi Pasteur	Diphtheria toxoid and tetanus toxoid adsorbed on aluminum hydroxide dihydrate
Infanrix®	Diphtheria and Tetanus Toxoids and Acellular Pertussis Vac-cine Adsorbed	GlaxoSmithKline	Diphtheria toxoid, tetanus toxoid, and 3 pertus-sis antigens (inactivated pertussis toxin, fila-mentous hemagglutinin, and pertactin); indi-vidually adsorbed onto aluminum hydroxide
Infanrix® IPV	Diphtheria, Tetanus, Pertussis (Acelluar, Component) and Poliomyelitis (Inactivated) Vaccine (Adsorbed)	GlaxoSmithKline	Diphtheria and tetanus toxoids and 3 pertussis antigens (inactivated pertussis toxin, fila-mentous hemagglutinin, and pertactin) ad-sorbed onto aluminum hydroxide; plus in-activated poliovirus types 1 (Mahoney), 2 (MEF-1), and 3 (Saukett) grown in VERO cell culture
Kinrix®	Diphtheria and Tetanus Toxoids and Acellular Pertussis Ad-sorbed and Inactivated Polio-virus Vaccine	GlaxoSmithKline	Diphtheria and tetanus toxoids and 3 pertussis antigens (inactivated pertussis toxin, fila-mentous hemagglutinin, and pertactin) ad-sorbed onto aluminum hydroxide; plus in-activated poliovirus types 1 (Mahoney), 2 (MEF-1), and 3 (Saukett) grown in VERO cell culture
Menitorix®	Haemophilus Type b and Menin-gococcal Group C Conjugate Vaccine	GlaxoSmithKline	*Haemophilus* type b and *Neisseria meningitides* group c polysaccharides conjugated to tetanus toxoid as a carrier protein
M-M-R® II	Measles, Mumps, and Rubella Virus Vaccine Live	Merck & Co.	Live, attenuated measles virus (Enders' attenuated Edmonston strain) propagated in chick embryo cell culture; plus live mumps virus (Jeryl Lynn™ (B level) strain) propagated in chick embryo cell culture; plus live, attenuated rubella virus (Wistar RA 27/3 strain) propagated in WI-38 human diploid lung fibroblasts
M-M Vax™	Measles and Mumps Virus Vac-cine Live	Merck & Co.	Live, attenuated measles virus (Enders' attenu-ated Edmonston strain) propagated in chick embryo cell culture; plus live mumps virus (Jeryl Lynn™ (B level) strain) propagated in chick embryo cell culture
Pediacel®	Diphtheria, Tetanus, 5 Compo-nent Acellular Pertussis, Inac-tivated Poliomyelitis, and Haemophilus Influenza Type B Conjugate Vaccine (Ad-sorbed)	Sanofi Pasteur	Tetanus and diphtheria toxoids, 5 pertussis an-tigens (pertussis toxoid, filamentous hemag-glutinin, pertactin, fimbriae types 2 and 3), plus inactivated poliovirus types 1 (Maho-ney), 2 (MEF-1), and 3 (Saukett) grown in VERO cell culture; plus *Haemophilus in-fluenzae* type b polysaccharide (polyribosyl-ribitol phosphate) conjugated to tetanus tox-oid, adsorbed onto aluminum phosphate
Pediarix®	Diphtheria and Tetanus Toxoids and Acellular Pertussis Ad-sorbed, Hepatitis B (Recombi-nant) and Inactivated Poliovi-rus Vaccine	GlaxoSmithKline	Diphtheria toxoid, tetanus toxoid, and 3 pertus-sis antigens (inactivated pertussis toxin, fila-mentous hemagglutinin, and pertactin) indi-vidually adsorbed onto aluminum hydroxide; plus purified hepatitis B surface antigen pro-duced by recombinant DNA technology in *Saccharomyces cerevisiae*, adsorbed onto aluminum phosphate; plus poliovirus Type 1 (Mahoney), Type 2 (MEF-1), and Type 3 (Saukett) grown in VERO cell culture and in-activated by formaldehyde

TABLE-18

Combination vaccines (continued)

Trade Name	Product Name	Company	Description
Pentacel®	Diphtheria and Tetanus Toxoids and Acellular Pertussis, Inactivated Poliovirus and Haemophilus b Conjugate (Tetanus Toxoid Conjugate) Vaccine	Sanofi Pasteur	Tetanus and diphtheria toxoids, 5 pertussis antigens (pertussis toxoid, filamentous hemagglutinin, pertactin, fimbriae types 2 and 3); plus inactivated poliovirus types 1 (Mahoney), 2 (MEF-1), and 3 (Saukett) grown in VERO cell culture; plus *Haemophilus influenzae* type b polysaccharide (polyribosylribitol phosphate) conjugated to tetanus toxoid; adjuvanted with aluminum phosphate
Pentavac® Pentaxim®	Adsorbed Diphtheria, Tetanus, Acellular Pertussis, Inactivated Poliomyelitis Vaccine and Conjugate *Haemophilus influenzae* type b Vaccine	Sanofi Pasteur	Diphtheria toxoid, tetanus toxoid, pertussis antigens (toxoid, filamentous hemagglutinin), inactivated poliomyelitis virus types 1, 2, and 3; plus Hib polysaccharide conjugated with tetanus protein
Priorix®	Combined Measles, Mumps, and Rubella Vaccine	GlaxoSmithKline	Live, attenuated measles virus (Schwarz strain), mumps virus (RIT 4385 strain) and rubella virus (Wistar RA 27/3 strain)
ProQuad®	Measles, Mumps, Rubella and Varicella Virus Vaccine Live	Merck & Co.	Live, attenuated measles virus (Enders' attenuated Edmonston strain) grown in chick embryo cell culture; mumps virus (Jeryl Lynn™) (B level) strain grown in chick embryo cell culture; rubella virus (Wistar RA 27/3 strain) grown in WI-38 human diploid lung fibroblasts; and varicella virus (Oka/Merck strain) grown in human diploid MRC-5 cells
Quadracel®	Diphtheria and Tetanus Toxoids and Acellular Pertussis Vaccine Adsorbed Combined with Inactivated Poliomyelitis Vaccine	Sanofi Pasteur	Tetanus and diphtheria toxoids, 5 pertussis antigens (pertussis toxoid, filamentous hemagglutinin, pertactin, fimbriae types 2 and 3); plus inactivated poliovirus types 1 (Mahoney), 2 (MEF-1), and 3 (Saukett) grown in VERO cell culture
Revaxis®	Diphtheria, Tetanus and Poliomyelitis (Inactivated) Vaccine (Adsorbed, Reduced Antigen Content)	Sanofi Pasteur	Diphtheria and tetanus toxoids; plus inactivated poliovirus types 1, 2, and 3; adsorbed on aluminum hydroxide
TriHIBit®	Haemophilus b Conjugate Vaccine (Tetanus Toxoid Conjugate) Reconstituted with Diphtheria and Tetanus Toxoids and Acellular Pertussis Vaccine Adsorbed	Sanofi Pasteur	Alum adsorbed diphtheria and tetanus toxoids combined with acellular pertussis concentrate (inactivated pertussis toxin and filamentous hemagglutinin); plus purified polyribose-ribitol-phosphate capsular polysaccharide of *Haemophilus influenzae* type b (strain 1482), covalently bound to tetanus toxoid
Trimovax®	Live Attenuated Virus Vaccine against Measles (Schwarz Stain), Mumps (Urabe AM-9 Strain) and Rubella (Wistar RA 27/3M Stain)	Sanofi Pasteur	Live, attenuated measles virus (Schwarz strain) grown in chick embryo cell culture, mumps virus (Urabe AM-9 strain) grown in embryonated hens eggs, and rubella virus (Wistar RA 27/3M strain) grown in human diploid cells
Tripedia®	Diphtheria and Tetanus Toxoids and Acellular Pertussis Vaccine Adsorbed	Sanofi Pasteur	Alum adsorbed diphtheria and tetanus toxoids combined with acellular pertussis concentrate (inactivated pertussis toxin and filamentous hemagglutinin)
Twinrix®	Hepatitis A and Hepatitis B (Recombinant) Vaccine	GlaxoSmithKline	Purified lysate of hepatitis A virus strain HM175 grown in human diploid MRC-5 cells, inactivated by formalin; plus purified hepatitis B surface antigen produced by recombinant DNA technology in *Saccharomyces cerevisiae*; individually adsorbed onto aluminum hydroxide

TABLE-19

Terms for Radicals and Groups Used for Nonproprietary Names

Listed by the United States Adopted Names (USAN) Council and the
World Health Organization (WHO) International Nonproprietary Names (INN) Programme

Name	Chemical Name	Molecular Formula	Graphic Formula
acefurate	acetate (ester) and furan-2-carboxylate (ester)	$C_2H_3O_2$ and $C_5H_3O_3$	
aceglumate	hydrogen *rac-N*-acetylgluta-mate	$C_7H_{10}NO_5$	and enantiomer
aceponate	acetate (ester) and propanoate (ester)	$C_2H_3O_2$ and $C_3H_5O_2$	
acetonide	isopropylidenedioxy [or propane-2,2-diylbis(oxy)]	$C_3H_6O_2$	
aceturate	*N*-acetylglycinate	$C_4H_6NO_3$	
acibutate	acetate (ester) and 2-methyl-propanoate (ester)	$C_2H_3O_2$ and $C_4H_7O_2$	
acistrate	acetate (ester) and octadeca-noate (salt)	$C_2H_3O_2$ and $C_{18}H_{35}O_2$	
acoxil	(acetyloxy)methyl	$C_3H_5O_2$	
alanetil	[(S)-1-ethoxy-1-oxopropan-2-yl]amino	$C_5H_{10}NO_2$	
alapivoxil	L-alanyl and [(2,2-dimethyl-propanoyl)oxy]methyl	C_3H_6NO and $C_6H_{11}O_2$	
alfoscerate	(2R)-2,3-dihydroxypropyl hy-drogen phosphate	$C_3H_8O_6P$	

TABLE-20

Terms for Radicals and Groups Used for Nonproprietary Names (Continued)

Name	Chemical Name	Molecular Formula	Graphic Formula
alideximer	poly[[oxy(2-hydroxyethane-1,1-diyl)]][oxy[1-(hydroxy-methyl)ethane-1,2-diyl]] partly *O*-etherified with carboxymethyl groups with some carboxy groups amide linked to the tetrapeptide residue (glycyl-glycyl-L-phenyl-alanylglycyl)		R = −H −CH$_2$COOH −CH$_2$CO-Gly-Gly-Phe-Gly
amsonate	2,2′-ethene-1,2-diylbis(5-aminobenzene-1-sulfonate)	$C_{14}H_{12}N_2O_6S_2$	
anisatil	2-(4-methoxyphenyl)-2-oxo-ethyl	$C_9H_9O_2$	
arbamel	2-(dimethylamino)-2-oxoethyl	C_4H_8NO	
axetil	*rac*-1-(acetyloxy)ethyl	$C_4H_7O_2$	and enantiomer
beloxil	benzyloxy	C_7H_7O	
benetonide	*rac*-*N*-benzoyl-2-methyl-β-alanine (ester) and propane-2,2-diyl-bis(oxy)	$C_{11}H_{12}NO_3$ and $C_3H_6O_2$	and enantiomer
besilate, besylate	benzenesulfonate	$C_6H_5O_3S$	
betadex	β-cyclodextrin	$C_{42}H_{70}O_{35}$	

TABLE-21

Name	Chemical Name	Molecular Formula	Graphic Formula
bezomil	(benzoyloxy)methyl	$C_8H_7O_2$	
buciclate	*trans*-4-butylcyclohexanecarboxylate	$C_{11}H_{19}O_2$	
bunapsilate	3,7-di-*tert*-butylnaphthalene-1,5-disulfonate	$C_{18}H_{22}O_6S_2$	
buteprate, probutate	butanoate (ester) and propanoate (ester) [or propionate (ester) and butyrate (ester)]	$C_4H_7O_2$ and $C_3H_5O_2$	
camsilate, camsylate	*rac*-(7,7-dimethyl-2-oxobicyclo[2.2.1]-heptan-1-yl)-methanesulfonate	$C_{10}H_{15}O_4S$	and enantiomer
caproate	hexanoate	$C_6H_{11}O_2$	
carbesilate	4-sulfobenzoate	$C_7H_5O_5S$	
ceribate	*rac*-2,3-dihydroxypropyl carbonate (ester)	$C_4H_7O_5$	and enantiomer
ciclotate, cyclotate	4-methylbicyclo[2.2.2]oct-2-ene-1-carboxylate	$C_{10}H_{13}O_2$	
cilexetil	*rac*-1-[[(cyclohexyloxy)carbonyl]oxy]ethyl	$C_9H_{15}O_3$	and enantiomer
cipecilate	cyclohexanecarboxylate (ester) and cyclopropanecarboxylate (ester)	$C_7H_{11}O_2$ and $C_4H_5O_2$	
cipionate, cypionate	cyclopentane propionoate [or 3-cyclopentylpropanoate]	$C_8H_{13}O_2$	

TABLE-22

Name	Chemical Name	Molecular Formula	Graphic Formula
citrate	2-hydroxypropane-1,2,3-tri-carboxylate	$C_6H_5O_7$	
cituxetan	*rac-N*-(4-[2-[bis(carboxy-methyl)amino]-3-[[2-[bis-(carboxymethyl)amino]eth-yl](carboxymethyl)amino-propyl]phenyl]thiocarbam-oyl	$C_{22}H_{29}N_4O_{10}S$	
clofibrol	2-(4-chlorophenoxy)-2-meth-ylpropyl	$C_{10}H_{12}ClO$	
closilate, closylate	4-chlorobenzene-1-sulfonate	$C_6H_4ClO_3S$	
crobefate	*rac*-(3*E*)-3-[(4-methoxyphen-yl)methylidene]-2-(4-meth-oxyphenyl)-4-oxochroman-6-yl phosphate(2−)	$C_{24}H_{19}O_8P$	
cromacate	2-[(6-hydroxy-4-methyl-2-oxo-2*H*-chromen-7-yl)-oxy]acetate	$C_{12}H_9O_6$	
cromesilate	(6,7-dihydroxy-2-oxo-2*H*-chromen-4-yl)methanesul-fonate	$C_{10}H_7O_7S$	
crosfumaril	(2*E*)-but-2-enedioyl	$C_4H_2O_2$	
cyclamate	cyclohexylsulfamate	$C_6H_{12}NO_3S$	
cyclotate	*See* ciclotate		
daloxate	L-alaninate (ester) and (5-methyl-2-oxo-1,3-dioxol-4-yl)methyl	$C_3H_6NO_2$ and $C_5H_5O_3$	

TABLE-23

Name	Chemical Name	Molecular Formula	Graphic Formula
dapropate, daropate	*N,N*-dimethyl-*β*-alaninate [or 3-(dimethylamino)pro-panoate]	$C_5H_{10}NO_2$	
deanil	2-(dimethylamino)ethyl	$C_4H_{10}N$	
detemir	tetradecanoyl	$C_{14}H_{27}O$	
dibudinate	2,6-di-*tert*-butylnaphthalene-1,5-disulfonate	$C_{18}H_{22}O_6S_2$	
dibunate	2,6-di-*tert*-butylnaphthalene-1-sulfonate	$C_{18}H_{23}O_3S$	
dicibate	dicyclohexylmethyl carbonate	$C_{14}H_{23}O_3$	
dicloacetate	2,2′-dichloroacetate (ester)	$C_2HCl_2O_2$	
digolil	2-(2-hydroxyethoxy)ethyl	$C_4H_9O_2$	
diolamine	diethanolamine [or 2,2′-azanediyldiethanol]	$C_4H_{11}NO_2$	
dofosfate	octadecyl hydrogen phosphate	$C_{18}H_{38}O_4P$	
ecamate	*N*-ethylcarbamate (ester)	$C_3H_6NO_2$	
edamine	ethane-1,2-diamine [or ethylenediamine]	$C_2H_8N_2$	
edetate (formerly edathamil)	ethylenediaminetetraacetate, and all anions derived from edetic acid (EDTA)	$C_{10}H_{12}N_2O_8$	

TABLE-24

Terms for Radicals and Groups Used for Nonproprietary Names (Continued)

Name	Chemical Name	Molecular Formula	Graphic Formula
edisilate, edisylate	1,2-ethanedisulfonate [or ethane-1,2-disulfonate]	$C_2H_4O_6S_2$	
embonate, pamoate	4,4-methylenebis(3-hydroxy-naphthalene-2-carboxylate	$C_{23}H_{14}O_6$	
enacarbil	[*rac*-1-[(2-methylpropanoyl)-oxy]ethoxy]carbonyl	$C_7H_{11}O_4$	and enantiomer
enantate, enanthate	heptanoate	$C_7H_{13}O_2$	
enbutate	acetate (ester) and butanoate (ester)	$C_2H_3O_2$ and $C_4H_7O_2$	
epolamine	2-(pyrrolidin-1-yl)ethanol	$C_6H_{13}NO$	
erbumine	*tert*-butylamine [or 2-methylpropan-2-amine]	$C_4H_{11}N$	
esilate, esylate	ethanesulfonate	$C_2H_5O_3S$	
estolate	propanoate (ester) and dodecyl sulfate (salt)	$C_3H_5O_2$ and $C_{12}H_{25}O_4S$	
etabonate	ethyl carbonate	$C_3H_5O_3$	
etemesil	2-(methylsulfonyl)ethanolate	$C_3H_7O_3S$	
etexilate	ethyl and (hexyloxy)carbonyl	C_2H_5 and $C_7H_{13}O_2$	
etibutil	2-ethylbutan-1-ol ester	$C_6H_{13}O$	
etiprate	ethyl ester and 2-benzamido-acetic acid	C_2H_5O and $C_9H_9NO_3$	

TABLE-25

Name	Chemical Name	Molecular Formula	Graphic Formula
etzadroxil	(2-ethylbutanoyloxy)metha-nolate	$C_7H_{13}O_3$	
farnesil	(2E,6E)-3,7,11-trimethyl-dodeca-2,6,10-trien-1-yl	$C_{15}H_{25}$	
fendizoate	2-(6-hydroxybiphenyl-3-car-bonyl)benzoate	$C_{20}H_{13}O_4$	
fosamil	phosphono	H_2O_3P	
fostedate	tetradecyl hydrogen phos-phate	$C_{14}H_{30}O_4P$	
fumarate	(2E)-2-butenedioate	$C_4H_2O_4$	
furetonide	1-benzofuran-2-carboxylate (ester) and propane-2,2-diylbis(oxy)	$C_9H_5O_3$ and $C_3H_6O_2$	
furoate	furan-2-carboxylate (ester)	$C_5H_3O_3$	
gamolenate	(6Z,9Z,12Z)-octadeca-6,9,12-trienoate (ester)	$C_{18}H_{29}O_2$	
gluceptate	D-glycero-D-gulo-heptanoate	$C_7H_{13}O_8$	
gluconate	(2R,3S,4R,5R)-2,3,4,5,6-pen-tahydroxyhexanoate	$C_6H_{11}O_7$	
glucuronide	β-D-glucopyranosiduronic acid [oside]	$C_6H_9O_7$	
glutamer	glutaraldehyde polymer	$(C_5H_7O_2)_n$	
guacil	2-methoxyphenyl	C_7H_7O	

TABLE-26

Name	Chemical Name	Molecular Formula	Graphic Formula
hemisuccinate	3-carboxypropanoate	$C_4H_5O_4$	
hexacetonide	3,3-dimethylbutanoate (ester) and propan-2,2-diylbis-(oxy) [or 3,3-dimethylbutyrate (ester) and acetonide]	$C_6H_{11}O_2$ and $C_3H_6O_2$	
hibenzate, hybenzate	2-(4-hydroxybenzoyl)benzo-ate	$C_{14}H_9O_4$	
hyclate	monohydrochloride hemi-ethanolate hemihydrate		$HCl \cdot \frac{1}{2}C_2H_5OH \cdot \frac{1}{2}H_2O$
hydroxynaphthoate	3-hydroxynaphthalene-2-car-boxylate	$C_{11}H_7O_3$	
isethionate, isetionate	2-hydroxyethane-1-sulfonate	$C_2H_5O_4S$	
isoproxil	hydroxymethyl isopropyl car-bonate	$C_5H_9O_4$	
kamedoxomil	(5-methyl-2-oxo-1,3-dioxol-4-yl)methyl (ester) and potas-sium salt	$C_5H_5O_3$ and K	
ketolaurate	3-oxododecanoate (ester)	$C_{12}H_{21}O_3$	
lactate	2-hydroxypropanoate	$C_3H_5O_3$	
laurate	dodecanoate	$C_{12}H_{23}O_2$	
lauril	dodecyl	$C_{12}H_{25}$	
laurilsulfate, lauryl sulfate	dodecyl sulfate	$C_{12}H_{25}O_4S$	
lisetil	L-lysinate (ester) and diethyl (ester)	$C_6H_{13}N_2O$ and $(C_2H_5)_2$	

TABLE-27

Terms for Radicals and Groups Used for Nonproprietary Names (Continued)

Name	Chemical Name	Molecular Formula	Graphic Formula
lisicol	[N-[(5S)-5-carboxy-5-(3α,7α,-12α-trihydroxy-5β-cholan-24-amido)pentyl]carbamo-thioyl]amino	$C_{31}H_{52}N_3O_6S$	
L-malate	(2S)-2-hydroxybutanedioic acid	$C_4H_4O_5$	
maleate	(Z)-but-2-enedioate	$C_4H_2O_4$	
mebutate	(2Z)-2-methylbut-2-enoate (ester)	$C_5H_7O_2$	
mecarbil	methoxycarbonyl	$C_2H_3O_2$	
medocaril	[(5-methyl-2-oxo-1,3-dioxol-4-yl)methoxy]carbonyl	$C_6H_5O_5$	
medoxomil	(5-methyl-2-oxo-1,3-dioxol-4-yl)methyl	$C_5H_5O_3$	
megallate	3,4,5-trimethoxybenzoate	$C_{10}H_{11}O_5$	
meglumine	N-methylglucamine [or 1-deoxy-1-(methylamino)-D-glucitol]	$C_7H_{17}NO_5$	
mepesuccinate	(2R)-2,6-dihydroxy-2-(2-methoxy-2-oxoethyl)-6-methylheptanoate (ester)	$C_{11}H_{19}O_6$	

TABLE-28

Name	Chemical Name	Molecular Formula	Graphic Formula
merpentan	4,5-bis(2-mercaptoacetamido)valeric acid and anions derived from this acid [or [N,N'-(5-oxopentane-1,2-diyl)bis(2-sulfanylacetamidato)](4−)]	$C_9H_{15}N_2O_3S_2$	
mertansine	(4RS)-4[[3-[[(1S)-2-[[(1S,2R,3S,5S,6S,16E,18E,20R,21S)-11-chloro-21-hydroxy-12,20-dimethoxy-2,5,9,16-tetramethyl-8,23-dioxo-4,24-dioxa-9,22-diazatetracyclo[19.3.1.1^{10,14}.0^{3,5}]hexacosa-10,12,14(26),16,18-pentaen-6-yl]oxy]-1-methyl-2-oxoethyl]methylamino]-3-oxo-propyl]disulfanyl]pentanoyl [or $N^{2'}$-deacetyl-$N^{2'}$-(3-sulfanylpropanoyl)maytansine]	$(C_{40}H_{55}ClN_3-O_{11}S_2)_n$	
mesilate, mesylate	methanesulfonate	CH_3O_3S	
metazoate	sodium 3-carboxybenzenesulfonate	$C_7H_5NaO_5S$	
metembonate	4,4'-methylenebis(3-methoxynaphthalene-2-carboxylate)	$C_{25}H_{18}O_6$	
mofetil	2-(morpholin-4-yl)ethyl	$C_6H_{12}NO$	
napadisilate	naphthalene-1,5-disulfonate	$C_{10}H_6O_6S_2$	
napsilate, napsylate	naphthalene-2-sulfonate	$C_{10}H_7O_3S$	
nicontinate	pyridine-3-carboxylate	$C_6H_4NO_2$	

TABLE-29

Name	Chemical Name	Molecular Formula	Graphic Formula
olamine	ethanolamine [or 2-aminoethanol]	C_2H_7NO	
oleate	(9Z)-octadec-9-enoate	$C_{18}H_{33}O_2$	
oxoglurate	hydrogen 2-oxopentanedioate	$C_5H_5O_5$	
palmitate	hexadecanoate	$C_{16}H_{31}O_2$	
pamoate	*See* embonate		
pentexil	*rac*-1-[(2,2-dimethylpropa-noyl)oxy]ethyl	$C_7H_{13}O_2$	and enantiomer
phenpropionate	3-phenylpropionate	$C_9H_9O_2$	
pivalate	trimethylacetate (ester) [or 2,2-dimethylpropanoate (ester)]	$C_5H_9O_2$	
pivoxetil	*rac*-1-[(2-methoxy-2-methyl-propanoyl)oxy]ethyl	$C_7H_{13}O_2$	and enantiomer
pivoxil	[(2,2-dimethylpropanoyl)-oxy]methyl	$C_6H_{11}O_2$	
placarbil	[[[(*R*)-2-methyl-1-[(2-methyl-propanoyl)oxy]propoxy]car-bonyl]amino]methyl	$C_{10}H_{18}NO_4$	
poliglumex	L-glutamic acid homopolymer [or poly[poly(L-glutamic acid)$_z$—(L-glutamate γ-ester)—poly (L-glutamic acid)$_y$]$_n$]		
probutate	*See* buteprate		

TABLE-30

Terms for Radicals and Groups Used for Nonproprietary Names (Continued)

Name	Chemical Name	Molecular Formula	Graphic Formula
proxetil	*rac*-1-[[(propan-2-yloxy)carbonyl]oxy]ethyl	$C_6H_{11}O_3$	
salicylate	2-hydroxybenzoate	$C_7H_5O_3$	
soproxil	[[(propan-2-yloxy)carbonyl]oxy]methyl	$C_5H_9O_3$	
steaglate	2-(octadecanoyloxy)acetate (ester)	$C_{20}H_{37}O_4$	
stearate	octadecanoate	$C_{18}H_{35}O_2$	
stinoprate	*N*-acetyl-L-cysteinate (salt) and propanoate (ester)	$C_5H_8NO_3S$ and $C_3H_5O_2$	
succinil	3-carboxypropanoyl	$C_4H_5O_3$	
suleptanate	8-[methyl-(2-sulfoethyl)amino]-8-oxooctanoate (ester), monosodium salt [or 7-methyl[(2-sulfoethyl)-carbamoyl]heptanoate ester, monosodium salt]	$C_{11}H_{19}NNaO_6$	
sulfoxylate	sulfinomethyl, monosodium salt	CH_2NaO_2S	
tartrate	(2*R*,3*R*)-2,3-dihydroxysuccinate	$C_4H_4O_6$	
tebutate	*tert*-butylacetate [or 3,3-dimethylbutanoate]	$C_6H_{11}O_2$	
tenoate	thiophene-2-carboxylate	$C_5H_3O_2S$	
teoclate	8-chloro-1,3-dimethyl-2,6-dioxo-3,6-dihydro-1*H*-purin-7-(2*H*)-ide	$C_7H_6ClN_4O_2$	

TABLE-31

Name	Chemical Name	Molecular Formula	Graphic Formula
teprosilate	3-(1,3-dimethyl-2,6-dioxo-1,-2,3,6-tetrahydro-7*H*-purin-7-yl)propane-1-sulfonate	$C_{10}H_{13}N_4O_5S$	
tetraxetan	2-[4,7,10-tris(carboxymethyl)-1,4,7,10-tetraazacyclodec-an-1-yl]-acetyl	$C_{16}H_{27}N_4O_7$	
tidoxil	*rac*-2-(decyloxy)-3-(dodecyl-sulfanyl)propyl	$C_{25}H_{51}OS$	and enantiomer
tiuxetan	*N*-[4-[(2*S*)-2-[bis(carboxy-methyl)amino]-3-[(2*RS*)-[2-[bis(carboxymethyl)ami-no]propyl](carboxymethyl)-amino]propyl]phenyl]thio-carbamoyl	$C_{23}H_{31}N_4O_{10}S$	
tocoferil	*rac*-(2*R*)-2,5,7,8-tetramethyl-2-[(4*R*,8*R*)-4,8,12-trimeth-yltridecyl]chroman-6-yl	$C_{29}H_{49}O$	and enantiomer
tofesilate	2-(1,3-dimethyl-2,6-dioxo-1,-2,3,6-tetrahydro-7*H*-purin-7-yl)ethane-1-sulfonate	$C_9H_{11}N_4O_5S$	
tosilate, tosylate	*p*-toluenesulfonate [or 4-methylbenzene-1-sulfo-nate]	$C_7H_7O_3S$	
triclofenate	2,4,5-trichlorophenolate	$C_6H_2Cl_3O$	
trifenatate	2,2,2-triphenylacetate	$C_{20}H_{16}O_2$	
triflutate	trifluoroacetate	$C_2F_3O_2$	

TABLE-32

Terms for Radicals and Groups Used for Nonproprietary Names (Continued)

Name	Chemical Name	Molecular Formula	Graphic Formula
trolamine	triethanolamine [or 2,2′,2″-nitrolotriethanol]	$C_6H_{15}NO_3$	
tromethamine	2-amino-2-(hydroxymethyl) propane-1,3-diol	$C_4H_{11}NO_3$	
troxundate	[2-(2-ethoxyethoxy)ethoxy]-acetate	$C_8H_{15}O_5$	
valactate	(2S)-2-[(2S)-2-amino-3-meth-ylbutanoyloxy]propanoate	$C_8H_{14}NO_4$	
valerate	pentanoate	$C_5H_9O_2$	
xinafoate	1-hydroxy-2-naphthoate [or 1-hydroxynaphthalene-2-carboxylate]	$C_{11}H_7O_3$	

TABLE-33

Nonproprietary Name Stems

The following table lists selected stems, prefixes, and infixes used in coining nonproprietary names to denote chemical structure, biological activity, or therapeutic use.

Term	Description	Term	Description
-ac	anti-inflammatory agents; acetic acid derivatives	-arsin	organoarsenic compounds
-actide	synthetic corticotropins (*See also* -tide)	arte-	antimalarials; artemisin derivatives
-adol or -adol-	analgesics; mixed opiate receptor agonists/antagonists	-ase	enzymes (*See subgroups: -dismase, -teplase, -uplase*)
-adom	analgesics; tifluadom derivatives	-ast	antiasthmatics and antiallergics not acting as antihistaminics; (*See subgroups: -lukast, -milast, -trodast, -zolast, -tegr-*)
-adox	antibacterials; quinoline dioxide derivatives		
-afenone	antiarrhythmics; propafenone derivatives	-astine	histamine-H_1 receptor antagonists
-afil	phosphodiesterase 5 (PDE 5) inhibitors	-(a)tadine	tricyclic histamine-H_1 receptor antagonist; loratadine derivatives
-aj-	antiarrhythmics; ajmaline derivatives	-axine or -faxine	antianxiety agents, antidepressants; norepinephrine and dopamine re-uptake inhibitors
-aldrate	antacids; aluminum salts		
-algron	alpha$_1$- and alpha$_2$- adrenoreceptor agonists	-azam, -azepam, or -azolam	antianxiety agents; diazepam derivatives
-alol	combined alpha and beta blockers	-azenil	benzodiazepine receptor agonists/antagonists (*See also -carnil, -nil, -quinil*)
-alox or -ox	antacids; aluminium derivatives		
-amidis	antimyloidotics	-azepide	cholecystokinin receptor antagonists; benzodiazepine derivatives
-amivir	antivirals; neuraminidase inhibitors (*See also -vir*)	-azocine	narcotic antagonists/agonists; 6,7-benzo-morphan derivatives
-(a)mostat	proteolytic enzyme inhibitors (*See also -stat*)	-azoline	antihistamines, local vasoconstrictors; antazoline derivatives
-ampanel	ionotropic non-NMDA glutamate receptor antagonists; *i.e.* aminohydroxymethyl-isoxazole-propionic acid (AMPA) and/or kainite antagonist (KA) receptor antagonists	-azosin	antihypertensives; prazosin derivatives
		-bacept	B-cell activating factor receptor molecules (*See also -cept*)
		-bactam	beta-lactamase inhibitors
-ampator	ionotropic non-NMDA glutamate receptor modulators; *i.e.* aminohydroxymethyl-isoxazole-propionic acid (AMPA) and/or kainite antagonist (KA) receptor modulators	-bamate	tranquilizers, antiepileptics; propanediol and pentanediol derivatives
		-barb or -barb-	barbituric acid derivatives
		-basib	sheddase inhibitors, ADAM (A disintegrin and metalloprotease) type
-andr- or andr-	androgens	-begron	beta-3-adrenoreceptor agonist
-anib	angiogenesis inhibitors (*See subgroup: -siranib*)	-benakin	interleukin-1β (IL-1β) analogs and derivatives (*See also -nakin*)
-anserin	serotonin 5-HT_2 receptor antagonists	-bendan	cardiac stimulants; pimobendan derivatives
-antel	anthelmintics (undefined group) (*See subgroup: -quantel*)		
-antrone	antineoplastics; anthraquinone derivatives	-bendazole	anthelmintics; thiabendazole derivatives
-apsel	P-selectin antagonists	-bercept	vascular endothelial growth factor (VEGF) receptor molecules (*See also -cept*)
-arabine	antineoplastics; arabinofuranosyl derivatives		
		-berel	beta estrogen receptor agonist
-aril, -aril-, or aril-	antiviral; arildone derivatives	-bermin	vascular endothelial growth factors (*See also -ermin*)
-arit	antirheumatics; clobuzarit and lobenzarit type	-bersat	anticonvulsants, antimigraine; benzoylamino-benzpyran derivatives
-arol	anticoagulants; dicumarol type		
-arone	antiarrhythmics (*See subgroup: -iodarone*)	-betasol	prednisone and prednisolone derivatives
-arotene	arotinoid derivatives	-bol- or bol-	anabolic steroids

TABLE-34

Term	Description
-bradine	sinus node inhibitors
-bufen	nonsteroidal anti-inflammatory agents (NSAIDs); fenbufen derivatives
-bulin	antineoplastics; mitotic inhibitors, tubulin binders
-butan	antiseptics (dapabutan type)
-butazone or -buzone	anti-inflammatory analgesics; phenyl-butazone derivatives
-buvir	antivirals; RNA polymerase (NS5B) inhibitors (See also -vir)
-caftor	cystic fibrosis transmembrane regulator (CFTR) protein modulators
-cain-	class I antiarrhythmics; procainamide and lidocaine derivatives
-caine	local anesthetics
-calcet	calcium receptor agonists (See also -cet)
-calci- or calci-	vitamin D analogs
-caleret	calcium receptor antagonists
-camra	antivirals; intracellular adhesion molecules-1 (ICAM-1) derivatives
-camsule	camphorsulfonic acid derivatives used as UVA sunscreens
-capoc	Gardos channel blockers (calcium activated potassium channel of intermediate conductance)
-capone	catechol-O-methyltransferase (COMT) inhibitors
-carbef	antibiotics; carbacephem derivatives
-carnil	benzodiazepine receptor antagonists/agonists; carboline derivatives (See also -azenil,-quinil)
-casan	caspase (interleukin-1β (IL-1β) converting enzyme) inhibitors
-caserin	serotonin receptor agonists, primarily 5-HT$_2$
-castat	dopamine β-hydrolase (DBH) inhibitors (See also -stat)
-catib	cathepsin inhibitors
-cavir	antivirals; carbocyclic nucleosides (See also -vir)
cef-	antibiotics; cephalosporins
cell-	cellulose derivatives
cell-ate	cellulose ester derivatives for substances containing acidic residues (See also cell-)
-cept	receptor molecules, native or modified (See subgroups: -bacept, -bercept, -cocept, -facept, -farcept, -lefacept, -nacept, -nercept, -tacept, -tercept, -vircept)
-cerfont	corticotropin releasing factor-1 (CRF-1) receptor antagonists
-cet	receptors (small molecule) (See subgroup: -calcet)

Term	Description
-cetrapib	cholesterol ester transfer protein (CETP) inhibitors
-cic	hepatoprotectives; timonacic derivatives
-ciclib	cyclin dependent kinase inhibitors (formerly -cidib)
-ciclovir or -cyclovir	antivirals; acyclovir derivatives (See also -vir)
-cidin	natural antibiotics (undefined group)
-ciguat	guanylate cyclase activators
-cillide, -cillin, or -cillinam	antibiotics; penicillins
-citabine	antivirals, antineoplastics, nucleoside type; cytarabine or azarabine derivatives
-clidine or -clidinium	muscarinic receptor agonists/antagonists
-clomol	heat-shock protein inducers; bimoclomal derivatives
-clone	hypnotics, tranquilizers; zopiclone derivatives
-closporin	cyclosporine derivatives
-cocept	complement receptor molecules (See also -cept)
-cog	blood coagulation factors (See subgroups: -eptacog, -nonacog, -octocog)
-cogin	blood coagulation cascade inhibitors
-conazole	systemic antifungals; miconazole derivatives
-corat	glucocorticoid receptor agonists (not glucocorticoids)
-cort-	cortisone derivatives
-coxib	cyclooxygenase-2 (COX-2) inhibitors
-cridar	multidrug resistance inhibitors; acridine carboxamide derivatives (See also -dar)
-crinat	diuretics; ethacrynic acid derivatives
-crine	acridine derivatives
-criviroc	CCR5 antagonists, immunomodulators (See also -viroc)
-cromil	antiallergics; cromoglicic acid derivatives
-curium or -curonium	neuromuscular blocking agents; quaternary ammonium compounds
-cycline	antibiotics; tetracycline derivatives
-dan	positive inotropic agents, pimobendan derivatives
-dapsone	antimycobacterials; diaminodiphenyl-sulfone derivatives
-dar	multidrug resistance inhibitors (See subgroups: -cridar, -icodar, -quidar, -spodar)

TABLE-35

Term	Description	Term	Description
-decakin	interleukin-10 (IL-10) analogs and derivatives (See also -kin)	-erg-	ergot alkaloid derivatives
-degib	hedgehog signaling inhibitors	-eridine or -ethidine	analgesics; meperidine (pethidine) derivatives
-denant	adenosine 2A antagonists	-ermin	growth factors (See subgroups: -bermin, -dermin, -fermin, -filermin, -nermin, -plermin, -otermin, -sermin, -termin)
-denoson	adenosine A receptor agonists		
-depsin	depsipeptide derivatives		
-dermin	epidermal growth factors (EGF) (See also -ermin)	-espib	heat shock protein inhibitors
		-estr- or estr-	estrogens
-dil, -dil-, dil-, -dilol, or -dyl	vasodilators	-estrant	estrogen antagonists
		-etanide	diuretics; piretanide derivatives
-dipine	phenylpyridine vasodilators; calcium channel blockers, nifedipine type	-exakin	interleukin-6 (IL-6) analogs and derivatives (See also -kin)
-dismase	enzymes with superoxide dismutase (SOD) activity (See also -ase)	-exine	mucolytics; bromhexine derivatives
-distim	conjugates of two different types of colony-stimulating factors (CSF) (See also -stim)	-facept or -lefacept	lymphocyte function-associated antigen 3 (LFA-3) receptors (See also -cept)
-ditan	antimigraine (5-HT$_1$ receptor agonists)	-farcept	interferon receptor molecules (See also -cept)
-ditraz	acaricide, tick repellent	-farin	warfarin analogs
-dodekin	interleukin-12 (IL-12) analogs and derivatives (See also -kin)	-farnib	farnesyl transferase inhibitors
		-fenamate	"fenamic acid" ester or salt derivatives (See also -fenamic acid)
-domide	thalidomide derivatives		
-dopa	dopamine receptor agonists, dopamine derivatives	-fenamic acid	anti-inflammatory agents; anthranilic acid derivatives (See also -fenamate)
-dopidine	dopamine D$_2$ receptor modulators	-fenin	diagnostic aids; (phenylcarbamoyl)-methyliminodiacetic acid derivatives
-dore	dopamine D$_2$D$_3$ receptor modulators		
-dotin	synthetic analogs of the dolastatin series	-fenine	analgesics; glafenine derivatives (fenamic acid subgroup)
-dox	antibacterials; quinazoline dioxide derivatives	-fentanil	narcotic analgesics; fentanyl derivatives
-dralazine	antihypertensives; hydrazine-phthalazine derivatives	-fentrine	phosphodiesterase inhibitors
-drine	sympathomimetics	-fermin	fibroblast growth factors (See also -ermin)
-dronate or -dronic acid	calcium metabolism regulators	-fetamin(e)	amfetamine derivatives
-dutant	neurokinin NK$_2$ receptor antagonists (See also -tant)	-fexor- or -fexorate	farnesoid X receptor agonists
		-fiban	fibrinogen receptor antagonists (glycoprotein II$_b$/III$_a$ receptor antagonists)
-ectedin	ecteinascodin derivatives		
-ectin	antiparasitics; ivermectin derivatives	-fibatide	peptide platelet aggregation inhibitors (glycoprotein II$_b$/III$_a$ receptor antagonists) (See also -tide)
-elestat	elastase inhibitors (See also -stat)		
-elvekin	interleukin-11 (IL-11) analogs and derivatives (See also -kin)	-fibrate	antihyperlipidemics; clofibrate derivatives
-emcinal	erythromycin derivatives without antibiotic activity; motilin agonists	-filcon	hydrophilic contact lens materials
-enicokin	interleukin-21 (IL-21) analogs and derivatives (See also -kin)	-filermin	leukemia-inhibitory factor (LIF) growth factors (See also -ermin)
-entan	endothelin receptor antagonists	-fingol	sphingosine derivatives
-eptacog	blood coagulation factor VII (See also -cog)	-flapon	5-lipoxygenase-activating protein (FLAP) inhibitors
-eptakin	interleukin-7 (IL-7) analogs and derivatives (See also -kin)	-flurane	general inhalation anesthetics; halogenated alkane derivatives

TABLE-36

Term	Description	Term	Description
-focon	hydrophobic contact lens materials	-glinide	antidiabetics; sodium-dependent glucose cotransporters (SGLT2) inhibitors, not phlorozin derivatives *(See also -gli- or gli-)*
-folastat	folate hydrolase (PSMA) inhibitors *(See also -stat)*		
-formin	hypoglycemics; phenformin derivatives	-gliptin	antidiabetics; dipeptidyl aminopeptidase-IV (DPP-IV) inhibitors *(See also -gli- or gli-)*
-fos-	phosphoro-derivatives		
-fosfamide	alkylating agents; isophosphoramide mustard derivatives	-glitazar	antidiabetics; peroxisome proliferator activating receptor (PPAR) agonists (not thiazolidene (TZD) derivatives) *(See also -gli- or gli-)*
-fosine	cytostatic phosphoro derivatives		
-fovir	antiviral phosphonic acid derivatives *(See also -vir)*	-glitazone	antidiabetics; peroxisome proliferator activating receptor (PPAR) agonists; thiazolidinedione (TZD) derivatives *(See also -gli- or gli-)*
-fradil	vasodilators; calcium channel blockers		
-frine	sympathomimetic phenethyl derivatives	-glumide	antiulcer, anxiolytics; cholecystokinin (CCK) antagonists
-fulven	antineoplastic acylfulven derivatives	-glustat	glucosyltransferase inhibitors *(See also -stat)*
-fungin	antifungal antibiotics	-glutide	glucagon-like peptide (GLP) peptide analogs *(See also -tide)*
-fylline	*N*-methylated xanthine (theophylline) derivatives		
-gab-	gabamimetics	-golide	dopamine receptor agonists, ergoline derivatives
-gacestat	gamma secretase inhibitors *(See also -stat)*	-golix	gonadotropin-releasing hormone (GnRH) receptor antagonists (nonpeptide)
gado-	diagnostic aids; gadolinium derivatives		
-ganan	antimicrobial, bactericidal permeability increasing polypeptides	-gosivir	antivirals; glucosidase inhibitors *(See also -vir)*
		-gramostim	granulocyte macrophage colony-stimulating factors (GM-CSF) *(See also -stim)*
-gapil	neuronal apoptosis inhibitors; glyceraldehyde 3-phosphate dehydrogenase (GAPDH) modulators		
		-grastim	granulocyte colony-stimulating factors (G-CSF) *(See also -stim)*
-gaptide	peptides acting as gap junction protein channel modulators *(See also -tide)*	-grel or -grel-	platelet aggregation inhibitors, primarily platelet P2Y12 receptor antagonists
-gatran	thrombin inhibitors; argatroban derivatives		
		guan-	antihypertensives; guanidine derivatives
-gene	gene therapy products	-ibat	ileal bile acid transport inhibitor
-gepant	calcitonin gene-related peptide (CGRP) receptor antagonists	-icam	anti-inflammatory agents; isoxicam derivatives
-gest-	progestins	-icodar	multidrug resistance inhibitors; pipecolic acid derivatives *(See also -dar)*
-gestr-	estrogens		
-giline	monoamine oxidase (MAO) inhibitors, type B	-ifen(e)	antiestrogens; clomifene and tamoxifen derivatives
-gillin	antibiotics obtained from *Aspergillus* strains	-ilide	class III antiarrhythmic agents
-gil- or gli-	antihyperglycemics, antidiabetics *(See subgroups: -gliatin, -gliflozin, -glinide, -gliptin, -glitazar, -glitazone)*	-imepodib	inosine monophosphate dehydrogenase inhibitors
		-imex	immunostimulants
		-imibe-	antihyperlipidemics; acyl CoA: cholesterol acyltransferase (ACAT) inhibitors
-gliatin	antidiabetics; glucokinase activators *(See also -gli- or gli-)*		
-gliflozin	antidiabetics; sodium glucose cotransporter (SGLT2) inhibitors, phlorozin derivatives (phenolic glycosides) *(See also -gli- or gli-)*	-imod	immunomodulators *(See subgroups: -mapimod, -tirimod, -tolimod)*
		-imus	immunosuppressants *(See subgroup: -rolimus)*

TABLE-37

Term	Description	Term	Description
-ine	alkaloids and organic bases	-locib	antineoplastics that inhibit the formation of 5 lipoxygenase activating protein (5-LO), leukotriene B4 (LTB$_4$), leukotriene C4 (LTC$_4$), and thromboxane B$_2$ (TxB$_2$) and activate peroxisome proliferator activating receptor gamma (PPARγ) nuclear receptors
-inostat	histone deacetylase (HDAC) inhibitors *(See also -stat)*		
io-	iodine-containing contrast media		
iod- or -io-	iodine-containing compounds other than contrast media		
-iodarone	antiarrhythmics with high iodine content *(See also -arone)*	-lubant	leukotriene B$_4$ receptor antagonists (used in treatment of inflammatory skin disorders)
-irudin	anticoagulants; hirudin derivatives		
-isant	histamine H$_3$ receptor antagonists	-lukast	antiasthmatics, antiallergics; leukotriene receptor antagonists *(See also -ast)*
-isomide	antiarrhythmics; disopyramide derivatives		
-ium or -onium	quaternary ammonium derivatives	-luren	inducers of ribosomal readthrough of nonsense mutation mRNA stop codons
-ivon	Von Willebrand factor inhibitors		
-ixafor	CXC chemokine receptor 4 (CXCR4) antagonists	-lutamide	nonsteroidal antiandrogens
-izine or -yzine	diphenylmethyl piperazine derivatives	-lutril	neutral endopeptidase inhibitors possessing additional endothelin converting enzyme inhibitory activity
-kacin	antibiotics obtained from *Streptomyces kanamyceticus* (related to kanamycin)		
-kalant	potassium channel antagonists	-mab	monoclonal antibodies *(follows an appropriate infix: -axo-, -kin-, -les-, -ne(r)-, -os-, -toxa-, -xizu-)*
-kalim	potassium channel agonists		
-kalner	opener of large conductance calcium-activated (maxi-k) K+ channels		
		-manid	mycolic acid inhibitors
-kef-	enkephalin agonists	-mantadine, -mantine, or -mantone	antivirals, antiparkinsonians; adamantane derivatives
-kin	interleukins *(See subgroups: -decakin, -dodekin, -elvekin, -enicokin, -eptakin, -exakin, -kinra, -leukin, -nakin, -nakinra, -nonakin, -octakin, -octadekin, -penkin, -trakin, -tredekin; see also -mab, -pab)*	-mapimod	immunomodulators; mitogen-activated protein (MAP) kinase inhibitors *(See also -imod)*
		-maprazole	antiulcer agents; acid pump inhibitors *(See also -prazole)*
		-mastat	matrix metalloprotease inhibitors *(See also -stat)*
-kinra	interleukin receptor antagonists *(See subgroup: -nakinra, -trakinra; see also -kin)*	-meline	cholinergic agents, arecoline derivatives (used in treatment of Alzheimer's disease)
-kiren	renin inhibitors	-melteon	selective melatonin receptor agonist
-lanstat	lanosterol 14α-demethylase inhibitors *(See also -stat)*	-mer	polymers
		-mesine	sigma receptor ligands
-lazad	lipid peroxidation inhibitors	-mestane	antineoplastics; aromatase inhibitors
-leptin	leptin derivatives	-metacin	anti-inflammatory agents; indomethacin derivatives
-leukin	interleukin-2 (IL-2) analogs and derivatives *(See also -kin)*		
-leuton	5-lipoxgenase (5-LO) inhibitors	-micin	antibiotics obtained from *Micromonospora* strains
-lind	pro-apoptotic cGMP phosphodiesterase inhibitors *(See subgroup: -sulind)*	-milast	antiasthmatics, antiallergics; phosphodiesterase IV (PDE IV) inhibitors *(See also -ast)*
-lintide	amylin derivatives or mimics (peptides) *(See also -tide)*	-monam	monobactam antibiotics
		-morelin	growth hormone-release stimulating peptides *(See also -relin)*
-lipim	lipoprotein lipase activators		
-lisib	phosphatidylinositol 3-kinase inhibitors	-moren	nonpeptidic growth hormone secretagogues
-listat	gastrointestinal lipase inhibitors *(See also -stat)*	-mostim	macrophage colony-stimulating factors (M-CSF) *(See also -stim)*

TABLE-38

Term	Description	Term	Description
-motide	peptides used as immunological agents for active immunization (*See also -tide*)	-nidap	nonsteroidal anti-inflammatory agents; tenidap derivatives
-motine	antivirals; quinoline derivatives	-nidazole	antiprotozoal substances; metronidazole derivatives
-moxin	monoamine oxidase inhibitors, hydrazine derivatives	nifur-	5-nitrofuran derivatives
-mulin	antibacterials; pleuromulin derivatives	-nil	benzodiazepine receptor antagonists/agonists (*See subgroups: -punil, -quinil; See also -azenil*)
-multin	mucosal tolerance inductors		
-murtide	peptides with muramic acid present (*See also -tide*)	nitro-, nitr-, nit-, ni-, or -ni-	NO$_2$ - derivatives
-mustine	antineoplastics; chloroethylamine derivatives	-nixin	anti-inflammatory agents; anilino-nicotinic acid derivatives
-mycin	antibiotics obtained from *Streptomyces* strains	-nonacog	blood coagulation factor IX (*See also -cog*)
-nab or nab-	cannabinol derivatives (*See subgroup: -nabant*)	-nonakin	interleukin-9 (IL-9) analogs and derivatives (*See also -kin*)
-nabant	CB cannabinoid receptor antagonists (*See also nab- or -nab-*)	-octadekin	interleukin-18 (IL-18) analogs and derivatives (*See also -kin*)
-nacept	interleukin-1 (IL-1) receptor molecules (*See also -cept*)	-octakin	interleukin-8 (IL-8) analogs and derivatives (*See also -kin*)
-nakalant	mixed sodium/potassium channel blockers	-octocog	blood coagulation factor VIII (*See also -cog*)
-nakin	interleukin-1 (IL-1) analogs and derivatives (*See subgroup: -benakin, -onakin; see also -kin*)	-olol	beta-adrenoreceptor antagonists (beta-blockers), propranolol derivatives
-nakinra	interleukin-1 (IL-1) receptor antagonists (*See also -kinra*)	-olone	steroids (not prednisolone derivatives)
nal-	narcotic agonists/antagonists; normorphine derivatives	-onakin	interleukin-1α (IL-1α) analogs and derivatives (*See also -kin, -nakin*)
-navir	antivirals; HIV protease inhibitors; saquinavir derivatives (*See also vir-*)	-one	ketones
		-onide	topical steroids; acetal derivatives
-nepag	nonprostanoid prostaglandin receptor agonists	-onidine	antihypertensives; clonidine derivatives
-neperit	neuropeptide Y5 (NPY5) receptor modulators (*See also -perit*)	-opamine	cardiac stimulants, antihypertensives, diuretics; dopaminergic agents, dopamine derivatives
-nercept	tumor necrosis factor (TNF) receptor molecules (*See also -cept*)	-opilone	epothilones
-nermin	tumor necrosis factors (TNF) (*See also -ermin*)	-orex	anorexiants
		-orexant	orexin antagonists
-nertant	neurotensin receptor antagonists	-orph- or -orphan-	narcotic antagonists/agonists, morphinan derivatives
-nesib	kinesin inhibitors	-osuran	urotensin receptor antagonists
-netant	neurokinin NK$_3$ receptor antagonists (*See also -tant*)	-otermin	bone morphogenetic proteins (*See also -ermin*)
-netide	peptides with neurologic indications (*See also -tide*)	-otilate	hepatoprotectants; di-isopropyl-1,3-dithiol-malonate derivatives
-neurin	neurotensin receptor antagonists; neurotropins	-oxacin	antibacterials; nalidixic acid (quinolone) derivatives
-nicate	antihypercholesterolaemics, vasodilators; nicotinic acid esters	-oxan or -oxane	alpha-adrenoceptor antagonists; benzodioxane derivatives
-nicline	nicotinic acetylcholine receptor partial agonists/agonists	-oxanide	antiparasitics; salicylanilide derivatives
nico-, nic, or ni-	nicotinic acid or nicotinoyl alcohol derivatives	-oxef	antibiotics; oxacefalosporanic acid derivatives

TABLE-39

Term	Description	Term	Description
-oxetine	antidepressants; fluoxetine type (See subgroup: -tioxetine)	-pin(e)	tricyclic compounds
-oxin	fluoroquinolone derivatives (not antibacterials)	-piprant	prostaglandin receptor antagonists, non prostinoid structure
-pab	polyclonal antibodies	-piprazole	psychotropics; phenylpiperazine derivatives
-pafant	platelet-activating factor antagonists	-pirdine	5-HT$_6$ inhibitors
-pamide	diuretics; sulfamoylbenzoic acid derivatives	-pirox	antimycotics; pyridone derivatives
-pamil	coronary vasodilators; verapamil derivatives	-pitant	neurokinin NK$_1$ (substance P) receptor antagonists (See also -tant)
-pamine	dopaminergics; butopamine derivatives	-plact	platelet factor 4 analogs and derivatives
-panel	AMPA receptor antagonists	-pladib	phospholipase A$_2$ inhibitors
-paratide	parathyroid hormone related peptides (See also -tide)	-planin	antibacterials obtained from Actinoplanes strains
-parcil	antithrombotics	-plasinin	plasminogen activator inhibitors - type 1
-parcin	glycopeptide antibiotics	-plasmid	gene therapy products
-parib	poly-ADP-ribose polymerase inhibitors	-plasmin	plasmin proteins and derivatives
-parin	heparin derivatives and low molecular weight (or depolymerized) heparins	-platin	antineoplastics; platinum derivatives
		-plermin	platelet derived growth factors (See also -ermin)
-parinux	antithrombotics; indirect selective synthetic factor Xa inhibitors	-plestim	interleukin 3 (IL-3) derivatives and analogs, pleiotropic colony-stimulating factors (See also -stim)
-paroid	antithrombotics; heparinoid derivatives		
-patril or -patrilat	angiotensin-1 converting enzyme (ACE) and neutral endopeptidase (NEP) inhibitors (See also -tril or -trilat)	-plon	anxiolytics, sedatives, hypnotics; non-benzodiazepine derivatives
		-poetin	erythropoietins
		-porfin	benzoporphyrin derivatives
		-poride	Na+/H+ antiport inhibitor
-paxar	protease-activated receptor 1 (PAR-1) antagonists	-potide	peptides with prostate cancer indications (See also -tide)
peg-	PEGylated compounds	-pramine	antidepressants; imipramine type
-penem	antibacterial antibiotics; carbapenem derivatives	-prazan	acid pump inhibitors, not dependent on acid activation
-penkin	interleukin-5 (IL-5) analogs and derivatives (See also -kin)	-prazole	antiulcer agents; benzimidazole derivatives (See subgroup: -maprazole)
perflu-	blood substitutes, diagnostics; perfluoro-chemicals		
-peridol	antipsychotics; haloperidol derivatives	pred-, -pred-, or -pred	prednisone and prednisolone derivatives
-peridone	antipsychotics; risperidone derivatives	-prenaline	bronchodilators; phenethylamine derivatives
-perit	neuropeptide Y (NPY) receptor modulators (See subgroup: -neperit)	-pressin	vasoconstrictors; vasopressin derivatives
-perone	antianxiety agents, neuroleptics; 4'-fluoro-4-piperidinobutyrophenone derivatives	-previr	antivirals; serine protease inhibitors (See also -vir)
		-pride	sulpiride derivatives
-pertin	glycine transporter inhibitors	-pril	antihypertensives; angiotensin-converting enzyme (ACE) inhibitors
-pezil	acetylcholinesterase inhibitors (used in treatment of Alzheimers disease)	-prilat	antihypertensives; angiotensin-converting enzyme (ACE) inhibitors, diacid analogs of the -pril entity
-pidem	hypnotics, sedatives; zolpidem derivatives	-prim	antibacterials; trimethoprim derivatives

TABLE-40

Term	Description	Term	Description
-prinim	nootropic agents, purine derivatives	-ritide	atrial natriuretic type peptides (See also -tide)
-pris-	steroidal compounds acting on pro-gesterone receptors (excluding -gest- compounds)	-rixin	CXC chemokine receptor 2 (CXCR2) modulators
-prisnil	selective progesterone receptor mod-ulators (SPRM)	-rizine	antihistaminics, cerebral (or periph-eral) vasodilators
-pristin	antibacterials; pristinamycin deriva-tives	-rocin	t-RNA synthetase inhibitors
-pristone	progesterone receptor antagonists	-rolimus	immunosuppressants; rapamycin derivatives (See also -imus)
-profen	anti-inflammatory, analgesic agents; ibuprofen derivatives	-rozole	aromatase inhibitors, imidazole/tria-zole derivatives
-proget	nonsteroidal ligand for the proges-terone receptor	-rsen	antisense oligonucleotides (See sub-group: -virsen)
-prost or -prost-	prostaglandins	-rubicin	antineoplastic antibiotics; daunorubi-cin derivatives
-prostil	anti-ulcer; prostaglandins	-sal, -sal-, or sal-	anti-inflammatory agents; salicylic acid derivatives
-protafib	protein tyrosine phosphatase 1B in-hibitors	-salan	disinfectants; brominated salicylam-ide derivatives
-pultide	peptides used as pulmonary surfac-tants (See also -tide)	salazo-	antibacterials; phenylazosalicylic acid derivatives
-punil	mitochondrial benzodiazepine recep-tor (MBR) selective, partial, or in-verse agonists (purine derivatives) (See also -nil)	-sarm	selective androgen receptor modula-tors (SARM), nonsteroidal
-quantel	anthelmintics; 2-deoxoparaherquam-ide A derivatives (See also -antel)	-sartan	angiotensin II receptor antagonists
-queside	cholesterol sequestrants, glycosides	-semide	diuretics; furosemide derivatives
-quidar	multidrug resistance inhibitors; quin-oline derivatives (See also -dar)	-sermin	insulin-like growth factors (See also -ermin)
-quiline	antibiotics; diarylquinoline derivatives	-serod	serotonin receptor antagonists and partial agonists
-quinil	benzodiazepine receptor selective, partial or inverse agonists; quino-line derivatives (See also -nil)	-serpine	Rauwolfia alkaloid derivatives
		-sertib	serine/threonine kinase inhibitors
-quistat	squalene synthase inhibitors (See also -stat)	-setrag	serotonin 5-HT receptor agonists, not principally 5-HT$_2$
-racetam	nootropic agents; piracetam deriva-tives	-setron	serotonin 5-HT$_3$ antagonists
-racil	antineoplastics; uracil derivatives	-siban	oxytocin antagonists
-rafenib	raf kinase inhibitors	-sidomine	antianginals; sydnone derivatives
-relin	prehormones or hormone-release stimulating peptides (See sub-groups: -morelin, -tirelin)	-sidone	antipsychotic with binding activity on serotonin (5-HT$_{2A}$) and dopa-mine (D$_2$) receptors
-relix	hormone-release inhibiting peptides	-siranib	angiogensis inhibitors; siRNA (See also -anib)
-renone	aldosterone antagonists, spironolac-tone type	som-	growth hormone derivatives
-restat or -restat-	aldose-reductase inhibitors (See also -stat)	-sonan	5-HT$_{1B}$ receptor antagonists
		-spirone	anxiolytics; buspirone type
-retin or -retin-	retinol derivatives	-spodar	multidrug resistance inhibitors; ciclo-sporin D derivatives (See also -dar)
-rev	therapeutic virus (See subgroup: -tu-cirev)	-sporivir	antivirals; cyclosporine derivatives (See also -vir)
-rian	ryanodine receptor modulators	-stat or -stat-	enzyme inhibitors (See subgroups: -(a)mostat, -castat, -elestat, -fola-stat, -gacestat, -glustat, -inostat, -lanstat, -listat, -mastat, -quistat, -restat-, -restat, -telstat, -tiostat, -tristat, -vastatin, -xostat)
-ribine	ribofuranil derivatives, pyrazofurin type		
rifa-	antibiotics; rifamycin derivatives		
-rinone	cardiotonics; amrinone derivatives		

TABLE-41

Term	Description	Term	Description
-steine	mucolytics (excluding bromhexine derivatives)	-thiazide	diuretics; thiazide derivatives
-ster-	steroids (androgens, anabolics)	-tiapine	antipsychotics; dibenzothiazepine derivatives
-steride	testosterone reductase inhibitors	-tiazem	calcium channel blockers, diltiazem derivatives
-stigmine	cholinesterase inhibitors, physostigmine type	-tibant	antiasthmatics; bradykinin antagonists
-stim	colony-stimulating factors (*See subgroups: -distim, -gramostim, -grastim, -mostim, -plestim*)	-tide	peptides (*See subgroups: -actide, -fibatide, -gaptide, -glutide, -lintide, -murtide, -netide, -paratide, -potide, -pultide, -ritide, -zotide*)
-stinel	*N*-methyl-d-asparate (NMDA) receptor antagonist, glycine recognition site	-tidine	histamine H_2-receptor antagonists; cimetidine derivatives
-sulam	antineoplastics, apoptosis inducing sulfonamide	-tinib	tyrosine kinase inhibitors
sulfa-	antimicrobials; sulfonamides derivatives	-tiostat	glutathione-*S*-transferase inhibitors (*See also -stat*)
-sulfan	antineoplastics, alkylating agents; methanesulfonate derivatives	-tioxetine	antidepressants, fluoxetine type; thioether derivatives (*See also -oxetine*)
-sulind	antineoplastics, sulindac metabolites (*See also -lind*)	-tirelin	thyrotropin releasing hormone analogs (*See also -relin*)
-tacept	cytotoxic T-lymphocyte-associated antigen 4 (CTLA-4) receptor molecules (*See also -cept*)	-tirimod	immunomodulators; napthyridine analogs (*See also -imod*)
-tadine	tricyclic histamine-H_1 receptor antagonists	-tirome	antihyperlipidemic; thyromimetic derivatives
-tansine	maytansinoid derivatives	-tizide	diuretics; chlorothiazide derivatives
-tant	tachykinin (neurokinin) receptor antagonists (*See subgroups: -dutant, -netant, -pitant*)	-tocin	oxytocin derivatives
		-toclax	B-cell lymphoma 2 (BCL-2) inhibitors
-tapide	microsomal triglyceride transfer protein (MTP) inhibitors	-toin	antiepileptics; hydantoin derivatives
-taxel	antineoplastics, taxane derivatives	-tolimod	immunomodulators; toll-like receptor agonists (*See also -imod*)
-tecan	antineoplastics; topoisomerase I inhibitors, camptothecine derivatives	-toran	toll-like receptor 4 (TLR-4) antagonists
-tecarin	antineoplastics; rebeccamycin derivatives	-tox(a)-	toxins
-tegrast	antiasthmatics/antiallergics; integrin antagonists (*See also -ast*)	-trakin	interleukin-4 (IL-4) analogs and derivatives (*See also -kin*)
-tegravir	antivirals; integrase inhibitors (*See also -vir*)	-trakinra	interleukin-4 receptor antagonists (*See also -kinra*)
-telstat	telomerase inhibitors (*See also -stat*)	-traline	selective serotonin reuptake inhibitors (SSRI)
-tepa	antineoplastics; thiotepa derivatives	-traposin	adipocyte lipid-binding protein (aP2) inhibitors
-teplase	tissue-type plasminogen activators (*See also -ase*)	-tredekin	interleukin-13 analogs and derivatives (*See also -kin*)
-tercept	transforming growth factor receptor molecules (*See also -cept*)	-trexate	antimetabolites; folic acid derivatives
-termin	transforming growth factors (*See also -ermin*)	-trexed	antineoplastics; thymidylate synthetase inhibitors
-terol	bronchodilators; phenethylamine derivatives	-tricin	antibiotics; polyene derivatives
-terone	antiandrogens	-tril or -trilat	endopeptidase inhibitors (*See subgroup: -patril/-patrilat*)
-tesind	thymidilate synthetase inhibitors; benzindole derivatives	-triptan	antimigraine agents, 5-HT_1 receptor agonists; sumatriptan derivatives
-texafin	texaphryn derivatives		

TABLE-42

Term	Description	Term	Description
-triptyline or -tiline	antidepressants; dibenzo[*a,d*]cyclo-heptane derivatives	-virenz	antivirals; non-nucleoside reverse transcriptase inhibitors (NNRTI), benzoxazinone derivatives *(See also -vir)*
-tristat	tryptophan hydroxylase inhibitors *(See also -stat)*		
		-virimat	antivirals that disrupt viral maturation
-troban	antithrombotics; thromboxane A_2 receptor antagonists	-virine	non-nucleoside reverse transcriptase inhibitors, not benzoxazinone derivatives
-trodast	antiasthmatics/antiallergics; thromboxane A_2 receptor antagonists *(See also -ast)*		
		-viroc	CCR5 antagonists *(See subgroup: -criviroc)*
-troline	antipsychotics; dopamine D_2 antagonists	-virsen	antivirals; antisense oligonucleotides *(See also -rsen)*
-trombopag	thrombopoetin agonists		
-trop- or trop-	atropine derivatives	-vudine	antineoplastics and antivirals; zido-vudine derivatives (exception: edoxudine)
-tucirev	tumoricidal therapeutic virus *(See also -rev)*		
		-xaban	antithrombotic; blood coagulation factor Xa inhibitors
-uclin	mucosal tolerance inhibitors		
-uplase	urokinase-type plasminogen activators *(See also -ase)*	-xanox	antiallergics; xanoxic acid derivatives
-ur, -(ur)amidine, or -uridine	antivirals and antineoplastics; uridine derivatives	-(x)antrone	antineoplastics; mitoxantrone derivatives, aza-anthracenedione class
-uracil	thyroid antagonists and antineoplastics; uracil derivatives	-xostat	xanthine oxidase and xanthine dehydrogenase inhibitors *(See also -stat)*
-urad	urate transporter inhibitors	-zafone	alozafone derivatives
-vaptan	vasopressin receptor antagonists	-zolamide	carbonic anhydrase inhibitors
-vastatin	antihyperlipidemics; HMG-CoA reductase inhibitors *(See also -stat)*	-zolast	antiasthmatics/antiallergics; leukotriene biosynthesis inhibitors, benzoxazole derivatives *(See also -ast)*
-vec	gene therapy product		
-verine	spasmolytic agents, papaverine derivatives	-zolid	oxazolidinone antibacterials
		-zomib	proteozome inhibitors
vin- or -vin-	vinca alkaloids	-zone	anti-inflammatory analgesics; phenylbutazone derivatives
-vir, -vir-, or vir-	antivirals *(See subgroups: -amivir, -buvir, -cavir, -ciclovir, -gosivir-, fovir, -navir, -previr, -sporivir, -tegravir, -virenz)*	-zotan	5-HT_{1A} receptor agonists/antagonists; primarily neuroprotective agents
-vircept	antiviral receptor molecules *(See also -cept)*	-zotide	zonulin antagonist peptides *(See also -tide)*

TABLE-43

Radioactive Isotopes Used in Medical Diagnosis and Therapy

Name and Symbol	Principal Nuclear Properties	Form	Use
Americium ^{241}Am	Half-life 432.7y α(5.49, 5.44) γ(0.060)	Encapsulated source	Diagnostic: External radiation source for bone mineral analyzer
			Therapeutic: Antineoplastic (intracavitary radiation source)
Calcium ^{47}Ca	Half-life 4.53d β^-(0.67, 1.98) γ(1.297)	Calcium chloride	Diagnostic: Calcium metabolism studies
Cesium ^{137}Cs	Half-life 30.0y β^-(1.176, 0.514)	Cesium chloride or cesium sulfate (encased in needles or applicator cells)	Therapeutic: Antineoplastic (teletherapy source, intracavitary or interstitial radiation source)
Daughter ^{137m}Ba	Half-life 2.552min γ(0.662)		
Californium ^{252}Cf	Half-life 2.645y α(6.217)	Sealed source	Therapeutic: Antineoplastic (intracavitary or interstitial radiation source)
Chromium ^{51}Cr	Half-life 27.704d K; γ(0.32)	Chromic chloride	Diagnostic: Determn of serum protein loss into the gastrointestinal tract
		Chromium disodium edetate	Diagnostic: Determn of glomerular filtration rate
		Labeled human serum albumin	Diagnostic: Placenta localization; gastrointestinal protein loss
		Sodium chromate labeled red blood cells	Diagnostic: Determn of red cell volume or mass; red cell survival time; evaluation of blood loss; spleen imaging; placenta localization
Cobalt ^{60}Co	Half-life 5.271y β^-(0.318, 1.48) γ(1.173, 1.332)	Metallic cobalt	Therapeutic: Antineoplastic (teletherapy source, intracavitary or interstitial radiation source)
		Radioactive vitamin B$_{12}$	Diagnostic: In Schilling test for absence of intrinsic factor (pernicious anemia) or other defects of intestinal vitamin B$_{12}$ absorption
^{57}Co	Half-life 271.77d K; γ(0.122)	Radioactive vitamin B$_{12}$	Diagnostic: In Schilling test for absence of intrinsic factor (pernicious anemia) or other defects of intestinal vitamin B$_{12}$ absorption
^{58}Co	Half-life 71.91d K; β^+(0.48) γ(0.811)	Radioactive vitamin B$_{12}$	Diagnostic: In Schilling test for absence of intrinsic factor (pernicious anemia) or other defects of intestinal vitamin B$_{12}$ absorption
Copper ^{64}Cu	Half-life 12.701h β^-(0.571), β^+(0.657), γ(1.34)	Copper versenate	Diagnostic: Brain scan
		Copper acetate	Diagnostic: Study of Wilson's disease
Fluorine ^{18}F	Half-life 1.8295h β^+(0.635)	Florbetapir	Diagnostic: Brain imaging for β-amyloid plaques
		Fludeoxyglucose (FDG)	Diagnostic: Functional brain imaging
		Sodium fluoride (reactor produced)	Diagnostic: Bone scan
Gadolinium ^{153}Gd	Half-life 241.6d K; γ(0.70, 0.097, 0.103)	Sealed source	Diagnostic: External radiation source for bone mineral analyzer
Gallium ^{67}Ga	Half-life 3.261d K; γ(0.093, 0.184, 0.300, 0.393)	Gallium citrate	Diagnostic: Detection of neoplastic and inflammatory lesions; tumor seeking agent
Gold ^{198}Au	Half-life 2.6935d β^-(1.371, 0.962) γ(0.412)	Colloidal gold	Diagnostic: Liver imaging
		Colloidal gold or seeds	Therapeutic: Antineoplastic (radiation source) in treatment of widespread abdominal carcinomatosis with ascites; carcinomatosis of pleura with effusion; lymphomas; interstitially in metastatic tumors

TABLE-44

Radioactive Isotopes Used in Medical Diagnosis and Therapy (Continued)

Name and Symbol	Principal Nuclear Properties	Form	Use
Indium ^{113m}In	Half-life 1.658h $\gamma(0.393)$	Indium colloid	Diagnostic: Liver and spleen imaging
		Indium Fe(OH)$_3$	Diagnostic: Pulmonary perfusion imaging; cardiac output
		Indium labeled red blood cells	Diagnostic: Determn of blood volume
		Indium pentetate (DTPA)	Diagnostic: Brain scan; renal function studies
		Indium transferrin	Diagnostic: Static cardiovascular blood pool imaging; hepatic and placenta blood pool imaging; placenta localization
^{111}In	Half-life 2.807d K; $\gamma(0.172, 0.247)$	Indium bleomycin	Diagnostic: Tumor detection
		Indium capromab pendetide	Diagnostic: Tumor detection for prostate cancer
		Indium chloride	Diagnostic: Hematopoietic bone marrow imaging; tumor detection
		Indium oxyquinoline (oxine) labeled leukocytes	Diagnostic: Detection of abscesses, infections and inflammation
		Indium oxyquinoline (oxine) labeled platelets	Diagnostic: Detection of deep vein thrombosis; cardiac thrombosis; renal transplant rejection
		Indium oxyquinoline (oxine) labeled red blood cells	Diagnostic: Detection of gastrointestinal bleeding
		Indium pentetate (DTPA)	Diagnostic: Gastric emptying studies; cardiac output; renal scintigraphy; cisternography
		Indium pentetreotide	Diagnostic: Neuroendocrine tumor detection
		Indium satumomab pendetide	Diagnostic: Tumor detection
Iodine ^{131}I	Half-life 8.040d $\beta^-(0.607, 0.81, 0.336)$ $\gamma(0.080, 0.284, 0.364, 0.637, 0.723)$	Diiodofluorescein	Diagnostic: Brain scan
		Iobenguane (MIBG)	Diagnostic: Adrenomedullary imaging and tumor detection
			Therapeutic: Antineoplastic (radiation source) in treatment of neuroendocrine tumors
		Iodinated fats and fatty acids, e.g. oleic acid, triolein	Diagnostic: Pancreatic function; intestinal fat absorption
		Iodinated fibrinogen	Diagnostic: *In vitro* determn of fibrinolytic enzymes
		Iodinated human serum albumin (IHSA)	Diagnostic: Plasma volume determn; peripheral vascular flow; cardiac output; circulation time; cerebral vascular flow. Brain scan; placenta localization. Cisternography
		Iodinated human serum albumin (macroaggregated)	Diagnostic: Pulmonary perfusion imaging
		Iodinated human serum albumin (microaggregated)	Diagnostic: Hepatic blood pool imaging
		Iodinated levothyroxine	Diagnostic: Metabolic study of endogenous thyroxine. *In vitro* determn of thyroid function
		Iodinated liothyronine	Diagnostic: *In vitro* determn of thyroid function
		Iodinated povidone	Diagnostic: Protein-loss enteropathy
		Iodinated rose bengal	Diagnostic: Liver function in hepatic excretion studies
		Iodinated tositumomab	Therapeutic: Antineoplastic (radiation source) in treatment of non-Hodgkin's lymphoma
		Iodohippurate sodium	Diagnostic: Determn of renal function, renal blood flow, urinary tract obstruction; renal imaging
		Sodium iodide	Diagnostic: Thyroid function studies; thyroid imaging
			Therapeutic: Hyperthyroidism; antineoplastic (radiation source) in treatment of thyroid cancer

TABLE-45

Radioactive Isotopes Used in Medical Diagnosis and Therapy (Continued)

Name and Symbol	Principal Nuclear Properties	Form	Use
Iodine ^{125}I	Half-life 60.14d K; $\gamma(0.035)$	Iodinated fats and fatty acids	Diagnostic: Pancreatic function; intestinal fat absorption
		Iodinated fibrinogen	Diagnostic: Localization of deep vein thrombosis; study of fibrinogen metabolism. *In vitro* determn of fibrinolytic enzymes
		Iodinated human serum albumin (IHSA)	Diagnostic: Determn of blood or plasma volume; circulation time; cardiac output
		Iodinated levothyroxine	Diagnostic: Metabolic study of endogenous thyroxine. *In vitro* determn of thyroid function
		Iodinated liothyronine	Diagnostic: *In vitro* determn of thyroid function
		Iodinated povidone	Diagnostic: Protein-loss enteropathy
		Iodinated rose bengal	Diagnostic: Liver function in hepatic excretion studies
		Sealed source	Diagnostic: External radiation source for bone mineral analyzer
		Sodium iodide	Diagnostic: Thyroid function studies; thyroid imaging
^{123}I	Half-life 13.2h K; $\gamma(0.159)$	IACFT	Diagnostic: SPECT cerebral imaging
		Iobenguane (MIBG)	Diagnostic: Adrenomedullary imaging and tumor detection
			Therapeutic: Antineoplastic (radiation source) in treatment of neuroendocrine tumors
		Iodohippurate sodium	Diagnostic: Determn of renal function, renal blood flow, urinary obstruction; renal imaging
		Iofetamine hydrochloride (IMP)	Diagnostic: Cerebral imaging
		Ioflupane (I-FP-CIT)	Diagnostic: SPECT cerebral imaging
		Sodium iodide	Diagnostic: Thyroid function studies; thyroid imaging
Iridium ^{192}Ir	Half-life 73.831d $\beta^-(0.67)$ $\gamma(0.296, 0.308, 0.317, 0.468, 0.589, 0.604, 0.612)$	Seed encased in nylon ribbon	Therapeutic: Antineoplastic (interstitial radiation source)
Iron ^{59}Fe	Half-life 44.496d $\beta^-(0.273, 0.475)$ $\gamma(1.095, 1.292)$	Ferric chloride Ferrous citrate Ferrous sulfate	Diagnostic: Ferrokinetics
		Labeled red blood cells	Diagnostic: Red cell maturation studies
^{55}Fe	Half-life 2.73y K	Labeled red blood cells	Diagnostic: Red cell maturation studies
Krypton ^{85}Kr	Half-life 10.72y $\beta^-(0.67)$ $\gamma(0.517)$	Gas	Diagnostic: Cardiac abnormalities; skeletal muscle, coronary or cerebral blood flow
^{81m}Kr	Half-life 13s $\gamma(0.19)$	Gas	Diagnostic: Lung ventilation studies
Lead RaD(^{210}Pb)	Half-life 22.3y $\beta^-(0.017)$ $\gamma(0.047)$	Beta ray applicator	Therapeutic: See Strontium (^{90}Sr)
Daughter RaE(^{210}Bi)	Half-life 5.013d $\beta^-(1.16)$		
Mercury ^{197}Hg	Half-life 2.6725d K; $\gamma(0.077)$	Chlormerodrin	Diagnostic: Brain scan; renal imaging
		Merisoprol	Diagnostic: Determn of renal function

TABLE-46

Radioactive Isotopes Used in Medical Diagnosis and Therapy (Continued)

Name and Symbol	Principal Nuclear Properties	Form	Use
Mercury ^{203}Hg	Half-life 46.60d β^-(0.214) γ(0.279)	Chlormerodrin	Diagnostic: Brain scan; renal imaging
Phosphorous ^{32}P	Half-life 14.282d β^-(1.71)	Chromic phosphate	Therapeutic: Antineoplastic (radiation source) in treatment of peritoneal or pleural effusions caused by metastatic disease
		Labeled red blood cells	Diagnostic: Blood volume determn
		Sodium phosphate	Diagnostic: Study of peripheral vascular disease; localization of ocular, brain and skin tumors; study of breast carcinomas
			Therapeutic: Polycythemia vera; chronic myelocytic leukemia; chronic lymphocytic leukemia; skeletal metastases; antineoplastic (radiation source)
Potassium ^{43}K	Half-life 22.3h β^-(0.83, 0.46, 1.22, 1.82) γ(0.618, 0.373, 0.39, 0.59, 0.22)	Potassium chloride	Diagnostic: Myocardial scan; determn of total exchangeable potassium
^{42}K	Half-life 12.360h β^-(3.52) γ(1.524)	Potassium carbonate	Diagnostic: Localization of brain tumors; determn of intracellular fluid space
		Potassium chloride	Diagnostic: Tumor detection; renal blood flow; determn of total exchangeable potassium
Radium ^{226}Ra	Half-life 1600y α(4.78, 4.60) γ(0.187)	Radium bromide; α and β particles filtered by platinum	Therapeutic: Antineoplastic (radiation source) in interstitial treatment of malignancies such as cancer of uterine cervix and fundus, oral pharynx, urinary bladder, skin and metastatic cancer of lymph nodes
Radon (Radium Emanation) ^{222}Rn (Daughter of ^{226}Ra)	Half-life 3.825d α(5.49) γ(0.510)	Gaseous radon; α and β particles filtered by 0.3 mm of gold	Therapeutic: See Radium (^{226}Ra)
Ruthenium ^{106}Ru	Half-life 1.020y β^-(0.039)	Beta ray applicator	Therapeutic: See Strontium (^{90}Sr)
Daughter (^{106}Rh)	Half-life 29.80s β^-(3.53, 3.1, 2.4) γ(0.512, 0.622, 1.128)		
Samarium ^{153}Sm	Half life 1.946d β^-(0.81, 0.71, 0.64) γ(0.103, 0.070)	Sm-lexidronam (EDTMP)	Therapeutic: Antineoplastic (radiation source) in treatment of bone cancer
Selenium ^{75}Se	Half-life 119.77d K; γ(0.265, 0.136, 0.121, 0.280, 0.401)	Selenomethionine	Diagnostic: Imaging of pancreas and parathyroid glands
Sodium ^{24}Na	Half-life 14.659h β^-(1.389, 4.17) γ(1.369, 2.754)	Sodium chloride	Diagnostic: Determn of circulation times, sodium space, total exchangeable sodium
^{22}Na	Half-life 2.602y β^+(0.545, 1.83) K; γ(1.275)	Sodium chloride	Diagnostic: Determn of sodium space and total exchangeable sodium
Strontium ^{85}Sr	Half-life 64.84d K; γ(0.514)	Strontium chloride Strontium nitrate	Diagnostic: Bone imaging
^{87m}Sr	Half-life 2.795h γ(0.388)		

TABLE-47

Radioactive Isotopes Used in Medical Diagnosis and Therapy (Continued)

Name and Symbol	Principal Nuclear Properties	Form	Use
Strontium ^{90}Sr	Half-life 28.5y β^-(0.546)	Beta ray applicator	Therapeutic: External irradiation for treatment of benign conditions of eye such as pterygia, traumatic corneal ulceration, corneal scars, vernal conjunctivitis, hemangioma of eyelid, vascularization of cornea and in preparation for a corneal transplant
Daughter ^{90}Y	Half-life 2.671d β^-(2.288) γ(2.186)		
Sulfur ^{35}S	Half-life 87.51d β^-(0.167)	Sodium sulfate	Diagnostic: Determn of extracellular fluid volume
Technetium ^{99m}Tc	Half-life 6.006h IT; γ(0.141)	Sodium pertechnetate	Diagnostic: Brain imaging. Cerebral angiography; thyroid imaging; salivary gland imaging; placenta localization; blood pool imaging; gastric mucosa imaging; cardiac function studies; renal blood flow studies. Urinary bladder imaging. Nasolacrimal drainage system imaging
		Sodium pertechnetate labeled red blood cells	Diagnostic: Determn of red blood cell volume, short-term survival studies. *In vitro* compatibility studies
		Tc-albumin	Diagnostic: Blood pool imaging; cardiovascular studies; placenta localization; determn of blood or plasma volumes
		Tc-albumin (aggregated)	Diagnostic: Pulmonary perfusion imaging
		Tc-albumin (microaggregated)	Diagnostic: Liver imaging
		Tc-apcitide	Diagnostic: Acute venous thrombosis imaging
		Tc-arcitumomab	Diagnostic: Tumor detection for colorectal cancer
		Tc-bicisate	Diagnostic: Brain imaging
		Tc-butedronate (DPD)	Diagnostic: Brain imaging
		Tc-depreotide	Diagnostic: Tumor detection for lung cancer
		Tc-disofenin (DISIDA)	Diagnostic: Hepatobiliary imaging
		Tc-etidronate (EHDP)	Diagnostic: Bone imaging
		Tc-exametazine (HM-PAO)	Diagnostic: Cerebral perfusion imaging
		Tc-fanolesomab	Diagnostic: Imaging and diagnosis of infections
		Tc-gluceptate	Diagnostic: Brain imaging; renal imaging; assess renal and brain perfusion
		Tc-labeled red blood cells	Diagnostic: Determn of red cell volume; short-term red cell survival studies
		Tc-lidofenin (HIDA)	Diagnostic: Hepatobiliary imaging
		Tc-mebrofenin	Diagnostic: Hepatobiliary imaging
		Tc-medronate (MDP)	Diagnostic: Bone imaging
		Tc-mertiatide (MAG3)	Diagnostic: Renal imaging
		Tc-oxidronate (HDP)	Diagnostic: Bone imaging
		Tc-pentetate (DTPA)	Diagnostic: Brain imaging; renal imaging; assess renal and brain perfusion; estimate glomerular filtration rate. Lung ventilation studies
		Tc-polyphosphates	Diagnostic: Bone imaging; myocardial imaging; blood pool imaging; detection of gastrointestinal bleeding
		Tc-pyrophosphate	Diagnostic: Bone imaging; cardiac imaging; blood pool imaging; detection of gastrointestinal bleeding
		Tc-sestamibi (HEXAMIBI)	Diagnostic: Myocardial perfusion imaging
		Tc-succimer	Diagnostic: Renal imaging
		Tc-sulesomab	Diagnostic: Detection of infections and inflammation

TABLE-48

Radioactive Isotopes Used in Medical Diagnosis and Therapy (Continued)

Name and Symbol	Principal Nuclear Properties	Form	Use
Technetium ^{99m}Tc (*Cont'd*)		Tc-sulfur colloid	Diagnostic: Liver, spleen and bone marrow imaging. Esophageal transit studies; gastro-esophageal reflux scintigraphy; determn of pulmonary aspiration of gastric contents; detection of intrapulmonary and lower gastrointestinal bleeding. Lung ventilation imaging
		Tc-teboroxime	Diagnostic: Myocardial perfusion imaging
		Tc-tetrofosmin	Diagnostic: Myocardial perfusion imaging
Thallium ^{201}Tl	Half-life 3.046d K; γ(0.135, 0.167)	Thallous chloride	Diagnostic: Myocardial perfusion imaging; localization of sites of parathyroid hyperactivity
Xenon ^{133}Xe	Half-life 5.245d β^-(0.346) γ(0.081)	Gas	Diagnostic: Pulmonary perfusion imaging; cerebral blood flow; lung ventilation studies
		Gas in saline solution	Diagnostic: Pulmonary perfusion imaging; regional blood flow; lung ventilation studies
^{127}Xe	Half-life 36.41d K; γ(0.172, 0.203)	Gas	Diagnostic: Pulmonary perfusion imaging; lung ventilation studies
Ytterbium ^{169}Yb	Half-life 32.022d K; γ(0.063, 0.100, 0.313, 0.177, 0.198, 0.308)	Yb-pentetate (DTPA)	Diagnostic: Cisternography; brain scan
Yttrium ^{90}Y	Half life 2.671d β^-(2.288) γ(2.186)	Y-edotreotide	Therapeutic: Antineoplastic (radiation source) in treatment of neuroendocrine tumors
		Y-epratuzumab	Therapeutic: Antineoplastic (radiation source) in treatment of non-Hodgkin's lymphoma
		Y-ibritumomab tiuxetan	Therapeutic: Antineoplastic (radiation source) in treatment of non-Hodgkin's lymphoma

References
1. S. Baum, R. Bramlet, *Basic Nuclear Medicine* (Appleton-Century-Crofts, New York, 1975).
2. C. Behrns *et al., Atomic Medicine* (Williams & Wilkins Co., Baltimore, 5th ed., 1969).
3. H.M. Chilton, R. L. Witcofski, *Nuclear Pharmacy* (Lea & Febiger, Philadelphia, 1986).
4. *CRC Handbook of Radioactive Nuclides,* Y. Wang, Ed. (Chemical Rubber Co., Cleveland, Ohio, 1969).
5. *Diagnostic and Investigational Uses of Radiolabeled Blood Elements,* R. J. Davey, M. E. Wallace, Eds. (Am. Assoc. of Blood Banks, Arlington, Va., 1987).
6. *Martindale: The Complete Drug Reference,* S. C. Sweetman, Ed. (Pharmaceutical Press, London, 34th ed., 2005) pp. 1522-1526.
7. *Physician's Desk Reference for Radiology and Nuclear Medicine,* L. M. Freeman, M. D. Blaufox, Editorial Consultants (Medical Economics Co., Oradell, N.J., 4th ed., 1974/75, 5th ed., 1975/76.).
8. *Remington: The Science and Practice of Pharmacy,* A. R. Gennaro, Ed. (Lippincott Williams & Wilkins, Baltimore, MD, 20th ed., 2000) pp. 469-482.
9. D. P. Swanson *et al., Pharmaceuticals in Medical Imaging* (Macmillan Publishing Co., Inc., New York, 1990).

Notes
1. The following abbreviations for time units are used: y = years, d = days, h = hours, min = minutes, s = seconds.
2. The column for Principal Nuclear Properties lists the principal modes of disintegration and energies of the radiation in million electronvolts (MeV) for the individual isotopes. Symbols used to represent the modes of decay are: α = alpha particle emission, β^- = beta particle, β^+ = positron, γ = gamma ray, K = electron capture.

TABLE-49

Amino Acids Found in Proteins

General structure of the 20 common amino acids found in proteins except for proline, which has a 3-carbon cyclic side chain bound to the nitrogen of the backbone.

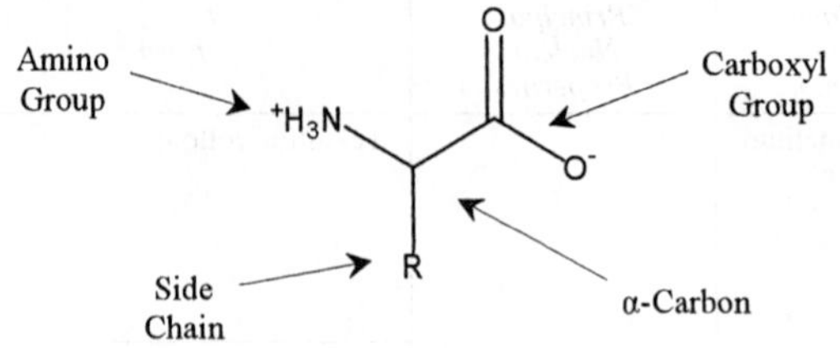

Amino Group

Carboxyl Group

^+H_3N

Side Chain

R

α-Carbon

Name	Abbreviations		pKa Values			R Group Structure
	3 Letter	1 Letter	–COOH	$-NH_3^+$	R Group	
Alanine	Ala	A	2.34	9.69		—CH_3
Arginine	Arg	R	2.17	9.04	12.48	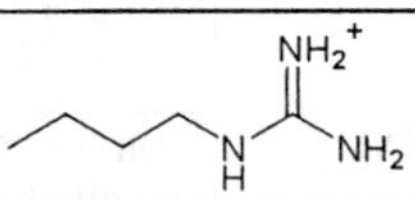
Asparagine	Asn	N	2.02	8.80		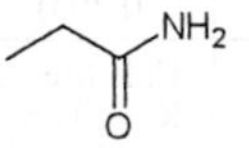
Aspartic Acid	Asp	D	1.88	9.60	3.65	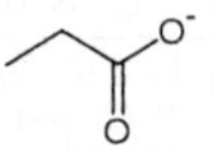
Cysteine	Cys	C	1.71	8.33	10.78	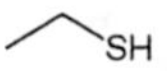
Glutamic Acid	Glu	E	2.19	9.67	4.25	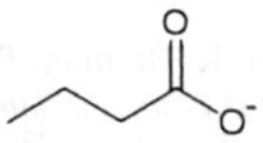
Glutamine	Gln	Q	2.17	9.13		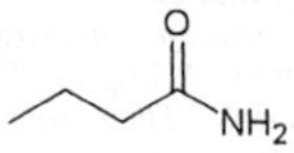
Glycine	Gly	G	2.34	9.60		—H
Histidine	His	H	1.82	9.17	6.00	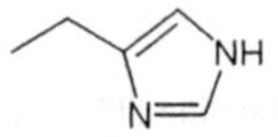
Isoleucine	Ile	I	2.36	9.68		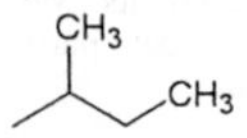
Leucine	Leu	L	2.36	9.60		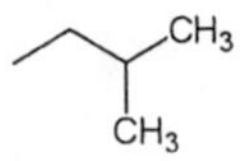
Lysine	Lys	K	2.18	8.95	10.53	

TABLE-50

Amino Acids Found in Proteins (Continued)

Name	Abbreviations		pKa Values			R Group Structure
	3 Letter	1 Letter	–COOH	–NH$_3^+$	R Group	
Methionine	Met	M	2.28	9.21		
Phenylalanine	Phe	F	1.83	9.13		
Proline	Pro	P	1.99	10.60		
Serine	Ser	S	2.21	9.15		
Threonine	Thr	T	2.63	10.43		
Tryptophan	Trp	W	2.38	9.39		
Tyrosine	Tyr	Y	2.20	9.11	10.07	
Valine	Val	V	2.32	9.62		

TABLE-51

Chemical Terms Translator

English	French	German	Russian	Japanese
acetic acid	acide acétique	Essigsäure	уксусная кислота	酢酸
acetone	acétone	Aceton	ацетон	アセトン
acidic	acide	sauer	кислотный	酸性の
alkaline	alcalin	alkalisch	щелочной	アルカリ性の
amorphous	amorphe	amorph	аморфный	無定形の
anhydrous	anhydre	wasserfrei	безводный	無水の
aqueous	aqueux	wässrig	водный	水性の
boiling point	point d'ébullition	Siedepunkt	точка кипения	沸点
chloroform	chloroforme	Chloroform	хлороформ	クロロホルム
chromatography	chromatographie	Chromatographie	хроматография	クロマトグラフィー
cold	froid	kalt	холодный	冷たい
colorless	incolore	farblos	бесцветный	無色の
compound	composé	Verbindung	соединение	化合物
condensation	condensation	Kondensation	конденсация	凝縮
cool	refroidir	kühlen	охлаждать	冷却する
corrosive	corrosif	korrosiv; ätzend	коррозионный	腐食性の
crude	brut	roh	неочищенный	粗製の
crystal	cristal	Kristall	кристалл	結晶
crystalline	cristallin	kristallin	кристаллический	結晶性の
decomposition	décomposition	Zersetzung	разложение	分解
dehydration	déshydratation	Dehydratisierung	дегидратация	脱水
deliquescense	déliquescence	Zerfließlichkeit	расплывание	潮解
density	densité	Dichte	плотность	密度
dextrorotatory	dextrogyre	rechtsdrehend	правовращающий	右施性の
dilute (adjective)	dilué	verdünnt	разбавленный	希釈した
dilute (verb)	diluer	verdünnen	разбавлять	希釈する
dissolve	dissoudre	lösen	растворять	溶解する
distill	distiller	destillieren	дистиллировать	蒸留する
dry	sécher	trocknen	сушить	乾燥する
easily soluble	facilement soluble	leicht löslich	легкорастворимый	溶解しやすい
empirical formula	formule empirique	Summenformel	эмпирическая формула	実験式
esterification	estérification	Veresterung	эстерификация	エステル化
ethanol	éthanol	Äthanol	этанол	エタノール
ether	éther	Äther	простой эфир	エーテル
evaporation	évaporation	Verdunstung	выпаривание	蒸発
explosive	explosif	explosiv	взрывчатый	爆発性の
flammable	inflammable	entflammbar	воспламеняемый	引火性の
flash point	point d'éclair	Flammpunkt	температура вспышки	引火点
fraction	fraction	Fraktion	фракция	分画
freezing point	point de congélation	Gefrierpunkt	точка замерзания	凝固点
gas	gaz	Gas	газ	気体
heat	chauffer	erhitzen	нагревать	加熱する
hot	chaud	heiß	горячий	熱い
hydrochloric acid	acide chlorhydrique	Salzsäure	соляная кислота	塩酸
hygroscopic	hygroscopique	hygroskopisch	гигроскопический	吸湿性の
immiscible	immiscible	unmischbar	несмешивающийся	混和しない
impure	impur	unrein	нечистый	不純な
IR (infrared)	IR (infrarouge)	IR (infrarot)	ИК (инфракрасный)	IR（赤外の）
insoluble	insoluble	unlöslich	нерастворимый	不溶性の
isolation	isolation	Isolierung	выделение	単離
isomer	isomère	Isomer	изомер	異性体

TABLE-52

Chemical Terms Translator (Continued)

English	French	German	Russian	Japanese
levorotatory	lévogyre	linksdrehend	левовращающий	左施性の
light	lumière	Licht	свет	光
liquid	liquide	Flüssigkeit	жидкость	液体
lustrous	lustré	glänzend	блестящий	光沢がある
MS (mass spectrometry)	spectrométrie de masse	MS (Massenspektrometrie)	МС (масс-спектрометрия)	MS (質量分析)
maximum	maximum	Maximum	максимум	最大の
melting point	point de fusion	Schmelzpunkt	точка плавления	融点
minimum	minimum	Minimum	минимум	最小の
miscible	miscible	mischbar	смешивающийся	混合できる
mix	mélanger	mischen	смешивать	混合する
mixture	mélange	Mischung	смесь	混合物
mole	mole	Mol	моль	モル
molecular weight	poids moléculaire	Molekulargewicht	молекулярный вес	分子量
nitric acid	acide nitrique	Salpetersäure	азотная кислота	硝酸
NMR (nuclear magnetic resonance)	résonance magnétique nucléaire	NMR (kernmagnetische Resonanz)	ЯМР (ядерный магнитный резонанс)	NMR （核磁気共鳴）
normal (N)	normal (N)	normal (N)	нормальный (о растворе)	規定 (N)
odorless	inodore	geruchlos	без запаха	無臭の
oil	huile	Öl	нефть	油
optical rotation	rotation optique	optische Drehung	оптическое вращение	旋光
phase	phase	Phase	фаза	相
powder	poudre	Pulver	порошок	粉末
practically insoluble	pratiquement insoluble	nahezu unlöslich	практически нерастворимый	ほとんど溶けない
precipitate (noun)	précipité	Präzipitat	осадок	沈殿
precipitate (verb)	précipiter	präzipitieren	осаждать	沈殿する
preparation	préparation	Präparat	препарат	調製
pure	pur	rein	чистый	純粋な
racemate	racémate	Razemat	рацемат	ラセミ体
recrystallize	recristalliser	umkristallisieren	перекристаллизовывать	再結晶する
residue	résidu	Rückstand	остаток	残留物
resolution	résolution	Auflösung	разрешающая способность	分解能
saturated	saturé	gesättigt	насыщенный	飽和した
slightly soluble	légèrement soluble	schwer löslich	слегка растворимый	溶けにくい
sodium hydroxide	hydroxyde de sodium	Natriumhydroxid	едкий натр	水酸化ナトリウム
solid	solide	Feststoff	твердое вещество	固体
solubility	solubilité	Löslichkeit	растворимость	溶解度
soluble	soluble	löslich	растворимый	可溶性の
solution	solution	Lösung	раствор	溶液
solvent	solvant	Lösungsmittel	растворитель	溶媒
somewhat soluble	quelque peu soluble	mäßig löslich	частично растворимый	若干可溶性の
sparingly soluble	difficilement soluble	wenig löslich	слаборастворимый	やや溶けにくい
sublime	sublimer	sublimieren	сублимировать	昇華する
synthesis	synthèse	Synthese	синтез	合成
uv (ultraviolet)	uv (ultraviolet)	UV (ultraviolett)	УФ (ультрафиолетовый)	uv (紫外線の)
vapor pressure	pression de vapeur	Dampfdruck	упругость пара	蒸気圧
very soluble	très soluble	sehr leicht löslich	высокорастворимый	非常に可溶性の
viscous	visqueux	viskos	вязкий	粘性の
volatile	volatil	flüchtig	летучий	揮発性の
warm	réchauffer	erwärmen	подогревать	温める
water	eau	Wasser	вода	水
yield	rendement	Ausbeute	выход	収量

TABLE-53

Selected Hexoses and Pentoses

Carbohydrates comprise one of the most abundant groups of naturally occurring organic compounds in living systems as well as being of great industrial importance. Presented below are various structural depictions of the common 5- and 6- carbon monosaccharides. Generally accepted carbohydrate nomenclature conventions include:

The use of D- and L-. The optical activity of the sugars is related to the reference compound glyceraldehyde which has only one D-form and one L-form. Since many of the aldoses have two or more chiral centers, the prefixes D- and L- refer to the configuration of the chiral carbon most distant from the carbonyl carbon. When the OH-group is drawn to the right in a Fischer projection, a D-sugar is designated.

Definition of the anomeric carbon. The new stereogenic center formed upon conversion from the Fischer projection to the cyclic hemiacetal is the anomeric carbon; the stereochemistry at this center is either α- or β-.

anomeric carbon

α–form   β–form

Sugar	Haworth Ring Pyranose	Haworth Ring Furanose	Chair Perspective β-form	Fischer Projection

Hexoses | | | | **Aldohexoses**

D-Allose	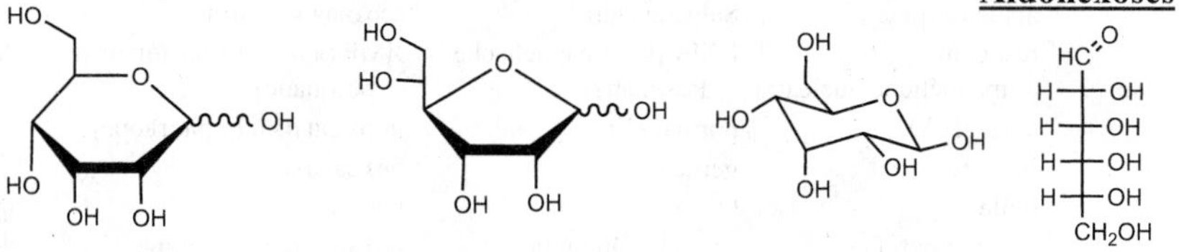			
D-Altrose	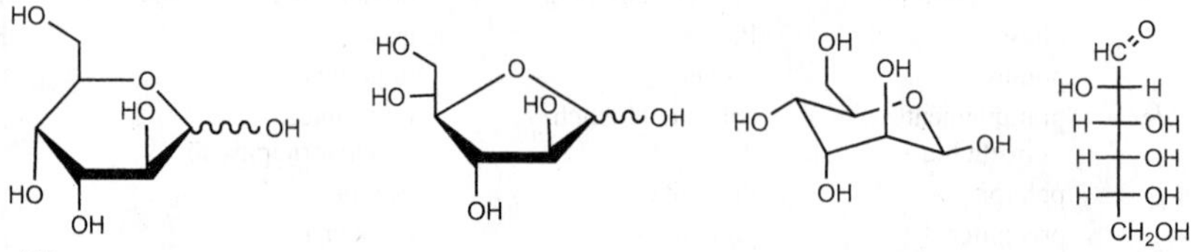			
D-Glucose	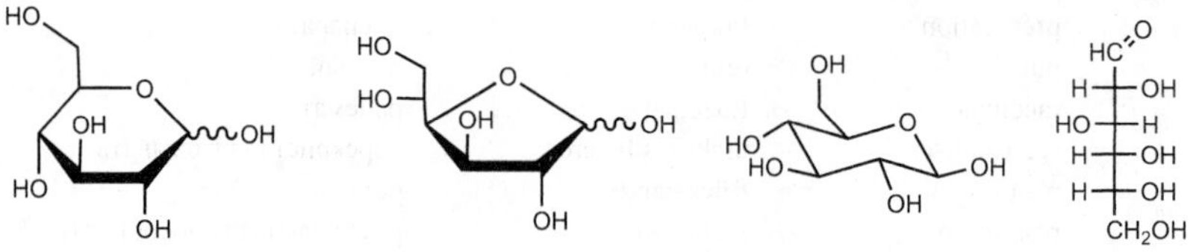			
D-Gulose	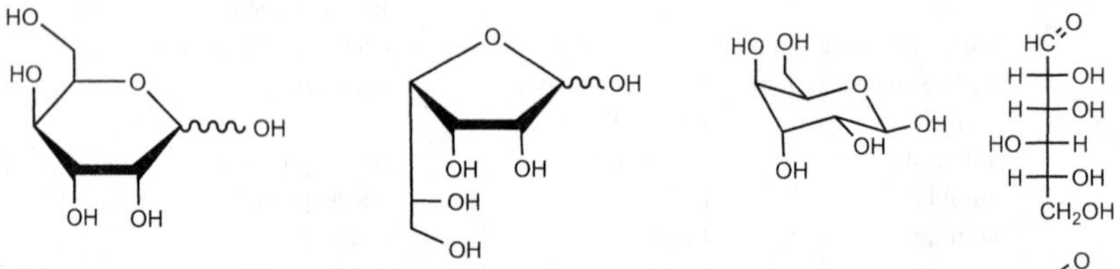			
D-Galactose	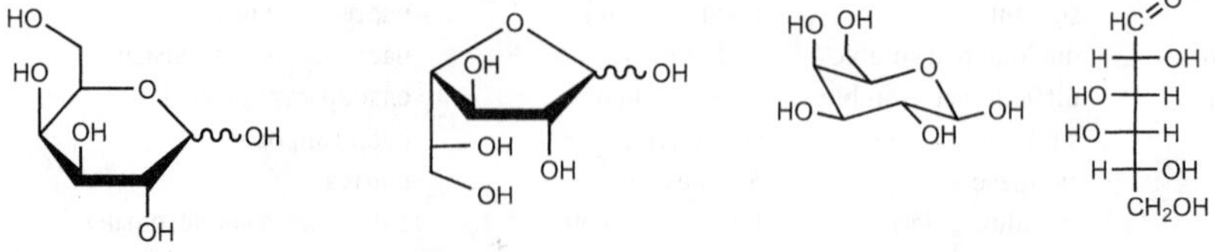			
D-Idose	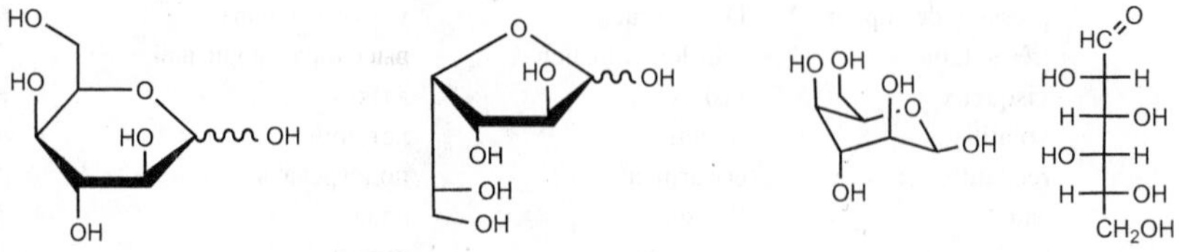			

TABLE-54

Selected Hexoses and Pentoses (Continued)

Sugar	Haworth Ring Pyranose	Haworth Ring Furanose	Chair Perspective β-form	Fischer Projection
D-Mannose				
D-Talose				
Ketoses				
D-Fructose β–form				
D-Sorbose β-form				
Pentoses			**Aldohexoses**	
D-Arabinose				
D-Lyxose				
D-Ribose				
D-Xylose				

TABLE-55

Common Heterocyclic Ring Systems

O - Containing Rings

  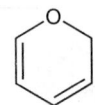 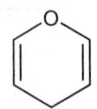 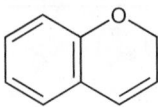 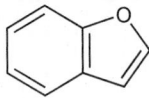

oxirane   furan   2*H*-pyran   4*H*-pyran   2*H*-chromene   benzo[*b*]furan

S - Containing Rings

 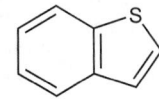 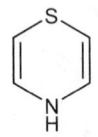

thiophene   benzo[*b*]thiophene   parathiazine

N - Containing Rings

 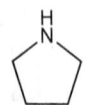 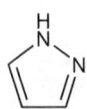 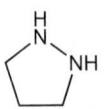 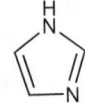

pyrrole   pyrrolidine   pyrazole   pyrazolidine   imidazole   imidazoline   imidazolidine

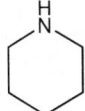 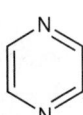 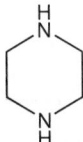 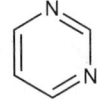

pyridine   piperidine   pyrazine   piperazine   pyrimidine   triazine

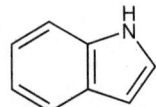 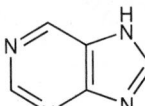 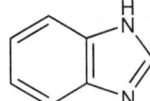 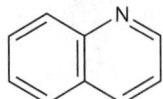 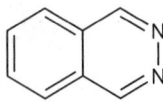

indole   purine   benzimidazole   quinoline   phthalazine

Other

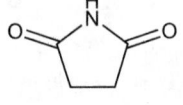 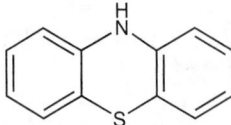 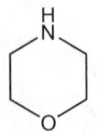 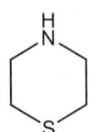

succinimide   thiazolidine   phenothiazine   morpholine   thiomorpholine

TABLE-56

International Patent Country Codes

Code	Country Name	Code	Country Name
AE	United Arab Emirates	KR	Republic of Korea
AG	Antigua and Barbuda	KZ	Kazakhstan
AL	Albania	LC	Saint Lucia
AM	Armenia	LI	Liechtenstein
AR	Argentina	LK	Sri Lanka
AT	Austria	LR	Liberia
AU	Australia	LS	Lesotho
AZ	Azerbaijan	LT	Lithuania
BA	Bosnia and Herzegovina	LU	Luxembourg
BB	Barbados	LV	Latvia
BE	Belgium	LY	Libyan Arab Jamahiriya
BF	Burkina Faso	MA	Morocco
BG	Bulgaria	MC	Monaco
BJ	Benin	MD	Republic of Moldova
BR	Brazil	MG	Madagascar
BW	Botswana	MK	The former Yugoslav Republic of Macedonia
BY	Belarus	ML	Mali
BZ	Belize	MN	Mongolia
CA	Canada	MR	Mauritania
CF	Central African Republic	MW	Malawi
CG	Congo	MX	Mexico
CH	Switzerland	MZ	Mozambique
CI	Côte d'Ivoire	NA	Namibia
CM	Cameroon	NE	Niger
CN	China	NG	Nigeria
CO	Colombia	NI	Nicaragua
CR	Costa Rica	NL	Netherlands
CS	Czechoslovakia	NO	Norway
CU	Cuba	NZ	New Zealand
CY	Cyprus	OM	Oman
CZ	Czech Republic	PG	Papua New Guinea
DD	German Democratic Republic	PH	Philippines
DE	Germany	PL	Poland
DK	Denmark	PT	Portugal
DM	Dominica	RO	Romania
DZ	Algeria	RU	Russian Federation
EC	Ecuador	SC	Seychelles
EE	Estonia	SD	Sudan
EG	Egypt	SE	Sweden
EP	European Patent Office	SG	Singapore
ES	Spain	SK	Slovakia
FI	Finland	SL	Sierra Leone
FR	France	SM	San Marino
GA	Gabon	SN	Senegal
GB	United Kingdom	SU	USSR
GD	Grenada	SY	Syrian Arab Republic
GE	Georgia	SZ	Swaziland
GH	Ghana	TD	Chad
GM	Gambia	TG	Togo
GN	Guinea	TJ	Tajikistan
GQ	Equatorial Guinea	TM	Turkmenistan
GR	Greece	TN	Tunisia
GW	Guinea-Bissau	TR	Turkey
HR	Croatia	TT	Trinidad and Tobago
HU	Hungary	TW	Taiwan, Province of China
ID	Indonesia	TZ	United Republic of Tanzania
IE	Ireland	UA	Ukraine
IL	Israel	UG	Uganda
IN	India	US	United States of America
IS	Iceland	UZ	Uzbekistan
IT	Italy	VC	Saint Vincent and the Grenadines
JP	Japan	VN	Viet Nam
KE	Kenya	WO	World Intellectual Property Organization (WIPO)
KG	Kyrgyzstan	YU	Serbia and Montenegro
KM	Comoros	ZA	South Africa
KN	Saint Kitts and Nevis	ZM	Zambia
KP	Democratic People's Republic of Korea	ZW	Zimbabwe

TABLE-57

INTERNATIONAL SYSTEM OF UNITS (SI)

SI Basic and Supplementary Units

Name	Symbol	Physical Quantity
SI BASE UNITS		
meter	m	length
kilogram	kg	mass
second	s	time
ampere	A	electric current
kelvin	K	thermodynamic temperature
mole	mol	amount of substance
candela	cd	luminous intensity
SI SUPPLEMENTARY UNITS		
radian	rad	plane angle
steradian	sr	solid angle

SI Derived Units with Special Names

Formula	Symbol	Name	Physical Quantity
$kg \cdot m/s^2$	N	Newton	force
N/m^2	Pa	Pascal	pressure or stress
$N \cdot m$	J	Joule	work, energy, or quantity of heat
J/s	W	watt	power or radiant energy flux
W/A	V	volt	electric potential, potential difference or electromotive force
A/V	S	siemens	electric conductance
V/A	Ω	ohm	electric resistance
$A \cdot s$	C	Coulomb	quantity of electricity or electric charge
C/V	F	Farad	electric capacitance
$V \cdot s$	Wb	Weber	magnetic flux
Wb/A	H	Henry	inductance
Wb/m^2	T	Tesla	magnetic flux density or magnetic induction
$cd \cdot sr$	lm	lumen	luminous flux
lm/m^2	lx	lux	illuminance
J/kg	Gy	gray	absorbed dose (of ionizing radiation)
1 (disintegrations)/s	Bq	Becquerel	activity (of a radionuclide)
1 (cycle)/s	Hz	hertz	frequency (of a periodic phenomenon)
$K - 273$	°C	degree Celsius	temperature

SI Prefixes

Factor	Prefix	Symbol	Factor	Prefix	Symbol
10^{-24}	yocto	y	10	deca	da
10^{-21}	zepto	z	10^2	hecto	h
10^{-18}	atto	a	10^3	kilo	k
10^{-15}	femto	f	10^6	mega	M
10^{-12}	pico	p	10^9	giga	G
10^{-9}	nano	n	10^{12}	tera	T
10^{-6}	micro	μ	10^{15}	peta	P
10^{-3}	milli	m	10^{18}	exa	E
10^{-2}	centi	c	10^{21}	zetta	Z
10^{-1}	deci	d	10^{24}	yotta	Y

TABLE-58

Numerical Prefixes Commonly Used in Forming Chemical Names

Numeral	Prefix	Numeral	Prefix	Numeral	Prefix
$\frac{1}{2}$	hemi-	13	trideca-	28	octacosa-
1	mono-	14	tetradeca-	29	nonacosa-
$1\frac{1}{2}$	sesqui-	15	pentadeca-	30	triaconta-
2	di-, bi-	16	hexadeca-	40	tetraconta-
$2\frac{1}{2}$	hemipenta-	17	heptadeca-	50	pentaconta-
3	tri-	18	octadeca-	60	hexaconta-
4	tetra-	19	nonadeca-	70	heptaconta-
5	penta-	20	eicosa-	80	octaconta-
6	hexa-	21	heneicosa-	90	nonaconta-
7	hepta-	22	docosa-	100	hecta-
8	octa-	23	tricosa-	101	henhecta-
9	ennea, nona-	24	tetracosa-	102	dohecta-
10	deca-	25	pentacosa-	110	decahecta-
11	hendeca-, undeca-	26	hexacosa-	120	eicosahecta-
12	dodeca-	27	heptacosa-	200	dicta-

Alchemical Symbols Used in Biology and Botany

⊙	Sun; gold; annual plant.
⊙ ⊙	Biennial plant.
O or ☽	Moon; silver; *sometimes* female sex.
☿	Mercury.
♀	Venus; copper; *female sex.*
♁	Earth; terra; *sometimes* male sex.
♁ ♀	Having male and female flowers separate.
♁ — ♀	Having male and female flowers on the same plant.
♁ : ♀	Having male and female flowers on different plants.
♂	Mars; iron; *male sex.*
☌	Conjunction; mating; mated.
♃	Jupiter; tin; perennial herb.
♄	Saturn; lead.
🜍	Sulfur.
🜺	Arsenic.

Prescription Notation

The units of quantities written on prescriptions in the Apothecaries System are usually designated by symbols. Occasionally abbreviations are used. The following list provides these symbols and abbreviations.

Apothecaries' Weight

Symbol	*Abbreviation*	*Unit*
—	gr.	grain
℈	sc.	scruple
ʒ	dr.	dram
℥	oz.	ounce
℔	lb.	pound

Apothecaries' Volume

Symbol	*Abbreviation*	*Unit*
♏	min.	minim
f ʒ	fl. dr.	fluid dram
f ℥	fl. oz.	fluid ounce
O.	pt.	pint
Cong.	gal.	gallon

Roman numerals (lower case) are always used following the symbol to designate the number of units required, but, if the abbreviation is used, arabic numerals are used and precede the appreviation. For example:

ʒ iv but 4 dr.

For less than one unit, one-half may be designated by ss following the symbol, but other fractions must be designated by arabic numeral fractions. For example:

ʒ ss but gr. $\frac{1}{8}$

Roman numerals are also usually used to designate the number of dosage forms required. For example:

Caps. No. xiv

Latin Terms

Abbreviation	Expanded form	Meaning	Abbreviation	Expanded form	Meaning
a.	ante	before	d.	dexter	right
a., aur.	auris	ear	d.	dies	a day
aa.	ana	of each	d.	dosis	a dose
a.c.	ante cibos	before meals	da.	da	give
ad	ad	to, up to	dent., d.	dentur	let be given, give
add.	adde, addantur, addendus, addendo	add, let them be added, to be added, by adding	det.	detur	let it be given
			dieb. alt.	diebus alternis	on alternate days
			dieb. secund.	diebus secundis	every second day
ad lib.	ad libitum	at pleasure	dil.	dilue	dilute
adm.	admove	apply	disp.	dispensa, dispensetur	dispense
ad man. med.	ad manus medici	(to be delivered) into the hands of the (pre-scribing) physician	div.	divide	divide
			div. in par. aeq.	dividatur in partes aequales	divide in equal parts
ad satur.	ad saturandum	to saturation			
agit.	agita	shake	d.t.d.	dentur tales doses	give such doses
alb.	albus	white	dulc.	dulcis	sweet
alt. hor.	alternis horis	every other hour	dur.	durus	hard
A.M.	ante meridiem	before noon	e.m.p.	ex modo prescripto	after the manner prescribed
ampl.	amplus	large			
ampul.	ampulla	ampul, ampule	et	et	and
applicand.	applicandus	to be applied	ex	ex	out of
aq.	aqua	water	ext.	extractum	an extract
aq. bull.	aqua bulliens	boiling water	f., ft.	fac, fiat, fiant	make, let be made
aq. dest.	aqua destillata	distilled water	ferv.	fervens	hot
aq. ferv.	aqua fervens	hot water	filt.	filtra	filter (imperative form)
aq. frig.	aqua frigida	cold water	fl.	fluidus	fluid
b.	bis	twice	flav.	flavus	yellow
ben.	bene	well	fldext.	fluidextractum	fluidextract
bib.	bibe	drink	fort.	fortis	strong
b.i.d.	bis in die	twice a day	frig.	frigidus	cold
bol.	bolus	a large pill	garg.	gargarisma	a gargle
brevis	brevis	short	Gm.	gramma	gram
bull.	bulliens, bulliat, bulliant	boiling, let boil	gr.	granum	grain
			gran.	granulatus	granulated
c̄	cum	with	gtt.	gutta	a drop
C	centum	a hundred	h.s.	hora somni	at the hour of sleep, at bedtime
cap.	capiat	let the patient take			
caps.	capsula	a capsule	hydrarg.	hydrargyrum.	mercury
cerat.	ceratum	wax ointment	i.c.	inter cibos	between meals
chart.	charta	paper, a powder in paper	juxt.	juxta	near
			Kal.	Kalium	potassium
chart. cerat.	charta cerata	waxed paper, parch-ment paper	l.a.	lege artis	according to the art
			laev.	laevus	left
chirurg.	chirurgicalis	surgical	lb.	libra	pound
chord chirurg.	chorda chirurgicalis	surgical "catgut"	lev.	levis	light
			liq.	liquor	a liquor, a solution
cito disp!	cito dispensetur!	let it be dispensed quickly	lot.	lotio	lotion
			m.	mane	in the morning
coch. amp.	cochleare amplum	a tablespoonful	M.	misce	mix
coch. mag.	cochleare magnum	a tablespoonful	m.	mitte	send
			mag.	magnus	large
coch. med.	cochleare medium	a dessertspoonful	m. dict.	more dicto	as directed
			min.	minimum	minim
coch. mod.	cochleare modicum	a dessertspoonful	mist.	mistura	mixture
			mixt.	mixtura	mixture
coch. parv.	cochleare parvum	a teaspoonful	m.p.	modo praescripto	in the manner prescribed
col.	cola	strain (imperative form)	m.t.d.	mitte tales doses	send such doses
			n.	naris	nostril
colet.	coletur	let it be strained	Natr.	Natrium	sodium
collun.	collunarium	a nose wash	nebul.	nebula	a spray
collut.	collutorium	a mouth wash	n. et m.	nocte maneque	night and morning
collyr.	collyrium	an eye wash	nig.	niger	black
comp.	compositus	compound, compounded	no.	numero	number
			noct.	nocte	at night
cong.	congius	a gallon	non rep.	non repetatur	do not repeat
consperg.	consperge, conspergetur	dust, sprinkle	O.	Octarius	a pint
			o.d.	oculus dexter	right eye
cont.	contra	against	o.l.	oculus laevus	left eye
contus.	contusus	bruised	omn. hor.	omni hora	at every hour
coq.	coque, coquatur	boil, let it boil	omn. man.	omni mane	on every morning

TABLE-60

Latin Terms (Continued)

Abbreviation	Expanded form	Meaning	Abbreviation	Expanded form	Meaning
opt.	optimus	best	s.a.	secundum artem	according to art
o.s.	oculus sinister	left eye	sat.	saturatus	saturated
o.u.	oculi uterque	both eyes	scat.	scatula	box
p. ae.	partes aequales	equal parts	scat. orig.	scatula	original package, man-
parv.	parvus	small		originalis	ufacturer's package
p.c.	post cibos	after meals			(and label)
phial.	phiala	bottle	sic.	siccus	dried
Plumb.	Plumbum	lead	s.n.	secundum	according to nature
P.M.	post meridiem	after noon		naturam	
p.o.	per os	by mouth	sol.	solubilis	soluble
pond.	ponderosus	heavy	sol.	solutio	solution
ppt.	praecipitatus	precipitated	solv.	solve	dissolve
p.r.n.	pro re nata	as occasion arises, as	s.o.s.	si opus sit	if there is need
		needed	spir.	spiritus	spirit
pro rect.	pro recto	rectal	spiss.	spissus	dried
pt.	perstetur	let it be continued	ss.	semis	one half
pulv.	pulvis, pulveres	powder, powders	stat.	statim	immediately
q., qq.	quodque,	each, every	suc.	succus	juice
	quaeque		sum.	sume, sumendus	take, to be taken
q.i.d.	quater in die	four times a day	S.V.R.	spiritus vini	alcohol
qq. hor.	quaque hora	every hour		rectificatus	
Q.R.	quantum	the quantity is correct	syr.	syrupus	syrup
	rectum		tab.	tabella	tablet
q.s.	quantum suffi-	a sufficient quantity	tal.	talis, tales, talia	such
	ciat, quantum		ter.	tere	rub
	satis		t.i.d.	ter in die	three times a day
quot. op. sit	quoties opus sit	as often as necessary	tinct., tr.	tinctura	tincture
q.v.	quantum voleris	as much as you wish	trit.	tritura	triturate
℞	recipe	you take	ult.	ultime	lastly
recen.	recens	fresh	unct.	unctus	smeared
rect.	rectificatus	rectified	ung., ungt.	unguentum	ointment
ren. sem.	renovetur semel	shall be renewed (only)	ust.	ustus	burnt
		once	ut dict.	ut dictum	as directed
rept.	repetatur	let it be repeated	v.	vel	or
rub.	ruber	red	vesp.	vesper	evening
S., Sig.	signa, signetur	write, it shall be written	vir.	viridis, viride	green
		(as instruction to the	vol.	volatilis, volatile	volatile
		patient)			

TABLE-61

Greek Alphabet

Name of Letter	Capital	Lower Case	Transliteration	Name of Letter	Capital	Lower Case	Transliteration
alpha	A	α	a	nu	N	ν	n
beta	B	β	b	xi	Ξ	ξ	x
gamma	Γ	γ	g	omicron	O	o	o short
delta	Δ	δ	d	pi	Π	π	p
epsilon	H	ε	e short	rho	P	ρ	r
zeta	Z	ζ	z	sigma	Σ	σ or ς	s
eta	H	η	e long	tau	T	τ	t
theta	Θ	θ	th	upsilon	Υ	υ	y
iota	I	ι	i	phi	Φ	φ or ϕ	f
kappa	K	κ	k, c	chi	X	χ	ch as in German echt
lambda	Λ	λ	l	psi	Ψ	ψ	ps
mu	M	μ	m	omega	Ω	ω	o long

Russian Alphabet

Cyrillic Print		Transliteration	Pronunciation	Cyrillic Print		Transliteration	Pronunciation
А	а	a	*a* in far	Р	р	r	*r*
Б	б	b	*b*	С	с	s	*s* in say
В	в	v	*v*	Т	т	t	*t*
Г	г	g (h)	*g* in gay	У	у	u	*oo* in boot
Д	д	d	*d*	Ф	ф	f	*f*
Е	е	e	*e* in fell; also *ye* in yell	Х	х	kh	like German *ch*
Ж	ж	zh	*z* in azure	Ц	ц	ts	*ts* in hoots
З	з	z	*z* in zeal	Ч	ч	ch	*ch* in church
И	и	i	*i* in meet	Ш	ш	sh	*sh*
Й	й	ĭ	*y* in boy	Щ	щ	shch	*shch* as in fre*sh ch*eese
К	к	k	*k*	Ъ	ъ	″	silent*
Л	л	l	*l*	Ы	ы	y	*y* in rhythm (hard)
М	м	m	*m*	Ь	ь	′	silent* (softens preceding consonant)
Н	н	n	*n*	Э	э	ė	*e* in met
О	о	o	*o* in or	Ю	ю	iu	*u* in union
П	п	p	*p*	Я	я	ia	*ya* in yard

* Hard sign; used to separate a consonant from a soft vowel especially in foreign words; frequently replaced by an apostrophe.

Roman Numerals

I 1	II 2	III 3	IV 4	V 5	VI 6	VII 7	VIII 8	IX 9	X 10
XX 20	XXX 30	XL 40	L 50	LX 60	LXX 70	LXXX 80	XC 90	IC 99	C 100
CC 200	CCC 300	CD 400	D 500	DC 600	DCC 700	DCCC 800	CM 900	XM 990	M 1000

TABLE-62

Fundamental Physical and Mathematical Constants

Physical Constants

QUANTITY	SYMBOL	VALUE
Atomic mass unit (Dalton)	amu	1.661×10^{-24} g
Avogadro constant	N_A, L	6.0221367×10^{23} mol^{-1}
Bohr magneton	μ_B	$9.2740154 \times 10^{-24}$ J T^{-1}
Boltzmann constant	k	1.380658×10^{-23} J K^{-1}
Electron mass	m_e	$9.1093897 \times 10^{-31}$ kg
Elementary charge	e	$1.60217653 \times 10^{-19}$ C
Faraday constant	F	9.6485309×10^4 C mol^{-1}
Gas constant	R	8.314510 J mol^{-1} K^{-1}
		8.314510×10^{-2} L bar mol^{-1} K^{-1}
		8.20578×10^{-2} L atm mol^{-1} K^{-1}
		6.2364 L Torr mol^{-1} K^{-1}
		1.9872 cal mol^{-1} K^{-1}
Gravitational constant	G	6.67259×10^{-11} m^3 kg^{-1} s^{-2}
Molar volume of ideal gas at STP	V_m	22.413996×10^{-3} m^3 mol^{-1}
Neutron mass	m_n	$1.6749286 \times 10^{-27}$ kg
Nuclear magneton	μ_N	$5.0507866 \times 10^{-27}$ J T^{-1}
Permeability of vacuum	μ_0	$4\pi \times 10^{-7}$ H m^{-1}
Permittivity of vacuum	ε_0	8.85419×10^{-12} J^{-1} C^2 m^{-1}
Planck constant	h	$6.6260755 \times 10^{-34}$ J s
	h	$4.13566743 \times 10^{-15}$ eV s
	$\hbar = h/2\pi$	$1.05457168 \times 10^{-34}$ J s
Proton mass	m_p	$1.6726231 \times 10^{-27}$ kg
Rydberg constant	R_∞	1.0973731534×10^7 m^{-1}
Speed of light in a vacuum	c, c_0	2.99792458×10^8 m s^{-1}
Standard acceleration of gravity	g	9.80665 m s^{-2}
Stefan-Boltzmann constant	σ	5.67051×10^{-8} W m^{-2} K^{-4}

Mathematical Constants

QUANTITY	SYMBOL	VALUE
Ratio of circumference to diameter of a circle	π	3.14159265359
Base of natural logarithms	e	2.71828182846
Natural logarithm of 10	ln 10	2.30258509299

Thermometric Equivalents

Temperature Scales

Symbol	Designation	Zero point	Freezing point of water	Boiling point of water at standard atm pressure
°C	degree Celsius	freezing point of water	0 °C	100 °C
K	Kelvin or thermodynamic temperature based on the Celsius scale	absolute zero	273.15 K	373.15 K
°F	degree Fahrenheit	-17.8 °C	32 °F	212 °F
°Rank	degree Rankine or thermodymanic temperature based on the Fahrenheit scale	absolute zero	491.4 °Rank	671.4 °Rank

The following formulas may be used to convert temperatures from one scale to another:

Temperature given in	Temperature wanted in			
	°C	K	°F	°Rank
°C	°C	°C + 273.15	1.8 °C + 32	1.8 °C + 491.4
K	K − 273.15	K	1.8 K − 459.4	1.8 K
°F	0.556 °F − 17.8	0.556 °F + 255.3	°F	°F + 459.4
°Rank	0.556 °Rank − 273.15	0.556 °Rank	°Rank − 459.4	°Rank

TABLE-63

Universal Conversion Factors

To convert units of one system into another, find the given units in the table and multiply by the appropriate conversion factor. For example:
To convert 26.5 centimeters into inches, find centimeters in the table, read across to inches and multiply by 0.3937.

$$26.5 \text{ cm} \times \frac{0.3937 \text{ inch}}{\text{cm}} = 10.4 \text{ inches}$$

To convert 8.75 inches into centimeters, find inches in the table, read across to centimeters and multiply by 2.540.

$$8.75 \text{ inches} \times \frac{2.540 \text{ cm}}{\text{inch}} = 22.2 \text{ cm}$$

TO CONVERT	INTO	MULTIPLY BY	TO CONVERT	INTO	MULTIPLY BY
A			baryes	dynes/sq cm	1.000
abcoulombs	statcoulombs	2.998×10^{10}	bolts (US cloth)	meters	36.576
acres	sq chains (Gunter's)	10	Btu	liter—atmosphere	10.409
acres	sq rods	160	Btu	ergs	1.0550×10^{10}
acres	square links		Btu	foot-lb	778.3
	(Gunter's)	1×10^5	Btu	gram-calories	252.0
acres	hectares or		Btu	horsepower-hr	3.931×10^{-4}
	sq hectometers	0.4047	Btu	joules	1,054.8
acres	sq ft	43,560.0	Btu	kilogram-calories	0.2520
acres	sq meters	4,047	Btu	kilogram-meters	107.5
acres	sq miles	1.562×10^{-3}	Btu	kilowatt-hr	2.928×10^{-4}
acres	sq yd	4,840	Btu/hr	foot-lb/sec	0.2162
acre-feet	cu ft	43,560.0	Btu/hr	gram-cal/sec	0.0700
acre-feet	gallons	3.259×10^5	Btu/hr	horsepower-hr	3.929×10^{-4}
amperes/sq cm	amps/sq in	6.452	Btu/hr	watts	0.2931
amperes/sq cm	amps/sq meter	10^4	Btu/min	foot-lb/sec	12.96
amperes/sq in	amps/sq cm	0.1550	Btu/min	horsepower	0.02356
amperes/sq in	amps/sq meter	1,550.0	Btu/min	kilowatts	0.01757
amperes/sq meter	amps/sq cm	10^{-4}	Btu/min	watts	17.57
amperes/sq meter	amps/sq in	6.452×10^{-4}	Btu/sq ft/min	watts/sq in	0.1221
ampere-hours	coulombs	3,600.0	bucket (UK dry)	cubic cm	1.818×10^4
ampere-hours	faradays	0.03731	bushels	cu ft	1.2445
ampere-turns	gilberts	1.257	bushels	cu in	2,150.42
ampere-turns/cm	amp-turns/in	2.540	bushels	cu meters	0.03524
ampere-turns/cm	amp-turns/meter	100.0	bushels	liters	35.24
ampere-turns/cm	gilberts/cm	1.257	bushels	pecks	4.0
ampere-turns/in	amp-turns/cm	0.3937	bushels	pints (dry)	64.0
ampere-turns/in	amp-turns/meter	39.37	bushels	quarts (dry)	32.0
ampere-turns/in	gilberts/cm	0.4950			
ampere-turns/meter	amp-turns/cm	0.01			
ampere-turns/meter	amp-turns/in	0.0254	**C**		
ampere-turns/meter	gilberts/cm	0.01257	calories, gram		
angstrom unit	inches	3937×10^{-9}	(mean)	Btu (mean)	3.9685×10^{-3}
angstrom unit	meters	1×10^{-10}	candle/sq cm	lamberts	3.142
angstrom unit	microns or (μ)	1×10^{-4}	candle/sq inch	lamberts	0.4870
ares	acres (US)	0.02471	Celsius	Fahrenheit	$1.8°C + 32$
ares	sq yd	119.60	centares (centiares)	sq meters	1.0
ares	acres	0.02471	centigrams	grams	0.01
ares	sq meters	100.0	centiliters	ounces (fl)	0.3382
astronomical units	kilometers	1.495×10^8	centiliters	cu in	0.6103
atmospheres	ton/sq in	0.007348	centiliters	drams	2.705
atmospheres	cm of mercury	76.0	centiliters	liters	0.01
atmospheres	mm of mercury	760.0	centimeters	feet	3.281×10^{-2}
atmospheres	torrs	760.0	centimeters	inches	0.3937
atmospheres	ft of water (at 4°C)	33.90	centimeters	kilometers	10^{-5}
atmospheres	in of mercury (at 0°C)	29.92	centimeters	meters	0.01
atmospheres	kg/sq cm	1.0333	centimeters	miles	6.214×10^{-6}
atmospheres	kg/sq meter	10,332	centimeters	millimeters	10.0
atmospheres	lb/sq in	14.70	centimeters	mils	393.7
atmospheres	tons/sq ft	1.058	centimeters	yards	1.094×10^{-2}
			centimeter-dynes	cm-grams	1.020×10^{-3}
B			centimeter-dynes	meter-kg	1.020×10^{-8}
barrels (US, dry)	cu in	7056	centimeter-dynes	lb-ft	7.376×10^{-8}
barrels (US, dry)	quarts (dry)	105.0	centimeter-grams	cm-dynes	980.7
barrels (US, liq)	gallons	31.5	centimeter-grams	meter-kg	10^{-5}
barrels (oil)	gallons (oil)	42.0	centimeter-grams	lb-ft	7.233×10^{-5}
bars	atmospheres	0.9869	centimeters of		
bars	dynes/sq cm	10^6	mercury	atmospheres	0.01316
bars	kg/sq meter	1.020×10^4	centimeters of		
bars	lb/sq ft	2,089	mercury	ft of water	0.4461
bars	lb/sq in	14.50			

TABLE-64

TO CONVERT	INTO	MULTIPLY BY	TO CONVERT	INTO	MULTIPLY BY
centimeters of mercury	kg/sq meter	136.0	cubic inches	bushels	4.650×10^{-4}
			cubic inches	cu cm	16.39
centimeters of mercury	lb/sq ft	27.85	cubic inches	cu ft	5.787×10^{-4}
			cubic inches	cu meters	1.639×10^{-5}
centimeters of mercury	lb/sq in	0.1934	cubic inches	cu yards	2.143×10^{-5}
			cubic inches	drams (fl)	4.4329
centimeters/sec	ft/min	1.1969	cubic inches	gallons (US)	4.329×10^{-3}
centimeters/sec	ft/sec	0.03281	cubic inches	gallons (UK)	3.605×10^{-3}
centimeters/sec	kilometers/hr	0.036	cubic inches	gills	0.1385
centimeters/sec	knots	0.1943	cubic inches	liters	0.01639
centimeters/sec	meters/min	0.6	cubic inches	milliliters	16.39
centimeters/sec	miles/hr	0.02237	cubic inches	mil-ft	1.061×10^5
centimeters/sec	miles/min	3.728×10^{-4}	cubic inches	minims	265.974
centimeters/sec/sec	ft/sec/sec	0.03281	cubic inches	ounces (fl)	0.5541
centimeters/sec/sec	km/hr/sec	0.036	cubic inches	pecks	1.860×10^{-3}
centimeters/sec/sec	meters/sec/sec	0.01	cubic inches	pints (dry)	0.0298
centimeters/sec/sec	miles/hr/sec	0.02237	cubic inches	pints (liq)	0.03463
chain	inches	792.00	cubic inches	quarts (dry)	0.0149
chain	meters	20.12	cubic inches	quarts (liq)	0.01732
chains (surveyors' or Gunter's)	yards	22.00	cubic meters	bushels	28.38
			cubic meters	cu cm	10^6
circular mils	sq cm	5.067×10^{-6}	cubic meters	cu ft	35.31
circular mils	sq mils	0.7854	cubic meters	cu in	61,023.74
circular mils	sq in	7.854×10^{-7}	cubic meters	cu yards	1.308
circumference	radians	6.283	cubic meters	gallons (US)	264.2
cords	cord ft	8	cubic meters	gallons (UK)	220.0
cord feet	cu ft	16	cubic meters	liters	1,000.0
coulomb	statcoulombs	2.998×10^9	cubic meters	milliliters	10^6
coulombs	faradays	1.036×10^{-5}	cubic meters	pecks	113.51
coulombs/sq cm	coulombs/sq in	64.52	cubic meters	pints (dry)	1,816.166
coulombs/sq cm	coulombs/sq meter	10^4	cubic meters	pints (liq)	2,113.0
coulombs/sq in	coulombs/sq cm	0.1550	cubic meters	quarts (dry)	908.083
coulombs/sq in	coulombs/sq meter	1,550	cubic meters	quarts (liq)	1,057
coulombs/sq meter	coulombs/sq cm	10^{-4}	cubic yards	cu cm	7.646×10^5
coulombs/sq/meter	coulombs/sq in	6.452×10^{-4}	cubic yards	cu ft	27.0
cubic centimeters	cu ft	3.531×10^{-5}	cubic yards	cu in	46,656.0
cubic centimeters	cu in	0.06102	cubic yards	cu meters	0.7646
cubic centimeters	cu meters	10^{-6}	cubic yards	gallons (US)	202.0
cubic centimeters	cu yards	1.308×10^{-4}	cubic yards	gallons (UK)	168.2
cubic centimeters	drams (fl)	0.2705	cubic yards	liters	764.6
cubic centimeters	gallons (US)	2.642×10^{-4}	cubic yards	milliliters	7.646×10^5
cubic centimeters	gills	8.454×10^{-3}	cubic yards	pints (liq)	1,615.9
cubic centimeters	liters	0.001	cubic yards	quarts (liq)	807.9
cubic centimeters	milliliters	1.0	cubic yards/min	cubic ft/sec	0.45
cubic centimeters	minims	16.231	cubic yards/min	gallons/sec	3.367
cubic centimeters	ounces (fl)	0.0338	cubic yards/min	liters/sec	12.74
cubic centimeters	pints (liq)	2.113×10^{-3}			
cubic centimeters	quarts (liq)	1.057×10^{-3}	**D**		
cubic feet	bushels	0.8036			
cubic feet	cu cm	28,316.85	Dalton	gram	1.650×10^{-24}
cubic feet	cu in	1,728.0	days	seconds	86,400.0
cubic feet	cu meters	0.02832	decigrams	grams	0.1
cubic feet	cu yards	0.03704	deciliters	liters	0.1
cubic feet	drams (fl)	7660.05	decimeters	meters	0.1
cubic feet	gallons (US)	7.48052	degrees (angle)	quadrants	0.01111
cubic feet	gallons (UK)	6.229	degrees (angle)	radians	0.01745
cubic feet	gills	239.38	degrees (angle)	seconds	3,600.0
cubic feet	liters	28.32	degrees/sec	radians/sec	0.01745
cubic feet	milliliters	28,316.85	degrees/sec	revolutions/min	0.1667
cubic feet	minims	459,603.1	degrees/sec	revolutions/sec	2.778×10^{-3}
cubic feet	ounces (fl)	957.51	dekagrams	grams	10.0
cubic feet	pecks	3.2143	dekaliters	liters	10.0
cubic feet	pints (dry)	51.428	dekameters	meters	10.0
cubic feet	pints (liq)	59.84	drams (avdp)	drams (apoth)	0.4557
cubic feet	quarts (dry)	25.714	drams (avdp)	grains	27.3437
cubic feet	quarts (liq)	29.92	drams (avdp)	grams	1.7718
cubic feet/min	cu cm/sec	472.0	drams (avdp)	kilograms	1.7718×10^{-3}
cubic feet/min	gallons/sec	0.1247	drams (avdp)	milligrams	1771.85
cubic feet/min	liters/sec	0.4720	drams (avdp)	ounces (apoth or troy)	0.0570
cubic feet/min	lb of water/min	62.43	drams (avdp)	ounces (avdp)	0.0625
cubic feet/sec	million gals/day	0.646317	drams (avdp)	pennyweights	1.139
cubic feet/sec	gallons/min	448.831			

TABLE-65

TO CONVERT	INTO	MULTIPLY BY	TO CONVERT	INTO	MULTIPLY BY
drams (avdp)	pounds (apoth or troy)	4.747×10^{-3}	faradays	coulombs	9.649×10^4
drams (avdp)	pounds (avdp)	3.906×10^{-3}	fathoms	meters	1.828804
drams (avdp)	scruples	1.367	fathoms	feet	6.0
drams (apoth)	drams (avdp)	2.1943	feet	centimeters	30.48
drams (apoth)	grains	60.0	feet	inches	12.0
drams (apoth)	grams	3.8879	feet	kilometers	3.048×10^{-4}
drams (apoth)	kilograms	3.888×10^{-3}	feet	meters	0.3048
drams (apoth)	milligrams	3887.93	feet	miles (nautical)	1.645×10^{-4}
drams (apoth)	ounces (apoth or troy)	0.125	feet	miles (statute)	1.894×10^{-4}
drams (apoth)	ounces (avdp)	0.1371429	feet	millimeters	304.8
drams (apoth)	pennyweights	2.5	feet	mils	1.2×10^4
drams (apoth)	pounds (apoth or troy)	0.0104	feet	yards	0.333
drams (apoth)	pounds (avdp)	8.571×10^{-3}	feet of water	atmospheres	0.02950
drams (apoth)	scruples	3.0	feet of water	in of mercury	0.8826
drams (fl)	cu cm	3.6967	feet of water	kg/sq cm	0.03048
drams (fl)	cu ft	1.3055×10^{-4}	feet of water	kg/sq meter	304.8
drams (fl)	cu in	0.2256	feet of water	lb/sq ft	62.43
drams (fl)	gallons (US)	9.7656×10^{-4}	feet of water	lb/sq in	0.4335
drams (fl)	gills	0.03125	feet/min	cm/sec	0.5080
drams (fl)	liters	3.6967×10^{-3}	feet/min	feet/sec	0.01667
drams (fl)	milliliters	3.6967	feet/min	km/hr	0.01829
drams (fl)	minims	60.0	feet/min	meters/min	0.3048
drams (fl)	ounces (fl)	0.125	feet/min	miles/hr	0.01136
drams (fl)	pints (liq)	7.8125×10^{-3}	feet/sec	cm/sec	30.48
drams (fl)	quarts (liq)	3.9063×10^{-3}	feet/sec	km/hr	1.097
dynes	grams	1.020×10^{-3}	feet/sec	knots	0.5921
dynes	joules/cm	10^{-7}	feet/sec	meters/min	18.29
dynes	joules/meter (newtons)	10^{-5}	feet/sec	miles/hr	0.6818
dynes	kilograms	1.020×10^{-6}	feet/sec	miles/min	0.01136
dynes	poundals	7.233×10^{-5}	feet/sec/sec	cm/sec/sec	30.48
dynes	pounds	2.248×10^{-6}	feet/sec/sec	km/hr/sec	1.097
dynes/cm	ergs/sq mm	.01	feet/sec/sec	meters/sec/sec	0.3048
dynes/sq cm	atmospheres	9.869×10^{-7}	feet/sec/sec	miles/hr/sec	0.6818
dynes/sq cm	bars	10^{-6}	feet/100 feet	percent grade	1.0
dynes/sq cm	in of mercury at 0°C	2.953×10^{-5}	foot-candles	lumens/sq meter	10.764
dynes/sq cm	in of water at 4°C	4.015×10^{-4}	foot-pounds	Btu	1.286×10^{-3}
			foot-pounds	ergs	1.356×10^7
E			foot-pounds	gram-calories	0.3238
ell	cm	114.30	foot-pounds	hp-hr	5.050×10^{-7}
ell	inches	45	foot-pounds	joules	1.356
em, pica	inches	.167	foot-pounds	kg-calories	3.24×10^{-4}
em, pica	cm	.4233	foot-pounds	kg-meters	0.1383
ergs	Btu	9.480×10^{-11}	foot-pounds	kilowatt-hr	3.766×10^{-7}
ergs	dyne-centimeters	1.0	foot-pounds/min	Btu/min	1.286×10^{-3}
ergs	foot-lb	7.367×10^{-8}	foot-pounds/min	foot-lb/sec	0.01667
ergs	gram-calories	0.2389×10^{-7}	foot-pounds/min	horsepower	3.030×10^{-5}
ergs	gram-cm	1.020×10^{-3}	foot-pounds/min	kg-calories/min	3.24×10^{-4}
ergs	horsepower-hr	3.7250×10^{-14}	foot-pounds/min	kilowatts	2.260×10^{-5}
ergs	joules	10^{-7}	foot-pounds/sec	Btu/hr	4.6263
ergs	kg-calories	2.389×10^{-11}	foot-pounds/sec	Btu/min	0.07717
ergs	kg-meters	1.020×10^{-8}	foot-pounds/sec	horsepower	1.818×10^{-3}
ergs	kilowatt-hr	0.2778×10^{-13}	foot-pounds/sec	kg-calories/min	0.01945
ergs	watt-hr	0.2778×10^{-10}	foot-pounds/sec	kilowatts	1.356×10^{-3}
ergs/sec	Btu/min	5.688×10^{-6}	furlongs	miles (US)	0.125
ergs/sec	dyne-cm/sec	1.000	furlongs	rods	40.0
ergs/sec	ft-lb/min	4.427×10^{-6}	furlongs	feet	660.0
ergs/sec	ft-lb/sec	7.3756×10^{-8}			
ergs/sec	horsepower	1.341×10^{-10}	**G**		
ergs/sec	kg-calories/min	1.433×10^{-9}	gallons (US)	cu cm	3,785.0
ergs/sec	kilowatts	10^{-10}	gallons (US)	cu ft	0.1337
			gallons (US)	cu in	231.0
F			gallons (US)	cu meters	3.785×10^{-3}
Fahrenheit	Celsius	$0.556°F - 17.8$	gallons (US)	cu yards	4.951×10^{-3}
farads	microfarads	10^6	gallons (US)	drams (fl)	1024.0
faradays/sec	ampere (abs)	9.6500×10^4	gallons (US)	gallons (UK)	0.83267
faradays	ampere-hr	26.80	gallons (US)	gills	32.0
			gallons (US)	liters	3.785
			gallons (US)	milliliters	3,785.0
			gallons (US)	minims	61,440.0
			gallons (US)	ounces (fl)	128.0
			gallons (US)	pints (liq)	8.0

TABLE-66

TO CONVERT	INTO	MULTIPLY BY	TO CONVERT	INTO	MULTIPLY BY
gallons (US)	quarts (liq)	4.0	grams	tons (short)	1.102×10^{-6}
gallons (UK)	cu ft	0.1605	grams/cm	lb/in	5.600×10^{-3}
gallons (UK)	cu in	277.4	grams/cu cm	lb/cu ft	62.43
gallons (UK)	cu meters	4.546×10^{-3}	grams/cu cm	lb/cu in	0.03613
gallons (UK)	cu yards	5.946×10^{-3}	grams/cu cm	lb/mil-ft	3.405×10^{-7}
gallons (UK)	gallons (US)	1.20095	grams/liter	grains/gal	58.417
gallons (UK)	liters	4.546	grams/liter	lb/1,000 gal	8.345
gallons of water	lb of water	8.3453	grams/liter	lb/cu ft	0.062427
gallons/min	cu ft/sec	2.228×10^{-3}	grams/liter	parts/million	1,000.0
gallons/min	liters/sec	0.06308	grams/sq cm	lb/sq ft	2.0481
gallons/min	cu ft/hr	8.0208	gram-calories	Btu	3.9683×10^{-3}
gauss	lines/sq in	6.452	gram-calories	ergs	4.1868×10^{7}
gauss	webers/sq cm	10^{-8}	gram-calories	ft-lb	3.0880
gauss	webers/sq in	6.452×10^{-8}	gram-calories	horsepower-hr	1.5596×10^{-6}
gauss	webers/sq meter	10^{-4}	gram-calories	kilowatt-hr	1.1630×10^{-6}
gilberts	ampere-turns	0.7958	gram-calories	watt-hr	1.1630×10^{-3}
gilberts/cm	amp-turns/cm	0.7958	gram-calories/sec	Btu/hr	14.286
gilberts/cm	amp-turns/in	2.021	gram-centimeters	Btu	9.297×10^{-8}
gilberts/cm	amp-turns/meter	79.58	gram-centimeters	ergs	980.7
gills	cu cm	118.2941	gram-centimeters	joules	9.807×10^{-5}
gills	cu ft	4.1775×10^{-3}	gram-centimeters	kg-cal	2.343×10^{-8}
gills	cu in	7.21875	gram-centimeters	kg-meters	10^{-5}
gills	drams (fl)	32.0			
gills	gallons (US)	0.03125			
gills	liters	0.1183		**H**	
gills	milliliters	118.2941	hand	cm	10.16
gills	minims	1920.0	hectares	acres	2.471
gills	ounces (fl)	4.0	hectares	sq ft	1.076×10^{5}
gills	pints (liq)	0.25	hectograms	grams	100.0
gills	quarts (liq)	0.125	hectoliters	liters	100.0
gills (UK)	cu cm	142.07	hectometers	meters	100.0
grade	radian	0.01571	hectowatts	watts	100.0
grains	drams (apoth)	0.0167	henries	millihenries	1,000.0
grains	drams (avdp)	0.03657143	hogsheads (UK)	cu ft	10.114
grains	grams	0.0648	hogsheads (US)	cu ft	8.42184
grains	kilograms	6.479×10^{-5}	hogsheads (US)	gallons (US)	63
grains	milligrams	64.799	horsepower	Btu/min	42.44
grains	ounces (apoth or		horsepower	ft-lb/min	33,000
	troy)	2.083×10^{-3}	horsepower	ft-lb/sec	550.0
grains	ounces (avdp)	2.286×10^{-3}	horsepower	horsepower (metric)	1.014
grains	pennyweights	0.04167	horsepower	kg-calories/min	10.68
grains	pounds (apoth or		horsepower	kilowatts	0.7457
	troy)	1.736×10^{-4}	horsepower	watts	745.7
grains	pounds (avdp)	1.423×10^{-4}	horsepower (boiler)	Btu/hr	33,479
grains	scruples	0.05	horsepower (boiler)	kilowatts	9.803
grains (troy)	grains (avdp)	1.0	horsepower (metric)	horsepower	0.9863
grains (troy)	grams	0.06480	horsepower (metric)	ft-lb/sec	542.5
grains (troy)	ounces (avdp)	2.286×10^{-3}	horsepower-hours	Btu	2,547
grains (troy)	pennyweight (troy)	0.04167	horsepower-hours	ergs	2.6845×10^{13}
grains/US gal	parts/million	17.118	horsepower-hours	ft-lb	1.98×10^{6}
grains/US gal	lb/million gal	142.86	horsepower-hours	gm-cal	641,190
grains/UK gal	parts/million	14.286	horsepower-hours	joules	2.684×10^{6}
grams	drams (apoth)	0.2572	horsepower-hours	kg-calories	641.1
grams	drams (avdp)	0.5644	horsepower-hours	kg-meters	2.737×10^{5}
grams	dynes	980.7	horsepower-hours	kilowatt-hr	0.7457
grams	grains	15.43	hours	days	4.167×10^{-2}
grams	joules/cm	9.807×10^{-5}	hours	weeks	5.952×10^{-3}
grams	joules/meter		hundredweights		
	(newtons)	9.807×10^{-3}	(long)	cwt (short)	1.12
grams	kilograms	0.001	hundredweights		
grams	milligrams	1,000.0	(long)	kilograms	50.802
grams	ounces (apoth or		hundredweights		
	troy)	0.03215	(long)	ounces	1792.0
grams	ounces (avdp)	0.03527	hundredweights		
grams	pennyweights	0.643	(long)	pounds	112
grams	pounds (apoth or		hundredweights		
	troy)	2.679×10^{-3}	(long)	slugs	3.4811
grams	pounds (avdp)	2.205×10^{-3}	hundredweights		
grams	poundals	0.07093	(long)	tons (long)	0.05
grams	scruples	0.7716	hundredweights		
grams	slugs	6.852×10^{-5}	(long)	tons (short)	0.056

TABLE-67

TO CONVERT	INTO	MULTIPLY BY	TO CONVERT	INTO	MULTIPLY BY
hundredweights (short)	cwt (short)	0.8929	kilograms	joules/meters (newtons)	9.807
hundredweights (short)	kilograms	45.359	kilograms	milligrams	10^6
hundredweights (short)	ounces	1600.0	kilograms	ounces (apoth or troy)	32.151
hundredweights (short)	pounds	100.0	kilograms	ounces (avdp)	35.274
hundredweights (short)	slugs	3.1081	kilograms	pennyweights	643.015
hundredweights (short)	tons (metric)	0.0453592	kilograms	pounds (apoth or troy)	2.679
hundredweights (short)	tons (long)	0.0446429	kilograms	pounds (avdp)	2.205
hundredweights (short)	tons (short)	0.05	kilograms	poundals	70.93
			kilograms	scruples	771.62
I			kilograms	slugs	0.0685
inches	centimeters	2.540	kilograms	tons (metric)	0.001
inches	feet	0.0833	kilograms	tons (long)	9.842×10^{-4}
inches	meters	2.540×10^{-2}	kilograms	tons (short)	1.102×10^{-3}
inches	miles	1.578×10^{-5}	kilograms/cu meter	gm/cu cm	0.001
inches	miles (naut)	1.3715×10^{-5}	kilograms/cu meter	lb/cu ft	0.06243
inches	millimeters	25.40	kilograms/cu meter	lb/cu in	3.613×10^{-5}
inches	mils	1,000.0	kilograms/cu meter	lb/mil-foot	3.405×10^{-10}
inches	yards	2.778×10^{-2}	kilograms/meter	lb/ft	0.6720
inches of mercury	atmospheres	0.03342	kilograms/sq cm	dynes	980,665
inches of mercury	ft of water	1.133	kilograms/sq cm	atmospheres	0.9678
inches of mercury	kg/sq cm	0.03453	kilograms/sq cm	ft of water	32.81
inches of mercury	kg/sq meter	345.3	kilograms/sq cm	in of mercury	28.96
inches of mercury	lb/sq ft	70.73	kilograms/sq cm	lb/sq ft	2,048
inches of mercury	lb/sq in	0.4912	kilograms/sq cm	lb/sq in	14.22
inches of water (at 4°C)	atmospheres	2.458×10^{-3}	kilograms/sq meter	atmospheres	9.678×10^{-5}
inches of water (at 4°C)	in of mercury	0.07355	kilograms/sq meter	bars	9.807×10^{-5}
			kilograms/sq meter	ft of water	3.281×10^{-3}
inches of water (at 4°C)	kg/sq cm	2.540×10^{-3}	kilograms/sq meter	in of mercury	2.896×10^{-3}
inches of water (at 4°C)	oz/sq in	0.5781	kilograms/sq meter	lb/sq ft	0.2048
			kilograms/sq meter	lb/sq in	1.422×10^{-3}
inches of water (at 4°C)	lb/sq ft	5.204	kilograms/sq mm	kg/sq meter	10^6
inches of water (at 4°C)	lb/sq in	0.03613	kilograms-calories	Btu	3.968
			kilogram-calories	ft-lb	3,088
international ampere	ampere (abs)	.9998	kilogram-calories	hp-hr	1.560×10^{-3}
international volt	volts (abs)	1.0003	kilogram-calories	joules	4,186
international volt	joules	9.654×10^4	kilogram-calories	kg-meters	426.9
			kilogram-calories	kilojoules	4.186
J			kilogram-calories	kilowatt-hr	1.163×10^{-3}
joules	Btu	9.480×10^{-4}	kilogram-meters	Btu	9.294×10^{-3}
joules	ergs	10^7	kilogram-meters	ergs	9.804×10^7
joules	ft-lb	0.7376	kilogram-meters	ft-lb	7.233
joules	kg-calories	2.389×10^{-4}	kilogram-meters	joules	9.804
joules	kg-meters	0.1020	kilogram-meters	kg-calories	2.342×10^{-3}
joules	watt-hr	2.778×10^{-4}	kilogram-meters	kilowatt-hr	2.723×10^{-6}
joules/cm	grams	1.020×10^4	kilolines	maxwells	1,000.0
joules/cm	dynes	10^7	kiloliters	liters	1,000.0
joules/cm	joules/meter (newtons)	100.0	kilometers	centimeters	10^5
			kilometers	feet	3,281
joules/cm	poundals	723.3	kilometers	inches	3.937×10^4
joules/cm	pounds	22.48	kilometers	meters	1,000.0
			kilometers	miles	0.6214
K			kilometers	millimeters	10^6
kilograms	cwt (long)	0.0197	kilometers	yards	1,094
kilograms	cwt (short)	0.022	kilometers/hr	cm/sec	27.78
kilograms	drams (apoth)	257.21	kilometers/hr	ft/min	54.68
kilograms	drams (avdp)	564.38	kilometers/hr	ft/sec	0.9113
kilograms	dynes	980,665	kilometers/hr	knots	0.5396
kilograms	grains	15,432.36	kilometers/hr	meters/min	16.67
kilograms	grams	1000.0	kilometers/hr	miles/hr	0.6214
kilograms	joules/cm	0.09807	kilometers/hr/sec	cm/sec/sec	27.78
			kilometers/hr/sec	ft/sec/sec	0.9113
			kilometers/hr/sec	meters/sec/sec	0.2778
			kilometers/hr/sec	miles/hr/sec	0.6214
			kilowatts	Btu/min	56.92
			kilowatts	foot-lb/min	4.426×10^4
			kilowatts	foot-lb/sec	737.6
			kilowatts	horsepower	1.341
			kilowatts	kg-calories/min	14.34

TABLE-68

TO CONVERT	INTO	MULTIPLY BY	TO CONVERT	INTO	MULTIPLY BY
kilowatts	watts	1,000.0	meters	millimeters	1,000.0
kilowatt-hours	Btu	3,413	meters	rods	0.1988
kilowatt-hours	ergs	3.600×10^{13}	meters	yards	1.094
kilowatt-hours	ft-lb	2.655×10^6	meters	varas	1.179
kilowatt-hours	gram-calories	859,850	meters/min	cm/sec	1.667
kilowatt-hours	horsepower-hr	1.341	meters/min	ft/min	3.281
kilowatt-hours	joules	3.6×10^6	meters/min	ft/sec	0.05468
kilowatt-hours	kg-calories	860.5	meters/min	km/hr	0.06
kilowatt-hours	kg-meters	3.671×10^5	meters/min	knots	0.03238
kilowatt-hours	lb of water evap-		meters/min	miles/hr	0.03728
	orated from and at		meters/sec	feet/min	196.8
	212°F.	3.53	meters/sec	feet/sec	3.281
kilowatt-hours	lb of water raised		meters/sec	kilometers/hr	3.6
	from 62° to 212°F.	22.75	meters/sec	kilometers/min	0.06
knots	ft/hr	6,080	meters/sec	miles/hr	2.237
knots	kilometers/hr	1.8532	meters/sec	miles/min	0.03728
knots	nautical miles/hr	1.0	meters/sec/sec	cm/sec/sec	100.0
knots	statute miles/hr	1.151	meters/sec/sec	ft/sec/sec	3.281
knots	yd/hr	2,027	meters/sec/sec	km/hr/sec	3.6
knots	ft/sec	1.689	meters/sec/sec	miles/hr/sec	2.237
			meter-kilograms	cm-dynes	9.807×10^7
L			meter-kilograms	cm-gm	10^5
league	miles (approx.)	3.0	meter-kilograms	lb-ft	7.233
light year	miles	5.9×10^{12}	microfarad	farads	10^{-6}
light year	kilometers	9.46091×10^{12}	micrograms	grams	10^{-6}
lines/sq cm	gauss	1.0	microhms	megohms	10^{-12}
lines/sq in	gauss	0.1550	microhms	ohms	10^{-6}
lines/sq in	webers/sq cm	1.550×10^{-9}	microliters	liters	10^{-6}
lines/sq in	webers/sq in	10^{-8}	microns	meters	1×10^{-6}
lines/sq in	webers/sq meter	1.550×10^{-5}	miles (naut)	feet	6,080.27
links(engineer's)	inches	12.0	miles (naut)	inches	7.2913×10^4
links(surveyor's)	inches	7.92	miles (naut)	kilometers	1.853
liters	bushels	0.02838	miles (naut)	meters	1,853
liters	cu cm	1,000.0	miles (naut)	miles (statute)	1.1516
liters	cu ft	0.03531	miles (naut)	yards	2,027
liters	cu in	61.024	miles (statute)	centimeters	1.609×10^5
liters	cu meters	0.001	miles (statute)	feet	5,280
liters	cu yards	1.308×10^{-3}	miles (statute)	inches	6.336×10^4
liters	drams (fl)	270.512	miles (statute)	kilometers	1.609
liters	gallons (US)	0.2642	miles (statute)	meters	1,609
liters	gallons (UK)	0.220	miles (statute)	miles (naut)	0.8684
liters	gills	8.454	miles (statute)	rods	320
liters	milliliters	1,000	miles (statute)	yards	1,760
liters	minims	16,230.73	miles/hr	cm/sec	44.70
liters	ounces (fl)	33.814	miles/hr	ft/min	88
liters	pecks	0.1135	miles/hr	ft/sec	1.467
liters	pints (dry)	1.8162	miles/hr	km/hr	1.609
liters	pints (liq)	2.113	miles/hr	km/min	0.02682
liters	quarts (dry)	0.9081	miles/hr	knots	0.8684
liters	quarts (liq)	1.057	miles/hr	meters/min	26.82
liters/min	cu ft/sec	5.886×10^{-4}	miles/hr	miles/min	0.01667
liters/min	gal/sec	4.403×10^{-3}	miles/hr/sec	cm/sec/sec	44.70
lumens/sq ft	foot-candles	1.0	miles/hr/sec	ft/sec/sec	1.467
lumen	spherical candle		miles/hr/sec	km/hr/sec	1.609
	power	0.07958	miles/hr/sec	meters/sec/sec	0.4470
lumen	watt	0.001496	miles/min	cm/sec	2,682
lumen/sq ft	lumen/sq meter	10.76	miles/min	ft/sec	88
lux	foot-candles	0.0929	miles/min	km/min	1.609
			miles/min	knots/min	0.8684
M			miles/min	miles/hr	60.0
maxwells	kilolines	0.001	mil-feet	cu in	9.425×10^{-6}
maxwells	webers	10^{-8}	milliers	kilograms	1,000
megalines	maxwells	10^6	milligrams	drams (apoth)	2.572×10^{-4}
megohms	microhms	10^{12}	milligrams	drams (avdp)	5.644×10^{-4}
megohms	ohms	10^6	milligrams	grains	0.01543236
meters	centimeters	100.0	milligrams	grams	0.001
meters	feet	3.281	milligrams	kilograms	10^{-6}
meters	inches	39.37	milligrams	ounces (apoth or	
meters	kilometers	0.001		troy)	3.215×10^{-5}
meters	miles (naut)	5.396×10^{-4}	milligrams	ounces (avdp)	3.527×10^{-5}
meters	miles (stat)	6.214×10^{-4}	milligrams	pennyweights	6.43×10^{-4}

TABLE-69

TO CONVERT	INTO	MULTIPLY BY	TO CONVERT	INTO	MULTIPLY BY
milligrams	pounds (apoth or troy)	2.679×10^{-6}	ounces (avdp)	milligrams	28,349.5
milligrams	pounds (avdp)	2.2046×10^{-6}	ounces (avdp)	ounces (apoth or troy)	0.9115
milligrams	scruples	7.7162×10^{-4}	ounces (avdp)	pennyweights	18.23
milligrams/liter	parts/million	1.0	ounces (avdp)	pounds (apoth or troy)	0.0759
millihenries	henries	0.001	ounces (avdp)	pounds (avdp)	0.0625
milliliters	cu cm	1.0	ounces (avdp)	scruples	21.875
milliliters	cu ft	3.531×10^{-5}	ounces (avdp)	slugs	1.9426×10^{-3}
milliliters	cu in	0.06102	ounces (avdp)	tons (long)	2.790×10^{-5}
milliliters	drams (fl)	0.2705	ounces (avdp)	tons (metric)	2.835×10^{-5}
milliliters	gallons (US)	2.642×10^{-4}	ounces (avdp)	tons (short)	3.125×10^{-5}
milliliters	gills	8.454×10^{-3}	ounces (apoth or troy)	drams (apoth)	8.0
milliliters	liters	0.001			
milliliters	minims	16.231	ounces (apoth or troy)	drams (avdp)	17.554
milliliters	ounces (fl)	0.0338			
milliliters	pints (liq)	2.113×10^{-3}	ounces (apoth or troy)	grains	480.0
milliliters	quarts (liq)	1.057×10^{-3}			
millimeters	centimeters	0.1	ounces (apoth or troy)	grams	31.103481
millimeters	feet	3.281×10^{-3}			
millimeters	inches	0.03937	ounces (apoth or troy)	kilograms	0.0311
millimeters	kilometers	10^{-6}			
millimeters	meters	0.001	ounces (apoth or troy)	milligrams	31,103.48
millimeters	miles	6.214×10^{-7}			
millimeters	mils	39.37	ounces (apoth or troy)	ounces (avdp)	1.09714
millimeters	yards	1.094×10^{-3}			
millimeters of Hg	atmospheres	1.316×10^{-3}	ounces (apoth or troy)	pennyweights	20.0
millimeters of Hg	torr	1.0			
millimicrons	meters	1×10^{-9}	ounces (apoth or troy)	pounds (apoth or troy)	0.08333
million gals/day	cu ft/sec	1.54723			
mils	centimeters	2.540×10^{-3}	ounces (apoth or troy)	pounds (avdp)	0.0686
mils	feet	8.333×10^{-5}			
mils	inches	0.001	ounces (apoth or troy)	scruples	24.0
mils	kilometers	2.540×10^{-8}	ounces (fl)	cu cm	29.5735
mils	yards	2.778×10^{-5}	ounces (fl)	cu ft	1.0444×10^{-3}
miner's inches	cu ft/min	1.5	ounces (fl)	cu in	1.805
minims (UK)	cu cm	0.059192	ounces (fl)	drams (fl)	8.0
minims	cu cm	0.061612	ounces (fl)	gallons (US)	7.8125×10^{-3}
minims	cu ft	2.176×10^{-6}	ounces (fl)	gills	0.25
minims	cu in	3.7598×10^{-3}	ounces (fl)	liters	0.02957
minims	drams (fl)	0.0167	ounces (fl)	milliliters	29.5735
minims	gallons (US)	1.628×10^{-5}	ounces (fl)	minims	480.0
minims	gills	5.208×10^{-4}	ounces (fl)	pints (liq)	0.0625
minims	liters	6.161×10^{-5}	ounces (fl)	quarts (liq)	0.03125
minims	milliliters	0.061612	ounces/sq in	dynes/sq cm	4309
minims	ounces (fl)	0.0021	ounces/sq in	pounds/sq in	0.0625
minims	pints (liq)	1.302×10^{-4}			
minims (liq)	quarts (liq)	6.51×10^{-5}		**P**	
minutes (angles)	degrees	0.01667	parsecs	miles	19×10^{12}
minutes (angles)	quadrants	1.852×10^{-4}	parsecs	kilometers	3.084×10^{13}
minutes (angles)	radians	2.909×10^{-4}	parts/million	grains/US gal	0.0584
minutes (angles)	seconds	60.0	parts/million	grains/UK gal	0.07016
myriagrams	kilograms	10.0	parts/million	lb/million gal	8.345
myriameters	kilometers	10.0	pecks (UK)	cu in	554.6
myriawatts	kilowatts	10.0	pecks (UK)	liters	9.091901
			pecks	bushels	0.25
	N		pecks	cu ft	0.3111
nepers	decibels	8.686	pecks	cu in	537.605
newtons	dynes	1×10^{5}	pecks	cu meters	8.8098×10^{-3}
			pecks	liters	8.8098
			pecks	pints (dry)	16.0
	O		pecks	quarts (dry)	8.0
ohm (Int)	ohm (abs)	1.0005	pennyweights	drams (apoth)	0.4
ohms	megohms	10^{-6}	pennyweights	drams (avdp)	0.8777
ohms	microhms	10^{6}	pennyweights	grains	24.0
ounces (avdp)	cwt (long)	5.5804×10^{-4}	pennyweights	grams	1.55517
ounces (avdp)	cwt (short)	6.25×10^{-4}	pennyweights	kilograms	1.555×10^{-3}
ounces (avdp)	drams (apoth)	7.292	pennyweights	milligrams	1,555.17
ounces (avdp)	drams (avdp)	16.0	pennyweights	ounces (apoth or troy)	0.05
ounces (avdp)	grains	437.5			
ounces (avdp)	grams	28.349527			
ounces (avdp)	kilograms	0.0283			

TABLE-70

TO CONVERT	INTO	MULTIPLY BY	TO CONVERT	INTO	MULTIPLY BY
pennyweights	ounces (avdp)	0.0549	pounds (apoth or troy)	ounces (apoth or troy)	12.0
pennyweights	pounds (apoth or troy)	4.1667×10^{-3}	pounds (apoth or troy)	ounces (avdp)	13.1657
pennyweights	pounds (avdp)	3.428×10^{-3}	pounds (apoth or troy)	pennyweights	240.0
pennyweights	scruples	1.2	pounds (apoth or troy)	pounds (avdp)	0.822857
pints (dry)	bushels	0.0156	pounds (apoth or troy)	scruples	288.0
pints (dry)	cu ft	0.0194	pounds (apoth or troy)	tons (long)	3.6753×10^{-4}
pints (dry)	cu in	33.60	pounds (apoth or troy)	tons (metric)	3.7324×10^{-4}
pints (dry)	cu meters	5.506×10^{-4}	pounds (apoth or troy)	tons (short)	4.1143×10^{-4}
pints (dry)	liters	0.5506	pounds of water	cu ft	0.01602
pints (dry)	pecks	0.0625	pounds of water	cu in	27.68
pints (dry)	quarts (dry)	0.5	pounds of water	gallons	0.1198
pints (liq)	cu cm	473.2	pounds of water/min	cu ft/sec	2.670×10^{-4}
pints (liq)	cu ft	0.01671	pound-feet	cm-dynes	1.356×10^{7}
pints (liq)	cu in	28.875	pound-feet	cm-gm	13,825
pints (liq)	cu meters	4.732×10^{-4}	pound-feet	meter-kg	0.1383
pints (liq)	cu yards	6.189×10^{-4}	pounds/cu ft	gm/cu cm	0.01602
pints (liq)	drams (fl)	128.0	pounds/cu ft	kg/cu meter	16.02
pints (liq)	gallons (US)	0.125	pounds/cu ft	lb/cu in	5.787×10^{-4}
pints (liq)	gills	4.0	pounds/cu ft	lb/mil-foot	5.456×10^{-9}
pints (liq)	liters	0.4732	pounds/cu in	gm/cu cm	27.68
pints (liq)	milliliters	473.2	pounds/cu in	kg/cu meter	2.768×10^{4}
pints (liq)	minims	7680.0	pounds/cu in	lb/cu ft	1,728
pints (liq)	ounces (fl)	16.0	pounds/cu in	lb/mil-foot	9.425×10^{-6}
pints (liq)	quarts (liq)	0.5	pounds/ft	kg/meter	1.488
Planck's quantum	erg-sec	6.624×10^{-27}	pounds/in	gm/cm	178.6
poise	gm/cm sec	1.00	pounds/mil-foot	gm/cu cm	2.306×10^{6}
poundals	dynes	13,826	pounds/sq ft	atmospheres	4.725×10^{-4}
poundals	grams	14.10	pounds/sq ft	ft of water	0.01602
poundals	joules/cm	1.383×10^{-3}	pounds/sq ft	in of mercury	0.01414
poundals	joules/meter (newtons)	0.1383	pounds/sq ft	kg/sq meter	4.882
poundals	kilograms	0.01410	pounds/sq ft	lb/sq in	6.944×10^{-3}
poundals	pounds	0.03108	pounds/sq in	atmospheres	0.06804
pounds (avdp)	cwt (long)	8.929×10^{-3}	pounds/sq in	ft of water	2.307
pounds (avdp)	cwt (short)	0.01	pounds/sq in	in of mercury	2.036
pounds (avdp)	drams (apoth)	116.67	pounds/sq in	kg/sq meter	703.1
pounds (avdp)	drams (avdp)	256.0	pounds/sq in	lb/sq ft	144.0
pounds (avdp)	dynes	44.4823×10^{4}			
pounds (avdp)	grains	7,000.0		**Q**	
pounds (avdp)	grams	453.5924	quadrants (angle)	degrees	90.0
pounds (avdp)	joules/cm	0.04448	quadrants (angle)	minutes	5,400.0
pounds (avdp)	joules/meter (newtons)	4.448	quadrants (angle)	radians	1.571
pounds (avdp)	kilograms	0.4536	quadrants (angle)	seconds	3.24×10^{5}
pounds (avdp)	milligrams	453,592.37	quarts (dry)	bushels	0.0313
pounds (avdp)	ounces (apoth or troy)	14.5833	quarts (dry)	cu ft	0.0389
			quarts (dry)	cu in	67.20
pounds (avdp)	ounces (avdp)	16.0	quarts (dry)	cu meters	1.101×10^{-3}
pounds (avdp)	pennyweights	291.667	quarts (dry)	liters	1.1012
pounds (avdp)	poundals	32.17	quarts (dry)	pecks	0.125
pounds (avdp)	pounds (apoth or troy)	1.21528	quarts (dry)	pints (dry)	2.0
pounds (avdp)	scruples	350.0	quarts (liq)	cu cm	946.4
pounds (avdp)	slugs	3.108×10^{-2}	quarts (liq)	cu ft	0.03342
pounds (avdp)	tons (long)	4.464×10^{-4}	quarts (liq)	cu in	57.75
pounds (avdp)	tons (metric)	4.536×10^{-4}	quarts (liq)	cu meters	9.464×10^{-4}
pounds (avdp)	tons (short)	5.0×10^{-4}	quarts (liq)	cu yards	1.238×10^{-3}
pounds (apoth or troy)	drams (apoth)	96.0	quarts (liq)	drams (fl)	256.0
			quarts (liq)	gallons (US)	0.25
pounds (apoth or troy)	drams (avdp)	210.65	quarts (liq)	gills	8.0
pounds (apoth or troy)	grains	5,760.0	quarts (liq)	liters	0.9464
			quarts (liq)	milliliters	946.4
pounds (apoth or troy)	grams	373.2417	quarts (liq)	minims	15,360.0
pounds (apoth or troy)	kilograms	0.3732	quarts (liq)	ounces (fl)	32.0
pounds (apoth or troy)	milligrams	373,241.72	quarts (liq)	pints (liq)	2.0

TABLE-71

TO CONVERT	INTO	MULTIPLY BY	TO CONVERT	INTO	MULTIPLY BY
R			square feet	sq miles	3.587×10^{-8}
radians	degrees	57.30	square feet	sq mm	9.290×10^4
radians	minutes	3,438	square feet	sq yd	0.1111
radians	quadrants	0.6366	square inches	circular mils	1.273×10^6
radians	seconds	2.063×10^5	square inches	sq cm	6.452
radians/sec	degrees/sec	57.30	square inches	sq ft	6.944×10^{-3}
radians/sec	rev/min	9.549	square inches	sq mm	645.2
radians/sec	rev/sec	0.1592	square inches	sq mils	10^6
radians/sec/sec	rev/min/min	573.0	square inches	sq yd	7.716×10^{-4}
radians/sec/sec	rev/min/sec	9.549	square kilometers	acres	247.1
radians/sec/sec	rev/sec/sec	0.1592	square kilometers	sq cm	10^{10}
revolutions	degrees	360.0	square kilometers	sq ft	1.076×10^7
revolutions	quadrants	4.0	square kilometers	sq in	1.550×10^9
revolutions	radians	6.283	square kilometers	sq meters	10^6
revolutions/min	degrees/sec	6.0	square kilometers	sq miles	0.3861
revolutions/min	radians/sec	0.1047	square kilometers	sq yd	1.196×10^6
revolutions/min	rev/sec	0.01667	square meters	acres	2.471×10^{-4}
revolutions/min/min	radians/sec/sec	1.745×10^{-3}	square meters	sq cm	10^4
revolutions/min/min	rev/min/sec	0.01667	square meters	sq ft	10.76
revolutions/min/min	rev/sec/sec	2.778×10^{-4}	square meters	sq in	1,550
revolutions/sec	degrees/sec	360.0	square meters	sq miles	3.861×10^{-7}
revolutions/sec	radians/sec	6.283	square meters	sq mm	10^6
revolutions/sec	rev/min	60.0	square meters	sq yd	1.196
revolutions/sec/sec	radians/sec/sec	6.283	square miles	acres	640.0
revolutions/sec/sec	rev/min/min	3,600.0	square miles	sq ft	2.788×10^7
revolutions/sec/sec	rev/min/sec	60.0	square miles	sq km	2.590
rods	chain (Gunter's)	0.25	square miles	sq meters	2.590×10^6
rods	meters	5.029	square miles	sq yd	3.098×10^6
rods (surveyors'			square millimeters	circular mils	1,973
measure)	yards	5.5	square millimeters	sq cm	0.01
rods	feet	16.5	square millimeters	sq ft	1.076×10^{-5}
			square millimeters	sq in	1.550×10^{-3}
S			square mils	circular mils	1.273
scruples	drams (apoth)	0.3333	square mils	sq cm	6.452×10^{-6}
scruples	drams (avdp)	0.7314	square mils	sq in	10^{-6}
scruples	grains	20.0	square yards	acres	2.066×10^{-4}
scruples	grams	1.296	square yards	sq cm	8,361
scruples	kilograms	1.296×10^{-3}	square yards	sq ft	9.0
scruples	milligrams	1295.97	square yards	sq in	1,296
scruples	ounces (apoth or		square yards	sq meters	0.8361
	troy)	0.0417	square yards	sq miles	3.228×10^{-7}
scruples	ounces (avdp)	0.0457	square yards	sq mm	8.361×10^5
scruples	pennyweights	0.8333			
scruples	pounds (apoth or		**T**		
	troy)	3.472×10^{-3}			
scruples	pounds (avdp)	2.857×10^{-3}	temperature (°C)	absolute tempera-	
seconds (angle)	degrees	2.778×10^{-4}	+273	ture (K)	1.0
seconds (angle)	minutes	0.01667	temperature (°C)		
seconds (angle)	quadrants	3.087×10^{-6}	+17.78	temperature (°F)	1.8
seconds (angle)	radians	4.848×10^{-6}	temperature (°F)	absolute tempera-	
slugs	cwt (long)	0.2873	+460	ture (°Rank)	1.0
slugs	cwt (short)	0.3217	temperature (°F)		
slugs	grams	1.459×10^4	−32	temperature (°C)	5/9
slugs	kilograms	14.59	tons (long)	cwt (long)	20
slugs	ounces (avdp)	514.79	tons (long)	cwt (short)	22.4
slugs	pounds (avdp)	32.17	tons (long)	kilograms	1,016
slugs	tons (long)	1.436×10^{-2}	tons (long)	ounces (avdp)	35,840.0
slugs	tons (short)	1.609×10^{-2}	tons (long)	pounds (avdp)	2,240.0
sphere	steradians	12.57	tons (long)	slugs	69.621
square centimeters	circular mils	1.973×10^5	tons (long)	tons (metric)	1.0160
square centimeters	sq ft	1.076×10^{-3}	tons (long)	tons (short)	1.120
square centimeters	sq in	0.1550	tons (metric)	cwt (short)	22.046
square centimeters	sq meters	0.0001	tons (metric)	kilograms	1,000.0
square centimeters	sq miles	3.861×10^{-11}	tons (metric)	ounces (avdp)	35,273.96
square centimeters	sq mm	100.0	tons (metric)	pounds (avdp)	2,205
square centimeters	sq yd	1.196×10^{-4}	tons (metric)	tons (long)	0.9842
square feet	acres	2.296×10^{-5}	tons (metric)	tons (short)	1.1023
square feet	circular mils	1.833×10^8	tons (short)	cwt (long)	17.857
square feet	sq cm	929.0	tons (short)	cwt (short)	20.0
square feet	sq in	144.0	tons (short)	grams	9.072×10^5
square feet	sq meters	0.09290	tons (short)	kilograms	907.1847

TABLE-72

TO CONVERT	INTO	MULTIPLY BY	TO CONVERT	INTO	MULTIPLY BY
tons (short)	ounces (apoth or troy)	29,166.66	watts (abs)	Btu (mean)/min	0.056884
tons (short)	ounces (avdp)	32,000.0	watts (abs)	joules/sec	1
tons (short)	pounds (apoth or		watt-hours	Btu	3.413
	troy)	2,430.56	watt-hours	ergs	3.60×10^{10}
tons (short)	pounds (avdp)	2,000.0	watt-hours	ft-lb	2,656
tons (short)	slugs	62.16	watt-hours	gm-cal	859.85
tons (short)	tons (long)	0.89286	watt-hours	horsepower-hr	1.341×10^{-3}
tons (short)	tons (metric)	0.9072	watt-hours	kg-cal	0.8605
tons (short)/sq ft	kg/sq meter	9,765	watt-hours	kg-meters	367.2
tons (short)/sq ft	lb/sq in	2,000	watt-hours	kilowatt-hr	0.001
tons of water/24 hr	lb of water/hr	83.333	watt (int)	watt (abs)	1.0002
tons of water/24 hr	gal/min	0.16643	webers	maxwells	10^8
tons of water/24 hr	cu ft/hr	1.3349	webers	kilolines	10^5
torr	mm of mercury	1.0	webers/sq in	gauss	1.550×10^7
torr	atmospheres	1.316×10^{-3}	webers/sq in	lines/sq in	10^8
			webers/sq in	webers/sq cm	0.1550
	V		webers/sq in	webers/sq meter	1,550
volt/inch	volt/cm	0.39370	webers/sq meter	gauss	10^4
volt (abs)	statvolts	0.003336	webers/sq meter	lines/sq in	6.452×10^4
			webers/sq meter	webers/sq cm	10^{-4}
	W		webers/sq meter	webers/sq in	6.452×10^{-4}
watts	Btu/hr	3.4129			
watts	Btu/min	0.05688		**Y**	
watts	ergs/sec	10^7	yards	centimeters	91.44
watts	ft-lb/min	44.27	yards	feet	3.0
watts	ft-lb/sec	0.7378	yards	inches	36.0
watts	horsepower	1.341×10^{-3}	yards	kilometers	9.144×10^{-4}
watts	horsepower (metric)	1.360×10^{-3}	yards	meters	0.9144
watts	kg-calories/min	0.01433	yards	miles (naut)	4.934×10^{-4}
watts	kilowatts	0.001	yards	miles (stat)	5.682×10^{-4}
			yards	millimeters	914.4

TABLE-73

Table of Minerals

Minerals are naturally occurring, homogeneous solids formed in the Earth's crust by the inorganic forces of nature, and possess definite, but generally not fixed, chemical compositions, physical characteristics, and highly ordered atomic arrangements. This table contains a group of common minerals selected for their relative importance and frequency of occurrence. Several physical properties commonly used to identify minerals have been included for each entry. The information provided here should be used with the understanding that there is a certain amount of variation in physical properties from specimen to specimen. For more information about the material presented here, consult the references cited at the end of the table.

Streak is the color of a finely powdered mineral obtained by rubbing the mineral on a streak plate, which is an unglazed piece of porcelain.

Since minerals are frequently polymorphic, the crystal system reported in the table is that of the most commonly observed form. The term amorphous (amor) is used to describe noncrystalline minerals that lack an ordered atomic arrangement. The crystal systems have been abbreviated in the table as follows:

hex	hexagonal	ortho	orthorhombic
iso	isometric	tet	tetragonal
mono	monoclinic	trig	trigonal

Hardness, which is the resistance of a smooth surface of a mineral to scratching, is expressed in terms of the Mohs' Hardness Scale. This relative scale spans from 1-9 and is defined as follows:

<2.5	can be scratched by a fingernail
>2.5-3	cannot be scratched by a fingernail; can be scratched by a copper cent
>3-5.5	cannot be scratched by a copper cent; can be scratched by a knife
>5.5-<7	cannot be scratched by a knife; can be scratched by quartz
>7	cannot be scratched by quartz

Mineral	Composition	Color	Streak	Crystal System	Hardness	Index of Refraction	Specific Gravity
Acanthite	Ag_2S	gray-black	black	mono	2-2.5	-	7.3
Actinolite	$Ca_2(Mg,Fe)_5Si_8O_{22}(OH)_2$	green	white	mono	6	1.64	3.0-3.4
Aegirine	$NaFeSi_2O_6$	dark green, brown, black	white	mono	6	1.82	3.6
Allanite	$(Ca,Ce)_3(Al,Fe)_2Si_3O_{12}(OH)$	brown to pitch-black	colorless	mono	5.5-6	1.70-1.81	3.5-4.2
Almandine	$Fe_3Al_2Si_3O_{12}$	red to black	white	iso	7-7.5	1.83	4.0-4.3
Alunite	$KAl_3(SO_4)_2(OH)_6$	white, red	colorless	hex	4	1.57	2.6-2.8
Amblygonite	$LiAlFPO_4$	white, pale green, blue	colorless	trig	6	1.60	3.0-3.1
Analcime	$NaAlSi_2O_6 \cdot H_2O$	colorless, white	colorless	iso	5-5.5	1.48-1.49	2.27
Anatase	TiO_2	brown	white	tet	5.5-6	2.6	3.9
Andalusite	Al_2SiO_5	reddish-brown, white, gray	colorless	ortho	7.5	1.64	3.16-3.20
Andradite	$Ca_3Fe_2Si_3O_{12}$	yellow, green, brown, black	white	iso	7-7.5	1.89	3.8-3.9
Anglesite	$PbSO_4$	colorless, white	colorless	ortho	3	1.88	6.2-6.4
Anhydrite	$CaSO_4$	colorless	colorless	ortho	3-3.5	1.58	2.89-2.98
Anthophyllite	$(Mg,Fe)_7Si_8O_{22}(OH)_2$	white, gray	colorless	ortho	5.5-6	1.61-1.71	2.85-3.2
Antimony	Sb	white to gray	gray	hex	3-3.5	-	6.7
Antlerite	$Cu_3(SO_4)(OH)_4$	dark emerald-green	light green	ortho	3.5-4	1.74	3.88
Apatite	$Ca_5(PO_4)_3(F,Cl,OH)$	green, blue, violet, brown, colorless	colorless	hex	5	1.63	3.15-3.20
Apophyllite	$KCa_4(Si_4O_{10})_2F \cdot 8H_2O$	colorless, white	colorless	tet	4.5-5	1.54	2.3-2.4
Aragonite	$CaCO_3$	colorless, white	colorless	ortho	3.5-4	1.68	2.95
Argentite	Ag_2S	black	dark gray-black	iso	2-2.5	-	7.2-7.4
Arsenic	As	tin-white	gray	hex	3.5	-	5.7

TABLE-74

Table of Minerals (Continued)

Mineral	Composition	Color	Streak	Crystal System	Hardness	Index of Refraction	Specific Gravity
Arsenopyrite	FeAsS	silver-white	black	mono	5.5-6	-	6.0-6.2
Atacamite	$Cu_2Cl(OH)_3$	emerald-green	light green	ortho	3-3.5	1.86	3.75-3.77
Augite	$(Ca,Na)(Mg,Fe,Al)(Si,Al)_2O_6$	black, dark green	white	mono	6	1.67-1.73	3.25-3.55
Autunite	$Ca(UO_2)_2(PO_4)_2.10-12H_2O$	lemon-yellow	yellow	tet	2-2.5	1.58	3.1-3.2
Axinite	$(Ca,Fe,Mn)_3Al_2(BO_3)(Si_4O_{12})(OH)$	clove-brown	colorless	trig	6.5-7	1.69	3.27-3.35
Azurite	$Cu_3(CO_3)_2(OH)_2$	azure-blue	light blue	mono	3.5-4	1.76	3.77
Barite	$BaSO_4$	colorless, white	colorless	ortho	3-3.5	1.64	4.5
Bauxite	A mixture of Al hydroxides	yellow, brown, gray, white	colorless	-	1-3	-	2.0-2.55
Beryl	$Be_3Al_2(Si_6O_{18})$	bluish-green, yellow, colorless	colorless	hex	7.5-8	1.57-1.61	2.65-2.8
Biotite	$K(Mg,Fe)_3(AlSi_3O_{10})(OH)_2$	dark brown, green to black	colorless	mono	2.5-3	1.61-1.70	2.95-3
Bismuth	Bi	silver-white	silver-white	hex	2-2.5	-	9.8
Boehmite	γ-AlO(OH)	white, yellow, brown	white	ortho	3	1.65	3.0-3.1
Borax	$Na_2B_4O_5(OH)_4.8H_2O$	colorless, white	colorless	mono	2-2.5	1.47	1.71
Bornite	Cu_5FeS_4	brownish-bronze	gray-black	tet	3	-	5.1
Boulangerite	$Pb_5Sb_4S_{11}$	bluish-lead-gray	brownish-gray	mono	2.5-3	-	6.23
Bournonite	$PbCuSbS_3$	steel-gray to black	steel-gray to black	ortho	2.5-3	-	5.8-5.9
Braunite	$3Mn_2O_3.MnSiO_3$	gray to black	gray to black	tet	6-6.5	-	4.7-4.8
Brochantite	$Cu_4SO_4(OH)_6$	green	green	mono	3.5-4	1.78	4.0
Brookite	TiO_2	brown to black	white to gray	ortho	5.5-6	2.6	4.1
Brucite	$Mg(OH)_2$	white, gray, green	colorless	hex	2.5	1.57	2.39
Bustamite	$(Mn,Ca,Fe)SiO_3$			trig	5.5-6.5	1.67-1.70	3.3-3.4
Calaverite	$AuTe_2$	tin-white to brass yellow	yellowish to green-gray	mono	2.5	-	9.4
Calcite	$CaCO_3$	colorless, white	colorless	hex	3	1.66	2.71
Cancrinite	$Na_6Ca_2(AlSiO_4)_6(CO_3)_2$	yellow, white, pink	white	hex	6	1.52	2.4-2.5
Carnallite	$KMgCl_3.6H_2O$	colorless, white	white	ortho	2.5	1.48	1.60
Carnotite	$K_2(UO_2)_2(VO_4)_2.3H_2O$	yellow	yellow	mono	2	1.93	4-5
Cassiterite	SnO_2	brown to black	white	tet	6-7	2.00	6.8-7.1
Celestite	$SrSO_4$	colorless, white, pale blue	colorless	ortho	3-3.5	1.62	3.95-3.97
Cerussite	$PbCO_3$	colorless, white	colorless	ortho	3-3.5	2.08	6.55
Chabazite	$Ca_2Al_2Si_4O_{12}.6H_2O$	white, flesh-red	colorless	hex	4-5	1.48	2.05-2.15
Chalcanthite	$CuSO_4.5H_2O$	blue	pale blue	trig	2.5	1.54	2.28
Chalcocite	Cu_2S	steel-gray	gray-black	ortho	2.5-3	-	5.7

TABLE-75

Table of Minerals (Continued)

Mineral	Composition	Color	Streak	Crystal System	Hardness	Index of Refraction	Specific Gravity
Chalcopyrite	$CuFeS_2$	brass-yellow	greenish-black	tet	3.5-4	-	4.1-4.3
Chlorargyrite	$AgCl$	pearl-gray, colorless	colorless	iso	2-3	2.07	~5.5
Chlorite	$Mg,Fe_3(Si,Al)_4O_{10}$ $(OH)_2.(Mg,Fe)_3(OH)_6$	green of various shades	colorless	mono	2-2.5	1.57-1.67	2.6-2.9
Chloritoid	$(Fe,Mg,Mn)_2(Al,Fe)$ $Al_3O_2(SiO_4)_2(OH)_4$	grassy green	white	mono	6.5	1.72-1.73	3.58-3.61
Chondrodite	$Mg_5(SiO_4)_2(F,OH)_2$	light yellow, brown	colorless	mono	6-6.5	1.60-1.63	3.1-3.2
Chromite	$FeCr_2O_4$	iron-black to brownish black	dark brown	iso	5.5	2.16	4.6
Chrysoberyl	$BeAl_2O_4$	yellowish to green	colorless	ortho	8.5	1.75	3.65-3.8
Chrysocolla	$(Cu,Al)_2H_2Si_2O_5$ $(OH)_4.nH_2O$	light green to turquoise-blue	light blue	amor	2-4	~4	2.0-2.4
Cinnabar	HgS	red to vermillion	bright red	hex	2-2.5	2.81	8.1
Clinozoisite	$Ca_2Al_3OSi_3O_{12}(OH)$	grayish-white, green	colorless	mono	6-6.5	1.67-1.72	3.25-3.37
Cobaltite	$CoAsS$	silver-white	gray-black	iso	5.5-6	-	6.3
Colemanite	$CaB_3O_4(OH)_3.H_2O$	colorless, white	colorless	mono	4-4.5	1.59	2.42
Columbite	$(Fe,Mn)Nb_2O_6$	black	dark brown to black	ortho	6	-	5.3-7.3
Copper	Cu	copper-red	copper-red	iso	2.5-3	-	8.9
Cordierite	$(Mg,Fe)_2Al_4Si_5$ $O_{18}.nH_2O$	blue, gray, brown	colorless	ortho	7-7.5	1.53-1.57	2.60-2.66
Corundum	Al_2O_3	gray, blue, pink, brown	white	trig	9	1.77	4.0
Covellite	CuS	indigo-blue to black	lead-gray to black	hex	1.5-2	-	4.6
Cristobalite	SiO_2	colorless, white	colorless	iso	7	1.48	2.32
Crocoite	$PbCrO_4$	red	orange-yellow	mono	2.5-3	2.36	5.9-6.1
Cryolite	Na_3AlF_6	colorless, white	colorless	mono	2.5	1.34	2.95-3.0
Cummingtonite	$(Fe,Mg,Mn)_7Si_8O_{22}$ $(OH)_2$	brown	white	mono	6	1.66-1.68	3.2-3.6
Cuprite	Cu_2O	red	brownish-red	iso	3.5-4	-	6.0
Danburite	$CaB_2(SiO_4)_2$	-	-	ortho	7	1.63	2.97-3.02
Datolite	$CaB(SiO_4)(OH)$	colorless, pale green	colorless	mono	5-5.5	1.65	2.8-3.0
Diamond	C	colorless, yellow, black	colorless	iso	10	2.42	3.5
Diaspore	$AlO(OH)$	white, gray	colorless	ortho	6.5-7	1.72	3.35-3.45
Diopside	$Ca(Mg,Fe)Si_2O_6$	white, green	white	mono	6	1.67	3.25-3.40
Dioptase	$CuSiO_2(OH)_2$	green		hex	5	1.65	3.3
Dolomite	$CaMg(CO_3)_2$	colorless, white, pink	colorless	hex	3.5-4	1.68	2.85
Enargite	Cu_3AsS_4	gray-black	gray-black	ortho	3	-	4.4
Enstatite	$MgSiO_3$	gray-brown, green, brown	colorless	ortho	5.5	1.65	3.2-3.5

TABLE-76

Table of Minerals (Continued)

Mineral	Composition	Color	Streak	Crystal System	Hardness	Index of Refraction	Specific Gravity
Epidote	$Ca_2(Al,Fe)Al_2(SiO_4)_3$ (OH)	yellowish to blackish green	colorless	mono	6-7	1.72-1.78	3.35-3.45
Epsomite	$MgSO_4.7H_2O$	white	white	ortho	2	1.46	1.67
Erythrite	$Co_3(AsO_4)_2.8H_2O$	red to pink	pink	mono	1.5-2.5	1.66	3.06
Euclase	$BeAl(SiO_4)(OH)$	-	-	mono	7.5	1.66	3.1
Eucryptite	$LiAlSiO_4$	-	-	hex	-	1.55	2.67
Fayalite	Fe_2SiO_4	-	-	ortho	6.5	1.86	4.39
Fluorite	CaF_2	colorless, violet, green	colorless	iso	4	1.43	3.18
Franklinite	$(Fe,Zn,Mn)(Fe,Mn)_2O_4$	iron-black	dark brown	iso	6	-	5.15
Gahnite	$ZnAl_2O_4$	dark green	colorless	iso	7.5-8	1.80	4.55
Galena	PbS	lead-gray	lead-gray	iso	2.5	-	7.6
Garnet	$(Ca,Mg,Fe,Mn)_3(Al,Fe,Cr)_2(SiO_4)_3$	brown to red	-	iso	6.5-7.5	1.71-1.88	3.5-4.3
Gersdorffite	$NiAsS$	silver-white	gray-black	iso	5.5	-	6.3
Gibbsite	$Al(OH)_3$	white, red, brown	white	mono	2.5-3.5	1.57	2.3-2.4
Glauconite	$(K,Na)(Fe,Mg,Al)_2(Si,Al)_4O_{10}(OH)_2$	dark green to black	pale green	mono	2	1.62	2.5-2.8
Glaucophane	$Na_2Mg_3Al_2Si_8O_{22}(OH)_2$	blue, blue-black	white	mono	6	1.62-1.67	3.0-3.3
Goethite	α-$FeO(OH)$	yellow-brown to dark brown	yellow-brown	ortho	5-5.5	2.39	4.37
Gold	Au	gold-yellow	gold-yellow	iso	2.5-3	-	15.0-19.3
Graphite	C	steel-gray to iron-black	black	hex	1-1.5	-	2.23
Grossular	$Ca_3Al_2Si_3O_{12}$	white, green, brown	white	iso	7-7.5	1.73	3.6
Grunerite	$Fe_7Si_8O_{22}(OH)_2$	light brown	colorless	mono	6	1.71	3.6
Gypsum	$CaSO_4.2H_2O$	colorless, white, gray	colorless	mono	2	1.52	2.32
Halite	$NaCl$	colorless, white	colorless	iso	2.5	1.54	2.1-2.3
Hausmannite	Mn_3O_4	brownish black	brown	tet	5.5-6	-	4.84
Hedenbergite	$CaFeSi_2O_6$	green to black	white	mono	5-6	1.73	3.40-3.55
Hematite	Fe_2O_3	red to vermillion	red-brown	hex	5.5-6.5	-	5.2
Hemimorphite	$Zn_4(Si_2O_7)(OH)_2.H_2O$	white	colorless	ortho	4.5-5	1.62	3.4-3.5
Heulandite	$CaAl_2Si_7O_{18}.6H_2O$	colorless, white	colorless	mono	3.5-4	1.48	2.18-2.20
Hornblende	$NaCa_2(Mg,Fe,Al)_5(Al,Si)_8O_{22}(OH)_2$	green, brown, black	white	mono	5-6	1.62-1.72	3.0-3.4
Humite	$(Mg,Fe)_7(SiO_4)_3(F,OH)_2$	yellow, brownish red	white	ortho	6-6.5	1.64	3.15-3.35
Hypersthene	$(Mg,Fe)SiO_3$	brown, black	white	ortho	5-6	1.68-1.73	3.4-4.0
Ilmenite	$FeTiO_3$	black	black	hex	5.5-6	-	4.7
Iron	Fe	gray	gray	iso	4	-	7.8
Jadeite	$NaAlSi_2O_6$	green, white	colorless	mono	6.5-7	1.66	3.3-3.5

TABLE-77

Table of Minerals (Continued)

Mineral	Composition	Color	Streak	Crystal System	Hardness	Index of Refraction	Specific Gravity
Jamesonite	$Pb_4FeSb_6S_{14}$	steel-gray to gray-black	steel-gray to gray-black	mono	2-3	-	5.5-6.0
Jarosite	$KFe_3(SO_4)_2(OH)_6$	brown	yellow	trig	3	1.82	2.9-3.3
Kainite	$KMg(Cl,SO_4).2\ 3/4\ H_2O$	-	-	mono	3	1.51	2.1
Kaolinite	$Al_2Si_2O_5(OH)_4$	white	colorless	trig	2-2.5	1.55	2.6-2.63
Kernite	$Na_2B_4O_6(OH)_2.3H_2O$	colorless, white	colorless	mono	3	1.47	1.95
Kyanite	Al_2SiO_5	blue	colorless	trig	5-7	1.72	3.56-3.66
Laumontite	$CaAl_2Si_4O_{12}.4H_2O$	white	white	mono	4	1.52	2.25-2.30
Lawsonite	$CaAl_2(Si_2O_7)(OH)_2.H_2O$	colorless, gray, pink	colorless	ortho	8	1.67	3.09
Lazulite	$(Mg,Fe)Al_2(PO_4)_2(OH)_2$	azure-blue	colorless	mono	5-5.5	1.64	3.0-3.1
Lazurite	$(Na,Ca)_8(AlSiO_4)_6(SO_4,S,Cl)_2$	azure-blue, greenish-blue	colorless	iso	5-5.5	1.50	2.4-2.45
Lepidolite	$K(Li,Al)_{2-3}(AlSi_3O_{10})(O,OH,F)_2$	pink to grayish white	colorless	mono	2.5-4	1.55-1.59	2.8-3.0
Leucite	$KAlSi_2O_6$	gray, white	colorless	iso	5.5-6	1.51	2.45-2.50
Lithiophilite	$Li(Mn,Fe)PO_4$	bluish-gray	colorless	ortho	4.5-5	1.67	3.5
Magnesite	$MgCO_3$	white, yellow	colorless	hex	3.5-5	1.70	3.0-3.2
Magnetite	Fe_3O_4	black	black	iso	6	-	5.18
Malachite	$Cu_2CO_3(OH)_2$	bright green	light green	mono	3.5-4	1.88	3.9-4.03
Manganite	$MnO(OH)$	steel-gray to iron-black	dark brown	mono	4	-	4.3
Marcasite	FeS_2	pale yellow	grayish-black	iso	6-6.5	-	4.9
Margarite	$CaAl_2(Al_2Si_2O_{10})(OH)_2$	pink, gray, white	colorless	mono	3.5-5	1.65	3.0-3.1
Melanterite	$FeSO_4.7H_2O$	pale green	white	mono	2	1.48	1.90
Melilite	$Ca_2Al(AlSiO_7) - Ca_2Mg(Si_2O_7)$	nearly color-less, reddish brown, gray	white	tet	5-6	-	2.94-3.05
Microcline	$KAlSi_3O_8$	colorless, white	colorless	trig	6	1.53	2.54-2.56
Microlite	$Ca_2Ta_2O_6(O,OH,F)$	yellow, brown, black	yellow to brown	iso	5.5	1.92-1.99	5.48-5.56
Millerite	NiS	brass-yellow	greenish-black	hex	3-3.5	-	5.5
Mimetite	$Pb_5(AsO_4)_3Cl$	white, yellow	white	hex	3.5-4	2.1-2.2	7.2-7.3
Molybdenite	MoS_2	lead-gray	grayish-black	hex	1-1.5	-	4.7
Monazite	$(Ce,La,Y,Th)PO_4$	yellowish-brown	colorless	mono	5-5.5	1.79	5.0-5.3
Monticellite	$CaMgSiO_4$			ortho	5	1.65	3.2
Montmorillonite	$(Al,Mg)_2Si_4O_{10}(OH)_2.nH_2O$	white, gray, green-gray	white	mono	2	1.50-1.64	2.0-2.7
Muscovite	$KAl_2(AlSi_3)O_{10}(OH)_2$	colorless, pale tints	colorless	mono	2-2.5	1.60	2.76-2.88
Natrolite	$Na_2Al_2Si_3O_{10}.2H_2O$	colorless, white	colorless	ortho	5-5.5	1.48	2.25
Nepheline	$(Na,K)AlSiO_4$	colorless, gray, brown	colorless	hex	5.5-6	1.54	2.55-2.65

TABLE-78

Table of Minerals (Continued)

Mineral	Composition	Color	Streak	Crystal System	Hardness	Index of Refraction	Specific Gravity
Nickel Skutterudite	$(Ni,Co)As_3$	silver-white	black	iso	5.5	-	6.1-6.9
Nickeline	NiAs	pale copper-red	brown-black	hex	5-5.5	-	7.8
Niter	KNO_3	white	colorless	ortho	2	1.50	2.09-2.14
Nitratite	$NaNO_3$	colorless, white	colorless	hex	1-2	1.59	2.29
Olivine	$(Mg,Fe)_2SiO_4$	olive to grayish-green	colorless	ortho	6.5-7	1.69	3.27-4.37
Opal	$SiO_2.nH_2O$	colorless, white	colorless	amor	5-6	1.44	1.9-2.2
Orpiment	As_2S_3	lemon-yellow	yellow	mono	1.5-2	2.8	3.49
Orthoclase	$KAlSi_3O_8$	colorless, white	colorless	mono	6	1.52	2.54-2.56
Pectolite	$Ca_2NaH(SiO_3)_3$	white	colorless	trig	5	1.60	2.7-2.8
Pentlandite	$(Fe,Ni)_9S_8$	yellowish-bronze	bronze-brown	iso	3.5-4	-	4.6-5.0
Periclase	MgO	white	white	iso	5	1.73	3.58
Phenacite	Be_2SiO_4	white, colorless	colorless	hex	7.5-8	1.65	2.97-3.0
Phlogopite	$KMg_3(AlSi_3O_{10})(OH)_2$	pale brown	colorless	mono	2.5-3	1.56-1.64	2.86
Plagioclase	$(Na,Ca)(Al,Si)_4O_8$	colorless, white, gray	colorless	trig	6	1.53-1.59	2.62
Platinum	Pt	steel-gray	steel-gray	iso	4-4.5	-	14-19
Prehnite	$Ca_2Al(AlSi_3O_{10})(OH)_2$	apple-green, white	colorless	ortho	6-6.5	1.63	2.8-2.95
Proustite	Ag_3AsS_3	ruby-red	bright red	hex	2-2.5	3.09	5.55
Pumpellyite	$Ca_2MgAl_2(SiO_4)(Si_2O_7)(OH)_2.H_2O$	bluish green to nearly white	white	mono	5.5-6	-	3.18-3.23
Pyrargyrite	Ag_3SbS_3	red	red	hex	2.5	3.08	5.85
Pyrite	FeS_2	pale brass-yellow	brownish-black	iso	6-6.5	-	5.0
Pyrochlore	$(Ca,Na)_2(Nb,Ta)_2O_6(OH,F)$	yellow, brown, black	white	iso	5-5.5	-	4.2-6.4
Pyrolusite	MnO_2	iron-black	black	tet	1-2	-	4.7
Pyromorphite	$Pb_5(PO_4)_3Cl$	green, brown, yellow, gray	colorless	hex	3.5-4	2.06	6.5-7.1
Pyrope	$Mg_3Al_2Si_3O_{12}$	red to black	white	iso	7-7.5	1.71	3.6
Pyrophyllite	$Al_2Si_4O_{10}(OH)_2$	white, gray, green	colorless	mono	1-2	1.59	2.8-2.9
Pyrrhotite	$Fe_{1-x}S$	brownish-bronze	black	mono	4	-	4.6
Quartz	SiO_2	colorless, white, various colors	colorless	hex	7	1.54	2.65
Realgar	AsS	red	red to orange	mono	1.5-2	2.60	3.48
Rhodochrosite	$MnCO_3$	pink, rose-red, brown	colorless	hex	3.5-4.5	1.82	3.45-3.6
Rhodonite	$MnSiO_3$	pink, brown	colorless	trig	5.5-6	1.73-1.75	3.58-3.70
Riebeckite	$Na_2(Mg,Fe)_3Fe_2Si_8O_{22}(OH)_2$	dark blue, black	blue-gray	mono	6	1.66-1.71	3.3-3.6
Romanechite	$BaMn_9O_{16}(OH)_4$	black	black	ortho	5-6	-	3.7-4.7

TABLE-79

Table of Minerals (Continued)

Mineral	Composition	Color	Streak	Crystal System	Hardness	Index of Refraction	Specific Gravity
Rutile	TiO_2	reddish-brown	pale brown	tet	6-6.5	2.61	4.18-4.25
Scapolite	$3NaAlSi_3O_8.NaCl - 3CaAlSi_2O_8.CaCO_3$	white, pink, gray	colorless	tet	5-6	1.55-1.60	2.65-2.74
Scheelite	$CaWO_4$	white, yellow, brown	colorless	tet	4.5-5	1.92	5.9-6.1
Serpentine	$Mg_3Si_2O_5(OH)_4$	shades of green	colorless	mono	2-5	1.55	2.3-2.66
Shattuckite	$Cu_5(SiO_3)_4(OH)_2$	blue		ortho	-	1.78	3.8
Siderite	$FeCO_3$	light to dark brown	yellow, brown	hex	3.5-4	1.88	3.83-3.88
Sillimanite	Al_2SiO_5	white, brown, gray	colorless	ortho	6-7	1.66	3.23
Silver	Ag	silver-white	silver-white	iso	2.5-3	-	10.5
Skutterudite	$(Co,Ni)As_3$	silver-white	black	iso	5.5-6	-	6.1-6.9
Smithsonite	$ZnCO_3$	green, blue, white	colorless	hex	5	1.85	4.35-4.40
Sodalite	$Na_8(AlSiO_4)_6Cl_2$	blue, white	colorless	iso	5.5-6	1.48	2.15-2.3
Sperrylite	$PtAs_2$	tin-white	black	iso	6-7	-	10.6
Spessartine	$Mn_3Al_2Si_3O_{12}$	orange, red, brown	white	iso	7-7.5	1.80	4.0-4.2
Sphalerite	ZnS	yellow-brown to black	white to yellow-brown	iso	3.5	2.37	3.9-4.1
Spinel	$MgAl_2O_4$	red, black, blue, green, brown	colorless	iso	8	1.72	3.6-4.0
Spodumene	$LiAlSi_2O_6$	white, gray, pink, green	colorless	mono	6.5-7	1.67	3.15-3.20
Staurolite	$Fe_2Al_9O_6(SiO_4)_4(O,OH)_2$	red-brown	colorless	ortho	7-7.5	1.75	3.65-3.75
Stibnite	Sb_2S_3	lead-gray to black	lead-gray to black	ortho	2	-	4.5
Stilbite	$NaCa_2Al_5Si_{13}O_{36}.14H_2O$	white	colorless	mono	3.5-4	1.50	2.1-2.2
Stilpnomelane	$K(Fe,Mg)_6Si_8Al(O,OH)_{27}.2-4H_2O$	black, yellowish to reddish brown	white	mono	3-4	1.58-1.74	2.59-2.96
Stromeyerite	$(Ag,Cu)_2S$	steel-gray	-	ortho	2.5-3	-	6.2-6.3
Strontianite	$SrCO_3$	colorless, white	colorless	ortho	3.5-4	1.67	3.7
Sulfur	S	yellow	colorless	ortho	1.5-2.5	2.04	2.05-2.09
Sylvanite	$(Au,Ag)Te_2$	tin-white	gray	mono	2	-	8-8.2
Sylvite	KCl	colorless, white	colorless	iso	2	1.49	1.99
Talc	$Mg_3Si_4O_{10}(OH)_2$	white, green, gray	colorless	mono	1	1.59	2.7-2.8
Tantalite	$(Fe,Mn)Ta_2O_6$	black	dark brown to black	ortho	6	-	5.3-7.3
Tennantite	$Cu_{12}As_4S_{13}$	black	brown to black	iso	3-4.5	-	4.6-5.1
Tetrahedrite	$Cu_{12}Sb_4S_{13}$	grayish black	black	iso	3-4.5	-	4.7-5.0
Thorite	$ThSiO_4$	brown to black	white	tet	5	1.8	5.3
Titanite	$CaTiO(SiO_4)$	brown, green, yellow	colorless	mono	5-5.5	1.91	3.4-3.55

TABLE-80

Table of Minerals (Continued)

Mineral	Composition	Color	Streak	Crystal System	Hardness	Index of Refraction	Specific Gravity
Topaz	$Al_2SiO_4(F,OH)_2$	colorless, yellow, blue	colorless	ortho	8	1.61-1.63	3.4-3.6
Torbernite	$Cu(UO_2)_2$ $(PO_4)_2.8-12H_2O$	emerald-green to apple-green	green	tet	2-2.5	1.59	3.2-3.7
Tourmaline	$(Na,Ca)(Li,Mg,Al)$ $(Al,Fe,Mn)_6(BO_3)_3$ $(Si_6O_{18})(OH)_4$	black, green, brown, blue, pink	colorless	hex	7-7.5	1.64-1.68	3.0-3.25
Tremolite	$Ca_2Mg_5Si_8O_{22}(OH)_2$	white	white	mono	6	1.61	3.0
Tridymite	SiO_2	colorless	colorless	hex	7	1.47	2.26
Triphylite	$Li(Fe,Mn)PO_4$	bluish-gray	colorless	ortho	4.5-5	1.69	3.42-3.56
Turquoise	$CuAl_6(PO_4)_4$ $(OH)_8.5H_2O$	blue, bluish-green	colorless	trig	6	1.62	2.6-2.8
Ulexite	$NaCaB_5O_6$ $(OH)_6.5 H_2O$	white	colorless	trig	1-2.5	1.50	1.96
Uraninite	UO_2	black	brownish-black	iso	5.5	-	9.0-9.7
Uvarovite	$Ca_3Cr_2Si_3O_{12}$	green	white	iso	7.5	1.87	3.90
Vanadinite	$Pb_5(VO_4)_3Cl$	ruby-red, brown, yellow	colorless	hex	3	2.25-2.42	6.7-7.1
Vermiculite	$(Mg,Fe,Al)_3(Al,Si)_4$ $O_{10}(OH)_2.4H_2O$	yellow, brown, green	white	mono	1.5	1.55-1.58	2.4
Vesuvianite	$Ca_{10}(Mg,Fe)_2Al_4Si_9$ $O_{34}(OH)_4$	green, brown, yellow, blue	colorless	tet	6.5	1.70-1.75	3.35-4.45
Vivianite	$Fe_3(PO_4)_2.8H_2O$	blue, blue-green	blue	mono	1-1.5	1.60	2.7
Wavellite	$Al_3(PO_4)_2$ $(OH)_3.5 H_2O$	green, white, brown	colorless	ortho	3.5-4	1.54	2.33
Willemite	Zn_2SiO_4	yellow-green, white, brown	colorless	hex	5.5	1.69	3.9-4.2
Witherite	$BaCO_3$	colorless, white	colorless	ortho	3.5	1.68	4.3
Wolframite	$(Fe,Mn)WO_4$	brown to black	black to brown	mono	4-4.5	-	7.0-7.5
Wollastonite	$CaSiO_3$	colorless, white	colorless	trig	5-5.5	1.63	2.8-2.9
Wulfenite	$PbMoO_4$	yellow, red	colorless	tet	3	2.40	6.8
Wurtzite	ZnS	brownish black	brown	hex	3.5-4	2.35	4.09
Xenotime	YPO_4	yellowish to reddish brown	white	tet	4-5	-	4.4-5.1
Zincite	ZnO	red to orange-yellow	orange-yellow	hex	4-4.5	2.01	5.68
Zircon	$ZrSiO_4$	brown, gray, colorless	colorless	tet	7.5	1.92-1.96	4.68
Zoisite	$Ca_2Al_3Si_3O_{12}(OH)$	gray, green, pink	white	ortho	7	1.69	3.3

References

1. L. G. Berry *et al.*, *Mineralogy: Concepts, Descriptions, Determinations* (W. H. Freeman and Co., San Francisco, 2nd ed., 1983).
2. A. M. Clark, *Hey's Mineral Index: Mineral Species, Varieties and Synonyms* (Chapman & Hall, London, 3rd ed., 1993).
3. C. Klein, C. S. Hurlbut, Jr., *Manual of Mineralogy* (John Wiley & Sons, New York, 20th ed., 1985).
4. P. G. Read, *Dictionary of Gemmology* (Butterworth, Heinemann, Oxford, 2nd ed., 1988).

TABLE-81

ATOMIC WEIGHTS
(Order of Atomic Number)

Atomic number	Element	Symbol	Atomic weight	Atomic number	Element	Symbol	Atomic weight
1	Hydrogen	H	[1.00784; 1.00811] 1.008*	55	Cesium	Cs	132.9054519
				56	Barium	Ba	137.327
2	Helium	He	4.002602	57	Lanthanum	La	138.90547
3	Lithium	Li	[6.938; 6.997] 6.94*	58	Cerium	Ce	140.116
				59	Praseodymium	Pr	140.90765
4	Beryllium	Be	9.012182	60	Neodymium	Nd	144.242
5	Boron	B	[10.806; 10.821] 10.81*	61	Promethium	Pm	144.9127**
				62	Samarium	Sm	150.36
6	Carbon	C	[12.0096; 12.0116] 12.011*	63	Europium	Eu	151.964
				64	Gadolinium	Gd	157.25
7	Nitrogen	N	[14.00643; 14.00728] 14.007*	65	Terbium	Tb	158.92535
				66	Dysprosium	Dy	162.500
8	Oxygen	O	[15.99903; 15.99977] 15.999*	67	Holmium	Ho	164.93032
				68	Erbium	Er	167.259
9	Fluorine	F	18.9984032	69	Thulium	Tm	168.93421
10	Neon	Ne	20.1797	70	Ytterbium	Yb	173.054
11	Sodium	Na	22.98976928	71	Lutetium	Lu	174.9668
12	Magnesium	Mg	24.3050	72	Hafnium	Hf	178.49
13	Aluminum	Al	26.9815386	73	Tantalum	Ta	180.94788
14	Silicon	Si	[28.084; 28.086] 28.085*	74	Tungsten	W	183.84
				75	Rhenium	Re	186.207
15	Phosphorus	P	30.973762	76	Osmium	Os	190.23
16	Sulfur	S	[32.059; 32.076] 32.06*	77	Iridium	Ir	192.217
				78	Platinum	Pt	195.084
17	Chlorine	Cl	[35.446; 35.457] 35.45*	79	Gold	Au	196.966569
				80	Mercury	Hg	200.59
18	Argon	Ar	39.948	81	Thallium	Tl	[204.382; 204.385] 204.38*
19	Potassium	K	39.0983				
20	Calcium	Ca	40.078	82	Lead	Pb	207.2
21	Scandium	Sc	44.955912	83	Bismuth	Bi	208.98040
22	Titanium	Ti	47.867	84	Polonium	Po	208.9824**
23	Vanadium	V	50.9415	85	Astatine	At	209.9871**
24	Chromium	Cr	51.9961	86	Radon	Rn	222.0176**
25	Manganese	Mn	54.938045	87	Francium	Fr	223.0197**
26	Iron	Fe	55.845	88	Radium	Ra	226.0254**
27	Cobalt	Co	58.933195	89	Actinium	Ac	227.0278**
28	Nickel	Ni	58.6934	90	Thorium	Th	232.03806
29	Copper	Cu	63.546	91	Protactinium	Pa	231.03588
30	Zinc	Zn	65.38	92	Uranium	U	238.02891
31	Gallium	Ga	69.723	93	Neptunium	Np	237.0482**
32	Germanium	Ge	72.63	94	Plutonium	Pu	244.0642**
33	Arsenic	As	74.92160	95	Americium	Am	243.0614**
34	Selenium	Se	78.96	96	Curium	Cm	247.0704**
35	Bromine	Br	79.904	97	Berkelium	Bk	247.0703**
36	Krypton	Kr	83.798	98	Californium	Cf	251.0796**
37	Rubidium	Rb	85.4678	99	Einsteinium	Es	252.0830**
38	Strontium	Sr	87.62	100	Fermium	Fm	257.0951**
39	Yttrium	Y	88.90585	101	Mendelevium	Md	258.0984**
40	Zirconium	Zr	91.224	102	Nobelium	No	259.1010**
41	Niobium	Nb	92.90638	103	Lawrencium	Lr	262.1096**
42	Molybdenum	Mo	95.96	104	Rutherfordium	Rf	265.1167**
43	Technetium	Tc	97.9072**	105	Dubnium	Db	268.125**
44	Ruthenium	Ru	101.07	106	Seaborgium	Sg	271.133**
45	Rhodium	Rh	102.90550	107	Bohrium	Bh	267.1277**
46	Palladium	Pd	106.42	108	Hassium	Hs	277.150**
47	Silver	Ag	107.8682	109	Meitnerium	Mt	276.151**
48	Cadmium	Cd	112.411	110	Darmstadtium	Ds	281.162**
49	Indium	In	114.818	111	Roentgenium	Rg	280.164**
50	Tin	Sn	118.710	112	Copernicium	Cn	285.174**
51	Antimony	Sb	121.760	113	Ununtrium	Uut	284.178**
52	Tellurium	Te	127.60	114	Flerovium	Fl	289.187**
53	Iodine	I	126.90447	115	Ununpentium	Uup	288.192**
54	Xenon	Xe	131.293	116	Livermorium	Lv	292.200**

Based on the 2009 IUPAC Atomic Weights of the Elements: M. E. Wieser, T. B. Coplen, *Pure Appl. Chem.* **83**, 359 (2011).
* Conventional atomic weight; representative value for an element having an atomic weight interval.
** Relative atomic mass of the isotope of that element of longest known half-life.

TABLE-82

ORGANIC NAME REACTIONS

ORGANIC NAME REACTIONS

The Organic Name Reactions (ONR) section is intended to serve the professional chemist and student by describing organic chemical reactions which have come to be recognized and referred to by name within the chemistry community. A select group has been chosen for addition to this section. Each reaction description is designed to be informative and representative of the pertinent literature; however, it is not meant to be comprehensive. The descriptions are composed of the following: (1) name(s) associated with the reaction, (2) the original and/or primary contributor(s) connected with the discovery and/or development of the reaction, (3) a concise description of the transformation, (4) a reaction scheme, (5) key references, and (6) cross references to other ONR based on commonalities. An index to this section follows the reactions monographs.

Abbreviations

Ac	acetyl	E	electrophile
AcOH	acetic acid	ee	enantiomeric excess
AIBN	2,2′-azobisisobutyronitrile	equiv	equivalents
Ar	aryl	Et	ethyl
aq	aqueous	Et$_2$O	diethyl ether
alc	alcohol	EtOH	ethanol
B	base	EWG	electron withdrawing group
BBN	borabicyclo[3.3.1]nonane	*hv*	irradiation with light
BINAP	2,2′-bis(diphenylphosphino)-1,1′-	HMPT	hexamethylphosphoric triamide
	binaphthyl	L	ligand
BINOL	2,2′-dihydroxy-1,1′-binaphthyl	liq	liquid
Bu	butyl	LDA	lithium diisopropylamide
cat	catalytic	LHMDS	lithium hexamethyldisilazide
conc	concentrated	mCPBA	3-chloroperoxybenzoic acid
Cp	cyclopentadienyl	Me	methyl
Δ	heat	MeOH	methanol
dba	dibenzylideneacetone	NBS	*N*-bromosuccinimide
dil	dilute	NuH	nucleophile
DCC	dicyclohexylcarbodiimide	Ph	phenyl
DMAP	4-(dimethylamino)pyridine	Pr	propyl
DMF	*N,N*-dimethylformamide	salen	*N,N*′-ethylenebis(salicylideneimine)
DMSO	dimethyl sulfoxide	Tf	trifluoromethanesulfonyl
dppf	1,1′-bis(diphenylphosphino)-	THF	tetrahydrofuran
	ferrocene	TMS	trimethylsilyl
dppp	1,3-bis(diphenylphosphino)propane	Ts	*p*-toluenesulfonyl
DTBMP	2,6-di-*tert*-butyl-4-methylpyridine		

The Merck Index Editorial Staff expresses gratitude to Professor David W. C. MacMillan of Princeton University for reviewing portions of this section and providing insightful comments and valuable suggestions.

Organic Name Reactions

1. Acetoacetic Ester Synthesis

Base-catalyzed alkylation or arylation of β-ketoesters. Subsequent mild hydrolysis and decarboxylation yield substituted ketones. Alternately, treatment with concentrated base produces substituted esters:

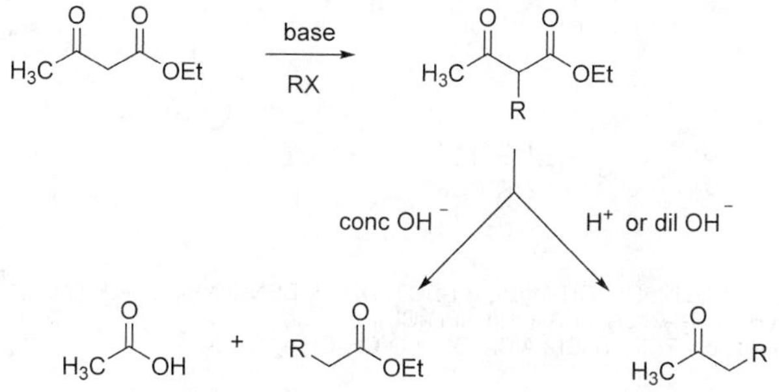

Synthetic applications: R. Kluger, M. Brandl, *J. Org. Chem.* **51**, 3964 (1986); T. Yamamitsu *et al.*, *J. Chem. Soc. Perkin Trans. 1* **1989**, 1811.

2. Acyloin Condensation

L. Bouveault, R. Loquin, *Compt. Rend.* **140**, 1593 (1905).

Reductive coupling of esters by sodium to yield acyloins (α-hydroxyketones). Yields are greatly improved in the presence of trimethylchlorosilane:

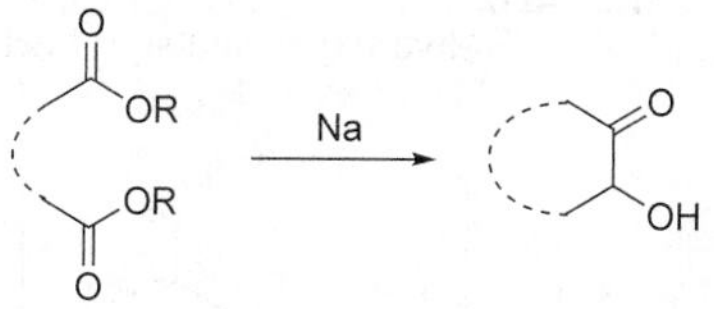

K. T. Finley, *Chem. Rev.* **64**, 573 (1964); K. Ziegler, *Houben-Weyl* **4/2**, 729-822 (1955); S. M. McElvain, *Org. React.* **4**, 256 (1948); J. J. Bloomfield *et al., ibid.* **23**, 259 (1976); R. Brettle, *Comp. Org. Syn.* **3**, 613-632 (1991). *Cf.* Benzoin Condensation.

3. Akabori Amino Acid Reactions

S. Akabori, *J. Chem. Soc. Jpn.* **52**, 606 (1931); *Ber.* **66**, 143, 151 (1933); *J. Chem. Soc.* **64**, 608 (1943).

1. Formation of aldehydes by oxidative decomposition of α-amino acids when heated with sugars according to the equation:

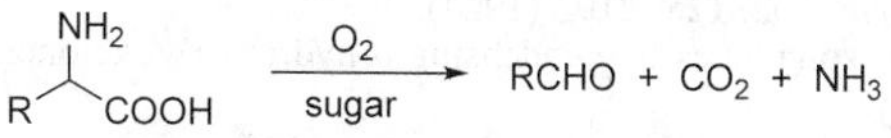

2. Reduction of α-amino acids and esters by sodium amalgam and ethanolic hydrogen chloride to the corresponding α-amino aldehydes:

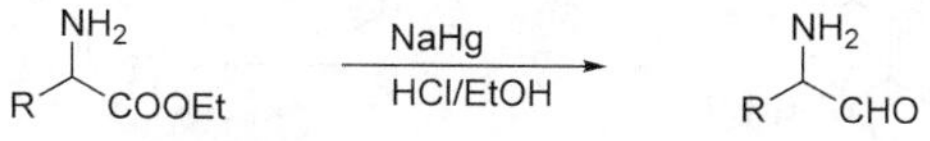

3. Formation of alkamines by heating mixtures of aromatic aldehydes and amino acids. No reaction was observed with tertiary amino groups.

E. Takagi *et al., J. Pharm. Soc. Jpn.* **71**, 648 (1951); **72**, 812 (1952); A. Lawson, H. V. Morley, *J. Chem. Soc.* **1955**, 1695; A. Lawson, *ibid.* **1956**, 307; K. Dose, *Ber.* **90**, 1251 (1957); V. N. Belikov *et al., Izv. Akad. Nauk SSSR Ser. Khim.* **1969**, 2536.

4. Aldol Reaction (Condensation)

R. Kane, *Ann. Phys. Chem., Ser. 2, **44**, 475 (1838); *idem, J. Prakt. Chem.* **15**, 129 (1838).

The acid- or base-catalyzed condensation of one carbonyl compound with the enolate/enol of another, which may or may not be the same, to generate a β-hydroxy carbonyl compound—an aldol. The method is compromised by self-condensation, polycondensation, generation of regioisomeric enols/enolates, and dehydration of the aldol followed by Michael addition, *q.v.* The development of methods for the preparation and use of preformed enolates or enol derivatives, that dictate specific carbon-carbon bond formation, have revolutionized the coupling of carbonyl compounds:

MX = LDA, LHMDS, CH$_3$MgBr, Bu$_2$BOTf/NR$_3$, 9-BBNCl/NR$_3$, LDA + ZnCl$_2$,
LDA + (C$_5$H$_5$)$_2$ZrCl$_2$, LDA + Ti(O-iPr)$_3$Cl
Lewis acid = TiCl$_4$, SnCl$_4$, AlCl$_3$, BF$_3$ • O(CH$_2$CH$_3$)$_2$, ZnCl$_2$

Historical perspective: C. H. Heathcock, *Comp. Org. Syn.* **2**, 133-179 (1991). General review: T. Mukaiyama, *Org. React.* **28**, 203-331 (1982). Application of lithium and magnesium enolates: C. H. Heathcock, *Comp. Org. Syn.* **2**, 181-238 (1991); of boron enolates: B. M. Kim *et al., ibid.* 239-275; of transition metal enolates: I. Paterson, *ibid.* 301-319. Stereoselective reactions of ester and thioester enolates: M. Braun, H. Sacha, *J. Prakt. Chem.* **335**, 653-668 (1993). Review of asymmetric methodology: A. S. Franklin, I. Paterson, *Contemp. Org. Synth.* **1**, 317-338 (1994). *Cf.* Claisen-Schmidt Condensation; Henry Reaction; Ivanov Reaction; Knoevenagel Condensation; Reformatsky Reaction; Robinson Annulation.

5. Algar-Flynn-Oyamada Reaction

J. Algar, J. P. Flynn, *Proc. R. Irish Acad.* **42B**, 1 (1934); B. Oyamada, *J. Chem. Soc. Jpn.* **55**, 1256 (1934).

Alkaline hydrogen peroxide oxidation of *o*-hydroxyphenyl styryl ketones (chalcones) to flavonols *via* the intermediate dihydroflavonols:

T. S. Wheeler, *Rec. Chem. Prog.* **18**, 133 (1957); W. P. Cullen *et al., J. Chem. Soc. C* **1971**, 2848. Mechanism: T. R. Gormley, *et al., Tetrahedron* **29**, 369 (1973); M. Bennett *et al., ibid.* **54**, 9911 (1998). Synthetic applications: H. Wagner *et al., ibid.* **33**, 1405 (1977); A. C. Jain *et al., Bull. Chem. Soc. Jpn.* **56**, 1267 (1983). *Cf.* Auwers Synthesis.

6. Allan-Robinson Reaction

J. Allan, R. Robinson, *J. Chem. Soc.* **125**, 2192 (1924).

Preparation of flavones or isoflavones by condensing *o*-hydroxyaryl ketones with anhydrides of aromatic acids and their sodium salts:

S. F. Dyke *et al., J. Org. Chem.* **26**, 2453 (1961); Seshandri in *The Chemistry of Flavonoid Compounds*, T. A. Geissman, Ed. (New York, 1962) p 182; Gripenberg, *ibid.* p 411; W. Rahman, K. T. Nasim, *J. Org. Chem.* **27**, 4215 (1962); D. L. Dreyer *et al., Tetrahedron* **20**, 2977 (1964). Synthesis applications: P. K. Dutta *et al., Indian J. Chem.* **21B**, 1037 (1982); T. Horie *et al., Chem. Pharm. Bull.* **37**, 1216 (1989); J. K. Makrandi *et al., Synth. Commun.* **19**, 1919 (1989); E. J. Corey *et al., Tetrahedron Lett.* **37**, 7162 (1996); B. P. Reddy *et al., J. Heterocycl. Chem.* **33**, 1561 (1996). *Cf.* Baker-Venkataraman Rearrangement; Kostanecki Acylation.

7. Allylic Rearrangements

L. Claisen, *Ber.* **45**, 3157 (1912).

Migration of a carbon-carbon double bond in a three carbon (allylic) system on treatment with nucleophiles under S$_N$1 conditions (or under S$_N$2 conditions when the nucleophilic attack takes place at the γ-carbon):

$$R-\underset{H}{\underset{|}{C}}=\underset{H}{\underset{|}{C}}-CH_2X \longrightarrow \left[R\overset{+}{-CH\cdots CH\cdots CH_2} \right] \longrightarrow R-\underset{H}{\underset{|}{C}}=\underset{H}{\underset{|}{C}}-CH_2Y + R-\underset{H}{\underset{|}{C}}-\underset{Y}{\underset{|}{C}}=CH_2$$

Reviews: J. R. DeWolfe, W. G. Young, *Chem. Rev.* **56**, 753 (1956); W. G. Young, *J. Chem. Educ.* **39**, 455 (1962); P. de la Mare in *Molecular Rearrangements* Part 1, P. de Mayo, Ed. (Wiley-Interscience, New York, 1963) pp 27-110; K. Mackenzie in *The Chemistry of Alkenes*, S. Patai, Ed. (Interscience, New York, 1964) pp 436-453; R. H. DeWolfe, W. G. Young in *ibid.* pp 681-738; J. March, *Advanced Organic Chemistry* (Wiley-Interscience, New York, 4th ed., 1992) pp 327-330.

8. Amadori Rearrangement
M. Amadori, *Atti Accad. Naz. Lincei* **2**(6), 337 (1925), *C.A.* **20**, 902 (1926); *ibid.* **9**(6), 68, 226 (1929), *C.A.* **23**, 3211, 3443 (1929).

Conversion of *N*-glycosides of aldoses to *N*-glycosides of the corresponding ketoses by acid or base catalysis:

$$\begin{array}{c} NHR \\ | \\ HC \\ | \\ HCOH \quad O \\ | \quad | \end{array} \longrightarrow \begin{array}{c} NHR \\ | \\ CH_2 \\ | \\ HOC \\ | \\ O \end{array}$$

J. E. Hodge, *Adv. Carbohydr. Chem.* **10**, 169 (1955); R. U. Lemieux in *Molecular Rearrangements* Part 2, P. de Mayo, Ed. (Wiley-Interscience, New York, 1964) p 753. ^{13}C-NMR studies: W. Funcke, *Ann.* **1978**, 2099. *Review:* K. Maruoka, H. Yamamoto, *Comp. Org. Syn.* **6**, 789-791 (1991).

9. Arens-van Dorp Synthesis; Isler Modification
D. A. van Dorp, J. F. Arens, *Nature* **160**, 189 (1947); J. F. Arens *et al., Rec. Trav. Chim.* **68**, 604, 609 (1949); O. Isler *et al., Helv. Chim. Acta* **39**, 259 (1956).

The preparation of alkoxyethynyl alcohols from ketones and ethoxyacetylene. In the **Isler modification** the tedious preparation of ethoxyacetylene is obviated by treating β-chlorovinyl ether with lithium amide to yield lithium ethoxyacetylene, which is then condensed with the ketone:

$$ClHC=CHOEt \xrightarrow{LiNH_2} Li-\!\!\equiv\!\!-O-Et \longrightarrow$$

$$\underset{OH}{\overset{C\equiv COEt}{\underset{|}{\underset{CCH_3}{|}}}}$$

H. Heusser *et al., Helv. Chim. Acta* **33**, 370 (1950); J. F. Arens, *Adv. Org. Chem.* **2**, 117-212 (1960); H. Meerwein, *Houben-Weyl* **6/3**, 189 (1965). *Cf.* Favorskii-Babayan Synthesis; Nef Synthesis.

10. Arndt-Eistert Synthesis
F. Arndt, B. Eistert, *Ber.* **68**, 200 (1935).

Homologation of carboxylic acids:

$$RCOOH \xrightarrow{SOCl_2} RCOCl \xrightarrow{2CH_2N_2} R-\overset{O}{\overset{||}{C}}-\overset{H}{\overset{|}{C}}-\overset{+}{N}\!\!\equiv\!\!N \xrightarrow[Ag_2O]{H_2O} RCH_2COOH$$

Alternative reagent for diazomethane: T. Aoyama, *Tetrahedron Lett.* **21**, 4461 (1980). Application to synthesis of unsaturated diazoketones: T. Hudlicky *et al., ibid.* **1979**, 2667; K. Gademann *et al., Angew. Chem. Int. Ed.* **38**, 1223 (1999); via ultrasonic activation: J-Y. Winum *et al., Tetrahedron Lett.* **37**, 1781 (1996); of amino acids: R. E. Marti *et al., ibid.* **38**, 6145 (1997); R. J. DeVita *et al., Bioorg. Med. Chem. Lett.* **9**, 2621 (1999). *Reviews:* W. E. Bachmann, W. S. Struve, *Org. React.* **1**, 38-62 (1942); B. Eistert in *Newer Methods of Preparative Organic Chemistry* vol. 1 (Interscience, New York, 1948) pp 513-570; G. B. Gill, *Comp. Org. Syn.* **3**, 888-889 (1991). *Cf.* Wolff Rearrangement.

11. Auwers Synthesis
K. v. Auwers *et al., Ber.* **41**, 4233 (1908); **48**, 85 (1915); **49**, 809 (1916); K. v. Auwers, P. Pohl, *Ann.* **405**, 243 (1914).

Expansion of coumarones to flavonols by treatment of 2-bromo-2-(α-bromobenzyl)coumarones with alcoholic alkali:

$$\xrightarrow[EtOH]{OH^-}$$

T. H. Minton, H. Stephen, *J. Chem. Soc.* **121**, 1598 (1922); J. Kalff, R. Robinson, *ibid.* **127**, 1968 (1925); B. H. Ingham *et al., ibid.* **1931**, 895; B. G. Acharya *et al., ibid.* **1940**, 817; S. Wawzonek, *Heterocycl. Compd.* **2**, 245 (1951). *Cf.* Algar-Flynn-Oyamada Reaction.

12. Baeyer-Drewson Indigo Synthesis
A. Baeyer, V. Drewson, *Ber.* **15**, 2856 (1882).
    Formation of indigos by an aldol reaction, *q.v.,* of *o*-nitrobenzaldehydes to acetone, pyruvic acid or acetaldehyde; of interest mainly as a method of protecting *o*-nitrobenzaldehydes:

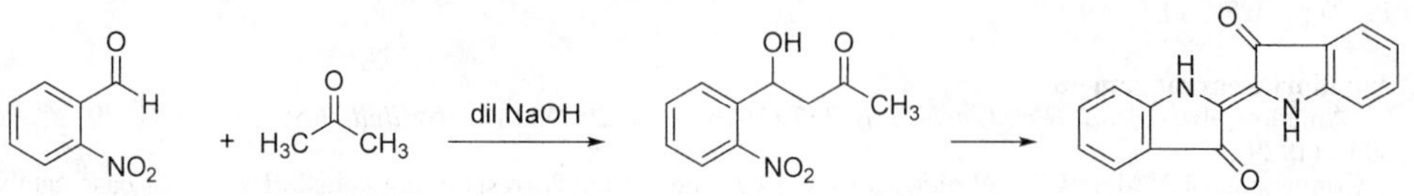

    K. Venkataraman, *Chemistry of Synthetic Dyes* **2**, 1008 (New York, 1952); M. Sainsbury, *Rodd's Chemistry of Carbon Compounds* **IVB**, 346, 353 (1977). Synthetic applications: J. R. Mckee *et al., J. Chem. Educ.* **68**, A242 (1991); L. Fitjer *et al., Tetrahedron* **55**, 14421 (1999).

13. Baeyer-Villiger Reaction
A. Baeyer, V. Villiger, *Ber.* **32**, 3625 (1899); **33**, 858 (1900).
    The oxidation of ketones to esters or lactones by peracids:

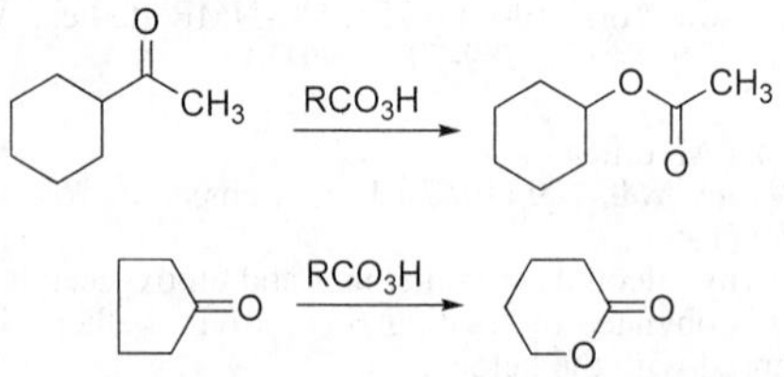

    Reviews: P. A. S. Smith in *Molecular Rearrangements* Part 1, P. de Mayo, Ed. (Wiley-Interscience, New York, 1963) pp 577-591; J. B. Lee, B. C. Uff, *Q. Rev. Chem. Soc.* **21**, 429-457 (1967); C. H. Hassall, *Org. React.* **9**, 73 (1957); G. R. Krow, *ibid.* **43**, 251-798 (1993); *idem, Comp. Org. Syn.* **7**, 671-688 (1991); G.-J. ten Brink *et al., Chem. Rev.* **104**, 4105-4123 (2004).

14. Baker-Venkataraman Rearrangement
W. Baker, *J. Chem. Soc.* **1933**, 1381; H. S. Mahal, K. Venkataraman, *ibid.* **1934**, 1767.
    Base-catalyzed rearrangement of *o*-acyloxyketones to β-diketones, important intermediates in the synthesis of chromones and flavones:

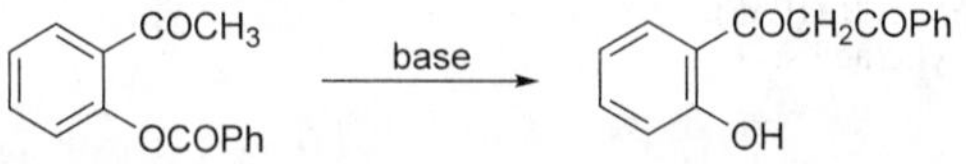

    Gripenberg in *The Chemistry of Flavonoid Compounds*, Geissman, Ed. (New York, 1962) p 410. Mechanistic studies: K. Bowden, M. Chehel-Amiran, *J. Chem. Soc. Perkin Trans. 2* **1986**, 2039. Synthetic applications: P. K. Jain *et al., Synthesis* **1982**, 221; J. Zhu *et al., Chem. Commun.* **1988**, 1549; A. V. Kalinin *et al., Tetrahedron Lett.* **39**, 4995 (1998); D. C. G. Pinto *et al., New J. Chem.* **24**, 85 (2000). *Cf.* Allan-Robinson Reaction; Kostanecki Acylation.

15. Bamberger Rearrangement
E. Bamberger, *Ber.* **27**, 1347, 1548 (1894).
    Intermolecular rearrangement of *N*-phenylhydroxylamines in aqueous acid to give the corresponding 4-aminophenols:

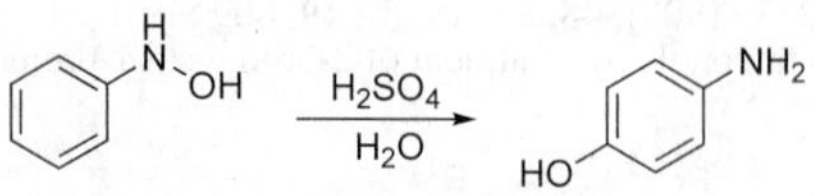

    Early review: H. J. Shine, *Aromatic Rearrangements* (Elsevier, New York, 1967) pp 182-190. Kinetic and mechanistic study: G. Kohnstam *et al., J. Chem. Soc. Perkin Trans. 2* **1984**, 423. Synthetic application: D. Johnston, D. Elder, *J. Labelled Compd. Radiopharm.* **25**, 1315 (1988). Modified conditions: A. Zoran *et al., Chem. Commun.* **1994**, 2239; M. Tordeux, C. Wakselman, *J. Fluorine Chem.* **74**, 251 (1995).

Organic Name Reactions

16. Bamford-Stevens Reaction; Shapiro Reaction

W. R. Bamford, T. S. Stevens, *J. Chem. Soc.* **1952**, 4735.

Formation of olefins by base-promoted decomposition of *p*-toluenesulfonylhydrazones of aldehydes and ketones:

$$R-\underset{\underset{H}{|}}{\overset{\overset{R^1}{|}}{C}}-\underset{\underset{R^2}{|}}{C}=N-NHTs \xrightarrow[\text{HOCH}_2\text{CH}_2\text{OH}]{\text{Na}} R-\underset{\underset{H}{|}}{\overset{\overset{R^1}{|}}{C}}=C-R^2$$

The formation of unrearranged alkenes, generally the less substituted isomers, by treatment of ketone derived *p*-toluenesulfonylhydrazones with alkyl lithium reagents is known as the **Shapiro reaction:** R. H. Shapiro, M. J. Heath, *J. Am. Chem. Soc.* **89,** 5734 (1967). Use of *N,N*-diethylaminosulfonylhydrazones: J. Kang *et al., Bull. Korean Chem. Soc.* **13,** 192 (1992).

Silicon directing effect: T. K. Sarkar, B. K. Ghorai, *Chem. Commun.* **1992**, 1184. *Reviews:* R. H. Shapiro, *Org. React.* **23,** 405-507 (1976); R. M. Adlington, A. G. M. Barrett, *Acc. Chem. Res.* **16,** 55-59 (1983); K. Maruka, H. Yamamoto, *Comp. Org. Syn.* **6,** 776-779 (1991); A. R. Chamberlin, D. J. Sall, *ibid.* **8,** 944-949.

17. Barbier(-type) Reaction

P. Barbier, *C. R. Hebd. Seances Acad. Sci.* **128,** 110 (1899).

One-step procedure for the preparation of alcohols from organic halides and aldehydes or ketones:

$$R-X \quad + \quad \overset{O}{\underset{}{\overset{||}{\diagup\!\!\diagdown}}} \quad \xrightarrow{M} \quad \overset{HO\quad R}{\underset{}{\diagup\!\!\diagdown}}$$

M = Mg, Li, Sm(II), Zn   R = alkyl, aryl, vinyl, allyl   X = Cl, Br, I

Review of mechanistic studies of Sm-mediated coupling: D. P. Curran *et al., Synlett* **1992**, 943-961. Book: C. Blomberg, *The Barbier Reaction and Related One-Step Processes*, K. Hafner *et al.,* Eds. (Springer-Verlag, New York, 1993) 183 pp. Zn-promoted coupling: F. Hong *et al., Chem. Commun.* **1994**, 289. Sm-mediated coupling: M. Kunishima *et al., Chem. Pharm. Bull.* **42,** 2190 (1994). Comparison with Ni(0) insertion chemistry for intramolecular cyclization: M. Kihara *et al., Tetrahedron* **48,** 67 (1992). *Cf.* Grignard Reaction.

18. Barbier-Wieland Degradation

H. Wieland, *Ber.* **45,** 484 (1912); P. Barbier, R. Locquin, *Compt. Rend.* **156,** 1443 (1913).

Stepwise carboxylic acid degradation of aliphatic acids (particularly in sterol side chains) to the next lower homolog. The ester is converted to a tertiary alcohol that is dehydrated with acetic anhydride, and the olefin oxidized with chromic acid to a lower homologous carboxylic acid:

$$\text{RCH}_2\text{COOCH}_3 \xrightarrow[\text{2. HX}]{\text{1. 2PhMgX}} \text{RCH}_2\text{COHPh}_2 \xrightarrow{\text{Ac}_2\text{O}} \text{RCH}=\text{CHPh}_2 \xrightarrow{\text{CrO}_3} \text{RCOOH} + \text{Ph}_2\text{CO}$$

H. Wieland *et al., Z. Physiol. Chem.* **161,** 80 (1926); C. W. Shoppee, *Ann. Repts.* (Chem. Soc., London) **44,** 184 (1947); W. Baker *et al., J. Chem. Soc.* **1958**, 1007; J. R. Dias, R. Ramachandra, *Tetrahedron Lett.* **1976**, 3685. Synthetic applications: S. C. Wilcox, J. J. Guadino, *J. Am. Chem. Soc.* **108,** 3102 (1986); C. D. Schteingart, A. E. Hofmann, *J. Lipid Res.* **29,** 1387 (1988). *Cf.* Krafft Degradation; Miescher Degradation.

19. Bart Reaction; Scheller Modification

H. Bart, **DE 250264** (1910); **DE 254092** (1910); **DE 264924** (1910); **DE 268172** (1912); *Ann.* **429,** 55 (1922); E. Scheller, **GB 261026**; A. W. Ruddy *et al., J. Am. Chem. Soc.* **64,** 828 (1942).

Formation of aromatic arsonic acids by treating aromatic diazonium compounds with alkali arsenites in the presence of cupric salts or powdered silver or copper; in the **Scheller modification** primary aromatic amines are diazotized in the presence of arsenious chloride and a trace of cuprous chloride:

$$\text{PhN}_2{}^+\text{Cl}^- + \text{Na}_3\text{AsO}_3 \longrightarrow \text{PhAsO}_3\text{Na}_2 + \text{NaCl} + \text{N}_2$$

$$\text{PhNH}_2 + \text{HNO}_2 + \text{AsCl}_3 + (\text{H}_2\text{O}) \longrightarrow \text{PhAsO}_3\text{H}_2 + \text{N}_2 + 3\text{HCl}$$

The modified Bart reaction can be applied to the formation of arylstibonic acids:

$$\text{ArNH}_2 + \text{HNO}_2 + \text{SbCl}_3 + (\text{H}_2\text{O}) \longrightarrow \text{ArSbO}_3\text{H}_2 + \text{N}_2 + 3\text{HCl}$$

C. F. Hamilton, J. F. Morgan, *Org. React.* **2,** 415 (1944); G. O. Doak, H. G. Steinman, *J. Am. Chem. Soc.* **68,** 1987 (1946); K. H. Saunders, *Aromatic Diazo-Compounds and Their Technical Applications* (London, 1949) p 330; W. A. Cowdry, D. S. Davies, *Q. Rev. Chem. Soc.* **6,** 363 (1952).

20. Bartoli Indole Synthesis

G. Bartoli *et al., Tetrahedron Lett.* **30,** 2129 (1989).

One-step reaction of *ortho*-substituted nitroarenes with vinyl Grignard reagents to yield 7-substituted indoles:

Mechanistic studies: M. Bosco et al., *J. Chem. Soc. Perkin Trans. 2* **1991**, 657. Improved protocols: A. Dobbs, *J. Org. Chem.* **66**, 638 (2001); M. C. Pirrung et al., *Synlett* **2002**, 143. Indole synthesis on solid supports: K. Knepper, S. Bräse, *Org. Lett.* **5**, 2829 (2003). *Review*: R. Dalpozzo, G. Bartoli, *Curr. Org. Synth.* **9**, 163-178 (2005).

21. Barton Decarboxylation

D. H. R. Barton et al., *Chem. Commun.* **1983**, 939; *eidem, Tetrahedron* **41**, 3901 (1985).

Radical decarboxylation of organic acids to the corresponding noralkane with tri-*n*-butyltin hydride or *t*-butylmercaptan:

Synthetic application: F. E. Ziegler, M. Belema, *J. Org. Chem.* **62**, 1083 (1997).

22. Barton Deoxygenation (Barton-McCombie Reaction)

D. H. R. Barton, S. W. McCombie, *J. Chem. Soc. Perkin Trans. 1* **1975**, 1574.

Deoxygenation of alcohols *via* their thiocarbonyl derivatives which undergo free radical scission upon treatment with tri-*n*-butyltin hydride:

R = H, CH$_3$, SCH$_3$, Ph, OPh, imidazolyl
R^2 = alkyl

Mechanistic study: J. E. Forbes, S. Z. Zard, *Tetrahedron Lett.* **30**, 4367 (1989). *Review:* M. Pereyre et al., *Tin in Organic Synthesis* (Butterworths, Boston, 1987) pp 84-96. Review of methodological improvements, particularly the replacement of tri-*n*-butyltin hydride with silicon hydrides: C. Chatgilialoglu, C. Ferreri, *Res. Chem. Intermed.* **19**, 755-775 (1993).

23. Barton Olefin Synthesis (Barton-Kellogg Reaction)

D. H. R. Barton et al., *Chem. Commun.* **1970**, 1226; R. M. Kellogg, S. Wassenaar, *Tetrahedron Lett.* **1970**, 1987; R. M. Kellogg et al., *ibid.* 4689.

Olefin synthesis by two-fold extrusion of nitrogen and sulfur from a Δ^3-1,3,4-thiadiazoline intermediate. Particularly applicable to the synthesis of moderately hindered *tetra*-substituted ethylenes:

R = Bu, Ph, OCH$_2$CH$_3$, N(CH$_2$CH$_3$)$_3$

Organic Name Reactions

Scope and limitations: D. H. R. Barton *et al., J. Chem. Soc. Perkin Trans. 1* **1974**, 1794. Synthetic applications: A. P. Schaap, G. R. Faler, *J. Org. Chem.* **38**, 3061 (1973); L. K. Bee *et al., ibid.* **40**, 2212 (1975); M. D. Bachi *et al., Tetrahedron Lett.* **1978**, 4167; J. E. McMurry *et al., J. Am. Chem. Soc.* **106**, 5018 (1984); F. J. Hoogesteger *et al., J. Org. Chem.* **60**, 4375 (1995). *Cf.* McMurry Reaction.

24. Barton Reaction

D. H. R. Barton *et al., J. Am. Chem. Soc.* **82**, 2640 (1960); **83**, 4076 (1961).

Conversion of a nitrite ester to a γ-oximino alcohol by photolysis involving the homolytic cleavage of a nitrogen-oxygen bond followed by hydrogen abstraction:

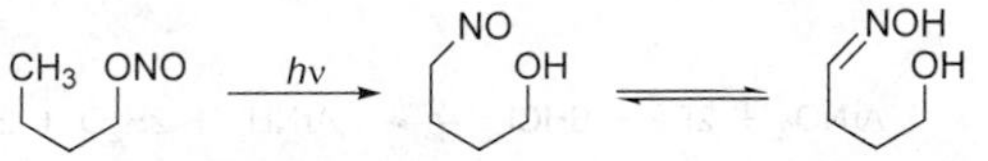

M. Akhtar, *Adv. Photochem.* **2**, 263 (1964); R. H. Hesse, *Adv. Free-Radical Chem.* **3**, 83 (1969); J. Kalvoda, *Angew. Chem. Int. Ed.* **8**, 525 (1964). Mechanism: D. H. R. Barton *et al., J. Chem. Soc. Perkin Trans. 1* **1979**, 1159. Synthetic application: A. Herzog *et al., Angew. Chem. Int. Ed.* **37**, 1552 (1998).

25. Barton-Zard Reaction

D. H. R. Barton *et al., Tetrahedron* **46**, 7587 (1990).

Formation of a pyrrole by condensation of a substituted nitroso-alkene with an isocyanoester:

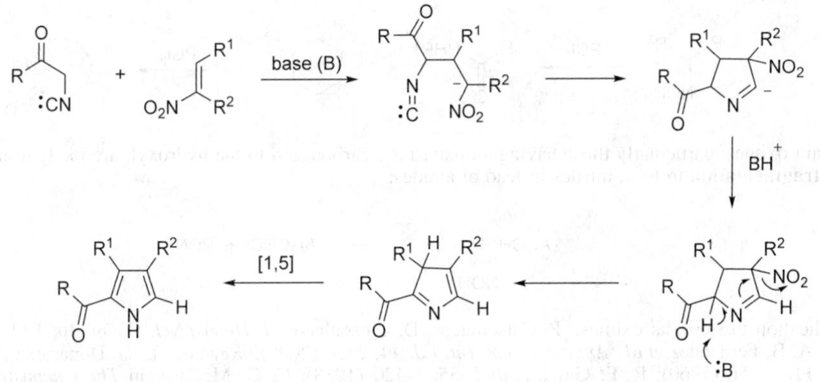

Synthetic applications: T. D. Lash *et al., Tetrahedron Lett.* **35**, 2493 (1994); *idem et al., ibid.* **38**, 2031 (1997); E. T. Pelkey *et al., Chem. Commun.* **1996**, 1909.

26. Baudisch Reaction

O. Baudisch *et al., Naturwissenschaften* **27**, 768, 769 (1939); *Science* **92**, 336 (1940); *J. Am. Chem. Soc.* **63**, 622 (1941).

Synthesis of *o*-nitrosophenols from benzene or substituted benzenes, hydroxylamine and hydrogen peroxide in the presence of copper salts:

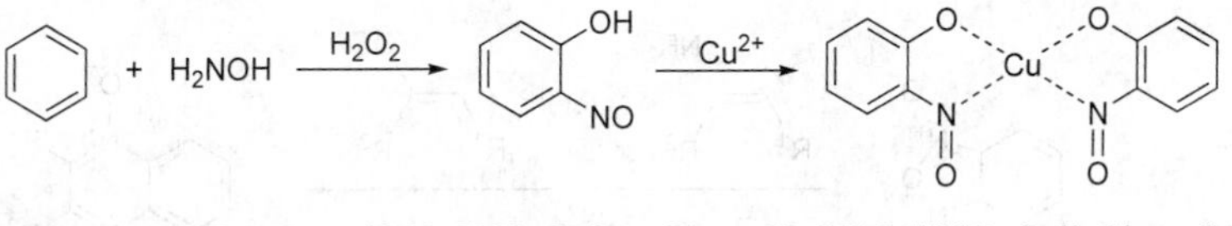

K. Maruyama *et al., Tetrahedron Lett.* **1966**, 5889; *J. Org. Chem.* **32**, 2516 (1967); *Bull. Chem. Soc. Jpn.* **44**, 3120 (1971); W. Seidenfaden, *Houben-Weyl* **10/1**, 1025, 1027 (1971).

27. Baylis-Hillman Reaction

A. B. Baylis, M. E. D. Hillman, **DE 2155113**; *eidem*, **US 3743669** (1972, 1973 both to Celanese).

Coupling of activated vinyl systems with aldehydes, catalyzed by 1,4-diazabicyclo[2.2.2]octane (DABCO), to yield α-hydroxyalkylated or -arylated products:

EWG = COOR, COR, CN, SO₂R, CONR₂

Mechanistic studies: Y. Fort *et al., Tetrahedron* **48,** 6371 (1992); E. L. M. van Rozendaal *et al., ibid.* **49,** 6931 (1993); K. E. Price *et al., J. Org. Chem.* **70,** 3980 (2005). Rate enhancement study: J. Augé *et al., Tetrahedron Lett.* **35,** 7947 (1994). Use of chiral auxillary: S. E. Drewes *et al., Synth. Commun.* **23,** 1215 (1993). Synthetic applications: *idem et al., ibid.* 2807; P. Perlmutter, T. D. McCarthy, *Aust. J. Chem.* **46,** 253 (1993). *Reviews:* S. E. Drewes, G. H. P. Roos, *Tetrahedron* **44,** 4653-4670 (1988); D. Basavaiah *et al., Chem. Rev.* **103,** 811-891 (2003).

28. Béchamp Reduction
A. J. Béchamp, *Ann. Chim. Phys.* **42**(3), 186, (1854).
   Reduction of aromatic nitro compounds to the corresponding amines by iron, ferrous salts or iron catalysts in aqueous acid:

$$ArNO_2 + 2Fe + 6HCl \longrightarrow ArNH_2 + 2H_2O + 2FeCl_3$$

J. Werner, *Ind. Eng. Chem.* **40,** 1575 (1948); **41,** 1841 (1949); S. Yagi *et al., Bull. Chem. Soc. Jpn.* **29,** 194 (1956); A. Courtin, *Helv. Chim. Acta* **62,** 2280 (1980). *Reviews:* C. S. Hamilton, J. F. Morgan, *Org. React.* **2,** 428 (1944); R. Schröter, *Houben-Weyl* **11/1,** 394-409 (1957).

29. Beckmann Rearrangement; Beckmann Fragmentation
E. Beckmann, *Ber.* **19,** 988 (1886).
   Acid-mediated isomerization of oximes to amides. Oximes of cyclic ketones give ring enlargements:

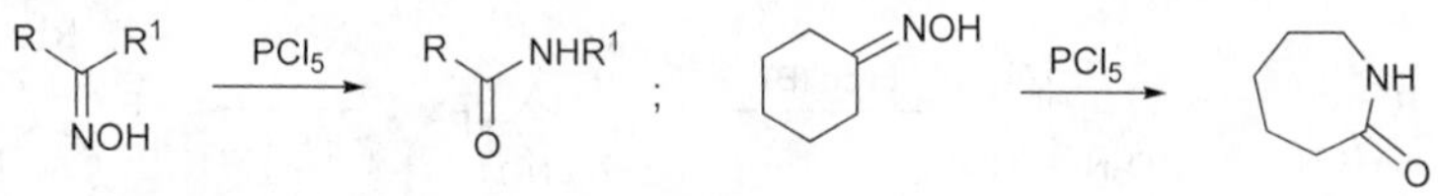

Certain oximes, particularly those having a quarternary carbon *anti* to the hydroxyl, are likely to undergo the **Beckmann fragmentation** to form nitriles instead of amides:

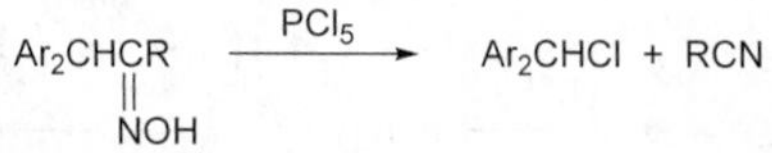

Application to steroidal oximes: P. Catsoulacos, D. Catsoulacos, *J. Heterocycl. Chem.* **30,** 1 (1993). Mechanistic study: A. B. Fernández *et al., Angew. Chem. Int. Ed.* **44,** 2370 (2005). *Reviews:* L. G. Donaruma, W. Z. Heldt, *Org. React.* **11,** 1-156 (1960); R. E. Gawley, *ibid.* **35,** 1-420 (1988); C. G. McCarty in *The Chemistry of the Carbon-Nitrogen Double Bond*, S. Patai, Ed. (Interscience, New York, 1970) pp 408-439; J. R. Hauske, *Comp. Org. Syn.* **1,** 98-100 (1991); K. Maruoka, H. Yamamoto, *ibid.* **6,** 763-775; D. Craig, *ibid.* **7,** 689-702. *Cf.* Schmidt Reaction; Tiemann Rearrangement.

30. Beirut Reaction
M. J. Haddadin, C. H. Issidorides, *Tetrahedron Lett.* **6,** 3253 (1965); C. H. Issidorides, M. J. Haddadin, *J. Org. Chem.* **31,** 4067 (1966).
   Benzofurazan 1-oxide condensation with an enamine or enolate anion to give the quinoxaline-1,4-dioxide:

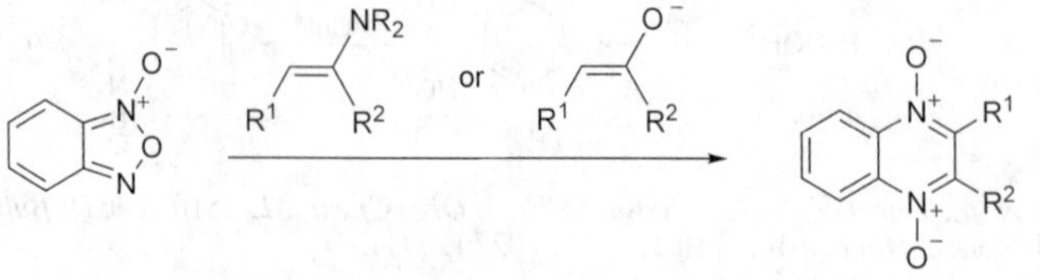

Reactivity studies with unsymmetrical 1,3-diketones: M. J. Haddadin *et al., Tetrahedron* **32,** 719 (1976); with phenolate anions: M. J. A. El-Haj *et al., J. Org. Chem.* **37,** 589 (1972); with substituted benzofurazan 1-oxides: N. A. Mufarrij *et al., J. Chem. Soc. Perkin Trans. 1* **1972,** 965; E. Abushanab, N. D. Alteri, Jr., *J. Org. Chem.* **40,** 157 (1975). Reaction catalysis with calcium salts: G. Stumm, H.-J. Niclas, *J. Prakt. Chem.* **331,** 736 (1989). *Reviews:* M. J. Haddadin, C. H. Issidorides, *Heterocycles* **4,** 767-816 (1976); *eidem, ibid.* **35,** 1503-1525 (1993).

31. Bénary Reaction
E. Bénary, *Ber.* **63,** 1573 (1930); **64,** 2543 (1931).
   Action of Grignard reagents on enamino ketones or aldehydes yields *β*-substituted *α,β*-unsaturated ketones or aldehydes:

$$RMgX \ + \ \diagdown NCH=CHCOR^1 \longrightarrow \left[\begin{array}{c} R \\ \diagdown CHCH=CR^1 \\ -N \diagdown \quad OMgX \end{array} \right] \longrightarrow RCH=CHCOR^1$$

T. Cuvigny, H. Normant, *Bull. Soc. Chim. Fr.* **1960**, 515. Use of lithio derivatives instead of Grignard reagents: C. Jutz, *Ber.* **91**, 1867 (1958). Mechanism: A. Pasteur *et al., Bull. Soc. Chim. Fr.* **1965**, 2328.

32. Benkeser Reduction
R. A. Benkeser *et al., J. Am. Chem. Soc.* **74**, 5699 (1952); **77**, 3230 (1955).

Reduction of aromatic and olefinic compounds with lithium or calcium and low molecular weight amines to mono-unsaturated olefins, as well as the fully reduced products. The extent of reduction and selectivity can be controlled by varying the reaction conditions:

$$\text{naphthalene} \xrightarrow[\text{H}_2\text{NR}]{\text{Li or Ca}} + \ +$$

$$R = CH_3, \ CH_2CH_3, \ CH_2CH_2CH_3, \ CH_2CH_2NH_2$$

Selectivity study: R. A. Benkeser *et al., Tetrahedron Lett.* no. 16, 1 (1960). Comparative review: E. M. Kaiser, *Synthesis* **1972**, 391-415 *passim.* Scope and limitations: R. A. Benkeser *et al., J. Org. Chem.* **48**, 2796 (1983). Synthetic applications: C. Eaborn *et al., J. Chem. Soc. Perkin Trans. 1* **1975**, 475; R. Eckrich, D. Kuck, *Synlett* **1993**, 344. *Cf.* Birch Reduction.

33. Benzidine Rearrangement; Semidine Rearrangement
A. W. Hofmann, *Proc. R. Soc. London* **12**, 576 (1863); P. Jacobson *et al., Ber.* **26**, 688 (1893).

Acid-catalyzed rearrangement of hydrazobenzenes to 4,4′-diaminobiphenyls. If the hydrazobenzene contains a *para* substituent, then the favored product is *p*-aminodiphenylamine (**Semidine rearrangement**):

$$Ph\overset{H}{\underset{}{N}}-\overset{H}{\underset{}{N}}Ph \xrightarrow{H^+} H_2N-\text{---}-NH_2$$

$$X-\overset{H}{\underset{}{N}}-\overset{H}{\underset{}{N}}- \xrightarrow{H^+} X-\text{---}-NH-\text{---}-NH_2$$

D. L. H. Williams, *Compr. Chem. Kinet.* vol. 13, C. H. Bamford, C. F. H. Tipper, Eds. (Elsevier, New York, 1972) pp 437-448; R. A. Cox, E. Buncel, *The Chemistry of Hydrazo, Azo and Azoxy Groups* pt. 2, S. Patai, Ed. (Wiley, New York, 1975) pp 775-807. Mechanistic studies: H. J. Shine *et al., J. Am. Chem. Soc.* **103**, 955 (1981); **104**, 5184 (1982); **106**, 7077 (1984). Synthetic applications: T. Nozoe *et al., Chem. Lett.* **1986**, 1577; K. H. Park, J. S. Kang, *J. Org. Chem.* **62**, 3794 (1997).

34. Benzilic Acid Rearrangement (Benzil-Benzilic Acid Rearrangement)
J. Liebig, *Ann.* **25**, 27 (1838); N. Zinin, *ibid.* **31**, 329 (1939).

Base-induced rearrangement of benzil to benzylic acid *via* phenyl group migration. More commonly perceived to include the migrations of other groups in α-dicarbonyl compounds:

$$\underset{\quad}{Ph-\overset{O}{\overset{||}{C}}-\overset{O}{\overset{||}{C}}-Ph} \xrightarrow{KOH} Ph-\overset{Ph}{\underset{OH}{\overset{|}{C}}}-COO^- \ K^+$$

Reviews: S. Selman, J. F. Eastham, *Q. Rev. Chem. Soc.* **14**, 221 (1960); D. J. Cram, *Fundamentals of Carbanion Chemistry* (Academic Press, 1965) pp 238-243; G. B. Gill, *Comp. Org. Syn.* **3**, 821-838 (1991).

35. Benzoin Condensation
A. J. Lapworth, *J. Chem. Soc.* **83**, 995 (1903); **85**, 1206 (1904).

Cyanide-catalyzed condensation of aromatic aldehydes to give benzoins (acyloins):

$$ArCHO \ + \ Ar^1CHO \xrightarrow{KCN} ArCHOHCOAr^1$$

H. Staudinger, *Ber.* **46**, 3530, 3535 (1913); W. S. Ide, J. S. Buck, *Org. React.* **4**, 269 (1948); H. Herlinger, *Houben-Weyl* **7/2a**, 653 (1973); A. Hassner, K. M. L. Rai, *Comp. Org. Syn.* **1**, 541-577 (1991). *Cf.* Acyloin Condensation.

Organic Name Reactions

36. Bergius Process
F. Bergius, *Gas World* **58**, 490 (1913); **GB 18232** (1914).

Formation of petroleum-like hydrocarbons by hydrogenation of coal at high temperatures and pressures (*e.g., 450°C* and 300 atm) with or without catalysts; production of toluene by subjecting aromatic naphthas to cracking temperatures at 100 atm with a low partial pressure of hydrogen in the presence of a catalyst.

B. T. Brooks, *The Chemistry of the Nonbenzenoid Hydrocarbons* (New York, 1950) p 115; *McGraw-Hill Encyclopedia of Science and Technology* **vol. 2** (New York, 1960) p 166; R. M. Baldwin in *Kirk-Othmer Encyclopedia of Chemical Technology* **vol. 6** (Wiley, New York, 4th ed., 1993) p 569. *Cf.* Fischer-Tropsch Syntheses.

37. Bergman Reaction
R. R. Jones, R. G. Bergman, *J. Am. Chem. Soc.* **94**, 660 (1972); R. G. Bergman, *Acc. Chem. Res.* **6**, 25 (1973).

The cyclization of an enediyne to generate a 1,4-benzenoid diradical species. Reaction of the radical with a hydrogen donor gives the arene:

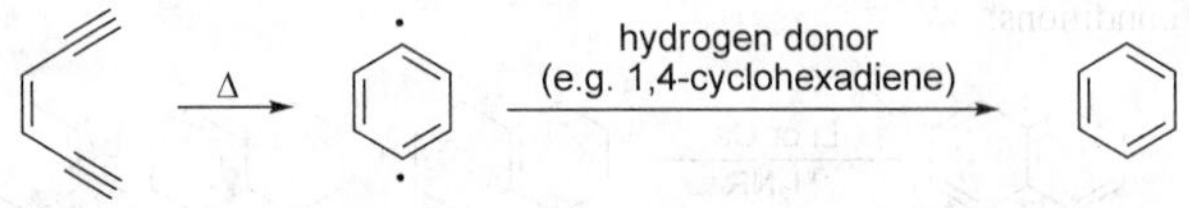

Application to ring annulation: J. W. Grissom *et al., Tetrahedron* **50**, 4635 (1994). Kinetic study: *idem et al., J. Chem.* **59**, 5833 (1994). Reaction energetics: E. Kraka, D. Cremer, *J. Am. Chem. Soc.* **116**, 4929 (1994). Reviews of enediyne chemistry and its application to the development of antitumor agents: K. C. Nicolaou *et al., Proc. Natl. Acad. Sci. USA* **90**, 5881-5888 (1993); K. Nicolaou, *Chem. Br.* **41**, 33-36, (1994).

38. Bergmann Azlactone Peptide Synthesis
M. Bergmann *et al., Ann.* **449**, 277 (1926).

Conversion of an acetylated amino acid and an aldehyde into an azlactone with an alkylene side chain, reaction with a second amino acid with ring opening and formation of an acylated unsaturated dipeptide, followed by catalytic hydrogenation and hydrolysis to the dipeptide:

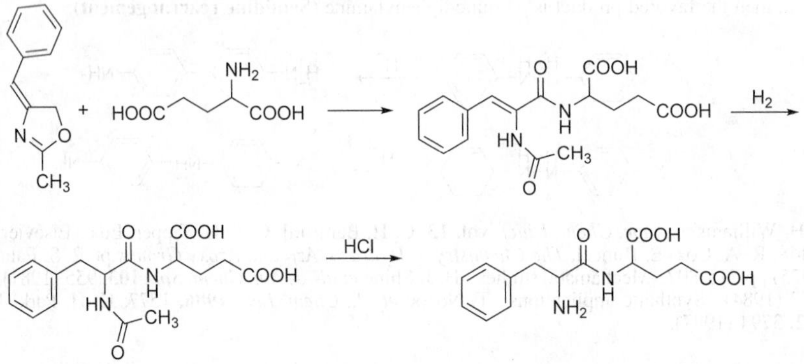

J. S. Fruton, *Adv. Protein Chem.* **V**, 15 (1949); S. Archer in *Amino Acids and Proteins*, D. M. Greenberg, Ed. (Thomas, Springfield, IL, 1951) p 181; H. D. Springall, *The Structural Chemistry of Proteins* (New York, 1954) p 29; E. Baltazzi, *Q. Rev. Chem. Soc.* **10**, 235 (1956). *Cf.* Erlenmeyer-Plöchl Azlactone and Amino Acid Synthesis.

39. Bergmann Degradation
M. Bergmann, *Science* **79**, 439 (1934).

Stepwise degradation of polypeptides involving benzoylation, conversion to azides and treatment of the azides with benzyl alcohol; this treatment yields, *via* rearrangement to isocyanates, carbobenzoxy compounds which undergo catalytic hydrogenation and hydrolysis to the amide of the degraded peptide:

$$R^1CONHCHRCON_3 \xrightarrow{C_6H_5CH_2OH} R^1CONHCHRNHCOOCH_2C_6H_5 \xrightarrow[H_2O]{H_2}$$

$$R^1CONH_2 + RCHO + CO_2 + C_6H_5CH_3 + NH_3$$

M. Bergmann, L. Zervas, *J. Biol. Chem.* **113**, 341 (1936); H. D. Springall, *The Structural Chemistry of Proteins* (New York, 1954) p 321. *Cf.* Curtius Rearrangement.

40. Bergmann-Zervas Carbobenzoxy Method
M. Bergmann, L. Zervas, *Ber.* **65**, 1192 (1932).

Formation of the *N*-carbobenzoxy derivative of an amino acid for use in peptide synthesis and liberation of the amino group at an appropriate stage of synthesis by hydrogenolysis of the labile carbon-oxygen bond:

C. L. A. Schmidt, *The Chemistry of the Amino Acids and Proteins* (Thomas, Springfield, IL, 1944) p 262; S. Archer in *Amino Acids and Proteins*, D. M. Greenberg, Ed. (Charles C. Thomas, Springfield, IL, 1951) p 177; G. W. Kenner, *J. Chem. Soc.* **1956**, 3689; T. W. Greene, *Protective Groups in Organic Synthesis* (Wiley, New York, 1981) p 239. *Cf.* Fischer Peptide Synthesis.

41. Bernthsen Acridine Synthesis
A. Bernthsen, *Ann.* **192**, 1 (1878); **224**, 1 (1884).

Formation of 5-substituted acridines by heating diarylamines in organic acids or anhydrides, usually in the presence of zinc chloride:

A. Albert, *The Acridines* (London, 1951) p 67; A. Albert, *Heterocycl. Compd.* **4**, 502 (1952); N. P. Buu-Hoi *et al.*, *J. Chem. Soc.* **1955**, 1082; R. M. Acheson in *The Chemistry of Heterocyclic Compounds*, A. Weissberger, Ed., *Acridines* (Interscience, New York, 1956) pp 19-25; F. D. Popp, *J. Org. Chem.* **27**, 2658 (1962). Alkyl migration: L. H. Klemm *et al.*, *Heterocyclic Chem.* **29**, 571 (1992).

42. Betti Reaction
M. Betti, *Gazz. Chim. Ital.* **30 II**, 301 (1900); **33 II**, 2 (1903); F. Pirrone, *ibid.* **66**, 518 (1936); **67**, 529 (1937).

The reaction of aromatic aldehydes, primary aromatic or heterocyclic amines and phenols leading to α-aminobenzylphenols:

R, R^1= aryl, heterocyclic

J. P. Phillips, *Chem. Rev.* **56**, 286 (1956); J. P. Phillips, E. M. Barrall, *J. Org. Chem.* **21**, 692 (1956). *Early review:* J. P. Phillips, Leach, *Trans. K. Acad. Sci.* **24**(3-4), 95 (1964). Mechanistic study: H. Möhrle *et al.*, *Ber.* **107**, 2675 (1974). Stereoselectivity: C. Cardellicchio *et al.*, *Tetrahedron: Asymmetry* **9**, 3667 (1998). *Cf.* Mannich Reaction.

43. Biginelli Reaction
P. Biginelli, *Ber.* **24**, 1317, 2962 (1891); **26**, 447 (1893).

Synthesis of tetrahydropyrimidinones by the acid-catalyzed condensation of an aldehyde, a β-keto ester and urea:

H. E. Zaugg, W. B. Martin, *Org. React.* **14**, 88 (1965); D. J. Brown, *The Pyrimidines* (Wiley, New York, 1962) p 440; *ibid.*, Suppl. I, **1970**, p 326, F. Sweet, Y. Fissekis, *J. Am. Chem. Soc.* **95**, 8741 (1973). Synthetic applications: M. V. Fernandez *et al.*, *Heterocycles* **27**, 2133 (1988); K. Singh *et al.*, *Tetrahedron* **55**, 12873 (1999); A. S. Franklin *et al., J. Org. Chem.* **64,** 1512 (1999). Modified conditions: C. O. Kappe *et al., Synthesis* **1999,** 1799; J. Lu, H. Ma. *Synlett* **2000,** 63. Use of cycloalkanones as starting material: Y.-L. Zhu *et al., Eur. J. Org. Chem.* **2005,** 2354.

44. Birch Reduction

A. J. Birch, *J. Chem. Soc.* **1944,** 430; **1945,** 809; **1946,** 593; **1947,** 102, 1642, **1949,** 2531.

Reduction of aromatic rings by means of alkali metals in liquid ammonia to give mainly unconjugated dihydro derivatives:

R = electron-donating   R = electron-withdrawing

Reviews: A. J. Birch, H. Smith, *Q. Rev. Chem. Soc.* **12,** 17 (1958); D. Caine, *Org. React.* **23,** 1-258 (1976); P. W. Rabideau, Z. Marcinow, *ibid.* **42,** 1-334 (1992); J. M. Hook, L. N. Mander, *Nat. Prod. Rep.* **3,** 35-85 (1986); L. N. Mander, *Comp. Org. Syn.* **8,** 489-521 (1991). *Cf.* Benkeser Reduction.

45. Bischler-Möhlau Indole Synthesis

A. Bischler *et al., Ber.* **25,** 2860 (1892); **26,** 1336 (1893); R. Möhlau, *ibid.* **14,** 171 (1881); **15,** 2480 (1882); E. Fischer, T. Schmitt, *ibid.* **21,** 1071 (1888).

Formation of 2-substituted indoles by heating ω-halogeno- or ω-hydroxy- ketones with excess aniline *via* cyclization of the intermediate 2-arylaminoketone:

P. L. Julian *et al., Heterocycl. Compd.* **3,** 22 (1952); R. J. Sundberg, *The Chemistry of Indoles* (Academic Press, New York, 1970) p 164; R. K. Brown in *The Chemistry of Heterocyclic Compounds,* A. Weissberger, E. C. Taylor, Eds., *Indoles* **Part I,** W. J. Houlihan, Ed. (Wiley, New York, 1972) p 317; J. R. Henry, J. H. Dodd, *Tetrahedron Lett.* **38,** 8763 (1998).

46. Bischler-Napieralski Reaction

A. Bischler, B. Napieralski, *Ber.* **26,** 1903 (1893).

Cyclodehydration of β-phenethylamides to 3,4-dihydroisoquinoline derivatives by means of condensing agents such as phosphorous pentoxide or zinc chloride:

W. M. Whaley, T. R. Govindachari, *Org. React.* **6,** 74 (1951); T. Kametani *et al., Tetrahedron* **27,** 5367 (1971); G. Fodor *et al., Angew. Chem. Int. Ed.* **11,** 919 (1972); G. Fodor, S. Nagubandi, *Tetrahedron* **36,** 1279 (1980); *eidem, Heterocycles* **15,** 165 (1981). Review of enantioselective modifications: M. O. Rozwadowska, *ibid.* **39,** 903-931 (1994). *Cf.* Bradsher Reaction; Pechmann Condensation; Pictet-Gams Isoquinoline Synthesis; Pictet-Hubert Reaction; Skraup Reaction.

47. Blaise Ketone Synthesis; Blaise-Maire Reaction

E. E. Blaise, A. Koehler, *Bull. Soc. Chim.* [4] **7,** 215 (1910); E. E. Blaise, M. Maire, *Compt. Rend.* **145,** 73 (1907); E. E. Blaise, *Bull. Soc. Chim.* [4] **9,** 1 (1911).

Formation of ketones by treatment of acid halides with organozinc compounds; the use of β-hydroxy carbonyl chlorides to give β-hydroxy ketones, convertible into α,β-unsaturated ketones in boiling dilute sulfuric acid, is known as the **Blaise-Maire reaction:**

$$RZnCl + R^1COCl \longrightarrow RCOR^1 + ZnCl_2$$

$$AcOCH_2CHRCOCl + R^1ZnCl \longrightarrow AcOCH_2CHRCOR^1 \longrightarrow HOCH_2CHRCOR^1 \longrightarrow H_2C{=}CRCOR^1$$

J. Cason, *Chem. Rev.* **40,** 17 (1947); D. A. Shirley, *Org. React.* **8,** 29 (1954).

48. Blaise Reaction

E. E. Blaise, *Compt. Rend.* **132,** 478 (1901).

Formation of β-oxoesters by treatment of α-bromocarboxylic esters with zinc in the presence of nitriles. The intermediate organozinc compound reacts with the nitrile and the complex is hydrolyzed with 30% potassium hydroxide:

A. Horeau, J. Jacques, *Bull. Soc. Chim.* **1947**, Mem. 58; J. Cason *et al., J. Org. Chem.* **18**, 1594 (1953); H. Henecka, *Houben-Weyl* **7/2a**, 518 (1973); K. Nützel, *ibid.* **13/2a**, 829. Modified conditions: S. M. Hannick, Y. Kishi, *J. Org. Chem.* **48**, 3833 (1983); N. Zylber *et al., J. Organomet. Chem.* **444**, 1 (1993); K. Narkunan, B.-J. Uang, *Synthesis* **1998**, 1713. Stereoselectivity: J. J. Duffield, A. C. Regan, *Tetrahedron: Asymmetry* **7**, 663 (1996); A. S.-Y. Lee *et al., Tetrahedron Lett.* **38**, 443 (1997); J. Syed *et al., Tetrahedron: Asymmetry* **9**, 805 (1998). *Cf.* Reformatsky Reaction.

49. Blanc Reaction (Chloromethylation)
G. Blanc, *Bull. Soc. Chim. Fr.* [4], **33**, 313 (1923).

Introduction of the chloromethyl group into aromatic rings on treatment with formaldehyde and hydrogen chloride in the presence of zinc chloride:

$$ArH \ + \ CH_2O \ + \ HCl \ \xrightarrow{\ ZnCl_2\ } \ ArCH_2Cl$$

Reviews: R. C. Fuson, C. H. McKeever, *Org. React.* **1**, 63 (1942); G. Olah, W. S. Tolgyesi, in *Friedel-Crafts and Related Reactions* **vol. II**, Part 2, G. Olah, Ed. (Interscience, New York, 1963) pp 659-784. *Cf.* Quelet Reaction.

50. Blanc Reaction-Blanc Rule
H. G. Blanc, *Compt. Rend.* **144**, 1356 (1907).

Cyclization of dicarboxylic acids on heating with acetic anhydride to give either cyclic anhydrides or ketones depending on the respective positions of the carboxyl groups; 1,4- and 1,5-diacids yield anhydrides, while diacids in which the carboxy groups are in 1,6 or further removed positions yield ketones:

H. Kwart, K. King in *The Chemistry of Carboxylic Acids and Esters*, J. Patai, Ed. (Interscience, London, 1969) p 362; K. D. Bode, *Houben-Weyl* **7/2**, 640 (1973). *Cf.* Ruzicka Large Ring Synthesis.

51. Bodroux-Chichibabin Aldehyde Synthesis
F. Bodroux, *Compt. Rend.* **138**, 92 (1904); A. E. Chichibabin, *Ber.* **37**, 186, 850 (1904).

Formation of aldehydes by treatment of orthoformates with Grignard reagents:

$$HC(OC_2H_5)_3 \ + \ RMgX \ \longrightarrow \ RCH(OC_2H_5)_2 \ \xrightarrow[H_2O]{H^+} \ RCHO$$

L. I. Smith *et al., J. Org. Chem.* **6**, 437, 489 (1941); H. W. Post, *The Chemistry of the Aliphatic Orthoesters* (New York, 1943) p 96; H. Meerwein, *Houben-Weyl* **6/3**, 243 (1965); R. H. DeWolfe, *Carboxylic Orthoacid Derivatives* (Academic Press, New York, 1970) p 224. *Cf.* Bouveault Aldehyde Synthesis.

52. Bodroux Reaction
F. Bodroux, *Bull. Soc. Chim. Fr.* **33**, 831 (1905), **35**, 519 (1906); **1**, 912 (1907); *Compt. Rend.* **138**, 1427 (1904); **140**, 1108 (1905); **142**, 401 (1906).

Formation of substituted amides by reaction of a simple aliphatic or aromatic ester with an aminomagnesium halide obtained by treatment of a primary or secondary amine with a Grignard reagent at room temperature:

$$RCOOR^1 \ + \ IMgNHR^2 \ \longrightarrow \ IMgOCR(OR^1)NHR^2 \ \xrightarrow{IMgNHR^2} \ RCONHR^2$$

H. L. Bassett, C. R. Thomas, *J. Chem. Soc.* **1954**, 1188; K. Nützel, *Houben-Weyl* **13/2a**, 278 (1973).

53. Bogert-Cook Synthesis
M. T. Bogert, *Science* **77**, 289 (1933); J. W. Cook, C. L. Hewett, *J. Chem. Soc.* **1933**, 1098.

Condensation of β-phenylethylmagnesium bromide with cyclohexanones followed by cyclodehydration of the tertiary alcohol with concentrated sulfuric acid with formation of octahydrophenanthrene derivatives and a small amount of spiran:

L. F. Fieser, M. Fieser, *Natural Products Related to Phenanthrene* (New York, 1949) p 90; C. Schmidt *et al., Can. J. Chem.* **51**, 3620 (1973). For a general approach to the synthesis of phenanthrenoid compounds, *see* D. A. Evans *et al., J. Am. Chem. Soc.* **99**, 7083 (1977).

54. Bohn-Schmidt Reaction
R. Bohn, **DE 46654** (1889); R. E. Schmidt, **DE 60855** (1891).

Hydroxylation of anthraquinones containing at least one hydroxyl group by treatment with fuming sulfuric acid or sulfuric acid and boric acid in the presence of a catalyst such as mercury:

Reviews: M. Phillips, *Chem. Rev.* **6**, 168 (1929); Fieser, Fieser, *Organic Chemistry* (New York, 1956) p 903. Studies and proposed mechanism: J. Winkler, W. Jenny, *Helv. Chim. Acta* **48**, 119 (1965); B. R. Dhruva *et al., Indian J. Chem.* **14 (B)**, 622 (1976).

55. Boord Olefin Synthesis
L. C. Swallen, C. E. Boord, *J. Am. Chem. Soc.* **52**, 651 (1930); **53**, 1505 (1931); **55**, 3293 (1933); H. B. Dykstra *et al., ibid.* **52**, 3396 (1930).

Regiospecific synthesis of olefins from aldehydes and Grignard reagents by zinc induced reductive elimination of halogen and alkoxy groups:

C. Niemann, C. D. Wagner, *J. Org. Chem.* **7**, 227 (1942); P. Bandart, *Bull. Soc. Chim.* **11**, 336 (1944); L. Crombie, *Q. Rev. Chem. Soc.* **6**, 131 (1952); M. Schlosser, *Houben-Weyl* **5/1b**, 213 (1972). Application to taxanes: J. S. Yadav *et al. Tetrahedron Lett.* **35**, 3617 (1994); P. H. Beusker *et al., Eur. J. Org. Chem.* **1998**, 2483.

56. Borsche-Drechsel Cyclization
E. Drechsel, *J. Prakt. Chem.* **38**(2), 69 (1858); W. Borsche, M. Feise, *Ber.* **20**, 378 (1904).

Formation of carbazole by acid-catalyzed rearrangement of cyclohexanone phenylhydrazone to tetrahydrocarbazole followed by oxidation:

N. Campbell, B. M. Barclay, *Chem. Rev.* **40**, 361 (1947); W. Freudenberg, *Heterocycl. Compd.* **3**, 298 (1952); P. Bruck, *J. Org. Chem.* **35**, 2222 (1970). *Cf.* Bucherer Carbazole Synthesis; Fischer Indole Synthesis; Piloty-Robinson Synthesis.

57. Bouveault Aldehyde Synthesis
L. Bouveault, *Bull. Soc. Chim. Fr.* **31**, 1306, 1322 (1904).

Action of Grignard or organic lithium reagents on *N,N*-disubstituted formamides yields the homologous aldehydes:

L. I. Smith, J. Nichols, *J. Org. Chem.* **6**, 489 (1941); J. Sicé *J. Am. Chem. Soc.* **75**, 3697 (1953); E. R. H. Jones *et al., J. Chem. Soc.* **1958**, 1054. Use of lithio derivatives instead of Grignard reagents: E. A. Evans, *Chem. Ind. (London)* **1957**, 1596. Synthetic applications using modified conditions: C. Pétrier *et al., Tetrahedron Lett.* **23**, 3361 (1982); J. Einhorn, J. L. Luche, *ibid.* **27**, 1791 (1986); H. Meier, H. Aust, *J. Prakt. Chem.* **341**, 466 (1999). *Cf.* Bodroux-Chichibabin Aldehyde Synthesis.

58. Bouveault-Blanc Reduction

L. Bouveault, G. Blanc, *Compt. Rend.* **136**, 1676 (1903); *Bull. Soc. Chim. Fr.* [3] **31**, 666 (1904).
Formation of alcohols by reduction of esters with sodium and an alcohol:

$$RCOOCH_3 \xrightarrow[R^1OH]{Na} RCH_2OH + CH_3OH$$

H. O. House, *Modern Synthetic Reactions* (W. A. Benjamin, Menlo Park, California, 2nd ed., 1972) p 150.

59. Boyland-Sims Oxidation

E. Boyland *et al., J. Chem. Soc.* **1953**, 3623; E. Boyland, P. Sims, *ibid.* **1954**, 980.
Alkaline persulfate oxidation of aromatic amines to yield predominantly the *o*-amino aryl sulfates. Acid-catalyzed hydrolysis generates the *o*-hydroxy aryl amines:

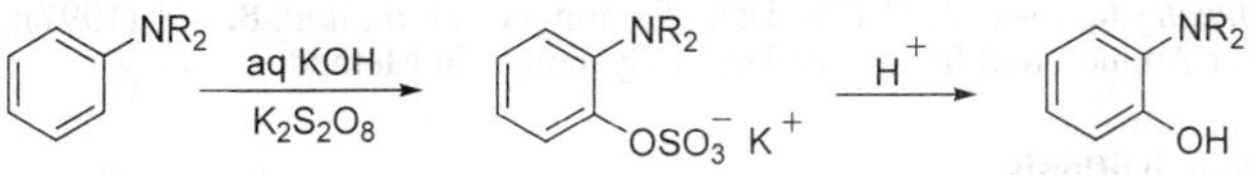

Regioselectivity/mechanistic study: E. J. Behrman, *J. Org. Chem.* **57**, 2266 (1992). *Review: idem, Org. React.* **35**, 421-511 (1988). *Cf.* Elbs Persulfate Oxidation.

60. Bradsher Cyclization (Bradsher Cycloaddition)

C. K. Bradsher, T. W. G. Solomons, *J. Am. Chem. Soc.* **80**, 933 (1958).
[4 + 2] addition of a common dienophile with cationic aromatic azadienes such as acridizinium or isoquinolinium:

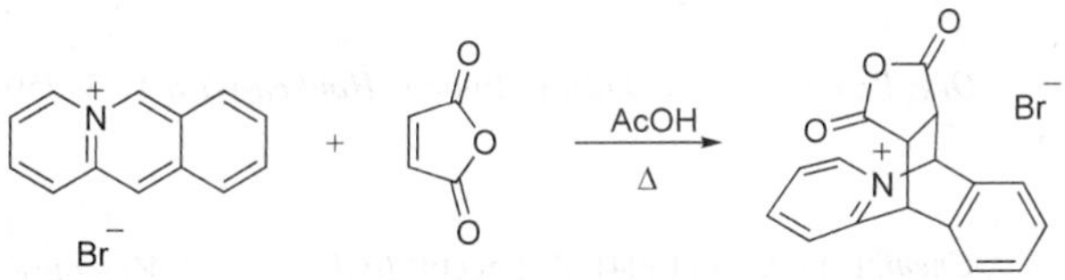

Mechanistic study: C. K. Bradsher, J. A. Stone, *J. Org. Chem.* **33**, 519 (1968). Synthetic applications: V. Bolitt *et al., J. Am. Chem. Soc.* **113**, 6320 (1991); H. Yin *et al., J. Org. Chem.* **57**, 644 (1992); T. E. Nicolas, R. W. Franck, *ibid.* **69**, 6904 (1995). *Review:* D. L. Boger, S. M. Weinreb, *Hetero Diels-Alder Methodology* in *Organic Synthesis* (Academic Press, NY, 1987) pp 239-299.

61. Bradsher Reaction

C. K. Bradsher, *J. Am. Chem. Soc.* **62**, 486 (1940).
Acid-catalyzed cyclodehydration of *o*-acyldiarylmethanes to anthracene derivatives:

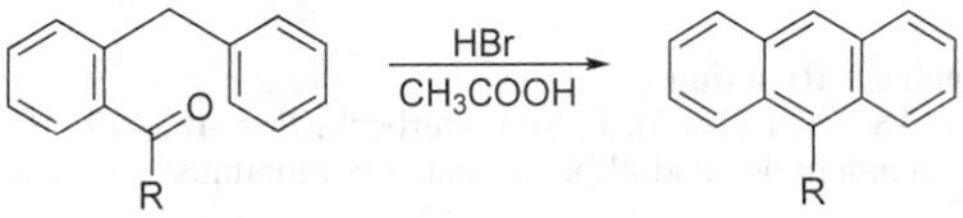

Extension to an *o*-acyldiaryl ether: H. Ishibashi *et al., Tetrahedron* **50**, 10215 (1994). Application: T. Yamato *et al., J. Chem. Soc. Perkin Trans. 1* **1997**, 1193. *Review:* C. K. Bradsher, *Chem. Rev.* **87**, 1277-1297 (1987). *Cf.* Bischler-Napieralski Reaction.

62. Brook Rearrangement

A. G. Brook, *J. Am. Chem. Soc.* **80**, 1886 (1958); *idem et al., ibid.* **81**, 981 (1959).
Base-catalyzed silicon migration from carbon to oxygen in α-, β- and γ-silyl alcohols, yielding silyl ethers:

n = 1,2,3
base = NaOH, Na/K, NaH, NR$_3$

Early review: A. G. Brook, *Acc. Chem. Res.* **7**, 77-84 (1974). Synthetic applications: H. J. Reich *et al., J. Am. Chem. Soc.* **112**, 5609 (1990); K. Takeda *et al., Synlett* **1993**, 841; I. Fleming, U. Ghosh *et al., J. Chem. Soc. Perkin Trans. 1* **1994**, 257.

63. Bucherer-Bergs Reaction

H. T. Bucherer, H. T. Fischbeck, *J. Prakt. Chem.* **140**, 69 (1934); H. T. Bucherer, W. Steiner, *ibid.* 291; H. Bergs, **DE 566094** (1929).

Preparation of hydantoin from carbonyl compounds by reaction with potassium cyanide and ammonium carbonate, or from the corresponding cyanohydrin and ammonium carbonate:

E. Ware, *Chem. Rev.* **46**, 422 (1950); A. Rousset *et al., Tetrahedron* **36**, 2649 (1980). Modified conditions: R. Sarges *et al., J. Med. Chem.* **33**, 1859 (1990). Synthetic applications to excitatory amino acids: K.-I. Tanaka *et al., Tetrahedron: Asymmetry* **6**, 1641, 2271 (1995); C. Domínguez *et al., ibid.* **8**, 511 (1997); J. Knabe, *Pharmazie* **52**, 912 (1997). *Cf.* Strecker Amino Acid Synthesis; Urech Cyanohydrin Method.

64. Bucherer Carbazole Synthesis

H. T. Bucherer, F. Seyde, *J. Prakt. Chem.* **77**(2), 403 (1908).

Formation of carbazoles from naphthols or naphthylamines, aryl hydrazines and sodium bisulfite:

Reviews: N. L. Drake, *Org. React.* **1**, 114 (1942); E. Enders, *Houben-Weyl* **10/2**, 250 (1967). *Cf.* Borsche-Drechsel Cyclization.

65. Bucherer Reaction

H. T. Bucherer, *J. Prakt. Chem.* [2] **69**, 49 (1904); R. Lepetit, *Bull. Soc. Ind. Mulhouse* **1903**, 326.

Reversible formation of β-naphthylamine from β-naphthol and aqueous ammonium sulfite or bisulfite *via* intermediate formation of tetralonesulfonic and tetraloneiminosulfonic acids:

N. L. Drake, *Org. React.* **1**, 105 (1942); H. Seeboth, *Angew. Chem. Int. Ed.* **6**, 307 (1967); M. S. Gibson in *The Chemistry of the Amino Group*, S. Patai, Ed. (Interscience, London, 1968) p 37; Z. Allan *et al., Tetrahedron Lett.* **1969**, 4855; W. H. Pirkle, T. C. Pochapsky, *J. Org. Chem.* **51**, 102 (1986); J. Bendig *et al., Tetrahedron* **48**, 9207 (1992).

66. Buchner-Curtius-Schlotterbeck Reaction

E. Buchner, T. Curtius, *Ber.* **18**, 2371 (1885); F. Schlotterbeck, *Ber.* **40**, 479 (1907); **42**, 2559 (1909).

Formation of ketones from aldehydes and aliphatic diazo compounds; ethylene oxides may also be formed:

$$RCHN_2 + R^1CHO \longrightarrow RCH_2COR^1 + N_2$$

B. Eistert in *Newer Methods of Preparative Organic Chemistry*, English Ed. (New York, 1948) p 521; C. D. Gutsche, *Org. React.* **8**, 364 (1954); J. B. Bastus, *Tetrahedron Lett.* **1963**, 955.

67. Buchner Method of Ring Enlargement

E. Buchner, *Ber.* **29**, 106 (1896); E. Buchner, K. Schottenhammer, *Ber.* **53**, 865 (1920).

Diazoacetic acid ester reacts with benzene and homologs to give the corresponding esters of noncaradienic acid, transformed at high temperatures to derivatives of cycloheptatriene and phenylacetic acid:

Organic Name Reactions

W. von F. Doering, L. H. Knox, *J. Am. Chem. Soc.* **79**, 352 (1957); W. Kirmse, *Carbene Chemistry* (Academic Press, New York, 2nd ed., 1971); A. F. Noels *et al., J. Org. Chem.* **46**, 873 (1981). *Cf.* Pfau-Plattner Azulene Synthesis.

68. Buchwald-Hartwig Cross Coupling Reaction

J. Louie, J. F. Hartwig, *Tetrahedron Lett.* **36**, 3609 (1995); A. S. Guram *et al., Angew. Chem. Int. Ed.* **34**, 1348 (1995).
Metal-catalyzed formation of an arylamine by the reaction of an aryl halide or triflate with a primary or secondary amine:

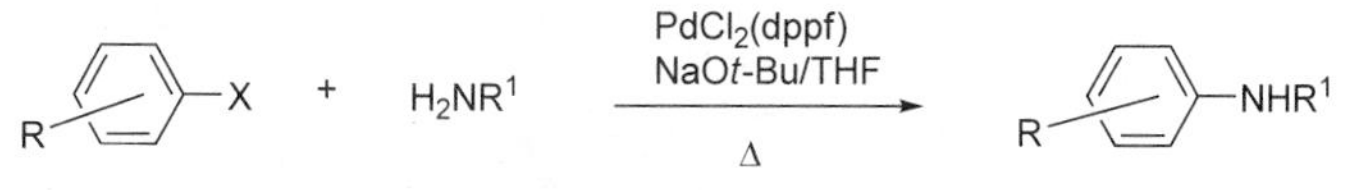

R = *o-, p-*alkyl, CN, C(O)Ph, C(O)NEt$_2$

R^1 = alkyl, aryl

X = Br, I, OTf

Application: S. L. MacNeil *et al., Synlett* **1998**, 419. *Review:* J. F. Hartwig, *Angew. Chem. Int. Ed.* **37**, 2046-2067 (1998).

69. Burgess Dehydration Reaction

E. M. Burgess *et al., J. Am. Chem. Soc.* **92**, 5224 (1970); *eidem, J. Org. Chem.* **38**, 26 (1973).
Dehydration of secondary and tertiary alcohols by treatment with Burgess reagent (inner salt of (methoxycarbonyl-sulfamoyl)triethylammonium hydroxide) to yield the corresponding olefins. Expansion modifications permit the preparation of carbamates from primary alcohols, nitriles from oximes and primary amides, isonitriles from formamides, nitrile oxides from primary nitroalkanes, oxazolines from β-hydroxyamides, and thiazolines from β-hydroxythioamides:

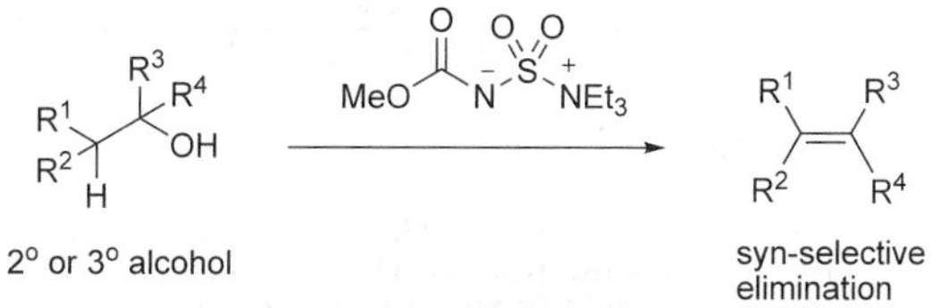

2° or 3° alcohol

syn-selective elimination

Prepn of carbamates: E. Burgess *et al., Org. Synth.* **coll. vol. VI**, 788 (1988); of nitriles: D. A. Claremon, B. T. Phillips, *Tetrahedron Lett.* **29**, 2155 (1988); of isonitriles: S. M. Creedon *et al., J. Chem. Soc. Perkin Trans. 1* **1998**, 1015; of nitrile oxides: N. Maugein *et al., Tetrahedron Lett.* **38**, 1547 (1997); of oxazolines: P. Wipf, C. P. Miller, *ibid.* **33**, 907 (1992); and thiazolines: *eidem, ibid.* 6267. *Reviews:* C. Lamberth, *J. Prakt. Chem.* **342**, 518-522 (2000); S. Khapli *et al., J. Indian Inst. Sci.* **81**, 461-476 (2001); D. D. Holsworth in *Name Reactions for Functional Group Transformations*, J. J. Li, E. J. Corey, Eds. (Wiley, Hoboken, NJ, 2007) pp 189-206. *See* monograph: Burgess Reagent.

70. Camps Quinoline Synthesis

R. Camps, *Ber.* **22**, 3228 (1899); *Arch. Pharm.* **237**, 659 (1899); **239**, 591 (1901).
Formation of hydroxyquinolines from *o*-acylaminoacetophenones in alcoholic sodium hydroxide. Two isomers are produced; the relative proportions are mainly determined by the residue on the amino nitrogen:

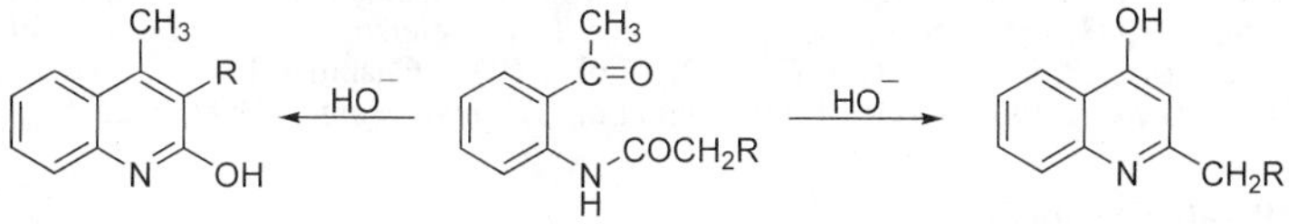

R. H. F. Manske, *Chem. Rev.* **30**, 127 (1942); B. Witkop *et al., J. Am. Chem. Soc.* **73**, 2641 (1951); J. Bornstein *et al., ibid.* **76**, 2760 (1954); R. C. Elderfield, *Heterocycl. Compd.* **4**, 60 (1952); H. Yanagisawa *et al., Chem. Pharm. Bull.* **21**, 1080 (1973).

71. Cannizzaro Reaction

S. Cannizzaro, *Ann.* **88**, 129 (1853); K. List, H. Limpricht, *Ann.* **90**, 180 (1854).
Base-catalyzed disproportionation reaction of aromatic or aliphatic aldehydes with no α-hydrogen to the corresponding acid and alcohol. If the aldehydes are different, the reaction is called the "crossed Cannizarro reaction":

$$2RCHO \xrightarrow{base} RCOO^- + RCH_2OH$$

Organic Name Reactions

T. A. Geissman, *Org. React.* **2,** 94 (1944); F. P. B. Van der Maeden *et al., Recl. Trav. Chim. Pays-Bas* **91**(2), 221 (1972); C. G. Swain *et al., J. Am. Chem. Soc.* **101,** 3576 (1979); R. S. McDonald, C. E. Sibley, *Can. J. Chem.* **59,** 1061 (1981). *Review:* T. Lane, A. Plagens, *Named Organic Reactions* (John Wiley & Sons, Chichester, 1998) p 40-42. *Cf.* Meerwein-Ponndorf-Verley Reduction; Oppenauer Oxidation; Tishchenko Reaction.

72. Carroll Rearrangement
M. F. Carroll, *J. Chem. Soc.* **1940,** 704; **1941,** 507; W. Kimel, A. C. Cope, *J. Am. Chem. Soc.* **65,** 1992 (1943).

Preparation of γ,δ-unsaturated ketones by base-catalyzed reaction of allylic alcohols with β-ketoesters or thermal rearrangement of allyl acetoacetates:

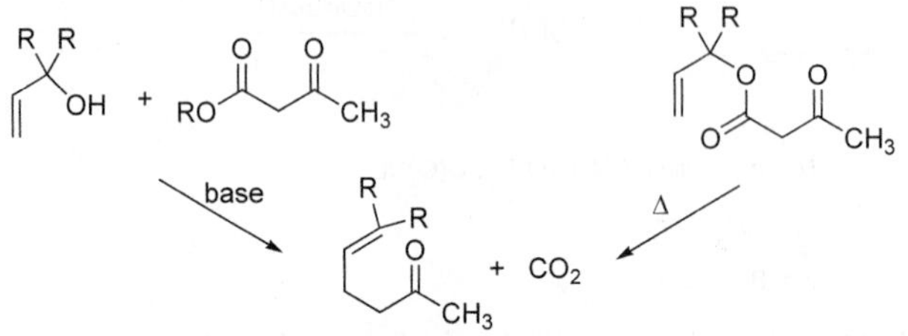

Detailed experimental: S. R. Wilson, C. E. Augelli, *Org. Synth.* **68,** 210 (1990). Synthetic applications: A. V. Echavarren *et al., Tetrahedron Lett.* **32,** 6421 (1991); N. Ouvrard *et al., ibid.* **34,** 1149 (1993). *Cf.* Claisen Rearrangement.

73. Castro-Stephens Coupling (Stephens-Castro Coupling) (Castro Reaction)
C. E. Castro, R. D. Stephens, *J. Org. Chem.* **28,** 2163 (1963); R. D. Stephens, C. E. Castro, *ibid.* 3313; A. M. Sladkov *et al., Bull. Acad. Sci. USSR Div. Chem. Sci.* **1963,** 2043.

The coupling of cuprous acetylides with aryl halides to yield arylacetylenes:

$$ ArX \ + \ Cu-C\equiv C-R \ \xrightarrow[\Delta]{Pd/pyridine} \ Ar-C\equiv C-R $$

X = I, Br, Cl
R = alkyl, aryl, vinyl

Synthetic applications: J. D. Kinder *et al., Synlett* **1993,** 149; J. Kabbara *et al., Synthesis* **1995,** 299; M. S. Yu *et al., Tetrahedron Lett.* **39,** 9347 (1998). Early reviews: G. H. Posner, *Org. React.* **22,** 253-400 *passim* (1975); A. M. Sladkov, I. R. Gol'ding, *Russ. Chem. Rev.* **48,** 868-896 (1979). *Cf.* Glaser Coupling.

74. Chapman Rearrangement
O. Mumm *et al., Ber.* **48,** 379 (1915); A. W. Chapman, *J. Chem. Soc.* **127,** 1992 (1925); **1927,** 174; **1929,** 569.

Thermal rearrangement of aryl imidates to *N,N*-diaryl amides:

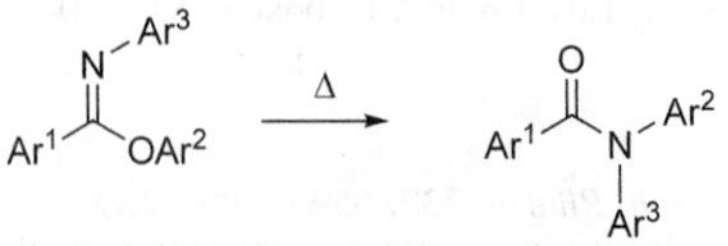

J. W. Schulenberg *et al., Org. React.* **14,** 1 (1965); C. G. McCarty, L. Garner in *The Chemistry of Amidines and Imidates* S. Patai, Ed. (Interscience, New York, 1975) p 189. Mechanistic study: N. A. Suttle, A. Williams, *J. Chem. Soc. Perkin Trans. 2* **1983,** 1369. Synthetic applications: L. H. Peterson *et al., J. Heterocycl. Chem.* **18,** 659 (1981); N. Dubau-Assibat *et al., Bull. Soc. Chim. Fr.* **132,** 1139 (1995). Chapman-like rearrangements: F. Esser *et al., J. Chem. Soc. Perkin Trans. 1* **1988,** 3311; M. Dessolin *et al., Chem. Commun.* **1992,** 132.

75. Chichibabin Pyridine Synthesis
A. E. Chichibabin, *J. Russ. Phys. Chem. Soc.* **37,** 1229 (1906); *J. Prakt. Chem.* **107,** 122 (1924).

Condensation of carbonyl compounds with ammonia or amines under pressure to form pyridine derivatives; the reaction is reversible and produces different pyridine derivatives along with byproducts:

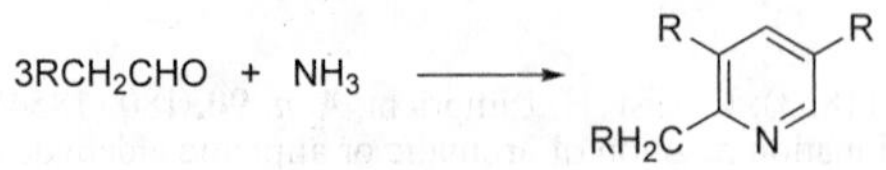

M. M. Sprung, *Chem. Rev.* **26,** 301 (1940); R. L. Frank, R. P. Seven, *J. Am. Chem. Soc.* **71,** 2629 (1949); H. S. Mosher, *Heterocycl. Compd.* **1,** 456 (1950); J. A. Gautier, J. Renault, *Bull. Soc. Chim. Fr.* **1955,** 588; C. P. Farley, E.

L. Eliel, *J. Am. Chem. Soc.* **78**, 3477 (1956); A. T. Soldatenkov, *Zh. Org. Khim.* **16**, 188 (1980). *Cf.* Hantzsch (Dihydro)Pyridine Synthesis; Kröhnke Pyridine Synthesis.

76. Chichibabin Reaction
A. E. Chichibabin, O. A. Zeide, *J. Russ. Phys. Chem. Soc.* **46**, 1216 (1914), *C.A.* **9**, 1901 (1915).
Amination of pyridines and other heterocyclic nitrogen compounds with alkali-metal amides:

H. S. Mosher, *Heterocycl. Compd.* **1**, 405 (1950); A. F. Pozharskii *et al., Russ. Chem. Rev.* **47**, 1042 (1978); H. J. W. van den Haak *et al., J. Org. Chem.* **46**, 2134 (1981). Applications: N. J. Kos *et al., ibid.* **44**, 3140 (1979); H. Tondys *et al., J. Heterocycl. Chem.* **22**, 353 (1985); E. Ciganek *et al., J. Org. Chem.* **57**, 4521 (1992). *Review:* H. C. van der Plas, M. Wozniak, *Croat. Chem. Acta* **59**, 33-49 (1986). *Cf.* Kröhnke Pyridine Synthesis.

77. Chugaev Reaction (Tschugaeff Olefin Synthesis)
L. Chugaev (Tschugaeff), *Ber.* **32**, 3332 (1899).
Formation of olefins from alcohols without rearrangement through pyrolysis of the corresponding xanthates *via cis-*elimination:

C. H. DePuy, R. W. King, *Chem. Rev.* **60**, 444 (1960); H. R. Nace, *Org. React.* **12**, 57 (1962); K. Harano, T. Taguchi, *Chem. Pharm. Bull.* **20**, 2357 (1972); J. March, *Advanced Organic Chemistry* (John Wiley & Sons, NY, 1992) 1014-1015. Synthetic applications: X. Fu, J. M. Cook, *Tetrahedron Lett.* **31**, 3409 (1990); P. S. Ray, M. J. Manning, *Heterocycles* **33**, 1361 (1994). *Cf.* Cope Elimination Reaction.

78. Ciamician-Dennstedt Rearrangement
G. L. Ciamician, M. Dennstedt, *Ber.* **14**, 1153 (1881).
Expansion of the pyrrole ring by heating with chloroform or other halogeno compounds in alkaline solution. The intermediate dichlorocarbene, by addition to the pyrrole, forms an unstable dihalogenocyclopropane which rearranges to a 3-halogenopyridine:

A. H. Corwin, *Heterocycl. Compd.* **1**, 309 (1950); H. S. Mosher, *ibid.* 475; P. S. Skell, R. S. Sandler, *J. Am. Chem. Soc.* **80**, 2024 (1958); E. Vogel, *Angew. Chem.* **72**, 8 (1960).

79. Claisen Condensation (Acetoacetic Ester Condensation)
L. Claisen, O. Lowman, *Ber.* **20**, 651 (1887).
Base-catalyzed condensation of an ester containing an α-hydrogen atom with a molecule of the same ester or a different one to give β-keto esters:

C. R. Hauser, B. E. Hudson, *Org. React.* **1**, 266-322 (1942); H. O. House, *Modern Synthetic Reactions* (W. A. Benjamin, Menlo Park, California, 2nd ed., 1972) pp 734-746; J. F. Garst, *J. Chem. Educ.* **56**, 721 (1979); J. E. Bartmess *et al., J. Am. Chem. Soc.* **103**, 1338 (1981); B. R. Davis, P. J. Garratt, *Comp. Org. Syn.* **2**, 795-805 (1991). *Cf.* Dieckmann Reaction.

80. Claisen Rearrangement; Eschenmoser-Claisen Rearrangement; Johnson-Claisen Rearrangement; Ireland-Claisen Rearrangement
L. Claisen, *Ber.* **45**, 3157 (1912); L. Claisen, E. Tietze, *ibid.* **58**, 275 (1925); **59**, 2344 (1926).
Highly stereoselective [3,3]-sigmatropic rearrangement of allyl vinyl or allyl aryl ethers to yield γ,δ-unsaturated carbonyl compounds or *o*-allyl substituted phenols, respectively:

When $R^1 = NR_2$, the reaction is referred to as the **Eschenmoser-Claisen rearrangement:** A. E. Wick *et al., Helv. Chim. Acta* **47**, 2425 (1964); M. Lautens *et al., Tetrahedron Lett.* **31**, 5829 (1990); B. Coates *et al., ibid.* **32**, 4199 (1991).

When $R^1 = OR$, the reaction is referred to as the **Johnson-Claisen rearrangement:** W. S. Johnson *et al., J. Am. Chem. Soc.* **92**, 741 (1970); R. Bao *et al., Synlett* **1992**, 217; D. Basavaiah, S. Pandiaraju, *Tetrahedron Lett.* **36**, 757 (1995).

When $R^1 = OSiR_3$ or OLi, the reaction is referred to as the **Ireland-Claisen rearrangement:** R. E. Ireland, R. H. Mueller, *J. Am. Chem. Soc.* **94**, 5897 (1972); R. E. Ireland *et al., J. Org. Chem.* **56**, 650 (1991); *idem et al., ibid.* 3572; K. Hattori, H. Yamamoto, *Tetrahedron* **50**, 3099 (1994).

Inclusive reviews: S. J. Rhoads, N. R. Raulins, *Org. React.* **22**, 1-252 (1975); F. E. Ziegler, *Chem. Rev.* **88**, 1423-1452 (1988); P. Wipf, *Comp. Org. Syn.* **5**, 827-873 (1991); A. M. M. Castro, *Chem. Rev.* **104**, 2939-3002 (2004). *Cf.* Carroll Rearrangement; Cope Rearrangement; Overman Rearrangement.

81. Claisen-Schmidt Condensation

L. Claisen, A. Claparède, *Ber.* **14**, 2460 (1881); J. G. Schmidt, *ibid.* 1459.

Condensation of an aromatic aldehyde with an aliphatic aldehyde or ketone in the presence of a relatively strong base (hydroxide or alkoxide ion) to form an α,β-unsaturated aldehyde or ketone:

A. T. Nielsen, W. J. Houlihan, *Org. React.* **16**, 1 (1968); H. O. House, *Modern Synthetic Reactions* (W. A. Benjamin, Menlo Park, California, 2nd ed., 1972) pp 632-639; J. A. Fine, P. Pulaski, *J. Org. Chem.* **38**, 1747 (1973). *Cf.* Aldol Reaction.

82. Clemmensen Reduction

E. Clemmensen, *Ber.* **46**, 1837 (1913); **47**, 51, 681 (1914).

Reduction of carbonyl groups of aldehydes and ketones to methylene groups with zinc amalgam and hydrochloric acid:

$$RCOR^1 \xrightarrow[HCl]{Zn(Hg)} RCH_2R^1$$

E. L. Martin, *Org. React.* **1**, 155 (1942); M. Smith in *Reduction*, R. L. Augustine, Ed. (M. Dekker, New York, 1968) pp 95-170; W. Reusch, *ibid.* pp 186-194; J. G. St. C. Buchanan, P. D. Woodgate, *Q. Rev. Chem. Soc.* **23**, 522 (1969); D. Straschewski, *Angew. Chem.* **71**, 726 (1959); E. Vedejs, *Org. React.* **22**, 401 (1975); S. Yamamura, S. Nishiyama, *Comp. Org. Syn.* **8**, 309-313 (1991). *Cf.* Haworth Phenanthrene Synthesis; Wolff-Kishner Reduction.

83. Combes Quinoline Synthesis

A. Combes, *Bull. Soc. Chim. Fr.* **49**, 89 (1888).

Formation of quinolines by condensation of β-diketones with primary arylamines followed by acid-catalyzed ring closure of the intermediate Schiff base:

W. S. Johnson, F. J. Matthews, *J. Am. Chem. Soc.* **66**, 210 (1944); F. W. Bergstrom, *Chem. Rev.* **35**, 156 (1944); J. C. Perche *et al., J. Chem. Soc. Perkin Trans. 1* **1972**, 260; J. Born, *J. Org. Chem.* **37**, 3952 (1972). *Cf.* Conrad-Limpach Reaction; Doebner Reaction.

84. Conrad-Limpach Cyclization

M. Conrad, L. Limpach, *Ber.* **20**, 944 (1887); **24**, 2990 (1891).

Thermal condensation of arylamines with β-ketoesters followed by cyclization of the intermediate Schiff bases to 4-hydroxyquinolines:

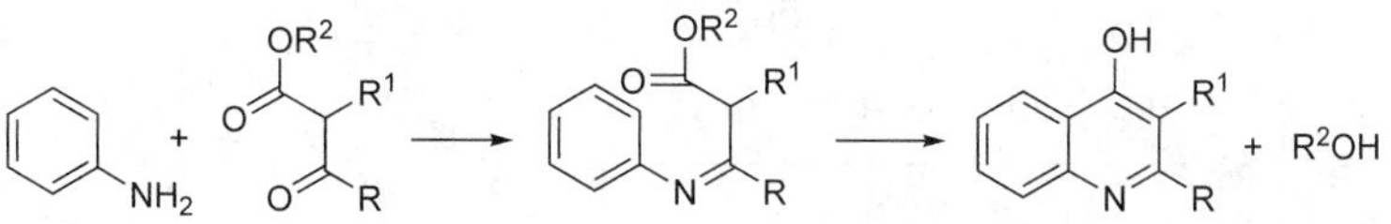

R. H. Manske, *Chem. Rev.* **30**, 121 (1942); R. H. Reitsema, *ibid.* **43**, 47 (1948); H. Henecka, *Chemie der Beta-Dicarbonylverbindungen* (Berlin, 1950) p 307; R. C. Elderfield, *Heterocycl. Compd.* **4**, 30 (1952); J.-C. Perche, G. Saint-Ruf, *J. Heterocycl. Chem.* **11**, 93 (1974); J. M. Barker *et al., J. Chem. Res. Synop.* **1980**, 4; J. A. Moore, T. D. Mitchell, *J. Polym. Chem.* **18**, 3029 (1980). *Cf.* Combes Quinoline Synthesis; Doebner Reaction.

85. Cope Elimination Reaction

A. C. Cope *et al., J. Am. Chem. Soc.* **71**, 3929 (1949); *idem et al., ibid.* **75**, 3212 (1953).
Formation of an olefin and a hydroxylamine by pyrolysis of an amine oxide:

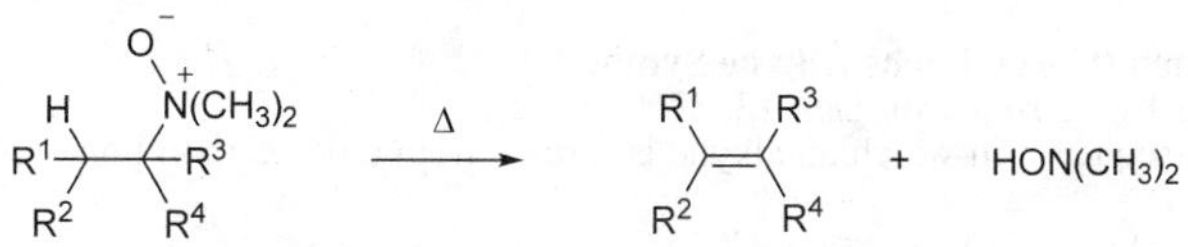

Early reviews: C. H. DePuy, R. W. King, *Chem. Rev.* **60**, 448 (1960); A. C. Cope, E. R. Trumbull, *Org. React.* **11**, 317-493 *passim* (1960). Synthetic application: E. Tojo *et al., Heterocycles* **27**, 2367 (1988). Mechanistic study: R. D. Bach, M. L. Braden, *J. Org. Chem.* **56**, 7194 (1991). Methods development: A. D. Woolhouse, *J. Heterocycl. Chem.* **30**, 873 (1993). Synthetic applications of the reverse reaction (retro-Cope elimination): E. Ciganek, *J. Org. Chem.* **55**, 3007 (1990); M. B. Gravestock *et al., Chem. Commun.* **1993**, 169. *Cf.* Chugaev Reaction; Hofmann Degradation.

86. Cope Rearrangement; Oxy-Cope Rearrangement

A. C. Cope *et al., J. Am. Chem. Soc.* **62**, 441 (1940).
Highly stereoselective [3,3]-sigmatropic rearrangement of 1,5-dienes; "all-carbon" equivalent of the Claisen rearrangement, *q.v.*:

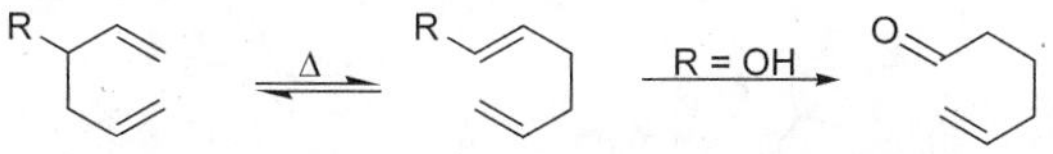

When R = OH, the transformation is referred to as the **oxy-Cope rearrangement:** J. Berson, M. Jones, *ibid.* **86**, 5019 (1964).
Reviews: S. J. Rhodds, N. R. Raulins, *Org. React.* **22**, 1-252 (1975); S. R. Wilson, *ibid.* **43**, 93-250 *passim* (1993); R. K. Hill, *Comp. Org. Syn.* **5**, 785-826 (1991). Review of hetero-Cope rearrangements: S. Blechert, *Synthesis* **1989**, 71-82. Brief review of synthetic applications: K. Durairaj, *Curr. Sci.* **66**, 917-922 (1994).

87. Corey-Bakshi-Shibata Reduction (CBS)

E. J. Corey *et al., J. Am. Chem. Soc.* **109**, 5551 (1987).
Enantioselective borane reduction of ketones catalyzed by chiral oxazaborolidines:

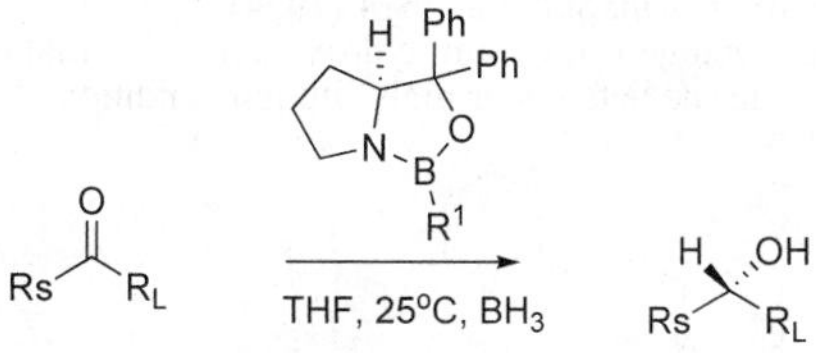

R_L = larger group
R_S = smaller group
R^1 = H, CH_3, C_2H_5, C_4H_9, etc.

Practical catalyst synthesis: D. J. Mathre *et al., J. Org. Chem.* **58**, 2880 (1993). Synthetic application: E. J. Corey *et al., J. Am. Chem. Soc.* **119**, 11769 (1997). *Reviews:* V. K. Singh, *Synthesis* **1992**, 605-617; L. Deloux, M. Srebnik, *Chem. Rev.* **93**, 763-784 (1993); E. J. Corey, C. J. Helal, *Angew. Chem. Int. Ed.* **37**, 1986-2012 (1998).

88. Corey-Chaykovsky Reaction

E. J. Corey, M. Chaykovsky, *J. Am. Chem. Soc.* **84**, 867 (1962); *eidem, ibid.* 3782; *eidem, ibid.* **87**, 1353 (1965).
Reaction of an electrophile with a sulfur ylide, either dimethylsulfoxonium methylide (Corey's reagent) or dimethylsulfonium methylide, leading to the formation of an epoxide, cyclopropane, aziridine, or thiirane:

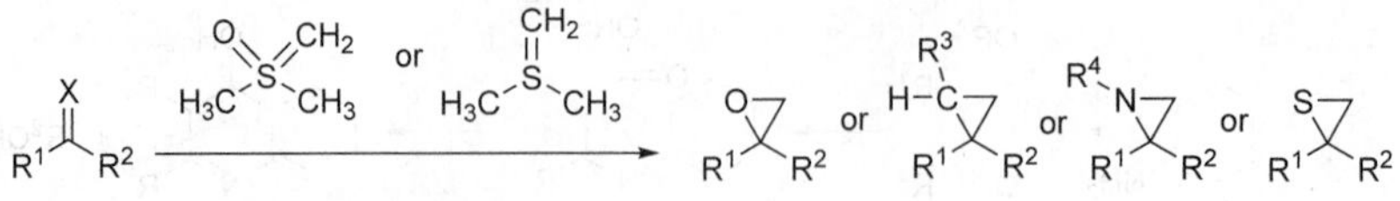

X = O, CHR³, NR⁴, S

Stereoselective epoxidation: T. Saito *et al.*, *Tetrahedron Lett.* **38**, 3755 (1997). Stereoselective cyclopropanation: M. Calmes *et al.*, *Tetrahedron: Asymmetry* **7**, 395 (1996). Reaction modifications utilizing phase transfer conditions: A. Merz, G. Märkl, *Angew. Chem. Int. Ed.* **12**, 845 (1973); ionic liquids: S. Chandrasekhar *et al.*, *Tetrahedron Lett.* **44**, 3629 (2003); organic bases: R. J. Paxton, R. J. K. Taylor, *Synlett* **2007**, 633; storable sulfoxonium salt/base mixtures. J. A. Ciaccio, C. E. Aman, *Synth. Commun.* **36**, 1333 (2006). *Review*: J. J. Li in *Name Reactions in Heterocyclic Chemistry*, J. J. Li, Ed. (Wiley, Hoboken, NJ, 2005) pp 2-14. *Cf.* Darzens Condensation. *See* monograph: Dimethylsulfoxonium Methylide.

89. Corey-Fuchs Reaction (Corey-Fuchs Alkyne Synthesis)
E. J. Corey, P. L. Fuchs, *Tetrahedron Lett.* **13**, 3769 (1972).
Two step reaction sequence in which an alkyne is formed from a one carbon homologation of an aldehyde:

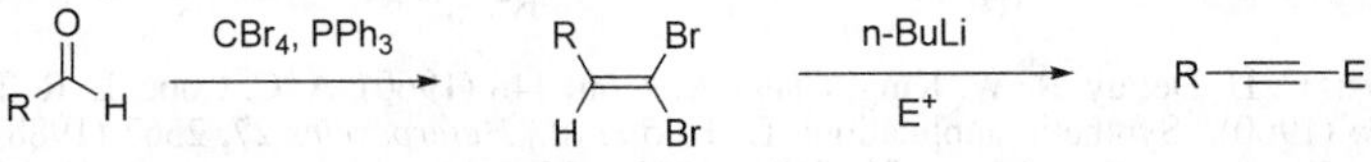

Reaction condition modifications: H. J. Bestmann, K. Li, *Ber.* **115**, 828 (1982); P. Michel *et al.*, *Tetrahedron Lett.* **40**, 8575 (1999); T. Nakahata *et al.*, *Chem. Eur. J.* **12**, 4584 (2006). *Review*: X. Han in *Name Reactions for Homologations, Part I* J. J. Li, Ed. (Wiley, Hoboken, NJ, 2009) pp 393-403. *Cf.* Wittig Reaction; Seyferth-Gilbert Homologation.

90. Corey-Kim Oxidation
E. J. Corey, C. U. Kim, *J. Am. Chem. Soc.* **94**, 7586 (1972).
Oxidation of primary and secondary alcohols *via* their alkoxysulfonium salts. Upon the addition of base, the salt rearranges intramolecularly to aldehydes and ketones, respectively:

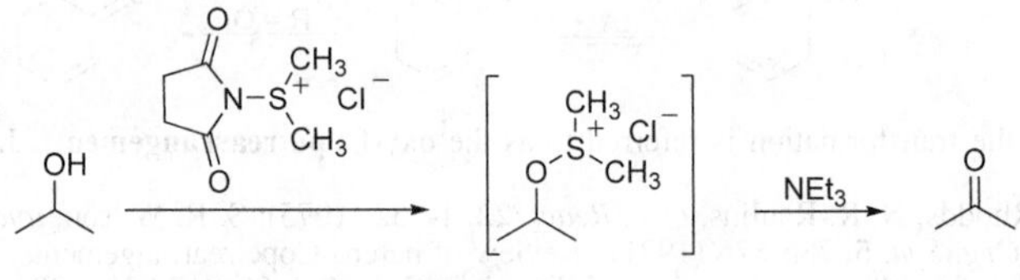

Application to the synthesis of α-hydroxy ketones: E. J. Corey, C. U. Kim, *Tetrahedron Lett.* **1974**, 287; of 1,3-dicarbonyl compounds: S. Katayama *et al.*, *Synthesis* **1988,** 178; J. T. Pulkkinen *et al., J. Org. Chem.* **61**, 8604 (1996). *Cf.* Pfitzner-Moffatt Oxidation; Swern Oxidation.

91. Corey-Nicolaou Macrolactonization (Corey-Nicolaou Double Activation Method)
E. J. Corey, K. C. Nicolaou, *J. Am. Chem. Soc.* **96**, 5614 (1974).
Preparation of small and large lactone rings from hydroxy acids; internal esterification is achieved through cyclization of the 2-pyridinethiol ester intermediate under high dilution conditions:

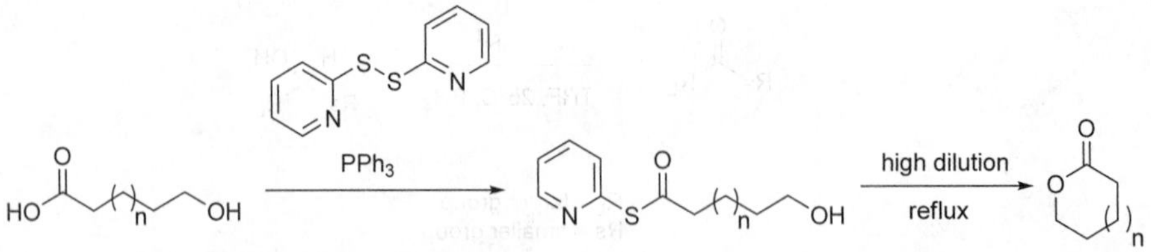

Original application: E. J. Corey *et al.*, *J. Am. Chem. Soc.* **97**, 653 (1975). Mechanistic studies: *idem et al.*, *Tetrahedron Lett.* **17**, 3405 (1976); K. Behinpour *et al.*, *ibid.* **22**, 275 (1981). Reaction modifications: H. Gerlach, A. Thalmann, *Helv. Chim. Acta* **57**, 2661 (1974); E. J. Corey, D. J. Brunelle, *Tetrahedron Lett.* **17**, 3409 (1976). *Reviews*: K. C. Nicolaou, *Tetrahedron* **33**, 683-710 (1977); T. G. Back, *ibid.* 3041-3059.

92. Corey-Suggs PCC Oxidation (PCC Oxidation); Corey-Schmidt PDC Oxidation
E. J. Corey, J. W. Suggs, *Tetrahedron Lett.* **16**, 2647 (1975).
Efficient oxidation of primary and secondary alcohols to aldehydes and ketones, respectively, through the use of pyridinium chlorochromate (PCC):

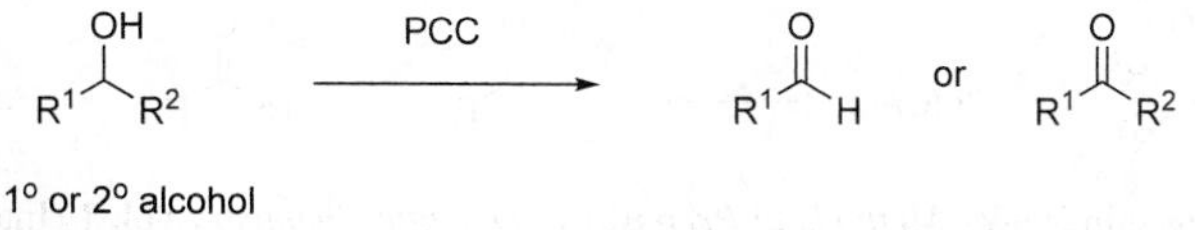

$$R^1 \overset{OH}{\underset{}{\diagup}} R^2 \xrightarrow{\text{PCC}} R^1 \overset{O}{\underset{}{\diagup}} H \quad \text{or} \quad R^1 \overset{O}{\underset{}{\diagup}} R^2$$

1° or 2° alcohol

Kinetic and mechanistic studies: K. K. Banerji, *Bull. Chem. Soc. Jpn.* **51**, 2732 (1978). Procedure modifications with molecular sieves: J. Herscovici *et al.*, *J. Chem. Soc. Perkin Trans. 1* **1982**, 1967; with silica gel and ultrasound conditions: L. L. Adams, F. A. Luzzio, *J. Org. Chem.* **54**, 5387 (1989); with catalytic acetic acid: S. Agarwal *et al.*, *Tetrahedron* **46**, 4417 (1990); with solvent-free conditions: S. Bhar, S. K. Chaudhuri, *ibid.* **59**, 3493 (2003); with catalytic PCC: M. Hunsen, *Tetrahedron Lett.* **46**, 1651 (2005). *Review*: G. Piancatelli *et al.*, *Synthesis* **1982**, 245-258.

When pyridinium dichromate (PDC) is utilized as the oxidizing agent, the reaction is known as the **Corey-Schmidt PDC Oxidation**. Use of PDC in alcohol oxidations: E. J. Corey, G. Schmidt, *Tetrahedron Lett.* **20**, 399 (1979); S. Czernecki *et al.*, *ibid.* **26**, 1699 (1985). *Cf.* Dess-Martin Oxidation; Swern Oxidaiton; Corey-Kim Oxidation; Pfitzner-Moffatt Oxidation; Griffith-Ley TPAP Oxidation. *See* monographs: Pyridinium Chlorochromate, Pyridinium Dichromate.

93. Corey-Winter Olefin Synthesis

E. J. Corey, R. A. E. Winter, *J. Am. Chem. Soc.* **85**, 2677 (1963).

Synthesis of olefins from 1,2-diols and thiocarbonyldiimidazole. Treatment of the intermediate cyclic thionocarbonate with trimethylphosphite yields the olefin by *cis*-elimination:

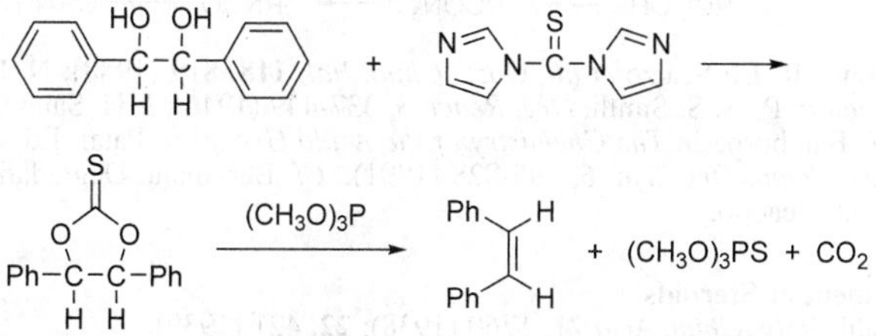

M. Tichy, J. Sicher, *Tetrahedron Lett.* **1969**, 4609; E. J. Corey, P. B. Hopkiss, *ibid.* **23**, 1797 (1982); S. Kaneko *et al.*, *Chem. Pharm. Bull.* **45**, 43 (1997). Applications in nucleotide synthesis: L. W. Dudycz, *Nucleosides Nucleotides* **8**, 35 (1989); R. L. K. Carr *et al.*, *Org. Prep. Proced. Int.* **22**, 245 (1990); in enediynes syntheses: M. F. Semmelhack, J. Gallagher, *Tetrahedron Lett.* **34**, 4121 (1993); D. Crich *et al.*, *Synth. Commun.* **29**, 359 (1999).

94. Cornforth Rearrangement

J. W. Cornforth, *The Chemistry of Penicillin* (Princeton University Press, New Jersey, 1949) p 700.

Thermal rearrangement of 4-carbonyl substituted oxazoles to their isomeric oxazoles *via* the postulated dicarbonyl nitrile ylides:

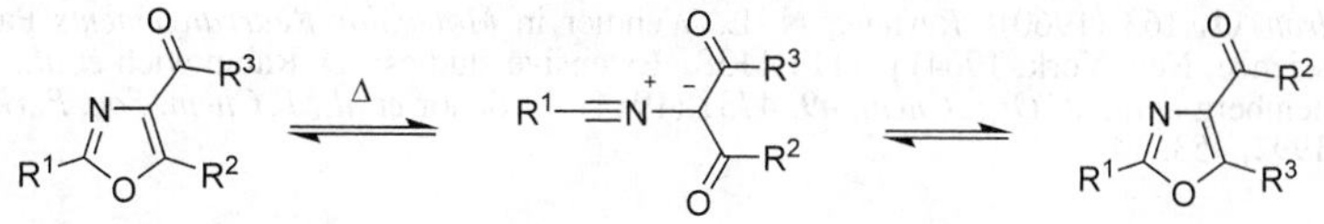

Mechanistic study: M. J. S. Dewar, I. J. Turchi, *J. Am. Chem. Soc.* **96**, 6148 (1974). Scope and limitations: *eidem*, *J. Org. Chem.* **40**, 1521 (1975). Extension to the synthesis of 5-aminothiazoles: S. L. Corrao *et al.*, *ibid.* **55**, 4484 (1990). Synthetic application: G. L'abbé *et al.*, *J. Chem. Soc. Perkin Trans. 1* **1993**, 2259.

95. Craig Method

L. C. Craig, *J. Am. Chem. Soc.* **56**, 231 (1934).

Introduction of a halogen into the α-position of aminopyridines by treatment with sodium nitrite in hydrohalic acid followed by warming:

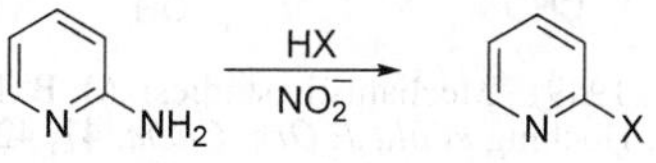

H. S. Mosher, *Heterocycl. Compd.* **1**, 515, 555 (1950); H. E. Mertel in *The Chemistry of Heterocyclic Compounds*, A. Weissberger, Ed., *Pyridine and Its Derivatives* **Pt. Two**, E. Klingsberg, Ed. (Interscience, New York, 1961) p 334.

96. Criegee Reaction

R. Criegee, *Ber.* **64**, 260 (1931).

Oxidative cleavage of vicinal glycols by lead tetraacetate:

$$R^1\text{–}C(OH)(R)\text{–}C(OH)(R^2)\text{–}R^3 + Pb(OAc)_4 \longrightarrow R\text{–CO–}R^1 + R^2\text{–CO–}R^3 + Pb(OAc)_2 + 2\,AcOH$$

Reviews: R. Criegee in *Newer Methods of Preparative Organic Chemistry* **vol. 1** (Interscience, New York, 1948) pp 12-20; H. O. House, *Modern Synthetic Reactions* (W. A. Benjamin, Menlo Park, California, 2nd ed., 1972) pp 359-387; K. W. Bentley in *Elucidation of Organic Structures by Physical and Chemical Methods* pt. 2, K. W. Bentley, G. W. Kirby, Eds. (Wiley, New York, 2nd ed., 1973) pp 169-177; S. Hatakeyama, H. Akimoto, *Res. Chem. Intermed.* **20**, 503-524 (1994). Mechanism: S. Chandrasekhar, C. D. Roy, *J. Chem. Soc. Perkin Trans. 2* **1994**, 2141; R. Ponec *et al., J. Org. Chem.* **62**, 2757 (1997); R. M. Goodman, Y. Kishi, *J. Am. Chem. Soc.* **120**, 9392 (1998). *Cf.* Malaprade Reaction.

97. Curtius Rearrangement; Curtius Reaction

T. Curtius, *Ber.* **23**, 3023 (1890); *idem, J. Prakt. Chem.* [2] **50**, 275 (1894).
   Formation of isocyanates by thermal decomposition of acyl azides:

$$RCON_3 \xrightarrow{\Delta} R\text{–N=C=O}$$

The stepwise conversion of a carboxylic acid to an amine having one fewer carbon unit, *via* the azide and isocyanate, is referred to as the **Curtius reaction:**

$$RCOOH \longrightarrow RCON_3 \longrightarrow RNCO \longrightarrow RNH_2$$

Synthetic applications: R. Lo Scalzo *et al., Gazz. Chim. Ital.* **118**, 819 (1988); N. De Kimpe *et al., J. Org. Chem.* **59**, 8215 (1994). *Reviews:* P. A. S. Smith, *Org. React.* **3**, 337-449 (1946); J. H. Saunders, R. J. Slocombe, *Chem. Rev.* **43**, 205 (1948); D. V. Banthorpe in *The Chemistry of the Azido Group*, S. Patai, Ed. (Interscience, New York, 1971) pp 397-405; T. Shioiri, *Comp. Org. Syn.* **6**, 795-828 (1991). *Cf.* Bergmann Degradation; Hofmann Reaction; Lossen Rearrangement; Schmidt Reaction.

98. D-Homo Rearrangement of Steroids

L. Ruzicka, H. Meldahl, *Helv. Chim. Acta* **21**, 1760 (1938); **22**, 421 (1939).
   Originally discovered in 17β-hydroxy-20-ketosteroids, but thoroughly studied in the 17α-hydroxy-20-keto series, this reaction involves an acid- or base-catalyzed acyloin rearrangement which yields a 6-membered D-ring:

R. B. Turner, *J. Am. Chem. Soc.* **75**, 3484 (1953); D. K. Fukushima *et al., ibid.* **77**, 6585 (1955); N. L. Wendler *et al., Tetrahedron* **11**, 163 (1960). *Review:* N. L. Wendler in *Molecular Rearrangements* Part 2, P. de Mayo, Ed. (Wiley-Interscience, New York, 1964) p 1114-1138. Extensive studies: D. Rabinovich *et al., Chem. Commun.* **1976**, 461; N. G. Steinberg *et al., J. Org. Chem.* **49**, 4731 (1984); L. Schor *et al., J. Chem. Soc. Perkin Trans. 1* **1990**, 163; *eidem, ibid.* **1992**, 453.

99. Dakin Reaction

H. D. Dakin, *Am. Chem. J.* **42**, 477 (1909).
   Replacement of the formyl or acetyl groups in phenolic aldehydes or ketones by a hydroxyl group by means of hydrogen peroxide:

J. E. Leffler, *Chem. Rev.* **45**, 385 (1949). Mechanistic studies: M. B. Hocking, *et al., Can. J. Chem.* **55**, 102 (1977); *eidem, ibid.* **56**, 2646 (1978); M. B. Hocking *et al., J. Org. Chem.* **47**, 4208 (1982). Sodium percarbonate as oxidizing reagent: G. W. Kabalka *et al., Tetrahedron Lett.* **33**, 865 (1992).

100. Dakin-West Reaction

H. D. Dakin, R. West, *J. Biol. Chem.* **78**, 91, 745, 757 (1928).
   Reaction of α-amino acids with acetic anhydride in the presence of base to give α-acetamido ketones. The reaction occurs *via* the intermediate azlactone:

Organic Name Reactions

$$RCHCOOH \xrightarrow{Ac_2O} RCHCOCH_3$$
$$\underset{NH_2}{|} \qquad\qquad \underset{NHCOCH_3}{|}$$

Mechanism: R. Knorr, R. Huisgen, *Ber.* **103**, 2598 (1970); W. Steglich, *et al., Ber.* **104**, 3644 (1971); G. Holfe *et al., Ber.* **105**, 1718 (1972); N. Allinger *et al., J. Org. Chem.* **39**, 1730 (1974); M. Kawase *et al., Chem. Pharm. Bull.* **48**, 114 (2000). Synthetic applications: J. R. Casimir *et al., Tetrahedron Lett.* **36**, 4797 (1995); T. T. Curran, *J. Fluorine Chem.* **74**, 107 (1995). *Review:* G. L. Buchanan, *Chem. Soc. Rev.* **17**, 91 (1988).

101. Darzens Condensation (Darzens-Claisen Reaction) (Glycidic Ester Condensation)
G. Darzens, *Compt. Rend.* **139**, 1214 (1904); **141**, 766 (1905); **142**, 214 (1906).

Formation of α,β-epoxy esters (glycidic esters) by the condensation of aldehydes or ketones with esters of α-halo-acids; the corresponding thermally unstable glycidic acids yield aldehydes or ketones on decarboxylation:

M. S. Newman, B. J. Magerlein, *Org. React.* **5**, 413 (1949); M. Ballester, *Chem. Rev.* **55**, 283 (1955); H. O. House, *Modern Synthetic Reactions* (W. Benjamin, Menlo Park, California, 2nd ed., 1972) pp 666-671. Intramolecular reaction: G. Fráter *et al., Tetrahedron Lett.* **34**, 2753 (1993). Enantioselectivity: D. Enders, R. Hett, *Synlett* **1998**, 961; S. Arai *et al., Tetrahedron* **55**, 6375 (1999). Modified conditions: R. F. Borch, *Tetrahedron Lett.* **1972**, 3761; I. Shibata *et al., J. Org. Chem.* **57**, 6909 (1992). Review: T. Rosen, *Comp. Org. Syn.* **2**, 409-439 (1991).

102. Darzens-Nenitzescu Synthesis of Ketones
G. Darzens, *Compt. Rend.* **150**, 707 (1910); C. D. Nenitzescu, I. P. Cantuniari, *Ann.* **510**, 269 (1934); C. D. Nenitzescu, C. Cioranescu, *Ber.* **69**, 1820 (1936).

Acylation of olefins with acid chlorides or anhydrides catalyzed by Lewis acids. When performed in the presence of a saturated hydrocarbon the product is the saturated ketone:

G. A. Olah, *Friedel-Crafts and Related Reactions* **vol. 1** (Interscience, New York, 1963) p 129; C. D. Nenitzescu, A. T. Balaban, *ibid.* **vol. 3,** Part 2, 1069 (1964); L. Ötvös *et al., Acta Chim. Acad. Sci. Hung.* **71**(2), 193 (1972); H. O. House, *Modern Synthetic Reactions* (W. A. Benjamin, Menlo Park, California, 2nd ed., 1972) p 786; J. K. Groves, *Chem. Soc. Rev.* **1**, 73 (1972). Synthetic applications: D. Villemin, D. Labiad, *Synth. Commun.* **22**, 3181 (1992); S. Nakanishi *et al., ibid.* **28**, 1967 (1998). *Cf.* Friedel-Crafts Reaction; Nencki Reaction; Nenitzescu Reductive Acylation.

103. Darzens Synthesis of Tetralin Derivatives
G. Darzens, *Compt. Rend.* **183**, 748 (1926).

Cyclization of α-benzyl-α-allylacetic acid type compounds by moderate heating in concentrated sulfuric acid to yield tetralin derivatives:

E. Bergmann, *Chem. Rev.* **29**, 536 (1941); J. N. Chatterjea *et al., Indian J. Chem.* **20B,** 264 (1981).

104. Delépine Reaction (Delépine Amine Synthesis)
M. Delépine, *Compt. Rend.* **120**, 501 (1895); *idem, ibid.* **124**, 292 (1897).

Preparation of primary amines by reaction of alkyl halides with hexamethylenetetramine followed by acid hydrolysis of the formed quaternary salts:

$$RCH_2X \; + \; C_6H_{12}N_4 \longrightarrow [RCH_2C_6H_{12}N_4]^+ \; X^- \xrightarrow{\;HX\;} RCH_2NH_3^+ \; X^-$$

S. J. Angyal, *Org. React.* **8**, 197 (1954); Y. Basace *et al., Bull. Soc. Chim. Fr.* **1971**, 1468. Synthetic applications: S. N. Quessy *et al., J. Chem. Soc. Perkin Trans. 1* **1979**, 512; S. Brandänge, B. Rodriguez, *Synthesis* **1988**, 347; R. A. Henry *et al., J. Org. Chem.* **55**, 1796 (1990).

105. de Mayo Reaction

P. de Mayo *et al., Proc. Chem. Soc. London* **1962**, 119; P. de Mayo, H. Takeshita, *Can. J. Chem.* **41**, 440 (1963).

Synthesis of 1,5-diketones by photoaddition of enol derivatives of 1,3-diketones to olefins, followed by a retro-aldol reaction, *q.v.*:

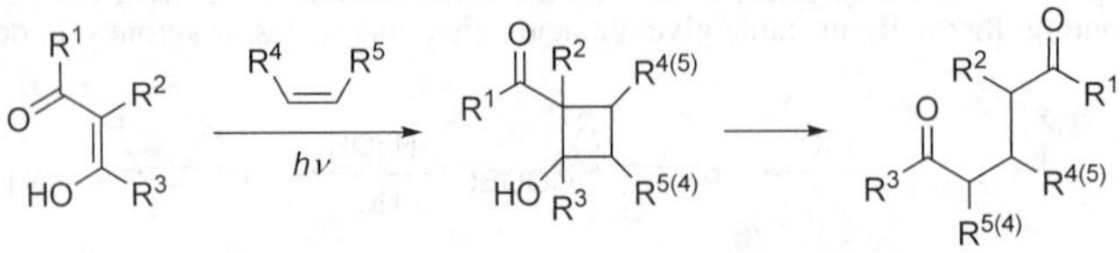

P. de Mayo, *Acc. Chem. Res.* **4**, 49 (1971); H. Meier, *Houben-Weyl* **4/5b**, 924 (1975); W. Oppolzer, *Pure Appl. Chem.* **53**, 1189 (1981). Intramolecular reactions: A. J. Barker, G. Pattenden, *Tetrahedron Lett.* **21**, 3513 (1980); *eidem, J. Chem. Soc. Perkin Trans. 1* **1983**, 1901. Intermolecular reactions: M. Sato *et al., Chem. Lett.* **1994**, 2191; P. Galatsis, J. J. Manwell, *Tetrahedron* **51**, 665 (1995); T. M. Quevillon, A. C. Weedon, *Tetrahedron Lett.* **37**, 3939 (1996).

106. Demjanov Rearrangement

N. J. Demjanov, M. Lushnikov, *J. Russ. Phys. Chem. Soc.* **35**, 26 (1903); *Chem. Zentralbl.* **1903**, 1, 828.

Deamination of primary amines by diazotization to give rearranged alcohols. In the case of alicyclic amines, ring enlargement or contraction occurs:

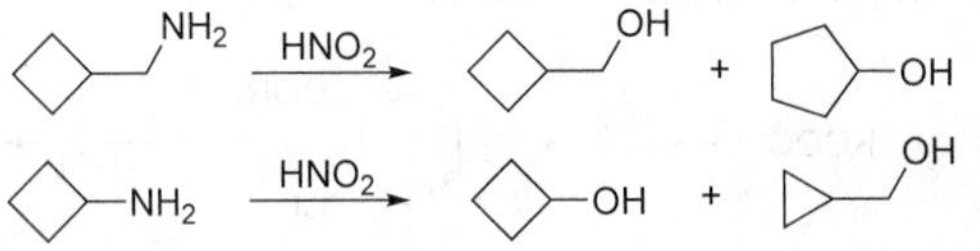

P. A. S. Smith, D. R. Baer, *Org. React.* **11**, 157 (1960); H. Stetter, P. Goebel, *Ber.* **96**, 550 (1963); R. Kotani, *J. Org. Chem.* **30**, 350 (1965); V. Dave *et al., Can. J. Chem.* **57**, 1557 (1979); R. K. Murray, Jr., T. M. Ford, *J. Org. Chem.* **44**, 3504 (1979); D. Fattori, *et al., Tetrahedron* **49**, 1649 (1993). *Cf.* Tiffeneau-Demjanov Rearrangement; Wagner-Meerwein Rearrangement.

107. Dess-Martin Oxidation

D. B. Dess, J. C. Martin, *J. Org. Chem.* **48**, 4155 (1983).

Mild oxidation of primary and secondary alcohols to aldehydes and ketones, respectively, employing the triace-toxyperiodinane (the "Dess-Martin Periodinane" reagent):

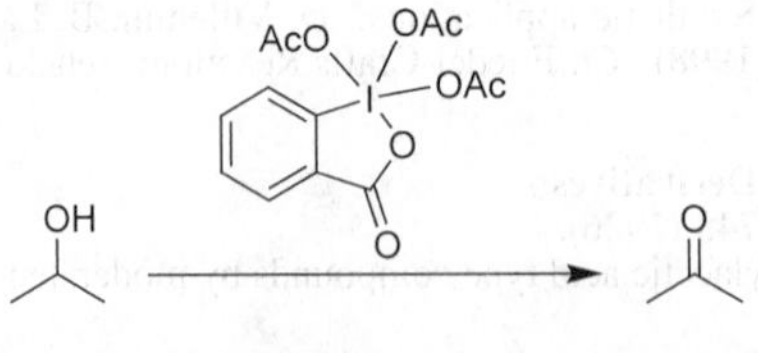

Scope and limitations of fluoroalkyl-substituted carbinols as substrates: R. J. Linderman, D. M. Graves, *J. Org. Chem.* **54**, 661 (1989). Methods development: D. B. Dess, J. C. Martin, *J. Am. Chem. Soc.* **113**, 7277 (1991). Application to the synthesis of 2'- and 3'-ketonucleosides: V. Samano, M. J. Robins, *J. Org. Chem.* **55**, 5186 (1990); of substituted oxazoles: P. Wipf, C. P. Miller, *ibid.* **58**, 3604 (1993). *See* monograph: Dess-Martin Periodinane.

108. Dieckmann Reaction

W. Dieckmann, *Ber.* **27**, 102, 965 (1894); **33**, 595, 2670, (1900); *Ann.* **317**, 51, 93, (1901).

Base-catalyzed cyclization of dicarboxylic acid esters to give β-ketoesters; the intramolecular equivalent of the Claisen condensation, *q.v.*:

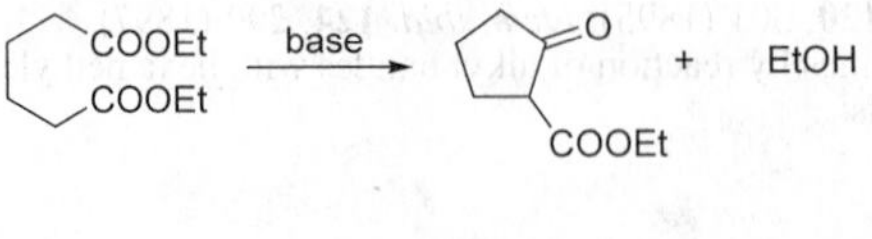

Organic Name Reactions

J. P. Schaefer, J. J. Bloomfield, *Org. React.* **15,** 1-203 (1967); H. O. House, *Modern Synthetic Reactions* (W. A. Benjamin, Menlo Park, California, 2nd ed., 1972) pp 740-743; H. Kwart, K. Sing in *The Chemistry of Carboxylic Acids and Esters,* S. Patai, Ed. (Interscience, New York, 1969) p 341; B. R. Davis, P. J. Garrett, *Comp. Org. Syn.* **2,** 806-829 (1991). *Cf.* Gabriel-Colman Rearrangement.

109. Diels-Alder Reaction

O. Diels, K. Alder, *Ann.* **460,** 98 (1928); **470,** 62 (1929); *Ber.* **62,** 2081, 2087 (1929).

The 1,4-addition of the double bond of a dienophile to a conjugated diene to generate a six-membered ring; up to four new stereocenters may be created simultaneously. The [4+2]-cycloaddition usually occurs with high regio- and stereoselectivity:

diene   dienophile

Heteroatomic analogs of the diene (e.g., $CHR{=}CR{-}CR{=}O$, $O{=}CR{-}CR{=}O$, and $RN{=}CR{-}CR{=}NR$) and dienophile (e.g., $RN{=}NR$, $R_2C{=}NR$, and $RN{=}O$) may also serve as reactants.

Early reviews: M. C. Kloetzel, *Org. React.* **4,** 1-59 (1948); H. L. Holmes *ibid.* 60-173; L. W. Butz, A. W. Rytina, *ibid.* **5,** 136-192 (1949). Intermolecular reactions: W. Oppolzer, *Comp. Org. Syn.* **5,** 315-399 (1991). Intramolecular reactions: E. Ciganek, *Org. React.* **32,** 1-374 (1984); W. R. Rousch, *Comp. Org. Syn.* **5,** 513-550 (1991). Use of heterodienophiles: S. M. Weinreb, *ibid.* 401-449. Use of nitroso dienophiles: J. Streith, A. DeFoin, *Synthesis* **1994,** 1107-1117. Use of heterodienes: D. L. Boger, *ibid,* 451-512. Review of diastereoselectivity: J. M. Coxon *et al.,* "Diastereofacial Selectivity in the Diels-Alder Reaction" in *Adv. Detailed React. Mech.* **3,** 131-166 (1994); T. Oh, M. Reilly, *Org. Prep. Proced. Int.* **26,** 131-158 (1994); H. Waldmann, *Synthesis* **1994,** 535-551. *Cf.* Wagner-Jauregg Reaction.

110. Dienone-Phenol Rearrangement

K. von Auwers, K. Ziegler, *Ann.* **425,** 217 (1921).

Transformation of a 4,4-disubstituted cyclohexadienone into a 3,4-disubstituted phenol upon acid treatment:

Reviews: C. J. Collins *et al.,* in *The Chemistry of the Carbonyl Group,* S. Patai, Ed. (Interscience, New York, 1966) pp 775-778; A. J. Waring, *Adv. Alicyclic Chem.* **1,** 207 (1967); B. Miller in *Mechanisms of Molecular Migrations* vol. **1,** B. S. Thyagarajan, Ed. (Interscience, New York, 1968) pp 275-285; B. Miller, *Acc. Chem. Res.* **8,** 277 (1975). Mechanism: G. Goodyear, A. J. Waring, *J. Chem. Soc. Perkin Trans. 2* **1990,** 103. Steric effects: A. G. Schultz, N. J. Green, *J. Am. Chem. Soc.* **114,** 1824 (1992); A. A. Frimer *et al., J. Org. Chem.* **59,** 1831 (1994). Synthetic applications: D. J. Hart *et al., Tetrahedron* **48,** 8179 (1992); R. W. Draper *et al., Steroids* **63,** 135 (1998).

111. Dimroth Rearrangement

O. Dimroth, *Ann.* **364,** 183 (1909); **459,** 39 (1927).

Rearrangement whereby exo- and endocyclic heteroatoms on a heterocyclic ring are translocated:

D. J. Brown, J. S. Harper in *Pteridine Chemistry,* W. Pfleiderer, E. C. Taylor, Ed. (Macmillan, New York, 1964) pp 219-230; D. J. Brown in *Mechanisms of Molecular Migrations* vol. **1,** B. S. Thyagarajan, Ed. (Wiley-Interscience, New York, 1968) p 209; D. J. Brown in *The Pyrimidines* Suppl. I (Interscience, New York, 1970) p 287; D. J. Brown, K. Lenega, *J. Chem. Soc. Perkin Trans. 1* **1974,** 372. Mechanism: K. Vaughan *et al., Heterocyclic Chem.* **28,** 1709 (1991); T. Itaya *et al., Chem. Pharm. Bull.* **45,** 832 (1997). Modified reaction: A. R. Katritzky *et al., J. Org. Chem.* **57,** 190 (1992); A. R. Pagano *et al., J. Org. Chem.* **63,** 3213 (1998). *Review:* E. S. H. El Ashry *et al., Adv. Heterocycl. Chem.* **75,** 79-167 (2000).

112. Directed *ortho* Metalation

H. Gilman, R. L. Bebb, *J. Am. Chem. Soc.* **61,** 109 (1939); G. Wittig, G. Fuhrmann, *Ber.* **73,** 1197 (1940).

Alkyllithium-based deprotonation of an arene's sp² hybridized carbon alpha to a directed metalation group; the resulting *ortho*-lithiated intermediate can then undergo an aromatic substitution reaction with an electrophile to yield a 1,2-disubstituted product:

Organic Name Reactions

Y = Directed Metalation Group (DMG)
e.g.

E⁺ = Electrophile
e.g. CH_3I, CO_2, Br_2, I_2, $B(OR)_3$
R_3SnCl, R_3SiCl, $ZnCl_2$

Mechanistic studies: M. Stratakis, *J. Org. Chem.* **62**, 3024 (1997); S. T. Chadwick *et al.*, *J. Am. Chem. Soc.* **122**, 8640 (2000). Anionic *ortho* Fries rearrangement: M. P. Sibi, V. Snieckus, *J. Org. Chem.* **48**, 1935 (1983). Use of cumyl groups as directed metalation groups: C. Metallinos *et al.*, *Org. Lett.* **1**, 1183 (1999). Review of metalations involving heterocycles: F. Mongin, G. Quéguiner, *Tetrahedron* **57**, 4059-4090 (2001); A. Turck *et al., ibid.* 4489-4505; of applications with cross-coupling chemistry: E. J.-G. Anctil, V. Snieckus, *J. Organomet. Chem.* **653**, 150-160 (2002). *Reviews:* V. Snieckus, *Chem. Rev.* **90**, 879-933 (1990); C. G. Hartung, V. Snieckus in *Modern Arene Chemistry*, D. Astruc, Ed. (Wiley, Weinheim, 2002) pp 330-367.

113. Doebner-Miller Reaction; Beyer Method for Quinolines
O. Doebner, W. v. Miller, *Ber.* **16**, 2464 (1883).

Acid-catalyzed synthesis of quinolines from primary aromatic amines and α,β-unsaturated carbonyl compounds. When the latter are prepared *in situ* from two molecules of aldehyde or an aldehyde and methyl ketone, the reaction is known as the **Beyer method for quinolines:**

F. W. Bergström, *Chem. Rev.* **35**, 153 (1944); Y. Ogata *et al., J. Chem. Soc. B* **1969**, 805; G. A. Dauphinee, T. P. Forrest, *J. Chem. Soc. D* **1969**, 327; *Can. J. Chem.* **56**, 632 (1978); C. M. Leir, *J. Org. Chem.* **42**, 911 (1977). Applications: G. K. Lund *et al., J. Chem. Eng. Data* **26**, 227 (1981); W. Buchowiecki *et al., J. Prakt. Chem.* **327**, 1015 (1985); T. Blitzke *et al., ibid.* **335**, 683 (1993). *Cf.* Gould-Jacobs Reaction; Knorr Quinoline Synthesis.

114. Doebner Reaction
O. Doebner, *Ann.* **242**, 265 (1887); *Ber.* **20**, 277 (1887); **27**, 352, 2020 (1894).

Formation of substituted cinchoninic acids from aromatic amines on heating with aldehydes and pyruvic acid:

F. W. Bergström, *Chem. Rev.* **35**, 156 (1944); R. C. Elderfield, *Heterocycl. Compd.* **4**, 25 (1952); C. Centini, *Rev. Soe. Venez. Quim.* **7**(5), 265 (1970), *C.A.* **74**, 76301x (1971); G. E. Gream, A. K. Serelis, *Aust. J. Chem.* **31**, 863 (1978). *Cf.* Combes Quinoline Synthesis; Conrad-Limpach Reaction.

115. Doering-LaFlamme Allene Synthesis
W. von E. Doering, P. M. LaFlamme, *Tetrahedron* **2**, 75 (1958); **US 2933544** (1960).

Treatment of an olefin with bromoform and an alkoxide to yield the 1,1-dibromocyclopropane which reacts with an active metal to produce an allene:

Reviews: M. Murray, *Houben-Weyl* **5/2a**, 985 (1977); V. Nair, *Comp. Org. Syn.* **4**, 1009-1012 (1991).

Organic Name Reactions

116. Dötz Reaction
K. H. Dötz, *Angew. Chem. Int. Ed.* **14**, 644 (1975).

Three component cyclization of an aromatic or vinylic alkoxy pentacarbonyl chromium carbene complex, an alkyne, and carbon monoxide, generating a $Cr(CO)_3$ coordinated phenol:

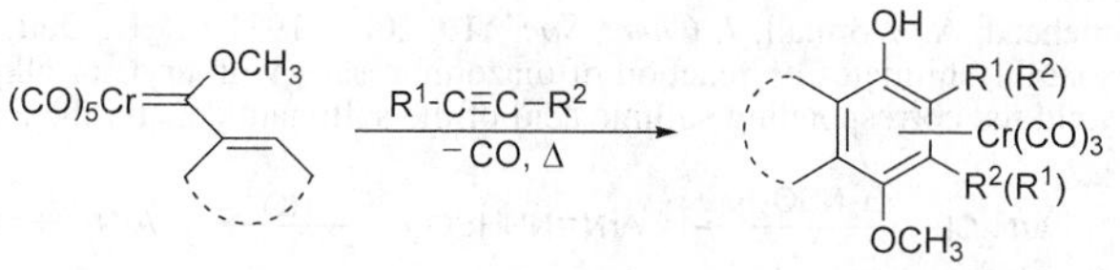

Solvent effects: K. S. Chan *et al., J. Organomet. Chem.* **334**, 9 (1987). Methods development: S. Chamberlin *et al., Tetrahedron* **49**, 5531 (1993); S. Chamberlin, W. D. Wulff, *J. Org. Chem.* **59**, 3047 (1994). Synthetic applications: W. D. Wulff *et al., J. Am. Chem. Soc.* **110**, 7419 (1988); D. L. Boger, I. C. Jacobson, *J. Org. Chem.* **56**, 2115 (1991). *Review:* K. H. Dötz, *New J. Chem.* **14**, 433-445 (1990).

117. Dowd-Beckwith Ring Expansion Reaction
A. L. J. Beckwith *et al., J. Am. Chem. Soc.* **110**, 2565 (1988); P. Dowd, S. C. Choi, *Tetrahedron* **45**, 77 (1989).

Free radical mediated ring expansions of haloalkyl β-ketoesters:

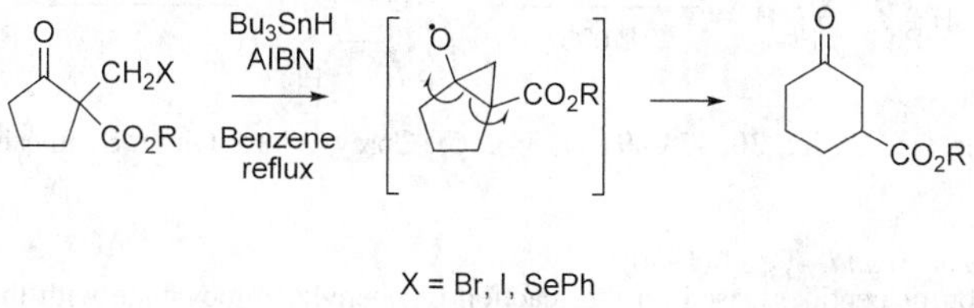

X = Br, I, SePh

Synthetic application: M. G. Banwell, J. M. Cameron, *Tetrahedron Lett.* **37**, 525 (1996); C. Wang *et al., Tetrahedron* **54**, 8355 (1998).

118. Doyle-Kirmse Reaction (Kirmse Reaction)
W. Kirmse, M. Kapps, *Ber.* **101**, 994 (1968); M. P. Doyle *et al., J. Org. Chem.* **46**, 5094 (1981).

Metal-catalyzed ylide formation and subsequent [2,3]-sigmatropic rearrangement of allyl substituted sulfides:

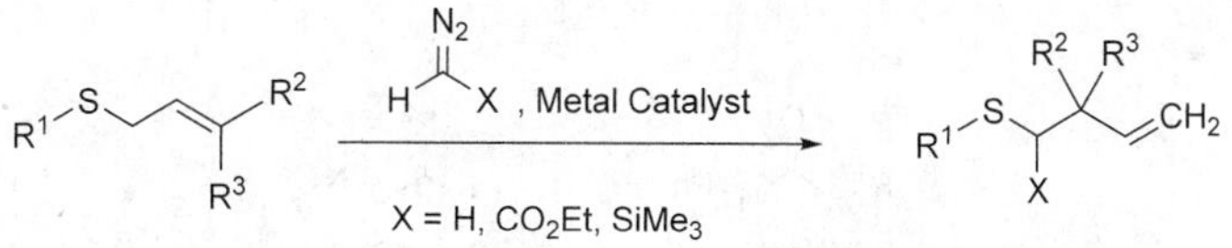

X = H, CO_2Et, $SiMe_3$

Comparison of catalysts and diazo compounds: D. S. Carter, D. L. Van Vranken, *Tetrahedron Lett.* **40**, 1617 (1999); V. K. Aggarwal *et al., ibid.* 8923. Variations involving Fe salt catalysis: D. S. Carter, D. L. Van Vranken, *Org. Lett.* **2**, 1303 (2000); Pd salt catalysis: K. L. Greenman *et al., Tetrahedron* **57**, 5219 (2001); diisopropyl diazomethylphosphonate: M. Gulea *et al., Synthesis* **1998**, 1635; diazoalkane free conditions: Y. Kato *et al., Org. Lett.* **5**, 2619 (2003). Synthetic applications: J. B. Perales *et al., J. Org. Chem.* **67**, 6711 (2002).

119. Duff Reaction
J. C. Duff, E. J. Bills, *J. Chem. Soc.* **1932**, 1987; **1934**, 1305; **1941**, 547; **1945**, 276.

Formylation of phenols or aromatic amines with hexamethylenetetramine in the presence of an acidic catalyst. *Ortho*-substitution is usual; however in the presence of anhydrous trifluoroacetic acid (TFA) regioselective *ortho* and *para* substitutions are observed:

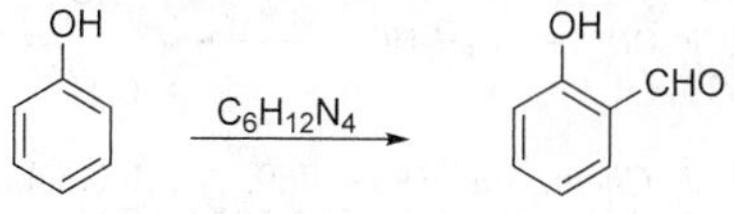

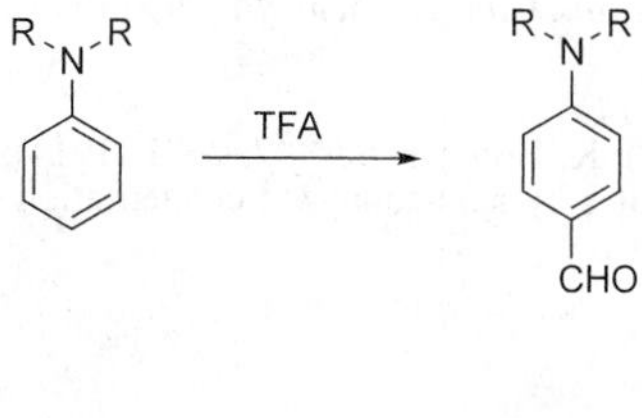

L. N. Ferguson, *Chem. Rev.* **38**, 230 (1946); Y. Ogata, F. Sugiura, *Tetrahedron* **24**, 5001 (1968); F. Wada *et al.*, *Bull. Chem. Soc. Jpn.* **53**, 1473 (1980). Use of TFA: W. E. Smith, *J. Org. Chem.* **37**, 3972 (1972); J. F. Larrow *et al.*, *ibid.* **59**, 1939 (1994); L. F. Lindoy *et al.*, *Synthesis* **1998**, 1029. *Cf.* Reimer-Tiemann Reaction.

120. Dutt-Wormall Reaction

P. K. Dutt, H. R. Whitehead, A. Wormall, *J. Chem. Soc.* **119**, 2088 (1921); P. K. Dutt, *ibid.* **125**, 1463 (1924).

Preparation of diazoaminosulfinates by reaction of diazonium salts with aryl- or alkylsulfonamides followed by alkaline hydrolysis to yield the corresponding sulfinic acid of the sulfonamide, and the azide:

$$ArN_2Cl \xrightarrow{H_2NSO_2R} ArN=NNHSO_2R \xrightarrow{HO^-} ArN_3 + HO_2SR$$

H. Bretschneider, H. Rager, *Monatsh. Chem.* **81.** 970 (1950); I. G. Laing, *Rodd's Chemistry of Carbon Compounds* **IIIC,** 107 (1973); C. Grundmann, *Houben-Weyl* **10/3,** 808 (1965).

121. Eastwood Reaction (Eastwood Deoxygenation)

G. Grank, F. W. Eastwood, *Aust. J. Chem.* **17**, 1392 (1964).

Stereospecific conversion of vicinal diols into olefins:

Review: E. Block, *Org. React.* **30,** 478-491 (1984). *Cf.* Corey-Winter Olefin Synthesis.

122. Edman Degradation

P. Edman, *Acta Chem. Scand.* **4**, 283 (1950).

Cyclic degradation of peptides based on the reaction of phenylisothiocyanate with the free amino group of the *N*-terminal residue such that amino acids are removed one at a time and identified as their phenylthiohydantoin derivatives:

S. Bösze *et al., J. Chromatogr. A* **668,** 345 (1994). *Reviews:* R. A. Laursen *et al., Methods Biochem. Anal.* **26,** 201-284 (1980); R. L. Heinrikson, "The Edman Degradation in Protein Sequence Analysis" in *Biochemical and Biophysical Studies of Proteins and Nucleic Acids*, T.-B. Lo *et al.*, Eds. (Elsevier, New York, 1984) pp 285-302; K.-K. Han *et al., Int. J. Biochem.* **17**, 429-445 (1985); C. G. Fields *et al., Pept. Res.* **6**, 39-47 (1993).

123. Ehrlich-Sachs Reaction

P. Ehrlich, F. Sachs, *Ber.* **32**, 2341 (1899).

Formation of *N*-phenylimines by the base-catalyzed condensation of compounds containing active methylene groups with aromatic nitroso compounds; nitrones also may be formed:

F. Barrow, F. J. Thorneycroft, *J. Chem. Soc.* **1939**, 769; A. McGookin, *J. Appl. Chem.* **5**, 65 (1955); F. Bell, *J. Chem. Soc.* **1957**, 516; D. M. W. Anderson, F. Bell, *ibid.* **1959**, 3708; D. M. W. Anderson, J. L. Duncan, *ibid.* **1961**, 1631; W. Seidenfaden, *Houben-Weyl* **10/1**, 1079 (1971). Applications: F. Millich, M. T. El-Shoubary, *Org. Prep. Proced. Int.* **28**, 366 (1996); S. K. De *et al., Can. J. Chem.* **76**, 199 (1998).

124. Einhorn-Brunner Reaction

A. Einhorn *et al., Ann.* **343**, 229 (1905); K. Brunner, *Ber.* **47**, 2671 (1914); *Monatsh. Chem.* **36**, 509 (1915).

Formation of substituted 1,2,4-triazoles by acid-catalyzed condensation of hydrazines or semicarbazides with diacylamines:

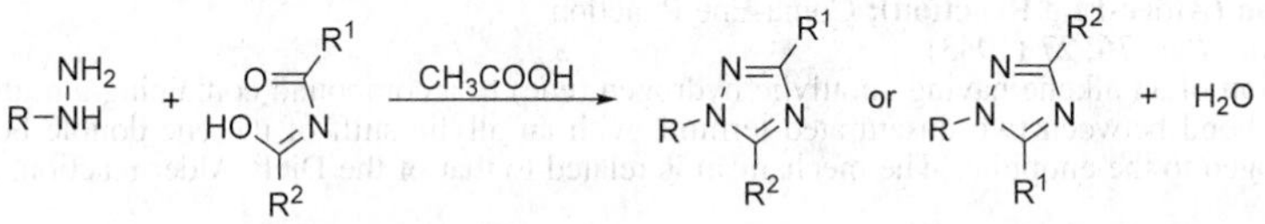

M. R. Atkinson, J. B. Polya, *J. Chem. Soc.* **1952**, 3418; **1954**, 141, 3319; Theilheimer, *Synthetic Methods* **9,** No. 449 (1955); K. T. Potts, *Chem. Rev.* **61,** 103 (1961); K. Hu *et al., J. Org. Chem.* **63,** 4786 (1998). *Cf.* Pellizzari Reaction.

125. Elbs Persulfate Oxidation

K. Elbs, *J. Prakt. Chem.* **48,** 179 (1893).

Hydroxylation of phenols to predominantely *p*-diphenols. If the phenol contains a *para* substituent, then the reaction occurs at the *ortho* position:

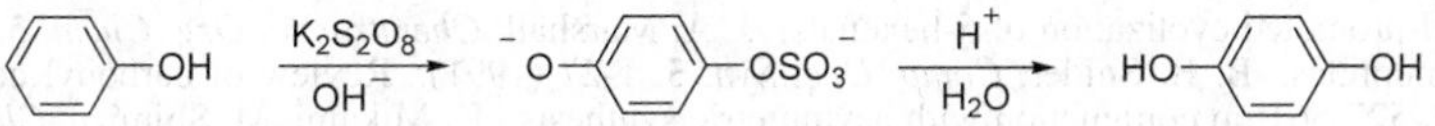

Evaluation in the synthesis of herbicides: K. G. Watson, A. Serban, *Aust. J. Chem.* **48,** 1503 (1995). *Reviews:* S. M. Sethna, *Chem. Rev.* **49,** 91-101 (1951); J. B. Lee, B. C. Uff, *Q. Rev. Chem. Soc.* **21,** 429-457 (1967); E. J. Behrman, *Org. React.* **35,** 421-511 (1988); *idem, Beilstein J. Org. Chem.* **2,** 22 (2006). *Cf.* Boyland-Sims Oxidation.

126. Elbs Reaction

K. Elbs, E. Larsen, *Ber.* **17,** 2847 (1884).

Formation of polyaromatics (*e.g.* anthracene) by intramolecular condensation of diaryl ketones containing a methyl or methylene substituent adjacent to the carbonyl group:

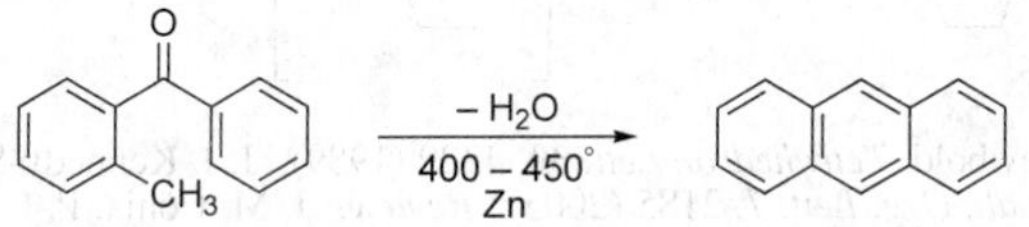

L. F. Fieser, *Org. React.* **1,** 129 (1942); G. N. Badger, B. J. Christie, *J. Chem. Soc.* **1956,** 3435; N. P. Buu-Hoi, D. Lavit, *Rec. Trav. Chim.* **76,** 419 (1957); Cl. Marie *et al., J. Chem. Soc.* **1971,** 431; M. S. Newman, V. K. Khanna, *J. Org. Chem.* **45,** 4507 (1980).

127. Emde Degradation

H. Emde, *Ber.* **42,** 2590 (1909); *Ann.* **391,** 88 (1912).

Modification of the Hofmann degradation, *q.v.,* method for reductive cleavage of the carbon-nitrogen bond by treatment of an alcoholic or aqueous solution of a quaternary ammonium halide with sodium amalgam. Also used as a catalytic method with palladium and platinum catalysts. The method succeeds with ring compounds not degraded by the Hofmann procedure:

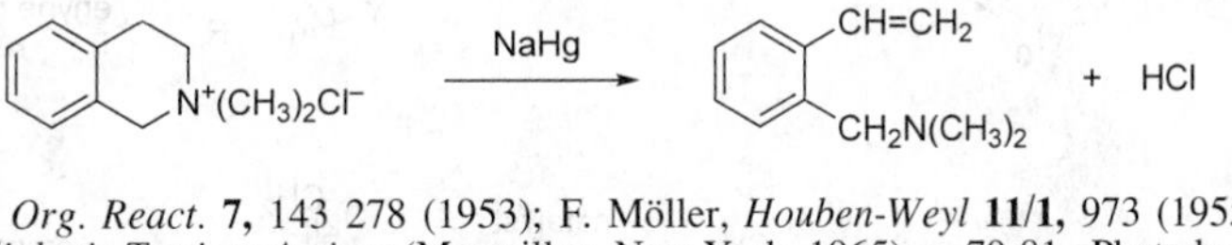

Reviews: A. Birch, *Org. React.* **7,** 143 278 (1953); F. Möller, *Houben-Weyl* **11/1,** 973 (1955); Z. Spialter, J. A. Pappalardo, *Acyclic Aliphatic Tertiary Amines* (Macmillan, New York, 1965) pp 79-81. Photodegradation: V. Partail, *Helv. Chim. Acta* **68,** 1952 (1985). Synthetic applications: J. G. Cannon *et al., J. Med. Chem.* **18,** 110 (1975); J. Lévy *et al., Tetrahedron: Asymmetry* **8,** 4127 (1997).

128. Emmert Reaction

B. Emmert, E. Asendorf, *Ber.* **72,** 1188 (1939); B. Emmert, E. Pirot, *ibid.* **74,** 714 (1941).

Formation of pyridyldialkylcarbinols by condensation of ketones with pyridine or its homologs in the presence of aluminum or magnesium amalgam:

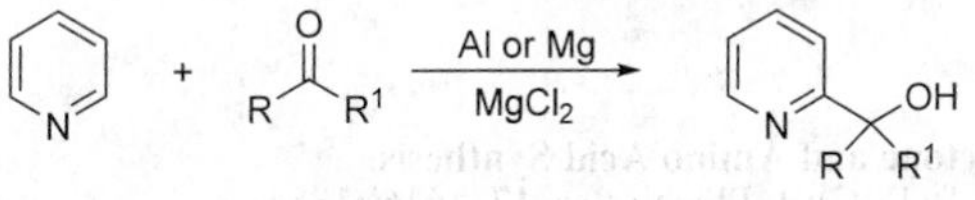

C. H. Tilford *et al., J. Am. Chem. Soc.* **70,** 4001 (1948); H. L. Lochti *et al., ibid.* **75,** 4477 (1953); R. Abramovitch, R. Vinutha, *J. Chem. Soc. C* **1969,** 2104; C. A. Russell *et al., J. Chem. Soc. D* **1970,** 1406; R. Tschesche, W. Führer, *Ber.* **111,** 3502 (1978).

Organic Name Reactions

129. Ene Reaction (Alder-Ene Reaction); Conia-Ene Reaction

K. Alder *et al., Ber.* **76**, 27 (1943).

The addition of an alkene having an allylic hydrogen (ene) to a compound containing a multiple bond (enophile) to form a new bond between two unsaturated termini, with an allylic shift of the ene double bond, and transfer of the allylic hydrogen to the enophile. The mechanism is related to that of the Diels-Alder reaction, *q.v.*:

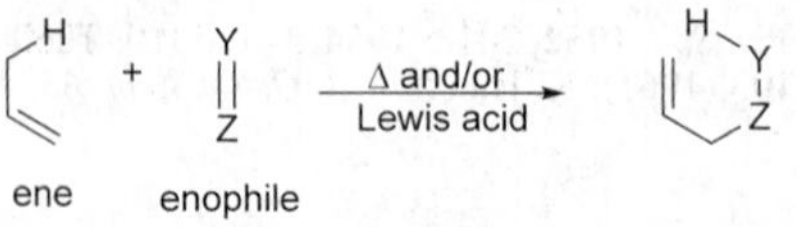

enophile = carbonyl and thiocarbonyl compounds, imines, alkenes, alkynes
Lewis acid = $BF_3 \bullet O(CH_2CH_3)_2$, $SnCl_4$, $Al(CH_2CH_3)Cl_2$, $Al(CH_3)_2Cl$

Lewis acid-promoted cyclization of 5-hexenals: J. A. Marshall, *Chemtracts: Org. Chem.* **5**, 1-7 (1992). Review of alkenes as enophiles: B. B. Snider, *Comp. Org. Syn.* **5**, 1-27 (1991). Review of carbonyl compounds as enophiles: *idem, ibid.* **2**, 527-561; in conjunction with asymmetric synthesis: K. Mikami, M. Shimizu, *Chem. Rev.* **92**, 1021-1050 (1992); K. Mikami *et al., Synlett* **1992**, 255-265.

The intramolecular Ene reaction of unsaturated ketones, in which the carbonyl functionality serves as the ene component, *via* its tautomer, and the olefinic moiety serves as the enophile, is known as the **Conia-Ene reaction**: F. Rouessac *et al., Tetrahedron Lett.* **6**, 3319 (1965).

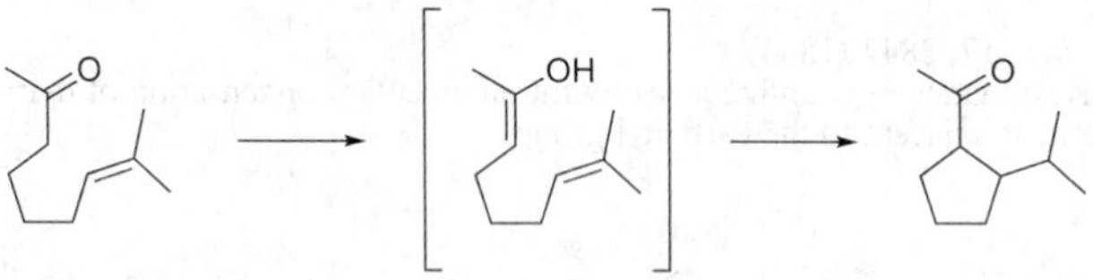

A. S. Kende, R. C. Newbold, *Tetrahedron Lett.* **30**, 4329 (1989). J. J. Kennedy-Smith *et al., J. Am. Chem. Soc.* **126**, 4526 (2004). Q. Gao *et al., Org. Lett.* **7**, 2185 (2005). *Review:* J. M. Conia, P. Le Perchec, *Synthesis* **1975**, 1-19.

130. Enyne Metathesis (Enyne Bond Reorganization)

T. J. Katz, T. M. Sivavec, *J. Am. Chem. Soc.* **107**, 737 (1985).

Carbon-carbon bond rearrangement of an alkene and alkyne to yield a 1,3-diene. Intramolecular cycloisomerization, referred to as ring-closing enyne metathesis, produces a cyclic 1,3-diene. The intermolecular version of the reaction, known as the cross enyne metathesis, results in an acyclic product. The metathesis reactions are catalyzed by either metal salts or metal carbenes:

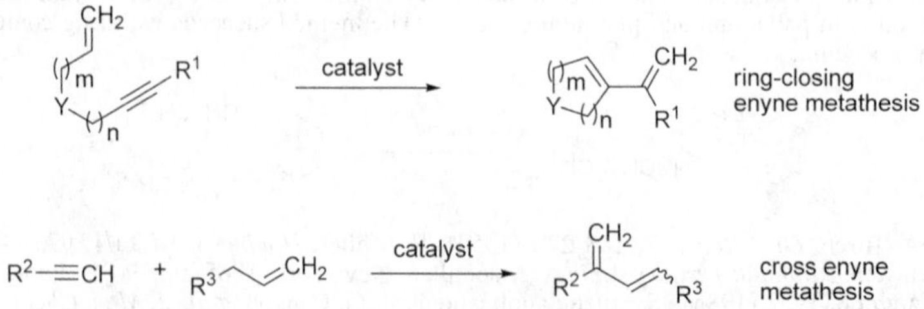

ring-closing enyne metathesis

cross enyne metathesis

Ring closing metathesis catalyzed by metal carbenes: P. F. Korkowski *et al., J. Am. Chem. Soc.* **110**, 2676 (1988); T. R. Hoye, J. A. Suriano, *Organometallics* **11**, 2044 (1992); by Grubbs' catalyst: S.-H. Kim *et al., J. Am. Chem. Soc.* **116**, 10801 (1994); A. Kinoshita, M. Mori, *Synlett* **1994**, 1020. Use of ethylene gas in intramolecular reactions: M. Mori *et al., J. Org. Chem.* **63**, 6082 (1998). Initial cross enyne metathesis reaction: R. Stragies *et al., Angew. Chem. Int. Ed.* **36**, 2518 (1997). Review of ring-closing metathesis reactions: H. Villar *et al., Chem. Soc. Rev.* **36**, 55-66 (2007). Review of cross enyne metathesis reactions: S. T. Diver, *J. Mol. Catal. A* **254**, 29-42 (2006). *Comprehensive review:* S. T. Diver, A. J. Giessert, *Chem. Rev.* **104**, 1317-1382 (2004). *Cf.* Olefin Metathesis. *See* monograph: Grubbs' Catalyst.

131. Erlenmeyer-Plöchl Azlactone and Amino Acid Synthesis

E. Erlenmeyer, *Ann.* **275**, 1 (1893); J. Plöchl, *Ber.* **17**, 1616 (1884).

Formation of azlactones by intramolecular condensation of acylglycines in the presence of acetic anhydride. The reaction of azlactones with carbonyl compounds followed by hydrolysis to the unsaturated α-acylamino acid and by reduction yields the amino acid; drastic hydrolysis gives the α-oxo acid:

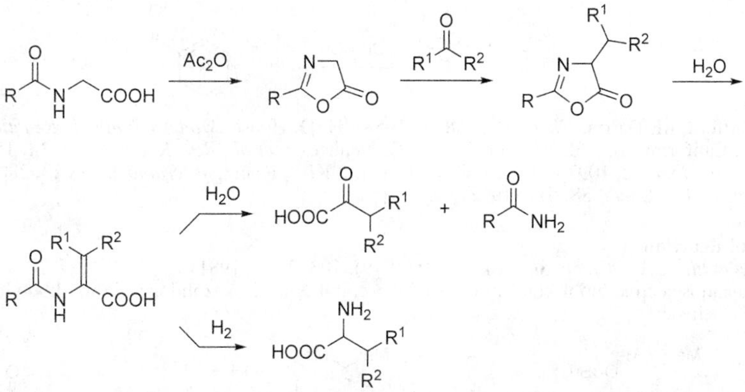

C. L. A. Schmidt, *The Chemistry of the Amino Acids and Proteins* (Springfield, IL, 1944) p 54; H. E. Carter, *Org. React.* **3,** 198 (1946); M. Crawford, W. T. Little, *J. Chem. Soc.* **1959,** 729; W. Steglich, *Fortschr. Chem. Forsch.* **12,** 84 (1969); J. Cornforth, D. Ming-hui, *J. Chem. Soc. Perkin Trans. 1* **1991,** 2183; A. P. Combs, R. W. Armstrong, *Tetrahedron Lett.* **33,** 6419 (1992). *Cf.* Bergmann Azlactone Peptide Synthesis; Perkin Reaction.

132. Eschenmoser Coupling Reaction (Sulfide Contraction)
A. Fischli, A. Eschenmoser, *Angew. Chem. Int. Ed.* **6,** 866 (1967); M. Roth *et al., Helv. Chim. Acta* **54,** 710 (1971).
Formation of vinylogous amides and urethanes by alkylation of secondary or tertiary thioamides with an electrophilic agent followed by elimination of sulfur:

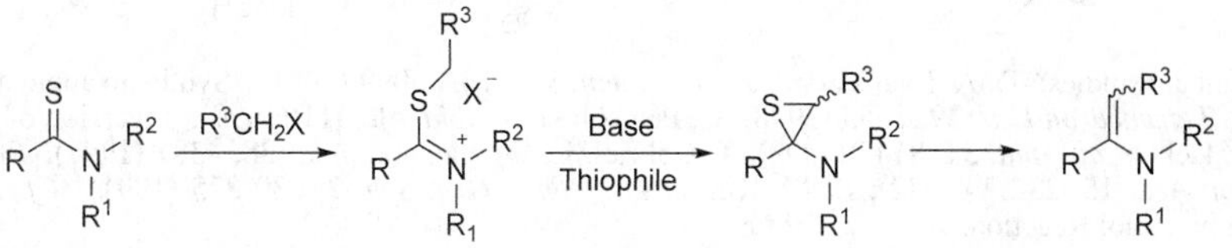

Synthetic applications: E. Götschi *et al., Angew. Chem. Int. Ed.* **12,** 910 (1973); O. Sakurai *et al., J. Org. Chem.* **61,** 7889 (1996), T. G. Minehan, Y. Kishi, *Tetrahedron Lett.* **38,** 6811 (1997). *Review:* K. Shiosaki, *Comp. Org. Syn.* **2,** 865-894 (1991).

133. Eschenmoser Fragmentation (Eschenmoser-Tanabe Fragmentation)
A. Eschenmoser *et al., Helv. Chim. Acta* **50,** 708 (1967); J. Schreiber *et al., ibid.* 2101; M. Tanabe *et al., Tetrahedron Lett.* **1967,** 3943.
Cleavage of α,β-epoxyketones under mild conditions, *via* sulfonylhydrazone intermediates, to yield acetylenic and carbonyl compounds:

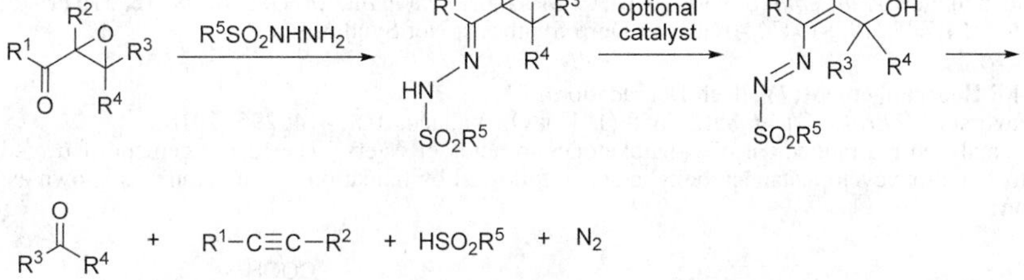

R^5 = *p*-toluene, 2,4,6-trimethylbenzene, 2,4-dinitrobenzene
optional catalyst = C_6H_5N, $NaHCO_3$, Na_2CO_3, silica gel

Early review: D. Felix *et al., Helv. Chim. Acta* **54,** 2896-2912 (1971). Synthetic applications: C. B. Reese, H. P. Sanders, *Synthesis* **1981,** 276; W. Dai, J. A. Katzenellenbogen, *J. Org. Chem.* **58,** 1900 (1993); A. Abad *et al., Synlett* **1991,** 787. *Cf.* Grob Fragmentation; Wharton Reaction.

134. Étard Reaction
A. L. Étard, *Compt. Rend.* **90,** 534 (1880); *Ann. Chim. Phys.* **22,** 218 (1881).
Oxidation of an aromatic methyl group to an aldehyde by treatment with chromyl chloride:

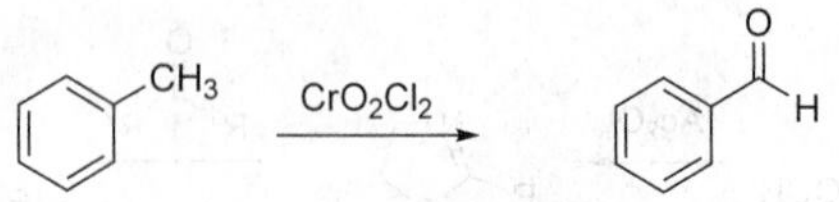

$$\text{(toluene)} \xrightarrow{\text{CrO}_2\text{Cl}_2} \text{(benzaldehyde)}$$

W. H. Hartford, M. Darrin, *Chem. Rev.* **58**, 1 (1958); H. O. House, *Modern Synthetic Reactions* (W. A. Benjamin, Menlo Park, California, 2nd ed., 1972) p 289; C. D. Nenitzescu *et al., Rev. Roum. Chim.* **14**, 1543, 1553 (1969); I. I. Schiketanz *et al., ibid.* **22**, 1097 (1977); J. C. W. Chien, J. K. Y. Kiang, *Macromolecules* **13**, 280 (1980); F. A. Luzzio, W. J. Moore, *J. Org. Chem.* **58**, 512 (1993).

135. Evans Aldol Reaction
D. A. Evans *et al., J. Am. Chem. Soc.* **101**, 6120 (1979); **103**, 2127 (1981).

Highly enantioselective aldol condensation of the chiral *N*-acyl-oxazolidone via its dibutylboryl enolate with the appropriate aldehyde:

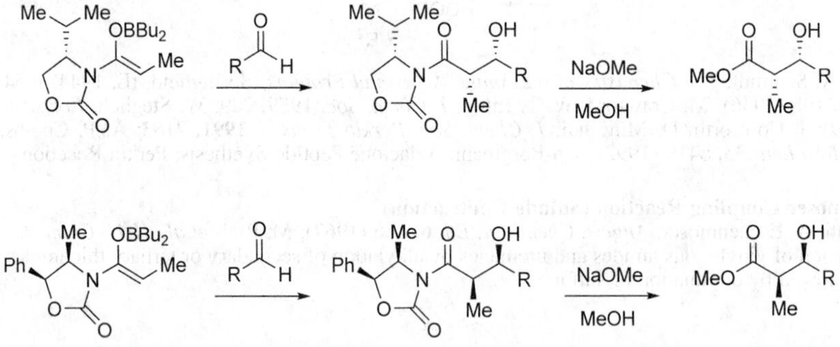

Mechanistic studies: D. A. Evans *et al., J. Am. Chem. Soc.* **103**, 3099 (1981). Synthetic applications: C. W. Phoon, C. Abell, *Tetrahedron Lett.* **39**, 2655 (1998); C. Pearson *et al., ibid.* **40**, 411 (1999). Inversion of product stereochemistry: K. Iseki *et al., ibid.* **34**, 8147 (1993); T. Gabriel, L. Wessjohann, *ibid.* **38**, 4387 (1997). *Review:* D. A. Evans, *Aldrichim. Acta* **15**, 23-32 (1982); B. M. Kim *et al., Comp. Org. Syn.* **2**, 239-275 (1991). *Cf.* Aldol Condensation, Mukaiyama Aldol Reaction.

136. Favorskii-Babayan Synthesis
A. E. Favorskii, *J. Russ. Phys. Chem. Soc.* **37**, 643 (1905); *Chem. Zentralbl.* **1905, II,** 1018; A. Babayan *et al., J. Gen. Chem. USSR* **9**, 1631 (1939).

Synthesis of acetylenic alcohols from ketones and terminal acetylenes in the presence of anhydrous alkali:

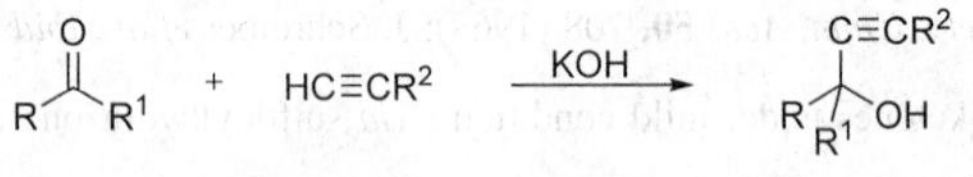

A. W. Johnson, *The Chemistry of Acetylenic Compounds* **vol. 1** (London, 1946) p 14; R. A. Raphael, *Acetylenic Compounds in Organic Synthesis* (New York, 1955) p 10; M. F. Shostakovskii *et al., Zh. Org. Khim.* **4**, 1747 (1968), A. V. Shchelkunov *et al., ibid.* **6**, 930 (1970); E. M. Glazunova *et al., Zh. Org. Khim.* **12**, 516 (1976); Y. M. Vilenchik *et al., ibid.* **14**, 447 (1978). *Cf.* Arens-van Dorp Synthesis; Nef Synthesis.

137. Favorskii Rearrangement; Wallach Degradation
A. E. Favorskii, *J. Prakt. Chem.* **88**(2), 658 (1913); O. Wallach, *Ann.* **414**, 296 (1918).

Base-catalyzed rearrangement of α-haloketones to acids or esters. The rearrangement of α,α′-dibromocyclohexanones to 1-hydroxycyclopentanecarboxylic acids, followed by oxidation to the ketones is known as the **Wallach degradation:**

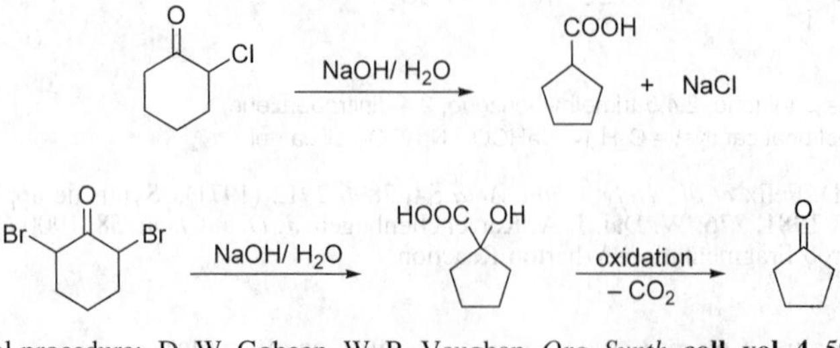

Detailed experimental procedure: D. W. Goheen, W. R. Vaughan, *Org. Synth.* **coll. vol. 4**, 594 (1963). Application to the synthesis of carboxylic acids: T. Satoh *et al., Bull. Chem. Soc. Jpn.* **66**, 2339 (1993). Applications to asymmetric synthesis: *idem et al., Tetrahedron Lett.* **34**, 4823 (1993); E. Lee, C. H. Yoon, *Chem. Commun.* **1994**, 479. *Reviews:*

A. S. Kende, *Org. React.* **11**, 261-316 (1960); P. J. Chenier, *J. Chem. Educ.* **55**, 286 (1978); A. Baretta, B. Waegill, "A Survey of Favorskii Rearrangement Mechanisms" in *Reactive Intermediates*, R. A. Abramovitch, Ed. (Plenum Press, New York, 1982) pp 527-585; J. Mann, *Comp. Org. Syn.* **3**, 839-859 (1991).

138. Feist-Bénary Synthesis

F. Feist, *Ber.* **35**, 1537, 1545 (1902); E. Bénary, *Ber.* **44**, 489, 493 (1911).

Formation of furans from α-halogenated ketones or ethers and 1,3-dicarbonyl compounds in the presence of pyridine. When ammonia is used as the condensing agent, pyrrole derivatives are always formed as secondary products:

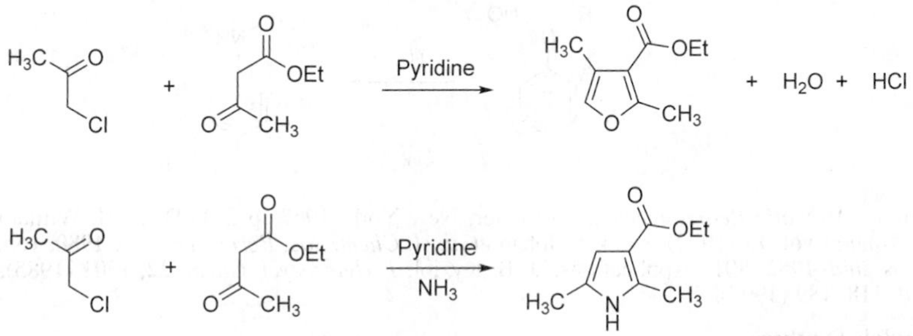

T. Reichstein, H. Zschokke, *Helv. Chim. Acta* **14**, 1270 (1931); **15**, 268, 1105, 1112 (1932); R. C. Elderfield, T. N. Dodd, *Heterocycl. Compd.* **1**, 132 (1950); J. Kagan, K. C. Mattes, *J. Org. Chem.* **45**, 1524 (1980). Alternative substrate: R. C. Cambie *et al.*, *Synth. Commun.* **20**, 1923 (1990). Catalytic, enantioselective "interrupted" reaction: M. A. Calter *et al.*, *J. Am. Chem. Soc.* **127**, 14566 (2005). *Cf.* Hantzsch Pyrrole Synthesis.

139. Fenton Reaction

H. J. H. Fenton, *Proc. Chem. Soc. London* **9**, 113 (1893); *J. Chem. Soc.* **65**, 899 (1894).

Oxidation of α-hydroxy acids with hydrogen peroxide and ferrous salts (Fenton's reagent) to α-keto acids or of 1,2-glycols to hydroxy aldehydes:

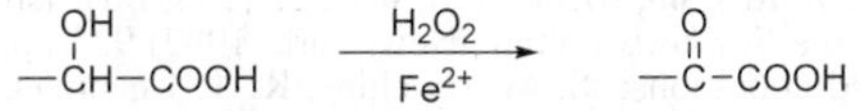

W. A. Waters in *Organic Chemistry* **vol. 4**, H. Gilman, Ed. (Wiley, New York, 1953) p 1157; G. Sosnovsky, D. Rawlinson in *Organic Peroxides* **vol. 2**, D. Swern, Ed. (Interscience, New York, 1970) pp 269-336; C. Walling, *Acc. Chem. Res.* **8**, 125 (1975); T. Tezuka *et al.*, *J. Am. Chem. Soc.* **103**, 3045 (1981); C. Walling, K. Amarnath, *ibid.* **104**, 1185 (1982). Extension to additional substrates: aromatic alcohols: F. J. Benitez *et al.*, *Ind. Eng. Chem. Res.* **38**, 1341 (1999); L. Lunar *et al.*, *Water Res.* **34**, 1791 (2000); *N*-heterocyclics: M. A. Oturan *et al.*, *New J. Chem.* **23**, 793 (1999); E. L. Bier *et al.*, *Environ. Toxicol. Chem.* **18**, 1078 (1999); organometals: K. Banerjee *et al.*, *Environ. Prog.* **18**, 280 (1999).

140. Ferrier Rearrangement

R. J. Ferrier, *J. Chem. Soc. Perkin Trans. 1* **1979**, 1455.

The stereochemically controlled conversion of hex-5-enopyranosides into cyclohexanones (inosose derivatives), catalyzed by mercury(II) salts, such that the 5-hydroxyl and the 3-substituent of the product are predominantly in a *trans* relationship:

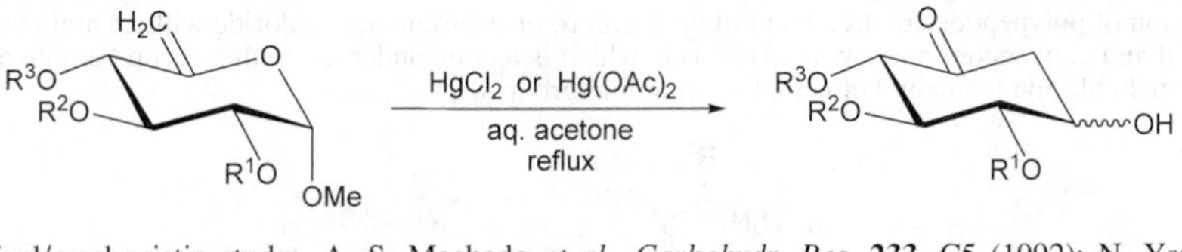

Stereochemical/mechanistic study: A. S. Machado *et al.*, *Carbohydr. Res.* **233**, C5 (1992); N. Yamauchi *et al.*, *Tetrahedron* **50**, 4125 (1994). Scope and limitations: N. Chida *et al.*, *Bull. Chem. Soc. Jpn.* **64**, 2118 (1991). Synthetic applications: D. H. R. Barton *et al.*, *Tetrahedron* **46**, 215 (1990); R. Chretien *et al.*, *Nat. Prod. Lett.* **2**, 69 (1993); A. B. Smith III *et al.*, *Org. Lett.* **1**, 909 (1999); *eidem, ibid.* 913. Modification of catalysis: J. C. López *et al.*, *J. Org. Chem.* **60**, 3851 (1995); T. Linker *et al.*, *Tetrahedron Lett.* **39**, 9637 (1998); B. S. Babu *et al.*, *Synth. Commun.* **29**, 4299 (1999); S. Hotha, A. Tripathi, *Tetrahedron Lett.* **46**, 4555 (2005).

141. Finkelstein Reaction

H. Finkelstein, *Ber.* **43**, 1528 (1910).

Reaction of alkyl halides with sodium iodide in acetone:

$$RBr + NaI \longrightarrow RI + NaBr$$

Organic Name Reactions

C. K. Ingold, *Structure and Mechanism in Organic Chemistry* (Cornell Univ. Press, London, 2nd ed., 1969) p 435; J. Hayami *et al., Tetrahedron Lett.* **1973**, 385; S. Samaan, F. Rolla, *Phosphorus Sulfur* **4**, 145 (1978); W. B. Smith, G. D. Branum, *Tetrahedron Lett.* **22**, 2055 (1981). Modified conditions: D. Landini *et al., J. Chem. Soc. Perkin Trans. 1* **1992**, 2309. Applications: A. J. Pearson, K. Lee, *J. Org. Chem.* **59**, 2304 (1994); A. Schmidt, M. K. Kindermann, *ibid.* **62**, 3910 (1997); T. Zoller *et al., Tetrahedron Lett.* **39**, 8089 (1998).

142. Fischer-Hepp Rearrangement (Nitrosamine Rearrangement)
O. Fischer, E. Hepp, *Ber.* **19**, 2991 (1886).

Rearrangement of secondary aromatic nitrosamines to *p*-nitrosoarylamines:

H. J. Shine, *Aromatic Rearrangements* (Elsevier, New York, 1967) p 231; D. L. H. Williams in *Comprehensive Chemical Kinetics* **vol. 13** (1972) p 454; S. Johan *et al., J. Chem. Soc. Perkin Trans. 2* **1980**, 165. Mechanism: D. L. H. Williams, *ibid.* **1982**, 801. Applications: J. B. Kyziol, *J. Heterocycl. Chem.* **22**, 1301 (1985); P. Kannan *et al., J. Mol. Catal.* **118**, 189 (1997).

143. Fischer Indole Synthesis
E. Fischer, F. Jourdan, *Ber.* **16**, 2241 (1883); E. Fischer, O. Hess, *ibid.* **17**, 559 (1884).

Formation of indoles on heating aryl hydrazones of aldehydes or ketones in the presence of catalysts such as Lewis or proton acids:

Reviews: B. Robinson, *Chem. Rev.* **63**, 373 (1963); **69**, 227 (1969); H. Ishii, *Acc. Chem. Res.* **14**, 233-247 (1981); B. Robinson, *The Fischer Indole Synthesis* (Wiley, New York, 1982) 923 pp.; D. L. Hughes, *Org. Prep. Proced. Int.* **25**, 607-632 (1993). Modified conditions: S. M. Hutchins, K. T. Chapman, *Tetrahedron Lett.* **37**, 4869 (1996); O. Miyata *et al., ibid.* **40**, 3601 (1999); S. Wagaw *et al., J. Am. Chem. Soc.* **121**, 10251 (1999). *Cf.* Borsche-Drechsel Cyclization; Piloty-Robinson Synthesis.

144. Fischer Oxazole Synthesis
E. Fischer, *Ber.* **29**, 205 (1896).

Condensation of equimolar amounts of aldehyde cyanohydrins and aromatic aldehydes in dry ether in the presence of dry hydrochloric acid:

R. H. Wiley, *Chem. Rev.* **37**, 410 (1945); J. W. Cornforth, R. H. Cornforth, *J. Chem. Soc.* **1949**, 1028; J. W. Cornforth, *Heterocycl. Compd.* **5**, 309 (1957); T. Onaka, *Tetrahedron Lett.* **1971**, 4391.

145. Fischer Peptide Synthesis
E. Fischer, *Ber.* **36**, 2982 (1903).

Formation of polypeptides by treatment of an α-chloro or α-bromo acyl chloride with an amino acid ester, hydrolysis to the acid and conversion to a new acid chloride which is again condensed with a second amino acid ester, and so on. The terminal chloride is finally converted to an amino group with ammonia:

(steps repeated)

C. L. A. Schmidt, *The Chemistry of the Amino Acids and Proteins* (Thomas, Springfield, IL, 1944) p 257; B. Rockland in *Amino Acids and Proteins*, D. M. Greenberg, Ed. (Charles C. Thomas, Springfield, IL, 1951) p 232; H. D. Springall, *The Structural Chemistry of Proteins* (New York, 1954) p 24. *Cf.* Bergmann-Zervas Carbobenzoxy Method.

146. Fischer Phenylhydrazine Synthesis
E. Fischer, *Ber.* **8**, 589 (1875).

Formation of arylhydrazines by reduction of diazo compounds with excess sodium sulfite and hydrolysis of the substituted hydrazine sulfonic acid salt with hydrochloric acid. The process is a standard industrial method for production of arylhydrazines:

$$Ar-N_2^+ \xrightarrow{NaSO_3^-} Ar-N=NSO_3Na \xrightarrow{NaHSO_3} \underset{Ar}{\overset{SO_3Na}{\underset{N}{\mid}}}\!NHSO_3Na \xrightarrow{HCl} Ar-NHNH_2$$

G. H. Colemann, *Org. Synth.* **coll. vol. I,** 432 (1932); K. H. Saunders, *The Aromatic Diazo-Compounds and Their Technical Applications* (London, 1949) p 183; R. Huisgen, R. Lux, *Ber.* **93**, 540 (1960).

147. Fischer Phenylhydrazone and Osazone Reaction
E. Fischer, *Ber.* **17**, 579 (1884).

Formation of phenylhydrazones and osazones by heating sugars with phenylhydrazine in dilute acetic acid:

$$\underset{HCOH}{\overset{CHO}{\mid}} \xrightarrow{C_6H_5NHNH_2} \underset{HCOH}{\overset{HC=NNHC_6H_5}{\mid}} \longrightarrow \underset{COH}{\overset{HCNHNC_6H_5}{\parallel}} \xrightarrow{-C_6H_5NH_2}$$

$$\underset{CO}{\overset{HC=NH}{\mid}} \xrightarrow{2\ equiv\ C_6H_5NHNH_2} \underset{C=NNHC_6H_5}{\overset{HC=NNHC_6H_5}{\mid}}$$

E. G. V. Percival, *Adv. Carbohydr. Chem.* **3**, 23 (1948); F. Micheel, *Chemie der Zucker und Polysaccharide* (Leipzig, 1956) p 54; W. Pigman, *The Carbohydrates* **1957**, 452, 455; H. Simon *et al., Fortschr. Chem. Forsch.* **14**, 451 (1970).

148. Fischer-Speier Esterification Method
E. Fischer, A. Speier, *Ber.* **28**, 3252 (1895).

Esterification of acids by refluxing with excess alcohol in the presence of hydrogen chloride or other acid catalysts:

$$CH_3COOH + CH_3CH_2OH \xrightarrow{3\%\ HCl} CH_3COOCH_2CH_3 + H_2O$$

E. D. Hughes, *Q. Rev. Chem. Soc.* **2**, 110 (1948); A. J. Kirby in *Comprehensive Chemical Kinetics* **vol. 10**, C. H. Bamford, C. F. H. Tipper, Eds. (Elsevier, New York, 1972) p 57; E. K. Euranto in *The Chemistry of Carboxylic Acids and Esters*, S. Patai, Ed. (Interscience, New York, 1969) p 505.

149. Fischer-Tropsch Syntheses; Synthol Process; Oxo Synthesis
F. Fischer, H. Tropsch, *Ber.* **56**, 2428 (1923).

Synthesis of hydrocarbons, aliphatic alcohols, aldehydes, and ketones by the catalytic hydrogenation of carbon monoxide using enriched synthesis gas from passage of steam over heated coke. The ratio of products varies with conditions. The high pressure **Synthol process** gives mainly oxygenated products and addition of olefins in the presence of cobalt catalyst, **Oxo synthesis,** produces aldehydes. Normal pressure synthesis leads mainly to petroleum-like hydrocarbons.

C. Masters, *Adv. Organomet. Chem.* **17**, 61 (1979); C. K. Rofer-DePoorter, *Chem. Rev.* **31**, 447 (1981); W. A. Herrmann, *Angew. Chem. Int. Ed.* **21**, 117 (1982). *Reviews:* P. M. Maitlis *et al., Chem. Commun.* **1996**, 1-8; H. Schulz, *Appl. Catal.* **186**, 3-12 (1999). *Cf.* Bergius Process; Oxo Process.

150. Fleming-Tamao Oxidation (Kumada-Fleming-Tamao Oxidation) (Fleming Oxidation) (Tamao-Kumada Oxidation)
K. Tamao *et al., Tetrahedron* **39**, 983 (1983); *idem et al., Organometallics* **2**, 1694 (1983); I. Fleming *et al., J. Chem. Soc. Chem. Commun.* **1984**, 29.

Mild, stereospecific oxidation of a silyl group to give the corresponding hydroxyl group. Fleming and Tamao independently reported that silyl functionalities can serve as masked alcohols. Fleming's reaction sequence requires conversion of the SiMe$_2$Ph group to the corresponding halosilane before oxidation while Tamao's procedure involves direct halosilane oxidation:

oxidation conditions
e.g. (H_2O_2) or
(m-CPBA with either Et_3N or KF)

$$\underset{R^1 \quad R^2}{SiR_3} \xrightarrow{\hspace{2cm}} \underset{R^1 \quad R^2}{OH}$$

$SiR_3 = SiCl_3, SiF_3, Si(OR)_3, SiR(OR)_2,$
$SiMe_2Ar$ where Ar = phenyl, furan
SiR_2X where X = H, Cl, F, OR, NR_2

Reaction substrate studies: K. Tamao *et al., J. Org. Chem.* **48**, 2120 (1983); *eidem, J. Organomet. Chem.* **254**, 13 (1983); K. Tamao, N. Ishida, *ibid.* **269**, C37 (1984); I. Fleming, P. E. J. Sanderson, *Tetrahedron Lett.* **28**, 4229 (1987). Scope expansion studies: S. S. Magar, P. L. Fuchs, *ibid.* **32**, 7513 (1991); K. Itami *et al., J. Org. Chem.* **64**, 8709 (1999); J. D. Sunderhaus *et al., Org. Lett.* **5**, 4571 (2003). *Reviews:* G. R. Jones, Y. Landais, *Tetrahedron* **52**, 7599-7662 (1996); I. Fleming, *Chemtracts Org. Chem.* **9**, 1-64 (1996); K. Tamao in *Advances in Silicon Chemistry* 3, G. L. Larson, Ed. (Jai Press Inc., Greenwich, CT, 1996) pp 1-62.

151. Flood Reaction
E. A. Flood, *J. Am. Chem. Soc.* **55**, 1735 (1933).
Formation of trialkylsilyl halides from hexaalkyldisiloxanes using concentrated sulfuric acid in the presence of ammonium chloride or fluoride, or by treatment of the intermediate silane sulfates with hydrogen chloride in the presence of ammonium sulfate:

$$R_3SiOSiR_3 \xrightarrow[H_2SO_4]{NH_4Cl} R_3SiCl$$

$$R_3SiOSiR_3 \xrightarrow{H_2SO_4} (R_3Si)_2SO_2 \xrightarrow[(NH_4)_2SO_4]{HCl} R_3SiCl$$

H. W. Post, *Silicones and Other Organic Compounds* (New York, 1949) p 64; E. G. Rochow *et al., The Chemistry of Organometallic Compounds* (New York, 1957) p 158, 159. Synthetic applications: L. Birkofer, O. Stuhl, *Top. Curr. Chem.* **88**, 33 (1980).

152. Forster Diazoketone Synthesis
M. O. J. Forster, *J. Chem. Soc.* **107**, 260 (1915).
Formation of diazoketones from α-oximinoketones by reaction with chloramine:

M. P. Cava, R. L. Litle, *Chem. Ind. (London)* **1957**, 367; W. Kirmse *et al., Angew. Chem.* **69**, 106 (1957). Mechanism: J. Meinwald *et al., J. Am. Chem. Soc.* **81**, 4751 (1959). Application to steroids: M. P. Cava, B. R. Vogt, *J. Org. Chem.* **30**, 3776 (1965). Review and applications: F. Weygand, H. J. Bestmann, *Angew. Chem.* **72**, 535 (1960); W. Rundel, *ibid.* **74**, 469 (1962).

153. Forster Reaction
M. O. J. Forster, *J. Chem. Soc.* **75**, 934 (1899); H. Decker, P. Becker, *Ann.* **395**, 362 (1913).
Formation of secondary amines by condensation of a primary amine with an aldehyde, addition of alkyl halide to the Schiff base, and subsequent hydrolysis:

H. Glaser, *Houben-Weyl* **11/1**, 108 (1957); F. Möller, *ibid.* p 956.

154. Franchimont Reaction
A. P. N. Franchimont, *Ber.* **5**, 1048 (1872).
Carboxylic acid dimerization to 1,2-dicarboxylic acids by treating α-bromocarboxylic acids with potassium cyanide followed by hydrolysis and decarboxylation:

N. Zelinsky, *Ber.* **21**, 3160 (1888); O. Poppe, *ibid.* **23**, 113 (1890); R. C. Fuson *et al., J. Am. Chem. Soc.* **51**, 1536 (1929); **52**, 4074 (1930); **60**, 1237 (1938); H. N. Rydon, *J. Chem. Soc.* **1936**, 593; H. Henecka, *Chemie der Beta-Dicarbonylverbindungen* (Berlin, 1950) p 176.

155. Frankland-Duppa Reaction
E. Frankland, *Ann.* **126**, 109 (1863); E. Frankland, B. F. Duppa, *Ann.* **135**, 25 (1865).

Formation of α-hydroxycarboxylic esters by reaction of dialkyl oxalates with alkyl halides in the presence of zinc, or amalgamated zinc, and acid:

$$ROOC-COOR + 2R^1I + 2Zn + 2HCl \longrightarrow R^1R^1C(OH)COOR + ROH + ZnI_2 + ZnCl_2$$

E. Krause, A. von Grosse, *Die Chemie der Metallorganischen Verbindungen* (Berlin, 1937) p 225; K. Nützel, *Houben-Weyl* **13/2a**, 741 (1973).

156. Frankland Synthesis
E. Frankland, *Ann.* **71**, 213 (1849); **85**, 3641 (1853).

Synthesis of zinc dialkyls from alkyl halides and zinc:

$$2C_2H_5I + 2Zn \longrightarrow (C_2H_5)_2Zn + ZnI_2$$

Reviews: K. Nützel, *Houben-Weyl* **13/2a**, 570 (1973); C. R. Noller, *Org. Synth.* **coll. vol. II**, 184 (1943).

157. Freund Reaction; Gustavson Reaction; Hass Cyclopropane Process
A. Freund, *Monatsh. Chem.* **3**, 625 (1882); G. Gustavson, *J. Prakt. Chem.* [2] **36**, 300 (1887); H. B. Hass *et al., Ind. Eng. Chem.* **28**, 1178 (1936).

Formation of alicyclic hydrocarbons by the action of sodium (Freund reaction) or zinc (**Gustavson reaction**) on open chain dihalo compounds; 1,3-dichloropropane derived from the chlorination of propane obtained from natural gas is cyclized in the **Hass cyclopropane process** by treating with zinc dust in aqueous alcohol in the presence of catalytic sodium iodide:

$$Cl\diagdown\diagup\diagdown Cl + Zn \longrightarrow \triangle + ZnCl_2$$

H. Gilman, *Organic Chemistry* **I** (New York, 1943) p 74; J. D. Bartleson *et al., J. Am. Chem. Soc.* **68**, 2513 (1946); R. N. Shortsidge *et al., ibid.* **70**, 946 (1948); B. T. Brooks, *The Chemistry of the Nonbenzenoid Hydrocarbons* (New York, 1950) p 88; H. F. Ebel, A. Lüttringhaus, *Houben-Weyl* **13/1**, 492 (1970).

158. Friedel-Crafts Reaction
C. Friedel, J. M. Crafts, *Compt. Rend.* **84**, 1392, 1450 (1877).

The alkylation or acylation of aromatic compounds catalyzed by aluminum chloride or other Lewis acids:

RCOX = acyl halides, anhydrides
RX = alkyl halides, alkenes, alkynes, alcohols

Reviews: C. C. Price, *Org. React.* **3**, 1 (1946); G. A. Olah, *Friedel-Crafts and Related Reactions* **vol. 1-4** (Interscience, New York, 1963-1965); J. K. Groves, *Chem. Soc. Rev.* **1**, 73 (1972); H. Heaney, *Comp. Org. Syn.* **2**, 733-752, 753-768 (1991); **3**, 293-339. Aliphatic version: S. C. Eyley, *ibid.* **2**, 707-731. Intramolecular reactions: H.-J. Knölker, *Angew. Chem. Int. Ed.* **38**, 2583 (1999); M.-C. P. Yeh *et al., J. Organomet. Chem.* **599**, 128 (2000); C.-L. Kao *et al., Tetrahedron Lett.* **41**, 2207 (2000). Modified conditions: U. Bierman, J. O. Metzger, *Angew. Chem. Int. Ed.* **38**, 3675 (1999). *Cf.* Darzens-Nenitzescu Synthesis of Ketones; Haworth Phenanthrene Synthesis; Nencki Reaction.

159. Friedlaender Synthesis
P. Friedlaender, *Ber.* **15**, 2572 (1882); P. Friedlaender, C. F. Gohring, *ibid.* **16**, 1833 (1883).

Base-catalyzed condensation of 2-aminobenzaldehydes with ketones to form quinoline derivatives:

Reviews: R. H. Manske, *Chem. Rev.* **30**, 124 (1942); C. C. Cheng, S. J. Yan, *Org. React.* **28**, 37 (1982). Cyclic ketones containing S, or N: G. Kempter, S. Hirschberg, *ibid.* **98**, 419 (1965); K. Rao *et al., J. Heterocycl. Chem.* **16**, 1241 (1979). Modified conditions: I.-S. Cho *et al., J. Org. Chem.* **56**, 7288 (1991); G. Sabitha *et al., Synth. Commun.*

29, 4403 (1999). Review: R. P. Thummel, *Synlett* **1992,** 1-12. *Cf.* Niementowski Quinoline Synthesis; Pfitzinger Reaction.

160. Fries Rearrangement
K. Fries, G. Fink, *Ber.* **41,** 4271 (1908); K. Fries, W. Pfaffendorf, *ibid.* **43,** 212 (1910).
Rearrangement of phenolic esters to *o*- and/or *p*-phenolic ketones with Lewis acid catalysts:

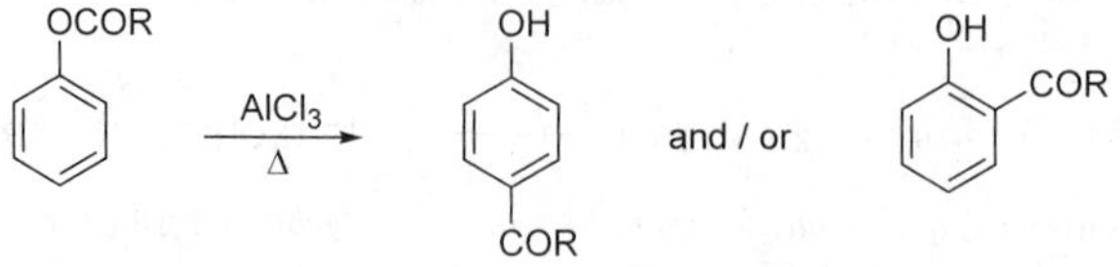

A. H. Blatt, *Org. React.* **1,** 342 (1942); A. Gerecs in *Friedel-Crafts and Related Reactions* in **vol. 3,** Part 1; G. Olah, Ed. (Interscience, New York, 1964) pp 499-533; F. R. Jensen, G. Goldman in *ibid.* Part 2, p 1349; R. Martin *et al., Monatsh. Chem.* **81,** 111 (1980); R. Martin, *Org. Prep. Proced. Int.* **24,** 369 (1992). Photo-rearrangement: J. C. Anderson, C. B. Reese, *Proc. Chem. Soc. London* **1960,** 217; D. Bellus, *Adv. Photochem.* **8,** 109 (1971); D. J. Crouse *et al., J. Org. Chem.* **46,** 374 (1981); W. Gu *et al., J. Am. Chem. Soc.* **121,** 9467 (1999). Modified conditions: K. J. Balkus, Jr. *et al., J. Mol. Catal. A* **134,** 137 (1998); B. Kaboudin, *Tetrahedron* **55,** 12865 (1999); B. M. Khadilkar, V. R. Madyar, *Synth. Commun.* **29,** 1195 (1999).

161. Fritsch-Buttenberg-Wiechell Rearrangement
P. Fritsch, *Ann.* **279,** 319 (1894); W. P. Buttenberg, *ibid.* 327; H. Wiechell, *ibid.* 332.
Carbene-mediated rearrangement of 1,1-diaryl-2-haloethylenes to diaryl acetylenes:

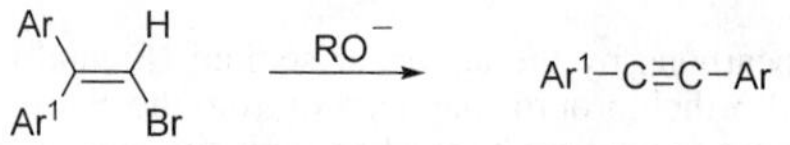

G. Köbrich, *Angew. Chem. Int. Ed.* **4,** 49 (1965); G. Köbrich, P. Buck in *Acetylenes,* H. G. Viehe, Ed. (Marcel Dekker, New York, 1969) pp 117, 131; G. Köbrich, *Angew. Chem. Int. Ed.* **11,** 473 (1972); P. J. Stang, D. P. Fox, *J. Org. Chem.* **43,** 364 (1978). Synthetic applications: V. Mouriès *et al., Synthesis* **1998,** 271; I. Creton *et al., Tetrahedron Lett.* **40,** 1899 (1999). Substituent effects: T. Kawase *et al., Chem. Lett.* **1995,** 499; H. Rezaei *et al., Org. Lett.* **2,** 419 (2000).

162. Fujimoto-Belleau Reaction
C. I. Fujimoto, *J. Am. Chem. Soc.* **73,** 1856 (1951); B. Belleau, *ibid.* 5441.
Synthesis of cyclic α-substituted α,β-unsaturated ketones from enol lactones and Grignard reagents prepared from primary halides:

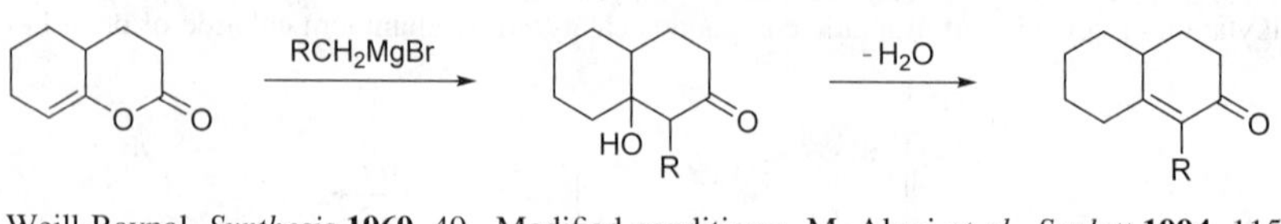

Review: J. Weill-Raynal, *Synthesis* **1969,** 49. Modified conditions: M. Aloui *et al., Synlett* **1994,** 115.

163. Fukuyama Reduction; Fukuyama Coupling Reaction
T. Fukuyama *et al., J. Am. Chem. Soc.* **112,** 7050 (1990).
Synthesis of a multifunctional aldehyde through a palladium-catalyzed thioester reduction with triethylsilane:

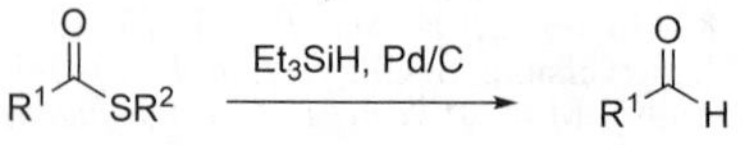

Synthesis of α-amino aldehydes: H. Tokuyama *et al., Synthesis* **2002,** 1121. Mechanistic studies: M. Kimura, M. Seki, *Tetrahedron Lett.* **45,** 3219 (2004). *Comprehensive review:* T. Fukuyama, H. Tokuyama, *Aldrichim. Acta* **37,** 87-96 (2004).
The preparation of a functionalized ketone by the treatment of a thioester with an organozinc reagent in the presence of a palladium-based catalyst is known as the **Fukuyama Coupling Reaction:**

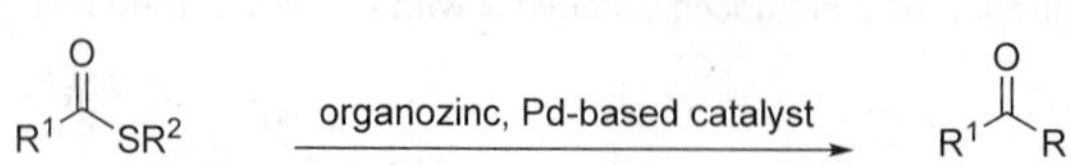

organozinc = $RZnI$, ZnR_2

Pd-based catalyst = $PdCl_2(PPh)_3$, Pd/C, $Pd(OH)_2$/C, $Pd(OAc)_2$

Ketone preparation: H. Tokuyama *et al.*, *Tetrahedron Lett.* **39**, 3189 (1998); *eidem, J. Braz. Chem. Soc.* **9**, 381 (1998). Catalytic procedures with palladium: T. Shimizu, M. Seki, *Tetrahedron Lett.* **42**, 429 (2001); with palladium hydroxide: Y. Mori, M. Seki, *J. Org. Chem.* **68**, 1571 (2003); with palladium acetate: *eidem, Synlett* **2005**, 2233. Reaction method utilizing dialkylzincs: *eidem, Tetrahedron Lett.* **45**, 7343 (2004). Study of reaction conditions and mechanisms: *eidem, Adv. Synth. Catal.* **349**, 2027 (2007). *Cf.* Suzuki Coupling; Heck Reaction; Sonogashira Coupling; Stille Coupling; Negishi Coupling.

164. Gabriel-Colman Rearrangement (Phthalimidoacetic Ester → Isoquinoline Rearrangement) (Gabriel Isoquinoline Synthesis)

S. Gabriel, J. Colman, *Ber.* **33**, 980, 996, 2630 (1900); **35**, 2421 (1902).

Formation of isoquinoline derivatives or substituted benzothiazines by the action of alkoxides on phthalimidoacetic or saccharin esters or ketones:

C. F. H. Allen, *Chem. Rev.* **47**, 284 (1950); H. Henecka, *Houben-Weyl* **8**, 578 (1952); J. H. M. Hill, *J. Org. Chem.* **30**, 620 (1965); W. C. Groutas *et al., Biochem. Biophys. Res. Commun.* **194**, 1491 (1993); *idem et al., Bioorg. Med. Chem.* **3**, 187 (1995); S.-K. Kwon, *J. Korean Chem. Soc.* **40**, 678 (1996). Mechanism: M. T. Ivery, J. E. Gready, *J. Chem. Res. Synop.* **9**, 349 (1993). *Cf.* Dieckmann Reaction.

165. Gabriel Ethylenimine Method (Gabriel-Marckwald Ethylenimine Synthesis)

S. Gabriel *et al., Ber.* **21**, 1049 (1888); W. Marckwald *et al., ibid.* **32**, 2036 (1899); **33**, 764 (1900); **34**, 3544 (1901).

Formation of ethylenimines (aziridines) by elimination of hydrogen halides from aliphatic vicinal haloamines with alkali. The method can be extended to the preparation of five- and six-membered cyclic amines:

O. C. Dermer, G. E. Ham, *Ethylenimine and Other Aziridines* (Academic Press, New York, 1969) pp 1-59; R. Bartnik *et al., Pol. J. Chem.* **53**, 537 (1979); K. H. Sunwoo *et al., Dyes Pigm.* **41**, 19 (1999).

166. Gabriel Synthesis

S. Gabriel, *Ber.* **20**, 2224 (1887).

Conversion of alkyl halides to primary amines by treatment with potassium phthalimide and subsequent hydrolysis:

M. S. Gibson, R. W. Bradshaw, *Angew. Chem. Int. Ed.* **7**, 919 (1968); B. Dietrich *et al., J. Am. Chem. Soc.* **103**, 1282 (1981); O. Mitsunobu, *Comp. Org. Syn.* **6**, 79-85 (1991). Modified conditions: S. E. Sen, S. L. Roach, *Synthesis* **1994**, 756; M. N. Khan, *J. Org. Chem.* **61**, 8063 (1996). Stereoselectivity: A. Kubo *et al., Tetrahedron Lett.* **37**, 4957 (1996).

167. Gattermann Aldehyde Synthesis

L. Gattermann, *Ber.* **31**, 1149 (1898); *Ann.* **313**, (1907).

Preparation of phenolic aldehydes, phenol ethers or heterocyclic compounds by treatment of the aromatic substrate with hydrogen cyanide and hydrogen chloride in the presence of Lewis acid catalysts:

W. E. Truce, *Org. React.* **9**, 37 (1957); E. Baltazzi, L. I. Krimen, *Chem. Rev.* **63**, 526 (1963); F. M. Aslam *et al., J. Chem. Soc. Perkin Trans. 1* **1972**, 892; Y. Sato *et al., J. Am. Chem. Soc.* **117**, 3037 (1995). *Cf.* Houben-Hoesch Reaction; Reimer-Tiemann Reaction.

168. Gattermann-Koch Reaction

L. Gattermann, J. A. Koch, *Ber.* **30**, 1622 (1897); L. Gattermann, *Ann.* **347**, 347 (1906).

Formylation of benzene, alkylbenzenes or polycyclic aromatic hydrocarbons with carbon monoxide and hydrogen chloride in the presence of aluminum chloride at high pressure. Addition of cuprous chloride allows the reaction to proceed at atmospheric pressure:

$$
\text{C}_6\text{H}_6 \ + \ \text{CO} \ + \ \text{HCl} \xrightarrow[\text{Cu}_2\text{Cl}_2]{\text{AlCl}_3} \text{C}_6\text{H}_5\text{CHO}
$$

N. N. Crounse, *Org. React.* **5**, 290 (1949); G. A. Olah, S. J. Kuhn in *Friedel-Crafts and Related Reactions* vol. **3**, Part 2, G. Olah, Ed. (Interscience, New York, 1964) pp 1153-1156. Use of CuCl(PPh$_3$)$_n$: L. Toniolo, M. Graziani, *J. Organomet. Chem.* **194**, 221 (1980).

169. Glaser Coupling; Eglinton Reaction; Cadiot-Chodkiewicz Coupling

C. Glaser, *Ber.* **2**, 422 (1869).

Homocoupling of terminal alkynes catalyzed by cuprous salts in the presence of an oxidant and ammonium chloride:

$$
\text{R-C}\equiv\text{CH} \xrightarrow[\text{NH}_4\text{Cl}]{\text{CuCl, O}_2} \text{R-C}\equiv\text{C-C}\equiv\text{C-R}
$$

Heterocoupling may be accomplished *via* the **Cadiot-Chodkiewicz coupling** of terminal alkynes with haloalkynes, catalyzed by cuprous salts in the presence of aliphatic amines:

$$
\text{HC-C}\equiv\text{CH} \ + \ \text{Br-C}\equiv\text{C-R}^1 \xrightarrow[\substack{\text{CH}_3\text{CH}_2\text{NH}_2 \\ \text{NH}_2\text{OH}\cdot\text{HCl}}]{\text{CuCl}} \text{R-C}\equiv\text{C-C}\equiv\text{C-R}^1
$$

Synthetic applications: F. M. Menger *et al., J. Am. Chem. Soc.* **115**, 6600 (1993); L. Guo *et al., Chem. Commun.* **1994**, 243.

This coupling may also be effected by cupric salts in pyridine and is often referred to as the **Eglinton reaction.** It is particularly applicable to cyclizations: G. Eglinton, A. R. Galbraith, *Chem. Ind. (London)* **1956**, 737; N. Hébert *et al., J. Org. Chem.* **57**, 1777 (1992).

W. Chodkiewicz *et al., Compt. Rend.* **245**, 322 (1957); B. N. Ghose, *Synth. React. Inorg. Met.-Org. Chem.* **24**, 29 (1994); with supercritical CO$_2$ as solvent: J. Li, H. Jiang, *Chem. Commun.* **1999**, 2369.

Inclusive reviews: P. Cadiot, W. Chodkiewicz, "Couplings of Acetylenes" in *Chemistry of Acetylenes*, H. G. Viehe, Ed. (Marcel Dekker, New York, 1969) pp 597-647; K. Sonogashira, *Comp. Org. Syn.* **3**, 551-561 (1991). *Cf.* Castro-Stephens Coupling; Ullmann Reaction.

170. Gomberg-Bachmann Reaction

M. Gomberg, W. E. Bachmann, *J. Am. Chem. Soc.* **46**, 2339 (1924).

Alkali dependent formation of diaryl compounds from aryl diazonium salts and aromatic compounds:

$$
\underset{\text{X}^-}{\text{Ar-N}\overset{+}{\equiv}\text{N}} \xrightarrow{\text{HO}^-} \text{Ar-N=N-OH} \xrightarrow{-\text{N}_2} \text{Ar}^\bullet \ + \ \text{HO}^\bullet
$$

$$
\text{Ar}^\bullet \ + \ \text{C}_6\text{H}_6 \longrightarrow \underset{\text{H}}{\overset{\text{Ar}}{\bigcirc}} \xrightarrow[-\text{H}_2\text{O}]{\text{HO}^\bullet} \overset{\text{Ar}}{\bigcirc}
$$

W. E. Bachmann, R. A. Hoffman, *Org. React.* **2**, 224 (1944); O. C. Dermer, M. T. Edmison, *Chem. Rev.* **57**, 77 (1957); D. H. Hey, *Adv. Free-Radical Chem.* **2**, 47 (1966); D. E. Rosenberg, *et al., Tetrahedron Lett.* **21**, 4141 (1980); J. R. Beadle *et al., J. Org. Chem.* **49**, 1594 (1984); T. C. McKenzie, S. M. Rolfes, *J. Heterocycl. Chem.* **24**, 859 (1987); M. Gurczynski, P. Tomasik, *Org. Prep. Proced. Int.* **23**, 438 (1991). For intramolecular version, *see* Pschorr Reaction.

171. Gomberg Free Radical Reaction

M. Gomberg, *J. Am. Chem. Soc.* **22**, 757 (1900).

Formation of free radicals by abstraction of the halogen from triarylmethyl halides with metals:

$$
2(\text{C}_6\text{H}_5)_3\text{CCl} \ + \ \text{Zn} \longrightarrow 2(\text{C}_6\text{H}_5)_3\text{C}\overset{\bullet-}{-}\text{Cl} \longrightarrow 2(\text{C}_6\text{H}_5)_3\text{C}^\bullet + \ \text{ZnCl}_2
$$

A. R. Forrester *et al.*, in *Organic Chemistry of Stable Free Radicals* (Academic Press, New York, 1968); Scholle, Rozantsev, *Russ. Chem. Rev.* **42**, 1101 (1973); J. M. McBride, *Tetrahedron* **30**, 2009 (1974).

172. Gould-Jacobs Reaction

R. G. Gould, W. A. Jacobs, *J. Am. Chem. Soc.* **61**, 2890 (1939).

Synthesis of 4-hydroxyquinolines from anilines and diethyl ethoxymalonate *via* cyclization of the intermediate anilinomethylenemalonate followed by hydrolysis and decarboxylation:

R. H. Reitsema, *Chem. Rev.* **43**, 53 (1948); R. C. Elderfield, *Heterocycl. Compd.* **4**, 38 (1952); C. C. Price, R. N. Roberts, *Org. Synth.* **coll. vol. III**, 272 (New York, 1955); D. G. Markees, L. S. Schwab, *Helv. Chim. Acta* **55**, 1319 (1972); R. Albrecht, G. A. Hoyer, *Ber.* **105**, 3118 (1972); J. M. Barker *et al., J. Chem. Res. Synop.* **1980**, 4; A. Pipaud *et al., Synth. Commun.* **27**, 1727 (1997); C. G. Dave, R. D. Shah, *Heterocycles* **51**, 1819 (1999). *Cf.* Doebner-Miller Reaction; Knorr Quinoline Synthesis.

173. Graebe-Ullmann Synthesis

C. Graebe, F. Ullmann, *Ann.* **291**, 16 (1896); F. Ullmann, *ibid.* **332**, 82 (1904).

Formation of carbazoles by the action of nitrous acid on 2-aminodiphenylamines, followed by thermal decomposition of the resulting benzotriazoles:

O. Bremer, *Ann.* **514**, 279 (1934); S. H. Tucker *et al., J. Chem. Soc.* **1942**, 500; N. Campbell, B. Barclay, *Chem. Rev.* **40**, 360 (1947); C. C. Colser *et al., J. Chem. Soc.* **1951**, 110; B. W. Ashton, H. Suschitzky, *ibid.* **1957**, 4559; R. A. Abramovitch, I. D. Spenser, *Adv. Heterocycl. Chem.* **3**, 128 (1964). Photo-decomposition: L. K. Mehta *et al., J. Chem. Soc. Perkin Trans. 1* **1993**, 1261. Synthetic applications: A. Molina *et al., J. Org. Chem.* **61**, 5587 (1996); D. J. Hagan *et al., J. Chem. Soc. Perkin Trans. 1* **1998**, 915.

174. Griess Diazo Reaction (Witt and Knoevenagel Diazotization Methods)

P. Griess, *Ann.* **106**, 123 (1858); **121**, 257 (1862); E. Knoevenagel, *Ber.* **23**, 2994 (1890); O. N. Witt, *ibid.* **42**, 2953 (1909).

Formation of aromatic diazonium salts from primary aromatic amines and nitrous acid or other nitrosating agents:

$$ArNH_2 + NaNO_2 + 2HX \longrightarrow ArN_2^+ X^- + NaX + 2H_2O$$

$$2ArNH_2 + N_2O_3 + 2HNO_3 + H_2O \longrightarrow 2ArN_2^+NO_3^- + 4H_2O \text{ (Griess reaction)}$$

$$2ArNH_2 + Na_2S_2O_5 + 4HNO_3 \longrightarrow 2ArN_2^+NO_3^- + Na_2S_2O_7 + 4H_2O \text{ (Witt method)}$$

$$ArNH_2 + RONO + HX \longrightarrow ArN_2^+ X^- + ROH + H_2O \text{ (Knoevenagel method)}$$

N. Kornblum, *Org. React.* **2**, 264 (1944); W. A. Cowdry, D. S. Davies, *Q. Rev. Chem. Soc.* **6**, 358 (1952); Ridd, *ibid.* **15**, 418 (1961); B. I. Belov, V. V. Kozlov, *Russ. Chem. Rev.* **32**, 59 (1963); K. Schank in *The Chemistry of Diazonium and Diazo Groups*, S. Patai, Ed. (Wiley, New York, 1978) p 645; J. B. Fox, Jr., *Anal. Chem.* **51**, 1493 (1979). Evaluation in determination of biological nitrogen: I. Guevara *et al., Clin. Chim. Acta* **274**, 177 (1998); K. Schulz *et al., Nitric Oxide: Biology & Chemistry* **3**, 225 (1999).

175. Griffith-Ley TPAP Oxidation (Ley-Griffith Oxidation) (TPAP Oxidation)

W. P. Griffith *et al., J. Chem. Soc. Chem. Commun.* **1987**, 1625.

Mild oxidation of primary and secondary alcohols to aldehydes and ketones, respectively, through the use of catalytic tetrapropylammonium perruthenate (TPAP) and the co-oxidant, *N*-methylmorpholine *N*-oxide (NMO). Other functional groups are typically unaffected by the reaction conditions:

Kinetic autocatalytic studies: D. G. Lee *et al.*, *J. Org. Chem.* **57**, 3276 (1992). Reaction condition modifications involving molecular oxygen: I. E. Markó *et al.*, *J. Am. Chem. Soc.* **119**, 12661 (1997); R. Lenz, S. V. Ley, *J. Chem. Soc. Perkin Trans. 1* **1997**, 3291; and polymer-supported perruthenate: B. Hinzen *et al.*, *Synthesis* **1998**, 977; involving polymer supported NMO: D. S. Brown *et al.*, *Synlett* **2001**, 1257. *Reviews*: W. P. Griffith, S. P. Ley, *Aldrichim. Acta* **23**, 13-19 (1990); S. V. Ley *et al.*, *Synthesis* **1994**, 639-666; P. Langer, *J. Prakt. Chem.* **342**, 728-730 (2000). *Cf.* Dess-Martin Oxidation; Swern Oxidation; Corey-Kim Oxidation; Pfitzner-Moffatt Oxidation. *See* monographs: Tetrapropylammonium Perruthenate, *N*-Methylmorpholine *N*-Oxide.

176. Grignard Degradation
W. Steinkopf *et al.*, *Ann.* **512**, 136 (1934); **543**, 128 (1940).

Stepwise dehalogenation of a polyhalo compound through its Grignard reagent which on treatment with water yields a product containing one halogen atom less:

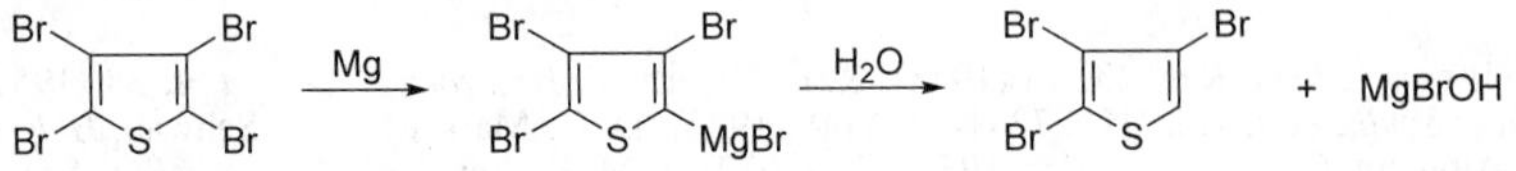

V. Grignard, *Compt. Rend.* **130**, 1322 (1900); F. F. Blicke, *Heterocycl. Compd.* **1**, 222 (1950); K. Nützel, *Houben-Weyl* **13/2a**, 128 (1973).

177. Grignard Reaction
V. Grignard, *Compt. Rend.* **130**, 1322 (1900).

Addition of organomagnesium compounds (Grignard reagents) to carbonyl compounds to generate alcohols. A more modern interpretation extends the scope of the reaction to include the addition of Grignard reagents to a wide variety of electrophilic substrates:

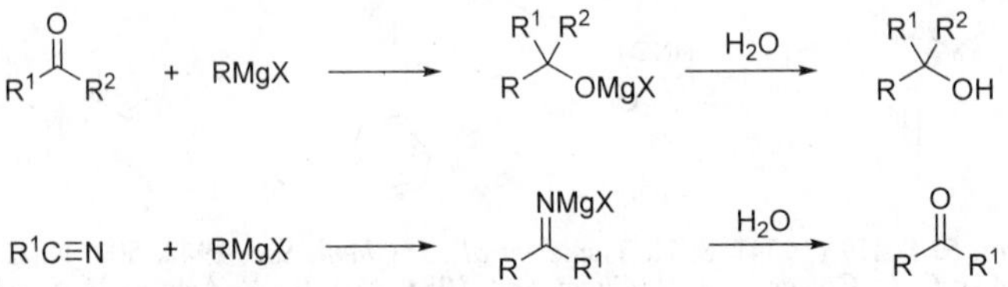

Early review: D. A. Shirley, *Org. React.* **8**, 28-58 (1954). Preparation of Grignard reagents: Y. H. Lai, *Synthesis* **1981**, 585-604. Mechanistic study: K. Maruyama, T. Katagiri, *J. Phys. Org. Chem.* **2**, 205 (1989). Review of stereoselective addition of carbonyl compounds: D. M. Huryn, *Comp. Org. Syn.* **1**, 49-75 (1991). General review: G. S. Silverman, P. E. Rakita in *Kirk-Othmer Encyclopedia of Chemical Technology* **vol. 12** (Wiley-Interscience, New York, 4th ed., 1994) pp 768-786. *Cf.* Barbier(-type) Reaction.

178. Grob Fragmentation
C. A. Grob, W. Baumann, *Helv. Chim. Acta* **38**, 594 (1955).

Carbon-carbon bond cleavage primarily *via* a concerted process involving a five atom system:

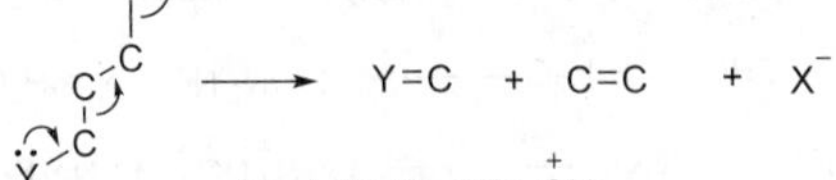

X = I, Br, Cl, OTs, $\overset{+}{O}H_2$
Y = NR$_2$, O$^-$

The intramolecular version is useful for the preparation of medium-size rings:

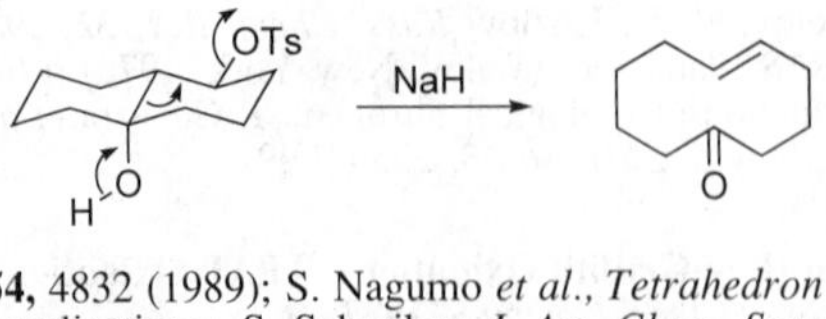

M. Ochiai *et al.*, *J. Org. Chem.* **54**, 4832 (1989); S. Nagumo *et al.*, *Tetrahedron* **49**, 10501 (1993); J.-J. Wang *et al.* *ibid.* **54**, 13149 (1998). Synthetic applications: S. Schreiber, *J. Am. Chem. Soc.* **102**, 6163 (1980); J. Boivin *et al.*, *Tetrahedron Lett.* **40**, 9239 (1999); A. Krief *et al.*, *ibid.* **41**, 3871 (2000). *Reviews:* C. A. Grob, *Angew. Chem. Int. Ed.* **8**, 535-546 (1969); P. Weyerstahl, H. Marschall, *Comp. Org. Syn.* **6**, 1044-1065 (1991). *Cf.* Eschenmoser Fragmentation; Wharton Reaction.

179. Grundmann Aldehyde Synthesis
C. Grundmann, *Ann.* **524**, 31 (1936).

Transformation of an acid into an aldehyde of the same chain length by conversion of the acid chloride, *via* the diazo ketone, to the acetoxy ketone, reduction with aluminum isopropoxide and hydrolysis to the glycol, and cleavage with lead tetraacetete:

$$RCOCl \xrightarrow{CH_2N_2} RCOCHN_2 \xrightarrow{AcOH} RCOCH_2OAc$$

$$\xrightarrow[H_2O]{Al(O\text{-}i\ Pr)_3} RCHOHCH_2OH \xrightarrow{Pb(OAc)_4} RCHO$$

E. Mosetting, *Org. React.* **8**, 225 (1954); O. Bayer, *Houben-Weyl* **7/1**, 239 (1954); H. K. Mangold, *J. Org. Chem.* **24**, 405 (1959). *Cf.* Sonn-Müller Method; Stephen Aldehyde Synthesis.

180. Guareschi-Thorpe Condensation
I. Guareschi, *Mem. Reale Accad. Sci. Torino* **II**, 46, 7, 11, 25 (1896); H. Baron, *et al., J. Chem. Soc.* **85**, 1726 (1904).
Synthesis of pyridine derivatives by condensation of cyanoacetic ester with acetoacetic ester in the presence of ammonia. In a second type of synthesis a mixture of cyanoacetic ester and a ketone is treated with alcoholic ammonia:

C. Hollins, *The Synthesis of Nitrogen Ring Compounds* (New York, 1924) p 197; V. Migrdichian, *The Chemistry of Organic Cyanogen Compounds* (New York, 1947) p 322; H. S. Mosher, *Heterocycl. Compd.* **1**, 466 (1950); R. W. Holder *et al., J. Org. Chem.* **47**, 1445 (1982); D. J. Collins, A. M. James, *Aust. J. Chem.* **42**, 215 (1989). *Cf.* Hantzsch (Dihydro)Pyridine Synthesis; Kröhnke Pyridine Synthesis.

181. Guerbet Reaction
M. Guerbet, *Compt. Rend.* **128**, 511 (1899).
Condensation of 1° or 2° alcohols at high temperature and pressure in the presence of alkali metal hydroxide or alkoxide by a dehydrogenation, aldol condensation, *q.v.,* and hydrogenation sequence:

$$2\ RCH_2CH_2OH \xrightarrow{NaOEt} 2\ RCH_2CHO \longrightarrow RCH_2CHOHCHRCHO \xrightarrow{-H_2O}$$

$$RCH_2CH=CRCHO \xrightarrow{H_2} RCH_2CH_2CHRCH_2OH$$

H. Machemer, *Angew. Chem.* **64**, 213 (1952); S. Veibel, J. T. Nielsen, *Tetrahedron* **23**, 1723 (1967); G. Gregorio *et al., J. Organomet. Chem.* **37**, 385 (1972); E. Klein, *et al., Ann.* **1973**, 1004. Rhodium-promoted reaction: P. L. Burk *et al., J. Mol. Catal.* **33**, 1 (1985).

182. Gutknecht Pyrazine Synthesis
H. Gutknecht, *Ber.* **12**, 2290 (1879); **13**, 1116 (1880).
Cyclization of α-amino ketones, produced by reduction of isonitroso ketones to yield the dihydropyrazines which are dehydrogenated with mercury(I) oxide or copper(II) sulfate, or sometimes with atmospheric oxygen:

$$2RCOCH_2R^1 \xrightarrow{2HNO_2} 2RCOC(=NOH)R^1 \xrightarrow{4H_2} 2RCOCH(NH_2)R^1 \longrightarrow$$

I. J. Krems, P. E. Spoerri, *Chem. Rev.* **40**, 291 (1947); Y. T. Pratt, *Heterocycl. Compd.* **6**, 379, 385 (1957).

183. Hajos-Parrish Reaction (Hajos-Parrish-Wiechert Reaction) (Hajos-Parrish-Eder-Sauer-Wiechert Reaction)
GB 1325631 (1971 to Hoffmann-La Roche); Z. G. Hajos, D. R. Parrish, *J. Org. Chem.* **39**, 1615 (1974); **GB 1352637** (1971 to Schering); U. Eder *et al., Angew. Chem. Int. Ed.* **10**, 496 (1971).
Proline-catalyzed, intramolecular aldol reaction of a cyclopentane or cyclohexane-1,3-dione. The resulting bicyclic diketone can be further dehydrated under acidic conditions to yield the corresponding enedione:

n = 1, 2

Organic Name Reactions

Enantioselective organocatalysis studies with aldolase antibodies: G. Zhong *et al.*, *J. Am. Chem. Soc.* **119**, 8131 (1997); with phenylalanine: S. Danishefsky, P. Cain, *ibid.* **98**, 4975 (1976); with β-amino acids: S. G. Davies *et al.*, *Org. Biomol. Chem.* **5**, 3190 (2007). Single step, proline-catalyzed reaction from 1,3-dione precursors: T. Bui, C. F. Barbas III, *Tetrahedron Lett.* **41**, 6951 (2000). Theoretical and mechanistic studies: C. Allemann *et al.*, *Acc. Chem. Res.* **37**, 558 (2004). Review of proline-catalyzed asymmetric reactions: B. List, *Tetrahedron* **58**, 5573-5590 (2002). *Cf.* Aldol Reaction, Robinson Annulation.

184. Haller-Bauer Reaction
A. Haller, E. Bauer, *Compt. Rend.* **148,** 70, 127 (1909); **149,** 5 (1909).
Cleavage of non-enolizable ketones with sodium amide; frequently applied to formation of trisubstituted acetic acid:

K. E. Hamlin, A. W. Weston, *Org. React.* **9,** 1 (1957); H. M. Walborsky *et al.*, *J. Org. Chem.* **36,** 2937 (1971); E. M. Kaiser, C. O. Warner, *Synthesis* **1975,** 395. Applications: G. Mehta, M. Praveen, *J. Org. Chem.* **60,** 279 (1995); *idem et al., Tetrahedron Lett.* **37,** 2289 (1996); A. Mittra *et al., J. Org. Chem.* **63,** 9555 (1998). Reviews and extension to amide formation: J. P. Gilday, L. A. Paquette, *Org. Prep. Proced. Int.* **22,** 167-201 (1990); G. Mahta, R. V. Venkateswaran, *Tetrahedron* **56,** 1399 (2000).

185. Hammick Reaction
P. Dyson, D. L. Hammick, *J. Chem. Soc.* **1937,** 1724.
Decarboxylation of α-picolinic or related acids in the presence of carbonyl compounds accompanied by the formation of a new carbon-carbon bond:

D. L. Hammick *et al., J. Chem. Soc.* **1939,** 809; **1949,** 659; N. H. Cantwell, E. V. Brown, *J. Am. Chem. Soc.* **75,** 1489 (1953); M. J. Betts, B. R. Brown, *J. Chem. Soc.* **1967,** 1730; E. V. Brown, M. B. Shambhu, *J. Org. Chem.* **36,** 2002 (1971). Effect of conditions on yield and products: V. P. Karandikar *et al., Indian J. Technol.* **23,** 28 (1985). Mechanism: R. Grigg *et al., J. Chem. Soc. Perkin Trans. 2* **1990,** 51; B. Bohn *et al., Heterocycles* **37,** 1731 (1994).

186. Hantzsch Dihydropyridine Synthesis (Pyridine Synthesis)
A. Hantzsch, *Ann.* **215,** 1, 72 (1882); *Ber.* **18,** 1744 (1885); **19,** 289 (1886).
Synthesis of dihydropyridines by condensation of two moles of a β-dicarbonyl compound with one mole of an aldehyde in the presence of ammonia. Dehydrogenation to the corresponding pyridine is accomplished with an oxidizing agent:

Note: if R at C-4 is benzyl then during oxidation cleavage will occur

H. S. Mosher, *Heterocycl. Compd.* **1,** 462 (1950); R. M. Kellog *et al., J. Org. Chem.* **45,** 2854 (1980); Y. Watanabe *et al., Synthesis* **1983,** 761. Mechanistic study: A. R. Katritzky *et al., Tetrahedron* **42,** 5729 (1986); **43,** 5171 (1987). Extension to the synthesis of unsymmetrical dihydropyridines: J. B. Sainani *et al., Indian J. Chem.* **34B,** 17 (1995); S. Visentin *et al., J. Med. Chem.* **42,** 1422 (1999). *Cf.* Chichibabin Pyridine Synthesis; Guareschi-Thorpe Condensation; Kröhnke Pyridine Synthesis.

187. Hantzsch Pyrrole Synthesis
A. Hantzsch, *Ber.* **23,** 1474 (1890).
Formation of pyrrole derivatives from α-chloromethyl ketones, β-keto esters and ammonia or amines:

R. Elderfield, T. N. Dodd, Jr., *Heterocycl. Compd.* **1**, 132 (1950); A. H. Corwin, *ibid.* 290; M. W. Roomi, S. F. MacDonald, *Can. J. Chem.* **48**, 1689 (1970); K. Kirschke *et al., J. Prakt. Chem.* **332**, 143 (1990); A. W. Trautwein *et al., Bioorg. Med. Chem. Lett.* **8**, 2381 (1998). *Cf.* Feist-Bénary Synthesis; Knorr Pyrrole Synthesis; Paal-Knorr Pyrrole Synthesis.

188. Harries Ozonide Reaction (Ozonolysis)
C. Harries, *Ann.* **343**, 311 (1905).

Treatment of olefins with ozone as a method of cleaving olefinic linkages. On hydrolysis or catalytic hydrogenation the initially formed ozonide yields two molecules of carbonyl compounds:

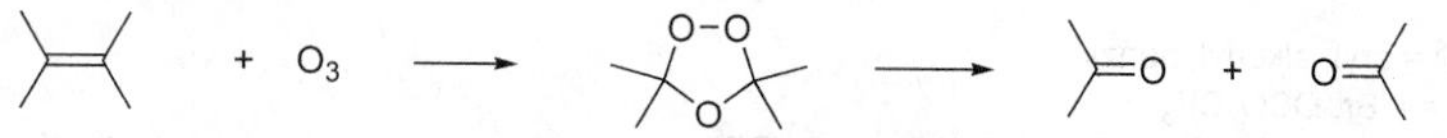

Reviews: P. S. Bailey, *Chem. Rev.* **58**, 925 (1958); L. J. Chinn, *Selection of Oxidants in Synthesis: Oxidation at the Carbon Atom* (Dekker, New York, 1971) pp 151-160; P. S. Bailey, *Ozonation in Organic Chemistry* **vols. 1 and 2** (Academic Press, New York, 1978, 1982). Mechanism: R. Criegee, *Rec. Chem. Prog.* **18**, 111 (1957); R. W. Murray, *Acc. Chem. Res.* **1**, 313 (1968); M. Miura *et al., J. Org. Chem.* **50**, 1504 (1985). Applications: J. Z. Gillies *et al., J. Am. Chem. Soc.* **110**, 7991 (1988); K. Griesbaum, V. Ball, *Tetrahedron Lett.* **35**, 1163 (1994).

189. Haworth Methylation
W. N. Haworth, *J. Chem. Soc.* **107**, 13 (1915).

Formation of methylated methyl glycosides from monosaccharides with dimethyl sulfate and 30% sodium hydroxide. The glycosidic methyl group is hydrolyzed with acid to yield the free methylated sugar:

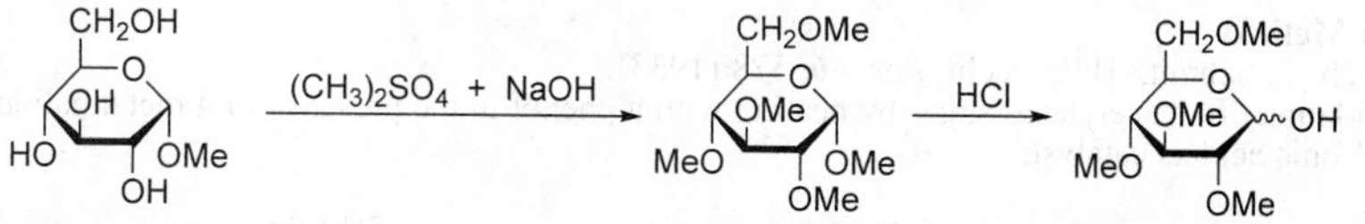

W. N. Haworth, H. Machemer, *J. Chem. Soc.* **1932**, 2270; C. C. Barker *et al., ibid.* **1946**, 783; E. J. Bourne, S. Peat, *Adv. Carbohydr. Chem.* **5**, 146 (1950); W. Pigman, *The Carbohydrates* **1957**, 369. *Cf.* Purdie Methylation.

190. Haworth Phenanthrene Synthesis
R. D. Haworth, *J. Chem. Soc.* **1932**, 1125, 2717, *idem et al., ibid.* 1784, 2248, 2720; **1934**, 454.

Preparation of phenanthrenes from naphthalenes *via* a series of steps including a Friedel-Crafts acylation and two Clemmensen or Wolff-Kishner reductions, *q.q.v.*:

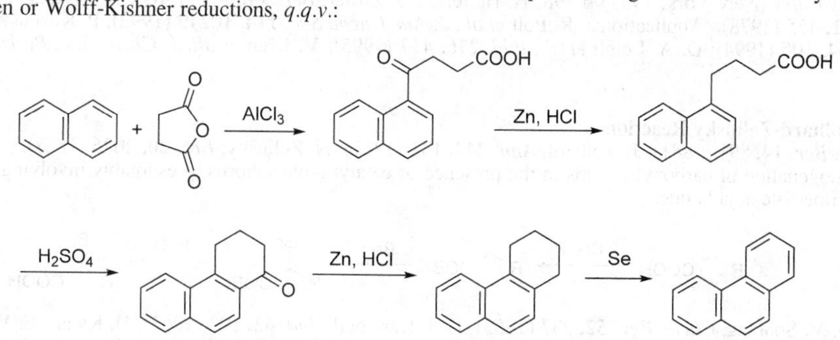

E. Berliner, *Org. React.* **5**, 229 (1949); I. Agranat, Y. S. Shih, *Synthesis* **1974**, 865; R. Menicagli, O. Piccolo, *J. Org. Chem.* **45**, 2581 (1980). *Cf.* Friedel-Crafts Reaction.

191. Hayashi Rearrangement
M. Hayashi, *J. Chem. Soc.* **1927**, 2516; **1930**, 1513, 1520, 1524.

Rearrangement of *o*-benzoylbenzoic acids in the presence of sulfuric acids or phosphorous pentoxide:

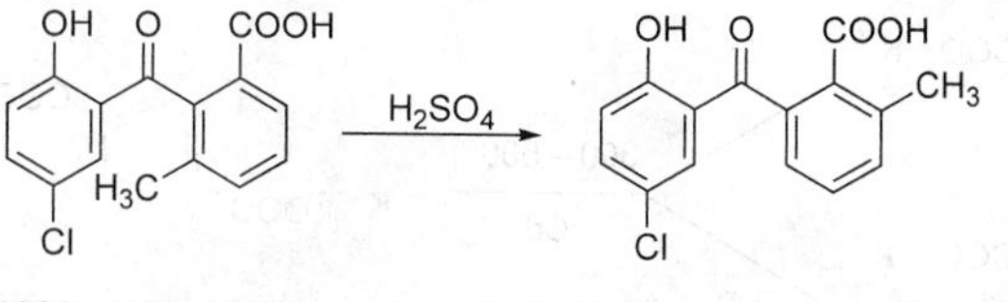

J. W. Cook, *J. Chem. Soc.* **1932**, 1472; M. Hayashi *et al., Bull. Chem. Soc. Jpn.* **11**, 184 (1936); R. B. Sandin, L. F. Fieser, *J. Am. Chem. Soc.* **62**, 3098 (1940); R. B. Sandin *et al., ibid.* **78**, 3817 (1956); R. Goncalves *et al., J. Org. Chem.* **17**, 705 (1952); S. Cristol, M. L. Caspar, *ibid.* **33**, 2020 (1968); M. Cushman *et al., ibid.* **45**, 5067 (1980).

192. Heck Reaction

R. F. Heck, J. P. Nolley, Jr., *J. Org. Chem.* **37**, 2320 (1972).

Stereospecific palladium-catalyzed coupling of alkenes with organic halides or triflates lacking sp^3-hybridized β-hydrogens:

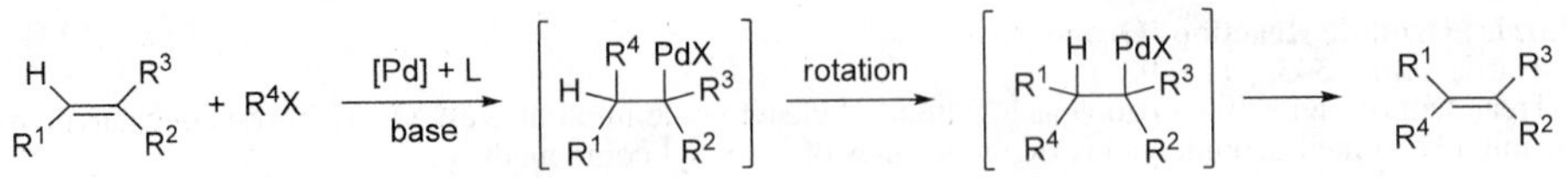

R^4 = aryl, alkenyl, benzyl
X = I, Br, OSO_2CF_3
[Pd] = $Pd(OCOCH_3)_2$, $PdCl_2$, $Pd(dba)_3$, $Pd(PPh_3)_4$
L = PAr_3, dppp, binap
base = $N(CH_2CH_3)_3$, K_2CO_3, $NaOCOCH_3$

Reviews: R. F. Heck, *Org. React.* **27**, 345-390 (1982); A. de Meijere, F. E. Meyer, *Angew. Chem. Int. Ed.* **33**, 2379-2411 (1994); W. Cabri, I. Candiani, *Acc. Chem. Res.* **28**, 2-7 (1995). Review of intramolecular reactions: L. E. Overman, *Pure Appl. Chem.* **66**, 1423-1430 (1994); S. E. Gibson *et al.*, *Contemp. Org. Synth.* **3**, 447-471 (1996); J. T. Link, L. E. Overman, *Met.-Catal. Cross-Coupling React.* **1998**, 231-269; of enantioselective syntheses: M. Shibasaki, E. M. Vogl, *J. Organomet. Chem.* **576**, 1-15 (1999); O. Loiseleur *et al., ibid.* 16-22; U. Iserloh, D. P. Curran, *Chemtracts* **12**, 289-296 (1999); of mechanism: G. T. Crisp, *Chem. Soc. Rev.* **27**, 427-436 (1998); M. Shibasaki *et al., Adv. Synth. Catal.* **346**, 1533-1552 (2004). *Cf.* Stille Coupling; Suzuki Coupling.

193. Helferich Method

B. Helferich, E. Schmitz-Hillebrecht, *Ber.* **66**, 378 (1933).

Glycosidation of an acetylated sugar by heating with a phenol in the presence of a metal halide ($ZnCl_2$, $FeCl_3$) or *p*-toluenesulfonic acid as catalyst:

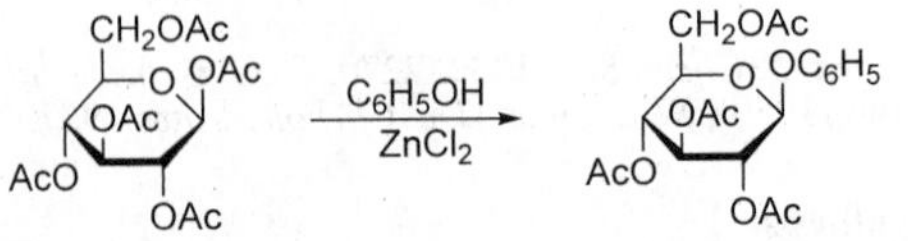

W. W. Pigman, R. M. Goepp, *Chemistry of the Carbohydrates* (New York, 1948) p 194; W. W. Pigman, *The Carbohydrates* (New York, 1957) p 198; B. Helferich, J. Zirner, *Ber.* **96**, 385 (1963); A. Piskala *et al., Nucleic Acid Chem.* **1**, 455 (1978). Applications: R. Polt *et al., J. Am. Chem. Soc.* **114**, 10249 (1992); P. Kosma *et al., Carbohydr. Res.* **254**, 105 (1994); D. A. Leigh *et al., ibid.* **276**, 417 (1995); V. Křen *et al., J. Chem. Soc. Perkin Trans. 1* **1997**, 2467.

194. Hell-Volhard-Zelinsky Reaction

C. Hell, *Ber.* **14**, 891 (1881); J. Volhard, *Ann.* **242**, 141 (1887); N. Zelinsky, *Ber.* **20**, 2026 (1887).

α-Halogenation of carboxylic acids in the presence of catalytic phosphorus, presumably involving the enol form of the intermediate acyl halide:

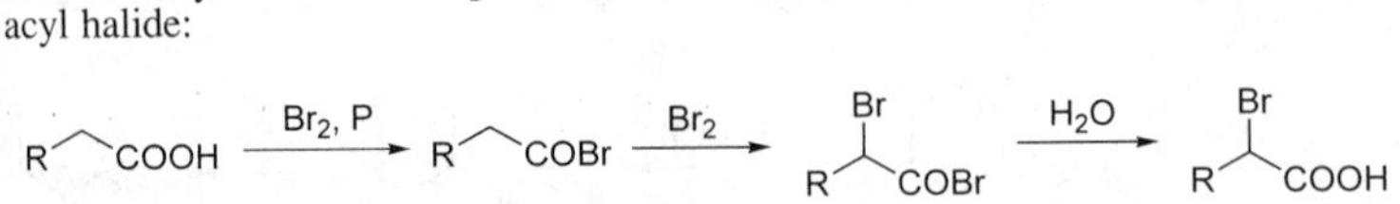

N. O. V. Sonntag, *Chem. Rev.* **52**, 237 (1953); H. J. Harwood, *ibid.* **62**, 102 (1962); H. Kwart, E. V. Scalzi, *ibid.* **86**, 5496 (1964); A. R. Sexton *et al., J. Am. Chem. Soc.* **91**, 7098 (1969); G. L. Lange, J. A. Otulakowski, *J. Org. Chem.* **47**, 5093 (1982); R. J. Crawford, *ibid.* **48**, 1364 (1983); H.-J. Liu, W. Luo, *Synth. Commun.* **21**, 2097 (1991).

195. Henkel Reaction (Raecke Process) (Henkel Process)

B. Raecke, **DE 936036** (1952) and **DE 958920** (1952) to Henkel & Co.

Industrial scale thermal rearrangement or disproportionation of alkaline salts of aromatic acids to symmetrical diacids in the presence of cadmium or other metallic salts:

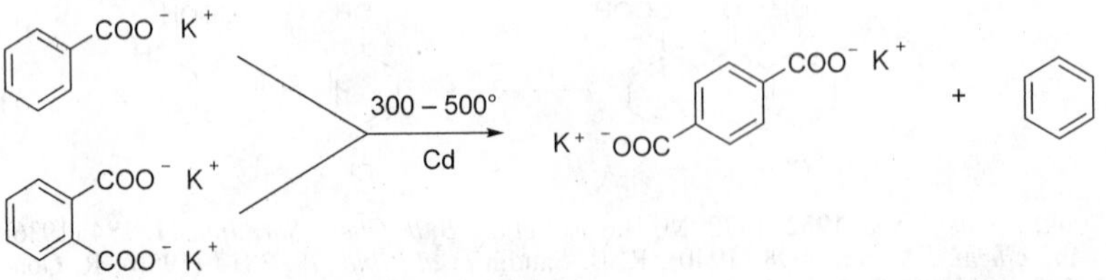

Review: B. Raecke, *Angew. Chem.* **70**, 1 (1958); Y. Ogata *et al., J. Org. Chem.* **25**, 2082 (1960); E. McNelis, *ibid.* **30**, 1209 (1965); J. Szammer, L. Otvos, *Radiochem. Radioanal. Lett.* **45**, 359 (1980).

196. Henry Reaction (Nitroaldol Reaction)

L. Henry, *Compt. Rend.* **120**, 1265 (1895); J. Kamlet, **US 2151517** (1939).

Base-catalyzed aldol-type condensation, *q.v.*, of nitroalkanes with aldehydes or ketones:

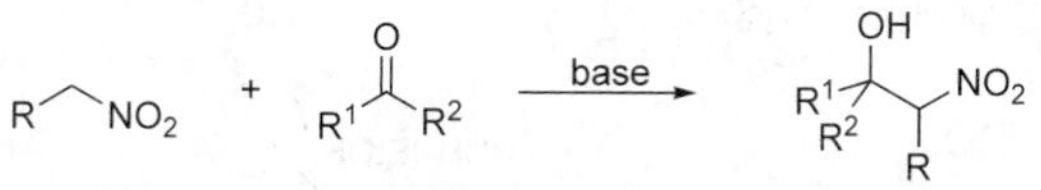

Application to sugars: R. Fernández *et al., Carbohydr. Res.* **247**, 239 (1993). Reagent controlled asymmetric induction: H. Sasai *et al., Tetrahedron Lett.* **34**, 855 (1993); R. Chinchilla *et al., Tetrahedron: Asymmetry* **5**, 1393 (1994); R. S. Varma *et al., Tetrahedron Lett.* **38**, 5131 (1997); R. Ballini, G. Bosica, *J. Org. Chem.* **62**, 425 (1997); V. J. Bulbule *et al., Tetrahedron* **55**, 9325 (1999). Catalyst effects: I. Morao, F. P. Cossio, *Tetrahedron Lett.* **38**, 6461 (1997); P. B. Kisanga, J. G. Verkade, *J. Org. Chem.* **64**, 4298 (1999); D. Simoni *et al., Tetrahedron Lett.* **41**, 1607 (2000); C. Palomo *et al, Angew. Chem. Int. Ed.* **43**, 5442 (2004). *Review:* G. Rosini, *Comp. Org. Syn.* **2**, 321-340 (1991). *Cf.* Knoevenagel Condensation.

197. HERON Rearrangement (Heteroatom Rearrangements on Nitrogen)

J. M. Buccigross *et al., Aust. J. Chem.* **48**, 353 (1995); J. M. Buccigross, S. A. Glover, *J. Chem. Soc. Perkin Trans. 2* **1995**, 595.

Rearrangement of *N*-alkoxy-*N*-aminoamides to esters and 1,1-diazenes via migration of oxygen from the nitrogen to the carbonyl carbon. Analogues of *N,N'*-diacyl-*N,N'*-dialkoxyhydrazines thermally decompose to esters and N$_2$ through two consecutive rearrangements:

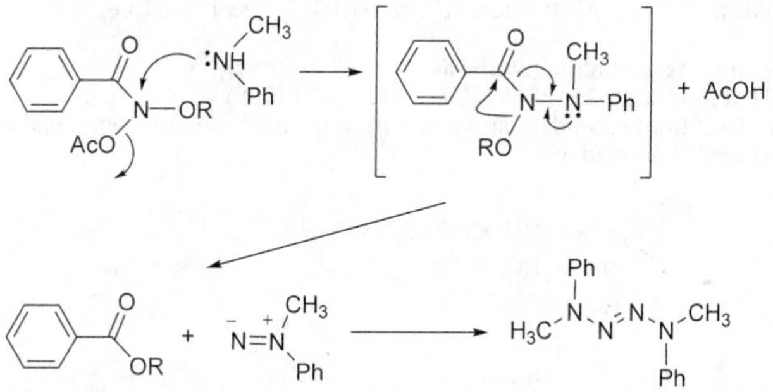

Application to *N,N'*-diacyl-*N,N'*-dialkoxyhydrazines: S. A. Glover *et al., J. Chem. Soc. Perkin Trans. 2* **1999**, 2053; to mutagenic *N*-acyloxy-*N*-alkoxybenzamides: J. J. Campbell, S. A. Glover, *J. Chem. Res.* **1999**, 474. Stereochemistry and computational studies: A. Rauk, S. A. Glover, *J. Org. Chem.* **61**, 2337 (1996); *eidem, ibid.* **64**, 2340 (1999). *Review:* S. A. Glover, *Tetrahedron* **54**, 7229-7272 (1998); *idem et al., Can. J. Chem.* **83**, 1492-1509 (2005).

198. Herz Reaction

R. Herz, **DE 360690** (1914 to Cassella & Co.); **US 1637023** (1928); **US 1699432** (1929).

Formation of *o*-aminothiophenols by heating aromatic amines with excess sulfur monochloride. The initial products are thiazothionium halides (Herz compounds) which will undergo chlorination if the position *para* to the amino group is unsubstituted:

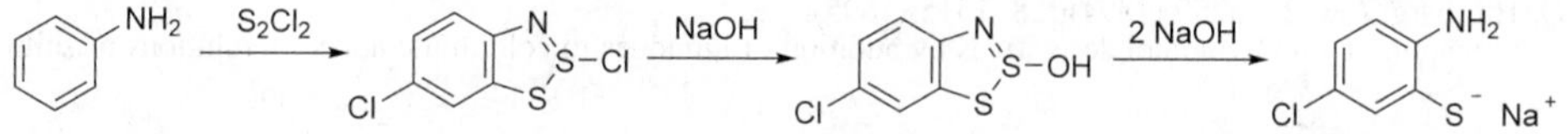

W. K. Warburton, *Chem. Rev.* **57**, 1011 (1957); L. D. Huestis *et al., J. Org. Chem.* **30**, 2763 (1965); P. Hope, L. A. Wiles, *J. Chem. Soc. C* **1967**, 1642; B. K. Strelets, L. S. Efros, *Zh. Org. Khim.* **1969**, 153; S. W. Schneller, *Int. J. Sulfur Chem.* **8**, 579 (1976); B. L. Chenard, *J. Org. Chem.* **49**, 1224 (1984).

199. Hilbert-Johnson Reaction

T. B. Johnson, G. E. Hilbert, *Science* **69**, 579 (1929); G. E. Hilbert, T. B. Johnson, *J. Am. Chem. Soc.* **52**, 2001, 4489 (1930).

Reaction of 2,4-dialkoxypyrimidines with halogenated sugar to yield pyrimidine nucleosides:

W. Zorbach, *Methods Carbohydr. Chem.* **6,** 445 (1972); T. Ueda, H. Ohtsuka, *Chem. Pharm. Bull.* **21,** 1451, 1530 (1973); C.-H. Kim *et al., J. Med. Chem.* **29,** 1374 (1986); A. A. Mourabit, *Tetrahedron: Asymmetry* **7,** 3455 (1996). Modified conditions: U. Neidballa, H. Vorbrüggen, *Angew. Chem. Int. Ed.* **9,** 469 (1970); H. Vorbrüggen, *et al., Ber.* **114,** 1279 (1981); H. Kristinsson *et al., Tetrahedron* **50,** 6825 (1994); G. Liu *et al., Synth. Commun.* **26,** 2681 (1996). Review of early studies: J. Pliml, M. Prystas, *Adv. Heterocycl. Chem.* **8,** 115 (1967).

200. Hinsberg Oxindole and Oxiquinoline Synthesis
O. Hinsberg, *Ber.* **21,** 110 (1888); **25,** 2545 (1892); **41,** 1367 (1908).

Formation of oxindoles from secondary aryl amines and the acid addition compound of glyoxal; primary aryl amines give glycine or glycinamide derivatives:

O. Hinsberg, J. Rosenzweig, *ibid.* **27,** 3253 (1894); C. Hollins, *Synthesis of Nitrogen Ring Compounds* (London, 1924) p 112; H. Burton, *J. Chem. Soc.* **1932,** 546; P. L. Julian *et al., Heterocycl. Compd.* **3,** 139 (1952). Mechanistic study: M. I. Abasolo *et al., J. Heterocycl. Chem.* **29,** 1279 (1992). Applications: M. I. Abasolo *et al., ibid.* **27,** 157 (1990); G. A. Rodrigo *et al., ibid.* **34,** 505 (1997). *Cf.* Stollé Synthesis.

201. Hinsberg Sulfone Synthesis
O. Hinsberg, *Ber.* **27,** 3259 (1894); **28,** 1315 (1895).

Formation of sulfonylquinol derivatives by addition of quinones to cold dilute aqueous solutions of sulfinic acids:

R. M. Scribner, *J. Org. Chem.* **31,** 3671 (1966); H. Ulrich *et al., Houben-Weyl* **7/3a,** 661 (1977). *Cf.* Thiele Reaction.

202. Hinsberg Synthesis of Thiophene Derivatives
O. Hinsberg, *Ber.* **43,** 901 (1910).

Formation of thiophene carboxylic acids from α-diketones and dialkyl thiodiacetates:

$$R \overset{O}{\underset{O}{\|}}R + R^1OOC\sim S\sim COOR^1 \xrightarrow{\text{KO-tBu}} \underset{R^1OOC}{\overset{R\quad R}{\bigodot_S}}COOR^1$$

H. Wynberg, D. J. Zwanenburg, *J. Org. Chem.* **29**, 1919 (1964); H. Wynberg, H. J. Kooreman, *J. Am. Chem. Soc.* **87**, 1739 (1965); A. Birch, D. A. Crombie, *Chem. Ind.* **1971**, 177; D. J. Chadwick *et al., J. Chem. Soc. Perkin Trans. 1* **1972**, 2079.

203. Hiyama Cross-Coupling Reaction; Hiyama-Denmark Cross-Coupling Reaction
Y. Hatanaka, T. Hiyama, *J. Org. Chem.* **53**, 918 (1988); *eidem, ibid.* **54**, 268 (1989).
   Palladium-catalyzed cross-coupling of organosilanes with organic halides or triflates in the presence of a fluoride or hydroxide activating agent:

$$R^1-X + R^2-SiR_3 \xrightarrow[\text{base activator or } F^-]{\text{PdL}_n \text{ Catalyst}} R^1-R^2$$

R^1 = alkenyl, allyl, vinyl, aryl, alkyl

R^2 = alkynyl, alkenyl, allyl, vinyl, aryl, alkyl

X = Cl, Br, I, OTf, OCO$_2$Et

SiR$_3$ = SiMe$_3$, SiMe$_2$F, SiMeF$_2$, SiF$_3$, Si(OMe)$_3$

Stereochemistry control studies: Y. Hatanaka, T. Hiyama, *J. Am. Chem. Soc.* **112**, 7793 (1990). Mechanistic studies with alkenylsilacyclobutanes and alkenylsilanols: S. E. Denmark *et al., Org. Lett.* **2**, 2491 (2000). *Reviews:* Y. Hatanaka, T. Hiyama, *Synlett* **1991** 845-853; T. Hiyama in *Metal-catalyzed Cross-coupling Reactions,* F. Diederich, P. J. Stang, Eds. (Wiley-VCH, New York, 1998) pp 421-453.
   When SiR$_3$ = SiMe$_2$OH and no fluoride activator is used, the modified reaction is referred to as the **Hiyama-Denmark Cross-Coupling Reaction:** S. E. Denmark, R. F. Sweis, *J. Am. Chem. Soc.* **123**, 6439 (2001). *Review:* S. E. Denmark, C. S. Regens, *Acc. Chem. Res.* **41**, 1486-1499 (2008). *Cf.* Heck Reaction; Negishi Cross Coupling; Stille Coupling; Suzuki Coupling.

204. Hoch-Campbell Aziridine Synthesis
J. Hoch, *Compt. Rend.* **198**, 1865 (1934); K. N. Campbell, J. F. McKenna, *J. Org. Chem.* **4**, 198 (1939).
   Formation of aziridines by treatment of ketoximes with Grignard reagents and subsequent hydrolysis of the organometallic complex:

$$\underset{Ph}{\overset{NOH}{\|}}CH_3 + 2PhMgX + 2HCl \longrightarrow \underset{Ph}{\overset{H}{\underset{N}{\bigtriangleup}}}CH_3 + C_6H_6 + 2MgXCl + H_2O$$

K. N. Campbell *et al., J. Org. Chem.* **8**, 99, 103 (1943); **9**, 184 (1944); J. P. Freeman, *Chem. Rev.* **73**, 283 (1973); O. C. Dermer, G. E. Ham, *Ethylenimine and Other Aziridines* (Academic Press, New York, 1969) pp 65-68; E. Y. Takehisa *et al., Chem. Pharm. Bull.* **24**, 1691 (1976); T. Sasaki *et al., Heterocycles* **11**, 235 (1978); G. Alvernhe, A. Laurent, *J. Chem. Res. Synop.* **1978**, 28.

205. Hofmann Degradation (Exhaustive Methylation)
A. W. Hofmann, *Ber.* **14**, 659 (1881).
   Formation of an olefin and a tertiary amine by pyrolysis of a quaternary ammonium hydroxide:

$$\underset{H}{\overset{\bigcirc}{N}}CH_3 \xrightarrow[\text{2) Ag}_2\text{O}]{\text{1) CH}_3\text{I}} \underset{H_3C \quad CH_3}{\overset{+}{N}}CH_3 \longrightarrow \sim\sim\sim\underset{CH_3}{\overset{CH_3}{N}}$$

$$\longrightarrow \sim\sim\sim\overset{+}{N}(CH_3)_3 \longrightarrow \sim\sim\sim + N(CH_3)_3$$

A. C. Cope, E. R. Trumbull, *Org. React.* **11**, 317-493 *passim* (1960); K. W. Bentley, G. W. Kirby in *Techniques of Organic Chemistry* **vol. IV,** Pt. 2, A. Weissberger, Ed., *Elucidation of Organic Structures by Physical and Chemical Methods* (Wiley, New York, 2nd ed., 1973) pp 255-289. Isotope effects: R. D. Bach, M. L. Braden, *J. Org. Chem.* **56**, 7194 (1991). Synthetic applications: A. D. Woolhouse *et al., J. Heterocycl. Chem.* **30**, 873 (1993); D. Berkes *et al., Synth. Commun.* **28**, 949 (1998). *Cf.* Cope Elimination Reaction; Emde Degradation.

206. Hofmann Isonitrile Synthesis (Carbylamine Reaction)
A. W. Hofmann, *Ann.* **146**, 107 (1868); *Ber.* **3**, 767 (1870).

Formation of isonitriles by the reaction of primary amines with chloroform in the presence of alkali; the odor of the isocyanide is a test for a primary amine:

$$C_2H_5NH_2 + CHCl_3 + 3NaOH \longrightarrow C_2H_5NC + 3NaCl + 3H_2O$$

P. A. S. Smith, N. W. Kalenda, *J. Org. Chem.* **23**, 1599 (1958); M. B. Frankel *et al., Tetrahedron Lett.* **1959**, 5; H. L. Jackson, B. C. McKusick, *Org. Synth.* **coll. vol. IV**, 438 (1963); W. P. Weber, G. W. Gokel, *Tetrahedron Lett.* **1972**, 1637.

207. Hofmann-Löffler-Freytag Reaction
A. W. Hofmann, *Ber.* **16**, 558 (1883); **18**, 5, 109 (1885); K. Löffler, C. Freytag, *ibid.* **42**, 3427 (1909).

Formation of pyrrolidines or piperidines by thermal or photochemical decomposition of protonated *N*-haloamines:

$$RCH_2(CH_2)_nNClR^1 \xrightarrow[\Delta \text{ or } h\nu]{H^+} RCHCl(CH_2)_n\overset{+}{N}H_2R^1 \longrightarrow$$

n = 3 or 4

M. E. Wolff, *Chem. Rev.* **63**, 55 (1963); E. J. Corey, W. R. Hertler, *J. Am. Chem. Soc.* **82**, 1657 (1960); R. Furstoss *et al., Tetrahedron Lett.* **1970**, 1263; S. Titouani *et al., Tetrahedron* **36**, 2961 (1980).

208. Hofmann-Martius Rearrangement (Aniline Rearrangement)
A. W. Hofmann, C. A. Martius, *Ber.* **4**, 742 (1871); A. W. Hofmann, *ibid.* **5**, 720 (1872).

Thermal conversion of *N*-alkylaniline hydrohalides to *o*- and *p*-alkylanilines:

H. Hart, J. R. Kosak, *J. Org. Chem.* **27**, 116 (1962); Y. Ogata *et al., Tetrahedron* **20**, 2717 (1964); *J. Org. Chem.* **35**, 1642 (1970); G. F. Grillot in *Mechanisms of Molecular Migrations* **vol. 3**, B. S. Thyagarajan, Ed. (Wiley, New York, 1971) p 237; A. G. Giumanini *et al., J. Org. Chem.* **40**, 1677 (1975); W. F. Burgoyne, D. D. Dixon, *J. Mol. Catal.* **62**, 61 (1990); M. G. Siskos *et al., Bull. Soc. Chim. Belg.* **105**, 759 (1996).

209. Hofmann Reaction
A. W. Hofmann, *Ber.* **14**, 2725 (1881).

Conversion of primary carboxylic amides to primary amines with one fewer carbon atom upon treatment with hypohalites or hydroxide *via* the intermediate isocyanate:

$$RCONH_2 \xrightarrow[NaOH]{Br_2} [R-N=C=O] \xrightarrow{H_2O} RNH_2$$

Early review: E. S. Wallis, J. F. Lane, *Org. React.* **3**, 267-306 (1949). Alternative reagents/strategies: S. Kajigaeshi *et al., Chem. Lett.* **1989**, 463; S. Jew *et al., Arch. Pharmacal Res.* **15**, 333 (1992); D. S. Rane, M. M. Sharma, *J. Chem. Technol. Biotechnol.* **59**, 271 (1994); H. Moustafa *et al., Tetrahedron* **53**, 625 (1997); Y. Matsumura *et al., J. Chem. Soc. Perkin Trans. 1* **1999**, 2057. *Review:* T. Shioiri, *Comp. Org. Syn.* **6**, 800-806 (1991). *Cf.* Curtius Rearrangement; Lossen Rearrangement; Schmidt Reaction; Weerman Degradation.

210. Hofmann-Sand Reactions
K. A. Hofmann, J. Sand, *Ber.* **33**, 1340, 1353 (1900).

Olefin mercuration with mercuric salts (halides, acetates, nitrates, or sulfates) in aqueous solution. In alcoholic solutions the accelerated reaction produces alkoxyalkyl compounds:

$$H_2C=CH_2 + HgX_2 + NaOH \xrightarrow{-NaX} HOCH_2CH_2HgX \xrightarrow[NaOH]{} XHgCH_2CH_2OCH_2CH_2HgX$$

$$H_2C=CH_2 + Hg(OCOCH_3)_2 + ROH \longrightarrow ROCH_2CH_2HgOCOCH_3 + CH_3COOH$$

Organic Name Reactions

J. Sand, *Ber.* **34,** 1385, 2906, 2910 (1901); *Ann.* **329,** 135 (1903); J. Chatt, *Chem. Rev.* **48,** 7 (1951); E. R. Rochow *et al., Chemistry of Organometallic Compounds* (New York, 1957) p 109; W. Kitching, *Organomet. Chem. Rev.* **3,** 35 (1968); K. P. Geller, H. Straub, *Houben-Weyl* **13/2b,** 130 (1974).

211. Hooker Reaction

S. C. Hooker, *J. Am. Chem. Soc.* **58,** 1174 (1936).

Oxidation of 2-hydroxy-3-alkyl-1,4-quinones with dilute alkaline permanganate with shortening of the alkyl side chain by a methylene group and simultaneous exchange of hydroxyl and alkyl or alkenyl group positions:

S. C. Hooker, A. Steyermark, *J. Am. Chem. Soc.* **58,** 1179 (1936); L. F. Fieser, M. Fieser, *ibid.* **70,** 3215 (1948); L. F. Fieser, A. R. Bader, *ibid.* **73,** 681 (1951); L. F. Fieser, M. Fieser, *Advanced Organic Chemistry* (New York, 1961) p 870.

212. Houben-Fischer Synthesis

J. Houben, W. Fischer, *J. Prakt. Chem.* [2] **123,** 89, 262, 313 (1929).

Formation of aromatic nitriles by basic hydrolysis of trichloromethyl aryl ketimines. Acidic hydrolysis yields ketones:

J. Houben, W. Fischer, *Ber.* **63,** 2464 (1930); **64,** 240, 2636, 2645 (1931); **66,** 339 (1933); D. T. Mowry, *Chem. Rev.* **42,** 221 (1948); P. E. Spoerri, A. S. DuBois, *Org. React.* **5,** 390 (1949); G. Hesse, *Houben-Weyl* **4/2** 103 (1955); W. Ruske in *Friedel-Crafts and Related Reactions* **vol. III,** Part 1, G. A. Olah, Ed. (Interscience, New York, 1964) p 407. *Cf.* Houben-Hoesch Reaction.

213. Houben-Hoesch Reaction

K. Hoesch, *Ber.* **48,** 1122 (1915); J. Houben, *ibid.* **59,** 2878 (1926).

Synthesis of acylphenols from phenols or phenolic ethers by the action of organic nitriles in the presence of hydrochloric acid and aluminum chloride as catalyst:

Reviews: P. E. Spoerri, A. S. DuBois, *Org. React.* **5,** 387 (1949); Thomas, *Anhydrous Aluminum Chloride in Organic Chemistry* (New York, 1941) p 504; W. Ruske in *Friedel-Crafts and Related Reactions* **vol. III,** Part 1, G. A. Olah, Ed. (Interscience, New York, 1964) p 383; M. I. Amer *et al., J. Chem. Soc. Perkin Trans. 1* **1983,** 1075; V. V. Arkhipov *et al., Chem. Heterocycl. Compd.* **33,** 515 (1997); R. Kawecki *et al., Synthesis* **1999,** 751. *Cf.* Gatterman Aldehyde Synthesis; Houben-Fischer Synthesis.

214. Houdry Cracking Process

E. Houdry, **US 1957648** and **US 1957649** (1934).

Decomposition of petroleum or heavy petroleum fractions into more useful lower boiling materials by heating at 500° and 30 psi, over a silica-alumina-magnanese oxide catalyst.

E. Houdry *et al., Oil Gas J.* **37,** 40 (1938); A. N. Sachanen, *Chemical Constituents of Petroleum* (New York, 1945) p 260; V. Haensel, M. J. Sterba, *Ind. Eng. Chem.* **40,** 1662 (1948); *Kirk-Othmer Encyclopedia of Chemical Technology* **4,** 323, 357 (New York, 1979); E. Boye, *Chem. Ztg.* **81,** 341 (1957); S. Gussow *et al., Oil Gas J.* **78,** 96 (1980); C. G. Mosley, *J. Chem. Educ.* **61,** 655 (1984); G. A. Mills, *Chemtech* **1986,** 72; Y. Nishimura, *Petrotech* **21,** 605 (1998).

215. Hunsdiecker Reaction (Borodine Reaction)

C. Hunsdiecker *et al.,* **US 2176181** (1939); H. Hunsdiecker, C. Hunsdiecker, *Ber.* **75,** 291 (1942); A. Borodine, *Ann.* **119,** 121 (1861).

Organic Name Reactions

Synthesis of organic halides by thermal decarboxylation of silver salts of the corresponding carboxylic acids in the presence of halogens:

$$RCOOAg + X_2 \xrightarrow{\Delta} RX + CO_2 + AgX$$

R. G. Johnson, R. K. Ingham, *Chem. Rev.* **56**, 219 (1956); C. V. Wilson, *Org. React.* **9**, 341 (1957); S. J. Cristol, W. C. Firth, Jr., *J. Org. Chem.* **26**, 280 (1961); F. F. Knapp, Jr., *Steroids* **33**, 245 (1979); A. I. Meyers, M. P. Fleming, *J. Org. Chem.* **44**, 3405 (1979). Modified catalysis by metal salt pool: S. Chowdhury, S. Roy, *Tetrahedron Lett.* **37**, 2623 (1996); D. Naskar, S. Roy, *J. Chem. Soc. Perkin Trans. 1* **1999**, 2436; *eidem, Tetrahedron* **56**, 1369 (2000). *Cf.* Kochi Reaction; Simonini Reaction.

216. Hydroboration Reaction

H. C. Brown, B. C. Subba Rao, *J. Am. Chem. Soc.* **78**, 5694 (1956); *J. Org. Chem.* **22**, 1135, 1136 (1957).

Addition of boron hydrides to alkenes, allenes, and alkynes to form organoboranes, such that boron adds to the less substituted carbon. Attack usually takes place on the less hindered side in a *cis* fashion:

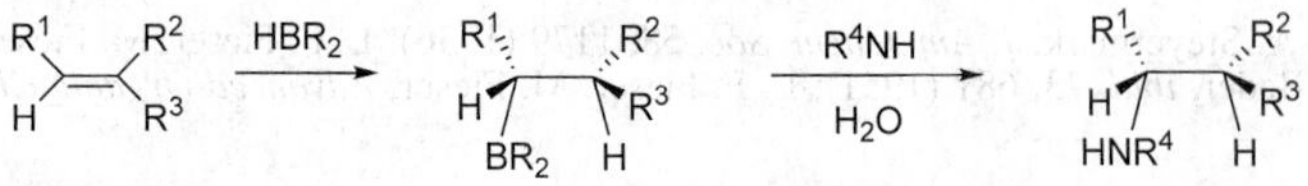

Diastereofacial and regioselectivity study: B. W. Gung *et al., Synth. Commun.* **24**, 167 (1994). Methods development for asymmetric synthesis: U. P. Dhokte, H. C. Brown, *Tetrahedron Lett.* **35**, 4715 (1994). Application to hydration: G. Zweifel, H. C. Brown, *Org. React.* **13**, 1-54 (1963). General reviews: H. O. House, *Modern Synthetic Reactions* (W. A. Benjamin, Menlo Park, California, 2nd ed., 1972) pp 106-130; K. Smith, A. Pelter, *Comp. Org. Syn.* **8**, 703-731 (1991). Reviews of asymmetric synthesis: H. C. Brown, *Tetrahedron* **37**, 3547-3587 (1981); K. Burgess, M. J. Ohlmeyer in *Adv. Chem. Ser.* **230**, entitled "Homogeneous Transition Metal Catalyzed Reactions" (ACS, Washington DC, 1992) pp 163-177. *Cf.* Suzuki Coupling.

217. Ivanov Reaction

D. Ivanov, A. Spassoff, *Bull. Soc. Chim. Fr.* **49**, 19, 375 (1931); D. Ivanov *et al., ibid.* **51**, 1321, 1325, 1331 (1932).

The addition of enediolates of aryl acetic acids (Ivanov reagents) to electrophiles, particularly carbonyl compounds:

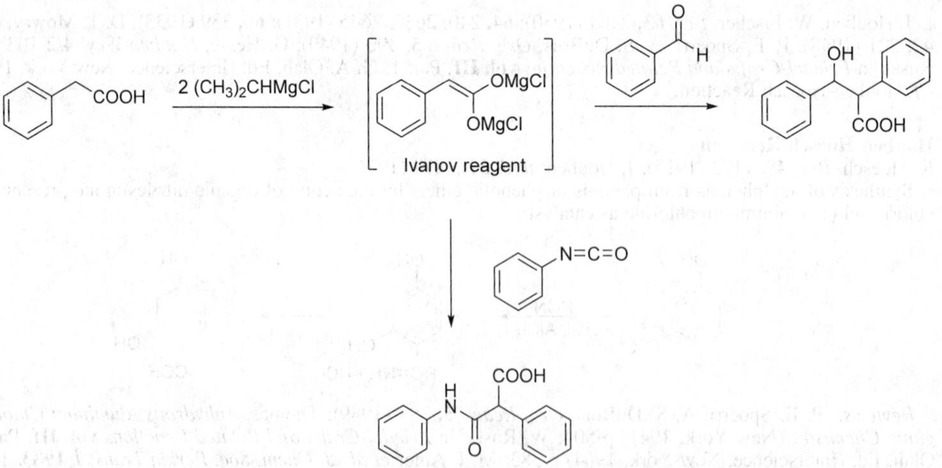

Early reviews: B. Blagoev, D. Ivanov, *Synthesis* **1970**, 615; D. Ivanov *et al., ibid.* **1975**, 83. Synthetic application: Y. A. Zhdanov *et al., Carbohydr. Res.* **29**, 274 (1973). Kinetic and mechanistic study: J. Toullec *et al., J. Org. Chem.* **50**, 2563 (1985). Stereoselectivity: M. Mladenova *et al., Tetrahedron* **37**, 2157 (1981); M. Momtchev *et al., Bull. Soc. Chim. Fr.* **5**, 844 (1985). *Cf.* Aldol Reaction; Knoevenagel Condensation.

218. Jacobsen Epoxidation

W. Zhang *et al., J. Am. Chem. Soc.* **112**, 2801 (1990); E. N. Jacobsen *et al. ibid.* **113**, 7063 (1991).

Chiral (salen)manganese(III)-catalyzed asymmetric epoxidation of alkenes. Enantio- and diastereo- selectivity depend strongly on the nature of the substrate:

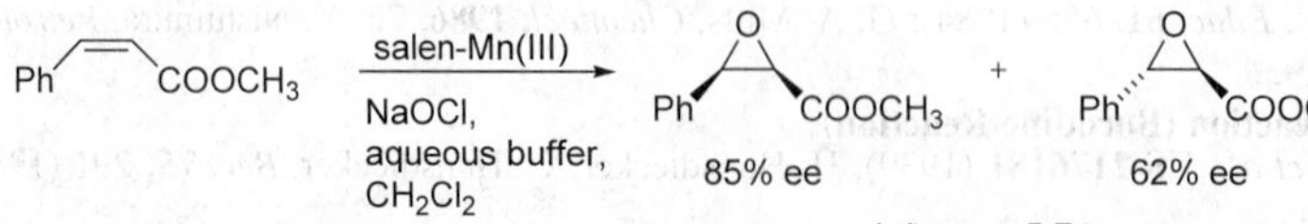

Methods development: E. N. Jacobsen *et al., Tetrahedron* **50**, 4323 (1994); S. Chang *et al., J. Am. Chem. Soc.* **116**, 6937 (1994); B. D. Brandes, E. N. Jacobsen, *J. Org. Chem.* **59**, 4378 (1994). Large-scale preparation of ligand: J. F. Larrow *et al., ibid.* 1939. *Review:* E. N. Jacobsen, "Asymmetric Catalytic Epoxidation of Unfunctionalized Olefins" in *Catalytic Asymmetric Synthesis*, I. Ojima, Ed. (VCH, New York, 1993) pp 159-202. For parallel studies, *see* N. Hosoya *et al., Synlett* **1993**, 641; H. Sasaki *et al., ibid.* **1994**, 356. Mechanistic study: D. L. Hughes *et al., J. Org. Chem.* **62**, 2222 (1997). Application: P. S. Savle *et al., Tetrahedron: Asymmetry* **9**, 1843 (1998). *Review:* T. Flessner *et al., J. Prakt. Chem.* **341**, 436-444 (1999).

219. Jacobsen Rearrangement

O. Jacobsen, *Ber.* **19**, 1209 (1886); **20**, 901 (1887).

Reaction of polymethylbenzenes with concentrated sulfuric acid to give rearranged polymethylbenzenesulfonic acids. Under identical conditions halogenated polymethylbenzenes undergo disproportionation:

L. I. Smith, *Org. React.* **1**, 370 (1942); H. Suzuki *et al., Bull. Chem. Soc. Jpn.* **36**, 1642 (1963); A. Koeberg-Telder, H. Cerfontain, *J. Chem. Soc. Perkin Trans. 2* **1977**, 717; M. Nakada *et al., Bull. Chem. Soc. Jpn.* **52**, 3671 (1979). Mechanism: J. L. Norula, R. P. Gupta, *Chem. Era* **10**, 7 (1974). ZrCl$_4$ catalysis: E. Solari *et al., Angew. Chem. Int. Ed.* **34**, 1510 (1995).

220. Janovsky Reaction

J. V. Janovsky, L. Erb, *Ber.* **19**, 2155 (1886).

Reaction of aldehydes and ketones containing α-methylene groups with *m*-dinitrobenzenes in the presence of a strong base resulting in the formation of an intense purple coloration, used for the detection of carbonyl compounds:

Reviews: Akatsuka, *J. Pharm. Soc. Jpn.* **80**, 389 (1960); Foster, Mackie, *Tetrahedron* **18**, 1131 (1962); Pollitt, Saunders, *J. Chem. Soc.* **1965**, 4615; M. Kimura *et al., Chem. Pharm. Bull.* **17**, 531 (1969); K. Kohashi *et al., ibid.* **25**, 50 (1977). Applications: R. G. Sutherland *et al., Can. J. Chem.* **64**, 2031 (1986); J. D. Artiss *et al., Microchem. J.* **65**, 277 (2000). *Cf.* Zimmermann Reaction.

221. Japp-Klingemann Reaction

F. R. Japp, F. Klingemann, *Ann.* **247**, 190 (1888); *Ber.* **20**, 2942, 3284, 3398 (1887).

Formation of hydrazones by coupling of aryldiazonium salts with active methylene compounds in which at least one of the activating groups is acyl or carboxyl. This group usually cleaves during the process:

Review: R. R. Phillips, *Org. React.* **10**, 143 (1959); H. C. Yao, P. Resnick, *J. Am. Chem. Soc.* **84**, 3504 (1962); M. O. Lozinskii, A. A. Gershkovich, *ibid.* **8**, 785 (1972); A. Kozikowski, W. C. Floyd, *Tetrahedron Lett.* **1978**, 19. Use of brominium ion as leaving group: G. Cirrincione *et al., J. Heterocycl. Chem.* **27**, 983 (1990). Synthetic applications:

F. Chetoni *et al., ibid.* **30**, 1481 (1993); B. Loubinoux *et al., J. Org. Chem.* **60**, 953 (1995); B. Pete *et al., Heterocycles* **53**, 665 (2000).

222. Jones Oxidation
K. Bowden *et al., J. Chem. Soc.* **1946**, 39.
    The oxidation of primary and secondary alcohols to acids and ketones, respectively, in the presence of chromic acid, aqueous sulfuric acid, and acetone. Isolated multiple bonds are not disturbed under these conditions:

$$ R\!\!-\!\!OH \xrightarrow[\text{acetone}]{CrO_3 \,/\, aq\; H_2SO_4} R\!\!-\!\!COOH $$

$$ R\!\!-\!\!CH(OH)\!\!-\!\!R^1 \xrightarrow[\text{acetone}]{CrO_3 \,/\, aq\; H_2SO_4} R\!\!-\!\!CO\!\!-\!\!R^1 $$

    P. Bladon *et al., J. Chem. Soc.* **1951**, 2402; E. R. H. Jones *et al., ibid.* **1953**, 457, 2548, 3019; C. Djerassi *et al., J. Org. Chem.* **21**, 1547 (1956); R. N. Warriner *et al., Aust. J. Chem.* **31**, 1113 (1978); S. V. Ley, A. Madin, *Comp. Org. Syn.* **7**, 253-256 (1991). Extensive synthetic applications: R. A. Epifanio *et al., Tetrahedron Lett.* **29**, 6403 (1988); P. A. Evans *et al., Synth. Commun.* **26**, 4685 (1996); N. M. Allanson *et al., Tetrahedron Lett.* **39**, 1889 (1998); Y. Watanabe *et al., ibid.* **40**, 3411 (1999). *Cf.* Sarett Oxidation.

223. Jourdan-Ullmann-Goldberg Synthesis
F. Jourdan, *Ber.* **18**, 1444 (1885); F. Ullmann, *ibid.* **36**, 2382 (1903); I. Goldberg, *ibid.* **39**, 1691 (1906); **40**, 4541 (1907).
    Synthesis of substituted diphenylamines, useful as intermediates in the synthesis of acridones:

    Reviews: R. M. Acheson, *Acridines* (Interscience, New York, 1956) p 148; Schulenberg, Archer, *Org. React.* **14**, 19 (1965).

224. Julia Olefination (Julia-Lythgoe Olefination)
M. Julia, M.-M. Paris, *Tetrahedron Lett.* **1973**, 4833.
    The formation of predominantly *trans*-olefins *via* the addition of phenyl sulfones to aldehydes or ketones, followed by alcohol functionalization and subsequent reductive elimination with sodium amalgam:

$$ R^3 = COCH_3, X = OCOCH_3; $$
$$ R^3 = COPh, SO_2CH_3, X = Cl $$

    Reviews: P. Kocienski, *Phosphorus Sulfur* **24**, 97-127 (1985); S. E. Kelly, *Comp. Org. Syn.* **1**, 792-806. Synthetic applications: R. Bellingham *et al., Synthesis* **1996**, 285; I. E. Markú *et al., Tetrahedron Lett.* **37**, 2089 (1996); T. Satoh *et al., ibid.* **39**, 6935 (1998); C. Charrier *et al., ibid.* **40**, 5705 (1999). Modified conditions: P. R. Blakemore *et al., Synthesis* **7**, 1209 (1999); P. J. Kocienski *et al., Synlett* **2000**, 365.

225. **Kahne Glycoslyation (Sulfoxide Glycosylation)**
   D. Kahne *et al.*, *J. Am. Chem. Soc.* **111**, 6881 (1989).
   Glycosidic bond formation between the anomeric carbon of the donor and the hydroxyl group of the acceptor. The reaction can be performed with a range of carbohydrate systems, and modified conditions allow for stereospecific glycosidic linkages and more functionalized glycosyl donor/acceptor pairs:

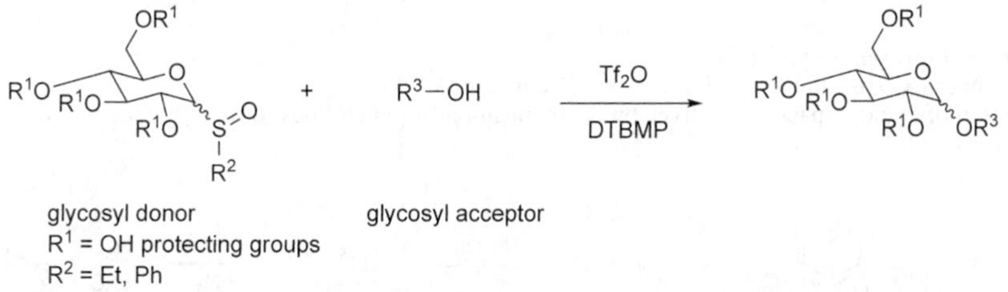

   glycosyl donor   glycosyl acceptor
   R^1 = OH protecting groups
   R^2 = Et, Ph

   Application to solid phase synthesis: L. Yan *et al.*, *J. Am. Chem. Soc.* **116**, 6953 (1994); and library synthesis: R. Liang *et al.*, *Science* **274**, 1520 (1996). Uniform reaction conditions: L. Yan, D. Kahne, *J. Am. Chem. Soc.* **118**, 9239 (1996). Mechanistic studies: D. Crich, S. Sun, *ibid.* **119**, 11217 (1997); J. Gildersleeve *et al.*, *ibid.* **120**, 5961 (1998). Reaction modifications involving by-product scavenging agents: *idem et al.*, *ibid.* **121**, 6176 (1999); aglycon sulfoxides: D. B. Berkowitz *et al.*, *Org. Lett.* **2**, 1149 (2000). *Review*: C. M. Taylor in *Solid Support Oligosaccharide Synthesis and Combinatorial Carbohydrate Libraries*, P. H. Seeberger, Ed. (Wiley, New York, 2001) pp 41-65.

226. **Keck-Mikami Allylation Reaction (Keck Allylation Reaction)**
   S. Aoki *et al.*, *Tetrahedron* **49**, 1783 (1993); G. E. Keck *et al.*, *J. Am. Chem. Soc.* **115**, 8467 (1993); G. E. Keck, L. S. Geraci, *Tetrahedron Lett.* **34**, 7827 (1993); G. E. Keck *et al.*, *J. Org. Chem.* **58**, 6543 (1993).
   Asymmetric allylation of an aldehyde, catalyzed by a Ti(IV)-1,1'-binaphthalene-2,2'-diol (BINOL) complex, to yield a chiral homoallylic alcohol:

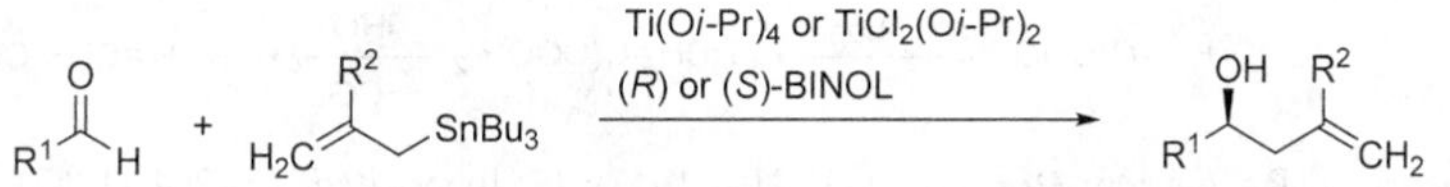

   Independent discovery of asymmetric induction with $TiCl_2(Oi\text{-}Pr)_2$ as the Ti(IV) source: A. L. Costa *et al.*, *J. Am. Chem. Soc.* **115**, 7001 (1993). Reaction rate studies: C.-M. Yu *et al.*, *Tetrahedron Lett.* **37**, 7095 (1996); *idem et al.*, *Chem. Commun.* **1997**, 761. Use of zirconium in catalyst complexes: S. Casolari *et al.*, *ibid.* 2123. Scope expansion studies: S. Weigand, R. Brückner, *Chem. Eur. J.* **2**, 1077 (1996); M. Bandin *et al.*, *Eur. J. Org. Chem.* **2000**, 491; S. Kii, K. Maruoka, *Chirality* **15**, 68 (2003). Methodology application to ketones: J. G. Kim *et al.*, *J. Am. Chem. Soc.* **126**, 12580 (2004). *Reviews*: S. E. Denmark, J. Fu, *Chem. Rev.* **103**, 2763-2793 (2003); M. A. Biamonte, T. A. Wynn in *Name Reactions for Homologations, Part II*, J. J. Li, Ed. (Wiley, Hoboken, NJ, 2009) pp 583-612. *Cf.* Sakurai Reaction. *See* monograph: BINOL.

227. **Kendall-Mattox Reaction**
   V. R. Mattox, E. C. Kendall, *J. Am. Chem. Soc.* **70**, 882 (1948); **72**, 2290 (1950); *J. Biol. Chem.* **188**, 287 (1951); E. C. Kendall, W. F. McGuckin, *J. Am. Chem. Soc.* **74**, 5811 (1952).
   Formation of a conjugated ketone from an α-bromoketone *via* a phenylhydrazone or semicarbazone:

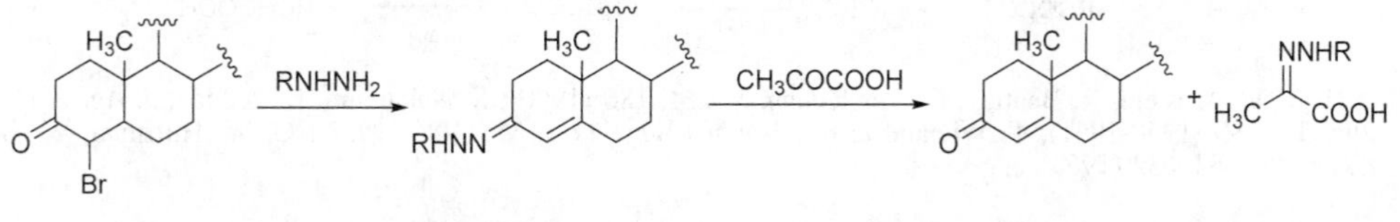

   C. Djerassi, *J. Am. Chem. Soc.* **71**, 1003 (1949); B. A. Koechlin *et al.*, *J. Biol. Chem.* **184**, 393 (1950); N. L. Wendler *et al.*, *J. Am. Chem. Soc.* **73**, 3818 (1951); J. J. Beereboom *et al.*, *ibid.* **75**, 3500 (1953); C. R. Engel, *ibid.* **78**, 4727 (1956); E. W. Warnhoff, *J. Org. Chem.* **28**, 887 (1963).

228. **Kiliani-Fischer Synthesis**
   H. Kiliani, *Ber.* **18**, 3066 (1885); E. Fischer, *ibid.* **22**, 2204 (1889).
   Extension of the carbon atom chain of aldoses by treatment with cyanide. Hydrolysis of the cyanohydrins followed by reduction of the lactone yields the homologous aldose:

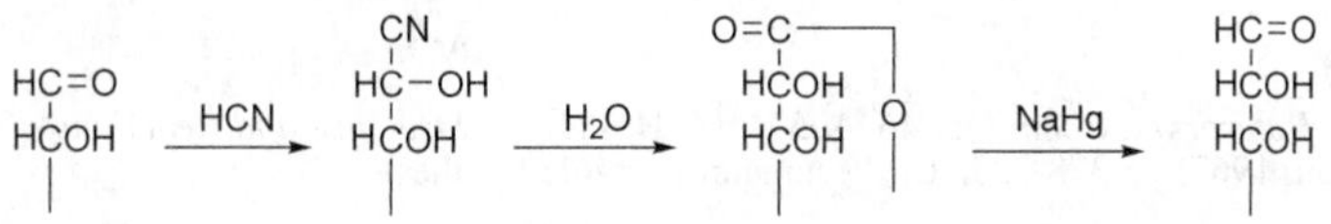

Organic Name Reactions

Reviews: C. S. Hudson, *Adv. Carbohydr. Chem.* **1**, 2 (1945); T. Moury, *Chem. Rev.* **42**, 239 (1948); L. Hough, A. C. Richardson, *The Carbohydrates* **1A**, 118 (1972); R. Kuhn, P. Klesse, *Ber.* **91**, 1989 (1958); R. Varma, D. French, *Carbohydr. Res.* **25**, 71 (1972); R. Blazer, T. W. Whalen, *J. Am. Chem. Soc.* **102**, 5082 (1980). Mechanistic study: A. S. Serianni *et al., J. Org. Chem.* **45**, 3329 (1980). Modified conditions: N. Adjé *et al., Tetrahedron Lett.* **37**, 5893 (1996). Stereoselective synthesis: J. Roos, F. Effenberger, *Tetrahedron: Asymmetry* **10**, 2817 (1999). *Cf.* Urech Cyanohydrin Method.

229. Kishner Cyclopropane Synthesis
N. M. Kishner, A. Zavadovskii, *J. Russ. Phys. Chem. Soc.* **43**, 1132 (1911).

Formation of cyclopropane derivatives by decomposition of pyrazolines formed by reacting α,β-unsaturated ketones or aldehydes with hydrazine:

L. I. Smith, E. R. Rogier, *J. Am. Chem. Soc.* **73**, 3840 (1951); G. S. Hammond, R. W. Todd, *ibid.* **76**, 4081 (1954); T. L. Jacobs, *Heterocycl. Compd.* **5**, 109 (1957). Mechanistic aspects of pyrazoline decomposition to cyclopropanes: R. G. Bergman in *Free Radicals* **vol. 1**, J. Kochi, Ed. (Wiley, New York, 1973) p 191; R. J. Crawford, M. Ohno, *Can. J. Chem.* **52**, 3134 (1974); R. J. Crawford, H. Tokunaga, *ibid.* 4033; J. A. Berson in *Rearrangements in Ground and Excited States* **vol. 1**, P. de Mayo, Ed. (Academic Press, New York, 1980) p 326.

230. Knoevenagel Condensation; Doebner Modification
E. Knoevenagel *Ber.* **31**, 2596 (1898); O. Doebner, *Ber.* **33**, 2140 (1900).

Condensation of aldehydes or ketones with active methylene compounds in the presence of ammonia or amines; the use of malonic acid and pyridine is known as the **Doebner modification:**

$$RCHO + H_2C(COOR)_2 \xrightarrow{\text{base}} RCH=C(COOR)_2 \xrightarrow[-CO_2]{HO^-} RCH=CH-COOH$$

Early reviews: J. R. Johnson, *Org. React.* **1**, 210 (1942); G. Jones, *ibid.* **15**, 204 (1967); H. O. House, *Modern Synthetic Reactions* (W. A. Benjamin, Menlo Park, California, 2nd ed., 1972) pp 646-653. Development of enantioselective methods: L. F. Tietze, P. Saling, *Chirality* **5**, 329 (1993). Application to the synthesis of indole alkaloids: L. F. Tietze *et al., Synthesis* **1994**, 1185. Modified conditions: J. McNulty *et al., Tetrahedron Lett.* **39**, 8013 (1998); B. M. Choudary *et al., J. Mol. Catal. A* **142**, 361 (1991). Synthetic applications: B. T. Watson, G. E. Christiansen, *Tetrahedron Lett.* **39**, 6087 (1998); R. W. Draper *et al., Tetrahedron* **56**, 1811 (2000). *Review:* L. F. Tietze, U. Beifuss, *Comp. Org. Syn.* **2**, 341-394 (1991). *Cf.* Aldol Reaction; Henry Reaction; Ivanov Reaction.

231. Knoop-Oesterlin Amino Acid Synthesis
F. Knoop, H. Oesterlin, *Z. Physiol. Chem.* **148**, 294 (1925).

Preparation of α-amino acids by catalytic hydrogenation of α-oxo acids in aqueous ammonia in the presence of platinum, palladium or Raney nickel catalysts, probably *via* an unstable iminocarboxylate ion intermediate:

H. R. V. Arnstein, R. Bentley, *Q. Rev. Chem. Soc.* **4**, 186 (1950); S. Nakamura, K. Ashida, *J. Agric. Chem. Soc. Jpn.* **24**, 185 (1950-1951); T. Wieland *et al., Houben-Weyl* **11/2**, 311, 482 (1958); C. W. Huffman, W. G. Skelly, *Chem. Rev.* **63**, 632 (1963).

232. Knorr Pyrazole Synthesis
L. Knorr, *Ber.* **16**, 2587 (1883).

Formation of pyrazole derivatives from hydrazines, hydrazides, semicarbazides, and aminoguanidines by condensation with 1,3-dicarbonyl compounds; substituted hydrazines yield two structurally isomeric pyrazoles:

T. J. Jacobs, *Heterocycl. Compd.* **5**, 46 (1957); M. H. Palmer, *Structure and Reactions of Heterocyclic Compounds* (Arnold, London, 1967) pp 378-385. *Cf.* Pechmann Pyrazole Synthesis.

Organic Name Reactions

233. Knorr Pyrrole Synthesis

L. Knorr, *Ber.* **17,** 1635 (1884); *Ann.* **236,** 290 (1886); L. Knorr, H. Lange, *Ber.* **35,** 2998 (1902).

Formation of pyrrole derivatives by condensation of α-amino ketones as such or generated *in situ* from isonitroso-ketones with carbonyl compounds containing active α-methylene groups:

A. H. Corwin, *Heterocycl. Compd.* **1,** 287 (1950); H. Fischer, *Org. Synth.* **coll. vol. III,** 573 (1955); S. Hauptmann, M. Martin, *Z. Chem.* **8,** 333 (1968); A. J. Castro *et al., J. Org. Chem.* **35,** 2815 (1970); Y. Tamura *et al., Chem. Ind. (London)* **1971,** 767; H. Rapoport, J. Harbuck, *J. Org. Chem.* **36,** 853 (1971); E. Fabiano, B. T. Golding, *J. Chem. Soc. Perkin Trans. 1* **1991,** 3371; A. Alberola *et al., Tetrahedron* **55,** 6555 (1999). Synthetic applications: J. A. Bastian, T. D. Lash, *ibid.* **54,** 6299 (1998); P. E. Harrington, M. A. Tius, *Org. Lett.* **1,** 649 (1999); L. Cheng, D. A. Lightner, *Synthesis* **1999,** 46. *Cf.* Hantzsch Pyrrole Synthesis; Paal-Knorr Pyrrole Synthesis.

234. Knorr Quinoline Synthesis

L. Knorr, *Ann.* **236,** 69 (1886); **245,** 357, 378 (1888).

Formation of α-hydroxyquinolines from β-ketoesters and arylamines above 100°. The intermediate anilide under-goes cyclization by dehydration with concentrated sulfuric acid:

F. W. Bergstrom, *Chem. Rev.* **35,** 157 (1944); C. R. Hauser, G. A. Reynolds, *J. Am. Chem. Soc.* **70,** 2402 (1948); *Org. Synth.* **coll. vol. III,** 593 (1955); R. C. Elderfield, *Heterocycl. Compd.* **4,** 30 (1952); A. J. Hodgkinson, B. Staskum, *J. Org. Chem.* **34,** 1709 (1969). Synthetic application: P. López-Alvarado *et al., Synthesis* **1998,** 186. *Cf.* Doebner-Miller Reaction; Gould-Jacobs Reaction.

235. Koch-Haaf Carboxylations

H. Koch, *Brennst.-Chem.* **36,** 321 (1955); H. Koch, W. Haaf, *Ann.* **618,** 251 (1958).

Formation of tertiary carboxylic acids by treating alcohols with carbon monoxide in strong acid:

Extension to olefins:

H. Langhals *et al., Tetrahedron Lett.* **22,** 2365 (1981); R. R. Rao, J. Bhattacharya, *Indian J. Chem.* **20B,** 207 (1981); *eidem, ibid.* **21B,** 405 (1982); O. Farooq *et al., J. Am. Chem. Soc.* **110,** 864 (1988). *Reviews:* K. E. Möller, *Brennst.-Chem.* **47,** 10 (1966); Y. T. Eidus, *et al., Russ. Chem. Rev.* **42,** 199 (1973); H. Bahrmann, "Koch Reactions" in *New Syntheses with Carbon Monoxide,* J. Falbe, Ed. (Springer-Verlag, New York, 1980) pp 372-413.

G. Olah, J. Olah in *Friedel-Crafts and Related Reactions* **vol. 3,** Part 2, G. A. Olah, Ed. (Interscience, New York, 1964) pp 1272-1296; C. W. Bird, *Chem. Rev.* **62,** 283 (1962). Extension to amides: C. Leonte, E. Carp, *Rev. Roum. Chim.* **34,** 1241 (1989).

236. Kochi Reaction

J. K. Kochi, *J. Am. Chem. Soc.* **87,** 2500 (1965).

Synthesis of organic chlorides by decarboxylation of carboxylic acids in the presence of lead tetraacetate and lithium chloride:

$$RCOOH + Pb(OAc)_4 \longrightarrow RCOOPb(OAc)_3 + AcOH \xrightarrow[\Delta]{LiCl} RCl + CO_2$$

R. A. Sheldon, J. K. Kochi, *Org. React.* **19,** 279 (1972); M. Mannier, J. P. Aycard, *Can. J. Chem.* **57,** 1257 (1979). *Cf.* Hunsdiecker Reaction.

Organic Name Reactions

237. Koenigs-Knorr Synthesis
W. Koenigs, E. Knorr, *Ber.* **34**, 957 (1901).

Formation of glycosides from acetylated glycosyl halides and alcohols or phenols in the presence of silver salts. The reaction proceeds with inversion of configuration:

Reviews: Evans *et al., Adv. Carbohydr. Chem.* **6**, 41-52 (1951); K. Igarashi, *ibid.* **34**, 243 (1977); H. M. Flowers, *Methods Carbohydr. Chem.* **6**, 474-480 (1972); R. R. Schmidt, *Comp. Org. Syn.* **6**, 33-64 (1991). Stereoselectivity: J.-I. Tamaru *et al., J. Carbohydr. Chem.* **12**, 893 (1993). Applications: A. Milius *et al., New J. Chem.* **15**, 337 (1991); F. W. Lichtenthaler, T. W. Metz, *Tetrahedron Lett.* **38**, 5477 (1997); S. Laszlo *et al., Chem. Commun.* **1999**, 591.

238. Kolbe Electrolytic Synthesis; Crum Brown-Walker Reaction
H. Kolbe, *Ann.* **69**, 257 (1849).

Formation of symmetrical dimers by the electrolysis of carboxylates (decarboxylative dimerization). The coupling of two distinct carboxylates yields unsymmetrical products:

$$2RCOO^- \xrightarrow{\text{electrolysis}} R-R + 2CO_2$$

The dimerization of half-esters is known as the **Crum Brown-Walker reaction:** A. Crum Brown, J. Walker, *ibid.* **261**, 107 (1891).

Reviews: B. C. L. Weedon, *Q. Rev. Chem. Soc.* **6**, 380 (1952); A. K. Vijh, B. E. Conway, *Chem. Rev.* **67**, 623 (1967); L. Eberson in *Organic Electrochemistry*, M. M. Baizer, Ed. (M. Dekker, New York, 1973) pp 469-507; H. J. Schäfer, *Comp. Org. Syn.* **3**, 633-658 (1991); J. Weiguny, H. J. Schäfer, *Ann.* **1994**, 225; G. Nuding *et al., Synthesis* **1996**, 71; J. Hiebl *et al., Tetrahedron* **54**, 2059 (1998); M. Sugiya, H. Noshira, *Chem. Lett.* **1998**, 479; *eidem, Bull. Chem. Soc. Jpn.* **73**, 705 (2000).

239. Kolbe-Schmitt Reaction
H. Kolbe, *Ann.* **113**, 125 (1860); R. Schmitt, *J. Prakt. Chem.* [2] **31**, 397 (1885).

Formation of aromatic hydroxy acids by carboxylation of phenolates, mostly in the *ortho* position, by carbon dioxide:

Reviews: A. S. Lindsey, H. Jeskey, *Chem. Rev.* **57**, 583 (1957); D. C. Ayres, *Carbanions in Synthesis* **1966**, 168-173; J. L. Hales *et al., J. Chem. Soc.* **1954**, 3145; J. March, *Advanced Organic Chemistry* (Wiley-Interscience, New York, 4th ed., 1992) p 546.

240. Kostanecki Acylation
S. von Kostanecki, A. Rozycki, *Ber.* **34**, 102 (1901).

Formation of chromones or coumarins by acylation of *o*-hydroxyaryl ketones with aliphatic acid anhydrides, followed by cyclization:

W. Baker, *J. Chem. Soc.* **1933**, 1381; C. R. Hauser, *Org. React.* **8**, 91 (1954); T. Szell *et al., Tetrahedron* **25**, 715 (1969); *idem et al., Helv. Chim. Acta* **52**, 2636 (1969); S. R. Save *et al., J. Indian Chem. Soc.* **48**, 675 (1971); Y. A. Shaikh, K. N. Trivedi, *ibid.* **49**, 599, 713 (1972); S. R. Save *et al., ibid.* **49**, 25 (1972). *Cf.* Allan-Robinson Reaction; Baker-Venkataraman Rearrangement.

241. Krafft Degradation
F. Krafft, *Ber.* **12**, 1664 (1879).

Conversion of carboxylic acids, especially of high molecular weight, into the next lower homolog by dry distillation of the alkaline earth salt with the corresponding acetate, followed by chromic acid oxidation of the methyl ketone:

Organic Name Reactions

$$(RCH_2COO)_2M \ + \ (AcO)_2M \longrightarrow RCH_2COCH_3 \xrightarrow{CrO_3} RCOOH$$

F. C. Whitmore, *Organic Chemistry* (New York, 1951) p 255; F. Klages, *Lehrbuch der organischen Chemie* **I** (Berlin, 1952) pp 262, 266, 368. *Cf.* Barbier-Wieland Degradation.

242. Krapcho Decarbalkoxylation
A. P. Krapcho *et al., Tetrahedron Lett.* **1967,** 215.

The decarbalkoxylation of malonate esters, β-keto esters, α-cyano esters and α-sulfonyl esters in dipolar aprotic solvents, at high temperatures, in the presence of water and/or salt, to yield esters, ketones, nitriles and sulfonyl derivatives, respectively:

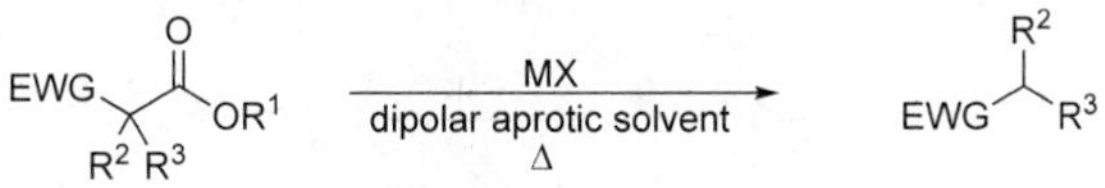

$$R^1 = CH_3, \ CH_2CH_3$$
$$EWG = COOR, \ COR, \ CN, \ SO_2R$$
solvent (possibly wet) = DMSO, DMF, HMPT
$$MX = LiCl, \ NaCl, \ LiI, \ NaCN, \ KCN, \ n\text{-}Bu_4NOAc$$

Scope and limitations: A. P. Krapcho *et al., J. Org. Chem.* **43,** 138 (1978). Mechanistic studies: A. M. Bernard *et al., Tetrahedron* **46,** 3929 (1990); P. J. Gilligan, P. J. Krenitsky, *Tetrahedron Lett.* **35,** 3441 (1994). Review of synthetic applications: A. P. Krapcho, *Synthesis* **1982,** 805-822, 893-914.

243. Kröhnke Oxidation
F. Kröhnke *et al., Ber.* **69,** 2006 (1936); **71,** 2583 (1938); **72,** 440 (1939).

Transformation of activated halides into aldehydes *via* their pyridinium salts, which yield nitrones upon treatment with *p*-nitrosodimethylaniline. Aldehydes or ketones are generated upon hydrolysis:

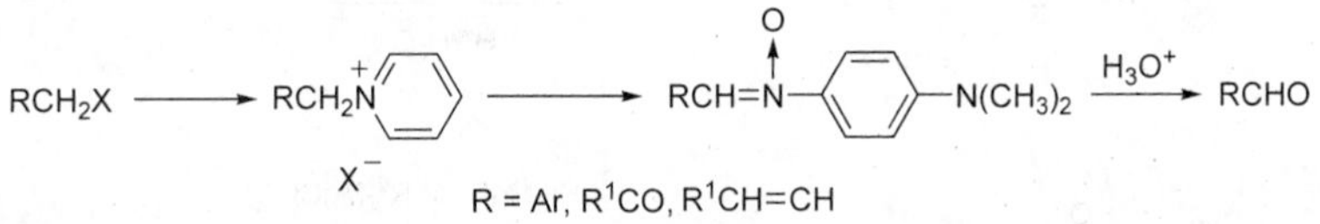

$$R = Ar, \ R^1CO, \ R^1CH=CH$$

A. A. Goldberg, H. A. Walker, *J. Chem. Soc.* **1954,** 2540; F. Kröhnke, *Angew. Chem. Int. Ed.* **2,** 380 (1963); A. Markovac *et al., Heterocyclic Chem.* **14,** 19 (1977); I. Maeba *et al., J. Chem. Soc. Perkin Trans. 1* **1991,** 939; S. N. Kilenyi, *Comp. Org. Syn.* **7,** 657-659 (1991). *Cf.* Sommelet Reaction.

244. Kröhnke Pyridine Synthesis
W. Zecher, F. Kröhnke, *Ber.* **94,** 690, 698 (1961); *eidem, Angew. Chem. Int. Ed.* **1,** 626 (1962).

1,4-Michael addition, *q.v.*, of α-pyridinium methyl ketone salts to α,β-unsaturated ketones, generating the 1,5-dicarbonyl compounds which undergo ammonium acetate-promoted ring closure, to yield substituted pyridines:

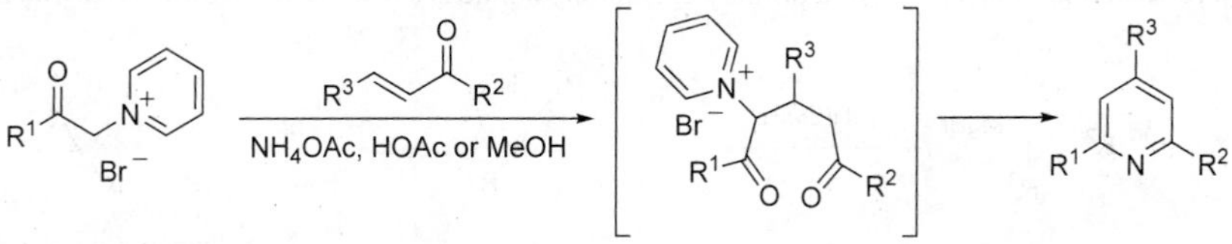

Early review: F. Kröhnke, *Synthesis* **1976,** 1-24. Synthetic applications: J. N. Chatterjea *et al., Indian J. Chem.* **15B,** 430 (1977); G. R. Newkome *et al., J. Org. Chem.* **51,** 850 (1986); P. Lhoták, A. Kurfürst, *Collect. Czech. Chem. Commun.* **57,** 1937 (1992); T. R. Kelly *et al., J. Org. Chem.* **62,** 2774 (1997). *Cf.* Chichibabin Pyridine Synthesis; Guareschi-Thorpe Condensation; Hantzsh (Dihydro)Pyridine Synthesis.

245. Kucherov Reaction
M. Kucherov, *Ber.* **14,** 1540 (1881).

Hydration of acetylenic hydrocarbons with dilute sulfuric acid in the presence of mercuric sulfate or boron trifluoride as catalyst:

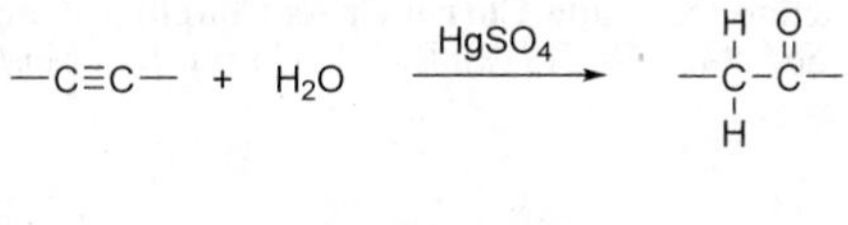

Organic Name Reactions

Reviews: A. D. Petrov. *Usp. Khim.* **21,** 250 (1952); M. Miocque *et al., Ann. Chim. (Paris)* **8,** 157 (1963); M. M. Khan, A. E. Martell, *Homogeneous Catalysis by Metal Complexes* **vol. 2** (Academic Press, New York, 1974) p 1974; B. S. Krupin, A. A. Petrov, *J. Gen. Chem. USSR* **33,** 3799 (1963); W. L. Budde, R. E. Dessy, *Tetrahedron Lett.* **1963,** 651; *J. Am. Chem. Soc.* **85,** 3964 (1963); K. G. Golodova, S. I. Yakimovich, *Zh. Org. Khim.* **8,** 2015 (1972). Extension to allenes: A. V. Fedorova, A. A. Petrov, *J. Gen. Chem. USSR* **32,** 1740 (1962).

246. Kuhn-Winterstein Reaction
R. Kuhn, A. Winterstein, *Helv. Chim. Acta* **11,** 87 (1928).

Conversion of 1,2-glycols into *trans* olefins by reaction with diphosphotetraiodide (P_2I_4) or other halogenated reagents. This reaction is useful in the preparation of polyenes:

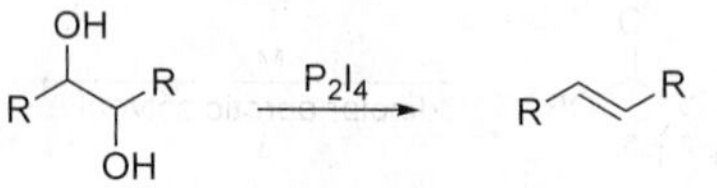

Kuhn *et al., Ber.* **71,** 1510 (1938); **84,** 566 (1961); **88,** 309 (1965); Inhoffen *et al., Ann.* **684,** 24 (1965); H. Kessler, W. Ott, *Tetrahedron Lett.* **1974,** 1383; W. W. Win *et al., J. Org. Chem.* **59,** 2803 (1994).

247. Kulinkovich Reaction; Kulinkovich-de Meijere Reaction; Kulinkovich-Szymoniak Reaction
O. G. Kulinkovich *et al., J. Org. Chem. USSR (Engl. Transl.)* **25,** 2027 (1989); *eidem, ibid.* **27,** 1249 (1991) *eidem, Synthesis* **1991,** 234.

Formation of a substituted cyclopropanol by the reaction of an ester with a Grignard reagent in the presence of titanium tetraisopropoxide. Alternately, a cyclopropanol product can be generated from a modification involving a terminal alkene:

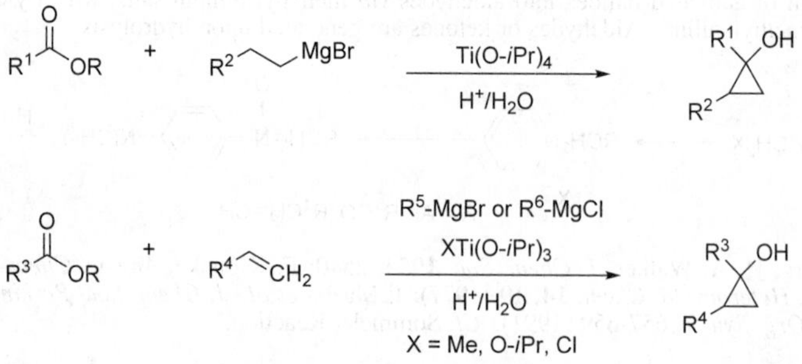

Mechanistic studies: Y.-D. Wu, Z.-X. Yu, *J. Am. Chem. Soc.* **123,** 5777 (2001). Ligand exchange modifications involving terminal alkenes: J. C. Lee *et al., Tetrahedron Lett.* **42,** 2059 (2001). *Comprehensive review*: O. G. Kulinkovich, A. de Meijere, *Chem. Rev.* **100,** 2789-2834 (2000).

The reaction of an amide under Kulinkovich conditions to yield a cyclopropyl amine is known as the **Kulinkovich-de Meijere Reaction**. Preparation of a primary cyclopropylamine is known as the **Kulinkovich-Szymoniak Reaction**:

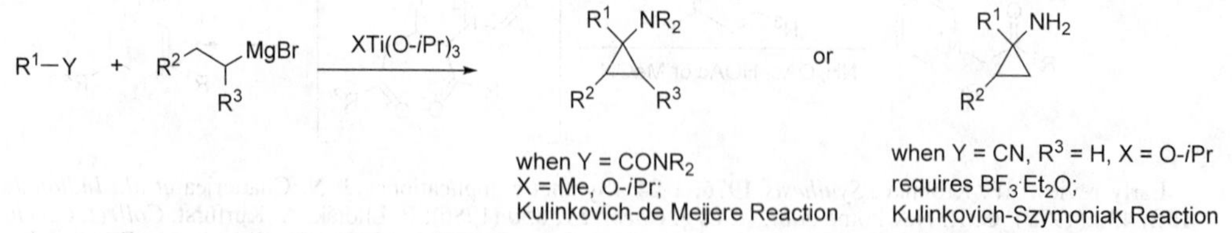

Kulinkovich-de Meijere synthetic applications: V. Chaplinski, A. de Meijere, *Angew. Chem. Int. Ed. Engl.* **35,** 413 (1996). Ligand exhange modifications involving terminal alkenes: J. Lee, J. K. Cha, *J. Org. Chem.* **62,** 1584 (1997). *Review*: A. de Meijere *et al., J. Organomet. Chem.* **689,** 2033-2055 (2004). Kulinkovich-Szymoniak synthetic applications: P. Bertus, J. Szymoniak, *Chem. Commun.* **2001,** 1792; *eidem, J. Org. Chem.* **68,** 7133 (2003). *Review: eidem, Synlett* **2007,** 1346-1356.

248. Kumada Cross-Coupling Reaction (Kumada-Corriu Cross-Coupling Reaction)
K. Tamao *et al., J. Am. Chem. Soc.* **94,** 4374 (1972); R. J. P. Corriu, J. P. Masse, *J. Chem. Soc. Chem. Commun.* **1972,** 144.

Organic Name Reactions

Nickel or palladium catalyzed cross-coupling reaction between an alkenyl, vinyl, or aryl halide and either an organomagnesium halide or an organolithium reagent:

$$R^1-X \ + \ R^2-Y \xrightarrow{\text{NiL}_n \text{ or PdL}_n \text{ Catalyst}} R^1-R^2$$

R^1 = alkenyl, vinyl, aryl

R^2 = alkenyl, vinyl, aryl, alkyl

X = Br, I, Cl, F, OTf

Y = MgBr, MgI, Li

Nickel catalyzed Grignard couplings: K. Tamao *et al.*, *Bull. Chem. Soc. Jpn.* **49**, 1958 (1976); *idem et al.*, *Tetrahedron* **38**, 3347 (1982). Mechanism of nickel catalysis: D. G. Morrell, J. K. Kochi, *J. Am. Chem. Soc.* **97**, 7262 (1975). Reaction modifications to include palladium catalysis: M. Yamamura *et al.*, *J. Organomet. Chem.* **91**, C39 (1975); S. Murahashi *et al.*, *J. Org. Chem.* **44**, 2408 (1979). *Review:* M. Kumada, *Pure Appl. Chem.* **52**, 669-679 (1980); of nickel catalyzed reactions: T. Takahashi, K. Kanno in *Modern Organonickel Chemistry*, Y. Tamaru, Ed. (Wiley-VCH, Weinheim, 2005) pp 41-55; of palladium catalyzed reactions: S. Huo, E. Negishi in *Handbook of Organopalladium Chemistry for Organic Synthesis*, E. Negishi, Ed. (Wiley, Hoboken, NJ, 2002) pp 335-408. *Cf.* Heck Reaction; Negishi Cross Coupling; Stille Coupling; Suzuki Coupling; Hiyama Cross-Coupling.

249. Ladenburg Rearrangement

A. Ladenburg, *Ber.* **16**, 410 (1883); *Ann.* **247**, 1 (1888).

Thermal rearrangement of an alkyl- or benzylpyridinium halide to an alkyl- or benzylpyridine:

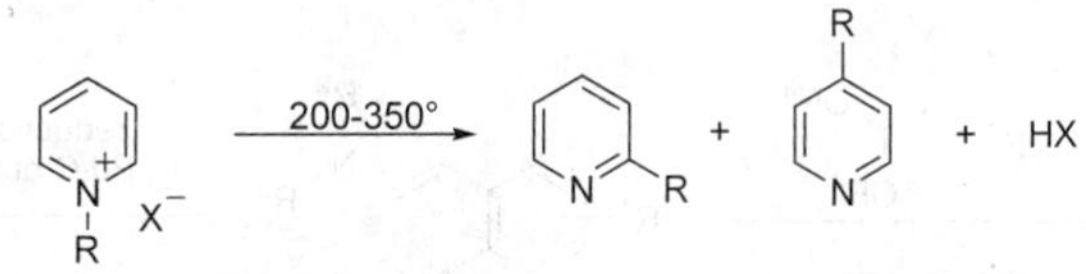

J. H. Brewster, E. L. Eliel, *Org. React.* **7**, 135 (1953); L. E. Tenenbau in *Pyridine and Its Derivatives* **Pt. 2**, E. Klingsberg, Ed. (Interscience, New York, 1961) p 163.

250. Larock Indole Synthesis (Larock Heteroannulation)

R. C. Larock, E. K. Yum, *J. Am. Chem. Soc.* **113**, 6689 (1991).

Palladium-catalyzed coupling reaction of *o*-iodoaniline derivatives with internal alkynes to prepare 2,3-disubstituted indoles:

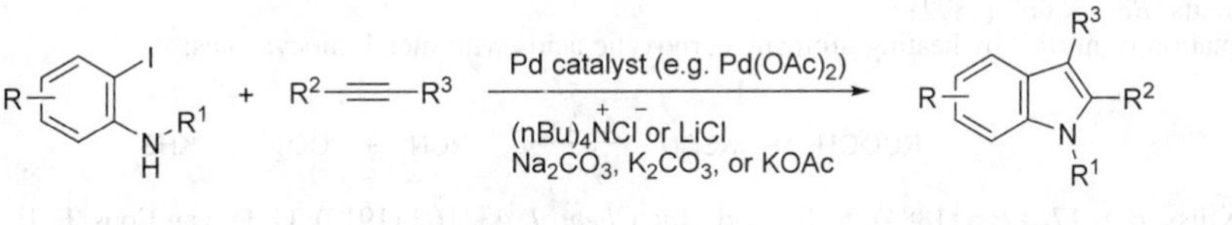

R^1 = alkyl, acyl, tosyl

R^2, R^3 = alkyl, aryl, alkenyl, CH$_2$OH, SiR$_3$

regioselectivity when R^2 is larger than R^3

Synthetic applications and reaction mechanism: R. C. Larock *et al.*, *J. Org. Chem.* **63**, 7652 (1998). Reaction procedure modifications with 2-bromo and 2-chloroanilines: M. Shen *et al.*, *Org. Lett.* **6**, 4129 (2004); with ligand- and salt-free conditions: N. Batail *et al.*, *Adv. Synth. Catal.* **351**, 2055 (2009); with silica immobilized palladium complexes: *eidem*, *Appl. Catal. A* **388**, 179 (2010). Reaction scope expansion to form isoindolo[2,1-*a*]indoles: K. R. Roesch, R. C. Larock, *J. Org. Chem.* **66**, 412 (2001); to form isoquinolines and pyridines: K. R. Roesch *et al.*, *ibid.* 8042. Review of indole syntheses: G. R. Humphrey, J. T. Kuethe, *Chem. Rev.* **106**, 2875-2911 (2006). *Cf.* Bartoli Indole Synthesis; Fischer Indole Synthesis; Leimgruber-Batcho Indole Synthesis; Madelung Synthesis.

251. Lebedev Process

S. V. Lebedev, *Zh. Obshch. Khim.* **3**, 698 (1933).

Formation of butadiene from ethanol by catalytic pyrolysis. The catalysts used are mixtures of sililcates and aluminum and zinc oxides:

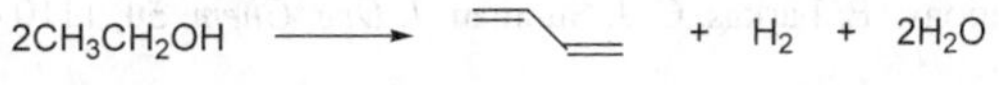

$$2CH_3CH_2OH \longrightarrow \text{/==\textbackslash} + H_2 + 2H_2O$$

Organic Name Reactions

S. V. Lebedev, **FR 665917** (1928); **GB 331482** (1929); **RU 24393** (1931); C. Ellis, *The Chemistry of Petroleum Derivatives* **II** (New York, 1937) p 173; G. Egloff, G. Hulla, *Chem. Rev.* **36**, 67 (1945); Y. A. Gorin, *Zh. Obshch. Khim.* **20**, 1596 (1950); *Kirk-Othmer Encyclopedia of Chemical Technology* **vol. 4** (New York, 3rd ed., 1978) p 322.

252. Lehmstedt-Tanasescu Reaction

K. Lehmstedt, *Ber.* **65**, 834 (1932); I. Tanasescu, *Bull. Soc. Chim. Fr.* **41**, 528 (1927).

Preparation of acridones (and 10-hydroxyacridones) from *o*-nitrobenzaldehyde and a halobenzene in the presence of concentrated sulfuric acid containing nitrous acid as catalyst:

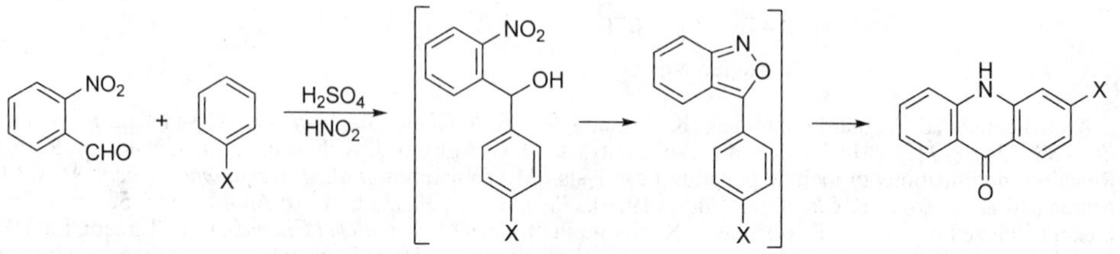

I. Tanasescu, Z. Frenkel, *ibid.* **1960**, 693. Mechanism: Silberg, Frenkel, *Rev. Roum. Chim.* **10**, 1035 (1965).

253. Leimgruber-Batcho Indole Synthesis (Batcho-Leimgruber Indole Synthesis)

A. D. Batcho, W. Leimgruber, **US 3732245** (1973 to Hoffman-La Roche).

Condensation of a formamide acetal with a derivative of *o*-nitrotoluene to yield an enamine intermediate that subsequently undergoes reductive cyclization to give the corresponding indole:

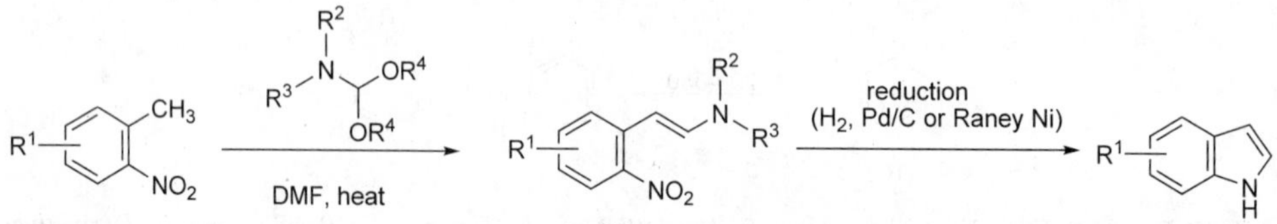

Reaction condition modifications involving use of amines during condensation: P. L. Feldman, H. Rapoport, *Synthesis* **1986**, 735; substituted formamide acetals: J. M. Bentley *et al.*, *Synth. Commun.* **34**, 2295 (2004). Prepn of 2-substituted indoles: E. E. Garcia, R. I. Fryer, *J. Heterocycl. Chem.* **11**, 219 (1974). *Reviews*: R. D. Clark, D. B. Repke, *Heterocycles* **22**, 195-221 (1984); J. Li, J. M. Cook in *Name Reactions in Heterocyclic Chemistry*, J. J. Li, E. J. Corey, Eds. (Wiley, Hoboken, NJ, 2005) pp 104-109.

254. Letts Nitrile Synthesis

E. A. Letts, *Ber.* **5**, 669 (1872).

Formation of nitriles by heating aromatic carboxylic acids with metal thiocyanates:

$$RCOOH \;+\; KSCN \xrightarrow{\Delta} RCN \;+\; CO_2 \;+\; KHS$$

G. Krüss, *Ber.* **17**, 1766 (1884); E. E. Reid, *Am. Chem. J.* **43**, 162 (1910); G. D. van Epps, E. E. Reid, *J. Am. Chem. Soc.* **38**, 2120 (1916); D. T. Mowry, *Chem. Rev.* **42**, 264 (1948); F. Klages, *Lehrbuch der organischen Chemie* **I** (Berlin, 1959) p 362.

255. Leuckart (Leukart) Reaction; Leuckart-Wallach Reaction; Eschweiler-Clarke Reaction

R. Leuckart, *Ber.* **18**, 2341 (1885).

Reductive alkylation of ammonium (or amine) salts of formic acid or formamides by aldehydes or ketones:

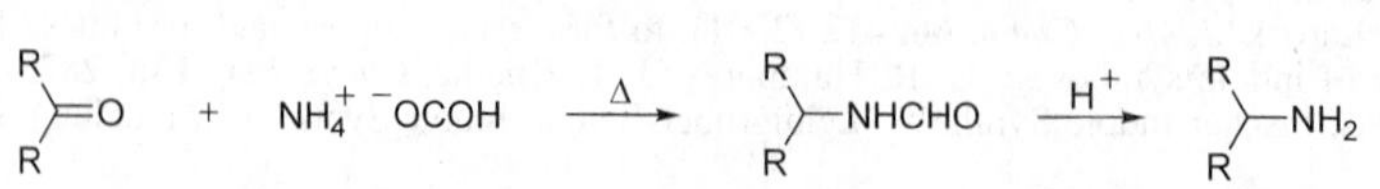

When the reaction is performed in the presence of excess formic acid it is referred to as the **Leuckart-Wallach reaction:** O. Wallach, *Ann.* **272**, 99 (1892). Application to steroids: W. E. Solomons, N. J. Doorenbos, *J. Pharm. Sci.* **63**, 19 (1974); A. M. Bellini *et al.*, *Steroids* **56**, 395 (1991).

The reductive methylation of primary or secondary amines employing formaldehyde and formic acid is known as the **Eschweiler-Clarke reaction:** W. Eschweiler, *Ber.* **38**, 880 (1905); H. T. Clarke *et al.*, *J. Am. Chem. Soc.* **55**, 4571 (1933). Synthetic applications: E. Farkas, C. J. Sunman, *J. Org. Chem.* **50**, 1110 (1985); J. Casanova, P. Devi, *Synth. Commun.* **23**, 245 (1993).

Organic Name Reactions

Early reviews: M. L. Moore, *Org. React.* **5,** 301-330 (1949); F. Möller, R. Schröter, *Houben-Weyl* **11/1,** 648-664 (1957). Application to deoxybenzoins: M. J. Villa *et al., Heterocycles* **24,** 1943 (1986). Mechanistic study: P. I. Awachie, V. C. Agwada, *Tetrahedron* **46,** 1899 (1990); A. G. Martinez *et al., Tetrahedron: Asymmetry* **10,** 1499 (1999). Optimized procedure: R. Carlson *et al., Acta Chem. Scand.* **47,** 1046 (1993). Modified conditions: A. Loupy *et al., Tetrahedron Lett.* **37,** 8177 (1996); I. Helland, T. Lejon, *Heterocycles* **51,** 611 (1999).

256. Leuckart Thiophenol Reaction

R. Leuckart, *J. Prakt. Chem.* [2] **41,** 179 (1890).

Decomposition of diazoxanthates, by warming gently in faintly acidic cuprous media, to the corresponding aryl xanthates which afford aryl thiols on alkaline hydrolysis and aryl thioethers on warming:

$$ArN_2Cl + KSCSOR \xrightarrow{70°} ArSCSOR \begin{cases} \xrightarrow{\Delta} ArSR + COS \\ \xrightarrow[H_2O]{alkali} ArSH + HSCOOR \end{cases}$$

D. S. Tarbell, D. K. Fukushima, *Org. Synth.* **coll. vol. III,** 809 (1955); K. H. Saunders, *The Aromatic Diazo-Compounds and Their Technical Applications* (London, 1949) p 325; D. S. Tarbell, M. A. McCall, *J. Am. Chem. Soc.* **74,** 48 (1952); A. R. Forrester, J. L. Wardell, *Rodd's Chemistry of Carbon Compounds* **IIIA,** 422 (1971); A. Schöberl, A. Wagner, *Houben-Weyl* **9,** 12 (1955); L. Tournier, S. Z. Zard, *Tetrahedron Lett.* **46,** 971 (2005).

257. Lieben Iodoform Reaction (Haloform Reaction)

A. Lieben, *Ann.* (Suppl.) **7,** 218 (1870).

Cleavage of methyl ketones with halogens (mostly iodine) and base to carboxylic acids and haloform:

$$RCOCH_3 + NaOI \longrightarrow RCOCl_3 \xrightarrow{NaOH} CHI_3 + RCOO^- Na^+$$

R. C. Fuson, B. A. Bull, *Chem. Rev.* **15,** 275 (1934); R. N. Seelye, T. A. Turney, *J. Chem. Educ.* **36,** 572 (1959); H. O. House, *Modern Synthetic Reactions* (W. A. Benjamin, Menlo Park, California, 2nd ed., 1972) pp 464-465; J. March, *Advanced Organic Chemistry* (Wiley-Interscience, New York, 4th ed., 1992) p 632.

258. Lobry de Bruyn-van Ekenstein Transformation

C. A. Lobry de Bruyn, *Rec. Trav. Chim.* **14,** 150 (1895); C. A. Lobry de Bruyn, W. A. van Ekenstein, *ibid.* 195, 203; **16,** 262 (1897).

Isomerization of carbohydrates in alkaline media, considered to embrace both epimerization of aldoses and ketoses and aldose-ketose interconversion:

Reviews: Evans, *Chem. Rev.* **31,** 544 (1942); Sattler, *Adv. Carbohydr. Chem.* **3,** 113 (1948); Pigman, *The Carbohydrates* (Academic Press, New York, 1957) p 60; Speck, *Adv. Carbohydr. Chem.* **13,** 63 (1958); Schaffer, *J. Org. Chem.* **29,** 1473 (1964); M. H. Johansson, O. Samuelson, *Chem. Scr.* **9,** 151 (1976). Synthetic applications: P. Köll, G. Papert, *Ann.* **1986,** 1568; B. Sauerbrei *et al., Carbohydr. Res.* **280,** 223 (1996); P. Sedmera *et al., J. Carbohydr. Chem.* **17,** 1351 (1998). Mechanistic study: B. M. Kabyemela *et al., Ind. Eng. Chem. Res.* **38,** 2888 (1999).

259. Lossen Rearrangement

W. Lossen, *Ann.* **161,** 347 (1872); **175,** 271, 313 (1874).

Conversion of a hydroxamic acid to an isocyanate *via* the intermediacy of its *O*-acyl, sulfonyl, or phosphoryl derivative. In the presence of amines, ureas are formed; in the presence of water, amines containing one less carbon than the starting material, are generated:

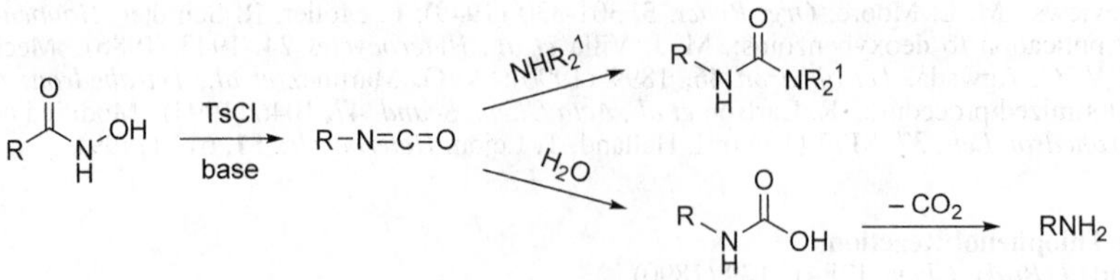

Reviews: H. L. Yale, *Chem. Rev.* **33,** 209 (1943); L. Bauer, O. Exner, *Angew. Chem. Int. Ed.* **13,** 376 (1974); T. Shiori, *Comp. Org. Syn.* **6,** 821-825 (1991). Reaction conditions leading to the formation of ureas: J. Pihuleac, L. Bauer, *Synthesis* **1989,** 61; extention to *N*-phosphinoylhydroxylamines: J. Fawcett *et al., Chem. Commun.* **1992,** 227; C. J. Salomon, E. Breuer, *J. Org. Chem.* **62,** 3858 (1997); to sulfonyloxy imides: D. A. Casteel *et al., Heterocycles* **36,** 485 (1993). Modifications: J. A. Stafford *et al., J. Org. Chem.* **63,** 10040 (1998); R. Anilkumar *et al., Tetrahedron Lett.* **41,** 5291 (2000). *Cf.* Curtius Rearrangement; Hofmann Reaction; Schmidt Reaction.

260. Luche Reduction

J.-L. Luche, *J. Am. Chem. Soc.* **100,** 2226 (1978); J.-L. Luche *et al., J. Chem. Soc. Chem. Commun.* **1978,** 601.

Selective 1,2 reduction of both substituted and unsubstituted α,β-unsaturated aldehydes and ketones to give the corresponding allylic alcohols. The Lewis acid conditions avoid epimerization, and steric hindrance has little to no effect on regioselectivity. In the case of cyclohexenones, axial hydride attack results in equitorial alcohols:

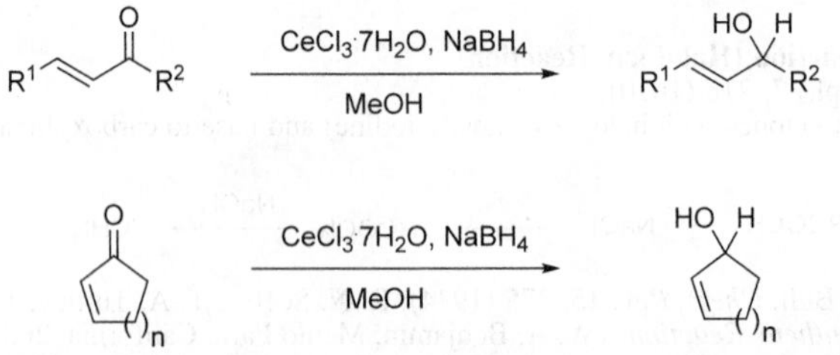

Synthetic and mechanistic studies: A. L. Gemal, J.-L. Luche, *J. Am. Chem. Soc.* **103,** 5454 (1981). Expansion of reaction substrates: K. Li *et al., Tetrahedron Lett.* **33,** 6569 (1992); G. Hutton *et al., ibid.* **36,** 7905 (1995); C. Liu, D. J. Burnell, *ibid.* **38,** 6573 (1997). *Review:* R. J. Mullins *et al.* in *Name Reactions for Functional Group Transformations,* J. J. Li, E. J. Corey, Eds. (Wiley, Hoboken, NJ, 2007) pp 112-122.

261. Madelung Synthesis

W. Madelung, *Ber.* **45,** 1128 (1912).

Formation of indole derivatives by intramolecular cyclization of an *N*-(2-alkylphenyl)alkanamide by a strong base at high temperature:

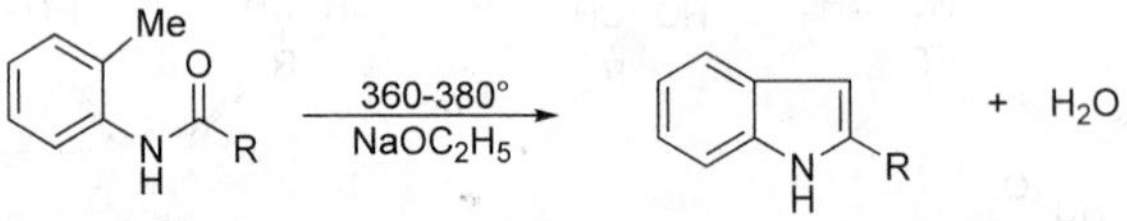

R. K. Brown in *The Chemistry of Heterocyclic Compounds,* A. Weissberger, Ed., *Indoles* **Part I,** W. J. Houlihan, Ed. (Wiley, New York, 1972) pp 385-396; W. J. Houlihan *et al., J. Org. Chem.* **46,** 4511, 4515 (1981). Under mild conditions: W. Verboom *et al., Tetrahedron Lett.* **26,** 685 (1985); *eidem, Tetrahedron* **42,** 5053 (1986); E. O. M. Orlemans *et al., ibid.* **43,** 3817 (1987).

262. Maillard Reaction ("Browning" Reaction)

L. C. Maillard, *Compt. Rend.* **154,** 66 (1912); *Ann. Chim.* **9,** 5, 258 (1916).

The reactions of amino groups of amino acids, peptides or proteins with the "glycosidic" hydroxyl group of sugars ultimately resulting in the formation of brown pigments.

G. P. Ellis, *Adv. Carbohydr. Chem.* **14,** 63 (1959); E. F. L. Anet, *ibid.* **19,** 181 (1964). Mechanism: M. Amrani-Hemaimi *et al., J. Agric. Food Chem.* **43,** 2818 (1995); high pressure effects: M. Bristow, N. S. Isaacs, *J. Chem. Soc. Perkin Trans. 2* **1999,** 221. Crosslinking in proteins: K. J. Wells-Knecht *et al., J. Org. Chem.* **60,** 6246 (1995); M. O. Lederer, R. G. Klaiber, *Bioorg. Med. Chem.* **7,** 2499 (1999). *Reviews:* C. Eriksson, *Prog. Food Nutr. Sci.* **5,** 159-176 (1981); *The Maillard Reaction in Foods and Medicine,* J. O. O'Brien *et al.,* Eds. (Royal Soc. Chem., Cambridge, U.K., 1998) 464 pp; S. Horvat, A. Jakas, *J. Pept. Sci.* **10,** 119-137 (2004).

263. Malaprade Reaction (Periodic Acid Oxidation)

L. Malaprade, *Bull. Soc. Chim. Fr.* [4] **43**, 683 (1928); *Compt. Rend.* **186**, 382 (1928).

Compounds containing two hydroxyl groups, or a hydroxyl and an amino group, attached to adjacent carbon atoms, undergo cleavage of the carbon-carbon bond when treated with periodic acid to yield aldehydes:

$$RCHOHCHOHR^1 + HIO_4 \longrightarrow RCHO + R^1CHO + H_2O + HIO_3$$
$$RCHOHCHNH_2R^1 + HIO_4 \longrightarrow RCHO + R^1CHO + NH_3 + HIO_3$$

H. O. House, *Modern Synthetic Reactions* (W. A. Benjamin, Menlo Park, California, 2nd ed., 1972) pp 353-359; K. W. Bentley in *Elucidation of Organic Structures by Physical and Chemical Methods* **Pt. 2**, K. W. Bentley, G. W. Kirby, Eds. (Wiley, New York, 2nd ed., 1973) pp 177-185. *Cf.* Criegee Reaction.

264. Malonic Ester Syntheses

Syntheses based on the strongly activated methylene group of malonic esters. Upon deprotonation with sodium ethoxide, the resulting resonance-stabilized ion is alkylated or acylated. After ester hydrolysis, the free alkylmalonic acid decarboxylates to yield a mono- or disubstituted monocarboxylic acid:

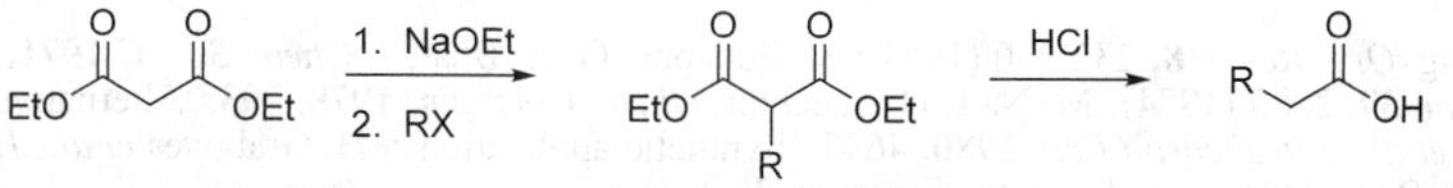

H. O. House, *Modern Synthetic Reactions* (W. A. Benjamin, Menlo Park, California, 2nd ed., 1972) pp 510-518, 756-761. Use of crown ethers as catalysts: D. H. Hunter, *et al., Synthesis* **1977**, 37. Modified conditions: M. A. Casadei *et al., J. Org. Chem.* **46**, 3127 (1981); B. K. Wilk, *Synth. Commun.* **26**, 3859 (1996). Stereoselectivity: T. Sato, J. Otera, *J. Org. Chem.* **60**, 2627 (1995); B. Klotz-Berendes *et al., Tetrahedron: Asymmetry* **8**, 1821 (1997). *Cf.* Perkin Alicyclic Synthesis.

265. Mannich Reaction

C. Mannich, W. Krosche, *Arch. Pharm.* **250**, 647 (1912).

Reaction of compounds having an active hydrogen with non-enolizable aldehydes and ammonia or primary or secondary amines to give aminomethylated products (Mannich bases):

$$(CH_3)_2NH + HCHO + CH_3COCH_3 \longrightarrow (CH_3)_2NCH_2CH_2COCH_3 + H_2O$$

Early reviews: F. F. Blicke, *Org. React.* **1**, 303 (1942); H. O. House, *Modern Synthetic Reactions* (W. A. Benjamin, Menlo Park, California, 2nd ed., 1972) pp 654-660. *p*-Substituted phenols as substrates: D. A. Leigh, P. Linnane, *Tetrahedron Lett.* **34**, 5639 (1993). In synthesis of vinylphosphonates: H. Krawezyk, *Synth. Commun.* **24**, 2263 (1994). Diastereoselectivity: P. C. B. Page *et al., J. Org. Chem.* **58**, 6902 (1993); enantioselectivity: H. Ishitani *et al., J. Am. Chem. Soc.* **119**, 7153 (1997); *eidem, Tetrahedron Lett.* **40**, 2161 (1999); K. Yamada, *Angew. Chem. Int. Ed.* **38**, 3504 (1999). *Reviews:* M. Tramontini *et al., Tetrahedron* **46**, 1791-1837 (1990); E. F. Kleinman, *Comp. Org. Syn.* **2**, 893-951 (1991); H. Heane, *ibid.* 953-973; L. E. Overman, D. J. Ricca, *ibid.* 1007-1046; M. Arend *et al., Angew. Chem. Int. Ed.* **37**, 1044-1070 (1998). *Cf.* Betti Reaction; Robinson-Schöpf Reaction.

266. Marschalk Reaction

C. Marschalk *et al., Bull. Soc. Chim. Fr.* **3**, 1545 (1936).

Sodium dithionite reduction of 1-hydroxy- or aminoanthraquinones to their leuco-forms, followed by condensation with aldehydes to yield the 2-alkylated anthraquinones. 2-Hydroxyanthraquinones yield 1-alkylated products:

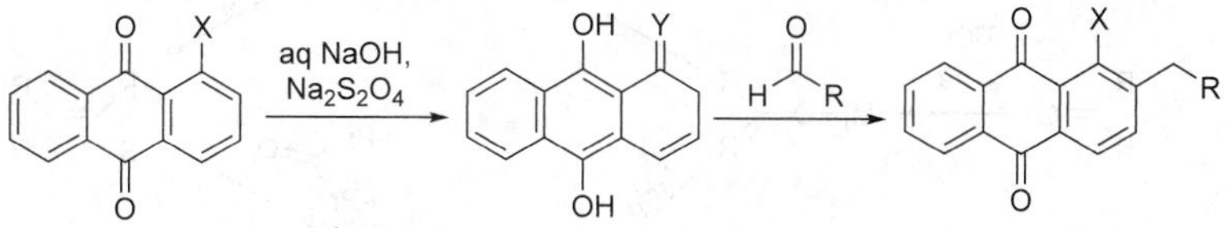

X = OH, Y = O
X = NH₂, Y = NH

Scope and limitations: K. Krohn, W. Baltus, *Tetrahedron* **44**, 49 (1988). Synthetic applications: F. Suzuki *et al., J. Am. Chem. Soc.* **100**, 2272 (1978); L. M. Harwood *et al., Can. J. Chem.* **62**, 1922 (1984); M. T. Furlong *et al., Synth. Commun.* **20**, 2691 (1990); N. R. Ayyangar *et al., Indian J. Chem.* **31B**, 3 (1992); K. Krohn, S. Bernhard, *J. Prakt. Chem.* **340**, 26 (1998).

267. Martinet Dioxindole Synthesis

A. Guyot, J. Martinet, *Compt. Rend.* **156**, 1625 (1913).

Formation of derivatives of dioxindole from esters of mesoxalic acid and aromatic amines or amino quinolines:

J. Martinet, *ibid.* **166,** 851, 998 (1918); *Ann. Chim.* [9] **11,** 85 (1919); W. Langenbeck *et al., Ann.* **499,** 201 (1932); **512,** 276 (1934); W. C. Sumpter, *Chem. Rev.* **37,** 472 (1945); P. L. Julian *et al., Heterocycl. Compd.* **3,** 239 (1952).

268. McFadyen-Stevens Reaction
J. S. McFadyen, T. S. Stevens, *J. Chem. Soc.* **1936,** 584.
    Base-catalyzed thermal decomposition of acylbenzenesulfonylhydrazines to aldehydes:

E. Mosettig, *Org. React.* **8,** 232-240 (1954); S. Siddappa, G. A. Bhat, *J. Chem. Soc. C* **1971,** 178; S. B. Matin *et al., J. Org. Chem.* **39,** 2285 (1974); M. Nair, H. Shechter, *Chem. Commun.* **1978,** 793. Alternative hydrazide reagent: C. C. Dudman *et al., Tetrahedron Lett.* **1980,** 4645. Synthetic applications: H. Graboyes *et al., J. Heterocycl. Chem.* **12,** 1225 (1975); R. K. Manna *et al., Synth. Commun.* **28,** 9 (1998).

269. McLafferty Rearrangement
F. W. McLafferty, *Anal. Chem.* **31,** 82 (1959).
    Electron-impact-induced cleavage of carbonyl compounds having a hydrogen in the γ-position, to an enolic fragment and an olefin:

D. G. I. Kingston *et al., Chem. Rev.* **74,** 215 (1974); K. Biemann, *Mass Spectrometry* (New York, 1962) p 119; Djerassi *et al., J. Am. Chem. Soc.* **87,** 817 (1965); **91,** 2069 (1969); **94,** 473 (1972); M. J. Lacey *et al., Org. Mass Spectrom.* **5,** 1391 (1971); G. Eadon, *J. Am. Chem. Soc.* **94,** 8938 (1972); F. Turecek, V. Hanus, *Org. Mass Spectrom.* **15,** 8 (1980). *Cf.* Norrish Type Cleavage.

270. McMurry Coupling Reaction
J. E. McMurry, M. P. Fleming, *J. Am. Chem. Soc.* **96,** 4708 (1974); S. Tyrlik, I. Wolochowicz, *Bull. Soc. Chim. Fr.* **1973,** 2147; T. Mukaiyama *et al., Chem. Lett.* **1973,** 1041.
    Deoxygenative coupling of carbonyl compounds to alkenes induced by low-valent titanium:

Synthetic application: A. Fürstner, D. N. Jumbam, *Tetrahedron* **48,** 5991 (1992); M. Rucker, R. Brückner, *Tetrahedron Lett.* **38,** 7353 (1997); P. Harter *et al., Polyhedron* **17,** 1141 (1998). Modified conditions: T. A. Lipski *et al., J. Org. Chem.* **62,** 4566 (1997); S. Talukdar *et al., ibid.* **63,** 4925 (1998). *Reviews:* J. E. McMurry, *Chem. Rev.* **89,** 1513-1524 (1989); G. M. Robertson, *Comp. Org. Syn.* **3,** 583-595 (1991); T. Lectka, *Act. Met.* **1996,** 85-131; M. Ephritikhine, *Chem. Commun.* **23,** 2549-2554 (1998). *Cf.* Barton Olefin Synthesis.

Organic Name Reactions

271. Meerwein Arylation
H. Meerwein *et al., J. Prakt. Chem.* **152**, 237 (1939).
Formation of arylated olefins on treatment of olefins with diazonium salts in the presence of cupric salts:

$$Z-C=C- \quad \xrightarrow[\text{CuCl}_2]{\text{ArN}_2^+ \text{ Cl}^-} \quad Z-C=C-$$

Z = C=C, C=O, Ar, CN, H

Synthetic applications: P. Sutter, C. D. Weis, *J. Heterocycl. Chem.* **24**, 69 (1987); G. Wurm, H. J. Gurka, *Pharmazie* **52**, 739 (1997); enhanced stereoselectivity: H. Brunner *et al., J. Organomet. Chem.* **541**, 89 (1997). Modified conditions: M. D. Obushak *et al., Tetrahedron Lett.* **39**, 9567 (1998). *Reviews:* C. S. Rondestvedt, Jr., *Org. React.* **11**, 189 (1960); *ibid.* **24**, 225-259 (1976); C. D. Weis, *Dyes Pigm.* **9**, 1-20 (1988). *Cf.* Pschorr Reaction.

272. Meerwein-Ponndorf-Verley Reduction (Aluminum Alkoxide Reduction)
H. Meerwein, R. Schmidt, *Ann.* **444**, 221 (1925); W. Ponndorf, *Angew. Chem.* **39**, 138 (1926); A. Verley, *Bull. Soc. Chim. Fr.* **37**, 537, 871 (1925).
Reduction of aldehydes or ketones to the corresponding alcohols with aluminum alkoxides (the reverse of the Oppenauer oxidation, *q.v.*):

$$\underset{R}{\overset{O}{\|}}\underset{R^1}{} + \underset{H_3C}{\overset{OH}{|}}\underset{CH_3}{} \xrightarrow{\text{Al[OCH(CH}_3)_2]_3} \underset{R}{\overset{OH}{|}}\underset{R^1}{} + \underset{H_3C}{\overset{O}{\|}}\underset{CH_3}{}$$

Reviews: A. L. Wilds, *Org. React.* **2**, 178-202 (1944); R. M. Kellogg, *Comp. Org. Syn.* **8**, 88-91 (1991); C. F. de Graauw *et al., Synthesis* **10**, 1007-1017 (1994). Enantioselectivity: D. A. Evans *et al., J. Am. Chem. Soc.* **115**, 9800 (1993); M. Node *et al., ibid.* **122**, 1927 (2000). Modified conditions: P. S. Kumbhar *et al., Chem. Commun.* **1998**, 535; T. Ooi *et al., J. Am. Chem. Soc.* **120**, 10790 (1998); Y. Nakano *et al., Tetrahedron Lett.* **41**, 1565 (2000). *Cf.* Cannizzaro Reaction; Tischenko Reaction.

273. Meinwald Rearrangement
J. Meinwald *et al., J. Am. Chem. Soc.* **85**, 582 (1963); *idem et al., Tetrahedron Lett.* **6**, 1789 (1965).
Acid-catalyzed epoxide rearrangement to yield carbonyl compounds. The ratio of the resulting products depends upon the catalyst and solvent as well as the relative migratory aptitude of the R^2 and R^3 substituents:

$$\underset{R^3}{\overset{R^1 \quad O \quad R^2}{\triangle}} \quad \xrightarrow{H^+} \quad \underset{R^3}{\overset{R^1 \quad \overset{O}{\|} \quad R^2}{}} \quad + \quad \underset{R^2}{\overset{R^1 \quad \overset{O}{\|} \quad R^3}{}}$$

Relative migratory aptitude studies: K. Maruoka *et al., Tetrahedron* **48**, 3303 (1992). Catalyst and product distribution studies: B. C. Ranu, U. Jana, *J. Org. Chem.* **63**, 8212 (1998); K. A. Bhatia *et al., Tetrahedron Lett.* **42**, 8129 (2001); I. Karamé *et al., ibid.* **44**, 7687 (2003).

274. Meisenheimer Rearrangements
J. Meisenheimer, *Ber.* **52**, 1667 (1919).
Formation of *O,N,N*-trisubstituted hydroxylamines from tertiary amine oxides *via* [1,2]-R group migration, or [2,3]-sigmatropic rearrangement when R′ = allyl:

$$\underset{R}{\overset{R^1}{\underset{R}{\overset{|}{N^+-O^-}}}} \quad \longrightarrow \quad \underset{R}{\overset{R}{\underset{R}{\overset{|}{N-OR^1}}}}$$

$$\underset{R \quad R}{\overset{+}{N}-O^-} \quad \longrightarrow \quad \underset{R \quad R}{\overset{O}{N}}$$

[1,2]-Rearrangements: N. Castagnoli, Jr. *et al., Tetrahedron* **26**, 4319 (1970); J. B. Bremner *et al., Aust. J. Chem.* **41**, 293 (1988); R. Yoneda *et al., Tetrahedron Lett.* **35**, 3749 (1994); *eidem, Tetrahedron* **52**, 14563 (1996). *Cf.* Stevens Rearrangement; [1,2]-Wittig Rearrangement.

[2,3]-Rearrangements: V. Rautenstrauch, *Helv. Chim. Acta* **56**, 2492 (1973); Y. Yamamato *et al., J. Org. Chem.* **41**, 303 (1976); or [1,2]: T. Kurihara *et al., Chem. Pharm. Bull.* **42**, 475 (1994). Asymmetric syntheses: D. Enders, H. Kempen, *Synlett* **1994**, 969; S. G. Davies, G. D. Smyth, *Tetrahedron: Asymmetry* **7**, 1001 (1996); J. E. H. Buston *et*

al., ibid. **9,** 1995 (1998). *Cf.* Mislow-Evans Rearrangement; Sommelet-Hauser Rearrangement; [2,3]-Wittig Rearrangement.

275. Menschutkin Reaction

N. Menschutkin, *Z. Phys. Chem.* **5,** 589 (1890); **6,** 41 (1890).

Reaction of tertiary amines with alkyl halides to form quaternary salts:

$$R_3N \ + \ R^1X \longrightarrow R_3R^1N^+ X^-$$

Mechanistic studies: C. K. Ingold, *Structure and Mechanism in Organic Chemistry* (Cornell Univ. Press, New York, 2nd ed., 1969) p 435; M. H. Abraham, *Prog. Phys. Org. Chem.* **11,** 1 (1974); E. M. Arnett, R. Reich, *J. Am. Chem. Soc.* **102,** 5892 (1980); S. Shaik *et al., ibid.* **116,** 262 (1994); S. H. Kim *et al., J. Phys. Org. Chem.* **11,** 254 (1998). Solvent effects: J.-L. M. Abboud *et al., J. Phys. Chem.* **93,** 214 (1989); S.-G. Kang *et al., Bull. Chem. Soc. Jpn.* **66,** 972 (1993).

276. Merrifield Solid-Phase Peptide Synthesis (SPPS)

R. B. Merrifield, *J. Am. Chem. Soc.* **85,** 2149 (1963).

Synthesis of long peptides involving the following steps: (1) attachment of the C-terminal amino acid to an insoluble polymeric support resin, (2) elongation of the peptide chain, and (3) cleavage of the peptide from the resin:

Elongation step (P = support resin):

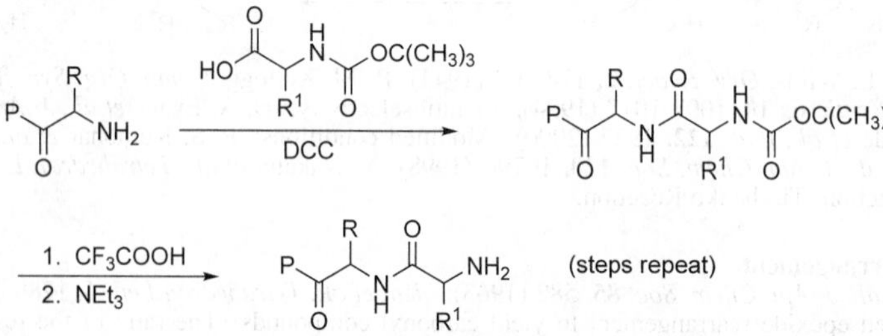

Method for monitoring synthesis: B. D. Larsen *et al., J. Am. Chem. Soc.* **115,** 6247 (1993). Synthetic applications: D. D. Smith *et al., J. Peptide Protein Res.* **44,** 183 (1994); M. J. O'Donnell *et al., J. Am. Chem. Soc.* **118,** 6070 (1996); R. Léger *et al., Tetrahedron Lett.* **39,** 4171 (1998). *Review:* C. Birr, *Aspects of Merrifield Peptide Synthesis,* K. Hafner *et al.,* Eds. (Springer-Verlag, New York, 1978) pp 102; B. Merrifield, *Science* **232,** 341-347 (1986); G. B. Wisdom *et al., Peptide Antigens* (Oxford University Press, 1994) pp 27-81. Autobiographical account: B. Merrifield, *Life During a Golden Age of Peptide Chemistry,* J. I. Seeman, Ed. (ACS, Washington, D.C., 1993) pp 54-118.

277. Meyer Reaction

G. Meyer, *Ber.* **16,** 1439 (1883).

Preparation of alkylstannonic acids by reacting alkali stannite with an alkyl iodide. When applied to alkali arsenites or plumbites the reaction yields alkylarsonic and alkylplumbonic acids, respectively:

$$Na_2SnO_2 \ + \ RX \longrightarrow RSnO_2Na \ + \ NaX$$

$$Na_3AsO_3 \ + \ RX \longrightarrow RAsO_3Na_2 \ + \ NaX$$

W. R. Cullen, *Adv. Organomet. Chem.* **4,** 148 (1966).

278. Meyer-Schuster Rearrangement; Rupe Rearrangement

K. H. Meyer, K. Schuster, *Ber.* **55,** 819 (1922); H. Rupe, E. Kambli, *Helv. Chim. Acta* **9,** 672 (1926).

Acid-catalyzed rearrangement of secondary and tertiary α-acetylenic alcohols to α,β-unsaturated carbonyl compounds; aldehydes result when the acetylenic group is terminal, ketones when it is internal:

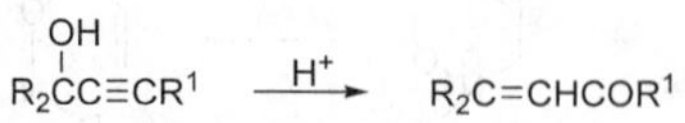

$$\underset{\displaystyle}{R_2\overset{\displaystyle OH}{\overset{|}{C}}C\equiv CR^1} \xrightarrow{\ H^+\ } R_2C=CHCOR^1$$

The conversion of tertiary alkylacetylenic carbinols with a terminal acetylenic group to predominantly α,β-unsaturated ketones and not the expected aldehydes, is referred to as the **Rupe rearrangement:**

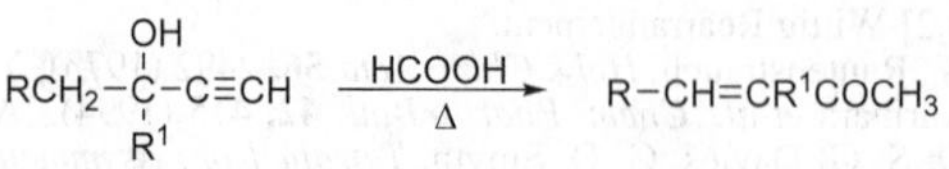

$$RCH_2-\overset{\displaystyle OH}{\underset{\displaystyle R^1}{\overset{|}{\underset{|}{C}}}}-C\equiv CH \xrightarrow[\Delta]{HCOOH} R-CH=CR^1COCH_3$$

Metal-based catalysis: P. Chabardes, *Tetrahedron Lett.* **29**, 6253 (1988); C. Y. Lorber, J. A. Osborn, *ibid.* **37**, 853 (1996). Mechanism studies: M. Edens *et al.*, *J. Org. Chem.* **42**, 3403 (1977); J. Andres *et al.*, *J. Am. Chem. Soc.* **110**, 666 (1988). Applications: E. A. Omar *et al.*, *J. Heterocycl. Chem.* **29**, 947 (1992); M. Yoshimatsu *et al.*, *J. Org. Chem.* **60**, 4798 (1995). Early reviews: R. Heilmann, R. Glenat, *Ann. Chim. (Paris)* **8**, 178 (1963); S. Swaminathan, K. V. Narayanan, *Chem. Rev.* **71**, 429 (1971).

279. Meyer Synthesis (Victor Meyer Synthesis)

V. Meyer, O. Stuber, *Ber.* **5**, 203 (1872).

Formation of aliphatic nitrites and nitro derivatives by the reaction of aliphatic halides with metal nitrites:

$$RX \ + \ MNO_2 \ \longrightarrow \ RONO \ + \ RNO_2 \ + \ MX$$

R. B. Reynolds, H. Adkins, *J. Am. Chem. Soc.* **51**, 279 (1929). *Reviews:* H. B. Hass, E. F. Riley, *Chem. Rev.* **32**, 373 (1943); N. Kornblum, *Org. React.* **12**, 101-156 (1962). Application to the synthesis of α,ω-dinitroalkanes: J. K. Stille, E. D. Vessel, *J. Org. Chem.* **25**, 478 (1960); G. Leston, *Org. Synth.* **4**, 368 (1963).

280. Meyers Aldehyde Synthesis

A. I. Meyers *et al.*, *J. Am. Chem. Soc.* **91**, 763 (1969); *eidem*, *J. Org. Chem.* **38**, 36 (1973).

Synthesis of aldehydes from alkylhalides and 2-lithiomethyltetrahydro-3-oxazine:

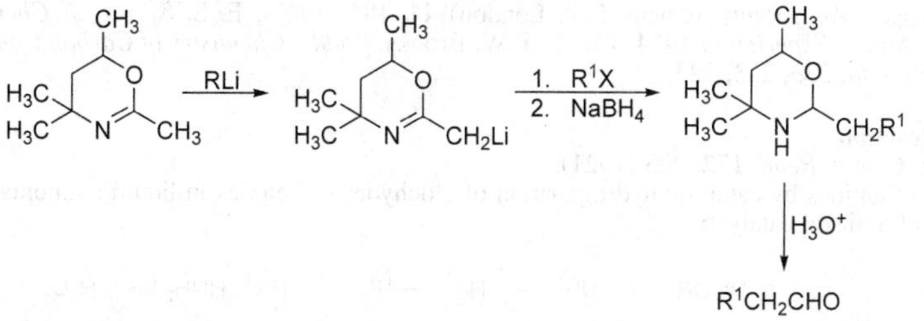

J. March, *Advanced Organic Chemistry* (Wiley-Interscience, New York, 4th ed., 1992) pp 478-479; A. I. Meyers *et al.*, *J. Org. Chem.* **46**, 783 (1981).

281. Michael Reaction (Addition) (Condensation)

A. Michael, *J. Prakt. Chem.* [2] **35**, 349 (1887).

Base-promoted conjugate addition of carbon nucleophiles (donors) to activated unsaturated systems (acceptors):

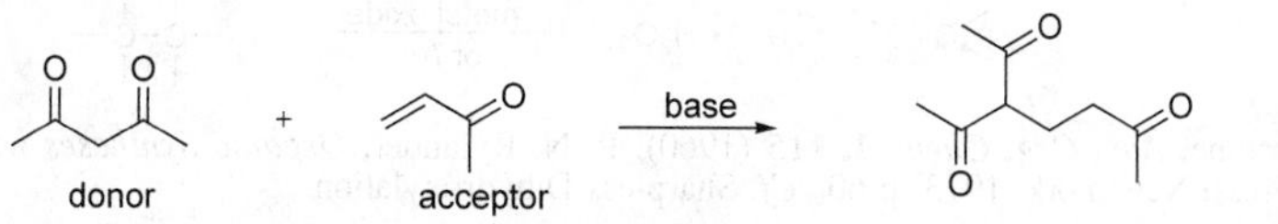

donor = malonates, cyanoacetates, acetoacetates, carboxylic esters, ketones, aldehydes, nitriles, nitro compounds, sulfones

acceptor = α,β-unsaturated ketones, esters, aldehydes, amides, carboxylic acids, nitriles, sulfoxides, sulfones, nitro compounds, phosphonates, phosphoranes

base = $NaOCH_2CH_3$, $NH(CH_2CH_3)_2$, KOH, $KOC(CH_3)_3$, $N(CH_2CH_3)_3$, NaH, BuLi, LDA

Reviews: E. D. Bergmann *et al.*, *Org. React.* **10**, 179-555 (1959); H. O. House, *Modern Synthetic Reactions* (W. A. Benjamin, Menlo Park, California, 2nd ed., 1972) pp 595-623; M. E. Jung, *Comp. Org. Syn.* **4**, 1-67 (1991). Review of organometallic nucleophiles: D. A. Hunt *et al.*, *Org. Prep. Proced. Int.* **21**, 705-749 (1989); V. J. Lee, *Comp. Org. Syn.* **4**, 69-137, 139-168 (1991); J. A. Kozlowski, *ibid.* 169-198. Reviews of stereoselective synthesis: H.-G. Schmalz, *ibid.* 199-236; D. A. Oare, C. H. Heathcock, *Top. Stereochem.* **20**, 87-170 (1991); J. d'Angelo *et al.*, *Tetrahedron: Asymmetry* **3**, 459-505 (1992); J. Leonard *et al.*, *Eur. J. Org. Chem.* **1998**, 2051-2061. *Cf.* Nagata Hydrocyanation; Robinson Annulation.

282. Michaelis-Arbuzov Reaction

A. Michaelis, R. Kaehne, *Ber.* **31**, 1048 (1898); A. E. Arbuzov, *J. Russ. Phys. Chem. Soc.* **38**, 687 (1906); *Chem. Zentralbl.* **1906**, II, 1639.

Formation of monoalkylphosphonic esters from alkyl halides and trialkyl phosphites, *via* the intermediate phosphonium salt:

$$P(OR)_3 \ + \ R^1X \ \longrightarrow \ [(RO)_3PR^1]X \ \xrightarrow{\Delta} \ (RO)_2P(O)R^1$$

Organic Name Reactions

K. Sasse, *Houben-Weyl* **12/1**, 433 (1963); B. A. Arbuzov, *Pure Appl. Chem.* **9**, 307 (1964); G. M. Kosolapoff, *Org. React.* **6**, 276 (1951); D. Redmore, *Chem. Rev.* **71**, 317 (1971); G. Bauer, G. Haegele, *Angew. Chem. Int. Ed.* **16**, 477 (1977); A. K. Bhattacharya, G. Thyagarajan, *Chem. Rev.* **81**, 415 (1981); B. Faure *et al.*, *Chem. Commun.* **1989**, 805; V. K. Yadav, *Synth. Commun.* **20**, 239 (1990).

283. Miescher Degradation
C. Meystre *et al.*, *Helv. Chim. Acta* **27**, 1815 (1944).

Adaptation of the Barbier-Wieland degradation, *q.v.*, to permit simultaneous elimination of three carbon atoms, as in degradation of the bile acid side chain to the methyl ketone stage. Conversion of the methyl ester of the bile acid to the tertiary alcohol, followed by dehydration, bromination, dehydrohalogenation and oxidation of the diene yields the chain-shortened ketone:

$$RR^1CHCH_2CH_2COOCH_3 \xrightarrow{PhMgBr} RR^1CHCH_2CH_2C(OH)Ph_2 \xrightarrow{-H_2O}$$

$$RR^1CHCH_2CH=CPh_2 \xrightarrow{(CH_2CO)_2NBr} RR^1CHCHBrCH=CPh_2 \xrightarrow{-HBr}$$

$$RR^1C=CH-CH=CPh_2 \xrightarrow{CrO_3} RCOR^1 + OHCCH=CPh_2$$

C. W. Shoppee, *Ann. Repts.* (Chem. Soc. London) **44**, 184 (1947); F. S. Spring, *J. Chem. Soc.* **1950**, 3355; A. Wettstein, G. Anner, *Experientia* **1954**, 407; C. J. W. Brooks, *Rodd's Chemistry of Carbon Compounds* **IID**, 26 (1970); P. G. Marshall, *ibid.* 233, 253, 323.

284. Mignonac Reaction
G. Mignonac, *Compt. Rend.* **172**, 223 (1921).

Formation of amines by catalytic hydrogenation of aldehydes or ketones in liquid ammonia and absolute ethanol in the presence of a nickel catalyst:

$$RCOR^1 + NH_3 + H_2 \xrightarrow{Ni} RR^1CHNH_2 + H_2O$$

F. Randvere, *An. Farm. Bioquim.* **18**, 81 (1948); *Houben-Weyl* **4/2**, 51 (1955).

285. Milas Hydroxylation of Olefins
N. A. Milas *et al.*, *J. Am. Chem. Soc.* **58**, 1302 (1936); **59**, 543, 2342, 2345 (1937); **61**, 1844 (1939); **62**, 1841 (1940).

Formation of *cis*-glycols by reaction of alkenes with hydrogen peroxide and either ultraviolet light or a catalytic amount of osmium, vanadium, or chromium oxide:

$$\text{>C=C<} + H_2O_2 \xrightarrow[\text{or } h\nu]{\text{metal oxide}} \begin{array}{c} HO \ \ OH \\ | \ \ \ | \\ -C-C- \\ | \ \ \ | \end{array}$$

F. D. Gunstone, *Adv. Org. Chem.* **1**, 115 (1960); P. N. Rylander, *Organic Syntheses with Noble Metal Catalysts* (Academic Press, New York, 1973) p 60. *Cf.* Sharpless Dihydroxylation.

286. Mislow-Evans Rearrangement
P. Bickart *et al.*, *J. Am. Chem. Soc.* **90**, 4869 (1968); D. A. Evans *et al.*, *ibid.* **93**, 4956 (1971).

[2,3]-Sigmatropic rearrangement of allylic sulfoxides to allylic sulfenates which are captured by thiophiles to generate the allylic alcohols, thereby effecting the 1,3-transposition of sulfoxide and alcohol functions. The reverse process is accomplished by treating the alcohol with arylsulfenyl chloride, followed by thermal rearrangement of the sulfenate to generate the allylic sulfoxide:

thiophile = $P(OCH_3)_3$, $(CH_3CH_2)_2NH$, piperidine, PhS^-

Early review: D. A. Evans, G. C. Andrews, *Acc. Chem. Res.* **7**, 147 (1974). Acid-catalyzed modification: Y. Masaki *et al.*, *Chem. Pharm. Bull.* **33**, 2531 (1985). Synthetic applications: H. J. Reich, S. Wollowitz, *J. Am. Chem. Soc.* **104**, 7051 (1982); G. H. Posner *et al.*, *J. Org. Chem.* **52**, 4836 (1987); A. Padwa *et al.*, *ibid.* **56**, 4252 (1991). Mechanistic studies: D. K. Jones-Hertzog, W. L. Jorgensen, *ibid.* **60**, 6682 (1995); *eidem, J. Am. Chem. Soc.* **117**, 9077 (1995). *Cf.* Meisenheimer Rearrangements; [2,3]-Wittig Rearrangement.

287. Mitsunobu Reaction
O. Mitsunobu *et al.*, *Bull. Chem. Soc. Jpn.* **40**, 935 (1967); O. Mitsunobu, Y. Yamada, *ibid.* 2380.

Condensation of alcohols and acidic components (NuH) on treatment with dialkyl azodicarboxylates and trialkyl- or triarylphosphines occurring primarily with inversion of configuration *via* the proposed intermediary oxyphosphonium salts:

NuH = phosphoric mono- and diesters, carboxylic acids, phenols, imides, oximes, hydroxymates, heterocycles, thiols, thioamides, β-keto esters

Methods development: R. F. C. Brown *et al., Tetrahedron* **50**, 5469 (1994); T. Tsunoda *et al., Tetrahedron Lett.* **40**, 7355 (1999); J. C. Pelletier, S. Kincaid, *ibid.* **41**, 797 (2000). Synthetic applications: M. A. Poelert *et al., Rec. Trav. Chim.* **113**, 355 (1994); A. Viso *et al., Tetrahedron Lett.* **41**, 407 (2000); H. Schedel *et al., Tetrahedron: Asymmetry* **11**, 2125 (2000). Solid-phase synthesis: S. R. Chhabra *et al., Tetrahedron Lett.* **41**, 1099 (2000); F. Zaragoza, H. Stephensen, *ibid.* 2015; P.-P. Kung, E. Swayze, *ibid.* **40**, 5651 (1999). Mechanism: T. Watanabe *et al., Chirality* **12**, 346 (2000). *Reviews:* O. Mitsunobu, *Synthesis* **1981**, 1-28; D. L. Hughes, *Org. React.* **29**, 1-162 (1983); D. L. Hughes, *Org. Prep. Proced. Int.* **28**, 127-164 (1996); J. A. Dodge, S. A. Jones, *Recent Res. Dev. Org. Chem.* **1**, 273-283 (1997); S. Dandapani, D. P. Curran, *Chem. Eur. J.* **10**, 3130-3138 (2004); R. Dembinski, *Eur. J. Org. Chem.* **2004**, 2763-2772.

288. Miyaura Boration
T. Ishiyama *et al., J. Org. Chem.* **60**, 7508 (1995).

Formation of an arylboronic, heteroarylboronic, or alkenylboronic ester from the palladium-catalyzed cross-coupling reaction of bis(pinacolato)diboron with the corresponding halide or triflate. Expansion of this method has also allowed for the preparation of allylboronic esters:

X = Cl, I, Br, OTf
Pd-based catalyst (e.g. PdCl₂(dppf), Pd(PPh₃)₂)
base (e.g. KOAc, KOPh, Et₃N)

Synthesis of arylboronic esters: M. Murata *et al., J. Org. Chem.* **62**, 6458 (1997); T. Ishiyama *et al., Tetrahedron Lett.* **38**, 3447 (1997); of alkenylboronic esters: J. Takagi *et al., J. Am. Chem. Soc.* **124**, 8001 (2002); T. Ishiyama *et al., J. Organomet. Chem.* **687**, 284 (2003). Method expansion to synthesize allylboronates: *idem et al., Tetrahedron Lett.* **37**, 6889 (1996); F.-Y. Yang *et al., J. Am. Chem. Soc.* **122**, 7122 (2000). Borylation with aryldiazonium tetrafluoroborate salts: D. M. Willis, R. M. Strongin, *Tetrahedron Lett.* **41**, 8683 (2000). Mechanistic studies: M. Sumimoto *et al., J. Am. Chem. Soc.* **126**, 10457 (2004). *Review:* K. Matos, E. Burkhardt, *Spec. Chem.* **25** (5), 40-44 (2005).

289. Moore Cyclization
J. O. Karlsson *et al., J. Am. Chem. Soc.* **107**, 3392 (1985).

Thermal generation of a biradical species by cyclization of an enyne-ketene. Reaction of the radical with a hydrogen donor gives the arene:

A. Rahm, W. D. Wulff, *J. Am. Chem. Soc.* **118**, 1807 (1996). *Reviews:* H. W. Moore, B. R. Yerxa, *Chemtracts* **1992**, 273-313; M. E. Maier, *Synlett* **1995**, 13-26; K. K. Wang, *Chem. Rev.* **96**, 207-222 (1996). *Cf.* Bergman Reaction.

290. Mukaiyama Aldol Reaction
T. Mukaiyama *et al., Chem. Lett.* **1973**, 1011; *idem et al., ibid.* **1974**, 323; *eidem, J. Am. Chem. Soc.* **96**, 7503 (1974).

Formation of β-hydroxy ketones via reaction of silyl enol ethers or ketene silyl acetals with aldehydes in presence of a Lewis acid, such as titanium tetrachloride, tin tetrachloride or boron trifluoride etherate:

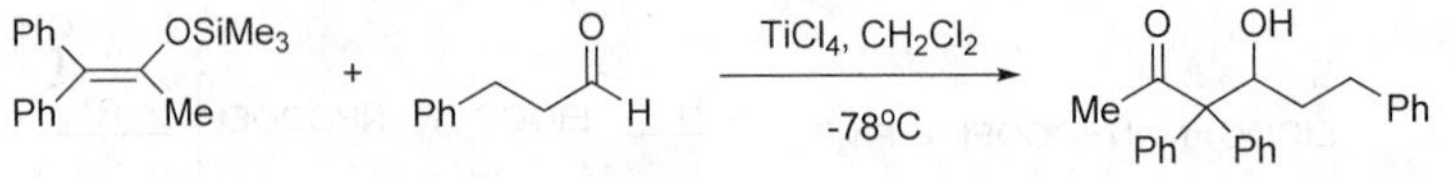

Enantioselectivity: E. M. Carreira *et al., J. Am. Chem. Soc.* **116**, 8837 (1994). Diastereoselectivity: S. E. Denmark *et al., Tetrahedron* **54**, 10389 (1998). *Reviews:* H. Gröger *et al., Chem. Eur. J.* **4**, 1137-1141 (1998); E. M. Carreira in *Comprehensive Asymmetric Catalysis I-III* **vol. 3**, E. N. Jacobsen *et al.,* Eds. (Springer-Verlag, Berlin, Germany, 1999) 997-1065; K. Iseki, *ACS Symp. Ser.* **746**, 38-51 (2000). *Cf.* Aldol Reaction; Evans Aldol Reaction.

291. Mukaiyama-Michael Reaction

K. Narasaka *et al., Bull. Chem. Soc. Jpn.* **49**, 779 (1976).

Formation of 1,5-dicarbonyl compounds by reaction of ketene silyl acetals with α,β-unsaturated ketones and esters:

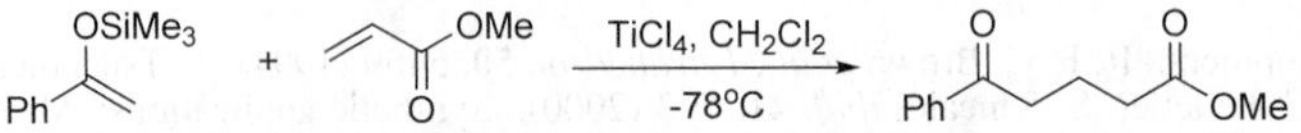

T. Mukaiyama, S. Kobayashi, *Heterocycles* **25**, 245 (1987). Enhanced diastereoselectivity: J. Otera *et al., Tetrahedron* **52**, 9409 (1996); in tandem-aldol reaction: N. Giuseppone *et al., Tetrahedron Lett.* **39**, 7874 (1998). Synthetic application: H. Paulsen *et al., Angew. Chem. Int. Ed.* **38**, 3373 (1999).

292. Myers-Saito Cyclization (Myers Cyclization) (Saito-Myers Cycloaromatization)

A. G. Myers *et al., J. Am. Chem. Soc.* **111**, 8057 (1989); R. Nagata *et al., Tetrahedron Lett.* **30**, 4995 (1989).

Allenyl enyne cyclization leading to a biradical species that subsequently reacts with a hydrogen donor to give the corresponding arene:

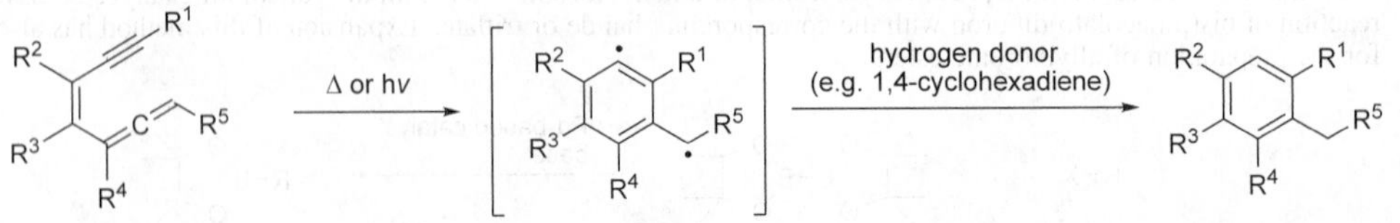

Mechanism and reactivity studies: A. G. Myers *et al., J. Am. Chem. Soc.* **114**, 9369 (1992). Reaction modifications involving cyclization at subambient temperatures: A. G. Myers, P. S. Dragovich, *ibid.* **111**, 9130 (1989); azaenyne allene cyclization: L. Feng *et al., Org. Lett.* **6**, 2059 (2004). *Review*: K. K. Wang, *Chem. Rev.* **96**, 207-222 (1996).

293. Nagata Hydrocyanation

W. Nagata *et al., Tetrahedron Lett.* **1962**, 461.

Alkylaluminum-mediated 1,4-addition of hydrogen cyanide to α,β-unsaturated carbonyl compounds:

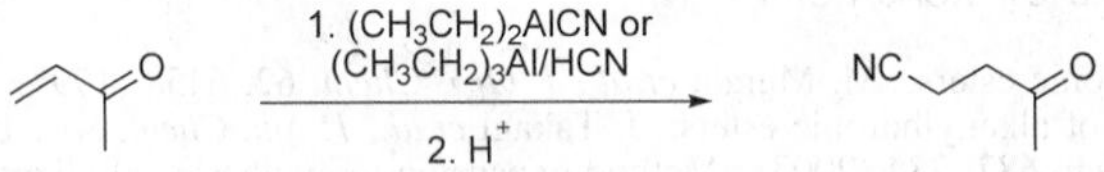

Early review: W. Nagata, M. Yoshioka, *Org. React.* **25**, 255-476 (1977). Synthetic application: T. F. Gallagher, J. L. Adams, *J. Org. Chem.* **57**, 3347 (1992). *Cf.* Michael Reaction.

294. Nametkin Rearrangement

S. S. Nametkin, *Ann.* **432**, 207 (1923).

A special case of carbonium ion rearrangement in camphene hydrochloride derivatives involving the migration of a methyl group:

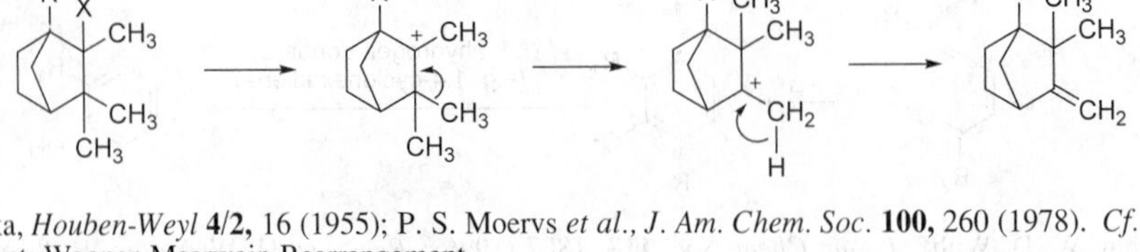

H. Henecka, *Houben-Weyl* **4/2**, 16 (1955); P. S. Moervs *et al., J. Am. Chem. Soc.* **100**, 260 (1978). *Cf.* Retropinacol Rearrangement; Wagner-Meerwein Rearrangement.

295. Nazarov Cyclization Reaction

I. N. Nazarov *et al., Izv. Akad. Nauk SSSR Otd. Khim. Nauk* **1942**, 200.

Protic or Lewis acid-catalyzed electrocyclic ring closure of divinyl ketones, or their equivalents, to yield 2-cyclopentenones:

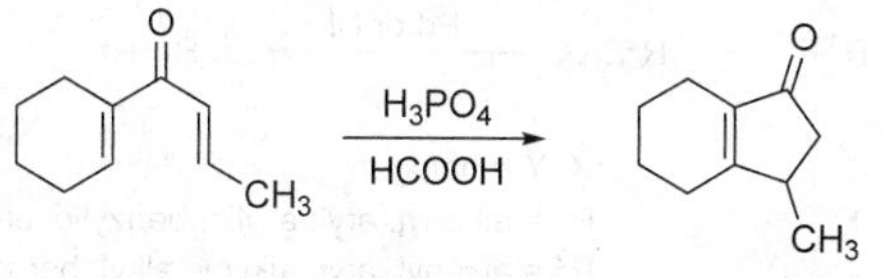

Silicon-directed cyclizations to α-methylenecyclopentanones: H. T. Kang *et al., Tetrahedron Lett.* **33**, 3495 (1992). Diastereoselectivity of interrupted reaction: J. A. Bender *et al., J. Org. Chem.* **63**, 2430 (1998); *idem et al., J. Am. Chem. Soc.* **121**, 7443 (1999); H. Hu *et al., ibid.* **121**, 9895 (1999). Lewis acid catalyzed reactions: C. Kuroda *et al., Chem. Commun.* **1997**, 1177; H. A. Buchholz, A. de Meijere, *Eur. J. Org. Chem.* **1998**, 2301. *Reviews:* S. E. Denmark, *Comp. Org. Syn.* **5**, 751-784 (1991); K. L. Habermas *et al., Org. React.* **45**, 1-158 (1994); S. Giese, F. G. West, *Tetrahedron Lett.* **39**, 8393 (1998); of interrupted reaction: D. Zuev, L. A. Paquette, *Chemtracts* **12**, 1019-1025 (1999); of methodological advances: M. A. Tius, *Eur. J. Org. Chem.* **2005**, 2193-2206.

296. Neber Rearrangement
P. W. Neber, A. v. Friedolsheim, *Ann.* **449**, 109 (1926); P. W. Neber, G. Huh, *ibid.* **515**, 283 (1935).

Formation of α-amino ketones by treatment of sulfonic esters of ketoximes with potassium ethoxide, followed by hydrolysis:

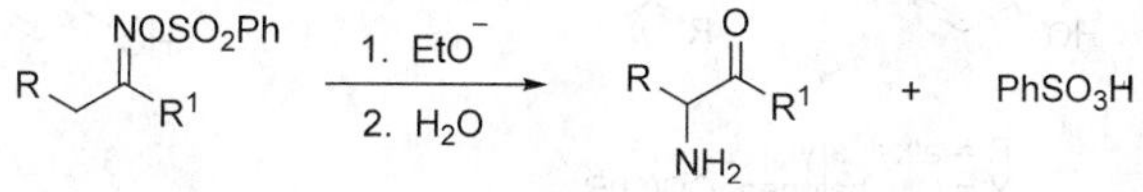

Reviews: C. O'Brien, *Chem. Rev.* **64**, 81 (1964); C. G. McCarty in *The Chemistry of the Carbon-Nitrogen Double Bond*, S. Patai, Ed. (Interscience, New York, 1970) p 447; Y. Tamura *et al., Synthesis* **1973**, 215; R. F. Parcell, J. C. Sanchez, *J. Org. Chem.* **46**, 5229 (1981); K. Maruoka, H. Yamamoto, *Comp. Org. Syn.* **6**, 786-789 (1991). Synthetic applications: I. Moldvai *et al., Heterocycles* **43**, 2377 (1996); M. J. Mphahlele, T. A. Modro, *Phosphorus Sulfur Silicon Relat. Elem.* **127**, 131 (1997); J. Y. L. Chung *et al., Tetrahedron Lett.* **40**, 6739 (1999).

297. Nef Reaction
J. U. Nef, *Ann.* **280**, 263 (1894).

Formation of aldehydes and ketones from primary and secondary nitroalkanes, respectively, by treatment of their salts with sulfuric acid:

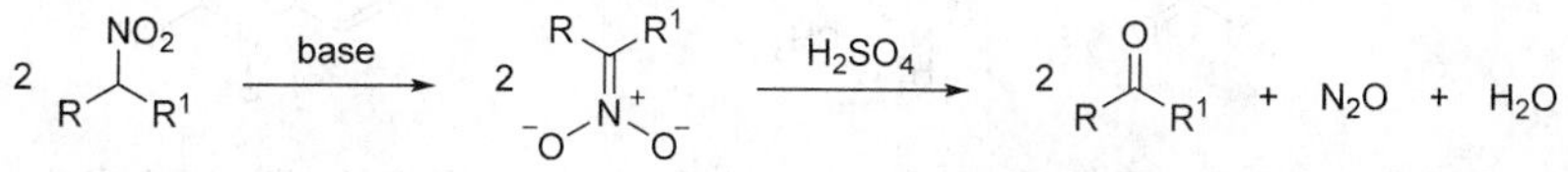

Modified conditions: W. Adam *et al., Synlett* **1998**, 1335; P. Ceccherelli *et al., Synth. Commun.* **28**, 3057 (1998). Application to spiroketals: T. Capecchi *et al., Tetrahedron Lett.* **39**, 5429 (1998). *Reviews:* P. Salomaa in *The Chemistry of the Carbonyl Group*, S. Patai, Ed. (Interscience, N.Y., 1966) pp 177-210; H. W. Pinnick, *Org. React.* **38**, 655-792 (1990); D. S. Grierson, H.-P. Husson, *Comp. Org. Syn.* **6**, 937-944 (1991); R. Ballini, M. Petrini, *Tetrahedron* **60**, 1017-1047 (2004).

298. Nef Synthesis
J. U. Nef, *Ann.* **308**, 281 (1899).

Addition of sodium acetylides to aldehydes and ketones to yield acetylenic carbinols; occasionally and erroneously referred to as the Nef reaction, *q.v.*:

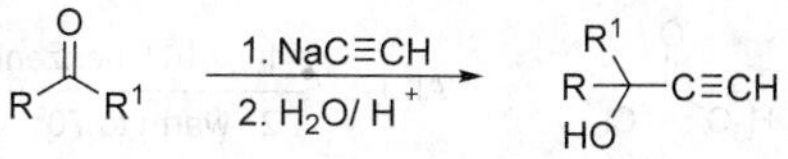

Farbenfabriken Bayer, **DE 280226; DE 285770** (1913); J. H. Saunders, *Org. Synth.* **20**, 40 (1940); A. W. Johnson, *The Chemistry of the Acetylenic Compounds* (London, 1946) p 11; C. D. Hurd, W. D. McPhee, *J. Am. Chem. Soc.* **69**, 239 (1947); W. Oroschnik, A. O. Mebane, *ibid.* **71**, 2062 (1949); R. A. Raphael, *Acetylenic Compounds in Organic Synthesis* (London, 1955) p 10. *Cf.* Arens-van Dorp Synthesis; Favorskii-Babayan Synthesis.

299. Negishi Cross Coupling
E. Negishi *et al., J. Org. Chem.* **42**, 1821 (1977).

Formation of unsymmetric biaryls by cross coupling arylhalides with arylzinc reagents in presence of catalytic Ni or Pd:

Organic Name Reactions

$$R^1Y \quad + \quad R^2ZnX \quad \xrightarrow{\text{Pd or Ni}} \quad R^1-R^2 \quad + \quad ZnXY$$

X, Y = halogen
R^1 = alkenyl, aryl, allylic, benzylic, propargylic
R^2 = alkenyl, aryl, alkyny, alkyl, benzylic, allylic

Synthetic application: S. Superchi *et al., Tetrahedron Lett.* **37,** 6057 (1996); J. A. Miller, R. P. Farrell, *ibid.* **39,** 6441 (1998). Extension to additional functional groups: E. Negishi, *Acc. Chem. Res.* **15,** 340 (1982). *Review:* P. Knochel, R. D. Singer, *Chem. Rev.* **93,** 2117-2188 (1993); *idem et al., Ber.* **130,** 1021-1027 (1997).

300. Nencki Reaction
M. Nencki, N. Sieber, *J. Prakt. Chem.* (2) **23,** 147 (1881).
The ring acylation of phenols with acids in the presence of zinc chloride, or the modification of the Friedel-Crafts reaction, *q.v.,* by substitution of ferric chloride for aluminum chloride.

R = alkyl, aryl
Y = OH, halogen, OC(O)R

M. Nencki, W. Schmid, *ibid.* 546; M. Nencki, *ibid.* **25,** 273 (1882); U. S. Chiema, K. Venkataraman, *J. Chem. Soc.* **1932,** 918; C. W. Schellhammer, *Houben-Weyl* **7/2a,** 284 (1973); A. S. Anjaneyulu *et al., Indian J. Chem.* **33B,** 847 (1994). *Cf.* Darzens-Nenitzescu Synthesis of Ketones.

301. Nenitzescu Indole Synthesis
C. D. Nenitzescu, *Bull. Soc. Chim. Romania* **11,** 37 (1929).
Synthesis of 5-hydroxyindole derivatives by condensation of *p*-benzoquinone with *β*-aminocrotonic esters:

Reviews: R. K. Brown in *The Chemistry of Heterocyclic Compounds,* W. J. Houlihan, Ed. (Wiley, New York, 1972) p 413; G. R. Allen, Jr., *Org. React.* **20,** 337 (1973). Synthetic applications: U. Kuecklander, W. Huehnermann, *Arch. Pharm.* **312,** 515 (1979); J. L. Bernier, J. P., Henichart, *J. Org. Chem.* **46,** 4197 (1981). M. Kinugawa *et al., J. Chem. Soc. Perkin Trans. 1* **1995,** 2677; J. M. Pawlak *et al., J. Org. Chem.* **61,** 9055 (1996).

302. Nenitzescu Reductive Acylation
C. D. Nenitzescu, E. Cioranescu, *Ber.* **69,** 1820 (1936).
Hydrogenative acylation of cycloolefins with acid chlorides in the presence of aluminum chloride; with five- and six-membered rings no change in ring size occurs but with seven-membered rings rearrangement takes place with formation of a cyclohexane derivative:

C. Nenitzescu, C. N. Ionescu, *Ann.* **491,** 189 (1931); C. D. Nenitzescu, J. P. Cantuniari, *ibid.* **510,** 269 (1934); C. D. Nenitzescu, I. Chicos, *Ber.* **68,** 1584 (1935); C. A. Thomas, *Anhydrous Aluminum Chloride in Organic Chemistry* (New York, 1941) p 759; S. L. Friess, R. Pinson, *J. Am. Chem. Soc.* **73,** 3512 (1951); Olah, *Friedel-Crafts and Related Reactions* **vol. III,** Part 2 (New York, 1964) p 1069. *Cf.* Darzens-Nenitzescu Synthesis of Ketones.

303. Nicholas Reaction
R. F. Lockwood, K. M. Nicholas, *Tetrahedron Lett.* **18,** 4163 (1977).
The reaction of dicobalthexacarbonyl-stabilized propargyl cations with nucleophiles, followed by oxidative demetalation to yield propargylated products:

R^1 = H, SiR_3, alkyl, aryl; R^2 = H, CH_3
R^3 = H, Ph, alkyl, substituted alkyl; R^4 = alkyl, substituted alkyl
H^+ = $HBF_4 \cdot O(CH_3)_2$, CF_3COOH;
Lewis acid = $BF_3 \cdot OEt_2$, $EtAlCl_2$, $TiCl_4$, $Bu_2BOSO_2CF_3$
NuH = electron rich aromatics, β-dicarbonyl compounds, ketones,
enolates, allyl silanes, hydride, amines, enamines
[O] = $Fe(NO_3)_3$, $Ce(NH_4)_2(NO_3)_6$, $(CH_3)_3NO$

Stereochemistry: S. L. Schreiber *et al.*, *J. Am. Chem. Soc.* **109**, 5749 (1987); A. V. Muehldorf *et al.*, *Tetrahedron Lett.* **35**, 8755 (1994). Scope and limitations: K. D. Roth, *Synlett* **1992**, 435; K. D. Roth, U. Müller, *Tetrahedron Lett.* **34**, 2919 (1993). Synthetic applications: P. A. Jacobi, W. Zheng, *ibid.* 2581, 2585; E. Tyrrell *et al.*, *Synlett* **1993**, 769. Diastereoselective applications: J. Berge *et al.*, *Tetrahedron Lett.* **38**, 685 (1997); A. Mann *et al.*, *J. Chem. Soc. Perkin Trans. 1* **1998**, 1427. Enantioselective applications: S. Tanaka *et al.*, *Tetrahedron* **50**, 12883 (1994); A. M. Montana *et al.*, *Tetrahedron Lett.* **40**, 6499 (1999). *Reviews:* K. M. Nicholas, *Acc. Chem. Res.* **20**, 207-214 (1987); B. J. Teobald, *Tetrahedron* **58**, 4133-4170 (2002).

304. Niementowski Quinazoline Synthesis
S. v. Niementowski, *J. Prakt. Chem.* [2] **51**, 564 (1895).
Formation of 4-oxo-3,4-dihydroquinazolines by cyclization of the reaction products of anthranilic acid and amides:

Reviews: T. A. Williamson, *Heterocycl. Compd.* **6**, 331 (1957); W. L. F. Armarego, *Adv. Heterocycl. Chem.* **1**, 253 (1963); E. Cuny *et al.*, *Tetrahedron Lett.* **21**, 3029 (1980).

305. Niementowski Quinoline Synthesis
S. v. Niementowski, *Ber.* **27**, 1394 (1894); **28**, 2809 (1895); **38**, 2044 (1905); **40**, 4285 (1907).
Formation of γ-hydroxyquinoline derivatives from anthranilic acids and carbonyl compounds:

R. H. Manske, *Chem. Rev.* **30**, 127 (1942); T. A. Williamson, *Heterocycl. Compd.* **6**, 331 (1957); W. L. F. Armarego, *Quinazolines* (Interscience, New York, 1967) p 74; E. Cuny *et al.*, *Tetrahedron Lett.* **21**, 3029 (1980). Synthetic applications: B. P. Suthar, *Indian J. Chem.* **21B**, 588 (1982); R. J. Chong *et al.*, *Tetrahedron Lett.* **27**, 5323 (1986); M. S. Khajavi *et al.*, *Iran. J. Chem. Chem. Eng.* **17**, 29 (1988). *Review:* T. Hisano, *Org. Prep. Proced. Int.* **5**, 145-193 (1973). *Cf.* Friedlaender Synthesis; Pfitzinger Reaction.

306. Nierenstein Reaction
D. A. Clibbens, M. Nierenstein, *J. Chem. Soc.* **107**, 1491 (1915).
Formation of ω-chloroacetophenones by reaction of diazomethane in dry ether with aroyl chlorides. Coumarano-nones are obtained if an *ortho*-hydroxy group is present:

$$ArCOCl + CH_2N_2 \longrightarrow ArCOCH_2Cl + N_2$$

W. E. Bachman, W. S. Struve, *Org. React.* **1**, 38 (1942); Y. Miyahara, *J. Heterocycl. Chem.* **16**, 1147 (1979).

307. Norrish Type Cleavage

R. G. W. Norrish, C. H. Bamford, *Nature* **138**, 1016 (1936); **140**, 195 (1937).

Norrish Type I Cleavage: Homolytic cleavage of aldehydes and ketones originating from their excited $n\pi^*$ state. Synthetically useful for the ring cleavage of cyclic ketones:

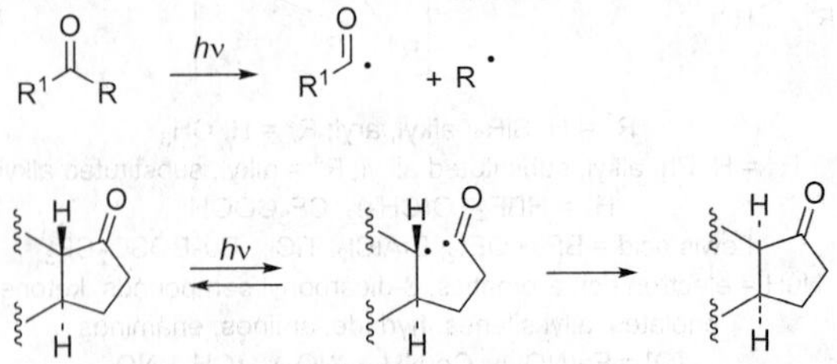

Norrish Type II Cleavage: Reaction originating from the $n\pi^*$ excited state of aldehydes and ketones that involves intramolecular γ-hydrogen abstraction followed by cleavage of the resulting diradical to an olefin and an enol which tautomerizes to the carbonyl compound:

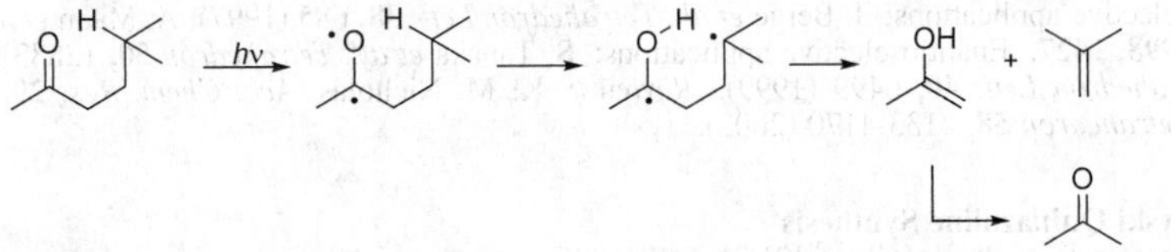

Norrish Type I: D. H. R. Barton *et al., J. Am. Chem. Soc.* **107**, 3607 (1985); J. R. Hwu *et al., Chem. Commun.* **1990**, 161. Norrish Type II: J. M. Nuss, M. M. Murphy, *Tetrahedron Lett.* **35**, 37 (1994); F. Hénin *et al., Tetrahedron* **50**, 2849 (1994). *Reviews:* J. D. Coyle, H. A. J. Carless, *Chem. Soc. Rev.* **1**, 465 (1972); O. L. Chapman, D. S. Weiss, *Org. Photochem.* **3**, 197-277 (1973); J. March, *Advanced Organic Chemistry* (Wiley-Interscience, New York, 4th ed., 1992) p 242; W. M. Horspool, *Photochemistry* **25**, 67-100 (1994). *Cf.* McLafferty Rearrangement.

308. Noyori Hydrogenation

T. Ikariya *et al., Chem. Commun.* **1985**, 922; R. Noyori *et al., J. Am. Chem. Soc.* **108**, 7117 (1986).

Homogeneous asymmetric catalytic hydrogenation of olefinic and carbonyl bonds mediated by enantiopure ruthenium(II) BINAP complexes. The substrates must have coordinating functionalities in neighboring positions which serve as directing groups during the transformation:

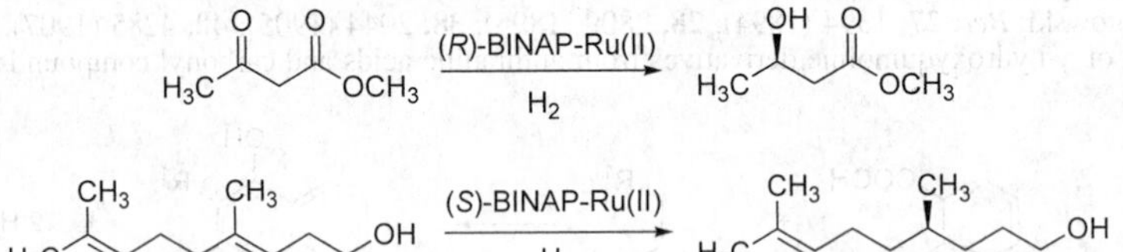

Detailed experimental procedure: M. Kitamura *et al., Org. Synth.* **71**, 1 (1993). Methods development for enamide substrates: *idem et al., J. Org. Chem.* **59**, 297 (1994); E. Vedejs *et al., ibid.* **64**, 6724 (1999). Development and use of arene substituted BINAP catalysts: K. Mashima *et al., ibid.* **59**, 3064 (1994). *Reviews:* H. Takaya *et al.* in *Adv. Chem. Ser.* **230**, entitled "Homogeneous Transition Metal Catalyzed Reactions" (ACS, Washington DC, 1992) pp 123-142; R. Noyori, *Asymmetric Catalysis in Organic Synthesis* (John Wiley & Sons, New York, 1994) pp 16-94.

309. Nozaki-Hiyama Coupling Reaction (Nozaki-Hiyama-Kishi Reaction)

Y. Okude *et al., J. Am. Chem. Soc.* **99**, 3179 (1977); K. Takai *et al., Tetrahedron Lett.* **24**, 5281 (1983).

Chromium chloride catalyzed redox additions of organic halides to aldehydes:

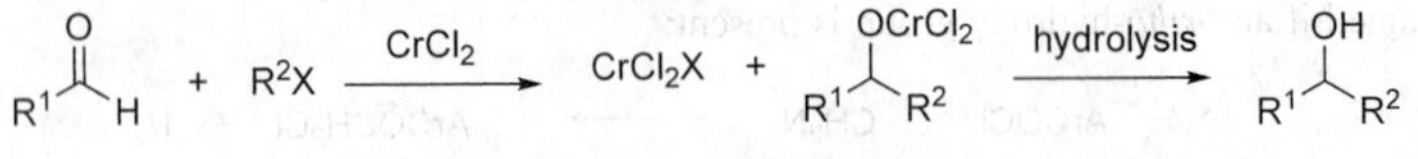

R_2 = aryl, alkynyl, alkenyl, allyl, propargyl

Catalysis with nickel salts: H. Jin *et al., J. Am. Chem. Soc.* **108**, 5644 (1986); K. Takai *et al., ibid.* 6048; with chromium salts: A. Furstner, N. Shi, *ibid.* **118**, 12349 (1996). Enantioselective reactions: K. Sugimoto *et al., J. Org. Chem.* **62**, 2322 (1997); M. Bandini *et al., Angew. Chem. Int. Ed.* **38**, 3357 (1999). Synthetic applications: Y. Kishi, *Pure Appl. Chem.* **64**, 343 (1992); D. P. Stamos *et al., J. Org. Chem.* **62**, 7552 (1997). *Review:* N. A. Saccomano, *Comp. Org. Syn.* **1**, 173-207 (1991).

310. Olefin Metathesis

Carbon-carbon bond rearrangements in presence of metal carbene catalyst complexes especially those of molybdenum and ruthenium:

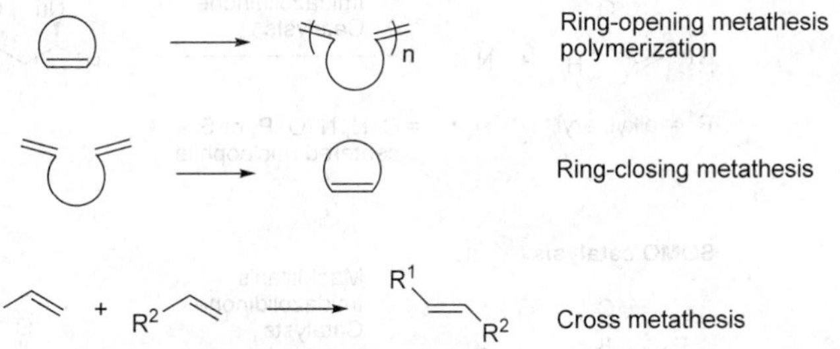

Ring-opening metathesis polymerization

Ring-closing metathesis

R^1 + R^2 ⟶ R^1 R^2 Cross metathesis

Comprehensive accounts: R. H. Grubbs, *Comp. Organomet. Chem.* **8**, 499 (1982); R. H. Grubbs, S. Chang, *Tetrahedron* **54**, 4413-4450 (1998). Synthetic applications: A. K. Chatterjee *et al.*, *J. Am. Chem. Soc.* **122**, 3728 (2000); C. W. Lee, R. H. Grubbs, *Org. Lett.* **2**, 2145 (2000). Series of articles on syntheses, polymerizations and catalysts: *J. Mol. Catal. A* **133**, 1-274 (1998). *Reviews:* A. Furstner, *Angew. Chem. Int. Ed.* **39**, 3012-3043 (2000); R. H. Grubbs, *Tetrahedron* **60**, 7117-7140 (2004). *See* monograph: Grubbs' Catalyst.

311. Oppenauer Oxidation

R. V. Oppenauer, *Rec. Trav. Chim.* **56**, 137 (1937).

The aluminum or potassium alkoxide-catalyzed oxidation of a secondary alcohol to the corresponding ketone (the reverse of the Meerwein-Ponndorf-Verley reduction, *q.v.*):

$$RCHOHR^1 + CH_3COCH_3 \underset{M = Al \; or \; K}{\overset{M(OC_3H_7)_3}{\rightleftharpoons}} RCOR^1 + CH_3CHOHCH_3$$

T. Beresin in *Newer Methods of Preparative Organic Chemistry* English Ed. (Interscience, New York, 1948) p 125; C. Djerassi, *Org. React.* **6**, 207 (1951); L. Horner, U. B. Kaps, *Ann.* **1980**, 192. Intramolecular reactions: B. B. Snider, B. E. Goldman, *Tetrahedron* **42**, 2951 (1986); G. A. Molander, J. A. McKie, *J. Am. Chem. Soc.* **115**, 5821 (1993). Alternate metals: B. Byrne, M. Karras, *Tetrahedron Lett.* **28**, 769 (1987); M. L. S. Almeida *et al.*, *J. Org. Chem.* **61**, 6587 (1996); K. Krohn *et al.*, *Synthesis* **1996**, 1341; K. Ishihara *et al.*, *J. Org. Chem.* **62**, 5664 (1997). *Review:* C. F. de Graauw *et al.*, *Synthesis* **1994**, 1007-1017. *Cf.* Cannizzaro Reaction.

312. Organocatalytic Diels-Alder Reaction (Iminium Diels-Alder Reaction); Enantioselective Imidazolidinone Organocatalysis

K. A. Ahrendt *et al.*, *J. Am. Chem. Soc.* **122**, 4243 (2000).

Highly enantioselective Diels-Alder reaction between an α,β-unsaturated carbonyl dienophile and a conjugated diene catalyzed by MacMillan's Imidazolidinone Catalysts:

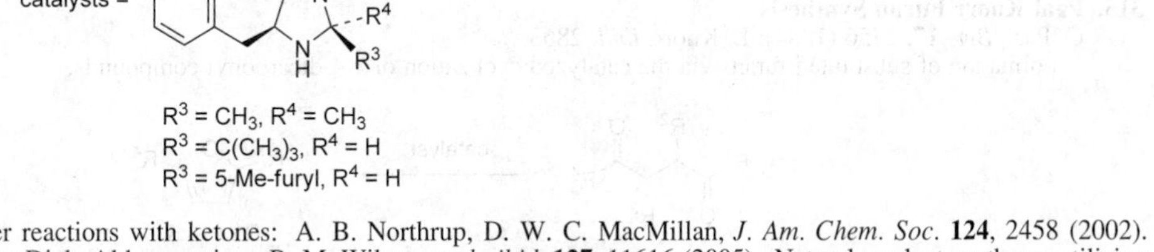

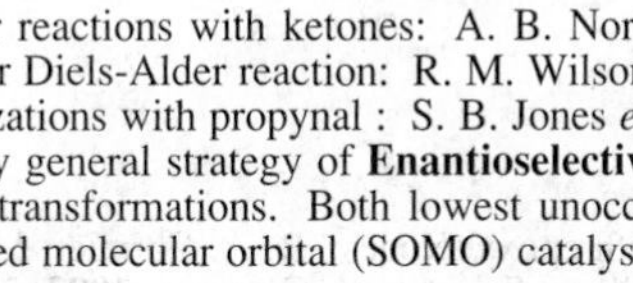

R^1 = alkyl, aryl   n = 0, 1, 2
R^2 = H, alkyl

endo adduct

catalysts =

R^3 = CH$_3$, R^4 = CH$_3$
R^3 = C(CH$_3$)$_3$, R^4 = H
R^3 = 5-Me-furyl, R^4 = H

Diels-Alder reactions with ketones: A. B. Northrup, D. W. C. MacMillan, *J. Am. Chem. Soc.* **124**, 2458 (2002). Intramolecular Diels-Alder reaction: R. M. Wilson *et al.*, *ibid.* **127**, 11616 (2005). Natural product syntheses utilizing cascade cyclizations with propynal : S. B. Jones *et al.*, *ibid.* **131**, 13606 (2009); *eidem*, *Nature* **475**, 183 (2011).

The broadly general strategy of **Enantioselective Imidazolidinone Organocatalysis** utilizes MacMillan's catalysts for aldehyde transformations. Both lowest unoccupied molecular orbital (LUMO)-lowering iminium activation and singly occupied molecular orbital (SOMO) catalysis enable a wide array of asymmetric reactions:

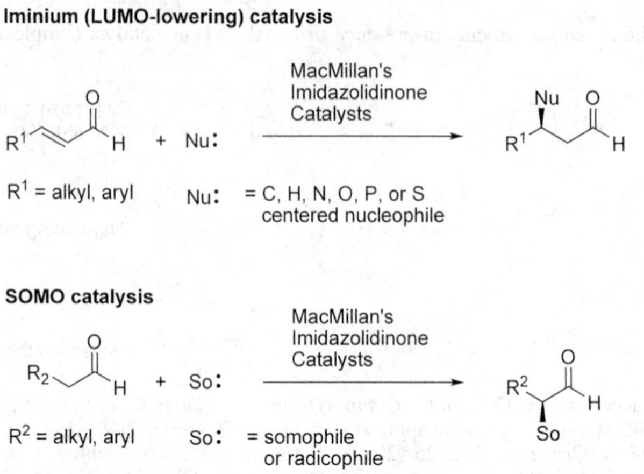
Iminium (LUMO-lowering) catalysis
MacMillan's
Imidazolidinone
Catalysts
R1 = alkyl, aryl
Nu: = C, H, N, O, P, or S
centered nucleophile
SOMO catalysis
R2 = alkyl, aryl
So: = somophile
or radicophile

Synthetic utility in Friedel-Crafts reactions: N. A. Paras, D. W. C. MacMillan, *J. Am. Chem. Soc.* **123**, 4370 (2001); in indole alkylations: J. F. Austin, D. W. C. MacMillan, *ibid.* **124**, 1172 (2002); in benzene alkylations: N. A. Paras, D. W. C. MacMillan, *ibid.* 7894; in Mukaiyama-Michael reactions: S. P. Brown *et al., ibid.* **125**, 1192 (2003); in aldehyde α-additions: T. D. Beeson *et al., Science* **316**, 582 (2007); H.-Y. Jang *et al., J. Am. Chem. Soc.*, **129**, 7004 (2007). Review of iminium activation: G. Lelais, D. W. C. MacMillan, *Aldrichim. Acta* **39**, 79-87 (2006); of organocatalysis: D. W. C. MacMillan, *Nature* **455**, 304-308 (2008). *Cf.* Diels-Alder Reaction. *See* monograph: MacMillan's Imidazolidinone Catalysts.

313. Overman Rearrangement

L. E. Overman, *J. Am. Chem. Soc.* **96,** 597 (1974); **98,** 2901 (1976).

Formal [3,3]-sigmatropic rearrangement of the trichloroacetimidate of allylic alcohols to allylic trichloroacetamides, thereby transposing the hydroxyl and amino functions with good chirality transfer:

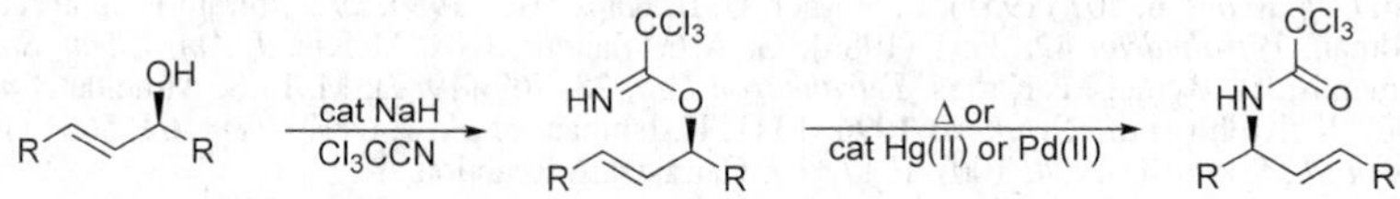
cat NaH
Cl3CCN
Δ or
cat Hg(II) or Pd(II)

Early review: L. E. Overman, *Acc. Chem. Res.* **13,** 218-224 (1980). Synthetic applications: M. Isobe *et al., Tetrahedron Lett.* **31,** 3327 (1990); T. Allmendinger *et al., ibid.* 7301; J. Gonda *et al., Synthesis* **1993,** 729; C. G. Cho *et al., Synth. Commun.* **30,** 1643 (2000); A. Montero *et al., Tetrahedron Lett.* **46,** 401 (2005). Use of a chiral template and mechanistic studies: T. Eguchi *et al., Tetrahedron* **49,** 4527 (1993). Modification of reaction conditions: T. Nishikawa *et al., J. Org. Chem.* **63,** 188 (1998). *Cf.* Claisen Rearrangement.

314. Oxo Process (Hydroformylation Reaction)

O. Roelen, **US 2327066** (1943); R. H. Hasek (Eastman), *Org. Chem. Bull.* **27,** No. 1 (1955).

Formation of alcohols from olefins, carbon monoxide and hydrogen in the liquid phase in the presence of catalysts (metallic cobalt compounds such as Raney cobalt or cobalt carbonyls) at 115-190° and high pressures (100-200 atmospheres) in a Fischer-Tropsch-type reaction, *q.v.* The process is sometimes carried out in two stages, the initial stage giving largely aldehydes which are then reduced to the alcohols.

B. Cornils, "Hydroformylation. Oxo Synthesis, Roelen Reaction" in *New Syntheses with Carbon Monoxide*, J. Falbe, Ed. (Springer-Verlag, Berlin, 1980) pp 1-225. Reppe modification (olefin + CO + H_2O + $Fe(CO)_5$): R. Massoudi *et al., J. Am. Chem. Soc.* **109,** 7428 (1987).

315. Paal-Knorr Furan Synthesis

C. Paal, *Ber.* **17**, 2756 (1884); L. Knorr, *ibid.* 2863.

Formation of substituted furans via the catalyzed cyclization of 1,4-dicarbonyl compounds:

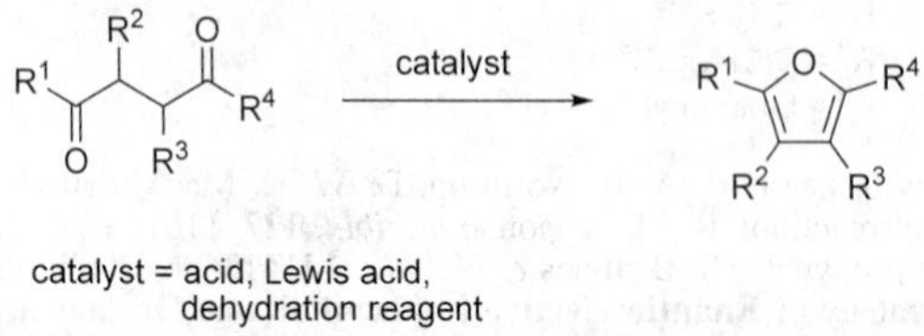
catalyst
catalyst = acid, Lewis acid,
dehydration reagent

Substrate expansion studies to substitute 1,4-dicarbonyls with epoxyketones: R. A. Cormier, M. D. Francis, *Synth. Commun.* **11**, 365 (1981); with alkynediols: J. Ji, X. Lu, *J. Chem. Soc. Chem. Commun.* **1993**, 764; with ene-diones and yne-diones: H. S. P. Rao, S. Jothilingam, *J. Org. Chem.* **68**, 5392 (2003). Solid phase synthesis applications: S. Raghavan, K. Anuradha, *Synlett* **2003**, 711. Use of an ionic liquid as a catalyst and reaction medium: G. Wang *et al.*, *Synth. Commun.* **40**, 370 (2010). Reaction mechanism studies: V. Amarnath, K. Amarnath, *J. Org. Chem.* **60**, 301 (1995). *Review:* K. M. Shea in *Name Reactions in Heterocyclic Chemistry*, J. J. Li, Ed. (Wiley, Hoboken, NJ, 2005) pp 168-181. *Cf.* Paal-Knorr Pyrrole Synthesis.

316. Paal-Knorr Pyrrole Synthesis
C. Paal, *Ber.* **18**, 367 (1885); L. Knorr, *ibid.* 299.
Formation of pyrroles via cyclization of 1,4-dicarbonyl compounds with ammonia or primary amines:

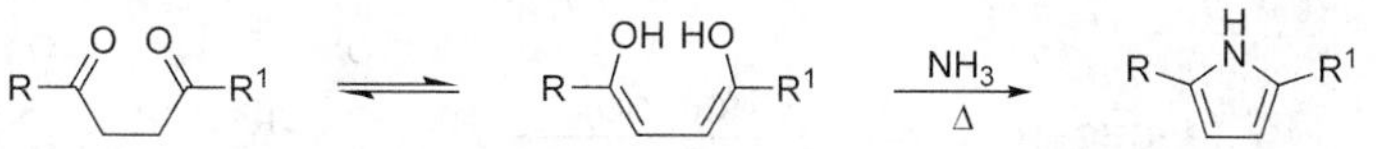

H. Fischer, H. Orth, *Die Chemie des Pyrrols* **1** (Leipzig, 1934) p 34; D. M. Young, C. F. H. Allen, *Org. Synth.* **16**, 25 (1936); A. H. Corwin, *Heterocycl. Compd.* **1**, 290 (1950); N. P. Buu-Hoi *et al.*, *J. Org. Chem.* **20**, 639, 850 (1955); H. H. Wassermann *et al.*, *Tetrahedron* **32**, 1863 (1976). Mechanistic studies: V. Amarnath *et al.*, *J. Org. Chem.* **56**, 6924 (1991); *idem*, K. Amarnath, *ibid.* **60**, 301 (1995). Applications: S.-X. Yu, P. W. Le Quesne, *Tetrahedron Lett.* **36**, 6205 (1995); R. Ballini *et al.*, *Synlett* **3**, 391 (2000); G. Minetto *et al.*, *Eur. J. Org. Chem.* **2005**, 5277 (2005). *Review:* S. E. Korostova *et al.*, *Russ. J. Org. Chem.* **34**, 1691 (1998). *Cf.* Hantzsch Pyrrole Synthesis; Knorr Pyrrole Synthesis.

317. Parham Cyclization
W. E. Parham *et al.*, *J. Org. Chem.* **40**, 2394 (1975).
Four- to seven-membered ring annulation of aryl bromides bearing *ortho* side chains having an electrophilic moiety, accomplished by halogen-metal exchange and subsequent nucleophilic ring closure:

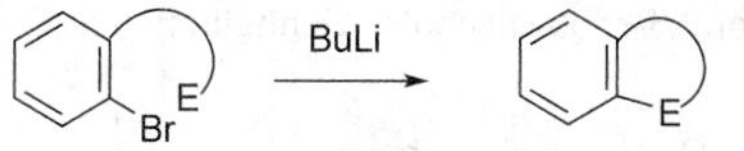

E = COOH, CONR$_2$, epoxide, CH$_2$Br, CH$_2$Cl,
OCONR$_2$, NCHAr, CONRCOCH$_2$R, POPh$_2$

Synthetic applications: M. R. Paleo *et al.*, *J. Org. Chem.* **40**, 2029 (1975); A. Couture *et al.*, *Chem. Commun.* **1994**; 1329; M. I. Collado *et al.*, *Tetrahedron Lett.* **37**, 6193 (1996); S. D. Larsen, *Synlett* **1997**; 1013; A. Ardeo *et al.*, *Tetrahedron Lett.* **41**, 5211 (2000); J. Ruiz *et al.*, *Tetrahedron* **61**, 3311 (2005). *Review:* W. E. Parham, C. K. Bradsher, *Acc. Chem. Res.* **15**, 300-305 (1982).

318. Passerini Reaction
M. Passerini, *Gazz. Chim. Ital.* **51**, 126, 181 (1921).
Formation of α-hydroxycarboxamides on treatment of an isonitrile with a carboxylic acid and an aldehyde or ketone:

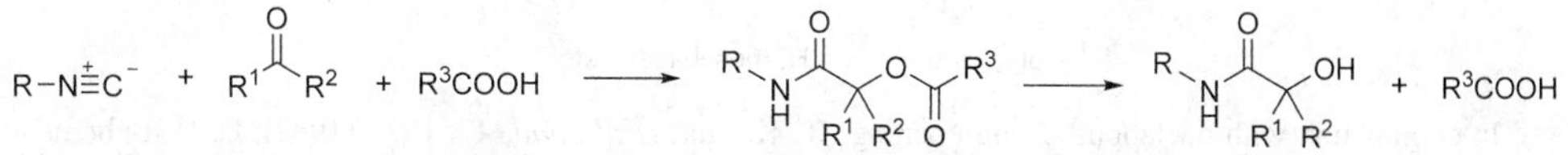

Synthetic applications: J. R. Falck, S. Manna, *Tetrahedron Lett.* **22**, 619 (1981); R. Bossio *et al.*, *Synthesis* **1993**, 783. Modifications: W. E. Lumma, *J. Org. Chem.* **46**, 3668 (1981); T. Carofiglio *et al.*, *Organometallics* **12**, 2726 (1993); Q. Xia, B. Ganem, *Org. Lett.* **4**, 1631 (2002); for stereoselectivity: H. Bock, I. Ugi, *J. Prakt. Chem.* **339**, 385 (1997) P. R. Andreana *et al.*, *Org. Lett.* **6**, 4231 (2004); for combinatorial chemistry; H. Bienaymé, *Tetrahedron Lett.* **39**, 4255 (1998); S. W. Kim *Tetrahedron Lett.* **39**, 7031 (1998). Catalytic, asymmetric reactions: S. E. Denmark, Y. Fan, *J. Am. Chem. Soc.* **125**, 7825 (2003). *Reviews:* I. Ugi, *Angew. Chem. Int. Ed.* **1**, 8 (1962); I. Ugi *et al.*, in *Isonitrile Chemistry* (Academic Press, New York, 1971) pp 133-143; I. Ugi *et al.*, *Comp. Org. Syn.* **2**, 1083-1087 (1991). *Cf.* Ugi Reaction.

319. Paterno-Büchi Reaction
E. Paterno, G. Chieffi, *Gazz. Chim. Ital.* **39**, 341 (1909); G. Büchi *et al.*, *J. Am. Chem. Soc.* **76**, 4327 (1954).
Formation of oxetanes by photochemical cycloaddition of carbonyl compounds to olefins:

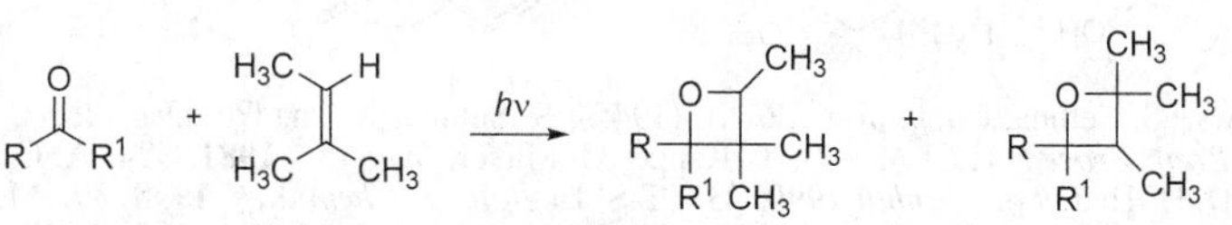

D. R. Arnold, *Adv. Photochem.* **6**, 301 (1968); G. Jones, II, *Org. Photochem.* **5**, 1 (1981); S. C. Freilich, K. S. Peters, *J. Am. Chem. Soc.* **103**, 6255 (1981); J. A. Porco, Jr., S. L. Schreiber, *Comp. Org. Syn.* **5**, 151-192 (1991). Mechanistic

Organic Name Reactions

studies: D. Sun *et al., J. Org. Chem.* **64**, 2250 (1999); *eidem, J. Chem. Soc. Perkin Trans. 2* **4**, 781 (1999). Stereo-controlled cycloadditions: S. A. Fleming, J. J. Gao, *Tetrahedron Lett.* **38**, 5407 (1997); G. Kollenz *et al., Tetrahedron* **55**, 2973 (1999); followed by oxetane ring opening: T. Bach *et al., Ann.* **1997**, 1529; S. R. Thopate *et al., Angew. Chem. Int. Ed.* **37**, 110 (1998). Regio- and stereoselectivity analyses: A. G. Griesbeck *et al., Acc. Chem. Res.* **37**, 919 (2004).

320. Pauson-Khand Reaction
I. U. Khand *et al., J. Chem. Soc. Perkin Trans. 1* **1973**, 977.

The formal [2+2+1] cycloaddition of an alkene, alkyne, and carbon monoxide to form cyclopentenones:

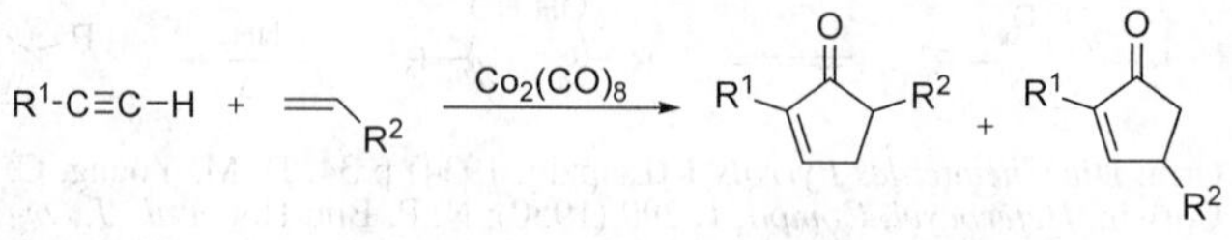

Use of a chiral auxillary: X. Verdaguer *et al., J. Am. Chem. Soc.* **116**, 2153 (1994); V. Bernardes *et al., J. Org. Chem.* **60**, 6670 (1995); J. Adrio, J. C. Carretero, *J. Am. Chem. Soc.* **121**, 7411 (1999). Catalytic version: N. Jeong *et al., ibid.* **116**, 3159 (1994). Intramolecular cyclizations: Y.-T. Shiu *et al., ibid.* **121**, 4066 (1999); F. A. Hicks *et al., ibid.* 5881; P. M. Breczinski *et al., Tetrahedron* **55**, 6797 (1999). Reactions of alkenyl sulfones and sulfoxides: M. R. Rivero *et al., Synlett* **2005**, 26. *Reviews:* N. E. Schore, *Org. React.* **40**, 1-90 (1991); *idem, Comp. Org. Syn.* **5**, 1037-1064 (1991); S. T. Ingate, J. Marco-Contelles, *Org. Prep. Proced. Int.* **30**, 123-143 (1998); O. Geis, H.-G. Schmalz, *Angew. Chem. Int. Ed.* **37**, 911-914 (1998); J. Blanco-Urgoiti *et al., Chem. Soc. Rev.* **33**, 32-42 (2004).

321. Payne Rearrangement
G. B. Payne, *J. Org. Chem.* **27**, 3819 (1962).

Base-promoted isomerization of 2,3-epoxyalcohols. Configuration at C-2 is inverted:

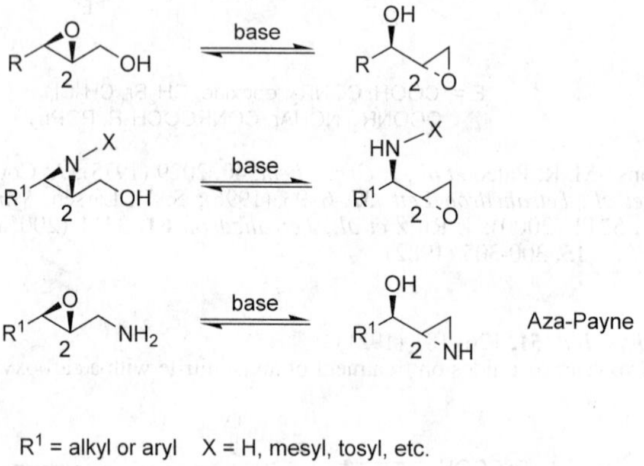

R^1 = alkyl or aryl   X = H, mesyl, tosyl, etc.

In conjunction with nucleophilic ring opening: T. Katsuki *et al., ibid.* **47**, 1373 (1982); C. H. Behrens *et al., ibid.* **50**, 5687 (1985); P. C. B. Page *et al., J. Chem. Soc. Perkin Trans. 1* **1990**, 1375; T. Konosu *et al., Chem. Pharm. Bull.* **40**, 562 (1992). *Aza-Payne rearrangements:* T. Ibuka *et al., J. Org. Chem.* **60**, 2044 (1995); K. Nakai *et al., Tetrahedron Lett.* **36**, 6247 (1995). Enhanced stereoselectivity: W. C. Frank, *Tetrahedron: Asymmetry* **9**, 3745 (1998). Review of aza-Payne: T. Ibuka *et al., Chem. Soc. Rev.* **27**, 145-154 (1998).

322. Pechmann Condensation
H. v. Pechmann, C. Duisberg, *Ber.* **16**, 2119 (1883).

Synthesis of coumarins by condensation of phenols with β-keto esters in the presence of Lewis acid catalysts:

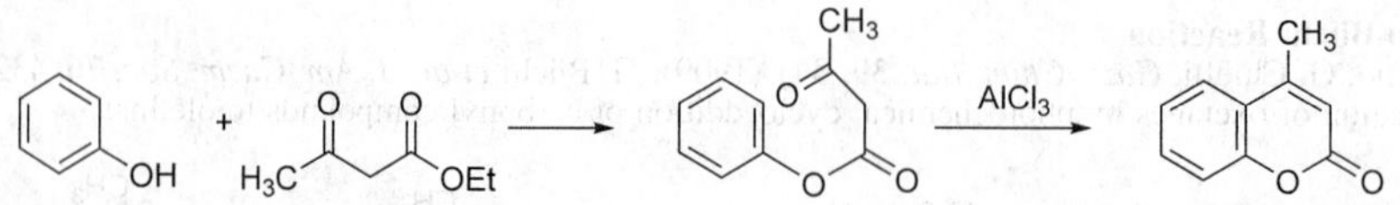

Early reviews: S. Sethna, *Chem. Rev.* **36**, 10 (1945); S. Sethna, R. Phadke, *Org. React.* **7**, 1 (1953). T. Kappe, E. Ziegler, *Org. Prep. Proced.* **1**, 61 (1969); T. Kappe, C. Mayer, *Synthesis* **1981**, 524; A. G. Osborne, *Tetrahedron* **37**, 2021 (1981); D. H. Hau *et al., Synlett* **1990**, 233; T-S. Li *et al., J. Chem. Res.* **1998**, 39. Modified conditions: J. E. T. Corrie, *J. Chem. Soc. Perkin Trans. 1* **1990**, 2151; D. H. Hua *et al., J. Org. Chem.* **57**, 399 (1992). *Cf.* Bischler-Napieralski Reaction; Simonis Chromone Cyclization.

323. Pechmann Pyrazole Synthesis
H. v. Pechmann, *Ber.* **31**, 2950 (1898).

Formation of pyrazoles from acetylenes and diazomethane. The analogous addition of diazoacetic esters to the triple bond yields pyrazolecarboxylic acid derivatives:

$$HC\equiv CH \quad + \quad H_2C=\overset{+}{N}=N^- \quad \longrightarrow$$

R. A. Raphael, *Acetylenic Compounds in Organic Synthesis* (London, 1955) p 179; T. L. Jacobs, *Heterocycl. Compd.* **5**, 70 (1957); B. Eistert *et al.*, *Houben-Weyl* **10/4**, 840 (1968). *Cf.* Knorr Pyrazole Synthesis.

324. Pellizzari Reaction
G. Pellizzari, *Gazz. Chim. Ital.* **41**, II, 20 (1911).

Formation of substituted 1,2,4-triazoles by the condensation of acyl hydrazines and amides. When the acyl groups of the amide and the acylhydrazine are different, interchange of acyl groups may occur with formation of a mixture of triazoles:

M. R. Atkinson, J. B. Polya, *J. Chem. Soc.* **1952**, 3418; P. Karrer, *Organic Chemistry* (New York, 4th ed., 1950) p 802; C. W. Bird, C. K. Wong, *Tetrahedron Lett.* **1974**, 1251. *Cf.* Einhorn-Brunner Reaction.

325. Pelouze Synthesis
J. Pelouze, *Ann.* **10**, 249 (1834).

Formation of nitriles from alkali cyanides by alkylation with alkyl sulfates or alkyl phosphates:

$$ROSO_2OK \quad + \quad KCN \quad \longrightarrow \quad RCN \quad + \quad K_2SO_4$$

V. Migrdichian, *Chemistry of Organic Cyanogen Compounds* (New York, 1947) p 6; D. T. Mowry, *Chem. Rev.* **42**, 192 (1948); P. Kurtz, *Houben-Weyl* **8**, 306 (1952).

326. Perkin Alicyclic Synthesis
W. H. Perkin, Jr., *Ber.* **16**, 1793 (1883).

Synthesis of alicyclic compounds from α,ω-dihaloalkanes and compounds containing active methylene groups in the presence of sodium ethoxide:

H. O. House, *Modern Synthetic Reactions* (W. A. Benjamin, Inc., Menlo Park, California, 2nd ed, 1972) pp 492-570. *Cf.* Malonic Ester Syntheses.

327. Perkin Reaction
W. H. Perkin, *J. Chem. Soc.* **21**, 53, 181 (1868); *idem, ibid.* **31**, 388 (1877).

Formation of α,β-unsaturated carboxylic acids by aldol condensation, *q.v.*, of aromatic aldehydes and acid anhydrides in the presence of an alkali salt of the acid:

Organic Name Reactions

Reviews: J. R. Johnson, *Org. React.* **1**, 210 (1942); H. O. House, *Modern Synthetic Reactions* (W. A. Benjamin, Menlo Park, California, 2nd ed, 1972) pp 660-663; N. Poonia *et al., Bull. Chem. Soc. Jpn.* **53**, 3338 (1980); T. Rosen, *Comp. Org. Syn.* **2**, 395-408 (1991). Applications: S. Kinastowski, A. Nowacki, *Tetrahedron Lett.* **23**, 3723 (1982); W. T. Brady *et al., J. Heterocycl. Chem.* **25**, 969 (1988). *Cf.* Erlenmeyer-Plöchl Azlactone and Amino Acid Synthesis; Stobbe Condensation.

328. Perkin Rearrangement (Coumarin-Benzofuran Ring Contraction)
W. H. Perkin, *J. Chem. Soc.* **23**, 368 (1870).
Formation of benzofuran-2-carboxylic acids and benzofurans by heating 3-halocoumarins with alkali:

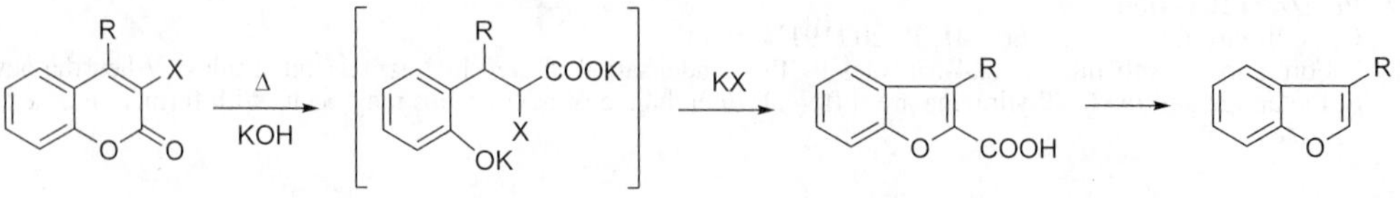

R. C. Elderfield, V. B. Meyer, *Heterocycl. Compd.* **2**, 2, 5 (1951); K. Bowden, S. Battah, *J. Chem. Soc. Perkin Trans. 2* **1998**, 1604.

329. Perkow Reaction
W. Perkow *et al., Naturwissenschaften* **39**, 353 (1952).
Formation of enol phosphates on treatment of α-halocarbonyl compounds with trialkyl phosphites:

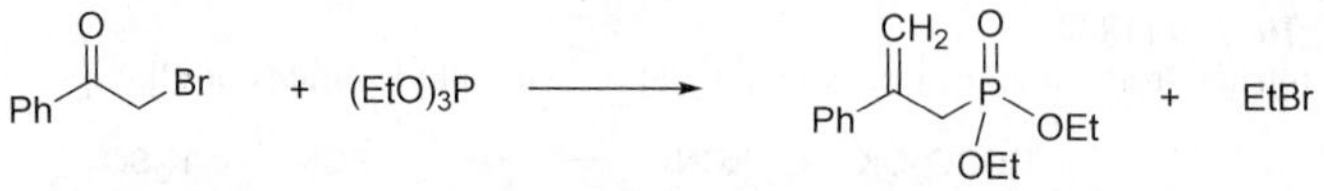

F. W. Lichtenthaler, *Chem. Rev.* **61**, 607 (1961); B. Miller in *Topics in Phosphorus Chemistry* **vol. 2**, M. Grayson, E. J. Griffith, Eds. (John Wiley, New York, 1965) p 178; K. Sasse, *Houben-Weyl* **12/1**, 423 (1963); A. J. Kirby, S. G. Warren, *The Organic Chemistry of Phosphorus* (Elsevier, Amsterdam, 1967) p 123; B. A. Arbuzow, *Pure Appl. Chem.* **9**, 306 (1964); I. J. Borowitz *et al., J. Org. Chem.* **38**, 1713 (1973); T. Winkler, W. L. Bencze, *Helv. Chim. Acta* **63**, 402 (1980); M. Sekine *et al., J. Org. Chem.* **46**, 4030 (1981).

330. Petasis Reaction (Petasis Boronic Acid-Mannich Reaction) (Boronic Acid Mannich Reaction)
N. A. Petasis, I. Akritopoulou, *Tetrahedron Lett.* **34**, 583 (1993).
Multicomponent reaction involving an amine, carbonyl, and boronic acid; utilized to prepare amine derivatives:

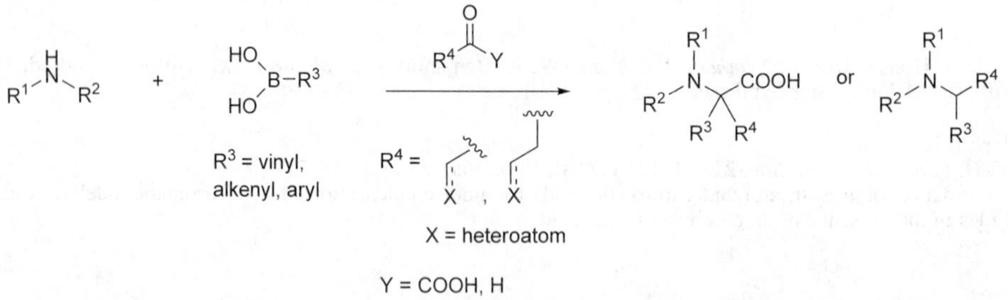

Prepn of α-amino acids: N. Petasis, I. A. Zavialov, *J. Am. Chem. Soc.* **119**, 445 (1997); *eidem, Tetrahedron* **53**, 16463 (1997). Reaction modifications involving solid-phase synthesis: N. Schlienger *et al., ibid.* **56**, 10023 (2000); tertiary aromatic amines: D. Naskar *et al., Tetrahedron Lett.* **44**, 5819 (2003); microwave conditions: M. Follmann *et al., Synlett* **2005**, 1009; *in situ* organoborane generation: N. Selander *et al., J. Am. Chem. Soc.* **129**, 13723 (2007); water as the reaction solvent: N. R. Candeias *et al., Eur. J. Org. Chem.* **2009**, 1859. *Review*: M. D. McReynolds, P. R. Hanson, *Chemtracts Org. Chem.* **14**, 796-801 (2001).

331. Peterson Reaction (Olefination)
D. J. Peterson, *J. Org. Chem.* **33**, 780 (1968).
Reaction of α-silyl carbanions with carbonyl compounds yielding β-silylalkoxides which undergo instantaneous elimination to afford olefins:

L. Birkofer, O. Stiehl, *Top. Curr. Chem.* **88,** 58 (1980); E. Colvin, *Silicon in Organic Synthesis* (Butterworth, London, 1981) p 143; D. J. Ager, *Synthesis* **1984,** 384-398; *idem, Org. React.* **38,** 1-223 (1990); S. E. Kelly, *Comp. Org. Syn.* **1,** 731-737, 782-783 (1991). *Review:* L. F. van Staden *et al., Chem. Soc. Rev.* **31,** 195-200 (2002). *Cf.* Tebbe Olefination; Wittig Reaction.

332. Petrenko-Kritschenko Piperidone Synthesis

P. Petrenko-Kritschenko *et al., Ber.* **39,** 1358 (1906); **40,** 2882 (1907); **41,** 1692 (1908); **42,** 2020, 3683 (1909).

Formation of piperidones *via* cyclization of two moles of aldehyde and one mole each of acetonedicarboxylic ester and ammonia or a primary amine:

R. Robinson, *J. Chem. Soc.* **111,** 762, 876, (1917); C. Mannich, O. Hieronimus, *Ber.* **75,** 49 (1942); H. S. Mosher, *Heterocyclic Compounds* **1,** 659 (New York, 1950). *Cf.* Robinson-Schöpf Reaction.

333. Pfau-Plattner Azulene Synthesis

A. St. Pfau, P. A. Plattner, *Helv. Chim. Acta* **22,** 202 (1939).

Formation of azulenes by ring enlargement of indanes on addition of diazoacetic ester, hydrolysis, dehydrogenation and decarboxylation of the resulting acid:

P. A. Plattner *et al., ibid.* **23,** 907 (1940); **24,** 483 (1941); **25,** 590 (1942); B. Eistert, *Newer Methods of Preparative Organic Chemistry* (Interscience, New York, 1948) p 555; D. H. Reid, *Chem. Soc. Spec. Publ.* **12,** 69 (1958); K. Hafner, *Angew. Chem.* **70,** 419 (1958). *Cf.* Buchner Method of Ring Enlargement.

334. Pfitzinger Reaction

W. Pfitzinger, *J. Prakt. Chem.* [2] **33,** 100 (1886); **38,** 582 (1888).

Formation of quinoline-4-carboxylic acids by condensation of isatic acids from isatin with α-methylene carbonyl compounds:

C. Hollins, *The Synthesis of Nitrogen Ring Compounds* (London, 1924) p 286; R. H. Manske, *Chem. Rev.* **30,** 126 (1942); F. W. Bergstrom, *ibid.* **35,** 152 (1944); R. C. Elderfield, *Heterocycl. Compd.* **4,** 47 (1952); N. P. Buu-Hoï *et al., Bull. Soc. Chim. Fr.* **1968,** 2476; M. H. Palmer, P. S. McIntyre, *J. Chem. Soc. B* **1969,** 539. *Review:* M. G.-A. Shvekhgeimer, *Chem. Heterocycl. Compd.* **40,** 257-294 (2004). *Cf.* Friedlaender Synthesis; Niementowski Quinoline Synthesis.

335. Pfitzner-Moffatt Oxidation (Moffatt Oxidation)

K. E. Pfitzner, J. G. Moffatt, *J. Am. Chem. Soc.* **85,** 3027 (1963).

Mild oxidation of primary and secondary alcohols, promoted by dicyclohexylcarbodiimide activation of dimethyl sulfoxide, evidently involving the alkoxysulfonium ylides, which rearrange intramolecularly to generate aldehydes and ketones, respectively:

Reviews: J. G. Moffatt, "Sulfoxide-Carbodiimide and Related Oxidations" in *Oxidation* **vol. 2,** R. L. Augustine, D. J. Trecker, Eds. (Dekker, New York, 1971) pp 1-64; T. T. Tidwell, *Org. React.* **39,** 297-572 *passim* (1990); T. V. Lee, *Comp. Org. Syn.* **7,** 291-303 *passim* (1991). *Cf.* Corey-Kim Oxidation; Swern Oxidation.

336. Pictet-Gams Isoquinoline Synthesis

A. Pictet, A. Gams, *Ber.* **43,** 2384 (1910).

Formation of isoquinolines by cyclization of acylated aminomethyl phenyl carbinols or their ethers with phosphorus pentoxide in toluene or xylene:

Reviews: W. M. Whaley, T. R. Govindachari, *Org. React.* **6,** 151 (1951); W. Y. Gensler, *Heterocycl. Compd.* **4,** 361 (1952); W. Herz, L. Tsai, *J. Am. Chem. Soc.* **77,** 3529 (1955); A. A. Bindra *et al., Tetrahedron Lett.* **1968,** 2677; N. Ardabilchi *et al., J. Chem. Soc. Perkin Trans. 1* **1979,** 539. *Cf.* Bischler-Napieralski Reaction.

337. Pictet-Hubert Reaction (Morgan-Walls Reaction)

A. Pictet, A. Hubert, *Ber.* **29,** 1182 (1896); C. T. Morgan, L. P. Walls, *J. Chem. Soc.* **1931,** 2447; **1932,** 2225.

Phenanthridine cyclization by dehydrative ring closure of acyl-*o*-aminobiphenyls on heating with zinc chloride at 250-300° (Pictet-Hubert), or with phosphorus oxychloride in boiling nitrobenzene (Morgan-Walls):

L. P. Walls, *J. Chem. Soc.* **1945,** 294; J. Cymerman, W. F. Short, *ibid.* **1949,** 703; R. S. Theobald, K. Schofield, *Chem. Rev.* **46,** 175 (1950); L. P. Walls, *Heterocycl. Compd.* **4,** 574 (1952); J. Eisch, H. Gilman, *Chem. Rev.* **57,** 525 (1957); N. Campbell, *Chemistry of Carbon Compounds* **IVA,** 691 (1957). *Cf.* Bischler-Napieralski Reaction.

338. Pictet-Spengler Isoquinoline Synthesis

A. Pictet, T. Spengler, *Ber.* **44,** 2030 (1911).

Formation of tetrahydroisoquinoline derivatives by condensation of β-arylethylamines with carbonyl compounds and cyclization of the Schiff bases formed:

Reviews: W. M. Whaley, T. R. Govindachari, *Org. React.* **6,** 151 (1951); R. A. Abramovitch, I. D. Spenser, *Adv. Heterocycl. Chem.* **3,** 79 (1964); K. Stuart, R. Woo-Ming, *Heterocycles* **3,** 223 (1975); D. Soerens *et al., J. Org. Chem.* **44,** 535 (1979); H. Ernst *et al., Ber.* **114,** 1894 (1981). Stereochemical study: E. Dominguez *et al., Tetrahedron* **43,** 1943 (1987). Review of enantioselective modifications: M. D. Rozwadowski, *Heterocycles* **39,** 903-931 (1994).

339. Piloty-Robinson Synthesis

O. Piloty, *Ber.* **43,** 489 (1910); G. M. Robinson, R. Robinson, *J. Chem. Soc.* **43,** 639 (1918).

Formation of pyrroles by heating azines of enolizable ketones with acid catalysts, usually zinc chloride or hydrogen chloride:

Review: N. V. Sidgwick, *Organic Chemistry of Nitrogen Compounds* (Oxford, 3rd ed., 1966) pp 619-641; H. Posvic *et al., J. Org. Chem.* **39,** 2575 (1974). *Cf.* Borsche-Drechsel Cyclization; Fischer Indole Synthesis.

340. Pinacol Coupling Reaction

R. Fittig, *Ann.* **110,** 17 (1859).

Formation of pinacols by a reductive radical-radical coupling of carbonyl compounds, especially ketones:

Reviews: G. M. Robertson, *Comp. Org. Syn.* **3,** 563 (1991); H. Jendralla *et al.,* in *Transition Metals for Organic Synthesis* (Wiley-VCH, Weinheim, 1998) pp. 403-417. *Cf.* McMurry Reaction.

341. Pinacol Rearrangement

R. Fittig, *Ann.* **114,** 54 (1860).

Acid-catalyzed rearrangement of vicinal diols to aldehydes or ketones:

Reviews: C. J. Collins, *Q. Rev. Chem. Soc.* **14,** 357 (1960); C. J. Collins, J. F. Eastham in *Chemistry of the Carbonyl Group,* S. Patai, Ed. (Interscience, New York, 1966) pp 762-767; B. Rickborn, *Comp. Org. Syn.* **3,** 721-732 (1991). *Cf.* Tiffeneau-Demjanov Rearrangement.

342. Pinner Reaction (Amidine and Ortho Ester Synthesis)

A. Pinner, F. Klein, *Ber.* **10,** 1889 (1877); **11,** 4, 1475 (1878); **16,** 352, 1643 (1883).

Formation of imino esters (alkyl imidates) by addition of dry hydrogen chloride to a mixture of a nitrile and an alcohol. Treatment of alkyl imidates with ammonia or primary or secondary amines affords amidines, while treatment with alcohols yields ortho-esters:

Reviews: R. Roger, D. Neilson, *Chem. Rev.* **61,** 179 (1961); E. N. Zil'berman, *Russ. Chem. Rev.* **31,** 615 (1962); P. L. Compagnon, M. Moeque, *Ann. Chim. (Paris)* **5,** 23 (1970); B. Decroix *et al., J. Chem. Res.* **1978,** 134.

343. Pinner Triazine Synthesis

A. Pinner, *Ber.* **23,** 2919 (1890).

Preparation of 2-hydroxy-4,6-diaryl-*s*-triazines by reaction of aryl amidines and phosgene. The reaction may be extended to halogenated aliphatic amidines:

A. Pinner, *Ber.* **25,** 1414 (1892); **28,** 483 (1895); J. Ephraim, *ibid.* **26,** 2226 (1893); P. Flatow, *ibid.* **30,** 2006 (1897); T. Rappaport, *ibid.* **34,** 1990 (1901); H. Schroeder, C. Grundmann, *J. Am. Chem. Soc.* **78,** 2447 (1956); E. M. Smolin, L. Rapoport, *The Chemistry of Heterocyclic Compounds,* A. Weissberger, Ed., *s-Triazines and Derivatives* (Interscience, New York, 1959) p 186.

344. Pinnick Oxidation

B. O. Lindgren, T. Nilsson, *Acta Chem. Scand.* **27,** 888 (1973).

Conversion of an aldehyde to the corresponding carboxylic acid; oxidative conditions utilize a dihydrogen phosphate buffered aqueous solution of sodium chlorite as well as a scavenger for the hypochlorous acid side product:

R^1 = alkyl, aryl, heteroaryl, allyl,

HClO scavenger = 2-methyl-2-butene, resorcinol, DMSO, sulfamic acid, H_2O_2

Expansion of hypochlorous acid scavengers to include 2-methyl-2-butene: G. A. Kraus, M. J. Taschner, *J. Org. Chem.* **45,** 1175 (1980); G. A. Kraus, B. Roth, *ibid.* 4825; B. S. Bal *et al., Tetrahedron* **37,** 2091 (1981); to include hydrogen peroxide and DMSO: E. Dalcanale, F. Montanari, *J. Org. Chem.* **51,** 567 (1986). Applications to solid-phase synthesis: T. Takemoto *et al., Synlett* **2001,** 1555. Utility in presence of tertiary disulfide functionalities: X. Fang *et al., ibid.* **2003,** 489. *Review:* A. Raach, O. Reiser, *J. Prakt. Chem.* **342,** 605-608 (2000).

345. Piria Reaction

R. Piria, *Ann.* **78,** 31 (1851).

Formation of arylsulfamic acids or sulfonation products or both by refluxing aromatic nitro compounds with a metal sulfite and boiling the mixture with dilute acid to yield the amines and sulfamic acids:

J. F. Bunnett, R. E. Zahler, *Chem. Rev.* **49,** 398 (1951); R. Schroter, *Houben-Weyl* **11/1,** 457 (1957); R. Budziarek, *Chem. Ind. (London)* **1978,** 583.

346. Polonovski Reaction; Potier-Polonovski Reaction

M. Polonovski, M. Polonovski, *Bull. Soc. Chim. Fr.* **41,** 1190 (1927).

Rearrangement of tertiary amine oxides upon treatment with acetic anhydride or acetyl chloride, in which one of the alkyl groups attached to the nitrogen is cleaved, generating the *N,N*-disubstituted acetamide and aldehyde:

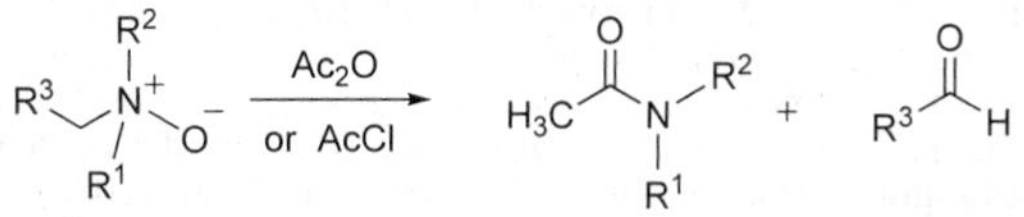

Reviews: A. R. Katritzky, J. N. Lagowski, *Chemistry of Heterocyclic N-Oxides* (Academic Press, New York, 1971) p 279, 362; D. Grierson, *Org. React.* **39**, 85-295 (1990); D. S. Grierson, H.-P. Husson, *Comp. Org. Syn.* **6**, 909-924 (1991).

The reaction proceeds *via* an iminium ion intermediate which becomes the stable reaction product when trifluoroacetic anhydride is employed. This modified procedure is commonly referred to as the **Potier-Polonovski reaction:** A. Cave *et al., Tetrahedron* **23**, 4681 (1967); T. Tamminen *et al., ibid.* **45**, 2683 (1989); R. J. Sundberg, *et al., ibid.* **48**, 277 (1992).

347. Pomeranz-Fritsch Reaction (Schlittler-Müller Modification)

C. Pomeranz, *Monatsh. Chem.* **14**, 116 (1893); P. Fritsch, *Ber.* **26**, 419 (1893); E. Schlittler, J. Müller, *Helv. Chim. Acta* **31**, 914, 1119 (1948).

Formation of isoquinolines by the acid-catalyzed cyclization of benzalaminoacetals prepared from aromatic aldehydes and aminoacetal; in the **Schlittler-Müller modification** the starting materials are benzyl amines and glyoxal semiacetal:

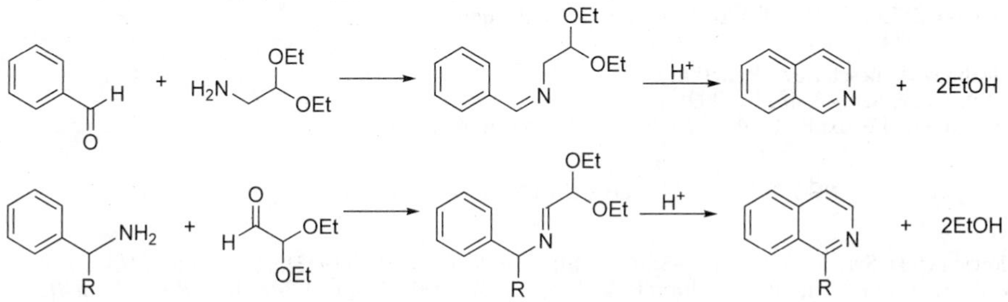

M. J. Bevis *et al., Tetrahedron* **27**, 1253 (1971); E. V. Brown, *J. Org. Chem.* **42**, 3208 (1977); R. Hirsenkorn, *Tetrahedron Lett.* **32**, 1775 (1991). *Reviews:* W. J. Gensler, *Org. React.* **6**, 191 (1951); *idem, Heterocycl. Compd.* **4**, 368 (1952); J. M. Bobbit, A. J. Bourque, *Heterocycles* **25**, 601-614 (1987).

348. Ponzio Reaction

G. Ponzio, *Gazz. Chim. Ital.* **27, I,** 171 (1897).

Formation of dinitrophenylmethanes from benzaldoximes by oxidation with nitrogen dioxide in ether:

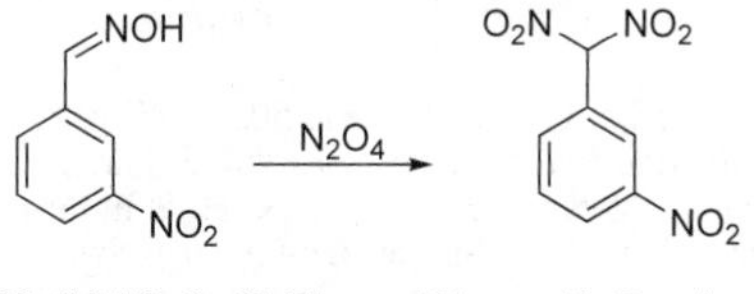

J. L. Riebsomer, *Chem. Rev.* **36**, 183 (1945); L. F. Fieser, L. von E. Doering, *J. Am. Chem. Soc.* **68**, 2252 (1946); L. F. Fieser, M. Fieser, *Reagents for Organic Synthesis* (New York, 1967) p 325; H. G. Padeken *et al., Houben-Weyl* **10/1**, 113 (1971). Improved procedure: H. Suzuki *et al., Bull. Chem. Soc. Jpn.* **61**, 2929 (1988).

349. Povarov Reaction (Aza-Diels-Alder Reaction) (Imino Diels-Alder Reaction)

L. S. Povarov, B. M. Mikhailov, *Russ. Chem. Bull.* **12**, 871 (1963); L. S. Povarov *et al., ibid.* 1878.

Lewis acid-catalyzed [4+2] cycloaddition reaction of electron rich alkene dienophiles and *N*-aryl imines to form 1,2,3,4-tetrahydroquinolines. The resulting products may also be further transformed into fully aromatized quinolines upon exposure to an oxidizing agent (*e.g.* atmospheric oxygen, potassium permanganate, *p*-toluenesulfonic acid):

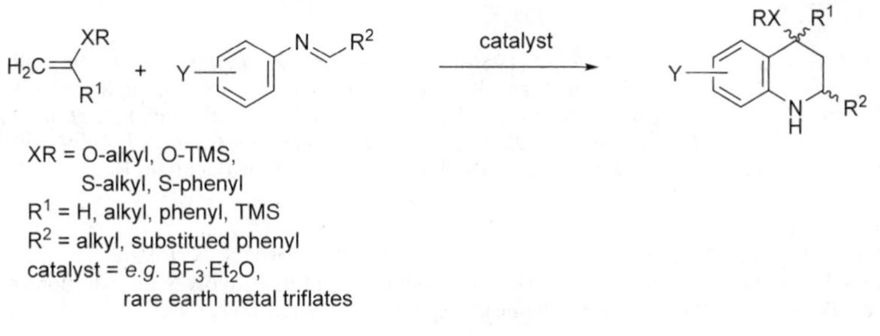

XR = O-alkyl, O-TMS,
S-alkyl, S-phenyl
R^1 = H, alkyl, phenyl, TMS
R^2 = alkyl, substitued phenyl
catalyst = *e.g.* $BF_3 \cdot Et_2O$,
rare earth metal triflates

Intramolecular reaction: S. Laschat, J. Lauterwein, *J. Org. Chem.* **58**, 2856 (1993). Multicomponent reaction conditions: Y. Ma *et al.*, *ibid.* **64**, 6462 (1999). Use of enamides as the alkene dienophile: M. Hadden *et al.*, *Tetrahedron* **57**, 5615 (2001). One-pot cyclization and oxidation reaction sequence: N. Shindoh *et al.*, *J. Org. Chem.* **73**, 7451 (2008); P. H. Dobbelaar, C. H. Marzabadi, *Tetrahedron Lett.* **51**, 201 (2010). Asymmetric cooperative catalysis with strong Bronsted acids and chiral ureas: H. Xu *et al.*, *Science* **327**, 986 (2010). Review of mechanistic studies: D. Bello *et al.*, *Curr. Org. Chem.* **14**, 332-356 (2010). *Reviews*: L. S. Povarov, *Russ. Chem. Rev.* **36**, 656-670 (1967); V. A. Glushkov, A. G. Tolstikov, *ibid.* **77**, 137-159 (2008); V. V. Kouznetsov, *Tetrahedron* **65**, 2721-2750 (2009). *Cf.* Combes Quinoline Synthesis; Doebner-Miller Reaction; Friedlaender Synthesis; Riehm Quinoline Synthesis; Skraup Reaction.

350. Prévost Reaction
C. Prévost, *Compt. Rend.* **196**, 1129 (1933), *C.A.* **27**, 3195 (1933).
Hydroxylation of olefins with iodine and silver benzoate in an anhydrous solvent to give *trans*-glycols:

Reviews: C. V. Wilson, *Org. React.* **9**, 350 (1957); F. D. Gunstone, *Adv. Org. Chem.* **1**, 117 (1960); H. O. House, *Modern Synthetic Reactions* (W. A. Benjamin, Menlo Park, California, 2nd ed., 1972) p 438; S. Amin *et al.*, *J. Org. Chem.* **46**, 2573 (1981). *Cf.* Woodward *cis*-Hydroxylation.

351. Prilezhaev (Prileschajew) Reaction
N. Prilezhaev, *Ber.* **42**, 4811 (1909).
Formation of epoxides by the reaction of alkenes with peracids:

Reviews: D. Swern, *Chem. Rev.* **45**, 16 (1949); *Org. React.* **7**, 378 (1953); H. O. House, *Modern Synthetic Reactions* (W. A. Benjamin, Menlo Park, California, 2nd ed., 1972) pp 302-319; D. I. Metelitra, *Russ. Chem. Rev.* **41**, 807 (1972); D. Schnurgfeil, *Z. Chem.* **20**, 445 (1980).

352. Prins Reaction
H. J. Prins, *Chem. Weekbl.* **16**, 64, 1072, 1510 (1919), *C.A.* **13**, 3155 (1919).
Acid-catalyzed addition of olefins to formaldehyde to give 1,3-diols, allylic alcohols or *meta*-dioxanes:

Reviews: R. Arundale, L. A. Mikeska, *Chem. Rev.* **51**, 505 (1952); V. I. Isagulyants *et al.*, *Russ. Chem. Rev.* **1968**, 17; C. W. Roberts in *Friedel-Crafts and Related Reactions* vol. **2**, Part 2, G. A. Olah, Ed. (Interscience, 1964) pp 1175-1210; D. R. Adams, S. P. Bhatnagar, *Synthesis* **1977**, 661; K. H. Schulte-Elte *et al.*, *Helv. Chim. Acta* **62**, 2673 (1979); R. El Gharbi *et al.*, *Synthesis* **1981**, 361; B. B. Snider, *Comp. Org. Syn.* **2**, 527-561 (1991).

353. Pschorr Reaction
R. Pschorr, *Ber.* **29**, 496 (1896).
Synthesis of phenanthrene derivatives from diazotized α-aryl-*o*-aminocinnamic acids by intramolecular arylation:

Reviews: P. H. Leake, *Chem. Rev.* **56**, 27 (1956); D. F. De Tar, *Org. React.* **9**, 409 (1957); R. A. Abramovitch, *Adv. Free-Radical Chem.* **2**, 88 (1967); T. Kametani, K. Fukumoto, *J. Heterocycl. Chem.* **8**, 341 (1971); S. Foldeak, *Tetrahedron* **27**, 3465 (1971); T. S. Kametani *et al.*, *ibid.* **27**, 5367 (1971); F. F. Gadallah *et al.*, *J. Org. Chem.* **38**, 2386 (1973); S. M. Kupchan *et al.*, *ibid.* 405; G. Daidone *et al.*, *J. Heterocycl. Chem.* **17**, 1409 (1980). Mechanistic study: P. Hanson *et al.*, *J. Chem. Soc. Perkin Trans. 2* **1999**, 49. *Cf.* Gomberg-Bachman Reaction; Meerwein Arylation.

354. Pudovik Reaction
A. N. Pudovik, B. A. Arbuzov, *Dokl. Akad. Nauk SSSR* **73**, 327 (1950); A. N. Pudovik, *ibid.* 499.
General method for the formation of a carbon-phosphorus bond that involves the addition of an unsaturated system (e.g. alkene, alkyne, carbonyl, imine) to a labile phosphorus-hydrogen bond:

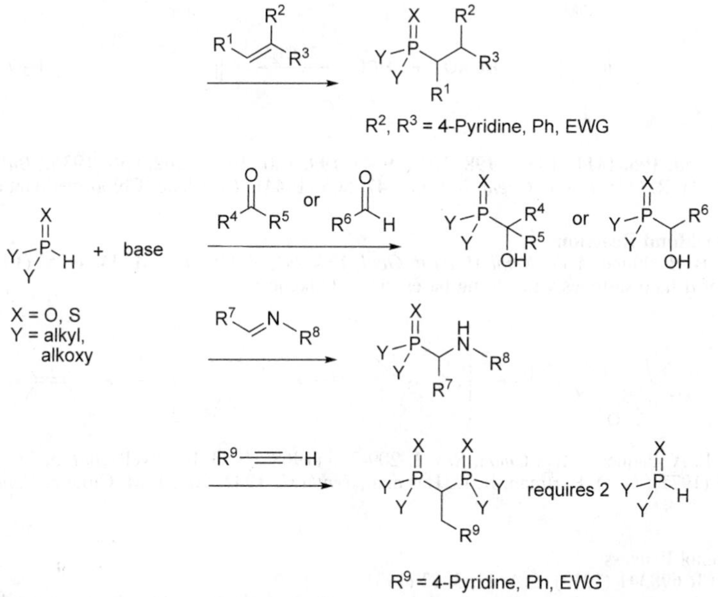

R^2, R^3 = 4-Pyridine, Ph, EWG

X = O, S
Y = alkyl, alkoxy

R^9 = 4-Pyridine, Ph, EWG

Kinetics and reaction mechanism: R. A. Cherkasov *et al.*, *Phosphorus Sulfur Silicon Relat. Elem.* **49-50**, 61 (1990). Reaction activation methodology studies: D. Semenzin *et al.*, *J. Org. Chem.* **62**, 2414 (1997). Stereoselective reaction applications: V. A. Alfonsov, *Phosphorus Sulfur Silicon Relat. Elem.* **183**, 2637 (2008); J. P. Abell, H. Yamamoto, *J. Am. Chem. Soc.* **130**, 10521 (2008). *Review*: A. N. Pudovik, I. V. Konovalova, *Synthesis* **1979**, 81-96.

355. Pummerer Rearrangement
R. Pummerer, *Ber.* **43**, 1401 (1910).

Rearrangement of sulfoxides to α-acyloxythioethers in the presence of acyclic anhydrides. When nucleophiles other than those derived from the anhydride are present, different functionalization is achieved:

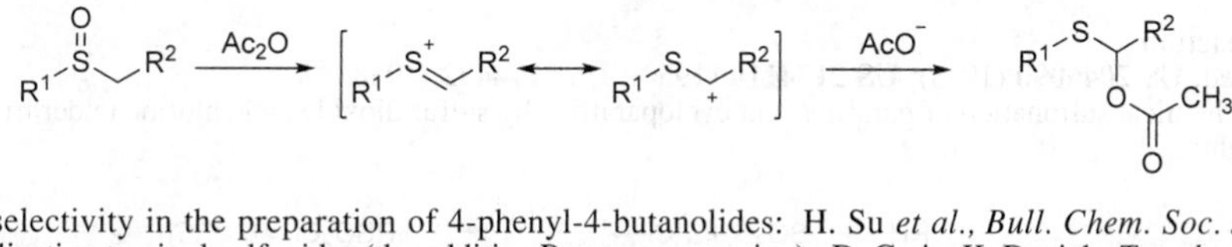

Diastereoselectivity in the preparation of 4-phenyl-4-butanolides: H. Su *et al.*, *Bull. Chem. Soc. Jpn.* **66**, 2603 (1993). Application to vinyl sulfoxides (the additive Pummerer reaction): D. Craig, K. Daniels, *Tetrahedron* **49**, 11263 (1993); to oxidative cyclization of C(3) indole derivatives: K. S. Feldman *et al.*, *J. Org. Chem.* **70**, 6429 (2005). Regiospecific cyclization: G. Majumdar, D. Mal, *Indian J. Chem.* **33B**, 700 (1994). Asymmetric synthesis: Y. Kita *et al.*, *Tetrahedron Lett.* **35**, 3575 (1994). *Reviews:* O. DeLucchi *et al.*, *Org. React.* **40**, 157-405 (1991); D. S. Grierson, H.-P. Husson, *Comp. Org. Syn.* **6**, 924-937 (1991); S. K. Bur, A. Padwa, *Chem. Rev.* **104**, 2401-2432 (2004).

356. Purdie Methylation (Irvine-Purdie Methylation)
T. Purdie, J. C. Irvine, *J. Chem. Soc.* **83**, 1021 (1903).

Exhaustive methylation of a methyl glycoside by repeated treatment with methyl iodide and silver oxide, followed by hydrolysis of the pentamethyl ether with dilute acid to yield the anomeric hydroxyl group:

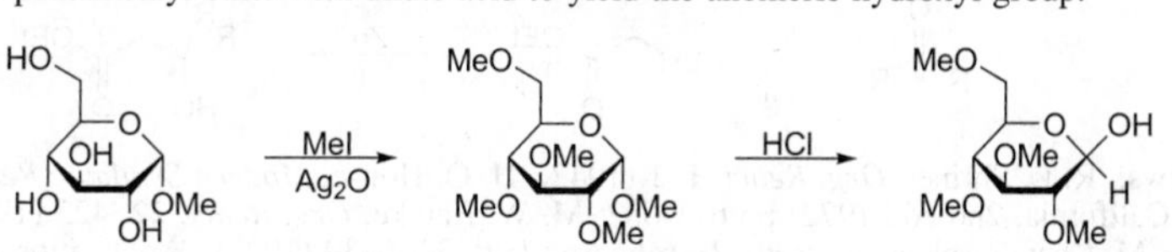

C. C. Barker, *et al., ibid.* **1946**, 753; E. J. Bourne, S. Peat, *Adv. Carbohydr. Chem.* **5**, 146 (1950); W. Pigman, *The Carbohydrates* (New York, 1957) p 370; P. V. Kovac *et al., Carbohydr. Res.* **58**, 327 (1977). *Cf.* Haworth Methylation.

357. Quelet Reaction
R. Quelet, *Compt. Rend.* **195**, 155 (1932).

Passage of dry hydrochloric acid through a solution in ligroin of a phenolic ether and an aliphatic aldehyde in the presence or absence of a dehydration catalyst to yield α-chloroalkyl derivatives by substitution in the *para* position to the ether group or in the *ortho* position in *para*-substituted phenolic ethers:

$$\text{(p-bromoanisole)} + RCHO + HCl \xrightarrow{ZnCl_2} \text{(substituted anisole with CHRCl)} + H_2O$$

R. Quelet, *ibid.* **196**, 1411 (1933); **198**, 102 (1934); **199**, 150 (1934); **202**, 956 (1936); *Bull. Soc. Chim. Fr.* **7**, 196 (1940); U. Neda, R. Oda, *J. Soc. Chem. Ind. Jpn.* **47**, 565 (1944). *Cf.* Blanc (Chloromethylation) Reaction.

358. Ramberg-Bäcklund Reaction

L. Ramberg, B. Bäcklund, *Arkiv Kemi Mineral. Geol.* **13A**(27), 50 (1940), *C.A.* **34**, 4725[5] (1940).

Reaction of α-halo sulfones with strong bases to yield alkenes:

$$\text{(α-halo sulfone)} \xrightarrow{B^-} \left[\text{(carbanion)} \longrightarrow \text{(episulfone)} \right] \longrightarrow \text{(alkene)} + SO_2$$

Reviews: L. A. Paquette, *Acc. Chem. Res.* **1**, 209-216 (1968); F. G. Bordwell, *ibid.* **3**, 28 (1970); L. Paquette, *Org. React.* **25**, 1 (1977); G. D. Hartman, R. D. Hartman, *Synthesis* **1982**, 504; J. M. Clough, *Comp. Org. Syn.* **3**, 861-886 (1991).

359. Raschig Phenol Process

F. Raschig, **FR 698341** (1930), *C.A.* **25**, 3012 (1931).

Commercial process for the production of phenol by the hydrolysis of chlorobenzene, produced by the chlorination of benzene with hydrochloric acid and air:

$$C_6H_6 + HCl \xrightarrow[\text{Cu -Fe}]{230°/\text{air}} C_6H_5Cl + H_2O \xrightarrow[\text{SiO}_2]{425°} C_6H_5OH + HCl$$

W. H. Prahl, **US 1963761** (1934); **US 2156402** (1939); J. A. Kent, *Riegel's Industrial Chemistry* (New York, 1962) p 339; W. L. Faith, D. B. Keyes, R. L. Clark, *Industrial Chemistry* (New York, 3rd ed., 1965) p 586; R. N. Shreve, *Chemical Process Industries* (New York, 3rd ed., 1967) p 105; *Kirk-Othmer Encyclopedia of Chemical Technology* **vol. 17** (New York, 1982) p 378.

360. Reed Reaction

C. F. Reed, **US 2046090** (1933); **US 2174110** (1934); **US 2174492** (1938).

Photochemical sulfonation of paraffins and cycloparaffins by sulfur dioxide and chlorine under irradiation with ultraviolet light:

$$RH + SO_2 + Cl_2 \xrightarrow{h\nu} RSO_2Cl + HCl$$

F. Asinger *et al., Ber.* **75**, 34, 42, 344 (1942); J. H. Helberger *et al., Ann.* **562**, 23 (1949); H. Eckoldt, *Houben-Weyl* **9**, 407-427 (1955); A. Schönberg, *Präparative Organische Photochemie* (Berlin, 1958) p 201; G. Sosnovsky, *Free Radical Reactions in Preparative Organic Chemistry* (New York, 1964) p 105.

361. Reformatsky (Reformatskii) Reaction

S. Reformatskii, *Ber.* **20**, 1210 (1887); *J. Russ. Phys. Chem. Soc.* **22**, 44 (1890).

Condensation of aldehydes or ketones with organozinc derivatives of α-halo esters to yield β-hydroxy esters:

$$\underset{R}{\overset{O}{\underset{}{\parallel}}}R^1 + Br\underset{O}{\overset{}{\diagup}}OEt \xrightarrow{Zn} R\underset{HO}{\overset{R^1}{\underset{}{}}}\underset{O}{\overset{}{}}OEt$$

Early reviews: R. L. Shriner, *Org. React.* **1**, 1 (1942); H. O. House, *Modern Synthetic Reactions* (W. A. Benjamin, Menlo Park, California, 2nd ed., 1972) pp 671-682; M. W. Rathke, *Org. React.* **22**, 423 (1975). Use of thiocarbonyl electrophiles: M. Chandrasekharam *et al., Tetrahedron Lett.* **34**, 6439 (1993). Application to the synthesis of β-keto esters: C. Kashima *et al., J. Org. Chem.* **58**, 793 (1993). Asymmetric synthesis: D. Pini *et al., Tetrahedron: Asymmetry* **5**, 1875 (1994). Methods development for the synthesis of β-lactones: H. Schick *et al., Tetrahedron* **51**, 2939 (1995). *Reviews:* A. Fürstner, *Synthesis* **1989**, 571-590; M. W. Rathke, P. Weipert, *Comp. Org. Syn.* **2**, 277-299 (1991); R. Ocampo, W. R. Dolbier, Jr., *Tetrahedron* **60**, 9325-9374 (2004). *Cf.* Aldol Reaction; Blaise Reaction.

362. Reimer-Tiemann Reaction

K. Reimer, F. Tiemann, *Ber.* **9**, 824, 1268, 1285 (1876).

Formation of phenolic aldehydes from phenols, chloroform and alkali:

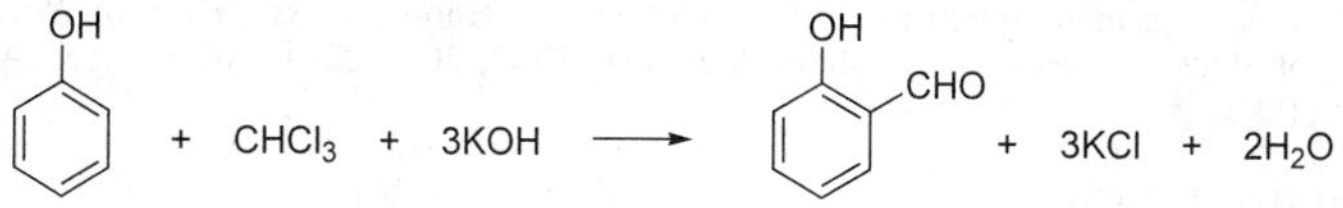

$$\text{phenol} + CHCl_3 + 3KOH \longrightarrow \text{salicylaldehyde} + 3KCl + 2H_2O$$

Review: H. Wynberg, *Chem. Rev.* **60,** 169 (1960); H. Wynberg, E. W. Meijer, *Org. React.* **28,** 2 (1982); H. Wynberg, *Comp. Org. Syn.* **2,** 769-775 (1991). *Cf.* Duff Reaction; Gattermann Aldehyde Synthesis.

363. Reissert Indole Synthesis
A. Reissert, *Ber.* **30,** 1030 (1897).
Condensation of an *o*-nitrotoluene with oxalic ester, reduction to the amine, and cyclization to the indole:

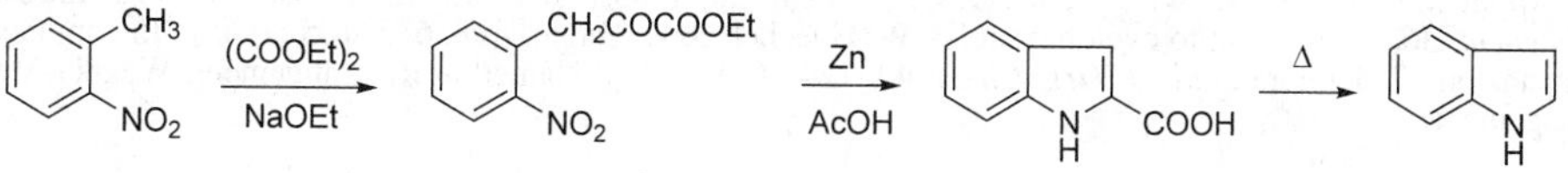

W. O. Kermack *et al., J. Chem. Soc.* **119,** 1602 (1921); P. C. Julian *et al., Heterocycl. Compd.* **3,** 18 (1962); J. G. Cannon *et al., J. Med. Chem.* **24,** 238 (1981).

364. Reissert Reaction (Grosheintz-Fischer-Reissert Aldehyde Synthesis)
A. Reissert, *Ber.* **38,** 1603, 3415 (1905); J. M. Grosheintz, H. O. L. Fischer, *J. Am. Chem. Soc.* **63,** 2021 (1941).
Formation of 1-acyl-2-cyano-1,2-dihydroquinoline derivatives (Reissert compounds) by reaction of acid chlorides with quinoline and potassium cyanide; hydrolysis of these compounds yields aldehydes and quinaldic acid:

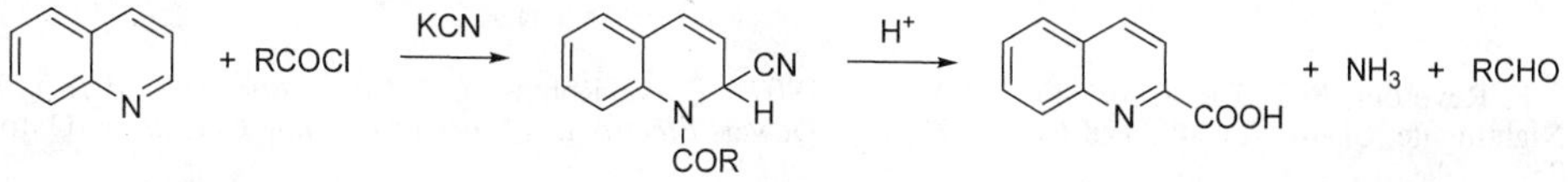

Reviews: E. Mosettig, *Org. React.* **8,** 220 (1954); W. E. McEwen, R. L. Cobb, *Chem. Rev.* **55,** 511 (1955); F. D. Popp, *Adv. Heterocycl. Chem.* **9,** 1 (1968); *idem, ibid.* **24,** 187 (1979); *idem. Bull. Soc. Chim. Belg.* **90,** 609 (1981); *idem* in *The Chemistry of Heterocyclic Compounds* **vol. 32,** Part 2, G. Jones, Ed. (Wiley, New York, 1982) p 353.

365. Reppe Chemistry
The term designates that phase of acetylene chemistry involving the use of acetylene at high pressures in the presence of suitable catalysts to carry out the fundamental reactions of vinylation, ethynylation, cyclopolymerization and carbonylation as developed from 1928 onward by Walter Reppe and associates in the I. G. Farbenindustrie laboratories in Ludwigshafen:

$$H_2C=CHOR \xleftarrow{ROH} HC\equiv CH \xrightarrow{RSH} H_2C=CHSR$$

$$RR^1CO + HC\equiv CH \longrightarrow RR^1C(OH)C\equiv CH \xrightarrow{RR^1CO} RR^1C(OH)C\equiv CC(OH)RR^1$$

$$4HC\equiv CH \longrightarrow \text{cyclooctatetraene}$$

$$4HC\equiv CH + 4C_2H_4OH + Ni(CO)_4 + 2HCl \longrightarrow 4H_2C=CHCOOC_2H_5 + NiCl_2 + H_2$$

J. W. Copenhaver, M. H. Bigelow, *Acetylene and Carbon Monoxide Chemistry* (New York, 1949) p 246; W. Reppe, *Acetylene Chemistry*, U.S. Dept. Commerce PB 18852-S (1949); *Neue Entwicklungen auf dem Gebiet des Acetylens und Kohlenoxyds* (Berlin 1949); H. Kröper, *Houben-Weyl* **4/II,** 413-422 (1955); D. W. F. Hardie, *Acetylene, Manufacture and Uses* (New York, 1965) p 67; L. F. Fieser, M. Fieser, *Reagents for Organic Synthesis* (New York, 1967) pp 61, 183, 185, 190, 519, 720, 722, 723. Review of carbonylations: A. Mullen, "Carbonylations Catalyzed by Metal Carbonyls-Reppe Reactions" in *New Syntheses with Carbon Monoxide*, J. Falbe, Ed. (Springer-Verlag, Berlin, 1980) pp 243-308. Mechanistic study of cyclooctatetraene synthesis: R. E. Colborn, K. P. C. Vollhardt, *J. Am. Chem. Soc.* **108,** 5470 (1986); C. J. Lawrie *et al., Organometallics* **8,** 2274 (1989).

366. Retro-Diels-Alder Reaction
Thermal dissociation of Diels-Alder adducts, occurring most readily when one or both fragments are particularly stable:

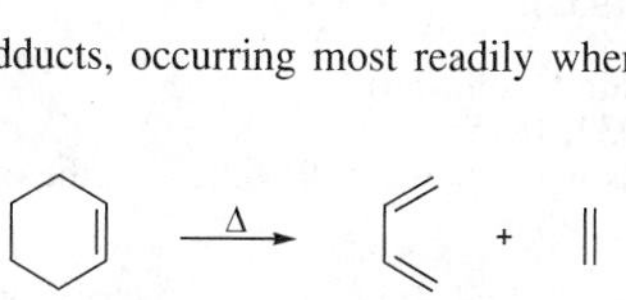

Early review: H. Kwart, K. King, *Chem. Rev.* **68,** 415-447 (1968). Acceleration by alkoxide substituent: O. Papies, W. Grimme, *Tetrahedron Lett.* **21,** 2799 (1980). Application to the synthesis of enethiols: Y. Vallée *et al., Synth. Commun.* **23,** 1267 (1993); of cyclopentadienyl ligands: B. Y. Lee *et al., J. Am. Chem. Soc.* **116,** 2163 (1994). Role in structure elucidation *via* mass spectrometry: F. Turecek, V. Hanus, *Mass Spectrom. Rev.* **3**, 85-152 (1984). Application to natural product synthesis: A. Ichihara, *Synthesis* **1987**, 207-222; R. W. Sweger, A. W. Czarnik, *Comp. Org. Syn.* **5**, 551-592 (1991).

367. Retropinacol Rearrangement
N. Zelinsky, J. Zelikow, *Ber.* **34**, 3249 (1901).
Conversion of an alcohol to the rearranged olefin on treatment with acid:

$$H_3C-\overset{\underset{\displaystyle CH_3}{|}}{\overset{\displaystyle CH_3}{|}}C-\overset{\underset{\displaystyle OH}{|}}{C}\begin{matrix}CH_3\\CH_3\end{matrix} \quad \xrightarrow[-H_2O]{H^+} \quad \begin{matrix}H_3C\\H_3C\end{matrix}C=C\begin{matrix}CH_3\\CH_3\end{matrix}$$

Application to sterols: W. F. Johns, *J. Org. Chem.* **26**, 4583 (1961); L. M. Harrison, P. V. Fennessey, *J. Steroid Biochem.* **36**, 407 (1990); to cyclohexanols: W. Hueckel, S. K. Gupte, *Ann.* **685**, 105 (1965). In conjunction with ring expansion: T. Kimura *et al., J. Org. Chem.* **43**, 1247 (1978). *Cf.* Nametkin Rearrangement; Wagner-Meerwein Rearrangement.

368. Reverdin Reaction
F. Reverdin, *Ber.* **29**, 997, 2595 (1896).
Migration of iodine during nitration of iodophenolic ethers:

$$\underset{I}{\overset{OR}{\bigcirc}} \quad \xrightarrow{HNO_3} \quad \underset{NO_2}{\overset{OR}{\bigcirc}}I$$

F. Reverdin, *Bull. Soc. Chim. Fr.* [4] **1**, 618 (1907); G. M. Robinson, *J. Chem. Soc.* **109**, 1078 (1916); D. V. Nightingale, *Chem. Rev.* **40**, 128 (1947); M. J. S. Dewar, *Electronic Theory of Organic Chemistry* (Oxford, 1949) p 232.

369. Riehm Quinoline Synthesis
P. Riehm *et al., Ber.* **18**, 2245 (1885); **19**, 1394 (1886); *idem, Ann.* **238**, 9 (1887).
Formation of quinoline derivatives by prolonged heating of arylamine hydrochlorides with ketones with or without use of aluminum chloride or phosphorus pentachloride:

$$\underset{NH_2 \cdot HCl}{\bigcirc} + \; 2CH_3COCH_3 \quad \xrightarrow[AlCl_3 \text{ or } PCl_5]{\Delta} \quad \underset{N}{\overset{CH_3}{\bigcirc\bigcirc}}CH_3 \; + \; CH_4 + 2H_2O$$

E. Knoevenagel *et al., Ann.* **55**, 1923, 1934 (1922); **56**, 2414 (1923); C. Hollins, *The Synthesis of Nitrogen Ring Compounds* (London, 1924) p 263; D. J. Craig, *J. Am. Chem. Soc.* **60**, 1458 (1938); R. C. Elderfield, J. R. McCarthy, *ibid.* **73**, 975 (1951); R. C. Elderfield, *Heterocycl. Compd.* **4**, 16 (1952).

370. Riemschneider Thiocarbamate Synthesis
R. Riemschneider, F. Wojahn, *Pharmazie* **4**, 460 (1949); *Chim. Ind. (Paris)* **64**, 99 (1950); *Pharm. Zentralhalle* **89**, 118 (1950).
The action of concentrated sulfuric acid followed by treatment with ice-water serves to transform arylthiocyanates into the corresponding thiocarbamates:

$$ArSCN \quad \xrightarrow[H_2O]{conc\; H_2SO_4} \quad ArS-\overset{\underset{\displaystyle NH}{||}}{C}-OH \quad \rightleftharpoons \quad ArSCONH_2$$

R. Riemschneider, *Chim. Ind. (Milan)* **33**, 483 (1951); *idem et al., J. Am. Chem. Soc.* **73**, 5905 (1951); R. Riemschneider, G. Orlick, *Angew. Chem.* **64**, 420 (1952); R. Riemschneider, *Chim. Ind. (Milan)* **34**, 353 (1952); *idem, Z. Naturforsch.* **7b**, 277 (1952); R. Riemschneider, G. Orlick, *Monatsh. Chem.* **84**, 316 (1953); K. Schmidt, P. Kolleck-Bös, *J. Am. Chem. Soc.* **75**, 6067 (1953).

371. Riley Oxidations (Selenium Dioxide Oxidation)
H. L. Riley *et al., J. Chem. Soc.* **1932**, 1875.
Oxidations of organic compounds with selenium dioxide; *e.g.*, the oxidation of active methylene groups to carbonyl groups:

$$CH_3CH_2CHO \; + \; SeO_2 \quad \longrightarrow \quad CH_3COCHO \; + \; H_2O \; + \; Se$$

N. Rabjohn, *Org. React.* **5**, 331 (1949); *Oxidation*, E. N. Trachtenberg, R. L. Augustine, Eds. (Marcel Dekker, New York, 1969) pp 119-187; H. O. House, *Modern Synthetic Reactions* (W. A. Benjamin, Menlo Park, California, 2nd ed., 1972) pp 407-411.

Organic Name Reactions

372. Ritter Reaction
J. J. Ritter, P. P. Minieri, *J. Am. Chem. Soc.* **70**, 4045 (1948); J. J. Ritter, J. Kalish, *ibid.* 4048.
Synthesis of amides from nitriles and alcohols or alkenes in strongly acidic media:

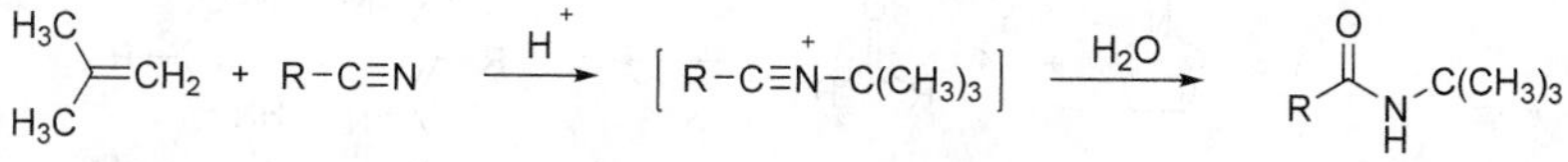

Reviews: L. I. Krimen, D. J. Cota, *Org. React.* **17**, 213-325 (1969); R. C. Larock, W. W. Leong, *Comp. Org. Syn.* **4**, 292-294 (1991); R. Bishop, *ibid.* **6**, 261-300 (1991). Synthetic applications: S. Top, G. Jaouen, *J. Org. Chem.* **46**, 78 (1981); D. M. Fink, R. C. Effland, *Synth. Commun.* **24**, 2793 (1994); W. M. Samaniego *et al., Tetrahedron Lett.* **35**, 6967 (1994).

373. Robinson Annulation
W. S. Rapson, R. Robinson, *J. Chem. Soc.* **1935**, 1285.
Formation of six-membered ring α,β-unsaturated ketones by the addition of cyclohexanones to methyl vinyl ketone (or simple derivatives of methyl vinyl ketone) or its equivalents, followed by an intramolecular aldol condensation, *q.v.*:

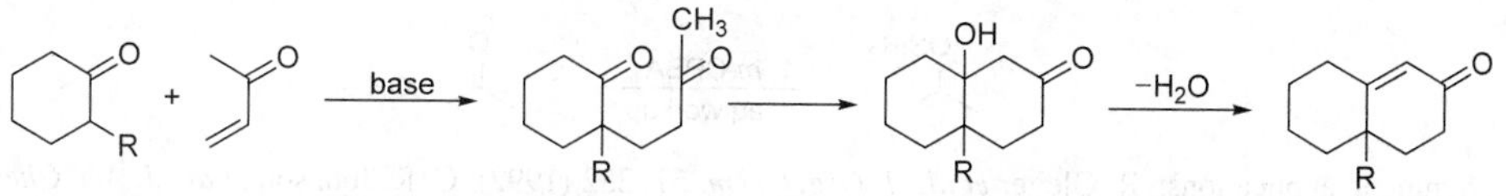

Early review: R. E. Gawley, *Synthesis* **1976**, 777-794. Improved methodology: T. Sato *et al., Tetrahedron Lett.* **31**, 1581 (1990). Stereochemical study: C. Nussbaumer, *Helv. Chim. Acta* **73**, 1621 (1990). Synthetic applications: R. V. Bonnert *et al., J. Chem. Soc. Perkin Trans. 1* **1991**, 1225; S. Kim, P. L. Fuchs, *J. Am. Chem. Soc.* **115**, 5934 (1993). *Cf.* Michael Reaction; Wichterle Reaction.

374. Robinson-Schöpf Reaction
R. Robinson, *J. Chem. Soc.* **111**, 762, 876 (1917); C. Schöpf, *Angew. Chem.* **50**, 779, 797 (1937).
Synthesis of tropinones from a dialdehyde, acetonedicarboxylic acid, and methylamine:

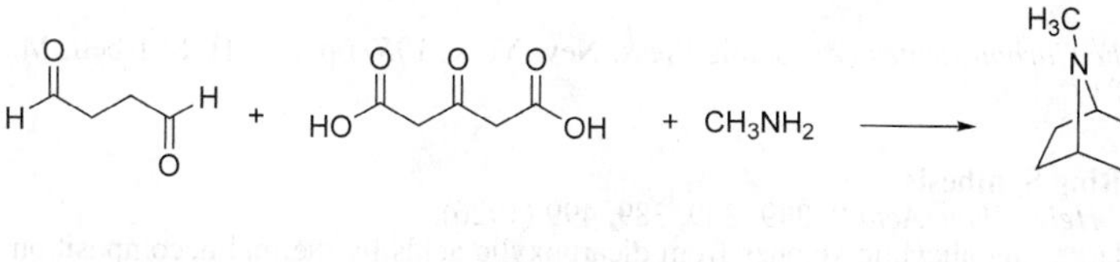

K. Alder *et al., Ann.* **601**, 147 (1956); R. D. Guthrie, J. F. McCarthy, *J. Chem. Soc. C* **1967**, 62; R. V. Stevens, A. W. M. Lee, *J. Am. Chem. Soc.* **101**, 7032 (1979); M. Langlois *et al., Synth. Commun.* **22**, 3115 (1992); T. Jarevang *et al., Acta Chem. Scand.* **52**, 1350 (1998). *Cf.* Mannich Reaction; Petrenko-Kritschenko Piperidone Synthesis.

375. Rosenmund Reduction
K. W. Rosenmund, *Ber.* **51**, 585 (1918); K. W. Rosenmund, F. Zetzsche, *ibid.* **54**, 425 (1921).
Catalytic reduction of acid chlorides to aldehydes. To prevent further hydrogenation a poison is added to the catalyst:

$$RCOCl \ + \ H_2 \ \xrightarrow{Pd-BaSO_4} \ RCHO \ + \ HCl$$

Reviews: E. Mosettig, R. Mozingo, *Org. React.* **4**, 362 (1948); A. Rachlin *et al., Org. Synth.* **51**, 8 (1971); J. A. Peters, H. Van Bekkum, *Rec. Trav. Chim.* **100**, 21 (1981). Investigation of reaction parameters: W. F. Maier *et al., J. Am. Chem. Soc.* **108**, 2608 (1986). Modified procedure applied to the synthesis of esters: V. V. Grushin, H. Alper, *J. Org. Chem.* **56**, 5159 (1991).

376. Rosenmund-von Braun Synthesis
K. W. Rosenmund, E. Struck, *Ber.* **52**, 1749 (1916); J. von Braun, G. Manz, *Ann.* **488**, 111 (1931).
Conversion of aryl halides to aromatic nitriles in the presence of cuprous cyanide:

$$ArX \ + \ CuCN \ \xrightarrow{\Delta} \ ArCN \ + \ CuX$$

Reviews: D. T. Moury, *Chem. Rev.* **42**, 207 (1948); J. E. Callen *et al., Org. Synth.* **3**, 212 (1955); M. S. Newman, *ibid.* 631; K. Takagi *et al., Bull. Chem. Soc. Jpn.* **48**, 3298 (1975); P. Bouyssou *et al., J. Heterocycl. Chem.* **29**, 895 (1992).

377. Rothemund Reaction
P. Rothemund, *J. Am. Chem. Soc.* **57**, 2010 (1935); **61**, 2912 (1939).
Preparation of *meso*-tetrasubstituted porphyrins by condensation of pyrrole with an aldehyde:

$$4 \quad \text{(pyrrole, } \overset{H}{N}\text{)} \quad + \quad 4 \quad \underset{R}{\overset{O}{\parallel}}\overset{}{C}H \quad \longrightarrow \quad \text{(porphyrin)}$$

Ball *et al.*, *J. Am. Chem. Soc.* **68**, 2278 (1946); Thomas, Martell, *ibid.* **78**, 1335 (1956). Mechanism: Badger *et al.*, *Aust. J. Chem.* **17**, 1028 (1964); R. G. Little, *J. Heterocycl. Chem.* **18**, 833 (1981).

378. Rubottom Oxidation
A. G. Brook, D. M. Macrae, *J. Organomet. Chem.* **77**, C19 (1974); A. Hassner *et al.*, *J. Org. Chem.* **40**, 3427 (1975); G. M. Rubottom *et al.*, *Tetrahedron Lett.* **1974**, 4319.
Oxidation of enolsilanes with *m*-chloroperbenzoic acid (*m*-CPBA) to afford α-hydroxy ketones:

$$\underset{}{\overset{OSiR_3}{\diagdown}} \quad \xrightarrow[\text{2. aq workup}]{\text{1. } m\text{-CPBA}} \quad \underset{}{\overset{O}{\parallel}}\text{OH}$$

Synthetic applications: R. Gleiter *et al.*, *J. Org. Chem.* **57**, 252 (1992); C. R. Johnson *et al.*, *J. Am. Chem. Soc.* **114**, 9414 (1992); M. T. Crimmins *et al.*, *ibid.* **115**, 3146 (1993).

379. Ruff-Fenton Degradation
O. Ruff, *Ber.* **31**, 1573 (1898); **32**, 550 (1899); H. J. H. Fenton, *Proc. Chem. Soc. London* **9**, 113 (1893).
Shortening of the carbon chain of sugars by the oxidation of aldonic acids (as calcium salts) with hydrogen peroxide and ferric salts:

$$RCH(OH)COOH \; + \; H_2O_2 \; \xrightarrow{Fe^{3+}} \; RCHO \; + \; CO_2 \; + \; 2H_2O$$

W. Pigman, *The Carbohydrates* (Academic Press, New York, 1957) p 118; H. S. Isbell, M. A. Salam, *Carbohydr. Res.* **90**, 123 (1981).

380. Ruzicka Large Ring Synthesis
L. Ruzicka *et al.*, *Helv. Chim. Acta* **9**, 249, 339, 389, 499 (1926).
Formation of large ring alicyclic ketones from dicarboxylic acids by thermal decomposition of salts with metals of the second and fourth groups of the periodic table (Ca, Th, Ce):

$$HO\underset{n}{\overset{O}{\parallel}}\cdots\overset{O}{\parallel}OH \quad \xrightarrow{\text{metal salt}} \quad \text{(cyclic ketone)}_n$$

L. Ruzicka, *Chem. Ind. (London)* **54**, 2 (1935); H. Gilman, *Organic Chemistry* **vol. 1** (New York, 1943) p 78; K. Ziegler, *Houben-Weyl* **4/2**, 754 (1955). *Cf.* Blanc Reaction.

381. Sabatier-Senderens Reduction
P. Sabatier, J. B. Senderens, *Compt. Rend.* **128**, 1173 (1899).
Catalytic hydrogenation of organic compounds in the vapor phase by passage over hot, finely divided nickel (the oldest of all hydrogenation methods).
E. B. Maxted in *Handbuch der Katalyse* **vol. 7**, G. M. Schwab, Ed. (Vienna, 1943) p 624; H. Roth *et al.*, *Houben-Weyl* **2**, 288 (1953); G. Schiller, *ibid.* **IV/2**, 284 (1955); H. O. House, *Modern Synthetic Reactions* (W. A. Benjamin, Menlo Park, California, 2nd ed., 1972) Chapter 1.

382. Saegusa Oxidation
Y. Ito *et al.*, *J. Org. Chem.* **43**, 1011 (1978).
Conversion of silyl enol ethers into corresponding α,β-enones using stoichiometric amounts of palladium acetate:

$$\underset{RCH_2 \quad OSiMe_3}{\overset{H \qquad R^1}{\diagup\diagdown}} \quad \xrightarrow[CH_3CN]{Pd(OAc)_2} \quad \underset{RCH_2 \quad O}{\overset{H \qquad R^1}{\diagup\diagdown}}$$

Application: M. Kim *et al.*, *Synth. Commun.* **20**, 989 (1990). Mechanism: S. Porth *et al.*, *Angew. Chem. Int. Ed.* **38**, 2015 (1999).

383. Sakurai Reaction (Hosomi-Sakurai Reaction)

A. Hosomi, H. Sakurai, *Tetrahedron Lett.* **1976**, 1295; A. Hosomi *et al., Chem. Lett.* **1976,** 941.

Lewis acid-promoted nucleophilic addition of allylic silanes to carbon electrophiles accompanied by regiospecific transposition of the allylic moiety:

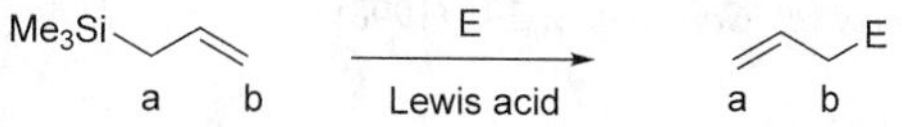

E = aldehydes, ketones, enones, acid chlorides, acetals, ketals, epoxides, iminium salts

Lewis acid = $TiCl_4$, $AlCl_3$, $BF_3 \cdot O(CH_2CH_3)_2$, $SnCl_4$, $(CH_3CH_2)_2AlCl$, cat $(CH_3)_3SiOSO_2CF_3$

Synthetic applications: I. E. Markó, D. J. Bayston, *Tetrahedron Lett.* **34,** 6595 (1993); H. Hioki *et al., ibid.* 6131. [$TiCp_2(OSO_2CF_3)_2$] as catalyst: T. K. Hollis *et al., ibid.* 4309. *Reviews:* I. Fleming *et al., Org. React.* **37,** 57-575 (1989); Y. Yamamoto, N. Sasaki, "The Stereochemistry of the Sakurai Reaction" in *Stereochemistry of Organometallic and Inorganic Compounds* **vol. 3,** I. Bernal, Ed. (Elsevier, New York, 1989) pp 363-437; I. Fleming, *Comp. Org. Syn.* **2,** 563-593 (1991).

384. Sandmeyer Diphenylurea Isatin Synthesis

T. Sandmeyer, *Z. Farben Text. Chem.* **2,** 129 (1903).

Formation of a cyanoformamidine by treatment of a symmetrical diphenylthiourea with potassium cyanide in alcohol containing lead carbonate, reduction with ammonium sulfide and ring-closure with concentrated sulfuric acid to isatin-2-anil; also formed smoothly by ring closure of the cyanoformamidine with aluminum chloride in benzene or carbon disulfide:

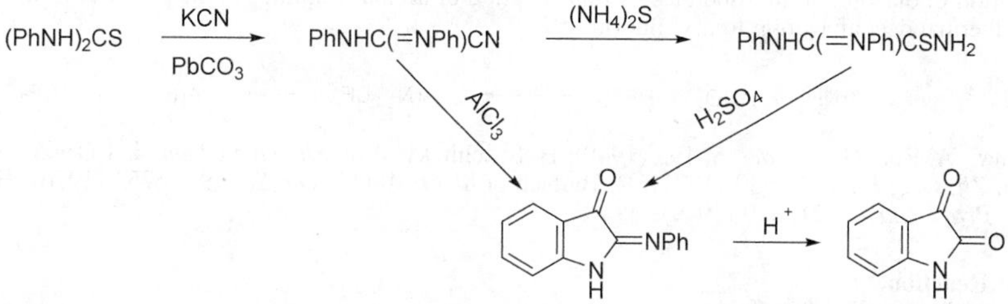

DE 115169, DE 116563 (both 1900 to J. R. Geigy & Co.); *Frdl.* **6,** 574, 575 (1900-1902); A. Reissert, *Ber.* **37,** 3708 (1904); G. Schultz *et al., J. Prakt. Chem.* [2] **74,** 74, 76 (1906); C. Hollins, *The Synthesis of Nitrogen Ring Compounds* (London, 1924) p 102; C. S. Marvel, G. S. Hiers, *Org. Synth.* **coll. vol. I,** 327 (1943); P. L. Julian *et al., Heterocycl. Compd.* **3,** 207 (1952); O. Bayer, W. Eckert, *Houben-Weyl* 7/4, 11 (1968). *Cf.* Sandmeyer Isonitrosoacetanilide Isatin Synthesis.

385. Sandmeyer Isonitrosoacetanilide Isatin Synthesis

T. Sandmeyer, *Helv. Chim. Acta* **2,** 234 (1919).

Formation of isonitrosoacetodiphenylamidine by condensation of chloral hydrate, hydroxylamine and aniline, cyclization with concentrated sulfuric acid, and quantitative hydrolysis to isatin on dilution:

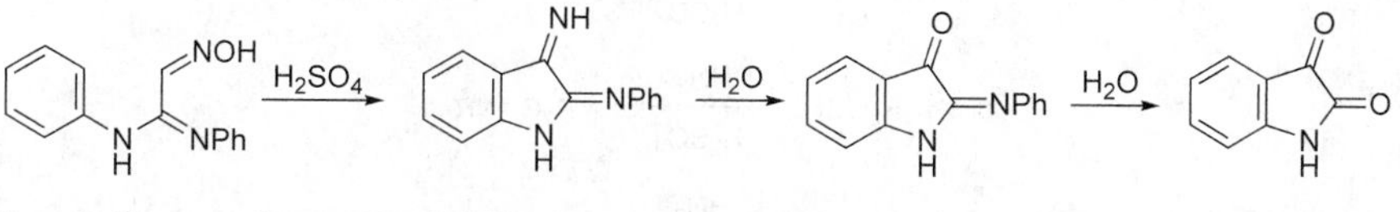

J. Martinet, P. Cousset, *Compt. Rend.* **172,** 1234 (1921); C. Hollins, *The Synthesis of Nitrogen Ring Compounds* (London, 1924) p 103; C. S. Marvel, G. S. Hiers, *Org. Synth.* **coll. vol. I,** 327 (1943); P. L. Julian *et al., Heterocycl. Compd.* **3,** 208 (1952); F. E. Sheibley, J. S. McNulty, *J. Org. Chem.* **21,** 171 (1956); O. Bayer, W. Eckert, *Houben-Weyl* 7/4, 14 (1968); S. J. Garden *et al., Tetrahedron Lett.* **38,** 1501 (1997). *Cf.* Sandmeyer Diphenylurea Isatin Synthesis.

386. Sandmeyer Reaction (Gattermann Reaction) (Körner-Contardi Reaction)

T. Sandmeyer, *Ber.* **17,** 1633, 2650 (1884); L. Gattermann, *Ber.* **23,** 1218 (1890); G. Körner, A. Contardi, *Atti Accad. Naz. Lincei* **23 II,** 464 (1914), *C.A.* **9,** 1478 (1915).

Substitution of diazonium groups in aromatic compounds by halo or cyano groups in the presence of cuprous salts (Sandmeyer reaction), copper powder and hydrochloric or hydrobromic acid (Gattermann reaction) or cupric salts (Körner-Contardi reaction):

Organic Name Reactions

Early reviews: H. H. Hodgson, *Chem. Rev.* **40**, 251-277 (1947); W. A. Coudrey, D. S. Davies, *Q. Rev. Chem. Soc.* **6**, 358-379 (1952); A. Roedig, *Houben-Weyl* **5/4**, 438 (1960); R. Stroh, *ibid.* **5/3**, 846 (1962). Direct conversion of aryl amines to aryl halides: M. P. Doyle, *J. Org. Chem.* **42**, 2426 (1977). Mechanistic studies: J. K. Kochi, *J. Am. Chem. Soc.* **79**, 2942 (1957); C. Galli, *J. Chem. Soc. Perkin Trans. 2* **1984**, 897. Synthetic application: C. Corral *et al., Heterocycles* **23**, 1431 (1985). Improved methodology: N. Suzuki *et al., J. Chem. Soc. Perkin Trans. 1* **1987**, 645; A. P. Krapcho, S. N. Haydar, *Heterocycl. Commun.* **4**, 291 (1998).

387. Sarett Oxidation; Collins Oxidation
G. I. Poos, G. E. Arth, R. E. Beyler, L. H. Sarett, *J. Am. Chem. Soc.* **75**, 422 (1953).

Oxidation of primary and secondary alcohols to aldehydes (and/or carboxylic acids) and ketones by means of CrO_3-pyridine complex:

$$\underset{OH}{\overset{H}{>}C} \quad \xrightarrow[\text{pyridine}]{CrO_3} \quad >C=O$$

J. R. Holum, *J. Org. Chem.* **26**, 4814 (1961); E. J. Kris, *Chem. Ind. (London)* **1961**, 1834; V. I. Stenberg, R. J. Perkins, *J. Org. Chem.* **28**, 323 (1963); P. G. Gassman, P. G. Pape, *J. Org. Chem.* **29**, 160 (1964); H. O. House, *Modern Synthetic Reactions* (W. A. Benjamin, Menlo Park, California, 2nd ed., 1972) pp 264-273. Mechanistic studies: F. Hasan, J. Rocek, *J. Am. Chem. Soc.* **97**, 1444, 3762 (1975).

The **Collins oxidation** is characterized by a modified procedure (dichloromethane as solvent) that reliably oxidizes primary alcohols to aldehydes: J. C. Collins, *Tetrahedron Lett.* **1968**, 3363; J. C. Collins, W. W. Hess, *Org. Synth.* **52**, 5 (1972); R. W. Ratcliffe, *ibid.* **55**, 84 (1976). *Cf.* Jones Oxidation.

388. Schiemann Reaction (Balz-Schiemann Reaction)
G. Balz, G. Schiemann, *Ber.* **60**, 1186 (1927).

Formation of diazonium fluoroborates by diazotization of aromatic amines in the presence of fluoroborates, followed by their thermal decomposition to aryl fluorides:

$$ArNH_2 + HNO_2 + HBF_4 \longrightarrow ArN_2^+ BF_4^- \xrightarrow{\Delta} ArF + N_2 + BF_3$$

Reviews: A. Roe, *Org. React.* **5**, 193 (1949); H. Suschitzky, *Adv. Fluorine Chem.* **4**, 1 (1965); T. K. Al'sing, E. G. Sochilin, *Zh. Org. Khim.* **7**, 530 (1971); R. Bartsch *et al., J. Am. Chem. Soc.* **98**, 6753 (1976); H. G. O. Becker, G. Israel, *J. Prakt. Chem.* **321**, 579 (1979).

389. Schmidt Reaction
R. F. Schmidt, *Ber.* **57**, 704 (1924).

Acid-catalyzed addition of hydrazoic acid to carboxylic acids, aldehydes and ketones to give amines, nitriles and amides, respectively. Tertiary alcohols and substituted alkenes yield imines upon treatment with hydrazoic acid:

$$RCOOH \xrightarrow[H_2SO_4]{HN_3} RNH_2$$

$$RCHO \xrightarrow[H_2SO_4]{HN_3} RCN + RNHCHO$$

$$RCOR \xrightarrow[H_2SO_4]{HN_3} RCONHR$$

$$R_3COH \xrightarrow[H_2SO_4]{HN_3} R_2C=NR$$

$$R_2C=CR_2 \xrightarrow[H_2SO_4]{HN_3} \underset{R}{R_2CHC=NR}$$

Early reviews: H. Wolff, *Org. React.* **3**, 307-336 (1946); P. A. S. Smith in *Molecular Rearrangements* Part 1, P. de Mayo, Ed. (Wiley-Interscience, New York, 1963) pp 507-558; D. V. Banthorpe, *The Chemistry of the Azido Group*, S. Patai, Ed. (Interscience, New York, 1971) pp 405-421; G. I. Koldobskii, *Russ. Chem. Rev.* **47**, 1084 (1978). Application to cyclic ketones: A. Lévai *et al., Heterocycles* **34**, 1523 (1992); J.-Y. Mérour *et al., J. Heterocycl. Chem.* **31**, 87 (1994); to alcohols and alkenes: W. H. Pearson *et al., J. Am. Chem. Soc.* **115**, 10183 (1993). Extension to dialkyl acylphosphonates: M. Sprecher, D. Kost, *ibid.* **116**, 1016 (1994). *Review:* T. Shioiri, *Comp. Org. Syn.* **6**, 817-821 (1991). *Cf.* Beckmann Rearrangement; Curtius Rearrangement; Hofmann Reaction; Lossen Rearrangement.

390. Scholl Reaction
R. Scholl, C. Seer, *Ann.* **394**, 111 (1912).

Coupling of aromatic molecules by treatment with Lewis acid catalysts:

$$2ArH \xrightarrow[H^+]{AlCl_3} Ar-Ar + H_2$$

$$\xrightarrow[AlCl_3]{100°} + H_2$$

Review: C. D. Nenitzescu, A. T. Balaban in *Friedel-Crafts and Related Reactions* **vol. 2,** part 2, G. Olah, Ed. (Wiley, New York, 1964) pp 979-1048; G. A. Clowes, *J. Chem. Soc. C* **1968,** 2519; A. C. Buchanan *et al., J. Am. Chem. Soc.* **102,** 5262 (1980).

391. Schöllkopf Bis-Lactim Amino Acid Synthesis
U. Schöllkopf *et al., Angew. Chem. Int. Ed.* **18,** 863 (1979); **20,** 798 (1981).

Asymmetric amino acid synthesis *via* diastereoselective alkylation of the lithiated bis-lactim ether (derived from L-Val and Gly or Ala) by an electrophile. Subsequent acid hydrolysis liberates L-Val-OCH$_3$ and the *(R)*-α-substituted amino acid ester. When the bis-lactim is generated from D-Val, the *(S)*-enantiomer forms:

$$\xrightarrow[\text{or LDA}]{n\text{-BuLi}}$$

$$\xrightarrow[\text{aq HCl}]{} H_2N \underset{O}{\overset{E^1 \quad R}{|}} OCH_3 + \text{L-Val-OCH}_3$$

R = H, CH$_3$
E = alkyl halides, aldehydes, ketones, thioketones, acid chlorides,
epoxides, acrylates

Synthetic applications: S. Kotha, A. Kuki, *Chem. Commun.* **1992,** 404; M. S. Allen *et al., Synth. Commun.* **22,** 2077 (1992). Isotopic labeling: N. R. Thomas, D. Gani, *Tetrahedron* **47,** 497 (1991). *Reviews:* U. Schöllkopf, *Top. Curr. Chem.* **109,** 65-84 (1983); *idem, Pure Appl. Chem.* **55,** 1799-1806 (1983); R. M. Williams, *Synthesis of Optically Active α-Amino Acids* (Pergamon, New York, 1989) pp 1-33.

392. Schotten-Baumann Reaction
C. Schotten, *Ber.* **17,** 2544 (1884); E. Baumann, *ibid.* **19,** 3218 (1886).

Acylation of alcohols or amines with acid chlorides in aqueous alkaline solution:

$$ROH + \underset{R^1 \quad X}{\overset{O}{||}} \xrightarrow{NaOH} \underset{R^1 \quad OR}{\overset{O}{||}}$$

Review: N. O. V. Sonntag, *Chem. Rev.* **52,** 272-273 (1953). Synthetic applications: M. Tsuchiya *et al., Bull. Chem. Soc. Jpn.* **42,** 1756 (1969); G. I. Georg, *Bioorg. Med. Chem. Lett.* **4,** 335 (1994).

393. Semmler-Wolff Reaction (Wolff-Semmler Aromatization) (Wolff Aromatization)
W. Semmler, *Ber.* **25,** 3352 (1892); L. Wolff, *Ann.* **322,** 351 (1902).

Rearrangement of α,β-unsaturated cyclohexenyl ketoximes into aromatic amines under acidic conditions:

$$\xrightarrow{}$$

Review: R. T. Conley, S. Ghosh in *Mechanisms of Molecular Migrations* **vol. 4,** B. S. Thyagarajan, Ed., (Interscience, New York, 1971) p 251; M. I. El-Sheikh, J. M. Cook, *J. Org. Chem.* **45,** 2585 (1980); Y. Tamura *et al., Synthesis* **1980,** 483.

394. Serini Reaction
  A. Serini *et al., Ber.* **72**, 391 (1939).
  Zinc-promoted rearrangement of 17-hydroxy-20-acetoxysterol derivatives into C-20 ketones; the reaction is applicable to other cyclic, as well as open-chain alcohols:

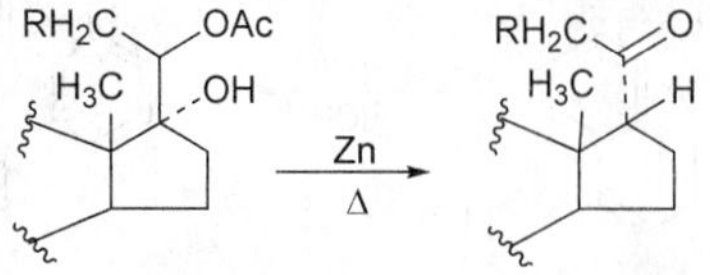

  Reviews: C. W. Shoppe, *Chimia* **4**, 418 (1948); L. F. Fieser, M. Fieser, *Steroids* (Reinhold Publishing Corp., New York, 1959) p 628; N. L. Wendler in *Molecular Rearrangements* Part 2, P. de Mayo, Ed. (Wiley-Interscience, New York, 1964) p 1038; E. Ghera, *Chem. Commun.* **1968**, 1639; *J. Org. Chem.* **35**, 660 (1970).

395. Seyferth-Gilbert Homologation; Ohira-Bestmann Modification
  E. W. Colvin, B. J. Hamill, *J. Chem. Soc. Chem. Commun.* **1973**, 151; *eidem, J. Chem. Soc. Perkin Trans. 1* **1977**, 869.
  Preparation of an alkyne by the reaction between an aldehyde or ketone with dimethyl diazomethyl phosphonate (Seyferth-Gilbert reagent) under basic conditions. In the **Ohira-Bestmann modification**, Seyferth-Gilbert reagent is generated *in situ*, leading to the formation of a terminal alkyne from an aldehyde:

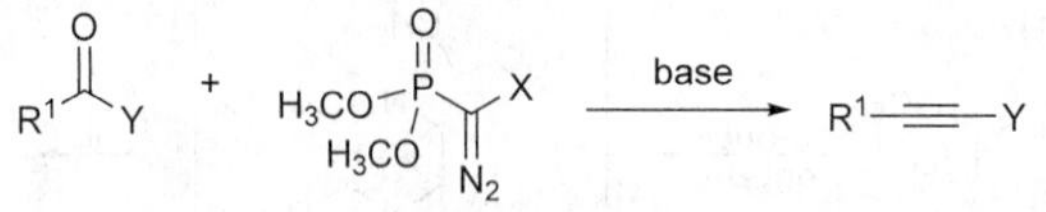

when X = H, Y = H, aryl, heteroaryl, base = *n*-BuLi, KO-*t*Bu
when X= Ac, Y = H, base = K_2CO_3 (Ohira-Bestmann modification)

  Methodology improvements and reaction mechanism: J. C. Gilbert, U. Weerasooriya, *J. Org. Chem.* **44**, 4997 (1979); *eidem, ibid.* **47**, 1837 (1982). Ohira-Bestmann modification: S. Ohira, *Synth. Commun.* **19**, 561 (1989); S. Müller *et al., Synlett* **1996**, 521; G. J. Roth *et al., Synthesis* **2004**, 59. One-pot applications: H. D. Dickson *et al., Tetrahedron Lett.* **45**, 5597 (2004); E. Quesada, R. J. K. Taylor, *ibid.* **46**, 6473 (2005); D. Luvino *et al., Synlett* **2007**, 3037. *Cf.* Seyferth-Gilbert Reagent; Bestmann-Ohira Reagent.

396. Sharpless Dihydroxylation
  E. N. Jacobsen *et al., J. Am. Chem. Soc.* **110**, 1968 (1988).
  Osmium-catalyzed asymmetric *cis*-dihydroxylation of olefins:

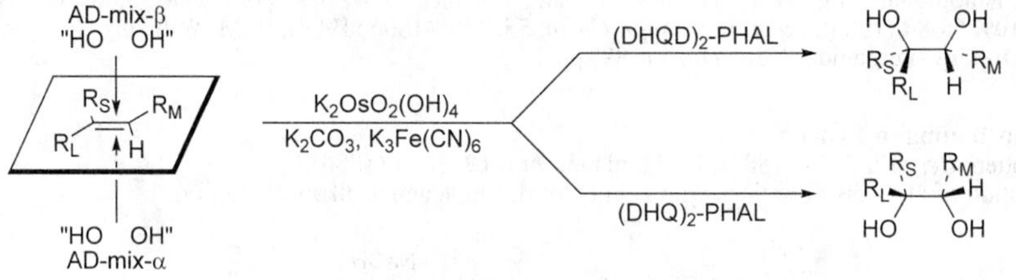

A premix of the four reagent components is commercially available. The
composition containing (DHQD)$_2$-PHAL is termed AD-mix-β; the composition
containing (DHQ)$_2$-PHAL is termed AD-mix-α.
(DHQD)$_2$-PHAL = 1,4-bis(9-O-dihydroquinidine)phthalazine; (DHQ)$_2$-PHAL =1,4-
bis(9-O-dihydroquinine)phthalazine.
R_L = largest substituent; R_M = medium-sized substituent; R_S = smallest substituent.

  Note: The scheme shown is an empirical mnemonic indicating olefin orientation and face selectivity. It is not to be considered an absolute predictor of new diol configurations.
  Allyl and vinyl silanes as substrates: A. R. Bassindale *et al., J. Chem. Soc. Perkin Trans. 1* **1994**, 1061. Chemoselective dihydroxylation of a polyene: S. C. Sinha, E. Keinan, *J. Org. Chem.* **59**, 949 (1994). *Reviews:* R. A. Johnson, K. B. Sharpless, "Catalytic Asymmetric Dihydroxylation" in *Catalytic Asymmetric Synthesis*, I. Ojima, Ed. (VCH, New York, 1993) pp 227-272; H. C. Kolb *et al., Chem. Rev.* **94**, 2483-2547 (1994). *Cf.* Milas Hydroxylation of Olefins.

397. Sharpless Epoxidation
  T. Katsuki, K. B. Sharpless, *J. Am. Chem. Soc.* **102**, 5974 (1980).
  Titanium-catalyzed asymmetric epoxidation of allylic alcohols employing titanium alkoxide, an optically active tartrate ester and an alkyl hydroperoxide. A high degree of enantiomeric purity is attainable having predictable absolute stereochemistry:

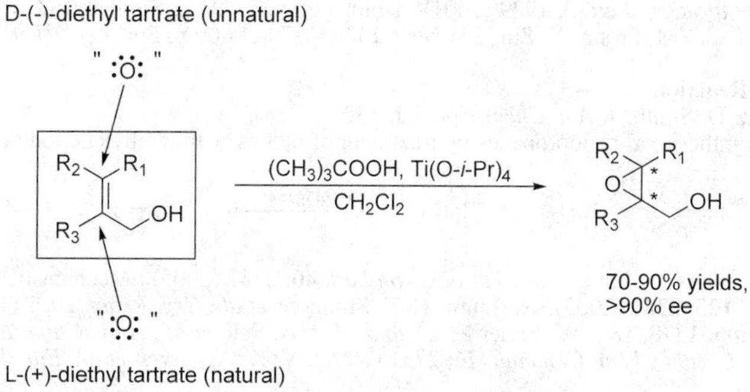

D-(-)-diethyl tartrate (unnatural)

" :Ö: "

$$R_2 \quad R_1$$
$$\quad \quad OH$$
$$R_3$$

$$\xrightarrow[\text{CH}_2\text{Cl}_2]{(\text{CH}_3)_3\text{COOH, Ti}(O\text{-}i\text{-Pr})_4}$$

$$R_2 \quad R_1$$
$$O \quad * \quad OH$$
$$R_3 \quad *$$

70-90% yields,
>90% ee

" :Ö: "

L-(+)-diethyl tartrate (natural)

Note: The asterisk at a chiral center denotes a preponderance of either the *R* or *S* configuration.

Mechanistic studies: S. S. Woodward *et al., J. Am. Chem. Soc.* **113**, 106 (1991); M. G. Finn, K. B. Sharpless, *ibid.* **113**. Methods development for the synthesis of enantiopure allylic alcohols: D. C. Dittmer *et al., J. Org. Chem.* **58**, 718 (1993). Alkenylsilanols as substrates: T. H. Chan *et al., Can. J. Chem.* **71**, 60 (1993). *Reviews:* R. A. Johnson, K. B. Sharpless, *Comp. Org. Syn.* **7**, 389-436 (1991); E. Höft, *Top. Curr. Chem.* **164**, 63-77 (1993).

398. Sharpless Oxyamination

K. B. Sharpless *et al., J. Am. Chem. Soc.* **97**, 2305 (1975).

Osmium-mediated *cis*-addition of nitrogen and oxygen moieties to mono-, di- and tri-substituted olefins to yield vicinal amino or amido alcohols:

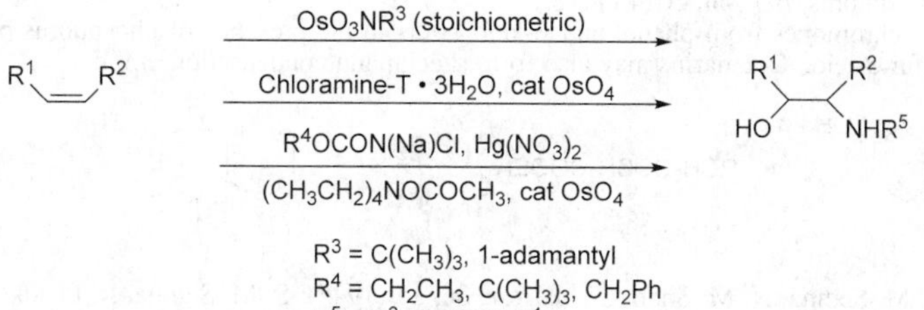

$$R^1 \quad R^2$$

$$\xrightarrow{\text{OsO}_3\text{NR}^3 \text{ (stoichiometric)}}$$

$$\xrightarrow{\text{Chloramine-T} \cdot 3\text{H}_2\text{O, cat OsO}_4}$$

$$\xrightarrow[(\text{CH}_3\text{CH}_2)_4\text{NOCOCH}_3, \text{ cat OsO}_4]{\text{R}^4\text{OCON(Na)Cl, Hg(NO}_3)_2}$$

$$R^1 \quad R^2$$
$$HO \quad NHR^5$$

$R^3 = C(CH_3)_3$, 1-adamantyl
$R^4 = CH_2CH_3$, $C(CH_3)_3$, CH_2Ph
$R^5 = R^3$, Ts, $COOR^4$

Methods development in the context of taxol synthesis: L. Mangatal *et al., Tetrahedron* **45**, 4177 (1989). Synthetic applications: S. K. Dubey, E. E. Knaus, *Can. J. Chem.* **61**, 565 (1983); M. Lemaire *et al., Synlett* **1990**, 615. Brief review: *Organic Syntheses by Oxidation with Metal Compounds*, W. J. Mijs, C. R. H. I. de Jonge, Eds. (Plenum Press, New York, 1986) pp 642-645.

399. Shi Epoxidation

Y. Tu *et al., J. Am. Chem. Soc.* **118**, 9806 (1996); Z.-X. Wang *et al., ibid.* **119**, 11224 (1997).

Enantioselective epoxidation of *trans*- and tri-substituted olefins using a fructose derived chiral ketone catalyst and potassium peroxomonosulfate or H_2O_2 as an oxidant:

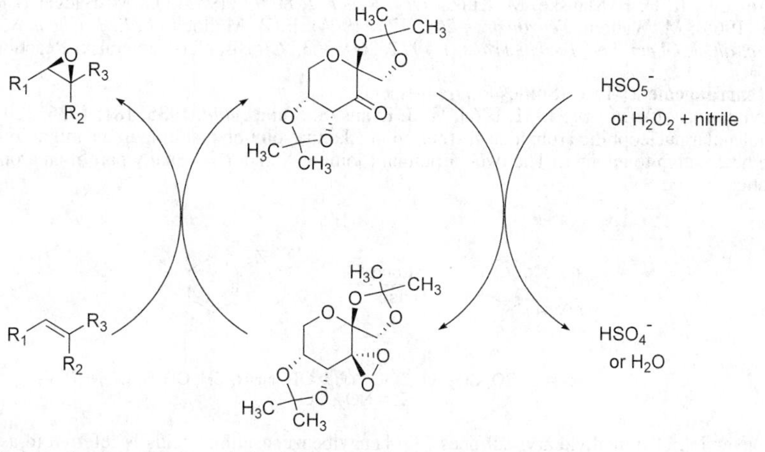

Catalyst modification for epoxidation of *cis*-olefins: H. Tian *et al., J. Am. Chem. Soc.* **122**, 11551 (2000). Use of hydrogen peroxide as the primary oxidant: L. Shu, Y. Shi, *Tetrahedron* **57**, 5213 (2001). Epoxidation of terminal olefins: H. Tian *et al.*, *Org. Lett.* **3**, 1929 (2001). Isotope effects: D. A. Singleton, Z. Wang, *J. Am. Chem. Soc.* **127**, 6679 (2005). *Reviews*: M. Frohn, Y. Shi, *Synthesis* **2000**, 1979-2000; Y. Shi, *Acc. Chem. Res.* **37**, 488-496 (2004).

400. Simmons-Smith Reaction

H. E. Simmons, R. D. Smith, *J. Am. Chem. Soc.* **80**, 5323 (1958).

Stereospecific synthesis of cyclopropanes by treatment of olefins with methylene iodide and zinc-copper couple:

$$\text{>=<} \quad + \quad CH_2I_2 \quad + \quad Zn(Cu) \quad \xrightarrow{(C_2H_5)_2O} \quad \text{>\hspace{-4pt}<} \quad + \quad ZnI_2 \quad + \quad Cu$$

Synthetic applications: J. Long *et al.*, *Tetrahedron Lett.* **46**, 2737 (2005). Mechanistic studies: M. Nakamura *et al.*, *J. Am. Chem. Soc.* **125**, 2341 (2003). *Reviews:* H. E. Simmons *et al.*, *Org. React.* **20**, 1 (1973); C. Girard, J. M. Conia, *J. Chem. Res. Synop.* **1978**, 182; W. Ratier *et al., ibid.* 179; A. Sele *et al.*, *Helv. Chim. Acta* **62**, 866 (1979); J. Joska, J. Fajkos, *Collect. Czech. Chem. Commun.* **46**, 2751 (1981); V. K. Aggarwal *et al.*, *Eur. J. Org. Chem.* **2002**, 319-326.

401. Simonini Reaction

A. Simonini, *Monatsh. Chem.* **13**, 320 (1892); **14**, 81 (1893).

The preparation of aliphatic esters by the reaction of the silver salt of a carboxylic acid with iodine:

$$2RCOOAg \quad + \quad I_2 \quad \xrightarrow{\Delta} \quad RCOOR \quad + \quad 2AgI \quad + \quad CO_2$$

H. Wieland, F. G. Fischer, *Ann.* **446**, 49 (1926); J. Kleinberg, *Chem. Rev.* **40**, 381 (1947); R. G. Johnson, R. K. Ingham, *ibid.* **56**, 259 (1956); C. V. Wilson, *Org. React.* **9**, 332 (1957); N. J. Bunce, M. Hadley, *Can. J. Chem.* **54**, 2612 (1976). *Cf.* Hunsdiecker Reaction.

402. Simonis Chromone Cyclization

E. Petschek, H. Simonis, *Ber.* **46**, 2014 (1913).

Formation of chromones from phenol and β-keto esters in the presence of phosphorus pentoxide, phosphorus oxy-chloride or sulfuric acid. Coumarins may also form (Pechmann condensation, *q.v.*):

$$\text{(phenol-OH)} \quad + \quad RCH_2COCH_2COOEt \quad \xrightarrow{P_2O_5} \quad \text{(chromone-CH}_2\text{R)} \quad + \quad EtOH \quad + \quad H_2O$$

Reviews: S. M. Sethna, N. M. Shah, *Chem. Rev.* **36**, 14 (1945); S. M. Sethna, R. Phadke, *Org. React.* **7**, 15 (1953); R. N. Lacey, *J. Chem. Soc.* **1954**, 854; O. Dann, G. Illing, *Ann.* **605**, 158 (1957); S. F. Tan, *Aust. J. Chem.* **25**, 1367 (1972).

403. Skraup Reaction

Z. H. Skraup, *Ber.* **13**, 2086 (1880).

Synthesis of quinolines from aromatic amines, glycerol, an oxidizing agent and sulfuric acid:

$$\text{(aniline-NH}_2\text{)} \quad + \quad \begin{matrix} CH_2OH \\ CHOH \\ CH_2OH \end{matrix} \quad \xrightarrow[C_6H_5NO_2]{H_2SO_4} \quad \text{(quinoline)}$$

Early review: R. H. F. Manske, M. Kulka, *Org. React.* **7**, 80-99 (1953). G. M. Badger *et al.*, *Aust. J. Chem.* **16**, 814, 828 (1963); M. Wahren, *Tetrahedron* **20**, 2773 (1964); E. B. Mullock *et al.*, *J. Chem. Soc. C* **1970**, 829; N. P. Buu-Hoi *et al.*, *J. Chem. Soc. Perkin Trans. 1* **1972**, 260, 263. *Cf.* Bischler-Napieralski Reaction.

404. Smiles Rearrangement; Truce-Smiles Rearrangement

A. A. Levi *et al.*, *J. Chem. Soc.* **1931**, 3264; W. J. Evans, S. Smiles, *ibid.* **1935**, 181; **1936**, 329.

Intramolecular nucleophilic aromatic substitution in alkaline solution resulting in the migration of an aromatic system from one heteroatom to another. The two-carbon unit joining X and Y is usually part of an aromatic ring but may also be aliphatic:

$$\xrightarrow[\text{base}]{\text{strong}}$$

X = S, SO, SO₂, O, COO   YH = OH, NHR, SH, CH₂R, CONHR
Z = NO₂, SO₂R

The conversion of *o*-methyldiaryl sulfones to *o*-benzylbenzenesulfinic acids is referred to as the **Truce-Smiles rearrangement:** W. E. Truce *et al., J. Am. Chem. Soc.* **80**, 3625 (1958); G. P. Crowther, C. R. Hauser, *J. Org. Chem.* **33**, 2228 (1968).

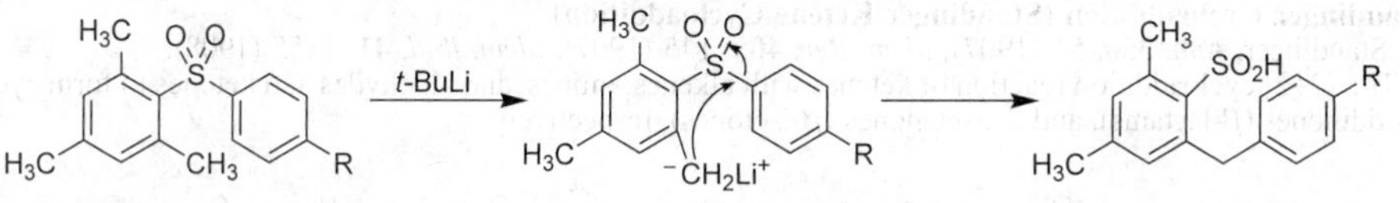

Early reviews: J. F. Bunnett, R. E. Zahler, *Chem. Rev.* **49**, 362 (1951); H. J. Shine, *Aromatic Rearrangements* (Elsevier, New York, 1967) pp 307-316; W. E. Truce *et al., Org. React.* **18**, 99-215 (1970). Conversion of phenols to anilines: I. G. C. Coutts, M. R. Southcott, *J. Chem. Soc. Perkin Trans. 1* **1990**, 767. Kinetic study: K. Bowden, P. R. Williams, *J. Chem. Soc. Perkin Trans. 2* **1991**, 215. Methods development for aliphatic substrates: M. Sako *et al., Chem. Pharm. Bull.* **42**, 806 (1994). Application to the synthesis of phenothiazines: S. K. Mukherjee *et al., Pharmazie* **49**, 453 (1994); J. Mukesh *et al., ibid.* 689.

405. Sommelet-Hauser Rearrangement
M. Sommelet, *Compt. Rend.* **205**, 56 (1937).

Rearrangement of benzyl quaternary ammonium salts to *ortho* substituted benzyldialkylamines on treatment with alkali metal amides:

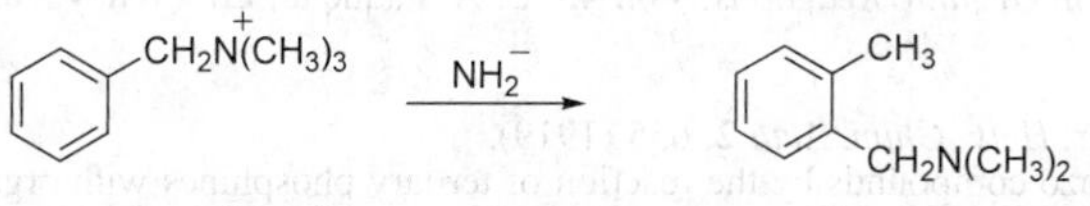

Early reviews: H. E. Zimmerman in *Molecular Rearrangements* Part 1, P. de Mayo, Ed. (Wiley-Interscience, New York, 1963) pp 382-391; S. H. Pine, *Org. React.* **18**, 403-464 (1970). Extension to sulfur ylides: M. Yamamoto *et al., Bull. Chem. Soc. Jpn.* **62**, 958 (1989); H. Ishibashi *et al., Chem. Pharm. Bull.* **39**, 2878 (1991). Effects of aromatic substitution: T. Tanaka *et al., ibid.* **40**, 518 (1992). Selectivity studies (Sommelet-Hauser rearrangement vs Stevens rearrangement, *q.v.*): T. Kitano *et al., J. Chem. Soc. Perkin Trans. 1* **1992**, 2851; T. Tanaka *et al., J. Org. Chem.* **57**, 5034 (1992). *Cf.* Meisenheimer Rearrangements; [2,3]-Wittig Rearrangement.

406. Sommelet Reaction
M. Sommelet, *Compt. Rend.* **157**, 852 (1913); *Bull. Soc. Chim. Fr.* [4] **23**, 95 (1918).

Preparation of aldehydes from aralkyl halides by treatment with hexamethylenetetramine to yield the quaternary salt, followed by mild hydrolysis:

$$ArCH_2X + C_6H_{12}N_4 \longrightarrow [ArCH_2-C_6H_{12}N_4]^+ X^- \xrightarrow[H_2O]{\Delta} ArCHO$$

Early reviews: S. J. Angyal, *Org. React.* **8**, 197-217 (1954); Bayer, *Houben-Weyl* **7/1**, 194 (1954). Synthetic applications: S. Miyano *et al., Bull. Chem. Soc. Jpn.* **59**, 3285 (1986); D. Evans *et al., Heterocycles* **26**, 1569 (1987). *Cf.* Kröhnke Oxidation.

407. Sonn-Müller Method
A. Sonn, E. Müller, *Ber.* **52**, 1927 (1919).

Reaction sequence employed to convert aromatic anilides to aldehydes. Treatment of the anilide with phosphorus pentachloride generates the imidoyl chloride, which is reduced to the imine with a mixture of stannous chloride and hydrochloric acid. Subsequent hydrolysis yields the aldehyde:

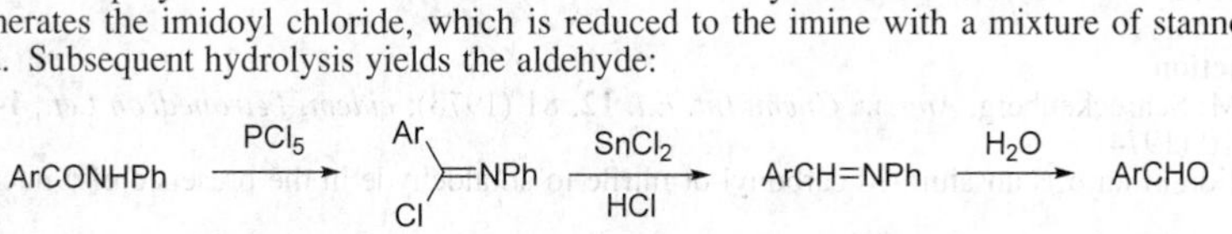

T. Reichstein, H. Zschokke, *Helv. Chim. Acta* **15**, 1105 (1932); W. E. Bachmann, *J. Am. Chem. Soc.* **57**, 1381 (1935); T. S. Work, *J. Chem. Soc.* **1942**, 429; L. N. Ferguson, *Chem. Rev.* **38**, 244 (1946); E. Mosettig, *Org. React.* **8**, 240 (1954); L. F. Fieser, M. Fieser, *Advanced Organic Chemistry* (New York, 1961) p 832. *Cf.* Grundmann Aldehyde Synthesis; Stephen Aldehyde Synthesis.

408. Sonogashira Coupling (Sonogashira-Hagihara Coupling)
K. Sonogashira *et al., Tetrahedron Lett.* **16**, 4467 (1975).

Palladium(0)-copper(I) catalyzed cross-coupling between sp²-hybridized organic halides and terminal acetylenes:

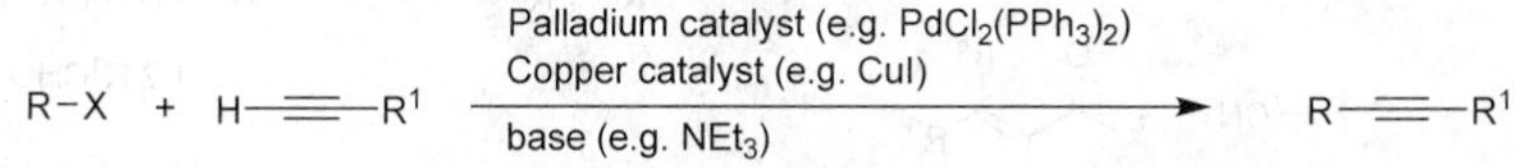

Synthetic applications: A. S. Karpov *et al., J. Org. Chem.* **68**, 1503 (2003); M. Erdélyi, A. Gogoll, *ibid.* 6431; A. Elangovan *et al., Org. Lett.* **5**, 1841 (2003); M. S. M. Ahmed, A. Mori, *ibid.* 3057; M. Eckhardt, G. C. Fu, *J. Am. Chem. Soc.* **125**, 13642 (2003); E. Mas-Marzá *et al., Tetrahedron Lett.* **44**, 6595 (2003). Nickel catalyzed reaction: I. P. Beletskaya *et al., ibid.* 5011. Copper-free modifications: A. Soheili *et al., Org. Lett.* **5**, 4191 (2003); L. Djakovitch, P. Rollet, *Adv. Synth. Catal.* **346**, 1782 (2004). *Reviews:* K. Sonogashira, *J. Organomet. Chem.* **653**, 46-49 (2002); R. R. Tykwinski, *Angew. Chem. Int. Ed.* **42**, 1566-1568 (2003).

Organic Name Reactions

409. Staudinger Cycloaddition (Staudinger Ketene Cycloaddition)

H. Staudinger, *Ann.* **356**, 51 (1907); *idem, Ber.* **40**, 1145 (1907); *idem, ibid.* **41**, 1355 (1908).

The [2+2] cycloaddition reaction of ketenes with alkenes, imines, and aldehydes and ketones to form cyclobutanones, azetidinones (β-lactams), and 2-oxetanones (β-lactones), respectively:

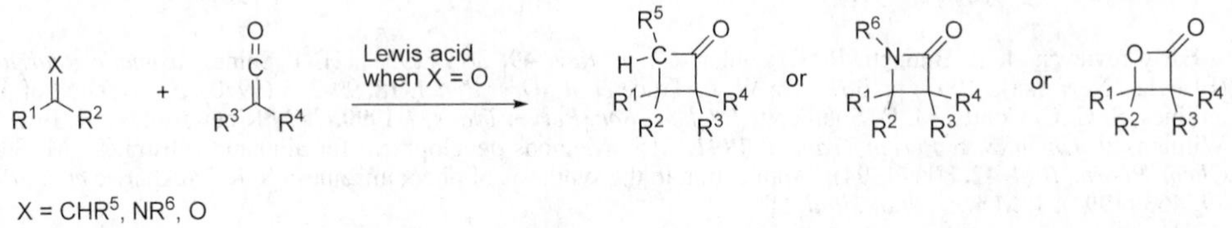

$$X = CHR^5, NR^6, O$$

Methodology application with thiocarbonyls: H. Kohn *et al., J. Org. Chem.* **43**, 4961 (1978); with alkynes: A. Hassner, J. L. Dillon, Jr., *ibid.* **48**, 3382 (1983). Review of reaction mechanism: A. Venturini, J. González, *Mini Rev. Org. Chem.* **3**, 185-194 (2006); F. P. Cossío *et al., Acc. Chem. Res.* **41**, 925-936 (2008). Review of asymmetric synthesis applications: R. K. Orr, M. A. Calter, *Tetrahedron* **59**, 3545-3565 (2003). Comprehensive review: J. A. Hyatt, P. W. Raynolds in *Organic Reactions*, **Vol. 45**, L. A. Paquette, Ed. (Wiley, New York, 1994) pp 159-646.

410. Staudinger Reaction

H. Staudinger, J. Meyer, *Helv. Chim. Acta* **2**, 635 (1919).

Synthesis of phosphazo compounds by the reaction of tertiary phosphines with organic azides:

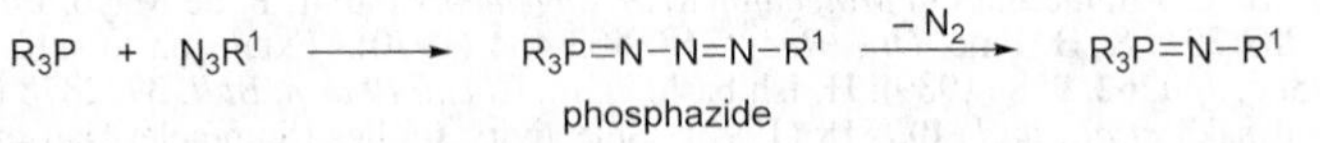

$$R_3P \; + \; N_3R^1 \longrightarrow R_3P=N-N=N-R^1 \xrightarrow{-N_2} R_3P=N-R^1$$
$$\text{phosphazide}$$

Synthetic applications: M. Taillefer *et al., Chem. Commun.* **6**, 565 (1999); M. D. Velasco *et al., Tetrahedron* **56**, 4079 (2000); P. Vanek, P. Klán, *Synth. Commun.* **30**, 1503 (2000). Cell surface engineering: E. Saxon, C. R. Bertozzi, *Science* **287**, 2007 (2000). *Reviews*: Y. G. Gololobov *et al., Tetrahedron* **37**, 437-472 (1981); Y. G. Gololobov, L. F. Kasukhin, *ibid.* 1353-1406 (1992); M. Köhn, R. Breinbauer, *Angew. Chem. Int. Ed.* **43**, 3106-3116 (2004).

411. Stephen Aldehyde Synthesis

H. Stephen, *J. Chem. Soc.* **127**, 1874 (1925); T. Stephen, H. Stephen, *ibid.* **1956**, 4695.

Reaction sequence employed to convert nitriles to aldehydes. Treatment of the nitrile with a mixture of stannous chloride and hydrochloric acid yields the imine salt complex which is subsequently hydrolyzed to the aldehyde. Practically applied only to aromatic aldehydes:

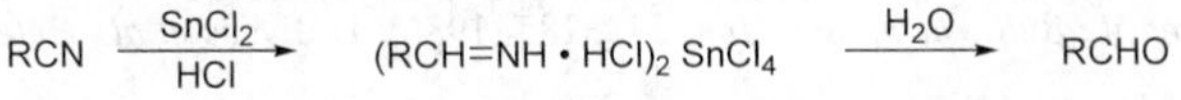

$$RCN \xrightarrow[HCl]{SnCl_2} (RCH=NH \cdot HCl)_2 \, SnCl_4 \xrightarrow{H_2O} RCHO$$

L. N. Ferguson, *Chem. Rev.* **38**, 243 (1946); E. Mosettig, *Org. React.* **8**, 246 (1954); O. Bayer, *Houben-Weyl* **7/1**, 299 (1954); E. N. Zilberman, P. S. Pyryalova, *J. Gen. Chem. USSR* (Engl. trans.) **33**, 3348 (1963); C. G. Stuckwisch, *J. Org. Chem.* **37**, 318 (1972). *Cf.* Grundmann Aldehyde Synthesis; Sonn-Müller Method.

412. Stetter Reaction

H. Stetter, M. Schreckenberg, *Angew. Chem. Int. Ed.* **12**, 81 (1973); *eidem, Tetrahedron Lett.* **14**, 1461 (1973); *eidem, Ber.* **107**, 210 (1974).

1,4-Addition of an α,β-unsaturated carbonyl or nitrile to an aldehyde in the presence of a nucleophilic catalyst:

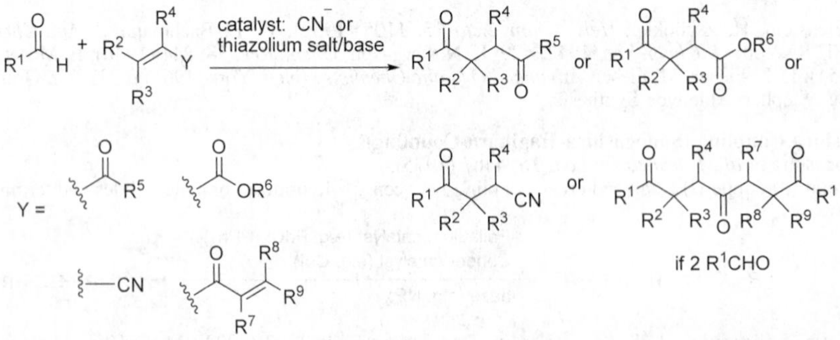

Asymmetric intramolecular reactions: D. Enders *et al., Helv. Chim. Acta* **79**, 1899 (1996); J. Read de Alaniz, T. Rovis, *J. Am. Chem. Soc.* **127**, 6284 (2005). Review of asymmetric reactions: M. Christmann, *Angew. Chem. Int. Ed.* **44**, 2632-2634 (2005). *Reviews*: H. Stetter, *Angew. Chem. Int. Ed. Engl.* **15**, 639-647 (1976); H. Stetter, H. Kuhlmann in *Organic Reactions* **vol. 40**, L. A. Paquette, Ed. (Wiley, New York, 1991) pp 407-496. *Cf.* Benzoin Condensation; Michael Condensation.

Organic Name Reactions

413. Stevens Rearrangement
T. S. Stevens *et al., J. Chem. Soc.* **1928**, 3193; **1930**, 2107, 2119; **1932**, 55, 1926, 1932.
Migration of an alkyl group from a sulfonium or quaternary ammonium salt to an adjacent carbanionic center on treatment with strong base. The product is a rearranged tertiary amine or sulfide:

$$R^1 \diagdown Y^+ \diagdown R^2 \quad \xrightarrow{NaNH_2} \quad R^1 \diagdown Y^+ \diagdown R^2 \quad \longrightarrow \quad R^1 \diagup Y \diagdown R^3$$

Y = NR or S

Early reviews: H. E. Zimmerman in *Molecular Rearrangements* Part 1, P. de Mayo, Ed. (Wiley-Interscience, New York, 1963) pp 345-406; D. J. Cram, *Fundamentals of Carbanion Chemistry* (Academic Press, New York, 1965) pp 223-229; S. M. Pine, *Org. React.* **18**, 403-464 (1970). Selectivity studies vs Sommelet-Hauser rearrangement, *q.v.*: T. Kitano *et al., J. Chem. Soc. Perkin Trans. 1* **1992**, 2851; T. Tanaka *et al., J. Org. Chem.* **57**, 5034 (1992). *Review:* I. E. Markó, *Comp. Org. Syn.* **3**, 913-932 (1991). Comprehensive review and applications in natural product synthesis: J. A. Vanecko *et al., Tetrahedron* **62**, 1043-1062 (2006). *Cf.* Meisenheimer Rearrangements; [1,2]-Wittig Rearrangement.

414. Stieglitz Rearrangement
J. Stieglitz, P. N. Leech, *Ber.* **46**, 2147 (1913); *J. Am. Chem. Soc.* **36**, 272 (1914).
Rearrangement of trityl hydroxylamines to Schiff bases on treatment with phosphorus pentachloride:

$$Ar_3CNHOH \quad \xrightarrow{PCl_5} \quad Ar_2C=NAr \ + \ POCl_3 \ + \ 2HCl$$

Reviews: P. A. S. Smith in *Molecular Rearrangements* Part 1, P. de Mayo, Ed. (Wiley-Interscience, New York, 1963) p 479; *Trans. N.Y. Acad. Sci.* **31**, 504 (1969); N. Koga, J. P. Anselme, *Tetrahedron Lett.* **1969**, 4773; R. V. Hoffman, D. J. Poelker, *J. Org. Chem.* **44**, 2364 (1979).

415. Stille Coupling
M. Kosugi *et al., Chem. Lett.* **1977**, 301 (1977); D. Milstein, J. K. Stille, *J. Am. Chem. Soc.* **100**, 3636 (1978).
Palladium-catalyzed cross coupling reaction of organostannanes with organic halides, acetates or perfluorinated sulfonates lacking a sp^3-hybridized β-hydrogen:

$$R_3Sn-R^1 \ + \ R^2-X \quad \xrightarrow{[Pd]} \quad R^1-R^2 \ + \ R_3SnX$$

$[Pd] = Pd(PPh_3)_4, PhCH_2Pd(PPh_3)_2Cl$
R^1 = alkynyl, alkenyl, aryl, allyl, benzyl, alkyl
R^2 = acyl, alkenyl, allyl, benzyl, aryl
X = Cl, Br, I, $OCOCH_3$, $OSO_2(C_nF_{2n+1})$, n = 0,1,4

Allylic acetates as substrates: L. Del Valle *et al., J. Org. Chem.* **55**, 3019 (1990). Effect of additives: S. Gronowitz *et al., J. Organomet. Chem.* **460**, 127 (1993); V. Farina *et al., J. Org. Chem.* **59**, 5905 (1994). Synthesis of α-methylene lactones: R. M. Adlington *et al., J. Chem. Soc. Perkin Trans. 1* **1994**, 1697. Solid-phase synthesis of 1,4-benzodiazepines: M. J. Plunkett, J. A. Ellman, *J. Am. Chem. Soc.* **117**, 3306 (1995). *Reviews:* J. K. Stille, *Angew. Chem. Int. Ed.* **25**, 508-524 (1986); M. Pereyre *et al., Tin in Organic Synthesis* (Butterworths, Boston, 1987) pp 185-207 *passim*. Review of synthetic applications: T. N. Mitchell, *Synthesis* **1992**, 803-815; of mechanisms: P. Espinet, A. M. Echavarren, *Angew. Chem. Int. Ed.* **43**, 4704-4734 (2004). *Cf.* Heck Reaction; Suzuki Coupling.

416. Stobbe Condensation
H. Stobbe, *Ber.* **26**, 2312 (1893); *Ann.* **282**, 280 (1894).
Condensation of aldehydes or ketones with diethyl succinate in the presence of a strong base to form monoesters of α-alkylidene (or arylidene) succinic acids:

Organic Name Reactions

Reviews: W. S. Johnson, G. H. Daub, *Org. React.* **6**, 1 (1951); H. O. House, *Modern Synthetic Reactions* (W. A. Benjamin, Menlo Park, California, 2nd ed., 1972) pp 663-666; R. J. Hart, H. G. Heller, *J. Chem. Soc. Perkin Trans. 1* **1972**, 1321; N. R. El-Rayyes, *J. Prakt. Chem.* **315**, 295 (1973); V. B. Bagos *et al., Helv. Chim. Acta* **62**, 90 (1979). *Cf.* Perkin Reaction.

417. Stollé Synthesis

R. Stollé, *Ber.* **46**, 3915 (1913); **47**, 2120 (1914); *J. Prakt. Chem.* **105**, 137 (1923); **128**, 1 (1930).

Formation of indole derivatives by the reaction of arylamines with α-haloacid chlorides or oxalyl chloride, followed by cyclization of the resulting amides with aluminum chloride:

W. C. Sumpter, *Chem. Rev.* **34**, 396 (1944); **37**, 446 (1945); P. L. Julian *et al., Heterocycl. Compd.* **3**, 142, 209 (1952); A. H. Beckett *et al., Tetrahedron* **24**, 6093 (1968). *Cf.* Hinsberg Oxindole Synthesis.

418. Stork Enamine Reaction

G. Stork *et al., J. Am. Chem. Soc.* **76**, 2029 (1954); G. Stork, H. Landesman, *ibid.* **78**, 5128 (1956).

Synthesis of α-alkyl or α-acyl carbonyl compounds from enamines and alkyl or acyl halides:

Reviews: J. Szmuszkovicz, *Adv. Org. Chem.* **4**, 1 (1963); A. G. Cook, Ed., *Enamines* (Marcel Dekker, New York, 1969); H. O. House, *Modern Synthetic Reactions* (W. A. Benjamin, Menlo Park, California, 2nd ed., 1972) pp 570-580, 766-772; S. F. Dyke, *Chemistry of Enamines* (Cambridge University Press, New York, 1973); P. W. Hickmott, *Tetrahedron* **38**, 1975 (1982). Synthetic applications: C. F. Bridge, D. O'Hagan, *J. Fluorine Chem.* **82**, 21 (1997); J. J. Li *et al., Tetrahedron Lett.* **39**, 6111 (1998).

419. Strecker Amino Acid Synthesis

A. Strecker, *Ann.* **75**, 27 (1850); **91**, 349 (1854).

Synthesis of α-amino acids by reaction of aldehydes with ammonia and hydrogen cyanide followed by hydrolysis of the resulting α-aminonitriles. Safer, milder, and more selective reaction conditions have been developed, esp in regard to asymmetric synthesis. The scope of the reaction has been extended to include primary and secondary amines:

Reviews: J. P. Greenstein, M. Winitz, *Chemistry of the Amino Acids* vol. 3 (New York, 1961) pp 698-700; G. C. Barrett, *Chemistry and Biochemistry of the Amino Acids* (Chapman and Hall, New York, 1985) pp 251, 261. Asymmetric synthesis using enantiopure sulfinimines: F. A. Davis *et al., Tetrahedron Lett.* **35**, 9351 (1994); *idem et al., J. Org. Chem.* **61**, 440 (1996). Asymmetric syntheses: M. S. Sigman, E. N. Jacobsen, *J. Am. Chem. Soc.* **120**, 4901 (1998); E. J. Corey, M. J. Grogan, *Org. Lett.* **1**, 157 (1999). Review of catalytic enantioselective syntheses: H. Gröger, *Chem. Rev.* **103**, 2795-2827 (2003). *Cf.* Bucherer-Bergs Reaction.

420. Strecker Degradation
A. Strecker, *Ann.* **123**, 363 (1862).

Interaction of an α-amino acid with a carbonyl compound in aqueous solution or suspension to give carbon dioxide and an aldehyde or ketone containing one less carbon atom. Inorganic oxidizing agents can also be used to bring about the reaction:

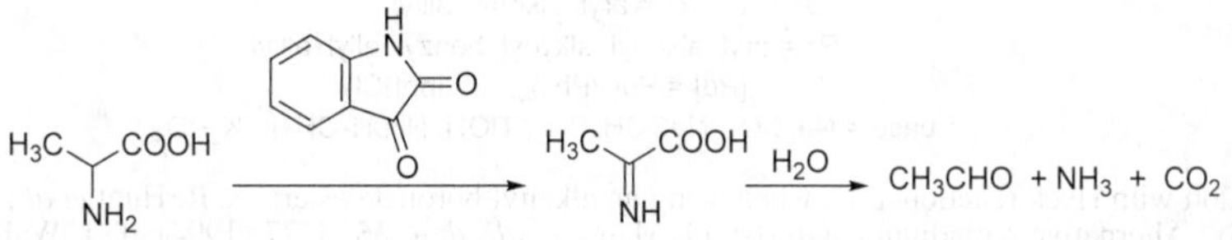

Early review: A. Schönberg, R. Moubacher, *Chem. Rev.* **50**, 261 (1952). Photo-promoted oxidation: Y. Ogata *et al.*, *Bull. Chem. Soc. Jpn.* **54**, 2057 (1981). Synthetic studies: A. Schönberg *et al.*, *J. Chem. Soc.* **1948**, 176; C.-T. Ho, G. J. Hartman, *J. Agric. Food Chem.* **1982**, 793; A. F. Ghiron *et al.*, *ibid.* **36**, 677 (1988); J. Koch *et al.*, *Carbohydr. Res.* **313**, 117 (1998).

421. Strecker Sulfite Alkylation
A. Strecker, *Ann.* **148**, 90 (1868).

Formation of alkyl sulfonates by reaction of alkyl halides with alkali or ammonium sulfites in aqueous solution in the presence of iodide:

$$RX \ + \ M_2SO_3 \ \xrightarrow{\text{NaI}} \ RSO_3M \ + \ MX$$

A. Collmann, *ibid.* 101; W. Hemilian, *ibid.* **168**, 145 (1873); *Ber.* **6**, 562 (1873); **CH 105845**; **CH 104907** (both 1925); F. C. Wagner, E. E. Reid, *J. Am. Chem. Soc.* **53**, 3409 (1931); C. Weygand, *Organic Preparations* (New York, 1945) p 306; M. Quaedvlieg, *Houben-Weyl* **9**, 372 (1955).

422. Suarez Reaction (Suarez Fragmentation)
J. I. Concepion *et al.*, *Tetrahedron Lett.* **25**, 1953 (1984); *eidem, J. Org. Chem.* **51**, 402 (1986).

Photoinduced conversion of hydroxyl-containing substrates with hypervalent iodine I(III)I$_2$ to the corresponding oxygen-centered radical:

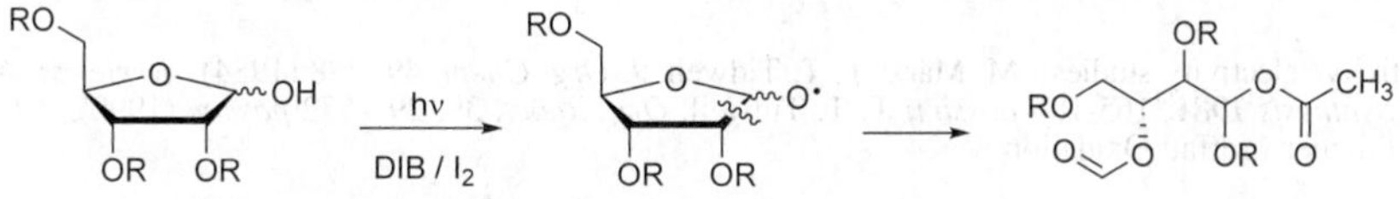

DIB = (diacetoxyiodo)benzene

P. De Armas *et al.*, *Angew. Chem. Int. Ed.* **31**, 772 (1992). Mechanistic studies: J. L. Courtneidge *et al.*, *Tetrahedron Lett.* **35**, 1003 (1994); T. Muraki *et al.*, *J. Chem. Soc. Perkin Trans. 1* **1999**, 1713. Synthetic applications: C. M. Hayward *et al.*, *Tetrahedron Lett.* **34**, 3989 (1993); A. Kittaka *et al.*, *Tetrahedron* **55**, 5319 (1999).

423. Sugasawa Reaction
T. Sugasawa *et al.*, *J. Am. Chem. Soc.* **100**, 4842 (1978); M. Adachi *et al.*, *Chem. Pharm. Bull.* **33**, 1826 (1985).

Ortho acylation of anilines by nitriles in the presence of BCl$_3$ and an auxillary Lewis acid:

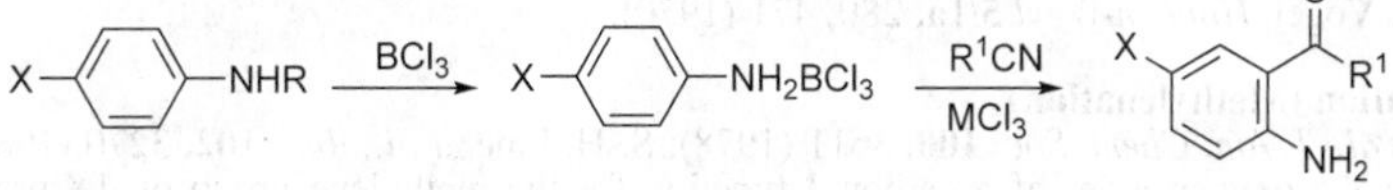

Mechanistic study: A. W. Douglas *et al.*, *Tetrahedron Lett.* **35**, 6807 (1994). Synthetic application: J. N. Houpis *et al.*, *ibid.* 6811.

424. Suzuki Coupling (Suzuki-Miyaura Cross-Coupling)
N. Miyaura *et al.*, *Tetrahedron Lett.* **1979**, 3437; N. Miyaura, A. Suzuki, *Chem. Commun.* **1979**, 866.

Palladium-catalyzed cross coupling of organic halides or perfluorinated sulfonates with organoboron derivatives proceeding with high stereo- and regioselectivity:

$$R^1-BY_2 \quad + \quad R^2-X \quad \xrightarrow[\text{base}]{[Pd]} \quad R^1-R^2$$

$BY_2 = B(OR)_2$, 9–BBN, $B(CHCH_3CH(CH_3)_2)_2$
X = I, Br, Cl, $OSO_2(C_nF_{2n+1})$, n = 0,1,4
R^1 = aryl, alkenyl, alkyl
R^2 = aryl, alkenyl, alkynyl, benzyl, allyl, alkyl
[Pd] = $Pd(PPh_3)_4$, $Pd(dppf)Cl_2$
base = Na_2CO_3, $NaOCH_2CH_3$, TlOH, $N(CH_2CH_3)_3$, K_3PO_4

Competition with Heck reaction, *q.v.*, when using an alkenyl boronate ester: A. R. Hunt *et al., Tetrahedron Lett.* **34,** 3599 (1993). Alternative palladium catalysts: G. Marck *et al., ibid.* **35,** 3277 (1994); T. I. Wallow, B. M. Novak, *J. Org. Chem.* **59,** 5034 (1994). *Reviews:* A. Suzuki, *Pure Appl. Chem.* **63,** 419-422 (1991); A. R. Martin, Y. Yang, *Acta Chem. Scand.* **47,** 221-230 (1993); S. Kotha *et al., Tetrahedron* **58,** 9633-9695 (2002). *Cf.* Hydroboration Reaction; Stille Coupling.

425. Swarts Reaction
F. Swartx, *Bull. Acad. R. Belg.* **24,** 309 (1892).
Fluorination of organic polyhalides with antimony trifluoride (or zinc and mercury fluorides) in the presence of a trace of a pentavalent antimony salt:

$$ClHC=CClCHCl_2 \quad + \quad SbF_3 \quad \xrightarrow{SbCl_5} \quad ClHC=CClCF_3 \quad + \quad SbCl_3$$

A. L. Henne, *Org. React.* **2,** 49 (1944); M. Hudlicky, *Chemistry of Organic Fluorine Compounds* (MacMillan, New York, 1962) pp 93-98.

426. Swern Oxidation (Moffatt-Swern Oxidation)
K. Omura, D. Swern, *Tetrahedron* **34,** 1651 (1978).
Mild oxidation of primary and secondary alcohols, promoted by oxalyl chloride activation of dimethyl sulfoxide, evidently involving the dimethyl alkoxysulfonium salts. Upon the addition of base, the intermediates rearrange intramolecularly to generate aldehydes or ketones, respectively:

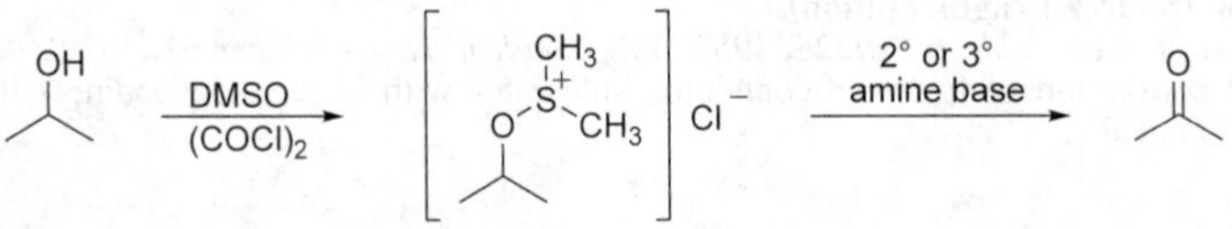

Reactivity/selectivity studies: M. Marx, T. T. Tidwell, *J. Org. Chem.* **49,** 788 (1984). *Reviews:* A. J. Mancuso, D. Swern, *Synthesis* **1981,** 165-185 *passim;* T. T. Tidwell, *Org. React.* **39,** 297-572 *passim* (1990). *Cf.* Corey-Kim Oxidation; Pfitzner-Moffatt Oxidation.

427. Tafel Rearrangement
J. Tafel, H. Hahl, *Ber.* **40,** 3312 (1907).
Rearrangement of the carbon skeleton of substituted acetoacetic esters to hydrocarbons with the same number of carbon atoms by electrolytic reduction at a lead cathode in alcoholic sulfuric acid:

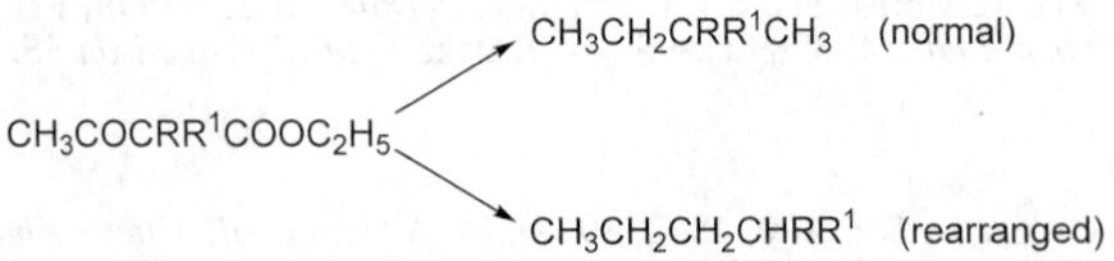

$$CH_3COCRR^1COOC_2H_5 \Big\langle {}^{\displaystyle CH_3CH_2CRR^1CH_3 \quad \text{(normal)}}_{\displaystyle CH_3CH_2CH_2CHRR^1 \quad \text{(rearranged)}}$$

J. Tafel, W. Jürgen, *ibid.* **42,** 2548 (1909); J. Tafel, *ibid.* **45,** 437 (1912); C. J. Brockman, *Electro-organic Chemistry* (New York, 1926) p 321; H. Stenzl, F. Fichter, *Helv. Chim. Acta* **17,** 669 (1934); **19,** 392 (1936); **20,** 846 (1937); F. Asinger, H. H. Vogel, *Houben-Weyl* **5/1a,** 280, 471 (1970).

428. Tebbe Olefination (Methylenation)
F. N. Tebbe *et al., J. Am. Chem. Soc.* **100,** 3611 (1978); S. H. Pine *et al., ibid.* **102,** 3270 (1980).
Exchange of the oxygen atom of a carbonyl function for the methylene group of the proposed titanium carbene complex (the Tebbe reagent) to yield terminal alkenes:

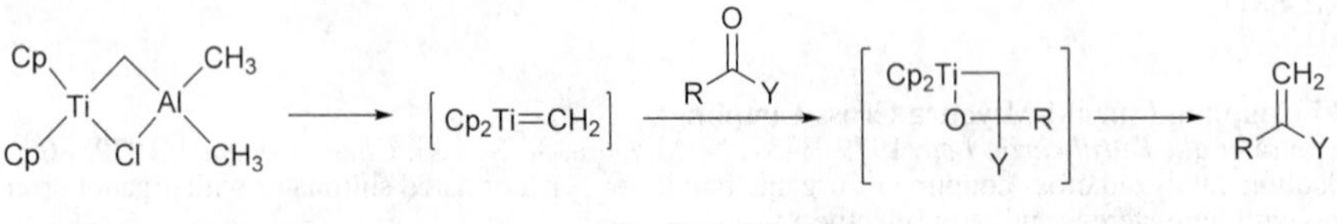

Y = H, R, OR, NR_2

Comparative study with Wittig reaction, *q.v.*: S. H. Pines *et al., Synthesis* **1991,** 165. *Reviews:* K. A. Brown-Wensley *et al., Pure Appl. Chem.* **55,** 1733-1744 (1983); S. E. Kelly, *Comp. Org. Syn.* **1,** 743-746 (1991); S. H. Pines, *Org. React.* **43,** 1-91 (1993). *Cf.* Peterson Reaction.

429. Thiele Reaction (Thiele-Winter Acetoxylation)

J. Thiele, *Ber.* **31,** 1247 (1898).

Formation of triacetoxy aromatic compounds by the reaction of quinones with acetic anhydride catalyzed by sulfuric acid or boron trifluoride:

Review: J. F. W. McOmie, J. N. Blatchly, *Org. React.* **19,** 199 (1972). J. M. Blatchly *et al., J. Chem. Soc. Perkin Trans. 1* **1972,** 2286; J. F. W. McOmie, S. A. Saleh, *ibid.* **1974,** 384; M. Hirama, S. Ito, *Chem. Lett.* **1977,** 627. *Cf.* Hinsberg Sulfone Synthesis.

430. Thorpe Reaction

H. Baron *et al., J. Chem. Soc.* **85,** 1726 (1904); K. Ziegler *et al., Ann.* **504,** 94 (1933).

Base-catalyzed self-condensation of nitriles to yield imines which tautomerize to enamines:

Reviews: J. P. Schaefer, J. J. Bloomfield, *Org. React.* **15,** 1 (1967); H. O. House, *Modern Synthetic Reactions* (W. Benjamin, Menlo Park, California, 2nd ed., 1972) p 742; E. C. Taylor, A. McKillop, *Chemistry of Enaminonitriles and o-Aminonitriles* (Wiley-Interscience, N.Y., 1970) pp 1-58; *eidem, Adv. Org. Chem.* **7,** 1 (1970). *Cf.* Ziegler Method.

431. Tiemann Rearrangement

F. Tiemann, *Ber.* **24,** 4162 (1891).

Rearrangement of amide oximes (available from nitriles and hydroxylamine) to monosubstituted ureas by treatment with benzenesulfonyl chloride and water:

P. A. S. Smith, *Org. React.* **3,** 366 (1946); M. W. Partridge, H. A. Turner, *J. Pharm. Pharmacol.* **5,** 103 (1953); R. F. Plapinger. O. O. Owens, *J. Org. Chem.* **21,** 1186 (1956); J. Garapon *et al., Tetrahedron Lett.* **1970,** 4905. *Cf.* Beckmann Rearrangement.

432. Tiffeneau-Demjanov Rearrangement

M. Tiffeneau *et al., Compt. Rend.* **205,** 54 (1937).

Rearrangement of β-amino alcohols upon diazotization with nitrous acid to give carbonyl compounds. Cyclic alcohols yield ring expanded or contracted products:

Reviews: P. A. S. Smith, D. R. Baer, *Org. React.* **11,** 157-188 (1960); H. Metzger, *Houben-Weyl* **10/4,** 233 (1968); D. J. Coveney, *Comp. Org. Syn.* **3,** 781-782 (1991). W. E. Parham, C. S. Roosevelt, *J. Org. Chem.* **37,** 1975 (1972); D. Fattori *et al., Tetrahedron* **49,** 1649 (1993). *Cf.* Demjanov Rearrangement; Pinacol Rearrangement.

433. Tishchenko Reaction

L. Claisen, *Ber.* **20,** 646 (1887); V. Tishchenko, *J. Russ. Phys. Chem. Soc.* **38,** 355, 482, 540, 547 (1906); *Chem. Zentralbl.* **1906 II,** 1309, 1552, 1555, 1556.

Organic Name Reactions

Formation of esters from aldehydes by an oxidation-reduction process that utilizes aluminum or sodium alkoxides:

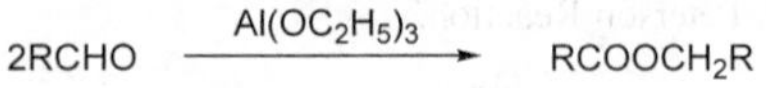

$$2RCHO \xrightarrow{Al(OC_2H_5)_3} RCOOCH_2R$$

O. Kamm, W. F. Kamm, *Org. Synth. coll. vol.* **I**, 104 (1941); Y. Ogata, A. Kawasaki, *Tetrahedron* **25**, 929, 2845 (1969); P. R. Stapp, *J. Org. Chem.* **38**, 1433 (1973); G. Fouquet *et al., Ann.* **1979**, 1591. *Review:* L. Cichon, *Wiad. Chem.* **20**, 641, 783 (1966), *C.A.* **66**, 54672b, 94408b (1967). *Cf.* Cannizzaro Reaction; Meerwein-Pondorf-Verley Reduction.

434. Traube Purine Synthesis
W. Traube, *Ber.* **33**, 1371, 3035 (1900).

Preparation of 4,5-diaminopyrimidines by introduction of the amino group into the 5-position of 4-amino-6-hydroxy- or 4,6-diaminopyrimidines by nitrosation and ammonium sulfide reduction, followed by ring closure with formic acid or chlorocarbonic ester:

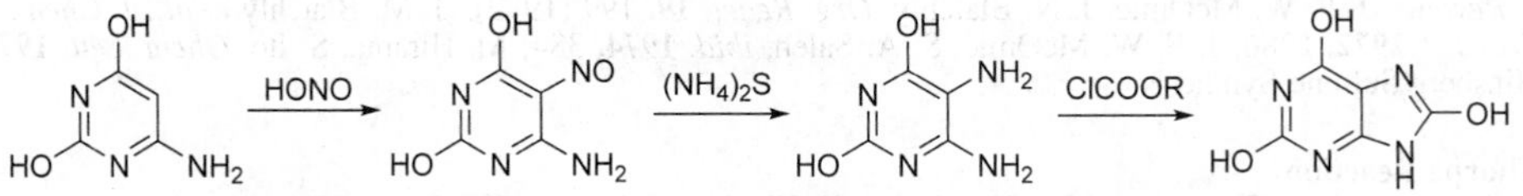

J. H. Davidson, *The Nucleic Acids* **I** (New York, 1955) p 131; A. R. Katritzky, *Q. Rev. Chem. Soc.* **10**, 397 (1956); *idem, Rev. Pure Appl. Chem.* **11**, 178 (1961); J. H. Lister, *Purines* (Wiley, New York, 1971) pp 31-90.

435. Trost Allylation (Tsuji-Trost Reaction)
J. Tsuji *et al., Tetrahedron Lett.* **1965**, 4387; B. M. Trost, T. J. Fullerton, *J. Am. Chem. Soc.* **95**, 292 (1973).

Palladium-catalyzed allylation of nucleophiles proceeding in an S$_N$2 or S$_N$2' fashion depending on the catalyst, nucleophile, and substituents on the substrate:

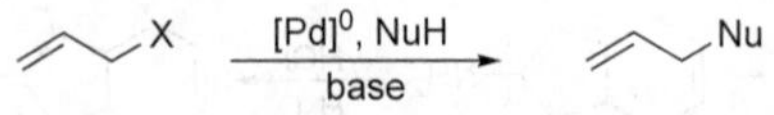

NuH = malonates, β-diketones, β-keto esters, enamines, β-keto sulfones, bis-sulfones
X = Br, Cl, OCOOR, SO$_2$R, OCOR, OCONR$_2$, OPO(OR)$_2$, NO$_2$
[Pd]^0 = Pd(PPh$_3$)$_4$, Pd$_2$(dba)$_3$ · CHCl$_3$, + PPh$_3$, Pd(OCOCH$_3$)$_2$ + PPh$_3$
base = NaH, if necessary

Scope and limitations under neutral conditions: J. Tsuji *et al., J. Org. Chem.* **50**, 1523 (1985); in biphasic media: C. de Bellefon *et al., J. Mol. Catal. A* **145**, 121 (1999). Application to the synthesis of polyprenoids: E. Keinan, D. Eren, *Pure Appl. Chem.* **60**, 89 (1988). Review of intramolecular applications: B. M. Trost in *Adv. Chem. Ser.* **230**, entitled "Homogeneous Transition Metal Catalyzed Reactions" (ACS, Washington DC, 1992) pp 463-478. *Review:* C. G. Frost *et al., Tetrahedron: Asymmetry* **3**, 1089-1122 (1992).

436. Trost Desymmetrization
B. M. Trost *et al., J. Am. Chem. Soc.* **114**, 9333 (1992).

Formation of an enantiomerically pure, azide or amine containing, five or six membered ring by a pallidium catalyzed desymmetrization using a nitrogen nucleophile, where the palladium complex is derived from a chiral ligand and π-allylpalladium chloride:

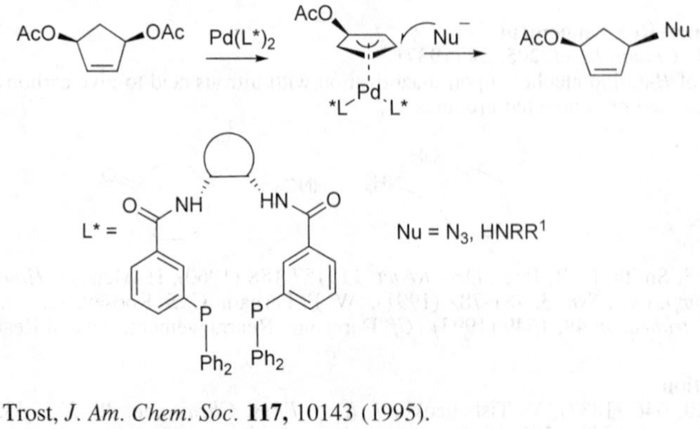

Nu = N$_3$, HNRR^1

S. R. Pulley, B. M. Trost, *J. Am. Chem. Soc.* **117**, 10143 (1995).

437. Tscherniac-Einhorn Reaction

J. Tscherniac, **DE 134979**; A. Einhorn *et al., Ann.* **343,** 207 (1905); **361,** 113 (1908).

Introduction of the amidomethyl group into aromatic rings or activated methylene groups in the presence of sulfuric acid:

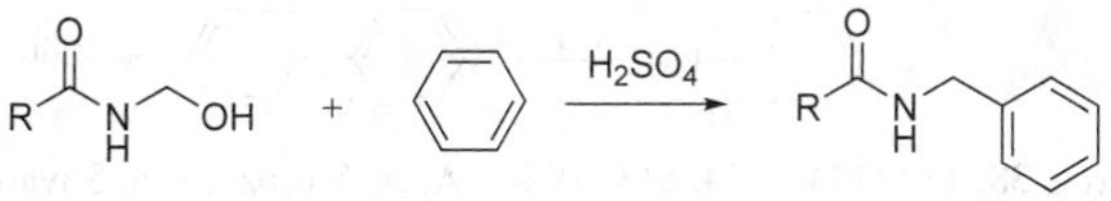

Reviews: R. Schröter, *Houben-Weyl* **11/1,** 795 (1957); Hellman *Angew. Chem.* **69,** 463 (1957); H. E. Zaugg, W. B. Martin, *Org. React.* **14,** 52 (1965); H. E. Zaugg *et al., J. Org. Chem.* **34,** 11, 14 (1969); K. Bott, *Ber.* **106,** 2513 (1973); A. R. Mitchell *et al., Tetrahedron Lett.* **1976,** 3795.

438. Twitchell Process

E. Twitchell, **US 601603; US 628503** (1898); **DE 365522; DE 385074.**

Commercial process for splitting fats to glycerol and fatty acids by heating the sulfuric-acid-washed fat 20-48 hours in an open tank with steam in a mixture of 25-50% water, 0.5% sulfuric acid and 0.75-1.25% Twitchell reagent (sulfonated petroleum products):

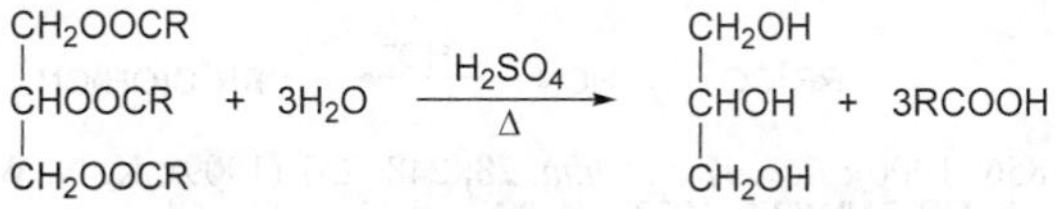

E. Twitchell, *J. Am. Chem. Soc.* **22,** 22 (1900); **28,** 196 (1906); J. W. Lawrie, *Glycerol and the Glycols* (New York, 1928) p 32; R. B. Trusler, *J. Oil Fat Ind.* **8,** 141 (1931); A. F. Bailey, *Industrial Oil and Fat Products* (New York, 1945) p 668; C. J. Marsel, H. D. Allen, *Chem. Eng.* **54**(6), 104 (1947); V. Mills, H. K. McClain, *Ind. Eng. Chem.* **41,** 1982 (1949); L. Lascaray, *J. Am. Oil Chem. Soc.* **29,** 362 (1952); *Faith, Keyes & Clark's Industrial Chemicals* (Wiley-Interscience, New York, 4th ed., 1975) p 431.

439. Ueno-Stork Cyclization

Y. Ueno *et al., J. Am. Chem. Soc.* **104,** 5564 (1982); G. Stork *et al., ibid.* **105,** 3741 (1983).

Radical cyclization of haloacetals. The cyclic acetal product can be converted to a lactone via Jones Oxidation, *q.v.*:

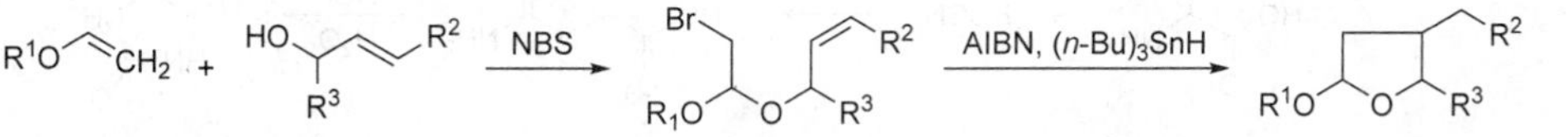

Synthetic applications: Y. Ueno *et al., J. Chem. Soc. Perkin Trans. 1* **1986,** 1351; G. Stork *et al., J. Am. Chem. Soc.* **108,** 6384 (1986); F. Villar *et al., Org. Lett.* **2,** 1061 (2000); G. K. Friestad, G. M. Fioroni, *ibid.* **7,** 2393 (2005). Stereochemistry studies: F. Villar *et al., Chem. Eur. J.* **9,** 1566 (2003); O. Corminboeuf *et al., ibid.* 1578. *Review*: X. J. Salom-Roig *et al., Synthesis* **2004,** 1903-1928.

440. Ugi Reaction (Four-Component Condensation) (4CC)

I. Ugi, *Angew. Chem. Int. Ed.* **1,** 8 (1962).

The α-addition of an iminium ion and the conjugate base of a carboxylic acid to an isocyanide, followed by spontaneous rearrangement of the α-adduct to yield an α-aminocarboxamide derivative. Carbonyl compounds and amines, or their condensation products, serve as precursors to the iminium ion. The nature of the product depends primarily on the acid component:

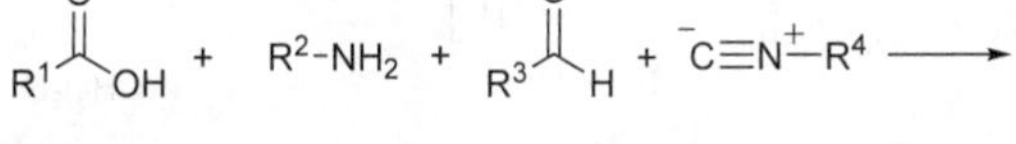

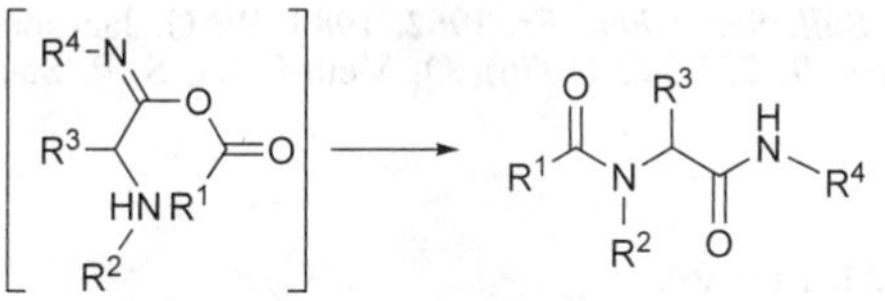

When four discrete reactants are used, the reaction is often referred to as the four-component condensation (4CC). Diastereoselective methods development: H. Kunz *et al., Synthesis* **1991,** 1039; M. Goebel, I. Ugi, *ibid.* 1095. Synthetic applications: T. Ziegler *et al., Tetrahedron Lett.* **39,** 5957 (1998); *eidem, Tetrahedron* **55,** 8397 (1999). *Reviews:* I. Ugi, *Proc. Est. Acad. Sci. Chem.* **40,** 1-13 (1991); I. Ugi *et al., Comp. Org. Syn.* **2,** 1083-1109 (1991).

Organic Name Reactions

441. Ullmann Reaction
F. Ullmann, *Ann.* **332**, 38 (1904); F. Ullmann, P. Sponagel, *Ber.* **38**, 2211 (1905).
  Copper-mediated coupling of aryl halides. Biaryl ether synthesis is similarly accomplished with aryl halides and phenols:

$$2 \; \text{C}_6\text{H}_5\text{I} \xrightarrow{\text{Cu}} \text{C}_6\text{H}_5\text{—C}_6\text{H}_5 \; + \; \text{CuI}_2$$

  P. E. Fanta, *Chem. Rev.* **38**, 139 (1946); **64**, 613 (1964); A. A. Moroz, M. S. Shvartsberg, *Russ. Chem. Rev.* **43**, 679 (1974); P. E. Fanta, *Synthesis* **1974**, 9; M. F. Semmelhack *et al., J. Am. Chem. Soc.* **103**, 6460 (1981); D. W. Knight, *Comp. Org. Syn.* **3**, 499-507 (1991). *Cf.* Glaser Coupling.

442. Urech Cyanohydrin Method (Ultee Cyanohydrin Method)
F. Urech, *Ann.* **164**, 225 (1872); A. J. Ultee, *Rec. Trav. Chim.* **28**, 1 (1909).
  Cyanohydrin formation by addition of alkali cyanide to the carbonyl group in the presence of acetic acid (Urech method) or by reaction of the carbonyl compound with anhydrous hydrogen cyanide in the presence of a basic catalyst (Ultee cyanohydrin method):

$$RR^1CO \; + \; KCN \; + \; CH_3COOH \longrightarrow RR^1C(OH)CN \; + \; CH_3COO^- \, K^+$$

$$RR^1CO \; + \; HCN \xrightarrow{HO^-} RR^1C(OH)CN$$

  A. J. Ultee, *Ber.* **39**, 1856 (1906); *Rec. Trav. Chim.* **28**, 248, 257 (1909); K. N. Welch, G. R. Clemo, *J. Chem. Soc.* **1928**, 2629; H. R. Dittmar, **US 2101823** (1937); V. Migrdichian, *The Chemistry of Organic Cyanogen Compounds* (New York, 1947) p 173; D. T. Mowry, *Chem. Rev.* **42**, 231 (1948); P. Kurz, *Houben-Weyl* **8**, 274 (1952); R. F. B. Cox, R. T. Stormont, *Org. Synth.* **coll. vol. 2**, 7 (1955). *Cf.* Bucherer-Bergs Reaction; Kiliani-Fischer Synthesis.

443. Urech Hydantoin Synthesis
F. Urech, *Ann.* **165**, 99 (1873).
  Formation of hydantoins from α-amino acids by treatment with potassium cyanate in aqueous solution and heating the salt of the intermediate hydantoic acid with 25% hydrochloric acid:

  H. D. Dakin, *Am. Chem. J.* **44**, 48 (1910); T. B. Johnson, *J. Am. Chem. Soc.* **35**, 780 (1913); W. J. Boyd, W. Robson, *Biochem. J.* **29**, 542, 546, 2256 (1935); E. Ware, *Chem. Rev.* **46**, 407 (1950); M. Sainsbury, R. S. Theobald, *Rodd's Chemistry of Carbon Compounds* **IVC**, 185 (1986).

444. Vilsmeier-Haack Reaction
A. Vilsmeier, A. Haack, *Ber.* **60**, 119 (1927).
  Formylation of activated aromatic or heterocyclic compounds with disubstituted formamides and phosphorus oxychloride:

  Reviews: M. R. de Maheas, *Bull. Soc. Chim. Fr.* **1962**, 1989; W. G. Jackson *et al., J. Am. Chem. Soc.* **103**, 533 (1981); C. Jutz, *Adv. Org. Chem.* **9**, 225-342 (1976); O. Meth-Cohn, S. P. Stanforth, *Comp. Org. Syn.* **2**, 777-794 (1991).

445. Voight Amination
K. Voight, *J. Prakt. Chem.* [2] **34**, 1 (1886).
  Amination of benzoins with amines in the presence of phosphorus pentoxide or hydrochloric acid:

H. H. Strain, *J. Am. Chem. Soc.* **51**, 269 (1929); R. M. Cowper, T. S. Stevens, *J. Chem. Soc.* **1940**, 347; P. L. Julian *et al., J. Am. Chem. Soc.* **67**, 1203 (1945); R. E. Lutz *et al., ibid.* **70**, 2016 (1948); I. A. Kaye *et al., ibid.* **75**, 746 (1953); J. Iwao *et al., J. Pharm. Soc. Jpn.* **74**, 551 (1954); R. E. Lutz, J. W. Baker, *J. Org. Chem.* **21**, 49 (1956).

446. Volhard-Erdmann Cyclization
J. Volhard, H. Erdmann, *Ber.* **18**, 454 (1885).

Synthesis of alkyl and aryl thiophenes by cyclization of disodium succinate or other 1,4-difunctional compounds (γ-oxo acids, 1,4-diketones, chloroacetyl-substituted esters) with phosphorus heptasulfide:

L. H. Friedburg, *J. Am. Chem. Soc.* **12**, 83 (1890); *J. Chem. Soc.* **58**, 1400 (1890); R. Phillips, *Org. Synth.* **coll. vol. II,** 578 (1943); F. F. Blicke, *Heterocycl. Compd.* **1**, 212 (1950); D. E. Wolf, K. Folkers, *Org. React.* **4**, 412 (1951); R. F. Feldkamp, B. F. Tullar, *Org. Synth.* **coll. vol. IV,** 671 (1963).

447. von Braun Amide Degradation
H. von Pechmann, *Ber.* **33**, 611 (1900); J. von Braun, *ibid.* **37**, 3210 (1904).

Formation of a nitrile and alkyl halide from the phosphorus pentahalide promoted degradation of an amide:

X = Cl or Br

Mechanistic study: B. A. Phillips *et al., Tetrahedron* **29**, 3309 (1973). Application to *N-t*-butylamides: R. B. Perni, G. W. Gribble, *Org. Prep. Proced. Int.* **15**, 297 (1983).

448. von Braun Reaction
J. von Braun, *Ber.* **40**, 3914 (1907); **42**, 2219 (1909); **44**, 1250 (1911).

Reaction of tertiary amines with cyanogen bromide to form disubstituted cyanamides and an alkyl halide:

$$R_2NR^1 + BrCN \longrightarrow R_2NCN + R^1Br$$

Mechanistic correlation with Ritter, Bischler-Napieralski, Beckmann and Schmidt reactions, *q.q.v.*: G. Fodor, S. Nagubandi, *Tetrahedron* **36**, 1279 (1980). Synthetic applications: S. Siddiqui *et al., Z. Naturforsch.* **37b**, 1481 (1982); *idem et al., Pak. J. Sci. Ind. Res.* **30**, 163 (1987). *Reviews:* H. A. Hageman, *Org. React.* **7**, 198-262 (1953); J. H. Cooley, E. J. Evain, *Synthesis* **1989**, 1-7.

449. von Richter (Cinnoline) Synthesis
V. von Richter, *Ber.* **16**, 677 (1883).

Formation of cinnoline derivatives by diazotization of *o*-aminoarylpropiolic acids or *o*-aminoarylacetylenes followed by hydration and cyclization:

M. Busch, M. Klett, *Ber.* **25**, 2847 (1892); N. J. Leonard, *Chem. Rev.* **37**, 270 (1945); K. Schofield, J. C. E. Simpson, *J. Chem. Soc.* **1945**, 512, K. Schofield, T. Swain, *ibid.* **1949**, 2393; J. C. E. Simpson, *Condensed Pyridazine and Pyrazine Rings* (New York, 1953) p 16; G. R. Ramage, J. K. Landquist, *Chemistry of Carbon Compounds* **IVB**, 1217 (1959); G. T. Rogere *et al., Tetrahedron Lett.* **9**, 1028 (1968); A. C. Ellis *et al., J. Chem. Soc. Chem. Commun.* **1977**, 152. *Cf.* Widman-Stoermer Synthesis.

450. von Richter Rearrangement
V. von Richter, *Ber.* **4**, 21, 459, 553 (1871).

Carboxylation of *para-* or *meta*-substituted aromatic nitro compounds with cyanide at 120-270°. The carboxyl group enters with cine substitution in a position *ortho* to the eliminated nitro group:

Organic Name Reactions

J. F. Bunnett, *Q. Rev. Chem. Soc.* **12**, 15 (1958); D. Samuel, *J. Chem. Soc.* **1960**, 1318; J. Sauer, R. Huisgen, *Angew. Chem.* **72**, 314 (1960); M. Rosenblum, *J. Am. Chem. Soc.* **82**, 3796 (1960); E. Cullen, P. L'Ecuyer, *Can. J. Chem.* **39**, 144, 154, 382, 862 (1961); E. F. Ullman, E. A. Bartkus, *Chem. Ind. (London)* **1962**, 93; K. M. Ibne-Rasa, E. Koubak, *J. Org. Chem.* **28**, 3240 (1963); G. T. Rogers, T. L. V. Ulbricht, *Tetrahedron Lett.* **9**, 1029 (1968); A. C. Ellis, I. D. Rae, *Chem. Commun.* **1977**, 152; E. Tomitori *et al., Yakugaku Zasshi* **103**, 601 (1983).

451. Vorbrüggen Glycosylation
U. Niedballa, H. Vorbrüggen, *Angew. Chem. Int. Ed.* **9**, 461 (1970).

The reaction of silylated heterocyclic bases with peracylated sugars in the presence of Lewis acids to yield natural β-nucleosides. If the sugar lacks a 2α-acyloxy substituent, an anomeric mixture forms:

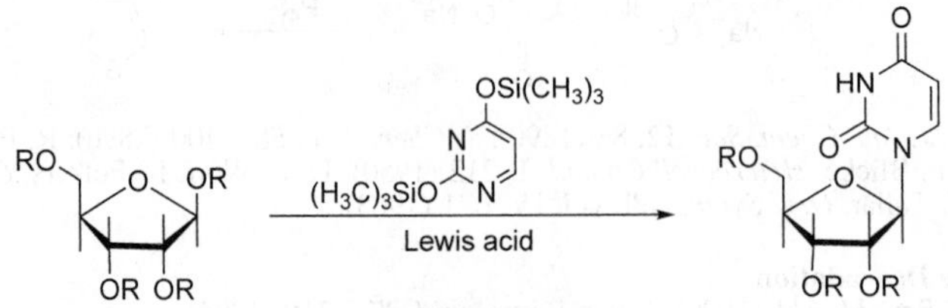

R = COCH₃, COPh
Lewis acid = BF₃ · O(CH₂CH₃)₂, SnCl₄, (CH₃)₃SiOSO₂CF₃, (CH₃)₃SiClO₄, TiCl₄, AlCl₃

Scope and limitations: H. Vorbrüggen *et al., Ber.* **114**, 1234 (1981). Mechanistic study: H. Vorbrüggen, G. Höfle, *ibid.* 1256. Synthetic applications: U. Niedballa, H. Vorbrüggen, *J. Org. Chem.* **39**, 3654, 3660, 3664, 3668, 3672 (1974); R. O. Dempcy, E. B. Skibo, *ibid.* **56**, 776 (1991); S. H. Kawai, G. Just, *Nucleosides Nucleotides* **10**, 1485 (1991). *Cf.* Hilbert-Johnson Reaction.

452. Wacker Oxidation
J. Smidt *et al., Angew. Chem. Int. Ed.* **1**, 176 (1959).

The oxidation of ethylene to acetaldehyde employing palladium chloride and cupric chloride as catalysts and molecular oxygen as oxidant. The reaction has been extensively developed for the oxidation of terminal alkenes to methyl ketones:

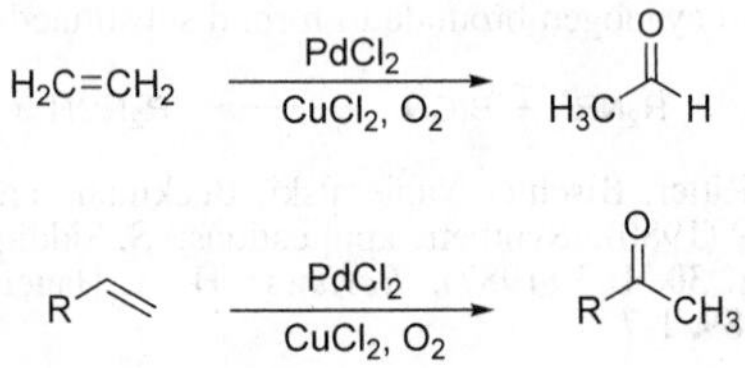

Application to hydroxy-α,β-unsaturated esters: S. X. Auclair *et al., Tetrahedron Lett.* **33**, 7739 (1992). Use of a multicomponent catalytic system: E. Monflier *et al., ibid.* **36**, 387 (1995). Synthetic applications: M. Romero *et al., ibid.* **35**, 3255 (1994); L. A. Paquette, X. Wang, *J. Org. Chem.* **59**, 2052 (1994). *Reviews:* L. S. Hegedus, *Comp. Org. Syn.* **4**, 552-559 (1991); J. Tsuji, *ibid.* **6**, 449-468.

453. Wagner-Jauregg Reaction
T. Wagner-Jauregg, *Ber.* **63**, 3213 (1930); *Ann.* **491**, 1 (1931).

Addition of maleic anhydride to diarylethylenes with formation of *bis* adducts which can be converted to aromatic ring systems:

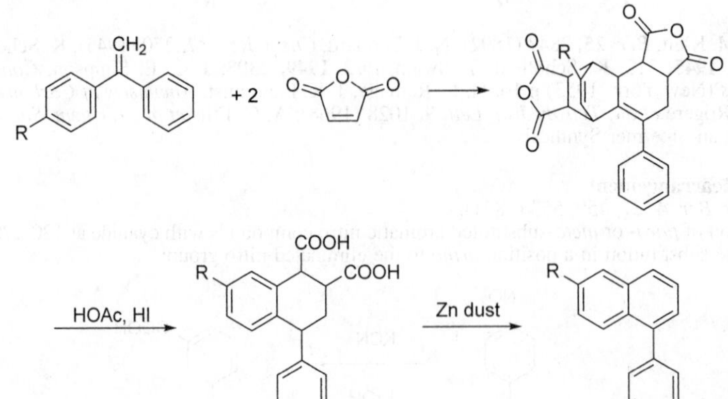

Organic Name Reactions

F. Bergmann *et al., J. Am. Chem. Soc.* **69**, 1773, 1777, 1779 (1947); K. Alder in *Newer Methods of Preparative Organic Chemistry*, English Ed. (Interscience, New York, 1948) p 425; M. C. Kloetzel, *Org. React.* **4**, 32 (1948). *Cf.* Diels-Alder Reaction.

454. Wagner-Meerwein Rearrangement

G. Wagner, *J. Russ. Phys. Chem. Soc.* **31**, 690 (1899); H. Meerwein, *Ann.* **405**, 129 (1914).

Carbon-to-carbon migration of alkyl, aryl or hydride ions. The original example is the acid-catalyzed rearrangement of camphene hydrochloride to isobornyl chloride:

C. Le Drian, P. Vogel, *Helv. Chim. Acta* **70**, 1703 (1987); M. Asaoka, H. Takei, *Tetrahedron Lett.* **28**, 6343 (1987); L. U. Román *et al., J. Org. Chem.* **56**, 1938 (1991). Review of applications to alcohols: Y. Pocker in *Molecular Rearrangements* Part 1, P. de Mayo, Ed. (Wiley-Interscience, New York, 1963) pp 6-15; to bicyclic systems: J. Berson, *ibid.* 111-231; to terpenes: J. F. King, P. de Mayo, *ibid.* 813-840; to alkaloids: E. W. Warnhof, *ibid.* 842-879; to steroids: N. L. Wendler, *ibid.* 1020-1028. *Reviews:* R. L. Cargill *et al., Acc. Chem. Res.* **7**, 106-113 (1974); H. Hogeveen, E. M. G. A. Van Kruchten, *Top. Curr. Chem.* **80**, 89-124 (1979); J. R. Hanson, *Comp. Org. Syn.* **3**, 705-719 (1991). *Cf.* Demjanov Rearrangement; Nametkin Rearrangement; Retropinacol Rearrangement.

455. Walden Inversion

P. Walden, *Ber.* **28**, 1287, 2766 (1895).

Inversion of configuration of a chiral center in bimolecular nucleophilic substitution (S$_N$2) reactions:

H. A. Bent, *Chem. Rev.* **68**, 587 (1968); D. P. G. Harmon, *J. Chem. Educ.* **47**, 398 (1970); L. Kryger *et al., Acta Chem. Scand.* **26**, 2339, 2349 (1972); C. W. Shoppee, J. Nemorin, *J. Chem. Soc. Perkin Trans. 1* **1973**, 542; K.-C. To *et al., J. Chem. Phys.* **74**, 1499 (1981).

456. Wallach Rearrangement

O. Wallach, L. Belli, *Ber.* **13**, 525 (1880).

Acid-catalyzed rearrangement of azoxybenzenes to *p*-hydroxyazobenzenes:

Reviews: K. H. Schündehütte, *Houben-Weyl* **10/3**, 771-773 (1965); E. Buncel in *Mechanisms of Molecular Migrations* **vol. 1**, B. S. Thyagarajan, Ed. (Interscience, New York, 1968) p 61; R. A. Cox, E. Buncel in *The Chemistry of Hydrazo, Azo and Azoxy Groups* pt. 2, S. Patai, Ed. (Wiley, New York, 1975) pp 808-837; J. Yamamoto *et al., Tetrahedron* **36**, 3177 (1980).

457. Weerman Degradation

R. A. Weerman, *Rec. Trav. Chim.* **37**, 1, 16 (1918).

Formation of an aldose with one less carbon from an aldonic acid by a Hofmann-type reaction, *q.v.*, of the corresponding amide. This is a general reaction of α-hydroxy carboxylic acids:

W. N. Haworth *et al., J. Chem. Soc.* **1934**, 1722; **1938**, 1975; E. S. Wallis, J. F. Lane, *Org. React.* **3**, 275 (1946); J. C. Sowden in *The Carbohydrates*, W. Pigman, Ed. (New York, 1957) p 120; L. F. Fieser, M. Fieser, *Advanced Organic Chemistry* (New York, 1961) p 945. *Cf.* Hofmann Reaction.

458. Weinreb Ketone Synthesis

S. Nahm, S. M. Weinreb, *Tetrahedron Lett.* **22**, 3815 (1981).

Preparation of a functionalized ketone by the addition of an organometallic reagent to the corresponding *N*-methoxy-*N*-methylamide (Weinreb amide). Alternately, an aldehyde can be synthesized via Weinreb amide reduction with lithium aluminum hydride or diisobutylaluminum hydride:

Weinreb amide

R^1, R^2 = alkyl, alkenyl, alkynyl,
aryl, or heteroaryl

Synthetic applications involving solid phase synthesis: O. B. Wallace, *Tetrahedron Lett.* **38**, 4939 (1997); involving modified Weinreb amides: O. Labeeuw *et al., ibid.* **45**, 7107 (2004). Review of prepn and synthetic utility of Weinreb amides: M. P. Sibi, *Org. Prep. Proced. Int.* **25**, 15-40 (1993); J. Singh *et al., J. Prakt. Chem.* **342**, 340-347 (2000); S. Balasubramaniam, I. S. Aidhen, *Synthesis* **2008**, 3707-3738.

459. Weiss Reaction

U. Weiss, J. M. Edwards, *Tetrahedron Lett.* **1968**, 4885.

Reaction of 1,2-dicarbonyl compounds with 3-oxoglutarates to yield *cis*-bicyclo[3.3.0]octane-3,7-dione or [n.3.3]propellanedione (n > 2) tetracarboxylates. Subsequent acid-catalyzed hydrolysis and decarboxylation yield the respective 2,4,6,8-unsubstituted diones:

E = COOCH$_3$; R^1 or R^2 = H, alkyl, aryl or R^1—R^2 = (CH$_2$)$_n$, n > 2

Experimental procedure: S. H. Bertz *et al., Org. Synth.* **coll. vol. VII**, 50 (1990). Review of synthetic applications: A. K. Gupta *et al., Tetrahedron* **47**, 3665-3710 (1991). *Review:* H.-U. Reissig, "The Weiss Reaction" in *Organic Synthesis Highlights*, J. Mulzer *et al.*, Eds. (VCH, New York, 1991) pp 121-125.

460. Wessely-Moser Rearrangement

F. Wessely, G. H. Moser, *Monatsh. Chem.* **56**, 97 (1930).

Rearrangement of flavones and flavanones possessing a 5-hydroxyl group, through fission of the heterocyclic ring and reclosure of the intermediate diaroylmethanes in the alternate direction:

alternate substrates = chromones, isoflavones, flavonols,
xanthones, furanochromones

Reviews: Wheeler, *Rec. Chem. Prog.* **18**, 133 (1957); T. R. Seshadri, *Tetrahedron* **6**, 169 (1959); H. D. Locksley, *Fortschr. Chem. Org. Naturst.* **30**, 292 (1973).

461. Westphalen-Lettré Rearrangement

T. Westphalen, *Ber.* **48**, 1064 (1915); H. Lettré, M. Müller, *ibid.* **70**, 1947 (1937).

Dehydration of 5-hydroxycholesterol derivatives accompanied by C-10 to C-5 methyl migration in compounds with a *β*-substituent at C-6:

Organic Name Reactions

Early review: N. L. Wendler in *Molecular Rearrangements* Part 2, P. de Mayo, Ed. (Wiley-Interscience, New York, 1964) p 1027. A. T. Rowland, *J. Org. Chem.* **29**, 222 (1964); J. W. Blunt *et al., Tetrahedron* **21**, 1567 (1965); K. Kieslich, G. Schulz, *Ann.* **726**, 152 (1969); B. Marples, J. G. L. Jones, *J. Chem. Soc. C* **1970**, 2273; J. Wicha, *Tetrahedron Lett.* **1972**, 2877; P. Kocovsky *et al., Collect. Czech. Chem. Commun.* **44**, 234 (1979).

462. Wharton Reaction

P. S. Wharton, D. H. Bohlen, *J. Org. Chem.* **26**, 3615 (1961); P. S. Wharton, *ibid.* 4781.
Reduction of α,β-epoxy ketones by hydrazine to allylic alcohols:

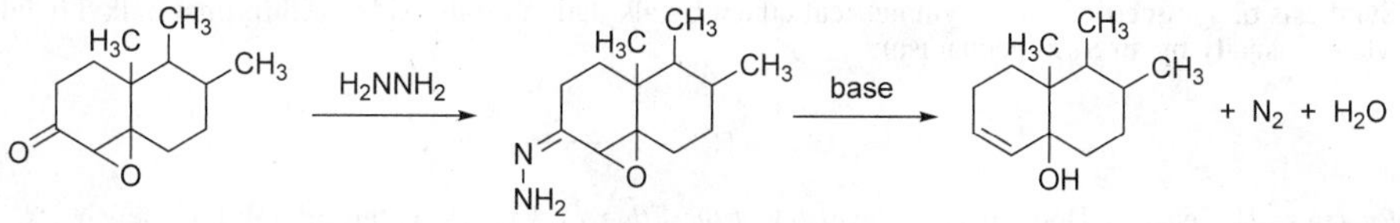

Improved procedure: C. Dupuy, J. L. Luche, *Tetrahedron* **45**, 3437 (1989). Synthetic applications: S. Takano *et al., Synlett* **1991**, 636; T. Yoshimitsu *et al., Synthesis* **1994**, 1029; K. Yamada *et al., J. Org. Chem.* **63**, 3666 (1998). *Review:* D. Caine, *Org. Prep. Proced. Int.* **20**, 3-8 (1988); A. R. Chamberlin, D. J. Sall, *Comp. Org. Syn.* **8**, 927-929 (1991). *Cf.* Eschenmoser Fragmentation; Grob Fragmentation.

463. Whiting Reaction

P. Nayler, M. C. Whiting, *J. Chem. Soc.* **1954**, 4006.
Alkynediols are reduced by lithium aluminum hydride in ether or tertiary amines to dienes:

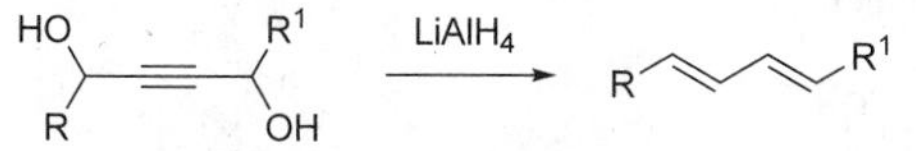

R. A. Raphael, *Acetylene Compounds in Organic Synthesis* (New York, 1955) p 114; O. Isler, *et al., Helv. Chim. Acta* **39**, 454 (1956); L. F. Fieser, M. Fieser, *Reagents for Organic Synthesis* (New York, 1967) p 385.

464. Wichterle Reaction

O. Wichterle *et al., Collect. Czech. Chem. Commun.* **13**, 300 (1948).
Modification of the Robinson annulation, *q.v.*, in which 1,3-dichloro-*cis*-2-butene is used instead of methyl vinyl ketone:

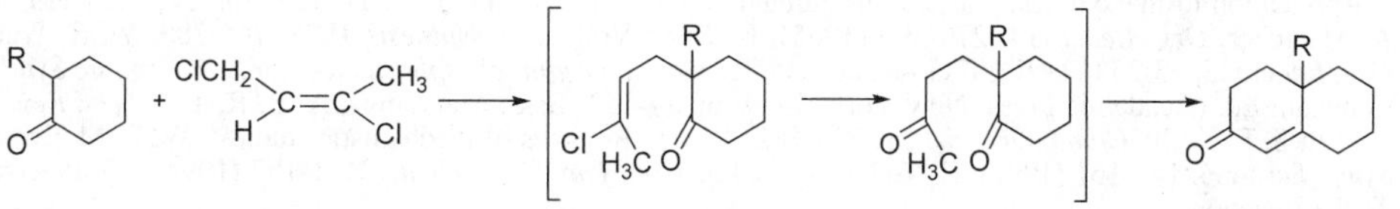

M. Kobayashi, T. Matsumoto, *Chem. Lett.* **1973**, 957; H. Yoshioka *et al., Tetrahedron Lett.* **1979**, 3489. *Review:* M. Hudlicky, *Collect. Czech. Chem. Commun.* **58**, 2229-2244 (1993).

465. Widman-Stoermer Synthesis

O. Widman, *Ber.* **17**, 722 (1884); R. Stoermer, H. Fincke, *Ber.* **42**, 3115 (1909).
Synthesis of cinnolines by cyclization of diazotized *o*-aminoarylethylenes at room temperature:

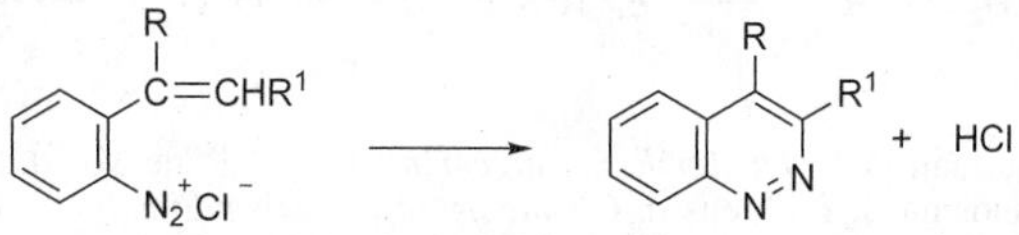

N. J. Leonard, *Chem. Rev.* **37**, 270 (1945); J. C. E. Simpson, *Condensed Pyridazine and Pyrazine Rings* (New York, 1953) p 6; T. L. Jacobs, *Heterocycl. Compd.* **6**, 137 (1957); G. R. Ramage, J. K. Landquist, *Chemistry of Carbon Compounds* **IVB**, 1217 (1959). *Cf.* von Richter (Cinnoline) Synthesis.

466. Willgerodt-Kindler Reaction

C. Willgerodt, *Ber.* **20**, 2467 (1887); **21**, 534 (1888); K. Kindler, *Ann.* **431**, 193 (1923).
Conversion of aryl alkyl ketones to amides and/or the ammonium salts of the corresponding acids by aqueous ammonium polysulfide or by sulfur and a primary or secondary amine:

Organic Name Reactions

Reviews: M. Carmack, M. A. Spielman, *Org. React.* **3,** 83 (1946); R. Wegler *et al., Newer Methods of Preparative Organic Chemistry* **vol. 3** (Academic Press, New York, 1964) pp 1-51; E. E. Campaigne in *The Chemistry of the Carbonyl Group,* S. Patai, Ed. (Wiley, New York, 1966) p 954; A. L. J. Beckwith, *The Chemistry of Amides,* J. Zabicky, Ed. (Interscience, London, 1970) pp 145-147; S. W. Schneller, *Int. J. Sulfur Chem.* **8,** 591 (1976); G. Purrello, *Heterocycles* **65,** 411-449 (2005).

467. Williamson Synthesis
A. W. Williamson, *J. Chem. Soc.* **4,** 229 (1852).

Synthesis of symmetrical or unsymmetrical ethers by alkylation of alkoxides with primary alkyl halides, sulfates or tosylates; usually by an S_N2 mechanism:

$$OR^- + R^1X \longrightarrow ROR^1$$

Reviews: H. Feuer, J. Hooz in *The Chemistry of the Ether Linkage,* S. Patai, Ed. (Wiley, New York, 1967) pp 446-460; H. O. Kalinowski *et al., Ber.* **114,** 477 (1981). Applications on solid support: A. Weissberg *et al., J. Comb. Chem.* **3,** 154 (2001). Catalytic synthesis of alkyl aryl ethers: E. Fuhrmann, J. Talbiersky, *Org. Process Res. Dev.* **9,** 206 (2005).

468. Wittig Reaction; Horner Reaction; Horner-Wadsworth-Emmons Reaction
G. Wittig, U. Schöllkopf, *Ber.* **87,** 1318 (1954); G. Wittig, W. Haag, *ibid.* **88,** 1654 (1955).

Alkene formation from carbonyl compounds and phosphonium ylides, proceeding primarily through the proposed betaine and/or oxaphosphetane intermediates. The stereoselectivity can be controlled by the choice of ylide, carbonyl compound, and reaction conditions:

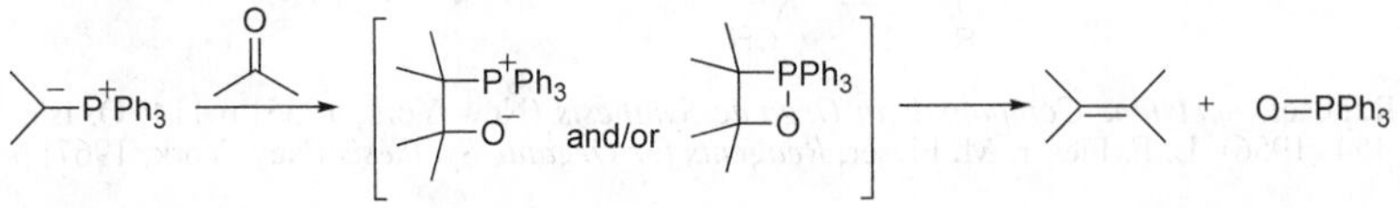

When the ylide is replaced with a phosphine oxide carbanion, the reaction is referred to as the **Horner reaction:** L. Horner *et al., Ber.* **91,** 61 (1958); *idem et al., ibid.* **92,** 2499 (1959).

When the ylide is replaced with a phosphonate carbanion, the reaction is referred to as the **Horner-Emmons-Wadsworth reaction:** W. S. Wadsworth, Jr., W. D. Emmons, *J. Am. Chem. Soc.* **83,** 1733 (1961).

Application to the synthesis of β,γ-unsaturated amides: T. Janecki *et al., Tetrahedron* **51,** 1721 (1995). *Reviews:* A. Maercker, *Org. React.* **14,** 270-490 (1965); K. P. C. Vollhardt, *Synthesis* **1975,** 765-780; W. S. Wadsworth, Jr., *Org. React.* **25,** 73-253 (1977); I. Gosney, A. G. Rowley in *Organophosphorus Reagents in Organic Synthesis,* J. I. G. Cadogan, Ed. (Academic Press, New York, 1979) pp 17-153; B. E. Maryanoff, A. B. Reitz, *Chem. Rev.* **89,** 863-927 (1989); S. E. Kelly, *Comp. Org. Syn.* **1,** 755-782 (1991). Reviews of mechanistic studies: W. E. McEwen *et al., ACS Symp. Ser.* **486,** 149-161 (1992); E. Vedejs, M. J. Peterson, *Top. Stereochem.* **21,** 1-157 (1994). *Cf.* Peterson Reaction; Tebbe Reaction.

469. [1,2]-Wittig Rearrangement
G. Wittig, L. Löhmann, *Ann.* **550,** 260 (1942); G. Wittig, *Experientia* **14,** 389 (1958).

Rearrangement of ethers with alkyl lithiums to yield alcohols *via* a [1,2]-shift:

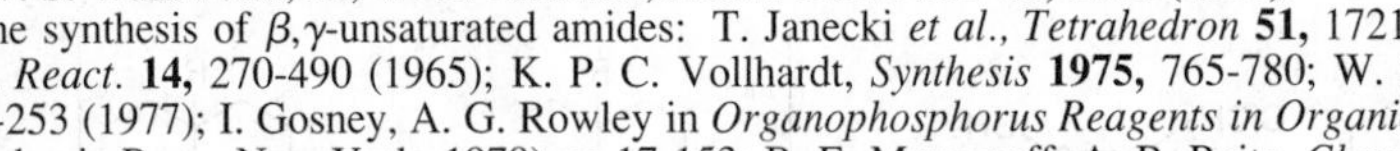

Reviews: H. E. Zimmerman in *Molecular Rearrangements* Part 1, P. de Mayo, Ed. (Wiley-Interscience, New York, 1963) p 372-377; L. Brandsma, J. F. Arens in *Chemistry of the Ether Linkage* S. Patai, Ed. (Interscience, New York, 1967) pp 570-580; U. Schöllkopf, *Angew. Chem.* **82,** 795 (1970); A. R. Lepley, A. G. Giumanini in *Mechanisms of Molecular Migrations* **vol. 3,** B. S. Thyagarajan, Ed. (Interscience, New York, 1971); U. Schöllkopf, *Ind. Chim. Belge* **36,** 1057 (1971); G. Tennant, *Annu. Rep. Prog. Chem. Sect. B* **68,** 241 (1972); R. W. Hoffmann, *Angew. Chem.* **91,** 625 (1979); *idem, Nachr. Chem. Tech. Lab.* **30,** 483 (1982). *Cf.* Meisenheimer Rearrangements; Stevens Rearrangement.

470. [2,3]-Wittig Rearrangement
J. Cast *et al., J. Chem. Soc.* **1960,** 3521; U. Schöllkopf, K. Fellenberger, *Ber.* **698,** 80 (1966); Y. Makisumi, S. Notzumoto, *Tetrahedron Lett.* **1966,** 6393.

[2,3]-Sigmatropic rearrangement of the conjugate bases of allylic ethers with high regioselectivity. The stereoselectivity is highly dependent on the nature of the substrate:

Y = alkynyl, alkenyl, Ph, COR, CN
base = LDA, *n*-BuLi, NaNH$_2$

Methods development for ring contractions generating enediynes: H. Audrain *et al., Tetrahedron* **50,** 1469 (1994). Review of stereoselectivity: K. Mikami, T. Nakai, *Synthesis* **1994,** 594. *Reviews:* J. A. Marshall, *Comp. Org. Syn.* **3,** 975-1014 (1991); T. Nakai, K. Mikami, *Org. React.* **46,** 105-209 (1994). *Cf.* Meisenheimer Rearrangements; Mislow-Evans Rearrangement; Sommelet-Hauser Rearrangement.

471. Wohl Degradation; Zemplén Modification

A. Wohl, *Ber.* **26,** 730 (1893); **32,** 3666 (1899); G. Zemplén, *Ber.* **59,** 1254, 2402 (1926).

Method for the conversion of an aldose into an aldose with one less carbon atom by the reversal of the cyanohydrin synthesis. In the Wohl method the nitrile group is eliminated by treatment with ammoniacal silver oxide; in the **Zemplén modification** sodium alkoxide is used in the elimination of the nitrile:

Reviews: V. Deulofeu, *Adv. Carbohydr. Chem.* **4,** 129, 138 (1949); R. Bognár *et al., Ann.* **680,** 118 (1964); W. W. Wendall, *Tetrahedron Lett.* **1970,** 3439; L. Hough, A. C. Richardson, *The Carbohydrates* **1A,** 128 (1972).

472. Wohl-Ziegler Reaction

A. Wohl, *Ber.* **52,** 51 (1919); K. Ziegler *et al., Ann.* **551,** 30 (1942).

Allylic bromination of olefins with *N*-bromosuccinimide. Peroxides or ultraviolet light are used as initiators:

Reviews: C. Djerassi, *Chem. Rev.* **43,** 271 (1948); L. Horner, E. M. Winkelman, *Angew. Chem.* **71,** 349 (1959); S. S. Novikov *et al., Russ. Chem. Rev.* **31,** 671 (1962); A. Nechvatal, *Adv. Free-Radical Chem.* **4,** 175-201 (1972).

473. Wolff-Kishner Reduction; Huang-Minlon Modification

N. Kishner, *J. Russ. Phys. Chem. Soc.* **43,** 582 (1911), *C.A.* **6,** 347 (1912); L. Wolff, *Ann.* **394,** 86 (1912); Huang-Minlon, *J. Am. Chem. Soc.* **68,** 2487 (1946).

Complete reduction of carbonyl compounds to methyl or methylene groups on heating with hydrazine hydrate and a base. In the **Huang-Minlon modification** diethylene glycol is used as a solvent:

Reviews: D. Todd, *Org. React.* **4,** 378 (1948); H. H. Szmant, *Angew. Chem. Int. Ed.* **7,** 120 (1968); F. Asinger, H. H. Vogel, *Houben-Weyl* **5/1a,** 251, 456 (1970); H. Balli, *ibid.* **5/1b,** 629 (1972); R. O. Hutchins, M. K. Hutchins, *Comp. Org. Syn.* **8,** 327-343 (1991). Bond cleavage: R. P. Lemieux, P. Beak, *Tetrahedron Lett.* **30,** 1353 (1989). Synthetic application: A. Srikrishna, D. Vijaykumuv, *J. Chem. Soc. Perkin Trans. 1* **1999,** 1265. *Cf.* Clemmensen Reduction; Haworth Phenanthrene Synthesis.

474. Wolff Rearrangement

L. Wolff *Ann.* **394,** 25 (1912).

Rearrangement of diazoketones to ketenes thermally, photochemically or catalytically. The rearrangement is the key step in the Arndt-Eistert synthesis, *q.v.*:

Reviews: P. A. S. Smith in *Molecular Rearrangements* Part 1, Ed. (Wiley-Interscience, New York, 1963) pp 528-550, 558-568; W. Kirmse, *Carbene Chemistry* (Academic Press, New York, 2nd ed., 1971) pp 475-492; H. Meier, K. P. Zeller, *Angew. Chem. Int. Ed.* **14**, 32 (1975); M. Torres, *Pure Appl. Chem.* **52**, 1623 (1980); C. B. Gill, *Comp. Org. Syn.* **3**, 887-912 (1991). Photo-induced mechanistic studies: T. Lippert *et al., J. Am. Chem. Soc.* **118**, 1551 (1996); Y. Chiang *et al., ibid.* **121**, 5930 (1999). Synthetic application: Y. R. Lee *et al., Tetrahedron Lett.* **40**, 8219 (1999).

475. Wolffenstein-Böters Reaction
O. Böters, R. Wolffenstein, **DE 194883** (1906); **FR 380121** (1907); **GB 17521** (1907); **US 923761** (1909).
Simultaneous oxidation and nitration of aromatic compounds to nitrophenols with nitric acid or the higher oxides of nitrogen in the presence of a mercury salt as catalyst:

R. Wolffenstein, O. Böters, *Ber.* **46**, 586 (1913); F. H. Westheimer *et al., J. Am. Chem. Soc.* **69**, 773 (1947); M. Carmack *et al., ibid.* 785; E. E. Aristoff *et al., Ind. Eng. Chem.* **40**, 1281 (1948); W. Seidenfaden, W. Pawellek, *Houben-Weyl* **10/1**, 815 (1971).

476. Woodward *cis*-Hydroxylation
R. B. Woodward, **US 2687435** (1954); R. B. Woodward, F. V. Brutcher, *J. Am. Chem. Soc.* **80**, 209 (1958).
The hydroxylation of an olefin with iodine and silver acetate in wet acetic acid to give *cis*-glycols:

L. B. Barkley, M. W. Farrar, *J. Am. Chem. Soc.* **76**, 5014, (1954); W. S. Knowles, Q. E. Thompson, *ibid.* **79**, 3212 (1957); W. F. Forbes, R. Shelton, *J. Org. Chem.* **24**, 436 (1959); F. D. Gunstone, *Adv. Org. Chem.* **1**, 117 (1960). Application to steroids: L. Mangoni, V. Dovinola, *Tetrahedron Lett.* **1969**, 5235; P. Kocovsky, V. Cerny, *Collect. Czech. Chem. Commun.* **42**, 163 (1977). Modification: L. Mangoni *et al., Gazz. Chim. Ital.* **105**, 377 (1975). *Cf.* Prévost Reaction.

477. Wurtz-Fittig Reaction
B. Tollens, R. Fittig, *Ann.* **131**, 303 (1864); R. Fittig, J. König, *ibid.* **144**, 277 (1867).
Formation of alkylated aromatic hydrocarbons on coupling of an alkyl and an aryl halide with sodium:

T. L. Kwa, C. Boelhouwer, *Tetrahedron* **25**, 5771 (1969); B. J. Wakefield, *Comp. Organomet. Chem.* **7**, 45 (1982); K. Miyoshi *et al., Chemosphere* **41**, 819 (2000).

478. Wurtz Reaction
A. Wurtz, *Ann. Chim. Phys.* [3] **44**, 275 (1855); *Ann.* **96**, 364 (1855).
Coupling of two alkyl radicals by treating two moles of alkyl halides with two moles of sodium:

$$2RX + 2Na \longrightarrow R-R + 2NaX$$

J. L. Wardell, *Comp. Organomet. Chem.* **1**, 52 (1982); W. E. Lindsell, *ibid.* 193; B. J. Wakefield, *ibid.* **7**, 45; D. C. Billington, *Comp. Org. Syn.* **3**, 413-423 (1991).

Organic Name Reactions

479. Yamaguchi Esterification; Yamaguchi Macrolactonization

J. Inanaga *et al.*, *Bull. Chem. Soc. Jpn.* **52**, 1989 (1979).

Synthesis of functionalized esters from carboxylic acids and alcohols via the use of 2,4,6-trichlorobenzoyl chloride and DMAP. In the **Yamaguchi macrolactonization**, the procedure is applied to the intramolecular cyclization of hydroxy acids to prepare the corresponding lactones:

R^1, R^2 = alkyl, aryl

Mechanistic studies: I. Dhimitruka, J. SantaLucia, Jr., *Org. Lett.* **8**, 47 (2006). Application to the prepn of lactone rings: P. A. Wender *et al.*, *J. Am. Chem. Soc.* **124**, 13648 (2002). Lactonization procedural modifications: H. Tone *et al.*, *Tetrahedron Lett.* **28**, 4569 (1987); M. Hikota *et al.*, *Tetrahedron* **46**, 4613 (1990). *Review*: N. M. Ahmad in *Name Reactions for Functional Group Transformations*, J. J. Li, E. J. Corey, Eds. (Wiley, Hoboken, NJ, 2007) pp 545-550.

480. Ziegler Method (Thorpe-Ziegler Method)

K. Ziegler *et al.*, *Ann.* **504**, 94 (1933).

Cyclization of dinitriles at high dilution in dialkyl ether in the presence of ether-soluble metal alkylanilide and hydrolysis of the resultant imino-nitrile with formation of macrocyclic ketones (yield is dependent on ring size):

K. Ziegler *et al.*, *ibid.* **511**, 1 (1933); **512**, 164; **513**, 43 (1934); *idem Ber.* **67**, 139 (1934); *idem et al.*, *Ann.* **528**, 114, 143 (1937); R. C. Fuson in *Organic Chemistry* vol. I, H. Gilman, Ed. (New York, 1943) p 89; V. Migrdichian, *The Chemistry of Organic Cyanogen Compounds* (New York, 1947) p 288; K. Ziegler, *Houben-Weyl* **4/2**, 758 (1955). *Review:* J. P. Schaefer, J. J. Bloomfield, *Org. React.* **15**, 1-203 (1967). *Cf.* Thorpe Reaction.

481. Ziegler-Natta Polymerization

K. Ziegler *et al.*, *Angew. Chem.* **67**, 426, 541 (1955); G. Natta, *ibid.* **68**, 393 (1956).

Polymerization of vinyl monomers under mild conditions using Lewis acid catalysts to give a stereoregulated, or tactic, polymer.

K. Ziegler, *ibid.* **71**, 623 (1959); **72**, 829 (1960); C. L. Arcus in *Progress in Stereochemistry* vol. 3, P. B. D. de la Mare, W. Klyne, Eds. (Butterworth Inc., Washington, D.C., 1962) pp 269-288; M. N. Berger *et al.*, *Adv. Catal.* **19**, 211 (1969); T. Keii, *Kinetics of Ziegler-Natta Polymerization* (Halsted Press, New York, 1973) pp 129-162; *Developments in Polymerization* vol. 2, R. N. Haward, Ed. (Burgess-Intl., Philadelphia, 1979) pp 81-148; H. J. Sinn, W. Kaminsky, *Adv. Organomet. Chem.* **18**, 207 (1980); D. M. P. Mingos, *Comp. Organomet. Chem.* **3**, 72-75 (1982); P. D. Gavens *et al.*, *ibid.* 475-547.

482. Zimmermann Reaction

W. Zimmermann, *Z. Physiol. Chem.* **233**, 257 (1935).

The reaction that occurs between methylene ketones and aromatic polynitro compounds in the presence of alkali. When applied to 17-oxosteroids, the colored compounds formed can be used for the quantitative determination of 17-oxosteroids:

W. Zimmerman *et al.*, *ibid.* **289**, 91 (1952); *idem. ibid.* **300**, 141 (1955). Studies on mechanism: Neunhoffer *et al.*, *ibid.* **323**, 116 (1961); Foster, Mackie, *Tetrahedron* **18**, 1131 (1962); H. Hoffmeister, C. Rufer, *Ber.* **98**, 2376 (1965); B. T. Rudd, O. M. Galal, *Proc. Assoc. Clin. Biochem.* **4**, 175 (1967); C. S. Feldkamp *et al.*, *Microchem. J.* **22**, 201 (1977). *Cf.* Janovsky Reaction.

483. Zincke Disulfide Cleavage

T. Zincke, *Ber.* **44**, 769 (1911).

Formation of sulfenyl halides by three essentially similar methods involving the action of chlorine or bromine on aryl disulfides, thiophenols, or arylbenzyl sulfides:

$$ArSSAr + X_2 \longrightarrow 2ArSX$$

$$ArSH + X_2 \longrightarrow ArSX + HX$$

$$ArSCH_2C_6H_5 + 2X_2 \longrightarrow ArSX + C_6H_5CHX_2 + HX$$

T. Zincke *et al., ibid.* **45**, 471 (1912); **51**, 751 (1918); *Ann.* **391**, 55 (1912); **400**, 1 (1913); **406**, 103 (1914); **416**, 86 (1918); M. H. Hubacher, *Org. Synth.* **coll. II**, 455 (1943); N. Kharasch *et al., Chem. Rev.* **39**, 283 (1946); A. Schöberl, A. Wagner, *Houben-Weyl* **9**, 268 (1955); E. Kühle, *Synthesis* **1970**, 561.

484. Zincke Nitration
T. Zincke, *J. Prakt. Chem.* **61**, 561 (1900).

Replacement of *ortho-* or *para*-bromine or iodine atoms (but not fluorine or chlorine atoms) in phenols by a nitro group on treatment with nitrous acid or a nitrite in acetic acid:

L. C. Raiford, W. Heyl, *Am. Chem. J.* **43**, 393 (1910); **44**, 209 (1911); H. H. Hodgson, J. Nixon, *J. Chem. Soc.* **1932**, 273; L. C. Raiford, G. R. Miller, *J. Am. Chem. Soc.* **55**, 2125 (1933); L. C. Raiford, A. L. LeRosen, *ibid.* **66**, 1872 (1944); W. Seidenfaden, D. Pawellek, *Houben-Weyl* **10/1**, 821 (1971).

485. Zincke-Suhl Reaction
T. Zincke, R. Suhl, *Ber.* **39**, 4148 (1906).

Phenol-dienone rearrangement of *p*-cresols by addition of carbon tetrachloride in the presence of aluminum chloride with formation of 4-methyl-4-trichloromethylcyclohexa-2,5-dienone:

M. S. Newman, A. G. Pinkus, *J. Org. Chem.* **19**, 978, 985, 992, 997 (1954); M. S. Newman, L. L. Wood, Jr., *J. Am. Chem. Soc.* **81**, 6450 (1959); G. A. Olah, *Friedel-Crafts and Related Reactions* **vol. I** (Interscience, New York, 1963) p 128.

Organic Name Reactions

Organic Name Reactions

Organic Name Reactions

CHEMICAL ABSTRACTS SERVICE
REGISTRY NUMBERS

CHEMICAL ABSTRACTS SERVICE REGISTRY NUMBERS

The following table provides the Chemical Abstracts Service (CAS) registry numbers for the monograph title compounds and selected derivatives. The table is ordered numerically by CAS registry number.

CHEMICAL ABSTRACTS REGISTRY NUMBERS

[50-00-0] Formaldehyde, *4263*
[50-01-1] Guanidine Hydrochloride, *4597*
[50-02-2] Dexamethasone, *2945*
[50-03-3] Hydrocortisone 21-Acetate, *4824*
[50-04-4] Cortisone 21-Acetate, *2525*
[50-06-6] Phenobarbital, *7352*
[50-07-7] Mitomycin C, *6300*
[50-09-9] Hexobarbital Sodium Salt, *4740*
[50-10-2] Oxyphenonium Bromide, *7074*
[50-11-3] Metharbital, *6033*
[50-12-4] Mephenytoin, *5920*
[50-13-5] Meperidine Hydrochloride, *5916*
[50-14-6] Vitamin D₂, *10215*
[50-18-0] Cyclophosphamide (anhydrous), *2743*
[50-21-5] DL-Lactic Acid, *5384*
[50-22-6] Corticosterone, *2524*
[50-23-7] Hydrocortisone, *4824*
[50-24-8] Prednisolone, *7841*
[50-27-1] Estriol, *3762*
[50-28-2] Estradiol, *3758*
[50-29-3] DDT, *2843*
[50-32-8] Benzo[a]pyrene, *1106*
[50-33-9] Phenylbutazone, *7390*
[50-34-0] Propantheline Bromide, *7919*
[50-35-1] Thalidomide, *9403*
[50-36-2] Cocaine, *2440*
[50-37-3] Lysergide, *5695*
[50-41-9] Clomiphene Citrate, *2377*
[50-42-0] Adiphenine Hydrochloride, *149*
[50-44-2] 6-Mercaptopurine, *5938*
[50-47-5] Desipramine, *2921*
[50-48-6] Amitriptyline, *483*
[50-49-7] Imipramine, *4962*
[50-50-0] Estradiol Benzoate, *3758*
[50-52-2] Thioridazine, *9512*
[50-53-3] Chlorpromazine, *2191*
[50-54-4] Quinidine Sulfate, *8176*
[50-55-5] Reserpine, *8265*
[50-56-6] Oxytocin, *7078*
[50-57-7] Lypressin, *5691*
[50-58-8] Phendimetrazine Bitartrate, *7334*
[50-60-2] Phentolamine, *7376*
[50-62-4] Hexobendine Dihydrochloride, *4741*
[50-63-5] Chloroquine Diphosphate, *2165*
[50-65-7] Niclosamide, *6604*
[50-67-9] Serotonin, *8601*
[50-69-1] D-Ribose, *8328*
[50-70-4] Sorbitol, *8851*
[50-71-5] Alloxan, *273*
[50-76-0] Dactinomycin, *2797*
[50-78-2] Aspirin, *841*
[50-81-7] Ascorbic Acid, *819*
[50-85-1] *m*-Cresotic Acid, *2569*
[50-89-5] Thymidine, *9552*
[50-91-9] Floxuridine, *4144*

[50-98-6] Ephedrine Hydrochloride, *3663*
[50-99-7] Glucose, *4495*
[51-03-6] Piperonyl Butoxide, *7589*
[51-05-8] Procaine Hydrochloride, *7875*
[51-12-7] Nialamide, *6576*
[51-15-0] Pralidoxime Chloride, *7822*
[51-17-2] Benzimidazole, *1083*
[51-18-3] Triethylenemelamine, *9836*
[51-20-7] 5-Bromouracil, *1450*
[51-21-8] Fluorouracil, *4213*
[51-24-1] Tiratricol, *9618*
[51-26-3] Thyropropic Acid, *9567*
[51-28-5] 2,4-Dinitrophenol, *3308*
[51-34-3] Scopolamine, *8543*
[51-35-4] Hydroxyproline, *4878*
[51-41-2] Norepinephrine, *6784*
[51-42-3] Epinephrine Bitartrate, *3674*
[51-43-4] Epinephrine, *3674*
[51-45-6] Histamine, *4756*
[51-48-9] Thyroxine, *9570*
[51-49-0] Dextrothyroxine, *9570*
[51-50-3] Dibenamine, *3009*
[51-52-5] Propylthiouracil, *7980*
[51-55-8] Atropine, *866*
[51-56-9] Homatropine Hydrobromide, *4768*
[51-57-0] Methamphetamine Hydrochloride, *6020*
[51-60-5] Neostigmine Methyl Sulfate, *6549*
[51-61-6] Dopamine, *3470*
[51-63-8] Dextroamphetamine Sulfate, *2956*
[51-64-9] Dextroamphetamine, *2956*
[51-66-1] *p*-Acetanisidine, *50*
[51-67-2] Tyramine, *10016*
[51-68-3] Meclofenoxate, *5851*
[51-71-8] Phenelzine, *7335*
[51-73-0] Tremorine, *9744*
[51-74-1] Histamine Phosphate, *4756*
[51-75-2] Mechlorethamine, *5846*
[51-77-4] Gefarnate, *4411*
[51-79-6] Urethan, *10059*
[51-83-2] Carbachol, *1781*
[51-85-4] Cystamine, *2773*
[51-98-9] Norethindrone Acetate, *6786*
[52-01-7] Spironolactone, *8887*
[52-21-1] Prednisolone 21-Acetate, *7841*
[52-24-4] Thiotepa, *9518*
[52-26-6] Morphine Hydrochloride, *6361*
[52-28-8] Codeine Phosphate, *2448*
[52-31-3] Cyclobarbital, *2704*
[52-39-1] Aldosterone, *218*
[52-43-7] Allobarbital, *255*
[52-46-0] Apholate, *715*
[52-49-3] Trihexyphenidyl Hydrochloride, *9864*
[52-51-7] Bronopol, *1457*
[52-52-8] Cycloleucine, *2727*
[52-53-9] Verapamil, *10144*
[52-62-0] Pentolinium Tartrate, *7245*
[52-66-4] Penicillamine DL-Form, *7197*

[52-67-5] Penicillamine, *7197*
[52-68-6] Trichlorfon, *9788*
[52-76-6] Lynestrenol, *5690*
[52-78-8] Norethandrolone, *6785*
[52-86-8] Haloperidol, *4634*
[52-88-0] Atropine Methylnitrate, *866*
[52-89-1] Cysteine Hydrochloride, *2778*
[52-90-4] Cysteine, *2778*
[53-03-2] Prednisone, *7842*
[53-05-4] Tetrahydrocortisone, *9355*
[53-06-5] Cortisone, *2525*
[53-16-7] Estrone, *3763*
[53-19-0] Mitotane, *6301*
[53-21-4] Cocaine Hydrochloride, *2440*
[53-33-8] Paramethasone, *7133*
[53-34-9] Fluprednisolone, *4223*
[53-36-1] Methylprednisolone 21-Acetate, *6184*
[53-39-4] Oxandrolone, *7020*
[53-41-8] Androsterone, *633*
[53-43-0] Dehydroepiandrosterone, *2875*
[53-46-3] Methantheline Bromide, *6030*
[53-57-6] NADPH, *6432*
[53-59-8] NADP, *6432*
[53-60-1] Promazine Hydrochloride, *7898*
[53-70-3] 1,2:5,6-Dibenzanthracene, *3011*
[53-73-6] Angiotensin Amide, *644*
[53-79-2] Puromycin, *8054*
[53-84-9] NAD, *6429*
[53-86-1] Indomethacin, *5009*
[53-96-3] *N*-2-Fluorenylacetamide, *4189*
[54-03-5] Hexobendine, *4741*
[54-04-6] Mescaline, *5975*
[54-05-7] Chloroquine, *2165*
[54-06-8] Adrenochrome, *161*
[54-11-5] Nicotine, *6609*
[54-21-7] Salicylic Acid Sodium Salt, *8469*
[54-25-1] 6-Azauridine, *897*
[54-28-4] γ-Tocopherol, *9655*
[54-31-9] Furosemide, *4338*
[54-32-0] Moxisylyte, *6378*
[54-36-4] Metyrapone, *6238*
[54-42-2] Idoxuridine, *4931*
[54-47-7] Pyridoxal 5-Phosphate, *8092*
[54-49-9] Metaraminol, *6003*
[54-62-6] Aminopterin, *468*
[54-64-8] Thimerosal, *9470*
[54-71-7] Pilocarpine Hydrochloride, *7536*
[54-85-3] Isoniazid, *5232*
[54-91-1] Pipobroman, *7593*
[54-92-2] Iproniazid, *5122*
[54-95-5] Pentylenetetrazole, *7254*
[54-96-6] Amifampridine, *398*
[55-03-8] Levothyroxine Sodium, *9570*
[55-06-1] Liothyronine Sodium Salt, *5565*
[55-10-7] Vanilmandelic Acid, *10121*
[55-16-3] Scopolamine Hydrochloride, *8543*

[60-26-4] Hexamethonium, *4724*
[60-27-5] Creatinine, *2557*
[60-29-7] Ethyl Ether, *3861*
[60-31-1] Acetylcholine Chloride, *82*
[60-32-2] ε-Aminocaproic Acid, *427*
[60-33-3] Linoleic Acid, *5560*
[60-34-4] Methylhydrazine, *6158*
[60-35-5] Acetamide, *44*
[60-39-9] Mazipredone Hydrochloride, *5832*
[60-40-2] Mecamylamine, *5843*
[60-41-3] Strychnine Sulfate, *8985*
[60-46-8] Aminopentamide, *455*
[60-51-5] Dimethoate, *3241*
[60-54-8] Tetracycline, *9341*
[60-56-0] Methimazole, *6043*
[60-57-1] Dieldrin, *3114*
[60-70-8] Veratramine, *10146*
[60-79-7] Ergonovine, *3712*
[60-80-0] Antipyrine, *703*
[60-81-1] Phloridzin, *7438*
[60-82-2] Phloretin, *7437*
[60-87-7] Promethazine, *7901*
[60-91-3] Diethazine, *3119*
[60-92-4] Cyclic AMP, *2701*
[60-93-5] Quinine Dihydrochloride, *8177*
[60-99-1] Levomepromazine, *5518*
[61-00-7] Acepromazine, *33*
[61-12-1] Dibucaine Hydrochloride, *3036*
[61-16-5] Methoxamine Hydrochloride, *6058*
[61-19-8] 5'-Adenylic Acid, *146*
[61-24-5] Cephalosporin C, *1978*
[61-25-6] Papaverine Hydrochloride, *7122*
[61-33-6] Penicillin G, *7203*
[61-50-7] N,N-Dimethyltryptamine, *3288*
[61-54-1] Tryptamine, *9976*
[61-56-3] Sulthiame, *9122*
[61-57-4] Niridazole, *6649*
[61-68-7] Mefenamic Acid, *5869*
[61-72-3] Cloxacillin, *2403*
[61-73-4] Methylene Blue, *6132*
[61-75-6] Bretylium Tosylate, *1377*
[61-76-7] Phenylephrine Hydrochloride, *7398*
[61-78-9] p-Aminohippuric Acid, *439*
[61-80-3] Zoxazolamine, *10399*
[61-82-5] Amitrole, *485*
[61-90-5] Leucine, *5503*
[61-96-1] Nordefrin Hydrochloride, *6781*
[62-13-5] Adrenalone Hydrochloride, *160*
[62-31-7] Dopamine Hydrochloride, *3470*
[62-33-9] EDTA Calcium Disodium Salt, *3565*
[62-37-1] Chlormerodrin, *2105*
[62-38-4] Phenylmercuric Acetate, *7411*
[62-44-2] Phenacetin, *7319*
[62-46-4] Thioctic Acid, *9479*
[62-49-7] Choline, *2211*
[62-50-0] Ethyl Methanesulfonate, *3882*
[62-51-1] Methacholine Chloride, *6012*
[62-53-3] Aniline, *652*
[62-54-4] Calcium Acetate, *1649*
[62-55-5] Thioacetamide, *9472*
[62-56-6] Thiourea, *9521*
[62-57-7] α-Aminoisobutyric Acid, *442*

[62-59-9] Cevadine, *2032*
[62-67-9] Nalorphine, *6446*
[62-73-7] Dichlorvos, *3089*
[62-74-8] Fluoroacetic Acid Sodium Salt, *4199*
[62-75-9] N-Nitrosodimethylamine, *6724*
[62-76-0] Sodium Oxalate, *8781*
[62-90-8] Nandrolone Phenpropionate, *6450*
[62-97-5] Diphemanil Methylsulfate, *3336*
[63-05-8] Androstenedione, *631*
[63-12-7] Benzquinamide, *1122*
[63-25-2] Carbaryl, *1788*
[63-39-8] Uridine 5'-Triphosphate, *10065*
[63-42-3] Lactose, *5392*
[63-45-6] Primaquine Diphosphate, *7863*
[63-56-9] Thonzylamine Hydrochloride, *9527*
[63-68-3] Methionine, *6047*
[63-74-1] Sulfanilamide, *9057*
[63-75-2] Arecoline, *768*
[63-84-3] Dopa DL-Form, *3469*
[63-89-8] Colfosceril Palmitate, *2461*
[63-91-2] Phenylalanine, *7382*
[63-92-3] Phenoxybenzamine Hydrochloride, *7368*
[63-98-9] Phenacemide, *7318*
[64-02-8] EDTA Tetrasodium Salt, *3565*
[64-04-0] Phenethylamine, *7339*
[64-10-8] N-Phenylurea, *7429*
[64-17-5] Ethyl Alcohol, *3814*
[64-18-6] Formic Acid, *4269*
[64-19-7] Acetic Acid, *54*
[64-31-3] Morphine Sulfate, *6361*
[64-39-1] Promedol, *7899*
[64-43-7] Amobarbital Sodium Salt, *567*
[64-47-1] Physostigmine Sulfate, *7496*
[64-55-1] Mebutamate, *5842*
[64-65-3] Bemegride, *1029*
[64-67-5] Diethyl Sulfate, *3149*
[64-69-7] Iodoacetic Acid, *5068*
[64-72-2] Chlortetracycline Hydrochloride, *2198*
[64-73-3] Demeclocycline Hydrochloride, *2890*
[64-75-5] Tetracycline Hydrochloride, *9341*
[64-77-7] Tolbutamide, *9667*
[64-85-7] Deoxycorticosterone, *2902*
[64-86-8] Colchicine, *2455*
[64-95-9] Adiphenine, *149*
[65-19-0] Yohimbine Hydrochloride, *10300*
[65-22-5] Pyridoxal Hydrochloride, *8091*
[65-23-6] Pyridoxine, *8095*
[65-28-1] Phentolamine Methanesulfonate, *7376*
[65-29-2] Gallamine Triethiodide, *4373*
[65-30-5] Nicotine Sulfate, *6609*
[65-31-6] Nicotine Bitartrate, *6609*
[65-45-2] Salicylamide, *8465*
[65-46-3] Cytidine, *2783*
[65-49-6] p-Aminosalicylic Acid, *473*
[65-71-4] Thymine, *9553*
[65-82-7] N-Acetylmethionine, *92*
[65-85-0] Benzoic Acid, *1093*
[65-86-1] Orotic Acid, *6972*

[66-02-4] 3,5-Diiodotyrosine, *3210*
[66-22-8] Uracil, *10035*
[66-23-9] Acetylcholine Bromide, *81*
[66-25-1] Caproic Aldehyde, *1762*
[66-27-3] Methyl Methanesulfonate, *6169*
[66-28-4] Strophanthidin, *8981*
[66-32-0] Strychnine Nitrate, *8985*
[66-71-7] o-Phenanthroline, *7328*
[66-72-8] Pyridoxal, *8091*
[66-75-1] Uracil Mustard, *10036*
[66-76-2] Dicumarol, *3100*
[66-79-5] Oxacillin, *7003*
[66-81-9] Cycloheximide, *2721*
[66-86-4] Neomycin C, *6539*
[66-93-3] 3-Quinuclidinol *dl*-Form Benzoate (Ester), *8199*
[66-97-7] Psoralen, *8037*
[67-03-8] Thiamine Hydrochloride, *9447*
[67-07-2] Phosphocreatine, *7452*
[67-16-3] Thiamine Disulfide, *9449*
[67-20-9] Nitrofurantoin, *6686*
[67-21-0] Ethionine DL-Form, *3792*
[67-28-7] Nihydrazone, *6626*
[67-42-5] EGTA, *3576*
[67-43-6] Pentetic Acid, *7237*
[67-45-8] Furazolidone, *4330*
[67-47-0] 5-(Hydroxymethyl)-2-furaldehyde, *4871*
[67-48-1] Choline Chloride, *2211*
[67-52-7] Barbituric Acid, *958*
[67-56-1] Methanol, *6029*
[67-62-9] Methoxyamine, *6060*
[67-63-0] Isopropyl Alcohol, *5254*
[67-64-1] Acetone, *65*
[67-66-3] Chloroform, *2142*
[67-68-5] Dimethyl Sulfoxide, *3285*
[67-71-0] Dimethyl Sulfone, *3283*
[67-72-1] Hexachloroethane, *4715*
[67-73-2] Fluocinolone Acetonide, *4181*
[67-78-7] Triamcinolone 16,21-Diacetate, *9757*
[67-92-5] Dicyclomine Hydrochloride, *3108*
[67-96-9] Dihydrotachysterol, *3198*
[67-97-0] Vitamin D_3, *10216*
[67-99-2] Gliotoxin, *4476*
[68-04-2] Sodium Citrate, *8737*
[68-11-1] Thioglycolic Acid, *9489*
[68-12-2] N,N-Dimethylformamide, *3267*
[68-19-9] Vitamin B_{12}, *10212*
[68-22-4] Norethindrone, *6786*
[68-23-5] Norethynodrel, *6787*
[68-26-8] Vitamin A, *10211*
[68-35-9] Sulfadiazine, *9035*
[68-36-0] 1,4-Bis(trichloromethyl)benzene, *1304*
[68-41-7] Cycloserine, *2749*
[68-76-8] Triaziquone, *9764*
[68-81-5] Iodinin, *5064*
[68-88-2] Hydroxyzine, *4889*
[68-90-6] Benziodarone, *1085*
[68-91-7] Trimethaphan Camsylate, *9876*
[68-94-0] Hypoxanthine, *4908*
[68-96-2] 17α-Hydroxyprogesterone, *4877*
[69-09-0] Chlorpromazine Hydrochloride, *2191*
[69-14-7] Hexestrol Bis(β-diethylaminoethyl ether) Dihydrochloride, *4738*

[69-22-7] Caffeine Citrate, *1639*
[69-23-8] Fluphenazine, *4220*
[69-24-9] Cinchotoxine, *2292*
[69-25-0] Eledoisin, *3588*
[69-27-2] Chlorisondamine Chloride, *2101*
[69-33-0] Tubercidin, *9984*
[69-43-2] Prenylamine Lactate, *7855*
[69-46-5] Calcium Acetylsalicylate, *1650*
[69-52-3] Ampicillin Sodium Salt, *583*
[69-53-4] Ampicillin, *583*
[69-57-8] Penicillin G Sodium Salt, *7203*
[69-65-8] Mannitol, *5810*
[69-72-7] Salicylic Acid, *8469*
[69-78-3] Ellman's Reagent, *3606*
[69-79-4] Maltose, *5778*
[69-81-8] Carbazochrome, *1789*
[69-89-6] Xanthine, *10257*
[69-93-2] Uric Acid, *10060*
[70-00-8] Trifluridine, *9855*
[70-07-5] Mephenoxalone, *5918*
[70-11-1] ω-Bromoacetophenone, *1409*
[70-18-8] Glutathione, *4511*
[70-19-9] Thurfyl Nicotinate, *9550*
[70-22-4] Oxotremorine, *7050*
[70-25-7] N-Methyl-N'-nitro-N-nitrosoguanidine, *6176*
[70-26-8] Ornithine, *6969*
[70-30-4] Hexachlorophene, *4716*
[70-34-8] 1-Fluoro-2,4-dinitrobenzene, *4204*
[70-43-9] Barthrin, *1000*
[70-47-3] Asparagine, *827*
[70-49-5] Thiomalic Acid, *9496*
[70-51-9] Deferoxamine, *2863*
[70-69-9] p-Aminopropiophenone, *466*
[70-70-2] Paroxypropione, *7149*
[70-78-0] 3-Iodotyrosine, *5090*
[71-00-1] Histidine, *4758*
[71-23-8] Propyl Alcohol, *7955*
[71-27-2] Succinylcholine Chloride, *9006*
[71-30-7] Cytosine, *2792*
[71-36-3] n-Butyl Alcohol, *1542*
[71-41-0] 1-Pentanol, *7230*
[71-43-2] Benzene, *1068*
[71-44-3] Spermine, *8870*
[71-48-7] Cobaltous Acetate, *2417*
[71-55-6] 1,1,1-Trichloroethane, *9801*
[71-58-9] Medroxyprogesterone 17-Acetate, *5864*
[71-62-5] Veratridine, *10148*
[71-63-6] Digitoxin, *3182*
[71-67-0] Sulfobromophthalein Sodium, *9084*
[71-68-1] Hydromorphone Hydrochloride, *4840*
[71-73-8] Thiopental Sodium, *9503*
[71-78-3] Pipradrol Hydrochloride, *7597*
[71-81-8] Isopropamide Iodide, *5248*
[71-82-9] Levallorphan Tartrate, *5510*
[71-91-0] Tetraethylammonium Bromide, *9344*
[72-14-0] Sulfathiazole, *9074*
[72-17-3] Sodium Lactate, *8767*
[72-18-4] Valine, *10095*
[72-19-5] Threonine, *9534*
[72-20-8] Endrin, *3633*
[72-23-1] 11-Dehydrocorticosterone, *2873*
[72-33-3] Mestranol, *5987*

[72-43-5] Methoxychlor, *6061*
[72-44-6] Methaqualone, *6032*
[72-48-0] Alizarin, *243*
[72-54-8] 1,1-Dichloro-2,2-bis(p-chlorophenyl)ethane, *3070*
[72-57-1] Trypan Blue, *9972*
[72-63-9] Methandrostenolone, *6023*
[72-69-5] Nortriptyline, *6804*
[72-80-0] Chlorquinaldol, *2196*
[73-03-0] Cordycepin, *2514*
[73-05-2] Phentolamine Hydrochloride, *7376*
[73-09-6] Etozolin, *3934*
[73-22-3] Tryptophan, *9977*
[73-24-5] Adenine, *140*
[73-31-4] Melatonin, *5884*
[73-32-5] Isoleucine, *5225*
[73-40-5] Guanine, *4599*
[73-48-3] Bendroflumethiazide, *1039*
[73-49-4] Quinethazone, *8172*
[73-67-6] Toxopyrimidine, *9720*
[73-78-9] Lidocaine Hydrochloride, *5535*
[74-11-3] p-Chlorobenzoic Acid, *2126*
[74-31-7] N,N'-Diphenyl-p-phenylenediamine, *3363*
[74-39-5] Magneson, *5759*
[74-55-5] Ethambutol, *3775*
[74-61-3] 2,3-Dimercapto-1-propanesulfonic Acid, *3232*
[74-79-3] Arginine, *770*
[74-82-8] Methane, *6024*
[74-83-9] Methyl Bromide, *6103*
[74-84-0] Ethane, *3778*
[74-85-1] Ethylene, *3845*
[74-86-2] Acetylene, *85*
[74-87-3] Methyl Chloride, *6112*
[74-88-4] Methyl Iodide, *6159*
[74-89-5] Methylamine, *6088*
[74-90-8] Hydrogen Cyanide, *4832*
[74-93-1] Methanethiol, *6028*
[74-95-3] Methylene Bromide, *6133*
[74-96-4] Ethyl Bromide, *3826*
[74-98-6] Propane, *7913*
[75-00-3] Ethyl Chloride, *3837*
[75-01-4] Vinyl Chloride, *10191*
[75-03-6] Ethyl Iodide, *3868*
[75-04-7] Ethylamine, *3817*
[75-05-8] Acetonitrile, *69*
[75-07-0] Acetaldehyde, *40*
[75-08-1] Ethanethiol, *3781*
[75-09-2] Methylene Chloride, *6135*
[75-10-5] Difluoromethane, *3170*
[75-11-6] Methylene Iodide, *6137*
[75-12-7] Formamide, *4265*
[75-13-8] Isocyanic Acid, *5210*
[75-15-0] Carbon Disulfide, *1811*
[75-18-3] Dimethyl Sulfide, *3281*
[75-19-4] Cyclopropane, *2744*
[75-20-7] Calcium Carbide, *1658*
[75-21-8] Ethylene Oxide, *3856*
[75-24-1] Trimethylaluminum, *9879*
[75-25-2] Bromoform, *1430*
[75-26-3] Isopropyl Bromide, *5256*
[75-27-4] Bromodichloromethane, *1425*
[75-29-6] Isopropyl Chloride, *5257*
[75-30-9] Isopropyl Iodide, *5260*
[75-31-0] Isopropylamine, *5255*
[75-34-3] Ethylidene Chloride, *3865*
[75-35-4] Vinylidene Chloride, *10192*
[75-36-5] Acetyl Chloride, *80*
[75-39-8] Acetaldehyde Ammonia, *41*
[75-44-5] Phosgene, *7447*
[75-46-7] Fluoroform, *4205*

[75-47-8] Iodoform, *5076*
[75-50-3] Trimethylamine, *9880*
[75-52-5] Nitromethane, *6698*
[75-56-9] Propylene Oxide, *7969*
[75-58-1] Tetramethylammonium Iodide, *9372*
[75-59-2] Tetramethylammonium Hydroxide, *9371*
[75-60-5] Cacodylic Acid, *1609*
[75-64-9] tert-Butylamine, *1547*
[75-65-0] tert-Butyl Alcohol, *1544*
[75-66-1] tert-Butyl Mercaptan, *1582*
[75-69-4] Trichlorofluoromethane, *9805*
[75-71-8] Dichlorodifluoromethane, *3073*
[75-73-0] Carbon Tetrafluoride, *1817*
[75-75-2] Methanesulfonic Acid, *6026*
[75-76-3] Tetramethylsilane, *9377*
[75-77-4] Chlorotrimethylsilane, *2180*
[75-84-3] Neopentyl Alcohol, *6542*
[75-85-4] tert-Pentyl Alcohol, *7253*
[75-86-5] Acetone Cyanohydrin, *66*
[75-87-6] Trichloroacetaldehyde, *9791*
[75-89-8] Trifluoroethanol, *9848*
[75-91-2] tert-Butyl Hydroperoxide, *1573*
[75-93-4] Methyl Sulfate, *6195*
[75-96-7] Tribromoacetic Acid, *9770*
[75-97-8] Pinacolone, *7551*
[75-98-9] Pivalic Acid, *7625*
[75-99-0] Dalapon, *2799*
[76-00-6] 1,1,1-Trichloro-2-propanol, *9810*
[76-01-7] Pentachloroethane, *7217*
[76-03-9] Trichloroacetic Acid, *9792*
[76-05-1] Trifluoroacetic Acid, *9847*
[76-06-2] Chloropicrin, *2159*
[76-08-4] Tribromo-tert-butyl Alcohol, *9773*
[76-09-5] Pinacol, *7550*
[76-14-2] Cryofluorane, *2594*
[76-19-7] Perfluoropropane, *7273*
[76-20-0] Sulfonethylmethane, *9088*
[76-22-2] Camphor, *1734*
[76-24-4] Alloxantin, *274*
[76-25-5] Triamcinolone Acetonide, *9758*
[76-29-9] 3-Bromo-d-camphor, *1421*
[76-30-2] Dihydroxytartaric Acid, *3205*
[76-36-8] α,α,β-Trichloro-n-butyraldehyde, *9799*
[76-37-9] 2,2,3,3-Tetrafluoro-1-propanol, *9351*
[76-38-0] Methoxyflurane, *6067*
[76-41-5] Oxymorphone, *7068*
[76-42-6] Oxycodone, *7058*
[76-43-7] Fluoxymesterone, *4218*
[76-44-8] Heptachlor, *4691*
[76-45-9] Protoverine, *8009*
[76-49-3] Bornyl Acetate, *1341*
[76-53-9] Veatchine, *10131*
[76-54-0] 2',7'-Dichlorofluorescein, *3076*
[76-57-3] Codeine, *2448*
[76-58-4] Ethylmorphine, *3884*
[76-59-5] Bromthymol Blue, *1455*
[76-60-8] Bromcresol Green, *1393*
[76-61-9] Thymol Blue, *9555*
[76-62-0] Tetrabromophenolphthalein, *9331*
[76-66-4] Rhynchophylline, *8322*
[76-74-4] Pentobarbital, *7243*
[76-77-7] Neoquassin, *6548*
[76-78-8] Quassin, *8142*

[76-80-2] Tephrosin, 9293
[76-83-5] Trityl Chloride, 9944
[76-84-6] Triphenylcarbinol, 9912
[76-87-9] Triphenyltin Hydroxide, 9920
[76-90-4] Mepenzolate Bromide, 5915
[76-93-7] Benzilic Acid, 1082
[76-99-3] Methadone, 6016
[77-02-1] Aprobarbital, 745
[77-03-2] Piperidione, 7581
[77-04-3] Pyrithyldione, 8108
[77-06-5] Gibberellic Acid, 4454
[77-07-6] Levorphanol, 5524
[77-09-8] Phenolphthalein, 7356
[77-10-1] Phencyclidine, 7333
[77-16-7] Plumericin, 7656
[77-19-0] Dicyclomine, 3108
[77-20-3] Alphaprodine, 306
[77-21-4] Glutethimide, 4513
[77-22-5] Caramiphen, 1778
[77-23-6] Carbetapentane, 1795
[77-25-8] Diethyl Diethylmalonate, 3130
[77-26-9] Butalbital, 1514
[77-27-0] Thiamylal, 9454
[77-28-1] Butethal, 1524
[77-32-7] Thiobarbital, 9474
[77-36-1] Chlorthalidone, 2200
[77-37-2] Procyclidine, 7881
[77-38-3] Chlorphenoxamine, 2187
[77-40-7] Bisphenol B, 1299
[77-41-8] Methsuximide, 6078
[77-42-9] β-Santalol, 8495
[77-48-5] Dibromantin, 3022
[77-51-0] Isoaminile, 5155
[77-52-1] Ursolic Acid, 10075
[77-53-2] Cedrol, 1913
[77-58-7] Dibutyltin Dilaurate, 3047
[77-59-8] Tomatidine, 9703
[77-60-1] Tigogenin, 9591
[77-66-7] Acecarbromal, 22
[77-67-8] Ethosuximide, 3801
[77-75-8] Meparfynol, 5913
[77-78-1] Dimethyl Sulfate, 3280
[77-79-2] 3-Sulfolene, 9086
[77-81-6] Tabun, 9151
[77-86-1] Tromethamine, 9951
[77-91-8] Choline Dihydrogen Citrate, 2211
[77-92-9] Citric Acid, 2325
[77-93-0] Citric Acid Ethyl Ester, 2325
[77-94-1] Butyl Citrate, 1565
[77-95-2] Quinic Acid, 8175
[77-98-5] Tetraethylammonium Hydroxide, 9346
[78-00-2] Tetraethyllead, 9347
[78-05-7] Phenoctide, 7353
[78-10-4] Ethyl Silicate, 3903
[78-11-5] Pentaerythritol Tetranitrate, 7222
[78-30-8] Tri-o-tolyl Phosphate, 9941
[78-34-2] Dioxathion, 3330
[78-40-0] Triethyl Phosphate, 9838
[78-44-4] Carisoprodol, 1842
[78-48-8] Tribufos, 9779
[78-59-1] Isophorone, 5242
[78-67-1] 2,2'-Azobisisobutyronitrile, 911
[78-70-6] Linalool, 5550
[78-75-1] Propylene Dibromide, 7966
[78-76-2] sec-Butyl Bromide, 1556
[78-77-3] Isobutyl Bromide, 5180
[78-79-5] Isoprene, 5247
[78-81-9] Isobutylamine, 5177
[78-82-0] Isobutyronitrile, 5201

[78-83-1] Isobutyl Alcohol, 5176
[78-84-2] Isobutyraldehyde, 5199
[78-86-4] sec-Butyl Chloride, 1562
[78-87-5] Propylene Dichloride, 7967
[78-89-7] Propylene Chlorohydrin, 7963
[78-90-0] Propylenediamine, 7965
[78-91-1] 2-Aminopropanol, 464
[78-92-2] sec-Butyl Alcohol, 1543
[78-93-3] Methyl Ethyl Ketone, 6143
[78-94-4] Methyl Vinyl Ketone, 6210
[78-95-5] Chloroacetone, 2114
[78-98-8] Methylglyoxal, 6153
[78-99-9] Propylidene Chloride, 7973
[79-00-5] 1,1,2-Trichloroethane, 9802
[79-01-6] Trichloroethylene, 9804
[79-03-8] Propionyl Chloride, 7942
[79-04-9] Chloroacetyl Chloride, 2067
[79-05-0] Propionamide, 7938
[79-06-1] Acrylamide, 120
[79-07-2] Chloroacetamide, 2110
[79-08-3] Bromoacetic Acid, 1407
[79-09-4] Propionic Acid, 7939
[79-10-7] Acrylic Acid, 121
[79-11-8] Chloroacetic Acid, 2112
[79-14-1] Glycolic Acid, 4534
[79-15-2] N-Bromoacetamide, 1405
[79-17-4] Aminoguanidine, 438
[79-19-6] Thiosemicarbazide, 9514
[79-20-9] Methyl Acetate, 6082
[79-21-0] Peracetic Acid, 7263
[79-22-1] Methyl Chlorocarbonate, 6114
[79-24-3] Nitroethane, 6684
[79-27-6] sym-Tetrabromoethane, 9330
[79-31-2] Isobutyric Acid, 5200
[79-33-4] L-Lactic Acid, 5385
[79-34-5] Tetrachloroethane, 9334
[79-36-7] 2,2-Dichloroacetyl Chloride, 3062
[79-37-8] Oxalyl Chloride, 7013
[79-40-3] Rubeanic Acid, 8412
[79-41-4] Methacrylic Acid, 6013
[79-42-5] Thiolactic Acid, 9493
[79-43-6] Dichloroacetic Acid, 3059
[79-46-9] 2-Nitropropane, 6715
[79-49-2] Pentabromoacetone, 7215
[79-54-9] Levopimaric Acid, 5522
[79-55-0] Pempidine, 7189
[79-57-2] Oxytetracycline, 7075
[79-58-3] Rubijervine, 8420
[79-61-8] Dichlorisone 21-Acetate, 3057
[79-63-0] Lanosterol, 5411
[79-64-1] Dimethisterone, 3240
[79-68-5] γ-Irone, 5142
[79-69-6] α-Irone, 5140
[79-70-9] β-Irone, 5141
[79-74-3] 2,5-Di-tert-pentylhydroquinone, 3334
[79-77-6] β-Ionone, 5099
[79-80-1] 3-Dehydroretinol, 2879
[79-81-2] Vitamin A Palmitate, 10211
[79-83-4] Pantothenic Acid, 7118
[79-85-6] Quatrimycin, 8144
[79-90-3] Triclobisonium Chloride, 9818
[79-91-4] Terebic Acid, 9304
[79-92-5] Camphene, 1732
[79-94-7] Tetrabromobisphenol A, 9328
[80-00-2] Sulphenone, 9118
[80-03-5] Acediasulfone, 24
[80-05-7] Bisphenol A, 1298
[80-08-0] Dapsone, 2822

[80-11-5] p-Tolylsulfonylmethylnitrosamide, 9702
[80-12-6] Tetramethylenedisulfotetramine, 9374
[80-13-7] Halazone, 4625
[80-17-1] Benzenesulfonyl Hydrazide, 1074
[80-32-0] Sulfachlorpyridazine, 9032
[80-35-3] Sulfamethoxypyridazine, 9051
[80-40-0] Ethyl p-Toluenesulfonate, 3907
[80-46-6] p-tert-Pentylphenol, 7255
[80-49-9] Homatropine Methylbromide, 4768
[80-53-5] Terpin, 9311
[80-56-8] α-Pinene, 7556
[80-58-0] α-Bromobutyric Acid, 1420
[80-59-1] Tiglic Acid, 9589
[80-62-6] Methacrylic Acid Methyl Ester, 6013
[80-69-3] Tartronic Acid, 9207
[80-70-6] 1,1,3,3-Tetramethylguanidine, 9375
[80-72-8] Reductic Acid, 8246
[80-73-9] 1,3-Dimethyl-2-imidazolidinone, 3273
[80-74-0] Acetyl Sulfisoxazole, 9082
[80-77-3] Chlormezanone, 2106
[80-78-4] Solanidine, 8831
[80-92-2] Pregnanediol, 7845
[80-97-7] Cholestanol, 2207
[81-04-9] Armstrong's Acid, 779
[81-05-0] 2-Naphthylamine-5-sulfonic Acid, 6489
[81-07-2] Saccharin, 8445
[81-11-8] Amsonic Acid, 590
[81-13-0] Dexpanthenol, 2949
[81-16-3] 2-Naphthylamine-1-sulfonic Acid, 6488
[81-23-2] Dehydrocholic Acid, 2872
[81-24-3] Taurocholic Acid, 9212
[81-25-4] Cholic Acid, 2210
[81-27-6] Sennoside A, 8594
[81-38-9] Laureline, 5438
[81-54-9] Purpurin, 8056
[81-61-8] Quinalizarin, 8166
[81-64-1] Quinizarin, 8180
[81-67-4] Pukateine, 8045
[81-77-6] Indanthrene, 4974
[81-81-2] Warfarin, 10236
[81-82-3] Coumachlor, 2544
[81-88-9] Rhodamine B, 8307
[81-90-3] Phenolphthalin, 7357
[81-92-5] Phenolphthalol, 7358
[81-93-6] Phenosafranin, 7363
[82-02-0] Khellin, 5356
[82-05-3] Benzanthrone, 1065
[82-08-6] Rottlerin, 8405
[82-12-2] Rufigallol, 8425
[82-22-4] Anthrimide, 681
[82-24-6] 1-Aminoanthraquinone-2-carboxylic Acid, 412
[82-40-6] Lunacrine, 5664
[82-45-1] 1-Aminoanthraquinone, 411
[82-54-2] Cotarnine, 2540
[82-57-5] Visnagin, 10209
[82-58-6] Lysergic Acid, 5694
[82-62-2] 3,4,6-Trichloro-2-nitrophenol, 9807
[82-66-6] Diphacinone, 3335
[82-68-8] Quintozene, 8197
[82-71-3] Styphnic Acid, 8989
[82-75-7] 1-Naphthylamine-8-sulfonic Acid, 6487

[82-76-8] 1-Anilino-8-naphthalenesul-fonate, *654*
[82-82-6] 4-Pyridoxic Acid, *8094*
[82-88-2] Phenindamine, *7347*
[82-89-3] Prodigiosin, *7884*
[82-92-8] Cyclizine, *2703*
[82-93-9] Chlorcyclizine, *2082*
[82-95-1] Buclizine, *1474*
[82-98-4] Piperidolate, *7582*
[82-99-5] Thiphenamil, *9524*
[83-01-2] Diphenylcarbamoyl Chloride, *3351*
[83-07-8] 4-Aminoantipyrine, *413*
[83-12-5] Phenindione, *7348*
[83-14-7] Pellotine, *7183*
[83-25-0] Succinanil, *8997*
[83-26-1] Pindone, *7555*
[83-28-3] 2-Isovalerylindane-1,3-dione, *5279*
[83-32-9] Acenaphthene, *29*
[83-34-1] Skatole, *8698*
[83-40-9] *o*-Cresotic Acid, *2570*
[83-43-2] Methylprednisolone, *6184*
[83-44-3] Deoxycholic Acid, *2901*
[83-45-4] Stigmastanol, *8943*
[83-46-5] β-Sitosterol, *8694*
[83-47-6] γ-Sitosterol, *8695*
[83-48-7] Stigmasterol, *8944*
[83-49-8] Hyodeoxycholic Acid, *4896*
[83-54-5] Echinopsine, *3545*
[83-58-9] Lunacridine, *5663*
[83-60-3] Reserpic Acid, *8263*
[83-67-0] Theobromine, *9433*
[83-70-5] Vitamin K$_5$, *10220*
[83-72-7] Lawsone, *5448*
[83-73-8] Iodoquinol, *5085*
[83-74-9] Ibogaine, *4914*
[83-79-4] Rotenone, *8403*
[83-81-8] *N,N,N',N'*-Tetraethylphthal-amide, *9348*
[83-86-3] Phytic Acid, *7499*
[83-88-5] Riboflavin, *8325*
[83-89-6] Quinacrine, *8160*
[83-94-3] Tabernanthine, *9150*
[83-95-4] Skimmianine, *8699*
[83-98-7] Orphenadrine, *6975*
[84-01-5] Chlorproethazine, *2189*
[84-02-6] Prochlorperazine Dimaleate, *7879*
[84-04-8] Pipamazine, *7567*
[84-06-0] Thiopropazate, *9509*
[84-11-7] Phenanthrenequinone, *7327*
[84-12-8] Phanquinone, *7312*
[84-16-2] Hexestrol, *4737*
[84-17-3] Dienestrol, *3115*
[84-19-5] Dienestrol Diacetate, *3115*
[84-21-9] 3'-Adenylic Acid, *145*
[84-22-0] Tetrahydrozoline, *9364*
[84-24-2] Physodic Acid, *7494*
[84-26-4] Rutecarpine, *8432*
[84-31-1] Quininone, *8179*
[84-36-6] Syrosingopine, *9147*
[84-37-7] Pseudoyohimbine, *8031*
[84-52-6] 3'-Cytidylic Acid, *2785*
[84-54-8] 2-Methylanthraquinone, *6095*
[84-55-9] Viquidil, *10199*
[84-58-2] 2,3-Dichloro-5,6-dicyanoben-zoquinone, *3072*
[84-62-8] Diphenyl Phthalate, *3368*
[84-65-1] Anthraquinone, *679*
[84-66-2] Phthalic Acid Ethyl Ester, *7483*
[84-68-4] 2,2'-Dichlorobenzidine, *3067*
[84-74-2] Dibutyl Phthalate, *3044*

[84-79-7] Lapachol, *5417*
[84-80-0] Phylloquinone, *7492*
[84-81-1] Menaquinone 6, *5901*
[84-82-2] Toxoflavin, *9718*
[84-86-6] 1-Naphthylamine-4-sulfonic Acid, *6485*
[84-87-7] 1-Naphthol-4-sulfonic Acid, *6476*
[84-88-8] 8-Hydroxy-5-quinolinesul-fonic Acid, *4882*
[84-89-9] 1-Naphthylamine-5-sulfonic Acid, *6486*
[84-96-8] Trimeprazine, *9873*
[84-97-9] Perazine, *7265*
[84-99-1] Xanthoxyletin, *10264*
[85-00-7] Diquat Dibromide, *3392*
[85-01-8] Phenanthrene, *7326*
[85-02-9] Benzo[*f*]quinoline, *1108*
[85-17-6] Streptidine, *8952*
[85-31-4] Thioguanosine, *9491*
[85-32-5] 5'-Guanylic Acid, *4605*
[85-38-1] 3-Nitrosalicylic Acid, *6717*
[85-41-6] Phthalimide, *7485*
[85-44-9] Phthalic Anhydride, *7484*
[85-47-2] 1-Naphthalenesulfonic Acid, *6461*
[85-55-2] 2-(*p*-Toluyl)benzoic Acid, *10002*
[85-61-0] Coenzyme A, *2450*
[85-64-3] Laudanine, *5435*
[85-66-5] Pyocyanine, *8062*
[85-73-4] Phthalylsulfathiazole, *7489*
[85-79-0] Dibucaine, *3036*
[85-82-5] 1-Xylylazo-2-naphthol, *10287*
[85-83-6] Scarlet Red, *8530*
[85-84-7] Yellow AB, *10293*
[85-85-8] PAN, *7107*
[85-86-9] Sudan III, *9015*
[85-87-0] Pyridoxamine, *8093*
[85-90-5] Tricromyl, *9823*
[85-94-9] 2'-Cytidylic Acid, *2784*
[85-97-2] 2-Phenyl-6-chlorophenol, *7391*
[85-98-3] *N,N'*-Diethylcarbanilide, *3129*
[86-00-0] *o*-Nitrobiphenyl, *6679*
[86-12-4] Thenaldine, *9429*
[86-13-5] Benztropine, *1124*
[86-14-6] Thiambutene, *9445*
[86-21-5] Pheniramine, *7349*
[86-22-6] Brompheniramine, *1453*
[86-29-3] Diphenylacetic Acid Nitrile, *3346*
[86-34-0] Phensuximide, *7373*
[86-35-1] Ethotoin, *3802*
[86-36-2] Mercumallylic Acid, *5939*
[86-40-8] Acriflavine, *115*
[86-42-0] Amodiaquin, *569*
[86-48-6] 1-Hydroxy-2-naphthoic Acid, *4873*
[86-50-0] Azinphos-methyl, *906*
[86-54-4] Hydralazine, *4802*
[86-55-5] 1-Naphthoic Acid, *6466*
[86-56-6] *N,N*-Dimethyl-1-naphthyl-amine, *3276*
[86-57-7] 1-Nitronaphthalene, *6701*
[86-58-8] 8-Quinolineboronic Acid, *8186*
[86-59-9] 8-Quinolinecarboxylic Acid, *8187*
[86-60-2] Badische Acid, *933*
[86-65-7] Amido-G-Acid, *395*
[86-68-0] Quininic Acid, *8178*

[86-73-7] 9*H*-Fluorene, *4187*
[86-74-8] Carbazole, *1791*
[86-84-0] 1-Naphthylisocyanate, *6492*
[86-87-3] 1-Naphthaleneacetic Acid, *6456*
[86-88-4] ANTU, *708*
[87-00-3] Homatropine, *4768*
[87-08-1] Penicillin V, *7209*
[87-09-2] Penicillin O, *7208*
[87-10-5] Tribromsalan, *9778*
[87-11-6] Thiolutin, *9495*
[87-17-2] Salicylanilide, *8467*
[87-18-3] 4-*tert*-Butylphenyl Salicylate, *1589*
[87-20-7] Isoamyl Salicylate, *5170*
[87-28-5] Glycol Salicylate, *4535*
[87-29-6] Cinnamyl Anthranilate, *2304*
[87-33-2] Isosorbide Dinitrate, *5271*
[87-39-8] Violuric Acid, *10196*
[87-40-1] 2,4,6-Trichloroanisole, *9794*
[87-42-3] 6-Chloropurine, *2163*
[87-44-5] Caryophyllene, *1875*
[87-45-6] 4-(5-Isopropenyl-2-methyl-1-cyclopenten-1-yl)-2-butanone, *5250*
[87-47-8] Pyrolan, *8114*
[87-51-4] Indoleacetic Acid, *5004*
[87-52-5] Gramine, *4569*
[87-53-6] Penicillanic Acid, *7200*
[87-56-9] Mucochloric Acid, *6383*
[87-58-1] Iodopyrrole, *5084*
[87-59-2] 2,3-Dimethylbenzenamine, *10282*
[87-61-6] 1,2,3-Trichlorobenzene, *9795*
[87-62-7] 2,6-Dimethylbenzenamine, *10282*
[87-65-0] 2,6-Dichlorophenol, *3083*
[87-66-1] Pyrogallol, *8112*
[87-67-2] Choline Bitartrate, *2211*
[87-68-3] Hexachlorobutadiene, *4714*
[87-69-4] L-Tartaric Acid, *9205*
[87-73-0] D-Glucaric Acid, *4485*
[87-74-1] Glucoheptonic Acid, *4491*
[87-79-6] Sorbose, *8852*
[87-81-0] D-Tagatose, *9161*
[87-86-5] Pentachlorophenol, *7218*
[87-88-7] Chloranilic Acid, *2079*
[87-89-8] Inositol, *5021*
[87-90-1] Trichloroisocyanuric Acid, *9806*
[87-91-2] Diethyl Tartrate, *3150*
[87-99-0] Xylitol, *10283*
[88-04-0] Chloroxylenol, *2182*
[88-06-2] 2,4,6-Trichlorophenol, *9809*
[88-09-5] Diethylacetic Acid, *3121*
[88-12-0] Povidone Monomer, *7814*
[88-13-1] 3-Thenoic Acid, *9430*
[88-14-2] 2-Furoic Acid, *4336*
[88-21-1] Orthanilic Acid, *6978*
[88-29-9] Versalide®, *10161*
[88-30-2] TFM, *9401*
[88-67-5] *o*-Iodobenzoic Acid, *5074*
[88-72-2] *o*-Nitrotoluene, *6737*
[88-73-3] *o*-Chloronitrobenzene, *2153*
[88-74-4] *o*-Nitroaniline, *6669*
[88-75-5] *o*-Nitrophenol, *6707*
[88-85-7] Dinoseb, *3315*
[88-88-0] Picryl Chloride, *7529*
[88-89-1] Picric Acid, *7522*
[88-95-9] Phthaloyl Chloride, *7487*
[88-96-0] Phthalamide, *7481*
[88-99-3] Phthalic Acid, *7483*
[89-00-9] Quinolinic Acid, *8190*
[89-01-0] 2,3-Pyrazinedicarboxylic Acid, *8069*

[108-87-2] Methylcyclohexane, *6118*
[108-88-3] Toluene, *9688*
[108-89-4] γ-Picoline, *7514*
[108-90-7] Chlorobenzene, *2122*
[108-91-8] Cyclohexylamine, *2722*
[108-93-0] Cyclohexanol, *2718*
[108-94-1] Cyclohexanone, *2719*
[108-95-2] Phenol, *7354*
[108-98-5] Thiophenol, *9508*
[108-99-6] β-Picoline, *7513*
[109-02-4] *N*-Methylmorpholine, *6362*
[109-06-8] α-Picoline, *7512*
[109-21-7] *n*-Butyl *n*-Butyrate, *1558*
[109-52-4] *n*-Valeric Acid, *10090*
[109-57-9] Thiosinamine, *9515*
[109-60-4] Propyl Acetate, *7954*
[109-61-5] Propyl Chloroformate, *7961*
[109-63-7] Boron Trifluoride Etherate, *1352*
[109-64-8] Trimethylene Bromide, *9884*
[109-65-9] *n*-Butyl Bromide, *1555*
[109-66-0] Pentane, *7228*
[109-67-1] 1-Pentene, *7234*
[109-68-2] 2-Pentene, *7235*
[109-69-3] *n*-Butyl Chloride, *1561*
[109-73-9] *n*-Butylamine, *1545*
[109-74-0] Butyronitrile, *1601*
[109-75-1] Allyl Cyanide, *281*
[109-77-3] Malononitrile, *5775*
[109-78-4] Ethylene Cyanohydrin, *3848*
[109-79-5] *n*-Butyl Mercaptan, *1580*
[109-80-8] 1,3-Propanedithiol, *7915*
[109-82-0] 2-(Methyleneamino)acetonitrile, *6129*
[109-83-1] 2-(Methylamino)ethanol, *6089*
[109-86-4] 2-Methoxyethanol, *6063*
[109-87-5] Methylal, *6086*
[109-89-7] Diethylamine, *3122*
[109-93-3] Divinyl Ether, *3427*
[109-94-4] Ethyl Formate, *3862*
[109-95-5] Ethyl Nitrite, *3885*
[109-96-6] 3-Pyrroline, *8131*
[109-97-7] 1*H*-Pyrrole, *8128*
[109-98-8] 2-Pyrazoline, *8072*
[109-99-9] Tetrahydrofuran, *9356*
[110-00-9] Furan, *4325*
[110-01-0] Tetrahydrothiophene, *9363*
[110-02-1] Thiophene, *9506*
[110-04-3] 1,2-Ethanedisulfonic Acid, *3779*
[110-05-4] DTBP, *3508*
[110-13-4] Acetonylacetone, *70*
[110-14-5] Succinamide, *8996*
[110-15-6] Succinic Acid, *8999*
[110-16-7] Maleic Acid, *5767*
[110-17-8] Fumaric Acid, *4316*
[110-18-9] TEMED, *9277*
[110-19-0] Isobutyl Acetate, *5175*
[110-27-0] Isopropyl Myristate, *5261*
[110-38-3] Ethyl Caprate, *3831*
[110-40-7] Sebacic Acid Diethyl Ester, *8552*
[110-43-0] 2-Heptanone, *4698*
[110-44-1] Sorbic Acid, *8847*
[110-45-2] Isoamyl Formate, *5164*
[110-46-3] Isoamyl Nitrite, *5168*
[110-49-6] 2-Methoxyethyl Acetate, *6064*
[110-53-2] *n*-Amyl Bromide, *596*
[110-54-3] *n*-Hexane, *4732*
[110-58-7] *n*-Amylamine, *593*
[110-59-8] Valeronitrile, *10091*

[110-60-1] Putrescine, *8058*
[110-61-2] Succinonitrile, *9003*
[110-62-3] *n*-Valeraldehyde, *10088*
[110-63-4] 1,4-Butanediol, *1518*
[110-66-7] *n*-Amyl Mercaptan, *604*
[110-71-4] 1,2-Dimethoxyethane, *3245*
[110-74-7] Propyl Formate, *7970*
[110-75-8] 2-Chloroethyl Vinyl Ether, *2140*
[110-77-0] 2-(Ethylthio)ethanol, *3906*
[110-80-5] 2-Ethoxyethanol, *3803*
[110-82-7] Cyclohexane, *2716*
[110-83-8] Cyclohexene, *2720*
[110-85-0] Piperazine, *7576*
[110-86-1] Pyridine, *8081*
[110-87-2] Dihydropyran, *3195*
[110-88-3] *s*-Trioxane, *9906*
[110-89-4] Piperidine, *7580*
[110-91-8] Morpholine, *6362*
[110-94-1] Glutaric Acid, *4509*
[110-97-4] Diisopropanolamine, *3214*
[111-01-3] Squalane, *8895*
[111-02-4] Squalene, *8896*
[111-13-7] Hexyl Methyl Ketone, *4747*
[111-14-8] Heptanoic Acid, *4695*
[111-15-9] 2-Ethoxyethyl Acetate, *3804*
[111-16-0] Pimelic Acid, *7543*
[111-17-1] 3,3′-Thiodipropionic Acid, *9485*
[111-20-6] Sebacic Acid, *8552*
[111-27-3] 1-Hexanol, *4734*
[111-28-4] Sorbic Alcohol (unspecified stereo), *8848*
[111-29-5] 1,5-Pentanediol, *7229*
[111-30-8] Glutaraldehyde, *4508*
[111-36-4] Butyl Isocyanate, *1578*
[111-42-2] Diethanolamine, *3118*
[111-43-3] Dipropyl Ether, *3385*
[111-44-4] *sym*-Dichloroethyl Ether, *3075*
[111-46-6] Diethylene Glycol, *3131*
[111-47-7] Dipropyl Sulfide, *3387*
[111-48-8] 2,2′-Thiodiethanol, *9483*
[111-51-3] Tetramethyldiaminobutane, *9373*
[111-55-7] Ethylene Glycol Diacetate, *3853*
[111-61-5] Stearic Acid Ethyl Ester, *8930*
[111-65-9] Octane, *6838*
[111-66-0] Caprylene, *1766*
[111-69-3] Adiponitrile, *152*
[111-70-6] 1-Heptanol, *4696*
[111-71-7] Heptanal, *4693*
[111-76-2] Butyl Cellosolve®, *1560*
[111-77-3] Diethylene Glycol Monomethyl Ether, *3136*
[111-78-4] 1,5-Cyclooctadiene, *2730*
[111-81-9] Undecylenic Acid Methyl Ester, *10033*
[111-83-1] *n*-Octyl Bromide, *6853*
[111-87-5] 1-Octanol, *6840*
[111-90-0] Diethylene Glycol Monoethyl Ether, *3134*
[111-92-2] *n*-Dibutylamine, *3037*
[111-96-6] Diglyme, *3184*
[112-05-0] Pelargonic Acid, *7179*
[112-12-9] Methyl Nonyl Ketone, *6177*
[112-15-2] Diethylene Glycol Monoethyl Ether Acetate, *3134*
[112-24-3] Trientine, *9828*
[112-27-6] Triethylene Glycol, *9835*
[112-30-1] *n*-Decyl Alcohol, *2858*
[112-34-5] Diethylene Glycol Monobutyl Ether, *3133*

[112-36-7] Diethylene Glycol Diethyl Ether, *3132*
[112-38-9] Undecylenic Acid, *10033*
[112-47-0] Decamethylene Glycol, *2853*
[112-49-2] Triglyme, *9861*
[112-53-8] 1-Dodecanol, *3451*
[112-61-8] Stearic Acid Methyl Ester, *8930*
[112-63-0] Methyl Linoleate, *6167*
[112-72-1] Myristyl Alcohol, *6421*
[112-79-8] Elaidic Acid, *3581*
[112-80-1] Oleic Acid, *6921*
[112-85-6] Behenic Acid, *1022*
[112-86-7] Erucic Acid, *3732*
[112-92-5] Stearyl Alcohol, *8932*
[113-00-8] Guanidine, *4597*
[113-15-5] Ergotamine, *3718*
[113-18-8] Ethchlorvynol, *3784*
[113-22-4] Estriol 16,17-Bis(sodium Hemisuccinate), *3762*
[113-38-2] Estradiol Dipropionate, *3758*
[113-42-8] Methylergonovine, *6142*
[113-45-1] Methylphenidate, *6183*
[113-52-0] Imipramine Hydrochloride, *4962*
[113-53-1] Dothiepin, *3478*
[113-59-7] Chlorprothixene, *2194*
[113-73-5] Gramicidin S, *4567*
[113-78-0] Deaminooxytocin, *2844*
[113-79-1] Arginine Vasopressin, *10129*
[113-80-4] Arginine Vasotocin, *10129*
[113-92-8] Chlorpheniramine Maleate, *2186*
[113-98-4] Penicillin G Potassium Salt, *7203*
[114-04-5] Neuraminic Acid, *6568*
[114-07-8] Erythromycin, *3739*
[114-21-6] Mandelic Acid Sodium Salt, *5781*
[114-25-0] Biliverdine, *1223*
[114-26-1] Propoxur, *7949*
[114-42-1] Flavaspidic Acid, *4121*
[114-43-2] Desaspidin BB, *2918*
[114-49-8] Scopolamine Hydrobromide Trihydrate, *8543*
[114-80-7] Neostigmine Bromide, *6549*
[114-85-2] Bethanidine Sulfate, *1188*
[114-86-3] Phenformin, *7345*
[114-90-9] Obidoxime Chloride, *6827*
[114-91-0] Metyridine, *6239*
[115-02-6] Azaserine, *892*
[115-07-1] Propylene, *7962*
[115-08-2] Thioformamide, *9486*
[115-10-6] Dimethyl Ether, *3266*
[115-11-7] Isobutylene, *5186*
[115-17-3] Bromal, *1390*
[115-19-5] 2-Methyl-3-butyn-2-ol, *6108*
[115-20-8] 2,2,2-Trichloroethanol, *9803*
[115-24-2] Sulfonmethane, *9090*
[115-25-3] Octafluorocyclobutane, *6834*
[115-26-4] Dimefox, *3228*
[115-27-5] Chlorendic Anhydride, *2087*
[115-29-7] Endosulfan, *3629*
[115-31-1] Isobornyl Thiocyanoacetate, *5174*
[115-32-2] Dicofol, *3096*
[115-33-3] Oxyphenisatin Acetate, *7073*
[115-37-7] Thebaine, *9427*
[115-38-8] Mephobarbital, *5921*
[115-39-9] Bromphenol Blue, *1454*

[123-47-7] Prolonium Iodide, 7897
[123-51-3] Isopentyl Alcohol, 5241
[123-54-6] Acetylacetone, 76
[123-56-8] Succinimide, 9001
[123-62-6] Propionic Anhydride, 7940
[123-63-7] Paraldehyde, 7131
[123-66-0] Ethyl Caproate, 3832
[123-72-8] Butyraldehyde, 1595
[123-73-9] Crotonaldehyde (E-form), 2587
[123-75-1] Pyrrolidine, 8129
[123-76-2] Levulinic Acid, 5526
[123-77-3] Azodicarbonamide, 912
[123-78-4] Sphingosine, 8874
[123-82-0] Tuaminoheptane, 9982
[123-84-2] 1-[(2-Aminoethyl)amino]-2-propanol, 435
[123-86-4] n-Butyl Acetate, 1537
[123-90-0] Thiamorpholine, 9452
[123-91-1] Dioxane, 3328
[123-92-2] Isoamyl Acetate, 5156
[123-93-3] Thiodiglycolic Acid, 9484
[123-95-5] Butyl Stearate, 1592
[123-96-6] 2-Octanol, 6841
[123-99-9] Azelaic Acid, 898
[124-02-7] Diallylamine, 2974
[124-03-8] Cetyldimethylethylammonium Bromide, 2028
[124-04-9] Adipic Acid, 150
[124-06-1] Myristic Acid Ethyl Ester, 6419
[124-07-2] Caprylic Acid, 1767
[124-09-4] Hexamethylenediamine, 4727
[124-13-0] Caprylic Aldehyde, 1768
[124-20-9] Spermidine, 8869
[124-28-7] Dymanthine, 3522
[124-38-9] Carbon Dioxide, 1809
[124-40-3] Dimethylamine, 3250
[124-41-4] Sodium Methoxide, 8775
[124-42-5] Acetamidine Hydrochloride, 45
[124-43-6] Carbamide Peroxide, 1784
[124-47-0] Urea Nitrate, 10054
[124-48-1] Chlorodibromomethane, 2135
[124-58-3] Methanearsonic Acid, 6025
[124-63-0] Methanesulfonyl Chloride, 6027
[124-65-2] Sodium Cacodylate, 8730
[124-68-5] 2-Amino-2-methyl-1-propanol, 446
[124-72-1] Teflurane, 9250
[124-76-5] Isoborneol, 5173
[124-83-4] Camphoric Acid d-Form, 1735
[124-87-8] Picrotoxin, 7527
[124-90-3] Oxycodone Hydrochloride, 7058
[124-92-5] Metopon Hydrochloride, 6227
[124-94-7] Triamcinolone, 9757
[124-97-0] Protoveratrine B, 8008
[124-98-1] Cevine, 2034
[124-99-2] Scillaren A, 8536
[125-02-0] Prednisolone 21-Disodium Phosphate, 7841
[125-04-2] Hydrocortisone 21-Sodium Succinate, 4824
[125-10-0] Prednisone 21-Acetate, 7842
[125-15-5] Vomicine, 10228
[125-20-2] Thymolphthalein, 9556
[125-24-6] Pseudomorphine, 8027
[125-28-0] Dihydrocodeine, 3189

[125-29-1] Hydrocodone, 4823
[125-30-4] Ethylmorphine Hydrochloride, 3884
[125-31-5] Xylenol Blue, 10280
[125-33-7] Primidone, 7865
[125-40-6] Butabarbital, 1509
[125-46-2] Usnic Acid, 10078
[125-51-9] Pipenzolate Bromide, 7573
[125-52-0] Oxyphencyclimine Hydrochloride, 7072
[125-53-1] Oxyphencyclimine, 7072
[125-64-4] Methyprylon, 6212
[125-65-5] Pleuromutilin, 7650
[125-68-8] Levomethorphan l-Form Hydrobromide, 5520
[125-69-9] Dextromethorphan Hydrobromide, 2957
[125-70-2] Levomethorphan, 5520
[125-71-3] Dextromethorphan, 2957
[125-73-5] Levorphanol d-Form, 5524
[125-84-8] Aminoglutethimide, 437
[125-85-9] Caramiphen Hydrochloride, 1778
[125-86-0] Caramiphen Ethanedisulfonate, 1778
[125-88-2] Aprobarbital Sodium Salt, 745
[125-99-5] Tridihexethyl Iodide, 9826
[126-00-1] Diphenolic Acid, 3342
[126-07-8] Griseofulvin, 4584
[126-11-4] Tris(hydroxymethyl)nitromethane, 9929
[126-12-5] Anileridine Dihydrochloride, 651
[126-14-7] Sucrose Octaacetate, 9013
[126-15-8] R-11, 8205
[126-17-0] Solasodine, 8836
[126-18-1] Smilagenin, 8703
[126-19-2] Sarsasapogenin, 8517
[126-22-7] Butanone, 9788
[126-27-2] Oxethazaine, 7032
[126-29-4] Violaxanthin, 10194
[126-30-7] Neopentyl Glycol, 6543
[126-31-8] Methiodal Sodium, 6045
[126-33-0] Sulfolane, 9085
[126-45-4] Silver Citrate, 8650
[126-49-8] Prephenic Acid, 7856
[126-72-7] Tris-BP, 9927
[126-73-8] Tributyl Phosphate, 9781
[126-81-8] 5,5-Dimethyl-1,3-cyclohexanedione, 3264
[126-85-2] Mechlorethamine Oxide, 5847
[126-90-9] Linalool S-(+)-Form, 5550
[126-91-0] Linalool R-(−)-Form, 5550
[126-96-5] Sodium Diacetate, 8743
[126-98-7] Methacrylonitrile, 6014
[127-00-4] sec-Propylene Chlorohydrin, 7964
[127-06-0] Acetoxime, 73
[127-07-1] Hydroxyurea, 4888
[127-08-2] Potassium Acetate, 7726
[127-09-3] Sodium Acetate, 8709
[127-17-3] Pyruvic Acid, 8135
[127-18-4] Tetrachloroethylene, 9335
[127-19-5] N,N-Dimethylacetamide, 3248
[127-20-8] Dalapon Sodium Salt, 2799
[127-22-0] Taraxerol, 9195
[127-25-3] Methyl Abietate, 6080
[127-27-5] Pimaric Acid, 7540
[127-31-1] Fludrocortisone, 4161
[127-33-3] Demeclocycline, 2890

[127-35-5] Phenazocine, 7332
[127-39-9] Diisobutyl Sodium Sulfosuccinate, 3213
[127-40-2] Xanthophyll, 10261
[127-41-3] α-Ionone, 5099
[127-47-9] Vitamin A Acetate, 10211
[127-48-0] Trimethadione, 9875
[127-52-6] Chloramine-B, 2074
[127-56-0] Sulfacetamide Sodium Salt, 9031
[127-57-1] Sulfapyridine Sodium Salt Monohydrate, 9069
[127-58-2] Sulfamerazine Monosodium Salt, 9045
[127-60-6] Acediasulfone Sodium Salt, 24
[127-63-9] Diphenyl Sulfone, 3372
[127-65-1] Chloramine-T, 2075
[127-69-5] Sulfisoxazole, 9082
[127-71-9] Sulfabenzamide, 9029
[127-77-5] Sulfabenz, 9028
[127-79-7] Sulfamerazine, 9045
[127-82-2] Zinc p-Phenolsulfonate, 10350
[127-83-3] Calcium Phenolsulfonate, 1692
[127-85-5] Arsanilic Acid Sodium Salt, 783
[127-91-3] β-Pinene, 7557
[127-95-7] Potassium Binoxalate, 7732
[127-96-8] Potassium Tetroxalate, 7806
[128-08-5] N-Bromosuccinimide, 1448
[128-09-6] N-Chlorosuccinimide, 2167
[128-13-2] Ursodiol, 10074
[128-19-8] 4-Pregnene-17α,20β,21-triol-3-one, 7850
[128-20-1] Pregnan-3α-ol-20-one, 7847
[128-23-4] 3,20-Pregnanedione, 7846
[128-27-8] Pyrocalciferol, 8110
[128-37-0] Butylated Hydroxytoluene, 1550
[128-46-1] Dihydrostreptomycin, 3197
[128-49-4] Docusate Calcium, 3445
[128-50-7] Nopol ((±)-form), 6771
[128-53-0] N-Ethylmaleimide, 3877
[128-57-4] Sennoside B, 8594
[128-62-1] Noscapine, 6807
[128-68-7] Phenicin, 7346
[128-71-2] Phenoltetrachlorophthalein Disodium Salt, 7361
[128-76-7] Laurotetanine, 5443
[128-80-3] Quinizarin Green SS, 8181
[128-87-0] Bis(4-amino-1-anthraquinonyl)amine, 1247
[129-00-0] Pyrene, 8074
[129-03-3] Cyproheptadine, 2770
[129-06-6] Warfarin Sodium Salt, 10236
[129-16-8] Merbromin, 5934
[129-17-9] Sulphan Blue, 9117
[129-18-0] Phenylbutazone Sodium Salt, 7390
[129-20-4] Oxyphenbutazone, 7071
[129-24-8] Viridicatin, 10201
[129-46-4] Suramin Sodium, 9135
[129-49-7] Methysergide Hydrogen Maleate, 6213
[129-51-1] Ergonovine Maleate, 3712
[129-64-6] Carbic Anhydride, 1797
[129-66-8] 2,4,6-Trinitrobenzoic Acid, 9899
[129-73-7] Leucomalachite Green, 5763
[129-74-8] Buclizine Dihydrochloride, 1474

[129-77-1] Piperidolate Hydrochloride, 7582

[129-79-3] 2,4,7-Trinitrofluorenone, 9900

[129-83-9] Phenampromide, 7323

[130-01-8] Senecionine, 8590

[130-15-4] 1,4-Naphthoquinone, 6480

[130-16-5] Cloxyquin, 2405

[130-22-3] Sodium Alizarinesulfonate, 8711

[130-24-5] Vitamin K₅ Hydrochloride, 10220

[130-26-7] Iodochlorhydroxyquin, 5075

[130-37-0] Menadione Sodium Bisulfite, 5899

[130-40-5] Flavin Mononucleotide Monosodium Salt, 4123

[130-61-0] Thioridazine Hydrochloride, 9512

[130-73-4] Methestrol, 6040

[130-79-0] Dimestrol, 3234

[130-80-3] Diethylstilbestrol Dipropionate, 3148

[130-85-8] Pamoic Acid, 7106

[130-86-9] Protopine, 8005

[130-89-2] Quinine Hydrochloride, 8177

[130-95-0] Quinine, 8177

[131-01-1] Deserpidine, 2919

[131-02-2] Reserpiline, 8264

[131-03-3] α-Yohimbine, 10301

[131-11-3] Dimethyl Phthalate, 3279

[131-12-4] Pimpinellin, 7548

[131-13-5] Menadiol Sodium Diphosphate, 5898

[131-28-2] Narceine, 6504

[131-48-6] N-Acetylneuraminic Acid, 8623

[131-49-7] Diatrizoic Acid Meglumine Salt, 2993

[131-52-2] Pentachlorophenol Sodium Salt, 7218

[131-53-3] Dioxybenzone, 3332

[131-54-4] Benzophenone-6, 1101

[131-56-6] Benzoresorcinol, 1109

[131-57-7] Oxybenzone, 7053

[131-67-9] Phthalofyne, 7486

[131-69-1] Phthalylsulfacetamide, 7488

[131-73-7] Dipicrylamine, 3375

[131-74-8] Ammonium Picrate, 544

[131-79-3] Yellow OB, 10294

[131-89-5] 2-Cyclohexyl-4,6-dinitrophenol, 2726

[131-91-9] 1-Nitroso-2-naphthol, 6728

[131-99-7] Inosinic Acid, 5020

[132-13-8] α-Ribazole, 8324

[132-17-2] Benztropine Methanesulfonate, 1124

[132-18-3] Diphenylpyraline Hydrochloride, 3370

[132-20-7] Pheniramine Maleate, 7349

[132-21-8] Brompheniramine d-Form, 1453

[132-22-9] Chlorpheniramine, 2186

[132-27-4] o-Phenylphenol Sodium Salt, 7415

[132-35-4] Proxazole Citrate, 8012

[132-40-1] Papaverine Nitrite, 7122

[132-49-0] Magnesium Acetylsalicylate, 5719

[132-57-0] Croceic Acid, 2578

[132-58-1] Cinchophen Hydrochloride, 2291

[132-60-5] Cinchophen, 2291

[132-64-9] Dibenzofuran, 3013

[132-66-1] Naptalam, 6500

[132-69-4] Benzydamine Hydrochloride, 1125

[132-73-0] Chloroquine Sulfate, 2165

[132-86-5] Naphthoresorcinol, 6481

[132-92-3] Methicillin Sodium, 6041

[132-93-4] Phenethicillin Potassium, 7337

[132-98-9] Penicillin V Potassium Salt, 7209

[133-04-0] Asarinin, 812

[133-06-2] Captan, 1774

[133-07-3] Folpet, 4251

[133-08-4] n-Butylmalonic Acid Diethyl Ester, 1579

[133-09-5] p-Aminosalicylic Acid Potassium Salt, 473

[133-11-9] Phenyl Aminosalicylate, 7383

[133-15-3] p-Aminosalicylic Acid Calcium Salt, 473

[133-16-4] Chloroprocaine, 2160

[133-17-5] o-Iodohippurate Sodium, 5077

[133-26-6] Peucedanin, 7306

[133-32-4] Indolebutyric Acid, 5005

[133-36-8] Piperazine Tartrate, 7576

[133-37-9] DL-Tartaric Acid, 9204

[133-51-7] N-Methylglucamine Antimonate, 6150

[133-53-9] Dichloroxylenol, 3086

[133-58-4] Nitromersol, 6697

[133-67-5] Trichlormethiazide, 9789

[133-89-1] Uridine Diphosphate Glucose, 10064

[133-91-5] 3,5-Diiodosalicylic Acid, 3208

[134-01-0] Peonidin, 7256

[134-03-2] Sodium Ascorbate, 819

[134-04-3] Pelargonidin, 7180

[134-09-8] Menthyl Anthranilate, 5908

[134-20-3] Methyl Anthranilate, 6094

[134-31-6] 8-Hydroxyquinoline Sulfate, 4881

[134-32-7] 1-Naphthylamine, 6483

[134-36-1] Erythromycin Propionate, 3742

[134-49-6] Phenmetrazine, 7351

[134-50-9] Aminacrine Hydrochloride, 402

[134-58-7] 8-Azaguanine, 889

[134-62-3] Deet, 2859

[134-63-4] Lobeline Hydrochloride, 5609

[134-64-5] Lobeline Sulfate, 5609

[134-65-6] Lobeline (±)-Form, 5609

[134-71-4] Ephedrine dl-Form Hydrochloride, 3663

[134-72-5] Ephedrine Sulfate, 3663

[134-80-5] Diethylpropion Hydrochloride, 3145

[134-81-6] Benzil, 1080

[134-95-2] Mandelic Acid Calcium Salt, 5781

[134-96-3] Syringaldehyde, 9145

[135-07-9] Methyclothiazide, 6079

[135-09-1] Hydroflumethiazide, 4826

[135-19-3] 2-Naphthol, 6469

[135-20-6] Cupferron, 2612

[135-23-9] Methapyrilene Hydrochloride, 6031

[135-31-9] Pyrrobutamine Diphosphate, 8126

[135-43-3] Lauroguadine, 5441

[135-48-8] Pentacene, 7216

[135-52-4] Zincon, 10345

[135-58-0] Mesulfen, 5988

[135-67-1] Phenoxazine, 7365

[135-68-2] 4-Amino-4'-chlorodiphenyl, 430

[135-97-7] Pseudotropine, 8030

[135-98-8] sec-Butylbenzene, 1552

[136-25-4] Erbon, 2799

[136-35-6] Diazoaminobenzene, 3001

[136-40-3] Phenazopyridine Hydrochloride, 7324

[136-44-7] Glyceryl p-Aminobenzoate, 4523

[136-47-0] Tetracaine Hydrochloride, 9333

[136-55-0] Procaine Butyrate, 7875

[136-60-7] n-Butyl Benzoate, 1554

[136-69-6] Protokylol Hydrochloride, 8004

[136-70-9] Protokylol, 8004

[136-72-1] Piperinic Acid, 7579

[136-77-6] 4-Hexylresorcinol, 4748

[136-79-8] 5-Nitro-o-phenetidine, 6704

[136-82-3] Piperocaine, 7587

[136-95-8] 2-Aminobenzothiazole, 419

[137-00-8] 4-Methyl-5-thiazoleethanol, 6199

[137-05-3] Methyl Cyanoacrylate, 6117

[137-06-4] o-Thiocresol, 9478

[137-08-6] Pantothenic Acid Calcium Salt, 7118

[137-16-6] Gardol®, 4401

[137-26-8] Thiram, 9525

[137-30-4] Ziram, 10372

[137-32-6] 2-Methyl-1-butanol, 6104

[137-40-6] Sodium Propionate, 8798

[137-42-8] Metham Sodium, 6021

[137-53-1] Dextrothyroxine Sodium, 9570

[137-58-6] Lidocaine, 5535

[137-76-8] Cetotiamine, 2022

[137-86-0] Octotiamine, 6849

[138-14-7] Deferoxamine Methanesulfonate, 2863

[138-15-8] Glutamic Acid Hydrochloride, 4505

[138-32-9] Cetrimonium Tosylate, 2024

[138-37-4] Mafenide Hydrochloride, 5712

[138-39-6] Mafenide, 5712

[138-41-0] Carzenide, 1876

[138-52-3] Salicin, 8460

[138-53-4] Mandelonitrile Glucoside, 5784

[138-55-6] Picrocrocin, 7523

[138-56-7] Trimethobenzamide, 9877

[138-59-0] Shikimic Acid, 8619

[138-61-4] Nordefrin Hydrochloride (unspecified stereo), 6781

[138-84-1] Potassium Aminobenzoate, 7727

[138-86-3] Limonene, 5546

[138-89-6] p-Nitroso-N,N-dimethylaniline, 6725

[138-91-0] Perillaldehyde Oxime, 7282

[138-92-1] Betazole Dihydrochloride, 1185

[139-02-6] Phenol Sodium Salt, 7354

[139-06-0] Calcium Cyclamate, 1667

[139-10-6] Amphetamine Phosphate, 579

[139-12-8] Aluminum Acetate, 320

[139-13-9] Nitrilotriacetic Acid, *6665*
[139-33-3] EDTA Disodium Salt, *3565*
[139-40-2] Propazine, *7926*
[139-42-4] Cerous Oxalate, *2000*
[139-66-2] Diphenyl Sulfide, *3371*
[139-68-4] Phenatine, *7330*
[139-85-5] Protocatechualdehyde, *8002*
[139-88-8] Sodium Tetradecyl Sulfate, *8817*
[139-89-9] Versenol®, *10162*
[139-91-3] Furaltadone, *4323*
[139-93-5] Arsphenamine, *801*
[139-94-6] Nithiazide, *6655*
[140-11-4] Benzyl Acetate, *1126*
[140-22-7] *sym*-Diphenylcarbazide, *3352*
[140-28-3] Benzathine, *1066*
[140-29-4] Benzyl Cyanide, *1134*
[140-40-9] Aminitrozole, *404*
[140-55-6] Pseudoconhydrine, *8022*
[140-59-0] Stilbamidine Isethionate, *8945*
[140-63-6] Propamidine Isethionate, *7911*
[140-64-7] Pentamidine Isethionate, *7227*
[140-65-8] Pramoxine, *7829*
[140-67-0] Estragole, *3760*
[140-80-7] Novoldiamine, *6812*
[140-87-4] Cyacetacide, *2670*
[140-88-5] Ethyl Acrylate, *3813*
[140-89-6] Potassium Xanthogenate, *7813*
[140-93-2] Sodium Isopropyl Xanthate, *8766*
[140-95-4] Oxymethurea, *7067*
[140-99-8] Calcium Succinate, *1707*
[141-01-5] Ferrous Fumarate, *4073*
[141-05-9] Diethyl Maleate, *3139*
[141-10-6] Pseudoionone, *8025*
[141-20-8] Diethylene Glycol Monolaurate, *3135*
[141-22-0] Ricinoleic Acid, *8335*
[141-26-4] α-Citronellal, *2328*
[141-27-5] Geranial, *2321*
[141-32-2] *n*-Butyl Acrylate, *1541*
[141-43-5] Ethanolamine, *3782*
[141-46-8] Glycolaldehyde, *4532*
[141-52-6] Sodium Ethoxide, *8749*
[141-53-7] Sodium Formate, *8756*
[141-62-8] Decamethyltetrasiloxane, *2854*
[141-63-9] Dodecamethylpentasiloxane, *3450*
[141-66-2] Dicrotophos, *3097*
[141-75-3] *n*-Butyryl Chloride, *1602*
[141-78-6] Ethyl Acetate, *3811*
[141-79-7] Mesityl Oxide, *5978*
[141-82-2] Malonic Acid, *5774*
[141-84-4] Rhodanine, *8308*
[141-90-2] 2-Thiouracil, *9520*
[141-94-6] Hexetidine, *4739*
[141-97-9] Ethyl Acetoacetate, *3812*
[142-03-0] Aluminum Subacetate, *359*
[142-04-1] Aniline Hydrochloride, *652*
[142-17-6] Calcium Oleate, *1686*
[142-47-2] Monosodium Glutamate, *6342*
[142-59-6] Nabam, *6426*
[142-61-0] Caproyl Chloride, *1765*
[142-62-1] *n*-Caproic Acid, *1761*
[142-68-7] Tetrahydropyran, *9362*
[142-71-2] Cupric Acetate, *2614*
[142-72-3] Magnesium Acetate, *5718*

[142-73-4] Iminodiacetic Acid, *4959*
[142-82-5] *n*-Heptane, *4694*
[142-84-7] *n*-Dipropylamine, *3383*
[142-88-1] Piperazine Adipate, *7577*
[142-96-1] *n*-Butyl Ether, *1572*
[143-07-7] Lauric Acid, *5439*
[143-08-8] *n*-Nonyl Alcohol, *6766*
[143-13-5] *n*-Nonyl Acetate, *6765*
[143-15-7] Lauryl Bromide, *5444*
[143-18-0] Potassium Oleate, *7771*
[143-19-1] Oleic Acid Sodium Salt, *6921*
[143-24-8] Tetraglyme, *9353*
[143-28-2] Oleyl Alcohol, *6924*
[143-33-9] Sodium Cyanide, *8741*
[143-37-3] Acetamidine, *45*
[143-50-0] Chlordecone, *2084*
[143-52-2] Metopon, *6227*
[143-57-7] Protoveratrine A, *8008*
[143-62-4] Digitoxigenin, *3181*
[143-66-8] Sodium Tetraphenylborate, *8818*
[143-67-9] Vinblastine Sulfate, *10179*
[143-71-5] Hydrocodone Bitartrate, *4823*
[143-74-8] Phenolsulfonphthalein, *7360*
[143-76-0] Cyclobarbital Calcium Salt, *2704*
[143-81-7] Butabarbital Sodium Salt, *1509*
[144-02-5] Barbital Sodium Salt, *957*
[144-12-7] Tiemonium Iodide, *9587*
[144-14-9] Anileridine, *651*
[144-21-8] Methanearsonic Acid Disodium Salt, *6025*
[144-23-0] Magnesium Citrate Dibasic, *5727*
[144-29-6] Piperazine Citrate, *7576*
[144-33-2] Citric Acid Disodium Salt, *2325*
[144-49-0] Fluoroacetic Acid, *4199*
[144-55-8] Sodium Bicarbonate, *8719*
[144-62-7] Oxalic Acid, *7010*
[144-68-3] Zeaxanthin, *10316*
[144-75-2] Sulfoxone Sodium, *9098*
[144-80-9] Sulfacetamide, *9031*
[144-82-1] Sulfamethizole, *9048*
[144-83-2] Sulfapyridine, *9069*
[145-12-0] Oxymesterone, *7064*
[145-13-1] Pregnenolone, *7851*
[145-41-5] Dehydrocholic Acid Sodium Salt, *2872*
[145-73-3] Endothall, *3630*
[146-14-5] Flavin-Adenine Dinucleotide, *4122*
[146-17-8] Flavin Mononucleotide, *4123*
[146-22-5] Nitrazepam, *6661*
[146-28-1] Thiopropazate Dihydrochloride, *9509*
[146-37-2] Laurolinium Acetate, *5442*
[146-48-5] Yohimbine, *10300*
[146-54-3] Triflupromazine, *9853*
[146-56-5] Fluphenazine Dihydrochloride, *4220*
[146-80-5] Xanthosine, *10263*
[146-84-9] Silver Picrate, *8665*
[146-90-7] Cinnabarine, *2297*
[147-14-8] Copper Phthalocyanine, *2505*
[147-20-6] Diphenylpyraline, *3370*
[147-24-0] Diphenhydramine Hydrochloride, *3339*

[147-48-8] Penicillin V Calcium Salt, *7209*
[147-52-4] Nafcillin, *6436*
[147-61-5] Pentoxyl, *7250*
[147-71-7] D-Tartaric Acid, *9203*
[147-73-9] *meso*-Tartaric Acid, *9202*
[147-81-9] Arabinose, *750*
[147-82-0] 2,4,6-Tribromoaniline, *9771*
[147-85-3] Proline, *7895*
[147-90-0] Morpholine Salicylate, *6362*
[147-93-3] Thiosalicylic Acid, *9513*
[147-94-4] Cytarabine, *2781*
[148-01-6] Dinitolmide, *3295*
[148-03-8] β-Tocopherol (*dl*-form), *9654*
[148-18-5] Ditiocarb Sodium, *3421*
[148-24-3] 8-Hydroxyquinoline, *4881*
[148-25-4] Chromotropic Acid, *2245*
[148-56-1] Flumethiazide, *4170*
[148-64-1] Chlorothen Citrate, *2170*
[148-65-2] Chlorothen, *2170*
[148-72-1] Pilocarpine Nitrate, *7536*
[148-75-4] 2-Naphthol-3,6-disulfonic Acid, *6472*
[148-79-8] Thiabendazole, *9440*
[148-82-3] Melphalan, *5896*
[148-83-4] Ostruthin, *6994*
[149-15-5] Butacaine Sulfate, *1510*
[149-16-6] Butacaine, *1510*
[149-29-1] Patulin, *7158*
[149-30-4] 2-Mercaptobenzothiazole, *5935*
[149-32-6] Erythritol, *3733*
[149-44-0] Sodium Formaldehyde Sulfoxylate, *8755*
[149-45-1] Tiron, *9621*
[149-64-4] *N*-Butylscopolammonium Bromide, *1591*
[149-73-5] Orthoformic Acid Trimethyl Ester, *6980*
[149-90-6] Acetylleucine Monoethanolamine Salt, *91*
[149-91-7] Gallic Acid, *4375*
[150-13-0] *p*-Aminobenzoic Acid, *418*
[150-25-4] Bicine, *1205*
[150-30-1] Phenylalanine DL-Form, *7382*
[150-38-9] EDTA Trisodium Salt, *3565*
[150-59-4] Alverine, *365*
[150-60-7] Dibenzyl Disulfide, *3018*
[150-61-8] 1,2-Dianilinoethane, *2990*
[150-68-5] Monuron, *6349*
[150-69-6] Dulcin, *3511*
[150-86-7] Phytol, *7502*
[150-90-3] Sodium Succinate, *8808*
[150-97-0] Mevalonic Acid, *6242*
[151-06-4] Chlorphentermine Hydrochloride, *2188*
[151-07-5] Chloroarsenol, *2119*
[151-18-8] 3-Aminopropionitrile, *465*
[151-21-3] Sodium Lauryl Sulfate, *8768*
[151-50-8] Potassium Cyanide, *7747*
[151-56-4] Ethylenimine, *3859*
[151-63-3] Aminoacetonitrile Sulfate (1:1), *406*
[151-67-7] Halothane, *4637*
[151-73-5] Betamethasone 21-Phosphate Disodium Salt, *1182*
[151-83-7] Methohexital, *6053*
[152-02-3] Levallorphan, *5510*
[152-11-4] Verapamil Hydrochloride, *10144*
[152-16-9] Schradan, *8532*

[152-43-2] Quinestrol, *8171*
[152-47-6] Sulfalene, *9042*
[152-58-9] 11-Desoxy-17-hydroxycorticosterone, *2933*
[152-62-5] Dydrogesterone, *3521*
[152-72-7] Acenocoumarol, *30*
[152-84-1] Ruberythric Acid, *8413*
[152-93-2] Vicine, *10171*
[152-95-4] Sophoricoside, *4426*
[152-97-6] Fluocortolone, *4184*
[153-00-4] Methenolone, *6039*
[153-18-4] Rutin, *8438*
[153-61-7] Cephalothin, *1981*
[153-77-5] 8-Hydroxyquinoline Aluminum Sulfate, *4881*
[153-87-7] Oxypertine, *7070*
[154-17-6] 2-Deoxy-D-glucose, *2905*
[154-21-2] Lincomycin, *5555*
[154-23-4] Catechin, *1902*
[154-41-6] Phenylpropanolamine Hydrochloride, *7417*
[154-42-7] Thioguanine, *9490*
[154-61-0] Primulaverin, *7867*
[154-68-7] Antazoline Phosphate, *672*
[154-69-8] Tripelennamine Hydrochloride, *9910*
[154-87-0] Thiamine Diphosphate, *9448*
[154-93-8] Carmustine, *1845*
[154-97-2] Pralidoxime Mesylate, *7822*
[155-09-9] Tranylcypromine, *9734*
[155-38-4] Noformicin, *6757*
[155-41-9] Methscopolamine Bromide, *6077*
[155-54-4] Hydroorotic Acid, *4841*
[155-58-8] Rhapontin, *8297*
[156-08-1] Benzphetamine, *1121*
[156-10-5] 4-Nitrosodiphenylamine, *6726*
[156-28-5] Phenethylamine Hydrochloride, *7339*
[156-34-3] Amphetamine *l*-Form, *579*
[156-43-4] *p*-Phenetidine, *7342*
[156-51-4] Phenelzine Acid Sulfate, *7335*
[156-56-9] Hypoglycine A, *4905*
[156-57-0] Cysteamine Hydrochloride, *2776*
[156-59-2] Acetylene Dichloride *cis*-Form, *87*
[156-60-5] Acetylene Dichloride *trans*-Form, *87*
[156-62-7] Calcium Cyanamide, *1664*
[157-03-9] 6-Diazo-5-oxo-L-norleucine, *3005*
[191-07-1] Coronene, *2523*
[192-97-2] Benzo[*e*]pyrene, *1107*
[195-19-7] 3,4-Benzphenanthrene, *1120*
[198-55-0] Perylene, *7299*
[213-46-7] Picene, *7508*
[217-59-4] Triphenylene, *9913*
[218-01-9] Chrysene, *2259*
[222-93-5] 2,3:6,7-Dibenzphenanthrene, *3015*
[244-69-9] γ-Carboline, *1805*
[253-14-5] Twistane, *10009*
[253-52-1] Phthalazine, *7482*
[253-66-7] Cinnoline, *2307*
[253-82-7] Quinazoline, *8169*
[254-04-6] 1,2-Benzopyran, *1105*
[260-94-6] Acridine, *114*
[261-31-4] Thioxanthene, *9522*
[271-44-3] 1*H*-Indazole, *4976*
[271-89-6] Benzofuran, *1090*

[274-07-7] Catecholborane, *1903*
[275-51-4] Azulene, *919*
[277-10-1] Cubane, *2601*
[278-06-8] Quadricyclane, *8138*
[280-57-9] Triethylenediamine, *9834*
[281-23-2] Adamantane, *137*
[286-08-8] Norcarane, *6777*
[287-23-0] Cyclobutane, *2708*
[287-92-3] Cyclopentane, *2736*
[288-13-1] Pyrazole, *8071*
[288-32-4] Imidazole, *4953*
[288-47-1] Thiazole, *9459*
[288-88-0] 1*H*-1,2,4-Triazole, *9766*
[288-99-3] 1,3,4-Oxadiazole, *7005*
[289-80-5] Pyridazine, *8080*
[289-95-2] Pyrimidine, *8100*
[290-37-9] Pyrazine, *8068*
[290-87-9] *s*-Triazine, *9763*
[292-46-6] Lenthionine, *5492*
[294-90-6] Cyclen, *2699*
[297-76-7] Ethynodiol Diacetate, *3909*
[297-78-9] Isobenzan, *5172*
[297-90-5] Levorphanol *dl*-Form, *5524*
[298-00-0] Methyl Parathion, *6181*
[298-02-2] Phorate, *7443*
[298-04-4] Disulfoton, *3408*
[298-12-4] Glyoxylic Acid, *4546*
[298-14-6] Potassium Bicarbonate, *7730*
[298-18-0] DL-Diepoxybutane, *3734*
[298-45-3] Bulbocapnine, *1486*
[298-46-4] Carbamazepine, *1783*
[298-55-5] Clocinizine, *2362*
[298-57-7] Cinnarizine, *2306*
[298-59-9] Methylphenidate Hydrochloride, *6183*
[298-81-7] Methoxsalen, *6059*
[298-95-3] Neotetrazolium Chloride, *6551*
[298-96-4] Triphenyltetrazolium Chloride, *9918*
[299-11-6] *N*-Methylphenazonium Methosulfate, *6182*
[299-26-3] α-Methyltryptamine, *6207*
[299-27-4] Potassium Gluconate, *4492*
[299-28-5] Calcium Gluconate, *1671*
[299-29-6] Ferrous Gluconate, *4074*
[299-42-3] Ephedrine, *3663*
[299-45-6] Potasan, *7724*
[299-84-3] Ronnel, *8387*
[299-85-4] DMPA, *3436*
[299-89-8] Acetiamine, *53*
[300-08-3] Arecoline Hydrobromide, *768*
[300-30-1] DL-Thyroxine, *9570*
[300-37-6] Iodopyracet, *5083*
[300-38-9] 3,5-Dibromo-L-tyrosine, *3035*
[300-48-1] 3-*O*-Methyldopa, *6128*
[300-54-9] Muscarine, *6395*
[300-62-9] Amphetamine, *579*
[300-76-5] Naled, *6442*
[300-85-6] β-Hydroxybutyric Acid, *4855*
[301-04-2] Lead Acetate, *5452*
[301-16-6] Scoparin, *8539*
[301-19-9] Robinin, *8372*
[301-21-3] Laudanidine, *5434*
[302-01-2] Hydrazine, *4809*
[302-15-8] Methylhydrazine Sulfate, *6158*
[302-17-0] Chloral Hydrate, *2071*
[302-22-7] Chlormadinone Acetate, *2102*

[302-27-2] Aconitine, *110*
[302-40-9] Benactyzine, *1031*
[302-41-0] Piritramide, *7611*
[302-49-8] Uredepa, *10057*
[302-53-4] Carbidopa DL-Form, *1798*
[302-66-9] Meparfynol Carbamate, *5913*
[302-70-5] Mechlorethamine Oxide Hydrochloride, *5847*
[302-72-7] DL-Alanine, *197*
[302-79-4] Retinoic Acid, *8288*
[302-83-0] Edrophonium Bromide, *3564*
[302-84-1] Serine DL-Form, *8599*
[302-91-0] Allopregnane-3α,11β,-17α,21-tetrol-20-one, *265*
[302-96-5] Stanozolol (1′*H* form), *8921*
[303-25-3] Cyclizine Hydrochloride, *2703*
[303-34-4] Lasiocarpine, *5429*
[303-40-2] Fluocortolone 21-Hexanoate, *4184*
[303-42-4] Methenolone 17-Enanthate, *6039*
[303-45-7] Gossypol, *4564*
[303-47-9] Ochratoxin A, *6829*
[303-49-1] Clomipramine, *2378*
[303-53-7] Cyclobenzaprine, *2706*
[303-69-5] Prothipendyl, *7998*
[303-81-1] Novobiocin, *6811*
[303-98-0] Ubiquinone 50, *10025*
[304-20-1] Hydralazine Hydrochloride, *4802*
[304-21-2] Harmaline, *4646*
[304-55-2] Succimer, *8995*
[304-59-6] Potassium Sodium Tartrate, *7790*
[304-84-7] Ethamivan, *3776*
[305-01-1] Esculetin, *3752*
[305-03-3] Chlorambucil, *2073*
[305-33-9] Iproniazid Phosphate, *5122*
[305-80-6] *p*-Diazobenzenesulfonic Acid, *3002*
[305-84-0] Carnosine, *1851*
[305-85-1] Disophenol, *3401*
[305-96-4] α-Methyl-*m*-tyrosine, *6208*
[306-03-6] Hyoscyamine Hydrobromide, *4897*
[306-07-0] Pargyline Hydrochloride, *7142*
[306-08-1] Homovanillic Acid, *4779*
[306-21-8] Hydroxyamphetamine Hydrobromide, *4847*
[306-44-5] Isonitrosoacetone, *5236*
[306-52-5] Triclofos, *9820*
[306-60-5] Agmatine, *181*
[306-61-6] Magnesium Thiocyanate, *5757*
[306-67-2] Spermine Tetrahydrochloride, *8870*
[306-94-5] Perfluorodecalin, *4215*
[309-00-2] Aldrin, *219*
[309-29-5] Doxapram, *3481*
[309-36-4] Methohexital Sodium Salt, *6053*
[309-43-3] Secobarbital Sodium, *8557*
[309-88-6] Ishikawa Reagent, *5153*
[311-45-5] Paraoxon, *7134*
[312-10-7] Penicillamine Disulfide, *7199*
[312-45-8] Hemicholinium Dibromide, *4678*
[312-84-5] Serine D-Form, *8599*
[313-04-2] Desmosterol, *2927*

[313-06-4] Estradiol Cypionate, *3758*
[313-67-7] Aristolochic Acid, *777*
[314-13-6] Evan's Blue, *3948*
[314-19-2] Apomorphine Hydrochloride, *733*
[314-35-2] Etamiphyllin, *3766*
[314-40-9] Bromacil, *1388*
[314-50-1] Orotidine, *6973*
[315-22-0] Monocrotaline, *6336*
[315-30-0] Allopurinol, *271*
[315-37-7] Testosterone Enanthate, *9324*
[315-72-0] Opipramol, *6948*
[315-80-0] Dibenzepin Hydrochloride, *3012*
[316-42-7] Emetine Dihydrochloride, *3616*
[316-81-4] Thioproperazine, *9510*
[317-34-0] Aminophylline, *461*
[317-52-2] Hexafluorenium Bromide, *4720*
[318-98-9] Propranolol Hydrochloride, *7953*
[319-89-1] Tetroquinone, *9397*
[320-67-2] Azacitidine, *884*
[321-64-2] Tacrine, *9154*
[322-35-0] Benserazide, *1052*
[322-79-2] Triflusal, *9856*
[326-43-2] Phenyramidol Hydrochloride, *7432*
[327-56-0] Norleucine D(−)-Form, *6795*
[327-57-1] Norleucine, *6795*
[327-97-9] Chlorogenic Acid, *2143*
[328-42-7] Oxalacetic Acid, *7008*
[328-50-7] α-Ketoglutaric Acid, *5350*
[329-56-6] Norepinephrine Hydrochloride, *6784*
[329-63-5] Epinephrine dl-Form Hydrochloride, *3674*
[329-65-7] Epinephrine dl-Form, *3674*
[329-71-5] 2,5-Dinitrophenol, *3309*
[329-98-6] PMSF, *7659*
[330-54-1] Diuron, *3425*
[330-55-2] Linuron, *5564*
[330-95-0] Nicarbazin, *6579*
[331-39-5] Caffeic Acid, *1638*
[333-18-6] Ethylenediamine Dihydrochloride, *3849*
[333-20-0] Potassium Thiocyanate, *7807*
[333-27-7] Methyl Trifluoromethanesulfonate, *6204*
[333-36-8] Flurothyl, *4231*
[333-41-5] Diazinon, *2998*
[333-93-7] Putrescine Dihydrochloride, *8058*
[334-20-3] Queen Substance, *8148*
[334-48-5] n-Capric Acid, *1760*
[334-50-9] Spermidine Trihydrochloride, *8869*
[334-88-3] Diazomethane, *3004*
[335-67-1] Perfluorooctanoic Acid, *7272*
[337-47-3] Thiamylal Sodium Salt, *9454*
[338-69-2] D-Alanine, *197*
[338-83-0] Perfluorotripropylamine, *4215*
[338-95-4] Isoflupredone, *5220*
[338-98-7] Isoflupredone 21-Acetate, *5220*
[339-43-5] Carbutamide, *1831*
[339-44-6] Glymidine, *4543*

[340-56-7] Methaqualone Hydrochloride, *6032*
[340-57-8] Mecloqualone, *5852*
[341-70-8] Diethazine Hydrochloride, *3119*
[341-88-8] HQNO, *4790*
[342-10-9] Kallidin, *5326*
[343-65-7] Kynurenine, *5376*
[345-78-8] Pseudoephedrine Hydrochloride, *8024*
[346-18-9] Polythiazide, *7705*
[348-67-4] Methionine D-Form, *6047*
[349-46-2] Cystine D-Form, *2779*
[350-03-8] Methyl Pyridyl Ketone, *6189*
[350-12-9] Sulbentine, *9024*
[352-21-6] 4-Amino-3-hydroxybutyric Acid, *441*
[352-32-9] p-Fluorotoluene, *4212*
[352-70-5] m-Fluorotoluene, *4212*
[352-93-2] Ethyl Sulfide, *3905*
[352-97-6] Glycocyamine, *4530*
[353-50-4] Carbonyl Fluoride, *1821*
[355-42-0] Perfluorohexane, *7271*
[355-80-6] 2,2,3,3,4,4,5,5-Octafluoro-1-pentanol, *6835*
[356-12-7] Fluocinonide, *4182*
[357-07-3] Oxymorphone Hydrochloride, *7068*
[357-08-4] Naloxone Hydrochloride, *6447*
[357-56-2] Dextromoramide, *2958*
[357-57-3] Brucine, *1464*
[357-67-5] Phenetharbital, *7336*
[357-70-0] Galantamine, *4370*
[358-23-6] Trifluoromethanesulfonic Anhydride, *9850*
[359-83-1] Pentazocine, *7233*
[360-68-9] Coprosterol, *2509*
[360-70-3] Nandrolone Decanoate, *6450*
[361-37-5] Methysergide, *6213*
[362-29-8] Propiomazine, *7936*
[362-74-3] Bucladesine, *1473*
[363-24-6] Prostaglandin E₂, *7988*
[364-62-5] Metoclopramide, *6219*
[364-98-7] Diazoxide, *3007*
[365-26-4] p-Hydroxyephedrine, *4863*
[366-13-2] Neutral Red (free base), *6573*
[366-18-7] 2,2'-Bipyridine, *1241*
[366-70-1] Procarbazine Hydrochloride, *7876*
[367-25-9] 2,4-Difluoroaniline, *3166*
[367-51-1] Sodium Thioglycolate, *8819*
[367-93-1] IPTG, *5126*
[368-39-8] Meerwein's Reagent, *5867*
[369-77-7] Cloflucarban, *2372*
[370-14-9] Pholedrine, *7442*
[371-11-9] Difluoroiodotoluene, *3169*
[371-40-4] 4-Fluoroaniline, *4200*
[371-86-8] Mipafox, *6287*
[372-09-8] Cyanoacetic Acid, *2679*
[372-66-7] Heptaminol, *4692*
[372-75-8] Citrulline, *2330*
[373-02-4] Nickel Acetate, *6584*
[378-44-9] Betamethasone, *1182*
[379-79-3] Ergotamine Tartrate, *3718*
[382-45-6] Adrenosterone, *165*
[382-67-2] Desoximetasone, *2931*
[386-17-4] Iodophthalein, *5081*
[388-51-2] Metofenazate, *6221*
[389-08-2] Nalidixic Acid, *6444*
[389-36-6] D-Glucaric Acid 1,4-Lactone, *4485*

[390-28-3] Methoxamine, *6058*
[390-64-7] Prenylamine, *7855*
[392-56-3] Hexafluorobenzene, *4722*
[395-28-8] Isoxsuprine, *5285*
[396-01-0] Triamterene, *9760*
[398-23-2] 4,4'-Difluorodiphenyl, *3168*
[402-61-9] 5-Methylpyrazole-3-carboxylic Acid, *6188*
[404-82-0] Fenfluramine Hydrochloride, *4004*
[404-86-4] Capsaicin, *1770*
[405-50-5] p-Fluorophenylacetic Acid, *4209*
[406-76-8] Carnitine DL-Form, *1849*
[406-90-6] Fluroxene, *4232*
[407-41-0] Phosphoserine, *7475*
[409-21-2] Silicon Carbide, *8631*
[420-04-2] Cyanamide, *2674*
[420-05-3] Cyanic Acid, *2676*
[420-12-2] Ethylene Sulfide, *3857*
[420-37-1] Trimethyloxonium Tetrafluoroborate, *9889*
[421-20-5] Methyl Fluorosulfonate, *6146*
[423-55-2] Perflubron, *7270*
[426-13-1] Fluorometholone, *4207*
[427-01-0] Geissospermine, *4414*
[427-51-0] Cyproterone Acetate, *2771*
[429-41-4] Tetrabutylammonium Fluoride, *9332*
[431-03-8] Diacetyl, *2969*
[432-60-0] Allylestrenol, *282*
[432-68-8] Echinenone, *3542*
[434-03-7] Ethisterone, *3795*
[434-05-9] Methenolone 17-Acetate, *6039*
[434-07-1] Oxymetholone, *7066*
[434-13-9] Lithocholic Acid, *5601*
[434-16-2] (3β)-7-Dehydrocholesterol, *2871*
[434-22-0] Nandrolone, *6450*
[434-43-5] Pentorex, *7246*
[435-97-2] Phenprocoumon, *7371*
[436-05-5] Curine, *2664*
[437-38-7] Fentanyl, *4028*
[437-50-3] Gentisin, *4434*
[437-74-1] Xanthinol Niacinate, *10258*
[438-08-4] Etiocholanic Acid, *3916*
[438-22-2] Androstane, *628*
[438-23-3] Etiocholane, *3915*
[438-41-5] Chlordiazepoxide Hydrochloride, *2085*
[438-60-8] Protriptyline, *8010*
[439-14-5] Diazepam, *2997*
[439-89-4] Gentianine, *4429*
[440-17-5] Trifluoperazine Dihydrochloride, *9846*
[440-58-4] Iodamide, *5055*
[441-38-3] Benzoin Oxime, *1096*
[442-16-0] Ethacridine, *3771*
[442-51-3] Harmine, *4649*
[442-52-4] Clemizole, *2344*
[443-48-1] Metronidazole, *6236*
[443-79-8] Isoleucine DL-Form, *5225*
[444-27-9] Timonacic, *9602*
[445-30-7] Homarine, *4767*
[446-71-9] Homoeriodictyol, *4774*
[446-72-0] Genistein, *4426*
[446-86-6] Azathioprine, *895*
[446-95-7] l-α-Isosparteine, *8864*
[447-41-6] Nylidrin, *6821*
[451-13-8] Homogentisic Acid, *4775*
[451-71-8] Glyhexamide, *4542*
[452-35-7] Ethoxzolamide, *3809*

[453-17-8] Glyceraldehyde D-Form, 4517
[453-18-9] Fluoroacetic Acid Methyl Ester, 4199
[454-14-8] Cuscohygrine, 2667
[454-29-5] DL-Homocysteine, 4772
[456-22-4] 4-Fluorobenzoic Acid, 4203
[456-59-7] Cyclandelate, 2697
[457-60-3] Neoarsphenamine, 6532
[457-87-4] N-Ethylamphetamine, 3818
[458-24-2] Fenfluramine, 4004
[458-35-5] Coniferyl Alcohol, 2488
[458-37-7] Curcumin, 2663
[458-88-8] Coniine, 2489
[459-67-6] Hydnocarpic Acid, 4800
[459-72-3] Fluoroacetic Acid Ethyl Ester, 4199
[459-86-9] Mitoguazone, 6298
[460-19-5] Cyanogen, 2681
[461-05-2] Carnitine DL-Form Hydrochloride, 1849
[461-58-5] Dicyanodiamide, 3103
[461-72-3] Hydantoin, 4799
[461-78-9] Chlorphentermine, 2188
[461-98-3] Kyanmethin, 5374
[462-02-2] Cyamelide, 2672
[462-06-6] Fluorobenzene, 4201
[462-08-8] β-Aminopyridine, 469
[462-10-2] Homocystine (unspecified stereo), 4773
[462-94-2] Cadaverine, 1614
[463-40-1] Linolenic Acid, 5561
[463-51-4] Ketene, 5346
[463-56-9] Thiocyanic Acid, 9481
[463-58-1] Carbonyl Sulfide, 1822
[463-72-9] Carbamyl Chloride, 1785
[463-78-5] Orthoformic Acid, 6980
[463-82-1] Neopentane, 6541
[463-88-7] Neurine, 6569
[464-20-0] Echitamine Hydroxide, 3547
[464-41-5] Bornyl Chloride, 1342
[464-43-7] Borneol (1R,2S,4R)-Form, 1340
[464-45-9] Borneol (1S,2R,4S)-Form, 1340
[464-48-2] Camphor l-Form, 1734
[464-49-3] Camphor d-Form, 1734
[464-72-2] Benzopinacol, 1103
[464-81-3] Bufotoxin, 1484
[465-11-2] Gamabufotalin, 4388
[465-16-7] Oleandrin, 6919
[465-21-4] Bufalin, 1478
[465-22-5] Scillarenin, 8537
[465-28-1] Carotol, 1858
[465-39-4] Resibufogenin, 8266
[465-42-9] Capsanthin, 1771
[465-65-6] Naloxone, 6447
[465-74-7] Quinovic Acid, 8192
[465-92-9] Marrubiin, 5820
[465-99-6] Hederagenin, 4659
[466-06-8] Proscillaridin, 7985
[466-07-9] Neriifolin, 6559
[466-09-1] Uzarigenin, 10083
[466-40-0] Isomethadone, 5230
[466-43-3] Atisine, 853
[466-49-9] Aspidospermine, 840
[466-61-5] Lycopodine, 5683
[466-80-8] α-Erythroidine, 3737
[466-81-9] β-Erythroidine, 3738
[466-85-3] Wieland-Gumlich Aldehyde, 10242
[466-90-0] Thebacon, 9426
[466-96-6] Pseudocodeine, 8021

[466-97-7] Normorphine, 6800
[466-99-9] Hydromorphone, 4840
[467-04-9] Oripavine, 6962
[467-14-1] Neopine, 6545
[467-15-2] Norcodeine, 6779
[467-36-7] Thialbarbital, 9444
[467-43-6] Methitural, 6050
[467-51-6] Samandarine, 8483
[467-53-8] Diginin, 3176
[467-55-0] Hecogenin, 4657
[467-60-7] Pipradrol, 7597
[467-83-4] Dipipanone, 3377
[467-85-6] Normethadone, 6798
[467-98-1] Thebainone, 9428
[468-10-0] Morphinan, 6360
[468-28-0] Lupulon, 5672
[468-61-1] Oxeladin, 7029
[468-76-8] Cassaine, 1891
[468-89-3] Nupharidine, 6816
[469-21-6] Doxylamine, 3489
[469-32-9] Hamamelitannin, 4640
[469-36-3] Oryzanol C, 6983
[469-59-0] Jervine, 5312
[469-62-5] Propoxyphene, 7952
[469-72-7] allo-Hydroxycitric Acid Lactone, 4861
[469-79-4] Ketobemidone, 5348
[469-83-0] Cafestol, 1637
[470-40-6] Thujopsene, 9547
[470-82-6] Eucalyptol, 3938
[470-90-6] Chlorfenvinphos, 2090
[471-09-0] Felinine, 3986
[471-15-8] β-Thujone, 9546
[471-25-0] Propiolic Acid, 7935
[471-29-4] Methylguanidine, 6155
[471-34-1] Calcium Carbonate, 1659
[471-35-2] Cacodyl, 1608
[471-46-5] Oxamide, 7017
[471-47-6] Oxamic Acid, 7016
[471-53-4] Enoxolone, 3644
[471-80-7] Steviol, 8938
[471-87-4] Stachydrine, 8899
[471-95-4] Bufotalin, 1482
[472-15-1] Betulinic Acid, 1193
[472-41-3] Dianin's Compound, 2991
[472-61-7] Astaxanthin, 845
[472-70-8] Cryptoxanthin, 2599
[472-87-7] 3-Dehydroretinal, 2878
[472-92-4] δ-Carotene, 1857
[472-93-5] γ-Carotene, 1856
[473-06-3] Chrysanthenone, 2256
[473-30-3] Thiazolsulfone, 9461
[473-34-7] Dichloramine T, 3056
[473-41-6] Tolbutamide Sodium Salt, 9667
[473-81-4] Glyceric Acid, 4519
[473-90-5] Mesoxalic Acid, 5983
[473-98-3] Betulin, 1192
[474-00-0] Eburnamonine, 3534
[474-07-7] Brazilin, 1372
[474-25-9] Chenodiol, 2054
[474-40-8] α₁-Sitosterol, 8693
[474-45-3] Funtumine, 4321
[474-62-4] Campesterol, 1731
[474-69-1] Lumisterol, 5662
[474-70-4] Isopyrocalciferol, 5265
[474-73-7] 24-Hydroxycholesterol, 4859
[474-77-1] Epicholesterol, 3668
[474-86-2] Equilin, 3700
[475-20-7] Longifolene, 5624
[475-25-2] Hematein, 4669
[475-26-3] DFDT, 2960
[475-31-0] Glycocholic Acid, 4529

[475-54-7] Oosporein, 6945
[475-67-2] Isocorydine, 5207
[475-81-0] Glaucine, 4471
[476-28-8] Lycorine, 5687
[476-32-4] Chelidonine (+)-Form, 2052
[476-33-5] Homochelidonine, 4770
[476-45-9] Javanicin, 5310
[476-66-4] Ellagic Acid, 3602
[476-69-7] Corydine, 2533
[476-70-0] Boldine, 1328
[476-71-1] Domesticine, 3462
[477-27-0] Colchiceine, 2454
[477-30-5] Demecolcine, 2891
[477-32-7] Visnadine, 10208
[477-47-4] Picropodophyllin, 7663
[477-58-7] Chondrocurine, 2217
[477-60-1] Bebeerine, 1014
[477-62-3] Isochondrodendrine, 5203
[477-67-8] Picropodophyllic Acid, 7662
[477-75-8] Triptycene, 9925
[478-10-4] Elliptone, 3605
[478-43-3] Rhein, 8299
[478-60-4] Citromycetin, 2327
[478-61-5] Berbamine, 1155
[478-73-9] Pseudococaine, 8020
[478-84-2] Bromolysergide, 1433
[478-85-3] Apo-β-erythroidine, 731
[478-94-4] Lysergamide, 5693
[478-95-5] Isolysergic Acid, 5226
[479-00-5] Ergometrinine, 3711
[479-13-0] Coumestrol, 2552
[479-18-5] Dyphylline, 3526
[479-20-9] Atranorin, 860
[479-23-2] Cholanthrene, 2204
[479-27-6] 1,8-Naphthalenediamine, 6457
[479-36-7] Daphnoline, 2819
[479-41-4] Indirubin (unspecified stereo), 4985
[479-45-8] Nitramine, 6659
[479-50-5] Lucanthone, 5650
[479-61-8] Chlorophyll A, 2158
[479-81-2] Bietamiverine, 1209
[479-92-5] Propyphenazone, 7982
[479-98-1] Aucubin, 868
[480-11-5] Oroxylin A, 6974
[480-15-9] Datiscetin, 2832
[480-16-0] Morin, 6355
[480-17-1] Leucocyanidin, 5504
[480-30-8] Dichloralphenazone, 3055
[480-36-4] Linarin, 5553
[480-40-0] Chrysin, 2260
[480-41-1] Naringenin, 6506
[480-44-4] Acacetin, 14
[480-49-9] Filipin III, 4110
[480-54-6] Retrorsine, 8290
[480-64-8] o-Orsellinic Acid, 6977
[480-68-2] 5-Nitrobarbituric Acid, 6672
[480-78-4] Platyphylline, 7646
[480-81-9] Seneciphylline, 8591
[480-85-3] Retronecine, 8289
[481-05-0] Artemisin, 806
[481-06-1] α-Santonin, 8498
[481-17-4] Chondrillasterol, 2216
[481-18-5] α-Spinasterol, 8876
[481-20-9] Coprostane, 2508
[481-21-0] (5α)-Cholestane, 2206
[481-26-5] Pregnane, 7844
[481-29-8] Epiandrosterone, 3664
[481-30-1] Epitestosterone, 3679
[481-37-8] Ecgonine, 3540
[481-39-0] Juglone, 5317
[481-42-5] Plumbagin, 7655
[481-49-2] Cepharanthine, 1984

[481-72-1] Aloe-Emodin, 300
[481-74-3] Chrysophanic Acid, 2262
[481-82-3] Betti Base, 1190
[481-85-6] Menadiol, 5898
[481-97-0] Estrone Sulfate, 3763
[482-05-3] Diphenic Acid, 3340
[482-15-5] Isothipendyl, 5273
[482-20-2] Tylophorine, 10012
[482-28-0] Cinchonamine, 2288
[482-35-9] Isoquercitrin, 5267
[482-44-0] Imperatorin, 4967
[482-66-6] 1,2-Cyclopentenophenanthrene, 2739
[482-68-8] Sarpagine, 8515
[482-70-2] Chimaphilin, 2058
[482-74-6] Cryptopine, 2598
[482-89-3] Indigo (unspecified stereo), 4981
[482-91-7] Aricine, 774
[483-04-5] Raubasine, 8237
[483-10-3] Corynanthine, 2535
[483-17-0] Cephaeline, 1973
[483-18-1] Emetine, 3616
[483-19-2] Emetamine, 3616
[483-55-6] Phthiocol, 7490
[483-57-8] Fervenulin, 4094
[483-63-6] Crotamiton, 2585
[483-64-7] Damascenine, 2807
[483-65-8] Retene, 8283
[483-78-3] Cadalene, 1613
[484-11-7] Neocuproine, 6534
[484-12-8] Osthole, 6993
[484-20-8] Bergapten, 1160
[484-23-1] Dihydralazine, 3187
[484-29-7] Dictamnine, 3099
[484-31-1] Apiole (Dill), 720
[484-49-1] Olivacine, 6928
[484-68-4] Pinitol (DL-form), 7561
[484-89-9] Fumigatin, 4317
[484-93-5] Ecgonidine, 3539
[485-19-8] Reticuline, 8286
[485-35-8] Cytisine, 2786
[485-41-6] Sulfachrysoidine, 9033
[485-43-8] Iridomyrmecin, 5133
[485-47-2] Ninhydrin, 6640
[485-49-4] Bicuculline, 1206
[485-50-7] Capnoidine, 154
[485-64-3] Hydrocinchonidine, 4820
[485-65-4] Hydrocinchonine, 4821
[485-70-1] Epicinchonine, 2290
[485-71-2] Cinchonidine, 2289
[485-72-3] Formononetin, 4271
[485-89-2] Oxycinchophen, 7056
[485-91-6] Allocryptopine, 257
[486-12-4] Triprolidine, 9922
[486-16-8] Carbinoxamine, 1800
[486-17-9] Captodiamine, 1775
[486-18-0] Flavopereirine, 4125
[486-35-1] Daphnetin, 2817
[486-39-5] Coclaurine, 2444
[486-47-5] Ethaverine, 3783
[486-55-5] Daphnin, 2818
[486-56-6] Cotinine, 2541
[486-62-4] Ononin, 4271
[486-66-8] Daidzein, 2798
[486-67-9] Mersalyl Acid, 5971
[486-70-4] Lupinine, 5669
[486-84-0] Harman, 4648
[486-86-2] Methylcytisine, 6121
[486-87-3] α-Isolupanine, 5667
[486-88-4] Lupanine l-Form, 5667
[486-89-5] Anagyrine, 617
[487-06-9] Limettin, 5545
[487-21-8] Lumazine, 5656

[487-27-4] Scopoline, 8547
[487-41-2] Phillyrin, 7434
[487-58-1] Hypaphorine, 4900
[487-60-5] Indican, 4980
[487-79-6] Kainic Acid, 5324
[487-90-1] Porphobilinogen, 7721
[487-93-4] Bufotenine, 1483
[487-94-5] Indoxyl Sulfate, 5013
[488-04-0] Holomycin, 4766
[488-10-8] Jasmone, 5307
[488-41-5] Mitobronitol, 6297
[488-43-7] Glucamine, 4484
[488-69-7] Fructose-1,6-diphosphate, 4304
[488-73-3] d-Quercitol, 8152
[488-81-3] Adonitol, 157
[488-84-6] D-Ribulose, 8331
[489-21-4] Sarkomycin A, 8513
[489-32-7] Icariin, 3672
[489-84-9] Guaiazulene, 4590
[489-86-1] Guaiol, 4592
[490-02-8] Aspergillic Acid, 831
[490-03-9] Diosphenol, 3326
[490-10-8] Nepetalactone, 6554
[490-11-9] Cinchomeronic Acid, 2286
[490-23-3] β-Tocotrienol, 9659
[490-53-9] Carnegine, 1847
[490-55-1] Amiphenazole, 479
[490-79-9] Gentisic Acid, 4433
[490-83-5] Dehydroascorbic Acid, 2870
[490-98-2] Hydroxytetracaine, 4885
[491-35-0] Lepidine, 5496
[491-50-9] Quercimeritrin, 8151
[491-58-7] Chrysarobin Pure Substance, 2257
[491-59-8] Chrysarobin 9-Hydroxy Analog, 2257
[491-67-8] Baicalein, 935
[491-70-3] Luteolin, 5675
[491-80-5] Biochanin A, 1231
[491-88-3] Isopilosine, 5245
[491-92-9] Pamaquine, 7102
[492-11-5] Leucopterin, 5506
[492-17-1] 2,4'-Biphenyldiamine, 1240
[492-18-2] Mersalyl, 5971
[492-22-8] Thioxanthone, 9523
[492-27-3] Kynurenic Acid, 5375
[492-38-6] Atropic Acid, 865
[492-39-7] Norpseudoephedrine, 6803
[492-70-6] Hydrobenzoin, 4815
[492-89-7] 3-Pentadecylcatechol, 7219
[492-99-9] Nioxime, 6646
[493-49-2] Corydaldine, 2530
[493-52-7] Methyl Red, 6192
[493-92-5] Prolintane, 7896
[494-03-1] Chlornaphazine, 2108
[494-08-6] Glucovanillin, 4500
[494-44-0] Cassella's Acid F, 1893
[494-47-3] Hydrofuramide, 4828
[494-52-0] Anabasine, 612
[494-55-3] Hydrohydrastinine, 4839
[494-79-1] Melarsoprol, 5883
[494-97-3] Nornicotine, 6801
[495-20-5] Conhydrine, 2484
[495-31-8] Nodakenin, 6755
[495-42-1] Halostachine, 4635
[495-45-4] Dypnone, 3527
[495-48-7] Azoxybenzene, 916
[495-54-5] Chrysoidine Free Base, 2261
[495-69-2] Hippuric Acid, 4753
[495-83-0] Tigloidine, 9590
[495-84-1] Salinazid, 8472

[495-85-2] Methysticin, 6214
[495-87-4] Isochavicinic Acid, 7579
[495-88-5] Isopiperinic Acid, 7579
[495-89-6] Chavicinic Acid, 7579
[495-91-0] Chavicine, 2048
[495-99-8] Hydroxystilbamidine, 4883
[496-00-4] Dibromopropamidine, 3029
[496-11-7] Indan, 4971
[496-16-2] Coumaran, 2547
[496-41-3] Coumarilic Acid, 2549
[496-46-8] Acetyleneurea, 88
[496-49-1] Hygrine, 4890
[496-65-1] Pantetheine, 7114
[496-67-3] Bromisovalum, 1404
[496-74-2] Toluene-3,4-dithiol, 9690
[496-77-5] Butyroin, 1599
[497-09-6] Glyceraldehyde L-Form, 4517
[497-18-7] Carbohydrazide, 1804
[497-19-8] Sodium Carbonate, 8731
[497-30-3] Ergothioneine, 3720
[497-39-2] DBMC, 2840
[497-59-6] Meconic Acid, 5853
[497-72-3] Methymycin, 6211
[497-73-4] Neomethymycin, 6538
[497-75-6] Dioxethedrine, 3331
[497-76-7] Arbutin, 763
[497-78-9] Rhododendrin, 8314
[497-92-7] Allethrin II, 252
[498-02-2] Apocynin, 728
[498-23-7] Citraconic Acid, 2320
[498-24-8] Mesaconic Acid, 5973
[498-40-8] Cysteic Acid L-Form, 2777
[498-59-9] Djenkolic Acid, 3431
[498-71-5] Sobrerol, 8705
[498-94-2] Isonipecotic Acid, 5235
[498-95-3] Nipecotic Acid, 6647
[498-96-4] Guvacine, 4619
[499-04-7] Arecaidine, 767
[499-12-7] Aconitic Acid, 109
[499-14-9] Chondrosine, 2220
[499-15-0] Hyalobiuronic Acid, 4795
[499-20-7] Dhurrin, 2961
[499-44-5] β-Thujaplicin, 9544
[499-67-2] Proparacaine, 7921
[499-75-2] Carvacrol, 1872
[499-89-8] Thujic Acid, 9545
[500-05-0] Coumalic Acid, 2545
[500-12-9] Goitrin, 4548
[500-24-3] Barrelene, 999
[500-34-5] β-Eucaine, 3937
[500-38-9] Nordihydroguaiaretic Acid (unspecified stereo), 6782
[500-44-7] Mimosine, 6280
[500-55-0] Apoatropine, 726
[500-62-9] Yangonin, 10291
[500-64-1] Kawain, 5335
[500-92-5] Chlorguanide, 2091
[501-15-5] Deoxyepinephrine, 2904
[501-30-4] Kojic Acid, 5365
[501-36-0] Resveratrol, 8279
[501-52-0] Hydrocinnamic Acid, 4822
[501-53-1] Carbobenzoxy Chloride, 1801
[501-65-5] Tolan, 9664
[501-68-8] Beclamide, 1016
[501-81-5] 3-Pyridineacetic Acid, 8082
[501-92-8] Chavicol, 2049
[501-94-0] Tyrosol, 10022
[502-37-4] Hypoglycine B, 4905
[502-42-1] Cycloheptanone, 2715
[502-54-5] Glycerol 1-Octanoate, 6338
[502-55-6] Dixanthogen, 3428
[502-61-4] α-Farnesene, 3973

[502-65-8] Lycopene, *5681*
[502-85-2] γ-Hydroxybutyrate Sodium Salt, *4854*
[502-98-7] Chloroazodin, *2121*
[503-01-5] Isometheptene, *5231*
[503-07-1] Vernolic Acid (+)-Form, *10159*
[503-30-0] Trimethylene Oxide, *9886*
[503-38-8] Diphosgene, *3374*
[503-40-2] Methionic Acid, *6046*
[503-49-1] Meglutol, *5877*
[503-64-0] Isocrotonic Acid, *5209*
[503-66-2] Hydracrylic Acid, *4801*
[503-74-2] Isovaleric Acid, *5277*
[504-02-9] Dihydroresorcinol, *3196*
[504-15-4] Orcinol, *6959*
[504-20-1] Phorone, *7445*
[504-24-5] Dalfampridine, *2802*
[504-29-0] α-Aminopyridine, *469*
[504-60-9] 1,3-Pentadiene, *7220*
[504-61-0] Crotyl Alcohol *trans*-Form, *2591*
[504-63-2] Trimethylene Glycol, *9885*
[504-64-3] Carbon Suboxide, *1815*
[505-32-8] Isophytol, *5244*
[505-48-6] Suberic Acid, *8992*
[505-60-2] Mustard Gas, *6401*
[505-70-4] Muconic Acid, *6385*
[505-75-9] Cicutoxin, *2270*
[505-94-2] Isomycomycin, *6411*
[506-03-6] Chimyl Alcohol, *2060*
[506-12-7] Margaric Acid, *5816*
[506-25-2] Isanic Acid, *5147*
[506-26-3] γ-Linolenic Acid, *5562*
[506-30-9] Arachidic Acid, *753*
[506-32-1] Arachidonic Acid, *754*
[506-33-2] Brassidic Acid, *1369*
[506-58-1] Ethylamine Hydriodide, *3817*
[506-59-2] Dimethylamine Hydrochloride, *3250*
[506-61-6] Potassium Silver Cyanide, *7789*
[506-64-9] Silver Cyanide, *8651*
[506-65-0] Gold Monocyanide, *4551*
[506-66-1] Beryllium Carbide, *1167*
[506-68-3] Cyanogen Bromide, *2683*
[506-77-4] Cyanogen Chloride, *2684*
[506-78-5] Cyanogen Iodide, *2685*
[506-80-9] Carbon Diselenide, *1810*
[506-82-1] Dimethylcadmium, *3261*
[506-87-6] Ammonium Carbonate, *505*
[506-89-8] Urea Hydrochloride, *10052*
[506-93-4] Guanidine Nitrate, *4597*
[506-96-7] Acetyl Bromide, *77*
[507-02-8] Acetyl Iodide, *90*
[507-09-5] Thioacetic Acid, *9473*
[507-16-4] Thionyl Bromide, *9500*
[507-19-7] *tert*-Butyl Bromide, *1557*
[507-20-0] *tert*-Butyl Chloride, *1563*
[507-25-5] Carbon Tetraiodide, *1818*
[507-27-7] Tetraphenylarsonium Bromide, *9383*
[507-28-8] Tetraphenylarsonium Chloride, *9383*
[507-36-8] *tert*-Amyl Bromide, *597*
[507-40-4] *tert*-Butyl Hypochlorite, *1574*
[507-42-6] Bromal Hydrate, *1391*
[507-60-8] Scilliroside, *8538*
[507-61-9] Azafrin, *888*
[507-70-0] Borneol, *1340*
[507-79-9] Tazettine, *9219*
[508-02-1] Oleanolic Acid, *6920*

[508-29-2] 3-Chloro-*d*-camphor, *2132*
[508-44-1] Anemonin, *635*
[508-52-1] Ouabagenin, *6999*
[508-54-3] Hydroxycodeinone, *4862*
[508-59-8] Parthenin, *7151*
[508-65-6] Germine, *4446*
[508-75-8] Convallatoxin, *2497*
[508-77-0] Cymarin, *2758*
[509-00-2] Cortisone 21-Cyclopentane-propionate, *2525*
[509-14-8] Tetranitromethane, *9381*
[509-15-9] Gelsemine, *4418*
[509-17-1] Ajacine, *5679*
[509-18-2] Delsoline, *2887*
[509-20-6] Aconine, *107*
[509-24-0] Songorine, *8843*
[509-36-4] β-Colubrine, *2474*
[509-40-0] Diaboline, *2962*
[509-44-4] α-Colubrine, *2474*
[509-60-4] Dihydromorphine, *3194*
[509-67-1] Pholcodine, *7441*
[509-71-7] Diacetyldihydromorphine, *2970*
[509-74-0] Methadyl Acetate, *6017*
[509-86-4] Heptabarbital, *4690*
[509-93-3] Ambrosin, *379*
[510-13-4] Malachite Green (carbinol base), *5763*
[510-20-3] Diethylmalonic Acid, *3140*
[510-35-0] Santonic Acid, *8497*
[510-53-2] Racemethorphan, *5520*
[511-07-9] Ergocristine 8α-Isomer, *3706*
[511-08-0] Ergocristine, *3706*
[511-09-1] α-Ergocryptine, *3707*
[511-10-4] Ergocryptinine, *3708*
[511-12-6] Dihydroergotamine, *3191*
[511-13-7] Chlophedianol Hydrochloride, *2066*
[511-18-2] Norcholanic Acid, *6778*
[511-20-6] Ergostane, *3714*
[511-26-2] Pregnenolone Methyl Ether, *7851*
[511-28-4] Vitamin D₄, *10217*
[511-34-2] Digitogenin, *3179*
[511-45-5] Pridinol, *7859*
[511-89-7] Plumieride, *7657*
[511-96-6] Gitogenin, *4466*
[512-04-9] Diosgenin, *3323*
[512-15-2] Cyclopentolate, *2741*
[512-29-8] Flavoxanthin, *4127*
[512-35-6] Benzenesulfonic Anhydride, *1072*
[512-48-1] Valdetamide, *10087*
[512-64-1] Echinomycin, *3544*
[512-69-6] Raffinose, *8213*
[512-85-6] Ascaridole, *816*
[513-10-0] Echothiophate Iodide, *3548*
[513-17-7] Methyl Sulfate Barium Salt, *6195*
[513-29-1] Glycine Sulfate, *4527*
[513-31-5] 2,3-Dibromopropene, *3030*
[513-35-9] Amylene, *601*
[513-36-0] Isobutyl Chloride, *5183*
[513-37-1] 1-Chloro-2-methyl-1-propene, *2149*
[513-38-2] Isobutyl Iodide, *5189*
[513-44-0] Isobutyl Mercaptan, *5192*
[513-48-4] *sec*-Butyl Iodide, *1577*
[513-49-5] *sec*-Butylamine *d*-Form, *1546*
[513-53-1] *sec*-Butyl Mercaptan, *1581*
[513-74-6] Ammonium Dithiocarbamate, *513*

[513-77-9] Barium Carbonate, *965*
[513-78-0] Cadmium Carbonate, *1621*
[513-79-1] Cobaltous Carbonate, *2420*
[513-81-5] 2,3-Dimethyl-1,3-butadiene, *3260*
[513-85-9] 2,3-Butylene Glycol, *1571*
[513-86-0] Acetoin, *63*
[513-88-2] 1,1-Dichloroacetone, *3060*
[513-92-8] Tetraiodoethylene, *9365*
[513-96-2] Chlorocyanohydrin, *2134*
[514-10-3] Abietic Acid, *7*
[514-17-0] Androstane-3β,11β-diol-17-one, *629*
[514-36-3] Fludrocortisone 21-Acetate, *4161*
[514-39-6] Periplogenin, *7291*
[514-61-4] Normethandrone, *6799*
[514-65-8] Biperiden, *1238*
[514-73-8] Dithiazanine Iodide, *3412*
[514-76-1] Astacin, *843*
[514-78-3] Canthaxanthin, *1756*
[514-92-1] Torularhodin, *9712*
[515-30-0] Atrolactic Acid, *864*
[515-40-2] Neophyl Chloride, *6544*
[515-42-4] Benzenesulfonic Acid Sodium Salt, *1071*
[515-46-8] Benzenesulfonic Acid Ethyl Ester, *1071*
[515-49-1] Sulfathiourea, *9075*
[515-64-0] Sulfisomidine, *9081*
[515-69-5] α-Bisabolol, *1245*
[515-72-0] Barium Benzenesulfonate, *962*
[515-82-2] Chloral Formamide, *2070*
[515-83-3] Chloral Alcoholate, *2069*
[515-84-4] Trichloroacetic Acid Ethyl Ester, *9792*
[515-94-6] 2,3-Diaminopropionic Acid, *2986*
[515-96-8] Semioxamazide, *8584*
[515-98-0] Ammonium Lactate, *527*
[516-00-7] Oxamic Acid Ammonium Salt, *7016*
[516-02-9] Barium Oxalate, *981*
[516-03-0] Ferrous Oxalate, *4078*
[516-12-1] *N*-Iodosuccinimide, *5088*
[516-16-5] Allopregnane-3β,11β,21-triol-20-one, *266*
[516-21-2] Cycloguanil, *2714*
[516-38-1] Cortol, *2528*
[516-41-6] Hydrallostane, *4803*
[516-42-7] Cortolone, *2529*
[516-51-8] Uranediol, *10038*
[516-53-0] Allopregnane-3β,20β-diol, *262*
[516-54-1] Allopregnan-3α-ol-20-one, *267*
[516-55-2] Allopregnan-3β-ol-20-one, *268*
[516-58-5] Allopregnan-20β-ol-3-one, *270*
[516-59-6] Allopregnan-20α-ol-3-one, *269*
[516-78-9] Fungisterol, *4320*
[516-85-8] Dehydroergosterol, *2877*
[516-95-0] Epicholestanol, *3667*
[516-98-3] Neoergosterol, *6536*
[517-04-4] Isoestradiol, *5212*
[517-06-6] 8-Isoestrone, *5213*
[517-09-9] Equilenin, *3699*
[517-10-2] Allocholesterol, *256*
[517-21-5] Glyoxal-Sodium Bisulfite, *4545*
[517-23-7] α-Acetylbutyrolactone, *78*

[540-88-5] *tert*-Butyl Acetate, *1539*
[540-92-1] Acetone Sodium Bisulfite, *68*
[540-97-6] Dodecamethylcyclohexasiloxane, *3449*
[541-02-6] Decamethylcyclopentasiloxane, *2852*
[541-07-1] Mevaldic Acid, *6241*
[541-09-3] Uranyl Acetate, *10046*
[541-14-0] Carnitine D-Form, *1849*
[541-15-1] Carnitine L-Form, *1849*
[541-16-2] Di-*tert*-butyl Malonate, *3041*
[541-19-5] Succinylcholine Iodide, *9007*
[541-22-0] Decamethonium Bromide, *2851*
[541-23-1] Isoamylamine Hydrochloride, *5157*
[541-25-3] Dichloro(2-chlorovinyl)arsine, *3071*
[541-28-6] Isoamyl Iodide, *5165*
[541-33-3] Butylidene Chloride, *1575*
[541-35-5] *n*-Butyramide, *1596*
[541-41-3] Ethyl Chloroformate, *3839*
[541-42-4] Isopropyl Nitrite, *5262*
[541-43-5] Barium Formate, *971*
[541-46-8] Isovaleramide, *5276*
[541-48-0] β-Aminobutyric Acid, *424*
[541-50-4] Acetoacetic Acid, *58*
[541-53-7] 2,4-Dithiobiuret, *3414*
[541-64-0] Furtrethonium Iodide, *4341*
[541-66-2] Oxapropanium Iodide, *7022*
[541-73-1] *m*-Dichlorobenzene, *3064*
[541-85-5] 5-Methyl-3-heptanone, *6156*
[541-88-8] Chloroacetic Anhydride, *2113*
[541-91-3] Muscone (±)-Form, *6398*
[542-05-2] Acetonedicarboxylic Acid, *67*
[542-08-5] Isopropyl Acetoacetate, *5252*
[542-10-9] Ethylidene Diacetate, *3866*
[542-11-0] Aniline Hydrobromide, *652*
[542-13-2] Aniline Hydrofluoride, *652*
[542-14-3] Aniline Acetate, *652*
[542-15-4] Aniline Nitrate, *652*
[542-16-5] Aniline Hemisulfate, *652*
[542-18-7] Cyclohexyl Chloride, *2725*
[542-32-5] α-Aminoadipic Acid, *410*
[542-40-5] β-Norbixin, *1312*
[542-42-7] Calcium Palmitate, *1689*
[542-46-1] Civetone, *2334*
[542-52-9] *n*-Butyl Carbonate, *1559*
[542-54-1] Isoamyl Cyanide, *5162*
[542-55-2] Isobutyl Formate, *5188*
[542-56-3] Isobutyl Nitrite, *5194*
[542-59-6] Ethylene Glycol Monoacetate, *3855*
[542-62-1] Barium Cyanide, *969*
[542-69-8] *n*-Butyl Iodide, *1576*
[542-75-6] 1,3-Dichloropropene, *3085*
[542-76-7] β-Chloropropionitrile, *2162*
[542-78-9] Malondialdehyde, *5773*
[542-83-6] Cadmium Cyanide, *1623*
[542-84-7] Cobaltous Cyanide, *2423*
[542-85-8] Ethyl Isothiocyanate, *3870*
[542-88-1] *sym*-Dichloromethyl Ether, *3079*
[542-91-6] Diethylsilane, *3147*
[542-92-7] Cyclopentadiene, *2734*
[543-15-7] Heptaminol Hydrochloride, *4692*
[543-20-4] Succinyl Chloride, *9004*

[543-21-5] Cellocidin, *1964*
[543-24-8] Aceturic Acid, *75*
[543-27-1] Isobutyl Chlorocarbonate, *5184*
[543-28-2] Isobutyl Carbamate, *5182*
[543-29-3] Isobutyl Nitrate, *5193*
[543-38-4] Canavanine, *1740*
[543-49-7] 2-Heptanol, *4697*
[543-59-9] Amyl Chloride, *600*
[543-63-5] *n*-Butylmercuric Chloride, *1583*
[543-67-9] Propyl Nitrite, *7976*
[543-80-6] Barium Acetate, *961*
[543-81-7] Beryllium Acetate, *1164*
[543-82-8] Octodrine, *6846*
[543-83-9] Galegine, *4371*
[543-87-3] Isoamyl Nitrate, *5167*
[543-90-8] Cadmium Acetate, *1619*
[543-94-2] Strontium Acetate, *8966*
[543-99-7] Diisoamylamine Hydrochloride, *3211*
[544-00-3] Diisoamylamine, *3211*
[544-01-4] Isoamyl Ether, *5163*
[544-10-5] 1-Chlorohexane, *2145*
[544-12-7] 3-Hexen-1-ol, *4736*
[544-13-8] Glutaronitrile, *4510*
[544-16-1] *n*-Butyl Nitrite, *1585*
[544-17-2] Calcium Formate, *1670*
[544-18-3] Cobaltous Formate, *2425*
[544-19-4] Cupric Formate, *2629*
[544-31-0] Palmidrol, *7096*
[544-35-4] Ethyl Linoleate, *3875*
[544-40-1] *n*-Butyl Sulfide, *1593*
[544-44-5] Agrocybin, *185*
[544-51-4] Mycomycin, *6411*
[544-60-5] Ammonium Oleate, *532*
[544-62-7] Batyl Alcohol, *1009*
[544-63-8] Myristic Acid, *6419*
[544-92-3] Cuprous Cyanide, *2651*
[544-97-8] Dimethylzinc, *3289*
[545-06-2] Trichloroacetonitrile, *9793*
[545-26-6] Gitoxigenin, *4468*
[545-47-1] Lupeol, *5668*
[545-55-1] Triethylenephosphoramide, *9837*
[545-56-2] Delcosine, *2887*
[545-61-9] Ajaconine, *186*
[545-80-2] Poldine Methylsulfate, *7671*
[545-93-7] Propallylonal, *7910*
[546-06-5] Conessine, *2480*
[546-43-0] Alantolactone, *200*
[546-46-3] Zinc Citrate, *10333*
[546-48-5] Pempidine Tartrate, *7189*
[546-62-3] Verbascose, *10152*
[546-67-8] Lead Tetraacetate, *5478*
[546-68-9] Titanium Tetraisopropoxide, *9636*
[546-74-7] Sodium Ethyl Sulfate, *8750*
[546-80-5] α-Thujone, *9546*
[546-88-3] Acetohydroxamic Acid, *62*
[546-89-4] Lithium Acetate, *5577*
[546-97-4] Columbin, *2476*
[547-25-1] Turanose, *10001*
[547-36-4] Succisulfone 2,2′-Iminodiethanol Salt, *9009*
[547-44-4] Sulfanilylurea, *9062*
[547-52-4] N^4-Sulfanilylsulfanilamide, *9061*
[547-57-9] Tropaeolin O, *9954*
[547-58-0] Methyl Orange, *6178*
[547-63-7] Methyl Isobutyrate, *6160*
[547-64-8] Methyl Lactate, *6166*
[547-65-9] α-Methylene Butyrolactone, *6134*

[547-66-0] Magnesium Oxalate, *5740*
[547-67-1] Nickel Oxalate, *6599*
[547-68-2] Zinc Oxalate, *10346*
[547-81-9] 16-Epiestriol, *3671*
[547-91-1] 8-Hydroxy-7-iodo-5-quinolinesulfonic Acid, *4865*
[548-00-5] Ethyl Biscoumacetate, *3825*
[548-04-9] Hypericin, *4902*
[548-24-3] Eosine I Bluish, *3657*
[548-37-8] Verbenalin, *10153*
[548-40-3] Oxyacanthine, *7052*
[548-42-5] Agroclavine, *184*
[548-43-6] Elymoclavine, *3611*
[548-51-6] *o*-Thymotic Acid, *9563*
[548-54-9] Violacein, *10193*
[548-57-2] Lucanthone Hydrochloride, *5650*
[548-62-9] Gentian Violet, *4430*
[548-66-3] Drofenine Hydrochloride, *3494*
[548-68-5] Thiphenamil Hydrochloride, *9524*
[548-73-2] Droperidol, *3499*
[548-76-5] Irigenin, *5134*
[548-77-6] Tectorigenin, *9245*
[548-80-1] Chromotrope 2B, *2244*
[548-83-4] Galangin, *4369*
[548-84-5] Pyrvinium Chloride, *8136*
[548-98-1] Cholane, *2202*
[549-18-8] Amitriptyline Hydrochloride, *483*
[549-28-0] Protostephanine, *8007*
[549-38-2] 3′-Methyl-1,2-cyclopentenophenanthrene, *6119*
[549-56-4] Quinine Bisulfate, *8177*
[550-10-7] Hydrocotarnine, *4825*
[550-24-3] Embelin, *3613*
[550-33-4] Nebularine, *6517*
[550-49-2] Adlumidine, *154*
[550-54-9] Epicinchonidine, *2289*
[550-60-7] 1-Nitro-2-naphthol, *6702*
[550-74-3] Picrolonic Acid, *7524*
[550-82-3] Resazurin, *8261*
[550-83-4] Propoxycaine Hydrochloride, *7950*
[550-90-3] Lupanine *d*-Form, *5667*
[550-97-0] 1-Naphthyl Salicylate, *6495*
[550-99-2] Naphazoline Hydrochloride, *6453*
[551-01-9] Plasmocid, *7636*
[551-06-4] 1-Naphthylisothiocyanate, *6493*
[551-09-7] *N*-(1-Naphthyl)ethylenediamine, *6491*
[551-11-1] Prostaglandin F$_{2α}$, *7989*
[551-16-6] 6-Aminopenicillanic Acid, *454*
[551-27-9] Propicillin, *7931*
[551-64-4] Zinc Tartrate, *10361*
[551-68-8] D-Psicose, *8033*
[551-76-8] 2,4,6-Trichloro-*m*-cresol, *9800*
[551-88-2] 3-Nitropentane, *6703*
[551-92-8] Dimetridazole, *3290*
[552-23-8] 2-Nitro-4-sulfobenzoic Acid, *6732*
[552-30-7] Trimellitic Anhydride, *9872*
[552-32-9] *o*-Nitroacetanilide, *6667*
[552-45-4] *o*-Xylyl Chloride, *10289*
[552-58-9] Eriodictyol, *3726*
[552-59-0] Prunetin, *8017*
[552-66-9] Daidzin, *2798*
[552-70-5] Pseudopelletierine, *8028*

[552-79-4] N-Methylephedrine, 6140
[552-82-9] Methyldiphenylamine, 6126
[552-93-2] Thymol Carbonate, 9554
[552-94-3] Salsalate, 8477
[553-03-7] Hydrocarbostyril, 4817
[553-12-8] Protoporphyrin IX, 8006
[553-19-5] Xanthyletin, 10267
[553-24-2] Neutral Red, 6573
[553-26-4] γ,γ'-Dipyridyl, 3389
[553-27-5] Aniline Mustard, 653
[553-30-0] Proflavine Sulfate, 7887
[553-39-9] Allenolic Acid, 251
[553-54-8] Lithium Benzoate, 5579
[553-58-2] Butethamine meta-Isomer Hydrochloride, 1526
[553-63-9] Dimethocaine Hydrochloride, 3242
[553-68-4] Butethamine Hydrochloride, 1526
[553-69-5] Phenyramidol, 7432
[553-70-8] Magnesium Benzoate, 5721
[553-74-2] Nitrin, 6666
[553-79-7] 5-Nitro-2-propoxyaniline, 6716
[553-84-4] Perilla Ketone, 7281
[553-90-2] Methyl Oxalate, 6179
[553-91-3] Lithium Oxalate, 5593
[554-01-8] 5-Methylcytosine, 6122
[554-12-1] Methyl Propanoate, 6185
[554-13-2] Lithium Carbonate, 5583
[554-15-4] N-Methylpyrroline, 6191
[554-18-7] Glucosulfone Sodium, 4499
[554-35-8] Linamarin, 5552
[554-57-4] Methazolamide, 6034
[554-70-1] Triethylphosphine, 9839
[554-72-3] Tryparsamide, 9974
[554-73-4] Tropaeolin OO, 9955
[554-84-7] m-Nitrophenol, 6706
[554-91-6] Gentiobiose, 4431
[554-92-7] Trimethobenzamide Hydrochloride, 9877
[554-99-4] N-Methylepinephrine, 6141
[555-03-3] m-Nitroanisole, 6671
[555-10-2] β-Phellandrene, 7316
[555-21-5] 4-Nitrobenzyl Cyanide, 6678
[555-30-6] Methyldopa, 6127
[555-31-7] Aluminum Isopropoxide, 342
[555-32-8] Aluminum Benzoate, 325
[555-35-1] Aluminum Palmitate, 351
[555-37-3] Neburon, 6519
[555-43-1] Tristearin, 9932
[555-44-2] Tripalmitin, 9908
[555-45-3] Trimyristin, 9896
[555-54-4] Diphenylmagnesium, 3361
[555-57-7] Pargyline, 7142
[555-59-9] Maleanilic Acid, 5766
[555-68-0] 3-Nitrocinnamic Acid, 6682
[555-75-9] Aluminum Ethoxide, 334
[555-76-0] Ferric Formate, 4050
[555-77-1] Trichlormethine, 9790
[555-89-5] Bis(p-chlorophenoxy)methane, 1252
[555-92-0] Coniine Hydrochloride, 2489
[556-10-5] (4-Nitrophenyl)urea, 6713
[556-24-1] Methyl Isovalerate, 6165
[556-27-4] Alliin, 254
[556-38-7] Zinc Valerate, 10364
[556-45-6] Butyric Acid Magnesium Salt, 1597
[556-50-3] Glycylglycine, 4539
[556-52-5] Glycidol, 4525

[556-53-6] Propylamine Hydrochloride, 7956
[556-56-9] Allyl Iodide, 285
[556-61-6] Methyl Isothiocyanate, 6164
[556-63-8] Lithium Formate, 5588
[556-64-9] Methyl Thiocyanate, 6200
[556-67-2] Octamethylcyclotetrasiloxane, 6836
[556-88-7] N-Nitroguanidine, 6695
[556-89-8] Nitrourea, 6738
[556-91-2] Aluminum tert-Butoxide, 329
[556-99-0] Triuret, 9945
[557-04-0] Magnesium Stearate, 5754
[557-05-1] Zinc Stearate, 10358
[557-07-3] Zinc Oleate, 10344
[557-08-4] Undecylenic Acid Zinc Salt, 10033
[557-09-5] Zinc Caprylate, 10329
[557-11-9] Allylurea, 291
[557-17-5] Methyl Propyl Ether, 6186
[557-18-6] Diethylmagnesium, 3138
[557-19-7] Nickel Cyanide, 6591
[557-20-0] Diethylzinc, 3151
[557-21-1] Zinc Cyanide, 10334
[557-24-4] Maleamic Acid, 5765
[557-25-5] Monobutyrin, 6335
[557-28-8] Zinc Propionate, 10353
[557-31-3] Allyl Ethyl Ether, 284
[557-34-6] Zinc Acetate, 10327
[557-35-7] sec-Octyl Bromide, 6854
[557-36-8] sec-Octyl Iodide, 6858
[557-39-1] Magnesium Formate, 5730
[557-40-4] Allyl Ether, 283
[557-41-5] Zinc Formate, 10336
[557-42-6] Zinc Thiocyanate, 10363
[557-59-5] Lignoceric Acid, 5541
[557-61-9] Octacosanol, 6833
[557-66-4] Ethylamine Hydrochloride, 3817
[559-48-8] Kopsine, 5368
[559-49-9] Annotinine, 669
[559-57-9] Diginatigenin, 3174
[559-70-6] β-Amyrin, 611
[559-74-0] Friedelin, 4300
[560-09-8] Camphoric Acid l-Form, 1735
[560-27-0] Mesoxalic Acid Hydrate, 5983
[561-07-9] Delphinine, 2886
[561-10-4] Isomethadone l-Form, 5230
[561-20-6] Cacotheline, 1610
[561-25-1] Dihydrothebaine, 3199
[561-27-3] Diacetylmorphine, 2971
[561-78-4] Alphaprodine Hydrochloride, 306
[561-86-4] Brallobarbital, 1366
[561-94-4] Ergosine, 3713
[561-95-5] Nimbiol, 6632
[562-09-4] Chlorphenoxamine Hydrochloride, 2187
[562-10-7] Doxylamine Succinate, 3489
[562-26-5] Phenoperidine, 7362
[562-34-5] Chlorogenin, 2144
[562-71-0] Suprasterol II, 9133
[562-76-5] Potassium Tetracyanoplatinate(II), 7803
[562-81-2] Barium Platinous Cyanide, 987
[562-90-3] Silicon Tetraacetate, 8637
[563-12-2] Ethion, 3790
[563-33-7] Methyl Sulfate Calcium Salt, 6195
[563-41-7] Semicarbazide Hydrochloride, 8583

[563-43-9] Ethylaluminum Dichloride, 3816
[563-47-3] 3-Chloro-2-methyl-1-propene, 2150
[563-52-0] 3-Chloro-1-butene, 2131
[563-63-3] Silver Acetate, 8644
[563-68-8] Thallium Acetate, 9405
[563-72-4] Calcium Oxalate, 1687
[563-80-4] Methyl Isopropyl Ketone, 6162
[564-00-1] Erythritol Anhydride, 3734
[564-25-0] Doxycycline (anhydrous), 3488
[564-36-3] Ergocornine, 3705
[564-37-4] Ergocorninine, 3705
[564-87-4] 11-cis-Retinal, 8287
[565-50-4] Terpin trans-Form, 9311
[565-63-9] Angelic Acid, 639
[565-74-2] α-Bromoisovaleric Acid, 1432
[566-48-3] Formestane, 4268
[566-56-3] Allopregnane-3β,20α-diol, 261
[566-57-4] Allopregnane-3α,20β-diol, 260
[566-58-5] Allopregnane-3α,20α-diol, 259
[566-65-4] 3,20-Allopregnanedione, 264
[567-18-0] 1-Naphthol-2-sulfonic Acid, 6475
[568-02-5] Alizarine Blue, 245
[568-21-8] Isothebaine, 5272
[568-53-6] α-Peltatin, 7184
[568-93-4] Alizarine Orange, 246
[569-31-3] Meconin, 5854
[569-57-3] Chlorotrianisene, 2178
[569-58-4] Aluminon, 318
[569-59-5] Phenindamine Hydrogen Tartrate, 7347
[569-64-2] Malachite Green, 5763
[569-65-3] Meclizine, 5848
[569-77-7] Purpurogallin, 8057
[569-84-6] Antipyrine Acetylsalicylate, 703
[570-54-7] Allopregnane-3β,17α-diol-20-one, 263
[572-09-8] Acetobromglucose, 59
[572-43-0] Benzil Dioxime γ-Form, 1081
[572-45-2] Benzil Dioxime β-Form, 1081
[572-59-8] Epiquinidine, 3675
[572-60-1] Epiquinine, 3676
[572-76-9] Nandinine, 6449
[572-96-3] Phytonadiol, 7492
[573-01-3] Menadoxime, 5900
[573-20-6] Menadiol Diacetate, 5898
[573-35-3] Inositol Monophosphate, 5021
[573-41-1] Theophylline Ethanolamine, 9436
[573-56-8] 2,6-Dinitrophenol, 3310
[573-58-0] Congo Red, 2482
[574-12-9] Isoflavone, 5219
[574-42-5] Diphenylmethyl Ether, 1092
[574-64-1] Trypan Red, 9973
[574-65-2] Vital Red, 10210
[574-66-3] Benzophenone Oxime, 1100
[574-84-5] Fraxetin, 4294
[574-95-8] Aureothricin, 871
[575-19-9] 6,7-Benzomorphan, 1097
[575-47-3] Amylpenicillin Sodium Salt, 609

[575-75-7] Toldimfos Sodium, *9672*
[576-19-2] Biocytin, *1232*
[576-26-1] Xylenol 2,6-Dimethylphenol, *10279*
[576-55-6] 3,4,5,6-Tetrabromo-*o*-cresol, *9329*
[576-97-6] Noprylsulfamide, *6772*
[577-11-7] Docusate Sodium, *3446*
[577-27-5] Ledol, *5485*
[577-37-7] Aphylline, *716*
[577-91-3] Iodoalphionic Acid, *5069*
[578-20-1] Thymyl *N*-Isoamylcarbamate, *9564*
[578-36-9] Potassium Salicylate, *7785*
[578-74-5] Apigetrin, *718*
[578-94-9] Phenarsazine Chloride, *7329*
[579-04-4] Cymarose, *2759*
[579-10-2] *N*-Methylacetanilide, *6081*
[579-13-5] Oligomycin A, *6927*
[579-38-4] Diloxanide, *3223*
[579-56-6] Isoxsuprine Hydrochloride, *5285*
[579-92-0] Diphenylamine-2,2'-dicarboxylic Acid, *3348*
[580-02-9] Methyl Acetylsalicylate, *6084*
[580-13-2] 2-Bromonaphthalene, *1437*
[580-52-9] Oxalenediuramidoxime, *7009*
[580-74-5] Xanthocillin X, *10259*
[581-49-7] Anatabine, *620*
[581-64-6] Thionine, *9499*
[581-75-9] 2,6-Naphthalenedisulfonic Acid, *6459*
[581-88-4] Debrisoquin Sulfate, *2846*
[582-36-5] Oxythiamine, *7076*
[582-52-5] Diacetoneglucose, *2968*
[582-60-5] 5,6-Dimethylbenzimidazole, *3258*
[583-03-9] Fenipentol, *4006*
[583-15-3] Mercuric Benzoate, *5941*
[583-39-1] 2-Benzimidazolethiol, *1084*
[583-50-6] D-Erythrose, *3746*
[583-52-8] Potassium Oxalate, *7772*
[583-91-5] Methionine Hydroxy Analog, *6048*
[584-02-1] 3-Pentanol, *7232*
[584-08-7] Potassium Carbonate, *7740*
[584-10-1] Potassium Trithiocarbonate, *7810*
[584-28-1] Aspidin, *836*
[584-42-9] Metachrome Yellow, *5989*
[584-79-2] Allethrin I, *252*
[584-84-9] Toluene 2,4-Diisocyanate, *9689*
[584-85-0] Anserine, *671*
[585-18-2] D-Erythrose 4-Phosphate, *3748*
[585-21-7] Glutamine DL-Form, *4507*
[585-29-5] Triethylammonium Formate, *9832*
[585-48-8] 2,6-Di-*tert*-butylpyridine, *3045*
[585-86-4] Lactitol, *5389*
[585-99-9] Melibiose, *5888*
[586-02-7] Phrenosin, *7479*
[586-06-1] Metaproterenol, *6002*
[586-60-7] Dyclonine, *3520*
[586-76-5] 4-Bromobenzoic Acid, *1415*
[586-96-9] Nitrosobenzene, *6721*
[587-23-5] Methenamine Mandelate, *6038*
[587-26-8] Lanthanum Carbonate, *5415*
[587-33-7] *m*-Tyrosine, *10021*

[587-63-3] Dihydrokawain, *5335*
[587-84-8] Diphenylamine Sulfate, *3347*
[587-90-6] 4,4'-Dinitrocarbanilide, *3304*
[587-98-4] Metanil Yellow, *5999*
[588-59-0] Stilbene, *8946*
[588-60-3] Levulinic Acid Phenylhydrazone, *5526*
[589-15-1] *p*-Bromobenzyl Bromide, *1416*
[589-17-3] *p*-Bromobenzyl Chloride, *1417*
[589-21-9] *p*-Bromophenylhydrazine, *1440*
[589-44-6] 3-Amino-4-hydroxybutyric Acid, *440*
[589-57-1] Diethyl Phosphorochloridite, *3144*
[589-59-3] Isobutyl Isovalerate, *5191*
[589-97-9] Potassium Percarbonate, *7773*
[590-01-2] *n*-Butyl Propionate, *1590*
[590-18-1] 2-Butene *cis*-Form, *1523*
[590-28-3] Potassium Cyanate, *7746*
[590-29-4] Potassium Formate, *7754*
[590-46-5] Betaine Hydrochloride, *1181*
[590-47-6] Betaine Monohydrate, *1181*
[590-63-6] Bethanechol Chloride, *1187*
[590-86-3] Isovaleraldehyde, *5275*
[590-92-1] β-Bromopropionic Acid, *1444*
[591-01-5] Dicyanodiamidine Sulfate, *3104*
[591-09-3] Acetyl Nitrate, *93*
[591-11-7] Angelica Lactone β-Form, *640*
[591-12-8] Angelica Lactone α-Form, *640*
[591-20-8] *m*-Bromophenol, *1439*
[591-27-5] *m*-Aminophenol, *456*
[591-43-5] Aniline Oxalate, *652*
[591-50-4] Iodobenzene, *5072*
[591-59-3] Glyceraldehyde 3-Phosphate, *4518*
[591-64-0] Calcium Levulinate, *1681*
[591-78-6] Methyl Butyl Ketone, *6107*
[591-81-1] γ-Hydroxybutyrate, *4854*
[591-87-7] Allyl Acetate, *276*
[591-89-9] Mercuric Potassium Cyanide, *5952*
[591-97-9] 1-Chloro-2-butene, *2130*
[592-01-8] Calcium Cyanide, *1666*
[592-04-1] Mercuric Cyanide, *5944*
[592-65-4] Isobutyl Sulfide, *5197*
[592-85-8] Mercuric Thiocyanate, *5958*
[592-87-0] Lead Thiocyanate, *5481*
[592-88-1] Allyl Sulfide, *288*
[593-26-0] Ammonium Palmitate, *535*
[593-29-3] Potassium Stearate, *7792*
[593-38-4] Parinaric Acid *cis*-Form, *7144*
[593-39-5] Petroselinic Acid, *7304*
[593-50-0] 1-Triacontanol, *9753*
[593-51-1] Methylamine Hydrochloride, *6088*
[593-53-3] Fluoromethane, *4206*
[593-56-6] Methoxyamine Hydrochloride, *6060*
[593-74-8] Dimethylmercury, *3274*
[593-81-7] Trimethylamine Hydrochloride, *9880*
[594-08-1] Trithiocarbonic Acid, *9936*

[595-05-1] Calycanthine, *1725*
[595-33-5] Megestrol Acetate, *5876*
[595-39-1] Isovaline, *5280*
[595-40-4] Isovaline L-Form, *5280*
[595-77-7] Algestone, *232*
[596-01-0] α-Naphtholphthalein, *6474*
[596-03-2] 4',5'-Dibromofluorescein, *3027*
[596-27-0] *o*-Cresolphthalein, *2565*
[596-51-0] Glycopyrrolate, *4537*
[597-12-6] Melezitose, *5887*
[597-44-4] Citramalic Acid, *2322*
[597-71-7] Pentaerythritol Tetraacetate, *7221*
[598-31-2] Bromoacetone, *1408*
[598-54-9] Cuprous Acetate, *2648*
[598-55-0] Methyl Carbamate, *6110*
[598-58-3] Methyl Nitrate, *6174*
[598-62-9] Manganese Carbonate, *5791*
[598-75-4] 3-Methyl-2-butanol, *6105*
[599-04-2] Pantolactone, *7116*
[599-71-3] Piloty's Acid, *7538*
[599-79-1] Sulfasalazine, *9073*
[599-88-2] Sulfaperine, *9065*
[600-22-6] Pyruvic Acid Methyl Ester, *8135*
[602-09-5] BINOL, *1230*
[602-41-5] Thiocolchicine 2-Glucoside Analog, *9477*
[602-64-2] Anthragallol, *675*
[603-00-9] Proxyphylline, *8014*
[603-33-8] Triphenylbismuth, *9911*
[603-35-0] Triphenylphosphine, *9916*
[603-45-2] Aurin, *872*
[603-50-9] Bisacodyl, *1246*
[603-57-6] Aminopyrine Salicylate, *470*
[603-59-8] Antipyrine Salicylacetate, *703*
[603-64-5] Antipyrine Mandelate, *703*
[604-51-3] Deptropine, *2913*
[604-75-1] Oxazepam, *7025*
[605-50-5] Isoamyl Phthalate, *5169*
[605-65-2] Dansyl Chloride, *2813*
[606-04-2] Pamabrom, *7101*
[606-17-1] Iodipamide, *5065*
[606-21-3] 2-Chloro-1,3-dinitrobenzene, *2138*
[606-22-4] 2,6-Dinitroaniline, *3297*
[606-58-6] Toyocamycin, *9721*
[606-90-6] Diphenylpyraline 8-Chlorotheophyllinate, *3370*
[607-80-7] Sesamin, *8609*
[607-91-0] Myristicin, *6420*
[607-99-8] 2,4,6-Tribromoanisole, *9772*
[608-07-1] 5-Methoxytryptamine, *6076*
[608-66-2] Galactitol, *4362*
[608-89-9] Monoethyl Tartrate, *6339*
[609-09-6] Mesoxalic Acid Diethyl Ester, *5983*
[609-36-9] Proline DL-Form, *7895*
[609-78-9] Cycloguanil Pamoate, *2714*
[610-60-6] Methyl Benzoylsalicylate, *6099*
[610-88-8] Tenuazonic Acid, *9292*
[611-13-2] Methyl 2-Furoate, *6149*
[611-40-5] Tectoridin, *9245*
[611-75-6] Bromhexine Hydrochloride, *1398*
[612-82-8] *o*-Tolidine Dihydrochloride, *9674*
[612-83-9] 3,3'-Dichlorobenzidine Dihydrochloride, *3068*
[613-78-5] 2-Naphthyl Salicylate, *6496*
[613-82-1] Phenylethanolamine Sulfate, *7399*

[693-30-1] Hemisulfur Mustard, *4681*
[693-65-2] *n*-Amyl Ether, *602*
[693-72-1] Vaccenic Acid, *10084*
[694-59-7] Pyridine 1-Oxide, *8084*
[695-34-1] 2-Amino-4-picoline, *462*
[695-53-4] Dimethadione, *3235*
[703-95-7] 5-Fluoroorotic Acid, *4208*
[705-16-8] Chrysanthemic Acid *dl-trans*-Form, *2255*
[705-86-2] δ-Decalactone, *2849*
[709-55-7] Etilefrin, *3914*
[709-98-8] Propanil, *7918*
[713-95-1] δ-Dodecalactone, *3448*
[716-79-0] 2-Phenyl-1*H*-benzimidazole, *7387*
[721-50-6] Prilocaine, *7862*
[723-46-6] Sulfamethoxazole, *9050*
[728-88-1] Tolperisone, *9681*
[729-99-7] Sulfamoxole, *9056*
[730-68-7] Methitural Sodium Salt, *6050*
[732-11-6] Phosmet, *7448*
[737-31-5] Diatrizoic Acid Sodium Salt, *2993*
[738-70-5] Trimethoprim, *9878*
[738-99-8] Dipin, *3376*
[739-71-9] Trimipramine, *9894*
[742-20-1] Cyclopenthiazide, *2740*
[745-65-3] Prostaglandin E₁, *7987*
[747-36-4] Hydroxychloroquine Sulfate, *4857*
[749-02-0] Spiperone, *8878*
[749-13-3] Trifluperidol, *9852*
[751-84-8] Penicillin G Benethamine, *7204*
[751-94-0] Fusidic Acid Sodium Salt, *4346*
[751-97-3] Rolitetracycline, *8381*
[752-56-7] Riboflavin 2′,3′,4′,5′-Tetrabutyrate, *8325*
[752-61-4] Digitalin, *3177*
[756-79-6] DMMP, *3435*
[759-73-9] *N*-Ethyl-*N*-nitrosourea, *3887*
[759-94-4] EPTC, *3697*
[760-78-1] Norvaline DL-Form, *6805*
[762-72-1] Allyltrimethylsilane, *290*
[764-05-6] Cyanogen Azide, *2682*
[767-21-5] Piperidine Phosphate, *7580*
[768-94-5] Amantadine, *368*
[770-05-8] Octopamine DL-Form Hydrochloride, *6848*
[770-12-7] Phenyl Dichlorophosphate, *7394*
[771-61-9] Pentafluorophenol, *7223*
[773-76-2] Chloroxine, *2181*
[790-43-2] 2,2,3,3-Tetrafluoro-1-propanol *p*-Nitrobenzoate, *9351*
[790-69-2] Loflucarban, *5616*
[791-35-5] Chlophedianol, *2066*
[795-38-0] Dihydroisocodeine, *3193*
[797-63-7] Levonorgestrel, *6793*
[797-64-8] Norgestrel (+)-Form, *6793*
[800-22-6] Chloracizine, *2068*
[800-24-8] 2,5-Bis(1-aziridinyl)-3,6-bis(2-methoxyethoxy)-1,4-benzoquinone, *1249*
[801-52-5] Porfiromycin, *7719*
[804-10-4] Chromonar, *2243*
[804-30-8] Fursultiamine, *4340*
[804-36-4] Nitrovin, *6741*
[804-53-5] Tetrabenazine Methanesulfonate, *9326*
[804-63-7] Quinine Sulfate, *8177*
[807-31-8] Aceperone, *31*

[808-71-9] Penethamate Hydriodide, *7194*
[811-54-1] Lead Formate, *5464*
[811-97-2] HFC-134a, *4750*
[813-94-5] Calcium Citrate, *1663*
[814-71-1] Calcium Thioglycollate, *1713*
[814-80-2] Calcium Lactate, *1680*
[814-87-9] Aluminum Oxalate, *349*
[814-89-1] Cobaltous Oxalate, *2429*
[814-91-5] Cupric Oxalate, *2635*
[814-93-7] Lead Oxalate, *5470*
[814-94-8] Stannous Oxalate, *8913*
[815-78-1] Aluminum Tartrate, *362*
[815-82-7] Cupric Tartrate, *2646*
[815-85-0] Stannous Tartrate, *8919*
[817-09-4] Trichlormethine Hydrochloride, *9790*
[818-08-6] Dibutyltin Oxide, *3048*
[821-33-0] 4,4′-Oxydi-2-butanol, *7059*
[822-16-2] Sodium Stearate, *8806*
[823-82-5] 2,5-Furandicarboxaldehyde, *4328*
[826-39-1] Mecamylamine Hydrochloride, *5843*
[828-00-2] Dimethoxane, *3244*
[829-74-3] Levonordefrin, *6781*
[829-85-6] Diphenylphosphine, *3364*
[830-89-7] Albutoin, *211*
[834-12-8] Ametryn, *387*
[834-28-6] Phenformin Hydrochloride, *7345*
[835-31-4] Naphazoline, *6453*
[837-27-4] Acefylline Sodium Salt, *25*
[841-73-6] Bucolome, *1475*
[845-10-3] Methyl Red Sodium Salt, *6192*
[846-48-0] Boldenone, *1327*
[846-49-1] Lorazepam, *5636*
[846-50-4] Temazepam, *9275*
[846-70-8] Naphthol Yellow S, *6478*
[847-25-6] Thiamphenicol DL-Form, *9453*
[847-84-7] Normethadone Hydrochloride, *6798*
[848-21-5] Norgestrienone, *6794*
[848-53-3] Homochlorcyclizine, *4771*
[848-75-9] Lormetazepam, *5639*
[849-55-8] Nylidrin Hydrochloride, *6821*
[850-52-2] Altrenogest, *315*
[852-19-7] Sulfazamet, *9076*
[853-34-9] Kebuzone, *5336*
[855-19-6] Clostebol Acetate, *2397*
[855-96-9] Eupatorin, *3943*
[859-18-7] Lincomycin Hydrochloride, *5555*
[860-22-0] Indigo Carmine, *4982*
[863-61-6] Menaquinone 4, *5901*
[865-04-3] 10-Methoxydeserpidine, *2919*
[865-21-4] Vinblastine, *10179*
[865-28-1] Tyrocidine B, *10017*
[865-44-1] Iodine Trichloride, *5063*
[866-82-0] Cupric Citrate, *2626*
[866-83-1] Potassium Citrate, Monobasic, *7745*
[866-84-2] Potassium Citrate, *7744*
[867-27-6] Demeton-*O*-methyl, *6123*
[867-81-2] Pantothenic Acid Sodium Salt, *7118*
[868-14-4] Potassium Bitartrate, *7736*
[868-18-8] Sodium Tartrate, *8812*

[868-54-2] 2-Amino-1,1,3-tricyanopropene, *476*
[870-09-7] Helenynolic Acid Methyl Ester, *4662*
[870-62-2] Hexamethonium Iodide, *4724*
[870-72-4] Formaldehyde Sodium Bisulfite, *4264*
[870-93-9] DL-Homocystine, *4773*
[872-50-4] 1-Methylpyrrolidone, *6190*
[875-74-1] α-Phenylglycine (*R*)-Form, *7402*
[875-83-2] 2-Naphthol Sodium Salt, *6469*
[876-04-0] Octopamine D(−)-Form, *6848*
[877-24-7] Potassium Biphthalate, *7733*
[881-17-4] Iodohippurate Sodium I 131, *5077*
[882-09-7] Clofibric Acid, *2371*
[882-33-7] Diphenyl Disulfide, *3355*
[886-38-4] Diphencyprone, *3338*
[886-65-7] 1,4-Diphenyl-1,3-butadiene, *3350*
[886-74-8] Chlorphenesin Carbamate, *2185*
[886-86-2] Tricaine, *9786*
[887-08-1] Daucol, *2833*
[892-20-6] Triphenyltin Hydride, *9919*
[894-71-3] Nortriptyline Hydrochloride, *6804*
[897-15-4] Dothiepin Hydrochloride, *3478*
[897-61-0] Penicillin O Potassium Salt, *7208*
[900-95-8] Triphenyltin Hydroxide Acetate, *9920*
[901-93-9] Estrone Acetate, *3763*
[904-04-1] Captodiamine Hydrochloride, *1775*
[908-54-3] Diminazene Aceturate, *3291*
[909-39-7] Opipramol Dihydrochloride, *6948*
[911-45-5] Clomiphene, *2377*
[911-65-9] Etonitazene, *3927*
[912-60-7] Noscapine Hydrochloride, *6807*
[914-00-1] Methacycline, *6015*
[915-30-0] Diphenoxylate, *3343*
[915-67-3] Amaranth (Dye), *369*
[917-13-5] Enniatin B, *3639*
[917-61-3] Sodium Cyanate, *8740*
[917-69-1] Cobaltic Acetate, *2412*
[918-04-7] Acetaldehyde Sodium Bisulfite, *42*
[919-16-4] Lithium Citrate, *5586*
[919-86-8] Demeton-*S*-methyl, *6123*
[921-53-9] Potassium Tartrate, *7796*
[922-55-4] Lanthionine, *5416*
[922-56-5] *meso*-Lanthionine, *5416*
[922-80-5] Diamyl Sodium Sulfosuccinate, *2989*
[923-06-8] Bromosuccinic Acid, *1447*
[923-32-0] Cystine DL-Form, *2779*
[926-26-1] Di-*tert*-butyl Succinate, *3046*
[926-93-2] Methallibure, *6018*
[929-77-1] Behenic Acid Methyl Ester, *1022*
[930-55-2] *N*-Nitrosopyrrolidine, *6730*
[932-53-6] 6-Azathymine, *896*
[932-72-9] Iodobenzene Dichloride, *5073*
[937-13-3] Oxonic Acid, *7046*

[1313-84-4] Sodium Sulfide Nonahydrate, *8810*
[1313-85-5] Sodium Selenide, *8801*
[1313-96-8] Niobium Pentoxide, *6645*
[1313-99-1] Nickel Monoxide, *6597*
[1314-06-3] Nickel Sesquioxide, *6601*
[1314-08-5] Palladium Oxide, *7094*
[1314-11-0] Strontium Oxide, *8976*
[1314-12-1] Thallium Oxide, *9415*
[1314-13-2] Zinc Oxide, *10347*
[1314-15-4] Platinic Oxide, *7642*
[1314-18-7] Strontium Peroxide, *8977*
[1314-20-1] Thorium Oxide, *9531*
[1314-22-3] Zinc Peroxide, *10349*
[1314-23-4] Zirconium Oxide, *10380*
[1314-24-5] Phosphorus Trioxide, *7472*
[1314-27-8] Lead Sesquioxide, *5472*
[1314-28-9] Rhenium Trioxide, *8303*
[1314-32-5] Thallium Sesquioxide, *9417*
[1314-34-7] Vanadium Trioxide, *10109*
[1314-35-8] Tungsten Trioxide, *9997*
[1314-41-6] Lead Tetroxide, *5480*
[1314-56-3] Phosphorus Pentoxide, *7467*
[1314-60-9] Antimony Pentoxide, *689*
[1314-61-0] Tantalum Pentoxide, *9188*
[1314-62-1] Vanadium Pentoxide, *10107*
[1314-64-3] Uranyl Sulfate, *10049*
[1314-68-7] Rhenium Heptoxide, *8301*
[1314-80-3] Phosphorus Pentasulfide, *7466*
[1314-82-5] Phosphorus Pentaselenide, *7465*
[1314-84-7] Zinc Phosphide, *10352*
[1314-85-8] Tetraphosphorus Trisulfide, *9385*
[1314-86-9] Phosphorus Triselenide, *7473*
[1314-87-0] Lead Sulfide, *5476*
[1314-91-6] Lead Telluride, *5477*
[1314-95-0] Stannous Sulfide, *8918*
[1314-96-1] Strontium Sulfide, *8979*
[1314-97-2] Thallium Sulfide, *9419*
[1314-98-3] Zinc Sulfide, *10360*
[1315-01-1] Stannic Sulfide, *8907*
[1315-03-3] Vanadium Trisulfide, *10110*
[1315-04-4] Antimony Pentasulfide, *688*
[1315-05-5] Antimony Triselenide, *700*
[1315-06-6] Stannous Selenide, *8916*
[1315-09-9] Zinc Selenide, *10356*
[1315-11-3] Zinc Telluride, *10362*
[1317-33-5] Molybdenum Disulfide, *6322*
[1317-34-6] Manganese Sesquioxide, *5802*
[1317-35-7] Manganese Oxide, *5800*
[1317-36-8] Lead Monoxide, *5468*
[1317-37-9] Ferrous Sulfide, *4085*
[1317-38-0] Cupric Oxide, *2636*
[1317-39-1] Cuprous Oxide, *2654*
[1317-40-4] Cupric Sulfide, *2645*
[1317-41-5] Cupric Selenide, *2641*
[1317-42-6] Cobaltous Sulfide, *2433*
[1317-54-0] Ferrite, *4062*
[1317-61-9] Ferrosoferric Oxide, *4067*
[1317-82-4] Sapphire, *8504*
[1318-00-9] Vermiculite, *10156*
[1318-93-0] Montmorillonite, *6348*
[1319-46-6] Basic Lead Carbonate, *1002*

[1319-77-3] Cresol, *2564*
[1320-44-1] Methylbenzethonium Chloride Monohydrate, *6097*
[1321-14-8] Potassium Guaiacolsulfonate, *7755*
[1327-39-5] Calcium Aluminosilicate, *1651*
[1327-41-9] Aluminum Hydroxychloride, *339*
[1327-53-3] Arsenic Trioxide, *795*
[1329-83-5] Tigonin, *9591*
[1330-20-7] Xylene, *10278*
[1330-43-4] Sodium Borate, *8726*
[1330-51-4] Pyrogallol Monoacetate, *8112*
[1330-78-5] Tritolyl Phosphate, *9940*
[1332-10-1] Potassium Arsenite Solution, *7729*
[1332-14-5] Cupric Sulfate, Basic, *2644*
[1332-77-0] Potassium Tetraborate, *7799*
[1333-22-8] Copper Sulfate Tribasic, *2644*
[1333-74-0] Hydrogen, *4829*
[1333-82-0] Chromium(VI) Oxide, *2239*
[1333-83-1] Sodium Bifluoride, *8720*
[1334-74-3] Sodium Glycerophosphate, *8757*
[1335-31-5] Mercuric Oxycyanide, *5951*
[1335-32-6] Lead Subacetate, *5474*
[1335-94-0] Irone, *5139*
[1336-20-5] Tetracycline Phosphate Complex, *9341*
[1336-21-6] Ammonia Water, *491*
[1336-80-7] Ferrocholinate, *4065*
[1338-39-2] Sorbitan Laurate, *8850*
[1338-41-6] Sorbitan Stearate, *8850*
[1338-43-8] Sorbitan Oleate, *8850*
[1339-92-0] Basic Aluminum Carbonate Gel, *1001*
[1340-26-7] Phenacetolin, *7320*
[1340-35-8] Urea Stibamine, *10056*
[1340-69-8] Bentoquatam, *1058*
[1341-49-7] Ammonium Bifluoride, *495*
[1343-78-8] Cochineal Extract, *2442*
[1343-98-2] Silicic Acid, *8629*
[1344-09-8] Sodium Silicate, *8804*
[1344-28-1] Aluminum Oxide, *350*
[1344-48-5] Mercuric Sulfide, Black, *5956*
[1344-48-5] Mercuric Sulfide, Red, *5957*
[1344-57-6] Uranium Dioxide, *10040*
[1344-58-7] Uranium Trioxide, *10045*
[1344-95-2] Calcium Silicate, *1704*
[1345-04-6] Antimony Trisulfide, *701*
[1345-05-7] Lithopone, *5602*
[1345-07-9] Bismuth Sulfide, *1287*
[1345-24-0] Gold Stannate, *4555*
[1345-25-1] Ferrous Oxide, *4079*
[1361-49-5] Taxine A, *9215*
[1362-42-1] Absinthin, *13*
[1390-65-4] Carmine, *2442*
[1390-93-8] Citrullol, *2331*
[1391-14-6] Coptine, *2510*
[1391-82-8] Grisein, *4583*
[1392-21-8] Leucomycin, *5505*
[1392-46-7] Micrococcin P, *6259*
[1392-81-0] Parotin, *7147*
[1392-87-6] Phasin, *7314*
[1392-97-8] Protokosin, *5370*

[1393-12-0] Rimocidin, *8354*
[1393-25-5] Secretin, *8558*
[1393-38-0] Subtilin, *8994*
[1393-48-2] Thiostrepton, *9517*
[1393-62-0] Abrin, *10*
[1393-63-1] Annatto, *667*
[1393-64-2] Araroba, *757*
[1393-68-6] Bottromycin, *1361*
[1393-87-9] Fusafungine, *4342*
[1393-92-6] Litmus, *5603*
[1394-02-1] Hachimycin, *4621*
[1395-18-2] Azolitmin, *913*
[1397-74-6] Acetyltannic Acid, *96*
[1397-77-9] Actinorhodine, *131*
[1397-84-8] Alazopeptin, *201*
[1397-89-3] Amphotericin B, *582*
[1397-94-0] Antimycin A, *702*
[1398-17-0] Bakankosin, *936*
[1398-61-4] Chitin, *2065*
[1398-78-3] Colocynthin, *2471*
[1399-64-0] Gymnemic Acid, *4620*
[1400-61-9] Nystatin, *6825*
[1400-62-0] Orcein, *6958*
[1401-54-3] Tanacetin, *9182*
[1401-55-4] Tannic Acid, *9184*
[1401-59-8] Granaticin Tetraacetylgranaticin, *4570*
[1401-69-0] Tylosin, *10013*
[1401-79-2] Viomycin Pantothenate Sulfate, *10197*
[1401-98-5] Condurangin, *2479*
[1402-10-4] Lichenin, *5531*
[1402-37-5] Actinomycetin, *130*
[1402-82-0] Amphomycin, *581*
[1403-17-4] Candicidin, *1743*
[1403-66-3] Gentamicin, *4427*
[1403-71-0] Hamycin, *4642*
[1404-00-8] Mitomycins, *6300*
[1404-04-2] Neomycin, *6539*
[1404-15-5] Nogalamycin, *6758*
[1404-23-5] Pleurotine, *7651*
[1404-26-8] Polymyxin B, *7693*
[1404-52-0] Rhodomycin B, *8315*
[1404-55-3] Ristocetin, *8362*
[1404-59-7] Oligomycin D, *6927*
[1404-64-4] Sparsomycin, *8863*
[1404-74-6] Streptovaricin, *8960*
[1404-88-2] Tyrothricin, *10023*
[1404-90-6] Vancomycin, *10116*
[1404-93-9] Vancomycin Monohydrochloride, *10116*
[1405-10-3] Neomycin B Sulfate, *6539*
[1405-20-5] Polymyxin B Sulfate, *7693*
[1405-36-3] Capreomycin Disulfate, *1759*
[1405-41-0] Gentamicin C Complex Sulfate, *4427*
[1405-59-0] Ristocetin B, *8362*
[1405-86-3] Glycyrrhizic Acid, *4541*
[1405-87-4] Bacitracin, *927*
[1405-89-6] Bacitracin Zinc, *929*
[1405-90-9] Candidin, *1744*
[1406-11-7] Polymyxin, *7693*
[1406-18-4] Vitamin E, *10218*
[1406-65-1] Chlorophyll, *2158*
[1407-47-2] Angiotensin, *644*
[1407-73-4] Vitamin T, *10221*
[1407-85-8] Dextroamphetamine Tannate, *2956*
[1407-93-8] Plantisul, *7632*
[1414-39-7] Albomycin, *207*
[1414-45-5] Nisin, *6650*
[1415-73-2] Aloin A, *303*
[1415-94-7] Iodopsin, *5082*

[1897-45-6] Chlorothalonil, *2169*
[1898-66-4] 1,1-Diphenyl-2-picrylhydrazyl (Free Radical), *3369*
[1904-98-9] 2,6-Diaminopurine, *2987*
[1910-42-5] Paraquat Dichloride, *7135*
[1910-68-5] Methisazone, *6049*
[1912-24-9] Atrazine, *862*
[1912-26-1] Trietazine, *9829*
[1917-65-3] Ethoxymethylfurfural, *3805*
[1918-00-9] Dicamba, *3050*
[1918-02-1] Picloram, *7509*
[1918-08-7] Dipropalin, *3381*
[1918-16-7] Propachlor, *7907*
[1921-70-6] Pristane, *7870*
[1923-76-8] Bunamiodyl Sodium Salt, *1490*
[1926-49-4] Clometocillin, *2376*
[1929-82-4] Nitrapyrin, *6660*
[1934-21-0] Tartrazine, *9206*
[1936-18-1] Salutaridine, *8479*
[1937-19-5] Aminoguanidine Hydrochloride, *438*
[1937-54-8] Solanone, *8834*
[1944-10-1] Fenoterol Hydrochloride, *4012*
[1944-12-3] Fenoterol Hydrobromide, *4012*
[1945-53-5] Palustric Acid, *7099*
[1945-77-3] Methylthymol Blue Tetrasodium Salt, *6202*
[1949-20-8] Oxolamine Citrate, *7043*
[1949-45-7] Metrizoic Acid, *6235*
[1950-39-6] Deferoxamine Hydrochloride, *2863*
[1951-25-3] Amiodarone, *478*
[1953-02-2] Tiopronin, *9609*
[1953-04-4] Galantamine Hydrobromide, *4370*
[1954-28-5] Etoglucid, *3925*
[1962-14-7] Acacic Acid, *16*
[1963-48-0] Anisomycin Hydrochloride, *663*
[1972-08-3] Δ⁹-Tetrahydrocannabinol, *9354*
[1972-28-7] Diethyl Azodicarboxylate, *3126*
[1977-10-2] Loxapine, *5645*
[1977-11-3] Perlapine, *7293*
[1981-58-4] Sulfamethazine Sodium Salt, *9047*
[1982-36-1] Homochlorcyclizine Dihydrochloride, *4771*
[1982-37-2] Methdilazine, *6036*
[1982-49-6] Siduron, *8626*
[1983-72-8] Medicagol, *5859*
[1984-15-2] Medronic Acid, *5863*
[1986-53-4] Bolandiol Dipropionate, *1325*
[1986-70-5] Calotropin, *1722*
[1987-71-9] Nicotinamide Ascorbate, *6608*
[1990-29-0] D-Altrose, *317*
[1997-15-5] Triamcinolone Acetonide 21-Disodium Phosphate, *9758*
[2001-95-8] Valinomycin, *10096*
[2002-24-6] Ethanolamine Hydrochloride, *3782*
[2002-29-1] Flumethasone 21-Pivalate, *4169*
[2002-44-0] MIF-I, *5882*
[2004-70-8] 1,3-Pentadiene (3*E*)-Form, *7220*
[2008-41-5] Butylate, *1548*

[2009-64-5] Neopterin D-*erythro*-Form, *6547*
[2011-67-8] Nimetazepam, *6634*
[2013-12-9] Norvaline D(−)-Form, *6805*
[2013-58-3] Meclocycline, *5849*
[2016-36-6] Choline Salicylate, *2213*
[2016-63-9] Bamifylline, *948*
[2016-88-8] Amiloride Hydrochloride, *401*
[2022-85-7] Flucytosine, *4156*
[2030-53-7] Aporeine, *735*
[2030-63-9] Clofazimine, *2367*
[2032-59-9] Aminocarb, *428*
[2032-65-7] Methiocarb, *6044*
[2033-24-1] Meldrum's Acid, *5885*
[2037-48-1] 2-Deoxystreptamine, *2910*
[2043-38-1] Buthiazide, *1527*
[2050-20-6] Pimelic Acid Diethyl Ester, *7543*
[2050-87-5] Diallyl Trisulfide, *2976*
[2052-01-9] α-Bromoisobutyric Acid, *1431*
[2052-63-3] Neovitamin A, *6552*
[2053-25-0] Etonitazene Hydrochloride, *3927*
[2058-46-0] Oxytetracycline Hydrochloride, *7075*
[2058-52-8] Clothiapine, *2399*
[2058-58-4] Asparagine D-Form, *827*
[2061-86-1] Methandriol Diacetate, *6022*
[2062-77-3] Trifluperidol Hydrochloride, *9852*
[2062-78-4] Pimozide, *7546*
[2062-84-2] Benperidol, *1050*
[2065-00-1] Filicinic Acid, *4109*
[2066-89-9] Isoniazid 4-Aminosalicylate, *5232*
[2068-78-2] Vincristine Sulfate, *10183*
[2071-20-7] DPPM, *1254*
[2074-50-2] Paraquat Bismethyl Sulfate, *7135*
[2078-54-8] Propofol, *7947*
[2079-00-7] Blasticidin S, *1317*
[2079-54-1] (±)-Deprenyl Hydrochloride, *8565*
[2079-78-9] Hexamethonium Tartrate, *4724*
[2085-33-8] Aluminum Tris(8-hydroxyquinoline), *364*
[2086-83-1] Berberine, *1156*
[2086-96-6] Canadine d-Form, *1738*
[2090-89-3] Butethamine, *1526*
[2092-16-2] Calcium Thiocyanate, *1712*
[2092-17-3] Barium Thiocyanate, *994*
[2094-98-6] 1,1'-Azobis(cyclohexanecarbonitrile), *910*
[2095-57-0] Thiobutabarbital, *9476*
[2096-42-6] Gougerotin, *4565*
[2098-66-0] Cyproterone, *2771*
[2099-26-5] Androstenediol Diacetate, *630*
[2103-64-2] Gallein, *4374*
[2104-64-5] EPN, *3682*
[2104-96-3] Bromophos, *1442*
[2105-47-7] Tetrabenazine Hydrochloride, *9326*
[2111-75-3] Perillaldehyde, *7282*
[2116-55-4] Mimosine *dl*-Form, *6280*
[2122-29-4] Retamine, *8280*
[2124-57-4] Menaquinone 7, *5901*
[2135-17-3] Flumethasone, *4169*
[2140-46-7] 25-Hydroxycholesterol, *4860*

[2141-09-5] Magnoflorine, *5760*
[2150-48-3] Pyronine B, *8118*
[2152-34-3] Pemoline, *7188*
[2152-44-5] Betamethasone 17-Valerate, *1182*
[2152-56-9] Arabitol, *751*
[2152-76-3] Idose, *4930*
[2153-98-2] Norpseudoephedrine Hydrochloride, *6803*
[2156-27-6] Benproperine, *1051*
[2163-80-6] Methanearsonic Acid Monosodium Salt, *6025*
[2164-08-1] Lenacil, *5489*
[2164-09-2] Dicryl, *3098*
[2164-17-2] Fluometuron, *4185*
[2165-19-7] Guanoxan, *4603*
[2167-85-3] Pipazethate, *7569*
[2169-64-4] 6-Azauridine 2',3',5'-Triacetate, *897*
[2169-75-7] Deptropine Citrate, *2913*
[2174-16-5] Triethanolamine Salicylate, *9830*
[2179-16-0] Ninopterin, *6641*
[2179-37-5] Bencyclane, *1035*
[2181-04-6] Canrenone Free Acid Potassium Salt, *1753*
[2182-14-1] Vindoline, *10185*
[2188-68-3] Lycorine Hydrochloride, *5687*
[2192-20-3] Hydroxyzine Dihydrochloride, *4889*
[2198-93-8] Trichodermol, *9812*
[2207-50-3] Aminorex, *472*
[2210-63-1] Mofebutazone, *6317*
[2210-64-2] Pyrrocaine Hydrochloride, *8127*
[2210-77-7] Pyrrocaine, *8127*
[2216-51-5] Menthol, *5905*
[2216-92-4] *N*-Phenylglycine Ethyl Ester, *7403*
[2217-44-9] Iodophthalein Sodium, *5081*
[2218-94-2] Nitron, *6700*
[2219-30-9] Penicillamine Hydrochloride, *7197*
[2219-31-0] Canavanine Sulfate, *1740*
[2221-95-6] Fichtelite, *4105*
[2226-96-2] TEMPOL, *9284*
[2227-17-0] Dienochlor, *3116*
[2227-29-4] Chlorodiisopropylsilane, *2136*
[2240-14-4] Fencamfamine Hydrochloride, *3998*
[2241-90-9] Cyclobuxine D, *2709*
[2243-33-6] α-Phellandrene d-Form, *7315*
[2243-76-7] Alizarine Yellow R, *247*
[2244-21-5] Troclosene Potassium, *9946*
[2255-39-2] Muscazone, *6396*
[2257-09-2] Phenethyl Isothiocyanate, *7340*
[2259-14-5] Chrysanthemic Acid *l*-*trans*-Form, *2255*
[2259-96-3] Cyclothiazide, *2752*
[2272-11-9] Ethanolamine Oleate, *3782*
[2276-90-6] Iothalamic Acid, *5107*
[2277-92-1] Oxyclozanide, *7057*
[2280-32-2] Kanamycin C, *5329*
[2284-31-3] Pratensein, *7834*
[2290-11-1] Verrucarin B, *10160*
[2292-79-7] Congressane, *2483*
[2295-58-1] Flopropione, *4138*
[2297-30-5] Androstenediol Dipropionate, *630*

[2898-11-5] Medazepam Hydrochloride, 5857
[2898-12-6] Medazepam, 5857
[2900-38-1] Sporidesmolide I, 8893
[2919-66-6] Melengestrol Acetate, 5886
[2921-88-2] Chlorpyrifos, 2195
[2921-92-8] Propatyl Nitrate, 7925
[2922-44-3] Dextromoramide Bitartrate, 2958
[2934-97-6] Tetrahydropalmatine, 9360
[2935-23-1] Chrysanthemic Acid dl-cis-Form, 2255
[2935-35-5] α-Phenylglycine (S)-Form, 7402
[2945-08-6] 4-Nitrophenylacetic Acid Methyl Ester, 6709
[2945-88-2] Sophorabioside, 8844
[2955-23-9] Olivil, 6931
[2955-38-6] Prazepam, 7836
[2957-21-3] Sakuranetin, 8455
[2970-95-8] Macromerine, 5709
[2971-90-6] Clopidol, 2386
[2998-57-4] Estramustine, 3761
[2998-94-9] p-Dimethylaminobenzophenone anti-Oxime, 3255
[2998-95-0] p-Dimethylaminobenzophenone syn-Oxime, 3255
[3001-72-7] 1,5-Diazabicyclo[4.3.0]non-5-ene, 2995
[3012-65-5] Ammonium Citrate, Dibasic, 509
[3017-60-5] Cobaltous Thiocyanate, 2434
[3031-95-6] SAICAR, 8454
[3031-98-9] 2-Keto-L-gulonic Acid Methyl Ester, 5351
[3040-38-8] Acetylcarnitine L-Form, 79
[3051-09-0] Murexide, 6391
[3054-35-1] Aspartic Acid Compound with L-Arginine, 830
[3056-17-5] Stavudine, 8929
[3059-97-0] Isovaline D-Form, 5280
[3060-41-1] 4-Amino-3-phenylbutyric Acid Hydrochloride, 460
[3060-89-7] Metobromuron, 6218
[3061-75-4] Behenic Acid Amide, 1022
[3064-61-7] Stibocaptate, 8941
[3081-61-6] Theanine, 9424
[3092-17-9] Midodrine Hydrochloride, 6263
[3093-35-4] Halcinonide, 4626
[3094-09-5] Doxifluridine, 3485
[3095-65-6] Ammonium Bitartrate, 500
[3100-04-7] 1-Methylcyclopropene, 6120
[3101-51-7] Ergoflavin, 3709
[3105-97-3] Hycanthone, 4798
[3115-49-9] (p-Nonylphenoxy)acetic Acid, 6769
[3116-76-5] Dicloxacillin, 3094
[3122-01-8] Thiazesim Hydrochloride, 9457
[3131-03-1] Pristinamycin IA, 7871
[3144-16-9] d-Camphorsulfonic Acid, 1736
[3147-14-6] Calmagite, 1720
[3147-55-5] 3,5-Dibromosalicylic Acid, 3033
[3148-09-2] Verrucarin A, 10160
[3150-28-5] D-Chalcose, 2038
[3155-48-4] (±)-Kawain, 5335
[3155-57-5] (±)-Dihydromethysticin, 6214
[3160-91-6] Moroxydine Hydrochloride, 6358

[3164-34-9] Calcium Tartrate, 1711
[3166-62-9] Benactyzine Methobromide, 1031
[3167-49-5] 6-Aminonicotinic Acid, 452
[3169-21-9] N-(p-Methoxyphenyl)-p-phenylenediamine Sulfate, 6072
[3183-08-2] DL-Lanthionine, 5416
[3194-55-6] Hexabromocyclododecane, 4712
[3198-07-0] Ethylnorepinephrine Hydrochloride, 3888
[3200-06-4] Nafronyl Acid Oxalate, 6437
[3200-75-7] Sporidesmolide II, 8893
[3202-84-4] 4-Salicyloylmorpholine, 8470
[3207-50-9] Linoleic Acid Cyclohexylamide, 5560
[3211-76-5] Selenomethionine, 8578
[3215-70-1] Hexoprenaline, 4743
[3230-75-9] Oxenin, 7031
[3230-94-2] L-Ornithine-L-aspartate, 6969
[3232-84-6] Urazole, 10051
[3239-44-9] Dexfenfluramine, 4004
[3239-45-0] Dexfenfluramine Hydrochloride, 4004
[3244-88-0] Acid Fuchsin, 99
[3251-23-8] Cupric Nitrate, 2633
[3252-29-7] Tyrocidine C, 10017
[3254-89-5] Diphenidol Hydrochloride, 3341
[3264-82-2] Nickel Acetylacetonate, 6585
[3269-83-8] Pheniramine p-Aminosalicylate, 7349
[3279-54-7] Disodium Phenyl Phosphate, 3399
[3286-46-2] Thiamine Disulfide O,O-Diisobutyrate, 9449
[3296-43-3] Petroselinic Acid Glyceryl Triester, 7304
[3306-52-3] Viridin, 10202
[3317-61-1] DMPO, 3437
[3326-32-7] Asenapine (5-isothiocyanate), 4120
[3329-91-7] Dioscorine, 3322
[3337-71-1] Asulam, 847
[3342-61-8] Deanol Aceglumate, 2845
[3344-18-1] Magnesium Citrate Tribasic, 5727
[3347-22-6] Dithianon, 3411
[3349-06-2] Nickel Formate, 6594
[3351-86-8] Fucoxanthin, 4310
[3363-58-4] Nifurfoline, 6617
[3366-95-8] Secnidazole, 8556
[3368-16-9] α-Phenylcinnamic Acid, 7393
[3371-85-5] Voacamine, 10223
[3372-02-9] Normorphine Hydrochloride, 6800
[3375-31-3] Palladium Diacetate, 7092
[3376-24-7] PBN, 7162
[3376-83-8] Biotin l-Sulfoxide, 1237
[3380-34-5] Triclosan, 9822
[3383-96-8] Temephos, 9278
[3385-03-3] Flunisolide, 4176
[3395-91-3] β-Bromopropionic Acid Methyl Ester, 1444
[3397-23-7] Ornipressin, 6968
[3416-24-8] Glucosamine, 4494
[3416-26-0] Lidoflazine, 5537
[3420-59-5] Isomaltol, 5227

[3424-98-4] Telbivudine, 9258
[3440-28-6] Betamipron, 1183
[3458-28-4] D-Mannose, 5812
[3459-06-1] Cyclopentamine Hydrochloride, 2735
[3459-20-9] Glymidine Sodium Salt, 4543
[3460-67-1] Furonazide, 4337
[3463-92-1] Carpaine, 1860
[3468-99-3] Diphenylacetic Acid Ethyl Ester, 3346
[3469-00-9] Diphenylacetic Acid Methyl Ester, 3346
[3475-65-8] Thiamine Triphosphate, 9450
[3483-12-3] 1,4-Dithiothreitol, 3419
[3485-14-1] Cyclacillin, 2695
[3485-62-9] Clidinium Bromide, 2352
[3486-35-9] Zinc Carbonate, 10330
[3486-66-6] Coptisine, 2512
[3486-67-7] Palmatine, 7095
[3487-44-3] Methylenetriphenylphosphorane, 6138
[3493-12-7] S-Methylmethionine DL-Form Chloride, 6170
[3505-38-2] Carbinoxamine Maleate, 1800
[3511-16-8] Hetacillin, 4707
[3521-62-8] Erythromycin Estolate, 3740
[3521-84-4] Iodipamide Bis[N-methylglucamine] Salt, 5065
[3522-50-7] Ferric Citrate (1:1 iron(III) citrate salt), 4048
[3543-75-7] Bendamustine Hydrochloride, 1036
[3544-35-2] Iproclozide, 5120
[3546-03-0] Cyamemazine, 2673
[3546-21-2] Homidium, 4769
[3546-41-6] Pyrvinium Pamoate, 8136
[3562-84-3] Benzbromarone, 1067
[3562-99-0] Menbutone, 5902
[3563-27-7] Dihydroequilin, 3190
[3563-49-3] Pyrovalerone, 8122
[3563-84-6] Dihydrostreptomycin Pantothenate, 3197
[3564-09-8] Ponceau 3R, 7712
[3566-44-7] N-(p-Methoxyphenyl)-p-phenylenediamine Hydrochloride, 6072
[3568-24-9] Propionylpromazine, 7943
[3568-94-3] 4-Methylaminorex, 6092
[3569-99-1] N-(Hydroxymethyl)nicotinamide, 4872
[3571-88-8] Platonin, 7645
[3572-06-3] Cuelure, 2606
[3572-43-8] Bromhexine, 1398
[3572-60-9] Amidinomycin, 394
[3572-74-5] Moxastine, 6373
[3572-80-3] Cyclazocine, 2698
[3573-82-8] Hygrophylline, 4893
[3575-80-2] Melperone, 5895
[3583-64-0] Bumadizone, 1488
[3588-17-8] Muconic Acid (E,E)-Form, 6385
[3589-73-9] 10-Methoxyharmalan, 6068
[3593-85-9] Methandriol Dipropionate, 6022
[3595-11-7] Propylhexedrine dl-Form, 7972
[3598-37-6] Acepromazine Maleate, 33
[3599-32-4] Indocyanine Green, 5002
[3600-95-1] Perezone, 7269

[3604-87-3] Ecdysone, *3538*
[3605-01-4] Piribedil, *7607*
[3607-18-9] Doxepin *trans*-Form Hydrochloride, *3483*
[3613-73-8] Dimebolin, *3225*
[3614-30-0] Emepronium Bromide, *3615*
[3614-69-5] Dimethindene Maleate, *3238*
[3615-24-5] Ramifenazone, *8220*
[3615-37-0] D-Fucose, *4307*
[3615-41-6] Rhamnose, *8295*
[3615-44-9] D-*manno*-Heptulose, *4701*
[3615-82-5] Phytic Acid Calcium Magnesium Salt, *7499*
[3616-78-2] Fenfluramine *l*-Form Hydrochloride, *4004*
[3618-43-7] Xylenol Orange Tetrasodium Salt, *10281*
[3621-36-1] Columbamine, *2475*
[3621-38-3] Jatrorrhizine, *5309*
[3625-06-7] Mebeverine, *5838*
[3627-49-4] Phenoperidine Hydrochloride, *7362*
[3632-91-5] Magnesium Gluconate, *4492*
[3635-74-3] Deanol Acetamidobenzoate, *2845*
[3639-12-1] Butethamate Citrate, *1525*
[3644-61-9] Tolperisone Hydrochloride, *9681*
[3650-09-7] Carnosic Acid, *1850*
[3650-17-7] Hirsutic Acid C, *4754*
[3653-48-3] MCPA Sodium Salt, *5833*
[3666-69-1] Dioxadrol Hydrochloride, *3327*
[3671-76-9] (+)-Anisomorphal, *3459*
[3682-35-7] 2,4,6-Tripyridyl-*s*-triazine, *9926*
[3684-26-2] 3-Quinuclidinol *dl*-Form, *8199*
[3685-84-5] Meclofenoxate Hydrochloride, *5851*
[3686-58-6] Tolycaine, *9700*
[3687-18-1] Tramiprosate, *9728*
[3688-62-8] Aminopromazine Fumarate, *463*
[3689-24-5] Sulfotep, *9096*
[3689-50-7] Oxomemazine, *7045*
[3689-76-7] Chlormidazole, *2107*
[3691-21-2] Buzepide, *1603*
[3691-35-8] Chlorophacinone, *2155*
[3691-74-5] Glyconiazide, *4536*
[3691-81-4] Sulfoniazide, *9089*
[3697-42-5] Chlorhexidine Dihydrochloride, *2092*
[3703-76-2] Cloperastine, *2384*
[3703-79-5] Bamethan, *947*
[3704-09-4] Mibolerone, *6253*
[3715-90-0] Tramazoline Hydrochloride Monohydrate, *9727*
[3717-88-2] Flavoxate Hydrochloride, *4128*
[3724-89-8] Phosphocysteamine Sodium Salt, *7453*
[3731-59-7] Moroxydine, *6358*
[3734-33-6] Denatonium Benzoate, *2893*
[3734-52-9] Metazocine, *6005*
[3735-45-3] Vetrabutine, *10168*
[3735-84-0] Reserpilic Acid Dimethylaminoethyl Ester Dihydrochloride, *8264*
[3736-08-1] Fenethylline, *4003*

[3736-36-5] Tropine Benzylate, *9961*
[3736-81-0] Diloxanide Furoate, *3223*
[3737-09-5] Disopyramide, *3402*
[3738-01-0] Otobain, *6998*
[3749-97-1] Laureline (±)-Form, *5438*
[3758-45-0] Mycarose 3-*O*-Methylmycarose, *6403*
[3763-55-1] Rubixanthin, *8422*
[3772-43-8] Butoxycaine, *1533*
[3772-76-7] Sulfamethomidine, *9049*
[3778-22-1] Methylthymol Blue, *6202*
[3778-73-2] Ifosfamide, *4937*
[3778-76-5] Todralazine Hydrochloride, *9660*
[3779-59-7] Cimigenol, *2284*
[3779-61-1] Ocimene *trans*-β-Form, *6830*
[3785-32-8] Salicylamide *O*-Acetic Acid Sodium Salt, *8466*
[3790-78-1] Nerolidol *cis*-Form, *6561*
[3801-06-7] Fluorometholone 17-Acetate, *4207*
[3804-89-5] Isoniazid Methanesulfonate Sodium, *5232*
[3810-35-3] Tenonitrozole, *9290*
[3810-74-0] Streptomycin Sesquisulfate, *8956*
[3810-80-8] Diphenoxylate Hydrochloride, *3343*
[3811-04-9] Potassium Chlorate, *7741*
[3811-20-9] Cuminaldehyde Thiosemicarbazone, *2611*
[3811-56-1] Aminoquinuride, *471*
[3811-75-4] Hexamidine, *4730*
[3818-50-6] Bephenium Hydroxynaphthoate, *1150*
[3818-62-0] Betoxycaine, *1189*
[3819-00-9] Piperacetazine, *7574*
[3825-26-1] Perfluorooctanoic Acid Ammonium Salt, *7272*
[3835-52-7] *O*-Cinnamoyltaxicin-II triacetate, *9214*
[3844-45-9] Brilliant Blue FCF, *1379*
[3844-67-5] Allantoin *S*-(+)-Form, *250*
[3847-29-8] Erythromycin Lactobionate, *3741*
[3856-25-5] Copaene, *2500*
[3858-89-7] Chloroprocaine Hydrochloride, *2160*
[3861-73-2] Anazolene Sodium, *622*
[3871-82-7] Moperone Hydrochloride, *6351*
[3882-38-0] Morphinan *N*-Methylmorphinan, *6360*
[3900-31-0] Fludiazepam, *4159*
[3902-71-4] Trioxsalen, *9907*
[3922-90-5] Oleandomycin, *6918*
[3926-62-3] Chloroacetic Acid Sodium Salt, *2112*
[3930-19-6] Streptonigrin, *8957*
[3930-20-9] Sotalol, *8854*
[3938-95-2] Pivalic Acid Ethyl Ester, *7625*
[3947-65-7] Neamine, *6514*
[3952-98-5] Sinigrin, *8683*
[3963-95-9] Methacycline Hydrochloride, *6015*
[3964-81-6] Azatadine, *894*
[3972-41-6] Bromosuccinic Acid *d*-Form, *1447*
[3978-86-7] Azatadine Dimaleate, *894*
[3982-91-0] Phosphorus Sulfochloride, *7468*
[4008-48-4] Nitroxoline, *6742*

[4021-34-5] 5-Methoxy-*N*,*N*-diisopropyltryptamine, *6062*
[4042-30-2] Parvaquone, *7154*
[4043-71-4] *trans*-Isosafrole, *5269*
[4044-65-9] Bitoscanate, *1309*
[4055-39-4] Mitomycin A, *6300*
[4055-40-7] Mitomycin B, *6300*
[4065-45-6] Sulisobenzone, *9113*
[4070-80-8] Sodium Stearyl Fumarate, *8807*
[4075-81-4] Calcium Propionate, *1700*
[4076-02-2] 2,3-Dimercapto-1-propanesulfonic Acid Sodium Salt, *3232*
[4080-31-3] Quaternium-15, *8143*
[4088-60-2] Crotyl Alcohol *cis*-Form, *2591*
[4093-35-0] Bromopride, *1443*
[4097-22-7] Dideoxyadenosine, *3112*
[4098-40-2] Mitragynine, *6303*
[4111-54-0] Diisopropylamine Lithium Salt, *3215*
[4112-89-4] Guaiacol Phenylacetate, *4589*
[4119-52-2] Ferric Thiocyanate, *4061*
[4135-11-9] Polymyxin B₁, *7693*
[4136-95-2] 2,4,6-Trichlorobenzoyl Chloride, *9798*
[4143-60-6] Diethyl Azodicarboxylate (Z-form), *3126*
[4143-61-7] Diethyl Azodicarboxylate (E-form), *3126*
[4147-51-7] Dipropetryn, *3382*
[4154-10-3] Rolitetracycline Compd with Chloramphenicol Succinate, *8381*
[4154-53-4] *d*-Camphocarboxylic Acid Basic Bismuth Salt, *1733*
[4156-16-5] Woodward's Reagent K, *10248*
[4163-15-9] Cyclorphan, *2748*
[4170-30-3] Crotonaldehyde, *2587*
[4171-13-5] Valnoctamide, *10098*
[4180-23-8] Anethole, *636*
[4197-25-5] Sudan Black B, *9016*
[4205-23-6] D-Gulose, *4613*
[4205-90-7] Clonidine, *2380*
[4205-91-8] Clonidine Hydrochloride, *2380*
[4221-98-1] α-Phellandrene *l*-Form, *7315*
[4221-99-2] *sec*-Butyl Alcohol *d*-Form, *1543*
[4245-41-4] Estradiol 3-Acetate, *3758*
[4254-14-2] Propylene Glycol *l*-Form, *7968*
[4254-15-3] Propylene Glycol *d*-Form, *7968*
[4258-85-9] Clocortolone 21-Acetate, *2363*
[4267-05-4] Teclothiazide, *9241*
[4268-36-4] Tybamate, *10010*
[4271-30-1] *N*-(*p*-Aminobenzoyl)glutamic Acid, *421*
[4276-74-8] Lactisole *S*-Isomer, *5388*
[4282-16-0] *N*,*N*'-Dimethylcyclobuxine D, *2709*
[4291-63-8] Cladribine, *2336*
[4299-57-4] Plastoquinone 9, *7637*
[4299-60-9] Sulfisoxazole Diethanolamine Salt, *9082*
[4300-28-1] D-Ribose-5-phosphoric Acid, *8329*
[4310-35-4] Tridihexethyl Chloride, *9826*

[4312-32-7] Stylopine, 8988
[4316-73-8] Sarcosine Sodium Salt, 8510
[4316-74-9] N-Methyltaurine Sodium Salt, 6196
[4317-14-0] Amitriptylinoxide, 484
[4320-30-3] Arginine Glutamate, 771
[4321-58-8] ε-Aminocaproic Acid Hydrochloride, 427
[4323-43-7] Hexoprenaline Dihydrochloride, 4743
[4330-99-8] Trimeprazine Tartrate, 9873
[4337-33-1] Dimethylsulfoniopropionate Chloride, 3284
[4342-03-4] Dacarbazine, 2795
[4345-03-3] α-Tocopherol Succinate, 9653
[4350-07-6] 5-Hydroxytryptophan D-Form, 4886
[4350-09-8] 5-Hydroxytryptophan L-Form, 4886
[4353-06-4] 2-Nonyldioxolane, 6767
[4356-43-8] Lupanine dl-Form, 5667
[4358-59-2] Isocrotonic Acid Methyl Ester, 5209
[4360-12-7] Ajmaline, 187
[4368-28-9] Tetrodotoxin, 9394
[4382-36-9] Fustin (2R,3R)-Form, 4347
[4386-35-0] Meralein Sodium, 5932
[4388-82-3] Propylhexedrine l-Form Ethylphenylbarbiturate, 7972
[4393-19-5] p-Sulfanilylbenzylamine, 9059
[4394-00-7] Niflumic Acid, 6615
[4396-01-4] Isopelletierine, 7181
[4403-90-1] Alizarin Cyanine Green F, 244
[4406-22-8] Cyprenorphine, 2766
[4412-09-3] Mucochloric Anhydride, 6384
[4419-39-0] Beclomethasone, 1017
[4432-31-9] MES, 5972
[4438-22-6] Atropine N-Oxide, 866
[4439-84-3] HON (unspecified stereo), 4780
[4448-95-7] Shellolic Acid, 8618
[4449-51-8] Cyclopamine, 2733
[4461-30-7] Chloroacetyl Isocyanate, 2117
[4462-95-7] Truxillic Acid, 9971
[4466-14-2] Jasmolin I, 5306
[4468-02-4] Zinc Gluconate, 4492
[4478-93-7] Sulforaphane (unspecified stereo), 9092
[4481-55-4] Pentryl, 7252
[4482-25-1] Bismark Brown R Free Base, 1256
[4482-55-7] Fenuron Trichloroacetate, 4036
[4482-83-1] BBB, 4111
[4493-18-9] Amylpenicillin, 609
[4493-23-6] Dodecahedrane, 3447
[4493-37-2] Chromous Formate, 2250
[4498-32-2] Dibenzepin, 3012
[4499-40-5] Choline Theophyllinate, 2215
[4508-48-9] Djenkolic Acid Monohydrochloride, 3431
[4533-39-5] Nitracrine, 6657
[4533-89-5] Flunisolide 21-Acetate, 4176
[4544-15-4] Piperilate Hydrochloride, 7583

[4546-39-8] Piperilate, 7583
[4548-53-2] Ponceau SX, 7713
[4556-55-2] Diethyl Oxalacetate Semicarbazone, 3141
[4562-36-1] Gitoxin, 4469
[4564-87-8] Carbomycin A, 1806
[4573-78-8] Hamamelose, 4641
[4579-64-0] Lysergic Acid Methyl Ester, 5694
[4589-33-7] Bostrycoidin, 1356
[4602-84-0] Farnesol, 3975
[4611-02-3] Chlorproethazine Hydrochloride, 2189
[4618-18-2] Lactulose, 5395
[4619-74-3] 2,4,6-Tribromo-m-cresol, 9774
[4626-00-0] Kitol, 5363
[4628-21-1] 1-Chloro-2-butene cis-Isomer, 2130
[4630-95-9] Prifinium Bromide, 7861
[4638-92-0] Chrysanthemic Acid d-trans-Form, 2255
[4643-58-7] Verrucarin J, 10160
[4670-05-7] Theaflavine, 9423
[4673-18-1] Buphanitine, 1497
[4674-50-4] Nootkatone, 6770
[4678-44-8] Grayanotoxin II, 4578
[4678-45-9] Grayanotoxin III, 4578
[4680-78-8] Guinea Green B, 4610
[4682-36-4] Orphenadrine Citrate, 6975
[4682-48-8] Cytolipin H, 2791
[4682-50-2] Trichodermin, 9812
[4685-14-7] Paraquat, 7135
[4695-13-0] Diphenylacetic Acid Amide, 3346
[4695-62-9] d-Fenchone, 4000
[4696-76-8] Kanamycin B, 5329
[4697-14-7] Ticarcillin Disodium Salt, 9584
[4697-36-3] Carbenicillin, 1793
[4705-64-0] Physalaemin Trifluoroacetate, 7493
[4719-75-9] Fosfestrol Tetrasodium Salt, 4280
[4720-09-6] Grayanotoxin I, 4578
[4727-40-6] S-Methylmethionine, 6170
[4733-39-5] Bathocuproine, 1005
[4747-99-3] Tetrahydropapaveroline, 9361
[4759-48-2] Isotretinoin, 5274
[4764-17-4] MDA, 5834
[4767-03-7] Dimethylolpropionic Acid, 3277
[4776-06-1] Fluorosalan, 4210
[4800-94-6] Carbenicillin Disodium Salt, 1793
[4803-27-4] Anthramycin, 677
[4825-86-9] Ochratoxin B, 6829
[4826-71-5] Phosphorylcholine Calcium Salt, 7474
[4828-27-7] Clocortolone, 2363
[4846-91-7] Fenoxazoline, 4016
[4850-21-9] Quebrachamine, 8146
[4857-44-7] Methionine Hydroxy Analog Calcium Salt, 6048
[4865-85-4] Ochratoxin C, 6829
[4880-88-0] Eburnamonine (−)-Form, 3534
[4891-15-0] Estramustine 17-(Dihydrogen Phosphate), 3761
[4894-61-5] 1-Chloro-2-butene trans-Isomer, 2130
[4914-30-1] Dehydroemetine, 2874
[4936-47-4] Nifuratel, 6616

[4940-11-8] Ethyl Maltol, 3879
[4940-39-0] Chromocarb, 2241
[4945-47-5] Bamipine, 949
[4955-90-2] Gentisic Acid Sodium Salt, 4433
[4956-37-0] Estradiol Enanthate, 3758
[4966-84-1] Vasicine Hydriodide, 10127
[4968-09-6] Algestone Cyclic Acetal with Acetone, 232
[4968-29-0] Guaiazulene Trinitrobenzene Deriv, 4590
[4969-02-2] Methixene, 6051
[4976-25-4] Tremetone, 9743
[4991-65-5] Tioxolone, 9612
[5001-33-2] Metanephrine, 5997
[5002-47-1] Fluphenazine Decanoate, 4220
[5003-47-4] Betoxycaine Monohydrochloride, 1189
[5003-48-5] Benorylate, 1047
[5008-52-6] Napelline, 6452
[5011-21-2] Potassium 5-Guaiacolsulfonate, 7755
[5011-34-7] Trimetazidine, 9874
[5036-02-2] Tetramisole, 5511
[5051-62-7] Guanabenz, 4593
[5053-06-5] Fenspiride, 4026
[5053-08-7] Fenspiride Hydrochloride, 4026
[5060-55-9] Prednisolone 21-Stearoylglycolate, 7841
[5064-31-3] Nitrilotriacetic Acid Trisodium Salt, 6665
[5072-26-4] Buthionine Sulfoximine, 1528
[5080-50-2] Acetylcarnitine L-Form Hydrochloride, 79
[5086-74-8] Tetramisole Hydrochloride, 5511
[5096-57-1] Canadine l-Form, 1738
[5103-42-4] Hydrindantin, 4813
[5104-49-4] Flurbiprofen, 4229
[5118-29-6] Melitracen, 5891
[5119-48-2] Withaferin A, 10246
[5141-20-8] Light Green SF Yellowish, 5538
[5142-76-7] Bismuth Subacetate, 1281
[5144-89-8] o-Phenanthroline Monohydrate, 7328
[5153-24-2] Zirconyl Acetate, 10383
[5158-50-9] Bromodimethylborane, 1426
[5169-78-8] Tipepidine, 9613
[5175-83-7] Bismuth Tribromophenate, 1291
[5189-11-7] Pizotyline Malate, 7629
[5192-84-7] Trengestone, 9746
[5197-58-0] Stenbolone, 8935
[5205-82-3] Bevonium Methyl Sulfate, 1197
[5221-53-4] Dimethirimol, 3239
[5231-60-7] Demethoxyvindoline, 10185
[5234-68-4] Carboxin, 1828
[5250-39-5] Floxacillin, 4143
[5251-34-3] Cloprednol, 2387
[5285-18-7] Piperic Acid, 7579
[5289-74-7] 20-Hydroxyecdysone, 3538
[5300-03-8] Alitretinoin, 241
[5306-80-9] Teclothiazide Potassium Salt, 9241
[5308-89-4] Taxicin-I, 9214
[5308-90-7] Taxicin-II, 9214

[5853-00-9] Carnosine D-Form, *1851*
[5853-25-8] Carnegine Hydrobromide Monohydrate, *1847*
[5854-93-3] L-Alanosine, *199*
[5862-27-1] (17:0)-Urushiol, *10076*
[5864-18-6] Carnegine Hydrochloride Monohydrate, *1847*
[5868-05-3] Niceritrol, *6582*
[5868-06-4] Fentonium Bromide, *4033*
[5870-29-1] Cyclopentolate Hydrochloride, *2741*
[5874-97-5] Metaproterenol Sulfate, *6002*
[5875-06-9] Proparacaine Hydrochloride, *7921*
[5892-10-4] Bismuth Subcarbonate, *1282*
[5892-11-5] Brucine Tetrahydrate, *1464*
[5893-64-1] DL-Lactic Acid Copper Salt Dihydrate, *5384*
[5895-59-0] Decamethylene Glycol Diethyl Ether, *2853*
[5895-86-3] Diacetonamine Acid Oxalate Monohydrate, *2965*
[5897-13-2] Cyacetacide Hydrochloride, *2670*
[5897-76-7] α-(α-Aminopropyl)benzyl Alcohol, *467*
[5902-51-2] Terbacil, *9298*
[5905-52-2] Ferrous Lactate, *4077*
[5907-38-0] Dipyrone, *3390*
[5908-63-4] Baptigenin, *953*
[5908-64-5] Barium Acetate Monohydrate, *961*
[5908-68-9] Benzoic Acid Barium Salt Dihydrate, *1093*
[5908-83-8] Batyl Alcohol Bis(*p*-nitrobenzoate), *1009*
[5908-87-2] Behenic Acid Ethyl Ester, *1022*
[5908-99-6] Atropine Sulfate Monohydrate, *866*
[5909-04-6] Chrysoidine Citrate, *2261*
[5911-58-0] Carnegine Methyliodide, *1847*
[5912-86-7] Isoeugenol *cis*-Form, *5216*
[5913-82-6] Conessine Dihydrobromide, *2480*
[5915-88-8] Propionic Acid Barium Salt Monohydrate, *7939*
[5928-26-7] Sissotrin, *1231*
[5928-72-3] Benzenesulfonic Acid Sesquihydrate, *1071*
[5928-84-7] Penicillin V Compd with Dibenzylethylenediamine, *7209*
[5932-68-3] Isoeugenol *trans*-Form, *5216*
[5934-14-5] Succisulfone, *9009*
[5936-28-7] Hydrastine Hydrochloride, *4806*
[5937-72-4] Androstenediol 17-Acetate, *630*
[5945-50-6] Monotropein, *6343*
[5945-86-8] Nimbin, *6631*
[5947-49-9] Podocarpic Acid, *7660*
[5949-29-1] Citric Acid Monohydrate, *2325*
[5949-44-0] Testosterone Undecanoate, *9324*
[5951-57-5] Dolichodial, *3459*
[5953-63-9] Androstenediol 3-Acetate-17-benzoate, *630*
[5957-75-5] Δ⁸-Tetrahydrocannabinol, *9354*

[5957-80-2] Carnosol, *1852*
[5959-35-3] γ-Aminobutyric Acid Hydrochloride, *425*
[5959-36-4] γ-Aminobutyric Acid Ethyl Ester, *425*
[5959-52-4] 3-Amino-2-naphthoic Acid, *448*
[5959-58-0] 1-Amino-2-naphthol-4-sulfonic Acid Sodium Salt, *450*
[5961-85-3] TCEP, *9222*
[5962-29-8] Xylulose, *10286*
[5965-09-3] Fosfestrol Disodium Salt, *4280*
[5965-13-9] Dihydrocodeine Tartrate, *3189*
[5965-33-3] Antimony Potassium Oxalate, *690*
[5965-49-1] Ketobemidone Hydrochloride, *5348*
[5965-53-7] α-Ketoglutaric Acid Diethyl Ester, *5350*
[5965-56-0] 2-Keto-L-gulonic Acid Ethyl Ester, *5351*
[5965-83-3] Sulfosalicylic Acid Dihydrate, *9094*
[5965-92-4] D-Lanthionine, *5416*
[5967-52-2] Metaraminol Hydrochloride, *6003*
[5967-53-3] Metaraminol Oxalate Dihydrate, *6003*
[5967-73-7] Methadone *l*-Form Hydrochloride, *6016*
[5967-97-5] Thioformamide Monohydrate, *9486*
[5968-11-6] Sodium Carbonate Monohydrate, *8731*
[5970-32-1] Mercuric Salicylate, *5953*
[5970-45-6] Zinc Acetate Dihydrate, *10327*
[5972-72-5] Ammonium Binoxalate, *496*
[5972-76-9] Ammonium Caprylate, *503*
[5973-17-1] Phenol Ammonium Salt, *7354*
[5974-09-4] Vetrabutine Hydrochloride, *10168*
[5978-55-2] *sec*-Octyl Bromide (*R*)-Form, *6854*
[5978-70-1] 2-Octanol *l*-Form, *6841*
[5978-87-0] D-Araboflavin, *752*
[5982-56-9] 1,2-Ethanedisulfonic Acid Dihydrate, *3779*
[5982-57-0] 1,2-Ethanedisulfonic Acid Sodium Salt Dihydrate, *3779*
[5984-50-9] Isometheptene Bitartrate (Acid Tartrate), *5231*
[5984-61-2] Isopelletierine Hydrochloride, *7181*
[5985-04-6] Hydrohydrastinine Hydrochloride, *4839*
[5985-05-7] Hydrohydrastinine Hydrobromide, *4839*
[5985-06-8] Hydrohydrastinine Platinichloride, *4839*
[5985-28-4] Synephrine Hydrochloride, *9142*
[5985-35-3] Levorphanol *dl*-Form Hydrobromide, *5524*
[5985-36-4] Morphinan Sulfate, *6360*
[5985-38-6] Levorphanol Tartrate Dihydrate, *5524*
[5986-55-0] Patchouli Alcohol, *7157*
[5987-82-6] Benoxinate Hydrochloride, *1049*

[5988-22-7] Phytonadiol Sodium Diphosphate, *7492*
[5988-39-6] *p*-Dimethylaminobenzaldehyde Hydrochloride, *3252*
[5988-51-2] Deanol Bitartrate, *2845*
[5989-43-5] Matricarin, *5826*
[5990-65-8] Tetraethylammonium Hydroxide Tetrahydrate, *9346*
[5990-94-3] *o*-Iodohippurate Sodium Dihydrate, *5077*
[5991-71-9] Clorazepic Acid Monopotassium Salt, *2392*
[5995-86-8] Gallic Acid Monohydrate, *4375*
[5996-14-5] α-Glucose-1-phosphate Dipotassium Salt Dihydrate, *4497*
[5997-53-5] Ensulizole Sodium Salt, *3647*
[6000-40-4] Glyceric Acid D(−)-Form, *4519*
[6000-41-5] Glyceric Acid D-Form Calcium Salt, *4519*
[6000-43-7] Glycine Hydrochloride, *4526*
[6000-74-4] Hydrocortisone 21-Phosphate Disodium Salt, *4824*
[6001-74-7] Clomethiazole Hydrochloride, *2375*
[6001-87-2] β-Chloropropionic Acid Methyl Ester, *2161*
[6001-97-4] Bis(1-methylamyl) Sodium Sulfosuccinate, *1259*
[6004-24-6] Cetylpyridinium Chloride Monohydrate, *2031*
[6009-70-7] Ammonium Oxalate Monohydrate, *534*
[6009-81-0] Morphine (monohydrate), *6361*
[6010-09-9] Ferrous Thiocyanate, *4086*
[6010-17-9] Harmalol *O*-Ethylharmalol, *4647*
[6011-14-9] Aminoacetonitrile Hydrochloride, *406*
[6011-32-1] Iminodiacetic Acid Sodium Salt Monohydrate, *4959*
[6011-39-8] Penicillin G Mixture with Clemizole, *7203*
[6018-19-5] *p*-Aminosalicylic Acid Sodium Salt Dihydrate, *473*
[6018-40-2] Corypalmine, *2536*
[6018-53-7] 6-Desoxy-D-glucosamine, *2932*
[6018-84-4] *p*-Aminobenzoic Acid Diethylamine Salt, *418*
[6018-91-3] Benzoic Acid Nickel Salt Trihydrate, *1093*
[6019-02-9] Nicotine Dihydrochloride, *6609*
[6020-39-9] *meso*-Cystine, *2779*
[6024-76-6] Tetraethylammonium Chloride Tetrahydrate, *9345*
[6024-77-7] Tetraethylammonium Hydroxide Hexahydrate, *9346*
[6027-13-0] L-Homocysteine, *4772*
[6027-14-1] D-Homocysteine, *4772*
[6027-15-2] D-Homocystine, *4773*
[6027-23-2] Hordenine Hydrochloride, *4785*
[6027-89-0] L-Gulose, *4614*
[6027-99-2] Harmalol Trihydrate, *4647*
[6028-01-9] Harmalol *O-n*-Propylharmalol, *4647*
[6028-08-6] Harmalol Lactate, *4647*
[6028-21-3] Thiolactic Acid Barium Salt, *9493*

[6028-35-9] Thibenzazoline, *9464*
[6031-86-3] Racemethorphan Hydrobromide, *5520*
[6032-06-0] Cyproheptadine Hydrochloride Monohydrate, *2770*
[6032-14-0] *N*-Methylepinephrine DL-Form, *6141*
[6032-29-7] 2-Pentanol, *7231*
[6032-32-2] Methylarbutin, *763*
[6032-74-2] Muconic Acid (*E,E*)-Form Diethyl Ester, *6385*
[6032-77-5] Muconic Acid (*Z,Z*)-Form Diethyl Ester, *6385*
[6032-80-0] Murexide Monohydrate, *6391*
[6032-82-2] Murexine Chloride Hydrochloride, *6392*
[6032-92-4] Mycarose, *6403*
[6033-08-5] Hematoporphyrin Hydrochloride Monohydrate, *4671*
[6033-23-4] 2-Heptanol (2*S*)-Form, *4697*
[6033-24-5] 2-Heptanol (2*R*)-Form, *4697*
[6033-69-8] Deserpidine Hydrochloride, *2919*
[6035-06-9] 4-Nitrophenylacetic Acid Benzyl Ester, *6709*
[6035-32-1] Normorphine (sesquihydrate), *6800*
[6035-40-1] Noscapine *dl*-Form, *6807*
[6035-45-6] Folinic Acid Calcium Salt Pentahydrate, *4249*
[6035-47-8] Sodium Formaldehyde Sulfoxylate Dihydrate, *8755*
[6036-25-5] Narbomycin, *6503*
[6038-19-3] DL-Homocysteine Thiolactone Hydrochloride, *4772*
[6046-93-1] Cupric Acetate Monohydrate, *2614*
[6046-97-5] Benzoic Acid Copper Salt Dihydrate, *1093*
[6047-24-1] Ferrous Lactate Trihydrate, *4077*
[6054-98-4] Olsalazine Disodium Salt, *6935*
[6055-19-2] Cyclophosphamide, *2743*
[6055-52-3] Hexamethylenediamine Dihydrochloride, *4727*
[6055-56-7] Ergocristine Ethanesulfonate, *3706*
[6055-73-8] Adrenochrome Oxime Sesquihydrate, *161*
[6056-11-7] Pipazethate Hydrochloride, *7569*
[6056-70-8] α-Glucose-1-phosphate Barium Salt Trihydrate, *4497*
[6057-35-8] Glyceric Acid L-Form Calcium Salt, *4519*
[6064-83-1] Fosfosal, *4282*
[6066-82-6] *N*-Hydroxysuccinimide, *4884*
[6078-26-8] Bikhaconitine, *1219*
[6080-56-4] Lead Acetate Trihydrate, *5452*
[6080-57-5] Benzoic Acid Lead Salt Monohydrate, *1093*
[6080-58-6] Lithium Citrate Tetrahydrate, *5586*
[6088-51-3] DDD (Analytical), *2842*
[6091-05-0] Physovenine, *7497*
[6097-58-1] α-Phenylglycine Ethyl Ester, *7402*
[6099-90-7] Phloroglucinol Dihydrate, *7439*

[6100-02-3] Benzoic Acid Potassium Salt Trihydrate, *1093*
[6100-03-4] Potassium Binoxalate Monohydrate, *7732*
[6100-05-6] Potassium Citrate Monohydrate, *7744*
[6100-18-1] Succinic Acid Potassium Salt Trihydrate, *8999*
[6100-20-5] Potassium Tetroxalate Dihydrate, *7806*
[6101-04-8] Strychnine Hydrochloride Dihydrate, *8985*
[6101-32-2] Sulfanilic Acid Monohydrate, *9058*
[6101-34-4] Sulfathiourea Sodium Salt, *9075*
[6106-02-1] Sodium Ethyl Sulfate Monohydrate, *8750*
[6106-20-3] Sodium Sesquicarbonate Dihydrate, *8803*
[6106-21-4] Sodium Succinate Hexahydrate, *8808*
[6106-22-5] Sulfanilic Acid Sodium Salt Dihydrate, *9058*
[6106-24-7] Sodium Tartrate Dihydrate, *8812*
[6106-46-3] Scopolamine Methyl Nitrate, *8543*
[6106-81-6] Scopolamine *N*-Oxide Hydrobromide, *8544*
[6108-05-0] Lidocaine Hydrochloride Monohydrate, *5535*
[6108-17-4] Lithium Acetate Dihydrate, *5577*
[6108-23-2] Lithium Formate Monohydrate, *5588*
[6109-70-2] 3-Quinuclidinol *dl*-Form Acetate Hydrochloride, *8199*
[6113-09-3] *l*-Lupinine Hydrochloride, *5669*
[6114-26-7] Pholedrine Sulfate, *7442*
[6117-91-5] Crotyl Alcohol, *2591*
[6119-47-7] Quinine Hydrochloride Dihydrate, *8177*
[6119-70-6] Quinine Sulfate Dihydrate, *8177*
[6130-64-9] Penicillin G Procaine, *7206*
[6131-90-4] Sodium Acetate Trihydrate, *8709*
[6131-98-2] Sodium Bitartrate Monohydrate, *8725*
[6131-99-3] Sodium Cacodylate Trihydrate, *8730*
[6132-02-1] Sodium Carbonate Decahydrate, *8731*
[6132-04-3] Sodium Citrate Dihydrate, *8737*
[6138-23-4] Trehalose Dihydrate, *9742*
[6138-26-7] Tricarballylic Acid Trimethyl Ester, *9787*
[6138-47-2] Liothyronine Hydrochloride, *5565*
[6138-56-3] Tripelennamine Citrate, *9910*
[6138-73-4] Trilobine, *9866*
[6138-79-0] Triprolidine Hydrochloride Monohydrate, *9922*
[6146-97-0] Benzoic Acid Manganese Salt Tetrahydrate, *1093*
[6147-11-1] Mangostin, *5807*
[6147-37-1] Menadione Sodium Bisulfite (trihydrate), *5899*
[6147-53-1] Cobaltous Acetate Tetrahydrate, *2417*

[6150-58-9] Lycorine Hydrochloride Monohydrate, *5687*
[6150-79-4] Magnesium Citrate Tribasic (tetradecahydrate), *5727*
[6150-97-6] Dipyrone Magnesium Salt, *3390*
[6151-25-3] Quercetin Dihydrate, *8150*
[6151-30-0] Quinacrine Dihydrochloride Dihydrate, *8160*
[6151-39-9] Quinidine Hydrogen Sulfate Tetrahydrate, *8176*
[6152-42-7] *N*-Phenylsulfanilic Acid Potassium Salt, *7423*
[6152-67-6] *N*-Phenylsulfanilic Acid Sodium Salt, *7423*
[6153-16-8] β-Phellandrene *d*-Form, *7316*
[6153-17-9] β-Phellandrene *l*-Form, *7316*
[6153-19-1] Phenacaine Hydrochloride Monohydrate, *7317*
[6153-33-9] Mebhydroline 1,5-Naphthalenedisulfonate Salt, *5839*
[6153-44-2] Orotic Acid Methyl Ester, *6972*
[6153-56-6] Oxalic Acid Dihydrate, *7010*
[6155-64-2] Salicylsulfuric Acid Monosodium Salt, *8471*
[6155-83-5] 2-Sulfoacetic Acid Monohydrate, *9083*
[6159-29-1] Tryparsamide Hemihydrate, *9974*
[6159-35-9] Tuaminoheptane Hydrochloride, *9982*
[6159-44-0] Uranyl Acetate Dihydrate, *10046*
[6159-55-3] Vasicine *S*-Form, *10127*
[6159-56-4] Vasicine (±)-Form, *10127*
[6160-12-9] Sparteine Sulfate Pentahydrate, *8864*
[6163-66-2] Di-*tert*-butyl Ether, *3040*
[6164-87-0] Nicotinyl Alcohol *d*-Tartrate, *6612*
[6164-98-3] Chlordimeform, *2086*
[6168-76-9] Crotethamide, *2586*
[6168-83-8] β-Hydroxybutyric Acid *d*-Form, *4855*
[6168-86-1] Isometheptene Hydrochloride, *5231*
[6169-06-8] 2-Octanol *d*-Form, *6841*
[6170-25-8] *p*-Aminopropiophenone Hydrochloride, *466*
[6170-29-2] Aminopyrine Hydrochloride, *470*
[6170-42-9] Chloropyramine Hydrochloride, *2164*
[6183-68-2] Quinine Bisulfate Heptahydrate, *8177*
[6190-36-9] Cotarnine Phthalate, *2540*
[6190-39-2] Dihydroergotamine Methanesulfonate, *3191*
[6190-43-8] Methenamine Anhydromethylenecitrate, *6038*
[6190-60-9] Mephentermine Sulfate Dihydrate, *5919*
[6191-56-6] Apomorphine Diacetate (Ester), *733*
[6192-44-5] 2-Phenoxyethanol Acetate, *7369*
[6192-52-5] *p*-Toluenesulfonic Acid Monohydrate, *9692*
[6192-89-8] Procaine Dihydrate, *7875*
[6192-92-3] Procaine Nitrate, *7875*

[6192-95-6] Propylhexedrine *dl*-Form Hydrochloride, *7972*

[6192-96-7] Propylhexedrine *d*-Form Hydrochloride, *7972*

[6192-97-8] Propylhexedrine *l*-Form, *7972*

[6192-98-9] Propylhexedrine *l*-Form Hydrochloride, *7972*

[6197-30-4] Octocrylene, *6845*

[6197-42-8] Harmalol *O-n*-Butylharmalol, *4647*

[6198-58-9] Vaccenic Acid Methyl Ester, *10084*

[6199-67-3] Cucurbitacin B, *2604*

[6202-23-9] Cyclobenzaprine Hydrochloride, *2706*

[6205-14-7] Hydroxycitric Acid (unspecified stereo), *4861*

[6208-97-5] D-*threo*-β-Hydroxyglutamic acid, *4864*

[6208-98-6] L-*threo*-β-Hydroxyglutamic acid, *4864*

[6208-99-7] D-*erythro*-β-Hydroxyglutamic acid, *4864*

[6209-00-3] L-*erythro*-β-Hydroxyglutamic acid, *4864*

[6209-17-2] Sulfacetamide Sodium Salt (monohydrate), *9031*

[6209-43-4] Murexine Chloride, *6392*

[6211-15-0] Morphine Sulfate Pentahydrate, *6361*

[6211-24-1] *N*-Phenylsulfanilic Acid Barium Salt, *7423*

[6214-20-6] Methyl *p*-Nitrobenzenesulfonate, *6175*

[6217-54-5] Docosahexaenoic Acid, *3443*

[6223-35-4] Guaiazulene 3-Sulfonate Sodium Salt, *4590*

[6227-35-6] Potassium Dicyanoaurate-(I) Dihydrate, *7749*

[6236-05-1] Nifuroxime, *6620*

[6254-89-3] *N*-Palmitoylsphingomyelin, *8873*

[6261-18-3] Jasmone *trans*-Form, *5307*

[6263-41-8] *p*-Toluenesulfonic Acid Sodium Salt, *9692*

[6272-74-8] Lapyrium Chloride, *5421*

[6277-14-1] Acetoxolone, *74*

[6284-40-8] *N*-Methylglucamine, *6150*

[6292-91-7] MDA Hydrochloride, *5834*

[6303-21-5] Hypophosphorous Acid, *4907*

[6314-28-9] Thionaphthene-2-carboxylic Acid, *9498*

[6325-93-5] 4-Nitrobenzenesulfonamide, *6675*

[6341-58-8] αMNP Hydroxy Acid (unspecified stereo), *6310*

[6358-53-8] Citrus Red 2, *2332*

[6358-69-6] Pyranine, *8065*

[6363-57-1] Pseudococaine Hydrochloride, *8020*

[6365-83-9] Dinoseb Ammonium Salt, *3315*

[6372-96-9] Polar® Yellow, *7670*

[6377-18-0] Chartreusin, *2044*

[6379-56-2] Hygromycin, *4891*

[6379-69-7] Trichothecin, *9815*

[6381-59-5] Potassium Sodium Tartrate Tetrahydrate, *7790*

[6381-79-9] Potassium Carbonate Sesquihydrate, *7740*

[6381-92-6] EDTA Disodium Salt (dihydrate), *3565*

[6384-92-5] NMDA, *6749*

[6385-58-6] Bithionol Sodium Salt, *1307*

[6385-62-2] Diquat Dibromide Monohydrate, *3392*

[6398-98-7] Amodiaquin Dihydrochloride Dihydrate, *569*

[6402-23-9] Ethacridine Lactate Monohydrate, *3771*

[6402-36-4] Traumatic Acid, *9737*

[6411-75-2] Tuaminoheptane Sulfate, *9982*

[6414-49-9] Synephrine Tartaric Acid Monoester, *9142*

[6416-10-0] Thiolactic Acid Copper Salt, *9493*

[6416-12-2] Thiolactic Acid Platinum Salt, *9493*

[6420-47-9] Dinoseb Triethanolamine Salt, *3315*

[6424-36-8] Ergocristine Phosphate, *3706*

[6452-71-7] Oxprenolol, *7051*

[6452-73-9] Oxprenolol Hydrochloride, *7051*

[6458-13-5] Ethyldimethyl-9-octadecenylammonium Bromide, *3844*

[6469-93-8] Chlorprothixene Hydrochloride, *2194*

[6484-52-2] Ammonium Nitrate, *531*

[6484-71-5] Phosphorylcholine Barium Salt, *7474*

[6484-74-8] Pteroylhexaglutamylglutamic Acid, *8044*

[6484-89-5] Folic Acid Sodium Salt, *4248*

[6487-29-2] Citric Acid Barium Salt Heptahydrate, *2325*

[6487-33-8] Isocorypalmine *dl*-Form, *5208*

[6487-38-3] Hydracrylic Acid Sodium Salt, *4801*

[6487-39-4] Lanthanum Carbonate Octahydrate, *5415*

[6487-47-4] Hippuric Acid Potassium Salt Monohydrate, *4753*

[6487-48-5] Potassium Oxalate Monohydrate, *7772*

[6489-61-8] Metampicillin Sodium Salt, *5996*

[6489-97-0] Metampicillin, *5996*

[6493-05-6] Pentoxifylline, *7249*

[6495-46-1] Dioxadrol, *3327*

[6500-81-8] Ethacrynic Acid Sodium Salt, *3772*

[6506-37-2] Nimorazole, *6637*

[6533-00-2] Norgestrel, *6793*

[6533-54-6] Dihydrostreptomycin Trihydrochloride, *3197*

[6533-73-9] Thallium Carbonate, *9408*

[6538-02-9] Ergostanol, *3715*

[6539-57-7] Nordefrin (unspecified stereo), *6781*

[6556-11-2] Inositol Niacinate, *5022*

[6556-12-3] D-Glucuronic Acid, *4501*

[6556-13-4] Hydracrylic Acid Calcium Salt Dihydrate, *4801*

[6556-29-2] Propylhexedrine *d*-Form, *7972*

[6556-35-0] Traumatic Acid *cis*-Form, *9737*

[6569-51-3] Borazine, *1337*

[6576-51-8] Distamycin A Hydrochloride, *3404*

[6577-41-9] Cyclonium Iodide, *2729*

[6591-55-5] Bismuth Oxalate, *1272*

[6591-63-5] Quinidine Sulfate Dihydrate, *8176*

[6592-85-4] Hydrastinine, *4807*

[6598-46-5] Quinacrine Methanesulfonate Monohydrate, *8160*

[6600-40-4] Norvaline, *6805*

[6606-65-1] *n*-Butyl Cyanoacrylate, *1566*

[6620-60-6] Proglumide, *7891*

[6621-47-2] Perhexiline, *7278*

[6645-46-1] Carnitine L-Form Hydrochloride, *1849*

[6673-35-4] Practolol, *7818*

[6674-22-2] 1,8-Diazabicyclo[5.4.0]undec-7-ene, *2996*

[6696-47-5] Oleandomycin Hydrochloride, *6918*

[6696-54-4] Arsanilic Acid Sodium Salt Tetrahydrate, *783*

[6700-39-6] Isoproterenol *dl*-Form Sulfate Dihydrate, *5263*

[6701-17-3] Octyl Cyanoacrylate, *6855*

[6724-53-4] Perhexiline Maleate, *7278*

[6736-85-2] Catalposide, *1901*

[6737-42-4] DPPP, *1254*

[6740-88-1] Ketamine, *5343*

[6745-93-3] Methotrexate Monohydrate, *6057*

[6746-59-4] Ethylmorphine Hydrochloride Dihydrate, *3884*

[6752-46-1] Mycarose D-Form, *6403*

[6753-98-6] Humulene, *4792*

[6754-13-8] Helenalin, *4661*

[6754-20-7] Polygodial, *7690*

[6780-13-8] Ethylenediamine Monohydrate, *3849*

[6793-24-4] Buphanamine, *1496*

[6795-23-9] Aflatoxin M₁, *173*

[6795-60-4] Norvinisterone, *6806*

[6799-26-4] *d*-Camphocarboxylic Acid Ammonium Salt, *1733*

[6804-07-5] Carbadox, *1782*

[6805-41-0] Escin, *3751*

[6809-52-5] Teprenone, *9296*

[6812-78-8] Rhodinol, *8309*

[6818-37-7] Olaflur, *6912*

[6823-79-6] Pentamidine Dimethanesulfonate, *7227*

[6829-98-7] Imipramine *N*-Oxide, *4963*

[6830-17-7] Oxamarin Dihydrochloride, *7014*

[6831-14-7] Arborescin, *761*

[6833-84-7] Nonactin, *6762*

[6834-92-0] Sodium Metasilicate, *8804*

[6834-98-6] Fungichromin, *4319*

[6835-16-1] Hyoscyamine Sulfate Dihydrate, *4897*

[6846-50-0] Texanol Isobutyrate, *9399*

[6856-31-1] Pridinol Methanesulfonate, *7859*

[6865-92-5] 2-(Methyleneamino)acetonitrile Trimer, *6129*

[6871-44-9] Echitamine, *3547*

[6873-15-0] Glycosine, *4538*

[6874-98-2] Vellosimine, *10137*

[6877-35-6] Digalogenin, *3173*

[6878-36-0] Echitamine Chloride, *3547*

[6878-83-7] Tecomanine, *9243*

[6878-98-4] Tropacine, *9952*

[6879-02-3] Tylocrebrine, *10011*

[6879-74-9] Himbacine, *4752*

[6882-99-1] Sempervirine, *8586*

[7440-69-9] Bismuth, *1261*
[7440-70-2] Calcium, *1648*
[7440-71-3] Californium, *1718*
[7440-72-4] Fermium, *4041*
[7440-73-5] Francium, *4290*
[7440-74-6] Indium, *4987*
[7446-07-3] Tellurium Dioxide, *9265*
[7446-08-4] Selenium Oxide, *8571*
[7446-09-5] Sulfur Dioxide, *9101*
[7446-11-9] Sulfur Trioxide, *9109*
[7446-14-2] Lead Sulfate, *5475*
[7446-18-6] Thallium Sulfate, *9418*
[7446-19-7] Zinc Sulfate Monohydrate, *10359*
[7446-20-0] Zinc Sulfate Heptahydrate, *10359*
[7446-26-6] Zinc Pyrophosphate, *10354*
[7446-27-7] Lead Phosphate, *5471*
[7446-32-4] Antimony Sulfate, *694*
[7446-70-0] Aluminum Chloride, *333*
[7447-39-4] Cupric Chloride, *2623*
[7447-40-7] Potassium Chloride, *7742*
[7447-41-8] Lithium Chloride, *5584*
[7449-03-8] Erythropterin, *3745*
[7455-39-2] Fonazine Methanesulfonate, *4258*
[7456-24-8] Fonazine, *4258*
[7460-12-0] Pseudoephedrine Sulfate, *8024*
[7481-89-2] Zalcitabine, *10307*
[7487-81-2] Isomethadone *l*-Form Hydrochloride, *5230*
[7487-88-9] Magnesium Sulfate, *5755*
[7487-94-7] Mercuric Chloride, *5943*
[7488-49-5] *cis*-(−)-Pentazocine, *7233*
[7488-55-3] Stannous Sulfate, *8917*
[7488-56-4] Selenium Disulfide, *8573*
[7488-99-5] α-Carotene, *1854*
[7491-09-0] Docusate Potassium, *3446*
[7491-74-9] Piracetam, *7600*
[7492-31-1] Isometheptene Mucate, *5231*
[7492-55-9] Sorbic Acid Calcium Salt, *8847*
[7512-17-6] *N*-Acetylglucosamine, *4494*
[7529-22-8] *N*-Methylmorpholine *N*-Oxide, *6171*
[7535-00-4] D-Galactosamine, *4364*
[7542-37-2] Paromomycin, *7146*
[7542-45-2] Astaxanthin (unspecified stereo), *845*
[7550-35-8] Lithium Bromide, *5582*
[7550-45-0] Titanium Tetrachloride, *9634*
[7553-56-2] Iodine, *5057*
[7554-65-6] Fomepizole, *4253*
[7558-79-4] Sodium Phosphate, Dibasic, *8789*
[7558-80-7] Sodium Phosphate, Monobasic, *8790*
[7562-61-0] Usnic Acid *d*-Form, *10078*
[7563-42-0] Pentobarbital Calcium Salt, *7243*
[7568-93-6] Phenylethanolamine, *7399*
[7580-67-8] Lithium Hydride, *5589*
[7585-39-9] β-Cyclodextrin, *2710*
[7601-54-9] Sodium Phosphate, Tribasic, *8792*
[7601-55-0] Metocurine Iodide, *6220*
[7601-89-0] Sodium Perchlorate, *8784*
[7601-90-3] Perchloric Acid, *7267*
[7616-22-0] γ-Tocopherol (*dl*-form), *9655*
[7616-94-6] Perchloryl Fluoride, *7268*

[7618-86-2] Furtrethonium, *4341*
[7622-06-2] Jervine Diacetyljervine, *5312*
[7631-86-9] Silicon Dioxide, *8632*
[7631-90-5] Sodium Bisulfite, *8724*
[7631-94-9] Sodium Dithionate, *8746*
[7631-95-0] Sodium Molybdate(VI), *8776*
[7631-99-4] Sodium Nitrate, *8778*
[7632-00-0] Sodium Nitrite, *8779*
[7632-04-4] Sodium Perborate, *8783*
[7633-29-6] Psychotrine, *8038*
[7635-46-3] Sodium Phosphate P 32, *8791*
[7635-51-0] Phendimetrazine Hydrochloride, *7334*
[7637-07-2] Boron Trifluoride, *1351*
[7646-69-7] Sodium Hydride, *8760*
[7646-78-8] Stannic Chloride, *8901*
[7646-79-9] Cobaltous Chloride, *2421*
[7646-85-7] Zinc Chloride, *10331*
[7646-93-7] Potassium Bisulfate, *7734*
[7647-01-0] Hydrochloric Acid, *4818*
[7647-01-0] Hydrogen Chloride, *4831*
[7647-10-1] Palladium Chloride, *7091*
[7647-14-5] Sodium Chloride, *8734*
[7647-15-6] Sodium Bromide, *8729*
[7647-17-8] Cesium Chloride, *2010*
[7647-18-9] Antimony Pentachloride, *686*
[7647-19-0] Phosphorus Pentafluoride, *7464*
[7658-08-4] Quinovose, *8194*
[7664-38-2] Phosphoric Acid, *7456*
[7664-39-3] Hydrofluoric Acid, *4827*
[7664-39-3] Hydrogen Fluoride, *4833*
[7664-41-7] Ammonia, *488*
[7664-93-9] Sulfuric Acid, *9104*
[7665-99-8] Cyclic GMP, *2702*
[7681-11-0] Potassium Iodide, *7764*
[7681-14-3] Prednisolone 2l-*tert*-Butylacetate, *7841*
[7681-28-9] Quinidine Polygalacturonate, *8176*
[7681-38-1] Sodium Bisulfate, *8722*
[7681-49-4] Sodium Fluoride, *8754*
[7681-52-9] Sodium Hypochlorite, *8762*
[7681-53-0] Sodium Hypophosphite, *8763*
[7681-55-2] Sodium Iodate, *8764*
[7681-57-4] Sodium Metabisulfite, *8770*
[7681-65-4] Cuprous Iodide, *2652*
[7681-76-7] Ronidazole, *8385*
[7681-79-0] Etafedrine, *3764*
[7681-82-5] Sodium Iodide, *8765*
[7681-93-8] Natamycin, *6509*
[7683-59-2] Isoproterenol, *5263*
[7688-25-7] DPPB, *1254*
[7688-64-4] Camptothecin Acetate, *1737*
[7688-65-5] Camptothecin Chloroacetate, *1737*
[7689-03-4] Camptothecin, *1737*
[7693-27-8] Magnesium Hydride, *5733*
[7696-12-0] Tetramethrin, *9370*
[7697-37-2] Nitric Acid, *6663*
[7699-41-4] Metasilicic Acid, *8629*
[7699-43-6] Zirconyl Chloride, *10384*
[7699-45-8] Zinc Bromide, *10328*
[7704-34-9] Sulfur, *9099*
[7704-34-9] Sulfur, Pharmaceutical, *9107*

[7704-98-5] Titanium Hydride, *9629*
[7704-99-6] Zirconium Hydride, *10376*
[7705-07-9] Titanium Trichloride, *9637*
[7705-08-0] Ferric Chloride, *4046*
[7706-67-4] Dimecrotic Acid, *3226*
[7718-54-9] Nickel Chloride, *6590*
[7719-09-7] Thionyl Chloride, *9501*
[7719-12-2] Phosphorus Trichloride, *7470*
[7720-78-7] Ferrous Sulfate, *4084*
[7721-01-9] Tantalum Pentachloride, *9186*
[7722-44-3] Colistin A, *2463*
[7722-64-7] Potassium Permanganate, *7776*
[7722-76-1] Ammonium Phosphate, Monobasic, *541*
[7722-84-1] Hydrogen Peroxide, *4835*
[7722-86-3] Caro's Acid, *1853*
[7722-88-5] Tetrasodium Pyrophosphate, *9388*
[7723-14-0] Phosphorus, *7459*
[7726-95-6] Bromine, *1401*
[7727-15-3] Aluminum Bromide, *328*
[7727-18-6] Vanadyl Trichloride, *10114*
[7727-21-1] Potassium Persulfate, *7777*
[7727-37-9] Nitrogen, *6688*
[7727-43-7] Barium Sulfate, *990*
[7727-54-0] Ammonium Peroxydisulfate, *539*
[7732-18-5] Water, *10237*
[7733-02-0] Zinc Sulfate, *10359*
[7753-60-8] Anecortave Acetate, *634*
[7757-79-1] Potassium Nitrate, *7769*
[7757-81-5] Sorbic Acid Sodium Salt, *8847*
[7757-82-6] Sodium Sulfate, *8809*
[7757-83-7] Sodium Sulfite, *8811*
[7757-86-0] Magnesium Phosphate, Dibasic, *5745*
[7757-87-1] Magnesium Phosphate, Tribasic, *5747*
[7757-88-2] Magnesium Sulfite, *5756*
[7757-93-9] Calcium Phosphate, Dibasic, *1694*
[7758-01-2] Potassium Bromate, *7738*
[7758-02-3] Potassium Bromide, *7739*
[7758-05-6] Potassium Iodate, *7763*
[7758-09-0] Potassium Nitrite, *7770*
[7758-11-4] Potassium Phosphate, Dibasic, *7779*
[7758-16-9] Sodium Acid Pyrophosphate, *8710*
[7758-19-2] Sodium Chlorite, *8735*
[7758-23-8] Calcium Phosphate, Monobasic, *1695*
[7758-29-4] Sodium Tripolyphosphate, *8824*
[7758-87-4] Calcium Phosphate, Tribasic, *1696*
[7758-88-5] Cerous Fluoride, *1997*
[7758-89-6] Cuprous Chloride, *2650*
[7758-94-3] Ferrous Chloride, *4070*
[7758-95-4] Lead Chloride, *5459*
[7758-97-6] Lead Chromate(VI), *5460*
[7758-98-7] Cupric Sulfate, *2643*
[7758-99-8] Cupric Sulfate Pentahydrate, *2643*
[7759-00-4] Manganese Silicate, *5803*
[7759-02-6] Strontium Sulfate, *8978*
[7759-35-5] Elcometrine, *3586*
[7761-88-8] Silver Nitrate, *8657*
[7772-98-7] Sodium Thiosulfate, *8821*

[7772-99-8] Stannous Chloride, *8910*
[7773-01-5] Manganese Chloride, *5793*
[7773-06-0] Ammonium Sulfamate, *551*
[7774-29-0] Mercuric Iodide, Red, *5946*
[7774-34-7] Calcium Chloride Hexahydrate, *1661*
[7775-09-9] Sodium Chlorate, *8733*
[7775-11-3] Sodium Chromate, *8736*
[7775-14-6] Sodium Dithionite, *8747*
[7775-19-1] Sodium Metaborate, *8771*
[7775-27-1] Sodium Persulfate, *8787*
[7775-41-9] Silver Fluoride, *8653*
[7778-18-9] Calcium Sulfate, *1708*
[7778-39-4] Arsenic Acid, *785*
[7778-43-0] Sodium Arsenate, Dibasic, *8715*
[7778-44-1] Calcium Arsenate, *1652*
[7778-50-9] Potassium Dichromate(VI), *7748*
[7778-53-2] Potassium Phosphate, Tribasic, *7781*
[7778-54-3] Calcium Hypochlorite, *1676*
[7778-74-7] Potassium Perchlorate, *7774*
[7778-77-0] Potassium Phosphate, Monobasic, *7780*
[7778-80-5] Potassium Sulfate, *7793*
[7779-25-1] Magnesium Citrate, *5727*
[7779-88-6] Zinc Nitrate, *10343*
[7779-90-0] Zinc Phosphate, *10351*
[7782-39-0] Deuterium, *2941*
[7782-40-3] Diamond, *2988*
[7782-41-4] Fluorine, *4194*
[7782-42-5] Graphite, *4575*
[7782-44-7] Oxygen, *7063*
[7782-49-2] Selenium, *8568*
[7782-50-5] Chlorine, *2095*
[7782-61-8] Ferric Nitrate Nonahydrate, *4054*
[7782-63-0] Ferrous Sulfate Heptahydrate, *4084*
[7782-64-1] Manganese Difluoride, *5794*
[7782-65-2] Germane, *4440*
[7782-68-5] Iodic Acid, *5056*
[7782-77-6] Nitrous Acid, *6739*
[7782-78-7] Nitrosylsulfuric Acid, *6735*
[7782-79-8] Hydrazoic Acid, *4812*
[7782-85-6] Sodium Phosphate, Dibasic Heptahydrate, *8789*
[7782-87-8] Potassium Hypophosphite, *7762*
[7782-89-0] Lithium Amide, *5578*
[7782-91-4] Molybdic(VI) Acid, *6326*
[7782-92-5] Sodium Amide, *8714*
[7782-94-7] Nitramide, *6658*
[7782-99-2] Sulfurous Acid, *9106*
[7783-00-8] Selenious Acid, *8567*
[7783-03-1] Tungstic(VI) Acid, *9998*
[7783-05-3] Pyrosulfuric Acid, *8121*
[7783-06-4] Hydrogen Sulfide, *4837*
[7783-07-5] Hydrogen Selenide, *4836*
[7783-08-6] Selenic Acid, *8566*
[7783-09-7] Hydrogen Telluride, *4838*
[7783-11-1] Ammonium Sulfite Monohydrate, *554*
[7783-18-8] Ammonium Thiosulfate, *559*
[7783-19-9] Ammonium Selenite, *549*
[7783-20-2] Ammonium Sulfate, *552*
[7783-21-3] Ammonium Selenate, *548*

[7783-22-4] Ammonium Uranate(VI), *562*
[7783-26-8] Trisilane, *9930*
[7783-28-0] Ammonium Phosphate, Dibasic, *540*
[7783-29-1] Tetrasilane, *9387*
[7783-33-7] Potassium Tetraiodomercurate(II), *7805*
[7783-35-9] Mercuric Sulfate, *5955*
[7783-36-0] Mercurous Sulfate, *5965*
[7783-39-3] Mercuric Fluoride, *5945*
[7783-40-6] Magnesium Fluoride, *5729*
[7783-41-7] Fluorine Monoxide, *4196*
[7783-42-8] Thionyl Fluoride, *9502*
[7783-44-0] Fluorine Dioxide, *4195*
[7783-46-2] Lead Fluoride, *5463*
[7783-47-3] Stannous Fluoride, *8911*
[7783-48-4] Strontium Fluoride, *8972*
[7783-49-5] Zinc Fluoride, *10335*
[7783-50-8] Ferric Fluoride, *4049*
[7783-51-9] Gallium Trifluoride, *4384*
[7783-52-0] Indium Trifluoride, *4999*
[7783-53-1] Manganese Trifluoride, *5806*
[7783-54-2] Nitrogen Fluoride, *6691*
[7783-55-3] Phosphorus Trifluoride, *7471*
[7783-56-4] Antimony Trifluoride, *697*
[7783-58-6] Germanium Tetrafluoride, *4445*
[7783-59-7] Lead Tetrafluoride, *5479*
[7783-60-0] Sulfur Tetrafluoride, *9108*
[7783-61-1] Silicon Tetrafluoride, *8640*
[7783-62-2] Stannic Fluoride, *8903*
[7783-63-3] Titanium Tetrafluoride, *9635*
[7783-64-4] Zirconium Fluoride, *10375*
[7783-66-6] Iodine Pentafluoride, *5061*
[7783-68-8] Niobium Pentafluoride, *6644*
[7783-70-2] Antimony Pentafluoride, *687*
[7783-71-3] Tantalum Pentafluoride, *9187*
[7783-72-4] Vanadium Pentafluoride, *10106*
[7783-75-7] Iridium Hexafluoride, *5130*
[7783-77-9] Molybdenum Hexafluoride, *6323*
[7783-79-1] Selenium Hexafluoride, *8570*
[7783-80-4] Tellurium Hexafluoride, *9266*
[7783-81-5] Uranium Hexafluoride, *10041*
[7783-82-6] Tungsten Hexafluoride, *9996*
[7783-83-7] Ammonium Ferric Sulfate Dodecahydrate, *515*
[7783-84-8] Ferric Hypophosphite, *4053*
[7783-85-9] Ammonium Ferrous Sulfate Hexahydrate, *518*
[7783-86-0] Ferrous Iodide, *4076*
[7783-90-6] Silver Chloride, *8648*
[7783-92-8] Silver Chlorate, *8647*
[7783-93-9] Silver Perchlorate, *8662*
[7783-95-1] Silver Difluoride, *8652*
[7783-96-2] Silver Iodide, *8655*
[7783-97-3] Silver Iodate, *8654*
[7783-98-4] Silver Permanganate, *8663*
[7783-99-5] Silver Nitrite, *8658*
[7784-01-2] Silver Chromate(VI), *8649*
[7784-03-4] Silver Tetraiodomercurate-(II), *8670*

[7784-09-0] Silver Phosphate, *8664*
[7784-13-6] Aluminum Chloride Hexahydrate, *333*
[7784-14-7] Ammonium Tetrachloroaluminate, *555*
[7784-15-8] Aluminum Chlorate Nonahydrate, *332*
[7784-16-9] Sodium Tetrachloroaluminate, *8815*
[7784-18-1] Aluminum Fluoride, *335*
[7784-19-2] Ammonium Hexafluoroaluminate, *521*
[7784-21-6] Aluminum Hydride, *337*
[7784-22-7] Aluminum Hypophosphite, *340*
[7784-23-8] Aluminum Iodide, *341*
[7784-24-9] Aluminum Potassium Sulfate Dodecahydrate, *354*
[7784-25-0] Aluminum Ammonium Sulfate, *323*
[7784-26-1] Aluminum Ammonium Sulfate Dodecahydrate, *323*
[7784-27-2] Aluminum Nitrate Nonahydrate, *347*
[7784-30-7] Aluminum Phosphate, *352*
[7784-31-8] Aluminum Sulfate Hydrate, *360*
[7784-33-0] Arsenic Tribromide, *791*
[7784-34-1] Arsenic Trichloride, *792*
[7784-35-2] Arsenic Trifluoride, *793*
[7784-36-3] Arsenic Pentafluoride, *787*
[7784-40-9] Lead Arsenate, *5454*
[7784-41-0] Potassium Arsenate, *7728*
[7784-42-1] Arsine, *799*
[7784-45-4] Arsenic Triiodide, *794*
[7784-46-5] Sodium Arsenite, *8716*
[7785-20-8] Nickel Ammonium Sulfate Hexahydrate, *6586*
[7785-23-1] Silver Bromide, *8645*
[7785-24-2] Cobaltous Arsenate Octahydrate, *2418*
[7785-84-4] Sodium Trimetaphosphate, *8823*
[7785-87-7] Manganese Sulfate, *5804*
[7786-30-3] Magnesium Chloride, *5726*
[7786-34-7] Mevinphos, *6244*
[7786-81-4] Nickel Sulfate, *6602*
[7787-32-8] Barium Fluoride, *970*
[7787-35-1] Barium Manganate(VI), *977*
[7787-36-2] Barium Permanganate, *984*
[7787-39-5] Barium Sulfite, *993*
[7787-46-4] Beryllium Bromide, *1166*
[7787-47-5] Beryllium Chloride, *1168*
[7787-49-7] Beryllium Fluoride, *1169*
[7787-52-2] Beryllium Hydride, *1170*
[7787-55-5] Beryllium Nitrate Trihydrate, *1172*
[7787-57-7] Bismuth Bromide Oxide, *1263*
[7787-58-8] Bismuth Bromide, *1262*
[7787-59-9] Bismuth Oxychloride, *1274*
[7787-60-2] Bismuth Chloride, *1264*
[7787-61-3] Bismuth Fluoride, *1265*
[7787-62-4] Bismuth Pentafluoride, *1275*
[7787-63-5] Bismuth Iodide Oxide, *1270*
[7787-64-6] Bismuth Iodide, *1269*
[7787-68-0] Bismuth Sulfate, *1286*
[7787-69-1] Cesium Bromide, *2008*
[7787-70-4] Cuprous Bromide, *2649*
[7787-71-5] Bromine Trifluoride, *1403*

[7788-96-7] Chromyl Fluoride, *2253*
[7788-97-8] Chromic Fluoride, *2227*
[7788-98-9] Ammonium Chromate(VI), *507*
[7788-99-0] Chromic Potassium Sulfate Dodecahydrate, *2233*
[7789-00-6] Potassium Chromate(VI), *7743*
[7789-01-7] Lithium Chromate(VI) Dihydrate, *5585*
[7789-04-0] Chromic Phosphate, *2231*
[7789-06-2] Strontium Chromate(VI), *8971*
[7789-09-5] Ammonium Dichromate-(VI), *512*
[7789-12-0] Sodium Dichromate(VI) Dihydrate, *8744*
[7789-17-5] Cesium Iodide, *2013*
[7789-18-6] Cesium Nitrate, *2014*
[7789-19-7] Cupric Fluoride, *2628*
[7789-20-0] Deuterium Oxide, *2942*
[7789-21-1] Fluorosulfonic Acid, *4211*
[7789-23-3] Potassium Fluoride, *7753*
[7789-24-4] Lithium Fluoride, *5587*
[7789-25-5] Nitrosyl Fluoride, *6734*
[7789-26-6] Fluorine Nitrate, *4197*
[7789-27-7] Thallium Fluoride, *9411*
[7789-28-8] Ferrous Fluoride, *4072*
[7789-29-9] Potassium Bifluoride, *7731*
[7789-30-2] Bromine Pentafluoride, *1402*
[7789-31-3] Bromic Acid, *1399*
[7789-33-5] Iodine Monobromide, *5059*
[7789-38-0] Sodium Bromate, *8728*
[7789-39-1] Rubidium Bromide, *8416*
[7789-40-4] Thallium Bromide, *9407*
[7789-41-5] Calcium Bromide, *1657*
[7789-42-6] Cadmium Bromide, *1620*
[7789-43-7] Cobaltous Bromide, *2419*
[7789-45-9] Cupric Bromide, *2619*
[7789-46-0] Ferrous Bromide, *4069*
[7789-47-1] Mercuric Bromide, *5942*
[7789-48-2] Magnesium Bromide, *5723*
[7789-57-3] Tribromosilane, *9777*
[7789-59-5] Phosphorus Oxybromide, *7460*
[7789-60-8] Phosphorus Tribromide, *7469*
[7789-61-9] Antimony Tribromide, *695*
[7789-65-3] Selenium Tetrabromide, *8574*
[7789-66-4] Silicon Tetrabromide, *8638*
[7789-67-5] Stannic Bromide, *8900*
[7789-68-6] Titanium Tetrabromide, *9633*
[7789-69-7] Phosphorus Pentabromide, *7462*
[7789-75-5] Calcium Fluoride, *1669*
[7789-78-8] Calcium Hydride, *1674*
[7789-79-9] Calcium Hypophosphite, *1677*
[7789-80-2] Calcium Iodate, *1678*
[7789-82-4] Calcium Molybdate(VI), *1683*
[7790-21-8] Potassium Periodate, *7775*
[7790-22-9] Lithium Iodide Trihydrate, *5591*
[7790-26-3] Sodium Iodide I 131, *8765*
[7790-28-5] Sodium Metaperiodate, *8772*
[7790-29-6] Rubidium Iodide, *8419*
[7790-30-9] Thallium Iodide, *9413*
[7790-33-2] Manganese Iodide, *5797*
[7790-37-6] Zinc Iodate, *10338*

[7790-44-5] Antimony Triiodide, *698*
[7790-47-8] Stannic Iodide, *8904*
[7790-48-9] Tellurium Tetraiodide, *9269*
[7790-53-6] Potassium Metaphosphate, *7767*
[7790-58-1] Potassium Tellurite, *7798*
[7790-59-2] Potassium Selenate, *7786*
[7790-60-5] Potassium Tungstate, *7811*
[7790-62-7] Potassium Pyrosulfate, *7784*
[7790-63-8] Potassium Uranate(VI), *7812*
[7790-69-4] Lithium Nitrate, *5592*
[7790-75-2] Calcium Tungstate(VI), *1714*
[7790-76-3] Calcium Pyrophosphate, *1701*
[7790-78-5] Cadmium Chloride Hemipentahydrate, *1622*
[7790-79-6] Cadmium Fluoride, *1624*
[7790-80-9] Cadmium Iodide, *1626*
[7790-84-3] Cadmium Sulfate Hydrate, *1632*
[7790-85-4] Cadmium Tungstate(VI), *1635*
[7790-86-5] Cerous Chloride, *1996*
[7790-87-6] Cerous Iodide, *1998*
[7790-89-8] Chlorine Monofluoride, *2098*
[7790-91-2] Chlorine Trifluoride, *2100*
[7790-92-3] Hypochlorous Acid, *4904*
[7790-93-4] Chloric Acid, *2093*
[7790-94-5] Chlorosulfonic Acid, *2168*
[7790-98-9] Ammonium Perchlorate, *538*
[7790-99-0] Iodine Monochloride, *5060*
[7791-03-9] Lithium Perchlorate, *5595*
[7791-07-3] Sodium Perchlorate Monohydrate, *8784*
[7791-08-4] Antimony Chloride Oxide, *684*
[7791-10-8] Strontium Chlorate, *8969*
[7791-11-9] Rubidium Chloride, *8417*
[7791-12-0] Thallium Chloride, *9409*
[7791-13-1] Cobaltous Chloride Hexahydrate, *2421*
[7791-16-4] Antimony Dichlorotrifluoride, *685*
[7791-18-6] Magnesium Chloride Hexahydrate, *5726*
[7791-20-0] Nickel Chloride Hexahydrate, *6590*
[7791-21-1] Chlorine Monoxide, *2099*
[7791-23-3] Selenium Oxychloride, *8572*
[7791-25-5] Sulfuryl Chloride, *9110*
[7791-26-6] Uranyl Chloride, *10047*
[7795-51-9] α-Methyltryptamine (*S*)-Form, *6207*
[7795-52-0] α-Methyltryptamine (*R*)-Form, *6207*
[7798-23-4] Cupric Phosphate, *2638*
[7803-49-8] Hydroxylamine, *4867*
[7803-51-2] Phosphine, *7450*
[7803-52-3] Stibine, *8940*
[7803-54-5] Magnesium Amide, *5720*
[7803-55-6] Ammonium Vanadate(V), *565*
[7803-57-8] Hydrazine Monohydrate, *4809*
[7803-58-9] Sulfamide, *9054*
[7803-60-3] Hypophosphoric Acid, *4906*

[7803-62-5] Silane, *8627*
[7803-63-6] Ammonium Bisulfate, *497*
[7803-65-8] Ammonium Hypophosphite, *525*
[7803-68-1] Telluric(VI) Acid, *9262*
[8000-10-0] Theophylline Sodium Glycinate, *9436*
[8000-25-7] Oil of Rosemary, *8395*
[8000-28-0] Oil of Lavender, *5445*
[8000-42-8] Oil of Caraway, *1779*
[8000-48-4] Oil of Eucalyptus, *3939*
[8000-66-6] Oil of Cardamom, *1833*
[8000-68-8] Parsley Seed Oil, *7150*
[8000-73-5] Ammonium Carbonate Carbamate, *505*
[8000-78-0] Oil of Garlic, *4403*
[8001-08-9] Acerin, *35*
[8001-27-2] Hirudin, *4755*
[8001-35-2] Toxaphene, *9716*
[8001-39-6] Japan Wax, *5304*
[8001-40-9] Iodized Oil, *5067*
[8001-54-5] Benzalkonium Chloride, *1061*
[8001-58-9] Creosote, Coal Tar, *2560*
[8001-61-4] Copaiba, *2501*
[8001-75-0] Ceresin, *1986*
[8001-95-4] Alseroxylon, *8238*
[8002-36-6] Ustilagic Acid, *10080*
[8002-41-3] Laurel Berry Oil, *5437*
[8002-43-5] Lecithins, *5483*
[8002-53-7] Montan Wax, *6345*
[8002-55-9] Myrtol, *6424*
[8002-65-1] Neem Fixed Oil, *6522*
[8002-66-2] German Chamomile Oil, *2040*
[8002-68-4] Oil of Juniper, *5318*
[8002-75-3] Palm Oil, *6910*
[8002-76-4] Papaveretum, *7121*
[8002-88-8] Theobromine Sodium Acetate, *9433*
[8002-89-9] Theophylline Sodium Acetate, *9436*
[8003-05-2] Phenylmercuric Nitrate, Basic, *7413*
[8003-22-3] Quinoline Yellow Spirit Soluble, *8189*
[8004-92-0] Quinoline Yellow, *8188*
[8005-78-5] Bismark Brown R (unspecified structure), *1256*
[8006-08-4] Ergotinine, *3721*
[8006-25-5] Ergotoxine, *3722*
[8006-28-8] Soda Lime, *8707*
[8006-44-8] Candelilla Wax, *1741*
[8006-54-0] Lanolin, *5409*
[8006-64-2] Turpentine Oil, *10007*
[8006-77-7] Oil of Pimenta, *7544*
[8006-80-2] Oil of Sassafras, *8518*
[8006-82-4] Black Pepper Oil, *1314*
[8006-84-6] Oil of Fennel, *4008*
[8006-87-9] Oil of Santal, *8490*
[8006-90-4] Oil of Peppermint, *7258*
[8006-91-5] Dextri-Maltose®, *2954*
[8007-04-3] Oil of Hops, *4784*
[8007-06-5] Oil of Cascarilla, *1881*
[8007-08-7] Oil of Ginger, *4458*
[8007-14-5] Argol, *772*
[8007-20-3] Oil of Cedar Leaf, *9543*
[8007-31-6] Toughened Silver Nitrate, *8657*
[8007-44-1] American Pennyroyal Oil, *7211*
[8007-56-5] Nitrohydrochloric Acid, *6696*
[8007-61-2] Verdigris, *2614*

[8007-70-3] Oil of Anise, *657*
[8007-80-5] Oil of Cassia, *2300*
[8007-87-2] Oil of Cubeb, *2602*
[8008-20-6] Kerosene, *5342*
[8008-45-5] Oil of Nutmeg, *6818*
[8008-52-4] Oil of Coriander, *2517*
[8008-53-5] Ethiodized Oil, *3789*
[8008-79-5] Oil of Spearmint, *8865*
[8008-88-6] Oil of Valerian, *10089*
[8008-93-3] Oil of Wormwood, *10250*
[8011-61-8] Tyrocidine, *10017*
[8011-62-9] Barium Sulfide, Black, *992*
[8011-63-0] Bordeaux Mixture, *1338*
[8012-89-3] Beeswax, *1019*
[8013-17-0] Invert Sugar, *5050*
[8013-43-2] Spirit of Ether, *8882*
[8013-44-3] Spirit of Ether Compound, *8883*
[8013-59-0] Spirit of Ammonia, Aromatic, *8881*
[8013-61-4] Ammonium Acetate Solution, *492*
[8013-75-0] Fusel Oil, *4345*
[8013-88-5] Calcium Cyanamide Citrated, *1665*
[8013-90-9] Ionone, *5099*
[8013-97-6] Oil of Copaiba, *2501*
[8013-99-8] Oil of Pennyroyal, *7211*
[8014-63-9] Seidlitz Mixture, *8560*
[8014-95-7] Sulfuric Acid, Fuming, *9104*
[8015-14-3] Lynestrenol Mixture with Mestranol, *5690*
[8015-18-7] Pfeiffer's Substance, *7309*
[8015-19-8] Dimethisterone Mixture with Ethinyl Estradiol, *3240*
[8015-30-3] Norethynodrel Mixture with Mestranol, *6787*
[8015-51-8] Cropropamide Combination with Crotethamide, *2582*
[8015-61-0] Aloin, *303*
[8015-64-3] Angelica Root Oil, *638*
[8015-79-0] Oil of Calamus, *1640*
[8015-86-9] Carnauba Wax, *1846*
[8015-91-6] Oil of Cinnamon, *2300*
[8015-92-7] Roman Chamomile Oil, *2040*
[8016-69-1] Oil of Asarum, *814*
[8016-78-2] Spike Lavender Oil, *5445*
[8018-01-7] Mancozeb, *5780*
[8018-15-3] Mercumallylic Acid Sodium Salt Compd with Theophylline, *5939*
[8021-39-4] Creosote, Wood, *2561*
[8021-82-7] Dobell's Solution, *3438*
[8022-00-2] Methyl Demeton, *6123*
[8022-07-9] Oil of Yarrow, *97*
[8022-15-9] Lavandin Oil, *5445*
[8023-77-6] Oleoresin Capsicum, *1772*
[8023-79-8] Palm Kernel Oil, *6910*
[8024-00-8] Oil of Savin, *8523*
[8025-81-8] Spiramycin, *8879*
[8027-33-6] Lanolin Alcohols, *5410*
[8029-68-3] Ichthammol, *4924*
[8030-30-6] Petroleum Benzin, *7303*
[8030-97-5] Pyroligneous Acids, *8115*
[8031-14-9] Oxychlorosene, *7055*
[8031-21-8] Napalm, *6451*
[8031-66-1] Osmaron B, *6987*
[8031-67-2] Royal Jelly, *8410*
[8031-76-3] Mistletoe Extract, *6294*
[8032-32-4] Ligroin, *5542*
[8038-65-1] Ambergris, *375*

[8046-19-3] Storax, *8949*
[8047-13-0] Ryania, *8441*
[8047-67-4] Iron Sucrose, *5145*
[8048-16-6] Streptodornase Mixture with Streptokinase, *8953*
[8048-31-5] Theobromine Sodium Salicylate, *9433*
[8048-52-0] Acriflavine Mixture with 3,6-Acridinediamine, *115*
[8049-11-4] Devarda's Metal, *2943*
[8049-20-5] Misch Metal, *6292*
[8049-47-6] Pancreatic Extract, *7108*
[8049-62-5] Zinc Insulin, *5023*
[8049-99-8] Milorganite®, *6276*
[8050-07-5] Olibanum, *6926*
[8050-81-5] Simethicone, *3237*
[8050-88-2] Celluloid, *1965*
[8051-02-3] Veratrine (Mixture), *10149*
[8052-16-2] Cactinomycin, *1611*
[8052-27-5] Kino, *5361*
[8052-41-3] Mineral Spirits Type I, *6282*
[8053-05-2] Yellow Phenolphthalein, *7356*
[8053-39-2] Horse Chestnut Seed Extract, *4787*
[8056-51-7] Norgestrel Mixture with Ethinyl Estradiol, *6793*
[8063-06-7] Curare, *2661*
[8063-07-8] Kanamycin, *5329*
[8063-16-9] Psyllium Seed Gum, *7631*
[8064-12-8] Diatrizoic Acid Mixture of Sodium And Meglumine Salts, *2993*
[8064-51-5] Megestrol Acetate Mixture with Mestranol, *5876*
[8064-66-2] Megestrol Acetate Mixture with Ethinyl Estradiol, *5876*
[8064-73-1] Tioxolone Combination with Hydrocortisone, *9612*
[8064-75-3] Norgesterone Mixture with Ethinyl Estradiol, *6791*
[8064-76-4] Lynestrenol Mixture with Ethinyl Estradiol, *5690*
[8064-90-2] Sulfamethoxazole Mixture with Trimethoprim, *9050*
[8065-48-3] Demeton, *2892*
[8065-51-8] Theobromine Calcium Salicylate, *9433*
[8065-91-6] Chlormadinone Acetate Mixture with Mestranol, *2102*
[8067-24-1] Ergoloid Mesylates, *3710*
[8067-82-1] Alfaxalone Mixture with Alfadolone Acetate, *228*
[8068-28-8] Colistin Sodium Methanesulfonate, *2463*
[8069-64-5] Meralluride, *5933*
[8075-95-4] Clofibrate Mixture with Androsterone, *2370*
[8075-98-7] Ascorbigen, *820*
[9000-01-5] Acacia, *15*
[9000-02-6] Amber, *374*
[9000-03-7] Ammoniacum, *490*
[9000-04-8] Asafetida, *811*
[9000-05-9] Gum Benzoin, *4615*
[9000-07-1] Carrageenan, *1864*
[9000-14-0] Copal, *2502*
[9000-16-2] Damar, *2806*
[9000-21-9] Furcellaran, *4331*
[9000-25-3] Gamboge, *4390*
[9000-28-6] Ghatti Gum, *4450*
[9000-29-7] Guaiac, *4588*
[9000-30-0] Guaran, *4606*
[9000-30-0] Guar Gum, *4607*
[9000-32-2] Gutta-Percha, *4618*

[9000-34-4] Resin Ipomea, *8268*
[9000-35-5] Resin Jalap, *8269*
[9000-36-6] Karaya Gum, *5332*
[9000-45-7] Myrrh, *6423*
[9000-47-9] Mesquite Gum, *5984*
[9000-55-9] Podophyllin, *7664*
[9000-58-2] Resin Scammony, *8270*
[9000-59-3] Shellac, *8617*
[9000-65-1] Gum Tragacanth, *4616*
[9000-69-5] Pectin, *7169*
[9000-70-8] Gelatin, *4415*
[9000-71-9] Casein, *1883*
[9000-81-1] Acetylcholinesterase, *2214*
[9000-85-5] α-Amylase (Bacterial), *594*
[9000-88-8] D-Amino Acid Oxidase, *408*
[9000-89-9] L-Amino Acid Oxidase, *409*
[9000-90-2] α-Amylase (Porcine), *594*
[9000-91-3] β-Amylase (Sweet Potato), *594*
[9000-92-4] Amylase, *594*
[9000-95-7] Apyrase, *748*
[9001-00-7] Bromelain, *1395*
[9001-01-8] Kallikrein, *5327*
[9001-03-0] Carbonic Anhydrase, *1812*
[9001-04-1] Pyruvate Decarboxylase, *8134*
[9001-05-2] Catalase, *1900*
[9001-08-5] Cholinesterase, *2214*
[9001-09-6] Chymopapain, *2263*
[9001-12-1] Collagenase, *2465*
[9001-19-8] Taka-Diastase, *9162*
[9001-24-5] Factor V, *3960*
[9001-25-6] Factor VII, *3961*
[9001-26-7] Prothrombin, *7999*
[9001-27-8] Factor VIII, *3962*
[9001-28-9] Factor IX, *3963*
[9001-29-0] Factor X, *3964*
[9001-30-3] Factor XII, *3966*
[9001-31-4] Fibrin, *4099*
[9001-32-5] Fibrinogen, *4100*
[9001-33-6] Ficin, *4106*
[9001-37-0] Glucose Oxidase, *4496*
[9001-45-0] β-Glucuronidase, *4502*
[9001-53-0] Diamine Oxidase, *2977*
[9001-54-1] Hyaluronidases, *4797*
[9001-57-4] Invertase, *5049*
[9001-60-9] Lactate Dehydrogenase, *5382*
[9001-62-1] Lipase, *5566*
[9001-63-2] Lysozyme, *5701*
[9001-66-5] Monoamine Oxidase, *6333*
[9001-68-7] Old Yellow Enzyme, *6917*
[9001-73-4] Papain, *7159*
[9001-75-6] Pepsin, *7219*
[9001-90-5] Plasmin, *7634*
[9001-91-6] Plasminogen, *7635*
[9001-97-2] Q-Enzyme, *8137*
[9001-98-8] Rennin, *8254*
[9001-99-4] Ribonuclease, *8326*
[9002-01-1] Streptokinase, *8955*
[9002-04-4] Thrombin, *9537*
[9002-07-7] Trypsin, *9975*
[9002-10-2] Tyrosinase, *10019*
[9002-12-4] Uricase, *10061*
[9002-13-5] Urease, *10055*
[9002-18-0] Agar, *176*
[9002-60-2] ACTH, *125*
[9002-61-3] Chorionic Gonadotropin (Human), *2221*
[9002-62-4] Prolactin, *7894*
[9002-64-6] Parathyroid Hormone, *7138*

[9002-67-9] Luteinizing Hormone, 5674
[9002-68-0] Follicle-Stimulating Hormone, 4250
[9002-69-1] Relaxin, 8249
[9002-70-4] Chorionic Gonadotropin (Equine), 2221
[9002-71-5] Thyrotropin, 9569
[9002-72-6] Somatotropin, 8842
[9002-84-0] Polytetrafluoroethylene, 7704
[9002-86-2] Polyvinyl Chloride, 7707
[9002-88-4] Polyethylene, 7687
[9002-89-5] Polyvinyl Alcohol, 7706
[9002-91-9] Metaldehyde, 5994
[9002-92-0] Polidocanol, 7675
[9002-93-1] Octoxynol, 6850
[9003-07-0] Polypropylene, 7701
[9003-11-6] Poloxalene, 7679
[9003-39-8] Povidone, 7814
[9003-98-9] Deoxyribonuclease I, 2907
[9004-02-8] Lipoprotein Lipase, 5567
[9004-07-3] α-Chymotrypsin, 2264
[9004-10-8] Insulin, 5023
[9004-17-5] Protamine Zinc Insulin, 5023
[9004-32-4] Carboxymethylcellulose Sodium, 1830
[9004-34-6] Cellulose, 1966
[9004-53-9] Dextrin, 2955
[9004-54-0] Dextran, 2950
[9004-57-3] Ethyl Cellulose, 3836
[9004-58-4] Cellulose Ethyl Hydroxyethyl Ether, 1968
[9004-61-9] Hyaluronic Acid, 4796
[9004-62-0] Hetastarch, 4708
[9004-64-2] Hydroxypropyl Cellulose, 4879
[9004-65-3] Hydroxypropyl Methylcellulose, 4880
[9004-66-4] Iron Dextran, 5138
[9004-67-5] Methylcellulose, 6111
[9004-70-0] Pyroxylin, 8125
[9005-32-7] Alginic Acid, 235
[9005-35-0] Alginic Acid Calcium Salt, 235
[9005-38-3] Algin, 234
[9005-48-5] Heparin Potassium Salt, 4688
[9005-49-6] Heparin, 4688
[9005-53-2] Lignin, 5540
[9005-65-6] Polysorbate 80, 7703
[9005-79-2] Glycogen, 4531
[9005-80-5] Inulin, 5048
[9005-84-9] Starch, Soluble, 8926
[9005-90-7] Turpentine, 10007
[9005-97-4] Thyroprotein, 9568
[9006-58-0] Lysalbinic Acid, 5692
[9006-65-9] Dimethicone, 3237
[9007-03-8] Bifidus Factor, 1215
[9007-12-9] Calcitonin, 1646
[9007-28-7] Chondroitin Sulfate, 2219
[9007-31-2] Clupeine, 2407
[9007-40-3] Crotoxin, 2590
[9007-41-4] CRP, 2593
[9007-43-6] Cytochrome c, 2788
[9007-57-2] Edestin, 3557
[9007-67-4] Enterogastrone, 3651
[9007-73-2] Ferritin, 4063
[9007-76-5] Fibroins, 4102
[9007-90-3] Gliadin, 4472
[9007-92-5] Glucagon, 4481
[9007-93-6] Glycinin, 4528
[9008-08-6] Hypalon®, 4899

[9008-19-9] Krebiozen, 5371
[9008-22-4] Laminaran, 5400
[9008-96-2] Phosvitins, 7477
[9008-99-5] Piromen, 7617
[9009-58-9] Porphyropsin, 7722
[9009-72-7] Purothionin, 8055
[9009-81-8] Rhodopsin, 8317
[9009-86-3] Ricin, 8333
[9010-30-4] Taraxein, 9194
[9010-71-3] Zinc Iodide-Starch, 10340
[9010-72-4] Zymosan, 10400
[9011-02-3] Policresulen, 7673
[9011-05-6] Ureaform, 10053
[9011-18-1] Dextran Sulfate Sodium, 2953
[9011-93-2] Lysostaphin, 5700
[9011-97-6] Cholecystokinin, 2205
[9012-94-4] Oxycodone Pectinate, 7058
[9013-55-2] Factor XI, 3965
[9013-56-3] Factor XIII, 3967
[9013-83-6] Insulinase, 5024
[9014-02-2] Zinostatin, 10367
[9014-44-2] Toxohormone, 9719
[9014-67-9] Aloxiprin, 305
[9015-13-8] ECTEOLA-Cellulose, 3551
[9015-51-4] Silver Protein, 8666
[9015-68-3] Asparaginase, 826
[9015-71-8] CRF, 2575
[9015-73-0] Detaxtran, 2938
[9015-94-5] Renin, 8253
[9016-72-2] Propineb (homopolymer), 7933
[9025-35-8] α-Galactosidase, 4366
[9025-70-1] Dextranase, 2951
[9025-75-6] Calcineurin, 1643
[9031-37-2] Ceruloplasmin, 2005
[9034-40-6] Gonadotropin-Releasing Hormone, 4561
[9035-55-6] Lipotropic Hormone, 5568
[9035-58-9] Thromboplastin, 9539
[9035-68-1] Proinsulin, 7893
[9037-22-3] Amioca, 477
[9037-24-5] Amberlyst 15®, 377
[9038-41-9] Sodium Cellulose Phosphate, 8732
[9039-53-6] Urokinase, 10072
[9039-61-6] Batroxobin, 1008
[9040-61-3] Staphylokinase, 8923
[9041-08-1] Heparin Sodium Salt, 4688
[9041-92-3] α₁-Antitrypsin, 705
[9041-93-4] Bleomycins Sulfate, 1319
[9044-70-6] Coherin, 2453
[9046-56-4] Ancrod, 624
[9048-46-8] Serum Albumin, 8607
[9050-67-3] Sizofiran, 8697
[9054-89-1] Superoxide Dismutase, 9131
[9057-02-7] Pullulan, 8047
[9061-61-4] Nerve Growth Factor, 6562
[9063-57-4] Tuftsin, 9989
[9064-91-9] Detaxtran Hydrochloride, 2938
[9064-92-0] Polidexide, 7674
[9066-59-5] Lysozyme Hydrochloride, 5701
[9067-32-7] Hyaluronic Acid Sodium Salt, 4796
[9072-41-7] Human Motilin, 6369
[9073-56-7] α-L-Iduronidase, 4935
[9073-60-3] Penicillinase, 7202
[9073-78-3] Thermolysin, 9437
[9079-25-8] Amberlite®, 376
[9080-79-9] Sodium Polystyrene Sulfonate, 8797

[9083-38-9] Melanostatin, 5882
[9087-70-1] Aprotinin, 746
[10001-43-1] Pimefylline, 7542
[10004-44-1] Hymexazol, 4895
[10008-73-8] Angelica Lactone γ-Form, 640
[10010-67-0] PIPES Monosodium Salt, 7592
[10016-20-3] α-Cyclodextrin, 2710
[10017-37-5] tert-Butylamine Hydrochloride, 1547
[10017-44-4] Carnitine D-Form Hydrochloride, 1849
[10018-19-6] Cotarnine Chloride, 2540
[10022-31-8] Barium Nitrate, 979
[10022-50-1] Nitryl Fluoride, 6745
[10022-68-1] Cadmium Nitrate Tetrahydrate, 1627
[10023-07-1] Frenolicin, 4298
[10024-89-2] Morpholine Hydrochloride, 6362
[10024-97-2] Nitrous Oxide, 6740
[10025-65-7] Platinous Chloride, 7643
[10025-66-8] Radium Chloride, 8210
[10025-67-9] Sulfur Chloride, 9100
[10025-68-0] Selenium Chloride, 8569
[10025-69-1] Stannous Chloride Dihydrate, 8910
[10025-70-4] Strontium Chloride Hexahydrate, 8970
[10025-71-5] Tellurium Dichloride, 9264
[10025-73-7] Chromic Chloride, 2226
[10025-77-1] Ferric Chloride Hexahydrate, 4046
[10025-78-2] Trichlorosilane, 9811
[10025-82-8] Indium Trichloride, 4998
[10025-83-9] Iridium Trichloride, 5132
[10025-84-0] Lanthanum Chloride (heptahydrate), 5414
[10025-85-1] Nitrogen Chloride, 6689
[10025-87-3] Phosphorus Oxychloride, 7461
[10025-91-9] Antimony Trichloride, 696
[10025-92-0] Thulium Chloride Heptahydrate, 9548
[10025-93-1] Uranium Trichloride, 10044
[10025-99-7] Potassium Tetrachloroplatinate(II), 7801
[10026-03-6] Selenium Tetrachloride, 8575
[10026-04-7] Silicon Tetrachloride, 8639
[10026-07-0] Tellurium Tetrachloride, 9268
[10026-08-1] Thorium Chloride, 9529
[10026-10-5] Uranium Tetrachloride, 10042
[10026-11-6] Zirconium Chloride, 10374
[10026-12-7] Niobium Pentachloride, 6643
[10026-13-8] Phosphorus Pentachloride, 7463
[10026-17-2] Cobaltous Fluoride, 2424
[10026-18-3] Cobaltic Fluoride, 2414
[10026-22-9] Cobaltous Nitrate Hexahydrate, 2428
[10026-24-1] Cobaltous Sulfate Heptahydrate, 2432
[10028-14-5] Nobelium, 6751
[10028-15-6] Ozone, 7080

[10028-17-8] Tritium, *9939*
[10028-18-9] Nickel Fluoride, *6593*
[10028-22-5] Ferric Sulfate, *4059*
[10028-24-7] Sodium Phosphate, Dibasic Dihydrate, *8789*
[10030-90-7] Ferrous Succinate, *4083*
[10031-13-7] Lead Arsenite, *5455*
[10031-18-2] Mercurous Bromide, *5961*
[10031-22-8] Lead Bromide, *5458*
[10031-23-9] Radium Bromide, *8210*
[10031-24-0] Stannous Bromide, *8909*
[10031-25-1] Chromic Bromide, *2225*
[10031-26-2] Ferric Bromide, *4045*
[10031-27-3] Tellurium Tetrabromide, *9267*
[10031-43-3] Cupric Nitrate Trihydrate, *2633*
[10034-81-8] Magnesium Perchlorate, *5742*
[10034-82-9] Sodium Chromate Tetrahydrate, *8736*
[10034-85-2] Hydriodic Acid, *4814*
[10034-85-2] Hydrogen Iodide, *4834*
[10034-88-5] Sodium Bisulfate Monohydrate, *8722*
[10034-93-2] Hydrazine Sulfate, *4810*
[10034-94-3] Magnesium Orthosilicate, *5751*
[10034-96-5] Manganese Sulfate Monohydrate, *5804*
[10034-99-8] Magnesium Sulfate Heptahydrate, *5755*
[10035-04-8] Calcium Chloride Dihydrate, *1661*
[10035-05-9] Calcium Chlorate Dihydrate, *1660*
[10035-06-0] Bismuth Nitrate Pentahydrate, *1271*
[10035-10-6] Hydrobromic Acid, *4816*
[10035-10-6] Hydrogen Bromide, *4830*
[10038-97-8] Palladium Chloride Dihydrate, *7091*
[10038-98-9] Germanium Tetrachloride, *4444*
[10039-32-4] Sodium Phosphate, Dibasic Dodecahydrate, *8789*
[10039-53-9] Sodium Chromate ^{51}Cr-Labeled Compound, *8736*
[10039-54-0] Hydroxylamine Sulfate, *4867*
[10039-56-2] Sodium Hypophosphite Monohydrate, *8763*
[10040-45-6] Picosulfate Sodium, *7518*
[10042-76-9] Strontium Nitrate, *8975*
[10043-01-3] Aluminum Sulfate, *360*
[10043-11-5] Boron Nitride, *1347*
[10043-35-3] Boric Acid, *1339*
[10043-49-9] Gold, Radioactive, Colloidal, *4552*
[10043-52-4] Calcium Chloride, *1661*
[10043-67-1] Aluminum Potassium Sulfate, *354*
[10043-84-2] Manganese Hypophosphite, *5796*
[10043-92-2] Radon, *8212*
[10045-25-7] Cyclen Tetrahydrochloride, *2699*
[10045-86-0] Ferric Phosphate, *4056*
[10045-89-3] Ammonium Ferrous Sulfate, *518*
[10045-94-0] Mercuric Nitrate, *5947*
[10048-32-5] Parasorbic Acid, *7136*
[10048-95-0] Sodium Arsenate, Dibasic Heptahydrate, *8715*

[10048-98-3] Barium Phosphate, Dibasic, *986*
[10048-99-4] Barium Mercuric Iodide, *978*
[10049-01-1] Bismuth Phosphate, *1276*
[10049-04-4] Chlorine Dioxide, *2096*
[10049-05-5] Chromous Chloride, *2248*
[10049-06-6] Titanium Dichloride, *9627*
[10049-07-7] Rhodium Chloride, *8313*
[10049-08-8] Ruthenium Trichloride, *8436*
[10049-10-2] Chromous Fluoride, *2249*
[10049-12-4] Vanadium Trifluoride, *10108*
[10049-14-6] Uranium Tetrafluoride, *10043*
[10049-17-9] Rhenium Hexafluoride, *8302*
[10049-18-0] Ferric Pyrophosphate Nonahydrate, *4057*
[10049-21-5] Sodium Phosphate, Monobasic Monohydrate, *8790*
[10049-23-7] Tellurous Acid, *9270*
[10049-25-9] Chromous Bromide, *2247*
[10058-07-8] Pimefylline Nicotinate, *7542*
[10058-44-3] Ferric Pyrophosphate, *4057*
[10060-10-3] Ceric Fluoride, *1987*
[10060-11-4] Germanium Dichloride, *4442*
[10061-01-5] 1,3-Dichloropropene *cis*-Form, *3085*
[10061-02-6] 1,3-Dichloropropene *trans*-Form, *3085*
[10072-50-1] Polymyxin D$_1$, *7693*
[10075-24-8] Imipramine Pamoate, *4962*
[10075-36-2] Isoaminile Cyclamate, *5155*
[10075-85-1] 9,10-Bis(2-phenylethynyl)anthracene, *1300*
[10085-81-1] Benzoctamine Hydrochloride, *1089*
[10088-95-6] Radicinin, *8209*
[10097-28-6] Silicon Monoxide, *8635*
[10099-58-8] Lanthanum Chloride, *5414*
[10099-59-9] Lanthanum Nitrate, *5414*
[10099-60-2] Lanthanum Sulfate, *5414*
[10099-74-8] Lead Nitrate, *5469*
[10099-79-3] Lead Vanadate(V), *5482*
[10101-50-5] Sodium Permanganate, *8785*
[10101-52-7] Zirconium Silicate, *10381*
[10101-53-8] Chromic Sulfate, *2234*
[10101-63-0] Lead Iodide, *5466*
[10101-83-4] Sodium Tellurate, *8813*
[10101-85-6] Sodium Dithionate Dihydrate, *8746*
[10101-88-9] Sodium Thiophosphate, *8820*
[10101-89-0] Sodium Phosphate, Tribasic Dodecahydrate, *8792*
[10101-97-0] Nickel Sulfate Hexahydrate, *6602*
[10101-98-1] Nickel Sulfate Heptahydrate, *6602*
[10102-03-1] Nitrogen Pentoxide, *6692*
[10102-05-3] Palladium Nitrate, *7093*
[10102-15-5] Sodium Sulfite Heptahydrate, *8811*
[10102-17-7] Sodium Thiosulfate Pentahydrate, *8821*

[10102-18-8] Sodium Selenite, *8802*
[10102-20-2] Sodium Tellurite, *8814*
[10102-23-5] Sodium Selenate Decahydrate, *8800*
[10102-24-6] Lithium Silicate, *5596*
[10102-25-7] Lithium Sulfate Monohydrate, *5597*
[10102-36-0] Sodium Permanganate Trihydrate, *8785*
[10102-40-6] Sodium Molybdate(VI) Dihydrate, *8776*
[10102-43-9] Nitric Oxide, *6664*
[10102-44-0] Nitrogen Dioxide, *6690*
[10102-45-1] Thallium Mononitrate, *9414*
[10102-68-8] Calcium Iodide, *1679*
[10102-71-3] Aluminum Sodium Sulfate, *357*
[10108-64-2] Cadmium Chloride, *1622*
[10108-73-3] Cerous Nitrate, *1999*
[10112-91-1] Mercurous Chloride, *5962*
[10114-58-6] Bismark Brown Y, *1257*
[10117-38-1] Potassium Sulfite, *7795*
[10118-56-6] Cascarillin, *1882*
[10118-76-0] Calcium Permanganate, *1690*
[10118-90-8] Minocycline, *6283*
[10124-36-4] Cadmium Sulfate, *1632*
[10124-37-5] Calcium Nitrate, *1684*
[10124-43-3] Cobaltous Sulfate, *2432*
[10124-53-5] Magnesium Thiosulfate, *5758*
[10125-13-0] Cupric Chloride Dihydrate, *2623*
[10129-92-7] Eledoisin Trifluoroacetate, *3588*
[10135-94-1] Sodium Metavanadate Tetrahydrate, *8774*
[10137-74-3] Calcium Chlorate, *1660*
[10138-04-2] Ammonium Ferric Sulfate, *515*
[10138-62-2] Holmium Chloride, *4765*
[10139-06-7] Linatine, *5554*
[10139-07-8] Linatine (−)-Isomer, *5554*
[10139-47-6] Zinc Iodide, *10339*
[10139-51-2] Nitric Acid Ammonium Cerium Salt, *1991*
[10140-70-2] Curvularin, *2666*
[10141-00-1] Chromic Potassium Sulfate, *2233*
[10141-05-6] Cobaltous Nitrate, *2428*
[10161-33-8] Trenbolone, *9745*
[10161-34-9] Trenbolone Acetate, *9745*
[10170-69-1] Manganese Carbonyl, *5792*
[10190-55-3] Lead Molybdate(VI), *5467*
[10191-41-0] *dl*-α-Tocopherol, *9653*
[10192-30-0] Ammonium Bisulfite, *499*
[10193-36-9] Orthosilicic Acid, *8629*
[10196-04-0] Ammonium Sulfite, *554*
[10199-21-0] Guanidinium Aluminum Sulfate Hexahydrate, *4598*
[10210-64-7] Beryllium Acetylacetonate, *1165*
[10210-68-1] Dicobalt Octacarbonyl, *3095*
[10212-25-6] Ancitabine Hydrochloride, *623*
[10213-09-9] Vanadyl Dichloride, *10112*
[10213-10-2] Sodium Tungstate(VI) Dihydrate, *8826*

[10214-39-8] Lead Borate (monohydrate), *5457*
[10236-47-2] Naringin, *6507*
[10236-58-5] Selenocysteine, *8577*
[10238-21-8] Glyburide, *4514*
[10246-75-0] Hydroxyzine Pamoate, *4889*
[10252-34-3] Sporidesmolide IV, *8893*
[10257-54-2] Cupric Sulfate Monohydrate, *2643*
[10257-55-3] Calcium Sulfite, *1710*
[10262-69-8] Maprotiline, *5813*
[10265-92-6] Methamidophos, *6019*
[10277-43-7] Lanthanum Nitrate (hexahydrate), *5414*
[10284-63-6] Pinitol, *7561*
[10290-12-7] Cupric Arsenite, *2616*
[10294-26-5] Silver Sulfate, *8668*
[10294-29-8] Gold Monochloride, *4550*
[10294-33-4] Boron Tribromide, *1349*
[10294-34-5] Boron Trichloride, *1350*
[10294-40-3] Barium Chromate(VI), *968*
[10294-52-7] Ferric Chromate(VI), *4047*
[10294-54-9] Cesium Sulfate, *2015*
[10294-62-9] Lanthanum Sulfate (nonahydrate), *5414*
[10294-64-1] Potassium Manganate, *7765*
[10294-66-3] Potassium Thiosulfate, *7808*
[10294-70-9] Stannous Iodide, *8912*
[10297-61-7] Propylure, *7981*
[10310-32-4] Tribenoside, *9769*
[10318-26-0] Mitolactol, *6299*
[10321-12-7] Propizepine, *7946*
[10323-20-3] Arabinose D-Form, *750*
[10325-94-7] Cadmium Nitrate, *1627*
[10326-21-3] Magnesium Chlorate, *5725*
[10326-24-6] Zinc Meta-arsenite, *10342*
[10326-27-9] Barium Chloride Dihydrate, *967*
[10326-41-7] D-Lactic Acid, *5383*
[10331-57-4] Niclofolan, *6603*
[10334-13-1] Isoborneol (1*R*,2*R*,4*R*)-Form, *5173*
[10343-61-0] Titanium Sesquisulfate, *9632*
[10347-81-6] Maprotiline Hydrochloride, *5813*
[10352-73-5] Pseudopederin, *7171*
[10361-03-2] Sodium Metaphosphate, *8773*
[10361-37-2] Barium Chloride, *967*
[10361-43-0] Bismuth Hydroxide, *1267*
[10361-44-1] Bismuth Nitrate, *1271*
[10373-81-6] 3-Hydroxycamphor, *4856*
[10376-48-4] Shionone, *8620*
[10377-48-7] Lithium Sulfate, *5597*
[10377-51-2] Lithium Iodide, *5591*
[10377-58-9] Magnesium Iodide, *5735*
[10377-60-3] Magnesium Nitrate, *5738*
[10377-62-5] Magnesium Permanganate, *5743*
[10377-66-9] Manganese Nitrate, *5798*
[10378-22-0] EDTA Trisodium Salt (monohydrate), *3565*
[10379-14-3] Tetrazepam, *9391*
[10380-31-1] Barium Uranium Oxide, *997*
[10381-36-9] Nickel Phosphate, *6600*
[10397-75-8] Iocarmic Acid, *5053*

[10402-53-6] Eprazinone Dihydrochloride, *3688*
[10402-90-1] Eprazinone, *3688*
[10403-00-6] Muscone, *6398*
[10405-02-4] Trospium Chloride, *9966*
[10415-75-5] Mercurous Nitrate, *5964*
[10417-94-4] Eicosapentaenoic Acid, *3578*
[10418-03-8] Stanozolol (2′*H* form), *8921*
[10421-48-4] Ferric Nitrate, *4054*
[10450-55-2] Ferric Acetate, Basic, *4043*
[10450-60-9] Periodic Acid, *7286*
[10453-86-8] Resmethrin, *8274*
[10453-89-1] Chrysanthemic Acid, *2255*
[10457-66-6] Geranylhydroquinone, *4439*
[10457-90-6] Bromperidol, *1452*
[10476-81-0] Strontium Bromide, *8967*
[10476-85-4] Strontium Chloride, *8970*
[10476-86-5] Strontium Iodide, *8974*
[10534-89-1] Hexaaminecobalt Trichloride, *4710*
[10539-19-2] Moxaverine, *6374*
[10540-29-1] Tamoxifen, *9180*
[10543-57-4] TAED, *9158*
[10544-72-6] Dinitrogen Tetroxide, *3306*
[10544-73-7] Dinitrogen Trioxide, *3307*
[10553-31-8] Barium Bromide, *964*
[10563-70-9] Melitracen Hydrochloride, *5891*
[10567-69-8] Barium Iodate, *975*
[10580-02-6] Ammonium Titanium Oxalate, *560*
[10588-01-9] Sodium Dichromate(VI), *8744*
[10592-03-7] Vincamine Hydrochloride, *10180*
[10596-23-3] Clodronic Acid, *2365*
[10597-60-1] Hydroxytyrosol, *4887*
[10597-89-4] *N*-Acetylmuramic Acid, *6389*
[10605-21-7] Carbendazim, *1792*
[11000-03-6] Strepogenin, *8951*
[11000-17-2] Vasopressin, *10129*
[11003-38-6] Capreomycin, *1759*
[11003-70-6] Scillaren, *8536*
[11005-20-2] Pseudohecogenin, *4657*
[11005-63-3] Strophanthin, *8982*
[11006-76-1] Virginiamycin, *10200*
[11006-83-0] Thermorubin, *9438*
[11015-37-5] Bambermycins, *945*
[11016-07-2] Perimycin, *7284*
[11016-39-0] Properdin, *7928*
[11017-56-4] Strigol, *8963*
[11021-66-2] Ristocetin A, *8362*
[11024-24-1] Digitonin, *3180*
[11028-71-0] Concanavalin A, *2478*
[11030-31-2] Lanatoside D, *5404*
[11030-71-0] Amanitin, *367*
[11031-48-4] Sarkomycin, *8513*
[11031-82-6] Streptovaricin B, *8960*
[11031-85-9] Streptovaricin G, *8960*
[11032-05-6] Teleocidin B₄, *9260*
[11034-45-0] *O*-Cinnamoyltaxicin-I, *9214*
[11041-12-6] Cholestyramine, *2209*
[11042-38-9] Xanthocillin, *10259*
[11042-64-1] γ-Oryzanol, *6983*
[11050-21-8] CTX-1, *2273*
[11050-94-5] Oligomycin B, *6927*

[11051-71-1] Avilamycin, *880*
[11052-72-5] Oligomycin C, *6927*
[11054-04-9] Isobotryococcene, *1360*
[11061-68-0] Insulin (human), *5023*
[11061-96-4] Cardiotoxin III, *1834*
[11070-73-8] Insulin (Bovine), *5023*
[11072-93-8] β-Escin, *3751*
[11075-36-8] Tuberactinomycin, *9983*
[11076-19-0] Bongkrekic Acid, *1334*
[11078-21-0] Filipin, *4110*
[11079-53-1] Hyperforin, *4901*
[11085-36-2] HCS, *4655*
[11088-20-3] Viscotoxin B, *10206*
[11089-65-9] Tunicamycin, *9999*
[11096-26-7] Erythropoietin, *3744*
[11096-49-4] Partricin, *7153*
[11096-82-5] Aroclor 1260, *7682*
[11097-69-1] Aroclor 1254, *7682*
[11103-72-3] Ruthenium Red, *8434*
[11104-88-4] Phosphomolybdic Acid, *7454*
[11110-52-4] Sodium Amalgam, *8713*
[11113-63-6] Graphite Fluoride, *4576*
[11113-80-7] Polyoxins, *7696*
[11115-82-5] Enduracidin, *3634*
[11116-31-7] Bleomycin A₂, *1319*
[11120-25-5] Ammonium Paratungstate, *536*
[11121-16-7] Aluminum Borate, *326*
[11121-32-7] Mepartricin, *5914*
[11125-96-5] Burgundy Mixture, *1338*
[11138-66-2] Xanthan Gum, *10255*
[11141-17-6] Azadirachtin, *886*
[12001-72-8] Mafenide Propionate, *5712*
[12002-03-8] Cupric Acetoarsenite, *2615*
[12002-30-1] Piperazine Edetate Calcium, *7576*
[12002-43-6] Gilsonite, *4457*
[12002-53-8] Trimedlure, *9870*
[12007-25-9] Magnesium Diboride, *5728*
[12007-56-6] Calcium Borate, *1655*
[12007-58-8] Ammonium Borate, *501*
[12007-60-2] Lithium Borate, *5580*
[12011-77-7] Dihydroxyaluminum Sodium Carbonate, *3202*
[12013-15-9] Copper Sulfate Dibasic, *2644*
[12015-53-1] Chlorine Heptoxide, *2097*
[12016-69-2] Cobaltous Chromate(III), *2422*
[12016-80-7] Cobaltic Oxide Monohydrate, *2415*
[12017-68-4] Samarium Cobalt, *8485*
[12018-01-8] Chromium Dioxide, *2237*
[12018-10-9] Cupric Chromite, *2625*
[12023-99-3] Gallium Hydroxide, *4376*
[12024-20-3] Gallium Suboxide, *4376*
[12024-21-4] Gallium Oxide, *4382*
[12026-06-1] Thallium Hydroxide, *9412*
[12026-66-3] Ammonium Phosphomolybdate, *543*
[12026-93-6] Ammonium 12-Tungstophosphate, *561*
[12027-06-4] Ammonium Iodide, *526*
[12027-38-2] Silicotungstic Acid, *8641*
[12027-67-7] Ammonium Molybdate(VI), *530*
[12029-98-0] Iodine Pentoxide, *5062*
[12033-88-4] Nitrogen Selenide, *6693*
[12033-89-5] Silicon Nitride, *8636*

[12036-44-1] Thulium Oxide, *9548*
[12040-41-4] Nicomorphine Hydrochloride, *6605*
[12040-44-7] Alcian Blue, *212*
[12041-72-4] Formosulfathiazole, *9074*
[12047-27-7] Barium Titanate(IV), *996*
[12048-50-9] Bismuth Tetroxide, *1290*
[12052-78-7] Samarium Cobalt, *8485*
[12054-48-7] Nickel Hydroxide, *6595*
[12054-85-2] Ammonium Molybdate-(VI) Tetrahydrate, *530*
[12055-62-8] Holmium Oxide, *4765*
[12057-24-8] Lithium Oxide, *5594*
[12058-66-1] Sodium Stannate(IV), *8805*
[12060-59-2] Strontium Titanate, *8980*
[12062-24-7] Cupric Hexafluorosilicate, *2631*
[12063-56-8] Yig, *10295*
[12063-98-8] Gallium Phosphide, *4383*
[12067-99-1] Phosphotungstic Acid, *7476*
[12068-69-8] Bismuth Selenide, *1279*
[12069-32-8] Boron Carbide, *1345*
[12069-69-1] Cupric Carbonate, Basic, *2621*
[12070-12-1] Tungsten Carbide, *9995*
[12071-83-9] Propineb (monomer), *7933*
[12108-13-3] MMT, *6309*
[12111-24-9] Pentetate Calcium Trisodium, *7236*
[12122-67-7] Zineb, *10365*
[12124-97-9] Ammonium Bromide, *502*
[12124-99-1] Ammonium Bisulfide, *498*
[12125-01-8] Ammonium Fluoride, *519*
[12125-02-9] Ammonium Chloride, *506*
[12125-03-0] Potassium Stannate Trihydrate, *7791*
[12125-08-5] Ammonium Osmium Chloride, *533*
[12125-09-6] Phosphonium Iodide, *7455*
[12126-59-9] Conjugated Estrogens, *2492*
[12135-05-6] Sodium Bisulfide Dihydrate, *8723*
[12135-06-7] Sodium Bisulfide Trihydrate, *8723*
[12135-22-7] Pearlman's Catalyst, *7166*
[12135-76-1] Ammonium Sulfide, *553*
[12141-46-7] Aluminum Silicate, *356*
[12142-33-5] Potassium Stannate, *7791*
[12150-46-8] DPPF, *3490*
[12173-47-6] Hectorite, *4658*
[12192-57-3] Aurothioglucose, *874*
[12209-98-2] Sodium Stannate(IV) Trihydrate, *8805*
[12230-71-6] Barium Hydroxide Octahydrate, *973*
[12232-99-4] Sodium Bismuthate(V), *8721*
[12233-56-6] Bismuth Germanate, *1266*
[12240-15-2] Prussian Blue (unspecified formula), *8018*
[12244-26-7] Kininogens, *5360*
[12244-57-4] Gold Sodium Thiomalate, *4553*
[12279-90-2] Arsenic Sulfide, *790*
[12286-76-9] Fructose Ferric Form, *4302*
[12304-65-3] Hydrotalcite, *5724*
[12354-84-6] Pentamethylcyclopentadienyliridium(III) Dichloride Dimer, *7226*

[12427-38-2] Maneb, *5786*
[12505-77-0] Boron Monoxide, *1346*
[12511-31-8] Aluminum Magnesium Silicate, *345*
[12542-36-8] Gossypol Acetic Acid, *4564*
[12550-17-3] Sodium Antimonylgluconate, *692*
[12584-58-6] Insulin (Porcine), *5023*
[12584-83-7] Cobrotoxin, *2437*
[12602-23-2] Cobaltous Carbonate Basic, *2420*
[12607-70-4] Nickel Carbonate Hydroxide, *6588*
[12607-92-0] Aceglutamide Aluminum Complex, *26*
[12607-93-1] Taxine, *9215*
[12609-84-6] Thiopeptin, *9504*
[12629-01-5] Human Growth Hormone, *8842*
[12650-69-0] Mupirocin, *6388*
[12663-44-4] Cardiotoxins, *1834*
[12698-96-3] Thearubigins, *9425*
[12704-90-4] Aurodox, *873*
[12740-44-2] Sodium-Lead Alloy, *8769*
[12751-04-1] Creolin, *2558*
[12758-40-6] Propagermanium, *7909*
[12771-68-5] Ancymidol, *625*
[12772-35-9] Butirosin, *1530*
[12772-57-5] Monorden, *6341*
[13002-65-8] Tamoxifen (*E*)-Form, *9180*
[13005-39-5] Potassium Tetrachloroaurate(III) Dihydrate, *7800*
[13007-85-7] Glucoheptonic Acid Sodium Salt, *4491*
[13007-90-4] Bis(triphenylphosphine)-dicarbonylnickel, *1305*
[13007-92-6] Chromium Carbonyl, *2236*
[13007-93-7] Cuproxoline, *2660*
[13008-73-6] L-Streptose, *8958*
[13009-99-9] Mafenide Acetate, *5712*
[13010-47-4] Lomustine, *5621*
[13018-10-5] Torilin, *9710*
[13042-18-7] Fendiline, *4002*
[13045-94-8] Melphalan D-Form, *5896*
[13051-01-9] Carbazochrome Salicylate, *1789*
[13055-82-8] Reproterol Hydrochloride, *8259*
[13058-67-8] Lucensomycin, *5651*
[13071-79-9] Terbufos, *9301*
[13073-35-3] Ethionine, *3792*
[13085-08-0] Mazipredone, *5832*
[13087-53-1] Iothalamic Acid *N*-Methylglucamine Salt, *5107*
[13092-66-5] Magnesium Phosphate, Monobasic, *5746*
[13093-88-4] Perimethazine, *7283*
[13100-82-8] Cysteic Acid, *2777*
[13103-34-9] Boldenone 10-Undecenoate, *1327*
[13115-03-2] Vitamin B_{12}-^{57}Co, *10213*
[13121-70-5] Cyhexatin, *2757*
[13137-64-9] Periplocin, *7289*
[13138-45-9] Nickel Nitrate, *6598*
[13157-90-9] Quercetin Pentabenzyl Ether, *8150*
[13171-21-6] Phosphamidon, *7449*
[13171-25-0] Trimetazidine Dihydrochloride, *9874*
[13190-97-1] Solanesol, *8830*
[13194-48-4] Ethoprop, *3799*

[13215-10-6] Demeclocycline Sesquihydrate, *2890*
[13250-12-9] *sec*-Butylamine *l*-Form, *1546*
[13258-72-5] Cephalosporin P_1, *1979*
[13269-28-8] Canavanine DL-Form, *1740*
[13291-74-2] Gentamicin A, *4427*
[13292-46-1] Rifampin, *8341*
[13292-87-0] Borane-dimethyl Sulfide Complex, *1336*
[13311-52-9] PAR Sodium Salt, *7125*
[13311-84-7] Flutamide, *4238*
[13327-32-7] Beryllium Hydroxide, *1171*
[13347-42-7] Dowicide 9®, *3479*
[13356-08-6] Fenbutatin Oxide, *3997*
[13364-32-4] Clobenzorex, *2357*
[13392-18-2] Fenoterol, *4012*
[13392-28-4] Rimantadine, *8351*
[13397-24-5] Calcium Sulfate Dihydrate, *1708*
[13400-13-0] Cesium Fluoride, *2011*
[13401-40-6] Phaseolin, *7313*
[13405-60-2] β-Glucogallin, *4490*
[13408-56-5] Ponasterone A, *7710*
[13410-01-0] Sodium Selenate, *8800*
[13411-16-0] Nifurpirinol, *6621*
[13412-64-1] Dicloxacillin Sodium Salt Monohydrate, *3094*
[13422-51-0] Hydroxocobalamin, *4846*
[13422-52-1] Aquacobalamin, *4846*
[13422-53-2] Vitamin B_{12}-^{60}Co, *10213*
[13422-55-4] Methylcobalamin, *6116*
[13424-46-9] Lead Azide, *5456*
[13425-22-4] Aminorex Fumarate, *472*
[13425-39-3] Etofylline Nicotinate, *3924*
[13444-24-1] 1-Ethyl-3-piperidinol, *3897*
[13444-90-1] Nitryl Chloride, *6744*
[13445-63-1] Itramin Tosylate, *5293*
[13446-03-2] Manganese Bromide, *5790*
[13446-18-9] Magnesium Nitrate Hexahydrate, *5738*
[13446-24-7] Magnesium Pyrophosphate, *5748*
[13446-34-9] Manganese Chloride Tetrahydrate, *5793*
[13446-49-6] Potassium Molybdate, *7768*
[13447-95-5] Isoniazid Methanesulfonate, *5232*
[13450-90-3] Gallium Chloride, *4378*
[13452-36-3] *O*-Cinnamoyltaxicin-I triacetate, *9214*
[13453-07-1] Gold Trichloride, *4556*
[13453-34-4] Thallium Cyanide, *9410*
[13453-78-6] Lithium Perchlorate Trihydrate, *5595*
[13454-94-9] Cerous Sulfate, *2001*
[13455-21-5] Potassium Fluoride Dihydrate, *7753*
[13455-36-2] Cobaltous Phosphate, *2431*
[13457-18-6] Pyrazophos, *8073*
[13462-88-9] Nickel Bromide, *6587*
[13462-90-3] Nickel Iodide, *6596*
[13463-39-3] Nickel Carbonyl, *6589*
[13463-40-6] Iron Pentacarbonyl, *5143*
[13463-41-7] Pyrithione Zinc Derivative, *8107*
[13463-67-7] Titanium Dioxide, *9628*

[14221-48-8] Ammonium Ferricyanide, 516

[14222-60-7] Protionamide, 8000

[14255-61-9] Oxidronic Acid Sodium Salt, 7039

[14255-87-9] Parbendazole, 7139

[14259-45-1] Asperuloside, 834

[14268-23-6] Trinitromethane Potassium Salt, 9901

[14277-97-5] Domoic Acid, 3466

[14283-05-7] Tetraamminecopper Sulfate, 9325

[14283-07-9] Lithium Tetrafluoroborate, 5599

[14286-84-1] Bencyclane Fumarate, 1035

[14293-44-8] Xipamide, 10276

[14293-70-0] Fremy's Salt, 4297

[14307-33-6] Calcium Dichromate(VI), 1668

[14307-35-8] Lithium Chromate(VI), 5585

[14317-18-1] Ioglycamic Acid Meglumine Salt, 5094

[14323-36-5] Potassium Tetracyanoplatinate(II) Trihydrate, 7803

[14334-40-8] Pramiverin, 7827

[14334-41-9] Pramiverin Hydrochloride, 7827

[14338-32-0] Mukaiyama Reagent, 6387

[14350-67-5] Ormosinine, 6966

[14357-78-9] Diprenorphine, 3380

[14358-43-1] Dehydroemetine (±)-Form Dihydrochloride, 2874

[14375-45-2] Abscisic Acid (±)-cis,-trans-Form, 12

[14376-16-0] Sulfaloxic Acid, 9044

[14398-53-9] Abscisic Acid (−)-cis,-trans-Form, 12

[14402-67-6] Potassium Titanyl Oxalate Dihydrate, 7809

[14402-73-4] Lithium Tetracyanoplatinate(II), 5598

[14402-75-6] Cadmium Potassium Cyanide, 1629

[14402-88-1] EDTA Magnesium Disodium Salt, 3565

[14402-89-2] Sodium Nitroprusside, 8780

[14417-88-0] Melinamide, 5890

[14429-30-2] 2-Deoxystreptamine Dihydrochloride, 2910

[14433-82-0] Thiacetarsamide Disodium Salt, 9441

[14434-22-1] Sodium Ferrocyanide Decahydrate, 8752

[14451-99-1] Menadione Dimethylpyrimidinol Bisulfite, 5899

[14452-57-4] Magnesium Peroxide, 5744

[14457-87-5] Cerous Bromide, 1994

[14459-29-1] Hematoporphyrin, 4671

[14459-95-1] Potassium Ferrocyanide Trihydrate, 7752

[14463-33-3] Cob(II)alamin, 2410

[14475-63-9] Zirconium Hydroxide, 10377

[14481-26-6] Potassium Titanyl Oxalate, 7809

[14481-29-9] Ammonium Ferrocyanide, 517

[14481-33-5] Thorium Tetracyanoplatinate(II), 9532

[14484-47-0] Deflazacort, 2865

[14484-64-1] Ferbam, 4040

[14487-05-9] Rifamycin O, 8342

[14492-68-3] Lapyrium Chloride Stearoyl Analog, 5421

[14504-73-5] Tritoqualine, 9942

[14507-19-8] Lanthanum Hydroxide, 5414

[14521-96-1] Etorphine, 3931

[14523-22-9] Rhodium Carbonyl Chloride, 8312

[14556-46-8] Bupranolol, 1500

[14559-79-6] Propizepine Hydrochloride, 7946

[14590-13-7] Ammonium Cobaltous Phosphate, 510

[14611-51-9] Selegiline, 8565

[14611-52-0] Selegiline Hydrochloride, 8565

[14613-30-0] Clofibric Acid Magnesium Salt, 2371

[14633-54-6] Cyclopropyl Phenyl Sulfide, 2747

[14635-75-7] Nitrosyl Tetrafluoroborate, 6736

[14636-12-5] Terlipressin, 9310

[14639-25-9] Chromium Picolinate, 2240

[14639-94-2] Ammonium Hexafluorogallate, 522

[14639-97-5] Ammonium Tetrachlorozincate, 556

[14639-98-6] Ammonium Pentachlorozincate, 537

[14644-61-2] Zirconium Sulfate, 10382

[14650-46-5] Tramiprosate Sodium Salt, 9728

[14663-23-1] Dantrolene Sodium Salt, 2815

[14679-73-3] Todralazine, 9660

[14694-95-2] Wilkinson's Catalyst, 10245

[14698-29-4] Oxolinic Acid, 7044

[14720-53-7] Lead Borate, 5457

[14721-21-2] Cupric Chlorate, 2622

[14758-11-3] Schultenite, 5454

[14759-06-9] Sulforidazine, 9093

[14769-73-4] Levamisole, 5511

[14807-96-6] Talc, 9168

[14816-18-3] Phoxim, 7478

[14816-67-2] Soterenol Monohydrochloride, 8855

[14836-73-8] Deferoxamine Iron Complex, 2863

[14838-15-4] Phenylpropanolamine, 7417

[14855-76-6] Methyl Green, 6154

[14860-49-2] Clobutinol, 2360

[14871-41-1] Vaska's Compound (unspecified stereo), 10128

[14871-79-5] Barium Hypophosphite, 974

[14874-24-9] Tris(ethylenediamine)cadmium Dihydroxide, 9928

[14875-96-8] Heme, 4674

[14882-18-9] Bismuth Subsalicylate, 1285

[14885-29-1] Ipronidazole, 5123

[14897-39-3] Rifamycin SV Sodium Salt, 8343

[14898-79-4] sec-Butyl Alcohol l-Form, 1543

[14901-08-7] Cycasin, 2694

[14912-44-8] Ylangene, 10297

[14918-35-5] Destomycin A, 2936

[14919-77-8] Benserazide Hydrochloride, 1052

[14929-11-4] Simfibrate, 8675

[14930-96-2] Cytochalasin B, 2787

[14932-42-4] Xenon-133, 10270

[14940-41-1] Ferrous Phosphate, 4080

[14959-86-5] Looplure, 5627

[14970-87-7] 3,6-Dioxaoctane-1,8-dithiol, 3329

[14976-57-9] Clemastine Hydrogen Fumarate, 2343

[14977-61-8] Chromyl Chloride, 2252

[14984-68-0] Cloperastine Hydrochloride, 2384

[14987-04-3] Magnesium Trisilicate, 5751

[14992-62-2] DL-Acetylcarnitine, 79

[15016-60-1] Chlorogenic Acid (E-form), 2143

[15060-55-6] Ammonium Tetrathiomolybdate, 557

[15069-99-5] Lycorine Methiodide, 5687

[15086-94-9] Eosin Y Free Acid, 3658

[15096-52-3] Cryolite, 2595

[15123-69-0] Cupric Selenate, 2640

[15139-76-1] Orange B, 6954

[15148-80-8] Bupranolol Hydrochloride, 1500

[15165-67-0] Dichlorprop (R)-(+)-Form, 3088

[15176-29-1] Edoxudine, 3563

[15180-03-7] Alcuronium Dichloride, 214

[15189-51-2] Sodium Tetrachloroaurate(III), 8816

[15238-00-3] Cobaltous Iodide, 2427

[15263-53-3] Cartap, 1865

[15265-28-8] Palitantin, 7088

[15275-09-9] Chromic Potassium Oxalate Trihydrate, 2232

[15280-09-8] Sodium Dicyanoaurate(I), 8745

[15283-45-1] Gold Sodium Thiosulfate, 4554

[15291-75-5] Ginkgolide A, 4461

[15291-76-6] Ginkgolide C, 4461

[15291-77-7] Ginkgolide B, 4461

[15299-99-7] Napropamide, 6497

[15301-48-1] Bezitramide, 1200

[15301-69-6] Flavoxate, 4128

[15301-80-1] Oxamarin, 7014

[15307-79-6] Diclofenac Sodium Salt, 3091

[15307-81-0] Diclofenac Potassium Salt, 3091

[15307-86-5] Diclofenac, 3091

[15308-34-6] Norfenefrine dl-Form Hydrochloride, 6788

[15318-31-7] Vaska's Compound, 10128

[15318-33-9] trans-Carbonylchlorobis-(triphenylphosphine)rhodium(I), 1819

[15318-45-3] Thiamphenicol, 9453

[15320-30-6] Magnesium Sulfate Trihydrate, 5755

[15328-32-2] 1-Cyanobenzotriazole, 2680

[15351-05-0] Buzepide Methiodide, 1603

[15356-74-8] Dihydroactinidiolide, 3188

[15358-48-2] Hydroxylupanine, *4868*
[15385-57-6] Mercurous Iodide, *5963*
[15421-84-8] Trapidil, *9735*
[15468-10-7] Oxidronic Acid, *7039*
[15477-33-5] Aluminum Chlorate, *332*
[15489-16-4] Stibophen, *8942*
[15489-90-4] Hematin, *4670*
[15490-91-2] Struvite, *8984*
[15500-66-0] Pancuronium Bromide, *7109*
[15537-71-0] *N*-Acetylpenicillamine D-Form, *94*
[15537-76-5] Chlorproguanil Hydrochloride, *2190*
[15545-48-9] Chlorotoluron, *2176*
[15571-91-2] Potassium Tellurate, *7797*
[15572-25-5] Thallium Selenide, *9416*
[15574-96-6] Pizotyline, *7629*
[15578-26-4] Stannous Pyrophosphate, *8915*
[15588-95-1] DOM, *3461*
[15589-00-1] DOM Hydrochloride, *3461*
[15590-23-5] Dimsyl Sodium, *3285*
[15599-39-0] Noxythiolin, *6813*
[15610-76-1] Ammonium Cupric Chloride, *511*
[15622-65-8] Molindone Hydrochloride, *6319*
[15656-28-7] Bis(pyridine)iodonium Tetrafluoroborate, *1303*
[15662-33-6] Ryanodine, *8442*
[15663-27-1] Cisplatin, *2316*
[15671-27-9] Sulfitocobalamin, *4846*
[15676-16-1] Sulpiride, *9119*
[15676-23-0] Norlevorphanol (+)-Form, *6796*
[15686-51-8] Clemastine, *2343*
[15686-71-2] Cephalexin, *1974*
[15686-83-6] Pyrantel, *8066*
[15687-27-1] Ibuprofen, *4919*
[15687-41-9] Oxyfedrine, *7061*
[15690-55-8] Clomiphene *cis*-Form, *2377*
[15690-57-0] Clomiphene *trans*-Form, *2377*
[15699-18-0] Nickel Ammonium Sulfate, *6586*
[15708-41-5] Ferric Sodium Edetate, *4058*
[15722-48-2] Olsalazine, *6935*
[15764-04-2] α-Vetivone, *10167*
[15768-18-0] Silver Lactate, *8656*
[15798-64-8] Crotonaldehyde (Z-form), *2587*
[15825-70-4] D-Mannitol Hexanitrate, *5811*
[15826-37-6] Cromolyn Disodium Salt, *2581*
[15845-98-4] Iothalamic Acid ^{131}I-Labeled Sodium Salt, *5107*
[15871-57-5] TOAC, *9645*
[15876-67-2] Distigmine Bromide, *3405*
[15879-93-3] α-Chloralose, *2072*
[15956-28-2] Rhodium(II) Acetate, *8311*
[15972-60-8] Alachlor, *194*
[16009-13-5] Hemin, *4679*
[16034-77-8] Iocetamic Acid, *5054*
[16037-91-5] Sodium Stibogluconate, *692*
[16039-53-5] Zinc Lactate, *10341*
[16048-96-7] Cupric Salicylate, *2639*

[16051-77-7] Isosorbide-5-mononitrate, *5271*
[16110-51-3] Cromolyn, *2581*
[16208-51-8] Dimesna, *3233*
[16241-25-1] Potassium 4-Guaiacolsulfonate, *7755*
[16283-36-6] Zinc Salicylate, *10355*
[16286-69-4] Cetiedil Citrate, *2019*
[16298-90-1] Jesaconitine, *5313*
[16320-04-0] Gestrinone, *4449*
[16353-77-8] *cis*-Carbonylchlorobis(triphenylphosphine)rhodium(I), *1819*
[16357-59-8] EEDQ, *3566*
[16377-00-7] Indospicine, *5011*
[16378-21-5] Piroheptine, *7616*
[16378-22-6] Piroheptine Hydrochloride, *7616*
[16397-28-7] Fenproporex, *4022*
[16409-46-4] Menthyl Isovalerate (unspecified stereo), *5909*
[16423-68-0] Erythrosine, *3749*
[16462-65-0] Plicatic Acid, *7653*
[16478-59-4] Hemicholinium, *4678*
[16506-27-7] Bendamustine, *1036*
[16509-46-9] Linalyl Acetate *R*-Form, *5551*
[16550-22-4] Cyprenorphine Hydrochloride, *2766*
[16589-24-5] Synephrine Tartrate, *9142*
[16590-41-3] Naltrexone, *6448*
[16593-81-0] PAR Sodium Salt (monohydrate), *7125*
[16595-80-5] Levamisole Hydrochloride, *5511*
[16604-45-8] Medifoxamine Fumarate, *5861*
[16625-20-0] Haplophytine, *4644*
[16643-66-6] Menbutone Magnesium Salt, *5902*
[16648-69-4] Oxyfedrine DL-Form Hydrochloride, *7061*
[16662-46-7] Gallopamil Hydrochloride, *4386*
[16662-47-8] Gallopamil, *4386*
[16672-87-0] Ethephon, *3787*
[16674-78-5] Magnesium Acetate Tetrahydrate, *5718*
[16676-29-2] Naltrexone Hydrochloride, *6448*
[16679-58-6] Desmopressin, *2926*
[16698-35-4] β-Tocopherol, *9654*
[16721-80-5] Sodium Bisulfide, *8723*
[16725-71-6] Isoborneol (1*S*,2*S*,4*S*)-Form, *5173*
[16731-55-8] Potassium Metabisulfite, *7766*
[16752-77-5] Methomyl, *6054*
[16755-07-0] Showdomycin, *8621*
[16773-42-5] Ornidazole, *6967*
[16774-21-3] Cerium(IV) Ammonium Nitrate, *1991*
[16777-42-7] Oxyfedrine L-Form Hydrochloride, *7061*
[16816-67-4] Pantethine, *7115*
[16830-15-2] Asiaticoside, *822*
[16833-54-8] Pinolenic Acid, *7562*
[16846-24-5] Josamycin, *5315*
[16846-34-7] Leucomycin A₁, *5505*
[16853-85-3] Aluminum Lithium Hydride, *344*
[16871-60-6] Potassium Hexachloroosmate(IV), *7756*
[16871-71-9] Zinc Hexafluorosilicate, *10337*

[16871-90-2] Potassium Hexafluorosilicate, *7759*
[16872-11-0] Fluoboric Acid, *4180*
[16893-85-9] Sodium Hexafluorosilicate, *8759*
[16903-35-8] Chloroauric Acid, *2120*
[16919-19-0] Ammonium Hexafluorosilicate, *524*
[16919-58-7] Ammonium Platinic Chloride, *545*
[16921-30-5] Potassium Hexachloroplatinate(IV), *7757*
[16921-96-3] Iodine Heptafluoride, *5058*
[16923-58-3] Sodium Hexachloroplatinate(IV), *8758*
[16923-95-8] Potassium Hexafluorozirconate, *7760*
[16925-39-6] Calcium Hexafluorosilicate, *1673*
[16940-66-2] Sodium Borohydride, *8727*
[16941-10-9] Aluminum Calcium Hydride, *330*
[16941-11-0] Ammonium Hexafluorophosphate, *523*
[16941-12-1] Platinic Chloride, *7641*
[16949-15-8] Lithium Borohydride, *5581*
[16949-65-8] Magnesium Hexafluorosilicate, *5732*
[16960-16-0] Cosyntropin, *2539*
[16961-25-4] Chloroauric Acid Trihydrate, *2120*
[16961-83-4] Fluosilicic Acid, *4214*
[16962-07-5] Aluminum Borohydride, *327*
[16980-89-5] Bucladesine Sodium Salt, *1473*
[17021-26-0] Calusterone, *1724*
[17032-39-2] Phosphorylcholine Magnesium Salt, *7474*
[17034-34-3] Secretin Porcine Secretin Acetate Salt, *8558*
[17034-35-4] Porcine Secretin, *8558*
[17035-90-4] *N*ᴳ-Methylarginine, *6096*
[17082-09-6] Cinnamoyl Chloride (2*E*)-Form, *2301*
[17086-28-1] Doxycycline (monohydrate), *3488*
[17088-37-8] Triacetone Triperoxide, *9752*
[17090-79-8] Monensin, *6332*
[17092-92-1] Dihydroactinidiolide (*R*)-Form, *3188*
[17093-76-4] Hydroxyglutamic Acid DL-*threo*-Form, *4864*
[17099-70-6] Aluminum Hexafluorosilicate, *336*
[17102-64-6] Sorbic Alcohol, *8848*
[17109-49-8] Edifenphos, *3558*
[17117-97-4] Dimethylglyoxime (2*E*,-3*E*-form), *3270*
[17125-80-3] Barium Hexafluorosilicate, *972*
[17140-60-2] Glucoheptonic Acid Calcium Salt, *4491*
[17140-68-0] Etamiphyllin Hydrochloride, *3766*
[17140-81-7] Nitrofurantoin Monohydrate, *6686*
[17146-95-1] Pentazocine Lactate, *7233*
[17155-31-6] Veralkamine, *10143*
[17169-60-7] Ferroglycine Sulfate, *4066*

[19246-24-3] Telomycin, *9273*
[19262-68-1] Methylphenidate *d-threo-*Form Hydrochloride, *6183*
[19286-37-4] Convicine, *2498*
[19287-45-7] Diborane(6), *3020*
[19295-31-9] Urothion, *10073*
[19326-29-5] Etamiphyllin Camphorsulfonate, *3766*
[19342-33-7] *N*-Acetyl-β-neuraminic Acid, *8623*
[19356-17-3] Calcifediol, *1641*
[19379-90-9] Benzoxonium Chloride, *1114*
[19387-91-8] Tinidazole, *9604*
[19388-87-5] Taurolidine, *9213*
[19396-03-3] Polyoxin A, *7696*
[19396-06-6] Polyoxin B, *7696*
[19428-14-9] Benproperine Trihydrogen Phosphate, *1051*
[19457-37-5] Batrachotoxinin A, *1007*
[19467-61-9] β-Ergocryptinine, *3708*
[19504-77-9] Pecilocin, *7168*
[19545-26-7] Wortmannin, *10251*
[19554-22-4] Cytohemin, *2790*
[19562-30-2] Piromidic Acid, *7618*
[19624-22-7] Pentaborane(9), *7213*
[19625-10-6] Lactobacillic Acid, *5390*
[19660-77-6] Phytochlorin, *7500*
[19666-30-9] Oxadiazon, *7006*
[19700-21-1] Geosmin, *4435*
[19721-56-3] Picromycin, *7525*
[19728-88-2] Metitepine Maleate, *6217*
[19767-45-4] Mesna, *5979*
[19771-63-2] L-2-Oxo-4-thiazolidinecarboxylic Acid, *7049*
[19774-82-4] Amiodarone Hydrochloride, *478*
[19793-20-5] Bolandiol, *1325*
[19794-93-5] Trazodone, *9740*
[19803-62-4] Picloxydine Dihydrochloride, *7510*
[19804-27-4] *p*-Methyldiphenhydramine, *6125*
[19809-78-0] Uliginosin A, *10028*
[19809-79-1] Uliginosin B, *10028*
[19875-60-6] Lisuride Maleate, *5574*
[19879-03-9] Granaticin B, *4570*
[19879-06-2] Granaticin, *4570*
[19881-18-6] Nitroscanate, *6719*
[19891-74-8] Lycoxanthin, *5688*
[19891-75-9] Lycophyll, *5682*
[19893-23-3] Enniatin C, *3639*
[19973-76-3] Estrone *dl*-Form, *3763*
[19982-08-2] Memantine, *5897*
[20004-62-0] Resistomycin, *8272*
[20039-37-6] Pyridinium Dichromate, *8087*
[20069-09-4] Piperlongumine, *7586*
[20123-80-2] Dobesilate Calcium, *3439*
[20150-34-9] Ferrous Bisglycinate, *4068*
[20153-98-4] Dilazep Dihydrochloride, *3222*
[20187-55-7] Bendazac, *1037*
[20196-67-2] Sinalbin, *8678*
[20229-30-5] Metitepine, *6217*
[20231-81-6] Uzarin, *10083*
[20236-82-2] Thebacon Hydrochloride, *9426*
[20246-33-7] D-Gulonic Acid, *4611*
[20267-87-2] Chelidonine (±)-Form, *2052*
[20283-48-1] Chalcomycin, *2035*

[20283-92-5] Rosmarinic Acid, *8399*
[20284-40-6] Murexine, *6392*
[20290-75-9] Stearidonic Acid, *8931*
[20300-26-9] Gossypol *S*-Form, *4564*
[20310-89-8] Saponarin, *8501*
[20311-78-8] Enanthotoxin, *3624*
[20315-46-2] β-Ergocryptine, *3707*
[20316-18-1] Lycodine, *5680*
[20338-08-3] Titanic(IV) Acid, *9625*
[20344-49-4] Ferric Hydroxide, *4052*
[20347-65-3] Bornyl Acetate *d*-Form, *1341*
[20350-15-6] Brefeldin A, *1374*
[20405-64-5] Cuprous Selenide, *2656*
[20408-97-3] 5-Thio-D-glucose, *9487*
[20427-56-9] Ruthenium Tetroxide, *8435*
[20427-59-2] Cupric Hydroxide, *2632*
[20432-64-8] Iprindole Hydrochloride, *5119*
[20438-98-6] Imipramine *N*-Oxide Hydrochloride, *4963*
[20445-31-2] Mosher's Reagent (*R*)-(+)-Form, *6365*
[20445-33-4] Mosher's Reagent (*S*)-(+)-Acid Chloride, *6365*
[20449-79-0] Melittin, *5892*
[20480-93-7] Methenamine Sulfosalicylate, *6038*
[20493-41-8] Quinapyramine, *8168*
[20537-88-6] Amifostine, *399*
[20548-54-3] Calcium Sulfide, *1709*
[20554-84-1] Parthenolide, *7152*
[20559-55-1] Oxibendazole, *7035*
[20562-02-1] Solanine, *8832*
[20562-03-2] α-Chaconine, *8832*
[20574-50-9] Morantel, *6353*
[20594-83-6] Nalbuphine, *6441*
[20623-13-6] Nitrocobalamin, *4846*
[20624-25-3] Ditiocarb Sodium Trihydrate, *3421*
[20638-18-0] Silicic Acid Pyrosilicic Acid, *8629*
[20642-05-1] Potassium Diformate, *7750*
[20644-87-5] Vanadium Carbonyl, *10105*
[20667-12-3] Silver Oxide, *8660*
[20684-06-4] Bamifylline Hydrochloride, *948*
[20702-77-6] Neohesperidin Dihydrochalcone, *6537*
[20717-86-6] Chlorotitanium Triisopropoxide, *2174*
[20725-03-5] Fustin, *4347*
[20734-58-1] DMAN, *3432*
[20770-09-6] Stannic Selenide, *8906*
[20788-07-2] Resorantel, *8275*
[20816-12-0] Osmium Tetroxide, *6990*
[20830-75-5] Digoxin, *3186*
[20830-81-3] Daunorubicin, *2834*
[20831-76-9] Gentiopicrin, *4432*
[20846-91-7] EDDS *S,S*-Form, *3556*
[20859-23-8] Bromosuccinic Acid *l*-Form, *1447*
[20859-73-8] Aluminum Phosphide, *353*
[20867-01-0] Fulvoplumierin, *4314*
[20902-45-8] Penicillamine Disulfide D-Form, *7199*
[20905-71-9] Pyridoxal 5-Phosphate *O*-Methyloxime, *8092*
[20972-43-4] Azoxybenzene *trans*-Form, *916*

[21013-96-7] Salicylaldoxime (*E*-form), *8464*
[21018-84-8] Amarogentin, *371*
[21040-45-9] Cinnamyl Acetate *trans*-Form, *2302*
[21040-59-5] Codamine, *2447*
[21041-93-0] Cobaltous Hydroxide, *2426*
[21041-95-2] Cadmium Hydroxide, *1625*
[21056-98-4] Calcium Phosphite, *1698*
[21087-64-9] Metribuzin, *6233*
[21109-95-5] Barium Sulfide, *991*
[21133-52-8] Lycoramine, *5686*
[21133-53-9] Glucofrangulin A, *4488*
[21150-22-1] β-Amanitin, *367*
[21174-80-1] Penicillamine Disulfide DL-Form, *7199*
[21187-73-5] Gardenin A, *4399*
[21187-98-4] Gliclazide, *4474*
[21200-24-8] Indolmycin, *5007*
[21215-62-3] Calcitonin (Human Synthetic), *1646*
[21238-30-2] Carbomycin B, *1806*
[21238-33-5] Oryzanol A, *6983*
[21245-01-2] 4-(Dimethylamino)benzoic Acid 3-Methylbutyl Ester, *3254*
[21245-02-3] 4-(Dimethylamino)benzoic Acid 2-Ethylhexyl Ester, *3254*
[21256-18-8] Oxaprozin, *7023*
[21259-20-1] T-2 Toxin, *9981*
[21259-76-7] Mercaptomerin Sodium, *5937*
[21286-57-7] *trans*-(−)-Metazocine, *6005*
[21286-60-2] *cis*-(−)-Metazocine, *6005*
[21293-20-9] Isoquassin, *5266*
[21293-29-8] Abscisic Acid, *12*
[21302-79-4] Ceanothic Acid, *1911*
[21312-10-7] *N*⁴-Acetylsulfamethoxazole, *9050*
[21351-79-1] Cesium Hydroxide, *2012*
[21361-93-3] Nicotine Hydrochloride, *6609*
[21362-69-6] Mepitiostane, *5925*
[21370-21-8] Fenoxazoline Hydrochloride, *4016*
[21411-53-0] Virginiamycin M₁, *10200*
[21416-53-5] Picrotin, *7526*
[21416-67-1] Razoxane, *8241*
[21462-39-5] Clindamycin Hydrochloride, *2354*
[21466-07-9] Bromofenofos, *1429*
[21498-08-8] Lofexidine Hydrochloride, *5615*
[21535-47-7] Mianserin Hydrochloride, *6251*
[21548-73-2] Silver Sulfide, *8669*
[21564-17-0] TCMTB, *9223*
[21593-23-7] Cephapirin, *1983*
[21645-51-2] Aluminum Hydroxide, *338*
[21649-57-0] Carbenicillin Phenyl Sodium, *1793*
[21650-42-0] Tretoquinol, *9749*
[21650-65-7] Azoxybenzene *cis*-Form, *916*
[21651-19-4] Stannous Oxide, *8914*
[21651-62-7] Nepetalactone *cis-trans*-Form, *6554*
[21652-27-7] Oleyl Hydroxyethyl Imidazoline, *6925*
[21679-14-1] Fludarabine, *4157*
[21715-46-8] Etifoxine, *3913*

[21725-46-2] Cyanazine, *2675*
[21730-16-5] Metapramine, *6001*
[21738-42-1] Oxamniquine, *7018*
[21755-66-8] Picoperine, *7516*
[21829-25-4] Nifedipine, *6613*
[21888-96-0] Dexetimide Hydrochloride, *2947*
[21888-98-2] Dexetimide, *2947*
[21898-19-1] Clenbuterol Hydrochloride, *2345*
[21908-53-2] Mercuric Oxide, Red, *5949*
[21908-53-2] Mercuric Oxide, Yellow, *5950*
[22059-60-5] Disopyramide Phosphate, *3402*
[22071-15-4] Ketoprofen, *5352*
[22089-22-1] Trofosfamide, *9947*
[22131-35-7] Butalamine, *1513*
[22139-77-1] Pinosylvin, *7563*
[22150-76-1] Biopterin, *1234*
[22156-91-8] *sec*-Butyl Chloride *d*-Form, *1562*
[22156-92-9] *sec*-Butyl Iodide *l*-Form, *1577*
[22157-31-9] *sec*-Butyl Chloride *l*-Form, *1562*
[22178-11-6] Pipotiazine Undecylenic Ester, *7595*
[22189-32-8] Spectinomycin Dihydrochloride Pentahydrate, *8866*
[22195-34-2] Guanadrel Sulfate, *4594*
[22199-08-2] Sulfadiazine Silver Salt, *9035*
[22204-24-6] Pyrantel Pamoate, *8066*
[22204-53-1] Naproxen, *6499*
[22205-45-4] Cuprous Sulfide, *2657*
[22224-92-6] Fenamiphos, *3991*
[22232-54-8] Carbimazole, *1799*
[22232-55-9] Pentorex Hydrogen D-Tartrate, *7246*
[22232-71-9] Mazindol, *5831*
[22248-79-9] Tetrachlorvinphos, *9337*
[22254-24-6] Ipratropium Bromide, *5117*
[22260-51-1] Bromocriptine Methanesulfonate, *1424*
[22263-79-2] Antheridiol, *673*
[22298-29-9] Betamethasone 17-Benzoate, *1182*
[22304-30-9] Apazone Dihydrate, *713*
[22316-47-8] Clobazam, *2356*
[22326-55-2] Barium Hydroxide Monohydrate, *973*
[22345-47-7] Tofisopam, *9662*
[22368-82-7] Pithecolobine, *7622*
[22398-80-7] Indium Phosphide, *4993*
[22457-89-2] Benfotiamine, *1043*
[22465-48-1] Acetatocobalamin, *4846*
[22494-42-4] Diflunisal, *3165*
[22535-46-2] Thiolactic Acid Sodium Salt, *9493*
[22537-19-5] Lawrencium, *5447*
[22560-16-3] Lithium Triethylborohydride, *5600*
[22560-50-5] Clodronic Acid Disodium Salt, *2365*
[22572-05-0] Penicillamine DL-Form Hydrochloride, *7197*
[22572-40-3] EDC Methiodide, *3555*
[22573-93-9] Alexidine, *226*
[22619-35-8] Tioclomarol, *9607*
[22633-88-1] Cosyntropin Hexaacetate, *2539*

[22662-39-1] Rafoxanide, *8214*
[22664-55-7] Metipranolol, *6216*
[22668-01-5] Etanidazole, *3768*
[22672-74-8] Isocorybulbine, *5206*
[22722-98-1] SMEAH, *8702*
[22733-60-4] Siccanin, *8625*
[22760-18-5] Proquazone, *7984*
[22763-02-6] Potassium Phosphate, Tribasic Heptahydrate, *7781*
[22775-52-6] Mycelianamide, *6404*
[22781-23-3] Bendiocarb, *1038*
[22831-39-6] Magnesium Silicide, *5752*
[22832-87-7] Miconazole Nitrate, *6257*
[22839-47-0] Aspartame, *829*
[22862-76-6] Anisomycin, *663*
[22888-70-6] Silybin, *8671*
[22916-47-8] Miconazole, *6257*
[22963-93-5] Juvenile Hormone III, *5321*
[22994-85-0] Benznidazole, *1086*
[23031-25-6] Terbutaline, *9302*
[23031-32-5] Terbutaline Sulfate, *9302*
[23031-36-9] Prallethrin, *7823*
[23047-25-8] Lofepramine, *5614*
[23076-35-9] Xylazine Hydrochloride, *10277*
[23089-26-1] α-Bisabolol (−)-Form, *1245*
[23092-17-3] Halazepam, *4624*
[23093-74-5] Bunitrolol Hydrochloride, *1493*
[23103-98-2] Pirimicarb, *7609*
[23109-05-9] α-Amanitin, *367*
[23110-15-8] Fumagillin, *4315*
[23135-22-0] Oxamyl, *7019*
[23142-01-0] Carbetapentane Citrate, *1795*
[23147-58-2] Glycolaldehyde Dimer, *4532*
[23152-29-6] Virginiamycin S₁, *10200*
[23155-02-4] Fosfomycin, *4281*
[23178-88-3] α-Bisabolol (+)-Form, *1245*
[23184-66-9] Butachlor, *1511*
[23192-42-9] Hexalure, *4723*
[23210-56-2] Ifenprodil, *4936*
[23214-92-8] Doxorubicin, *3487*
[23214-96-2] Alcuronium, *214*
[23239-37-4] Etoxadrol (+)-Hydrochloride, *3932*
[23239-51-2] Ritodrine Hydrochloride, *8365*
[23249-97-0] Procodazole, *7880*
[23255-54-1] Dihydro-β-erythroidine, *3192*
[23255-59-6] Lunularic Acid, *5666*
[23255-93-8] Hycanthone Mesylate, *4798*
[23255-99-4] Bendazac Sodium Salt, *1037*
[23256-30-6] Nifurtimox, *6622*
[23256-40-8] Guanoxabenz Hydrochloride, *4602*
[23256-50-0] Guanabenz Monoacetate, *4593*
[23257-56-9] Levophacetoperane Hydrochloride, *5521*
[23277-43-2] Nalbuphine Hydrochloride, *6441*
[23282-20-4] Nivalenol, *6746*
[23282-55-5] Sulfachlorpyridazine Sodium Salt, *9032*
[23288-49-5] Probucol, *7873*
[23288-60-0] Sodium Pertechnetate Tc 99m, *8788*

[23315-05-1] Elaiomycin, *3582*
[23319-63-3] Vedejs Reagent, *10134*
[23325-78-2] Cephalexin Monohydrate, *1974*
[23327-57-3] Nefopam Hydrochloride, *6525*
[23344-16-3] Streptovaricin A, *8960*
[23344-17-4] Streptovaricin C, *8960*
[23383-11-1] Ferrous Citrate, *4071*
[23454-33-3] Trenbolone Cyclohexylmethylcarbonate, *9745*
[23465-76-1] Caroverine, *1859*
[23491-44-3] Bisbenzimide, *1250*
[23491-45-4] Bisbenzimide Trihydrochloride, *1250*
[23491-52-3] Bisbenzimide Ethoxide, *1250*
[23496-41-5] Verticine, *10165*
[23505-41-1] Pirimiphos-ethyl, *7610*
[23509-16-2] Batrachotoxin, *1007*
[23509-17-3] Homobatrachotoxin, *1007*
[23513-14-6] [6]-Gingerol, *4459*
[23541-50-6] Daunorubicin Hydrochloride, *2834*
[23560-59-0] Heptenophos, *4699*
[23564-05-8] Thiophanate-methyl, *9505*
[23564-06-9] Thiophanate, *9505*
[23593-75-1] Clotrimazole, *2401*
[23602-78-0] Benfluorex, *1041*
[23642-66-2] Benfluorex Hydrochloride, *1041*
[23647-14-5] SPADNS, *8859*
[23651-95-8] Droxidopa, *3504*
[23666-50-4] Rhodomycin A, *8315*
[23668-11-3] Pactamycin, *7083*
[23672-07-3] Levosulpiride, *9119*
[23674-86-4] Difluprednate, *3171*
[23694-17-9] Sultopride Hydrochloride, *9123*
[23694-81-7] Mepindolol, *5923*
[23696-28-8] Olaquindox, *6916*
[23696-85-7] β-Damascenone (unspecified stereo), *2808*
[23726-93-4] β-Damascenone, *2808*
[23736-58-5] Cloxacillin Benzathine Salt, *2403*
[23769-39-3] α-Thioacetaldehyde, *9471*
[23769-40-6] β-Thioacetaldehyde, *9471*
[23777-80-2] Hexaborane(10), *4711*
[23779-99-9] Floctafenine, *4135*
[23795-03-1] Probenecid Sodium Salt, *7872*
[23828-92-4] Ambroxol Hydrochloride, *380*
[23843-52-9] Phloionic Acid, *7436*
[23873-81-6] Benzil Dioxime, *1081*
[23887-31-2] Clorazepic Acid, *2392*
[23915-80-2] Chromocarb Diethylamine, *2241*
[23918-98-1] Eritadenine, *3728*
[23930-19-0] Alfaxalone, *228*
[23930-37-2] Alfadolone Acetate, *228*
[23947-60-6] Ethirimol, *3794*
[23950-58-5] Propyzamide, *7983*
[23964-57-0] Carticaine Hydrochloride, *1869*
[23964-58-1] Carticaine, *1869*
[23978-09-8] Cryptand 222, *2596*
[24017-47-8] Triazophos, *9767*
[24027-80-3] Dideoxyadenosine 5′-Triphosphate, *3112*
[24047-25-4] Guanoxabenz, *4602*
[24147-36-2] Thiazole Orange Iodide, *9460*

Chemical Abstracts Registry Numbers

[28911-01-5] Triazolam, 9765
[28920-43-6] 9-Fluorenylmethyl Chloroformate, 4190
[28947-50-4] Fencamine, 3999
[28950-34-7] Tetrasulfur Tetranitride, 9389
[28981-97-7] Alprazolam, 308
[28983-56-4] Methyl Blue, 6102
[28994-41-4] o-Benzylphenol, 1144
[29025-14-7] Butropium Bromide, 1536
[29031-19-4] Glucosamine Sulfate Salt, 4494
[29074-38-2] Canadine dl-Form, 1738
[29091-21-2] Prodiamine, 7883
[29094-61-9] Glipizide, 4477
[29106-32-9] Chaulmoogric Acid, 2047
[29110-47-2] Guanfacine, 4596
[29110-48-3] Guanfacine Hydrochloride, 4596
[29117-48-4] sec-Octyl Iodide (R)-Form, 6858
[29119-03-7] Frequentin, 4299
[29122-68-7] Atenolol, 850
[29177-84-2] Ethyl Loflazepate, 3876
[29202-00-4] Gardenin D, 4399
[29216-28-2] Mequitazine, 5931
[29218-27-7] Toloxatone, 9680
[29232-93-7] Pirimiphos-methyl O,O-Dimethyl Analog, 7610
[29306-57-8] Trinitrotriazidobenzene, 9903
[29334-07-4] Sulmarin, 9114
[29342-05-0] Ciclopirox, 2269
[29342-22-1] Moore's Ketene, 6350
[29343-52-0] HNE (unspecified stereo), 4764
[29365-11-5] Giractide Hexaacetate Salt, 4463
[29393-20-2] Bacilysin, 926
[29400-42-8] Helvolic Acid, 4668
[29493-77-4] 4-Methylaminorex (±)-cis Form, 6092
[29550-05-8] Gardenin C, 4399
[29550-07-0] Gardenin E, 4399
[29560-58-5] Moricizine Hydrochloride, 6354
[29587-89-1] δ-Dodecalactone R-Form, 3448
[29608-49-9] Almitrine Dimethanesulfonate, 294
[29679-58-1] Fenoprofen, 4011
[29684-56-8] Burgess Reagent, 1505
[29701-07-3] Kanamycin B Sulfate, 5329
[29714-87-2] Ocimene, 6830
[29728-34-5] Acetoxolone Aluminum Salt, 74
[29767-20-2] Teniposide, 9288
[29790-52-1] Nicotine Salicylate, 6609
[29804-22-6] Disparlure, 3403
[29836-26-8] n-Octyl-β-D-glucoside, 6857
[29838-46-8] Roccellic Acid, 8373
[29856-33-5] Sodium Trimetaphosphate Hexahydrate, 8823
[29868-97-1] Pirenzepine Dihydrochloride, 7604
[29883-15-6] Amygdalin, 592
[29908-03-0] S-Adenosylmethionine, 144
[29913-86-8] Amarolide, 372
[29915-38-6] TAPS, 9191
[29952-87-2] Clofibric Acid Pyridoxine Salt, 2371

[29956-24-9] Pyridoxal 5-Phosphate Calcium Salt, 8092
[29975-16-4] Estazolam, 3757
[30009-42-8] Aleuritic Acid Methyl Ester (unspecified stereo), 225
[30042-37-6] Lankamycin, 5407
[30123-17-2] Tianeptine Sodium Salt, 9575
[30223-48-4] Fluacizine, 4145
[30237-26-4] DL-Fructose, 4303
[30272-08-3] Amineptine Hydrochloride, 403
[30286-75-0] Oxitropium Bromide, 7042
[30299-08-2] Clinofibrate, 2355
[30392-40-6] Bitolterol, 1308
[30392-41-7] Bitolterol Methanesulfonate, 1308
[30403-03-3] Gallium Citrate (unspecified stoichiometry), 4379
[30408-30-1] Nybomycin, 6820
[30452-69-8] L-Cystine S,S-Dioxide, 2780
[30484-77-6] Flunarizine Dihydrochloride, 4175
[30516-87-1] Zidovudine, 10322
[30525-89-4] Paraformaldehyde, 7129
[30544-47-9] Etofenamate, 3921
[30544-61-7] Clanobutin, 2337
[30560-19-1] Acephate, 32
[30562-34-6] Geldanamycin, 4416
[30578-37-1] Amezinium Methyl Sulfate, 388
[30652-11-0] Deferiprone, 2862
[30685-43-9] β-Methyldigoxin, 3186
[30748-29-9] Feprazone, 4039
[30751-05-4] Troxipide, 9970
[30861-27-9] Aloesin, 301
[30909-51-4] Flupentixol Decanoate, 4219
[30964-13-7] Cynarine, 2763
[31036-80-3] Lofexidine, 5615
[31112-62-6] Metrizamide, 6234
[31139-87-4] Tipepidine Hibenzate, 9613
[31218-83-4] Propetamphos, 7929
[31272-51-2] Cryptoxanthin (±)-Form, 2599
[31282-04-9] Hygromycin B, 4892
[31329-57-4] Nafronyl, 6437
[31342-36-6] Chloramphenicol Pantothenate Calcium Complex (4:1), 2077
[31352-82-6] Zolazepam, 10387
[31362-50-2] Bombesin, 1332
[31366-25-3] Tetrathiafulvalene, 9390
[31377-23-8] Amantadine Sulfate, 368
[31418-71-0] Coprogen, 2507
[31428-61-2] Tiamenidine, 9573
[31430-15-6] Flubendazole, 4149
[31431-39-7] Mebendazole, 5837
[31431-43-3] Cyclobendazole, 2705
[31477-60-8] Centchroman, 1972
[31566-31-1] Glyceryl Monostearate, 4524
[31570-39-5] Neocembrene, 6533
[31586-77-3] Bismuth Sodium Tartrate, 1280
[31637-97-5] Etofibrate, 3923
[31677-93-7] Bupropion Hydrochloride, 1503
[31698-14-3] Ancitabine, 623
[31721-17-2] Quinupramine, 8200
[31774-33-1] Abikoviromycin, 8
[31828-50-9] Cephradine Dihydrate, 1985

[31828-68-9] L-Homocysteine Thiolactone Hydrochloride, 4772
[31828-71-4] Mexiletine, 6248
[31842-61-2] Rimiterol Hydrobromide, 8353
[31853-38-0] 1,2-Naphthoquinone 2-Semicarbazone, 6479
[31868-18-5] Mexazolam, 6245
[31883-05-3] Moricizine, 6354
[31884-76-1] Sulfanilic Acid Zinc Salt Tetrahydrate, 9058
[31884-77-2] Meclizine Dihydrochloride Monohydrate, 5848
[31944-97-5] Chondrofoline, 2218
[32133-82-7] Martin Sulfurane, 5821
[32157-29-2] Propanethial S-Oxide, 7916
[32164-16-2] Verticillin A, 10164
[32164-26-4] Streptovaricin D, 8960
[32190-57-1] Wildfire Toxin, 10244
[32222-06-3] Calcitriol, 1647
[32245-40-2] 1-Theobromineacetic Acid Sodium Salt, 9434
[32248-43-4] Samarium Iodide, 8486
[32266-10-7] Hexoprenaline Sulfate, 4743
[32267-39-3] Divicine, 3426
[32289-58-0] Polihexanide, 7677
[32305-98-9] (R,R)-DIOP, 3319
[32315-10-9] Triphosgene, 9921
[32359-34-5] Medifoxamine, 5861
[32381-42-3] Vernolic Acid (−)-Form, 10159
[32383-76-9] Medicarpin, 5860
[32385-11-8] Sisomicin, 8688
[32434-42-7] Febrifugine Dihydrochloride, 3982
[32449-92-6] D-Glucuronolactone, 4503
[32467-88-2] ACV, 133
[32476-67-8] Periplocymarin, 7290
[32527-55-2] Tiaramide, 9579
[32534-81-9] Pentabromodiphenyl Ether, 7681
[32535-84-5] Ammonium Zirconyl Carbonate, 566
[32619-42-4] Oleuropein, 6923
[32630-75-4] Pododacric Acid, 7661
[32665-36-4] Eprozinol, 3693
[32672-69-8] Mesoridazine Benzenesulfonate, 5980
[32728-78-2] Heliosupine, 4664
[32780-64-6] Labetalol Hydrochloride, 5377
[32795-44-1] Acecainide, 21
[32795-47-4] Nomifensine Maleate, 6760
[32809-16-8] Procymidone, 7882
[32828-81-2] Picotamide, 7519
[32854-75-4] Lappaconitine, 5420
[32886-97-8] Amdinocillin Pivoxil, 384
[32887-01-7] Amdinocillin, 383
[32887-03-9] Amdinocillin Pivoxil Hydrochloride, 384
[32909-92-5] Sulfametrole, 9052
[32953-89-2] Rimiterol, 8353
[32986-56-4] Tobramycin, 9647
[32988-50-4] Viomycin, 10197
[33005-95-7] Tiaprofenic Acid, 9577
[33032-12-1] Methapyrilene Fumarate, 6031
[33069-62-4] Paclitaxel, 7081
[33089-61-1] Amitraz, 482
[33103-21-8] Tuberactinomycin A, 9983

[33103-22-9] Enviomycin, 3654
[33125-97-2] Etomidate, 3926
[33137-73-4] Tuberactinomycin O, 9983
[33159-27-2] Ecabet, 3535
[33178-86-8] Alinidine, 238
[33194-27-3] Alsactide Acetate, 310
[33237-74-0] Aprindine Hydrochloride, 743
[33239-19-9] 4′,5′-Diiodofluorescein Sodium Salt, 3207
[33245-39-5] Fluchloralin, 4152
[33286-22-5] Diltiazem Hydrochloride, 3224
[33320-16-0] Methyl Aminolevulinate, 6090
[33342-05-1] Gliquidone, 4478
[33368-20-6] Enduracidin A Hydrochloride, 3634
[33386-08-2] Buspirone Hydrochloride, 1507
[33396-37-1] Proscillaridin-4-methyl Ether, 7985
[33401-94-4] Pyrantel Tartrate, 8066
[33402-03-8] Metaraminol Bitartrate, 6003
[33404-78-3] Negamycin, 6526
[33419-42-0] Etoposide, 3929
[33445-35-1] Sulfamidochrysoidine Hydrochloride, 9055
[33490-33-4] Capreomycin IB, 1759
[33507-63-0] Substance P, 8993
[33515-09-2] Gonadorelin, 4561
[33515-32-1] S-Methylmethionine Bromide, 6170
[33538-71-5] Venturicidin A, 10141
[33538-72-6] Venturicidin B, 10141
[33564-30-6] Cefoxitin Sodium Salt, 1938
[33564-31-7] Diflorasone Diacetate, 3159
[33570-04-6] Bilobalide, 1224
[33579-45-2] Scotophobin, 8548
[33580-30-2] Tertatolol Hydrochloride, 9316
[33605-94-6] Pyrisuccideanol, 8104
[33614-49-2] Gold Sodium Thiosulfate Dihydrate, 4554
[33629-47-9] Butralin, 1534
[33636-93-0] Stryker's Reagent, 8987
[33665-90-6] Acesulfame, 38
[33671-46-4] Clotiazepam, 2400
[33754-49-3] Zolazepam Hydrochloride, 10387
[33765-68-3] Oxendolone, 7030
[33797-51-2] Dimethyl(methylene)ammonium Iodide, 3275
[33817-20-8] Pivampicillin, 7626
[33818-15-4] Citicoline Sodium Salt, 2318
[33842-02-3] Viehe's Salt, 10174
[33861-17-5] Phenylselenotrimethylsilane, 7421
[33864-99-2] Alcian Blue 8G, 212
[33993-35-0] Benzylideneaniline (Z-form), 1141
[33996-33-7] Oxaceprol, 7002
[34014-18-1] Tebuthiuron, 9232
[34031-32-8] Auranofin, 869
[34089-81-1] Sodium Ferric Gluconate, 8751
[34097-16-0] Clocortolone 21-Pivalate, 2363
[34118-92-8] Acecainide Hydrochloride, 21

[34123-59-6] Isoproturon, 5264
[34135-85-8] Methyl Allyl Trisulfide, 6087
[34140-59-5] Trimebutine Maleate, 9868
[34156-56-4] Foscarnet Sodium Hexahydrate, 4278
[34157-83-0] Celastrol, 1957
[34161-23-4] Fipexide Hydrochloride, 4115
[34161-24-5] Fipexide, 4115
[34167-45-8] Polymyxin D₂, 7693
[34183-22-7] Propafenone Hydrochloride, 7908
[34184-77-5] Promegestone, 7900
[34195-34-1] Hydrocodone Bitartrate Hemipentahydrate, 4823
[34214-51-2] Floxacillin Sodium Monohydrate, 4143
[34217-90-8] Stercobilin, 10068
[34218-61-6] Juvenile Hormone II, 5321
[34255-08-8] Rhodoviolascin, 8319
[34256-82-1] Acetochlor, 60
[34272-51-0] Histrionicotoxin, 4761
[34273-10-4] Saralasin, 8509
[34291-02-6] Butirosin A, 1530
[34291-03-7] Butirosin B, 1530
[34301-55-8] Isometamidium Chloride, 5229
[34302-69-7] Neocuproine Hemihydrate, 6534
[34302-70-0] Neocuproine Dihydrate, 6534
[34312-10-2] Citrulline Hydrochloride, 2330
[34316-15-9] Chelerythrine, 2050
[34327-18-9] Chloramphenicol Monosuccinate Arginine Salt, 2077
[34341-58-7] Potassium Fluoride Tetrahydrate, 7753
[34368-04-2] Dobutamine, 3440
[34381-68-5] Acebutolol Hydrochloride, 20
[34391-04-3] Albuterol (R)-Form, 210
[34433-31-3] Asoxime Chloride, 825
[34444-01-4] Cefamandole, 1916
[34461-22-8] Metformin Embonate, 6010
[34461-56-8] Dichloro(2-chlorovinyl)-arsine (Z-form), 3071
[34487-61-1] Phenolsulfonphthalein Sodium Salt, 7360
[34493-98-6] Dibekacin, 3008
[34503-87-2] Polymyxin B₂, 7693
[34521-09-0] Antimony Sodium Tartrate, 693
[34522-46-8] Oxetorone Fumarate, 7033
[34524-20-4] Boromycin, 1343
[34552-83-5] Loperamide Hydrochloride, 5628
[34580-13-7] Ketotifen, 5354
[34580-14-8] Ketotifen Fumarate, 5354
[34590-94-8] Dipropylene Glycol Monomethyl Ether, 3384
[34642-77-8] Amoxicillin Sodium Salt, 574
[34649-22-4] Oroidin, 6971
[34661-75-1] Urapidil, 10050
[34675-84-8] Cetraxate, 2023
[34707-92-1] Chlorothricin, 2172
[34722-90-2] Bromthymol Blue Sodium Salt, 1455

[34725-61-6] Bromphenol Blue Sodium Salt, 1454
[34758-83-3] Zipeprol, 10370
[34758-84-4] Zipeprol Dihydrochloride, 10370
[34765-96-3] Alsactide, 310
[34765-98-5] Enduracidin B Hydrochloride, 3634
[34786-70-4] Nystatin A₁, 6825
[34787-01-4] Ticarcillin, 9584
[34816-55-2] Moxestrol, 6375
[34866-46-1] Carbuterol Hydrochloride, 1832
[34866-47-2] Carbuterol, 1832
[34911-55-2] Bupropion, 1503
[34915-68-9] Bunitrolol, 1493
[34973-08-5] Gonadorelin Acetate, 4561
[34994-11-1] Hypusine, 4909
[35035-05-3] Timepidium Bromide, 9599
[35080-11-6] Prajmaline, 7820
[35112-53-9] Barium Thiosulfate, 995
[35121-78-9] Prostacyclin, 7986
[35124-13-1] d-cis-α-Irone, 5140
[35189-28-7] Norgestimate, 6792
[35212-22-7] Ipriflavone, 5118
[35225-79-7] Dibenzalacetone trans-trans-Form, 3010
[35237-02-6] (15:1)-Urushiol, 10076
[35274-05-6] Cetyl Lactate, 2029
[35287-72-0] Secalonic Acid A, 8555
[35334-12-4] Azidocillin Sodium Salt, 902
[35354-74-6] Honokiol, 4781
[35367-38-5] Diflubenzuron, 3161
[35380-71-3] Estradiol Hemihydrate, 3758
[35413-63-9] Streptovaricin E, 8960
[35440-78-9] Thiolactic Acid Calcium Salt, 9493
[35457-80-8] Midecamycin A₁, 6262
[35512-37-9] Streptovaricin F, 8960
[35523-89-8] Saxitoxin, 8526
[35531-88-5] Carindacillin, 1839
[35543-24-9] Buflomedil Hydrochloride, 1480
[35554-08-6] Saxitoxin Dihydrochloride, 8526
[35554-44-0] Enilconazole, 3637
[35595-03-0] Centaurein, 1969
[35597-44-5] Phosphinothricin, 7451
[35604-67-2] Viloxazine Hydrochloride, 10178
[35607-66-0] Cefoxitin, 1938
[35619-65-9] Trithiozine, 9937
[35691-65-7] 1,2-Dibromo-2,4-dicyanobutane, 3026
[35700-21-1] Carboprost Methyl Ester, 1824
[35700-23-3] Carboprost, 1824
[35729-37-4] 3-Chloro-1-butene L(+)-Form, 2131
[35788-00-2] Cuprous Sulfite, 2658
[35834-26-5] Rosaramicin, 8392
[35836-73-8] Nopol, 6771
[35846-53-8] Maytansine, 5829
[35891-70-4] Myriocin, 6418
[35898-87-4] Dilazep, 3222
[35941-65-2] Butriptyline, 1535
[35941-71-0] Tiaramide Hydrochloride, 9579
[36025-69-1] Midecamycin A₃, 6262
[36085-73-1] Talipexole Dihydrochloride, 9171

[36104-64-0] Micranthine, *6258*
[36104-80-0] Camazepam, *1727*
[36167-63-2] Halofantrine Hydrochloride, *4630*
[36190-93-9] Oleandrin Desacetyloleandrin, *6919*
[36222-39-6] Gallopamil *l*-Form Hydrochloride, *4386*
[36282-47-0] Tramadol Hydrochloride, *9726*
[36289-36-8] 1-Cyanoimidazole, *2686*
[36296-31-8] Oxiniacic Acid Ethanolamine Salt, *7040*
[36304-84-4] Dimemorfan Phosphate, *3229*
[36309-01-0] Dimemorfan, *3229*
[36322-90-4] Piroxicam, *7619*
[36330-85-5] Fenbufen, *3996*
[36338-96-2] Carthamin, *1867*
[36364-49-5] Imidazole Salicylate, *4954*
[36478-76-9] Uranyl Nitrate, *10048*
[36499-65-7] EDTA Cobalt Salt, *3565*
[36505-84-7] Buspirone, *1507*
[36508-71-1] Zorubicin Hydrochloride, *10395*
[36508-79-9] Nitroxynil D-*N*-Methylglucamine Salt, *6743*
[36531-26-7] Oxantel, *7021*
[36590-19-9] Amocarzine, *568*
[36592-77-5] Metipranolol Hydrochloride, *6216*
[36622-28-3] Verapamil (*S*)-Form Hydrochloride, *10144*
[36622-29-4] Verapamil (*S*)-Form, *10144*
[36634-48-7] Ricinolsulfuric Acid, *8335*
[36635-61-7] Tosylmethyl Isocyanide, *9715*
[36637-18-0] Etidocaine, *3911*
[36637-19-1] Etidocaine Hydrochloride, *3911*
[36647-02-6] Cotarnine Hydrochloride, *2540*
[36653-82-4] Cetyl Alcohol, *2027*
[36703-88-5] Inosine Pranobex, *5019*
[36734-19-7] Iprodione, *5121*
[36735-22-5] Quazepam, *8145*
[36791-04-5] Ribavirin, *8323*
[36861-47-9] 4-Methylbenzylidene Camphor, *6101*
[36894-69-6] Labetalol, *5377*
[36945-98-9] Icilin, *4925*
[36981-91-6] Fepradinol, *4038*
[37002-48-5] (*S*,*S*)-DIOP, *3319*
[37025-55-1] Carbetocin, *1796*
[37065-29-5] Miloxacin, *6277*
[37091-65-9] Azlocillin Sodium Salt, *908*
[37091-66-0] Azlocillin, *908*
[37091-73-9] DMC, *3434*
[37106-97-1] Bentiromide, *1056*
[37112-31-5] Levoglucosenone, *5517*
[37115-32-5] Adinazolam, *148*
[37134-40-0] Bicyclomycin Benzoate, *1207*
[37148-27-9] Clenbuterol, *2345*
[37189-34-7] Stem Bromelain, *1395*
[37196-57-9] Crotamine, *2584*
[37213-56-2] Adipsin, *153*
[37220-17-0] Konjac Mannan, *5367*
[37221-79-7] VIP, *10198*
[37228-64-1] Glucocerebrosidase, *4487*

[37248-47-8] Validamycin A, *10094*
[37267-86-0] Phosphoric Acid, Meta, *7457*
[37270-69-2] Benserazide Combination with Levodopa, *1052*
[37270-89-6] Heparin Calcium Salt, *4688*
[37280-35-6] Capreomycin IA, *1759*
[37288-97-4] Pinguinain, *7560*
[37291-07-9] Crilanomer, *2576*
[37296-80-3] Colestipol Hydrochloride, *2459*
[37300-21-3] Pentosan Polysulfate, *7247*
[37301-55-6] Chlormadinone Acetate Mixture with Ethinyl Estradiol, *2102*
[37319-17-8] Pentosan Polysulfate Sodium Salt, *7247*
[37321-09-8] Apramycin, *739*
[37332-99-3] Avoparcin, *882*
[37338-39-9] Pyrimethamine Combination with Sulfadoxine, *8098*
[37339-90-5] Lentinan, *5493*
[37340-82-2] Streptodornase, *8953*
[37341-53-0] Keratinase, *5340*
[37342-97-5] Schwartz's Reagent, *8533*
[37350-58-6] Metoprolol, *6228*
[37370-49-3] Alcian Blue 7G, *212*
[37415-62-6] Mycophenolic Acid Sodium Salt, *6412*
[37517-26-3] Pipotiazine Palmitic Ester, *7595*
[37517-28-5] Amikacin, *400*
[37517-30-9] Acebutolol, *20*
[37561-27-6] Fenoverine, *4013*
[37571-84-9] Amidephrine, *393*
[37577-24-5] Fenfluramine *l*-Form, *4004*
[37595-74-7] Phenyl Triflimide, *7427*
[37640-71-4] Aprindine, *743*
[37661-08-8] Bacampicillin Hydrochloride, *924*
[37671-82-2] Prenoxdiazine Hibenzate, *7854*
[37686-84-3] Terguride, *9307*
[37686-85-4] Terguride Hydrogen Maleate, *9307*
[37693-01-9] Clofoctol, *2373*
[37723-78-7] Iopronic Acid, *5106*
[37758-47-7] Ganglioside G_{m1}, *4394*
[37764-25-3] Dichlormid, *3058*
[37878-19-6] Detoxin D$_1$, *2940*
[37883-00-4] Viomycin Sulfate, *10197*
[37895-35-5] Albofungin, *206*
[37933-66-7] Thevetin A, *9439*
[38029-10-6] Pirbuterol Dihydrochloride, *7602*
[38078-09-0] Diethylaminosulfur Trifluoride, *3124*
[38129-37-2] Bicyclomycin, *1207*
[38142-58-4] *ar*-Turmerone (±)-Form, *10006*
[38176-02-2] Dexverapamil Hydrochloride, *10144*
[38176-09-9] Gallopamil *d*-Form Hydrochloride, *4386*
[38183-12-9] Fluorescamine, *4191*
[38194-50-2] Sulindac, *9112*
[38222-83-2] 2,6-Di-*tert*-butyl-4-methylpyridine, *3042*
[38234-21-8] Fertirelin, *4088*
[38260-01-4] Trientine Dihydrochloride, *9828*
[38270-90-5] Strontium Chloride ^{89}Sr-Labeled Form, *8970*

[38304-91-5] Minoxidil, *6285*
[38321-02-7] Verapamil (*R*)-Form Hydrochloride, *10144*
[38345-66-3] Chiral, *2062*
[38363-32-5] Penbutolol Sulfate, *7191*
[38363-40-5] Penbutolol, *7191*
[38396-39-3] Bupivacaine, *1499*
[38398-32-2] Ganaxolone, *4392*
[38455-77-5] Stannic Chromate(VI), *8902*
[38455-90-2] *N*-Methylephedrine Hydrochloride, *6140*
[38562-01-5] Prostaglandin F$_{2\alpha}$ Tromethamine Salt, *7989*
[38577-97-8] 4′,5′-Diiodofluorescein, *3207*
[38640-92-5] Rintatolimod, *8357*
[38641-94-0] Glyphosate Mono(isopropylamine) Salt, *4547*
[38677-81-5] Pirbuterol, *7602*
[38677-85-9] Flunixin, *4178*
[38721-52-7] L-Selectride, *8564*
[38748-32-2] Triptolide, *9923*
[38821-49-7] Carbidopa, *1798*
[38821-52-2] Indoramin Hydrochloride, *5010*
[38821-53-3] Cephradine, *1985*
[38916-34-6] Somatostatin, *8841*
[38932-40-0] Cephalexin Sodium Salt, *1974*
[38957-41-4] Emorfazone, *3619*
[38965-69-4] Xanthocillin Y$_1$, *10259*
[38965-70-7] Xanthocillin Y$_2$, *10259*
[38966-21-1] Aphidicolin, *714*
[39025-23-5] Z-Guggulsterone, *4608*
[39025-24-6] E-Guggulsterone, *4608*
[39133-31-8] Trimebutine, *9868*
[39148-24-8] Fosetyl Al, *4279*
[39156-41-7] 2,4-Diaminoanisole Sulfate, *2978*
[39236-46-9] Imidurea, *4957*
[39238-36-3] Myxin Copper(II) Complex, *6425*
[39290-85-2] Cupric Borate, *2618*
[39295-60-8] Sulfadoxine Mixture with Trimethoprim, *9038*
[39300-45-3] Dinocap, *3314*
[39366-37-5] Levonorgestrel Mixture with Ethinyl Estradiol, *6793*
[39379-15-2] Neurotensin, *6571*
[39391-39-4] Nocardicin A, *6753*
[39404-00-7] Hemozoin, *4684*
[39409-82-0] Magnesium Carbonate Hydroxide, *5724*
[39416-48-3] Pyridinium Bromide Perbromide, *8085*
[39455-18-0] Chondroitin 4-Sulfate Disodium Salt, *2219*
[39492-01-8] Gabexate, *4350*
[39515-40-7] Cyphenothrin, *2765*
[39515-41-8] Fenpropathrin, *4019*
[39543-79-8] Befunolol Hydrochloride, *1021*
[39552-01-7] Befunolol, *1021*
[39562-70-4] Nitrendipine, *6662*
[39577-19-0] Picumast, *7530*
[39577-20-3] Picumast Dihydrochloride, *7530*
[39589-98-5] Dimethyl Carbate, *3262*
[39637-99-5] Mosher's Reagent (*R*)-(−)-Acid Chloride, *6365*
[39640-15-8] Piberaline, *7504*
[39647-11-5] Methyl Dihydrojasmonate (+)-(1*R*,2*S*)-Form, *6124*

[39664-27-2] Ethylamine Oleate, 3817
[39698-78-7] Saralasin Hydrated Acetate, 8509
[39711-79-0] Ethyl Menthane Carboxamide, 3880
[39715-02-1] Endralazine, 3632
[39718-89-3] Alminoprofen, 293
[39733-35-2] Ammonium Magnesium Chloride, 528
[39800-16-3] Perfosfamide (unspecified stereo), 7276
[39807-15-3] Oxadiargyl, 7004
[39809-25-1] Penciclovir, 7192
[39831-55-5] Amikacin Sulfate, 400
[39860-99-6] Pipotiazine, 7595
[39878-70-1] Talampicillin Hydrochloride, 9165
[39937-23-0] Bixins trans-Form, 1312
[39968-33-7] 1-Hydroxy-7-azabenzotriazole, 4848
[39978-42-2] Nifurzide, 6624
[40034-42-2] Rosoxacin, 8400
[40054-69-1] Etizolam, 3919
[40077-57-4] Human VIP, 10198
[40180-04-9] Ticrynafen, 9586
[40372-00-7] Taribavirin Hydrochloride, 9197
[40391-99-9] Pamidronic Acid, 7104
[40431-64-9] Methylphenidate d-threo-Form, 6183
[40487-42-1] Pendimethalin, 7193
[40507-78-6] Indanazoline, 4972
[40507-80-0] Indanazoline Hydrochloride, 4972
[40516-48-1] Tocoretinate, 9657
[40580-59-4] Guanadrel, 4594
[40596-69-8] Methoprene, 6055
[40626-29-7] Phenylpropanolamine (+)-Form Hydrochloride, 7417
[40665-92-7] Cloprostenol, 2388
[40666-16-8] Fluprostenol, 4224
[40716-66-3] Nerolidol trans-Form, 6561
[40819-93-0] Lorajmine Hydrochloride, 5634
[40828-46-4] Suprofen, 9134
[40918-97-6] Cinnamyl Cinnamate trans-trans-Form, 2305
[41096-46-2] Hydroprene, 4842
[41100-52-1] Memantine Hydrochloride, 5897
[41114-59-4] Nysted Reagent, 6826
[41183-64-6] Gallium Citrate ⁶⁷Ga-Labeled Form, 4379
[41198-08-7] Profenofos, 7886
[41294-56-8] 1α-Hydroxycholecalciferol, 4858
[41340-25-4] Etodolac, 3920
[41342-53-4] Erythromycin Ethylsuccinate, 3739
[41354-29-4] Cyproheptadine Hydrochloride Sesquihydrate, 2770
[41372-08-1] Methyldopa (sesquihydrate), 6127
[41372-20-7] Apomorphine Hydrochloride (hemihydrate), 733
[41394-05-2] Metamitron, 5995
[41451-75-6] Bruceantin, 1463
[41470-05-7] 11-cis-3-Dehydroretinal, 2878
[41483-43-6] Bupirimate, 1498
[41544-24-5] Tromantadine Hydrochloride, 9950
[41567-78-6] Fumagillin Dicyclohexylamine Salt, 4315

[41570-61-0] Tulobuterol, 9991
[41575-94-4] Carboplatin, 1823
[41621-49-2] Ciclopirox Ethanolamine Salt (1:1), 2269
[41663-50-1] Isobutol, 3775
[41708-72-9] Tocainide, 9648
[41744-40-5] Sulbenicillin, 9022
[41767-29-7] Fluocortin Butyl, 4183
[41826-92-0] Trepibutone, 9747
[41859-67-0] Bezafibrate, 1199
[41906-86-9] Nitrocefin, 6681
[41944-01-8] Bismuth Potassium Iodide, 1277
[41992-22-7] Spirogermanium Dihydrochloride, 8885
[41992-23-8] Spirogermanium, 8885
[42017-89-0] Fenofibric Acid, 4009
[42116-76-7] Carnidazole, 1848
[42151-56-4] N-Methylephedrine d-Form, 6140
[42200-33-9] Nadolol, 6431
[42399-41-7] Diltiazem, 3224
[42408-82-2] Butorphanol, 1532
[42461-84-7] Flunixin Meglumine Salt, 4178
[42471-28-3] Nimustine, 6639
[42540-40-9] Cefamandole Nafate, 1916
[42542-10-9] MDMA, 5836
[42553-65-1] Crocetin Di-gentiobiose Ester, 2579
[42576-02-3] Bifenox, 1212
[42597-57-9] Ronifibrate, 8386
[42719-34-6] Botryococcene, 1360
[42794-76-3] Midodrine, 6263
[42835-25-6] Flumequine, 4168
[42852-95-9] N-Methyl-α-L-glucosamine, 6151
[42864-78-8] Bevantolol Hydrochloride, 1195
[42874-03-3] Oxyfluorfen, 7062
[42924-53-8] Nabumetone, 6428
[42971-09-5] Vinpocetine, 10188
[43021-26-7] Ecgonidine Methyl Ester, 3539
[43033-72-3] Levomethadyl Acetate Hydrochloride, 5519
[43121-43-3] Triadimefon, 9754
[43200-80-2] Zopiclone, 10393
[43210-67-9] Fenbendazole, 3994
[43222-48-6] Difenzoquat Methyl Sulfate, 3157
[43229-80-7] Formoterol Fumarate, 4272
[46464-11-3] Meobentine, 5911
[46817-91-8] Viloxazine, 10178
[47141-42-4] Levobunolol, 5514
[47543-65-7] Prenoxdiazine, 7854
[47562-08-3] Lorajmine, 5634
[47739-98-0] Clocapramine, 2361
[47747-56-8] Talampicillin, 9165
[47931-85-1] Calcitonin (Salmon Synthetic), 1646
[49557-75-7] GHK, 4452
[49562-28-9] Fenofibrate, 4009
[49564-56-9] Fazadinium Bromide, 3980
[49582-09-4] PBB, 4111
[49638-23-5] Betaprodine Hydrochloride, 306
[49642-07-1] Statine, 8927
[49656-78-2] Methcathinone Hydrochloride, 6035
[49658-21-1] Sodium Chlorite Trihydrate, 8735

[49669-74-1] Aminoethoxyvinylglycine, 434
[49697-38-3] Rimexolone, 8352
[49715-04-0] Chloromethyl Chlorosulfate, 2147
[49745-95-1] Dobutamine Hydrochloride, 3440
[49746-00-1] Carbinoxamine l-Form d-Tartrate, 1800
[49763-96-4] Stiripentol (unspecified stereo), 8948
[49866-87-7] Difenzoquat, 3157
[50264-69-2] Lonidamine, 5625
[50276-98-7] Erythrocentaurin, 3736
[50291-21-9] Rose Bengal ¹³¹I-Labeled Disodium Salt, 8393
[50293-90-8] Albuterol (R)-Form Hydrochloride, 210
[50327-22-5] Nylon 46, 6824
[50357-45-4] Pentamidine Dihydrochloride, 7227
[50361-05-2] Dichloro(2-chlorovinyl)-arsine (E-form), 3071
[50370-12-2] Cefadroxil, 1915
[50435-25-1] Nimidane, 6635
[50450-21-0] Bromodimethylsulfonium Bromide, 1427
[50471-44-8] Vinclozolin, 10181
[50594-66-6] Acifluorfen, 101
[50629-82-8] Halometasone, 4633
[50650-76-5] Piroctone, 7615
[50679-08-8] Terfenadine, 9306
[50700-49-7] Acetimidoquinone, 56
[50700-72-6] Vecuronium Bromide, 10132
[50767-79-8] Prodlure, 7885
[50782-69-9] VX, 10234
[50813-16-6] Insoluble Sodium Metaphosphate, 8773
[50838-36-3] Tolciclate, 9669
[50847-11-5] Ibudilast, 4918
[50865-01-5] Protoporphyrin IX Disodium Salt, 8006
[50887-69-9] Orotic Acid Monohydrate, 6972
[50896-27-0] Hydroxyglutamic Acid DL-erythro-Form, 4864
[50924-49-7] Mizoribine, 6308
[50927-09-8] Sydnones, 9141
[50933-06-7] Sulfalene Mixture with Trimethoprim, 9042
[50933-33-0] Gossyplure, 4563
[50935-04-1] Carubicin, 1870
[50935-71-2] Mocimycin, 6311
[50936-59-9] Iduronate-2-sulfatase, 4934
[50972-17-3] Bacampicillin, 924
[51005-85-7] PBP, 4111
[51012-32-9] Tiapride, 9576
[51012-33-0] Tiapride Hydrochloride, 9576
[51016-68-3] Josamycin Propionate, 5315
[51022-69-6] Amcinonide, 382
[51022-70-9] Albuterol Sulfate, 210
[51022-71-0] Nabilone, 6427
[51022-98-1] Butirosin Sulfate Dihydrate, 1530
[51023-56-4] Centchroman Hydrochloride, 1972
[51025-85-5] Arbekacin, 760
[51037-30-0] Acipimox, 102
[51115-67-4] Trimethyl Isopropyl Butanamide, 9887

[51207-31-9] 2,3,7,8-Tetrachlorodibenzofuran, *7684*

[51218-45-2] Metolachlor, *6223*

[51218-49-6] Pretilachlor, *7858*

[51229-78-8] Quaternium-15 *cis*-Form, *8143*

[51235-04-2] Hexazinone, *4735*

[51264-14-3] Amsacrine, *589*

[51274-83-0] Tiamenidine Hydrochloride, *9573*

[51276-47-2] Phosphinothricin DL-Form, *7451*

[51306-35-5] DTAF, *3507*

[51312-42-6] Sodium Phosphotungstate, *8795*

[51322-75-9] Tizanidine, *9642*

[51325-91-8] DCM, *2841*

[51333-22-3] Budesonide, *1476*

[51338-27-3] Diclofop-methyl, *3092*

[51395-42-7] Butedronic Acid, *1520*

[51410-30-1] Pirenoxine Sodium Salt, *7603*

[51424-33-0] Salsoline Hydrochloride, *8478*

[51460-26-5] Carbazochrome Sodium Sulfonate, *1790*

[51481-10-8] Vomitoxin, *10229*

[51481-60-8] Naloxone Hydrochloride (dihydrate), *6447*

[51481-61-9] Cimetidine, *2282*

[51481-65-3] Mezlocillin, *6250*

[51484-40-3] Difenpiramide, *3156*

[51503-61-8] Ammonium Phosphite, *542*

[51542-71-3] Magnalium (70:30), *5716*

[51543-40-9] Tarenflurbil, *9196*

[51570-36-6] Milbemycins, *6271*

[51579-82-9] Amfenac, *389*

[51596-10-2] Milbemycin A₃, *6269*

[51596-11-3] Milbemycin A₄, *6269*

[51596-16-8] Milbemycin β₁, *6271*

[51598-60-8] Cimetropium Bromide, *2283*

[51627-14-6] Cefatrizine, *1917*

[51630-58-1] Fenvalerate, *4037*

[51674-17-0] Sodium Thiophosphate Dodecahydrate, *8820*

[51707-55-2] Thidiazuron, *9465*

[51762-05-1] Cefroxadine, *1945*

[51773-92-3] Mefloquine Hydrochloride, *5872*

[51781-06-7] Carteolol, *1866*

[51781-21-6] Carteolol Hydrochloride, *1866*

[51803-78-2] Nimesulide, *6633*

[51805-45-9] TCEP Hydrochloride, *9222*

[51873-93-9] Azulene Sodium Sulfonate, *919*

[51888-09-6] Prochlorperazine Dimethanesulfonate, *7879*

[51901-85-0] *B*-Bromocatecholborane, *1423*

[51931-66-9] Tilidine, *9594*

[51938-42-2] Solanine, *8832*

[51940-44-4] Pipemidic Acid, *7572*

[51952-41-1] Gonadotropin-Releasing Hormone Hydrochloride, *4561*

[52093-21-7] Micronomicin, *6260*

[52094-70-9] Tetrantoin, *9382*

[52109-93-0] Cyclodrine, *2712*

[52110-55-1] Strobilurin A, *8964*

[52128-35-5] Trimetrexate, *9893*

[52152-93-9] Cefsulodin Sodium Salt, *1946*

[52205-73-9] Estramustine 17-(Dihydrogenphosphate) Disodium Salt, *3761*

[52207-99-5] Gossyplure (Z,Z)-Form (anhydrous), *4563*

[52209-35-5] Cryptoxanthin Monoacetate, *2599*

[52211-63-9] Viquidil Hydrochloride, *10199*

[52212-02-9] Pipecurium Bromide, *7571*

[52212-86-9] Verticillin B, *10164*

[52214-84-3] Ciprofibrate, *2312*

[52225-20-4] *dl*-α-Tocopherol Acetate, *9653*

[52239-63-1] Thiethylperazine Dimalate, *9467*

[52260-69-2] Dicyanine, *3101*

[52275-61-3] Streptovaricin J, *8960*

[52279-57-9] Nandrolone *p*-Hexyloxyphenylpropionate, *6450*

[52304-36-6] Ethyl Butylacetylaminopropionate, *3828*

[52306-35-1] Lunacridine (±)-Form, *5663*

[52315-07-8] Cypermethrin, *2764*

[52365-63-6] Dipivefrin, *3378*

[52423-56-0] Bromopride Hydrochloride, *1443*

[52432-72-1] Oxeladin Citrate, *7029*

[52441-47-1] Semtex, *2728*

[52443-21-7] Glucametacin, *4483*

[52463-83-9] Pinazepam, *7553*

[52468-60-7] Flunarizine, *4175*

[52485-79-7] Buprenorphine, *1501*

[52486-78-9] Mercurophen, *5959*

[52500-59-1] Mexican, *6247*

[52500-60-4] Thioredoxin, *9511*

[52500-61-5] β-Amino-α-methylphenethyl Alcohol, *444*

[52549-17-4] Pranoprofen, *7831*

[52589-12-5] Diginatin, *3175*

[52645-53-1] Permethrin, *7294*

[52665-69-7] Calcimycin, *1642*

[52667-15-9] Methionic Acid Aluminum Salt, *6046*

[52699-48-6] Gonadorelin Acetate (hydrate), *4561*

[52705-43-8] α-Hydroxybenzylphosphinic Acid, *4852*

[52712-76-2] Bunazosin Hydrochloride, *1491*

[52731-38-1] Glucofrangulin, *4488*

[52740-16-6] Calcium Arsenite, *1653*

[52757-95-6] Sevelamer, *8613*

[52775-76-5] Methylenomycin A, *6139*

[52775-77-6] Methylenomycin B, *6139*

[52794-97-5] Carubicin Hydrochloride, *1870*

[52809-07-1] Quisqualic Acid, *8202*

[52906-84-0] Oxychlorosene Sodium Salt, *7055*

[52906-92-0] Motilin, *6369*

[52918-63-5] Deltamethrin, *2888*

[53003-10-4] Salinomycin, *8473*

[53005-05-3] DIDS, *3113*

[53016-31-2] Norelgestromin, *6783*

[53042-79-8] Gossyplure (Z,E)-Form, *4563*

[53078-86-7] Arogenic Acid, *781*

[53092-86-7] Lindlar Catalyst, *5557*

[53112-28-0] Pyrimethanil, *8099*

[53123-88-9] Rapamycin, *8232*

[53152-21-9] Buprenorphine Hydrochloride, *1501*

[53164-05-9] Acemetacin, *28*

[53179-09-2] Sisomicin Sulfate, *8688*

[53179-11-6] Loperamide, *5628*

[53179-13-8] Pirfenidone, *7606*

[53230-10-7] Mefloquine, *5872*

[53237-59-5] Urushiol, *10076*

[53251-94-8] Pinaverium Bromide, *7552*

[53260-52-9] Heparamine, *4687*

[53267-01-9] Cifenline, *2272*

[53308-83-1] N^G-Methylarginine Acetate, *6096*

[53318-36-8] α-Glucogallin, *4489*

[53414-68-9] Tonin, *9705*

[53447-14-6] Isocorypalmine, *5208*

[53449-58-4] Ciclonicate, *2268*

[53450-33-2] Oxalomolybdic Acid, *7012*

[53469-21-9] Aroclor 1242, *7682*

[53516-73-7] Quinovin, *8193*

[53558-25-1] Pyriminil, *8101*

[53567-47-8] Tiropramide Hydrochloride, *9622*

[53583-79-2] Sultopride, *9123*

[53586-99-5] Periodyl, *7287*

[53597-25-4] Salmine Sulfate, *8476*

[53608-75-6] Pancrelipase, *7108*

[53643-48-4] Vindesine, *10184*

[53643-53-1] Tuberin (unspecified stereo), *9986*

[53648-05-8] Ibuproxam, *4920*

[53648-55-8] Dezocine, *2959*

[53659-00-0] Pyrisuccideanol Dimaleate, *8104*

[53678-77-6] Muramyl Dipeptide, *6390*

[53684-48-3] Beryllium Potassium Sulfate, *1175*

[53714-56-0] Leuprolide, *5509*

[53716-44-2] Rociverine, *8374*

[53716-49-7] Carprofen, *1862*

[53716-50-0] Oxfendazole, *7034*

[53734-79-5] Metralindole Hydrochloride, *6231*

[53746-45-5] Fenoprofen Calcium Salt Dihydrate, *4011*

[53772-83-1] Clopenthixol *cis(Z)*-Form, *2383*

[53780-34-0] Mefluidide, *5873*

[53783-83-8] Tromantadine, *9950*

[53797-35-6] Ribostamycin Sulfate, *8330*

[53808-87-0] Tetroxoprim, *9398*

[53808-88-1] Lonazolac, *5623*

[53850-34-3] Thaumatin, *9422*

[53850-35-4] Dubnium, *3509*

[53850-36-5] Rutherfordium, *8437*

[53861-57-7] γ-Carboxyglutamic Acid, *1829*

[53882-12-5] Lodoxamide, *5613*

[53882-13-6] Lodoxamide Diethyl Ester, *5613*

[53885-35-1] Ticlopidine Hydrochloride, *9585*

[53902-12-8] Tranilast, *9731*

[53910-25-1] Pentostatin, *7248*

[53984-74-0] Tocainide *R*-(−)-Form Hydrochloride, *9648*

[53984-76-2] Tocainide *S*-(+)-Form Hydrochloride, *9648*

[53994-73-3] Cefaclor, *1914*

[54024-22-5] Desogestrel, *2928*

[54029-12-8] Albendazole Sulfoxide, *203*

[54037-14-8] Bohrium, *1324*

[54037-57-9] Hassium, *4651*
[54038-01-6] Meitnerium, *5878*
[54038-81-2] Seaborgium, *8551*
[54048-10-1] Etonogestrel, *3928*
[54063-32-0] Clobetasone, *2359*
[54063-53-5] Propafenone, *7908*
[54063-54-6] Reproterol, *8259*
[54083-22-6] Zorubicin, *10395*
[54083-77-1] Darmstadtium, *2828*
[54084-26-3] Copernicium, *2503*
[54084-70-7] Element 113, *3589*
[54085-16-4] Element 114, *3590*
[54085-64-2] Element 115, *3591*
[54100-71-9] Element 116, *3592*
[54101-14-3] Element 117, *3593*
[54110-25-7] Pirozadil, *7620*
[54114-10-2] *N*-Methylephedrine *d*-Form Hydrochloride, *6140*
[54120-61-5] Prostalene, *7991*
[54125-02-9] 1-Methoxy-3-(trimethylsilyloxy)-1,3-butadiene (*E*-form), *6075*
[54143-55-4] Flecainide, *4130*
[54143-56-5] Flecainide Monoacetate, *4130*
[54143-57-6] Metoclopramide Monohydrochloride Monohydrate, *6219*
[54144-19-3] Element 118, *3594*
[54182-58-0] Sucralfate, *9010*
[54187-04-1] Rilmenidine, *8347*
[54188-38-4] Metralindole, *6231*
[54239-37-1] Cimaterol, *2281*
[54283-65-7] Dimecrotic Acid Magnesium Salt, *3226*
[54340-58-8] Meptazinol, *5930*
[54340-62-4] Bufuralol, *1485*
[54350-48-0] Etretinate, *3936*
[54386-24-2] Roentgenium, *8376*
[54397-85-2] Thromboxane B$_2$, *9542*
[54464-57-2] Iso E Super® (unspecified stereo), *5214*
[54479-70-8] Heparin Magnesium Salt, *4688*
[54504-70-0] Theofibrate, *9435*
[54527-84-3] Nicardipine Hydrochloride, *6580*
[54530-86-8] Clometocillin Sodium Salt, *2376*
[54533-85-6] Nizofenone, *6748*
[54533-86-7] Nizofenone Fumarate, *6748*
[54556-98-8] Propiverine Hydrochloride, *7945*
[54573-75-0] Doxercalciferol, *3484*
[54575-49-4] K-Selectride, *8564*
[54593-83-8] Chlorethoxyfos, *2088*
[54605-45-7] Iocarmic Acid Di-*N*-methylglucamine Salt, *5053*
[54651-39-7] Viscotoxin A$_2$, *10206*
[54657-08-8] *sec*-Butyl Acetate *l*-Form, *1538*
[54663-47-7] Tibezonium Iodide, *9581*
[54739-18-3] Fluvoxamine, *4246*
[54749-86-9] Phenylbutazone 2-Amino-2-thiazoline Salt, *7390*
[54749-90-5] Chlorozotocin, *2183*
[54761-04-5] Ytterbium Triflate, *10304*
[54783-95-8] Fortimicin B, *4275*
[54910-51-9] Disparlure, *3403*
[54910-89-3] Fluoxetine, *4217*
[54940-97-5] Citrulline Malate (Salt), *2330*
[54965-21-8] Albendazole, *203*
[54965-24-1] Tamoxifen Citrate, *9180*
[54974-54-8] TRH Tartrate, *9750*

[54986-75-3] Botrydial, *1359*
[55028-70-1] Arbaprostil, *759*
[55028-71-2] Fluprostenol Sodium Salt, *4224*
[55028-72-3] Cloprostenol Sodium Salt, *2388*
[55077-30-0] Aclatonium Napadisilate, *105*
[55079-83-9] Acitretin, *103*
[55096-26-9] Nalmefene, *6445*
[55134-13-9] Narasin, *6501*
[55142-85-3] Ticlopidine, *9585*
[55179-31-2] Bitertanol, *1306*
[55219-65-3] Triadimenol, *9755*
[55242-55-2] Propentofylline, *7927*
[55268-74-1] Praziquantel, *7837*
[55268-75-2] Cefuroxime, *1956*
[55283-68-6] Ethalfluralin, *3774*
[55285-14-8] Carbosulfan, *1827*
[55285-45-5] Pirifibrate, *7608*
[55290-64-7] Oxidimethiin, *7037*
[55297-95-5] Tiamulin, *9574*
[55297-96-6] Tiamulin Fumarate, *9574*
[55335-06-3] Triclopyr, *9821*
[55354-43-3] Arylsulfatase B, *809*
[55406-53-6] IPBC, *5112*
[55429-45-3] Nitracrine Dihydrochloride Monohydrate, *6657*
[55453-87-7] Isoxepac, *5284*
[55465-79-7] Viscotoxin A$_3$, *10206*
[55482-89-8] Aspirin Guaiacol Ester, *841*
[55512-33-9] Pyridate, *8079*
[55560-96-8] Tixocortol 21-Pivalate, *9641*
[55589-62-3] Acesulfame Potassium Salt, *38*
[55608-72-5] Pseudococaine *n*-Propyl Ester Analog, *8020*
[55661-38-6] Nimustine Hydrochloride, *6639*
[55720-26-8] Aminoethoxyvinylglycine Hydrochloride, *434*
[55721-31-8] Salinomycin Sodium Salt, *8473*
[55722-12-8] *S*-Adenosylmethionine Disulfate Tosylate, *144*
[55726-47-1] Enocitabine, *3640*
[55739-58-7] (*R*,*R*)-DIPAMP, *3333*
[55774-33-9] Azathioprine Sodium Salt, *895*
[55779-06-1] Fortimicin A, *4275*
[55837-18-8] Butibufen, *1529*
[55837-20-2] Halofuginone, *4632*
[55837-25-7] Buflomedil, *1480*
[55837-27-9] Piretanide, *7605*
[55837-29-1] Tiropramide, *9622*
[55837-30-4] Clonixin Lysine Salt, *2381*
[55852-84-1] Bacitracin Methylenedisalicylate, *928*
[55881-07-7] Miokamycin, *6286*
[55893-12-4] Gephyrotoxin, *4436*
[55905-53-8] Clebopride, *2342*
[55965-63-4] *O*-(β-Hydroxyethyl)rutinosides (mixture), *9969*
[55965-84-9] Methylisothiazolinone Mixture with Methylchloroisothiazolinone, *6163*
[55981-09-4] Nitazoxanide, *6653*
[55985-32-5] Nicardipine, *6580*
[56030-54-7] Sufentanil, *9018*
[56038-13-2] Sucralose, *9011*
[56073-07-5] Difenacoum, *3153*

[56073-10-0] Brodifacoum, *1386*
[56087-11-7] Dextranomer, *2952*
[56124-62-0] Valrubicin, *10101*
[56180-94-0] Acarbose, *19*
[56187-89-4] Ximoprofen, *10275*
[56208-01-6] Pifarnine, *7532*
[56211-40-6] Torsemide, *9711*
[56222-04-9] Femoxetine Hydrochloride, *3989*
[56238-63-2] Cefuroxime Sodium Salt, *1956*
[56254-07-0] Iodohippurate Sodium I 123, *5077*
[56281-36-8] Motretinide, *6370*
[56281-37-9] 2C-B Hydrochloride, *1910*
[56283-74-0] Laidlomycin, *5397*
[56287-74-2] Afloqualone, *175*
[56341-08-3] Mabuterol, *5703*
[56377-79-8] Nosiheptide, *6808*
[56390-09-1] Epirubicin Hydrochloride, *3678*
[56391-56-1] Netilmicin, *6564*
[56391-57-2] Netilmicin Sulfate, *6564*
[56392-17-7] Metoprolol Tartrate, *6228*
[56393-22-7] Pildralazine Dihydrochloride, *7535*
[56396-94-2] Mepindolol Sulfate Salt, *5923*
[56420-45-2] Epirubicin, *3678*
[56425-91-3] Flurprimidol, *4234*
[56430-99-0] Flumecinol, *4167*
[56518-41-3] Brodimoprim, *1387*
[56553-60-7] Sodium Triacetoxyborohydride, *8822*
[56592-32-6] Efrotomycin, *3572*
[56695-65-9] Rosaprostol, *8391*
[56695-66-0] Rosaprostol Sodium Salt, *8391*
[56775-88-3] Zimeldine, *10325*
[56776-01-3] Tulobuterol Hydrochloride, *9991*
[56776-32-0] Etifoxine Hydrochloride, *3913*
[56796-20-4] Cefmetazole, *1927*
[56796-39-5] Cefmetazole Sodium Salt, *1927*
[56824-20-5] Amiprilose, *480*
[56833-74-0] Streptovirudin, *8961*
[56839-43-1] Eperisone Hydrochloride, *3661*
[56974-46-0] Butalamine Hydrochloride, *1513*
[56974-61-9] Gabexate Methanesulfonate, *4350*
[56980-93-9] Celiprolol, *1962*
[56995-20-1] Flupirtine, *4221*
[57018-04-9] Tolclofos-methyl, *9670*
[57021-61-1] Isonixin, *5238*
[57041-67-5] Desflurane, *2920*
[57109-90-7] Clorazepic Acid Dipotassium Salt, *2392*
[57117-31-4] 2,3,4,7,8-Pentachlorodibenzofuran, *7684*
[57132-53-3] Proglumetacin, *7890*
[57149-07-2] Naftopidil, *6440*
[57149-08-3] Naftopidil Dihydrochloride, *6440*
[57197-43-0] Sulfamoxole Mixture with Trimethoprim, *9056*
[57213-69-1] Triclopyr Triethylamine Salt, *9821*
[57248-88-1] Pamidronic Acid Disodium Salt, *7104*

[57249-13-5] Heptaminol 5'-Adenylate, 4692
[57308-51-7] Carbidopa Combination with Levodopa, 1798
[57333-96-7] Tacalcitol, 9152
[57381-26-7] Irsogladine, 5146
[57432-61-8] Methylergonovine Maleate, 6142
[57455-37-5] Ultramarine, 10031
[57460-41-0] Talinolol, 9170
[57469-78-0] Ketoprofen Lysine Salt, 5352
[57470-78-7] Celiprolol Hydrochloride, 1962
[57474-29-0] Nifuroquine, 6618
[57475-17-9] Brovincamine, 1460
[57491-54-0] Aleuritic Acid Methyl Ester, 225
[57524-89-7] Hydrocortisone 17-Valerate, 4824
[57526-81-5] Prenalterol, 7853
[57574-09-1] Amineptine, 403
[57576-44-0] Aclacinomycin A, 104
[57576-52-0] Thromboxane A₂, 9542
[57596-79-9] Aclacinomycin B, 104
[57644-54-9] Colloidal Bismuth Subcitrate, 2469
[57645-91-7] Clebopride Malate, 2342
[57648-21-2] Timiperone, 9600
[57695-04-2] Sitamaquine, 8691
[57704-10-6] Bufuralol (−)-Hydrochloride, 1485
[57704-11-7] Bufuralol (+)-Hydrochloride, 1485
[57754-85-5] Clopyralid Monoethanolamine Salt, 2389
[57760-36-8] Alborixin, 208
[57773-63-4] Triptorelin, 9924
[57773-65-6] Deslorelin, 2924
[57775-26-5] Sultosilic Acid, 9124
[57775-27-6] Sultosilic Acid Piperazine Salt, 9124
[57775-29-8] Carazolol, 1780
[57801-81-7] Brotizolam, 1459
[57808-65-8] Closantel, 2396
[57808-66-9] Domperidone, 3467
[57817-89-7] Stevioside, 8939
[57821-29-1] Sulodexide, 9116
[57837-19-1] Metalaxyl, 5993
[57852-57-0] Idarubicin Hydrochloride, 4927
[57938-82-6] Adinazolam Methanesulfonate, 148
[57960-19-7] Acequinocyl, 34
[57966-95-7] Cymoxanil, 2762
[57982-77-1] Buserelin, 1506
[57982-78-2] Budipine, 1477
[57998-68-2] Diaziquone, 2999
[58001-44-8] Clavulanic Acid, 2339
[58024-13-8] Isomaltulose (monohydrate), 5228
[58045-23-1] Clopenthixol cis(Z)-Form Dihydrochloride, 2383
[58066-85-6] Miltefosine, 6279
[58073-76-0] Phenazocine cis-(±)-Form, 7332
[58095-31-1] Sulbenox, 9023
[58138-08-2] Tridiphane, 9827
[58152-03-7] Isepamicin, 5151
[58186-27-9] Idebenone, 4929
[58207-19-5] Clindamycin Hydrochloride (monohydrate), 2354
[58298-97-8] Spermine Diphosphate Hexahydrate, 8870

[58306-30-2] Febantel, 3981
[58321-78-1] Dibenzalacetone cis-cis-Form, 3010
[58337-35-2] Elliptinium Acetate, 3604
[58409-52-2] Albaspidin, 202
[58479-61-1] tert-Butyldiphenylchlorosilane, 1568
[58503-79-0] Meobentine Sulfate, 5911
[58543-16-1] Rebaudioside A, 8243
[58543-17-2] Rebaudioside B, 8243
[58551-69-2] Carboprost Tromethamine Salt, 1824
[58579-51-4] Anagrelide Hydrochloride, 616
[58580-55-5] Dibekacin Sulfate, 3008
[58581-89-8] Azelastine, 899
[58632-95-4] 2-(Boc-oxyimino)-2-phenylacetonitrile, 1323
[58640-87-2] Phenazocine cis-(−)-Form, 7332
[58670-63-6] Dinosterol, 3316
[58694-52-3] Tertiapin, 9318
[58712-69-9] Traxanox, 9739
[58749-22-7] Licochalcone A, 5532
[58749-23-8] Licochalcone B, 5532
[58785-63-0] Lonomycin A, 5626
[58786-99-5] Butorphanol Tartrate, 1532
[58795-03-2] Apalcillin Sodium Salt, 710
[58798-97-3] Berninamycin, 1162
[58845-80-0] Lonomycin A Sodium Salt, 5626
[58864-81-6] α-Tocotrienol, 9658
[58880-19-6] Trichostatin A, 9814
[58895-00-4] Trichostatin B, 9814
[58895-64-0] Nalmefene Hydrochloride, 6445
[58909-84-5] N-Stearoylsphingomyelin, 8873
[58934-46-6] Lorcainide Hydrochloride, 5637
[58944-73-3] Sinefungin, 8682
[58957-92-9] Idarubicin, 4927
[58970-76-6] Ubenimex, 10024
[58994-96-0] Ranimustine, 8227
[59017-64-0] Ioxaglic Acid, 5110
[59056-94-9] Iso E Super® (trans-form), 5214
[59092-07-8] Lycopene 15,15'-cis-Form, 5681
[59122-46-2] Misoprostol, 6293
[59128-97-1] Haloxazolam, 4638
[59160-29-1] Lidofenin, 5536
[59170-23-9] Bevantolol, 1195
[59198-70-8] Diflucortolone 21-Valerate, 3162
[59209-40-4] Proglumetacin Dimaleate, 7890
[59227-89-3] Laurocapram, 5440
[59263-76-2] Meptazinol Hydrochloride, 5930
[59277-89-3] Acyclovir, 134
[59285-67-5] δ-Decalactone S-Form, 2849
[59333-67-4] Fluoxetine Hydrochloride, 4217
[59338-87-3] Alizapride Hydrochloride, 242
[59338-93-1] Alizapride, 242
[59392-53-9] Maitotoxin, 5762
[59414-23-2] 1-Methoxy-3-(trimethylsilyloxy)-1,3-butadiene (unspecified stereo), 6075

[59467-70-8] Midazolam, 6261
[59467-94-6] Midazolam Maleate, 6261
[59467-96-8] Midazolam Hydrochloride, 6261
[59512-37-7] Piketoprofen Hydrochloride, 7534
[59536-65-1] Firemaster BP-6, 7680
[59598-29-7] Vernolepin (±)-Form, 10158
[59619-81-7] Etiproston, 3918
[59652-29-8] Bufuralol Hydrochloride, 1485
[59653-73-5] Triglycidyl Isocyanurate α-Form, 9860
[59653-74-6] Triglycidyl Isocyanurate β-Form, 9860
[59669-26-0] Thiodicarb, 9482
[59672-20-7] Sulfaloxic Acid Calcium Salt, 9044
[59703-84-3] Piperacillin Sodium Salt, 7575
[59721-28-7] Camostat, 1730
[59721-29-8] Camostat Methanesulfonate, 1730
[59727-70-7] Octacaine Hydrochloride, 6832
[59729-31-6] Lorcainide, 5637
[59729-32-7] Citalopram Hydrobromide, 2317
[59729-33-8] Citalopram, 2317
[59767-13-4] Setastine Hydrochloride, 8611
[59787-61-0] Cyclosporin C, 2750
[59798-30-0] Mezlocillin Sodium Salt Monohydrate, 6250
[59803-98-4] Brimonidine, 1381
[59804-37-4] Tenoxicam, 9291
[59859-58-4] Femoxetine, 3989
[59862-55-4] p-Bromoacetophenone Oxime (E-form), 1410
[59865-13-3] Cyclosporin A, 2750
[59865-23-5] MeBmt, 5840
[59917-39-4] Vindesine Sulfate Salt, 10184
[59937-28-9] Malotilate, 5776
[59973-80-7] Exisulind, 3956
[59985-21-6] Diquafosol, 3391
[59989-18-3] Eniluracil, 3638
[59995-64-1] Thienamycin, 9466
[60084-10-8] Tiazofurin, 9580
[60134-71-6] Nocardicin B, 6753
[60142-96-3] Gabapentin, 4348
[60166-93-0] Iopamidol, 5100
[60168-88-9] Fenarimol, 3992
[60200-06-8] Clorsulon, 2395
[60202-16-6] Protein C, 7995
[60207-31-0] Azaconazole, 885
[60207-90-1] Propiconazole, 7932
[60212-94-4] Xenon Hydrate, 10270
[60282-87-3] Gestodene, 4447
[60318-52-7] Trichosanthin, 9813
[60325-46-4] Sulprostone, 9120
[60414-06-4] Amiprilose Hydrochloride, 480
[60450-21-7] Morindin, 6356
[60477-34-1] Pifithrin-β, 7533
[60478-52-6] (−)-Dolichodial, 3459
[60529-33-1] Frangulin, 4292
[60539-20-0] Dixyrazine Dihydrochloride, 3429
[60560-33-0] Pinacidil, 7549
[60561-17-3] Sufentanil Citrate, 9018
[60569-19-9] Propiverine, 7945
[60576-13-8] Piketoprofen, 7534

[60583-39-3] Brucine Sulfate Heptahydrate, 1464
[60607-34-3] Oxatomide, 7024
[60607-68-3] Indenolol, 4979
[60628-96-8] Bifonazole, 1216
[60668-24-8] Alafosfalin, 195
[60719-84-8] Amrinone, 587
[60731-46-6] Elcatonin, 3585
[60761-10-6] TMD, 9643
[60789-62-0] Clocapramine Dihydrochloride Monohydrate, 2361
[60925-61-3] Ceforanide, 1932
[60940-34-3] Ebselen, 3533
[61036-64-4] Teicoplanin, 9255
[61129-30-4] Zimeldine Dihydrochloride Monohydrate, 10325
[61197-73-7] Loprazolam, 5631
[61228-92-0] Periplanone B, 7288
[61230-25-9] Aplasmomycin, 724
[61260-05-7] Prenalterol Hydrochloride, 7853
[61270-58-4] Cefonicid, 1930
[61270-78-8] Cefonicid Disodium Salt, 1930
[61318-90-9] Sulconazole, 9025
[61318-91-0] Sulconazole Nitrate, 9025
[61336-70-7] Amoxicillin Trihydrate, 574
[61337-67-5] Mirtazapine, 6291
[61379-65-5] Rifapentine, 8344
[61413-54-5] Rolipram, 8380
[61422-45-5] Carmofur, 1844
[61434-67-1] cis-Resveratrol, 8279
[61477-96-1] Piperacillin, 7575
[61512-20-7] Cord Factors, 2513
[61545-06-0] Temocillin Disodium Salt, 9280
[61570-90-9] Tioxidazole, 9611
[61618-27-7] Amfenac Sodium Salt Monohydrate, 389
[61622-34-2] Cefotiam, 1936
[61676-87-7] Cymiazole, 2761
[61717-82-6] o-Iodoxybenzoic Acid (cyclic tautomer), 5091
[61718-82-9] Fluvoxamine Maleate, 4246
[61788-48-5] Acetylated Lanolin, 5409
[61788-49-6] Acetylated Lanolin Alcohols, 5410
[61789-32-0] Sodium Cocoyl Isethionate, 8739
[61802-93-5] Metaclazepam Hydrochloride, 5990
[61825-94-3] Oxaliplatin, 7011
[61849-14-7] Prostacyclin Sodium Salt, 7986
[61869-07-6] Domiodol, 3463
[61869-08-7] Paroxetine, 7148
[61891-34-7] Flufenamic Acid Aluminum Salt, 4163
[61951-99-3] Tixocortol, 9641
[61954-97-0] Drotaverine Acephyllinate, 3503
[62013-04-1] Dirithromycin, 3394
[62031-54-3] Fibroblast Growth Factor, 4101
[62087-72-3] Pentigetide, 7241
[62232-46-6] Bifemelane Hydrochloride, 1210
[62265-68-3] Quinfamide, 8173
[62288-83-9] Desmopressin Monoacetate, 2926
[62304-98-7] Thymalfasin, 9560
[62327-61-1] Perimycin A, 7284

[62435-42-1] Perfosfamide, 7276
[62476-59-9] Acifluorfen Sodium Salt, 101
[62499-27-8] Gastrodin, 4408
[62534-68-3] Mepartricin A, 5914
[62534-69-4] Mepartricin B, 5914
[62568-57-4] DSIP, 3505
[62571-86-2] Captopril, 1776
[62587-73-9] Cefsulodin, 1946
[62613-82-5] Oxiracetam, 7041
[62625-22-3] Zincon Sodium Salt, 10345
[62625-30-3] Bromcresol Purple Sodium Salt, 1394
[62625-32-5] Bromcresol Green Sodium Salt, 1393
[62637-91-6] Tetrabromophenolphthalein Ethyl Ester Potassium Salt, 9331
[62637-93-8] Trimethylamine N-Oxide Dihydrate, 9881
[62640-05-5] Ajugarin-I, 190
[62640-06-6] Ajugarin-II, 190
[62640-07-7] Ajugarin-III, 190
[62658-63-3] Bopindolol, 1335
[62666-20-0] Progabide, 7888
[62669-70-9] Rhodamine 123, 8306
[62765-90-6] Iron Sorbitex, 5144
[62766-26-1] Mimosine Hydrochloride, 6280
[62774-96-3] Tilisolol Hydrochloride, 9595
[62778-11-4] Olah's Reagent, 6913
[62796-23-0] Merocyanine 540, 5969
[62893-19-0] Cefoperazone, 1931
[62893-20-3] Cefoperazone Sodium Salt, 1931
[62952-06-1] Lysine Acetylsalicylate, 5698
[62973-76-6] Azanidazole, 890
[62989-33-7] Sapropterin, 8505
[62994-47-2] Warburganal, 10235
[62996-74-1] Staurosporine, 8928
[63147-28-4] DBHBT, 2839
[63208-82-2] Pifithrin-α, 7533
[63279-13-0] Rebaudioside D, 8243
[63279-14-1] Rebaudioside E, 8243
[63329-53-3] Lobenzarit, 5610
[63333-35-7] Bromethalin, 1396
[63358-49-6] Aspoxicillin, 842
[63394-05-8] Plafibride, 7630
[63451-28-5] Methyl Red Hydrochloride, 6192
[63469-19-2] Apalcillin, 710
[63527-52-6] Cefotaxime, 1934
[63543-09-9] Thiarubrin(e) A, 9456
[63547-13-7] Adrafinil, 158
[63550-99-2] Rebaudioside C, 8243
[63585-09-1] Foscarnet Sodium, 4278
[63590-64-7] Terazosin, 9297
[63610-08-2] Indobufen, 5001
[63610-09-3] Lodoxamide Tromethamine Salt, 5613
[63612-50-0] Nilutamide, 6629
[63628-25-1] αMNP, 6310
[63628-26-2] αMNP (−)-(R)-Form, 6310
[63659-18-7] Betaxolol, 1184
[63659-19-8] Betaxolol Hydrochloride, 1184
[63661-61-0] Budipine Hydrochloride, 1477
[63675-72-9] Nisoldipine, 6651
[63701-55-3] Baclofen (R)-Form Hydrochloride, 930

[63717-27-1] Amifostine Monohydrate, 399
[63732-85-4] Norlevorphanol Hydrobromide, 6796
[63739-93-5] Verrucarin K, 10160
[63749-94-0] Sulfametrole Mixture with Trimethoprim, 9052
[63767-79-3] Ceforanide L-Lysine Salt, 1932
[63775-95-1] Cyclosporin B, 2750
[63775-96-2] Cyclosporin D, 2750
[63814-06-2] Isomethadone d-Form Hydrochloride, 5230
[63836-75-9] Pivcefalexin, 7627
[63918-50-3] Fencamine Hydrochloride, 3999
[63968-64-9] Artemisinin, 807
[64000-73-3] Pildralazine, 7535
[64006-44-6] Paroxetine Maleate, 7148
[64019-93-8] Dipivefrin Hydrochloride, 3378
[64023-94-5] trans-(+)-Metazocine, 6005
[64024-15-3] Pentazocine Hydrochloride, 7233
[64057-70-1] MDMA Hydrochloride, 5836
[64058-48-6] Spectinomycin Sulfate Tetrahydrate, 8866
[64211-45-6] Oxiconazole, 7036
[64211-46-7] Oxiconazole Nitrate, 7036
[64217-62-5] Cefatrizine Compd with Propylene Glycol, 1917
[64218-02-6] Plaunotol, 7647
[64221-86-9] Imipenem, 4961
[64228-81-5] Atracurium Besylate, 859
[64238-92-2] Benproperine Pamoate, 1051
[64285-06-9] Anatoxin a, 621
[64294-95-7] Setastine, 8611
[64297-64-9] o-Iodoxybenzoic Acid, 5091
[64318-79-2] Gemeprost, 4421
[64359-81-5] 4,5-Dichloro-2-octyl-3-isothiazolone, 3080
[64461-82-1] Tizanidine Hydrochloride, 9642
[64475-85-0] Mineral Spirits, 6282
[64485-93-4] Cefotaxime Sodium Salt (syn-Isomer), 1934
[64490-92-2] Tolmetin Sodium Salt Dihydrate, 9677
[64506-49-6] Sofalcone, 8828
[64512-28-3] Tetradecahexaenal, 8036
[64512-29-4] Hexadecaheptaenal, 8036
[64536-78-3] Crabtree's Catalyst, 2554
[64544-07-6] Cefuroxime 1-Acetoxyethyl Ester, 1956
[64603-91-4] Gaboxadol, 4351
[64628-44-0] Triflumuron, 9844
[64700-56-7] Triclopyr Butoxyethyl Ester, 9821
[64706-54-3] Bepridil, 1152
[64726-91-6] Japonilure, 5305
[64784-13-0] Indigo, 4981
[64808-48-6] Lobenzarit Disodium Salt, 5610
[64840-90-0] Eperisone, 3661
[64872-76-0] Butoconazole, 1531
[64872-77-1] Butoconazole Nitrate, 1531
[64896-28-2] (S,S)-Chiraphos, 2063
[64902-72-3] Chlorsulfuron, 2197
[64924-67-0] Halofuginone Hydrobromide, 4632

[69558-55-0] Thymopentin, *9558*
[69655-05-6] Didanosine, *3109*
[69657-51-8] Acyclovir Sodium Salt, *134*
[69712-56-7] Cefotetan, *1935*
[69739-16-8] Cefodizime, *1929*
[69756-53-2] Halofantrine, *4630*
[69770-45-2] Flumethrin, *4171*
[69787-79-7] Avilamycin A, *880*
[69787-80-0] Avilamycin C, *880*
[69806-50-4] Fluazifop-butyl, *4147*
[69815-49-2] Norepinephrine *d*-Bitartrate, *6784*
[69819-86-9] Darinaparsin, *2827*
[69819-87-0] Darinaparsin Pyridinium Hydrochloride, *2827*
[69900-72-7] Trimoprostil, *9895*
[69975-86-6] Doxofylline, *3486*
[70020-71-2] ε-Acetamidocaproic Acid Zinc Salt, *46*
[70024-40-7] Terazosin Hydrochloride Dihydrate, *9297*
[70052-12-9] Eflornithine, *3570*
[70059-30-2] Cimetidine Hydrochloride, *2282*
[70111-54-5] Loprazolam Methanesulfonate, *5631*
[70124-77-5] Flucythrinate, *4155*
[70187-32-5] *N*-Methylmorpholine *N*-Oxide Monohydrate, *6171*
[70197-13-6] Methyltrioxorhenium, *6206*
[70209-81-3] Ivermectin Component B$_{1b}$, *5296*
[70222-87-6] (±)-*cis*-Metazocine Hydrochloride Monohydrate, *6005*
[70280-88-5] Difemerine Hydrochloride, *3152*
[70288-86-7] Ivermectin, *5296*
[70356-03-5] Cefaclor Monohydrate, *1914*
[70356-09-1] Avobenzone, *881*
[70359-46-5] Brimonidine D-Tartrate, *1381*
[70369-47-0] Bucindolol Hydrochloride, *1471*
[70374-39-9] Lornoxicam, *5640*
[70384-29-1] Peplomycin Sulfate Salt, *7257*
[70458-92-3] Pefloxacin, *7172*
[70458-95-6] Perfloxacin Methanesulfonate, *7172*
[70458-96-7] Norfloxacin, *6789*
[70476-82-3] Mitoxantrone Dihydrochloride, *6302*
[70543-06-5] Iminodisuccinic Acid (unspecified stereo), *4960*
[70565-74-1] Propanethial *S*-Oxide (Z-form), *7916*
[70608-72-9] 5-HETE, *4709*
[70613-99-9] (±)-Periplanone B, *7288*
[70630-17-0] Metalaxyl *R*-Form, *5993*
[70667-26-4] Ornoprostil, *6970*
[70704-03-9] Vinconate, *10182*
[70775-75-6] Octenidine Dihydrochloride, *6843*
[70797-11-4] Cefpiramide, *1940*
[70878-79-4] Phenazocine *cis*-(±)-Form Hydrobromide, *7332*
[70879-28-6] Alfentanil Hydrochloride Monohydrate, *229*
[70981-66-7] Naproxen Piperazine Salt, *6499*

[71010-52-1] Gellan Gum, *4417*
[71013-43-9] Carminic Acid Methyl Tetra-*O*-methylcarminate, *1843*
[71031-15-7] Cathinone, *1907*
[71109-09-6] Vedaprofen, *10133*
[71116-82-0] Tiaprost, *9578*
[71117-51-6] Zoapatanol, *10385*
[71119-11-4] Bucindolol, *1471*
[71125-38-7] Meloxicam, *5894*
[71138-35-7] Desogestrel Mixture with Ethinyl Estradiol, *2928*
[71142-71-7] PPACK, *7816*
[71160-24-2] Leukotriene B$_4$, *5507*
[71195-58-9] Alfentanil, *229*
[71231-14-6] Lucifer Yellow VS, *5653*
[71247-25-1] Ceruletide Diethylamine Salt, *2004*
[71251-02-0] Octenidine, *6843*
[71283-80-2] Fenoxaprop-ethyl (*R*)-Form, *4015*
[71320-77-9] Moclobemide, *6312*
[71395-14-7] Tocainide (±)-Form Hydrochloride, *9648*
[71439-68-4] Bisantrene Dihydrochloride, *1248*
[71486-22-1] Vinorelbine, *10187*
[71539-72-5] Thiarubrin(e) B, *9456*
[71620-89-8] Reboxetine, *8244*
[71626-11-4] Benalaxyl, *1032*
[71675-85-9] Amisulpride, *481*
[71678-03-0] Ilimaquinone, *4939*
[71731-58-3] Tiquizium Bromide, *9616*
[71751-41-2] Abamectin, *2*
[71771-90-9] Denopamine, *2896*
[71789-16-7] Clopamide Mixture with Pindolol, *2382*
[71827-03-7] Ivermectin Component B$_{1a}$, *5296*
[71939-50-9] Dihydroartemisinin, *807*
[71963-77-4] Artemether, *805*
[71980-98-8] Pseudomonic Acid C, *8026*
[72025-60-6] Leukotriene C$_4$, *5507*
[72058-36-7] Lochnericine, *5611*
[72059-45-1] Leukotriene A$_4$, *5507*
[72178-02-0] Fomesafen, *4254*
[72301-79-2] Enviroxime, *3655*
[72324-18-6] Stepronin, *8936*
[72332-33-3] Procaterol, *7877*
[72432-03-2] Miglitol, *6267*
[72432-10-1] Aniracetam, *655*
[72450-51-2] Hendrickson's Reagent, *4685*
[72459-58-6] Triazoxide, *9768*
[72479-26-6] Fenticonazole, *4032*
[72490-01-8] Fenoxycarb, *4017*
[72496-41-4] Pirarubicin, *7601*
[72509-76-3] Felodipine, *3987*
[72522-13-5] Eptazocine, *3696*
[72558-82-8] Ceftazidime, *1948*
[72559-06-9] Rifabutin, *8338*
[72571-82-5] Pipemidic Acid Trihydrate, *7572*
[72599-27-0] Miglustat, *6268*
[72732-56-0] Piritrexim, *7612*
[72797-41-2] Tianeptine, *9575*
[72804-96-7] *O*-(Diphenylphosphinyl)-hydroxylamine, *3365*
[72822-12-9] Dapiprazole, *2820*
[72822-13-0] Dapiprazole Monohydrochloride, *2820*
[72869-16-0] Pramiracetam Sulfate, *7826*
[72945-61-0] Technetium Tc 99m Oxidronate, *7039*

[72956-09-3] Carvedilol, *1873*
[72957-37-0] GHK Acetate, *4452*
[72962-43-7] Brassinolide, *1370*
[73080-51-0] Repirinast, *8258*
[73121-56-9] Enprostil, *3645*
[73151-29-8] Fenticonazole Mononitrate, *4032*
[73173-12-3] Tetroxoprim Mixture with Sulfadiazine, *9398*
[73183-34-3] Bis(pinacolato)diborane, *1301*
[73210-73-8] Xamoterol Hemifumarate, *10253*
[73217-88-6] Apraclonidine Dihydrochloride, *738*
[73218-79-8] Apraclonidine Hydrochloride, *738*
[73220-03-8] Remoxipride Hydrochloride, *8252*
[73231-34-2] Florfenicol, *4141*
[73232-52-7] Methylnaltrexone Bromide, *6172*
[73250-68-7] Mefenacet, *5868*
[73270-47-0] L-Glutamic Acid 5-Ethyl Ester Hydrochloride, *4506*
[73301-54-9] Yingzhaosu A, *10296*
[73334-07-3] Iopromide, *5105*
[73360-54-0] Bupivacaine Hydrochloride (monohydrate), *1499*
[73384-59-5] Ceftriaxone, *1955*
[73384-60-8] Sulmazole, *9115*
[73391-87-4] Pizotyline Hydrochloride, *7629*
[73523-00-9] Luprostiol, *5670*
[73573-87-2] Formoterol, *4272*
[73573-88-3] Mevastatin, *6243*
[73590-58-6] Omeprazole, *6939*
[73630-23-6] Quin2 (salt), *8158*
[73679-07-9] Phosphinothricin D-Form, *7451*
[73681-12-6] Indecainide Hydrochloride, *4977*
[73744-33-9] *p*-Bromoacetophenone Oxime (Z-form), *1410*
[73771-04-7] Prednicarbate, *7840*
[73803-48-2] Tripamide, *9909*
[73816-42-9] Meclocycline 5-Sulfosalicylate, *5849*
[73836-78-9] Leukotriene D$_4$, *5507*
[73957-86-5] α-Avoparcin, *882*
[73957-87-6] β-Avoparcin, *882*
[73963-72-1] Cilostazol, *2280*
[74011-58-8] Enoxacin, *3641*
[74014-51-0] Rokitamycin, *8379*
[74050-97-8] Haloperidol Decanoate, *4634*
[74050-98-9] Ketanserin, *5344*
[74051-80-2] Sethoxydim, *8612*
[74087-85-7] Dimethyldioxirane, *3265*
[74103-06-3] Ketorolac, *5353*
[74103-07-4] Ketorolac (±)-Form Tromethamine Salt, *5353*
[74115-24-5] Clofentezine, *2369*
[74150-27-9] Pimobendan, *7545*
[74176-31-1] Alfaprostol, *227*
[74191-85-8] Doxazosin, *3482*
[74214-62-3] Ethyl β-Carboline-3-carboxylate, *3834*
[74219-54-8] Aurodox Sodium Salt, *873*
[74222-97-2] Sulfometuron-methyl, *9087*
[74223-64-6] Metsulfuron-methyl, *6237*
[74252-25-8] Indomethacin Sodium Salt Trihydrate, *5009*

[74258-86-9] Alacepril, *193*

[74272-66-5] Brassard's Diene (unspecified stereo), *1368*

[74298-63-8] Chlormidazole Hydrochloride, *2107*

[74341-78-9] MDE Hydrochloride, *5835*

[74347-32-3] Glucoheptonic Acid Magnesium Salt, *4491*

[74356-00-6] Cefotetan Disodium Salt, *1935*

[74381-53-6] Leuprolide Acetate, *5509*

[74397-12-9] Limaprost, *5543*

[74431-23-5] Imipenem, *4961*

[74432-68-1] Ractopamine *R,R*-Form Hydrochloride, *8208*

[74436-00-3] Cyclosporin G, *2750*

[74512-12-2] Omoconazole, *6941*

[74517-78-5] Indecainide, *4977*

[74545-79-2] Aloeresin A, *301*

[74563-64-7] Phytantriol, *7498*

[74578-69-1] Ceftriaxone Disodium Salt, *1955*

[74591-03-0] Endiandric Acid A, *3626*

[74639-40-0] Docarpamine, *3441*

[74697-28-2] Cloricromen Hydrochloride, *2393*

[74711-43-6] Zaltoprofen, *10310*

[74712-19-9] Bromobutide, *1419*

[74717-53-6] Detoxin C₁, *2940*

[74738-17-3] Fenpiclonil, *4018*

[74764-40-2] Bepridil Hydrochloride Monohydrate, *1152*

[74807-65-1] Quadrone (±)-Form, *8139*

[74811-65-7] Croscarmellose Sodium, *2583*

[74811-93-1] Dendrotoxins, *2894*

[74812-63-8] Nordefrin, *6781*

[74847-35-1] Pyronaridine, *8117*

[74849-93-7] Cefpiramide Sodium Salt, *1940*

[74863-84-6] Argatroban, *769*

[74913-18-1] Dynorphin, *3525*

[74978-16-8] Magaldrate, *5714*

[74984-58-0] Amidinomycin Sulfate, *394*

[75011-65-3] Ibopamine Hydrochloride, *4915*

[75139-06-9] Tetronasin, *9396*

[75216-20-5] Fluosol DA, *4215*

[75272-39-8] Nemonapride, *6531*

[75318-43-3] Appel's Salt, *736*

[75330-75-5] Lovastatin, *5644*

[75418-09-6] Tiletamine Combination of Hydrochloride with Zolazepam Hydrochloride, *9592*

[75438-57-2] Moxonidine, *6379*

[75438-58-3] Moxonidine Hydrochloride, *6379*

[75498-96-3] Cefminox Sodium Salt, *1928*

[75507-68-5] Flupirtine Maleate, *4221*

[75530-68-6] Nilvadipine, *6630*

[75558-90-6] Amperozide, *577*

[75607-67-9] Fludarabine 5'-Monophosphate, *4157*

[75621-03-3] CHAPS, *2043*

[75659-07-3] Dilevalol, *5377*

[75659-08-4] Dilevalol Hydrochloride, *5377*

[75695-93-1] Isradipine, *5287*

[75706-12-6] Leflunomide, *5486*

[75715-89-8] Leukotriene E₄, *5507*

[75738-58-8] Cefmenoxime Hydrochloride (*syn*-Isomer), *1926*

[75757-02-7] Osteocalcin (human reduced), *6992*

[75783-06-1] Dipipanone Hydrochloride, *3377*

[75821-71-5] Lonazolac Calcium Salt, *5623*

[75847-73-3] Enalapril, *3623*

[75881-23-1] Alcian Blue 8G (unspecified structure), *212*

[75887-54-6] Arteether, *803*

[75917-92-9] Iofetamine I 123, *5092*

[75921-69-6] Afamelanotide, *168*

[75975-70-1] Cephradine Monohydrate, *1985*

[76025-73-5] Carpetimycin A, *1861*

[76060-33-8] Endiandric Acid B, *3626*

[76060-34-9] Endiandric Acid C, *3626*

[76094-36-5] Carpetimycin B, *1861*

[76095-16-4] Enalapril Maleate, *3623*

[76120-72-4] Ramifenazone Mixture with Phenylbutazone, *8220*

[76123-46-1] Calcium Magnesium Acetate, *1682*

[76168-82-6] Ramoplanin, *8222*

[76189-55-4] BINAP (*R*)-(+)-Form, *1227*

[76189-56-5] BINAP (*S*)-(−)-Form, *1227*

[76333-53-4] Cathinone Hydrochloride, *1907*

[76343-93-6] Latrunculin A, *5433*

[76343-94-7] Latrunculin B, *5433*

[76420-72-9] Enalaprilat (anhydrous), *3623*

[76467-71-5] Indospicine Monohydrochloride Monohydrate, *5011*

[76470-66-1] Loracarbef, *5633*

[76497-13-7] Sultamicillin, *9121*

[76547-98-3] Lisinopril, *5572*

[76568-02-0] Flosequinan, *4142*

[76578-14-8] Quizalofop-ethyl, *8203*

[76584-70-8] Valproic Acid Sodium Salt (2:1), *10099*

[76610-84-9] Cefbuperazone, *1919*

[76631-46-4] Detomidine, *2939*

[76648-01-6] Cefbuperazone Sodium Salt, *1919*

[76674-21-0] Flutriafol, *4242*

[76703-62-3] Gamma-cyhalothrin, *5398*

[76712-82-8] Histrelin, *4760*

[76738-62-0] Paclobutrazol, *7082*

[76738-75-5] (+)-Epi-α-bisabolol, *1245*

[76748-86-2] Pyronaridine Tetraphosphate, *8117*

[76792-22-8] α-Bromoisovaleric Acid (*R*)-Form, *1432*

[76820-74-1] Ioxaglate Meglumine mixture with Ioxaglate Sodium, *5110*

[76822-96-3] Viscotoxin, *10206*

[76824-35-6] Famotidine, *3971*

[76836-02-7] PIPES Disodium Salt, *7592*

[76932-56-4] Nafarelin, *6435*

[76932-60-0] Nafarelin Acetate, *6435*

[76963-41-2] Nizatidine, *6747*

[77086-19-2] Dizocilpine (−)-Form, *3430*

[77086-21-6] Dizocilpine, *3430*

[77086-22-7] Dizocilpine Maleate, *3430*

[77107-46-1] Calmodulin, *1721*

[77134-01-1] Cinnamyl Acetate *cis*-Form, *2302*

[77174-66-4] Croconazole Monohydrochloride, *2580*

[77175-51-0] Croconazole, *2580*

[77181-69-2] Sorivudine, *8853*

[77182-82-2] Phosphinothricin DL-Form Monoammonium Salt, *7451*

[77191-36-7] Nefiracetam, *6524*

[77287-05-9] Rioprostil, *8359*

[77327-04-9] Didemnin A, *3111*

[77327-05-0] Didemnin B, *3111*

[77327-06-1] Didemnin C, *3111*

[77337-73-6] Acamprosate Calcium, *18*

[77392-58-6] Paraherquamide, *7130*

[77469-98-8] Pimobendan Hydrochloride, *7545*

[77521-29-0] AMPA, *575*

[77671-31-9] Enoximone, *3643*

[77732-09-3] Oxadixyl, *7007*

[77734-91-9] Palytoxin, *7100*

[77855-81-3] Milbemycin D, *6271*

[77874-90-9] Ellman's Anion, *3606*

[77883-43-3] Doxazosin Methanesulfonate, *3482*

[77907-69-8] Interferon Alfa-2a, *5034*

[77912-79-9] Sulfazecin, *9077*

[78110-38-0] Aztreonam, *918*

[78111-17-8] Okadaic Acid, *6911*

[78113-36-7] Romurtide, *8384*

[78126-10-0] Stepronin Sodium Salt, *8936*

[78148-59-1] (−)-Epi-α-bisabolol, *1245*

[78178-41-3] Podophyllinic Acid Hydrazide, *7662*

[78186-34-2] Bisantrene, *1248*

[78204-83-8] Tetramethyldiaminobutane Dihydrochloride, *9373*

[78213-16-8] Diclofenac Diethylammonium Salt, *3091*

[78246-49-8] Paroxetine Hydrochloride, *7148*

[78266-06-5] Mebrofenin, *5841*

[78277-23-3] Benzyl Cinnamate (*trans*-form), *1133*

[78281-72-8] Nepafenac, *6553*

[78415-72-2] Milrinone, *6278*

[78439-06-2] Ceftazidime Pentahydrate, *1948*

[78473-71-9] Enterolactone, *3652*

[78474-55-2] Teleocidins, *9260*

[78491-02-8] Diazolidinyl Urea, *3003*

[78587-05-0] Hexythiazox, *4749*

[78613-35-1] Amorolfine, *570*

[78613-38-4] Amorolfine Hydrochloride, *570*

[78628-28-1] Roxatidine Acetate, *8407*

[78628-80-5] Terbinafine Hydrochloride, *9299*

[78649-41-9] Iomeprol, *5097*

[78654-44-1] Amicoumacin A, *392*

[78697-56-0] Lophotoxin, *5629*

[78712-43-3] Ozagrel Hydrochloride, *7079*

[78718-25-9] Benexate Hydrochloride, *1040*

[78755-81-4] Flumazenil, *4166*

[78822-40-9] Pirlimycin Hydrochloride, *7613*

[78833-03-1] Pentisomide, *7242*
[78853-39-1] Cyclodrine Hydrochloride, *2712*
[78919-13-8] Iloprost, *4942*
[78948-87-5] Magnesium Monoperoxyphthalate, *5737*
[78964-85-9] Fosfomycin Tromethamine, *4281*
[78967-07-4] Mofezolac, *6318*
[78994-23-7] Centchroman (−)-Form, *1972*
[79094-20-5] Daltroban, *2805*
[79277-27-3] Thifensulfuron-methyl, *9468*
[79307-93-0] Azelastine Hydrochloride, *899*
[79350-37-1] Cefixime, *1925*
[79350-82-6] Cefixime Disodium Salt, *1925*
[79416-27-6] Methyl Aminolevulinate Hydrochloride, *6090*
[79455-30-4] Nicaraven, *6578*
[79483-69-5] Piritrexim Isethionate, *7612*
[79516-68-0] Levocabastine, *5515*
[79517-01-4] Octreotide Acetate, *6851*
[79538-32-2] Tefluthrin, *9251*
[79547-78-7] Levocabastine Hydrochloride, *5515*
[79548-73-5] Pirlimycin, *7613*
[79559-97-0] Sertraline Hydrochloride, *8606*
[79617-96-2] Sertraline, *8606*
[79622-59-6] Fluazinam, *4148*
[79645-27-5] Tobramycin Sulfate, *9647*
[79660-72-3] Fleroxacin, *4131*
[79704-88-4] α-Methylfentanyl, *6145*
[79770-24-4] Iotrolan, *5108*
[79778-41-9] Neridronic Acid, *6558*
[79794-75-5] Loratadine, *5635*
[79831-76-8] Castanospermine, *1896*
[79871-54-8] Norgestimate Mixture with Ethinyl Estradiol, *6792*
[79874-76-3] Delmopinol, *2884*
[79902-63-9] Simvastatin, *8677*
[79917-90-1] Butylmethylimidazolium Chloride, *1584*
[79944-56-2] Idazoxan Hydrochloride, *4928*
[79944-58-4] Idazoxan, *4928*
[79983-71-4] Hexaconazole, *4717*
[80012-43-7] Epinastine, *3673*
[80060-09-9] Diafenthiuron, *2972*
[80125-14-0] Remoxipride, *8252*
[80210-62-4] Cefpodoxime Free Acid, *1942*
[80214-83-1] Roxithromycin, *8409*
[80295-38-1] Conestat Alfa, *2481*
[80370-57-6] Ceftiofur, *1951*
[80382-23-6] Loxoprofen Sodium Salt, *5647*
[80387-96-8] Difemerine, *3152*
[80409-82-1] Crabtree's Catalyst Tetrafluoroborate, *2554*
[80432-08-2] Butylmethylimidazolium, *1584*
[80433-71-2] Folinic Acid *l*-Form Calcium Salt, *4249*
[80449-31-6] Urinastatin, *10067*
[80455-68-1] Fredericamycin A, *4296*
[80471-63-2] Epostane, *3683*
[80474-14-2] Fluticasone Propionate, *4240*
[80529-93-7] Gadopentetic Acid, *4357*

[80530-63-8] Picotamide Monohydrate, *7519*
[80573-04-2] Balsalazide, *938*
[80576-83-6] Edatrexate, *3554*
[80621-81-4] Rifaximin, *8345*
[80663-95-2] Iobenguane, *5051*
[80729-79-9] Neridronic Acid Monosodium Salt, *6558*
[80734-02-7] Lenampicillin Hydrochloride, *5491*
[80755-51-7] Bunazosin, *1491*
[80844-07-1] Etofenprox, *3922*
[80863-62-3] Alitame, *240*
[81025-03-8] Lactitol Dihydrate, *5389*
[81025-04-9] Lactitol Monohydrate, *5389*
[81098-60-4] Cisapride, *2315*
[81103-11-9] Clarithromycin, *2338*
[81110-73-8] Racecadotril, *8207*
[81129-83-1] Cilastatin Sodium Salt, *2274*
[81131-70-6] Pravastatin Sodium, *7835*
[81147-92-4] Esmolol, *3755*
[81161-17-3] Esmolol Hydrochloride, *3755*
[81262-93-3] Procaterol Hydrochloride Hemihydrate, *7877*
[81267-65-4] Dehydroequol, *2876*
[81290-20-2] Trifluoromethyltrimethylsilane, *9851*
[81334-34-1] Imazapyr, *4946*
[81335-37-7] Imazaquin, *4947*
[81335-77-5] Imazethapyr, *4948*
[81403-68-1] Alfuzosin Hydrochloride, *231*
[81403-80-7] Alfuzosin, *231*
[81405-85-8] Imazamethabenz, *4944*
[81406-37-3] Fluroxypyr 1-Methylheptyl Ester, *4233*
[81409-90-7] Cabergoline, *1606*
[81486-22-8] Nipradilol, *6648*
[81510-83-0] Imazapyr Isopropylamine Salt, *4946*
[81525-10-2] Nafamostat, *6434*
[81572-37-4] NB-Enantride, *6513*
[81591-81-3] Glyphosate Trimethylsulfonium Salt, *4547*
[81627-83-0] Macrophage Colony-Stimulating Factor, *5710*
[81655-41-6] Mosher's Reagent, *6365*
[81669-57-0] Anistreplase, *666*
[81732-46-9] Bambuterol Hydrochloride, *946*
[81732-65-2] Bambuterol, *946*
[81777-89-1] Clomazone, *2374*
[81789-85-7] Indenolol Hydrochloride, *4979*
[81800-41-1] Dihydroactinidiolide (S)-Form, *3188*
[81801-12-9] Xamoterol, *10253*
[81840-15-5] Vesnarinone, *10166*
[81846-19-7] Treprostinil, *9748*
[81858-94-8] Peptide YY (porcine), *7262*
[81872-10-8] Zofenopril, *10386*
[81907-61-1] Ganoderic Acid B, *4396*
[81907-62-2] Ganoderic Acid A, *4396*
[81919-14-4] Bendazac Lysine Salt, *1037*
[81938-43-4] Zofenopril Calcium Salt, *10386*
[81971-15-5] NB-Enantride NB-Enantrane, *6513*
[81982-32-3] Alpiropride, *307*

[81988-88-7] Ramoplanin A$_2$, *8222*
[82009-34-5] Cilastatin, *2274*
[82030-87-3] Methionyl Human Growth Hormone, *8842*
[82034-46-6] Loteprednol Etabonate, *5642*
[82097-50-5] Triasulfuron, *9761*
[82101-18-6] Balsalazide Disodium Salt (anhydrous), *938*
[82115-62-6] Interferon-γ, *5036*
[82159-09-9] Epalrestat, *3660*
[82186-77-4] Lumefantrine, *5657*
[82188-90-7] PPACK Dihydrochloride, *7816*
[82225-47-6] Ajugarin-IV, *190*
[82231-14-9] Ajugarin-V, *190*
[82248-59-7] Atomoxetine Hydrochloride, *854*
[82310-93-8] Hypusine Dihydrochloride, *4909*
[82318-06-7] Deslorelin Acetate, *2924*
[82410-32-0] Ganciclovir, *4393*
[82413-20-5] Droloxifene, *3495*
[82419-36-1] Ofloxacin, *6863*
[82473-24-3] CHAPS Hydroxy Analog, *2043*
[82504-20-9] Asoxime Chloride Monohydrate, *825*
[82547-58-8] Cefteram, *1949*
[82547-81-7] Cefteram Pivaloyloxymethyl Ester, *1949*
[82558-50-7] Isoxaben, *5281*
[82560-54-1] Benfuracarb, *1044*
[82571-53-7] Ozagrel, *7079*
[82586-52-5] Moexipril Hydrochloride, *6315*
[82586-55-8] Quinapril Hydrochloride, *8167*
[82586-57-0] Moexiprilat Hydrochloride, *6315*
[82619-04-3] Cefpodoxime Free Acid Sodium Salt, *1942*
[82626-48-0] Zolpidem, *10391*
[82640-04-8] Raloxifene Hydrochloride, *8215*
[82657-04-3] Bifenthrin, *1213*
[82657-92-9] Pro-Urokinase, *8011*
[82697-16-3] Clofencet Sodium Salt, *2368*
[82697-71-0] Clofencet Potassium Salt, *2368*
[82747-56-6] Cicletanine Hydrochloride, *2267*
[82752-99-6] Nefazodone Hydrochloride, *6523*
[82768-85-2] Quinaprilat, *8167*
[82785-45-3] Neuropeptide Y, *6570*
[82801-81-8] MDE, *5835*
[82834-16-0] Perindopril, *7285*
[82855-09-2] Combretastatin, *2477*
[82857-38-3] Bopindolol Malonate, *1335*
[82952-64-5] Trimetrexate D-Glucuronate, *9893*
[82956-11-4] Nafamostat Dimethanesulfonate, *6434*
[82964-04-3] Tolrestat, *9683*
[82989-25-1] Tazanolast, *9217*
[83014-44-2] Quin2 (acid), *8158*
[83015-26-3] Atomoxetine, *854*
[83055-99-6] Bensulfuron-methyl, *1053*
[83058-94-0] Terrecyclic Acid, *9314*
[83104-85-2] Quin2 Acetoxymethyl Ester, *8158*

[83105-70-8] Sultamicillin Tosylate, *9121*
[83121-18-0] Teflubenzuron, *9249*
[83150-76-9] Octreotide, *6851*
[83164-33-4] Diflufenican, *3163*
[83314-01-6] Bryostatin 1, *1466*
[83366-66-9] Nefazodone, *6523*
[83382-98-3] Synsorbs, *9144*
[83387-25-1] Methylnaltrexonium, *6172*
[83435-66-9] Delapril, *2881*
[83435-67-0] Delapril Hydrochloride, *2881*
[83461-56-7] Mifamurtide, *6265*
[83480-29-9] Voglibose, *10226*
[83602-05-5] Spiraprilat, *8880*
[83621-06-1] Omoconazole Nitrate, *6941*
[83643-88-3] L-AMPA, *575*
[83647-97-6] Spirapril, *8880*
[83657-18-5] Diniconazole (−)-Form, *3294*
[83657-19-6] Diniconazole (+)-Form, *3294*
[83657-24-3] Diniconazole, *3294*
[83688-84-0] Tertatolol, *9316*
[83712-60-1] Defibrotide, *2864*
[83729-01-5] Salvinorin A, *8481*
[83799-24-0] Fexofenadine, *4097*
[83846-83-7] Ketanserin Tartrate, *5344*
[83869-56-1] Granulocyte-Macrophage Colony-Stimulating Factor, *4574*
[83881-51-0] Cetirizine, *2021*
[83881-52-1] Cetirizine Dihydrochloride, *2021*
[83905-01-5] Azithromycin, *907*
[83915-83-7] Lisinopril, *5572*
[83919-23-7] Mometasone Furoate, *6327*
[83930-13-6] Somatoliberin, *8840*
[84031-17-4] Metaclazepam, *5990*
[84057-84-1] Lamotrigine, *5403*
[84057-95-4] Ropivacaine, *8389*
[84082-34-8] Bilberry Extract, *1221*
[84088-42-6] Roquinimex, *8390*
[84294-96-2] Enoxacin Sesquihydrate, *3641*
[84297-59-6] Cabenegrin A-I, *1605*
[84297-60-9] Cabenegrin A-II, *1605*
[84305-41-9] Cefminox, *1928*
[84371-65-3] Mifepristone, *6266*
[84449-90-1] Raloxifene, *8215*
[84504-69-8] Irsogladine Maleate, *5146*
[84611-23-4] Erdosteine, *3704*
[84625-59-2] Dotarizine, *3477*
[84625-61-6] Itraconazole, *5292*
[84665-66-7] Magnesium Monoperoxyphthalate Hexahydrate, *5737*
[84680-54-6] Enalaprilat (dihydrate), *3623*
[84799-02-0] Laidlomycin Propionate Potassium, *5397*
[84845-57-8] Ritipenem, *8364*
[84845-58-9] Ritipenem Sodium Salt, *8364*
[84878-61-5] Maduramicin, *5711*
[84937-45-1] Gusperimus (−)-Form Trihydrochloride, *4617*
[84957-29-9] Cefpirome, *1941*
[84957-30-2] Cefquinome, *1944*
[84964-12-5] Brovincamine Hydrogen Fumarate, *1460*
[85068-76-4] Iofetamine I 123 Hydrochloride, *5092*

[85118-33-8] Gaboxadol Hydrochloride, *4351*
[85136-71-6] Tilisolol, *9595*
[85175-67-3] Zatebradine, *10313*
[85233-19-8] BAPTA, *952*
[85248-93-7] Pseudomonic Acid D, *8026*
[85320-68-9] Amosulalol, *572*
[85329-89-1] Cabergoline Diphosphate, *1606*
[85371-64-8] Pinacidil Monohydrate, *7549*
[85441-61-8] Quinapril, *8167*
[85468-01-5] Gusperimus Trihydrochloride, *4617*
[85509-19-9] Flusilazole, *4236*
[85622-93-1] Temozolomide, *9282*
[85637-73-6] Atrial Natriuretic Peptide, *863*
[85649-12-3] Cyfluthrin *(1R,3R,αR)*-Form, *2754*
[85649-15-6] Cyfluthrin *(1R,3R,αS)*-Form, *2754*
[85649-19-0] Cyfluthrin *(1R,3S,αS)*-Form, *2754*
[85650-56-2] Asenapine Maleate, *821*
[85721-33-1] Ciprofloxacin, *2313*
[85856-54-8] Moveltipril, *6371*
[85865-74-3] Pygeum Africanum Extract, *8060*
[85897-35-4] Sacrosidase, *8447*
[85898-30-2] Interleukin-2, *5038*
[85921-53-5] Moveltipril Calcium Salt, *6371*
[85960-17-4] Imipenem Combination with Cilastatin Sodium, *4961*
[86050-77-3] Gadopentetic Acid Dimeglumine Salt, *4357*
[86168-78-7] Sermorelin, *8600*
[86197-47-9] Dopexamine, *3472*
[86209-51-0] Primisulfuron-methyl, *7866*
[86220-42-0] Nafarelin Acetate Hydrate, *6435*
[86227-47-6] Eicosapentaenoic Acid Ethyl Ester, *3578*
[86273-18-9] Lenampicillin, *5491*
[86303-22-2] BIGCHAP, *1217*
[86303-23-3] Deoxy-BIGCHAP, *1217*
[86329-79-5] Cefodizime Disodium Salt, *1929*
[86340-94-5] *N*-Iodosaccharin, *5086*
[86347-14-0] Medetomidine, *5858*
[86347-15-1] Medetomidine Hydrochloride, *5858*
[86386-73-4] Fluconazole, *4153*
[86393-32-0] Ciprofloxacin Monohydrochloride Monohydrate, *2313*
[86401-95-8] Methylprednisolone Aceponate, *6184*
[86408-72-2] Ecabet Sodium Salt, *3535*
[86479-06-3] Hexaflumuron, *4719*
[86484-91-5] Dopexamine Dihydrochloride, *3472*
[86541-74-4] Benazepril Hydrochloride, *1033*
[86541-75-5] Benazepril, *1033*
[86541-78-8] Benazeprilat, *1033*
[86598-92-7] Imibenconazole, *4950*
[86697-68-9] Fasciculins, *3977*
[86725-37-3] Amperozide Hydrochloride, *577*
[86767-75-1] Octenidine Disaccharin, *6843*

[86780-90-7] Aranidipine, *756*
[86832-68-0] Carumonam Disodium Salt, *1871*
[87041-58-5] Hementin, *4675*
[87051-43-2] Ritanserin, *8363*
[87056-78-8] Quinagolide, *8161*
[87081-35-4] Leptomycin B, *5499*
[87130-20-9] Diethofencarb, *3120*
[87233-61-2] Emedastine, *3614*
[87233-62-3] Emedastine Difumarate, *3614*
[87234-24-0] Piroxicam Cinnamic Acid Ester, *7619*
[87238-52-6] Ritipenem Acetoxymethyl Ester, *8364*
[87239-81-4] Cefpodoxime Proxetil, *1942*
[87333-19-5] Ramipril, *8221*
[87344-06-7] Amtolmetin Guacil, *591*
[87392-12-9] Metolachlor *S*-Form, *6223*
[87413-09-0] Dess-Martin Periodinane, *2934*
[87440-45-7] Taprostene Sodium Salt, *9190*
[87480-01-1] Quadrone (+)-Form, *8139*
[87546-18-7] Flumiclorac-pentyl, *4173*
[87547-04-4] Flumiclorac, *4173*
[87625-62-5] Ptaquiloside, *8039*
[87638-04-8] Carumonam, *1871*
[87674-68-8] Dimethenamid, *3236*
[87679-37-6] Trandolapril, *9729*
[87679-71-8] Trandolaprilat, *9729*
[87726-17-8] Panipenem, *7111*
[87760-53-0] Tandospirone, *9183*
[87771-40-2] Ioversol, *5109*
[87805-34-3] GLP-1 (human), *4482*
[87806-31-3] Porfimer Sodium, *7718*
[87818-31-3] Cinmethylin, *2296*
[87820-88-0] Tralkoxydim, *9724*
[87848-99-5] Acrivastine, *118*
[88040-23-7] Cefepime, *1923*
[88069-49-2] Pilsicainide Hydrochloride Hemihydrate, *7539*
[88069-67-4] Pilsicainide, *7539*
[88122-99-0] Ethylhexyl Triazone, *3864*
[88150-42-9] Amlodipine, *487*
[88150-47-4] Amlodipine Maleate, *487*
[88189-03-1] Bismuth Triflate, *1292*
[88200-01-5] Chelidonine (−)-Form, *2052*
[88255-01-0] Netobimin, *6565*
[88283-41-4] Pyrifenox, *8096*
[88416-50-6] Clodronic Acid Disodium Salt (tetrahydrate), *2365*
[88426-33-9] Buparvaquone, *1495*
[88430-50-6] Beraprost, *1154*
[88475-69-8] Beraprost Sodium Salt, *1154*
[88495-63-0] Artesunate, *808*
[88641-36-5] Cefminox Sodium Salt Heptahydrate, *1928*
[88671-89-0] Myclobutanil, *6406*
[88678-31-3] Liranaftate, *5570*
[88768-40-5] Cilazapril, *2275*
[88805-35-0] Prohexadione, *7892*
[88816-02-8] Belleau's Reagent, *1027*
[88889-14-9] Fosinopril Sodium Salt, *4283*
[89030-95-5] GHK Complex with Copper, *4452*
[89213-87-6] Carperitide, *863*
[89226-50-6] Manidipine, *5808*

Chemical Abstracts Registry Numbers

[114084-78-5] Ibandronic Acid, *4912*
[114311-32-9] Imazamox, *4945*
[114369-43-6] Fenbuconazole, *3995*
[114471-18-0] Brain Natriuretic Peptide, *1365*
[114560-48-4] Apaziquone, *712*
[114615-82-6] Tetrapropylammonium Perruthenate, *9386*
[114798-26-4] Losartan, *5641*
[114870-03-0] Fondaparinux Sodium, *4259*
[114899-77-3] Ecteinascidin 743, *3550*
[114921-13-0] Gamboge Pigment, *4390*
[114977-28-5] Docetaxel, *3442*
[115103-54-3] Tiagabine, *9572*
[115256-11-6] Dofetilide, *3456*
[115266-92-7] Domitroban (±)-Form, *3465*
[115436-72-1] Risedronic Acid Monosodium Salt, *8360*
[115457-83-5] Hecameg®, *4656*
[115464-59-0] Methyl(trifluoromethyl)-dioxirane, *6205*
[115550-35-1] Marbofloxacin, *5815*
[115575-11-6] Liarozole, *5528*
[115587-57-0] Dibenzalacetone *cis-trans*-Form, *3010*
[115850-27-6] Metconazole *cis*-Form, *6006*
[115850-28-7] Metconazole *trans*-Form, *6006*
[115852-48-7] Fenoxanil, *4014*
[115956-12-2] Dolasetron, *3458*
[115956-13-3] Dolasetron Methanesulfonate, *3458*
[115964-29-9] Tosufloxacin Tosylate, *9714*
[115966-23-9] Urodilatin, *10071*
[115966-66-0] Histatin 1 (unspecified), *4757*
[115966-67-1] Histatin 3 (unspecified), *4757*
[115966-68-2] Histatin 5, *4757*
[116036-70-5] Fibrolase, *4103*
[116078-65-0] Bidisomide, *1208*
[116094-23-6] Insulin Aspart, *5025*
[116095-17-1] Antimycin A$_{3b}$, *702*
[116255-48-2] Bromuconazole, *1456*
[116355-83-0] Fumonisin B$_1$, *4318*
[116476-13-2] Semotiadil, *8585*
[116476-14-3] Semotiadil Fumarate, *8585*
[116539-59-4] Duloxetine, *3512*
[116644-53-2] Mibefradil, *6252*
[116649-85-5] Ramatroban, *8218*
[116666-63-8] Mibefradil Dihydrochloride, *6252*
[116714-46-6] Novaluron, *6809*
[116836-09-0] Isomethadone *dl*-Form, *5230*
[116876-37-0] Clavulanic Acid Combination of Potassium Salt with Ticarcillin Disodium, *2339*
[117048-59-6] Combretastatin A-4, *2477*
[117060-71-6] Sulbactam Mixture of Sodium Salt with Ampicillin Sodium, *9021*
[117091-64-2] Etoposide Phosphate, *3929*
[117149-90-3] Thymostimulin, *9562*
[117276-75-2] 4-221-Colony-stimulating Factor 1 (Human Clone p3ACSF-69 Protein Moiety Reduced), *5710*

[117279-73-9] Israpafant, *5288*
[117337-19-6] Fluthiacet-methyl, *4239*
[117381-09-6] Trimetrexate Monoacetate Monohydrate, *9893*
[117399-94-7] Endothelin, *3631*
[117428-22-5] Picoxystrobin, *7520*
[117467-28-4] Cefditoren Pivaloyloxymethyl Ester, *1922*
[117591-79-4] Remoxipride Hydrochloride Monohydrate, *8252*
[117704-25-3] Doramectin, *3473*
[117718-60-2] Thiazopyr, *9463*
[117772-70-0] Azithromycin Dihydrate, *907*
[117976-89-3] Rabeprazole, *8206*
[117976-90-6] Rabeprazole Sodium Salt, *8206*
[118072-93-8] Zoledronic Acid, *10388*
[118081-34-8] Ceftibuten Dihydrate, *1950*
[118134-30-8] Spiroxamine, *8889*
[118248-94-5] Mangafodipir (free acid), *5787*
[118288-08-7] Lafutidine, *5396*
[118292-40-3] Tazarotene, *9218*
[118390-30-0] Interferon Alfacon-1, *5034*
[118443-89-3] Cefquinome Sulfate, *1944*
[118457-14-0] Nebivolol, *6516*
[118812-69-4] Ularitide, *10071*
[118949-22-7] Vorozole (±)-Form, *10233*
[118949-61-4] Pyridine Bis(oxazoline) Ligands (*S,S*)-*i*-Pr-pybox, *8083*
[118997-30-1] Peptide YY (human), *7262*
[119006-77-8] Flutrimazole, *4243*
[119068-77-8] Semduramicin Sodium Salt, *8582*
[119119-85-6] α-Glucose-1-phosphate Calcium Salt, *4497*
[119129-70-3] Ananain, *1395*
[119141-88-7] Omeprazole *S*-Form, *6939*
[119168-77-3] Tebufenpyrad, *9231*
[119169-78-7] Episteride, *3690*
[119295-40-8] Diphenylcarbazone (Z-form), *3353*
[119295-41-9] Diphenylcarbazone (*E*-form), *3353*
[119302-91-9] Rocuronium Bromide, *8375*
[119356-77-3] Dapoxetine, *2821*
[119413-54-6] Topotecan Hydrochloride, *9707*
[119446-68-3] Difenoconazole, *3154*
[119478-55-6] Danofloxacin Methanesulfonate, *2812*
[119478-56-7] Meropenem Trihydrate, *5970*
[119506-01-3] Deserpidine Oxalate, *2919*
[119515-38-7] Picaridin, *7506*
[119520-05-7] Zilpaterol, *10324*
[119520-06-8] Zilpaterol Hydrochloride, *10324*
[119567-79-2] Taribavirin, *9197*
[119600-43-0] Vinconate Monohydrochloride, *10182*
[119683-68-0] Ferumoxides, *4090*
[119742-13-1] Roxindole Methanesulfonate, *8408*
[119784-94-0] Tenidap Sodium Salt, *9287*

[119791-41-2] Emamectin, *3612*
[119817-90-2] Loxiglumide (*R*)-Form, *5646*
[119914-60-2] Grepafloxacin, *4580*
[120010-32-4] Park Nucleotide (pimelate form), *7145*
[120011-70-3] Donepezil Hydrochloride, *3468*
[120014-06-4] Donepezil, *3468*
[120066-54-8] Gadoteridol, *4358*
[120068-37-3] Fipronil, *4116*
[120103-35-7] HLö-7, *4762*
[120116-88-3] Cyazofamid, *2693*
[120138-50-3] Quinupristin, *8201*
[120162-55-2] Azimsulfuron, *905*
[120210-48-2] Tenidap, *9287*
[120279-96-1] Dorzolamide, *3475*
[120287-85-6] Cetrorelix, *2025*
[120373-24-2] Unoprostone Isopropyl Ester, *10034*
[120373-36-6] Unoprostone, *10034*
[120410-24-4] Biapenem, *1201*
[120444-71-5] Deramciclane, *2916*
[120444-72-6] Deramciclane Fumarate, *2916*
[120511-73-1] Anastrozole, *619*
[120608-46-0] Duteplase, *9624*
[120638-55-3] Bromfenac Monosodium Salt Sesquihydrate, *1397*
[120666-13-9] Trimethyl-Verkade's Superbase, *10155*
[120685-11-2] Midostaurin, *6264*
[120895-52-5] Amphotericin B Compd with Cholesteryl Sulfate, *582*
[120923-37-7] Amidosulfuron, *397*
[120928-09-8] Fenazaquin, *3993*
[120993-53-5] Desirudin, *4755*
[121011-80-1] Pseudopterosin E, *8029*
[121025-46-5] Sphingofungin C, *8872*
[121032-29-9] Nelarabine, *6527*
[121034-85-3] Cymiazole Hydrochloride, *2761*
[121067-52-5] Yingzhaosu C, *10296*
[121123-17-9] Cefprozil, *1943*
[121124-29-6] Emamectin, *3612*
[121181-53-1] Filgrastim, *4573*
[121268-17-5] Alendronic Acid Monosodium Salt Trihydrate, *223*
[121281-41-2] Technetium Tc 99m Bicisate, *9236*
[121412-77-9] Cefprozil Z-Form, *1943*
[121424-52-0] Emamectin, *3612*
[121451-02-3] Noviflumuron, *6810*
[121543-68-8] Lunasine (−)-Form Perchlorate, *5665*
[121547-04-4] Mirimostim, *5710*
[121552-61-2] Cyprodinil, *2769*
[121584-18-7] Valspodar, *10103*
[121652-76-4] Eptastigmine Tartrate, *3695*
[121679-13-8] Naratriptan, *6502*
[121696-62-6] Piroxicam Compd with β-Cyclodextrin, *7619*
[121808-62-6] Pidotimod, *7531*
[121961-22-6] Loracarbef Monohydrate, *5633*
[122007-85-6] Monteplase, *6347*
[122008-85-9] Cyhalofop-butyl, *2755*
[122039-27-4] Woollins' Reagent, *10249*
[122111-03-9] Gemcitabine Hydrochloride, *4420*
[122191-40-6] Caspase-1, *1884*
[122312-54-3] Epoetin Beta, *3744*

[142583-61-7] Policosanol, *7672*
[142796-21-2] Isosilybin A, *8671*
[142796-22-3] Isosilybin B, *8671*
[142797-34-0] Silybin B, *8671*
[142825-10-3] Sulforaphane, *9092*
[142891-20-1] Cinidon-ethyl, *2294*
[143003-46-7] Alglucerase, *4487*
[143011-72-7] Granulocyte Colony-Stimulating Factor, *4573*
[143090-92-0] Anakinra, *5045*
[143201-11-0] Cerivastatin Sodium Salt, *1992*
[143322-58-1] Eletriptan, *3597*
[143334-20-7] Carbon Nitride, *1814*
[143388-64-1] Naratriptan Hydrochloride, *6502*
[143390-89-0] Kresoxim-methyl, *5372*
[143413-84-7] Thiazole Orange Dimer, *9460*
[143491-57-0] Emtricitabine, *3622*
[143558-00-3] Rocuronium, *8375*
[143653-53-6] Abciximab, *5*
[143807-66-3] Chromafenozide, *2223*
[143831-71-4] Dornase Alfa, *2907*
[144017-65-2] Stiripentol (*R*)-Form, *8948*
[144017-66-3] Stiripentol (*S*)-Form, *8948*
[144034-80-0] Rizatriptan, *8370*
[144046-30-0] D-Galactose Transpulmonary Microparticulate Form, *4365*
[144058-40-2] Satumomab, *8521*
[144060-53-7] Febuxostat, *3983*
[144115-48-0] Dichlormid Combination with Acetochlor, *3058*
[144143-96-4] Eprosartan Mesylate, *3691*
[144171-61-9] Indoxacarb (*RS*)-Form, *5012*
[144189-66-2] 3-Nitrosobenzamide, *6720*
[144245-52-3] Fomivirsen, *4256*
[144252-71-1] Asoxime Dimethanesulfonate, *825*
[144348-08-3] Binodenoson, *1229*
[144412-49-7] Lamifiban, *5399*
[144412-50-0] Lamifiban Trifluoroacetate Salt, *5399*
[144481-98-1] Landiolol Hydrochloride, *5405*
[144494-65-5] Tirofiban, *9620*
[144506-14-9] Licochalcone C, *5532*
[144506-15-0] Licochalcone D, *5532*
[144510-96-3] Pixantrone, *7628*
[144538-83-0] Iminodisuccinic Acid Tetrasodium Salt, *4960*
[144550-36-7] Iodosulfuron-methyl-sodium, *5089*
[144598-75-4] Paliperidone, *7087*
[144665-07-6] Lubeluzole, *5648*
[144675-97-8] Pixantrone Dimalate Salt, *7628*
[144689-63-4] Olmesartan, *6933*
[144701-48-4] Telmisartan, *9271*
[144740-53-4] Flupyrsulfuron-methyl, *4225*
[144740-54-5] Flupyrsulfuron-methyl Sodium Salt, *4225*
[144875-48-9] Resiquimod, *8271*
[144913-72-4] Graphite Fluoride Poly-(dicarbon Monofluoride), *4576*
[144980-29-0] Repinotan, *8257*
[144980-77-8] Repinotan Hydrochloride, *8257*

[145026-81-9] Propoxycarbazone, *7951*
[145026-88-6] Flucarbazone, *4151*
[145040-37-5] Candesartan Cilexetil, *1742*
[145108-58-3] Dexmedetomidine Hydrochloride, *2948*
[145137-38-8] Desmoteplase, *3506*
[145155-23-3] Interferon Beta-1b, *5035*
[145158-71-0] Tegaserod, *9253*
[145202-66-0] Rizatriptan Benzoate, *8370*
[145258-61-3] Interferon Beta-1a, *5035*
[145375-43-5] Mitiglinide, *6296*
[145414-76-2] Dihydroxybergamottin, *1159*
[145464-28-4] Capromab Pendetide, *1764*
[145525-41-3] Mitiglinide Calcium, *6296*
[145599-86-6] Cerivastatin, *1992*
[145613-73-6] HLö-7 Dimethanesulfonate, *4762*
[145626-87-5] Bis(2-mercaptoethyl)sulfone, *1258*
[145672-81-7] Cetrorelix Acetate, *2025*
[145701-21-9] Diclosulam, *3093*
[145701-23-1] Florasulam, *4139*
[145808-47-5] Margatoxin, *5817*
[145821-59-6] Tiagabine Hydrochloride, *9572*
[145858-50-0] Liarozole Hydrochloride, *5528*
[145858-52-2] Liarozole Fumarate, *5528*
[145918-75-8] Troxacitabine, *9968*
[145941-26-0] Oprelvekin, *5044*
[146368-15-2] Cy 5, *2669*
[146374-27-8] Ellman's Sulfinamides, *3607*
[146389-61-9] Zaragozic Acid B, *10312*
[146389-62-0] Zaragozic Acid C, *10312*
[146397-20-8] Cy 3, *2668*
[146426-40-6] Flavopiridol, *4126*
[146464-95-1] Pralatrexate, *7821*
[146939-27-7] Ziprasidone, *10371*
[146961-34-4] Trovafloxacin Hydrochloride, *9967*
[147059-72-1] Trovafloxacin, *9967*
[147059-75-4] Trovafloxacin Methanesulfonate, *9967*
[147098-20-2] Rosuvastatin Calcium Salt, *8402*
[147116-67-4] Maropitant, *5819*
[147127-20-6] Tenofovir, *9289*
[147149-76-6] Nolatrexed, *6759*
[147150-35-4] Cloransulam-methyl, *2391*
[147221-93-0] Delavirdine Mesylate, *2882*
[147245-92-9] Glatiramer Acetate, *4470*
[147253-67-6] (*R,R*)-Me-DuPHOS, *3513*
[147359-76-0] Flibanserin Hydrochloride, *4132*
[147403-03-0] Azilsartan, *903*
[147413-41-0] Magnalium (50:50), *5716*
[147511-69-1] Pitavastatin, *7621*
[147526-32-7] Pitavastatin Calcium Salt, *7621*
[147536-97-8] Bosentan, *1355*

[147664-63-9] Pexiganan, *7308*
[147816-23-7] Cefcapene Pivoxil Hydrochloride, *1920*
[147816-24-8] Cefcapene Pivoxil Hydrochloride Hydrate, *1920*
[148016-81-3] Doripenem, *3474*
[148408-66-6] Docetaxel Trihydrate, *3442*
[148477-71-8] Spirodiclofen, *8884*
[148553-50-8] Pregabalin, *7843*
[148641-02-5] Muplestim, *5039*
[148717-58-2] Palau'amine, *7085*
[148717-90-2] Squalamine, *8894*
[148849-67-6] Ivabradine Hydrochloride, *5294*
[148883-56-1] Tifacogin, *9623*
[148893-10-1] HATU, *4653*
[149003-01-0] Razoxane (+)-Form Hydrochloride, *8241*
[149152-95-4] Splendipherin, *8890*
[149193-61-3] Xemilofiban Acid, *10269*
[149204-42-2] Kahalalide F, *5323*
[149470-60-0] Ivabradine (±)-Form, *5294*
[149508-90-7] Simeconazole, *8673*
[149647-78-9] Vorinostat, *10232*
[149656-63-3] Jacobsen's Catalyst, *5300*
[149676-40-4] Pefloxacin Methanesulfonate Dihydrate, *7172*
[149820-74-6] Xemilofiban, *10269*
[149824-15-7] Ilodecakin, *5043*
[149838-22-2] FM1-43, *4247*
[149845-06-7] Saquinavir Mesylate, *8506*
[149845-07-8] Tiludronic Acid Disodium Salt, *9598*
[149877-41-8] Bifenazate, *1211*
[149888-94-8] Azimilide Dihydrochloride, *904*
[149908-53-2] Azimilide, *904*
[149920-56-9] Idraparinux, *4932*
[149961-52-4] Dimoxystrobin, *3292*
[149979-41-9] Tepraloxydim, *9295*
[150103-82-5] Sonicated Human Serum Albumin, *8607*
[150322-43-3] Prasugrel, *7833*
[150374-95-1] Sivelestat Sodium, *8696*
[150378-17-9] Indinavir, *4983*
[150399-21-6] Balsalazide Disodium Salt Dihydrate, *938*
[150399-23-8] Pemetrexed Sodium Salt, *7186*
[150436-68-3] Lactisole Sodium Salt, *5388*
[150501-62-5] Tedisamil Sesquifumarate, *9246*
[150529-93-4] Bis(oxazoline) Ligands (*R,R*)-Ph-box, *2277*
[150683-30-0] Tolvaptan, *9699*
[150726-52-6] Bosentan Sodium Salt, *1355*
[150812-12-7] Ezogabine, *3959*
[150812-13-8] Ezogabine Dihydrochloride, *3959*
[150824-47-8] Nitenpyram, *6654*
[150915-40-5] Tirofiban Hydrochloride Monohydrate, *9620*
[150915-41-6] Perospirone, *7296*
[150937-43-2] Brinzolamide Hydrochloride, *1382*
[151031-37-7] Stellar®, *8933*
[151096-09-2] Moxifloxacin, *6377*

[151126-32-8] Pramlintide, 7828
[151319-34-5] Zaleplon, 10309
[151438-54-9] Harpin, 4650
[151606-29-0] Zinquin, 10369
[151716-36-8] Methyl Dihydrojasmonate (+)-(1S,2S)-Form, 6124
[151763-64-3] Capromab, 1764
[151767-02-1] Montelukast Monosodium Salt, 6346
[151879-73-1] Aprinocarsen, 744
[151912-11-7] Amediplase, 385
[152044-53-6] Epothilone A, 3684
[152044-54-7] Epothilone B, 3684
[152074-97-0] Dirucotide, 3396
[152459-95-5] Imatinib, 4943
[152520-56-4] Nebivolol Hydrochloride, 6516
[152657-84-6] Nalfurafine, 6443
[152658-17-8] Nalfurafine Hydrochloride, 6443
[152751-57-0] Sevelamer Hydrochloride, 8613
[152811-62-6] Piboserod, 7505
[152854-19-8] Xanomeline Tartrate, 10254
[152923-56-3] Daclizumab, 2796
[152923-57-4] Luteinizing Hormone Lutropin Alfa, 5674
[152946-68-4] Nolatrexed Dihydrochloride, 6759
[153168-05-9] Pleconaril, 7648
[153205-46-0] Asimadoline, 823
[153233-91-1] Etoxazole, 3933
[153259-65-5] Cilomilast, 2279
[153439-40-8] Fexofenadine Hydrochloride, 4097
[153504-70-2] Cevimeline Hydrochloride Hemihydrate, 2033
[153507-46-1] Technetium Tc 99m Apcitide Dimer, 9235
[153531-96-5] Magnesium Citrate Tribasic (nonahydrate), 5727
[153559-49-0] Bexarotene, 1198
[153719-23-4] Thiamethoxam, 9446
[153773-82-1] Ertapenem Sodium Salt, 3731
[153832-46-3] Ertapenem, 3731
[154039-60-8] Marimastat, 5818
[154229-18-2] Abiraterone Acetate, 9
[154229-19-3] Abiraterone, 9
[154248-97-2] Imiglucerase, 4958
[154323-57-6] Almotriptan, 298
[154361-48-5] Arcitumomab, 764
[154361-49-6] Technetium Tc 99m Arcitumomab, 764
[154361-50-9] Capecitabine, 1757
[154427-83-5] Samarium Sm 153 Lexidronam, 8487
[154598-52-4] Efavirenz, 3569
[155030-63-0] Emodepside, 3617
[155141-29-0] Rosiglitazone Maleate, 8396
[155148-31-5] Plerixafor Octahydrochloride, 7649
[155148-32-6] Plerixafor Octahydrobromide, 7649
[155206-00-1] Bimatoprost, 1225
[155213-67-5] Ritonavir, 8366
[155270-99-8] Istradefylline, 5289
[155319-91-8] Mangafodipir (hexahydrogen), 5787
[155453-10-4] Tiludronic Acid Disodium Salt (hemihydrate), 9598
[155648-60-5] Minodronic Acid Monohydrate, 6284

[155773-57-2] Pegorgotein, 9131
[155798-07-5] Ioflupane I 123, 5093
[155974-00-8] Ivabradine, 5294
[156052-68-5] Zoxamide, 10398
[156053-89-3] Alvimopan, 366
[156137-99-4] Rapacuronium Bromide, 8231
[156436-89-4] Motexafin Gadolinium (hydrate), 6367
[156436-90-7] Motexafin Lutetium (hydrate), 6368
[156586-91-3] Xemilofiban Hydrochloride, 10269
[156592-72-2] MacMillan's Enamine Catalyst, 5707
[156604-79-4] Dexketoprofen Trometamol, 5352
[156611-76-6] Rupatadine Trihydrochloride, 8430
[156897-06-2] Licofelone, 5533
[156910-61-1] Phenserine Tartrate, 7372
[157212-55-0] Bosentan Monohydrate, 1355
[157283-68-6] Travoprost, 9738
[157379-44-7] PPACK Trifluoroacetate Salt, 7816
[157716-52-4] Perifosine, 7280
[157810-81-6] Indinavir Sulfate, 4983
[157856-25-2] Technetium Tc 99m Sulesomab, 9027
[158062-67-0] Flonicamid, 4137
[158113-60-1] Opebecan, 931
[158237-07-1] Fentrazamide, 4034
[158382-37-7] Canfosfamide, 1748
[158440-71-2] Irofulven, 5136
[158681-13-1] Rimonabant Hydrochloride, 8355
[158747-02-5] Frovatriptan, 4301
[158827-34-0] Pralmorelin Dihydrochloride, 7824
[158861-67-7] Pralmorelin, 7824
[158876-82-5] Rupatadine, 8430
[158930-09-7] Frovatriptan Succinate (andydrous), 4301
[158930-17-7] Frovatriptan Succinate Monohydrate, 4301
[158966-92-8] Montelukast, 6346
[159138-80-4] Cariporide, 1840
[159138-81-5] Cariporide Methanesulfonate, 1840
[159351-69-6] Everolimus, 3950
[159518-97-5] Cloransulam, 2391
[159519-65-0] Enfuvirtide, 3636
[159652-53-6] Phenserine (racemate), 7372
[159752-39-3] Methoxyethylbenzeneboronic Acid (+)-R-Form, 6065
[159776-70-2] Melagatran, 5879
[159811-51-5] Ulipristal, 10030
[159912-53-5] Sabcomeline, 8444
[159912-58-0] Sabcomeline Hydrochloride, 8444
[159989-64-7] Nelfinavir, 6528
[159989-65-8] Nelfinavir Methanesulfonate, 6528
[159997-94-1] Biricodar, 1244
[160003-66-7] Iniparib, 5017
[160337-95-1] Insulin Glargine, 5027
[160369-77-7] Fomivirsen Sodium, 4256
[160369-78-8] Samarium Sm 153 Lexidronam Pentasodium Salt, 8487
[160430-64-8] Acetamiprid, 48

[160707-69-7] Apricitabine, 742
[160970-54-7] Silodosin, 8642
[161050-58-4] Methoxyfenozide, 6066
[161265-03-8] Xantphos, 10268
[161326-34-7] Fenamidone, 3990
[161735-79-1] Rasagiline Methanesulfonate, 8234
[161814-49-9] Amprenavir, 585
[161832-65-1] Talampanel, 9164
[161967-81-3] Grepafloxacin Hydrochloride, 4580
[161973-10-0] Esomeprazole Magnesium, 6939
[161982-62-3] Depreotide, 2912
[162011-90-7] Rofecoxib, 8377
[162359-55-9] Fingolimod, 4114
[162359-56-0] Fingolimod Hydrochloride, 4114
[162394-19-6] Palifermin, 7086
[162401-32-3] Roflumilast, 8378
[162610-17-5] Idraparinux, 4932
[162635-04-3] Temsirolimus, 9285
[162652-95-1] Vinflunine, 10186
[162808-62-0] Caspofungin, 1889
[163120-03-4] Nodulisporic Acid, 6756
[163133-43-5] Naproxcinod, 6498
[163222-33-1] Ezetimibe, 3958
[163252-36-6] Clevudine, 2351
[163253-35-8] Sitafloxacin Sesquihydrate, 8689
[163451-81-8] Teriflunomide, 9308
[163515-14-8] Dimethenamid P, 3236
[163515-35-3] Chlorotoxin, 2177
[163520-33-0] Isoxadifen-ethyl, 5282
[163521-08-2] Vilazodone Hydrochloride, 10176
[163521-12-8] Vilazodone, 10176
[163545-26-4] Ancestim, 499
[163680-77-1] Pazufloxacin Methanesulfonate, 7161
[163706-06-7] Cangrelor, 1749
[163706-36-3] Cangrelor Tetrasodium Salt, 1749
[164656-23-9] Dutasteride, 3518
[165101-51-9] Becaplermin, 7639
[165133-56-2] β-Cyclodextrin Sulfobutyl Ether, 2711
[165252-70-0] Dinotefuran, 3317
[165450-17-9] Neotame, 6550
[165668-41-7] Indisulam, 4986
[165800-03-3] Linezolid, 5559
[165800-06-6] Zoledronic Acid, 10388
[165800-07-7] Zoledronic Acid Disodium Salt Tetrahydrate, 10388
[165800-08-8] Zoledronic Acid Trisodium Salt Hydrate, 10388
[166191-23-7] Methoxyethylbenzeneboronic Acid (−)-S-Form, 6065
[166518-60-1] Avasimibe, 876
[166518-61-2] Avasimibe Sodium Salt, 876
[166663-25-8] Anidulafungin, 649
[166944-16-7] Saporin-6, 8503
[167221-71-8] Clevidipine, 2350
[167305-00-2] Omapatrilat, 6937
[167354-41-8] Zosuquidar, 10396
[167465-36-3] Zosuquidar Trihydrochloride, 10396
[167747-19-5] Sulesomab, 9027
[167933-07-5] Flibanserin, 4132
[168021-77-0] Disufenton, 3406
[168021-79-2] Disufenton Sodium Salt, 3406
[168079-32-1] Lixivaptan, 5606

[190977-41-4] Oblimersen Sodium, 6828
[191114-48-4] Telithromycin, 9261
[191217-81-9] Pramipexole Dihydrochloride Monohydrate, 7825
[191588-94-0] Tenecteplase, 9286
[191670-18-5] Volicitin, 10227
[191732-72-6] Lenalidomide, 5490
[192050-59-2] Ruboxistaurin Methanesulfonate, 8423
[192185-68-5] Tipifarnib (unspecified stereo), 9614
[192185-72-1] Tipifarnib, 9614
[192230-81-2] Synsorb Pk, 9144
[192314-93-5] Iclaprim, 4926
[192329-42-3] Prinomastat, 7869
[192391-48-3] Iodine I 131 Tositumomab, 9713
[192564-14-0] Oritavancin Diphosphate, 6963
[192725-17-0] Lopinavir, 5630
[192880-93-6] Benthiavalicarb-isopropyl Mixture with Mancozeb, 1055
[192939-46-1] Ximelagatran, 10274
[193153-04-7] Otamixaban, 6996
[193275-84-2] Lonafarnib, 5622
[193901-90-5] Gadofosveset Trisodium Salt, 4355
[194085-75-1] Carisbamate, 1841
[194100-83-9] Thyrotropin Alfa, 9569
[194468-36-5] Vinflunine Ditartrate, 10186
[194798-83-9] Fluconazole Dihydrogen Phosphate Ester, 4153
[194804-75-6] Garenoxacin, 4402
[194874-06-1] Apoptolidin, 734
[195532-12-8] Pradofloxacin, 7819
[195733-43-8] Atrasentan Hydrochloride, 861
[195875-84-4] Tesofensine, 9322
[195883-09-1] 14-Succinyl Triptolide Sodium Salt, 9923
[195962-23-3] Corifollitropin Alfa, 2518
[196597-26-9] Ramelteon, 8219
[196618-13-0] Oseltamivir, 6986
[196929-78-9] Ellman's Sulfinamides (R)-Form, 3607
[197922-42-2] Teduglutide, 9248
[198153-51-4] Peginterferon Alfa-2a, 7175
[198228-20-5] Viscotoxin A$_1$, 10206
[198283-73-7] Tebanicline, 9227
[198283-74-8] Tebanicline Tosylate, 9227
[198470-84-7] Parecoxib, 7140
[198470-85-8] Parecoxib Sodium Salt, 7140
[198481-32-2] Bazedoxifene, 1011
[198481-33-3] Bazedoxifene Acetate, 1011
[198821-22-6] Merimepodib, 5968
[198904-31-3] Atazanavir, 849
[199396-76-4] Asoprisnil, 824
[199463-33-7] Revaprazan, 8291
[199612-75-4] Py-Phe, 8063
[199739-10-1] Paliperidone Palmitate, 7087
[200393-05-1] S-Adenosylmethionine 1,4-Butanedisulfonate, 144
[200815-49-2] Formoterol R,R-Form L-Tartrate, 4272
[201341-05-1] Tenofovir Disoproxil, 9289

[201530-41-8] Deferasirox, 2861
[201677-61-4] Sivelestat Sodium Salt Tetrahydrate, 8696
[201677-75-0] Taltirelin Tetrahydrate, 9177
[201688-00-8] Gadofosveset, 4355
[202138-50-9] Tenofovir Disoproxil Fumarate, 9289
[202189-78-4] Bilastine, 1220
[202260-21-7] Ruboxistaurin Methanesulfonate Monohydrate, 8423
[202289-38-1] Deoxo-Fluor®, 2899
[202409-33-4] Etoricoxib, 3930
[202825-46-5] Safinamide Methanesulfonate, 8450
[203120-17-6] Laninamivir, 5406
[203120-18-7] Laninamivir Trifluoroacetate Salt, 5406
[203120-46-1] Laninamivir Octanoic Acid Ester, 5406
[203258-38-2] Brostallicin Hydrochloride, 1458
[203258-60-0] Brostallicin, 1458
[203313-25-1] Spirotetramat, 8888
[203399-38-6] Oroidin Hydrochloride, 6971
[203564-54-9] Tebanicline Hydrochloride, 9227
[203923-89-1] Cositecan, 2538
[204204-73-9] R-138727, 7833
[204248-78-2] Omiganan, 6940
[204255-11-8] Oseltamivir Phosphate, 6986
[204318-14-9] Edotreotide, 3561
[204512-90-3] Tecadenoson, 9233
[204656-20-2] Liraglutide, 5569
[204992-09-6] Nemifitide Ditriflutate, 6530
[205110-48-1] Cethromycin, 2018
[205599-45-7] Cameleons, 1729
[205599-75-3] Orexin A, 6961
[205599-76-4] Orexin B, 6961
[205923-56-4] Cetuximab, 2026
[205923-57-5] Epratuzumab, 3687
[206181-63-7] Ibritumomab Tiuxetan, 4917
[206361-99-1] Darunavir, 2829
[206873-63-4] Tariquidar, 9198
[207623-20-9] Agatolimod, 180
[207748-29-6] Insulin Glulisine, 5028
[207844-01-7] Mitiglinide Calcium Salt Dihydrate, 6296
[208265-92-3] Granulocyte Colony-Stimulating Factor Pegfilgrastim, 4573
[208465-21-8] Mesosulfuron-methyl, 5981
[208517-65-1] IACFT ^{123}I, 4910
[208538-73-2] Micafungin Sodium Salt, 6254
[208921-02-2] Tositumomab, 9713
[209216-23-9] Entecavir Monohydrate, 3649
[209342-40-5] Finafloxacin, 4112
[209342-41-6] Finafloxacin Hydrochloride, 4112
[209467-52-7] Ceftobiprole, 1953
[209810-58-2] Darbepoetin Alfa, 2825
[209860-87-7] Tafluprost, 9160
[210101-16-9] Conivaptan, 2491
[210353-53-0] Gemifloxacin Methanesulfonate, 4423
[210421-74-2] Sitaxsentan Sodium Salt, 8692

[210880-92-5] Clothianidin, 2398
[211050-90-7] Diflufenzopyr Mixture with Dicamba, 3164
[211427-08-6] Diquafosol Tetrasodium Salt, 3391
[211439-12-2] Davunetide, 2836
[211448-85-0] Denufosol, 2898
[211513-37-0] Dalcetrapib, 2801
[211914-51-1] Dabigatran, 2794
[211915-06-9] Dabigatran Etexilate, 2794
[212141-54-3] Vatalanib, 10130
[212142-18-2] Vatalanib Succinate, 10130
[213327-37-8] Oregovomab, 6960
[213464-77-8] Orthosulfamuron, 6981
[213697-53-1] DavePhos Phosphine Ligand, 1469
[213819-48-8] Belotecan Hydrochloride, 1028
[214745-43-4] Efalizumab, 3567
[214766-78-6] Degarelix, 2866
[215647-85-1] Peginterferon Alfa-2b, 7176
[216167-82-7] Succinobucol, 9002
[216503-57-0] Alemtuzumab, 222
[216586-46-8] Virulizin®, 10203
[216974-75-3] Bevacizumab, 1194
[217500-96-4] Tulathromycin A, 9990
[217797-14-3] Paroxetine Methanesulfonate, 7148
[218600-44-3] Bardoxolone, 959
[218600-53-4] Bardoxolone Methyl Ester, 959
[218620-50-9] Pegvisomant, 7178
[218791-21-0] Imisopasem Manganese, 4965
[218949-48-5] Tesamorelin, 9320
[219685-50-4] Eculizumab, 3553
[219685-93-5] Pexelizumab, 7307
[219714-96-2] Penoxsulam, 7212
[219861-08-2] Escitalopram Oxalate, 2317
[219897-32-2] Betti Base Hydrochloride, 1190
[219897-35-5] Betti Base (R)-Form, 1190
[219897-38-8] Betti Base (S)-Form, 1190
[219989-84-1] Ixabepilone, 5297
[220119-17-5] Selamectin, 8561
[220127-57-1] Imatinib Methanesulfonate, 4943
[220201-34-3] Talaporfin Tetrasodium Salt, 9166
[220578-59-6] Gemtuzumab Ozogamicin, 4424
[220620-09-7] Tigecycline, 9588
[220810-26-4] Histrelin Acetate, 4760
[220899-03-6] Metrafenone, 6230
[220991-20-8] Lumiracoxib, 5661
[221012-82-4] (R)-P-Phos, 7817
[221054-79-1] Isoequol, 3701
[221373-18-8] Olanzapine Pamoate Monohydrate, 6914
[222030-63-9] Fosbretabulin, 4277
[222535-22-0] Alefacept, 220
[222716-86-1] Pegaptanib Sodium, 7174
[223460-79-5] GLP-2 (human), 4482
[223577-45-5] Aviscumine, 6295
[223652-82-2] Garenoxacin Methanesulfonate, 4402
[223652-90-2] Garenoxacin Methanesulfonate (monohydrate), 4402

[835876-32-9] Platensimycin, *7640*
[838853-48-8] Mifamurtide Sodium Salt (hydrate), *6265*
[842131-33-3] Ulimorelin, *10029*
[843663-66-1] Bedaquiline, *1018*
[845264-92-8] Atacicept, *848*
[845273-93-0] Sevelamer Carbonate, *8613*
[845533-86-0] Bedaquiline Fumarate, *1018*
[846589-98-8] Lorcaserin Hydrochloride, *5638*
[847353-30-4] Arbaclofen Placarbil, *758*
[849479-74-9] Vapreotide Acetate, *10123*
[850140-72-6] Afatinib, *169*
[850649-61-5] Alogliptin, *302*
[850649-62-6] Alogliptin Benzoate, *302*
[851199-59-2] Linaclotide, *5548*
[851199-60-5] Linaclotide Acetate, *5548*
[856676-23-8] Fenofibric Acid Choline Salt, *4009*
[856866-72-3] Tedizolid, *9247*
[856867-39-5] Tedizolid Phosphate Disodium Salt, *9247*
[856867-55-5] Tedizolid Phosphate, *9247*
[857402-23-4] Retaspimycin, *8282*
[857402-63-2] Retaspimycin Hydrochloride, *8282*
[857876-30-3] Motesanib Diphosphate, *6366*
[857890-39-2] Lenvatinib Methanesulfonate, *5494*
[858954-83-3] Aminocyclopyrachlor Methyl Ester, *432*
[858956-08-8] Aminocyclopyrachlor, *432*
[858956-35-1] Aminocyclopyrachlor Potassium Salt, *432*
[861205-83-6] Tesofensine Citrate, *9322*
[862111-32-8] Aflibercept, *174*
[863031-21-4] Azilsartan Azilsartan Medoxomil, *903*
[863031-24-7] Azilsartan Azilsartan Kamedoxomil, *903*
[863126-95-8] Salinosporamide B, *8474*
[863126-96-9] Salinosporamide C, *8474*
[868540-17-4] Carfilzomib, *1837*
[869572-92-9] Tecovirimat, *9244*
[871038-72-1] Raltegravir Monopotassium Salt, *8216*

[871224-64-5] Almorexant, *297*
[873054-44-5] Ivacaftor, *5295*
[873857-62-6] Fidaxomicin, *4108*
[874819-74-6] Toceranib Phosphate, *9650*
[875148-45-1] Regadenoson Monohydrate, *8247*
[875446-37-0] Anacetrapib, *615*
[875455-82-6] Rusalatide Acetate (3:2), *8431*
[875756-97-1] Bisbenzimide Ethoxide Trihydrochloride, *1250*
[876657-17-9] Ecgonine (±)-Form, *3540*
[877399-52-5] Crizotinib, *2577*
[879085-55-9] Vismodegib, *10207*
[882737-42-0] Pridopidine Hydrochloride, *7860*
[882976-95-6] Cortistatin A, *2526*
[884595-19-1] *N*-Acetyl-α-neuraminic Acid, *8623*
[884604-91-5] Velaglucerase Alfa, *10135*
[885051-90-1] Pegloticase, *7177*
[887144-94-7] Togni's Reagents Carbonyl Compound, *9663*
[887144-97-0] Togni's Reagents Dimethyl Compound, *9663*
[894356-79-7] Trebananib, *9741*
[901119-35-5] Fostamatinib, *4287*
[901758-09-6] Tesamorelin Acetate, *9320*
[903508-18-9] Jervine Hydrochloride, *5312*
[905818-69-1] Bupropion Hydrobromide, *1503*
[905854-02-6] Tivantinib, *9640*
[906748-38-7] Indirubin, *4985*
[906805-06-9] Necitumumab, *6520*
[909260-86-2] Dipipanone Hydrobromide, *3377*
[914088-09-8] Brentuximab Vedotin, *1375*
[914295-16-2] Fostamatinib Disodium Salt (hexahydrate), *4287*
[914613-48-2] Canakinumab, *1739*
[915713-02-9] Elisidepsin Trifluoroacetate, *3601*
[915942-22-2] Neratinib Maleate, *6557*
[918504-65-1] Vemurafenib, *10138*
[919364-56-0] Isofagomine Tartrate, *5217*
[923564-51-6] Navitoclax, *6512*
[923590-37-8] Vaniprevir, *10122*
[925681-61-4] Trabedersen, *9723*
[925705-41-5] Arsenicin A, *786*
[928659-70-5] Eliglustat Tartrate, *3599*

[934246-14-7] Degarelix Acetate Hydrate, *2866*
[936111-69-2] Ceftolozane Sulfate, *1954*
[936500-94-6] Elinogrel, *3600*
[936501-01-8] Elinogrel Potassium Salt, *3600*
[936631-70-8] Afatinib Dimaleate, *169*
[941678-49-5] Ruxolitinib, *8440*
[944804-62-0] Cortistatin J, *2526*
[945667-22-1] Saxagliptin Monohydrate, *8525*
[951326-02-6] Ulimorelin Hydrochloride Monohydrate, *10029*
[953077-35-5] Telcagepant Potassium Salt Monoethanolate, *9259*
[956103-76-7] Florbetapir F 18, *4140*
[957136-80-0] 2,2,3-Trimethylcyclopentanebutanoic Acid, *9883*
[960404-48-2] Dapagliflozin Compound with (2*S*)-1,2-Propanediol Hydrate, *2816*
[1000120-98-8] Mipomersen, *6288*
[1005808-93-4] Taliglucerase Alfa, *9169*
[1025687-58-4] Fostamatinib Disodium Salt, *4287*
[1026016-83-0] Tetrabenazine *R,R*-Form, *9326*
[1041434-82-5] Peramivir Trihydrate, *7264*
[1046050-73-0] Laropiprant Combination with Nicotinic Acid, *5426*
[1048007-93-7] Masitinib Methanesulfonate, *5822*
[1051375-16-6] Dolutegravir, *3460*
[1051375-19-9] Dolutegravir Sodium Salt, *3460*
[1070663-78-3] BrettPhos Phosphine Ligand, *1469*
[1071557-77-1] Morphinan Hydrochloride, *6360*
[1092799-01-3] MacMillan's Enamine Catalyst Hydrochloride, *5707*
[1092939-17-7] Ruxolitinib Phosphate, *8440*
[1162664-19-8] Tecovirimat Monohydrate, *9244*
[1186486-62-3] Evacetrapib, *3947*
[1192491-61-4] Avibactam Sodium Salt, *878*
[1192500-31-4] Avibactam, *878*
[1227300-83-5] Estramustine 17-(Dihydrogenphosphate) Disodium Salt Monohydrate, *3761*
[1269195-08-5] Vaniprevir Potassium Salt, *10122*

THERAPEUTIC CATEGORY AND
BIOLOGICAL ACTIVITY
INDEX

THERAPEUTIC CATEGORY AND BIOLOGICAL ACTIVITY INDEX

This index is designed to serve as an additional entry point to the monographs of this edition for which a therapeutic category (THERAP CAT) has been given. It contains listings for discrete chemical entities. Plants and plant portions, *e.g.* seeds or leaves, although used medicinally, have generally been excluded. An attempt has been made to standardize the nomenclature used for index terms. Therefore, the heading under which a compound is listed may differ slightly from the therapeutic category found within the monograph.

Biological activities indicating mechanism of action have also been included as index headings. Whenever possible, cross references to appropriate therapeutic categories have been given for these listings. Cross references have also been provided for synonyms of preferred terms or, in some cases, for closely related entries. In addition, many therapeutic category headings have been sub-classified according to chemical or pharmacological features.

Monographs are listed alphabetically by title under each heading. Selected derivatives and isomers of title compounds have been listed by generic or trivial name and are referenced to the appropriate monograph by number.

Therapeutic categories have been assigned to reflect the major indications listed in manufacturer's product information, reported in the clinical literature, or published by the USAN Council. Inclusion of a drug in this index does not imply efficacy or endorsement, nor does it imply that the compound is *presently* being marketed. Omission of a compound from a particular category does not imply that the substance is not in use or under investigation for that indication. Additional information on each entry can be found in the cited monograph. Reference to primary sources is encouraged.

ABORTIFACIENT/INTERCEPTIVE

Gemeprost, *4421*
Mifepristone, *6266*
Prostaglandin E$_2$, *7988*
Prostaglandin F$_{2\alpha}$, *7989*
Sulprostone, *9120*
Trichosanthin, *9813*

ACE-INHIBITOR *see also Antihypertensive; Vasopeptidase Inhibitor*

Alacepril, *193*
Benazepril, *1033*
Captopril, *1776*
Ceronapril, *1993*
Cilazapril, *2275*
Delapril, *2881*
Enalapril, *3623*
Fosinopril, *4283*
Imidapril, *4952*
Lisinopril, *5572*
Moexipril, *6315*
Moveltipril, *6371*
Perindopril, *7285*
Quinapril, *8167*
Ramipril, *8221*
Spirapril, *8880*
Temocapril, *9279*
Trandolapril, *9729*
Zofenopril, *10386*

ACIDIFIER

Ammonium Chloride, *506*
Betaine Hydrochloride *see 1181*
Glutamic Acid Hydrochloride *see 4505*
Methionine, *6047*

ADENOSINE RECEPTOR AGONIST

Binodenoson, *1229*
Regadenoson, *8247*
Tecadenoson, *9233*

α-ADRENERGIC AGONIST *see also Antihypotensive; Antihypertensive; Antiulcerative; Decongestant; Mydriatic*

Adrafinil, *159*
Adrenalone, *160*
Amidephrine, *393*
Apraclonidine, *738*
Clonidine, *2380*
Cyclopentamine, *2735*
Dexmedetomidine, *2948*
Dipivefrin, *3378*
Ephedrine, *3663*
Epinephrine, *3674*
Fenoxazoline, *4016*
Guanabenz, *4593*
Guanfacine, *4596*
Ibopamine, *4915*
Indanazoline, *4972*
Isometheptene, *5231*
Mephentermine, *5919*
Metaraminol, *6003*
Methoxamine, *6058*
Methylhexaneamine, *6157*
Midodrine, *6263*
Mivazerol, *6306*

Modafinil, *6314*
Moxonidine, *6379*
Naphazoline, *6453*
Norepinephrine, *6784*
Norfenefrine, *6788*
Octodrine, *6846*
Octopamine, *6848*
Oxymetazoline, *7065*
Phenylephrine, *7398*
Phenylpropanolamine, *7417*
Phenylpropylmethylamine, *7418*
Pholedrine, *7442*
Propylhexedrine, *7972*
Pseudoephedrine, *8024*
Rilmenidine, *8347*
Synephrine, *9142*
Talipexole, *9171*
Tetrahydrozoline, *9364*
Tiamenidine, *9573*
Tramazoline, *9727*
Tuaminoheptane, *9982*
Tymazoline, *10015*
Tyramine, *10016*
Xylometazoline, *10284*

β-ADRENERGIC AGONIST *see also Bronchodilator; Cardiotonic; Tocolytic*

Albuterol, *210*
Bambuterol, *946*
Bitolterol, *1308*
Carbuterol, *1832*
Clenbuterol, *2345*
Denopamine, *2896*
Dioxethedrine, *3331*
Dobutamine, *3440*
Dopexamine, *3472*
Ephedrine, *3663*
Epinephrine, *3674*
Etafedrine, *3764*
Ethylnorepinephrine, *3888*
Fenoterol, *4012*
Formoterol, *4272*
Hexoprenaline, *4743*
Ibopamine, *4915*
Indacaterol, *4970*
Isoetharine, *5215*
Isoproterenol, *5263*
Mabuterol, *5703*
Metaproterenol, *6002*
Methoxyphenamine, *6071*
Oxyfedrine, *7061*
Pirbuterol, *7602*
Prenalterol, *7853*
Procaterol, *7877*
Protokylol, *8004*
Reproterol, *8259*
Rimiterol, *8353*
Ritodrine, *8365*
Salmeterol, *8475*
Soterenol, *8855*
Terbutaline, *9302*
Tretoquinol, *9749*
Tulobuterol, *9991*
Xamoterol, *10253*

α-ADRENERGIC BLOCKER *see also Antihypertensive*

Alfuzosin, *231*
Amosulalol, *572*
Arotinolol, *782*
Dapiprazole, *2820*

Doxazosin, *3482*
Ergoloid Mesylates, *3710*
Fenspiride, *4026*
Indoramin, *5010*
Labetalol, *5377*
Naftopidil, *6440*
Nicergoline, *6581*
Prazosin, *7838*
Silodosin, *8642*
Tamsulosin, *9181*
Terazosin, *9297*
Tolazoline, *9666*
Yohimbine, *10300*

β-ADRENERGIC BLOCKER *see also Antianginal; Antiarrhythmic; Antiglaucoma; Antihypertensive*

Acebutolol, *20*
Alprenolol, *309*
Amosulalol, *572*
Arotinolol, *782*
Atenolol, *850*
Befunolol, *1021*
Betaxolol, *1184*
Bevantolol, *1195*
Bisoprolol, *1295*
Bopindolol, *1335*
Bucindolol, *1471*
Bufuralol, *1485*
Bunitrolol, *1493*
Bupranolol, *1500*
Carazolol, *1780*
Carteolol, *1866*
Carvedilol, *1873*
Celiprolol, *1962*
Esmolol, *3755*
Indenolol, *4979*
Labetalol, *5377*
Landiolol, *5405*
Levobunolol, *5514*
Mepindolol, *5923*
Metipranolol, *6216*
Metoprolol, *6228*
Nadolol, *6431*
Nebivolol, *6516*
Nipradilol, *6648*
Oxprenolol, *7051*
Penbutolol, *7191*
Pindolol, *7554*
Practolol, *7818*
Propranolol, *7953*
Sotalol, *8854*
Talinolol, *9170*
Tertatolol, *9316*
Tilisolol, *9595*
Timolol, *9601*

ADRENOCORTICAL STEROID *see Glucocorticoid; Mineralocorticoid*

ADRENOCORTICAL SUPPRESSANT

Aminoglutethimide, *437*
Trilostane, *9867*

ADRENOCORTICOTROPIC HORMONE

ACTH, *125*
Giractide, *4463*

ALCOHOL DEPENDENCE TREATMENT

Acamprosate Calcium, *18*
Calcium Cyanamide Citrated, *1665*
Disulfiram, *3407*
Naltrexone, *6448*

ALDOSE REDUCTASE INHIBITOR

Epalrestat, *3660*
Fidarestat, *4107*
Tolrestat, *9683*
Zopolrestat, *10394*

ALDOSTERONE ANTAGONIST *see also* **Diuretic**

Canrenone, *1753*
Eplerenone, *3681*
Spironolactone, *8887*

ALKALINIZING AGENT

Potassium Citrate, *7744*
Sodium Bicarbonate, *8719*
Sodium Citrate, *8737*
Sodium Lactate, *8767*
Tromethamine, *9951*

5-ALPHA REDUCTASE INHIBITOR *see 5α-Reductase Inhibitor*

AMPA RECEPTOR ANTAGONIST

Talampanel, *9164*

ANABOLIC

Bolandiol, *1325*
Clostebol, *2397*
Ethylestrenol, *3860*
Formebolone, *4267*
Methandriol, *6022*
Methenolone, *6039*
Nandrolone, *6450*
Oxymesterone, *7064*
Pizotyline, *7629*

ANALEPTIC *see CNS Stimulant*

ANALGESIC *see also* **Anti-inflammatory (Nonsteroidal)**

Non-opioids

Acetic/Propionic Acid Derivatives
Aceclofenac, *23*
Etodolac, *3920*
Felbinac, *3985*
Fenoprofen, *4011*
Flurbiprofen, *4229*
Ibuprofen, *4919*
Indomethacin, *5009*
Ketoprofen, *5352*
Loxoprofen, *5647*
Mofezolac, *6318*
Naproxcinod, *6498*
Naproxen, *6499*
Zaltoprofen, *10310*

Anilides
Acetaminophen, *47*
Bucetin, *1467*
Isonixin, *5238*
Phenacetin, *7319*
Propacetamol, *7906*

Anthranilic Acid Derivatives
Floctafenine, *4135*
Flufenamic Acid, *4163*
Tolfenamic Acid, *9673*

Pyrazolones
Aminopyrine, *470*
Antipyrine, *703*
Dipyrone, *3390*
Propyphenazone, *7982*
Ramifenazone, *8220*

Salicylic Acid Derivatives
Aloxiprin, *305*
Ammonium Salicylate, *547*
Aspirin, *841*
Benorylate, *1047*
Calcium Acetylsalicylate, *1650*
Choline Salicylate, *2213*
Diflunisal, *3165*
Ethenzamide, *3786*
Fosfosal, *4282*
Gentisic Acid, *4433*
Imidazole Salicylate, *4954*
Lysine Acetylsalicylate, *5698*
Magnesium Acetylsalicylate, *5719*
Morpholine Salicylate *see 6362*
Salicin, *8460*
Salicylamide, *8465*
Salicylamide *O*-Acetic Acid, *8466*
Salsalate, *8477*
Sodium Salicylate *see 8469*
Sodium Salicylsulfate *see 8471*

Others
Benzydamine, *1125*
Cizolirtine, *2335*
Clonixin, *2381*
Emorfazone, *3619*
Flupirtine, *4221*
Gabapentin, *4348*
Ketorolac, *5353*
Lacosamide, *5381*
Lornoxicam, *5640*
Nefopam, *6525*
Parecoxib, *7140*
Phenazopyridine, *7324*
Phenyramidol, *7432*
Pregabalin, *7843*
Ziconotide, *10321*

Opioids

Benzomorphan Derivatives
Pentazocine, *7233*
Phenazocine, *7332*

Diphenylpropylamine Derivatives
Bezitramide, *1200*
Dextromoramide, *2958*
Dipipanone, *3377*
Methadone, *6016*
Piritramide, *7611*
Propoxyphene, *7952*

Morphinan Derivatives
Butorphanol, *1532*
Levorphanol, *5524*

Oripavine Derivatives
Buprenorphine, *1501*

Phenanthrenes
Codeine, *2448*
Diacetylmorphine, *2971*
Dihydrocodeine, *3189*
Hydrocodone, *4823*
Hydromorphone, *4840*
Morphine, *6361*
Nalbuphine, *6441*
Nicomorphine, *6605*
Opium, *6949*
Oxycodone, *7058*
Oxymorphone, *7068*
Papaveretum, *7121*

Phenylpiperidines
Alfentanil, *229*
Alphaprodine, *306*
Anileridine, *651*
Fentanyl, *4028*
Ketobemidone, *5348*
Meperidine, *5916*
Phenoperidine, *7362*
Promedol, *7899*
Remifentanil, *8251*
Sufentanil, *9018*

Others
Dezocine, *2959*
Eptazocine, *3696*
Meptazinol, *5930*
Tapentadol, *9189*
Tilidine, *9594*
Tramadol, *9726*

ANALGESIC (DENTAL)

Chlorobutanol, *2129*
Clove Oil *see 2402*
Eugenol, *3940*

ANDROGEN

Dehydroepiandrosterone, *2875*
Fluoxymesterone, *4218*
Mesterolone, *5986*
Methandrostenolone, *6023*
17-Methyltestosterone, *6197*
Norethandrolone, *6785*
Normethandrone, *6799*
Oxandrolone, *7020*
Oxymesterone, *7064*
Oxymetholone, *7066*
Stanolone, *8920*
Stanozolol, *8921*
Testosterone, *9324*

ANESTHETIC (INHALATION)

Cyclopropane, *2744*
Desflurane, *2920*
Divinyl Ether, *3427*
Enflurane, *3635*
Ethylene, *3845*
Ethyl Ether, *3861*
Halothane, *4637*
Isoflurane, *5221*
Methoxyflurane, *6067*
Methyl Propyl Ether, *6186*
Nitrous Oxide, *6740*
Sevoflurane, *8614*
Teflurane, *9250*

Trichloroethylene, *9804*
Xenon, *10270*

ANESTHETIC (INTRAVENOUS)

Fospropofol, *4286*
Hexobarbital, *4740*
Ketamine, *5343*
Methohexital, *6053*
Midazolam, *6261*
Phencyclidine, *7333*
Propanidid, *7917*
Propofol, *7947*
Sodium Oxybate *see 4854*
Thialbarbital, *9444*
Thiamylal, *9454*
Thiobutabarbital, *9476*
Thiopental Sodium, *9503*

ANESTHETIC (LOCAL)

Amylocaine, *607*
Benoxinate, *1049*
Benzocaine, *1088*
Betoxycaine, *1189*
Bupivacaine, *1499*
Butacaine, *1510*
Butamben, *1515*
Butethamine, *1526*
Butoxycaine, *1533*
Carticaine, *1869*
Chloroprocaine, *2160*
Cocaethylene, *2439*
Cocaine, *2440*
Dibucaine, *3036*
Dimethocaine, *3242*
Dixyrazine, *3429*
Dyclonine, *3520*
Ethyl Chloride, *3837*
Etidocaine, *3911*
β-Eucaine, *3937*
Fomocaine, *4257*
Hexylcaine Hydrochloride, *4744*
Hydroxytetracaine, *4885*
Isobutyl *p*-Aminobenzoate, *5178*
Lidocaine, *5535*
Mepivacaine, *5926*
Octacaine, *6832*
Orthocaine, *6979*
Oxethazaine, *7032*
Phenacaine Hydrochloride, *7317*
Phenol, *7354*
Piperocaine, *7587*
Pramoxine, *7829*
Prilocaine, *7862*
Procaine, *7875*
Proparacaine, *7921*
Propoxycaine Hydrochloride, *7950*
Pseudococaine, *8020*
Pyrrocaine, *8127*
Ropivacaine, *8389*
Salicyl Alcohol, *8461*
Tetracaine Hydrochloride, *9333*
Tolycaine, *9700*
Trimecaine, *9869*

ANGIOTENSIN CONVERTING ENZYME INHIBITOR *see ACE-Inhibitor*

ANGIOTENSIN II RECEPTOR ANTAGONIST *see also Antihypertensive*

Azilsartan, *903*
Candesartan, *1742*

Eprosartan, *3691*
Irbesartan, *5127*
Losartan, *5641*
Olmesartan, *6933*
Telmisartan, *9271*
Valsartan, *10102*

ANOREXIC

Aminorex, *472*
Amphetamine, *579*
Benzphetamine, *1121*
Chlorphentermine, *2188*
Clobenzorex, *2357*
Dextroamphetamine, *2956*
Diethylpropion, *3145*
N-Ethylamphetamine, *3818*
Fenfluramine, *4004*
Fenproporex, *4022*
Levophacetoperane, *5521*
Mazindol, *5831*
Mefenorex, *5870*
Methamphetamine, *6020*
Norpseudoephedrine, *6803*
Pentorex, *7246*
Phendimetrazine, *7334*
Phenmetrazine, *7351*
Phentermine, *7374*
Phenylpropanolamine, *7417*
Sibutramine, *8624*

ANTACID

Almagate, *292*
Aluminum Hydroxide, *338*
Aluminum Magnesium Silicate, *345*
Aluminum Phosphate, *352*
Azulene, *919*
Basic Aluminum Carbonate Gel, *1001*
Bismuth Phosphate, *1276*
Bismuth Subcarbonate, *1282*
Bismuth Subnitrate, *1284*
Calcium Carbonate, *1659*
Dihydroxyaluminum Aminoacetate, *3201*
Dihydroxyaluminum Sodium Carbonate, *3202*
Magaldrate, *5714*
Magnesium Carbonate Hydroxide, *5724*
Magnesium Hydroxide, *5734*
Magnesium Oxide, *5741*
Magnesium Peroxide, *5744*
Magnesium Phosphate, Tribasic, *5747*
Magnesium Silicates, *5751*
Sodium Bicarbonate, *8719*

ANTHELMINTIC (CESTODES)

Albendazole, *203*
Aspidin, *836*
Aspidinol, *837*
Dichlorophen, *3081*
Kosins, *5370*
Naphthalene, *6455*
Niclosamide, *6604*
Nitazoxanide, *6653*
Pelletierine, *7181*
Quinacrine, *8160*

ANTHELMINTIC (NEMATODES)

Amocarzine, *568*
Amoscanate, *571*

Ascaridole, *816*
Bephenium Hydroxynaphthoate, *1150*
Bitoscanate, *1309*
Carbon Tetrachloride, *1816*
Carvacrol, *1872*
Cyclobendazole, *2705*
Diethylcarbamazine, *3128*
Dithiazanine Iodide, *3412*
Dymanthine, *3522*
Gentian Violet, *4430*
4-Hexylresorcinol, *4748*
Ivermectin, *5296*
Kainic Acid, *5324*
Levamisole, *5511*
Mebendazole, *5837*
2-Naphthol, *6469*
Oxantel, *7021*
Papain, *7119*
Piperazine, *7576*
Piperazine Adipate, *7577*
Pyrantel, *8066*
Pyrvinium Pamoate, *8136*
α-Santonin, *8498*
Suramin Sodium, *9135*
Tetrachloroethylene, *9335*
Thiabendazole, *9440*
Thymol, *9554*
Thymyl *N*-Isoamylcarbamate, *9564*
Urea Stibamine, *10056*

ANTHELMINTIC (SCHISTOSOMA)

Amoscanate, *571*
Antimony Potassium Tartrate, *691*
Antimony Sodium Tartrate, *693*
Hycanthone, *4798*
Lucanthone, *5650*
Niridazole, *6649*
Oxamniquine, *7018*
Praziquantel, *7837*
Sodium Antimonylgluconate *see 692*
Stibocaptate, *8941*
Stibophen, *8942*
Trichlorfon, *9788*
Urea Stibamine, *10056*

ANTHELMINTIC (TREMATODES)

Tetrachloroethylene, *9335*

ANTIACNE *see also Keratolytic*

Adapalene, *138*
Azelaic Acid, *898*
Benzoyl Peroxide, *1119*
Cyproterone, *2771*
Isotretinoin, *5274*
Motretinide, *6370*
Resorcinol, *8276*
Retinoic Acid, *8288*
Tazarotene, *9218*
Tioxolone, *9612*

ANTIALLERGIC *see also Antihistaminic; Decongestant; Glucocorticoid*

Amlexanox, *486*
Cromolyn, *2581*
Ibudilast, *4918*
Ketotifen, *5354*
Lodoxamide, *5613*

ANTIALLERGIC (continued)

Montelukast, 6346
Nedocromil, 6521
Omalizumab, 6936
Oxatomide, 7024
Pemirolast, 7187
Pentigetide, 7241
Picumast, 7530
Ramatroban, 8218
Repirinast, 8258
Suplatast Tosylate, 9132
Tazanolast, 9217
Tranilast, 9731
Traxanox, 9739

ANTIALLERGIC (HYPOSENSITI-ZATION THERAPY)

Histamine, 4756
Poison Ivy Extract see 7666
Poison Oak Extract see 7667
Poison Sumac Extract see 7668
Urushiol, 10076

ANTIALLERGIC (STEROIDAL, NASAL) see also Glucocorticoid

Beclomethasone, 1017
Dexamethasone, 2945
Flunisolide, 4176
Fluticasone Propionate, 4240
Triamcinolone Acetonide, 9758

ANTIALOPECIA AGENT

Diphencyprone, 3338
Finasteride, 4113
Minoxidil, 6285
Squaric Acid Dibutylester see 8897

ANTIAMEBIC

Arsthinol, 802
Cephaeline, 1973
Chloroquine, 2165
Chlortetracycline, 2198
Dehydroemetine, 2874
Dibromopropamidine, 3029
Diloxanide, 3223
Emetine, 3616
Fumagillin, 4315
8-Hydroxy-7-iodo-5-quinoline-sulfonic Acid, 4865
Iodochlorhydroxyquin, 5075
Iodoquinol, 5085
Paromomycin, 7146
Phanquinone, 7312
Propamidine, 7911
Quinfamide, 8173
Secnidazole, 8556
Sulfarside, 9071
Teclozan, 9242
Tetracycline, 9341
Tinidazole, 9604

ANTIAMYLOIDOGENIC

Bapineuzumab, 951
Semagacestat, 8580
Tarenflurbil, 9196
Tramiprosate, 9728

ANTIANDROGEN see also Antiacne; Antialopecia Agent; Antineoplastic (Hormonal); Antiprostatic Hypertrophy

Bicalutamide, 1204
Cyproterone, 2771
Flutamide, 4238
Nilutamide, 6629
Osaterone, 6985
Oxendolone, 7030

ANTIANEMIC see Hematinic; Hematopoietic

ANTIANGINAL see also Vasodilator (Coronary)

Acebutolol, 20
Alprenolol, 309
Amlodipine, 487
Arotinolol, 782
Atenolol, 850
Barnidipine, 998
Bepridil, 1152
Bevantolol, 1195
Bufuralol, 1485
Bunitrolol, 1493
Bupranolol, 1500
Carazolol, 1780
Carteolol, 1866
Celiprolol, 1962
Diltiazem, 3224
Felodipine, 3987
Gallopamil, 4386
Indenolol, 4979
Isosorbide Dinitrate, 5271
Isradipine, 5287
Ivabradine, 5294
Limaprost, 5543
Mepindolol, 5923
Metoprolol, 6228
Molsidomine, 6320
Nadolol, 6431
Nicardipine, 6580
Nicorandil, 6606
Nifedipine, 6613
Nilvadipine, 6630
Nipradilol, 6648
Nisoldipine, 6651
Nitroglycerin, 6694
Oxprenolol, 7051
Oxyfedrine, 7061
Ozagrel, 7079
Penbutolol, 7191
Pentaerythritol Tetranitrate, 7222
Pindolol, 7554
Propranolol, 7953
Ranolazine, 8229
Semotiadil, 8585
Sotalol, 8854
Timolol, 9601
Verapamil, 10144
Zatebradine, 10313

ANTIARRHYTHMIC

Acebutolol, 20
Acecainide, 21
Adenosine, 141
Ajmaline, 187
Alprenolol, 309
Amiodarone, 478
Aprindine, 743
Arotinolol, 782

Atenolol, 850
Azimilide, 904
Bevantolol, 1195
Bidisomide, 1208
Bretylium Tosylate, 1377
Bunitrolol, 1493
Bupranolol, 1500
Carazolol, 1780
Carteolol, 1866
Cifenline, 2272
Disopyramide, 3402
Dofetilide, 3456
Dronedarone, 3498
Encainide, 3625
Esmolol, 3755
Flecainide, 4130
Hydroquinidine, 4843
Ibutilide, 4921
Indecainide, 4977
Indenolol, 4979
Ipratropium Bromide, 5117
Landiolol, 5405
Lidocaine, 5535
Lorajmine, 5634
Lorcainide, 5637
Meobentine, 5911
Mexiletine, 6248
Moricizine, 6354
Nifekalant, 6614
Oxprenolol, 7051
Penbutolol, 7191
Pentisomide, 7242
Pilsicainide, 7539
Pindolol, 7554
Pirmenol, 7614
Practolol, 7818
Prajmaline, 7820
Procainamide Hydrochloride, 7874
Propafenone, 7908
Propranolol, 7953
Pyrinoline, 8102
Quinidine, 8176
Sematilide, 8581
Sotalol, 8854
Talinolol, 9170
Tecadenoson, 9233
Tedisamil, 9246
Tilisolol, 9595
Timolol, 9601
Tocainide, 9648
Verapamil, 10144
Vernakalant, 10157

ANTIARTHRITIC/ANTIRHEU-MATIC see also Anti-inflammatory (Nonsteroidal); Glucocorticoid

Actarit, 124
Auranofin, 869
Aurothioglucose, 874
Azathioprine, 895
Bucillamine, 1470
Chloroquine, 2165
Cuproxoline, 2660
Diacerein, 2963
Fostamatinib, 4287
Glucosamine, 4494
Gold Sodium Thiomalate, 4553
Gold Sodium Thiosulfate, 4554
Hydroxychloroquine, 4857
Leflunomide, 5486
Lobenzarit, 5610
Melittin, 5892
Methotrexate, 6057
Penicillamine, 7197

ANTIASTHMATIC (NONBRON-CHODILATOR) see also Bronchodilator; Glucocorticoid

Altrakincept see 5046
Amlexanox, 486
Cilomilast, 2279
Cromolyn, 2581
Domitroban, 3465
Ibudilast, 4918
Israpafant, 5288
Ketotifen, 5354
Montelukast, 6346
Nedocromil, 6521
Omalizumab, 6936
Oxatomide, 7024
Pranlukast, 7830
Ramatroban, 8218
Roflumilast, 8378
Seratrodast, 8598
Suplatast Tosylate, 9132
Tiaramide, 9579
Traxanox, 9739
Zafirlukast, 10306
Zileuton, 10323

ANTIASTHMATIC (STEROIDAL, INHALANT) see also Glucocorticoid

Beclomethasone, 1017
Budesonide, 1476
Ciclesonide, 2266
Dexamethasone, 2945
Flunisolide, 4176
Fluticasone Propionate, 4240
Triamcinolone Acetonide, 9758

ANTIATHEROSCLEROTIC see also Antilipemic

Alagebrium Chloride, 196
Anacetrapib, 615
Dalcetrapib, 2801
Darapladib, 2824
Motexafin Lutetium, 6368
Pyridinol Carbamate, 8088
Succinobucol, 9002
Torcetrapib, 9708
Varespladib, 10126

ANTIBACTERIAL (ANTIBIOTICS)
Aminoglycosides
Amikacin, 400
Arbekacin, 760
Bambermycins, 945
Butirosin, 1530
Dibekacin, 3008
Dihydrostreptomycin, 3197
Fortimicins, 4275
Gentamicin, 4427
Isepamicin, 5151
Kanamycin, 5329
Micronomicin, 6260
Neomycin, 6539
Netilmicin, 6564
Paromomycin, 7146
Ribostamycin, 8330
Sisomicin, 8688
Spectinomycin, 8866
Streptomycin, 8956
Tobramycin, 9647

Amphenicols
Chloramphenicol, 2077
Thiamphenicol, 9453

Ansamycins
Rifamide, 8340
Rifampin, 8341
Rifamycin SV, 8343
Rifapentine, 8344
Rifaximin, 8345

β-Lactams
Carbacephems
Loracarbef, 5633

Carbapenems
Biapenem, 1201
Doripenem, 3474
Ertapenem, 3731
Imipenem, 4961
Meropenem, 5970
Panipenem, 7111

Cephalosporins
Cefaclor, 1914
Cefadroxil, 1915
Cefamandole, 1916
Cefatrizine, 1917
Cefazolin, 1918
Cefcapene, 1920
Cefdinir, 1921
Cefditoren, 1922
Cefepime, 1923
Cefetamet, 1924
Cefixime, 1925
Cefmenoxime, 1926
Cefodizime, 1929
Cefonicid, 1930
Cefoperazone, 1931
Ceforanide, 1932
Cefoselis, 1933
Cefotaxime, 1934
Cefotiam, 1936
Cefozopran, 1939
Cefpiramide, 1940
Cefpirome, 1941
Cefpodoxime, 1942
Cefprozil, 1943
Cefroxadine, 1945
Cefsulodin, 1946
Ceftaroline, 1947
Ceftazidime, 1948
Cefteram, 1949
Ceftibuten, 1950
Ceftizoxime, 1952
Ceftobiprole Medocaril, 1953
Ceftolozane, 1954
Ceftriaxone, 1955
Cefuroxime, 1956
Cephalexin, 1974
Cephalothin, 1981
Cephapirin, 1983
Cephradine, 1985
Pivcefalexin, 7627

Cephamycins
Cefbuperazone, 1919
Cefmetazole, 1927
Cefminox, 1928
Cefotetan, 1935
Cefoxitin, 1938

Monobactams
Aztreonam, 918
Carumonam, 1871

Oxacephems
Flomoxef, 4136
Moxalactam, 6372

Penems
Faropenem, 3976
Ritipenem, 8364

Penicillins
Amdinocillin, 383
Amdinocillin Pivoxil, 384
Amoxicillin, 574
Ampicillin, 583
Apalcillin, 710
Aspoxicillin, 842
Azidocillin, 902
Azlocillin, 908
Bacampicillin, 924
Carbenicillin, 1793
Carindacillin, 1839
Clometocillin, 2376
Cloxacillin, 2403
Cyclacillin, 2695
Dicloxacillin, 3094
Epicillin, 3669
Floxacillin, 4143
Hetacillin, 4707
Lenampicillin, 5491
Metampicillin, 5996
Methicillin Sodium, 6041
Mezlocillin, 6250
Nafcillin, 6436
Oxacillin, 7003
Penamecillin, 7190
Penethamate Hydriodide, 7194
Penicillin G, 7203
Penicillin G Benzathine, 7205
Penicillin G Procaine, 7206
Penicillin N, 7207
Penicillin O, 7208
Penicillin V, 7209
Phenethicillin Potassium, 7337
Piperacillin, 7575
Pivampicillin, 7626
Propicillin, 7931
Quinacillin, 8159
Sulbenicillin, 9022
Sultamicillin, 9121
Talampicillin, 9165
Temocillin, 9280
Ticarcillin, 9584

Lincosamides
Clindamycin, 2354
Lincomycin, 5555

Macrolides
Azithromycin, 907
Cethromycin, 2018
Clarithromycin, 2338
Dirithromycin, 3394
Erythromycin, 3739
Erythromycin Estolate, 3740
Erythromycin Lactobionate, 3741
Erythromycin Propionate, 3742
Erythromycin Stearate, 3743
Fidaxomicin, 4108
Josamycin, 5315
Leucomycin, 5505
Midecamycins, 6262
Miokamycin, 6286
Oleandomycin, 6918
Primycin, 7868
Rokitamycin, 8379
Rosaramicin, 8392
Roxithromycin, 8409
Spiramycin, 8879
Telithromycin, 9261
Troleandomycin, 9949

Sulbactam, *9021*
Sultamicillin, *9121*
Tazobactam, *9220*

Renal Dipeptidase Inhibitors
Cilastatin, *2274*

Renal Protectant
Betamipron, *1183*

ANTIBIOTIC *see Antibacterial (Antibiotics); Antifungal (Antibiotics); Antineoplastic*

ANTICANCER *see Antineoplastic*

ANTICHOLELITHOGENIC *see Cholelitholytic Agent*

ANTICHOLESTEREMIC *see Antilipemic*

ANTICHOLINERGIC *see Antimuscarinic*

ANTICOAGULANT *see also Antithrombotic; Thrombolytic*
Acenocoumarol, *30*
Ancrod, *624*
Anisindione, *661*
Bivalirudin, *1311*
Bromindione, *1400*
Dextran Sulfate Sodium, *2953*
Dicumarol, *3100*
Ethyl Biscoumacetate, *3825*
Ethylidene Dicoumarol, *3867*
Fluindione, *4165*
Heparin, *4688*
Hirudin, *4755*
Lyapolate Sodium, *5678*
Pentosan Polysulfate, *7247*
Phenindione, *7348*
Phenprocoumon, *7371*
Phosvitins, *7477*
Picotamide, *7519*
Tioclomarol, *9607*
Warfarin, *10236*

ANTICONVULSANT
Albutoin, *211*
Aminoglutethimide, *437*
4-Amino-3-hydroxybutyric Acid, *441*
Beclamide, *1016*
Brivaracetam, *1384*
Calcium Bromide, *1657*
Carbamazepine, *1783*
Carisbamate, *1841*
Clomethiazole, *2375*
Clonazepam, *2379*
Diazepam, *2997*
Dimethadione, *3235*
Eslicarbazepine, *3754*
Eterobarb, *3770*
Ethadione, *3773*
Ethosuximide, *3801*
Ethotoin, *3802*
Ezogabine, *3959*
Felbamate, *3984*

Fosphenytoin, *4285*
Gabapentin, *4348*
Ganaxolone, *4392*
Lacosamide, *5381*
Lamotrigine, *5403*
Levetiracetam, *5513*
Lorazepam, *5636*
Magnesium Bromide, *5723*
Magnesium Sulfate, *5755*
Mephenytoin, *5920*
Mephobarbital, *5921*
Metharbital, *6033*
Methsuximide, *6078*
Midazolam, *6261*
Nitrazepam, *6661*
Oxcarbazepine, *7028*
Paramethadione, *7132*
Phenacemide, *7318*
Phenetharbital, *7336*
Pheneturide, *7344*
Phenobarbital, *7352*
Phensuximide, *7373*
Phenytoin, *7433*
Potassium Bromide, *7739*
Pregabalin, *7843*
Primidone, *7865*
Progabide, *7888*
Rufinamide, *8426*
Safinamide, *8450*
Sodium Bromide, *8729*
Solanum, *8835*
Stiripentol, *8948*
Strontium Bromide, *8967*
Sulthiame, *9122*
Talampanel, *9164*
Tetrantoin, *9382*
Tiagabine, *9572*
Topiramate, *9706*
Trimethadione, *9875*
Valproic Acid, *10099*
Valpromide, *10100*
Vigabatrin, *10175*
Zonisamide, *10392*

ANTIDEPRESSANT *see also Antimanic*
Bicyclics
Citalopram, *2317*
Escitalopram *see 2317*
Fencamine, *3999*
Nefopam, *6525*
Nomifensine, *6760*
Oxitriptan *see 4886*
Oxypertine, *7070*
Paroxetine, *7148*
Sertraline, *8606*
Thiazesim, *9457*
Trazodone, *9740*

Hydrazides/Hydrazines
Iproclozide, *5120*
Iproniazid, *5122*
Isocarboxazid, *5202*
Nialamide, *6576*
Phenelzine, *7335*

Phenyloxazolidinones
Befloxatone, *1020*
Toloxatone, *9680*

Pyrrolidones
Cotinine, *2541*

Tetracyclics
Maprotiline, *5813*
Metralindole, *6231*

Mianserin, *6251*
Mirtazapine, *6291*

Tricyclics
Adinazolam, *148*
Amineptine, *403*
Amitriptyline, *483*
Amitriptylinoxide, *484*
Amoxapine, *573*
Butriptyline, *1535*
Clomipramine, *2378*
Desipramine, *2921*
Dibenzepin, *3012*
Dothiepin, *3478*
Doxepin, *3483*
Fluacizine, *4145*
Imipramine, *4962*
Imipramine *N*-Oxide, *4963*
Iprindole, *5119*
Lofepramine, *5614*
Melitracen, *5891*
Metapramine, *6001*
Nortriptyline, *6804*
Opipramol, *6948*
Pizotyline, *7629*
Propizepine, *7946*
Protriptyline, *8010*
Quinupramine, *8200*
Tianeptine, *9575*
Trimipramine, *9894*

Others
Agomelatine, *183*
Bupropion, *1503*
Desvenlafaxine, *2937*
Duloxetine, *3512*
Femoxetine, *3989*
Fluoxetine, *4217*
Fluvoxamine, *4246*
Hematoporphyrin, *4671*
Levophacetoperane, *5521*
Medifoxamine, *5861*
Milnacipran, *6275*
Minaprine, *6281*
Moclobemide, *6312*
Nefazodone, *6523*
Nemifitide, *6530*
Piberaline, *7504*
Prolintane, *7896*
Pyrisuccideanol, *8104*
Reboxetine, *8244*
Ritanserin, *8363*
Roxindole, *8408*
Rubidium Chloride, *8417*
Selegiline, *8565*
Sulpiride, *9119*
Tandospirone, *9183*
Thozalinone, *9533*
Tranylcypromine, *9734*
Tryptophan, *9977*
Venlafaxine, *10140*
Viloxazine, *10178*
Zimeldine, *10325*

ANTIDIABETIC
Biguanides
Buformin, *1481*
Metformin, *6010*
Phenformin, *7345*

Hormones/Analogs
Amylin, *603*
Insulin, *5023*
Insulin Aspart, *5025*
Insulin Detemir, *5026*
Insulin Glargine, *5027*

ANTIDIABETIC

Hormones/Analogs (*continued*)
Insulin Glulisine, *5028*
Insulin Lispro, *5030*
Liraglutide, *5569*
Lixisenatide, *5605*
Pramlintide, *7828*
Taspoglutide, *9209*

Sulfonylurea Derivatives
Acetohexamide, *61*
Carbutamide, *1831*
Chlorpropamide, *2192*
Glibornuride, *4473*
Gliclazide, *4474*
Glimepiride, *4475*
Glipizide, *4477*
Gliquidone, *4478*
Glisoxepid, *4479*
Glyburide, *4514*
Glybuthiazole, *4515*
Glybuzole, *4516*
Glyhexamide, *4542*
Glymidine, *4543*
Tolazamide, *9665*
Tolbutamide, *9667*
Tolcyclamide, *9671*

Thiazolidinediones
Pioglitazone, *7565*
Rosiglitazone, *8396*
Troglitazone, *9948*

Others
Acarbose, *19*
Albiglutide, *204*
Aleglitazar, *221*
Alogliptin, *302*
Bromocriptine, *1424*
Dapagliflozin, *2816*
Exenatide *see 3955*
Linagliptin, *5549*
Miglitol, *6267*
Mitiglinide, *6296*
Nateglinide, *6510*
Repaglinide, *8256*
Saxagliptin, *8525*
Sitagliptin, *8690*
Vildagliptin, *10177*
Voglibose, *10226*

ANTIDIARRHEAL

Acetyltannic Acid, *96*
Bismuth Subcarbonate, *1282*
Bismuth Subsalicylate, *1285*
Catechin, *1902*
Difenoxin, *3155*
Diphenoxylate, *3343*
Lidamidine, *5534*
Loperamide, *5628*
Opium, *6949*
Racecadotril, *8207*
Trillium, *9865*
Uzarin, *10083*
Zaldaride, *10308*

ANTIDIURETIC

Desmopressin, *2926*
Lypressin, *5691*
Oxycinchophen, *7056*
Pituitary, Posterior, *7623*
Vasopressin, *10129*

ANTIDOTE (ACETAMINOPHEN POISONING)

Acetylcysteine, *83*
Cysteamine, *2776*
Methionine, *6047*

ANTIDOTE (CURARE)

Edrophonium Chloride, *3564*
Neostigmine, *6549*
Tacrine, *9154*

ANTIDOTE (CYANIDE)

p-Aminopropiophenone, *466*
Dicobalt Edetate *see 3565*
Hydroxocobalamin, *4846*
Methylene Blue, *6132*
Potassium Nitrite, *7770*
Sodium Nitrite, *8779*
Sodium Thiosulfate, *8821*

ANTIDOTE (FOLIC ACID ANTAGONISTS)

Folinic Acid, *4249*

ANTIDOTE (HEAVY METAL POISONING)

N-Acetylpenicillamine, *94*
Albumen, *209*
Deferoxamine, *2863*
Dimercaprol, *3231*
2,3-Dimercapto-1-propanesulfonic Acid, *3232*
Ditiocarb Sodium, *3421*
Prussian Blue, *8018*
Succimer, *8995*
Tiopronin, *9609*

ANTIDOTE (METHANOL AND ETHYLENE GLYCOL POISONING)

Fomepizole, *4253*

ANTIDOTE (ORGANOPHOSPHATE POISONING)

Asoxime Chloride, *825*
HLö-7, *4762*
Obidoxime Chloride, *6827*
Pralidoxime Chloride, *7822*

ANTIDYSKINETIC *see also Antiparkinsonian*

Amantadine, *368*
Clonidine, *2380*
Haloperidol, *4634*
Pimozide, *7546*
Sarizotan, *8512*
Tetrabenazine, *9326*
Tiapride, *9576*

ANTIECZEMATIC

Alitretinoin, *241*
Evening Primrose Oil, *3949*
γ-Linolenic Acid, *5562*
Pimecrolimus, *7541*
Tacrolimus, *9155*

ANTIEMETIC

Alizapride, *242*
Aprepitant, *741*
Azasetron, *893*
Benzquinamide, *1122*
Bismuth Subsalicylate, *1285*
Bromopride, *1443*
Buclizine, *1474*
Chlorpromazine, *2191*
Clebopride, *2342*
Cyclizine, *2703*
Dexamethasone, *2945*
Dimenhydrinate, *3230*
Diphenhydramine, *3339*
Diphenidol, *3341*
Dixyrazine, *3429*
Dolasetron, *3458*
Domperidone, *3467*
Granisetron, *4572*
Meclizine, *5848*
Metoclopramide, *6219*
Metopimazine, *6226*
Nabilone, *6427*
Ondansetron, *6943*
Oxypendyl, *7069*
Palonosetron, *7098*
Pipamazine, *7567*
Prochlorperazine, *7879*
Promethazine, *7901*
Ramosetron, *8223*
Scopolamine, *8543*
Sulpiride, *9119*
Tetrahydrocannabinols, *9354*
Thiethylperazine, *9467*
Thioproperazine, *9510*
Trimethobenzamide, *9877*
Tropisetron, *9962*

ANTIEPILEPTIC *see Anticonvulsant*

ANTIFIBROTIC

Pirfenidone, *7606*
Potassium Aminobenzoate, *7727*

ANTIFLATULENT

Dimethicone, *3237*
α-Galactosidase, *4366*

ANTIFUNGAL (ANTIBIOTICS)

Echinocandins
Anidulafungin, *649*
Caspofungin, *1889*
Micafungin, *6254*

Polyenes
Amphotericin B, *582*
Candicidin, *1743*
Filipin, *4110*
Fungichromin, *4319*
Hachimycin, *4621*
Hamycin, *4642*
Lucensomycin, *5651*
Mepartricin, *5914*
Natamycin, *6509*
Nystatin, *6825*
Pecilocin, *7168*
Perimycin, *7284*

Others
Griseofulvin, *4584*
Oligomycins, *6927*

Pyrrolnitrin, *8132*
Siccanin, *8625*
Viridin, *10202*

ANTIFUNGAL (SYNTHETIC)
Allylamines
Butenafine, *1521*
Naftifine, *6439*
Terbinafine, *9299*

Imidazoles
Bifonazole, *1216*
Butoconazole, *1531*
Chlormidazole, *2107*
Clotrimazole, *2401*
Croconazole, *2580*
Econazole, *3549*
Fenticonazole, *4032*
Flutrimazole, *4243*
Isoconazole, *5205*
Ketoconazole, *5349*
Lanoconazole, *5408*
Luliconazole, *5655*
Miconazole, *6257*
Neticonazole, *6563*
Omoconazole, *6941*
Oxiconazole, *7036*
Sertaconazole, *8604*
Sulconazole, *9025*
Tioconazole, *9608*

Thiocarbamates
Liranaftate, *5570*
Tolciclate, *9669*
Tolindate, *9675*
Tolnaftate, *9678*

Triazoles
Fluconazole, *4153*
Isavuconazole, *5149*
Itraconazole, *5292*
Posaconazole, *7723*
Ravuconazole, *8239*
Saperconazole, *8499*
Terconazole, *9303*
Voriconazole, *10231*

Others
Amorolfine, *570*
Calcium Propionate, *1700*
Chlorphenesin, *2184*
Ciclopirox, *2269*
Cloxyquin, *2405*
Flucytosine, *4156*
Hexetidine, *4739*
Loflucarban, *5616*
Nifuratel, *6616*
Potassium Iodide, *7764*
Propionic Acid, *7939*
Pyrithione, *8107*
Salicylanilide, *8467*
Sodium Propionate, *8798*
Sulbentine, *9024*
Tenonitrozole, *9290*
Triacetin, *9751*
Undecylenic Acid, *10033*
Zinc Propionate, *10353*

ANTIGLAUCOMA
Acetazolamide, *52*
Apraclonidine, *738*
Befunolol, *1021*
Betaxolol, *1184*
Bimatoprost, *1225*
Brimonidine, *1381*
Brinzolamide, *1382*
Bupranolol, *1500*
Carteolol, *1866*
Dapiprazole, *2820*
Dichlorphenamide, *3087*
Dipivefrin, *3378*
Dorzolamide, *3475*
Epinephrine, *3674*
Latanoprost, *5431*
Levobunolol, *5514*
Methazolamide, *6034*
Metipranolol, *6216*
Pilocarpine, *7536*
Pindolol, *7554*
Tafluprost, *9160*
Timolol, *9601*
Travoprost, *9738*
Unoprostone, *10034*

ANTIGONADOTROPIN
Danazol, *2811*
Gestrinone, *4449*
Paroxypropione, *7149*

ANTIGOUT
Allopurinol, *271*
Carprofen, *1862*
Colchicine, *2455*
Febuxostat, *3983*
Pegloticase, *7177*
Probenecid, *7872*
Sulfinpyrazone, *9079*

ANTIHEMOPHILIC FACTOR
Factor VIII, *3962*
Moroctocog Alfa, *6357*

ANTIHEMORRHAGIC see *Hemostatic*

ANTIHISTAMINIC see also *Antiallergic*
Alkylamine Derivatives
Acrivastine, *118*
Bamipine, *949*
Bilastine, *1220*
Brompheniramine, *1453*
Chlorpheniramine, *2186*
Dimethindene, *3238*
Pheniramine, *7349*
Pyrrobutamine, *8126*
Thenaldine, *9429*
Tolpropamine, *9682*
Triprolidine, *9922*

Aminoalkylethers
Bromodiphenhydramine, *1428*
Carbinoxamine, *1800*
Chlorphenoxamine, *2187*
Clemastine, *2343*
Diphenhydramine, *3339*
Diphenylpyraline, *3370*
Doxylamine, *3489*
p-Methyldiphenhydramine, *6125*
Moxastine, *6373*
Orphenadrine, *6975*
Phenyltoloxamine, *7425*
Setastine, *8611*

Ethylenediamine Derivatives
Chloropyramine, *2164*
Chlorothen, *2170*
Methapyrilene, *6031*
Pyrilamine, *8097*
Thenyldiamine, *9431*
Thonzylamine Hydrochloride, *9527*
Tripelennamine, *9910*

Piperazines
Cetirizine, *2021*
Chlorcyclizine, *2082*
Cinnarizine, *2306*
Clocinizine, *2362*
Homochlorcyclizine, *4771*
Hydroxyzine, *4889*

Tricyclics
Phenothiazines
Mequitazine, *5931*
Promethazine, *7901*
Thiazinamium Methylsulfate, *9458*

Other Tricyclics
Azatadine, *894*
Cyproheptadine, *2770*
Deptropine, *2913*
Desloratadine, *2923*
Isothipendyl, *5273*
Loratadine, *5635*
Olopatadine, *6934*
Rupatadine, *8430*

Others
Antazoline, *672*
Astemizole, *846*
Azelastine, *899*
Bepotastine, *1151*
Clemizole, *2344*
Ebastine, *3531*
Emedastine, *3614*
Epinastine, *3673*
Fexofenadine, *4097*
Levocabastine, *5515*
Mebhydroline, *5839*
Mizolastine, *6307*
Phenindamine, *7347*
Terfenadine, *9306*
Tritoqualine, *9942*

ANTIHYPERCHOLESTEROLEMIC see *Antilipemic*

ANTIHYPERLIPIDEMIC see *Antilipemic*

ANTIHYPERPARATHYROID
Calcitriol, *1647*
Cinacalcet, *2285*
Doxercalciferol, *3484*
Maxacalcitol, *5828*
Paricalcitol, *7143*

ANTIHYPERPHOSPHATEMIC
Aluminum Hydroxide, *338*
Aluminum Hydroxychloride, *339*
Calcium Acetate, *1649*
Lanthanum Carbonate, *5415*
Sevelamer, *8613*

ANTIHYPERTENSIVE *see also Diuretic; Pulmonary Hypertension Treatment*

Arylethanolamine Derivatives
Amosulalol, *572*
Bufuralol, *1485*
Labetalol, *5377*
Sotalol, *8854*

Aryloxypropanolamine Derivatives
Acebutolol, *20*
Alprenolol, *309*
Arotinolol, *782*
Atenolol, *850*
Betaxolol, *1184*
Bevantolol, *1195*
Bisoprolol, *1295*
Bopindolol, *1335*
Bucindolol, *1471*
Bunitrolol, *1493*
Bupranolol, *1500*
Carazolol, *1780*
Carteolol, *1866*
Carvedilol, *1873*
Celiprolol, *1962*
Indenolol, *4979*
Mepindolol, *5923*
Metipranolol, *6216*
Metoprolol, *6228*
Nadolol, *6431*
Nebivolol, *6516*
Nipradilol, *6648*
Oxprenolol, *7051*
Penbutolol, *7191*
Pindolol, *7554*
Propranolol, *7953*
Talinolol, *9170*
Tertatolol, *9316*
Tilisolol, *9595*
Timolol, *9601*

Biphenyltetrazole Derivatives
Azilsartan, *903*
Candesartan, *1742*
Irbesartan, *5127*
Losartan, *5641*
Olmesartan, *6933*
Valsartan, *10102*

N-Carboxyalkyl (peptide/lactam) Derivatives
Alacepril, *193*
Benazepril, *1033*
Captopril, *1776*
Ceronapril, *1993*
Cilazapril, *2275*
Delapril, *2881*
Enalapril, *3623*
Fosinopril, *4283*
Imidapril, *4952*
Lisinopril, *5572*
Moexipril, *6315*
Moveltipril, *6371*
Perindopril, *7285*
Quinapril, *8167*
Ramipril, *8221*
Spirapril, *8880*
Temocapril, *9279*
Trandolapril, *9729*
Zofenopril, *10386*

Dihydropyridine Derivatives
Amlodipine, *487*
Aranidipine, *756*
Azelnidipine, *900*
Barnidipine, *998*
Benidipine, *1045*

Cilnidipine, *2278*
Clevidipine, *2350*
Efonidipine, *3571*
Felodipine, *3987*
Isradipine, *5287*
Lacidipine, *5379*
Lercanidipine, *5500*
Manidipine, *5808*
Nicardipine, *6580*
Nifedipine, *6613*
Nilvadipine, *6630*
Nisoldipine, *6651*
Nitrendipine, *6662*

Guanidine Derivatives
Bethanidine, *1188*
Debrisoquin, *2846*
Guanabenz, *4593*
Guanadrel, *4594*
Guanethidine, *4595*
Guanfacine, *4596*
Guanoxabenz, *4602*
Guanoxan, *4603*
Pinacidil, *7549*

Hydrazines/Phthalazines
Dihydralazine, *3187*
Endralazine, *3632*
Hydralazine, *4802*
Pildralazine, *7535*
Todralazine, *9660*

Imidazoleacrylic Acid Derivatives
Eprosartan, *3691*

Imidazole Derivatives
Clonidine, *2380*
Lofexidine, *5615*
Moxonidine, *6379*
Phentolamine, *7376*
Tiamenidine, *9573*

Quaternary Ammonium Compounds
Hexamethonium, *4724*
Pentolinium Tartrate, *7245*

Quinazoline Derivatives
Alfuzosin, *231*
Bunazosin, *1491*
Doxazosin, *3482*
Prazosin, *7838*
Terazosin, *9297*

Reserpine Derivatives
Deserpidine, *2919*
Rescinnamine, *8262*
Reserpine, *8265*
Syrosingopine, *9147*

Sulfonamide Derivatives
Clopamide, *2382*
Furosemide, *4338*
Tripamide, *9909*
Xipamide, *10276*

Thiazides and Analogs
Althiazide, *314*
Bendroflumethiazide, *1039*
Benzthiazide, *1123*
Benzylhydrochlorothiazide, *1139*
Buthiazide, *1527*
Chlorothiazide, *2171*
Chlorthalidone, *2200*
Cyclopenthiazide, *2740*
Cyclothiazide, *2752*
Diazoxide, *3007*
Hydrochlorothiazide, *4819*

Hydroflumethiazide, *4826*
Indapamide, *4975*
Methyclothiazide, *6079*
Metolazone, *6224*
Polythiazide, *7705*
Quinethazone, *8172*
Teclothiazide, *9241*
Trichlormethiazide, *9789*

Others
Ajmaline, *187*
Aliskiren, *239*
Atrial Natriuretic Peptide, *863*
Bosentan, *1355*
Cicletanine, *2267*
Clentiazem, *2346*
Darusentan, *2830*
Eplerenone, *3681*
Fenoldopam, *4010*
Flosequinan, *4142*
Indoramin, *5010*
Ketanserin, *5344*
Levcromakalim, *5512*
Mebutamate, *5842*
Mecamylamine, *5843*
Methyldopa, *6127*
Mibefradil, *6252*
Minoxidil, *6285*
Naftopidil, *6440*
Omapatrilat, *6937*
Pargyline, *7142*
Pempidine, *7189*
Piperoxan, *7591*
Protoveratrines, *8008*
Raubasine, *8237*
Rilmenidine, *8347*
Sampatrilat, *8489*
Saralasin, *8509*
Semotiadil, *8585*
Sodium Nitroprusside, *8780*
Telmisartan, *9271*
Ticrynafen, *9586*
Trimethaphan Camsylate, *9876*
Tyrosinase, *10019*
Urapidil, *10050*

ANTIHYPERTHYROID

2-Aminothiazole, *475*
Carbimazole, *1799*
3,5-Dibromo-L-tyrosine, *3035*
3,5-Diiodotyrosine, *3210*
Iodine, *5057*
Methimazole, *6043*
Methylthiouracil, *6201*
Propylthiouracil, *7980*
Sodium Perchlorate, *8784*
Thibenzazoline, *9464*
Thiobarbital, *9474*
2-Thiouracil, *9520*

ANTIHYPOGLYCEMIC

Diazoxide, *3007*
Glucagon, *4481*
Glucose, *4495*

ANTIHYPOTENSIVE

Amezinium Methyl Sulfate, *388*
Angiotensin Amide *see 644*
Dopamine, *3470*
Etilefrin, *3914*
Metaraminol, *6003*
Methoxamine, *6058*
Midodrine, *6263*

Norepinephrine, 6784
Pholedrine, 7442
Synephrine, 9142

ANTIHYPOTHYROID

Liothyronine, 5565
Thyroid, 9565
Thyroidin, 9566
Thyroxine, 9570
Tiratricol, 9618

ANTI-INFECTIVE *see Antiseptic/Disinfectant*

ANTI-INFLAMMATORY (BIOLOGICAL RESPONSE MODIFIER)

Abatacept, 4
Adalimumab, 136
Alicaforsen, 237
Anakinra *see 5045*
Bardoxolone, 959
Bimosiamose, 1226
Certolizumab Pegol, 2002
Eculizumab, 3553
Etanercept, 3767
Golimumab, 4560
Infliximab, 5014
Natalizumab, 6508
Pexelizumab, 7307
Reslizumab, 8273
Rilonacept, 8348
Rituximab, 8367
Tocilizumab, 9651

ANTI-INFLAMMATORY (GASTROINTESTINAL)

Alicaforsen, 237
Balsalazide, 938
Infliximab, 5014
Mesalamine, 5974
Natalizumab, 6508
Olsalazine, 6935
Sulfasalazine, 9073

ANTI-INFLAMMATORY (NONSTEROIDAL) *see also Antiarthritic/Antirheumatic*

Aminoarylcarboxylic Acid Derivatives

Etofenamate, 3921
Flufenamic Acid, 4163
Isonixin, 5238
Meclofenamic Acid, 5850
Mefenamic Acid, 5869
Niflumic Acid, 6615
Talniflumate, 9176
Tolfenamic Acid, 9673

Arylacetic Acid Derivatives

Aceclofenac, 23
Acemetacin, 28
Amfenac, 389
Amtolmetin Guacil, 591
Bromfenac, 1397
Bufexamac, 1479
Diclofenac, 3091
Etodolac, 3920
Felbinac, 3985
Fenclozic Acid, 4001
Fentiazac, 4030

Glucametacin, 4483
Indomethacin, 5009
Isoxepac, 5284
Lonazolac, 5623
Metiazinic Acid, 6215
Mofezolac, 6318
Nepafenac, 6553
Oxametacine, 7015
Proglumetacin, 7890
Sulindac, 9112
Tiaramide, 9579
Tolmetin, 9677
Tropesin, 9957

Arylbutyric Acid Derivatives

Bumadizone, 1488
Butibufen, 1529
Fenbufen, 3996

Arylcarboxylic Acids

Ketorolac, 5353

Arylpropionic Acid Derivatives

Alminoprofen, 293
Carprofen, 1862
Fenoprofen, 4011
Flunoxaprofen, 4179
Flurbiprofen, 4229
Ibuprofen, 4919
Ibuproxam, 4920
Ketoprofen, 5352
Loxoprofen, 5647
Naproxcinod, 6498
Naproxen, 6499
Oxaprozin, 7023
Piketoprofen, 7534
Pranoprofen, 7831
Suprofen, 9134
Tiaprofenic Acid, 9577
Ximoprofen, 10275
Zaltoprofen, 10310

Pyrazolones

Feprazone, 4039
Kebuzone, 5336
Mofebutazone, 6317
Oxyphenbutazone, 7071
Phenylbutazone, 7390
Propyphenazone, 7982
Ramifenazone, 8220
Suxibuzone, 9139

Salicylic Acid Derivatives

Aspirin, 841
Balsalazide, 938
Benorylate, 1047
Calcium Acetylsalicylate, 1650
Diflunisal, 3165
Gentisic Acid, 4433
Glycol Salicylate, 4535
Imidazole Salicylate, 4954
Lysine Acetylsalicylate, 5698
Mesalamine, 5974
Morpholine Salicylate *see 6362*
1-Naphthyl Salicylate, 6495
Olsalazine, 6935
Salicylamide *O*-Acetic Acid, 8466
Salicylsulfuric Acid, 8471
Salsalate, 8477
Sodium Salicylate *see 8469*
Sulfasalazine, 9073

Thiazinecarboxamides

Ampiroxicam, 584
Lornoxicam, 5640
Meloxicam, 5894

Piroxicam, 7619
Tenoxicam, 9291

Others

ε-Acetamidocaproic Acid, 46
S-Adenosylmethionine, 144
Ajulemic Acid, 191
3-Amino-4-hydroxybutyric Acid, 440
Bendazac, 1037
Benzydamine, 1125
α-Bisabolol, 1245
Bucolome, 1475
Celecoxib, 1958
Difenpiramide, 3156
Ditazol, 3410
Emorfazone, 3619
Epirizole, 3677
Etoricoxib, 3930
Fepradinol, 4038
Guaiazulene, 4590
Imisopasem Manganese, 4965
Lexipafant, 5527
Licofelone, 5533
Lumiracoxib, 5661
Nabumetone, 6428
Nimesulide, 6633
Oxaceprol, 7002
Parecoxib, 7140
Proquazone, 7984
Rofecoxib, 8377
Superoxide Dismutase, 9131
Tenidap, 9287
Valdecoxib, 10086

ANTI-INFLAMMATORY (STEROIDAL) *see Glucocorticoid*

ANTILEPROTIC *see Antibacterial (Leprostatic)*

ANTILEUKEMIC *see Antineoplastic*

ANTILIPEMIC

Bile Acid Sequesterants

Cholestyramine, 2209
Colesevelam Hydrochloride, 2457
Colestilan, 2458
Colestipol, 2459
Polidexide, 7674

CETP Inhibitors

Anacetrapib, 615
Dalcetrapib, 2801
Evacetrapib, 3947
Torcetrapib, 9708

Fibrates

Bezafibrate, 1199
Binifibrate, 1228
Ciprofibrate, 2312
Clinofibrate, 2355
Clofibrate, 2370
Clofibric Acid, 2371
Etofibrate, 3923
Fenofibrate, 4009
Gemfibrozil, 4422
Pirifibrate, 7608
Ronifibrate, 8386
Simfibrate, 8675
Theofibrate, 9435

ANTILIPEMIC (*continued*)

HMG CoA Reductase Inhibitors
Atorvastatin, 855
Cerivastatin, 1992
Fluvastatin, 4245
Lovastatin, 5644
Pitavastatin, 7621
Pravastatin Sodium, 7835
Rosuvastatin, 8402
Simvastatin, 8677

Nicotinic Acid Derivatives
Acipimox, 102
Aluminum Nicotinate, 346
Niceritrol, 6582
Oxiniacic Acid, 7040

Thyroid Hormones/Analogs
Dextrothyroxine *see 9570*
Eprotirome, 3692
Thyropropic Acid, 9567

Others
Avasimibe, 876
Benfluorex, 1041
Detaxtran, 2938
Eicosapentaenoic Acid, 3578
Ezetimibe, 3958
Meglutol, 5877
Melinamide, 5890
Mipomersen, 6288
Omega-3 Acid Ethyl Esters, 6938
γ-Oryzanol, 6983
Pantethine, 7115
Pirozadil, 7620
Policosanol, 7672
Probucol, 7873
β-Sitosterol, 8694
Sultosilic Acid, 9124
Tiadenol, 9571

ANTILIPIDEMIC *see Antilipemic*

ANTIMALARIAL
Amodiaquin, 569
Arteether, 803
Arteflene, 804
Artemether, 805
Artemisinin, 807
Artesunate, 808
Bebeerine, 1014
Berberine, 1156
Chirata, 2064
Chlorguanide, 2091
Chloroquine, 2165
Chlorproguanil, 2190
Cinchona, 2287
Cinchonidine, 2289
Cinchonine, 2290
Cycloguanil, 2714
Fosmidomycin, 4284
Gentiopicrin, 4432
Halofantrine, 4630
Hydroxychloroquine, 4857
Lumefantrine, 5657
Mefloquine, 5872
Pamaquine, 7102
Plasmocid, 7636
Primaquine, 7863
Pyrimethamine, 8098
Pyronaridine, 8117
Quinacrine, 8160
Quinidine, 8176

Quinine, 8177
Quinocide, 8183
Quinoline, 8185
Sulfadoxine, 9038
Tafenoquine, 9159

ANTIMANIC
Carbamazepine, 1783
Lithium Acetate, 5577
Lithium Carbonate, 5583
Lithium Chloride, 5584
Lithium Citrate, 5586
Lithium Sulfate, 5597
Valproic Acid, 10099

ANTIMETHEMOGLOBINEMIC
Methylene Blue, 6132

ANTIMIGRAINE
Almotriptan, 298
Alpiropride, 307
Dihydroergotamine, 3191
Dotarizine, 3477
Eletriptan, 3597
Ergocriptine, 3707
Ergot, 3717
Ergotamine, 3718
Fonazine, 4258
Frovatriptan, 4301
Lisuride, 5574
Lomerizine, 5620
Methysergide, 6213
Naratriptan, 6502
Oxetorone, 7033
Pizotyline, 7629
Rizatriptan, 8370
Sumatriptan, 9126
Telcagepant, 9259
Topiramate, 9706
Valproic Acid, 10099
Zolmitriptan, 10390

ANTIMUSCARINIC *see also Antiparkinsonian; Antispasmodic; Mydriatic*
Aclidinium Bromide, 106
Acotiamide, 113
Adiphenine, 149
Ambutonium Bromide, 381
Aminopentamide, 455
Atropine, 866
Benactyzine, 1031
Benzetimide, 1077
Benztropine, 1124
Bevonium Methyl Sulfate, 1197
Biperiden, 1238
Butropium Bromide, 1536
N-Butylscopolammonium Bromide, 1591
Buzepide, 1603
Chlorbenzoxamine, 2081
Cimetropium Bromide, 2283
Clidinium Bromide, 2352
Cyclodrine, 2712
Cyclopentolate, 2741
Darifenacin, 2826
Dexetimide, 2947
Dicyclomine, 3108
Diethazine, 3119
Difemerine, 3152
Diphemanil Methylsulfate, 3336
Emepronium Bromide, 3615

Ethopropazine, 3800
Ethybenztropine, 3810
Fentonium Bromide, 4033
Fesoterodine, 4095
Flavoxate, 4128
Glycopyrrolate, 4537
Hexocyclium Methyl Sulfate, 4742
Homatropine, 4768
Hyoscyamine, 4897
Ipratropium Bromide, 5117
Isopropamide Iodide, 5248
Mepenzolate Bromide, 5915
Methantheline Bromide, 6030
Methixene, 6051
Methscopolamine Bromide, 6077
Otilonium Bromide, 6997
Oxitropium Bromide, 7042
Oxybutynin, 7054
Oxyphencyclimine, 7072
Oxyphenonium Bromide, 7074
Pipenzolate Bromide, 7573
Piperidolate, 7582
Piperilate, 7583
Poldine Methylsulfate, 7671
Prifinium Bromide, 7861
Procyclidine, 7881
Propantheline Bromide, 7919
Propiverine, 7945
Scopolamine, 8543
Scopolamine N-Oxide, 8544
Solifenacin, 8838
Tiemonium Iodide, 9587
Timepidium Bromide, 9599
Tiotropium Bromide, 9610
Tiquizium Bromide, 9616
Tolterodine, 9684
Tridihexethyl Iodide, 9826
Trihexyphenidyl Hydrochloride, 9864
Tropicamide, 9959
Trospium Chloride, 9966
Valethamate Bromide, 10092

ANTIMYASTHENIC
Ambenonium Chloride, 373
Neostigmine, 6549
Pyridostigmine Bromide, 8090

ANTIMYCOTIC *see Antifungal (Antibiotics); Antifungal (Synthetic)*

ANTINAUSEANT *see Antiemetic*

ANTINEOPLASTIC
Alkaloids/Natural Products

Camptothecin Derivatives
9-Aminocamptothecin, 426
Belotecan, 1028
Cositecan, 2538
Irinotecan, 5135
Rubitecan, 8421
Topotecan, 9707

Podophyllum Derivatives
Etoposide, 3929
Teniposide, 9288

Staurosporine Derivatives
Enzastaurin, 3656
Midostaurin, 6264

Taxanes
    Cabazitaxel, *1604*
    Docetaxel, *3442*
    Paclitaxel, *7081*

Vinca Alkaloids
    Vinblastine, *10179*
    Vincristine, *10183*
    Vindesine, *10184*
    Vinflunine, *10186*
    Vinorelbine, *10187*

Others
    Aplidine, *725*
    Elisidepsin, *3601*
    Elliptinium Acetate, *3604*
    Fosbretabulin, *4277*
    Homoharringtonine, *4776*
    Irofulven, *5136*
    Ixabepilone, *5297*
    Kahalalide F, *5323*
    Trabectedin *see 3550*

Alkylating Agents

Alkyl Sulfonates
    Busulfan, *1508*
    Piposulfan, *7594*

Aziridines
    Carboquone, *1825*
    Diaziquone, *2999*
    Uredepa, *10057*

Ethylenimines and Methylmelamines
    Altretamine, *316*
    Thiotepa, *9518*
    Triethylenemelamine, *9836*
    Triethylenephosphoramide, *9837*

Nitrogen Mustards
    Bendamustine, *1036*
    Canfosfamide, *1748*
    Chlorambucil, *2073*
    Chlornaphazine, *2108*
    Cyclophosphamide, *2743*
    Estramustine, *3761*
    Glufosfamide, *4504*
    Ifosfamide, *4937*
    Mechlorethamine, *5846*
    Mechlorethamine Oxide, *5847*
    Melphalan, *5896*
    Perfosfamide, *7276*
    Trichlormethine, *9790*
    Trofosfamide, *9947*
    Uracil Mustard, *10036*

Nitrosoureas
    Carmustine, *1845*
    Chlorozotocin, *2183*
    Fotemustine, *4289*
    Lomustine, *5621*
    Nimustine, *6639*
    Ranimustine, *8227*

Others
    Apaziquone, *712*
    Dacarbazine, *2795*
    Etoglucid, *3925*
    Laromustine, *5425*
    Mitobronitol, *6297*
    Mitolactol, *6299*
    Pipobroman, *7593*
    Procarbazine, *7876*
    Temozolomide, *9282*

Antibiotics and Analogs

Actinomycins
    Cactinomycin, *1611*
    Dactinomycin, *2797*

Anthracyclines
    Aclacinomycins, *104*
    Amrubicin, *588*
    Carubicin, *1870*
    Daunorubicin, *2834*
    Doxorubicin, *3487*
    Epirubicin, *3678*
    Idarubicin, *4927*
    Pirarubicin, *7601*
    Valrubicin, *10101*
    Zorubicin, *10395*

Others
    Bleomycins, *1319*
    Eribulin, *3723*
    Maytansinoid DM1, *5830*
    Mitomycins, *6300*
    Peplomycin, *7257*
    Plicamycin, *7652*
    Porfiromycin, *7719*
    Retaspimycin, *8282*
    Streptozocin, *8962*
    Zinostatin, *10367*

Antimetabolites

Folic Acid Analogs/Antagonists
    Aminopterin, *468*
    Edatrexate, *3554*
    Methotrexate, *6057*
    Nolatrexed, *6759*
    Pemetrexed, *7186*
    Piritrexim, *7612*
    Pralatrexate, *7821*
    Pteropterin, *8043*
    Raltitrexed, *8217*
    Trimetrexate, *9893*

Purine Analogs
    Cladribine, *2336*
    Clofarabine, *2366*
    Fludarabine, *4157*
    6-Mercaptopurine, *5938*
    Nelarabine, *6527*
    Thiamiprine, *9451*
    Thioguanine, *9490*
    Tiazofurin, *9580*

Pyrimidine Analogs
    Ancitabine, *623*
    Azacitidine, *884*
    6-Azauridine, *897*
    Capecitabine, *1757*
    Carmofur, *1844*
    Cytarabine, *2781*
    Decitabine, *2856*
    Doxifluridine, *3485*
    Enocitabine, *3640*
    Floxuridine, *4144*
    Fluorouracil, *4213*
    Gemcitabine, *4420*
    Tegafur, *9252*
    Troxacitabine, *9968*

Cyclin-dependent Kinase Inhibitors
    Seliciclib, *8579*

Enzymes
    Asparaginase, *826*
    Ranpirnase, *8230*

Farnesyl Transferase Inhibitors
    Lonafarnib, *5622*
    Tipifarnib, *9614*

Histone Deacetylase Inhibitors
    Belinostat, *1025*
    Panobinostat, *7113*
    Romidepsin, *8382*
    Vorinostat, *10232*

Immunomodulators
    Aldesleukin *see 5038*
    Interferon-α, *5034*
    Lentinan, *5493*
    Mifamurtide, *6265*
    Oregovomab, *6960*
    Peginterferon Alfa-2b, *7176*
    Polysaccharide-K, *7702*
    Pomalidomide, *7708*
    Propagermanium, *7909*
    Roquinimex, *8390*
    Sipuleucel-T, *8685*
    Sizofiran, *8697*
    Teceleukin *see 5038*
    Ubenimex, *10024*
    Velimogene Aliplasmid, *10136*

Immunotoxins
    Cintredekin Besudotox, *2311*
    Denileukin Diftitox, *2895*

Monoclonal Antibodies
    Alemtuzumab, *222*
    Bevacizumab, *1194*
    Brentuximab Vedotin, *1375*
    Catumaxomab, *1909*
    Cetuximab, *2026*
    Epratuzumab, *3687*
    Gemtuzumab Ozogamicin, *4424*
    Iodine I 131 Tositumomab *see 9713*
    Ipilimumab, *5114*
    Necitumumab, *6520*
    Nimotuzumab, *6638*
    Ofatumumab, *6862*
    Oregovomab, *6960*
    Panitumumab, *7112*
    Pertuzumab, *7298*
    Rituximab, *8367*
    Trastuzumab, *9736*

mTOR Inhibitors
    Everolimus, *3950*
    Ridaforolimus, *8336*
    Temsirolimus, *9285*

Oligonucleotides
    Agatolimod, *180*
    Aprinocarsen, *744*
    Oblimersen Sodium, *6828*
    Trabedersen, *9723*

Platinum Complexes
    Carboplatin, *1823*
    Cisplatin, *2316*
    Lobaplatin, *5607*
    Oxaliplatin, *7011*
    Picoplatin, *7517*
    Satraplatin, *8520*

Poly(ADP-ribose) Polymerase Inhibitors
    Iniparib, *5017*
    Olaparib, *6915*

ANTINEOPLASTIC (*continued*)

Proteasome Inhibitors
Bortezomib, *1353*
Carfilzomib, *1837*

Raf Kinase Inhibitors
Sorafenib, *8846*
Vemurafenib, *10138*

Retinoids and Analogs
Alitretinoin, *241*
Bexarotene, *1198*
Fenretinide, *4025*
Mofarotene, *6316*
Tamibarotene, *9179*

Sulfonamides
Indisulam, *4986*

Tyrosine Kinase Inhibitors
Afatinib, *169*
Axitinib, *883*
Bosutinib, *1357*
Brivanib, *1383*
Canertinib, *1747*
Crizotinib, *2577*
Dasatinib, *2831*
Erlotinib, *3730*
Gefitinib, *4412*
Imatinib, *4943*
Intedanib, *5031*
Lapatinib, *5418*
Lenvatinib, *5494*
Masitinib, *5822*
Motesanib, *6366*
Neratinib, *6557*
Pazopanib, *7160*
Ruxolitinib, *8440*
Sorafenib, *8846*
Sunitinib, *9129*
Tivantinib, *9640*
Vandetanib, *10117*
Vatalanib, *10130*

Others
Amsacrine, *589*
Arsenic Trioxide, *795*
Atrasentan, *861*
Bisantrene, *1248*
Brostallicin, *1458*
Calcitriol, *1647*
Cilengitide, *2276*
Contusugene Ladenovec, *2495*
Darinaparsin, *2827*
Dehydroequol, *2876*
Edotecarin, *3560*
Eflornithine, *3570*
Elesclomol, *3596*
Flavopiridol, *4126*
Gallium Nitrate, *4380*
Hydrazine Sulfate, *4810*
Hydroxyurea, *4888*
Lonidamine, *5625*
Miltefosine, *6279*
Mitoguazone, *6298*
Mitoxantrone, *6302*
Navitoclax, *6512*
Nitracrine, *6657*
Pentostatin, *7248*
Perifosine, *7280*
Pixantrone, *7628*
Plinabulin, *7654*
Razoxane, *8241*
Seocalcitol, *8595*
Sobuzoxane, *8706*
Spirogermanium, *8885*
Tasquinimod, *9210*

Tirapazamine, *9617*
Trebananib, *9741*
Vismodegib, *10207*
Zibotentan, *10320*

ANTINEOPLASTIC (HORMONAL)

Androgens
Dromostanolone, *3497*
Epitiostanol, *3680*
Mepitiostane, *5925*
Testolactone, *9323*

Antiadrenals
Aminoglutethimide, *437*
Mitotane, *6301*
Trilostane, *9867*

Antiandrogens
Abiraterone, *9*
Bicalutamide, *1204*
Flutamide, *4238*
Nilutamide, *6629*

Antiestrogens
Arzoxifene, *810*
Droloxifene, *3495*
Fulvestrant, *4313*
Tamoxifen, *9180*
Toremifene, *9709*

Antiprogestins
Onapristone, *6942*

Aromatase Inhibitors
Aminoglutethimide, *437*
Anastrozole, *619*
Exemestane, *3954*
Fadrozole, *3968*
Formestane, *4268*
Letrozole, *5502*
Vorozole, *10233*

Estrogens
Diethylstilbestrol, *3148*
Fosfestrol, *4280*
Hexestrol, *4737*
Polyestradiol Phosphate, *7686*

GnRH Agonists
Buserelin, *1506*
Goserelin, *4562*
Histrelin, *4760*
Leuprolide, *5509*
Triptorelin, *9924*

GnRH Antagonists
Abarelix, *3*
Cetrorelix, *2025*
Degarelix, *2866*

Progestogens
Chlormadinone Acetate, *2102*
Medroxyprogesterone, *5864*
Megestrol Acetate, *5876*

Somatostatin Analog
Lanreotide, *5412*

ANTINEOPLASTIC (PHOTOSENSI-TIZER)

δ-Aminolevulinic Acid, *443*
Methyl Aminolevulinate, *6090*
Motexafin Lutetium, *6368*
Porfimer Sodium, *7718*
Talaporfin, *9166*
Temoporfin, *9281*

ANTINEOPLASTIC (RADIATION SOURCE)

Americium ²⁴¹Am *see 386*
Californium ²⁵²Cf *see 1718*
Cesium, *2007*
Chromic Phosphate P 32 *see 2231*
Cobalt ⁶⁰Co *see 2411*
Ethiodized Oil I 131 *see 3789*
Gold, Radioactive, Colloidal, *4552*
Iobenguane I 131 *see 5051*
Iodine I 131 Tositumomab *see 9713*
Radium, *8210*
Radium-223 Chloride, *8211*
Radon, *8212*
Samarium Sm 153 Lexidronam, *8487*
Sodium Iodide I 131 *see 8765*
Sodium Phosphate P 32, *8791*
Strontium Chloride Sr 89 *see 8970*
Strontium ⁸⁹Sr *see 8965*
Yttrium Y 90 Edotreotide *see 3561*
Yttrium Y 90 Epratuzumab Tetraxetan *see 3687*
Yttrium Y 90 Ibritumomab Tiuxetan *see 4917*

ANTINEOPLASTIC ADJUNCT

Antimetastatic Agent
Marimastat, *5818*
Prinomastat, *7869*

Chemomodulator
Eniluracil, *3638*

Chemosensitizer
Biricodar, *1244*
Tariquidar, *9198*
Tesmilifene, *9321*
Valspodar, *10103*
Zosuquidar, *10396*

Folic Acid Replenisher
Folinic Acid, *4249*

Radioprotective
Amifostine, *399*
Palifermin, *7086*

Radiosensitizer
Broxuridine, *1461*
Efaproxiral, *3568*
Etanidazole, *3768*
Motexafin Gadolinium, *6367*

Uroprotective
Dimesna, *3233*
Mesna, *5979*

ANTINEUTROPENIC

Granulocyte Colony-Stimulating Factor, *4573*
Granulocyte-Macrophage Colony-Stimulating Factor, *4574*
Interleukin-3, *5039*

ANTIOBESITY AGENT *see also Anorexic*

Cetilistat, *2020*
Lorcaserin, *5638*

Orlistat, *6964*
Rimonabant, *8355*
Sibutramine, *8624*
Tesofensine, *9322*

ANTIOBSESSIONAL

Clomipramine, *2378*
Fluoxetine, *4217*
Fluvoxamine, *4246*
Paroxetine, *7148*
Sertraline, *8606*

ANTIOSTEOPOROTIC *see also*
Bone Resorption Inhibitor; Calcium
Regulator

Alendronic Acid, *223*
Arzoxifene, *810*
Bazedoxifene, *1011*
Clodronic Acid, *2365*
Denosumab, *2897*
Eldecalcitol, *3587*
Etidronic Acid, *3912*
Ibandronic Acid, *4912*
Incadronic Acid, *4969*
Lasofoxifene, *5430*
Minodronic Acid, *6284*
Neridronic Acid, *6558*
Odanacatib, *6861*
Pamidronic Acid, *7104*
Raloxifene, *8215*
Risedronic Acid, *8360*
Sodium Fluoride, *8754*
Strontium Ranelate *see 8224*
Teriparatide Acetate, *9309*
Zoledronic Acid, *10388*

ANTIPAGETIC *see also **Bone Re-***
sorption Inhibitor; Calcium Regula-
tor

Alendronic Acid, *223*
Calcitonin, *1646*
Elcatonin, *3585*
Gallium Nitrate, *4380*
Minodronic Acid, *6284*
Neridronic Acid, *6558*
Pamidronic Acid, *7104*
Risedronic Acid, *8360*
Tiludronic Acid, *9598*
Zoledronic Acid, *10388*

ANTIPARKINSONIAN

Amantadine, *368*
Apomorphine, *733*
Benserazide, *1052*
Benzetimide, *1077*
Benztropine, *1124*
Biperiden, *1238*
Bromocriptine, *1424*
Budipine, *1477*
Cabergoline, *1606*
Carbidopa, *1798*
Dexetimide, *2947*
Diethazine, *3119*
Droxidopa, *3504*
Entacapone, *3648*
Ethopropazine, *3800*
Ethybenztropine, *3810*
Istradefylline, *5289*
Lazabemide, *5449*
Levodopa, *5516*
Memantine, *5897*

Methixene, *6051*
Pergolide, *7277*
Piroheptine, *7616*
Pramipexole, *7825*
Pridinol Hydrochloride *see 7859*
Procyclidine, *7881*
Rasagiline, *8234*
Ropinirole, *8388*
Safinamide, *8450*
Scopolamine *N*-Oxide, *8544*
Selegiline, *8565*
Talipexole, *9171*
Terguride, *9307*
Tolcapone, *9668*
Trihexyphenidyl Hydrochloride, *9864*
Tropacine, *9952*
Zonisamide, *10392*

ANTIPERISTALTIC *see Antidiar-*
rheal

ANTIPHEOCHROMOCYTOMA

Metyrosine, *6240*
Phenoxybenzamine, *7368*
Phentolamine, *7376*

ANTIPNEUMOCYSTIC

Atovaquone, *857*
Eflornithine, *3570*
Pentamidine, *7227*
Sulfamethoxazole, *9050*

ANTIPROGESTIN *see also Selective*
Progesterone Receptor Modulator

Onapristone, *6942*

ANTIPROSTATIC HYPERTROPHY

Alfuzosin, *231*
Doxazosin, *3482*
Dutasteride, *3518*
Episteride, *3690*
Finasteride, *4113*
Gestonorone Caproate, *4448*
Mepartricin, *5914*
Osaterone, *6985*
Oxendolone, *7030*
Silodosin, *8642*
Tamsulosin, *9181*
Terazosin, *9297*

ANTIPROTOZOAL (AMEBA) *see*
Antiamebic

ANTIPROTOZOAL (CRYPTO-
SPORIDIUM)

Nitazoxanide, *6653*

ANTIPROTOZOAL (GIARDIA)

Tinidazole, *9604*

ANTIPROTOZOAL (LEISHMANIA)

Hydroxystilbamidine, *4883*
N-Methylglucamine, *6150*
Miltefosine, *6279*

Pentamidine, *7227*
Sitamaquine, *8691*
Sodium Stibogluconate *see 692*
Stilbamidine, *8945*
Urea Stibamine, *10056*

ANTIPROTOZOAL (MALARIA) *see*
Antimalarial

ANTIPROTOZOAL (PNEUMOCYS-
TIS) *see Antipneumocystic*

ANTIPROTOZOAL (TOXOPLAS-
MA)

Pyrimethamine, *8098*

ANTIPROTOZOAL (TRICHOMO-
NAS)

Acetarsone, *51*
Aminitrozole, *404*
Azanidazole, *890*
Furazolidone, *4330*
Hachimycin, *4621*
Lauroguadine, *5441*
Mepartricin, *5914*
Metronidazole, *6236*
Nifuratel, *6616*
Nifuroxime, *6620*
Nimorazole, *6637*
Secnidazole, *8556*
Silver Picrate, *8665*
Tenonitrozole, *9290*
Tinidazole, *9604*

ANTIPROTOZOAL (TRYPANOSO-
MA)

Benznidazole, *1086*
Eflornithine, *3570*
Melarsoprol, *5883*
Nifurtimox, *6622*
Pentamidine, *7227*
Propamidine, *7911*
Quinapyramine, *8168*
Stilbamidine, *8945*
Suramin Sodium, *9135*
Trypan Red, *9973*
Tryparsamide, *9974*

ANTIPRURITIC

Camphor, *1734*
Cyproheptadine, *2770*
Dichlorisone, *3057*
3-Hydroxycamphor, *4856*
Menthol, *5905*
Mesulfen, *5988*
Methdilazine, *6036*
Nalfurafine, *6443*
Phenol, *7354*
Thenaldine, *9429*
Tolpropamine, *9682*
Trimeprazine, *9873*

ANTIPSORIATIC

Acitretin, *103*
Alefacept, *220*
Ammonium Salicylate, *547*
Anthralin, *676*
Apremilast, *740*

ANTIPSORIATIC (continued)

6-Azauridine, *897*
Bergapten, *1160*
Calcipotriene, *1644*
Calcitriol, *1647*
Clobetasol, *2358*
Efalizumab, *3567*
Etanercept, *3767*
Etretinate, *3936*
Flurandrenolide, *4227*
Halobetasol Propionate, *4629*
Infliximab, *5014*
Maxacalcitol, *5828*
Tacalcitol, *9152*
Tazarotene, *9218*
Ustekinumab, *10079*
Voclosporin, *10224*

ANTIPSYCHOTIC

Benzamides
Amisulpride, *481*
Nemonapride, *6531*
Remoxipride, *8252*
Sulpiride, *9119*
Sultopride, *9123*

Benzisoxazoles
Iloperidone, *4941*
Paliperidone, *7087*
Risperidone, *8361*

Butyrophenones
Benperidol, *1050*
Bromperidol, *1452*
Droperidol, *3499*
Fluanisone, *4146*
Haloperidol, *4634*
Melperone, *5895*
Moperone, *6351*
Pipamperone, *7568*
Spiperone, *8878*
Timiperone, *9600*
Trifluperidol, *9852*

Phenothiazines
Butaperazine, *1519*
Chlorpromazine, *2191*
Cyamemazine, *2673*
Dixyrazine, *3429*
Fluphenazine, *4220*
Levomepromazine, *5518*
Mesoridazine, *5980*
Metofenazate, *6221*
Perazine, *7265*
Pericyazine, *7279*
Perimethazine, *7283*
Perphenazine, *7297*
Piperacetazine, *7574*
Pipotiazine, *7595*
Prochlorperazine, *7879*
Promazine, *7898*
Sulforidazine, *9093*
Thiopropazate, *9509*
Thioproperazine, *9510*
Thioridazine, *9512*
Trifluoperazine, *9846*
Triflupromazine, *9853*

Thioxanthenes
Chlorprothixene, *2194*
Clopenthixol, *2383*
Flupentixol, *4219*
Thiothixene, *9519*

Other Tricyclics
Asenapine, *821*
Benzquinamide, *1122*

Clocapramine, *2361*
Clothiapine, *2399*
Clozapine, *2406*
Mosapramine, *6363*
Olanzapine, *6914*
Opipramol, *6948*
Prothipendyl, *7998*
Quetiapine, *8155*
Zotepine, *10397*

Others
Aripiprazole, *776*
Bifeprunox, *1214*
Blonanserin, *1320*
Fluspirilene, *4237*
Lurasidone, *5673*
Molindone, *6319*
Penfluridol, *7196*
Perospirone, *7296*
Pimozide, *7546*
Sertindole, *8605*
Ziprasidone, *10371*

ANTIPYRETIC

Acetaminophen, *47*
Aminopyrine, *470*
Aspirin, *841*
Benorylate, *1047*
Berberine, *1156*
Calcium Acetylsalicylate, *1650*
Choline Salicylate, *2213*
Dipyrone, *3390*
Ibuprofen, *4919*
Imidazole Salicylate, *4954*
Lysine Acetylsalicylate, *5698*
Magnesium Acetylsalicylate, *5719*
Meclofenamic Acid, *5850*
Morpholine Salicylate *see 6362*
Naproxen, *6499*
Phenacetin, *7319*
Propacetamol, *7906*
Propyphenazone, *7982*
Ramifenazone, *8220*
Salicylamide *O*-Acetic Acid, *8466*
Sodium Salicylate *see 8469*

ANTIRESTENOTIC

Paclitaxel, *7081*
Rapamycin, *8232*

ANTIRETROVIRAL *see also Antiviral*

Entry Inhibitors
Maraviroc, *5814*
Vicriviroc, *10172*

Fusion Inhibitors
Enfuvirtide, *3636*

Integrase Inhibitors
Dolutegravir, *3460*
Elvitegravir, *3610*
Raltegravir, *8216*

Maturation Inhibitors
Bevirimat, *1196*

Protease Inhibitors
Amprenavir, *585*
Atazanavir, *849*
Darunavir, *2829*
Fosamprenavir, *4276*

Indinavir, *4983*
Lopinavir, *5630*
Nelfinavir, *6528*
Ritonavir, *8366*
Saquinavir, *8506*
Tipranavir, *9615*

Reverse Transcriptase Inhibitors
Nonnucleosides
Delavirdine, *2882*
Efavirenz, *3569*
Etravirine, *3935*
Nevirapine, *6575*
Rilpivirine, *8349*

Nucleosides/Nucleotides
Abacavir, *1*
Apricitabine, *742*
Didanosine, *3109*
Dideoxyadenosine, *3112*
Emtricitabine, *3622*
Lamivudine, *5402*
Stavudine, *8929*
Tenofovir, *9289*
Zalcitabine, *10307*
Zidovudine, *10322*

ANTIRHEUMATIC *see Antiarthritic/ Antirheumatic*

ANTIRICKETTSIAL

p-Aminobenzoic Acid, *418*
Chloramphenicol, *2077*
Tetracycline, *9341*

ANTISEBORRHEIC

Chloroxine, *2181*
Piroctone, *7615*
Pyrithione, *8107*
Resorcinol, *8276*
Selenium Sulfides, *8573*
Tioxolone, *9612*

ANTISEPSIS

Drotrecogin Alfa *see 7995*
Eritoran, *3729*
Opebacan *see 931*

ANTISEPTIC/DISINFECTANT

Alcohols
Dichlorobenzyl Alcohol, *3069*
Ethyl Alcohol, *3814*
Isopropyl Alcohol, *5254*

Aldehydes
Formaldehyde Solution *see 4263*
Glutaraldehyde, *4508*

Dyes
Acriflavine, *115*
Aminacrine, *402*
Brilliant Green, *1380*
Ethacridine, *3771*
Gentian Violet, *4430*
Magenta I, *5715*
Methyl Blue, *6102*

Guanidines
Alexidine, *226*
Chlorhexidine, *2092*
Picloxydine, *7510*
Polihexanide, *7677*

Halogens/Halogen Containing Compounds
Bibrocathol, *1203*
Bismuth Iodide Oxide, *1270*
Bismuth Tribromophenate, *1291*
Bornyl Chloride, *1342*
Calcium Iodate, *1678*
Cloflucarban, *2372*
Fluorosalan, *4210*
Iodic Acid, *5056*
Iodine, *5057*
Iodine Monochloride, *5060*
Iodine Trichloride, *5063*
Iodoform, *5076*
Oxychlorosene, *7055*
Povidone-Iodine, *7815*
Sodium Hypochlorite, *8762*
Trichloroisocyanuric Acid, *9806*
Triclocarban, *9819*
Triclosan, *9822*
Troclosene Potassium, *9946*

Mercurial compounds
Meralein Sodium, *5932*
Merbromin, *5934*
Mercuric Chloride, *5943*
Mercuric Sodium *p*-Phenolsulfonate, *5954*
Mercuric Sulfide, Red, *5957*
Mercurophen, *5959*
Mercurous Acetate, *5960*
Mercurous Chloride, *5962*
Nitromersol, *6697*
Thimerosal, *9470*

Nitrofurans
Furazolidone, *4330*
Nifuroxime, *6620*
Nifurzide, *6624*
Nitrofurazone, *6687*

Peroxides/Permanganates
Calcium Peroxide, *1691*
Carbamide Peroxide, *1784*
Hydrogen Peroxide, *4835*
Magnesium Peroxide, *5744*
Potassium Permanganate, *7776*
Strontium Peroxide, *8977*

Phenols
Bithionol, *1307*
Cadmium Salicylate, *1630*
Carvacrol, *1872*
Chlorocresol, *2133*
Chloroxylenol, *2182*
Creosote, Wood, *2561*
Cresol, *2564*
Fenticlor, *4031*
Hexachlorophene, *4716*
1-Naphthyl Salicylate, *6495*
2-Naphthyl Salicylate, *6496*
2,4,6-Tribromo-*m*-cresol, *9774*

Quaternary Ammonium Compounds
Benzalkonium Chloride, *1061*
Benzethonium Chloride, *1076*
Benzoxonium Chloride, *1114*
Cetalkonium Chloride, *2016*
Cethexonium Bromide, *2017*
Cetrimonium Bromide, *2024*
Cetylpyridinium Chloride, *2031*
Dequalinium Chloride, *2914*
Domiphen Bromide, *3464*
Lapyrium Chloride, *5421*
Laurolinium Acetate, *5442*
Methylbenzethonium Chloride, *6097*

Phenoctide, *7353*
Tibezonium Iodide, *9581*
Triclobisonium Chloride, *9818*

Quinolines
Aminoquinuride, *471*
Broxyquinoline, *1462*
Chloroxine, *2181*
Chlorquinaldol, *2196*
Cloxyquin, *2405*
Hydrastine, *4806*
8-Hydroxyquinoline, *4881*
Iodochlorhydroxyquin, *5075*

Silver compounds
Silver Bromide, *8645*
Silver Fluoride, *8653*
Silver Lactate, *8656*
Silver Nitrate, *8657*
Silver Protein, *8666*

Others
Aluminum Acetate, *320*
Aluminum Acetotartrate, *321*
Aluminum Lactate, *343*
Aluminum Subacetate, *359*
3-Amino-4-hydroxybutyric Acid, *440*
Boric Acid, *1339*
Chloroazodin, *2121*
m-Cresyl Acetate, *2572*
Cupric Citrate, *2626*
Cupric Sulfate, *2643*
Dibromopropamidine, *3029*
Hexamidine, *4730*
Ichthammol, *4924*
Noxythiolin, *6813*
Octenidine, *6843*
Ornidazole, *6967*
Policresulen, *7673*
Polynoxylin, *7694*
β-Propiolactone, *7934*
α-Terpineol, *9313*

ANTISPASMODIC *see also* **Antimuscarinic**
Adiphenine, *149*
Alibendol, *236*
Alverine, *365*
Ambutonium Bromide, *381*
Aminopentamide, *455*
Aminopromazine, *463*
Atropine, *866*
Benactyzine, *1031*
Bevonium Methyl Sulfate, *1197*
Bietamiverine, *1209*
Butropium Bromide, *1536*
N-Butylscopolammonium Bromide, *1591*
Buzepide, *1603*
Caroverine, *1859*
Chlorbenzoxamine, *2081*
Cimetropium Bromide, *2283*
Clebopride, *2342*
Clidinium Bromide, *2352*
Cyclonium Iodide, *2729*
Darifenacin, *2826*
Dicyclomine, *3108*
Difemerine, *3152*
Drofenine, *3494*
Drotaverine, *3503*
Emepronium Bromide, *3615*
Ethaverine, *3783*
Fenoverine, *4013*
Fentonium Bromide, *4033*
Fesoterodine, *4095*

Flavoxate, *4128*
Flopropione, *4138*
Glycopyrrolate, *4537*
Hymecromone, *4894*
Hyoscyamine, *4897*
Isopropamide Iodide, *5248*
Mebeverine, *5838*
Memantine, *5897*
Mepenzolate Bromide, *5915*
Methantheline Bromide, *6030*
Methscopolamine Bromide, *6077*
Moxaverine, *6374*
Otilonium Bromide, *6997*
Oxyphencyclimine, *7072*
Oxyphenonium Bromide, *7074*
Phloroglucinol, *7439*
Pinaverium Bromide, *7552*
Pipenzolate Bromide, *7573*
Piperidolate, *7582*
Piperilate, *7583*
Pipoxolan Hydrochloride, *7596*
Poldine Methylsulfate, *7671*
Pramiverin, *7827*
Prifinium Bromide, *7861*
Propantheline Bromide, *7919*
Proxazole, *8012*
Rociverine, *8374*
Scopolamine, *8543*
Tiemonium Iodide, *9587*
Tigloidine, *9590*
Timepidium Bromide, *9599*
Tiquizium Bromide, *9616*
Tiropramide, *9622*
Trepibutone, *9747*
Tricromyl, *9823*
Tridihexethyl Iodide, *9826*
Trimebutine, *9868*
Trospium Chloride, *9966*
Valethamate Bromide, *10092*

ANTITHROMBOCYTHEMIC
Anagrelide, *616*

ANTITHROMBOCYTOPENIC
Eltrombopag, *3609*
Oprelvekin *see 5044*
Romiplostim, *8383*

ANTITHROMBOTIC *see also* **Anticoagulant; Thrombolytic**
Abciximab, *5*
Apixaban, *723*
Ardeparin, *765*
Argatroban, *769*
Aspirin, *841*
Beraprost, *1154*
Bivalirudin, *1311*
Cangrelor, *1749*
Cilostazol, *2280*
Clopidogrel, *2385*
Cloricromen, *2393*
Dabigatran, *2794*
Dalteparin, *2804*
Daltroban, *2805*
Danaparoid, *2810*
Defibrotide, *2864*
Dipyridamole, *3388*
Edoxaban, *3562*
Elinogrel, *3600*
Enoxaparin, *3642*
Eptifibatide, *3698*
Fondaparinux Sodium, *4259*
Idraparinux, *4932*

ANTITHROMBOTIC (continued)

Indobufen, 5001
Lamifiban, 5399
Melagatran, 5879
Nadroparin, 6433
Otamixaban, 6996
Ozagrel, 7079
Pamicogrel, 7103
Picotamide, 7519
Plafibride, 7630
Prasugrel, 7833
Reviparin Sodium, 8293
Ridogrel, 8337
Rivaroxaban, 8368
Sulfinpyrazone, 9079
Sulodexide, 9116
Taprostene, 9190
Ticagrelor, 9583
Ticlopidine, 9585
Tifacogin see 9623
Tinzaparin, 9606
Tirofiban, 9620
Triflusal, 9856
Vorapaxar, 10230
Xemilofiban, 10269
Ximelagatran, 10274

ANTITUBERCULAR see Antibacterial (Tuberculostatic)

ANTITUMOR see Antineoplastic

ANTITUSSIVE

Benproperine, 1051
Benzonatate, 1098
Bromoform, 1430
Butamirate, 1516
Butethamate, 1525
Caramiphen, 1778
Carbetapentane, 1795
Chlophedianol, 2066
Clobutinol, 2360
Cloperastine, 2384
Codeine, 2448
Dextromethorphan, 2957
Dihydrocodeine, 3189
Dimemorfan, 3229
Dropropizine, 3500
Eprazinone, 3688
Ethyl Dibunate, 3843
Ethylmorphine, 3884
Fominoben, 4255
Hydrocodone, 4823
Isoaminile, 5155
Levopropoxyphene, 5523
Narceine, 6504
Normethadone, 6798
Noscapine, 6807
Oxeladin, 7029
Oxolamine, 7043
Pholcodine, 7441
Picoperine, 7516
Pipazethate, 7569
Piperidione, 7581
Prenoxdiazine, 7854
Thebacon, 9426
Tipepidine, 9613
Zipeprol, 10370

ANTIULCERATIVE see also Antacid

Aceglutamide Aluminum Complex see 26

Acetoxolone, 74
Aldioxa, 216
Benexate Hydrochloride, 1040
Bismuth Subnitrate, 1284
Carbenoxolone, 1794
Cetraxate, 2023
Cimetidine, 2282
Colloidal Bismuth Subcitrate, 2469
Ebrotidine, 3532
Ecabet, 3535
Enprostil, 3645
Famotidine, 3971
Gefarnate, 4411
Guaiazulene, 4590
Irsogladine, 5146
Lafutidine, 5396
Lansoprazole, 5413
S-Methylmethionine, 6170
Misoprostol, 6293
Nizatidine, 6747
Omeprazole, 6939
Ornoprostil, 6970
γ-Oryzanol, 6983
Pantoprazole, 7117
Pifarnine, 7532
Pirenzepine, 7604
Plaunotol, 7647
Polaprezinc, 7669
Rabeprazole, 8206
Ranitidine, 8228
Rebamipide, 8242
Revaprazan, 8291
Rioprostil, 8359
Rosaprostol, 8391
Roxatidine Acetate, 8407
Sofalcone, 8828
Sucralfate, 9010
Teprenone, 9296
Trimoprostil, 9895
Trithiozine, 9937
Troxipide, 9970
Zinc Acexamate see 46
Zolimidine, 10389

ANTIUROLITHIC

Acetohydroxamic Acid, 62
Allopurinol, 271
Potassium Citrate, 7744
Succinimide, 9001

ANTIVENIN

Antivenin (Latrodectus mactans), 707

ANTIVERTIGO see also Antiemetic; Antihistaminic

Acetylleucine, 91
Dimenhydrinate, 3230
Meclizine, 5848

ANTIVIRAL see also Antiretroviral

Adamantanes
Amantadine, 368
Rimantadine, 8351
Tromantadine, 9950

Imidazoquinolines
Imiquimod, 4964
Resiquimod, 8271

Monoclonal Antibodies
Palivizumab, 7089

Nucleosides/Nucleotides
Acyclovir, 134
Adefovir, 139
Brivudine, 1385
Cidofovir, 2271
Clevudine, 2351
Edoxudine, 3563
Entecavir, 3649
Famciclovir, 3970
Floxuridine, 4144
Ganciclovir, 4393
Idoxuridine, 4931
Inosine Pranobex, 5019
Penciclovir, 7192
Ribavirin, 8323
Sorivudine, 8853
Taribavirin, 9197
Telbivudine, 9258
Tenofovir, 9289
Trifluridine, 9855
Valacyclovir, 10085
Valganciclovir, 10093
Vidarabine, 10173

Polynucleotides
Fomivirsen, 4256
Rintatolimod, 8357

Protease Inhibitors
Boceprevir, 1322
Telaprevir, 9256
Vaniprevir, 10122

Sialic Acid Analogs
Laninamivir, 5406
Oseltamivir, 6986
Peramivir, 7264
Zanamivir, 10311

Others
Acemannan, 27
n-Docosanol, 3444
Foscarnet Sodium, 4278
Interferon-α, 5034
Interferon-β, 5035
Lysozyme, 5701
Merimepodib, 5968
Methisazone, 6049
Moroxydine, 6358
Peginterferon Alfa-2a, 7175
Peginterferon Alfa-2b, 7176
Pleconaril, 7648
Podophyllotoxin, 7663
Tecovirimat, 9244

ANXIOLYTIC

Arylpiperazines
Buspirone, 1507
Ipsapirone, 5125
Lesopitron, 5501
Tandospirone, 9183

Benzodiazepine Derivatives
Alprazolam, 308
Bromazepam, 1392
Camazepam, 1727
Chlordiazepoxide, 2085
Clobazam, 2356
Clorazepic Acid, 2392
Clotiazepam, 2400
Cloxazolam, 2404
Diazepam, 2997
Ethyl Loflazepate, 3876
Etizolam, 3919
Fludiazepam, 4159
Halazepam, 4624

Ketazolam, *5345*
Lorazepam, *5636*
Medazepam, *5857*
Metaclazepam, *5990*
Mexazolam, *6245*
Nordazepam, *6780*
Oxazepam, *7025*
Oxazolam, *7026*
Pinazepam, *7553*
Prazepam, *7836*
Tofisopam, *9662*

Carbamates
Meprobamate, *5929*
Tybamate, *10010*

Others
Benzoctamine, *1089*
Captodiamine, *1775*
Chlormezanone, *2106*
Deramciclane, *2916*
Duloxetine, *3512*
Escitalopram *see 2317*
Etifoxine, *3913*
Hydroxyzine, *4889*
Loxapine, *5645*
Pagoclone, *7084*
Phenibut *see 460*
Pregabalin, *7843*
Valnoctamide, *10098*
Venlafaxine, *10140*

AROMATASE INHIBITORS *see also*
Antineoplastic (Hormonal)
Aminoglutethimide, *437*
Anastrozole, *619*
Exemestane, *3954*
Fadrozole, *3968*
Formestane, *4268*
Letrozole, *5502*
Vorozole, *10233*

ASTRINGENT
Alkannin, *249*
Aluminum Acetate, *320*
Aluminum Acetotartrate, *321*
Aluminum Chlorate, *332*
Aluminum Chloride, *333*
Aluminum Hydroxychloride, *339*
Aluminum Potassium Sulfate,
354
Aluminum Sodium Sulfate, *357*
Aluminum Subacetate, *359*
Aluminum Sulfate, *360*
Ammonium Alum *see 323*
Ammonium Ferric Sulfate, *515*
Baicalein, *935*
Bismuth Oxide, *1273*
Bismuth Tannate, *1288*
Boric Acid, *1339*
Calcium Hydroxide, *1675*
Cupric Citrate, *2626*
Dichloroacetic Acid, *3059*
Ferric Chloride, *4046*
Gallic Acid, *4375*
Iodic Acid, *5056*
Lead Acetate, *5452*
Methionic Acid, *6046*
Silver Bromide, *8645*
Silver Lactate, *8656*
Silver Nitrate, *8657*
Sodium Formate, *8756*
Tannic Acid, *9184*
Zinc Acetate, *10327*

Zinc Carbonate, *10330*
Zinc Chloride, *10331*
Zinc Iodide, *10339*
Zinc Oxide, *10347*
Zinc Peroxide, *10349*
Zinc *p*-Phenolsulfonate, *10350*
Zinc Salicylate, *10355*
Zinc Sulfate, *10359*

BENZODIAZEPINE ANTAGONIST
Flumazenil, *4166*

BETA-BLOCKER *see β-Adrenergic*
Blocker

BLOOD SUBSTITUTE
Fluosol DA, *4215*
Polymerized Pyridoxylated He-
moglobin, *7692*

BONE RESORPTION INHIBITOR
*see also **Calcium Regulator***
Alendronic Acid, *223*
Clodronic Acid, *2365*
Denosumab, *2897*
Eldecalcitol, *3587*
Etidronic Acid, *3912*
Gallium Nitrate, *4380*
Ibandronic Acid, *4912*
Incadronic Acid, *4969*
Minodronic Acid, *6284*
Neridronic Acid, *6558*
Odanacatib, *6861*
Pamidronic Acid, *7104*
Risedronic Acid, *8360*
Strontium Ranelate *see 8224*
Tiludronic Acid, *9598*
Zoledronic Acid, *10388*

BRADYCARDIC AGENT *see also*
Antianginal
Tedisamil, *9246*
Zatebradine, *10313*

BRADYKININ ANTAGONIST
Icatibant, *4922*

BRONCHODILATOR *see also Anti-*
asthmatic; Glucocorticoid
Ephedrine Derivatives
Albuterol, *210*
Bambuterol, *946*
Bitolterol, *1308*
Carbuterol, *1832*
Clenbuterol, *2345*
Dioxethedrine, *3331*
Ephedrine, *3663*
Epinephrine, *3674*
Eprozinol, *3693*
Etafedrine, *3764*
Ethylnorepinephrine, *3888*
Fenoterol, *4012*
Formoterol, *4272*
Hexoprenaline, *4743*
Isoetharine, *5215*
Isoproterenol, *5263*
Mabuterol, *5703*

Metaproterenol, *6002*
Pirbuterol, *7602*
Procaterol, *7877*
Protokylol, *8004*
Reproterol, *8259*
Rimiterol, *8353*
Salmeterol, *8475*
Soterenol, *8855*
Terbutaline, *9302*
Tulobuterol, *9991*

Quaternary Ammonium Com-
pounds
Aclidinium Bromide, *106*
Ipratropium Bromide, *5117*
Oxitropium Bromide, *7042*
Tiotropium Bromide, *9610*

Xanthine Derivatives
Acefylline, *25*
Aminophylline, *461*
Bamifylline, *948*
Choline Theophyllinate, *2215*
Doxofylline, *3486*
Dyphylline, *3526*
Etamiphylline, *3766*
Etofylline, *3924*
Guaithylline *see 4591*
Proxyphylline, *8014*
Theobromine, *9433*
1-Theobromineacetic Acid, *9434*
Theophylline, *9436*

Others
Fenspiride, *4026*
Indacaterol, *4970*
Methoxyphenamine, *6071*
Tretoquinol, *9749*

CALCINEURIN INHIBITOR *see also*
Immunosuppressant
Cyclosporine *see 2750*
Pimecrolimus, *7541*
Tacrolimus, *9155*
Voclosporin, *10224*

CALCIUM CHANNEL BLOCKER
*see also **Antianginal; Antihyperten-***
*sive; Vasodilator (Coronary)***
Arylalkylamines
Bepridil, *1152*
Fendiline, *4002*
Gallopamil, *4386*
Mibefradil, *6252*
Prenylamine, *7855*
Semotiadil, *8585*
Verapamil, *10144*

Benzothiazepines
Clentiazem, *2346*
Diltiazem, *3224*

Dihydropyridine Derivatives
Amlodipine, *487*
Aranidipine, *756*
Barnidipine, *998*
Benidipine, *1045*
Cilnidipine, *2278*
Clevidipine, *2350*
Efonidipine, *3571*
Felodipine, *3987*
Isradipine, *5287*
Lacidipine, *5379*
Lercanidipine, *5500*
Manidipine, *5808*

CALCIUM CHANNEL BLOCKER
Dihydropyridine Derivatives
(*continued*)

Nicardipine, *6580*
Nifedipine, *6613*
Nilvadipine, *6630*
Nimodipine, *6636*
Nisoldipine, *6651*
Nitrendipine, *6662*

Piperazine Derivatives
Cinnarizine, *2306*
Dotarizine, *3477*
Flunarizine, *4175*
Lidoflazine, *5537*
Lomerizine, *5620*

Others
Bencyclane, *1035*
Etafenone, *3765*
Perhexiline, *7278*

CALCIUM REGULATOR
Calcifediol, *1641*
Calcitonin, *1646*
Calcitriol, *1647*
Dihydrotachysterol, *3198*
Elcatonin, *3585*
Ipriflavone, *5118*
Parathyroid Hormone, *7138*
Teriparatide Acetate, *9309*

CALCIUM SUPPLEMENT *see Replenishers/Supplements*

CANCER CHEMOTHERAPY *see Antineoplastic*

CAPILLARY PROTECTANT *see Vasoprotectant*

CARBONIC ANHYDRASE INHIBITOR *see also Antiglaucoma; Diuretic*
Acetazolamide, *52*
Brinzolamide, *1382*
Dichlorphenamide, *3087*
Dorzolamide, *3475*
Ethoxzolamide, *3809*
Flumethiazide, *4170*
Methazolamide, *6034*

CARDIAC DEPRESSANT (ANTIARRHYTHMIC) *see Antiarrhythmic*

CARDIOPROTECTIVE
Acadesine, *17*
Cariporide, *1840*
Dexrazoxane *see 8241*

CARDIOTONIC
Acetyldigitoxins, *84*
2-Amino-4-picoline, *462*
Amrinone, *587*
Bucladesine, *1473*
Convallatoxin, *2497*
Cymarin, *2758*

Denopamine, *2896*
Deslanoside, *2922*
Digitalin, *3177*
Digitalis, *3178*
Digitoxin, *3182*
Digoxin, *3186*
Dobutamine, *3440*
Docarpamine, *3441*
Dopamine, *3470*
Dopexamine, *3472*
Enoximone, *3643*
Gitoxin, *4469*
Glycocyamine, *4530*
Heptaminol, *4692*
Hydrastinine, *4807*
Ibopamine, *4915*
Lanatosides, *5404*
Levosimendan, *5525*
Loprinone, *5632*
Milrinone, *6278*
Neriifolin, *6559*
Oleandrin, *6919*
Ouabain, *7000*
Oxyfedrine, *7061*
Pimobendan, *7545*
Prenalterol, *7853*
Proscillaridin, *7985*
Resibufogenin, *8266*
Scillaren, *8536*
Scillarenin, *8537*
Strophanthin, *8982*
Sulmazole, *9115*
Theobromine, *9433*
Vesnarinone, *10166*
Xamoterol, *10253*

CATHARTIC *see Laxative/Cathartic*

CATHEPSIN K INHIBITOR *see also Antiosteoporotic; Bone Resorption Inhibitor*
Odanacatib, *6861*

CATION-EXCHANGE RESIN *see Ion-exchange Resin*

CCK ANTAGONIST
Loxiglumide, *5646*
Proglumide, *7891*

CENTRAL STIMULANT *see CNS Stimulant*

CEREBRAL VASODILATOR *see Vasodilator (Cerebral)*

CETP INHIBITOR *see also Antiatherosclerotic; Antilipemic*
Anacetrapib, *615*
Dalcetrapib, *2801*
Evacetrapib, *3947*
Torcetrapib, *9708*

CHELATING AGENT
Ammonium Tetrathiomolybdate, *557*
Deferasirox, *2861*

Deferiprone, *2862*
Deferoxamine, *2863*
Ditiocarb Sodium, *3421*
EDTA, *3565*
Penicillamine, *7197*
Pentetate Calcium Trisodium, *7236*
Pentetic Acid, *7237*
Succimer, *8995*
Trientine, *9828*

CHOLECYSTOKININ ANTAGONIST *see CCK Antagonist*

CHOLELITHOLYTIC AGENT
Chenodiol, *2054*
Methyl *tert*-Butyl Ether, *6106*
Monoctanoin, *6338*
Ursodiol, *10074*

CHOLERETIC
Alibendol, *236*
Anethole Trithione, *637*
Cholic Acid, *2210*
Clanobutin, *2337*
Cynarine, *2763*
Dehydrocholic Acid, *2872*
Deoxycholic Acid, *2901*
Dimecrotic Acid, *3226*
α-Ethylbenzyl Alcohol, *3823*
Fenipentol, *4006*
Hymecromone, *4894*
Menbutone, *5902*
Osalmid, *6984*
Ox Bile Extract, *7027*
4,4'-Oxydi-2-butanol, *7059*
Piprozolin, *7598*
4-Salicyloylmorpholine, *8470*
Sincalide, *8681*
Taurocholic Acid, *9212*
Tocamphyl, *9649*
Trepibutone, *9747*

CHOLESTERYL ESTER TRANSFER PROTEIN INHIBITOR *see See CETP Inhibitor*

CHOLINERGIC *see also Cholinesterase Inhibitor*
Aceclidine *see 8199*
Acetylcholine Bromide, *81*
Acetylcholine Chloride, *82*
Aclatonium Napadisilate, *105*
Bethanechol Chloride, *1187*
Carbachol, *1781*
Demecarium Bromide, *2889*
Dexpanthenol, *2949*
Echothiophate Iodide, *3548*
Edrophonium Chloride, *3564*
Eptastigmine, *3695*
Furtrethonium, *4341*
Guanidine, *4597*
Isoflurophate, *5222*
Methacholine Chloride, *6012*
Muscarine, *6395*
Neostigmine, *6549*
Oxapropanium Iodide, *7022*
Physostigmine, *7496*
Pyridostigmine Bromide, *8090*
Xanomeline, *10254*

CHOLINESTERASE INHIBITOR
see also **Cholinergic**

Ambenonium Chloride, *373*
Distigmine Bromide, *3405*
Edrophonium Chloride, *3564*
Eptastigmine, *3695*
Galantamine, *4370*
Neostigmine, *6549*
Phenserine, *7372*
Pyridostigmine Bromide, *8090*
Rivastigmine, *8369*
Tacrine, *9154*

CHOLINESTERASE REACTIVA-TOR

Asoxime Chloride, *825*
HLö-7, *4762*
Obidoxime Chloride, *6827*
Pralidoxime Chloride, *7822*

CNS STIMULANT

Adrafinil, *159*
Amphetamine, *579*
Amphetaminil, *580*
Bemegride, *1029*
Benzphetamine, *1121*
Brucine, *1464*
Caffeine, *1639*
Chlorphentermine, *2188*
Coca, *2438*
Dextroamphetamine, *2956*
Diethylpropion, *3145*
Ethamivan, *3776*
N-Ethylamphetamine, *3818*
Fencamfamine, *3998*
Fenethylline, *4003*
Flurothyl, *4231*
Lisdexamfetamine, *5571*
Lobeline, *5609*
Mazindol, *5831*
Methamphetamine, *6020*
Methylphenidate, *6183*
Modafinil, *6314*
Nikethamide, *6627*
Pemoline, *7188*
Pentylenetetrazole, *7254*
Phendimetrazine, *7334*
Phenmetrazine, *7351*
Phentermine, *7374*
Picrotoxin, *7527*
Pipradrol, *7597*
Prolintane, *7896*
Pyrovalerone, *8122*

COGNITION ACTIVATOR see Noo-tropic

COMT INHIBITOR

Entacapone, *3648*
Tolcapone, *9668*

CONTRACEPTIVE (IMPLANT-ABLE)

Elcometrine, *3586*
Etonogestrel, *3928*
Levonorgestrel see *6793*

CONTRACEPTIVE (INJECTABLE)

Medroxyprogesterone, *5864*
Norethindrone, *6786*

CONTRACEPTIVE (ORAL)

Centchroman, *1972*
Desogestrel, *2928*
Dienogest, *3117*
Drospirenone, *3502*
Ethinyl Estradiol, *3788*
Ethynodiol, *3909*
Gestodene, *4447*
Lynestrenol, *5690*
Mestranol, *5987*
Norethindrone, *6786*
Norethynodrel, *6787*
Norgestimate, *6792*
Norgestrel, *6793*
Ulipristal Acetate, *10030*

CONTRACEPTIVE (TRANSDER-MAL)

Norelgestromin, *6783*

CONVERTING ENZYME INHIBI-TOR see ACE-Inhibitor

CORONARY VASODILATOR see Vasodilator (Coronary)

CYCLOOXYGENASE-2 SELEC-TIVE INHIBITOR see also Anti-inflammatory (Nonsteroidal)

Celecoxib, *1958*
Etoricoxib, *3930*
Lumiracoxib, *5661*
Parecoxib, *7140*
Rofecoxib, *8377*
Valdecoxib, *10086*

CYTOPROTECTANT (GASTRIC)
see also **Antiulcerative**

Aceglutamide Aluminum Com-plex see *26*
Acetoxolone, *74*
Benexate Hydrochloride, *1040*
Carbenoxolone, *1794*
Cetraxate, *2023*
Dosmalfate, *3476*
Guaiazulene, *4590*
Irsogladine, *5146*
Plaunotol, *7647*
Polaprezinc, *7669*
Rebamipide, *8242*
Sofalcone, *8828*
Sucralfate, *9010*
Teprenone, *9296*
Troxipide, *9970*
Zolimidine, *10389*

DEBRIDING AGENT

Bromelain, *1395*
Collagenase, *2465*
Deoxyribonuclease I, *2907*
Papain, *7119*
Streptodornase, *8953*

DECONGESTANT

Amidephrine, *393*
Cyclopentamine, *2735*
Ephedrine, *3663*
Epinephrine, *3674*
Fenoxazoline, *4016*
Indanazoline, *4972*
Naphazoline, *6453*
Nordefrin, *6781*
Octodrine, *6846*
Oxymetazoline, *7065*
Phenylephrine, *7398*
Phenylpropanolamine, *7417*
Phenylpropylmethylamine, *7418*
Propylhexedrine, *7972*
Pseudoephedrine, *8024*
Tetrahydrozoline, *9364*
Tramazoline, *9727*
Tuaminoheptane, *9982*
Tymazoline, *10015*
Xylometazoline, *10284*

DENTAL PLAQUE INHIBITOR

Delmopinol, *2884*

DEPIGMENTOR

Hydroquinine, *4844*
Hydroquinone, *4845*
Monobenzone, *6334*

DERMATITIS HERPETIFORMIS SUPPRESSANT

Dapsone, *2822*
Sulfapyridine, *9069*

DIAGNOSTIC AID

Alsactide, *310*
Americium ^{241}Am see *386*
p-Aminohippuric Acid, *439*
Anazolene Sodium, *622*
Antipyrine, *703*
Arbutamine, *762*
Arginine, *770*
Bentiromide, *1056*
Betazole, *1185*
Binodenoson, *1229*
Broxuridine, *1461*
Ceruletide, *2004*
Colfosceril Palmitate, *2461*
Congo Red, *2482*
Cosyntropin, *2539*
Cystatin C see *2775*
Dexamethasone, *2945*
Dipyridamole, *3388*
Dobutamine, *3440*
Edrophonium Chloride, *3564*
Evan's Blue, *3948*
Fluorescein, *4192*
D-Galactose, *4365*
Glycerol, *4520*
Gold Sodium Thiosulfate, *4554*
Hexaminolevulinate, *4731*
Histamine, *4756*
Hydroxyamphetamine, *4847*
Ibopamine, *4915*
Indocyanine Green, *5002*
Inulin, *5048*
Iodinated Serum Albumin see *8607*
Isosulphan Blue see *9117*
Mannitol, *5810*
Methacholine Chloride, *6012*
Metyrapone, *6238*
Nickel Sulfate, *6602*
Oleic Acid, *6921*

DIAGNOSTIC AID (*continued*)

Penicilloyl Polylysine, *7210*
3-Pentadecylcatechol, *7219*
Pentagastrin, *7224*
Phenolsulfonphthalein, *7360*
Phenoltetrachlorophthalein, *7361*
Phentolamine, *7376*
Piperoxan, *7591*
Porfimer Sodium, *7718*
Regadenoson, *8247*
Ristocetin, *8362*
Rose Bengal, *8393*
Saralasin, *8509*
Secretin, *8558*
Sodium Benzoate, *8718*
Sodium Chromate Cr 51 *see 8736*
Sodium Iodide I 131 *see 8765*
Sulfobromophthalein Sodium, *9084*
Teriparatide Acetate, *9309*
Thyrotropin, *9569*
Tolonium Chloride, *9679*
Tuberculin, *9985*
Vitamin B$_{12}$, Radioactive, *10213*
Xylose, *10285*

DIAGNOSTIC AID (MRI CONTRAST AGENT)

Ferumoxides, *4090*
Ferumoxsil, *4091*
Ferumoxtran 10, *4092*
Ferumoxytol, *4093*
Gadobenate Dimeglumine, *4352*
Gadobutrol, *4353*
Gadodiamide, *4354*
Gadofosveset, *4355*
Gadopentetic Acid, *4357*
Gadoteridol, *4358*
Gadoversetamide, *4359*
Gadoxetic Acid, *4360*
Mangafodipir, *5787*
Perflubron, *7270*

DIAGNOSTIC AID (RADIOACTIVE IMAGING AGENT)

Florbetapir F 18, *4140*
Fludeoxyglucose F 18, *4158*
Gallium Citrate Ga 67 *see 4379*
IACFT I 123 *see 4910*
Indium In 111 Capromab Pendetide *see 1764*
Indium In 111 Pentetreotide *see 7238*
Indium In 111 Satumomab Pendetide *see 8521*
Iobenguane I 123 *see 5051*
Iodohippurate Sodium I 123 *see 5077*
Iodohippurate Sodium I 131 *see 5077*
Iofetamine I 123, *5092*
Ioflupane I 123, *5093*
Krypton Kr 81m *see 5373*
Selenomethionine Se 75 *see 8578*
Sodium Pertechnetate Tc 99m, *8788*
Technetium Tc 99m Apcitide, *9235*
Technetium Tc 99m Arcitumomab *see 764*
Technetium Tc 99m Bicisate, *9236*

Technetium Tc 99m Butedronate *see 1520*
Technetium Tc 99m Depreotide *see 2912*
Technetium Tc 99m Disofenin *see 3400*
Technetium Tc 99m Exametazime *see 3953*
Technetium Tc 99m Fanolesomab, *9237*
Technetium Tc 99m Gluceptate *see 4491*
Technetium Tc 99m Lidofenin *see 5536*
Technetium Tc 99m Mebrofenin *see 5841*
Technetium Tc 99m Medronate *see 5863*
Technetium Tc 99m Mertiatide, *9238*
Technetium Tc 99m Oxidronate *see 7039*
Technetium Tc 99m Pyrophosphate *see 8915*
Technetium Tc 99m Sestamibi, *9239*
Technetium Tc 99m Succimer *see 8995*
Technetium Tc 99m Sulesomab *see 9027*
Technetium Tc 99m Teboroxime, *9240*
Technetium Tc 99m Tetrofosmin *see 9395*
Xenon Xe 133 *see 10270*

DIAGNOSTIC AID (RADIOPAQUE MEDIUM)

Barium Sulfate, *990*
Bunamiodyl, *1490*
Diatrizoic Acid, *2993*
Ethiodized Oil, *3789*
Iobitridol, *5052*
Iocarmic Acid, *5053*
Iocetamic Acid, *5054*
Iodamide, *5055*
Iodipamide, *5065*
Iodixanol, *5066*
Iodized Oil, *5067*
Iodoalphionic Acid, *5069*
o-Iodohippurate Sodium, *5077*
Iodophthalein Sodium, *5081*
Iodopyracet, *5083*
Ioglycamic Acid, *5094*
Iohexol, *5095*
Iomeglamic Acid, *5096*
Iomeprol, *5097*
Iopamidol, *5100*
Iopanoic Acid, *5101*
Iopentol, *5102*
Iophendylate, *5103*
Iophenoxic Acid, *5104*
Iopromide, *5105*
Iopronic Acid, *5106*
Iothalamic Acid, *5107*
Iotrolan, *5108*
Ioversol, *5109*
Ioxaglic Acid, *5110*
Ioxilan, *5111*
Ipodate, *5115*
Methiodal Sodium, *6045*
Metrizamide, *6234*
Metrizoic Acid, *6235*
Phentetiothalein Sodium, *7375*
Thorium Oxide, *9531*
Tyropanoate Sodium, *10018*

DIAGNOSTIC AID (ULTRASOUND CONTRAST AGENT)

Perfluorohexane, *7271*
Perfluoropropane, *7273*
Serum Albumin, *8607*

DIGESTIVE AID

Amylase, *594*
Bromelain, *1395*
α-Galactosidase, *4366*
Lipase, *5566*
Papain, *7119*
Pepsin, *7259*
Rennin, *8254*

DIPEPTIDYL PEPTIDASE IV INHIBITOR

Alogliptin, *302*
Linagliptin, *5549*
Saxagliptin, *8525*
Sitagliptin, *8690*
Vildagliptin, *10177*

DISINFECTANT *see Antiseptic/Disinfectant*

DIURETIC

Organomercurials
Chlormerodrin, *2105*
Meralluride, *5933*
Mercaptomerin Sodium, *5937*
Mercumatilin Sodium *see 5939*
Mercurous Chloride, *5962*
Mersalyl, *5971*

Purines
Pamabrom, *7101*
Theobromine, *9433*

Steroids
Canrenone, *1753*
Oleandrin, *6919*
Spironolactone, *8887*

Sulfonamide Derivatives
Acetazolamide, *52*
Azosemide, *915*
Bumetanide, *1489*
Chloraminophenamide, *2076*
Clopamide, *2382*
Ethoxzolamide, *3809*
Furosemide, *4338*
Mefruside, *5874*
Methazolamide, *6034*
Piretanide, *7605*
Torsemide, *9711*
Tripamide, *9909*
Xipamide, *10276*

Thiazides and Analogs
Althiazide, *314*
Bendroflumethiazide, *1039*
Benzthiazide, *1123*
Benzylhydrochlorothiazide, *1139*
Buthiazide, *1527*
Chlorothiazide, *2171*
Chlorthalidone, *2200*
Cyclopenthiazide, *2740*
Cyclothiazide, *2752*
Hydrochlorothiazide, *4819*
Hydroflumethiazide, *4826*
Indapamide, *4975*

Methyclothiazide, *6079*
Metolazone, *6224*
Polythiazide, *7705*
Quinethazone, *8172*
Teclothiazide, *9241*
Trichlormethiazide, *9789*

Others
Amiloride, *401*
Ethacrynic Acid, *3772*
Etozolin, *3934*
Isosorbide, *5270*
Mannitol, *5810*
Perhexiline, *7278*
Ticrynafen, *9586*
Triamterene, *9760*
Urea, *10052*

DOPAMINE RECEPTOR AGONIST
*see also **Antihypertensive; Antimigraine; Antiparkinsonian; Prolactin Inhibitor***
Apomorphine, *733*
Bromocriptine, *1424*
Cabergoline, *1606*
Dopexamine, *3472*
Fenoldopam, *4010*
Ibopamine, *4915*
Lisuride, *5574*
Pergolide, *7277*
Pramipexole, *7825*
Quinagolide, *8161*
Ropinirole, *8388*
Rotigotine, *8404*
Roxindole, *8408*
Talipexole, *9171*

DOPAMINE RECEPTOR ANTAGONIST *see also **Antiemetic; Antipsychotic***
Amisulpride, *481*
Clebopride, *2342*
Domperidone, *3467*
Metoclopramide, *6219*
Mosapramine, *6363*
Nemonapride, *6531*
Remoxipride, *8252*
Sarizotan, *8512*
Sulpiride, *9119*
Sultopride, *9123*
Veralipride, *10142*

DOPAMINE-SEROTONIN SYSTEM STABILIZER
Aripiprazole, *776*
Bifeprunox, *1214*

DOPAMINERGIC STABILIZER
Pridopidine, *7860*

ECTOPARASITICIDE
Benzyl Benzoate, *1130*
Carbaryl, *1788*
Crotamiton, *2585*
Dixanthogen, *3428*
Lindane, *5556*
Malathion, *5764*
Mercuric Oleate, *5948*
Mesulfen, *5988*
Permethrin, *7294*

Sulfiram, *9080*
Sulfur, Pharmaceutical, *9107*

ELECTROLYTE REPLENISHER
*see **Replenishers/Supplements***

EMETIC
Apocodeine, *727*
Apomorphine, *733*
Cephaeline, *1973*
Ipecac, *5113*
Sodium Chloride, *8734*
Zinc Acetate, *10327*

ENDOTHELIN RECEPTOR ANTAGONIST *see also **Antineoplastic; Vasodilator (Peripheral)***
Ambrisentan, *378*
Atrasentan, *861*
Bosentan, *1355*
Clazosentan, *2340*
Darusentan, *2830*
Tezosentan, *9400*
Zibotentan, *10320*

ENKEPHALINASE INHIBITOR *see **Neutral Endopeptidase Inhibitor***

ENZYME
Digestive
Amylase, *594*
Lipase, *5566*
Pepsin, *7259*
Rennin, *8254*

Mucolytic
Lysozyme, *5701*

Penicillin inactivating
Penicillinase, *7202*

Proteolytic
Chymopapain, *2263*
Chymotrypsins, *2264*
Collagenase, *2465*
Papain, *7119*
Trypsin, *9975*

ENZYME COFACTOR *see also **Vitamin***
Acetiamine, *53*
Benfotiamine, *1043*
Carnitine, *1849*
Cetotiamine, *2022*
Dexpanthenol, *2949*
Flavin-Adenine Dinucleotide, *4122*
Flavin Mononucleotide, *4123*
Fursultiamine, *4340*
Methylcobalamin, *6116*
Nicotinamide, *6608*
Nicotinic Acid, *6610*
Octotiamine, *6849*
Pantothenic Acid, *7118*
Prosultiamine, *7993*
Pyridoxal 5-Phosphate, *8092*
Pyridoxine, *8095*
Riboflavin, *8325*
Sapropterin, *8505*

Thiamine, *9447*
Thiamine Diphosphate, *9448*
Thiamine Disulfide, *9449*

ENZYME INDUCER (HEPATIC)
Flumecinol, *4167*

ENZYME REPLACEMENT THERAPY
Agalsidase *see 4366*
Alglucerase *see 4487*
Alglucosidase alfa *see 100*
α_1-Antitrypsin, *705*
Galsulfase *see 809*
Idursulfase *see 4934*
Imiglucerase, *4958*
Laronidase *see 4935*
Pancreatic Extract, *7108*
Sacrosidase, *8447*
Taliglucerase Alfa, *9169*
Velaglucerase Alfa, *10135*

ERECTILE DYSFUNCTION TREATMENT
Apomorphine, *733*
Avanafil, *875*
Sildenafil, *8628*
Tadalafil, *9157*
Udenafil, *10027*
Vardenafil, *10124*

ESTROGEN
Nonsteroidal
Chlorotrianisene, *2178*
Dienestrol, *3115*
Diethylstilbestrol, *3148*
Fosfestrol, *4280*
Hexestrol, *4737*
Methestrol, *6040*

Steroidal
Conjugated Estrogens, *2492*
Equilenin, *3699*
Equilin, *3700*
Estradiol, *3758*
Estriol, *3762*
Estrone, *3763*
Ethinyl Estradiol, *3788*
Mestranol, *5987*
Moxestrol, *6375*
Quinestrol, *8171*

ESTROGEN ANTAGONIST *see **Antineoplastic (Hormonal)***

EXPECTORANT
Ambroxol, *380*
Ammonium Bicarbonate, *494*
Bromhexine, *1398*
Calcium Iodide, *1679*
Carbocysteine, *1802*
Guaiacol, *4589*
Guaifenesin, *4591*
Guaithylline *see 4591*
Hydriodic Acid, *4814*
Potassium Guaiacolsulfonate, *7755*
Potassium Iodide, *7764*
Storax, *8949*
Terpin, *9311*

Gelatin, 4415
Monsel's Solution, 6344
1,2-Naphthoquinone, 6479
1-Naphthylamine-4-sulfonic
Acid, 6485
Oxamarin, 7014
Oxidized Cellulose, 7038
Styptic Collodion see 2468
Sulmarin, 9114
Thrombin, 9537
Thromboplastin, 9539
Tolonium Chloride, 9679
Tranexamic Acid, 9730
Vapreotide, 10123
Vasopressin, 10129

HEPARIN ANTAGONIST

Hexadimethrine Bromide, 4718

HEPATOPROTECTANT

S-Adenosylmethionine, 144
Betaine, 1181
Catechin, 1902
Citiolone, 2319
Malotilate, 5776
Methionine, 6047
Orazamide, 6956
Phosphorylcholine, 7474
Protoporphyrin IX, 8006
Silymarin see 6274
Thioctic Acid, 9479
Timonacic, 9602
Tiopronin, 9609

**HISTAMINE H₁-RECEPTOR AN-
TAGONIST** see Antihistaminic

**HISTAMINE H₂-RECEPTOR AN-
TAGONIST** see also Antiulcerative

Cimetidine, 2282
Ebrotidine, 3532
Famotidine, 3971
Lafutidine, 5396
Nizatidine, 6747
Ranitidine, 8228
Roxatidine Acetate, 8407

HIV FUSION INHIBITOR see also
Antiretroviral

Enfuvirtide, 3636

HIV INTEGRASE INHIBITOR see
also *Antiretroviral*

Dolutegravir, 3460
Elvitegravir, 3610
Raltegravir, 8216

HIV PROTEASE INHIBITOR see
also *Antiretroviral*

Amprenavir, 585
Atazanavir, 849
Darunavir, 2829
Fosamprenavir, 4276
Indinavir, 4983
Lopinavir, 5630
Nelfinavir, 6528
Ritonavir, 8366
Saquinavir, 8506

**HMG COA REDUCTASE INHIBI-
TOR** see also *Antilipemic*

Atorvastatin, 855
Cerivastatin, 1992
Fluvastatin, 4245
Lovastatin, 5644
Pitavastatin, 7621
Pravastatin Sodium, 7835
Rosuvastatin, 8402
Simvastatin, 8677

HYPNOTIC see *Sedative/Hypnotic*

HYPOCHOLESTEREMIC see *Antilipemic*

HYPOLIPIDEMIC see *Antilipemic*

HYPOTENSIVE see *Antihypertensive*

IMMUNOMODULATOR

Acemannan, 27
Agatolimod, 180
Aldesleukin see 5038
Amiprilose, 480
Bucillamine, 1470
Canakinumab, 1739
Dirucotide, 3396
Ditiocarb Sodium, 3421
Fingolimod, 4114
Glatiramer, 4470
Imiquimod, 4964
Inosine Pranobex, 5019
Interferon-β, 5035
Interferon-γ, 5036
Laquinimod, 5422
Leflunomide, 5486
Lenalidomide, 5490
Lentinan, 5493
Levamisole, 5511
Lisofylline, 5573
Macrophage Colony-Stimulating
Factor, 5710
Mitoxantrone, 6302
Pidotimod, 7531
Platonin, 7645
Polyoxidonium, 7695
Procodazole, 7880
Propagermanium, 7909
Resiquimod, 8271
Rintatolimod, 8357
Romurtide, 8384
Talactoferrin, 9163
Teriflunomide, 9308
Thalidomide, 9403
Thymalfasin see 9560
Thymomodulin, 9557
Thymopentin, 9558
Thymostimulin, 9562
Ubenimex, 10024
Ustekinumab, 10079
Virulizin®, 10203

IMMUNOSUPPRESSANT

Abetimus Sodium, 6
Alefacept, 220
Alemtuzumab, 222
Atacicept, 848
Azathioprine, 895
Basiliximab, 1003

Belatacept, 1023
Belimumab, 1024
Brequinar, 1376
Cyclosporine see 2750
Daclizumab, 2796
Everolimus, 3950
Gusperimus, 4617
6-Mercaptopurine, 5938
Mizoribine, 6308
Muromonab CD3, 6393
Mycophenolic Acid, 6412
Pimecrolimus, 7541
Rapamycin, 8232
Tacrolimus, 9155
Thioguanine, 9490
Tofacitinib, 9661
Voclosporin, 10224

IMPDH INHIBITOR

Mycophenolic Acid, 6412
Ribavirin, 8323
Tiazofurin, 9580

IMPOTENCE THERAPY see *Erectile
Dysfunction Treatment*

INOTROPIC AGENT see *Cardiotonic*

INSULIN SENSITIZER see also *Antidiabetic*

Pioglitazone, 7565
Rosiglitazone, 8396
Troglitazone, 9948

ION-EXCHANGE RESIN

Cholestyramine, 2209
Colestilan, 2458
Colestipol, 2459
Polidexide, 7674
Sodium Polystyrene Sulfonate,
8797

KERATOLYTIC see also *Antiacne*

Benzoyl Peroxide, 1119
Dichloroacetic Acid, 3059
Glutaraldehyde, 4508
Resorcinol, 8276
Retinoic Acid, 8288
Salicylic Acid, 8469
Sulfur, Pharmaceutical, 9107

**LACTATION STIMULATING HOR-
MONE**

Prolactin, 7894

LAXATIVE/CATHARTIC

Agar, 176
Aloe-Emodin, 300
Bisacodyl, 1246
Bisoxatin Acetate, 1296
Calcium Polycarbophil, 1699
Casanthranol, 1878
Castor Oil, 1898
Cellulose Ethyl Hydroxyethyl
Ether, 1968
Colocynthin, 2471

LAXATIVE/CATHARTIC
(continued)

Danthron, *2814*
Docusate Calcium, *3445*
Docusate Sodium, *3446*
Emodin, *3618*
Frangulin, *4292*
Glucofrangulin, *4488*
Lactitol, *5389*
Lactulose, *5395*
Lubiprostone, *5649*
Magnesium Carbonate Hydroxide, *5724*
Magnesium Chloride, *5726*
Magnesium Citrate, *5727*
Magnesium Hydroxide, *5734*
Magnesium Lactate, *5736*
Magnesium Phosphate, Dibasic, *5745*
Magnesium Sulfate, *5755*
Mercurous Chloride, *5962*
Methylcellulose, *6111*
Oxyphenisatin Acetate, *7073*
Petrolatum, Liquid, *7301*
Phenolphthalein, *7356*
Phenolphthalol, *7358*
Phenoltetrachlorophthalein, *7361*
Picosulfate Sodium, *7518*
Polyethylene Glycol, *7688*
Potassium Bisulfate, *7734*
Potassium Bitartrate, *7736*
Potassium Phosphate, Dibasic, *7779*
Potassium Sodium Tartrate, *7790*
Potassium Sulfate, *7793*
Potassium Tartrate, *7796*
Seidlitz Mixture, *8560*
Senna, *8593*
Sennosides, *8594*
Sodium Phosphate, Dibasic, *8789*
Sodium Sulfate, *8809*
Sodium Tartrate, *8812*
Sorbitol, *8851*

LEUKOTRIENE ANTAGONIST *see also Antiasthmatic*

Ibudilast, *4918*
Montelukast, *6346*
Pranlukast, *7830*
Zafirlukast, *10306*

LIPOTROPIC

Inositol, *5021*
Lecithins, *5483*

5-LIPOXYGENASE INHIBITOR *see also Antiallergic; Antiasthmatic; Anti-inflammatory; Antipsoriatic*

Amlexanox, *486*
Tenidap, *9287*
Zileuton, *10323*

LOCAL ANESTHETIC *see Anesthetic (Local)*

LUPUS ERYTHEMATOSUS SUPPRESSANT

Chloroquine, *2165*
Hydroxychloroquine, *4857*

MACULAR DEGENERATION TREATMENT

Aflibercept, *174*
Anecortave Acetate, *634*
Motexafin Lutetium, *6368*
Pegaptanib Sodium, *7174*
Ranibizumab, *8226*
Rostaporfin, *8401*
Verteporfin, *10163*

MAJOR TRANQUILIZER *see Antipsychotic*

MATRIX METALLOPROTEINASE INHIBITOR

Prinomastat, *7869*

MELATONIN RECEPTOR AGONIST

Ramelteon, *8219*
Tasimelteon, *9208*

MINERALOCORTICOID

Aldosterone, *218*
Deoxycorticosterone, *2902*
Deoxycorticosterone Acetate, *2903*
Fludrocortisone, *4161*

MINOR TRANQUILIZER *see Anxiolytic*

MIOTIC

Carbachol, *1781*
Physostigmine, *7496*
Pilocarpine, *7536*

MONOAMINE OXIDASE INHIBITOR *see also Antidepressant; Antihypertensive; Antiparkinsonian*

Befloxatone, *1020*
Iproclozide, *5120*
Iproniazid, *5122*
Isocarboxazid, *5202*
Lazabemide, *5449*
Moclobemide, *6312*
Pargyline, *7142*
Phenelzine, *7335*
Safinamide, *8450*
Selegiline, *8565*
Toloxatone, *9680*
Tranylcypromine, *9734*

MUCOLYTIC

Acetylcysteine, *83*
Bromhexine, *1398*
Carbocysteine, *1802*
Domiodol, *3463*
Erdosteine, *3704*
Lysozyme, *5701*
Mecysteine, *5856*
Mesna, *5979*
Sobrerol, *8705*
Stepronin, *8936*
Tiopronin, *9609*
Tyloxapol, *10014*

MUSCLE RELAXANT (SKELETAL) *see also Neuromuscular Blocking Agent*

Afloqualone, *175*
Arbaclofen Placarbil, *758*
Baclofen, *930*
Botulin Toxins, *1362*
Carisoprodol, *1842*
Chlormezanone, *2106*
Chlorphenesin Carbamate, *2185*
Chlorzoxazone, *2201*
Cyclobenzaprine, *2706*
Dantrolene, *2815*
Diazepam, *2997*
Eperisone, *3661*
Idrocilamide, *4933*
Mephenesin, *5917*
Mephenoxalone, *5918*
Methocarbamol, *6052*
Mivacurium Chloride, *6305*
Orphenadrine, *6975*
Phenprobamate, *7370*
Pridinol Mesylate *see 7859*
Quinine, *8177*
Tetrazepam, *9391*
Thiocolchicoside *see 9477*
Tizanidine, *9642*
Tolperisone, *9681*

MUSCLE RELAXANT (SMOOTH) *see Antimuscarinic*

MYDRIATIC

Atropine, *866*
Cyclodrine, *2712*
Cyclopentolate, *2741*
Epinephrine, *3674*
Homatropine, *4768*
Hydroxyamphetamine, *4847*
Phenylephrine, *7398*
Scopolamine, *8543*
Tropicamide, *9959*
Yohimbine, *10300*

NARCOTIC ANALGESIC *see Analgesic*

NARCOTIC ANTAGONIST

Amiphenazole, *479*
Cyclazocine, *2698*
Levallorphan, *5510*
Methylnaltrexone Bromide, *6172*
Nalmefene, *6445*
Nalorphine, *6446*
Naloxone, *6447*
Naltrexone, *6448*

NASAL DECONGESTANT *see Decongestant*

NEURAMINIDASE INHIBITOR *see also Antiviral*

Laninamivir, *5406*
Oseltamivir, *6986*
Peramivir, *7264*
Zanamivir, *10311*

**NEUROKININ RECEPTOR AN-
TAGONIST** see also **Antiemetic**

Aprepitant, *741*

NEUROLEPTIC see **Antipsychotic**

**NEUROMUSCULAR BLOCKING
AGENT**

Depolarizing Agents
Decamethonium Bromide, *2851*
Succinylcholine Bromide, *9005*
Succinylcholine Chloride, *9006*
Succinylcholine Iodide, *9007*

Nondepolarizing Agents
Alcuronium, *214*
Atracurium Besylate, *859*
Doxacurium Chloride, *3480*
Fazadinium Bromide, *3980*
Gallamine Triethiodide, *4373*
Metocurine Iodide, *6220*
Pancuronium Bromide, *7109*
Pipecurium Bromide, *7571*
Rapacuronium Bromide, *8231*
Rocuronium, *8375*
Tubocurarine Chloride, *9987*
Vecuronium Bromide, *10132*

**NEUROMUSCULAR BLOCK RE-
VERSANT**

Edrophonium Chloride, *3564*
Neostigmine, *6549*
Pyridostigmine Bromide, *8090*
Sugammadex, *9019*

**NEUTRAL ENDOPEPTIDASE
INHIBITOR** see also **Antihyperten-
sive; Antidiarrheal; Vasopeptidase
Inhibitor**

Candoxatril, *1745*
Racecadotril, *8207*

NEUROPROTECTIVE

Aptiganel, *747*
Arimoclomol, *775*
Citicoline, *2318*
Davunetide, *2836*
Dexanabinol, *2946*
Disufenton, *3406*
Lubeluzole, *5648*
Memantine, *5897*
Remacemide, *8250*
Repinotan, *8257*
Riluzole, *8350*
Xaliproden, *10252*
Ziconotide, *10321*

NMDA RECEPTOR ANTAGONIST
see also **Neuroprotective**

Aptiganel, *747*
Dexanabinol, *2946*
Memantine, *5897*
Remacemide, *8250*

NOOTROPIC

Acetylcarnitine, *79*
Adrafinil, *159*

Aniracetam, *655*
Bifemelane, *1210*
Choline Alfoscerate, *2212*
Donepezil, *3468*
Fipexide, *4115*
Idebenone, *4929*
Ispronicline, *5286*
Nefiracetam, *6524*
Nizofenone, *6748*
Oxiracetam, *7041*
Piracetam, *7600*
Pramiracetam, *7826*
Propentofylline, *7927*
Pyritinol, *8109*
Rivastigmine, *8369*
Sabcomeline, *8444*
Tacrine, *9154*
Vinconate, *10182*
Xanomeline, *10254*

NSAID see **Anti-inflammatory (Nonsteroidal)**

OPIOID ANALGESIC see **Analgesic**

ORAL CONTRACEPTIVE See **Contraceptive (Oral)**

OSTEOINDUCTIVE AGENT

Dibotermin alfa see *1333*

OVARIAN HORMONE

Relaxin, *8249*

OXYTOCIC

Carbetocin, *1796*
Carboprost, *1824*
Deaminooxytocin, *2844*
Ergonovine, *3712*
Gemeprost, *4421*
Methylergonovine, *6142*
Oxytocin, *7078*
Pituitary, Posterior, *7623*
Prostaglandin E_2, *7988*
Prostaglandin $F_{2\alpha}$, *7989*

PARASYMPATHOMIMETIC see
Cholinergic

PEDICULICIDE see **Ectoparasiticide**

PERIPHERAL VASODILATOR see
Vasodilator (Peripheral)

PERISTALTIC STIMULANT see
Gastroprokinetic

**PHOSPHODIESTERASE INHIBI-
TOR** see also **Antiasthmatic; Antithrombotic**

Apremilast, *740*
Avanafil, *875*
Cilomilast, *2279*
Cilostazol, *2280*

Dipyridamole, *3388*
Roflumilast, *8378*
Sildenafil, *8628*
Tadalafil, *9157*
Vardenafil, *10124*

PIGMENTATION AGENT

Methoxsalen, *6059*
Trioxsalen, *9907*

PLASMA VOLUME EXPANDER

Dextran, *2950*
Hetastarch, *4708*
Serum Albumin, *8607*

**PLATELET ACTIVATING FAC-
TOR ANTAGONIST**

Israpafant, *5288*
Lexipafant, *5527*
Rupatadine, *8430*

**POTASSIUM CHANNEL ACTIVA-
TOR/OPENER** see also **Antihypertensive; Antianginal**

Ezogabine, *3959*
Levcromakalim, *5512*
Nicorandil, *6606*
Pinacidil, *7549*

POTASSIUM CHANNEL BLOCKER
see also **Antiarrhythmic**

Azimilide, *904*
Dalfampridine, *2802*
Dofetilide, *3456*
Nifekalant, *6614*
Tedisamil, *9246*

PRESSOR AGENT see **Antihypotensive**

PROGESTOGEN

Algestone Acetophenide, *233*
Allylestrenol, *282*
Chlormadinone Acetate, *2102*
Desogestrel, *2928*
Dienogest, *3117*
Dimethisterone, *3240*
Drospirenone, *3502*
Dydrogesterone, *3521*
Elcometrine, *3586*
Ethynodiol, *3909*
Etonogestrel, *3928*
Flurogestone Acetate, *4230*
Gestodene, *4447*
Gestonorone Caproate, *4448*
17α-Hydroxyprogesterone, *4877*
Lynestrenol, *5690*
Medrogestone, *5862*
Medroxyprogesterone, *5864*
Megestrol Acetate, *5876*
Norelgestromin, *6783*
Norethindrone, *6786*
Norethynodrel, *6787*
Norgesterone, *6791*
Norgestimate, *6792*
Norgestrel, *6793*
Norgestrienone, *6794*

PROGESTOGEN (*continued*)
Norvinisterone, *6806*
Pentagestrone, *7225*
Progesterone, *7889*
Promegestone, *7900*
Trengestone, *9746*

PROLACTIN INHIBITOR
Bromocriptine, *1424*
Cabergoline, *1606*
Lisuride, *5574*
Metergoline, *6009*
Quinagolide, *8161*
Terguride, *9307*

**PROSTAGLANDIN/PROSTAGLAN-
DIN ANALOG** *see also Abortifa-
cient/Interceptive; Antiglaucoma;
Antithrombotic; Antiulcerative; Oxy-
tocic*
Beraprost, *1154*
Bimatoprost, *1225*
Carboprost, *1824*
Enprostil, *3645*
Gemeprost, *4421*
Latanoprost, *5431*
Limaprost, *5543*
Misoprostol, *6293*
Ornoprostil, *6970*
Prostacyclin, *7986*
Prostaglandin E$_1$, *7987*
Prostaglandin E$_2$, *7988*
Prostaglandin F$_{2\alpha}$, *7989*
Rioprostil, *8359*
Rosaprostol, *8391*
Sulprostone, *9120*
Trimoprostil, *9895*
Unoprostone, *10034*

PROTEASE INHIBITOR *see also
HIV Protease Inhibitor*
Aprotinin, *746*
Camostat, *1730*
Gabexate, *4350*
Nafamostat, *6434*
Tipranavir, *9615*
Urinastatin, *10067*

PROTON PUMP INHIBITOR *see
Gastric Proton Pump Inhibitor*

**PULMONARY HYPERTENSION
TREATMENT**
Ambrisentan, *378*
Bosentan, *1355*
Epoprostenol *see 7986*
Iloprost, *4942*
Riociguat, *8358*
Sildenafil, *8628*
Tadalafil, *9157*
Treprostinil, *9748*

PULMONARY SURFACTANT
Beractant, *1153*
Calfactant, *1716*
Colfosceril Palmitate, *2461*
Exosurf®, *3957*
Lucinactant *see 8680*

Poractant Alfa, *7717*
Pumactant, *8049*

PURINOCEPTOR P2Y AGONIST
Denufosol, *2898*
Diquafosol, *3391*

**PURINOCEPTOR P2Y ANTAGO-
NIST**
Cangrelor, *1749*
Clopidogrel, *2385*
Elinogrel, *3600*
Ticagrelor, *9583*
Ticlopidine, *9585*

5α-REDUCTASE INHIBITOR *see
also Antiprostatic Hypertrophy*
Dutasteride, *3518*
Episteride, *3690*
Finasteride, *4113*

RENIN INHIBITOR
Aliskiren, *239*

REPLENISHERS/SUPPLEMENTS
Calcium
Calcium Carbonate, *1659*
Calcium Citrate, *1663*
Calcium Gluconate, *1671*
Calcium Hypophosphite, *1677*
Calcium Lactate, *1680*
Calcium Levulinate, *1681*
Calcium Phosphate, Dibasic,
1694
Calcium Phosphate, Tribasic,
1696

Chromium
Chromium Picolinate, *2240*

Copper
Cupric Bisglycinate, *2617*
Cupric Gluconate, *2630*
Cupric Oxide, *2636*
Cupric Sulfate, *2643*

Electrolyte
Calcium Chloride, *1661*
Potassium Acetate, *7726*
Potassium Chloride, *7742*
Sodium Chloride, *8734*
Sodium Lactate, *8767*

Iodine
Periodyl, *7287*
Potassium Iodate, *7763*
Potassium Iodide, *7764*
Prolonium Iodide, *7897*
Rubidium Iodide, *8419*
Sodium Iodide, *8765*
Strontium Iodide, *8974*

Magnesium
Magnesium Gluconate *see 4492*
Magnesium Sulfate, *5755*

Nutrient
Aceglutamide, *26*
Glucose, *4495*
Glutamic Acid, *4505*
Protein Hydrolysates, *7996*

Phosphorus
Calcium Glycerophosphate, *1672*
Durapatite, *3514*

Potassium
Potassium Acetate, *7726*
Potassium Bicarbonate, *7730*
Potassium Gluconate *see 4492*

Selenium
Sodium Selenate, *8800*
Sodium Selenite, *8802*

Zinc
Zinc Gluconate *see 4492*
Zinc Sulfate, *10359*

RESPIRATORY STIMULANT
Almitrine, *294*
Bemegride, *1029*
Carbon Dioxide, *1809*
Cropropamide, *2582*
Crotethamide, *2586*
Dimefline, *3227*
Doxapram, *3481*
Ethamivan, *3776*
Fominoben, *4255*
Lobeline, *5609*
Mepixanox, *5927*
Nikethamide, *6627*
Picrotoxin, *7527*
Tacrine, *9154*

**RETROVIRAL PROTEASE INHIBI-
TOR** *see HIV Protease Inhibitor*

**REVERSE TRANSCRIPTASE IN-
HIBITOR** *see also Antiretroviral*
Abacavir, *1*
Delavirdine, *2882*
Didanosine, *3109*
Dideoxyadenosine, *3112*
Efavirenz, *3569*
Emtricitabine, *3622*
Etravirine, *3935*
Foscarnet Sodium, *4278*
Lamivudine, *5402*
Nevirapine, *6575*
Rilpivirine, *8349*
Stavudine, *8929*
Tenofovir, *9289*
Zalcitabine, *10307*
Zidovudine, *10322*

SCABICIDE *see Ectoparasiticide*

SCLEROSING AGENT
Ethanolamine, *3782*
Ethylamine, *3817*
2-Hexyldecanoic Acid, *4745*
Polidocanol, *7675*
Sodium Ricinoleate *see 8335*
Sodium Tetradecyl Sulfate, *8817*
Tribenoside, *9769*

SEDATIVE/HYPNOTIC *see also
Anxiolytic*
Acylic Ureides
Acecarbromal, *22*
Bromisovalum, *1404*

Alcohols
Ethchlorvynol, *3784*
Meparfynol, *5913*
tert-Pentyl Alcohol, *7253*
2,2,2-Trichloroethanol, *9803*

Amides
Niaprazine, *6577*
Trimetozine, *9892*
Valdetamide, *10087*

Barbituric Acid Derivatives
Allobarbital, *255*
Amobarbital, *567*
Aprobarbital, *745*
Barbital, *957*
Brallobarbital, *1366*
Butabarbital, *1509*
Butalbital, *1514*
Butethal, *1524*
Cyclobarbital, *2704*
Heptabarbital, *4690*
Hexobarbital, *4740*
Mephobarbital, *5921*
Methitural, *6050*
Pentobarbital, *7243*
Phenobarbital, *7352*
Propallylonal, *7910*
Proxibarbal, *8013*
Secobarbital Sodium, *8557*
Talbutal, *9167*
Vinylbital, *10190*

Benzodiazepine Derivatives
Brotizolam, *1459*
Estazolam, *3757*
Flunitrazepam, *4177*
Flurazepam, *4228*
Haloxazolam, *4638*
Loprazolam, *5631*
Lormetazepam, *5639*
Midazolam, *6261*
Nimetazepam, *6634*
Nitrazepam, *6661*
Quazepam, *8145*
Temazepam, *9275*
Triazolam, *9765*

Bromides
Ammonium Bromide, *502*
Calcium Bromide, *1657*
Lithium Bromide, *5582*
Magnesium Bromide, *5723*
Potassium Bromide, *7739*
Sodium Bromide, *8729*

Chloral Derivatives
Chloral Formamide, *2070*
Chloral Hydrate, *2071*
Dichloralphenazone, *3055*
Triclofos, *9820*

Imidazopyridines
Indiplon, *4984*
Zaleplon, *10309*
Zolpidem, *10391*

Piperidinediones
Glutethimide, *4513*
Methyprylon, *6212*
Piperidione, *7581*
Pyrithyldione, *8108*
Thalidomide, *9403*

Quinazolone Derivatives
Etaqualone, *3769*
Mecloqualone, *5852*

Others
Almorexant, *297*
Bromoform, *1430*
α-Chloralose, *2072*
Clomethiazole, *2375*
Cypripedium, *2767*
Dexmedetomidine, *2948*
Diphenhydramine, *3339*
Doxylamine, *3489*
Etomidate, *3926*
Gaboxadol, *4351*
Paraldehyde, *7131*
Perlapine, *7293*
Propiomazine, *7936*
Ramelteon, *8219*
Rilmazafone, *8346*
Sulfonethylmethane, *9088*
Sulfonmethane, *9090*
Tasimelteon, *9208*
Zopiclone, *10393*

**SELECTIVE ESTROGEN RECEP-
TOR MODULATOR (SERM)** *see
also **Antineoplastic (Hormonal);
Antiosteoporotic***
Arzoxifene, *810*
Bazedoxifene, *1011*
Droloxifene, *3495*
Lasofoxifene, *5430*
Ospemifene, *6991*
Raloxifene, *8215*
Tamoxifen, *9180*
Toremifene, *9709*

**SELECTIVE PROGESTERONE RE-
CEPTOR MODULATOR (SPRM)**
Asoprisnil, *824*
Ulipristal Acetate, *10030*

SERENIC
Eltoprazine, *3608*

**SEROTONIN-DOPAMINE ANTAG-
ONIST** *see also **Antipsychotic***
Asenapine, *821*
Blonanserin, *1320*
Clozapine, *2406*
Iloperidone, *4941*
Lurasidone, *5673*
Olanzapine, *6914*
Paliperidone, *7087*
Perospirone, *7296*
Quetiapine, *8155*
Risperidone, *8361*
Sertindole, *8605*
Ziprasidone, *10371*
Zotepine, *10397*

**SEROTONIN NORADRENALINE
REUPTAKE INHIBITOR (SNRI)**
*see also **Antidepressant; Antiobesity
Agent***
Desvenlafaxine, *2937*
Doxepin, *3483*
Duloxetine, *3512*
Milnacipran, *6275*
Sibutramine, *8624*
Venlafaxine, *10140*

**SEROTONIN RECEPTOR AGO-
NIST** *see also **Anxiolytic; Anti-
depressant; Antimigraine; Neuro-
protective; Serenic***
Almotriptan, *298*
Buspirone, *1507*
Dihydroergotamine, *3191*
Eletriptan, *3597*
Eltoprazine, *3608*
Ergotamine, *3718*
Flibanserin, *4132*
Frovatriptan, *4301*
Ipsapirone, *5125*
Lesopitron, *5501*
Lorcaserin, *5638*
Methysergide, *6213*
Naratriptan, *6502*
Repinotan, *8257*
Rizatriptan, *8370*
Sarizotan, *8512*
Sumatriptan, *9126*
Tandospirone, *9183*
Tegaserod, *9253*
Vilazodone, *10176*
Xaliproden, *10252*
Zolmitriptan, *10390*

**SEROTONIN RECEPTOR ANTAG-
ONIST** *see also **Antidepressant;
Antiemetic; Antihypertensive; Anti-
migraine; Anxiolytic***
Agomelatine, *183*
Alosetron, *304*
Azasetron, *893*
Deramciclane, *2916*
Dolasetron, *3458*
Dotarizine, *3477*
Flibanserin, *4132*
Granisetron, *4572*
Ketanserin, *5344*
Methysergide, *6213*
Nefazodone, *6523*
Ondansetron, *6943*
Oxetorone, *7033*
Palonosetron, *7098*
Piboserod, *7505*
Ramosetron, *8223*
Ritanserin, *8363*
Tropisetron, *9962*

SEROTONIN UPTAKE INHIBITOR
*see also **Antidepressant; Antiobses-
sional***
Citalopram, *2317*
Dapoxetine, *2821*
Escitalopram *see 2317*
Femoxetine, *3989*
Fluoxetine, *4217*
Fluvoxamine, *4246*
Milnacipran, *6275*
Paroxetine, *7148*
Sertraline, *8606*
Vilazodone, *10176*
Zimeldine, *10325*

SIALAGOGUE
Anethole Trithione, *637*
Cevimeline, *2033*
Pilocarpine, *7536*

SKELETAL MUSCLE RELAXANT
*see **Muscle Relaxant (Skeletal)***

SNRI *see Serotonin Noradrenaline Reuptake Inhibitor (SNRI)*

SOMATOSTATIN ANALOG
Edotreotide, *3561*
Lanreotide, *5412*
Octreotide, *6851*
Pasireotide, *7155*
Pentetreotide, *7238*
Vapreotide, *10123*

SPASMOLYTIC *see Antispasmodic*

STOOL SOFTENER *see Laxative/Cathartic*

SUCCINYLCHOLINE SYNERGIST
Hexafluorenium Bromide, *4720*

SUNSCREEN *see Ultraviolet Screen*

SYMPATHOMIMETIC *see α-Adrenergic Agonist; β-Adrenergic Agonist*

THROMBOLYTIC *see also Anticoagulant; Antithrombotic*
Alfimeprase, *230*
Alteplase *see 9624*
Amediplase, *385*
Anistreplase, *666*
Desmoteplase *see 3506*
Duteplase *see 9624*
Monteplase, *6347*
Pamiteplase, *7105*
Plasmin, *7634*
Pro-Urokinase, *8011*
Reteplase, *8284*
Staphylokinase, *8923*
Streptokinase, *8955*
Tenecteplase, *9286*
Tissue Plasminogen Activator, *9624*
Urokinase, *10072*

THROMBOXANE A$_2$-RECEPTOR ANTAGONIST *see also Antithrombotic; Antiasthmatic*
Daltroban, *2805*
Domitroban, *3465*
Ramatroban, *8218*
Ridogrel, *8337*
Seratrodast, *8598*

THROMBOXANE A$_2$-SYTHETASE INHIBITOR *see also Antithrombotic*
Ozagrel, *7079*
Ridogrel, *8337*

THYROID HORMONE *see also Antihypothyroid*
Liothyronine, *5565*
Thyroid, *9565*

Thyroidin, *9566*
Thyroxine, *9570*

THYROID INHIBITOR *see Antihyperthyroid*

THYROTROPIC HORMONE
TRH, *9750*

TISSUE ADHESIVE
n-Butyl Cyanoacrylate, *1566*
Isobutyl Cyanoacrylate, *5185*
Methyl Cyanoacrylate, *6117*
Octyl Cyanoacrylate, *6855*

TOCOLYTIC
Albuterol, *210*
Atosiban, *856*
Fenoterol, *4012*
Hexoprenaline, *4743*
Ritodrine, *8365*
Terbutaline, *9302*

TOPICAL PROTECTANT
Allantoin, *250*
Balsam Peru, *942*
Balsam Traumatic, *944*
Bismuth Phosphate, *1276*
Bismuth Tannate, *1288*
Collodion, *2468*
Gum Benzoin, *4615*
Hydroxypropyl Cellulose, *4879*
Pyroxylin, *8125*
Shark Liver Oil, *8616*
Storax, *8949*
Titanium Dioxide, *9628*
Zinc Oxide, *10347*

TOPOISOMERASE I INHIBITOR *see also Antineoplastic*
9-Aminocamptothecin, *426*
Belotecan, *1028*
Cositecan, *2538*
Edotecarin, *3560*
Irinotecan, *5135*
Rubitecan, *8421*
Topotecan, *9707*

TOPOISOMERASE II INHIBITOR *see also Antineoplastic*
Amrubicin, *588*
Amsacrine, *589*
Daunorubicin, *2834*
Doxorubicin, *3487*
Epirubicin, *3678*
Etoposide, *3929*
Garenoxacin, *4402*
Idarubicin, *4927*
Mitoxantrone, *6302*
Pirarubicin, *7601*
Pixantrone, *7628*
Razoxane, *8241*
Sobuzoxane, *8706*
Teniposide, *9288*

TRANQUILIZER *see Antipsychotic; Anxiolytic*

ULTRAVIOLET SCREEN
p-Aminobenzoic Acid, *418*
Avobenzone, *881*
Bemotrizinol, *1030*
Bisoctrizole, *1294*
β-Carotene, *1855*
Cinoxate, *2310*
4-(Dimethylamino)benzoic Acid, *3254*
Dioxybenzone, *3332*
Drometrizole, *3496*
Ecamsule, *3537*
Ensulizole, *3647*
Ethylhexyl Triazone, *3864*
Homosalate, *4777*
Lawsone, *5448*
Menthyl Anthranilate, *5908*
4-Methylbenzylidene Camphor, *6101*
Mexenone, *6246*
Octabenzone, *6831*
Octocrylene, *6845*
Octyl Methoxycinnamate, *6859*
Octyl Salicylate, *6860*
Oxybenzone, *7053*
Phenyl Salicylate, *7420*
Sulisobenzone, *9113*
Titanium Dioxide, *9628*
Zinc Oxide, *10347*

URICOSURIC *see also Antigout*
Benzbromarone, *1067*
Benziodarone, *1085*
Orotic Acid, *6972*
Oxycinchophen, *7056*
Probenecid, *7872*
Sulfinpyrazone, *9079*
Ticrynafen, *9586*

VASOCONSTRICTOR
Felypressin, *3988*
Ornipressin, *6968*
Terlipressin, *9310*

VASODILATOR (CEREBRAL)
Bencyclane, *1035*
Ciclonicate, *2268*
Cinnarizine, *2306*
Cyclandelate, *2697*
Diisopropylamine Dichloroacetate, *3216*
Eburnamonine, *3534*
Fasudil, *3979*
Flunarizine, *4175*
Ibudilast, *4918*
Ifenprodil, *4936*
Lomerizine, *5620*
Nafronyl, *6437*
Nicergoline, *6581*
Nimodipine, *6636*
Papaverine, *7122*
Pentifylline, *7240*
Vincamine, *10180*
Vinpocetine, *10188*
Viquidil, *10199*

VASODILATOR (CORONARY)

Binodenoson, *1229*
Chloracizine, *2068*
Chromonar, *2243*
Cloricromen, *2393*
Dilazep, *3222*
Erythrityl Tetranitrate, *3735*
Etafenone, *3765*
Fendiline, *4002*
Hexestrol Bis(β-diethylamino-
  ethyl ether), *4738*
Hexobendine, *4741*
Itramin Tosylate, *5293*
Khellin, *5356*
Lidoflazine, *5537*
D-Mannitol Hexanitrate, *5811*
Nitroglycerin, *6694*
Pentaerythritol Tetranitrate, *7222*
Pentrinitrol, *7251*
Perhexiline, *7278*
Pimefylline, *7542*
Prenylamine, *7855*
Propatyl Nitrate, *7925*
Regadenoson, *8247*
Trapidil, *9735*
Tricromyl, *9823*
Trimetazidine, *9874*
Visnadine, *10208*

VASODILATOR (PERIPHERAL)

Bamethan, *947*
Bencyclane, *1035*
Beraprost, *1154*
Betahistine, *1180*
Bradykinin, *1364*
Brovincamine, *1460*
Buflomedil, *1480*
Butalamine, *1513*
Cetiedil, *2019*
Ciclonicate, *2268*
Cinnarizine, *2306*
Cyclandelate, *2697*
Diisopropylamine Dichloroace-
  tate, *3216*
Flunarizine, *4175*
Ifenprodil, *4936*
Iloprost, *4942*
Inositol Niacinate, *5022*
Isoxsuprine, *5285*
Kallidin, *5326*
Kallikrein, *5327*
Moxisylyte, *6378*
Nafronyl, *6437*
Nebivolol, *6516*
Nicergoline, *6581*
Nicotinyl Alcohol, *6612*
Nylidrin, *6821*
Pentifylline, *7240*
Piribedil, *7607*
Prostaglandin E₁, *7987*
Tolazoline, *9666*
Xanthinol Niacinate, *10258*

VASOPEPTIDASE INHIBITOR *see also ACE-Inhibitor; Neutral Endopeptidase Inhibitor*

Omapatrilat, *6937*
Sampatrilat, *8489*

VASOPRESSIN RECEPTOR AGONIST

Desmopressin, *2926*
Felypressin, *3988*
Ornipressin, *6968*
Terlipressin, *9310*

VASOPRESSIN RECEPTOR ANTAGONIST

Conivaptan, *2491*
Lixivaptan, *5606*
Tolvaptan, *9699*

VASOPRESSOR *see Antihypotensive*

VASOPROTECTANT

Bioflavonoids, *1233*
Chromocarb, *2241*
Clazosentan, *2340*
Diosmin, *3325*
Dobesilate Calcium, *3439*
Escin, *3751*
Leucocyanidin, *5504*
Quercetin, *8150*
Rutin, *8438*
Troxerutin, *9969*

VITAMIN/VITAMIN SOURCE *see also Enzyme Cofactor*

Vitamin A
α-Carotene, *1854*
β-Carotene, *1855*
γ-Carotene, *1856*
Vitamin A, *10211*

Vitamin B₁
Acetiamine, *53*
Benfotiamine, *1043*
Cetotiamine, *2022*
Fursultiamine, *4340*
Octotiamine, *6849*
Prosultiamine, *7993*
Thiamine, *9447*
Thiamine Diphosphate, *9448*
Thiamine Disulfide, *9449*

Vitamin B₂
Flavin-Adenine Dinucleotide, *4122*
Flavin Mononucleotide, *4123*
Riboflavin, *8325*

Vitamin B₃ (Niacin)
Nicotinamide, *6608*
Nicotinic Acid, *6610*

Vitamin B₅
Dexpanthenol, *2949*
Pantothenic Acid, *7118*

Vitamin B₆
Pyridoxal 5-Phosphate, *8092*
Pyridoxine, *8095*

Vitamin B₁₂
Cobamamide, *2435*
Hydroxocobalamin, *4846*
Methylcobalamin, *6116*
Vitamin B₁₂, *10212*

Vitamin C
Ascorbic Acid, *819*

Vitamin D
Calcipotriene, *1644*
Calcitriol, *1647*
Ergosterol, *3716*
1α-Hydroxycholecalciferol, *4858*
Vitamin D₂, *10215*
Vitamin D₃, *10216*

Vitamin E
α-Tocopherol, *9653*

Vitamin K
Menadiol, *5898*
Menadione, *5899*
Menadoxime, *5900*
Menaquinones, *5901*
Phylloquinone, *7492*
Vitamin K₅, *10220*

Others
Biotin, *1236*
Carnitine, *1849*
Folic Acid, *4248*
Inositol, *5021*

VULNERARY

Acetylcysteine, *83*
Allantoin, *250*
Becaplermin *see 7639*
Cadexomer Iodine, *1615*
Chitin, *2065*
Dextranomer, *2952*
Oxaceprol, *7002*
Prezatide Copper Acetate *see 4452*
Rusalatide, *8431*
Tocoretinate, *9657*

WILSON'S DISEASE TREATMENT

Ammonium Tetrathiomolybdate, *557*
Ditiocarb Sodium, *3421*
Penicillamine, *7197*
Trientine, *9828*

XANTHINE OXIDASE INHIBITOR *see also Antigout*

Allopurinol, *271*
Febuxostat, *3983*

FORMULA INDEX

FORMULA INDEX

A

AgBr
Silver Bromide, 8645
AgCl
Silver Chloride, 8648
AgClO₃

$AgClO_3$
Silver Chlorate, 8647
AgClO₄

$AgClO_4$
Silver Perchlorate, 8662
AgF
Silver Fluoride, 8653
AgF₂

AgF_2
Silver Difluoride, 8652
AgI
Silver Iodide, 8655
AgIO₃

$AgIO_3$
Silver Iodate, 8654
AgMnO₄

$AgMnO_4$
Silver Permanganate, 8663
AgNO₂

$AgNO_2$
Silver Nitrite, 8658
AgNO₃

$AgNO_3$
Silver Nitrate, 8657
AgO
Silver(II) Oxide, 8661
Ag₂CrO₄

Ag_2CrO_4
Silver Chromate(VI), 8649
Ag₂HgI₄

Ag_2HgI_4
Silver Tetraiodomercurate(II), 8670
Ag₂O

Ag_2O
Silver Oxide, 8660
Ag₂O₄S

Ag_2O_4S
Silver Sulfate, 8668
Ag₂S

Ag_2S
Silver Sulfide, 8669
Ag₂Se

Ag_2Se
Silver Selenide, 8667
Ag₃O₄P

Ag_3O_4P
Silver Phosphate, 8664
AlB₃H₁₂

AlB_3H_{12}
Aluminum Borohydride, 327
AlBr₃

$AlBr_3$
Aluminum Bromide, 328
AlCl₃

$AlCl_3$
Aluminum Chloride, 333
AlCl₃O₉

$AlCl_3O_9$
Aluminum Chlorate, 332
AlCl₄H₄N

$AlCl_4H_4N$
Ammonium Tetrachloroaluminate, 555
AlCl₄Na

$AlCl_4Na$
Sodium Tetrachloroaluminate, 8815
AlF₃

AlF_3
Aluminum Fluoride, 335
AlF₆H₁₂N₃

$AlF_6H_{12}N_3$
Ammonium Hexafluoroaluminate, 521
AlF₆Na₃

AlF_6Na_3
Cryolite, 2595
AlH₃

AlH_3
Aluminum Hydride, 337
AlH₃O₃

AlH_3O_3
Aluminum Hydroxide, 338

AlH₄Li

AlH_4Li
Aluminum Lithium Hydride, 344
AlH₄NO₈S₂

$AlH_4NO_8S_2$
Aluminum Ammonium Sulfate, 323
AlH₆O₆P₃

$AlH_6O_6P_3$
Aluminum Hypophosphite, 340
AlI₃

AlI_3
Aluminum Iodide, 341
AlKO₈S₂

$AlKO_8S_2$
Aluminum Potassium Sulfate, 354
AlN
Aluminum Nitride, 348
AlN₃O₉

AlN_3O_9
Aluminum Nitrate, 347
AlNaO₂

$AlNaO_2$
Sodium Aluminate, 8712
AlNaO₈S₂

$AlNaO_8S_2$
Aluminum Sodium Sulfate, 357
AlO₄P

AlO_4P
Aluminum Phosphate, 352
AlP
Aluminum Phosphide, 353
AlSb
Aluminum Antimonide, 324
Al₂CaH₈

Al_2CaH_8
Aluminum Calcium Hydride, 330
Al₂F₁₈Si₃

$Al_2F_{18}Si_3$
Aluminum Hexafluorosilicate, 336
Al₂MgO₈Si₂

$Al_2MgO_8Si_2$
Aluminum Magnesium Silicate, 345
Al₂O₃

Al_2O_3
Aluminum Oxide, 350
Sapphire, 8504
Al₂O₅Si

Al_2O_5Si
Aluminum Silicate, 356
Al₂O₁₂S₃

$Al_2O_{12}S_3$
Aluminum Sulfate, 360
Al₂S₃

Al_2S_3
Aluminum Sulfide, 361
Al₂Se₃

Al_2Se_3
Aluminum Selenide, 355
AsBr₃

$AsBr_3$
Arsenic Tribromide, 791
AsCl₃

$AsCl_3$
Arsenic Trichloride, 792
AsF₃

AsF_3
Arsenic Trifluoride, 793
AsF₅

AsF_5
Arsenic Pentafluoride, 787
AsGa
Gallium Arsenide, 4377
AsHNa₂O₄

$AsHNa_2O_4$
Sodium Arsenate, Dibasic, 8715
AsH₂KO₄

AsH_2KO_4
Potassium Arsenate, 7728
AsH₃

AsH_3
Arsine, 799
AsH₃O₄

AsH_3O_4
Arsenic Acid, 785
AsI₃

AsI_3
Arsenic Triiodide, 794
AsIn
Indium Arsenide, 4989
AsKO₂

$AsKO_2$
Potassium Arsenite, 7729

AsNaO₂

$AsNaO_2$
Sodium Arsenite, 8716
As₂Ca₃O₈

$As_2Ca_3O_8$
Calcium Arsenate, 1652
As₂Co₃O₈

$As_2Co_3O_8$
Cobaltous Arsenate, 2418
As₂H₂O₄Pb

$As_2H_2O_4Pb$
Lead Arsenite, 5455
As₂O₃

As_2O_3
Arsenic Trioxide, 795
As₂O₄Zn

As_2O_4Zn
Zinc Meta-arsenite, 10342
As₂O₅

As_2O_5
Arsenic Pentoxide, 789
As₂S₃

As_2S_3
Arsenic Trisulfide, 797
As₂S₅

As_2S_5
Arsenic Pentasulfide, 788
As₂Se₃

As_2Se_3
Arsenic Triselenide, 796
As₄S₄

As_4S_4
Arsenic Sulfide, 790
AuCl
Gold Monochloride, 4550
AuCl₃

$AuCl_3$
Gold Trichloride, 4556
AuCl₄H

$AuCl_4H$
Chloroauric Acid, 2120
AuCl₄K

$AuCl_4K$
Potassium Tetrachloroaurate(III), 7800
AuCl₄Na

$AuCl_4Na$
Sodium Tetrachloroaurate(III), 8816
AuH₃O₃

AuH_3O_3
Gold Trihydroxide, 4557
AuNa₃O₆S₄

$AuNa_3O_6S_4$
Gold Sodium Thiosulfate, 4554
Au₂O₃

Au_2O_3
Gold Trioxide, 4558
Au₂S₃

Au_2S_3
Gold Trisulfide, 4559

B

BBr₃

BBr_3
Boron Tribromide, 1349
BCl₃

BCl_3
Boron Trichloride, 1350
BF₃

BF_3
Boron Trifluoride, 1351
BF₄H

BF_4H
Fluoboric Acid, 4180
BF₄K

BF_4K
Potassium Tetrafluoroborate, 7804
BF₄Li

BF_4Li
Lithium Tetrafluoroborate, 5599
BF₄NO

BF_4NO
Nitrosyl Tetrafluoroborate, 6736
BF₄Na

BF_4Na
Sodium Fluoborate, 8753
BH₃O₃

BH_3O_3
Boric Acid, 1339

BH₄K
Potassium Borohydride, 7737
BH₄Li
Lithium Borohydride, 5581
BH₄Na
Sodium Borohydride, 8727
BH₆N
Ammonia Borane, 489
BN
Boron Nitride, 1347
BNaO₂
Sodium Metaborate, 8771
BNaO₃
Sodium Perborate, 8783
BO
Boron Monoxide, 1346
B₂Cl₄
Diboron Tetrachloride, 3021
B₂H₆
Diborane(6), 3020
B₂Mg
Magnesium Diboride, 5728
B₂MgO₄
Magnesium Borate, 5722
B₂O₃
Boron Oxide, 1348
B₂O₄Pb
Lead Borate, 5457
B₃H₆N₃
Borazine, 1337
B₄CaO₇
Calcium Borate, 1655
B₄H₈N₂O₇
Ammonium Borate, 501
B₄H₁₀
Tetraborane(10), 9327
B₄K₂O₇
Potassium Tetraborate, 7799
B₄Li₂O₇
Lithium Borate, 5580
B₄Na₂O₇
Sodium Borate, 8726
B₅H₉
Pentaborane(9), 7213
B₅H₁₁
Pentaborane(11), 7214
B₆H₁₀
Hexaborane(10), 4711
B₁₀H₁₄
Decaborane(14), 2847
BaBr₂
Barium Bromide, 964
BaBr₂O₆
Barium Bromate, 963
BaCl₂
Barium Chloride, 967
BaCl₂O₆
Barium Chlorate, 966
BaCl₂O₈
Barium Perchlorate, 983
BaCrO₄
Barium Chromate(VI), 968
BaF₂
Barium Fluoride, 970
BaF₆Si
Barium Hexafluorosilicate, 972
BaHO₄P
Barium Phosphate, Dibasic, 986
BaH₂O₂
Barium Hydroxide, 973
BaH₄O₄P₂
Barium Hypophosphite, 974
BaHgI₄
Barium Mercuric Iodide, 978
BaI₂
Barium Iodide, 976

BaI₂O₆
Barium Iodate, 975
BaMnO₄
Barium Manganate(VI), 977
BaMn₂O₈
Barium Permanganate, 984
BaN₂O₄
Barium Nitrite, 980
BaN₂O₆
Barium Nitrate, 979
BaO
Barium Oxide, 982
BaO₂
Barium Peroxide, 985
BaO₃S
Barium Sulfite, 993
BaO₃S₂
Barium Thiosulfate, 995
BaO₃Ti
Barium Titanate(IV), 996
BaO₄S
Barium Sulfate, 990
BaO₇U₂
Barium Uranium Oxide, 997
BaS
Barium Sulfide, 991
BaSe
Barium Selenide, 988
BaSi₂
Barium Silicide, 989
BeBr₂
Beryllium Bromide, 1166
BeCl₂
Beryllium Chloride, 1168
BeF₂
Beryllium Fluoride, 1169
BeF₄Na₂
Beryllium Sodium Fluoride, 1176
BeH₂
Beryllium Hydride, 1170
BeH₂O₂
Beryllium Hydroxide, 1171
BeK₂O₈S₂
Beryllium Potassium Sulfate, 1175
BeN₂O₆
Beryllium Nitrate, 1172
BeO
Beryllium Oxide, 1174
BeO₄S
Beryllium Sulfate, 1177
Be₃N₂
Beryllium Nitride, 1173
BiBrO
Bismuth Bromide Oxide, 1263
BiBr₃
Bismuth Bromide, 1262
BiClO
Bismuth Oxychloride, 1274
BiCl₃
Bismuth Chloride, 1264
BiF₃
Bismuth Fluoride, 1265
BiF₅
Bismuth Pentafluoride, 1275
BiH₃
Bismuthine, 1268
BiH₃O₃
Bismuth Hydroxide, 1267
BiIO
Bismuth Iodide Oxide, 1270
BiI₃
Bismuth Iodide, 1269
BiI₇K₄
Bismuth Potassium Iodide, 1277
BiN₃O₉
Bismuth Nitrate, 1271

BiNaO₃
Sodium Bismuthate(V), 8721
BiO₄P
Bismuth Phosphate, 1276
Bi₂O₃
Bismuth Oxide, 1273
Bi₂O₄
Bismuth Tetroxide, 1290
Bi₂O₁₂S₃
Bismuth Sulfate, 1286
Bi₂S₃
Bismuth Sulfide, 1287
Bi₂Se₃
Bismuth Selenide, 1279
Bi₂Te₃
Bismuth Telluride, 1289
Bi₄Ge₃O₁₂
Bismuth Germanate, 1266
Bi₅H₉N₄O₂₂
Bismuth Subnitrate, 1284
BrCs
Cesium Bromide, 2008
BrCu
Cuprous Bromide, 2649
BrF₃
Bromine Trifluoride, 1403
BrF₅
Bromine Pentafluoride, 1402
BrH
Hydrogen Bromide, 4830
BrHO₃
Bromic Acid, 1399
BrH₄N
Ammonium Bromide, 502
BrI
Iodine Monobromide, 5059
BrK
Potassium Bromide, 7739
BrKO₃
Potassium Bromate, 7738
BrLi
Lithium Bromide, 5582
BrNa
Sodium Bromide, 8729
BrNaO₃
Sodium Bromate, 8728
BrRb
Rubidium Bromide, 8416
BrTl
Thallium Bromide, 9407
Br₂Ca
Calcium Bromide, 1657
Br₂Cd
Cadmium Bromide, 1620
Br₂Co
Cobaltous Bromide, 2419
Br₂Cr
Chromous Bromide, 2247
Br₂Cu
Cupric Bromide, 2619
Br₂Fe
Ferrous Bromide, 4069
Br₂Hg
Mercuric Bromide, 5942
Br₂Hg₂
Mercurous Bromide, 5961
Br₂Mg
Magnesium Bromide, 5723
Br₂Mn
Manganese Bromide, 5790
Br₂Ni
Nickel Bromide, 6587
Br₂OS
Thionyl Bromide, 9500
Br₂Pb
Lead Bromide, 5458

Br₂Sn
Stannous Bromide, 8909
Br₂Sr
Strontium Bromide, 8967
Br₂Zn
Zinc Bromide, 10328
Br₃Ce
Cerous Bromide, 1994
Br₃Cr
Chromic Bromide, 2225
Br₃Fe
Ferric Bromide, 4045
Br₃HSi
Tribromosilane, 9777
Br₃Ho
Holmium Bromide, 4765
Br₃In
Indium Tribromide, 4997
Br₃OP
Phosphorus Oxybromide, 7460
Br₃P
Phosphorus Tribromide, 7469
Br₃Sb
Antimony Tribromide, 695
Br₄Se
Selenium Tetrabromide, 8574
Br₄Si
Silicon Tetrabromide, 8638
Br₄Sn
Stannic Bromide, 8900
Br₄Te
Tellurium Tetrabromide, 9267
Br₄Ti
Titanium Tetrabromide, 9633
Br₅P
Phosphorus Pentabromide, 7462

C

CAgN
Silver Cyanide, 8651
CAg₂O₃
Silver Carbonate, 8646
CAuN
Gold Monocyanide, 4551
CB₄
Boron Carbide, 1345
CBaO₃
Barium Carbonate, 965
CBe₂
Beryllium Carbide, 1167
CBi₂O₅
Bismuth Subcarbonate, 1282
CBrN
Cyanogen Bromide, 2683
CCaN₂
Calcium Cyanamide, 1664
CCaO₃
Calcium Carbonate, 1659
CCdO₃
Cadmium Carbonate, 1621
CClN
Cyanogen Chloride, 2684
CCl₂F₂
Dichlorodifluoromethane, 3073
CCl₂O
Phosgene, 7447
CCl₃F
Trichlorofluoromethane, 9805
CCl₃NO₂
Chloropicrin, 2159
CCl₄
Carbon Tetrachloride, 1816

CCoO₃
Cobaltous Carbonate, 2420
CCs₂O₃
Cesium Carbonate, 2009
CCuN
Cuprous Cyanide, 2651
CCuNS
Cuprous Thiocyanate, 2659
CF₂O
Carbonyl Fluoride, 1821
CF₃I
Trifluoroiodomethane, 9849
CF₄
Carbon Tetrafluoride, 1817
CHBrCl₂
Bromodichloromethane, 1425
CHBr₂Cl
Chlorodibromomethane, 2135
CHBr₃
Bromoform, 1430
CHCl₃
Chloroform, 2142
CHFO
Formyl Fluoride, 4274
CHF₃
Fluoroform, 4205
CHF₃O₃S
Triflic Acid, 9842
CHI₃
Iodoform, 5076
CHKO₂
Potassium Formate, 7754
CHKO₃
Potassium Bicarbonate, 7730
CHLiO₂
Lithium Formate, 5588
CHN
Hydrogen Cyanide, 4832
CHNO
Cyanic Acid, 2676
Isocyanic Acid, 5210
CHNS
Thiocyanic Acid, 9481
CHN₃O₆
Trinitromethane, 9901
CHNaO₂
Sodium Formate, 8756
CHNaO₃
Sodium Bicarbonate, 8719
CH₂AlNaO₅
Dihydroxyaluminum Sodium Carbonate, 3202
CH₂Br₂
Methylene Bromide, 6133
CH₂ClNO
Carbamyl Chloride, 1785
CH₂Cl₂
Methylene Chloride, 6135
CH₂Cl₂Na₂O₆P₂
Clodronic Acid Disodium Salt, 2365
CH₂Cl₂O₃S
Chloromethyl Chlorosulfate, 2147
CH₂Cu₂O₅
Cupric Carbonate, Basic, 2621
CH₂F₂
Difluoromethane, 3170
CH₂INaO₃S
Methiodal Sodium, 6045
CH₂I₂
Methylene Iodide, 6137
CH₂N₂
Cyanamide, 2674
Diazomethane, 3004
CH₂O
Formaldehyde, 4263
CH₂O₂
Formic Acid, 4269

CH₂O₃
Performic Acid, 7275
CH₂S₃
Trithiocarbonic Acid, 9936
CH₃AsNa₂O₃
Methanearsonic Acid Disodium Salt, 6025
CH₃BNNa
Sodium Cyanoborohydride, 8742
CH₃Br
Methyl Bromide, 6103
CH₃Cl
Methyl Chloride, 6112
CH₃ClO₂S
Methanesulfonyl Chloride, 6027
CH₃F
Fluoromethane, 4206
CH₃FO₃S
Methyl Fluorosulfonate, 6146
CH₃I
Methyl Iodide, 6159
CH₃NO
Formamide, 4265
CH₃NO₂
Nitromethane, 6698
CH₃NO₃
Methyl Nitrate, 6174
CH₃NS
Thioformamide, 9486
CH₃N₃O₃
Nitrourea, 6738
CH₃NaO
Sodium Methoxide, 8775
CH₃NaO₃S
Sodium Formaldehyde Sulfoxylate, 8755
CH₃NaO₄S
Formaldehyde Sodium Bisulfite, 4264
CH₃O₃Re
Methyltrioxorhenium, 6206
CH₄
Methane, 6024
CH₄AsNaO₃
Methanearsonic Acid Monosodium Salt, 6025
CH₄Cl₂O₆P₂
Clodronic Acid, 2365
CH₄N₂O
Urea, 10052
CH₄N₂O₂
Hydroxyurea, 4888
CH₄N₂S
Ammonium Thiocyanate, 558
Thiourea, 9521
CH₄N₄O₂
N-Nitroguanidine, 6695
CH₄Ni₃O₇
Nickel Carbonate Hydroxide, 6588
CH₄O
Methanol, 6029
CH₄O₃
Orthoformic Acid, 6980
CH₄O₃S
Methanesulfonic Acid, 6026
CH₄O₄S
Methyl Sulfate, 6195
CH₄O₆S₂
Methionic Acid, 6046
CH₄S
Methanethiol, 6028
CH₅AsO₃
Methanearsonic Acid, 6025
CH₅N
Methylamine, 6088
CH₅NO
Methoxyamine, 6060

C₂H₃BrO₂
Bromoacetic Acid, 1407
C₂H₃Br₃O₂
Bromal Hydrate, 1391
C₂H₃Cl
Vinyl Chloride, 10191
C₂H₃ClO
Acetyl Chloride, 80
Chloroacetaldehyde, 2109
C₂H₃ClO₂
Chloroacetic Acid, 2112
Methyl Chlorocarbonate, 6114
C₂H₃Cl₃
1,1,1-Trichloroethane, 9801
1,1,2-Trichloroethane, 9802
C₂H₃Cl₃NaO₄P
Triclofos Monosodium Salt, 9820
C₂H₃Cl₃O
2,2,2-Trichloroethanol, 9803
C₂H₃Cl₃O₂
Chloral Hydrate, 2071
C₂H₃CuO₂
Cuprous Acetate, 2648
C₂H₃FO₂
Fluoroacetic Acid, 4199
C₂H₃F₃O
Trifluoroethanol, 9848
C₂H₃F₃O₃S
Methyl Trifluoromethanesulfonate, 6204
C₂H₃IO
Acetyl Iodide, 90
C₂H₃IO₂
Iodoacetic Acid, 5068
C₂H₃KO₂
Potassium Acetate, 7726
C₂H₃KO₄
Potassium Diformate, 7750
C₂H₃LiO₂
Lithium Acetate, 5577
C₂H₃N
Acetonitrile, 69
C₂H₃NO
Methyl Isocyanate, 6161
C₂H₃NO₃
Oxamic Acid, 7016
C₂H₃NO₄
Acetyl Nitrate, 93
C₂H₃NS
Methyl Isothiocyanate, 6164
Methyl Thiocyanate, 6200
C₂H₃N₃
1*H*-1,2,4-Triazole, 9766
C₂H₃N₃O₂
Urazole, 10051
C₂H₃NaO₂
Sodium Acetate, 8709
C₂H₃NaO₂S
Sodium Thioglycolate, 8819
C₂H₃O₂Tl
Thallium Acetate, 9405
C₂H₄
Ethylene, 3845
C₂H₄BrNO
N-Bromoacetamide, 1405
C₂H₄Br₂
Ethylene Dibromide, 3850
C₂H₄ClNO
Chloroacetamide, 2110
C₂H₄Cl₂
Ethylene Dichloride, 3851
Ethylidene Chloride, 3865
C₂H₄Cl₂N₆
Chloroazodin, 2121
C₂H₄Cl₂O
sym-Dichloromethyl Ether, 3079

C₂H₄Cl₃O₄P
Triclofos, 9820
C₂H₄FNO
Fluoroacetamide, 4198
C₂H₄NNaS₂
Metham Sodium, 6021
C₂H₄N₂
Aminoacetonitrile, 406
C₂H₄N₂O₂
Oxamide, 7017
C₂H₄N₂O₆
Ethylene Glycol Dinitrate, 3854
C₂H₄N₂S₂
Rubeanic Acid, 8412
C₂H₄N₄
Amitrole, 485
Dicyanodiamide, 3103
C₂H₄N₄O₂
Azodicarbonamide, 912
C₂H₄Na₂O₆S₂.2H₂O
1,2-Ethanedisulfonic Acid Sodium Salt Dihydrate, 3779
C₂H₄Na₂O₈S₂
Glyoxal-Sodium Bisulfite, 4545
C₂H₄O
Acetaldehyde, 40
Ethylene Oxide, 3856
C₂H₄OS
Thioacetic Acid, 9473
C₂H₄O₂
Acetic Acid, 54
Glycolaldehyde, 4532
Methyl Formate, 6148
C₂H₄O₂S
Thioglycolic Acid, 9489
C₂H₄O₃
Glycolic Acid, 4534
Peracetic Acid, 7263
C₂H₄O₅S
2-Sulfoacetic Acid, 9083
C₂H₄S
Ethylene Sulfide, 3857
C₂H₄S₅
Lenthionine, 5492
C₂H₅AlCl₂
Ethylaluminum Dichloride, 3816
C₂H₅AsO₅
Arsonoacetic Acid, 800
C₂H₅Br
Ethyl Bromide, 3826
C₂H₅BrO
Ethylene Bromohydrin, 3846
C₂H₅Cl
Ethyl Chloride, 3837
C₂H₅ClHg
Ethylmercuric Chloride, 3881
C₂H₅ClO
Chloromethyl Methyl Ether, 2148
Ethylene Chlorohydrin, 3847
C₂H₅FeNO₆S
Ferroglycine Sulfate, 4066
C₂H₅I
Ethyl Iodide, 3868
C₂H₅N
Ethylenimine, 3859
C₂H₅NO
Acetaldoxime, 43
Acetamide, 44
N-Methylformamide, 6147
C₂H₅NO₂
Acetohydroxamic Acid, 62
Ethyl Nitrite, 3885
Glycine, 4526
Methyl Carbamate, 6110
Nitroethane, 6684
C₂H₅NO₄
Ammonium Binoxalate, 496

C₂H₅NS
Thioacetamide, 9472
C₂H₅N₃O₂
Biuret, 1310
Semioxamazide, 8584
C₂H₅N₃S₂
2,4-Dithiobiuret, 3414
C₂H₅N₅O₃
N-Methyl-*N'*-nitro-*N*-nitrosoguanidine, 6176
C₂H₅NaO
Sodium Ethoxide, 8749
C₂H₅NaOS
Dimsyl Sodium, 3285
C₂H₅NaO₃S₂
Mesna, 5979
C₂H₅NaO₄S
Acetaldehyde Sodium Bisulfite, 42
Sodium Ethyl Sulfate, 8750
C₂H₆
Ethane, 3778
C₂H₆AlCl
Dimethylaluminum Chloride, 3249
C₂H₆AlNO₄
Dihydroxyaluminum Aminoacetate, 3201
C₂H₆AsNaO₂
Sodium Cacodylate, 8730
C₂H₆BBr
Bromodimethylborane, 1426
C₂H₆BaO₈S₂
Methyl Sulfate Barium Salt, 6195
C₂H₆Br₂S
Bromodimethylsulfonium Bromide, 1427
C₂H₆CaO₈S₂
Methyl Sulfate Calcium Salt, 6195
C₂H₆Cd
Dimethylcadmium, 3261
C₂H₆ClO₃P
Ethephon, 3787
C₂H₆Co₅O₁₂
Cobaltous Carbonate Basic, 2420
C₂H₆Hg
Dimethylmercury, 3274
C₂H₆N₂
Acetamidine, 45
C₂H₆N₂O
N-Nitrosodimethylamine, 6724
C₂H₆Na₂O₇P₂
Etidronic Acid Disodium Salt, 3912
C₂H₆O
Dimethyl Ether, 3266
Ethyl Alcohol, 3814
C₂H₆OS
Dimethyl Sulfoxide, 3285
2-Mercaptoethanol, 5936
C₂H₆O₂
Ethylene Glycol, 3852
C₂H₆O₂S
Dimethyl Sulfone, 3283
C₂H₆O₃S
Methyl Methanesulfonate, 6169
C₂H₆O₄S
Dimethyl Sulfate, 3280
Ethyl Sulfate, 3904
Isethionic Acid, 5152
C₂H₆O₆S₂
1,2-Ethanedisulfonic Acid, 3779
C₂H₆S
Dimethyl Sulfide, 3281
Ethanethiol, 3781
C₂H₆S₂
1,2-Ethanedithiol, 3780
C₂H₆Zn
Dimethylzinc, 3289

C₃H₅N
  Propionitrile, 7941
C₃H₅NO
  Acrylamide, 120
  Ethylene Cyanohydrin, 3848
C₃H₅NO₂
  Isonitrosoacetone, 5236
C₃H₅NO₄
  Hadacidin, 4622
C₃H₅NS
  Ethyl Isothiocyanate, 3870
  Thiocyanic Acid Ethyl Ester, 9481
C₃H₅N₃O
  Cyacetacide, 2670
C₃H₅N₃O₉
  Nitroglycerin, 6694
C₃H₅NaO₂
  Sodium Propionate, 8798
C₃H₅NaO₂S
  Thiolactic Acid Sodium Salt, 9493
C₃H₅NaO₃
  Hydracrylic Acid Sodium Salt, 4801
  Sodium Lactate, 8767
C₃H₆
  Cyclopropane, 2744
  Propylene, 7962
C₃H₆Al₂O₁₈S₆
  Methionic Acid Aluminum Salt, 6046
C₃H₆As₄O₃
  Arsenicin A, 786
C₃H₆BrNO₄
  Bronopol, 1457
C₃H₆Br₂
  Propylene Dibromide, 7966
  Trimethylene Bromide, 9884
C₃H₆Cl₂
  Propylene Dichloride, 7967
  Propylidene Chloride, 7973
C₃H₆Cl₂O
  1,3-Dichloro-2-propanol, 3084
C₃H₆Cl₃N
  Viehe's Salt, 10174
C₃H₆NNaO₂
  Sarcosine Sodium Salt, 8510
C₃H₆N₂
  3-Aminopropionitrile, 465
  2-Pyrazoline, 8072
C₃H₆N₂O
  2-Imidazolidinone, 4955
C₃H₆N₂O₂
  Cycloserine, 2749
C₃H₆N₂S
  Ethylene Thiourea, 3858
C₃H₆N₄O₃
  Triuret, 9945
C₃H₆N₆
  Melamine, 5880
C₃H₆N₆O₆
  Cyclonite, 2728
C₃H₆O
  Acetone, 65
  Allyl Alcohol, 277
  Propionaldehyde, 7937
  Propylene Oxide, 7969
  Trimethylene Oxide, 9886
C₃H₆OS
  Propanethial S-Oxide, 7916
C₃H₆O₂
  Acetol, 64
  Dimethyldioxirane, 3265
  Ethyl Formate, 3862
  Glycidol, 4525
  Methyl Acetate, 6082
  Propionic Acid, 7939
C₃H₆O₂S
  Thiolactic Acid, 9493

C₃H₆O₃
  Dihydroxyacetone, 3200
  Dimethyl Carbonate, 3263
  Glyceraldehyde, 4517
  Hydracrylic Acid, 4801
  D-Lactic Acid, 5383
  DL-Lactic Acid, 5384
  L-Lactic Acid, 5385
  s-Trioxane, 9906
C₃H₆O₄
  Glyceric Acid, 4519
C₃H₇Br
  Isopropyl Bromide, 5256
  Propyl Bromide, 7958
C₃H₇CaO₆P
  Calcium Glycerophosphate, 1672
C₃H₇Cl
  Isopropyl Chloride, 5257
  Propyl Chloride, 7960
C₃H₇ClO
  Propylene Chlorohydrin, 7963
  sec-Propylene Chlorohydrin, 7964
C₃H₇ClO₂
  α-Chlorohydrin, 2146
C₃H₇I
  Isopropyl Iodide, 5260
  Propyl Iodide, 7974
C₃H₇N
  Allylamine, 278
C₃H₇NO
  Acetoxime, 73
  N,N-Dimethylformamide, 3267
  Propionamide, 7938
C₃H₇NO₂
  Alanine, 197
  β-Alanine, 198
  Formicin, 4270
  Isopropyl Nitrite, 5262
  1-Nitropropane, 6714
  2-Nitropropane, 6715
  Propyl Nitrite, 7976
  Sarcosine, 8510
  Urethan, 10059
C₃H₇NO₂S
  Cysteine, 2778
C₃H₇NO₂Se
  Selenocysteine, 8577
C₃H₇NO₃
  Propyl Nitrate, 7975
  Serine, 8599
C₃H₇NO₅S
  Cysteic Acid, 2777
C₃H₇N₂O₂P
  Seyferth-Gilbert Reagent, 8615
C₃H₇N₃O₂
  N-Ethyl-N-nitrosourea, 3887
  Glycocyamine, 4530
C₃H₇N₃O₄
  L-Alanosine, 199
C₃H₇NaO₃S₃
  2,3-Dimercapto-1-propanesulfonic
    Acid Sodium Salt, 3232
C₃H₇NaO₄S
  Acetone Sodium Bisulfite, 68
C₃H₇Na₂O₆P
  Sodium Glycerophosphate, 8757
C₃H₇O₄P
  Fosfomycin, 4281
C₃H₇O₆P
  Glyceraldehyde 3-Phosphate, 4518
C₃H₈
  Propane, 7913
C₃H₈IN
  Dimethyl(methylene)ammonium Io-
    dide, 3275
C₃H₈NNaO₃S
  Tramiprosate Sodium Salt, 9728

C₃H₈NO₅P
  Glyphosate, 4547
C₃H₈NO₆P
  Phosphoserine, 7475
C₃H₈N₂OS
  Noxythiolin, 6813
C₃H₈N₂O₂
  2,3-Diaminopropionic Acid, 2986
C₃H₈N₂O₃
  Oxymethurea, 7067
C₃H₈N₂S
  N,N'-Dimethylthiourea, 3287
C₃H₈O
  Ethyl Methyl Ether, 3883
  Isopropyl Alcohol, 5254
  Propyl Alcohol, 7955
C₃H₈OS
  Dimethylsulfoxonium Methylide,
    3286
C₃H₈OS₂
  Dimercaprol, 3231
C₃H₈O₂
  2-Methoxyethanol, 6063
  Methylal, 6086
  Propylene Glycol, 7968
  Trimethylene Glycol, 9885
C₃H₈O₂S
  Thioglycerol, 9488
C₃H₈O₃
  Glycerol, 4520
C₃H₈O₃S
  Ethyl Methanesulfonate, 3882
C₃H₈O₃S₃
  2,3-Dimercapto-1-propanesulfonic
    Acid, 3232
C₃H₈S₂
  Bis(methylthio)methane, 1260
  1,3-Propanedithiol, 7915
C₃H₉Al
  Trimethylaluminum, 9879
C₃H₉AsO₃
  1-Propanearsonic Acid, 7914
C₃H₉BF₄O
  Trimethyloxonium Tetrafluoroborate,
    9889
C₃H₉BO₃
  Trimethyl Borate, 9882
C₃H₉ClSi
  Chlorotrimethylsilane, 2180
C₃H₉N
  Isopropylamine, 5255
  Propylamine, 7956
  Trimethylamine, 9880
C₃H₉NNa₂O₇P₂
  Pamidronic Acid Disodium Salt, 7104
C₃H₉NO₃
  Ammonium Lactate, 527
C₃H₉NO₃S
  N-Methyltaurine, 6196
  Tramiprosate, 9728
C₃H₉O₂PS
  EMPTA, 3621
C₃H₉O₃P
  DMMP, 3435
  EMPA, 3620
C₃H₉O₆P
  Glycerophosphoric Acid, 4522
C₃H₁₀N₂
  Propylenediamine, 7965
C₃H₁₁Br₂N₃S
  AET, 167
C₃H₁₁NO₇P₂
  Pamidronic Acid, 7104
C₃H₁₃N₃O₁₀Zr
  Ammonium Zirconyl Carbonate, 566
C₃H₁₆N₄O₁₁U
  Ammonium Uranium Carbonate, 563

Methacrylic Acid, 6013
Methyl Acrylate, 6085
Vinyl Acetate, 10189

C₄H₆O₂S
3-Sulfolene, 9086

C₄H₆O₃
Acetic Anhydride, 55
Acetoacetic Acid, 58

C₄H₆O₄
Methyl Oxalate, 6179
Succinic Acid, 8999

C₄H₆O₄Pb
Lead Acetate, 5452

C₄H₆O₄Pd
Palladium Diacetate, 7092

C₄H₆O₄S
Thiodiglycolic Acid, 9484
Thiomalic Acid, 9496

C₄H₆O₄S₂
Succimer, 8995

C₄H₆O₄Sn
Stannous Acetate, 8908

C₄H₆O₄Sr
Strontium Acetate, 8966

C₄H₆O₄Zn
Zinc Acetate, 10327

C₄H₆O₅
Malic Acid, 5771

C₄H₆O₆
D-Tartaric Acid, 9203
DL-Tartaric Acid, 9204
L-Tartaric Acid, 9205
meso-Tartaric Acid, 9202

C₄H₆O₆S
Sulfonyldiacetic Acid, 9091

C₄H₆O₆U
Uranyl Acetate, 10046

C₄H₆O₈
Dihydroxytartaric Acid, 3205

C₄H₇AlN₄O₅
Aldioxa, 216

C₄H₇AlO₅
Aluminum Subacetate, 359

C₄H₇BrO₂
α-Bromobutyric Acid, 1420
α-Bromoisobutyric Acid, 1431

C₄H₇Br₂Cl₂O₄P
Naled, 6442

C₄H₇Br₃O
Tribromo-tert-butyl Alcohol, 9773

C₄H₇Cl
1-Chloro-2-butene, 2130
3-Chloro-1-butene, 2131
1-Chloro-2-methyl-1-propene, 2149
3-Chloro-2-methyl-1-propene, 2150

C₄H₇ClO
n-Butyryl Chloride, 1602
2-Chloroethyl Vinyl Ether, 2140

C₄H₇ClO₂
β-Chloropropionic Acid Methyl Ester, 2161
Ethyl Chloroacetate, 3838
Propyl Chloroformate, 7961

C₄H₇Cl₂O₄P
Dichlorvos, 3089

C₄H₇Cl₃O
Chlorobutanol, 2129

C₄H₇Cl₃O₂
Chloral Alcoholate, 2069

C₄H₇FeO₅
Ferric Acetate, Basic, 4043

C₄H₇N
Butyronitrile, 1601
Isobutyronitrile, 5201
3-Pyrroline, 8131

C₄H₇NO
Acetone Cyanohydrin, 66
2-Pyrrolidone, 8130

C₄H₇NO₂
2-Azetidinecarboxylic Acid, 901

C₄H₇NO₂S
Timonacic, 9602

C₄H₇NO₃
Aceturic Acid, 75

C₄H₇NO₄
Aspartic Acid, 830
Iminodiacetic Acid, 4959

C₄H₇N₃O
Creatinine, 2557

C₄H₇NaOS₂
Sodium Isopropyl Xanthate, 8766

C₄H₇NaO₃
γ-Hydroxybutyrate Sodium Salt, 4854

C₄H₇NaO₄
Sodium Diacetate, 8743

C₄H₈
1-Butene, 1522
2-Butene, 1523
Cyclobutane, 2708
Isobutylene, 5186

C₄H₈Br₂
α-Butylene Dibromide, 1569

C₄H₈CaN₃O₅P.4H₂O
Phosphocreatine Calcium Salt Tetrahydrate, 7452

C₄H₈Cl₂
Butylidene Chloride, 1575

C₄H₈Cl₂O
sym-Dichloroethyl Ether, 3075

C₄H₈Cl₂S
Mustard Gas, 6401

C₄H₈Cl₃O₄P
Trichlorfon, 9788

C₄H₈CuN₂O₄
Cupric Bisglycinate, 2617

C₄H₈FeN₂O₄
Ferrous Bisglycinate, 4068

C₄H₈N₂
Lysidine, 5696

C₄H₈N₂O
Allylurea, 291
N-Nitrosopyrrolidine, 6730

C₄H₈N₂O₂
Dimethylglyoxime, 3270
N-Nitrosomorpholine, 6727
Succinamide, 8996

C₄H₈N₂O₃
Asparagine, 827
Glycylglycine, 4539

C₄H₈N₂O₉Ti
Ammonium Titanium Oxalate, 560

C₄H₈N₂S
Thiosinamine, 9515

C₄H₈N₃Na₂O₅P
Phosphocreatine Sodium Salt, 7452

C₄H₈N₄O₄S₂
Tetramethylenedisulfotetramine, 9374

C₄H₈N₆O₄
Oxalenediuramidoxime, 7009

C₄H₈N₈O₈
Octogen, 6847

C₄H₈Na₂O₆S₄
Dimesna, 3233

C₄H₈O
Butyraldehyde, 1595
Crotyl Alcohol, 2591
Cyclopropyl Methyl Ether, 2746
Isobutyraldehyde, 5199
2-Methoxypropene, 6073
Methyl Ethyl Ketone, 6143
Tetrahydrofuran, 9356

C₄H₈O₂
Acetoin, 63
Aldol, 217
Butyric Acid, 1597
Dioxane, 3328
Ethyl Acetate, 3811
Isobutyric Acid, 5200
Methyl Propanoate, 6185
Propyl Formate, 7970

C₄H₈O₂S
Sulfolane, 9085

C₄H₈O₃
Ethylene Glycol Monoacetate, 3855
Glycerol Formal, 4521
γ-Hydroxybutyrate, 4854
β-Hydroxybutyric Acid, 4855
Methyl Lactate, 6166

C₄H₈O₄
D-Erythrose, 3746
L-Erythrose, 3747
L-Erythrulose, 3750
Glycolaldehyde Dimer, 4532
D-Threose, 9535
L-Threose, 9536

C₄H₈S
Tetrahydrothiophene, 9363

C₄H₈S₃
Methyl Allyl Trisulfide, 6087

C₄H₉Br
n-Butyl Bromide, 1555
sec-Butyl Bromide, 1556
tert-Butyl Bromide, 1557
Isobutyl Bromide, 5180

C₄H₉Cl
n-Butyl Chloride, 1561
sec-Butyl Chloride, 1562
tert-Butyl Chloride, 1563
Isobutyl Chloride, 5183

C₄H₉ClHg
n-Butylmercuric Chloride, 1583

C₄H₉ClO
tert-Butyl Hypochlorite, 1574

C₄H₉ClOS
Hemisulfur Mustard, 4681

C₄H₉Cl₃Sn
Butyltin Trichloride, 1594

C₄H₉F₃O₃SSi
Trimethylsilyl Triflate, 9891

C₄H₉F₃Si
Trifluoromethyltrimethylsilane, 9851

C₄H₉I
n-Butyl Iodide, 1576
sec-Butyl Iodide, 1577
Isobutyl Iodide, 5189

C₄H₉N
Pyrrolidine, 8129

C₄H₉NO
n-Butyramide, 1596
N,N-Dimethylacetamide, 3248
Morpholine, 6362

C₄H₉NO₂
α-Aminobutyric Acid, 423
β-Aminobutyric Acid, 424
γ-Aminobutyric Acid, 425
α-Aminoisobutyric Acid, 442
n-Butyl Nitrite, 1585
tert-Butyl Nitrite, 1586
N,N-Dimethylglycine, 3269
Isobutyl Nitrite, 5194

C₄H₉NO₂S
Homocysteine, 4772
Mecysteine, 5856

C₄H₉NO₃
3-Amino-4-hydroxybutyric Acid, 440
4-Amino-3-hydroxybutyric Acid, 441
Homoserine, 4778

$C_5H_6Br_3N$
Pyridinium Bromide Perbromide, 8085
$C_5H_6ClCrNO_3$
Pyridinium Chlorochromate, 8086
$C_5H_6Cl_2N_2O_2$
1,3-Dichloro-5,5-dimethylhydantoin, 3074
$C_5H_6N_2$
α-Aminopyridine, 469
Dalfampridine, 2802
Glutaronitrile, 4510
$C_5H_6N_2OS$
Methylthiouracil, 6201
$C_5H_6N_2O_2$
5-Methylpyrazole-3-carboxylic Acid, 6188
6-Methyluracil, 6209
Thymine, 9553
$C_5H_6N_2O_4$
Hydroorotic Acid, 4841
Ibotenic Acid, 4916
Maleuric Acid, 5770
Muscazone, 6396
$C_5H_6N_6$
2,6-Diaminopurine, 2987
$C_5H_6Na_4O_{10}P_2$
Butedronic Acid Tetrasodium Salt, 1520
$C_5H_6O_2$
Angelica Lactone, 640
Ethyl Propiolate, 3898
Furfuryl Alcohol, 4334
α-Methylene Butyrolactone, 6134
$C_5H_6O_3$
Reductic Acid, 8246
$C_5H_6O_4$
Citraconic Acid, 2320
Itaconic Acid, 5290
Mesaconic Acid, 5973
$C_5H_6O_5$
Acetonedicarboxylic Acid, 67
α-Ketoglutaric Acid, 5350
C_5H_7NOS
Goitrin, 4548
$C_5H_7NO_2$
Ethyl Cyanoacetate, 3841
$C_5H_7NO_3$
Dimethadione, 3235
L-Pyroglutamic Acid, 8113
$C_5H_7N_3$
Amifampridine, 398
$C_5H_7N_3O$
5-Methylcytosine, 6122
$C_5H_7N_3O_2$
Dimetridazole, 3290
$C_5H_7N_3O_4$
Azaserine, 892
$C_5H_7N_3O_5$
Quisqualic Acid, 8202
C_5H_8
Isoprene, 5247
1,3-Pentadiene, 7220
$C_5H_8NNaO_4$
Monosodium Glutamate, 6342
$C_5H_8N_2S_4Zn$
Propineb, 7933
$C_5H_8N_4O_3S_2$
Methazolamide, 6034
$C_5H_8N_4O_{12}$
Pentaerythritol Tetranitrate, 7222
C_5H_8O
Cyclopentanone, 2738
Dihydropyran, 3195
2-Methyl-3-butyn-2-ol, 6108
Senecialdehyde, 8587

$C_5H_8O_2$
Acetylacetone, 76
Allyl Acetate, 276
Angelic Acid, 639
Ethyl Acrylate, 3813
Glutaraldehyde, 4508
Isopropenyl Acetate, 5249
Tiglic Acid, 9589
$C_5H_8O_3$
Levulinic Acid, 5526
Methyl Acetoacetate, 6083
$C_5H_8O_4$
Glutaric Acid, 4509
Methyl Malonate, 6168
$C_5H_8O_5$
Citramalic Acid, 2322
$C_5H_9BrO_2$
α-Bromoisovaleric Acid, 1432
Ethyl α-Bromopropionate, 3827
C_5H_9ClO
Isovaleryl Chloride, 5278
$C_5H_9ClO_2$
Ethyl α-Chloropropionate, 3840
Isobutyl Chlorocarbonate, 5184
$C_5H_9Cl_2N_3O_2$
Carmustine, 1845
$C_5H_9IO_3$
Domiodol, 3463
C_5H_9N
N-Methylpyrroline, 6191
Valeronitrile, 10091
C_5H_9NO
Butyl Isocyanate, 1578
1-Methylpyrrolidone, 6190
$C_5H_9NO_2$
Proline, 7895
$C_5H_9NO_3$
δ-Aminolevulinic Acid, 443
Hydroxyproline, 4878
$C_5H_9NO_3S$
Acetylcysteine, 83
Tiopronin, 9609
$C_5H_9NO_4$
Glutamic Acid, 4505
HON, 4780
NMDA, 6749
$C_5H_9NO_4S$
Carbocysteine, 1802
$C_5H_9NO_5$
Hydroxyglutamic Acid, 4864
$C_5H_9N_2O_7P$
Bestmann-Ohira Reagent, 1179
$C_5H_9N_3$
Betazole, 1185
Histamine, 4756
$C_5H_9N_3O_{10}$
Pentrinitrol, 7251
C_5H_{10}
Amylene, 601
Cyclopentane, 2736
1-Pentene, 7234
2-Pentene, 7235
$C_5H_{10}Cl_2N_2$
DMC, 3434
$C_5H_{10}NNaS_2$
Ditiocarb Sodium, 3421
$C_5H_{10}N_2O$
1,3-Dimethyl-2-imidazolidinone, 3273
$C_5H_{10}N_2O_2S$
Methomyl, 6054
$C_5H_{10}N_2O_3$
Glutamine, 4507
L-Isoglutamine, 5223
$C_5H_{10}N_2O_7P_2$
Zoledronic Acid, 10388

$C_5H_{10}N_2S_2$
Dazomet, 2837
$C_5H_{10}O$
Allyl Ethyl Ether, 284
Cyclopentanol, 2737
Diethyl Ketone, 3137
Isovaleraldehyde, 5275
Methyl Isopropyl Ketone, 6162
Methyl Propyl Ketone, 6187
2-Methyltetrahydrofuran, 6198
Pivalaldehyde, 7624
Tetrahydropyran, 9362
n-Valeraldehyde, 10088
$C_5H_{10}O_2$
Ethyl Propionate, 3899
Isobutyl Formate, 5188
Isopropyl Acetate, 5251
Isovaleric Acid, 5277
Methyl Butyrate, 6109
Methyl Isobutyrate, 6160
Pivalic Acid, 7625
Propyl Acetate, 7954
Tetrahydrofurfuryl Alcohol, 9358
n-Valeric Acid, 10090
$C_5H_{10}O_2S$
Dimethylsulfoniopropionate, 3284
$C_5H_{10}O_3$
Ethyl Carbonate, 3835
Ethyl Lactate, 3872
2-Methoxyethyl Acetate, 6064
$C_5H_{10}O_3S$
Methionine Hydroxy Analog, 6048
$C_5H_{10}O_4$
D-2-Deoxyribose, 2909
Dimethylolpropionic Acid, 3277
Monacetin, 6328
$C_5H_{10}O_5$
Apiose, 722
Arabinose, 750
D-Lyxose, 5702
D-Ribose, 8328
D-Ribulose, 8331
Xylose, 10285
Xylulose, 10286
$C_5H_{10}O_{10}P_2$
Butedronic Acid, 1520
$C_5H_{11}Br$
n-Amyl Bromide, 596
tert-Amyl Bromide, 597
(2S)-1-Bromo-2-methylbutane, 1435
Isoamyl Bromide, 5159
$C_5H_{11}Cl$
Amyl Chloride, 600
Isoamyl Chloride, 5161
$C_5H_{11}ClHgN_2O_2$
Chlormerodrin, 2105
$C_5H_{11}Cl_2N$
Mechlorethamine, 5846
$C_5H_{11}Cl_2NO$
Mechlorethamine Oxide, 5847
$C_5H_{11}I$
Isoamyl Iodide, 5165
$C_5H_{11}N$
Piperidine, 7580
$C_5H_{11}NO$
Isovaleramide, 5276
$C_5H_{11}NO_2$
Betaine, 1181
Isoamyl Nitrite, 5168
Isobutyl Carbamate, 5182
Isovaline, 5280
N-Methylmorpholine N-Oxide, 6171
3-Nitropentane, 6703
Norvaline, 6805
Valine, 10095

Pipecolic Acid, 7570
Vigabatrin, 10175
C₆H₁₁NO₃
Methyl Aminolevulinate, 6090
C₆H₁₁NO₃S
Alliin, 254
C₆H₁₁NO₄
α-Aminoadipic Acid, 410
C₆H₁₁N₂O₄PS₃
Methidathion, 6042
C₆H₁₁N₃O₉
Propatyl Nitrate, 7925
C₆H₁₂
Cyclohexane, 2716
C₆H₁₂Br₂OZn₃
Nysted Reagent, 6826
C₆H₁₂Br₂O₄
Mitobronitol, 6297
Mitolactol, 6299
C₆H₁₂Cl₃N
Trichlormethine, 9790
C₆H₁₂F₂N₂O₂
Eflornithine, 3570
C₆H₁₂FeN₃O₁₂
Ammonium Ferric Oxalate, 514
C₆H₁₂FeN₉
Ammonium Ferricyanide, 516
C₆H₁₂NNaO₃S
Cyclamic Acid Sodium Salt, 2696
C₆H₁₂NO₂P
TPMPA, 9722
C₆H₁₂NO₄PS₂
Formothion, 4273
C₆H₁₂N₂
Triethylenediamine, 9834
C₆H₁₂N₂O₃
Aminoethoxyvinylglycine, 434
Daminozide, 2809
C₆H₁₂N₂O₄Pt
Carboplatin, 1823
C₆H₁₂N₂O₄S
Lanthionine, 5416
C₆H₁₂N₂O₄S₂
Cystine, 2779
C₆H₁₂N₂O₆S₂
L-Cystine S,S-Dioxide, 2780
C₆H₁₂N₂O₁₂P₄¹⁵³Sm
Samarium Sm 153 Lexidronam, 8487
C₆H₁₂N₂S₄
Thiram, 9525
C₆H₁₂N₂S₄Zn
Ziram, 10372
C₆H₁₂N₃OP
Triethylenephosphoramide, 9837
C₆H₁₂N₃PS
Thiotepa, 9518
C₆H₁₂N₄
Methenamine, 6038
C₆H₁₂O
Caproic Aldehyde, 1762
Cyclohexanol, 2718
Cyclopentyl Methyl Ether, 2742
3-Hexen-1-ol, 4736
Isopropylacetone, 5253
Methyl Butyl Ketone, 6107
Pinacolone, 7551
C₆H₁₂O₂
n-Butyl Acetate, 1537
sec-Butyl Acetate, 1538
tert-Butyl Acetate, 1539
tert-Butylacetic Acid, 1540
n-Caproic Acid, 1761
Diacetone Alcohol, 2967
Diethylacetic Acid, 3121
Ethyl Butyrate, 3830
Ethyl Isobutyrate, 3869

Isoamyl Formate, 5164
Isobutyl Acetate, 5175
Methyl Isovalerate, 6165
Propyl Propanoate, 7979
C₆H₁₂O₂S
2,4-Dimethylsulfolane, 3282
C₆H₁₂O₃
2-Ethoxyethyl Acetate, 3804
Isopropylidene Glycerol, 5259
Paraldehyde, 7131
2,5-Tetrahydrofurandimethanol, 9357
C₆H₁₂O₄
Digitoxose, 3183
Kethoxal, 5347
Mevalonic Acid, 6242
C₆H₁₂O₅
2-Deoxy-D-glucose, 2905
D-Fucose, 4307
L-Fucose, 4308
d-Quercitol, 8152
Quinovose, 8194
Rhamnose, 8295
C₆H₁₂O₅S
5-Thio-D-glucose, 9487
C₆H₁₂O₆
D-Allose, 272
D-Altrose, 317
Fructose, 4302
DL-Fructose, 4303
D-Galactose, 4365
Glucose, 4495
D-Gulose, 4613
L-Gulose, 4614
Hamamelose, 4641
Idose, 4930
Inositol, 5021
D-Mannose, 5812
D-Psicose, 8033
Sorbose, 8852
D-Tagatose, 9161
C₆H₁₂O₇
Gluconic Acid, 4492
D-Gulonic Acid, 4611
L-Gulonic Acid, 4612
C₆H₁₂S₃
Thioacetaldehyde, 9471
C₆H₁₃Cl
1-Chlorohexane, 2145
C₆H₁₃N
Cyclohexylamine, 2722
C₆H₁₃NO
Diacetonamine, 2965
C₆H₁₃NO₂
ε-Aminocaproic Acid, 427
Isoleucine, 5225
Leucine, 5503
Norleucine, 6795
C₆H₁₃NO₂S
Ethionine, 3792
S-Methylmethionine, 6170
C₆H₁₃NO₃
Isofagomine, 5217
C₆H₁₃NO₃S
Cyclamic Acid, 2696
C₆H₁₃NO₄
Bicine, 1205
1-Deoxynojirimycin, 2906
6-Desoxy-D-glucosamine, 2932
Fucosamine, 4306
Mycosamine, 6413
C₆H₁₃NO₄S
MES, 5972
C₆H₁₃NO₅
D-Galactosamine, 4364
Glucosamine, 4494
Tricine, 9816

C₆H₁₃N₃
Galegine, 4371
C₆H₁₃N₃O₃
Citrulline, 2330
C₆H₁₃N₅O
Moroxydine, 6358
C₆H₁₃O₉P
Fructose-6-phosphate, 4305
α-Glucose-1-phosphate, 4497
Glucose-6-phosphate, 4498
Inositol Monophosphate, 5021
C₆H₁₄
n-Hexane, 4732
C₆H₁₄BrNO₂S
S-Methylmethionine Bromide, 6170
C₆H₁₄ClN₃O₅S₂
Laromustine, 5425
C₆H₁₄FO₃P
Isoflurophate, 5222
C₆H₁₄F₃NO₂S
Deoxo-Fluor®, 2899
C₆H₁₄N₂O₂
Lysine, 5697
C₆H₁₄N₂O₃
2-Deoxystreptamine, 2910
C₆H₁₄N₂O₇
Ammonium Citrate, Dibasic, 509
C₆H₁₄N₄O₂
Arginine, 770
C₆H₁₄O
tert-Amyl Methyl Ether, 605
Dipropyl Ether, 3385
Ethyl *tert*-Butyl Ether, 3829
1-Hexanol, 4734
Isopropyl Ether, 5258
C₆H₁₄O₂
Acetal, 39
Butyl Cellosolve®, 1560
Hexamethylene Glycol, 4728
Hexylene Glycol, 4746
Pinacol, 7550
C₆H₁₄O₂S₂
3,6-Dioxaoctane-1,8-dithiol, 3329
C₆H₁₄O₃
Diethylene Glycol Monoethyl Ether, 3134
Diglyme, 3184
C₆H₁₄O₄
Triethylene Glycol, 9835
C₆H₁₄O₅S
1-Thiosorbitol, 9516
C₆H₁₄O₆
Galactitol, 4362
Mannitol, 5810
Sorbitol, 8851
C₆H₁₄O₆S₂
Busulfan, 1508
C₆H₁₄O₁₂P₂
Fructose-1,6-diphosphate, 4304
C₆H₁₄S
Dipropyl Sulfide, 3387
C₆H₁₄Si
Allyltrimethylsilane, 290
C₆H₁₅AlO₃
Aluminum Ethoxide, 334
C₆H₁₅B
Triethylborane, 9833
C₆H₁₅BF₄O
Meerwein's Reagent, 5867
C₆H₁₅ClN₂O₂
Carbachol, 1781
C₆H₁₅ClSi
tert-Butyldimethylchlorosilane, 1567
Chlorodiisopropylsilane, 2136
Chlorotriethylsilane, 2179

C₆H₁₅N
Diisopropylamine, 3215
n-Dipropylamine, 3383
Triethylamine, 9831

C₆H₁₅NO
2-(Diethylamino)ethanol, 3123

C₆H₁₅NO₂
Diisopropanolamine, 3214

C₆H₁₅NO₃
Triethanolamine, 9830

C₆H₁₅NO₅
Glucamine, 4484

C₆H₁₅NO₆S
TES, 9319

C₆H₁₅N₅
Buformin, 1481

C₆H₁₅O₃P
Triethyl Phosphite, 9840

C₆H₁₅O₃PS₂
Methyl Demeton, 6123

C₆H₁₅O₄P
Triethyl Phosphate, 9838

C₆H₁₅P
Triethylphosphine, 9839

C₆H₁₆AlNaO₄
SMEAH, 8702

C₆H₁₆BLi
Lithium Triethylborohydride, 5600

C₆H₁₆Cl₂Si₂
1,2-Bis(chlorodimethylsilyl)ethane, 1251

C₆H₁₆FN₂OP
Mipafox, 6287

C₆H₁₆FeN₁₀
Ammonium Ferrocyanide, 517

C₆H₁₆NNaO₇P₂
Neridronic Acid Monosodium Salt, 6558

C₆H₁₆N₂
Hexamethylenediamine, 4727
TEMED, 9277

C₆H₁₆Si
Triethylsilane, 9841

C₆H₁₇NO₇P₂
Neridronic Acid, 6558

C₆H₁₇N₃O₁₀S
Glycine Sulfate, 4527

C₆H₁₈AlO₉P₃
Fosetyl Al, 4279

C₆H₁₈N₃OP
HMPA, 4763

C₆H₁₈N₄
Trientine, 9828

C₆H₁₈OSi₂
Hexamethyldisiloxane, 4726

C₆H₁₈O₂₄P₆
Phytic Acid, 7499

C₆H₁₉NSi₂
Hexamethyldisilazane, 4725

C₆H₂₆CdN₆O₂
Tris(ethylenediamine)cadmium Dihydroxide, 9928

C₆K₃O₁₂Sb
Antimony Potassium Oxalate, 690

C₆N₄
Tetracyanoethylene, 9340

C₆N₁₂O₆
Trinitrotriazidobenzene, 9903

C₆Na₂O₆
Sodium Rhodizonate, 8799

C₆O₆V
Vanadium Carbonyl, 10105

C₇H₂Cl₄O
2,4,6-Trichlorobenzoyl Chloride, 9798

C₇H₃Br₂NO
Bromoxynil, 1451

C₇H₃ClN₂O₅
3,5-Dinitrobenzoyl Chloride, 3303

C₇H₃Cl₂N
Dichlobenil, 3052

C₇H₃IN₂O₃
Nitroxynil, 6743

C₇H₃N₃O₈
2,4,6-Trinitrobenzoic Acid, 9899

C₇H₄BrNO
p-Bromophenyl Isocyanate, 1441

C₇H₄Br₂O₂
3,5-Dibromosalicylaldehyde, 3032

C₇H₄Br₂O₃
3,5-Dibromosalicylic Acid, 3033

C₇H₄Br₄O
3,4,5,6-Tetrabromo-*o*-cresol, 9329

C₇H₄ClF₃
p-Chlorobenzotrifluoride, 2127

C₇H₄ClNO₂
Chlorzoxazone, 2201

C₇H₄ClNO₃
4-Nitrobenzoyl Chloride, 6677

C₇H₄ClNO₃S
N-Chlorosaccharin, 2166

C₇H₄Cl₃NO₃
Triclopyr, 9821

C₇H₄F₃NO₃
TFM, 9401

C₇H₄HgO₃
Mercuric Salicylate, 5953

C₇H₄INO₃S
N-Iodosaccharin, 5086

C₇H₄I₂O₃
3,5-Diiodosalicylic Acid, 3208

C₇H₄N₂O₅
2,4-Dinitrobenzaldehyde, 3298

C₇H₄N₂O₆
3,4-Dinitrobenzoic Acid, 3301
3,5-Dinitrobenzoic Acid, 3302

C₇H₄N₄
1-Cyanobenzotriazole, 2680

C₇H₄O₃S
Tioxolone, 9612

C₇H₄O₆
Chelidonic Acid, 2051

C₇H₄O₇
Meconic Acid, 5853

C₇H₅AgO₂
Benzoic Acid Silver Salt, 1093

C₇H₅BiO₄
Bismuth Subsalicylate, 1285

C₇H₅BiO₆
Bismuth Subgallate, 1283

C₇H₅BrO₂
4-Bromobenzoic Acid, 1415

C₇H₅Br₃O
2,4,6-Tribromoanisole, 9772
2,4,6-Tribromo-*m*-cresol, 9774

C₇H₅ClN₂O
Zoxazolamine, 10399

C₇H₅ClN₃NaO₄S₂
Chlorothiazide Sodium Salt, 2171

C₇H₅ClO
Benzoyl Chloride, 1115

C₇H₅ClO₂
m-Chlorobenzoic Acid, 2124
o-Chlorobenzoic Acid, 2125
p-Chlorobenzoic Acid, 2126

C₇H₅ClO₃
3-Chloroperoxybenzoic Acid, 2154

C₇H₅Cl₂FN₂O₃
Fluroxypyr, 4233

C₇H₅Cl₂NO₄S
Halazone, 4625

C₇H₅Cl₃
Benzotrichloride, 1112

C₇H₅Cl₃O
2,4,6-Trichloroanisole, 9794
2,4,6-Trichloro-*m*-cresol, 9800

C₇H₅FO₂
4-Fluorobenzoic Acid, 4203

C₇H₅F₃
Benzotrifluoride, 1113

C₇H₅HgNO₃
Nitromersol, 6697

C₇H₅IN₂O₃
Iniparib, 5017

C₇H₅IO₂
o-Iodobenzoic Acid, 5074

C₇H₅IO₄
o-Iodoxybenzoic Acid, 5091

C₇H₅KO₂.3H₂O
Benzoic Acid Potassium Salt Trihydrate, 1093

C₇H₅KO₃
Potassium Salicylate, 7785

C₇H₅LiO₂
Lithium Benzoate, 5579

C₇H₅N
Benzonitrile, 1099

C₇H₅NO
2-Furanacrylonitrile, 4327
Phenyl Isocyanate, 7407

C₇H₅NO₃
Nitrobenzaldehyde, 6673

C₇H₅NO₃S
Saccharin, 8445

C₇H₅NO₄
Cinchomeronic Acid, 2286
Isocinchomeronic Acid, 5204
Nitrobenzoic Acid, 6676
Quinolinic Acid, 8190

C₇H₅NO₅
3-Nitrosalicylic Acid, 6717
5-Nitrosalicylic Acid, 6718

C₇H₅NO₇S
2-Nitro-4-sulfobenzoic Acid, 6732

C₇H₅NS
Benzothiazole, 1110
Phenyl Isothiocyanate, 7408

C₇H₅NS₂
2-Mercaptobenzothiazole, 5935

C₇H₅N₃O₅
Nitromide, 6699

C₇H₅N₃O₆
2,4,6-Trinitrotoluene, 9902

C₇H₅N₅O₈
Nitramine, 6659

C₇H₅NaO₂
Sodium Benzoate, 8718

C₇H₅NaO₂S
Thiosalicylic Acid Sodium Salt, 9513

C₇H₅NaO₄
Gentisic Acid Sodium Salt, 4433

C₇H₆BrCl
p-Bromobenzyl Chloride, 1417

C₇H₆BrNO₂
5-Bromoanthranilic Acid, 1412

C₇H₆BrNO₃
5-Bromosalicylhydroxamic Acid, 1446

C₇H₆Br₂
p-Bromobenzyl Bromide, 1416

C₇H₆ClN₃O₄S₂
Chlorothiazide, 2171

C₇H₆Cl₂
Benzal Chloride, 1059

C₇H₆Cl₂O
Dichlorobenzyl Alcohol, 3069

C₇H₆KNO₂
Potassium Aminobenzoate, 7727

$C_7H_6NNaO_3 \cdot 2H_2O$
p-Aminosalicylic Acid Sodium Salt
  Dihydrate, 473
$C_7H_6NO_3K$
p-Aminosalicylic Acid Potassium
  Salt, 473
$C_7H_6N_2$
Benzimidazole, 1083
1H-Indazole, 4976
$C_7H_6N_2O_2$
3-Nitrosobenzamide, 6720
$C_7H_6N_2O_2S_2$
Holomycin, 4766
$C_7H_6N_2O_5$
Dinitrocresol, 3305
$C_7H_6N_2S$
2-Aminobenzothiazole, 419
6-Aminobenzothiazole, 420
2-Benzimidazolethiol, 1084
$C_7H_6N_4O$
N,N'-Carbonyldiimidazole, 1820
$C_7H_6N_4O_2$
Tirapazamine, 9617
C_7H_6O
Benzaldehyde, 1060
$C_7H_6OS_2$
Dithiosalicylic Acid, 3418
$C_7H_6O_2$
Benzoic Acid, 1093
p-Hydroxybenzaldehyde, 4849
Salicylaldehyde, 8462
Tropolone, 9963
$C_7H_6O_2S$
Thiosalicylic Acid, 9513
$C_7H_6O_3$
2-Furanacrylic Acid, 4326
p-Hydroxybenzoic Acid, 4850
Perbenzoic Acid, 7266
Protocatechualdehyde, 8002
β-Resorcylaldehyde, 8277
Salicylic Acid, 8469
$C_7H_6O_4$
Gentisic Acid, 4433
Patulin, 7158
Protocatechuic Acid, 8003
β-Resorcylic Acid, 8278
Terreic Acid, 9315
$C_7H_6O_5$
Gallic Acid, 4375
$C_7H_6O_6S$
Salicylsulfuric Acid, 8471
Sulfosalicylic Acid, 9094
C_7H_7Br
Benzyl Bromide, 1131
Bromotoluene, 1449
Tropylium Bromide, 9965
C_7H_7BrO
Bromophenol p-Form Methyl Ether,
  1439
C_7H_7Cl
Benzyl Chloride, 1132
Chlorotoluene, 2175
$C_7H_7ClNNaO_2S$
Chloramine-T, 2075
$C_7H_7ClNa_2O_6P_2S$
Tiludronic Acid Disodium Salt, 9598
C_7H_7ClO
Chlorocresol, 2133
$C_7H_7ClO_2S$
p-Toluenesulfonyl Chloride, 9693
$C_7H_7Cl_2NO$
Clopidol, 2386
$C_7H_7Cl_2NO_2S$
Dichloramine T, 3056
$C_7H_7Cl_3NO_3PS$
Chlorpyrifos O,O-Dimethyl Analog,
  2195

C_7H_7F
Fluorotoluene, 4212
$C_7H_7FO_2S$
PMSF, 7659
$C_7H_7F_2I$
Difluoroiodotoluene, 3169
C_7H_7IO
o-Iodoanisole, 5071
$C_7H_7KO_5S$
Potassium Guaiacolsulfonate, 7755
C_7H_7N
3-Ethenylpyridine, 3785
C_7H_7NO
Benzamide, 1062
Formanilide, 4266
Methyl Pyridyl Ketone, 6189
$C_7H_7NO_2$
m-Aminobenzoic Acid, 416
o-Aminobenzoic Acid, 417
p-Aminobenzoic Acid, 418
Homarine, 4767
Methyl Nicotinate, 6173
Nitrotoluene, 6737
3-Pyridineacetic Acid, 8082
Salicylaldoxime, 8464
Salicylamide, 8465
Trigonelline, 9863
$C_7H_7NO_3$
p-Aminosalicylic Acid, 473
Mesalamine, 5974
Nitroanisole, 6671
Salicylhydroxamic Acid, 8468
$C_7H_7NO_4S$
Carzenide, 1876
$C_7H_7NO_5S$
Methyl p-Nitrobenzenesulfonate,
  6175
$C_7H_7N_3O_3$
(4-Nitrophenyl)urea, 6713
$C_7H_7N_3O_4$
Nihydrazone, 6626
$C_7H_7N_5O_2$
Fervenulin, 4094
Toxoflavin, 9718
$C_7H_7NaO_3S$
p-Toluenesulfonic Acid Sodium Salt,
  9692
$C_7H_7O_6P$
Fosfosal, 4282
C_7H_8
Norbornadiene, 6776
Quadricyclane, 8138
Toluene, 9688
$C_7H_8ClN_3O_4S_2$
Hydrochlorothiazide, 4819
$C_7H_8N_2O$
N-Phenylurea, 7429
$C_7H_8N_2O_2$
3,5-Diaminobenzoic Acid, 2980
N-(Hydroxymethyl)nicotinamide,
  4872
$C_7H_8N_2O_3S$
Saccharin Ammonium Salt, 8445
$C_7H_8N_2S$
N-Phenylthiourea, 7424
$C_7H_8N_3NaO_4S$
Isoniazid Methanesulfonate Sodium,
  5232
$C_7H_8N_4O_2$
Theobromine, 9433
Theophylline, 9436
C_7H_8O
Anisole, 662
Benzyl Alcohol, 1127
Cresol, 2564

$C_7H_8O_2$
Guaiacol, 4589
Orcinol, 6959
Salicyl Alcohol, 8461
$C_7H_8O_2S$
p-Toluenesulfinic Acid, 9691
$C_7H_8O_3$
Ethyl Maltol, 3879
2-Furoic Acid Ethyl Ester, 4336
Sarkomycin A, 8513
$C_7H_8O_3S$
p-Toluenesulfonic Acid, 9692
C_7H_8S
Thiobenzyl Alcohol, 9475
Thiocresol, 9478
$C_7H_8S_2$
Toluene-3,4-dithiol, 9690
$C_7H_9AsN_2O_4$
Carbarsone, 1787
$C_7H_9ClN_2O$
Pralidoxime Chloride, 7822
Trigonellamide Chloride, 9862
C_7H_9ClO
Ethchlorvynol, 3784
$C_7H_9ClO_6P_2S$
Tiludronic Acid, 9598
$C_7H_9IN_2O$
Pralidoxime Iodide, 7822
C_7H_9N
Benzylamine, 1128
3-Ethylpyridine, 3900
4-Ethylpyridine, 3901
2,6-Lutidine, 5677
Methylaniline, 6093
Toluidine, 9696
C_7H_9NO
Anisidine, 660
Moore's Ketene, 6350
$C_7H_9NO_2$
Ammonium Benzoate, 493
Deferiprone, 2862
$C_7H_9NO_3$
Ammonium Salicylate, 547
$C_7H_9N_3O$
4-Phenylsemicarbazide, 7422
$C_7H_9N_3O_2$
p-Aminosalicylic Acid Hydrazide,
  474
$C_7H_9N_3O_2S_2$
Sulfathiourea, 9075
$C_7H_9N_3O_3S$
Sulfanilylurea, 9062
$C_7H_9N_3O_4S$
Isoniazid Methanesulfonate, 5232
$C_7H_9O_3P$
α-Hydroxybenzylphosphinic Acid,
  4852
$C_7H_{10}ClN_3O$
Girard Reagent P, 4465
$C_7H_{10}ClN_3O_3$
Ornidazole, 6967
$C_7H_{10}NNaO_7P_2$
Risedronic Acid Monosodium Salt,
  8360
$C_7H_{10}N_2$
Diallylcyanamide, 2975
DMAP, 3433
Tolylhydrazine, 9701
$C_7H_{10}N_2O$
2,4-Diaminoanisole, 2978
$C_7H_{10}N_2OS$
Propylthiouracil, 7980
$C_7H_{10}N_2O_2S$
Carbimazole, 1799
Mafenide, 5712
p-Toluenesulfonylhydrazide, 9694

$C_7H_{10}N_2O_4$
AMPA, 575
$C_7H_{10}N_3NaO_6S$
Avibactam Sodium Salt, 878
$C_7H_{10}N_4O_2S$
Sulfaguanidine, 9040
$C_7H_{10}N_4O_3$
Cymoxanil, 2762
$C_7H_{10}N_4O_4$
Etanidazole, 3768
$C_7H_{10}O_4$
Terebic Acid, 9304
$C_7H_{10}O_5$
Mesoxalic Acid Diethyl Ester, 5983
Shikimic Acid, 8619
$C_7H_{10}O_7$
D-Glucoascorbic Acid, 4486
$C_7H_{11}F_6N$
Ishikawa Reagent, 5153
$C_7H_{11}NO_2$
Arecaidine, 767
Ethosuximide, 3801
Hypoglycine A, 4905
Meparfynol Carbamate, 5913
$C_7H_{11}NO_3$
Ethadione, 3773
Paramethadione, 7132
$C_7H_{11}NO_4$
Oxaceprol, 7002
$C_7H_{11}NO_7P_2$
Risedronic Acid, 8360
$C_7H_{11}N_3O_2$
Ipronidazole, 5123
$C_7H_{11}N_3O_3$
Secnidazole, 8556
$C_7H_{11}N_3O_6S$
Avibactam, 878
C_7H_{12}
Norcarane, 6777
$C_7H_{12}ClN_5$
Simazine, 8672
$C_7H_{12}N_2$
1,5-Diazabicyclo[4.3.0]non-5-ene,
2995
$C_7H_{12}N_2O_2$
Heptoxime, 4700
$C_7H_{12}N_2O_4$
Aceglutamide, 26
$C_7H_{12}O$
Cycloheptanone, 2715
$C_7H_{12}O_2$
n-Butyl Acrylate, 1541
Cyclohexanecarboxylic Acid, 2717
$C_7H_{12}O_3$
Ethyl Levulinate, 3874
Isopropyl Acetoacetate, 5252
$C_7H_{12}O_4$
n-Butylmalonic Acid, 1579
Diethylmalonic Acid, 3140
Ethyl Malonate, 3878
Pimelic Acid, 7543
$C_7H_{12}O_5$
Diacetin, 2964
$C_7H_{12}O_6$
Quinic Acid, 8175
$C_7H_{12}O_7$
2-Keto-L-gulonic Acid Methyl Ester,
5351
$C_7H_{13}N$
Quinuclidine, 8198
$C_7H_{13}NO$
Cycloheptanone Oxime, 2715
3-Quinuclidinol, 8199
$C_7H_{13}NO_2$
Stachydrine, 8899

$C_7H_{13}NO_3S$
N-Acetylmethionine, 92
N-Acetylpenicillamine, 94
$C_7H_{13}NO_3S_2$
Bucillamine, 1470
$C_7H_{13}NO_4$
L-Glutamic Acid 5-Ethyl Ester, 4506
$C_7H_{13}N_3O_3S$
Oxamyl, 7019
$C_7H_{13}NaO_8$
Glucoheptonic Acid Sodium Salt,
4491
$C_7H_{13}O_6P$
Mevinphos, 6244
C_7H_{14}
Methylcyclohexane, 6118
$C_7H_{14}AsClO_3$
Chloroarsenol, 2119
$C_7H_{14}B_2ClF_9N_2$
Selectfluor®, 8562
$C_7H_{14}NO_5P$
Monocrotophos, 6337
$C_7H_{14}N_2$
N,N'-Diisopropylcarbodiimide, 3218
$C_7H_{14}N_2O_2S$
Aldicarb, 215
$C_7H_{14}N_2O_3$
Theanine, 9424
$C_7H_{14}N_2O_4S$
L-Cystathionine, 2774
$C_7H_{14}N_2O_4S_2$
Djenkolic Acid, 3431
$C_7H_{14}N_3O_3P$
Uredepa, 10057
$C_7H_{14}N_4O_3$
Dinotefuran, 3317
$C_7H_{14}N_4S_2$
Methallibure, 6018
$C_7H_{14}O$
Cyclohexylcarbinol, 2724
Dipropyl Ketone, 3386
Heptanal, 4693
2-Heptanone, 4698
$C_7H_{14}O_2$
n-Butyl Propionate, 1590
Ethyl Isovalerate, 3871
Heptanoic Acid, 4695
Isoamyl Acetate, 5156
Isobutyl Propionate, 5195
Propyl Butyrate, 7959
n-Valeric Acid Ethyl Ester, 10090
$C_7H_{14}O_4$
D-Chalcose, 2038
Cymarose, 2759
Monobutyrin, 6335
Mycarose, 6403
Sarmentose, 8514
$C_7H_{14}O_6$
α-Methylglucoside, 6152
Pinitol, 7561
$C_7H_{14}O_7$
D-manno-Heptulose, 4701
$C_7H_{14}O_8$
Glucoheptonic Acid, 4491
$C_7H_{15}Cl_2N_2O_2P$
Ifosfamide, 4937
$C_7H_{15}Cl_2N_2O_2P\cdot H_2O$
Cyclophosphamide, 2743
$C_7H_{15}Cl_2N_2O_4P$
Perfosfamide, 7276
$C_7H_{15}NO$
1-Ethyl-3-piperidinol, 3897
$C_7H_{15}NO_2$
Isobutyl Urethane, 5198
$C_7H_{15}NO_3$
Carnitine, 1849

$C_7H_{15}NO_4S$
MOPS, 6352
$C_7H_{15}NO_5$
N-Methyl-α-L-glucosamine, 6151
$C_7H_{15}N_3O_2$
Indospicine, 5011
$C_7H_{15}N_3O_2S_2$
Cartap, 1865
C_7H_{16}
n-Heptane, 4694
$C_7H_{16}BrNO_2$
Acetylcholine Bromide, 81
$C_7H_{16}ClN$
Mepiquat Chloride, 5924
$C_7H_{16}ClNO_2$
Acetylcholine Chloride, 82
$C_7H_{16}ClN_3O_2S_2$
Cartap Monohydrochloride, 1865
$C_7H_{16}FO_2P$
Soman, 8839
$C_7H_{16}INO_2$
Oxapropanium Iodide, 7022
$C_7H_{16}N_4O_2$
N^G-Methylarginine, 6096
$C_7H_{16}N_4O_4S_2$
Taurolidine, 9213
$C_7H_{16}O$
1-Heptanol, 4696
2-Heptanol, 4697
$C_7H_{16}O_3$
Dipropylene Glycol Monomethyl
Ether, 3384
Orthoformic Acid Triethyl Ester,
6980
$C_7H_{16}O_4S_2$
Sulfonmethane, 9090
$C_7H_{17}AsClNO_3$
Chloroarsenol Ammonium Salt, 2119
$C_7H_{17}ClN_2O_2$
Bethanechol Chloride, 1187
$C_7H_{17}N$
Methylhexaneamine, 6157
Tuaminoheptane, 9982
$C_7H_{17}NO_2$
Triethylammonium Formate, 9832
$C_7H_{17}NO_5$
N-Methylglucamine, 6150
$C_7H_{17}NO_6S$
TAPS, 9191
$C_7H_{17}N_4O_4P$
Anatoxin A(s), 621
$C_7H_{17}O_2PS_3$
Phorate, 7443
$C_7H_{18}NO_7P$
Fosfomycin Tromethamine, 4281
$C_7H_{19}N_3$
Spermidine, 8869
C_8BrF_{17}
Perflubron, 7270
$C_8Cl_2N_2O_2$
2,3-Dichloro-5,6-dicyanobenzoqui-
none, 3072
$C_8Cl_4N_2$
Chlorothalonil, 2169
$C_8Co_2O_8$
Dicobalt Octacarbonyl, 3095
$C_8HF_{15}O_2$
Perfluorooctanoic Acid, 7272
$C_8H_2Cl_4O_5$
Mucochloric Anhydride, 6384
$C_8H_4Cl_2O_2$
Phthaloyl Chloride, 7487
$C_8H_4Cl_6$
1,4-Bis(trichloromethyl)benzene,
1304

$C_8H_4F_{15}NO_2$
Perfluorooctanoic Acid Ammonium
Salt, 7272

$C_8H_4K_2O_{12}Sb_2.3H_2O$
Antimony Potassium Tartrate, 691

$C_8H_4N_2S_2$
Bitoscanate, 1309

$C_8H_4Na_2O_{12}Sb_2$
Antimony Sodium Tartrate, 693

$C_8H_4O_3$
Phthalic Anhydride, 7484

$C_8H_5Cl_3O_3$
2,4,5-T, 9149

$C_8H_5F_3N_2OS$
Riluzole, 8350

$C_8H_5F_6NO_4S_2$
Phenyl Triflimide, 7427

$C_8H_5KO_4$
Potassium Biphthalate, 7733

C_8H_5NOS
Benzoyl Isothiocyanate, 1117

$C_8H_5NO_2$
Agrocybin, 185
Isatin, 5148
Phthalimide, 7485

$C_8H_5NO_3$
N-Hydroxyphthalimide, 4876

$C_8H_5N_3O_3S_2$
Tenonitrozole, 9290

C_8H_6
Ethynylbenzene, 3910

C_8H_6BrClO
p-Chlorophenacyl Bromide, 2156

C_8H_6BrN
α-Bromobenzyl Cyanide, 1418

$C_8H_6Br_2O$
p-Bromophenacyl Bromide, 1438

$C_8H_6Cl_2O_3$
2,4-D, 2793
Dicamba, 3050

$C_8H_6F_3N_3O_4S_2$
Flumethiazide, 4170

$C_8H_6KNO_4S$
Indoxyl Sulfate Potassium Salt, 5013

$C_8H_6N_2$
Cinnoline, 2307
Phthalazine, 7482
Quinazoline, 8169
Quinoxaline, 8195

$C_8H_6N_2OS_2$
Acibenzolar-S-methyl, 98

$C_8H_6N_2O_2$
4-Nitrobenzyl Cyanide, 6678

$C_8H_6N_4O_5$
Nitrofurantoin, 6686

$C_8H_6N_4O_8$
Alloxantin, 274

$C_8H_6N_6O_{11}$
Pentryl, 7252

C_8H_6O
Benzofuran, 1090

$C_8H_6O_2$
o-Phthalaldehyde, 7480

$C_8H_6O_3$
Piperonal, 7588

$C_8H_6O_4$
Isophthalic Acid, 5243
Phthalic Acid, 7483
Piperonylic Acid, 7590
Terephthalic Acid, 9305

$C_8H_6O_5$
Fomecin B, 4252
4-Hydroxyisophthalic Acid, 4866

C_8H_6S
Thianaphthene, 9455

C_8H_7BrO
ω-Bromoacetophenone, 1409
p-Bromoacetophenone, 1410

$C_8H_7BrO_3$
p-Bromomandelic Acid, 1434

$C_8H_7ClN_2O_2S$
Diazoxide, 3007

C_8H_7ClO
ω-Chloroacetophenone, 2115
p-Chloroacetophenone, 2116

$C_8H_7ClO_2$
p-Anisoyl Chloride, 665
Carbobenzoxy Chloride, 1801
m-Chlorobenzoic Acid Methyl Ester, 2124
p-Chlorobenzoic Acid Methyl Ester, 2126

$C_8H_7Cl_4N_3O_4S_2$
Teclothiazide, 9241

$C_8H_7FO_2$
p-Fluorophenylacetic Acid, 4209

C_8H_7N
Benzyl Cyanide, 1134
Indole, 5003
o-Tolunitrile, 9697
p-Tolunitrile, 9698

C_8H_7NO
Mandelonitrile, 5783

$C_8H_7NO_2$
Acetimidoquinone, 56
Isonitrosoacetophenone, 5237

$C_8H_7NO_3$
4-Pyridoxic Acid Lactone, 8094

$C_8H_7NO_4$
4-Nitrophenylacetic Acid, 6709

$C_8H_7NO_4S$
Indoxyl Sulfate, 5013

$C_8H_7N_3O_2$
Luminol, 5660

$C_8H_7N_3O_5$
Dinitolmide, 3295
Furazolidone, 4330

$C_8H_7NaO_3$
Mandelic Acid Sodium Salt, 5781

$C_8H_7NaO_4.H_2O$
Dehydroacetic Acid Sodium Salt Hydrate, 2869

C_8H_8
Barrelene, 999
Cubane, 2601
Cyclooctatetraene, 2731
Styrene, 8990

$C_8H_8BrCl_2O_3PS$
Bromophos, 1442

C_8H_8BrNO
p-Bromoacetanilide, 1406
p-Bromoacetophenone Oxime, 1410

C_8H_8ClNO
Chloroacetanilide, 2111

$C_8H_8ClNO_3S$
N-Acetylsulfanilyl Chloride, 95

$C_8H_8ClN_3O_2$
Aminocyclopyrachlor, 432

$C_8H_8Cl_2N_4$
Guanabenz, 4593

$C_8H_8Cl_2N_4O$
Guanoxabenz, 4602

$C_8H_8Cl_2O$
Dichloroxylenol, 3086

$C_8H_8Cl_3N_3O_4S_2$
Clorsulon, 2395
Trichlormethiazide, 9789

$C_8H_8Cl_3O_3PS$
Ronnel, 8387

$C_8H_8F_3N_3O_4S_2$
Hydroflumethiazide, 4826

$C_8H_8HgO_2$
Phenylmercuric Acetate, 7411

$C_8H_8K_2O_5$
Endothall Dipotassium Salt, 3630

$C_8H_8N_2O_2$
Phthalamide, 7481
Ricinine, 8334

$C_8H_8N_2O_2S_2$
Thiolutin, 9495

$C_8H_8N_2O_3$
Nitroacetanilide, 6667

$C_8H_8N_2O_3S$
Zonisamide, 10392

$C_8H_8N_3Na_2O_6S^{99m}Tc$
Technetium Tc 99m Mertiatide, 9238

$C_8H_8N_4$
Hydralazine, 4802

$C_8H_8N_6O_6$
Murexide, 6391

C_8H_8O
Acetophenone, 71
Coumaran, 2547
Phenylacetaldehyde, 7377
o-Tolualdehyde, 9686

$C_8H_8O_2$
p-Anisaldehyde, 656
Benzyl Formate, 1137
Methyl Benzoate, 6098
Phenyl Acetate, 7379
Phenylacetic Acid, 7380
Toluic Acid, 9695

$C_8H_8O_2S$
Phenyl Vinyl Sulfone, 7431

$C_8H_8O_3$
p-Anisic Acid, 659
m-Cresotic Acid, 2569
o-Cresotic Acid, 2570
p-Cresotic Acid, 2571
Mandelic Acid, 5781
Methylparaben, 6180
Methyl Salicylate, 6193
Phenoxyacetic Acid, 7366
Resacetophenone, 8260
Resorcinol Monoacetate, 8276
Vanillin, 10120

$C_8H_8O_4$
Dehydroacetic Acid, 2869
2,6-Dimethoxyquinone, 3246
Fumigatin, 4317
Gallacetophenone, 4372
Homogentisic Acid, 4775
o-Orsellinic Acid, 6977
Pyrogallol Monoacetate, 8112
Vanillic Acid, 10119

$C_8H_8O_5$
Fomecin A, 4252
Gallic Acid Methyl Ester, 4375

C_8H_9Br
Xylyl Bromide, 10288

C_8H_9BrO
Bromophenol p-Form Ethyl Ether, 1439

C_8H_9Cl
Xylyl Chloride, 10289

C_8H_9ClO
Chloroxylenol, 2182

$C_8H_9FN_2O_3$
Tegafur, 9252

C_8H_9NO
Acetanilide, 49
Aminoacetophenone, 407
Phenacylamine, 7321
α-Phenylacetamide, 7378
o-Toluamide, 9687

$C_8H_9NO_2$
Acetaminophen, 47
p-Aminophenylacetic Acid, 459
Methyl Anthranilate, 6094

α-Phenylglycine, 7402
N-Phenylglycine, 7403

C₈H₉NO₃
N-(4-Hydroxyphenyl)glycine, 4875
Orthocaine, 6979
Pyridoxal, 8091

C₈H₉NO₄
4-Pyridoxic Acid, 8094

C₈H₉NO₅
Clavulanic Acid, 2339

C₈H₉N₂NaO₃S
Sulfacetamide Sodium Salt, 9031

C₈H₉N₃O₄
Nicorandil, 6606

C₈H₁₀
Ethylbenzene, 3820
Xylene, 10278

C₈H₁₀AsNO₅
Acetarsone, 51

C₈H₁₀AsN₂NaO₄
Tryparsamide, 9974

C₈H₁₀ClN₃O
Lazabemide, 5449

C₈H₁₀ClN₃S
Tiamenidine, 9573

C₈H₁₀ClN₃O₃S
Thiamethoxam, 9446

C₈H₁₀FN₃O₃S
Emtricitabine, 3622

C₈H₁₀IN₃
Iobenguane, 5051

C₈H₁₀NNaO₅S
Sulbactam Sodium Salt, 9021

C₈H₁₀NO₅PS
Methyl Parathion, 6181

C₈H₁₀NO₆P
Pyridoxal 5-Phosphate, 8092

C₈H₁₀N₂O
p-Aminoacetanilide, 405
Benzylurea, 1149
p-Nitroso-N,N-dimethylaniline, 6725

C₈H₁₀N₂O₃
5-Nitro-o-phenetidine, 6704

C₈H₁₀N₂O₃S
N⁴-Acetylsulfanilamide, 9057
Sulfacetamide, 9031
p-Tolylsulfonylmethylnitrosamide, 9702

C₈H₁₀N₂O₄
Mimosine, 6280

C₈H₁₀N₂O₄S
Asulam, 847

C₈H₁₀N₂S
S-Benzylthiourea, 1148
Ethionamide, 3791

C₈H₁₀N₄O₂
Caffeine, 1639

C₈H₁₀N₆
Dihydralazine, 3187

C₈H₁₀O
Benzyl Methyl Ether, 1143
2-Ethylphenol, 3892
Methylenomycin B, 6139
Phenethyl Alcohol, 7338
Phenetole, 7343
Xylenol, 10279

C₈H₁₀O₂
Anise Alcohol, 658
Creosol, 2559
2-Phenoxyethanol, 7369
Styrene Glycol, 8991
Tyrosol, 10022
Veratrole, 10150

C₈H₁₀O₃
Ethoxymethylfurfural, 3805
Filicinic Acid, 4109
Hydroxytyrosol, 4887

C₈H₁₀O₃S
Benzenesulfonic Acid Ethyl Ester, 1071

C₈H₁₀O₄
Penicillic Acid, 7201

C₈H₁₀O₅
Endothall, 3630

C₈H₁₁Cl₂NO
Dichlormid, 3058

C₈H₁₁Cl₂N₃O₂
Uracil Mustard, 10036

C₈H₁₁Cl₃O₆
α-Chloralose, 2072

C₈H₁₁Cl₃O₇
Urochloralic Acid, 10069

C₈H₁₁N
N,N-Dimethylaniline, 3256
Ethylaniline, 3819
3-Ethyl-4-picoline, 3894
4-Ethyl-2-picoline, 3895
5-Ethyl-2-picoline, 3896
α-Methylbenzylamine, 6100
Phenethylamine, 7339
2,4,6-Trimethylpyridine, 9890
Xylidine, 10282

C₈H₁₁NO
Metyridine, 6239
o-Phenetidine, 7341
p-Phenetidine, 7342
Phenylethanolamine, 7399
Tyramine, 10016

C₈H₁₁NO₂
n-Butyl Cyanoacrylate, 1566
Dopamine, 3470
Isobutyl Cyanoacrylate, 5185
Norfenefrine, 6788
Octopamine, 6848

C₈H₁₁NO₃
Ammonium Mandelate, 529
Norepinephrine, 6784
Pyridoxine, 8095

C₈H₁₁NO₄S₂
Erdosteine, 3704

C₈H₁₁NO₅S
Sulbactam, 9021

C₈H₁₁NO₈
Iminodisuccinic Acid, 4960

C₈H₁₁N₂NaO₃
Barbital Sodium Salt, 957

C₈H₁₁N₃O₃S
Apricitabine, 742
Lamivudine, 5402

C₈H₁₁N₃O₄
Troxacitabine, 9968

C₈H₁₁N₃O₆
6-Azauridine, 897

C₈H₁₁N₅
Phenyl Biguanide, 7389

C₈H₁₁N₅O₃
Acyclovir, 134

C₈H₁₂
1,5-Cyclooctadiene, 2730
Cyclooctyne, 2732

C₈H₁₂INO₂
IPBC, 5112

C₈H₁₂NO₅PS₂
Cythioate, 2782

C₈H₁₂N₂
Betahistine, 1180
Dimethyl-p-phenylenediamine, 3278
Phenelzine, 7335

C₈H₁₂N₂O₂
Pyridoxamine, 8093

C₈H₁₂N₂O₂S
Thiobarbital, 9474

C₈H₁₂N₂O₃
Barbital, 957

C₈H₁₂N₂O₃S
6-Aminopenicillanic Acid, 454

C₈H₁₂N₂O₆
Kifunensine, 5357

C₈H₁₂N₄
2,2'-Azobisisobutyronitrile, 911

C₈H₁₂N₄O₃S
Carnidazole, 1848

C₈H₁₂N₄O₄
Decitabine, 2856

C₈H₁₂N₄O₅
Azacitidine, 884
Ribavirin, 8323

C₈H₁₂N₅O₄P
Adefovir, 139

C₈H₁₂O₂
5,5-Dimethyl-1,3-cyclohexanedione, 3264

C₈H₁₂O₄
Diethyl Maleate, 3139

C₈H₁₂O₅
Diethyl Oxalacetate, 3141

C₈H₁₂O₈Pb
Lead Tetraacetate, 5478

C₈H₁₂O₈Rh₂
Rhodium(II) Acetate, 8311

C₈H₁₂O₈Si
Silicon Tetraacetate, 8637

C₈H₁₃NO₂
Arecoline, 768
Bemegride, 1029
Retronecine, 8289
Scopoline, 8547

C₈H₁₃N₂O₅P
Pyridoxamine Phosphate, 8093

C₈H₁₃N₃O₄S
Tinidazole, 9604

C₈H₁₃N₅O₄
Taribavirin, 9197

C₈H₁₄ClNS₂
Sulfallate, 9043

C₈H₁₄ClN₅
Atrazine, 862

C₈H₁₄Cl₃O₂P
Trichlorfon Butanoic Acid Ester, 9788

C₈H₁₄CuO₄
Cupric Butyrate, 2620

C₈H₁₄INO
Furtrethonium Iodide, 4341

C₈H₁₄MgO₄
Butyric Acid Magnesium Salt, 1597

C₈H₁₄NNaO₃
ε-Acetamidocaproic Acid Sodium Salt, 46

C₈H₁₄NO⁺
Furtrethonium, 4341

C₈H₁₄N₂O
Loline, 5618

C₈H₁₄N₂O₂
Levetiracetam, 5513

C₈H₁₄N₂O₄
Diisopropyl Azodicarboxylate, 3217

C₈H₁₄N₂O₄Pt
Oxaliplatin, 7011

C₈H₁₄N₃O₇P
Cidofovir, 2271

C₈H₁₄N₄NiO₄
Nickel Dimethylglyoxime, 6592

C₈H₁₄N₄OS
Metribuzin, 6233

C₈H₁₄N₄O₇
Diazolidinyl Urea, 3003

C₈H₁₄O₂S₂
Thioctic Acid, 9479

C₈H₁₄O₃
Butyric Anhydride, 1598

C$_8$H$_{14}$O$_4$
Dimethoxane, 3244
Suberic Acid, 8992
Succinic Acid Diethyl Ester, 8999

C$_8$H$_{14}$O$_6$
Diethyl Tartrate, 3150

C$_8$H$_{14}$O$_7$
2-Keto-L-gulonic Acid Ethyl Ester, 5351

C$_8$H$_{15}$BF$_4$N$_2$
Butylmethylimidazolium Tetrafluoroborate, 1584

C$_8$H$_{15}$BrO$_2$
α-Bromo-n-caproic Acid Ethyl Ester, 1422

C$_8$H$_{15}$ClN$_2$
Butylmethylimidazolium Chloride, 1584

C$_8$H$_{15}$F$_6$N$_2$P
Butylmethylimidazolium, 1584

C$_8$H$_{15}$N
β-Coniceine, 2485
γ-Coniceine, 2486
Tropane, 9956

C$_8$H$_{15}$NO
Hygrine, 4890
Pelletierine, 7181
Pseudotropine, 8030
Tropine, 9960

C$_8$H$_{15}$NO$_2$
Tranexamic Acid, 9730

C$_8$H$_{15}$NO$_3$
ε-Acetamidocaproic Acid, 46
Acetylleucine, 91

C$_8$H$_{15}$NO$_4$
Castanospermine, 1896

C$_8$H$_{15}$N$_2$$^+$
Butylmethylimidazolium, 1584

C$_8$H$_{15}$N$_3$O$_7$
Streptozocin, 8962

C$_8$H$_{15}$N$_5$O
Noformicin, 6757
Pildralazine, 7535

C$_8$H$_{15}$N$_5$S
Simetryn, 8674

C$_8$H$_{15}$N$_7$O$_2$S$_3$
Famotidine, 3971

C$_8$H$_{15}$NaO$_2$
Valproic Acid Sodium Salt (1:1), 10099

C$_8$H$_{16}$
Caprylene, 1766

C$_8$H$_{16}$NO$_3$PS$_2$
Mephosfolan, 5922

C$_8$H$_{16}$NO$_5$P
Dicrotophos, 3097

C$_8$H$_{16}$N$_2$Na$_2$O$_6$S$_2$
PIPES Disodium Salt, 7592

C$_8$H$_{16}$N$_2$O$_2$S$_2$
4,4′-Dithiodimorpholine, 3415

C$_8$H$_{16}$N$_2$O$_4$S$_2$
L-Homocystine, 4773
Penicillamine Cysteine Disulfide, 7198

C$_8$H$_{16}$N$_2$O$_7$
Cycasin, 2694

C$_8$H$_{16}$N$_6$OS$_2$
Kethoxal Bis(thiosemicarbazone), 5347

C$_8$H$_{16}$O
Caprylic Aldehyde, 1768
Hexyl Methyl Ketone, 4747
5-Methyl-3-heptanone, 6156
3-Octanone, 6842

C$_8$H$_{16}$O$_2$
n-Butyl n-Butyrate, 1558
Butyroin, 1599
Caprylic Acid, 1767

Ethyl Caproate, 3832
Isobutyl n-Butyrate, 5181
Isobutyl Isobutyrate, 5190
Valproic Acid, 10099

C$_8$H$_{16}$O$_2$Si
1-Methoxy-3-(trimethylsilyloxy)-1,3-butadiene, 6075

C$_8$H$_{16}$O$_4$
Diethylene Glycol Monoethyl Ether Acetate, 3134
Mycarose 3-O-Methylmycarose, 6403

C$_8$H$_{17}$Br
n-Octyl Bromide, 6853
sec-Octyl Bromide, 6854

C$_8$H$_{17}$Cl$_2$NO$_2$
Diisopropylamine Dichloroacetate, 3216

C$_8$H$_{17}$I
sec-Octyl Iodide, 6858

C$_8$H$_{17}$N
Coniine, 2489

C$_8$H$_{17}$NNa$_2$O$_6$P$_2$
Incadronic Acid Disodium Salt, 4969

C$_8$H$_{17}$NO
Conhydrine, 2484
Pseudoconhydrine, 8022
Valnoctamide, 10098
Valpromide, 10100

C$_8$H$_{17}$NO$_2$
Butyroin Oxime, 1599
Octanohydroxamic Acid, 6839
Pregabalin, 7843

C$_8$H$_{17}$NO$_3$
Desosamine, 2930
Statine, 8927

C$_8$H$_{17}$NO$_3$S
CHES, 2055
Felinine, 3986

C$_8$H$_{17}$NO$_4$
Mycaminose, 6402

C$_8$H$_{17}$NO$_5$
Miglitol, 6267

C$_8$H$_{17}$N$_3$
EDC, 3555

C$_8$H$_{18}$
Isooctane, 5239
Octane, 6838

C$_8$H$_{18}$ClNO$_2$
Methacholine Chloride, 6012

C$_8$H$_{18}$Cl$_2$Si
Di-tert-butyldichlorosilane, 3039

C$_8$H$_{18}$N$_2$O$_3$S
Buthionine Sulfoximine, 1528

C$_8$H$_{18}$N$_2$O$_4$S
Burgess Reagent, 1505
HEPES, 4689

C$_8$H$_{18}$N$_2$O$_6$S$_2$
PIPES, 7592

C$_8$H$_{18}$N$_6$O$_4$
Streptidine, 8952

C$_8$H$_{18}$O
n-Butyl Ether, 1572
Di-tert-butyl Ether, 3040
2-Ethyl-1-hexanol, 3863
Isobutyl Ether, 5187
Isooctyl Alcohol, 5240
1-Octanol, 6840
2-Octanol, 6841

C$_8$H$_{18}$OSn
Dibutyltin Oxide, 3048

C$_8$H$_{18}$O$_2$
DTBP, 3508
Ethohexadiol, 3797

C$_8$H$_{18}$O$_3$
Diethylene Glycol Diethyl Ether, 3132
Diethylene Glycol Monobutyl Ether, 3133

4,4′-Oxydi-2-butanol, 7059
Trimethyl Orthovalerate, 9888

C$_8$H$_{18}$O$_4$
Triglyme, 9861

C$_8$H$_{18}$O$_4$S$_2$
Sulfonethylmethane, 9088

C$_8$H$_{18}$S
n-Butyl Sulfide, 1593
Isobutyl Sulfide, 5197

C$_8$H$_{19}$Al
Diisobutylaluminum Hydride, 3212

C$_8$H$_{19}$N
n-Dibutylamine, 3037
Diisopropylethylamine, 3219
Octodrine, 6846

C$_8$H$_{19}$NO
Heptaminol, 4692

C$_8$H$_{19}$NO$_2$
Ammonium Caprylate, 503

C$_8$H$_{19}$NO$_6$P$_2$
Incadronic Acid, 4969

C$_8$H$_{19}$O$_2$PS$_2$
Ethoprop, 3799

C$_8$H$_{19}$O$_2$PS$_3$
Disulfoton, 3408

C$_8$H$_{19}$O$_3$PS$_2$
Demeton, 2892

C$_8$H$_{20}$BrN
Tetraethylammonium Bromide, 9344

C$_8$H$_{20}$ClN
Tetraethylammonium Chloride, 9345

C$_8$H$_{20}$NO$_6$P
Choline Alfoscerate, 2212

C$_8$H$_{20}$N$_2$
Tetramethyldiaminobutane, 9373

C$_8$H$_{20}$N$_4$
Cyclen, 2699

C$_8$H$_{20}$O$_4$Si
Ethyl Silicate, 3903

C$_8$H$_{20}$O$_5$P$_2$S$_2$
Sulfotep, 9096

C$_8$H$_{20}$O$_7$P$_2$
Tetraethyl Pyrophosphate, 9349

C$_8$H$_{20}$Pb
Tetraethyllead, 9347

C$_8$H$_{21}$NO
Tetraethylammonium Hydroxide, 9346

C$_8$H$_{24}$N$_4$O$_3$P$_2$
Schradan, 8532

C$_8$H$_{24}$O$_2$Si$_3$
Octamethyltrisiloxane, 6837

C$_8$H$_{24}$O$_4$Si$_4$
Octamethylcyclotetrasiloxane, 6836

C$_8$N$_8$Pt$_2$Th
Thorium Tetracyanoplatinate(II), 9532

C$_9$F$_{21}$N
Perfluorotripropylamine, 4215

C$_9$H$_2$Cl$_6$O$_3$
Chlorendic Anhydride, 2087

C$_9$H$_4$Cl$_3$NO$_2$S
Folpet, 4251

C$_9$H$_4$Cl$_8$O
Isobenzan, 5172

C$_9$H$_4$O$_5$
Trimellitic Anhydride, 9872

C$_9$H$_5$Br$_2$NO
Broxyquinoline, 1462

C$_9$H$_5$ClINO
Iodochlorhydroxyquin, 5075

C$_9$H$_5$Cl$_2$NO
Chloroxine, 2181

C$_9$H$_5$Cl$_3$N$_4$
Anilazine, 650

C$_9$H$_5$I$_2$NO
Iodoquinol, 5085

C₉H₅NO₄
o-Nitrophenylpropiolic Acid, 6712

C₉H₆ClNO
Cloxyquin, 2405

C₉H₆Cl₆O₃S
Endosulfan, 3629

C₉H₆F₃N₃O
Flonicamid, 4137

C₉H₆INO₄S
8-Hydroxy-7-iodo-5-quinolinesulfonic
Acid, 4865

C₉H₆N₂O₂
Toluene 2,4-Diisocyanate, 9689

C₉H₆N₂O₃
Nitroxoline, 6742

C₉H₆N₂S₃
TCMTB, 9223

C₉H₆N₇NaO₂S
Azathioprine Sodium Salt, 895

C₉H₆O₂
Coumarin, 2550

C₉H₆O₂S
Thionaphthene-2-carboxylic Acid,
9498

C₉H₆O₃
Coumarilic Acid, 2549
Umbelliferone, 10032

C₉H₆O₄
Daphnetin, 2817
Esculetin, 3752
Ninhydrin, 6640

C₉H₆O₆
Trimellitic Acid, 9871

C₉H₇ClO
Cinnamoyl Chloride, 2301

C₉H₇Cl₂N₅
Irsogladine, 5146
Lamotrigine, 5403

C₉H₇INNaO₃
o-Iodohippurate Sodium, 5077

C₉H₇MnO₃
MMT, 6309

C₉H₇N
Isoquinoline, 5268
Quinoline, 8185

C₉H₇NO
Carbostyril, 1826
8-Hydroxyquinoline, 4881

C₉H₇NO₄
3-Nitrocinnamic Acid, 6682

C₉H₇NO₄S
8-Hydroxy-5-quinolinesulfonic Acid,
4882

C₉H₇N₅O₅
Erythropterin, 3745

C₉H₇N₇O₂S
Azathioprine, 895

C₉H₈
Indene, 4978

C₉H₈BNO₂
8-Quinolineboronic Acid, 8186

C₉H₈ClNS₂
Nimidane, 6635

C₉H₈ClN₅S
Tizanidine, 9642

C₉H₈ClNaO₃
MCPA Sodium Salt, 5833

C₉H₈Cl₂O₃
Dichlorprop, 3088

C₉H₈Cl₃NO₂S
Captan, 1774

C₉H₈I₂O₃
3,5-Diiodosalicylic Acid Ethyl Ester,
3208

C₉H₈NNaO₄
Salicylamide O-Acetic Acid Sodium
Salt, 8466

C₉H₈N₂O₂
Pemoline, 7188

C₉H₈N₃NaO₂S₂.1½H₂O
Sulfathiazole Sodium Salt Sesquihydrate, 9074

C₉H₈N₄OS
Thidiazuron, 9465

C₉H₈N₄O₆
Nifurtoinol, 6623

C₉H₈N₈O₂S
Thiamiprine, 9451

C₉H₈O
1,2-Benzopyran, 1105
Cinnamaldehyde, 2298

C₉H₈O₂
Atropic Acid, 865
Cinnamic Acid, 2299

C₉H₈O₃
Carbic Anhydride, 1797
p-Coumaric Acid, 2548

C₉H₈O₄
Aspirin, 841
Caffeic Acid, 1638

C₉H₉Br₂NO₃
3,5-Dibromo-L-tyrosine, 3035

C₉H₉ClO₃
MCPA, 5833

C₉H₉Cl₂NO
Propanil, 7918

C₉H₉Cl₂NO₂
Diloxanide, 3223

C₉H₉Cl₂N₃
Clonidine, 2380

C₉H₉Cl₂N₃O
Guanfacine, 4596

C₉H₉HgNaO₂S
Thimerosal, 9470

C₉H₉I₂NO₃
3,5-Diiodotyrosine, 3210

C₉H₉N
Skatole, 8698

C₉H₉NO
Hydrocarbostyril, 4817

C₉H₉NO₂S
Tosylmethyl Isocyanide, 9715

C₉H₉NO₃
Adrenochrome, 161
Adrenolutin, 163
Hippuric Acid, 4753

C₉H₉NO₄
Ethyl Nitrobenzoate, 3886
Salicylamide O-Acetic Acid, 8466

C₉H₉NS
Phenethyl Isothiocyanate, 7340

C₉H₉N₂NaO₃
p-Aminohippuric Acid Sodium Salt,
439

C₉H₉N₃O₂
Carbendazim, 1792

C₉H₉N₃O₂S₂
Sulfathiazole, 9074
Thiazolsulfone, 9461

C₉H₉N₃S
Amiphenazole, 479

C₉H₉N₄NaO₄
Acefylline Sodium Salt, 25
1-Theobromineacetic Acid Sodium
Salt, 9434

C₉H₉N₅
Benzoguanamine, 1091

C₉H₁₀
Indan, 4971

C₉H₁₀ClNO₃
Carisbamate, 1841

C₉H₁₀ClN₅O₂
Imidacloprid, 4951

C₉H₁₀Cl₂N₂O
Diuron, 3425

C₉H₁₀Cl₂N₂O₂
Linuron, 5564

C₉H₁₀Cl₂N₄
Apraclonidine, 738

C₉H₁₀FIN₂O₅
Fialuridine, 4098

C₉H₁₀INO₃
3-Iodotyrosine, 5090

C₉H₁₀NO₅PS
Cyanophos, 2687

C₉H₁₀N₂
5,6-Dimethylbenzimidazole, 3258

C₉H₁₀N₂O
Aminorex, 472
1-Phenyl-3-pyrazolidinone, 7419

C₉H₁₀N₂O₂
Phenacemide, 7318

C₉H₁₀N₂O₂S
Sulbenox, 9023
Thibenzazoline, 9464

C₉H₁₀N₂O₂S₂
Aureothricin, 871

C₉H₁₀N₂O₃
p-Aminohippuric Acid, 439

C₉H₁₀N₂O₃S₂
Ethoxzolamide, 3809

C₉H₁₀N₄O₂S₂
Sulfamethizole, 9048

C₉H₁₀N₄O₃S₂
Sulfametrole, 9052

C₉H₁₀N₄O₄
Acefylline, 25
1-Theobromineacetic Acid, 9434

C₉H₁₀N₆O₅
Orazamide, 6956

C₉H₁₀O
p-Anol, 670
Chavicol, 2049
Cinnamyl Alcohol, 2303
Phenylacetone, 7381
Propiophenone, 7944

C₉H₁₀O₂
Benzyl Acetate, 1126
m-Cresyl Acetate, 2572
o-Cresyl Acetate, 2573
Ethyl Benzoate, 3821
Hydrocinnamic Acid, 4822
Paroxypropione, 7149

C₉H₁₀O₃
Apocynin, 728
Atrolactic Acid, 864
Ethylparaben, 3890
Ethyl Salicylate, 3902
Ethyl Vanillin, 3908
Tropic Acid, 9958
Veratraldehyde, 10145

C₉H₁₀O₄
Flopropione, 4138
Glycol Salicylate, 4535
Homovanillic Acid, 4779
Methylenomycin A, 6139
Syringaldehyde, 9145
Veratric Acid, 10147

C₉H₁₀O₅
Vanilmandelic Acid, 10121

C₉H₁₀S
Cyclopropyl Phenyl Sulfide, 2747

C₉H₁₁BrN₂O₂
Metobromuron, 6218

C₉H₁₁BrN₂O₅
Broxuridine, 1461

C₉H₁₁ClN₂O
Monuron, 6349
Tebanicline, 9227

C₉H₁₁ClO₃
Chlorphenesin, 2184
C₉H₁₁Cl₂FN₂O₂S₂
Dichlofluanid, 3053
C₉H₁₁Cl₂N₃O₄S₂
Methyclothiazide, 6079
C₉H₁₁Cl₂O₃PS
Tolclofos-methyl, 9670
C₉H₁₁Cl₃NO₃PS
Chlorpyrifos, 2195
C₉H₁₁FN₂O₅
Doxifluridine, 3485
Floxuridine, 4144
C₉H₁₁F₂N₃O₄
Gemcitabine, 4420
C₉H₁₁IN₂O₅
Idoxuridine, 4931
C₉H₁₁N
Tranylcypromine, 9734
C₉H₁₁NO
Acetotoluide, 72
p-Aminopropiophenone, 466
Cathinone, 1907
p-Dimethylaminobenzaldehyde, 3252
N-Methylacetanilide, 6081
C₉H₁₁NO₂
p-Acetanisidine, 50
Benzocaine, 1088
4-(Dimethylamino)benzoic Acid, 3254
Ethenzamide, 3786
Phenylalanine, 7382
Phenylurethane, 7430
C₉H₁₁NO₃
Adrenalone, 160
Tyrosine, 10020
m-Tyrosine, 10021
C₉H₁₁NO₄
Dopa, 3469
Levodopa, 5516
C₉H₁₁NO₅
Droxidopa, 3504
C₉H₁₁NO₆
Showdomycin, 8621
C₉H₁₁N₃O₄
Ancitabine, 623
C₉H₁₁N₅O₃
Biopterin, 1234
C₉H₁₁N₅O₄
Eritadenine, 3728
Neopterin, 6547
C₉H₁₂
Cumene, 2607
Mesitylene, 5977
Propylbenzene, 7957
Pseudocumene, 8023
C₉H₁₂ClN₅O
Moxonidine, 6379
C₉H₁₂ClO₄P
Heptenophos, 4699
C₉H₁₂NO₅PS
Fenitrothion, 4007
C₉H₁₂N₂
Nornicotine, 6801
C₉H₁₂N₂Na₃O₁₅P₃.2H₂O
Uridine 5'-Triphosphate Trisodium
    Salt Dihydrate, 10065
C₉H₁₂N₂O
Fenuron, 4036
C₉H₁₂N₂O₂
Dulcin, 3511
C₉H₁₂N₂O₃
5-Nitro-2-propoxyaniline, 6716
C₉H₁₂N₂O₄S
Pidotimod, 7531
C₉H₁₂N₂O₅
Deoxyuridine, 2911

C₉H₁₂N₂O₅S
Tiazofurin, 9580
C₉H₁₂N₂O₆
Uridine, 10062
C₉H₁₂N₂O₇P₂
Minodronic Acid, 6284
C₉H₁₂N₂S
Protionamide, 8000
C₉H₁₂N₄O₃
Etofylline, 3924
C₉H₁₂N₅NaO₄
Ganciclovir Sodium Salt, 4393
C₉H₁₂N₆
Triethylenemelamine, 9836
C₉H₁₂O
Benzyl Ethyl Ether, 1136
p,α-Dimethylbenzyl Alcohol, 3259
α-Ethylbenzyl Alcohol, 3823
C₉H₁₂O₃
Phenylglyceryl Ether, 7401
C₉H₁₂O₃S
Ethyl p-Toluenesulfonate, 3907
C₉H₁₂O₄
3-Methoxy-4-hydroxyphenylglycol, 6069
C₉H₁₃BO₃
Methoxyethylbenzeneboronic Acid, 6065
C₉H₁₃BrN₂O₂
Bromacil, 1388
Pyridostigmine Bromide, 8090
C₉H₁₃ClN₂O₂
Terbacil, 9298
C₉H₁₃ClN₆
Cyanazine, 2675
C₉H₁₃ClN₆O₂
Nimustine, 6639
C₉H₁₃Cl₂N₃O₂
Dopan, 3471
C₉H₁₃N
Amphetamine, 579
Cumidine, 2610
Dextroamphetamine, 2956
C₉H₁₃NNaO₂P
Toldimfos Sodium, 9672
C₉H₁₃NO
β-Amino-α-methylphenethyl Alcohol, 444
Halostachine, 4635
Hydroxyamphetamine, 4847
Norpseudoephedrine, 6803
Phenylpropanolamine, 7417
C₉H₁₃NO₂
Deoxyepinephrine, 2904
Ecgonidine, 3539
Metaraminol, 6003
Phenylephrine, 7398
Pyrithyldione, 8108
Synephrine, 9142
C₉H₁₃NO₃
Epinephrine, 3674
Nordefrin, 6781
Normetanephrine, 6797
C₉H₁₃N₂O₉P
5'-Uridylic Acid, 10066
C₉H₁₃N₃O
Iproniazid, 5122
C₉H₁₃N₃O₃
Zalcitabine, 10307
C₉H₁₃N₃O₅
Cytarabine, 2781
Cytidine, 2783
C₉H₁₃N₃O₆
Mizoribine, 6308
C₉H₁₃N₅O₄
Ganciclovir, 4393

C₉H₁₄IN
Phenyltrimethylammonium Iodide, 7428
C₉H₁₄N₂O₃
Metharbital, 6033
C₉H₁₄N₂O₆S
Itramin Tosylate, 5293
C₉H₁₄N₂O₁₂P₂
Uridine 5'-Diphosphate, 10063
C₉H₁₄N₃O₈P
2'-Cytidylic Acid, 2784
3'-Cytidylic Acid, 2785
C₉H₁₄N₄O₃
Carnosine, 1851
Nimorazole, 6637
C₉H₁₄N₄O₄
Molsidomine, 6320
C₉H₁₄N₄O₅
Acadesine, 17
C₉H₁₄N₅O₄P
Tenofovir, 9289
C₉H₁₄O
Isophorone, 5242
Phorone, 7445
C₉H₁₄OS₃
Ajoene, 188
C₉H₁₄O₅
α-Ketoglutaric Acid Diethyl Ester, 5350
C₉H₁₄O₆
Triacetin, 9751
C₉H₁₄SeSi
Phenylselenotrimethylsilane, 7421
C₉H₁₅AlO₉
Aluminum Lactate, 343
C₉H₁₅BrN₂O₃
Acecarbromal, 22
C₉H₁₅Br₆O₄P
Tris-BP, 9927
C₉H₁₅Cl₆O₄P
TDCPP, 9224
C₉H₁₅NO
Pseudopelletierine, 8028
C₉H₁₅NO₂
Diacetone Acrylamide, 2966
Piperidione, 7581
3-Quinuclidinol dl-Form Acetate
    (Ester), 8199
C₉H₁₅NO₃
Ecgonine, 3540
C₉H₁₅NO₃S
Captopril, 1776
Mycobacidin, 6407
C₉H₁₅N₂O₁₅P₃
Uridine 5'-Triphosphate, 10065
C₉H₁₅N₃O₂
Hercynine, 4702
C₉H₁₅N₃O₂S
Ergothioneine, 3720
C₉H₁₅N₃O₅
Diethyl Oxalacetate Semicarbazone, 3141
C₉H₁₅N₅O
Minoxidil, 6285
C₉H₁₅N₅O₃
Sapropterin, 8505
Theophylline Ethanolamine, 9436
C₉H₁₅N₅O₇S₂
Amidosulfuron, 397
C₉H₁₅O₆P
TCEP, 9222
C₉H₁₆ClN₃O₂
Lomustine, 5621
C₉H₁₆ClN₃O₇
Chlorozotocin, 2183

$C_9H_{16}ClN_5$
Propazine, 7926
Trietazine, 9829
$C_9H_{16}Cl_2N_4$
Quaternium-15, 8143
$C_9H_{16}N_2$
1,8-Diazabicyclo[5.4.0]undec-7-ene, 2996
$C_9H_{16}N_4OS$
Tebuthiuron, 9232
$C_9H_{16}O$
2-Nonenal, 6763
$C_9H_{16}O_2$
HNE, 4764
$C_9H_{16}O_4$
Azelaic Acid, 898
$C_9H_{17}NO$
N-Methylpelletierine, 7181
Valdetamide, 10087
$C_9H_{17}NO_2$
Gabapentin, 4348
$C_9H_{17}NO_4$
Acetylcarnitine, 79
$C_9H_{17}NO_5$
Pantothenic Acid, 7118
$C_9H_{17}NO_7$
Muramic Acid, 6389
$C_9H_{17}NO_8$
Neuraminic Acid, 6568
$C_9H_{17}N_2NaO_6$
Betaine Sodium Aspartate, 1181
$C_9H_{17}N_5S$
Ametryn, 387
$C_9H_{18}Cl_3N_2O_2P$
Trofosfamide, 9947
$C_9H_{18}FeN_3S_6$
Ferbam, 4040
$C_9H_{18}NO$
TEMPO, 9283
$C_9H_{18}NO_2$
TEMPOL, 9284
$C_9H_{18}NO_3PS_2$
Fosthiazate, 4288
$C_9H_{18}N_2O$
MacMillan's Enamine Catalyst, 5707
$C_9H_{18}N_2O_3Pt$
Lobaplatin, 5607
$C_9H_{18}N_2O_4$
Meprobamate, 5929
$C_9H_{18}N_3OP$
Metepa, 6008
$C_9H_{18}N_4O$
Amidinomycin, 394
$C_9H_{18}N_6$
Altretamine, 316
$C_9H_{18}N_6O_6$
Hexamethylolmelamine, 4729
$C_9H_{18}O_2$
n-Amyl Butyrate, 598
Ethyl Oenanthate, 3889
Isoamyl Butyrate, 5160
Isobutyl Isovalerate, 5191
Pelargonic Acid, 7179
$C_9H_{18}O_3$
n-Butyl Carbonate, 1559
$C_9H_{18}O_3Si$
Brassard's Diene, 1368
$C_9H_{18}O_5S$
IPTG, 5126
$C_9H_{18}O_6$
Triacetone Triperoxide, 9752
$C_9H_{19}ClN_3O_5P$
Fotemustine, 4289
$C_9H_{19}N$
Cyclopentamine, 2735
Isometheptene, 5231

$C_9H_{19}NOS$
EPTC, 3697
$C_9H_{19}NO_3S$
CAPS, 1769
$C_9H_{19}NO_4$
Dexpanthenol, 2949
$C_9H_{20}NO_2^+$
Muscarine, 6395
$C_9H_{20}N_2O_2$
Propamocarb, 7912
$C_9H_{20}N_4O_4$
Negamycin, 6526
$C_9H_{20}O$
n-Nonyl Alcohol, 6766
$C_9H_{21}AlO_3$
Aluminum Isopropoxide, 342
$C_9H_{21}ClO_3Ti$
Chlorotitanium Triisopropoxide, 2174
$C_9H_{21}O_2PS_3$
Terbufos, 9301
$C_9H_{21}O_6P_3$
Propylphosphonic Anhydride, 7978
$C_9H_{22}N_2$
Novoldiamine, 6812
$C_9H_{22}N_2O$
Bredereck's Reagent, 1373
$C_9H_{22}O_4P_2S_4$
Ethion, 3790
$C_9H_{23}INO_3PS$
Echothiophate Iodide, 3548
$C_9H_{23}NO_7P_2$
Ibandronic Acid, 4912
$C_9H_{24}I_2N_2O$
Prolonium Iodide, 7897
$C_9H_{28}Si_4$
Tris(trimethylsilyl)silane, 9934
$C_{10}Cl_{10}$
Dienochlor, 3116
$C_{10}Cl_{10}O$
Chlordecone, 2084
$C_{10}Cl_{12}$
Mirex, 6290
$C_{10}F_{18}$
Perfluorodecalin, 4215
$C_{10}H_4Cl_2O_2$
Dichlone, 3054
$C_{10}H_4N_2Na_2O_8S$
Naphthol Yellow S, 6478
$C_{10}H_5ClN_2$
o-Chlorobenzylidenemalononitrile, 2128
$C_{10}H_5Cl_7$
Heptachlor, 4691
$C_{10}H_5F_6IO_4$
Phenyliodine(III) Bis(trifluoroacetate), 7406
$C_{10}H_5NNa_2O_8S_2$
Nitroso-R Salt, 6731
$C_{10}H_5NaO_5S$
Sodium β-Naphthoquinone-4-sulfonate, 8777
$C_{10}H_6ClNO_2$
Quinoclamine, 8184
$C_{10}H_6ClN_5O$
Triazoxide, 9768
$C_{10}H_6Cl_2N_2O_2$
Pyrrolnitrin, 8132
$C_{10}H_6Cl_4O_4$
Chlorthal-dimethyl, 2199
$C_{10}H_6Cl_8$
Chlordane, 2083
$C_{10}H_6N_2OS_2$
Oxythioquinox, 7077
$C_{10}H_6Na_2O_8S_2\cdot 2H_2O$
Chromotropic Acid Disodium Salt Dihydrate, 2245

$C_{10}H_6O_2$
1,2-Naphthoquinone, 6479
1,4-Naphthoquinone, 6480
$C_{10}H_6O_3$
Juglone, 5317
Lawsone, 5448
$C_{10}H_6O_4$
Chromocarb, 2241
Coumarin-3-carboxylic Acid, 2551
$C_{10}H_6O_8$
Pyromellitic Acid, 8116
$C_{10}H_7Br$
1-Bromonaphthalene, 1436
2-Bromonaphthalene, 1437
$C_{10}H_7Cl$
1-Chloronaphthalene, 2151
2-Chloronaphthalene, 2152
$C_{10}H_7Cl_2NO$
Chlorquinaldol, 2196
$C_{10}H_7Cl_2N_3O$
Anagrelide, 616
$C_{10}H_7Cl_5O$
Tridiphane, 9827
$C_{10}H_7F_3O_4$
Triflusal, 9856
$C_{10}H_7F_4NO_4$
2,2,3,3-Tetrafluoro-1-propanol p-Nitrobenzoate, 9351
$C_{10}H_7KN_2O$
Pemirolast Potassium Salt, 7187
$C_{10}H_7NO_2$
1-Nitronaphthalene, 6701
1-Nitroso-2-naphthol, 6728
N-Phenylmaleimide, 7410
Quinaldic Acid, 8162
8-Quinolinecarboxylic Acid, 8187
$C_{10}H_7NO_3$
Kynurenic Acid, 5375
1-Nitro-2-naphthol, 6702
$C_{10}H_7NO_4$
Xanthurenic Acid, 10266
$C_{10}H_7N_3S$
Thiabendazole, 9440
$C_{10}H_7NaO$
2-Naphthol Sodium Salt, 6469
$C_{10}H_7NaO_4S$
2-Naphthol-6-sulfonic Acid Sodium Salt, 6477
$C_{10}H_8$
Azulene, 919
Naphthalene, 6455
$C_{10}H_8ClN_4NaO_2S$
Sulfachlorpyridazine Sodium Salt, 9032
$C_{10}H_8F_2N_4O$
Rufinamide, 8426
$C_{10}H_8NNaO_4S$
1-Amino-2-naphthol-4-sulfonic Acid Sodium Salt, 450
$C_{10}H_8NNaO_4S\cdot\frac{1}{2}H_2O$
1-Amino-2-naphthol-6-sulfonic Acid Sodium Salt Hemipentahydrate, 451
$C_{10}H_8NNaO_3S\cdot 4H_2O$
1-Naphthylamine-4-sulfonic Acid Sodium Salt Tetrahydrate, 6485
$C_{10}H_8N_2$
2,2'-Bipyridine, 1241
γ,γ'-Dipyridyl, 3389
$C_{10}H_8N_2O_2S_2$
Pyrithione Dimer, 8107
$C_{10}H_8N_2O_2S_2Zn$
Pyrithione Zinc Derivative, 8107
$C_{10}H_8N_2O_4$
α-Furildioxime, 4335
$C_{10}H_8N_4O_3S_2$
Nitrodan, 6683

$C_{10}H_8N_4O_5$
Picrolonic Acid, 7524
$C_{10}H_8N_6O$
Pemirolast, 7187
$C_{10}H_8O$
1-Naphthol, 6468
2-Naphthol, 6469
$C_{10}H_8OS_3$
Anethole Trithione, 637
$C_{10}H_8O_2$
Naphthoresorcinol, 6481
Tricromyl, 9823
$C_{10}H_8O_3$
Erythrocentaurin, 3736
Hymecromone, 4894
$C_{10}H_8O_3S$
1-Naphthalenesulfonic Acid, 6461
2-Naphthalenesulfonic Acid, 6462
$C_{10}H_8O_4$
Anemonin, 635
Scopoletin, 8545
$C_{10}H_8O_4S$
Cassella's Acid, 1892
Croceic Acid, 2578
1-Naphthol-2-sulfonic Acid, 6475
1-Naphthol-4-sulfonic Acid, 6476
2-Naphthol-6-sulfonic Acid, 6477
$C_{10}H_8O_5$
Fraxetin, 4294
$C_{10}H_8O_6S_2$
Armstrong's Acid, 779
1,6-Naphthalenedisulfonic Acid, 6458
2,6-Naphthalenedisulfonic Acid, 6459
2,7-Naphthalenedisulfonic Acid, 6460
$C_{10}H_8O_7S_2$
1-Naphthol-4,8-disulfonic Acid, 6471
2-Naphthol-3,6-disulfonic Acid, 6472
2-Naphthol-6,8-disulfonic Acid, 6473
$C_{10}H_8O_8S_2$
Chromotropic Acid, 2245
$C_{10}H_8O_{10}S_2$
Sulmarin, 9114
$C_{10}H_8S$
1-Naphthalenethiol, 6463
2-Naphthalenethiol, 6464
$C_{10}H_9ClN_4O_2S$
Sulfachlorpyridazine, 9032
$C_{10}H_9ClN_4S$
Thiacloprid, 9443
$C_{10}H_9Cl_2NO$
Dicryl, 3098
$C_{10}H_9Cl_2NaO_3$
2,4-DB Sodium Salt, 2838
$C_{10}H_9Cl_4NO_2S$
Captafol, 1773
$C_{10}H_9Cl_4O_4P$
Tetrachlorvinphos, 9337
$C_{10}H_9F_3O_3$
Mosher's Reagent, 6365
$C_{10}H_9N$
Lepidine, 5496
1-Naphthylamine, 6483
2-Naphthylamine, 6484
Quinaldine, 8163
$C_{10}H_9NO$
4-Amino-1-naphthol, 449
Echinopsine, 3545
$C_{10}H_9NO_2$
Gentianine, 4429
Indoleacetic Acid, 5004
Succinanil, 8997
$C_{10}H_9NO_3$
Maleanilic Acid, 5766
$C_{10}H_9NO_3S$
Badische Acid, 933
Cassella's Acid F, 1893
1,6-Cleve's Acid, 2348

1,7-Cleve's Acid, 2349
1-Naphthylamine-4-sulfonic Acid, 6485
1-Naphthylamine-5-sulfonic Acid, 6486
1-Naphthylamine-8-sulfonic Acid, 6487
2-Naphthylamine-1-sulfonic Acid, 6488
2-Naphthylamine-5-sulfonic Acid, 6489
$C_{10}H_9NO_4S$
1-Amino-2-naphthol-4-sulfonic Acid, 450
1-Amino-2-naphthol-6-sulfonic Acid, 451
$C_{10}H_9NO_6S_2$
Amido-G-Acid, 395
Amido-R-Acid, 396
$C_{10}H_9NO_7S_2$
1-Naphthol-8-amino-3,6-disulfonic Acid, 6470
$C_{10}H_9N_3O$
Amrinone, 587
$C_{10}H_9N_4NaO_2S$
Sulfadiazine Sodium Salt, 9035
$C_{10}H_9N_5O$
Kinetin, 5359
$C_{10}H_{10}$
Bullvalene, 1487
$C_{10}H_{10}CaO_5$
Prohexadione Calcium Salt, 7892
$C_{10}H_{10}Cl_2N_2S_2$
3,3-Dithiodipyridine Dihydrochloride, 3416
$C_{10}H_{10}Cl_2O_3$
2,4-DB, 2838
$C_{10}H_{10}Cl_2Ti$
Titanocene Dichloride, 9638
$C_{10}H_{10}F_3IO$
Togni's Reagents Dimethyl Compound, 9663
$C_{10}H_{10}Fe$
Ferrocene, 4064
$C_{10}H_{10}IN_2 \cdot BF_4$
Bis(pyridine)iodonium Tetrafluoroborate, 1303
$C_{10}H_{10}NNaO_4S_2$
Stepronin Sodium Salt, 8936
$C_{10}H_{10}N_2$
1,8-Naphthalenediamine, 6457
$C_{10}H_{10}N_2O$
Norphenazone, 6802
$C_{10}H_{10}N_2O_2$
Procodazole, 7880
$C_{10}H_{10}N_2O_3$
Imidazole Salicylate, 4954
$C_{10}H_{10}N_4O$
Metamitron, 5995
$C_{10}H_{10}N_4OS$
Methisazone, 6049
$C_{10}H_{10}N_4O_2S$
Sulfadiazine, 9035
Sulfapyrazine, 9068
$C_{10}H_{10}N_6O_2$
Azanidazole, 890
$C_{10}H_{10}O$
Benzylideneacetone, 1140
$C_{10}H_{10}O_2$
Isosafrole, 5269
Safrole, 8453
$C_{10}H_{10}O_4$
Dimethyl Phthalate, 3279
Ferulic Acid, 4089
Meconin, 5854
Methyl Acetylsalicylate, 6084

Terephthalic Acid Dimethyl Ester, 9305
$C_{10}H_{10}O_5$
Opianic Acid, 6946
$C_{10}H_{10}O_6$
Chorismic Acid, 2222
Prephenic Acid, 7856
$C_{10}H_{10}V$
Vanadocene, 10111
$C_{10}H_{11}BrN_2O_3$
Brallobarbital, 1366
$C_{10}H_{11}ClFN_5O_3$
Clofarabine, 2366
$C_{10}H_{11}ClKO_3$
Mecoprop Potassium Salt, 5855
$C_{10}H_{11}ClN_4$
Acetamiprid, 48
$C_{10}H_{11}ClO_3$
Clofibric Acid, 2371
Mecoprop, 5855
$C_{10}H_{11}ClZr$
Schwartz's Reagent, 8533
$C_{10}H_{11}F_3N_2O$
Fluometuron, 4185
$C_{10}H_{11}F_3N_2O_5$
Trifluridine, 9855
$C_{10}H_{11}NO$
Abikoviromycin, 8
Tryptophol, 9978
$C_{10}H_{11}NO_2$
Acetoacetanilide, 57
Tuberin, 9986
$C_{10}H_{11}NO_3$
Actarit, 124
Betamipron, 1183
Succinanilic Acid, 8998
$C_{10}H_{11}NO_4S_2$
Stepronin, 8936
$C_{10}H_{11}N_3OS$
Methabenzthiazuron, 6011
$C_{10}H_{11}N_3O_3S$
Sulfamethoxazole, 9050
$C_{10}H_{11}N_3O_5S$
Nifuratel, 6616
$C_{10}H_{11}N_4NaO_5S$
Carbazochrome Sodium Sulfonate, 1790
Tazobactam Sodium Salt, 9220
$C_{10}H_{11}N_5O$
Pymetrozine, 8061
$C_{10}H_{12}$
Tetralin®, 9368
$C_{10}H_{12}BrCl_2O_3PS$
Bromophos O,O-Diethyl Analog, 1442
$C_{10}H_{12}CaN_2Na_2O_8$
EDTA Calcium Disodium Salt, 3565
$C_{10}H_{12}ClNO$
Beclamide, 1016
$C_{10}H_{12}ClNO_2$
Baclofen, 930
Chlorpropham, 2193
$C_{10}H_{12}ClNO_4$
Chlorphenesin Carbamate, 2185
$C_{10}H_{12}ClN_3O_3S$
Quinethazone, 8172
$C_{10}H_{12}ClN_5O_3$
Cladribine, 2336
$C_{10}H_{12}Cl_2N_4O_2$
Guanabenz Monoacetate, 4593
$C_{10}H_{12}Co_2N_2O_8$
EDTA Cobalt Salt, 3565
$C_{10}H_{12}Cr_2N_2O_7$
Pyridinium Dichromate, 8087
$C_{10}H_{12}FN_5O_4$
Fludarabine, 4157

$C_{10}H_{12}FeN_2NaO_8$
Ferric Sodium Edetate, 4058
$C_{10}H_{12}MgN_2Na_2O_8$
EDTA Magnesium Disodium Salt,
3565
$C_{10}H_{12}N_2$
Anatabine, 620
Tolazoline, 9666
Tryptamine, 9976
$C_{10}H_{12}N_2Na_4O_8$
EDTA Tetrasodium Salt, 3565
$C_{10}H_{12}N_2O$
Cotinine, 2541
4-Methylaminorex, 6092
Serotonin, 8601
$C_{10}H_{12}N_2O_3$
Allobarbital, 255
Kynurenine, 5376
$C_{10}H_{12}N_2O_3S$
Bentazon, 1054
$C_{10}H_{12}N_2O_4$
Stavudine, 8929
$C_{10}H_{12}N_2O_5$
Dinoseb, 3315
$C_{10}H_{12}N_2O_5S$
7-Aminocephalosporanic Acid, 429
$C_{10}H_{12}N_2O_6S$
Ritipenem, 8364
$C_{10}H_{12}N_2O_8$
Orotidine, 6973
$C_{10}H_{12}N_3O_3PS_2$
Azinphos-methyl, 906
$C_{10}H_{12}N_4OS$
Thiacetazone, 9442
$C_{10}H_{12}N_4O_2S_2$
Sulfaethidole, 9039
$C_{10}H_{12}N_4O_3$
Carbazochrome, 1789
$C_{10}H_{12}N_4O_4$
Nebularine, 6517
$C_{10}H_{12}N_4O_5$
Inosine, 5018
$C_{10}H_{12}N_4O_5S$
Tazobactam, 9220
$C_{10}H_{12}N_4O_6$
Xanthosine, 10263
$C_{10}H_{12}N_5O_6P$
Cyclic AMP, 2701
$C_{10}H_{12}N_5O_7P$
Cyclic GMP, 2702
$C_{10}H_{12}O$
Anethole, 636
Cuminaldehyde, 2611
Estragole, 3760
Tetralol, 9369
$C_{10}H_{12}O_2$
Cumic Acid, 2608
Duroquinone, 3517
Ethyl Phenylacetate, 3893
Eugenol, 3940
Isoeugenol, 5216
β-Thujaplicin, 9544
Thujic Acid, 9545
$C_{10}H_{12}O_3$
Coniferyl Alcohol, 2488
Propylparaben, 7977
$C_{10}H_{12}O_4$
Lactisole, 5388
$C_{10}H_{12}O_5$
Prohexadione, 7892
Propyl Gallate, 7971
$C_{10}H_{13}BrN_2O_3$
Propallylonal, 7910
$C_{10}H_{13}Cl$
Neophyl Chloride, 6544
$C_{10}H_{13}ClN_2$
Chlordimeform, 2086

$C_{10}H_{13}ClN_2O$
Chlorotoluron, 2176
$C_{10}H_{13}ClN_2O_3S$
Chlorpropamide, 2192
$C_{10}H_{13}ClO$
Chlorothymol, 2173
$C_{10}H_{13}Cl_2N$
Aniline Mustard, 653
$C_{10}H_{13}FN_2O_5$
Clevudine, 2351
$C_{10}H_{13}FN_5O_7P$
Fludarabine 5′-Monophosphate, 4157
$C_{10}H_{13}FN_6O_6S$
Nucleocidin, 6815
$C_{10}H_{13}NO$
Methcathinone, 6035
$C_{10}H_{13}NO_2$
4-Amino-3-phenylbutyric Acid, 460
Fusaric Acid, 4343
4′-Hydroxybutyranilide, 4853
MDA, 5834
Phenacetin, 7319
Phenprobamate, 7370
Propham, 7930
$C_{10}H_{13}NO_3$
Damascenine, 2807
α-Methyl-m-tyrosine, 6208
Metyrosine, 6240
Surinamine, 9136
$C_{10}H_{13}NO_4$
Glyceryl p-Aminobenzoate, 4523
Methyldopa, 6127
3-O-Methyldopa, 6128
$C_{10}H_{13}NO_5$
Arogenic Acid, 781
$C_{10}H_{13}N_2NaO_3$
Aprobarbital Sodium Salt, 745
$C_{10}H_{13}N_2Na_3O_8$
EDTA Trisodium Salt, 3565
$C_{10}H_{13}N_3$
Debrisoquin, 2846
$C_{10}H_{13}N_3O_2$
Guanoxan, 4603
5-(α-Phenylethyl)semioxamazide,
7400
$C_{10}H_{13}N_3O_5S$
Nifurtimox, 6622
$C_{10}H_{13}N_4O_8P$
Inosinic Acid, 5020
$C_{10}H_{13}N_5O$
Zeatin, 10315
$C_{10}H_{13}N_5O_2$
Dideoxyadenosine, 3112
$C_{10}H_{13}N_5O_3$
Cordycepin, 2514
$C_{10}H_{13}N_5O_4$
Adenosine, 141
Vidarabine, 10173
Zidovudine, 10322
$C_{10}H_{13}N_5O_4S$
Thioguanosine, 9491
$C_{10}H_{13}N_5O_5$
Guanosine, 4601
$C_{10}H_{13}N_5O_{12}P_3$
Cordycepin Triphosphate, 2514
$C_{10}H_{14}$
n-Butylbenzene, 1551
sec-Butylbenzene, 1552
tert-Butylbenzene, 1553
Cymene, 2760
Durene, 3515
Isobutylbenzene, 5179
Isodurene, 5211
$C_{10}H_{14}BeO_4$
Beryllium Acetylacetonate, 1165
$C_{10}H_{14}BrNO_2$
2C-B, 1910

$C_{10}H_{14}CaO_6$
Calcium Levulinate, 1681
$C_{10}H_{14}ClN$
Chlorphentermine, 2188
$C_{10}H_{14}ClNO_3$
Pyridoxal Monoethylacetal Hydro-
chloride, 8091
$C_{10}H_{14}Cl_2NO_2PS$
DMPA, 3436
$C_{10}H_{14}Cl_6N_4O_2$
Triforine, 9859
$C_{10}H_{14}NNaO_3$
Tenuazonic Acid Sodium Salt, 9292
$C_{10}H_{14}NO_5PS$
Parathion, 7137
$C_{10}H_{14}NO_6P$
Paraoxon, 7134
$C_{10}H_{14}N_2$
Anabasine, 612
Nicotine, 6609
$C_{10}H_{14}N_2Na_2O_8$
EDTA Disodium Salt, 3565
$C_{10}H_{14}N_2O$
Isonicotinic Acid Diethylamide, 5234
Nikethamide, 6627
$C_{10}H_{14}N_2O_3$
Aprobarbital, 745
$C_{10}H_{14}N_2O_4$
Porphobilinogen, 7721
Proxibarbal, 8013
$C_{10}H_{14}N_2O_4 \cdot H_2O$
Carbidopa, 1798
$C_{10}H_{14}N_2O_4S_2$
Sulthiame, 9122
$C_{10}H_{14}N_2O_5$
Telbivudine, 9258
Thymidine, 9552
$C_{10}H_{14}N_4O_2$
Morphazinamide, 6359
$C_{10}H_{14}N_4O_3$
Proxyphylline, 8014
$C_{10}H_{14}N_4O_4$
Dyphylline, 3526
$C_{10}H_{14}N_5O_7P$
3′-Adenylic Acid, 145
5′-Adenylic Acid, 146
$C_{10}H_{14}N_5O_8P$
3′-Guanylic Acid, 4604
5′-Guanylic Acid, 4605
$C_{10}H_{14}NiO_4$
Nickel Acetylacetonate, 6585
$C_{10}H_{14}O$
4-tert-Butylphenol, 1588
Carvacrol, 1872
Carvone, 1874
Chrysanthenone, 2256
Cumic Alcohol, 2609
Perillaldehyde, 7282
Safranal, 8452
Thymol, 9554
d-Verbenone, 10154
$C_{10}H_{14}O_2$
Dolichodial, 3459
Durohydroquinone, 3516
Nepetalactone, 6554
Perilla Ketone, 7281
$C_{10}H_{14}O_3$
Mephenesin, 5917
$C_{10}H_{14}O_4$
Guaifenesin, 4591
$C_{10}H_{15}BrO$
3-Bromo-d-camphor, 1421
$C_{10}H_{15}ClO$
3-Chloro-d-camphor, 2132
$C_{10}H_{15}F_6N_6OP$
HATU, 4653

C₁₀H₁₅N
Diethylaniline, 3125
Methamphetamine, 6020
Phentermine, 7374
Phenylpropylmethylamine, 7418

C₁₀H₁₅NO
α-(α-Aminopropyl)benzyl Alcohol, 467
Anatoxin A, 621
Ephedrine, 3663
Hordenine, 4785
Perillaldehyde Oxime, 7282
Pholedrine, 7442
Pseudoephedrine, 8024

C₁₀H₁₅NO₃
Ethylnorepinephrine, 3888
Metanephrine, 5997
N-Methylepinephrine, 6141
Tenuazonic Acid, 9292

C₁₀H₁₅NO₄
Kainic Acid, 5324

C₁₀H₁₅NO₅S
Tricaine, 9786

C₁₀H₁₅N₂NaO₂S
Thiobutabarbital Sodium Salt, 9476

C₁₀H₁₅N₂Na₃O₇
Versenol®, 10162

C₁₀H₁₅N₃
Bethanidine, 1188

C₁₀H₁₅N₃O
Sabcomeline, 8444

C₁₀H₁₅N₃O₅
Benserazide, 1052
Dinoseb Ammonium Salt, 3315

C₁₀H₁₅N₃O₈
Convicine, 2498

C₁₀H₁₅N₃S
Talipexole, 9171

C₁₀H₁₅N₅
Phenformin, 7345
Trapidil, 9735

C₁₀H₁₅N₅O₃
Penciclovir, 7192

C₁₀H₁₅N₅O₁₀P₂
Adenosine Diphosphate, 142

C₁₀H₁₅OPS₂
Fonofos, 4260

C₁₀H₁₅O₃PS₂
Fenthion, 4029

C₁₀H₁₆
Adamantane, 137
Camphene, 1732
3-Carene, 1835
Limonene, 5546
β-Myrcene, 6416
Ocimene, 6830
α-Phellandrene, 7315
β-Phellandrene, 7316
α-Pinene, 7556
β-Pinene, 7557
Terpinene, 9312
Twistane, 10009

C₁₀H₁₆BrNO
Edrophonium Bromide, 3564

C₁₀H₁₆Br₂N₂O₂
Pipobroman, 7593

C₁₀H₁₆ClNO
Edrophonium Chloride, 3564

C₁₀H₁₆Cl₃NOS
Triallate, 9756

C₁₀H₁₆KNO₉S₂
Sinigrin, 8683

C₁₀H₁₆N₂
Tetramethyl-p-phenylenediamine, 9376

C₁₀H₁₆N₂O
Rilmenidine, 8347

C₁₀H₁₆N₂OS
Albutoin, 211

C₁₀H₁₆N₂O₂S
Thiobutabarbital, 9476

C₁₀H₁₆N₂O₃
Butabarbital, 1509
Butethal, 1524

C₁₀H₁₆N₂O₃S
Amidephrine, 393
Biotin, 1236

C₁₀H₁₆N₂O₄
TAED, 9158

C₁₀H₁₆N₂O₄S
Biotin l-Sulfoxide, 1237

C₁₀H₁₆N₂O₄S₃
Dorzolamide, 3475

C₁₀H₁₆N₂O₈
EDDS, 3556
EDTA, 3565

C₁₀H₁₆N₄O₃
Anserine, 671

C₁₀H₁₆N₄O₇
Vicine, 10171

C₁₀H₁₆N₅O₁₃P₃
Adenosine Triphosphate, 143

C₁₀H₁₆N₆S
Cimetidine, 2282

C₁₀H₁₆O
Camphor, 1734
Citral, 2321
d-Fenchone, 4000
Piperitone, 7585
Pulegone, 8046
Thujone, 9546

C₁₀H₁₆O₂
Ascaridole, 816
Chrysanthemic Acid, 2255
Diosphenol, 3326
3-Hydroxycamphor, 4856
Iridomyrmecin, 5133
Lineatin, 5558
Sebacil, 8553

C₁₀H₁₆O₃
Queen Substance, 8148

C₁₀H₁₆O₄
Camphoric Acid, 1735
Prelog-Djerassi Lactone, 7852

C₁₀H₁₆O₄S
d-Camphorsulfonic Acid, 1736

C₁₀H₁₆O₅
Senecic Acid, 8588

C₁₀H₁₇Cl
Bornyl Chloride, 1342

C₁₀H₁₇Cl₂NOS
Diallate, 2973

C₁₀H₁₇N
Amantadine, 368

C₁₀H₁₇NOS
Cevimeline, 2033

C₁₀H₁₇NO₂
Methyprylon, 6212

C₁₀H₁₇NO₅S
Ethamsylate, 3777

C₁₀H₁₇NO₆
Linamarin, 5552

C₁₀H₁₇N₃O₂
Isolan, 5224

C₁₀H₁₇N₃O₅
Linatine, 5554

C₁₀H₁₇N₃O₆S
Glutathione, 4511

C₁₀H₁₇N₃S
Pramipexole, 7825

C₁₀H₁₇N₅O₃
Theophylline Isopropanolamine, 9436

C₁₀H₁₇N₇O₄²⁺
Saxitoxin, 8526

C₁₀H₁₈
Decalin®, 2850

C₁₀H₁₈CaO₆S₂
Methionine Hydroxy Analog Calcium Salt, 6048

C₁₀H₁₈ClN₃O₇
Ranimustine, 8227

C₁₀H₁₈N₂O₃
Desthiobiotin, 2935

C₁₀H₁₈N₂O₇
α-Ketoglutaric Acid Compound with L(+)-Ornithine, 5350

C₁₀H₁₈N₄O₄S₃
Thiodicarb, 9482

C₁₀H₁₈O
Borneol, 1340
Citronellal, 2328
Eucalyptol, 3938
Geraniol, 4437
Grandisol, 4571
Isoborneol, 5173
Linalool, 5550
l-Menthone, 5906
Nerol, 6560
α-Terpineol, 9313

C₁₀H₁₈O₂
δ-Decalactone, 2849
Sebacoin, 8554
Sobrerol, 8705

C₁₀H₁₈O₃
Royal Jelly Acid, 8410

C₁₀H₁₈O₄
Adipic Acid Diethyl Ester, 150
Sebacic Acid, 8552

C₁₀H₁₈O₄Zn
Zinc Valerate, 10364

C₁₀H₁₈O₅
Di-tert-butyl Dicarbonate, 3038

C₁₀H₁₈O₆
Di-tert-butyl Peroxyoxalate, 3043

C₁₀H₁₉ClNO₅P
Phosphamidon, 7449

C₁₀H₁₉NO
Lupinine, 5669

C₁₀H₁₉NO₃
MeBmt, 5840

C₁₀H₁₉NO₅
Hopantenic Acid, 4783

C₁₀H₁₉N₂O₃
TOAC, 9645

C₁₀H₁₉N₃O₂
Guanadrel, 4594

C₁₀H₁₉N₅O
Prometon, 7903

C₁₀H₁₉N₅O₂
Amicarbazone, 390

C₁₀H₁₉N₅S
Prometryn, 7904

C₁₀H₁₉O₆PS₂
Malathion, 5764

C₁₀H₂₀CaN₂O₈S₂
Acamprosate Calcium, 18

C₁₀H₂₀NO₄PS
Propetamphos, 7929

C₁₀H₂₀N₂O₄
Mebutamate, 5842
Piperazine Adipate, 7577

C₁₀H₂₀N₂O₄S₂
Penicillamine Disulfide, 7199

C₁₀H₂₀N₂S₃
Sulfiram, 9080

C₁₀H₂₀N₂S₄
Disulfiram, 3407

C₁₀H₂₀O
β-Citronellol, 2329
Menthol, 5905
Rhodinol, 8309

$C_{10}H_{20}O_2$
n-Capric Acid, 1760
Ethyl Caprylate, 3833
Isoamyl Isovalerate, 5166
Octyl Acetate, 6852
Terpin, 9311
$C_{10}H_{21}Cl_2N_2O_7P$
Glufosfamide, 4504
$C_{10}H_{21}N$
Pempidine, 7189
Propylhexedrine, 7972
$C_{10}H_{21}NO$
Trimethyl Isopropyl Butanamide,
9887
$C_{10}H_{21}NOS$
Pebulate, 7167
$C_{10}H_{21}NO_4$
Miglustat, 6268
$C_{10}H_{21}NO_7$
Voglibose, 10226
$C_{10}H_{21}N_3O$
Diethylcarbamazine, 3128
$C_{10}H_{21}N_5O_6$
Aspartic Acid Compound with L-Ar-
ginine, 830
$C_{10}H_{22}Cl_2N_2O_4Pt$
Satraplatin, 8520
$C_{10}H_{22}N_4$
Guanethidine, 4595
$C_{10}H_{22}O$
n-Amyl Ether, 602
n-Decyl Alcohol, 2858
Isoamyl Ether, 5163
$C_{10}H_{22}O_2$
Decamethylene Glycol, 2853
$C_{10}H_{22}O_5$
Tetraglyme, 9353
$C_{10}H_{23}B$
Bis(1,2-dimethylpropyl)borane, 1253
$C_{10}H_{23}N$
Diisoamylamine, 3211
$C_{10}H_{23}N_3O_3$
Hypusine, 4909
$C_{10}H_{23}O_2PS_2$
Cadusafos, 1636
$C_{10}H_{24}N_2O_2$
Ethambutol, 3775
$C_{10}H_{24}N_4$
Tetrakis(dimethylamino)ethylene,
9366
$C_{10}H_{26}N_4$
Spermine, 8870
$C_{10}H_{30}O_3Si_4$
Decamethyltetrasiloxane, 2854
$C_{10}H_{30}O_5Si_5$
Decamethylcyclopentasiloxane, 2852
$C_{10}Mn_2O_{10}$
Manganese Carbonyl, 5792
$C_{11}H_6ClN_3O_6$
Lodoxamide, 5613
$C_{11}H_6Cl_2N_2$
Fenpiclonil, 4018
$C_{11}H_6O_3$
Psoralen, 8037
$C_{11}H_7NO$
1-Naphthylisocyanate, 6492
$C_{11}H_7NS$
1-Naphthylisothiocyanate, 6493
$C_{11}H_8ClNO_2$
Cloxyquin Acetate (Ester), 2405
Quinmerac, 8182
$C_{11}H_8ClNO_2S$
Fenclozic Acid, 4001
$C_{11}H_8N_2$
γ-Carboline, 1805
$C_{11}H_8N_2O_3S_2$
Firefly Luciferin, 4117

$C_{11}H_8N_3NaO_2$
PAR Sodium Salt, 7125
$C_{11}H_8Na_4O_8P_2$
Menadiol Sodium Diphosphate, 5898
$C_{11}H_8O_2$
Menadione, 5899
1-Naphthoic Acid, 6466
2-Naphthoic Acid, 6467
$C_{11}H_8O_3$
1-Hydroxy-2-naphthoic Acid, 4873
3-Hydroxy-2-naphthoic Acid, 4874
Phthiocol, 7490
Plumbagin, 7655
$C_{11}H_8O_5$
Purpurogallin, 8057
$C_{11}H_9FN_2O_3$
Sorbinil, 8849
$C_{11}H_9I_3N_2O_4$
Diatrizoic Acid, 2993
Iothalamic Acid, 5107
$C_{11}H_9NO_2$
3-Amino-2-naphthoic Acid, 448
$C_{11}H_9NO_3$
Quininic Acid, 8178
$C_{11}H_9N_3O_2$
1,2-Naphthoquinone 2-Semicarba-
zone, 6479
PAR, 7125
$C_{11}H_9NaO_5S$
Menadione Sodium Bisulfite, 5899
$C_{11}H_{10}BrN_5$
Brimonidine, 1381
$C_{11}H_{10}Cl_2F_2N_4O_3S$
Sulfentrazone, 9078
$C_{11}H_{10}FNO_2S$
Flosequinan, 4142
$C_{11}H_{10}N_2S$
ANTU, 708
$C_{11}H_{10}N_3NaO_2S.H_2O$
Sulfapyridine Sodium Salt Monohy-
drate, 9069
$C_{11}H_{10}N_4O_4$
Carbadox, 1782
$C_{11}H_{10}O$
2-Methoxynaphthalene, 6070
$C_{11}H_{10}O_2$
Menadiol, 5898
$C_{11}H_{10}O_4$
Limettin, 5545
Scoparone, 8541
$C_{11}H_{11}Br_2N_5O$
Oroidin, 6971
$C_{11}H_{11}Cl_2NO_2$
Benoxacor, 1048
$C_{11}H_{11}F_3N_2O_3$
Flutamide, 4238
$C_{11}H_{11}I_3O_3$
Iophenoxic Acid, 5104
$C_{11}H_{11}NO$
3-Indolylacetone, 5008
Vitamin K_5, 10220
$C_{11}H_{11}NO_2$
Phensuximide, 7373
$C_{11}H_{11}NO_4S$
Woodward's Reagent K, 10248
$C_{11}H_{11}N_3O_2$
Mivazerol, 6306
$C_{11}H_{11}N_3O_2S$
Sulfapyridine, 9069
$C_{11}H_{11}N_4NaO_2S$
Sulfamerazine Monosodium Salt,
9045
$C_{11}H_{11}N_4NaO_3S$
Sulfamethoxypyridazine Sodium Salt,
9051
$C_{11}H_{11}N_5$
Phenazopyridine, 7324

$C_{11}H_{11}N_5O_3S_2$
Urothion, 10073
$C_{11}H_{12}AsNO_5S_2$
Thiacetarsamide, 9441
$C_{11}H_{12}ClNO_3S$
Chlormezanone, 2106
$C_{11}H_{12}Cl_2N_2O$
Lofexidine, 5615
$C_{11}H_{12}Cl_2N_2O_5$
Chloramphenicol, 2077
$C_{11}H_{12}Cl_2O_3$
2,4-D Isopropyl Ester, 2793
$C_{11}H_{12}I_3NO_2$
Iopanoic Acid, 5101
$C_{11}H_{12}NO_4PS_2$
Phosmet, 7448
$C_{11}H_{12}N_2O$
Antipyrine, 703
Vasicine, 10127
$C_{11}H_{12}N_2O_2$
Ethotoin, 3802
Idazoxan, 4928
Thozalinone, 9533
Tryptophan, 9977
$C_{11}H_{12}N_2O_2S$
Zileuton, 10323
$C_{11}H_{12}N_2O_3$
5-Hydroxytryptophan, 4886
$C_{11}H_{12}N_2S$
Levamisole, 5511
$C_{11}H_{12}N_4O_2$
Todralazine, 9660
$C_{11}H_{12}N_4O_2S$
Sulfamerazine, 9045
Sulfaperine, 9065
$C_{11}H_{12}N_4O_3S$
Sulfalene, 9042
Sulfameter, 9046
Sulfamethoxypyridazine, 9051
$C_{11}H_{12}O_2$
Cinnamic Acid Ethyl Ester, 2299
Cinnamyl Acetate, 2302
6-Methoxy-α-tetralone, 6074
$C_{11}H_{12}O_3$
Ethyl Benzoylacetate, 3822
Myristicin, 6420
$C_{11}H_{12}O_5$
Elenolide, 3595
$C_{11}H_{13}BrN_2O_5$
Brivudine, 1385
$C_{11}H_{13}BrN_2O_6$
Sorivudine, 8853
$C_{11}H_{13}ClF_3N_3O_4S_3$
Polythiazide, 7705
$C_{11}H_{13}ClN_2$
Epibatidine, 3665
$C_{11}H_{13}ClO$
Dowicide 9®, 3479
$C_{11}H_{13}Cl_3N_2O_3$
Fenuron Trichloroacetate, 4036
$C_{11}H_{13}F_3N_2O_3S$
Mefluidide, 5873
$C_{11}H_{13}N$
Pargyline, 7142
$C_{11}H_{13}NNa_2O_7S_2$
Disufenton Sodium Salt, 3406
$C_{11}H_{13}NO_2$
Hydrohydrastinine, 4839
Idrocilamide, 4933
$C_{11}H_{13}NO_3$
Corydaldine, 2530
Hydrastinine, 4807
4-Salicyloylmorpholine, 8470
Thurfyl Nicotinate, 9550
Toloxatone, 9680

C$_{11}$H$_{13}$NO$_4$
　Bendiocarb, 1038
　Mephenoxalone, 5918
C$_{11}$H$_{13}$N$_3$O
　4-Aminoantipyrine, 413
C$_{11}$H$_{13}$N$_3$O$_3$S
　Sulfamoxole, 9056
　Sulfisoxazole, 9082
C$_{11}$H$_{14}$AsNO$_3$S$_2$
　Arsthinol, 802
C$_{11}$H$_{14}$ClN
　Lorcaserin, 5638
C$_{11}$H$_{14}$ClNO
　Propachlor, 7907
C$_{11}$H$_{14}$ClN$_3$O$_4$S$_3$
　Althiazide, 314
C$_{11}$H$_{14}$ClN$_5$
　Cycloguanil, 2714
C$_{11}$H$_{14}$N$_2$
　Gramine, 4569
　α-Methyltryptamine, 6207
C$_{11}$H$_{14}$N$_2$O
　Cytisine, 2786
　5-Methoxytryptamine, 6076
C$_{11}$H$_{14}$N$_2$O$_2$
　Levulinic Acid Phenylhydrazone,
　　5526
　Pheneturide, 7344
C$_{11}$H$_{14}$N$_2$O$_3$S
　Sulfadicramide, 9036
C$_{11}$H$_{14}$N$_2$O$_4$
　Felbamate, 3984
C$_{11}$H$_{14}$N$_2$S
　Pyrantel, 8066
C$_{11}$H$_{14}$N$_4$O$_2$
　Epirizole, 3677
C$_{11}$H$_{14}$N$_4$O$_4$
　Doxofylline, 3486
　Tubercidin, 9984
C$_{11}$H$_{14}$O$_2$
　n-Butyl Benzoate, 1554
　Methyleugenol, 6144
C$_{11}$H$_{14}$O$_3$
　Butylparaben, 1587
　o-Thymotic Acid, 9563
　Zingerone, 10366
C$_{11}$H$_{14}$O$_4$
　Dimethyl Carbate, 3262
C$_{11}$H$_{15}$BrClO$_3$PS
　Profenofos, 7886
C$_{11}$H$_{15}$ClN$_2$O$_2$
　Iproclozide, 5120
C$_{11}$H$_{15}$ClN$_4$O$_2$
　Nitenpyram, 6654
C$_{11}$H$_{15}$Cl$_2$N$_5$
　Chlorproguanil, 2190
C$_{11}$H$_{15}$NO
　PBN, 7162
　Phenmetrazine, 7351
C$_{11}$H$_{15}$NO$_2$
　Butamben, 1515
　Isobutyl p-Aminobenzoate, 5178
　MDMA, 5836
　Salsoline, 8478
C$_{11}$H$_{15}$NO$_2$S
　Methiocarb, 6044
C$_{11}$H$_{15}$NO$_3$
　Anhalamine, 646
　Propoxur, 7949
C$_{11}$H$_{15}$NO$_4$
　Mephenesin Carbamate, 5917
C$_{11}$H$_{15}$NO$_5$
　Methocarbamol, 6052
C$_{11}$H$_{15}$NO$_7$S$_2$
　Disufenton, 3406
C$_{11}$H$_{15}$N$_2$O$_8$P
　NMN, 6750

C$_{11}$H$_{15}$N$_3$O$_4$
　Pyridinol Carbamate, 8088
C$_{11}$H$_{15}$N$_5$O$_5$
　Nelarabine, 6527
　Psicofuranine, 8032
C$_{11}$H$_{16}$
　Amylbenzene, 595
C$_{11}$H$_{16}$ClN$_3$O$_4$S$_2$
　Buthiazide, 1527
C$_{11}$H$_{16}$ClN$_5$
　Chlorguanide, 2091
C$_{11}$H$_{16}$FN$_3$O$_3$
　Carmofur, 1844
C$_{11}$H$_{16}$F$_6$N$_5$OP
　HBTU, 4654
C$_{11}$H$_{16}$I$_2$N$_2$O$_5$
　Iodopyracet, 5083
C$_{11}$H$_{16}$N$_2$O
　Tocainide, 9648
C$_{11}$H$_{16}$N$_2$O$_2$
　Aminocarb, 428
　Pilocarpine, 7536
C$_{11}$H$_{16}$N$_2$O$_3$
　Butalbital, 1514
　Talbutal, 9167
　Vinylbital, 10190
C$_{11}$H$_{16}$N$_2$O$_3$S
　Etozolin Free Acid, 3934
C$_{11}$H$_{16}$N$_2$O$_4$S
　Thienamycin, 9466
C$_{11}$H$_{16}$N$_2$O$_5$
　Edoxudine, 3563
C$_{11}$H$_{16}$N$_4$O
　Lidamidine, 5534
C$_{11}$H$_{16}$N$_4$O$_4$
　Pentostatin, 7248
　Razoxane, 8241
C$_{11}$H$_{16}$N$_8$O$_8$
　Imidurea, 4957
C$_{11}$H$_{16}$O
　Fenipentol, 4006
　Jasmone, 5307
　p-$tert$-Pentylphenol, 7255
C$_{11}$H$_{16}$O$_2$
　Butylated Hydroxyanisole, 1549
　Dihydroactinidiolide, 3188
C$_{11}$H$_{16}$O$_3$
　d-Camphocarboxylic Acid, 1733
C$_{11}$H$_{17}$Cl$_2$NOS
　4,5-Dichloro-2-octyl-3-isothiazolone,
　　3080
C$_{11}$H$_{17}$N
　N-Ethylamphetamine, 3818
　Mephentermine, 5919
　Pentorex, 7246
C$_{11}$H$_{17}$NO
　Methoxyphenamine, 6071
　N-Methylephedrine, 6140
　Mexiletine, 6248
　Tecomanine, 9243
C$_{11}$H$_{17}$NO$_2$
　Dioxethedrine, 3331
　Isoproterenol, 5263
　Mescaline, 5975
　Metaproterenol, 6002
　Methoxamine, 6058
C$_{11}$H$_{17}$N$_2$NaO$_2$S
　Thiopental Sodium, 9503
C$_{11}$H$_{17}$N$_2$NaO$_3$
　Amobarbital Sodium Salt, 567
　Pentobarbital Sodium Salt, 7243
C$_{11}$H$_{17}$N$_3$O
　Meobentine, 5911
C$_{11}$H$_{17}$N$_3$O$_3$
　Emorfazone, 3619
C$_{11}$H$_{17}$N$_3$O$_3$S
　Carbutamide, 1831

C$_{11}$H$_{17}$N$_3$O$_8$
　Tetrodotoxin, 9394
C$_{11}$H$_{17}$O$_4$PS$_2$
　Fensulfothion, 4027
C$_{11}$H$_{18}$BrN$_5$O$_3$
　Pamabrom, 7101
C$_{11}$H$_{18}$ClN$_3$O$_2$
　Murexine Chloride, 6392
C$_{11}$H$_{18}$N$_2$O$_2$
　p-Aminobenzoic Acid Diethylamine
　　Salt, 418
C$_{11}$H$_{18}$N$_2$O$_3$
　Amobarbital, 567
　Pentobarbital, 7243
C$_{11}$H$_{18}$N$_3$O$_2$$^+$
　Murexine, 6392
C$_{11}$H$_{18}$N$_4$O$_2$
　Pirimicarb, 7609
C$_{11}$H$_{18}$N$_4$O$_6$
　Lysine Monoorotate, 5697
C$_{11}$H$_{18}$O
　Nopol, 6771
C$_{11}$H$_{18}$O$_5$
　D-Apiose Di-O-isopropylidene, 722
C$_{11}$H$_{18}$O$_7$
　Di-$tert$-butyl Tricarbonate, 3049
C$_{11}$H$_{19}$NOS
　Octhilinone, 6844
C$_{11}$H$_{19}$NO$_3$
　d-Camphocarboxylic Acid Ammo-
　　nium Salt, 1733
C$_{11}$H$_{19}$NO$_4$
　Garner's Aldehyde, 4404
C$_{11}$H$_{19}$NO$_9$
　N-Acetylneuraminic Acid, 8623
C$_{11}$H$_{19}$N$_3$O
　Dimethirimol, 3239
　Ethirimol, 3794
C$_{11}$H$_{19}$N$_3$O$_6$
　Wildfire Toxin, 10244
C$_{11}$H$_{19}$N$_5$S
　Irgarol®, 5128
C$_{11}$H$_{20}$N$_2$O$_2$
　Brivaracetam, 1384
C$_{11}$H$_{20}$N$_2$O$_6$
　L-Saccharopine, 8446
C$_{11}$H$_{20}$N$_3$O$_3$PS
　Pirimiphos-methyl O,O-Dimethyl An-
　　alog, 7610
C$_{11}$H$_{20}$O$_2$
　Undecylenic Acid, 10033
C$_{11}$H$_{20}$O$_4$
　Di-$tert$-butyl Malonate, 3041
　Diethyl Diethylmalonate, 3130
C$_{11}$H$_{20}$O$_{10}$
　Primeverose, 7864
C$_{11}$H$_{21}$N
　Mecamylamine, 5843
C$_{11}$H$_{21}$NO$_3$
　Ethyl Butylacetylaminopropionate,
　　3828
　Hexaminolevulinate, 4731
C$_{11}$H$_{21}$NO$_8$
　Choline Dihydrogen Citrate, 2211
C$_{11}$H$_{21}$N$_5$S
　Dipropetryn, 3382
C$_{11}$H$_{22}$N$_2$O$_4$S
　Pantetheine, 7114
C$_{11}$H$_{22}$O
　Methyl Nonyl Ketone, 6177
C$_{11}$H$_{22}$O$_2$
　n-Amyl Caproate, 599
　Ethyl Pelargonate, 3891
　n-Nonyl Acetate, 6765
C$_{11}$H$_{22}$O$_4$
　Glycerol 1-Octanoate, 6338

$C_{11}H_{23}MoN_4O_6P$
Vedejs Reagent, 10134
$C_{11}H_{23}NOS$
Butylate, 1548
$C_{11}H_{23}N_3O_6$
L-Lysine L-Glutamate, 5699
$C_{11}H_{23}N_5O_6$
Arginine Glutamate, 771
$C_{11}H_{24}FeNO_{11}$
Ferrocholinate, 4065
$C_{11}H_{26}NO_2PS$
VX, 10234
$C_{12}Br_{10}O$
Decabromodiphenyl Ether, 2848
$C_{12}H_4Cl_2F_6N_4OS$
Fipronil, 4116
$C_{12}H_4Cl_4Na_2O_2S$
Bithionol Sodium Salt, 1307
$C_{12}H_4Cl_4O$
2,3,7,8-Tetrachlorodibenzofuran, 7684
$C_{12}H_4Cl_4O_2$
TCDD, 9221
$C_{12}H_4Na_5O_{16}S_4Sb.7H_2O$
Stibophen, 8942
$C_{12}H_5N_7O_{12}$
Dipicrylamine, 3375
$C_{12}H_6Cl_2NNaO_2$
2,6-Dichloroindophenol Sodium, 3078
$C_{12}H_6Cl_2N_2O_6$
Niclofolan, 6603
$C_{12}H_6Cl_4O_2S$
Bithionol, 1307
Tetradifon, 9343
$C_{12}H_6Cl_4O_3S$
Bithionol Sulfoxide, 1307
$C_{12}H_6F_2N_2O_2$
Fludioxonil, 4160
$C_{12}H_6N_2O_2$
Phanquinone, 7312
$C_{12}H_6N_2O_5S$
Berninamycinic Acid, 1162
$C_{12}H_6N_2O_8SSr_2$
Ranelic Acid Strontium Salt, 8224
$C_{12}H_6Na_6O_{12}S_6Sb_2$
Stibocaptate, 8941
$C_{12}H_6O_{12}$
Mellitic Acid, 5893
$C_{12}H_7Br_4O_5P$
Bromofenofos, 1429
$C_{12}H_7Cl_2NO_3$
Nitrofen, 6685
$C_{12}H_7Cl_3O_2$
Triclosan, 9822
$C_{12}H_7NO_4$
Resazurin, 8261
$C_{12}H_8Cl_2O_2S$
Fenticlor, 4031
$C_{12}H_8Cl_6$
Aldrin, 219
$C_{12}H_8Cl_6O$
Dieldrin, 3114
Endrin, 3633
$C_{12}H_8F_2$
4,4'-Difluorodiphenyl, 3168
$C_{12}H_8F_3N_5O_3S$
Florasulam, 4139
$C_{12}H_8HgNa_2O_8S_2$
Mercuric Sodium p-Phenolsulfonate, 5954
$C_{12}H_8N_2$
o-Phenanthroline, 7328
Phenazine, 7331
$C_{12}H_8N_2O$
Hemipyocyanine, 4680

$C_{12}H_8N_2O_4$
Iodinin, 5064
$C_{12}H_8N_2O_4S_2$
Nitrophenide, 6705
$C_{12}H_8N_4O_6S$
Nifurzide, 6624
$C_{12}H_8O$
Dibenzofuran, 3013
$C_{12}H_8O_4$
Bergapten, 1160
Methoxsalen, 6059
Naphthalic Acid, 6465
$C_{12}H_8S_3$
α-Terthienyl, 9317
$C_{12}H_9AsClN$
Phenarsazine Chloride, 7329
$C_{12}H_9ClF_3N_3O$
Norflurazon, 6790
$C_{12}H_9ClO$
2-Phenyl-6-chlorophenol, 7391
4-Phenyl-2-chlorophenol, 7392
$C_{12}H_9ClO_2S$
Sulphenone, 9118
$C_{12}H_9Cl_2NO_3$
Vinclozolin, 10181
$C_{12}H_9F_2N_5O_2S$
Flumetsulam, 4172
$C_{12}H_9F_3N_2O_2$
Leflunomide, 5486
Teriflunomide, 9308
$C_{12}H_9N$
Carbazole, 1791
$C_{12}H_9NO$
Phenoxazine, 7365
$C_{12}H_9NO_2$
Dictamnine, 3099
o-Nitrobiphenyl, 6679
p-Nitrobiphenyl, 6680
$C_{12}H_9NO_6$
Miloxacin, 6277
$C_{12}H_9NS$
Phenothiazine, 7364
$C_{12}H_9N_2NaO_5S$
Tropaeolin O, 9954
$C_{12}H_9N_3O$
Milrinone, 6278
$C_{12}H_9N_3O_4$
Magneson, 5759
$C_{12}H_9N_3O_5$
Nifuroxazide, 6619
$C_{12}H_9N_3O_5S$
Nitazoxanide, 6653
$C_{12}H_9N_5O_3$
Cyadox, 2671
$C_{12}H_9NaO$
o-Phenylphenol Sodium Salt, 7415
$C_{12}H_{10}$
Acenaphthene, 29
Diphenyl, 3344
$C_{12}H_{10}BaO_6S_2$
Barium Benzenesulfonate, 962
$C_{12}H_{10}BaO_8S_2$
p-Phenolsulfonic Acid Barium Salt, 7359
$C_{12}H_{10}Ba_3O_{14}.7H_2O$
Citric Acid Barium Salt Heptahydrate, 2325
$C_{12}H_{10}CaO_2$
Calcium Phenoxide, 1693
$C_{12}H_{10}CaO_8S_2$
Calcium Phenolsulfonate, 1692
$C_{12}H_{10}CaO_{10}S_2$
Dobesilate Calcium, 3439
$C_{12}H_{10}Ca_3O_{14}$
Calcium Citrate, 1663
$C_{12}H_{10}ClN$
4-Amino-4'-chlorodiphenyl, 430

$C_{12}H_{10}ClN_3O$
Forchlorfenuron, 4262
$C_{12}H_{10}ClN_3S$
Thionine, 9499
$C_{12}H_{10}ClO_3P$
Diphenyl Phosphoryl Chloride, 3367
$C_{12}H_{10}ClP$
Chlorodiphenylphosphine, 2139
$C_{12}H_{10}Cl_2N_2$
2,2'-Dichlorobenzidine, 3067
3,3'-Dichlorobenzidine, 3068
$C_{12}H_{10}FNO_4S_2$
N-Fluorobenzenesulfonimide, 4202
$C_{12}H_{10}FN_3O_4$
Fidarestat, 4107
$C_{12}H_{10}F_3N_3O_4$
Nilutamide, 6629
$C_{12}H_{10}F_3N_4NaO_6S$
Flucarbazone Sodium Salt, 4151
$C_{12}H_{10}Mg$
Diphenylmagnesium, 3361
$C_{12}H_{10}Mg_3O_{14}$
Magnesium Citrate, 5727
$C_{12}H_{10}N_2$
Azobenzene, 909
Harman, 4648
$C_{12}H_{10}N_2O$
Azoxybenzene, 916
3-Hydroxymethyl-β-carboline, 3834
4-Nitrosodiphenylamine, 6726
$C_{12}H_{10}N_2O_4$
Nifurpirinol, 6621
$C_{12}H_{10}N_2O_5$
Cinoxacin, 2309
$C_{12}H_{10}N_2O_8S$
Ranelic Acid, 8224
$C_{12}H_{10}N_3O_3P$
Diphenylphosphoryl Azide, 3366
$C_{12}H_{10}N_4$
2,3-Diaminophenazine, 2984
$C_{12}H_{10}N_4O_2$
Lumichrome, 5658
$C_{12}H_{10}O$
Diphenyl Ether, 3357
o-Phenylphenol, 7415
p-Phenylphenol, 7416
$C_{12}H_{10}O_2$
Chimaphilin, 2058
1-Naphthaleneacetic Acid, 6456
$C_{12}H_{10}O_2S$
Diphenyl Sulfone, 3372
$C_{12}H_{10}O_2S_2$
Phenyl Benzenethiosulfonate, 7386
$C_{12}H_{10}O_3$
2-Naphthoxyacetic Acid, 6482
$C_{12}H_{10}O_4$
Piperic Acid, 7579
Quinhydrone, 8174
$C_{12}H_{10}O_5S_2$
Benzenesulfonic Anhydride, 1072
$C_{12}H_{10}O_7$
Echinochrome A, 3543
$C_{12}H_{10}O_8S_2Zn$
Zinc p-Phenolsulfonate, 10350
$C_{12}H_{10}O_{14}Zn_3$
Zinc Citrate, 10333
$C_{12}H_{10}P_2Se_4$
Woollins' Reagent, 10249
$C_{12}H_{10}S$
Diphenyl Sulfide, 3371
$C_{12}H_{10}S_2$
Diphenyl Disulfide, 3355
$C_{12}H_{11}ClN_2O_5S$
Furosemide, 4338
$C_{12}H_{11}ClN_6O_2S_2$
Azosemide, 915

C$_{12}$H$_{11}$Cl$_2$NO
Propyzamide, 7983

C$_{12}$H$_{11}$Cl$_2$N$_3$O$_2$
Anecortave Acetate, 885

C$_{12}$H$_{11}$F$_3$N$_4$O$_6$S
Flucarbazone, 4151

C$_{12}$H$_{11}$I$_3$N$_2$O$_4$
Iodamide, 5055
Metrizoic Acid, 6235

C$_{12}$H$_{11}$N
p-Biphenylamine, 1239
Diphenylamine, 3347

C$_{12}$H$_{11}$NO
Pirfenidone, 7606

C$_{12}$H$_{11}$NOS
Thionalide, 9497

C$_{12}$H$_{11}$NO$_2$
4-Amino-1-naphthol N-Acetyl Deriv,
449
Carbaryl, 1788

C$_{12}$H$_{11}$NO$_3$S
N-Phenylsulfanilic Acid, 7423

C$_{12}$H$_{11}$N$_2$NaO$_3$
Phenobarbital Sodium Salt, 7352

C$_{12}$H$_{11}$N$_3$
p-Aminoazobenzene, 414
Diazoaminobenzene, 3001

C$_{12}$H$_{11}$N$_3$O$_2$
Furonazide, 4337

C$_{12}$H$_{11}$N$_7$
Triamterene, 9760

C$_{12}$H$_{11}$P
Diphenylphosphine, 3364

C$_{12}$H$_{12}$Al$_2$O$_{18}$
Aluminum Tartrate, 362

C$_{12}$H$_{12}$BrN$_4$NaO$_2$S.H$_2$O
Sulfabromomethazine Sodium Salt
Monohydrate, 9030

C$_{12}$H$_{12}$Br$_2$N$_2$
Diquat Dibromide, 3392

C$_{12}$H$_{12}$ClNO$_2$S
Dansyl Chloride, 2813

C$_{12}$H$_{12}$ClN$_5$O$_4$S
Chlorsulfuron, 2197

C$_{12}$H$_{12}$I$_3$N$_2$NaO$_2$
Ipodate Sodium Salt, 5115

C$_{12}$H$_{12}$NO$_2$P
O-(Diphenylphosphinyl)hydroxyl-
amine, 3365

C$_{12}$H$_{12}$N$_2$
Benzidine, 1079
2,4'-Biphenyldiamine, 1240
1,1-Diphenylhydrazine, 3359

C$_{12}$H$_{12}$N$_2$O
Harmalol, 4647

C$_{12}$H$_{12}$N$_2$OS$_2$
p-Dimethylaminobenzalrhodanine,
3253

C$_{12}$H$_{12}$N$_2$O$_2$
Tetrantoin, 9382

C$_{12}$H$_{12}$N$_2$O$_2$S
Dapsone, 2822
Enoximone, 3643
Sulfabenz, 9028

C$_{12}$H$_{12}$N$_2$O$_3$
Nalidixic Acid, 6444
Phenobarbital, 7352

C$_{12}$H$_{12}$N$_4$
Chrysoidine Free Base, 2261
p-Diaminoazobenzene, 2979

C$_{12}$H$_{12}$N$_4$O$_3$
Benznidazole, 1086

C$_{12}$H$_{12}$N$_6$Na$_2$O$_{10}$S$_2$
Carumonam Disodium Salt, 1871

C$_{12}$H$_{12}$O
2-Ethoxynaphthalene, 3806

C$_{12}$H$_{12}$O$_2$
2-(2-Naphthyloxy)ethanol, 6494

C$_{12}$H$_{12}$O$_5$
Fraxetin Dimethyl Ether, 4294
Radicinin, 8209

C$_{12}$H$_{12}$O$_6$
Pyrogallol Triacetate, 8112

C$_{12}$H$_{13}$BrN$_4$O$_2$S
Sulfabromomethazine, 9030

C$_{12}$H$_{13}$ClF$_3$N$_3$O$_4$
Fluchloralin, 4152

C$_{12}$H$_{13}$ClN$_4$
Chrysoidine, 2261
Pyrimethamine, 8098

C$_{12}$H$_{13}$Cl$_2$N$_3$
Alinidine, 238

C$_{12}$H$_{13}$I$_3$N$_2$O$_2$
Ipodate, 5115

C$_{12}$H$_{13}$I$_3$N$_2$O$_3$
Iocetamic Acid, 5054
Iomeglamic Acid, 5096

C$_{12}$H$_{13}$N
N,N-Dimethyl-1-naphthylamine, 3276
Rasagiline, 8234

C$_{12}$H$_{13}$NO$_2$
Indolebutyric Acid, 5005
Methsuximide, 6078

C$_{12}$H$_{13}$NO$_2$S
Carboxin, 1828

C$_{12}$H$_{13}$NO$_3$
Aniracetam, 655

C$_{12}$H$_{13}$N$_2$NaO$_5$
2-Cyclohexyl-4,6-dinitrophenol
Sodium Salt, 2726

C$_{12}$H$_{13}$N$_3$
4,4'-Diaminodiphenylamine, 2981
Pyrimethanil, 8099

C$_{12}$H$_{13}$N$_3$O$_2$
Isocarboxazid, 5202
Triaziquone, 9764

C$_{12}$H$_{13}$N$_3$O$_4$
Olaquindox, 6916

C$_{12}$H$_{13}$N$_3$O$_4$S$_2$
N^4-Sulfanilylsulfanilamide, 9061

C$_{12}$H$_{13}$N$_3$O$_6$
Glyconiazide, 4536

C$_{12}$H$_{13}$N$_4$NaO$_2$S
Sulfamethazine Sodium Salt, 9047

C$_{12}$H$_{13}$N$_5$O$_2$S
Sulfamidochrysoidine, 9055

C$_{12}$H$_{13}$N$_5$O$_4$
Toyocamycin, 9721

C$_{12}$H$_{13}$N$_5$O$_6$S$_2$
Thifensulfuron-methyl, 9468

C$_{12}$H$_{14}$ClNO$_2$
Clomazone, 2374

C$_{12}$H$_{14}$ClNO$_3$
Cotarnine Chloride, 2540

C$_{12}$H$_{14}$Cl$_2$FNO$_4$S
Florfenicol, 4141

C$_{12}$H$_{14}$Cl$_2$N$_2$
Paraquat Dichloride, 7135

C$_{12}$H$_{14}$Cl$_2$O$_3$
2,4-D Butyl Ester, 2793

C$_{12}$H$_{14}$Cl$_3$O$_4$P
Chlorfenvinphos, 2090

C$_{12}$H$_{14}$F$_5$N$_3$O$_4$S
Pyroxasulfone, 8123

C$_{12}$H$_{14}$N$_2$
Detomidine, 2939
N-(1-Naphthyl)ethylenediamine, 6491

C$_{12}$H$_{14}$N$_2$$^{2+}$
Paraquat, 7135

C$_{12}$H$_{14}$N$_2$O$_2$
Abrine, 11
Mephenytoin, 5920
Primidone, 7865

C$_{12}$H$_{14}$N$_2$O$_3$S
Tioxidazole, 9611

C$_{12}$H$_{14}$N$_2$O$_5$
N-(p-Aminobenzoyl)glutamic Acid,
421
2-Cyclohexyl-4,6-dinitrophenol, 2726

C$_{12}$H$_{14}$N$_2$O$_6$
Dinoseb Acetate, 3315

C$_{12}$H$_{14}$N$_2$S
Cymiazole, 2761

C$_{12}$H$_{14}$N$_4$O$_2$S
Sulfamethazine, 9047
Sulfisomidine, 9081

C$_{12}$H$_{14}$N$_4$O$_3$S
Sulfacytine, 9034
Sulfamethomidine, 9049

C$_{12}$H$_{14}$N$_4$O$_4$
Cyclohexanone 2,4-Dinitrophenylhy-
drazone, 2719

C$_{12}$H$_{14}$N$_4$O$_4$S
Sulfadimethoxine, 9037
Sulfadoxine, 9038

C$_{12}$H$_{14}$N$_4$O$_4$S$_2$
Thiophanate O,O-Dimethyl Analog,
9505

C$_{12}$H$_{14}$N$_6$O$_{10}$S$_2$
Carumonam, 1871

C$_{12}$H$_{14}$O$_2$
Precocene I, 7839

C$_{12}$H$_{14}$O$_3$
Cuelure, 2606

C$_{12}$H$_{14}$O$_4$
Apiole (Dill), 720
Apiole (Parsley), 721
Dimecrotic Acid, 3226
Phthalic Acid Ethyl Ester, 7483

C$_{12}$H$_{15}$AsN$_6$OS$_2$
Melarsoprol, 5883

C$_{12}$H$_{15}$ClNO$_4$PS$_2$
Phosalone, 7446

C$_{12}$H$_{15}$ClO$_3$
Clofibrate, 2370

C$_{12}$H$_{15}$Cl$_2$NO
Karsil, 5333

C$_{12}$H$_{15}$Cl$_2$NO$_5$S
Thiamphenicol, 9453

C$_{12}$H$_{15}$N
6,7-Benzomorphan, 1097
MPTP, 6380

C$_{12}$H$_{15}$NNaO$_5$S
Faropenem Sodium Salt, 3976

C$_{12}$H$_{15}$NO$_3$
Anhalonine, 648
Carbofuran, 1803
Hydrocotarnine, 4825

C$_{12}$H$_{15}$NO$_4$
Cotarnine, 2540
Ethopabate, 3798

C$_{12}$H$_{15}$NO$_5$S
Faropenem, 3976

C$_{12}$H$_{15}$N$_2$NaO$_3$
Hexobarbital Sodium Salt, 4740

C$_{12}$H$_{15}$N$_2$O$_3$PS
Phoxim, 7478

C$_{12}$H$_{15}$N$_3$
Indanazoline, 4972

C$_{12}$H$_{15}$N$_3$O$_2$S
Albendazole, 203

C$_{12}$H$_{15}$N$_3$O$_2$S$_2$
Glybuzole, 4516

C$_{12}$H$_{15}$N$_3$O$_3$
Oxibendazole, 7035

C$_{12}$H$_{15}$N$_3$O$_5$S
Amezinium Methyl Sulfate, 388

C$_{12}$H$_{15}$N$_3$O$_6$
Triglycidyl Isocyanurate, 9860

$C_{12}H_{15}N_5O_3$
Entecavir, 3649
$C_{12}H_{15}N_5O_3S$
Sulfaguanole, 9041
$C_{12}H_{16}BiK_3O_{14}$
Colloidal Bismuth Subcitrate, 2469
$C_{12}H_{16}ClNO_3$
Meclofenoxate, 5851
$C_{12}H_{16}ClN_3O_2S$
Oxythiamine, 7076
$C_{12}H_{16}Cl_2N_2O$
Neburon, 6519
$C_{12}H_{16}F_3N$
Fenfluramine, 4004
$C_{12}H_{16}N_2$
N,N-Dimethyltryptamine, 3288
Fenproporex, 4022
$C_{12}H_{16}N_2O$
Bufotenine, 1483
Methylcytisine, 6121
Psilocin, 8034
$C_{12}H_{16}N_2O_2$
Eltoprazine, 3608
$C_{12}H_{16}N_2O_3$
Cyclobarbital, 2704
Hexobarbital, 4740
$C_{12}H_{16}N_2S$
Morantel, 6353
Xylazine, 10277
$C_{12}H_{16}N_3O_3PS$
Triazophos, 9767
$C_{12}H_{16}N_3O_3PS_2$
Azinphos-methyl O,O-Diethyl Analog, 906
$C_{12}H_{16}N_4O_2S_2$
Glybuthiazole, 4515
$C_{12}H_{16}O_2$
Isoamyl Benzoate, 5158
Thymol Acetate, 9554
$C_{12}H_{16}O_3$
Asarones, 813
Isoamyl Salicylate, 5170
$C_{12}H_{16}O_4$
Aspidinol, 837
$C_{12}H_{16}O_4S_2$
Malotilate, 5776
$C_{12}H_{16}O_7$
Arbutin, 763
$C_{12}H_{17}ClN_4OS$
Thiamine, 9447
$C_{12}H_{17}Cl_2NO_3$
2,4-DB Dimethylamine Salt, 2838
$C_{12}H_{17}NO$
Deet, 2859
N,N-Diethylphenylacetamide, 3143
Phendimetrazine, 7334
$C_{12}H_{17}NOS$
Tiletamine, 9592
$C_{12}H_{17}NO_2$
Ciclopirox, 2269
MDE, 5835
$C_{12}H_{17}NO_3$
Anhalonidine, 647
Bucetin, 1467
Bufexamac, 1479
Cerulenin, 2003
Deacetylanisomycin, 663
Ethamivan, 3776
Rimiterol, 8353
$C_{12}H_{17}N_2NaO_2S$
Thiamylal Sodium Salt, 9454
$C_{12}H_{17}N_2NaO_3$
Secobarbital Sodium, 8557
$C_{12}H_{17}N_2O_4P$
Psilocybin, 8035
$C_{12}H_{17}N_3O$
Cimaterol, 2281

$C_{12}H_{17}N_3O_3S$
Cariporide, 1840
$C_{12}H_{17}N_3O_4$
Agaritine, 179
$C_{12}H_{17}N_3O_4S$
Imipenem, 4961
$C_{12}H_{17}N_5O_4S$
Thiamine Mononitrate, 9447
$C_{12}H_{17}NaO_7$
Dikegulac Sodium Salt, 3220
$C_{12}H_{17}Na_3O_{17}Sb_2.9H_2O$
Sodium Stibogluconate, 692
$C_{12}H_{18}Br_6$
Hexabromocyclododecane, 4712
$C_{12}H_{18}ClN$
Mefenorex, 5870
$C_{12}H_{18}ClNO$
Tulobuterol, 9991
$C_{12}H_{18}ClNO_2S$
Dimethenamid, 3236
$C_{12}H_{18}ClNO_4$
Methyldopa Ethyl Ester Hydrochloride, 6127
$C_{12}H_{18}ClN_4O_4PS$
Thiamine Monophosphate, 9448
$C_{12}H_{18}Cl_2N_2O$
Clenbuterol, 2345
$C_{12}H_{18}{}^{123}IN$
Iofetamine I 123, 5092
$C_{12}H_{18}N_2O$
Isoproturon, 5264
Oxotremorine, 7050
$C_{12}H_{18}N_2O_2S$
Thiamylal, 9454
$C_{12}H_{18}N_2O_3S$
Tolbutamide, 9667
$C_{12}H_{18}N_2O_4$
Midodrine, 6263
$C_{12}H_{18}N_2O_5$
Bacilysin, 926
Hypoglycine B, 4905
$C_{12}H_{18}N_2O_7$
Bicyclomycin, 1207
$C_{12}H_{18}N_4O_6S$
Oryzalin, 6982
$C_{12}H_{18}O$
Propofol, 7947
$C_{12}H_{18}O_2$
4-Hexylresorcinol, 4748
$C_{12}H_{18}O_3$
Jasmonic Acid, 5308
$C_{12}H_{18}O_7$
Dikegulac, 3220
$C_{12}H_{19}BrN_2O_2$
Neostigmine Bromide, 6549
$C_{12}H_{19}ClN_4O_7P_2S$
Thiamine Diphosphate, 9448
$C_{12}H_{19}Cl_3O_8$
Sucralose, 9011
$C_{12}H_{19}NO$
Etafedrine, 3764
$C_{12}H_{19}NO_2$
Bamethan, 947
DOM, 3461
Fepradinol, 4038
Octyl Cyanoacrylate, 6855
$C_{12}H_{19}NO_3$
Macromerine, 5709
Prenalterol, 7853
Terbutaline, 9302
$C_{12}H_{19}NO_4$
Choline Salicylate, 2213
$C_{12}H_{19}N_2NaO_2S_2$
Methitural Sodium Salt, 6050
$C_{12}H_{19}N_2O_2{}^+$
Neostigmine, 6549

$C_{12}H_{19}N_3O$
Procarbazine, 7876
$C_{12}H_{19}N_4O_{10}P_3S$
Thiamine Triphosphate, 9450
$C_{12}H_{20}B_2CaO_{16}$
Calcium Borogluconate, 1656
$C_{12}H_{20}Cl_4N_2O_2$
Fertilysin, 4087
$C_{12}H_{20}LiN_3O_3$
Pyrrolysine Lithium Salt, 8133
$C_{12}H_{20}N_2$
Tremorine, 9744
$C_{12}H_{20}N_2O_2$
Aspergillic Acid, 831
$C_{12}H_{20}N_2O_2S_2$
Methitural, 6050
$C_{12}H_{20}N_2O_3$
Pirbuterol, 7602
$C_{12}H_{20}N_2O_3S$
Sotalol, 8854
$C_{12}H_{20}N_2O_4S$
Soterenol, 8855
$C_{12}H_{20}N_4O_2$
Hexazinone, 4735
$C_{12}H_{20}N_4O_7$
Zanamivir, 10311
$C_{12}H_{20}N_4O_6S$
Sulfazecin, 9077
$C_{12}H_{20}O_2$
Bornyl Acetate, 1341
Linalyl Acetate, 5551
$C_{12}H_{20}O_4$
Traumatic Acid, 9737
$C_{12}H_{20}O_6$
Diacetoneglucose, 2968
$C_{12}H_{20}O_7$
Citric Acid Ethyl Ester, 2325
$C_{12}H_{21}ClO_2$
Trimedlure, 9870
$C_{12}H_{21}N$
Memantine, 5897
Rimantadine, 8351
$C_{12}H_{21}NO_8S$
Topiramate, 9706
$C_{12}H_{21}NO_{11}$
Chondrosine, 2220
Hyalobiuronic Acid, 4795
$C_{12}H_{21}N_2O_3PS$
Diazinon, 2998
$C_{12}H_{21}N_2O_8S_2{}^{99m}Tc$
Technetium Tc 99m Bicisate, 9236
$C_{12}H_{21}N_3O_3$
Pyrrolysine, 8133
$C_{12}H_{21}N_3O_5S_3$
Brinzolamide, 1382
$C_{12}H_{21}N_5O_2S_2$
Nizatidine, 6747
$C_{12}H_{21}N_5O_3$
Choline Theophyllinate, 2215
$C_{12}H_{21}NaO_7S$
Diisobutyl Sodium Sulfosuccinate, 3213
$C_{12}H_{22}AsN_3O_6S$
Darinaparsin, 2827
$C_{12}H_{22}CaO_{14}$
Calcium Gluconate, 1671
$C_{12}H_{22}Cl_2N_2O_3$
Pirbuterol Dihydrochloride, 7602
$C_{12}H_{22}CuO_{14}$
Cupric Gluconate, 2630
$C_{12}H_{22}FeO_{14}$
Ferrous Gluconate, 4074
$C_{12}H_{22}MgO_{14}$
Gluconic Acid Magnesium Salt, 4492
$C_{12}H_{22}N_2O_2$
Crotethamide, 2586

C₁₃H₁₁NO
Benzanilide, 1063
Benzophenone Oxime, 1100
C₁₃H₁₁NO₂
Nicotinic Acid Benzyl Ester, 6611
N-Phenylanthranilic Acid, 7384
Salicylanilide, 8467
C₁₃H₁₁NO₃
Fagarine, 3969
Osalmid, 6984
Phenyl Aminosalicylate, 7383
C₁₃H₁₁NO₅
Oxolinic Acid, 7044
C₁₃H₁₁N₂NaO₂
Ozagrel Sodium Salt, 7079
C₁₃H₁₁N₃
Proflavine, 7887
C₁₃H₁₁N₃O
Drometrizole, 3496
C₁₃H₁₁N₃O₂
Salinazid, 8472
C₁₃H₁₁N₃O₄
Pomalidomide, 10690
C₁₃H₁₁N₃O₄S
Sulfoniazide, 9089
C₁₃H₁₁N₃O₄S₂
Tenoxicam, 9291
C₁₃H₁₂
Diphenylmethane, 3362
C₁₃H₁₂BrCl₂N₃O
Bromuconazole, 1456
C₁₃H₁₂ClN₃S
Azure C, 922
C₁₃H₁₂Cl₂N₂
4,4′-Methylenebis[2-chloroaniline], 6131
C₁₃H₁₂Cl₂O₄
Ethacrynic Acid, 3772
C₁₃H₁₂F₂N₆O
Fluconazole, 4153
C₁₃H₁₂F₃N₅O₅S
Flazasulfuron, 4129
C₁₃H₁₂N₂
9H-Fluorene-2,7-diamine, 4188
C₁₃H₁₂N₂O
Carbanilide, 1786
Harmine, 4649
Salicylaldehyde Phenylhydrazone, 8463
C₁₃H₁₂N₂O₂
Ozagrel, 7079
C₁₃H₁₂N₂O₃S
Sulfabenzamide, 9029
C₁₃H₁₂N₂O₅S
Nimesulide, 6633
C₁₃H₁₂N₂S
sym-Diphenylthiourea, 3373
C₁₃H₁₂N₄
PhIP, 7435
C₁₃H₁₂N₄O
Diphenylcarbazone, 3353
C₁₃H₁₂N₄O₂
Lumiflavine, 5659
C₁₃H₁₂N₄O₃
Pyriminil, 8101
C₁₃H₁₂N₄S
Dithizone, 3420
C₁₃H₁₂N₅NaO₅S₂
Ceftizoxime Sodium Salt, 1952
C₁₃H₁₂O
Benzohydrol, 1092
o-Benzylphenol, 1144
p-Benzylphenol, 1145
C₁₃H₁₂O₂
Monobenzone, 6334

C₁₃H₁₂O₃
Allenolic Acid, 251
Euparin, 3942
C₁₃H₁₂O₇S₂
Sultosilic Acid, 9124
C₁₃H₁₃As₂N₂NaO₄S
Neoarsphenamine, 6532
C₁₃H₁₃ClN₄O₂S
Cyazofamid, 2693
C₁₃H₁₃Cl₂N₃O₃
Iprodione, 5121
C₁₃H₁₃IO₄S
Koser's Reagent, 5369
C₁₃H₁₃IO₈
Dess-Martin Periodinane, 2934
C₁₃H₁₃N
Benzhydrylamine, 1078
Benzylaniline, 1129
Methyldiphenylamine, 6126
C₁₃H₁₃N₃
1,3-Diphenylguanidine, 3358
Nitrin, 6666
Varenicline, 10125
C₁₃H₁₃N₃O₃
Cyclobendazole, 2705
Lenalidomide, 5490
C₁₃H₁₃N₃O₅S₂
Succinylsulfathiazole, 9008
C₁₃H₁₃N₅O₄S
Sulfachrysoidine, 9033
C₁₃H₁₃N₅O₅S₂
Ceftizoxime, 1952
C₁₃H₁₄ClNOS
Alagebrium Chloride, 196
C₁₃H₁₄Cl₂O₂
Ciprofibrate, 2312
C₁₃H₁₄F₃N₃O₄
Ethalfluralin, 3774
C₁₃H₁₄N₂
p,p′-Diaminodiphenylmethane, 2982
Tacrine, 9154
C₁₃H₁₄N₂O
Harmaline, 4646
10-Methoxyharmalan, 6068
N-(p-Methoxyphenyl)-p-phenylenedi-amine, 6072
Phenyramidol, 7432
C₁₃H₁₄N₂O₂
Metomidate, 6225
C₁₃H₁₄N₂O₂S
p-Sulfanilylbenzylamine, 9059
C₁₃H₁₄N₂O₃
2-(Boc-oxyimino)-2-phenylacetoni-trile, 1323
Mephobarbital, 5921
C₁₃H₁₄N₂O₄
Menadoxime, 5900
C₁₃H₁₄N₂O₄S₂
Gliotoxin, 4476
C₁₃H₁₄N₃NaO₄S
Glymidine Sodium Salt, 4543
C₁₃H₁₄N₄O
sym-Diphenylcarbazide, 3352
C₁₃H₁₄O₂
Tremetone, 9743
C₁₃H₁₄O₅
Citrinin, 2326
C₁₃H₁₅BrN₄O₂
Brodimoprim, 1387
C₁₃H₁₅ClN₆O₇S
Halosulfuron-methyl, 4636
C₁₃H₁₅NO₂
Glutethimide, 4513
Securinine, 8559
C₁₃H₁₅N₃O₂
Pyrolan, 8114

C₁₃H₁₅N₃O₃
Imazapyr, 4946
C₁₃H₁₅N₃O₄S
Acetyl Sulfisoxazole, 9082
Glymidine, 4543
C₁₃H₁₅N₅O₃
Tazanolast, 9217
C₁₃H₁₅N₅O₆
Nifurfoline, 6617
C₁₃H₁₆ClNO
Ketamine, 5343
C₁₃H₁₆F₃N₃O₄
Benfluralin, 1042
Trifluralin, 9854
C₁₃H₁₆HgNNaO₆
Mersalyl, 5971
C₁₃H₁₆N₂
Dexmedetomidine, 2948
Medetomidine, 5858
Tetrahydrozoline, 9364
C₁₃H₁₆N₂O
Adrenoglomerulotropin, 162
Oxantel, 7021
C₁₃H₁₆N₂O₂
Aminoglutethimide, 437
Melatonin, 5884
Mofebutazone, 6317
C₁₃H₁₆N₂O₂S
Thialbarbital, 9444
C₁₃H₁₆N₃NaO₄S.H₂O
Dipyrone, 3390
C₁₃H₁₆N₄O₂
Diaveridine, 2994
C₁₃H₁₆N₄O₆
Furaltadone, 4323
C₁₃H₁₆N₁₀O₅S
Azimsulfuron, 905
C₁₃H₁₆O₂
Cinnamic Acid n-Butyl Ester, 2299
Dihydrotremetone, 9743
R-11, 8205
C₁₃H₁₆O₃
Precocene II, 7839
C₁₃H₁₆O₅
Trinexapac-ethyl, 9897
C₁₃H₁₆O₇
Helicin, 4663
C₁₃H₁₆O₁₀
α-Glucogallin, 4489
β-Glucogallin, 4490
C₁₃H₁₇ClN₂O₂
Moclobemide, 6312
C₁₃H₁₇F₃N₄O₄
Prodiamine, 7883
C₁₃H₁₇HgNO₆
Mersalyl Acid, 5971
C₁₃H₁₇N
Selegiline, 8565
C₁₃H₁₇NO
Crotamiton, 2585
C₁₃H₁₇NO₂
Alminoprofen, 293
C₁₃H₁₇NO₄
Alibendol, 236
C₁₃H₁₇NO₇
Synephrine Tartaric Acid Monoester, 9142
C₁₃H₁₇N₂NaO₄S₂
Penicillin O Sodium Salt, 7208
C₁₃H₁₇N₃
Tramazoline, 9727
C₁₃H₁₇N₃O
Aminopyrine, 470
C₁₃H₁₇N₃O₂
Parbendazole, 7139
C₁₃H₁₇N₅O₈S₂
Aztreonam, 918

C$_{14}$H$_{10}$BaO$_4$.2H$_2$O
Benzoic Acid Barium Salt Dihydrate, 1093

C$_{14}$H$_{10}$BrN$_3$O
Bromazepam, 1392

C$_{14}$H$_{10}$CaO$_4$.3H$_2$O
Benzoic Acid Calcium Salt Trihydrate, 1093

C$_{14}$H$_{10}$CdO$_6$
Cadmium Salicylate, 1630

C$_{14}$H$_{10}$ClNO$_4$
Lobenzarit, 5610

C$_{14}$H$_{10}$ClN$_3$S$_2$
Lanoconazole, 5408

C$_{14}$H$_{10}$Cl$_2$KNO$_2$
Diclofenac Potassium Salt, 3091

C$_{14}$H$_{10}$Cl$_2$NNaO$_2$
Diclofenac Sodium Salt, 3091

C$_{14}$H$_{10}$Cl$_4$
1,1-Dichloro-2,2-bis(p-chlorophenyl)-ethane, 3070
Mitotane, 6301

C$_{14}$H$_{10}$CuO$_4$.2H$_2$O
Benzoic Acid Copper Salt Dihydrate, 1093

C$_{14}$H$_{10}$CuO$_6$
Cupric Salicylate, 2639

C$_{14}$H$_{10}$F$_3$NO$_2$
Flufenamic Acid, 4163

C$_{14}$H$_{10}$F$_3$NO$_5$
Nitisinone, 6656

C$_{14}$H$_{10}$HgO$_4$
Mercuric Benzoate, 5941

C$_{14}$H$_{10}$MgO$_4$
Magnesium Benzoate, 5721

C$_{14}$H$_{10}$MgO$_6$
Magnesium Salicylate, 5749

C$_{14}$H$_{10}$MnO$_4$.4H$_2$O
Benzoic Acid Manganese Salt Tetrahydrate, 1093

C$_{14}$H$_{10}$N$_2$O$_5$
Cinnabarine, 2297

C$_{14}$H$_{10}$N$_2$O$_6$
Olsalazine, 6935

C$_{14}$H$_{10}$N$_4$O
Loprinone, 5632

C$_{14}$H$_{10}$N$_4$O$_5$
Dantrolene, 2815

C$_{14}$H$_{10}$NiO$_4$.3H$_2$O
Benzoic Acid Nickel Salt Trihydrate, 1093

C$_{14}$H$_{10}$O
Anthranol, 678
Anthrone, 682
Diphenylketene, 3360

C$_{14}$H$_{10}$O$_2$
Benzil, 1080

C$_{14}$H$_{10}$O$_3$
Anthralin, 676
Benzoic Anhydride, 1094

C$_{14}$H$_{10}$O$_4$
Benzoyl Peroxide, 1119
Diphenic Acid, 3340

C$_{14}$H$_{10}$O$_5$
Gentisin, 4434
Salsalate, 8477

C$_{14}$H$_{10}$O$_6$
Phenicin, 7346

C$_{14}$H$_{10}$O$_6$Zn
Zinc Salicylate, 10355

C$_{14}$H$_{10}$O$_7$
Citromycetin, 2327

C$_{14}$H$_{10}$O$_8$
Oosporein, 6945

C$_{14}$H$_{10}$O$_9$
Digallic Acid, 3172

C$_{14}$H$_{11}$ClFN$_5$O$_5$S
Cloransulam-methyl Free Acid, 2391

C$_{14}$H$_{11}$ClN$_2$O$_4$S
Chlorthalidone, 2200

C$_{14}$H$_{11}$ClO
Diphenylacetic Acid Chloride, 3346

C$_{14}$H$_{11}$Cl$_2$NO$_2$
Diclofenac, 3091
Meclofenamic Acid, 5850

C$_{14}$H$_{11}$F$_3$N$_2$O$_2$
Flunixin, 4178

C$_{14}$H$_{11}$N
Diphenylacetic Acid Nitrile, 3346

C$_{14}$H$_{11}$NO$_4$
Benzoylpas, 1118
Diphenylamine-2,2'-dicarboxylic Acid, 3348

C$_{14}$H$_{11}$NO$_5$
Tolcapone, 9668

C$_{14}$H$_{11}$N$_4$NaO$_2$S
Sulfaquinoxaline Sodium Salt, 9070

C$_{14}$H$_{12}$
1,1-Diphenylethene, 3356
Stilbene, 8946

C$_{14}$H$_{12}$CaN$_2$O$_6$
p-Aminosalicylic Acid Calcium Salt, 473

C$_{14}$H$_{12}$ClNO$_2$
Cicletanine, 2267
Tolfenamic Acid, 9673

C$_{14}$H$_{12}$ClN$_3$O$_4$S$_2$
Indisulam, 4986

C$_{14}$H$_{12}$Cl$_2$N$_2$O
Pyrifenox, 8096

C$_{14}$H$_{12}$FNO$_3$
Flumequine, 4168

C$_{14}$H$_{12}$N$_2$
Neocuproine, 6534

C$_{14}$H$_{12}$N$_2$O$_2$
Benzil Dioxime, 1081
Ethyl β-Carboline-3-carboxylate, 3834

C$_{14}$H$_{12}$N$_2$O$_2$S
Zolimidine, 10389

C$_{14}$H$_{12}$N$_4$OS
Nolatrexed, 6759

C$_{14}$H$_{12}$N$_4$O$_2$S
Sulfaquinoxaline, 9070

C$_{14}$H$_{12}$N$_6$O
Levosimendan, 5525

C$_{14}$H$_{12}$N$_6$O$_3$
Pteroic Acid, 8042

C$_{14}$H$_{12}$N$_6$O$_6$
Nitrovin, 6741

C$_{14}$H$_{12}$O
Phenyl Tolyl Ketone, 7426

C$_{14}$H$_{12}$O$_2$
Benzoin, 1095
Benzyl Benzoate, 1130
Diphenylacetic Acid, 3346
Felbinac, 3985
Pinosylvin, 7563

C$_{14}$H$_{12}$O$_3$
Benzilic Acid, 1082
Benzyl Salicylate, 1146
Oxybenzone, 7053
Resveratrol, 8279
Trioxsalen, 9907
Xanthyletin, 10267

C$_{14}$H$_{12}$O$_3$S
Suprofen, 9134
Tiaprofenic Acid, 9577

C$_{14}$H$_{12}$O$_4$
Dioxybenzone, 3332
Fulvoplumierin, 4314

C$_{14}$H$_{12}$O$_5$
Khellin, 5356

C$_{14}$H$_{12}$O$_6$S
Sulisobenzone, 9113

C$_{14}$H$_{12}$S$_2$
Mesulfen, 5988

C$_{14}$H$_{13}$ClN$_6$O$_5$S
Imazosulfuron, 4949

C$_{14}$H$_{13}$ClO$_5$S
Sulcotrione, 9026

C$_{14}$H$_{13}$F$_3$N$_6$O$_5$S
Pyroxsulam, 8124

C$_{14}$H$_{13}$F$_4$N$_3$O$_2$S
Flufenacet, 4162

C$_{14}$H$_{13}$IN$_5$NaO$_6$S
Iodosulfuron-methyl-sodium, 5089

C$_{14}$H$_{13}$NO
Diphenylacetamide, 3345
Diphenylacetic Acid Amide, 3346

C$_{14}$H$_{13}$NO$_2$
Benzoin Oxime, 1096
Flindersine, 4133

C$_{14}$H$_{13}$NO$_4$
Skimmianine, 8699

C$_{14}$H$_{13}$NO$_7$S
Mesotrione, 5982

C$_{14}$H$_{13}$N$_2$NaO$_4$S
Acediasulfone Sodium Salt, 24

C$_{14}$H$_{13}$N$_3$
Fadrozole, 3968
Mepanipyrim, 5912

C$_{14}$H$_{13}$N$_3$O$_2$S
Sulmazole, 9115

C$_{14}$H$_{13}$N$_3$O$_4$S$_2$
Meloxicam, 5894

C$_{14}$H$_{13}$N$_3$O$_5$S
Sulfanitran, 9064

C$_{14}$H$_{13}$N$_5$O$_5$S$_2$
Cefdinir, 1921

C$_{14}$H$_{13}$N$_8$NaO$_4$S$_3$
Cefazolin Sodium Salt, 1918

C$_{14}$H$_{13}$NaO$_3$
Naproxen Sodium Salt, 6499

C$_{14}$H$_{14}$
Bibenzyl, 1202

C$_{14}$H$_{14}$As$_2$N$_2$Na$_2$O$_8$S$_2$
Sulfarsphenamine, 9072

C$_{14}$H$_{14}$ClNS
Ticlopidine, 9585

C$_{14}$H$_{14}$ClN$_3$
Acriflavine, 115

C$_{14}$H$_{14}$ClN$_3$O$_4$S$_2$
Benzylhydrochlorothiazide, 1139

C$_{14}$H$_{14}$ClN$_3$S
Azure A, 920

C$_{14}$H$_{14}$ClO$_3$P
Dibenzyl Chlorophosphonate, 3017

C$_{14}$H$_{14}$Cl$_2$N$_2$O
Enilconazole, 3637

C$_{14}$H$_{14}$Hg
p-Ditolylmercury, 3424

C$_{14}$H$_{14}$HgO$_6$
Mercumallylic Acid, 5939

C$_{14}$H$_{14}$NO$_4$PS
EPN, 3682

C$_{14}$H$_{14}$N$_2$
Naphazoline, 6453

C$_{14}$H$_{14}$N$_2$Na$_2$O$_6$S$_3$
Sulfoxone Sodium, 9098

C$_{14}$H$_{14}$N$_2$O
Metyrapone, 6238

C$_{14}$H$_{14}$N$_2$O$_2$
Isonixin, 5238

C$_{14}$H$_{14}$N$_2$O$_4$S
Acediasulfone, 24
N-Methylphenazonium Methosulfate, 6182

C$_{14}$H$_{14}$N$_2$O$_6$S$_2$
Amsonic Acid, 590

$C_{14}H_{14}N_3NaO_3S$
Methyl Orange, 6178
$C_{14}H_{14}N_4O$
Phenamidine, 7322
$C_{14}H_{14}N_4O_2S$
Cambendazole, 1728
$C_{14}H_{14}N_8O_4S_3$
Cefazolin, 1918
$C_{14}H_{14}O$
Benzyl Ether, 1135
$C_{14}H_{14}O_2$
Hydrobenzoin, 4815
$C_{14}H_{14}O_2P_2S_4$
Lawesson's Reagent, 5446
$C_{14}H_{14}O_3$
Euparin O-Methyl Deriv, 3942
2-Isovalerylindane-1,3-dione, 5279
Kawain, 5335
αMNP, 6310
Naproxen, 6499
Pindone, 7555
$C_{14}H_{14}O_4$
Phthalofyne, 7486
$C_{14}H_{14}O_6$
Radicinin Monoacetate, 8209
$C_{14}H_{14}S$
Benzyl Sulfide, 1147
$C_{14}H_{14}S_2$
Dibenzyl Disulfide, 3018
$C_{14}H_{15}BrClNO_6$
X-gal, 10272
$C_{14}H_{15}Cl_2N$
Chlornaphazine, 2108
$C_{14}H_{15}Cl_2N_3OS$
Prothioconazole, 7997
$C_{14}H_{15}N$
Dibenzylamine, 3016
$C_{14}H_{15}N_3$
o-Aminoazotoluene, 415
Cyprodinil, 2769
p-Dimethylaminoazobenzene, 3251
$C_{14}H_{15}N_3O_2$
Indolmycin, 5007
$C_{14}H_{15}N_3O_5$
Entacapone, 3648
$C_{14}H_{15}N_5O$
Endralazine, 3632
$C_{14}H_{15}N_5O_5S_2$
Cefetamet, 1924
$C_{14}H_{15}N_5O_6S$
Metsulfuron-methyl, 6237
$C_{14}H_{15}O_2PS_2$
Edifenphos, 3558
$C_{14}H_{15}O_3P$
Dibenzyl Phosphite, 3019
$C_{14}H_{16}$
Chamazulene, 2039
$C_{14}H_{16}ClN_3O$
Metazachlor, 6004
$C_{14}H_{16}ClN_3O_2$
Triadimefon, 9754
$C_{14}H_{16}ClN_3O_4S_2$
Cyclothiazide, 2752
$C_{14}H_{16}ClN_5O_5S$
Triasulfuron, 9761
$C_{14}H_{16}ClO_5PS$
Coumaphos, 2546
$C_{14}H_{16}Cl_2N_4O_3$
Asoxime Chloride, 825
Obidoxime Chloride, 6827
$C_{14}H_{16}Cl_3NO_2$
Zoxamide, 10398
$C_{14}H_{16}N_2$
Atipamezole, 852
1,2-Dianilinoethane, 2990
o-Tolidine, 9674

$C_{14}H_{16}N_2O$
Harmalol O-Ethylharmalol, 4647
$C_{14}H_{16}N_2O_2$
Dianisidine, 2992
Etomidate, 3926
$C_{14}H_{16}N_2O_3$
Phenetharbital, 7336
$C_{14}H_{16}N_2O_3S$
Monastrol, 6330
$C_{14}H_{16}N_4$
Imiquimod, 4964
$C_{14}H_{16}N_4O_3$
Piromidic Acid, 7618
$C_{14}H_{16}O_3$
Dihydrokawain, 5335
$C_{14}H_{17}BrN_6O_2S_3$
Ebrotidine, 3532
$C_{14}H_{17}Cl_2NO_2$
Fenhexamid, 4005
$C_{14}H_{17}Cl_2N_3O$
Hexaconazole, 4717
$C_{14}H_{17}I_3N_2O_2$
Ipodate Ethyl Ester, 5115
$C_{14}H_{17}NO_2$
3-Quinuclidinol dl-Form Benzoate
(Ester), 8199
$C_{14}H_{17}NO_3$
EEDQ, 3566
$C_{14}H_{17}NO_4$
Chromocarb Diethylamine, 2241
$C_{14}H_{17}NO_6$
Indican, 4980
Mandelonitrile Glucoside, 5784
$C_{14}H_{17}NO_7$
Dhurrin, 2961
$C_{14}H_{17}N_2NaO_3$
Methohexital Sodium Salt, 6053
$C_{14}H_{17}N_3O$
Frovatriptan, 4301
$C_{14}H_{17}N_3O_2S$
Fasudil, 3979
$C_{14}H_{17}N_3O_9$
6-Azauridine 2',3',5'-Triacetate, 897
$C_{14}H_{17}N_5O_3$
Pipemidic Acid, 7572
$C_{14}H_{17}N_5O_7S_2$
Rimsulfuron, 8356
$C_{14}H_{17}O_5PS$
Potasan, 7724
$C_{14}H_{18}CaN_3Na_3O_{10}$
Pentetate Calcium Trisodium, 7236
$C_{14}H_{18}ClN_3O_2$
Triadimenol, 9755
$C_{14}H_{18}ClN_3S$
Chlorothen, 2170
$C_{14}H_{18}N_2$
DMAN, 3432
$C_{14}H_{18}N_2O$
Ibudilast, 4918
Propyphenazone, 7982
$C_{14}H_{18}N_2O_2$
Hypaphorine, 4900
Nefiracetam, 6524
$C_{14}H_{18}N_2O_3$
Methohexital, 6053
Physovenine, 7497
$C_{14}H_{18}N_2O_4$
Oxadixyl, 7007
α-Ribazole, 8324
$C_{14}H_{18}N_2O_5$
Aspartame, 829
Lidofenin, 5536
$C_{14}H_{18}N_2O_6S$
Carpetimycin A, 1861
$C_{14}H_{18}N_2O_7$
Dinobuton, 3313

$C_{14}H_{18}N_2O_9S_2$
Carpetimycin B, 1861
$C_{14}H_{18}N_4O_3$
Benomyl, 1046
Trimethoprim, 9878
$C_{14}H_{18}N_4O_4S_2$
Thiophanate, 9505
$C_{14}H_{18}N_6O$
Abacavir, 1
$C_{14}H_{18}O_3$
Stiripentol, 8948
$C_{14}H_{18}O_4$
Cinoxate, 2310
$C_{14}H_{18}O_7$
Picein, 7507
$C_{14}H_{18}O_8$
Glucovanillin, 4500
$C_{14}H_{19}BrO_9$
Acetobromglucose, 59
$C_{14}H_{19}ClN_4$
Amprolium, 586
$C_{14}H_{19}Cl_2NO_2$
Chlorambucil, 2073
$C_{14}H_{19}NO$
Ethoxyquin, 3807
$C_{14}H_{19}NO_2$
Levophacetoperane, 5521
Methylphenidate, 6183
Piperoxan, 7591
$C_{14}H_{19}NO_4$
Anisomycin, 663
$C_{14}H_{19}NO_4S$
Furtrethonium Benzenesulfonate,
4341
Trithiozine, 9937
$C_{14}H_{19}NO_5$
Trimetozine, 9892
$C_{14}H_{19}N_3O$
Oxolamine, 7043
Ramifenazone, 8220
$C_{14}H_{19}N_3O_2$
Zilpaterol, 10324
$C_{14}H_{19}N_3S$
Methapyrilene, 6031
Thenyldiamine, 9431
$C_{14}H_{19}N_5O_4$
Famciclovir, 3970
$C_{14}H_{19}N_5O_5$
Tecadenoson, 9233
$C_{14}H_{20}$
Congressane, 2483
$C_{14}H_{20}Br_2N_2$
Bromhexine, 1398
$C_{14}H_{20}Br_2N_4O$
Pyrithiamine, 8105
$C_{14}H_{20}ClNO_2$
Acetochlor, 60
Alachlor, 194
$C_{14}H_{20}ClN_3O_3$
Arimoclomol, 775
$C_{14}H_{20}ClN_3O_3S$
Clopamide, 2382
$C_{14}H_{20}Cl_6N_2$
Chlorisondamine Chloride, 2101
$C_{14}H_{20}FN_3OSi$
Simeconazole, 8673
$C_{14}H_{20}GdN_3O_{10}$
Gadopentetic Acid, 4357
$C_{14}H_{20}IN_3O_8$
Nitroxynil D-N-Methylglucamine Salt,
6743
$C_{14}H_{20}N_2O$
Pyrrocaine, 8127
Siduron, 8626
Tymazoline, 10015

$C_{14}H_{20}N_2O_2$
Bunitrolol, 1493
Pindolol, 7554
$C_{14}H_{20}N_2O_3$
Propacetamol, 7906
Vorinostat, 10232
$C_{14}H_{20}N_2O_3S$
Tolcyclamide, 9671
$C_{14}H_{20}N_2O_6S$
p-Methylaminophenol Sulfate, 6091
$C_{14}H_{20}N_2O_8S_2$
Paraquat Bismethyl Sulfate, 7135
$C_{14}H_{20}N_3O_5PS$
Pyrazophos, 8073
$C_{14}H_{20}N_4$
1,1'-Azobis(cyclohexanecarbonitrile), 910
$C_{14}H_{20}N_4O_7S_2$
Netobimin, 6565
$C_{14}H_{20}O_2$
Butibufen, 1529
$C_{14}H_{20}O_4$
Frequentin, 4299
$C_{14}H_{21}NO_2$
Amylocaine, 607
4-(Dimethylamino)benzoic Acid 3-Methylbutyl Ester, 3254
$C_{14}H_{21}NO_4$
Diethofencarb, 3120
$C_{14}H_{21}NO_8$
1,3,4-Tri-O-acetyl-N-acetyl-6-desoxy-β-D-glucosamine, 2932
$C_{14}H_{21}N_3O_2S$
Sumatriptan, 9126
$C_{14}H_{21}N_3O_3$
Oxamniquine, 7018
$C_{14}H_{21}N_3O_3S$
Tolazamide, 9665
$C_{14}H_{21}N_3O_4$
Butralin, 1534
$C_{14}H_{21}N_3O_5$
Leonurine, 5495
$C_{14}H_{21}N_3O_6S$
Penicillin N, 7207
$C_{14}H_{21}N_5O_6S.H_2O$
Serotonin Complex with Creatinine Sulfate, Monohydrate, 8601
$C_{14}H_{22}BrN_3O_2$
Bromopride, 1443
$C_{14}H_{22}ClNO$
Clobutinol, 2360
$C_{14}H_{22}ClNO_2$
Bupranolol, 1500
$C_{14}H_{22}ClNO_3$
Mecoprop Diethylamine Salt, 5855
$C_{14}H_{22}ClN_3O_2$
Metoclopramide, 6219
$C_{14}H_{22}Cl_2N_2O_6S_4$
Clomethiazole Ethanedisulfonate, 2375
$C_{14}H_{22}CuN_6O_4$
GHK Complex with Copper, 4452
$C_{14}H_{22}N_2O$
Ispronicline, 5286
Lidocaine, 5535
Octacaine, 6832
$C_{14}H_{22}N_2O_2$
Rivastigmine, 8369
$C_{14}H_{22}N_2O_3$
Atenolol, 850
Bucolome, 1475
Practolol, 7818
Trimetazidine, 9874
$C_{14}H_{22}N_2O_3S$
Piprozolin, 7598
$C_{14}H_{22}N_2O_4S$
Amylpenicillin, 609

$C_{14}H_{22}N_6O_5$
Valganciclovir, 10093
$C_{14}H_{22}O$
Irone, 5139
α-Irone, 5140
β-Irone, 5141
γ-Irone, 5142
$C_{14}H_{22}O_4$
Palitantin, 7088
$C_{14}H_{23}N$
2,6-Di-tert-butyl-4-methylpyridine, 3042
$C_{14}H_{23}NO$
Affinin, 170
Tapentadol, 9189
$C_{14}H_{23}NO_2$
Piroctone, 7615
$C_{14}H_{23}N_3OS$
Xanomeline, 10254
$C_{14}H_{23}N_3O_3S$
Sematilide, 8581
$C_{14}H_{23}N_3O_{10}$
Pentetic Acid, 7237
$C_{14}H_{24}CaN_4O_8$
Piperazine Edetate Calcium, 7576
$C_{14}H_{24}N_2O_3$
Ciclopirox Ethanolamine Salt (1:1), 2269
$C_{14}H_{24}N_2O_5$
Pirbuterol Monoacetate, 7602
$C_{14}H_{24}N_2O_7$
Spectinomycin, 8866
$C_{14}H_{24}N_2O_{10}$
EGTA, 3576
$C_{14}H_{24}N_6O_4$
GHK, 4452
$C_{14}H_{24}O_2$
Japonilure, 5305
$C_{14}H_{25}NO$
Pellitorine, 7182
$C_{14}H_{25}N_3O_4S$
Alitame, 240
$C_{14}H_{25}N_3O_6S$
ACV, 133
$C_{14}H_{25}N_4NaO_{11}P_2$
Citicoline Sodium Salt, 2318
$C_{14}H_{25}NaO_7S$
Diamyl Sodium Sulfosuccinate, 2989
$C_{14}H_{26}CaO_{16}$
Glucoheptonic Acid Calcium Salt, 4491
$C_{14}H_{26}MgO_{16}$
Glucoheptonic Acid Magnesium Salt, 4491
$C_{14}H_{26}N_4O_{11}P_2$
Citicoline, 2318
$C_{14}H_{26}O_2$
Looplure, 5627
$C_{14}H_{26}O_4$
Sebacic Acid Ethyl Ester, 8552
$C_{14}H_{27}NO_6$
Amiprilose, 480
$C_{14}H_{27}N_3O_2$
Pramiracetam, 7826
$C_{14}H_{28}O_2$
Ethyl Laurate, 3873
Myristic Acid, 6419
$C_{14}H_{28}O_6$
n-Octyl-β-D-glucoside, 6857
$C_{14}H_{28}P_2$
(R,R)-Me-BPE, 1363
$C_{14}H_{29}NaSO_4$
Sodium Tetradecyl Sulfate, 8817
$C_{14}H_{30}Br_2N_2O_4$
Succinylcholine Bromide, 9005
$C_{14}H_{30}Cl_2N_2O_4$
Succinylcholine Chloride, 9006

$C_{14}H_{30}I_2N_2O_4$
Succinylcholine Iodide, 9007
$C_{14}H_{30}O$
Myristyl Alcohol, 6421
$C_{14}H_{30}O_2$
Decamethylene Glycol Diethyl Ether, 2853
$C_{14}H_{30}O_2S_2$
Tiadenol, 9571
$C_{14}H_{32}N_2O_4$
Entprol, 3653
$C_{14}H_{42}O_5Si_6$
Tetradecamethylhexasiloxane, 9342
$C_{15}H_8Cl_2FNO$
Quinoxyfen, 8196
$C_{15}H_8O_5$
Coumestrol, 2552
$C_{15}H_8O_6$
Rhein, 8299
$C_{15}H_9BrO_2$
Bromindione, 1400
$C_{15}H_9FO_2$
Fluindione, 4165
$C_{15}H_9NO_4$
1-Aminoanthraquinone-2-carboxylic Acid, 412
$C_{15}H_{10}BrClN_4S$
Brotizolam, 1459
$C_{15}H_{10}ClF_3N_2NaO_6S$
Fomesafen Sodium Salt, 4254
$C_{15}H_{10}ClF_3N_2O_3$
Triflumuron, 9844
$C_{15}H_{10}ClF_3N_2O_6S$
Fomesafen, 4254
$C_{15}H_{10}ClN_3O_3$
Clonazepam, 2379
$C_{15}H_{10}Cl_2N_2O_2$
Lonidamine, 5625
Lorazepam, 5636
$C_{15}H_{10}I_4NNaO_4$
Thyroxine Sodium Salt, 9570
$C_{15}H_{10}O$
Diphencyprone, 3338
$C_{15}H_{10}O_4$
Chrysin, 2260
Chrysophanic Acid, 2262
Daidzein, 2798
Rubiadin, 8414
$C_{15}H_{10}O_5$
Aloe-Emodin, 300
Apigenin, 717
Baicalein, 935
Emodin, 3618
Galangin, 4369
Genistein, 4426
Sulfuretin, 9102
$C_{15}H_{10}O_6$
Baptigenin, 953
Datiscetin, 2832
Fisetin, 4119
Kaempferol, 5322
Luteolin, 5675
Scutelarein, 8549
$C_{15}H_{10}O_7$
Morin, 6355
Quercetin, 8150
$C_{15}H_{10}O_8$
Myricetin, 6417
Quercetagetin, 8149
$C_{15}H_{11}BrClF_3N_2O$
Chlorfenapyr, 2089
$C_{15}H_{11}BrNNaO_3.1\frac{1}{2}H_2O$
Bromfenac Monosodium Salt Sesquihydrate, 1397
$C_{15}H_{11}ClF_3NO_4$
Oxyfluorfen, 7062

C$_{15}$H$_{11}$ClN$_2$O
  Mecloqualone, 5852
  Nordazepam, 6780
C$_{15}$H$_{11}$ClN$_2$O$_2$
  Oxazepam, 7025
C$_{15}$H$_{11}$ClO$_2$
  9-Fluorenylmethyl Chloroformate, 4190
C$_{15}$H$_{11}$ClO$_5$
  Pelargonidin, 7180
C$_{15}$H$_{11}$ClO$_6$
  Cyanidin Chloride, 2677
C$_{15}$H$_{11}$ClO$_7$
  Delphinidin, 2885
C$_{15}$H$_{11}$I$_3$NNaO$_4$
  Liothyronine Sodium Salt, 5565
C$_{15}$H$_{11}$I$_3$O$_4$
  Thyropropic Acid, 9567
C$_{15}$H$_{11}$I$_4$NO$_4$
  Thyroxine, 9570
C$_{15}$H$_{11}$NO$_2$
  Viridicatin, 10201
C$_{15}$H$_{11}$NO$_5$
  Bostrycoidin, 1356
C$_{15}$H$_{11}$N$_2$NaO$_2$
  Phenytoin Sodium Salt, 7433
C$_{15}$H$_{11}$N$_3$O
  PAN, 7107
C$_{15}$H$_{11}$N$_3$O$_3$
  Nitrazepam, 6661
C$_{15}$H$_{12}$BrNO$_3$
  Bromfenac, 1397
C$_{15}$H$_{12}$Br$_4$O$_2$
  Tetrabromobisphenol A, 9328
C$_{15}$H$_{12}$ClNO$_2$
  Carprofen, 1862
C$_{15}$H$_{12}$F$_2$N$_4$O$_3$
  Diflufenzopyr, 3164
C$_{15}$H$_{12}$F$_3$NO$_4$S
  Isoxaflutole, 5283
C$_{15}$H$_{12}$F$_4$N$_4$O$_7$S
  Primisulfuron-methyl, 7866
C$_{15}$H$_{12}$I$_2$O$_3$
  Iodoalphionic Acid, 5069
C$_{15}$H$_{12}$I$_3$NO$_4$
  Liothyronine, 5565
C$_{15}$H$_{12}$NNaO$_3$.H$_2$O
  Amfenac Sodium Salt Monohydrate, 389
C$_{15}$H$_{12}$N$_2$O
  Carbamazepine, 1783
C$_{15}$H$_{12}$N$_2$O$_2$
  Oxcarbazepine, 7028
  Phenytoin, 7433
C$_{15}$H$_{12}$N$_2$O$_3$
  Hydrofuramide, 4828
C$_{15}$H$_{12}$O
  Chalcone, 2036
C$_{15}$H$_{12}$O$_2$
  Dibenzoylmethane, 3014
  α-Phenylcinnamic Acid, 7393
C$_{15}$H$_{12}$O$_3$
  Chrysarobin, 2257
  Dehydroequol, 2876
  2-(p-Toluyl)benzoic Acid, 10002
C$_{15}$H$_{12}$O$_4$
  o-(p-Anisoyl)benzoic Acid, 664
  Methyl Benzoylsalicylate, 6099
C$_{15}$H$_{12}$O$_5$
  Naringenin, 6506
C$_{15}$H$_{12}$O$_6$
  Eriodictyol, 3726
  Fustin, 4347
  5,5′-Methylenedisalicylic Acid, 6136
C$_{15}$H$_{12}$O$_8$
  Ampelopsin, 576

C$_{15}$H$_{13}$ClFNO$_2$
  Lumiracoxib, 5661
C$_{15}$H$_{13}$ClFN$_5$O$_5$S
  Cloransulam-methyl, 2391
C$_{15}$H$_{13}$ClN$_2$
  Chlormidazole, 2107
C$_{15}$H$_{13}$ClN$_2$O$_5$
  Gallocyanine, 4385
C$_{15}$H$_{13}$Cl$_2$N$_3$O$_2$
  Azafenidin, 887
C$_{15}$H$_{13}$Cl$_2$N$_5$
  Robenidine, 8371
C$_{15}$H$_{13}$FO$_2$
  Flurbiprofen, 4229
  Tarenflurbil, 9196
C$_{15}$H$_{13}$I$_2$NO$_4$
  3,5-Diiodothyronine, 3209
C$_{15}$H$_{13}$NO
  N-2-Fluorenylacetamide, 4189
C$_{15}$H$_{13}$NO$_2$
  Dibenzoylmethane Monoxime, 3014
C$_{15}$H$_{13}$NO$_2$S
  Metiazinic Acid, 6215
C$_{15}$H$_{13}$NO$_3$
  Amfenac, 389
  Ketorolac, 5353
  Pranoprofen, 7831
C$_{15}$H$_{13}$NO$_3$S$_2$
  Epalrestat, 3660
C$_{15}$H$_{13}$N$_3$O$_2$S
  Fenbendazole, 3994
C$_{15}$H$_{13}$N$_3$O$_3$S
  Oxfendazole, 7034
C$_{15}$H$_{13}$N$_3$O$_4$S
  Piroxicam, 7619
C$_{15}$H$_{13}$N$_4$NaO$_2$S.H$_2$O
  Sulfaphenazole Sodium Salt Monohydrate, 9066
C$_{15}$H$_{14}$ClN$_3$O$_4$S
  Cefaclor, 1914
C$_{15}$H$_{14}$ClN$_3$O$_4$S$_3$
  Benzthiazide, 1123
C$_{15}$H$_{14}$ClN$_3$O$_6$
  Lodoxamide Diethyl Ester, 5613
C$_{15}$H$_{14}$Cl$_2$F$_3$N$_3$O$_3$
  Carfentrazone-ethyl, 1836
C$_{15}$H$_{14}$Cl$_2$N$_2$O$_3$
  Oxadiargyl, 7004
C$_{15}$H$_{14}$FNO$_3$
  Ibafloxacin, 4911
C$_{15}$H$_{14}$FN$_3$O$_3$
  Flumazenil, 4166
C$_{15}$H$_{14}$F$_3$N$_3$O$_4$S$_2$
  Bendroflumethiazide, 1039
C$_{15}$H$_{14}$F$_3$N$_5$O$_7$S
  Flupyrsulfuron-methyl, 4225
C$_{15}$H$_{14}$NNaO$_3$.2H$_2$O
  Tolmetin Sodium Salt Dihydrate, 9677
C$_{15}$H$_{14}$N$_2$Na$_2$O$_6$S$_2$
  Ticarcillin Disodium Salt, 9584
C$_{15}$H$_{14}$N$_2$O$_2$
  Eslicarbazepine, 3754
  Nepafenac, 6553
C$_{15}$H$_{14}$N$_2$O$_4$S
  Belinostat, 1025
C$_{15}$H$_{14}$N$_3$NaO$_2$
  Methyl Red Sodium Salt, 6192
C$_{15}$H$_{14}$N$_4$O
  Nevirapine, 6575
C$_{15}$H$_{14}$N$_4$O$_2$S
  Sulfaphenazole, 9066
C$_{15}$H$_{14}$N$_4$O$_6$S$_2$
  Ceftibuten, 1950
C$_{15}$H$_{14}$O$_2$
  p-Cresyl Phenylacetate, 2574
  Diphenylacetic Acid Methyl Ester, 3346

C$_{15}$H$_{14}$O$_3$
  Equol, 3701
  Fenoprofen, 4011
  Guaiacol Phenylacetate, 4589
  Lapachol, 5417
  Mexenone, 6246
C$_{15}$H$_{14}$O$_4$
  Lunularic Acid, 5666
  Menadiol Diacetate, 5898
  Menbutone, 5902
  Peucedanin, 7306
  Xanthoxyletin, 10264
  Yangonin, 10291
C$_{15}$H$_{14}$O$_5$
  Benzophenone-6, 1101
  Methysticin, 6214
  Phloretin, 7437
C$_{15}$H$_{14}$O$_6$
  Catechin, 1902
  Javanicin, 5310
  Plumericin, 7656
C$_{15}$H$_{14}$O$_7$
  Fusarubin, 4344
  Leucocyanidin, 5504
C$_{15}$H$_{15}$ClFN$_3$O$_3$S$_2$
  Fluthiacet-methyl, 4239
C$_{15}$H$_{15}$ClN$_2$O$_4$S
  Xipamide, 10276
C$_{15}$H$_{15}$ClN$_4$O$_6$S
  Chlorimuron-ethyl, 2094
C$_{15}$H$_{15}$Cl$_2$N$_2$NaO$_8$
  Chloramphenicol Monosuccinate Sodium Salt, 2077
C$_{15}$H$_{15}$FN$_4$O
  Zolazepam, 10387
C$_{15}$H$_{15}$F$_3$N$_2$O$_2$
  Flurprimidol, 4234
C$_{15}$H$_{15}$I$_3$NNaO$_3$
  Bunamiodyl Sodium Salt, 1490
C$_{15}$H$_{15}$NO
  p-Dimethylaminobenzophenone, 3255
C$_{15}$H$_{15}$NO$_2$
  Apo-β-erythroidine, 731
  Mefenamic Acid, 5869
C$_{15}$H$_{15}$NO$_2$S
  Modafinil, 6314
C$_{15}$H$_{15}$NO$_3$
  Tolmetin, 9677
C$_{15}$H$_{15}$NO$_3$S
  Adrafinil, 159
C$_{15}$H$_{15}$NO$_6$
  Ascorbigen, 820
C$_{15}$H$_{15}$N$_3$O
  Ethacridine, 3771
C$_{15}$H$_{15}$N$_3$O$_2$
  Methyl Red, 6192
C$_{15}$H$_{15}$N$_5$O$_4$
  Etofylline Nicotinate, 3924
C$_{15}$H$_{16}$ClN$_3$S
  Azure B, 921
  Tolonium Chloride, 9679
C$_{15}$H$_{16}$Cl$_3$N$_3$O$_2$
  Prochloraz, 7878
C$_{15}$H$_{16}$F$_3$N$_5$O$_4$S
  Prosulfuron, 7992
C$_{15}$H$_{16}$F$_5$NO$_2$S$_2$
  Dithiopyr, 3417
C$_{15}$H$_{16}$I$_3$NO$_3$
  Bunamiodyl, 1490
C$_{15}$H$_{16}$N$_2$Na$_2$O$_8$S$_3$
  Noprylsulfamide, 6772
C$_{15}$H$_{16}$N$_2$O
  Phenatine, 7330
C$_{15}$H$_{16}$N$_2$O$_2$
  Ancymidol, 625
C$_{15}$H$_{16}$N$_2$O$_3$S
  N-Sulfanilyl-3,4-xylamide, 9063

$C_{15}H_{16}N_2O_6S_2$
Ticarcillin, 9584
$C_{15}H_{16}N_4O_2$
Nicaraven, 6578
$C_{15}H_{16}N_4O_5S$
Sulfometuron-methyl, 9087
$C_{15}H_{16}N_7NaO_5S_3$
Cefmetazole Sodium Salt, 1927
$C_{15}H_{16}O_2$
Bisphenol A, 1298
Nabumetone, 6428
$C_{15}H_{16}O_3$
Osthole, 6993
$C_{15}H_{16}O_5$
Lactucin, 5394
Vernolepin, 10158
$C_{15}H_{16}O_6$
Picrotoxinin, 7528
$C_{15}H_{16}O_8$
Skimmin, 8700
$C_{15}H_{16}O_9$
Cichoriin, 2265
Daphnin, 2818
Esculin, 3753
$C_{15}H_{17}Br_2NO_2$
Bromoxynil Octanoate Ester, 1451
$C_{15}H_{17}ClN_4$
Myclobutanil, 6406
Neutral Red, 6573
$C_{15}H_{17}Cl_2N_3O$
Diniconazole, 3294
$C_{15}H_{17}Cl_2N_3O_2$
Propiconazole, 7932
$C_{15}H_{17}FN_4O_2$
Flupirtine, 4221
$C_{15}H_{17}FN_4O_3$
Enoxacin, 3641
$C_{15}H_{17}F_2N_6NaO_7S_2$
Flomoxef Sodium Salt, 4136
$C_{15}H_{17}I_2N_5O_4$
HLö-7, 4762
$C_{15}H_{17}I_3NNaO_3$
Tyropanoate Sodium, 10018
$C_{15}H_{17}N$
Ethylbenzylaniline, 3824
$C_{15}H_{17}NO_2$
Agomelatine, 183
$C_{15}H_{17}NS_2$
Tipepidine, 9613
$C_{15}H_{17}N_3O$
Metralindole, 6231
$C_{15}H_{17}N_4NaO_7S$
Propoxycarbazone Sodium Salt, 7951
$C_{15}H_{17}N_5O_6S_2$
Cefpodoxime Free Acid, 1942
$C_{15}H_{17}N_7O_5S_3$
Cefmetazole, 1927
$C_{15}H_{17}NaO_3$
Loxoprofen Sodium Salt, 5647
$C_{15}H_{17}NaO_3S$
Guaiazulene 3-Sulfonate Sodium Salt, 4590
$C_{15}H_{18}$
Cadalene, 1613
Guaiazulene, 4590
$C_{15}H_{18}ClN_3O$
Cyproconazole, 2768
$C_{15}H_{18}ClN_3O_3S$
Tiaramide, 9579
$C_{15}H_{18}Cl_2N_2O_2$
Fenoxanil, 4014
$C_{15}H_{18}Cl_2N_2O_3$
Oxadiazon, 7006
$C_{15}H_{18}Cl_3NO$
Carpropamid, 1863
$C_{15}H_{18}Cl_6N_2O_5$
Dichloralphenazone, 3055

$C_{15}H_{18}F_2N_6O_7S_2$
Flomoxef, 4136
$C_{15}H_{18}F_3NO_5$
Befloxatone, 1020
$C_{15}H_{18}I_3NO_5$
Iopronic Acid, 5106
$C_{15}H_{18}N_2O$
Harmalol O-n-Propylharmalol, 4647
Huperzine A, 4794
$C_{15}H_{18}N_4O_4S$
Biapenem, 1201
$C_{15}H_{18}N_4O_5$
Mitomycin C, 6300
$C_{15}H_{18}N_4O_7S$
Ethoxysulfuron, 3808
Propoxycarbazone, 7951
$C_{15}H_{18}N_6O_2$
Pimefylline, 7542
$C_{15}H_{18}N_6O_6S$
Nicosulfuron, 6607
$C_{15}H_{18}N_8O_5$
Regadenoson, 8247
$C_{15}H_{18}O_3$
Ambrosin, 379
Irofulven, 5136
Loxoprofen, 5647
α-Santonin, 8498
Xanthatin, 10256
$C_{15}H_{18}O_4$
Artemisin, 806
Helenalin, 4661
Parthenin, 7151
$C_{15}H_{18}O_5$
Coriamyrtin, 2516
$C_{15}H_{18}O_6$
Tutin, 10008
$C_{15}H_{18}O_7$
Picrotin, 7526
$C_{15}H_{18}O_8$
Bilobalide, 1224
$C_{15}H_{19}BrN_2O_5$
Mebrofenin, 5841
$C_{15}H_{19}ClN_4$
Toluylene Blue, 10003
$C_{15}H_{19}NO_2$
Tasimelteon, 9208
Tropacocaine, 9953
$C_{15}H_{19}NO_3$
Ximoprofen, 10275
$C_{15}H_{19}N_3O_3$
Imazethapyr, 4948
$C_{15}H_{19}N_3O_4$
Imazamox, 4945
$C_{15}H_{19}N_3O_5$
Carboquone, 1825
$C_{15}H_{19}N_5$
Rizatriptan, 8370
$C_{15}H_{20}ClN_3O$
Paclobutrazol, 7082
$C_{15}H_{20}N_2O$
Anagyrine, 617
$C_{15}H_{20}N_2O_2$
Fenspiride, 4026
$C_{15}H_{20}N_2O_4S$
Acetohexamide, 61
$C_{15}H_{20}N_6O_5$
Alazopeptin, 201
$C_{15}H_{20}O$
ar-Turmerone, 10006
$C_{15}H_{20}O_2$
Alantolactone, 200
Periplanone A, 7288
$C_{15}H_{20}O_3$
Arborescin, 761
Illudin M, 4940
Parthenolide, 7152

Perezone, 7269
Periplanone B, 7288
Quadrone, 8139
Terrecyclic Acid, 9314
$C_{15}H_{20}O_4$
Abscisic Acid, 12
Hirsutic Acid C, 4754
Illudin S, 4940
Santonic Acid, 8497
Tanacetin, 9182
$C_{15}H_{20}O_6$
Shellolic Acid, 8618
Vomitoxin, 10229
$C_{15}H_{20}O_7$
Nivalenol, 6746
$C_{15}H_{21}ClN_6$
Lesopitron, 5501
$C_{15}H_{21}Cl_2FN_2O_3$
Fluroxypyr 1-Methylheptyl Ester, 4233
$C_{15}H_{21}F_3N_2O_2$
Fluvoxamine, 4246
$C_{15}H_{21}N$
Fencamfamine, 3998
$C_{15}H_{21}NO$
Eptazocine, 3696
Metazocine, 6005
$C_{15}H_{21}NO_2$
Ciclonicate, 2268
β-Eucaine, 3937
Indenolol, 4979
Ketobemidone, 5348
Meperidine, 5916
$C_{15}H_{21}NO_4S$
Furtrethonium p-Toluenesulfonate, 4341
$C_{15}H_{21}NO_6$
Domoic Acid, 3466
$C_{15}H_{21}N_3O$
Cizolirtine, 2335
Primaquine, 7863
Quinocide, 8183
$C_{15}H_{21}N_3O_2$
Physostigmine, 7496
$C_{15}H_{21}N_3O_2S_3$
Arotinolol, 782
$C_{15}H_{21}N_3O_3$
Geneserine, 4425
$C_{15}H_{21}N_3O_3S$
Gliclazide, 4474
$C_{15}H_{21}N_3O_4S$
Panipenem, 7111
$C_{15}H_{22}BrNO$
Bromobutide, 1419
$C_{15}H_{22}ClNO_2$
Metolachlor, 6223
$C_{15}H_{22}FN_3O_6$
Capecitabine, 1757
$C_{15}H_{22}IN_3O_8$
Nitroxynil N-Ethylglucamine, 6743
$C_{15}H_{22}N_2O$
MacMillan's Imidazolidinone Second Generation Catalyst, 5708
$C_{15}H_{22}N_2O_2$
Mepindolol, 5923
$C_{15}H_{22}N_2O_3$
Tolycaine, 9700
$C_{15}H_{22}N_2O_4$
Troxipide, 9970
$C_{15}H_{22}N_2O_6$
Lysine Acetylsalicylate, 5698
Nipradilol, 6648
$C_{15}H_{22}N_4O_3$
Propentofylline, 7927
$C_{15}H_{22}N_6O_5S$
S-Adenosylmethionine, 144

$C_{15}H_{22}O$
Nootkatone, 6770
Vetivones, 10167
$C_{15}H_{22}O_2$
Drimenin, 3493
Polygodial, 7690
$C_{15}H_{22}O_3$
[S-(all-E)]-3-(1,3,5,7,9-Dodecapen-
taenyloxy)-1,2-propanediol, 3452
Gemfibrozil, 4422
Octyl Salicylate, 6860
Trichodermol, 9812
Warburganal, 10235
Yingzhaosu C, 10296
$C_{15}H_{22}O_5$
Artemisinin, 807
$C_{15}H_{22}O_9$
Aucubin, 868
$C_{15}H_{23}N$
Prolintane, 7896
$C_{15}H_{23}NO$
Meptazinol, 5930
$C_{15}H_{23}NO_2$
Alprenolol, 309
Nupharidine, 6816
$C_{15}H_{23}NO_2S$
Pridopidine, 7860
$C_{15}H_{23}NO_3$
Oxprenolol, 7051
$C_{15}H_{23}NO_4$
Cycloheximide, 2721
$C_{15}H_{23}N_3O_2$
Acecainide, 21
$C_{15}H_{23}N_3O_3S$
Amdinocillin, 383
$C_{15}H_{23}N_3O_4S$
Cyclacillin, 2695
Sulpiride, 9119
$C_{15}H_{23}N_7O_5$
Sinefungin, 8682
$C_{15}H_{24}$
Cadinenes, 1617
Caryophyllene, 1875
Copaene, 2500
α-Farnesene, 3973
β-Farnesene, 3974
Humulene, 4792
Longifolene, 5624
Thujopsene, 9547
Ylangene, 10297
$C_{15}H_{24}NO_4PS$
Isofenphos, 5218
$C_{15}H_{24}N_2O$
Aphylline, 716
Lupanine, 5667
Matrine, 5827
Trimecaine, 9869
$C_{15}H_{24}N_2O_3$
Hydroxytetracaine, 4885
$C_{15}H_{24}N_2O_4S$
Tiapride, 9576
$C_{15}H_{24}N_2O_{17}P_2$
Uridine Diphosphate Glucose, 10064
$C_{15}H_{24}N_4O_2S_2$
Prosultiamine, 7993
$C_{15}H_{24}N_4O_6S_2$
Doripenem, 3474
$C_{15}H_{24}O$
Butylated Hydroxytoluene, 1550
DBMC, 2840
α-Santalol, 8494
β-Santalol, 8495
$C_{15}H_{24}O_5$
Artemisinin Dihydro Derivative, 807
$C_{15}H_{25}ClN_2O_2$
Tetracaine Hydrochloride, 9333

$C_{15}H_{25}NO_3$
Metoprolol, 6228
$C_{15}H_{25}N_3O$
Lisdexamfetamine, 5571
$C_{15}H_{26}N_2$
Sparteine, 8864
$C_{15}H_{26}N_2O$
Retamine, 8280
$C_{15}H_{26}O$
α-Bisabolol, 1245
Carotol, 1858
Cedrol, 1913
Farnesol, 3975
Guaiol, 4592
Ledol, 5485
Nerolidol, 6561
Patchouli Alcohol, 7157
$C_{15}H_{26}O_2$
Daucol, 2833
$C_{15}H_{26}O_4$
Yingzhaosu A, 10296
$C_{15}H_{26}O_6$
Tributyrin, 9785
$C_{15}H_{28}NNaO_3$
Gardol®, 4401
$C_{15}H_{28}N_4O_4$
Peramivir, 7264
$C_{15}H_{28}O_2$
Exaltolide®, 3952
Menthyl Isovalerate, 5909
$C_{15}H_{29}NO_7$
Hecameg®, 4656
$C_{15}H_{29}N_3O_5$
Marimastat, 5818
$C_{15}H_{32}N_4O_5$
Fortimicin B, 4275
$C_{15}H_{32}Sn$
Allyltributylstannane, 289
$C_{15}H_{33}N_3O_2$
Dodine, 3455
$C_{15}N_{16}N_2O_4$
Apaziquone, 712
$C_{16}H_5N_5O_6$
9-Dicyanomethylene-2,4,7-trinitro-
fluorene, 3105
$C_{16}H_7N_2NaO_5$
Pirenoxine Sodium Salt, 7603
$C_{16}H_7Na_3O_{10}S_3$
Pyranine, 8065
$C_{16}H_8Br_2N_2O_2$
6,6'-Dibromoindigo, 3028
$C_{16}H_8Cl_2FN_5O$
Fluquinconazole, 4226
$C_{16}H_8Cl_2F_6N_2O_3$
Hexaflumuron, 4719
$C_{16}H_8N_2Na_2O_8S_2$
Indigo Carmine, 4982
$C_{16}H_8N_2O_5$
Pirenoxine, 7603
$C_{16}H_8O_6$
Medicagol, 5859
$C_{16}H_9N_2Na_3O_{11}S_3$
SPADNS, 8859
$C_{16}H_9N_3Na_2O_{10}S_2$
Chromotrope 2B, 2244
$C_{16}H_9N_4Na_3O_9S_2$
Tartrazine, 9206
$C_{16}H_{10}$
Pyrene, 8074
$C_{16}H_{10}ClKN_2O_4$
Clorazepic Acid Monopotassium Salt,
2392
$C_{16}H_{10}ClN_3$
Lotrifen, 5643
$C_{16}H_{10}Cl_4O_5$
Diploicin, 3379

$C_{16}H_{10}MgO_{10}$
Magnesium Monoperoxyphthalate,
5737
$C_{16}H_{10}N_2Na_2O_7S_2$
Sunset Yellow FCF, 9130
$C_{16}H_{10}N_2O_2$
Indigo, 4981
Indirubin, 4985
$C_{16}H_{10}N_2O_6S_4$
DIDS, 3113
$C_{16}H_{10}O_5$
Pseudobaptigenin, 8019
$C_{16}H_{10}O_7$
Laccaic Acid D, 5378
$C_{16}H_{10}O_8$
Kermesic Acid, 5341
$C_{16}H_{11}ClK_2N_2O_4$
Clorazepic Acid Dipotassium Salt,
2392
$C_{16}H_{11}ClN_2O_3$
Clorazepic Acid, 2392
$C_{16}H_{11}ClN_4$
Estazolam, 3757
$C_{16}H_{11}NO_2$
Cinchophen, 2291
$C_{16}H_{11}NO_3$
Oxycinchophen, 7056
$C_{16}H_{11}N_2NaO_4S$
Orange I, 6952
Orange II, 6953
$C_{16}H_{12}ClFN_2O$
Fludiazepam, 4159
$C_{16}H_{12}Cl_2N_2O_2$
Lormetazepam, 5639
$C_{16}H_{12}FNO_3$
Flunoxaprofen, 4179
$C_{16}H_{12}FN_3O_3$
Flubendazole, 4149
Flunitrazepam, 4177
$C_{16}H_{12}O_3$
Anisindione, 661
$C_{16}H_{12}O_4$
Chrysin Methylchrysin, 2260
Formononetin, 4271
Isoxepac, 5284
$C_{16}H_{12}O_5$
Biochanin A, 1231
$C_{16}H_{12}O_6$
Diosmetin, 3324
Hematein, 4669
Pratensein, 7834
Tectorigenin, 9245
$C_{16}H_{12}O_7$
Rhamnetin, 8294
$C_{16}H_{13}ClFNO_5$
Flumiclorac, 4173
$C_{16}H_{13}ClN_2O$
Diazepam, 2997
Mazindol, 5831
$C_{16}H_{13}ClN_2O_2$
Clobazam, 2356
Temazepam, 9275
$C_{16}H_{13}ClN_6$
Vorozole, 10233
$C_{16}H_{13}ClO_6$
Peonidin, 7256
$C_{16}H_{13}ClO_7$
Petunidin, 7305
$C_{16}H_{13}Cl_2NO_4$
Aceclofenac, 23
Quinfamide, 8173
$C_{16}H_{13}Cl_3N_2OS$
Tioconazole, 9608
$C_{16}H_{13}F_2N_3O$
Flutriafol, 4242

C$_{16}$H$_{13}$NO$_3$S
1-Anilino-8-naphthalenesulfonate, 654

C$_{16}$H$_{13}$N$_2$NaO$_3$
Bendazac Sodium Salt, 1037

C$_{16}$H$_{13}$N$_2$Na$_2$O$_6$P
Fosphenytoin Disodium Salt, 4285

C$_{16}$H$_{13}$N$_3$
Yellow AB, 10293

C$_{16}$H$_{13}$N$_3$O$_3$
Mebendazole, 5837
Nimetazepam, 6634

C$_{16}$H$_{13}$N$_3$O$_4$
Icilin, 4925

C$_{16}$H$_{14}$
1,4-Diphenyl-1,3-butadiene, 3350

C$_{16}$H$_{14}$CaO$_6$
Mandelic Acid Calcium Salt, 5781

C$_{16}$H$_{14}$ClN$_3$O
Chlordiazepoxide, 2085

C$_{16}$H$_{14}$Cl$_2$O$_4$
Diclofop-methyl, 3092

C$_{16}$H$_{14}$FN$_3$O
Afloqualone, 175

C$_{16}$H$_{14}$F$_2$N$_3$NaO$_4$S
Pantoprazole Sodium Salt, 7117

C$_{16}$H$_{14}$F$_3$NO$_3$S
Tolrestat, 9683

C$_{16}$H$_{14}$F$_3$N$_3$O$_2$S
Lansoprazole, 5413

C$_{16}$H$_{14}$F$_3$N$_5$O
Voriconazole, 10231

C$_{16}$H$_{14}$F$_5$N$_5$O$_5$S
Penoxsulam, 7212

C$_{16}$H$_{14}$N$_2$O
Glycosine, 4538
Methaqualone, 6032

C$_{16}$H$_{14}$N$_2$O$_2$S
Mefenacet, 5868

C$_{16}$H$_{14}$N$_2$O$_3$
Bendazac, 1037

C$_{16}$H$_{14}$N$_2$O$_3$S
Valdecoxib, 10086

C$_{16}$H$_{14}$N$_2$O$_4$
Amlexanox, 486
Nybomycin, 6820

C$_{16}$H$_{14}$N$_2$O$_6$S
Phthalylsulfacetamide, 7488

C$_{16}$H$_{14}$O
Dypnone, 3527

C$_{16}$H$_{14}$O$_2$
Benzyl Cinnamate, 1133

C$_{16}$H$_{14}$O$_3$
Fenbufen, 3996
Ketoprofen, 5352

C$_{16}$H$_{14}$O$_4$
Imperatorin, 4967
Medicarpin, 5860

C$_{16}$H$_{14}$O$_5$
Aspirin Guaiacol Ester, 841
Brazilin, 1372
Licochalcone B, 5532
Sakuranetin, 8455

C$_{16}$H$_{14}$O$_6$
Hematoxylin, 4672
Hesperetin, 4705
Homoeriodictyol, 4774

C$_{16}$H$_{15}$ClN$_2$
Medazepam, 5857

C$_{16}$H$_{15}$ClN$_2$OS
Clotiazepam, 2400

C$_{16}$H$_{15}$Cl$_3$O$_2$
Methoxychlor, 6061

C$_{16}$H$_{15}$FN$_2$O$_4$
Pazufloxacin, 7161

C$_{16}$H$_{15}$F$_2$N$_3$O$_4$S
Pantoprazole, 7117

C$_{16}$H$_{15}$F$_2$N$_3$Si
Flusilazole, 4236

C$_{16}$H$_{15}$F$_3$O
Flumecinol, 4167

C$_{16}$H$_{15}$F$_6$N$_5$O
Sitagliptin, 8690

C$_{16}$H$_{15}$N
Dizocilpine, 3430

C$_{16}$H$_{15}$N$_2$NaO$_6$S$_2$
Cephalothin Sodium Salt, 1981

C$_{16}$H$_{15}$N$_2$O$_6$P
Fosphenytoin, 4285

C$_{16}$H$_{15}$N$_3$
Epinastine, 3673

C$_{16}$H$_{15}$N$_3$O$_5$
Opiniazide, 6947

C$_{16}$H$_{15}$N$_3$O$_7$S
Sulfaloxic Acid, 9044

C$_{16}$H$_{15}$N$_4$NaO$_8$S
Cefuroxime Sodium Salt, 1956

C$_{16}$H$_{15}$N$_5$O$_7$S$_2$
Cefixime, 1925

C$_{16}$H$_{16}$
[2.2]Metacyclophane, 5991

C$_{16}$H$_{16}$ClNO$_2$S
Clopidogrel, 2385

C$_{16}$H$_{16}$ClNO$_3$
Fenoldopam, 4010

C$_{16}$H$_{16}$ClNO$_4$S
Daltroban, 2805

C$_{16}$H$_{16}$ClN$_3$O$_3$S
Indapamide, 4975
Metolazone, 6224

C$_{16}$H$_{16}$ClN$_3$O$_4$
Loracarbef, 5633

C$_{16}$H$_{16}$NO$_6$P
Naftalofos, 6438

C$_{16}$H$_{16}$N$_2$Na$_2$O$_7$S$_2$
Sulbenicillin Disodium Salt, 9022
Temocillin Disodium Salt, 9280

C$_{16}$H$_{16}$N$_2$O$_2$
Isolysergic Acid, 5226
Lysergic Acid, 5694
Rugulovasines, 8428
Salen, 8458

C$_{16}$H$_{16}$N$_2$O$_3$S
Febuxostat, 3983

C$_{16}$H$_{16}$N$_2$O$_4$
Desmedipham, 2925
NPPB, 6814
Phenmedipham, 7350

C$_{16}$H$_{16}$N$_2$O$_5$S
Succisulfone, 9009

C$_{16}$H$_{16}$N$_2$O$_6$S$_2$
Cephalothin, 1981

C$_{16}$H$_{16}$N$_3$NaO$_4$S
Cephalexin Sodium Salt, 1974

C$_{16}$H$_{16}$N$_3$NaO$_7$S$_2$
Cefoxitin Sodium Salt, 1938

C$_{16}$H$_{16}$N$_4$
Stilbamidine, 8945

C$_{16}$H$_{16}$N$_4$O
Hydroxystilbamidine, 4883

C$_{16}$H$_{16}$N$_4$O$_2$S
Sulfazamet, 9076

C$_{16}$H$_{16}$N$_4$O$_8$S
Cefuroxime, 1956

C$_{16}$H$_{16}$N$_5$NaO$_4$S
Azidocillin Sodium Salt, 902

C$_{16}$H$_{16}$N$_5$NaO$_7$S$_2$
Cefotaxime Sodium Salt (*syn*-Isomer), 1934

C$_{16}$H$_{16}$O$_2$
Diphenylacetic Acid Ethyl Ester, 3346
Felbinac Ethyl Ester, 3985

C$_{16}$H$_{16}$O$_3$
Parvaquone, 7154

C$_{16}$H$_{16}$O$_5$
Alkannin, 249

C$_{16}$H$_{17}$BrClN$_3$O$_3$
Halofuginone, 4632

C$_{16}$H$_{17}$BrN$_2$
Zimeldine, 10325

C$_{16}$H$_{17}$ClN$_2$O
Tetrazepam, 9391

C$_{16}$H$_{17}$F$_5$N$_2$O$_2$S
Thiazopyr, 9463

C$_{16}$H$_{17}$KN$_2$O$_4$S
Penicillin G Potassium Salt, 7203

C$_{16}$H$_{17}$KN$_2$O$_5$S
Penicillin V Potassium Salt, 7209

C$_{16}$H$_{17}$NO
Diphenamid, 3337

C$_{16}$H$_{17}$NO$_3$
Normorphine, 6800

C$_{16}$H$_{17}$NO$_4$
Lycorine, 5687
Tetrahydropapaveroline, 9361

C$_{16}$H$_{17}$N$_2$NaO$_4$S
Penicillin G Sodium Salt, 7203

C$_{16}$H$_{17}$N$_3$O
Lysergamide, 5693

C$_{16}$H$_{17}$N$_3$O$_4$
Anthramycin, 677

C$_{16}$H$_{17}$N$_3$O$_4$S
Cephalexin, 1974

C$_{16}$H$_{17}$N$_3$O$_5$S
Cefadroxil, 1915

C$_{16}$H$_{17}$N$_3$O$_7$S$_2$
Cefoxitin, 1938

C$_{16}$H$_{17}$N$_5$O$_4$S
Azidocillin, 902

C$_{16}$H$_{17}$N$_5$O$_7$S$_2$
Cefotaxime, 1934

C$_{16}$H$_{17}$N$_9$O$_5$S$_2$
Cefteram, 1949

C$_{16}$H$_{17}$N$_9$O$_5$S$_3$
Cefmenoxime, 1926

C$_{16}$H$_{18}$
1,1-Di-*p*-tolylethane, 3422
1,2-Di-*p*-tolylethane, 3423

C$_{16}$H$_{18}$ClN
Clobenzorex, 2357
Dibenamine, 3009

C$_{16}$H$_{18}$ClN$_3$S
Methylene Blue, 6132

C$_{16}$H$_{18}$Cl$_2$N$_2$O$_4$
Mefenpyr-diethyl, 5871

C$_{16}$H$_{18}$FN$_3$O$_2$
Ezogabine, 3959

C$_{16}$H$_{18}$FN$_3$O$_3$
Norfloxacin, 6789

C$_{16}$H$_{18}$N$_2$
Agroclavine, 184
Metapramine, 6001
Nomifensine, 6760

C$_{16}$H$_{18}$N$_2$O
Amphenone B, 578
Elymoclavine, 3611

C$_{16}$H$_{18}$N$_2$OS.HBr
Pifithrin-α, 7533

C$_{16}$H$_{18}$N$_2$O$_3$
Isopilosine, 5245
Levcromakalim, 5512

C$_{16}$H$_{18}$N$_2$O$_4$S
Penicillin G, 7203
Sulfaproxyline, 9067

C$_{16}$H$_{18}$N$_2$O$_5$S
Penicillin V, 7209

C$_{16}$H$_{18}$N$_2$O$_7$S$_2$
Sulbenicillin, 9022
Temocillin, 9280

$C_{16}H_{18}N_3NaO_5S$
Amoxicillin Sodium Salt, 574
$C_{16}H_{18}N_4O_2$
Nialamide, 6576
Piribedil, 7607
$C_{16}H_{18}N_4O_4$
Xemilofiban Acid, 10269
$C_{16}H_{18}N_4O_7S$
Bensulfuron-methyl, 1053
$C_{16}H_{18}N_6O_7S_2$
Sulfosulfuron, 9095
$C_{16}H_{18}O_2$
Bisphenol B, 1299
$C_{16}H_{18}O_3$
Strobilurin A, 8964
$C_{16}H_{18}O_9$
Chlorogenic Acid, 2143
Scopolin, 8546
$C_{16}H_{18}O_{10}$
Fraxin, 4295
$C_{16}H_{19}BrN_2$
Brompheniramine, 1453
$C_{16}H_{19}ClN_2$
Chlorpheniramine, 2186
$C_{16}H_{19}ClN_2O$
Carbinoxamine, 1800
$C_{16}H_{19}ClSi$
tert-Butyldiphenylchlorosilane, 1568
$C_{16}H_{19}NO_2$
Medifoxamine, 5861
$C_{16}H_{19}NO_3$
α-Erythroidine, 3737
β-Erythroidine, 3738
Lunacrine, 5664
$C_{16}H_{19}N_3O_3$
Febrifugine, 3982
$C_{16}H_{19}N_3O_4S$
Ampicillin, 583
Cephradine, 1985
$C_{16}H_{19}N_3O_5S$
Amoxicillin, 574
Cefroxadine, 1945
$C_{16}H_{19}N_3O_6$
Mitomycin B, 6300
$C_{16}H_{19}N_3S$
Isothipendyl, 5273
Prothipendyl, 7998
$C_{16}H_{20}ClN_3$
Chloropyramine, 2164
$C_{16}H_{20}ClN_3O_3S$
Tripamide, 9909
$C_{16}H_{20}ClN_5O_2$
Fentrazamide, 4034
$C_{16}H_{20}FN_3O_4$
Linezolid, 5559
$C_{16}H_{20}F_3N_3OS$
Penthiopyrad, 7239
$C_{16}H_{20}NNaO_8S$
1-Naphthylamine-4-sulfonic Acid
Sodium Salt Glucoside, 6485
$C_{16}H_{20}N_2$
Benzathine, 1066
Pheniramine, 7349
$C_{16}H_{20}N_2O$
Chanoclavine, 2041
Harmalol O-n-Butylharmalol, 4647
$C_{16}H_{20}N_2O_3$
Imazamethabenz, 4944
$C_{16}H_{20}N_2O_4S_2$
Pyritinol, 8109
$C_{16}H_{20}N_2O_5$
Eterobarb, 3770
$C_{16}H_{20}N_4O_2$
Apazone, 713
$C_{16}H_{20}N_4O_3S$
Torsemide, 9711

$C_{16}H_{20}N_4O_5$
Porfiromycin, 7719
$C_{16}H_{20}N_4O_6$
Diaziquone, 2999
$C_{16}H_{20}N_6O$
Tofacitinib, 9661
$C_{16}H_{20}N_6O_6S$
Orthosulfamuron, 6981
$C_{16}H_{20}N_7NaO_7S_3.7H_2O$
Cefminox Sodium Salt Heptahydrate,
1928
$C_{16}H_{20}O_5$
Curvularin, 2666
$C_{16}H_{20}O_6P_2S_3$
Temephos, 9278
$C_{16}H_{20}O_9$
Gentiopicrin, 4432
$C_{16}H_{21}Cl_2N_3O_2$
Bendamustine, 1036
$C_{16}H_{21}N$
Morphinan, 6360
$C_{16}H_{21}NO$
Norlevorphanol, 6796
$C_{16}H_{21}NO_2$
HQNO, 4790
Propranolol, 7953
Ramelteon, 8219
$C_{16}H_{21}NO_3$
Annotinine, 669
Dihydro-β-erythroidine, 3192
Homatropine, 4768
Rolipram, 8380
$C_{16}H_{21}NO_4$
Befunolol, 1021
$C_{16}H_{21}NS_2$
Thiambutene, 9445
$C_{16}H_{21}N_3$
Tripelennamine, 9910
$C_{16}H_{21}N_3O_2$
Zolmitriptan, 10390
$C_{16}H_{21}N_3O_4S$
Epicillin, 3669
$C_{16}H_{21}N_3O_8S$
Cephalosporin C, 1978
$C_{16}H_{21}N_5O_2$
Alizapride, 242
$C_{16}H_{21}N_7O_7S_3$
Cefminox, 1928
$C_{16}H_{22}ClN_3O$
Tebuconazole, 9229
$C_{16}H_{22}ClN_3O_2$
Renzapride, 8255
$C_{16}H_{22}ClN_3O_4$
Plafibride, 7630
$C_{16}H_{22}FNO$
Melperone, 5895
$C_{16}H_{22}N_2$
Lycodine, 5680
$C_{16}H_{22}N_2O_3$
Procaterol, 7877
$C_{16}H_{22}N_2O_3S$
Glyhexamide, 4542
$C_{16}H_{22}N_2O_4$
Antipyrine Methylethylglycolate, 703
$C_{16}H_{22}N_2O_6$
2,5-Bis(1-aziridinyl)-3,6-bis(2-meth-
oxyethoxy)-1,4-benzoquinone,
1249
$C_{16}H_{22}N_4O_4$
Tetroxoprim, 9398
$C_{16}H_{22}N_4O_4S$
Acetiamine, 53
$C_{16}H_{22}N_6O_4$
TRH, 9750
$C_{16}H_{22}O_2$
Geranylhydroquinone, 4439

$C_{16}H_{22}O_3$
Homosalate, 4777
$C_{16}H_{22}O_4$
Dibutyl Phthalate, 3044
$C_{16}H_{22}O_6$
Trepibutone, 9747
$C_{16}H_{22}O_8$
Coniferin, 2487
$C_{16}H_{22}O_{10}$
Swertiamarin, 9140
$C_{16}H_{22}O_{11}$
Monotropein, 6343
$C_{16}H_{23}BrN_2O_3$
Remoxipride, 8252
$C_{16}H_{23}ClN_4O$
Thonzylamine Hydrochloride, 9527
$C_{16}H_{23}HgN_6NaO_8$
Meralluride, 5933
$C_{16}H_{23}NO$
Dezocine, 2959
Pyrovalerone, 8122
Tolperisone, 9681
$C_{16}H_{23}NO_2$
Alphaprodine, 306
Bufuralol, 1485
Etoxadrol, 3932
Piperocaine, 7587
$C_{16}H_{23}NO_6$
Monocrotaline, 6336
$C_{16}H_{23}NO_8$
Bakankosin, 936
$C_{16}H_{23}N_3OS$
Buprofezin, 1502
$C_{16}H_{23}N_3O_4$
Gabexate, 4350
$C_{16}H_{23}N_5O$
Tegaserod, 9253
$C_{16}H_{24}ClNO_2$
Hexylcaine Hydrochloride, 4744
$C_{16}H_{24}NO_5^+$
Sinapine, 8679
$C_{16}H_{24}N_2$
Isoaminile, 5155
Xylometazoline, 10284
$C_{16}H_{24}N_2O$
Oxymetazoline, 7065
Ropinirole, 8388
$C_{16}H_{24}N_2O_2$
Molindone, 6319
N,N,N',N'-Tetraethylphthalamide,
9348
$C_{16}H_{24}N_2O_3$
Carteolol, 1866
$C_{16}H_{24}N_2O_4$
Ubenimex, 10024
$C_{16}H_{24}N_2O_6$
Pyrisuccideanol, 8104
$C_{16}H_{24}N_{10}O_6$
Aminophylline, 461
$C_{16}H_{24}O_4$
Brefeldin A, 1374
$C_{16}H_{24}O_7$
Rhododendrin, 8314
$C_{16}H_{25}HgNNa_2O_6S$
Mercaptomerin Sodium, 5937
$C_{16}H_{25}NO$
Lycopodine, 5683
$C_{16}H_{25}NO_2$
Butethamate, 1525
Desvenlafaxine, 2937
Thymyl N-Isoamylcarbamate, 9564
Tramadol, 9726
$C_{16}H_{25}NO_2S$
Tertatolol, 9316
$C_{16}H_{25}NO_3$
Moxisylyte, 6378

$C_{16}H_{25}NO_4$
Esmolol, 3755
$C_{16}H_{25}N_2NaO_5S$
Cilastatin Sodium Salt, 2274
$C_{16}H_{25}N_3O_5$
Xamoterol, 10253
$C_{16}H_{25}N_7O_8$
Gougerotin, 4565
$C_{16}H_{26}GdN_5O_8$
Gadodiamide, 4354
$C_{16}H_{26}NO_4S.C_7H_7O_3S$
Suplatast Tosylate, 9132
$C_{16}H_{26}N_2O_2$
Dimethocaine, 3242
$C_{16}H_{26}N_2O_3$
Proparacaine, 7921
$C_{16}H_{26}N_2O_5S$
Cilastatin, 2274
$C_{16}H_{26}O$
Iso E Super®, 5214
$C_{16}H_{26}O_2$
2,5-Di-*tert*-pentylhydroquinone, 3334
$C_{16}H_{26}O_3$
Juvenile Hormone III, 5321
$C_{16}H_{26}O_5$
Artemether, 805
$C_{16}H_{26}O_7$
Picrocrocin, 7523
$C_{16}H_{27}ClN_2O_3$
Propoxycaine Hydrochloride, 7950
$C_{16}H_{27}NO_6$
Gabapentin Encarbil, 4349
$C_{16}H_{27}N_3O_8$
Dinoseb Triethanolamine Salt, 3315
$C_{16}H_{28}N_2O_2$
Tromantadine, 9950
$C_{16}H_{28}N_2O_4$
Oseltamivir, 6986
$C_{16}H_{28}N_2O_6Zn$
ε-Acetamidocaproic Acid Zinc Salt, 46
$C_{16}H_{28}N_4O_4S$
Biocytin, 1232
$C_{16}H_{28}O_2$
Hydnocarpic Acid, 4800
Prodlure, 7885
$C_{16}H_{29}NaO_7S$
Bis(1-methylamyl) Sodium Sulfosuccinate, 1259
$C_{16}H_{30}N_2O_3$
Piroctone Ethanolamine Salt (1:1), 7615
$C_{16}H_{30}O$
Muscone, 6398
$C_{16}H_{30}O_4$
Texanol Isobutyrate, 9399
$C_{16}H_{30}O_4Zn$
Zinc Caprylate, 10329
$C_{16}H_{32}O_2$
2-Hexyldecanoic Acid, 4745
Myristic Acid Ethyl Ester, 6419
Palmitic Acid, 7097
$C_{16}H_{32}O_4$
Diethylene Glycol Monolaurate, 3135
$C_{16}H_{32}O_5$
Aleuritic Acid, 225
$C_{16}H_{33}NO_2$
Delmopinol, 2884
Dodecylbetaine, 3453
$C_{16}H_{34}O$
Cetyl Alcohol, 2027
$C_{16}H_{35}NO_2$
Ammonium Palmitate, 535
$C_{16}H_{36}FN$
Tetrabutylammonium Fluoride, 9332
$C_{16}H_{38}Br_2N_2$
Decamethonium Bromide, 2851

$C_{16}H_{42}I_7N_8O_{16}$
Tetraglycine Hydroperiodide, 9352
$C_{17}H_7Cl_2F_9N_2O_3$
Noviflumuron, 6810
$C_{17}H_8Cl_2F_8N_2O_3$
Lufenuron, 5654
$C_{17}H_9ClF_8N_2O_4$
Novaluron, 6809
$C_{17}H_9Cl_3N_4O_2$
Diclazuril, 3090
$C_{17}H_9NO_4$
Alizarine Blue, 245
$C_{17}H_{10}Cl_2N_4O_2$
Clazuril, 2341
$C_{17}H_{10}O$
Benzanthrone, 1065
$C_{17}H_{10}O_4$
Fluorescamine, 4191
$C_{17}H_{11}ClF_4N_2S$
Quazepam, 8145
$C_{17}H_{11}NO_7$
Aristolochic Acid, 777
$C_{17}H_{11}N_5$
Letrozole, 5502
$C_{17}H_{12}Br_2O_3$
Benzbromarone, 1067
$C_{17}H_{12}ClF_3N_2O$
Halazepam, 4624
$C_{17}H_{12}ClNO_2S$
Fentiazac, 4030
$C_{17}H_{12}Cl_2N_2O$
Fenarimol, 3992
$C_{17}H_{12}Cl_2N_4$
Triazolam, 9765
$C_{17}H_{12}I_2O_3$
Benziodarone, 1085
$C_{17}H_{12}O_2$
2-Naphthyl Benzoate, 6490
$C_{17}H_{12}O_3$
1-Naphthyl Salicylate, 6495
2-Naphthyl Salicylate, 6496
$C_{17}H_{12}O_6$
Aflatoxin B₁, 171
$C_{17}H_{12}O_7$
Aflatoxin G₁, 172
Aflatoxin M₁, 173
$C_{17}H_{13}ClFNO_4$
Clodinafop-propargyl, 2364
$C_{17}H_{13}ClFN_3O$
Epoxiconazole, 3685
$C_{17}H_{13}ClN_2O_2$
Lonazolac, 5623
$C_{17}H_{13}ClN_4$
Alprazolam, 308
Liarozole, 5528
$C_{17}H_{13}Cl_3N_4S$
Imibenconazole, 4950
$C_{17}H_{13}N_3Na_2O_6.2H_2O$
Balsalazide Disodium Salt Dihydrate, 938
$C_{17}H_{13}N_3O_5S_2$
Phthalylsulfathiazole, 7489
$C_{17}H_{14}$
1,2-Cyclopentenophenanthrene, 2739
$C_{17}H_{14}BrFN_2O_2$
Haloxazolam, 4638
$C_{17}H_{14}ClF_7O_2$
Tefluthrin, 9251
$C_{17}H_{14}Cl_2F_2N_2O_3$
Roflumilast, 8378
$C_{17}H_{14}Cl_2N_2O_2$
Cloxazolam, 2404
$C_{17}H_{14}F_3N_3O_2S$
Celecoxib, 1958
$C_{17}H_{14}F_3N_3O_3S$
Deracoxib, 2915

$C_{17}H_{14}N_2$
Ellipticine, 3603
Flavopereirine, 4125
Olivacine, 6928
$C_{17}H_{14}N_2O_3$
Rosoxacin, 8400
$C_{17}H_{14}N_2O_5S$
Calmagite, 1720
$C_{17}H_{14}O$
Dibenzalacetone, 3010
$C_{17}H_{14}O_3S$
Zaltoprofen, 10310
$C_{17}H_{14}O_4S$
Rofecoxib, 8377
$C_{17}H_{14}O_5$
Pterocarpin, 8041
$C_{17}H_{14}O_6$
Aflatoxin B₂, 171
Pectolinarigenin, 7170
$C_{17}H_{14}O_7$
Aflatoxin G₂, 172
Aflatoxin M₂, 173
$C_{17}H_{15}ClN_4S$
Etizolam, 3919
$C_{17}H_{15}ClO_7$
Malvidin Chloride, 5779
$C_{17}H_{15}FN_6O_3$
Tedizolid, 9247
$C_{17}H_{15}NO$
Betti Base, 1190
$C_{17}H_{15}NO_5$
Benorylate, 1047
$C_{17}H_{15}N_3$
Yellow OB, 10294
$C_{17}H_{15}N_3O_6$
Balsalazide, 938
$C_{17}H_{15}N_5O$
Zaleplon, 10309
$C_{17}H_{15}N_7Na_2O_8S_4$
Cefotetan Disodium Salt, 1935
$C_{17}H_{16}Br_2O_3$
Bromopropylate, 1445
$C_{17}H_{16}ClFN_2O_2$
Progabide, 7888
$C_{17}H_{16}ClF_3O_6S$
Tembotrione, 9276
$C_{17}H_{16}ClN_3O$
Amoxapine, 573
$C_{17}H_{16}F_3NO_2$
Flutolanil, 4241
$C_{17}H_{16}F_6N_2O$
Mefloquine, 5872
$C_{17}H_{16}N_2Na_2O_6S$
Carbenicillin Disodium Salt, 1793
$C_{17}H_{16}N_2O$
Etaqualone, 3769
$C_{17}H_{16}N_2O_3$
Eslicarbazepine Acetate, 3754
$C_{17}H_{16}N_2O_3S$
TSQ, 9979
$C_{17}H_{16}O_3$
Eugenol Benzoate, 3940
$C_{17}H_{17}ClFN_3O_3$
Clinafloxacin, 2353
$C_{17}H_{17}ClF_4N_4O_5S$
Saflufenacil, 8451
$C_{17}H_{17}ClN_2O$
Etifoxine, 3913
$C_{17}H_{17}ClN_6O_3$
Zopiclone, 10393
$C_{17}H_{17}ClO_6$
Griseofulvin, 4584
$C_{17}H_{17}Cl_2N$
Sertraline, 8606
$C_{17}H_{17}Cl_2N_2NaO_5S$
Clometocillin Sodium Salt, 2376

$C_{17}H_{17}NO_2$
Apomorphine, 733
$C_{17}H_{17}N_2^+$
Difenzoquat, 3157
$C_{17}H_{17}N_2NaO_5S$
Piretanide Monosodium Salt, 7605
$C_{17}H_{17}N_3O$
Ramosetron, 8223
$C_{17}H_{17}N_3OS$
Fenamidone, 3990
$C_{17}H_{17}N_3O_3$
Imazaquin, 4947
$C_{17}H_{17}N_3O_6S_2$
Cephapirin, 1983
$C_{17}H_{17}N_7O_8S_4$
Cefotetan, 1935
$C_{17}H_{18}Br_2N_4O_2$
Dibromopropamidine, 3029
$C_{17}H_{18}ClNO_4$
Pirifibrate, 7608
$C_{17}H_{18}ClN_3O_4$
Celestin Blue, 1961
$C_{17}H_{18}Cl_2N_2O_5S$
Clometocillin, 2376
$C_{17}H_{18}FN_3O_3$
Ciprofloxacin, 2313
$C_{17}H_{18}FN_3O_3S$
Rufloxacin, 8427
$C_{17}H_{18}F_3NO$
Fluoxetine, 4217
$C_{17}H_{18}F_3N_3O_3$
Fleroxacin, 4131
$C_{17}H_{18}N_2$
Amphetaminil, 580
$C_{17}H_{18}N_2O_5S$
Piretanide, 7605
$C_{17}H_{18}N_2O_6$
Nifedipine, 6613
$C_{17}H_{18}N_2O_6S$
Carbenicillin, 1793
Menadione Dimethylpyrimidinol Bi-
sulfite, 5899
$C_{17}H_{18}N_2S_2$
Sulbentine, 9024
$C_{17}H_{18}N_3NaO_4S$
Metampicillin Sodium Salt, 5996
$C_{17}H_{18}N_4O$
Alosetron, 304
$C_{17}H_{18}N_4O_3S$
Enviroxime, 3655
$C_{17}H_{18}N_6$
Ruxolitinib, 8440
$C_{17}H_{18}O_3$
4-tert-Butylphenyl Salicylate, 1589
$C_{17}H_{18}O_4$
Diphenolic Acid, 3342
$C_{17}H_{19}ClN_2O$
Pyronine Y, 8119
$C_{17}H_{19}ClN_2OS$
Opromazine, 6950
$C_{17}H_{19}ClN_2S$
Chlorpromazine, 2191
$C_{17}H_{19}FN_2O_2$
Safinamide, 8450
$C_{17}H_{19}FN_4O_4$
Marbofloxacin, 5815
$C_{17}H_{19}F_2N_3O_3$
Lomefloxacin, 5619
$C_{17}H_{19}F_3N_6O_6S$
Triflusulfuron-methyl, 9857
$C_{17}H_{19}KN_2O_5S$
Phenethicillin Potassium, 7337
$C_{17}H_{19}NO$
Nefopam, 6525
$C_{17}H_{19}NO_3$
Chavicine, 2048
Coclaurine, 2444
Hydromorphone, 4840

Morphine, 6361
Norcodeine, 6779
Piperine, 7584
$C_{17}H_{19}NO_4$
Buphanamine, 1496
Fenoxycarb, 4017
Oxymorphone, 7068
$C_{17}H_{19}N_2NaO_6S$
Methicillin Sodium, 6041
$C_{17}H_{19}N_3$
Antazoline, 672
Mirtazapine, 6291
$C_{17}H_{19}N_3O$
Phentolamine, 7376
Piberaline, 7504
$C_{17}H_{19}N_3O_3S$
Omeprazole, 6939
$C_{17}H_{19}N_3O_4S$
Metampicillin, 5996
$C_{17}H_{19}N_5$
Anastrozole, 619
$C_{17}H_{19}N_5O_2$
Piritrexim, 7612
Pixantrone, 7628
$C_{17}H_{19}N_5O_6S_2$
Cefcapene, 1920
Cefovecin, 1937
$C_{17}H_{20}BrNO$
Bromodiphenhydramine, 1428
$C_{17}H_{20}ClNO$
Chlophedianol, 2066
$C_{17}H_{20}ClN_3O_2$
Furametpyr, 4324
$C_{17}H_{20}ClN_3O_3$
Azasetron, 893
$C_{17}H_{20}FN_3O_3$
Pefloxacin, 7172
$C_{17}H_{20}F_6N_2O_3$
Flecainide, 4130
$C_{17}H_{20}N_2O$
N,N'-Diethylcarbanilide, 3129
Michler's Ketone, 6256
Remacemide, 8250
$C_{17}H_{20}N_2O_2$
Tropicamide, 9959
Tropisetron, 9962
$C_{17}H_{20}N_2O_3$
Bifenazate, 1211
Nicotine Salicylate, 6609
$C_{17}H_{20}N_2O_5S$
Bumetanide, 1489
$C_{17}H_{20}N_2S$
Promazine, 7898
Promethazine, 7901
$C_{17}H_{20}N_4O$
Propizepine, 7946
$C_{17}H_{20}N_4O_2$
Propamidine, 7911
$C_{17}H_{20}N_4O_3$
Imazaquin Ammonium Salt, 4947
$C_{17}H_{20}N_4O_6$
D-Araboflavin, 752
Riboflavin, 8325
$C_{17}H_{20}N_4S$
Olanzapine, 6914
$C_{17}H_{20}N_6O_7S$
Foramsulfuron, 4261
$C_{17}H_{20}O_5$
Matricarin, 5826
Oleocanthal, 6922
$C_{17}H_{20}O_5S$
Firocoxib, 4118
$C_{17}H_{20}O_6$
Mycophenolic Acid, 6412
$C_{17}H_{20}O_9$
Chlorogenic Acid 3'-Methyl Ether,
2143

$C_{17}H_{21}ClN_2O_2S$
Hexythiazox, 4749
$C_{17}H_{21}Cl_2F_3N_5Na_4O_{12}P_3S_2$
Cangrelor Tetrasodium Salt, 1749
$C_{17}H_{21}N$
Benzphetamine, 1121
N,N-Diethylbenzhydrylamine, 3127
$C_{17}H_{21}NO$
Atomoxetine, 854
Diphenhydramine, 3339
Phenyltoloxamine, 7425
$C_{17}H_{21}NO_2$
Apoatropine, 726
Napropamide, 6497
$C_{17}H_{21}NO_3$
Dihydromorphine, 3194
Etodolac, 3920
Galantamine, 4370
Ritodrine, 8365
$C_{17}H_{21}NO_4$
Cocaine, 2440
Fenoterol, 4012
Pseudococaine, 8020
Scopolamine, 8543
$C_{17}H_{21}NO_5$
Buphanitine, 1497
Scopolamine N-Oxide, 8544
$C_{17}H_{21}N_4O_9P$
Flavin Mononucleotide, 4123
$C_{17}H_{21}N_5O_9S_2$
Mesosulfuron-methyl, 5981
$C_{17}H_{22}BrNOS_2$
Timepidium Bromide, 9599
$C_{17}H_{22}ClN_3O$
Metconazole, 6006
$C_{17}H_{22}ClN_9O_2$
Palau'amine, 7085
$C_{17}H_{22}I_3N_3O_8$
Iomeprol, 5097
Iopamidol, 5100
$C_{17}H_{22}NO_3^+$
Lunasine, 5665
$C_{17}H_{22}N_2$
Michler's Base, 6255
$C_{17}H_{22}N_2O$
Doxylamine, 3489
$C_{17}H_{22}N_2OS$
Neticonazole, 6563
$C_{17}H_{22}N_2O_3$
Trichostatin A, 9814
$C_{17}H_{22}N_2S$
Thenaldine, 9429
$C_{17}H_{22}N_4O$
Minaprine, 6281
$C_{17}H_{22}N_4O_2$
Resiquimod, 8271
$C_{17}H_{22}O_2$
Cicutoxin, 2270
Enanthotoxin, 3624
$C_{17}H_{22}O_3$
Podocarpic Acid, 7660
$C_{17}H_{22}O_5$
Pyrethrosin, 8076
$C_{17}H_{23}Cl_2NO$
Tesofensine, 9322
$C_{17}H_{23}NO$
Levorphanol, 5524
$C_{17}H_{23}NO_2$
Tilidine, 9594
$C_{17}H_{23}NO_3$
Atropine, 866
Hyoscyamine, 4897
Lycoramine, 5686
Mesembrine, 5976
$C_{17}H_{23}NO_4$
Atropine N-Oxide, 866
Cetraxate, 2023
Lunacridine, 5663

C₁₇H₂₃N₃O
Pyrilamine, 8097

$C_{17}H_{23}N_7O_5$
Taltirelin, 9177

$C_{17}H_{24}ClNO_4$
Tepraloxydim, 9295

$C_{17}H_{24}N_2O$
Pilsicainide, 7539

$C_{17}H_{24}N_2O_3$
Tilisolol, 9595

$C_{17}H_{24}O_3$
Cyclandelate, 2697

$C_{17}H_{24}O_4$
Trichodermin, 9812

$C_{17}H_{24}O_9$
Syringin, 9146

$C_{17}H_{24}O_{10}$
Verbenalin, 10153

$C_{17}H_{25}Cl_2F_3N_5O_{12}P_3S_2$
Cangrelor, 1749

$C_{17}H_{25}N$
Phencyclidine, 7333

$C_{17}H_{25}NO$
Eperisone, 3661

$C_{17}H_{25}NO_2$
Menthyl Anthranilate, 5908
Promedol, 7899

$C_{17}H_{25}NO_3$
Cyclopentolate, 2741
Levobunolol, 5514
Pecilocin, 7168

$C_{17}H_{25}NO_4$
Buflomedil, 1480
Ibopamine, 4915

$C_{17}H_{25}N_3O$
Plasmocid, 7636
Proxazole, 8012

$C_{17}H_{25}N_3O_2$
Vildagliptin, 10177

$C_{17}H_{25}N_3O_2S$
Almotriptan, 298
Naratriptan, 6502

$C_{17}H_{25}N_3O_5S$
Meropenem, 5970
Veralipride, 10142

$C_{17}H_{25}N_5O_{13}$
Polyoxin B, 7696

$C_{17}H_{25}N_7O_4$
Binodenoson, 1229

$C_{17}H_{26}ClN$
Sibutramine, 8624

$C_{17}H_{26}ClNO_3S$
Clethodim, 2347

$C_{17}H_{26}N_2O$
5-Methoxy-N,N-diisopropyltryptamine, 6062
Phenampromide, 7323
Ropivacaine, 8389

$C_{17}H_{26}N_2O_4S$
Sultopride, 9123

$C_{17}H_{26}N_4O$
Emedastine, 3614

$C_{17}H_{26}N_4O_3S_2$
Fursultiamine, 4340

$C_{17}H_{26}N_4O_4S$
Alpiropride, 307

$C_{17}H_{26}N_8O_5$
Blasticidin S, 1317

$C_{17}H_{26}O_3$
(p-Nonylphenoxy)acetic Acid, 6769

$C_{17}H_{26}O_3S$
DBHBT, 2839

$C_{17}H_{26}O_4$
Embelin, 3613
[6]-Gingerol, 4459

$C_{17}H_{26}O_5$
Botrydial, 1359

$C_{17}H_{26}O_{10}$
Loganin, 5617

$C_{17}H_{27}NO_2$
4-(Dimethylamino)benzoic Acid 2-Ethylhexyl Ester, 3254
Venlafaxine, 10140

$C_{17}H_{27}NO_3$
Butoxycaine, 1533
Pramoxine, 7829

$C_{17}H_{27}NO_4$
Metipranolol, 6216
Nadolol, 6431

$C_{17}H_{27}N_3O_4S$
Amisulpride, 481

$C_{17}H_{28}N_2O$
Etidocaine, 3911

$C_{17}H_{28}N_2O_3$
Benoxinate, 1049

$C_{17}H_{28}N_2O_4$
Embelin Dioxime, 3613

$C_{17}H_{28}N_2O_5$
Perindopril Diacid Form, 7285

$C_{17}H_{28}O_3$
Juvenile Hormone II, 5321

$C_{17}H_{28}O_5$
Arteether, 803

$C_{17}H_{29}GdN_4O_7$
Gadoteridol, 4358

$C_{17}H_{29}NO_3S$
Sethoxydim, 8612

$C_{17}H_{30}O$
Civetone, 2334

$C_{17}H_{30}O_2$
Hydroprene, 4842

$C_{17}H_{31}ClN_2O_5S$
Pirlimycin, 7613

$C_{17}H_{32}O_4$
Roccellic Acid, 8373

$C_{17}H_{34}N_4O_{10}$
Ribostamycin, 8330

$C_{17}H_{34}O_2$
Isopropyl Myristate, 5261
Margaric Acid, 5816

$C_{17}H_{35}N_5O_6$
Fortimicin A, 4275

$C_{17}H_{36}GeN_2$
Spirogermanium, 8885

$C_{17}H_{37}N_7O_3$
Gusperimus, 4617

$C_{17}H_{38}BrN$
Myristyltrimethylammonium Bromide, 6422

$C_{18}BF_{15}$
Tris(pentafluorophenyl)boron, 9931

$C_{18}Fe_7N_{18}$
Prussian Blue, 8018

$C_{18}H_6BiBr_9O_3$
Bismuth Tribromophenate, 1291

$C_{18}H_{10}I_6N_2O_7$
Ioglycamic Acid, 5094

$C_{18}H_{10}O_6$
Hydrindantin, 4813

$C_{18}H_{12}$
1,2-Benzanthracene, 1064
3,4-Benzphenanthrene, 1120
Chrysene, 2259
Naphthacene, 6454
Triphenylene, 9913

$C_{18}H_{12}Cl_2N_2O$
Boscalid, 1354

$C_{18}H_{12}Cl_{12}$
Dechlorane® Plus, 2855

$C_{18}H_{12}CrN_3O_6$
Chromium Picolinate, 2240

$C_{18}H_{12}NNaO_3$
Naptalam Sodium Salt, 6500

$C_{18}H_{12}N_2O_2$
Xanthocillin X, 10259

$C_{18}H_{12}N_2O_3$
Xanthocillin Y₁, 10259

$C_{18}H_{12}N_2O_4$
Xanthocillin Y₂, 10259

$C_{18}H_{12}N_5O_6$
1,1-Diphenyl-2-picrylhydrazyl (Free Radical), 3369

$C_{18}H_{12}N_6$
2,4,6-Tripyridyl-s-triazine, 9926

$C_{18}H_{12}O_4$
Karanjin, 5331

$C_{18}H_{13}ClFN_3$
Midazolam, 6261

$C_{18}H_{13}ClN_2O$
Pinazepam, 7553

$C_{18}H_{13}Cl_4N_3O$
Oxiconazole, 7036

$C_{18}H_{13}N$
6-Chrysenamine, 2258

$C_{18}H_{13}NNa_2O_8S_2$
Picosulfate Sodium, 7518

$C_{18}H_{13}NO_3$
Naptalam, 6500

$C_{18}H_{13}N_3O$
Rutecarpine, 8432

$C_{18}H_{14}BrCl_2N_5O_2$
Chlorantraniliprole, 2080

$C_{18}H_{14}CaO_8$
Calcium Acetylsalicylate, 1650

$C_{18}H_{14}ClFN_2O_3$
Ethyl Loflazepate, 3876

$C_{18}H_{14}ClNaN_2O_6S_2$
Sitaxsentan Sodium Salt, 8692

$C_{18}H_{14}Cl_4N_2O$
Isoconazole, 5205
Miconazole, 6257

$C_{18}H_{14}F_3NO_2$
Flurtamone, 4235

$C_{18}H_{14}F_3N_3O_4S$
Toltrazuril, 9685

$C_{18}H_{14}F_3N_3O_6S$
Ponazuril, 7711

$C_{18}H_{14}F_4N_2O_4S$
Bicalutamide, 1204

$C_{18}H_{14}MgO_8$
Magnesium Acetylsalicylate, 5719

$C_{18}H_{14}N_2Na_2O_7S_2$
Ponceau SX, 7713

$C_{18}H_{14}N_2Na_2O_8S_2$
Allura Red AC, 275

$C_{18}H_{14}N_3NaO_3S$
Metanil Yellow, 5999
Tropaeolin OO, 9955

$C_{18}H_{14}N_4O_5S$
Sulfasalazine, 9073

$C_{18}H_{14}O_3$
Cinnamic Acid Anhydride, 2299

$C_{18}H_{15}Bi$
Triphenylbismuth, 9911

$C_{18}H_{15}Br_2P$
Triphenylphosphine Dibromide, 9917

$C_{18}H_{15}ClN_2O$
Croconazole, 2580

$C_{18}H_{15}ClN_2O_2S$
Etoricoxib, 3930

$C_{18}H_{15}ClN_2O_6S_2$
Sitaxsentan, 8692

$C_{18}H_{15}ClN_4$
Phenosafranin, 7363

$C_{18}H_{15}Cl_3N_2O$
Econazole, 3549

$C_{18}H_{15}Cl_3N_2S$
Sulconazole, 9025

$C_{18}H_{15}NO_3$
Oxaprozin, 7023

C$_{18}$H$_{15}$O$_4$P
  Triphenyl Phosphate, 9915
C$_{18}$H$_{15}$P
  Triphenylphosphine, 9916
C$_{18}$H$_{16}$
  3'-Methyl-1,2-cyclopentenophenan-
    threne, 6119
C$_{18}$H$_{16}$ClNO$_5$
  Fenoxaprop-ethyl, 4015
C$_{18}$H$_{16}$Cl$_2$N$_2$O$_2$
  Mexazolam, 6245
C$_{18}$H$_{16}$F$_3$NO$_4$
  Picoxystrobin, 7520
C$_{18}$H$_{16}$N$_2$
  N,N'-Diphenyl-p-phenylenediamine,
    3363
C$_{18}$H$_{16}$N$_2$O
  1-Xylylazo-2-naphthol, 10287
C$_{18}$H$_{16}$N$_2$O$_2$
  Blebbistatin, 1318
C$_{18}$H$_{16}$N$_2$O$_3$
  Citrus Red 2, 2332
  Roquinimex, 8390
C$_{18}$H$_{16}$N$_4$O$_6$S
  Quinacillin, 8159
C$_{18}$H$_{16}$N$_6$Na$_2$O$_8$S$_3$
  Cefonicid Disodium Salt, 1930
C$_{18}$H$_{16}$N$_8$Na$_2$O$_7$S$_3$.3½H$_2$O
  Ceftriaxone Disodium Salt Hemihep-
    tahydrate, 1955
C$_{18}$H$_{16}$OSn
  Triphenyltin Hydroxide, 9920
C$_{18}$H$_{16}$O$_2$
  Cinnamyl Cinnamate, 2305
C$_{18}$H$_{16}$O$_3$
  Ipriflavone, 5118
  Phenprocoumon, 7371
C$_{18}$H$_{16}$O$_4$
  Benzyl Fumarate, 1138
  Truxillic Acid, 9971
C$_{18}$H$_{16}$O$_7$
  Eupatorin, 3943
  Usnic Acid, 10078
C$_{18}$H$_{16}$O$_8$
  Irigenin, 5134
  Rosmarinic Acid, 8399
C$_{18}$H$_{16}$Sn
  Triphenyltin Hydride, 9919
C$_{18}$H$_{17}$Br$_2$NO$_5$
  Eprotirome, 3692
C$_{18}$H$_{17}$ClN$_2$O$_2$
  Oxazolam, 7026
C$_{18}$H$_{17}$ClO$_6$
  Monorden, 6341
C$_{18}$H$_{17}$F$_3$N$_2$O$_3$
  Ridogrel, 8337
C$_{18}$H$_{17}$NO$_2$
  Aporeine, 735
C$_{18}$H$_{17}$NO$_3$
  Indobufen, 5001
  Isoxadifen-ethyl, 5282
  Pukateine, 8045
C$_{18}$H$_{17}$NO$_4$
  Actinodaphnine, 129
C$_{18}$H$_{17}$NO$_5$
  Tranilast, 9731
C$_{18}$H$_{18}$
  Retene, 8283
C$_{18}$H$_{18}$BrClN$_2$O
  Metaclazepam, 5990
C$_{18}$H$_{18}$ClNOS
  Zotepine, 10397
C$_{18}$H$_{18}$ClNO$_4$
  Clanobutin, 2337
C$_{18}$H$_{18}$ClNO$_5$
  Etofibrate, 3923

C$_{18}$H$_{18}$ClNS
  Chlorprothixene, 2194
C$_{18}$H$_{18}$ClN$_3$O
  Loxapine, 5645
C$_{18}$H$_{18}$ClN$_3$S
  Clothiapine, 2399
C$_{18}$H$_{18}$F$_3$NO$_2$
  Flufenamic Acid Butyl Ester, 4163
C$_{18}$H$_{18}$F$_3$NO$_4$
  Etofenamate, 3921
C$_{18}$H$_{18}$F$_3$N$_3$O$_3$
  Pleconaril, 7648
C$_{18}$H$_{18}$N$_2$
  Cifenline, 2272
C$_{18}$H$_{18}$N$_2$O
  Proquazone, 7984
C$_{18}$H$_{18}$N$_2$O$_4$
  Antipyrine Salicylate, 703
C$_{18}$H$_{18}$N$_6$O$_5$S$_2$
  Cefamandole, 1916
  Cefatrizine, 1917
C$_{18}$H$_{18}$N$_6$O$_8$S$_3$
  Cefonicid, 1930
C$_{18}$H$_{18}$N$_8$O$_7$S$_3$
  Ceftriaxone, 1955
C$_{18}$H$_{18}$Na$_4$O$_8$P$_2$
  Fosfestrol Tetrasodium Salt, 4280
C$_{18}$H$_{18}$O$_2$
  Dienestrol, 3115
  Equilenin, 3699
  Honokiol, 4781
  Magnolol, 5761
C$_{18}$H$_{18}$O$_4$
  Enterolactone, 3652
C$_{18}$H$_{18}$O$_6$
  Samaderin A, 8482
C$_{18}$H$_{18}$O$_7$
  Frenolicin, 4298
C$_{18}$H$_{19}$ClN$_2$O$_2$
  Halofenozide, 4631
C$_{18}$H$_{19}$ClN$_4$
  Clozapine, 2406
C$_{18}$H$_{19}$Cl$_2$NO$_4$
  Felodipine, 3987
C$_{18}$H$_{19}$F$_3$N$_2$S
  Trifluromazine, 9853
C$_{18}$H$_{19}$N
  Benzoctamine, 1089
C$_{18}$H$_{19}$NOS
  Duloxetine, 3512
  Tolindate, 9675
C$_{18}$H$_{19}$NO$_2$
  Apocodeine, 727
C$_{18}$H$_{19}$NO$_3$
  Oripavine, 6962
  Tsuduranine, 9980
C$_{18}$H$_{19}$NO$_4$
  Hydroxycodeinone, 4862
  Kresoxim-methyl, 5372
C$_{18}$H$_{19}$N$_3$O
  Ondansetron, 6943
C$_{18}$H$_{19}$N$_3$O$_3$S
  Rosiglitazone, 8396
C$_{18}$H$_{19}$N$_3$O$_5$S
  Cefprozil, 1943
C$_{18}$H$_{20}$ClN$_3$O$_6$S$_2$
  Sporidesmin A, 8892
C$_{18}$H$_{20}$Cl$_2$N$_8$
  Bismark Brown Y, 1257
C$_{18}$H$_{20}$FN$_3$O$_4$
  Ofloxacin, 6863
C$_{18}$H$_{20}$FN$_5$O$_4$
  Gemifloxacin, 4423
C$_{18}$H$_{20}$F$_4$O$_3$
  Metofluthrin, 6222
C$_{18}$H$_{20}$N$_2$
  Mianserin, 6251

C$_{18}$H$_{20}$N$_2$O$_2$
  Vinconate, 10182
C$_{18}$H$_{20}$N$_2$O$_2$S
  Liranaftate, 5570
C$_{18}$H$_{20}$N$_2$O$_6$
  Nitrendipine, 6662
C$_{18}$H$_{20}$N$_2$S
  Methdilazine, 6036
C$_{18}$H$_{20}$N$_4$O$_2$
  Nitracrine, 6657
C$_{18}$H$_{20}$O$_2$
  Dianin's Compound, 2991
  Diethylstilbestrol, 3148
  Equilin, 3700
C$_{18}$H$_{20}$O$_5$
  Combretastatin A-4, 2477
C$_{18}$H$_{21}$ClN$_2$
  Chlorcyclizine, 2082
C$_{18}$H$_{21}$ClO$_4$
  Barthrin, 1000
C$_{18}$H$_{21}$FINO$_2$
  IACFT, 4910
C$_{18}$H$_{21}$KN$_2$O$_5$S
  Propicillin Potassium Salt, 7931
C$_{18}$H$_{21}$NO
  Pipradrol, 7597
C$_{18}$H$_{21}$NO$_4$
  Cephalotaxine, 1980
  Oxycodone, 7058
C$_{18}$H$_{21}$NO$_5$
  Protokylol, 8004
  Tazettine, 9219
C$_{18}$H$_{21}$NO$_6$
  Naproxcinod, 6498
C$_{18}$H$_{21}$N$_3$O
  Dibenzepin, 3012
C$_{18}$H$_{21}$N$_3$O$_3$S
  Rabeprazole, 8206
C$_{18}$H$_{21}$N$_3$O$_5$S$_2$
  Prinomastat, 7869
C$_{18}$H$_{21}$N$_5$O$_2$
  Alogliptin, 302
C$_{18}$H$_{21}$N$_5$O$_2$S
  Amocarzine, 568
C$_{18}$H$_{21}$O$_8$P
  Fosbretabulin, 4277
C$_{18}$H$_{22}$ClNO
  Chlorphenoxamine, 2187
  Phenoxybenzamine, 7368
C$_{18}$H$_{22}$ClNO$_3$
  Cloquintocet-mexyl, 2390
C$_{18}$H$_{22}$I$_3$N$_3$O$_8$
  Metrizamide, 6234
C$_{18}$H$_{22}$N$_2$
  Cyclizine, 2703
  Desipramine, 2921
C$_{18}$H$_{22}$N$_2$O$_2$
  Carazolol, 1780
C$_{18}$H$_{22}$N$_2$O$_2$S
  Oxomemazine, 7045
C$_{18}$H$_{22}$N$_2$O$_5$S
  Propicillin, 7931
C$_{18}$H$_{22}$N$_2$S
  Diethazine, 3119
  Trimeprazine, 9873
C$_{18}$H$_{22}$N$_4$Na$_4$O$_{23}$P$_4$
  Diquafosol Tetrasodium Salt, 3391
C$_{18}$H$_{22}$N$_4$O$_7$
  Galactoflavin, 4363
C$_{18}$H$_{22}$O
  4-Methylbenzylidene Camphor, 6101
C$_{18}$H$_{22}$O$_2$
  Dihydroequilin, 3190
  Estrone, 3763
  Hexestrol, 4737

8-Isoestrone, 5213
Trenbolone, 9745

C₁₈H₂₂O₄
Nordihydroguaiaretic Acid, 6782

C₁₈H₂₂O₅
Zearalenone, 10314

C₁₈H₂₂O₆
Combretastatin, 2477

C₁₈H₂₂O₈P₂
Fosfestrol, 4280

C₁₈H₂₂O₁₁
Asperuloside, 834

C₁₈H₂₃ClN₂O₂
Phenacaine Hydrochloride, 7317

C₁₈H₂₃FINO₂
Ioflupane I 123, 5093

C₁₈H₂₃N
Tolpropamine, 9682

C₁₈H₂₃NO
Bifemelane, 1210
p-Methyldiphenhydramine, 6125
Moxastine, 6373
Orphenadrine, 6975

C₁₈H₂₃NO₃
Dihydrocodeine, 3189
Dihydroisocodeine, 3193
Dobutamine, 3440
Isoxsuprine, 5285
Ractopamine, 8208

C₁₈H₂₃NO₄
Arbutamine, 762
Cocaethylene, 2439
Denopamine, 2896

C₁₈H₂₃NO₅
Seneciphylline, 8591

C₁₈H₂₃N₃O₄
Pefurazoate, 7173

C₁₈H₂₃N₃O₅S
Acediasulfone Morpholine Salt, 24

C₁₈H₂₃N₅NaO₈P
Bucladesine Sodium Salt, 1473

C₁₈H₂₃N₅Na₄O₂₁P₄
Denufosol Tetrasodium Salt, 2898

C₁₈H₂₃N₅O₂
Fenethylline, 4003

C₁₈H₂₃N₅O₅
Reproterol, 8259

C₁₈H₂₃N₉O₄S₃
Cefotiam, 1936

C₁₈H₂₄BrNO₃S
Bretylium Tosylate, 1377

C₁₈H₂₄BrNO₄
Methscopolamine Bromide, 6077

C₁₈H₂₄ClN₃O
Tebufenpyrad, 9231

C₁₈H₂₄FN₃O₃S
Benthiavalicarb-isopropyl, 1055

C₁₈H₂₄INO₂S
Tiemonium Iodide, 9587

C₁₈H₂₄I₃N₃O₈
Iopromide, 5105
Ioxilan, 5111

C₁₈H₂₄I₃N₃O₉
Ioversol, 5109

C₁₈H₂₄N₂O₄
Isoxaben, 5281

C₁₈H₂₄N₂O₅·2H₂O
Enalapril Diacid Dihydrate, 3623

C₁₈H₂₄N₂O₅S
Amosulalol, 572

C₁₈H₂₄N₂O₆
Dinocap, 3314

C₁₈H₂₄N₄O
Granisetron, 4572

C₁₈H₂₄N₅O₈P
Bucladesine, 1473

C₁₈H₂₄O₂
Estradiol, 3758
α-Estradiol, 3759
Isoestradiol, 5212
Nimbiol, 6632

C₁₈H₂₄O₃
Doisynolic Acid, 3457
16-Epiestriol, 3671
Estriol, 3762

C₁₈H₂₄O₈P₂
Hexestrol Diphosphate, 4737

C₁₈H₂₅N
Dimemorfan, 3229

C₁₈H₂₅NO
Cyclazocine, 2698
Dextromethorphan, 2957
Levomethorphan, 5520

C₁₈H₂₅NO₂
Allylprodine, 287

C₁₈H₂₅NO₅
Senecionine, 8590

C₁₈H₂₅NO₆
Retrorsine, 8290

C₁₈H₂₅NO₇
Retrorsine *N*-Oxide, 8290

C₁₈H₂₅N₃O₂
Saxagliptin, 8525

C₁₈H₂₆ClN₃
Chloroquine, 2165

C₁₈H₂₆ClN₃O
Hydroxychloroquine, 4857

C₁₈H₂₆ClN₃O₃
Prucalopride, 8015

C₁₈H₂₆I₃N₃O₉
Iothalamic Acid *N*-Methylglucamine
Salt, 5107

C₁₈H₂₆N₂O₂
DPQ, 3491

C₁₈H₂₆N₂O₄
Proglumide, 7891

C₁₈H₂₆N₂O₄S
Glibornuride, 4473

C₁₈H₂₆N₂O₅
Disofenin, 3400

C₁₈H₂₆N₄O₆S
Cetotiamine, 2022

C₁₈H₂₆N₄O₂₃P₄
Diquafosol, 3391

C₁₈H₂₆N₁₀O₃
Netropsin, 6566

C₁₈H₂₆O
HHCB, 4751
Versalide®, 10161
Xibornol, 10273

C₁₈H₂₆O₂
Cinmethylin, 2296
Isanic Acid, 5147
Nandrolone, 6450

C₁₈H₂₆O₃
Octyl Methoxycinnamate, 6859

C₁₈H₂₆O₄
Isoamyl Phthalate, 5169

C₁₈H₂₆O₅
Zeranol, 10319

C₁₈H₂₇NO₂
Caramiphen, 1778
Dyclonine, 3520

C₁₈H₂₇NO₃
Capsaicin, 1770

C₁₈H₂₇NO₅
Platyphylline, 7646
Propanidid, 7917

C₁₈H₂₇NO₆
Hygrophylline, 4893

C₁₈H₂₇N₅O₂₁P₄
Denufosol, 2898

C₁₈H₂₈N₂O
Bupivacaine, 1499

C₁₈H₂₈N₂O₃
Iprovalicarb, 5124

C₁₈H₂₈N₂O₄
Acebutolol, 20

C₁₈H₂₈N₄O
Butalamine, 1513

C₁₈H₂₈O₂
Bolandiol, 1325
Parinaric Acid, 7144
Stearidonic Acid, 8931

C₁₈H₂₈O₃S
Sulfoxide, 9097

C₁₈H₂₈P₂
(*R*,*R*)-Me-DuPHOS, 3513

C₁₈H₂₉NO₂
Penbutolol, 7191

C₁₈H₂₉NO₃
Betaxolol, 1184
Butamirate, 1516

C₁₈H₂₉N₃O₅
Bambuterol, 946

C₁₈H₂₉NaO₃S
Sodium Dodecylbenzenesulfonate,
8748

C₁₈H₃₀N₂O₂
Butacaine, 1510

C₁₈H₃₀O₂
Linolenic Acid, 5561
γ-Linolenic Acid, 5562
Pinolenic Acid, 7562

C₁₈H₃₀O₃
Helenynolic Acid, 4662
Juvenile Hormone I, 5321

C₁₈H₃₁GdN₄O₉
Gadobutrol, 4353

C₁₈H₃₁NO₄
Bisoprolol, 1295

C₁₈H₃₂CaN₂O₁₀
Pantothenic Acid Calcium Salt, 7118

C₁₈H₃₂I₂O₃
Periodyl, 7287

C₁₈H₃₂O
Grundmann's Ketone, 4587

C₁₈H₃₂O₂
Chaulmoogric Acid, 2047
Gossyplure, 4563
Linoleic Acid, 5560
Propylure, 7981

C₁₈H₃₂O₃
Vernolic Acid, 10159

C₁₈H₃₂O₇
Butyl Citrate, 1565

C₁₈H₃₂O₁₆
Melezitose, 5887
Raffinose, 8213

C₁₈H₃₃ClN₂O₅S
Clindamycin, 2354

C₁₈H₃₃NaO₂
Oleic Acid Sodium Salt, 6921

C₁₈H₃₃NaO₃
Ricinoleic Acid Sodium Salt, 8335

C₁₈H₃₄ClN₂O₈PS
Clindamycin 2-Dihydrogen Phosphate, 2354

C₁₈H₃₄N₂O₆S
Lincomycin, 5555

C₁₈H₃₄OSn
Cyhexatin, 2757

C₁₈H₃₄O₂
Elaidic Acid, 3581
Hexalure, 4723
Oleic Acid, 6921
Petroselinic Acid, 7304
Vaccenic Acid, 10084

$C_{18}H_{34}O_3$
  Ricinoleic Acid, 8335
  Rosaprostol, 8391

$C_{18}H_{34}O_6$
  Phloionic Acid, 7436
  Sorbitan Laurate, 8850

$C_{18}H_{34}O_6S$
  Ricinoleic Acid Acid Sulfate, 8335

$C_{18}H_{35}KO_2$
  Potassium Stearate, 7792

$C_{18}H_{35}NO$
  Dodemorph, 3454
  Laurocapram, 5440

$C_{18}H_{35}NO_2$
  Spiroxamine, 8889

$C_{18}H_{35}NaO_2$
  Sodium Stearate, 8806

$C_{18}H_{36}N_2O_6$
  Cryptand 222, 2596

$C_{18}H_{36}N_4O_{10}$
  Gentamicin A, 4427

$C_{18}H_{36}N_4O_{11}$
  Kanamycin A, 5329

$C_{18}H_{36}O$
  Oleyl Alcohol, 6924

$C_{18}H_{36}O_2$
  Stearic Acid, 8930

$C_{18}H_{36}O_4$
  9,10-Dihydroxystearic Acid, 3204

$C_{18}H_{36}P_2$
  (R,R)-Et-BPE, 1363

$C_{18}H_{37}NO_2$
  Ammonium Oleate, 532
  Palmidrol, 7096
  Sphingosine, 8874

$C_{18}H_{37}N_5O_8$
  Dibekacin, 3008

$C_{18}H_{37}N_5O_9$
  Tobramycin, 9647

$C_{18}H_{38}O$
  Stearyl Alcohol, 8932

$C_{18}H_{39}NO_2$
  Ammonium Stearate, 550

$C_{18}H_{39}N_4P$
  Triisobutyl-Verkade's Superbase, 10155

$C_{18}H_{40}O_4P_2$
  Tetrofosmin, 9395

$C_{19}H_9Br_4NaO_5S$
  Bromphenol Blue Sodium Salt, 1454

$C_{19}H_9HgI_2NaO_7S$
  Meralein Sodium, 5932

$C_{19}H_{10}Br_4O_5S$
  Bromphenol Blue, 1454

$C_{19}H_{11}Cl_2I_2NO_3$
  Rafoxanide, 8214

$C_{19}H_{11}F_5N_2O_2$
  Diflufenican, 3163

$C_{19}H_{12}F_3N_3O_3S$
  Zopolrestat, 10394

$C_{19}H_{12}F_4N_2O_2$
  Picolinafen, 7511

$C_{19}H_{12}O_6$
  Dicumarol, 3100

$C_{19}H_{12}O_8$
  Diacerein, 2963

$C_{19}H_{13}NaO_5S$
  Phenolsulfonphthalein Sodium Salt, 7360

$C_{19}H_{14}BrClN_6O_2$
  Cyantraniliprole, 2689

$C_{19}H_{14}Cl_2N_2O_3S$
  Vismodegib, 10207

$C_{19}H_{14}NO_4^+$
  Coptisine, 2512

$C_{19}H_{14}N_4O$
  1,3-Di-6-quinolylurea, 3393

$C_{19}H_{14}O_3$
  Aurin, 872

$C_{19}H_{14}O_5S$
  Phenolsulfonphthalein, 7360

$C_{19}H_{15}Cl$
  Trityl Chloride, 9944

$C_{19}H_{15}ClNNaO_4.3H_2O$
  Indomethacin Sodium Salt Trihydrate, 5009

$C_{19}H_{15}ClN_2O_4$
  Rebamipide, 8242

$C_{19}H_{15}ClN_4$
  Triphenyltetrazolium Chloride, 9918

$C_{19}H_{15}ClO_4$
  Coumachlor, 2544

$C_{19}H_{15}FN_2O_4$
  Flumioxazin, 4174

$C_{19}H_{15}F_3N_2O_3$
  Tecovirimat, 9244

$C_{19}H_{15}F_3N_4O_3$
  Tosufloxacin, 9714

$C_{19}H_{15}NNa_2O_7$
  Nedocromil Disodium Salt, 6521

$C_{19}H_{15}NO_2$
  Cinchophen Allyl Ester, 2291

$C_{19}H_{15}NO_6$
  Acenocoumarol, 30

$C_{19}H_{15}NaO_4$
  Warfarin Sodium Salt, 10236

$C_{19}H_{16}$
  Triphenylmethane, 9914

$C_{19}H_{16}ClFN_3NaO_5S.H_2O$
  Floxacillin Sodium Monohydrate, 4143

$C_{19}H_{16}ClNO_4$
  Indomethacin, 5009

$C_{19}H_{16}Cl_2N_3NaO_5S.H_2O$
  Dicloxacillin Sodium Salt Monohydrate, 3094

$C_{19}H_{16}N_2$
  Sempervirine, 8586

$C_{19}H_{16}N_2Na_2O_7S_2$
  Ponceau 3R, 7712

$C_{19}H_{16}N_2O$
  Difenpiramide, 3156

$C_{19}H_{16}N_5NaO_7S_3$
  Ceftiofur Monosodium Salt, 1951

$C_{19}H_{16}N_6O_4S$
  Zibotentan, 10320

$C_{19}H_{16}O$
  Triphenylcarbinol, 9912

$C_{19}H_{16}O_4$
  Warfarin, 10236

$C_{19}H_{17}ClFN_3O_5S$
  Floxacillin, 4143

$C_{19}H_{17}ClN_2NaO_3$
  Laquinimod Sodium Salt, 5422

$C_{19}H_{17}ClN_2O$
  Prazepam, 7836

$C_{19}H_{17}ClN_2O_3$
  Laquinimod, 5422

$C_{19}H_{17}ClN_2O_4$
  Oxametacine, 7015
  Quizalofop-ethyl, 8203

$C_{19}H_{17}ClN_3NaO_5S.H_2O$
  Cloxacillin Sodium Monohydrate, 2403

$C_{19}H_{17}ClN_4$
  Fenbuconazole, 3995

$C_{19}H_{17}Cl_2NO_4$
  Cinidon-ethyl, 2294

$C_{19}H_{17}Cl_2N_3O_3$
  Difenoconazole, 3154

$C_{19}H_{17}Cl_2N_3O_5S$
  Dicloxacillin, 3094

$C_{19}H_{17}Cl_3N_2S$
  Butoconazole, 1531

$C_{19}H_{17}NOS$
  Tolnaftate, 9678

$C_{19}H_{17}NO_4$
  Stylopine, 8988

$C_{19}H_{17}NO_5$
  Mofezolac, 6318

$C_{19}H_{17}NO_7$
  Nedocromil, 6521

$C_{19}H_{17}N_2NaO_4S$
  Parecoxib Sodium Salt, 7140

$C_{19}H_{17}N_2S^+$
  Thiazole Orange, 9460

$C_{19}H_{17}N_3O$
  DCM, 2841
  Evodiamine, 3951

$C_{19}H_{17}N_5O_2$
  Nafamostat, 6434

$C_{19}H_{17}N_5O_5S$
  Salazosulfadimidine, 8457

$C_{19}H_{17}N_5O_7S_3$
  Ceftiofur, 1951

$C_{19}H_{17}N_6NaO_5S_3.1\frac{1}{2}H_2O$
  Cefditoren Sodium Salt, Sesquihydrate, 1922

$C_{19}H_{17}N_6NaO_6S_2$
  Cefamandole Nafate, 1916

$C_{19}H_{17}N_9O_5S_2$
  Cefozopran, 1939

$C_{19}H_{17}P$
  Methylenetriphenylphosphorane, 6138

$C_{19}H_{18}CaN_2O_9$
  Calcium Acetylsalicylate Complex with Urea, 1650

$C_{19}H_{18}ClF_2N_3O_3$
  Sitafloxacin, 8689

$C_{19}H_{18}ClN_3O_3$
  Camazepam, 1727

$C_{19}H_{18}ClN_3O_4$
  Pyraclostrobin, 8064

$C_{19}H_{18}ClN_3O_5S$
  Cloxacillin, 2403
  Rivaroxaban, 8368

$C_{19}H_{18}ClN_5$
  Adinazolam, 148

$C_{19}H_{18}N_2O_3$
  Kebuzone, 5336

$C_{19}H_{18}N_2O_4S$
  Parecoxib, 7140

$C_{19}H_{18}N_2O_5S$
  Zinquin, 10369

$C_{19}H_{18}N_3NaO_5S.H_2O$
  Oxacillin Sodium Salt Monohydrate, 7003

$C_{19}H_{18}N_4O_2$
  Pimobendan, 7545

$C_{19}H_{18}N_4O_6$
  Bispyribac, 1302

$C_{19}H_{18}N_6O_5S_3$
  Cefditoren, 1922

$C_{19}H_{18}N_6O_6$
  Nicarbazin, 6579

$C_{19}H_{18}N_7NaO_6$
  Folic Acid Sodium Salt, 4248

$C_{19}H_{18}O_7$
  Gardenin B, 4399

$C_{19}H_{18}O_8$
  Atranorin, 860
  Gardenin D, 4399

$C_{19}H_{18}O_9$
  Gardenin E, 4399

$C_{19}H_{19}ClN_2$
  Desloratadine, 2923

$C_{19}H_{19}FN_4O_3$
  Brivanib, 1383

$C_{19}H_{19}N$
  Phenindamine, 7347

$C_{19}H_{19}NOS$
Ketotifen, 5354
$C_{19}H_{19}NO_3$
Laureline, 5438
$C_{19}H_{19}NO_4$
Bulbocapnine, 1486
Domesticine, 3462
Nandinine, 6449
$C_{19}H_{19}N_2NaO_2$
Phenylbutazone Sodium Salt, 7390
$C_{19}H_{19}N_3O_3$
Talampanel, 9164
$C_{19}H_{19}N_3O_5S$
Oxacillin, 7003
$C_{19}H_{19}N_3O_6$
Nilvadipine, 6630
$C_{19}H_{19}N_7O_6$
Folic Acid, 4248
$C_{19}H_{20}Br_2N_6O_4S$
Macitentan, 5705
$C_{19}H_{20}ClNO_4$
Bezafibrate, 1199
$C_{19}H_{20}ClNO_5$
Romifibrate, 8386
$C_{19}H_{20}ClN_3$
Clemizole, 2344
$C_{19}H_{20}FNO_3$
Paroxetine, 7148
$C_{19}H_{20}FN_3O_3$
Danofloxacin, 2812
$C_{19}H_{20}F_3NO_2$
Benfluorex, 1041
$C_{19}H_{20}F_3NO_4$
Fluazifop-butyl, 4147
$C_{19}H_{20}F_3N_3O_3$
Niflumic Acid β-Morpholinoethyl Es-
ter, 6615
Orbifloxacin, 6957
$C_{19}H_{20}N_2$
Mebhydroline, 5839
$C_{19}H_{20}N_2O$
Vellosimine, 10137
$C_{19}H_{20}N_2O_2$
Phenylbutazone, 7390
$C_{19}H_{20}N_2O_3$
Ditazol, 3410
Dolasetron, 3458
Oxyphenbutazone, 7071
$C_{19}H_{20}N_2O_3S$
Pioglitazone, 7565
$C_{19}H_{20}N_2O_7$
Aranidipine, 756
$C_{19}H_{20}N_4O_2$
Plinabulin, 7654
$C_{19}H_{20}N_4O_2S_2$
Elesclomol, 3596
$C_{19}H_{20}N_6O$
Imidocarb, 4956
$C_{19}H_{20}N_8O_5$
Aminopterin, 468
$C_{19}H_{20}O_2$
Cyclofenil Free Diol, 2713
$C_{19}H_{20}O_9$
Cervicarcin, 2006
$C_{19}H_{21}BrO_5$
Metrafenone, 6230
$C_{19}H_{21}ClFN_3O_3$
Besifloxacin, 1178
$C_{19}H_{21}ClN_2OS$
Chloracizine, 2068
$C_{19}H_{21}ClN_4O_5$
Theofibrate, 9435
$C_{19}H_{21}FN_2O_4$
Nadifloxacin, 6430
$C_{19}H_{21}F_3N_2S$
Trifluomeprazine, 9845

$C_{19}H_{21}N$
Nortriptyline, 6804
Protriptyline, 8010
$C_{19}H_{21}NO$
Doxepin, 3483
$C_{19}H_{21}NO_3$
Isothebaine, 5272
Nalorphine, 6446
Thebaine, 9427
$C_{19}H_{21}NO_4$
Boldine, 1328
Corytuberine, 2537
Laurotetanine, 5443
Naloxone, 6447
Salutaridine, 8479
$C_{19}H_{21}NS$
Dothiepin, 3478
Pizotyline, 7629
$C_{19}H_{21}N_3$
Perlapine, 7293
$C_{19}H_{21}N_3O$
Zolpidem, 10391
$C_{19}H_{21}N_3O_5$
Isradipine, 5287
$C_{19}H_{21}N_3S$
Cyamemazine, 2673
$C_{19}H_{21}N_5O_2$
Pirenzepine, 7604
$C_{19}H_{21}N_5O_4$
Prazosin, 7838
$C_{19}H_{22}BrNO_4S_2$
Tiotropium Bromide, 9610
$C_{19}H_{22}ClN_5O$
Trazodone, 9740
$C_{19}H_{22}FN_3O$
Azaperone, 891
$C_{19}H_{22}FN_3O_3$
Enrofloxacin, 3646
Grepafloxacin, 4580
$C_{19}H_{22}FN_3O_4$
Gatifloxacin, 4409
$C_{19}H_{22}F_2N_4O_3$
Sparfloxacin, 8862
$C_{19}H_{22}KN_3O_4S$
Hetacillin Potassium Salt, 4707
$C_{19}H_{22}N_2$
Triprolidine, 9922
$C_{19}H_{22}N_2O$
Cinchonidine, 2289
Cinchonine, 2290
Cinchotoxine, 2292
Eburnamonine, 3534
$C_{19}H_{22}N_2OS$
Acepromazine, 33
Thiazesim, 9457
$C_{19}H_{22}N_2O_2$
Cupreine, 2613
Sarpagine, 8515
Wieland-Gumlich Aldehyde, 10242
$C_{19}H_{22}N_2O_3$
Bumadizone, 1488
Dimoxystrobin, 3292
$C_{19}H_{22}N_2O_6S$
Penamecillin, 7190
$C_{19}H_{22}N_4O_3$
Iclaprim, 4926
$C_{19}H_{22}N_8O_6S_2$
Cefoselis, 1933
$C_{19}H_{22}O_2$
Diethylstilbestrol Monomethyl Ether,
3148
Vedaprofen, 10133
$C_{19}H_{22}O_3$
Ostruthin, 6994
$C_{19}H_{22}O_6$
Gibberellic Acid, 4454
Strigol, 8963

$C_{19}H_{22}O_7$
Samaderin B, 8482
$C_{19}H_{23}ClN_2$
Clomipramine, 2378
Homochlorcyclizine, 4771
$C_{19}H_{23}ClN_2O_2S$
Pyridate, 8079
$C_{19}H_{23}ClN_2S$
Chlorproethazine, 2189
$C_{19}H_{23}NO$
Diphenylpyraline, 3370
$C_{19}H_{23}NO_3$
Armepavine, 778
Dihydrothebaine, 3199
Ethylmorphine, 3884
Oxyfedrine, 7061
Reboxetine, 8244
$C_{19}H_{23}NO_4$
Reticuline, 8286
Sinomenine, 8684
$C_{19}H_{23}NO_5$
Metaphanine, 6000
Tretoquinol, 9749
$C_{19}H_{23}N_3$
Amitraz, 482
$C_{19}H_{23}N_3O$
Benzydamine, 1125
$C_{19}H_{23}N_3O_2$
Ergometrinine, 3711
Ergonovine, 3712
$C_{19}H_{23}N_3O_4S$
Hetacillin, 4707
$C_{19}H_{23}N_4O_6PS$
Benfotiamine, 1043
$C_{19}H_{23}N_5O_3$
Trimetrexate, 9893
$C_{19}H_{23}N_5O_3S$
Ipsapirone, 5125
$C_{19}H_{24}BrNS_2$
Tiquizium Bromide, 9616
$C_{19}H_{24}N_2$
Bamipine, 949
Imipramine, 4962
$C_{19}H_{24}N_2O$
Aminopentamide, 455
Cinchonamine, 2288
Hydrocinchonidine, 4820
Hydrocinchonine, 4821
Imipramine N-Oxide, 4963
Palonosetron, 7098
$C_{19}H_{24}N_2OS$
Levomepromazine, 5518
$C_{19}H_{24}N_2O_2$
Praziquantel, 7837
$C_{19}H_{24}N_2O_3$
Labetalol, 5377
$C_{19}H_{24}N_2O_4$
Formoterol, 4272
$C_{19}H_{24}N_2O_4S_2$
Omapatrilat, 6937
$C_{19}H_{24}N_2O_6$
Ketorolac (±)-Form Tromethamine
Salt, 5353
$C_{19}H_{24}N_2S$
Ethopropazine, 3800
$C_{19}H_{24}N_4O_2$
Pentamidine, 7227
$C_{19}H_{24}N_6O_5S_2$
Cefepime, 1923
$C_{19}H_{24}O_2$
Methyltrienolone, 6203
$C_{19}H_{24}O_3$
Adrenosterone, 165
Prallethrin, 7823
Testolactone, 9323
$C_{19}H_{24}O_5$
Trichothecin, 9815

$C_{20}H_{10}Br_4O_4$
  Tetrabromophenolphthalein, 9331
$C_{20}H_{10}Cl_2O_5$
  2′,7′-Dichlorofluorescein, 3076
  4′,5′-Dichlorofluorescein, 3077
$C_{20}H_{10}Cl_4O_4$
  Phenoltetrachlorophthalein, 7361
$C_{20}H_{10}I_2O_5$
  4′,5′-Diiodofluorescein, 3207
$C_{20}H_{10}I_4O_4$
  Iodophthalein Sodium Free Acid,
    5081
$C_{20}H_{10}Na_2O_5$
  Fluorescein Disodium Salt, 4192
$C_{20}H_{10}Na_2O_7$
  Gallein Disodium Salt, 4374
$C_{20}H_{11}N_2Na_3O_{10}S_3$
  Amaranth (Dye), 369
$C_{20}H_{12}$
  Benzo[a]pyrene, 1106
  Benzo[e]pyrene, 1107
  Perylene, 7299
$C_{20}H_{12}Li_2N_2O_{10}S_3$
  Lucifer Yellow VS, 5653
$C_{20}H_{12}N_3NaO_7S$
  Eriochrome® Black T, 3725
$C_{20}H_{12}O_5$
  Fluorescein, 4192
$C_{20}H_{12}O_7$
  Gallein, 4374
$C_{20}H_{13}N_3O_3$
  Violacein, 10193
$C_{20}H_{14}$
  Cholanthrene, 2204
  Triptycene, 9925
$C_{20}H_{14}ClNO_4$
  Sanguinarine Chloride, 8493
$C_{20}H_{14}I_6N_2O_6$
  Iodipamide, 5065
$C_{20}H_{14}NO_4^+$
  Sanguinarine, 8493
$C_{20}H_{14}N_4$
  Porphine, 7720
$C_{20}H_{14}O_2$
  BINOL, 1230
$C_{20}H_{14}O_2S_2$
  DDD (Analytical), 2842
$C_{20}H_{14}O_4$
  Diphenyl Phthalate, 3368
  Phenolphthalein, 7356
$C_{20}H_{14}O_5$
  Fluorescin, 4193
$C_{20}H_{14}O_6$
  Ethylidene Dicoumarol, 3867
$C_{20}H_{15}BrN_6O$
  Etravirine, 3935
$C_{20}H_{15}ClFN_5O_5S_2$
  Elinogrel, 3600
$C_{20}H_{15}ClN_4$
  Vatalanib, 10130
$C_{20}H_{15}Cl_3N_2OS$
  Sertaconazole, 8604
$C_{20}H_{15}F_3N_4O_3$
  Trovafloxacin, 9967
$C_{20}H_{15}N$
  β,β′-Dinaphthylamine, 3293
$C_{20}H_{15}N_3O_6$
  Rubitecan, 8421
$C_{20}H_{15}N_4NaO_6S$
  Zincon Sodium Salt, 10345
$C_{20}H_{16}$
  9,10-Dimethyl-1,2-benzanthracene,
    3257
$C_{20}H_{16}N_2O_4$
  Camptothecin, 1737
$C_{20}H_{16}N_4$
  Nitron, 6700

$C_{20}H_{16}N_4O_2S$
  Indiplon, 4984
$C_{20}H_{16}N_4O_6S$
  Zincon, 10345
$C_{20}H_{16}O_6$
  Elliptone, 3605
  Viridin, 10202
$C_{20}H_{17}ClN_2O_3$
  Ketazolam, 5345
$C_{20}H_{17}ClO_3$
  Indanofan, 4973
$C_{20}H_{17}Cl_3N_2O_2$
  Omoconazole, 6941
$C_{20}H_{17}FO_3S$
  Sulindac, 9112
$C_{20}H_{17}FO_4S$
  Exisulind, 3956
$C_{20}H_{17}F_2N_3O_3$
  Sarafloxacin, 8507
$C_{20}H_{17}F_3N_2O_4$
  Floctafenine, 4135
  Tasquinimod, 9210
$C_{20}H_{17}F_5N_2O_2$
  Cyflufenamid, 2753
$C_{20}H_{17}NO_6$
  Adlumidine, 154
  Bicuculline, 1206
$C_{20}H_{17}N_3Na_2O_9S_3$
  Acid Fuchsin, 99
$C_{20}H_{17}N_3O_4$
  9-Aminocamptothecin, 426
$C_{20}H_{18}ClNO_6$
  Ochratoxin A, 6829
$C_{20}H_{18}NO_4^+$
  Berberine, 1156
$C_{20}H_{18}N_2O_2$
  Nile Red, 6628
$C_{20}H_{18}N_4O_5S_2$
  Cephalonium, 1977
$C_{20}H_{18}N_6Na_2O_7S_4$
  Cefodizime Disodium Salt, 1929
$C_{20}H_{18}N_6Na_2O_9S$
  Moxalactam Disodium Salt, 6372
$C_{20}H_{18}O_2Sn$
  Triphenyltin Hydroxide Acetate, 9920
$C_{20}H_{18}O_3$
  Phenolphthalol, 7358
$C_{20}H_{18}O_4$
  Phaseolin, 7313
$C_{20}H_{18}O_6$
  Asarinin, 812
  Sesamin, 8609
$C_{20}H_{18}O_7$
  Sesamolin, 8610
$C_{20}H_{18}O_9$
  Frangulin B, 4292
$C_{20}H_{19}FN_4O_4$
  Finafloxacin, 4112
$C_{20}H_{19}FN_8O_2$
  Riociguat, 8358
$C_{20}H_{19}F_2N_3O_5$
  Dolutegravir, 3460
$C_{20}H_{19}F_3N_2O_4$
  Trifloxystrobin, 9843
$C_{20}H_{19}NO_3$
  Pyriproxyfen, 8103
$C_{20}H_{19}NO_5$
  Chelidonine, 2052
  Papaveraldine, 7120
  Protopine, 8005
$C_{20}H_{19}NO_6$
  Ochratoxin B, 6829
$C_{20}H_{19}NO_8S$
  Berberine Acid Sulfate, 1156
$C_{20}H_{20}$
  Dodecahedrane, 3447

$C_{20}H_{20}ClN_3$
  Magenta I, 5715
$C_{20}H_{20}ClN_3O_3$
  Tepoxalin, 9294
$C_{20}H_{20}Cl_2MgO_6$
  Clofibric Acid Magnesium Salt, 2371
$C_{20}H_{20}FKN_6O_5$
  Raltegravir Monopotassium Salt,
    8216
$C_{20}H_{20}FNO_3S$
  Prasugrel, 7833
$C_{20}H_{20}FNO_4$
  Cyhalofop-butyl, 2755
$C_{20}H_{20}NO_4^+$
  Columbamine, 2475
  Jatrorrhizine, 5309
$C_{20}H_{20}N_2O_2$
  Feprazone, 4039
$C_{20}H_{20}N_2O_3$
  Elliptinium Acetate, 3604
$C_{20}H_{20}N_2O_5$
  Antipyrine Acetylsalicylate, 703
$C_{20}H_{20}N_2O_6$
  Antipyrine Salicylacetate, 703
$C_{20}H_{20}N_6O_7S_4$
  Cefodizime, 1929
$C_{20}H_{20}N_6O_9S$
  Moxalactam, 6372
$C_{20}H_{20}N_8Na_2O_5$
  Methotrexate Disodium Salt, 6057
$C_{20}H_{20}O_4$
  Otobain, 6998
$C_{20}H_{20}O_6$
  Cubebin, 2603
$C_{20}H_{20}O_7$
  Herqueinone, 4703
$C_{20}H_{20}O_9$
  Gardenin C, 4399
$C_{20}H_{20}O_{14}$
  Hamamelitannin, 4640
$C_{20}H_{21}AlCl_2O_7$
  Clofibric Acid Basic Aluminum Salt,
    2371
$C_{20}H_{21}CaN_7O_7.5H_2O$
  Folinic Acid Calcium Salt Pentahy-
    drate, 4249
$C_{20}H_{21}ClN_2O_4$
  Fipexide, 4115
$C_{20}H_{21}ClO_4$
  Fenofibrate, 4009
$C_{20}H_{21}FN_2O$
  Citalopram, 2317
$C_{20}H_{21}FN_6O_5$
  Raltegravir, 8216
$C_{20}H_{21}F_3N_2OS$
  Fluacizine, 4145
$C_{20}H_{21}F_3N_4O$
  Flibanserin, 4132
$C_{20}H_{21}N$
  Cyclobenzaprine, 2706
$C_{20}H_{21}NOS$
  Tolciclate, 9669
$C_{20}H_{21}NO_2$
  Moxaverine, 6374
$C_{20}H_{21}NO_3$
  Dimefline, 3227
  Mepixanox, 5927
$C_{20}H_{21}NO_4$
  Canadine, 1738
  Dicentrine, 3051
  Papaverine, 7122
$C_{20}H_{21}NO_5$
  Repirinast, 8258
$C_{20}H_{21}N_3$
  Aptiganel, 747
$C_{20}H_{21}N_3O_7S$
  Ampiroxicam, 584

$C_{20}H_{21}N_5O_6$
Pemetrexed, 7186
$C_{20}H_{21}N_7O_6$
Methopterin, 6056
Ninopterin, 6641
$C_{20}H_{21}N_7O_6S_2$
Ceforanide, 1932
$C_{20}H_{22}ClN$
Pyrrobutamine, 8126
$C_{20}H_{22}ClN_3O$
Amodiaquin, 569
$C_{20}H_{22}N_2$
Azatadine, 894
$C_{20}H_{22}N_2O$
Fenazaquin, 3993
$C_{20}H_{22}N_2O_2$
Akuammicine, 192
Gelsemine, 4418
Quininone, 8179
$C_{20}H_{22}N_2O_3$
Perivine, 7292
$C_{20}H_{22}N_2O_7S$
Sivelestat, 8696
$C_{20}H_{22}N_2S$
Mequitazine, 5931
$C_{20}H_{22}N_4O_5$
Camostat, 1730
$C_{20}H_{22}N_4O_6S$
Febantel, 3981
$C_{20}H_{22}N_4O_{10}S$
Cefuroxime 1-Acetoxyethyl Ester, 1956
$C_{20}H_{22}N_5NaO_6S$
Azlocillin Sodium Salt, 908
$C_{20}H_{22}N_8O_5$
Methotrexate, 6057
$C_{20}H_{22}N_8O_6S_2$
Ceftobiprole, 1953
$C_{20}H_{22}O_2$
Norgestrienone, 6794
$C_{20}H_{22}O_3$
Avobenzone, 881
$C_{20}H_{22}O_6$
Columbin, 2476
$C_{20}H_{22}O_8$
Populin, 7716
$C_{20}H_{22}O_{10}$
Plicatic Acid, 7653
$C_{20}H_{23}N$
Amitriptyline, 483
Maprotiline, 5813
$C_{20}H_{23}NO$
Amitriptylinoxide, 484
$C_{20}H_{23}NO_2$
Dioxadrol, 3327
$C_{20}H_{23}NO_3$
Benalaxyl, 1032
$C_{20}H_{23}NO_4$
Corydine, 2533
Corypalmine, 2536
Efaproxiral, 3568
Isocorydine, 5207
Isocorypalmine, 5208
Naltrexone, 6448
Thebacon, 9426
$C_{20}H_{23}NS$
Methixene, 6051
$C_{20}H_{23}N_3O_2$
Bitertanol, 1306
Phenserine, 7372
$C_{20}H_{23}N_5O_6$
Azlocillin, 908
$C_{20}H_{23}N_7O_7$
Folinic Acid, 4249
$C_{20}H_{24}BrN_3O$
Bromolysergide, 1433

$C_{20}H_{24}ClNO$
Cloperastine, 2384
$C_{20}H_{24}ClN_3O_2$
Clebopride, 2342
$C_{20}H_{24}ClN_3S$
Prochlorperazine, 7879
$C_{20}H_{24}Cl_2N_{10}$
Picloxydine, 7510
$C_{20}H_{24}FN_3O_4$
Balofloxacin, 937
$C_{20}H_{24}NO_4^+$
Magnoflorine, 5760
$C_{20}H_{24}N_2$
Dimethindene, 3238
$C_{20}H_{24}N_2O$
Indecainide, 4977
$C_{20}H_{24}N_2OS$
Lucanthone, 5650
Propiomazine, 7936
Propionylpromazine, 7943
$C_{20}H_{24}N_2O_2$
Epiquinidine, 3675
Epiquinine, 3676
Quinidine, 8176
Quinine, 8177
Viquidil, 10199
$C_{20}H_{24}N_2O_2S$
Hycanthone, 4798
$C_{20}H_{24}N_2O_4$
Metoquinone, 6229
$C_{20}H_{24}N_2O_6$
Nisoldipine, 6651
$C_{20}H_{24}N_2S_2$
Metitepine, 6217
$C_{20}H_{24}N_4O_4$
Istradefylline, 5289
$C_{20}H_{24}O_2$
Dimestrol, 3234
Ethinyl Estradiol, 3788
Exemestane, 3954
$C_{20}H_{24}O_3$
Estrone Acetate, 3763
Trenbolone Acetate, 9745
$C_{20}H_{24}O_4$
Crocetin, 2579
$C_{20}H_{24}O_6$
Dibenzo-18-crown-6, 2592
Triptolide, 9923
$C_{20}H_{24}O_7$
Olivil, 6931
$C_{20}H_{24}O_9$
Ginkgolide A, 4461
Nodakenin, 6755
$C_{20}H_{24}O_{10}$
Ginkgolide B, 4461
$C_{20}H_{24}O_{11}$
Ginkgolide C, 4461
$C_{20}H_{25}ClN_2O_5$
Amlodipine, 487
$C_{20}H_{25}ClO_4$
Osaterone, 6985
$C_{20}H_{25}FN_4O$
Niaprazine, 6577
$C_{20}H_{25}{}^{18}FN_2O_3$
Florbetapir F 18, 4140
$C_{20}H_{25}NO$
Normethadone, 6798
Pridinol, 7859
$C_{20}H_{25}NOS$
Thiphenamil, 9524
$C_{20}H_{25}NO_2$
Adiphenine, 149
Dienogest, 3117
Femoxetine, 3989
Fomocaine, 4257
$C_{20}H_{25}NO_2S_2$
Tiagabine, 9572

$C_{20}H_{25}NO_3$
Benactyzine, 1031
Difemerine, 3152
$C_{20}H_{25}NO_4$
Cilomilast, 2279
Codamine, 2447
Laudanidine, 5434
Laudanine, 5435
$C_{20}H_{25}N_3O$
Lysergide, 5695
Prodigiosin, 7884
$C_{20}H_{25}N_3O_2$
Methylergonovine, 6142
$C_{20}H_{25}N_3S$
Perazine, 7265
$C_{20}H_{25}N_5O_7S_2$
Cefetamet Pivaloyloxymethyl Ester, 1924
$C_{20}H_{26}ClNO_5$
Cloricromen, 2393
$C_{20}H_{26}N_2$
Trimipramine, 9894
$C_{20}H_{26}N_2O$
Ibogaine, 4914
Tabernanthine, 9150
$C_{20}H_{26}N_2O_2$
Ajmaline, 187
Epsiprantel, 3694
Hydroquinidine, 4843
Hydroquinine, 4844
$C_{20}H_{26}N_2O_4$
Itopride, 5291
$C_{20}H_{26}N_2O_5S$
Alacepril, 193
$C_{20}H_{26}N_2O_5S_2$
Spirapril Diacid, 8880
$C_{20}H_{26}N_4O$
Lisuride, 5574
$C_{20}H_{26}N_4OS$
Oxypendyl, 7069
$C_{20}H_{26}N_4O_2$
Hexamidine, 4730
$C_{20}H_{26}O$
3-Dehydroretinal, 2878
$C_{20}H_{26}O_2$
Methestrol, 6040
Norethindrone, 6786
Norethynodrel, 6787
$C_{20}H_{26}O_4$
Carnosol, 1852
$C_{20}H_{26}O_5$
Allethrin II, 252
$C_{20}H_{26}O_7$
Cnicin, 2408
$C_{20}H_{27}BrN_2O$
Ambutonium Bromide, 381
$C_{20}H_{27}N$
Alverine, 365
$C_{20}H_{27}NO$
Cyclorphan, 2748
$C_{20}H_{27}NO_2$
Vetrabutine, 10168
$C_{20}H_{27}NO_3$
Tralkoxydim, 9724
Trilostane, 9867
$C_{20}H_{27}NO_4$
Bevantolol, 1195
$C_{20}H_{27}NO_4S$
Domitroban, 3465
$C_{20}H_{27}NO_5$
Buquinolate, 1504
Chromonar, 2243
$C_{20}H_{27}NO_{11}$
Amygdalin, 592
$C_{20}H_{27}N_3O_5$
Cilazapril Diacid, 2275

$C_{20}H_{27}N_3O_6$
Imidapril, 4952
$C_{20}H_{27}N_3O_7S$
Succisulfone 2,2'-Iminodiethanol
Salt, 9009
$C_{20}H_{27}N_5O_2$
Cilostazol, 2280
$C_{20}H_{27}N_5O_3$
Bamifylline, 948
$C_{20}H_{27}N_5O_5S$
Glisoxepid, 4479
$C_{20}H_{27}P$
JohnPhos Phosphine Ligand, 1469
$C_{20}H_{28}BrN$
Emepronium Bromide, 3615
$C_{20}H_{28}Cl_4N_2O_4$
Teclozan, 9242
$C_{20}H_{28}I_3N_3O_9$
Iobitridol, 5052
Iopentol, 5102
$C_{20}H_{28}N_2O_3$
Oxyphencyclimine, 7072
$C_{20}H_{28}N_2O_5$
Enalapril, 3623
Remifentanil, 8251
$C_{20}H_{28}N_2O_5S$
Tamsulosin, 9181
$C_{20}H_{28}N_4O$
Terguride, 9307
$C_{20}H_{28}N_4O_8S_2$
Stilbamidine Isethionate, 8945
$C_{20}H_{28}N_6O_2$
Fencamine, 3999
$C_{20}H_{28}O$
3-Dehydroretinol, 2879
Lynestrenol, 5690
Retinal, 8287
$C_{20}H_{28}O_2$
Alitretinoin, 241
Ethynodiol, 3909
Isotretinoin, 5274
Methandrostenolone, 6023
Norgesterone, 6791
Norvinisterone, 6806
Retinoic Acid, 8288
$C_{20}H_{28}O_3$
Cafestol, 1637
Cinerin I, 2293
$C_{20}H_{28}O_3S$
Ethyl Dibunate, 3843
$C_{20}H_{28}O_4$
Carnosic Acid, 1850
Marrubiin, 5820
$C_{20}H_{28}O_5$
Pododacric Acid, 7661
$C_{20}H_{28}O_5S$
Ecabet, 3535
$C_{20}H_{28}O_6$
Amarolide, 372
Phorbol, 7444
$C_{20}H_{28}O_6S$
Tiaprost, 9578
$C_{20}H_{28}O_{13}$
Primulaverin, 7867
$C_{20}H_{29}FO_3$
Fluoxymesterone, 4218
$C_{20}H_{29}NO_5S$
Latrunculin B, 5433
$C_{20}H_{29}N_3O_2$
Dibucaine, 3036
$C_{20}H_{29}N_3O_7$
Amicoumacin A, 392
$C_{20}H_{29}N_5O_3$
Urapidil, 10050
$C_{20}H_{30}BrNO_3$
Ipratropium Bromide, 5117

$C_{20}H_{30}Cl_4Ir_2$
Pentamethylcyclopentadienyliridium-
(III) Dichloride Dimer, 7226
$C_{20}H_{30}N_2O_2$
Furazabol, 4329
$C_{20}H_{30}N_2O_2.H_2SO_4$
Pseudoephedrine Sulfate, 8024
$C_{20}H_{30}N_2O_5$
Neotame, 6550
$C_{20}H_{30}N_2O_5S$
Benfuracarb, 1044
$C_{20}H_{30}O$
Neovitamin A, 6552
Vitamin A, 10211
$C_{20}H_{30}O_2$
Abietic Acid, 7
Eicosapentaenoic Acid, 3578
Isopimaric Acid, 5246
Levopimaric Acid, 5522
Methenolone, 6039
17-Methyltestosterone, 6197
Mibolerone, 6253
Norethandrolone, 6785
Oxendolone, 7030
Oxenin, 7031
Palustric Acid, 7099
Pimaric Acid, 7540
Stenbolone, 8935
$C_{20}H_{30}O_3$
Leukotriene A_4, 5507
Oxymesterone, 7064
Steviol, 8938
$C_{20}H_{30}O_5$
Andrographolide, 627
Taxicin-II, 9214
$C_{20}H_{30}O_6$
Taxicin-I, 9214
$C_{20}H_{30}O_8$
Ptaquiloside, 8039
$C_{20}H_{31}NO$
Deramciclane, 2916
$C_{20}H_{31}NO_2$
Drofenine, 3494
$C_{20}H_{31}NO_2S$
Cetiedil, 2019
$C_{20}H_{31}NO_3$
Carbetapentane, 1795
$C_{20}H_{31}NO_4$
Vernakalant, 10157
$C_{20}H_{31}NO_7$
Heliosupine, 4664
$C_{20}H_{31}NaO_5$
Prostacyclin Sodium Salt, 7986
$C_{20}H_{32}$
Neocembrene, 6533
$C_{20}H_{32}ClNO$
Trihexyphenidyl Hydrochloride, 9864
$C_{20}H_{32}F_2O_5$
Lubiprostone, 5649
$C_{20}H_{32}N_2O_3S$
Carbosulfan, 1827
$C_{20}H_{32}N_5O_8P$
Adefovir Di(pivaloyloxymethyl) Es-
ter, 139
$C_{20}H_{32}N_6O_{12}S_2$
Glutathione Disulfide, 4511
$C_{20}H_{32}O$
Ethylestrenol, 3860
$C_{20}H_{32}O_2$
Arachidonic Acid, 754
Dromostanolone, 3497
Etiocholanic Acid, 3916
Mestanolone, 5985
Mesterolone, 5986
Methandriol, 6022

$C_{20}H_{32}O_3$
Grindelic Acid, 4582
5-HETE, 4709
$C_{20}H_{32}O_4$
Leukotriene B_4, 5507
$C_{20}H_{32}O_5$
Grayanotoxin II, 4578
Prostacyclin, 7986
Prostaglandin E_2, 7988
Thromboxane A_2, 9542
$C_{20}H_{33}NO$
Fenpropimorph, 4021
$C_{20}H_{33}NO_3$
Oxeladin, 7029
$C_{20}H_{33}NO_3S$
N-Acetyl-L-farnesylcysteine, 89
$C_{20}H_{33}NO_7$
Candoxatril Diacid, 1745
$C_{20}H_{33}N_3O_3$
Talinolol, 9170
$C_{20}H_{33}N_3O_3S$
Quinagolide, 8161
$C_{20}H_{33}N_3O_4$
Celiprolol, 1962
$C_{20}H_{34}AuO_9PS$
Auranofin, 869
$C_{20}H_{34}GdN_5O_{10}$
Gadoversetamide, 4359
$C_{20}H_{34}O_2$
Plaunotol, 7647
$C_{20}H_{34}O_4$
Aphidicolin, 714
Zoapatanol, 10385
$C_{20}H_{34}O_5$
Prostaglandin E_1, 7987
Prostaglandin $F_{2\alpha}$, 7989
$C_{20}H_{34}O_6$
Grayanotoxin III, 4578
Thromboxane B_2, 9542
$C_{20}H_{35}ClO_4S$
Oxychlorosene, 7055
$C_{20}H_{35}NO_{13}$
Validamycin A, 10094
$C_{20}H_{36}CaNO_5.\frac{1}{2}H_2O$
Hopantenic Acid Calcium Salt Hemi-
hydrate, 4783
$C_{20}H_{36}N_2O_3S$
Ibutilide, 4921
$C_{20}H_{36}N_6O$
Lauroguadine, 5441
$C_{20}H_{36}O_2$
Ethyl Linoleate, 3875
$C_{20}H_{36}O_6$
Dicyclohexano-18-crown-6, 2592
$C_{20}H_{37}KO_7S$
Docusate Potassium, 3446
$C_{20}H_{37}NO_3$
Rociverine, 8374
$C_{20}H_{37}N_3O_{13}$
Destomycin A, 2936
Hygromycin B, 4892
$C_{20}H_{37}NaO_7S$
Docusate Sodium, 3446
$C_{20}H_{38}O_2$
Oleic Acid Ethyl Ester, 6921
$C_{20}H_{40}N_2O_{12}$
Hexamethonium Tartrate, 4724
$C_{20}H_{40}O$
Isophytol, 5244
Phytol, 7502
$C_{20}H_{40}O_2$
Arachidic Acid, 753
Stearic Acid Ethyl Ester, 8930
$C_{20}H_{41}N_5O_7$
Gentamicin C_2, 4427
Micronomicin, 6260

$C_{20}H_{42}O$
Octyldodecanol, 6856
$C_{20}H_{42}O_3$
Phytantriol, 7498
$C_{20}H_{43}N$
Dymanthine, 3522
$C_{20}H_{44}BrN$
Cetyldimethylethylammonium Bromide, 2028
$C_{20}H_{60}O_8Si_9$
Eicosamethylnonasiloxane, 3577
$C_{21}H_6Br_9N_3O_3$
Tris(tribromophenoxy)triazine, 9933
$C_{21}H_{11}ClF_6N_2O_3$
Flufenoxuron, 4164
$C_{21}H_{11}NO_5S$
Asenapine, 4120
$C_{21}H_{13}Br_4NaO_5S$
Bromcresol Green Sodium Salt, 1393
$C_{21}H_{13}F_3N_2O_4$
Talniflumate, 9176
$C_{21}H_{14}Br_4O_5S$
Bromcresol Green, 1393
$C_{21}H_{15}AlO_6$
Aluminum Benzoate, 325
$C_{21}H_{15}Br_2NaO_5S$
Bromcresol Purple Sodium Salt, 1394
$C_{21}H_{15}CeO_6.3H_2O$
Benzoic Acid Cerium Salt Trihydrate, 1093
$C_{21}H_{15}N_3O_4$
Deferasirox, 2861
$C_{21}H_{16}$
3-Methylcholanthrene, 6115
$C_{21}H_{16}Br_2O_5S$
Bromcresol Purple, 1394
$C_{21}H_{16}ClFN_4O_5$
Fluoxastrobin, 4216
$C_{21}H_{16}ClF_3N_4O_3$
Sorafenib, 8846
$C_{21}H_{16}N_4O_8S_2$
Nitrocefin, 6681
$C_{21}H_{16}O_6$
Justicidin B, 5320
$C_{21}H_{17}ClN_2O_3$
Rhodamine 123, 8306
$C_{21}H_{18}ClNO_6$
Acemetacin, 28
$C_{21}H_{18}F_3N_3O_3$
Temafloxacin, 9274
$C_{21}H_{18}NO_4^+$
Chelerythrine, 2050
$C_{21}H_{18}O_5S$
Cresol Purple, 2567
Cresol Red, 2568
$C_{21}H_{18}O_{12}$
Scutellarein Glucuronide, 8549
$C_{21}H_{19}ClFNO_4S$
Laropiprant, 5426
$C_{21}H_{19}ClN_4O_4$
Lenvatinib, 5494
$C_{21}H_{19}F_2N_3O_3$
Difloxacin, 3160
$C_{21}H_{19}N_3O_3S$
Amsacrine, 589
$C_{21}H_{20}BrN_3$
Homidium Bromide, 4769
$C_{21}H_{20}ClNO_5$
Flavopiridol, 4126
$C_{21}H_{20}Cl_2N_6O_3$
Rilmazafone, 8346
$C_{21}H_{20}Cl_2O_3$
Permethrin, 7294
$C_{21}H_{20}FN_3O_6S$
Prulifloxacin, 8016

$C_{21}H_{20}N_2O_3$
Alstonine, 312
Serpentine (Alkaloid), 8603
$C_{21}H_{20}N_2O_5$
Varespladib, 10126
$C_{21}H_{20}N_3^+$
Homidium, 4769
$C_{21}H_{20}N_4O_3$
Picotamide, 7519
$C_{21}H_{20}N_6O$
Aminoquinuride, 471
$C_{21}H_{20}O_6$
Cabenegrin A-I, 1605
Curcumin, 2663
$C_{21}H_{20}O_8$
α-Peltatin, 7184
$C_{21}H_{20}O_9$
Chrysophanic Acid Glucoside, 2262
Daidzein 7-Glucoside, 2798
Frangulin A, 4292
$C_{21}H_{20}O_{10}$
Apigetrin, 718
Genistin, 4426
$C_{21}H_{20}O_{11}$
Kaempferol 3-Glucoside, 5322
Luteolin 5-Glucoside, 5675
Quercitrin, 8153
$C_{21}H_{20}O_{12}$
Isoquercitrin, 5267
Myricetin 3-Rhamnoside, 6417
Quercimeritrin, 8151
$C_{21}H_{20}O_{12}.2½H_2O$
Quercetin 3-β-D-Galactoside Hemipentahydrate, 8150
$C_{21}H_{20}O_{13}$
Gossypitrin, 8151
Quercetagetin 7-Glucoside, 8149
$C_{21}H_{21}ClN_2O_8$
Demeclocycline, 2890
$C_{21}H_{21}ClN_4OS$
Ziprasidone, 10371
$C_{21}H_{21}ClN_4O_3$
Nizofenone, 6748
$C_{21}H_{21}ClO_{10}$
Pelargonidin 3-Glucoside, 7180
$C_{21}H_{21}ClO_{11}$
Cyanidin Chloride 3-Glucoside, 2677
$C_{21}H_{21}ClO_{12}$
Delphinidin 3-Glucoside, 2885
$C_{21}H_{21}FN_2O_4S$
Ramatroban, 8218
$C_{21}H_{21}FN_4O_3$
Pradofloxacin, 7819
$C_{21}H_{21}HgN_4NaO_8$
Mercumallylic Acid Sodium Salt Compd with Theophylline, 5939
$C_{21}H_{21}N$
Cyproheptadine, 2770
Naftifine, 6439
$C_{21}H_{21}NO_2$
Oxetorone, 7033
$C_{21}H_{21}NO_2S$
Tazarotene, 9218
$C_{21}H_{21}NO_6$
Adlumine, 155
Hydrastine, 4806
Rheadine, 8298
$C_{21}H_{21}NO_7$
Narcotoline, 6505
$C_{21}H_{21}N_2NaO_5S.H_2O$
Nafcillin Sodium Salt Monohydrate, 6436
$C_{21}H_{21}N_3O_7$
Cacotheline, 1610
$C_{21}H_{21}O_4P$
Tritolyl Phosphate, 9940
Tri-o-tolyl Phosphate, 9941

$C_{21}H_{22}ClNO_4$
Dimethomorph, 3243
$C_{21}H_{22}Cl_2FN_5O$
Crizotinib, 2577
$[C_{21}H_{22}NO_4]^+$
Palmatine, 7095
$C_{21}H_{22}N_2O_3$
Strychnine N^6-Oxide, 8986
$C_{21}H_{22}N_2O_5S$
Nafcillin, 6436
$C_{21}H_{22}N_2O_6S_2$
Dibenzoyldjenkolic Acid, 3431
$C_{21}H_{22}N_4O_6S$
Raltitrexed, 8217
$C_{21}H_{22}O_2$
Endiandric Acid A, 3626
$C_{21}H_{22}O_4$
Bergamottin, 1159
Licochalcone C, 5532
$C_{21}H_{22}O_5$
Licochalcone D, 5532
Pleurotine, 7651
$C_{21}H_{22}O_6$
Cabenegrin A-II, 1605
$C_{21}H_{22}O_7$
Ostruthol, 6995
$C_{21}H_{22}O_9$
Aloin, 303
Gardenin A, 4399
$C_{21}H_{22}O_{10}$
Eriodictyol 7-L-Rhamnoside, 3726
$C_{21}H_{23}BrFNO_2$
Bromperidol, 1452
$C_{21}H_{23}ClFNO_2$
Haloperidol, 4634
$C_{21}H_{23}ClFNO_5$
Flumiclorac Pentyl Ester, 4173
$C_{21}H_{23}ClFN_3O$
Flurazepam, 4228
$C_{21}H_{23}Cl_2NO_6$
Clevidipine, 2350
$C_{21}H_{23}F_2NO_2$
Etoxazole, 3933
$C_{21}H_{23}IN_2$
Quinaldine Red, 8165
$C_{21}H_{23}NO$
Dapoxetine, 2821
$C_{21}H_{23}NO_3$
Olopatadine, 6934
3-Quinuclidinol dl-Form Benzilate (Ester), 8199
$C_{21}H_{23}NO_4S$
Racecadotril, 8207
$C_{21}H_{23}NO_5$
Allocryptopine, 257
Cryptopine, 2598
Diacetylmorphine, 2971
Homochelidonine, 4770
$C_{21}H_{23}NO_6$
Colchiceine, 2454
$C_{21}H_{23}N_3OS$
Pericyazine, 7279
$C_{21}H_{23}N_3O_2$
Panobinostat, 7113
$C_{21}H_{23}N_3O_7S$
Lenampicillin, 5491
$C_{21}H_{23}N_7O_2S$
Pazopanib, 7160
$C_{21}H_{24}ClN_2NaO_4S$
Tianeptine Sodium Salt, 9575
$C_{21}H_{24}ClN_3OS$
Pipamazine, 7567
$C_{21}H_{24}ClN_3O_3$
Fominoben, 4255
$C_{21}H_{24}Cl_2O_4$
Spirodiclofen, 8884

$C_{21}H_{24}FN_3O_4$
Moxifloxacin, 6377
$C_{21}H_{24}F_3N_3S$
Trifluoperazine, 9846
$C_{21}H_{24}N_2$
Quinupramine, 8200
$C_{21}H_{24}N_2O_2$
Catharanthine, 1905
$C_{21}H_{24}N_2O_3$
Diaboline, 2962
Lochnericine, 5611
Raubasine, 8237
$C_{21}H_{24}N_2O_4S$
Repinotan, 8257
$C_{21}H_{24}N_5NaO_5S_2.H_2O$
Mezlocillin Sodium Salt Monohydrate, 6250
$C_{21}H_{24}N_8$
Bismark Brown R Free Base, 1256
$C_{21}H_{24}O_2$
Gestrinone, 4449
$C_{21}H_{24}O_4$
Bisphenol A Diglycidyl Ether, 1298
$C_{21}H_{24}O_7$
Visnadine, 10208
$C_{21}H_{24}O_9$
Rhapontin, 8297
$C_{21}H_{24}O_{10}$
Phloridzin, 7438
$C_{21}H_{24}O_{11}$
Helicin Tetraacetate, 4663
$C_{21}H_{25}BrN_2O_3$
Brovincamine, 1460
$C_{21}H_{25}ClFN_3O_3$
Mosapride, 6364
$C_{21}H_{25}ClN_2O_3$
Bepotastine, 1151
Cetirizine, 2021
$C_{21}H_{25}ClN_2O_4S$
Tianeptine, 9575
$C_{21}H_{25}ClO_2$
Trengestone, 9746
$C_{21}H_{25}ClO_5$
Cloprednol, 2387
$C_{21}H_{25}ClO_6$
Dapagliflozin, 2816
$C_{21}H_{25}FN_2O_2$
Fluanisone, 4146
$C_{21}H_{25}N$
Melitracen, 5891
Terbinafine, 9299
$C_{21}H_{25}NO$
Benztropine, 1124
$C_{21}H_{25}NO_2$
Piperidolate, 7582
$C_{21}H_{25}NO_3$
Nalmefene, 6445
Piperilate, 7583
$C_{21}H_{25}NO_4$
Glaucine, 4471
Isocorybulbine, 5206
Tetrahydropalmatine, 9360
$C_{21}H_{25}NO_5$
Demecolcine, 2891
Diacetyldihydromorphine, 2970
$C_{21}H_{25}N_3$
Dimebolin, 3225
$C_{21}H_{25}N_3O_2S$
Quetiapine, 8155
$C_{21}H_{25}N_3O_3S$
Pipazethate, 7569
$C_{21}H_{25}N_5O_8S_2$
Mezlocillin, 6250
$C_{21}H_{26}BrNO_3$
Mepenzolate Bromide, 5915
Methantheline Bromide, 6030

$C_{21}H_{26}BrNO_4$
Methylnaltrexone Bromide, 6172
$C_{21}H_{26}ClNO$
Clemastine, 2343
$C_{21}H_{26}ClN_3OS$
Perphenazine, 7297
$C_{21}H_{26}ClN_3O_2$
Nemonapride, 6531
$C_{21}H_{26}Cl_2N_8$
Bismark Brown R, 1256
$C_{21}H_{26}Cl_2O$
Clofoctol, 2373
$C_{21}H_{26}Cl_2O_4$
Dichlorisone, 3057
$C_{21}H_{26}N_2OS_2$
Mesoridazine, 5980
$C_{21}H_{26}N_2O_2S_2$
Sulforidazine, 9093
$C_{21}H_{26}N_2O_3$
Corynanthine, 2535
Pseudoyohimbine, 8031
Vincamine, 10180
Yohimbine, 10300
α-Yohimbine, 10301
$allo$-Yohimbine, 10302
$C_{21}H_{26}N_2O_7$
Nimodipine, 6636
$C_{21}H_{26}N_2S_2$
Thioridazine, 9512
$C_{21}H_{26}O_2$
Altrenogest, 315
Cannabinol, 1751
Gestodene, 4447
Mestranol, 5987
$C_{21}H_{26}O_3$
Acitretin, 103
Buparvaquone, 1495
Moxestrol, 6375
Octabenzone, 6831
$C_{21}H_{26}O_5$
Prednisone, 7842
$C_{21}H_{26}O_{12}$
Plumieride, 7657
$C_{21}H_{27}ClN_2O$
Pyronine B, 8118
$C_{21}H_{27}ClN_2O_2$
Hydroxyzine, 4889
$C_{21}H_{27}FO_5$
Fluprednisolone, 4223
Isoflupredone, 5220
$C_{21}H_{27}FO_6$
Triamcinolone, 9757
$C_{21}H_{27}N$
Budipine, 1477
Butriptyline, 1535
Pramiverin, 7827
$C_{21}H_{27}NO$
Benproperine, 1051
Diphenidol, 3341
Isomethadone, 5230
Methadone, 6016
$C_{21}H_{27}NO_2$
Etafenone, 3765
Ifenprodil, 4936
$C_{21}H_{27}NO_3$
Propafenone, 7908
$C_{21}H_{27}NO_4$
Laudanosine, 5436
Nalbuphine, 6441
Protostephanine, 8007
$C_{21}H_{27}NO_4S$
Diphemanil Methylsulfate, 3336
$C_{21}H_{27}NO_5$
Hasubanonine, 4652
Spirotetramat, 8888
$C_{21}H_{27}N_3O_2$
Methysergide, 6213

$C_{21}H_{27}N_3O_7S$
Bacampicillin, 924
$C_{21}H_{27}N_5O_4S$
Glipizide, 4477
$C_{21}H_{27}N_5O_7S$
Aspoxicillin, 842
$C_{21}H_{27}N_5O_9S_2$
Cefpodoxime 1-(Isopropoxycarbonyl-oxy)ethyl Ester, 1942
$C_{21}H_{27}N_7O$
Reversine, 8292
$C_{21}H_{27}N_7O_{14}P_2$
NAD, 6429
$C_{21}H_{27}Na_2O_8P$
Prednisolone 21-Disodium Phosphate, 7841
$C_{21}H_{28}BrNO_4$
Cimetropium Bromide, 2283
$C_{21}H_{28}N_2O_2$
Aspidospermine N-Formyl-N-deacetylaspidospermine, 840
$C_{21}H_{28}N_2O_5$
Trimethobenzamide, 9877
$C_{21}H_{28}N_7O_{17}P_3$
NADP, 6432
$C_{21}H_{28}O_2$
Dydrogesterone, 3521
Ethisterone, 3795
Guggulsterone, 4608
Norgestrel, 6793
Tetrahydrogestrinone, 9359
Tibolone, 9582
$C_{21}H_{28}O_3$
Cyclethrin, 2700
Pyrethrin I, 8075
$C_{21}H_{28}O_4$
11-Dehydrocorticosterone, 2873
Formebolone, 4267
$C_{21}H_{28}O_5$
Aldosterone, 218
Cinerin II, 2293
Cortisone, 2525
Prednisolone, 7841
$C_{21}H_{29}ClO_3$
Clostebol Acetate, 2397
$C_{21}H_{29}ClO_6S$
Luprostiol, 5670
$C_{21}H_{29}FO_5$
Fludrocortisone, 4161
$C_{21}H_{29}NO$
Biperiden, 1238
$C_{21}H_{29}NO_2$
Butorphanol, 1532
Norelgestromin, 6783
$C_{21}H_{29}NS_2$
Captodiamine, 1775
$C_{21}H_{29}N_3O$
Disopyramide, 3402
$C_{21}H_{29}N_5O_2$
Tandospirone, 9183
$C_{21}H_{29}N_5O_4$
Pfeiffer's Substance, 7309
$C_{21}H_{29}N_7O_{14}P_2$
NAD Reduced Form, 6429
$C_{21}H_{29}Na_2O_8P$
Hydrocortisone 21-Phosphate Disodium Salt, 4824
$C_{21}H_{30}BrNO_4$
N-Butylscopolammonium Bromide, 1591
$C_{21}H_{30}Br_2N_4O_{10}S_2$
Dibromopropamidine Isethionate, 3029
$C_{21}H_{30}Cl_2N_2O_5$
Loxiglumide, 5646

Formula Index

C₂₁H₃₀Cl₂N₆O₁₀
Chloramphenicol Monosuccinate Arginine Salt, 2077
C₂₁H₃₀FN₃O₂
Pipamperone, 7568
C₂₁H₃₀N₂O₈S
Docarpamine, 3441
C₂₁H₃₀N₄O₄
Cinitapride, 2295
C₂₁H₃₀N₄O₅S
Acotiamide, 113
C₂₁H₃₀N₇O₁₇P₃
NADP Reduced Form, 6432
C₂₁H₃₀O₂
Cannabidiol, 1750
Progesterone, 7889
Tetrahydrocannabinols, 9354
C₂₁H₃₀O₃
Deoxycorticosterone, 2902
17α-Hydroxyprogesterone, 4877
Jasmolin I, 5306
Nandrolone Propionate, 6450
C₂₁H₃₀O₄
Algestone, 232
Corticosterone, 2524
11-Desoxy-17-hydroxycorticosterone, 2933
C₂₁H₃₀O₄S
Tixocortol, 9641
C₂₁H₃₀O₅
Humulon, 4793
Hydrocortisone, 4824
4-Pregnene-17α,20β,21-triol-3,11-di-one, 7849
C₂₁H₃₁ClN₆O₃
PPACK, 7816
C₂₁H₃₁NO₄
Furethidine, 4332
C₂₁H₃₁N₃O₅
Lisinopril, 5572
C₂₁H₃₁N₅O₂
Buspirone, 1507
C₂₁H₃₂N₂O
Stanozolol, 8921
C₂₁H₃₂N₄O₁₀S₂
Propamidine Isethionate, 7911
C₂₁H₃₂N₆O₃
Alfentanil, 229
C₂₁H₃₂O
Allylestrenol, 282
C₂₁H₃₂O₂
3,20-Allopregnanedione, 264
Bolasterone, 1326
Calusterone, 1724
Methyl Abietate, 6080
Norbolethone, 6774
3,20-Pregnanedione, 7846
Pregnenolone, 7851
C₂₁H₃₂O₃
Alfaxalone, 228
Oxymetholone, 7066
C₂₁H₃₂O₄
4-Pregnene-17α,20β,21-triol-3-one, 7850
C₂₁H₃₂O₅
Hydrallostane, 4803
4-Pregnene-11β,17α,20β,21-tetrol-3-one, 7848
Tetrahydrocortisone, 9355
C₂₁H₃₃NO₇
Lasiocarpine, 5429
C₂₁H₃₃N₂O₆P
Ceronapril, 1993
C₂₁H₃₃N₃O
Sitamaquine, 8691
C₂₁H₃₃N₃O₂
Eptastigmine, 3695

C₂₁H₃₃N₃O₅S
Amdinocillin Pivoxil, 384
C₂₁H₃₄BrNO₃
Oxyphenonium Bromide, 7074
C₂₁H₃₄O₂
Allopregnan-3α-ol-20-one, 267
Allopregnan-3β-ol-20-one, 268
Allopregnan-20α-ol-3-one, 269
Allopregnan-20β-ol-3-one, 270
Pregnan-3α-ol-20-one, 7847
(15:1)-Urushiol, 10076
C₂₁H₃₄O₃
Allopregnane-3β,17α-diol-20-one, 263
C₂₁H₃₄O₅
Allopregnane-3α,11β,17α,21-tetrol-20-one, 265
Arbaprostil, 759
Cortolone, 2529
C₂₁H₃₅ClN₂O₃
Lapyrium Chloride, 5421
C₂₁H₃₅Cl₂MnN₅
Imisopasem Manganese, 4965
C₂₁H₃₅NO
Amorolfine, 570
Funtumine, 4321
C₂₁H₃₅NO₅S
Procyclidine Methosulfate, 7881
C₂₁H₃₆
Allopregnane, 258
Pregnane, 7844
C₂₁H₃₆ClNO
Tridihexethyl Chloride, 9826
C₂₁H₃₆INO
Tridihexethyl Iodide, 9826
C₂₁H₃₆N₂O₅S
Hexocyclium Methyl Sulfate, 4742
C₂₁H₃₆N₇O₁₆P₃S
Coenzyme A, 2450
C₂₁H₃₆O₂
Allopregnane-3α,20α-diol, 259
Allopregnane-3α,20β-diol, 260
Allopregnane-3β,20α-diol, 261
Allopregnane-3β,20β-diol, 262
3-Pentadecylcatechol, 7219
Pregnanediol, 7845
Uranediol, 10038
(15:0)-Urushiol, 10076
C₂₁H₃₆O₅
Carboprost, 1824
Cortol, 2528
C₂₁H₃₈ClN
Cetylpyridinium Chloride, 2031
C₂₁H₃₈N₂OS
Kalkitoxin, 5325
C₂₁H₃₈O₄
Rioprostil, 8359
C₂₁H₃₉NO₆
Myriocin, 6418
C₂₁H₃₉N₇O₁₂
Streptomycin, 8956
C₂₁H₄₀N₈O₆
Tuftsin, 9989
C₂₁H₄₁N₅O₇
Netilmicin, 6564
C₂₁H₄₁N₅O₁₁
Apramycin, 739
C₂₁H₄₁N₅O₂
Butirosin, 1530
C₂₁H₄₁N₇O₁₂
Dihydrostreptomycin, 3197
C₂₁H₄₃N₅O₇
Gentamicin C₁, 4427
C₂₁H₄₄O₃
Batyl Alcohol, 1009
C₂₁H₄₅N₃
Hexetidine, 4739

C₂₁H₄₆NO₄P
Miltefosine, 6279
C₂₂H₁₄
1,2:5,6-Dibenzanthracene, 3011
2,3:6,7-Dibenzphenanthrene, 3015
Pentacene, 7216
Picene, 7508
C₂₂H₁₄Cl₂I₂N₂O₂
Closantel, 2396
C₂₂H₁₆Cl₂O₄S
Tioclomarol, 9607
C₂₂H₁₆F₂N₂
Flutrimazole, 4243
C₂₂H₁₆N₄Na₂O₉S₂
Orange B, 6954
C₂₂H₁₆N₄O
Sudan III, 9015
C₂₂H₁₆O₆
Resistomycin, 8272
C₂₂H₁₆O₈
Ethyl Biscoumacetate, 3825
C₂₂H₁₇ClF₃N₃O₇
Indoxacarb, 5012
C₂₂H₁₇ClN₂
Clotrimazole, 2401
C₂₂H₁₇ClN₂O₅
Camptothecin Chloroacetate, 1737
C₂₂H₁₇F₂N₅OS
Isavuconazole, 5149
Ravuconazole, 8239
C₂₂H₁₇N₃O₅
Azoxystrobin, 917
C₂₂H₁₈Cl₂FNO₃
Cyfluthrin, 2754
C₂₂H₁₈N₂
Bifonazole, 1216
C₂₂H₁₈N₂O₄
Famoxadone, 3972
C₂₂H₁₈N₂O₅
Camptothecin Acetate, 1737
C₂₂H₁₈N₄OS
Axitinib, 883
C₂₂H₁₈N₆
Rilpivirine, 8349
C₂₂H₁₈O₄
o-Cresolphthalein, 2565
C₂₂H₁₈O₇
Justicidin A, 5320
C₂₂H₁₈O₁₁
EGCG, 3573
C₂₂H₁₉Br₂NO₃
Deltamethrin, 2888
C₂₂H₁₉Br₄NO₃
Tralomethrin, 9725
C₂₂H₁₉ClO₃
Atovaquone, 857
C₂₂H₁₉Cl₂NO₃
Cypermethrin, 2764
C₂₂H₁₉N
Bis(1-naphthylmethyl)amine, 1293
C₂₂H₁₉NO₄
Bisacodyl, 1246
C₂₂H₁₉N₃O₄
Tadalafil, 9157
C₂₂H₁₉N₄NaO₈S₂
Cefsulodin Sodium Salt, 1946
C₂₂H₂₀N₂O₂
Piketoprofen, 7534
C₂₂H₂₀N₄O₈S₂
Cefsulodin, 1946
C₂₂H₂₀N₈O₅S₄
Ceftaroline, 1947
C₂₂H₂₀O₁₀
Granaticin, 4570
C₂₂H₂₀O₁₃
Carminic Acid, 1843

$C_{22}H_{21}CIN_2O_8$
Meclocycline, 5849
$C_{22}H_{21}Cl_3N_4O$
Rimonabant, 8355
$C_{22}H_{21}FN_2O$
Sarizotan, 8512
$C_{22}H_{21}NO_2$
Benzylimidobis(p-methoxyphenyl)-
methane, 1142
$C_{22}H_{21}N_8O_8PS_4$
Ceftaroline Fosamil, 1947
$C_{22}H_{22}CINO_6$
Ochratoxin C, 6829
$C_{22}H_{22}CIN_3O_5$
Propaquizafop, 7920
$C_{22}H_{22}CIN_5O_2S$
Apafant, 709
$C_{22}H_{22}FN_3O_2$
Droperidol, 3499
$C_{22}H_{22}FN_3O_3$
Ketanserin, 5344
$C_{22}H_{22}F_3N$
Cinacalcet, 2285
$C_{22}H_{22}N_2O_4$
Ambrisentan, 378
$C_{22}H_{22}N_2O_6$
Darusentan, 2830
$C_{22}H_{22}N_2O_8$
Methacycline, 6015
$C_{22}H_{22}N_6O_5S_2$
Cefpirome, 1941
$C_{22}H_{22}N_6O_7S_2$
Ceftazidime, 1948
$C_{22}H_{22}N_8$
Bisantrene, 1248
$C_{22}H_{22}O_4$
Dienestrol Diacetate, 3115
$C_{22}H_{22}O_8$
β-Peltatin, 7185
Podophyllotoxin, 7663
$C_{22}H_{22}O_9$
Formononetin 7-Glucoside, 4271
$C_{22}H_{22}O_{10}$
Sissotrin, 1231
Prunetin 4'-Glucoside, 8017
$C_{22}H_{22}O_{11}$
Scoparin, 8539
Tectoridin, 9245
$C_{22}H_{23}CIN_2O_2$
Loratadine, 5635
$C_{22}H_{23}CIN_2O_8$
Chlortetracycline, 2198
$C_{22}H_{23}CIN_6O$
Losartan, 5641
$C_{22}H_{23}FN_4$
Revaprazan, 8291
$C_{22}H_{23}F_4NO_2$
Trifluperidol, 9852
$C_{22}H_{23}NO_3$
Fenpropathrin, 4019
$C_{22}H_{23}NO_4$
Nequinate, 6556
$C_{22}H_{23}NO_4S_2$
Zofenopril, 10386
$C_{22}H_{23}NO_6$
Aureothin, 870
$C_{22}H_{23}NO_7$
Noscapine, 6807
$C_{22}H_{23}N_3O_4$
Erlotinib, 3730
$C_{22}H_{23}N_3O_9$
Aluminon, 318
$C_{22}H_{23}N_5O$
Motesanib, 6366
$C_{22}H_{24}BrFN_4O_2$
Vandetanib, 10117

$C_{22}H_{24}CIFN_4O_3$
Gefitinib, 4412
$C_{22}H_{24}CIN_3O$
Azelastine, 899
$C_{22}H_{24}CIN_5O_2$
Domperidone, 3467
$C_{22}H_{24}FN_3OS$
Timiperone, 9600
$C_{22}H_{24}FN_3O_2$
Benperidol, 1050
$C_{22}H_{24}MnN_4O_{14}P_2$
Mangafodipir, 5787
$C_{22}H_{24}N_2O_2$
Acrivastine, 118
$C_{22}H_{24}N_2O_3$
Colubrines, 2474
$C_{22}H_{24}N_2O_4$
Alstonidine, 311
Kopsine, 5368
Vomicine, 10228
$C_{22}H_{24}N_2O_7S$
Apremilast, 740
$C_{22}H_{24}N_2O_8$
Quatrimycin, 8144
Tetracycline, 9341
$C_{22}H_{24}N_2O_8 \cdot H_2O$
Doxycycline, 3488
$C_{22}H_{24}N_2O_9$
Oxytetracycline, 7075
$C_{22}H_{24}N_2O_{10}$
BAPTA, 952
$C_{22}H_{24}O_8$
Lophotoxin, 5629
$C_{22}H_{24}O_9$
Podophyllic Acids, 7662
$C_{22}H_{24}O_{10}$
Sakuranetin 5-Glucoside, 8455
$C_{22}H_{25}CIN_2OS$
Clopenthixol, 2383
$C_{22}H_{25}CIN_2O_4S$
Clentiazem, 2346
$C_{22}H_{25}FN_4O_2$
Toceranib, 9650
$C_{22}H_{25}F_2NO_4$
Nebivolol, 6516
$C_{22}H_{25}F_2N_3O_2S$
Lubeluzole, 5648
$C_{22}H_{25}N$
Piroheptine, 7616
$C_{22}H_{25}NO_2$
Tropacine, 9952
$C_{22}H_{25}NO_3$
Tamibarotene, 9179
Tropine Benzylate, 9961
$C_{22}H_{25}NO_5S$
Thiocolchicine, 9477
$C_{22}H_{25}NO_6$
Colchicine, 2455
$C_{22}H_{25}N_3O$
Indoramin, 5010
$C_{22}H_{25}N_3O_2$
Bucindolol, 1471
$C_{22}H_{25}N_3O_4$
Vesnarinone, 10166
$C_{22}H_{25}N_3O_4S$
Moricizine, 6354
$C_{22}H_{25}N_3O_7S$
Ertapenem, 3731
$C_{22}H_{25}N_7O_5$
Edatrexate, 3554
$C_{22}H_{26}BrNO_3$
Clidinium Bromide, 2352
$C_{22}H_{26}CIFO_4$
Clobetasone, 2359
$C_{22}H_{26}CINO_3$
Pipoxolan Hydrochloride, 7596

$C_{22}H_{26}CIN_7O_2S$
Dasatinib, 2831
$C_{22}H_{26}FNO_2$
Moperone, 6351
$C_{22}H_{26}F_3N_3OS$
Fluphenazine, 4220
$C_{22}H_{26}N_2O_2$
Vinpocetine, 10188
$C_{22}H_{26}N_2O_2S$
Eletriptan, 3597
$C_{22}H_{26}N_2O_3$
Corynantheine, 2534
$C_{22}H_{26}N_2O_4$
Aricine, 774
Tofisopam, 9662
$C_{22}H_{26}N_2O_4S$
Diltiazem, 3224
$C_{22}H_{26}N_4$
Calycanthine, 1725
Chimonanthine, 2059
$C_{22}H_{26}N_4O_2S$
Phenylbutazone 2-Amino-2-thiazoline
Salt, 7390
$C_{22}H_{26}O_3$
Bioresmethrin, 1235
Resmethrin, 8274
$C_{22}H_{26}O_4$
Seratrodast, 8598
$C_{22}H_{26}O_{12}$
Catalposide, 1901
$C_{22}H_{27}CIF_2O_5$
Halometasone, 4633
$C_{22}H_{27}CIN_2O$
Lorcainide, 5637
$C_{22}H_{27}CIN_2O_3$
Lorajmine, 5634
$C_{22}H_{27}CIO_3$
Cyproterone, 2771
$C_{22}H_{27}CIO_5$
Osaterone Acetate Ester, 6985
$C_{22}H_{27}FN_4O_2$
Sunitinib, 9129
$C_{22}H_{27}MnN_4Na_3O_{14}P_2$
Mangafodipir Trisodium Salt, 5787
$C_{22}H_{27}NO$
Ethybenztropine, 3810
Phenazocine, 7332
$C_{22}H_{27}NO_2$
Amineptine, 403
Danazol, 2811
Lobeline, 5609
$C_{22}H_{27}NO_4$
Corydaline, 2531
$C_{22}H_{27}N_3O_2$
Caroverine, 1859
$C_{22}H_{27}N_3O_3S_2$
Metopimazine, 6226
$C_{22}H_{27}N_3O_5$
Physostigmine Salicylate, 7496
$C_{22}H_{27}N_3O_6S$
Pivcefalexin, 7627
$C_{22}H_{27}N_9O_4$
Distamycin A, 3404
$C_{22}H_{27}N_9O_7S_2$
Cefteram Pivaloyloxymethyl Ester,
1949
$C_{22}H_{28}BrN$
Prifinium Bromide, 7861
$C_{22}H_{28}BrNO_3$
Pipenzolate Bromide, 7573
$C_{22}H_{28}CIFO_4$
Clobetasol, 2358
Clocortolone, 2363
$C_{22}H_{28}CINO$
Setastine, 8611
$C_{22}H_{28}CINaO_6$
Cloprostenol Sodium Salt, 2388

$C_{22}H_{28}FN_3O_6S$
Rosuvastatin, 8402
$C_{22}H_{28}FNa_2O_8P$
Betamethasone 21-Phosphate Di-
sodium Salt, 1182
Dexamethasone 21-Phosphate Di-
sodium Salt, 2945
$C_{22}H_{28}F_2O_4$
Diflucortolone, 3162
$C_{22}H_{28}F_2O_5$
Diflorasone, 3159
Flumethasone, 4169
$C_{22}H_{28}N_2O$
Buzepide, 1603
Fentanyl, 4028
$C_{22}H_{28}N_2O_2$
Anileridine, 651
Encainide, 3625
Tebufenozide, 9230
$C_{22}H_{28}N_2O_2S$
Perimethazine, 7283
$C_{22}H_{28}N_2O_3$
Methoxyfenozide, 6066
Voacangine, 10223
$C_{22}H_{28}N_2O_4$
Rhynchophylline, 8322
$C_{22}H_{28}N_2O_5$
Mycelianamide, 6404
Reserpic Acid, 8263
$C_{22}H_{28}N_4O_3$
Etonitazene, 3927
$C_{22}H_{28}N_4O_6$
Mitoxantrone, 6302
$C_{22}H_{28}N_6O_3S$
Delavirdine, 2882
$C_{22}H_{28}N_9NaO_9S_2$
Cefbuperazone Sodium Salt, 1919
$C_{22}H_{28}O_2$
Etonogestrel, 3928
$C_{22}H_{28}O_3$
Canrenone, 1753
Norethindrone Acetate, 6786
$C_{22}H_{28}O_6$
Isoquassin, 5266
Quassin, 8142
$C_{22}H_{29}ClN_2O_4$
Echitamine Chloride, 3547
$C_{22}H_{29}ClO_5$
Alclometasone, 213
Beclomethasone, 1017
$C_{22}H_{29}ClO_6$
Cloprostenol, 2388
$C_{22}H_{29}FO_4$
Desoximetasone, 2931
Fluocortolone, 4184
Fluorometholone, 4207
$C_{22}H_{29}FO_5$
Betamethasone, 1182
Dexamethasone, 2945
Paramethasone, 7133
$C_{22}H_{29}KO_4$
Canrenone Free Acid Potassium Salt,
1753
$C_{22}H_{29}NO_2$
Levopropoxyphene, 5523
Noracymethadol, 6773
Propoxyphene, 7952
$C_{22}H_{29}NO_5$
Trimebutine, 9868
$C_{22}H_{29}NO_7S$
Poldine Methylsulfate, 7671
$C_{22}H_{29}N_2O_4^+$
Echitamine, 3547
$C_{22}H_{29}N_3O_4S$
Lafutidine, 5396
$C_{22}H_{29}N_3O_6S$
Pivampicillin, 7626

$C_{22}H_{29}N_3S_2$
Thiethylperazine, 9467
$C_{22}H_{29}N_7O_5$
Puromycin, 8054
$C_{22}H_{29}N_9O_6$
Diminazene Aceturate, 3291
$C_{22}H_{29}N_9O_9S_2$
Cefbuperazone, 1919
$C_{22}H_{30}Cl_2N_{10}$
Chlorhexidine, 2092
$C_{22}H_{30}N_2$
Aprindine, 743
$C_{22}H_{30}N_2O$
Pirmenol, 7614
$C_{22}H_{30}N_2O_2$
Aspidospermine, 840
Eprozinol, 3693
$C_{22}H_{30}N_2O_2S$
Sufentanil, 9018
$C_{22}H_{30}N_2O_5$
Echitamine Hydroxide, 3547
Trandolapril Diacid, 9729
$C_{22}H_{30}N_2O_5S_2$
Spirapril, 8880
$C_{22}H_{30}N_4O_2S_2$
Thioproperazine, 9510
$C_{22}H_{30}N_6O_4S$
Sildenafil, 8628
$C_{22}H_{30}O$
Desogestrel, 2928
$C_{22}H_{30}O_2$
Promegestone, 7900
$C_{22}H_{30}O_3$
Siccanin, 8625
$C_{22}H_{30}O_4$
Ilimaquinone, 4939
$C_{22}H_{30}O_5$
Jasmolin II, 5306
Methylprednisolone, 6184
$C_{22}H_{30}O_6$
Neoquassin, 6548
$C_{22}H_{31}NO$
Tolterodine, 9684
$C_{22}H_{31}NO_3$
Epostane, 3683
Oxybutynin, 7054
Songorine, 8843
$C_{22}H_{31}NO_5S$
Latrunculin A, 5433
$C_{22}H_{31}N_3O_2$
Piboserod, 7505
$C_{22}H_{31}N_3O_5$
Cilazapril, 2275
$C_{22}H_{31}N_5O_4$
Melagatran, 5879
$C_{22}H_{32}Br_2N_4O_4$
Distigmine Bromide, 3405
$C_{22}H_{32}IN_3O_4S$
Penethamate Hydriodide, 7194
$C_{22}H_{32}N_2O_2$
Dopexamine, 3472
$C_{22}H_{32}N_2O_5$
Benzquinamide, 1122
$C_{22}H_{32}N_2O_6$
Hexoprenaline, 4743
$C_{22}H_{32}O_2$
Docosahexaenoic Acid, 3443
Synhexyl, 9143
$C_{22}H_{32}O_3$
Medroxyprogesterone, 5864
Medrysone, 5865
Methenolone 17-Acetate, 6039
Stenbolone Acetate, 8935
Testosterone Propionate, 9324
$C_{22}H_{32}O_4$
Algestone 16α-Methyl Ether, 232
Iloprost, 4942

$C_{22}H_{32}O_5$
Ajugarin-V, 190
Torilin, 9710
$C_{22}H_{32}O_6$
Ajugarin-II, 190
$C_{22}H_{32}O_7$
Cascarillin, 1882
$C_{22}H_{33}NO_2$
Atisine, 853
Veatchine, 10131
$C_{22}H_{33}NO_3$
Ajaconine, 186
Napelline, 6452
$C_{22}H_{33}N_3O_3$
Propylhexedrine l-Form Ethylphenyl-
barbiturate, 7972
$C_{22}H_{34}CaN_4O_6$
Pentobarbital Calcium Salt, 7243
$C_{22}H_{34}ClN_3O_2$
Bidisomide, 1208
$C_{22}H_{34}INO_2$
Cyclonium Iodide, 2729
$C_{22}H_{34}N_2O_4$
Oxamarin, 7014
$C_{22}H_{34}N_4O_{10}$
Sobuzoxane, 8706
$C_{22}H_{34}O_2$
Eicosapentaenoic Acid Ethyl Ester,
3578
$C_{22}H_{34}O_3$
Mesterolone Acetate, 5986
$C_{22}H_{34}O_5$
Pleuromutilin, 7650
$C_{22}H_{34}O_7$
Colforsin, 2460
$C_{22}H_{35}NO_2$
Himbacine, 4752
$C_{22}H_{36}N_2O_5S$
Tirofiban, 9620
$C_{22}H_{36}N_8O_{11}$
Pentigetide, 7241
$C_{22}H_{36}O_2$
Ganaxolone, 4392
$C_{22}H_{36}O_5$
Limaprost, 5543
Prostalene, 7991
$C_{22}H_{36}O_7$
Grayanotoxin I, 4578
$C_{22}H_{36}P_2$
(R,R)-Et-DuPHOS, 3513
$C_{22}H_{37}FO_5$
Fluocortolone 21-Pivalate, 4184
$C_{22}H_{37}NO_2$
Anandamide, 618
$C_{22}H_{37}N_3O_3S$
Isoaminile Cyclamate, 5155
$C_{22}H_{38}O_4$
Citrullol, 2331
$C_{22}H_{38}O_4Zn$
Undecylenic Acid Zinc Salt, 10033
$C_{22}H_{38}O_5$
Carboprost Methyl Ester, 1824
Misoprostol, 6293
Unoprostone, 10034
$C_{22}H_{39}NaO_4$
Sodium Stearyl Fumarate, 8807
$C_{22}H_{40}BrNO$
Domiphen Bromide, 3464
$C_{22}H_{40}O_7$
Agaricic Acid, 178
$C_{22}H_{41}NO_7$
Sphingofungin C, 8872
$C_{22}H_{42}N_2O$
Oleyl Hydroxyethyl Imidazoline,
6925
$C_{22}H_{42}N_4O_8S_2$
Pantethine, 7115

$C_{22}H_{42}O_2$
Brassidic Acid, 1369
Erucic Acid, 3732
$C_{22}H_{43}N_5O_{12}$
Isepamicin, 5151
$C_{22}H_{43}N_5O_{13}$
Amikacin, 400
$C_{22}H_{44}N_6O_{10}$
Arbekacin, 760
$C_{22}H_{44}O_2$
Behenic Acid, 1022
Butyl Stearate, 1592
Isobutyl Stearate, 5196
$C_{22}H_{45}NO$
Behenic Acid Amide, 1022
$C_{22}H_{46}BrN$
Ethyldimethyl-9-octadecenylammon-
ium Bromide, 3844
$C_{22}H_{46}N_4O_2$
Pithecolobine, 7622
$C_{22}H_{46}O$
n-Docosanol, 3444
$C_{22}H_{48}ClN$
Didecyldimethylammonium Chloride,
3110
$C_{22}H_{63}N_{13}P_4$
Schwesinger P$_4$ Base, 8535
$C_{23}H_{12}Cl_2N_4O_5$
DTAF, 3507
$C_{23}H_{14}F_2NNaO_2$
Brequinar Sodium Salt, 1376
$C_{23}H_{14}Na_2O_{11}$
Cromolyn Disodium Salt, 2581
$C_{23}H_{15}ClO_3$
Chlorophacinone, 2155
$C_{23}H_{15}F_2NO_2$
Brequinar, 1376
$C_{23}H_{16}O_3$
Diphacinone, 3335
$C_{23}H_{16}O_6$
Pamoic Acid, 7106
$C_{23}H_{16}O_{11}$
Cromolyn, 2581
$C_{23}H_{18}ClF_2N_3O_3S$
Vemurafenib, 10138
$C_{23}H_{18}ClN_4NaO_7S_2$
Polar® Yellow, 7670
$C_{23}H_{19}ClF_3NO_3$
Cyhalothrin, 2756
Lambda-cyhalothrin, 5398
$C_{23}H_{19}N_3O_2$
Tivantinib, 9640
$C_{23}H_{20}F_2N_2O_4$
Garenoxacin, 4402
$C_{23}H_{20}N_2O_3S$
Sulfinpyrazone, 9079
$C_{23}H_{20}N_2O_5$
Bentiromide, 1056
$C_{23}H_{20}O_{10}$
Fustin Tetraacetate, 4347
$C_{23}H_{21}ClN_6O_3$
Loprazolam, 5631
$C_{23}H_{21}ClO_3$
Chlorotrianisene, 2178
$C_{23}H_{21}F_7N_4O_3$
Aprepitant, 741
$C_{23}H_{21}N_2NaO_6S$
Carbenicillin Phenyl Sodium, 1793
$C_{23}H_{22}ClF_3O_2$
Bifenthrin, 1213
$C_{23}H_{22}ClNO_2$
Licofelone, 5533
$C_{23}H_{22}ClNO_4$
Mandipropamid, 5785
$C_{23}H_{22}ClN_3O_2$
Pagoclone, 7084

$C_{23}H_{22}F_7IN_2O_4S$
Flubendiamide, 4150
$C_{23}H_{22}O_5S$
Xylenol Blue, 10280
$C_{23}H_{22}O_6$
Deguelin, 2867
Rotenone, 8403
$C_{23}H_{22}O_7$
Epicocconone, 3670
Lactucin p-Hydroxyphenylacetate
Hydrate, 5394
Tephrosin, 9293
$C_{23}H_{23}ClFNO_5$
Elvitegravir, 3610
$C_{23}H_{23}IN_2S_2$
Dithiazanine Iodide, 3412
$C_{23}H_{23}N_3O_5$
Topotecan, 9707
$C_{23}H_{23}N_7O_5$
Pralatrexate, 7821
$C_{23}H_{24}N_2O_4S$
Eprosartan, 3691
$C_{23}H_{24}N_4O_6$
Merimepodib, 5968
$C_{23}H_{24}N_4O_9$
Nocardicin A, 6753
$C_{23}H_{24}N_6O_5S_2$
Cefquinome, 1944
$C_{23}H_{24}O_2$
Endiandric Acid B, 3626
$C_{23}H_{24}O_4$
Cyclofenil, 2713
$C_{23}H_{24}O_8$
Wortmannin, 10251
$C_{23}H_{25}ClN_2$
Malachite Green, 5763
$C_{23}H_{25}ClO_{12} \cdot 4H_2O$
Malvidin Chloride 3-β-Glucoside
Tetrahydrate, 5779
$C_{23}H_{25}ClO_{12} \cdot 5\frac{1}{2}H_2O$
Malvidin Chloride 3-Galactoside He-
mihendecahydrate, 5779
$C_{23}H_{25}F_3N_2OS$
Flupentixol, 4219
$C_{23}H_{25}N$
Fendiline, 4002
$C_{23}H_{25}N_5O_5$
Doxazosin, 3482
$C_{23}H_{26}ClN_7O_3$
Avanafil, 875
$C_{23}H_{26}Cl_2O_6$
Simfibrate, 8675
$C_{23}H_{26}FN_3O_2$
Spiperone, 8878
$C_{23}H_{26}FN_6O_9P$
Fostamatinib, 4287
$C_{23}H_{26}N_2$
Leucomalachite Green, 5763
$C_{23}H_{26}N_2O$
Roxindole, 8408
$C_{23}H_{26}N_2O_2$
Benzetimide, 1077
Dexetimide, 2947
Solifenacin, 8838
$C_{23}H_{26}N_2O_5$
Quinapril Diacid, 8167
$C_{23}H_{26}N_4O_9S_2$
1,3-Di-6-quinolylurea Bismethosul-
fate, 3393
$C_{23}H_{26}N_5NaO_7S$
Piperacillin Sodium Salt, 7575
$C_{23}H_{26}O_3$
Phenothrin, 7325
$C_{23}H_{27}ClO_4$
Delmadinone Acetate, 2883
$C_{23}H_{27}Cl_2N_3O_2$
Aripiprazole, 776

$C_{23}H_{27}FN_4O_2$
Risperidone, 8361
$C_{23}H_{27}FN_4O_3$
Paliperidone, 7087
$C_{23}H_{27}N$
Butenafine, 1521
$C_{23}H_{27}NO$
Deptropine, 2913
$C_{23}H_{27}NO_8$
Narceine, 6504
$C_{23}H_{27}N_3O$
Prenoxdiazine, 7854
$C_{23}H_{27}N_3O_7$
Minocycline, 6283
$C_{23}H_{27}N_5O_7S$
Piperacillin, 7575
$C_{23}H_{28}ClN_3O_2S$
Thiopropazate, 9509
$C_{23}H_{28}ClN_3O_4$
Benexate Hydrochloride, 1040
$C_{23}H_{28}ClN_3O_5S$
Glyburide, 4514
$C_{23}H_{28}ClN_5O_3$
Azimilide, 904
$C_{23}H_{28}Cl_2O_5$
Dichlorisone 21-Acetate, 3057
$C_{23}H_{28}F_2N_6O_4S$
Ticagrelor, 9583
$C_{23}H_{28}F_3NaO_6$
Fluprostenol Sodium Salt, 4224
$C_{23}H_{28}GdN_3Na_2O_{11}$
Gadoxetic Acid Sodium Salt, 4360
$C_{23}H_{28}N_2O_3$
Bopindolol, 1335
$C_{23}H_{28}N_2O_5$
Reserpiline, 8264
$C_{23}H_{28}N_2O_5S_2$
Temocapril, 9279
$C_{23}H_{28}O_6$
Enprostil, 3645
Prednisone 21-Acetate, 7842
$C_{23}H_{28}O_{15}$
Purpurogallin Diglucoside, 8057
$C_{23}H_{29}ClFN_3O_4$
Cisapride, 2315
$C_{23}H_{29}ClO_4$
Chlormadinone Acetate, 2102
$C_{23}H_{29}FO_6$
Isoflupredone 21-Acetate, 5220
$C_{23}H_{29}F_2N_3O$
Amperozide, 577
$C_{23}H_{29}F_3O_6$
Fluprostenol, 4224
$C_{23}H_{29}NO_3$
Phenoperidine, 7362
Propiverine, 7945
$C_{23}H_{29}NO_{12}$
Hygromycin, 4891
$C_{23}H_{29}N_3O$
Opipramol, 6948
$C_{23}H_{29}N_3O_2$
Oxypertine, 7070
$C_{23}H_{29}N_3O_2S_2$
Thiothixene, 9519
$C_{23}H_{29}N_5O_8S_2$
Cefcapene Pivaloyloxymethyl Ester,
1920
$C_{23}H_{30}BrNO_3$
Piperilate Ethyl Bromide, 7583
Propantheline Bromide, 7919
$C_{23}H_{30}ClN_3O$
Quinacrine, 8160
$C_{23}H_{30}Cl_2NNa_2O_6P \cdot H_2O$
Estramustine 17-(Dihydrogenphos-
phate) Disodium Salt Monohydrate,
3761

$C_{23}H_{30}FN_3$
Blonanserin, 1320
$C_{23}H_{30}GdN_3O_{11}$
Gadoxetic Acid, 4360
$C_{23}H_{30}N_2O$
α-Methylfentanyl, 6145
$C_{23}H_{30}N_2O_4$
Mitragynine, 6303
Pholcodine, 7441
$C_{23}H_{30}N_4O_2S$
Perospirone, 7296
$C_{23}H_{30}N_4O_4S$
Lexipafant, 5527
$C_{23}H_{30}N_{12}O_8S_2$
Ceftolozane, 1954
$C_{23}H_{30}O_3$
Etretinate, 3936
$C_{23}H_{30}O_4$
Elcometrine, 3586
Spiromesifen, 8886
$C_{23}H_{30}O_6$
Cortisone 21-Acetate, 2525
Fenprostalene, 4023
Prednisolone 21-Acetate, 7841
$C_{23}H_{31}Cl_2NO_3$
Estramustine, 3761
$C_{23}H_{31}FO_5$
Flurogestone Acetate, 4230
$C_{23}H_{31}FO_6$
Fludrocortisone 21-Acetate, 4161
$C_{23}H_{31}NO_2$
Levomethadyl Acetate, 5519
Methadyl Acetate, 6017
Motretinide, 6370
$C_{23}H_{31}NO_3$
Norgestimate, 6792
$C_{23}H_{31}NO_7S$
Bevonium Methyl Sulfate, 1197
Sulprostone, 9120
$C_{23}H_{31}N_3O_4S_2$
Batimastat, 1006
$C_{23}H_{32}Cl_2NO_6P$
Estramustine 17-Phosphate, 3761
$C_{23}H_{32}N_2OS$
Diafenthiuron, 2972
$C_{23}H_{32}N_2O_3$
Zipeprol, 10370
$C_{23}H_{32}N_2O_4$
Pinoxaden, 7564
$C_{23}H_{32}N_2O_5$
Ramipril, 8221
$C_{23}H_{32}N_2O_8$
Trichostatin C, 9814
$C_{23}H_{32}N_6O_4S$
Vardenafil, 10124
$C_{23}H_{32}N_6O_{14}$
Polyoxin A, 7696
$C_{23}H_{32}O_2$
Dimethisterone, 3240
Medrogestone, 5862
$C_{23}H_{32}O_3$
Estradiol 17-Valerate, 3758
$C_{23}H_{32}O_4$
Deoxycorticosterone Acetate, 2903
17α-Hydroxyprogesterone Acetate,
4877
$C_{23}H_{32}O_6$
Strophanthidin, 8981
$C_{23}H_{32}O_7$
16-Hydroxystrophanthidin, 8981
$C_{23}H_{33}IN_2O$
Isopropamide Iodide, 5248
$C_{23}H_{33}N_2O_2^+$
Prajmaline, 7820
$C_{23}H_{33}NaO_5$
Treprostinil Sodium Salt, 9748

$C_{23}H_{34}NO_5P$
Fosinopril Diacid, 4283
$C_{23}H_{34}O_4$
Digitoxigenin, 3181
Uzarigenin, 10083
$C_{23}H_{34}O_5$
Alfadolone Acetate, 228
Digoxigenin, 3185
Gitoxigenin, 4468
Mevastatin, 6243
Periplogenin, 7291
Treprostinil, 9748
$C_{23}H_{34}O_6$
Ajugarin-IV, 190
Diginatigenin, 3174
Dihydrostrophanthidin, 8981
$C_{23}H_{34}O_8$
Ouabagenin, 6999
$C_{23}H_{35}NO_2S$
Dalcetrapib, 2801
$C_{23}H_{35}NaO_7$
Pravastatin Sodium, 7835
$C_{23}H_{36}N_2O_2$
Finasteride, 4113
$C_{23}H_{36}N_2O_4$
Eliglustat, 3599
$C_{23}H_{36}N_4O_5S_3$
Octotiamine, 6849
$C_{23}H_{36}N_6O_5S$
Argatroban, 769
$C_{23}H_{36}O_3$
Dromostanolone Propionate, 3497
$C_{23}H_{37}NO_5S$
Leukotriene E$_4$, 5507
$C_{23}H_{37}NO_6$
Tocamphyl, 9649
$C_{23}H_{37}N_5O_6S$
Etamiphyllin Camphorsulfonate, 3766
$C_{23}H_{38}N_2O_5$
Volicitin, 10227
$C_{23}H_{38}O$
Teprenone, 9296
$C_{23}H_{38}O_2$
Norcholanic Acid, 6778
$C_{23}H_{38}O_3$
Pregnanediol 3-Acetate, 7845
$C_{23}H_{38}O_4$
Trimoprostil, 9895
$C_{23}H_{38}O_5$
Gemeprost, 4421
$C_{23}H_{38}O_6$
Ornoprostil, 6970
$C_{23}H_{40}O_2$
(17:0)-Urushiol, 10076
$C_{23}H_{42}ClNO_2$
Benzoxonium Chloride, 1114
$C_{23}H_{42}N_2O_{12}$
Pentolinium Tartrate, 7245
$C_{23}H_{45}N_5O_{14}$
Paromomycin, 7146
$C_{23}H_{46}$
Muscalure, 6394
$C_{23}H_{46}N_6O_{13}$
Neomycin B, 6539
$C_{23}H_{46}O_2$
Behenic Acid Methyl Ester, 1022
$C_{24}H_{12}$
Coronene, 2523
$C_{24}H_{16}As_2O_3$
10,10'-Oxydiphenoxarsine, 7060
$C_{24}H_{16}F_6N_4O_2$
Metaflumizone, 5992
$C_{24}H_{16}O_{12}$
Laccaic Acid B, 5378
$C_{24}H_{18}O_2P_2S_4$
Belleau's Reagent, 1027

$C_{24}H_{19}NO_5$
Oxyphenisatin Acetate, 7073
$C_{24}H_{19}NO_6$
Bisoxatin Acetate, 1296
$C_{24}H_{19}N_3O_5S$
Piroxicam Cinnamic Acid Ester, 7619
$C_{24}H_{20}AsBr$
Tetraphenylarsonium Bromide, 9383
$C_{24}H_{20}AsCl$
Tetraphenylarsonium Chloride, 9383
$C_{24}H_{20}BNa$
Sodium Tetraphenylborate, 8818
$C_{24}H_{20}Cl_2N_2OS$
Fenticonazole, 4032
$C_{24}H_{20}I_6N_4O_8$
Iocarmic Acid, 5053
$C_{24}H_{20}N_2$
N,N'-Diphenylbenzidine, 3349
$C_{24}H_{20}N_4O$
Scarlet Red, 8530
$C_{24}H_{20}N_6O_3$
Candesartan, 1742
$C_{24}H_{20}P^+$
Tetraphenylphosphonium, 9384
$C_{24}H_{21}F_2NO_3$
Ezetimibe, 3958
$C_{24}H_{21}I_6N_5O_8$
Ioxaglic Acid, 5110
$C_{24}H_{22}F_3N$
Xaliproden, 10252
$C_{24}H_{23}ClO_2$
Ospemifene, 6991
$C_{24}H_{23}FN_4O_3$
Olaparib, 6915
$C_{24}H_{23}NO_5S$
Aleglitazar, 221
$C_{24}H_{23}N_3O_2$
Bifeprunox, 1214
$C_{24}H_{23}N_3O_6S$
Talampicillin, 9165
$C_{24}H_{24}CaI_6N_4O_4$
Ipodate Calcium Salt, 5115
$C_{24}H_{24}N_2O_5$
Amtolmetin Guacil, 591
$C_{24}H_{25}ClFN_5O_3$
Afatinib, 169
Canertinib, 1747
$C_{24}H_{25}FNNaO_4$
Fluvastatin Sodium Salt, 4245
$C_{24}H_{25}FN_6O$
Mizolastine, 6307
$C_{24}H_{25}NO_3$
Cyphenothrin, 2765
$C_{24}H_{25}NO_4$
Flavoxate, 4128
$C_{24}H_{26}BrN_3O_3$
Nicergoline, 6581
$C_{24}H_{26}ClFN_4O$
Sertindole, 8605
$C_{24}H_{26}FNO_4$
Fluvastatin, 4245
$C_{24}H_{26}MgO_8$
Dimecrotic Acid Magnesium Salt,
3226
$C_{24}H_{26}N_2O_4$
Carvedilol, 1873
$C_{24}H_{26}N_2O_6$
Suxibuzone, 9139
$C_{24}H_{26}O_6$
Mangostin, 5807
$C_{24}H_{26}O_{13}$
Centaurein, 1969
Irigenin 7-Glucoside, 5134
$C_{24}H_{27}FN_2O_4$
Iloperidone, 4941
$C_{24}H_{27}N$
Prenylamine, 7855

$C_{24}H_{27}NO_2$
Octocrylene, 6845
$C_{24}H_{27}NO_4$
Tylocrebrine, 10011
Tylophorine, 10012
$C_{24}H_{27}NO_5S$
Troglitazone, 9948
$C_{24}H_{27}NO_7$
Platensimycin, 7640
$C_{24}H_{27}N_3O_4$
Fenpyroximate, 4024
$C_{24}H_{28}ClN_5O_3$
Dimenhydrinate, 3230
$C_{24}H_{28}F_3N_3O_3$
Tafenoquine, 9159
$C_{24}H_{28}N_2O_3$
Indacaterol, 4970
Ivacaftor, 5295
Naftopidil, 6440
$C_{24}H_{28}N_2O_5$
Benazepril, 1033
$C_{24}H_{28}N_4O_6$
Lamifiban, 5399
$C_{24}H_{28}O_2$
Bexarotene, 1198
$C_{24}H_{28}O_4$
Diethylstilbestrol Dipropionate, 3148
$C_{24}H_{29}ClO_4$
Cyproterone Acetate, 2771
$C_{24}H_{29}FN_2O_2$
Aceperone, 31
$C_{24}H_{29}FO_6$
Fluprednidene Acetate, 4222
$C_{24}H_{29}NO_3$
Donepezil, 3468
$C_{24}H_{29}NO_4$
Ethaverine, 3783
$C_{24}H_{29}N_5O_3$
Valsartan, 10102
$C_{24}H_{29}NaO_5$
Beraprost Sodium Salt, 1154
Taprostene Sodium Salt, 9190
$C_{24}H_{30}CaN_4O_6$
Cyclobarbital Calcium Salt, 2704
$C_{24}H_{30}ClFO_5$
Clocortolone 21-Acetate, 2363
$C_{24}H_{30}ClN_7O_4S$
Edoxaban, 3562
$C_{24}H_{30}FNa_2O_9P$
Triamcinolone Acetonide 21-Di-
sodium Phosphate, 9758
$C_{24}H_{30}F_2O_6$
Fluocinolone Acetonide, 4181
$C_{24}H_{30}N_2O_2$
Doxapram, 3481
$C_{24}H_{30}N_2O_2S$
Piperacetazine, 7574
$C_{24}H_{30}N_2O_3$
Chromafenozide, 2223
$C_{24}H_{30}N_2O_5$
Demethoxyvindoline, 10185
$C_{24}H_{30}O_3$
Drospirenone, 3502
$C_{24}H_{30}O_5$
Beraprost, 1154
Taprostene, 9190
$C_{24}H_{30}O_6$
Eplerenone, 3681
$C_{24}H_{30}O_7$
Athamantin, 851
$C_{24}H_{30}O_8$
Desaspidin BB, 2918
Flavaspidic Acid, 4121
$C_{24}H_{31}ClO_7$
Loteprednol Etabonate, 5642

$C_{24}H_{31}FO_6$
Betamethasone 21-Acetate, 1182
Dexamethasone 21-Acetate, 2945
Flunisolide, 4176
Paramethasone 21-Acetate, 7133
Triamcinolone Acetonide, 9758
$C_{24}H_{31}NO$
Abiraterone, 9
Dipipanone, 3377
$C_{24}H_{31}NO_4$
Drotaverine, 3503
$C_{24}H_{31}N_3OS$
Butaperazine, 1519
$C_{24}H_{32}ClFO_5$
Halcinonide, 4626
$C_{24}H_{32}N_2O_2$
Eprazinone, 3688
$C_{24}H_{32}O_4$
Acequinocyl, 34
Estradiol Dipropionate, 3758
Ethynodiol Diacetate, 3909
Megestrol Acetate, 5876
Resibufogenin, 8266
Scillarenin, 8537
$C_{24}H_{32}O_4S$
Spironolactone, 8887
$C_{24}H_{32}O_6$
Desonide, 2929
Methylprednisolone 21-Acetate, 6184
$C_{24}H_{32}O_7$
Etiproston, 3918
$C_{24}H_{33}FO_6$
Flurandrenolide, 4227
$C_{24}H_{33}NO_3$
Nafronyl, 6437
$C_{24}H_{33}N_3O_2S$
Dixyrazine, 3429
$C_{24}H_{33}N_3O_3S_2$
Pipotiazine, 7595
$C_{24}H_{33}N_3O_4$
Ranolazine, 8229
$C_{24}H_{33}NaO_5$
Dehydrocholic Acid Sodium Salt,
2872
$C_{24}H_{33}NaO_7S$
Bufalin Sodium Sulfate, 1478
$C_{24}H_{34}Br_2N_2O_4$
Hemicholinium Dibromide, 4678
$C_{24}H_{34}N_2Na_2O_{18}S_3$
Glucosulfone Sodium, 4499
$C_{24}H_{34}N_2O$
Bepridil, 1152
$C_{24}H_{34}N_2O_4^{2+}$
Hemicholinium, 4678
$C_{24}H_{34}N_2O_5$
Trandolapril, 9729
$C_{24}H_{34}N_4O_5S$
Glimepiride, 4475
$C_{24}H_{34}N_8O_4S_2$
Thiamine Disulfide, 9449
$C_{24}H_{34}O_3$
Rimexolone, 8352
$C_{24}H_{34}O_4$
Algestone Cyclic Acetal with Ace-
tone, 232
Bufalin, 1478
Medroxyprogesterone 17-Acetate,
5864
$C_{24}H_{34}O_5$
Dehydrocholic Acid, 2872
Gamabufotalin, 4388
$C_{24}H_{34}O_7$
Ajugarin-I, 190
$C_{24}H_{34}O_9$
T-2 Toxin, 9981
$C_{24}H_{35}NO_5$
Decoquinate, 2857

$C_{24}H_{35}N_5O_5$
Ximelagatran, 10274
$C_{24}H_{36}N_2O_9S$
Celesticetin, 1960
$C_{24}H_{36}N_4O_6S_2$
Romidepsin, 8382
$C_{24}H_{36}N_8O_{10}P_2S_2$
Thiamine Disulfide Phosphate, 9449
$C_{24}H_{36}O_4$
Bolandiol Dipropionate, 1325
$C_{24}H_{36}O_5$
Lovastatin, 5644
$C_{24}H_{36}O_8$
Ajugarin-III, 190
$C_{24}H_{38}N_2O_2$
Laurolinium Acetate, 5442
$C_{24}H_{38}N_4O_{10}S_2$
Hexamidine Isethionate, 4730
$C_{24}H_{38}O_4$
DEHP, 2868
$C_{24}H_{38}O_5$
Alfaprostol, 227
$C_{24}H_{39}NO_4$
Cassaine, 1891
$C_{24}H_{39}NO_9S$
Tiaprost Tromethamine Salt, 9578
$C_{24}H_{39}NaO_4$
Deoxycholic Acid Sodium Salt, 2901
$C_{24}H_{39}NaO_5$
Cholic Acid Sodium Salt, 2210
$C_{24}H_{40}N_2$
Conessine, 2480
$C_{24}H_{40}N_8O_4$
Dipyridamole, 3388
$C_{24}H_{40}O_2$
Cholanic Acid, 2203
$C_{24}H_{40}O_3$
Lithocholic Acid, 5601
$C_{24}H_{40}O_4$
Chenodiol, 2054
Deoxycholic Acid, 2901
Hyodeoxycholic Acid, 4896
Ursodiol, 10074
$C_{24}H_{40}O_5$
Cholic Acid, 2210
$C_{24}H_{41}NO_4$
Cassaidine, 1890
$C_{24}H_{42}$
Cholane, 2202
$C_{24}H_{42}CaO_{24}$
Lactobionic Acid Calcium Salt, 5391
$C_{24}H_{43}NO$
Linoleic Acid Cyclohexylamide, 5560
$C_{24}H_{43}NO_9$
Pseudopederin, 7171
$C_{24}H_{44}O_6$
Sorbitan Oleate, 8850
$C_{24}H_{46}O_6$
Sorbitan Stearate, 8850
$C_{24}H_{48}O_2$
Behenic Acid Ethyl Ester, 1022
Lignoceric Acid, 5541
$C_{24}H_{50}BrNO$
Cethexonium Bromide, 2017
$C_{24}H_{72}O_{10}Si_{11}$
Tetracosamethylhendecasiloxane,
9339
$C_{25}H_{17}NO_{13}$
Laccaic Acid C, 5378
$C_{25}H_{18}N_4O_2$
Asperlicin D, 833
$C_{25}H_{18}N_4O_3$
Asperlicin E, 833
$C_{25}H_{20}ClF_2N_3O_3$
Flucycloxuron, 4154
$C_{25}H_{20}N_4O_5$
Azilsartan, 903

C$_{26}$H$_{22}$O
Diphenylmethyl Ether, 1092
C$_{26}$H$_{22}$O$_2$
Benzopinacol, 1103
C$_{26}$H$_{23}$F$_2$NO$_4$
Flucythrinate, 4155
C$_{26}$H$_{23}$K$_4$N$_3$O$_{10}$
Quin2, 8158
C$_{26}$H$_{24}$P$_2$
DPPE, 1254
C$_{26}$H$_{25}$ClN$_2$O$_3$
Tolvaptan, 9699
C$_{26}$H$_{25}$F$_9$N$_2$O$_4$
Torcetrapib, 9708
C$_{26}$H$_{25}$N$_2$NaO$_6$S
Carindacillin Sodium Salt, 1839
C$_{26}$H$_{25}$N$_3$O$_3$S
Fenoverine, 4013
C$_{26}$H$_{25}$N$_3$O$_{11}$
Tunichrome B-1, 10000
C$_{26}$H$_{25}$N$_8$NaO$_{11}$S$_2$
Ceftobiprole Medocaril Sodium Salt, 1953
C$_{26}$H$_{26}$ClN$_3$
Rupatadine, 8430
C$_{26}$H$_{26}$F$_2$N$_2$
Flunarizine, 4175
C$_{26}$H$_{26}$N$_2$O$_2$S
BBOT, 1012
C$_{26}$H$_{26}$N$_2$O$_6$S
Carindacillin, 1839
C$_{26}$H$_{26}$N$_8$O$_{11}$S$_2$
Ceftobiprole Medocaril, 1953
C$_{26}$H$_{27}$ClN$_2$
Clocinizine, 2362
C$_{26}$H$_{27}$ClN$_2$O
Lofepramine, 5614
C$_{26}$H$_{27}$F$_5$N$_6$O$_3$
Telcagepant, 9259
C$_{26}$H$_{27}$NO$_9$
Idarubicin, 4927
C$_{26}$H$_{27}$NO$_{10}$
Carubicin, 1870
C$_{26}$H$_{27}$N$_5$O$_2$
Vilazodone, 10176
C$_{26}$H$_{28}$ClNO
Clomiphene, 2377
Toremifene, 9709
C$_{26}$H$_{28}$ClN$_3$
Pyrvinium Chloride, 8136
C$_{26}$H$_{28}$Cl$_2$N$_4$O$_4$
Ketoconazole, 5349
C$_{26}$H$_{28}$N$_2$
Cinnarizine, 2306
C$_{26}$H$_{28}$N$_4$O$_2$
Zaldaride, 10308
C$_{26}$H$_{28}$O$_{14}$
Apiin, 719
C$_{26}$H$_{29}$Cl$_2$N$_5$O$_3$
Bosutinib, 1357
C$_{26}$H$_{29}$FN$_2$O$_2$
Levocabastine, 5515
C$_{26}$H$_{29}$F$_2$N$_7$
Almitrine, 294
C$_{26}$H$_{29}$NO
Tamoxifen, 9180
C$_{26}$H$_{29}$NO$_2$
Droloxifene, 3495
C$_{26}$H$_{29}$N$_3$O$_6$
Nicardipine, 6580
C$_{26}$H$_{30}$BrNO$_4$S$_2$
Aclidinium Bromide, 106
C$_{26}$H$_{30}$Cl$_2$F$_3$NO
Halofantrine, 4630
C$_{26}$H$_{30}$Na$_2$O$_9$
Estriol 16,17-Bis(sodium Hemisuccinate), 3762

C$_{26}$H$_{30}$O$_8$
Limonin, 5547
Physodic Acid, 7494
C$_{26}$H$_{31}$Cl$_2$N$_5$O$_3$
Terconazole, 9303
C$_{26}$H$_{31}$N$_5$O$_2$
Physostigmine, 7496
C$_{26}$H$_{32}$ClFO$_5$
Clobetasone 17-Butyrate, 2359
C$_{26}$H$_{32}$F$_2$O$_7$
Diflorasone Diacetate, 3159
Fluocinonide, 4182
C$_{26}$H$_{32}$MgN$_6$O$_8$S$_2$
Dipyrone Magnesium Salt, 3390
C$_{26}$H$_{32}$N$_2$O$_5$
Delapril, 2881
C$_{26}$H$_{32}$N$_2$O$_8$
Tritoqualine, 9942
C$_{26}$H$_{32}$N$_3$NaO$_6$S$_2$
Merocyanine 540, 5969
C$_{26}$H$_{33}$FNNaO$_5$
Cerivastatin Sodium Salt, 1992
C$_{26}$H$_{33}$FO$_7$
Flunisolide 21-Acetate, 4176
C$_{26}$H$_{33}$NO$_2$
Abiraterone Acetate, 9
Fenretinide, 4025
C$_{26}$H$_{33}$NO$_4$
Cyprenorphine, 2766
C$_{26}$H$_{33}$NO$_6$
Lacidipine, 5379
C$_{26}$H$_{33}$NaO$_8$
Methylprednisolone 21-Succinate Sodium Salt, 6184
C$_{26}$H$_{34}$FNO$_5$
Cerivastatin, 1992
C$_{26}$H$_{34}$O$_4$
Methestrol Dipropionate, 6040
C$_{26}$H$_{34}$O$_7$
Cinobufotalin, 2308
Fumagillin, 4315
C$_{26}$H$_{35}$FO$_5$
Fluocortin Butyl, 4183
C$_{26}$H$_{35}$F$_3$O$_6$
Travoprost, 9738
C$_{26}$H$_{35}$NO$_4$
Diprenorphine, 3380
C$_{26}$H$_{35}$O$_2$P
SPhos Phosphine Ligand, 1469
C$_{26}$H$_{36}$NP
DavePhos Phosphine Ligand, 1469
C$_{26}$H$_{36}$N$_2$O$_5$
Zatebradine, 10313
C$_{26}$H$_{36}$N$_2$O$_9$
Antimycin A$_{3b}$, 702
C$_{26}$H$_{36}$O$_3$
Estradiol 17β-Cyclopentanepropanoate, 3758
C$_{26}$H$_{36}$O$_6$
Bufotalin, 1482
Prednisolone 21-Trimethylacetate, 7841
C$_{26}$H$_{37}$ClN$_4$O$_6$S$_2$.H$_2$O
Penicillin O 2-Chloroprocaine Salt Monohydrate, 7208
C$_{26}$H$_{37}$NO$_3$
Fesoterodine, 4095
C$_{26}$H$_{37}$N$_5$O$_2$
Cabergoline, 1606
C$_{26}$H$_{38}$N$_2$O$_4$
Mazipredone, 5832
C$_{26}$H$_{38}$O$_3$
Pentagestrone, 7225
C$_{26}$H$_{38}$O$_4$
Gestonorone Caproate, 4448
Lupulon, 5672

C$_{26}$H$_{38}$O$_5$S
Tixocortol 21-Pivalate, 9641
C$_{26}$H$_{39}$NO$_5$S
Epothilone A, 3684
C$_{26}$H$_{40}$BLiO
NB-Enantride, 6513
C$_{26}$H$_{40}$Cl$_4$N$_5$O$_{10}$PS
Canfosfamide, 1748
C$_{26}$H$_{40}$N$_4$O$_9$S
Sampatrilat, 8489
C$_{26}$H$_{40}$O$_3$
Testosterone Enanthate, 9324
C$_{26}$H$_{40}$O$_4$
Methandriol Dipropionate, 6022
C$_{26}$H$_{40}$O$_5$
Latanoprost, 5431
C$_{26}$H$_{41}$Br$_2$NO$_4$
Pinaverium Bromide, 7552
C$_{26}$H$_{41}$NO
Melinamide, 5890
C$_{26}$H$_{42}$O$_4$
DINP, 3318
Maxacalcitol, 5828
C$_{26}$H$_{42}$O$_9$
Pseudomonic Acid D, 8026
C$_{26}$H$_{43}$NO$_6$
Glycocholic Acid, 4529
C$_{26}$H$_{44}$NNaO$_7$S
Taurocholic Acid Sodium Salt, 9212
C$_{26}$H$_{44}$O$_2$
Tocol, 9652
C$_{26}$H$_{44}$O$_8$
Pseudomonic Acid C, 8026
C$_{26}$H$_{44}$O$_9$
Mupirocin, 6388
C$_{26}$H$_{45}$NO$_7$S
Taurocholic Acid, 9212
C$_{26}$H$_{49}$NO$_3$S
Cetrimonium Bromide p-Toluenesulfonate Analog, 2024
C$_{26}$H$_{50}$O$_4$
Bis(2-ethylhexyl) Sebacate, 1255
Glycol Dilaurate, 4533
C$_{26}$H$_{54}$NO$_7$P
1-O-Hexadecyl PAF, 7638
C$_{26}$H$_{56}$N$_{10}$
Alexidine, 226
C$_{27}$H$_{18}$AlN$_3$O$_3$
Aluminum Tris(8-hydroxyquinoline), 364
C$_{27}$H$_{20}$N$_4$O
Pyrinoline, 8102
C$_{27}$H$_{20}$O$_{12}$
Collinomycin, 2466
C$_{27}$H$_{21}$ClFN$_3$O$_2$
Lixivaptan, 5606
C$_{27}$H$_{22}$Cl$_2$N$_4$
Clofazimine, 2367
C$_{27}$H$_{22}$Cl$_2$N$_4$O
Tipifarnib, 9614
C$_{27}$H$_{22}$O$_{18}$
Tannic Acid, 9184
C$_{27}$H$_{23}$N$_5$O$_4$
Pranlukast, 7830
C$_{27}$H$_{24}$N$_2$O$_9$
Albofungin, 206
C$_{27}$H$_{25}$F$_2$N$_3$OS
Ritanserin, 8363
C$_{27}$H$_{25}$N$_9$Na$_2$O$_6$S
Tezosentan Sodium Salt, 9400
C$_{27}$H$_{26}$P$_2$
DPPP, 1254
C$_{27}$H$_{27}$Br$_2$NaO$_5$S
Bromthymol Blue Sodium Salt, 1455
C$_{27}$H$_{27}$N$_9$O$_6$S
Tezosentan, 9400

$C_{27}H_{28}Br_2O_5S$
Bromthymol Blue, 1455
$C_{27}H_{28}N_2O_7$
Cilnidipine, 2278
$C_{27}H_{28}N_5NaO_6S$
Bosentan Sodium Salt, 1355
$C_{27}H_{29}IN_2$
Dicyanine, 3101
$C_{27}H_{29}NO_{10}$
Daunorubicin, 2834
Pirozadil, 7620
$C_{27}H_{29}NO_{11}$
Doxorubicin, 3487
Epirubicin, 3678
$C_{27}H_{29}N_3O_6$
Barnidipine, 998
$C_{27}H_{29}N_5O_6S$
Bosentan, 1355
$C_{27}H_{30}Cl_2O_6$
Mometasone Furoate, 6327
$C_{27}H_{30}F_2N_2O_3$
Lomerizine, 5620
$C_{27}H_{30}F_6N_2O_2$
Dutasteride, 3518
$C_{27}H_{30}N_2O_2$
Asimadoline, 823
$C_{27}H_{30}N_4O$
Oxatomide, 7024
$C_{27}H_{30}O_5S$
Thymol Blue, 9555
$C_{27}H_{30}O_6$
Sofalcone, 8828
$C_{27}H_{30}O_{14}$
Glucofrangulin, 4488
Kaempferol 3,7-Dirhamnoside, 5322
Morindin, 6356
Sophorabioside, 8844
$C_{27}H_{30}O_{15}$
Saponarin, 8501
$C_{27}H_{30}O_{16}$
Rutin, 8438
$C_{27}H_{31}Br_2ClN_4O_2$
Lonafarnib, 5622
$C_{27}H_{31}ClN_2O$
Chlorbenzoxamine, 2081
$C_{27}H_{31}ClO_{15}$
Cyanidin Chloride 3-Rhamnogluco-
side, 2677
Pelargonidin 3,5-Diglucoside, 7180
$C_{27}H_{31}ClO_{16}$
Cyanidin Chloride 3-Sophoroside,
2677
$C_{27}H_{31}ClO_{17}$
Delphinidin 3,5-Diglucoside, 2885
$C_{27}H_{31}N_2NaO_6S_2$
Sulphan Blue, 9117
$C_{27}H_{32}N_2O_4S_3$
Mesoridazine Benzenesulfonate, 5980
$C_{27}H_{32}N_4O_8$
Pyridomycin, 8089
$C_{27}H_{32}O_8$
Verrucarin J, 10160
$C_{27}H_{32}O_9$
Verrucarin B, 10160
$C_{27}H_{32}O_{14}$
Naringin, 6507
$C_{27}H_{33}NO_3$
Prenylamine Lactate, 7855
$C_{27}H_{33}NO_{10}S$
Thiocolchicine 2-Glucoside Analog,
9477
$C_{27}H_{33}N_3O_6S$
Gliquidone, 4478
$C_{27}H_{33}N_3O_8$
Rolitetracycline, 8381
$C_{27}H_{33}N_9O_{15}P_2$
Flavin-Adenine Dinucleotide, 4122

$C_{27}H_{34}F_2O_7$
Difluprednate, 3171
$C_{27}H_{34}N_2O_4S$
Brilliant Green, 1380
$C_{27}H_{34}N_4O$
Piritramide, 7611
$C_{27}H_{34}O_3$
Nandrolone Phenpropionate, 6450
$C_{27}H_{34}O_8$
Verrucarin K, 10160
$C_{27}H_{34}O_9$
Verrucarin A, 10160
$C_{27}H_{34}O_{11}$
Phillyrin, 7434
$C_{27}H_{35}BrClN_3$
Methyl Green, 6154
$C_{27}H_{35}N_5O_7S$
Met^5-Enkephalin, 3627
$C_{27}H_{36}ClFO_5$
Clocortolone 21-Pivalate, 2363
$C_{27}H_{36}F_2O_5$
Diflucortolone 21-Valerate, 3162
$C_{27}H_{36}F_2O_6$
Flumethasone 21-Pivalate, 4169
$C_{27}H_{36}N_2O_4$
Repaglinide, 8256
$C_{27}H_{36}N_2O_5$
Ivabradine, 5294
$C_{27}H_{36}O_8$
Prednicarbate, 7840
$C_{27}H_{37}FO_6$
Betamethasone 17-Valerate, 1182
$C_{27}H_{37}N_3O_7S$
Darunavir, 2829
$C_{27}H_{38}N_2O_4$
Verapamil, 10144
$C_{27}H_{38}O_3$
Norethindrone Enanthate, 6786
$C_{27}H_{38}O_4$
Azafrin, 888
$C_{27}H_{38}O_6$
Prednisolone 2l-tert-Butylacetate,
7841
$C_{27}H_{39}NO_2$
Veratramine, 10146
$C_{27}H_{39}NO_3$
Jervine, 5312
$C_{27}H_{40}N_2O_2$
Pifarnine, 7532
$C_{27}H_{40}N_8O_7$
Cilengitide, 2276
$C_{27}H_{40}O$
Neoergosterol, 6536
$C_{27}H_{40}O_4$
17α-Hydroxyprogesterone Caproate,
4877
$C_{27}H_{41}NO_2$
Cyclopamine, 2733
$C_{27}H_{41}NO_5S$
Epothilone D, 3684
$C_{27}H_{41}NO_6S$
Epothilone B, 3684
$C_{27}H_{42}ClNO$
Phenoctide, 7353
$C_{27}H_{42}ClNO_2$
Benzethonium Chloride, 1076
$C_{27}H_{42}FeN_9O_{12}$
Ferrichrome, 4051
$C_{27}H_{42}N_2O_5S$
Ixabepilone, 5297
$C_{27}H_{42}O_3$
Diosgenin, 3323
Methenolone 17-Enanthate, 6039
$C_{27}H_{42}O_4$
Hecogenin, 4657
$C_{27}H_{43}NO$
Solanidine, 8831

$C_{27}H_{43}NO_2$
Rubijervine, 8420
Solasodine, 8836
Veralkamine, 10143
$C_{27}H_{43}NO_3$
Imperialine, 4968
$C_{27}H_{43}NO_8$
Cevine, 2034
Germine, 4446
$C_{27}H_{43}NO_9$
Protoverine, 8009
$C_{27}H_{44}O$
(3β)-7-Dehydrocholesterol, 2871
Desmosterol, 2927
Vitamin D$_3$, 10216
$C_{27}H_{44}O_2$
Calcifediol, 1641
Gefarnate, 4411
1α-Hydroxycholecalciferol, 4858
$C_{27}H_{44}O_3$
Calcitriol, 1647
Paricalcitol, 7143
Sarsasapogenin, 8517
Smilagenin, 8703
Tacalcitol, 9152
Tigogenin, 9591
$C_{27}H_{44}O_4$
Chlorogenin, 2144
Digalogenin, 3173
Gitogenin, 4466
$C_{27}H_{44}O_5$
Digitogenin, 3179
$C_{27}H_{44}O_6$
Ecdysone, 3538
Ponasterone A, 7710
$C_{27}H_{44}O_7$
20-Hydroxyecdysone, 3538
$C_{27}H_{45}NO_2$
Tomatidine, 9703
$C_{27}H_{45}NO_3$
Verticine, 10165
$C_{27}H_{45}N_5O_5$
Boceprevir, 1322
$C_{27}H_{46}N_2O_2$
Solanocapsine, 8833
$C_{27}H_{46}O$
Allocholesterol, 256
Cholesterol, 2208
Epicholesterol, 3668
$C_{27}H_{46}O_2$
24-Hydroxycholesterol, 4859
25-Hydroxycholesterol, 4860
δ-Tocopherol, 9656
$C_{27}H_{47}ClN_2O_3$
Lapyrium Chloride Stearoyl Analog,
5421
$C_{27}H_{48}$
(5α)-Cholestane, 2206
Coprostane, 2508
$C_{27}H_{48}N_6O_9$
Nocardamin, 6752
$C_{27}H_{48}O$
Cholestanol, 2207
Coprosterol, 2509
Epicholestanol, 3667
$C_{27}H_{50}N_6O_9$
Deferoxamine N-Acetyl Derivative,
2863
$C_{27}H_{58}N_2O_3 \cdot 2HF$
Olaflur, 6912
$C_{28}H_{14}N_2O_4$
Indanthrene, 4974
$C_{28}H_{15}NO_4$
Anthrimide, 681
$C_{28}H_{17}N_3O_4$
Bis(4-amino-1-anthraquinonyl)amine,
1247

$C_{28}H_{18}O_4$
α-Naphtholphthalein, 6474

$C_{28}H_{19}N_5Na_2O_6S_4$
Thiazol Yellow G, 9462

$C_{28}H_{20}CaN_2O_8.5H_2O$
Benzoylpas Calcium Salt Pentahydrate, 1118

$C_{28}H_{20}N_2Na_2O_8S_2$
Alizarin Cyanine Green F, 244

$C_{28}H_{22}Cl_2FNO_3$
Flumethrin, 4171

$C_{28}H_{22}N_2O_2$
Quinizarin Green SS, 8181

$C_{28}H_{22}N_4Na_2O_8S_2$
Blankophor® R, 1316

$C_{28}H_{22}O_3$
Diphenylacetic Acid Anhydride, 3346

$C_{28}H_{24}Br_2N_6$
Fazadinium Bromide, 3980

$C_{28}H_{24}ClNO_6$
Tropesin, 9957

$C_{28}H_{26}ClN_7$
Isometamidium Chloride, 5229

$C_{28}H_{26}N_4O_3$
Staurosporine, 8928

$C_{28}H_{27}ClF_5NO$
Penfluridol, 7196

$C_{28}H_{27}NO_4S$
Raloxifene, 8215

$C_{28}H_{28}N_2O_2$
Difenoxin, 3155

$C_{28}H_{28}N_4O_3$
Ruboxistaurin, 8423

$C_{28}H_{28}O_2P_2$
(R,R)-DIPAMP, 3333

$C_{28}H_{28}O_3$
Adapalene, 138

$C_{28}H_{28}P_2$
DPPB, 1254
(S,S)-Chiraphos, 2063

$C_{28}H_{29}ClN_4S$
Israpafant, 5288

$C_{28}H_{29}F_2N_3O$
Pimozide, 7546

$C_{28}H_{29}NO_4$
Bephenium Hydroxynaphthoate, 1150

$C_{28}H_{29}NO_4S$
Arzoxifene, 810

$C_{28}H_{30}N_2O_2$
Darifenacin, 2826

$C_{28}H_{30}N_6OS$
Masitinib, 5822

$C_{28}H_{30}O_4$
Thymolphthalein, 9556

$C_{28}H_{30}O_{14}.3H_2O$
Pseudobaptigenin 7-Rhamnoglucoside Trihydrate, 8019

$C_{28}H_{31}ClN_2O_3$
Rhodamine B, 8307

$C_{28}H_{31}FN_4O$
Astemizole, 846

$C_{28}H_{31}NO_2$
Lasofoxifene, 5430

$C_{28}H_{31}NO_5$
Bitolterol, 1308

$C_{28}H_{31}NO_9$
Rhodomycin B, 8315

$C_{28}H_{31}N_3O_6$
Benidipine, 1045

$C_{28}H_{31}NaO_9S$
Prednisolone 21-m-Sulfobenzoate Sodium Salt, 7841

$C_{28}H_{32}FNO_6$
Dexamethasone 21-Isonicotinate, 2945

$C_{28}H_{32}IN_3S_2$
Tibezonium Iodide, 9581

$C_{28}H_{32}N_2O_5$
Nalfurafine, 6443

$C_{28}H_{32}O_{14}$
Acacetin 7-Rhamnoglucoside, 14
Linarin, 5553

$C_{28}H_{32}O_{15}$
Diosmin, 3325

$C_{28}H_{33}ClN_2$
Buclizine, 1474

$C_{28}H_{33}ClO_{16}$
Peonidin 3,5-Diglucoside, 7256

$C_{28}H_{33}ClO_{17}$
Petunidin 3,5-Diglucoside, 7305

$C_{28}H_{33}N_3O_5$
Haplophytine O-Methylhaplophytine, 4644

$C_{28}H_{34}N_2O_3$
Denatonium Benzoate, 2893

$C_{28}H_{34}O_8$
Uliginosin B, 10028

$C_{28}H_{34}O_8S_2$
Ecamsule, 3537

$C_{28}H_{34}O_9$
Nomilin, 6761

$C_{28}H_{34}O_{15}$
Hesperidin, 4706

$C_{28}H_{35}ClN_4O$
Mosapramine, 6363

$C_{28}H_{35}FO_7$
Amcinonide, 382

$C_{28}H_{35}FO_9$
Triamcinolone Acetonide 21-Hemisuccinate, 9758

$C_{28}H_{35}NO_3$
Ulipristal, 10030

$C_{28}H_{35}NO_4$
Asoprisnil, 824

$C_{28}H_{35}N_3O_5$
Paraherquamide, 7130

$C_{28}H_{35}N_3O_7$
Virginiamycin M_1, 10200

$C_{28}H_{36}N_2O_4$
Psychotrine, 8038

$C_{28}H_{36}N_4O_2S$
Lurasidone, 5673

$C_{28}H_{36}O_6$
Clinofibrate, 2355

$C_{28}H_{36}O_8$
Uliginosin A, 10028

$C_{28}H_{36}O_{11}$
Bruceantin, 1463

$C_{28}H_{36}O_{15}$
Neohesperidin Dihydrochalcone, 6537

$C_{28}H_{37}ClN_4O$
Clocapramine, 2361

$C_{28}H_{37}ClO_7$
Alclometasone 17,21-Dipropionate, 213
Beclomethasone Dipropionate, 1017

$C_{28}H_{37}FO_7$
Betamethasone 17,21-Dipropionate, 1182
Dexamethasone 17,21-Dipropionate, 2945

$C_{28}H_{37}N_3O_3$
Bilastine, 1220

$C_{28}H_{37}N_5O_7$
Leu⁵-Enkephalin, 3627

$C_{28}H_{38}BrNO_4$
Butropium Bromide, 1536

$C_{28}H_{38}F_3N_5O_2$
Vicriviroc, 10172

$C_{28}H_{38}N_2O_4$
Cephaeline, 1973

$C_{28}H_{38}N_4O_8$
Pactamycin, 7083

$C_{28}H_{38}O_6$
Withaferin A, 10246

$C_{28}H_{38}O_{19}$
Octaacetyl-aldehydo-cellobiose, 1963
Sucrose Octaacetate, 9013

$C_{28}H_{39}FO_5$
Fluocortolone 21-Hexanoate, 4184

$C_{28}H_{40}N_2O_5$
Gallopamil, 4386

$C_{28}H_{40}N_2O_9$
Antimycin A, 702

$C_{28}H_{40}O_4$
Pentagestrone Acetate, 7225

$C_{28}H_{40}O_7$
Diginin, 3176

$C_{28}H_{41}N_3O_2$
Teleocidin B_4, 9260

$C_{28}H_{41}N_3O_3$
Oxethazaine, 7032
Tiropramide, 9622

$C_{28}H_{41}N_3O_8$
Detoxin D_1, 2940

$C_{28}H_{42}Cl_4N_4O_2$
Ambenonium Chloride, 373

$C_{28}H_{42}O$
Dehydroergosterol, 2877

$C_{28}H_{42}O_2$
β-Tocotrienol, 9659

$C_{28}H_{44}ClNO_2$
Methylbenzethonium Chloride, 6097

$C_{28}H_{44}O$
Ergosterol, 3716
Isopyrocalciferol, 5265
Lumisterol, 5662
Pyrocalciferol, 8110
Suprasterol II, 9133
Tachysterol, 9153
Vitamin D_2, 10215

$C_{28}H_{44}O_2$
Doxercalciferol, 3484

$C_{28}H_{44}O_3$
Nandrolone Decanoate, 6450

$C_{28}H_{46}CuN_{12}O_8$
GHK Bis Complex with Copper, 4452

$C_{28}H_{46}O$
Dihydrotachysterol, 3198
Ergosterol 22,23-Dihydro Analog, 3716
Vitamin D_4, 10217

$C_{28}H_{47}NO_4S$
Tiamulin, 9574

$C_{28}H_{47}NO_7$
Narbomycin, 6503

$C_{28}H_{47}NO_8$
Picromycin, 7525

$C_{28}H_{48}O$
Campesterol, 1731
Fungisterol, 4320

$C_{28}H_{48}O_2$
β-Tocopherol, 9654
γ-Tocopherol, 9655

$C_{28}H_{48}O_6$
Brassinolide, 1370

$C_{28}H_{50}$
Ergostane, 3714

$C_{28}H_{50}N_2O_4$
Carpaine, 1860

$C_{28}H_{50}N_4O_7$
Epoxomicin, 3686

$C_{28}H_{50}O$
Ergostanol, 3715

$C_{28}H_{54}N_8$
Plerixafor, 7649

$C_{28}H_{58}NO_7P$
1-O-Octadecyl-PAF, 7638

C$_{28}$H$_{58}$O
  Octacosanol, 6833
C$_{28}$H$_{60}$Al$_{14}$O$_{71}$S$_7$
  Dosmalfate, 3476
C$_{29}$H$_{24}$N$_4$O$_8$
  Niceritrol, 6582
C$_{29}$H$_{24}$N$_6$
  Sudan Black B, 9016
C$_{29}$H$_{24}$O$_{12}$
  Theaflavine, 9423
C$_{29}$H$_{25}$NO$_3$
  Py-Phe, 8063
C$_{29}$H$_{25}$N$_3$O$_5$
  Nicomorphine, 6605
C$_{29}$H$_{26}$ClFN$_4$O$_4$S
  Lapatinib, 5418
C$_{29}$H$_{27}$ClN$_2$O$_{14}$S
  Meclocycline 5-Sulfosalicylate, 5849
C$_{29}$H$_{27}$NO$_4$S$_2$
  Tipepidine Hibenzate, 9613
C$_{29}$H$_{27}$N$_3$O$_{14}$
  Fura-2, 4322
C$_{29}$H$_{28}$N$_2$O$_5$
  Thymidine Monotrityl Thymidine, 9552
C$_{29}$H$_{28}$N$_4$O$_{11}$
  Edotecarin, 3560
C$_{29}$H$_{30}$N$_6$O$_6$
  Olmesartan, 6933
C$_{29}$H$_{30}$O$_{13}$
  Amarogentin, 371
C$_{29}$H$_{31}$F$_2$N$_3$O
  Fluspirilene, 4237
C$_{29}$H$_{31}$F$_3$N$_2$O$_3$
  Almorexant, 297
C$_{29}$H$_{31}$N$_7$O
  Imatinib, 4943
C$_{29}$H$_{32}$ClN$_5$O$_2$
  Pyronaridine, 8117
C$_{29}$H$_{32}$N$_2$O$_6$S
  Semotiadil, 8585
C$_{29}$H$_{32}$O$_{13}$
  Etoposide, 3929
C$_{29}$H$_{33}$ClN$_2$O$_2$
  Loperamide, 5628
C$_{29}$H$_{33}$FN$_2$O$_4$
  Vorapaxar, 10230
C$_{29}$H$_{33}$FO$_6$
  Betamethasone 17-Benzoate, 1182
C$_{29}$H$_{33}$N$_9$O$_{12}$
  Pteropterin, 8043
C$_{29}$H$_{33}$O$_{16}$P
  Etoposide Phosphate, 3929
C$_{29}$H$_{34}$N$_2$O$_2$
  Dotarizine, 3477
C$_{29}$H$_{34}$O$_6$
  Tribenoside, 9769
C$_{29}$H$_{34}$O$_{15}$
  Pectolinarigenin 7-Rutinoside, 7170
C$_{29}$H$_{35}$ClO$_{17}$
  Malvidin Chloride 3,5-Diglucoside, 5779
C$_{29}$H$_{35}$NO$_2$
  Mifepristone, 6266
C$_{29}$H$_{36}$N$_2$O$_4$
  Emetamine, 3616
C$_{29}$H$_{36}$O$_4$
  Algestone Acetophenide, 233
C$_{29}$H$_{36}$O$_7$
  O-Cinnamoyltaxicin-I, 9214
C$_{29}$H$_{37}$NO$_5$
  Cytochalasin B, 2787
C$_{29}$H$_{37}$N$_3$O$_6$
  Calcimycin, 1642
C$_{29}$H$_{38}$FN$_3$O$_3$
  Mibefradil, 6252

C$_{29}$H$_{38}$F$_3$N$_3$O$_2$S
  Fluphenazine Enanthate, 4220
C$_{29}$H$_{38}$N$_2$O$_4$
  Dehydroemetine, 2874
C$_{29}$H$_{38}$N$_2$O$_6$
  Atrasentan, 861
C$_{29}$H$_{38}$N$_4$O$_6$S.H$_2$O
  Penicillin G Procaine, 7206
C$_{29}$H$_{38}$N$_4$O$_9$
  Pipacycline, 7566
C$_{29}$H$_{38}$N$_4$O$_{10}$
  Lymecycline, 5689
C$_{29}$H$_{38}$O$_4$
  Celastrol, 1957
C$_{29}$H$_{38}$O$_9$
  Uscharidin, 10077
C$_{29}$H$_{39}$NO$_2$
  Mofarotene, 6316
C$_{29}$H$_{39}$NO$_3$
  Onapristone, 6942
C$_{29}$H$_{39}$N$_9$
  Homoharringtonine, 4776
C$_{29}$H$_{39}$N$_3$O$_2$
  Echinuline, 3546
C$_{29}$H$_{39}$N$_5$O$_8$
  Tigecycline, 9588
C$_{29}$H$_{40}$N$_2$O$_4$
  Emetine, 3616
C$_{29}$H$_{40}$N$_2$O$_9$
  Geldanamycin, 4416
C$_{29}$H$_{40}$O$_6$
  Cortisone 21-Cyclopentanepropionate, 2525
C$_{29}$H$_{40}$O$_9$
  Calotropin, 1722
C$_{29}$H$_{41}$F$_2$N$_5$O
  Maraviroc, 5814
C$_{29}$H$_{41}$NO$_4$
  Buprenorphine, 1501
C$_{29}$H$_{41}$NO$_7$
  Candoxatril, 1745
C$_{29}$H$_{41}$O$_2$
  Galvinoxyl, 4387
C$_{29}$H$_{42}$N$_6$O$_9$
  Amicetin, 391
C$_{29}$H$_{42}$O$_5$
  Antheridiol, 673
C$_{29}$H$_{42}$O$_9$
  Helveticoside, 4667
C$_{29}$H$_{42}$O$_{10}$
  Adonitoxin, 158
  Convallatoxin, 2497
C$_{29}$H$_{43}$BrN$_2$O$_4$
  Otilonium Bromide, 6997
C$_{29}$H$_{43}$NO$_4$S
  Avasimibe, 876
C$_{29}$H$_{44}$O$_2$
  α-Tocotrienol, 9658
C$_{29}$H$_{44}$O$_{12}$
  Ouabain, 7000
C$_{29}$H$_{47}$NO$_6$
  Coumingine, 2553
C$_{29}$H$_{48}$O
  Chondrillasterol, 2216
  7-Dehydrositosterol, 2880
  Fucosterol, 4309
  α-Spinasterol, 8876
  Stigmasterol, 8944
C$_{29}$H$_{50}$O$_2$
  α-Tocopherol, 9653
C$_{29}$H$_{52}$O
  Stigmastanol, 8943
C$_{29}$H$_{53}$NO$_5$
  Orlistat, 6964
C$_{30}$H$_{16}$O$_8$
  Hypericin, 4902

C$_{30}$H$_{18}$
  9,10-Bis(2-phenylethynyl)anthracene, 1300
C$_{30}$H$_{20}$F$_{12}$O$_2$S
  Martin Sulfurane, 5821
C$_{30}$H$_{21}$FeN$_3$O$_{15}$$^{3-}$
  Ferric Enterobactin, 3650
C$_{30}$H$_{21}$NO$_9$
  Fredericamycin A, 4296
C$_{30}$H$_{23}$BrO$_4$
  Bromadiolone, 1389
C$_{30}$H$_{25}$F$_{10}$NO$_3$
  Anacetrapib, 615
C$_{30}$H$_{26}$CaO$_6$.2H$_2$O
  Fenoprofen Calcium Salt Dihydrate, 4011
C$_{30}$H$_{26}$MgO$_8$
  Menbutone Magnesium Salt, 5902
C$_{30}$H$_{26}$O$_{14}$
  Ergoflavin, 3709
C$_{30}$H$_{27}$N$_3$O$_{15}$
  Enterobactin, 3650
C$_{30}$H$_{28}$N$_6$O$_6$S$_4$
  Verticillin A, 10164
C$_{30}$H$_{28}$N$_6$O$_7$S$_4$
  Verticillin B, 10164
C$_{30}$H$_{28}$N$_6$O$_7$S$_5$
  Verticillin C, 10164
C$_{30}$H$_{28}$O$_8$
  Rottlerin, 8405
C$_{30}$H$_{29}$ClN$_6$O$_3$
  Neratinib, 6557
C$_{30}$H$_{30}$O$_8$
  Gossypol, 4564
C$_{30}$H$_{31}$ClN$_6$
  Janus Green B, 5303
C$_{30}$H$_{32}$Cl$_3$NO
  Lumefantrine, 5657
C$_{30}$H$_{32}$N$_2$O$_2$
  Diphenoxylate, 3343
C$_{30}$H$_{34}$N$_2$O$_3$
  Bazedoxifene, 1011
C$_{30}$H$_{34}$O$_{13}$
  Picrotoxin, 7527
C$_{30}$H$_{35}$BrN$_{12}$O$_5$
  Brostallicin, 1458
C$_{30}$H$_{35}$BrO$_7$
  3-(p-Bromobenzoyl)strophanthidin, 8981
C$_{30}$H$_{35}$F$_2$N$_3$O
  Lidoflazine, 5537
C$_{30}$H$_{35}$NO$_3$
  Centchroman, 1972
C$_{30}$H$_{36}$O$_7$
  3-Benzoylstrophanthidin, 8981
C$_{30}$H$_{36}$O$_9$
  Nimbin, 6631
C$_{30}$H$_{37}$NO$_4$
  Ulipristal Acetate, 10030
C$_{30}$H$_{37}$N$_5$O$_5$
  Ergosine, 3713
C$_{30}$H$_{39}$FN$_4$O$_4$
  Ulimorelin, 10029
C$_{30}$H$_{40}$Cl$_2$N$_4$
  Dequalinium Chloride, 2914
C$_{30}$H$_{40}$O$_2$
  β-Citraurin, 2323
C$_{30}$H$_{40}$O$_6$
  Absinthin, 13
  Anabsinthin, 613
C$_{30}$H$_{41}$FO$_7$
  Triamcinolone Hexacetonide, 9759
C$_{30}$H$_{42}$N$_2$O$_{15}$S$_2$
  Sinalbin, 8678
C$_{30}$H$_{42}$O$_8$
  Proscillaridin, 7985

C$_{30}$H$_{43}$O$_2$P
RuPhos Phosphine Ligand, 1469
C$_{30}$H$_{44}$K$_2$O$_{16}$S$_2$
Atractyloside, 858
C$_{30}$H$_{44}$N$_2$O$_{10}$
Hexobendine, 4741
C$_{30}$H$_{44}$O$_3$
Boldenone 10-Undecenoate, 1327
C$_{30}$H$_{44}$O$_9$
Cymarin, 2758
C$_{30}$H$_{45}$NNaO$_7$P
Fosinopril Sodium Salt, 4283
C$_{30}$H$_{46}$NO$_7$P
Fosinopril, 4283
C$_{30}$H$_{46}$N$_2$O$_{14}$S$_2$
Aclatonium Napadisilate, 105
C$_{30}$H$_{46}$O$_3$
Seocalcitol, 8595
C$_{30}$H$_{46}$O$_4$
Enoxolone, 3644
C$_{30}$H$_{46}$O$_5$
Ceanothic Acid, 1911
Quinovic Acid, 8192
C$_{30}$H$_{46}$O$_8$
Neriifolin, 6559
Periplocymarin, 7290
C$_{30}$H$_{47}$NO$_4$S
Retapamulin, 8281
C$_{30}$H$_{47}$N$_3$O$_9$S
Leukotriene C$_4$, 5507
C$_{30}$H$_{48}$N$_2$O$_2$
Hexestrol Bis(β-diethylaminoethyl ether), 4738
C$_{30}$H$_{48}$O$_3$
Betulinic Acid, 1193
β-Boswellic Acid, 1358
Nandrolone Dodecanoate, 6450
Oleanolic Acid, 6920
Testosterone Undecanoate, 9324
Ursolic Acid, 10075
C$_{30}$H$_{48}$O$_4$
Hederagenin, 4659
C$_{30}$H$_{48}$O$_5$
Acacic Acid, 16
Cimigenol, 2284
C$_{30}$H$_{49}$Br$_2$N$_3$
FM1-43, 4247
C$_{30}$H$_{49}$N$_9$O$_9$
Thymopentin, 9558
C$_{30}$H$_{50}$
Squalene, 8896
C$_{30}$H$_{50}$O
α-Amyrin, 610
β-Amyrin, 611
Friedelin, 4300
Lanosterol, 5411
Lupeol, 5668
Shionone, 8620
α_1-Sitosterol, 8693
Taraxasterol, 9193
Taraxerol, 9195
C$_{30}$H$_{50}$O$_5$
Eldecalcitol, 3587
C$_{30}$H$_{51}$B
Longifolene Borane Deriv, 5624
C$_{30}$H$_{52}$O
Dinosterol, 3316
C$_{30}$H$_{52}$O$_{26}$
Verbascose, 10152
C$_{30}$H$_{53}$NO$_{11}$
Benzonatate, 1098
C$_{30}$H$_{53}$N$_3$O$_6$
Aliskiren, 239
C$_{30}$H$_{60}$I$_3$N$_3$O$_3$
Gallamine Triethiodide, 4373
C$_{30}$H$_{62}$
Squalane, 8895

C$_{30}$H$_{62}$O
1-Triacontanol, 9753
C$_{31}$H$_{23}$BrO$_2$S
Difethialone, 3158
C$_{31}$H$_{23}$BrO$_3$
Brodifacoum, 1386
C$_{31}$H$_{24}$O$_3$
Difenacoum, 3153
C$_{31}$H$_{28}$N$_2$Na$_4$O$_{13}$S
Xylenol Orange Tetrasodium Salt, 10281
C$_{31}$H$_{29}$N$_5$O$_4$
Asperlicin, 833
C$_{31}$H$_{29}$N$_5$O$_5$
Asperlicin B, 833
C$_{31}$H$_{32}$N$_2$O$_{13}$S
Xylenol Orange, 10281
C$_{31}$H$_{32}$N$_4$O$_2$
Bezitramide, 1200
C$_{31}$H$_{32}$N$_6$O$_2$
Purmorphamine, 8053
C$_{31}$H$_{32}$O$_2$P$_2$
(R,R)-DIOP, 3319
C$_{31}$H$_{33}$F$_3$N$_2$O$_5$S
Tipranavir, 9615
C$_{31}$H$_{33}$N$_3$O$_6$S
Zafirlukast, 10306
C$_{31}$H$_{33}$N$_3$O$_{11}$
Carzinophilin A, 1877
C$_{31}$H$_{33}$N$_5$O$_4$
Intedanib, 5031
C$_{31}$H$_{34}$BrNO$_4$
Fentonium Bromide, 4033
C$_{31}$H$_{35}$N$_2$NaO$_{11}$
Novobiocin Monosodium Salt, 6811
C$_{31}$H$_{35}$N$_3$O$_4$S
Penicillin G Benethamine, 7204
C$_{31}$H$_{36}$ClN$_3$O$_5$S
Metofenazate, 6221
C$_{31}$H$_{36}$F$_6$N$_6$O$_2$
Evacetrapib, 3947
C$_{31}$H$_{36}$N$_2$O$_{11}$
Novobiocin, 6811
C$_{31}$H$_{39}$N$_5$O$_5$
Ergocornine, 3705
C$_{31}$H$_{40}$O$_2$
Menaquinone 4, 5901
C$_{31}$H$_{41}$ClFNO$_3$
Haloperidol Decanoate, 4634
C$_{31}$H$_{41}$NO$_4$
Bardoxolone, 959
C$_{31}$H$_{42}$N$_2$O$_6$
Batrachotoxin, 1007
C$_{31}$H$_{43}$NO$_7$
Milbemycin Oxime, 6270
C$_{31}$H$_{43}$N$_3$Na$_{10}$O$_{49}$S$_8$
Fondaparinux Sodium, 4259
C$_{31}$H$_{44}$N$_2$O$_5$S
Dronedarone, 3498
C$_{31}$H$_{44}$N$_2$O$_{10}$
Dilazep, 3222
C$_{31}$H$_{44}$O$_7$
Milbemycin A$_3$, 6269
C$_{31}$H$_{44}$O$_8$
Proscillaridin-4-methyl Ether, 7985
C$_{31}$H$_{45}$N$_3$O$_8$
Retaspimycin, 8282
C$_{31}$H$_{46}$O$_2$
Phylloquinone, 7492
C$_{31}$H$_{47}$NaO$_6$
Fusidic Acid Sodium Salt, 4346
C$_{31}$H$_{48}$O$_2$S$_2$
Probucol, 7873
C$_{31}$H$_{48}$O$_6$
Fusidic Acid, 4346
C$_{31}$H$_{50}$IrNP.F$_6$P
Crabtree's Catalyst, 2554

C$_{31}$H$_{51}$NO$_9$
Rosaramicin, 8392
C$_{31}$H$_{52}$N$_2$O$_5$S
Valnemulin, 10097
C$_{31}$H$_{55}$N$_3$O$_6$
Enocitabine, 3640
C$_{32}$H$_{16}$CuN$_8$
Copper Phthalocyanine, 2505
C$_{32}$H$_{19}$N$_6$Na$_5$O$_{15}$S$_5$
Trypan Red, 9973
C$_{32}$H$_{22}$N$_6$Na$_2$O$_6$S$_2$
Congo Red, 2482
C$_{32}$H$_{24}$O$_{12}$
Thermorubin A, 9438
C$_{32}$H$_{26}$N$_4$O$_2$
Conivaptan, 2491
C$_{32}$H$_{26}$O$_{14}$
Actinorhodine, 131
C$_{32}$H$_{28}$CaN$_6$O$_{14}$S$_2$
Sulfaloxic Acid Calcium Salt, 9044
C$_{32}$H$_{29}$N$_5$O$_2$
Enzastaurin, 3656
C$_{32}$H$_{30}$F$_5$N$_3$O$_5$
Elagolix, 3580
C$_{32}$H$_{30}$O$_{14}$
Secalonic Acids, 8555
C$_{32}$H$_{31}$BrN$_2$O$_2$
Bedaquiline, 1018
C$_{32}$H$_{31}$F$_2$N$_3$O$_2$
Zosuquidar, 10396
C$_{32}$H$_{31}$N$_3$O$_{12}$
Indo-1, 5000
C$_{32}$H$_{32}$N$_2$O$_{10}$
Cotarnine Phthalate, 2540
C$_{32}$H$_{32}$N$_2$O$_{12}$
o-Cresolphthalein Complexone, 2566
C$_{32}$H$_{32}$O$_{13}$S
Teniposide, 9288
C$_{32}$H$_{32}$O$_{14}$
Chartreusin, 2044
C$_{32}$H$_{34}$CaN$_4$O$_{10}$S$_2$
Penicillin V Calcium Salt, 7209
C$_{32}$H$_{37}$NO$_{12}$
Pirarubicin, 7601
C$_{32}$H$_{38}$N$_2$O$_5$
Cortivazol, 2527
C$_{32}$H$_{38}$N$_2$O$_8$
Deserpidine, 2919
C$_{32}$H$_{38}$N$_4$
Etioporphyrin, 3917
C$_{32}$H$_{39}$NO$_2$
Ebastine, 3531
C$_{32}$H$_{39}$NO$_4$
Fexofenadine, 4097
C$_{32}$H$_{40}$BrN$_5$O$_5$
Bromocriptine, 1424
C$_{32}$H$_{40}$N$_2$O
Maropitant, 5819
C$_{32}$H$_{40}$N$_2$O$_5$S$_2$
Trimethaphan Camsylate, 9876
C$_{32}$H$_{41}$NO$_2$
Terfenadine, 9306
C$_{32}$H$_{41}$N$_5$O$_5$
Ergocryptine, 3707
Ergocryptinine, 3708
C$_{32}$H$_{44}$F$_3$N$_3$O$_2$S
Fluphenazine Decanoate, 4220
C$_{32}$H$_{44}$N$_2$O$_6$
Homobatrachotoxin, 1007
C$_{32}$H$_{44}$N$_2$O$_8$
Lappaconitine, 5420
C$_{32}$H$_{44}$O$_7$
Ciclesonide, 2266
C$_{32}$H$_{44}$O$_8$
Cucurbitacin E, 2604
C$_{32}$H$_{44}$O$_{12}$
Scilliroside, 8538

$C_{32}H_{45}NO_7$
Milbemycin Oxime, 6270

$C_{32}H_{45}N_3O_4S$
Nelfinavir, 6528

$C_{32}H_{46}N_8O_6S_2$
Thiamine Disulfide O,O-Diisobutyrate, 9449

$C_{32}H_{46}O_7$
Milbemycin A$_4$, 6269

$C_{32}H_{46}O_8$
Cucurbitacin B, 2604

$C_{32}H_{47}F_5O_3S$
Fulvestrant, 4313

$C_{32}H_{48}O_5$
Acetoxolone, 74

$C_{32}H_{48}O_7$
Milbemycin β_1, 6271

$C_{32}H_{48}O_9$
Neriifolin 2'-Acetate, 6559
Oleandrin, 6919

$C_{32}H_{49}NO_9$
Cevadine, 2032

$C_{32}H_{51}NO_8S$
Tiamulin Fumarate, 9574

$C_{32}H_{57}Br_2N_4O_4$
Demecarium Bromide, 2889

$C_{32}H_{52}O_2$
Taraxasterol Acetate, 9193

$C_{32}H_{53}N_2O_4^+$
Rocuronium, 8375

$C_{32}H_{56}N_4O_8$
Sporidesmolide III, 8893

$C_{32}H_{58}N_2O_7S$
CHAPS, 2043

$C_{32}H_{58}N_2O_8S$
CHAPS Hydroxy Analog, 2043

$C_{32}H_{62}CaO_4$
Calcium Palmitate, 1689

$C_{32}H_{64}O_2$
Cetyl Palmitate, 2030

$C_{32}H_{64}O_4Sn$
Dibutyltin Dilaurate, 3047

$C_{33}H_{25}F_3O_4$
Flocoumafen, 4134

$C_{33}H_{25}N_3O_3$
Norbormide, 6775

$C_{33}H_{30}N_4O_2$
Telmisartan, 9271

$C_{33}H_{34}N_4O_6$
Azelnidipine, 900
Biliverdine, 1223

$C_{33}H_{34}N_6O_6$
Candesartan 1-[[(Cyclohexyloxy)carbonyl]oxy]ethyl Ester, 1742

$C_{33}H_{35}FN_2O_5$
Atorvastatin, 855

$C_{33}H_{35}N_5O_5$
Ergotamine, 3718
Ergotaminine, 3719

$C_{33}H_{36}N_4O_6$
Bilirubin, 1222

$C_{33}H_{37}N_5O_5$
Dihydroergotamine, 3191

$C_{33}H_{38}GdN_3Na_3O_{14}P$
Gadofosveset Trisodium Salt, 4355

$C_{33}H_{38}N_4O_6$
Irinotecan, 5135

$C_{33}H_{40}O_{15}$
Icariin, 3672

$C_{33}H_{40}O_{19}$
Robinin, 8372

$C_{33}H_{41}GdN_3O_{14}P$
Gadofosveset, 4355

$C_{33}H_{42}N_4O_6$
Urobilin, 10068

$C_{33}H_{42}O_{19}$
Troxerutin, 9969

$C_{33}H_{43}FN_{10}O_6$
Nemifitide, 6530

$C_{33}H_{44}N_4O_{10}$
Riboflavin 2',3',4',5'-Tetrabutyrate, 8325

$C_{33}H_{44}O_8$
Helvolic Acid, 4668

$C_{33}H_{45}NO_9$
Delphinine, 2886

$C_{33}H_{46}Bi_2O_{11}$
d-Camphocarboxylic Acid Basic Bismuth Salt, 1733

$C_{33}H_{46}N_4O_6$
Stercobilin, 10068

$C_{33}H_{46}O_4$
Nandrolone p-Hexyloxyphenylpropionate, 6450

$C_{33}H_{47}NO_{13}$
Natamycin, 6509

$C_{33}H_{48}O_6$
Leptomycin B, 5499

$C_{33}H_{48}O_7$
Milbemycin D, 6271

$C_{33}H_{49}P$
XPhos Phosphine Ligand, 1469

$C_{33}H_{50}O_8$
Cephalosporin P$_1$, 1979

$C_{33}H_{53}NO_8$
Imperialine β-D-Glucopyranoside, 4968

$C_{33}H_{55}NO_8$
Discodermolide, 3397
Verticine 3-Glucoside, 10165

$C_{33}H_{57}N_3O_9$
Enniatin B, 3639

$C_{33}H_{58}N_4O_8$
Sporidesmolide I, 8893

$C_{34}H_{24}CaCl_2N_4O_4$
Lonazolac Calcium Salt, 5623

$C_{34}H_{24}N_4Na_2O_{10}S_2$
Benzo Azurine G, 1087

$C_{34}H_{24}N_6Na_4O_{14}S_4$
Evan's Blue, 3948
Trypan Blue, 9972

$C_{34}H_{25}N_6Na_3O_9S_3$
Vital Red, 10210

$C_{34}H_{26}N_6Na_2O_6S_2$
Benzopurpurine 4B, 1104

$C_{34}H_{28}FeP_2$
DPPF, 3490

$C_{34}H_{32}ClFeN_4O_4$
Hemin, 4679

$C_{34}H_{32}FeN_4O_4$
Heme, 4674

$C_{34}H_{32}N_2O_5$
Micranthine, 6258

$C_{34}H_{32}N_4Na_2O_4$
Protoporphyrin IX Disodium Salt, 8006

$C_{34}H_{32}N_6O_4$
1,3-Diphenylguanidine Phthalate, 3358

$C_{34}H_{33}FeN_4O_5$
Hematin, 4670

$C_{34}H_{34}N_4O_4$
Protoporphyrin IX, 8006

$C_{34}H_{35}N_3O_{10}$
Zorubicin, 10395

$C_{34}H_{36}F_3NO_{13}$
Valrubicin, 10101

$C_{34}H_{36}N_2O_6$
Pseudomorphine, 8027

$C_{34}H_{36}N_4O_6$
Phytochlorin, 7500

$C_{34}H_{38}N_3O_7P$
Efonidipine, 3571

$C_{34}H_{38}N_4O_6$
Hematoporphyrin, 4671

$C_{34}H_{40}O_{12}$
Filixic Acid PBP, 4111

$C_{34}H_{41}N_3O_7$
Biricodar, 1244

$C_{34}H_{46}ClN_3O_{10}$
Maytansine, 5829

$C_{34}H_{46}O_8$
Helvolic Acid Methyl Ester, 4668

$C_{34}H_{47}NO_{11}$
Aconitine, 110

$C_{34}H_{48}I_6N_4O_{16}$
Iodipamide Bis[N-methylglucamine] Salt, 5065

$C_{34}H_{48}N_2O_9$
Ajacine, 5679

$C_{34}H_{48}Na_2O_7$
Carbenoxolone Disodium Salt, 1794

$C_{34}H_{50}N_4O_9S$
Dalfopristin, 2803

$C_{34}H_{50}O_7$
Carbenoxolone, 1794

$C_{34}H_{50}O_{12}$
Thapsigargin, 9421

$C_{34}H_{51}NO_{13}$
Tetrin A, 9393

$C_{34}H_{51}NO_{14}$
Tetrin B, 9393

$C_{34}H_{53}NaO_8$
Lasalocid A Sodium Salt, 5427

$C_{34}H_{54}O_8$
Lasalocid A, 5427

$C_{34}H_{56}CuN_6O_{14}S_4$
Cuproxoline, 2660

$C_{34}H_{57}BrN_2O_4$
Vecuronium Bromide, 10132

$C_{34}H_{58}$
Botryococcene, 1360

$C_{34}H_{59}NO_{15}$
Fumonisin B$_1$, 4318

$C_{34}H_{60}N_4O_8$
Sporidesmolide II, 8893

$C_{34}H_{63}N_5O_9$
Pepstatin, 7260

$C_{34}H_{67}N_3O_5S$
Squalamine, 8894

$C_{35}H_{30}N_4O_4$
Midostaurin, 6264

$C_{35}H_{34}N_2O_5$
Trilobine, 9866

$C_{35}H_{35}ClNNaO_3S$
Montelukast Monosodium Salt, 6346

$C_{35}H_{36}ClNO_3S$
Montelukast, 6346

$C_{35}H_{36}N_2O_6$
Daphnoline, 2819

$C_{35}H_{38}Cl_2N_8O_4$
Itraconazole, 5292

$C_{35}H_{38}F_2N_8O_4$
Saperconazole, 8499

$C_{35}H_{38}N_4O_6$
Manidipine, 5808

$C_{35}H_{39}N_5O_5$
Ergocristine, 3706

$C_{35}H_{41}N_9O_9$
Diphenhydramine Di(acefyllinate), 3339

$C_{35}H_{42}N_2O_9$
Rescinnamine, 8262

$C_{35}H_{42}N_2O_{11}$
Syrosingopine, 9147

$C_{38}H_{30}NiO_2P_2$
Bis(triphenylphosphine)dicarbonyl-
nickel, 1305

$C_{38}H_{34}N_2O_4P_2$
(R)-P-Phos, 7817

$C_{38}H_{36}CaN_4O_4$
Phenylbutazone Calcium Salt, 7390

$C_{38}H_{36}EuN_8S_2$
Quantum Dye, 8140

$C_{38}H_{37}N_5Na_4O_9$
Talaporfin Tetrasodium Salt, 9166

$C_{38}H_{38}N_4O_6$
Tariquidar, 9198

$C_{38}H_{41}N_5O_9$
Talaporfin, 9166

$C_{38}H_{42}CaN_4O_6\cdot\frac{1}{2}H_2O$
Bumadizone Calcium Salt Hemihydrate, 1488

$C_{38}H_{42}N_2O_6$
Tetrandrine, 9379

$C_{38}H_{44}N_2O_6$
Dauricine, 2835

$C_{38}H_{44}O_8$
Gambogic Acid, 4391

$C_{38}H_{48}CaN_2O_6\cdot2H_2O$
Mitiglinide Calcium Salt Dihydrate,
6296

$C_{38}H_{49}N_3O_5$
Bemotrizinol, 1030

$C_{38}H_{50}N_6O_5$
Saquinavir, 8506

$C_{38}H_{52}N_6O_2$
Tirilazad, 9619

$C_{38}H_{52}N_6O_7$
Atazanavir, 849

$C_{38}H_{54}I_6N_6O_{18}$
Iocarmic Acid Di-N-methylglucamine
Salt, 5053

$C_{38}H_{54}O_{13}$
Colocynthin, 2471

$C_{38}H_{55}N_5O_9S$
Vaniprevir, 10122

$C_{38}H_{55}Na_9O_{49}S_7$
Idraparinux Nonasodium Salt, 4932

$C_{38}H_{58}CaN_4O_{10}S_2$
Moveltipril Calcium Salt, 6371

$C_{38}H_{60}O_{18}$
Rebaudioside B, 8243
Stevioside, 8939

$C_{38}H_{61}I_2N_3S_3$
Platonin, 7645

$C_{38}H_{64}O_{49}S_7$
Idraparinux, 4932

$C_{38}H_{69}NO_{13}$
Clarithromycin, 2338

$C_{38}H_{72}N_2O_{12}$
Azithromycin, 907

$C_{39}H_{32}OP_2$
Xantphos, 10268

$C_{39}H_{43}N_3O_{11}S$
Ecteinascidin 743, 3550

$C_{39}H_{47}NO_{14}$
Rifamycin O, 8342
Streptovaricin F, 8960

$C_{39}H_{49}NO_{14}$
Rifamycin B, 8342

$C_{39}H_{49}NO_{16}$
Nogalamycin, 6758

$C_{39}H_{53}N_9O_{15}S$
β-Amanitin, 367

$C_{39}H_{54}N_{10}O_{14}S$
α-Amanitin, 367

$C_{39}H_{54}O_{13}$
Zaragozic Acid B, 10312

$C_{39}H_{61}NO_{14}$
Rimocidin, 8354

$C_{39}H_{67}N_5O_7$
Monomethylauristatin E, 6340

$C_{40}H_{32}Cl_2N_8O_2$
Tetrazolium Blue, 9392

$C_{40}H_{33}F_3N_4O_3$
Dirlotapide, 3395

$C_{40}H_{43}ClN_4O_{11}$
Rhod-2, 8305

$C_{40}H_{44}N_4O^{2+}$
C-Curarine I, 2662

$C_{40}H_{46}N_4O_2^{2+}$
Toxiferine I, 9717

$C_{40}H_{48}I_2N_2O_6$
Metocurine Iodide, 6220

$C_{40}H_{48}N_4O_2^{2+}$
C-Toxiferine II, 9717

$C_{40}H_{48}N_4O_3$
Geissospermine, 4414

$C_{40}H_{48}O_4$
Astacin, 843

$C_{40}H_{49}NO_{14}$
Streptovaricin E, 8960

$C_{40}H_{50}O_2$
Rhodoxanthin, 8320

$C_{40}H_{50}O_{14}$
Zaragozic Acid C, 10312

$C_{40}H_{51}NO_{13}$
Streptovaricin D, 8960

$C_{40}H_{51}NO_{14}$
Streptovaricin C, 8960

$C_{40}H_{51}NO_{15}$
Streptovaricin G, 8960

$C_{40}H_{52}CaN_2O_8S_2$
Domitroban Calcium Salt, 3465

$C_{40}H_{52}O_2$
Canthaxanthin, 1756
Torularhodin, 9712

$C_{40}H_{52}O_4$
Astaxanthin, 845

$C_{40}H_{53}N_7O_5S_2$
Cobicistat, 2436

$C_{40}H_{54}O$
Echinenone, 3542

$C_{40}H_{56}$
α-Carotene, 1854
β-Carotene, 1855
γ-Carotene, 1856
δ-Carotene, 1857
Lycopene, 5681

$C_{40}H_{56}O$
Cryptoxanthin, 2599
Lycoxanthin, 5688
Rubixanthin, 8422

$C_{40}H_{56}O_2$
Lycophyll, 5682
Xanthophyll, 10261
Zeaxanthin, 10316

$C_{40}H_{56}O_3$
Capsanthin, 1771
Chrysanthemaxanthin, 2254
Flavoxanthin, 4127

$C_{40}H_{56}O_4$
Violaxanthin, 10194

$C_{40}H_{57}N_5O_7$
Carfilzomib, 1837

$C_{40}H_{58}O$
Rhodopin, 8316

$C_{40}H_{58}O_4$
Oryzanol A, 6983

$C_{40}H_{59}NO_{11}$
Eribulin, 3723

$C_{40}H_{60}BNaO_{14}$
Aplasmomycin, 724

$C_{40}H_{60}N_4O_{10}$
Bufotoxin, 1484

$C_{40}H_{60}O_2$
Kitol, 5363

$C_{40}H_{63}N_3O_4S_2$
Pipotiazine Palmitic Ester, 7595

$C_{40}H_{64}O_{12}$
Nonactin, 6762

$C_{40}H_{66}N_6$
Ormosinine, 6966

$C_{40}H_{66}O_{10}$
Venturicidin B, 10141

$C_{40}H_{67}NO_{14}$
Leucomycin A_1, 5505

$C_{40}H_{68}$
Phytofluene, 7501

$C_{40}H_{68}O_{11}$
Nigericin, 6625

$C_{40}H_{71}NO_{14}$
Erythromycin Propionate, 3742

$C_{40}H_{74}CaO_{14}S_2$
Docusate Calcium, 3445

$C_{40}H_{80}NO_8P$
Colfosceril Palmitate, 2461

$C_{41}H_{39}ClO_{21}$
Delphinidin Compound with Glucose
+ Hydroxybenzoic Acid, 2885

$C_{41}H_{42}N_4O_8$
Verteporfin, 10163

$C_{41}H_{48}N_2O_8$
Thalicarpine, 9402

$C_{41}H_{50}N_6O_2$
Bisoctrizole, 1294

$C_{41}H_{52}N_6O_7$
Ergonovine Hydracrylate, 3712

$C_{41}H_{56}O_2$
Menaquinone 6, 5901

$C_{41}H_{58}FeN_9O_{20}$
Ferrichrome A, 4051

$C_{41}H_{60}O_4$
Oryzanol C, 6983

$C_{41}H_{63}NO_{14}$
Protoveratrine A, 8008

$C_{41}H_{63}NO_{15}$
Protoveratrine B, 8008

$C_{41}H_{64}O_8$
Prednisolone 21-Stearoylglycolate,
7841

$C_{41}H_{64}O_{13}$
Digitoxin, 3182

$C_{41}H_{64}O_{14}$
Digoxin, 3186
Gitoxin, 4469

$C_{41}H_{64}O_{15}$
Diginatin, 3175

$C_{41}H_{65}NO_{10}$
Spinosyn A, 8877

$C_{41}H_{65}NO_{15}$
Midecamycin A_3, 6262

$C_{41}H_{66}O_{12}$
α-Hederin, 4660

$C_{41}H_{67}NO_{11}$
Venturicidin A, 10141

$C_{41}H_{67}NO_{15}$
Midecamycin A_1, 6262
Troleandomycin, 9949

$C_{41}H_{76}N_2O_{15}$
Roxithromycin, 8409

$C_{41}H_{79}N_3O_{12}$
Tulathromycin, 9990

$C_{42}H_{27}AlF_9N_3O_6$
Flufenamic Acid Aluminum Salt,
4163

$C_{42}H_{30}N_6O_{12}$
Inositol Niacinate, 5022

$C_{42}H_{38}O_{20}$
Sennosides, 8594

$C_{42}H_{42}N_2O_4P_2$
(R)-Tol-P-Phos, 7817

$C_{42}H_{51}NO_{15}$
Aclacinomycin B, 104

$C_{42}H_{53}NO_{15}$
Aclacinomycin A, 104
Streptovaricin J, 8960
$C_{42}H_{53}NO_{16}$
Streptovaricin A, 8960
$C_{42}H_{58}O_6$
Fucoxanthin, 4310
$C_{42}H_{59}N_3O_{10}$
Cethromycin, 2018
$C_{42}H_{60}K_2O_{16}$
Glycyrrhizic Acid Dipotassium Salt, 4541
$C_{42}H_{60}O_2$
Rhodoviolascin, 8319
$C_{42}H_{62}N_8O_7S$
Bottromycin A$_2$, 1361
$C_{42}H_{62}O_{16}$
Glycyrrhizic Acid, 4541
$C_{42}H_{64}O_{19}$
Thevetin A, 9439
$C_{42}H_{65}N_{13}O_{10}$
Saralasin, 8509
$C_{42}H_{66}O_{14}$
β-Methyldigoxin, 3186
$C_{42}H_{67}NO_{10}$
Spinosyn D, 8877
$C_{42}H_{67}NO_{15}$
Carbomycin B, 1806
$C_{42}H_{67}NO_{16}$
Carbomycin A, 1806
$C_{42}H_{69}NO_{15}$
Josamycin, 5315
Rokitamycin, 8379
$C_{42}H_{69}NaO_{11}$
Salinomycin Sodium Salt, 8473
$C_{42}H_{70}O_{11}$
Salinomycin, 8473
$C_{42}H_{70}O_{35}$
β-Cyclodextrin, 2710
$C_{42}H_{72}O_{16}$
Lankamycin, 5407
$C_{42}H_{75}N_3O_{16}$
BIGCHAP, 1217
$C_{42}H_{78}N_2O_{14}$
Dirithromycin, 3394
$C_{43}H_{42}O_{22}$
Carthamin, 1867
$C_{43}H_{47}N_2NaO_6S_2$
Indocyanine Green, 5002
$C_{43}H_{49}N_7O_{10}$
Virginiamycin S$_1$, 10200
$C_{43}H_{50}N_4O_{14}S_2$
Cy 3, 2668
$C_{43}H_{51}N_3O_{11}$
Rifaximin, 8345
$C_{43}H_{52}N_4O_5$
Voacamine, 10223
$C_{43}H_{53}NO_6$
Nodulisporic Acid, 6756
$C_{43}H_{53}NO_{14}$
Docetaxel, 3442
$C_{43}H_{55}N_5O_7$
Vindesine, 10184
$C_{43}H_{58}N_2O_{13}$
Rifamide, 8340
$C_{43}H_{58}N_4O_{12}$
Rifampin, 8341
$C_{43}H_{60}N_2O_{12}$
Mocimycin, 6311
$C_{43}H_{62}N_2O_{12}$
5,6-Dihydromocimycin, 6311
$C_{43}H_{63}NO_{11}$
Selamectin, 8561
$C_{43}H_{65}N_5O_{10}$
Telithromycin, 9261
$C_{43}H_{65}N_{11}O_{12}S_2$
Deaminooxytocin, 2844

$C_{43}H_{66}N_{12}O_{12}S_2$
Oxytocin, 7078
$C_{43}H_{66}O_{14}$
Acetyldigitoxins, 84
$C_{43}H_{66}O_{15}$
α-Acetyldigoxin, 3186
$C_{43}H_{67}N_{11}O_{12}S_2$
Atosiban, 856
$C_{43}H_{67}N_{15}O_{12}S_2$
Arginine Vasotocin, 10129
$C_{43}H_{68}ClNO_{11}$
Pimecrolimus, 7541
$C_{43}H_{72}Cl_2P_2Ru$
Grubbs' Catalyst, 4585
$C_{43}H_{72}O_{11}$
Narasin, 6501
$C_{43}H_{73}NaO_{14}$
Lonomycin C Sodium Salt, 5626
$C_{43}H_{74}N_2O_{14}$
Spiramycin I, 8879
$C_{43}H_{75}NO_{16}$
Erythromycin Ethylsuccinate, 3739
$C_{43}H_{78}N_6O_{13}$
Romurtide, 8384
$C_{44}CaH_{54}F_2N_6O_{12}S_2$
Rosuvastatin Calcium Salt, 8402
$C_{44}H_{32}N_4O_4$
Temoporfin, 9281
$C_{44}H_{32}P_2$
BINAP, 1227
$C_{44}H_{44}CaN_2O_8S_4$
Zofenopril Calcium Salt, 10386
$C_{44}H_{50}Cl_2N_4O_2$
Alcuronium Dichloride, 214
$C_{44}H_{50}N_4O_2^{2+}$
Alcuronium, 214
$C_{44}H_{62}N_2O_{12}$
Aurodox, 873
$C_{44}H_{68}O_{13}$
Okadaic Acid, 6911
$C_{44}H_{69}NO_{12}$
Tacrolimus, 9155
$C_{44}H_{70}O_{22}$
Rebaudioside C, 8243
$C_{44}H_{70}O_{23}$
Rebaudioside A, 8243
$C_{44}H_{72}O_{11}$
Oligomycin D, 6927
$C_{44}H_{75}NaO_{14}$
Lonomycin A Sodium Salt, 5626
$C_{44}H_{76}O_{14}$
Lonomycin A, 5626
$C_{45}H_{26}N_{10}Na_6O_{21}S_6$
Sirius Red, 8686
$C_{45}H_{52}N_4O_{14}S_2$
Cy 5, 2669
$C_{45}H_{54}F_2N_4O_8$
Vinflunine, 10186
$C_{45}H_{54}N_4O_8$
Vinorelbine, 10187
$C_{45}H_{54}N_8O_{10}$
Pristinamycin IA, 7871
$C_{45}H_{55}N_9O_6$
Pralmorelin, 7824
$C_{45}H_{57}NO_{14}$
Cabazitaxel, 1604
$C_{45}H_{63}N_{13}O_{12}S_2$
Ornipressin, 6968
$C_{45}H_{69}N_{11}O_{12}S$
Carbetocin, 1796
$C_{45}H_{71}NO_{17}$
Miokamycin, 6286
$C_{45}H_{72}O_{12}$
Oligomycin B, 6927
$C_{45}H_{72}O_{16}$
Dioscin, 3320

$C_{45}H_{73}NO_{14}$
α-Chaconine, 8832
$C_{45}H_{73}NO_{15}$
α-Solanine, 8832
$C_{45}H_{73}NO_{16}$
Josamycin Propionate, 5315
Solasonine, 8837
$C_{45}H_{74}BNO_{15}$
Boromycin, 1343
$C_{45}H_{74}O$
Solanesol, 8830
$C_{45}H_{74}O_{10}$
Oligomycin C, 6927
$C_{45}H_{74}O_{11}$
Oligomycin A, 6927
$C_{45}H_{76}N_2O_{15}$
Spiramycin II, 8879
$C_{45}H_{76}O_{16}$
Semduramicin, 8582
$C_{45}H_{86}O_6$
Trimyristin, 9896
$C_{46}H_{36}Cl_2N_6O_5$
Zinpyr-1, 10368
$C_{46}H_{38}N_6O_5$
Zinpyr-2, 10368
$C_{46}H_{46}O_4P_2Pd_2$
Herrmann-Beller Catalyst, 4704
$C_{46}H_{50}N_2O_4P_2$
(R)-Xyl-P-Phos, 7817
$C_{46}H_{52}Na_2O_{16}$
Bimosiamose Disodium Salt, 1226
$C_{46}H_{54}O_{16}$
Bimosiamose, 1226
$C_{46}H_{56}N_4O_{10}$
Vincristine, 10183
$C_{46}H_{58}ClN_5O_8$
Proglumetacin, 7890
$C_{46}H_{58}N_4O_9$
Vinblastine, 10179
$C_{46}H_{62}N_4O_{11}$
Rifabutin, 8338
$C_{46}H_{64}N_6O_{10}$
Lysergide D-Tartrate, 5695
$C_{46}H_{64}N_{14}O_{12}S_2$
Desmopressin, 2926
$C_{46}H_{64}O_2$
Menaquinone 7, 5901
$C_{46}H_{65}Cl_2N_2PRu$
Grubbs' Second Generation Catalyst, 4586
$C_{46}H_{65}N_{13}O_{11}S_2$
Felypressin, 3988
$C_{46}H_{65}N_{13}O_{12}S_2$
Lypressin, 5691
$C_{46}H_{65}N_{15}O_{12}S_2$
Arginine Vasopressin, 10129
$C_{46}H_{77}NO_{17}$
Tylosin, 10013
$C_{46}H_{78}N_2O_{15}$
Spiramycin III, 8879
$C_{46}H_{80}N_2O_{13}$
Tilmicosin, 9596
$C_{46}H_{80}O_2$
β-Amyrin Palmitate, 611
$C_{47}H_{51}NO_{14}$
Paclitaxel, 7081
$C_{47}H_{55}ClF_3N_5O_6S_3$
Navitoclax, 6512
$C_{47}H_{64}N_4O_{12}$
Rifapentine, 8344
$C_{47}H_{68}O_{17}$
Bryostatin 1, 1466
$C_{47}H_{70}O_{14}$
Abamectin, 2
$C_{47}H_{71}NO_{17}$
Candidin, 1744

$C_{47}H_{72}O_{14}$
Ivermectin Component B_{1b}, 5296
$C_{47}H_{73}NO_{17}$
Amphotericin B, 582
$C_{47}H_{74}O_{19}$
Deslanoside, 2922
$C_{47}H_{75}NO_{17}$
Nystatin A_1, 6825
$C_{47}H_{79}NaO_{17}$
Maduramicin Sodium Salt, 5711
$C_{47}H_{83}NO_{17}$
Maduramicin, 5711
$C_{48}H_{56}N_6O_8S_2$
Penicillin G Benzathine, 7205
$C_{48}H_{66}N_6O_6$
Ethylhexyl Triazone, 3864
$C_{48}H_{72}O_{14}$
Abamectin, 2
$C_{48}H_{74}O_{14}$
Ivermectin Component B_{1a}, 5296
$C_{48}H_{78}O_{19}$
Asiaticoside, 822
$C_{48}H_{80}O_{40}$
γ-Cyclodextrin, 2710
$C_{48}H_{84}O_{14}$
Alborixin, 208
$C_{48}H_{86}CaO_{12}$
Calcium Stearyl-2 Lactylate, 1706
$C_{48}H_{93}AlO_6$
Aluminum Palmitate, 351
$C_{48}H_{93}NO_9$
Phrenosin, 7479
$C_{49}H_{56}ClFeN_4O_6$
Cytohemin, 2790
$C_{49}H_{61}N_{13}O_{24}$
Pteroylhexaglutamylglutamic Acid, 8044
$C_{49}H_{62}N_{10}O_{16}S_3$
Sincalide, 8681
$C_{49}H_{66}N_{10}O_{10}S_2$
Octreotide, 6851
$C_{49}H_{69}ClO_{14}$
Brevetoxin C, 1378
$C_{49}H_{70}N_{14}O_{11}$
Angiotensin Amide, 644
$C_{49}H_{70}O_{13}$
Brevetoxin A, 1378
$C_{49}H_{73}NO_{14}$
Eprinomectin Component B_{1b}, 3689
$C_{49}H_{74}O_{14}$
Avermectins, 877
$C_{49}H_{76}O_3$
Tocoretinate, 9657
$C_{49}H_{76}O_{19}$
Lanatoside A, 5404
$C_{49}H_{76}O_{20}$
Lanatoside B, 5404
$C_{49}H_{76}O_{21}$
Lanatoside D, 5404
$C_{49}H_{78}N_6O_{12}$
Didemnin A, 3111
$C_{49}H_{89}NO_{25}$
Erythromycin Lactobionate, 3741
$C_{49}H_{91}N_{15}O_{15}$
Polymyxin D_2, 7693
$C_{50}H_{40}O_7$
Quercetin Pentabenzyl Ether, 8150
$C_{50}H_{46}CaF_2N_2O_8$
Pitavastatin Calcium Salt, 7621
$C_{50}H_{62}N_{12}O_{12}S_2$
Triostin A, 9905
$C_{50}H_{63}ClO_{16}$
Chlorothricin, 2172
$C_{50}H_{70}O_{14}$
Brevetoxin B, 1378
$C_{50}H_{72}CoN_{13}O_8$
Dicyanocobinamide, 3102

$C_{50}H_{73}N_{15}O_{11}$
Bradykinin, 1364
$C_{50}H_{74}O_{14}$
Doramectin, 3473
$C_{50}H_{75}NO_{14}$
Eprinomectin Component B_{1a}, 3689
$C_{50}H_{80}O_{28}$
Rebaudioside D, 8243
$C_{50}H_{81}N_4O_{15}P$
Calyculin A, 1726
$C_{50}H_{82}O_{23}$
F-Gitonin, 4467
$C_{50}H_{83}NO_{21}$
Tomatine, 9704
$C_{50}H_{93}N_{15}O_{15}$
Polymyxin D_1, 7693
$C_{51}H_{34}N_6Na_6O_{23}S_6$
Suramin Sodium, 9135
$C_{51}H_{43}N_{13}O_{12}S_6$
Nosiheptide, 6808
$C_{51}H_{51}N_{15}O_{14}S$
Berninamycin, 1162
$C_{51}H_{51}N_{15}O_{15}S$
Berninamycin, 1162
$C_{51}H_{63}FeN_6O_9$
Trichostatin B, 9814
$C_{51}H_{64}N_4O_{13}$
Rifalazil, 8339
$C_{51}H_{64}N_{12}O_{12}S_2$
Echinomycin, 3544
$C_{51}H_{74}O_{19}$
Gitoxin Pentaacetate, 4469
$C_{51}H_{79}NO_{13}$
Rapamycin, 8232
$C_{51}H_{98}O_6$
Tripalmitin, 9908
$C_{52}H_{72}GdN_5O_{14}$
Motexafin Gadolinium, 6367
$C_{52}H_{72}LuN_5O_{14}$
Motexafin Lutetium, 6368
$C_{52}H_{74}Cl_2O_{18}$
Fidaxomicin, 4108
$C_{52}H_{74}N_{16}O_{15}S_2$
Terlipressin, 9310
$C_{52}H_{76}O_{24}$
Plicamycin, 7652
$C_{52}H_{78}N_{10}O_{17}$
Inosine Pranobex, 5019
$C_{52}H_{82}N_6O_{14}$
Didemnin C, 3111
$C_{52}H_{88}BaN_2O_{14}S_2$
Taurocholic Acid Barium Salt, 9212
$C_{52}H_{88}N_{10}O_{15}$
Caspofungin, 1889
$C_{52}H_{97}NO_{18}S$
Erythromycin Estolate, 3740
$C_{52}H_{98}N_{16}O_{13}$
Circulin B, 2314
Polymyxin E_2, 2463
$C_{53}H_{67}N_9O_{10}S$
Quinupristin, 8201
$C_{53}H_{80}O_2$
Plastoquinone 9, 7637
$C_{53}H_{83}NO_{14}$
Everolimus, 3950
$C_{53}H_{84}NO_{14}P$
Ridaforolimus, 8336
$C_{53}H_{100}N_{16}O_{13}$
Circulin A, 2314
Colistin A, 2463
$C_{54}H_{45}ClP_3Rh$
Wilkinson's Catalyst, 10245
$C_{54}H_{56}Cl_2N_8O_{10}S_2$
Cloxacillin Benzathine Salt, 2403
$C_{54}H_{69}N_{11}O_{10}S_2$
Lanreotide, 5412

$C_{54}H_{70}MgN_4O_6$
Chlorophyll D, 2158
$C_{54}H_{70}N_{12}O_{12}S_2$
Triostin C, 9905
$C_{54}H_{85}N_{13}O_{15}S$
Eledoisin, 3588
$C_{54}H_{90}N_6O_{18}$
Valinomycin, 10096
$C_{54}H_{95}N_9O_{16}$
Viscosin, 10205
$C_{54}H_{105}AlO_6$
Aluminum Stearate, 358
$C_{55}H_{46}OP_3Rh$
Tristriphenylphosphine Rhodium Carbonyl Hydride, 9935
$C_{55}H_{70}MgN_4O_6$
Chlorophyll B, 2158
$C_{55}H_{72}MgN_4O_5$
Chlorophyll A, 2158
$C_{55}H_{74}IN_3O_{21}S_4$
Calicheamicin γ_1^I, 1717
$C_{55}H_{75}N_{17}O_{13}$
Gonadotropin-Releasing Hormone, 4561
$C_{55}H_{76}N_{16}O_{12}$
Fertirelin, 4088
$C_{55}H_{84}N_{17}O_{21}S_3$
Bleomycin A_2, 1319
$C_{55}H_{96}N_{16}O_{13}$
Polymyxin B_2, 7693
$C_{55}H_{103}NO_{15}$
Erythromycin Stearate, 3743
$C_{56}H_{71}N_9O_{23}S$
Micafungin, 6254
$C_{56}H_{78}Cl_2N_2O_{16}$
Doxacurium Chloride, 3480
$C_{56}H_{82}N_{16}O_{19}S_2 \cdot 5H_2O$
Terlipressin Diacetate Pentahydrate, 9310
$C_{56}H_{85}N_{17}O_{12}$
Kallidin, 5326
$C_{56}H_{87}NO_{16}$
Temsirolimus, 9285
$C_{56}H_{88}O_2$
Vitamin D_1, 10214
$C_{56}H_{92}O_{27}$
Tigonin, 9591
$C_{56}H_{92}O_{29}$
Digitonin, 3180
$C_{56}H_{98}N_{16}O_{13}$
Polymyxin B_1, 7693
$C_{57}H_{70}N_{12}O_9S_2$
Vapreotide, 10123
$C_{57}H_{82}O_{26}$
Chromomycin A_3, 2242
$C_{57}H_{83}N_3O_{39}S$
Piroxicam Compd with β-Cyclodextrin, 7619
$C_{57}H_{87}N_7O_{15}$
Aplidine, 725
$C_{57}H_{89}N_7O_{15}$
Didemnin B, 3111
$C_{57}H_{104}O_6$
Petroselinic Acid Glyceryl Triester, 7304
Triolein, 9904
$C_{57}H_{110}O_6$
Tristearin, 9932
$C_{58}H_{66}N_{10}O_9$
Pasireotide, 7155
$C_{58}H_{73}N_7O_{17}$
Anidulafungin, 649
$C_{58}H_{73}N_{13}O_{21}S_2$
Ceruletide, 2004
$C_{58}H_{80}Cl_2N_2O_{14}$
Mivacurium Chloride, 6305

$C_{98}H_{138}N_{24}O_{33}$
Bivalirudin, 1311
$C_{99}H_{155}N_{29}O_{21}S$
Alsactide, 310
$C_{100}H_{132}N_{26}O_{13}$
Indolicidin, 5006
$C_{100}H_{156}N_{34}O_{22}S$
Giractide, 4463
$C_{102}H_{172}N_{36}O_{32}S_7$
Ziconotide, 10321
$C_{104}H_{168}N_{34}O_{20}S$
Penetratin, 7195
$C_{106}H_{180}N_{34}O_{23}S_5$
Tertiapin, 9318
$C_{108}H_{96}Cu_6P_6$
Stryker's Reagent, 8987
$C_{112}H_{162}N_{36}O_{43}S_{10}$
Technetium Tc 99m Apcitide Dimer, 9235
$C_{119}H_{154}ClN_{21}O_{40}$
Ramoplanin A$_2$, 8222
$C_{122}H_{210}N_{32}O_{22}$
Pexiganan, 7308
$C_{126}H_{238}N_{26}O_{22}$
Sinapultide, 8680
$C_{129}H_{223}N_3O_{54}$
Palytoxin, 7100
$C_{130}H_{220}N_{44}O_{41}$
Porcine Secretin, 8558
$C_{131}H_{229}N_{39}O_{31}$
Melittin, 5892
$C_{133}H_{195}N_{51}O_{33}$
Histatin 5, 4757
$C_{136}H_{210}N_{40}O_{31}S$
Cosyntropin, 2539
$C_{143}H_{230}N_{42}O_{37}S_7$
Nisin, 6650
$C_{145}H_{234}N_{52}O_{44}S_3$
Ularitide, 10071
$C_{148}H_{227}N_{39}O_{38}S_5$
Subtilin, 8994
$C_{148}H_{244}N_{42}O_{47}$
Elcatonin, 3585
$C_{149}H_{246}N_{44}O_{42}S$
Sermorelin, 8600
$C_{152}H_{232}N_{40}O_{45}$
Taspoglutide, 9209
$C_{153}H_{225}N_{43}O_{49}S$
Glucagon, 4481
$C_{164}H_{252}N_{44}O_{55}S$
Teduglutide, 9248
$C_{164}H_{256}Na_2O_{68}S_2$
Maitotoxin, 5762
$C_{171}H_{267}N_{51}O_{53}S_2$
Pramlintide, 7828
$C_{172}H_{265}N_{43}O_{51}$
Liraglutide, 5569
$C_{176}H_{277}N_{57}O_{55}S_7$
Charybdotoxin, 2045
$C_{178}H_{258}N_{64}O_{48}$
Histatin 3, 4757
$C_{178}H_{286}N_{52}O_{50}S_7$
Margatoxin, 5817
$C_{192}H_{225}N_{75}Na_{19}O_{98}P_{19}S_{19}$
Alicaforsen Nonadecasodium Salt, 237
$C_{196}H_{230}N_{68}Na_{19}O_{105}P_{19}S_{19}$
Aprinocarsen Nonadecasodium Salt, 744
$C_{196}H_{249}N_{68}O_{105}P_{19}S_{19}$
Aprinocarsen, 744
$C_{197}H_{328}N_{64}O_{67}S_6$
Viscotoxin B, 10206
$C_{198}H_{340}N_{68}O_{56}S_8$
α_1-Purothionin, 8055
$C_{199}H_{318}N_{62}O_{66}S_6$
Viscotoxin A$_2$, 10206

$C_{200}H_{330}N_{62}O_{68}S_6$
Viscotoxin A$_1$, 10206
$C_{201}H_{328}N_{62}O_{64}S_6$
Viscotoxin A$_3$, 10206
$C_{202}H_{348}N_{68}O_{59}S_8$
α_2-Purothionin, 8055
$C_{203}H_{339}N_{67}O_{59}S_8$
β-Purothionin, 8055
$C_{204}H_{243}N_{63}Na_{20}O_{114}P_{20}S_{20}$
Fomivirsen Sodium Salt, 4256
$C_{204}H_{263}N_{63}O_{114}P_{20}S_{20}$
Fomivirsen, 4256
$C_{204}H_{301}N_{51}O_{64}$
Enfuvirtide, 3636
$C_{215}H_{347}N_{61}O_{65}S$
Lixisenatide, 5605
$C_{217}H_{298}N_{69}O_{64}P$
Histatin 1, 4757
$C_{221}H_{366}N_{72}O_{67}S$
Tesamorelin, 9320
$C_{236}H_{303}N_{70}Na_{23}O_{133}P_{23}S_{23}$
Agatolimod Tricosasodium Salt, 180
$C_{236}H_{326}N_{70}O_{133}P_{23}S_{23}$
Agatolimod, 180
$C_{258}H_{384}N_{64}O_{78}S_6$
Insulin Glulisine, 5028
$C_{267}H_{402}N_{64}O_{76}S_6$
Insulin Detemir, 5026
$C_{272}H_{318}N_{106}Na_{26}O_{138}P_{26}S_{26}$
Edifoligide Sodium, 3559
$C_{284}H_{432}N_{84}O_{79}S_7$
Aprotinin, 746
$C_{305}H_{442}N_{88}O_{91}S_8$
Ecallantide, 3536
$C_{938}H_{1465}N_{257}O_{278}S_6$
Somatotropin, 8842
$C_{938}H_{1469}N_{255}O_{275}S_7$
Somatotropin, 8842
$C_{976}H_{1533}N_{265}O_{286}S_8$
Somatotropin, 8842
$C_{977}H_{1527}N_{265}O_{287}S_7$
Somatotropin, 8842
$C_{978}H_{1537}N_{265}O_{286}S_9$
Somatotropin, 8842
$C_{979}H_{1527}N_{265}O_{287}S_8$
Somatotropin, 8842
$C_{987}H_{1550}N_{268}O_{291}S_9$
Somatotropin, 8842
$C_{990}H_{1529}N_{263}O_{299}S_7$
Human Growth Hormone, 8842
$C_{995}H_{1537}N_{263}O_{301}S_8$
Methionyl Human Growth Hormone, 8842
$C_{1020}H_{1596}N_{274}O_{302}S_9$
Somatotropin, 8842
$CaCl_2$
Calcium Chloride, 1661
$CaCl_2O_2$
Calcium Hypochlorite, 1676
$CaCl_2O_6$
Calcium Chlorate, 1660
$CaCrO_4$
Calcium Chromate(VI), 1662
$CaCr_2O_7$
Calcium Dichromate(VI), 1668
CaF_2
Calcium Fluoride, 1669
CaF_6Si
Calcium Hexafluorosilicate, 1673
$CaHO_3P$
Calcium Phosphite, 1698
$CaHO_4P$
Calcium Phosphate, Dibasic, 1694
CaH_2
Calcium Hydride, 1674
CaH_2O_2
Calcium Hydroxide, 1675

$CaH_4O_4P_2$
Calcium Hypophosphite, 1677
$CaH_4O_8P_2$
Calcium Phosphate, Monobasic, 1695
CaI_2
Calcium Iodide, 1679
CaI_2O_6
Calcium Iodate, 1678
$CaMn_2O_8$
Calcium Permanganate, 1690
$CaMoO_4$
Calcium Molybdate(VI), 1683
CaN_2O_4
Calcium Nitrite, 1685
CaN_2O_6
Calcium Nitrate, 1684
CaO
Calcium Oxide, 1688
CaO_2
Calcium Peroxide, 1691
CaO_3S
Calcium Sulfite, 1710
CaO_4S
Calcium Sulfate, 1708
CaO_4W
Calcium Tungstate(VI), 1714
CaS
Calcium Sulfide, 1709
$CaSe$
Calcium Selenide, 1703
$Ca_2O_7P_2$
Calcium Pyrophosphate, 1701
$Ca(C_3H_5O_4)_2.2H_2O$
Glyceric Acid L-Form Calcium Salt, 4519
$Ca_3O_8P_2$
Calcium Phosphate, Tribasic, 1696
Ca_3P_2
Calcium Phosphide, 1697
$CdCl_2$
Cadmium Chloride, 1622
CdF_2
Cadmium Fluoride, 1624
CdH_2O_2
Cadmium Hydroxide, 1625
CdI_2
Cadmium Iodide, 1626
CdN_2O_6
Cadmium Nitrate, 1627
CdO
Cadmium Oxide, 1628
CdO_4S
Cadmium Sulfate, 1632
CdO_4W
Cadmium Tungstate(VI), 1635
CdS
Cadmium Sulfide, 1633
$CdSe$
Cadmium Selenide, 1631
$CdTe$
Cadmium Telluride, 1634
$CeCl_3$
Cerous Chloride, 1996
CeF_3
Cerous Fluoride, 1997
CeF_4
Ceric Fluoride, 1987
$CeH_8N_8O_{18}$
Cerium(IV) Ammonium Nitrate, 1991
CeI_3
Cerous Iodide, 1998
CeN_3O_9
Cerous Nitrate, 1999
CeO_2
Ceric Oxide, 1988
CeO_8S_2
Ceric Sulfate, 1989

Ce₂O₁₂S₃
Cerous Sulfate, 2001
ClCs
Cesium Chloride, 2010
ClCu
Cuprous Chloride, 2650
ClF
Chlorine Monofluoride, 2098
ClFO₃
Perchloryl Fluoride, 7268
ClF₃
Chlorine Trifluoride, 2100
ClH
Hydrogen Chloride, 4831
ClHO
Hypochlorous Acid, 4904
ClHO₃
Chloric Acid, 2093
ClHO₃S
Chlorosulfonic Acid, 2168
ClHO₄
Perchloric Acid, 7267
ClH₄N
Ammonium Chloride, 506
ClH₄NO₄
Ammonium Perchlorate, 538
ClI
Iodine Monochloride, 5060
ClK
Potassium Chloride, 7742
ClKO₃
Potassium Chlorate, 7741
ClKO₄
Potassium Perchlorate, 7774
ClLi
Lithium Chloride, 5584
ClLiO₄
Lithium Perchlorate, 5595
ClNO
Nitrosyl Chloride, 6733
ClNO₂
Nitryl Chloride, 6744
ClNa
Sodium Chloride, 8734
ClNaO
Sodium Hypochlorite, 8762
ClNaO₂
Sodium Chlorite, 8735
ClNaO₃
Sodium Chlorate, 8733
ClNaO₄
Sodium Perchlorate, 8784
ClOSb
Antimony Chloride Oxide, 684
ClO₂
Chlorine Dioxide, 2096
ClRb
Rubidium Chloride, 8417
ClTl
Thallium Chloride, 9409
Cl₂²²³Ra
Radium-223 Chloride, 8211
Cl₂Co
Cobaltous Chloride, 2421
Cl₂Cr
Chromous Chloride, 2248
Cl₂CrO₂
Chromyl Chloride, 2252
Cl₂Cu
Cupric Chloride, 2623
Cl₂CuO₆
Cupric Chlorate, 2622
Cl₂CuO₈
Cupric Perchlorate, 2637
Cl₂F₃Sb
Antimony Dichlorotrifluoride, 685

Cl₂Fe
Ferrous Chloride, 4070
Cl₂Ge
Germanium Dichloride, 4442
Cl₂H₆N₂Pt
Cisplatin, 2316
Cl₂Hg
Mercuric Chloride, 5943
Cl₂Hg₂
Mercurous Chloride, 5962
Cl₂Mg
Magnesium Chloride, 5726
Cl₂MgO₆
Magnesium Chlorate, 5725
Cl₂MgO₈
Magnesium Perchlorate, 5742
Cl₂Mn
Manganese Chloride, 5793
Cl₂Ni
Nickel Chloride, 6590
Cl₂O
Chlorine Monoxide, 2099
Cl₂OS
Thionyl Chloride, 9501
Cl₂OSe
Selenium Oxychloride, 8572
Cl₂OV
Vanadyl Dichloride, 10112
Cl₂OZr
Zirconyl Chloride, 10384
Cl₂O₂S
Sulfuryl Chloride, 9110
Cl₂O₂U
Uranyl Chloride, 10047
Cl₂O₆Sr
Strontium Chlorate, 8969
Cl₂O₇
Chlorine Heptoxide, 2097
Cl₂O₈Zn
Zinc Perchlorate, 10348
Cl₂Pb
Lead Chloride, 5459
Cl₂Pd
Palladium Chloride, 7091
Cl₂Pt
Platinous Chloride, 7643
Cl₂S₂
Sulfur Chloride, 9100
Cl₂Se₂
Selenium Chloride, 8569
Cl₂Sn
Stannous Chloride, 8910
Cl₂Sr
Strontium Chloride, 8970
Cl₂Te
Tellurium Dichloride, 9264
Cl₂Ti
Titanium Dichloride, 9627
Cl₂Zn
Zinc Chloride, 10331
Cl₃CoH₁₈N₆
Hexaaminecobalt Trichloride, 4710
Cl₃Cr
Chromic Chloride, 2226
Cl₃Fe
Ferric Chloride, 4046
Cl₃Ga
Gallium Chloride, 4378
Cl₃HSi
Trichlorosilane, 9811
Cl₃H₄MgN
Ammonium Magnesium Chloride, 528
Cl₃Ho
Holmium Chloride, 4765
Cl₃I
Iodine Trichloride, 5063

Cl₃In
Indium Trichloride, 4998
Cl₃Ir
Iridium Trichloride, 5132
Cl₃La
Lanthanum Chloride, 5414
Cl₃N
Nitrogen Chloride, 6689
Cl₃OP
Phosphorus Oxychloride, 7461
Cl₃OV
Vanadyl Trichloride, 10114
Cl₃P
Phosphorus Trichloride, 7470
Cl₃PS
Phosphorus Sulfochloride, 7468
Cl₃Rh
Rhodium Chloride, 8313
Cl₃Ru
Ruthenium Trichloride, 8436
Cl₃Sb
Antimony Trichloride, 696
Cl₃Ti
Titanium Trichloride, 9637
Cl₃U
Uranium Trichloride, 10044
Cl₄CuH₈N₂
Ammonium Cupric Chloride, 511
Cl₄Ge
Germanium Tetrachloride, 4444
Cl₄H₈N₂Pt
Ammonium Platinous Chloride, 546
Cl₄H₈N₂Zn
Ammonium Tetrachlorozincate, 556
Cl₄K₂Pt
Potassium Tetrachloroplatinate(II), 7801
Cl₄Se
Selenium Tetrachloride, 8575
Cl₄Si
Silicon Tetrachloride, 8639
Cl₄Sn
Stannic Chloride, 8901
Cl₄Te
Tellurium Tetrachloride, 9268
Cl₄Th
Thorium Chloride, 9529
Cl₄Ti
Titanium Tetrachloride, 9634
Cl₄U
Uranium Tetrachloride, 10042
Cl₄Zr
Zirconium Chloride, 10374
Cl₅H₁₂N₃Zn
Ammonium Pentachlorozincate, 537
Cl₅Nb
Niobium Pentachloride, 6643
Cl₅P
Phosphorus Pentachloride, 7463
Cl₅Sb
Antimony Pentachloride, 686
Cl₅Ta
Tantalum Pentachloride, 9186
Cl₆H₂Pt
Platinic Chloride, 7641
Cl₆H₈N₂Os
Ammonium Osmium Chloride, 533
Cl₆H₈N₂Pt
Ammonium Platinic Chloride, 545
Cl₆H₄₂N₁₄O₂Ru₃
Ruthenium Red, 8434
Cl₆K₂Os
Potassium Hexachloroosmate(IV), 7756
Cl₆K₂Pt
Potassium Hexachloroplatinate(IV), 7757
Cl₆Na₂Pt
Sodium Hexachloroplatinate(IV), 8758

Formula Index

CoCr₂O₄
Cobaltous Chromate(III), 2422
CoF₂
Cobaltous Fluoride, 2424
CoF₃
Cobaltic Fluoride, 2414
CoHO₂
Cobaltic Oxide Monohydrate, 2415
CoH₂O₂
Cobaltous Hydroxide, 2426
CoH₄NO₄P
Ammonium Cobaltous Phosphate, 510
CoI₂
Cobaltous Iodide, 2427
CoK₃N₆O₁₂
Cobaltic Potassium Nitrite, 2416
CoN₂O₆
Cobaltous Nitrate, 2428
CoN₆Na₃O₁₂
Sodium Cobaltinitrite, 8738
CoO
Cobaltous Oxide, 2430
CoO₄S
Cobaltous Sulfate, 2432
CoS
Cobaltous Sulfide, 2433
Co₃O₄
Cobaltic-Cobaltous Oxide, 2413
Co₃O₈P₂
Cobaltous Phosphate, 2431
Co₅Sm
Samarium Cobalt, 8485
Co₁₇Sm₂
Samarium Cobalt, 8485
CrCuO₄
Cupric Chromate(VI), 2624
CrF₂
Chromous Fluoride, 2249
CrF₂O₂
Chromyl Fluoride, 2253
CrF₃
Chromic Fluoride, 2227
CrH₃O₃
Chromic Hydroxide, 2229
CrH₄NO₈S₂
Ammonium Chromic Sulfate, 508
CrH₈N₂O₄
Ammonium Chromate(VI), 507
CrKO₈S₂
Chromic Potassium Sulfate, 2233
CrK₂O₄
Potassium Chromate(VI), 7743
CrLi₂O₄
Lithium Chromate(VI), 5585
CrN₃O₉
Chromic Nitrate, 2230
CrNa₂O₄
Sodium Chromate, 8736
CrO₂
Chromium Dioxide, 2237
CrO₃
Chromium(VI) Oxide, 2239
CrO₄P
Chromic Phosphate, 2231
CrO₄Pb
Lead Chromate(VI), 5460
CrO₄S
Chromous Sulfate, 2251
CrO₄Sr
Strontium Chromate(VI), 8971
CrPb₂O₅
Lead Chromate(VI) Oxide, 5461
Cr₂CuO₄
Cupric Chromite, 2625
Cr₂H₈N₂O₇
Ammonium Dichromate(VI), 512

Cr₂K₂O₇
Potassium Dichromate(VI), 7748
Cr₂Na₂O₇
Sodium Dichromate(VI), 8744
Cr₂O₃
Chromium(III) Oxide, 2238
Cr₂O₈Sn
Stannic Chromate(VI), 8902
Cr₂O₁₂S₃
Chromic Sulfate, 2234
Cr₃Fe₂O₁₂
Ferric Chromate(VI), 4047
CsF
Cesium Fluoride, 2011
CsHO
Cesium Hydroxide, 2012
CsI
Cesium Iodide, 2013
CsNO₃
Cesium Nitrate, 2014
Cs₂O₄S
Cesium Sulfate, 2015
CuF₂
Cupric Fluoride, 2628
CuF₆Si
Cupric Hexafluorosilicate, 2631
CuFeS₂
Chalcopyrite, 2037
CuH₂O₂
Cupric Hydroxide, 2632
CuH₁₂N₄O₄S
Tetraamminecopper Sulfate, 9325
CuH₁₄N₄O₂
Schweizer's Reagent, 8534
CuI
Cuprous Iodide, 2652
CuN₂O₆
Cupric Nitrate, 2633
CuO
Cupric Oxide, 2636
CuO₄S
Cupric Sulfate, 2643
CuO₄Se
Cupric Selenate, 2640
CuO₄W
Cupric Tungstate(VI), 2647
CuS
Cupric Sulfide, 2645
CuSe
Cupric Selenide, 2641
Cu₂HgI₄
Cuprous Mercuric Iodide, 2653
Cu₂O
Cuprous Oxide, 2654
Cu₂O₃S
Cuprous Sulfite, 2658
Cu₂S
Cuprous Sulfide, 2657
Cu₂Se
Cuprous Selenide, 2656
Cu₃H₄O₈S
Copper Sulfate Dibasic, 2644
Cu₃O₆S
Cupro-cupric Sulfate, 2658
Cu₃O₈P₂
Cupric Phosphate, 2638
Cu₄H₆O₁₀S
Copper Sulfate Tribasic, 2644

D

Dy₂O₃
Dysprosium Oxide, 3529

E

Er₂O₃
Erbium Oxide, 3703
Eu₂O₃
Europium Sesquioxide, 3946

F

FH
Hydrogen Fluoride, 4833
FHO₃S
Fluorosulfonic Acid, 4211
FH₄N
Ammonium Fluoride, 519
FK
Potassium Fluoride, 7753
FLi
Lithium Fluoride, 5587
FNO
Nitrosyl Fluoride, 6734
FNO₂
Nitryl Fluoride, 6745
FNO₃
Fluorine Nitrate, 4197
FNa
Sodium Fluoride, 8754
FTl
Thallium Fluoride, 9411
F₂Fe
Ferrous Fluoride, 4072
F₂HK
Potassium Bifluoride, 7731
F₂HNa
Sodium Bifluoride, 8720
F₂H₅N
Ammonium Bifluoride, 495
F₂Hg
Mercuric Fluoride, 5945
F₂Kr
Krypton Difluoride, 5373
F₂Mg
Magnesium Fluoride, 5729
F₂Mn
Manganese Difluoride, 5794
F₂Ni
Nickel Fluoride, 6593
F₂O
Fluorine Monoxide, 4196
F₂OS
Thionyl Fluoride, 9502
F₂O₂
Fluorine Dioxide, 4195
F₂O₂S
Sulfuryl Fluoride, 9111
F₂Pb
Lead Fluoride, 5463
F₂Sn
Stannous Fluoride, 8911
F₂Sr
Strontium Fluoride, 8972
F₂Zn
Zinc Fluoride, 10335
F₃Fe
Ferric Fluoride, 4049
F₃Ga
Gallium Trifluoride, 4384
F₃In
Indium Trifluoride, 4999
F₃Mn
Manganese Trifluoride, 5806
F₃N
Nitrogen Fluoride, 6691

F₃P
Phosphorus Trifluoride, 7471
F₃Sb
Antimony Trifluoride, 697
F₃V
Vanadium Trifluoride, 10108
F₄Ge
Germanium Tetrafluoride, 4445
F₄Pb
Lead Tetrafluoride, 5479
F₄S
Sulfur Tetrafluoride, 9108
F₄Se
Selenium Tetrafluoride, 8576
F₄Si
Silicon Tetrafluoride, 8640
F₄Sn
Stannic Fluoride, 8903
F₄Ti
Titanium Tetrafluoride, 9635
F₄U
Uranium Tetrafluoride, 10043
F₄Zr
Zirconium Fluoride, 10375
F₅H₁₂N₃O₂U
Ammonium Uranium Fluoride, 564
F₅I
Iodine Pentafluoride, 5061
F₅Nb
Niobium Pentafluoride, 6644
F₅P
Phosphorus Pentafluoride, 7464
F₅Sb
Antimony Pentafluoride, 687
F₅Ta
Tantalum Pentafluoride, 9187
F₅V
Vanadium Pentafluoride, 10106
F₆GaH₁₂N₃
Ammonium Hexafluorogallate, 522
F₆H₂Si
Fluosilicic Acid, 4214
F₆H₄NP
Ammonium Hexafluorophosphate,
523
F₆H₈N₂Si
Ammonium Hexafluorosilicate, 524
F₆Ir
Iridium Hexafluoride, 5130
F₆K₂Si
Potassium Hexafluorosilicate, 7759
F₆K₂Zr
Potassium Hexafluorozirconate, 7760
F₆MgSi
Magnesium Hexafluorosilicate, 5732
F₆Mo
Molybdenum Hexafluoride, 6323
F₆Na₂Si
Sodium Hexafluorosilicate, 8759
F₆Os
Osmium Hexafluoride, 6989
F₆PbSi
Lead Hexafluorosilicate, 5465
F₆Re
Rhenium Hexafluoride, 8302
F₆S
Sulfur Hexafluoride, 9103
F₆Se
Selenium Hexafluoride, 8570
F₆SiZn
Zinc Hexafluorosilicate, 10337
F₆Te
Tellurium Hexafluoride, 9266
F₆U
Uranium Hexafluoride, 10041
F₆W
Tungsten Hexafluoride, 9996

F₇I
Iodine Heptafluoride, 5058
FeHO₂
Ferric Hydroxide, 4052
FeH₂O₂
Ferrous Hydroxide, 4075
FeH₄NO₈S₂
Ammonium Ferric Sulfate, 515
FeH₆O₆P₃
Ferric Hypophosphite, 4053
FeH₈N₂O₈S₂
Ammonium Ferrous Sulfate, 518
FeI₂
Ferrous Iodide, 4076
FeN₃O₉
Ferric Nitrate, 4054
FeO
Ferrous Oxide, 4079
FeO₄P
Ferric Phosphate, 4056
FeO₄S
Ferrous Sulfate, 4084
FeS
Ferrous Sulfide, 4085
FeSe
Ferrous Selenide, 4082
Fe₂O₃
Ferric Oxide, 4055
Fe₂O₁₂S₃
Ferric Sulfate, 4059
Fe₂P
Ferrous Phosphide, 4081
Fe₃O₄
Ferrosoferric Oxide, 4067
Fe₃O₈P₂
Ferrous Phosphate, 4080
Fe₄O₂₁P₆
Ferric Pyrophosphate, 4057
Fe₅O₁₂Y₃
Yig, 10295

G

GaN
Gallium Nitride, 4381
GaN₃O₉
Gallium Nitrate, 4380
GaP
Gallium Phosphide, 4383
Gd₂O₃
Gadolinium Oxide, 4356
GeH₄
Germane, 4440
GeMg₂
Magnesium Germanide, 5731
GeO₂
Germanium Dioxide, 4443

H

HAsO₄Pb
Lead Arsenate, 5454
HBr
Hydrobromic Acid, 4816
HCl
Hydrochloric Acid, 4818
HF
Hydrofluoric Acid, 4827
HI
Hydriodic Acid, 4814
Hydrogen Iodide, 4834

HIO₃
Iodic Acid, 5056
HKO
Potassium Hydroxide, 7761
HKO₄S
Potassium Bisulfate, 7734
HKS
Potassium Bisulfide, 7735
HK₂O₃P
Potassium Phosphite, 7782
HK₂O₄P
Potassium Phosphate, Dibasic, 7779
HLi
Lithium Hydride, 5589
HLiO
Lithium Hydroxide, 5590
HMgO₄P
Magnesium Phosphate, Dibasic, 5745
HNO₂
Nitrous Acid, 6739
HNO₃
Nitric Acid, 6663
HNO₅S
Nitrosylsulfuric Acid, 6735
HN₃
Hydrazoic Acid, 4812
HNa
Sodium Hydride, 8760
HNaO
Sodium Hydroxide, 8761
HNaO₃S
Sodium Bisulfite, 8724
HNaO₄S
Sodium Bisulfate, 8722
HNaS
Sodium Bisulfide, 8723
HNa₂O₃P
Sodium Phosphite, 8793
HNa₂O₄³²P
Sodium Phosphate P 32, 8791
HNa₂O₄P
Sodium Phosphate, Dibasic, 8789
HORb
Rubidium Hydroxide, 8418
HOTl
Thallium Hydroxide, 9412
HO₃P
Phosphoric Acid, Meta, 7457
H₂KO₂P
Potassium Hypophosphite, 7762
H₂KO₄P
Potassium Phosphate, Monobasic,
7780
H₂LiN
Lithium Amide, 5578
H₂Mg
Magnesium Hydride, 5733
H₂MgO₂
Magnesium Hydroxide, 5734
H₂Mg₃O₁₂Si₄
Talc, 9168
H₂MoO₄
Molybdic(VI) Acid, 6326
H₂NNa
Sodium Amide, 8714
H₂N₂O₂
Nitramide, 6658
H₂NaO₂P
Sodium Hypophosphite, 8763
H₂NaO₄P
Sodium Phosphate, Monobasic, 8790
H₂Na₂O₇P₂
Sodium Acid Pyrophosphate, 8710
H₂NiO₂
Nickel Hydroxide, 6595
H₂O
Water, 10237

K

KMnO₄
Potassium Permanganate, 7776
KNO₂
Potassium Nitrite, 7770
KNO₃
Potassium Nitrate, 7769
K₂MnO₄
Potassium Manganate, 7765
K₂MoO₄
Potassium Molybdate, 7768
K₂NO₇S₂
Fremy's Salt, 4297
K₂O₃S
Potassium Sulfite, 7795
K₂O₃S₂
Potassium Thiosulfate, 7808
K₂O₃Sn
Potassium Stannate, 7791
K₂O₃Te
Potassium Tellurite, 7798
K₂O₄S
Potassium Sulfate, 7793
K₂O₄Se
Potassium Selenate, 7786
K₂O₄Te
Potassium Tellurate, 7797
K₂O₄W
Potassium Tungstate, 7811
K₂O₅S₂
Potassium Metabisulfite, 7766
K₂O₇S₂
Potassium Pyrosulfate, 7784
K₂O₇U₂
Potassium Uranate(VI), 7812
K₂O₈S₂
Potassium Persulfate, 7777
K₂S
Potassium Sulfide, 7794
K₂Se
Potassium Selenide, 7787
K₃O₄P
Potassium Phosphate, Tribasic, 7781
K₄O₇P₂
Potassium Pyrophosphate, 7783

L

LaN₃O₉
Lanthanum Nitrate, 5414
La₂O₃
Lanthanum Oxide, 5414
La₂O₁₂S₃
Lanthanum Sulfate, 5414
LiNO₃
Lithium Nitrate, 5592
Li₂O
Lithium Oxide, 5594
Li₂O₃Si
Lithium Silicate, 5596
Li₂O₄S
Lithium Sulfate, 5597

M

MgMn₂O₈
Magnesium Permanganate, 5743

MgN₂O₆
Magnesium Nitrate, 5738
MgO
Magnesium Oxide, 5741
MgO₂
Magnesium Peroxide, 5744
MgO₃S
Magnesium Sulfite, 5756
MgO₃S₂
Magnesium Thiosulfate, 5758
MgO₄S
Magnesium Sulfate, 5755
MgSe
Magnesium Selenide, 5750
MgSiO₃
Magnesium Metasilicate, 5751
Mg₂O₇P₂
Magnesium Pyrophosphate, 5748
Mg₂Si
Magnesium Silicide, 5752
Mg₂SiO₄
Magnesium Orthosilicate, 5751
Mg₂Si₃O₈
Magnesium Trisilicate, 5751
Mg₂Sn
Magnesium Stannide, 5753
Mg₃O₇Si₂
Serpentine (Mineral), 5751
Mg₃O₈P₂
Magnesium Phosphate, Tribasic, 5747
MnN₂O₆
Manganese Nitrate, 5798
MnNaO₄
Sodium Permanganate, 8785
MnO₂
Manganese Dioxide, 5795
MnO₄S
Manganese Sulfate, 5804
MnS
Manganese Sulfide, 5805
MnSe
Manganese Selenide, 5801
Mn₂O₃
Manganese Sesquioxide, 5802
Mn₃O₄
Manganese Oxide, 5800
MoNa₂O₄
Sodium Molybdate(VI), 8776
MoO₃
Molybdenum Trioxide, 6325
MoO₄Pb
Lead Molybdate(VI), 5467
MoS₂
Molybdenum Disulfide, 6322
Mo₂O₃
Molybdenum Sesquioxide, 6324
Mo₁₂Na₃O₄₀P
Sodium Phosphomolybdate, 8794

N

NNaO₂
Sodium Nitrite, 8779
NNaO₃
Sodium Nitrate, 8778
NO
Nitric Oxide, 6664
NO₂
Nitrogen Dioxide, 6690
NO₃Tl
Thallium Mononitrate, 9414
NTi
Titanium Nitride, 9630

N₂Na₂O₃
Angeli's Salt, 641
N₂NiO₆
Nickel Nitrate, 6598
N₂O
Nitrous Oxide, 6740
N₂O₃
Dinitrogen Trioxide, 3307
N₂O₄
Dinitrogen Tetroxide, 3306
N₂O₅
Nitrogen Pentoxide, 6692
N₂O₆Pb
Lead Nitrate, 5469
N₂O₆Pd
Palladium Nitrate, 7093
N₂O₆Sr
Strontium Nitrate, 8975
N₂O₆Zn
Zinc Nitrate, 10343
N₂O₈U
Uranyl Nitrate, 10048
N₃Na
Sodium Azide, 8717
N₄O₁₂Th
Thorium Nitrate, 9530
N₄O₁₂Zr
Zirconium Nitrate, 10379
N₄S₄
Tetrasulfur Tetranitride, 9389
N₄Se₄
Nitrogen Selenide, 6693
N₄Si₃
Silicon Nitride, 8636
N₆Pb
Lead Azide, 5456
NaO₃V
Sodium Metavanadate, 8774
NaO₄⁹⁹ᵐTc
Sodium Pertechnetate Tc 99m, 8788
Na₂O
Sodium Oxide, 8782
Na₂O₂
Sodium Peroxide, 8786
Na₂O₃S
Sodium Sulfite, 8811
Na₂O₃S₂
Sodium Thiosulfate, 8821
Na₂O₃Se
Sodium Selenite, 8802
Na₂O₃Si
Sodium Metasilicate, 8804
Na₂O₃Sn
Sodium Stannate(IV), 8805
Na₂O₃Te
Sodium Tellurite, 8814
Na₂O₄S
Sodium Sulfate, 8809
Na₂O₄S₂
Sodium Dithionite, 8747
Na₂O₄Se
Sodium Selenate, 8800
Na₂O₄Te
Sodium Tellurate, 8813
Na₂O₄W
Sodium Tungstate(VI), 8826
Na₂O₅S₂
Sodium Metabisulfite, 8770
Na₂O₆S₂
Sodium Dithionate, 8746
Na₂O₇U₂
Sodium Uranate(VI), 8827
Na₂O₈S₂
Sodium Persulfate, 8787
Na₂S
Sodium Sulfide, 8810

Na₂Se
Sodium Selenide, 8801
Na₃O₃PS
Sodium Thiophosphate, 8820
Na₃O₄P
Sodium Phosphate, Tribasic, 8792
Na₃O₉P₃
Sodium Trimetaphosphate, 8823
Na₄O₇P₂
Tetrasodium Pyrophosphate, 9388
Na₅O₁₀P₃
Sodium Tripolyphosphate, 8824
Nb₂O₅
Niobium Pentoxide, 6645
NiO
Nickel Monoxide, 6597
NiO₄S
Nickel Sulfate, 6602
Ni₂O₃
Nickel Sesquioxide, 6601
Ni₃O₈P₂
Nickel Phosphate, 6600

O

OPb
Lead Monoxide, 5468
OPd
Palladium Oxide, 7094
OSi
Silicon Monoxide, 8635
OSn
Stannous Oxide, 8914
OSr
Strontium Oxide, 8976
OTl₂
Thallium Oxide, 9415
OZn
Zinc Oxide, 10347
O₂Pb
Lead Dioxide, 5462
O₂Pt
Platinic Oxide, 7642
O₂S
Sulfur Dioxide, 9101
O₂Se
Selenium Oxide, 8571
O₂Si
Silicon Dioxide, 8632
O₂Sn
Stannic Oxide, 8905
O₂Sr
Strontium Peroxide, 8977
O₂Te
Tellurium Dioxide, 9265
O₂Th
Thorium Oxide, 9531
O₂Ti
Titanium Dioxide, 9628
O₂U
Uranium Dioxide, 10040
O₂Zn
Zinc Peroxide, 10349
O₂Zr
Zirconium Oxide, 10380
O₃
Ozone, 7080
O₃P₂
Phosphorus Trioxide, 7472
O₃Pb₂
Lead Sesquioxide, 5472
O₃Re
Rhenium Trioxide, 8303

O₃S
Sulfur Trioxide, 9109
O₃Sb₂
Antimony Trioxide, 699
O₃Sc₂
Scandium Oxide, 8529
O₃SrTi
Strontium Titanate, 8980
O₃Tb₂
Terbium Oxide, 9300
O₃Tl₂
Thallium Sesquioxide, 9417
O₃Tm₂
Thulium Oxide, 9548
O₃U
Uranium Trioxide, 10045
O₃V₂
Vanadium Trioxide, 10109
O₃W
Tungsten Trioxide, 9997
O₃Y₂
Yttrium Oxide, 10305
O₃Yb₂
Ytterbium Oxide, 10303
O₄Os
Osmium Tetroxide, 6990
O₄PbS
Lead Sulfate, 5475
O₄Pb₃
Lead Tetroxide, 5480
O₄Ru
Ruthenium Tetroxide, 8435
O₄SSn
Stannous Sulfate, 8917
O₄SSr
Strontium Sulfate, 8978
O₄STl₂
Thallium Sulfate, 9418
O₄SZn
Zinc Sulfate, 10359
O₄SiZn₂
Zinc Silicate, 10357
O₄SiZr
Zirconium Silicate, 10381
O₅P₂
Phosphorus Pentoxide, 7467
O₅STi
Titanium Oxysulfate, 9631
O₅SV
Vanadyl Sulfate, 10113
O₅Sb₂
Antimony Pentoxide, 689
O₅Ta₂
Tantalum Pentoxide, 9188
O₅V₂
Vanadium Pentoxide, 10107
O₆PbV₂
Lead Vanadate(V), 5482
O₆SU
Uranyl Sulfate, 10049
O₇P₂Sn₂
Stannous Pyrophosphate, 8915
O₇P₂Zn₂
Zinc Pyrophosphate, 10354
O₇Re₂
Rhenium Heptoxide, 8301
O₈P₂Pb₃
Lead Phosphate, 5471
O₈P₂Zn₃
Zinc Phosphate, 10351
O₈S₂Zr
Zirconium Sulfate, 10382
O₁₂S₃Sb₂
Antimony Sulfate, 694
O₁₂S₃Ti₂
Titanium Sesquisulfate, 9632

P

P₂S₅
Phosphorus Pentasulfide, 7466
P₂Se₅
Phosphorus Pentaselenide, 7465
P₂Zn₃
Zinc Phosphide, 10352
P₄S₃
Tetraphosphorus Trisulfide, 9385
P₄Se₃
Phosphorus Triselenide, 7473
PbS
Lead Sulfide, 5476
PbTe
Lead Telluride, 5477

S

SSn
Stannous Sulfide, 8918
SSr
Strontium Sulfide, 8979
STl₂
Thallium Sulfide, 9419
SZn
Zinc Sulfide, 10360
S₂Si
Silicon Disulfide, 8633
S₂Sn
Stannic Sulfide, 8907
S₃Sb₂
Antimony Trisulfide, 701
S₃V₂
Vanadium Trisulfide, 10110
S₅Sb₂
Antimony Pentasulfide, 688
Sb₂Se₃
Antimony Triselenide, 700
SeS₂
Selenium Disulfide, 8573
SeSn
Stannous Selenide, 8916
SeTl₂
Thallium Selenide, 9416
SeZn
Zinc Selenide, 10356
Se₂Sn
Stannic Selenide, 8906

T

TeZn
Zinc Telluride, 10362
TlN₃O₉
Thallium Trinitrate, 9420
TmCl₃.7H₂O
Thulium Chloride Heptahydrate, 9548
Tm₂(C₂O₄)₃.6H₂O
Thulium Oxalate Hexahydrate, 9548
Tm₂(SO₄)₃.8H₂O
Thulium Sulfate Octahydrate, 9548
Tm(OH)₃
Thulium Hydroxide, 9548

NAME INDEX

NAME INDEX

The Name Index refers the reader to monograph numbers with two types of citations:

Ethacrynic Acid, 3772 (format for titles)
Ethacrynate Sodium *see* 3772 (format for all synonyms)

Company name included in citation denotes trademark owner or its subsidiary, affiliate, licensee or distributor.

Amistar [Syngenta] *see* 917

02-115 *see* 9170
2-5410-3A *see* 5066
2F2 *see* 6862
3C Antibiotic *see* 6425
3PAA *see* 8082
4A65 *see* 4956
4-C-32 *see* 9585
6-12 [Union Carbide] *see* 3797
6-H.E.S. [Morishita] *see* 4708
7-ACA *see* 429
7E3 *see* 5
**7E11-C5.3-glycyl-tyrosyl-(*N*-ε-diethyl-
   enetriaminepentaacetic Acid)-lysine**
   see 1764
7E11-C5.3-GYK-DTPA *see* 1764
8-AL *see* 6582
9-AC *see* 426
15AU81 *see* 9748
17-AAGH₂ *see* 8282
18:4(n-3) *see* 8931
27-400 *see* 2750
31K-Precursor *see* 7905
51W89 *see* 859
53-32 C *see* 9585
84L *see* 3128
87/312 *see* 6681
90CE *see* 5425
101-E *see* 6485
101-G *see* 6485
101M *see* 5425
106-7 *see* 2749
141W94 *see* 585
191C49 *see* 9445
217-MI *see* 3548
256U *see* 10085
264-EUP-99 *see* 5283
295C51 *see* 9922
311C90 *see* 10390
501-P *see* 4499
506U *see* 6527
516-MD *see* 2306
524W91 *see* 3622
566C80 *see* 857
683-M *see* 7043
688-A *see* 7368
710-F *see* 7636
722 D *see* 7620
746-CE *see* 3688
776C85 *see* 3638
921 C *see* 9826
1081 *see* 4198
1162-F *see* 9057
1314-Th *see* 3791
1358F *see* 2822
1489-RB *see* 7572
1497-CB *see* 7943
1522-CB *see* 33
1589mRB *see* 7172
1589RB *see* 7172
1592U89 *see* 1
1600 Antibiotic *see* 7146
1609-RB *see* 7601
1665-RB *see* 760
1703-18B *see* 4891

2249-F *see* 7022
2329 Labaz *see* 1085
2601-A *see* 2191
3123-L *see* 8054
5512-M *see* 672
6063 *see* 52
6315-S *see* 4136
6720 CDRI *see* 1972
7228 *see* 4093
7432-S *see* 1950
8053CB *see* 9453
8102 CB *see* 948
10040 *see* 2092
10275-S *see* 3680
10364-S *see* 5925
13832-JL *see* 10275
29060-LE *see* 10179
47657 *see* 739
81723 hfu *see* 9574
450191-S *see* 8346
710674-S *see* 2580
A-002 *see* 10126
A-21 *see* 5263
A-41-304 *see* 2931
A-65 *see* 7611
A-66 *see* 7351
A-101 *see* 6780
A-145 *see* 7898
A-272 *see* 6927
A-300-I *see* 9814
A-363 *see* 428
A-585 *see* 9124
A-820 *see* 1534
A-3823A *see* 6332
A-4166 *see* 6510
A-4696 *see* 123
A-4942 *see* 4937
A-5610 *see* 899
A-8103 *see* 7593
A-9145 *see* 8682
A-16686 *see* 8222
A-19120 *see* 7142
A-20968 *see* 7594
A-23187 *see* 1642
A-29622 *see* 4278
A-33547 *see* 8252
A-35957 *see* 315
A-38414 *see* 624
A-40926 *see* 2800
A-43818 *see* 5509
A-46745 *see* 4449
A-56268 *see* 2338
A-56619 *see* 3160
A-56620 *see* 8507
A-60386X *see* 1153
A-60969 *see* 9714
A-61827 *see* 9714
A-62254 *see* 9274
A-63004 *see* 9274
A-64077 *see* 10323
A-64730 *see* 9714
A-65006 *see* 5413
A-70569 *see* 9572
A-83543A *see* 8877

A-83543D *see* 8877
A-84538 *see* 8366
(+)-A-127722 *see* 861
A-147627.1 *see* 861
A-157378 *see* 5630
A-195773 *see* 2018
A-771726 *see* 9308
Aα *see* 4037
Aβ *see* 608
β/A4 *see* 608
AA-149 *see* 9747
AA-673 *see* 486
AA-2414 *see* 8598
Aa-5648 *see* 9087
AAB-001 *see* 951
Aacifemine [Aaciphar] *see* 3762
AAF *see* 4189
AAT *see* 705
A1AT *see* 705
AAtrex [Syngenta] *see* 862
AB-021 *see* 9642
AB-100 *see* 10057
AB-206 *see* 6277
AB-1010 *see* 5822
ABA *see* 12
AB-A-663 *see* 2281
Abacavir, 1
Abacin [Benedetti] *see* 9050
Abadol [Specia] *see* 475
Abalyn *see* 6080
Abamectin, 2
Abarelix, 3
Abasin [Bayer] *see* 22
Abatacept, 4
Abate [BASF] *see* 9278
Abbocillin [Abbott] *see* 7206
Abboflox [Abbott] *see* 4143
Abbokinase [Abbott] *see* 10072
Abboticine [Abbott] *see* 3743
Abbott 16900 *see* 9250
Abbott 30400 *see* 7188
Abbott 35616 *see* 2392
Abbott 36581 *see* 1516
Abbott 43326 *see* 1866
Abbott 43818 *see* 5509
Abbott 44747 *see* 4275
Abbott 45975 *see* 9297
Abbott 46811 *see* 1946
Abbott 48999 *see* 1936
Abbott 50192 *see* 1926
Abbott 50711 *see* 10099
Abbott 56619 *see* 3160
Abbott 56620 *see* 8507
Abbott 64077 *see* 10323
Abbott 73001 *see* 8598
Abbott 84538 *see* 8366
Abbsa [Sanko] *see* 9681
ABC-12/3 *see* 3486
ABCD *see* 582
Abciximab, 5
ABDP *see* 223
Abelcet [Liposome Co.] *see* 582
Abequito [Am. Cyanamid] *see* 6635
Aberel [Janssen-Cilag] *see* 8288

Abetimus Sodium, 6
ABG-6215 see 4017
Abicoviromycin see 8
Δ^6,8(14)-Abietadienoic Acid see 5522
Abietic Acid, 7
Abietic Anhydride see 8397
Abietin see 2487
Abikoviromycin, 8
Abil [Goldschmidt Chem.] see 8634
Abilify [Otsuka] see 776
Abilit [Sumitomo] see 9119
Abimasten 100 [ABIC] see 6618
Abiocine [Lepetit] see 3197
Abiol [3V Inc.] see 4957
Abiraterone, 9
Abitilguanide see 6358
Abitylguanide see 6358
ABLC see 582
Abminthic [Pfizer] see 3412
ABOB see 6358
Abomasal Enzyme see 8254
Aboren [Promeco] see 6262
Abovis [Toyama] see 105
ABR-215050 see 9210
ABR-215062 see 5422
Abramen [Mepha] see 380
Abraxane [Abraxis Biosci.] see 7081
Abreva [GSK] see 3444
Abrial Lavandin Oil see 5445
Abrifam [Abbott] see 8341
Abrin, 10
Abrine, 11
Abrodil [Schering] see 6045
Abrus Agglutinin see 10
Abscisic Acid, 12
(R)-(−)-Abscisic Acid see 12
Abscisin II see 12
Absentol [Nourypharma] see 9875
Absinthiin see 13
Absinthin, 13
Absinthium see 10250
Absolute Alcohol see 3814
Absonal [Boehringer, Ing.] see 1114
Absorbable Cellulose see 7038
Abstem see 1665
Abstral [ProStrakan] see 4028
Absynthin see 13
ABT-263 see 6512
ABT-335 see 4009
ABT-378 see 5630
ABT-538 see 8366
ABT-594 see 9227
ABT-627 see 861
ABT-773 see 2018
Abufène [Théraplix] see 198
Abuphenine see 1525
Aburamycin B see 2242
ABX-EGF see 7112
AC see 4832
AC-001 see 603
AC-17 see 1790
AC-137 see 7828
AC-145 see 34
AC-223 see 5890
AC-293 see 4944
AC-528 see 3330
AC-820 see 6443
AC-1198 see 3235
AC-1802 see 743
AC-2993 see 3955
AC-3422 see 7137
AC-3810 see 948
AC-3911 see 7443
AC-4464 see 9711
AC-5223 see 3455

AC-38555 see 2104
AC-47300 see 4007
AC-52160 see 9278
AC-84633 see 6635
AC-84777 see 3157
AC-92100 see 9301
AC-92553 see 7193
AC-217300 see 4804
AC-222293 see 4944
AC-222705 see 4155
AC-252214 see 4947
AC-252925 see 4946
AC-263499 see 4948
AC-263780 see 2281
AC-299263 see 4945
AC-303630 see 2089
AC-375839 see 6230
AC-382042 see 4014
AC-900001 see 7511
Acabel [Grünenthal] see 1197
Acacetin, 14
Acacetin-β-rutinoside see 5553
Acacia, 15
Acacic Acid, 16
Acaciin see 14
Acadesine, 17
Acadione [Sanofi-Aventis] see 9609
Ac-D-2Nal-D-4Cpa-D-3Pal-Ser-4Aph(L-
   hydroorotyl)-D-4Aph(carbamoyl)-Leu-
   ILys-Pro-D-Ala-NH₂ see 2866
Acamol [Volta] see 47
Acamprosate Calcium, 18
Acanto [Syngenta] see 7520
Acapodene [GTx] see 9709
Acaprin see 3393
Acarbose, 19
Acardi [Boehringer, Ing.] see 7545
Acarexx [Idexx] see 5296
Acariflor [Sipcam] see 4749
Acarin [Makhteshim-Agan] see 3096
Acarol [Syngenta] see 1445
Acarosan [Allergopharma] see 1130
ACC-9653 see 4285
Accelerator Globulin see 3960
Accelerin see 3960
Accelerine see 6725
Accent [Underwood] see 6342
Accent [DuPont] see 6607
Acclaim see 4015
Accolate [AstraZeneca] see 10306
Accord [Dow AgroSci.] see 4547
Accupril [Parke-Davis] see 8167
Accuprin [Pfizer] see 8167
Accupro [Pfizer] see 8167
Accutane [Roche] see 5274
Acdeam [Grelan] see 5701
Ac-Di-Sol [FMC] see 2583
Acea Gel [Ferndale] see 6236
Acebutolol, 20
Acecainide, 21
Acecarbromal, 22
Aceclidine see 8199
Aceclofenac, 23
Acecol [Sankyo] see 9279
Acecoline [Lematte et Boinot] see 82
Acecor [SPA] see 20
Acediamine see 45
Acediasulfone, 24
Acedicone [Boehringer, Ing.] see 9426
Acef [KG Italia] see 1918
Acefylline, 25
Acefylline Piperazine see 25
Aceglutamide, 26
Aceglutamide Aluminum see 26
Acemannan, 27

Acemetacin, 28
Acemethadone see 6017
Acemix [Bioprogress] see 28
Acenalin [Kyowa] see 2315
Acenaphthene, 29
Acenocoumarin see 30
Acenocoumarol, 30
Aceon [Solvay] see 7285
Aceperone, 31
Acephate, 32
Acephen see 5851
Acepifylline see 25
Acepress [BMS] see 1776
Acepril [BMS] see 1776
Acepromazine, 33
Acequin [Recordati] see 8167
Acequinocyl, 34
Acerbon [AstraZeneca] see 5572
Acerin, 35
Acerola, 36
Aceroxatidine see 8407
ACES, 37
Acesulfame, 38
Acesulfame-K see 38
Acetabutone see 31
Acetadote [Cumberland] see 83
Acetal, 39
Acetaldehyde, 40
Acetaldehyde Ammonia, 41
Acetaldehyde Diethyl Acetal see 39
Acetaldehyde Dimethyl Acetal see 3247
Acetaldehyde Homopolymer see 5994
Acetaldehyde Oxime see 43
Acetaldehyde Sodium Bisulfite, 42
Acetaldol see 217
Acetaldoxime, 43
Acetamide, 44
Acetamidine, 45
3-Acetamido-5-(acetamidomethyl)-2,4,6-
   triiodobenzoic Acid see 5055
Acetamidoacetic Acid see 75
N-(2-Acetamido)-2-aminoethanesulfonic
   Acid see 37
p-Acetamidobenzaldehyde Thiosemicarbazone see 9442
p-Acetamidobenzenesulfonyl Chloride
   see 95
ε-Acetamidocaproic Acid, 46
2-Acetamido-2-deoxy-3-O-(D-2-propion-
   yl-L-alanyl-D-isoglutamine)-D-glucopy-
   ranose see 6390
(2R,3R,4S)-3-Acetamido-2-[(1R,2R)-2,3-
   dihydroxy-1-methoxypropyl]-4-guani-
   dino-3,4-dihydro-2H-pyran-6-carbox-
   ylic Acid see 5406
(−)-(S)-2-Acetamido-N-(3,4-dihydroxy-
   phenethyl)-4-(methylthio)butyramide
   Bis(ethylcarbonate) (Ester) see 3441
5-Acetamido-2,4-dimethyltrifluoro-
   methanesulfonanilide see 5873
6-Acetamido-1,2-dithiolo[4,3-b]pyrrol-
   5(4H)-one see 4766
(4S,5R,6R)-5-Acetamido-4-guanidino-6-
   [(1R,2R)-2-hydroxy-1-methoxy-3-(octanoyloxy)propyl]-5,6-dihydro-4H-py-
   ran-2-carboxylic Acid see 5406
5-Acetamido-4-guanidino-2,3,4,5,7-pen-
   tadeoxy-7-methoxy-D-glycero-D-galac-
   to-non-2-enopyranosoic Acid see 5406
5-Acetamido-4-guanidino-2,3,4,5-tetra-
   deoxy-D-glycero-D-galacto-non-2-eno-
   pyranosonic Acid see 10311
6-Acetamidohexanoic Acid see 46
3-Acetamido-4-hydroxybenzenearsonic
   Acid see 51

Acetone Bis(3,5-di-*tert*-butyl-4-hydroxy-phenyl)mercaptole *see* 7873
Acetone-bromoform *see* 9773
Acetonecarboxylic Acid *see* 58
Acetone Chloroform *see* 2129
Acetone Cyanohydrin, 66
Acetonedicarboxylic Acid, 67
Acetone Glycerol *see* 5259
Acetone Oxime *see* 73
Acetone Sodium Bisulfite, 68
Acetone Sulfite *see* 68
Acetonitrile, 69
Acetonylacetone, 70
3-(α-Acetonylbenzyl)-4-hydroxycoumarin *see* 10236
Acetonyl Chloride *see* 2114
3-(α-Acetonyl-*p*-chlorobenzyl)-4-hydroxycoumarin *see* 2544
3-Acetonyl-5,8-dihydroxy-2-(hydroxymethyl)-6-methoxy-1,4-naphthoquinone *see* 4344
3-Acetonyl-5,8-dihydroxy-6-methoxy-2-methyl-1,4-naphthoquinone *see* 5310
3-Acetonylindole *see* 5008
(±)-Acetonyl Methyl 1,4-Dihydro-2,6-dimethyl-4-(*o*-nitrophenyl)-3,5-pyridinedicarboxylate *see* 756
3-(α-Acetonyl-*p*-nitrobenzyl)-4-hydroxycoumarin *see* 30
2-Acetonylpiperidine *see* 7181
p-Acetophenetide *see* 7319
p-Acetophenetidide *see* 7319
Acetophenetidin *see* 7319
para-Acetophenetidin *see* 7319
Acetophenone, 71
Acetopromazine *see* 33
Acetopyrrothine *see* 9495
Acetoquat CDAC *see* 2016
Acetorphan *see* 8207
Acetorphine *see* 3931
Acetosulfam *see* 38
Acetothioamide *see* 9472
Acetotoluide, 72
m-Acetotoluide *see* 72
o-Acetotoluide *see* 72
p-Acetotoluide *see* 72
m-Acetotoluidide *see* 72
Acetovanillon *see* 728
Acetoxime, 73
Acetoxolone, 74
2-Acetoxy-4'-(acetamino)phenylbenzoate *see* 1047
9-[4-Acetoxy-3-(acetoxymethyl)but-1-yl]-2-aminopurine *see* 3970
5-Acetoxy-3-acetylthio-2-[(4-amino-2-methyl-5-pyrimidinyl)methyl-*N*-formylamino]-2-pentene *see* 53
2-(2'-Acetoxy)benzamido-5-nitrothiazole *see* 6653
Acetoxybenzene *see* 7379
2-Acetoxybenzoic Acid *see* 841
17α-Acetoxy-6-chloro-6,7-dehydroprogesterone *see* 2102
2-Acetoxy-5-(α-cyclopropylcarbonyl-2-fluorobenzyl)-4,5,6,7-tetrahydro-thieno[3,2-*c*]pyridine *see* 7833
2-Acetoxy-3-(*N*,*N*-diethylcarboxamido)-9,10-dimethoxy-1,2,3,4,6,7-hexahydro-11b*H*-benzopyridocoline *see* 1122
21-Acetoxy-6α,9α-difluoro-11β-hydroxy-16α,17α-isopropylidenedioxy-1,4-pregnadiene-3,20-dione *see* 4182
3-Acetoxy-6-dimethylamino-4,4-diphenylheptane *see* 6017
5-Acetoxy-2-dimethylamino-4,4-diphenylheptane *see* 6017

17α-Acetoxy-11β-(4-dimethylamino-phenyl)-19-norpregna-4,9-dien-3,20-dione *see* 10030
6-Acetoxy-2,4-dimethyl-*m*-dioxane *see* 3244
α-*dl*-3-Acetoxy-4,4-diphenyl-6-methyl-aminoheptane *see* 6773
cis-1-Acetoxy-7-dodecene *see* 5627
7β-Acetoxy-8,13-epoxy-1α,6β,9α-trihydroxylabd-14-en-11-one *see* 2460
17β-Acetoxyestra-4,9,11-trien-3-one *see* 9745
Acetoxyethane *see* 3811
Acetoxyethene *see* 10189
1-(2-Acetoxyethyl)-4-[3-(2-chloro-10-phenothiazinyl)propyl]piperazine *see* 9509
N-(β-Acetoxyethyl)-*N*'-[γ-(2'-chloro-10'-phenothiazinyl)propyl]piperazine *see* 9509
Acetoxyethylene *see* 10189
17α-Acetoxy-13-ethyl-17-ethynylgon-4-en-3-one Oxime *see* 6792
10-[3-[1-(2-Acetoxyethyl)-4-piperazinyl]-propyl]-2-chlorophenothiazine *see* 9509
17α-Acetoxy-9α-fluoro-11β-hydroxyprogesterone *see* 4230
cis-1-Acetoxy-7-hexadecene *see* 4723
Acetoxyl [Stiefel] *see* 1119
1-Acetoxy-2-methoxyethane *see* 6064
3-Acetoxymethyl-7-aminoceph-3-em-4-oic Acid *see* 429
α-*dl*-3-Acetoxy-6-methylamino-4,4-di-phenylheptane *see* 6773
17β-Acetoxy-1α-methyl-5α-androstan-3-one *see* 5986
Acetoxymethyl Benzylpenicillinate *see* 7190
4'-Acetoxy-3'-(α-methylbutyryloxy)-2',2'-dimethyldihydropyrano(7,8:6',-5')coumarin *see* 10208
17α-Acetoxy-6-methylpregna-4,6-diene-3,20-dione *see* 5876
17α-Acetoxy-6α-methylprogesterone *see* 5864
(7*R*)-3-(Acetoxymethyl)-7-[2-(2-thienyl)-acetamido]-3-cephem-4-carboxylic Acid *see* 1981
17β-Acetoxy-3-oxoestra-4,9,11-triene *see* 9745
Acetoxyphenylmercury *see* 7411
threo-1-Acetoxy-1-phenyl-1-(2-piperidyl)-methane *see* 5521
3-Acetoxy-1-phenylpropene *see* 2302
3α-Acetoxypregnan-20α-ol *see* 7845
20α-Acetoxypregnan-3α-ol *see* 7845
17α-Acetoxyprogesterone *see* 4877
17α-Acetoxyprogesterone 3-Cyclopentyl Enol Ether *see* 7225
1-Acetoxypropane *see* 7954
2-Acetoxypropane *see* 5251
3-Acetoxy-1-propene *see* 276
trans-1-Acetoxy-10-(*n*-propyl)trideca-5,9-diene *see* 7981
(2-Acetoxypropyl)trimethylammonium Chloride *see* 6012
(3β)-Acetoxy-17-(3-pyridyl)androsta-5,16-diene *see* 9
3-Acetoxyquinuclidine *see* 8199
6-Acetoxythymol 2-(Dimethylamino)-ethyl Ether *see* 6378
(6-Acetoxythymoxy)ethyldimethylamine *see* 6378
Acetoxytriphenyltin *see* 9920

Acetphenarsine *see* 51
Acetphenolisatin *see* 7073
Acetpyrogall *see* 8112
Acetulan [Amerchol] *see* 5410
Aceturic Acid, 75
α-Acetylacetanilide *see* 57
Acetylacetic Acid *see* 58
Acetylacetone, 76
Acetylamidine *see* 45
Acetylaminoacetic Acid *see* 75
3-(Acetylamino)-5-[(acetylamino)meth-yl]-2,4,6-triiodobenzoic Acid *see* 5055
3-(Acetylamino)-5-(acetylmethylamino)-2,4,6-triiodobenzoic Acid *see* 6235
2-[[3-(Acetylamino)-5-(acetylmethylami-no)-2,4,6-triiodobenzoyl]amino]-2-de-oxy-D-glucose *see* 6234
(3*R*,4*R*,5*S*)-4-(Acetylamino)-5-amino-3-(1-ethylpropoxy)-1-cyclohexene-1-car-boxylic Acid Ethyl Ester *see* 6986
5-(Acetylamino)-4-[(aminoiminomethyl)-amino]-2,6-anhydro-3,4,5-trideoxy-7-*O*-methyl-D-*glycero*-D-*galacto*-non-2-enonic Acid *see* 5406
5-(Acetylamino)-4-[(aminoiminomethyl)-amino]-2,6-anhydro-3,4,5-trideoxy-D-*glycero*-D-*galacto*-non-2-enonic Acid *see* 10311
p-Acetylaminobenzaldehyde Thiosemi-carbazone *see* 9442
Acetylaminobenzene *see* 49
4-(Acetylamino)benzeneacetic Acid *see* 124
4-(Acetylamino)benzenesulfonyl Chloride *see* 95
N-(*p*-Acetylaminobenzene Sulfonyl)-*p*-nitroaniline *see* 9064
N-[(Acetylamino)carbonyl]-2-bromo-2-ethylbutanamide *see* 22
(*R*)-2-(Acetylamino)-3-*O*-(1-carboxyeth-yl)-2-deoxy-D-glucose *see* 5055
(4''*R*)-4''-(Acetylamino)-5-*O*-demethyl-25-de(1-methylpropyl)-4''-deoxy-25-(1-methylethyl)avermectin A$_{1a}$ *see* 3689
(4''*R*)-4''-(Acetylamino)-5-*O*-demethyl-4''-deoxyavermectin A$_{1a}$ *see* 3689
(4''*R*)-4''-(Acetylamino)-4''-deoxyavermec-tin B$_1$ *see* 3689
(7α,25*R*)-7-[[2-(Acetylamino)-2-deoxy-β-D-glucopyranosyl]oxy]-26-(acetyloxy)-cholest-4-en-3-one *see* 7159
(7*S*,9*S*)-9-Acetyl-9-amino-7-[(2-deoxy-β-D-*erythro*-pentopyranosyl)oxy]-7,8,-9,10-tetrahydro-6,11-dihydroxy-5,12-naphthacenedione *see* 588
5-(Acetylamino)-3,5-dideoxy-D-*glycero*-D-*galacto*-2-nonulosonic Acid *see* 8623
4-(Acetylamino)-*N*-[2-(diethylamino)eth-yl]benzamide *see* 21
(5*R*,6*R*)-3-[(*R*)-[(1*E*)-2-(Acetylamino)eth-enyl]sulfinyl]-6-(1-hydroxy-1-methyl-ethyl)-7-oxo-1-azabicyclo[3.2.0]hept-2-ene-2-carboxylic Acid *see* 1861
4-(Acetylamino)-2-ethoxybenzoic Acid Methyl Ester *see* 3798
(1*S*,2*S*,3*R*,4*R*)-3-[(1*S*)-1-(Acetylamino)-2-ethylbutyl]-4-[(aminoiminomethyl)-amino]-2-hydroxycyclopentanecarbox-ylic Acid *see* 7264
2-Acetylaminofluorene *see* 4189
6-(Acetylamino)hexanoic Acid *see* 46
N-Acetyl-6-aminohexanoic Acid *see* 46
As-[3-(Acetylamino)-4-hydroxyphenyl]-arsonic Acid *see* 51
(2*R*)-2-(Acetylamino)-3-methoxy-*N*-(phenylmethyl)propanamide *see* 5381

2-Acetylamino-4-(methylolcycloethylene-dimercaptoarsine)phenol *see* 802

5-Acetylamino-4-methyl-Δ^2-1,3,4-thiadiazoline-2-sulfonamide *see* 6034

C,C'-[4-[2-[[(2*S*)-2-(Acetylamino)-4-(methylthio)-1-oxobutyl]amino]ethyl]-1,2-phenylene]carbonic Acid *C,C'*-Diethyl Ester *see* 3441

2-Acetylamino-5-nitrothiazole *see* 404

N-Acetyl-*p*-aminophenol *see* 47

p-Acetylaminophenol *see* 47

(8*R*)-7-Acetyl-5-(4-aminophenyl)-8,9-dihydro-8-methyl-7*H*-1,3-dioxolo[4,5-*h*]-[2,3]benzodiazepine *see* 9164

2-[[4-(Acetylamino)phenyl]methylene]hydrazinecarbothioamide *see* 9442

3-(Acetylamino)-1-propanesulfonic Acid Calcium Salt (2:1) *see* 18

2-[[[4-[(Acetylamino)sulfonyl]phenyl]amino]carbonyl]benzoic Acid *see* 7488

2-Acetylamino-1,3,4-thiadiazole-5-sulfonamide *see* 52

(8*S*,10*S*)-8-Acetyl-10-[(3-amino-2,3,6-trideoxy-α-L-*lyxo*-hexopyranosyl)oxy]-7,-8,9,10-tetrahydro-1,6,8,11-tetrahydroxy-5,12-naphthacenedione *see* 1870

(8*S*,10*S*)-8-Acetyl-10-[(3-amino-2,3,6-trideoxy-α-L-*lyxo*-hexopyranosyl)oxy]-7,-8,9,10-tetrahydro-6,8,11-trihydroxy-1-methoxy-5,12-naphthacenedione *see* 2834

(7*S*,9*S*)-9-Acetyl-7-[(3-amino-2,3,6-trideoxy-α-L-*lyxo*-hexopyranosyl)oxy]-7,-8,9,10-tetrahydro-6,9,11-trihydroxy-5,12-naphthacenedione *see* 4927

3-(Acetylamino)-2,4,6-triiodo-5-[(methylamino)carbonyl]benzoic Acid *see* 5107

3-[2-(3-Acetylamino-2,4,6-triiodophenoxy)ethoxy]-2-ethylpropionic Acid *see* 5106

2-[[2-[3-(Acetylamino)-2,4,6-triiodophenoxy]ethoxy]methyl]butanoic Acid *see* 5106

3-[Acetyl-(3-amino-2,4,6-triiodophenyl)amino]-2-methylpropanoic Acid *see* 5054

N-Acetyl-*N*-(3-amino-2,4,6-triiodophenyl)-2-methyl-β-alanine *see* 5054

Acetylandromedol *see* 4578

Acetylaniline *see* 49

N-Acetylanthranilic Acid Ester *see* 5679

N-Acetyl-L-aspartic Acid *see* 830

Acetylated Lanolin *see* 5409

Acetylated Lanolin Alcohols *see* 5410

2-Acetyl-9-azabicyclo[4.2.1]non-2-ene *see* 621

Acetylbenzene *see* 71

N-(*p*-Acetylbenzenesulfonyl)-*N'*-cyclohexylurea *see* 61

N-Acetyl-*p*-benzoquinonimine *see* 56

Acetyl Bromide, 77

1-Acetyl-4-bromobenzene *see* 1410

Acetylbromodiethylacetylcarbamide *see* 22

N-Acetyl-*N*-bromodiethylacetylurea *see* 22

N-Acetyl-*N'*-α-bromo-α-ethylbutyrylcarbamide *see* 22

N-Acetyl-*N*-butyl-β-alanine Ethyl Ester *see* 3828

N-[3-Acetyl-4-(3'-*tert*-butylamino-2'-hydroxy)propoxy]phenyl-*N'*-diethylurea *see* 1962

L-3-Acetyl-5-*sec*-butyl-4-hydroxy-3-pyrrolin-2-one *see* 9292

L-3-Acetyl-5-*sec*-butyltetramic Acid *see* 9292

1-(2-Acetyl-4-*n*-butyramidophenoxy)-2-hydroxy-3-isopropylaminopropane *see* 20

α-Acetylbutyrolactone, 78

Acetylcarbinol *see* 64

Acetylcarbromal *see* 22

Acetylcarnitine, 79

Acetyl-L-carnitine *see* 79

Acetyl-L-carnitine Chloride *see* 79

Acetyl Chloride, 80

(4a*R*,4b*S*,6a*S*,7*R*,9a*S*,9b*R*)-7-Acetyl-11-chloro-4a,4b,5,6,6a,7,8,9,9a,9b-decahydro-7-hydroxy-4a,6a-dimethyl-cyclopenta[5,6]naphtho[1,2-*c*]pyran-2(4*H*)-one *see* 6985

Acetylcholine Acetylhydrolase *see* 2214

Acetylcholine Bromide, 81

Acetylcholine Chloride, 82

Acetyl Choline Esterase *see* 2214

Acetylcholinesterase *see* 2214

Acetyl-*o*-cresol *see* 2573

4-Acetyl-*N*-[(cyclohexylamino)carbonyl]-benzenesulfonamide *see* 61

Acetylcysteine, 83

N-Acetyl-L-cysteine *see* 83

N-Acetyldeferoxamine *see* 2863

Acetyldemethyldihydrothebaine *see* 9426

N-Acetyldesferrioxamine B *see* 2863

cis-1-Acetyl-4-[4-[[2-(2,4-dichlorophenyl)-2-(1*H*-imidazol-1-ylmethyl)-1,3-dioxolan-4-yl]methoxy]phenyl]piperazine *see* 5349

(17*R*)-1-Acetyl-19,20-didehydro-17,18-epoxycuran-17-ol *see* 2962

(1*S*)-1-*C*-[(2*S*,3*S*)-7-[[4-*O*-Acetyl-2,6-dideoxy-3-*O*-(2,6-dideoxy-4-*O*-methyl-α-D-*lyxo*-hexopyranosyl)-β-D-*lyxo*-hexopyranosyl]oxy]-3-[[*O*-4-*O*-acetyl-2,6-dideoxy-3-*C*-methyl-α-L-*arabino*-hexopyranosyl-(1 → 3)-*O*-2,6-dideoxy-β-D-*arabino*-hexopyranosyl-(1 → 3)-2,6-dideoxy-β-D-*arabino*-hexopyranosyl]oxy]-1,2,3,4-tetrahydro-5,10-dihydroxy-6-methyl-4-oxo-2-anthracenyl]-5-deoxy-1-*O*-methyl-D-*threo*-2-pentulose *see* 2242

3^D -*O*-(4-*O*-Acetyl-2,6-dideoxy-3-*C*-methyl-α-L-*arabino*-hexopyranosyl)-7-methyllolivomycin D *see* 2242

(3*R*,4*S*,5*R*,6*S*,7*S*,9*S*,11*R*,12*S*,13*S*,14*R*)-4-[(4-*O*-Acetyl-2,6-dideoxy-3-*C*-methyl-3-*O*-methyl-α-L-*xylo*-hexopyranosyl)oxy]-12-(acetyloxy)-6-[(4,6-dideoxy-3-*O*-methyl-β-D-*xylo*-hexopyranosyl)oxy]-9-hydroxy-14-[(1*S*,2*S*)-2-hydroxy-1-methylpropyl]-3,5,7,9,11,13-hexamethyloxacyclotetradecane-2,10-dione *see* 5407

Acetyldigitoxins, 84

α-Acetyldigoxin *see* 3186

β-Acetyldigoxin *see* 3186

Acetyldihydrocodeinone *see* 9426

3-Acetyldihydro-2(3*H*)-furanone *see* 78

*O*³-Acetyl-7,8-dihydro-7α-[1(*R*)-hydroxy-1-methylbutyl]-*O*⁶-methyl-6,14-*endo*-ethenomorphine *see* 3931

(5*S*)-3-Acetyl-1,5-dihydro-4-hydroxy-5-[(1*S*)-1-methylpropyl]-2*H*-pyrrol-2-one *see* 9292

5-[Acetyl(2,3-dihydroxypropyl)amino]-*N*¹,*N*³-bis(2,3-dihydroxypropyl)-2,4,6-triiodo-1,3-benzenedicarboxamide *see* 5095

5-[Acetyl(2,3-dihydroxypropyl)amino]-*N*¹-(2,3-dihydroxypropyl)-*N*³-(2-hydroxyethyl)-2,4,6-triiodo-1,3-benzenedicarboxamide *see* 5111

O-Acetyl-6-dimethylamino-4,4-diphenyl-3-heptanol *see* 6017

2-Acetyl-10-(3-dimethylaminopropyl)-phenothiazine *see* 33

3-Acetyl-10-(3-dimethylaminopropyl)-phenothiazine *see* 33

N'-[3-Acetyl-4-[3-[(1,1-dimethylethyl)-amino]-2-hydroxypropoxy]phenyl]-*N,N*-diethylurea *see* 1962

Acetyldiphenylamine *see* 3345

Acetylene, 85

Acetylene Carbamide *see* 88

Acetylenecarboxylic Acid *see* 7935

Acetylene Dibromide, 86

Acetylenedicarboxamide *see* 1964

Acetylenedicarboxylic Acid Diamide *see* 1964

Acetylene Dichloride, 87

Acetylenediureine *see* 88

Acetylene Tetrabromide *see* 9330

Acetylene Tetrachloride *see* 9334

Acetyleneurea, 88

Acetylenogen *see* 1658

Acetyl Ethylene *see* 6210

N-Acetyl-*S*-*trans,trans*-farnesyl-L-cysteine *see* 89

N-Acetyl-L-farnesylcysteine, 89

Acetylformaldehyde *see* 6153

Acetylformic Acid *see* 8135

Acetylgalactosamine 4-Sulfatase *see* 809

Acetylgalactosamine 4-Sulfatase (Human CSL4S-342 Cell) *see* 809

α-*N*-Acetyl-L-glutamine *see* 26

*N*²-Acetyl-L-glutamine *see* 26

Acetyl Glyceryl Ether Phosphorylcholine *see* 7638

N-Acetylglycine *see* 75

N-Acetylglycine Compd with 4,4'-(Diazoamino)dibenzamidine *see* 3291

N-Acetylglycine Compd with 4,4'-(1-Triazene-1,3-diyl)bis(benzenecarboximidamide) (2:1) *see* 3291

Acetylglycocoll *see* 75

3-*O*-Acetyl-18β-glycyrrhetic Acid *see* 74

Acetylglycyrrhetinic Acid *see* 74

(1*S*,3*S*)-3-Acetyl-1,2,3,4,6,11-hexahydro-3,5,10,12-tetrahydroxy-6,11-dioxo-1-naphthacenyl 3-Amino-2,3,6-trideoxy-α-L-*lyxo*-hexopyranoside *see* 1870

(1*S*,3*S*)-3-Acetyl-1,2,3,4,6,11-hexahydro-3,5,12-trihydroxy-6,11-dioxo-1-naphthacenyl-3-amino-2,3,6-trideoxy-α-L-*lyxo*-hexopyranoside *see* 4927

N-Acetylhomocysteinethiolactone *see* 2319

Acetyl Hydroperoxide *see* 7263

*N*⁴-[5-[[4-[[5-(Acetylhydroxyamino)pentyl]amino]-1,4-dioxobutyl]hydroxyamino]pentyl]-*N*¹-(5-aminopentyl)-*N*¹-hydroxybutanediamide *see* 2863

N-Acetyl-4-hydroxy-*m*-arsanilic Acid *see* 51

α-Acetyl-γ-hydroxybutyric Acid γ-Lactone *see* 78

17β-Acetyl-17-hydroxyestr-4-ene-3-one Hexanoate *see* 4448

2-Acetyl-10-[3-[γ-(2-hydroxyethyl)piperidino]propyl]phenothiazine *see* 7574

2-Acetyl-10-[3-[4-(β-hydroxyethyl)piperidino]propyl]phenothiazine *see* 7574

5-Acetyl-6-hydroxy-2-isopropenylbenzofuran *see* 3942

Actemra [Chugai] see 9651
ACTH, 125
Acthar [RPR] see 125
ACTH-β-lipotropin Common Precursor see 7905
Acthormon [Shionogi] see 4463
Acticolin [Upsamedica] see 2318
Actidil [Wellcome] see 9922
Actidilon [Wellcome] see 9922
Actidione see 2721
Actigall [Novartis] see 10074
Actigam [Daltan] see 4497
Actilin see 6539
Actilyse [Boehringer, Ing.] see 9624
Actimmune [Genentech] see 5036
Actin, 126
α-Actin see 126
β-Actin see 126
γ-Actin see 126
F-Actin see 126
G-Actin see 126
Actinex [Schwarz] see 6782
Actinium, 127
Actinium X see 8210
Actinobolin, 128
Actinodaphnine, 129
Actinolite see 815
Actinomycetin, 130
Actinomycin I₁ see 2797
Actinomycin IV see 2797
Actinomycin X₁ see 2797
Actinomycin A_IV see 2797
Actinomycin C see 1611
Actinomycin C₁ see 2797
Actinomycin C₂ see 1611
Actinomycin C₃ see 1611
Actinomycin D see 2797
Actinomycin-[thr-val-pro-sar-meval] see 2797
Actinon see 8212
Actinorhodine, 131
Actinospectacin see 8866
Actinum K see 4290
Action [Kumiai] see 4239
Actiplas [Dompé] see 9624
Actiq [Cephalon] see 4028
Actira [Bayer] see 6377
Actithiazic Acid see 6407
Activase [Genentech] see 9624
Activated Barium Hydroxide see 973
Activated Carbon see 1808
Activated Charcoal see 1808
Activated Dimethicone see 3237
Active Amyl Alcohol see 6104
Active Methionine see 144
Activin, 132
Activin A see 132
Activin AB see 132
Activol [Zeneca] see 4454
Actol [Fournier] see 6615
Actomyosin see 126
Acton [Ferring] see 125
Actonel [Procter & Gamble] see 8360
Actos [Takeda] see 7565
Actosin [Daiichi Seiyaku] see 1473
Actosolv [HMR] see 10072
Actrapid [Novo] see 5023
Actriol [Organon] see 3671
Acuatim [Otsuka] see 6430
Acuitel [Pfizer] see 8167
Acular [Allergan] see 5353
Acupan [Meda] see 6525
AcuTect [Diatide] see 9235
Acuvail [Allergan] see 5353
ACV, 133

Acycloguanosine see 134
Acyclovir, 134
Acylan [Croda] see 5409
Acylanid [Sandoz] see 84
Acylcholine Acylhydrolase see 2214
N-(Acylcolaminoformylmethyl)pyridinium Chloride see 5421
Acylpyrin [Slovakofarma] see 841
ACZ-885 see 1739
AD₆ see 2393
AD-32 see 10101
AD-810 see 10392
AD-4833 see 7565
AD-5423 see 1320
ADA, 135
Ada see 3201
Adage [Syngenta] see 9446
ADAH see 1248
Adalat(e) [Bayer] see 6613
Adalimumab, 136
"Adam" see 5836
1-Adamantanamine see 368
Adamantane, 137
N-1-Adamantyl-N-[2-(dimethylamino)ethoxy]acetamide see 9950
6-[3-(1-Adamantyl)-4-methoxyphenyl]-2-naphthoic Acid see 138
Adams' Catalyst see 7642
Adamsite see 7329
Adancor [Merck-Clévenot] see 6606
Adapalene, 138
Adapin [Fisons] see 3483
Adaptinol [Bayer] see 10261
Adartrel [GSK] see 8388
ADCA see 1248
Adcetris [Seattle Genetics] see 1375
Adcirca [Lilly] see 9157
Ad5CMV-p53 see 2495
Adcortyl [BMS] see 9758
ADD-03055 see 3984
ADD-234037 see 5381
Adecut [Takeda] see 2881
Adefovir, 139
Adefovir Dipivoxil see 139
Adekin [Desitin] see 368
Adelir [Merck KGaA] see 10025
Ademetione see 144
Ademide [Asahi] see 2435
Ademin(e) see 9760
Ademol [BMS] see 4170
Adenine, 140
Adenine Arabinoside see 10173
Adenine Riboside see 141
Adenocard [Fujisawa] see 141
Adenock [Tanabe Mitsubishi] see 271
Adenocor [Sanofi Winthrop] see 141
Adenohypophyseal Growth Hormone see 8842
Adenohypophysial Luteotropin see 7894
Adenoscan [Fujisawa] see 141
Adenosine, 141
Adenosine Cyclic 3',5'-(Hydrogen Phosphate) see 2701
Adenosine 3',5'-Cyclic Monophosphate see 2701
Adenosine 3',5'-Cyclic Phosphate see 2701
Adenosine Diphosphatase see 748
Adenosine Diphosphate, 142
Adenosinediphosphoric Acid see 142
Adenosine 3'-Monophosphate see 145
Adenosine 3',5'-Monophosphate see 2701
Adenosine 5'-Monophosphate see 146
Adenosine-3'-monophosphoric Acid see 145

Adenosine-5'-monophosphoric Acid see 146
Adenosine Phosphate see 146
Adenosine 3',5'-Phosphate see 2701
Adenosine-3'-phosphoric Acid see 145
Adenosine-5'-phosphoric Acid see 146
Adenosine 5'-Pyrophosphoric Acid see 142
Adenosine 5'-(Tetrahydrogen Triphosphate) see 143
Adenosine 5'-(Trihydrogen Diphosphate) see 142
Adenosine 5'-(Trihydrogen Diphosphate) 2'-(Dihydrogen Phosphate) P' → 5'-Ester with 3-(Aminocarbonyl)-1-β-D-ribofuranosylpyridinium Inner Salt see 6432
Adenosine 5'-(Trihydrogen Diphosphate) 3'-(Dihydrogen Phosphate) P'-[(R)-3-Hydroxy-4-[[3-[(2-mercaptoethyl)amino]-3-oxopropyl]amino]-2,2-dimethyl-4-oxobutyl]ester see 2450
Adenosine 5'-(Trihydrogen Diphosphate) P' → 5'-Ester with 3-(Aminocarbonyl)-1-β-D-ribofuranosylpyridinium Inner Salt see 6429
Adenosine 5'-(Trihydrogen Diphosphate) 5' → 5'-Ester with Riboflavine see 4122
Adenosine Triphosphate, 143
Adenosine 5'-Triphosphoric Acid see 143
Adenosyl-B₁₂ see 2435
Adenosylcobalamin see 2435
S-Adenosylmethionine, 144
Adényl [Am. Home] see 146
Adenylate Cyclase see 2701
3'-Adenylic Acid, 145
5'-Adenylic Acid, 146
Adenylic Acid b see 145
5'-Adenylphosphoric Acid see 142
Adenylpyrophosphatase see 748
Adepril [Teofarma] see 483
Adeps see 5423
Adepsine Oil see 7301
Adermykon see 2184
Aderoxal see 8092
Adesitrin [Pharmacia] see 6694
Adetol [Kyowa] see 143
Adetphos [Kowa] see 143
Adfeed [Lead Chem.] see 4229
ADH see 10129
Adhatoda, 147
Adhatodai see 147
Adiazine [Recordati] see 9035
Adicillin see 7207
Adifax [Servier] see 4004
Adigal [Beiersdorf] see 5404
Adinazolam, 148
Adinazolam Mesylate see 148
Adipex-P [Gate] see 7374
Adiphenine, 149
Adipic Acid, 150
Adipic Acid Di(3-carboxy-2,4,6-triiodoanilide) see 5065
Adipic Acid Dinitrile see 152
Adipic Ketone see 2738
Adipine [Chiesi] see 6613
Adipiodone see 5065
Adipogenesis Inhibitory Factor see 5044
Adipokinetic Hormone see 5568
Adipomin [Streuli] see 4004
Adiponectin, 151
Adiponitrile, 152
AdipoQ see 151

Adiposetten [Reiss] *see* 6803
3,3′-(Adipoyldiimino)bis[2,4,6-triiodo-
  benzoic Acid] *see* 5065
5,5′-(Adipoyldiimino)bis[2,4,6-triiodo-*N*-
  methylisophthalamic Acid] *see* 5053
Adipsin, 153
N,*N*′-Adipylbis(3-amino-2,4,6-triiodo-
  benzoic Acid) *see* 5065
Adiuretin SD *see* 2926
Adizem [Napp] *see* 3224
ADL-8-2698 *see* 366
Adlumidine, 154
Adlumine, 155
Admire [Bayer CropSci.] *see* 4951
Admitt [Dow AgroSci.] *see* 8124
Admon [Esteve] *see* 6636
ADN *see* 152
A-DNA *see* 2908
Adnexol [Protina] *see* 4924
AdoCbl *see* 2435
Adofen [Ferrer] *see* 4217
AdoMet *see* 144
Adona [Tanabe Seiyaku] *see* 1790
Adonis Vernalis, 156
Adonite *see* 157
Adonitol, 157
Adonitoxin, 158
D-Adonose *see* 8331
ADP *see* 142, 7104
Adphen [Ferndale] *see* 7334
ADR-033 *see* 9909
ADR-529 *see* 8241
Adrafinil, 159
Adral [Lifepharma] *see* 6619
Adrenalin [Monarch] *see* 3674
Adrenaline *see* 3674
Adrenaline Oxidase *see* 6333
Adrenalone, 160
Adrenochrome, 161
Adrenochrome Monosemicarbazone *see*
  1789
Adrenocorticotrop(h)ic Hormone of the
  Pituitary Gland *see* 125
Adrenocorticotrop(h)in *see* 125
Adrenoglomerulotropin, 162
Adrenolutin, 163
Adrenomedullin, 164
Adrenone *see* 160
Adrenor *see* 6784
Adrenosterone, 165
Adrenoxyl [Sanofi-Aventis] *see* 1789
Adrevil [Zyma] *see* 1513
Adriacin [Pfizer] *see* 3487
Adriamicina [Farmochim. Ital.] *see* 6015
Adriamycin [Pfizer] *see* 3487
Adriamycinone *see* 3487
Adriblastina [Pfizer] *see* 3487
Adronat [Neopharmed] *see* 223
Adroyd [Parke-Davis] *see* 7066
Adrucil [Sicor] *see* 4213
ADT *see* 637
Adulsa *see* 147
Adumbran [Boehringer, Ing.] *see* 7025
Adurix [Nycomed] *see* 2382
Advantage [Bayer] *see* 4951
Advantan [Schering AG] *see* 6184
Advate [Baxter] *see* 3962
Advexin [Introgen] *see* 2495
Advil [Wyeth] *see* 4919
Advion [DuPont] *see* 5012
Advocin [Pfizer] *see* 2812
AE-17 *see* 9139
AE-0172747 *see* 9276
AE-B066752 *see* 7912
AE-C638206 *see* 4186

AE-F107892 *see* 5871
AE-F115008 *see* 5089
AE-F122006 *see* 5282
AE-F130060 *see* 5981
AE-F130360 *see* 4261
Aedurid [Robugen] *see* 3563
Aequamen [Promonta] *see* 1180
Aequorin, 166
Aerobec [3M Pharma] *see* 1017
Aerobid [Forest] *see* 4176
Aerobin [Farmasan] *see* 9436
Aerolate [Fleming] *see* 9436
Aerolin [3M Pharma] *see* 210
Aerolone [Lilly] *see* 5263
Aeromatt *see* 1659
Aeroseb-Dex [Allergan] *see* 2945
Aerosol AY *see* 2989
Aerosol IB *see* 3213
Aerosol MA *see* 1259
Aerosol OT [Cyanamid] *see* 3446
Aerosporin [Burroughs Wellcome] *see*
  7693
Aerotrol [Abbott] *see* 5263
Aerrane [Baxter] *see* 5221
Aescin *see* 3751
Aescorin [Steigerwald] *see* 4787
Aescusan [Jenapharm] *see* 4787
AET, 167
Aether Oenanthicus *see* 3889
Aethone [Sopar] *see* 6980
Aethoxysklerol [Kreussler] *see* 7675
Aethylis Chloridum *see* 3837
Aetina *see* 3791
Aetiocholanic Acid *see* 3916
Aetoxisclerol [Kreussler] *see* 7675
AF-102B *see* 2033
AF-438 *see* 7043
AF-634 *see* 8012
AF-983 *see* 1037
AF-1161 *see* 9740
AF-1890 *see* 5625
AF-2139 *see* 2820
Afalon [Hoechst] *see* 5564
Afamelanotide, 168
Afatinib, 169
AFC *see* 89
Afema [Novartis] *see* 3968
Affinin, 170
Affinitac [Lilly] *see* 744
Affinitak [Lilly] *see* 744
Affinitins *see* 5484
Affirm [Syngenta] *see* 3612
Affront [Wellmark] *see* 2195
aFGF *see* 4101
aFGF-1 *see* 4101
aFGF-2 *see* 4101
Afinitor [Novartis] *see* 3950
AFI-phyllin *see* 3526
AFI-Tiazin *see* 7364
Aflatoxicol *see* 171
Aflatoxin B₁ *see* 171
Aflatoxin B₂ *see* 171
Aflatoxin G₁ *see* 172
Aflatoxin G₂ *see* 172
Aflatoxin M₁ *see* 173
Aflatoxin M₂ *see* 173
Aflatoxins B, 171
Aflatoxins G, 172
Aflatoxins M, 173
Aflibercept, 174
Aflix *see* 4273
Afloben [Esseti] *see* 1125
Afloqualone, 175
Afloxan [Rottapharm] *see* 7890
Afluon [Viatris] *see* 899

Afongan [Galderma] *see* 6941
Afonilum [Knoll] *see* 461
Afonilum *see* 9436
Afos [Salus] *see* 4281
AFP-168 *see* 9160
Afrazine [Schering-Plough] *see* 7065
African Oil Palm *see* 6910
African Saffron *see* 1868
Afrin [Schering-Plough] *see* 7065
Aftate [Schering-Plough] *see* 9678
Afugan *see* 8073
Afwillite *see* 1704
Afzelechin *see* 1902
AG-3-5 *see* 4925
AG-337 *see* 6759
AG-1343 *see* 6528
AG-1346 *see* 6528
AG-1749 *see* 5413
AG-3340 *see* 7869
AG-013736 *see* 883
Agar, 176
Agar-agar *see* 176
Agaric, 177
Agaric Acid *see* 178
Agaricic Acid, 178
Agaricin *see* 178
Agaricinic Acid *see* 178
Agarin *see* 6397
Agaritine, 179
Agaropectin *see* 176
Agarose *see* 176
Agate *see* 8632
Agatolimod, 180
Agatolimod Sodium *see* 180
Agbayun *see* 6289
Agedoite *see* 827
AG-EE-623-ZW *see* 8256
Agene *see* 6689
Agenerase [GSK] *see* 585
Agent Orange *see* 9221
AGEPC *see* 7638
Ageratochromene *see* 7839
Agerite *see* 6334
Ageroplas [Serono] *see* 3410
Agglutinin *see* 10
Agglutinins *see* 5484
Aggrastat [Merck & Co.] *see* 9620
Aggrenox [Boehringer, Ing.] *see* 3388
AGI-1067 *see* 9002
AGIF *see* 5044
Agifutol S [Kyorin] *see* 4511
Agilect [Teva] *see* 8234
Agilene [Am. Agile] *see* 7687
Agit [Schwarz] *see* 3191
Agkistrodon rhodostoma Venom Protein-
  ase *see* 624
Agkistrodon Serine Proteinase *see* 624
Aglumin *see* 3777
Agmatine, 181
AGN-1135 *see* 8234
AGN-190168 *see* 9218
AGN-190342 *see* 1381
AGN-190342-LF *see* 1381
AGN-191622 *see* 1362
AGN-192024 *see* 1225
Agnolyt [Madaus] *see* 182
Agnus Castus, 182
Agofollin [Biotika] *see* 3758
Agomelatine, 183
Agon [TheraPharm] *see* 3987
Agoniadin *see* 7657
Agontan [Knoll] *see* 3210
Agopton [Takeda] *see* 5413
Agr-1240 *see* 6281
Agradil [Sanofi-Aventis] *see* 10142

Agram [Fabre] *see* 574
Agréal [Sanofi-Aventis] *see* 10142
Agricultural Limestone *see* 5544
Agri-Mek [Syngenta] *see* 2
AgriStrep [Merck & Co.] *see* 8956
Agritox [Nufarm] *see* 5833
Agroclavine, 184
Agrocybin, 185
Agropyrum *see* 9938
Agrosan [Zeneca] *see* 7411
Agroxone [Zeneca] *see* 5833
Agrylin [Shire] *see* 616
Agrypnal [Eggochemia] *see* 7352
Agstone *see* 5544
Ague Tree *see* 8518
Agyrax [Vedim] *see* 5848
AH-42 *see* 6031
AH-289 *see* 2082
AH-2250 *see* 1499
AH-3365 *see* 210
AH-5158A *see* 5377
AH-8165 *see* 3980
AH-19065 *see* 8228
AHA *see* 62
AHCTL *see* 2319
AHF *see* 3962
AHG *see* 3962
AHMHA *see* 8927
AHP-200 *see* 7002
AHPrBP *see* 7104
AHR-85 *see* 6052
AHR-233 *see* 5918
AHR-504 *see* 4537
AHR-619 *see* 3481
AHR-857 *see* 9046
AHR-3018 *see* 713
AHR-3053 *see* 1802
AHR-3070-C *see* 6219
AHR-5850D *see* 389
AHR-9434 *see* 6553
AHR-10282 *see* 1397
AHR-10282B *see* 1397
[Aib^{8,35}]hGLP-1(7-36)NH$_2$ *see* 9209
AIBN *see* 911
Aicamin [Crinos] *see* 6956
AICA Orotate *see* 6956
AICA-riboside *see* 17
Aicorat [Mack, Illert.] *see* 6956
Aida [Paraphar] *see* 4845
Aiglonyl [Fumouze] *see* 9119
Aimax [Ayerst] *see* 6018
Airol [Pierre Fabre] *see* 8288
Airomir [IVAX] *see* 210
Airtal [Prodes] *see* 23
Airum [Promeco] *see* 4012
Aisemide *see* 4338
Aizumycin *see* 1207
Ajacine *see* 5679
Ajaconine, 186
Ajan [3M] *see* 6525
Aji-no-moto [Suzuki] *see* 6342
(17R,21α)-Ajmalan-17,21-diol *see* 187
(17R,21α)-Ajmalan-17,21-diol 17-(2-Chloroacetate) *see* 5634
Ajmalicine *see* 8237
Ajmaline, 187
Ajoene, 188
Ajowan Oil, 189
Ajugarin-I *see* 190
Ajugarin-II *see* 190
Ajugarin-III *see* 190
Ajugarin-IV *see* 190
Ajugarin-V *see* 190
Ajugarins, 190
Ajulemic Acid, 191

AK-33X *see* 6309
Akarpine [Akorn] *see* 7536
Akatinol [Merz] *see* 5897
Ak-Con [Akorn] *see* 6453
AKD-2023 *see* 34
Ak-Dex [Akorn] *see* 2945
Ak-Dilate [Akorn] *see* 7398
Akermanite *see* 1704
Ak-Fluor [Akorn] *see* 4192
Akineton [Knoll] *see* 1238
Akinophyl [Knoll] *see* 1238
Akistin [Nycomed] *see* 8466
Aklavinone *see* 104
Ak-Nefrin [Akorn] *see* 7398
Aknemycin [Hermal] *see* 3739
Akneroxid L [Lubapharm] *see* 1119
Aknin [Lichtenstein] *see* 3739
Ak-Pentolate [Akorn] *see* 2741
Akrochem ETU-22 [Akron Chem.] *see* 3858
Ak-Sulf [Akorn] *see* 9031
Ak-Taine [Akorn] *see* 7921
Aktamin *see* 6784
Ak-Tate [Akorn] *see* 7841
Aktedron [Chinoin] *see* 579
Aktivin [Kobayashi] *see* 9449
Ak-Tracin [Akorn] *see* 927
Akuammicine, 192
Akwa Tears [Akorn] *see* 7706
AL-108 *see* 2836
AL-208 *see* 2836
AL-1241 *see* 7880
AL-4682 *see* 1382
AL-4943A *see* 6934
AL-6221 *see* 9738
AL-6515 *see* 6553
Ala *see* 197
ALA *see* 443
Alabandite *see* 5805
Alabaster *see* 1708
Alacepril, 193
Alachlor, 194
Ala-Cort [Del-Ray] *see* 4824
Ala²-cyclosporine *see* 2750
Alafosfalin, 195
Alagebrium Chloride, 196
[β-Ala¹,lys¹⁷]ACTH¹⁻¹⁷-4-amino-N-butylamide *see* 310
[β-Ala¹,lys¹⁷]corticotropin-(1-17)-heptadecapeptide-4-amino-N-butylamide *see* 310
Alamine *see* 3201
Alamon [Grelan] *see* 4889
Alamycin [Norbrook] *see* 7075
Alanap *see* 6500
D-**Ala-D-(β-naphthyl)-Ala-Ala-Trp-D-Phe-Lys-NH₂** *see* 7824
Alanetorin [Takata; Nippon Kayaku] *see* 216
Alanex [Makhteshim-Agan] *see* 194
Alanine, 197
L-**Alanine** *see* 197
L-**α-Alanine** *see* 197
β-**Alanine**, 198
1-β-Alanine-17-[N-(4-aminobutyl)-L-lysinamide]-α¹⁻¹⁷-corticotropin *see* 310
1-β-Alanine-17-[L-2,6-diamino-N-(4-aminobutyl)hexanamide]-α¹⁻¹⁷-corticotropin *see* 310
L-**Alanine (1R)-2-[[4-[(4-Fluoro-2-methyl-1H-indol-5-yl)oxy]-5-methylpyrrolo[2,1-f][1,2,4]triazin-6-yl]oxy]-1-methylethyl Ester** *see* 1383
Alanine Nitrogen Mustard *see* 5896
L-**Alanosine**, 199

Alant Camphor *see* 200
Alantin *see* 5048
Alanto [Bayer CropSci.] *see* 9443
Alantolactone, 200
Alant Starch *see* 5048
1R-1-(L-Alanylamino)ethylphosphonic Acid *see* 195
N-L-**Alanyl-blood Coagulation Factor LACI (Human Clone AP9 Protein Moiety Reduced)** *see* 9623
L-**Alanyl-(6-diazo-5-oxo)-L-norleucine-(6-diazo-5-oxo)-L-norleucine** *see* 201
L-**Alanylglycyl-L-tyrosyl-L-lysyl-L-prolyl-L-α-aspartyl-L-α-glutamylglycyl-L-L-lysyl-L-arginylglycyl-L-α-aspartyl-L-alanyl-L-cysteinyl-L-α-glutamylglycyl-L-α-aspartyl-L-serylglycylglycyl-L-prolyl-L-phenylalanyl-L-valinamide** *see* 8431
β-**Alanyl-L-histidinato Zinc** *see* 7669
[N-β-**Alanyl-L-histidinato(2−)-N,N^N,-O^α]zinc** *see* 7669
β-**Alanyl-L-histidine** *see* 1851
L-**Alanyl-L-leucyl-L-cysteinyl-L-asparaginyl-L-cysteinyl-L-asparaginyl-L-arginyl-L-isoleucyl-L-isoleucyl-L-isoleucyl-L-prolyl-L-histidyl-L-methionyl-L-cysteinyl-L-tryptophyl-L-lysyl-L-lysyl-L-cysteinylglycyl-L-lysyl-L-lysinamide** *see* 9318
β-**Alanyl-3-methyl-L-histidine** *see* 671
D-**Alanyl-3-(2-naphthalenyl)-D-alanyl-L-alanyl-L-tryptophyl-D-phenylalanyl-L-lysinamide** *see* 7824
[1R-(1α,2β,6α)]-N-L-**Alanyl-3-(5-oxo-7-oxabicyclo[4.1.0]hept-2-yl)-L-alanine** *see* 926
L-**Alanyl-3-[(1R,2S,6R)-5-oxo-7-oxabicyclo[4.1.0]hept-2-yl]-L-alanine** *see* 926
L-**Alanyl-L-seryl-L-threonyl-L-threonyl-L-threonyl-L-asparaginyl-L-tyrosyl-L-threonine** *see* 7261
L-**Alanyl-L-valyl-L-glutaminyl-L-seryl-L-lysyl-L-prolyl-L-prolyl-L-seryl-L-lysyl-L-argininyl-L-α-aspartyl-L-prolyl-L-prolyl-L-lysyl-L-methionyl-L-glutaminyl-L-threonyl-L-aspartic Acid** *see* 9148
[D-**Ala¹] Peptide T Amide** *see* 7261
Alaphosphin *see* 195
Alapril [Mediolanum] *see* 5572
Alar [Uniroyal] *see* 2809
Alathon [DuPont] *see* 7687
Alavert [Wyeth] *see* 5635
Alazopeptin, 201
Albac [Alpharma] *see* 929
Albacar *see* 1659
Albalon [Allergan] *see* 6453
Albamycin [Pharmacia & Upjohn] *see* 6811
Albasil C *see* 9061
Albaspidin, 202
Albaspidin-BB *see* 202
Albego [Boehringer, Ing.] *see* 1727
Albendazole, 203
Albendazole Oxide *see* 203
Albenza [GSK] *see* 203
Albert-285 *see* 7927
Albiglutide, 204
Albiotic [Pharmacia] *see* 5555
Albipen [Intervet] *see* 583
Albistat [Cilag] *see* 6257
Albizziin, 205
Albocresil [Altana] *see* 7673
Albofungin, 206

Alboline *see* 7301
Albomycetin *see* 7525
Albomycin, 207
Albon [Pfizer] *see* 9037
Albone *see* 4835
Alborixin, 208
Albothyl [Altana] *see* 7673
Albucid [Ankerpharm] *see* 9031
Albugon *see* 204
Albumen, 209
Albuminar [ZLB Behring] *see* 8607
Albumose Silver *see* 8666
Albumotope-131I [Bracco] *see* 8607
Albunex [Molecular Biosystems] *see* 8607
Albuterol, 210
Albutest [Miles] *see* 1454
Albutoin, 211
Alcacyl [Novartis] *see* 1650
Alcaine [Alcon] *see* 7921
ALCAR *see* 79
Alchloquin *see* 5075
Alcian Blue, 212
Alcian Blue 7G *see* 212
Alcian Blue 8G *see* 212
Alcian Blue 8GS *see* 212
Alcian Blue 8GX *see* 212
Alclometasone, 213
Alcodin [Alcon] *see* 1221
Alcohol *see* 3814
95% Alcohol *see* 3814
Alcohol Ethoxylates *see* 7697
Alcohol, USP *see* 3814
Alcomicin [Alcon] *see* 4427
Alconefrin [Polymedica] *see* 7398
Alcopan-250 *see* 2949
Alcuronium, 214
Aldactazine *see* 314
Aldactone [Pharmacia & Upjohn] *see* 8887
Aldara [3M Pharma] *see* 4964
Aldazine [Alphapharm] *see* 9512
Aldebaranium *see* 10303
Aldecin [Schering-Plough] *see* 1017
"Aldehyde" *see* 40
Aldehyde Ammonia *see* 41
Aldehyde C-7 *see* 4693
Aldehyde-collidine *see* 3896
Aldehydine *see* 3896
Alder Buckthorn *see* 4291
Aldesleukin *see* 5038
Aldesulfone Sodium *see* 9098
Aldicarb, 215
Aldifen *see* 3308
Aldioxa, 216
Aldipin [Helvepharm] *see* 6613
Aldocorten [Novartis] *see* 218
Aldol, 217
Aldomet [Merck & Co.] *see* 6127
Aldomycin *see* 6687
Aldosterone, 218
Aldosterone-stimulating Hormone *see* 162
Aldoxime *see* 43
Aldrin, 219
Aldrox [Am. Home] *see* 338
Aldurazyme [BioMarin] *see* 4935
ALEC *see* 8049
Alecor [Andromaco] *see* 3693
Alefacept, 220
Aleglitazar, 221
Alemtuzumab, 222
Alendronate Sodium *see* 223
Alendronic Acid, 223
Alendros [Abiogen] *see* 223
Aleppo-galls *see* 6817

Alergicide *see* 2082
Alerlisin [Prodes] *see* 2021
Alesion [Boehringer, Ing.] *see* 3673
Aletris, 224
Aleudrin [Lewis] *see* 5263
Aleuritic Acid, 225
α-Aleuritic Acid *see* 225
β-Aleuritic Acid *see* 225
Alevaire [Sanofi Winthrop] *see* 10014
Aleve [Procter & Gamble] *see* 6499
Alexan [Ebewe] *see* 2781
Alexidine, 226
Alfacalcidol *see* 4858
Alfacet [Galenika] *see* 1914
Alfadil [Pfizer] *see* 3482
Alfadione *see* 228
Alfadolone Acetate *see* 228
Alfadryl [Leciva] *see* 6373
Alfa-interferon *see* 5034
Alfamedin [Kade] *see* 3482
Alfamox [Teofarma] *see* 574
Alfaprostol, 227
Alfarol [Chugai] *see* 4858
Alfason [Astellas] *see* 4824
Alfaspoven [Alfa] *see* 1974
Alfatil [Dexo] *see* 1914
Alfavet [Vetem] *see* 227
Alfaxalone, 228
Alfaxan [Vetoquinol] *see* 228
Alfenamin *see* 4163
Alfenta *see* 229
Alfentanil, 229
Alferon [Hemispherx] *see* 5034
Alficetin [BMS] *see* 2463
Alfimeprase, 230
Alfonic [Vista] *see* 7697
Alfospas [Rorer] *see* 9622
Alfuzosin, 231
Alganex [Roche] *see* 9291
Algestone, 232
Algestone Acetonide *see* 232
Algestone Acetophenide, 233
Algil [Maggioni] *see* 5916
Algin, 234
Alginic Acid, 235
Alginic Acid Sodium Salt *see* 234
Alginodia [Beta] *see* 3390
Alginor [Boehringer, Ing.] *see* 2283
Alglucosidase Alfa *see* 100
Algobaz [Labaz] *see* 6116
Algocalmin *see* 3390
Algocor [Ravizza] *see* 4386
Algylen *see* 9804
Alibendol, 236
Alicaforsen, 237
Alicaforsen Sodium *see* 237
Alidase [Searle] *see* 4797
Aliette [Bayer CropSci.] *see* 4279
Alimemazine *see* 9873
Alimet [Novus] *see* 6048
Alimix [Janssen] *see* 2315
Alimta [Lilly] *see* 7186
Alinam [Lucien] *see* 2106
Alinamin [Takeda] *see* 7993
Alinamin F [Takeda] *see* 4340
Alindapril *see* 2881
Alinia [Romark] *see* 6653
Alinidine, 238
Alisactide *see* 310
Aliskiren, 239
Alisobumal *see* 1514
Alitame, 240
Alitretinoin, 241
Alival [Hoechst] *see* 6760
Alizapride, 242

Alizarin, 243
Alizarin Blue R *see* 245
Alizarin Bordeaux B *see* 8166
Alizarin Cyanine Green F, 244
Alizarine Blue, 245
Alizarine Carmine *see* 8711
Alizarine Orange, 246
Alizarine S *see* 8711
Alizarine Yellow C *see* 4372
Alizarine Yellow GG *see* 5989
Alizarine Yellow R, 247
Alizarine Yellow RW *see* 247
β-2-Alizarin Primeveroside *see* 8413
Alizarin Violet *see* 4374
Alkacitron [PDH] *see* 2325
Alkalovert [Klein] *see* 7499
Alkamuls SML [Alkaril] *see* 8850
Alkamuls SMO [Alkaril] *see* 8850
Alkamuls SMS [Alkaril] *see* 8850
Alkanet, 248
Alkanet Extract *see* 249
Alkanna *see* 248
Alkanna Red *see* 249
Alkannin, 249
Alkannin Paper *see* 248
Alkathene [ICI] *see* 7687
Alkavervir *see* 10151
Alkeran [GSK] *see* 5896
Alkiron *see* 6201
Alkron *see* 7137
1-O-Alkyl-2-acetyl-sn-glycero-3-phos-phorylcholine *see* 7638
Alkylaluminum Halide *see* 322
Alkylaluminum Sesquihalide *see* 322
Alkylbenzyldimethyl Quaternary Ammonium Compounds Chlorides *see* 1061
Alkyldimethyl(phenylmethyl)ammonium Chloride *see* 1061
Al(lact)₃ *see* 343
Allantoin, 250
(Allantoinato)dihydroxyaluminum *see* 216
Allantoxanic Acid *see* 7046
Allegra [Aventis] *see* 4097
Allegron [Lilly] *see* 6804
Allenolic Acid, 251
Alleoside A *see* 4667
Allercur [Schering AG] *see* 2344
Allergefon [Lafon] *see* 1800
Allergisan [Pharmacia] *see* 2186
Allergocrom [Ursapharm] *see* 2581
Allergodil [Viatris] *see* 899
Allerplus [AstraZeneca] *see* 9306
Allethrin I *see* 252
Allethrin II *see* 252
Allethrins, 252
Allethrolone Ester of Chrysanthemum-dicarboxylic Acid Monomethyl Ester *see* 252
Allethrolone Ester of Chrysanthemum-monocarboxylic Acid *see* 252
alli [GSK] *see* 6964
Allicin, 253
Allie [DuPont] *see* 6237
Alliin, 254
Allitridium *see* 2976
Allium *see* 4403
Allobarbital, 255
Allobarbitone *see* 255
Alloca [Pharmacia & Upjohn] *see* 6970
Allocholesterol, 256
Allocor [Kwizda] *see* 5404
Allocryptopine, 257
Allodihydro F *see* 4803

Allodihydrohydrocortisone *see* 4803
Alloferin [Roche] *see* 214
Allohydroxycitric Acid *see* 4861
Allohydroxyproline *see* 4878
Alloid [Kyosei] *see* 234
L-(+)-Alloisoleucine *see* 5225
2-D-Alloisoleucine Sporidesmolide I *see* 8893
Allomaleic Acid *see* 4316
Allomelanins *see* 5881
Allophycocyanins *see* 7491
Allopregnane, 258
Allopregnane-3α,20α-diol, 259
Allopregnane-3α,20β-diol, 260
Allopregnane-3β,20α-diol, 261
Allopregnane-3β,20β-diol, 262
Allopregnane-3β,17α-diol-20-one, 263
3,20-Allopregnanedione, 264
Allopregnane-3α,11β,17α,21-tetrol-20-one, 265
Allopregnane-11β,17α,21-triol-3,20-dione *see* 4803
Allopregnane-3β,11β,21-triol-20-one, 266
Allopregnan-3α-ol-20-one, 267
Allopregnan-3β-ol-20-one, 268
Allopregnan-20α-ol-3-one, 269
Allopregnan-20β-ol-3-one, 270
Alloptrex [Fabre] *see* 2581
Allopurinol, 271
β-D-Allopyranose *see* 272
Allorphine *see* 6446
Allose [Kyosei] *see* 234
D-Allose, 272
Allotelluric Acid *see* 9262
3α-Allotetrahydrocortisol *see* 265
Allotetrahydroprogesterone *see* 267
Allotropal [Heyl] *see* 5913
Allovectin [Vical] *see* 10136
Allovectin-7 [Vical] *see* 10136
Alloxan, 273
Alloxan 5-Oxime *see* 10196
Alloxantin, 274
Alloyohimbine *see* 10302
Allspice *see* 7544
Allspice Berry Oil *see* 7544
Alltox *see* 9716
D-Allulose *see* 8033
Allura Red AC, 275
Allurit [Teofarma] *see* 271
Alluval [Berlin-Chemie] *see* 1404
Allvoran [TAD] *see* 3091
Ally [DuPont] *see* 6237
Allyl Acetate, 276
Allyl Alcohol, 277
Allylamine, 278
Allylamine-epichlorohydrin Copolymer *see* 8613
Allylamine Polymer with 1-Chloro-2,3-epoxypropane *see* 8613
17-Allylamino-17-demethoxygeldanamycin Hydroquinone *see* 8282
6-Allyl-2-amino-5,6,7,8-tetrahydro-4*H*-thiazolo[4,5-*d*]azepine *see* 9171
p-Allylanisole *see* 3760
Allylbarbital *see* 1514
Allyl Bromide, 279
5-Allyl-5-(2-bromoallyl)barbituric Acid *see* 1366
5-Allyl-5-*sec*-butylbarbituric Acid *see* 9167
3-Allyl-5-*sec*-butyl-2-thiohydantoin *see* 211
Allylcarbamide *see* 291
Allylcatechol Methylene Ether *see* 8453

Allyl Chloride, 280
Allyl Cinerins *see* 252
Allyl Cyanide, 281
5-Allyl-5-(2-cyclohexen-1-yl)-2-thiobarbituric Acid *see* 9444
S-Allyl-L-cysteine Sulfoxide *see* 254
2-(*N*-Allyl-2,6-dichloroanilino)-2-imidazoline *see* 238
2-[*N*-Allyl-*N*-(2,6-dichlorophenyl)amino]-2-imidazoline *see* 238
17-Allyl-1,14-dihydroxy-12-[2-(4-hydroxy-3-methoxycyclohexyl)-1-methylvinyl]-23,25-dimethoxy-13,19,21,27-tetramethyl-11,28-dioxa-4-azatricyclo[22.3.1.0^{4,9}]octacos-18-ene-2,3,10,16-tetraone *see* 9155
1-Allyl-1-(3α,17β-dihydroxy-2β-morpholino-5α-androstan-16β-yl)pyrrolidinium 17-Acetate *see* 8375
1-Allyl-2-(3α,17β-dihydroxy-2β-piperidino-5α-androstan-16β-yl)piperidinium Bromide 3-Acetate 17-Propionate *see* 8231
1-Allyl-2,3-dimethoxy-4,5-(methylenedioxy)benzene *see* 720
1-Allyl-2,5-dimethoxy-3,4-methylenedioxybenzene *see* 721
4-Allyl-1,2-dimethoxybenzene *see* 6144
Allyldioxybenzene Methylene Ether *see* 8453
17-Allyl-4,5α-epoxy-3,14-dihydroxymorphinan-6-one *see* 6447
1-[(6-Allylergoline-8β-yl)carbonyl]-1-[3-(dimethylamino)propyl]-3-ethylurea *see* 1606
Allylestrenol, 282
17-Allylestr-4-en-17β-ol *see* 282
Allyl Ether, 283
Allyl Ethyl Ether, 284
Allyl Glucosinolate *see* 8683
Allylguaiacol *see* 3940
Allyl 1,5-Hexadienyltrisulfide *see* 188
17α-Allyl-17-hydroxyestra-4,9,11-trien-3-one *see* 315
5-Allyl-2-hydroxy-*N*-(2-hydroxyethyl)-*m*-anisamide *see* 236
l-*N*-Allyl-3-hydroxymorphinan *see* 5510
17α-Allyl-17-hydroxy-19-nor-4-androstene *see* 282
(−)-*N*-Allyl-14-hydroxynordihydromorphinone *see* 6447
5-Allyl-5-(β-hydroxypropyl)barbituric Acid *see* 8013
5-Allyl-5-(2-hydroxypropyl)barbituric Acid *see* 8013
5-Allyl-5-(β-hydroxypropyl)malonylurea *see* 8013
Allyl Iodide, 285
5-Allyl-5-isobutylbarbituric Acid *see* 1514
3-Allyl-5-isobutyl-2-thiohydantoin *see* 211
5-Allyl-5-isopropylbarbituric Acid *see* 745
Allyl Isosulfocyanate *see* 286
Allyl Isothiocyanate, 286
Allylmercaptomethylpenicillin *see* 7208
Allylmercaptomethylpenicillinic Acid *see* 7208
5-Allyl-1-methoxy-2,3-(methylenedioxy)-benzene *see* 6420
4-Allyl-2-methoxyphenol *see* 3940
5-Allyl-5-(1-methylbutyl)barbituric Acid Sodium Salt *see* 8557
5-Allyl-5-(1-methylbutyl)malonylurea Sodium Salt *see* 8557

5-Allyl-5-(1-methylbutyl)-2-thiobarbituric Acid *see* 9454
4-Allyl-1,2-methylenedioxybenzene *see* 8453
α-*dl*-5-Allyl-1-methyl-5-(1-methyl-2-pentynyl)barbituric Acid *see* 6053
α-3-Allyl-1-methyl-4-phenyl-4-piperidinol Propionate *see* 287
α-3-Allyl-1-methyl-4-phenyl-4-propionoxypiperidine *see* 287
5-Allyl-5-(1-methylpropyl)barbituric Acid *see* 9167
5-Allyl-5-(2-methylpropyl)barbituric Acid *see* 1514
Allyl Methyl Trisulfide *see* 6087
Allyl Mustard Oil *see* 286
N-Allylnormorphine *see* 6446
N-Allylnoroxymorphone *see* 6447
(±)-1-[β-(Allyloxy)-2,4-dichlorophenethyl]imidazole *see* 3637
1-[*o*-(Allyloxy)phenoxy]-3-(isopropylamino)-2-propanol *see* 7051
p-Allylphenol *see* 2049
1-(*o*-Allylphenoxy)-3-(isopropylamino)-2-propanol *see* 309
Allyl 2-Phenylcinchoninate *see* 2291
α-Allylpiperidine *see* 2485
Allylprodine, 287
m-Allylpyrocatechin Methylene Ether *see* 8453
(±)-*N*-[(1-Allyl-2-pyrrolidinyl)methyl]-4-amino-5-(methylsulfamoyl)-*o*-anisamide *see* 307
N-(1-Allyl-2-pyrrolidinylmethyl)-2-methoxy-4-amino-5-methylsulfamoylbenzamide *see* 307
N-[(1-Allyl-2-pyrrolidinyl)methyl]-6-methoxy-1*H*-benzotriazole-5-carboxamide *see* 242
N-[(1-Allyl-2-pyrrolidinyl)methyl]-5-sulfamoyl-*o*-veratramide *see* 10142
Allyl Sulfide, 288
3-((*S*)-Allylsulfinyl)-L-alanine *see* 254
Allylthiocarbamide *see* 9515
3-[(Allylthio)methyl]-6-chloro-3,4-dihydro-2*H*-1,2,4-benzothiadiazine-7-sulfonamide 1,1-Dioxide *see* 314
3-Allylthiomethyl-6-chloro-7-sulfamoyl-3,4-dihydrobenzothiadiazine 1,1-Dioxide *see* 314
[(Allylthio)methyl] Penicillin *see* 7208
Allyl Thiourea *see* 9515
Allyl Tribromide *see* 9776
Allyltributylstannane, 289
Allyltributyltin *see* 289
Allyltrimethylsilane, 290
Allyl Trisulfide *see* 2976
Allylurea, 291
4-Allylveratrole *see* 6144
Allypropymal *see* 745
Allytrenbolone *see* 315
Almacet [Celtia] *see* 74
Almagate, 292
Alman [Kyosei] *see* 234
Almarl [Sumitomo] *see* 782
Almarytm [Synthelabo] *see* 4130
Almax [Almirall] *see* 292
Almeta [Shionogi] *see* 213
Alminoprofen, 293
Almitrine, 294
Almitrine Dimesylate *see* 294
Almocarpine [Ayerst] *see* 7536
Almogran [Almirall] *see* 298
Almond, Bitter, 295
Almond, Sweet, 296

Almora [Forest] see 4492
Almorexant, 297
Almotriptan, 298
Alnert [Fujisawa] see 1210
Alnide [Cusi] see 2741
Alnulin see 9195
ALO-2145 see 738
Alodan [Gerot] see 5916
Aloe, 299
Aloe-Emodin, 300
Aloeresin B see 301
Aloeresins see 301
Aloesin, 301
Aloe Sugar see 750
Alogliptin, 302
Aloin, 303
Aloin A see 303
Alomide [Alcon] see 5613
Aloperidin see 4634
Alopexil [Fabre] see 6285
Aloprim [Nabi] see 271
Alosetron, 304
Alositol [Tanabe Mitsubishi] see 271
Alostil [Pfizer] see 6285
Alotec [Boehringer, Ing.] see 6002
Aloxi [Helsinn] see 7098
Aloxiprin, 305
Aloxyn see 4881
ALP-201 see 3976
Alpha₁-antitrypsin see 705
Alphacetylmethadol see 6017
Alpha-Chymocutan [Hasenclever] see 2264
Alpha D₃ [Teva] see 4858
Alphadione see 228
Alphagan [Allergan] see 1381
Alpha-hypophamine see 7078
Alpha-interferon see 5034
Alphakil [Rentokil] see 2072
Alphamide see 9059
Alphamin [SS Pharm.] see 2343
Alpha-naphthol see 6468
Alphanate [Alpha Ther.] see 3962
AlphaNine [Alpha Ther.] see 3963
Alpha-pipradrol see 7597
Alphapress [Alphapharm] see 4802
Alphaprodine, 306
Alpharadin [Algeta] see 8211
Alphasol AY see 2989
Alphasol IB see 3213
Alphasol MA [259]
Alphasone see 232
Alphasone Acetonide see 232
Alphasone Acetophenide see 233
Alphaxalone see 228
Alphol see 6495
Alpiny [SS Pharm.] see 47
Alpiropride, 307
Alplax [Gador] see 308
Alprazolam, 308
Alprenolol, 309
Alpress LP [Pfizer] see 7838
Alprostadil see 7987
Alq₃ see 364
Alredase [Am. Home] see 9683
Alreumat [Bayer] see 5352
Alrex [Bausch & Lomb] see 5642
Alrheumun [Bayer] see 5352
Alronidase see 4935
ALRT-1057 see 241
ALS-357 see 1193
Alsactide, 310
Alsanate [Dainippon] see 4411
Alseroxylon see 8238
Alsol [Athenstaedt] see 321

Alstonia Cortex see 3409
Alstonidine, 311
Alstonine, 312
Alsystin [Bayer] see 9844
ALT-711 see 196
Altabactina see 4323
Altabax [GSK] see 8281
Altace [Aventis] see 8221
Altacite [HMR] see 5724
Altacor [DuPont] see 2080
Altafur [Norwich] see 4323
Altaite see 5477
Altan see 2814
Altargo [GSK] see 8281
Altat [Teikoku Hormone] see 8407
Alteplase see 9624
ALternaGEL [J & J-Merck] see 338
Althea, 313
Altheine see 827
Althesin [Glaxo] see 228
Althiazide, 314
Altiazem [Lusofarmaco] see 3224
Altim [Roussel Diamant] see 2527
Altimina [Miquel] see 3999
Altinil [BMS] see 9457
Altiopril see 6371
Altizide see 314
Alto [Kyosei] see 234
Alto [Syngenta] see 2768
Altodor [Synthelabo] see 3777
Altosid [Wellmark] see 6055
Altrenogest, 315
Altretamine, 316
Altropane [Organix] see 4910
D-Altropyranose see 317
D-Altrose, 317
Aludrine see 5263
Alubasine see 3201
Aludex [Intervet] see 482
Aludrin [Boehringer, Ing.] see 5263
Aludyal [Sandoz] see 338
Alufibrate see 2371
Alum see 354
Alum Flour see 354
Alumina see 350
Aluminate (AlO₂¹⁻) Sodium (1:1) see 8712
Aluminim Hydride (AlH₃) see 337
Aluminium see 319
Aluminon, 318
Aluminum, 319
Aluminum Acetate, 320
Aluminum Acetate, Basic see 359
Aluminum Acetate Solution see 320
Aluminum Acetotartrate, 321
Aluminum Alkyls, 322
Aluminum Alloy, Base, Al,Mg (Magnalium) see 5716
Aluminum Aminoacetate (Basic) see 3201
Aluminum Ammonium Chloride see 555
Aluminum Ammonium Sulfate, 323
Aluminum Antimonide, 324
Aluminum Benzoate, 325
Aluminum Borate, 326
Aluminum Borohydride, 327
Aluminum Bromide, 328
Aluminum Bromide (AlBr₃) see 328
Aluminum tert-Butoxide, 329
Aluminum Calcium Hydride, 330
Aluminum Calcium Silicate see 1651
Aluminum Carbide, 331
Aluminum Carbide (Al₄C₃) see 331
Aluminum Chlorate, 332
Aluminum Chloride, 333

Aluminum Chloride (AlCl₃) see 333
Aluminum Chlorohydrate see 339
Aluminum Chlorohydroxide see 339
Aluminum Compd. with Antimony (1:1) see 324
Aluminum Diacetate see 359
Aluminum Dihydroxy Allantoinate see 216
Aluminum Dihydroxyaminoacetate see 3201
Aluminum Ethoxide, 334
Aluminum Ethylate see 334
Aluminum Flufenamate see 4163
Aluminum Fluoride, 335
Aluminum Fluoride (AlF₃) see 335
Aluminum Fluosilicate see 336
Aluminum Glycinate see 3201
Aluminum Hexafluorosilicate, 336
Aluminum Hydrate see 338
Aluminum Hydride, 337
Aluminum Hydroxide, 338
Aluminum Hydroxide (Al(OH)₃) see 338
Aluminum Hydroxyacetate see 359
Aluminum Hydroxychloride, 339
Aluminum Hydroxydiacetate see 359
Aluminum Hypophosphite, 340
Aluminum Iodide, 341
Aluminum Iodide (AlI₃) see 341
Aluminum Isopropoxide, 342
Aluminum Isopropylate see 342
Aluminum Lactate, 343
Aluminum Lithium Hydride, 344
Aluminum Magnesium Carbonate Hydroxide (AlMg₃(CO₃)(OH)₇) Hydrate (1:2) see 292
Aluminum Magnesium Hydroxide Sulfate (Al₅Mg₁₀(OH)₃₁(SO₄)₂) Hydrate see 5714
Aluminum Magnesium Silicate, 345
Aluminum Methanedisulfonate see 6046
Aluminum Nicotinate, 346
Aluminum Nitrate, 347
Aluminum Nitride, 348
Aluminum Nitride (AlN) see 348
Aluminum Orthophosphate see 352
Aluminum Oxalate, 349
Aluminum Oxide, 350
Aluminum Oxide (Al₂O₃) see 350
Aluminum Oxide Silicate [Al₂O(SiO₄)] see 356
Aluminum Palmitate, 351
Aluminum Phosphate, 352
Aluminum Phosphide, 353
Aluminum Phosphide (AlP) see 353
Aluminum Potassium Sulfate, 354
Aluminum Selenide, 355
Aluminum Selenide (Al₂Se₃) see 355
Aluminum Silicate, 356
Aluminum Silicofluoride see 336
Aluminum Sodium Carbonate Hydroxide see 3202
Aluminum Sodium Chloride see 8815
Aluminum Sodium Oxide see 8712
Aluminum Sodium Sulfate, 357
Aluminum Stearate, 358
Aluminum Subacetate, 359
Aluminum Sulfate, 360
Aluminum(III) Sulfate see 360
Aluminum Sulfide, 361
Aluminum Sulfide (Al₂S₃) see 361
Aluminum Sulfocyanate see 363
Aluminum Tartrate, 362
Aluminum Tetrahydroborate see 327
Aluminum Thiocyanate, 363
Aluminum Triacetate see 320

Aluminum Tribromide *see* 328
Aluminum Trichloride *see* 333
Aluminum Trifluoride *see* 335
Aluminum Trihydrate *see* 338
Aluminum Trihydride *see* 337
Aluminum Triiodide *see* 341
Aluminum Trimethanide *see* 9879
Aluminum Trimethyl *see* 9879
Aluminum Tris(*tert*-butylate) *see* 329
Aluminum Tris(ethyl Phosphite) *see* 4279
Aluminum Tris(*O*-ethylphosphonate) *see* 4279
Aluminum Tris(8-hydroxyquinoline), 364
Aluminum Tristearate *see* 358
Alum Meal *see* 354
Alum Root *see* 4438
Alundum *see* 350
Alunogen *see* 360
Alupent [Boehringer, Ing.] *see* 6002
Alurate [Roche] *see* 745
Alusa [Towa Yakuhin] *see* 216
"Alutiae" *see* 7644
Alvedon [AstraZeneca] *see* 47
Alven [Firma] *see* 9769
Alveograf *see* 3514
Alverde [BASF] *see* 5992
Alverine, 365
Alvesco [Altana] *see* 2266
Alvimopan, 366
Alvo [Taisho] *see* 7023
Alvocidib *see* 4126
Alvonal MR [Gödecke] *see* 2758
ALX-0600 *see* 9248
Alyrane [Baxter] *see* 3635
Alzhemed *see* 9728
Alzinox [Berlex] *see* 3201
AM-80 *see* 9179
AM-109 *see* 6050
AM-715 *see* 6789
AM-725 *see* 7172
AM-833 *see* 4131
AM-1091 *see* 2353
AM-1155 *see* 4409
AM-2282 *see* 8928
AMA-1080 *see* 1871
Amacetam *see* 7826
Amadol [TAD] *see* 9726
Amadou *see* 177
Amal [Knoll] *see* 567
Amanin *see* 367
Amanitin, 367
α-Amanitin *see* 367
β-Amanitin *see* 367
Amantadine, 368
Amaranth (Dye), 369
Amaranth (Plant), 370
Amarel [Sanofi-Aventis] *see* 4475
Amarogentin, 371
Amarolide, 372
Amaromycin *see* 7525
Amaryl [Sanofi-Aventis] *see* 4475
Amarylline *see* 5687
Amasulin [Takeda] *see* 1871
Amasust [Knoll] *see* 567
Amatine [Roberts] *see* 6263
Amaze *see* 5218
Ambacamp [Pharmacia & Upjohn] *see* 924
Ambamide *see* 5712
Ambaxin [Pharmacia] *see* 924
Ambene [Merckle] *see* 7390
Ambenonium Chloride, 373
Amber, 374

Amber [Syngenta] *see* 9761
Amber Acid *see* 8999
Ambergris, 375
Amberlite [Rohm & Haas], 376
Amberlyst 15 [Rohm & Haas], 377
Ambien [Sanofi-Aventis] *see* 10391
Ambilhar [Novartis] *see* 6649
AmBisome [Nexstar] *see* 582
Ambivalon [Nattermann] *see* 484
Amblosin [Hoechst] *see* 583
Amblygonite *see* 5576
Ambodryl [Parke-Davis] *see* 1428
Ambramicina [Lepetit] *see* 9341
Ambra-Vena [Lepetit] *see* 7566
Ambril [Cascan] *see* 380
Ambrisentan, 378
Ambrosia *see* 6877
Ambrosin, 379
Ambroxol, 380
Ambush [Syngenta] *see* 7294
Ambuterol *see* 5703
Ambutonium Bromide, 381
Ambutyrosin *see* 1530
AMCHA *see* 9730
Amchem 70-25 *see* 1534
Amciderm [Hermal] *see* 382
Amcinonide, 382
Amco [Am. Mfg.] *see* 7701
AMD-473 *see* 7517
AMD-3100 *see* 7649
Amdinocillin, 383
Amdinocillin Pivoxil, 384
Amdray [Novartis] *see* 10103
Amdro [BASF] *see* 4804
Amebil *see* 5075
Amechol [Moorsfields] *see* 6012
Amedel [Dainippon] *see* 7593
Amediplase, 385
Ameisensäure (German) *see* 4269
Amekrin [Warner-Lambert] *see* 589
Ameliaroside *see* 7507
Amen [Elan] *see* 5864
Amenide [Sterling Winthrop] *see* 8173
Amenox [Sanofi Winthrop] *see* 8173
Amenyl *see* 2102
Amerchol 400 [Amerchol] *see* 5410
Amerge [GSK] *see* 6502
Americaine [Medeva] *see* 1088
American Cyanamid 12880 *see* 3241
American Hellebore *see* 10151
American Horsemint *see* 6329
American Indian Hemp *see* 730
American Mandrake *see* 7664
American Nightshade Root *see* 7503
American Oil Palm *see* 6910
American Pennyroyal *see* 7211
American Saffron *see* 1868
American Spikenard *see* 755
American Valerian *see* 2767
American Veratrum *see* 10151
American Wormseed *see* 6877
Americium, 386
Amerscan MDP [Amersham] *see* 5863
Ametazole *see* 1185
Amethocaine Hydrochloride *see* 9333
Amethopterin *see* 6057
Amethyst *see* 8632
Ametox *see* 8821
Ametrex [Makhteshim-Agan] *see* 387
Ametriodinic Acid *see* 5055
Ametryn, 387
Ametryne *see* 387
Ametycine [Sanofi Winthrop] *see* 6300
Amevive [Biogen] *see* 220

Amexine [CFPI] *see* 1534
Amezinium Methyl Sulfate, 388
AmF 297 *see* 6912
Amfamox [Merck & Co.] *see* 3971
Amfebutamone *see* 1503
Amfenac, 389
Amfepramone *see* 3145
Amfipen [Yamanouchi] *see* 583
Amfomycin *see* 581
AMG-073 *see* 2285
AMG-162 *see* 2897
AMG-386 *see* 9741
AMG-412 *see* 3687
AMG-531 *see* 8383
AMG-706 *see* 6366
AMI-25 *see* 4090
AMI-121 *see* 4091
AMI-227 *see* 4092
Amianthus *see* 815
Amias [Takeda] *see* 1742
Amicar [Wyeth] *see* 427
Amicarbazone, 390
Amicetin, 391
Amicos [Banyu] *see* 2342
Amicoumacin A, 392
Amidate [Hospira] *see* 3926
Amidazine *see* 3791
Amidazophen *see* 470
Amidephrine, 393
Amidephrine Mesylate *see* 393
[[1-[*N*-(*p*-Amidinobenzoyl)-L-tyrosyl]-4-piperidinyl]oxy]acetic Acid *see* 5399
N-[(*R*)-[[(2*S*)-2-[[*p*-Amidinobenzyl]carbamoyl]-1-azetidinyl]carbonyl]cyclohexylmethyl]glycine *see* 5879
N-Amidino-3,5-diamino-6-chloropyrazinamide *see* 401
N-Amidino-3,5-diamino-6-chloropyrazinecarboxamide *see* 401
N-Amidino-2-(2,6-dichlorophenyl)acetamide *see* 4596
(−)-*cis*-*N*-(2-Amidinoethyl)-3-aminocyclopentanecarboxamide *see* 394
2-[*N*-(2-Amidinoethyl)carbamoyl]-5-iminopyrrolidine *see* 6757
N′-(2-Amidinoethyl)-4-(2-guanidinoacetamido)-1,1′-dimethyl-*N*,4′-bi[pyrrole-2-carboxamide] *see* 6566
N-(2-Amidinoethyl)-5-imino-2-pyrrolidinecarboxamide *see* 6757
N-Amidinoglycine *see* 4530
Amidinoguanidine *see* 1218
Amidinomycin, 394
6-Amidino-2-naphthyl-4-guanidinobenzoate *see* 6434
L-6-Amidinonorleucine *see* 5011
7-*m*-Amidinophenyldiazoamino-2-amino-10-ethyl-9-phenylphenanthridinium Chloride *see* 5229
8-[3-(*m*-Amidinophenyl)-2-triazeno]-3-amino-5-ethyl-6-phenylphenanthridinium Chloride *see* 5229
N-Amidinosarcosine *see* 2556
*N*¹-Amidinosulfanilamide *see* 9040
2-Amidino-1,2,3,4-tetrahydroisoquinoline *see* 2846
Amidocyanogen *see* 2674
Amido-G-Acid, 395
Amidol *see* 2985
Amidolacetate *see* 6017
Amidonal [PCR Arzneimittel] *see* 743
Amidopyrine *see* 470
Amido-R-Acid, 396
Amidosulfonic Acid *see* 9053
1-(*p*-Amidosulfonylphenyl)-2-thiapiperidine 2,2-Dioxide *see* 9122

Aminobenzene *see* 652
α-Aminobenzeneacetic Acid *see* 7402
4-Aminobenzeneacetic Acid *see* 459
p-Aminobenzenearsonic Acid *see* 783
3-(p-Aminobenzenesulfamido)-6-methoxypyridazine *see* 9051
2-(p-Aminobenzenesulfamido)-5-tertiobutyl-1,3,4-thiadiazole *see* 4515
4-Aminobenzenesulfonamide *see* 9057
4-(4'-Aminobenzenesulfonamido)benzenesulfonamide *see* 9061
6-(4-Aminobenzenesulfonamido)-4,5-dimethoxypyrimidine *see* 9038
5-(p-Aminobenzenesulfonamido)-3,4-dimethylisooxazole *see* 9082
2-(p-Aminobenzenesulfonamido)-4,5-dimethyloxazole *see* 9056
2-(p-Aminobenzenesulfonamido)-5-ethylthiadiazole *see* 9039
2-(p-Aminobenzenesulfonamido)-3-methoxypyrazine *see* 9042
2-(p-Aminobenzenesulfonamido)-5-methoxypyrimidine *see* 9046
5-(p-Aminobenzenesulfonamido)-3-methyl-1-phenylpyrazole *see* 9076
2-(p-Aminobenzenesulfonamido)-5-methylthiadiazole *see* 9048
3-(p-Aminobenzenesulfonamido)-2-phenyl-5-methylpyrazole *see* 9076
3-(p-Aminobenzenesulfonamido)-2-phenylpyrazole *see* 9066
2-(p-Aminobenzenesulfonamido)pyridine *see* 9069
2-(p-Aminobenzenesulfonamido)thiazole *see* 9074
p-Aminobenzenesulfonanilide *see* 9028
2-Aminobenzenesulfonic Acid *see* 6978
3-Aminobenzenesulfonic Acid *see* 5998
4-Aminobenzenesulfonic Acid *see* 9058
p-Aminobenzenesulfonoacetamide *see* 9031
4-Aminobenzenesulfono-p-sulfamoylanilide *see* 9061
p-Aminobenzenesulfonyl-2-amino-4,5-dimethyloxazole *see* 9056
6-(p-Aminobenzenesulfonyl)amino-2,4-dimethylpyrimidine *see* 9081
4-(p-Aminobenzenesulfonyl)amino-2-methyl-6-methoxypyrimidine *see* 9049
N-(p-Aminobenzenesulfonyl)benzamide *see* 9029
N-(4-Aminobenzenesulfonyl)-N'-butylurea *see* 1831
p-Aminobenzenesulfonyl Fluoride *see* 9060
p-Aminobenzenesulfonylguanidine *see* 9040
p-Aminobenzoate of Diethylaminoneopentyl Alcohol *see* 3242
1-(2-Aminobenzoate) 3-Phenyl-2-propen-1-ol *see* 2304
Aminobenzoate Potassium *see* 7727
4-Amino-2-[4-(1,4-benzodioxan-2-carbonyl)piperazin-1-yl]-6,7-dimethoxyquinazoline *see* 3482
m-Aminobenzoic Acid, 416
o-Aminobenzoic Acid, 417
p-Aminobenzoic Acid, 418
2-Aminobenzoic Acid *see* 417
3-Aminobenzoic Acid *see* 416
4-Aminobenzoic Acid *see* 418
4-Aminobenzoic Acid Butyl Ester *see* 1515
4-Aminobenzoic Acid 2-(Diethylamino)ethyl Ester *see* 7875

4-Aminobenzoic Acid Ethyl Ester *see* 1088
3-Aminobenzoic Acid Ethyl Ester Methanesulfonate (1:1) *see* 9786
2-Aminobenzoic Acid Methyl Ester *see* 6094
4-Aminobenzoic Acid 2-Methylpropyl Ester *see* 5178
p-Aminobenzoic Acid Monoglyceryl Ester *see* 4523
4-Aminobenzoic Acid Potassium Salt (1:1) *see* 7727
2-Aminobenzothiazole, 419
6-Aminobenzothiazole, 420
3-Amino-1,2,4-benzotriazine 1,4-Dioxide *see* 9617
3-(p-Aminobenzoyl)-1-di-n-butylaminopropane *see* 1510
N-(p-Aminobenzoyl)aminoacetic Acid *see* 439
2-Amino-3-benzoylbenzeneacetamide *see* 6553
2-Amino-3-benzoylbenzeneacetic Acid *see* 389
p-Aminobenzoyldibutylaminopropanol *see* 1510
p-Aminobenzoyldiethylaminoethanol *see* 7875
1-Aminobenzoyl-2,2-dimethyl-3-diethylaminopropanol *see* 3242
N-(p-Aminobenzoyl)glutamic Acid *see* 421
N-(4-Aminobenzoyl)-L-glutamic Acid, 421
N-(4-Aminobenzoyl)glycine *see* 439
(E)-2-Amino-6-benzoyl-1-(isopropylsulfonyl)benzimidazole Oxime *see* 3655
2-Amino-3-benzoylphenylacetic Acid *see* 389
1-(α-Aminobenzyl)-2-naphthol *see* 1190
D(−)-α-Aminobenzylpenicillin *see* 583
4-Amino-N-(1-benzyl-4-piperidyl)-5-chloro-o-anisamide *see* 2342
4-Aminobiphenyl *see* 1239
5-Amino-[3,4'-bipyridin]-6(1H)-one *see* 587
5-Amino-1,3-bis(2-ethylhexyl)hexahydro-5-methylpyrimidine *see* 4739
2-Amino-5-bromobenzoic Acid *see* 1412
2-Amino-3-(4-bromobenzoyl)benzeneacetic Acid *see* 1397
4-[[4-Amino-5-bromo-6-(4-cyano-2,6-dimethylphenyloxy)-2-pyrimidinyl]amino]benzonitrile *see* 3935
4-[[6-Amino-5-bromo-2-[(4-cyanophenyl)amino]-4-pyrimidinyl]oxy]-3,5-dimethylbenzonitrile *see* 3935
4-Amino-5-bromo-N-[2-(diethylamino)-ethyl]-o-anisamide *see* 1443
4-Amino-5-bromo-N-[2-(diethylamino)-ethyl]-2-methoxybenzamide *see* 1443
4-Amino-N-(5-bromo-4,6-dimethyl-2-pyrimidinyl)benzenesulfonamide *see* 9030
1-Aminobutane *see* 1545
2-Aminobutane *see* 1546
(S)-Aminobutanedioic Acid *see* 830
2-Aminobutanoic Acid *see* 423
3-Aminobutanoic Acid *see* 424
4-Aminobutanoic Acid *see* 425
2-Amino-1-butanol, 422
3-Amino-4-butoxybenzoic Acid 2-[2-(Diethylamino)ethoxy]ethyl Ester *see* 1189
4-Amino-3-butoxybenzoic Acid 2-(Diethylamino)ethyl Ester *see* 1049

2-Amino-n-butyl Alcohol *see* 422
4-Amino-N-[(butylamino)carbonyl]benzenesulfonamide *see* 1831
4-Amino-α-[(tert-butylamino)methyl]-3-chloro-5-(trifluoromethyl)benzyl Alcohol *see* 5703
4-Amino-α-[(tert-butylamino)methyl]-3,5-dichlorobenzyl Alcohol *see* 2345
(3β,5α,7α,24R)-3-[[3-[(4-Aminobutyl)amino]propyl]amino]-cholestane-7,24-diol 24-(Hydrogen Sulfate) *see* 8894
N^1-[3-[(4-Aminobutyl)amino]propyl]-13-[(4-amino-4,6-dideoxy-α-L-talopyranosyl)oxy]-19-demethyl-12-hydroxybleomycinamide *see* 9175
4-Amino-1-(N-t-butylcarbamoyl)-3-isopropyl-1,2,4-triazolin-5-one *see* 390
N-(4-Aminobutyl)-1,3-diaminopropane *see* 8869
3β-N-1-[N-[3-(4-Aminobutyl)]-1,3-diaminopropane]-7α,24ζ-dihydroxy-5α-cholestane 24-Sulfate *see* 8894
N-(4-Aminobutyl)guanidine *see* 181
4-Amino-N-tert-butyl-3-isopropyl-5-oxo-Δ^2-1,2,4-triazoline-1-carboxamide *see* 390
4-Amino-6-tert-butyl-3-(methylthio)-as-triazin-5(4H)-one *see* 6233
2-Amino-4-(S-butylsulfonimidoyl)butanoic Acid *see* 1528
α-Aminobutyric Acid, 423
β-Aminobutyric Acid, 424
γ-Aminobutyric Acid, 425
α-Amino-n-butyric Acid *see* 423
β-Amino-n-butyric Acid *see* 424
γ-Amino-n-butyric Acid *see* 425
9-Aminocamptothecin, 426
9-Amino-20(S)-camptothecin *see* 426
α-Aminocaproic Acid *see* 6795
ε-Aminocaproic Acid, 427
Aminocaproic Lactam *see* 1763
Aminocarb, 428
(2S,4S,5S,7S)-5-Amino-N-(2-carbamoyl-2-methylpropyl)-4-hydroxy-2-isopropyl-7-[4-methoxy-3-(3-methoxypropoxy)benzyl]-8-methylnonanamide *see* 239
3-[(Aminocarbonyl)amino]-L-alanine *see* 205
[3-[(Aminocarbonyl-κO)amino]-2-methoxypropyl-κC]chloromercury *see* 2105
(2Z)-4-[(Aminocarbonyl)amino]-4-oxo-2-butenoic Acid *see* 5770
[4-[(Aminocarbonyl)amino]phenyl]arsonic Acid *see* 1787
N-(Aminocarbonyl)benzeneacetamide *see* 7318
N-(Aminocarbonyl)-2-bromo-3-methylbutanamide *see* 1404
4-(Aminocarbonyl)-1-[[(6R,7R)-2-carboxy-8-oxo-7-[[(2R)-2-phenyl-2-sulfoacetyl]amino]-5-thia-1-azabicyclo-[4.2.0]oct-2-en-3-yl]methyl]pyridinium Inner Salt *see* 1946
4-(Aminocarbonyl)-1-[[(6R,7R)-2-carboxy-8-oxo-7-[[2-(2-thienyl)acetyl]amino]-5-thia-1-azabicyclo[4.2.0]oct-2-en-3-yl]methyl]pyridinium Inner Salt *see* 1977
3-(Aminocarbonyl)-O^4-deacetyl-3-de-(methoxycarbonyl)vincaleukoblastine *see* 10184
N-[7-[3-O-(Aminocarbonyl)-6-deoxy-5-C-methyl-4-O-methyl-α-L-lyxo-hexopyranosyl]oxy]-4-hydroxy-8-methyl-2-

4-Amino-*N*-[[(4,5-dimethyl-2-oxazolyl)-amino]iminomethyl]benzenesulfon-amide *see* 9041

4-Amino-*N*-(4,5-dimethyl-2-oxazolyl)ben-zenesulfonamide *see* 9056

(2*S*,5*R*,6*R*)-6-Amino-3,3-dimethyl-7-oxo-4-thia-1-azabicyclo[3.2.0]heptane-2-carboxylic Acid *see* 454

2-Amino-*N*-(2,6-dimethylphenyl)propan-amide *see* 9648

4-Amino-2,3-dimethyl-1-phenyl-3-pyra-zolin-5-one *see* 413

4-Amino-2,6-dimethylpyrimidine *see* 5374

6-Amino-2,4-dimethylpyrimidine *see* 5374

4-Amino-*N*-(2,6-dimethyl-4-pyrimidinyl)-benzenesulfonamide *see* 9081

4-Amino-*N*-(4,6-dimethyl-2-pyrimidinyl)-benzenesulfonamide *see* 9047

2-Amino-4,6-dinitrophenol *see* 7521

2-[[3-(2-Amino-1,2-dioxoethyl)-2-ethyl-1-(phenylmethyl)-1*H*-indol-4-yl]oxy]ace-tic Acid *see* 10126

(α*S*)-α-Amino-3,5-dioxo-1,2,4-oxadiazoli-dine-2-propanoic Acid *see* 8202

4-Amino-2-(2,6-dioxo-3-piperidinyl)-1*H*-isoindole-1,3(2*H*)-dione *see* 7708

3-Amino-*N*-(2,6-dioxo-3-piperidyl)-phthalamide *see* 7708

p-Aminodiphenyl *see* 1239

p-Aminodiphenylimide *see* 414

α-Aminodiphenylmethane *see* 1078

(2*S*,4a*S*,4b*S*,6a*S*,6b*R*,7*S*,7a*R*,10*R*,11a*S*,-12a*R*,13a*S*,13b*R*,15a*S*)-2-Aminodoco-sahydro-4a,6a,7,10-tetramethyl-naphth[2″,1″:4′,5′]indeno[1′,2′:5,6]py-rano[3,2-*b*]pyridin-11a(1*H*)-ol *see* 8833

4-Amino-1-dodecyl-2-methylquinolinium Acetate (1:1) *see* 5442

4-Amino-1-dodecylquinaldinium Acetate *see* 5442

Aminodur [Berlex] *see* 461

(3β,5α,16α,22α,23β,25β)-3-Amino-16,23-epoxy-16,28-secosolanidan-23-ol *see* 8833

Aminoethane *see* 3817

2-Aminoethanesulfonic Acid *see* 9211

2-Aminoethanethiol *see* 2776

2-Aminoethanethiol 1-Dihydrogen Phos-phate *see* 7453

Aminoethanoic Acid *see* 4526

1-Aminoethanol *see* 41

2-Aminoethanol *see* 3782

2-Aminoethanol 1-Nitrate 4-Methylben-zenesulfonate (1:1) *see* 5293

2-Aminoethanol Nitrate Mono-*p*-toluene-sulfonate *see* 5293

6-Amino-2-[3-(ethenylsulfonyl)phenyl]-2,3-dihydro-1,3-dioxo-1*H*-benz[*de*]iso-quinoline-5,8-disulfonic Acid Dilith-ium Salt *see* 5653

2-[(2-Aminoethoxy)methyl]-4-(2-chloro-phenyl)-1,4-dihydro-6-methyl-3,5-pyr-idinedicarboxylic Acid 3-Ethyl 5-Methyl Ester *see* 487

(±)-2-[(2-Aminoethoxy)methyl]-4-(2-chlorophenyl)-3-ethoxycarbonyl-5-methoxycarbonyl-6-methyl-1,4-dihy-dropyridine *see* 487

4-Amino-2-(ethoxymethyl)-α,α-dimethyl-1*H*-imidazo[4,5-*c*]quinoline-1-ethanol *see* 8271

Aminoethoxyvinylglycine, 434

α-Aminoethyl Alcohol *see* 41

β-Aminoethyl Alcohol *see* 3782

(6*R*,7*R*)-3-[[4-[[[(2-Aminoethyl)amino]-carbonyl]amino]-2,3-dihydro-3-imino-2-methyl-1*H*-pyrazol-1-yl]methyl]-7-[[(2*Z*)-2-(5-amino-1,2,4-thiadiazol-3-yl)-2-[(1-carboxy-1-methylethoxy)imi-no]acetyl]amino]-8-oxo-5-thia-1-azabi-cyclo[4.2.0]oct-2-ene-2-carboxylic Acid *see* 1954

1-[(4*R*,5*S*)-5-[(2-Aminoethyl)amino]-*N* ²-[(10*R*,12*S*)-10,12-dimethyl-1-oxotetra-decyl]-4-hydroxy-L-ornithine]-5-[(3*R*)-3-hydroxy-L-ornithine]pneumocandin B₀ *see* 1889

N ¹-(2-Aminoethyl)-*N* ²-[2-[(2-aminoeth-yl)amino]ethyl]-1,2-ethanediamine Polymer with 2-(Chloromethyl)oxi-rane *see* 2459

1-[(2-Aminoethyl)amino]-2-propanol, 435

α-Aminoethylbenzene *see* 6100

β-Aminoethylbenzene *see* 7339

4-(2-Aminoethyl)-1,2-benzenediol *see* 3470

rel-(α*S*)-α-[(1*R*)-1-Aminoethyl]benzene-methanol *see* 7417

(α*S*)-α-[(1*S*)-1-Aminoethyl]benzenemeth-anol *see* 6803

N-(2-Aminoethyl)-5-chloropicolinamide *see* 5449

N-(2-Aminoethyl)-5-chloro-2-pyridine-carboxamide *see* 5449

4-Amino-*N*-(1-ethyl-1,2-dihydro-2-oxo-4-pyrimidinyl)benzenesulfonamide *see* 9034

(*erythro*)-(±)-α-(1-Aminoethyl)-3,4-dihy-droxybenzyl Alcohol *see* 6781

α-(1-Aminoethyl)-2,5-dimethoxybenzene-methanol *see* 6058

α-(1-Aminoethyl)-2,5-dimethoxybenzyl Alcohol *see* 6058

β-Aminoethylglyoxaline *see* 4756

(α*R*)-α-[(1*S*)-1-Aminoethyl]-3-hydroxy-benzenemethanol *see* 6003

(−)-α-(1-Aminoethyl)-*m*-hydroxybenzyl Alcohol *see* 6003

3-(β-Aminoethyl)-5-hydroxyindole *see* 8601

(4*S*)-10-Amino-4-ethyl-4-hydroxy-1*H*-py-rano[3′,4′:6,7]indolizino[1,2-*b*]quino-line-3,14(4*H*,12*H*)-dione *see* 426

(5*R*,9*R*,11*E*)-5-Amino-11-ethylidene-5,6,-9,10-tetrahydro-7-methyl-5,9-metha-nocycloocta[*b*]pyridin-2(1*H*)-one *see* 4794

β-Aminoethylimidazole *see* 4756

3-(2-Aminoethyl)indole *see* 9976

3-(2-Aminoethyl)-1*H*-indol-5-ol *see* 8601

β-Aminoethylisothiuronium Bromide Hydrobromide *see* 167

S-(2-Aminoethyl)isothiuronium Bromide Hydrobromide *see* 167

2-Aminoethyl Mercaptan *see* 2776

α-Amino-γ-(ethylmercapto)butyric Acid *see* 3792

3-(2-Aminoethyl)-5-methoxyindole *see* 6076

p-β-Aminoethylphenol *see* 10016

4-(2-Aminoethyl)phenol *see* 10016

2-Amino-2-ethyl-1,3-propanediol, 436

3-(β-Aminoethyl)pyrazole *see* 1185

3-(2-Aminoethyl)pyrazole *see* 1185

4-(2-Aminoethyl)pyrocatechol *see* 3470

4-Amino-*N*-[(1-ethyl-2-pyrrolidinyl)-methyl]-5-(ethylsulfonyl)-*o*-anisamide *see* 481

4-Amino-*N*-[(1-ethyl-2-pyrrolidinyl)-methyl]-5-(ethylsulfonyl)-2-methoxy-benzamide *see* 481

4-Amino-*N*-(5-ethyl-1,3,4-thiadiazol-2-yl)benzenesulfonamide *see* 9039

2-Amino-4-(ethylthio)butyric Acid *see* 3792

(5*R*,6*S*)-3-[(2-Aminoethyl)thio]-6-[(1*R*)-1-hydroxyethyl]-7-oxo-1-azabicyclo-[3.2.0]hept-2-ene-2-carboxylic Acid *see* 9466

2-(2-Aminoethyl)-2-thiopseudourea Di-hydrobromide *see* 167

3-Amino-α-ethyl-2,4,6-triiodobenzene-propanoic Acid *see* 5101

3-Amino-α-ethyl-2,4,6-triiodohydrocin-namic Acid *see* 5101

2-Amino-4-(4-fluorobenzylamino)-1-eth-oxycarbonylaminobenzene *see* 3959

N-[2-Amino-4-(4-fluorobenzylamino)-phenyl] Carbamic Acid Ethyl Ester *see* 3959

2-Amino-6-[(*p*-fluorobenzyl)amino]-3-pyridinecarbamic Acid Ethyl Ester *see* 4221

(−)-*cis*-4-Amino-5-fluoro-1-(2-hydroxy-methyl-1,3-oxathiolan-5-yl)-(1*H*)-pyr-imidin-2-one *see* 3622

4-Amino-5-fluoro-1-[(2*R*,5*S*)-2-(hydroxy-methyl)-1,3-oxathiolan-5-yl]-2(1*H*)-pyrimidinone *see* 3622

6-Amino-2-(fluoromethyl)-3-(2-methyl-phenyl)-4(3*H*)-quinazolinone *see* 175

6-Amino-2-fluoromethyl-3-(*o*-tolyl)-4(3*H*)-quinazolinone *see* 175

N-[2-Amino-4-[[(4-fluorophenyl)methyl]-amino]phenyl]carbamic Acid Ethyl Ester *see* 3959

N-[2-Amino-6-[[(4-fluorophenyl)methyl]-amino]-3-pyridinyl]carbamic Acid Ethyl Ester *see* 4221

6-Amino-5-fluoro-2(1*H*)-pyrimidinone *see* 4156

4-Aminofolic Acid *see* 468

Aminoformamidine *see* 4597

N-[4-[[(2-Amino-5-formyl-3,4,5,6,7,8-hexahydro-4-oxo-6-pteridinyl)methyl]-amino]benzoyl]-L-glutamic Acid *see* 4249

N-[*p*-[[(2-Amino-5-formyl-5,6,7,8-tetra-hydro-4-hydroxy-6-pteridinyl)methyl]-amino]benzoyl]glutamic Acid *see* 4249

Amino-G-acid *see* 395

6-Amino-5-(β-D-glucopyranosyloxy)-2,4-(1*H*,3*H*)-pyrimidinedione *see* 2498

2-Aminoglutaramic Acid *see* 4507

4-Amino-L-glutaramic Acid *see* 5223

α-Aminoglutaric Acid *see* 4505

α-Aminoglutaric Acid Lactam *see* 8113

Aminoglutethimide, 437

Aminoguanidine, 438

1-Amino-19-guanidino-11-hydroxy-4,-9,12-triazanonadecane-10,13-dione *see* 4617

L-2-Amino-4-(guanidinooxy)butyric Acid *see* 1740

2-Amino-5-guanidinovaleric Acid *see* 770

1-Amino-4-guanidobutane *see* 181

2-Aminoheptane *see* 9982

7-[(3*R*)-3-Aminohexahydro-1*H*-azepin-1-yl]-8-chloro-1-cyclopropyl-6-fluoro-

α-Amino-β-thiolpropionic Acid *see* 2778
1-Aminothiourea *see* 9514
N-[4-[[(2-Aminothioxomethyl)hydrazin-
ylidene]methyl]phenyl]acetamide *see*
9442
Aminotoluene *see* 1128
2-Aminotoluene *see* 9696
3-Aminotoluene *see* 9696
4-Aminotoluene *see* 9696
α-Amino-*p*-toluenesulfonamide *see* 5712
α-Amino-α-toluic Acid *see* 7402
p-Amino-α-toluic Acid *see* 459
(6*R*,7*R*)-7-[2-(α-Amino-*o*-tolyl)acetami-
do]-3-[[[1-(carboxymethyl)-1*H*-tetra-
zol-5-yl]thio]methyl]-8-oxo-5-thia-1-
azabicyclo[4.2.0]oct-2-ene-2-carboxylic
Acid *see* 1932
Aminotriazole *see* 485
3-Amino-*s*-triazole *see* 485
3-Amino-1*H*-1,2,4-triazole *see* 485
5-Amino-1*H*-*v*-triazolo[*d*]pyrimidin-7-ol
see 889
4-Amino-6-(1,2,2-trichloroethenyl)-1,3-
benzenedisulfonamide *see* 2395
4-Amino-3,5,6-trichloropicolinic Acid *see*
7509
4-Amino-3,5,6-trichloro-2-pyridinecar-
boxylic Acid *see* 7509
4-Amino-6-(trichlorovinyl)-*m*-benzenedi-
sulfonamide *see* 2395
2-Amino-1,1,3-tricyanopropene, 476
2-Amino-1,3,3-tricyano-2-propene *see*
476
1-Aminotricyclo[3.3.1.1³,⁷]decane *see* 368
(8*S*,10*S*)-10-[(3-Amino-2,3,6-trideoxy-α-
L-*arabino*-hexopyranosyl)oxy]-7,8,-
9,10-tetrahydro-6,8,11-trihydroxy-8-
(2-hydroxyacetyl)-1-methoxy-5,12-
naphthacenedione *see* 3678
(8*S*,10*S*)-10-[(3-Amino-2,3,6-trideoxy-α-
L-*lyxo*-hexopyranosyl)oxy]-7,8,9,10-
tetrahydro-6,8,11-trihydroxy-8-(2-hy-
droxyacetyl)-1-methoxy-5,12-naphtha-
cenedione *see* 3487
(4″*R*)-22-*O*-(3-Amino-2,3,6-trideoxy-3-*C*-
methyl-α-L-*arabino*-hexopyranosyl)-
*N*³″-[(4′-chloro[1,1′-biphenyl]-4-yl)-
methyl]vancomycin *see* 6963
(8*S*,10*S*)-10-[[3-Amino-2,3,6-trideoxy-4-
O-[(2*R*)-tetrahydro-2*H*-pyran-2-yl]-α-
L-*lyxo*-hexopyranosyl]oxy]-7,8,9,10-
tetrahydro-6,8,11-trihydroxy-8-(2-hy-
droxyacetyl)-1-methoxy-5,12-naphtha-
cenedione *see* 7601
7-Amino-4,5,6-triethoxy-3-(5,6,7,8-tetra-
hydro-4-methoxy-6-methyl-1,3-diox-
olo[4,5-*g*]isoquinolin-5-yl)-1(3*H*)-iso-
benzofuranone *see* 9942
7-Amino-4,5,6-triethoxy-3-(5,6,7,8-tetra-
hydro-4-methoxy-6-methyl-1,3-diox-
olo[4,5-*g*]isoquinolin-5-yl)phthalide
see 9942
2-Amino-6-(trifluoromethoxy)benzothia-
zole *see* 8350
2-Amino-6-(1,2,3-trihydroxypropyl)-
4(3*H*)-pteridinone *see* 6547
2-Amino-4,6,7-trihydroxypteridine *see*
5506
2-Amino-4,6,7-trihydroxypyrimido[4,5-
b]pyrazine *see* 5506
3′-Amino-2′,4′,6′-triiodo-*N*-methylglu-
taranilic Acid *see* 5096
3-(3-Amino-2,4,6-triiodophenyl)-2-ethyl-
propanoic Acid *see* 5101
3-(3-Amino-2,4,6-triiodophenyl)-2-ethyl-
propionic Acid *see* 5101

5-[(3-Amino-2,4,6-triiodophenyl)methyl-
amino]-5-oxopentanoic Acid *see* 5096
Aminourea Hydrochloride *see* 8583
2-Amino-3-ureidopropionic Acid *see* 205
α-Amino-δ-ureidovaleric Acid *see* 2330
L-α-Aminovaleric Acid *see* 6805
(*S*)-2-Aminovaleric Acid *see* 6805
Aminoxafen *see* 472
Aminoxaphen *see* 472
Aminoxidin [Farmitalia] *see* 7146
2-Amino-1,3-xylene *see* 10282
2-Amino-1,4-xylene *see* 10282
3-Amino-*o*-xylene *see* 10282
4-Amino-*o*-xylene *see* 10282
4-Amino-1,3-xylene *see* 10282
5-Amino-1,3-xylene *see* 10282
Aminoxytropine Tropate *see* 866
Amioca, 477
Amiodar [Ecosol] *see* 478
Amiodarone, 478
Amipaque [Winthrop] *see* 6234
Amipenix [Toyo Jozo] *see* 583
Amiphenazole, 479
Amipramidin *see* 401
Amipramizide *see* 401
Amipress [Dox-Al] *see* 5377
Amiprilose, 480
Amisalin [Daiichi] *see* 7874
Amistar [Syngenta] *see* 917
Amisulpride, 481
Amithiozone *see* 9442
Amitiza [Sucampo] *see* 5649
Amitraz, 482
Amitriptyline, 483
Amitriptyline *N*-Oxide *see* 484
Amitriptylinoxide, 484
Amitrole, 485
Amizol [CFPI] *see* 485
Amlexanox, 486
Amlintide *see* 603
Amlodipine, 487
Amlodipine Besylate *see* 487
Amlor [Pfizer] *see* 487
Ammate [DuPont] *see* 551
Ammidin *see* 4967
(*SP*-4-3)-Amminedichloro(2-methylpyri-
dine)platinum *see* 7517
cis-Amminedichloro(2-methylpyridine)-
platinum(II) *see* 7517
(*T*-4)-Amminetrihydroboron *see* 489
Amminosidin *see* 7146
Ammivin *see* 5356
Ammo [FMC] *see* 2764
Ammoidin *see* 6059
Ammonia, 488
Ammonia Borane, 489
Ammoniacum, 490
Ammonia Soap *see* 532
Ammonia Water, 491
Ammonia Water—10% *see* 491
2-Ammonio-4-methylphosphinicobutyr-
ate *see* 7451
Ammonium Acetate, 492
Ammonium Acetate Solution *see* 492
Ammonium Acid Oxalate *see* 496
Ammonium Acid Sulfite *see* 499
Ammonium Acid Tartrate *see* 500
Ammonium Alum *see* 323
Ammonium Aluminum Fluoride *see* 521
Ammonium Aluminum Sulfate *see* 323
Ammonium Aminoformate *see* 504
Ammonium Antimony Sodium Tungsten
Oxide ((NH₄)₁₇Sb₉Na₂W₂₁O₈₆) *see*
4789
Ammonium Antimony Tungsten Oxide
see 4789

Ammonium Benzoate, 493
Ammonium Biborate *see* 501
Ammonium Bicarbonate, 494
Ammonium Bichromate *see* 512
Ammonium Bifluoride, 495
Ammonium Binoxalate, 496
Ammonium Biphosphate *see* 541
Ammonium Bis(oxalato)oxotitanate *see*
560
Ammonium Bisulfate, 497
Ammonium Bisulfide, 498
Ammonium Bisulfite, 499
Ammonium Bitartrate, 500
Ammonium Bithiolicum *see* 4924
Ammonium Bituminosulfonate *see* 4924
Ammonium Borate, 501
Ammonium Boron Oxide ((NH₄)₂B₄O₇)
see 501
Ammonium Bromide, 502
Ammonium Bromide ((NH₄)Br) *see* 502
Ammonium Caprylate, 503
Ammonium Carbamate, 504
Ammonium Carbazoate *see* 544
Ammonium Carbolate *see* 7354
Ammonium Carbonate, 505
Ammonium Ceric Nitrate *see* 1991
Ammonium Chloride, 506
Ammonium Chloride ((NH₄)Cl) *see* 506
Ammonium Chloroaluminate *see* 555
Ammonium Chlorocuprate *see* 511
Ammonium Chloroheptenearsonate *see*
2119
Ammonium Chloroosmate *see* 533
Ammonium Chloroplatinate *see* 545
Ammonium Chloroplatinite *see* 546
Ammonium Chromate(VI), 507
Ammonium Chromic Sulfate, 508
Ammonium Citrate, Dibasic, 509
Ammonium Cobaltous Phosphate, 510
Ammonium Cocoyl Isethionate *see* 8739
Ammonium Cryolite *see* 521
Ammonium Cupric Chloride, 511
Ammonium Cupric Sulfate *see* 9325
Ammonium (*OC*-6-11)-Diammine-
tetrakis(thiocyanato-κ*N*)chromate(1−)
1:1) *see* 8248
Ammonium Dichromate(VI), 512
Ammonium Dihydrogen Phosphate *see*
541
Ammonium Dioxopentafluorouranate-
(VI) *see* 564
Ammonium Dioxotricarbonatouranate-
(VI) *see* 563
Ammonium Disulfatochromate(III) *see*
508
Ammonium Dithiocarbamate, 513
Ammonium Diuranate *see* 562
Ammonium Ferric Citrate *see* 4044
Ammonium Ferric Oxalate, 514
Ammonium Ferric Sulfate, 515
Ammonium Ferricyanide, 516
Ammonium Ferrocyanide, 517
Ammonium Ferrous Sulfate, 518
Ammonium Fluoaluminate *see* 521
Ammonium Fluoride, 519
Ammonium Fluoride ((NH₄)F) *see* 519
Ammonium Fluoride ((NH₄)(HF₂)) *see*
495
Ammonium Fluosilicate *see* 524
Ammonium Formate, 520
Ammonium Heptamolybdate *see* 530
Ammonium Hexachloroosmate(IV) *see*
533
Ammonium Hexachloroplatinate(IV) *see*
545

Amudane *see* 4584
Amycor [Lipha] *see* 1216
Amygdalic Acid *see* 5781
Amygdalin, 592
Amygdalinic Acid *see* 5781
Amygdalose *see* 4431
Amygdaloside *see* 592
Amylacetic Ester *see* 5156
dl-sec-Amyl Alcohol *see* 7231
n-Amyl Alcohol *see* 7230
tert-Amyl Alcohol *see* 7253
n-Amylamine, 593
Amylase, 594
α-Amylase (*Aspergillus oryzae*) *see* 9162
γ-Amylases *see* 594
Amylbenzene, 595
n-Amylbenzene *see* 595
d-Amyl Bromide *see* 1435
n-Amyl Bromide, 596
tert-Amyl Bromide, 597
n-Amyl Butyrate, 598
n-Amyl Caproate, 599
Amylcarbinol *see* 4734
Amyl Chloride, 600
n-Amyl Chloride *see* 600
Amyleine *see* 607
Amylene, 601
α-*n*-Amylene *see* 7234
β-*n*-Amylene *see* 7235
Amylene Hydrate *see* 7253
n-Amyl Ether, 602
Amyl Ethyl Ketone *see* 6156
3-Amyl-1-hydroxy-6,6,9-trimethyl-6*H*-di-
   benzo[*b,d*]pyran *see* 1751
Amylin, 603
Amyl Isovalerate *see* 5166
n-Amyl Mercaptan, 604
Amylmethylcarbinol *see* 4697
tert-Amyl Methyl Ether, 605
Amyl Nitrite, 606
Amylobarbitone *see* 567
Amylocaine, 607
Amylodextrin *see* 8926
Amylogen *see* 8926
Amyloid β Peptide, 608
Amyloid Precursor Protein *see* 608, 7857
β-Amyloid Protein *see* 608
Amylopectin *see* 477
Amylose *see* 8925
Amyl Oxide *see* 602
Amylpenicillin, 609
p-tert-Amylphenol *see* 7255
cis-4-[3-(4-*tert*-Amylphenyl)-2-methyl-
   propyl]-2,6-dimethylmorpholine *see*
   570
Amyl Phthalate *see* 5169
Amyl Thioalcohol *see* 604
Amylum *see* 8925
Amyl Valerate *see* 5166
α-Amyrenol *see* 610
β-Amyrenol *see* 611
α-Amyrin, 610
β-Amyrin, 611
Amytal [Flynn] *see* 567
Amyvid [Avid Radiopharma] *see* 4140
AN 1 [Krugmann] *see* 580
AN-148 *see* 6016
AN-1317 *see* 7283
AN-1324 *see* 4516
Anabact [Cambridge Healthcare] *see* 6236
Anabasine, 612
Anabiol [Searle] *see* 1325
Anabo [Hokuriku] *see* 5985
Anabolex [Samil] *see* 8920
Anabsinthin, 613

Anabsynthin *see* 613
Anacardic Acid, 614
Anacardone *see* 6627
Anacetrapib, 615
Anacobin *see* 10212
Anadonis Green *see* 2238
Anadrol [Syntex] *see* 7066
Anaesthesin [Ritsert] *see* 1088
Anaflex [Geistlich] *see* 7694
Anafranil [Novartis] *see* 2378
Anagregal [Gentili] *see* 9585
Anagrelide, 616
Anagyrine, 617
Anahist [Warner-Lambert] *see* 9527
Analate [Winston] *see* 5749
Analeptin *see* 9142
Analetil *see* 9348
Analgin *see* 3390
Analud [Unifa] *see* 4039
Analux [Polfa] *see* 5851
Anamidol [Iwaki] *see* 7064
Anamycin [Chephasaar] *see* 3739
Ananain *see* 1395
Ananase [Rottapharm] *see* 1395
Anandamide, 618
Anandron [Cassenne] *see* 6629
Anapen [Celltech] *see* 3674
Anapolon [Syntex] *see* 7066
Anaprox [Syntex] *see* 6499
Anaptivan [Help] *see* 1956
Anaritide *see* 863
Anasterone *see* 7066
Anastil [Eberth] *see* 4589
Anastrozole, 619
Anatabine, 620
Anatase *see* 9628, 9631
Anatensol [BMS] *see* 4220
Anatoxin a *see* 621
Anatoxin a(s) *see* 621
Anatoxins, 621
Anatran [Tobishi] *see* 9789
Anaus [Molteni] *see* 9877
Anauxite *see* 356
Anavar [Searle] *see* 7020
Anazolene Sodium, 622
Anboxan [Shionogi] *see* 3465
Ancaron [Sanofi-Aventis] *see* 478
Ancef [GSK] *see* 1918
Anceron [Schering-Plough] *see* 1017
Ancestim *see* 8934
Anchoic Acid *see* 898
Anchorin *see* 668
Anchovyxanthin *see* 10316
Anchusa *see* 248
Anchusa Acid *see* 249
Anchusin *see* 249
Anchusin Paper *see* 248
Ancitabine, 623
Ancobon [Valeant] *see* 4156
Ancolan [Boots] *see* 5848
Ancotil [Valeant] *see* 4156
Ancrod, 624
Ancylol [Am. Cyanamid] *see* 3401
Ancymidol, 625
Ancytabine *see* 623
Ancyte [Abbott] *see* 7594
Andalin [Solvay Duphar] *see* 4154
Andalusite *see* 356
Andantol [Sumitomo] *see* 5273
Andanton [Lacer] *see* 5273
Andaxin [EGYT] *see* 5929
Andere [Toyama] *see* 1481
Andiamine [Polfa] *see* 4741
Andion [GEA] *see* 1017

Andirine *see* 9136
Andolex [3M Pharma] *see* 1125
Andractim [Besins-Iscovesco] *see* 8920
Andrade Indicator *see* 99
Andriol [Organon] *see* 9324
Andro [Forest] *see* 9324
Andro LA [Forest] *see* 9324
Androcur [Schering AG] *see* 2771
Androderm [GSK] *see* 9324
Androdiol *see* 6022
Androfurazanol *see* 4329
Andrographis, 626
Andrographolide, 627
Android [Valeant] *see* 6197
Androlone [Orma] *see* 8920
Andromedotoxin *see* 4578
Andropatch [GSK] *see* 9324
1,4-Androstadien-17β-ol-3-one *see* 1327
Androstanazole *see* 8921
Androstane, 628
(5α)-Androstane *see* 628
(5β)-Androstane *see* 3915
(5β,17β)-Androstane-17-carboxylic Acid
   see 3916
Androstane-3β,11β-diol-17-one, 629
Androstanolone *see* 8920
Androstan-3α-ol-17-one *see* 633
3β-Androstanol-17-one *see* 3664
Androstan-17β-ol-3-one *see* 8920
Androstenediol, 630
Δ⁵-Androstene-3β,17β-diol *see* 630
(3β,17β)-Androst-5-ene-3,17-diol *see* 630
Androstenedione, 631
Δ⁴-Androstenedione *see* 631
Androst-4-ene-3,17-dione *see* 631
Androst-4-ene-3,11,17-trione *see* 165
Δ¹⁶-Androsten-3-ol *see* 632
(3α,5α)-Androst-16-en-3-ol, 632
Δ⁵-Androsten-3β-ol-17-one *see* 2875
Δ⁴-Androsten-17β-ol-3-one *see* 9324
Androsterone, 633
cis-Androsterone *see* 633
Androtardyl [Schering AG] *see* 9324
Androteston-M *see* 6022
Androviron [Schering AG] *see* 5986
Anecortave Acetate, 634
Anectine [GSK] *see* 9006
Anelda [Sagrochem] *see* 1548
Anelmid [Lilly] *see* 3412
Anemone Camphor *see* 635
Anemonin, 635
Anergex [Mulford] *see* 7667
Anertan [Boehringer, Mann.] *see* 9324
Anesthetic Ether *see* 3861
Anestil *see* 7875
Anethaine [Glaxo] *see* 9333
Anethole, 636
cis-Anethole *see* 636
Anethole Dithiolthione *see* 637
Anetholesulfonic Acid Sodium Salt Poly-
   mer *see* 8796
Anethole Trithione, 637
Aneurin *see* 9447
Aneurin Disulfide *see* 9449
Anexate [Roche] *see* 4166
ANF *see* 863
AN Factor *see* 1237
Angel Dust *see* 7333
Angelica, 638
Angelic Acid, 639
Angelica Lactone, 640
Δ¹-Angelica Lactone *see* 640
Δ²-Angelica Lactone *see* 640
Angeline *see* 9136
Angeliq [Schering AG] *see* 3502

2-*p*-Anisyl-1,3-indandione *see* 661
trans-1-*p*-Anisylpropene *see* 636
3-(*p*-Anisyl)trithione *see* 637
ANIT *see* 6493
Ankilostin *see* 9335
Ankyrin *see* 8867
Annalin *see* 1708
Annatto, 667
Annexins, 668
Annotinine, 669
Annotta *see* 667
[4]Annulene *see* 2707
[8]Annulene *see* 2731
Anobesina *see* 2726
Anodynon *see* 3837
p-Anol, 670
Anone *see* 2719
Anoprolin [Azwell] *see* 271
Anorex [Crinex] *see* 3145
Anorthite *see* 1651
Anovlar [Schering AG] *see* 6786
ANP *see* 863
ANP 95-126 *see* 10071
ANP-235 *see* 5851
ANP-3624 *see* 9586
Anparton [Sanwa] *see* 2370
Anprolene *see* 3856
Anquil [Janssen-Cilag] *see* 1050
ANRL *see* 5866
ANS *see* 654
ANS-118 *see* 2223
Ansaid [Pfizer] *see* 4229
Ansar 6.6 [ISK Biosci] *see* 6025
Ansar 8100 [ISK Biosci] *see* 6025
Ansatin [Ono] *see* 4163
Ansatipine [Farmitalia] *see* 8338
Anseren [Novartis] *see* 5345
Anserine, 671
Ansiacal [Beolet] *see* 2085
Ansial [Searle] *see* 1507
Ansilan [Lek] *see* 5857
Ansimar [ABC] *see* 3486
Ansiven [Abbott] *see* 7947
Ansmin [SS Pharm.] *see* 3341
Ansolysen Bitartrate [Wyeth] *see* 7245
Ansolysen Tartrate [Wyeth] *see* 7245
Antabuse [Odyssey] *see* 3407
Antacal [Errekappa] *see* 487
Antadys [Théramex] *see* 4229
Antagonil [Novartis] *see* 6580
Antagonil [Wildlife Pharm.] *see* 10300
Antagosan [HMR] *see* 746
Antalgin [Syntex] *see* 6499
Antallin *see* 3565
Antalone [Kobayashi] *see* 5985
Antalzyme *see* 5701
Antamine [Grove] *see* 8097
Antapentan [Ayerst] *see* 7334
Antasten [Novartis] *see* 672
Antaxone [Zambon] *see* 6448
Antazoline, 672
Antébor [Biologiques] *see* 9031
Antees [Oriental] *see* 1795
Antegren [Elan] *see* 6508
Antemin [Streuli] *see* 3230
Antemovis [Vister] *see* 8601
Antepan [Henning] *see* 9750
Antepar [Glaxo Wellcome] *see* 7576
Antepsin [Wyeth] *see* 9010
Anterior Pituitary Growth Hormone *see* 8842
Anterior Pituitary Luteotropin *see* 7894
Antevas [Chugai] *see* 6578
Anthelcide EQ [Pfizer] *see* 7035
Anthelone *see* 3574

Anthelone E *see* 3651
Anthelone U *see* 3574
Antheridiol, 673
Anthesterin *see* 9193
Anthio *see* 4273
Anthion *see* 7777
Anthiphen *see* 3081
Anthisan [M & B] *see* 8097
Anthophyllite *see* 815
Anthorine *see* 853
Anthracene, 674
Anthracene Brown *see* 675
9,10-Anthracenedicarboxaldehyde Bis-[(4,5-dihydro-1*H*-imidazol-2-yl)hydrazone] *see* 1248
9,10-Anthracenedicarboxaldehyde Bis(2-imidazolin-2-ylhydrazone) *see* 1248
9,10-Anthracenedione *see* 679
9-Anthracenol *see* 678
9(10*H*)-Anthracenone *see* 682
Anthraforte [Stiefel] *see* 676
Anthragallic Acid *see* 675
Anthragallol, 675
Anthralin, 676
Anthramycin, 677
Anthranilic Acid *see* 417
Anthranilic Acid Cinnamyl Ester *see* 2304
Anthranilic Acid *p*-Menth-3-yl Ester *see* 5908
3-Anthraniloylalanine *see* 5376
Anthranol [Stiefel] *see* 676
Anthranol, 678
9,10-Anthraquinone *see* 679
Anthraquinone, 679
6,6′-(1,4-Anthraquinonylenediimino)di-*m*-toluenesulfonic Acid Disodium Salt *see* 244
Anthrarufin, 680
Anthrascalp [Stiefel] *see* 676
Anthrimide, 681
9-Anthrol *see* 678
Anthrone, 682
Anthropodesoxycholic Acid *see* 2054
Antiangor [I.S.O.M.] *see* 2243
Anti-B1 Antibody *see* 9713
Antibason *see* 6201
Antibiocin [Orion] *see* 7209
Antibiotic 67-694 *see* 8392
Antibiotic 1037 *see* 9721
Antibiotic 5879 *see* 1207
Antibiotic 6640 *see* 8688
Antibiotic 32232RP *see* 8682
Antibiotic A 246 *see* 4319
Antibiotic A-300 *see* 9814
Antibiotic A 5283 *see* 6509
Antibiotic A-23187 *see* 1642
Antibiotic A-28086 Factor A *see* 6501
Antibiotic AB 206 *see* 6277
Antibiotic Ab 651 *see* 1861
Antibiotic B-41 *see* 6271
Antibiotic B-41A1 *see* 6271
Antibiotic B-41A3 *see* 6269
Antibiotic B-41A4 *see* 6269
Antibiotic B-41D *see* 6271
Antibiotic C-19393-H₂ *see* 1861
Antibiotic C-19393-S₂ *see* 1861
Antibiotic DE-3936 *see* 5626
Antibiotic FI 1163 *see* 5651
Antibiotic G-6302 *see* 9077
Antibiotic K 178 *see* 6625
Antibiotic KA-6643-A *see* 1861
Antibiotic KA-6643-B *see* 1861
Antibiotic KW-1062 *see* 6260
Antibiotic LA-7017 *see* 7652

Antibiotic M139603 *see* 9396
Antibiotic MA 144A1 *see* 104
Antibiotic MA 144B1 *see* 104
Antibiotic MYC 8003 *see* 6311
Antibiotic S-541A *see* 6529
Antibiotic S 14750A *see* 208
Antibiotic SF-837A3 *see* 6262
Antibiotic WR 141 *see* 4570
Antibiotic X-464 *see* 6625
Antibiotic X-465A *see* 2044
Antibiotic X-537A *see* 5427
Antibiotic X-5108 *see* 873
Antibiotic X-14868A Ammonium Salt *see* 5711
Antibiotique EF 185 *see* 6539
Anticanitic Vitamin *see* 418
Anticarie *see* 4713
Antichlor *see* 8821
Antichoc Hipmag *see* 5758
Anti-chromotrichia Factor *see* 418
Antidiuretic Hormone *see* 10129
Antidrasi *see* 3087
Antifebrin *see* 49
Antifoam A [Dow] *see* 3237
"Antifungin" *see* 5722
Antigorite *see* 5751
Anti-gray-hair Factor *see* 418
Antihemophilic Factor A *see* 3962
Antihemophilic Factor B *see* 3963
Antihemophilic Globulin *see* 3962
Anti-(human β-Amyloid) Immunoglobulin G1 (Human-mouse Monoclonal Heavy Chain) Disulfide with Human-mouse Monoclonal Light Chain, Dimer *see* 951
Anti-(human Angiopoietin-2/angiopoietin-1) Peptibody (Synthetic Clone 2xCon4 (C)) *see* 9741
Anti-(human Antigen 17-1A) Immunoglobulin G2a (Mouse Monoclonal Ho-3/TP-A-01/TPBs01 Heavy Chain) Disulfide with Mouse Monoclonal Ho-3/TP-A-01/TPBs01 Light Chain Disulfide with Anti-(human CD3 (Antigen)) Immunoglobulin G2b (Rat Monoclonal 26/II/6-1.2/TPBs01 Heavychain) Disulfide with Rat Monoclonal 26/II/6-1.2/TPBs01 Light Chain *see* 1909
Anti-(human CA125(carbohydrate Antigen)) Immunoglobulin G1 (Mouse Monoclonal B43.13 γ₁-Chain) Disulfide with Mouse Monoclonal B43.13 κ-Chain, Dimer *see* 6960
Anti-(human Carcinoembryonic Antigen) Immunoglobulin G1 Fab′ Fragment (Mouse Monoclonal IMMU-4 γ₁-Chain) Disulfide with Mouse Monoclonal IMMU-4 Light Chain *see* 764
Anti-(human CD11a (Antigen)) Immunoglobulin G1 (Human-mouse Monoclonal Hu1124 γ₁-Chain) Disulfide with Human-mouse Monoclonal Hu1124 Light Chain, Dimer *see* 3567
Anti-(human CD15 (Antigen)) Immunoglobulin M (Mouse Monoclonal RB5 μ-Chain) Disulfide with Mouse Monoclonal RB5 Light Chain, Pentamer, Technetium-⁹⁹ᵐTc Salt *see* 9237
Anti-(human CD20 (Antigen)) Immunoglobulin G1 (Human-mouse Monoclonal IDEC-C2B8 γ₁-Chain) Disulfide with Human-mouse Monoclonal IDEC-C2B8 κ-Chain, Dimer *see* 8367
Anti-(human CD20 (Antigen)) Immunoglobulin G1 (Mouse Monoclonal

Name Index

Monoclonal cA2 Heavy Chain) Disulfide with Human-Mouse Monoclonal cA2 Light Chain, Dimer *see* 5014
Anti-(human Vascular Endothelial Growth Factor) Immunoglobulin G1 Fab Fragment (Human-mouse Monoclonal RhuFAB V2 γ_1-Chain) Disulfide with Human-mouse Monoclonal RhuFAB V2 Light Chain *see* 8226
Anti-(human Vascular Endothelial Growth Factor) Immunoglobulin G1 (Human-mouse Monoclonal Rhu-MAb-VEGF γ_1-Chain) Disulfide with Human-mouse Monoclonal RhuMAb-VEGF Light Chain, Dimer *see* 1194
Antihypertensive Neutral Renomedullary Lipid *see* 5866
Antihypertensive Polar Renomedullary Lipid *see* 7638
Anti-hypo *see* 7773
Anti-infective Vitamin *see* 10211
Anti-inflammatory Hormone *see* 4824
Anti-JH *see* 7839
Antikrein [Teikoku Zoki] *see* 746
Antilirium [Forest] *see* 7496
Antilon *see* 3770
Antimalarine *see* 7636
Antimonial Saffron *see* 688
Antimonic Acid (H_3SbO_4) Lead(2+) Salt (2:3) *see* 5453
Antimonic Oxide *see* 689
Antimonic Sulfide *see* 688
Antimonous Fluoride *see* 697
Antimonous Sulfate *see* 694
Antimonous Sulfide *see* 701
Antimony, 683
Antimony(III) Chloride *see* 696
Antimony Chloride ($SbCl_5$) *see* 686
Antimony Chloride Fluoride *see* 685
Antimony Chloride Oxide, 684
Antimony Chloride Solution *see* 696
Antimony Compd with Indium (1:1) *see* 4988
Antimony Dichlorofluoride *see* 685
Antimony Dichlorotrifluoride, 685
"Antimony Dimercaptosuccinate" *see* 8941
Antimony Fluoride *see* 697
Antimony Fluoride (SbF_5) *see* 687
Antimony Glance *see* 701
Antimony Hydride *see* 8940
Antimony Lead Oxide ($Sb_2Pb_3O_8$) *see* 5453
Antimony Oxide (Sb_2O_3) *see* 699
Antimony Oxide (Sb_2O_5) *see* 689
Antimony Oxychloride *see* 684
Antimony Pentachloride, 686
Antimony Pentafluoride, 687
Antimony Pentasulfide, 688
Antimony Pentoxide, 689
Antimony Perchloride *see* 686
Antimony Potassium Oxalate, 690
Antimony Potassium Tartrate, 691
Antimony Pyrocatechol Sodium Disulfonate Heptahydrate *see* 8942
Antimony Red *see* 688
"Antimony Salt" *see* 690
Antimony Selenide (Sb_2Se_3) *see* 700
Antimony Sodium Gluconate, 692
Antimony Sodium Oxide L-(+)-Tartrate *see* 693
Antimony Sodium Tartrate, 693
Antimony Sulfate, 694
Antimony Sulfide (Sb_2S_3) *see* 701
Antimony Sulfide (Sb_2S_5) *see* 688

Antimony Tribromide, 695
Antimony Trichloride, 696
Antimony Trichloride Solution *see* 696
Antimony Trifluoride, 697
Antimony Trifluorodichloride *see* 685
Antimony Triiodide, 698
Antimony Trioxide, 699
Antimony Triselenide, 700
Antimony Trisulfate *see* 694
Antimony Trisulfide, 701
Antimosan *see* 8942
Antimycin *see* 2326
Antimycin A, 702
Antin *see* 7425
Antinonnin *see* 3305
Antinosin *see* 5081
Antipar [Farmitalia] *see* 3119
Antiparkin [Viatris] *see* 8565
Antipellagra Vitamin *see* 6610
Antipernicious Anemia Principle *see* 10212
α_2-Antiplasmin *see* 7634
Antipyrine, 703
Antipyrine Amygdalate *see* 703
Antipyrine 2-Hydroxy-2-methylbutyrate *see* 703
(Antipyrinylmethylamino)methanesulfonic Acid Sodium Salt *see* 3390
Antiradon *see* 167
Anti-(respiratory Syncytial Virus Protein F) Immunoglobulin G1 (Human-Mouse Monoclonal MEDI-493 γ_1-Chain) Disulfide with Human-Mouse Monoclonal MEDI-493 κ-Chain, Dimer *see* 7089
Antireticular Cytotoxic Serum, 704
Antirex [Kyorin] *see* 3564
Antirobe *see* 2354
Antirrhinin *see* 2677
Antiscabiosum [Strathmann] *see* 1130
Antiscorbutic Vitamin *see* 819
Antisedan [Pfizer] *see* 852
Antistine [Novartis] *see* 672
Anti-Stress [Sintyal] *see* 5913
Antitanil *see* 3198
Anti-tetany Substance 10 *see* 3198
Antithermin *see* 5526
Antithrombin III *see* 4688
α_1-Antitrypsin, 705
Antituxil-Z [Ghimas] *see* 10370
Antivariz *see* 3782
Antivenin [Merck & Co.] *see* 707
Antivenin (Crotalidae) Polyvalent, 706
Antivenin (Latrodectus Mactans), 707
Antiverm *see* 7364
Antivert [Pfizer] *see* 5848
Antiweinsäure (German) *see* 9202
Antixerophthalmic Vitamin *see* 10211
Antizol [Orphan Med.] *see* 4253
Antlerite *see* 2504, 2644
Antocin *see* 856
Antodyne *see* 7401
Antoral [Recordati] *see* 9581
Antorphine *see* 6446
Antotalgin [Farmina] *see* 703
Antra [AstraZeneca] *see* 6939
Antracol [Bayer] *see* 7933
Antraderm [Protea] *see* 676
Antrancine 8 [Jandekker] *see* 1550
Antrancine 12 [Jandekker] *see* 1549
Antrapurol *see* 2814
Antrenyl [Novartis] *see* 7074
Antrin [Pharmacyclics] *see* 6368
Antrycide *see* 8168
Antrypol [AstraZeneca] *see* 9135

ANTU, 708
Antulcus *see* 7072
Anturan [Novartis] *see* 9079
Anturane [Novartis] *see* 9079
Anturano [Novartis] *see* 9079
Anturat *see* 708
(+)-ANTX-a *see* 621
ANTX-a(s) *see* 621
ANTXS *see* 621
Anvil [Syngenta] *see* 4717
Anvirzel *see* 6919
Anvitoff [Abbott] *see* 9730
Anxut [Eisai] *see* 1507
Anydrous Lanolin *see* 5409
Anymol *see* 1245
Anzatax [Hospira] *see* 7081
Anzemet [Sanofi-Aventis] *see* 3458
Anzief [Nippon Chemiphar] *see* 271
AO-128 *see* 10226
Aolan [Beiersdorf] *see* 3769
Aolept [Bayer] *see* 7279
Aotal [Merck KGaA] *see* 18
4-AP *see* 2802
AP-12009 *see* 9723
AP-23573 *see* 8336
6-APA *see* 454
Apache [FMC] *see* 1636
Apacizin *see* 474
Apacizina [Farmitalia] *see* 474
Apafant, 709
Apalcillin, 710
Apamin, 711
Aparkane [ICN] *see* 9864
Aparkazin *see* 3119
Apatef *see* 1935
Apaurin [Krka] *see* 2997
Apaziquone, 712
Apazone, 713
APC *see* 7319
APC-8015 *see* 8685
APD-356 *see* 5638
Apegmone [Lipha] *see* 9607
Aperdor [Tokyo Tanabe] *see* 9660
Apertase *see* 4797
Apetinil-Depo [Drossapharm] *see* 3818
Apex 462-5 *see* 9927
APF-1 *see* 10026
APFO *see* 7272
Aphanin *see* 3542
Aphet [Mitsui] *see* 7239
Aphidicolin, 714
Aphilan [UCB] *see* 1474
Aphistar [Dow AgroSci.] *see* 9762
Apholate, 715
Aphox [Zeneca] *see* 7609
Aphoxide *see* 9837
Aphrodine *see* 10300
Aphrodyne [Star] *see* 10300
Aphthasol [Block] *see* 486
Aphylline, 716
A1PI *see* 705
Apidra [Sanofi-Aventis] *see* 5028
Apigenin, 717
Apigenin-7-apiosylglucoside *see* 719
Apigenin-7-D-glucoside *see* 718
Apigenin-4'-methyl Ether *see* 14
Apigetrin, 718
Apihepar [Meda] *see* 6274
Apiin, 719
Apilak *see* 8410
Apinol *see* 5906
7-[(2-O-D-Apio-β-D-furanosyl-β-D-glucopyranosyl)oxy]-5-hydroxy-2-(4-hydroxyphenyl)-4H-1-benzopyran-4-one *see* 719

Arcitumomab, 764
Arcomonol Tablets see 6317
Arcosolv DPM [Arco] see 3384
Arcoxia [Merck & Co.] see 3930
Arcton [ICI] see 2141
Arcton 11 [ICI] see 9805
Arcton 12 [ICI] see 3073
Arcton 114 [ICI] see 2594
Arctuvan [Klinge] see 10082
Ardeparin, 765
Ardeytropin [Ardeypharm] see 9977
AR-DF 26 see 4478
Ardine [Antibioticos] see 574
Arduan [Organon] see 7571
Areca, 766
Arecaidine, 767
Arecaine see 767
Arecaline see 768
Arechin [Polfa] see 2165
Arecholine see 768
Arecoline, 768
Aredia [Novartis] see 7104
Arelix [HMR] see 7605
Arelon [Griffin] see 5264
Arena [Arvesta] see 2398
A-Rest [SePRO] see 625
Arestin [OroPharma] see 6283
Aretit [Hoechst] see 3315
Arfen [Lisapharma] see 4919
Arfonad [Roche] see 9876
Arformoterol see 4272
Arformoterol Tartrate see 4272
Arg see 770
Argatroban, 769
Argentic Fluoride see 8652
Argentic Oxide see 8661
Argentite see 8643, 8669
Argentoproteinum see 8666
Argentous Chlorate see 8647
Argentous Fluoride see 8653
Argentous Oxide see 8660
Argentous Sulfide see 8669
Argentum Vitellinum see 8666
D-Arg-[Hyp³,Thi⁵,D-Tic⁷,Oic⁸]bradykinin
    see 4922
Argil see 772
Argilla see 5330
Argilla Vini see 772
29-L-Argininamide-1-29-somatoliberin
    (Human Pancreatic Islet) see 8600
Arginine, 770
L-Arginine see 770
Arginine L-Aspartate see 830
Arginine Glutamate, 771
5-L-Arginine-27-glycine-33-L-isoleucine-
    34-L-serine-42-glycine-purothionin A I
    (Reduced) see 8055
[199-Arginine, 223-Histidine]prepro-α-
    glucosidase (Human) see 100
8-D-Arginine-1-(3-mercaptopropanoic
    Acid) Vasopressin see 2926
8-L-Arginineoxytocin see 10129
5-L-Arginine-6-L-threonine-18-L-serine-
    26-L-serine-27-L-threonine-42-glycine-
    purothionin A I (Reduced) see 8055
Arginine Vasopressin see 10129
8-L-Argininevasopressin see 10129
Arginine Vasotocin see 10129
8-L-Argininevasotocin see 10129
D-Arginyl-L-arginyl-L-prolyl-(4R)-4-hy-
    droxy-L-prolylglycyl-3-(2-thienyl)-L-
    alanyl-L-seryl-(3R)-1,2,3,4-tetrahydro-
    3-isoquinolinecarbonyl-(2S,3aS,7aS)-
    octahydro-1H-indole-2-carbonyl-L-ar-
    ginine see 4922

L-Arginyl-L-lysyl-L-α-aspartyl-L-valyl-L-
    tyrosine see 9558
Argipidine see 769
Argipressin see 10129
Argobyl [Specia] see 4138
Argol, 772
Argon, 773
Argun L [Merckle] see 5623
Argyrol [J & J] see 8666
Aria [FMC] see 4137
Aricept [Eisai] see 3468
Aricine, 774
Ariclaim [Lilly] see 3512
Aricyl [Winthrop] see 800
Ariflo [GSK] see 2279
Arilin [Wolff] see 6236
Arimidex [AstraZeneca] see 619
Arimoclomol, 775
Aripiprazole, 776
Aristamid [Nordmark] see 9081
Aristocort [Lederle] see 9757
Aristolochic Acid, 777
Aristolochic Acid-I see 777
Aristolochine see 777
Aristololactams see 777
Aristosol [Lederle] see 9758
Aristospan [Fujisawa] see 9759
Aritmina [Altana] see 187
Arixtra [GSK] see 4259
Arkin [Otsuka] see 10166
Arkitropin see 4768
ARL see 782
AR-L 115BS see 9115
ARL-16556 see 3406
Arlacel 20 [ICI] see 8850
Arlacel 60 [ICI] see 8850
Arlacel 80 [ICI] see 8850
Arlanto [Nichiiko] see 216
Arlef [Parke-Davis] see 4163
Arlidin [USV] see 6821
Arliflav see 1233
Arlitene [Viatris] see 6378
Armazal see 2670
Armepavine, 778
Armol see 6091
Armstrong's Acid, 779
Arnaudon's Green see 2231
Arnica, 780
Arnica Flowers see 780
Arnold's Base see 6255
Arnotta see 667
Arobon see 5612
Aroclor [Monsanto] see 7682
Aroclor 1242 see 7682
Aroclor 1254 see 7682
Aroclor 1260 see 7682
Arofuto [Mitsubishi] see 175
Arogenate see 781
Arogenic Acid, 781
Aromasin [Pharmacia & Upjohn] see 3954
Aropax [GSK] see 7148
Arotinolol, 782
Arovit [Roche] see 10211
AR-P900758XX see 6498
Arpicolin [Rosemont] see 7881
Arpimycin [Rosemont] see 3739
ARQ-197 see 9640
Arquel [Parke-Davis] see 5850
Arranon [GSK] see 6527
Arret [Janssen] see 5628
Arrivo [FMC] see 2764
Arrow Wood see 4291

Arsacol [Zambon] see 10074
Arsanilic Acid, 783
Arsanyl [Taiyo] see 4411
Arsecodile see 8730
Arsenal [BASF] see 4946
Arsenamide see 9441
Arsenenous Acid Potassium Salt (1:1)
    see 7729
Arsenenous Acid Sodium Salt (1:1) see
    8716
Arsenenous Acid Zinc Salt see 10342
Arsen (German) see 784
Arsenic, 784
Arsenic Acid, 785
Arsenic Acid Anhydride see 789
Arsenic Acid (AsH₃O₄) see 785
Arsenic Acid (H₃AsO₄) Calcium Salt
    (2:3) see 1652
Arsenic Acid (H₃AsO₄) Cobalt(2+) Salt
    (2:3) see 2418
Arsenic Acid (H₃AsO₄) Lead(2+) Salt
    (1:1) see 5454
Arsenic Acid (H₃AsO₄) Potassium Salt
    (1:1) see 7728
Arsenic Acid (H₃AsO₄) Sodium Salt
    (1:2) see 8715
Arsenical Solution see 7729
Arsenic Chloride Solution see 798
Arsenicin A, 786
Arsenic(III) Oxide see 795
Arsenic(V) Oxide see 789
Arsenic Oxide (As₂O₃) see 795
Arsenic Oxide (As₂O₅) see 789
Arsenic Pentafluoride, 787
Arsenic Pentasulfide, 788
Arsenic Pentoxide, 789
Arsenic Selenide (As₂Se₃) see 796
Arsenic Sesquioxide see 795
Arsenic Sulfide, 790
Arsenic Sulfide (As₂S₃) see 797
Arsenic Sulfide (As₂S₅) see 788
Arsenic Tribromide, 791
Arsenic Trichloride, 792
Arsenic Trifluoride, 793
Arsenic Trihydride see 799
Arsenic Triiodide, 794
Arsenic Trioxide, 795
Arsenic Triselenide, 796
Arsenic Trisulfide, 797
Arsenic Yellow see 797
Arsenik see 784
Arsenious Acid Copper(2+) Salt (1:1) see
    2616
Arsenious Acid Solution, 798
Arsenious Selenide see 796
Arsenolite see 795
Arsenosobenzene see 7048
Arsenous Acid see 795
Arsenous Acid Anhydride see 795
Arsenous Acid Lead(2+) Salt see 5455
Arsenous Oxide see 795
Arsenous Selenide see 796
Arsenous Tribromide see 791
Arsenous Trichloride see 792
Arsenous Trifluoride see 793
Arsenous Triiodide see 794
Arsenoxide see 7047
Arsevan see 6532
Arsicodile see 8730
Arsine, 799
Arsinyl [Cleary] see 6025
Arsobal [Specia] see 5883
Arsonic Acid Calcium Salt (1:1) see 1653
Arsonic Acid Copper(2+) Salt (1:1) see
    2616

Arsonoacetic Acid, 800
p-Arsonophenylurea *see* 1787
2-Arsonoylacetic Acid *see* 800
Arsphenamine, 801
Arsphenamine Methylenesulfoxylic Acid
  Sodium Salt *see* 6532
Arsthinol, 802
Arsycodile *see* 8730
Artabotrine *see* 5207
Artane [Wyeth] *see* 9864
Artate [Nippon Chemiphar] *see* 2306
Arteannuin *see* 807
Arteether, 803
Arteflene, 804
Artegodan [Artesan] *see* 7122
Artelac [Mann] *see* 4880
Artemether, 805
Artemin *see* 4410
Artemisia Oil *see* 10250
Artemisin, 806
Artemisine *see* 807
Artemisinin, 807
Artemotil *see* 803
Arteoptic [Novartis] *see* 1866
Arterenol [HMR] *see* 6784
Arteriohom [Weisskopf] *see* 2371
Arteriovinca [Farma-Lepori] *see* 10180
Arterocoline *see* 82
Artes [Sumitomo] *see* 5890
Artesunate, 808
Artesunic Acid *see* 808
Arteven [Boehringer, Ing. Italia] *see* 4688
Artevil *see* 2370
Artex [Servier] *see* 9316
Artexal [Servier] *see* 9316
Arthaxan [GSK] *see* 6428
Arthrocine [Substipharm] *see* 9112
Arthrodont [Veyron-Froment] *see* 3644
Arthropan [Purdue Frederick] *see* 2213
Arthrosetten [Hotz] *see* 2944
Arthrotabs [Duopharm] *see* 2944
Articaine *see* 1869
Artificial Cognac Essence *see* 3889
Artificial Essential Oil of Almond *see*
  1060
Artificial Ethyl Oenanthate *see* 3889
Artificial Lanthanite *see* 5415
Artificial Oil of Ants *see* 4333
Artil [Hoechst] *see* 5284
Artolon [Roter] *see* 5929
Artosin [Roche] *see* 9667
Artribid [Merck & Co.] *see* 9112
Artrizin [Leo Pharm] *see* 7390
Artrobione *see* 2213
Artrodar [Proter] *see* 2963
Artrolasi [Lenza] *see* 9131
Artrosilene [Dompé] *see* 5352
Artroxicam *see* 7619
Artume *see* 1707
ARTZ [Seikagaku] *see* 4796
Arumel [Hisamitsu] *see* 4213
Arusa *see* 147
Arvin [Knoll] *see* 624
Arvynol [Pfizer] *see* 3784
Arwin [Knoll] *see* 624
Arxxant [Lilly] *see* 8423
Arylsulfatase B, 809
Aryplase *see* 809
Arythmol [Abbott] *see* 7908
Arzene *see* 7048
Arzerra [GSK] *see* 6862
Arzoxifene, 810
AS-101 *see* 783
AS-4370 *see* 6364
5-ASA *see* 5974

Asabaine *see* 6030
Asacol [Tillotts] *see* 5974
Asacolitin [Tillotts] *see* 5974
Asafetida, 811
Asafoetida *see* 811
Asamedol [Maruko] *see* 3765
Asant *see* 811
Asarabacca Camphor *see* 813
Asarin *see* 813
Asarinin, 812
l-Asarinin *see* 812
α-Asarone *see* 813
β-Asarone *see* 813
Asarones, 813
Asarum, 814
Asarum Camphor *see* 813
Asatard [Boehringer, Ing.] *see* 841
Asbestos, 815
ASC *see* 95
ASC-66824 *see* 4288
Ascabiol [RPR] *see* 1130
Ascal [Viatris] *see* 1650
Ascaridil [Cilag] *see* 5511
Ascaridole, 816
Ascarite [Baker] *see* 815
Ascaryl [Wander] *see* 4748
Asceptichrome *see* 5934
Ascharite *see* 5722
Asclepias, 817
Asclepias Syriaca, 818
Asclera [Kreussler] *see* 7675
Ascomp [Nippon Chemiphar] *see* 216
Ascorbic Acid, 819
L-Ascorbic Acid *see* 819
L-*xylo*-Ascorbic Acid *see* 819
Ascorbicap [Valeant] *see* 819
Ascorbigen, 820
Ascorbigen A *see* 820
Ascorbigen B *see* 820
Ascorbin [Montavit] *see* 819
Ascorvit [Jenapharm] *see* 819
Ascotoxin *see* 1374
AS-E 136 *see* 3394
Asebotoxin *see* 4578
Asenapine, 821
Asendin [Lederle] *see* 573
Asendis [Wyeth] *see* 573
ASH *see* 162
Asiatic Acid *see* 822
Asiaticoside, 822
Asil *see* 4924
Asimadoline, 823
Asiprenol *see* 5263
Aska-Rid *see* 1628
Askensil [Atmos] *see* 462
ASL-279 *see* 3470
ASL-601 *see* 21
ASL-8052 *see* 3755
Asmalar *see* 5263
Asmanex [Schering-Plough] *see* 6327
Asmasal [UCB] *see* 210
Asmaten [3M Pharma] *see* 8353
Asmaterolo [Lusofarmaco] *see* 8259
Asmaven [APS] *see* 210
Asn *see* 827
L-Asnase *see* 826
Asoprisnil, 824
Asoxime Chloride, 825
Asp *see* 830
D-Asp *see* 830
ASP-47 *see* 9096
ASP-2314 *see* 7860
ASP-8825 *see* 4349
Asparagic Acid *see* 830
Asparaginase, 826

L-Asparaginase *see* 826
Asparagine, 827
L-Asparagine *see* 827
L-β-Asparagine *see* 827
L-Asparagine Amidohydrolase *see* 826
103-L-Asparagine-117-L-glutamine-296-
  L-alanine-297-L-alanine-298-L-alanine-
  299-L-alanineplasminogen Activator
  (Human Tissue-type) *see* 9286
27-L-Asparagine-nisin *see* 6650
[30-Asparagine,32-threonine,87-val-
  ine,88-asparagine,90-threonine]eryth-
  ropoietin (Human) *see* 2825
1-Asparagine-5-valine-angiotensin II *see*
  644
Asparaginic Acid *see* 830
Asparaginsäure (German) *see* 830
L-Asparaginyl-L-alanyl-L-prolyl-L-valyl-
  L-seryl-L-isoleucyl-L-prolyl-L-gluta-
  mine *see* 2836
Asparagus, 828
Asparamide *see* 827
Aspartame, 829
Aspartic Acid, 830
L-Aspartic Acid *see* 830
Aspartic Acid β-Amide *see* 827
28^B-L-Aspartic Acid-insulin (Human) *see*
  5025
3-(L-Aspartyl-D-alaninamido)-2,2,4,4-te-
  tramethylthietane *see* 240
L-Aspartyl-D-alanine-*N*-(2,2,4,4-tetra-
  methylthietan-3-yl)amide *see* 240
L-α-Aspartyl-L-α-glutamyl-L-asparagi-
  nyl-L-prolyl-L-valyl-L-valyl-L-histidyl-
  L-phenylalanyl-L-phenylalanyl-L-lysyl-
  L-asparaginyl-L-isoleucyl-L-valyl-L-
  threonyl-L-prolyl-L-arginyl-L-threo-
  nine *see* 3396
L-α-Aspartyl-L-phenylalanine 2-Methyl
  Ester *see* 829
L-α-Aspartyl-L-seryl-L-α-aspartyl-L-pro-
  lyl-L-arginine *see* 7241
(3*R*)-L-β-Aspartyl-L-seryl-L-threonyl-L-
  allothreonyl-L-alanylglycyl-(3*S*)-3-hy-
  droxy-L-prolyl-α,β-didehydrotrypto-
  phyl-β-methyl-L-tryptophyl-(3*S*)-3-hy-
  droxy-L-leucyl-3-hydroxy-L-proline
  (11 → 3)-Lactone *see* 9273
L-α-Aspartyl-*N*-(2,2,4,4-tetramethyl-3-
  thietanyl)-D-alaninamide *see* 240
AspB28-insulin (Human) *see* 5025
ASPC *see* 842
Aspenon [Mitsui] *see* 743
Aspergillic Acid, 831
Aspergillin, 832
Aspergillus Diastase *see* 9162
Asperlicin, 833
Asperlicin B *see* 833
Asperlicin C *see* 833
Asperlicin D *see* 833
Asperlicin E *see* 833
Asperule Absolute *see* 10247
Asperuloside, 834
Asphalt, 835
Asphaltum *see* 835
Asphocalcium *see* 819
Aspiculamycin *see* 4565
Aspidin, 836
Aspidinol, 837
Aspidium, 838
Aspidol [Piam] *see* 5698
Aspidosperma, 839
Aspidospermine, 840
Aspirin, 841
Aspirin Lysine Salt *see* 5698

Aspisol [Bayer] see 5698
Aspogen [Eaton] see 3201
Aspoxicillin, 842
Aspro [Roche] see 841
Assail [Nippon Soda] see 48
Assert [BASF] see 4944
Assiprenol see 5263
Assoral [Glaxo] see 8409
Assugrin see 2696
Assure [DuPont] see 8203
Asta C 4898 see 3222
Asta Z-4942 see 4937
Astacene see 843
Astacin, 843
AstaReal [Fuji] see 845
Astat see 5408
Astatine, 844
Astaxanthin, 845
rac-Astaxanthin see 845
Astelin [Wallace] see 899
Astemisan [Zdravlje] see 846
Astemizole, 846
Astenile [Recordati] see 2875
Asterin see 2677
Asteromycin see 4565
Asthmahaler [Menley & James] see 3674
Asthma Weed see 5608
Asthmolysin [Kade] see 3526
Astiban [Roche] see 8941
Astix CMPP [Agrotec] see 5855
Astmamasit [Showa Yakuhin] see 3526
Astomin [Astellas] see 3229
Astonin H [Merck KGaA] see 4161
Astragalin see 5322
Astrocar see 6627
Astroderm [Medix] see 3057
Astromicin see 4275
Astrophyllin see 3526
Asturidon [Cilag] see 1509
Asuccin see 8999
[ASU^{1,7}]-ECT see 3585
Asulam, 847
Asulox [RP Agr.] see 847
Asuntol [Bayer] see 2546
Asuro [Nippon Kayaku] see 2953
Asverin [Tanabe Seiyaku] see 9613
Asymmetrical Trimethylbenzene see
    8023
AT-7 see 4716
AT 10 see 3198
AT-17 see 3229
AT-101 see 5270
AT-327 see 9613
AT-877 see 3979
AT-2101 see 5217
AT-2266 see 3641
AT-4140 see 8862
ATA see 485
Atabrine Hydrochloride [Sanofi Win-
    throp] see 8160
Atacand [AstraZeneca] see 1742
Atacicept, 848
Atapren [Sumitomo] see 9660
Atarax [Pfizer] see 4889
Atazanavir, 849
ATC see 9602
Ateculon [Nippon Chemiphar] see 2370
Atehexal [Hexal] see 850
Atelec [Morishita] see 2278
Atem [Chiesi] see 5117
Atemi [Syngenta] see 2768
Atemorin see 5913
Atempol [Norgine] see 5913
Atenen [Tsuruhara] see 143
Atenezol [Tsuruhara] see 52

Ateno Basan [Schoenenberger] see 850
Atenol [AstraZeneca] see 850
Atenolol, 850
Atenos [UCB] see 9991
Atensil see 7535
Atensin see 5917
Aterax see 4889
Ateriosan see 2370
Ateromixol [Dalmer] see 7672
Aterosan [Aanderson] see 8088
Atgard [Boehringer, Ing.] see 3089
Athamantin, 851
Atheroitin see 2219
Atherolip see 2371
Atherolipin [Schwarz] see 2371
Atheropront [Mack, Illert.] see 2370
Athymil [Organon] see 6251
Athyromazole see 1799
ATI-BC-1 see 8211
Atipam [Dechra] see 852
Atipamezole, 852
Atisine, 853
Ativan [Wyeth] see 5636
ATL-962 see 2020
ATL-1251 see 8255
Atlacide [Chipman] see 8733
Atlas G-4829 [ICI] see 7675
Atlizumab see 9651
Atmosgen [Maruko] see 9681
Atock [Yamanouchi] see 4272
Atolant [Takeda] see 6563
Atomoxetine, 854
Atopica [Novartis] see 2750
Atorvastatin, 855
Atosiban, 856
Atosil [Bayer] see 7901
Atovaquone, 857
Atover [Recordati] see 8088
Atoxyl see 783
Atoxylic Acid see 783
ATP see 143
ATP-dependent Proteolytic Factor see
    10026
ATP-diphosphatase see 748
ATP Diphosphohydrolase (Phosphate-
    forming) see 748
Atractyl see 5782
Atractylin (C$_{30}$ Glucoside) see 858
Atractyloside, 858
Atracurium Besylate, 859
Atral see 3393
Atranex [Makhteshim-Agan] see 862
Atranoric Acid see 860
Atranorin, 860
Atrasentan, 861
Atravet [Wyeth-Ayerst] see 33
Atraxin [Daiichi] see 5929
Atrazine, 862
Atrial Natriuretic Factor see 863
Atrial Natriuretic Peptide, 863
Atriance [GSK] see 6527
Atrican [Innotech] see 9290
Atridox [Atrix] see 3488
Atrilon 5 [Winthrop] see 7925
Atrinal [Roche] see 3220
Atriopeptidase see 863
Atriopeptin see 863
Atriphos [Biochimica] see 143
Atrolactic Acid, 864
Atromid [Ayerst] see 2370
Atromidin see 2370
Atromid-S [Ayerst] see 2370
Atropamine see 726
Atropic Acid, 865
Atropine, 866

Atropine Aminoxide see 866
Atropisol [Novartis] see 866
Atropt [Sigma] see 866
Atropyltropeine see 726
Atroscine see 8543
Atrovent [Boehringer, Ing.] see 5117
Attacins, 867
Attar of Rose see 6903
Attenace [Cephalon] see 6314
Attentil [Ravizza] see 4115
Attenzo [BASF] see 6230
Attribut [Bayer CropSci.] see 7951
Aturgyl [Sanofi-Aventis] see 4016
Atysmal [Schaefer] see 3801
Aubagio [Sanofi] see 9308
Aubépine see 2555
Aucubin, 868
Aucuboside see 868
Audax [Napp] see 2213
Augmentin [SKB] see 2339
Augmerosen see 6828
Auligen see 3428
Aulin [Boehringer, Mann.] see 6633
Aulinogen [Boehringer, Ing.] see 3428
Auranofin, 869
Aurantiin see 6507
Aurantin [Pfizer] see 7433
Aurcoloid-198 [Abbott] see 4552
Aurcoscan-198 [Abbott] see 4552
Aureine see 8590
Aurelic Acid see 7652
Aureociclina see 2198
Aureocina see 2198
Aureolic Acid see 7652
Aureomycin [Lederle] see 2198
Aureotan [Byk Gulden] see 874
Aureothin, 870
Aureothricin, 871
Aureotope [BMS] see 4552
Auric Chloride see 4556
Auric Hydroxide see 4557
Auric Oxide see 4558
Auric Sulfide see 4559
Auriculin [Scios] see 863
Aurin, 872
Aurinol see 4400
Auripigment see 797
Aurochlorohydric Acid see 2120
Aurocidin see 4554
Aurodox, 873
Aurolin [UCB] see 4554
Aurone [Lennon] see 703
Aurora [FMC] see 1836
Aurorix [Roche] see 6312
Aurothioglucose, 874
Aurous Chloride see 4550
Aurous Cyanide see 4551
Aurous Stannate see 4555
Aurum Paradoxum see 9263
Australian Fever Bark see 3409
Austrawolf [Commonwealth] see 8238
Austrian Cinnabar see 5461
Auteral [Boehringer, Mann.] see 7530
Authority [FMC] see 9078
Autoprothrombin II see 3963
Autoprothrombin II-A see 7995
Autoworm [Schering-Plough] see 7034
Autumn Crocus see 2456
Auxisone [Boehringer, Ing.] see 2945
Auxit [BMS] see 1398
Auxura [Merz] see 5897
AV-290 see 882
Ava-ava see 5334
Avadex see 2973
Avadex BW [Monsanto] see 9756

Avagal *see* 6030
Avage [Allergan] *see* 9218
Avalox [Bayer] *see* 6377
Avan [Takeda] *see* 4929
Avanafil, 875
Avandia [GSK] *see* 8396
Avans-330 *see* 4547
Avantrin [Inverni] *see* 9735
Avapro [Sanofi-Synthelabo] *see* 5127
Avasimibe, 876
Avastin [Genentech] *see* 1194
Avatec [Roche] *see* 5427
Avaunt [DuPont] *see* 5012
Avazyme *see* 2264
AVE-0010 *see* 5605
AVE-1330A *see* 878
Avelox [Bayer] *see* 6377
Avenein *see* 4500
Avenge [BASF] *see* 3157
Avenin *see* 1883
Aventox [Dow AgroSci.] *see* 9829
Aventyl [Lilly] *see* 6804
Avermectin A$_{1a/b}$ *see* 877
Avermectin A$_{2a/b}$ *see* 877
Avermectin B$_1$ *see* 2
Avermectin B$_{1a/b}$ *see* 2
Avermectin B$_{2a/b}$ *see* 877
Avermectins, 877
Aviax [Pfizer] *see* 8582
Avibactam, 878
Avibon [Aventis] *see* 10211
Avicel *see* 1966
Avid [Syngenta] *see* 2
Avidin, 879
Avigilen [Riemser] *see* 7600
Avil *see* 7349
Avilamycin, 880
Avilamycin A *see* 880
Avilamycin C *see* 880
Avinar [Armour Pharm.] *see* 10057
Avinza [Ligand] *see* 6361
Aviochina [Agrimont-Vetem] *see* 9070
Aviral [Medici Domus] *see* 5019
Avirase [Lampugnani] *see* 134
Avisco Rayon [FMC] *see* 8240
Avishot [Kanebo] *see* 6440
Avita [Bertek] *see* 8288
Avitol [Lannacher] *see* 10211
Avlocardyl [AstraZeneca] *see* 7953
Avloclor [AstraZeneca] *see* 2165
Avlosulfon [Wyeth-Ayerst] *see* 2822
AVM *see* 877
Avobenzone, 881
Avodart [GSK] *see* 3518
Avogadrite *see* 7804
Avolin *see* 3279
Avolve [GSK] *see* 3518
Avomine [Sanofi-Aventis] *see* 7901
Avonex [Biogen] *see* 5035
Avoparcin, 882
Avornin *see* 4292
Avosyl *see* 5917
Avotan [Roche] *see* 882
Avoxyl *see* 5917
AVP *see* 10129
AVS *see* 6578
AVT *see* 10129
AVX-754 *see* 742
AW-14'2333 *see* 7293
AW-105-843 *see* 6439
AWP *see* 561
Axeen [Hommel] *see* 8013
Axepim [BMS] *see* 1923
Axer Alfa [Alfa] *see* 6499
Axerophthal *see* 8287

Axerophthol *see* 10211
Axert [Pharmacia & Upjohn] *see* 298
Axial [Syngenta] *see* 7564
Axid [Lilly] *see* 6747
Axil [Boehringer, Mann.] *see* 7531
Axiquel [McNeil] *see* 10098
Axiten *see* 5842
Axitinib, 883
Axonyl [Pfizer] *see* 7600
Axsain [Rodlen] *see* 1770
Axungia Porci *see* 5423
Axuris [Fournier] *see* 4430
AXV-034343 *see* 2340
AY-4166 *see* 6510
AY-5406-1 *see* 1031
AY-5710 *see* 5714
AY-6108 *see* 583
AY-6608 *see* 7224
AY-21011 *see* 7818
AY-22989 *see* 8232
AY-24031 *see* 4561
AY-24236 *see* 3920
AY-25650 *see* 9924
AY-27773 *see* 9683
AY-62014 *see* 1535
AY-62021 *see* 2383
AY-62022 *see* 5862
AY-64043 *see* 7953
Ayermicina [Ayerst] *see* 5505
Ayfactin *see* 7153
Aygestin [Barr] *see* 6786
AZ 8 [Millet Roux] *see* 4590
AZ 8 Beris [Weimer] *see* 4590
10-Azaanthracene *see* 114
5-Aza-10-arsenaanthracene Chloride *see* 7329
O-(7-Azabenzotriazol-1-yl)-*N*,*N*,*N*′*N*′-tetramethyluronium Hexafluorophosphate *see* 4653
1-(1*R*,6*R*)-9-Azabicyclo[4.2.1]non-2-en-2-ylethanone *see* 621
1-Azabicyclo[2.2.2]octane *see* 8198
1-Azabicyclo[2.2.2]octan-3-ol *see* 8199
N-1-Azabicyclo[2.2.2]oct-3-yl-6-chloro-3,4-dihydro-4-methyl-3-oxo-2*H*-1,4-benzoxazine-8-carboxamide *see* 893
5-(1-Azabicyclo[2.2.2]oct-3-yl)-10,11-dihydro-5*H*-dibenz[*b*,*f*]azepine *see* 8200
(3a*S*)-2-(3*S*)-1-Azabicyclo[2.2.2]oct-3-yl-2,3,3a,4,5,6-hexahydro-1*H*-benz[*de*]-isoquinolin-1-one *see* 7098
10-(1-Azabicyclo[2.2.2]oct-3-ylmethyl)-10*H*-phenothiazine *see* 5931
1-(3-Azabicyclo[3.3.0]oct-3-yl)-3-(*p*-tolylsulfonyl)urea *see* 4474
5-AzaC *see* 884
Azacitidine, 884
Azacort *see* 2865
Azactam [BMS] *see* 918
Azacyclohexane *see* 7580
2-(1′-Azacyclooctyl)ethylguanidine *see* 4595
Azacyclopropane *see* 3859
5-Azacytidine *see* 884
5-Aza-2′-deoxycytidine *see* 2856
Azadirachtin, 886
Azaepothilone B *see* 5297
Azafenidin, 887
9-Azafluorene *see* 1791
Azafrin, 888
8-Azaguanine, 889
7-Aza-1-hydroxybenzotriazole *see* 4848
1-Azaindene *see* 5003
Azalomycin M *see* 6625
Azameno [Wyeth] *see* 6941

4-Aza-5-(*N*-methyl-4-piperidinylidene)-10,11-dihydro-5*H*-dibenzo[*a*,*d*]cycloheptene *see* 894
6-Azamianserin *see* 6291
Azamin 4B *see* 1104
2-Azanaphthalene *see* 5268
Azanidazole, 890
Azanin [Tanabe] *see* 895
Azantac [GSK] *see* 8228
5-Azaorotic Acid *see* 7046
Azapen [Pfizer] *see* 6041
Azaperone, 891
1-Azaphenothiazine-10-carboxylic Acid 2-(2-Piperidinoethoxy)ethyl Ester *see* 7569
Azapropazone *see* 713
Azarek [Ecosol] *see* 895
Azaribine *see* 897
Azaron [Chefaro] *see* 9910
Azasan [Salix] *see* 895
Azaserine, 892
Azasetron, 893
5α,20β$_F$,22α$_F$,25β$_F$,27-Azaspirostan-3β-ol *see* 9703
Δ^5-20β$_F$,22α$_F$,25α$_F$,27-Azaspirosten-3β-ol *see* 8836
Azatadine, 894
Azathioprine, 895
6-Azathymine, 896
6-Azauracil Riboside *see* 897
6-Azauridine, 897
AZD-2281 *see* 6915
AZD-3582 *see* 6498
AZD-6140 *see* 9583
Azelaic Acid, 898
Azelastine, 899
Azelex [Allergan] *see* 898
Azelnidipine, 900
Azene [Endo] *see* 2392
1-[4-(2-Azepan-1-ylethoxy)benzyl]-2-(4-hydroxyphenyl)-3-methyl-1*H*-indol-5-ol *see* 1011
Azeptin [Eisai] *see* 899
2-Azetidinecarboxylic Acid, 901
5-[(2*R*)-2-Azetidinylmethoxy]-2-chloropyridine *see* 9227
Azidic Acid *see* 4812
Azidobenzene *see* 7385
α-Azidobenzylpenicillin *see* 902
Azidocillin, 902
3′-Azido-3′-deoxythymidine *see* 10322
6-[D-α-Azidophenylacetamido]penicillanic Acid *see* 902
(2*S*,5*R*,6*R*)-6-[[(2*R*)-Azidophenylacetyl]-amino]-3,3-dimethyl-7-oxo-4-thia-1-azabicyclo[3.2.0]heptane-2-carboxylic Acid *see* 902
Azidothymidine *see* 10322
Azilect [Teva] *see* 8234
Azilsartan, 903
Azilsartan Kamedoxomil *see* 903
Azilsartan Medoxomil *see* 903
Azilsartan Medoxomil Potassium Salt *see* 903
Azimethylene *see* 3004
Azimidobenzene *see* 1111
Azimilide, 904
Azimsulfuron, 905
Azindole *see* 1083
Azinepurine *see* 8040
Azinomycin B *see* 1877
Azinphos-ethyl *see* 906
Azinphos-methyl, 906
Aziridine *see* 3859
Aziridinylbenzoquinone *see* 2999

5-(1-Aziridinyl)-3-(hydroxymethyl)-2-
[(1*E*)-3-hydroxy-1-propen-1-yl]-1-
methyl-1*H*-indole-4,7-dione *see* 712
1-Aziridinylphosphonitrile Trimer *see* 715
Azithromycin, 907
Azitrocin *see* 907
Azlaire [Schering-Plough] *see* 7830
Azlin [Bayer] *see* 908
Azlocillin, 908
Azmacort [Aventis] *see* 9758
Azoangin [Ursapharm] *see* 2261
p-Azoaniline *see* 2979
Azobenzene, 909
Azobenzide *see* 909
Azobenzol *see* 909
4,4'-Azobisbenzenamine *see* 2979
1,1'-Azobiscarbamide *see* 912
Azobiscarbonamide *see* 912
Azobiscarboxamide *see* 912
α,α'-Azobis[chloroformamidine] *see* 2121
1,1'-Azobis[1-cyanocyclohexane] *see* 910
1,1'-Azobis(cyclohexanecarbonitrile), 910
1,1'-Azobis(cyclohexane-1-carbonitrile) *see* 910
1,1'-Azobis(cyclohexanenitrile) *see* 910
Azobiscyclohexylnitrile *see* 910
1,1'-Azobisformamide *see* 912
3,3'-Azobis(6-hydroxybenzoic Acid) *see* 6935
2,2'-Azobisisobutyronitrile, 911
1,1'-Azobis[3-methyl-2-phenylimidazo-[1,2-*a*]pyridinium] Dibromide *see* 3980
5,5'-Azobis(salicylic Acid) *see* 6935
Azochloramide *see* 2121
Azo Compd No. 4 *see* 9033
4,4'-Azodianiline *see* 2979
Azodicarbonamide, 912
Azodicarboxamide *see* 912
Azodiformic Acid Diethyl Ester *see* 3126
α,α'-Azodiisobutyronitrile *see* 911
Azodisal *see* 6935
Azodisal Sodium *see* 6935
Azodrin *see* 6337
Azohel [Ursapharm] *see* 2261
Azol *see* 458
Azole *see* 8128
Azolid [USV] *see* 7390
Azolitmin, 913
Azolmen [Menarini] *see* 1216
Azomycin, 914
Azone [Nelson] *see* 5440
Azoniaspiro(3α-benziloyloxynortropane-8,1'-pyrrolidine) Chloride *see* 9966
Azoniaspiro(3α-diphenylglycoloyloxy-nortropan-8,1'-pyrrolidine) Chloride *see* 9966
Azophenylene *see* 7331
Azopt [Alcon] *see* 1382
Azosemide, 915
Azothioprine *see* 895
Azotic Acid *see* 6663
Azovan Blue *see* 3948
Azoxodone *see* 7188
Azoxybenzene, 916
Azoxybenzide *see* 916
Azoxystrobin, 917
AZQ *see* 2999
AZT *see* 10322
Aztec [BASF] *see* 9762
Azthreonam *see* 918
Aztreonam, 918

Aztreonam Lysinate *see* 918
Aztreonam Lysine *see* 918
Azubromaron [Azupharma] *see* 1067
Azudimidine [Pharmacia] *see* 8457
Azudoxat [Azupharma] *see* 3488
Azuglucon [Azupharma] *see* 4514
Azulene, 919
Azulfidine [Pfizer] *see* 9073
Azulon [Rorer] *see* 4590
Azunol [Nippon Shinyaku] *see* 919
AzUR *see* 897
Azure I *see* 6130
Azure II *see* 6130
Azure II Eosin *see* 6130
Azure A, 920
Azure B, 921
Azure Blue II *see* 6130
Azure C, 922
Azurene [Cilag-Chemie] *see* 1452
Azurite *see* 2621
Azusalen [Ohta] *see* 919
Azusaleon [Towa Yakuhin] *see* 3673
Azutranquil [Azupharma] *see* 7025
B7-1001 *see* 9716
B8-1412 *see* 9716
B-9 *see* 2809
B72.3 *see* 8521
B72.3-Glycyl-tyrosyl-*N*-ε-diethylenetri-aminepentaacetic Acid-In-111 *see* 8521
B-360 *see* 7149
B-436 *see* 7855
B-518 *see* 2743
B-577 *see* 3921
B-663 *see* 2367
B-995 *see* 2809
B-1287 *see* 3729
B-1500 *see* 5874
B-1776 *see* 9779
B2036-PEG *see* 7178
B-2310 *see* 6359
B-2311 *see* 6359
B-11420 *see* 5106
B-15000 *see* 5100
B-16880 *see* 5097
B-19036/7 *see* 4352
B-66256 *see* 10050
B-68138 *see* 3991
BA *see* 1334
BA-59P [Chemtura] *see* 9328
Ba-168 *see* 5615
Ba-253 *see* 7042
Ba-253-BR-L *see* 7042
Ba-679 *see* 9610
Ba-679 BR *see* 9610
Ba-2758 *see* 2905
Ba-5473 *see* 7074
Ba-7205 *see* 2357
BA-7602-06 *see* 9176
Ba-10370 *see* 9032
Ba-21401 *see* 9769
Ba-29038 *see* 1327
Ba-29837 *see* 2863
Ba-30803 *see* 1089
Ba-32644 *see* 6649
Ba-34276 *see* 5813
Ba-34647 *see* 930
Ba-39089 *see* 7051
Ba-180265 *see* 206
Babbitt Metal, 923
Babesan *see* 3393
Babidium *see* 4769
Baburan *see* 3393
Bacacil [Rottapharm] *see* 924
Bacampicillin, 924

Bacampicine [Pharmacia & Upjohn] *see* 924
Bacancosin *see* 936
Bacarate [Tutag] *see* 7334
Baccara [Aventis] *see* 4235
Baccidal [Kyorin] *see* 6789
Bachelor's Buttons *see* 6819
Baciferm [Roche] *see* 929
Bacifurane [Meram] *see* 6619
Baciim [Pharma-Tek] *see* 927
Bacillin *see* 926
Bacillus Calmette-Guérin *see* 1013
Bacillus Thermoproteolyticus Neutral Proteinase *see* 9437
Bacillus Thuringiensis, 925
Bacilysin, 926
Bacimex [Alter] *see* 9121
Bacitracin, 927
Bacitracin A *see* 927
Bacitracin Methylenebis[2-hydroxyben-zoate] (Salt) *see* 928
Bacitracin Methylenedisalicylate, 928
Bacitracin Methylenedisalicylic Acid *see* 928
Bacitracin Zinc, 929
Baclofen, 930
l-Baclofen *see* 930
Bacterial Vitamin H[1] *see* 418
Bactericidal Permeability-Increasing Protein, 931
1-193-Bactericidal/permeability-increas-ing Protein [132-Alanine] (Human Re-duced) *see* 931
Bacterio-opsin *see* 932
Bacteriopsins *see* 6951
Bacteriorhodopsin, 932
Bactidan *see* 3641
Bactocill [SKB] *see* 7003
Bactoderm [Pfizer] *see* 6388
Bactofen [HMR] *see* 1114
Bactol *see* 5075
Bactox [Innotech] *see* 574
Bactramin [Roche] *see* 9050
Bactramyl [Carrion] *see* 7618
Bactrim [Roche] *see* 9050
Bactroban [GSK] *see* 6388
Baddeleyite *see* 10380
BADGE *see* 1298
Badil [Bayer] *see* 4430
Badional *see* 9075
Badische Acid, 933
BAF *see* 5037
Bagasse, 934
Bahama White Wood *see* 1746
Bahia Powder *see* 757
Baicalein, 935
BAJ2740 *see* 8884
Bajaten [Merck KGaA] *see* 4975
Bakankoside *see* 936
Bakankosin, 936
Bakash *see* 147
Baker's Ammonia *see* 505
Baking Soda *see* 8719
Baktar [Shionogi] *see* 9050
Baktonium [Bode] *see* 1061
BAL *see* 3231
BAL-4815 *see* 5149
BAL-5788 *see* 1953
BAL-8557 *see* 5149
BAL-9141 *see* 1953
Balan [Dow AgroSci.] *see* 1042
Balance [Yamanouchi] *see* 2085
Balance [Bayer CropSci.] *see* 5283
Balanophorin *see* 611
Balarsen *see* 802

Balfin [Dow AgroSci.] *see* 1042
Balminil DM [Rougier] *see* 2957
"Balm of Gilead" *see* 939
Balm of Gilead *see* 941
Balmox [GSK] *see* 6428
Balofloxacin, 937
Baloxin [Chugai] *see* 937
Balsalazide, 938
Balsam Canada, 939
Balsam Capivi *see* 2501
Balsam Gurjun, 940
Balsam Mecca, 941
Balsam of Fir *see* 939
Balsam of Gilead *see* 941
Balsam Peru, 942
Balsams Copaiba *see* 2501
Balsam Tolu, 943
Balsam Traumatic, 944
Balsam Tree *see* 5823
Balsán-Katél *see* 941
Baltic Amber *see* 374
Bambec [AstraZeneca] *see* 946
Bambermycin *see* 945
Bambermycins, 945
Bamboo Curare *see* 2661
Bambuterol, 946
Bamethan, 947
Bamifylline, 948
Bamiphylline *see* 948
Bamipine, 949
Banabins *see* 7632
Banabin-Sintyal *see* 2106
Banamine [Schering-Plough] *see* 4178
Banana Oil *see* 5156
Banewort *see* 1026
Banflex [Forest] *see* 6975
Banicol *see* 2016
Banisterine *see* 4649
Banistyl [M & B] *see* 4258
Ban-kakari *see* 7664
Banminth [Pfizer] *see* 8066
Bannal *see* 8540
Banner [Syngenta] *see* 7932
Banocide [Wellcome] *see* 3128
Banol [Bayer CropSci.] *see* 7912
Bantenol [Janssen] *see* 5837
Bantex *see* 5935
Banthine Bromide [Searle] *see* 6030
Banthionine [Banyu] *see* 6047
Bantu *see* 708
Banvel [BASF] *see* 3050
Banzel [Eisai] *see* 8426
Baobab, 950
Bapineuzumab, 951
BAPTA, 952
Baptigenin, 953
Baptisia, 954
Baptitoxine *see* 2786
BAQD-10 *see* 2914
Baraclude [BMS] *see* 3649
Barakshin *see* 4924
Baratol [Wyeth] *see* 5010
Barazan [Teofarma] *see* 6789
Barbaloin *see* 303
Barbamil *see* 567
Barbasco, 955
Barberry Bark, 956
Barbexaclone *see* 7972
Barbiphenyl [Lääkefarmos] *see* 7352
Barbipil [Conal] *see* 7352
Barbital, 957
Barbital Sodium *see* 957
Barbitone *see* 957
Barbitone Sodium *see* 957
Barbituric Acid, 958

Barbonin [Knoll] *see* 3783
Barbosec [Rowell] *see* 8557
Barbourin *see* 3698
Bardac 2250/2280 [Lonza] *see* 3110
Bardana *see* 5419
Bardoxolone, 959
Bareon [Hokuriku] *see* 5619
Bariard [Bayer CropSci.] *see* 9443
Barite *see* 990
Baritop [Sakai] *see* 990
Barium, 960
Barium Acetate, 961
Barium Benzenesulfonate, 962
Barium Benzoate *see* 1093
Barium Bromate, 963
Barium Bromide, 964
Barium Carbonate, 965
Barium Chlorate, 966
Barium Chloride, 967
Barium Chloride (BaCl$_2$) *see* 967
Barium Chromate(VI), 968
Barium Citrate *see* 2325
Barium Cyanide, 969
Barium Cyanoplatinate(II) *see* 987
Barium Dihydroxide *see* 973
Barium Dioxide *see* 985
Barium Diuranate *see* 997
Barium Ethyl Sulfate *see* 3904
Barium Fluoride, 970
Barium Fluosilicate *see* 972
Barium Formate, 971
Barium Hexafluorosilicate, 972
Barium Hydroxide, 973
Barium Hydroxide (Ba(OH)$_2$) *see* 973
Barium Hypophosphite, 974
Barium Hyposulfite *see* 995
Barium Iodate, 975
Barium Iodide, 976
Barium Lactate *see* 5384
Barium Manganate(VI), 977
Barium Manganate(VII) *see* 984
Barium Mercuric Iodide, 978
Barium Metatitanate *see* 996
Barium Methyl Sulfate *see* 6195
Barium Monoxide *see* 982
Barium Nitrate, 979
Barium Nitrite, 980
Barium Oleate *see* 6921
Barium Oxalate, 981
Barium Oxide, 982
Barium Perchlorate, 983
Barium Permanganate, 984
Barium Peroxide, 985
Barium *p*-Phenolsulfonate *see* 7359
Barium Phosphate, Dibasic, 986
Barium Platinocyanide *see* 987
Barium Platinous Cyanide, 987
Barium Propionate *see* 7939
Barium Protoxide *see* 982
Barium Rhodanide *see* 994
Barium Selenide, 988
Barium Silicide, 989
Barium Silicofluoride *see* 972
Barium Sulfate, 990
Barium Sulfide, 991
Barium Sulfide (BaS), Black *see* 992
Barium Sulfide, Black, 992
Barium Sulfite, 993
Barium Sulfocyanate *see* 994
Barium Sulfocyanide *see* 994
Barium Superoxide *see* 985
Barium Taurocholate *see* 9212
Barium Tetracyanoplatinate(II) *see* 987
Barium Tetraiodomercurate(II) *see* 978
Barium Thiocyanate, 994

Barium Thiosulfate, 995
Barium Titanate(IV), 996
Barium Titanium Oxide (BaTiO$_3$) *see* 996
Barium Uranate(VI) *see* 997
Barium Uranium Oxide, 997
Barium Uranium Oxide (BaU$_2$O$_7$) *see* 997
Barizin [Lek] *see* 6580
Barluenga's Reagent *see* 1303
Barnetil [Delagrange] *see* 9123
Barnidipine, 998
Barnotil [Vita] *see* 9123
Baron *see* 2799
Baroque [Yashima] *see* 3933
Baros [Horii] *see* 3237
Baros Camphor *see* 1340
Barosma Camphor *see* 3326
Barosmin *see* 3325
Barosperse [Mallinckrodt] *see* 990
Barquinol *see* 5075
Barrelene, 999
Barricade [BASF] *see* 2764
Barricade [Syngenta] *see* 7883
Barthrin, 1000
Baryta Yellow *see* 968
Barytes *see* 990
BAS-083 *see* 5924
BAS-238F *see* 3454
BAS-320I *see* 5992
BAS-352F *see* 10181
BAS-392H *see* 4152
BAS-480-F *see* 3685
BAS-490 F *see* 5372
BAS-490-02F *see* 5372
BAS-500F *see* 8064
BAS-505 *see* 3292
BAS-510F *see* 1354
BAS-518 *see* 8182
BAS-518H *see* 8182
BAS-560F *see* 6230
BAS-560-02F *see* 6230
BAS-615H *see* 2294
BAS-620H *see* 9295
BAS-635 *see* 9943
BAS-656H *see* 3236
BAS-662 *see* 3164
BAS-800H *see* 8451
BAS-3460 *see* 1792
BAS-9052 *see* 8612
BAS-42100F *see* 4021
BAS-67054 *see* 1792
BAS-85559X *see* 5924
Basagran [BASF] *see* 1054
Basalin [BASF] *see* 4152
Basaljel [Wyeth] *see* 1001
Basamid [Certis] *see* 2837
Basanite [BASF] *see* 3315
Basathrin [BASF] *see* 2764
Basecil *see* 6201
Basedol *see* 475
Basen [Takeda] *see* 10226
Basergin *see* 3712
Basethyrin *see* 6201
Basfapon *see* 2799
Basforin [BASF] *see* 9859
Basicaina [Galenica] *see* 5535
Basic Aluminum Acetate *see* 359
Basic Aluminum Carbonate Gel, 1001
Basic Aluminum Chloride *see* 339
Basic Antimony Chloride *see* 684
Basic Bismuth Bromide *see* 1263
Basic Bismuth Chloride *see* 1274
Basic Bismuth Iodide *see* 1270
Basic Bismuth Salicylate *see* 1285

Basic Chromic Sulfates *see* 2234
Basic Cupric Acetate *see* 2614
Basic Cupric Chromates *see* 2624
Basic Cupric Sulfate *see* 2644
Basic Dextran *see* 2938
Basic Fibroblast Growth Factor *see* 4101
Basic Iron(III) Acetate *see* 4043
Basic Lead Carbonate, 1002
Basic Lead Chromate *see* 5461
Basic Nickel Carbonate *see* 6588
Basi-Cop *see* 2644
Basic Phenylmercuric Nitrate *see* 7413
Basic Zirconium Chloride *see* 10384
Basiliximab, 1003
Basilixerit [Bastian-Werk] *see* 4514
Basodexan [Hermal] *see* 10052
Basofortina [Novartis] *see* 6142
B28-asp-insulin *see* 5025
Bassado [Pharmacia] *see* 3488
Basswood, 1004
Basta [Bayer CropSci.] *see* 7451
Bastard Saffron *see* 1868
Bastiverit [Bastian-Werk] *see* 4514
Bastnaesite *see* 6292
Basudin [Zagro] *see* 2998
BAT *see* 9961
Bathocuproine, 1005
Batimastat, 1006
Batixim [So.Se.] *see* 5535
BatPA *see* 3506
Batrachotoxin, 1007
Batrachotoxinin A *see* 1007
Batrachotoxinin A 20-(2,4-Dimethyl-1*H*-
   pyrrole-3-carboxylate) *see* 1007
Batrafen [HMR] *see* 2269
Batroxobin, 1008
Batyl Alcohol, 1009
Baum's Acid *see* 6475
Bauxite *see* 350
Bavenite *see* 1651
Bavistin [BASF] *see* 1792
BAX-1400Z *see* 3235
BAX-1526 *see* 2263
BAX-2739Z *see* 948
Baxacor [Mack, Illert.] *see* 3765
Baxan [BMS] *see* 1915
Baxo [Toyama] *see* 7619
Bay-12-8039 *see* 6377
BAY-35-3377 *see* 4112
Bay-38-9456 *see* 10124
Bay-43-9006 *see* 8846
Bay-44-4400 *see* 3617
Bay-56-6854 *see* 3976
BAY-59-7939 *see* 8368
BAY 63-2521 *see* 8358
Bay 1470 *see* 10277
Bay 2502 *see* 6622
Bay 5621 *see* 7478
Bay 9002 *see* 6438
Bay 9010 *see* 7949
Bay 19639 *see* 3408
Bay 25141 *see* 4027
Bay 34727 *see* 2687
Bay 39007 *see* 7949
Bay 44646 *see* 428
Bay 68138 *see* 3991
Bay 70143 *see* 1803
Bay 77488 *see* 7478
Bay 92114 *see* 5218
Bay 94337 *see* 6233
Bay a 1040 *see* 6613
Bay b 5097 *see* 2401
Bay e 5009 *see* 6662
Bay e 6905 *see* 908
Bay e 9736 *see* 6636
Bay f 1353 *see* 6250

Bay g 5421 *see* 19
Bay h 4502 *see* 1216
Bay h 5757 *see* 3981
Bay k 5552 *see* 6651
Bay m 1099 *see* 6267
Bay n 5595 *see* 2906
Bay o 6893 *see* 8359
Bay o 9867 *see* 2313
Bay q 3939 *see* 2313
Bay q 7821 *see* 5125
Bay u 3405 *see* 8218
Bay Va 1470 *see* 10277
Bay Va 9391 *see* 6916
Bay Vh 5757 *see* 3981
Bay Vi 9142 *see* 9685
Bay Vi 9143 *see* 7711
Bay Vp 2674 *see* 3646
Bay w 6628 *see* 1992
Bay x 3702 *see* 8257
Bay-y-7432 *see* 8207
BAY DRW 1139 *see* 5995
BAY FCR 1272 *see* 2754
BAY FOE-5043 *see* 4162
BAY HWG 1608 *see* 9229
BAY KWG 0519 *see* 9755
BAY KWG 0599 *see* 1306
BAY MEB 6447 *see* 9754
BAY MKH 3586 *see* 390
BAY MKH 6561 *see* 7951
BAY MKH 6562 *see* 4151
BAY NTN 33893 *see* 4951
BAY SAS 9244 *see* 9768
BAY SIR 8514 *see* 9844
BAY-YRC-2388 *see* 4034
BAY YRC 2894 *see* 9443
BAY-314666 *see* 390
Bayberry Bark, 1010
Baycain [Bayer] *see* 9700
Baycaron [Bayer] *see* 5874
Baycid [Bayer CropSci.] *see* 4029
Baycidal [Bayer] *see* 9844
Baycillin [Bayer] *see* 7931
Baycipen [Bayer] *see* 6250
Baycol [Bayer] *see* 1992
Baycor [Bayer] *see* 1306
Baycovin *see* 3146
Baycox [Bayer] *see* 9685
Bayer 21/116 *see* 6123
Bayer 21/199 *see* 2546
Bayer 205 *see* 9135
Bayer 1219 *see* 9873
Bayer 1362 *see* 1519
Bayer 1420 *see* 7917
Bayer 2353 *see* 6604
Bayer 3231 *see* 9764
Bayer 5312 *see* 3791
Bayer 5630 *see* 6236
Bayer 5633 *see* 6011
Bayer 8169 *see* 2892
Bayer 9015 *see* 6603
Bayer 9051 *see* 5511
Bayer 16259 *see* 906
Bayer 17147 *see* 906
Bayer 25820 *see* 6438
Bayer 29493 *see* 4029
Bayer 36205 *see* 7077
Bayer 37344 *see* 6044
Bayer 41831 *see* 4007
Bayer 46131 *see* 7933
Bayer 47531 *see* 3053
Bayer 68138 *see* 3991
Bayer 71628 *see* 6019
Bayer 74283 *see* 6011
Bayer 78418 *see* 3558
Bayer A 128 *see* 746

Bayer A 173 *see* 7879
Bayer E 39 Soluble *see* 1249
Bayer E 393 *see* 9096
Bayer L 1359 *see* 9788
Bayer S 5660 *see* 4007
Bayerite *see* 350
Bayer's Acid *see* 2578
Bayfidan [Bayer] *see* 9755
Baygon [Bayer] *see* 7949
Bay Laurel *see* 5437
Bay Leaf Oil *see* 5437
Bayleton [Bayer CropSci.] *see* 9754
Bayluscid [Bayer] *see* 6604
Baymicin [Bayer] *see* 8688
Baymix [Bayer] *see* 2546
Baymycard [Bayer] *see* 6651
Baynas [Bayer] *see* 8218
Bay-o-nox [Bayer] *see* 6916
Bayotensin [Bayer] *see* 6662
Baypen [Bayer] *see* 6250
Baypress [Bayer] *see* 6662
Baypure CX 100 [Bayer] *see* 4960
Bayrena [Bayer] *see* 9046
Bayrepel [Bayer] *see* 7506
Bayrogel [Bayropharm] *see* 3921
Baysiston [Bayer CropSci.] *see* 3408
Baytan [Bayer] *see* 9755
Baytex [Bayer CropSci.] *see* 4029
Baythion [Bayer] *see* 7478
Baythroid [Bayer CropSci.] *see* 2754
Bayticol [Bayer] *see* 4171
Baytril [Bayer] *see* 3646
Bayvarol [Provet] *see* 4171
Bazedoxifene, 1011
Bazoton [Abbott] *see* 6567
BB-94 *see* 1006
BB-882 *see* 5527
BB-2516 *see* 5818
B.B.C. *see* 1418
BB-K8 [BMS] *see* 400
9-BBN-nopol Benzyl Ether Adduct *see*
   6513
BBOT, 1012
BBR-2778 *see* 7628
BBS-1067 *see* 1943
BC-48 *see* 2889
BC-51 *see* 3405
BC-105 *see* 7629
BC-681 *see* 3566
BC-757 *see* 7650
levo-BC-2627 *see* 1532
Bc Conjugate *see* 8044
B-cell Activating Factor *see* 5037
B-cell Growth Factor *see* 5040
B-cell Growth Factor II *see* 5041
B-cell Stimulatory Factor-1 *see* 5040
B-cell Stimulatory Factor-2 *see* 5042
BCG, 1013
BCGF-1 *see* 5040
BCGFII *see* 5041
(−)-BCH-189 *see* 5402
(−)-BCH-204 *see* 9968
BCH-4556 *see* 9968
BCH-10618 *see* 742
BChE *see* 2214
BCM *see* 1792
BCME *see* 3079
BCNU *see* 1845
BCP *see* 1005, 1475
BCX-1812 *see* 7264
BCX-2600 *see* 8948
BD-40A *see* 4272
BDCM *see* 1425
BDE-47 *see* 7681
BDE-209 *see* 2848

BDF-5895 see 6379
BDFP see 950
BDH-312 see 5917
BDMPEA see 1910
BDMS see 1427
B-DNA see 2908
BDNF see 6572
BDO see 1518
Beacillin [EGYT] see 7205
Beacon [Syngenta] see 7866
Beamette [Dawbarn] see 7701
Bearberry see 10082
Bearberry Bark see 1880
Bear's Weed see 3727
Bearwood see 1880
Bebate see 1182
Bebeerine, 1014
d-Bebeerine see 1014
l-Bebeerine see 2664
Bebeeru Bark, 1015
Beben [Pfizer] see 1182
Becaplermin see 7639
Becenun [BMS] see 1845
Beclacin [Kaigai] see 1017
Beclamide, 1016
Becloforte [GSK] see 1017
Beclomet [Orion] see 1017
Beclomethasone, 1017
Beclorhinol [Lindopharm] see 1017
Beclovent [GSK] see 1017
Becodisks [GSK] see 1017
Beconase [GSK] see 1017
Beconasol [GSK] see 1017
Becort [Rachelle] see 1182
Becotide [GSK] see 1017
Bedaquiline, 1018
Bedermin [Damor] see 1182
Bedoz see 10212
Bedranol [Lagap] see 7953
Bedriol [Andromaco] see 1216
Beechwood Creosote see 2561
Bee Glue see 7948
Beeswax, 1019
Beet Sugar see 9012
Befaprost see 1154
Befizal [Obverval] see 1199
Beflavin [Roche] see 8325
Befloxatone, 1020
Befunolol, 1021
Behenic Acid, 1022
Behenic Alcohol see 3444
N⁴-Behenoyl-1-β-D-arabinofuranosylcy-
   tosine see 3640
Behenoylcytosine Arabinoside see 3640
Behenyl Alcohol see 3444
Behepan [Pfizer] see 10212
Behyd [Kyorin] see 1139
Bei-1293 see 10276
Bekanamycin see 5329
Belamarine see 5687
Belatacept, 1023
Belcomycine [Sanofi-Aventis] see 2463
Beleaf [FMC] see 4137
Belfacillin see 6041
Belimumab, 1024
Belinostat, 1025
Belivon [Organon] see 8361
Belladonna, 1026
Bellaradine see 2667
Bellasthman see 5263
Belleau's Reagent, 1027
Bell Pepper see 1772
Belmark [Shell] see 4037
Beloc [AstraZeneca] see 6228
Belotecan, 1028

Belseren [BMS] see 2392
Belt see 2083
Belt [Bayer CropSci.] see 4150
Bemarside see 9071
Bemegride, 1029
Bemotrizinol, 1030
BEMT see 1030
Benactyzine, 1031
Benadon [Roche] see 8095
Benadryl [Pfizer] see 3339
Benalaxyl, 1032
Benazepril, 1033
Benazeprilat see 1033
Bence-Jones Proteins, 1034
Bencyclane, 1035
Bendalina [Angelini] see 1037
Bendaline see 1037
Bendamustine, 1036
Bendazac, 1037
Bendazac Lysine see 1037
Bendazolic Acid see 1037
Bendectin [Duchesnay] see 3489
Bendiocarb, 1038
Bendioxide see 1054
Bendogen [Lagap] see 1188
Bendopa [ICN] see 5516
Bendrofluazide see 1039
Bendroflumethiazide, 1039
Benecel M [Hercules] see 6111
Benecel MP [Hercules] see 4880
Benedorm see 8108
Benefex [Makhteshim-Agan] see 1042
Benefin see 1042
Benemid [Merck & Co.] see 7872
Benerva [Roche] see 9447
Benethamine Penicillin G see 7204
Benexate-CD see 1040
Benexate Hydrochloride, 1040
dl-Benflumelol see 5657
Benflumetol see 5657
Benfluorex, 1041
Benfluralin, 1042
Benfluramate see 1041
Benfofen [Sanofi-Synthelabo] see 3091
Benfotiamine, 1043
Benfuracarb, 1044
Benfuran [Kaken] see 1021
Bengal Isinglass see 176
Benicar [Sankyo] see 6933
Benidipine, 1045
Benirol [Linobion] see 1061
Benisone [Warner-Chilcott] see 1182
Benitoite see 9626
Benlate [DuPont] see 1046
Benlysta [Human Genome Sciences] see
   1024
Benne Oil see 8608
Benocten [Medinova] see 3339
Benodaine [Merck & Co.] see 7591
Benomyl, 1046
Benoquin see 6334
Benoral [Sanofi Winthrop] see 1047
Benortan [Sanofi Winthrop] see 1047
Benorylate, 1047
Benoxacor, 1048
Benoxil [Santen] see 1049
Benoxinate, 1049
Benoxyl [Stiefel] see 1119
Benozil [Kyowa] see 4228
Benperidol, 1050
Benproperine, 1051
Benserazide, 1052
Bensulfuron-methyl, 1053
Bensylyt see 7368
Bentazon, 1054

Bentazone see 1054
Bentelan [GSK] see 1182
Benthiavalicarb-isopropyl, 1055
Bentiromide, 1056
Bentone 34 [Rheox] see 1058
Bentonite, 1057
Bentoquatam, 1058
Bentos [Kaken] see 1021
Bentrofene [Cilag-Chemie] see 24
Bentyl [Aventis] see 3108
Bentylol [Aventis] see 3108
Benur [Pfizer] see 3482
Benuride [Bengue] see 7344
Ben-u-ron [Novartis] see 47
Benylin [J & J] see 2957
Benzac [Galderma] see 1119
Benzagel [RPR] see 1119
Benzaidin see 1188
Benzaknen [Galderma] see 1119
Benzalacetone see 1140
Benzalacetophenone see 2036
Benzalaniline see 1141
Benzal Chloride, 1059
Benzaldehyde, 1060
Benzaldehyde Cyanohydrin see 5783
Benzaldehyde FFC see 1060
Benzaldehyde Green see 5763
Benzaldehyde N-Phenylimine see 1141
Benzal Green see 5763
Benzalin [Shionogi] see 6661
Benzalkonium Chloride, 1061
Benzamide, 1062
Benzamidoacetic Acid see 4753
DL-α-Benzamido-p-[2-(diethylamino)eth-
   oxy]-N,N-dipropylhydrocinnamamide
   see 9622
DL-4-Benzamido-N,N-dipropylglutar-
   amic Acid see 7891
(S)-p-(α-Benzamido-p-hydroxyhydrocin-
   namamido)benzoic Acid see 1056
3-[2-(4-Benzamidopiperidino)ethyl]in-
   dole see 5010
β-Benzamidopropionic Acid see 1183
4-Benzamidosalicylic Acid see 1118
Benzamine 3937
Benzamine Blue see 9972
Benzamizole see 5281
Benzamon see 4341
Benzanilide, 1063
Benz[a]anthracene see 1064
1,2-Benzanthracene, 1064
2,3-Benzanthracene see 6454
7H-Benz[de]anthracen-7-one see 1065
Benzanthrene see 1064
Benzanthrone, 1065
Benzathine, 1066
Benzathine Benzylpenicillin see 7209
Benzathine Penicillin G see 7205
Benzathine Penicillin V see 7209
Benzazimidol see 4851
1-Benzazine see 8185
2-Benzazine see 5268
1-Benzazole see 5003
Benzazoline see 9666
Benzbromarone, 1067
Benzchlorpropamide see 1016
3,4-Benzchrysene see 7508
Benzcurine Iodide see 4373
Benzcyclan see 1035
Benzedrex [SK & F] see 7972
Benzedrine [SKB] see 579
Benzelmin [Fort Dodge] see 7034
Benzenamine see 652
Benzene, 1068
Benzeneacetaldehyde see 7377

Benzodiapin [Lisapharma] see 2085
7,8-Benzo-1,3-diazaspiro[4.5]decane-2,4-dione see 9382
1,2-Benzodiazine see 2307
1,3-Benzodiazine see 8169
1,4-Benzodiazine see 8195
2,3-Benzodiazine see 7482
1,3-Benzodiazole see 1083
1,3,2-Benzodioxaborolane see 1903
1,3,2-Benzodioxaborole see 1903
Benzodioxane see 7591
2-[2-(1,4-Benzodioxanyl)]-2-imidazoline see 4928
(1,4-Benzodioxan-2-ylmethyl)guanidine see 4603
1-(1,4-Benzodioxan-5-yl)piperazine see 3608
1,3-Benzodioxole-5-carboxaldehyde see 7588
1,3-Benzodioxole-5-carboxylic Acid see 7590
5-[(1S,3aR,4R,6aR)-4-(1,3-Benzodioxolol-5-yloxy)tetrahydro-1H,3H-furo[3,4-c]-furan-1-yl]-1,3-benzodioxole see 8610
(2R,3R,4S)-4-(1,3-Benzodioxol-5-yl)-1-[2-(dibutylamino)-2-oxoethyl]-2-(4-meth-oxyphenyl)-3-pyrrolidinecarboxylic Acid see 861
9-(1,3-Benzodioxol-5-yl)-6,7-dimethoxy-naphtho[2,3-c]furan-1(3H)-one see 5320
(1E)-1-(1,3-Benzodioxol-5-yl)-4,4-dimeth-yl-1-penten-3-ol see 8948
(6R)-6-[(1E)-2-(1,3-Benzodioxol-5-yl)eth-enyl]-5,6-dihydro-4-methoxy-2H-py-ran-2-one see 6214
(6R,12aR)-6-(1,3-Benzodioxol-5-yl)-2,3,-6,7,12,12a-hexahydro-2-methylpyra-zino[1',2':1,6]pyrido[3,4-b]indole-1,4-dione see 9157
3-(1,3-Benzodioxol-5-yl)-7-hydroxy-4H-1-benzopyran-4-one see 8019
1-(1,3-Benzodioxol-5-ylmethyl)-4-[(4-chlorophenoxy)acetyl]piperazine see 4115
4-[2-[[2-(1,3-Benzodioxol-5-yl)-1-methyl-ethyl]amino]-1-hydroxyethyl]-1,2-ben-zenediol see 8004
10-[[4-(1,3-Benzodioxol-5-ylmethyl)-1-piperazinyl]acetyl]-10H-phenothiazine see 4013
1-[4-(1,3-Benzodioxol-5-ylmethyl)-1-piperazinyl]-2-(4-chlorophenoxy)etha-none see 4115
2-[4-(1,3-Benzodioxol-5-ylmethyl)-1-piperazinyl]-1-(10H-phenothiazin-10-yl)ethanone see 4013
2-[4-(1,3-Benzodioxol-5-ylmethyl)-1-piperazinyl]pyrimidine see 7607
1-(1,3-Benzodioxol-5-ylmethyl)-4-(3,7,11-trimethyl-2,6,10-dodecatrien-1-yl)-piperazine see 7532
1-[(2E,4E)-5-(1,3-Benzodioxol-5-yl)-1-oxo-2,4-pentadienyl]piperidine see 7584
(Z,Z)-1-[5-(1,3-Benzodioxol-5-yl)-1-oxo-2,4-pentadienyl]piperidine see 2048
(2R)-2-[2-[3-[[2-(1,3-Benzodioxol-5-yl-oxy)ethyl]methylamino]propoxy]-5-methoxyphenyl]-4-methyl-2H-1,4-ben-zothiazin-3(4H)-one see 8585
(3S,4R)-3-[(1,3-Benzodioxol-5-yloxy)-methyl]-4-(4-fluorophenyl)piperidine see 7148
5-(1,3-Benzodioxol-5-yl)-2,4-pentadienoic Acid see 7579

(2E,4E)-5-(1,3-Benzodioxol-5-yl)-1-(1-pi-peridinyl)-2,4-pentadien-1-one see 7584
(2Z,4Z)-5-(1,3-Benzodioxol-5-yl)-1-(1-pi-peridinyl)-2,4-pentadien-1-one see 2048
(7S,8R,9R)-9-(1,3-Benzodioxol-5-yl)-6,7,-8,9-tetrahydro-7,8-dimethylnaphtho-[1,2-d]-1,3-dioxole see 6998
9-(1,3-Benzodioxol-5-yl)-4,6,7-trimeth-oxynaphtho[2,3-c]furan-1(3H)-one see 5320
Benzofuran, 1090
2-Benzofurancarboxylic Acid see 2549
4-(4-Benzofurazanyl)-1,4-dihydro-2,6-di-methyl-3,5-pyridinedicarboxylic Acid Methyl 1-Methylethyl Ester see 5287
3-Benzofuro[3,2-c][1]benzoxepin-6(12H)-ylidene-N,N-dimethyl-1-propanamine see 7033
Benzoglyoxaline see 1083
Benzoguanamine, 1091
Benzohydrol, 1092
Benzoic Acid, 1093
Benzoic Acid, Aluminum Salt (3:1) see 325
Benzoic Acid 2-[1-[(2S,4S)-4-[(3-Amino-2,3,6-trideoxy-α-L-lyxo-hexopyrano-syl)oxy]-1,2,3,4,6,11-hexahydro-2,5,12-trihydroxy-7-methoxy-6,11-dioxo-2-napthacenyl]ethylidene]hydrazide see 10395
Benzoic Acid Ammonium Salt (1:1) see 493
Benzoic Acid 1,1'-Anhydride see 1094
Benzoic Acid Benzyl Ester see 1130
Benzoic Acid Butyl Ester see 1554
Benzoic Acid Ethyl Ester see 3821
Benzoic Acid Hydrazide 3-Hydrazone with Daunorubicin see 10395
Benzoic Acid Lithium Salt (1:1) see 5579
Benzoic Acid Magnesium Salt (2:1) see 5721
Benzoic Acid Mercury(2+) Salt (2:1) see 5941
Benzoic Acid Methyl Ester see 6098
Benzoic Acid Phenyl Ester see 7388
Benzoic Acid Phenylmethyl Ester see 1130
Benzoic Acid Sodium Salt (1:1) see 8718
Benzoic Acid 4-Sulfamide see 1876
Benzoic Aldehyde see 1060
Benzoic Anhydride, 1094
o-Benzoic N-Chlorosulphinide see 2166
Benzoic Sulfimide see 8445
Benzoin, 1095
Benzoinated Lard see 5423
Benzoin Oxime, 1096
Benzol see 1068
Benzoline see 5542
6,7-Benzomorphan, 1097
Benzo[b]naphthacene see 7216
Benzonaphthol see 6490
Benzonatate, 1098
Benzonitrile, 1099
Benzononatine see 1098
Benzoparadiazine see 8195
Benzo[c]phenanthrene see 1120
Benzo[def]phenanthrene see 8074
Benzophenone, 1100
Benzophenone-1 see 1109
Benzophenone-3 see 7053
Benzophenone-4 see 9113
Benzophenone-6, 1101
Benzophenone-8 see 3332

Benzophenone-10 see 6246
Benzophenone-12 see 6831
Benzophenone Imine, 1102
Benzopinacol, 1103
Benzopurpurine 4B, 1104
1,2-Benzopyran, 1105
2H-1-Benzopyran see 1105
α-5:6-Benzopyran see 1105
2H-1-Benzopyran-2-one see 2550
2-(5H-[1]Benzopyrano[2,3-b]pyridin-7-yl)propionic Acid see 7831
Benzo[a]pyrazine see 8195
Benzopyrazole see 4976
Benzo[a]pyrene, 1106
Benzo[e]pyrene, 1107
Benzo[a]pyrene-7,8-diol 9,10-Epoxide see 1106
Benzo[c]pyridazine see 2307
Benzo[d]pyridazine see 7482
Benzo[b]pyridine see 8185
Benzo[c]pyridine see 5268
Benzo[a]pyrimidine see 8169
5,6-Benzopyrimidine see 8169
1,2-Benzopyrone see 2550
Benzo-γ-pyronecarboxylic Acid see 2241
Benzo[b]pyrrole see 5003
2,3-Benzopyrrole see 5003
Benzoquin see 6334
Benzo[f]quinoline, 1108
5,6-Benzoquinoline see 1108
1,4-Benzoquinone see 8191
Benzoresorcinol, 1109
Benzosalin see 6099
Benzosulfimide see 8445
Benzosulfonazole see 1110
Benzotetrazine see 8040
1,2,3-Benzothiadiazole-7-carbothioic Acid S-Methyl Ester see 98
Benzothiazide see 1123
2-Benzothiazolamine see 419
6-Benzothiazolamine see 420
Benzothiazole, 1110
2-Benzothiazolethiol see 5935
2(3H)-Benzothiazolethione see 5935
1-(2-Benzothiazolyl)-1,3-dimethylurea see 6011
N-2-Benzothiazolyl-N,N'-dimethylurea see 6011
(αS)-4-(2-Benzothiazolylmethylamino)-α-[(3,4-difluorophenoxy)methyl]-1-piperidineethanol see 5648
2-(1,3-Benzothiazol-2-yloxy)-N-methyl-acetanilide see 5868
2-(2-Benzothiazolyloxy)-N-methyl-N-phenylacetamide see 5868
Benzothiazyl Disulfide see 3413
N-(1-Benzo[b]thien-2-ylethyl)-N-hy-droxyurea see 10323
Benzothiofuran see 9455
Benzo[b]thiophene see 9455
2-Benzothiophenecarboxylic Acid see 9498
Benzo[b]thiophene-2-carboxylic Acid see 9498
1,2,4-Benzotriazin-3-amine 1,4-Dioxide see 9617
1H-Benzotriazole, 1111
1,2,3-Benzotriazole see 1111
1H-Benzotriazole-1-carbonitrile see 2680
2-(2H-Benzotriazol-2-yl)-p-cresol see 3496
2-(2H-Benzotriazol-2-yl)-4-methylphenol see 3496
O-(Benzotriazol-1-yl)-N,N,N',N'-tetra-methyluronium Hexafluorophosphate see 4654

Bergamotine *see* 1159
Bergamottin, 1159
Bergaptan *see* 1160
Bergapten, 1160
Bergaptene *see* 1160
Bergaptin *see* 1159
Berkatens [Berk] *see* 10144
Berkelium, 1161
Berkfurin [Berk] *see* 6686
Berkmycen [APS] *see* 7075
Berkolol [APS] *see* 7953
Berkozide [APS] *see* 1039
Berlin Blue *see* 8018
Berninamycin, 1162
Berninamycin A *see* 1162
Berninamycin B *see* 1162
Bernstein *see* 374
Bernsteinsäure (German) *see* 8999
Bernsteinsäureanhydrid (German) *see* 9000
Berolase [Roche] *see* 9448
Beronald [Kowa] *see* 4338
Berotec [Boehringer, Ing.] *see* 4012
Berry Alder *see* 4291
Berry Wax *see* 1010
Bertholite *see* 2095
Bertrandite *see* 1163
Berubi [Cesra] *see* 10212
Beryl *see* 1163
Beryllia *see* 1174
Beryllium, 1163
Beryllium Acetate, 1164
Beryllium Acetylacetonate, 1165
Beryllium Bromide, 1166
Beryllium Bromide (BeBr$_2$) *see* 1166
Beryllium Carbide, 1167
Beryllium Carbide (Be$_2$C) *see* 1167
Beryllium Chloride, 1168
Beryllium Chloride (BeCl$_2$) *see* 1168
Beryllium Dibromide *see* 1166
Beryllium Dichloride *see* 1168
Beryllium Difluoride *see* 1169
Beryllium Dihydride *see* 1170
Beryllium Dihydroxide *see* 1171
Beryllium Fluoride, 1169
Beryllium Fluoride (BeF$_2$) *see* 1169
Beryllium Hydride, 1170
Beryllium Hydride (BeH$_2$) *see* 1170
Beryllium Hydroxide, 1171
Beryllium Hydroxide (Be(OH)$_2$) *see* 1171
Beryllium Nitrate, 1172
Beryllium Nitride, 1173
Beryllium Nitride (Be$_3$N$_2$) *see* 1173
Beryllium Oxide, 1174
Beryllium Oxide (BeO) *see* 1174
Beryllium Potassium Sulfate, 1175
Beryllium Sodium Fluoride, 1176
Beryllium Sulfate, 1177
Berzelianite *see* 2656
Berzeline *see* 2656
Besivance [Bausch & Lomb] *see* 1178
Besifloxacin, 1178
Besnoline [Kotobuki] *see* 9681
Bespar [BMS] *see* 1507
Bessisterol *see* 8876
Bestatin [Nippon Kayaku] *see* 10024
Bestcall [Takeda] *see* 1926
Bestguard [Takeda] *see* 6654
Bestmann-Ohira Reagent, 1179
Betabactyl [SKB] *see* 2339
Betabion [Merck KGaA] *see* 9447
Betacaine *see* 3937
Beta-Cardone [UCB] *see* 8854
Betacetylmethadol *see* 6017

Beta-Corlan *see* 1182
Betadexamethasone *see* 1182
Betadine [Purdue Frederick] *see* 7815
Betadran [Logeais] *see* 1500
Betadrenol [Schwarz] *see* 1500
Betaferon [Bayer Schering] *see* 5035
Betafluorene [Lepetit] *see* 1182
Betagan [Allergan] *see* 5514
Betagon [Schering AG] *see* 5923
Betahistine, 1180
Betaine, 1181
Betaine Hydrazide Hydrochloride *see* 4465
Betaine Viologen *see* 10195
Beta-interferon *see* 5035
Betaisodona [Mundipharma] *see* 7815
Betalin S [Lilly] *see* 9447
Betalin-12 [Lilly] *see* 10212
Betaling [Tanabe] *see* 1188
Betaloc [AstraZeneca] *see* 6228
Betalone [Lepetit] *see* 5928
Betamann [Mann] *see* 6216
Betamaze [Pfizer] *see* 9021
Betamet [Pharmaquest] *see* 6216
Betamethasone, 1182
Betamethasone 21-(Dihydrogen Phosphate) Disodium Salt *see* 1182
Betamipron, 1183
Betamox [Norbrook] *see* 574
Betanal [Bayer CropSci.] *see* 7350
Beta-naphthol *see* 6469
Betanaphthol Benzoate *see* 6490
Betanaphthoxyethanol *see* 6494
Betanex [Bayer CropSci.] *see* 2925
Betanidol [Tanabe Seiyaku] *see* 1188
Betanol [Dulcis] *see* 6216
Betapace [Schering AG] *see* 8854
Betapar [Parke-Davis] *see* 5928
Betapindol [Helvepharm] *see* 7554
Betapred [Schering] *see* 5928
Betapressin [HMR] *see* 7191
Betaprodine *see* 306
Betaprodine Hydrochloride *see* 306
Betaprone [Forest] *see* 7934
Betaserc [Duphar] *see* 1180
Betaseron [Bayer Schering] *see* 5035
Betasolon *see* 1182
Beta-Tablinen [Winthrop] *see* 7953
Betavel [Sanofi Winthrop] *see* 2404
Betaxin [Sterling Winthrop] *see* 9447
Betaxina [Terap. M. R.] *see* 6444
Betaxolol, 1184
Betazole, 1185
BETE *see* 9961
Betel, 1186
Betel Nuts *see* 766
Bethanechol Chloride, 1187
Bethanidine, 1188
Bethrodine *see* 1042
Beth Root *see* 9865
Betim [Valeant] *see* 9601
Betimol [Vistakon] *see* 9601
Betnelan [GSK] *see* 1182
Betnesol [GSK] *see* 1182
Betnesol-V [GSK] *see* 1182
Betneval [GSK] *see* 1182
Betnovate [GSK] *see* 1182
Betol *see* 6496
Betolvex [Dumex] *see* 10212
Betoptic [Alcon] *see* 1184
Betoptima [Alcon] *see* 1184
Betotastine *see* 1151
Betoxycaine, 1189
Betriol [Boehringer, Ing.] *see* 1493
Betti Base, 1190

Betula, 1191
Betula Oil *see* 6193
Betulic Acid *see* 1193
Betulin, 1192
Betulinic Acid, 1193
Betulinol *see* 1192
Betulol *see* 1192
Betuloside *see* 8314
Bevacizumab, 1194
Bevantolol, 1195
Bevirimat, 1196
Bevonium Methyl Sulfate, 1197
Bewon [Wyeth] *see* 9447
Bexarotene, 1198
Bexedan [Smit] *see* 2082
Bexide *see* 3428
Bexol *see* 3331
Bextasol [GSK] *see* 1182
Bextra [Pharmacia & Upjohn] *see* 10086
Bexxar [Coulter] *see* 9713
Bezafibrate, 1199
Bezalip [Boehringer, Mann.] *see* 1199
Bezatol [Kissei] *see* 1199
Bezitramide, 1200
BFE-60 *see* 1021
bFGF *see* 4101
des 1-15 bFGF *see* 4101
BFL *see* 5657
BG-8301 *see* 5038
BG-8967 *see* 1311
BG-9273 *see* 220
BG-9712 *see* 220
BGO *see* 1266
BGP *see* 6992
α-Bgt *see* 1492
6R-BH$_4$ *see* 8505
BH-6 *see* 6827
BHA *see* 1549
BH-AC *see* 3640
Bhang *see* 1752
BHB *see* 1536
γ-BHC *see* 5556
Bhimsaim Camphor *see* 1340
B Hormone *see* 6381
BHT *see* 1550
B-HT-920 *see* 9171
BI 61.012 *see* 4574
BI-397 *see* 2800
BI-1356 *see* 5549
BIA 2-093 *see* 3754
BIA 2-194 *see* 3754
Biacetyl *see* 2969
Bialaphos *see* 7451
Bialcol [Novartis] *see* 1114
Bialzepam [Bial] *see* 2997
Biapenem, 1201
Biarison [Sandoz-Wander] *see* 7984
Biaryl Phosphane Ligands *see* 1469
Biathlon [BASF] *see* 9943
Biaxin [Abbott] *see* 2338
Bibapcitide *see* 9235
Bibenzal *see* 8946
Bibenzene *see* 3344
o,o'-Bibenzoic Acid *see* 3340
Bibenzoyl *see* 1080
Bibenzyl, 1202
Bibenzylidene *see* 8946
BIBF-1120 *see* 5031
Bibiru Bark *see* 1015
BIBR 277 *see* 9271
BIBR-953 *see* 2794
BIBR-1048 *see* 2794
Bibrocathin *see* 1203
Bibrocathol, 1203
BIBW-2992 *see* 169

BIC see 1578
Bicalutamide, 1204
Bicarbamimide see 10051
2,2'-Bichavicol see 5761
Bichloracetic Acid see 3059
Bichromate of Soda see 8744
Bicillin L-A [Wyeth-Ayerst] see 7205
Bicine, 1205
Biciron [Basotherm] see 9727
Bicnu [BMS] see 1845
Bicolorin see 3753
Bicomarin [Schering-Plough] see 1207
Bicozamycin see 1207
Bicuculline, 1206
Bicyclo[4.4.0]decane see 2850
Bicyclo-[0.3.5]-deca-1,3,5,7,9-pentaene
    see 919
Bicyclo-[5.3.0]-deca-2,4,6,8,10-pentaene
    see 919
Bicyclo[2.2.1]hepta-2,5-diene see 6776
Bicyclo[4.1.0]heptane see 6777
rel-(1R,2S,3R,4S)-Bicyclo[2.2.1]hept-5-
    ene-2,3-dicarboxylic Acid Dimethyl
    Ester see 3262
endo-cis-Bicyclo[2.2.1]hept-5-ene-2,3-di-
    carboxylic Anhydride see 1797
3-Bicyclo[2.2.1]hept-5-en-2-yl-6-chloro-
    3,4-dihydro-2H-1,2,4-benzothiadi-
    azine-7-sulfonamide 1,1-Dioxide see
    2752
α-Bicyclo[2.2.1]hept-5-en-2-yl-α-phenyl-
    1-piperidinepropanol see 1238
1-Bicycloheptenyl-1-phenyl-3-piperidino-
    propanol see 1238
Bicyclohexylammonium Fumagillin see
    4315
[1,1'-Bicyclohexyl]-1-carboxylic Acid 2-
    (Diethylamino)ethyl Ester see 3108
Bicyclomycin see 1207, 1207
Bicyclo[2.2.2]octa-2,5,7-triene see 999
BID see 1886
Bidien [IDI] see 1476
Bidisomide, 1208
Δ 2,2'-Bi-1,3-dithiole see 9390
Bidrin [Amvac] see 3097
Bieberite see 2432
Biebrich Scarlet Red see 8530
Bietamiverine, 1209
Bietanautine see 3339
Biethylene see 1512
Bifemelane, 1210
Bifenazate, 1211
Bifenox, 1212
Bifenthrin, 1213
Bifeprunox, 1214
Bifeprunox Mesylate see 1214
Bifex [Provet] see 7949
Bifidus Factor, 1215
Bifonazole, 1216
Biformyl see 4544
Biforon [Meiji] see 1481
Bifril [Menarini] see 10386
BIGCHAP, 1217
Bigitalin see 4469
Biguanide, 1218
Bigunal [Nikken] see 1481
Bijosal see 3208
Bi-Ketolan see 7027
Bikhaconitine, 1219
Bikhroot see 112
Biklin [BMS] see 400
Bikunin Trypsin Inhibitor see 10067
Bilamid(e) [Cilag-Chemie] see 4872
Bilastine, 1220
Bilberry, 1221

Bilein see 7027
Biletan [Gador] see 9479
Bilevon see 4716
Bilevon-M see 6603
Bilicholan see 7027
Bilidren see 2872
Biligrafin [Schering AG] see 5065
Biligram [Schering AG] see 5094
Bilijodon-Natrium see 5101
Bilimiro [Bracco] see 5106
Bilimiron [Bracco] see 5106
Bilineurine see 2211
Biliognost see 5069
Biliphorine [Couvreur] see 9649
Bilirubin, 1222
Bilirubin IXα see 1222
Biliselectan see 5069
Bilitrast see 5081
Biliverdine, 1223
Bilobalide, 1224
Bilopaque [Sterling Winthrop] see 10018
Biloptin [Schering AG] see 5115
Biloxazol see 1306
Biltricide [Bayer] see 7837
BIM-23014 see 5412
BIM-23014C see 5412
BIM-51077 see 9209
Δ2,2'-Bimalononitrile see 9340
Bimatoprost, 1225
Bimethyl see 3778
2,2'-Bimorphine see 8027
Bimosiamose, 1226
Bimotrim [Vetpharma] see 9038
BIMT-17 see 4132
BINAP, 1227
[1,1'-Binaphthalene]-2,2'-diol see 1230
1,1'-[1,1'-Binaphthalene]-2,2'-diylbis[1,1-
    diphenylphosphine] see 1227
β-Binaphthol see 1230
β,β-Binaphthyleneethene see 7508
Binazin [Polfa] see 9660
Bindan see 7348
Bindazac see 1037
Bingo [BASF] see 2294
Binifibrate, 1228
Biniwas [Chiesi] see 1228
Binixin [Bayer] see 4178
Binocrit [Sandoz] see 3744
Binodenoson, 1229
Binodrenal see 6784
BINOL, 1230
Binotal [Grünenthal] see 583
Binova [Gentili] see 7993
Binovum [Ortho-Cilag] see 6786
BioAstin [Cyanotech] see 845
Biobenzyfuroline [Sumitomo] see 1235
Biocefalin [Benvegna] see 8109
Biochanin A, 1231
Biochanin A 7-O-Glucoside see 1231
Biochanin B see 4271
Biociclin [Francia] see 1956
Biocidan [Clin-Comar-Byla] see 2017
Biocin [Ibirn] see 4281
Bio-Cox [Alpharma] see 8473
Biocytin, 1232
Biodermatin [Lafare] see 1236
Biofanal [Pfleger] see 6825
Bioflavonoids, 1233
Biofos [Leben's] see 4281
Biofusal see 4342
BIOLF-62 see 4393
Biomioran see 2201
Biomitsin see 2198
Biomoduline see 5493
Bion [Syngenta] see 98

Bionicard [Bioindustria] see 6580
Biopal CVL-10 see 5080
Bioperazone [Biopharma] see 1931
Bioperidolo [Firma] see 4634
Bioplex [Thames] see 1794
Biopoin [CT Arzneimittel] see 3744
Biopten [Maruho] see 8505
Biopterin, 1234
Biopure 100 [Biophil] see 4957
Bioral [SKB] see 1794
Bioresmethrin, 1235
Biorgasept [Biorga] see 2092
Biorphen see 6975
Bioscleran [Pfleger] see 2370
Bios I see 5021
Bios II see 1236
Biosinon [Farmakhim] see 3777
Bio Soft [Stepan] see 7697
Biosol [Biokema] see 6539
Biosone [Berk] see 3644
Biosynthetic Human Insulin see 5023
Biotamin [Sankyo] see 1043
Bioterciclin see 2890
Biotertussin [Lisapharma] see 2360
Bi(OTf)₃ see 1292
Biotin, 1236
Biotin Complex of Yeast see 1232
Biotinin [Sankyo] see 9449
Biotin l-Sulfoxide, 1237
ε-N-Biotinyl-L-lysine see 1232
Biotiren [Benvegna] see 3035
rel-(2R,2'R)-2,2'-Bioxirane see 3734
rel-(2R,2'S)-2,2'-Bioxirane see 3734
Biozate [Ind. Bioquim] see 2762
Biperiden, 1238
Bipéridys [Fabre] see 3467
Biphenate see 1213
Biphenthrin see 1213
Biphentrin see 1213
1,1'-Biphenyl see 3344
[1,1'-Biphenyl]-4-acetic Acid see 3985
4-Biphenylacetic Acid see 3985
p-Biphenylamine, 1239
[1,1'-Biphenyl]-4-amine see 1239
[1,1'-Biphenyl]-2,4'-diamine see 1240
[1,1'-Biphenyl]-4,4'-diamine see 1079
2,4'-Biphenyldiamine, 1240
(2-Biphenyl)di-tert-butylphosphine see
    1469
[1,1'-Biphenyl]-2,2'-dicarboxylic Acid
    see 3340
3,3'-[[1,1'-Biphenyl]-4,4'-diylbis(azo)]-
    bis[4-amino-1-naphthalenesulfonic
    Acid] Disodium Salt see 2482
3,3'-[[1,1'-Biphenyl]-4,4'-diylbis(2,1-di-
    azenediyl)]bis[4-amino-1-naphthalene-
    sulfonic Acid] Sodium Salt (1:2)
    2482
3,3'-[1,1'-Biphenyl]-4,4'-diylbis[2,5-di-
    phenyl-2H-tetrazolium] Dichloride see
    6551
2,2'-[[1,1'-Biphenyl]-4,4'-diylbis[2-hy-
    droxy-4,4-dimethylmorpholinium] see
    4678
3,3'-[4,4'-Biphenylene]bis[2,5-diphenyl-
    2H-tetrazolium] Dichloride see 6551
o-Biphenylenemethane see 4187
2,2'-Biphenylene Oxide see 3013
(1,1'-Biphenyl)-2-ol see 7415
[1,1'-Biphenyl]-4-ol see 7416
2-Biphenylol see 7415
3-(3-p-Biphenyl-1,2,3,4-tetrahydro-
    naphth-1-yl)-4-hydroxycoumarin see
    3153
[1,1'-Biphenyl]-2-ylbis(1,1-dimethyleth-
    yl)phosphine see 1469

Bis(cyano-κC)cuprate(1−) Potassium (1:1) see 2655

1-[Bis(4-cyanophenyl)methyl]-1,2,4-triazole see 5502

(TPS-7-1-232′4′54)-[Bis[(1,2-cyclohexanedione Dioximato)(1−)-O] [(1,2-Cyclohexanedione Dioximato)(2−)-O]-methylborato(2−)-N,N′,N″,N‴,N⁗,-N⁗′]chlorotechnetium-⁹⁹ᵐTc see 9240

Bis(cyclohexyl)carboxylic Acid Diethylaminoethyl Ester see 3108

Biscyclopentadienyliron see 4064

Biscyclopentadienyltitanium(IV) Dichloride see 9638

Bis(cyclopentadienyl)vanadium see 10111

3′,7′-Bis(cyclopropylmethyl)spiro[cyclopentane-1,9′-[3,7]diazabicyclo[3.3.1]-nonane] see 9246

Bis-DEAE-fluorenone see 9597

1,6-Bisdehydro-6-chloro-17α-acetoxyprogesterone see 2883

α,α-Bis(3,5-dibromo-4-hydroxyphenyl)-α-hydroxy-o-toluenesulfonic Acid γ-Sultone see 1454

3,3-Bis(3,5-dibromo-4-hydroxyphenyl)-1(3H)-isobenzofuranone see 9331

2,2-Bis(3,5-dibromo-4-hydroxyphenyl)-propane see 9328

α,α-Bis(3,5-dibromo-4-hydroxy-o-tolyl)-α-hydroxytoluenesulfonic Acid, γ-Sultone see 1393

(R,R)-[N,N′-Bis(3,5-di-tert-butylsalicylidene)-1,2-cyclohexanediamine]manganese(III) Chloride see 5300

3,3′-Bis[N,N-di(carboxymethyl)aminomethyl]thymolsulfonephthalein see 6202

N,N′-Bis(dichloroacetyl)-N,N′-bis(2-ethoxyethyl)-1,4-bis(aminomethyl)benzene see 9242

N,N′-Bis(dichloroacetyl)-1,8-octamethylenediamine see 4087

Bis[3,4-dichloro-2(5)-furanonyl] Ether see 6384

Bis(dichloro(η⁵-pentamethylcyclopentadienyl)iridium) see 7226

2,6-Bis(diethanolamino)-4,8-dipiperidinopyrimido-[5,4-d]pyrimidine see 3388

Bis[S-(diethoxyphosphinothioyl)mercapto]methane see 3790

4,4′-Bis(β-diethylaminoethoxy)-α,β-diethyldiphenylethane see 4738

2,7-Bis[2-(diethylamino)ethoxy]-9H-fluoren-9-one see 9597

6,7-Bis[2-(diethylamino)ethoxy]-4-methyl-2H-1-benzopyran-2-one see 7014

6,7-Bis[2-(diethylamino)ethoxy]-4-methylcoumarin see 7014

3,4-Bis[p-(β-diethylaminoethoxy)phenyl]hexane see 4738

N,N′-Bis[2-diethylaminoethyl]oxamide Bis[2-chlorobenzyl Chloride] see 373

3,6-Bis(diethylamino)xanthylium Chloride (1:1) see 8118

(−)-1,2-Bis[(2R,5R)-2,5-diethylphospholano]benzene see 3513

(+)-1,2-Bis[(2R,5R)-2,5-diethylphospholano]ethane see 1363

Bis-O,O-diethylphosphoric Anhydride see 9349

Bis(diethylthiocarbamoyl) Disulfide see 3407

Bis(diethylthiocarbamoyl)sulfide see 9080

Bis(diethylthiocarbamyl) Disulfide see 3407

2-[[[[[4,6-Bis(difluoromethoxy)-2-pyrimidinyl]amino]carbonyl]amino]sulfonyl]-benzoic Acid Methyl Ester see 7866

2-[3-(4,6-Bis(difluoromethoxy)pyrimidin-2-yl)ureidosulfonyl]benzoic Acid Methyl Ester see 7866

N,N′-Bis[3-(4,5-dihydro-1H-imidazol-2-yl)phenyl]urea see 4956

2,6-Bis[(4S)-4,5-dihydro-4-(1-methylethyl)-2-oxazolyl]pyridine see 8083

2,6-Bis[(4S)-4,5-dihydro-4-phenyl-2-oxazolyl]pyridine see 8083

1,8-Bis[(3,5-dibromo-4-hydroxyphenyl)...

1,8-Bis[(4S)-4,5-dihydro-3,5,7-trihydroxy-2H-1-benzopyran-2-yl]-3,4,6-trihydroxy-5H-benzocyclohepten-5-one see 9423

(T-4)-Bis[4,5-di(hydroxy-κO)-1,3-benzenedisulfonato(4−)]antimonate(5−) Sodium Hydrate (1:5:7) see 8942

2,3-Bis(3,4-dihydroxybenzyl)butane see 6782

Bis[μ-[2,3-di(hydroxy-κO)butanedioato(4−)-κO¹:κO⁴]]diantimonate(2−) Potassium Hydrate (1:2:3) Stereoisomer see 691

Bis[μ-[(2R,3R)-2,3-di(hydroxy-κO)butanedioato(4−)-κO¹:κO⁴]]diantimonate(2−) Sodium (1:2) see 693

2,2-Bisdihydroxymethyl-1,3-propanediol Tetranitrate see 7222

N,N′-Bis[2-(3,4-dihydroxyphenyl)-2-hydroxyethyl]hexamethylenediamine see 4743

(1R,3R,4S,5R)-1,3-Bis[[3-(3,4-dihydroxyphenyl)-1-oxo-2-propen-1-yl]oxy]-4,5-dihydroxycyclohexanecarboxylic Acid see 2763

N,N′-Bis(2,3-dihydroxypropyl)-5-[N-(2,3-dihydroxypropyl)acetamido]-2,-4,6-triiodoisophthalamide see 5095

N¹,N³-Bis(2,3-dihydroxypropyl)-5-[(2-hydroxyacetyl)(2-hydroxyethyl)amino]-2,4,6-triiodo-1,3-benzenedicarboxamide see 5109

N¹,N³-Bis(2,3-dihydroxypropyl)-5-[(2-hydroxyacetyl)methylamino]-2,4,6-triiodo-1,3-benzenedicarboxamide see 5097

N,N′-Bis(2,3-dihydroxypropyl)-5-[N-(2-hydroxyethyl)glycolamido]-2,4,6-triiodoisophthalamide see 5109

N¹,N³-Bis(2,3-dihydroxypropyl)-5-[[3-hydroxy-2-(hydroxymethyl)-1-oxopropyl]amino]-2,4,6-triiodo-N¹,N³-dimethyl-1,3-benzenedicarboxamide see 5052

N,N′-Bis(2,3-dihydroxypropyl)-5-[N-(2-hydroxy-3-methoxypropyl)acetamido]-2,4,6-triiodoisophthalamide see 5102

N,N′-Bis(2,3-dihydroxypropyl)-2,4,6-triiodo-5-(2-methoxyacetamido)-N-methylisophthalamide see 5105

N¹,N³-Bis(2,3-dihydroxypropyl)-2,4,6-triiodo-5-[(2-methoxyacetyl)amino]-N¹-methyl-1,3-benzenedicarboxamide see 5105

N,N′-Bis(2,3-dihydroxypropyl)-2,4,6-triiodo-5-(N-methylglycolamido)isophthalamide see 5097

2,6-Bis[(4,6-dimethoxy-2-pyrimidinyl)oxy]benzoic Acid see 1302

Bis(dimethylamido)fluorophosphate see 3228

Bis(dimethylamido)phosphoryl Fluoride see 3228

4,4′-Bis(dimethylamino)benzophenone see 6256

1,4-Bis(dimethylamino)butane see 9373

Bis(dimethylamino)-tert-butyloxymethane see 1373

(4S,4aS,5aR,12aS)-4,7-Bis(dimethylamino)-9-[[2-[(1,1-dimethylethyl)amino]acetyl]amino]-1,4,4a,5,5a,6,11,12a-octahydro-3,10,12,12a-tetrahydroxy-1,11-dioxo-2-naphthacenecarboxamide see 9588

Bis[6-(dimethylamino)-2-[2-(2,5-dimethyl-1-phenylpyrrol-3-yl)vinyl]-1-methylquinolinium] 4,4′-Methylenebis(3-hydroxy-2-naphthoate) see 8136

Bis[2-dimethylaminoethyl]succinate Bis-[methobromide] see 9005

Bis[2-dimethylaminoethyl]succinate Bis-[methochloride] see 9006

Bis(β-dimethylaminoethyl)succinate Bis-(methyl Iodide) see 9007

Bisdimethylaminofluorophosphine Oxide see 3228

1-[Bis(dimethylamino)methylene]-1H-benzotriazolium 3-Oxide Hexafluorophosphate(1−) (1:1) see 4654

1-[Bis(dimethylamino)methylene]-1H-1,-2,3-triazolo[4,5-b]pyridinium 3-Oxide Hexafluorophosphate(1−) (1:1) see 4653

1,8-Bis(dimethylamino)naphthalene see 3432

(4S,4aS,5aR,12aS)-4,7-Bis(dimethylamino)-1,4,4a,5,5a,6,11,12a-octahydro-3,-10,12,12a-tetrahydroxy-1,11-dioxo-2-naphthacenecarboxamide see 6283

3,7-Bis(dimethylamino)phenazathionium Chloride see 6132

3,7-Bis(dimethylamino)phenothiazin-5-ium Chloride (1:1) see 6132

Bis(p-dimethylaminophenyl)methane see 6255

Bis[4-(dimethylamino)phenyl]methanone see 6256

N-[4-[Bis[4-(dimethylamino)phenyl]-methylene]-2,5-cyclohexadien-1-ylidene]-N-methylmethanaminium Chloride (1:1) see 4430

Bis(p-dimethylaminophenyl)phenylmethane see 5763

Bis[p-(dimethylamino)phenyl]phenylmethylium Chloride see 5763

10-[2,3-Bis(dimethylamino)propyl]phenothiazine see 463

3,6-Bis(dimethylamino)xanthylium Chloride (1:1) see 8119

(T-4)-Bis(N,N-dimethylcarbamodithioato-κS,κS′)zinc see 10372

1-[Bis(3′,5′-N,N-dimethylcarbamoyloxy)phenyl]-2-N-tert-butylaminoethanol see 946

Bis(dimethyldithiocarbamato)zinc see 10372

(3S,8S,9S,12S)-3,12-Bis(1,1-dimethylethyl)-8-hydroxy-4,11-dioxo-9-(phenylmethyl)-6-[[4-(2-pyridinyl)phenyl]-methyl]-2,5,6,10,13-pentaazatetradecanedioic Acid Dimethyl Ester see 849

N-[2,4-Bis(1,1-dimethylethyl)-5-hydroxyphenyl]-1,4-dihydro-4-oxo-3-quinolinecarboxamide see 5295

2,4-Bis(1,1-dimethylethyl)-5-methylphenol see 2840

3,4-Bis(3-methyl-4-hydroxyphenyl)hex-
ane *see* 6040
Bis[methylmercapto]methane *see* 1260
1,4-Bis[(4-methylphenyl)amino]-9,10-an-
thracenedione *see* 8181
N,N-Bis[*N*-methyl-*N*-phenyl-*tert*-butyl-
acetamido]-β-hydroxyethylamine *see*
7032
Bis(4-methylphenyl)mercury *see* 3424
N,N-Bis(2-methylpropyl)carbamothioic
Acid *S*-Ethyl Ester *see* 1548
(*R**,*S**)-1,3-Bis(1-methyl-2-pyrrolidinyl)-
2-propanone *see* 2667
1,2-Bis(methylsulfonyl)-1-(2-chloro-
ethyl)hydrazine *see* 5425
1,2-Bis(methylsulfonyl)-1-(2-chloro-
ethyl)-2-[(methylamino)carbonyl]-
hydrazine *see* 5425
1,4-Bis[3-[(methylsulfonyl)oxy]-1-oxo-
propyl]piperazine *see* 7594
N,N′-Bis(3-methylsulfonyloxypropion-
yl)piperazine *see* 7594
(−)-(*R*)-1-[4,4-Bis(3-methyl-2-thienyl)-3-
butenyl]nipecotic Acid *see* 9572
(3*R*)-1-[4,4-Bis(3-methyl-2-thienyl)-3-bu-
ten-1-yl]-3-piperidinecarboxylic Acid
see 9572
N,N′-Bis[1-methylthioacetaldehyde *O*-
(*N*-Methylcarbamoyl)oxime]sulfide *see*
9482
Bis-[*O*-(1-methylthioethylimino)-*N*-meth-
ylcarbamic Acid]-*N,N*′-sulfide *see*
9482
Bis(methylthio)methane, 1260
N,N′-Bis[1-methyl-3-(2,2,6-trimethyl-
cyclohexyl)propyl]-*N,N*′-dimethyl-1,6-
hexanediamine Bis(methochloride) *see*
9818
Bismetin [Mect] *see* 7855
Bismite *see* 1273
(*T*-4)-Bis[monothiosulfato(2−)-κ*O*,κ*S*]-
aurate(3−) Sodium (1:3) *see* 4554
Bismugel [Biothrax] *see* 1276
Bismutan *see* 1268
Bismuth, 1261
Bismuth Acetate Basic *see* 1281
Bismuth Acetate Oxide *see* 1281
Bismuth Bromide, 1262
Bismuth Bromide Oxide, 1263
Bismuth Carbonate Oxide (Bi₂(CO₃)O₂)
see 1282
Bismuth Chloride, 1264
Bismuth Chloride Oxide *see* 1274
Bismuth Fluoride, 1265
Bismuth Fluoride (BiF₅) *see* 1275
Bismuth Gallate, Basic *see* 1283
Bismuth Germanate, 1266
Bismuth Germanium Oxide (Bi₄Ge₃O₁₂)
see 1266
Bismuth Hydrate *see* 1267
Bismuth Hydroxide, 1267
Bismuth Hydroxide (Bi(OH)₃) *see* 1267
Bismuth Hydroxide Nitrate Oxide (Bi₅-
(OH)₉(NO₃)₄O) *see* 1284
Bismuthine, 1268
Bismuthinite *see* 1287
Bismuth Iodide, 1269
Bismuth Iodide Oxide, 1270
Bismuth Nitrate, 1271
Bismuth Nitrate Basic *see* 1284
Bismuthous Oxide *see* 1273
Bismuth Oxalate, 1272
Bismuth Oxide, 1273
Bismuth(3+) Oxide *see* 1273
Bismuth Oxide Acetate *see* 1281

Bismuth Oxide (Bi₂O₃) *see* 1273
Bismuth Oxybromide *see* 1263
Bismuth Oxycarbonate *see* 1282
Bismuth Oxychloride, 1274
Bismuth Oxyiodide *see* 1270
Bismuth Oxynitrate *see* 1284
Bismuth Oxysalicylate *see* 1285
Bismuth Pentafluoride, 1275
"Bismuth Peroxide" *see* 1290
Bismuth Phosphate, 1276
Bismuth Potassium Iodide, 1277
Bismuth Potassium Tartrate, 1278
Bismuth Selenide, 1279
Bismuth Selenide (Bi₂Se₃) *see* 1279
Bismuth Sodium Oxide (BiNaO₃) *see*
8721
Bismuth Sodium Tartrate, 1280
Bismuth Subacetate, 1281
Bismuth "Subbromide" *see* 1263
Bismuth Subcarbonate, 1282
Bismuth Subchloride *see* 1274
Bismuth Subgallate, 1283
Bismuth "Subiodide" *see* 1270
Bismuth Subnitrate, 1284
Bismuth Subnitricum *see* 1284
Bismuth Subsalicylate, 1285
Bismuth Sulfate, 1286
Bismuth Sulfide, 1287
Bismuth(3+) Sulfide *see* 1287
Bismuth Sulfide (Bi₂S₃) *see* 1287
Bismuth Tannate, 1288
Bismuth Telluride, 1289
Bismuth(3+) Telluride *see* 1289
Bismuth Telluride (Bi₂Te₃) *see* 1289
Bismuth Tetrabromopyrocatechol *see*
1203
Bismuth Tetroxide, 1290
Bismuth Tribromide *see* 1262
Bismuth Tribromophenate, 1291
Bismuth Tribromophenol *see* 1291
Bismuth Trichloride *see* 1264
Bismuth Triflate, 1292
Bismuth Trifluoride *see* 1265
Bismuth Trihydride *see* 1268
Bismuth Trihydroxide *see* 1267
Bismuth Triiodide *see* 1269
Bismuth Trinitrate *see* 1271
Bismuth Trioxide *see* 1273
Bismuth Triselenide *see* 1279
Bismuth(III) Tris(trifluoromethanesul-
fonate) *see* 1292
Bismuth White *see* 1284
Bismuth Yellow *see* 1273
Bismuthyl Acetate *see* 1281
Bismuthyl Bromide *see* 1263
Bismuthyl Chloride *see* 1274
Bismuthyl Iodide *see* 1270
Bismuthyl Nitrate *see* 1284
Bismutite *see* 1282
Bis(1-naphthylmethyl)amine, 1293
1,2-Bis(nicotinamido)propane *see* 6578
(*OC*-6-11)-Bis(nitrato-κ*O*,κ*O*′)dioxoura-
nium *see* 10048
sym-Bis(5-nitro-2-furfurylidene)acetone
Guanylhydrazone *see* 6741
1,5-Bis(5-nitro-2-furyl)-3-pentadienone
Amidinohydrazone *see* 6741
1,5-Bis(5-nitro-2-furyl)-3-pentadienone
Guanylhydrazone *see* 6741
2,2-Bis[(nitrooxy)methyl]-1,3-propane-
diol 1,3-Dinitrate *see* 7222
2,2-Bis[(nitrooxy)methyl]-1,3-propane-
diol 1-Nitrate *see* 7251
Bis(3-nitrophenyl)disulfide *see* 6705
N,N′-Bis(4-nitrophenyl)urea *see* 3304

N,N′-Bis(4-nitrophenyl)urea, Compd
with 4,6-Dimethyl-2(1*H*)-pyrimidinone
(1:1) *see* 6579
Bisoctrizole, 1294
1,10-Bis-[4-(octylamino)-1-pyridinium]-
decane *see* 6843
Bisolvon [Thomae] *see* 1398
Bisoprolol, 1295
Bis(oxalato)oxotitanate(2−) Diammon-
ium *see* 560
Bisoxatin Acetate, 1296
Bis(oxazoline) Ligands, 1297
ᴅ-Bis(*N*-pantothenyl-β-aminoethyl) Di-
sulfide *see* 7115
Bis(pentabromophenyl) Ether *see* 2848
Bis(pentachloro-2,4-cyclopentadien-1-yl)
see 3116
(*T*-4)-Bis(2,4-pentanedionato-κ*O*²,κ*O*⁴)-
beryllium *see* 1165
(*SP*-4-1)-Bis(2,4-pentanedionato-κ*O*²,-
κ*O*⁴)nickel *see* 6585
Bis(2,4-pentanediono)nickel(II) *see* 6585
p-Bis(perchloromethyl)benzene *see* 1304
Bisphenol A, 1298
Bisphenol B, 1299
Bis(phenoxarsin-10-yl)ether *see* 7060
Bis(10-phenoxarsinyl)oxide *see* 7060
Bis(10-phenoxarsyl)oxide *see* 7060
2,4-Bis(4-phenoxyphenyl)-1,3,2,4-dithia-
diphosphetane 2,4-Disulfide *see* 1027
4,4′-Bis(phenylamino)-1,1′-biphenyl *see*
3349
9,10-Bis(2-phenylethynyl)anthracene,
1300
1,1′-Bis(phenylmethyl)-4,4′-bipyridinium
see 10195
Bis(phenylmethyl) Disulfide *see* 3018
*N*¹,*N*²-Bis(phenylmethyl)-1,2-ethanedi-
amine *see* 1066
2,2-Bis[2-[4(*S*)-phenyl-1,3-oxazolinyl]]-
propane *see* 1297
2,6-Bis[(4*S*)-4-phenyl-2-oxazolin-2-yl]-
pyridine *see* 8083
Bis(γ-phenylpropyl)ethylamine *see* 365
4,4′-Bis(3-phenylureido)-2,2′-stilbenedi-
sulfonic Acid Disodium Salt *see* 1316
Bisphonal [Yamanouchi] *see* 4969
N,N′-Bis(3-picolyl)-4-methoxyisophthal-
amide *see* 7519
Bis(pinacolato)diborane, 1301
(*R*)-Bis(POC)PMPA *see* 9289
Bis(POM)-PMEA *see* 139
Bispyribac, 1302
Bis(pyridine)iodine(1+) Tetrafluorobo-
rate(1−) (1:1) *see* 1303
Bis(pyridine)iodonium Tetrafluorobo-
rate, 1303
Bis(2-pyridylthio)zinc 1,1′-Dioxide *see*
8107
Bis(6-quinolyl)urea *see* 3393
Bis(rhodium Diacetate) *see* 8311
[[4-[Bis[4-[(sulfophenyl)amino]phenyl]-
methylene]-2,5-cyclohexadien-1-yli-
dene]amino]benzenesulfonic Acid
Sodium Salt (1:2) *see* 6102
Bissy Nuts *see* 5366
Bis-*N,N,N*′,*N*′-tetramethylphosphorodi-
amidic Anhydride *see* 8532
Biston [Zentiva] *see* 1783
Bis(3,5,6-trichloro-2-hydroxyphenyl)-
methane *see* 4716
1,4-Bis(trichloromethyl)benzene, 1304
Bis(trichloromethyl) Carbonate *see* 9921
Bis(trifluoroacetoxy)iodobenzene *see*
7406

Blood-coagulation Factor X *see* 3964
Blood-coagulation Factor XI *see* 3965
Blood-coagulation Factor XII *see* 3966
Blood-coagulation Factor XIII *see* 3967
Blood-coagulation Factor XIV *see* 7995
Blood Platelet-activating Factor *see* 7638
Bloodroot *see* 8492
Blood Serum Albumin *see* 8607
Blood Sugar *see* 4495
Biopress [Takeda] *see* 1742
Blox [Biomed. Foscama] *see* 5628
BL-P-1322 *see* 1983
BL-S578 *see* 1915
BL-S640 *see* 1917
BL-S786 *see* 1932
Blue Asbestos *see* 815
Blue Cohosh, 1321
Bluestone *see* 2643
Blue Tetrazolium *see* 9392
Blue Vitriol *see* 2643
Blusalt *see* 8734
Blutene [Abbott] *see* 9679
BM-21.0955 *see* 4912
BM-102 *see* 809
BM-02015 *see* 9711
BM-06022 *see* 8284
BM-13505 *see* 2805
BM-14190 *see* 1873
BM-15075 *see* 1199
BM-15100 *see* 7530
BM-51052 *see* 1780
BMC *see* 1792
BMD [Alpharma] *see* 928
BMIM *see* 1584
BMP-2A *see* 1333
BMP-7 *see* 1333
BMPs *see* 1333
BMS *see* 1258, 1336
BMS-180549 *see* 4092
BMS-181173 *see* 4617
BMS-182751 *see* 8520
BMS-186295 *see* 5127
BMS-186716 *see* 6937
BMS-188667 *see* 4
BMS-200475 *see* 3649
BMS-207147 *see* 8239
BMS-214778 *see* 9208
BMS-217380-01 *see* 9321
BMS-224818 *see* 1023
BMS-232632 *see* 849
BMS-247550 *see* 5297
BMS-284756 *see* 4402
BMS-354825 *see* 2831
BMS-477118 *see* 8525
BMS-540215 *see* 1383
BMS-562247 *see* 723
BMS-582664 *see* 1383
BMS-477118-11 *see* 8525
BMS-512148-05 *see* 2816
BMY-13754 *see* 6523
BMY-26517 *see* 7187
BMY-26538-01 *see* 616
BMY-27857 *see* 8929
BMY-28100 *see* 1943
BMY-28100-03-800 *see* 1943
BMY-28142 *see* 1923
BMY-28167 *see* 1943
BMY-28488 *see* 1921
BMY-30056 *see* 4629
BMY-33419 *see* 9321
BMY-40481 *see* 3929
BMY-40900 *see* 3109
BMY-41606 *see* 10123
BMY-45594 *see* 8520
B-mycin *see* 1361

BN-1270 *see* 2267
BN-52020 *see* 4461
BN-52021 *see* 4461
BN-52022 *see* 4461
B-Nine [Uniroyal] *see* 2809
BNP *see* 1365
BNP-1350 *see* 2538
BNP-7787 *see* 3233
BO-714 *see* 2319
Bobierrite *see* 5747
Boc-β-Ala-Trp-Met-Asp-Phe(NH₂) *see* 7224
Boc Anhydride *see* 3038
Boceprevir, 1322
Boc₂O *see* 3038
Boc-ON *see* 1323
2-(Boc-oxyimino)-2-phenylacetonitrile, 1323
B.O.E.A. *see* 3825
Boehmite *see* 350
Boettger's Paper *see* 248
Boforsin (obsolete) *see* 2460
Bog Bean *see* 5910
Bogomolets' Serum *see* 704
Bohrium, 1324
BOL-148 *see* 1433
BOL-303224-A *see* 1178
Bolandiol, 1325
Bolasterone, 1326
Boldea *see* 1329
Boldenone, 1327
Boldenone Undecylenate *see* 1327
Boldine, 1328
Boldine Dimethyl Ether *see* 4471
Boldo, 1329
Boldoa *see* 1329
Boldu *see* 1329
Boldus *see* 1329
Bole, Armenian, 1330
Bolecic Acid *see* 5147
Boleko Oil, 1331
Boletic Acid *see* 4316
Bolus Alba *see* 5330
Bolus Armena *see* 1330
Bolus Rubra *see* 1330
Bolvidon [Organon] *see* 6251
Bombesin, 1332
Bonadorm [Smith & Nephew] *see* 3055
Bonaid [Norwich] *see* 1504
Bonalon [Merck & Co.] *see* 223
Bonamid [Heilmittelwerke] *see* 894
Bonamine [Pfizer] *see* 5848
Bonanza [Novartis Agro] *see* 2768
Bonasanit [Weimer] *see* 8095
Bondou *see* 8519
Bondronat [Roche] *see* 4912
Bonealfa [Teijin] *see* 9152
"Bone Ash" *see* 1696
Bonefos [Schering AG] *see* 2365
Bone Gla Protein *see* 6992
Bone Morphogenetic Protein 2 (Human Recombinant RhBMP-2) *see* 1333
Bone Morphogenetic Proteins, 1333
Bone Morphogenic Factors *see* 1333
Bone Morphogenic Proteins *see* 1333
Boneset *see* 3944
Bongkrekic Acid, 1334
Bonidon [Mepha] *see* 5009
Bonifen [Merck KGaA] *see* 8109
Bonine [Insight] *see* 5848
Boniva [Roche] *see* 4912
Bonoform *see* 9334
Bonopen [Belupo] *see* 5996
Bonoteo [Astellas] *see* 6284
Bonpac [Hoei] *see* 4258

Bonpyrin [Takeda] *see* 3390
Bonviva [Roche] *see* 4912
Bonyl [Erco] *see* 6499
Bonzi [Syngenta] *see* 7082
Bonzol [Tokyo Tanabe] *see* 2811
Bopindolol, 1335
Boracic Acid *see* 1339
Borane Ammonia Complex *see* 489
Borane Complexes, 1336
Borane-dimethyl Sulfide Complex *see* 1336
Borane-DMS *see* 1336
Borane-methyl Sulfide *see* 1336
Borane-tetrahydrofuran Complex *see* 1336
Borane-THF *see* 1336
Borax *see* 8726
Borax Glass *see* 8726
Borazane *see* 489, 1337
Borazine, 1337
Borazole *see* 1337
Borazon *see* 1347
Bordeaux Mixture, 1338
Borgal [Intervet] *see* 9038
Boric Acid, 1339
Boric Acid Aluminum Salt *see* 326
Boric Acid Copper Salt *see* 2618
Boric Acid (H₃BO₃) *see* 1339
Boric Acid (H₂B₄O₇) Dipotassium Salt *see* 7799
Boric Acid (HBO₂) Lead(2+) Salt (2:1) *see* 5457
Boric Acid (HBO₂) Magnesium Salt (2:1) *see* 5722
Boric Acid (HBO₂) Sodium Salt (1:1) *see* 8771
Boric Acid Trimethyl Ester (H₃BO₃) *see* 9882
Boric Anhydride *see* 1348
Boric Oxide *see* 1348
endo-2-Bornanol *see* 1340
exo-2-Bornanol *see* 5173
2-Bornanone *see* 1734
Borneo Camphor *see* 1340
Borneol, 1340
(±)-Borneol *see* 1340
dl-Borneol *see* 1340
(1R,2S,4R)-Borneol *see* 1340
(1S,2R,4S)-Borneol *see* 1340
Borneol Acetate *see* 1341
Bornite *see* 2037, 2504
Bornyl Acetate, 1341
(±)-Bornyl Acetate *see* 1341
dl-Bornyl Acetate *see* 1341
(1R,2S,4R)-Bornyl Acetate *see* 1341
(1S,2R,4S)-Bornyl Acetate *see* 1341
Bornyl Alcohol *see* 1340
Bornyl Chloride, 1342
Borobutane *see* 9327
Boroethane *see* 3020
Borofax [Burroughs Wellcome] *see* 1339
Borofluoric Acid *see* 4180
Borohexane *see* 4711
Boromycin, 1343
Boron, 1344
Boron Calcium Oxide (B₄CaO₇) *see* 1655
Boron Carbide, 1345
Boron Carbide (B₄C) *see* 1345
Boron Chloride *see* 3021
Boron Fluoride Etherate *see* 1352
Boron Fluoride Ethyl Ether *see* 1352
Boron Lithium Oxide (B₄Li₂O₇) *see* 5580
Boron Monoxide, 1346

Boron Nitride, 1347
Boron Nitride (BN) see 1347
Boron Oxide, 1348
Boron Oxide (BO) see 1346
Boron Oxide (B₂O₃) see 1348
Boron Potassium Oxide (B₄K₂O₇) see 7799
Boron Sesquioxide see 1348
Boron Sodium Oxide (B₄Na₂O₇) see 8726
Boron Tribromide, 1349
Boron Trichloride, 1350
Boron Trifluoride, 1351
Boron Trifluoride Etherate, 1352
Boron Trimethoxide see 9882
Boron Trioxide see 1348
Bortezomib, 1353
Boscalid, 1354
Bosentan, 1355
Bostrycoidin, 1356
Bosutinib, 1357
β-Boswellic Acid, 1358
Bothrops atrox serine proteinase see 1008
Bothrops Venom Proteinase see 1008
BoTox [Allergan] see 1362
Botrilex [Bayer] see 8197
Botropase [Ravizza] see 1008
Botrydial, 1359
Botryococcene, 1360
Bottromycic A₂ Acid Methyl Ester see 1361
Bottromycin, 1361
Bottromycin A₂ see 1361
Botulin A see 1362
Botulin B see 1362
Botulin Toxins, 1362
Botulinum Toxins see 1362
Botulinum Toxin Type A see 1362
Botulinum Toxin Type B see 1362
Botulinus Toxins see 1362
Bouncing Bet see 8500
Bourbonal see 3908
Bovatec see 5427
Bovilene [Syntex] see 4023
Bovine Somatotropin see 8842
Bovine Thymosin α1 see 9560
Bo-Xan see 10261
Boxer [Dow AgroSci.] see 4139
Box Ligands see 1297
Boxogetten [Cheplapharm] see 7417
Boxol [Pharmacia & Upjohn] see 2804
BP-662 see 4115
BPAA see 3985
BPD-MA see 10163
BPEA see 1300
BPE Ligands, 1363
BPI see 931
B₂pin₂ see 1301
BPPS see 7922
BQ 22-708 see 3632
BQC Reagent see 3031
BR see 932
BR-18 see 7596
BR-700 see 4030
BRA see 8278
Bradosol [Novartis] see 1061, 3464
Bradykinin, 1364
Brain-derived Neurotrophic Factor see 6572
Brain Natriuretic Factor see 1365
Brain Natriuretic Peptide, 1365
Brain Sugar see 4365
Brallobarbital, 1366
Branching Factor see 8137

Brandy, 1367
Brandy Mint see 7258
Branigen [GSK] see 79
Brantur [Dong-A] see 6281
Brasilin see 1372
Brassard's Diene, 1368
Brassel [Searle] see 2318
Brassicol [Hoechst] see 8197
Brassidic Acid, 1369
Brassinolide, 1370
Bratenol [Elmu] see 7608
Bratton-Marshall Reagent see 6491
Braunite see 5788
Braunol [Braun Melsungen] see 7815
Braunosan H [Braun Melsungen] see 7815
Bravo [Syngenta] see 2169
Braxin C see 8039
Brayera, 1371
Brazilin, 1372
Brazil Powder see 757
Brazil Wax see 1846
Brazil Wood see 7295
Brazzos [Spiess-Urania] see 4949
BrBcat see 1423
BrdUrd see 1461
Bredereck's Reagent, 1373
Bredinin [Toyo Jozo] see 6308
Bredon [Organon] see 7043
Brefeldin A, 1374
Brek [Irbi] see 5628
Brelomax [Abbott] see 9991
Bremax [Abbott] see 9991
Bremen Blue see 2621
Bremen Green see 2621
Bremfol see 6641
Brenal [Tanabe Seiyaku] see 5851
Brentuximab Vedotin, 1375
Brenzschleimsäure (German) see 4336
Brenztraubensäure (German) see 8135
Breokinase [Sanofi Winthrop] see 10072
Breon see 7707
Brequinar, 1376
Brestan [Hoechst] see 9920
Brethaire [Theragenex] see 9302
Brethine [Novartis] see 9302
BrettPhos Phosphine Ligand see 1469
Bretylate [GSK] see 1377
Bretylium Tosylate, 1377
Bretylol [GSK] see 1377
Brevafen [Janssen-Cilag] see 229
Brevetoxin A see 1378
Brevetoxin B see 1378
Brevetoxin C see 1378
Brevetoxins, 1378
Brevibloc [Baxter] see 3755
Brevicid [AWD] see 10259
Brevicidin see 10017
Brevicon [Watson] see 6786
Brevidil M [M & B] see 9005
Brevifolin see 10265
Brevimytal [Lilly] see 6053
Brevinarcon [Heyden] see 9476
Brevital see 6053
Brevium see 7994
Brevoxyl [Stiefel] see 1119
Brexidol [Chiesi] see 7619
Brexin [Chiesi] see 7619
Brezal [Novartis] see 2212
Bricanyl [AstraZeneca] see 9302
Bricef [BMS] see 1917
Bridion [Schering-Plough] see 9019
Briem [Fabre] see 1033
Brietal [Lilly] see 6053
Brigade [FMC] see 1213

Brigham Tea see 3662
Brij [ICI] see 7697
Brij 58 [ICI] see 7697
Brilinta [AstraZeneca] see 9583
Brilique [AstraZeneca] see 9583
Brilliant Blue FCF, 1379
Brilliant Congo R see 10210
Brilliant Cotton Blue see 6102
Brilliant Green, 1380
Brilliant Vital Red see 10210
Brimonidine, 1381
Brimstone see 9099
Brinaldix [Novartis] see 2382
Brinavess [Merck & Co., Inc.] see 10157
Brinzolamide, 1382
Briofil [Alfa] see 948
Briplatin [BMS] see 2316
Brisfirina [BMS] see 1983
Brispen see 3094
Bristacin [BMS] see 8381
Bristamox [BMS] see 574
Bristamycin [BMS] see 3743
Bristopen [BMS] see 7003
Britaject [Britannia] see 733
British Anti-lewisite see 3231
British Gum see 2955
BritLofex [Britannia] see 5615
Brivanib, 1383
Brivanib Alaninate see 1383
Brivaracetam, 1384
Brivex [Menarini] see 1385
Brivirac [Menarini] see 1385
Brivudin see 1385
Brivudine, 1385
BRL-284 see 7931
BRL-804 see 4707
BRL-1241 see 6041
BRL-1341 see 583
BRL-1400 see 7003
BRL-1621 see 2403
BRL-1702 see 3094
BRL-2039 see 4143
BRL-2064 see 1793
BRL-2288 see 9584
BRL-2333 see 574
BRL-2351 see 902
BRL-3475 see 1793
BRL-4910A see 6388
BRL-8988 see 9165
BRL-14777 see 6428
BRL-17421 see 9280
BRL-24924 see 8255
BRL-26921 see 666
BRL-29060 see 7148
BRL-34915 see 5512
BRL-38227 see 5512
BRL-38705 see 3694
BRL-39123 see 7192
BRL-39123A see 7192
BRL-42810 see 3970
BRL-43694A see 4572
BRL-49653 see 8396
BRL-49653c see 8396
Broadstrike [Dow AgroSci.] see 4172
Brocasipal see 6975
Brochantite see 2644
Brocide see 3851
Brocillin see 7931
Brockmann Aluminum Oxide see 350
Brodiar [Intervet] see 1462
Brodifacoum, 1386
Brodimoprim, 1387
Broflex [Bioglan] see 9864
Brolene Drops [RPR] see 7911
Brolene Ointment [RPR] see 3029

Brolitène [Lab. Medicia] see 4933
Bromacil, 1388
Bromadiolone, 1389
Bromal, 1390
Bromal Hydrate, 1391
Bromallylene see 279
Bromanautine see 1428
Bromat see 2024
Bromated Camphor see 1421
Bromazepam, 1392
Bromazine see 1428
Bromchlophos see 6442
Bromcholitin see 4471
Bromcresol Green, 1393
Bromcresol Purple, 1394
Bromday [Ista] see 1397
Bromdiphenhydramine see 1428
Bromelain, 1395
Bromelin see 1395
Brometazepam see 5990
Bromethalin, 1396
Brometone see 9773
Bromex see 6442
Bromfenac, 1397
Bromhexine, 1398
Bromic Acid, 1399
Bromic Acid Barium Salt (2:1) see 963
Bromic Acid Potassium Salt (1:1) see 7738
Bromic Acid Sodium Salt (1:1) see 8728
Bromic Ether see 3826
Brominal see 1451
Brominated Biphenyls see 7680
Bromindione, 1400
Bromine, 1401
Bromine Cyanide see 2683
Bromine Fluoride (BrF$_3$) see 1403
Bromine Fluoride (BrF$_5$) see 1402
Bromine Monoiodide see 5059
Bromine Pentafluoride, 1402
Bromine Trifluoride, 1403
Bromisoval see 1404
Bromisovalum, 1404
N-Bromoacetamide, 1405
p-Bromoacetanilide, 1406
4'-Bromoacetanilide see 1406
Bromoacetic Acid, 1407
2-Bromoacetic Acid see 1407
Bromoacetone, 1408
ω-Bromoacetophenone, 1409
p-Bromoacetophenone, 1410
4'-Bromoacetophenone see 1410
α-Bromoallyl Bromide see 3030
2-Bromoallyl Bromide see 3030
5-(2-Bromoallyl)-5-isopropylbarbituric Acid see 7910
5-Bromo-2-aminobenzoic Acid see 1412
Bromoanilide see 1406
p-Bromoaniline, 1411
4-Bromoaniline see 1411
p-Bromoanisole see 1439
5-Bromoanthranilic Acid, 1412
Bromo-Benadryl see 1428
Bromobenzene, 1413
α-Bromobenzeneacetonitrile see 1418
4-Bromobenzeneamine see 1411
(E)-N-p-Bromobenzenesulfonyl-N'-[2-[[[2-[(aminoiminomethyl)amino]-4-thiazolyl]methyl]thio]ethyl]formamidine see 3532
4-Bromobenzenesulfonyl Chloride, 1414
β-(p-Bromobenzhydryloxy)ethyldimethylamine see 1428
2-Bromo-1,3-2-benzodioxaborole see 1423

4-Bromobenzoic Acid, 1415
p-Bromobenzyl Bromide, 1416
p-Bromobenzyl Chloride, 1417
α-Bromobenzyl Cyanide, 1418
(o-Bromobenzyl)ethyldimethylammonium p-Toluenesulfonate see 1377
3-[3-(4'-Bromo[1,1'-biphenyl]-4-yl)-3-hydroxy-1-phenylpropyl]-4-hydroxy-2H-1-benzopyran-2-one see 1389
3-[3-(4'-Bromo[1,1'-biphenyl]-4-yl)-1,2,-3,4-tetrahydro-1-naphthalenyl]-4-hydroxy-2H-1-benzopyran-2-one see 1386
3-[3-(4'-Bromo[1,1'-biphenyl]-4-yl)-1,2,-3,4-tetrahydro-1-naphthalenyl]-4-hydroxy-2H-1-benzothiopyran-2-one see 3158
3-[3-(4'-Bromobiphenyl-4-yl)-1,2,3,4-tetrahydro-1-naphthyl]-4-hydroxycoumarin see 1386
3-Bromo-d-2-bornanone see 1421
(2E)-6-Bromo-2-(6-bromo-1,3-dihydro-3-oxo-2H-indol-2-ylidene)-1,2-dihydro-3H-indol-3-one see 3028
1-Bromo-4-(bromomethyl)benzene see 1416
2-Bromo-2-(bromomethyl)glutaronitrile see 3026
2-Bromo-2-(bromomethyl)pentanedinitrile see 3026
2-Bromo-1-(4-bromophenyl)ethanone see 1438
4-Bromo-α-(4-bromophenyl)-α-hydroxybenzeneacetic Acid 1-Methyl Ethyl Ester see 1445
1-Bromobutane see 1555
2-Bromobutane see 1556
(2R)-2-Bromobutane see 1556
(2S)-2-Bromobutane see 1556
2-Bromobutanedioic Acid see 1447
(2R)-2-Bromobutanedioic Acid see 1447
(2S)-2-Bromobutanedioic Acid see 1447
2-Bromobutanoic Acid see 1420
Bromobutide, 1419
5-Bromo-3-sec-butyl-6-methyluracil see 1388
α-Bromobutyric Acid, 1420
dl-2-Bromobutyric Acid see 1420
α-Bromo-d-camphor see 1421
α'-Bromo-d-camphor see 1421
3-Bromo-d-camphor, 1421
3α-Bromo-d-camphor see 1421
3β-Bromo-d-camphor see 1421
α-Bromo-n-caproic Acid, 1422
p-Bromocarbanil see 1441
B-Bromocatecholborane, 1423
α-Bromo-p-chloroacetophenone see 2156
2-Bromo-4'-chloroacetophenone see 2156
7-Bromo-6-chlorofebrifugine see 4632
rel-7-Bromo-6-chloro-3-[3-[(2R,3S)-3-hydroxy-2-piperidinyl]-2-oxopropyl]-4(3H)-quinazolinone see 4632
(±)-trans-7-Bromo-6-chloro-3-[3-(3-hydroxy-2-piperidyl)acetonyl]-4(3H)-quinazolinone see 4632
5-Bromo-4-chloro-1H-indol-3-yl β-D-Galactopyranoside see 10272
1-Bromo-4-(chloromethyl)benzene see 1417
3-Bromo-N-[4-chloro-2-methyl-6-[(methylamino)carbonyl]phenyl]-1-(3-chloro-2-pyridinyl)-1H-pyrazole-5-carboxamide see 2080

7-Bromo-5-(2-chlorophenyl)-2,3-dihydro-2-(methoxymethyl)-1-methyl-1H-1,4-benzodiazepine see 5990
2-Bromo-1-(4-chlorophenyl)ethanone see 2156
4-Bromo-2-(4-chlorophenyl)-1-(ethoxymethyl)-5-(trifluoromethyl)-1H-pyrrole-3-carbonitrile see 2089
O-(4-Bromo-2-chlorophenyl) O-Ethyl S-Propyl Phosphorothioate see 7886
2-Bromo-4-(2-chlorophenyl)-9-methyl-6H-thieno[3,2-f][1,2,4]triazolo[4,3-a]-[1,4]diazepine see 1459
2-Bromo-4-(o-chlorophenyl)-9-methyl-6H-thieno[3,2-f]-s-triazolo[4,3-a][1,4]-diazepine see 1459
8-Bromo-6-(o-chlorophenyl)-1-methyl-4H-s-triazolo[3,4-c]thieno[2,3-e]-1,4-diazepine see 1459
3-Bromo-1-(3-chloro-2-pyridinyl)-N-[4-cyano-2-methyl-6-[(methylamino)carbonyl]phenyl]-1H-pyrazole-5-carboxamide see 2689
3-Bromo-1-(3-chloro-2-pyridyl)-4'-cyano-2'-methyl-6'-(methylcarbamoyl)-pyrazole-5-carboxanilide see 2689
p-Bromo-α-chlorotoluene see 1417
Bromochlorotrifluoroethane see 4637
2-Bromo-2-chloro-1,1,1-trifluoroethane see 4637
Bromocresol Green see 1393
Bromocresol Purple see 1394
Bromocriptine, 1424
Bromocyanogen see 2683
Bromocyclohexane see 2723
Bromocyl see 1446
5-Bromo-2'-deoxyuridine see 1461
p-Bromo-N-[(E)-[[2-[[[2-[(diaminomethylene)amino]-4-thiazolyl]methyl]thio]ethyl]amino]methylene]benzenesulfonamide see 3532
Bromodichloromethane, 1425
O-(4-Bromo-2,5-dichlorophenyl)-O,O-dimethylphosphorothioate see 1442
1-[[4-Bromo-2-(2,4-dichlorophenyl)tetrahydro-2-furanyl]methyl]-1H-1,2,4-triazole see 1456
1-[(2RS,4RS;2RS,4SR)-4-Bromo-2-(2,4-dichlorophenyl)tetrahydrofurfuryl]-1H-1,2,4-triazole see 1456
(8β)-2-Bromo-9,10-didehydro-N,N-diethyl-6-methylergoline-8-carboxamide see 1433
2-Bromo-N,N-diethyl-D-lysergamide see 1433
8-Bromo-3,9-dihydro-1,3-dimethyl-1H-purine-2,6-dione Compd with 2-Amino-2-methyl-1-propanol (1:1) see 7101
(3α,14β,16α)-11-Bromo-14,15-dihydro-14-hydroxyeburnamenine-14-carboxylic Acid Methyl Ester see 1460
5-Bromo-N-(4,5-dihydro-1H-imidazol-2-yl)-6-quinoxalinamine see 1381
7-Bromo-1,3-dihydro-5-(2-pyridinyl)-2H-1,4-benzodiazepin-2-one see 1392
5-Bromo-N,2-dihydroxybenzamide see 1446
4'-Bromo-2,6-dihydroxybenzanilide see 8275
4-Bromo-2,5-dimethoxybenzeneethanamine see 1910
4-Bromo-2,5-dimethoxyphenethylamine see 1910
4-[(2-Bromo-4,5-dimethoxyphenyl)methyl]-4-[2-[2-(6,6-dimethylbicyclo[3.1.1]-

hept-2-yl)ethoxy]ethyl]morpholinium
Bromide (1:1) *see* 7552
5-[(4-Bromo-3,5-dimethoxyphenyl)meth-
yl]-2,4-pyrimidinediamine *see* 1387
2-[*p*-Bromo-α-(2-dimethylaminoethyl)-
benzyl]pyridine *see* 1453
(α*S*,β*R*)-6-Bromo-α-[2-(dimethylamino)-
ethyl]-2-methoxy-α-1-naphthalenyl-β-
phenyl-3-quinolineethanol *see* 1018
2-Bromo-*N*-(α,α-dimethylbenzyl)-3,3-di-
methylbutyramide *see* 1419
Bromodimethylborane, 1426
2-Bromo-3,3-dimethyl-*N*-(1-methyl-1-
phenylethyl)butanamide *see* 1419
α-Bromo-β-dimethylpropanoylurea *see*
1404
*N*¹-(5-Bromo-4,6-dimethyl-2-pyrimidin-
yl)sulfanilamide *see* 9030
5-Bromo-4,6-dimethyl-2-sulfanilamido-
pyrimidine *see* 9030
Bromodimethylsulfonium Bromide, 1427
5-Bromo-6,6′-dimethyl-2,2′,3′,4′-tetra-
methoxybenzophenone *see* 6230
Bromodiphenhydramine, 1428
1-Bromododecane *see* 5444
Bromoeosine *see* 3658
2-Bromoergocryptine *see* 1424
2-Bromo-α-ergokryptin *see* 1424
Bromoethane *see* 3826
2-Bromoethanol *see* 3846
5-[(1*E*)-2-Bromoethenyl]-2′-deoxyuridine
see 1385
β-Bromoethyl Alcohol *see* 3846
2-Bromo-*N*-ethyl-*N*,*N*-dimethylbenzene-
methanaminium 4-Methylbenzenesul-
fonate (1:1) *see* 1377
3-Bromo-*N*-[[(2*S*)-1-ethyl-2-pyrrolidin-
yl]methyl]-2,6-dimethoxybenzamide
see 8252
Bromofenofos, 1429
Bromofluoresceic Acid *see* 3658
4-(4-Bromo-2-fluoroanilino)-6-methoxy-
7-(1-methylpiperidin-4-ylmethoxy)-
quinazoline *see* 10117
N-(4-Bromo-2-fluorophenyl)-6-methoxy-
7-[(1-methylpiperidin-4-yl)methoxy]-
quinazolin-4-amine *see* 10117
N-(4-Bromo-2-fluorophenyl)-6-methoxy-
7-[(1-methyl-4-piperidinyl)methoxy]-4-
quinazolinamine *see* 10117
10-Bromo-11b-(2-fluorophenyl)-2,3,7,-
11b-tetrahydrooxazolo[3,2-*d*][1,4]ben-
zodiazepin-6(5*H*)-one *see* 4638
Bromoform, 1430
Bromofos *see* 1442
1-Bromo-1,1,2,2,3,3,4,4,5,5,6,6,7,7,8,8,8-
heptadecafluorooctane *see* 7270
2-Bromohexanoic Acid *see* 1422
4-Bromo-α-hydroxybenzeneacetic Acid
see 1434
(5′α)-2-Bromo-12′-hydroxy-2′-(1-meth-
ylethyl)-5′-(2-methylpropyl)ergota-
man-3′,6′,18-trione *see* 1424
α-(5-Bromo-4-hydroxy-*m*-tolyl)-α-(3-
bromo-5-methyl-4-oxo-2,5-cyclohexa-
dien-1-ylidene)-*o*-toluenesulfonic Acid
see 1394
5-Bromo-6-(2-imidazolin-2-ylamino)qui-
noxaline *see* 1381
2-Bromoisobutane *see* 1557
α-Bromoisobutyric Acid, 1431
1-Bromo-4-isocyanatobenzene *see* 1441
2-Bromoisopentane *see* 597
α-Bromoisovaleric Acid, 1432
(α-Bromoisovaleryl)urea *see* 1404

Bromo-Kin [Sanofi-Aventis] *see* 1424
Bromol *see* 9775
Bromo-LSD *see* 1433
D-2-Bromolysergic Acid Diethylamide
see 1433
Bromolysergide, 1433
p-Bromomandelic Acid, 1434
[[[(3-Bromomesityl)carbamoyl]methyl]-
imino]diacetic Acid *see* 5841
Bromomethane *see* 6103
(3-Bromo-6-methoxy-2-methylphenyl)(2,-
3,4-trimethoxy-6-methylphenyl)metha-
none *see* 6230
(1*R*,2*S*)-1-(6-Bromo-2-methoxyquinolin-
3-yl)-4-(dimethylamino)-2-(napthalen-
1-yl)-1-phenylbutan-2-ol *see* 1018
(Bromomethyl)benzene *see* 1131
1-Bromo-3-methylbenzene *see* 1449
(*S*)-(+)-1-Bromo-2-methylbutane *see*
1435
1-Bromo-3-methylbutane *see* 5159
(2*S*)-1-Bromo-2-methylbutane, 1435
2-Bromo-2-methylbutane *see* 597
2-Bromo-3-methylbutanoic Acid *see*
1432
1-(Bromomethyl)-3-methylbenzene *see*
10288
5-Bromo-6-methyl-3-(1-methylpropyl)-
2,4(1*H*,3*H*)-pyrimidinedione *see* 1388
5-Bromo-6-methyl-3-(1-methylpropyl)-
uracil *see* 1388
1-Bromo-2-methylpropane *see* 5180
2-Bromo-2-methylpropane *see* 1557
2-Bromo-2-methylpropanoic Acid *see*
1431
α-Bromonaphthalene *see* 1436
β-Bromonaphthalene *see* 1437
1-Bromonaphthalene, 1436
2-Bromonaphthalene, 1437
Bromone *see* 1389
8β-[(5-Bromonicotinoyloxy)methyl]-1,6-
dimethyl-10α-methoxyergoline *see*
6581
2-Bromo-2-nitro-1,3-propanediol *see*
1457
β-Bromo-β-nitrotrimethyleneglycol *see*
1457
1-Bromooctane *see* 6853
2-Bromooctane *see* 6854
Bromooxobismuthine *see* 1263
1-Bromopentane *see* 596
Bromoperfluorooctane *see* 7270
p-Bromophenacyl Bromide, 1438
p-Bromophenetole *see* 1439
Bromophenol, 1439
m-Bromophenol *see* 1439
o-Bromophenol *see* 1439
p-Bromophenol *see* 1439
Bromophenol Blue *see* 1454
Bromophenophos *see* 1429
N-(4-Bromophenyl)acetamide *see* 1406
α-Bromophenylacetonitrile *see* 1418
2-(*p*-Bromo-α-phenylbenzyloxy)-*N*,*N*-di-
methylethylamine *see* 1428
N-[5-(4-Bromophenyl)-6-[2-[(5-bromo-2-
pyrimidinyl)oxy]ethoxy]-4-pyrimidin-
yl]-*N*′-propylsulfamide *see* 5705
4-Bromophenylcarbimide *see* 1441
p-Bromophenylcarbonimide *see* 1441
N-(4-Bromophenyl)-2,6-dihydroxybenz-
amide *see* 8275
γ-(4-Bromophenyl)-*N*,*N*-dimethyl-2-pyr-
idinepropanamine *see* 1453
(2*Z*)-3-(4-Bromophenyl)-*N*,*N*-dimethyl-3-
(3-pyridinyl)-2-propen-1-amine *see*
10325

2-(4-Bromophenyl)-1,3-dioxohydrindene
see 1400
1-(4-Bromophenyl)ethanone *see* 1410
2-Bromo-1-phenylethanone *see* 1409
p-Bromophenylglycolic Acid *see* 1434
p-Bromophenylhydrazine, 1440
(4-Bromophenyl)hydrazine *see* 1440
3-[α-[*p*-(*p*-Bromophenyl)-β-hydroxy-
phenethyl]benzyl]-4-hydroxycoumarin
see 1389
4-[4-(*p*-Bromophenyl)-4-hydroxypi-
peridino]-4′-fluorobutyrophenone *see*
1452
4-[4-(4-Bromophenyl)-4-hydroxy-1-pi-
peridinyl]-1-(4-fluorophenyl)-1-buta-
none *see* 1452
2-(4-Bromophenyl)-1*H*-indene-1,3(2*H*)-
dione *see* 1400
p-Bromophenyl Isocyanate, 1441
3-(*p*-Bromophenyl)-1-methoxy-1-meth-
ylurea *see* 6218
N′-(4-Bromophenyl)-*N*-methoxy-*N*-
methylurea *see* 6218
2-[(4-Bromophenyl)phenylmethoxy]-
N,*N*-dimethylethanamine *see* 1428
(*Z*)-3-(4′-Bromophenyl)-3-(3″-pyridyl)di-
methylallylamine *see* 10325
1-(*p*-Bromophenyl)-1-(2-pyridyl)-3-di-
methylaminopropane *see* 1453
3-(*p*-Bromophenyl)-3-(2-pyridyl)-*N*,*N*-
dimethylpropylamine *see* 1453
Bromophos, 1442
Bromophos-ethyl *see* 1442
Bromophos-methyl *see* 1442
Bromophthalein Magenta E *see* 9331
Bromopride, 1443
1-Bromopropane *see* 7958
2-Bromopropane *see* 5256
3-Bromopropanoic Acid *see* 1444
2-Bromopropanoic Acid Ethyl Ester *see*
3827
1-Bromo-2-propanone *see* 1408
3-Bromo-1-propene *see* 279
5-(2-Bromo-2-propen-1-yl)-5-(1-methyl-
ethyl)-2,4,6(1*H*,3*H*,5*H*)-pyrimidinetri-
one *see* 7910
5-(2-Bromo-2-propen-1-yl)-5-(2-propen-
1-yl)-2,4,6(1*H*,3*H*,5*H*)-pyrimidinetri-
one *see* 1366
β-Bromopropionic Acid, 1444
Bromopropylate, 1445
3-Bromopropylene *see* 279
7-Bromo-5-(2-pyridyl)-3*H*-1,4-benzodi-
azepin-2(1*H*)-one *see* 1392
5-Bromo-2,4(1*H*,3*H*)-pyrimidinedione
see 1450
1-Bromo-2,5-pyrrolidinedione *see* 1448
4′-Bromo-γ-resorcylanilide *see* 8275
5-Bromosalicylhydroxamic Acid, 1446
Bromosuccinic Acid, 1447
N-Bromosuccinimide, 1448
5-Bromosulfamethazine *see* 9030
Bromosulfophthalein *see* 9084
2-Bromo-1,1,1,2-tetrafluoroethane *see*
9250
3-Bromo-2′,3′,4′,6-tetramethoxy-2,6′-di-
methylbenzophenone *see* 6230
8-Bromotheophylline Compd with 2-
Amino-2-methyl-1-propanol (1:1) *see*
7101
Bromothymol Blue *see* 1455
Bromotiren *see* 3035
Bromotoluene, 1449
α-Bromotoluene *see* 1131
ω-Bromotoluene *see* 1131

m-Bromotoluene *see* 1449
o-Bromotoluene *see* 1449
p-Bromotoluene *see* 1449
3-Bromotoluene *see* 1449
α-Bromo-α-tolunitrile *see* 1418
N-(3-Bromo-2,4,6-trimethylacetanilide)-iminodiacetic Acid *see* 5841
3-Bromo-1,7,7-trimethylbicyclo[2.2.1]-heptan-2-one *see* 1421
N-[2-[(3-Bromo-2,4,6-trimethylphenyl)-amino]-2-oxoethyl]-*N*-(carboxymeth-yl)glycine *see* 5841
Bromotriphenylphosphonium Bromide *see* 9917
5-Bromouracil, 1450
5-Bromouracil Deoxyriboside *see* 1461
4-(6-Bromoveratryl)-4-[2-[2-(6,6-dimeth-yl-2-norpinyl)ethoxy]ethyl]morpholi-nium Bromide *see* 7552
cis-11-Bromovincamine *see* 1460
5-Bromovinyl-araU *see* 8853
(*E*)-5-(2-Bromovinyl)-2′-deoxyuridine *see* 5
α-Bromo-*m*-xylene *see* 10288
ω-Bromo-*m*-xylene *see* 10288
Bromoxynil, 1451
Bromperidol, 1452
Brompheniramine, 1453
Bromphenol Blue, 1454
Bromphenphos *see* 1429
Bromsulphalein [Becton Dickinson] *see* 9084
Bromthalein [Merck KGaA] *see* 9084
Bromthymol Blue, 1455
Bromuconazole, 1456
Bromural [Knoll] *see* 1404
Bromuvan *see* 1404
Bromvaletone *see* 1404
Bronalide [Syntex] *see* 4176
Bronalin [Byk-Liprandi] *see* 4743
Broncatar [Pulitzer] *see* 7043
Bronchocillin [Bellon] *see* 7194
Bronchodil [Viatris] *see* 8259
Broncholin [Kaken] *see* 5703
Bronchopront [Mack, Illert.] *see* 380
Bronchoretard [Astellas] *see* 9436
Bronchospasmin [Viatris] *see* 8259
Broncoplus [Sigma-Tau] *see* 8936
Bronica [Takeda] *see* 8598
Bronkephrine [Sterling Winthrop] *see* 3888
Bronner's Acid *see* 1893
Bronopol, 1457
Bronosol [Green Cross] *see* 1457
Bronsecur [SK & F] *see* 1832
Brontine [Yamanouchi] *see* 2913
Brontyl [Reckitt Benckiser] *see* 8014
Brookite *see* 9628
Broom *see* 8540
Brosalamid *see* 1446
Brostallicin, 1458
Brotizolam, 1459
Brotopon [Pfizer Taito] *see* 4634
Brovalurea *see* 1404
Brovana [Sepracor] *see* 4272
Brovavir *see* 8853
Brovel [Lepetit] *see* 3693
Brovincamine, 1460
Brown Acetate of Lime *see* 1649
Brown Mustard *see* 6400
Broxine [NeoPharm] *see* 1461
Broxuridine, 1461
Broxykinolin *see* 1462
Broxynil *see* 1451
Broxyquinoline, 1462

BRS-640 *see* 1364
Bruceantin, 1463
Brucine, 1464
Brucite *see* 5734
Brufen [Abbott] *see* 4919
Brufort [Lampugnani] *see* 4919
Bruggolite [Brueggemann] *see* 8755
Bruisewort *see* 8500
Brulidine [M & B] *see* 3029
Brumetidina [Bruschettini] *see* 2282
Brumixol [Bruschettini] *see* 2269
Brunac [Bruschettini] *see* 83
Bruneomycin *see* 8957
Brunet Saxifrage *see* 7547
Brushite *see* 1694
Bruxicam [Bruschettini] *see* 7619
Bryamycin [BMS] *see* 9517
Bryonia, 1465
Bryony *see* 1465
Bryostatins, 1466
BS-100-141 *see* 4596
BS-572 *see* 2697
BS-4231 *see* 4479
BS-5930 *see* 6975
BS-6987 *see* 2913
BSF-208075 *see* 378
BSF-1 *see* 5040
BSF-2 *see* 5042
B.S.G. [Lemmon] *see* 1283
BSI-201 *see* 5017
BSN-2060 *see* 8886
BSO *see* 1528
BSP *see* 9084
Bt *see* 925
BT *see* 9392
BT-621 *see* 9660
BTC *see* 9921
BTC-1010 [Onyx] *see* 3110
BTE *see* 9961
BTHF *see* 1336
BTMP *see* 1043
BTPABA *see* 1056
BTS *see* 1307
BTS-18322 *see* 4229
BTS-27419 *see* 482
BTS-40542 *see* 7878
BTS-49465 *see* 4142
BTS-54524 *see* 8624
BTX *see* 1378
BTX-B *see* 1378
BTX-C *see* 1378
B-type Natriuretic Peptide *see* 1365
BU-2231 *see* 9175
BU-2231A *see* 9175
BU-2231B *see* 9175
BU-4061T *see* 3686
i-Bu₂AlH *see* 3212
Buccastem [Reckitt Benckiser] *see* 7879
Bucco *see* 1468
Bucetalon *see* 1467
Bucetin, 1467
Buchu, 1468
Buchu Camphor *see* 3326
Buchu Resin *see* 3325
Buchwald Phosphine Ligands, 1469
Bucillamine, 1470
Bucindolol, 1471
Buck Bean *see* 5910
Buckminsterfullerene, 1472
Buckthorn Bark *see* 4291, 8296
Bucku *see* 1468
Buckyball *see* 1472
Bucladesine, 1473
Buclina [Sanofi-Synthelabo] *see* 1474
Buclizine, 1474

Bucolome, 1475
Bucrilate *see* 5185
Bucrylate *see* 5185
Buctril [Bayer CropSci.] *see* 1451
Buddleoflavonoloside *see* 5553
Budeprion [Teva] *see* 1503
Budeson [AstraZeneca] *see* 1476
Budesonide, 1476
Budipine, 1477
Budirol [Fher] *see* 7982
Budoform [Dolder] *see* 5075
BUdR *see* 1461
Bueno [ISK Biosci] *see* 6025
Bufalin, 1478
Bufedil [Abbott] *see* 1480
Bufedon [Cosmopharma] *see* 6821
Bufemid [Lederle] *see* 3996
Bufexamac, 1479
Buflan [Fournier] *see* 1480
Buflocit [CT] *see* 1480
Buflomedil, 1480
Buflonat [Nattermann] *see* 1480
Bufonamin [Kaken] *see* 1481
Bufopto Homatrocel [Softcon] *see* 4768
Bufor [Bofors] *see* 6582
Buformin, 1481
Bufotalin, 1482
Bufotalin 3-Suberoylarginine Ester *see* 1484
Bufotenine, 1483
Bufotoxin, 1484
Bufuralol, 1485
Bugbane *see* 1313
Bugleweed *see* 5685
Bugwort *see* 1313
Buksamin *see* 441
Buku *see* 1468
Bulab 6009 [Bulab] *see* 9223
Bulbocapnine, 1486
Bulbold [Sandoz-Wander] *see* 4520
Bulbonin [Sankyo] *see* 1481
Bulbus Scillae *see* 8898
Bull Nettle *see* 8835
Bullvalene, 1487
Bumadizone, 1488
t-BuMe₂SiCl *see* 1567
Bumetanide, 1489
Bumex [Roche] *see* 1489
Buminate [Baxter] *see* 8607
Bunamiodyl, 1490
Buna S [I. G. Farben] *see* 8527
Bunazosin, 1491
α-Bungarotoxin *see* 1492
β-Bungarotoxin *see* 1492
Bungarotoxins, 1492
Buniodyl *see* 1490
Bunitrolol, 1493
l-Bunolol *see* 5514
Bunsenite *see* 6597
Bunt-cure *see* 4713
Bunte Salts, 1494
Bunt-no-more *see* 4713
t-Bu-P₄ *see* 8535
Buparvaquone, 1495
Bupatol [Gedeon Richter] *see* 947
Buphanamine, 1496
Buphanitine, 1497
Buphedrin [Tatsumi] *see* 6821
Buphenine *see* 6821
t-BuPh₂SiCl *see* 1568
Bupirimate, 1498
Bupivacaine, 1499
(*S*)-Bupivacaine *see* 1499
Bupranol *see* 1500
Bupranolol, 1500

Butopamine Hydrochloride *see* 8208
Butorphanol, 1532
Butox [Intervet] *see* 2888
Butoxone *see* 2838
3-Butoxy-4-aminobenzoic Acid 2-(Diethylamino)ethyl Ester *see* 1049
4-Butoxybenzoic Acid 2-(Diethylamino)-ethyl Ester *see* 1533
8-(*p*-Butoxybenzyl)-3α-hydroxy-1α*H*,-5α*H*-tropanium Bromide (−)-Tropate *see* 1536
l-[1-(*p-n*-Butoxybenzyl)hyoscyaminium] Bromide *see* 1536
tert-Butoxybis(dimethylamino)methane *see* 1373
Butoxycaine, 1533
N-[*N*-[*N*-[*N*-(*N*-*tert*-Butoxycarbonyl-β-alanyl)-L-tryptophanyl]-L-methionyl]-L-aspartyl]-L-phenylalaninamide *see* 7224
2-(*tert*-Butoxycarbonyloxyimino)-2-phenylacetonitrile *see* 1323
2-Butoxy-*N*-(2-diethylaminoethyl)cinchoninamide *see* 3036
2-Butoxy-*N*-[2-(diethylamino)ethyl]-4-quinolinecarboxamide *see* 3036
2-Butoxyethanol *see* 1560
2-(2-Butoxyethoxy)ethanol *see* 3133
α-[2-(2-Butoxyethoxy)ethoxy]-4,5-methylenedioxy-2-propyltoluene *see* 7589
5-[[2-(2-Butoxyethoxy)ethoxy]methyl]-6-propyl-1,3-benzodioxole *see* 7589
4-Butoxy-*N*-hydroxybenzeneacetamide *see* 1479
N-(Butoxymethyl)-2-chloro-2′,6′-diethylacetanilide *see* 1511
N-(Butoxymethyl)-2-chloro-*N*-(2,6-diethylphenyl)acetamide *see* 1511
4-[3-(4-Butoxyphenoxy)propyl]morpholine *see* 7829
2-(*p*-Butoxyphenyl)acetohydroxamic Acid *see* 1479
[3(*S*)-*endo*]-8-[(4-Butoxyphenyl)methyl]-3-(3-hydroxy-1-oxo-2-phenylpropoxy)-8-methyl-8-azoniabicyclo[3.2.1]octane Bromide *see* 1536
p-Butoxyphenyl γ-Morpholinopropyl Ether *see* 7829
4-Butoxyphenyl Piperidineethyl Ketone *see* 3520
1-(4-Butoxyphenyl)-3-(1-piperidinyl)-1-propanone *see* 3520
4-Butoxy-β-piperidinopropiophenone *see* 3520
4-*n*-Butoxy-β-(1-piperidyl)propiophenone *see* 3520
tert-Butoxy-*N*,*N*,*N'*,*N'*-tetramethylmethanediamine *see* 1373
Butoz [Hamilton] *see* 7390
Butralin, 1534
Butriptylene *see* 1535
Butriptyline, 1535
Butropium Bromide, 1536
Buttercup Yellow *see* 10332
Butterfly Weed *see* 817
Butternut *see* 5316
Butter of Antimony *see* 696
Butter of Arsenic *see* 792
Butter of Zinc *see* 10331
Butter Yellow *see* 3251
β-BuTX *see* 1492
Butyl Carbitol [Dow] *see* 3133
Butyl Chemosept *see* 1587
Butyl Diicinol [ICI] *see* 3133
Butyl Parasept *see* 1587

n-Butyl Acetate, 1537
sec-Butyl Acetate, 1538
tert-Butyl Acetate, 1539
21-*tert*-Butylacetate-9α-fluoro-11β-hydroxy-16α,17α-(isopropylidenedioxy)-pregna-1,4-diene-3,20-dione *see* 9759
tert-Butylacetic Acid, 1540
3-(*N*-Butyl-*N*-acetyl)-aminopropionic Acid Ethyl Ester *see* 3828
n-Butyl Acrylate, 1541
Butyl Alcohol *see* 1542
n-Butyl Alcohol, 1542
sec-Butyl Alcohol, 1543
tert-Butyl Alcohol, 1544
tert-Butyl Alcohol Aluminum Salt *see* 329
n-Butylamine, 1545
sec-Butylamine, 1546
tert-Butylamine, 1547
Butyl Aminobenzoate *see* 1515
n-Butyl *p*-Aminobenzoate *see* 1515
4-(Butylamino)benzoic Acid 2-(Dimethylamino)ethyl Ester Hydrochloride (1:1) *see* 9333
p-Butylaminobenzoic Acid ω-*O*-Methylnonaethyleneglycol Ester *see* 1098
4-(Butylamino)benzoic Acid 3,6,9,12,15,-18,21,24,27-Nonaoxaoctacos-1-yl Ester *see* 1098
p-Butylaminobenzoyl-2-dimethylaminoethanol Hydrochloride *see* 9333
N-[1-[(Butylamino)carbonyl]-1*H*-benzimidazol-2-yl]carbamic Acid Methyl Ester *see* 1046
N-[(Butylamino)carbonyl]-4-methylbenzenesulfonamide *see* 9667
1-*tert*-Butylamino-3-(2-chloro-5-methylphenoxy)-2-propanol *see* 1500
(±)-2-(*tert*-Butylamino)-3′-chloropropiophenone *see* 1503
1-(*tert*-Butylamino)-3-[(6-chloro-*m*-tolyl)-oxy]-2-propanol *see* 1500
(*S*)-1-(*tert*-Butylamino)-3-(*o*-cyclopentylphenoxy)-2-propanol *see* 7191
2-(*tert*-Butylamino)-4-(cyclopropylamino)-6-(methylthio)-*s*-triazine *see* 5128
4-(Butylamino)-2-hydroxybenzoic Acid 2-(Dimethylamino)ethyl Ester *see* 4885
(−)-1-*tert*-Butylamino-2-hydroxy-3-(2′-cyclopentylphenoxy)propane *see* 7191
2-(2-*tert*-Butylamino-1-hydroxyethyl)-7-ethylbenzofuran *see* 1485
(±)-5-[2-(*tert*-Butylamino)-1-hydroxyethyl]-*m*-phenylene Bis(dimethylcarbamate) *see* 946
2-(*tert*-Butylamino)-1-(4-hydroxy-3-hydroxymethylphenyl)ethanol *see* 210
2-Butylamino-1-*p*-hydroxyphenylethanol *see* 947
o-[3-(*tert*-Butylamino)-2-hydroxypropoxy]benzonitrile *see* 1493
5-[3-(*tert*-Butylamino)-2-hydroxypropoxy]-3,4-dihydrocarbostyril *see* 1866
(−)-5-[3-(*tert*-Butylamino)-2-hydroxypropoxy]-3,4-dihydro-1(2*H*)-naphthalenone *see* 5514
4-[3-(*tert*-Butylamino)-2-hydroxypropoxy]-*N*-methylisocarbostyril *see* 9595
S-(−)-3-(3-*tert*-Butylamino-2-hydroxypropoxy)-4-morpholino-1,2,5-thiadiazole *see* 9601
(±)-1-[*p*-[3-(*tert*-Butylamino)-2-hydroxypropoxy]phenyl]-3-cyclohexylurea *see* 9170

(2*R*,3*S*)-5-[3-(*tert*-Butylamino)-2-hydroxypropoxy]-1,2,3,4-tetrahydronaphthalene-2,3-diol *see* 6431
2-(3′-*tert*-Butylamino-2′-hydroxypropylthio)-4-(5′-carbamoyl-2′-thienyl)thiazole *see* 782
α-[(*tert*-Butylamino)methyl]-*o*-chlorobenzyl Alcohol *see* 9991
α-[(*tert*-Butylamino)methyl]-3,5-dihydroxybenzyl Alcohol *see* 9302
α-[(*tert*-Butylamino)methyl]-3,4-dihydroxybenzyl Alcohol 3,4-Di-*p*-toluate *see* 1308
α-[(*tert*-Butylamino)methyl]-7-ethyl-2-benzofuranmethanol *see* 1485
α-[(Butylamino)methyl]-4-hydroxybenzenemethanol *see* 947
α-[(Butylamino)methyl]-*p*-hydroxybenzyl Alcohol *see* 947
α-(*t*-Butylaminomethyl)-4-hydroxy-3-ureidobenzyl Alcohol *see* 1832
α¹-[(*tert*-Butylamino)methyl]-4-hydroxy-*m*-xylene-α,α′-diol *see* 210
(±)-1-(*tert*-Butylamino)-3-[(2-methylindol-4-yl)oxy]-2-propanol Benzoate (Ester) *see* 1335
3-(Butylamino)-4-phenoxy-5-sulfamoylbenzoic Acid *see* 1489
p-Butylaminosalicylic Acid 2-Dimethylaminoethyl Ester *see* 4885
1-(*tert*-Butylamino)-3-[(5,6,7,8-tetrahydro-*cis*-6,7-dihydroxy-1-naphthyl)-oxy]-2-propanol *see* 6431
(±)-1-*tert*-Butylamino-3-(1-thiachroman-8-yloxy)-2-propanol *see* 9316
Butylate, 1548
Butylated Hydroxyanisole, 1549
Butylated Hydroxytoluene, 1550
Butylbenzene *see* 1551
n-Butylbenzene, 1551
sec-Butylbenzene, 1552
tert-Butylbenzene, 1553
α-Butylbenzenemethanol *see* 4006
N-(6-Butyl-1*H*-benzimidazol-2-yl)carbamic Acid Methyl Ester *see* 7139
n-Butyl Benzoate, 1554
2-Butyl-3-benzofuranyl-4-[2-(diethylamino)ethoxy]-3,5-diiodophenyl Ketone *see* 478
(2-Butyl-3-benzofuranyl)[4-[2-(diethylamino)ethoxy]-3,5-diiodophenyl]methanone *see* 478
α-Butylbenzyl Alcohol *see* 4006
1-(*p*-*tert*-Butylbenzyl)-4-(*p*-chlorodiphenylmethyl)piperazine *see* 1474
N-(4-*t*-Butylbenzyl)-4-chloro-3-ethyl-1-methylpyrazole-5-carboxamide *see* 9231
1-(*p*-*tert*-Butylbenzyl)-4-(*p*-chloro-α-phenylbenzyl)piperazine *see* 1474
N-(*p*-*tert*-Butylbenzyl)-*N*-methyl-1-naphthalenemethylamine *see* 1521
n-Butylbiguanide *see* 1481
1-Butylbiguanide *see* 1481
n-Butyl Bromide, 1555
sec-Butyl Bromide, 1556
tert-Butyl Bromide, 1557
N-Butyl-1-butanamine *see* 3037
2-*tert*-Butyl-5-(4-*tert*-butylbenzylthio)-4-chloropyridazin-3(2*H*)-one *see* 8078
N-*sec*-Butyl-4-*tert*-butyl-2,6-dinitroaniline *see* 1534
n-Butyl *n*-Butyrate, 1558
N-Butylcarbamic Acid 2-[[(Aminocarbonyl)oxy]methyl]-2-methylpentyl Ester *see* 10010

tert-Butyl Hypochlorite, 1574

(11β,16α)-16,17-[Butylidenebis(oxy)]-11,21-dihydroxypregna-1,4-diene-3,20-dione see 1476

Butylidene Chloride, 1575

p,p'-sec-Butylidenediphenol see 1299

6-tert-Butyl-3-(2-imidazolin-2-ylmethyl)-2,4-dimethylphenol see 7065

N-Butylimidodicarbonimidic Diamide see 1481

2-tert-Butylimino-3-isopropyl-5-phenyl-perhydro-1,3,5-thiadiazin-4-one see 1502

n-Butyl Iodide, 1576

sec-Butyl Iodide, 1577

6-sec-Butyl-3-isobutylpyrazinol 1-Oxide see 831

Butyl Isocyanate, 1578

n-Butyl Isocyanate see 1578

n-Butylmalonic Acid, 1579

Butylmalonic Acid Mono(1,2-diphen-ylhydrazide) see 1488

n-Butyl Mercaptan, 1580

sec-Butyl Mercaptan, 1581

tert-Butyl Mercaptan, 1582

p-Butylmercaptobenzhydryl β-Dimethyl-aminoethyl Sulfide see 1775

n-Butylmercuric Chloride, 1583

Butyl Methoxydibenzoylmethane see 881

4-tert-Butyl-4'-methoxydibenzoylmeth-ane see 881

2-tert-Butyl-4-methoxyphenol see 1549

3-tert-Butyl-4-methoxyphenol see 1549

N-tert-Butyl-N'-(3-methoxy-o-toluoyl)-3,5-xylohydrazide see 6066

(2S,5S)-(−)-2-tert-Butyl-3-methyl-5-ben-zyl-4-imidazolidinone see 5708

tert-Butyl Methyl Ether see 6106

Butylmethylimidazolium, 1584

1-Butyl-3-methyl-1H-imidazolium see 1584

3-Butyl-1-methyl-1H-imidazolium see 1584

Butyl Methyl Ketone see 6107

tert-Butyl Methyl Ketone see 7551

2-sec-Butyl-2-methyl-1,3-propanediol Di-carbamate see 5842

N-Butyl-2-methyl-2-propyl-1,3-propane-diol Dicarbamate see 10010

2-sec-Butyl-2-methyltrimethylenecarba-mate see 5842

2'-tert-Butyl-5-methyl-2'-(3,5-xyloyl)-chromane-6-carbohydrazide see 2223

N-Butylmoranoline see 6268

n-Butyl Nitrite, 1585

tert-Butyl Nitrite, 1586

Butylon see 1467

2-[p-(Butyloxy)phenyl]acetohydroxamic Acid see 1479

Butylparaben, 1587

tert-Butyl Peroxalate see 3043

Butylphen see 1588

4-tert-Butylphenethyl Quinazolin-4-yl Ether see 3993

4-tert-Butylphenol, 1588

2-(p-tert-Butylphenoxy)cyclohexyl Pro-pargyl Sulfite see 7922

4-Butyl-1-phenyl-3,5-dioxopyrazolidine see 6317

4-[2-[4-(tert-Butyl)phenyl]ethoxy]quin-azoline see 3993

1-(p-tert-Butylphenyl)-4-[4'-(α-hydroxy-diphenylmethyl)-1'-piperidyl]butanol see 9306

α-(p-tert-Butylphenyl)-4-(α-hydroxy-α-phenylbenzyl)-1-piperidinebutanol see 9306

(RS)-1-[3-(4-tert-Butylphenyl)-2-methyl-propyl]piperidine see 4020

N-tert-Butyl-α-phenylnitrone see 7162

4-Butyl-1-phenyl-3,5-pyrazolidinedione see 6317

4-tert-Butylphenyl Salicylate, 1589

n-Butyl Phthalate see 3044

5-Butylpicolinic Acid see 4343

dl-N-n-Butylpipecolic Acid 2,6-Xylidide see 1499

dl-1-Butyl-2',6'-pipecoloxylidide see 1499

dl-1-n-Butylpiperidine-2-carboxylic Acid 2,6-Dimethylanilide see 1499

N-[(1-Butyl-4-piperidinyl)methyl]-3,4-di-hydro-2H-[1,3]oxazino[3,2-a]indole-10-carboxamide see 7505

2-Butylpropanedioic Acid see 1579

2-Butylpropanedioic Acid-1-(1,2-diphen-ylhydrazide) see 1488

n-Butyl Propionate, 1590

5-Butyl-2-pyridinecarboxylic Acid see 4343

Butylscopolamine Bromide see 1591

N-Butylscopolammonium Bromide, 1591

2-n-Butyl-4-spirocyclopentane-1-[(2'-(tet-razol-5-yl)biphenyl-4-yl)methyl]-2-imidazolin-5-one see 5127

Butyl Stearate, 1592

1-Butyl-3-sulfanilylurea see 1831

n-Butyl Sulfide, 1593

tert-Butylsulfinamide see 3607

2-S-(n-Butylsulfonylamino)-3-[4-(piperi-din-4-yl)butyloxyphenyl]propionic Acid see 9620

N-(Butylsulfonyl)-O-[4-(4-piperidinyl)bu-tyl]-L-tyrosine see 9620

N-(Butylsulfonyl)-4-[4-(4-piperidyl)bu-toxy]-L-phenylalanine see 9620

2-Butyl-3-[[2'-(2H-tetrazol-5-yl)[1,1'-bi-phenyl]-4-yl]methyl]-1,3-diazaspiro-[4.4]non-1-en-4-one see 5127

Butyl 3'-(1H-Tetrazol-5-yl)oxanilate see 9217

2-Butyl-3-[p-(o-1H-tetrazol-5-ylphenyl)-benzyl]-1,3-diazaspiro[4.4]non-1-en-4-one see 5127

N-(5-tert-Butyl-1,3,4-thiadiazol-2-yl)ben-zenesulfonamide see 4516

1-(5-tert-Butyl-1,3,4-thiadiazol-2-yl)-1,3-dimethylurea see 9232

N¹-(5-tert-Butyl-1,3,4-thiadiazol-2-yl)sul-fanilamide see 4515

sec-Butyl Thioalcohol see 1581

Butylthiobutane see 1593

p-Butylthiodiphenylmethyl 2-Dimethyl-aminoethyl Sulfide see 1775

S-tert-Butylthiomethyl O,O-Diethyl Phos-phorodithioate see 9301

2-[p-(Butylthio)-α-phenylbenzylthio]-N,N-dimethylethylamine see 1775

2-[[[4-(Butylthio)phenyl]phenylmethyl]-thio]-N,N-dimethylethanamine see 1775

Butyltin Trichloride, 1594

1-Butyl-3-(p-tolylsulfonyl)urea see 9667

N-n-Butyl-N'-tosylurea see 9667

Butyltrichlorostannane see 1594

Butyltrichlorotin see 1594

Butyl 2-[4-(5-Trifluoromethyl-2-pyrid-yloxy)phenoxy]propionate see 4147

1-Butyl-2-(2,6-xylylcarbamoyl)piperidine see 1499

Butyn Sulfate [Abbott] see 1510

2-Butynediamide see 1964

1,1'-(2-Butyne-1,4-diyl)bispyrrolidine see 9744

Butynorate see 3047

1,1'-(2-Butynylene)dipyrrolidine see 9744

Butyrac [Union Carbide] see 2838

Butyraldehyde, 1595

n-Butyramide, 1596

3-Butyramido-α-ethyl-2,4,6-triiodocinna-mic Acid see 1490

3-Butyramido-α-ethyl-2,4,6-triiodohy-drocinnamic Acid Sodium Salt see 10018

5'-Butyramido-2'-(2-hydroxy-3-isopro-pylaminopropoxy)acetophenone see 20

Butyrchloral see 9799

Butyric Acid, 1597

n-Butyric Acid see 1597

Butyric Acid Butyl Ester see 1558

Butyric Acid Ethyl Ester see 3830

1-Butyric Acid-2-[3-(p-methoxyphenyl)-L-alanine]oxytocin see 1796

Butyric Acid Nitrile see 1601

Butyric Anhydride, 1598

Butyroin, 1599

Butyrolactone, 1600

γ-Butyrolactone see 1600

Butyrone see 3386

Butyronitrile, 1601

Butyroxymethyl Methyl 4-(2',3'-Dichlo-rophenyl)-2,6-dimethyl-1,4-dihydro-pyridine-3,5-dicarboxylate see 2350

N-Butyroyl-p-aminophenol see 4853

3-(3-Butyrylamino-2,4,6-triiodophenyl)-2-ethylacrylic Acid see 1490

Butyryl Chloride see 1602

n-Butyryl Chloride, 1602

Butyrylcholinesterase see 2214

3'-[(5-Butyryl-2,4-dihydroxy-3,3-dimeth-yl-6-oxo-1,4-cyclohexadien-1-yl)meth-yl]-2',6'-dihydroxy-4'-methoxybutyro-phenone see 2918

3'-[(5-Butyryl-2,4-dihydroxy-3,3-dimeth-yl-6-oxo-1,4-cyclohexadien-1-yl)meth-yl]-2',4'-dihydroxy-6'-methoxy-5'-methylbutyrophenone see 836

3'-[(5-Butyryl-2,4-dihydroxy-3,3-dimeth-yl-6-oxo-1,4-cyclohexadien-1-yl)meth-yl]-5'-methylphlorobutyrophenone see 4121

4-Butyryl-3,5-dihydroxy-1-methoxy-2-methylbenzene see 837

1-Butyrylglycerol see 6335

3-Butyryl-sn-glycerol see 6335

4-Butyryl-2-methylphloroglucinol 1-Methyl Ether see 837

2-Butyryl-10-[3-(4-methyl-1-piperazinyl)-propyl]phenothiazine see 1519

Butyryl Oxide see 1598

Butyrylperazine see 1519

Butyryl Triglyceride see 9785

Butyvinal see 10190

Buventol [HRA-Pharma] see 210

Buzepide, 1603

Buzon [Knoll] see 7390

BV-26-723 see 1460

BV-araU see 8853

BVAU see 8853

BVDU see 1385

B.V.U. see 1404

BW15AU see 9748

BW-33A see 859

BW-33-T-57 see 6049

BW-57-322 *see* 895
BW-248U *see* 134
BW-256 *see* 10085
BW-256U87 *see* 10085
BW-270C *see* 118
BW-301U *see* 7612
BW-311C90 *see* 10390
BW-356-C-61 *see* 5347
BW-430C *see* 5403
BW-467C60 *see* 1188
BW-524W91 *see* 3622
BW-566C *see* 857
BW-566C-80 *see* 857
BW-720C *see* 1495
BW-759 *see* 4393
BW-759U *see* 4393
BW-825C *see* 118
BW-993C *see* 7154
BW-56158 *see* 271
BW-A509U *see* 10322
BW-A825C *see* 118
BW-A938U *see* 3480
BW-B759U *see* 4393
BW-B1090U *see* 6305
BX-112 *see* 7892
BX-661A *see* 938
BXR-220 *see* 775
BY-217 *see* 8378
BY-1023 *see* 7117
BY-9010 *see* 2266
Bydolax *see* 7073
Byemite [Bayer] *see* 7478
Byetta [Amylin] *see* 3955
BYI-8330 *see* 8888
BYK 1512 *see* 4743
Bykahepar [Byk Gulden] *see* 2337
Bykomycin [Byk Gulden] *see* 6539
Bylotensin [Yoshitomi] *see* 6662
Bystolic [Forest] *see* 6516
Byzantine Purple *see* 3028
BZ *see* 8199
BZ-55 *see* 1831
BZQ *see* 1122
C-6 *see* 4724
c7E3 Fab *see* 5
C-10 *see* 2851
C-16 JH *see* 5321
C-17 JH *see* 5321
C-18 JH *see* 5321
C-78 *see* 9991
C-83 *see* 8391
C-238 *see* 7859
C-247 *see* 3331
C-254 *see* 882
C-283 *see* 6657
C-637 *see* 5505
C-709 *see* 3097
C-1414 *see* 6337
C-5473 *see* 7074
C-5581H *see* 7425
C-5720 *see* 1862
C-5968 *see* 4802
C-7337 *see* 7376
C-7819B *see* 6501
C-9211 *see* 3080
C-9333-Go *see* 571
C-19393-H$_2$ *see* 1861
C-19393-S$_2$ *see* 1861
C-48401-Ba *see* 4633
C225 *see* 2026
C.A. *see* 1418
CA1 *see* 2477
cA2 *see* 5014
CA4 *see* 2477
CA4P *see* 4277

CA4P Mono-TRIS Salt *see* 4277
Ca-AOTA *see* 18
Cabagin-U [Kowa] *see* 6170
(−)-Cabastine *see* 5515
Cabazitaxel, 1604
Cabenegrin A-I *see* 1605
Cabenegrin A-II *see* 1605
Cabenegrins, 1605
Cabergoline, 1606
Cabeza De Negro *see* 10290
Cabral [Solvay] *see* 7432
Cabrio [BASF] *see* 8064
Cacao Butter *see* 9432
Cacao Shell, 1607
Cachectin *see* 9992
Cachou *see* 1904
Cacit [Norwich] *see* 1659
Cacodyl, 1608
Cacodylic Acid, 1609
Cacodylic Acid Sodium Salt *see* 8730
Cacotheline, 1610
Cacoxenite *see* 4056
CACP *see* 2316
Cactinomycin, 1611
Cactoid *see* 1612
Cactus Grandiflorus, 1612
CAC10-vcMMAE *see* 1375
CAD *see* 1885
Cadalene, 1613
Cadalin *see* 1613
Cadaverine, 1614
Caddy [Cleary] *see* 1622
Cadexomer Iodine, 1615
Cadherins, 1616
Cadillo *see* 10058
α-Cadinene *see* 1617
β-Cadinene *see* 1617
Cadinenes, 1617
Cadmium, 1618
Cadmium Acetate, 1619
Cadmium Bromide, 1620
Cadmium Carbonate, 1621
Cadmium Chloride, 1622
Cadmium Chloride (CdCl$_2$) *see* 1622
Cadmium Cyanide, 1623
Cadmium Dimethyl *see* 3261
Cadmium Fluoride, 1624
Cadmium Hydrate *see* 1625
Cadmium Hydroxide, 1625
Cadmium Iodide, 1626
Cadmium Nitrate, 1627
Cadmium Oxide, 1628
Cadmium Potassium Cyanide, 1629
Cadmium Salicylate, 1630
Cadmium Selenide, 1631
Cadmium Sulfate, 1632
Cadmium Sulfide, 1633
Cadmium Telluride, 1634
Cadmium Tungstate(VI), 1635
Cadmium Tungsten Oxide (CdWO$_4$) *see* 1635
Cadmium Yellow *see* 1633
Cadoxen *see* 9928
Cadusafos, 1636
Caelyx [Schering-Plough] *see* 3487
Caerin 2.3 *see* 8890
Caerulein *see* 2004
Caeruloplasmin *see* 2005
Caesarweed *see* 10058
Caesium *see* 2007
CAF *see* 4104
Cafcit [Roxane] *see* 1639
Cafesterol *see* 1637
Cafestol, 1637

Caffearine *see* 9863
Caffeic Acid, 1638
Caffeic Acid 1-Carboxy-4,5-dihydroxy-
  1,3-cyclohexylene Ester *see* 2763
Caffeic Acid 3-Methyl Ether *see* 4089
Caffeine, 1639
Caffeine Citrate *see* 1639
α-O-Caffeoyl-(3,4-dihydroxyphenyl)lac-
  tic Acid *see* 8399
3-Caffeoylquinic Acid *see* 2143
N-8-Caffeyl-N′-methyl-N′-(α-methyl-
  phenethyl)ethylenediamine *see* 3999
Cafrolicycline *see* 8381
Caid [Lipha] *see* 2155
Cajeputene *see* 5546
Cajeputi Oil *see* 6871
Cajeputol *see* 3938
Cajuput Oil *see* 6871
Cake Alum *see* 360
Calabar Bean *see* 7495
Calabash Curare *see* 2661
Calamus, 1640
Calan [Searle] *see* 10144
Cal-Aspirin [Winthrop] *see* 1650
Calavarite *see* 4549
Calblock [Sankyo] *see* 900
Calcamine [Wander] *see* 3198
Calcibind [Mission Pharmacal] *see* 8732
Calcichew [Shire] *see* 1659
Calcic Liver of Sulfur *see* 1709
Calcidia [Roche] *see* 1659
Calcifediol, 1641
Calciferol *see* 10215
Calciforte [Serolam] *see* 4491
Calcigenol Simple *see* 1696
Calcijex [Abbott] *see* 1647
Calcimar [Sanofi-Aventis] *see* 1646
Calcimax [Gador] *see* 1663
Calcimycin, 1642
Calcined Baryta *see* 982
Calcined Magnesia *see* 5741
Calcineurin, 1643
Calcineurin A *see* 1643
Calcineurin B *see* 1643
Calcinil [Sclavo] *see* 3585
Calciofon *see* 1671
Calciopen [AstraZeneca] *see* 7209
Calciparine [Sanofi Winthrop] *see* 4688
Calcipen [Leo Pharm] *see* 7209
Calcipotriene, 1644
Calcipotriol *see* 1644
Calciseptin (*Dendroaspis polylepis poly-
  lepis*) *see* 1645
Calciseptine, 1645
Calcisorb [3M Pharma] *see* 8732
Calcite *see* 1659
Calcitonin, 1646
Calcitonin Gene Related Peptide *see* 603
Calcitriol, 1647
Calcium, 1648
Calcium Acetate, 1649
Calcium Acetyl Homotaurinate *see* 18
Calcium Acetylsalicylate, 1650
Calcium Acetylsalicylate Carbamide *see*
  1650
Calcium Aluminosilicate, 1651
Calcium Aluminum Hydride *see* 330
Calcium Arsenate, 1652
Calcium Arsenite, 1653
Calcium Ascorbate *see* 819
Calcium Aspirin *see* 1650
Calcium Benzoate *see* 1093
Calcium Bichromate *see* 1668
Calcium Biphosphate *see* 1695
Calcium Bisacetyl Homotaurine *see* 18

(1*S*,3*R*)-Camphoric Acid *see* 1735
Camphoric Acid 1-(*p*,α-Dimethylbenzyl)
  Ester, Compd with 2,2′-Imidodi-
  ethanol (1:1) *see* 9649
Camphor Monobromated *see* 1421
Camphor Monochlorated *see* 2132
Camphor Solubilized *see* 1733
(+)-β-Camphorsulfonic Acid *see* 1736
d-Camphorsulfonic Acid, 1736
d-10-Camphorsulfonic Acid *see* 1736
Camphydryl *see* 1733
Camposan *see* 3787
Campral [Merck Santé] *see* 18
Campto [Sanofi-Aventis] *see* 5135
Camptosar [Pharmacia & Upjohn] *see*
  5135
Camptothecin, 1737
Camsellite *see* 5722
Camtobell [Chong Kun Dang] *see* 1028
Camvet [Merck & Co.] *see* 1728
Camyna [Nycomed] *see* 9612
Canada Snakeroot *see* 814
Canada Turpentine *see* 939
Canadian Hemp *see* 730
Canadine, 1738
Canadol *see* 5542
Canakinumab, 1739
Cananga Oil *see* 10298
Canary Dextrin *see* 2955
Cancer Bush *see* 9138
Canchalagua *see* 1970
Cancidas [Merck & Co.] *see* 1889
Candelilla Wax, 1741
Candesartan, 1742
Candesartan Cilexetil *see* 1742
Candex [Miles] *see* 6825
Candicidin, 1743
Candicidin D *see* 1743
Candidin, 1744
Candidinin *see* 1744
Candidoin *see* 1744
Candio-Hermal [Hermal] *see* 6825
Candit [BASF] *see* 5372
Candleberry Bark *see* 1010
Candoxatril, 1745
Candoxatrilat *see* 1745
Candyleaf *see* 8937
Canella, 1746
Canertinib, 1747
Canescine *see* 2919
Canesten [Bayer] *see* 2401
Cane Sugar *see* 9012
Canex [Mallinckrodt Vet.] *see* 8403
Canferon [Takeda] *see* 5034
Canfosfamide, 1748
Canfoxil [Gentili] *see* 1733
Cangrelor, 1749
Canifug [Wolff] *see* 2401
Caniphedrin [Streuli] *see* 3663
Cannabidiol, 1750
Cannabinol, 1751
Cannabis, 1752
Cannabiscetin *see* 6417
Cannoc [Takeda] *see* 3757
Canocenta [Byk Gulden] *see* 5643
Canrenone, 1753
Cantabilin [Formenti] *see* 4894
Cantabiline [Merck KGaA] *see* 4894
Cantan [Aventis] *see* 819
Cantharides, 1754
Cantharides Camphor *see* 1755
Cantharidin, 1755
Canthaxanthin, 1756
Cantil [Aventis] *see* 5915

Canton's Phosphorus *see* 1709
Cantor [Sanofi Winthrop] *see* 6281
Cantralax [Ferrosan] *see* 1878
Cantrex *see* 5329
Cantrodifene (obsolete) *see* 6719
Cantus [BASF] *see* 1354
Caomet [AstraZeneca] *see* 10025
Caoutchouc *see* 8411
CAP-440 *see* 856
Caparlem *see* 5319
Caparol [Syngenta] *see* 7904
Caparsolate *see* 9441
Capastat [Lilly] *see* 1759
Capecitabine, 1757
Caperase *see* 1900
Capistan [Pierre Fabre] *see* 8524
Capisten [Kissei] *see* 5352
Capitol [Dermal] *see* 1061
Capitrol [Westwood] *see* 2181
Capitus [Berk] *see* 3801
Capla [Wallace Labs.] *see* 5842
Caplenal [Teva] *see* 271
Capmul 8210 [Stokely-Van Camp] *see*
  6338
Capmul O [Capital City Prod.] *see* 8850
Capnoidine *see* 154
Capoten [BMS] *see* 1776
Capotena [BMS] *see* 1776
Caprelsa [AstraZeneca] *see* 10117
Caprenin, 1758
Capreomycin, 1759
Capreomycin IA *see* 1759
Capreomycin IB *see* 1759
Capreomycin IIA *see* 1759
Capreomycin IIB *see* 1759
n-Capric Acid, 1760
Caprine *see* 6795
n-Caproic Acid, 1761
n-Caproic Acid *n*-Amyl Ester *see* 599
Caproaldehyde *see* 1762
Caprocin [Lilly] *see* 1759
Caprokol [Merck & Co.] *see* 4748
Caprolactam, 1763
ε-Caprolactam *see* 1763
Caprolan [AlliedSignal] *see* 6823
Capromab, 1764
Capromab Pendetide *see* 1764
Caproyl Chloride, 1765
Caprylaldehyde *see* 1768
Capryl α-Cyanoacrylate *see* 6855
Caprylene, 1766
Caprylic Acid, 1767
Caprylic Acid Ammonium Salt *see* 503
Caprylic Acid α-Monoglyceride *see* 6338
Caprylic Alcohol *see* 6840
Caprylic Aldehyde, 1768
Caprylohydroxamic Acid *see* 6839
CAPS, 1769
Capsaicin, 1770
Capsanthin, 1771
Capsebon [Pitman-Moore] *see* 1633
Capsicum, 1772
Capstar [Novartis] *see* 6654
Captafol, 1773
Captagon [Viatris] *see* 4003
Captan, 1774
Captax *see* 5935
Captimer [MIT] *see* 9609
Captin [Krewel] *see* 47
Captisol [CyDex] *see* 2711
Captodiam *see* 1775
Captodiamine, 1775
Captodramin *see* 1775
Captolane [Sanofi-Aventis] *see* 1776

Captopril, 1776
Captoril [Sankyo] *see* 1776
Capture [FMC] *see* 1213
Capval [Dreluso] *see* 6807
Carac [Dermik] *see* 4213
Carace *see* 5572
Carachol *see* 2872
Caracurine VII *see* 10242
Caradrin [AWD] *see* 7985
Carafate [Aventis] *see* 9010
Caramba *see* 6006
Caramel, 1777
Caramiphen, 1778
Caraway, 1779
Carazolol, 1780
Carbacefaclor *see* 5633
Carbachol, 1781
Carbadox, 1782
Carbaglu [Orphan Europe] *see* 1838
Carbamaldehyde *see* 4265
Carbamamidine *see* 4597
Carbamate of 2-(Hydroxymethyl)-2-
  methylpentyl Ester of Butylcarbamic
  Acid *see* 10010
Carbamazepine, 1783
Carbamazine *see* 3128
Carbamic Acid Ammonium Salt (1:1)
  see 504
Carbamic Acid 2-*sec*-Butyl-2-methyltri-
  methylene Ester *see* 5842
Carbamic Acid 3-(*p*-Chlorophenoxy)-2-
  hydroxypropyl Ester *see* 2185
Carbamic Acid Ethyl Ester *see* 10059
Carbamic Acid Methyl Ester *see* 6110
Carbamic Acid (1-Methyl-5-nitroimid-
  azol-2-yl)methyl Ester *see* 8385
Carbamic Acid 2-Methylpropyl Ester
  see 5182
Carbamic Acid 2-Methyl-2-propyltri-
  methylene Ester *see* 5929
Carbamic Acid 3-Phenylpropyl Ester *see*
  7370
Carbamic Acid 2-Phenyltrimethylene
  Ester *see* 3984
Carbamic Chloride *see* 1785
Carbamidal *see* 6627
Carbamide *see* 10052
Carbamide Peroxide, 1784
Carbamidine *see* 4597
p-Carbamidobenzenearsonic Acid *see*
  1787
Carbamimidothioic Acid 2-Aminoethyl
  Ester Hydrobromide (1:2) *see* 167
Carbamimidothioic Acid Phenylmethyl
  Ester *see* 1148
3-[[2-[(4-Carbamimidoylphenylamino)-
  methyl]-1-methyl-1*H*-benzoimidazole-
  5-carbonyl]pyridin-2-ylamino]pro-
  pionic Acid *see* 2794
Carbamodithioic Acid Ammonium Salt
  (1:1) *see* 513
Carbamothioic Acid *S*^C,*S*^C′-[2-(Dimethylamino)-1,3-propanediyl] Ester *see*
  1865
N-Carbamoylarsanilic Acid *see* 1787
(6*R*,7*S*)-7-[4-(Carbamoylcarboxymethyl-
  ene)-1,3-dithietane-2-carboxamido]-7-
  methoxy-3-[[(1-methyl-1*H*-tetrazol-5-
  yl)thio]methyl]-8-oxo-5-thia-1-azabicy-
  clo[4.2.0]oct-2-ene-2-carboxylic Acid
  see 1935
(*S*)-2-*O*-Carbamoyl-1-*o*-chlorophenyl-
  ethanol *see* 1841
5-Carbamoyl-5*H*-dibenz[*b*,*f*]azepine
  1783

Carindacillin, 1839
Carindapen [Pfizer] *see* 1839
Cariporide, 1840
Cariporide Mesylate *see* 1840
Carisbamate, 1841
Carisoma [Pharmax] *see* 1842
Carisoprodate *see* 1842
Carisoprodol, 1842
Carlytene [Viatris] *see* 6378
Carmilax [Pfizer] *see* 5734
Carmine *see* 2442
Carminic Acid, 1843
Carminomycin *see* 1870
Carminomycin I *see* 1870
Carmofur, 1844
Carmubris [BMS] *see* 1845
Carmustine, 1845
Carnacid-Cor [TAD] *see* 4469
Carnallite *see* 5717, 7725
Carnauba Wax, 1846
Carnegine, 1847
Carnicor [Sigma-Tau] *see* 1849
Carnidazole, 1848
Carnigen [Albert Roussel] *see* 4863
Carnitine, 1849
L-Carnitine *see* 1849
Carnitor [Sigma-Tau] *see* 1849
Carnosic Acid, 1850
Carnosine, 1851
Carnosol, 1852
Carnotite *see* 10104
Carob Flour *see* 5612
Carolina Pink *see* 8875
Carony Bark *see* 645
Carophyll Pink [DSM] *see* 845
Carophyll Red [Roche] *see* 1756
Caro's Acid, 1853
Carotaben [Hermal] *see* 1855
Carotaben Plus *see* 1756
α-Carotene, 1854
β-Carotene, 1855
β,β-Carotene *see* 1855
β,ψ-Carotene *see* 1856
γ-Carotene, 1856
δ-Carotene, 1857
ε,ψ-Carotene *see* 1857
ψ,ψ-Carotene *see* 5681
(6'R)-β,ε-Carotene *see* 1854
all-trans-β-Carotene-3,3'-diol *see* 10316
(3R,3'R)-β,β-Carotene-3,3'-diol *see*
   10316
(3R,3'R,6'R)-β,ε-Carotene-3,3'-diol *see*
   10261
ψ,ψ-Carotene-16,16'-diol *see* 5682
β,β-Carotene-4,4'-dione *see* 1756
ψ,ψ-Carotene-16-ol *see* 5688
β,β-Carotene-3,3',4,4'-tetrone *see* 843
(R)-*all-trans*-β-Caroten-3-ol *see* 2599
(3R)-β,β-Caroten-3-ol *see* 2599
(3R)-β,ψ-Caroten-3-ol *see* 8422
β,β-Caroten-4-one *see* 3542
Carotol, 1858
Caroverine, 1859
Carpaine, 1860
Carpene [Isagro] *see* 3455
Carperitide *see* 863
Carpetimycin A *see* 1861
Carpetimycin B *see* 1861
Carpetimycins, 1861
Carpidine *see* 5245
Carpiline *see* 5245
Carprofen, 1862
Carpropamid, 1863
Carrageen *see* 1864
Carrageenan, 1864

Carrageenin *see* 1864
Carrisyn [Carrington] *see* 27
Cartap, 1865
Carteol [Otsuka] *see* 1866
Carteolol, 1866
Carthamic Acid *see* 1867
Carthamin, 1867
Carthamus, 1868
Carthamus Red *see* 1867
Carticaine, 1869
Cartonic [Yamanouchi] *see* 587
Cartric [Sanwa] *see* 8262
Cartrophen [Univet] *see* 7247
Carubicin, 1870
Carubinose *see* 5812
Carudol [Fher] *see* 7390
Carumonam, 1871
Carvacrol, 1872
Carvacron [Taiyo] *see* 9789
Carvasin [Teofarma] *see* 5271
Carvedilol, 1873
Carvol *see* 1874
6,8-Carvomenthenediol *see* 8705
Carvone, 1874
Carylderm [Mundipharma] *see* 1788
Caryolysine [Sanofi-Synthelabo] *see* 5846
Caryophyllene, 1875
α-Caryophyllene *see* 4792
β-Caryophyllene *see* 1875
γ-Caryophyllene *see* 1875
cis-Caryophyllene *see* 1875
trans-Caryophyllene *see* 1875
Caryophyllic Acid *see* 3940
Caryophyllin *see* 6920
Caryophyllus *see* 2402
Carzenide, 1876
Carzinophilin A, 1877
Casakol [Upjohn] *see* 1878
Casamino Acids *see* 1883
Casanthranol, 1878
Casanthranol A *see* 1878
Casanthranol B *see* 1878
Casca Bark *see* 8519
Cascade [BASF] *see* 4164
Cascapride [Cascan] *see* 1443
Cascara Amara, 1879
Cascara Sagrada, 1880
Cascarilla, 1881
Cascarillin, 1882
Cascarin *see* 4292
Cascarosides *see* 1880
Caseanine *see* 9360
Casein, 1883
Cas-Evac *see* 1880
Cashoo *see* 1904
Casodex [AstraZeneca] *see* 1204
Casoron [Solvay Duphar] *see* 3052
Caspase-1, 1884
Caspase-3, 1885
Caspase-8, 1886
Caspase-9, 1887
Caspase-activated DNase *see* 1885
Caspases, 1888
Caspofungin, 1889
Cassadan [Temmler] *see* 308
Cassaidine, 1890
Cassaine, 1891
Cassella 4489 *see* 2243
Cassella's Acid, 1892
Cassella's Acid F, 1893
Cassel's Green *see* 977
Cassia *see* 2300
Cassia Fistula, 1894
Cassia Pods *see* 1894
Cassia Pulp *see* 1894

Cassic Acid *see* 8299
Cassiopeium *see* 5676
Cassiterite *see* 8905
Castanea, 1895
Castanospermine, 1896
Castellan [AgrEvo] *see* 4226
Castle's Intrinsic Factor, 1897
Castor Oil, 1898
Castor Oil, Hydrogenated, 1899
Castorwax *see* 1899
Cat *see* 6035
Cataclot [Ono] *see* 7079
Cataflam [Novartis] *see* 3091
Catalase, 1900
Catalin [Senju] *see* 7603
Catalpin *see* 1901
Catalposide, 1901
Catalyst [Wellmark] *see* 7929
Catapres [Boehringer, Ing.] *see* 2380
Catapressan [Boehringer, Ing.] *see* 2380
Catarase [Cooper] *see* 2264
Cataria [Santhera] *see* 1908
Catechin, 1902
Catechinic Acid *see* 1902
Catechol *see* 1902, 8111
Catecholase *see* 10019
Catecholborane, 1903
Catechol Boron Bromide *see* 1423
Catechu *see* 4389
Catechu Black, 1904
Catechuic Acid *see* 1902
Catena [Santhera] *see* 4929
Catenins *see* 1616
Catenulin *see* 7146
Catergen [Zyma] *see* 1902
Catharanthine, 1905
Cathepsin C *see* 1906
Cathepsin D *see* 1906
Cathepsin G *see* 1906
Cathepsins, 1906
Cathine *see* 6803
Cathinone, 1907
Catlep [Sumitomo] *see* 5009
Catmint *see* 1908
Catnep, 1908
Catnip *see* 1908
Cat's Hair *see* 3945
Cattlyst [Syntex] *see* 5397
Catumaxomab, 1909
Caudaline [Exa] *see* 9585
Caulophylline *see* 6121
Caulophyllum *see* 1321
Caulosapogenin *see* 4659
Caustic Alcohol *see* 8749
Caustic Barley *see* 8443
Caustic Baryta *see* 973
Caustic Potash *see* 7761
Caustic Soda *see* 8761
Causyth [Inverni] *see* 7982
Cav-Ecol *see* 10239
Caverject [Pharmacia & Upjohn] *see* 7987
Cavinton [Thiemann] *see* 10188
Cavonyl *see* 2704
Cayenne Pepper *see* 1772
Cayston [Gilead] *see* 918
Caytine *see* 8004
CB *see* 2707
CB-154 *see* 1424
CB-154 Mesylate *see* 1424
CB-302 *see* 4302
CB-304 *see* 897
CB-311 *see* 8842
CB-313 *see* 6301
CB-337 *see* 5877
CB-1048 *see* 2108

CB-1348 *see* 2073
CB-1639 *see* 2727
CB-1678 *see* 7936
CB-2041 *see* 1508
CB-3025 *see* 5896
CB-3026 *see* 5896
CB-4261 *see* 9391
CB-4306 *see* 2392
CB-4311 *see* 2392
CB-7598 *see* 9
CB-7630 *see* 9
CB-8061 *see* 7247
CB-30038 *see* 6281
2C-B, 1910
CBA-93626 *see* 2381
CBD *see* 2707
CBDCA *see* 1823
CBN *see* 1751
(−)-C$_{34}$-botryococcene *see* 1360
CBS *see* 2469
CC-2481 *see* 2705
CC-4047 *see* 10690
CC-5013 *see* 5490
CC-10004 *see* 740
CCA *see* 5610
CCC *see* 1665, 2104
CCD-1042 *see* 4392
β-CCE *see* 3834
CCI-779 *see* 9285
CCI-15641 *see* 1956
CCI-18781 *see* 4240
CCK-179 *see* 3710
CCK-PZ *see* 2205
CCK *C*-Terminal Octapeptide *see* 8681
CCNU *see* 5621
CCP *see* 2361
CCRG-81045 *see* 9282
CD-68 *see* 2083
CD-271 *see* 138
CDA *see* 2028
2-CdA *see* 2336
CDB-2914 *see* 10030
CDC *see* 2054
CDD 95-126 *see* 10071
CDDD-5604 *see* 5642
CDDO *see* 959
CDDO-Me *see* 959
CDEC *see* 9043
CDFs *see* 7684
CDM *see* 2086
CDO-MeB *see* 9240
CDP-771 *see* 4424
CDP-870 *see* 2002
CDP-choline *see* 2318
CDR *see* 1721
CE-3624 *see* 9586
Ceanothic Acid, 1911
CEA-Scan [Immunomedics] *see* 764
Cebera [Irex] *see* 236
Cebesine [Chauvin] *see* 1049
Cebion [Merck KGaA] *see* 819
Cebrogen [Walker] *see* 4507
Cebrum [SIFA] *see* 2085
Cebutid [Boots] *see* 4229
Cecenu [Medac] *see* 5621
Ceclor [Lilly] *see* 1914
Cecon [Abbott] *see* 819
Cecropins, 1912
CED-3 *see* 1888
Cedad [Recordati] *see* 1031
Cedar Camphor *see* 1913
Cedar Leaf Oil *see* 9543
Cedax [Shionogi] *see* 1950
Cedilanid [Sandoz] *see* 5404
Cedilanide [Novartis] *see* 2922

Cedocard [Pharmacia] *see* 5271
8β*H*-Cedran-8-ol *see* 1913
Cedrol, 1913
Cedro Oil *see* 6890
Cedur [Boehringer, Mann.] *see* 1199
CEE *see* 2492
CeeNU [BMS] *see* 5621
Ceepryn [J. B. Williams] *see* 2031
Cefizox [Astellas] *see* 1952
Cefacidal [BMS] *see* 1918
Cefaclor, 1914
Cefadol [Nippon Shinyaku] *see* 3341
Cefa-Dri [Fort Dodge] *see* 1983
Cefadril [AGIPS] *see* 1915
Cefa-Drops [Fort Dodge] *see* 1915
Cefadroxil, 1915
Cefadyl *see* 1983
Cefa-Lak [Fort Dodge] *see* 1983
Cefalen [Lenza] *see* 7627
Cefalexin *see* 1974
Cefaloject [BMS] *see* 1983
Cefalonium *see* 1977
Cefalotin *see* 1981
Cefam [Magis] *see* 1916
Cefamandole, 1916
Cefamar [Firma] *see* 1956
Cefamedin [Astellas] *see* 1918
Cefamezin [Pfizer] *see* 1918
Cefamox [BMS] *see* 1915
Cefaperos *see* 1917
Cefapirin *see* 1983
Cefatamet Pivoxil *see* 1924
Cefatrex [BMS] *see* 1983
Cefatrexyl [BMS] *see* 1983
Cefatriaxone *see* 1955
Cefatrizine, 1917
Cefazil [Italfarmaco] *see* 1918
Cefazolin, 1918
Cefazone [Firma] *see* 1931
Cefbon-C$_2$F [Central Glass] *see* 4576
Cefbon-CF [Central Glass] *see* 4576
Cefbuperazone, 1919
Cefcapene, 1920
Cefcapene Pivoxil *see* 1920
Cefdinir, 1921
Cefditoren, 1922
Cefditoren Pivoxil *see* 1922
Cefepime, 1923
Cefepime Hydrochloride *see* 1923
Cefetamet, 1924
Cefetamet Pivaloyloxymethyl Ester *see* 1924
Cefetamet Pivoxil Hydrochloride *see* 1924
Cefixime, 1925
Cefixoral [Menarini] *see* 1925
Ceflatonin [ChemGenex] *see* 4776
Cefmenoxime, 1926
Cefmetazole, 1927
Cefmetazon [Sankyo] *see* 1927
Cefminox, 1928
Cefobid [Pfizer] *see* 1931
Cefobis [Pfizer] *see* 1931
Cefodie [Cristalfarma] *see* 1930
Cefodizime, 1929
Cefodox [Scharper] *see* 1942
Cefoneg [Tosi] *see* 1931
Cefonicid, 1930
Cefoperazone, 1931
Cefoprim [Esseti] *see* 1956
Ceforal [Grünenthal] *see* 1915
Ceforanide, 1932
Cefoselis, 1933
Cefotan *see* 1935
Cefotax [Sanofi-Aventis] *see* 1934

Cefotaxime, 1934
Cefotetan, 1935
Cefotiam, 1936
Cefotiam Hexetil Hydrochloride *see* 1936
Cefovecin, 1937
Cefoxitin, 1938
Cefozopran, 1939
Cefpiramide, 1940
Cefpiran [Dainippon Sumitomo] *see* 1940
Cefpirome, 1941
Cefpodoxime *see* 1942, 1942
Cefpodoxime Proxetil *see* 1942
Cefprozil, 1943
Cefquinome, 1944
Cefraden [Merck KGaA] *see* 1955
Cefradine *see* 1985
Cefrom [Sanofi-Aventis] *see* 1941
Cefroxadine, 1945
Cefspan [Astellas] *see* 1925
Cefsulodin, 1946
Ceftaroline, 1947
Ceftaroline Fosamil *see* 1947
Ceftaroline Fosamil Monoacetate *see* 1947
Ceftazidime, 1948
Cefteram, 1949
Cefteram Pivoxil *see* 1949
Ceftetrame, 1949
Ceftibuten, 1950
Ceftin [GSK] *see* 1956
Ceftiofur, 1951
Ceftix [Astellas] *see* 1952
Ceftizoxime, 1952
Ceftobiprole *see* 1953
Ceftobiprole Medocaril, 1953
Ceftolozane, 1954
Ceftriaxone, 1955
Ceftriaxone Sodium *see* 1955
Cefurax [Lindopharm] *see* 1956
Cefurin [Magis] *see* 1956
Cefuroxime, 1956
Cefuroxime Axetil *see* 1956
Cefyl [Takeda] *see* 1924
Cefzil [BMS] *see* 1943
Ceglunat [DDR] *see* 5404
Cekufon [United Phosphorus] *see* 9788
Cela W-524 *see* 9859
Celadigal [Beiersdorf] *see* 5404
Celance [Lilly] *see* 7277
Celaskon [Leciva] *see* 819
Celastrol, 1957
Celbenin [SKB] *see* 6041
Celebrex [Pfizer] *see* 1958
Celecoxib, 1958
Celectol [Sanofi-Aventis] *see* 1962
Celeport [Eisai] *see* 1210
Celery Seed, 1959
Celest [Syngenta] *see* 4160
Celestan [Essex] *see* 1182
Celestan-V [Essex] *see* 1182
Celestene [Schering-Plough] *see* 1182
Celesticetin, 1960
Celestin Blue, 1961
Celestine *see* 8965, 8978
Celestite *see* 8978
Celestoderm-V [Schering-Plough] *see* 1182
Celestone [Schering-Plough] *see* 1182
Celestovet [Biokema] *see* 1182
Celevac [Shire] *see* 6111
Celexa [Forest] *see* 2317
Celin [GSK] *see* 819
Célio [Novartis] *see* 2390
Celiomycin *see* 10197

Cesium Iodide (CsI) *see* 2013
Cesium Nitrate, 2014
Cesium Sulfate, 2015
Cesol [Merck KGaA] *see* 7837
Cesplon [Esteve] *see* 1776
Cestex [Pfizer] *see* 3694
Cestocide [Bayer] *see* 6604
Cetab *see* 2024
Cetaceum *see* 8868
Cetacillin *see* 7931
Cetacort [Coria] *see* 4824
Cetal [Warner-Lambert] *see* 10180
Cetalkonium Chloride, 2016
Cetamide [Alcon] *see* 9031
Cetapril [Dainippon] *see* 193
Cetats [Zeeland] *see* 2024
Cetavlon [Ayerst] *see* 2024
Ceteareth *see* 7697
Cetebe [GSK] *see* 819
Ceteth *see* 7697
Ceteth 20 *see* 7697
Cethexonium Bromide, 2017
C.E.T. HEXtra [Virbac] *see* 2092
Cethromycin, 2018
Cethylose [Ascher] *see* 1830
Cetiedil, 2019
Cetilistat, 2020
Cetiprin [Kabi] *see* 3615
Cetirizine, 2021
(−)-Cetirizine *see* 2021
Cetol *see* 2016
Cetosanol [Sanol] *see* 5404
Cetotiamine, 2022
Cetoxil [Ivax] *see* 1956
Cetraxate, 2023
Cetrexin [Leciva] *see* 5851
Cetrimide *see* 2024
Cetrimonium Bromide, 2024
Cetrimonium Tosylate *see* 2024
Cetrorelix, 2025
Cetrotide [Serono] *see* 2025
Cetuximab, 2026
Cetyl Alcohol, 2027
Cetylamine *see* 2024
α-Cetylcitric Acid *see* 178
Cetyldimethylbenzylammonium Chloride *see* 2016
N-Cetyl-*N*,*N*-dimethyl-2-cyclohexanolammonium Bromide *see* 2017
Cetyldimethylethylammonium Bromide, 2028
Cetyldimethyl(2-hydroxycyclohexyl)ammonium Bromide *see* 2017
Cetylic Acid *see* 7097
Cetyl Lactate, 2029
Cetyl Palmitate, 2030
Cetylpyridinium Chloride, 2031
Cetyltrimethylammonium Bromide *see* 2024
Cevadilla *see* 8443
Cevadine, 2032
Cevalin [Lilly] *see* 819
(3β,5α,6α)-Cevane-3,6,20-triol *see* 10165
Cevimeline, 2033
Cevine, 2034
Ce-Vi-Sol [BMS] *see* 819
Cevitamic Acid *see* 819
Cevitan [Aventis] *see* 819
Cewin [Sanofi-Synthelabo] *see* 819
Ceylon Cinnamon *see* 2300
Ceylon Isinglass *see* 176
CEZ *see* 1918
CF *see* 4249
CFCs *see* 2141
C-Film [Arun] *see* 6764

C₃-Fluoroalcohol *see* 9351
CFPQ *see* 3646
CG *see* 2221
CG-113 *see* 7858
CG-201 *see* 1197
CG-315E *see* 9726
CG-4203 *see* 9190
CG-4509 *see* 8011
CG-5503 *see* 9189
CGA-15324 *see* 7886
CGA-18731 *see* 5264
CGA-23654 *see* 6719
CGA-24705 *see* 6223
CGA-26423 *see* 7858
CGA-45156 *see* 9482
CGA-48988 *see* 5993
CGA-50439 *see* 2761
CGA-64250 *see* 7932
CGA-72662 *see* 2772
CGA-89317 *see* 9817
CGA-106630 *see* 2972
CGA-131036 *see* 9761
CGA-136872 *see* 7866
CGA-142705 *see* 4018
CGA-152005 *see* 7992
CGA-154281 *see* 1048
CGA-163935 *see* 9897
CGA-169374 *see* 3154
CGA-173506 *see* 4160
CGA-179246 *see* 6270
CGA-184699 *see* 5654
CGA-184927 *see* 2364
CGA-185072 *see* 2390
CGA-215944 *see* 8061
CGA-219417 *see* 2769
CGA-245704 *see* 98
CGA-248757 *see* 4239
CGA-279202 *see* 9843
CGA-293343 *see* 9446
CGA-329351 *see* 5993
cGMP *see* 2702
CGP *see* 4655
CGP 19835A *see* 6265
CGP-2175 *see* 6228
CGP-4540 *see* 571
CGP-6140 *see* 568
CGP-7174/E *see* 1946
CGP-7760B *see* 7853
CGP-9000 *see* 1945
CGP-14221/E *see* 1936
CGP-14458 *see* 4629
CGP-19835 *see* 6265
CGP-30083 *see* 3681
CGP-30694 *see* 3554
CGP-32349 *see* 4268
CGP-33101 *see* 8426
CGP-39360 *see* 8928
CGP-39393 *see* 4755
CGP-41251 *see* 6264
CGP-42446 *see* 10388
CGP-42446A *see* 10388
CGP-42446B *see* 10388
CGP-42935 *see* 385
CGP-45840B *see* 3091
CGP-48933 *see* 10102
CGP-51901 *see* 6936
CGP-57148B *see* 4943
CGP-60536 *see* 239
CGP-60536B *see* 239
CGP-64128A *see* 744
CGP-72670 *see* 2861
CGP-73547 *see* 849
CGP-79787 *see* 10130
CGP-79787D *see* 10130
CGS-9343B *see* 10308

CGS-14824A *see* 1033
CGS-14831 *see* 1033
CGS-16949A *see* 3968
CGS-20267 *see* 5502
CGT-003 *see* 3559
CGX-635 *see* 4776
CH-800 *see* 4030
CH-846 *see* 3534
CH-3635 *see* 9822
α-Chaconine *see* 8832
Chaetomidin *see* 6945
Chalcanthite *see* 2643
Chalcedony *see* 8632
Chalcocite *see* 2504, 2657
Chalcomycin, 2035
Chalcone, 2036
cis-Chalcone *see* 2036
trans-Chalcone *see* 2036
Chalcopyrite, 2037
Chalcopyrite (CuFeS₂) *see* 2037
D-Chalcose, 2038
Chalkone *see* 2036
Chalkopyrite *see* 2037
Chamazulene, 2039
Chamber Crystals *see* 6735
Chameleon Mineral *see* 7776
Chamomile, 2040
Champaca Camphor *see* 4592
Champacol *see* 4592
Champix [Pfizer] *see* 10125
Chanchalagua *see* 1970
Channel Black *see* 1808
Channel Forming Integral Protein *see* 749
Chanoclavin-I *see* 2041
Chanoclavine, 2041
Chanoclavine II *see* 2041
Chanodesethylapovincamine *see* 10182
Chantix [Pfizer] *see* 10125
Chaperonins, 2042
CHAPS, 2043
CHAPSO *see* 2043
Charas *see* 1752
Charcot-Neumann Crystals *see* 8870
Chartarin *see* 2044
Chartreusin, 2044
Charybdotoxin, 2045
Chasteberry *see* 182
Chaste Tree *see* 182
Chat *see* 5355
Chaulmestrol *see* 2047
Chaulmoogra Oil, 2046
Chaulmoogric Acid, 2047
Chavicine, 2048
Chavicinic Acid *see* 7579
Chavicol, 2049
Chavicol Methyl Ether *see* 3760
Cheeseflower *see* 5772
Chefir [Pfizer] *see* 1930
Chelafer *see* 4065
Chelaplex III *see* 3565
Chelatran [Serb] *see* 3565
Chelen *see* 3837
Chelerythrine, 2050
ψ-Chelerythrine *see* 8493
Chelidonic Acid, 2051
Chelidonine, 2052
Chel-Iron *see* 4065
Chemerin, 2053
Chemet [McNeil] *see* 8995
Chem-Fish [Tifa] *see* 7589
Chemical 109 *see* 708
Chemical Mace *see* 2115
Chemifluor [Chemipharm] *see* 8754
Chemiofuran [Italfarmaco] *see* 6686

Chemofuran *see* 6687
Chemotrim [Rosemont] *see* 9050
Chemox DN [Blue Spruce] *see* 3315
Chemox PE [Blue Spruce] *see* 3315
Chemox Selective [Blue Spruce] *see* 3315
Chendol [CP Pharm.] *see* 2054
Chenic Acid *see* 2054
Chenocol [Astellas] *see* 2054
Chenodeoxycholic Acid *see* 2054
Chenodiol, 2054
Chenofalk [Falk] *see* 2054
Chenopodiol [Pohl] *see* 6877
Chenoposan *see* 6877
Chenoposetten *see* 6877
Chenossil [Sanofi-Aventis] *see* 2054
Cheque *see* 6253
CHES, 2055
Chessylite *see* 2621
Chestnut *see* 1895
Chevreul's Salt *see* 2658
Chibro Pilocarpine [Chibret] *see* 7536
Chibro-Proscar [Chibret] *see* 4113
Chibroxin [Merck & Co.] *see* 6789
Chibroxol [Merck & Co.] *see* 6789
Chick Antidermatitis Factor *see* 7118
Chicle, 2056
Chi-jian *see* 4407
Chilean Centaury *see* 1970
Chile Saltpeter *see* 8778
Chilli Pepper *see* 1772
Chimaphila, 2057
Chimaphilin, 2058
Chimeric A2 Antibody *see* 5014
Chimonanthine, 2059
Chimono [Lusofarmaco] *see* 5619
Chimyl Alcohol, 2060
(*S*)-(+)-Chimyl Alcohol *see* 2060
China Bark *see* 8156
China Clay *see* 5330
China Green *see* 5763
China Oil *see* 942
"China White" *see* 2971
China Wood Oil *see* 9993
Chinese Anise *see* 8924
Chinese Blistering Flies *see* 6414
Chinese Blue *see* 8018
Chinese Cantharides *see* 6414
Chinese Caterpillar Fungus *see* 2515
Chinese Cinnamon *see* 2300
Chinese Ginger *see* 4368
Chinese Isinglass *see* 176
Chinese Pea *see* 8856
Chinese Red *see* 5957
Chinese Wax, 2061
Chinic Acid *see* 8175
Chinicine *see* 10199
Chinidin-Duriles [Astra] *see* 8176
Chiniofon *see* 4865
Chinocide *see* 8183
Chinoform *see* 5075
Chinofungin [Chinoin] *see* 9678
Chinoleine *see* 8185
Chinomethionat(e) *see* 7077
Chinosol [Chinosolfabrik] *see* 4881
Chinova Acid *see* 8192
Chinovic Acid *see* 8192
Chinovin *see* 8193
Chinovose *see* 8194
CHIP28 *see* 749
Chipco Signature [Bayer CropSci.] *see* 4279
CHIPCO-26019 [Bayer CropSci.] *see* 5121
Chiptox [Rhône-Poulenc] *see* 5833
Chirald, 2062

(*S,S*)-Chiraphos, 2063
Chirata, 2064
Chirayita *see* 2064
Chiretta *see* 2064
Chirocaine [Abbott] *see* 1499
Chitin, 2065
Chitosamine *see* 4494
Chitosan *see* 2065
Chittem Bark *see* 1880
Chittim Bark *see* 1880
Chlochinate *see* 4865
Chlokale [Sawai] *see* 216
Chloor-hexaviet *see* 4724
Chlophedianol, 2066
Chloquinate *see* 4865
Chloroptic [Allergan] *see* 2077
Chloracetone *see* 2114
Chloracetyl Chloride, 2067
Chloracizine, 2068
Chloractil [DDSA] *see* 2191
Chloracysin *see* 2068
Chloral *see* 9791
Chloral Acholate, 2069
Chloralamide *see* 2070
Chloraldurat [Pohl] *see* 2071
Chloral Ethylalcoholate *see* 2069
Chloral Formamide, 2070
Chloral Hydrate, 2071
Chloral Hydrocyanide *see* 2134
Chlorallylene *see* 280
Chloralosane *see* 2072
α-Chloralose, 2072
β-Chloralose *see* 2072
Chlorambucil, 2073
Chloramide *see* 2070
Chloramine *see* 2075
Chloramine-B, 2074
Chloramine-T, 2075
Chloraminophenamide, 2076
Chloraminophene *see* 2073
Chloramiphene *see* 2377
Chloramphenicol, 2077
Chloramphenicol Arginine Succinate *see* 2077
Chloramphenicol Pantothenate *see* 2077
Chloramphenicol Sodium Succinate *see* 2077
Chloranautine *see* 3230
Chloranil, 2078
Chloranilic Acid, 2079
Chloranocryl *see* 3098
Chlorantraniliprole, 2080
Chlorapatite *see* 7459
Chlorasept 2000 [Baxter] *see* 2092
Chlorazene [WPC] *see* 2075
Chlorazin [Streuli] *see* 2191
Chlorazine (Russian) *see* 2714
Chlorazodin *see* 2121
Chlorazol Yellow 2G *see* 9462
Chlorbenzoxamine, 2081
5-Chlorbenzoxazolin-2-one *see* 2201
Chlorbenzoxyethamine *see* 2081
Chlorbismol *see* 1274
Chlorbutol *see* 2129
Chlorcosane *see* 7128
Chlorcyclizine, 2082
Chlordan *see* 2083
Chlordane, 2083
Chlordecone, 2084
Chlordiazepoxide, 2085
Chlordimeform, 2086
Chloreal *see* 9806
Chlorendic Anhydride, 2087
Chloresium [Rystan] *see* 2158
Chlorethiazol *see* 2375

Chlorethoxyfos, 2088
Chlorethyl *see* 3837
Chloretone [Parke-Davis] *see* 2129
Chlorex *see* 3075
Chlorfenapyr, 2089
Chlorfenvinphos, 2090
Chlorguanide, 2091
Chlorguanide Triazine *see* 2714
Chlorhexamed [GSK] *see* 2092
Chlorhexidine, 2092
Chlorhydrol [Reheis] *see* 339
Chloric Acid, 2093
Chloric Acid, Aluminum Salt *see* 332
Chloric Acid Barium Salt (2:1) *see* 966
Chloric Acid Calcium Salt (2:1) *see* 1660
Chloric Acid Copper(2+) Salt *see* 2622
Chloric Acid Magnesium Salt (2:1) *see* 5725
Chloric Acid Potassium Salt (1:1) *see* 7741
Chloric Acid Silver(1+) Salt (1:1) *see* 8647
Chloric Acid Sodium Salt (1:1) *see* 8733
Chloric Acid Strontium Salt (2:1) *see* 8969
Chlor-IFC *see* 2193
Chlorimipramine *see* 2378
Chlorimuron-ethyl, 2094
Chlorin e₆ *see* 7500
Chlorinated Biphenyls *see* 7682
Chlorinated Camphene *see* 9716
Chlorinated Lime *see* 1676
Chlorine, 2095
Chlorine Cyanide *see* 2684
Chlorine Dioxide, 2096
Chlorine Fluoride *see* 2098
Chlorine Fluoride (ClF₃) *see* 2100
Chlorine Heptoxide, 2097
Chlorine Monofluoride, 2098
Chlorine Monoxide, 2099
Chlorine Nitride *see* 6689
Chlorine Oxide (ClO₂) *see* 2096
Chlorine Oxide (Cl₂O) *see* 2099
Chlorine Peroxide *see* 2096
Chlorine Trifluoride, 2100
Chlorins *see* 7720
Chlorisept [Chinosolfabrik] *see* 2405
Chlorisondamine Chloride, 2101
Chlorisondamine Dimethochloride *see* 2101
Chlormadinone Acetate, 2102
Δ¹-Chlormadinone Acetate *see* 2883
Chlormephos, 2103
Chlormeprazine *see* 7879
Chlormequat Chloride, 2104
Chlormerodrin, 2105
Chlormeroprin *see* 2105
Chlormethazanone *see* 2106
Chlormethiazole *see* 2375
Chlormethine *see* 5846
Chlormethylfos *see* 2103
Chlormezanone, 2106
Chlormidazole, 2107
Chlornaphazine, 2108
Chloroacetaldehyde, 2109
2-Chloroacetaldehyde *see* 2109
Chloroacetamide, 2110
2-Chloroacetamide *see* 2110
Chloroacetanilide, 2111
m-Chloroacetanilide *see* 2111
o-Chloroacetanilide *see* 2111
p-Chloroacetanilide *see* 2111
3′-Chloroacetanilide *see* 2111
Chloroacetic Acid, 2112
2-Chloroacetic Acid *see* 2112

(+)-N-(o-Chlorobenzyl)-α-methylphen-
ethylamine see 2357
1-[1-[2-[(3-Chlorobenzyl)oxy]phenyl]vi-
nyl]-1H-imidazole see 2580
7-[3-[4-(p-Chlorobenzyl)-1-piperazinyl]-
propoxy]-3,4-dimethylcoumarin see
7530
1-(p-Chlorobenzyl)-2-(1-pyrrolidinyl-
methyl)benzimidazole see 2344
1-(p-Chlorobenzyl)-2-pyrrolidylmethyl-
enebenzimidazole see 2344
5-(o-Chlorobenzyl)-4,5,6,7-tetrahydro-
thieno[3,2-c]pyridine see 9585
7-Chlorobicyclo[3.2.0]hepta-2,6-dien-6-yl
Dimethylphosphate see 4699
4'-Chloro[1,1'-biphenyl]-4-amine see 430
3-Chloro[1,1'-biphenyl]-2-ol see 7391
3-Chloro-(1,1'-biphenyl)-4-ol see 7392
Chlorobiphenyls see 7682
2-Chloro-N,N-bis(2-chloroethyl)ethan-
amine see 9790
(TPS-7-1-232'4'54)-Chloro[bis[(1,2-
cyclohexanedione 1-Oxime 2-Oximate-
κO) (1−)][[1,2-Cyclohexanedione 1-
Oxime-κO) 2-Oximato] (2−)]Methyl-
borato(2−)-κN,κN',κN'',κN''',κN'''',-
κN'''']technetium-⁹⁹Tc see 9240
Chlorobis(η⁵-2,4-cyclopentadien-1-yl)hy-
drozirconium see 8533
6-Chloro-1,6-bisdehydroretroprogester-
one see 9746
2-Chloro-4,6-bis(ethylamino)-s-triazine
see 8672
2-Chloro-4,6-bis(isopropylamino)-s-tri-
azine see 7926
Chlorobis(1-methylethyl)silane see 2136
6-Chloro-N²,N⁴-bis(1-methylethyl)-1,3,5-
triazine-2,4-diamine see 7926
5-Chloro-2,4-bis(sulfonamido)aniline see
2076
2-Chlorobornane see 1342
3-Chloro-d-2-bornanone see 2132
4-Chloro-ω-bromoacetophenone see
2156
Chlorobromoform see 2135
α-Chloro-4-bromotoluene see 1417
1-Chlorobutane see 1561
2-Chlorobutane see 1562
(2R)-2-Chlorobutane see 1562
(2S)-2-Chlorobutane see 1562
Chlorobutanol, 2129
1-Chloro-2-butene, 2130
3-Chloro-1-butene, 2131
(3R)-3-Chloro-1-butene see 2131
(3S)-3-Chloro-1-butene see 2131
m-Chloro-α-(tert-butylamino)propio-
phenone see 1503
(β-Chloro-tert-butyl)benzene see 6544
α-Chloro-β-butylene see 2130
γ-Chloro-α-butylene see 2131
Chlorocain [Pharm. Mfg.] see 5926
2-Chlorocamphane see 1342
α'-Chloro-d-camphor see 2132
α-Chloro-d-camphor see 2132
3-Chloro-d-camphor, 2132
3α-Chloro-d-camphor see 2132
3β-Chloro-d-camphor see 2132
2-Chloro-10-[3-(4-carbamoylpiperidin-
yl)propyl]phenothiazine see 7567
m-Chlorocarbanilic Acid Isopropyl Ester
see 2193
8-Chlorocarbochromen see 2393
trans-Chlorocarbonylbis(triphenylphosphine)iridium(I) see 10128
trans-Chlorocarbonylbis(triphenylphosphine)rhodium(I) see 1819

2-(Chlorocarbonyl)furan see 4339
3-Chlorocarpipramine see 2361
2-Chloro-N-(4'-chlorobiphenyl-2-yl)nico-
tinamide see 1354
2-Chloro-N-(4'-chloro[1,1'-biphenyl]-2-
yl)-3-pyridinecarboxamide see 1354
3-Chloro-N-[3-chloro-2,6-dinitro-4-(trifluoromethyl)phenyl]-5-(trifluoromethyl)-2-pyridinamine see 4148
2-Chloro-N-(2-chloroethyl)-N-methyl-
ethanamine see 5846
2-Chloro-N-(2-chloroethyl)-N-methyl-
ethanamine N-Oxide see 5847
1-Chloro-2-(β-chloroethylthio)ethane see
6401
(2Z)-2-Chloro-3-[2-chloro-5-(1,3,4,5,6,7-
hexahydro-1,3-dioxo-2H-isoindol-2-
yl)phenyl]-2-propenoic Acid Ethyl Ester see 2294
3-[5-Chloro-α-(p-chloro-β-hydroxy-
phenethyl)-2-thenyl]-4-hydroxycouma-
rin see 9607
6-Chloro-3-(chloromethyl)-3,4-dihydro-
2-methyl-2H-1,2,4-benzothiadiazine-7-
sulfonamide 1,1-Dioxide see 6079
6-Chloro-3-chloromethyl-3,4-dihydro-2-
methyl-7-sulfamoyl-1,2,4-benzothiadi-
azine 1,1-Dioxide see 6079
6-Chloro-3-chloromethyl-2-methyl-7-sul-
famyl-3,4-dihydro-1,2,4-benzothiadi-
azine 1,1-Dioxide see 6079
5-Chloro-N-(2-chloro-4-nitrophenyl)-2-
hydroxybenzamide see 6604
3-Chloro-4-(3-chloro-2-nitrophenyl)pyr-
role see 8132
5-Chloro-N-(2'-chloro-4'-nitrophenyl)-
salicylamide see 6604
3'-Chloro-4'-(p-chlorophenoxy)-3,5-diio-
dosalicylanilide see 8214
N-[3-Chloro-4-(4-chlorophenoxy)phen-
yl]-2-hydroxy-3,5-diiodobenzamide see
8214
1-[[2-[2-Chloro-4-(4-chlorophenoxy)-
phenyl]-4-methyl-1,3-dioxolan-2-yl]-
methyl]-1H-1,2,4-triazole see 3154
N-[5-Chloro-4-[(4-chlorophenyl)cyano-
methyl]-2-methylphenyl]-2-hydroxy-
3,5-diiodobenzamide see 2396
2-Chloro-α-(4-chlorophenyl)-4-(4,5-dihy-
dro-3,5-dioxo-1,2,4-triazin-2(3H)-yl)-
benzeneacetonitrile see 2341
7-Chloro-5-(2-chlorophenyl)-1,3-dihy-
dro-3-hydroxy-2H-1,4-benzodiazepin-
2-one see 5636
7-Chloro-5-(2-chlorophenyl)-1,3-dihy-
dro-3-hydroxy-1-methyl-2H-1,4-ben-
zodiazepin-2-one see 5639
3-[2-Chloro-5-(2-chlorophenyl)ethenyl]-
2,2-dimethylcyclopropanecarboxylic
Acid Cyano(4-fluoro-3-phenoxyphen-
yl)methyl Ester see 4171
10-Chloro-11b-(2-chlorophenyl)-2,3,5,6,-
7,11b-hexahydrobenzo[6,7]-1,4-diaze-
pino[5,4-b]oxazol-6-one see 2404
8-Chloro-6-(o-chlorophenyl)-1-methyl-
4H-s-triazolo[4,3-a][1,4]benzodiaze-
pine see 9765
8-Chloro-6-(2-chlorophenyl)-1-methyl-
4H-[1,2,4]triazolo[4,3-a][1,4]benzod-
iazepine see 9765
10-Chloro-11b-(2-chlorophenyl)-6-oxo-
2,3,5,6,7,11b-hexahydrooxazolo[3,2-d]-
[1,4]benzodiazepine see 2404
10-Chloro-11b-(2-chlorophenyl)-2,3,7,-
11b-tetrahydro-3-methyloxazolo[3,2-

d][1,4]benzodiazepin-6(5H)-one see
6245
10-Chloro-11b-(o-chlorophenyl)-2,3,7,-
11b-tetrahydro-3-methyloxazolo[3,2-
d][1,4]benzodiazepin-6(5H)-one see
6245
10-Chloro-11b-(2-chlorophenyl)-2,3,7,-
11b-tetrahydrooxazolo[3,2-d][1,4]ben-
zodiazepin-6(5H)-one see 2404
7-Chloro-5-(2-chlorophenyl)tetrahy-
drooxazolo[5,4-b]-2,3,4,5-tetrahydro-
1H-1,4-benzodiazepin-2-one see 2404
4-Chloro-α-(4-chlorophenyl)-α-(trichloromethyl)benzenemethanol see 3096
2-Chloro-N-[4-chloro-3-(2-pyridinyl)-
phenyl]-4-(methylsulfonyl)benzamide
see 10207
3-Chloro-N-(3-chloro-5-trifluoromethyl-
2-pyridinyl)-α,α,α-trifluoro-2,6-dini-
tro-p-toluidine see 4148
Chlorocholine Chloride see 2104
Chlorocizin [Egis] see 2077
Chlorocizin see 2068
Chlorocresol, 2133
4-Chloro-m-cresol see 2133
6-Chloro-m-cresol see 2133
Chlorocyanide see 2684
4-Chloro-2-cyano-N,N-dimethyl-5-(4-
methylphenyl)-1H-imidazole-1-sulfon-
amide see 2693
4-Chloro-2-cyano-N,N-dimethyl-5-p-tol-
ylimidazole-1-sulfonamide see 2693
Chlorocyanogen see 2684
Chlorocyanohydrin, 2134
2,2'-[(2-Chloro-5-cyano-1,3-phenylene)-
diimino]bis(2-oxoacetic Acid) see 5613
N,N'-(2-Chloro-5-cyano-m-phenylene)-
dioxamic Acid see 5613
Chlorocyclizine see 2082
Chlorocyclohexane see 2725
(SP-5-13)-Chloro[[rel-2,2'-[(1R,2R)-1,2-
cyclohexanediylbis[(nitrilo-κN)methyl-
idyne]]bis[4,6-bis(1,1'-dimethylethyl)-
phenolato-κO]](2−)]manganese see
5300
[2-Chloro-5-(cyclohex-1-ene-1,2-dicar-
boximido)-4-fluorophenoxy]acetic
Acid see 4173
7-Chloro-5-(1-cyclohexen-1-yl)-1,3-dihy-
dro-1-methyl-2H-1,4-benzodiazepin-2-
one see 9391
7-Chloro-5-(1-cyclohexenyl)-1-methyl-2-
oxo-2,3-dihydro-1H-[1,4]benzo[f]di-
azepine see 9391
5-Chloro-N-[2-[4-[[[(cyclohexylamino)-
carbonyl]amino]sulfonyl]phenyl]eth-
yl]-2-methoxybenzamide see 4514
Chloro(η⁵-2,4-cyclopentadien-1-yl)[(4R,-
5R)-2,2-dimethyl-α4,α4,α5,α5-tetra-
phenyl-1,3-dioxolane-4,5-dimethano-
lato(2−)-κO⁴,κO⁵]titanium see 3519
6-Chloro-3-(cyclopentylmethyl)-3,4-dihy-
dro-2H-1,2,4-benzothiadiazine-7-sul-
fonamide 1,1-Dioxide see 2740
6-Chloro-3-cyclopentylmethyl-3,4-dihy-
dro-7-sulfamoyl-2H-1,2,4-benzothiadi-
azine 1,1-Dioxide see 2740
4-Chloro-2-cyclopentylphenol see 3479
4-[3-Chloro-4-[[[(cyclopropylamino)car-
bonyl]amino]phenoxy]-7-methoxy-6-
quinolinecarboxamide see 5494
1-[α-(4-Chloro-α-cyclopropylbenzyli-
deneaminooxy)-p-tolyl]-3-(2,6-difluo-
robenzoyl)urea see 4154
2-[2-(1-Chlorocyclopropyl)-3-(2-chloro-
phenyl)-2-hydroxypropyl]-1,2-dihy-

dro-3*H*-1,2,4-triazole-3-thione *see* 7997

2-(1-Chlorocyclopropyl)-1-(2-chloro-phenyl)-3-(5-mercapto-1,2,4-triazol-1-yl)propan-2-ol *see* 7997

(4*S*)-6-Chloro-4-(2-cyclopropylethynyl)-1,4-dihydro-4-(trifluoromethyl)-2*H*-3,1-benzoxazin-2-one *see* 3569

7-Chloro-1-(cyclopropylmethyl)-1,3-di-hydro-5-phenyl-2*H*-1,4-benzodiazepin-2-one *see* 7836

Chlorocyzine *see* 2068

6-Chloro-6-dehydro-17α-acetoxyproges-terone *see* 2102

21-Chloro-11-dehydrobetamethasone *see* 2359

6-Chloro-6-dehydro-17α-hydroxy-1,2α-methyleneprogesterone *see* 2771

6-Chloro-6-dehydro-17α-hydroxyproges-terone Acetate *see* 2102

7-Chloro-6-demethyltetracycline *see* 2890

2-Chlorodeoxyadenosine *see* 2336

2-Chloro-2'-deoxyadenosine *see* 2336

2-Chloro-9-(2-deoxy-2-fluoro-β-D-arabi-nofuranosyl)adenine *see* 2366

2-Chloro-9-(2-deoxy-2-fluoro-β-D-arabi-nofuranosyl)-9*H*-purin-6-amine *see* 2366

7(*S*)-Chloro-7-deoxylincomycin *see* 2354

33-*epi*-Chloro-33-desoxyascomycin *see* 7541

2-[(8-Chlorodibenzo[*b,f*]thiepin-10-yl)-oxy]-*N,N*-dimethylethanamine *see* 10397

Chlorodibromomethane, 2135

3-Chloro-1,2-dibromopropane *see* 3025

Chlorodicarbonylrhodium(I) Dimer *see* 8312

1-Chloro-2-[2,2-dichloro-1-(4-chloro-phenyl)ethyl]benzene *see* 6301

6-Chloro-3-(dichloromethyl)-3,4-dihy-dro-2*H*-1,2,4-benzothiadiazine-7-sul-fonamide 1,1-Dioxide *see* 9789

6-Chloro-3-dichloromethyl-7-sulfamyl-3,4-dihydro-1,2,4-benzothiadiazine 1,1-Dioxide *see* 9789

6-Chloro-5-(2,3-dichlorophenoxy)-2-(methylthio)-1*H*-benzimidazole *see* 9817

5-Chloro-2-(2,4-dichlorophenoxy)phenol *see* 9822

7-Chloro-3-[1-(2,4-dichlorophenyl)-2-(1*H*-imidazol-1-yl)ethoxy-methyl]-benzo[*b*]thiophene *see* 8604

2-Chloro-1-(2,4-dichlorophenyl)vinyl Di-ethyl Phosphate *see* 2090

6-Chloro-1,6-didehydroretroprogester-one *see* 9746

(*SP*-5-13)-Chloro[7,12-diethenyl-3,8,-13,17-tetramethyl-21*H*,23*H*-porphine-2,18-dipropanoato(4−)-κ*N*²¹,κ*N*²²,-κ*N*²³,κ*N*²⁴]ferrate(2−) Hydrogen (1:2) *see* 4679

Chlorodiethoxyphosphine *see* 3144

3-Chloro-7-diethoxyphosphinothioyloxy-4-methylcoumarin *see* 2546

2-Chloro-10-(*N,N*-diethyl-β-alanyl)phe-nothiazine *see* 2068

Chlorodiethylaluminum *see* 322

2-Chloro-4-diethylamino-6-ethylamino-*s*-triazine *see* 9829

7-Chloro-1-[2-(diethylamino)ethyl]-5-(2-fluorophenyl)-1,3-dihydro-2*H*-1,4-ben-zodiazepin-2-one *see* 4228

2-[[8-Chloro-3-[2-(diethylamino)ethyl]-4-methyl-2-oxo-2*H*-1-benzopyran-7-yl]-oxy]acetic Acid Ethyl Ester *see* 2393

2-Chloro-5-(ω-diethylamino-α-methyl-butylamino)-7-methoxyacridine *see* 8160

6-Chloro-9-[[4-(diethylamino)-1-methyl-butyl]amino]-2-methoxyacridine *see* 8160

7-Chloro-4-(4-diethylamino-1-methylbu-tylamino)quinoline *see* 2165

7-Chloro-4-(3-diethylaminomethyl-4-hy-droxyanilino)quinoline *see* 569

7-Chloro-4-(3-diethylaminomethyl-4-hy-droxyphenylamino)quinoline *see* 569

2'-Chloro-2-[2-[(diethylamino)methyl]-imidazol-1-yl]-5-nitrobenzophenone *see* 6748

2-Chloro-10-[3-(diethylamino)-1-oxopro-pyl]-10*H*-phenothiazine *see* 2068

2-Chloro-10-(β-diethylaminopropionyl)-phenothiazine *see* 2068

2-Chloro-10-(3-diethylaminopropyl)phe-nothiazine *see* 2189

3-Chloro-10-(3-diethylaminopropyl)phe-nothiazine *see* 2189

2-Chloro-2,6-diethyl-*N*-(butoxymeth-yl)acetanilide *see* 1511

2-Chloro-2-diethylcarbamoyl-1-methylvi-nyl Dimethyl Phosphate *see* 7449

6-Chloro-3-(*O,O*-diethyldithiophosphor-ylmethyl)benzoxazolone *see* 7446

2-Chloro-2',6'-diethyl-*N*-(methoxymeth-yl)acetanilide *see* 194

2-Chloro-*N,N*-diethyl-10*H*-phenothia-zine-10-propanamine *see* 2189

2-Chloro-*N*-(2,6-diethylphenyl)-*N*-(meth-oxymethyl)acetamide *see* 194

2-Chloro-*N*-(2,6-diethylphenyl)-*N*-(2-propoxyethyl)acetamide *see* 7858

6-Chloro-*N*²,*N*⁴-diethyl-1,3,5-triazine-2,4-diamine *see* 8672

21-Chloro-6α,9-difluoro-11β,17-dihy-droxy-16β-methylpregna-1,4-diene-3,20-dione 17-Propionate *see* 4629

(6α,11β,16β)-21-Chloro-6,9-difluoro-11-hydroxy-16-methyl-17-(1-oxopro-poxy)pregna-1,4-diene-3,20-dione *see* 4629

2-Chloro-1-(difluoromethoxy)-1,1,2-trifluoroethane *see* 3635

2-Chloro-2-(difluoromethoxy)-1,1,1-trifluoroethane *see* 5221

(6α,11β,16α)-2-Chloro-6,9-difluoro-11,-17,21-trihydroxy-16-methylpregna-1,4-diene-3,20-dione *see* 4633

10-Chloro-5,10-dihydroarsacridine *see* 7329

4-(8-Chloro-5,6-dihydro-11*H*-benzo[5,6]-cyclohepta[1,2-*b*]pyridin-11-ylidene)-1-piperidinecarboxylic Acid Ethyl Ester *see* 5635

6-Chloro-3,4-dihydro-2*H*-1,2,4-benzo-thiadiazine-7-sulfonamide 1,1-Dioxide *see* 4819

1'-[3-(3-Chloro-10,11-dihydro-5*H*-di-benz[*b,f*]azepin-5-yl)propyl][1,4'-bipi-peridine]-4'-carboxamide *see* 2361

1'-[3-(3-Chloro-10,11-dihydro-5*H*-di-benz[*b,f*]azepin-5-yl)propyl]hexahy-drospiro[imidazo[1,2-*a*]pyridine-3(2*H*),4'-piperidin]-2-one *see* 6363

4'-[3-(3-Chloro-10,11-dihydro-5*H*-di-benz[*b,f*]azepin-5-yl)propyl]methyl-amino]acetophenone *see* 5614

1'-[3-(3-Chloro-10,11-dihydro-5*H*-di-benz[*b,f*]azepin-5-yl)propyl]-1,2,3,5,-6,7,8,8a-octahydro-2-oxoimidazo[1,2-*a*]-pyridine-3-spiro-4'-piperidine *see* 6363

7-Chloro-1,3-dihydro-3-(*N,N*-dimethyl-carbamoyl)-1-methyl-5-phenyl-2*H*-1,4-benzodiazepin-2-one *see* 1727

3-Chloro-10,11-dihydro-*N,N*-dimethyl-5*H*-dibenz[*b,f*]azepine-5-propan-amine *see* 2378

2-Chloro-4,5-dihydro-1,3-dimethyl-1*H*-imidazolium Chloride (1:1) *see* 3434

11-Chloro-8,12b-dihydro-2,8-dimethyl-12b-phenyl-4*H*-[1,3]oxazino[3,2-*d*][1,-4]benzodiazepine-4,7(6*H*)-dione *see* 5345

8-Chloro-3,7-dihydro-1,3-dimethyl-1*H*-purine-2,6-dione Compd with 2-(Di-phenylmethoxy)-*N,N*-dimethylethan-amine (1:1) *see* 3230

(±)-[2-Chloro-4-(4,5-dihydro-3,5-dioxo-*as*-triazin-2(3*H*)-yl)phenyl](*p*-chloro-phenyl)acetonitrile *see* 2341

Chloro[dihydrogen 3,7,12,17-Tetrameth-yl-8,13-divinyl-2,18-porphinedipro-pionato(2−)]iron *see* 4679

(1β,2β)-6-Chloro-1,2-dihydro-17-hy-droxy-3'*H*-cyclopropa[1,2]pregna-1,-4,6-triene-3,20-dione *see* 2771

(*R*)-*N*-[(5-Chloro-3,4-dihydro-8-hydroxy-3-methyl-1-oxo-1*H*-2-benzopyran-7-yl)carbonyl]-L-phenylalanine *see* 6829

(*R*)-*N*-[(5-Chloro-3,4-dihydro-8-hydroxy-3-methyl-1-oxo-1*H*-2-benzopyran-7-yl)carbonyl]-L-phenylalanine Ethyl Ester *see* 6829

7-Chloro-1,3-dihydro-3-hydroxy-1-meth-yl-5-phenyl-2*H*-1,4-benzodiazepin-2-one *see* 9275

7-Chloro-1,3-dihydro-3-hydroxy-1-meth-yl-5-phenyl-2*H*-1,4-benzodiazepin-2-one Dimethylcarbamate (Ester) *see* 1727

2-Chloro-5-(2,3-dihydro-1-hydroxy-3-oxo-1*H*-isoindol-1-yl)benzenesulfon-amide *see* 2200

7-Chloro-1,3-dihydro-3-hydroxy-5-phen-yl-2*H*-1,4-benzodiazepin-2-one *see* 7025

(3*Z*)-5-Chloro-2,3-dihydro-3-(hydroxy-2-thienylmethylene)-2-oxo-1*H*-indole-1-carboxamide *see* 9287

5-Chloro-*N*-(4,5-dihydro-1*H*-imidazol-2-yl)-2,1,3-benzothiadiazol-4-amine *see* 9642

4-Chloro-*N*-(4,5-dihydro-1*H*-imidazol-2-yl)-6-methoxy-2-methyl-5-pyrimidin-amine *see* 6379

6-Chloro-3,4-dihydro-3-isobutyl-7-sul-famoyl-1,2,4-benzothiadiazine 1,1-Di-oxide *see* 1527

(4a*S*)-7-Chloro-2,5-dihydro-2-[[(meth-oxycarbonyl)[4-(trifluoromethoxy)-phenyl]amino]carbonyl]indeno[1,2-*e*]-[1,3,4]oxadiazine-4a(3*H*)-carboxylic Acid Methyl Ester *see* 5012

7-[(3-Chloro-6,11-dihydro-6-methyldi-benzo[*c,f*][1,2]thiazepin-11-yl)amino]-heptanoic Acid S,S-Dioxide *see* 9575

7-[(3-Chloro-6,11-dihydro-6-methyl-5,5-dioxidodibenzo[*c,f*][1,2]thiazepin-11-yl)amino]heptanoic Acid *see* 9575

2-Chloro-5-[3,6-dihydro-3-methyl-2,6-di-oxo-4-(trifluoromethyl)-1(2*H*)-pyrimi-

ethyl]amino]methyl]-2-furanyl]-4-
quinazolinamine *see* 5418
6-[(3-Chloro-2-fluorophenyl)methyl]-1,4-
dihydro-1-[(1*S*)-1-(hydroxymethyl)-2-
methylpropyl]-7-methoxy-4-oxo-3-
quinolinecarboxylic Acid *see* 3610
8-Chloro-6-(2-fluorophenyl)-1-methyl-
4*H*-imidazo[1,5-*a*][1,4]benzodiazepine
see 6261
6-[3-(2-Chloro-6-fluorophenyl)-5-methyl-
4-isoxazolecarboxamido]penicillanic
Acid *see* 4143
(2*S*,5*R*,6*R*)-6-[[[3-(2-Chloro-6-fluoro-
phenyl)-5-methyl-4-isoxazolyl]carbon-
yl]amino]-3,3-dimethyl-7-oxo-4-thia-1-
azabicyclo[3.2.0]heptane-2-carboxylic
Acid *see* 4143
3-(2-Chloro-6-fluorophenyl)-5-methyl-4-
isoxazolylpenicillin *see* 4143
(2*R*)-2-[4-[(5-Chloro-3-fluoro-2-pyridin-
yl)oxy]phenoxy]propanoic Acid 2-Pro-
pyn-1-yl Ester *see* 2364
N′-{2-Chloro-4-fluoro-5-[1,2,3,6-tetrahy-
dro-3-methyl-2,6-dioxo-4-(trifluoromethyl)pyrimidin-1-yl]benzoyl}-*N*-iso-
propyl-*N*-methylsulfamide *see* 8451
2-[[2-Chloro-4-fluoro-5-[(tetrahydro-3-
oxo-1*H*,3*H*-[1,3,4]thiadiazolo[3,4-*a*]-
pyridazin-1-ylidene)amino]phenyl]-
thio]acetic Acid Methyl Ester *see* 4239
2-Chloro-4-fluoro-5-[(3,4,5,6-tetrahy-
dro)phthalimido]phenoxyacetic Acid
see 4173
21-Chloro-9-fluoro-11β,16α,17-trihy-
droxypregn-4-ene-3,20-dione Cyclic
16,17-Acetal with Acetone *see* 4626
Chlorofolin *see* 2158
Chloroform, 2142
Chloroformamide *see* 1785
Chloroformic Acid Benzyl Ester *see*
1801
Chloroformic Acid Ethyl Ester *see* 3839
Chloroformic Acid Isobutyl Ester *see*
5184
Chloroformic Acid Propyl Ester *see*
7961
Chloroformic Acid Trichloromethyl
Ester *see* 3374
Chloroformyl Chloride *see* 7447
Chlorofos *see* 9788
4-Chloro-*N*-furfuryl-5-sulfamoylanthra-
nilic Acid *see* 4338
4-Chloro-*N*-(2-furylmethyl)-5-sulfamoyl-
anthranilic Acid *see* 4338
Chlorogenic Acid, 2143
Chlorogenin, 2144
Chloroguanide *see* 2091
Chlorohemin *see* 4679
2-Chloro-1-heptene-1-arsonic Acid *see*
2119
(2-Chloro-1-heptenyl)arsonic Acid *see*
2119
(3α,5aα,10bα,11β,11aα)-9-Chloro-2,3,-
5a,6,10b,11-hexahydro-10b,11-dihy-
droxy-7,8-dimethoxy-2,3,6-trimethyl-
3,11a-epidithio-11a*H*-pyrazino[1′,-
2′:1,5]pyrrolo[2,3-*b*]indole-1,4-dione
see 8892
4-Chloro-*N*-(*endo*-hexahydro-4,7-meth-
anoisoindolin-2-yl)-3-sulfamoylbenz-
amide *see* 9909
10-Chloro-2,3,5,6,7,11b-hexahydro-2-
methyl-11b-phenylbenzo[6,7]-1,4-di-
azepino[5,4-*b*]oxazol-6-one *see* 7026
1-Chlorohexane, 2145

α-Chlorohydrin, 2146
(17β)-4-Chloro-17-hydroxyandrost-4-en-
3-one *see* 2397
5-Chloro-2-hydroxybenzoxazole *see* 2201
2-Chloro-2′-hydroxydiethyl Sulfide *see*
4681
2-Chloro-5-hydroxy-1,3-dimethylbenzene
see 2182
5-Chloro-2-hydroxydiphenylmethane *see*
2394
5-Chloro-3-[4-(2-hydroxyethyl)-1-pipera-
zinyl]carbonylmethyl-2-benzothiazoli-
none *see* 9579
5-Chloro-3-[2-[4-(2-hydroxyethyl)-1-
piperazinyl]-2-oxoethyl]-2(3*H*)-benzo-
thiazolone *see* 9579
2-Chloro-9-[3-[4-(2-hydroxyethyl)-1-
piperazinyl]propylidene]thiaxanthene
see 2383
2-Chloro-10-[3-[1-(2-hydroxyethyl)-4-
piperazinyl]propyl]phenothiazine *see*
7297
5-Chloro-8-hydroxy-7-iodoquinoline *see*
5075
2-Chloro-4-(hydroxymercuriphenol) *see*
4869
6-Chloro-17-hydroxy-1α,2α-methylene-
pregna-4,6-diene-3,20-dione *see* 2771
7-Chloro-5-hydroxy-1-[2-methyl-4-(2-
methylbenzoylamino)benzoyl]-2,3,4,5-
tetrahydro-1*H*-1-benzazepine *see* 9699
6-Chloro-4-hydroxy-2-methyl-*N*-2-pyri-
dinyl-2*H*-thieno[2,3-*e*]-1,2-thiazine-3-
carboxamide 1,1-Dioxide *see* 5640
6-Chloro-4-hydroxy-2-methyl-3-(2-pyri-
dylcarbamoyl)-2*H*-thieno[2,3-*e*]-1,2-
thiazine-1,1-dioxide *see* 5640
6-Chloro-17-hydroxy-2-oxapregna-4,6-
diene-3,20-dione *see* 6985
2-Chloro-5-(1-hydroxy-3-oxo-1-isoindol-
inyl)benzenesulfonamide *see* 2200
2-Chloro-5-(3-hydroxy-1-oxoisoindolin-
3-yl)benzenesulfonamide *see* 2200
7-Chloro-3-hydroxy-5-phenyl-1,3-dihy-
dro-2*H*-1,4-benzodiazepin-2-one *see*
7025
(3-Chloro-4-hydroxyphenyl)hydroxy-
mercury *see* 4869
6-Chloro-17-hydroxypregna-4,6-diene-
3,20-dione Acetate *see* 2102
6-Chloro-17-hydroxypregna-1,4,6-triene-
3,20-dione Acetate *see* 2883
1-Chloro-2-hydroxypropane *see* 7964
1-(3-Chloro-2-hydroxypropyl)-2-methyl-
5-nitroimidazole *see* 6967
5-Chloro-8-hydroxyquinoline *see* 2405
2-Chloro-5-hydroxytoluene *see* 2133
6-Chloro-3-hydroxytoluene *see* 2133
2-Chloro-5-hydroxy-*m*-xylene *see* 2182
5-Chloro-4-(2-imidazolin-2-ylamino)-2,-
1,3-benzothiadiazole *see* 9642
7-Chloro-3-(1*H*-imidazol-1-yl)-1,2,4-ben-
zotriazine 1-Oxide *see* 9768
(±)-5-(*m*-Chloro-α-imidazol-1-ylbenzyl)-
benzimidazole *see* 5528
1-(2-Chloroimidazo[1,2-*a*]pyridin-3-
ylsulfonyl)-3-(4,6-dimethoxypyrimi-
din-2-yl)urea *see* 4949
4-Chloro-2,2′-iminodibenzoic Acid *see*
5610
*N*¹-(5-Chloro-1*H*-indol-7-yl)-1,4-ben-
zenedisulfonamide *see* 4986
N-(3-Chloro-7-indolyl)-1,4-benzenedisul-
fonamide *see* 4986
N-(3-Chloro-1*H*-indol-7-yl)-4-sulfamoyl-
benzenesulfonamide *see* 4986

Chloroiodoquin *see* 5075
5-Chloro-7-iodo-8-quinolinol *see* 5075
Chloro-IPC *see* 2193
2-Chloroisobutane *see* 1563
α-Chloroisobutylene *see* 2149
γ-Chloroisobutylene *see* 2150
2-Chloro-*N*-isopropylacetanilide *see*
7907
1-Chloroisopropyl Alcohol *see* 7964
6-Chloro-4-isopropyl-1-methyl-3-phenol
see 2173
4′-Chloro-*N*-(1-isopropyl-4-piperidyl)-2-
phenylacetanilide *see* 5637
1-Chloro-2-ketopropane *see* 2114
1-[3-(Chloromercuri)-2-methoxypropyl]-
urea *see* 2105
2-[2-Chloro-4-mesyl-3-[(2,2,2-trifluoro-
ethoxy)methyl]benzoyl]cyclohexane-
1,3-dione *see* 9276
Chloromethane *see* 6112
Chloromethapyrilene *see* 2170
*N*⁴-(6-Chloro-2-methoxy-9-acridinyl)-
*N*¹,*N*¹-diethyl-1,4-pentanediamine *see*
8160
4-[(7-Chloro-2-methoxybenzo[*b*]-1,5-
naphthyridin-10-yl)amino]-2,6-bis(1-
pyrrolidinylmethyl)phenol *see* 8117
(3*S*,4*R*,5*S*,8*R*,9*E*,12*S*,14*S*,15*R*,16*S*,18*R*,-
19*R*,26a*S*)-3-[(1*E*)-2-[(1*R*,3*R*,4*S*)-4-
Chloro-3-methoxycyclohexyl]-1-meth-
ylethenyl]-8-ethyl-5,6,8,11,12,13,14,-
15,16,17,18,19,24,25,26,26a-hexadeca-
hydro-5,19-dihydroxy-14,16-dimeth-
oxy-4,10,12,18-tetramethyl-15,19-ep-
oxy-3*H*-pyrido[2,1-*c*][1,4]oxaazacyclo-
tricosine-1,7,20,21(4*H*,23*H*)-tetrone
see 7541
Chloro-(2-methoxyethyl)mercury *see*
3881
Chloromethoxymethane *see* 2148
rel-5-Chloro-2-methoxy-4-(methylamino)-*N*-[(2*R*,3*R*)-2-methyl-1-(phenylmethyl)-3-pyrrolidinyl]benzamide *see*
6531
(4*S*,4a*S*,6a*R*,11*E*,12a*R*,15*R*,16a*S*,21a*R*,-
21b*R*)-4-[[4-*O*-[3-*O*-(3-Chloro-6-meth-
oxy-2-methylbenzoyl)-2,6-dideoxy-β-
D-*arabino*-hexopyranosyl]-2,6-dideoxy-
β-D-*arabino*-hexopyranosyl]oxy]-1,2,-
3,4,4a,6a,7,8,9,10,12a,15,16,21,21a,-
21b-hexadecahydro-22-hydroxy-15,-
21a-dimethyl-18,21-dioxo-18*H*-16a,19-
metheno-16a*H*-naphtho[*e*]naphtho[2,1-
m][1,4]dioxacyclopentadecin-14-car-
boxylic Acid *see* 2172
3-Chloro-7-methoxy-9-(1-methyl-4-di-
ethyl Aminobutylamino)acridine *see*
8160
4-Chloro-6-methoxy-2-methyl-5-(2-imid-
azolin-2-yl)aminopyrimidine *see* 6379
2-Chloro-*N*-[[(4-methoxy-6-methyl-1,3,5-
triazin-2-yl)amino]carbonyl]benzene-
sulfonamide *see* 2197
4-[*p*-Chloro-*N*-(*p*-methoxyphenyl)benz-
amido]butyric Acid *see* 2337
4-[[(3-Chloro-4-methoxyphenyl)methyl]-
amino]-2-[(2*S*)-2-(hydroxymethyl)-1-
pyrrolidinyl]-*N*-(2-pyrimidinylmeth-
yl)-5-pyrimidinecarboxamide *see* 875
4-Chloro-*N*-[2-[3-methoxy-4-(2-propyn-
1-yloxy)phenyl]ethyl]-α-(2-propyn-1-
yloxy)benzeneacetamide *see* 5785
2-[[[[(4-Chloro-6-methoxy-2-pyrimidin-
yl)amino]carbonyl]amino]sulfonyl]ben-
zoic Acid Ethyl Ester *see* 2094

Chloromethoxysulfonyl Chloride *see* 2147

7-Chloro-2-methylamino-5-phenyl-3*H*-1,4-benzodiazepine 4-Oxide *see* 2085

4-Chloro-5-(methylamino)-2-[3-(trifluoromethyl)phenyl]-3(2*H*)-pyridazinone *see* 6790

4-Chloro-5-(methylamino)-2-(α,α,α-trifluoro-*m*-tolyl)-3(2*H*)-pyridazinone *see* 6790

2-(3-Chloro-2-methylanilino)nicotinic Acid *see* 2381

(Chloromethyl)benzene *see* 1132

1-Chloro-2-methylbenzene *see* 2175

1-Chloro-3-methylbenzene *see* 2175

1-Chloro-4-methylbenzene *see* 2175

N-Chloro-4-methylbenzenesulfonamide Sodium Salt (1:1) *see* 2075

7-Chloro-3-methyl-2*H*-1,2,4-benzothiadiazine 1,1-Dioxide *see* 3007

1-Chloro-3-methylbutane *see* 5161

6-Chloro-α-methyl-9*H*-carbazole-2-acetic Acid *see* 1862

Chloromethyl Chlorosulfate, 2147

4(or 5)-Chloro-2-methylcyclohexanecarboxylic acid 1,1-dimethylethyl ester *see* 9870

(μ-Chloro)(μ-methylene)bis(cyclopentadienyl)(dimethylaluminum)titanium *see* 9228

7-Chloro-6-methylene-5-hydroxytetracycline *see* 5849

6-Chloro-1,2α-methylene-4,6-pregnadien-17α-ol-3,20-dione *see* 2771

Chloromethyl 17α-Ethoxycarbonyloxy-11β-hydroxyandrosta-1,4-diene-3-one-17β-carboxylate *see* 5642

4-Chloro-α-(1-methylethyl)benzeneacetic Acid Cyano(3-phenoxyphenyl)methyl Ester *see* 4037

2-Chloro-*N*-(1-methylethyl)-*N*-phenylacetamide *see* 7907

1-(Chloromethyl)-4-fluoro-1,4-diazoniabicyclo[2.2.2]octane Tetrafluoroborate(1−) (1:2) *see* 8562

1,1′,1″-(Chloromethylidyne)trisbenzene *see* 9944

2-Chloro-4-methyl-3-(2′-imidazolin-2′-ylamino)thiophene *see* 9573

4-Chloro-*N*-(2-methyl-1-indolinyl)-3-sulfamoylbenzamide *see* 4975

5-Chloro-2-methyl-3(2*H*)-isothiazolone *see* 6163

N-(4-Chloro-3-methyl-5-isoxazolyl)-2-[2-(6-methyl-1,3-benzodioxol-5-yl)acetyl]-3-thiophenesulfonamide *see* 8692

N-(4-Chloro-3-methyl-5-isoxazolyl)-2-[3,4-(methylenedioxy)-6-methyl]phenylacetyl-3-thiophenesulfonamide *see* 8692

1-(Chloromethyl)-3-methylbenzene *see* 10289

Chloromethyl Methyl Ether, 2148

4-Chloro-5-methyl-2-(1-methylethyl)phenol *see* 2173

α-(Chloromethyl)-2-methyl-5-nitro-1*H*-imidazole-1-ethanol *see* 6967

(βR)-4-Chloro-β-[[[[(1S)-2-methyl-1-(2-methyl-1-oxopropoxy)propoxy]carbonyl]amino]methyl]benzenepropanoic Acid *see* 758

5-Chloro-6′-methyl-3-[4-(methylsulfonyl)phenyl]-2,3′-bipyridine *see* 3930

3′-Chloro-α-[methyl[(morpholinocarbonyl)methyl]amino]-*o*-benzotoluidide *see* 4255

3′-Chloro-β-[*N*-methyl-*N*-[(morpholinocarbonyl)methyl]aminomethyl]benzanilide *see* 4255

N-[3-Chloro-2-[[methyl-[2-(4-morpholinyl)-2-oxoethyl]amino]methyl]phenyl]benzamide *see* 4255

2-(Chloromethyl)oxirane *see* 3666

O-3-Chloro-4-methyl-2-oxo-2*H*-chromen-7-yl *O,O*-Diethyl Phosphorothioate *see* 2546

N-[3′-(1′-Chloro-3′-methyl-2′-oxopentan)]-3,5-dichloro-4-methylbenzamide *see* 10398

4-Chloro-3-methylphenol *see* 2133

2-(4-Chloro-2-methylphenoxy)acetic Acid *see* 5833

1-(6-Chloro-3-methylphenoxy)-3-*tert*-butylaminopropan-2-ol *see* 1500

1-(2-Chloro-5-methylphenoxy)-3-[(1,1-dimethylethyl)amino]-2-propanol *see* 1500

2-(4-Chloro-2-methylphenoxy)propanoic Acid *see* 5855

2-[(3-Chloro-2-methylphenyl)amino]benzoic Acid *see* 9673

2-[(3-Chloro-2-methylphenyl)amino]-3-pyridinecarboxylic Acid *see* 2381

7-Chloro-*N*-methyl-5-phenyl-3*H*-1,4-benzodiazepin-2-amine 4-Oxide *see* 2085

7-Chloro-1-methyl-5-phenyl-1*H*-1,5-benzodiazepine-2,4(3*H,5H*)-dione *see* 2356

7-Chloro-1-methyl-5-phenyl-3*H*-1,4-benzodiazepin-2(1*H*)-one *see* 2997

2-[(*p*-Chloro-α-methyl-α-phenylbenzyl)oxy]-*N,N*-dimethylethylamine *see* 2187

1-[2-(4-Chloro-α-methyl-α-phenylbenzyloxy)ethyl]azepam *see* 8611

1-[2-[(*p*-Chloro-α-methyl-α-phenylbenzyl)oxy]ethyl]hexahydro-1*H*-azepine *see* 8611

N-[2-[((4-Chloro-α-methyl-α-phenylbenzyl)oxy]ethyl]hexamethylenimine *see* 8611

(+)-2-[2-[(*p*-Chloro-α-methyl-α-phenylbenzyl)oxy]ethyl]-1-methylpyrrolidine *see* 2343

N′-(4-Chloro-2-methylphenyl)-*N,N*-dimethylmethanimidamide *see* 2086

N′-(3-Chloro-4-methylphenyl)-*N,N*-dimethylurea *see* 2176

N-(2-Chloro-6-methylphenyl)-2-[[6-[4-(2-hydroxyethyl)-1-piperazinyl]-2-methyl-4-pyrimidinyl]amino]-5-thiazolecarboxamide *see* 2831

N-(3-Chloro-4-methylphenyl)-2-methylpentanamide *see* 8829

1-Chloro-2-methyl-2-phenylpropane *see* 6544

8-Chloro-1-methyl-6-phenyl-4*H*-[1,2,4]triazolo[4,3-*a*][1,4]benzodiazepine *see* 308

8-Chloro-1-methyl-6-phenyl-4*H*-*s*-triazolo[4,3-*a*][1,4]benzodiazepine *see* 308

S-(Chloromethyl)phosphorodithioic Acid *O,O*-Diethyl Ester *see* 2103

8-Chloro-11-(4-methyl-1-piperazinyl)-5*H*-dibenzo[b,e][1,4]diazepine *see* 2406

2-Chloro-11-(4-methyl-1-piperazinyl)dibenzo[b,f][1,4]thiazepine *see* 2399

2-Chloro-11-(4-methyl-1-piperazinyl)dibenz[b,f][1,4]oxazepine *see* 5645

2-Chloro-10-[3-(1-methyl-4-piperazinyl)propyl]phenothiazine *see* 7879

2-Chloro-10-[3-(4-methyl-1-piperazinyl)propyl]-10*H*-phenothiazine *see* 7879

3-Chloro-10-[3-(4-methyl-1-piperazinyl)propyl]phenothiazine *see* 7879

7α-Chloro-16α-methylprednisolone *see* 213

9α-Chloro-16β-methylprednisolone *see* 1017

9α-Chloro-16β-methyl-1,4-pregnadiene-11β,17α,21-triol-3,20-dione *see* 1017

1-Chloro-2-methylpropane *see* 5183

2-Chloro-2-methylpropane *see* 1563

2-Chloro-3-methylpropane *see* 1562

1-Chloro-2-methyl-1-propene, 2149

3-Chloro-2-methyl-1-propene, 2150

2-Chloro-1-methylpyridinium Iodide (1:1) *see* 6387

7-Chloro-3-methyl-8-quinolinecarboxylic Acid *see* 8182

2-[(4-Chloro-*N*¹-methyl-3-sulfamoylbenzenesulfonamido)methyl]-2-methyltetrahydrofuran *see* 5874

2-[2-Chloro-4-(methylsulfonyl)benzoyl]-1,3-cyclohexanedione *see* 9026

5-Chloro-3-(4-methylsulfonyl)phenyl-2-(2-methyl-5-pyridinyl)pyridine *see* 3930

2-[2-Chloro-4-(methylsulfonyl)-3-[(2,2,2-trifluoroethoxy)methyl]benzoyl]-1,3-cyclohexanedione *see* 9276

4-Chloro-*N*¹-methyl-*N*¹-[(tetrahydro-2-methyl-2-furanyl)methyl]-1,3-benzenedisulfonamide *see* 5874

4-Chloro-*N*¹-methyl-*N*¹-(tetrahydro-2-methylfurfuryl)-*m*-benzenedisulfonamide *see* 5874

2-[(2-Chloro-4-methyl-3-thienyl)amino]-2-imidazoline *see* 9573

N-(2-Chloro-4-methyl-3-thienyl)-4,5-dihydro-1*H*-imidazol-2-amine *see* 9573

cis,trans-3-Chloro-4-[4-methyl-2-(1*H*-1,2,4-triazol-1-ylmethyl)-1,3-dioxolan-2-yl]phenyl 4-Chlorophenyl Ether *see* 3154

3-Chloro-4-methylumbelliferone *O,O*-Diethyl Phosphorothioate *see* 2546

3′-Chloro-2-methyl-*p*-valerotoluidide *see* 8829

p-Chloro-*N*-(2-morpholinoethyl)benzamide *see* 6312

4-Chloro-*N*-[2-(4-morpholinyl)ethyl]benzamide *see* 6312

Chloromycetin [Warner-Lambert] *see* 2077

α-Chloronaphthalene *see* 2151

β-Chloronaphthalene *see* 2152

1-Chloronaphthalene, 2151

2-Chloronaphthalene, 2152

(+)-2-(7-Chloro-1,8-naphthyridin-2-yl)-2,3-dihydro-3-(5-methyl-2-oxohexyl)-1*H*-isoindol-1-one *see* 7084

(+)-2-(7-Chloro-1,8-naphthyridin-2-yl)-3-(5-methyl-2-oxohexyl)-isoindolinone *see* 7084

Chloronitrobenzene, 2153

m-Chloronitrobenzene *see* 2153

o-Chloronitrobenzene *see* 2153

p-Chloronitrobenzene *see* 2153

1-Chloro-3-nitrobenzene *see* 2153

3-Chloro-4-(2′-nitro-3′-chlorophenyl)-pyrrole *see* 8132

N-(2′-Chloro-4′-nitrophenyl)-5-chlorosalicylamide *see* 6604

Chloronitrous Acid *see* 6696

Choline Orotate *see* 6972
Choline Phosphate Chloride *see* 7474
Choline Phosphate Hexadecyl Ester, Hydroxide, Inner Salt *see* 6279
Choline Phosphoric Acid Ester (Chloride) *see* 7474
"Choline Plasmalogen" *see* 7633
Choline Salicylate, 2213
Choline Salicylic Acid Salt *see* 2213
Cholinesterase, 2214
Choline Succinate Dichloride *see* 9006
Choline Theophyllinate, 2215
Cholinfall [Tanabe Seiyaku] *see* 6051
Cholit Ursan [Stada] *see* 10074
Chologon *see* 2872
Cholografin [BMS] *see* 5065
Cholonerton [Dolorgiet] *see* 4894
Cholospect *see* 5065
Choloxin [Knoll] *see* 9570
N-Choloyltaurine *see* 9212
Cholspasmin [Dolorgiet] *see* 4894
Cholumbrin *see* 5081
Cholybar [Warner-Lambert] *see* 2209
N-Cholylglycine *see* 4529
Cholyltaurine *see* 9212
d-Chondocurine *see* 2217
Chondodendrine *see* 1014
Chondrillasterol, 2216
Chondrocurine, 2217
Chondrofoline, 2218
Chondroitin Hydrogen Sulfate *see* 2219
Chondroitin Sulfate, 2219
Chondroitin 4-Sulfate *see* 2219
Chondroitin 6-Sulfate *see* 2219
Chondroitin Sulfate A *see* 2219
Chondroitin Sulfate B *see* 2219
Chondroitin Sulfate C *see* 2219
Chondroitin 4-Sulfate Disodium Salt *see* 2219
Chondroitinsulfuric Acid *see* 2219
Chondrosamine *see* 4364
Chondrosine, 2220
Chonsurid *see* 2219
Chop Nut *see* 7495
Chopper [BASF] *see* 4946
Choragon [Ferring] *see* 2221
Choriogonadotropin *see* 2221
Choriogonadotropin Alfa *see* 2221
Chorionic Gonadotropin, 2221
Chorionic Gonadotropin (Human α-Subunit Protein Moiety Reduced) Complex with Chorionic Gonadotropin (Human β-Subunit Protein Moiety Reduced) *see* 2221
Chorionic Growth Hormone-prolactin *see* 4655
Chorismic Acid, 2222
Chorus [Syngenta] *see* 2769
chRFT5 *see* 1003
Christmas Factor *see* 3963
Chromadax [Gloucester] *see* 8382
Chromafenozide, 2223
R-(−)-2-[4-[(Chroman-2-ylmethyl)amino]butyl]-1,1-dioxobenzo[d]isothiazolone *see* 8257
Chromargyre *see* 5934
Chromatophorotropic Hormone *see* 6381
Chromax [Nutrition 21] *see* 2240
Chrome Alum *see* 2233
Chrome Alum Ammonium *see* 508
Chrome Green *see* 2238
1,2-Chromene *see* 1105
3-Chromene *see* 1105
Chrome Ocher *see* 2238

Chrome Oxide Green *see* 2238
Chrome Red *see* 5461
Chrome Yellow *see* 5460
Chromia *see* 2238
Chromic Acetate, 2224
Chromic Acid ($H_2Cr_2O_7$) Ammonium Salt (1:2) *see* 512
Chromic Acid (H_2CrO_4) Ammonium Salt (1:2) *see* 507
Chromic Acid (H_2CrO_4) Barium Salt (1:1) *see* 968
Chromic Acid ($H_2Cr_2O_7$) Calcium Salt (1:1) *see* 1668
Chromic Acid (H_2CrO_4) Calcium Salt (1:1) *see* 1662
Chromic Acid ($H_2Cr_2O_7$) Compound with Pyridine (1:2) *see* 8087
Chromic Acid (H_2CrO_4) Copper(2+) Salt (1:1) *see* 2624
Chromic Acid (H_2CrO_4) Iron(3+) Salt (3:2) *see* 4047
Chromic Acid (H_2CrO_4) Lead(2+) Salt (1:1) *see* 5460
Chromic Acid (H_2CrO_4) Lithium Salt (1:2) *see* 5585
Chromic Acid ($H_2Cr_2O_7$) Potassium Salt (1:2) *see* 7748
Chromic Acid (H_2CrO_4) Potassium Salt (1:2) *see* 7743
Chromic Acid (H_2CrO_4) Silver(1+) Salt (1:2) *see* 8649
Chromic Acid ($H_2Cr_2O_7$) Sodium Salt (1:2) *see* 8744
Chromic Acid (H_2CrO_4) Sodium Salt (1:2) *see* 8736
Chromic Acid (H_2CrO_4) Strontium Salt (1:1) *see* 8971
Chromic Acid (H_2CrO_4) Tin(2+) Salt (2:1) *see* 8902
Chromic Acid (H_2CrO_4) Zinc Salt (1:1) *see* 10332
Chromic Acid Lead(2+) Salt (1:2) *see* 5461
Chromic Ammonium Sulfate *see* 508
Chromic Anhydride *see* 2239
Chromic Bromide, 2225
Chromic Chloride, 2226
Chromic Fluoride, 2227
Chromic Formate, 2228
Chromic Hydroxide, 2229
Chromic Nitrate, 2230
Chromic Oxide *see* 2238
Chromic Oxide Gel *see* 2229
Chromic Oxide, Hydrous *see* 2229
Chromic Phosphate, 2231
Chromic Phosphate P 32 *see* 2231
Chromic Potassium Oxalate, 2232
Chromic Potassium Sulfate, 2233
Chromic Sulfate, 2234
Chromite *see* 2235
Chromitope Sodium [Bracco] *see* 8736
Chromium, 2235
Chromium Bromide (CrB₃) *see* 2225
Chromium Carbonyl, 2236
(*OC*-6-11)-Chromium Carbonyl (Cr(CO)₆) *see* 2236
Chromium Chloride (CrCl₂) *see* 2248
Chromium Chloride (CrCl₃) *see* 2226
Chromium Cobalt Oxide (Cr₂CoO₄) *see* 2422
Chromium Copper Oxide (Cr₂CuO₄) *see* 2625
Chromium Dioxide, 2237
Chromium Dioxychloride *see* 2252
Chromium Fluoride (CrF₃) *see* 2227

Chromium Hexacarbonyl *see* 2236
Chromium Hydroxide (Cr(OH)₃) *see* 2229
Chromium(III) Potassium Sulfate *see* 2233
Chromium Lead Oxide *see* 5461
Chromium(III) Nitrate *see* 2230
Chromium(III) Oxide, 2238
Chromium(VI) Oxide, 2239
Chromium Oxide (CrO₃) *see* 2239
Chromium Oxide (CrO₂) *see* 2237
Chromium Oxide (Cr₂O₃) *see* 2238
Chromium Oxyfluoride *see* 2253
Chromium Picolinate, 2240
Chromium Sesquioxide *see* 2238
Chromium(III) Sulfate 2234
Chromium Trihydroxide *see* 2229
Chromium Trioxide *see* 2239
Chromium Tripicolinate 2240
Chromium(III) Trispicolinate *see* 2240
Chromocarb, 2241
Chromomycin A₃, 2242
Chromomycinone *see* 2242, 7652
Chromonar, 2243
2-Chromonecarboxylic Acid *see* 2241
Chromosorb P [Celite Corp.] *see* 9144
Chromotrichia Factor *see* 418
Chromotrope 2B, 2244
Chromotropic Acid, 2245
Chromous Acetate, 2246
Chromous Bromide, 2247
Chromous Chloride, 2248
Chromous Fluoride, 2249
Chromous Formate, 2250
Chromous Sulfate, 2251
Chromyl Chloride, 2252
Chromyl Fluoride, 2253
Chronadalate [Bayer] *see* 6613
Chronexan [Viatris] *see* 10276
Chronogest [Intervet] *see* 4230
Chrono-Indocid [Merck & Co.] *see* 5009
Chronosyn [Veterinaria] *see* 2102
Chrysalin [OrthoLogic] *see* 8431
Chrysanthemaxanthin, 2254
Chrysanthemic Acid, 2255
Chrysanthemin *see* 2677
N-(Chrysanthemoxymethyl)-1-cyclohexene-1,2-dicarboximide *see* 9370
Chrysanthemumdicarboxylic Acid Monomethyl Ester Pyrethrolone Ester *see* 8075
Chrysanthemumic Acid *see* 2255
Chrysanthemummonocarboxylic Acid *see* 2255
Chrysanthemummonocarboxylic Acid 6-Chloropiperonyl Ester *see* 1000
Chrysanthemummonocarboxylic Acid Ester with 3-(2-Cyclopenten-1-yl)-2-methyl-4-oxo-2-cyclopenten-1-ol *see* 2700
Chrysanthemummonocarboxylic Acid Pyrethrolone Ester *see* 8075
Chrysanthenone, 2256
Chrysarobin, 2257
Chrysatropic Acid *see* 8545
Chrysazin *see* 2814
Chrysazin-3-carboxylic Acid *see* 8299
6-Chrysenamine, 2258
Chrysene, 2259
Chrysenex *see* 2258
6-Chrysenylamine *see* 2258
Chrysidenon 1438 *see* 2260
Chrysin, 2260
Chrysoberyl *see* 1163
Chrysogen *see* 6454

C.I. Acid Blue 1 *see* 9117
C.I. Acid Blue 9 *see* 1379
C.I. Acid Blue 74 *see* 4982
C.I. Acid Blue 92 *see* 622
C.I. Acid Blue 93 *see* 6102
C.I. Acid Green 3 *see* 4610
C.I. Acid Green 5 *see* 5538
C.I. Acid Green 25 *see* 244
C.I. Acid Orange 5 *see* 9955
C.I. Acid Orange 6 *see* 9954
C.I. Acid Orange 7 *see* 6953
C.I. Acid Orange 11 *see* 3027
C.I. Acid Orange 20 *see* 6952
C.I. Acid Orange 52 *see* 6178
C.I. Acid Orange 137 *see* 6954
C.I. Acid Red 2 *see* 6192
C.I. Acid Red 27 *see* 369
C.I. Acid Red 51 *see* 3749
C.I. Acid Red 87 *see* 3658
C.I. Acid Red 91 *see* 3657
C.I. Acid Red 92 *see* 7440
C.I. Acid Red 94 *see* 8393
C.I. Acid Red 95 *see* 3207
C.I. Acid Red 176 *see* 2244
C.I. Acid Violet 19 *see* 99
C.I. Acid Yellow 1 *see* 6478
C.I. Acid Yellow 3 *see* 8188
C.I. Acid Yellow 23 *see* 9206
C.I. Acid Yellow 36 *see* 5999
C.I. Acid Yellow 40 *see* 7670
C.I. Acid Yellow 73 *see* 4192
C.I. Azoic Coupling Component 1 *see* 6469
C.I. Basic Blue 9 *see* 6132
C.I. Basic Blue 17 *see* 9679
C.I. Basic Brown 1 *see* 1257
C.I. Basic Brown 4 *see* 1256
C.I. Basic Green 1 *see* 1380
C.I. Basic Green 4 *see* 5763
C.I. Basic Orange 2 *see* 2261
C.I. Basic Red 5 *see* 6573
C.I. Basic Violet 3 *see* 4430
C.I. Basic Violet 10 *see* 8307
C.I. Basic Violet 14 *see* 5715
C.I. Developer 1 *see* 6802
C.I. Developer 5 *see* 6469
C.I. Direct Blue 8 *see* 1087
C.I. Direct Blue 14 *see* 9972
C.I. Direct Blue 53 *see* 3948
C.I. Direct Red 2 *see* 1104
C.I. Direct Red 28 *see* 2482
C.I. Direct Red 34 *see* 10210
C.I. Direct Red 80 *see* 8686
C.I. Direct Yellow 9 *see* 9462
C.I. Fluorescent Brightener 30 *see* 1316
C.I. Fluorescent Brightener 184 *see* 1012
C.I. Food Blue 1 *see* 4982
C.I. Food Blue 2 *see* 1379
C.I. Food Blue 3 *see* 9117
C.I. Food Green 1 *see* 4610
C.I. Food Green 3 *see* 3978
C.I. Food Red 1 *see* 7713
C.I. Food Red 6 *see* 7712
C.I. Food Red 14 *see* 3749
C.I. Food Red 17 *see* 275
C.I. Food Yellow 3 *see* 9130
C.I. Food Yellow 4 *see* 9206
C.I. Food Yellow 8 *see* 9954
C.I. Food Yellow 15 *see* 8325
C.I. Ingrain Blue 1 *see* 212
C.I. Ingrain Blue 3 *see* 212
C.I. Mordant Black 11 *see* 3725
C.I. Mordant Blue 10 *see* 4385
C.I. Mordant Blue 14 *see* 1961
C.I. Mordant Orange 1 *see* 247

C.I. Mordant Orange 14 *see* 246
C.I. Mordant Red 3 *see* 8711
C.I. Mordant Red 11 *see* 243
C.I. Mordant Violet 26 *see* 8166
C.I. Mordant Yellow 1 *see* 5989
C.I. Mordant Yellow 5 *see* 6935
C.I. Natural Brown 1 *see* 4119
C.I. Natural Brown 7 *see* 5317
C.I. Natural Orange 4 *see* 667
C.I. Natural Orange 6 *see* 5448
C.I. Natural Red 3 *see* 5341
C.I. Natural Red 4 *see* 1843
C.I. Natural Red 8 *see* 8056
C.I. Natural Red 16 *see* 8056
C.I. Natural Red 20 *see* 249
C.I. Natural Red 24 *see* 1372
C.I. Natural Red 25 *see* 5378
C.I. Natural Red 26 *see* 1867
C.I. Natural Red 34 *see* 1772
C.I. Natural Violet 1 *see* 3028
C.I. Natural Yellow 3 *see* 2663
C.I. Natural Yellow 8 *see* 6355
C.I. Natural Yellow 11 *see* 5706, 6355
C.I. Natural Yellow 24 *see* 4390
C.I. Oxidation Base 12 *see* 2978
C.I. Pigment Black 15 *see* 2636
C.I. Pigment Blue 15 *see* 2505
C.I. Pigment Blue 27 *see* 8018
C.I. Pigment Blue 29 *see* 10031
C.I. Pigment Blue 34 *see* 2645
C.I. Pigment Blue 66 *see* 4981
C.I. Pigment Brown 9 *see* 2627
C.I. Pigment Green 17 *see* 2238
C.I. Pigment Green 20 *see* 2614
C.I. Pigment Green 21 *see* 2615
C.I. Pigment Red 83 *see* 243
C.I. Pigment Red 105 *see* 5480
C.I. Pigment Red 106 *see* 5957
C.I. Pigment Red 109 *see* 4555
C.I. Pigment Violet 14 *see* 2431
C.I. Pigment White 1 *see* 1002
C.I. Pigment White 4 *see* 10347
C.I. Pigment White 5 *see* 5602
C.I. Pigment White 6 *see* 9628
C.I. Pigment White 14 *see* 1274
C.I. Pigment Yellow 31 *see* 968
C.I. Pigment Yellow 33 *see* 1662
C.I. Pigment Yellow 34 *see* 5460
C.I. Pigment Yellow 36 *see* 10332
C.I. Pigment Yellow 39 *see* 790
C.I. Pigment Yellow 40 *see* 2416
C.I. Pigment Yellow 45 *see* 4047
C.I. Solvent Black 3 *see* 9016
C.I. Solvent Blue 8 *see* 6132
C.I. Solvent Brown 12 *see* 1256
C.I. Solvent Green 3 *see* 8181
C.I. Solvent Green 7 *see* 8065
C.I. Solvent Orange 3 *see* 2261
C.I. Solvent Orange 7 *see* 10287
C.I. Solvent Orange 32 *see* 3077
C.I. Solvent Red 23 *see* 9015
C.I. Solvent Red 24 *see* 8530
C.I. Solvent Red 43 *see* 3658
C.I. Solvent Red 48 *see* 7440
C.I. Solvent Red 72 *see* 3027
C.I. Solvent Red 73 *see* 3207
C.I. Solvent Red 80 *see* 2332
C.I. Solvent Yellow 1 *see* 414
C.I. Solvent Yellow 2 *see* 3251
C.I. Solvent Yellow 3 *see* 415
C.I. Solvent Yellow 5 *see* 10293
C.I. Solvent Yellow 6 *see* 10294
C.I. Solvent Yellow 33 *see* 8189
C.I. Solvent Yellow 94 *see* 4192
C.I. Vat Blue 1 *see* 4981

Ciafos *see* 2687
Cialis [Lilly] *see* 9157
Cianatil *see* 2673
Cianurina *see* 5944
Ciatyl [Troponwerke] *see* 2383
Ciba 570 *see* 7449
Ciba 2059 *see* 4185
Ciba 3126 *see* 6218
Ciba 4311b *see* 6183
Ciba 5968 *see* 4802
Ciba 8514 *see* 2086
Ciba 10370 *see* 9032
Ciba 11925 *see* 7312
Ciba 32644-Ba *see* 6649
Cibacalcin [Novartis] *see* 1646
Cibacen [Novartis] *see* 1033
Cibazol [Novartis] *see* 9074
Cibenol [Fujisawa] *see* 2272
Cibenzoline *see* 2272
Ciberon [Taisho] *see* 1800
Ciblor [Inava] *see* 2339
Cichorigenin *see* 3752
Cichoriin, 2265
Ciclacillin *see* 2695
Cicladol [Master] *see* 7619
Ciclesonide, 2266
Cicletanide *see* 2267
Cicletanine, 2267
Ciclobendazole *see* 2705
Ciclobiotic [Beta] *see* 6015
Ciclochem [Novag] *see* 2269
Ciclonicate, 2268
Ciclonium Iodide *see* 2729
Ciclopirox, 2269
Ciclopirox Olamine *see* 2269
Ciclosporin *see* 2750
Cicloven [AGIPS] *see* 8088
Cicutine *see* 2489
Cicutoxin, 2270
Cidal [CID] *see* 8465
Cidandopa [Cidán] *see* 5516
Cidecin [Cubist] *see* 2823
Cidex [J & J] *see* 4508
Cidex OPA [J & J] *see* 7480
Cidine [Almirall] *see* 2295
Cidofovir, 2271
Cidomycin [Aventis] *see* 4427
Cidoxepin Hydrochloride *see* 3483
Cifenline, 2272
Ciflox [Bayer] *see* 2313
CIFN *see* 5034
CIG *see* 4104
Cignolin [Bayer] *see* 676
Ciguatoxins, 2273
Cilantro *see* 2517
Cilastatin, 2274
Cilastatin Sodium *see* 2274
Cilazapril, 2275
Cilazaprilat *see* 2275
Cilengitide, 2276
Cilest [Cilag] *see* 6792
Ciliary Neurotrophic Factor, 2277
Cilicaine [Sigma] *see* 7206
Cilleral [BMS] *see* 583
Cilligen [Sigma] *see* 7210
Cilnidipine, 2278
Cilomilast, 2279
Ciloprost *see* 4942
Cilostazol, 2280
Ciloxan [Alcon] *see* 2313
Cimzia [UCB] *see* 2002
Cimadronate Sodium *see* 4969
Cimal [Alpharma] *see* 2282
Cimaterol, 2281
Cimectacarb *see* 9897

Cimetidine, 2282
Cimetropium Bromide, 2283
Cimicifuga see 1313
Cimicifugol see 2284
Cimigenol, 2284
Cinacalcet, 2285
Cinalong [Fujirebio] see 2278
Cinaperazine [Yakult] see 2306
Cinarine see 2763
Cinazyn [Italchimici] see 2306
Cinch [DuPont] see 2296
Cinchamidine see 4820
Cinchocaine see 3036
Cinchol see 8694
Cincholepidine see 5496
Cincholic Acid β-D-Quinovoside see 8193
Cinchomeronic Acid, 2286
Cinchona, 2287
Cinchona Bark see 2287
Cinchonamine, 2288
(8α,9R)-Cinchonan-6′,9-diol see 2613
(8α,9R)-Cinchonan-9-ol see 2289
(8α,9S)-Cinchonan-9-ol see 2289
(9R)-Cinchonan-9-ol see 2290
(9S)-Cinchonan-9-ol see 2290
Cinchonicine see 2292
Cinchonidine, 2289
Cinchonine, 2290
Cinchophen, 2291
Cinchotine see 4821
Cinchotoxine, 2292
Cinchovatine see 774, 2289
Cincofarm [Lepori] see 4886
Cinconifine see 4821
Cinene see 5546
Cineole see 3938
1,8-Cineole see 3938
Cinerin I see 2293
Cinerin II see 2293
Cinerins, 2293
Cinidon-ethyl, 2294
Cinitapride, 2295
Cinmethylin, 2296
Cinnabar see 5957
Cinnabarine, 2297
Cinnacet [Schwarzhaupt] see 2306
Cinnageron [Streuli] see 2306
Cinnaloid [Pfizer] see 8262
Cinnamal see 2298
Cinnamaldehyde, 2298
Cinnamein see 1133
Cinnamene see 8990
Cinnamic Acid, 2299
Cinnamic Acid Chloride see 2301
Cinnamic Acid Cinnamyl Ester see 2305
Cinnamic Alcohol see 2303
Cinnamic Aldehyde see 2298
Cinnamic Anhydride see 2299
Cinnamol see 8990
Cinnamon, 2300
Cinnamonin see 6407
Cinnamon Wood see 8518
Cinnamoyl Chloride, 2301
trans-Cinnamoyl Chloride see 2301
Cinnamyl Acetate, 2302
(E)-Cinnamyl Acetate see 2302
(Z)-Cinnamyl Acetate see 2302
Cinnamyl Alcohol, 2303
Cinnamyl o-Aminobenzoate see 2304
Cinnamyl Anthranilate, 2304
1-Cinnamyl-4-benzhydrylpiperazine see 2306
1-Cinnamyl-4-(4-chlorobenzhydryl)-piperazine see 2362

Cinnamyl Cinnamate, 2305
1-Cinnamyl-4-(di-p-fluorobenzhydryl)-piperazine see 4175
1-Cinnamyl-4-diphenylmethylpiperazine see 2306
1-trans-Cinnamyl-4-diphenylmethyl-piperazine see 2306
(±)-(E)-Cinnamyl 2-Methoxyethyl 1,4-Dihydro-2,6-dimethyl-4-(m-nitrophen-yl)-3,5-pyridinedicarboxylate see 2278
Cinnamyl Methyl Ketone see 1140
(E)-N-Cinnamyl-N-methyl-1-naphtha-lenemethylamine see 6439
Cinnarizine, 2306
Cinnipirine see 2306
Cinnoline, 2307
Cinnoxicam see 7619
Cinnyl Cinnamate see 2305
CINO-40 [Tutag] see 9757
Cinobac [Lilly] see 2309
Cinobufotalin, 2308
Cinopal [Lederle] see 3996
Cinopenil [Hoechst] see 6041
Cinopol [Lederle] see 3996
Cinoxacin, 2309
Cinoxate, 2310
Cinquil [Cephalon] see 8273
Cin-Quin [Rowell] see 8176
Cintredekin Besudotox, 2311
CiNU [BMS] see 5621
CIPC see 2193
Cipralan [Roche] see 2272
Cipralex [Lundbeck] see 2317
Cipramil [Lundbeck] see 2317
Cipro [Bayer] see 2313
Ciprobay [Bayer] see 2313
Ciprofibrate, 2312
Ciprofloxacin, 2313
Ciprol [Sterling] see 2312
Ciproxan [Bayer] see 2313
Ciproxin [Bayer] see 2313
Cirantin see 4706
Circair [Circle] see 3526
Circanol [3M] see 3710
Circolene [Inverni] see 8237
Circo-Maren [Krewel] see 6581
Circubid [W. F. Merchant] see 3783
Circulen [Kyorin] see 143
Circuletin see 5327
Circulin A see 2314
Circulin B see 2314
Circulins, 2314
Circupon [Gepepharm] see 3914
Cirotyl [Parke-Davis] see 7073
Cisapride, 2315
Cisatracurium Besylate see 859
Cisordinol [Lundbeck] see 2383
Cisplatin, 2316
Cisplatyl [Sanofi-Aventis] see 2316
Cistobil [Bracco] see 5101
Citalopram, 2317
S-(+)-Citalopram see 2317
Citanest [AstraZeneca] see 7862
Citicoline, 2318
Citiolase [HMR] see 2319
Citiolone, 2319
Citizeta [CT] see 10370
Citodon see 4740
Citofur [Lusofarmaco] see 9252
Citogel [Zyma] see 9010
Citopan see 4740
Citosarin [Toyo Jozo] see 2695
Citoxid [Disprovent] see 6437
Citracal [Mission Pharmacal] see 1663
Citraconic Acid, 2320

Citral, 2321
cis-Citral see 2321
Citral A see 2321
Citral B see 2321
Citramag [Sanochemia] see 5727
Citramalic Acid, 2322
Citrated Caffeine see 1639
Citrated Calcium Carbimide see 1665
Citraurin see 2323
β-Citraurin, 2323
Citrazinic Acid, 2324
Citric Acid, 2325
Citric Acid Diammonium Salt see 509
Citric Acid Iron(2+) Salt see 4071
Citric Acid Tributyl Ester see 1565
Citric Acid Trilithium Salt see 5586
Citrical [Shire] see 1659
Citridic Acid see 109
Citrinin, 2326
Citromycetin, 2327
Citronellal, 2328
α-Citronellal see 2328
Citronellol, 2329
α-Citronellol see 8309
β-Citronellol, 2329
l-Citronellol see 8309
Citronin A see 6478
Citrophen see 7342
Citropten see 5545
Citrosodine [GSK] see 8737
Citrostadienol see 8693
Citrovorum Factor see 4249
Citrucel [GSK] see 6111
Citrulline, 2330
Citrullol, 2331
Citrus Flavonoid Compounds see 1233
Citrus Red 2, 2332
Citrylideneacetone see 8025
Civet, 2333
Civetone, 2334
Cizolirtine, 2335
CJ-91B see 6935
CJ-11972 see 5819
CK-1752 see 8581
CK-1752A see 8581
CKD-602 see 1028
CL-20 see 4733
CL-68 see 2363
CL-337 see 892
CL-636 see 9034
Cl-1025 see 2491
CL-1388R see 4594
CL-1848C see 3932
CL-1957A see 5499
CL-11344 see 8079
CL-12625 see 6509
CL-13494 see 9051
CL-13900 see 8054
CL-14377 see 6057
CL-19823 see 9757
CL-26691 see 2782
CL-34433 see 9759
CL-34699 see 382
CL-36010 see 8172
CL-62362 see 5645
CL-67772 see 573
CL-71563 see 5645
CL-81588 see 882
CL-82204 see 3996
CL-84777 see 3157
CL-112302 see 1501
CL-184116 see 7718
CL-186815 see 1201
CL-206576 see 9023
CL-216942 see 1248

Coelentrazine *see* 166
Coenzile [Fatro] *see* 2435
Coenzyme I *see* 6429
Coenzyme II *see* 6432
Coenzyme A, 2450
Coenzyme B$_{12}$ *see* 2435
Coenzyme Q$_{10}$ *see* 10025
Coenzyme Q-199 *see* 10025
Coenzyme R *see* 1236
Coenzymes Q *see* 10025
Coeruleolactite *see* 352
Co-Ervonum [Glaxo] *see* 5876
Coffearine *see* 9863
Coffee Bean Oil *see* 2452
Coffee Beans *see* 2451
Coffee, Green, 2451
Coffee Oil, 2452
Coffeine *see* 1639
Cofosfolactamines *see* 4281
Co-Fram [Abbott] *see* 9056
Co-galactoisomerase *see* 10064
Cogentin [Merck & Co.] *see* 1124
Cognac Oil, Synthetic *see* 3889
Cognex [Warner-Lambert] *see* 9154
Cogomycin *see* 4319
Coherin, 2453
Co I *see* 6429
Co II *see* 6432
Cola *see* 5366
Colace [Roberts] *see* 3446
Colalin *see* 2210
Colamine *see* 3782
Colaspase *see* 826
Colazal [Salix] *see* 938
Colazide [Shire] *see* 938
Colcemid *see* 2891
Colchamine *see* 2891
Colchiceine, 2454
Colchicine, 2455
Colchicine-binding Protein *see* 9988
Colchicum, 2456
Colcrys [Mutual Pharma] *see* 2455
Coldan *see* 6453
Coldrin [Nippon Shinyaku] *see* 2066
Colecalciferol *see* 10216
Coleflux [Finadiet] *see* 7598
Colemanite *see* 1655
Colenormol *see* 7059
Colepax *see* 5101
Colepur [Draco] *see* 1462
Colerainite *see* 345
Colesevelam Hydrochloride, 2457
Colesterinex [Galenica] *see* 8088
Colestid [Pfizer] *see* 2459
Colestilan, 2458
Colestimide *see* 2458
Colestipol, 2459
Colestyramin *see* 2209
Coletrast *see* 5069
Colforsin, 2460
Colfosceril Palmitate, 2461
Colicine *see* 2462
Colicins, 2462
Colicon [Resse] *see* 3237
Colicort [Merck & Co.] *see* 7841
Colic Root *see* 224, 3321, 4368
Colidosan *see* 8335
Colifoam [Meda] *see* 4824
Colifos *see* 7474
Colimicina [UCB] *see* 2463
Colimune [Sanofi-Aventis] *see* 2581
Colimycine [Sanofi-Aventis] *see* 2463
Coliopan [Eisai] *see* 1536
Coliquifilm [Allergan] *see* 2129
Colisone [Frosst] *see* 7842

Colistimethate Sodium *see* 2463
Colistin, 2463
Colistin B *see* 2463
Collagen, 2464
Collagenase, 2465
Collazin [Vitaplex] *see* 10359
α-Collidine *see* 3895
α,γ,α′-Collidine *see* 9890
β-Collidine *see* 3894
γ-Collidine *see* 9890
sym-Collidine *see* 9890
2,4,6-Collidine *see* 9890
Collinomycin, 2466
Collinsonia, 2467
Collodion, 2468
Collodion Cotton *see* 8125
Collodion Wool *see* 8125
Colloidal Bismuth Subcitrate, 2469
Colloidal Gold *see* 5692
Colloidal Gold ^{198}Au *see* 4552
Colloxylin *see* 8125
Collunosol *see* 9808
Collunovar [Dexo] *see* 2092
Collunovar *see* 6532
Colme [Lasa] *see* 1665
Colocynth, 2470
Colocynthin, 2471
Colofac [Duphar] *see* 5838
Colofoam [Meda] *see* 4824
Cologel [Lilly] *see* 6111
Cologne Yellow *see* 5460
Colombo *see* 1723
Colomycin [Forest] *see* 2463
Colomycin (Tabl.) [Forest] *see* 2463
Colonorm [Mundipharma] *see* 8594
Colony-stimulating Factor 1 *see* 5710
1-214-Colony-stimulating Factor 1 (Human Clone p3ACSF-69 Protein Moiety Reduced) Homodimer *see* 5710
4-221-Colony-stimulating Factor 1 (Human Clone p3ACSF-69 Protein Moiety Reduced) *see* 5710
Colony-stimulating Factor-2 *see* 4574
Colony-stimulating Factor 2 (Human Clone pCSF-1 Protein Moiety Reduced) Glycoform GMC 89-107 *see* 4574
Colony-stimulating Factor 2 (Human Clone pHG25 Protein Moiety Reduced) *see* 4574
Colony Stimulating Factors, 2472
Colophony *see* 8397
Colo-Pleon [Henning] *see* 9073
Colpermin [Pharmacia] *see* 7258
Colpogyn [Angelini] *see* 3762
Colpro [Dagra] *see* 5862
Colprone [Ayerst] *see* 5862
Colrex [Solvay] *see* 4591
Coltericin [Quimica Argentia] *see* 5329
Coltramyl [Roussel Diamant] *see* 9477
Coltrax [HMR] *see* 9477
Coltsfoot, 2473
α-Colubrine *see* 2474
β-Colubrine *see* 2474
Colubrines, 2474
Columbamine, 2475
Columbin, 2476
Columbite *see* 6642, 9185
Columbium *see* 6642
Columbium Pentachloride *see* 6643
Columbium Pentafluoride *see* 6644
Columbium Pentoxide *see* 6645
Coly-Mycin M [Monarch] *see* 2463
Colyonal [Mochida] *see* 2953
Colza Oil *see* 8233

Combantrin [Pfizer] *see* 8066
Combat *see* 4804
Combec [Tokyo Tanabe] *see* 4163
Combidex [Advanced Magnetics] *see* 4092
Combretastatin *see* 2477
(−)-Combretastatin *see* 2477
Combretastatin A-1 *see* 2477
Combretastatin A-4 *see* 2477
Combretastatin A-4 Disodium Phosphate *see* 4277
Combretastatin A-4 Phosphate *see* 4277
Combretastatins, 2477
Comelian [Kowa] *see* 3222
Comet [BASF] *see* 8064
Comfolax [Searle] *see* 3446
Comfortis [Elanco] *see* 8877
Comite [Uniroyal] *see* 7922
Command [FMC] *see* 2374
Commando [Boehringer, Ing.] *see* 3790
Common Mallow *see* 5772
Common Sage *see* 8480
Common Salt *see* 8734
Common Sundew *see* 3501
Compactin *see* 6243
Compass [Bayer] *see* 9843
Compazine [GSK] *see* 7879
Compd 3-120 *see* 9070
Compd 43-663 *see* 4602
Compd 47-83 *see* 2703
Compd 47-282 *see* 2082
Compd 48/268 *see* 9005
Compd 118 *see* 219
Compd 180/442 *see* 3652
Compd 269 *see* 3633
Compd 347 *see* 3635
Compd 469 *see* 5221
Compd 497 *see* 3114
Compd 545 *see* 5095
Compd 711 *see* 219
Compd 933F *see* 7591
Compd 57926 *see* 8682
Compd 64716 *see* 2309
Compd 72500 *see* 4234
Compd 79891 *see* 6501
Compd 83405 *see* 1916
Compd 83846 *see* 743
Compd 99638 *see* 1914
Compd 99170 *see* 743
Compd B *see* 2524
Compd F-2 *see* 10314
Compendium [Polifarma] *see* 1392
Compitox [M & B] *see* 5855
Compitox Plus [M & B] *see* 5855
Complamin [3M Pharma] *see* 10258
Complement Factor D *see* 153
Complement Factor P *see* 7928
Complexone *see* 3565
Comply [Syngenta] *see* 4017
Component B$_{1a}$ *see* 5296
Component B$_{1b}$ *see* 5296
Compound 1080 *see* 4199
Compound 81929 *see* 3440
Compound 112531 *see* 10184
Comprecin [Parke-Davis] *see* 3641
Comtan [Orion] *see* 3648
Comtess [Orion] *see* 3648
ConA *see* 2478
Conalbumin *see* 9732
Conbriza [Wyeth] *see* 1011
Concanavalin A, 2478
Conceptrol [Ortho] *see* 6764
Conceral [Takeda] *see* 4088
Concerta [Alza] *see* 6183
Conco NI [Continental Chem.] *see* 6764

Conco NI-90 [Continental Chem.] *see* 6764

Conco NIX-100 [Continental Chem.] *see* 6850

Concor [Merck KGaA] *see* 1295

Concordin [Merck & Co.] *see* 8010

Condor Vine *see* 2479

Condrosulf [IBSA] *see* 2219

Conducton [Klinge] *see* 1780

Condurangin, 2479

Condurangogenin A *see* 2479

Condyline [Nycomed] *see* 7663

Condylox [Oclassen] *see* 7663

Conessine, 2480

Conestat Alfa, 2481

Confidor [Bayer CropSci.] *see* 4951

Confirm [Dow AgroSci.] *see* 9230

Confortid [Alpharma] *see* 5009

Congo Blue *see* 9972

Congocidine *see* 6566

Congo Red, 2482

Congressane, 2483

Conhydrine, 2484

β-Coniceine, 2485

γ-Coniceine, 2486

Conicine *see* 2489

Coniel [Kyowa] *see* 1045

Coniferin, 2487

Coniferyl Alcohol, 2488

Coniine, 2489

Conium Fruit, 2490

Conivaptan, 2491

Conjugated Equine Estrogens *see* 2492

Conjugated Estrogens, 2492

Conjugol *see* 3763

Conjuncain [Mann] *see* 1049

Conludag [Searle] *see* 6786

Conmel [Winthrop] *see* 3390

Connectin *see* 9639

Connettivina [Fidia] *see* 4796

Connexins, 2493

Connexon *see* 2493

Conoco C-50 *see* 8748

Conoco C-60 *see* 8748

Conoco SD 40 *see* 8748

Conofite [Schering-Plough] *see* 6257

ω-Conopeptide MVIIA *see* 10321

Conotoxins, 2494

Conova [Gold Cross] *see* 3909

Conovid E [Searle] *see* 6787

Conquinine *see* 8176

Conray [Mallinckrodt] *see* 5107

Conray-400 [Mallinckrodt] *see* 5107

Conserve [Dow AgroSci.] *see* 8877

Conseus Interferon *see* 5034

Consolan [Novo] *see* 6428

Constilac [Alra] *see* 5395

Consult [Dow AgroSci.] *see* 4719

Consumptive's Weed *see* 3727

Contalax [3M Pharma] *see* 1246

Contax *see* 7073

Conteben [Bayer] *see* 9442

Contergan *see* 9403

Contomin [Takeda] *see* 2191

Contracne [Biorga] *see* 5274

Contralac [Virbac] *see* 6009

Contramal [Grünenthal] *see* 9726

Contrathion [Sanofi-Aventis] *see* 7822

Contratuss *see* 5523

Contrix "28" *see* 5107

Control-Om [OM] *see* 5918

Contusugene Ladenovec, 2495

Convacard [Madaus] *see* 2496

Convallan [Gödecke] *see* 2496

Convallaria, 2496

Convallaria Glycosides *see* 2496

Convallaton *see* 2497

Convallatoxigenin *see* 8981

Convallatoxin, 2497

Convallen *see* 2496

Convalyt *see* 2496

Convasid *see* 2496

Convenia [Pfizer] *see* 1937

Convenil [Hommel] *see* 1525

Convicine, 2498

Convulex [Pharmacia] *see* 10099

Coolspan [Hishiyama] *see* 9119

Coomassie Blue RL [BASF] *see* 622

Coopex [Bayer CropSci.] *see* 7294

CO-ORD [Baxter] *see* 211

COP, 2499

COP 1 *see* 4470

COP 1 (Polyamide) *see* 4470

Copaene, 2500

α-Copaene *see* 2500

Copaiba, 2501

Copal, 2502

Coparogin [Nippon Chemiphar] *see* 9252

Copaxone [Teva] *see* 4470

Copegus [Roche] *see* 8323

Copernicium, 2503

Copinal [Vinas] *see* 46

Copolymer 1 *see* 4470

Copper, 2504

Copper(II) Acetate *see* 2614

Copper Acetate Arsenite *see* 2615

Copperas *see* 4084

Copper Bis(octadecanoate) *see* 2642

Copper Bis(perchlorate) *see* 2637

Copper Bromide (CuBr) *see* 2649

Copper Bromide (CuBr$_2$) *see* 2619

Copper Carbonate Hydroxide *see* 2621

Copper Chelazome [Albion] *see* 2617

Copper(I) Chloride *see* 2650

Copper(II) Chloride *see* 2623

Copper Chloride (CuCl) *see* 2650

Copper Chloride (CuCl$_2$) *see* 2623

Copper Chromate *see* 2624

Copper(2+) Chromite *see* 2625

Copper-chromium Oxide *see* 2625

Copper Cyanide (Cu(CN)) *see* 2651

Copper Dibromide *see* 2619

Copper Dibutyrate *see* 2620

Copper Difluoride *see* 2628

Copper Diformate *see* 2629

Copper Dihydroxide *see* 2632

Copper Dioleate *see* 2634

Copper Diperchlorate *see* 2637

Copper Disalicylate *see* 2639

Copper Distearate *see* 2642

Copper DOS *see* 2660

Copper Fluoride (CuF$_2$) *see* 2628

Copper Glance *see* 2657

Copper(II) Gluconate *see* 2630

Copper Hydrate *see* 2632

Copper Hydroxide (Cu(OH)$_2$) *see* 2632

Copper Hydroxide Sulfate *see* 2644

Copper Hydroxide Sulfate (Cu$_3$(OH)$_4$-(SO$_4$)) *see* 2644

Copper Hydroxide Sulfate (Cu$_4$(OH)$_6$-(SO$_4$)) *see* 2644

Copper Iodide (CuI) *see* 2652

Copper Monoacetate *see* 2648

Copper Monobromide *see* 2649

Copper Monocyanide *see* 2651

Copper Monoiodide *see* 2652

Copper Monoselenide *see* 2641

Copper Monosulfide *see* 2645

Copper Monothiocyanate *see* 2659

Copper(II) Nitrate *see* 2633

Copper(II) Oxide *see* 2636

Copper Oxide (CuO) *see* 2636

Copper Oxide (Cu$_2$O) *see* 2654

Copper Oxysulfate *see* 2644

Copper Phosphate (3:2) *see* 2638

Copper Phthalocyanine, 2505

Copper Pyrites *see* 2037

Copper(2+) Selenide *see* 2641

Copper Selenide (CuSe) *see* 2641

Copper Selenide (Cu$_2$Se) *see* 2656

Copper(II) Sulfate *see* 2643

Copper Sulfate Dibasic *see* 2644

Copper Sulfate Tribasic *see* 2644

Copper Sulfide (CuS) *see* 2645

Copper Sulfide (Cu$_2$S) *see* 2657

Copper(I) Thiophene-2-carboxylate, 2506

Copper Tungstate *see* 2647

Copper Tungsten Oxide (CuWO$_4$) *see* 2647

Coppinger's Radical *see* 4387

Copra Oil *see* 2446

Copren [Fuso] *see* 4511

COP Resist *see* 2499

Coprogen, 2507

Coprola [Dunster] *see* 3446

Coprostane, 2508

3β-Coprostanol *see* 2509

Coprostenol *see* 256

4:5-Coprosten-3-ol *see* 256

Coprosterol, 2509

Coptine, 2510

Coptis, 2511

Coptisine, 2512

CoQ$_{10}$ *see* 10025

Coracon *see* 6627

Coracten [UCB] *see* 6613

Coractiv N [Phyteia] *see* 6627

Coragen [DuPont] *see* 2080

Corail [Bayer CropSci.] *see* 9229

Co-ral [Bayer] *see* 2546

Coralan [Servier] *see* 5294

Coralgil *see* 4738

Coralgina *see* 4738

Corallin *see* 872

Coramedan [Medice] *see* 3182

Coramine [Novartis] *see* 6627

Corangin [Novartis] *see* 5271

Coranormol *see* 7254

Corasil [Nufarm] *see* 3088

Corathiem [Ohta] *see* 2306

Corax [Empire] *see* 2085

Coraxam [Servier] *see* 5294

Corazole *see* 7254

Corbadrine *see* 6781

Corbasil *see* 6781

Corbel [BASF] *see* 4021

Corbionax [Sanofi-Aventis] *see* 478

Corchorgenin *see* 8981

Corchorin *see* 8981

Corchsularin *see* 8981

Cordan [IBI] *see* 4002

Cordanum [AWD] *see* 9170

Cordarex [Sanofi-Aventis] *see* 478

Cordarone [Sanofi-Aventis] *see* 478

Cord Factors, 2513

Cordiamin *see* 6627

Cordianine *see* 250

Cordicant [Mundipharma] *see* 6613

Cordilan [Roche] *see* 6613

Cordipatch [Schwarz] *see* 6694

Corditrine [Aventis] *see* 6694

Cordium [Riom] *see* 1152

Cordoval [Tempelhof] *see* 4469

Cordran [Watson] *see* 4227

Cordycepic Acid *see* 5810
Cordycepin, 2514
Cordycepin-5′-triphosphate *see* 2514
9-Cordyceposidoadenine *see* 2514
Cordyceps, 2515
Coredamin [Meiji] *see* 7855
Corediol [Meiji] *see* 6627
Coreg [GSK] *see* 1873
Corein 2R *see* 1961
Coretal [Polfa] *see* 7051
Corey-Chaykovsky Reagent *see* 3286
Corey's Reagent *see* 3286
Corflazine [Daltan] *see* 5537
Corgard [King] *see* 6431
Corglykon *see* 2497
Coriamyrtin, 2516
Coriander, 2517
Coriandrol *see* 5550
Coriantin [Lepetit] *see* 2221
Coric [BMS] *see* 5572
Corid [Merck & Co.] *see* 586
Coridil [Delalande] *see* 3388
Cori Ester *see* 4497
Corifollitropin Alfa, 2518
Corilagin *see* 9184
Corindolan [Schering AG] *see* 5923
Coriphate [Tokyo Tanabe] *see* 4181
Coristin [San Carlo] *see* 3710
Coritat [Green Cross] *see* 6788
Cork, 2519
Corlan [UCB] *see* 4824
Corlentor [Servier] *see* 5294
Corliprol [Pharmacia & Upjohn] *see* 1962
Corlopam [SKB] *see* 4010
Corlux [Corcept] *see* 6266
Cormed [Reiss] *see* 6627
Cormelian [Viatris] *see* 3222
Cormid *see* 6627
Cornaron [TAD] *see* 478
Cornetite *see* 2638
Cornin *see* 10153
Cornocentin *see* 3712
Corn Oil, 2520
Cornox [Boots] *see* 5833
Corn Steep Liquor, 2521
Corn Steep Water *see* 2521
Corn Sugar *see* 4495
Cornus, 2522
Corodane 2083
Corodilan [Meiji] *see* 3765
Coronarin *see* 3526
Coronarine [NEGMA] *see* 3388
Coronene, 2523
Corontin [Chinoin] *see* 7855
Corotrend [Siegfried] *see* 6613
Corotrope [Sanofi-Synthelabo] *see* 6278
Corovliss [Roche] *see* 5271
Corprilin [Streuli] *see* 5572
Corps de Pfeiffer *see* 7309
Corpus Luteum Hormone *see* 7889
Correctol [Schering-Plough] *see* 1246
Corrigast [Pfizer] *see* 7919
Corrigen *see* 6919
Corrinoids *see* 10212
Corrosive Mercury Chloride *see* 5943
Corrosive Sublimate *see* 5943
Corsodyl [GSK] *see* 2092
Cortaid [J & J] *see* 4824
Cortancyl [Aventis] *see* 7842
Cortate [Schering] *see* 2903
Cort-Dome [Bayer] *see* 4824
Cortef [Pharmacia & Upjohn] *see* 4824
Cortenema [Solvay] *see* 4824
Cortensor *see* 4692
Cortex Angosturae see 645

Cortex Cuspariae see 645
Cortexilar [Veterinaria] *see* 4169
Cortexolone *see* 2933
Cortexone *see* 2902
Cortexone Acetate *see* 2903
Cor-Theophyllin [Paramedical] *see* 3526
Corticoliberin(e) *see* 2575
Corticosteroid-binding Globulin *see* 4824
Corticosterone, 2524
Corticotrophin *see* 125
Corticotrophin Zinc Hydroxide Suspension *see* 125
Corticotropin *see* 125
$α^{1\text{-}24}$-Corticotropin *see* 2539
$α^{1\text{-}24}$-Corticotropin Hexaacetate (Salt) *see* 2539
$β^{1\text{-}24}$-Corticotropin *see* 2539
Corticotropin-like Intermediate Lobe Peptide *see* 7905
Corticotropin-releasing Factor *see* 2575
Corticotropin-releasing Hormone *see* 2575
Cortidene [Syntex Latino] *see* 7133
Cortifoam [Schwarz] *see* 4824
Cortiphyson [Promonta] *see* 125
Cortiplastol [Medici] *see* 4181
Cortiron [Schering AG] *see* 2903
Cortisol *see* 4824
Cortisone, 2525
$Δ^1$-Cortisone *see* 7842
Cortisone Acetate *see* 2525
Cortistab [Abbott] *see* 2525
Cortistatin A *see* 2526
Cortistatin J *see* 2526
Cortistatins, 2526
Cortisyl *see* 2525
Cortivazol, 2527
Cortivent [AstraZeneca] *see* 1476
Cortogen *see* 2525
Cortol, 2528
α-Cortol *see* 2528
β-Cortol *see* 2528
Cortolone, 2529
Cortone [Merck & Co.] *see* 2525
Cortril [Pfizer] *see* 4824
Cortrophin [Organon] *see* 125
Cortrophine-Z [Organon] *see* 125
Cortrosyn [Amphastar] *see* 2539
Corulon [Intervet] *see* 2221
Corundum *see* 350
Corvasal [Aventis] *see* 6320
Corvasol [Rieswerke] *see* 7254
Corvasymton *see* 9142
Corvaton [Aventis] *see* 6320
Corvert [Pfizer] *see* 4921
Corvitol *see* 6627
Corvotone *see* 6627
CorVue [King] *see* 1229
Corwin [Zeneca] *see* 10253
Corydaldine, 2530
Corydaline, 2531
d-Corydaline *see* 2531
Corydalis, 2532
Corydine, 2533
Corylophyline *see* 4496
Corynantheine, 2534
Corynanthidine *see* 10301
Corynanthine, 2535
Corynine *see* 10300
Corynomycolic Acids *see* 6410
Corypalmine, 2536
Corytuberine, 2537
Cosaldon [HMR] *see* 7240
Cosbiol [Laserson & Sabetay] *see* 8895

CoSept 200 [RDT HallStar] *see* 8143
Cositecan, 2538
Cosmegen [Merck & Co.] *see* 2797
Cosmocil CQ [Arch] *see* 7677
Cosmoline [E. F. Houghton] *see* 7300
Cosmos *see* 6035
Cosopt [Merck & Co.] *see* 3475
Cospanon [Eisai] *see* 4138
Cossmetin *see* 718
Cossym [Egnaro] *see* 1795
Costilate [Green Cross] *see* 5710
Costimulator *see* 5038
Cosumix [Novartis] *see* 9032
Cosylan [Recip] *see* 3884
Cosyntropin, 2539
COT *see* 2731
Cotarnine, 2540
Cotarnine Chloride *see* 2540
Cotarninium Chloride *see* 2540
Co-tetroxazine *see* 9398
Co-thromboplastin *see* 3961
Cotinine, 2541
Cotnion [Bayer CropSci.] *see* 906
Cotoran [Griffin] *see* 4185
Co-trifamole *see* 9056
Co-trimazine *see* 9035
Co-trimoxazole *see* 9050
Cotton Red 4B *see* 1104
Cottonex [Makhteshim-Agan] *see* 4185
Cotton-root Bark, 2542
Cottonseed Oil, 2543
Cotunnite *see* 5459
Couch Grass *see* 9938
Coughwort *see* 2473
Coumachlor, 2544
Coumadin [BMS] *see* 10236
Coumafos *see* 2546
Coumalic Acid, 2545
Coumaphos, 2546
Coumaran, 2547
p-Coumaric Acid, 2548
Coumarilic Acid, 2549
Coumarin, 2550
Coumarin-3-carboxylic Acid, 2551
cis-o-Coumarinic Acid Lactone *see* 2550
Coumarinic Anhydride *see* 2550
Coumarone *see* 1090
Coumarone-2-carboxylic Acid *see* 2549
Coumestrol, 2552
Coumingine, 2553
Counter [AMVAC] *see* 9301
Courlene [Courtaulds] *see* 7687
Courlene PY [Courtaulds] *see* 7701
Cousso *see* 1371
Covatine [Bailly] *see* 1775
Covatix [Gödecke] *see* 1775
Covellite *see* 2645
Covera-HS [Searle] *see* 10144
Coverine [Astier] *see* 2267
Coversum [Servier] *see* 7285
Coversyl [Servier] *see* 7285
Coviracil [Triangle] *see* 3622
Co-waldenase *see* 10064
Cow Clover *see* 9858
Cowrie *see* 2502
Cow's Milk *see* 6273
COX-189 *see* 5661
Coxistat *see* 6687
Coyden [Dow] *see* 2386
Cozaar [DuPont] *see* 5641
Cozymase *see* 6429
Cozyme [Travenol] *see* 2949
CP-20 *see* 2862
CP-1001 *see* 2258
CP-1044-J3 *see* 1479

Crown Compounds *see* 2592
Crown Ethers, 2592
Croysulfone *see* 2822
CRP, 2593
Crude Chrysarobin *see* 757
Crude Cream of Tartar *see* 772
Crude Oil *see* 7302
Crude Opium *see* 6949
Crude Potassium Bitartrate *see* 772
Cruex [Novartis] *see* 10033
Cruiser [Syngenta] *see* 9446
β-Crustacyanin *see* 845
Crustecdysone *see* 3538
Cryofluorane, 2594
Cryolite, 2595
Cryptand 222, 2596
Cryptates *see* 2592
Cryptocavine *see* 2598
Cryptocillin [Hoechst] *see* 7003
Cryptohalite *see* 524
Cryptophycin-1 *see* 2597
Cryptophycin-52 *see* 2597
Cryptophycin A *see* 2597
Cryptophycins, 2597
Cryptopine, 2598
Cryptoxanthin, 2599
β-Cryptoxanthin *see* 2599
all-*trans*-Cryptoxanthin *see* 2599
Cryptoxanthol *see* 2599
Crystallinic Acid *see* 6811
Crystallins, 2600
Crystallose [Tenneco] *see* 8445
Crystal Violet *see* 4430
Crystamine *see* 10212
Crystapen [Britannia] *see* 7203
Crystex *see* 9099
Crysticillin [BMS] *see* 7206
Crystodigin [Lilly] *see* 3182
Crystoids [Merck & Co.] *see* 4748
CS *see* 2128
CS-045 *see* 9948
CS-370 *see* 2404
CS-386 *see* 6245
CS-430 *see* 4638
CS-443 *see* 1183
CS-500 *see* 6243
CS-514 *see* 7835
CS-533 *see* 7111
CS-600 *see* 5647
CS-622 *see* 9279
CS-684 *see* 7647
CS-747 *see* 7833
CS-807 *see* 1942
CS-866 *see* 6933
CS-905 *see* 900
CS-976 *see* 7111
CS-1170 *see* 1927
CS-8958 *see* 5406
CSA *see* 2219
CSFα *see* 4574
CSF-β *see* 4573
γ-CSF *see* 2775
CSF-1 *see* 5710
CSF-2 *see* 4574
CSFs *see* 2472
CSIF *see* 5043
CSP *see* 4104
CT-3 *see* 191
CT-848 *see* 7041
CT-1341 *see* 228
CT-1501R *see* 5573
CT-2103 *see* 7081
C.T.A.B. *see* 2024
CTAC *see* 8143
CTLA-4Ig *see* 4

CTR-6110 *see* 6683
CTR-6669 *see* 1792
CTX *see* 2045, 2273
Ctx *see* 2177, 2494
CTX-1 *see* 2273
CTX-2 *see* 2273
CTX-3 *see* 2273
CTX-III *see* 1834
Cubane, 2601
Cube Alum *see* 354
Cubeb, 2602
Cubebin, 2603
α-Cubebin *see* 2603
β-Cubebin *see* 2603
(−)-(8*R*,8′*R*,9*R*)-Cubebin *see* 2603
(−)-(8*R*,8′*R*,9*S*)-Cubebin *see* 2603
Cubicin [Cubist] *see* 2823
Cubic Niter *see* 8778
Cuca *see* 2438
Cucoline *see* 8684
Cucurbitacin B *see* 2604
Cucurbitacin E *see* 2604
Cucurbitacins, 2604
Cudbear, 2605
Cuelure, 2606
Cuemid [Merck & Co.] *see* 2209
Cuivasil [IDC] *see* 5535
Cujec [Mallinckrodt] *see* 2660
Culpen [Fujisawa] *see* 4143
Cultar [Zeneca Ag Prod] *see* 7082
Culver's Root *see* 5497
Cu-lyt *see* 2650
Cumaldehyde *see* 2611
Cumaran *see* 2547
Cumarin *see* 2550
Cumarone *see* 1090
Cumene, 2607
3-*p*-Cumenyl-1,1-dimethylurea *see* 5264
2-[(*o*-Cumenyloxy)methyl]-2-imidazoline
  see 4016
Cumertilin Sodium [Endo] *see* 5939
Cumic Acid, 2608
Cumic Alcohol, 2609
Cumidine, 2610
Cuminal *see* 2611
Cuminaldehyde, 2611
Cuminic Acid *see* 2608
Cuminol *see* 2609
Cuminyl Alcohol *see* 2609
Cumol *see* 2607
Cumotocopherol *see* 9654
Cupferron, 2612
Cuprammonium Sulfate *see* 9325
Cupreine, 2613
Cupreol *see* 8694
Cupressin [Takeda] *see* 2881
Cuprex [Merck KGaA] *see* 2634
Cupric Acetate, 2614
Cupric Acetoarsenite, 2615
Cupric Aminoacetate *see* 2617
Cupric Ammonium Chloride *see* 511
Cupric Arsenite, 2616
Cupric Benzoate *see* 1093
Cupric Bisglycinate, 2617
Cupric Bis[8-hydroxyquinoline Di(dieth-
  ylammonium Sulfonate)] *see* 2660
Cupric Borate, 2618
Cupric Bromide, 2619
Cupric Butyrate, 2620
Cupric Carbonate, Basic, 2621
Cupric Chlorate, 2622
Cupric Chloride, 2623
Cupric Chromate(III) *see* 2625
Cupric Chromate(VI), 2624
Cupric Chromite, 2625

Cupric Citrate, 2626
Cupric Ferrocyanide, 2627
Cupric Ferrous Sulfide *see* 2037
Cupric Fluoride, 2628
Cupric Fluosilicate *see* 2631
Cupric Formate, 2629
Cupric Gluconate, 2630
Cupric Hexacyanoferrate(II) *see* 2627
Cupric Hexafluorosilicate, 2631
Cupric Hydroxide, 2632
Cupricin *see* 2651
Cupric Lactate *see* 5384
Cupric Nitrate, 2633
Cupric Oleate, 2634
Cupric Oxalate, 2635
Cupric Oxide, 2636
Cupric Perchlorate, 2637
Cupric Phosphate, 2638
Cupric Salicylate, 2639
Cupric Selenate, 2640
Cupric Selenide, 2641
Cupric Silicofluoride *see* 2631
Cupric Stearate, 2642
Cupric Subcarbonate *see* 2621
Cupric Subsulfate *see* 2644
Cupric Sulfate, 2643
Cupric Sulfate, Ammoniated *see* 9325
Cupric Sulfate, Basic, 2644
Cupric Sulfide, 2645
Cupric Tartrate, 2646
Cupric Tungstate(VI), 2647
Cupric Wolframate *see* 2647
Cuprimine [Merck & Co.] *see* 7197
Cuprimyl [M & B] *see* 2660
Cuprimyxin *see* 6425
Cupripen [Rubio] *see* 7197
Cuprite *see* 2654
Cuprocitrol *see* 2626
Cupro-cupric Sulfate *see* 2658
Cupron [Ayerst] *see* 1096
Cuprous Acetate, 2648
Cuprous Bromide, 2649
Cuprous Chloride, 2650
Cuprous Cyanide, 2651
Cuprous Iodide, 2652
Cuprous Mercuric Iodide, 2653
Cuprous Oxide, 2654
Cuprous Potassium Cyanide, 2655
Cuprous Selenide, 2656
Cuprous Sulfide, 2657
Cuprous Sulfite, 2658
Cuprous Sulfocyanate *see* 2659
Cuprous Tetraiodomercurate(2−) *see*
  2653
Cuprous Thiocyanate, 2659
Cuproxat [Agrolinz] *see* 2644
Cuproxoline, 2660
Curacne [Fabre] *see* 5274
Curacron [Syngenta] *see* 7886
Curamil *see* 8073
(16α)-Curan-17-ol *see* 4413
Curantyl [Berlin-Chemie] *see* 3388
Curare, 2661
C-Curarine I, 2662
Curatoderm [Merck KGaA] *see* 9152
Curatrem [Merial] *see* 2395
Curbix [Bayer CropSci.] *see* 3793
Curcuma Longa *see* 10005
Curcumin, 2663
Curfew [Dow AgroSci.] *see* 3085
Curine, 2664
Curium, 2665
Curled Dock *see* 8429
Curled Mint *see* 8865
Curling Factor *see* 4584

Cyanogen Iodide, 2685
Cyanogran *see* 8741
N-Cyanoguanidine *see* 3103
2-Cyano-10-[3-(4-hydroxypiperidino)-
    propyl]phenothiazine *see* 7279
(2*Z*)-2-Cyano-3-hydroxy-*N*-[4-(trifluoro-
    methyl)phenyl]-2-butenamide *see* 9308
N-Cyanoimidazole *see* 2686
1-Cyanoimidazole, 2686
1-[4-(5-Cyanoindol-3-yl)butyl]-4-(2-car-
    bamoylbenzofuran-5-yl)piperazine *see*
    10176
5-[4-[4-(5-Cyano-1*H*-indol-3-yl)butyl]-1-
    piperazinyl]-2-benzofurancarbox-
    amide *see* 10176
2-(3-Cyano-4-isobutyloxyphenyl)-4-
    methyl-5-thiazolecarboxylic Acid *see*
    3983
Cyanomethane *see* 69
1-(2-Cyano-2-methoxyiminoacetyl)-3-eth-
    ylurea *see* 2762
Cyanomethylamine *see* 406
17α-Cyanomethyl-17β-hydroxy-4,9-es-
    tradien-3-one *see* 3117
17α-Cyanomethyl-17β-hydroxy-13β-
    methylgona-4,9-dien-3-one *see* 3117
N-Cyano-*N'*-methyl-*N''*-[2-[[(4-methyl-
    1*H*-imidazol-5-yl)methyl]thio]ethyl]-
    guanidine *see* 2282
2-[3-Cyano-4-(2-methylpropoxy)phenyl]-
    4-methyl-5-thiazolecarboxylic Acid *see*
    3983
(6*R*,7*S*)-7-[[2-[(Cyanomethyl)thio]acetyl]-
    amino]-7-methoxy-3-[[(1-methyl-1*H*-
    tetrazol-5-yl)thio]methyl]-8-oxo-5-thia-
    1-azabicyclo[4.2.0]oct-2-ene-2-carbox-
    ylic Acid *see* 1927
N-(Cyanomethyl)-4-(trifluoromethyl)nic-
    otinamide *see* 4137
N-(Cyanomethyl)-4-(trifluoromethyl)-3-
    pyridinecarboxamide *see* 4137
α-Cyano-3-phenoxybenzyl α-(4-Chloro-
    phenyl)isovalerate *see* 4037
α-Cyano-3-phenoxybenzyl-2-(4-chloro-
    phenyl)-3-methylbutyrate *see* 4037
(*S*)-α-Cyano-3-phenoxybenzyl (1*R*,3*R*)-3-
    [(*Z*)-2-Chloro-3,3,3-trifluoropropenyl]-
    2,2-dimethylcyclopropanecarboxylate
    see 5398
α-Cyano-3-phenoxybenzyl 3-(2-Chloro-
    3,3,3-trifluoroprop-1-en-1-yl)-2,2-di-
    methylcyclopropanecarboxylate *see*
    2756
(*RS*)-α-Cyano-3-phenoxybenzyl (1*R*)-
    cis,trans-Chrysanthemate *see* 2765
(*S*)-α-Cyano-3-phenoxybenzyl (1*R*)-*cis*-3-
    (2,2-Dibromovinyl)-2,2-dimethylcyclo-
    propane Carboxylate *see* 2888
(±)-α-Cyano-3-phenoxybenzyl (±)-*cis*,-
    trans-3-(2,2-Dichlorovinyl)-2,2-dimeth-
    ylcyclopropane Carboxylate *see* 2764
α-Cyano-*m*-phenoxybenzyl 2,2-Dimeth-
    yl-3-(2-methylpropenyl)cyclopropane-
    carboxylate *see* 2765
(*S*)-α-Cyano-3-phenoxybenzyl (1*R*-*cis*)-
    2,2-Dimethyl-3-[(*Z*)-3-oxo-3-[2-(1,1,1,-
    3,3,3-hexafluoropropoxy)-1-propen-
    yl]cyclopropanecarboxylate *see* 117
α-Cyano-3-phenoxybenzyl 2,2,3,3-Tetra-
    methylcyclopropanecarboxylate *see*
    4019
1-(2-Cyanophenoxy)-2-hydroxy-3-*tert*-
    butylaminopropane *see* 1493
Cyano(3-phenoxyphenyl)methyl 4-Chlo-
    ro-α-(1-methylethyl)benzeneacetate
    see 4037

(±)-Cyano-(3-phenoxyphenyl)methyl (+)-
    4-(Difluoromethoxy)-α-(1-methyleth-
    yl)benzeneacetate *see* 4155
(α*E*)-2-[[6-(2-Cyanophenoxy)-4-pyrimi-
    dinyl]oxy]-α-(methoxymethylene)ben-
    zeneacetic Acid Methyl Ester *see* 917
O-(4-Cyanophenyl) *O*,*O*-Dimethyl Phos-
    phorothioate *see* 2687
5-*p*-Cyanophenyl-5,6,7,8-tetrahydroimid-
    azo[1,5-*a*]pyridine *see* 3968
(2*R*,3*R*)-3-[4-(4-Cyanophenyl)thiazol-2-
    yl]-2-(2,4-difluorophenyl)-1-(1*H*-1,2,4-
    triazol-1-yl)-2-butanol *see* 8239
(2*R*,3*R*)-3-[4-(4-Cyanophenyl)thiazol-2-
    yl]-2-(2,5-difluorophenyl)-1-(1*H*-1,2,4-
    triazol-1-yl)butan-2-ol *see* 5149
4-[1-(4-Cyanophenyl)-1-(1,2,4-triazol-1-
    yl)methyl]benzonitrile *see* 5502
2-[2-(4-Cyanophenyl)-1-[3-(trifluoro-
    methyl)phenyl]ethylidene]-*N*-[4-(tri-
    fluoromethoxy)phenyl]hydrazinecar-
    boxamide *see* 5992
(*EZ*)-2'-[2-(4-Cyanophenyl)-1-(α,α,α-tri-
    fluoro-*m*-tolyl)ethylidene]-4-(tri-
    fluoromethoxy)carbanilohydrazide *see*
    5992
Cyanophos, 2687
2-Cyanopropane *see* 5201
2-Cyano-1-propene *see* 6014
2-Cyano-2-propenoic Acid Butyl Ester
    see 1566
2-Cyano-2-propenoic Acid Ethyl Ester
    see 3842
2-Cyano-2-propenoic Acid Methyl Ester
    see 6117
2-Cyano-2-propenoic Acid 2-Methylpro-
    pyl Ester *see* 5185
2-Cyano-2-propenoic Acid Octyl Ester
    see 6855
Cyanopsin, 2688
N-[3-(3-Cyanopyrazolo[1,5-*a*]pyrimidin-
    7-yl)phenyl]-*N*-ethylacetamide *see*
    10309
N-Cyano-*N'*-4-pyridinyl-*N''*-(1,2,2-tri-
    methylpropyl)guanidine *see* 7549
Cyanosin *see* 7440
ω-Cyanotoluene *see* 1134
o-Cyanotoluene *see* 9697
Cyanotrichloromethane *see* 9793
(±)-4'-Cyano-α,α,α-trifluoro-3-[(*p*-fluo-
    rophenyl)sulfonyl]-2-methyl-*m*-lacto-
    toluidide *see* 1204
4-Cyano-3-trifluoromethyl-*N*-(3-*p*-fluoro-
    phenylsulfonyl-2-hydroxy-2-methyl-
    propionyl)aniline *see* 1204
N-[4-Cyano-3-(trifluoromethyl)phenyl]-
    3-[(4-fluorophenyl)sulfonyl]-2-hy-
    droxy-2-methylpropanamide *see* 1204
(*T*-4)-(Cyano-*κC*)trihydroborate(1−)
    Sodium (1:1) *see* 8742
Cyanox [Sumitomo] *see* 2687
Cyantin [Lederle] *see* 6686
Cyantraniliprole, 2689
Cyanuric Acid, 2690
Cyanuric Chloride, 2691
Cyanuric Fluoride, 2692
Cyanuric Trifluoride *see* 2692
Cyanurotriamide *see* 5880
Cyasorb UV 9 [Cytec] *see* 7053
Cyasorb UV 24 (obsolete) [Am. Cyanam-
    id] *see* 3332
Cyater [Lederle] *see* 9306
Cyazofamid, 2693
Cyazypyr [DuPont] *see* 2689
Cybolt [BASF] *see* 4155

CYC-202 *see* 8579
Cycasin, 2694
Cyclacillin, 2695
Cycladiene [Bruneau] *see* 3115
Cycladol [Kiesi] *see* 7619
Cyclaine [Merck & Co.] *see* 4744
Cyclamate Calcium *see* 1667
Cyclamate Sodium *see* 2696
Cyclamic Acid, 2696
Cyclamide *see* 61, 9671
Cyclamin *see* 5779
Cyclamycin [Wyeth] *see* 9949
Cyclan *see* 1667
Cyclandelate, 2697
Cyclanon *see* 4400
Cyclapen [Wyeth] *see* 2695
Cyclazocine, 2698
Cyclen, 2699
Cyclergine [Byk Gulden] *see* 2697
Cycletanide *see* 2267
Cyclethrin, 2700
Cyclic Adenosine 3',5'-Monophosphate
    see 2701
Cyclic AMP, 2701
Cyclic (5 → 14),(7 → 12)-L-Arginylgly-
    cylglycyl-L-leucyl-L-cysteinyl-L-tyro-
    syl-L-cysteinyl-L-arginylglycyl-L-argi-
    nyl-L-phenylalanyl-L-cysteinyl-L-valyl-
    L-cysteinyl-L-valylglycyl-L-arginin-
    amide *see* 5150
Cyclic GMP, 2702
Cyclic Guanosine 3',5'-Monophosphate
    see 2702
Cyclic Methylene (4-Chloro-*o*-tolyl)di-
    thioimidocarbonate *see* 6635
Cyclic Nucleotide Phosphodiesterases *see*
    2701
Cyclic Pifithrin-α *see* 7533
Cyclic Propylene (Diethoxyphosphinyl)-
    dithioimidocarbonate *see* 5922
Cyclic Sodium Trimetaphosphate *see*
    8823
Cyclizine, 2703
Cyclo[L-alanyl-D-alanyl-*N*-methyl-L-leu-
    cyl-*N*-methyl-L-leucyl-*N*-methyl-L-va-
    lyl-(3*R*,4*R*,6*E*)-6,7-didehydro-3-hy-
    droxy-*N*,4-dimethyl-L-2-aminooctan-
    oyl-L-2-aminobutanoyl-*N*-methylgly-
    cyl-*N*-methyl-L-leucyl-L-valyl-*N*-meth-
    yl-L-leucyl] *see* 2750
Cyclo[L-alanyl-D-alanyl-*N*-methyl-L-leu-
    cyl-*N*-methyl-L-leucyl-*N*-methyl-L-va-
    lyl-[(2*S*,3*R*,4*R*,6*E*)-3-hydroxy-4-meth-
    yl-2-(methylamino)nona-6,8-dienoyl]-
    (2*S*)-2-aminobutanoyl-*N*-methylglycyl-
    N-methyl-L-leucyl-L-valyl-*N*-methyl-L-
    leucyl] *see* 10224
Cyclo[(2*Z*)-2-amino-2-butenoyl-L-valyl-
    (3*S*,4*E*)-3-hydroxy-7-mercapto-4-hep-
    tenoyl-D-valyl-D-cysteinyl] Cyclic
    (3 → 5)-Disulfide *see* 8382
Cyclo[(2*Z*)-3-[(aminocarbonyl)amino]-
    2,3-didehydroalanyl-(2*S*)-2-[(4*R*)-2-
    amino-3,4,5,6-tetrahydro-4-pyrimidin-
    yl]glycyl]-2-[[(3*R*,4*R*)-3,6-diamino-
    4-hydroxy-1-oxohexyl]amino]-β-alan-
    yl-L-seryl-L-seryl] *see* 3654
Cyclo[(4*R*)-4-(2-aminoethylcarbamoyl-
    oxy)-L-prolyl-L-phenylglycyl-D-trypto-
    phyl-L-lysyl-4-*O*-benzyl-L-tyrosyl-L-
    phenylalanyl-] *see* 7515
Cycloamyloses *see* 2710
Cyclo(L-arginylglycyl-L-α-aspartyl-D-
    phenylalanyl-*N*-methyl-L-valyl) *see*
    2276

Cycloartenyl Ferulate *see* 6983
Cyclobarbital, 2704
Cyclobarbitone *see* 2704
Cyclobendazole, 2705
Cyclobenzaprine, 2706
Cycloblastin [Pharmacia] *see* 2743
Cyclobral [Norgine] *see* 2697
Cyclobutadiene, 2707
1,3-Cyclobutadiene *see* 2707
Cyclobutane, 2708
cis-[*trans*-1,2-Cyclobutanebis(methylamine)][(*S*)-lactato-*O*¹,*O*²]platinum *see* 5607
1,1-Cyclobutanedicarboxylic Acid Platinum Complex *see* 1823
(*SP*-4-3)-[*rel*-(1*R*,2*R*)-1,2-Cyclobutanedimethanamine-κ*N*,κ*N*′][(2*S*)-2-(hydroxy-κ*O*)propanoato(2−)-κ*O*]platinum *see* 5607
(−)-*N*-Cyclobutylmethyl-3,14-dihydroxymorphinan *see* 1532
(5α,6α)-17-(Cyclobutylmethyl)-4,5-epoxymorphinan-3,6,14-triol *see* 6441
N-Cyclobutylmethyl-14-hydroxydihydronormorphine *see* 6441
17-(Cyclobutylmethyl)morphinan-3,14-diol *see* 1532
Cyclobuxine *see* 2709
Cyclobuxine D, 2709
Cyclo-C [Kohjin] *see* 623
α-Cyclocitrylideneacetone *see* 5099
Cyclocort [Fujisawa] *see* 382
*O*²,²′-Cyclocytidine *see* 623
2,2′-*O*-Cyclocytidine *see* 623
1,2-Cyclodecanedione *see* 8553
1-Cyclodecanol-2-one *see* 8554
α-Cyclodextrin *see* 2710
β-Cyclodextrin *see* 2710
γ-Cyclodextrin *see* 2710
Cyclodextrins, 2710
β-Cyclodextrin Sulfobutyl Ether, 2711
Cyclo-dibromodi-μ-methylene[μ-(tetrahydrofuran)]trizinc *see* 6826
Cyclo[2,2-dimethyl-β-alanyl-(2*S*)-2-hydroxy-4-methylpentanoyl-(2*E*,5*S*,6*S*)-5-hydroxy-6-[(2*R*,3*R*)-3-phenyloxiranyl]-2-heptenoyl-3-chloro-*o*-methyl-D-tyrosyl] *see* 2597
4-Cyclododecyl-2,6-dimethylmorpholine *see* 3454
Cyclodol *see* 9864
Cyclodorm *see* 2704
Cyclodrine, 2712
Cycloestrol *see* 4737
Cyclofenil, 2713
Cycloform *see* 5178
Cyclogest [Alpharma] *see* 7889
Cycloglucans *see* 2710
Cycloglycylglycine *see* 7578
Cycloguanil, 2714
Cycloguanil Embonate *see* 2714
Cyclogyl [Alcon] *see* 2741
Cycloheptaamylose *see* 2710
(9*Z*)-9-Cycloheptadecen-1-one *see* 2334
Cycloheptaglucan *see* 2710
1,2-Cycloheptanedione 1,2-Dioxime *see* 4700
Cycloheptanone, 2715
Cycloheptatrienocarbonium Bromide *see* 9965
Cycloheptatrienylium Bromide (1:1) *see* 9965
5-(1-Cyclohepten-1-yl)-5-ethylbarbituric Acid *see* 4690
5-(1-Cyclohepten-1-yl)-5-ethyl-2,4,6(1*H*,-3*H*,5*H*)-pyrimidinetrione *see* 4690

P,*P*′-[(Cycloheptylamino)methylene]bisphosphonic Acid *see* 4969
Cyclohexaamylose *see* 2710
Cyclohexadextrin *see* 2710
1,4-Cyclohexadienedione *see* 8191
2,5-Cyclohexadiene-1,4-dione *see* 8191
2,5-Cyclohexadiene-1,4-dione Compd with 1,4-Benzenediol (1:1) *see* 8174
Cyclohexanamine *see* 2722
Cyclohexane, 2716
Cyclohexanecarbinol *see* 2724
N-[3-*(N*-Cyclohexanecarbonyl-D-alanylthio)-2-methylpropanoyl]-L-proline *see* 6371
Cyclohexanecarboxylic Acid, 2717
(*SP*-4-2)-[(1*R*,2*R*)-1,2-Cyclohexanediamine-κ*N*,κ*N*′][ethanedioato(2−)-κ*O*¹,κ*O*²]platinum *see* 7011
[(1*R*,2*R*)-1,2-Cyclohexanediamine-*N*,*N*′]-[oxalato(2−)-*O*,*O*′]platinum *see* 7011
1,3-Cyclohexanedione *see* 3196
1,2-Cyclohexanedione 1,2-Dioxime *see* 6646
Cyclohexanehexol *see* 5021
1,2,3,4,5,6-Cyclohexanehexolphosphoric Acid *see* 7499
Cyclohexanehexyl Hexaphosphate *see* 7499
Cyclohexanemethanol *see* 2724
1,2,3,4,5-Cyclohexanepentol *see* 8152
Cyclohexanesulfamic Acid *see* 2696
Cyclohexanesulfamic Acid Calcium Salt *see* 1667
Cyclohexanol, 2718
Cyclohexanone, 2719
Cyclohexatriene *see* 1068
Cyclohexene, 2720
5-(2-Cyclohexen-1-yl)-5-allyl-2-thiobarbituric Acid *see* 9444
5-(2-Cyclohexen-1-yl)dihydro-5-(2-propen-1-yl)-2-thioxo-4,6(1*H*,5*H*)-pyrimidinedione *see* 9444
5-(1-Cyclohexen-1-yl)-1,5-dimethylbarbituric Acid *see* 4740
5-Cyclohexenyl-3,5-dimethylbarbituric Acid *see* 4740
5-(1-Cyclohexen-1-yl)-1,5-dimethyl-2,4,6-(1*H*,3*H*,5*H*)-pyrimidinetrione *see* 4740
5-(1-Cyclohexen-1-yl)-5-ethylbarbituric Acid *see* 2704
5-(1-Cyclohexen-1-yl)-5-ethyl-2,4,6(1*H*,-3*H*,5*H*)-pyrimidinetrione *see* 2704
(1*R*,4*R*,5*S*)-1-[(*S*)-(1*S*)-2-Cyclohexen-1-yl-hydroxymethyl]-4-ethyl-5-methyl-6-oxa-2-azabicyclo[3.2.0]heptane-3,7-dione *see* 8474
Cycloheximide, 2721
Cyclohexitol *see* 5021
3-Cyclohexyl-1-(*p*-acetylphenylsulfonyl)urea *see* 61
Cyclohexylamine, 2722
N-[2-[[[[(Cyclohexylamino)carbonyl]-amino]sulfonyl]phenyl]ethyl]-5-methyl-2-pyrazinecarboxamide *see* 4477
N-[(Cyclohexylamino)carbonyl]-2,3-dihydro-1*H*-indene-5-sulfonamide *see* 4542
N-[(Cyclohexylamino)carbonyl]-4-[2-(3,4-dihydro-7-methoxy-4,4-dimethyl-1,3-dioxo-2(1*H*)-isoquinolinyl)ethyl]-benzenesulfonamide *see* 4478
N-[(Cyclohexylamino)carbonyl]-4-methylbenzenesulfonamide *see* 9671
2-(Cyclohexylamino)ethanesulfonic Acid *see* 2055
3-(Cyclohexylamino)-1-propanesulfonic Acid *see* 1769

1-(Cyclohexylamino)-2-propanol 2-Benzoate Hydrochloride (1:1) *see* 4744
25-Cyclohexylavermectin B₁ *see* 3473
α-Cyclohexylbenzeneacetic Acid 2-(Diethylamino)ethyl Ester *see* 3494
Cyclohexyl Bromide, 2723
Cyclohexylcarbinol, 2724
N-(Cyclohexylcarbonyl)-D-alanyl-(2*S*)-3-mercapto-2-methylpropanoyl-L-proline *see* 6371
(−)-*N*-[(*S*)-[3-(*N*-Cyclohexylcarbonyl-D-alanyl)thio]-2-methylpropionyl]-L-proline *see* 6371
2-(Cyclohexylcarbonyl)-2,3,6,7,8,12b-hexahydropyrazino[2,1-*a*][2]benzazepin-4(1*H*)-one *see* 3694
2-(Cyclohexylcarbonyl)-1,2,3,6,7,11b-hexahydro-4*H*-pyrazino[2,1-*a*]isoquinolin-4-one *see* 7837
2-Cyclohexylcarbonyl-4-oxo-1,2,3,4,6,7,-12b-octahydropyrazino[2,1-*a*][2]benzazepine *see* 3694
Cyclohexyl Chloride, 2725
N-Cyclohexylcyclohexanamine *see* 3106
(5*Z*)-25-Cyclohexyl-4′-*O*-de(2,6-dideoxy-3-*O*-methyl-α-L-*arabino*-hexopyranosyl)-5-demethoxy-25-de(1-methylpropyl)-22,23-dihydro-5-(hydroxyimino)-avermectin A₁ₐ *see* 8561
25-Cyclohexyl-5-*O*-demethyl-25-de(1-methylpropyl)avermectin A₁ₐ *see* 3473
3-Cyclohexyl-6,7-dihydro-1*H*-cyclopenta-pyrimidine-2,4-(3*H*,5*H*)-dione *see* 5489
1-Cyclohexyl-3-[[*p*-[2-(3,4-dihydro-7-methoxy-4,4-dimethyl-1,3-dioxo-2(1*H*)-isoquinolyl)ethyl]phenyl]sulfonyl]urea *see* 4478
3-Cyclohexyl-6-(dimethylamino)-1-methyl-1,3,5-triazine-2,4(1*H*,3*H*)-dione *see* 4735
N-Cyclohexyl-*N*′-[4-[3-[(1,1-dimethylethyl)amino]-2-hydroxypropoxy]phenyl]-urea *see* 9170
2-Cyclohexyl-4,6-dinitrophenate *see* 2726
2-Cyclohexyl-4,6-dinitrophenol, 2726
1-Cyclohexyl-3-(5-hydrindenylsulfonyl)urea *see* 4542
N-[(1*R*)-1-Cyclohexyl-2-[(2*S*)-2-[[[[4-[(hydroxyamino)iminomethyl]phenyl]methyl]amino]carbonyl]-1-azetidinyl]-2-oxoethyl]glycine Ethyl Ester *see* 10274
α-Cyclohexyl-α-hydroxybenzeneacetic Acid 4-(Diethylamino)-2-butyn-1-yl Ester *see* 7054
α-Cyclohexyl-α-hydroxybenzeneacetic Acid (1,4,5,6-Tetrahydro-1-methyl-2-pyrimidinyl)methyl Ester *see* 7072
(4*S*)-4-Cyclohexyl-1-[[[(*RS*)-1-hydroxy-2-methylpropoxy](4-phenylbutyl)phosphinyl]acetyl]-L-proline Propionate (Ester) *see* 4283
6-Cyclohexyl-1-hydroxy-4-methyl-2(1*H*)-pyridinone *see* 2269
2-Cyclohexyl-3-hydroxy-1,4-naphthalenedione *see* 7154
2-Cyclohexyl-3-hydroxy-1,4-naphthoquinone *see* 7154
(5*Z*)-7-[(1*R*,2*S*,3*R*,5*S*)-2-[(3*S*)-5-Cyclohexyl-3-hydroxy-1-pentyn-1-yl]-3,5-dihydroxycyclopentyl]-5-heptenoic Acid Methyl Ester *see* 227
4-(β-Cyclohexyl-β-hydroxy-β-phenethyl)-1,1-dimethylpiperazinium Methyl Sulfate *see* 4742

Name Index

Cyclohexylhydroxyphenylacetic Acid Diethylmethylaminoethyl Ester Bromide *see* 7074

2-[(2-Cyclohexyl-2-hydroxy-2-phenylacetyl)oxy]-*N*,*N*-diethyl-*N*-methylethanaminium Bromide (1:1) *see* 7074

4-(2-Cyclohexyl-2-hydroxy-2-phenylethyl)-1,1-dimethylpiperazinium Methyl Sulfate (1:1) *see* 4742

N-(*β*-Cyclohexyl-*β*-hydroxy-*β*-phenylethyl)-*N*¹-methylpiperazine Dimethylsulfate *see* 4742

N-(*β*-Cyclohexyl-*β*-hydroxy-*β*-phenylethyl)-*N*¹-methylpiperazine Methosulfate *see* 4742

1-(3-Cyclohexyl-3-hydroxy-3-phenylpropyl)-1-methylpyrrolidinium Chloride *see* 7881

(3-Cyclohexyl-3-hydroxy-3-phenylpropyl)triethylammonium Iodide *see* 9826

3-[(Z)-[(3a*R*,4*R*,5*R*,6a*S*)-4-[(1*E*,3*S*)-3-Cyclohexyl-3-hydroxy-1-propen-1-yl]-hexahydro-5-hydroxy-2*H*-cyclopenta[*b*]furan-2-ylidene]methyl]benzoic Acid *see* 9190

α-[(2*Z*,3a*R*,4*R*,5*R*,6a*S*)-4-[(1*E*,3*S*)-3-Cyclohexyl-3-hydroxypropenyl]hexahydro-5-hydroxy-2*H*-cyclopenta[*b*]furan-2-ylidene]-*m*-toluic Acid *see* 9190

2,2′-[Cyclohexylidenebis(4,1-phenyleneoxy)]bis[2-methylbutanoic Acid] *see* 2355

2,2′-(4,4′-Cyclohexylidenediphenoxy)-2,2′-dimethyldibutyric Acid *see* 2355

α-Cyclohexylidene-*α*-(*p*-hydroxyphenyl)-*p*-cresol Diacetate *see* 2713

4,4′-(Cyclohexylidenemethylene)bisphenol 1,1′-Diacetate *see* 2713

1-Cyclohexyl-3-(5-indanylsulfonyl)urea *see* 4542

N-Cyclohexyllinoleamide *see* 5560

Cyclohexylmethane *see* 6118

(17*β*)-17-[(Cyclohexylmethoxy)carbonyl]oxyestra-4,9,11-trien-3-one *see* 9745

N-Cyclohexyl-*N*-methyl-2-(2-amino-3,5-dibromo)benzylammonium *see* 1398

1-Cyclohexyl-2-methylaminopropane *see* 7972

l-1-Cyclohexyl-2-(methylamino)propane Ethylphenylbarbiturate *see* 7972

(11*β*,16*α*)-16,17-[[(*R*)-Cyclohexylmethylene]bis[oxy]]-11-hydroxy-21-(2-methyl-1-oxopropoxy)pregna-1,4-diene-3,20-dione *see* 2266

2-[2-(Cyclohexylmethylene)hydrazinyl]-adenosine *see* 1229

4-Cyclohexyl-*α*-methyl-1-naphthaleneacetic Acid *see* 10133

(4*S*)-4-Cyclohexyl-1-[2-[(*R*)-[(1*S*)-2-methyl-1-(1-oxopropoxy)propoxy](4-phenylbutyl)phosphinyl]acetyl]-L-proline *see* 4283

1-Cyclohexyl-3-[[*p*-[2-(5-methylpyrazinecarboxamido)ethyl]phenyl]sulfonyl]urea *see* 4477

9-Cyclohexyl-*N*-[4-(4-morpholinyl)phenyl]-2-(1-naphthalenyloxy)-9*H*-purin-6-amine *see* 8053

*N*⁶-Cyclohexyl-*N*²-[4-(4-morpholinyl)phenyl]-9*H*-purine-2,6-diamine *see* 8292

2-(4-Cyclohexyl-1-naphthyl)propionic Acid *see* 10133

1-[(2-Cyclohexyl-2-phenyl-1,3-dioxolan-4-yl)methyl]-1-methylpiperidinium Iodide (1:1) *see* 2729

α-Cyclohexyl-*α*-phenyl-1-piperidinepropanol Hydrochloride (1:1) *see* 9864

α-Cyclohexyl-*α*-phenyl-1-pyrrolidinepropanol *see* 7881

1-Cyclohexyl-1-phenyl-3-pyrrolidino-1-propanol *see* 7881

1-Cyclohexyl-1-phenyl-3-(1-pyrrolidinyl)-1-propanol *see* 7881

(1*S*,3a*R*,6a*S*)-2-[(2*S*)-2-[[(2*S*)-Cyclohexyl[(pyrazinylcarbonyl)amino]acetyl]-amino]-3,3-dimethylbutanoyl]-*N*-[(1*S*)-1-[2-(cyclopropylamino)oxoacetyl]butyl]octahydrocyclopenta[*c*]pyrrole-1-carboxamide *see* 9256

(1*S*,3a*R*,6a*S*)-(2*S*)-2-Cyclohexyl-*N*-(2-pyrazinylcarbonyl)glycyl-3-methyl-L-valyl-*N*-[(1*S*)-1-[2-(cyclopropylamino)-2-oxoacetyl]butyl]octahydrocyclopenta[*c*]pyrrole-1-carboxamide *see* 9256

N-Cyclohexylsulfamic Acid *see* 2696

N-Cyclohexylsulfamic Acid Calcium Salt (2:1) *see* 1667

N-Cyclohexyltaurine *see* 2055

3-Cyclohexyl-1,5,6,7-tetrahydro-2*H*-cyclopentapyrimidine-2,4(3*H*)-dione *see* 5489

6-[4-(1-Cyclohexyl-1*H*-tetrazol-5-yl)butoxy]-3,4-dihydrocarbostyril *see* 2280

6-[4-(1-Cyclohexyl-1*H*-tetrazol-5-yl)butoxy]-3,4-dihydro-2(1*H*)-quinolinone *see* 2280

6-[4-(1-Cyclohexyl-5-tetrazolyl)butoxy]-1,2,3,4-tetrahydro-2-oxoquinoline *see* 2280

α-Cyclohexyl-*α*-(3-thienyl)acetic Acid 2-Hexamethyleneiminoethyl Ester *see* 2019

α-Cyclohexyl-3-thiopheneacetic Acid 2-(Hexahydro-1*H*-azepin-1-yl)ethyl Ester *see* 2019

1-Cyclohexyl-3-*p*-tolylsulfonylurea *see* 9671

γ-Cyclohexyl-*N*,*N*,*N*-triethyl-*γ*-hydroxybenzenepropanaminium Iodide (1:1) *see* 9826

3-Cyclohexyl-5,6-trimethyleneuracil *see* 5489

1-(4-Cyclohexylureidophenoxy)-2-hydroxy-3-*tert*-butylaminopropane *see* 9170

Cyclo(L-homocysteinyl-*N*-methyl-L-phenylalanyl-L-tyrosyl-D-tryptophyl-L-lysyl-L-valyl), (1 → 1′)-thioether with 3-[(mercaptoacetyl)amino]-L-alanyl-L-lysyl-L-cysteinyl-L-lysinamide *see* 2912

Cyclo[(*αR*)-*α*-hydroxy-4-(4-morpholinyl)benzenepropanoyl-*N*-methyl-L-leucyl-(2*R*)-2-hydroxypropanoyl-*N*-methyl-L-leucyl-(*αR*)-*α*-hydroxy-4-(4-morpholinyl)benzenepropanoyl-*N*-methyl-L-leucyl-(2*R*)-2-hydroxypropanoyl-*N*-methyl-L-leucyl] *see* 3617

(3*β*)-9,19-Cyclolanost-24-en-3-ol 3-(4-Hydroxy-3-methoxyphenyl)-2-propenoate *see* 6983

Cycloleucine, 2727

Cyclolyt *see* 2697

Cyclomaltoheptaose *see* 2710

Cyclomaltohexaose *see* 2710

Cyclomaltooctaose *see* 2710

Cyclomandol [Yamanouchi] *see* 2697

Cyclomen [Sanofi-Synthelabo] *see* 2811

Cyclo-(*N*-Me)Phe-Tyr-(D-Trp)-Lys-Val-Hcy(CH₂CO-(*β*-Dap)-Lys-Cys-Lys-NH₂) *see* 2912

Cyclomethiazide *see* 2740

Cyclo[(2*R*)-2-methyl-*β*-alanyl-(2*S*)-2-hydroxy-4-methylpentanoyl-(2*E*,5*S*,6*S*)-5-hydroxy-6-[(2*R*,3*R*)-3-phenyloxiranyl]-2-heptenoyl-3-chloro-*o*-methyl-D-tyrosyl] *see* 2597

Cyclomorph *see* 3454

Cyclonal *see* 4740

Cyclonal Sodium *see* 4740

Cyclonamine *see* 3777

Cyclonarol [Hépatrol] *see* 2735

Cyclonicate *see* 2268

Cyclonite, 2728

Cyclonium Iodide, 2729

Cyclonorm [Streuli] *see* 2102

Cyclooctaamylose *see* 2710

1,5-Cyclooctadiene, 2730

[(1,2,5,6-*η*)-1,5-Cyclooctadiene](pyridine)(tricyclohexylphosphine)iridium-(1+) Hexafluorophosphate(1−) (1:1) *see* 2554

(*T*-4)-1,5-Cyclooctanediyl[(1*S*,2*S*,3*S*,5*R*)-6,6-dimethyl-2-[2-(phenylmethoxy)ethyl]bicyclo[3.1.1]hept-3-yl]hydroborate-(1−) Lithium *see* 6513

Cyclooctatetraene, 2731

1,3,5,7-Cyclooctatetraene *see* 2731

Cyclooctyne, 2732

Cyclopamine, 2733

Cyclopent [Jenapharm] *see* 2712

Cyclopentacycloheptene *see* 919

Cyclopentadiene, 2734

1,3-Cyclopentadiene *see* 2734

Cyclopentadrine *see* 2735

α,*β*-Cyclopentamethylenetetrazole *see* 7254

Cyclopentamine, 2735

Cyclopentane, 2736

Cyclopentanol, 2737

Cyclopentanone, 2738

Cyclopentanone 2*α*,3*α*-Epithio-5*α*-androstan-17*β*-yl Methyl Acetal *see* 5925

2-Cyclopenten-2,3-diol-1-one *see* 8246

(1*S*)-2-Cyclopentene-1-tridecanoic Acid *see* 2047

(1*R*)-2-Cyclopentene-1-undecanoic Acid *see* 4800

1,2-Cyclopentenophenanthrene, 2739

3-(2-Cyclopentenyl)-2-methyl-4-oxo-2-cyclopentenyl Ester of Chrysanthemummonocarboxylic Acid *see* 2700

D-13-(2-Cyclopenten-1-yl)tridecanoic Acid *see* 2047

11-(2-Cyclopenten-1-yl)undecanoic Acid *see* 4800

Cyclopenthiazide, 2740

Cyclopentolate, 2741

Cyclopentyl Alcohol *see* 2737

3-[(2-Cyclopentyl-2-hydroxy-2-phenylacetyl)oxy]-1,1-dimethylpyrrolidinium Bromide (1:1) *see* 4537

α-Cyclopentylmandelic Acid Ester with 3-Hydroxy-1,1-dimethylpyrrolidinium Bromide *see* 4537

Cyclopentyl 3-[2-Methoxy-4-[(*o*-tolylsulfonyl)carbamoyl]benzyl]-1-methylindole-5-carbamate *see* 10306

1-Cyclopentyl-2-methylaminopropane *see* 2735

3-Cyclopentylmethyl-6-chloro-7-sulfamyl-3,4-dihydro-1,2,4-benzothiadiazine 1,1-Dioxide *see* 2740

Cyclopentyl Methyl Ether, 2742

N-[4-[5-(Cyclopentyloxycarbonyl)amino-1-methylindol-3-ylmethyl]-3-methoxy-

(Cyclopropylthio)benzene *see* 2747
N^2-Cyclopropyl-1,3,5-triazine-2,4,6-tri-amine *see* 2772
1-Cyclopropyl-5,6,8-trifluoro-(*cis*-3,5-di-methyl-1-piperazinyl)-1,4-dihydro-4-oxoquinoline-3-carboxylic Acid *see* 6957
Cyclo(RGDf-N(Me)V-) *see* 2276
Cyclorphan, 2748
Cyclosa [Nourypharma] *see* 2928
Cycloserine, 2749
Cycloset [VeroScience] *see* 1424
Cyclospasmol [Wyeth-Ayerst] *see* 2697
Cyclosporin A *see* 2750
Cyclosporin B *see* 2750
Cyclosporin C *see* 2750
Cyclosporin D *see* 2750
Cyclosporin G *see* 2750
Cyclosporine *see* 2750
Cyclosporins, 2750
Cyclostin [Pharmacia] *see* 2743
Cyclosulfyne *see* 7922
Cyclotetramethylene Tetranitramine *see* 6847
Cyclotheonamide A *see* 2751
Cyclotheonamide B *see* 2751
Cyclotheonamides, 2751
Cyclothiazide, 2752
Cyclotrimethylenetrinitramine *see* 2728
Cycloviran [Sigma-Tau] *see* 134
Cycocel [BASF] *see* 2104
Cycogan [Makhteshim-Agan] *see* 2104
Cycostat [Am. Cyanamid] *see* 8371
Cycrin [Wyeth] *see* 5864
Cydectin [Fort Dodge] *see* 6376
Cydonia Seed *see* 8170
Cyflufenamid, 2753
Cyfluthrin, 2754
Cyfoxylate *see* 2754
Cygro [Am. Cyanamid] *see* 5711
Cyhalofop-butyl, 2755
Cyhalothrin, 2756
λ-Cyhalothrin *see* 5398
Cyhexatin, 2757
Cykill [Hacco] *see* 1396
Cyklodorm *see* 2704
Cyklokapron [Pfizer] *see* 9730
Cyklosal *see* 2735
Cylert [Abbott] *see* 7188
Cyllind [Abbott] *see* 2338
Cymarigenin *see* 8981
Cymarin, 2758
Cymarose, 2759
Cymbalta [Lilly] *see* 3512
Cymbi [Dolorgiet] *see* 583
Cymbush [Syngenta] *see* 2764
Cymene, 2760
m-Cymene *see* 2760
o-Cymene *see* 2760
p-Cymene *see* 2760
2-*p*-Cymenol *see* 1872
3-*p*-Cymenol *see* 9554
Cymerin [Tokyo Tanabe] *see* 8227
Cymeval [Roche] *see* 10093
Cymevan [Roche] *see* 4393
Cymevene [Roche] *see* 4393
Cymiazole, 2761
Cymonic Acid *see* 4199
Cymoxanil, 2762
Cynarine, 2763
Cynaroside *see* 5675
Cynisin *see* 2408
Cynoff [FMC] *see* 2764
Cynoglossophine *see* 4664
Cynomel [GSK] *see* 5565

Cynotoxin *see* 8981
Cynt [Beiersdorf] *see* 6379
Cypermethrin, 2764
Cyphenothrin, 2765
Cypip [Am. Cyanamid] *see* 3128
Cyprenorphine, 2766
Cypress Camphor *see* 1913
Cypripedium, 2767
Cyproconazole, 2768
Cyprodinil, 2769
Cyproheptadine, 2770
Cyprome Ether *see* 2746
Cyprostat [Schering AG] *see* 2771
Cyproterone, 2771
Cyren A [Bayer] *see* 3148
Cyren B [Bayer] *see* 3148
Cyromazine, 2772
Cyrpon [Bayer] *see* 5929
Cys *see* 2778
Cystadane [Orphan Med.] *see* 1181
Cystafos *see* 7453
Cystagon [Mylan] *see* 2776
Cystamine, 2773
Cystaphos *see* 7453
L-Cystathionine, 2774
Cystatin A *see* 2775
Cystatin B *see* 2775
Cystatin C *see* 2775
Cystatins, 2775
Cysteamine, 2776
Cysteamine-S-phosphate *see* 7453
Cysteic Acid, 2777
DL-Cysteic Acid *see* 2777
Cysteine, 2778
L-Cysteine *see* 2778
L-Cysteine Methyl Ester *see* 5856
Cysteine Protease P32 *see* 1885
α_1-Cysteine Proteinase Inhibitor *see* 5360
α_2-Cysteine Proteinase Inhibitor *see* 5360
L-Cysteine Thioacetal of Formaldehyde *see* 3431
L-Cysteinyl-L-cysteinyl-L-α-glutamyl-L-tyrosyl-L-cysteinyl-L-cysteinyl-L-aspar-aginyl-L-prolyl-L-alanyl-L-cysteinyl-L-threonylglycyl-L-cysteinyl-L-tyrosine Cyclic (1 → 6),(2 → 10),(5 → 13)-Tris-(disulfide) *see* 5548
L-Cysteinyl-L-lysylglycyl-L-lysylglycyl-L-alanyl-L-lysyl-L-cysteinyl-L-seryl-L-ar-ginyl-L-leucyl-L-methionyl-L-tyrosyl-L-α-aspartyl-L-cysteinyl-L-cysteinyl-L-threonylglycyl-L-seryl-L-cysteinyl-L-ar-ginyl-L-serylglycyl-L-lysyl-L-cystein-amide Cyclic (1 → 16), (8 → 20), (15 → 25)-Tris(disulfide) *see* 10321
N^2-[N-(N-L-Cysteinyl-L-tyrosyl)-L-cystei-nyl]-interferon γ (Human Lympho-cyte Protein Moiety Reduced) *see* 5036
Cystemme [Abbott] *see* 8737
Cystine, 2779
L-Cystine *see* 2779
Cystine S,S'-Dioxide *see* 2780
L-Cystine S,S-Dioxide, 2780
L-Cystine Thiosulfonate *see* 2780
Cystit [BMS] *see* 6686
Cystospaz [Alcon] *see* 4897
Cysto-Conray [Mallinckrodt] *see* 5107
Cystografin [Bracco] *see* 2993
Cystorelin [Merial] *see* 4561
Cystrin [Sanofi-Aventis] *see* 7054
Cys-Tyr-Cys-interferon γ *see* 5036
CYT-099 *see* 8521

CYT-356 *see* 1764
CYT-424 *see* 8487
Cytacon [Goldshield] *see* 10212
Cytadren [Novartis] *see* 437
Cytamen [UCB] *see* 10212
Cytarabine, 2781
Cythioate, 2782
Cythion *see* 5764
Cytidine, 2783
Cytidine Diphosphate Choline Ester *see* 2318
Cytidine-2'-monophosphate *see* 2784
Cytidine-3'-monophosphate *see* 2785
Cytidine-2'-phosphate *see* 2784
Cytidine-3'-phosphate *see* 2785
2'-Cytidinephosphoric Acid *see* 2784
3'-Cytidinephosphoric Acid *see* 2785
Cytidine 5'-(Trihydrogen Diphosphate) P'-[2-(Trimethylammonio)ethyl] Ester Inner Salt *see* 2318
Cytidylic Acid *a* *see* 2784
Cytidylic Acid *b* *see* 2785
2'-Cytidylic Acid, 2784
3'-Cytidylic Acid, 2785
Cytisine, 2786
Cytiton [USSR] *see* 2786
Cytoadhesins *see* 5032
Cytobin [Pfizer] *see* 5565
Cytobion [Merck KGaA] *see* 10212
Cytochalasin B *see* 2787
Cytochalasins, 2787
Cytochrome *c*, 2788
Cytochromes P_{450}, 2789
Cytohemin, 2790
Cytokine Receptor TACI (Synthetic Hu-man Extracellular Domain Fragment) Fusion Protein with Immunoglobulin G1 (Synthetic Human γ_1-Chain Fc Fragment), Dimer *see* 848
Cytokine Synthesis Inhibitory Factor *see* 5043
Cytolipin H, 2791
Cytolysin *see* 7274
Cytomel [GSK] *see* 5565
Cytophosphane *see* 2743
Cytorest [Mochida] *see* 2788
Cytosar [Pharmacia & Upjohn] *see* 2781
Cytosine, 2792
β-Cytosine Arabinoside *see* 2781
Cytosine Riboside *see* 2783
1-(1'-Cytosinyl)-4-[L-3'-amino-5'-(1''-N-methylguanidino)-valerylamino]-1,2,-3,4-tetradeoxy-β-D-*erythro*-hex-2-enu-ronic Acid *see* 1317
Cytostasan *see* 1036
Cytostatin *see* 4737
2'-Cytosylic Acid *see* 2784
3'-Cytosylic Acid *see* 2785
Cytotec [Pharmacia] *see* 6293
Cytotoxin D1 (Naja Naja Atra) *see* 1834
Cytovene [Roche] *see* 4393
Cytoxan [BMS] *see* 2743
Cytoxin-anticytotoxin Serum *see* 704
Cytozyme *see* 9539
Cytrol *see* 485
Cytrolane [BASF] *see* 5922
CZ *see* 1877
D$_s$ *see* 2852
1,3-D *see* 3085
2,4-D, 2793
D & C Blue No. 6 *see* 4981
D & C Green No. 5 *see* 244
D & C Green No. 6 *see* 8181
D & C Green No. 8 *see* 8065
D & C Orange No. 4 *see* 6953

D & C Orange No. 5 *see* 3027
D & C Orange No. 8 *see* 3077
D & C Orange No. 10 *see* 3207
D & C Orange No. 11 *see* 3207
D & C Red No. 17 *see* 9015
D & C Red No. 19 *see* 8307
D & C Red No. 21 *see* 3658
D & C Red No. 22 *see* 3658
D & C Red No. 27 *see* 7440
D & C Red No. 28 *see* 7440
D & C Yellow No. 7 *see* 4192
D & C Yellow No. 8 *see* 4192
D & C Yellow No. 10 *see* 8188
D & C Yellow No. 11 *see* 8189
D2E7 *see* 136
D4T *see* 8929
D-20 *see* 2714
D-40TA *see* 3757
D-41 *see* 9950
D-50 *see* 8067
D-65MT *see* 308
D-109 *see* 4744
D-138 *see* 6792
D-145 *see* 5897
D-201 *see* 5273
D-206 *see* 7998
D-254 *see* 7569
D-301 *see* 1114
D-306 *see* 8948
D-365 *see* 10144
D-563 *see* 7061
D-600 *see* 4386
D-706 *see* 7069
D-860 *see* 9667
D-1959 *see* 8259
D-2083 *see* 2929
D-2341 *see* 1211
D-7093 *see* 5979
D-9998 *see* 4221
D-18506 *see* 6279
D-19466 *see* 5607
D-19575 *see* 4504
D-20443 *see* 3959
D-20761 *see* 2025
D-21266 *see* 7280
D-23129 *see* 3959
DA-398 *see* 3677
DA-688 *see* 4411
DA-708 *see* 9250
DA-759 *see* 6067
DA-2370 *see* 4039
DA-3177 *see* 2283
DA-7157 *see* 9247
DA-7218 *see* 9247
DA-8159 *see* 10027
2,4-DAA *see* 2978
DAAO *see* 408
Dabco [Air Prods.] *see* 9834
Dabigatran, 2794
DAB$_{389}$IL2 *see* 2895
DABS *see* 3002
DAC-2787 *see* 2169
Dacarbazine, 2795
Dacarel [Roemmers] *see* 6580
Dacatic [Orion] *see* 2795
Dacin [Lipomed] *see* 2795
Dacliximab *see* 2796
Daclizumab, 2796
Dacogen [SuperGen] *see* 2856
Daconate [ISK Biosci] *see* 6025
Daconil 2787 [Syngenta] *see* 2169
Dacoren [Nattermann] *see* 3710
Dacortin [Merck KGaA] *see* 7842
DACPM *see* 6131
Dacron [DuPont] *see* 7689

Dacthal [Amvac] *see* 2199
Dactil [RPR] *see* 7582
Dactin *see* 3074
Dactinomycin, 2797
DADA *see* 3216
DADPS *see* 2822
Dafalgan [BMS] *see* 47
Daflon [Servier] *see* 3325
Dafnegin [Poli] *see* 2269
Dagenan *see* 9069
Dagger [BASF] *see* 4944
Dagutan *see* 8445
Dahi *see* 10299
Dahlin *see* 5048
Dahl's Acid *see* 6489
Daidzein, 2798
Daidzein 7-Glucoside *see* 2798
Daidzin *see* 2798
Daidzoside *see* 2798
Daipisate [Taisho] *see* 7583
Daitec [Lederle] *see* 3985
Daivonex [Leo Pharm] *see* 1644
Dakin's Solution *see* 8762
Daktar [Cilag] *see* 6257
Daktarin [Janssen] *see* 6257
Dalacin C [Pharmacia & Upjohn] *see* 2354
Dalacin T [Pharmacia & Upjohn] *see* 2354
Dalalone [Forest] *see* 2945
Dalapon, 2799
Dalbavancin, 2800
Dalcetrapib, 2801
Dal-E-Rad [Vineland] *see* 6025
Dalfampridine, 2802
Dalfopristin, 2803
Dalgan [Wyeth] *see* 2959
Dalgen [Elmu] *see* 4038
Dalgol *see* 5913
Daliresp [Nycomed] *see* 8378
Dalmadorm [Roche] *see* 4228
Dalmane [Roche] *see* 4228
Dalmate [Roche] *see* 4228
Dalmatian Insect Powder *see* 8077
Dalmatian Sage *see* 8480
Dalnate [USV] *see* 9675
Dalpac *see* 1550
Dalteparin, 2804
Daltroban, 2805
Dalysep [Syntex] *see* 9042
Dalzic [Zeneca] *see* 7818
Damar, 2806
Damascenine, 2807
β-Damascenone, 2808
Dambose *see* 5021
Damide *see* 4975
Daminozide, 2809
Dammar *see* 2806
DAN-2163 *see* 481
Danabol *see* 6023
Danadim [Cheminova] *see* 3241
Danantizol [Gador] *see* 6043
Danaparoid, 2810
Danatrol [Sanofi-Synthelabo] *see* 2811
Danazol, 2811
Dandelion *see* 9192
Daneral [Aventis] *see* 7349
Danilon [Esteve] *see* 9139
Danish Agar *see* 4331
Danishefsky's Diene *see* 6075
Danitol [Sumitomo] *see* 4019
Danka [Mediolanum] *see* 3500
Danocrine [Sanofi-Synthelabo] *see* 2811
Danofloxacin, 2812
Danol [Sanofi-Synthelabo] *see* 2811
Danoval [Krka] *see* 2811

Dansyl Chloride, 2813
Dantamacrin [Procter & Gamble] *see* 2815
Danthron, 2814
Dantrium [Bago] *see* 1035
Dantrium [Procter & Gamble] *see* 2815
Dantrolen [Procter & Gamble] *see* 2815
Dantrolene, 2815
Dantron *see* 2814
DAO *see* 2977
Daonil [Sanofi-Aventis] *see* 4514
3,4-DAP *see* 398
Dapagliflozin, 2816
Dapagliflozin Propylene Glycol Hydrate *see* 2816
Daphnandrine *see* 2819
Daphnetin, 2817
Daphnetin 7-β-D-Glucoside *see* 2818
Daphnin, 2818
Daphnoline, 2819
Dapiprazole, 2820
Dapotum [BMS] *see* 4220
Dapoxetine, 2821
Dapropterin *see* 8505
Dapsone, 2822
DAPT *see* 479
Daptazile *see* 479
Daptazole [Nicholas] *see* 479
Daptomycin, 2823
DAR *see* 2963
Daramin *see* 8445
Daranide [Merck & Co.] *see* 3087
Darapladib, 2824
Daraprim [GSK] *see* 8098
Darbepoetin Alfa, 2825
Darbid [SK & F] *see* 5248
D'Arcet Metal-fusible *see* 1261
Dardum [Lisapharma] *see* 1931
Daricol *see* 7072
Daricon [Pfizer] *see* 7072
Darifenacin, 2826
Darilin [Recordati] *see* 10093
Darinaparsin, 2827
Darmstadtium, 2828
Darob [Abbott] *see* 8854
Darotol *see* 2158
Dartal [Searle] *see* 9509
Dartalan [Searle] *see* 9509
Daruma [Takeda] *see* 4929
Darunavir, 2829
Darusentan, 2830
Darvisul [Burroughs Wellcome] *see* 2994
Darvon [AAI Pharma] *see* 7952
Darvon Alcohol *see* 2062
Darvon-N [AAI Pharma] *see* 7952
Darvyl *see* 5697
Dasatinib, 2831
Daserol *see* 5917
Dasovas [Erba] *see* 2864
DAST *see* 3124
Dasten [Duncan] *see* 7024
Dastosin [Astellas] *see* 3229
D.A.T. *see* 53
DATC *see* 2973
Datiscetin, 2832
Datril [BMS] *see* 47
DATS *see* 2976
DaTSCAN [GE Healthcare] *see* 5093
Daturine *see* 4897
Daucol, 2833
Daunoblastina [Farmitalia] *see* 2834
Daunomycin *see* 2834
Daunomycinone *see* 2834
Daunorubicin, 2834
Daunosamine *see* 2834, 3487

Dehydol [Henkel] *see* 7697
Dehydrated Alcohol *see* 3814
Dehydrite *see* 5742
Dehydroacetic Acid, 2869
trans-Dehydroandrosterone *see* 2875
Dehydroascorbic Acid, 2870
Dehydrobenzperidol [Janssen] *see* 3499
Dehydrobilirubin *see* 1223
(3β)-7-Dehydrocholesterol, 2871
24-Dehydrocholesterol *see* 2927
Dehydrocholic Acid, 2872
Dehydrocollinusin *see* 5320
11-Dehydrocorticosterone, 2873
Δ¹-Dehydrocortisol *see* 7841
Δ¹-Dehydrocortisone *see* 7842
Dehydrodichavicol *see* 5761
Dehydrodidemnin B *see* 725
Dehydroemetine, 2874
2-Dehydroemetine *see* 2874
2,3-Dehydroemetine *see* 2874
Dehydroepiandrosterone, 2875
Dehydroequol, 2876
Dehydroergosterol, 2877
1-Dehydro-9α-fluorohydrocortisone *see* 5220
Δ¹-Dehydrohydrocortisone *see* 7841
11-Dehydro-17-hydroxycorticosterone *see* 2525
Dehydroisoandrosterone *see* 2875
2,3-Dehydroisoemetine *see* 2874
6-Dehydro-6-methyl-17α-acetoxyprogesterone *see* 5876
1-Dehydro-16α-methyl-9α-fluorohydrocortisone *see* 2945
1-Dehydro-6α-methylhydrocortisone *see* 6184
1-Dehydro-17α-methyltestosterone *see* 6023
Dehydronivalenol *see* 10229
Dehydroperillic Acid *see* 9545
6-Dehydro-*retro*-progesterone *see* 3521
3-Dehydroretinal, 2878
all-trans-3,4-Dehydroretinal *see* 2878
3-Dehydroretinol, 2879
all-trans-3-Dehydroretinol *see* 2879
7-Dehydrositosterol, 2880
Dehydro-6-sulfoabietic Acid *see* 3535
1-Dehydrotestololactone *see* 9323
Dehydrotestosterone *see* 1327
6-Dehydrotestosterone-17α-propionic Acid γ-Lactone *see* 1753
Deidrocolico Vita *see* 2872
Deiten [ABC] *see* 6662
Dejo *see* 3412
Dekadin *see* 2914
DeKalin *see* 2850
Dekamin *see* 2914
Dekamin [Monico] *see* 7996
Dekelmin *see* 6239
Dekrysil *see* 3305
DEL-1267 *see* 6219
Delan [BASF] *see* 3411
Delaprem [Savage] *see* 4743
Delapril, 2881
Delatestryl [BMS] *see* 9324
Delavirdine, 2882
Delavirdine Mesylate *see* 2882
Delaxin [Ferndale] *see* 6052
Delcaine *see* 8020
Delcosine *see* 2887
Delecit [LPB] *see* 2212
Delestrogen [Monarch] *see* 3758
Delfen [Ortho] *see* 6764
Delgesic [Karlspharma] *see* 5698
Delimmun [Newport] *see* 5019

Delipid [Coop. Farm.] *see* 9571
Deliver [Certis] *see* 925
Delix [Aventis] *see* 8221
Delmadinone Acetate, 2883
Delmeson [Aventis] *see* 4207
Delmopinol, 2884
Delnav *see* 3330
Delonal [Essex] *see* 213
m-Delphene *see* 2859
Delphicort [Lederle] *see* 9758
Delphidenolon 1575 *see* 6417
Delphin *see* 2885
Delphinac [Riemser] *see* 3091
Delphinic Acid *see* 5277
Delphinidin, 2885
Delphinidol *see* 2885
Delphinin *see* 2885
Delphinine, 2886
Delphinium *see* 5424
Delphoside *see* 2885
Delpregnin [Novo] *see* 5876
Delsene [DuPont] *see* 1792
Delsoline, 2887
Delta Goal [Dow AgroSci.] *see* 7062
Deltacortene [Bruno] *see* 7842
Deltacortisone *see* 7842
Deltacortone [Merck & Co.] *see* 7842
Deltacortril [Pfizer] *see* 7841
Delta E *see* 7842
Delta F *see* 7841
Delta Gluconolactone *see* 4493
Deltalin [Lilly] *see* 10215
Deltamethrin, 2888
Deltamycin A₄ *see* 1806
Deltar [Osiris] *see* 2381
Delta Sleep Factor *see* 3505
Delta Sleep-inducing Peptide (Rabbit) *see* 3505
Delta Sleep Peptide *see* 3505
Deltasone [Upjohn] *see* 7842
Deltastab *see* 7841
Delta-1-testololactone *see* 9323
Delthione [Tobishi] *see* 4511
Deltazen [Sanofi-Aventis] *see* 3224
Deltison [Ferring] *see* 7842
Deltisona B *see* 5928
Delursan [Axcan] *see* 10074
Delvex [Lilly] *see* 3412
Delvomycin *see* 6311
Demadex [Roche] *see* 9711
Demecarium Bromide, 2889
Demeclocycline, 2890
Demecolcine, 2891
Demerol Hydrochloride [Sanofi Winthrop] *see* 5916
6-Demethoxyageratochromene *see* 7839
4-Demethoxydaunomycin *see* 4927
4-Demethoxydaunorubicin *see* 4927
2-Demethoxy-2-glucosidoxythiocolchicine *see* 9477
Demethoxyvindoline *see* 10185
5-*O*-Demethylavermectin A₁ₐ and 5-*O*-Demethyl-25-de(1-methylpropyl)-25-(1-methylethyl)avermectin A₁ₐ (4:1) *see* 2
1-Demethylcalycanthidine *see* 2059
Demethylchlortetracycline *see* 2890
4-*O*-Demethyldaunorubicin *see* 1870
(4″*R*)-5-*O*-Demethyl-25-de(1-methylpropyl)-4″-deoxy-4″-(methylamino)-25-(1-methylethyl)avermectin A₁ₐ *see* 3612
5-*O*-Demethyl-25-de(1-methylpropyl)-22,23-dihydro-25-(1-methylethyl)avermectin A₁ₐ *see* 5296
(6*R*,23*S*)-5-*O*-Demethyl-28-deoxy-25-[(1*E*)-1,3-dimethyl-1-buten-1-yl]-6,28-

epoxy-23-hydroxymilbemycin B *see* 6529
(6*R*,23*E*,25*S*)-5-*O*-Demethyl-28-deoxy-25-(1,3-dimethyl-1-buten-1-yl)-6,28-epoxy-23-(methoxyimino)milbemycin B *see* 6376
(6*R*,25*R*)-5-*O*-Demethyl-28-deoxy-6,28-epoxy-25-(1-methylethyl)milbemycin B *see* 6271
(6*R*,25*R*)-5-*O*-Demethyl-28-deoxy-6,28-epoxy-25-methylmilbemycin *see* 6269
(4″*R*)-5-*O*-Demethyl-4″-deoxy-4″-(methylamino)avermectin A₁ₐ *see* 3612
S-Demethyl-3′-depropyl-*S*-[2-[(2-hydroxybenzoyl)oxy]ethyl]-7-*O*-methyllincomycin *see* 1960
40-Demethyl-3,7-dideoxo-3,7-dihydroxy-*N*⁴⁷-methyl-5-oxocandicidin D Methyl Ester Cyclic 15,19-Hemiacetal *see* 5914
5-*O*-Demethyl-22,23-dihydroavermectin A₁ₐ *see* 5296
Demethyldihydrothebaine Acetate (Ester) *see* 9426
Demethyldopan *see* 10036
4′-Demethylepipodophyllotoxin 9-[4,6-*O*-Ethylidene-β-D-glucopyranoside] *see* 3929
4′-Demethylepipodophyllotoxin 9-(4,6-*O*-2-Thenylidene-β-D-glucopyranoside) *see* 9288
4′-Demethylepipodophyllotoxin-β-D-thenylidine Glucoside *see* 9288
N-Demethyl-*N*-heptylphysostigmine *see* 3695
Demethylhomopterocarpin *see* 5860
6-Demethylmevinolin *see* 6243
26-Demethyloligomycin A *see* 6927
4-Demethyl-3-oxovobasan-17-oic Acid Methyl Ester *see* 7292
*O*³-Demethylthebaine *see* 6962
N-Demethylthiolutin *see* 4766
Demeton, 2892
Demeton-methyl *see* 6123
Demetraciclina [Bios] *see* 2890
Demetrin [Parke-Davis] *see* 7836
Demolox [Am. Cyanamid] *see* 573
Demon [Syngenta] *see* 2764
Demoxytocin *see* 2844
Demser [Merck & Co.] *see* 6240
Demulen [Searle] *see* 3909
DEN *see* 6723
DENA *see* 6723
Denagard [Novartis] *see* 9574
Denamone *see* 10239
Denan [Boehringer, Ing.] *see* 8677
Denapol [Nagase] *see* 2306
Denarcotized Opium *see* 6949
Denatonium Benzoate, 2893
Denatured Alcohol *see* 3815
Denavir [SKB] *see* 7192
Dendrid [Alcon] *see* 4931
Dendrotoxins, 2894
Deniban [Synthelabo] *see* 481
Denileukin Diftitox, 2895
Denim [Syngenta] *see* 3612
De-Nol [Yamanouchi] *see* 2469
De-Noltab [Yamanouchi] *see* 2469
Denopamine, 2896
Denosine [Tanabe Seiyaku] *see* 4393
Denosumab, 2897
Denufosol, 2898
Denvar [Merck KGaA] *see* 1925
[9*S*(*R*)]-9-Deoxy-11-deoxy-9,11-[imino[2-(2-methoxyethoxy)ethylidene]oxy]erythromycin *see* 3394

Dequadin Chloride [Allen & Hanburys] *see* 2914
Dequafungan [Kreussler] *see* 2914
Dequalinium Chloride, 2914
Dequavagyn [Kreussler] *see* 2914
Dequavet *see* 2914
Dequest 2010 [Solutia] *see* 3912
Deracil *see* 9520
Deracoxib, 2915
Deracyn [Upjohn] *see* 148
Deralbine [Andromaco] *see* 6257
Deralin [Alphapharm] *see* 7953
Deramaxx [Novartis] *see* 2915
Deramciclane, 2916
Derbac [Mundipharma] *see* 1788
Derbac-M [Seaton] *see* 5764
Dermabet [Taro] *see* 1182
Dermabond [Ethicon] *see* 6855
Dermacid *see* 5935
Dermacor [DuPont] *see* 2080
Dermacort [Melaleuca] *see* 4824
Dermadex [Alconox] *see* 4716
Dermaflor [Brocchieri] *see* 3159
Dermalar *see* 4181
Dermaplus [Ripari-Gero] *see* 4182
Dermatan Sulfate *see* 2219
Dermatol [Hoechst] *see* 1283
Dermatop [Sanofi-Aventis] *see* 7840
Dermevan *see* 5080
Dermofix [Ferrer] *see* 8604
Dermonistat [Cilag] *see* 6257
Dermosol [Iwaki] *see* 1182
Dermotricine [Bellon] *see* 10023
Dermovaleas [Valeas] *see* 1182
Dermovate [GSK] *see* 2358
Derosal [Hoechst] *see* 1792
Deroxat [GSK] *see* 7148
Derride *see* 3605
Derringer [Roussel Uclaf] *see* 7589
Derris Root, 2917
Dervan [Caffaro] *see* 8733
DES *see* 3148
Destonate [Meiji] *see* 2936
Desacchromin Dispersion *see* 7617
Desace [Dexo] *see* 2922
Desacetyldigilanide C *see* 2922
Desacetylvinblastine Amide *see* 10184
Desaci [Simes] *see* 2922
Desaglybuzole *see* 4516
1-Desamino-8-D-arginine Vasopressin *see* 2926
1-Desamino-1-monocarba-[2-tyr(OMe)]-OT *see* 1796
Desaminooxytocin *see* 2844
Desanden [Nycomed] *see* 1119
Desaspidin *see* 2918
Desaspidin BB, 2918
Descarboethoxyloratadine *see* 2923
Deschlorobiomycin *see* 9341
Desclidium [Nattermann] *see* 10199
Desdemin [Vitacain] *see* 4338
Déselmine *see* 3412
Desenex [Novartis] *see* 10033
Deseril [Novartis] *see* 6213
Désernil [Novartis] *see* 6213
Deserol *see* 1428
Deserpidine, 2919
Desfedrin *see* 6020
Desferal [Novartis] *see* 2863
Desferricoprogen *see* 2507
Desferrioxamine B *see* 2863
Desferrioxamine Mesylate *see* 2863
Desflam [Merck KGaA] *see* 1488
Desflurane, 2920
Desglucotransvaaline *see* 7985

Des-Gly[10]-NH$_2$-LH-RH-ethylamide *see* 4088
Desicol [Parke-Davis] *see* 7027
Desipramine, 2921
Desirudin *see* 4755
Desitriptilina *see* 6804
Deskar [Marnac] *see* 7606
Deslanoside, 2922
Desloratadine, 2923
Deslorelin, 2924
Desmedipham, 2925
Desmesterol *see* 2927
11-Desmethoxyreserpine *see* 2919
Desmethylamitriptyline *see* 6804
N-Desmethylcodeine *see* 6779
Desmethyldiazepam *see* 6780
α-Desmethyl DOB *see* 1910
Desmethyldopan *see* 10036
Desmethylemetine *see* 1973
Desmethylimipramine *see* 2921
Desmethylmethadone *see* 6798
Desmethylmorphine *see* 6800
Desmethylnarcotine *see* 6505
5-*O*-Desmethyltangeretin *see* 4399
O-Desmethylvenlafaxine *see* 2937
Desmospray [Ferring] *see* 2926
Desmocollins *see* 1616
Desmogleins *see* 1616
Desmopressin, 2926
Desmosterol, 2927
Desmoteplase *see* 3506
Desmycosin *see* 10013
Desocort [Bausch & Lomb] *see* 2945
Desogen [Organon] *see* 2928
Desogestrel, 2928
Desol [CT] *see* 10074
Desomedine [Chauvin] *see* 4730
Desonax [Pulitzer] *see* 1476
Desonide, 2929
Desopan [Mochida] *see* 9867
Desosamine, 2930
Desoximetasone, 2931
Desoxiribon *see* 2908
Desoxostrophanthidin *see* 7291
S-(5'-Desoxyadenosin-5'-yl)-L-methionine *see* 144
2-Desoxy-D-altromethylose *see* 3183
Desoxycholic Acid *see* 2901
Desoxycorticosterone *see* 2902
Desoxycorticosterone Acetate *see* 2903
11-Desoxycortisone *see* 2933
Desoxycortone *see* 2902
Desoxycortone Acetate *see* 2903
14-Desoxy-14-[(2-diethylaminoethyl)mercaptoacetoxy]mutilin *see* 9574
d-Desoxyephedrine *see* 6020
Desoxyepothilone B *see* 3684
Desoxyfed *see* 6020
21-Desoxy-9α-fluoro-6α-methylprednisolone *see* 4207
6-Desoxy-D-glucosamine, 2932
D-*arabino*-2-Desoxyhexose *see* 2905
11-Desoxy-17-hydroxycorticosterone, 2933
Desoxymethasone *see* 2931
6-Desoxy-6-methylenenaltrexone *see* 6445
21-Desoxy-6α-methyl-9α-fluoroprednisolone *see* 4207
Desoxyn [Abbott] *see* 6020
(±)-Desoxynorephedrine *see* 579
3-Desoxynorlutin *see* 5690
[3'-Desoxy-3'-oxo-MeBmt]^1-[Val]^2-cyclosporin *see* 10103
2-Desoxyphenobarbital *see* 7865

5-Desoxyquercetin *see* 4119
Desoxyribonucleic Acid *see* 2908
Desoxyribose *see* 2909
DESP *see* 4026
Des-38-proline-exendine-4 (*Heloderma suspectum*)-(1-39)-peptidylpenta-L-lysyl-L-lysinamide *see* 5605
Desquaman [Hermal] *see* 8107
Dessin [Union Carbide] *see* 3313
Dess-Martin Periodinane, 2934
Desthiobiotin, 2935
Des-(1-6)-[7-threonine,46-threonine,291-lysine,301-serine]uricase (EC 1.7.3.3, Urate Oxidase) *Sus scrofa* (Pig) Tetramer, Non Acetylated, Carbamates with α-Carboxy-ω-methoxypoly(oxyethylene) *see* 7177
Destim [Central Pharm.] *see* 6020
Destolit [Norgine] *see* 10074
Destomycin A, 2936
Destriol *see* 3762
Destrone *see* 3763
N-Desulfoheparin *see* 4687
63-Desulfohirudin (*Hirudo medicinalis* Isoform HV1) *see* 4755
Desuric [Sanofi Winthrop] *see* 1067
Desvenlafaxine, 2937
Desyrel [BMS] *see* 9740
m-DETA *see* 2859
Detal *see* 3305
Detamide *see* 2859
Detantol [Eisai] *see* 1491
Detaxtran, 2938
Detensiel [Merck-Clévenot] *see* 1295
Detergent Alkylate #5 *see* 9824
Dethylandiamine *see* 9431
Dethyrona [Abbott] *see* 9570
Deticene [Sanofi-Aventis] *see* 2795
Detomidine, 2939
Detoxepa [Wyeth-Ayerst] *see* 9602
Detoxin C$_1$ *see* 2940
Detoxin C$_{α1}$ *see* 2940
Detoxin Complex, 2940
Detoxin D$_1$ *see* 2940
Detoxinine *see* 2940
Detravis [Vis] *see* 2890
Detrol [Pharmacia & Upjohn] *see* 9684
Detrunorm [Schering] *see* 7945
Detrusitol [Pharmacia & Upjohn] *see* 9684
Détryptoréline *see* 9924
Dettol [Reckitt & Colman] *see* 2182
Detulin [Woelm] *see* 9653
Deumacard [Jenapharm] *see* 7254
Deursil [Sanofi-Aventis] *see* 10074
Deuterium, 2941
Deuterium Oxide, 2942
Devarda's Alloy *see* 2943
Devarda's Metal, 2943
Devaricin *see* 4745
Develin [Gödecke] *see* 7952
Developer Z *see* 6802
Devil's Apple *see* 8950
Devil's Claw, 2944
Devil's Dung *see* 811
Devitol [Orion] *see* 10215
Devrinol [Stauffer] *see* 6497
Dewaxed Lanolin *see* 5409
Dexabene [Merckle] *see* 2945
Dexacilina [BMS] *see* 3669
Dexacillin [BMS] *see* 3669
Dexacortal [Organon] *see* 2945
Dexacortin [Streuli] *see* 2945
Dexambutol [L'Arguenon] *see* 3775
Dexamethasone, 2945

Dibenzalacetone, 3010
Dibenz[*a,h*]anthracene *see* 3011
Dibenz[*de,kl*]anthracene *see* 7299
1,2:5,6-Dibenzanthracene, 3011
5*H*-Dibenz[*b,f*]azepine-5-carboxamide *see* 1783
4-[3-(5*H*-Dibenz[*b,f*]azepin-5-yl)propyl]-1-piperazineethanol *see* 6948
Dibenzcozamide *see* 2435
Dibenzepin, 3012
Dibenzheptropine *see* 2913
1,2,3,4-Dibenznaphthalene *see* 9913
2,3,6,7-Dibenzoanthracene *see* 7216
4-[3-(5*H*-Dibenzo[*b,f*]azepin-5-yl)propyl]-1-(2-hydroxyethyl)piperazine *see* 6948
Dibenzo-18-crown-6 *see* 2592
3-(5*H*-Dibenzo[*a,d*]cyclohepten-5-ylidene)-*N,N*-dimethyl-1-propanamine *see* 2706
4-(5*H*-Dibenzo[*a,d*]cyclohepten-5-ylidene)-1-methylpiperidine *see* 2770
Dibenzofuran, 3013
Dibenzo[*b,d*]furan *see* 3013
2,3,11,12-Dibenzo-1,4,7,10,13,16-hexaoxacyclooctadeca-2,11-diene *see* 2592
Dibenzoparadiazine *see* 7331
Dibenzopenthiophene *see* 9522
Dibenzo[*a,i*]phenanthrene *see* 7508
Dibenzo[*b,h*]phenanthrene *see* 3015
Dibenzopyrazine *see* 7331
Dibenzo[*b,e*]pyridine *see* 114
Dibenzo-γ-pyrone *see* 10260
Dibenzopyrrole *see* 1791
2-[2-(4-Dibenzo[*b,f*][1,4]thiazepin-11-yl-1-piperazinyl)ethoxy]ethanol *see* 8155
Dibenzothiazine *see* 7364
2,2′-Dibenzothiazyl Disulfide *see* 3413
3-Dibenzo[*b,e*]thiepin-11(6*H*)-ylidene-*N,N*-dimethyl-1-propanamine *see* 3478
Dibenzothiopyran *see* 9522
3-Dibenz[*b,e*]oxepin-11(6*H*)-ylidene-*N,N*-dimethyl-1-propanamine *see* 3483
Dibenzoyl *see* 1080
Dibenzoyldjenkolic Acid *see* 3431
Dibenzoylmethane, 3014
Dibenzoyl Peroxide *see* 1119
β,β′-Dibenzphenanthrene *see* 3015
1,2,7,8-Dibenzphenanthrene *see* 7508
2,3:6,7-Dibenzphenanthrene, 3015
Dibenzthiazyl Disulfide *see* 3413
Dibenzthione *see* 9024
Dibenzyl *see* 1202
Dibenzylamine, 3016
N,N-Dibenzylaminoethyl Chloride *see* 3009
N,N-Dibenzyl-β-chloroethylamine *see* 3009
Dibenzylchlorophosphate *see* 3017
Dibenzyl Chlorophosphonate, 3017
N,N′-Dibenzyl-γ,γ′-dipyridylium *see* 10195
Dibenzyl Disulfide, 3018
Dibenzyl Ether *see* 1135
N,N′-Dibenzylethylenediamine *see* 1066
N,N′-Dibenzylethylenediamine Bis[benzylpenicillin] *see* 7205
Dibenzylethylenediamine Dipenicillin G *see* 7205
Dibenzyl Fumarate *see* 1138
Dibenzyl Hydrogen Phosphite *see* 3019
Dibenzylidene Acetone *see* 3010
Dibenzyline [Wellspring] *see* 7368

d-3,4-(1′,3′-Dibenzyl-2′-ketoimidazo-lido)-1,2-trimethylenethiophanium *d*-Camphorsulfonate *see* 9876
4,6-Dibenzyl-5-oxo-1-thia-4,6-diazatricyclo[6.3.0.0^{3,7}]undecanium (+)-β-Camphorsulfonate *see* 9876
Dibenzyl Phosphite, 3019
Dibenzyl Phosphonate *see* 3019
Dibenzylphosphoryl Chloride *see* 3017
Dibenzylsulfide *see* 1147
3,5-Dibenzyltetrahydro-2*H*-1,3,5-thiadiazine-2-thione *see* 9024
Dibenzyran [Procter & Gamble] *see* 7368
Dibestil [Sterling Winthrop] *see* 3148
N,N′-Dibetaine-γ,γ′-dipyridylium *see* 10195
Dibetos [Kodama] *see* 1481
Dibicor [Pik] *see* 9211
Dibismuth Triselenide *see* 1279
Dibismuth Tritelluride *see* 1289
Diblocin [AstraZeneca] *see* 3482
Diborane(6), 3020
Diboron Hexahydride *see* 3020
Diboron Tetrachloride, 3021
Dibotermin Alfa *see* 1333
Dibrom [Amvac] *see* 6442
Dibromantin, 3022
p,α-Dibromoacetophenone *see* 1438
2,4′-Dibromoacetophenone *see* 1438
9,10-Dibromoanthracene, 3023
p-Dibromobenzene, 3024
1,4-Dibromobenzene *see* 3024
4,4′-Dibromobenzilic Acid Isopropyl Ester *see* 1445
3,5-Dibromo-*N*-(4-bromophenyl)-2-hydroxybenzamide *see* 9778
1,2-Dibromobutane *see* 1569
2,3-Dibromobutanedioic Acid *see* 3034
[2,7-Dibromo-9-(*o*-carboxyphenyl)-6-hydroxy-3-oxo-3*H*-xanthen-4-yl]hydroxymercury Disodium Salt *see* 5934
2,6-Dibromo-*N*-chloro-*p*-benzoquinoneimine *see* 3031
4-[2-[4-[(11*R*)-3,10-Dibromo-8-chloro-6,11-dihydro-5*H*-benzo[5,6]cyclohepta[1,2-*b*]pyridin-11-yl]-1-piperidinyl]-2-oxoethyl]-1-piperidinecarboxamide *see* 5622
2,6-Dibromo-4-(chloroimino)-2,5-cyclohexadien-1-one *see* 3031
Dibromochloromethane *see* 2135
Dibromochloropropane, 3025
1,2-Dibromo-3-chloropropane *see* 3025
2,6-Dibromo-*N*-chloroquinonimine *see* 3031
5,5′-Dibromo-*o*-cresolsulfonphthalein *see* 1394
2,6-Dibromo-4-cyanophenol *see* 1451
3,5-Dibromo-*N*^α-cyclohexyl-*N*^α-methyltoluene-α,2-diamine *see* 1398
2′,2″-Dibromo-4′,4″-diamidino-1,3-diphenoxypropane *see* 3029
2,4-Dibromo-1-(2,4-dibromophenoxy)-benzene *see* 7681
1,2-Dibromo-2,2-dichloroethyl Dimethyl Phosphate *see* 6442
1,2-Dibromo-2,4-dicyanobutane, 3026
1,6-Dibromo-1,6-dideoxydulcitol *see* 6299
1,6-Dibromo-1,6-dideoxygalactitol *see* 6299
1,6-Dibromo-1,6-dideoxy-D-mannitol *see* 6297
4′,5′-Dibromo-3′,6′-dihydroxy-2′,7′-dinitrospiro[isobenzofuran-1(3*H*),9′-[9*H*]-

xanthen]-3-one Sodium Salt (1:2) *see* 3657
(2′,7′-Dibromo-3′,6′-dihydroxy-3-oxospiro[isobenzofuran-1(3*H*),9′-[9*H*]xanthen]-4′-yl)hydroxymercury Sodium Salt (1:2) *see* 5934
4′,5′-Dibromo-3′,6′-dihydroxyspiro[isobenzofuran-1(3*H*),9′-[9*H*]xanthen]-3-one *see* 3027
1,3-Dibromo-5,5-dimethylhydantoin *see* 3022
1,3-Dibromo-5,5-dimethyl-2,4-imidazolidinedione *see* 3022
4′,5′-Dibromo-2′,7′-dinitrofluorescein Disodium Salt *see* 3657
Dibromodulcit *see* 6299
Dibromodulcitol *see* 6299
sym-Dibromoethane *see* 3850
1,2-Dibromoethane *see* 3850
1,2-Dibromoethene *see* 86
(1*R*,3*R*)-3-(2,2-Dibromoethenyl)-2,2-dimethylcyclopropanecarboxylic Acid (*S*)-Cyano(3-phenoxyphenyl)methyl Ester *see* 2888
sym-Dibromoethylene *see* 86
1,2-Dibromoethylene *see* 86
4,5-Dibromo-3,6-fluorandiol *see* 3027
4′,5′-Dibromofluorescein, 3027
3,5-Dibromo-2-hydroxybenzaldehyde *see* 3032
3,5-Dibromo-2-hydroxybenzoic Acid *see* 3033
3,5-Dibromo-4-hydroxybenzonitrile *see* 1451
3-(3,5-Dibromo-4-hydroxybenzoyl)-2-ethylbenzofuran *see* 1067
N-[3,5-Dibromo-4-(4-hydroxy-3-isopropylphenoxy)phenyl]malonamic Acid *see* 3692
Dibromohydroxymercurifluorescein Disodium Salt *see* 5934
3-[[3,5-Dibromo-4-[4-hydroxy-3-(1-methylethyl)phenoxy]phenyl]amino]-3-oxopropanoic Acid *see* 3692
β-(3,5-Dibromo-4-hydroxyphenyl)alanine *see* 3035
3,5-Dibromo-4-hydroxyphenyl Cyanide *see* 1451
2-[(3,5-Dibromo-4-hydroxyphenyl)(3,5-dibromo-4-oxo-2,5-cyclohexadien-1-ylidene)methyl]benzoic Acid Ethyl Ester *see* 9331
3,5-Dibromo-4-hydroxyphenyl 2-Ethyl-3-benzofuranyl Ketone *see* 1067
(3,5-Dibromo-4-hydroxyphenyl)(2-ethyl-3-benzofuranyl)methanone *see* 1067
5,7-Dibromo-8-hydroxyquinoline *see* 1462
3,5-Dibromo-2-hydroxy-*N*-[3-(trifluoromethyl)phenyl]benzamide *see* 4210
6,6′-Dibromoindigo, 3028
6,6′-Dibromoindigotin *see* 3028
Dibromomannitol *see* 6297
Dibromomethane *see* 6133
2′,6′-Dibromo-2-methyl-4′-trifluoromethoxy-4′-trifluoromethyl-1,3-thiazole-5-carboxanilide *see* 9469
Dibromopropamidine, 3029
α,γ-Dibromopropane *see* 9884
ω,ω′-Dibromopropane *see* 9884
1,2-Dibromopropane *see* 7966
1,3-Dibromopropane *see* 9884
2,3-Dibromo-1-propanol 1,1′,1″-Phosphate *see* 9927
2,3-Dibromopropene, 3030

ano(3-phenoxyphenyl)methyl Ester
see 2764

3-(2,2-Dichloroethenyl)-2,2-dimethyl-cyclopropanecarboxylic Acid (3-Phen-oxyphenyl)methyl Ester *see* 7294

Dichloroethylaluminum *see* 3816

p-Di(2-chloroethyl)amino-L-phenylala-nine *see* 5896

N,N-Di-2-chloroethyl-γ-*p*-aminophenyl-butyric Acid *see* 2073

γ-[*p*-Di(2-chloroethyl)aminophenyl]bu-tyric Acid *see* 2073

5-[Di(β-chloroethyl)amino]uracil *see* 10036

asym-Dichloroethylene *see* 10192

sym-Dichloroethylene *see* 87

1,1-Dichloroethylene *see* 10192

1,2-Dichloroethylene *see* 87

β,β′-Dichloroethyl Ether *see* 3075

sym-Dichloroethyl Ether, 3075

1,1′-(2,2-Dichloroethylidene)bis[4-chloro-benzene] *see* 3070

Di(chloroethyl)methylamine *see* 5846

Dichloroethyl-β-naphthylamine *see* 2108

Di(2-chloroethyl)-β-naphthylamine *see* 2108

β,β′-Dichloroethyl Sulfide *see* 6401

(*OC*-6-13)-Dichloro[*rel*-ethyl (18*R*,19*S*)-3,4,20,21-Tetradehydro-4,9,14,19-tetraethyl-18,19-dihydro-3,8,13,18-tetramethyl-20-phorbinecarboxylato-(2−)-κN²³,κN²⁴,κN²⁵,κN²⁶]tin *see* 8401

2′,7′-Dichloro-3,6-fluorandiol *see* 3076

4,5-Dichloro-3,6-fluorandiol *see* 3077

Dichlorofluorescein *see* 3076

2′,7′-Dichlorofluorescein, 3076

4′,5′-Dichlorofluorescein, 3077

N-[[[3,5-Dichloro-2-fluoro-4-(1,1,2,3,3,3-hexafluoropropoxy)phenyl]amino]car-bonyl]-2,6-difluorobenzamide *see* 6810

N-[3,5-Dichloro-2-fluoro-4-(1,1,2,3,3,3-hexafluoropropoxy)phenyl]-*N*′-(2,6-di-fluorobenzoyl)urea *see* 6810

2,2-Dichloro-*N*-[(1*S*,2*R*)-1-(fluorometh-yl)-2-hydroxy-2-[4-(methylsulfonyl)-phenyl]ethyl]acetamide *see* 4141

N-[(Dichlorofluoromethyl)thio]-*N*′,*N*′-di-methyl-*N*-phenylsulfamide *see* 3053

5,7-Dichloro-4-(4-fluorophenoxy)quino-line *see* 8196

3-[(1*R*)-1-(2,6-Dichloro-3-fluorophenyl)-ethoxy]-5-[1-(4-piperidinyl)-1*H*-pyra-zol-4-yl]-2-pyridinamine *see* 2577

3,5-Dichloro-4′-fluorothiocarbanilide *see* 5616

α,β-Dichloro-β-formylacrylic Acid *see* 6383

(11β,16α)-9,21-Dichloro-17-[(2-furanyl-carbonyl)oxy]-11-hydroxy-16-methyl-pregna-1,4-diene-3,20-dione *see* 6327

N-[[[2,5-Dichloro-4-(1,1,2,3,3,3-hexafluo-ropropoxy)phenyl]amino]carbonyl]-2,6-difluorobenzamide *see* 5654

1-[2,5-Dichloro-4-(1,1,2,3,3,3-hexafluoro-propoxy)phenyl]-3-(2,6-difluorobenzo-yl)urea *see* 5654

α-Dichlorohydrin *see* 3084

2,2-Dichloro-*N*-[(1*R*,2*R*)-2-hydroxy-1-(hydroxymethyl)-2-[4-(methylsulfo-nyl)phenyl]ethyl]acetamide *see* 9453

2,2-Dichloro-*N*-[(1*R*,2*R*)-2-hydroxy-1-(hydroxymethyl)-2-(4-nitrophenyl)eth-yl]acetamide *see* 2077

2,2-Dichloro-4′-hydroxy-*N*-methylacet-anilide *see* 3223

2,6-Dichloro-4-[(4-hydroxyphenyl)imi-no]-2,5-cyclohexadien-1-one Sodium Salt (1:1) *see* 3078

Di[5-chloro-2-hydroxyphenyl]methane *see* 3081

2,2-Dichloro-*N*-(4-hydroxyphenyl)-*N*-methylacetamide *see* 3223

N-(2,3-Dichloro-4-hydroxyphenyl)-1-methylcyclohexanecarboxamide *see* 4005

5,7-Dichloro-8-hydroxyquinaldine *see* 2196

5,7-Dichloro-8-hydroxyquinoline *see* 2181

2,6-Dichloro-*N*-2-imidazolidinylidene-benzenamine *see* 2380

2,6-Dichloro-*N*′-2-imidazolidinylidene-1,4-benzenediamine *see* 738

2′,4′-Dichloro-2-(imidazol-1-yl)aceto-phenone *O*-(2,4-Dichlorobenzyl)oxime *see* 7036

2,6-Dichloroindophenol Sodium, 3078

(Dichloroiodo)benzene *see* 5073

sym-Dichloroisopropyl Alcohol *see* 3084

Dichloromalealdehydic Acid *see* 6383

2,3-Dichloromaleic Aldehyde Acid *see* 6383

Dichloromethane *see* 6135

Dichloromethanediphosphonic Acid *see* 2365

Dichloromethazanone *see* 2106

3,6-Dichloro-2-methoxybenzoic Acid *see* 3050

3,4-Dichloro-α-methoxybenzylpenicillin *see* 2376

4-[(2,4-Dichloro-5-methoxyphenyl)-amino]-6-methoxy-7-[(4-methyl-1-piperazinyl)propoxy]-3-quinoline-carbonitrile *see* 1357

4-[(2,4-Dichloro-5-methoxyphenyl)-amino]-6-methoxy-7-[3-(4-methyl-piperazin-1-yl)propoxy]quinoline-3-carbonitrile *see* 1357

3′,4′-Dichloro-2-methylacrylanilide *see* 3098

(Dichloromethyl)benzene *see* 1059

N,N-Dichloro-4-methylbenzenesulfon-amide *see* 3056

3-Dichloromethyl-6-chloro-7-sulfamyl-3,4-dihydro-1,2,4-benzothiadiazine 1,1-Dioxide *see* 9789

2,2′-Dichloro-*N*-methyldiethylamine *see* 5846

2,2′-Dichloro-*N*-methyldiethylamine *N*-Oxide *see* 5847

P,*P*′-(Dichloromethylene)bisphosphonic Acid *see* 2365

[2,3-Dichloro-4-(2-methylenebutyryl)-phenoxy]acetic Acid *see* 3772

N-(Dichloromethylene)-*N*,*N*-dimethyli-minium Chloride *see* 10174

(Dichloromethylene)diphosphonic Acid *see* 2365

N-(Dichloromethylene)-*N*-methylmethan-aminium Chloride *see* 10174

2-[2,3-Dichloro-4-(2-methylene-1-oxobu-tyl)phenoxy]acetic Acid *see* 3772

sym-Dichloromethyl Ether, 3079

3-[2,4-Dichloro-5-(1-methylethoxy)phen-yl]-5-(1,1-dimethylethyl)-1,3,4-oxadi-azol-2(3*H*)-one *see* 7006

3-Dichloromethylhydrochlorothiazide *see* 9789

5,7-Dichloro-2-methyl-8-hydroxyquino-line *see* 2196

Dichloromethylmethanaminium Chlo-ride (1:1) *see* 10174

Dichloromethylmethane *see* 3865

Dichloromethyl Methyl Ketone *see* 3060

2-[(2,6-Dichloro-3-methylphenyl)amino]-benzoic Acid *see* 5850

O-(2,6-Dichloro-4-methylphenyl)phos-phorothioic Acid *O*,*O*-Dimethyl Ester *see* 9670

5,7-Dichloro-2-methyl-8-quinolinol *see* 2196

3′,4′-Dichloro-2-methylvaleranilide *see* 5333

Dichloromonoxide *see* 2099

2,3-Dichloro-1,4-naphthalenedione *see* 3054

2,3-Dichloro-1,4-naphthoquinone *see* 3054

2,4-Dichloro-1-(4-nitrophenoxy)benzene *see* 6685

2′,5-Dichloro-4′-nitrosalicylanilide *see* 6604

4,5-Dichloro-2-*n*-octyl-4-isothiazolin-3-one *see* 3080

4,5-Dichloro-2-octyl-3-isothiazolone, 3080

4,5-Dichloro-2-octyl-3(2*H*)-isothiazolone *see* 3080

(2*Z*)-2,3-Dichloro-4-oxo-2-butenoic Acid *see* 6383

Dichlorooxovanadium *see* 10112

Dichlorooxozirconium *see* 10384

Dichloropentamethylcyclopenta-dienyliridium Dimer *see* 7226

Dichlorophen, 3081

Dichlorophene *see* 3081

2,4-Dichlorophenol, 3082

2,6-Dichlorophenol, 3083

2,6-Dichlorophenol-indophenol Sodium *see* 3078

2-(2,4-Dichlorophenoxy)acetic Acid *see* 2793

4-(2,4-Dichlorophenoxy)butanoic Acid *see* 2838

4-(2,4-Dichlorophenoxy)butyric Acid *see* 2838

2-[1-(2,6-Dichlorophenoxy)ethyl]-4,5-di-hydro-1*H*-imidazole *see* 5615

2-[1-(2,6-Dichlorophenoxy)ethyl]-2-imid-azoline *see* 5615

Di-(*p*-chlorophenoxy)methane *see* 1252

5-(2,4-Dichlorophenoxy)-2-nitrobenzoic Acid Methyl Ester *see* 1212

2-[4-(2,4-Dichlorophenoxy)phenoxy]pro-panoic Acid Methyl Ester *see* 3092

2-(2,4-Dichlorophenoxy)propanoic Acid *see* 3088

(±)-2-(2,4-Dichlorophenoxy)propionic Acid *see* 3088

[(2,6-Dichlorophenyl)acetyl]guanidine *see* 4596

2-[(2,6-Dichlorophenyl)amino]benzene-acetic Acid *see* 3091

2-[(2,6-Dichlorophenyl)amino]benzene-acetic Acid Carboxymethyl Ester *see* 23

2-[(2,6-Dichlorophenyl)amino]phenyl-acetoxyacetic Acid *see* 23

1-(3′,4′-Dichlorophenyl)-3-(4′-chloro-phenyl)urea *see* 9819

3-(2,3-Dichlorophenyl)-4-cyanopyrrole *see* 4018

1-[2-(2,4-Dichlorophenyl)-2-[(2,4-dichlo-rophenyl)methoxy]ethyl]-1*H*-imid-azole *see* 6257

N-[[[3,5-Dichloro-4-(1,1,2,2-tetrafluoro-
ethoxy)phenyl]amino]carbonyl]-2,6-di-
fluorobenzamide *see* 4719
1-[3,5-Dichloro-4-(1,1,2,2-tetrafluoroeth-
oxy)phenyl]-(2,6-difluorobenzoyl)urea
see 4719
6,7-Dichloro-1,2,3,5-tetrahydroimidazo-
[2,1-*b*]quinazolin-2-one *see* 616
3,5-Dichlorotetrahydro-2,4,6-trioxo-*s*-
triazin-1(2*H*)-yl Potassium *see* 9946
[2,3-Dichloro-4-(2-thenoyl)phenoxy]ace-
tic Acid *see* 9586
2-[2,3-Dichloro-4-(2-thienylcarbonyl)-
phenoxy]acetic Acid *see* 9586
[2,3-Dichloro-4-(2-thiophenecarbonyl)-
phenoxy]acetic Acid *see* 9586
Dichlorotitanium *see* 9627
α,α-Dichlorotoluene *see* 1059
N,N-Dichloro-*p*-toluenesulfonamide *see*
3056
N-(2,6-Dichloro-*m*-tolyl)anthranilic Acid
see 5850
O-(2,6-Dichloro-*p*-tolyl) *O,O*-Dimethyl
Phosphorothioate *see* 9670
1,3-Dichloro-1,3,5-triazine-2,4,6(1*H*,3*H*,-
5*H*)-trione Potassium Salt (1:1) *see*
9946
5-[(4,6-Dichloro-1,3,5-triazin-2-yl)ami-
no]-3′,6′-dihydroxyspiro[isobenzofu-
ran-1(3*H*),9′-[9*H*]xanthen]-3-one *see*
3507
5-(4,6-Dichlorotriazinyl)aminofluorescein
see 3507
4,4′-Dichloro-*α*-(trichloromethyl)-
benzhydrol *see* 3096
4,4′-Dichloro-3-(trifluoromethyl)carbani-
lide *see* 2372
1-(1,3-Dichloro-6-trifluoromethyl-9-
phenanthryl)-3-di(*n*-butyl)aminopro-
panol *see* 4630
Dichlorovos *see* 3089
Dichloroxide *see* 2099
Dichloroxylenol, 3086
2,4-Dichloro-3,5-xylenol *see* 3086
Dichlorphenamide, 3087
Dichlorprop, 3088
Dichlorprop-P *see* 3088
Dichlorprop-P-potassium *see* 3088
Dichlor-Stapenor [Bayer] *see* 3094
Dichlorvos, 3089
Dichlotride [Merck & Co.] *see* 4819
β-Dichroine *see* 3982
Dichystrolum *see* 3198
Dickite *see* 356
Diclazuril, 3090
Diclectin [Duchesnay] *see* 3489
Diclocil [BMS] *see* 3094
Diclofenac, 3091
Dicloflex [Dexcel] *see* 3091
Diclofop-methyl, 3092
Dicloguamine *see* 5146
Diclomax [Provalis] *see* 3091
Diclondazolic Acid *see* 5625
Diclophlogont [Azupharma] *see* 3091
Dicloreum [Alfa] *see* 3091
Diclosulam, 3093
Dicloxacillin, 3094
Dicobalt Edetate *see* 3565
Dicobalt Octacarbonyl, 3095
Dicodid (Tabl.) [Abbott] *see* 4823
Dicodin [Mundipharma] *see* 3189
Dicoferin *see* 6619
Dicofol, 3096
Diconal [Amdipharm] *see* 3377
Dicopac [Amersham] *see* 10213

Dicophane *see* 2843
Dicopper Monoselenide *see* 2656
Dicopper Monosulfide *see* 2657
Dicopper Monoxide *see* 2654
Dicopper(1+) Tetraiodomercurate(2−)
see 2653
Dicopur DP [Nufarm] *see* 3088
Dicoumarin *see* 3100
Dicoumarol *see* 3100
Dicromil [Organon] *see* 2928
Dicrotalic Acid *see* 5877
Dicrotophos, 3097
Dicryl, 3098
Dictamine *see* 3099
Dictamnine, 3099
Dictycide [Fort Dodge] *see* 2670
Dictyzide *see* 2670
Dicumacyl *see* 3825
Dicumarol, 3100
Dicumol *see* 3100
Dicuprene [UCB] *see* 2660
Dicural [Fort Dodge] *see* 3160
Dicuran [Syngenta] *see* 2176
Dicurone *see* 4503
Dicyan *see* 2681
Dicyanine, 3101
2,2′-Dicyano-2,2′-azopropane *see* 911
1,4-Dicyanobutane *see* 152
β,β-Dicyano-*o*-chlorostyrene *see* 2128
Dicyanocobinamide, 3102
Dicyanocobyrinic Acid *a,b,c,d,e,g*-Hex-
amide *f*-D-2-Hydroxypropylamide *see*
3102
Dicyanodiamide, 3103
Dicyanodiamidine Sulfate, 3104
2,3-Dicyano-1,4-dithiaanthraquinone *see*
3411
sym-Dicyanoethane *see* 9003
Dicyanomercury *see* 5944
Dicyanomethane *see* 5775
4-Dicyanomethylene-2-methyl-6-*p*-di-
methylaminostyryl-4*H*-pyran *see* 2841
9-Dicyanomethylene-2,4,7-trinitro-
fluorene, 3105
Dicyanonickel *see* 6591
1,3-Dicyano-2,4,5,6-tetrachlorobenzene
see 2169
Dicyclidine *see* 7882
Dicyclohexano-18-crown-6 *see* 2592
Dicyclohexylamine, 3106
Dicyclohexylcarbodiimide, 3107
Dicyclohexyl-18-crown-6 *see* 2592
Dicyclohexyl(2′,6′-dimethoxy[1,1′-bi-
phenyl]-2-yl)phosphine *see* 1469
Dicyclohexyl[3,6-dimethoxy-2′,4′,6′-
tris(1-methylethyl)[1,1′-biphenyl]-2-
yl]phosphine *see* 1469
2-(2,2-Dicyclohexylethyl)piperidine *see*
7278
2-Dicyclohexylphosphino-2′,6′-diisopro-
poxybiphenyl *see* 1469
2-Dicyclohexylphosphino-2′,6′-dimeth-
oxybiphenyl *see* 1469
2-(Dicyclohexylphosphino)-3,6-dimeth-
oxy-2′,4′,6′-triisopropyl-1,1′-biphenyl
see 1469
2-Dicyclohexylphosphino-2′-(*N,N*-di-
methylamino)biphenyl *see* 1469
2′-(Dicyclohexylphosphino)-*N,N*-dimeth-
yl-[1,1′-biphenyl]-2-amine *see* 1469
2-Dicyclohexylphosphino-2′,4′,6′-triiso-
propylbiphenyl *see* 1469
1,1-Dicyclohexyl-2-(2-piperidyl)ethane
see 7278

Dicyclohexyl[2′,4′,6′-tris(1-methylethyl)-
[1,1′-biphenyl]-2-yl]phosphine *see*
1469
Dicyclomine, 3108
Dicyclopentadiene *see* 2734
Dicyclopentadienyliron *see* 4064
Dicyclopentadienylvanadium *see* 10111
2-[*N*-(Dicyclopropylmethyl)amino]oxazo-
line *see* 8347
N-(Dicyclopropylmethyl)-4,5-dihydro-2-
oxazolamine *see* 8347
(2*α*,7*α*,8*R*,19*α*,22*E*)-7,19:8,19-Dicyclo-
9,10-secoergosta-5(10),22-dien-2-ol *see*
9133
Dicycloverin *see* 3108
Dicynene [Lorex] *see* 3777
Dicynone [OM] *see* 3777
Dicysteine *see* 2779
Didakene *see* 9335
Didanosine, 3109
Didecyldimethylammonium Chloride,
3110
4′,5′-Didehydro-4,5′-*retro*-*β,β*-carotene-
3,3′-dione *see* 8320
3′,4′-Didehydro-*β,ψ*-caroten-16′-oic
Acid *see* 9712
18,21-Didehydro-17-demethoxy-18,21-di-
deoxo-18,21-dihydroxy-17-(2-propen-
1-ylamino)geldanamycin *see* 8282
1′,4-Didehydro-1-deoxy-1,4-dihydro-3′-
hydroxy-5′-[4-(2-methylpropyl)-1-
piperazinyl]-1-oxorifamycin VIII *see*
8339
1′,4-Didehydro-1-deoxy-1,4-dihydro-5′-
(2-methylpropyl)-1-oxorifamycin XIV
see 8338
(16*R*,19*E*)-19,20-Didehydro-16-[(10*β*,-
13*β*,21*β*)-23-deoxy-21,22-dihydro-11-
oxa-1,2-secostrychnidin-10-yl]cory-
nan-17-oic Acid Methyl Ester *see* 4414
3′,4′-Didehydro-4′-deoxy-*C*′-norvinca-
leukoblastine *see* 10187
2′,3′-Didehydro-3′-deoxythymidine *see*
8929
(2′*E*)-2′,3′-Didehydro-2′-deoxyverru-
carin A *see* 10160
(8*β*)-9,10-Didehydro-*N,N*-diethyl-6-
methylergoline-8-carboxamide *see*
5695
12,13-Didehydro-13,14-dihydro-*α*-ery-
throidine *see* 3738
11,12-Didehydro-7,10-dihydro-10-hy-
droxyretinol *see* 7031
(3*β*)-1,6-Didehydro-14,17-dihydro-3-
methoxy-16(15*H*)-oxaerythrinan-15-
one *see* 3192
(16*E*,20*β*)-16,17-Didehydro-9,17-dimeth-
oxycorynan-16-carboxylic Acid
Methyl Ester *see* 6303
(3*β*,19*α*,20*α*)-16,17-Didehydro-10,11-di-
methoxy-19-methyloxayohimban-16-
carboxylic Acid Methyl Ester *see* 8264
(*E*)-16,17-Didehydro-9,17-dimethoxy-
17,18-seco-20*α*-yohimban-16-carbox-
ylic Acid Methyl Ester *see* 6303
8,9-Didehydro-6,8-dimethylergoline *see*
184
(5*α*,6*α*,7*α*,12*β*,19*α*)-2,3-Didehydro-6,7-
epoxyaspidospermidine-3-carboxylic
Acid Methyl Ester *see* 5611
(4*S*,17*R*)-19,20-Didehydro-17,18-epoxy-
curan-17-ol *see* 10242
(*βS*,16*β*)-19,20-Didehydro-*β*,18-epoxy-
10,11-dihydro-12-nitro-10,11-dioxo-
17-norcuran-16-propanoic Acid *see*
1610

Difluprednate, 3171
Diflupyl see 5222
Diflurex [Anphar-Rolland] see 9586
Difluron see 3161
Difolatan see 1773
Diformyl see 4544
2,5-Diformylfuran see 4328
1,3-Diformylpropane see 4508
Difosfocin [Magis] see 2318
Difosfonal [SPA] see 2365
Difrarel [Sigma-Tau] see 1221
1,2-Di-2-furanyl-1,2-ethanedione 1,2-
    Dioxime see 4335
N,N'-Difurfurylidene-2-furanmethanedi-
    amine see 4828
Digacin [Beiersdorf-Lilly] see 3186
Digallic Acid, 3172
"Digallic Acid" see 9184
m-Digallic Acid see 3172
Digalogenin, 3173
Digenic Acid see 5324
Digenin see 5324
Digerent [Polifarma] see 9868
Digicor [Hennig] see 3182
Digilanide A see 5404
Digilanide B see 5404
Digilanide C see 5404
Digilanide D see 5404
Digilong [Boehringer, Mann.] see 3182
Digimed [Trommsdorff] see 3182
Digimerck [Merck KGaA] see 3182
Digin see 4466
Diginatigenin, 3174
Diginatin, 3175
Diginigenin see 3176
Diginin, 3176
Diginorgin see 3177
Diginose see 3176
3β-(Diginosyloxy)-12α,20α-epoxy-14β,-
    17α-pregn-5-ene-11,15-dione see 3176
Digipural [Schaper & Brümmer] see 3182
Digitalin, 3177
Digitalin, Crystalline see 3182
Digitalinum True see 3177
Digitalinum Verum see 3177
Digitalis, 3178
Digitin see 3180
Digitoflavone see 5675
Digitogenin, 3179
Digitonin, 3180
Digitophyllin see 3182
Digitoxigenin, 3181
Digitoxin, 3182
Digitoxose, 3183
Diglycine see 4959
Diglycolic Acid Bis(2,4,6-triiodo-3-car-
    boxanilide) see 5094
Diglycol Laurate see 3135
3,3'-(Diglycoloyldiimino)bis(2,4,6-triio-
    dobenzoic Acid) see 5094
Diglycolyldiamide see 7578
Diglyme, 3184
Dignonitrat [Sankyo] see 5271
Dignover [Sankyo] see 10144
Digold Trioxide see 4558
Digoxigenin, 3185
Digoxin, 3186
Diguanide see 1218
2,4-Diguanidino-1-dodecyloxybenzene
    see 5441
2,4-Diguanidino-1-lauryloxybenzene see
    5441
2,4-Diguanidinophenyl Dodecyl Ether
    see 5441
2,4-Diguanidinophenyl Lauryl Ether see
    5441

1,3-Diguanido-2,4,5,6-cyclohexanetetrol
    see 8952
p,p'-Diguanyldiazoaminobenzene Diacet-
    urate see 3291
Dihexyl Sodium Sulfosuccinate see 1259
Di-Hydan see 7433
Dihydergot [Novartis] see 3191
1,4-Dihydracryloylpiperazine Dimeth-
    anesulfonate see 7594
Dihydral [Duphar] see 3198
Dihydralazine, 3187
Dihydralazine Mesylate see 3187
1,4-Dihydrazinophthalazine see 3187
1,4-Dihydrazinylphthalazine see 3187
o-Dihydric Phenolase see 10019
Dihydrite see 2638
22,23-Dihydroabamectin see 5296
1,2-Dihydroacenaphthylene see 29
Dihydroactinidiolide, 3188
N,N'-Dihydro-1,2,1',2'-anthraquinona-
    zine see 4974
6,15-Dihydro-5,9,14,18-anthrazinetet-
    rone see 4974
Dihydroartemisinin see 807
Dihydroartemisinin Ethyl Ether see 803
Dihydroartemisinin Methyl Ether see
    805
22,23-Dihydroavermectin B₁ see 5296
22,23-Dihydroavermectin B₁ₐ see 5296
22,23-Dihydroavermectin B₁ᵦ see 5296
Dihydrobaikiaine see 7570
1,2-Dihydrobenz[j]aceanthrylene see
    2204
9,10-Dihydro-9,10-[1',2']-benzenoanthra-
    cene see 9925
(S)-2-[1-[2-(2,3-Dihydrobenzfuran-5-yl)-
    ethyl]-3-pyrrolidinyl]-2,2-diphenyl-
    acetamide see 2826
1,3-Dihydro-2H-benzimidazole-2-thione
    see 1084
(6aS,11bR)-7,11b-Dihydrobenz[b]indeno-
    [1,2-d]pyran-3,4,6a,9,10(6H)-pentol
    see 4672
(6aS,11bR)-7,11b-Dihydrobenz[b]indeno-
    [1,2-d]pyran-3,6a,9,10(6H)-tetrol see
    1372
4-(9,10-Dihydro-4H-benzo[4,5]cyclo-
    hepta[1,2-b]thien-4-ylidene)-1-methyl-
    piperidine see 7629
2-(2,3-Dihydro-1,4-benzodioxin-2-yl)-4,5-
    dihydro-1H-imidazole see 4928
N-[(1R,2R)-2-(2,3-Dihydro-1,4-benzo-
    dioxin-6-yl)-2-hydroxy-1-(1-pyrrolidi-
    nylmethyl)ethyl]octanamide see 3599
N-[(1R,2R)-1-(2,3-Dihydro-1,4-benzo-
    dioxin-6-yl)-1-hydroxy-3-(pyrrolidin-
    1-yl)propan-2-yl]octanamide see 3599
N-[(2,3-Dihydro-1,4-benzodioxin-2-yl)-
    methyl]guanidine see 4603
1-[(2,3-Dihydro-1,4-benzodioxin-2-yl)-
    methyl]piperidine see 7591
1-(2,3-Dihydro-1,4-benzodioxin-5-yl)-
    piperazine see 3608
2,3-Dihydrobenzofuran see 2547
N-[[(1R,2R)-2-(2,3-Dihydro-4-benzofu-
    ranyl)cyclopropyl]methyl]propanam-
    ide see 9208
(1R-trans)-N-[[2-(2,3-Dihydrobenzofu-
    ran-4-yl)cycloprop-1-yl]methyl]pro-
    panamide see 9208
(3S)-1-[2-(2,3-Dihydro-5-benzofuranyl)-
    ethyl]-α,α-diphenyl-3-pyrrolidineacet-
    amide see 2826
2-[4-[[[(2R)-3,4-Dihydro-2H-1-benzopy-
    ran-2-yl]methyl]amino]butyl]-1,2-ben-

zisothiazol-3(2H)-one 1,1-Dioxide see
    8257
N-[[(2R)-3,4-Dihydro-2H-1-benzopyran-
    2-yl]methyl]-5-(4-fluorophenyl)-3-pyri-
    dinemethanamine see 8512
1-[(3,4-Dihydro-2H-1-benzothiopyran-8-
    yl)oxy]-3-[(1,1-dimethylethyl)amino]-
    2-propanol see 9316
6,7-Dihydrobis[1,3]benzodioxolo[5,6-
    a:4',5'-g]quinolizinium see 2512
1,3-Dihydro-1,3-bis(hydroxymethyl)-2H-
    benzimidazole-2-thione see 9464
rel-(3R,4R)-Dihydro-3,4-bis[(3-hydroxy-
    phenyl)methyl]-2(3H)-furanone see
    3652
Dihydrobis[2-(methoxy-κO)ethanolato-
    κO]aluminate(1−) Sodium (1:1) see
    8702
22,23-Dihydro C-076B₁ see 5296
22,23-Dihydro C-076B₁ₐ see 5296
3,4-Dihydrocarbostyril see 4817
2,3-Dihydro-5-carboxanilido-6-methyl-
    1,4-oxathiin see 1828
1,3-Dihydro-5-(4-chlorophenyl)-7-hy-
    droxy-6-methylfuro[3,4-c]pyridine see
    2267
3,4-Dihydrochlorothiazide see 4819
Dihydrocholesterol see 2207
(8α,9R)-10,11-Dihydrocinchonan-9-ol see
    4820
(9S)-10,11-Dihydrocinchonan-9-ol see
    4821
(−)-Dihydrocinchonidine see 4820
(+)-Dihydrocinchonine see 4821
Dihydrocodehydrogenase II see 6432
Dihydrocodeine, 3189
Dihydrocodeine Bitartrate see 3189
Dihydrocodeinone see 4823
Dihydrocodeinone Enol Acetate see 9426
4,5α-Dihydrocortisol see 4803
Dihydrocoumarone see 2547
16,17-Dihydro-15H-cyclopenta[a]phen-
    anthrene see 2739
9,13b-Dihydro-1H-dibenz[c,f]imidazo-
    [1,5-a]azepin-3-amine see 3673
7-[(10,11-Dihydro-5H-dibenzo[a,d]cyclo-
    hepten-5-yl)amino]heptanoic Acid see
    403
3-(10,11-Dihydro-5H-dibenzo[a,d]cyclo-
    hepten-5-ylidene)-N,N-dimethyl-1-
    propanamine see 483
3-(10,11-Dihydro-5H-dibenzo[a,d]cyclo-
    hepten-5-ylidene)-N,N-dimethyl-1-
    propanamine N-Oxide see 484
3-(10,11-Dihydro-5H-dibenzo[a,d]cyclo-
    hepten-5-ylidene)-1-ethyl-2-methylpyr-
    rolidine see 7616
3-(10,11-Dihydro-5H-dibenzo[a,d]cyclo-
    hepten-5-ylidene)-N-methyl-1-propan-
    amine see 6804
3-(10,11-Dihydro-5H-dibenzo[a,d]cyclo-
    hepten-5-ylidene)-N-methylpropylam-
    ine see 6804
(3-endo)-3-[(10,11-Dihydro-5H-dibenzo-
    [a,d]cyclohepten-5-yl)oxy]-8-methyl-8-
    azabicyclo[3.2.1]octane see 2913
3α-[(10,11-Dihydro-5H-dibenzo[a,d]cy-
    clohepten-5-yl)oxy]-1αH,5αH-tropane
    see 2913
Dihydrodiethylstilbestrol see 4737
(5R,6R)-5,6-Dihydro-5,6-dihydroxy-10'-
    apo-β,ψ-carotenoic Acid see 888
9,10-Dihydro-4,5-dihydroxy-9,10-dioxo-
    2-anthracenecarboxylic Acid see 8299

oxy-17,18,19,20-tetranorprostadienoic Acid *see* 3918

Di(hydroxyethyl)glycine *see* 1205

1(*S*),3(*R*)-Dihydroxy-20(*R*)-(5′-ethyl-5′-hydroxyhepta-1′(*E*),3′(*E*)-dien-1′-yl)-9,10-secopregna-5(*Z*),7(*E*),10(19)-triene *see* 8595

Di-(2-hydroxyethyl)nitrosamine *see* 6722

5,7-Dihydroxyflavone *see* 2260

3′,6′-Dihydroxyfluoran *see* 4192

11β,17β-Dihydroxy-9α-fluoro-17α-methyl-4-androsten-3-one *see* 4218

(3β,5α,25*R*)-3,26-Dihydroxyfurost-20-(22)-en-12-one *see* 4657

(1α,7β)-1,7-Dihydroxyginkgolide A *see* 4461

4,6-Dihydroxy-2-(β-D-glucosido)-β-(*p*-hydroxyphenyl)propiophenone *see* 7438

Dihydroxy(glycinato)aluminum *see* 3201

(*Z*)-8β,11-Dihydroxy-1β-guai-4-en-3-one 11-Acetate 2-Methylcrotonate *see* 9710

2,2′-Dihydroxy-3,3′,5,5′,6,6′-hexachlorodiphenylmethane *see* 4716

1,6-Dihydroxyhexane *see* 4728

5,7-Dihydroxy-3-(3-hydroxy-4,5-dimethoxyphenyl)-6-methoxy-4*H*-1-benzopyran-4-one *see* 5134

γ,4-Dihydroxy-2-(6-hydroxy-1-heptenyl)-4-cyclopentanecrotonic Acid λ-Lactone *see* 1374

3,5-Dihydroxy-4-(3-hydroxy-4-methoxy-hydrocinnamoyl)phenyl-2-*O*-(6-deoxy-α-L-mannopyranosyl)-β-D-gluco-pyranoside *see* 6537

5,7-Dihydroxy-2-(3-hydroxy-4-methoxyphenyl)-4*H*-1-benzopyran-4-one *see* 3324

5,7-Dihydroxy-3-(3-hydroxy-4-methoxyphenyl)-4*H*-1-benzopyran-4-one *see* 7834

l-3,4-Dihydroxy-1-[1-hydroxy-2-(methylamino)ethyl]benzene *see* 3674

1,8-Dihydroxy-3-(hydroxymethyl)-9,10-anthracenedione *see* 300

1,8-Dihydroxy-3-(hydroxymethyl)anthraquinone *see* 300

1,8-Dihydroxy-3-hydroxymethyl-10-(6-hydroxymethyl-3,4,5-trihydroxy-2-pyranyl)anthrone *see* 303

5,8-Dihydroxy-2-(hydroxymethyl)-6-methoxy-3-(2-oxopropyl)-1,4-naphthalenedione *see* 4344

7-[3,5-Dihydroxy-2-(3-hydroxy-3-methyl-1-octenyl)cyclopentyl]-5-heptenoic Acid *see* 1824

[1-[9-[(4*S*,2*R*,3*R*,5*R*)-3,4-Dihydroxy-5-(hydroxymethyl)oxolan-2-yl]-6-amino-purin-2-yl]pyrazol-4-yl]-*N*-methylcarboxamide *see* 8247

5,8-Dihydroxy-2-[(1*S*)-1-hydroxy-4-methyl-3-penten-1-yl]-1,4-naphthalenedione *see* 249

(−)-5,8-Dihydroxy-2-(1-hydroxy-4-methyl-3-pentenyl)-1,4-naphthoquinone *see* 249

3,5-Dihydroxy-α-[[(*p*-hydroxy-α-methylphenethyl)amino]methyl]benzyl Alcohol *see* 4012

(3*R*,4*R*,5*R*)-3,4-Dihydroxy-5-hydroxymethylpiperidine *see* 5217

[10-[2,3-Dihydroxy-1-(hydroxymethyl)-propyl]-1,4,7,10-tetraazacyclododecane-1,4,7-triacetato(3−)-κ*N*¹,κ*N*⁴,-

κ*N*⁷,κ*N*¹⁰,κ*O*¹,κ*O*⁴,κ*O*⁷]gadolinium *see* 4353

7-[3,5-Dihydroxy-2-(3-hydroxy-1-octenyl)cyclopentyl]-5-heptenoic Acid *see* 7989

rel-(4*S*)-7-[(1*R*,2*R*,3*R*,5*S*)-3,5-Dihydroxy-2-[(1*E*,3*R*)-3-hydroxy-4-phenoxy-1-buten-1-yl]cyclopentyl]-4,5-heptadienoic Acid Methyl Ester *see* 4023

5,7-Dihydroxy-2-(4-hydroxyphenyl)-4*H*-1-benzopyran-4-one *see* 717

5,7-Dihydroxy-3-(4-hydroxyphenyl)-4*H*-1-benzopyran-4-one *see* 4426

(*R*)-3,4-Dihydroxy-α-[[[4-(*p*-hydroxyphenyl)butyl]amino]methyl]benzyl Alcohol *see* 762

5,7-Dihydroxy-3-(4-hydroxyphenyl)-6-methoxy-4*H*-1-benzopyran-4-one *see* 9245

3,4-Dihydroxy-*N*-[3-(4-hydroxyphenyl)-1-methylpropyl]-β-phenylethylamine *see* 3440

(5*Z*)-7-[(1*R*,2*R*,3*R*,5*S*)-3,5-Dihydroxy-2-[(1*E*,3*S*)-3-hydroxy-5-phenyl-1-penten-1-yl]cyclopentyl]-*N*-ethyl-5-heptenamide *see* 1225

(5*Z*)-7-[(1*R*,2*R*,3*R*,5*S*)-3,5-Dihydroxy-2-[(3*R*)-3-hydroxy-5-phenylpentyl]cyclopentyl]-5-heptenoic Acid 1-Methylethyl Ester *see* 5431

(2*R*)-2,4-Dihydroxy-*N*-(3-hydroxypropyl)-3,3-dimethylbutanamide *see* 2949

D(+)-α,γ-Dihydroxy-*N*-(3-hydroxypropyl)-β,β-dimethylbutyramide *see* 2949

7-[3,5-Dihydroxy-2-[3-hydroxy-4-(3-thienyloxy)-1-butenyl]cyclopentyl]-5-heptenoic Acid *see* 9578

rel-(5*Z*)-7-[(1*R*,2*R*,3*R*,5*S*)-3,5-Dihydroxy-2-[(1*E*,3*R*)-3-hydroxy-4-[3-(trifluoromethyl)phenoxy]-1-buten-1-yl]cyclopentyl]-5-heptenoic Acid *see* 4224

(5*Z*)-7-[(1*E*,3*R*)-3-hydroxy-4-[3-(trifluoromethyl)phenoxy]-1-buten-1-yl]cyclopentyl]-5-heptenoic Acid 1-Methylethyl Ester *see* 9738

2,2-Dihydroxy-1,3-indanedione *see* 6640

2,2-Dihydroxy-1*H*-indene-1,3(2*H*)-dione *see* 6640

(*S*)-4′,7-Dihydroxyisoflavane *see* 3701

4′,7-Dihydroxyisoflavone *see* 2798

2,6-Dihydroxyisonicotinic Acid *see* 2324

3,4-Dihydroxy-α-[(isopropylamino)methyl]benzyl Alcohol *see* 5263

3,5-Dihydroxy-α-[(isopropylamino)methyl]benzyl Alcohol *see* 6002

3,4-Dihydroxy-α-[1-(isopropylamino)-propyl]benzyl Alcohol *see* 5215

β,β′-Dihydroxyisopropyl Chloride *see* 2146

11β,21-Dihydroxy-16α,17-isopropylidenedioxy-1,4-pregnadiene-3,20-dione *see* 2929

11,12-Dihydroxy-13-isopropylpodocarpa-8,11,13-trien-17-oic Acid *see* 1850

5,6-Dihydroxy-2-(2,6-xylidino)-4*H*-1,3-thiazine *see* 10277

Dihydroxymaleic Acid, 3203

Dihydroxymalonic Acid *see* 5983

(2*R*)-2,4-Dihydroxy-*N*-[3-[(2-mercaptoethyl)amino]-3-oxopropyl]-3,3-dimethylbutanamide *see* 7114

2,4-Dihydroxy-*N*-[2-[(2-mercaptoethyl)carbamoyl]ethyl]-3,3-dimethylbutyramide *see* 7114

(11β)-11,17-Dihydroxy-21-mercapto-pregn-4-ene-3,20-dione *see* 9641

α,4-Dihydroxy-3-methoxybenzeneacetic Acid *see* 10121

2,2′-Dihydroxy-4-methoxybenzophenone *see* 3332

7,8-Dihydroxy-6-methoxy-2*H*-1-benzopyran-2-one *see* 4294

(3β,16*R*)-3,17-Dihydroxy-16-(methoxycarbonyl)-4-methyl-2,4(1*H*)-cyclo-3,4-secoakuammilanium *see* 3547

7,8-Dihydroxy-6-methoxycoumarin *see* 4294

7,8-Dihydroxy-6-methoxycoumarin-8-β-D-glucoside *see* 4295

4′,5-Dihydroxy-7-methoxyflavanone *see* 8455

4′,5-Dihydroxy-7-methoxyflavanone 5-Glucoside *see* 8455

5,7-Dihydroxy-4′-methoxyflavone *see* 14

5,7-Dihydroxy-6-methoxyflavone *see* 6974

5,7-Dihydroxy-4′-methoxyflavone-D-glucosido-L-rhamnoside *see* 5553

4′,5-Dihydroxy-7-methoxyisoflavone *see* 8017

5,7-Dihydroxy-4′-methoxyisoflavone *see* 1231

5,7-Dihydroxy-6-methoxy-2-(4-methoxyphenyl)-4*H*-1-benzopyran-4-one *see* 7170

1,8-Dihydroxy-3-methoxy-6-methylanthraquinone *see* 3618

5,8-Dihydroxy-6-methoxy-3-methyl-2-aza-9,10-anthraquinone *see* 1356

6,9-Dihydroxy-7-methoxy-3-methylbenz-[g]isoquinoline-5,10-dione *see* 1356

2′,6′-Dihydroxy-4′-methoxy-3′-methyl-1-butyrophenone *see* 837

2-[[2,6-Dihydroxy-4-methoxy-3-methyl-5-(1-oxobutyl)phenyl]methyl]-3,5-dihydroxy-4,4-dimethyl-6-(1-oxobutyl)-2,5-cyclohexadien-1-one *see* 836

5,8-Dihydroxy-6-methoxy-2-methyl-3-(2-oxopropyl)-1,4-naphthalenedione *see* 5310

1-(2,6-Dihydroxy-4-methoxy-3-methyl-phenyl)-1-butanone *see* 837

3,17-Dihydroxy-11β-methoxy-19-nor-17α-pregna-1,3,5-trien-20-yne *see* 6375

2-[[2,4-Dihydroxy-6-methoxy-3-(1-oxobutyl)phenyl]methyl]-3,5-dihydroxy-4,4-dimethyl-6-(1-oxobutyl)-2,5-cyclohexadien-1-one *see* 2918

1,2-Dihydroxy-3-(2-methoxyphenoxy)-propane *see* 4591

5,7-Dihydroxy-2-(4-methoxyphenyl)-4*H*-1-benzopyran-4-one *see* 14

5,7-Dihydroxy-3-(4-methoxyphenyl)-4*H*-1-benzopyran-4-one *see* 1231

5,7-Dihydroxy-6-methoxy-2-phenyl-4*H*-1-benzopyran-4-one *see* 6974

2-(3,4-Dihydroxy-5-methoxyphenyl)-3,5-bis(β-D-glucopyranosyloxy)-7-hydroxy-1-benzopyrylium Chloride *see* 7305

(2*E*)-3-(3,4-Dihydroxy-2-methoxyphenyl)-1-[4-hydroxy-3-methyl-2-butenyl)phenyl]-2-propen-1-one *see* 5532

(2*E*)-3-(3,4-Dihydroxy-2-methoxyphenyl)-1-(4-hydroxyphenyl)-2-propen-1-one *see* 5532

(1*S*,2*S*,3*R*)-1-(3,4-Dihydroxy-5-methoxyphenyl)-1,2,3,4-tetrahydro-2,3,7-trihydroxy-3-(hydroxymethyl)-6-methoxy-2-naphthalenecarboxylic Acid *see* 7653

2-(3,4-Dihydroxy-5-methoxyphenyl)-3,-5,7-trihydroxy-1-benzopyrylium Chloride (1:1) *see* 7305

1,7-Dihydroxy-3-methoxy-9*H*-xanthen-9-one *see* 4434

4,7-Dihydroxy-2-methoxyxanthone *see* 4434

3,4-Dihydroxy-α-methylaminoaceto-phenone *see* 160

3′,4′-Dihydroxy-2-(methylamino)aceto-phenone *see* 160

(−)-3,4-Dihydroxy-α-[(methylamino)-methyl]benzyl Alcohol *see* 3674

(±)-3,4-Dihydroxy-α-[(methylamino)-methyl]benzyl Alcohol 3,4-Dipivalate *see* 3378

(17β)-4,17-Dihydroxy-17-methylandrost-4-en-3-one *see* 7064

1,3-Dihydroxy-2-methyl-9,10-anthra-cenedione *see* 8414

1,8-Dihydroxy-3-methyl-9,10-anthra-cenedione *see* 2262

1,8-Dihydroxy-3-methyl-9(10*H*)-anthra-cenone *see* 2257

1,3-Dihydroxy-2-methylanthraquinone *see* 8414

1,8-Dihydroxy-3-methylanthraquinone *see* 2262

2,4-Dihydroxy-6-methylbenzenecarbox-ylic Acid *see* 6977

2,4-Dihydroxy-6-methylbenzoic Acid *see* 6977

2,6-Dihydroxy-4-methyl-5-bis[2-chloro-ethyl]aminopyrimidine *see* 3471

6,7-Dihydroxy-4-methylcoumarin Disul-fate *see* 9114

(3α,4α,5α,17α)-3,17-Dihydroxy-4-meth-yl-9,15-cyclo-*C*,18-dinor-14,15-secoan-drostane-4,17-dimethanol *see* 714

(3*R*)-3,4-Dihydroxy-2-methylenebutanoic Acid (3a*R*,4*S*,6*E*,10*Z*,11a*R*)-2,3,3a,4,5,-8,9,11a-Octahydro-10-(hydroxymeth-yl)-6-methyl-3-methylene-2-oxocyclo-deca[*b*]furan-4-yl Ester *see* 2408

1,1-Di(hydroxymethyl)ethylamine *see* 445

(11β,16α)-11,21-Dihydroxy-16,17-[(1-methylethylidene)bis(oxy)]pregna-1,4-diene-3,20-dione *see* 2929

3,4-Dihydroxy-4′-methyl-5-nitrobenzo-phenone *see* 9668

(11α,17β)-11,17-Dihydroxy-17-methyl-3-oxoandrosta-1,4-diene-2-carboxalde-hyde *see* 4267

4,17β-Dihydroxy-17α-methyl-3-oxoan-drost-4-ene *see* 7064

(5*Z*,11α,13*E*,15*R*)-11,15-Dihydroxy-15-methyl-9-oxoprosta-5,13-dien-1-oic Acid *see* 759

(11α,13*E*)-(±)-11,16-Dihydroxy-16-meth-yl-9-oxoprost-13-en-1-oic Acid Methyl Ester *see* 6293

8,9-Dihydroxy-2-methyl-4-oxo-4*H*,5*H*-pyrano[3,2-*c*][1]benzopyran-10-car-boxylic Acid *see* 2327

3,5-Dihydroxy-3-methylpentanoic Acid *see* 6242

α,β-Dihydroxy-γ-(2-methylphenoxy)pro-pane *see* 5917

1,2-Dihydroxy-3-(2-methylphenoxy)pro-pane *see* 5917

(7*S*,13*E*,16*R*,20*R*,21*E*)-7,20-Dihydroxy-16-methyl-10-phenyl-24-oxo[14]cyto-

chalasa-6(12),13,21-triene-1,23-dione *see* 2787

(11β)-11,17-Dihydroxy-21-(4-methyl-1-piperazinyl)pregna-1,4-diene-3,20-di-one *see* 5832

11β,17-Dihydroxy-21-(4-methyl-1-piper-azinyl)-Δ¹-progesterone *see* 5832

(*E*)-3β,7β-Dihydroxy-14α-methyl-8β-po-docarpane-Δ¹³,α-acetic Acid 2-(Di-methylamino)ethyl Ester *see* 1890

(16β)-17,21-Dihydroxy-16-methylpreg-na-1,4-diene-3,11,20-trione *see* 5928

11β,21-Dihydroxy-2′-methyl-5′β*H*-preg-na-1,4-dieno[17,16-*d*]oxazole-3,20-di-one 2l-Acetate *see* 2865

1,3-Dihydroxy-2-methyl-2-propylamine *see* 445

2,4-Dihydroxy-5-methylpyrimidine *see* 9553

3,5-Dihydroxy-6-methyl-1,2,4-triazine *see* 896

N,*N*′-Dihydroxymethylurea *see* 7067

β,δ-Dihydroxy-β-methylvaleric Acid *see* 6242

3,5-Dihydroxy-3-methylvaleric Acid *see* 6242

β,δ-Dihydroxy-β-methyl-δ-valerolactone *see* 6241

1,5-Dihydroxy-2-methyl-6-[(6-*O*-β-D-xylopyranosyl-β-D-glucopyranosyl)-oxy]-9,10-anthracenedione *see* 6356

1,3-Dihydroxynaphthalene *see* 6481

1,8-Dihydroxynaphthalene-3,6-disulfonic Acid *see* 2245

4,5-Dihydroxy-2,7-naphthalenedisulfonic Acid *see* 2245

5,6-Dihydroxynaphtho[2,3-*f*]quinoline-7,12-dione *see* 245

1,2-Dihydroxy-3-nitro-9,10-anthracene-dione *see* 246

1,2-Dihydroxy-3-nitroanthraquinone *see* 246

2,4-Dihydroxy-4′-nitroazobenzene *see* 5759

4,5-Dihydroxy-3-[2-(4-nitrophenyl)di-azenyl]-2,7-naphthalenedisulfonic Acid Sodium Salt (1:2) *see* 2244

D-*threo*-*N*-(1,1′-Dihydroxy-1-*p*-nitro-phenylisopropyl)dichloroacetamide *see* 2077

(3,4-Dihydroxy-5-nitrophenyl)(4-methyl-phenyl)methanone *see* 9668

(±)-3,4-Dihydroxynorephedrine *see* 6781

rel-(9*R*,10*R*)-9,10-Dihydroxyoctadecane-dioic Acid *see* 7436

9,10-Dihydroxyoctadecanoic Acid *see* 3204

(3β,4α)-3,23-Dihydroxyolean-12-en-28-oic Acid *see* 4659

3′,6′-Dihydroxy-5(or 6)-Isothiocyanato-spiro[isobenzofuran-1(3*H*),9′-[9*H*]-xanthen]-3-one *see* 4120

1α,25-Dihydroxy-22-oxavitamin D₃ *see* 5828

1,6β-Dihydroxy-4-oxo-10α*H*-ambrosa-2,11(13)-dien-12-oic Acid γ-Lactone *see* 7151

6α,8β-Dihydroxy-4-oxoambrosa-2,11-(13)-dien-12-oic Acid 12,8-Lactone *see* 4661

N,*N*′-[[[(3α,5β,12α)-3,12-Dihydroxy-24-oxocholan-24-yl]imino]di-3,1-propan-diyl]bis-D-gluconamide *see* 1217

(5*Z*)-7-[(1*R*,2*R*,3*R*,5*S*)-3,5-Dihydroxy-2-(3-oxodecyl)cyclopentyl]-5-heptenoic Acid *see* 10034

4β,27-Dihydroxy-1-oxo-5β,6β-epoxywi-tha-2,24-dienolide *see* 10246

6α,8α-Dihydroxy-3-oxoeudesma-1,4-dien-12-oic Acid 12,6-Lactone *see* 806

3,8-Dihydroxy-11-oxo-1-(2-oxoheptyl)-6-pentyl-11*H*-dibenzo[*b*,*e*][1,4]dioxepin-7-carboxylic Acid *see* 7494

3β,17-Dihydroxy-20-oxo-5α-pregnane *see* 263

(5*Z*,11α,13*E*,15*S*)-11,15-Dihydroxy-9-ox-oprosta-5,13-dien-1-oic Acid *see* 7988

(11α,13*E*,15*S*)-11,15-Dihydroxy-9-oxo-prost-13-en-1-oic Acid *see* 7987

(1*S*,3*S*,7*S*,10*R*,11*S*,12*S*,16*R*)-7,11-Dihy-droxy-8,8,10,12,16-pentamethyl-3-[(1*E*)-1-methyl-2-(2-methyl-4-thiazol-yl)ethenyl]-17-oxa-4-azabicyclo[14.1.-0]heptadecane-5,9-dione *see* 5297

(4*S*,7*R*,8*S*,9*S*,13*Z*,16*S*)-4,8-Dihydroxy-5,-5,7,9,13-pentamethyl-16-[(1*E*)-1-meth-yl-2-(2-methyl-4-thiazolyl)ethenyl]oxa-cyclohexadec-13-ene-2,6-dione *see* 3684

1,5-Dihydroxypentane *see* 7229

1-[(4*R*,5*R*)-4,5-Dihydroxy-*N*²-[4-[5-[4-(pentyloxy)phenyl]-3-isoxazolyl]benzo-yl]-L-ornithine]-4-[(4*S*)-4-hydroxy-4-[4-hydroxy-3-(sulfooxy)phenyl]-L-thre-onine]pneumocandin A0 *see* 6254

1-[(4*R*,5*R*)-4,5-Dihydroxy-*N*²-[[4″-(pent-yloxy)[1,1′:4′,1″-terphenyl]-4-yl]car-bonyl]-L-ornithine]echinocandin B *see* 649

1,5-Dihydroxyphenazine *N*,*N*′-Dioxide *see* 5064

3,4-Dihydroxyphenethylamine *see* 3470

(5*Z*)-7-[(1*R*,2*R*,3*R*,5*S*)-3,5-Dihydroxy-2-[(1*E*)-2-[2-(phenoxymethyl)-1,3-dioxo-lan-2-yl]ethenyl]cyclopentyl]-5-hepten-oic Acid *see* 3918

1,2-Dihydroxy-3-phenoxypropane *see* 7401

2,5-Dihydroxyphenylacetic Acid *see* 4775

3-(3,4-Dihydroxyphenyl)alanine *see* 3469

(−)-3-(3,4-Dihydroxyphenyl)-L-alanine *see* 5516

β-(3,4-Dihydroxyphenyl)-L-alanine *see* 5516

β-(3,4-Dihydroxyphenyl)-α-alanine *see* 3469

1-(3,4-Dihydroxyphenyl)-2-aminobutanol *see* 3888

α-(3,4-Dihydroxyphenyl)-β-aminoethane *see* 3470

5,7-Dihydroxy-2-phenyl-4*H*-1-benzopy-ran-4-one *see* 2260

1-(3,5-Dihydroxyphenyl)-2-(*tert*-butyl-amino)ethanol *see* 9302

4-[2-(2,4-Dihydroxyphenyl)diazenyl]ben-zenesulfonic Acid Sodium Salt (1:1) *see* 9954

(2*R*,3*S*)-2-(3,4-Dihydroxyphenyl)-3,4-di-hydro-2*H*-1-benzopyran-3,4,5,7-tetrol *see* 5504

(2*R*,3*S*)-2-(3,4-Dihydroxyphenyl)-3,4-di-hydro-2*H*-1-benzopyran-3,5,7-triol *see* 1902

(2*S*)-2-(3,4-Dihydroxyphenyl)-2,3-dihy-dro-5,7-dihydroxy-4*H*-1-benzopyran-4-one *see* 3726

rel-(2*R*,3*R*)-2-(3,4-Dihydroxyphenyl)-2,3-dihydro-3,7-dihydroxy-4*H*-1-benzopy-ran-4-one *see* 4347

2-(3,4-Dihydroxyphenyl)-3,7-dihydroxy-4*H*-1-benzopyran-4-one *see* 4119

2-(3,4-Dihydroxyphenyl)-5,7-dihydroxy-4*H*-1-benzopyran-4-one *see* 5675

2-(3,4-Dihydroxyphenyl)-3,5-dihydroxy-7-methoxy-4*H*-1-benzopyran-4-one *see* 8294

α-(3,4-Dihydroxyphenyl)-2-dimethylaminoethanol *see* 6141

2-(3,4-Dihydroxyphenyl)ethanol *see* 4887

l-3,4-Dihydroxyphenylethanolamine *see* 6784

Dihydroxyphenylethanolisopropylamine *see* 5263

1-(2,4-Dihydroxyphenyl)ethanone *see* 8260

3,4-Dihydroxyphenylethyl Acetate *see* 4887

3,4-Dihydroxyphenylethylamine *see* 2904

1-(3',4'-Dihydroxyphenyl)-2-ethylamino-1-propanol *see* 3331

2-(3,4-Dihydroxyphenyl)-3-(β-D-glucofuranosyloxy)-5,7-dihydroxy-4*H*-1-benzopyran-4-one *see* 5267

2-(3,4-Dihydroxyphenyl)-7-(β-D-glucopyranosyloxy)-3,5-dihydroxy-4*H*-1-benzopyran-4-one *see* 8151

1-(3,4-Dihydroxyphenyl)-1-hydroxy-2-aminobutane *see* 3888

2-(2,4-Dihydroxyphenyl)-6-hydroxy-3-benzofurancarboxylic Acid δ-Lactone *see* 2552

α-(3,4-Dihydroxyphenyl)-α-hydroxy-β-dimethylaminoethane *see* 6141

7-[3-[[2-(3,5-Dihydroxyphenyl)-2-hydroxyethyl]amino]propyl]-3,7-dihydro-1,3-dimethyl-1*H*-purine-2,6-dione *see* 8259

1-(3,5-Dihydroxyphenyl)-1-hydroxy-2-[(4-hydroxyphenyl)isopropylamino]ethane *see* 4012

1-(3,5-Dihydroxyphenyl)-1-hydroxy-2-isopropylaminoethane *see* 6002

1-(3,4-Dihydroxyphenyl)-2-(4-(4-hydroxyphenyl)butylamino)ethanol *see* 762

1-(3,4-Dihydroxyphenyl)-2-isopropylamino-1-butanol *see* 5215

1-(3,4-Dihydroxyphenyl)-2-isopropylaminoethanol *see* 5263

1-(3,5-Dihydroxyphenyl)-2-isopropylaminoethanol *see* 6002

Dihydroxyphenylmethenylbenzyl Alcohol *see* 7358

L-(3,4-Dihydroxyphenyl)-2-methylalanine *see* 6127

l-1-(3,4-Dihydroxyphenyl)-2-(methylamino)ethanol *see* 3674

1-(3,4-Dihydroxyphenyl)-2-(methylamino)ethanone *see* 160

α-(3,4-Dihydroxyphenyl)-β-[2-(3,4-methylenedioxyphenyl)isopropylamino]ethanol *see* 8004

(2*Z*)-2-[(3,4-Dihydroxyphenyl)methylene]-6-hydroxy-3(2*H*)-benzofuranone *see* 9102

1-(3,4-Dihydroxyphenyl)-2-(α-methyl-3,4-methylenedioxyphenethylamino)ethanol *see* 8004

1-[(3,4-Dihydroxyphenyl)methyl]-1,2,3,4-tetrahydro-6,7-isoquinolinediol *see* 9361

(α*R*)-α-[[(2*E*)-3-(3,4-Dihydroxyphenyl)-1-oxo-2-propen-1-yl]oxy]-3,4-dihydroxybenzenepropanoic Acid *see* 8399

(1*S*,3*R*,4*R*,5*R*)-3-[[3-(3,4-Dihydroxyphenyl)-1-oxo-2-propen-1-yl]oxy]-1,4,5-trihydroxycyclohexanecarboxylic Acid *see* 2143

(2,4-Dihydroxyphenyl)phenylmethanone *see* 1109

erythro-α-(3,4-Dihydroxyphenyl)-2-piperidinemethanol *see* 8353

erythro-3,4-Dihydroxyphenyl-2-piperidinylcarbinol *see* 8353

3-(3,4-Dihydroxyphenyl)-2-propenoic Acid *see* 1638

2,3-Dihydroxy-4-phenylquinoline *see* 10201

L-*threo*-3-(3,4-Dihydroxyphenyl)serine *see* 3504

2-(3,4-Dihydroxyphenyl)-3,5,6,7-tetrahydroxy-4*H*-1-benzopyran-4-one *see* 8149

2-(2,4-Dihydroxyphenyl)-3,5,7-trihydroxy-4*H*-1-benzopyran-4-one *see* 6355

2-(3,4-Dihydroxyphenyl)-3,5,7-trihydroxy-4*H*-1-benzopyran-4-one *see* 8150

2-(3,4-Dihydroxyphenyl)-3,5,7-trihydroxy-1-benzopyrylium Chloride (1:1) *see* 2677

(3,4-Dihydroxyphenyl)(2,4,6-trihydroxyphenyl)methanone *see* 5706

1,1-Dihydroxyphosphinecarboxylic Acid 1-Oxide Sodium Salt (1:3) *see* 4278

Dihydroxyphosphinecarboxylic Acid Oxide Trisodium Salt *see* 4278

(2*R*,3*R*)-2,3-Dihydroxy-4-(phosphonooxy)butanal *see* 3748

7,8-Dihydroxy-5,6-phthalylquinoline *see* 245

(2α,11α)-2,11-Dihydroxypicrasane-1,-12,16-trione *see* 372

Dihydroxypivalic Acid *see* 3277

17,21-Dihydroxypregna-1,4-diene-3,-11,20-trione *see* 7842

3α,20α-Dihydroxy-5α-pregnane *see* 259

3α,20β-Dihydroxy-5α-pregnane *see* 260

3β,20α-Dihydroxy-5α-pregnane *see* 261

3β,20β-Dihydroxy-5α-pregnane *see* 262

(3β,5α)-3,17-Dihydroxypregnan-20-one *see* 263

3β,17-Dihydroxy-5α-pregnan-20-one *see* 263

(11β)-11,21-Dihydroxypregn-4-ene-3,20-dione *see* 2524

(16α)-16,17-Dihydroxypregn-4-ene-3,20-dione *see* 232

17,21-Dihydroxypregn-4-ene-3,20-dione *see* 2933

16α,17-Dihydroxypregn-4-ene-3,20-dione Cyclic Acetal with Acetophenone *see* 233

17,21-Dihydroxypregn-4-ene-3,11,20-trione *see* 2525

16α,17-Dihydroxyprogesterone *see* 232

16α,17α-Dihydroxyprogesterone Acetophenide *see* 233

2,3-Dihydroxypropanal *see* 4517

1,2-Dihydroxypropane *see* 7968

1,3-Dihydroxypropane *see* 9885

(1*S*,2*S*)-1,2-Dihydroxy-1,2,3-propanetricarboxylic Acid *see* 4861

(2*R*)-2,3-Dihydroxypropanoic Acid *see* 4519

(2*S*)-2,3-Dihydroxypropanoic Acid *see* 4519

2,3-Dihydroxypropanoic Acid *see* 4519

(*S*)-(−)-2,3-Dihydroxypropanol *see* 4517

1,3-Dihydroxy-2-propanone *see* 3200

α,β-Dihydroxypropionaldehyde *see* 4517

α,β-Dihydroxypropionic Acid *see* 4519

2-[[[(2*R*)-2,3-Dihydroxypropoxy]hydroxyphosphinyl]oxy]-*N*,*N*,*N*-trimethylethanaminium Inner Salt *see* 2212

9-[(1,3-Dihydroxy-2-propoxy)methyl]guanine *see* 4393

(17*R*,21α)-17,21-Dihydroxy-4-propylajmalanium *see* 7820

7-(2,3-Dihydroxypropyl)-3,7-dihydro-1,3-dimethyl-1*H*-purine-2,6-dione *see* 3526

N-(2,3-Dihydroxypropyl)-5-[*N*-(2,3-dihydroxypropyl)acetamido]-*N*'-(2-hydroxyethyl)-2,4,6-triiodoisophthalamide *see* 5111

4-[*o*-(2',3'-Dihydroxypropyloxycarbonyl)phenyl]amino-8-trifluoromethylquinoline *see* 4135

1-(2,3-Dihydroxypropyl)-4-phenylpiperazine *see* 3500

(1,2-Dihydroxy-3-propyl)theophylline *see* 3526

7-(2,3-Dihydroxypropyl)theophylline *see* 3526

N-(1,3-Dihydroxy-2-propyl)valiolamine *see* 10226

2,6-Dihydroxy-4-pyridinecarboxylic Acid *see* 2324

β,3-Dihydroxy-2-pyrrolidinepropanoic Acid *see* 2940

4,8-Dihydroxyquinaldic Acid *see* 10266

4,8-Dihydroxy-2-quinolinecarboxylic Acid *see* 10266

1β,5α-Dihydroxy-6β,7α*H*-selina-4(15),11(13)-dien-6,12-olide *see* 9182

12,18-Dihydroxysenecionan-11,16-dione *see* 8290

12,18-Dihydroxysenecionan-11,16-dione-4-oxide *see* 8290

Δ⁵-3β,12α-Dihydroxysolanidene *see* 8420

3',6'-Dihydroxyspiro[isobenzofuran-1(3*H*),9'-[9*H*]xanthen]-3-one *see* 4192

(2α,3β,5α,15β,25*R*)-2,15-Dihydroxyspirostan-3-yl *O*-β-D-Glycopyranosyl-(1→3)-*O*-β-D-galactopyranosyl-(1→2)-*O*-[β-D-xylopyranosyl-(1→3)]-*O*-β-D-glucopyranosyl-(1→4)-β-D-galactopyranoside *see* 3180

9,10-Dihydroxystearic Acid, 3204

trans-3,5-Dihydroxystilbene *see* 7563

d-α,β-Dihydroxysuccinic Acid *see* 9205

D-*threo*-2,3-Dihydroxysuccinic Acid *see* 9203

(11β)-11,17-Dihydroxy-21-[(3-sulfobenzoyl)oxy]pregna-1,4-diene-3,20-dione Monosodium Salt *see* 7841

4,5-Dihydroxy-3-[(4-sulfophenyl)azo]-2,7-naphthalenedisulfonic Acid Trisodium Salt *see* 8859

4,5-Dihydroxy-3-[2-(4-sulfophenyl)diazenyl]-2,7-naphthalenedisulfonic Acid Sodium Salt (1:3) *see* 8859

Dihydroxytartaric Acid, 3205

3,3'-Dihydroxy-2,3,2',3'-tetradehydro-β,β-carotene-4,4'-dione *see* 843

3',6'-Dihydroxy-2',4',5',7'-tetraiodospiro[isobenzofuran-1-(3*H*),9'-[9*H*]xanthen]-3-one Sodium Salt (1:2) *see* 3749

(1*S*,3*S*,7*S*,10*R*,11*S*,12*S*,16*R*)-7,11-Dihydroxy-8,8,10,12-tetramethyl-3-[(1*E*)-1-methyl-2-(2-methyl-4-thiazolyl)ethenyl]-4,17-dioxabicyclo[14.1.0]heptadecane-5,9-dione *see* 3684

11β,17α-Dihydroxy-21-thio-3,20-dioxo-4-pregnene *see* 9641
6,7-Dihydroxy-3-tiglyloxytropane *see* 6007
3,5-Dihydroxytoluene *see* 6959
2,5-Dihydroxy-α-toluic Acid *see* 4775
4,6-Dihydroxy-*o*-toluic Acid *see* 6977
3,4-Dihydroxy-5-[(3,4,5-trihydroxybenzoyl)oxy]benzoic Acid *see* 3172
3′,5-Dihydroxy-4′,6,7-trimethoxyflavone *see* 3943
(2*E*,4*E*,6*E*,8*E*,10*E*,12*E*,14*E*)-15-[(1*R*,2*R*)-1,2-Dihydroxy-2,6,6-trimethylcyclohexyl]-4,9,13-trimethyl-2,4,6,8,10,-12,14-pentadecaheptaenoic Acid *see* 888
(1*R*,3a*S*,3b*S*,6a*S*,8*S*,9*R*,10a*R*,10b*S*,12a*S*)-1-[(1*S*,2*R*,3*R*,4*S*)-2,3,Dihydroxy-1,4,5-trimethylhexyl]hexadecahydro-8,9-dihydroxy-10a,12a-dimethyl-6*H*-benz[*c*]-indeno[5,4-*e*]oxepin-6-one *see* 1370
2,4-Dihydroxy-1,3,5-trinitrobenzene *see* 8989
(3β,7β,25*R*)-3,7-Dihydroxy-11,15,23-trioxolanost-8-en-26-oic Acid *see* 4396
(7β,15α,25*R*)-7,15-Dihydroxy-3,11,23-trioxolanost-8-en-26-oic Acid *see* 4396
4′,4″-Dihydroxytriphenylmethane-2-carboxylic Acid *see* 7357
3,5-Dihydroxy-2,6,6-tris(3-methyl-2-buten-1-yl)-4-(3-methyl-1-oxobutyl)-2,4-cyclohexadien-1-one *see* 5672
6,7-Dihydroxytropinetiglic Acid Ester *see* 6007
(β*R*)-β,3-Dihydroxy-L-tyrosine *see* 3504
(α*E*)-3,5-Dihydroxy-L-tyrosyl-α,β-didehydro-3,5-dihydroxy-*N*-[(1*Z*)-2-(3,4,5-trihydroxyphenyl)ethenyl]tyrosinamide *see* 10000
2,5-Dihydroxy-3-undecyl-*p*-benzoquinone *see* 3613
2,5-Dihydroxy-3-undecyl-2,5-cyclohexadiene-1,4-dione *see* 3613
6-(6,10-Dihydroxyundecyl)-β-resorcylic Acid μ-Lactone *see* 10319
1α,24*R*-Dihydroxyvitamin D$_3$ *see* 9152
1α,25-Dihydroxyvitamin D$_3$ *see* 1647
2-(3,6-Dihydroxy-9*H*-xanthen-9-yl)benzoic Acid *see* 4193
Dihydroyohimbine *see* 10302
Dihyprylone *see* 7581
Dihyzin [Henning] *see* 3187
3,3′-Di-2-imidazolin-2-ylcarbanilide *see* 4956
Diimidazol-1-yl Ketone *see* 1820
Di-1*H*-imidazol-1-ylmethanone *see* 1820
Diimidotricarbonic Diamide *see* 9945
Diindium Trioxide *see* 4992
Diindium Trisulfate *see* 4995
Diindium Tritelluride *see* 4996
Diiodine Pentoxide *see* 5062
2,4-Diiodoaniline, 3206
2,4-Diiodobenzenamine *see* 3206
4′,5′-Diiodofluorescein, 3207
Diiodoform *see* 9365
Di(iodohexamethyl)diaminoisopropanol *see* 7897
8,9-Diiodo-11-hydroxy-8-heptadecene-1-carboxylic Acid *see* 7287
2,7-Diiodo-4-hydroxymercuriresorcinsulfonphthalein Monosodium Salt *see* 5932
3,5-Diiodo-4-hydroxy-β-phenylalanine *see* 3210
Diiodohydroxyquin *see* 5085

5,7-Diiodo-8-hydroxyquinoline *see* 5085
3,5-Diiodo-4-(3-iodo-4-hydroxyphenoxy)-phenylacetic Acid *see* 9618
Diiodomercury *see* 5946
Diiodomethane *see* 6137
2,6-Diiodo-4-nitrophenol *see* 3401
3,5-Diiodo-4-oxo-1(4*H*)-pyridineacetic Acid Compound with 2,2′-Iminobis-[ethanol] (1:1) *see* 5083
Diiodo-oxyquinoline *see* 5085
3,5-Diiodo-α-phenylphloretic Acid *see* 5069
3,5-Diiodo-4-pyridone-*N*-acetic Acid Diethanolamine Salt *see* 5083
5,7-Diiodo-8-quinolinol *see* 5085
Diiodoricinstearolic Acid *see* 7287
3,5-Diiodosalicylic Acid, 3208
Diiodosamarium *see* 8486
3,5-Diiodothyronine, 3209
3,5-Diiodotyrosine, 3210
Diiodyl *see* 7287
Diiridium Trioxide *see* 5131
Diiron Phosphide *see* 4081
Diiron Trisulfate *see* 4059
Diisoamylamine, 3211
Di-*sec*-isoamylborine *see* 1253
Diisoamyl Ether *see* 5163
Diisoamyl Phthalate *see* 5169
Diisobutenyl Ketone *see* 7445
Diisobutylaluminum Hydride, 3212
Diisobutyl Ether *see* 5187
Diisobutylphenoxyethoxyethyl Dimethyl Benzyl Ammonium Chloride *see* 1076
Diisobutyl Sodium Sulfosuccinate, 3213
Diisobutyl Sulfide *see* 5197
Diisobutylthiocarbamic Acid *S*-Ethyl Ester *see* 1548
3,4-Di-*o*-isobutyryl Epinine *see* 4915
Diisocarb *see* 1548
2,4-Diisocyanato-1-methylbenzene *see* 9689
2,4-Diisocyanatotoluene *see* 9689
4,4′-[(1*Z*,3*Z*)-2,3-Diisocyano-1,3-butadiene-1,4-diyl]bis-1,2-benzenediol *see* 10259
4,4′-[(1*Z*,3*Z*)-2,3-Diisocyano-1,3-butadiene-1,4-diyl]bisphenol *see* 10259
Diisononyl Phthalate *see* 3318
Diisopentyl Ether *see* 5163
Diisopropanolamine, 3214
Diisopropenyl *see* 3260
2-(2′,6′-Diisopropoxyphenyl)phenyldicyclohexylphosphine *see* 1469
(2,6-Diisopropylacetanilido)iminodiacetic Acid *see* 3400
Diisopropylamine, 3215
Diisopropylamine Dichloroacetate, 3216
Diisopropylamine Dichloroethanoate *see* 3216
(+)-(*R*)-2-[α-[2-(Diisopropylamino)ethyl]-benzyl]-*p*-cresol *see* 9684
N-(*N*′,*N*′-Diisopropylaminoethyl)-[2-(2-hydroxy-4,5-dimethoxybenzoylamino)-1,3-thiazole-4-yl]carboxyamide *see* 113
3-[2-(Diisopropylamino)ethyl]-5-methoxyindole *see* 6062
2-[2-(Diisopropylamino)ethyl]-4-methyl-2-(2-pyridyl)pentanamide *see* 7242
N-[2-(Diisopropylamino)ethyl]-2-oxo-1-pyrrolidineacetamide *see* 7826
α-[2-(Diisopropylamino)ethyl]-α-phenyl-2-pyridineacetamide *see* 3402
β-Diisopropylaminoethyl 9-Xanthenecarboxylate Methobromide *see* 7919

(*R*)-(+)-2-(3-Diisopropylamino-1-phenylpropyl)-4-hydroxymethylphenylisobutyrate Ester *see* 4095
4-(Diisopropylamino)-2-phenyl-2-(2-pyridyl)butyramide *see* 3402
Diisopropylammonium Dichloroacetate *see* 3216
Diisopropyl Azodicarboxylate, 3217
Diisopropyl Azodiformate *see* 3217
N,*N*′-Diisopropylcarbodiimide, 3218
1,3-Diisopropylcarbodiimide *see* 3218
Diisopropylchlorosilane *see* 2136
5′,5″-Diisopropyl-2′,2″-dimethylphenolphthalein *see* 9556
Diisopropyl 1,3-Dithiol-2-ylidenemalonate *see* 5776
Diisopropyl Ether *see* 5258
Diisopropylethylamine, 3219
N,*N*-Diisopropylethylamine *see* 3219
Diisopropyl Fluorophosphate *see* 5222
Diisopropyl Fluorophosphonate *see* 5222
(+)-*N*,*N*-Diisopropyl-3-(2-hydroxy-5-methylphenyl)-3-phenylpropylamine *see* 9684
Diisopropylidene Acetone *see* 7445
2,3:4,5-Di-*O*-isopropylidene-β-D-fructopyranose Sulfamate *see* 9706
1,2:5,6-Diisopropylidene-D-glucofuranose *see* 2968
1,2:5,6-Diisopropylidene-D-glucose *see* 2968
α-2,3:4,6-Di-*O*-isopropylidene-L-*xylo*-hexulofuranosonic Acid *see* 3220
Di-*O*-isopropylidene-2-keto-L-gulonic Acid *see* 3220
2,6-Diisopropylphenol *see* 7947
[[[(2,6-Diisopropylphenyl)carbamoyl]-methyl]imino]diacetic Acid *see* 3400
2,6-Diisopropylphenyl[(2,4,6-triisopropylphenyl)acetyl]sulfamate *see* 876
N,*N*′-Diisopropylphosphorodiamidic Fluoride *see* 6287
Diisopropylphosphorofluoridate *see* 5222
Diisopropylthiocarbamic Acid *S*-2,3-Dichloroallyl Ester *see* 2973
Diisopropylthiocarbamic Acid *S*-(2,3,3-Trichloroallyl) Ester *see* 9756
1,4-Diisothiocyanatobenzene *see* 1309
4,4′-Diisothiocyanato-2,2′-stilbenedisulfonic Acid *see* 3113
Dikalium Phosphate *see* 7779
Dikegulac, 3220
Dikegulac Sodium *see* 3220
Diketene, 3221
2,3-Diketobutane *see* 2969
Diketocyclobutenediol *see* 8897
2,3-Diketoindoline *see* 5148
3,20-Diketo-11β,18-oxido-4-pregnene-18,21-diol *see* 218
2,5-Diketopiperazine *see* 7578
2,5-Diketopyrrolidine *see* 9001
Diketosuccinic Acid *see* 3205
2,5-Diketotetrahydrofuran *see* 9000
3,5-Diketotriazolidine *see* 10051
Dikol *see* 5069
Dilabron [Sterling Winthrop] *see* 5215
Dilacor XR [Watson] *see* 3224
Diladel [Polifarma] *see* 3224
Dilafurane *see* 1085
Dilanacin [AWD] *see* 3186
Dilangil *see* 5811
Dilangio [Fournier] *see* 1035
Dilantin [Pfizer] *see* 7433
Dilar [Cassenne] *see* 7133
Dilasenil [Celtia] *see* 6581

Name Index

(E)-1-[4'-(2-Dimethylaminoethoxy)phenyl]-1-(3-hydroxyphenyl)-2-phenylbut-1-ene *see* 3495

N-[[4-[2-(Dimethylamino)ethoxy]phenyl]methyl]-3,4-dimethoxy Benzamide *see* 5291

2-Dimethylaminoethoxyphenylmethyl-2-picoline *see* 3489

N-[[4-[2-(Dimethylamino)ethoxy]phenyl]methyl]-3,4,5-trimethoxybenzamide *see* 9877

3-[(1E)-1-[4-[2-(Dimethylamino)ethoxy]phenyl]-2-phenyl-1-buten-1-yl]phenol *see* 3495

(−)-[1R,2S,4R]-2-(2-Dimethylaminoethoxy)-2-phenyl-1,7,7-trimethylbicyclo-[2.2.1]heptane *see* 2916

2-[2-(Dimethylamino)ethoxy]-N-tricyclo-[3.3.1.1³,⁷]dec-1-ylacetamide *see* 9950

4-(2-Dimethylaminoethoxy)-N-(3,4,5-trimethoxybenzoyl)benzylamine *see* 9877

β-Dimethylaminoethyl Alcohol *see* 2845

(αR)-α-[(1S)-1-(Dimethylamino)ethyl]-benzenemethanol *see* 6140

β-Dimethylaminoethyl Benzhydryl Ether *see* 3339

β-Dimethylaminoethyl Benzhydryl Ether 8-Chlorotheophyllinate *see* 3230

erythro-α-[1-(Dimethylamino)ethyl]benzyl Alcohol *see* 6140

2-[α-(2-Dimethylaminoethyl)benzyl]pyridine *see* 7349

β-Dimethylaminoethyl *p*-Bromobenzhydryl Ether *see* 1428

2-Dimethylaminoethyl 4-*n*-Butylaminobenzoate Hydrochloride *see* 9333

2-Dimethylaminoethyl *p*-Butylaminosalicylate *see* 4885

α-(Dimethylaminoethyl)-*o*-chlorobenzhydrol *see* 2066

N-Dimethylaminoethyl-N-*p*-chlorobenzyl-α-aminopyridine *see* 2164

β-Dimethylaminoethyl (*p*-Chloro-α-methylbenzhydryl) Ether *see* 2187

Dimethylaminoethyl *p*-Chlorophenoxyacetate *see* 5851

S-β-Dimethylaminoethyl-O,O-diethylthionophosphate Methiodide *see* 3548

(+)-*cis*-5-[2-(Dimethylamino)ethyl]-2,3-dihydro-3-hydroxy-2-(*p*-methoxyphenyl)-1,5-benzothiazepin-4(5H)-one Acetate (Ester) *see* 3224

10-[2-(Dimethylamino)ethyl]-5,10-dihydro-5-methyl-11H-dibenzo[b,e][1,4]diazepin-11-one *see* 3012

5-[2-(Dimethylamino)ethyl]-2,3-dihydro-2-phenyl-1,5-benzothiazepin-4(5H)-one *see* 9457

β-Dimethylaminoethyl (1-Hydroxycyclopentyl)phenylacetate *see* 2741

2-(Dimethylamino)ethyl [5-Hydroxy-4-(hydroxymethyl)-6-methyl-3-pyridyl]succinic Acid Methyl Ester *see* 8104

3-(β-Dimethylaminoethyl)-5-hydroxyindole *see* 1483

2-Dimethylaminoethyl 1-Hydroxy-α-phenylcyclopentaneacetate *see* 2741

2-[1-[2-[2-(Dimethylamino)ethyl]inden-3-yl]ethyl]pyridine *see* 3238

3-[2-(Dimethylamino)ethyl]indole *see* 3288

3-[2-(Dimethylamino)ethyl]-1H-indol-4-ol *see* 8034

3-[2-(Dimethylamino)ethyl]-1H-indol-5-ol *see* 1483

3-(2-Dimethylaminoethyl)-5-indolol *see* 1483

3-[2-(Dimethylamino)ethyl]-1H-indol-4-ol 4-(Dihydrogen Phosphate) *see* 8035

1-[[3-(2-Dimethylaminoethyl)-5-indolyl]methanesulfonyl]pyrrolidine *see* 298

(4S)-4-[[3-[2-(Dimethylamino)ethyl]-1H-indol-5-yl]methyl]-2-oxazolidinone *see* 10390

1-[[[3-[2-(Dimethylamino)ethyl]-1H-indol-5-yl]methyl]sulfonyl]pyrrolidine *see* 298

6-[2-[6-[2-(Dimethylamino)ethyl]-4-methoxy-1,3-benzodioxol-5-yl]acetyl]-2,3-dimethoxybenzoic Acid *see* 6504

2-[(2-Dimethylaminoethyl)(*p*-methoxybenzyl)amino]pyridine *see* 8097

2-[(2-Dimethylaminoethyl)(*p*-methoxybenzyl)amino]pyrimidine Hydrochloride *see* 9527

6-[[6-[2-(Dimethylamino)ethyl]-2-methoxy-3,4-(methylenedioxy)phenyl]acetyl]-*o*-veratric Acid *see* 6504

β-Dimethylaminoethyl 2-Methylbenzhydryl Ether *see* 6975

β-Dimethylaminoethyl 4-Methylbenzhydryl Ether *see* 6125

3-[2-(Dimethylamino)ethyl]-N-methyl-1H-indole-5-methanesulfonamide *see* 9126

3-[(1R,2R)-3-(Dimethylamino)-1-ethyl-2-methylpropyl]phenol *see* 9189

4-[2-(Dimethylamino)ethyl]phenol *see* 4785

β-Dimethylaminoethyl-2-pyridylaminotoluene *see* 9910

β-Dimethylaminoethyl-2-pyridylbenzylamine *see* 9910

N-(2-Dimethylaminoethyl)-N-2-pyridyl-3-thenylamine *see* 9431

2-Dimethylaminoethyl Succinate Dimethochloride *see* 9006

2-[(2-Dimethylaminoethyl)-2-thenylamino]pyridine *see* 6031

2-[(2-Dimethylaminoethyl)-3-thenylamino]pyridine *see* 9431

3-[2-(Dimethylamino)ethyl]-5-(1H-1,2,4-triazol-1-ylmethyl)indole *see* 8370

(αR,βS)-β-(Dimethylamino)-α-hydroxy Benzenepropanoic Acid (1S,2R,3E,5S,-7S,8S,10R,13S)-2,13-Bis(acetyloxy)-7,10-dihydroxy-8,12,15,15-tetramethyl-9-oxotricyclo[9.3.1.1⁴,⁸]hexadeca-3,11-dien-5-yl Ester *see* 9215

4-[2-(Dimethylamino)-1-(1-hydroxycyclohexyl)ethyl]phenol *see* 2937

4-[2-(Dimethylamino)-1-hydroxyethyl]-1,2-benzenediol *see* 6141

1-[2-(Dimethylamino)-1-(4-hydroxyphenyl)ethyl]cyclohexanol *see* 2937

7-(Dimethylamino)-3-imino-3H-phenothiazine Hydrochloride *see* 920

4-(Dimethylamino)-2-isopropyl-2-phenylvaleronitrile *see* 5155

10-Dimethylaminoisopropyl-2-propionylphenothiazine *see* 7936

N-Dimethylaminoisopropylthiophenylpyridylamine *see* 7936

6-Dimethylamino-9-[3'-(*p*-methoxy-L-phenylalanylamino)-3'-deoxy-β-D-ribofuranosyl]purine *see* 8054

1-[2-(Dimethylamino)-1-(4-methoxyphenyl)ethyl]cyclohexanol *see* 10140

2-(Dimethylamino)-N-[[(methylamino)-carbonyl]oxy]-2-oxoethanimidothioic Acid Methyl Ester *see* 7019

3-(Dimethylamino)-7-(methylamino)phenothiazin-5-ium Chloride (1:1) *see* 921

1-(Dimethylamino)-2-methyl-2-butanol 2-Benzoate *see* 607

2-Dimethylamino-4-methyl-5-*n*-butyl-6-hydroxypyrimidine *see* 3239

α-[(Dimethylamino)methyl]-3,4-dihydroxybenzyl Alcohol *see* 6141

Dimethylaminomethyl-(3,4-dihydroxyphenyl) Carbinol *see* 6141

α-[(Dimethylamino)methyl]-3,4-dimethoxybenzenemethanol *see* 5709

Dimethylaminomethyl 3,4-Dimethoxyphenyl Carbinol *see* 5709

4-(Dimethylaminomethyl)-1,3-dioxacyclopentane Methiodide *see* 7022

(2S,3R)-(+)-4-Dimethylamino-3-methyl-1,2-diphenyl-2-butanol *see* 2062

α-d-4-Dimethylamino-3-methyl-1,2-diphenyl-2-butanol Propionate *see* 7952

α-l-4-Dimethylamino-3-methyl-1,2-diphenyl-2-butanol Propionate *see* 5523

1-Dimethylamino-2-methyl-3,3-diphenyl-4-hexanone *see* 5230

6-(Dimethylamino)-5-methyl-4,4-diphenyl-3-hexanone *see* 5230

3-[[(Dimethylamino)methyl]amino]-2,4,6-triiodobenzenepropanoic Acid *see* 5115

3-[(Dimethylaminomethylene)amino]-2,-4,6-triiodohydrocinnamic Acid *see* 5115

β-(3-Dimethylaminomethyleneamino-2,-4,6-triiodophenyl)propionic Acid *see* 5115

(4S)-10-[(Dimethylamino)methyl]-4-ethyl-4,9-dihydroxy-1H-pyrano[3',4':6,7]-indolizino[1,2-b]quinoline-3,14(4H,-12H)-dione *see* 9707

N-(2'-Dimethylamino-2'-methyl)ethylphenothiazine *see* 7901

(αS)-α-[(1R)-2-(Dimethylamino)-1-methylethyl]-α-phenylbenzeneethanol *see* 2062

(αR)-α-[(1S)-2-(Dimethylamino)-1-methylethyl]-α-phenylbenzeneethanol 1-Propanoate *see* 5523

(αS)-α-[(1R)-2-(Dimethylamino)-1-methylethyl]-α-phenylbenzeneethanol 1-Propanoate *see* 7952

10-(2-Dimethylamino-2-methylethyl)-10H-pyrido[3,2-b][1,4]benzothiazine *see* 5273

N'-[2-[[[-5-[(Dimethylamino)methyl]-2-furanyl]methyl]thio]ethyl]-N-methyl-2-nitro-1,1-ethenediamine *see* 8228

9-[(Dimethylamino)methyl]-10-hydroxy-(20S)-camptothecin *see* 9707

7-(Dimethylamino)-3-(methylimino)-3H-phenothiazine Hydrochloride *see* 921

3-(Dimethylaminomethyl)indole *see* 4569

(±)-1-[α-[(Dimethylamino)methyl]-*p*-methoxybenzyl]cyclohexanol *see* 10140

8-[(Dimethylamino)methyl]-7-methoxy-3-methylflavone *see* 3227

8-[(Dimethylamino)methyl]-7-methoxy-3-methyl-2-phenyl-4H-1-benzopyran-4-one *see* 3227

8-Dimethylaminomethyl-7-methoxy-3-methyl-2-phenylchromone *see* 3227

rel-(1R,2R)-2-[(Dimethylamino)methyl]-1-(3-methoxyphenyl)cyclohexanol *see* 9726

1-(Dimethylaminomethyl)-1-methylpropyl Benzoate *see* 607

4-(Dimethylamino)-3-methylphenol 1-(*N*-Methylcarbamate) *see* 428

4-[4-(Dimethylamino)-*N*-methyl-L-phenylalanine]-5-[(2*S*,5*R*)-5-[[[(3*S*)-1-azabicyclo[2.2.2]oct-3-yl]thio]methyl]-4-oxo-2-piperidinecarboxylic Acid]virginiamycin S1 *see* 8201

4-[4-(Dimethylamino)-*N*-methyl-L-phenylalanine]virginiamycin S₁ *see* 7871

P-[4-(Dimethylamino)-2-methylphenyl]-phosphinic Acid Sodium Salt (1:1) *see* 9672

10-(3-Dimethylamino-2-methylpropyl)-3-cyanophenothiazine *see* 2673

5-[3-(Dimethylamino)-2-methylpropyl]-10,11-dihydro-5*H*-dibenz[*b*,*f*]azepine *see* 9894

5-(3-Dimethylamino-2-methylpropyl)-10,11-dihydro-5*H*-dibenzo[*a*,*d*]cycloheptene *see* 1535

5-(3-Dimethylamino-2-methylpropyl)iminodibenzyl *see* 9894

3-Dimethylamino-7-methyl-1,2-(*n*-propylmalonyl)-1,2-dihydro-1,2,4-benzotriazine *see* 713

(−)-10-(3-Dimethylamino-2-methylpropyl)-2-methoxyphenothiazine *see* 5518

10-[3-(Dimethylamino)-2-methylpropyl]-phenothiazine *see* 9873

10-[3-(Dimethylamino)-2-methylpropyl]-10*H*-phenothiazine-2-carbonitrile *see* 2673

10-(3-Dimethylamino-2-methylpropyl)-phenothiazine 5,5-Dioxide *see* 7045

5-(Dimethylamino)-9-methyl-8-propyl-1*H*-pyrazolo[1,2-*a*][1,2,4]benzotriazine-1,3(2*H*)-dione *see* 713

10-[3-(Dimethylamino)-2-methylpropyl]-2-(trifluoromethyl)phenothiazine *see* 9845

α-(Dimethylaminomethyl)protocatechuyl Alcohol *see* 6141

(9*S*)-9-[(Dimethylamino)methyl]-6,7,-10,11-tetrahydro-9*H*,18*H*-5,21:12,17-dimethenodibenzo[*e*,*k*]pyrrolo[3,4*h*][1,-4,13]oxadiazacyclohexadecine-18,-2019*H*)-dione *see* 8423

N'-[2-[[[2-[(Dimethylamino)methyl]-4-thiazolyl]methyl]thio]ethyl]-*N*-methyl-2-nitro-1,1-ethenediamine *see* 6747

α-[(Dimethylamino)methyl]veratryl Alcohol *see* 5709

1-Dimethylaminonaphthalene *see* 3276

5-(Dimethylamino)-1-naphthalenesulfonyl Chloride *see* 2813

(4*S*,4a*R*,5*S*,5a*R*,6*S*,12a*S*)-4-(Dimethylamino)-1,4,4a,5,5a,6,11,12a-octahydro-3,5,6,10,12,12a-hexahydroxy-6-methyl-1,11-dioxo-2-naphthacenecarboxamide *see* 7075

(4*S*,4a*S*,5a*S*,6*S*,12a*S*)-4-(Dimethylamino)-1,4,4a,5,5a,6,11,12a-octahydro-3,-6,10,12,12a-pentahydroxy-*N*-[4-(2-hydroxyethyl)-1-piperazinyl]methyl]-6-methyl-1,11-dioxo-2-naphthacenecarboxamide *see* 7566

(4*R*,4a*S*,5a*S*,6*S*,12a*S*)-4-(Dimethylamino)-1,4,4a,5,5a,6,11,12a-octahydro-3,-6,10,12,12a-pentahydroxy-6-methyl-1,11-dioxo-2-naphthacenecarboxamide *see* 8144

(4*S*,4a*S*,5a*S*,6*S*,12a*S*)-4-(Dimethylamino)-1,4,4a,5,5a,6,11,12a-octahydro-3,-6,10,12,12a-pentahydroxy-6-methyl-1,11-dioxo-2-naphthacenecarboxamide *see* 9341

(4*S*,4a*R*,5*S*,5a*R*,6*R*,12a*S*)-4-(Dimethylamino)-1,4,4a,5,5a,6,11,12a-octahydro-3,5,10,12,12a-pentahydroxy-6-methyl-1,11-dioxo-2-naphthacenecarboxamide Hydrate (1:1) *see* 3488

*N*⁶-[[[[(4*S*,4a*S*,5a*S*,6*S*,12a*S*)-4-(Dimethylamino)-1,4,4a,5,5a,6,11,12a-octahydro-3,6,10,12,12a-pentahydroxy-6-methyl-1,11-dioxo-2-naphthacenyl]carbonyl]-amino]methyl]-L-lysine *see* 5689

(4*S*,4a*S*,5a*S*,6*S*,12a*S*)-4-(Dimethylamino)-1,4,4a,5,5a,6,11,12a-octahydro-3,-6,10,12,12a-pentahydroxy-6-methyl-1,11-dioxo-*N*-(1-pyrrolidinylmethyl)-2-naphthacenecarboxamide *see* 8381

(4*S*,4a*R*,5*S*,5a*R*,12a*S*)-4-(Dimethylamino)-1,4,4a,5,5a,6,11,12a-octahydro-3,-5,10,12,12a-pentahydroxy-6-methyl-ene-1,11-dioxo-2-naphthacenecarboxamide *see* 6015

Dimethylaminophenazone *see* 470

4-[[(4-Dimethylamino)phenyl]azo]benzenesulfonic Acid Sodium Salt *see* 6178

2-[[4-(Dimethylamino)phenyl]azo]benzoic Acid *see* 6192

3-*trans*-Dimethylamino-4-phenyl-4-*trans*-carbethoxy-Δ¹-cyclohexene *see* 9594

rel-(1*R*,2*S*)-2-(Dimethylamino)-1-phenyl-3-cyclohexene-1-carboxylic Acid Ethyl Ester *see* 9594

4-[2-[4-(Dimethylamino)phenyl]diazenyl]benzenesulfonic Acid Sodium Salt (1:1) *see* 6178

2-[2-[4-(Dimethylamino)phenyl]diazenyl]benzoic Acid *see* 6192

4-[[4-(Dimethylamino)phenyl][4-(dimethylimino)-2,5-cyclohexadien-1-ylidene]-methyl]-*N*-ethyl-*N*,*N*-dimethylbenzenaminium Bromide Chloride (1:1:1) *see* 6154

Dimethylaminophenyldimethylpyrazolone *see* 470

2-[2-[4-(Dimethylamino)phenyl]ethenyl]-1-ethylquinolinium Iodide (1:1) *see* 8165

2-[2-[2-[4-(Dimethylamino)phenyl]ethenyl]-6-methyl-4*H*-pyran-4-ylidene]-propanedinitrile *see* 2841

α-(*p*-Dimethylaminophenylethylene)-quinoline Ethiodide *see* 8165

7-[4-(Dimethylamino)phenyl]-*N*-(β-D-glucopyranosyloxy)-4,6-dimethyl-7-oxo-2,4-heptadienamide *see* 9814

[*R*-(*E*,*E*)]-7-[4-(Dimethylamino)phenyl]-*N*-hydroxy-4,6-dimethyl-7-oxo-2,4-heptadienamide *see* 9814

(11β,13α,17α)-11-[4-(Dimethylamino)-phenyl]-17-hydroxy-17-(3-hydroxy-propyl)estra-4,9-dien-3-one *see* 6942

11β-(4-Dimethylaminophenyl)-17α-hydroxy-17β-(3-hydroxypropyl)-13α-methyl-4,9-gonadien-3-one *see* 6942

(11β)-11-[4-(Dimethylamino)phenyl]-17-hydroxy-19-norpregna-4,9-diene-3,20-dione *see* 10030

(11β,17β)-11-[4-(Dimethylamino)phenyl]-17-hydroxy-17-(1-propyn-1-yl)estra-4,9-dien-3-one *see* 6266

5-[[4-(Dimethylamino)phenyl]methylene]-2-thioxo-4-thiazolidinone *see* 3253

2-(Dimethylamino)-5-phenyl-2-oxazolin-4-one *see* 9533

2-(Dimethylamino)-5-phenyl-4(5*H*)-oxazolone *see* 9533

[4-(Dimethylamino)phenyl]phenylmethanone *see* 3255

(*E*)-[4-(Dimethylamino)phenyl]phenylmethanone Oxime *see* 3255

(*Z*)-[4-(Dimethylamino)phenyl]phenylmethanone Oxime *see* 3255

N-[4-[[4-(Dimethylamino)phenyl]phenylmethylene]-2,5-cyclohexadien-1-ylidene]-*N*-methylmethanaminium Chloride (1:1) *see* 5763

(1*R*,2*S*)-2-Dimethylamino-1-phenylpropanol *see* 6140

11β-[4-(*N*,*N*-Dimethylamino)phenyl]-17α-(prop-1-ynyl)-Δ⁴,⁹-estradiene-17β-ol-3-one *see* 6266

3-Dimethylamino-1-phenyl-1-*p*-tolylpropane *see* 9682

9-[[3-(Dimethylamino)propyl]amino]-1-nitroacridine *see* 6657

10-(γ-Dimethylaminopropyl)-1-azaphenothiazine *see* 7998

10-(2-Dimethylaminopropyl)-1-azaphenothiazine *see* 5273

N-[3-(Dimethylamino)propyl]carbamic Acid Propyl Ester *see* 7912

5-(γ-Dimethylaminopropyl)-3-chloroiminodibenzyl *see* 2378

N-(3-Dimethylaminopropyl)-3-chlorophenothiazine *see* 2191

10-(γ-Dimethylaminopropyl)-3-chlorphenothiazine 9-Oxide *see* 6950

2-[3-(Dimethylamino)propyl]-8,8-diethyl-2-aza-8-germaspiro[4.5]decane *see* 8885

5-(3-Dimethylaminopropyl)-10,11-dihydro-5*H*-dibenz[*b*,*f*]azepine *see* 4962

5-[3-(Dimethylamino)propyl]-10,11-dihydro-5*H*-dibenz[*b*,*f*]azepine *N*-Oxide *see* 4963

6-[2-(Dimethylamino)propyl]-6,11-dihydro-5*H*-pyrido[2,3-*b*][1,5]benzodiazepin-5-one *see* 7946

10-[2-(Dimethylamino)propyl]-*N*,*N*-dimethyl-10*H*-phenothiazine-2-sulfonamide *see* 4258

10-(2-Dimethylaminopropyl)-3-dimethylsulfamidophenothiazine *see* 4258

(8β)-*N*-[3-(Dimethylamino)propyl]-*N*-[(ethylamino)carbonyl]-6-(2-propen-1-yl)ergoline-8-carboxamide *see* 1606

N-(3-Dimethylaminopropyl)-*N'*-ethylcarbodiimide *see* 3555

β-[2-(Dimethylamino)propyl]-α-ethyl-β-phenylbenzeneethanol 1-Acetate *see* 6017

(α*S*)-β-[(2*S*)-2-(Dimethylamino)propyl]-α-ethyl-β-phenylbenzeneethanol 1-Acetate *see* 5519

1-[3-(Dimethylamino)propyl]-1-(4-fluorophenyl)-1,3-dihydro-5-isobenzofurancarbonitrile *see* 2317

1-[3-(Dimethylamino)propyl]-1-(4-fluorophenyl)-5-phthalancarbonitrile *see* 2317

5-[3-(Dimethylamino)propyl]-6,7,8,9,-10,11-hexahydro-5*H*-cyclooct[*b*]indole *see* 5119

1-(3-Dimethylaminopropyl)-2,3-hexamethyleneindole *see* 5119

6-(3-Dimethylaminopropylidene)benzo-[*b*]benzofurano[2,3-*e*]oxepine *see* 7033

5-(3-Dimethylaminopropylidene)dibenzo-[*a*,*d*][1,4]cycloheptadiene *see* 483

5-(3-Dimethylaminopropylidene)dibenzo-[*a*,*e*]cycloheptatriene *see* 2706

N,N'-Dimethyl-2,2-diphenoxyethylamine *see* 5861

N,N-Dimethyl-α,α-diphenyl Acetamide *see* 3337

N,N-Dimethyl-2,2-diphenylacetamide *see* 3337

N'^1,N'^3-Dimethyl-*N'^1,N'^3*-di(phenylcarbonothioyl)malonohydrazide *see* 3596

2,6-Dimethyldiphenylene Disulfide *see* 5988

2,9-Dimethyl-4,7-diphenyl-1,10-phenanthroline *see* 1005

1,2-Dimethyl-3,5-diphenyl-1*H*-pyrazolium *see* 3157

N,N'-Dimethyl-γ,γ'-dipyridylium *see* 7135

Dimethyldithiocarbamic Acid Iron Salt *see* 4040

Dimethyldithiocarbamic Acid Zinc Salt *see* 10372

3,3-Dimethyl-3-3'-dithiodialanine *see* 7198

N,N-Dimethyldodecylbetaine *see* 3453

6,7-Dimethyl-9-(*d*-1'-dulcityl)isoalloxazine *see* 4363

7,8-Dimethyl-10-(*d*-1'-dulcityl)isoalloxazine *see* 4363

trans,trans-2,2'-[Dimethylenebis(carbonyloxytrimethylene)]bis[1,2,3,4-tetrahydro-6,7,8-trimethoxy-2-methyl-1-(3,4,5-trimethoxybenzyl)isoquinolinium] Dichloride *see* 3480

N,N-Dimethyleneoxidebis(pyridinium-4-aldoxime) Dichloride *see* 6827

6β,7β,15β,16β-Dimethylene-3-oxo-4-androstene-[17(β-1')-spiro-5']perhydrofuran-2'-one *see* 3502

Dimethylenesulfone-α,α'-dicarboxylic Acid *see* 9091

Dimethylenimine *see* 3859

6β,7β,15β,16β-Dimethylen-3-oxo-17α-pregn-4-ene-21,17-carbolactone *see* 3502

exo-1,2-*cis*-Dimethyl-3,6-epoxyhexahydrophthalic Anhydride *see* 1755

N-[[(8β)-1,6-Dimethylergolin-8-yl]methyl]carbamic Acid Phenylmethyl Ester *see* 6009

(3β,4α,5α,22*E*)-4,23-Dimethylergost-22-en-3-ol *see* 3316

1-1'-[(1*S*,2*S*)-1,2-Dimethyl-1,2-ethanediyl]bis[1,1-diphenylphosphine] *see* 2063

N,N-Dimethylethanolamine *see* 2845

Dimethyl Ether, 3266

6α,21-Dimethylethisterone *see* 3240

N-[(1,1-Dimethylethoxy)carbonyl]-β-alanyl-L-tryptophyl-L-methionyl-L-α-aspartyl-L-phenylalaninamide *see* 7224

(α*R*,β*S*)-β-[[(1,1-Dimethylethoxy)carbonyl]amino]-α-hydroxybenzenepropanoic Acid (2a*R*,4*S*,4a*S*,6*R*,9*S*,11*S*,12*S*,12a*R*,12b*S*)-12b-(Acetyloxy)-12-(benzoyloxy)-2a,3,4,4a,5,6,9,10,11,12,12a,12b-dodecahydro-11-hydroxy-4,6-dimethoxy-4a,8,13,13-tetramethyl-5-oxo-7,11-methano-1*H*-cyclodeca[3,4]benz[1,2-*b*]oxet-9-yl Ester *see* 1604

(α*R*,β*S*)-β-[[(1,1-Dimethylethoxy)carbonyl]amino]-α-hydroxybenzenepropanoic Acid (2a*R*,4*S*,4a*S*,6*R*,9*S*,11*S*,12*S*,12a*R*,12b*S*)-12b-(Acetyloxy)-12-(benzoyloxy)-2a,3,4,4a,5,6,9,10,11,12,12a,12b-dodecahydro-4,6,11-trihy-

droxy-4a,8,13,13-tetramethyl-5-oxo-7,11-methano-1*H*-cyclodeca[3,4]benz-[1,2-*b*]oxet-9-yl Ester *see* 3442

α-[[[(1,1-Dimethylethoxy)carbonyl]oxy]-imino]benzeneacetonitrile *see* 1323

2-(1,1-Dimethylethoxy)-2-methylpropane *see* 3040

4-[2-[(1*E*)-3-(1,1-Dimethylethoxy)-3-oxo-1-propen-1-yl]phenyl]-1,4-dihydro-2,6-dimethyl-3,5-pyridinedicarboxylic Acid 3,5-Diethyl Ester *see* 5379

1-(1,1-Dimethylethoxy)-*N,N,N',N'*-tetramethylmethanediamine *see* 1373

(17β)-17-[[(1,1-Dimethylethyl)amino]carbonyl]androsta-3,5-diene-3-carboxylic Acid *see* 3690

(2*S*)-*N*^1[(1*S*,2*R*)-3-[(3*S*,4a*S*,8a*S*)-3-[[(1,1-Dimethylethyl)amino]carbonyl]octahydro-2(1*H*)-isoquinolinyl]-2-hydroxy-1-phenylmethyl]propyl]-2-[(2-quinolinylcarbonyl)amino]butanediamide *see* 8506

5-[2-[(1,1-Dimethylethyl)amino]-1-hydroxyethyl]-1,3-benzenediol *see* 9302

N-[5-[2-[(1,1-Dimethylethyl)amino]-1-hydroxyethyl]-2-hydroxyphenyl]urea *see* 1832

2-[3-[(1,1-Dimethylethyl)amino]-2-hydroxypropoxy]benzonitrile *see* 1493

5-[(2*S*)-3-[(1,1-Dimethylethyl)amino]-2-hydroxypropoxy]-3,4-dihydro-1(2*H*)-naphthalenone *see* 5514

5-[3-[(1,1-Dimethylethyl)amino]-2-hydroxypropoxy]-3,4-dihydro-2(1*H*)-quinolinone *see* 1866

4-[3-[(1,1-Dimethylethyl)amino]-2-hydroxypropoxy]-2-methyl-1(2*H*)-isoquinolinone *see* 9595

5-[3-[(1,1-Dimethylethyl)amino]-2-hydroxypropoxy]-1,2,3,4-tetrahydro-2,3-naphthalenediol *see* 6431

5-[2-[[3-[(1,1-Dimethylethyl)amino]-2-hydroxypropyl]thio]-4-thiazolyl]-2-thiophenecarboxamide *see* 782

α-[[[(1,1-Dimethylethyl)amino]methyl]-7-ethyl-2-benzofuranmethanol *see* 1485

α^1-[[(1,1-Dimethylethyl)amino]methyl]-4-hydroxy-1,3-benzenedimethanol *see* 210

α^6-[[(1,1-Dimethylethyl)amino]methyl]-3-hydroxy-2,6-pyridinedimethanol *see* 7602

1-[[(1,1-Dimethylethyl)amino]-3-[(2-methyl-1*H*-indol-4-yl)oxy]-2-propanol 2-Benzoate *see* 1335

(2*S*)-1-[(1,1-Dimethylethyl)amino]-3-[[4-(4-morpholinyl)-1,2,5-thiadiazol-3-yl]oxy]-2-propanol *see* 9601

1,4-Dimethyl-7-ethylazulene *see* 2039

(1,1-Dimethylethyl)benzene *see* 1553

Dimethyl Ethyl Carbinol *see* 7253

2-[[4-(1,1-Dimethylethyl)cyclohexyl]-methyl]-3-hydroxy-1,4-naphthalenedione *see* 1495

(3*S*,4a*S*,8a*S*)-*N*-(1,1-Dimethylethyl)decahydro-2-[(2*R*,3*R*)-2-hydroxy-3-[(3-hydroxy-2-methylbenzoyl)amino]-4-(phenylthio)butyl]-3-isoquinolinecarboxamide *see* 6528

(3a*S*,5a*R*,8*R*,8a*S*,9*R*,10a*S*)-9-(1,1-Dimethylethyl)-10,10a-dihydro-8,9-dihydroxy-4*H*,5a*H*,9*H*-furo[2,3-*b*]furo[3',-2':2,3]cyclopenta[1,2-*c*]furan-2,4,7-(3*H*,8*H*)trione *see* 1224

(2*R*,5*S*)-2-(1,1-Dimethylethyl)-3,5-dimethyl-4-imidazolidinone *see* 5707

2-[[4-(1,1-Dimethylethyl)-2,6-dimethyl-phenyl]methyl]-4,5-dihydro-1*H*-imidazole *see* 10284

1-(1,1-Dimethylethyl)-4,4-diphenylpiperidine *see* 1477

sym-Dimethylethylene *see* 1523

Dimethylethylene Glycol *see* 1571

N,N'-Dimethylethyleneurea *see* 3273

1,1-Dimethylethyl Ethyl Ether *see* 3829

8-(1,1-Dimethylethyl)-*N*-ethyl-*N*-propyl-1,4-dioxaspiro[4.5]decane-2-methamine *see* 8889

1,1-Dimethylethyl 4-Formyl-2,2-dimethyloxazolidine-3-carboxylate *see* 4404

1,1-Dimethylethylhydroperoxide *see* 1573

4-(1,1-Dimethylethyl)-*N*-[6-(2-hydroxyethoxy)-5-(2-methoxyphenoxy)[2,2'-bipyrimidin]-4-yl]benzenesulfonamide *see* 1355

(3*Z*,6*Z*)-3-[[5-(1,1-Dimethylethyl)-1*H*-imidazol-4-yl]methylene]-6-(phenylmethylene)-2,5-piperazinedione *see* 7654

2-[(1,1-Dimethylethyl)imino]tetrahydro-3-(1-methylethyl)-5-phenyl-4*H*-1,3,5-thiadiazin-4-one *see* 1502

(1,1-Dimethylethyl)-4-methoxyphenol *see* 1549

(2*S*,3*S*)-*N*-[[5-(1,1-Dimethylethyl)-2-methoxyphenyl]methyl]-2-(diphenylmethyl)-1-azabicyclo[2.2.2]octan-3-amine *see* 5819

(2*S*,5*S*)-2-(1,1-Dimethylethyl)-3-methyl-5-(phenylmethyl)-4-imidazolidinone *see* 5708

4-(1,1-Dimethylethyl)-*N*-(1-methylpropyl)-2,6-dinitrobenzenamine *see* 1534

α,α-Dimethylethyl Nitrite *see* 1586

3,5-Dimethyl-5-ethyloxazolidine-2,4-dione *see* 7132

4-[[(1,1-Dimethylethyl)oxidoimino]methyl]-1,3-benzenedisulfonic Acid *see* 3406

(5α,17β)-*N*-(1,1-Dimethylethyl)-3-oxo-4-azaandrost-1-ene-17-carboxamide *see* 4113

4-(1,1-Dimethylethyl)phenol *see* 1588

1-[4-(1,1-Dimethylethyl)phenyl]-4-[4-(diphenylmethoxy)-1-piperidinyl]-1-butanone *see* 3531

4-[2-[4-(1,1-Dimethylethyl)phenyl]ethoxy]quinazoline *see* 3993

α-[4-(1,1-Dimethylethyl)phenyl]-4-(hydroxydiphenylmethyl)-1-piperidinebutanol *see* 9306

1-[4-(1,1-Dimethylethyl)phenyl]-3-(4-methoxyphenyl)-1,3-propanedione *see* 881

N-[[4-(1,1-Dimethylethyl)phenyl]methyl]-*N*-methyl-1-naphthalenemethanamine *see* 1521

rel-(2*R*,6*S*)-4-[3-[4-(1,1-Dimethylethyl)-phenyl]-2-methylpropyl]-2,6-dimethylmorpholine *see* 4021

1-[3-[4-(1,1-Dimethylethyl)phenyl]-2-methylpropyl]piperidine *see* 4020

6-[*O*-(1,1-Dimethylethyl)-D-serine]-10-deglycinamideluteinizing Hormone-releasing Factor (Pig) 2-(Aminocarbonyl)hydrazide *see* 4562

6-[*O*-(1,1-Dimethylethyl)-D-serine]-9-(*N*-ethyl-L-prolinamide)-10-deglycinamideluteinizing Hormone-releasing Factor (Pig) *see* 1506

7-oxo-4-thia-1-azabicyclo[3.2.0]heptane-2-carboxylic Acid *see* 7003

N,N-Dimethyl-2-[(2-methylphenyl)phenylmethoxy]ethanamine *see* 6975

N,N-Dimethyl-2-[(4-methylphenyl)phenylmethoxy]ethanamine *see* 6125

Dimethyl Methylphosphonate *see* 3435

N,N-Dimethyl-9-[3-(4-methyl-1-piperazinyl)propylidene]-9H-thioxanthene-2-sulfonamide *see* 9519

N,N-Dimethyl-10-[3-(4-methyl-1-piperazinyl)propyl]-10H-phenothiazine-2-sulfonamide *see* 9510

2,2-Dimethyl-3-(2-methyl-1-propen-1-yl)cyclopropanecarboxylic Acid *see* 2255

trans-(+)-2,2-Dimethyl-3-(2-methylpropenyl)cyclopropanecarboxylic Acid (5-Benzyl-3-furyl)methyl Ester *see* 1235

(1R,3R)-2,2-Dimethyl-3-(2-methyl-1-propenyl)cyclopropanecarboxylic Acid (1S)-3-(2Z)-(2-Butenyl)-2-methyl-4-oxo-2-cyclopenten-1-yl-ester *see* 2293

2,2-Dimethyl-3-(2-methyl-1-propen-1-yl)cyclopropanecarboxylic Acid (6-Chloro-1,3-benzodioxol-5-yl) Methyl Ester *see* 1000

2,2-Dimethyl-3-(2-methylpropenyl)cyclopropanecarboxylic Acid 6-Chloropiperonyl Ester *see* 1000

2,2-Dimethyl-3-(2-methyl-1-propen-1-yl)cyclopropanecarboxylic Acid Cyano-(3-phenoxyphenyl)methyl Ester *see* 2765

2,2-Dimethyl-3-(2-methyl-1-propen-1-yl)cyclopropanecarboxylic Acid 3-(2-Cyclopenten-1-yl)-2-methyl-4-oxo-2-cyclopenten-1-yl Ester *see* 2700

2,2-Dimethyl-3-(2-methylpropenyl)cyclopropanecarboxylic Acid Ester with N-(Hydroxymethyl)-1-cyclohexene-1,2-dicarboximide *see* 9370

2,2-Dimethyl-3-(2-methyl-1-propen-1-yl)cyclopropanecarboxylic Acid (1,3,4,5,-6,7-Hexahydro-1,3-dioxo-2H-isoindol-2-yl)methyl Ester *see* 9370

(1R,3R)-2,2-Dimethyl-3-(2-methyl-1-propenyl)cyclopropanecarboxylic Acid (1S)-2-Methyl-4-oxo-3-(2Z)-2,4-pentadienyl-2-cyclopenten-1-yl Ester *see* 8075

(1R,3R)-2,2-Dimethyl-3-(2-methyl-1-propenyl)cyclopropanecarboxylic Acid (1S)-2-Methyl-4-oxo-3-(2Z)-2-pentenyl-2-cyclopenten-1-yl Ester *see* 5306

2,2-Dimethyl-3-(2-methyl-1-propenyl)cyclopropanecarboxylic Acid 2-Methyl-4-oxo-3-(2-propenyl)-2-cyclopenten-1-yl Ester *see* 252

2,2-Dimethyl-3-(2-methyl-1-propenyl)cyclopropanecarboxylic Acid 2-Methyl-4-oxo-3-(2-propynyl)-2-cyclopenten-1-yl Ester *see* 7823

2,2-Dimethyl-3-(2-methylpropenyl)cyclopropanecarboxylic Acid *m*-Phenoxybenzyl Ester *see* 7325

2,2-Dimethyl-3-(2-methyl-1-propen-1-yl)cyclopropanecarboxylic Acid (3-Phenoxyphenyl)methyl Ester *see* 7325

(1R,3R)-2,2-Dimethyl-3-(2-methyl-1-propen-1-yl)cyclopropanecarboxylic Acid [5-(Phenylmethyl)-3-furanyl]methyl Ester *see* 1235

2,2-Dimethyl-3-(2-methyl-1-propen-1-yl)cyclopropanecarboxylic Acid [5-(Phenylmethyl)-3-furanyl]methyl Ester *see* 8274

N,N-Dimethyl-2-[(1-methyl-1H-pyrazol-5-yl)phenylmethoxy]ethanamine *see* 2335

N²-[1,1-Dimethyl-2-(methylsulfonyl)ethyl]-3-iodo-N¹-[2-methyl-4-[1,2,2,2-tetrafluoro-1-(trifluoromethyl)ethyl]phenyl]-1,2-benzenedicarboxamide *see* 4150

(2S,5R,6R)-3,3-Dimethyl-6-[[(2R)-2-[[[3-methylsulfonyl)-2-oxo-1-imidazolidinyl]carbonyl]amino]-2-phenylacetyl]amino]-7-oxo-4-thia-1-azabicyclo[3.2.-0]heptane-2-carboxylic Acid *see* 6250

N,N-Dimethyl-N-[2-[2-[methyl-4-(1,1,3,3-tetramethylbutyl)phenoxy]ethoxy]ethyl]benzenemethanaminium Chloride (1:1) *see* 6097

2,4-Dimethyl-N-(3-methyl-2(3H)-thiazolylidene)benzenamine *see* 2761

3,5-Dimethyl-4-(methylthio)phenol 1-(N-Methylcarbamate) *see* 6044

O,O-Dimethyl O-4-(Methylthio)-*m*-tolyl Phosphorothioate *see* 4029

[1α,2α,5α(1E,3E),6β(1E,3E,5E)]-3,6-Dimethyl-5-[2-methyl-4-(2,6,6-trimethyl-1-cyclohexen-1-yl)-1,3-butadienyl]-6-[4-methyl-6-(2,6,6-trimethyl-1-cyclohexen-1-yl)-1,3,5-hexatrienyl]-3-cyclohexene-1,2-dimethanol *see* 5363

(9α,13α,14α)-3,17-Dimethylmorphinan *see* 3229

N,N-Dimethyl-1-naphthalenamine *see* 3276

2,7-Dimethyl-1,4-naphthalenedione *see* 2058

(αS)-N,N-Dimethyl-α-[2-(1-naphthalenyloxy)ethyl]benzenemethanamine *see* 2821

2,7-Dimethyl-1,4-naphthoquinone *see* 2058

N,N-Dimethyl-1-naphthylamine, 3276

N¹,N¹-Dimethyl-N³-(1-nitro-9-acridinyl)-1,3-propanediamine *see* 6657

1,2-Dimethyl-5-nitro-1H-imidazole *see* 3290

α,2-Dimethyl-5-nitro-1H-imidazole-1-ethanol *see* 8556

Dimethylnitromethane *see* 6715

(±)-2,6-Dimethyl-4-(3-nitrophenyl)-1,4-dihydropyridine-3,5-dicarboxylic Acid-3-(1-benzyl-3-piperidyl) Ester-5-methyl Ester *see* 1045

2,6-Dimethyl-4-(3'-nitrophenyl)-1,4-dihydropyridine-3,5-dicarboxylic Acid 3-β-Methoxyethyl Ester 5-Isopropyl Ester *see* 6636

O,O-Dimethyl O-*p*-Nitrophenyl Phosphorothioate *see* 6181

O,O-Dimethyl O-*p*-Nitrophenyl Thiophosphate *see* 6181

Dimethylnitrosamine *see* 6724

N,N-Dimethyl-4-nitrosobenzenamine *see* 6725

N,4-Dimethyl-N-nitrosobenzenesulfonamide *see* 9702

O,O-Dimethyl O-4-Nitro-*m*-tolyl Phosphorothioate *see* 4007

O,O-Dimethyl O-4-Nitro-*m*-tolyl Thiophosphate *see* 4007

5,5-Dimethyl-3-[4-nitro-3-(trifluoromethyl)phenyl]-2,4-imidazolidinedione *see* 6629

(2Z,6E)-2-[(3E)-4,8-Dimethyl-3,7-nonadienyl]-6-methyl-2,6-octadiene-1,8-diol *see* 7647

(*all*-E)-2,3-Dimethyl-5-(3,7,11,15,19,23,-27,31,35-nonamethyl-2,6,10,14,18,22,-26,30,34-hexatriacontanonaenyl)-2,5-cyclohexadiene-1,4-dione *see* 7637

L-N,N-Dimethylnorephedrine *see* 6140

17α,21-Dimethyl-19-norpregna-4,9-diene-3,20-dione *see* 7900

7α,17α-Dimethyl-19-nortestosterone *see* 6253

N,N-Dimethyl-1-octadecanamine *see* 3522

N,N-Dimethyloctadecylamine *see* 3522

3,7-Dimethyl-2,6-octadienal *see* 2321

(2E)-3,7-Dimethyl-2,6-octadien-1-ol *see* 4437

(2Z)-3,7-Dimethyl-2,6-octadien-1-ol *see* 6560

cis-2,6-Dimethyl-2,6-octadien-8-ol *see* 6560

2,6-Dimethyl-2,7-octadien-6-ol *see* 5550

3,7-Dimethyl-1,6-octadien-3-ol *see* 5550

trans-3,7-Dimethyl-2,6-octadien-8-ol *see* 4437

3,7-Dimethyl-1,6-octadien-3-ol 3-Acetate *see* 5551

2-[(2E)-3,7-Dimethyl-2,6-octadien-1-yl]-1,4-benzenediol *see* 4439

trans-(3,7-Dimethyl-2,6-octadienyl)hydroquinone *see* 4439

6-[(2E)-3,7-Dimethyl-2,6-octadien-1-yl]-7-hydroxy-2H-1-benzopyran-2-one *see* 6994

(E)-6-(3,7-Dimethyl-2,6-octadienyl)-7-hydroxycoumarin *see* 6994

4-[[(2E)-3,7-Dimethyl-2,6-octadien-1-yl]oxy]-7H-furo[3,2-g][1]benzopyran-7-one *see* 1159

(6S)-3-[[4-[[(2E)-3,7-Dimethyl-2,6-octadien-1-yl]oxy]phenyl]methylene]-1,4-dihydroxy-6-methyl-2,5-piperazinedione *see* 6404

6-(3,7-Dimethyl-2,6-octadienyl)umbelliferone *see* 6994

3,7-Dimethyl-1,3,?-octatriene *see* 6830

3,7-Dimethyl-1,3,6-octatriene *see* 6830

3,7-Dimethyl-1,3,7-octatriene *see* 6830

3,7-Dimethyl-6-octenal *see* 2328

3,7-Dimethyl-7-octenal *see* 2328

2,6-Dimethyl-2-octen-8-ol *see* 2329

(3S)-3,7-Dimethyl-7-octen-1-ol *see* 8309

3,7-Dimethyl-6-octen-1-ol *see* 2329

Dimethylolpropionic Acid, 3277

Dimethyl Oxalate *see* 6179

5,5-Dimethyl-2,4-oxazolidinedione *see* 3235

N¹-[(4,5-Dimethyl-2-oxazolyl)amidino]-sulfanilamide *see* 9041

N¹-(4,5-Dimethyl-2-oxazolyl)sulfanilamide *see* 9056

1-(4,5-Dimethyloxazol-2-yl)-3-sulfanilylguanidine *see* 9041

5-(5,5-Dimethyl-2-oxido-1,3,2-dioxaphosphorinan-2-yl)-1,4-dihydro-2,6-dimethyl-4-(3-nitrophenyl)-3-pyridinecarboxylic Acid 2-[Phenyl(phenylmethyl)amino]ethyl Ester *see* 3571

[R-(E,E)]-9-(3,3-Dimethyloxiranyl)-3,7-dimethyl-2,6-nonadienoic Acid Methyl Ester *see* 5321

(1S,4R)-7,7-Dimethyl-2-oxobicyclo[2.2.1]-heptane-1-methanesulfonic Acid *see* 1736

(11β,16α)-21-(3,3-Dimethyl-1-oxobutoxy)-9-fluoro-11-hydroxy-16,17-[(1-methylethylidene)bis(oxy)]pregna-1,4-diene-3,20-dione *see* 9759

4,4'-(1,1-Dioxido-3*H*-2,1-benzoxathiol-3-ylidene)bis[2,5-dimethylphenol] *see* 10280

N,*N*'-[(1,1-Dioxido-3*H*-2,1-benzoxathiol-3-ylidene)bis[(6-hydroxy-2-methyl-5-(1-methylethyl)-3,1-phenylene]methylene]]bis[*N*-(carboxymethyl)glycine] *see* 6202

N,*N*'-[(1,1-Dioxido-3*H*-2,1-benzoxathiol-3-ylidene)bis[(6-hydroxy-5-methyl-3,1-phenylene)methylene]]bis[*N*-(carboxymethyl)glycine] *see* 10281

4,4'-(1,1-Dioxido-3*H*-2,1-benzoxathiol-3-ylidene)bis[5-methyl-2-(1-methylethyl)phenol] *see* 9555

4,4'-(1,1-Dioxido-3*H*-2,1-benzoxathiol-3-ylidene)bis[2-methylphenol] *see* 2568

4,4'-(1,1-Dioxido-3*H*-2,1-benzoxathiol-3-ylidene)bis[3-methylphenol] *see* 2567

4,4'-(1,1-Dioxido-3*H*-2,1-benzoxathiol-3-ylidene)bisphenol *see* 7360

(−)-2-[*N*-[4-(1,1-Dioxido-3-oxo-2,3-dihydrobenzisothiazol-2-yl)butyl]aminomethyl]chroman *see* 8257

2-[(1,4-Dioxido-2-quioxalinyl)methylene]hydrazinecarboxylic Acid Methyl Ester *see* 1782

Dioxin *see* 9221

Dioxin (obsolete) *see* 3244

9,10-Dioxoanthracene *see* 679

3,3'-Dioxo-[Δ²,²'-biindoline]-5,5'-disulfonic Acid Disodium Salt *see* 4982

2,2'-[(1,4-Dioxo-1,4-butanediyl)bis(oxy)]-bis[*N*,*N*,*N*-trimethylethanaminium] Bromide (1:2) *see* 9005

2,2'-[(1,4-Dioxo-1,4-butanediyl)bis(oxy)]-bis[*N*,*N*,*N*-trimethylethanaminium] Chloride (1:2) *see* 9006

2,2'-[(1,4-Dioxo-1,4-butanediyl)bis(oxy)]-bis[*N*,*N*,*N*-trimethylethanaminium] Iodide (1:2) *see* 9007

rel-(1*R*,1'*S*,2*S*,2'*R*)-2,2'-[(1,4-Dioxo-1,4-butanediyl)bis(oxy-3,1-propanediyl)]-bis[1,2,3,4-tetrahydro-6,7,8-trimethoxy-2-methyl-1-[(3,4,5-trimethoxyphenyl)methyl]isoquinolinium] Chloride (1:2) *see* 3480

4,4'-Dioxo-*β*-carotene *see* 1756

2,4-Dioxo-5-diazopyrimidine *see* 3006

2,4-Dioxo-3,3-diethyl-5-methylpiperidine *see* 6212

2,4-Dioxo-3,3-diethyltetrahydropyridine *see* 8108

2,4-Dioxo-3,3-diethyl-1,2,3,4-tetrahydropyridine *see* 8108

3,3'-(1,3-Dioxo-1,3-digermoxanediyl)bispropanioic Acid *see* 7909

5,6-Dioxo-5,6-dihydro-4,7-phenanthroline *see* 7312

1,3-Dioxo-2-(2,6-dioxopiperidin-3-yl)-4-aminoisoindoline *see* 7708

3,5-Dioxo-1,2-diphenyl-4-*n*-butylpyrazolidine *see* 3342

N,*N*'-[(1,2-Dioxo-1,2-ethanediyl)bis(imino-2,1-ethanediyl)]bis[2-chloro-*N*,*N*-diethylbenzenemethanaminium] Chloride (1:2) *see* 373

4,6-Dioxo-1-ethyl-10-propyl-4*H*,6*H*-pyrano[3,2-g]quinoline-2,8-dicarboxylic Acid *see* 6521

2,4-Dioxo-5-fluoropyrimidine *see* 4213

3,3'-[(1,6-Dioxo-1,6-hexanediyl)diimino]-bis[2,4,6-triiodobenzoic Acid] *see* 5065

3,3'-[(1,6-Dioxo-1,6-hexanediyl)diimino]-bis[2,4,6-triiodo-5-[(methylamino)carbonyl]benzoic Acid] *see* 5053

N-(2,5-Dioxo-4-imidazolidinyl)urea *see* 250

[*N*-(2,5-Dioxo-4-imidazolidinyl)ureato]-dihydroxyaluminum *see* 216

2,3-Dioxoindoline *see* 5148

β-L-(−) Dioxolane Cytidine *see* 9968

1,3-Dioxolane-4-methanol *see* 4521

7-(1,3-Dioxolan-2-ylmethyl)-3,7-dihydro-1,3-dimethyl-1*H*-purine-2,6-dione *see* 3486

7-(1,3-Dioxolan-2-ylmethyl)theophylline *see* 3486

(1,3-Dioxolan-4-ylmethyl)trimethylammonium Iodide *see* 7022

2,6-Dioxo-4-methyl-4-ethylpiperidine *see* 1029

(1*R*,1'*R*)-2,2'-[[(4*E*)-1,8-Dioxo-4-octene-1,8-diyl]bis[oxy-3,1-propanediyl)]-bis[1,2,3,4-tetrahydro-6,7-dimethoxy-2-methyl-1-[(3,4,5-trimethoxyphenyl)-methyl]]isoquinolinium Chloride (1:2) *see* 6305

β-(3,5-Dioxo-1,2,4-oxodiazolidin-2-yl)-L-alanine *see* 8202

3,5-Dioxo-4-(1-oxopropyl)cyclohexanecarboxylic Acid *see* 7892

1,1-Dioxopenicillanoyloxymethyl 6-(D-*α*-Amino-*α*-phenylacetamido)penicillanate *see* 9121

3-(9,9-Dioxo-10-phenothiazinyl)-2-methyl-1-dimethylaminopropane *see* 7045

1,3-Dioxo-5-phthalancarboxylic Acid *see* 9872

2,6-Dioxo-3-phthalimidopiperidine *see* 9403

2,5-Dioxopiperazine *see* 7578

2-(2,6-Dioxo-3-piperidinyl)-1*H*-isoindole-1,3(2*H*)-dione *see* 9403

N-(2,6-Dioxo-3-piperidyl)phthalimide *see* 9403

3,20-Dioxo-5*α*-pregnane *see* 264

5,5'-[(1,3-Dioxo-1,3-propanediyl)bis(methylimino)]bis[*N*,*N*'-bis[2,3-dihydroxy-1-(hydroxymethyl)propyl]-2,4,6-triiodo-1,3-benzenedicarboxamide] *see* 5108

3,5-Dioxo-4-propionylcyclohexanecarboxylic Acid *see* 7892

N-[1-(1,2-Dioxopropyl)-L-prolyl]didemnin A *see* 725

2,6-Dioxopurine *see* 10257

2,4-Dioxopyrimidine *see* 10035

2,5-Dioxopyrrolidine *see* 9001

1-[6-[-(2,5-Dioxo-1-pyrrolidinyl)oxy]-6-oxohexyl]-2-[5-[1-[6-[(2,5-dioxo-1-pyrrolidinyl)oxy]-6-oxohexyl]-1,3-dihydro-3,3-dimethyl-5-sulfo-2*H*-indol-2-ylidene]-1,3-pentadien-1-yl]-3,3-dimethyl-5-sulfo-3*H*-indolium Inner Salt *see* 2669

1-[6-[(2,5-Dioxo-1-pyrrolidinyl)oxy]-6-oxohexyl]-2-[3-[1-[6-[(2,5-dioxo-1-pyrrolidinyl)oxy]-6-oxohexyl]-1,3-dihydro-3,3-dimethyl-5-sulfo-2*H*-indol-2-ylidene]-1-propen-1-yl]-3,3-dimethyl-5-sulfo-3*H*-indolium Inner Salt *see* 2668

Dioxo[sulfato(2−)-*κO*]uranium *see* 10049

2,4-Dioxo-1,2,3,4-tetrahydro-1,3,5-triazine-6-carboxylic Acid *see* 7046

3,5-Dioxo-2,3,4,5-tetrahydro-1,2,4-triazine Riboside *see* 897

3,20-Dioxo-11*β*,17*α*,21-trihydroxy-1,4-pregnadiene *see* 7841

Dioxybenzone, 3332

3*α*,7*β*-Dioxycholanic Acid *see* 10074

Dioxyfilline *see* 3486

Dioxygen Difluoride *see* 4195

10-[4-(3,4-Dioxymethylenebenzyl)-1-piperazinylacetyl]phenothiazine *see* 4013

Dioxymethyleneprotocatechuic Aldehyde *see* 7588

Di(*p*-oxyphenyl)-2,4-hexadiene *see* 3115

DIPA *see* 3214

DIPA-DCA *see* 3216

Dipalmitoyl-L-*α*-glycerylphosphorylcholine *see* 2461

L-*α*-Dipalmitoyl Lecithin *see* 2461

Dipalmitoyl Phosphatidylcholine *see* 2461

Dipalmitoylphosphatidylcholine *see* 8049

(*R*,*R*)-DIPAMP, 3333

(*S*,*S*)-DIPAMP *see* 3333

Dipan *see* 3346

Di-Paralene [Abbott] *see* 2082

Diparcol [Specia] *see* 3119

DIPC *see* 3218

DIPE *see* 5258

DIPEA *see* 3214

DiPel [Valent] *see* 925

Dipentene *see* 5546

Dipenteneglycol *see* 9311

Dipentum [Pfizer] *see* 6935

2,5-Di-*tert*-pentylhydroquinone, 3334

Dipeptidyl Amino Peptidase I *see* 1906

Dipeptidyl Transferase *see* 1906

Diphacil *see* 149

Diphacin [Hacco] *see* 3335

Diphacinone, 3335

Diphebuzol *see* 7390

Diphemanil Methylsulfate, 3336

Diphemin [Alcon] *see* 3378

Diphenadione *see* 3335

Diphenamid, 3337

Diphenasone *see* 2822

Diphenatrile *see* 3346

Diphencyprone, 3338

Diphenhydramine, 3339

Diphenhydramine Bis(theophyllin-7-yl-acetate) *see* 3339

Diphenhydramine Teoclate *see* 3230

Diphenic Acid, 3340

Diphenidol, 3341

Diphenolic Acid, 3342

Diphenoxylate, 3343

Diphenpyramide *see* 3156

Di-phenthane-70 *see* 3081

Diphenyl, 3344

Diphenylacetamide, 3345

N,*N*-Diphenylacetamide *see* 3345

Diphenylacetic Acid, 3346

Diphenylacetic Acid 1-Ethyl-3-piperidyl Ester *see* 7582

Diphenylacetic Acid 3*α*-Tropanyl Ester *see* 9952

Diphenylacetonitrile *see* 3346

Diphenylacetyldiethylaminoethanol *see* 149

2-Diphenylacetyl-1,3-diketohydrindene *see* 3335

Diphenylacetylene *see* 9664

2-Diphenylacetyl-1,3-indandione *see* 3335

2-(2,2-Diphenylacetyl)-1*H*-indene-1,3-(2*H*)-dione *see* 3335

2,3-Diphenylacrylic Acid *see* 7393

Diphenylamine, 3347

Diphenylamine-2-carboxylic Acid *see* 7384

Di-2-pyridyl-(6,6-di-2-pyridylfulven-2-yl)methanol *see* 8102
3,3-Dipyridyl Disulfide Dihydrochloride *see* 3416
3-(Di-2-pyridylmethylene)-α,α-di-2-pyridyl-1,4-cyclopentadiene-1-methanol *see* 8102
Dipyrine *see* 470
Dipyrithione *see* 8107
Dipyrone, 3390
1,4-Dipyrrolidino-2-butyne *see* 9744
(16α)-21-[4-(2,6-Di-1-pyrrolidinyl-4-pyrimidinyl)-1-piperazinyl]-16-methylpregna-1,4,9(11)-triene-3,20-dione *see* 9619
Diquafosol, 3391
Diquat Dibromide, 3392
Diquinol [Parke-Davis] *see* 3783
N,N'-Di-6-quinolinylurea *see* 3393
1,3-Di-6-quinolylurea, 3393
sym-Di-(6-quinolyl)urea *see* 3393
6,6'-D nolylurea *see* 3393
Dirax ichi] *see* 5053
Direktan [Gerot] *see* 6610
Direxiode [Delalande] *see* 5085
Dirhenium Heptoxide *see* 8301
Dirhodium Tetraacetate *see* 8311
Dirithromycin, 3394
Dirlotapide, 3395
Dirocide [Solvay] *see* 3128
Dirox [Gramon] *see* 47
Dirucotide, 3396
Dirythmin SA [AstraZeneca] *see* 3402
Disadine D.P. [Stuart] *see* 7815
Disalcid [3M] *see* 8477
Disalgesic [3M] *see* 8477
Disalicylic Acid *see* 8477
Disalunil [Berlin-Chemie] *see* 4819
Disarim [Bial] *see* 2085
Disarm [Arysta LifeScience] *see* 4216
Discase [Boots] *see* 2263
Dis-Cinil [Lusofarmaco] *see* 7059
Discodermolide, 3397
Discoid [Sagitta] *see* 4338
Discotrine [3M Pharma] *see* 6694
Disdolen [Uriach] *see* 4282
Diseptal C *see* 9061
Disgren [Uriach] *see* 9856
Disiamylborane *see* 1253
DISIDA *see* 3400
Disilane, 3398
Disilicane *see* 3398
Disilicic Acid *see* 8629
Disilicoethane *see* 3398
Disilicon Hexahydride *see* 3398
(*T*-4)-Disilver(1+) Tetraiodomercurate-(2−) *see* 8670
Disintegrins *see* 5362
Disipal [Doetsch, Grether] *see* 6975
Di-Sipidin *see* 7623
Diskin *see* 7059
Dismutec [Sterling Winthrop] *see* 9131
Disoderm [Schering] *see* 3057
Disodium Acetarsenate *see* 800
Disodium Arsenate *see* 8715
Disodium Arsonoacetate *see* 800
Disodium Azodisalicylate *see* 6935
Disodium 4-[(*tert*-Butylimino)methyl]-benzene-1,3-disulfonate *N*-Oxide *see* 3406
Disodium Chromate *see* 8736
Disodium Combretastatin A-4 3-*O*-Phosphate *see* 4277
Disodium Cromoglycate *see* 2581

Disodium 3,3'-Diamino-4,4'-dihydroxyarsenobenzene *N*-Dimethylenesulfonate *see* 9072
Disodium *p,p'*-Diaminodiphenylsulfone-*N,N'*-diglucose Sulfonate *see* 4499
Disodium *o*-Dianisidinediazobis(1-naphthol-4-sulfonate) *see* 1087
Disodium Dihydrogen (1-Hydroxyethylidene)bis[phosphonate] *see* 3912
Disodium Dihydrogen Pyrophosphate *see* 8710
Disodium-1,2-dihydroxybenzene-3,5-disulfonate *see* 9621
*N*⁴-(Disodium 1,3-Disulfo-3-phenylpropyl)sulfanilamide *see* 6772
Disodium 4,4'-Disulfoxydiphenyl-(2-pyridyl)methane *see* 7518
Disodium 2,2'-Dithiobis(ethanesulfonate) *see* 3233
Disodium Ethylenebis[dithiocarbamate] *see* 6426
Disodium Formaldehydesulfoxylate-diaminodiphenylsulfone *see* 9098
Disodium Hexachloroplatinate(2−) *see* 8758
Disodium Hydrogen Citrate *see* 2325
Disodium Hydrogen Phosphate *see* 8789
Disodium Hydrogen Phosphate ³²P *see* 8791
Disodium Hydrogen Phosphite *see* 8793
Disodium 5,5'-Indigotin Disulfonate *see* 4982
Disodium Methylarsonate *see* 6025
Disodium Monohydroxy Methylene Diphosphonate *see* 7039
Disodium Monoselenide *see* 8801
Disodium Orthophosphate *see* 8789
Disodium Phenoltetrabromophthalein Sulfonate *see* 9084
Disodium Phenyl Phosphate, 3399
Disodium *p*-(γ-Phenylpropylamino)benzenesulfonamide-α,γ-disulfonate *see* 6772
Disodium 1-Phenyl-3-*p*-sulfamoylanilino-1,3-propanedisulfonate *see* 6772
Disodium Phosphate *see* 8789
Disodium Phosphite *see* 8793
Disodium Pyrocatechol-3,5-disulfonate *see* 9621
Disodium Selenate *see* 8800
Disodium Selenide *see* 8801
Disodium Stannate *see* 8805
Disodium Succinate *see* 8808
Disodium[sulfonylbis(*p*-phenylenimino)]dimethanesulfinate *see* 9098
Disodium Tellurite *see* 8814
Disodium Tetraoxotellurate *see* 8813
Disodium *o*-Tolidinediazobis(1-naphthylamine-4-sulfonate) *see* 1104
Disodium Trioxotellurate *see* 8814
Disofenin, 3400
Disomer [Schering] *see* 1453
Disonate [Lannett] *see* 3446
Disopain [Mitsubishi] *see* 6318
Disophenol, 3401
Disoprivan [AstraZeneca] *see* 7947
Disoprofol *see* 7947
Disopyramide, 3402
Disoquin *see* 5085
Disorat [Boehringer, Mann.] *see* 6216
Distat [Isis] *see* 3216
Dispadol [Specia] *see* 5916
Disparlure, 3403
(2*S-cis*)-Disparlure *see* 3403
(7*R,8S*)-Disparlure *see* 3403

Dispril *see* 1650
Disprin *see* 1650
Disprol [Reckitt Benckiser] *see* 47
Disrupt II GM [Hercon] *see* 3403
Dissenten [SPA] *see* 5628
Distaclor [Flynn] *see* 1914
Distamine [Alliance] *see* 7197
Distamycin A, 3404
Distaxid *see* 6747
Distigmine Bromide, 3405
Distilbène [Gerda] *see* 3148
Distinct [BASF] *see* 3164
Disto-5 [Cogla] *see* 1307
Distol 8 *see* 3565
Distraneurin [AstraZeneca] *see* 2375
Distyryl *see* 3350
Distyryl Ketone *see* 3010
Disufenton, 3406
1,3-Disulfamyl-4,5-dichlorobenzene *see* 3087
Disulfan *see* 9061
Disulfatozirconic Acid *see* 10382
Disulfiram, 3407
1,6-Disulfonaphthalene *see* 6458
2,7-Disulfonaphthalene *see* 6460
2,4-Disulfonyl-α-phenyl-*tert*-butylnitrone *see* 3406
Disulfoton, 3408
Disulfur Dichloride *see* 9100
Disulfuric Acid *see* 8121
Disulfuric Acid Potassium Salt (1:2) *see* 7784
Disulfurous Acid Dipotassium Salt (1:2) *see* 7766
Disulfurous Acid Sodium Salt (1:2) *see* 8770
Disulon [Specia] *see* 9061
Disulone [Specia] *see* 2822
Disulphine Blue [Zeneca] *see* 9117
Disyncran *see* 6036
Disynformon *see* 3763
Di-Syston [Bayer CropSci.] *see* 3408
Dita Bark, 3409
Ditaine *see* 3547
Ditan *see* 3362
Ditaven [Cascan] *see* 3182
Ditazol, 3410
Diteftin *see* 4340
Ditetrazolium Chloride *see* 9392
Dithane D-14 [Rohm & Haas] *see* 6426
Dithane M-22 [Dow AgroSci.] *see* 5786
Dithane M-45 [Dow AgroSci.] *see* 5780
1,4-Dithiaanthraquinone-2,3-dicarbonitrile *see* 3411
δ-[3-(1,2-Dithiacyclopentyl)]pentanoic Acid *see* 9479
Dithianon, 3411
1,2-Dithia-5,8,11,14,17-pentaazacycloeicosane Cyclic Peptide Deriv *see* 6851
2,6-Dithia-1,3,5,7-tetraazaadamantane 2,2,6,6-Tetraoxide *see* 9374
2,6-Dithia-1,3,5,7-tetraazatricyclo[3.3.1.-1³,⁷]decane 2,2,6,6-Tetraoxide *see* 9374
Dithiazanine Iodide, 3412
3-(Di-2-thienylmethylene)-5-methoxy-1,1-dimethylpiperidinium Bromide (1:1) *see* 9599
3-(Di-2-thienylmethylene)-1-methylpiperidine *see* 9613
3-(Di-2-thienylmethylene)-5-methyl-*trans*-quinolizidinium Bromide *see* 9616
rel-(5*R*,9*aR*)-3-(Di-2-thienylmethylene)-octahydro-5-methyl-2*H*-quinolizinium Bromide (1:1) *see* 9616

Dithio *see* 9096
(2S,2′S)-4,4′-Dithiobis[2-aminobutanoic
Acid] *see* 4773
L-4,4′-Dithiobis[2-aminobutyric Acid]
see 4773
[R-(R*,R*)]-3,3′-Dithiobis[2-aminopro-
panoic Acid] *see* 2779
2,2′-Dithiobisbenzothiazole, 3413
N,N′-[Dithiobis[1-[(carboxymethyl)car-
bamoyl]ethylene]]diglutamine *see*
4511
2,2′-Dithiobisethanamine *see* 2773
(2R,2′R)-N,N′-[Dithiobis[2,1-ethanediyl-
imino(3-oxo-3,1-propanediyl)]]bis-
[2,4-dihydroxy-3,3-dimethylbutanam-
ide] *see* 7115
2,2′-Dithiobisethanesulfonic Acid
Sodium Salt (1:2) *see* 3233
2,2′-Dithiobis[ethylamine] *see* 2773
N,N′-[Dithiobis(ethyleneiminocarbonylethylene)]bis(2,4-dihydroxy-3,3-di-
methylbutyramide) *see* 7115
N,N′-[Dithiobis[2-(2-hydroxyethyl)-1-
methyl-2,1-ethenediyl]]bis[N-[(4-ami-
no-2-methyl-5-pyrimidinyl)methyl]-
formamide] *see* 9449
3,3′-[Dithiobis(methylene)]bis[5-hy-
droxy-6-methyl-4-pyridinemethanol]
see 8109
4,4′-Dithiobismorpholine *see* 3415
6,6′-Dithiobis-2-naphthalenol *see* 2842
6,6′-Dithiobis(2-naphthol) *see* 2842
3,3′-Dithiobis[6-nitrobenzoic Acid] *see*
3606
5,5′-Dithiobis[2-nitrobenzoic Acid] *see*
3606
3,3′-Dithiobis[pyridine] Dihydrochloride
see 3416
2,2′-Dithiobispyridine 1,1′-Dioxide *see*
8107
Dithiobis[thioformic Acid] O,O-Diethyl
Ester *see* 3428
3,3′-Dithiobisvaline *see* 7199
2,4-Dithiobiuret, 3414
Dithiocarb *see* 3421
Dithiocarbamic Acid Monoammonium
Salt *see* 513
Dithiocarbonic Acid Cyclic S,S-(6-Meth-
yl-2,3-quinoxalinediyl) Ester *see* 7077
Dithiocarbonic Anhydride *see* 1811
[N-[2-[(Dithiocarboxy)amino]ethyl]car-
bamodithioato(2−)-κS,κS′]manganese
see 5786
[N-[2-[(Dithiocarboxy)amino]ethyl]car-
bamodithioato(2−)-κS,κS′]manganese
Mixt. with [N-[2-[(Dithiocarboxy)ami-
no]ethyl]carbamodithioato(2−)-κS,-
κS′]zinc *see* 5780
[N-[2-[(Dithiocarboxy)amino]ethyl]car-
bamodithioato(2−)-κS, κS′]Zinc *see*
10365
[N-[2-[(Dithiocarboxy)amino]-1methyl-
ethyl]carbamodithioato(2−)-κS,κS′]-
zinc *see* 7933
Dithiodemeton *see* 3408
β,β′-Dithiodialanine *see* 2779
4,4′-Dithiodimorpholine, 3415
6,6′-Dithiodi-2-naphthol *see* 2842
3,3-Dithiodipyridine Dihydrochloride,
3416
3,3′-Dithiodivaline *see* 7199
Dithioethyleneglycol *see* 3780
1,2-Dithioglycerol *see* 3231
Dithioglycolyl p-Arsenobenzamide *see*
9441

"Dithiol" *see* 9690
1,2-Dithiolane-3-pentanoic Acid *see* 9479
1,2-Dithiolane-3-valeric Acid *see* 9479
5-[3-(1,2-Dithiolanyl)]pentanoic Acid *see*
9479
5-(1,2-Dithiolan-3-yl)valeric Acid *see*
9479
2,3-Dithiolpropanesulfonic Acid *see* 3232
2-(1,3-Dithiol-2-ylidene)-1,3-dithiole *see*
9390
2-(1,3-Dithiol-2-ylidene)propanedioic
Acid 1,3-Bis(1-methylethyl) Ester *see*
5776
Dithione *see* 9096
Dithionic Acid Sodium Salt (1:2) *see*
8746
Dithionous Acid Sodium Salt (1:2) *see*
8747
Dithiooxamide *see* 8412
Dithiophos *see* 9096
Dithiophosphoric Acid O,O′-Dimethyl-S-
[(5-methoxy-1,3,4-thiadiazol-2(3H)-
one-3-yl)methyl] Ester *see* 6042
Dithiopropylthiamine *see* 7993
Dithiopyr, 3417
Dithiosalicylic Acid, 3418
Dithiosteine *see* 3704
1,4-Dithiothreitol, 3419
Dithiotrimethyleneglycol *see* 7915
Dithizone, 3420
Dithranol *see* 676
Dithrocream [Dermal] *see* 676
Ditin Diphosphate *see* 8915
Ditin Pyrophosphate *see* 8915
Ditiocarb Sodium, 3421
Dititanium Trisulfate *see* 9632
1,4-Di-p-toluidinoanthraquinone *see*
8181
Ditolyldiazo-3,6-disulfo-β-naphthylam-
ine-β-naphthylamine-6-sulfonic Acid
Sodium Salt *see* 10210
α,α-Di-p-tolylethane *see* 3422
α,β-Di-p-tolylethane *see* 3423
asym-Di-p-tolylethane *see* 3422
sym-Di-p-tolylethane *see* 3423
1,1-Di-p-tolylethane, 3422
1,2-Di-p-tolylethane, 3423
p-Ditolylmercury, 3424
Ditrac [Bell] *see* 3335
Ditrazin *see* 3128
1,4-Di(2,2,2-trichloro-1-formamidoeth-
yl)piperazine *see* 9859
Di[tri-(2-methyl-2-phenylpropyl)tin]ox-
ide *see* 3997
Ditripentat [Heyl] *see* 7236
Ditropan [Sanofi-Aventis] *see* 7054
Ditrosol *see* 3305
Diucardin [Ayerst] *see* 4826
Diucardyn Sodium [Ayerst] *see* 5937
Diumax [Cusi-Norte] *see* 7605
Diural [Apothokermes] *see* 4338
Diurese [Am. Urologicals] *see* 9789
Diuretin [Knoll] *see* 9433
Diurex [Makhteshim-Agan] *see* 3425
Diurexan [Viatris] *see* 10276
P¹,P⁴-Diuridine 5′-Tetraphosphate *see*
3391
Diuril [Merck & Co.] *see* 2171
Diuriwas [IFI] *see* 52
Diuron, 3425
Diurone *see* 2105
Divalproex Sodium *see* 10099
Divarine [Tutag] *see* 3390
Divarius [Chiesi] *see* 7148
Divasil *see* 8661

Divergan [BASF] *see* 7814
Divicine, 3426
Divicine-β-glucoside *see* 10171
Divicine 5-Glucoside *see* 10171
Dividend [Syngenta] *see* 3154
Divinorin A *see* 8481
Divinorins *see* 8481
Divinyl *see* 1512
Divinylene Oxide *see* 4325
Divinylene Sulfide *see* 9506
Divinylenimine *see* 8128
Divinyl Ether, 3427
Divinyl Oxide *see* 3427
Divipan [Makhteshim-Agan] *see* 3089
Dixanthogen, 3428
Dixarit [Boehringer, Ing.] *see* 2380
Dixeran [Lundbeck] *see* 5891
Dixnalate [Sana] *see* 4411
Di-m-xylylene *see* 5991
m-Dixylylene *see* 5991
N,N-Di-(2,4-xylyliminomethyl)methyl-
amine *see* 482
Dixyrazine, 3429
Dizan [Elanco] *see* 3412
Dizocilpine, 3430
Dizol *see* 479
DJ-7041 *see* 8384
Djenkolic Acid, 3431
DK-7419 *see* 769
DKB *see* 3008
DKP *see* 7779
DL-458-IT *see* 2865
DL-473 *see* 8344
DL-717-IT *see* 5643
DL-832 *see* 4471
DL-3117 *see* 5108
DL-8280 *see* 6863
DLP [Dawbarn] *see* 7701
DLP-787 *see* 8101
DM *see* 7329
DM-9384 *see* 6524
DM1 *see* 5830
DMAA *see* 5897
DMAC *see* 3248
DMAE *see* 2845
DMAN, 3432
DMAP, 3433
DMC, 3434, 7197
DMD *see* 3265
DMDL Protein *see* 10081
DMDO *see* 3265
DMDP *see* 2365
DMDR *see* 4927
DMDT *see* 6061
DMDZ *see* 6780
DMF *see* 3267
DMFA *see* 3267
DMG *see* 3269
DMGG *see* 6010
DMHP *see* 2862
DMI *see* 3273
DMMP, 3435
DMMPA *see* 6524
DMN *see* 6724
DMNA *see* 6724
DMO *see* 3235
DMP *see* 3279
DMP-115 *see* 7273
DMP-266 *see* 3569
DMPA [Geo] *see* 3277
DMPA, 3436
DMPO, 3437
DMPS *see* 3232
DMS *see* 3280, 8995
DMSA *see* 8995

DMSO *see* 3285
DMSO₂ *see* 3283
DMSP *see* 3284, 4027
DMSY *see* 3286
DMT *see* 3288, 9305
DMTA-P *see* 3236
DMTT *see* 2837
DMX-115 *see* 7273
DN *see* 3305
DN-289 *see* 3315
DNA *see* 2908
DNase I *see* 2907
DNBP *see* 3315
DNC *see* 3304, 3305
DNFB *see* 4204
DNOC *see* 3305
DNOCHP *see* 2726
DNOCP *see* 3314
DNP *see* 3401
DNTP *see* 7137
DO-14 *see* 7922
Dobell's Solution, 3438
Dobendan [Cassella-med] *see* 2031
Dobesilate Calcium, 3439
Dobesin [Abigo] *see* 3145
Dobren [Ravizza] *see* 9119
Dobuject [Pisa] *see* 3440
Doburil [Boehringer, Ing.] *see* 2752
Dobutamine, 3440
Dobutrex [Lilly] *see* 3440
Doca [Organon] *see* 2903
Docarpamine, 3441
Docelan [Aventis] *see* 4846
Docémine [Roussel Uclaf] *see* 10212
Docetaxel, 3442
Docigram [UCB] *see* 10212
Dociton [MIBE] *see* 7953
Doconexent *see* 3443
Docosahexaenoic Acid, 3443
(4Z,7Z,10Z,13Z,16Z,19Z)-4,7,10,13,16,19-
  Docosahexaenoic Acid *see* 3443
(4a*S*,5a*R*,10a*R*,11a*S*,15a*R*,15b*S*,16*E*,18*E*,-
  19a*S*,19b*R*)-2,3,4,4a,5,5a,6,7,8,9,10,-
  10a,11,11a,12,13,14,15,15a,15b,19a,-
  19b-Docosahydro-1*H*-diindeno[2,1-
  f:1′,2′-1][1,5]diazacyclotridecine *see*
  7124
Docosanoic Acid *see* 1022
n-Docosanol, 3444
1-Docosanol *see* 3444
Δ¹³-*cis*-Docosenoic Acid *see* 3732
(13*E*)-13-Docosenoic Acid *see* 1369
(13*Z*)-13-Docosenoic Acid *see* 3732
Docosyl Alcohol *see* 3444
Docusate Calcium, 3445
Docusate Potassium *see* 3446
Docusate Sodium, 3446
1,2,3,4,7,8,9,10,13,13,14,14-Dodecachlo-
  ro-1,4,4a,5,6,6a,7,10,10a,11,12,12a-do-
  decahydro-1,4:7,10-dimethanodi-
  benzo[*a,e*]cyclooctene *see* 2855
1,1a,2,2,3,3a,4,5,5,5a,5b,6-Dodecachloro-
  octahydro-1,3,4-metheno-1*H*-cyclo-
  buta[*cd*]pentalene *see* 6290
Dodecahedrane, 3447
(2*E*)-2-[(1*R*,4a*S*,4b*R*,7*S*,8a*R*,10*S*,10a*S*)-
  Dodecahydro-7,10-dihydroxy-1,4b,8,8-
  tetramethyl-2(1*H*)-phenanthrenyl-
  idene]acetic Acid 2-(Dimethylamino)-
  ethyl Ester *see* 1890
1,2,3,3a,3b,7,10,10a,10b,11,12,12a-Do-
  decahydro-1,11-dihydroxy-2,5,10a,-
  12a-tetramethyl-7-phenylcyclopenta-
  [7,8]phenanthro[2,3-*c*]pyrazol-1-yl
  Hydroxymethyl Ketone Acetate *see*
  2527

Dodecahydrodiphenylamine *see* 3106
(7a*S*,13a*R*,13b*R*,13c*S*)-Dodecahydro-1*H*,-
  5*H*,10*H*-dipyrido[2,1-*f*:3′,2′,1′-*ij*][1,6]-
  naphthyridin-10-one *see* 5827
Dodecahydro-3-hydroxy-6-(hydroxymethyl)-3,3a,6-trimethyl-1*H*-benz[*e*]in-
  dene-7-acetic Acid δ-Lactone *see* 7020
(2*S*,7*S*,7a*R*,14*S*,14a*S*)-Dodecahydro-2-hy-
  droxy-7,14-methano-2*H*,11*H*-dipyrido-
  [1,2-*a*:1′,2′-*e*][1,5]diazocin-11-one *see*
  4868
(3b*S*,5a*S*,7*R*,8*R*,10a*R*,10b*S*)-3b,4,5,6,7,8,-
  9,10,10a,10b,11,12-Dodecahydro-7-hy-
  droxy-10b-methyl-5a,8-methano-5a*H*-
  cyclohepta[5,6]naphtho[2,1-*b*]furan-7-
  methanol *see* 1637
(2*E*)-[(1*R*,4a*S*,4b*R*,7*S*,8a*R*,10a*S*)-Dodeca-
  hydro-7-hydroxy-1,4b,8,8-tetramethyl-
  10-oxo-2(1*H*)-phenanthrenylidene]ace-
  tic Acid 2-(Dimethylamino)ethyl Ester
  see 1891
5,6,7,8,9,10,10′,9′,8′,7′,6′,5′-Dodecahy-
  drolycopene *see* 7501
(7*S*,7a*R*,14*S*,14a*S*)-Dodecahydro-7,14-
  methano-2*H*,6*H*-dipyrido[1,2-*a*:1′,2′-
  e][1,5]diazocine *see* 8864
(1*S*,7*R*,7a*S*,14*R*,14a*S*)-Dodecahydro-7,14-
  methano-2*H*,6*H*-dipyrido[1,2-*a*:1′,2′-
  e][1,5]diazocin-1-ol *see* 8280
(7*R*,7a*S*,14*S*,14a*R*)-Dodecahydro-7,14-
  methano-2*H*,6*H*-dipyrido[1,2-*a*:1′,2′-
  e][1,5]diazocin-6-one *see* 716
(7α,7aα,14α,14aβ)-Dodecahydro-7,14-
  methano-2*H*,11*H*-dipyrido[1,2-*a*:1′,2′-
  e][1,5]diazocin-11-one *see* 5667
Dodecahydro-7,14-methano-4*H*,6*H*-di-
  pyrido[1,2-*a*:1′,2′-*e*][1,5]diazocin-4-
  one *see* 5667
(1*S*,8a*R*,9*S*,11*R*,12a*R*)-Dodecahydro-11-
  methyl-1,9-ethanobenzo[*i*]quinolizin-
  14-one *see* 5683
(1*R*,3a*R*,5a*R*,6*R*,9a*S*)-Dodecahydro-6-
  (2*Z*)-2-penten-4-ynylpyrrolo[1,2-*a*]-
  quinoline-1-ethanol *see* 4436
Dodecahydrosqualene *see* 8895
δ-Dodecalactone, 3448
Dodecamethylcyclohexasiloxane, 3449
2,2,4,4,6,6,8,8,10,10,12,12-Dodecamethyl-
  cyclohexasiloxane *see* 3449
Dodecamethylpentasiloxane, 3450
1,1,1,3,3,5,5,7,7,9,9,9-Dodecamethylpen-
  tasiloxane *see* 3450
Dodecamolybdophosphoric Acid *see*
  7454
Dodecanoic Acid *see* 5439
Dodecanoic Acid 1,1′-(Dibutylstannylene) Ester *see* 3047
Dodecanoic Acid 1,1′-(1,2-Ethanediyl)
  Ester *see* 4533
Dodecanoic Acid Ethyl Ester *see* 3873
Dodecanoic Acid 2-(2-Hydroxyethoxy)-
  ethyl Ester *see* 3135
Dodecanoic Acid 2-Sulfoethyl Ester
  Sodium Salt (1:1) *see* 8739
1-Dodecanol, 3451
5-Dodecanolide *see* 3448
1-[(2-Dodecanoyloxyethylcarbamoyl)-
  methyl]pyridinium Chloride *see* 5421
(2*S*)-3-[(1*E*,3*E*,5*E*,7*E*,9*E*)-1,3,5,7,9-Dode-
  capentaen-1-yloxy]-1,2-propanediol
  see 3452
[*S*-(*all-E*)]-3-(1,3,5,7,9-Dodecapentaen-
  yloxy)-1,2-propanediol, 3452
Dodécavit [Serb] *see* 4846
trans-2-Dodecenedioic Acid *see* 9737

(2*E*)-2-Dodecenedioic Acid *see* 9737
(2*Z*)-2-Dodecenedioic Acid *see* 9737
(7*Z*)-7-Dodecen-1-ol Acetate *see* 5627
cis-7-Dodecenyl Acetate *see* 5627
Dodecoic Acid *see* 5439
Dodecyl Alcohol *see* 3451
Dodecyl Alcohol Polyoxyethylene Ether
  see 7675
1-Dodecyl-4-aminoquinaldinium Acetate
  see 5442
1-Dodecylazacycloheptan-2-one *see* 5440
Dodecylbenzene Sodium Sulfonate *see*
  8748
Dodecylbenzenesulfonic Acid Sodium
  Salt (1:1) *see* 8748
Dodecylbetaine, 3453
N-Dodecyl-*N*,*N*-bis(2-hydroxyethyl)ben-
  zenemethanaminium Chloride (1:1)
  see 1114
Dodecyl Bromide *see* 5444
N-Dodecyl-ε-caprolactam *see* 5440
Dodecyl 2,4-Diguanidinophenyl Ether
  see 5441
Dodecyldi(β-hydroxyethyl)benzylammo-
  nium Chloride *see* 1114
(Dodecyldimethylammonio)acetate *see*
  3453
n-Dodecyl-*N*,*N*-dimethylglycine *see* 3453
Dodecyldimethyl(2-phenoxyethyl)ammo-
  nium Bromide *see* 3464
N-Dodecylguanidine Acetate (1:1) *see*
  3455
1-Dodecylhexahydro-2*H*-azepin-2-one
  see 5440
α-Dodecyl-ω-hydroxypoly(oxy-1,2-eth-
  anediyl) *see* 7675
(2*R*,3*S*)-2-Dodecyl-3-methylbutanedioic
  Acid *see* 8373
d-α-Dodecyl-β-methylsuccinic Acid *see*
  8373
(2*R*,3*S*)-2-Dodecyl-3-methylsuccinic Acid
  see 8373
N,*N*‴-[4-(Dodecyloxy)-1,3-phenylene]-
  bisguanidine *see* 5441
1,1′-[4-(Dodecyloxy)-*m*-phenylene]di-
  guanidine *see* 5441
Dodemorph, 3454
Dodigen 1881 [Hoechst] *see* 3110
Dodine, 3455
DODT *see* 3329
Dofetilide, 3456
Dogalina [Angelini] *see* 1037
Dogbane *see* 729
Dog Buttons *see* 6819
Dog Grass *see* 9938
Dogmatil [Synthelabo] *see* 9119
Dogmatyl [Fujisawa] *see* 9119
Dogwood *see* 2522
Doisynolic Acid, 3457
Doktacillin *see* 583
Dolac [Syntex] *see* 5353
Doladene *see* 6030
Dolalgial [Pharmainvesti] *see* 2381
Dolantin [HMR] *see* 5916
Dolasetron, 3458
Dolasetron Mesylate *see* 3458
Dolazon [Graeub] *see* 3390
Dolcidium [Galephar] *see* 5009
Dolcol [Dainippon] *see* 7572
Dolcontin *see* 6361
Dolcymene *see* 2760
Dolerophane *see* 2644
Dolestine [Teva] *see* 5916
Dolgit [Dolorgiet] *see* 4919
Dolichodial, 3459

(−)-Dolichodial see 3459
(cis,trans)-Dolichodial see 3459
Dolipol [Aventis] see 9667
Doliprane [Sanofi-Aventis] see 47
Dolisina see 3119
Dolitabs [Sanofi-Aventis] see 47
Dolmatil [Lorex] see 9119
Dolmen [Sigma-Tau] see 9291
Dolobid [Merck & Co.] see 3165
Dolobis [Merck & Co.] see 3165
Dolocyl [Novartis] see 4919
Dolomite see 5717
Dolophine [Lilly] see 6016
Dolosal [Specia] see 5916
Doloteffin [Ardeypharm] see 2944
Dolutegravir, 3460
Dolviran [Bayer] see 47
DOM, 3461
Domain see 9081
Domar [Zambeletti] see 7553
Domark [Isagro] see 9338
Domeboro [Bayer] see 320
Domesticine, 3462
Domestine see 3462
D.O.M.F. see 5934
Domical [Berk] see 483
Domin [Boehringer, Ing.] see 9171
Dominal [AWD] see 7998
Dominator see 7610
Domiodol, 3463
Domiphen Bromide, 3464
Domitor [Orion] see 5858
Domitroban, 3465
Domoic Acid, 3466
Domoso [Pfizer Animal Health] see 3285
Domperidone, 3467
DON see 3005
Dona [Rottapharm] see 4494
Donamet see 144
Donaxine see 4569
Donepezil, 3468
DongChongXiaCao see 2515
Donmox [Horita] see 52
Donorest [Wyeth] see 4030
Doom [United Phosphorus] see 3089
Doom Bark see 8519
DOP see 2868
Dopa, 3469
L-Dopa see 5516
Dopacard see 3472
Dopaflex [Medimpex] see 5516
Dopaidan [DeAngeli] see 5516
Dopal [Kyowa] see 5516
Dopalina [Lepetit] see 5516
Dopamet [Actavis] see 6127
Dopamine, 3470
Dopan, 3471
Dopar see 5516
Doparkine [Armstrong] see 5516
Doparl [Kyowa] see 5516
threo-Dopaserine see 3504
Dopasol [Daiichi] see 5516
Dopastat [Parke-Davis] see 3470
Dopaston [Sankyo] see 5516
Dopastral [Astra] see 5516
Dopegyt [Egis] see 6127
Dopergin [Schering AG] see 5574
Dopexamine, 3472
Dopexamine Hydrochloride see 3472
Dopram [Baxter] see 3481
Doprin [SK & F] see 5516
Dops [Sumitomo] see 3504
L-DOPS see 3504
L-threo-DOPS see 3504
Dorado [Syngenta] see 8096

Doral [Wallace] see 8145
Doralese [GSK] see 5010
Doralin [Menarini] see 6997
Doramectin, 3473
Dorantamin [Dorsey] see 8097
DoraQuest [Agrovet] see 3473
Doraxamin [Dorsey] see 3201
Dorcalm [Frere] see 2072
Dorcostrin [Dorsey] see 2903
Doricenone [Firmenich] see 2808
Dorico see 4740
Dorico Soluble see 4740
Doridamina [Angelini] see 5625
Doriden [Novartis] see 4513
Doripenem, 3474
Dorison see 5913
Dorixina [Roemmers] see 2381
Dormalest see 5913
Dormalin [Schering-Plough] see 8145
Dormate [Wallace] see 5842
Dormicum [Roche] see 6261
Dormidin see 5913
Dormigen see 5913
Dormigene [Pharmacobel] see 1404
Dormin see 12
Dormiphen see 5913
Dormison [Schering] see 5913
Dormodor [Roche] see 4228
Dormonoct [Hoechst] see 5631
Dormosan see 5913
Dormosedan [Orion] see 2939
Dormwell [Smith & Nephew] see 3055
Dornase Alfa see 2907
Dornavac [Merck & Co.] see 2907
Dorner [Toray] see 1154
Dorsiflex [Drossapharm] see 5918
Doryl [Merck KGaA] see 1781
Doryx [Faulding] see 3488
Dorzolamide, 3475
Dosberotec [Boehringer, Ing.] see 4012
Dosiseptive [Gifrer Barbezat] see 2092
Dosmalfate, 3476
Dostinex [Pharmacia & Upjohn] see 1606
Dosulepin see 3478
Dotan see 2103
(DOTA D-Phe¹,Tyr³)octreotide see 3561
Dotarizine, 3477
DOTATOC see 3561
(−)-DOTC see 742
Dothiepin, 3478
Double-mycin [Heyl] see 3197
Dovenix [Specia] see 6743
Dovonex [Westwood-Squibb] see 1644
Dow 1329 see 3436
Dow General [Dow] see 3315
Dow Selective [Dow] see 3315
Dow Shield [Dow AgroSci.] see 2389
Dowanol DPM [Dow] see 3384
Dowco 118 see 3436
Dowco 179 see 2195
Dowco 213 [Dow] see 2757
Dowco 214 see 2195
Dowco 233 see 9821
Dowco 290 see 2389
Dowco 356 see 9827
Dowco 433 see 4233
Dowex 1-X2-Cl see 2209
Dowfax 9N [Dow] see 6764
Dowfax 9N9 [Dow] see 6764
Dowfume W 85 see 3850
Dowicide 1 [Dow] see 7415
Dowicide 2 [Dow] see 9808
Dowicide 2S [Dow] see 9809
Dowicide 9 [Dow], 3479
Dowicide A [Dow] see 7415

Dowicide B [Dow] see 9808
Dowicil 75 [Dow] see 8143
Dowicil 200 [Dow] see 8143
Dowlap see 9807
Dowmycin E [Dow] see 3743
Dowpon see 2799
Doxacurium Chloride, 3480
Doxamax [Esparma] see 3482
Doxapram, 3481
Doxazosin, 3482
Doxazosin Mesylate see 3482
Doxephrin see 6020
Doxepin, 3483
Doxercalciferol, 3484
Doxergan see 7045
Doxicrisol [Quimifar] see 3488
Doxidan [HMR] see 3445
Doxifluridine, 3485
Doxil [Ortho Biotech] see 3487
Doxinate [Hoechst] see 3446
Doxirobe [Pharmacia & Upjohn] see 3488
Doxium [Carrion] see 3439
Doxofylline, 3486
Doxol [Blair] see 3446
Doxophylline see 3486
Doxorubicin, 3487
Doxychol see 7027
Doxycycline, 3488
Doxycycline Hyclate see 3488
Doxylamine, 3489
Doxylar [Sandoz] see 3488
Doyle [Tanabe Seiyaku] see 842
Dozic [Rosemont] see 4634
2,4-DP see 3088
DP-178 see 3636
DPA see 3342, 7918
DPC see 3338
DPCC see 3351
DPC-Cl see 3351
DPD see 1520
DPE see 3378
D-Penamine [Alphapharm] see 7197
DPGME see 3384
DPM see 3384
DPN see 6429
DPNH see 6429
DPPA see 3366
DPPB see 1254
DPPC see 2461
DPPD see 3363
DPPE see 1254, 9321
DPPF, 3490
DPPH see 3369
DPPM see 1254
DPPP see 1254
DPQ, 3491
D-Pron [Minnesota Pharm.] see 3390
DPX-1410 see 7019
DPX-3217 see 2762
DPX-3674 see 4735
DPX-3792 see 34
DPX-4189 see 2197
DPX-5648 see 9087
DPX-43898 see 2088
DPX-66037 see 9857
DPX-A8947 see 905
DPX-E2Y45 see 2080
DPX-E9636 see 8356
DPX EY-059 see 4164
DPX-F5384 see 1053
DPX-F6025 see 2094
DPX-H6573 see 4236
DPX-HGW86 see 2689
DPX-JE874 see 3972
DPX-JW062 see 5012

E-Mycin *see* 3739
EN-141 *see* 5315
EN-313 *see* 6354
EN-1010 *see* 8127
EN-1639A *see* 6448
EN-1733A *see* 6319
EN-2234A *see* 6441
EN-15304 *see* 6447
ENA-713 *see* 8369
Enable [Dow AgroSci.] *see* 3995
Enablex [Novartis] *see* 2826
Enacard [Merial] *see* 3623
Enadel [Pfizer] *see* 2404
Enalapril, 3623
Enalaprilic Acid *see* 3623
Enallachrome *see* 3753
Enanthal *see* 4693
Enanthaldehyde *see* 4693
Enanthic Acid *see* 4695
Enanthic Alcohol *see* 4696
Enanthotoxin, 3624
Enantone [Takeda] *see* 5509
Enantyum [Menarini] *see* 5352
Enapren [Merck & Co.] *see* 3623
Enargite *see* 2504
Enarmon [Teikoku Zoki] *see* 9324
Enavid [Searle] *see* 6787
Enbol [Chugai] *see* 8109
Enbrel [Wyeth] *see* 3767
Enbucrilate *see* 1566
Encainide, 3625
Encaprin [Procter & Gamble] *see* 841
Encare [Norwich] *see* 6764
Encephabol [Merck KGaA] *see* 8109
Encetrop [Alpharma] *see* 7600
Enclomiphene *see* 2377
Encorton [Polfa] *see* 7842
Endak [Madaus] *see* 1866
Endep [Roche] *see* 483
Endiandric Acid A *see* 3626
Endiandric Acid B *see* 3626
Endiandric Acid C *see* 3626
Endiandric Acids, 3626
Endoamylases *see* 594
Endocaine [Endo] *see* 8127
Endociclina [Del Saz & Filippini] *see* 4281
Endocistobil [Bracco] *see* 5065
Endocorion [ELEA] *see* 2221
Endodextranases *see* 2951
Endofolliculina *see* 3763
Endogenous Pyrogen *see* 5037
Endografin [Schering AG] *see* 5065
Endojodin [Bayer] *see* 7897
Endometrion [Jenapharm] *see* 3117
Endomixin [Lusofarmaco] *see* 6539
Endo-Paractol [Temmler] *see* 3237
Endophenolphthalein *see* 7073
Endoprost [Italfarmaco] *see* 4942
Endorem [Guerbet] *see* 4090
α-Endorphin *see* 3627
β-Endorphin *see* 3627
γ-Endorphin *see* 3627
α-*neo*-Endorphin *see* 3627
β-*neo*-Endorphin *see* 3627
Endorphins, 3627
Endostatin, 3628
Endosulfan, 3629
Endothal *see* 3630
Endothall, 3630
Endothelial Cell Growth Factor *see* 4101
Endothelial-Leukocyte Adhesion Molecule-1 *see* 8563
Endothelin, 3631
Endothelin-1 *see* 3631

Endothelin-2 *see* 3631
Endothelin-3 *see* 3631
Endothium-derived Relaxing Factor *see* 6664
Endoxan [Baxter] *see* 2743
3,6-Endoxohexahydrophthalic Acid *see* 3630
Endralazine, 3632
Endralazine Mesylate *see* 3632
Endrate Disodium [Abbott] *see* 3565
Endrate Tetrasodium [Abbott] *see* 3565
Endrin, 3633
Endura [BASF] *see* 1354
Enduracidin, 3634
Enduracidin A Hydrochloride *see* 3634
Enduracidin B Hydrochloride *see* 3634
Endurance [Syngenta] *see* 7883
Enduron [Abbott] *see* 6079
Enduronum [Abbott] *see* 6079
E.N.E. *see* 3888
Enelfa [Dolorgiet] *see* 47
Enerbol [Polfa] *see* 8109
Energona [Maurer] *see* 6788
Enflurane, 3635
Enfuvirtide, 3636
Engemycin [Intervet] *see* 7075
English Chamomile *see* 2040
English Hawthorn *see* 2555
English White *see* 1659
Enheptin [Am. Cyanamid] *see* 453
Enheptin-A [Am. Cyanamid] *see* 404
Enhexymal *see* 4740
Enide [Tuco] *see* 3337
Enidin *see* 5779
Enilconazole, 3637
Eniluracil, 3638
Enin *see* 5779
Enirant [Desitin] *see* 3710
Enisyl [Person & Covey] *see* 5697
Enkaid [Mead Johnson] *see* 3625
Enkalene [Kunstzijde] *see* 7689
Enkalon [Am. Enka] *see* 6823
Enkephalinase *see* 863
Enkephalins *see* 3627
Enlon [Baxter] *see* 3564
Ennds *see* 2158
Enniatin A *see* 3639
Enniatin B *see* 3639
Enniatin C *see* 3639
Enniatins, 3639
Enocitabine, 3640
3-Enolpyruvic Ether of *trans*-3,4-Dihydroxycyclohexa-1,5-diene Carboxylic Acid *see* 2222
Enovid *see* 6787
Enoxacin, 3641
Enoxaparin, 3642
Enoxen [Dainippon] *see* 3641
Enoximone, 3643
Enoxolone, 3644
Enoxor [Fabre] *see* 3641
Enpro [Kobayashi] *see* 7875
Enprostil, 3645
Enradin [Takeda] *see* 3634
Enramycin *see* 3634
Enrofloxacin, 3646
Enrumay [Cooper] *see* 8097
E.N.S. *see* 3888
Ensidon [Novartis] *see* 6948
Enstatite *see* 5751
Ensulizole, 3647
ENT-987 *see* 9525
ENT-1122 *see* 3315
ENT-7796 *see* 5556
ENT-14250 *see* 7589

ENT-14874 *see* 10365
ENT-15108 *see* 7137
ENT-16225 *see* 3114
ENT-16273 *see* 9096
ENT-17034 *see* 5764
ENT-17251 *see* 3633
ENT-17292 *see* 6181
ENT-17510 *see* 252
ENT-17588 *see* 8114
ENT-20218 *see* 2859
ENT-20852 *see* 1451
ENT-22014 *see* 906
ENT-22374 *see* 6244
ENT-22879 *see* 3330
ENT-23233 *see* 906
ENT-23347 *see* 3408
ENT-23648 *see* 3096
ENT-23969 *see* 1788
ENT-24042 *see* 7443
ENT-24105 *see* 3790
ENT-24482 *see* 3097
ENT-24727 *see* 3314
ENT-24988 *see* 6442
ENT-25445 *see* 485
ENT-25515 *see* 7449
ENT-25540 *see* 4029
ENT-25567 *see* 6438
ENT-25604 *see* 2782
ENT-25647 *see* 3436
ENT-25705 *see* 7448
ENT-25715 *see* 4007
ENT-025719 *see* 6290
ENT-25784 *see* 428
ENT-25841 *see* 9337
ENT-25991 *see* 5922
ENT-26398 *see* 4208
ENT-26538 *see* 1774
ENT-27129 *see* 6337
ENT-27162 *see* 1442
ENT-27165 *see* 9278
ENT-27226 *see* 7922
ENT-27258 *see* 1442
ENT-27311 *see* 2195
ENT-27395 *see* 2757
ENT-27396 *see* 6019
ENT-27520 *see* 2195
ENT-27552 *see* 1445
ENT-27766 *see* 7609
ENT-27967 *see* 482
ENT-29054 *see* 3161
ENT-29106 *see* 6635
ENT-33266 *see* 5627
ENT-34886 *see* 3403
ENT-50852 *see* 316
ENT-50882 *see* 4763
Entacapone, 3648
Entacyl [BDH] *see* 7577
Entecavir, 3649
Enteramine *see* 8601
Entereg [Adolor] *see* 366
Enterfram *see* 6539
Enterosalil [Farmabraz] *see* 8469
Enteroanthelone *see* 3651
Enterobactin, 3650
Enterobiocine [Rousselot] *see* 9074
Enterochelin *see* 3650
Enterocid [Schering] *see* 7488
Enterocol *see* 6742
Enterocura [Nordmark] *see* 9041
Enterogastrone, 3651
Enterokanacin [Sidus] *see* 5329
Enterol [Lenza] *see* 7618
Enterolactone, 3652
Enteromide [Consolidated Chem.] *see* 9044

Enteroquinol *see* 5075
Enterosalicyl [Delalande] *see* 8469
Enterosept *see* 5085
Entero-Septol *see* 5075
Enterosulfamid [Katwijk] *see* 7488
Enterosulfon [CMC] *see* 7488
Enterozol *see* 5075
Entobex *see* 7312
Entocort [AstraZeneca] *see* 1476
Entothein *see* 8555
Entprol, 3653
Entrokin *see* 5075
Entrust [Dow AgroSci.] *see* 8877
Entumine [Novartis] *see* 2399
Enturen [Novartis] *see* 9079
Entyderma [Taiyo] *see* 1017
ENU *see* 3887
Envacar [Pfizer] *see* 4603
Envidor [Bayer] *see* 8884
Envilon [Toyo] *see* 7707
Enviomycin, 3654
Enviroxime, 3655
Enzactin [Ayerst] *see* 9751
Enzamin [Kowa] *see* 1125
Enzaprost F [Ceva] *see* 7989
Enzastaurin, 3656
Enzicoba [Farmasa] *see* 2435
Enzodase [Squibb] *see* 4797
Enzorb-A [Regis] *see* 7367
Enzyme Q *see* 8137
EO9 *see* 712
Eoden [Woelm] *see* 4692
EOquin [Spectrum] *see* 712
Eosine *see* 3658
Eosin Blue *see* 7440
Eosine *see* 3658
Eosine Acid *see* 3658
Eosine I Bluish, 3657
Eosinophil Differentiation Factor *see* 5041
Eosin Y, 3658
Eosin Yellowish *see* 3658
Eosote *see* 2561
Eotaxin, 3659
Eovist [Schering AG] *see* 4360
Ep *see* 3744
EP-475 *see* 2925
3-EP *see* 3785
EPA *see* 3578, 7344
Epadel [Mochida] *see* 3578
Epafosforil [Coli] *see* 7474
Epalons *see* 267
Epalrestat, 3660
Epanutin [Pfizer] *see* 7433
Epatec [Nissan] *see* 5352
EPEG *see* 3929
Eperisone, 3661
Ephedra, 3662
Ephedrine, 3663
d-*ψ*-Ephedrine *see* 8024
l-Ephedrine *see* 3663
Ephedrone *see* 6035
Ephicillin Hydriodide *see* 7194
Ephynal [Roche] *see* 9653
Epi-Aberel [Janssen-Cilag] *see* 8288
4′-Epiadriamycin *see* 3678
Epiallopregnanolone *see* 267
5-Epiandrostane *see* 3915
Epiandrosterone, 3664
Epibatidine, 3665
Epibloc [Pestcon] *see* 2146
(−)-Epicatechin *see* 1902
Epichlorohydrin, 3666
dl-*α*-Epichlorohydrin *see* 3666
Epicholestanol, 3667

Epicholesterol, 3668
Epicillin, 3669
Epicinchonidine *see* 2289
Epicinchonine *see* 2290
Epiclase [Bellon] *see* 7318
Epicocconone, 3670
Epicure [Novartis Agro] *see* 2768
Epidermal Growth Factor *see* 3574
α-Epidibromohydrin *see* 3030
Epidicentrine *see* 3462
3,17-Epidihydroxyestratriene *see* 3758
Epidione [Bellon] *see* 9875
Epidosin [Solvay] *see* 10092
4′-Epidoxorubicin *see* 3678
Epidropal [Teofarma] *see* 271
4′-Epi-DX *see* 3678
Epi-echinuline *see* 3546
8-Epiestradiol *see* 5212
Δ^{1,3,5}-8*α*-Epiestratriene-3,17*β*-diol *see* 5212
Δ^{1,3,5}-8-Epiestratrien-3-ol-17-one *see* 5213
16-Epiestriol, 3671
8-Epiestrone *see* 5213
Epifrin [Allergan] *see* 3674
Epifucose *see* 8194
(−)-Epigallocatechin 3-*O*-Gallate *see* 3573
(−)-Epigallocatechol Gallate *see* 3573
Epihetacillin *see* 4707
3-Epihydroxyetioallocholan-17-one *see* 633
Epileo Petitmal [Eisai] *see* 3801
Epilim [Sanofi-Aventis] *see* 10099
Epimedii Herba, 3672
(17*S*:12*S*:25*S*)-22,26-Epimino-18-(13 → 17)-*abeo*-cholesta-5,12-diene-3*β*,16*β*-diol *see* 10143
(22*S*:25*S*)-12,26-Epimino-17*β*-methyl-18-norcholesta-5,12-diene-3*β*,16*β*-diol *see* 10143
Epi-Monistat [Cilag] *see* 6257
Epinastine, 3673
Epinephrine, 3674
Epinephrine Isopropyl Homolog *see* 5263
Epinine *see* 2904
Epinyl [Nordic] *see* 3773
16-Epioestriol *see* 3671
EpiPen [Dey] *see* 3674
Epi-Pevaryl [Cilag-Chemie] *see* 3549
Epiphenethicillin Potassium *see* 7337
Epiquinidine, 3675
Epiquinine, 3676
Epirenor [Byk Gulden] *see* 6848
D-Epirhamnose *see* 8194
Epirizole, 3677
Epirubicin, 3678
Episesamin *see* 812
(−)-Episesamin *see* 812
(+)-Episesamin *see* 812
Epistilbite *see* 1651
Epitestosterone, 3679
Epitetracycline *see* 8144
Epithienamycins *see* 6929
(2*α*,3*α*,5*α*,17*β*)-2,3-Epithioandrostan-17-ol *see* 3680
(2*α*,3*α*,5*α*,17*β*)-2,3-Epithio-17-[(1-methoxycyclopentyl)oxy]androstane *see* 5925
Epitiostanol, 3680
Epitol [Teva] *see* 1783
Epitopic [Gerda] *see* 3171
Epivir [GSK] *see* 5402
Epizizanal *see* 6908

Eplerenone, 3681
EPN, 3682
Epo *see* 3744
EPO *see* 3949
EPO-906 *see* 3684
Epobron [Ono] *see* 4919
Epocan [Merck KGaA] *see* 8109
Epocelin [Astellas] *see* 1952
Epodyl [Ayerst] *see* 3925
Epoetin *see* 3744
Epoetin Alfa *see* 3744
Epoetin Beta *see* 3744
Epoetin Theta *see* 3744
Epoetin Zeta *see* 3744
Epogam [Scotia] *see* 3949
Epogen [Amgen] *see* 3744
Epogin [Chugai] *see* 3744
Epontol [Bayer] *see* 7917
Epoprostenol *see* 7986
Epoprostenol Sodium *see* 7986
Eporal *see* 2822
Eporatio [Ratiopharm] *see* 3744
Eposerin [Pfizer] *see* 1952
Epostane, 3683
Epothilone A *see* 3684
Epothilone B *see* 3684
Epothilone D *see* 3684
Epothilones, 3684
Epoxiconazole, 3685
Epoxomicin, 3686
(1*α*,4*α*,5*β*,16*β*)-1,4-Epoxy-3-aza-A-homoandrostan-16-ol *see* 8483
(3*β*,4*α*,7*α*,15*α*,16*β*)-4,9-Epoxycevane-3,-4,7,14,15,16,20-heptol *see* 4446
(3*α*,4*α*,16*β*)-4,9-Epoxycevane-3,4,12,14,-16,17,20-heptol *see* 2034
(3*β*,4*α*,16*β*)-4,9-Epoxycevane-3,4,12,14,-16,17,20-heptol 3-(3,4-Dimethoxybenzoate) *see* 10148
[3*β*,4*α*,16*β*]-4,9-Epoxycevane-3,4,12,14,-16,17,20-heptol 3-[(2*Z*)-2-Methyl-2-butenoate] *see* 2032
(3*β*,4*α*,6*α*,7*α*,15*α*,16*β*)-4,9-Epoxycevane-3,4,6,7,14,15,16,20-octol *see* 8009
[3*β*(2*R*,3*R*),4*α*,6*α*,7*α*,15*α*(*R*),16*β*]-4,9-Epoxycevane-3,4,6,7,14,15,16,20-octol 6,7-Diacetate 3-(2,3-Dihydroxy-2-methylbutanoate) 15-(2-Methylbutanoate) *see* 8008
[3*β*(*S*),4*α*,6*α*,7*α*,15*α*(*R*),16*β*]-4,9-Epoxycevane-3,4,6,7,14,15,16,20-octol 6,7-Diacetate 3-(2-Hydroxy-2-methylbutanoate) 15-(2-Methylbutanoate) *see* 8008
6*β*,6a*β*-Epoxy-2,3,3a*α*,3b,4,5,6,6a,7,7a*α*-decahydro-5*β*-hydroxy-2*β*,3b*β*-dimethyl-4-methylene-1*H*-cyclopenta[*a*]pentalene-2-carboxylic Acid *see* 4754
(11*α*,15*R*)-11,15-Epoxy-16,16-difluoro-15-hydroxy-9-oxoprostan-1-oic Acid *see* 5649
(3*S*,3′*R*,5*R*,6′*R*,8*R*)-5,8-Epoxy-5,8-dihydro-*β*,*ε*-carotene-3,3′-diol *see* 4127
(3*S*,3′*R*,5*R*,6′*R*,8*S*)-5,8-Epoxy-5,8-dihydro-*β*,*ε*-carotene-3,3′-diol *see* 2254
(4*α*,5*α*,17*β*)-4,5-Epoxy-3,17-dihydroxyandrost-2-ene-2-carbonitrile *see* 9867
(4*α*,5*α*,17*β*)-4,5-Epoxy-3,17-dihydroxy-4,17-dimethylandrost-2-ene-2-carbonitrile *see* 3683
1,10-Epoxy-6,8-dihydroxygermacra-4,11-(13)-dien-12-oic Acid 12,8-Lactone Acetate *see* 8076
15,16-Epoxy-6*β*,9-dihydroxy-8*βH*-labda-13(16),14-dien-19-oic Acid *γ*-Lactone *see* 5820

(5α)-4,5-Epoxy-3,14-dihydroxy-17-meth-ylmorphinan-6-one *see* 7068

(5α)-4,5-Epoxy-3,14-dihydroxy-17-(2-propen-1-yl)morphinan-6-one *see* 6447

(5Z,9α,11α,13E,15S)-6,9-Epoxy-11,15-dihydroxyprosta-5,13-dien-1-oic Acid *see* 7986

(1′α)-6′,7-Epoxy-6,12′-dimethoxy-2′-methyloxyacanthan *see* 9866

4a,9a-Epoxy-3-(2,3-epoxybutyryl)-1,2,3,-4,4a,9a-hexahydro-1,3,4,5,10-pentahy-droxy-2-methylanthrone *see* 2006

3α,9α-Epoxy-14β,18β-(epoxyethano-*N*-methylimino)-5β-pregna-7,16-diene-3β,11α,20α-triol 20α-Ester with 2,4-Dimethylpyrrole-3-carboxylic Acid *see* 1007

(8R)-6β,7β-Epoxy-8-ethyl-3α-hydroxy-1αH,5αH-tropanium Bromide (−)-Tropate *see* 7042

β,17-Epoxy-α-(3-ethylidene-1,2,3,4,6,7,-12,12b-octahydroindole[2,3-*a*]quinoli-zin-2-yl)-curan-1-propanoic Acid Methyl Ester *see* 4414

4,4a-Epoxy-5-ethylidene-2,3,4,4a-tetra-hydro-5*H*-1-pyridine *see* 8

5β,20-Epoxy-1,2α,4,7β,10β,13α-hexahy-droxytax-11-en-9-one 4,10-Diacetate 2-Benzoate 13-Ester with (2R,3S)-*N*-Benzoyl-3-phenylisoserine *see* 7081

(3β,5β,15β)-14,15-Epoxy-3-hydroxy-bufa-20,22-dienolide *see* 8266

(8β,10β)-8,10-Epoxy-8-hydroxy-3,4-di-methoxy-17-methylhasubanan-7-one *see* 6000

(5α)-4,5-Epoxy-3-hydroxy-5,17-dimeth-ylmorphinan-6-one *see* 6227

(2α,4α,5α,17β)-4,5-Epoxy-17-hydroxy-4,17-dimethyl-3-oxoandrostane-2-carbonitrile *see* 3683

4,5α-Epoxy-6β-hydroxygermacra-1(10),11(13)-dien-12-oic Acid γ-Lac-tone *see* 7152

1,10-Epoxy-6β-hydroxy-1β,5β,7α-guaii-3-en-12-oic Acid γ-Lactone *see* 761

(αR,5α,7α)-4,5-Epoxy-3-hydroxy-6-methoxy-α,17-dimethyl-α-propyl-6,14-ethenomorphinan-7-methanol *see* 3931

(5α)-4,5-Epoxy-14-hydroxy-3-methoxy-17-methylmorphinan-6-one *see* 7058

(5α)-4,5-Epoxy-3-hydroxy-17-methyl-morphinan-6-one *see* 4840

6β,7β-Epoxy-3β-hydroxy-8-methyl-1αH,5αH-tropanium Bromide Di-2-thienylglycolate *see* 9610

6β,7β-Epoxy-3α-hydroxy-8-methyl-1αH,5αH-tropanium Bromide Tro-pate (Ester) *see* 6077

4,5-Epoxy-17-hydroxy-3-oxoandrostane-2-carbonitrile *see* 9867

(7α,11α,17α)-9,11-Epoxy-17-hydroxy-3-oxopregn-4-ene-7,21-dicarboxylic Acid γ-Lactone 7-Methyl Ester *see* 3681

(5Z,9α,11α,13E,15S)-9,11-Epoxy-15-hy-droxythromboxa-5,13-dien-1-oic Acid *see* 9542

5,6-Epoxy-3-hydroxy-*p*-toluquinone *see* 9315

12,13-Epoxy-4-hydroxytrichothec-9-en-8-one Crotonate *see* 9815

3α,6α-Epoxy-7β-hydroxytropane *see* 8547

(3β,23β)-17,23-Epoxy-3-hydroxyveratra-man-11-one *see* 5312

9,13-Epoxylabd-7-en-15-oic Acid *see* 4582

1,8-Epoxy-*p*-menthane *see* 3938

Epoxymethamine Bromide *see* 6077

9α,11-Epoxy-7α-(methoxycarbonyl)-3-oxo-17α-pregn-4-ene-21,17-carbolac-tone *see* 3681

9α,11α-Epoxy-7α-methoxycarbonyl-20-spirox-4-ene-3,21-dione *see* 3681

(5α,6α)-4,5-Epoxy-3-methoxy-17-methyl-morphinan-6-ol *see* 3189

(5α,6β)-4,5-Epoxy-3-methoxy-17-methyl-morphinan-6-ol *see* 3193

(5α)-4,5-Epoxy-3-methoxy-17-methyl-morphinan-6-one *see* 4823

6′,7-Epoxy-6-methoxy-2-methyl-oxy-acanthan-12′-ol *see* 6258

(5α,6α)-4,5-Epoxy-17-methylmorphinan-3,6-diol *see* 3194

(5α,6α)-4,5-Epoxy-17-methylmorphinan-3,6-diol Diacetate (Ester) *see* 2970

cis-7,8-Epoxy-2-methyloctadecane *see* 3403

Epoxymexrenone *see* 3681

cis-12,13-Epoxyoctadec-*cis*-9-enoic Acid *see* 10159

12,13-Epoxyoleic Acid *see* 10159

(12R,13S)-(−)-12,13-Epoxyoleic Acid *see* 10159

(12S,13R)-(+)-12,13-Epoxyoleic Acid *see* 10159

(4β)-12,13-Epoxy-4-[[(2Z)-1-oxo-2-buten-yl]oxy]trichothec-9-en-8-one *see* 9815

2,3-Epoxy-4-oxo-7,10-dodecadienamide *see* 2003

2,3-Epoxy-4-oxo-7,10-dodecadienoylam-ide *see* 2003

1,3-Epoxypropane *see* 9886

2,3-Epoxy-1-propanol *see* 4525

(−)-(1R,2S)-(1,2-Epoxypropyl)phos-phonic Acid *see* 4281

(3α,4β,7α)-12,13-Epoxy-3,4,7,15-tetrahy-droxytrichothec-9-en-8-one *see* 6746

(3α,4β,8α)-12,13-Epoxytrichothec-9-ene-3,4,8,15-tetrol 4,15-Diacetate 8-(3-Methylbutanoate) *see* 9981

(4β)-12,13-Epoxytrichothec-9-en-4-ol Acetate *see* 9812

14,15-Epoxy-3β,5,16β-trihydroxy-5β-bufa-20,22-dienolide 16-Acetate *see* 2308

15,16-Epoxy-1β,4,12-trihydroxy-5,9-di-methyl-17,18-dinor-8βH,9βH,10α-lab-da-2,13(16),14-triene-19,20-dioic Acid 19,1:20,12-Dilactone *see* 2476

(4β,5β,6β,22R)-5,6-Epoxy-4,22,27-trihy-droxy-1-oxoergosta-2,24-dien-26-oic Acid δ-Lactone *see* 10246

(3α,7α)-12,13-Epoxy-3,7,15-trihydroxy-trichothec-9-en-8-one *see* 10229

3β,7β-Epoxy-1βH,5βH-tropan-6α-ol *see* 8547

6β,7β-Epoxy-1αH,5αH-tropan-3α-ol (−)-Tropate *see* 8543

6β,7β-Epoxy-1αH,5αH-tropan-3α-ol (−)-Tropate 8-Oxide *see* 8544

6β,7β-Epoxy-3α-tropanyl S-(−)-Tropate *see* 8543

6,7-Epoxytropine Tropate *see* 8543

Epratuzumab, 3687

Epratuzumab, 1,4,7,10-Tetraazacyclodo-decane-1,4,7,10-tetraacetic Acid Con-jugate, Yttrium-[90]Y Chelate *see* 3687

Eprazinone, 3688

Eprex [Janssen-Cilag] *see* 3744

Eprinex [Merck & Co.] *see* 3689

Eprinomectin, 3689

Epristeride, 3690

Eprosartan, 3691

Eprosartan Mesylate *see* 3691

Eprotirome, 3692

Eprozinol, 3693

Epsikapron [Pharmacia] *see* 427

Epsilcapramin *see* 427

Epsilon-aminocaproic Acid *see* 427

Epsiprantel, 3694

Epsomite *see* 5717, 5755

Epsom Salts *see* 5755

Epsyl [Exa] *see* 7598

Eptam [Syngenta] *see* 3697

Eptastatin Sodium *see* 7835

Eptastigmine, 3695

Eptazocine, 3696

EPTC, 3697

Eptifibatide, 3698

Equa [Novartis] *see* 10177

Equal [NutraSweet Co.] *see* 829

Equalactin [Numark] *see* 1699

Equanil [Wyeth-Ayerst] *see* 5929

Equasym [UCB] *see* 6183

Equest [Fort Dodge] *see* 6376

Equetro [Validus] *see* 1783

Equibral [Ravizza] *see* 2085

Equilase *see* 1900

Equilenin, 3699

Equilibrin [RPR] *see* 484

Equilin, 3700

Equimate *see* 4224

Equine Chorionic Gonadotropin *see* 2221

Equinox [BASF] *see* 9295

Equipalazone [NCN] *see* 7390

Equipertine *see* 7070

Equipoise [Fort Dodge] *see* 1327

Equipose *see* 4889

Equiproxen [Syntex] *see* 6499

Equisetic Acid *see* 109

Equitac [Pfizer] *see* 7035

Equitrol [Farnam] *see* 9337

Equivurm Plus [Janssen] *see* 5837

Equizole [Merial] *see* 9440

Equol, 3701

Equron [Solvay] *see* 4796

Eqvalan [Merial] *see* 5296

ER-115 *see* 9275

ER-30346 *see* 8239

ER-086526 *see* 3723

ER-203492-00 *see* 5494

Erabutoxin A *see* 3702

Erabutoxin B *see* 3702

Erabutoxin C *see* 3702

Erabutoxins, 3702

Eracine [Sanofi Winthrop] *see* 8400

Eradacil [Winthrop] *see* 8400

Eradacin [Sanofi Winthrop] *see* 8400

Eraldin [Zeneca] *see* 7818

Eraxis [Pfizer] *see* 649

Eraze [Motomco] *see* 10352

Erazon [Krka] *see* 7619

Erbalax-N [Erba] *see* 4488

Erbia *see* 3703

Erbitux [ImClone] *see* 2026

Erbium, 3703

Erbocain [Heilit] *see* 4257

Ercefurol *see* 6619

Ercefuryl [Sanofi-Synthelabo] *see* 6619

Erco-Fer [Orion] *see* 4073

Ercoquin *see* 4857

Erdosteine, 3704

Erdotin [Galen] *see* 3704

Erebile *see* 2872
Erecnos [Fournier] *see* 6378
Erelib [Bial-Portela] *see* 3754
Eremeyevite *see* 326
Eremfat [Fatol] *see* 8341
4*βH*,5*α*-Eremophila-1(10),7(11)-dien-2-one *see* 10167
Eremursine *see* 4785
Erex [Ion] *see* 10300
Ergadenylic Acid *see* 146
Ergamisol [Janssen] *see* 5511
Ergenyl [Sanofi-Synthelabo] *see* 10099
Ergine *see* 5693
Ergobasine *see* 3712
Ergobel [Hormosan] *see* 6581
Ergocalciferol *see* 10215
Ergocalm [Teofarma] *see* 5639
Ergochrome AA(2,2′) *see* 8555
Ergochrome CC(2,2′) *see* 3709
Ergoclavine *see* 3713
Ergoclavinine *see* 3713
Ergocornine, 3705
Ergocristine, 3706
Ergocryptine, 3707
α-Ergocryptine *see* 3707
β-Ergocryptine *see* 3707
Ergocryptinine, 3708
α-Ergocryptinine *see* 3708
β-Ergocryptinine *see* 3708
Ergodesit [Graf] *see* 3710
Ergoflavin, 3709
Ergohydrin [Streuli] *see* 3710
Ergoklinine *see* 3712
Ergokryptine *see* 3707
α-Ergokryptinine *see* 3708
Ergoloid Mesylates, 3710
Ergomar [Lotus] *see* 3718
Ergometrine *see* 3712
Ergometrinine, 3711
Ergomimet [Astellas] *see* 3191
Ergonovine, 3712
Ergonovinine *see* 3711
Ergont [Desitin] *see* 3191
Ergoplus [Formosan] *see* 3710
Ergosine, 3713
Ergosinine *see* 3713
(3*β*,5*α*)-Ergosta-7-en-3-ol *see* 4320
Ergostane, 3714
(5*α*)-Ergostane *see* 3714
Ergostanol, 3715
(3*β*,5*α*)-Ergostan-3-ol *see* 3715
Ergostat [Warner-Lambert] *see* 3718
(3*β*,22*E*)-Ergosta-5,7,9(11),22-tetraen-3-ol *see* 2877
Ergosta-5:6,7:8,22:23-trien-3-ol *see* 3716
(3*β*,9*β*,10*α*,22*E*)-Ergosta-5,7,22-trien-3-ol *see* 5662
(3*β*,9*β*,22*E*)-Ergosta-5,7,22-trien-3-ol *see* 5265
(3*β*,10*α*,22*E*)-Ergosta-5,7,22-trien-3-ol *see* 8110
(3*β*,22*E*)-Ergosta-5,7,22-trien-3-ol *see* 3716
(3*β*,9*β*,10*α*,22*E*)-Ergosta-5,7,22-trien-3-ol Compd with (1*S*,3*Z*)-4-Methylene-3-[(2*E*)-2-[(1*R*,3a*S*,7a*R*)-octahydro-7a-methyl-1-[(1*R*,2*E*,4*R*)-1,4,5-trimethyl-2-hexen-1-yl]-4*H*-inden-4-ylidene]ethylidene]cyclohexanol (1:1) *see* 10214
(3*β*,24*R*)-Ergost-5-en-3-ol *see* 1731
Ergosterin *see* 3716
Ergosterol, 3716
9*β*-Ergosterol *see* 5265
Ergostetrine *see* 3712
Ergot, 3717

Ergotamine, 3718
Ergotaminine, 3719
Ergothioneine, 3720
L(+)-Ergothioneine *see* 3720
Ergotinine, 3721
Ergotocine *see* 3712
Ergotonin [Streuli] *see* 3191
Ergotop [Kwizda] *see* 6581
Ergotoxine, 3722
Ergotoxinine *see* 3721
Ergotrate [Lilly] *see* 3712
Ergotrate-H [Lilly] *see* 3712
Ergotrate Maleate [Lilly] *see* 3712
Eribulin, 3723
Eribulin Mesylate *see* 3723
Erigeron, 3724
Eril [Asahi] *see* 3979
Erimin [Sumitomo] *see* 6634
Erinitrit *see* 8779
Eriochrome Black T [Ciba], 3725
Eriodictin *see* 3726
Eriodictyol, 3726
Eriodictyon, 3727
Eriodictyonone *see* 4774
Eriosept [Kreussler] *see* 2914
Eriscel [Rachelle] *see* 3740
Erisimin *see* 4667
Erispan [Sumitomo] *see* 4159
Eritadenine, 3728
Eritoran, 3729
Eritrocina [Abbott] *see* 3739
Eritroger [Isnardi] *see* 3740
Erlosamide *see* 5381
Erlotinib, 3730
Ermetrine [Organon] *see* 3712
Ermysin [Orion] *see* 3739
Ernodasa [ERN] *see* 8953
Eromycin *see* 3740
Errolon *see* 4338
Ertapenem, 3731
Erucic Acid, 3732
Ervasil [Gödecke] *see* 345
ERYC [Warner-Chilcott] *see* 3739
Erycen [APS] *see* 3739
Erycin [Nycomed] *see* 3739
Erycinum [Cytochemia] *see* 3739
Erycorbin *see* 5171
Ery-Derm [Abbott] *see* 3739
Eryliquid [Linden] *see* 3739
Erymax [Merz] *see* 3739
Erypar [Parke-Davis] *see* 3743
Eryped [Abbott] *see* 3739
Erypo [Janssen-Cilag] *see* 3744
Eryprim [Scarium] *see* 3743
Erysan *see* 2108
Erysimin *see* 4667
Ery-Tab [Abbott] *see* 3739
Erythorbic Acid *see* 5171
Erythrene *see* 1512
Erythricine *see* 4429
Erythrite *see* 2418
Erythritol, 3733
meso-Erythritol *see* 3733
Erythritol Anhydride, 3734
Erythritol Tetranitrate *see* 3735
Erythrityl Tetranitrate, 3735
Erythro S [Sanko] *see* 3743
Erythrocentaurin, 3736
Erythrocin [Abbott] *see* 3743
Erythrocin Lactobionate [Abbott] *see* 3741
Erythrocuprein *see* 9131
Erythrogenic Acid *see* 5147
Erythroglucin *see* 3733
D-Erythrohexulose *see* 8033

α-Erythroidine, 3737
β-Erythroidine, 3738
Erythrol *see* 3733
Erythrol Tetranitrate *see* 3735
Erythromid [Abbott] *see* 3739
Erythromycin, 3739
Erythromycin A *see* 3739
Erythromycin Estolate, 3740
Erythromycin 4-*O*-*β*-D-Galactopyranosyl-D-gluconate (1:1) *see* 3741
Erythromycin Lactobionate, 3741
Erythromycin 9-[*O*-[(2-Methoxyethoxy)methyl]oxime] *see* 8409
Erythromycin Octadecanoate (1:1) *see* 3743
Erythromycin 2′-Propanoate *see* 3742
Erythromycin 2′-Propanoate Dodecyl Sulfate (1:1) *see* 3740
Erythromycin Propionate, 3742
Erythromycin Propionate Lauryl Sulfate *see* 3740
Erythromycin Stearate, 3743
Erythroped [Abbott] *see* 3739
Erythropoiesis Stimulating Factor *see* 3744
Erythropoietin, 3744
1-165 Erythropoietin (Human Clone B03XA01) *see* 3744
1-165-Erythropoietin (Human Clone λHEPOFL13 Protein Moiety), Glycoform *α see* 3744
1-165-Erythropoietin (Human Clone λHEPOFL13 Protein Moiety), Glycoform *β see* 3744
1-165 Erythropoietin (Human), Glycoform *θ see* 3744
Erythropterin, 3745
D-Erythrose, 3746
L-Erythrose, 3747
D-Erythrose 4-Phosphate, 3748
4-D-Erythrosephosphoric Acid *see* 3748
Erythrosine, 3749
Erythrosine B *see* 3749
Erythrosine BS *see* 3749
Erythrosine Extra Yellowish *see* 3207
Erythroxylon *see* 2438
L-Erythrulose, 3750
Erytrarsin *see* 1608
ES-771 *see* 7542
ES-902 *see* 7542
Esafosfina [Foscama] *see* 4304
Esametina *see* 4724
Esantene *see* 5022
Esbatal [Burroughs Wellcome] *see* 1188
Esbecythrin *see* 2888
Esbericard [Schaper & Brümmer] *see* 2555
Esberidin [Schaper & Brümmer] *see* 10180
Esberiven [Schaper & Brümmer] *see* 5889
Esbriet [Intermune] *see* 7606
Esbuphon [Schaper & Brümmer] *see* 6788
Escalol 106 [Van Dyk] *see* 4523
Escalol 506 [Van Dyk] *see* 3254
Escalol 507 [Van Dyk] *see* 3254
Eschenmoser's Salt *see* 3275
Escin, 3751
α-Escin *see* 3751
β-Escin *see* 3751
Escitalopram *see* 2317
Escitalopram Oxalate *see* 2317
Esclama [Montedison] *see* 6637
Esclerosina *see* 3782
Escobedin *see* 888
Escodarone [Streuli] *see* 478

3-[(3R,4R)-3-Ethenyl-4-piperidinyl]-1-(6-methoxy-4-quinolinyl)-1-propanone *see* 10199

3-[(3R,4R)-3-Ethenyl-4-piperidinyl]-1-(4-quinolinyl)-1-propanone *see* 2292

3-Ethenylpyridine, 3785

4-Ethenylpyridine 1-Oxide Homopolymer *see* 8059

1-Ethenyl-2-pyrrolidinone *see* 7814

1-Ethenyl-2-pyrrolidinone Homopolymer *see* 7814

1-Ethenyl-2-pyrrolidinone Homopolymer Compd with Iodine *see* 7815

(Ethenylsulfonyl)benzene *see* 7431

(4aS,5R,6S)-5-Ethenyl-4,4a,5,6-tetrahydro-6-[[2-O-[(3,3',5-trihydroxy[1,1'-biphenyl]-2-yl)carbonyl]-β-D-glucopyranosyl]oxy]-1H,3H-pyrano[3,4-c]-pyran-1-one *see* 371

Ethenzamide, 3786

Etheophyl [Lindopharm] *see* 9436

Ethephon, 3787

Ether *see* 3861

Ether Chloratus *see* 3837

Ether Hydrochloric *see* 3837

Ether Muriatic *see* 3837

Ethidium *see* 4769

Ethimide *see* 3791

Ethinazone *see* 3769

Ethine *see* 85

Ethinyl Estradiol, 3788

17-Ethinylestradiol *see* 3788

17α-Ethinylestradiol 3-Cyclopentyl Ether *see* 8171

Ethinylestrenol *see* 5690

17α-Ethinyl-17β-hydroxyestr-4-ene *see* 5690

17α-Ethinyltestosterone *see* 3795

Ethinyl Trichloride *see* 9804

Ethiodan *see* 5103

Ethiodized Oil, 3789

Ethiodol [Savage] *see* 3789

Ethiofos *see* 399

Ethion, 3790

Ethionamide, 3791

Ethioniamide *see* 3791

Ethionine, 3792

Ethiops Iron *see* 4067

Ethiops Mineral *see* 5956

Ethiprole, 3793

Ethirimol, 3794

Ethisterone, 3795

Ethmosine *see* 6354

Ethmozin *see* 6354

Ethmozine [Roberts] *see* 6354

Ethnine [Allen & Hanburys] *see* 7441

Ethocel *see* 3836

Ethodin [Winthrop] *see* 3771

Ethodryl [BDH] *see* 3128

Ethofat [Akzo] *see* 7698

Ethofenprox *see* 3922

Ethofibrate *see* 3923

Ethofumesate, 3796

Ethoglucid *see* 3925

Ethohexadiol, 3797

Ethol *see* 2027

Ethopabate, 3798

Ethoprop, 3799

Ethopropazine, 3800

Ethoprophos *see* 3799

Ethoproxyfen *see* 3922

Ethosperse [Lonza] *see* 7697

Ethosuximide, 3801

Ethotoin, 3802

Ethovan *see* 3908

2-Ethoxy-4-acetamidobenzoic Acid Methyl Ester *see* 3798

p-Ethoxyacetanilide *see* 7319

7-Ethoxy-3,9-acridinediamine *see* 3771

1-Ethoxy-2-amino-4-nitrobenzene *see* 6704

2-Ethoxyaniline *see* 7341

4-Ethoxyaniline *see* 7342

2-Ethoxybenzamide *see* 3786

2-Ethoxybenzenamine *see* 7341

4-Ethoxybenzenamine *see* 7342

Ethoxybenzene *see* 7343

2-Ethoxybenzenecarbonamide *see* 3786

6-Ethoxy-2-benzothiazolesulfonamide *see* 3809

(2S,3R,4R,5S,6R)-2-(3-(4-Ethoxybenzyl)-4-chlorophenyl)-6-hydroxymethyltetrahydro-2H-pyran-3,4,5-triol *see* 2816

2-*p*-Ethoxybenzyl-1-(2-diethylaminoethyl)-5-nitrobenzimidazole *see* 3927

5-[(Ethoxycarbonyl)amino]-3-(4-morpholinyl)-1,2,3-oxadiazolium Inner Salt *see* 6320

(2S,3aS,7aS)-1-[(2S)-2-[[(1S)-1-(Ethoxycarbonyl)butyl]amino]-1-oxopropyl]-octahydro-1H-indole-2-carboxylic Acid *see* 7285

N-Ethoxycarbonyl-2-ethoxy-1,2-dihydroquinoline *see* 3566

1-Ethoxycarbonyl-3-methyl-2-thio-4-imidazoline *see* 1799

N-(Ethoxycarbonyl)-3-(4-morpholinyl)-sydnone Imine *see* 6320

17α-Ethoxycarbonyloxy-Δ'-cortienic Acid Chloromethyl Ester *see* 5642

(3β,16β,17α,18β,20α)-18-[[4-[(Ethoxycarbonyl)oxy]-3,5-dimethoxybenzoyl]-oxy]-11,17-dimethoxyyohimban-16-carboxylic Acid Methyl Ester *see* 9147

4-[1-(Ethoxycarbonyloxy)ethoxy]-2-methyl-*N*-(2-pyridyl)-2H-1,2-benzothiazine-3-carboxamide 1,1-Dioxide *see* 584

1'-Ethoxycarbonyloxyethyl 6-(D-α-Aminophenylacetamido)penicillanate *see* 924

(11β,17α)-17-[(Ethoxycarbonyl)oxy]-11-hydroxy-3-oxoandrosta-1,4-diene-17-carboxylic Acid Chloromethyl Ester *see* 5642

(11β)-17-[(Ethoxycarbonyl)oxy]-11-hydroxy-21-(1-oxopropoxy)pregna-1,4-diene-3,20-dione *see* 7840

cis-4-(Ethoxycarbonyloxy)-8-methoxy-3-(2,5-xylyl)-1-azaspiro[4.5]dec-3-en-2-one *see* 8888

N-[(1S)-1-(Ethoxycarbonyl)-3-phenylpropyl]-L-alanyl-*N*-(2,3-dihydro-1H-inden-2-yl)glycine *see* 2881

N-[*N*-[(S)-1-(Ethoxycarbonyl)-3-phenylpropyl]-L-alanyl]-*N*-(indan-2-yl)glycine *see* 2881

(3aR,7aS)-1-[*N*-[(S)-(Ethoxycarbonyl)-3-phenylpropyl]-(S)-alanyl]octahydroindole-2(S)-carboxylic Acid *see* 9729

N-[(1S)-1-(Ethoxycarbonyl)-3-phenylpropyl]-L-alanyl-L-proline *see* 3623

(1S,9S)-9-[[(1S)-1-(Ethoxycarbonyl)-3-phenylpropyl]amino]octahydro-10-oxo-6H-pyridazino[1,2-a][1,2]diazepine-1-carboxylic Acid *see* 2275

(8S)-7-[(2S)-2-[[(1S)-1-(Ethoxycarbonyl)-3-phenylpropyl]amino]-1-oxopropyl]-1,4-dithia-7-azaspiro[4.4]nonane-8-carboxylic Acid *see* 8880

(4S)-3-[(2S)-2-[[(1S)-1-(Ethoxycarbonyl)-3-phenylpropyl]amino]-1-oxopropyl]-1-methyl-2-oxo-4-imidazolidinecarboxylic Acid *see* 4952

(2S,3aS,6aS)-1-[(2S)-2-[[(1S)-1-(Ethoxycarbonyl)-3-phenylpropyl]amino]-1-oxopropyl]octahydrocyclopenta[b]pyrrole-2-carboxylic Acid *see* 8221

(2S,3aR,7aS)-1-[(2S)-2-[[(1S)-1-(Ethoxycarbonyl)-3-phenylpropyl]amino]-1-oxopropyl]octahydro-1H-indole-2-carboxylic Acid *see* 9729

(3S)-2-[(2S)-2-[[(1S)-1-(Ethoxycarbonyl)-3-phenylpropyl]amino]-1-oxopropyl]-1,2,3,4-tetrahydro-6,7-dimethoxy-3-isoquinolinecarboxylic Acid *see* 6315

(3S)-2-[(2S)-2-[[(1S)-1-(Ethoxycarbonyl)-3-phenylpropyl]amino]-1-oxopropyl]-1,2,3,4-tetrahydro-3-isoquinolinecarboxylic Acid *see* 6315

(4S)-3-[(2S)-2-[*N*-[(1S)-1-(Ethoxycarbonyl)-3-phenylpropyl]amino]propionyl]-1-methyl-2-oxoimidazolidine-4-carboxylic Acid *see* 4952

(3S)-3-[[(1S)-1-(Ethoxycarbonyl)-3-phenylpropyl]amino]-2,3,4,5-tetrahydro-2-oxo-1H-1-benzazepine-1-acetic Acid *see* 1033

(2S,6R)-6-[[(1S)-1-(Ethoxycarbonyl)-3-phenylpropyl]amino]tetrahydro-5-oxo-2-(2-thienyl)-1,4-thiazepine-4(5H)-acetic Acid *see* 9279

11-[*N*-(Ethoxycarbonyl)-4-piperidylidene]-8-chloro-6,11-dihydro-5H-benzo[5,6]cyclohepta[1,2-b]pyridine *see* 5635

(3R,5aS,6R,8aS,9R,10S,12R,12aR)-10-Ethoxydecahydro-3,6,9-trimethyl-3,12-epoxy-12H-pyrano[4,3-j]-1,2-benzodioxepin *see* 803

2-Ethoxy-6,9-diaminoacridine *see* 3771

2-Ethoxy-2,3-dihydro-3,3-dimethyl-5-benzofuranol 5-Methanesulfonate *see* 3796

6-Ethoxy-1,2-dihydro-2,2,4-trimethylquinoline *see* 3807

3-Ethoxy-1,1-dihydroxy-2-butanone *see* 5347

(S)-3,4-[(N,N'-1,1'-((2''-Ethoxy)-3'''(O)-4'''-(N,N-dimethylamino)butane)-bis-3,3'-indolyl)]-1(H)-pyrrole-2,3-dione *see* 8423

Ethoxyethane *see* 3861

2-Ethoxyethanol, 3803

2-Ethoxyethanol 1-Acetate *see* 3804

2-(2-Ethoxyethoxy)ethanol *see* 3134

2-Ethoxyethyl Acetate, 3804

4-[2-[4-[1-(2-Ethoxyethyl)-1H-benzimidazol-2-yl]-1-piperidinyl]ethyl]-α,α-dimethylbenzeneacetic Acid *see* 1220

2-(4-(2-(4-(1-(2-Ethoxyethyl)-1H-benzo[d]imidazol-2-yl)piperidin-1-yl)ethyl)phenyl)-2-methylpropanoic Acid *see* 1220

2,2'-[1-(1-Ethoxyethyl)-1,2-ethanediylidene]bishydrazinecarbothioamide *see* 5347

1-(2-Ethoxyethyl)-2-(hexahydro-4-methyl-1H-1,4-diazepin-1-yl)-1H-benzimidazole *see* 3614

2-Ethoxyethyl *p*-Methoxycinnamate *see* 2310

1-[2-(Ethoxy)ethyl]-2-(4-methyl-1-homopiperazinyl)benzimidazole *see* 3614

2-[2-Ethoxy-5-[(4-ethyl-1-piperazinyl)sulfonyl]phenyl]-5-methyl-7-propylimidazo[5,1-*f*][1,2,4]triazin-4(1*H*)-one *see* 10124

3-Ethoxy-4-hydroxybenzaldehyde *see* 3908

p-Ethoxy-*N*-(β-hydroxybutyryl)aniline *see* 1467

[3-[[4-(3-Ethoxy-2-hydroxypropoxy)-phenyl]amino]-3-oxopropyl]dimethylsulfonium 4-Methylbenzenesulfonate (1:1) *see* 9132

[3-[[4-(3-Ethoxy-2-hydroxypropoxy)-phenyl]amino]-3-oxopropyl]dimethylsulfonium Salt with 4-Methylbenzenesulfonic Acid *see* 9132

(±)-[2-[[*p*-(3-Ethoxy-2-hydroxypropoxy)-phenyl]carbamoyl]ethyl]dimethylsulfonium *p*-Toluenesulfonate *see* 9132

2-[1-(Ethoxyimino)butyl]-5-[2-(ethylthio)propyl]-3-hydroxy-2-cyclohexen-1-one *see* 8612

2-[1-(Ethoxyimino)propyl]-3-hydroxy-5-mesitylcyclohex-2-en-1-one *see* 9724

2-[1-(Ethoxyimino)propyl]-3-hydroxy-5-(2,4,6-trimethylphenyl)-2-cyclohexen-1-one *see* 9724

(+)-2-Ethoxy-α-[[(*S*)-α-isobutyl-*o*-piperidinobenzyl]carbamoyl]-*p*-toluic Acid *see* 8256

β-Ethoxy-α-ketobutyraldehyde Monohydrate *see* 5347

Ethoxylated Fatty Acid Esters *see* 7698

Ethoxylated Fatty Alcohols *see* 7697

N-[2-[(1*S*)-1-(3-Ethoxy-4-methoxyphenyl)-2-(methylsulfonyl)ethyl]-2,3-dihydro-1,3-dioxo-1*H*-isoindol-4-yl]acetamide *see* 740

(Ethoxymethyl)benzene *see* 1136

(1*R*,2*R*,3*S*)-2-Ethoxymethyl-3-(3,4-dichlorophenyl)tropane *see* 9322

2-[[Ethoxy[(1-methylethyl)amino]phosphinothioyl]oxy]benzoic Acid 1-Methylethyl Ester *see* 5218

5-(Ethoxymethyl)-2-furaldehyde *see* 3805

5-(Ethoxymethyl)-2-furancarboxaldehyde *see* 3805

Ethoxymethylfurfural, 3805

ω-Ethoxymethylfurfuraldehyde *see* 3805

4-Ethoxy-2-methyl-5-(4-morpholinyl)-3(2*H*)-pyridazinone *see* 3619

5-[2-Ethoxy-5-[(4-methyl-1-piperazinyl)-sulfonyl]phenyl]-1,6-dihydro-1-methyl-3-propyl-7*H*-pyrazolo[4,3-*d*]pyrimidin-7-one *see* 8628

2-Ethoxy-4-[2-[[(1*S*)-3-methyl-1-[2-(1-piperidinyl)phenyl]butyl]amino]-2-oxoethyl]benzoic Acid *see* 8256

2-Ethoxy-2-methylpropane *see* 3829

2-Ethoxynaphthalene, 3806

(2*S*,5*R*,6*R*)-6-[[(2-Ethoxy-1-naphthalenyl)carbonyl]amino]-3,3-dimethyl-7-oxo-4-thia-1-azabicyclo[3.2.0]heptane-2-carboxylic Acid *see* 6436

6-(2-Ethoxy-1-naphthamido)penicillin *see* 6436

2-Ethoxy-5-nitrobenzenamine *see* 6704

3-Ethoxy-2-oxobutyraldehyde Hydrate *see* 5347

2-Ethoxy-1-[[2'-(5-oxo-4,5-dihydro-1,2,4-oxadiazol-3-yl)biphenyl-4-yl]methyl]-1*H*-benzimidazole-7-carboxylic Acid *see* 903

3-[4-(β-Ethoxyphenethyl)-1-piperazinyl]-2-methylpropiophenone *see* 3688

(±)-(2*R**)-2-[(α*R**)-α-(*o*-Ethoxyphenoxy)benzyl]morpholine *see* 8244

5-[(2*R*)-2-[[2-(2-Ethoxyphenoxy)ethyl]-amino]propyl]-2-methoxybenzenesulfonamide *see* 9181

2-[(2-Ethoxyphenoxy)methyl]morpholine *see* 10178

2-(2-Ethoxyphenoxymethyl)tetrahydro-1,4-oxazine *see* 10178

rel-(2*R*)-2-[(*R*)-2-(2-Ethoxyphenoxy)phenylmethyl]morpholine *see* 8244

N-(4-Ethoxyphenyl)acetamide *see* 7319

4-Ethoxy-7-phenyl-3,5-dioxa-6-aza-4-phosphaoct-6-ene-8-nitrile 4-Sulfide *see* 7478

5-Ethoxy-2-phenyl-4,6-dioxa-3-aza-5-phosphaoct-2-enenitrile 5-Sulfide *see* 7478

3-[4-(2-Ethoxy-2-phenylethyl)-1-piperazinyl]-2-methyl-1-phenyl-1-propanone *see* 3688

N-(4-Ethoxyphenyl)-3-hydroxybutanamide *see* 1467

2-[(4-Ethoxyphenyl)methyl]-*N*,*N*-diethyl-5-nitro-1*H*-benzimidazole-1-ethanamine *see* 3927

2'-(4-Ethoxyphenyl)-5-(4-methyl-1-piperazinyl)-2,5'-bi-1*H*-benzimidazole *see* 1250

1-[[2-(4-Ethoxyphenyl)-2-methylpropoxy]methyl]-3-phenoxybenzene *see* 3922

2-(4-Ethoxyphenyl)-2-methylpropyl-3-phenoxybenzyl Ether *see* 3922

N-(4-Ethoxyphenyl)urea *see* 3511

(*S*)-(+)-2-Ethoxy-4-[*N*-[1-(2-piperidinophenyl)-3-methyl-1-butyl]aminocarbonylmethyl]benzoic Acid *see* 8256

3-Ethoxy-1-propene *see* 284

Ethoxyquin, 3807

2-Ethoxy-1-(2*H*)-quinolinecarboxylic Acid Ethyl Ester *see* 3566

Ethoxysulfuron, 3808

2-Ethoxy-1-[[2'-(2*H*-tetrazol-5-yl)[1,1'-biphenyl]-4-yl]methyl]-1*H*-benzimidazole-7-carboxylic Acid *see* 1742

2-Ethoxy-1-[4-[2-(1*H*-tetrazol-5-yl)phenyl]benzyl]-7-benzimidazolecarboxylic Acid *see* 1742

Ethoxzolamide *see* 3809

Ethoxzolamide, 3809

Ethrane [Baxter] *see* 3635

Ethrel [Bayer CropSci.] *see* 3787

Ethril [BMS] *see* 3743

Ethryn [Faulding] *see* 3743

Ethybenztropine, 3810

Ethyl Acetate, 3811

Ethylacetic Acid *see* 1597

Ethyl Acetoacetate, 3812

Ethyl Acetone *see* 6187

Ethyl Acetylenecarboxylate *see* 3898

Ethyl Acid Tartrate *see* 6339

Ethyl Acrylate, 3813

Ethyl Adipate *see* 150

Ethyl Alcohol, 3814

Ethyl Alcohol, Denatured, 3815

Ethylaldehyde *see* 40

Ethylaluminum Dichloride, 3816

Ethyl (3*S*)-3-[3-[(*p*-Amidinophenyl)carbamoyl]propionamido]-4-pentynoate *see* 10269

Ethylamine, 3817

Ethyl Aminobenzoate *see* 1088

Ethyl *m*-Aminobenzoate Methanesulfonate *see* 9786

2-Ethylamino-4-diethylamino-6-chloro-*s*-triazine *see* 9829

(4*R*)-4-(Ethylamino)-3,4-dihydro-2-(3-methoxypropyl)-2*H*-thieno[3,2-*e*]-1,2-thiazine-6-sulfonamide 1,1-Dioxide *see* 1382

(4*S*,6*S*)-4-(Ethylamino)-5,6-dihydro-6-methyl-4*H*-thieno[2,3-*b*]thiopyran-2-sulfonamide 7,7-Dioxide *see* 3475

2-Ethylamino-1-(3',4'-dihydroxyphenyl)-1-propanol *see* 3331

α-(1-Ethylaminoethyl)protocatechuyl Alcohol *see* 3331

Ethyl Aminoformate *see* 10059

4-[2-(Ethylamino)-1-hydroxypropyl]-1,2-benzenediol *see* 3331

2-Ethylamino-4-isopropylamino-6-methylmercapto-*s*-triazine *see* 387

2-(Ethylamino)-4-(isopropylamino)-6-(methylthio)-*s*-triazine *see* 387

(2*E*)-3-[[(Ethylamino)methoxyphosphinothioyl]oxy]-2-butenoic Acid 1-Methylethyl Ester *see* 7929

2-Ethylamino-4-methyl-5-*n*-butyl-6-hydroxypyrimidine *see* 3794

α-[(Ethylamino)methyl]-3-hydroxybenzenemethanol *see* 3914

α-[(Ethylamino)methyl]-*m*-hydroxybenzyl Alcohol *see* 3914

Ethyl 1-(4-Aminophenethyl)-4-phenylisonipecotate *see* 651

2-Ethylamino-3-phenylbicyclo[2.2.1]heptane *see* 3998

3-Ethyl-3-(*p*-aminophenyl)-2,6-dioxopiperidine *see* 437

Ethyl *p*-Aminophenyl Ketone *see* 466

2-Ethylamino-3-phenylnorbornane *see* 3998

2-Ethylamino-3-phenylnorcamphane *see* 3998

2-Ethylamino-1-phenylpropane *see* 3818

2-(Ethylamino)-2-(2-thienyl)cyclohexanone *see* 9592

2-Ethylamino-1-(3-trifluoromethylphenyl)propane *see* 4004

N-Ethylamphetamine, 3818

Ethyl Amyl Ketone *see* 6156

Ethyl *n*-Amyl Ketone *see* 6842

Ethylaniline, 3819

Ethyl Apovincamin-22-oate *see* 10188

N-Ethyl-8-aza-3-bicyclo[3.2.1]octyl Benzhydryl Ether *see* 3810

Ethyl Azodicarboxylate *see* 3126

Ethylbenzatropine *see* 3810

N-Ethylbenzenamine *see* 3819

Ethylbenzene, 3820

α-Ethylbenzeneacetic Acid 2-[2-(Diethylamino)ethyl]ethyl Ester *see* 1516

α-Ethylbenzeneacetic Acid 2-(Diethylamino)ethyl Ester *see* 1525

α-Ethylbenzenemethanol *see* 3823

Ethyl Benzenesulfonate *see* 1071

Ethyl Benzoate, 3821

1-(7-Ethylbenzofuran-2-yl)-2-*tert*-butylamino-1-hydroxyethane *see* 1485

2-Ethyl-3-benzofuranyl 4-Hydroxy-3,5-dibromophenyl Ketone *see* 1067

(2-Ethyl-3-benzofuranyl) (4-Hydroxy-3,5-diiodophenyl) Methanone *see* 1085

2-Ethyl-3-benzofuryl 3,5-Dibromo-4-hydroxyphenyl Ketone *see* 1067

2-Ethyl-3-benzofuryl 3',5'-Diiodo-4'-hydroxyphenyl Ketone *see* 1085

Ethyl Benzoylacetate, 3822

Ethylbenzoylecgonine *see* 2439

N-Ethylbenztropine *see* 3810

α-Ethylbenzyl Alcohol, 3823

ω-Ethylbenzyl Alcohol *see* 3823
Ethylbenzylaniline, 3824
3-(α-Ethylbenzyl)-4-hydroxycoumarin *see* 7371
Ethyl [Bis(1-aziridinyl)phosphinyl]carbamate *see* 10057
Ethyl Biscoumacetate, 3825
Ethyl *N*-[Bis(ethyleneimido)phosphoro]carbamate *see* 10057
Ethyl Bis(4-hydroxycoumarinyl)acetate *see* 3825
5-Ethyl-1,3-bis(methoxymethyl)-5-phenylbarbituric Acid *see* 3770
5-Ethyl-1,3-bis(methoxymethyl)-5-phenyl-2,4,6(1*H*,3*H*,5*H*)-pyrimidinetrione *see* 3770
Ethyl 2-[4,5-Bis(*p*-methoxyphenyl)-2-thiazolyl]pyrrole-1-acetate *see* 7103
EthylBloc [AgroFresh] *see* 6120
Ethyl Bromide, 3826
N-Ethyl-*N*-*o*-bromobenzyl-*N*,*N*-dimethylammonium Tosylate *see* 1377
Ethyl α-Bromobutyrate *see* 1420
(−)-*N*-Ethyl-2-(3-bromo-2,6-dimethoxybenzamidomethyl)pyrrolidine *see* 8252
Ethyl α-Bromopropionate, 3827
2-Ethylbutanoic Acid *see* 3121
Ethyl Butylacetylaminopropionate, 3828
Ethyl-3-(*N*-*n*-butyl-*N*-acetyl)aminopropionate *see* 3828
S-[2-[[1-(2-Ethylbutyl)cyclohexane]carbonylamino]phenyl]-2-methylpropanethioate *see* 2801
Ethyl (3-*tert*-Butyl-1-dimethylcarbamoyl-1*H*-1,2,4-triazol-5-ylthio)acetate *see* 9762
Ethyl *tert*-Butyl Ether, 3829
Ethyl *tert*-Butyl Oxide *see* 3829
Ethyl Butyrate, 3830
Ethyl *n*-Butyrate *see* 3830
Ethyl Caprate, 3831
Ethyl Caproate, 3832
Ethyl Caprylate, 3833
Ethyl Carbamate *see* 10059
Ethyl Carbanilate *see* 7430
Ethyl β-Carboline-3-carboxylate, 3834
Ethyl Carbonate, 3835
*N*³-(Ethylcarbonimidoyl)-*N*¹,*N*¹-dimethyl-1,3-propanediamine *see* 3555
Ethyl (*S*)-2-[[(*S*)-1-[(Carboxymethyl)-2-indanylcarbamoyl]ethyl]amino]-4-phenylbutyrate *see* 2881
Ethyl Cellulose, 3836
Ethyl Centralite *see* 3129
Ethyl Cetab *see* 2028
Ethyl Chloride, 3837
Ethyl Chloroacetate, 3838
Ethyl Chlorocarbonate *see* 3839
Ethyl (*Z*)-2-Chloro-3-[2-chloro-5-(1,3-dioxo-4,5,6,7-tetrahydroisoindol-2-yl)phenyl]acrylate *see* 2294
Ethyl 2-Chloro-3-[2-chloro-4-fluoro-5-(4-difluoromethyl-4,5-dihydro-3-methyl-5-oxo-1*H*-1,2,4-triazol-1-yl)phenyl]propionate *see* 1836
Ethyl [[8-Chloro-3,2-[2-(diethylamino)ethyl]-4-methyl-2-oxo-2*H*-1-benzopyran-7-yl]oxy]acetate *see* 2393
Ethyl 7-Chloro-5-(*o*-fluorophenyl)-2,3-dihydro-2-oxo-1*H*-1,4-benzodiazepine-3-carboxylate *see* 3876
Ethyl Chloroformate, 3839
Ethyl 2-[[[[[(4-Chloro-6-methoxypyrimidin-2-yl)amino]carbonyl]amino]sulfonyl]benzoate *see* 2094

Ethyl *p*-Chlorophenoxyisobutyrate *see* 2370
Ethyl 2-(*p*-Chlorophenoxy)-2-methylpropionate *see* 2370
Ethyl Chlorophosphite *see* 3144
Ethyl α-Chloropropionate, 3840
Ethyl 2-[4-(6-Chloro-2-quinoxalinyloxy)phenoxy]propionate *see* 8203
Ethyl β-Chlorovinyl Ethynyl Carbinol *see* 3784
24α-Ethylcholestanol *see* 8943
24β-Ethyl-Δ⁵-cholesten-3β-ol *see* 8694
Ethyl Cinnamate *see* 2299
Ethyl Citrate *see* 2325
N-Ethyl-*o*-crotonotoluidide *see* 2585
Ethyl Cyanide *see* 7941
Ethyl Cyanoacetate, 3841
Ethyl Cyanoacrylate, 3842
Ethyl α-Cyanoacrylate *see* 3842
Ethyl 2-Cyanoacrylate *see* 3842
Ethyl 1-(3-Cyano-3,3-diphenylpropyl)-4-phenylisonipecotate *see* 3343
Ethyl 1-(3-Cyano-3,3-diphenylpropyl)-4-phenyl-4-piperidinecarboxylate *see* 3343
Ethyl Cyanoethanoate *see* 3841
5-Ethyl-5-cycloheptenylbarbituric Acid *see* 4690
10-Ethyl-10-deazaaminopterin *see* 3554
Ethyl Decanoate *see* 3831
Ethyl 6-(*n*-Decyloxy)-7-ethoxy-4-hydroxyquinoline-3-carboxylate *see* 2857
5-Ethyl-1-(2′-deoxy-β-D-ribofuranosyl)uracil *see* 3563
5-Ethyl-2′-deoxyuridine *see* 3563
Ethyl Diazoacetate *see* 3000
Ethyl Diazoethanoate *see* 3000
2-Ethyl-3-(3,5-dibromo-4-hydroxybenzoyl)benzofuran *see* 1067
2-Ethyl-3-(3,5-dibromo-4-hydroxybenzoyl)oxaindene *see* 1067
Ethyl Dibunate, 3843
Ethyl 2,7-Di-*tert*-butylnaphthalene-4-sulfonate *see* 3843
Ethyl 3,6-Di-*tert*-butyl-1-naphthalenesulfonate *see* 3843
Ethyldicoumarol Acetate *see* 3825
Ethyl [[3-[2-(Diethylamino)ethyl]-4-methyl-2-oxo-2*H*-1-benzopyran-7-yl]oxy]acetate *see* 2243
1-Ethyl-6,8-difluoro-1,4-dihydro-7-(3-methyl-1-piperazinyl)-4-oxo-3-quinolinecarboxylic Acid *see* 5619
Ethyl Digol *see* 3134
Ethyl 1,4-Dihydro-5-(acetoxycarbonyl)-2,6-dimethyl-4-(3-nitrophenyl)-3-pyridinecarboxylate *see* 6662
Ethyl *N*-[2,3-Dihydro-2,2-dimethylbenzofuran-7-yloxycarbonyl(methyl)amino]thio]-*N*-isopropyl-β-alaninate *see* 1044
9-Ethyl-6,9-dihydro-4,6-dioxo-10-propyl-4*H*-pyrano[3,2-*g*]quinoline-2,8-dicarboxylic Acid *see* 6521
Ethyl 4,5-Dihydro-5,5-diphenyl-1,2-oxazole-3-carboxylate *see* 5282
5-(2-Ethyl-2,3-dihydro-1*H*-inden-2-yl)-1*H*-imidazole *see* 7536
5-Ethyldihydro-5-(1-methylbutyl)-2-thioxo-4,6(1*H*,5*H*)-pyrimidinedione Sodium Salt (1:1) *see* 9503
1-Ethyl-1,4-dihydro-6,7-methylenedioxy-4-oxo-3-quinolinecarboxylic Acid *see* 7044
(3*S*,4*R*)-3-Ethyldihydro-4-[(1-methyl-1*H*-imidazol-5-yl)methyl]-2(3*H*)-furanone *see* 7536

3-Ethyl-2,5-dihydro-4-methyl-*N*-[2-[4-[[[[(*trans*-4-methylcyclohexyl)amino]carbonyl]amino]sulfonyl]phenyl]ethyl]-2-oxo-1*H*-pyrrole-1-carboxamide *see* 4475
3-Ethyl-6,7-dihydro-2-methyl-5-(morpholinomethyl)indol-4(5*H*)-one *see* 6319
1-Ethyl-1,4-dihydro-7-methyl-4-oxo-1,8-naphthyridine-3-carboxylic Acid *see* 6444
5-Ethyldihydro-5-(1-methylpropyl)-2-thioxo-4,6(1*H*,5*H*)-pyrimidinedione *see* 9476
1-Ethyl-1,4-dihydro-4-oxo[1,3]dioxolo-[4,5-*g*]cinnoline-3-carboxylic Acid *see* 2309
5-Ethyl-5,8-dihydro-8-oxo-1,3-dioxolo-[4,5-*g*]quinoline-7-carboxylic Acid *see* 7044
8-Ethyl-5,8-dihydro-5-oxo-2-(1-piperazinyl)pyrido[2,3-*d*]pyrimidine-6-carboxylic Acid *see* 7572
1-Ethyl-1,4-dihydro-4-oxo-7-(4-pyridinyl)-3-quinolinecarboxylic Acid *see* 8400
N′-(1-Ethyl-1,2-dihydro-2-oxo-4-pyrimidinyl)sulfanilamide *see* 9034
8-Ethyl-5,8-dihydro-5-oxo-2-(1-pyrrolidinyl)pyrido[2,3-*d*]pyrimidine-6-carboxylic Acid *see* 7618
N-[1-[2-(4-Ethyl-4,5-dihydro-5-oxo-1*H*-tetrazol-1-yl)ethyl]-4-(methoxymethyl)-4-piperidinyl]-*N*-phenylpropanamide *see* 229
5-Ethyldihydro-5-phenyl-4,6(1*H*,5*H*)-pyrimidinedione *see* 7865
Ethyl 4,4′-Dihydroxydicoumarinyl-3,3′-acetate *see* 3825
(1α,15β)-21-Ethyl-1,15-dihydroxy-4-methyl-16-methylene-7,20-cycloveatchan-12-one *see* 8843
N-Ethyl-3,4-dihydroxynorephedrine *see* 3331
2-Ethyl-3-(3′,5′-diiodo-4′-hydroxybenzoyl)benzofuran *see* 1085
2-Ethyl-3-(3,5-diiodo-4-hydroxybenzoyl)coumarone *see* 1085
2-Ethyl-3-(3′,5′-diiodo-4′-hydroxybenzoyl)oxaindene *see* 1085
Ethyl 3,5-Diiodosalicylate *see* 3208
Ethyl 6,7-Diisobutoxy-4-hydroxyquinoline-3-carboxylate *see* 1504
S-Ethyl *N*,*N*-Diisobutylthiocarbamate *see* 1548
Ethyldiisopropylamine *see* 3219
O-Ethyl *S*-[2-(Diisopropylamino)ethyl]-methylphosphonothioate *see* 10234
3-Ethyl-6,7-dimethoxy-1-(phenylmethyl)isoquinoline *see* 6374
(*S*)-*N*-Ethyl-3-[(1-dimethylamino)ethyl]-*N*-methylphenylcarbamate *see* 8369
Ethyl 2-(Dimethylamino)-1-phenyl-2-cyclohexene-1-carboxylate *see* 9594
1-Ethyl-3-(3′-dimethylaminopropyl)-3-(6′-allylergoline-8′β-carbonyl)urea *see* 1606
1-Ethyl-3-(3′-dimethylaminopropyl)-carbodiimide *see* 3555
N-Ethyl-*N*′-(3-dimethylaminopropyl)-carbodiimide *see* 3555
7-Ethyl-1,4-dimethylazulene *see* 2039
Ethyl Dimethyl Carbinol *see* 7253
3-Ethyl-5,5-dimethyl-2,4-diketooxazolidine *see* 3773

Name Index

cyclohexadien-1-ylidene](*m*-sulfoben-zyl)ammonium Hydroxide Inner Salt Disodium Salt *see* 5538

N-Ethyl-*N*-[4-[[4-[ethyl[(3-sulfophenyl)-methyl]amino]phenyl](4-hydroxy-2-sulfophenyl)methylene]-2,5-cyclohexa-dien-1-ylidene]-3-sulfobenzenemethan-aminium Inner Salt Sodium Salt (1:2) *see* 3978

N-Ethyl-*N*-[4-[[4-[ethyl[(3-sulfophenyl)-methyl]amino]phenyl]phenylmethyl-ene]-2,5-cyclohexadien-1-ylidene]-3-sulfobenzenemethanaminium Inner Salt Sodium Salt (1:1) *see* 4610

N-Ethyl-*N*-[4-[[4-[ethyl[(3-sulfophenyl)-methyl]amino]phenyl](2-sulfophenyl)-methylene]-2,5-cyclohexadien-1-yli-dene]-3-sulfobenzenemethananimium Inner Salt Sodium Salt (1:2) *see* 1379

N-Ethyl-*N*-[4-[[4-[ethyl[(3-sulfophenyl)-methyl]amino]phenyl](4-sulfophenyl)-methylene]-2,5-cyclohexadien-1-yli-dene]-3-sulfobenzenemethanaminium Inner Salt Sodium Salt (1:2) *see* 5538

6-[(3*R*,4*S*,5*S*,7*R*)-7-[(2*S*,3*S*,5*S*)-5-Ethyl-5-[(2*R*,5*R*,6*S*)-5-ethyltetrahydro-5-hy-droxy-6-methyl-2*H*-pyran-2-yl]tetra-hydro-3-methyl-2-furanyl]-4-hydroxy-3,5-dimethyl-6-oxononyl]-2-hydroxy-3-methylbenzoic Acid *see* 5427

13β-Ethyl-17α-ethynyl-Δ⁴,⁹,¹¹-gonatri-ene-17β-ol-3-one *see* 4449

13β-Ethyl-17α-ethynyl-17β-hydroxy-4,-9,11-gonatrien-3-one *see* 4449

13β-Ethyl-17α-ethynyl-17β-hydroxygon-4-en-3-one *see* 6793

Ethyl Ethynyl Methyl Carbinol *see* 5913

17β-Ethyletiocholane *see* 7844

Ethyl-8-fluoro-5,6-dihydro-5-methyl-6-oxo-4*H*-imidazo[1,5-*a*][1,4]benzodiaze-pine-3-carboxylate *see* 4166

1-Ethyl-6-fluoro-1,4-dihydro-7-(4-meth-yl-1-piperazinyl)-4-oxo-3-quinoline-carboxylic Acid *see* 7172

1-Ethyl-6-fluoro-1,4-dihydro-4-oxo-7-(1-piperazinyl)-1,8-naphthyridine-3-car-boxylic Acid *see* 3641

1-Ethyl-6-fluoro-1,4-dihydro-4-oxo-7-(1-piperazinyl)-3-quinolinecarboxylic Acid *see* 6789

Ethyl [(3a*R*,4a*R*,8a*R*,9a*S*)-9(*S*)-[(*E*)-2-[5-(3-Fluorophenyl)-2-pyridinyl]ethenyl]-dodecahydro-1(*R*)-methyl-3-oxonaph-tho[2,3-*c*]furan-6(*R*)-yl]carbamate *see* 10230

Ethyl Formate, 3862

Ethylformic Acid *see* 7939

Ethyl 2-Furoate *see* 4336

γ-Ethyl L-Glutamate *see* 4506

N-Ethyl-L-glutamine *see* 9424

Ethyl Green *see* 6154

Ethyl Heptanoate *see* 3889

Ethyl *n*-Heptoate *see* 3889

N-[4-[4-(Ethylheptylamino)-1-hydroxy-butyl]phenyl]methanesulfonamide *see* 4921

Ethylhexadecyldimethylammonium Bro-mide *see* 2028

1-[[(2*R*,3*R*,11b*S*)-3-Ethyl-1,3,4,6,7,11b-hexahydro-9,10-dimethoxy-2*H*-benzo-[*a*]quinolizin-2-yl]methyl]-3,4-dihydro-7-methoxy-6-isoquinolinol *see* 8038

(1*R*)-1-[[(2*S*,3*R*,11b*S*)-3-Ethyl-1,3,4,6,7,-11b-hexahydro-9,10-dimethoxy-2*H*-benzo[*a*]quinolizin-2-yl]methyl]-1,2,-

3,4-tetrahydro-7-methoxy-6-isoquino-linol *see* 1973

(2*S*,3*R*,11b*S*)-3-Ethyl-1,3,4,6,7,11b-hexa-hydro-9,10-dimethoxy-2-[[(1*R*)-1,2,3,4-tetrahydro-6,7-dimethoxy-1-isoquino-linyl]methyl]-2*H*-benzo[*a*]quinolizine *see* 3616

3-(3-Ethylhexahydro-1-methyl-1*H*-aze-pin-3-yl)phenol *see* 5930

[1*R*-(1α,2β,4β)]-2-Ethyl-1,2,3,4,6,11-hexahydro-2,5,7-trihydroxy-6,11-di-oxo-4-[[2,3,6-trideoxy-4-*O*-[2,6-dide-oxy-4-*O*-[(2*R-trans*)-tetrahydro-6-methyl-5-oxo-2*H*-pyran-2-yl]-α-L-*lyxo*-hexopyranosyl]-3-(dimethylamino)-α-L-*lyxo*-hexopyranosyl]oxy]-1-naphtha-cenecarboxylic Acid Methyl Ester *see* 104

2-Ethyl-1,3-hexanediol *see* 3797

Ethyl Hexanoate *see* 3832

2-Ethyl-1-hexanol, 3863

2-Ethylhexyl Acetate *see* 6852

2-Ethylhexyl Alcohol *see* 3863

2-Ethylhexyl-2-cyano-3,3-diphenylacry-late *see* 6845

2-Ethylhexyl *p*-Methoxycinnamate *see* 6859

Ethyl 3-[[[2-[[[4-[[[(Hexyloxy)carbonyl]-amino]iminomethyl]phenyl]amino]-methyl]-1-methyl-1*H*-benzimidazol-5-yl]carbonyl](pyridin-2-yl)amino]pro-panoate *see* 2794

2-Ethylhexyl Salicylate *see* 6860

Ethylhexyl Triazone, 3864

S-Ethyl-L-homocysteine *see* 3792

Ethyl Hydrate *see* 3814

Ethyl Hydride *see* 3778

Ethyl Hydrogen Sulfate *see* 3904

Ethyl Hydroxide *see* 3814

1-Ethyl-2-hydroxybenzene *see* 3892

Ethyl *o*-Hydroxybenzoate *see* 3902

Ethyl *p*-Hydroxybenzoate *see* 3890

7-Ethyl-10-hydroxycamptothecin *see* 5135

(+)-7-Ethyl-10-hydroxycamptothecine 10-[1,4′-Bipiperidine]-1′-carboxylate *see* 5135

Ethyl *m*-Hydroxycarbanilate Carbani-late *see* 2925

2-Ethyl-3-(4-hydroxy-3,5-diiodobenzoyl)-benzofuran *see* 1085

N-Ethyl-3-hydroxy-*N*,*N*-dimethylbenzen-aminium Chloride (1:1) *see* 3564

(17α)-13-Ethyl-17-hydroxy-18,19-dinor-pregna-4,15-dien-20-yn-3-one *see* 4447

(17α)-13-Ethyl-17-hydroxy-18,19-dinor-pregna-4,9,11-trien-3-one *see* 9359

(17α)-13-Ethyl-17-hydroxy-18,19-dinor-pregna-4,9,11-trien-20-yn-3-one *see* 4449

(17α)-(±)-13-Ethyl-17-hydroxy-18,19-di-norpregn-4-en-3-one *see* 6774

(8α,9β,10α,13α,14β)-13-Ethyl-17-hy-droxy-18,19-dinorpregn-4-en-20-yn-3-one *see* 6793

(17α)-(±)-13-Ethyl-17-hydroxy-18,19-di-norpregn-4-en-20-yn-3-one *see* 6793

(17α)-13-Ethyl-17-hydroxy-18,19-dinor-pregn-4-en-20-yn-3-one Oxime *see* 6783

(+)-13-Ethyl-17-hydroxy-18,19-dinor-17α-pregn-4-en-20-yn-3-one Oxime Acetate (Ester) *see* 6792

1-Ethyl-3-[(2-hydroxy-2,2-diphenylace-tyl)oxy]-1-methylpiperidinium Bro-mide (1:1) *see* 7573

17α-Ethyl-17β-hydroxy-4-estrene *see* 3860

(16β,17β)-16-Ethyl-17-hydroxyestr-4-en-3-one *see* 7030

2-(*R*)-(1-Ethyl-2-hydroxyethylamino)-6-benzylamino-9-isopropylpurine *see* 8579

7-[2-[Ethyl(2-hydroxyethyl)amino]ethyl]-3,7-dihydro-1,3-dimethyl-8-(phenyl-methyl)-1*H*-purine-2,6-dione *see* 948

Ethyl Hydroxyethyl Cellulose *see* 1968

Ethyl 2-Hydroxyethyl Ether Cellulose *see* 1968

(17α)-13-Ethyl-17-hydroxy-11-methyl-ene-18,19-dinorpregn-4-en-20-yn-3-one *see* 3928

(4*S*)-4-Ethyl-4-hydroxy-11-[2-[(1-methyl-ethyl)amino]ethyl]-1*H*-pyrano[3′,4′:6,-7]indolizino[1,2-*b*]quinoline-3,14-(4*H*,12*H*)-dione *see* 1028

(1*R*,3*S*,5*Z*)-5-[(2*E*)-2-[(1*R*,3a*S*,7a*R*)-1-[(1*R*,2*E*,4*E*)-6-Ethyl-6-hydroxy-1-methyl-2,4-octadien-1-yl]octahydro-7a-methyl-4*H*-inden-4-ylidene]ethyli-dene]-4-methylene-1,3-cyclohexanediol *see* 8595

1-Ethyl-3-hydroxy-1-methylpiperidinium Bromide Benzilate *see* 7573

2-Ethyl-2-(hydroxymethyl)-1,3-propane-diol Trinitrate *see* 7925

N-Ethyl-α-(hydroxymethyl)-*N*-(4-pyri-dinylmethyl)benzeneacetamide *see* 9959

(1α,6β,14α,16β)-20-Ethyl-4-(hydroxy-methyl)-1,6,14,16-tetramethoxyaconi-tane-7,8-diol *see* 5679

(4*S*)-4-Ethyl-4-hydroxy-10-nitro-1*H*-py-rano[3′,4′:6,7]indolizino[1,2-*b*]quino-line-3,14(4*H*,12*H*)-dione *see* 8421

17α-Ethyl-17-hydroxy-4-norandrosten-3-one *see* 6785

17α-Ethyl-17-hydroxy-19-norandrost-4-en-3-one *see* 6785

(1α,2β,4β,5α,7β)-9-Ethyl-7-[2(*S*)-3-hy-droxy-1-oxo-2-phenylpropoxy]-9-methyl-3-oxa-9-azoniatricyclo[3.3.1.-0²,⁴]nonane Bromide (1:1) *see* 7042

(3*R*,5*R*,6*S*,7*S*,9*R*,11*E*,13*S*,14*R*)-14-Ethyl-13-hydroxy-3,5,7,9,13-pentamethyl-6-[[3,4,6-trideoxy-3-(dimethylamino)-β-D-*xylo*-hexopyranosyl]oxy]oxacyclo-tetradec-11-ene-2,4,10-trione *see* 7525

Ethyl(*m*-hydroxyphenyl)dimethylammo-nium Chloride *see* 3564

Ethyl *p*-Hydroxyphenyl Ketone *see* 7149

1-Ethyl-3-hydroxypiperidine *see* 3877

Ethyl α-Hydroxypropionate *see* 3872

(4*S*)-4-Ethyl-4-hydroxy-1*H*-pyrano[3′,-4′:6,7]indolizino[1,2-*b*]quinoline-3,14-(4*H*,12*H*)-dione *see* 1737

2-Ethyl-3-hydroxy-4*H*-pyran-4-one *see* 3879

3-Ethyl-7-hydroxy-2,8,12,16-tetramethyl-5,13-dioxo-9-[[3,4,6-trideoxy-3-(di-methylamino)-β-D-*xylo*-hexopyrano-syl]oxy]-4,17-dioxabicyclo[14.1.0]hep-tadec-14-ene-10-acetaldehyde *see* 8392

(3*R*,4*S*,5*S*,7*R*,9*E*,11*S*,12*R*)-12-Ethyl-11-hydroxy-3,5,7,11-tetramethyl-4-[[3,4,6-trideoxy-3-(dimethylamino)-β-D-*xylo*-hexopyranosyl]oxy]oxacyclododec-9-ene-2,8-dione *see* 6211

α-Ethyl-3-hydroxy-2,4,6-triiodobenzene-propanoic Acid *see* 5104

α-Ethyl-3-hydroxy-2,4,6-triiodohydro-cinnamic Acid *see* 5104

4'-Ethyl-2-methyl-3-piperidinopropio-
phenone *see* 3661
N-Ethyl-*N*-(2-methyl-2-propen-1-yl)-2,6-
dinitro-4-(trifluoromethyl)benzenam-
ine *see* 3774
5-Ethyl-5-(1-methylpropyl)barbituric
Acid *see* 1509
α-Ethyl-4-(2-methylpropyl)benzeneacetic
Acid *see* 1529
N-[3-(1-Ethyl-1-methylpropyl)-5-isox-
azolyl]-2,6-dimethoxybenzamide *see*
5281
5-Ethyl-5-(1-methylpropyl)-2,4,6(1*H*,3*H*,-
5*H*)-pyrimidinetrione *see* 1509
5-Ethyl-5-(1-methylpropyl)-2-thiobarbi-
turic Acid *see* 9476
1-Ethyl-1-methyl-2-propynyl Acid
Phthalate *see* 7486
β-Ethyl-γ-methylpyridine *see* 3894
3-Ethyl-4-methylpyridine *see* 3894
3-Ethyl-6-methylpyridine *see* 3896
4-Ethyl-2-methylpyridine *see* 3895
5-Ethyl-2-methylpyridine *see* 3896
3-Ethyl-3-methyl-2,5-pyrrolidinedione
see 3801
α-Ethyl-α-methylsuccinimide *see* 3801
2-Ethyl-2-methylsuccinimide *see* 3801
Ethyl 3-Methyl-2-thioimidazoline-1-car-
boxylate *see* 1799
Ethyl 4-(Methylthio)-*m*-tolyl Isopropylphosphoramidate *see* 3991
N-Ethyl-α-methyl-3-(trifluoromethyl)-
benzeneethanamine *see* 4004
N-Ethyl-α-methyl-*m*-(trifluoromethyl)-
phenethylamine *see* 4004
7-Ethyl-2-methyl-4-undecanol 4-(Hydrogen Sulfate) Sodium Salt (1:1) *see*
8817
α-Ethyl-β-methylvaleramide *see* 10098
2-Ethyl-3-methylvaleramide *see* 10098
Ethylmorphine, 3884
1-Ethyl-4-[2-(4-morpholinyl)ethyl]-3,3-
diphenyl-2-pyrrolidinone *see* 3481
Ethyl Mustard Oil *see* 3870
Ethyl Myristate *see* 6419
Ethyl β-Naphtholate *see* 3806
Ethyl β-Naphthyl Ether *see* 3806
3-Ethylnirvanol *see* 5920
Ethyl Nitrite, 3885
Ethyl Nitrobenzoate, 3886
2-Ethyl-2-[(nitrooxy)methyl]-1,3-pro-
panediol 1,3-Dinitrate *see* 7925
Ethyl *p*-Nitrophenyl Benzenethiophos-
phonate *see* 3682
O-Ethyl *O*-4-Nitrophenyl Phenylphos-
phonothioate *see* 3682
N-Ethyl-*N*-nitrosoethanamine *see* 6723
N-Ethyl-*N*-nitrosourea, 3887
N-Ethyl-*N*'-(5-nitro-2-thiazolyl)urea *see*
6655
Ethyl Nonanoate *see* 3891
Ethylnoradrenaline *see* 3888
Ethylnorepinephrine, 3888
Ethylnorgestrienone *see* 4449
Ethyl Norharmancarboxylate *see* 3834
(−)-*N*-Ethylnorscopolamine Metho-
bromide *see* 7042
Ethylnorsuprarenin *see* 3888
16β-Ethyl-19-nortestosterone *see* 7030
17α-Ethyl-19-nortestosterone *see* 6785
N-Ethylnortropine Benzhydryl Ether *see*
3810
(3a*S*,4*R*,7*R*,9*R*,10*R*,11*R*,13*R*,15*R*,15a*R*)-
4-Ethyloctahydro-3a,7,9,11,13,15-
hexamethyl-11-[[3-(3-quinolinyl)-2-

propenyl]oxy]-10-[[3,4,6-trideoxy-3-
(dimethylamino)-β-D-*xylo*-hexopyran-
osyl]oxy]-2*H*-oxacyclotetradecino[4,3-
d]oxazole-2,6,8,14(1*H*,7*H*,9*H*)-tetrone
see 2018
13a-Ethyl-2,3,5,6,12,13,13a,13b-octahy-
dro-12-hydroxy-1*H*-indolo[3,2,1-*de*]-
pyrido[3,2,1-*ij*][1,5]naphthyridine-12-
carboxylic Acid Methyl Ester *see*
10180
1-Ethyl-1,2,3,4,4a,9,10,10a-octahydro-7-
hydroxy-2-methyl-2-phenanthrenecar-
boxylic Acid *see* 3457
7-Ethyl-1,4,5,6,7,8,9,10-octahydro-2*H*-
3,7-methanoazacycloundecino[5,4-*b*]-
indole *see* 8146
(3a*S*,4*R*,7*R*,9*R*,10*R*,11*R*,13*R*,15*R*,15a*R*)-
4-Ethyloctahydro-11-methoxy-3a,7,9,-
11,13,15-hexamethyl-1-[4-[4-(3-pyri-
dinyl)-1*H*-imidazol-1-yl]butyl]-10-[[3,-
4,6-trideoxy-3-(dimethylamino)-β-D-
xylo-hexopyranosyl]oxy]-2*H*-oxacyclo-
tetradecino[4,3-*d*]oxazole-2,6,8,14(1*H*,-
7*H*,9*H*)-tetrone *see* 9261
(α*E*,2*S*,3*S*,12β*S*)-3-Ethyl-1,2,3,4,6,7,12,-
12b-octahydro-8-methoxy-α-(methoxymethylene)-indolo[2,3-*a*]quinoli-
zine-2-acetic Acid Methyl Ester *see*
6303
(α*E*,1'*R*,6'*R*,7'*S*,8'a*S*)-6'-Ethyl-1,2,2',3',-
6',7',8',8'a-octahydro-α-(methoxymethylene)-2-oxo-spiro[3*H*-indole-3,-
1'(5'*H*)-indolizine]-7'-acetic Acid
Methyl Ester *see* 8322
Ethyl Octanoate *see* 3833
Ethyl Octylate *see* 3833
Ethyl Oenanthate, 3889
Ethylolamine *see* 3782
Ethyl Oleate *see* 6921
1-(β-Ethylol)-2-methyl-5-nitro-3-azapyr-
role *see* 6236
Ethyl Orthoformate *see* 6980
Ethyl Oxalacetate *see* 3141
Ethyl Oxalate *see* 3142
Ethyl Oxide *see* 3861
Ethyl 3-Oxobutanoate *see* 3812
Ethyl Oxomalonate *see* 5983
3-Ethyl-4-oxo-5-piperidino-Δ²,α-thiazoli-
dineacetic Acid Ethyl Ester *see* 7598
2-[3-Ethyl-4-oxo-5-(1-piperidinyl)-2-thia-
zolidinylidene]acetic Acid Ethyl Ester
see 7598
(α*S*,4*R*)-α-Ethyl-2-oxo-4-propyl-1-pyr-
rolidineacetamide *see* 1384
(α*S*)-α-Ethyl-2-oxo-1-pyrrolidineacetam-
ide *see* 5513
N-[1-[2-(4-Ethyl-5-oxo-2-tetrazolin-1-yl)-
ethyl]-4-(methoxymethyl)-4-piperidyl]-
propionanilide *see* 229
Ethyl Pabate *see* 3798
Ethylpapaverine *see* 3783
Ethylparaben, 3890
Ethyl Parasept [Tenneco] *see* 3890
Ethyl Pelargonate, 3891
2-Ethyl-3,5,6,7,8-pentahydroxy-1,4-naph-
thalenedione *see* 3543
2-Ethyl-3,5,6,7,8-pentahydroxy-1,4-naph-
thoquinone *see* 3543
Ethylphenacemide *see* 7344
o-Ethylphenol *see* 3892
2-Ethylphenol, 3892
Ethyl [2-(4-Phenoxyphenoxy)ethyl]-
carbamate *see* 4017
Ethyl Phenylacetate, 3893
α-Ethyl-α-phenylacetylurea *see* 7344

Ethyl Phenylacrylate *see* 2299
Ethylphenylamine *see* 3819
5-Ethyl-5-phenylbarbituric Acid *see*
7352
N-Ethyl-*N*-phenylbenzenemethanamine
see 3824
N-Ethyl-*N*-phenylbenzylamine *see* 3824
N-Ethyl-3-phenylbicyclo[2.2.1]heptan-2-
amine *see* 3998
2-Ethyl-2-phenylbutyric Acid 2-(2-Dieth-
ylaminoethoxy)ethyl Ester *see* 7029
Ethyl Phenylcarbamate *see* 7430
Ethyl 3-Phenylcarbamoyloxycarbanilate
see 2925
Ethyl Phenyl Carbinol *see* 3823
N-Ethyl-*N*-phenyl-1,2-dihydro-4-hy-
droxy-5-chloro-1-methyl-2-oxo-quino-
line-3-carboxamide *see* 5422
3-Ethyl-3-phenyl-2,6-diketopiperidine
see 4513
(+)-2-(2-Ethyl-2-phenyl-1,3-dioxolan-4-
yl)piperidine *see* 3932
(2*S*)-2-[(2*S*,4*S*)-2-Ethyl-2-phenyl-1,3-
dioxolan-4-yl]piperidine *see* 3932
3-Ethyl-3-phenyl-2,6-dioxopiperidine *see*
4513
Ethylphenylephrine *see* 3914
Ethyl Phenyl Ether *see* 7343
O-Ethyl *S*-Phenyl Ethylphosphonothiolothionate *see* 4260
α-Ethyl-α-phenylglutarimide *see* 4513
2-Ethyl-2-phenylglutarimide *see* 4513
Ethyl Phenylglycinate *see* 7403
Ethyl DL-Phenylglycinate *see* 7402
5-Ethyl-5-phenylhexahydropyrimidine-
4,6-dione *see* 7865
3-Ethyl-5-phenylhydantoin *see* 3802
3-Ethyl-5-phenyl-2,4-imidazolidinedione
see 3802
N-Ethyl-ω-phenylisopropylamine *see*
3818
N-Ethyl-5-phenylisoxazolium-3'-sulfo-
nate *see* 10248
Ethyl Phenyl Ketone *see* 7944
N-(3-Ethylphenyl)-*N*-methyl-*N*'-1-naph-
thalenylguanidine *see* 747
1-(4-Ethylphenyl)-2-methyl-3-(1-piperi-
dinyl)-1-propanone *see* 3661
3-(2-Ethylphenyl)-2-methyl-4(3*H*)-quina-
zolinone *see* 3769
N-Ethyl-3-phenyl-2-norbornanamine *see*
3998
3-Ethyl-3-phenyl-2,6-piperidinedione *see*
4513
2-Ethyl-2-phenyl-4-(2-piperidyl)-1,3-di-
oxolane *see* 3932
N-Ethyl-*N*-(3-phenylpropyl)benzenepro-
panamine *see* 365
N-Ethyl-2-phenyl-*N*-(4-pyridylmethyl)-
hydracrylamide *see* 9959
5-Ethyl-5-phenyl-2,4,6(1*H*,3*H*,5*H*)-pyr-
imidinetrione *see* 7352
α-Ethyl-α-phenyl-3-(trifluoromethyl)-
benzenemethanol *see* 4167
Ethyl Phosphate *see* 9838
Ethyl Phosphite *see* 9840
P-Ethylphosphonodithioic Acid *O*-Ethyl
S-Phenyl Ester *see* 4260
Ethyl Phosphorochloridite *see* 3144
Ethyl Phthalate *see* 7483
3-Ethyl-γ-picoline *see* 3894
3-Ethyl-4-picoline, 3894
4-Ethyl-α-picoline *see* 3895
4-Ethyl-2-picoline, 3895
5-Ethyl-α-picoline *see* 3896

Ethynylestradiol *see* 3788
17α-Ethynylestradiol 3-Methyl Ether *see* 5987
17α-Ethynyl-1,3,5(10)-estratriene-3,17β-diol *see* 3788
17α-Ethynyl-4,9,11-estratrien-17β-ol-3-one *see* 6794
17α-Ethynyl-4-estrene-3β,17β-diol *see* 3909
17α-Ethynylestr-4-en-17β-ol *see* 5690
17α-Ethynyl-13-ethyl-17β-hydroxy-4,15-gonadien-3-one *see* 4447
17α-Ethynyl-17β-hydroxy-4-androsteno-[2,3-*d*]isoxazole *see* 2811
17α-Ethynyl-17β-hydroxy-4-androsten-3-one *see* 3795
17α-Ethynyl-17-hydroxy-6α,21-dimethylandrost-4-en-3-one *see* 3240
17α-Ethynyl-17-hydroxy-5(10)-estren-3-one *see* 6787
17α-Ethynyl-17β-hydroxy-18-methyl-4,15-estradien-3-one *see* 4447
17-Ethynyl-17β-hydroxy-18-methyl-11-methyleneestr-4-en-3-one *see* 3928
17α-Ethynyl-17β-hydroxy-3-oxo-4,9,11-estratriene *see* 6794
17α-Ethynyl-11β-methoxyestra-1,3,5-(10)-triene-3,17β-diol *see* 6375
17α-Ethynyl-3-methoxy-1,3,5(10)-estratrien-17β-ol *see* 5987
17α-Ethynyl-13β-methyl-Δ⁴,⁹,¹¹-gonatriene-17β-ol-3-one *see* 6794
17α-Ethynyl-18-methyl-11-methylene-Δ⁴-estren-17β-ol *see* 2928
17-Ethynyl-18-methyl-19-nortestosterone *see* 6793
17α-Ethynyl-19-norandrost-4-ene-3β,-17β-diol *see* 3909
17α-Ethynyl-19-nortestosterone *see* 6786
N-(3-Ethynylphenyl)-6,7-bis(2-methoxyethoxy)-4-quinazolinamine *see* 3730
5-Ethynyl-2,4(1*H*,3*H*)-pyrimidinedione *see* 3638
17α-Ethynyltestosterone *see* 3795
5-Ethynyluracil *see* 3638
Ethyol [U.S. Bioscience] *see* 399
Etianic Acid *see* 3916
Etibi [Gerot] *see* 3775
Eticol *see* 7134
Etidocaine, 3911
Etidron [Abiogen] *see* 3912
Etidronate Disodium *see* 3912
Etidronic Acid, 3912
Etifoxine, 3913
Etiladrianol *see* 3914
Etilefrin, 3914
Etioallocholane *see* 628
Etiocholane, 3915
Etiocholane-17β-carboxylic Acid *see* 3916
Etiocholanic Acid, 3916
Δ⁴-Etiocholendione-3,17 *see* 631
Etiocobalamin *see* 3102
Etioporphyrin, 3917
Etioporphyrin III *see* 3917
Etiproston, 3918
Etizolam, 3919
Etoc [Sumitomo] *see* 7823
Etodolac, 3920
Etodolic Acid *see* 3920
Etofenamate, 3921
Etofenprox, 3922
Etofibrate, 3923
Etofylline, 3924
Etofylline Clofibrate *see* 9435

Etogesic [Fort Dodge] *see* 3920
Etoglucid, 3925
Etomidate, 3926
Etonitazene, 3927
Etonogestrel, 3928
Etopophos [BMS] *see* 3929
Etoposide, 3929
Etoricoxib, 3930
Etorphine, 3931
Etoscol [Byk Gulden] *see* 4743
Etoval *see* 1524
Etoxadrol, 3932
Etoxazole, 3933
Etozolin, 3934
ETP *see* 9288
Etravirine, 3935
Etrenol [Winthrop] *see* 4798
Etretin *see* 103
Etretinate, 3936
Etroflex *see* 6052
Etrolene *see* 8387
Etruscomicina *see* 5651
Etruscomycin *see* 5651
Etrynit [Bofors] *see* 7925
Et₃SiCl *see* 2179
ETTN *see* 7925
Ettriol Trinitrate *see* 7925
ETU *see* 3858
Etulos *see* 1968
Etumine [Novartis] *see* 2399
Etybenzatropine *see* 3810
Etydion [Nordic] *see* 3773
Etylon *see* 9344
Etyprenaline *see* 5215
5-EU *see* 3638
EU-1806 *see* 6437
EU-4200 *see* 7607
EU-5306 *see* 7172
EU4EA *see* 6092
Eubine [Chiesi] *see* 7058
β-Eucaine, 3937
Eucaine B *see* 3937
Eucalyptol, 3938
Eucalyptus, 3939
Eucardic [Roche] *see* 1873
Eucazulen *see* 4590
Eucerin [Beiersdorf] *see* 5410
Eucil [Farmasa] *see* 6219
Eucistin [San Carlo] *see* 6444
Euclorina [Bracco] *see* 2075
Eucoran *see* 6627
Eucytol *see* 2908
Eudatin [Abbott] *see* 7142
Eudemine [UCB] *see* 3007
Eudermol *see* 6609
Eudigox [Simes] *see* 3186
EUDR *see* 3563
Eudyna [BASF] *see* 8288
Eufans [Sigma-Tau] *see* 591
Eufibron [Berlin-Chemie] *see* 7982
Eufin *see* 3835
Euflavine *see* 115
Euflex [Schering] *see* 4238
Eugallol [Knoll] *see* 8112
Eugenic Acid *see* 3940
Eugenol, 3940
Eugenol Methyl Ether *see* 6144
Euglucon [Sanofi-Aventis] *see* 4514
Euhypnos [Sigma] *see* 9275
Euipnos [Farmitalia] *see* 9275
Eukodal [Merck KGaA] *see* 7058
Eukraton [Nordmark] *see* 1029
Eukystol [Merckle] *see* 4634
Eulexin [Schering] *see* 4238
Eulip [SIT] *see* 9571

Eulipos [Boehringer, Mann.] *see* 9570
Eumelanins *see* 5881
Euminex [Lacer] *see* 4483
Eumotol [Byk Gulden] *see* 1488
Eumovate [GSK] *see* 2359
Eumulgin [Henkel] *see* 7697
Eumydrin [SKB] *see* 866
Eunades [Pfizer] *see* 10187
Eunal [Schering] *see* 5574
Eunatrol *see* 6921
Eunephran [Servier] *see* 1527
Eunerpan [Nordmark] *see* 5895
Eunoctal [Houdé] *see* 567
Eunoctin [Gedeon Richter] *see* 6661
Euonymit *see* 4362
Euonymus, 3941
Eupantol [Altana] *see* 7117
Euparen(e) [Bayer] *see* 3053
Euparin, 3942
Eupatorin, 3943
Eupatorium, 3944
Eupaverin [Merck KGaA] *see* 6374
Eupaverina [Bracco] *see* 6374
Euphorbia, 3945
Euphyllina [Byk Gulden] *see* 461
Euphylline [Altana] *see* 9436
Euphylong [Altana] *see* 9436
Euplit [Desitin] *see* 483
Euplix [Desitin] *see* 7148
Eupnéron [Lyocentre] *see* 3693
Eupractone [Baxter] *see* 3235
Eupragin [Alcon-Thilo] *see* 3740
Euprax [Baxter] *see* 211
Eupressyl [Byk Gulden] *see* 10050
Euradal [Lacer] *see* 1295
Eurax [Ranbaxy] *see* 2585
Euraxil [Novartis] *see* 2585
Eurelix [Hoechst] *see* 7605
Euresol [Knoll] *see* 8276
Eurex [Sanofi-Aventis] *see* 7838
Eurodin [Takeda] *see* 3757
Eurodopa [Europharma] *see* 5516
European Blueberry *see* 1221
European Cranberrybush *see* 10169
European Pennyroyal *see* 7211
European White Birch *see* 1191
Europia *see* 3946
Europium, 3946
Eurosan [Mepha] *see* 2997
Eusaprim [GSK] *see* 9050
Eusmanid *see* 1188
Eusolex 232 [Merck KGaA] *see* 3647
Eusolex 2292 [Merck KGaA] *see* 6859
Eusolex 4360 [Merck KGaA] *see* 7053
Eusolex 6300 [Merck KGaA] *see* 6101
Eusolex 9020 [Merck KGaA] *see* 881
Eusolex HMS [Merck KGaA] *see* 4777
Eusolex OCR [Merck KGaA] *see* 6845
Eusolex OS [Merck KGaA] *see* 6860
Eusovit [Strathmann] *see* 9653
Euspiran *see* 5263
Eutanol G [Cognis] *see* 6856
Eutensin [Aventis] *see* 4338
Euthyrox [Merck KGaA] *see* 9570
Eutonyl [Abbott] *see* 7142
Eutrit [Takeda] *see* 10283
Euvaderm [Gödecke] *see* 1182
Euvernil [Heyden] *see* 9062
Euvitol [Allen & Hanburys] *see* 3998
E-VA-16 *see* 4972
Evacetrapib, 3947
Evadene [Wyeth] *see* 1535
Evadyne [Wyeth] *see* 1535
Evan's Blue, 3948
Evansite *see* 352

Evastel [Almirall] see 3531
Evazol [Ravensberg] see 2914
"Eve" see 5835
Evening Primrose Oil, 3949
Everest [Bayer] see 4151
Everfree [Cooper] see 6683
Everninic Acid Methyl Ester see 8861
Everolimus, 3950
Everone [Hyrex] see 9324
Evik [Syngenta] see 387
Evion [Merck KGaA] see 9653
Evipal [Winthrop] see 4740
Evipal Sodium [Winthrop] see 4740
Evipan [Bayer] see 4740
Evipan Sodium [Bayer] see 4740
Evista [Lilly] see 8215
Evodiamine, 3951
Evolus [DuPont] see 887
Evonogenin see 3181
Evorel [Janssen-Cilag] see 3758
Evoxac [Daiichi] see 2033
Evra [Janssen-Cilag] see 6783
Evramycin [Wyeth] see 9949
Ewer-Pick Acid see 6458
Ex-12-095 see 3770
Exacin [Schering] see 5151
Exacor [Searle] see 2272
Exacyl [Sanofi-Aventis] see 9730
Exal [Nippon Kayaku] see 10179
Exalgin see 6081
Exaltolide [Firmenich], 3952
Exametazime, 3953
Exangit see 1098
Exanta [AstraZeneca] see 10274
Excede [Pfizer] see 1951
Excegran [Dainippon] see 10392
Excenel [Pfizer] see 1951
EXD see 3428
Exelderm [Syntex] see 9025
Exelon [Novartis] see 8369
Exelon Patch [Novartis] see 8369
Exemestane, 3954
Exenatide see 3955
Exendin-3 see 3955
Exendin-4 see 3955
Exendin (9-39) see 3955
Exendin (9-39) Amide see 3955
9-39-Exendin 3 (Heloderma horridum) see 3955
9-39-Exendin 4 see 3955
Exendins, 3955
Exhirud [Sanofi Winthrop] see 4755
Exisulind, 3956
Exitelite see 699
Exjade [Novartis] see 2861
Exlan see 2563
Ex-Lax [Novartis] see 1246
Exlutena [Organon] see 5690
Exluton(a) [Organon] see 5690
Exna [Wyeth] see 1123
Exoamylases see 594
Exocin [Allergan] see 6863
Exocorpol [Green Cross] see 7679
Exoderil [Rentschler] see 6439
Exodextranases see 2951
Exodor-Grun see 2158
Exofene see 4716
(±)-2-Exo-(2-methylbenzyloxy)-1-methyl-4-isopropyl-7-oxabicyclo[2.2.1]heptane see 2296
Exonal [Toyama] see 9252
Exoseptoplix [Tonipharm] see 2092
Exosulfonyl [Théraplix] see 9009
Exosurf [Glaxo Wellcome], 3957
Exosurf Neonatal [GSK] see 2461

EXP-105-1 see 368
EXP-126 see 8351
EXP-999 see 6226
EXP-3174 see 5641
Expandox [Expanpharm] see 47
Expansine see 7158
Experimental Insecticide No. 269 see 3633
Exponcit N [Salutas Fahlberg-List] see 6803
Expressed Almond Oil see 6905
Exrheudon N [Optimed] see 7390
Exsel [Herbert] see 8573
Exsiccated Alum see 354
Exsiccated Ammonium Alum see 323
Exsiccated Ferrous Sulfate see 4084
Exsiccated Sodium Phosphate see 8789
Extacol [Nikken] see 7370
Extavia [Novartis] see 5035
Ext. D & C Orange 3 see 6952
Ext. D & C Red No. 14 see 10287
Ext. D & C Red No. 15 see 7712
Ext. D & C Yellow No. 1 see 5999
Ext. D & C Yellow No. 7 see 6478
Ext. D & C Yellow No. 9 see 10293
Ext. D & C Yellow No. 10 see 10294
Extencilline [RPR] see 7205
Exterol [Dermal] see 1784
Extinguish [Wellmark] see 6055
Extracort [Basotherm] see 9758
Extractum Ginkgo biloba 761 see 4460
Extramycin [Bayer] see 8688
Extranase [Rottapharm] see 1395
Extrinsic Factor see 10212
Extrinsic Plasminogen Activator see 9624
Exubera [Pfizer] see 5023
EYE-001 see 7174
Eye-derived Growth Factor-II see 4101
Eyekas [Showa Shinyaku] see 8325
Eylea [Regeneron] see 174
Ezetimibe, 3958
Ezetrol [Merck/Schering-Plough] see 3958
Ezogabine, 3959
E-Z-Paque [E-Z-Em] see 990
F-4 see 890
F-155 see 8673
F-190 see 51
F-314 see 856
F-319 see 4895
F-440 see 2815
F-1500 see 9009
F-1785 see 4137
F-1983 see 8122
F-1991 see 1046
F-2207 see 6275
F-2559 see 4373
F-3616 see 3476
F-6060 see 2713
F-6066 see 2713
F-6285 see 9078
F-8426 see 9078
F-28249α see 6529
FA-402 see 4033
FA-2071 see 5121
FAA see 4199
2-FAA see 4189
Fabahistin [Bayer] see 5839
Fablyn [Pfizer] see 5430
Fabrazyme [Genzyme] see 4366
Fabrol [Zyma] see 83
F Acid see 1892
Factitious Air see 6740
Factive [Oscient] see 4423
Factor [Syngenta] see 7883

Factor I see 4100
Factor II see 7999
Factor IIa see 9537
Factor III see 9539
Factor V, 3960
Factor VII, 3961
Factor VIII, 3962
Factor IX, 3963
Factor IX Complex (Human) see 3963
Factor X, 3964
Factor XI, 3965
Factor XII, 3966
Factor XIII, 3967
Factor B see 3102
Factor I see 425
Factor T see 10221
Factrel [Wyeth] see 4561
FAD see 4122
FADD-like ICE see 1886
Fademin [Chugai] see 4122
Fadrozole, 3968
Fadul [Hexal] see 3971
Faecla see 7736
Faecula see 7736
Fagarasterol see 5668
Fagarine, 3969
α-Fagarine see 257
β-Fagarine see 8699
γ-Fagarine see 3969
Fagarol see 8609
Fagine see 2211
Fairy Gloves see 3178
Falapen [Frosst] see 7203
Falcol [Bayer] see 23
Falicard [AWD] see 10144
Falignost [Hexal] see 5096
Falithrom [Hexal] see 7371
Falmonox [Winthrop] see 9242
False Hellebore see 156
False Indigo see 954
False Saffron see 1868
False Unicorn see 4666
False Winter's Bark see 1746
Falvin [Recordati] see 4032
Famciclovir, 3970
Fame [Bayer CropSci.] see 4150
Famodil [Sigma-Tau] see 3971
Famosan [ProMed] see 3971
Famosept [Spofa] see 7414
Famotidine, 3971
Famoxadone, 3972
Famoxal [Silanes] see 3971
Famoxate [DuPont] see 3972
Fampridine see 2802
Famvir [SKB] see 3970
Fanapt [Vanda] see 4941
Fanasil [Roche] see 9038
Fancol ACEL [Fanning] see 5409
Fancol ALA [Fanning] see 5410
Fancol LA [Fanning] see 5410
Fanodormo see 2704
Fansidar [Roche] see 8098
Fansidol [Brocchieri] see 6633
Fantacin [Kyowa Hakko] see 4168
Fantorin [Glaxo] see 8942
2-F-araA see 4157
2-F-ara-AMP see 4157
Farecef [Lafare] see 1931
Faremicin [Lafare] see 4281
Fareston [GTx] see 9709
Faretrizin [Lafare] see 1917
Fargan [Pfizer] see 7901
Farganesse [Pfizer] see 7901
Far-Go [Monsanto] see 9756
Farial [Riemser] see 4972

Faridak [Novartis] *see* 7113
Farlutal [Pharmacia] *see* 5864
Farmaproina [CEPA] *see* 7206
Farmidril [Farmitalia] *see* 5441
Farmiglucin [Farmitalia] *see* 7146
Farminosidin [Farmitalia] *see* 7146
Farmiserina [Farmitalia] *see* 2749
Farmitalia 204/122 *see* 9042
Farmolisina [Centralvet] *see* 3390
Farmorubicina [Pharmacia] *see* 3678
Farnesene *see* 3973
α-Farnesene, 3973
β-Farnesene, 3974
Farnesol, 3975
(*E,E*)-Farnesol *see* 3975
trans,trans-Farnesol *see* 3975
Farnesyl Alcohol *see* 3975
Farnoquinone *see* 5901
Farom [Daiichi] *see* 3976
Faropenem, 3976
Faropenem Medoxomil *see* 3976
FAS *see* 3977
Fasciculins, 3977
Fasigin [Pfizer] *see* 9604
Fasigyn [Pfizer] *see* 9604
Fasinex [Novartis] *see* 9817
Faslodex [AstraZeneca] *see* 4313
Fast Green *see* 5763
Fast Green FCF, 3978
Fast Green J *see* 1380
Fastic [Ajinomoto] *see* 6510
Fastin [SKB] *see* 7374
Fastrac [Bell] *see* 1396
Fast Scarlet *see* 1104
Fastum [Menarini] *see* 5352
Fasturtec [Sanofi-Aventis] *see* 8235
Fast Yellow *see* 9955
Fasudil, 3979
Fasupond [Eu Rho Pharma] *see* 6803
Fat Ponceau R *see* 8530
all-cis-Fatty Acid 20:5 Omega-3 *see* 3578
Faustan [Temmler] *see* 2997
[¹⁸F]AV-45 *see* 4140
Faverin [Solvay] *see* 4246
Favistan [Temmler] *see* 6043
Faximin [Novartis] *see* 4494
Fazadinium Bromide, 3980
Fazadon [Glaxo Wellcome] *see* 3980
Fazol [Schering] *see* 5205
Fazor *see* 5769
FB₁ *see* 4318
FB/2 *see* 3392
FB-5097 *see* 2401
FBA-1420 *see* 7917
5F-BAPTA *see* 952
FBS *see* 4202
5-FC *see* 4156
FC 11 *see* 9805
FC-54 *see* 236
FC-1157a *see* 9709
FC-1271a *see* 6991
FC-3001 *see* 9577
FCCs *see* 2141
FCE-20124 *see* 8244
FCE-21336 *see* 1606
FCE-22101 *see* 8364
FCE-22891 *see* 8364
FCE-24304 *see* 3954
FCE-26743 *see* 8450
FCR-1272 *see* 2754
FCRC-A48 *see* 4296
FCV *see* 3970
FDA-20 *see* 4215
FDA-1541 *see* 3697
FDC *see* 4215

FD & C Blue No. 1 *see* 1379
FD & C Blue No. 2 *see* 4982
FD & C Green 1 *see* 4610
FD & C Green No. 2 *see* 5538
FD & C Green No. 3 *see* 3978
FD & C Orange I *see* 6952
FD & C Red No. 1 *see* 7712
FD & C Red No. 2 *see* 369
FD & C Red No. 3 *see* 3749
FD & C Red No. 4 *see* 7713
FD & C Red No. 32 *see* 10287
FD & C Red No. 40 *see* 275
FD & C Yellow No. 1 *see* 6478
FD & C Yellow No. 3 *see* 10293
FD & C Yellow No. 4 *see* 10294
FD & C Yellow No. 5 *see* 9206
FD & C Yellow No. 6 *see* 9130
¹⁸FDG *see* 4158
FDNB *see* 4204
F3TDR *see* 9855
Fe-200486 *see* 2866
Featherfew *see* 4096
Featherfoil *see* 4096
Febantel, 3981
Febrifugine, 3982
Febuxostat, 3983
Fecapentaene 12 *see* 3452
Fectrim [DDSA] *see* 9050
Fedacilina [Fedal] *see* 5996
Fedan [ICI] *see* 6916
Fedibaretta *see* 7336
Felacrinos *see* 2872
Felbamate, 3984
Felbatol [MedPointe] *see* 3984
Felbinac, 3985
Felden [Pfizer] *see* 7619
Feldene [Pfizer] *see* 7619
Felicur [Asche] *see* 3823
Felimazole [Dechra] *see* 6043
Felinine, 3986
(−)-Felinine *see* 3986
Felison [Sigurta] *see* 4228
Felitrope *see* 3823
Felkin-Crabtree Catalyst *see* 2554
Felmane *see* 4228
Feloday [Novartis] *see* 3987
Felodipine, 3987
Felosan *see* 4872
Felviten [Grünenthal] *see* 637
Felypressin, 3988
Femara [Novartis] *see* 5502
Femestrone Inj *see* 3763
Femidyn *see* 3763
FemiFect [Lumavita] *see* 4319
Feminone [Pharmacia & Upjohn] *see* 3788
Femodene [Schering AG] *see* 4447
Femovan [Schering AG] *see* 4447
Femoxetine, 3989
Fempress [Schwarz] *see* 6315
Femring [Warner-Chilcott] *see* 3758
Femstat [Syntex] *see* 1531
Femtrace [Warner-Chilcott] *see* 3758
Femulen [Searle] *see* 3909
Fenactil [Jelfa] *see* 2191
Fenadone *see* 6016
Fenamidone, 3990
Fenamiphos, 3991
Fenamisal *see* 7383
Fenamizol *see* 479
Fenarimol, 3992
Fenarol [Winthrop] *see* 2106
Fenasprate *see* 1047
Fenatin *see* 7330
Fenazaquin, 3993
Fenazil [Sella] *see* 7901

Fenazol [Hokuriku Seiyaku] *see* 4163
Fenazox [Meiji] *see* 389
Fenazoxine *see* 6525
Fenbendazole, 3994
Fenbid [Goldshield] *see* 4919
Fenbuconazole, 3995
Fenbufen, 3996
Fenbutatin Oxide, 3997
Fencamfamine, 3998
Fencamine, 3999
Fenchlorphos *see* 8387
d-Fenchone, 4000
Fenclor [Caffaro] *see* 7682
Fenclozic Acid, 4001
Fendilar [SPA] *see* 4002
Fendiline, 4002
Fenethanil *see* 3995
Fenethylline, 4003
Fenetrazole *see* 9229
Fenfluramine, 4004
Fenformin *see* 7345
Fenhexamid, 4005
Fenhydren *see* 7348
Fenibut *see* 460
Fenibutazona *see* 7390
Fenibutol [Atral] *see* 7390
Fenican [Rhone-Poulenc] *see* 3163
Fenicol [Alcon] *see* 2077
Fenidrone *see* 7056
Fenilin *see* 7348
Fenilor [UCB] *see* 1462
Fenipentol, 4006
Fenistil [Novartis] *see* 3238
Fenitrothion, 4007
Fenizolan [Recordati] *see* 4032
Fennel, 4008
Fenofibrate, 4009
Fenofibric Acid *see* 4009
Fenofibric Acid Choline Salt *see* 4009
Fenofibric Acid Isopropyl Ester *see* 4009
Fenoldopam, 4010
Fenoldopam Mesylate *see* 4010
Fenolovo *see* 9920
Fenoprofen, 4011
Fenopron [Dista] *see* 4011
Fenormin *see* 7345
Fenos [Bayer CropSci.] *see* 4150
Fenospen [Pharmacia] *see* 7209
Fenoterol, 4012
Fenoverine, 4013
Fenoxanil, 4014
Fenoxaprop-ethyl, 4015
Fenoxaprop-P-ethyl *see* 4015
Fenoxazoline, 4016
Fenoximone *see* 3643
Fenoxycarb, 4017
Fenoxypen [Novo] *see* 7209
Fenpiclonil, 4018
Fenpropanate *see* 4019
Fenpropathrin, 4019
Fenpropidin, 4020
Fenpropimorph, 4021
Fenproporex, 4022
Fenprostalene, 4023
Fenpyrate *see* 8079
Fenpyroximate, 4024
Fenretinide, 4025
Fenspiride, 4026
Fensulfothion, 4027
Fensum [Merckle] *see* 47
Fental [Kanebo] *see* 9252
Fentalim [Angelini] *see* 229
Fentanest [Pharmacia] *see* 4028
Fentanyl, 4028
Fentatienil [Angelini] *see* 9018

Fentazin [Allen & Hanburys] see 7297
Fenthion, 4029
Fentiazac, 4030
Fentiazin see 7364
Fenticlor, 4031
Fenticonazole, 4032
Fentigyn [Novartis] see 4032
Fentin Acetate see 9920
Fentin Hydroxide see 9920
Fentonium Bromide, 4033
Fentrazamide, 4034
Fentrinol [Frika] see 393
Fentrol [Cheminova] see 5398
Fenugreek, 4035
Fenuron, 4036
Fenuron TCA see 4036
Fenvalerate, 4037
Fenyramidol see 7432
Feosol [SK & F] see 4084
Feospan [SK & F] see 4084
Fepradinol, 4038
Feprazone, 4039
Fepron [Lilly] see 4011
Feprona [Lilly] see 4011
Feraheme [AMAG Pharm] see 4093
Ferbam, 4040
Ferbeck see 4040
Fergon [Miles] see 4074
Feridex [Advanced Magnetics] see 4090
Fer-in-Sol [Mead Johnson] see 4084
Ferlixit [Aventis] see 8751
Ferlucon see 4074
Fermate see 4040
Fermcozyme see 4496
Fermentation Amyl Alcohol see 5241
Fermentation Butyl Alcohol see 5176
Fermentation L. casei Factor see 8043
Fermine see 3279
Fermium, 4041
Fernambuco see 7295
Fernos [Zeneca] see 7609
Fero-Gradumet [Abbott] see 4084
Ferolactan [Bioindustria] see 7894
Feromax see 4084
Feron [Toray] see 5035
Feroritard [Nikken] see 4084
Ferradow see 4040
Ferrazone [Akzo-Nobel] see 4058
Ferredoxins, 4042
Ferric Acetate, Basic, 4043
Ferric Alum see 515
Ferric Ammonium Citrate, 4044
Ferric Ammonium Ferrocyanide see 8018
Ferric Ammonium Oxalate see 514
Ferric Ammonium Sulfate see 515
Ferric Bromide, 4045
Ferric Chloride, 4046
Ferric Chromate(VI), 4047
Ferric Citrate, 4048
Ferric Dimethyldithiocarbamate see 4040
Ferric Enterobactin see 3650
Ferric Ferrocyanide see 8018
Ferric Fluoride, 4049
Ferric Formate, 4050
Ferric Gallotannate see 4060
Ferric Hexacyanoferrate(II) see 8018
Ferrichrome see 4051
Ferrichrome A see 4051
Ferrichromes, 4051
Ferrichrysin see 4051
Ferric Hydroxide, 4052
Ferric Hydroxide Oxide see 4052
Ferric Hypophosphite, 4053

Ferric Monosodium Ethylenediamine-tetraacetate see 4058
Ferric Nitrate, 4054
Ferric Oxide, 4055
Ferric Persulfate see 4059
Ferric Phosphate, 4056
Ferric Pyrophosphate, 4057
Ferricrocin see 4051
Ferric Sesquioxide see 4055
Ferric Sesquisulfate see 4059
Ferric Sodium Edetate, 4058
Ferric Subsulfate Solution see 6344
Ferric Sulfate, 4059
Ferric Sulfocyanate see 4061
Ferric Sulfocyanide see 4061
Ferric Tannate, 4060
Ferric Thiocyanate, 4061
Ferricytochrome c see 2788
Ferriferrous Oxide see 4067
Ferriheme Chloride see 4679
Ferriheme Hydroxide see 4670
Ferrihemoglobin see 6037
Ferrimycins see 2863
Ferrioxamine B see 2863
Ferriporphyrin Chloride see 4679
Ferriporphyrin Hydroxide see 4670
Ferriprotoporphyrin Basic see 4670
Ferriprotoporphyrin Chloride see 4679
Ferriprox [Apotex] see 2862
Ferrirhodin see 4051
Ferrirubin see 4051
FerriSeltz [Otsuka] see 4044
Ferrite, 4062
Ferrite (Ferrospinel) see 4062
Ferritin, 4063
Ferritose [Calbiochem] see 4302
Ferrivenin [Fresenius] see 5145
Ferrlecit [Watson] see 8751
Ferro Angelini [Angelini] see 4057
Ferrosprint [Pfizer] see 8751
Ferrocene, 4064
1,1'-Ferrocenediylbis(diphenylphos-phine) see 3490
Ferrochel [Albion] see 4068
Ferrocholinate, 4065
Ferrocontin [Napp] see 4066
Ferrocytochrome c see 2788
Ferrofolin [Schering AG] see 4063
Ferroglycine Sulfate, 4066
Ferro-Gradumet [Abbott] see 4084
Ferroheme see 4674
Ferroin see 7328
Ferrol [UCB] see 4063
Ferrolip see 4065
Ferromyn [Calmic] see 4083
Ferron see 4865
Ferronat [Galena] see 4073
Ferronicum [Sandoz] see 4074
Ferrophosphate see 4080
Ferroprotoporphyrin see 4674
Ferrosanol [Schwarz] see 4066
Ferrosoferric Oxide, 4067
Ferrospinel see 4062
Ferrosprint [Pfizer] see 4063
Ferrostar [Mediolanum] see 4063
Ferrostrane [Teofarma] see 4058
Ferrostrene see 4058
Ferrotemp [Medix] see 4073
Ferrous Aminoacetosulfate see 4066
Ferrous Ammonium Sulfate see 518
Ferrous Bisglycinate, 4068
Ferrous Bromide, 4069
Ferrous Chloride, 4070
Ferrous Citrate, 4071
Ferrous Fluoride, 4072

Ferrous Fumarate, 4073
Ferrous Gluconate, 4074
Ferrous Hydroxide, 4075
Ferrous Iodide, 4076
Ferrous Lactate, 4077
Ferrous Oxalate, 4078
Ferrous Oxide, 4079
Ferrous Phosphate, 4080
Ferrous Phosphide, 4081
Ferrous Selenide, 4082
Ferrous Succinate, 4083
Ferrous Sulfate, 4084
Ferrous Sulfide, 4085
Ferrous Sulfocyanate see 4086
Ferrous Sulfocyanide see 4086
Ferrous Thiocyanate, 4086
Ferrox see 4078
Ferroxidase see 2005
Fersaday [Goldshield] see 4073
Fersamal [Goldshield] see 4073
Fertagyl [Intervet] see 4561
Fertiletten [Chassot] see 2102
Fertilysin, 4087
Fertinorm [Serono] see 4250
Fertiral see 4561
Fertirelin, 4088
Fertodur [Schering AG] see 2713
Ferulic Acid, 4089
3-Feruloylquinic Acid see 2143
Ferumoxides, 4090
Ferumoxsil, 4091
Ferumoxtran 10, 4092
Ferumoxytol, 4093
Fervenulin, 4094
FES see 10314
Fesofor [SK & F] see 4084
Fesoterodine, 4095
Fespan [SK & F] see 4084
Festucine see 5618
Fétizon's Reagent see 8646
Fevarin [Solvay] see 4246
Feverall see 3390
Feverfew, 4096
Fevonil [Century] see 3390
Feximac [Nicholas] see 1479
Fexofenadine, 4097
(E)-2-(2-(2-(2-¹⁸F-fluoroethoxy)ethoxy)-ethoxy)-5-(4-methylaminostyryl)pyri-dine see 4140
(E)-4-(2-(6-(2-(2-(2-¹⁸F-fluoroethoxy)eth-oxy)ethoxy)pyridin-3-yl)vinyl)-N-methylbenzenamine see 4140
FG-4963 see 3989
FG-5111 see 5895
FG-5606 see 577
FG-7051 see 7148
FGF see 4101
FGN-1 see 3956
FI-106 see 3487
FI-1163 see 5651
FI-3542 see 3532
FI-5853 see 7146
FI-6026 see 3477
F.I. 6426 see 3404
FI-6714 see 6581
FI-6934 see 2004
FI-7045 see 8604
FI-7056 see 8604
Fialuridine, 4098
FIAU see 4098
Fiber A see 6965
Fiber V [DuPont] see 7689
Fiber X-51 see 2563
Fiber X-54 see 2563
Fiberall [Novartis] see 7631

FiberCon [Pfizer] *see* 1699
Fiber-Lax [Rugby] *see* 1699
Fiblaferon [Rentschler] *see* 5035
Fiboran [Christiaens] *see* 743
Fibrase [SK & F] *see* 7247
Fibravyl [Rhovyl] *see* 7707
Fibrezym [Bene-Arzneimittel] *see* 7247
Fibrin, 4099
Fibrinase *see* 3967
Fibrin-i *see* 4099
Fibrinogen, 4100
Fibrinogenase *see* 9537
Fibrinokinase *see* 9624
Fibrinolysin *see* 7634
Fibrin-s *see* 4099
Fibrin-stabilizing Factor *see* 3967
Fibroblast Growth Factor, 4101
24-163-Fibroblast Growth Factor 7 (Human) *see* 7086
Fibroblast Interferon *see* 5035
Fibrogammin [Centeon] *see* 3967
Fibroins, 4102
Fibrolase, 4103
3-203-Fibrolase [3-Serine] (*Agkistrodon contortrix contortrix* Recombinant) *see* 230
Fibrolax [Gipharmex] *see* 7631
Fibronectins, 4104
Ficain *see* 4106
Ficam [Bayer CropSci.] *see* 1038
Fichtelite, 4105
Ficin, 4106
Ficoid [Fisons] *see* 4184
Ficortril [Pfizer] *see* 4824
Ficusin *see* 8037
Ficus Protease *see* 4106
Ficus Proteinase *see* 4106
Fidarestat, 4107
Fidaxomicin, 4108
FIF *see* 5035
Filaribits [SKB] *see* 3128
Filazine *see* 3128
Fildesin [Shionogi] *see* 10184
Filgrastim *see* 4573
Filicic Acid *see* 4111
Filicin *see* 4111
Filicinic Acid, 4109
Filicinsäure (German) *see* 4109
Filimarisin *see* 4110
Filipin, 4110
Filipin III *see* 4110
Filixic Acid BBB *see* 4111
Filixic Acid PBB *see* 4111
Filixic Acid PBP *see* 4111
Filixic Acids, 4111
Filix Mas (B.P.) *see* 838
Filixsäure (German) *see* 4111
Filtrax [Biomed. Foscama] *see* 7572
Finaceae [Schering AG] *see* 898
Finadyne [Fisons] *see* 4178
Finafloxacin, 4112
Finaplix [Intervet] *see* 9745
Finasteride, 4113
Finastid [Neopharmed] *see* 4113
Finaten [Finadiet] *see* 4255
Finaven [BASF] *see* 3157
Fingolimod, 4114
Finibax [Shionogi] *see* 3474
Finlepsin [AWD] *see* 1783
Finquel [Argent] *see* 9786
Finuret *see* 4826
Fipexide, 4115
Fiproles *see* 4116
Fipronil, 4116
Firazyr [Jerini] *see* 4922

Firdapse [Biomarin] *see* 398
Fire Guard 2000 [Teijin] *see* 9328
Firefly Luciferin, 4117
Firemaster BP-6 *see* 7680
Firemaster LV-T 23P [Michigan Chem.] *see* 9927
Firemaster T 23P [Michigan Chem.] *see* 9927
Firmagon [Ferring] *see* 2866
Firocoxib, 4118
First Rate [Dow AgroSci.] *see* 2391
Firstcin [Takeda] *see* 1939
Fir-wood Oil *see* 6902
Fisalamine *see* 5974
Fischer's Yellow *see* 2416
Fisetin, 4119
Fish-berry *see* 2441
Fish Glue *see* 5154
Fisidenolon 1521 *see* 4119
Fisiobil [Salvat] *see* 3226
Fisiodar [Abiogen] *see* 2963
Fisostina *see* 4341
Fixed Oil *see* 6400
Fixical [Pharmascience] *see* 1659
FK-027 *see* 1925
FK-037 *see* 1933
FK-228 *see* 8382
FK-235 *see* 6630
FK-435 *see* 3495
FK-463 *see* 6254
FK-482 *see* 1921
FK-506 *see* 9155
FK-749 *see* 1952
FK-1160 *see* 9579
FL-113 *see* 5118
FL-1039 *see* 384
FL-1060 *see* 383
FLA-731 *see* 8252
Flabelline *see* 6041
Flagentyl [Rhône-Poulenc] *see* 8556
Flagyl [Searle] *see* 6236
Flake Lead *see* 1002
Flamazine [Smith & Nephew] *see* 9035
Flammazine [Solvay] *see* 9035
Flammex AP *see* 9927
Flammex T 23P *see* 9927
Flamon [Mepha] *see* 10144
Flanax [Syntex] *see* 6499
Flantadin [Lepetit] *see* 2865
Flarex [Alcon] *see* 4207
Flavacidin *see* 609
Flavalfate *see* 3476
Flavamed [Berlin-Chemie] *see* 1088
Flavan [Pharmafarm] *see* 5504
3,3′,4,4′,5,7-Flavanhexol *see* 5504
trans-(+)-3,3′,4′,5,7-Flavanpentol *see* 1902
Flavaspidic Acid, 4121
Flavaxin [Winthrop] *see* 8325
Flavianic Acid *see* 6478
Flavicin *see* 609
Flavin-Adenine Dinucleotide, 4122
Flavine Yellow Shade *see* 8153
Flavin Mononucleotide, 4123
Flavitan [Toa Eiyo] *see* 4122
Flavomycin [Hoechst] *see* 945
Flavone, 4124
Flavopereirine, 4125
Flavophospholipol *see* 945
Flavopiridol, 4126
Flavoquine [Aventis] *see* 569
Flavoxanthin, 4127
Flavoxate, 4128
Flavurol *see* 5934
Flaxedil *see* 4373

Flaxseed *see* 5563
Flazasulfuron, 4129
FLC-1374 *see* 9576
Fleabane *see* 3724
Flea Seed *see* 7631
Flebopex [Profarma] *see* 3325
Flebosan [Dukron] *see* 9769
Flebosmil [Bouchara] *see* 3325
Flebosten [Bonomelli] *see* 3325
Flebotropin [Bago] *see* 3325
Flécaine [3M Pharma] *see* 4130
Flecainide, 4130
Flectadol [Maggioni] *see* 5698
Flectron [BASF] *see* 2764
Fleishmilchsäure *see* 5385
Flemoxine [Yamanouchi] *see* 574
Flerovium *see* 3590
Fleroxacin, 4131
Flex [Syngenta] *see* 4254
Flexartal [Sanofi Winthrop] *see* 1842
Flexase [TAD] *see* 7619
Flexazone *see* 7390
Flexeril [Merck & Co.] *see* 2706
Flexiban [Merck & Co.] *see* 2706
Flexible Collodion *see* 2468
Flexidol [Daker-Farmasimes] *see* 4038
Flexidor [Dow AgroSci.] *see* 5281
Flexin [McNeil] *see* 10399
Flexity [BASF] *see* 6230
Flexstar [Syngenta] *see* 4254
Flibanserin, 4132
FLICE *see* 1886
FLICE Proteinase *see* 1886
Flight Control [DCV] *see* 679
Flindersine, 4133
Flint *see* 8632
Flint [Bayer CropSci.] *see* 9843
Flivas [Asahi] *see* 6440
Flixonase [GSK] *see* 4240
Flixotide [GSK] *see* 4240
Flobacin [Sigma-Tau] *see* 6863
Flociprin [IBI] *see* 2313
Flocor [CytRx] *see* 7679
Flocoumafen, 4134
Floctafenine, 4135
Flodil [AstraZeneca] *see* 3987
Flogar [UCB] *see* 7015
Flogencyl [Expanpharm] *see* 3751
Flogene [Polifarma] *see* 4030
Floginax [Lifepharma] *see* 6499
Flogitolo *see* 7071
Flogobron [Intersint] *see* 7043
Flogoril *see* 7071
Flogos [Gentili] *see* 9139
Flogovital [Bago] *see* 6633
Flogozen [SPA] *see* 4954
Floionic Acid *see* 7436
Flolan [GSK] *see* 7986
Flomax [Chiesi] *see* 6615
Flomax [Astellas] *see* 9181
Flomaxtra [Astellas] *see* 9181
Flomox [Shionogi] *see* 1920
Flomoxef, 4136
Flonase [GSK] *see* 4240
Flonicamid, 4137
Flonorm [Alfa Wassermann] *see* 8345
Flopion [Kyoritsu] *see* 4138
Flopropione, 4138
Floramite [Uniroyal] *see* 1211
Floraquin [Searle] *see* 5085
Florasulam, 4139
Florbetapir F 18, 4140
Flores Martis *see* 4046
Florfenicol, 4141
Florid [Mochida] *see* 6257

1-(3′-p-Fluorobenzoylpropyl)-4-hydroxy-4-(3″-trifluoromethylphenyl)piperidine *see* 9852

N-[1-[3-(p-Fluorobenzoyl)propyl]-4-phenylpiperidin-4-ylmethyl]acetamide *see* 31

N-[[1-[3-(p-Fluorobenzoyl)propyl]-4-phenyl-4-piperidyl]methyl]acetamide *see* 31

8-[3-(p-Fluorobenzoyl)propyl]-1-phenyl-1,3,8-triazaspiro[4.5]decan-4-one *see* 8878

1-[γ-(4-Fluorobenzoyl)propyl]-4-piperidinopiperidine-4-carboxamide *see* 7568

1-[1-[3-(p-Fluorobenzoyl)propyl]-4-piperidyl]-2-benzimidazolinone *see* 1050

1-[1-[3-(4-Fluorobenzoyl)propyl]-4-piperidyl]-2,3-dihydrobenzimidazole-2-thione *see* 9600

1-[1-[3-(4-Fluorobenzoyl)propyl]-4-piperidyl]-2-mercaptobenzimidazole *see* 9600

1-[3-(4-Fluorobenzoyl)propyl]-4-(2-pyridyl)piperazine *see* 891

1-[1-[3-(p-Fluorobenzoyl)propyl]-1,2,3,6-tetrahydro-4-pyridyl]-2-benzimidazolinone *see* 3499

2-[[1-[1-(p-Fluorobenzyl)-2-benzimidazolyl]-4-piperidinyl]methylamino]-4(3H)-pyrimidinone *see* 6307

N-(4-Fluorobenzyl)-5-hydroxy-1-methyl-2-[1-methyl-1-[[(5-methyl-1,3,4-oxadiazol-2-yl)carbonyl]amino]ethyl]-6-oxo-1,6-dihydropyrimidine-4-carboxamide *see* 8216

1-(p-Fluorobenzyl)-2-[[1-(p-methoxyphenethyl)-4-piperidyl]amino]benzimidazole *see* 846

2-(2-Fluoro-4-biphenylyl)propionic Acid *see* 4229

(−)-2-(2-Fluoro-4-biphenylyl)propionic Acid *see* 9196

N-Fluorobis(phenylsulfonyl)amine *see* 4202

Fluorocarbon 11 *see* 9805

5-Fluoro-6-carboxyuracil *see* 4208

9α-Fluoro-21-chloro-11β,16α,17α-trihydroxypregn-4-ene-3,20-dione 16,17-Acetonide *see* 4626

9α-Fluorocortisol *see* 4161

5-Fluorocytosine *see* 4156

6α-Fluoro-1-dehydrohydrocortisone *see* 4223

2-[¹⁸F]Fluoro-2-deoxy-D-glucose *see* 4158

5-Fluoro-2′-deoxy-β-uridine *see* 4144

6α-Fluorodexamethasone *see* 4169

9-Fluoro-6,7-dihydro-5,8-dimethyl-1-oxo-1H,5H-benzo[ij]quinolizine-2-carboxylic Acid *see* 4911

(2S,4S)-6-Fluoro-2,3-dihydro-2′,5′-dioxospiro[4H-1-benzopyran-4,4′-imidazolidine]-2-carboxamide *see* 4107

9-Fluoro-6,7-dihydro-8-(4-hydroxy-1-piperidinyl)-5-methyl-1-oxo-1H,5H-benzo[ij]quinolizine-2-carboxylic Acid *see* 6430

9-Fluoro-2,3-dihydro-3-methyl-10-(4-methyl-1-piperazinyl)-7-oxo-7H-pyrido[3,2,1-ij][4,1,2]benzoxadiazine-6-carboxylic Acid *see* 5815

9-Fluoro-2,3-dihydro-3-methyl-10-(4-methyl-1-piperazinyl)-7-oxo-7H-pyrido[1,2,3-de]-1,4-benzoxazine-6-carboxylic Acid *see* 6863

9-Fluoro-6,7-dihydro-5-methyl-1-oxo-1H,5H-benzo[ij]quinolizine-2-carboxylic Acid *see* 4168

8-Fluoro-5,6-dihydro-5-methyl-6-oxo-4H-imidazo[1,5-a][1,4]benzodiazepine-3-carboxylic Acid Ethyl Ester *see* 4166

9-Fluoro-2,3-dihydro-10-(4-methyl-1-piperazinyl)-7-oxo-7H-pyrido[1,2,3-de]-1,4-benzothiazine-6-carboxylic Acid *see* 8427

5-[(Z)-(5-Fluoro-1,2-dihydro-2-oxo-3H-indol-3-ylidene)methyl]-2,4-dimethyl-N-[2-(1-pyrrolidinyl)ethyl]-1H-pyrrole-3-carboxamide *see* 9650

N-(7-Fluoro-3,4-dihydro-3-oxo-4-prop-2-ynyl-2H-1,4-benzoxazin-6-yl)cyclohex-1-ene-1,2-dicarboximide *see* 4174

2-[7-Fluoro-3,4-dihydro-3-oxo-4-(2-propyn-1-yl)-2H-1,4-benzoxazin-6-yl]-4,5,6,7-tetrahydro-1H-isoindole-1,3(2H)-dione *see* 4174

(4S)-6-Fluoro-2,3-dihydrospiro[4H-1-benzopyran-4,4′-imidazolidine]-2′,5′-dione *see* 8849

6α-Fluoro-11β,21-dihydroxy-16α,17-isopropylidenedioxy-Δ¹,⁴-pregnadiene-3,20-dione *see* 4176

9α-Fluoro-11β,21-dihydroxy-16α,17α-isopropylidenedioxy-1,4-pregnadiene-3,20-dione *see* 9758

(11β,17β)-9-Fluoro-11,17-dihydroxy-17-methylandrost-4-en-3-one *see* 4218

(6α,11β,16α)-6-Fluoro-11,21-dihydroxy-16,17-[(1-methylethylidene)bis(oxy)]-pregna-1,4-diene-3,20-dione *see* 4176

(11β,16α)-9-Fluoro-11,21-dihydroxy-16,17-[(1-methylethylidene)bis(oxy)]-pregna-1,4-diene-3,20-dione *see* 9758

(6α,11β,16α)-6-Fluoro-11,21-dihydroxy-16,17-[(1-methylethylidene)bis(oxy)]-pregn-4-ene-3,20-dione *see* 4227

(6α,11β)-9-Fluoro-11,17-dihydroxy-6-methylpregna-1,4-diene-3,20-dione *see* 4207

(6α,11β,16α)-6-Fluoro-11,21-dihydroxy-16-methylpregna-1,4-diene-3,20-dione *see* 4184

(11β,16α)-9-Fluoro-11,21-dihydroxy-16-methylpregna-1,4-diene-3,20-dione *see* 2931

9-Fluoro-11β,17-dihydroxypregn-4-ene-3,20-dione-17-acetate *see* 4230

9-Fluoro-11β,17-dihydroxyprogesterone 17-Acetate *see* 4230

1-Fluoro-2,4-dinitrobenzene, 4204

(2S,4S)-6-Fluoro-2′,5′-dioxospiro[chroman-4,4′-imidazolidine]-2-carboxamide *see* 4107

Fluoroethanoic Acid *see* 4199

4-[(1E)-2-[6-[2-[2-[2-(Fluoro-¹⁸F)ethoxy]ethoxy]ethoxy]-3-pyridinyl]ethenyl]-N-methylbenzenamine *see* 4140

6-Fluoro-2-(2′-fluoro[1,1′-biphenyl]-4-yl)-3-methyl-4-quinolinecarboxylic Acid *see* 1376

6-Fluoro-1-(4-fluorophenyl)-1,4-dihydro-7-(4-methyl-1-piperazinyl)-4-oxo-3-quinolinecarboxylic Acid *see* 3160

6-Fluoro-1-(4-fluorophenyl)-1,4-dihydro-4-oxo-7-(1-piperazinyl)-3-quinolinecarboxylic Acid *see* 8507

1-[o-Fluoro-α-(p-fluorophenyl)-α-phenylbenzyl]imidazole *see* 4243

Fluoroform, 4205

Fluorogesarol *see* 2960

5-Fluoro-N-hexyl-3,4-dihydro-2,4-dioxo-1(2H)-pyrimidinecarboxamide *see* 1844

9α-Fluorohydrocortisone *see* 4161

4′-Fluoro-4-(4-hydroxy-4-p-chlorophenylpiperidino)butyrophenone *see* 4634

9α-Fluoro-17-hydroxycorticosterone *see* 4161

Δ¹-9α-Fluoro-16α-hydroxyhydrocortisone *see* 9757

(6α,11β,16α)-6-Fluoro-11-hydroxy-16-methyl-3,20-dioxopregna-1,4-dien-21-oic Acid Butyl Ester *see* 4183

(−)-(2R,5S)-5-Fluoro-1-[2-(hydroxymethyl)-1,3-oxathiolan-5-yl]cytosine *see* 3622

p-Fluoro-4-(4′-hydroxy-4′-p-methylphenylpiperidino)butyrophenone *see* 6351

9α-Fluoro-11β-hydroxy-17α-methyltestosterone *see* 4218

9α-Fluoro-16α-hydroxyprednisolone *see* 9757

9α-Fluoro-16α-hydroxyprednisolone Acetonide *see* 9758

p-Fluoro-4-(4′-hydroxy-4′-p-tolylpiperidino)butyrophenone *see* 6351

p-Fluoro-4-[4′-hydroxy-4′-(3″-trifluoromethyl)phenyl]piperidinobutyrophenone *see* 9852

4′-Fluoro-4-[4-hydroxy-4-(α,α,α-trifluoro-m-tolyl)piperidino]butyrophenone *see* 9852

9α-Fluoro-16α,17-isopropylidenedioxyprednisolone *see* 9758

6α-Fluoro-16α,17α-isopropylidenedioxy-4-pregnene-11β,21-diol-3,20-dione *see* 4227

Fluoromar (formerly) [Ohio Med.] *see* 4232

Fluoromethane, 4206

Fluoromethcholone, 4207

4-[[(1R)-2-[5-(2-Fluoro-3-methoxyphenyl)-3-[[2-fluoro-6-(trifluoromethyl)phenyl]methyl]-3,6-dihydro-4-methyl-2,6-dioxo-1(2H)-pyrimidinyl]-1-phenylethyl]amino]butanoic Acid *see* 3580

4′-Fluoro-4-[4-(o-methoxyphenyl)-1-piperazinyl]butyrophenone *see* 4146

[4-(6-Fluoro-7-methylamino-2,4-dioxo-1,4-dihydro-2H-quinazolin-3-yl)-phenyl]-5-chlorothiophen-2-yl-sulfonylurea *see* 3600

2′-Fluoro-5-methyl-β-L-arabinofuranosyluridine *see* 2351

Fluoromethylbenzene *see* 4212

1-Fluoro-2-methylbenzene *see* 4212

2-Fluoro-α-methyl[1,1′-biphenyl]-4-acetic Acid *see* 4229

(αR)-2-Fluoro-α-methyl-[1,1′-biphenyl]-4-acetic acid *see* 9196

9α-Fluoro-16α-methyl-Δ¹-corticosterone *see* 2931

6α-Fluoro-16α-methyl-1-dehydrocorticosterone *see* 4184

9α-Fluoro-16α-methyl-17-desoxyprednisolone *see* 2931

S-Fluoromethyl 6α,9α-Difluoro-11β-hydroxy-16α-methyl-17α-propionyloxy-3-oxoandrosta-1,4-diene-17β-carbothioate *see* 4240

9α-Fluoro-16-methyleneprednisolone 21-Acetate *see* 4222

9α-Fluoro-16-methylene-Δ¹,⁴-pregnadiene-11β,17,21-triol-3,20-dione 21-Acetate *see* 4222

Fluorosilicic Acid see 4214
(+)-(4S)-6-Fluorospiro[chroman-4,4'-imidazolidine]-2',5'-dione see 8849
4'-Fluoro-5'-O-sulfamoyladenosine see 6815
9-(4-Fluoro-5-O-sulfamoylpentofuranosyl)adenine see 6815
Fluorosulfonic Acid, 4211
Fluorosulfuric Acid see 4211
Fluorosulfuric Acid Methyl Ester see 6146
5-Fluoro-1,2,3,6-tetrahydro-2,6-dioxo-4-pyrimidinecarboxylic Acid see 4208
5-Fluoro-1-(tetrahydro-2-furanyl)-2,4-(1H,3H)-pyrimidinedione see 9252
5-Fluoro-1-(tetrahydro-2-furyl)uracil see 9252
(11β, 16α)-9-Fluoro-11,16,17,21-tetrahydroxypregna-1,4-diene-3,20-dione see 9757
6α-Fluoro-11β,16α,17,21-tetrahydroxypregna-1,4-diene-3,20-dione Cyclic 16,17-Acetal with Acetone see 4176
9α-Fluoro-11β,16α,17,21-tetrahydroxypregna-1,4-diene-3,20-dione Cyclic 16,17-Acetal with Acetone see 9758
9-Fluoro-11β,16α,17,21-tetrahydroxypregna-1,4-diene-3,20-dione Cyclic 16,17-Acetal with Acetone, 21-(3,3-Dimethylbutyrate) see 9759
9-Fluoro-11β,16α,17,21-tetrahydroxypregna-1,4-diene-3,20-dione Cyclic 16,17-Acetal with Cyclopentanone 21-Acetate see 382
6α-Fluoro-11β,16α,17,21-tetrahydroxypregn-4-ene-3,20-dione Cyclic 16,17-Acetal with Acetone see 4227
6α-Fluoro-11β,16α,17,21-tetrahydroxyprogesterone Cyclic 16,17-Acetal with Acetone see 4227
Fluorothiamphenicol see 4141
4-Fluoro-4-[4-(2-thioxo-1-benzimidazolinyl)piperidino]butyrophenone see 9600
Fluorotoluene, 4212
m-Fluorotoluene see 4212
o-Fluorotoluene see 4212
p-Fluorotoluene see 4212
Fluorotrichloromethane see 9805
4'-Fluoro-6-[(α,α,α-trifluoro-m-tolyl)oxy]picolinanilide see 7511
9α-Fluoro-11β,17,21-trihydroxy-16-methylenepregna-1,4-diene-3,20-dione 21-Acetate see 4222
(6α,11β,16α)-6-Fluoro-11,17,21-trihydroxy-16-methylpregna-1,4-diene-3,20-dione see 7133
(11β,16α)-9-Fluoro-11,17,21-trihydroxy-16-methylpregna-1,4-diene-3,20-dione see 2945
(11β,16β)-9-Fluoro-11,17,21-trihydroxy-16-methylpregna-1,4-diene-3,20-dione see 1182
(6α,11β)-6-Fluoro-11,17,21-trihydroxypregna-1,4-diene-3,20-dione see 4223
(11β)-9-Fluoro-11,17,21-trihydroxypregna-1,4-diene-3,20-dione see 5220
(11β)-9-Fluoro-11,17,21-trihydroxypregn-4-ene-3,20-dione see 4161
6-[[5-Fluoro-2-[(3,4,5-trimethoxyphenyl)amino]-4-pyrimidinyl]amino]-2,2-dimethyl-4-[(phosphonooxy)methyl]-2H-pyrido[3,2-b]-1,4-oxazin-3-(4H)-one see 4287
Fluorouracil, 4213

2-Fluorovidarabine see 4157
Fluorspar see 1669
Fluosilicic Acid, 4214
Fluosol DA, 4215
Fluostigmine see 5222
Fluosulfonic Acid see 4211
Fluothane [Ayerst] see 4637
Fluovitef [Italfarmaco] see 4181
Fluoxastrobin, 4216
Fluoxeren [Menarini] see 4217
Fluoxetine, 4217
Fluox-Puren [Alpharma] see 4217
Fluoxymesterone, 4218
Flupenthixol see 4219
Flupentixol, 4219
Fluphenacur see 5654
Fluphenazine, 4220
Flupirtine, 4221
Fluprednidene Acetate, 4222
Fluprednisolone, 4223
Fluprednylidene 21-Acetate see 4222
Fluprostenol, 4224
Fluprowit [Thiemann] see 83
Flupyrazapon see 10387
Flupyrsulfuron-methyl, 4225
Fluquinconazole, 4226
Flura-Drops see 8754
Fluram [Roche] see 4191
Flurandrenolide, 4227
Flurandrenolone see 4227
Flurandrenolone Acetonide see 4227
Flurazepam, 4228
Flurazepam Hydrochloride see 4228
Flurbiprofen, 4229
(R)-Flurbiprofen see 9196
Flurinol [Boehringer, Ing.] see 3673
Flurizan [Myriad] see 9196
Fluroblastin [Pharmacia] see 4213
Flurobate [Texas Pharmacal] see 1182
Flurofen [Boots] see 4229
Flurogestone Acetate, 4230
Fluropryl see 5222
Flurothyl, 4231
Fluroxene, 4232
Fluroxypyr, 4233
Fluroxypyr-meptyl see 4233
Flurprimidol, 4234
Flurtamone, 4235
Flusalan see 4210
Flusilazole, 4236
Flusol [Ecosol] see 4217
Fluspirilene, 4237
Flutamide, 4238
Flutamin [Merck KGaA] see 4238
Fluthiacet-methyl, 4239
Fluticasone Propionate, 4240
Flutolanil, 4241
Flutra [Schering] see 9789
Flutriafen see 4242
Flutriafol, 4242
Flutrimazole, 4243
Flutron [Roche] see 3485
Fluvalinate, 4244
Fluvastatin, 4245
Fluvean [Kowa] see 4181
Fluvermal [Janssen] see 4149
Fluvoxamine, 4246
Fluxarten [SKB] see 4175
Fluxema [Italfarmaco] see 1035
Fluxet [Krewel] see 4217
Fluzilazol see 4236
Fluzon [Taisho] see 4181
Flypel see 2859
FM1-43, 4247
L-FMAU see 2351

L-FMAU-TP see 2351
FMC-1240 see 3790
FMC-9260 see 9370
FMC-17370 see 8274
FMC-30980 see 2764
FMC-33297 see 7294
FMC-35001 see 1827
FMC-45498 see 2888
FMC-54800 see 1213
FMC-57020 see 2374
FMC-67825 see 1636
FMC-97285 see 9078
FML [Allergan] see 4207
FMN see 4123
Fmoc Chloride see 4190
Fmoc-Cl see 4190
5-FOA see 4208
Focalin [Novartis] see 6183
Fodrin see 8867
FOE-1976 see 5868
FOE-5043 see 4162
Foipan [Ono] see 1730
Folacin see 4248
Folaren [Ist. Chim. Inter.] see 4249
Folbex VA [Syngenta] see 1445
Folcodal [Syncro] see 2306
Foldine [Specia] see 4248
Folettes [Fawns & McAllan] see 4248
Folex [Pfizer] see 6057
Foliamin see 4248
Foliben [Firma] see 4249
Folic Acid, 4248
Folicet [Mission Pharmacal] see 4248
Folicur [Bayer CropSci.] see 9229
Folidol see 7137
Folidol-M [Bayer CropSci.] see 6181
Foligan [DESMA] see 271
Folikrin see 3763
Folinerin see 6919
Folinic Acid, 4249
Folipac see 4248
Folipex see 3763
Folisan see 3763
Follestrine see 3763
Follicle-Stimulating Hormone, 4250
Follicle-stimulating Hormone (Human α-Subunit Reduced) Complex with Follicle-stimulating Hormone (Human β-Subunit Reduced) Fusion Protein with 118-145-Chorionic Gonadotropin (Human β-Subunit) see 2518
Follicular Hormone see 3763
Follicular Hormone Hydrate see 3762
Folliculin see 3763
Follicunodis see 3763
Follidrin (Tablets) see 3763
Follistim [Organon] see 4250
Follitropin see 4250
Follitropin Alfa see 4250
Follitropin Beta see 4250
Folosan [Uniroyal] see 8197
Folotyn [Allos] see 7821
Folpet, 4251
Folsan [Kali-Chemie] see 4248
Folsäure see 4248
Folvite [Lederle] see 4248
Fomecin A see 4252
Fomecin B see 4252
Fomecins, 4252
Fomepizole, 4253
Fomesafen, 4254
Fominoben, 4255
Fomivirsen, 4256
Fomocaine, 4257
Fonazine, 4258

Fortress [Dow AgroSci.] *see* 8196
Fortrol [Shell] *see* 2675
Fortum [GSK] *see* 1948
Fortumset [GSK] *see* 1948
Fortzaar [Merck & Co.] *see* 5641
Forum [BASF] *see* 3243
Forza [Zeneca Ag Prod] *see* 9251
Fosamac [Merck & Co.] *see* 223
Fosamax [Merck & Co.] *see* 223
Fosamprenavir, 4276
Fosbretabulin, 4277
Foscan [Scotia] *see* 9281
Foscarnet Sodium, 4278
Foscavir [AstraZeneca] *see* 4278
Fosdan [General Quimica] *see* 7448
Fosenopril *see* 4283
Fosetyl Al, 4279
Fosfalugel [Yamanouchi] *see* 352
Fosfestrol, 4280
Fosfluconazole *see* 4153
Fosfobiotic [Bergamon] *see* 4281
Fosfocin [Crinos] *see* 4281
Fosfocina [C.E.P.A.] *see* 4281
Fosfocolina [Afarit] *see* 7474
Fosfocrisolo [Zambon] *see* 4554
Fosfogram [Firma] *see* 4281
Fosfomycin, 4281
Fosfomycin Trometamol *see* 4281
Fosfonomycin *see* 4281
Fosforal [Fournier] *see* 4281
Fosforina B$_{12}$ [Francia] *see* 7475
Fosfosal, 4282
Fosfotricina [Italfarmaco] *see* 4281
Foshagite *see* 1704
Foshallasite *see* 1704
Fosinopril, 4283
Fosinorm [BMS] *see* 4283
Fositen [BMS] *see* 4283
Fosmicin [Meiji] *see* 4281
Fosmidomycin, 4284
Fosphenytoin, 4285
Fosphenytoin Sodium *see* 4285
Fospropofol, 4286
Fospropofol Disodium *see* 4286
Fosrenol [Shire] *see* 5415
Fossil Flour *see* 5015
Fossyol [Merckle] *see* 6236
Fostamatinib, 4287
Fosten [Serono] *see* 746
Fostex P [Henkel] *see* 3912
Fosthiazate, 4288
Foston [Hoechst] *see* 9672
Fostox E *see* 7137
Fotemustine, 4289
Fouadin *see* 8942
Fourneau 190 *see* 51
Fourneau 309 *see* 9135
Fourneau 710 *see* 7636
Fourneau 933 *see* 7591
Fourneau 1500 *see* 9009
Fovane [Pfizer] *see* 1123
Fowler's Solution *see* 7729
Foxetin [Gador] *see* 4217
Foxglove *see* 3178
Fox Green *see* 5002
Foximin [Caber] *see* 4281
"Foxy" *see* 6062
FOY [Ono] *see* 4350
FOY-305 *see* 1730
Fozitec [Lipha] *see* 4283
Foznol *see* 5415
4-F-Phe-4-OH-Pro-Arg-Gly-Trp-NH$_2$ *see* 6530
FPL-670 *see* 2581
FPL-12924AA *see* 8250

FPL-59002 *see* 6521
FPL-59002KP *see* 6521
FPL-60278 *see* 3472
FPL-60278AR *see* 3472
FPRMeCl *see* 7816
FR-02A *see* 3572
FR-245 *see* 9933
FR-613 *see* 9775
FR-860 *see* 2804
FR-1206 *see* 4712
FR-1210 [ICL] *see* 2848
FR-13479 *see* 1952
FR-17027 *see* 1925
FR-31564 *see* 4284
FR-34235 *see* 6630
FR-264205 *see* 1954
FR-310826 *see* 7860
FR-900494 *see* 5357
FR-900506 *see* 9155
FR-901228 *see* 8382
Frabel *see* 7071
Fraction P *see* 2864
Fradiomycin *see* 6539
Fragmin [Pharmacia & Upjohn] *see* 2804
Framycetin *see* 6539
Framygen *see* 6539
Francital [Francia] *see* 4281
Francium, 4290
Frandol [Toa Eiyo] *see* 5271
Frangula, 4291
Frangula Emodin *see* 3618
Frangulic Acid *see* 3618
Frangulin, 4292
Frangulin A *see* 4292
Frangulin B *see* 4292
Franguloside *see* 4292
Franidipine *see* 5808
Frankincense *see* 6926
Franklinite *see* 10326
Franocide [Burroughs Wellcome] *see* 3128
Franrose [Hishiyama] *see* 9252
Fraquinol *see* 6539
Frataxin, 4293
Fraxetin, 4294
Fraxetin-8-glucoside *see* 4295
Fraxin, 4295
Fraxiparine [Sanofi Winthrop] *see* 6433
Fraxoside *see* 4295
FRC-8653 *see* 2278
Fredericamycin A, 4296
Freedox [Pharmacia & Upjohn] *see* 9619
Fremy's Salt, 4297
Frénactil [Janssen-Cilag] *see* 1050
Frenantol [Syntex] *see* 7149
French Chalk *see* 9168
French Green *see* 2615
French Purple *see* 2605
French Saffron *see* 8449
Frenohypon *see* 7149
Frenolicin, 4298
Frenolon [EGYT] *see* 6221
Frenopect [Hefa] *see* 380
Freon [DuPont] *see* 2141
Freon 11 [DuPont] *see* 9805
Freon 12 [DuPont] *see* 3073
Freon 14 [DuPont] *see* 1817
Freon 40 *see* 6112
Freon 114 [DuPont] *see* 2594
Freon C318 [DuPont] *see* 6834
Frequentic Acid *see* 2327
Frequentin, 4299
Fresmin [Takeda] *see* 10212
FRG-8813 *see* 5396
Friar's Balsam *see* 944

Friar's Cowl *see* 108
Friedelan-3-one *see* 4300
Friedelin, 4300
D:A-Friedooleanan-3-one *see* 4300
(3β)-D-Friedoolean-14-en-3-ol *see* 9195
D:A-Friedo-18,19-secolup-19-en-3-one *see* 8620
Frigen [Hoechst] *see* 2141
Frigen 11 [Hoechst] *see* 9805
Frigen 12 [Hoechst] *see* 3073
Frigen 114 [Hoechst] *see* 2594
Fringanor [Sobio] *see* 7334
Frisium [HMR] *see* 2356
Froben [Boots] *see* 4229
Frone [Serono] *see* 5035
Frontier [BASF] *see* 3236
Frontline [Merial] *see* 4116
Fropenem *see* 3976
Frova [Vernalis] *see* 4301
Frovatriptan, 4301
Frovelan [Elan] *see* 4301
Frowncide [ISK] *see* 4148
Frucote *see* 1546
β-Fructofuranosidase *see* 5049
β-Fructofuranosidase (*Saccharomyces cerevisiae* Clone F14 Protein Moiety Reduced) *see* 8447
β-Fructofuranosidases *see* 5049
β-D-Fructofuranoside Fructohydrolase *see* 8447
β-D-Fructofuranosyl-*O*-α-D-galacto-pyranosyl-(1 → 6)-*O*-α-D-galacto-pyranosyl-(1 → 6)-*O*-α-D-galacto-pyranosyl-(1 → 6)-α-D-glucopyrano-side *see* 10152
β-D-Fructofuranosyl-*O*-α-D-galacto-pyranosyl-(1 → 6)-α-D-glucopyrano-side *see* 8213
β-D-Fructofuranosyl-α-D-glucopyrano-side *see* 9012
β-D-Fructofuranosyl-α-D-glucopyrano-side Octakis(hydrogen Sulfate) Alumi-num Complex *see* 9010
Fructose, 4302
D-Fructose *see* 4302
DL-Fructose, 4303
β-D-Fructose *see* 4302
D-Fructose 1,6-Bis(dihydrogen Phos-phate) *see* 4304
D-Fructose 6-(Dihydrogen Phosphate) *see* 4305
Fructose-1,6-diphosphate, 4304
1,6-D-Fructosediphosphoric Acid *see* 4304
Fructose Monophosphate *see* 4305
Fructose-6-phosphate, 4305
D-Fructose-6-phosphoric Acid *see* 4305
β-*h*-Fructosidases *see* 5049
Fructosteril *see* 4302
Fruit Bromelain *see* 1395
Fruitone-N [Rhone-Poulenc] *see* 6456
Fruit Sugar *see* 4302
Frupica [Kumiai] *see* 5912
Frusemide *see* 4338
Frusetic [Unimed] *see* 4338
Frusid [DDSA] *see* 4338
FS-069 *see* 7273
FSF *see* 3967
FSH *see* 4250
FSH-CTP *see* 2518
FSR-3 *see* 5232
FT-207 *see* 9252
Ftalicetimida *see* 7488
Ftalofyne *see* 7486
(−)-FTC *see* 3622

F-TEDA-BF$_4$ *see* 8562
Ftoracizine *see* 4145
Ftorafur [Grünenthal] *see* 9252
Ftorocort [Gedeon Richter] *see* 9758
FTPA *see* 4215
FTY-720 *see* 4114
5-FU *see* 4213
FU-40A *see* 734
Fuadin [Winthrop] *see* 8942
Fua-Med [Med. Fabrik] *see* 6686
Fuchsine *see* 5715
Fuchsin(e) Acid *see* 99
Fucidine [Leo Pharm] *see* 4346
Fucidin [Leo Pharm] *see* 4346
Fucithalmic [Leo Pharm] *see* 4346
Fucosamine, 4306
DL-Fucosamine *see* 4306
D-Fucose, 4307
L-Fucose, 4308
Fucostanol *see* 8943
Fucosterol, 4309
Fucoxanthin, 4310
all-trans-Fucoxanthin *see* 4310
Fucus, 4311
FUDR [Roche] *see* 4144
Fugerel [Schering] *see* 4238
Fugu Poison *see* 9394
Fujiglucon [Fujisawa] *see* 4493
FujiMite [Nichino] *see* 4024
Fulaid [Takeda] *see* 9252
Fulcin [Zeneca] *see* 4584
Fuldazin [Sumitomo] *see* 6277
Fulfeel [Kyorin] *see* 9252
Fulfill [Syngenta] *see* 8061
Ful-Glo [Sola-Barnes-Hinds] *see* 4192
[5]Fullerane-C$_{20}$-*I$_h$* *see* 3447
[5,6]Fullerene-C$_{60}$-Ih *see* 1472
Fullerenes *see* 1472
Fuller's Earth, 4312
Fuller's Herb *see* 8500
Fullsafe [Ohta] *see* 4163
Fulminic Acid Mercury(2+) Salt *see* 5967
Fulsix [Tatsumi] *see* 4338
Fuluvamide [Mect] *see* 4338
Fulvestrant, 4313
Fulvicin [Schering] *see* 4584
Fulvoplumierin, 4314
Fumafer [Sanofi-Synthelabo] *see* 4073
Fumagillin, 4315
Fumaric Acid, 4316
Fumaric Acid Dibenzyl Ester *see* 1138
Fumaric Acid Monooctadecyl Ester Sodium Salt *see* 8807
Fumarine *see* 8005
Fumazone *see* 3025
Fumidil B [Sanofi] *see* 4315
Fumigacin *see* 4668
Fumigatin, 4317
Fuming Liquid Arsenic *see* 792
Fuming Spirit of Libavius *see* 8901
Fumiron [Sanofi-Synthelabo] *see* 4073
Fumonisin B$_1$, 4318
Fundal [Schering] *see* 2086
Funduscein [IOLAB] *see* 4192
Fungacetin [Harvey] *see* 9751
Fungex [Streuli] *see* 10033
Fungibacid [Asche] *see* 9608
Fungichromin, 4319
Fungicidin *see* 6825
Fungifos [S & K Pharma] *see* 9669
Fungilin [BMS] *see* 582
Fungimycin *see* 7284
Funginex [Shell] *see* 9859
Fungiplex [Hermal] *see* 9024

Fungisan [Galderma] *see* 6941
Fungisdin [Isdin] *see* 6257
Fungistat [Cilag] *see* 9303
Fungisterol, 4320
Fungizone [BMS] *see* 582
Fungoid Solution [Pedinol] *see* 10033
Fungoral [Janssen-Cilag] *see* 5349
Funtumine, 4321
Fura-2, 4322
Fura-2/AM *see* 4322
Furachel [Rachelle] *see* 6686
Furacin [Eaton] *see* 6687
Furacinetten *see* 6687
Furacoccid *see* 6687
Furacrylic Acid *see* 4326
Furadan [FMC] *see* 1803
Furadantin [Dura Pharm.] *see* 6686
Furadantine MC [Procter & Gamble] *see* 6686
Furadoin [Eaton] *see* 6686
Furafluor [Green Cross] *see* 9252
2-Furalacetic Acid *see* 4326
Furalan [Lannett] *see* 6686
2-Furaldehyde *see* 4333
Furaltadone, 4323
Furametpyr, 4324
Furamide [Abbott] *see* 3223
Furamon *see* 4341
Furan, 4325
Furanace [Dainippon] *see* 6621
2-Furanacrylic Acid, 4326
2-Furanacrylonitrile, 4327
2-Furancarbinol *see* 4334
2-Furancarbonyl Chloride *see* 4339
2-Furancarboxaldehyde *see* 4333
2-Furancarboxylic Acid *see* 4336
2-Furancarboxylic Acid 1-(Dichloroacetyl)-1,2,3,4-tetrahydro-6-quinolinyl Ester *see* 8173
2-Furancarboxylic Acid Methyl Ester *see* 6149
Furan-2,5-dialdehyde *see* 4328
2,5-Furandicarbaldehyde *see* 4328
2,5-Furandicarboxaldehyde, 4328
2,5-Furandione *see* 5768
N$_1$-(2'-Furanidyl)-5-fluorouracil *see* 9252
2-Furanmethanol *see* 4334
Furanol *see* 4341
Furantoina [Uriach] *see* 6686
1-(2-Furanyl)-N,N'-bis(2-furanylmethylene)methanediamine *see* 4828
(1R,4R,4aR,6aR,9S,10aS,10bS)-9-(3-Furanyl)-1,4,4a,5,6,6a,9,10,10a,10b-decahydro-4-hydroxy-4a,10a-dimethyl-1,4-etheno-3H,7H-benzo[1,2-*c*:3,4-*c'*]dipyran-3,7-dione *see* 2476
(2aR,4aR,4bR,5aS,8S,8aS,10aR,10bR,-14aS)-8-(3-Furanyl)decahydro-2,2,4a,-8a-tetramethyl-11H,13H-oxireno[d]-pyrano[4',3':3,3a]isobenzofuro[5,4-*f*]-[2]benzopyran-4,6,13(2H,5aH)-trione *see* 5547
(2aS,5aS,5bR,7R,8aR,8bR)-6-[2-(3-Furanyl)ethyl]decahydro-6-hydroxy-2a,5a,7-trimethyl-2H-naphtho[1,8-*bc*]furan-2-one *see* 5820
2-[(2-Furanylmethyl)(1H-imidazol-1-ylcarbonyl)amino]butanoic Acid 4-Penten-1-yl Ester *see* 7173
1-(3-Furanyl)-4-methyl-1-pentanone *see* 7281
N-(2-Furanylmethyl)-9H-purin-6-amine *see* 5359
(+)-2-[[(2-Furanylmethyl)sulfinyl]-N-[(2Z)-4-[[4-(1-piperidinylmethyl)-2-

pyridinyl]oxy]-2-buten-1-yl]acetamide *see* 5396
(1R,4S,5R,7S,9aS)-4-(3-Furanyl)octahydro-1,7-dimethyl-2H-quinolizine 5-Oxide *see* 6816
3-(2-Furanyl)-2-propenenitrile *see* 4327
3-(2-Furanyl)-2-propenoic Acid *see* 4326
Furaplast *see* 6687
Furazabol, 4329
Furazalon *see* 4329
Furazol W *see* 6687
Furazolidone, 4330
Furazolin [USSR] *see* 4323
Furazosin *see* 7838
Furcellaran, 4331
Furcellaria Gum *see* 4331
Furesis [BMS] *see* 4338
Furesol *see* 6687
Furethidine, 4332
Furfural, 4333
Furfuralcohol *see* 4334
Furfuramide *see* 4828
Furfuran *see* 4325
"Furfurol" *see* 4333
N^6-Furfuryladenine *see* 5359
6-Furfurylaminopurine *see* 5359
Furfurylidene Acetic Acid *see* 4326
2-(Furfurylsulfinyl)-N-[(Z)-4-[[4-(piperidinomethyl)-2-pyridyl]oxy]-2-buten-yl]acetamide *see* 5396
Furfuryltrimethylammonium *see* 4341
Furilazone *see* 4337
α-Furildioxime, 4335
Furiton *see* 6626
Furloe [PPG] *see* 2193
Furmethide [SK & F] *see* 4341
Furmethonol *see* 4323
Furnace Black *see* 1808
Furobactina [Esteve] *see* 6686
7H-Furo[3,2-g][1]benzopyran-7-one *see* 8037
Furo[3,2-g]coumarin *see* 8037
Furofutran [Taiyo] *see* 9252
α-Furoic Acid *see* 4336
2-Furoic Acid, 4336
2-Furoic Acid Ester with 1-(Dichloroacetyl)-1,2,3,4-tetrahydro-6-quinolinol *see* 8173
α-Furoic Chloride *see* 4339
Furonazide, 4337
Furopenem *see* 3976
Furophen T-Caps [Krewel] *see* 6686
Furo-Puren [Klinge-Nattermann] *see* 4338
Furore Super [Bayer CropSci.] *see* 4015
Furosedon [Santen] *see* 4338
Furosemide, 4338
Furovag *see* 4330
Furoxane *see* 4330
Furoxone [Eaton] *see* 4330
2-Furoyl Chloride, 4339
2-[4-(2-Furoyl)piperazin-1-yl]-4-amino-6,7-dimethoxyquinazoline *see* 7838
Furpirinol *see* 6621
Furpyrinol *see* 6621
Fursemide *see* 4338
Fursultiamine, 4340
Furtrethonium, 4341
Furtrimethonium *see* 4341
Furtulon [Roche] *see* 3485
β-2-Furylacrylic Acid *see* 4326
3-(2-Furyl)acrylonitrile *see* 4327
α-Furylcarbinol *see* 4334
2-Furylcarbinol *see* 4334
1-(3-Furyl)decahydro-11-hydroxy-4b,7,7-11a,13a-pentamethyloxireno(4,4a)-2-

benzopyrano[6,5-g](2)benzoxepin-3,5,-9(3a*H*,4b*H*,6*H*)-trione Acetate *see* 6761
5-[2-(3-Furyl)ethyl]decahydro-5,8-dihydroxy-1,4a,6-trimethyl-1-naphthoic Acid γ-Lactone *see* 5820
β-Furyl Isoamyl Ketone *see* 7281
2-Furyl Methyl Ketone Isonicotinoylhydrazone *see* 4337
1-(3-Furyl)-4-methyl-1-pentanone *see* 7281
Fusafungine, 4342
Fusaloyos *see* 4342
Fusaric Acid, 4343
Fusarine *see* 4342
Fusariotoxin T-2 *see* 9981
Fusarubin, 4344
Fused Borax *see* 8726
Fused Boric Acid *see* 1348
Fusel Alcohols *see* 4345
Fusel Oil, 4345
Fusidic Acid, 4346
Fusilade [Syngenta] *see* 4147
Fussol [Sankyo] *see* 4198
Fustin, 4347
(−)-Fustin *see* 4347
(+)-Fustin *see* 4347
Fustpentane *see* 1795
FUT-175 *see* 6434
Futhan [Torii] *see* 6434
Futraful [Taiho] *see* 9252
Futrican [Astra] *see* 2107
Fuzeon [Roche] *see* 3636
FW-293 *see* 3096
FW-734 *see* 7918
FW-925 *see* 6685
FXV-673 *see* 6996
Fybogel [Reckitt Benckiser] *see* 7631
Fybozest [Reckitt Benckiser] *see* 7631
Fyfanon [Cheminova] *see* 5764
Fyracyl *see* 5719
Fyrex *see* 540
Fyrol FR-2, 9224
Fyrol HB 32 [Stauffer] *see* 9927
Fysionorm [Nourypharma] *see* 5690
G-I *see* 4578
G-II *see* 4578
G-III *see* 4578
G-4 *see* 3081
G-11 *see* 4716
G-020 *see* 2068
G-137 *see* 7519
G-277 *see* 4920
G-605 *see* 9089
G-3139 *see* 6828
G-13871 *see* 7390
G-21133 *see* 2544
G-22008 *see* 8114
G-22355 *see* 4962
G-23350 *see* 30
G-23611 *see* 5224
G-24480 *see* 2998
G-27202 *see* 7071
G-27692 *see* 8672
G-27901 *see* 9829
G-28315 *see* 9079
G-30027 *see* 862
G-30028 *see* 7926
G-30320 *see* 2367
G-31435 *see* 7903
G-32883 *see* 1783
G-32911 *see* 8674
G-33182 *see* 2200
G-34161 *see* 7904
G-34162 *see* 387

G-34586 *see* 2378
G-35020 *see* 2921
G-704650 *see* 223
GA *see* 9151
GAA *see* 100
GABA *see* 425
Gabacet [Sanofi-Synthelabo] *see* 7600
Gabapentin, 4348
Gabapentin Encarbil, 4349
Gabbromicina [Farmitalia] *see* 7146
Gabbromycin [Farmitalia] *see* 7146
Gabbroral [Farmitalia] *see* 7146
Gabbrostim [Vetem] *see* 227
Gabexate, 4350
Gabexate Mesylate *see* 4350
Gabitril [Cephalon] *see* 9572
GABOB *see* 441
Gabomade [Knoll] *see* 441
Gaboxadol, 4351
Gabren(e) [Synthelabo] *see* 7888
G Acid *see* 6473
Gacilin [Andromaco] *see* 4022
Gadobenate Dimeglumine, 4352
Gadobutrol, 4353
Gadodiamide, 4354
Gadofosveset, 4355
Gadolinia *see* 4356
Gadolinium, 4356
Gadolinium Benzyloxypropionictetraacetate Dimeglumine *see* 4352
Gadolinium Diethylenetriaminepentaacetic Acid *see* 4357
Gadolinium Diethylenetriamine Pentaacetic Acid Bismethylamide *see* 4354
Gadolinium Diethylenetriamine Pentaacetic Acid Bis(methyoxyethylamide) *see* 4359
Gadolinium Ethoxybenzyldiethylenetriaminepentaacetic Acid *see* 4360
Gadolinium Texaphyrin *see* 6367
Gadolinium(III) 1,4,7-Tris(carboxymethyl)-10-(2'-hydroxypropyl)-1,4,7,10-tetraazacyclododecane *see* 4358
Gadopentetic Acid, 4357
Gadoteridol, 4358
Gadoversetamide, 4359
Gadovist [Schering AG] *see* 4353
Gadoxetate *see* 4360
Gadoxetic Acid, 4360
Gaduol *see* 2449
GA-GCB *see* 10135
Gahnite *see* 10326
Gainex *see* 7387
α-Gal *see* 4366
Galactaric Acid, 4361
Galactin *see* 7894
Galactitol, 4362
Galactoflavin, 4363
D-Galactomethylose *see* 4307
L-Galactomethylose *see* 4308
4-D-Galactopyranosyl-4-D-fructofuranose *see* 5395
4-*O*-β-D-Galactopyranosyl-D-fructose *see* 5395
O-α-D-Galactopyranosyl-(1 → 6)-[*O*-α-D-galactopyranosyl-(1 → 6)-]₂-*O*-α-D-glucopyranosyl-(1 → 2)-β-D-fructofuranoside *see* 10152
4-*O*-β-D-Galactopyranosyl-D-glucitol *see* 5389
4-*O*-β-D-Galactopyranosyl-D-gluconic Acid *see* 5391
1-*O*-(4-*O*-β-D-Galactopyranosyl-β-D-glucopyranosyl)ceramide *see* 2791
4-*O*-β-D-Galactopyranosyl-D-glucose *see* 5392

6-*O*-α-D-Galactopyranosyl-D-glucose *see* 5888
(2*R*)-*N*-[(1*S*,2*R*,3*E*)-1-[(β-D-Galactopyranosyloxy)methyl]-2-hydroxy-3-heptadecen-1-yl]-2-hydroxytetracosanamide *see* 7479
Galactoquin [Mundipharma] *see* 8176
Galactosaccharic Acid *see* 4361
D-Galactosamine, 4364
D-Galactose, 4365
α-Galactosidase, 4366
α-Galactosidase A *see* 4366
α-Galactosidase (Human Clone λAG18 Isoenzyme A Subunit Protein Moiety Reduced) *see* 4366
α-D-Galactoside Galactohydrolase *see* 4366
β-Galactoside Sorbitol *see* 5389
4-β-D-Galactosido-D-fructose *see* 5395
4-(β-D-Galactosido)-D-gluconic Acid *see* 5391
4-(β-D-Galactosido)-D-glucose *see* 5392
6-(α-D-Galactosido)-D-glucose *see* 5888
4-*O*-β-D-Galactosyl-D-fructose *see* 5395
D-Galacturonic Acid, 4367
Galanga, 4368
Galangal *see* 4368
Galangin, 4369
Galantamine, 4370
Galanthamine *see* 4370
Galanthidine *see* 5687
Galatone *see* 4536
Galatur [Wyeth] *see* 5119
Galaxolide [Int. Flavors & Fragrances] *see* 4751
Galbas *see* 4018
Galben [Isagro] *see* 1032
Galcodine [Galen] *see* 2448
Galecron *see* 2086
Galegine, 4371
Galena *see* 5451, 5476
Galenphol [Galen] *see* 7441
Galfer [Thornton & Ross] *see* 4073
Galβ1 → 4Glcβ1 → 1cer *see* 2791
Galla *see* 6817
Gallacetophenone, 4372
Gallaldehyde 3,5-Dimethyl Ether *see* 9145
Gallamine Triethiodide, 4373
Gallein, 4374
Gallepronin [Taiyo] *see* 4138
Gallery [Dow AgroSci.] *see* 5281
Gallia *see* 4382
Gallic Acid, 4375
Gallic Acid Bismuth Basic Salt *see* 1283
Gallic Acid 5,6-Dihydroxy-3-carboxyphenyl Ester *see* 3172
Gallic Acid 3-Monogallate *see* 3172
Gallic Acid Propyl Ester *see* 7971
Gallicin *see* 4375
Gallimycin [Bimeda] *see* 3739
Gallimycin [Abbott] *see* 3743
Gallium, 4376
Gallium Arsenide, 4377
Gallium Chloride, 4378
Gallium Chloride (GaCl₃) *see* 4378
Gallium Citrate, 4379
Gallium Citrate Ga 67 *see* 4379
Gallium Fluoride (GaF₃) *see* 4384
Gallium Mononitride *see* 4381
Gallium Nitrate, 4380
Gallium Nitride, 4381
Gallium Oxide, 4382
Gallium(III) Oxide *see* 4382
Gallium Phosphide, 4383

Gallium Sesquioxide see 4382
Gallium Trichloride see 4378
Gallium Trifluoride, 4384
Gallium Trinitrate see 4380
Gallocatechin see 1902
Gallochrome see 5934
Gallocyanine, 4385
Gallodesoxycholic Acid see 2054
Gallopamil, 4386
Gallotannic Acid see 9184
Gallotannin see 9184
m-Galloylgallic Acid see 3172
1-Galloyl-α-D-glucose see 4489
1-Galloyl-β-D-glucose see 4490
Galls see 6817
GalN see 4364
Galphol [Galen] see 7441
Galpseud [Thorton & Ross] see 8024
Galsud [Thorton & Ross] see 8024
Galsulfase see 809
Galuteolin see 5675
Galvinoxyl, 4387
Galvus [Novartis] see 10177
Galzin [Lemmon] see 10327
Gamabufagin see 4388
Gamabufogenin see 4388
Gamabufotalin, 4388
Gamanil [Lundbeck] see 5614
Gamaquil [Siegfried] see 7370
Gambir, 4389
Gambir Catechu see 4389
Gambit [Syngenta] see 4018
Gamboge, 4390
Gambogic Acid, 4391
Gamibetal [SIT] see 441
Gamma Benzene Hexachloride see 5556
Gamma-cyhalothrin see 5398
Gamma Hexachlor see 5556
Gamma-hydroxybutyrate see 4854
Gamma-interferon see 5036
Gammajust 50 [Horita] see 6983
Gamma-OH [Clintec Nutrition] see 4854
Gamma-OZ [Kanebo] see 6983
Gammaphos see 399
Gammariza [Toyo] see 6983
Gammatsul [Nippon Chemiphar] see 6983
Gamma-vinyl GABA see 10175
Gamolenic Acid see 5562
Gamonil [Merck KGaA] see 5614
Gamophen [J & J] see 9822
Ganaton [Abbott] see 5291
Ganaxolone, 4392
Ganciclovir, 4393
Gangliosides, 4394
Gangliostat see 4724
Gangliovet [Fidia] see 4394
Ganidan [Specia] see 9040
Ganirelix, 4395
Ganite [Genta] see 4380
Ganja see 1752
Ganoderic Acids, 4396
Ganoderma, 4397
Ganor [Boehringer, Ing.] see 3971
Gansies see 9138
Gantanol [Roche] see 9050
Gantaprim [Lenza] see 9050
Gantrim [Geymonat] see 9050
Gantrisin [Roche] see 9082
GAP A see 4104
GAR-936 see 9588
Garamycin [Schering-Plough] see 4427
Garantose see 8445
Garasol [Schering-Plough] see 4427
Garcinia, 4398
Garcinia Acid see 4861

Garcinia Lactone see 4861
Garcinol see 4398
Gardenal [Specia] see 7352
Garden Chamomile see 2040
Gardenin A see 4399
Gardenin B see 4399
Gardenin C see 4399
Gardenin D see 4399
Gardenin E see 4399
Gardenins, 4399
Garden Lavender see 5445
Garden Sage see 8480
Gardinol see 4400
Gardinol Type Detergents, 4400
Gardol, 4401
Gardona [BASF] see 9337
Gardrin(e) [Syntex] see 3645
Garenoxacin, 4402
Garenoxacin Mesylate see 4402
Garget see 7503
Gargilon see 2914
Garlic, 4403
Garlon [Dow AgroSci.] see 9821
Garmian [Fuso] see 947
(*R*)-Garner Aldehyde see 4404
(*S*)-Garner Aldehyde see 4404
Garner's Aldehyde, 4404
Garnierite see 6583
Garosamine see 8688
GAs see 4455
Gas Black see 1808
GASH see 4598
Gaslon [Nippon Shinyaku] see 5146
Gasmotin [Dainippon] see 6364
Gasoline, 4405
Gasstenon [Tatsumi] see 4138
Gaster [Yamanouchi] see 3971
Gasteril [Ripari-Gero] see 7604
Gastomax [Brocchieri] see 2081
Gastracid [AWD] see 6939
Gastralgin [De Angeli] see 8407
Gastramine see 1185
Gastrax [Asche] see 6747
Gastrese [Wyeth] see 6219
Gastrhéma see 1897
Gastridin [Merck & Co.] see 3971
Gastrins, 4406
Gastrobid [Napp] see 6219
Gastrodia, 4407
Gastrodiagnost [Merck KGaA] see 7224
Gastrodin, 4408
Gastrofrenal [Sigma-Tau] see 2581
Gastrogard [Merial] see 6939
Gastrografin [Bracco] see 2993
Gastroloc see 6939
GastroMARK [Mallinckrodt] see 4091
Gastromax [Farmitalia] see 6219
Gastron [Tanabe] see 3535
Gastron [Winthrop] see 6030
Gastronerton [Dolorgiet] see 6219
Gastronorm [Janssen] see 3467
Gastropen [Schwarz] see 3971
Gastrosedan see 6030
Gastrósil [Heumann] see 6219
Gastro-Tablinen [Sanorania] see 6219
Gastrozepin [Thomae] see 7604
Gastrurol [Gibipharma] see 7618
Gas-X [Sandoz] see 3237
Gatalone [Barnes-Hind] see 4536
Gatifloxacin, 4409
Gaucho [Gustafson] see 4951
Gaudium [Recordati] see 2317
GB see 8511
GB-1 see 1378
GB-2 see 1378

GB-94 see 6251
GBP28 see 151
GC-1189 see 2084
GC-1283 see 6290
GC-2466 see 6384
GC-4072 see 2090
GCP-23339A see 7104
G-CSF see 4573
GD see 8839
Gd-BOPTA/Dimeg see 4352
GDC-0449 see 10207
Gd-DO3A-butrol see 4353
Gd-DTPA see 4357
Gd-DTPA-BMA see 4354
Gd-DTPA-BMEA see 4359
Gd-EOB-DTPA see 4360
Gd(HP-DO3A) see 4358
GDNF, 4410
Gd-T2BET see 6367
Gd-Tex see 6367
GE-132 see 7909
GEA-6414 see 9673
Geangin [GEA] see 10144
Gebutox [Hoechst] see 3315
Gedamycin Methyl Ester see 5914
Gefanil [Sumitomo] see 4411
Gefarnate, 4411
Gefarnil see 4411
Gefarnyl see 4411
Gefitinib, 4412
Gefulcer [Ohta] see 4411
Gehlenite see 1651
Geissoschizoline, 4413
Geissospermine, 4414
Gelatin, 4415
Gelbin see 1662
Geldanamycin, 4416
Geldene [Pfizer] see 7619
Gelée Royale (French) see 8410
Gelfilm [Pfizer] see 4415
Gelfoam [Pfizer] see 4415
Gellan Gum, 4417
Gelled Gasoline see 6451
Gelomyrtol [Pohl] see 6424
Gelose see 176
Gelovermin see 4748
Gelrite [CP Kelco] see 4417
Gelsemine, 4418
Gelseminic Acid see 8545
Gelsemium, 4419
Gelthix [Kerfoot] see 9275
Gelucystine [Parke-Davis] see 2779
Geluprane [Sanofi-Aventis] see 47
Gelvatol [Monsanto] see 7706
Gem [Bayer CropSci.] see 9843
Gemalgene see 9804
Gemcitabine, 4420
Gemeprost, 4421
Gemfibrozil, 4422
Gemifloxacin, 4423
Gemifloxacin Mesylate see 4423
Gemonil [Abbott] see 6033
Gemtuzumab Ozogamicin, 4424
Gemzar [Lilly] see 4420
Genasense [Genta] see 6828
Genatropine see 866
Gencaro [Arca Biopharma] see 1471
Gencor [Zoecon] see 4842
Gendon [Organon] see 8238
Generation [LiphaTech] see 3158
Generlac [Morton Grove] see 5395
GenESA [Gensia] see 762
Geneserine, 4425
Genesis [Monsanto] see 2368

Genestein 4′-Methyl Ether *see* 1231
Genetron [AlliedSignal] *see* 2141
Genetron 12 [AlliedSignal] *see* 3073
Genetron 32 [Honeywell] *see* 3170
Genicide *see* 10260
Geniphene *see* 9716
Genistein, 4426
Genisteine (Alkaloid) *see* 8864
Genistein-4′-glucoside *see* 4426
Genistein-7-*O*-β-D-glucoside *see* 4426
Genistein-4′-glucosidorhamnoside *see* 8844
Genisteol *see* 4426
Genistin *see* 4426
Genlip [Teofarma] *see* 4422
Genol *see* 6091
Genoptic [Allergan] *see* 4427
Genoscopolamine [Amido] *see* 8544
Genostrychnine *see* 8986
Genotropin [Pharmacia & Upjohn] *see* 8842
Gentacin [Schering-Plough] *see* 4427
Gent-Ak [Akorn] *see* 4427
Gentalline [Schering-Plough] *see* 4427
Gentalyn [Essex] *see* 4427
Gentamicin, 4427
Gentamicin A *see* 4427
Gentamicin C₁ *see* 4427
Gentamicin C₁ₐ *see* 4427
Gentamicin C₂ *see* 4427
Gentamicin C₂ᵦ *see* 6260
Gentamicin D *see* 4427
Gentamycin *see* 4427
GenTeal [Novartis] *see* 4880
Gentersal [Ortho] *see* 6764
Gentian, 4428
Gentianic Acid *see* 4434
Gentianin *see* 4434
Gentianine, 4429
Gentian Violet, 4430
Gentiaverm [Berlin-Chemie] *see* 4430
Gentibioptal [Farmila] *see* 4427
Genticin [Roche] *see* 4427
Gentiin *see* 4434
Gentinatre *see* 4433
Gentiobiose, 4431
Gentiopicrin, 4432
Gentiopicroside *see* 4432
Gentisic Acid, 4433
Gentisin, 4434
Gentisine U.C.B. *see* 4433
Gentisod *see* 4433
Gentlax [Purdue Frederick] *see* 1246
Gentocin [Schering-Plough] *see* 4427
Gentogram [Merck-Sante] *see* 4427
Gent-Ophtal [Winzer] *see* 4427
Gentran [Baxter] *see* 2950
Gentran 40 [Baxter] *see* 2950
Gentran 70 [Baxter] *see* 2950
Gentrol [Wellmark] *see* 4842
Genurin [Recordati] *see* 4128
Genz-99067 *see* 3599
Genz-112638 *see* 3599
Geocillin [Roerig] *see* 1839
Geodon [Pfizer] *see* 10371
Geoffroyine *see* 9136
Geomycin [Pliva] *see* 7075
Geon [Goodrich] *see* 7707
Geopen [Pfizer] *see* 1793
Geosmin, 4435
Gephyrotoxin, 4436
GER-11 *see* 438
Geram [UCB] *see* 7600
Geranial *see* 2321
Geraniol, 4437

Geranium, 4438
5-Geranoxypsoralen *see* 1159
Geranyl-1,4-benzenediol *see* 4439
Geranyl Farnesylacetate *see* 4411
Geranylgeranylacetone *see* 9296
Geranylhydroquinone, 4439
Gerbin [Sanofi Winthrop] *see* 23
Gerdaxyl [Gerda] *see* 5861
Geref [Serono] *see* 8600
Gerfil [G. F. Chem.] *see* 7701
Gerhardite *see* 2633
Germalgene *see* 9804
Germall 115 [Sutton] *see* 4957
Germall II [Sutton] *see* 3003
German Chamomile *see* 2040
German Chamomile Oil *see* 2040
Germane, 4440
German Fungus *see* 177
Germanin [Bayer] *see* 9135
Germanium, 4441
Germanium(IV) Chloride *see* 4444
Germanium Chloride (GeCl₂) *see* 4442
Germanium Compd. with Magnesium (1:2) *see* 5731
Germanium Dichloride, 4442
Germanium Dioxide, 4443
Germanium Hydride (GeH₄) *see* 4440
Germanium(IV) Oxide *see* 4443
Germanium Oxide (GeO₂) *see* 4443
Germanium Tetrachloride, 4444
Germanium Tetrafluoride, 4445
Germanium Tetrahydride *see* 4440
Germapect [Thiemann] *see* 1795
Germiciclin [Mendelejeff] *see* 6015
Germine, 4446
Gernebcin [Lilly] *see* 9647
Gero [URPAC] *see* 7875
Gerobit [Gerot] *see* 6020
Gerontine *see* 8870
Geropon AS-200 [Rhodia] *see* 8739
Geroquinol *see* 4439
Gerostop [Merckle] *see* 6849
Gerot-Epilan [Gerot] *see* 5920
Gesafram [Syngenta] *see* 7903
Gesagard [Syngenta] *see* 7904
Gesapax [Syngenta] *see* 387
Gesaprim [Syngenta] *see* 862
Gesatop [Syngenta] *see* 8672
Gestafortin [Merck KGaA] *see* 2102
Gestamestrol [Hermal] *see* 2102
Gestanin [Organon] *see* 282
Gestanon [Organon] *see* 282
Gestanyn [Organon] *see* 282
Gestodene, 4447
Gestone [Nordic] *see* 7889
Gestonorone Caproate, 4448
Gestoral [Novartis] *see* 5864
Gestovis *see* 7225
Gestrinone, 4449
Gestronol Caproate *see* 4448
Gevatran [Anphar-Rolland] *see* 6437
Gevilon [Pfizer] *see* 4422
Gewacalm [Nycomed] *see* 2997
Gewazol *see* 7254
GEWO-399 *see* 5232
GF-196960 *see* 9157
GFP *see* 4579
GG-167 *see* 10311
GG-745 *see* 3518
GGA *see* 9296
GH *see* 8842
Ghatti Gum, 4450
GHB *see* 4854
Ghee *see* 4451
Ghi, 4451

GHK, 4452
GHK-Cu *see* 4452
GHL *see* 4452
Ghrelin, 4453
Ghrelin (Human Clone CTB-187P1 Gene GHRELIN) *see* 4453
GH-RF *see* 8840
GH-RH *see* 8840
GH-RIF *see* 8841
GHRP-2 *see* 7824
GI-87084 *see* 8251
GI-87084B *see* 8251
GI-198745 *see* 3518
Giant Yam *see* 10290
Giardil *see* 4330
Giarlam *see* 4330
Giasion [Zambon] *see* 1922
Gibbane *see* 4455
Gibberellane *see* 4455
Gibberellic Acid, 4454
Gibberellin A₃ *see* 4454
Gibberellins, 4455
Gibberellin X *see* 4454
Gibbsite *see* 350
Gibbs Reagent, 4456
Gibrel [AgriDyne] *see* 4454
Gifblaar Poison *see* 4199
Giganten [Troponwerke] *see* 2306
Gilemal [Chinoin] *see* 4514
Gilenya [Novartis] *see* 4114
Gilsonite, 4457
Gilurytmal [Solvay] *see* 187
Gina [Winthrop] *see* 7925
Ginapect [Winthrop] *see* 7925
Gindarine *see* 9360
Ginger, 4458
[6]-Gingerol, 4459
Gingilli Oil *see* 8608
Ginkgo, 4460
Ginkgoic Acid *see* 614
Ginkgolide A *see* 4461
Ginkgolide B *see* 4461
Ginkgolide C *see* 4461
Ginkgolides, 4461
Ginoden [Schering AG] *see* 4447
Ginorite *see* 1655
Ginseng, 4462
Ginsenosides *see* 4462
Ginstar [Bayer CropSci.] *see* 9465
Giractide, 4463
Girard Reagent D, 4464
Girard Reagent P *see* 4465
Girard Reagents, 4465
Girard Reagent T *see* 4465
Gismondite *see* 1651
Gitogenin, 4466
Gitogenin β-Lycotetraoside *see* 4467
Gitonin *see* 4467
F-Gitonin, 4467
Gitoxigenin, 4468
Gitoxin, 4469
Gittalun [Boehringer, Ing.] *see* 3489
GIV-3727 *see* 9883
Give-Tan [Givaudan] *see* 2310
Giv-Gard DXN [Givaudan] *see* 3244
Gix *see* 2960
Gjellebaekite *see* 1704
GL-7 *see* 4535
Gla *see* 1829
GLA *see* 5562
Glacial Acetic Acid *see* 54
Glacial Phosphoric Acid *see* 7457
Gladiator [Agrisel] *see* 1396
Gladium [Bayer CropSci.] *see* 3808
Glakay [Eisai] *see* 5901

Glamidolo [Angelini] *see* 2820
Glandubolin *see* 3763
Glanil [Janssen] *see* 2306
Glasel [Troponwerke] *see* 3921
Glassy Sodium Metaphosphate *see* 8773
Glassy Sodium Polyphosphate *see* 8773
Glatiramer, 4470
Glauber's Salt *see* 8809
Glaucine, 4471
Glaucon [Alcon] *see* 3674
Glauconex [Alcon] *see* 1021
Glaucostat [Merck & Co.] *see* 8199
Glaucotat [Chibret] *see* 8199
Glaucotensil [Farmila] *see* 3809
Glaudin [Sifi] *see* 8199
Glauline [Smith & Nephew] *see* 6216
Glaupax [Erco] *see* 52
Glaurin *see* 3135
Glausyn [Novartis] *see* 6216
Glauvent *see* 4471
Glazidim [GSK] *see* 1948
GlcCer-β-glucosidase *see* 4487
Glean [DuPont] *see* 2197
Gleevec [Novartis] *see* 4943
Gliadel *see* 7676
Gliadin, 4472
Glial Cell Line-derived Neurotrophic
    Factor *see* 4410
Glianimon [Bayer] *see* 1050
Gliatilin [Italfarmaco] *see* 2212
Glibenclamide *see* 4514
Glibenese [Pfizer] *see* 4477
Glibornuride, 4473
Glicerinformal [Calipe] *see* 4521
Gliclazide, 4474
Glimepiride, 4475
Glimicron [Dainippon] *see* 4474
Glimidstada [Stada] *see* 4514
Glioten [Armstrong] *see* 3623
Gliotoxin, 4476
Glipasol *see* 4515
Glipizide, 4477
Gliporal [Grossmann] *see* 1481
Gliquidone, 4478
Glisoxepid, 4479
Glivec [Novartis] *see* 4943
Gln *see* 4507
Globacillin [Astra] *see* 902
Globaline [Arapahoe] *see* 9352
Globenicol [Yamanouchi] *see* 2077
Globin, 4480
Globocef [Roche] *see* 1924
Globularicitrin *see* 8438
Globulin G₁ *see* 5701
Glofil-131 [Abbott] *see* 5107
Glomycin *see* 7075
Glonoin *see* 6694
Gloxazone *see* 5347
GLP-1 *see* 4482
GLP-2 *see* 4482
GLQ223 [Genelabs] *see* 9813
Glu *see* 4505
Gluborid [Grünenthal] *see* 4473
GlucaGen [Novo] *see* 4481
Glucagon, 4481
Glucagon-Like Peptide I *see* 4482
7-36-Glucagon-Like Peptide I [8-Gly-
    cine] (Human) Fusion Protein with 7-
    36-Glucagon-Like Peptide I [8-Gly-
    cine] (Human) Fusion Protein with
    Serum Albumin (Human) *see* 204
Glucagon-Like Peptide II *see* 4482
Glucagon-Like Peptide II [2-Glycine]
    (Human) *see* 9248
Glucagon-Like Peptides, 4482

Glucagon-related Peptide I *see* 4482
Glucagon-related Peptide II *see* 4482
Glucal *see* 1671
Glucametacin, 4483
Glucametacine *see* 4483
Glucamethacin *see* 4483
Glucamine, 4484
D-Glucamine *see* 4484
α-Glucan Branching Glycosyltransferase
    see 8137
α-1,6-Glucan 6-Glucanohydrolase *see*
    2951
α-1,4-Glucan 4-Glucanohydrolases *see*
    594
α-1,4-Glucan Maltohydrolases *see* 594
Glucantim [Farmitalia] *see* 6150
Glucantime [Specia] *see* 6150
D-Glucaric Acid, 4485
D-Glucaric Acid Calcium Salt (1:1) *see*
    1702
Glucazide *see* 4536
Gluceptate Calcium *see* 4491
Gluceptate Sodium *see* 4491
Glucid *see* 8445
Glucidoral [Servier] *see* 1831
Glucinan [Merck KGaA] *see* 6010
Glucinium *see* 1163
D-Glucitol *see* 8851
Glucitol Iron Complex, Compd with Cit-
    ric Acid *see* 5144
Glucoamylases *see* 594
D-Glucoascorbic Acid, 4486
Glucobay [Bayer] *see* 19
Glucobiogen *see* 1671
Glucocerebrosidase, 4487
β-Glucocerebrosidase *see* 4487
Glucochloral *see* 2072
α-D-Glucochloralose *see* 2072
Glucofrangulin, 4488
Glucofrangulin A *see* 4488
D-Glucofuranurono-6,3-lactone *see* 4503
Glucogallic Acid *see* 4490
α-Glucogallin, 4489
β-Glucogallin, 4490
Glucoheptonic Acid, 4491
α-Glucoheptonic Acid *see* 4491
Glucolin *see* 4495
D-Glucomethylose *see* 8194
Glucomonocarbonic Acid *see* 4491
Gluconasturtiin *see* 7340
Gluconiazide *see* 4536
D-Gluconic Acid *see* 4492
Gluconic Acid, 4492
D-Gluconic Acid Calcium Salt (2:1) *see*
    1671
D-Gluconic Acid Cyclic 4,5-Ester with
    Boric Acid (H₃BO₃) Calcium Salt
    (2:1) *see* 1656
D-Gluconic Acid Iron(3+) Sodium Salt
    (1:?:?) *see* 8751
D-Gluconic Acid δ-Lactone *see* 4493
Gluconic Acid Quinidine Salt *see* 8176
Glucono Delta Lactone *see* 4493
Gluconolactone, 4493
Gluconsan K [Kayaku] *see* 4492
Glucoperiplocymarin *see* 7289
Glucophage [Merck KGaA] *see* 6010
Glucoproscillaridin A *see* 8536
β-D-Glucopyranose Aerodehydrogenase
    see 4496
α-D-Glucopyranose 1-(Dihydrogen Phos-
    phate) *see* 4497
α-D-Glucopyranose-1-gallate *see* 4489
β-D-Glucopyranose-1-gallate *see* 4490
α-D-Glucopyranose-1-phosphate *see*
    4497

α-D-Glucopyranose 1-(3,4,5-Trihydroxy-
    benzoate) *see* 4489
β-D-Glucopyranose 1-(3,4,5-Trihydroxy-
    benzoate) *see* 4490
(3β,5β,12β)-3-[(O-β-D-Glucopyranosyl-
    (1 → 4)-O-3-O-acetyl-2,6-dideoxy-β-D-
    ribo-hexopyranosyl-(1 → 4)-O-2,6-di-
    deoxy-β-D-*ribo*-hexopyranosyl-
    (1 → 4)-2,6-dideoxy-β-D-*ribo*-hexo-
    pyranosyl)oxy]-12,14-dihydroxycard-
    20(22)-enolide *see* 5404
(3β,5β,16β)-3-[(O-β-D-Glucopyranosyl-
    (1 → 4)-O-3-O-acetyl-2,6-dideoxy-β-D-
    ribo-hexopyranosyl-(1 → 4)-O-2,6-di-
    deoxy-β-D-*ribo*-hexopyranosyl-
    (1 → 4)-2,6-dideoxy-β-D-*ribo*-hexo-
    pyranosyl)oxy]-14,16-dihydroxycard-
    20(22)-enolide *see* 5404
(3β,5β)-3-[(O-β-D-Glucopyranosyl-
    (1 → 4)-O-3-O-acetyl-2,6-dideoxy-β-D-
    ribo-hexopyranosyl-(1 → 4)-O-2,6-di-
    deoxy-β-D-*ribo*-hexopyranosyl-
    (1 → 4)-2,6-dideoxy-β-D-*ribo*-hexo-
    pyranosyl)oxy]-14-hydroxycard-20-
    (22)-enolide *see* 5404
(3β,5β,12β,16β)-3-[(O-β-D-Glucopyran-
    osyl-(1 → 4)-O-3-O-acetyl-2,6-dideoxy-
    β-D-*ribo*-hexopyranosyl-(1 → 4)-O-2,6-
    dideoxy-β-D-*ribo*-hexopyranosyl-
    (1 → 4)-2,6-dideoxy-β-D-*ribo*-hexo-
    pyranosyl)oxy]-12,14,16-trihydroxy-
    card-20(22)-enolide *see* 5404
α-D-Glucopyranosyl Bromide 2,3,4,6-
    Tetraacetate *see* 59
2-O-β-D-Glucopyranosylcucurbitacin E
    see 2471
β-D-Glucopyranosyl N,N'-Di(2-chloro-
    ethyl) Phosphoric Acid Diamide *see*
    4504
(3β,5β,12β)-3-[(O-β-D-Glucopyranosyl-
    (1 → 4)-O-2,6-dideoxy-β-D-*ribo*-hexo-
    pyranosyl-(1 → 4)-O-2,6-dideoxy-β-D-
    ribo-hexopyranosyl-(1 → 4)-2,6-dide-
    oxy-β-D-*ribo*-hexopyranosyl)oxy]-
    12,14-dihydroxycard-20(22)-enolide
    see 2922
12-β-D-Glucopyranosyl-12,13-dihydro-
    2,10-dihydroxy-6-[[2-hydroxy-1-(hy-
    droxymethyl)ethyl]amino]-5H-indolo-
    [2,3-a]pyrrolo[3,4-c]carbazole-5,7(6H)-
    dione *see* 3560
7-β-D-Glucopyranosyl-9,10-dihydro-3,5,-
    6,8-tetrahydroxy-1-methyl-9,10-dioxo-
    2-anthracenecarboxylic Acid *see* 1843
8-β-D-Glucopyranosyl-5,7-dihydroxy-2-
    (4-hydroxy-3-methoxyphenyl)-4H-1-
    benzopyran-4-one *see* 8539
(10R)-10-β-D-Glucopyranosyl-1,8-dihy-
    droxy-3-(hydroxymethyl)-9(10H)-an-
    thracenone *see* 303
(10S)-10-β-D-Glucopyranosyl-1,8-dihy-
    droxy-3-(hydroxymethyl)-9(10H)-an-
    thracenone *see* 303
6-O-α-D-Glucopyranosyl-D-fructofura-
    nose *see* 5228
α-D-Glucopyranosyl-β-D-fructofurano-
    side *see* 9012
O-α-D-Glucopyranosyl-(1 → 3)-β-D-fruc-
    tofuranosyl-α-D-glucopyranoside *see*
    5887
3-O-α-D-Glucopyranosyl-D-fructose *see*
    10001
6-O-α-D-Glucopyranosyl-D-fructose *see*
    5228
α-D-Glucopyranosyl-α-D-glucopyrano-
    side *see* 9742

L-Glutamic Acid 5-[2-[4-(Hydroxymethyl)phenyl]hydrazide] see 179
Glutamic Acid Lactam see 8113
275-L-Glutamic Acid (1-91)-(174-527)-Plasminogen Activator (Human Tissue-type Protein Moiety) see 7105
L-Glutamic Acid Polymer with L-Alanine, L-Lysine and L-Tyrosine see 4470
L-Glutamic Acid Sodium Salt (1:1) see 6342
Glutamine, 4507
L-Glutamine see 4507
2-L-Glutamine-6-L-asparaginealytesin see 1332
Glutaminic Acid see 4505
Glutaminol [Syntex] see 4505
3-γ-D-Glutamyl-D-alanylamino-3-methoxyazetidin-2-one-1-sulfonic Acid see 9077
(3R)-3-(γ-D-Glutamyl-D-alanylamino)-3-methoxy-2-oxoazetidine-1-sulfonic Acid see 9077
γ-Glutamyl-α-amino-β-((2-ethyl-N,N,-N',N'-tetrakis(2-chloro)ethylphosphoramidate)sulfonyl)propionyl-(R)-(−)phenylglycine see 1748
1-[N-(γ-L-Glutamyl)amino]-D-proline see 5554
(2R)-L-γ-Glutamyl-3-[[2-[[bis[bis(2-chloroethyl)amino]phosphinyl]oxy]ethyl]sulfonyl]-L-alanyl-2-phenylglycine see 1748
L-γ-Glutamyl-S-[(1R,2E,4E,6Z,9Z)-1-[(1S)-4-carboxy-1-hydroxybutyl]-2,4,-6,9-pentadecatetraenyl]-L-cysteinylglycine see 5507
L-γ-Glutamyl-L-cysteinylglycine see 4511
L-γ-Glutamyl-L-cysteinylglycine Disulfide see 4511
L-γ-Glutamyl-S-(dimethylarsino)-L-cysteinylglycine see 2827
β-N-[γ-L(+)-Glutamyl]-4-hydroxymethylphenylhydrazine see 179
γ-L-Glutamylhypoglycine see 4905
D-γ-Glutamyl-N-[(3R)-3-methoxy-2-oxo-1-sulfo-3-azetidinyl]-D-alaninamide see 9077
N-L-γ-Glutamyl-3-(2-methylenecyclopropyl)alanine see 4905
L-Glutamyl-L-phenylalanyl-[495(497)-L-histidine (R → H)]human Glucosylceramidase (β-Glucocerebrosidase) Peptide with L-Aspartyl-L-leucyl-L-leucyl-L-valyl-L-aspartyl-L-threonyl-L-methionine Glycosylated Peptide 1-506 see 9169
Glutaral see 4508
Glutaraldehyde, 4508
Glutaric Acid, 4509
Glutaric Acid Dinitrile see 4510
Glutaric Dialdehyde see 4508
Glutarol [Dermal] see 4508
Glutaronitrile, 4510
ε-N-(L-Glutar-2-yl)-L-lysine see 8446
Glutasan [Sancarlo] see 4511
Glutathin [Mochida] see 4511
Glutathiol see 4511
Glutathion [Nichiiko] see 4511
L-Glutathione see 4511
Glutathione, 4511
Glutathione-SH see 4511
Gluten, 4512
Glutestere see 4506
Glutethimide, 4513

Glutimic Acid see 8113
Glutiminic Acid see 8113
Glutinal [Sankyo] see 4511
Glutril [Roche] see 4473
Gly see 4526
[Gly(A21),Arg(B31),Arg(B32)]insulin (Human) see 5027
Glybenzcyclamide see 4514
Glyburide, 4514
Glybuthiazole, 4515
Glybuthizol see 4515
Glybuzole, 4516
Glycacil [Lonza] see 5112
Glycamine see 4484
Glyceol [Chugai] see 4520
DL-Glyceraldehyde see 4517
Glyceraldehyde, 4517
Glyceraldehyde 3-Phosphate, 4518
Glyceric Acid, 4519
DL-Glyceric Acid see 4519
Glyceric Aldehyde see 4517
Glycerin see 4520
Glycerine see 4520
Glycerol, 4520
Glycerol Diacetate see 2964
Glycerol α,γ-Dichlorohydrin see 3084
sym-Glycerol Dichlorohydrin see 3084
Glycerol Dimethylketal see 5259
Glycerol Formal, 4521
Glycerol Guaiacolate see 4591
Glycerol α-(2-Methoxyphenyl) Ether see 4591
Glycerol α-Monobutyrate see 6335
Glycerol α-Monochlorohydrin see 2146
Glycerol Mono(2-methoxyphenyl) Ether see 4591
Glycerol α-Monophenyl Ether see 7401
Glycerol Nitric Acid Triester see 6694
Glycerol 1-Octanoate see 6338
DL-α-Glycerol Phosphate see 4522
Glycerol Tribromohydrin see 9776
Glycerol Tributanoate see 9785
Glycerophosphoric Acid, 4522
α-Glycerophosphoric Acid Calcium Salt see 1672
β-Glycerophosphoric Acid Calcium Salt see 1672
sn-Glycero-3-phosphorylcholine see 2212
Glycerose see 4517
Glyceryl p-Aminobenzoate, 4523
Glyceryl 2-p-Chlorophenoxyisobutyrate-1,3-dinicotinate see 1228
Glyceryl Diacetate see 2964
α-Glyceryl Guaiacol Ether see 4591
Glyceryl Guaiacyl Ether see 4591
Glyceryl Monoacetate see 6328
Glyceryl 3-Monobutyrate see 6335
Glyceryl Monostearate, 4524
L-α-Glycerylphosphorylcholine see 2212
Glyceryl o-Tolyl Ether see 5917
Glyceryl Triacetate see 9751
Glyceryl Tributyrate see 9785
Glyceryl Trihexadecanoate see 9908
Glyceryl Trimyristate see 9896
Glyceryl Trinitrate see 6694
Glyceryl Trioleate see 9904
Glyceryl Tripalmitate see 9908
Glyceryl Tripetroselinate see 7304
Glyceryl Tristearate see 9932
Glycidol, 4525
Glycidyl Methacrylate-ethyl Acrylate Polymer see 2499
Glycin see 4875
(Glycinato-N,O)dihydroxyaluminum see 3201

(T-4)-(Glycinato-κN,κO)dihydroxyaluminum see 3201
(T-4)-(Glycinato-κN,κO) [Sulfato(2−)-κO,κO']ferrate(1−) Hydrogen (1:1) see 4066
Glycine, 4526
Glycine, Aluminum Salt see 3201
Glycine Anhydride see 7578
1-Glycine-18-L-argininamide-α^(1-18)-corticotropin see 4463
21^A-Glycine-30^B a-L-arginine-30^B b-L-arginine-insulin (Human) see 5027
Glycine Betaine see 1181
[653-Glycine][Human Interleukin-1 Receptor Accessory Protein-(1-339)-peptide (Extracellular Domain Fragment) Fusion Protein with Human Type I Interleukin-1 Receptor-(5-316)-peptide (Extracellular Domain Fragment) Fusion Protein with Human Immunoglobulin G1-(229 C-Terminal Residues)-peptide (Fc Fragment)] Dimer see 8348
Glycine Hydriodide Compd with Iodine (16:4:5) see 9352
Glycine Nitrile see 406
Glycine Sulfate, 4527
Glycinin, 4528
Glycinonitrile see 406
Glycocholic Acid, 4529
Glycocoll see 4526
Glycocoll Betaine see 1181
Glycocoll-copper see 2617
Glycocyamine, 4530
Glycodiazine see 4543
Glycogen, 4531
Glycogenic Acid see 4492
Glycolaldehyde, 4532
Glycolande [Sanofi-Synthelabo] see 4514
Glycol Bromohydrin see 3846
Glycol Chlorohydrin see 3847
Glycol Cyanohydrin see 3848
Glycol Diacetate see 3853
Glycol Dilaurate, 4533
Glycol Dinitrate see 3854
Glycoleucine see 6795
Glycolic Acid, 4534
Glycolic Acid [o-(2,6-Dichloroanilino)phenyl]acetate see 23
Glycolic Acid 8-Ester with Octahydro-5,8-dihydroxy-4,6,9,10-tetramethyl-6-vinyl-3a,9-propano-3aH-cyclopentacycloocten-1(4H)-one see 7650
Glycolic Aldehyde see 4532
Glycol-monoacetin see 3855
3-Glycoloyl-1,2,3,4,6,11-hexahydro-3,-5,12-trihydroxy-10-methoxy-6,11-dioxo-1-naphthacenyl-3-amino-2,3,6-tridioxy-α-L-arabino-hexopyranoside see 3678
Glycol Salicylate, 4535
Glycoluril see 88
Glycolylurea see 4799
Glycomul L [Lonza] see 8850
Glycomul O [Lonza] see 8850
Glycomul S [Lonza] see 8850
Glyconiazide, 4536
Glyconic Acid see 4492
Glyconormal see 4543
Glycophene see 5121
Glycoprotein G see 9541
Glycoprotein-P see 7310
Glycopyrrolate, 4537
Glycopyrronium Bromide see 4537
Glycosine, 4538

7-D-Glycosylapigenin *see* 718
8-Glycosyl-4′,5,7-trihydroxy-3′-meth-
oxyflavone *see* 8539
Glycyclamide *see* 9671
Glycylglycine, 4539
Glycylglycine Lactam *see* 7578
N-(Glycylglycylglycyl)-8-L-lysinevaso-
pressin *see* 9310
N^α-Glycylglycylglycyl-8-lysine-vasopres-
sin *see* 9310
[Glycyl-*κN*-L-histidyl-*κN*,*κN*³-L-lysinato-
(2−)] Copper *see* 4452
Glycyl-L-histidyl-L-lysine *see* 4452
Glycyl-L-isoleucylglycyl-L-lysyl-L-phen-
ylalanyl-L-leucyl-L-lysyl-L-lysyl-L-alan-
yl-Llysyl-L-lysyl-L-phenylalanylglycyl-
L-lysyl-L-alanyl-L-phenylalanyl-Lvalyl-
L-lysyl-Lisoleucyl-L-leucyl-L-lysyl-L-ly-
sinamide *see* 7308
Glycylpressin [Ferring] *see* 9310
Glycyl-L-seryl-*O*-(1-oxooctyl)-L-seryl-L-
phenylalanyl-L-leucyl-L-seryl-L-prolyl-
L-α-glutamyl-L-histidyl-L-glutaminyl-
L-arginyl-L-valyl-L-glutaminyl-L-glu-
taminyl-L-arginyl-L-lysyl-L-α-gluta-
myl-L-seryl-L-lysyl-L-lysyl-L-prolyl-L-
prolyl-L-alanyl-L-lysyl-L-leucyl-L-glu-
taminyl-L-prolyl-L-arginine *see* 4453
Glycyrrhetic Acid *see* 3644
Glycyrrhetic Acid Acetate *see* 74
18β-Glycyrrhetic Acid Hydrogen Suc-
cinate *see* 1794
18β-Glycyrrhetinic Acid *see* 3644
Glycyrrhetinic Acid Glycoside *see* 4541
Glycyrrhiza, 4540
Glycyrrhizic Acid, 4541
Glycyrrhizin *see* 4541
Glycyrrhizinic Acid *see* 4541
Glydiazinamide *see* 4477
Glyfyllin *see* 3526
Glyhexamide, 4542
Glykin *see* 9961
Glykocellon *see* 1830
Glykresin *see* 5917
Glyme *see* 3245
Glymidine, 4543
Glymol *see* 7301
Glyotol *see* 5917
Glyoxal, 4544
Glyoxal Compd with Sodium Bisulfite
see 4545
Glyoxaldiureine *see* 88
Glyoxalic Acid *see* 4546
Glyoxaline *see* 4953
Glyoxaline-5-alanine *see* 4758
Glyoxal-Sodium Bisulfite, 4545
Gly-Oxide [SKB] *see* 1784
Glyoxyldiureide *see* 250
Glyoxylic Acid, 4546
Glypesin [Stada] *see* 4739
Glyphenarsine *see* 9974
Glyphomax [Dow AgroSci.] *see* 4547
Glyphosate, 4547
Glyphosate-trimesium *see* 4547
Glyphylline *see* 3526
Glypolix [Stroder] *see* 4004
Glypressin [Ferring] *see* 9310
Glypro [Dow AgroSci.] *see* 4547
Glysal *see* 4535
Glysennid [Dorsey] *see* 8594
Glyset [Pfizer] *see* 6267
Glyvenol [Novartis] *see* 9769
GM-CSF *see* 4574
GM-CSF/IL-3 Fusion Protein *see* 5039
GM-DF *see* 4573

GMP *see* 4605
3′,5′-GMP *see* 2702
GMP-140 *see* 8563
Gnoscopine *see* 6807
GnRH *see* 4561
Gö-687 *see* 3934
Gö-919 *see* 7598
Gö-1261C *see* 9594
Gö-3450 *see* 4348
Goal [Dow AgroSci.] *see* 7062
Goa Powder *see* 757
GOBAB *see* 440
Gödecke 3282 *see* 3934
GOE-3450 *see* 4348
Goethite *see* 4052
Goetsch's Vitamin *see* 10221
Goitrin, 4548
Gokilaht [Sumitomo] *see* 2765
Gold, 4549
Gold-bloom *see* 1715
Gold Chloride (AuCl) *see* 4550
Gold Chloride (AuCl₃) *see* 4556
Gold Colloid ¹⁹⁸Au *see* 4552
Gold Cyanide (Au(CN)) *see* 4551
Golden Antimony Sulfide *see* 688
Golden Apple Seed *see* 8170
Golden Ragwort *see* 8589
Golden Seal *see* 4808
Gold Hydroxide (Au(OH)₃) *see* 4557
Goldinodox *see* 873
Goldinomycin *see* 873
Gold Isotope of Mass 198 *see* 4552
Gold Monochloride, 4550
Gold Monocyanide, 4551
Gold Orange *see* 6178
Gold Oxide *see* 4558
Gold Oxide (Au₂O₃) *see* 4558
Gold Potassium Chloride *see* 7800
Gold Potassium Cyanide *see* 7749
Gold, Radioactive, Colloidal, 4552
Gold Sesquioxide *see* 4558
Gold Sodium Chloride *see* 8816
Gold Sodium Cyanide *see* 8745
Gold Sodium Thiomalate, 4553
Gold Sodium Thiosulfate, 4554
Gold Stannate, 4555
Gold Sulfide (Au₂S₃) *see* 4559
Gold Thioglucose *see* 874
Goldthread *see* 2511
Gold-tin Precipitate *see* 4555
Gold-tin Purple *see* 4555
Gold Trichloride, 4556
Gold Trichloride Hydrochloride *see*
2120
Gold Trihydroxide, 4557
Gold Trioxide, 4558
Gold Trisulfide, 4559
Gold Yellow *see* 9954
Golimumab, 4560
Goltix [Bayer] *see* 5995
Gonacrine *see* 115
Gonadoliberin *see* 4561
Gonadorelin *see* 4561
Gonadotropin-Releasing Factor *see* 4561
Gonadotropin-Releasing Hormone, 4561
Gona-1,3,5,7,9,11,13-heptaene *see* 2739
Gonak [Akorn] *see* 4880
Gonal-F [Serono] *see* 4250
Gonapeptyl [Ferring] *see* 9924
Gondafon [Schering] *see* 4543
Goniosol [Novartis] *see* 4880
Gonosan *see* 5335
Gonyaulax Toxin *see* 8526
Good Buffers *see* 4689, 5972
Gooroo Nuts *see* 5366

Gopten [Abbott] *see* 9729
Goserelin, 4562
Gossypine *see* 2211
Gossypitrin *see* 8151
Gossyplure, 4563
Gossypol, 4564
(−)-Gossypol *see* 4564
(+)-Gossypol *see* 4564
Gossypose *see* 8213
Gougerotin, 4565
Gourd Curare *see* 2661
Govern [Dow AgroSci.] *see* 3995
Government Rubber Styrene *see* 8527
GP-2-121-3 *see* 762
GP-26872 *see* 7390
GP-45840 *see* 3091
GP-47680 *see* 7028
GPIIb-IIIa *see* 5032
L-α-GPC *see* 2212
GPI-15715 *see* 4286
G-Proteins, 4566
GR-2/234 *see* 228
GR-2/925 *see* 2358
GR-2/1214 *see* 2359
GR-2/1574 *see* 228
GR-20263 *see* 1948
GR-33343G *see* 8475
GR-33343X *see* 8475
GR-38032F *see* 6943
GR-43175 *see* 9126
GR-43175C *see* 9126
GR-43659X *see* 5379
GR-68755 *see* 304
GR-68755C *see* 304
GR-85548A *see* 6502
GR-109714X *see* 5402
GR-121167X *see* 10311
GR-122311X *see* 8228
Gracevit [Daiichi] *see* 8689
Gradient [Polifarma] *see* 4175
Gradocycline *see* 8381
Graham's Salt *see* 8773
Gralise [Abbott] *see* 4348
Gramalil [Fujisawa] *see* 9576
Gramaxin [Astellas] *see* 1918
Gramicidin C (Soviet) *see* 4567
Gramicidin D (Dubos) *see* 4568
Gramicidin S, 4567
Gramicidins, 4568
Gramicidin S (Soviet) *see* 4567
Gramidil [Dexo] *see* 574
Gramine, 4569
Graminis *see* 9938
Gramipan [Mayoly-Spindler] *see* 2914
Gram-Micina [Lagap] *see* 4281
Grammite *see* 1704
Gramoderm [Schering] *see* 4568
Gramoxone [Syngenta] *see* 7135
Gramplus [Chiesi] *see* 2373
Granaticin, 4570
Granaticin B *see* 4570
Granatum *see* 7709
Grandaxin [EGYT] *see* 9662
Grandisol, 4571
Grandlure *see* 4571
Grandstand [Dow AgroSci.] *see* 9821
Granisetron, 4572
Granit [Rhône-Poulenc] *see* 1456
Granite [Dow AgroSci.] *see* 7212
Granocyte [Chugai] *see* 4573
Granosan *see* 3881
Granudoxy [Pierre Fabre] *see* 3488
α-Granule Membrane Protein 140 *see*
8563
Granulestin *see* 5483

Granulocyte Colony-Stimulating Factor, 4573
Granulocyte Elastase *see* 3583
Granulocyte-Macrophage Colony-Stimulating Factor, 4574
Granzymes *see* 7274
Grape Sugar *see* 4495
Graphite, 4575
Graphite Fluoride, 4576
Graphite Oxide *see* 4577
Graphitic Acid, 4577
Graphitic Oxide *see* 4577
Graphol *see* 6091
Graslan [Dow AgroSci.] *see* 9232
Grasp [Dow AgroSci.] *see* 7212
Grasp [Syngenta] *see* 9724
Gratil [Aventis] *see* 397
Gratus Strophanthin *see* 7000
Gravol [Church & Dwight] *see* 3230
Gray Acetate of Lime *see* 1649
(3β,6β,14R)-Grayanotoxane-3,5,6,10,-14,16-hexol *see* 4578
(3β,6β,14R)-Grayanotoxane-3,5,6,10,-14,16-hexol 14-Acetate *see* 4578
(3β,6β,14R)-Grayanotox-10(20)-ene-3,5,-6,14,16-pentol *see* 4578
Grayanotoxin I *see* 4578
Grayanotoxin II *see* 4578
Grayanotoxin III *see* 4578
Grayanotoxins, 4578
Grazie [Bayer CropSci.] *see* 3808
GR-C507/75 *see* 6943
Greek Hay *see* 4035
Green Broom *see* 8540
Green Cinnabar *see* 2238
Green Fluorescent Protein, 4579
Green Ginger *see* 10250
Greenhartin *see* 5417
Greenheart *see* 1015
Green Hellebore *see* 10151
Green Hydroquinone *see* 8174
Green Mint *see* 8865
"Green Nickel Oxide" *see* 6595
Greenockite *see* 1633
Green Oxide of Chromium *see* 2238
Green Rouge *see* 2238
Green Tea *see* 9225
Green Vitriol *see* 4084
Grenade [Zeneca] *see* 2756
Grepafloxacin, 4580
Grey Arsenic *see* 784
GRF *see* 8840
GRF(1-29)NH₂ *see* 8600
Griffith's Zinc White *see* 5602
Grifulvin [McNeil] *see* 4584
Grilon [Fibron] *see* 6823
Grindelia, 4581
Grindelic Acid, 4582
Grisactin [Ayerst] *see* 4584
Griséfuline [Sanofi Winthrop] *see* 4584
Grisein, 4583
Griseofulvin, 4584
Grisovin [GSK] *see* 4584
Gris-PEG [Herbert] *see* 4584
GR-M *see* 6546
Grocreme [Grossmann] *see* 2914
Grolean [Pitman-Moore] *see* 8842
Groliberin [Kabi] *see* 8600
Grorm [Serono] *see* 8842
Grossularite *see* 1651
Ground Holly *see* 2057
Ground Lily *see* 9865
Groundnut *see* 7164
Groundnut Oil *see* 7165
Grovex [Groves] *see* 7589

Growth Hormone *see* 8842
Growth Hormone-Release Inhibiting Factor *see* 8841
Growth Hormone-releasing Factor *see* 8840
Growth Hormone-Releasing Hormone *see* 8840
Growth Hormone-releasing Peptide-2 *see* 7824
GR-S *see* 8527
Grubbs' Catalyst, 4585
Grubbs II Catalyst *see* 4586
Grubbs' First Generation Catalyst *see* 4585
Grubbs' Second Generation Catalyst, 4586
Grundmann's Ketone, 4587
Grysio [Ayerst] *see* 4584
GS-95 *see* 9467
GS-393 *see* 139
GS-504 *see* 2271
GS-840 *see* 139
GS-1278 *see* 9289
GS-1339 *see* 3522
GS-2989 *see* 5849
GS-3065 *see* 3488
GS-4071 *see* 6986
GS-4104 *see* 6986
GS-4331-05 *see* 9289
GS-6244 *see* 1782
GS-9137 *see* 3610
GS-9350 *see* 2436
GS-13005 *see* 6042
GS-16068 *see* 3382
GS-19851 *see* 1445
GS-23654 *see* 6719
G Salt *see* 6473
GSH *see* 4511
GSK-716155 *see* 204
GSK-1349572 *see* 3460
GSK-1349572A *see* 3460
GSK-1838262 *see* 4349
GSK-1841157 *see* 6862
GSSG *see* 4511
G-strophanthidin *see* 6999
GT16-026A *see* 8613
GT-31-104 *see* 2457
GT-41 *see* 1508
GT-92 *see* 2712
GT-1012 *see* 7820
GTP Binding Proteins *see* 4566
Guacetisal *see* 841
Guaiac, 4588
Guaiac Alcohol *see* 4592
Guaiacol, 4589
Guaiacol Glyceryl Ether *see* 4591
Guaiacol Glyceryl Ether Carbamate *see* 6052
Guaiacum *see* 4588
Guaiacum (Resin) *see* 4588
Guaiacuran *see* 4591
Guaiacyl Glyceryl Ether *see* 4591
Guaiaspir [Lampugnani] *see* 841
Guaiazulene, 4590
S-Guaiazulene *see* 4590
Guaiazulene Soluble *see* 4590
Guaifenesin, 4591
Guaiol, 4592
Guaiphenesin *see* 4591
Guajol *see* 4592
Guanabenz, 4593
Guanadrel, 4594
Guanamprazine *see* 401
Guanazolo *see* 889
Guaneran [Burroughs Wellcome] *see* 9451

Guanethidine, 4595
Guanfacine, 4596
Guanicil [Cilag] *see* 9040
Guanidine, 4597
Guanidineacetic Acid *see* 4530
Guanidinium Aluminum Sulfate Hexahydrate, 4598
4-Guanidino-1-butanol Syringate *see* 5495
4-Guanidino-2,4-dideoxy-2,3-dehydro-N-acetylneuraminic Acid *see* 10311
1-(2-Guanidinoethyl)octahydroazocine *see* 4595
2-Guanidinomethyl-1,4-benzodioxan *see* 4603
4-Guanidino-Neu5Ac2en *see* 10311
Guanidoacetic Acid *see* 4530
Guanine, 4599
Guanine Nucleotide Binding Proteins *see* 4566
Guanine Riboside *see* 4601
Guanine Riboside-3-phosphoric Acid *see* 4604
Guanine Riboside-5-phosphoric Acid *see* 4605
Guanite *see* 8984
Guano, 4600
Guanosine, 4601
Guanosine Cyclic 3′,5′-(Hydrogen Phosphate) *see* 2702
Guanosine 3′,5′-Cyclic Monophosphate *see* 2702
Guanosine 3′,5′-Cyclic Phosphate *see* 2702
Guanosine 3′-Monophosphate *see* 4604
Guanosine 3′,5′-Monophosphate *see* 2702
Guanosine 5′-Monophosphate *see* 4605
Guanosine 5′-Phosphate *see* 4605
Guanoxabenz, 4602
Guanoxan, 4603
Guantal *see* 3358
Guanylate Cyclase *see* 2702
Guanylguanidine *see* 1218
Guanylhydrazine *see* 438
3′-Guanylic Acid, 4604
5′-Guanylic Acid, 4605
Guanylic Acid B *see* 4604
N¹-Guanylsulfanilamide *see* 9040
Guanylurea Sulfate *see* 3104
Guaran, 4606
Guaranine *see* 1639
Guardsman [BASF] *see* 3236
Guarem [Shire] *see* 4607
Guar Flour *see* 4607
Guar Gum, 4607
Guarina [Norgine] *see* 4607
Guastil [Uriach] *see* 9119
Guatambuinine *see* 6928
Guayule *see* 8411
Gubernal [Novartis] *see* 309
Guelder Rose *see* 10169
Guggul *see* 4609
Guggulsterone, 4608
Guggulu, 4609
Guglip [Cipla] *see* 4609
Gugulipid *see* 4609
Guidazide *see* 4536
Guinea Green B, 4610
Guinea-pig-anti-stiffness Factor *see* 8944
Gujaphenyl [DDD] *see* 4589
L-Gulitol *see* 8851
Gulliostin [Taiyo] *see* 3388
D-Gulonic Acid, 4611
L-Gulonic Acid, 4612

D-Gulose, 4613
L-Gulose, 4614
Gum Ammoniac see 490
Gum Arabic see 15
Gumbaral [AWD] see 144
Gum Benjamin see 4615
Gum Benzoin, 4615
Gum Camphor see 1734
Gum Copal see 2502
Gum Cyamopsis see 4607
Gum Damar see 2806
Gum Ghatti see 4450
Gum Guaiac see 4588
Gum Guggul see 4609
Gum Guggulu see 4609
Gum Karaya see 5332
Gum Kino see 5361
Gummigutt see 4390
Gum Opium see 6949
Gum Plant see 3727
Gum-plant (of California) see 4581
Gum Quince Seed see 8170
Gum-resin Myrrh see 6423
Gum Thus see 6926, 10007
Gum Tragacanth, 4616
Guntrin [Zensei] see 6983
Gunyl [Hogapharm] see 4589
G.U.-Pen [Pfizer] see 1839
Guronsan see 4503
Guru Nuts see 5366
Gusathion M [Bayer CropSci.] see 906
Gusperimus, 4617
Gutron [Nycomed] see 6263
Guttalax-Fher [Boehringer, Ing., Italia] see 7518
Gutta-Percha, 4618
β-Guttiferin see 4391
Guvacine, 4619
GVG see 10175
GW-506U78 see 6527
GW-1000 see 1752
GW-433908G see 4276
GW-572016 see 5418
GW-572016F see 5418
GW-786034 see 7160
GW-786034B see 7160
GX-1048 see 5379
Gy-bon [Ciba-Geigy] see 8674
Gymiso [HRA-Pharma] see 6293
Gymnemic Acid, 4620
Gymnemin see 4620
Gynäsan [Bastian] see 3762
Gynefollin see 3115
Gyne-Lotrimin [Schering-Plough] see 2401
Gynera [Schering AG] see 4447
Gynergen [Novartis] see 3718
Gynesine see 9863
Gynestrel [Recordati] see 6499
Gyn-Hydralin [Lefrancq] see 4526
Gynipral [Chemie Linz] see 4743
Gynocardia Oil see 2046
Gynochrome see 5934
Gyno-Daktarin [Janssen-Cilag] see 6257
Gynofug [Wolff] see 4919
Gynol II [Ortho-Cilag] see 6764
Gyno-Myfungar [Klinge] see 7036
Gynomyk [Cassenne] see 1531
Gyno-Pevaryl [Cilag-Chemie] see 3549
Gynoplix [Théraplix] see 51
Gynorest [Mead Johnson] see 3521
Gyno-Terazol [Cilag] see 9303
Gynotherax [Pfizer] see 2196
Gyno-Travogen [Schering AG] see 5205
Gyno-Trosyd [Pfizer] see 9608

Gynovlar [Schering AG] see 6786
Gynoxin [Recordati] see 4032
Gypsum see 1708
Gyramid [Warner-Lambert] see 3641
Gyrolite see 1704
h5G1.1 Fab see 3553
h5G1.1-scFv see 7307
h5G1.1VHC + h5G1.1VLC see 3553
H48 see 5737
H-56/28 see 309
(−)-H-80/62 see 7853
H-93/26 see 6228
cis-H-102/09 see 10325
H-115 see 2107
H-133 see 3052
H-133/22 see 7853
H-154/82 see 3987
H-168/68 see 6939
H-199/18 see 6939
H-319/68 see 5879
H-321 see 6044
H-365 see 7149
H-376/95 see 10274
H-610 see 3998
H-814 see 4003
H-990 see 7065
H-3292 see 3402
H-3452 see 2713
H-3625 see 2527
H-3749 see 8088
H-3774 see 236
H-4723 see 2356
HA-1077 see 3979
Haarlem Oil see 10007
Habekacin [Meiji Seika] see 760
Habitrol [Novartis] see 6609
Hachimycin, 4621
H Acid see 6470
Hadacidin, 4622
h-Adenylic Acid see 145
Haemocomplettan P [CSL Behring] see 4100
Haemofort see 4084
Haemostop Injection see 6479
Haemozoin see 4684
Haertolan see 6490
Hafnium, 4623
Hageman Factor see 3966
Haginin E see 2876
Hahnium see 3509
HAL see 4632
Halamid [Akzo] see 2075
Halamid [Viatris] see 6521
Halane see 3074
Halaven [Eisai] see 3723
Halazepam, 4624
Halazone, 4625
Halciderm [BMS] see 4626
Halcimat [BMS] see 4626
Halcinonide, 4626
Halcion [Pharmacia & Upjohn] see 9765
Haldol [McNeil] see 4634
Haldol Decanoate [Janssen] see 4634
Haldrate [Lilly] see 7133
Haldrone [Lilly] see 7133
Halfan [SKB] see 4630
Halfa Wax see 3756
Half-cystine see 2778
Halibut Liver Oil, 4627
Halichondrins, 4628
Halidor [Sumitomo] see 1035
Halinone [USV] see 1400
Halite see 8734
Halloysite see 4312
Haloanisone see 4146

Halobetasol Propionate, 4629
Halochondrine A see 6911
Halofantrine, 4630
Halofenozide, 4631
Halofuginone, 4632
Halog [BMS] see 4626
Halogabide see 7888
Halometasone, 4633
Halomonth [Dainippon] see 4634
Halon see 3073
Haloperidol, 4634
Halopyramine see 2164
Halostachine, 4635
Halosten [Shionogi] see 4634
Halosulfuron-methyl, 4636
Halotestin [Pfizer] see 4218
Halothane, 4637
Haloxazolam, 4638
Halset [Strohschein] see 2031
Hamaméliode P [Oberlin] see 5504
Hamamelis, 4639
Hamamelitannin, 4640
Hamamelose, 4641
Hamburg Blue see 8018
Hämovannid [Bastian] see 5022
Hamycin, 4642
Hamycin A see 4642
Hanatoxins, 4643
Hanp [Zeria] see 863
α-HANP see 863
HAPA-B see 5151
Hapadex [Schering] see 6565
Haplophytine, 4644
Haptocil [Cilag-Chemie] see 2705
Haptocorrin see 10212
Haptoglobins, 4645
Harden-Young Ester see 4304
Hard Paraffin see 7127
H-D-Arg-Arg-Pro-Hyp-Gly-Thia-Ser-D-Tic-Oic-Arg-OH see 4922
Harkoseride see 5381
Harlem Oil see 5319
Harmaline, 4646
Harmalol, 4647
Harmalol Methyl Ether see 4646
Harman, 4648
Harmidine see 4646
Harmine, 4649
Harmogen [Abbott] see 3763
Harmonet [Wyeth] see 4447
Harmony [DuPont] see 9468
Harmonyl [Abbott] see 2919
Harnal [Astellas] see 9181
Harness [Monsanto] see 60
Harodase see 4797
Harpin, 4650
Harpin Ea see 4650
Hartolan [Croda] see 5410
Hartshorn Salt see 505
Harvade [Uniroyal] see 7037
Harzol [Hoyer] see 8694
Hasethrol see 7222
Hashish see 1752
Hassium, 4651
Hasubanan see 4652
Hasubanonine, 4652
HAT see 2796
Hatchett's Brown see 2627
HATU, 4653
HaTx see 4643
HaTx₁ see 4643
HaTx₂ see 4643
Hauser's Salt see 10382
Hausmannite see 5800
Havapen [Wyeth] see 7190

Havlane [Roussel Diamant] *see* 5631
Haw Apple *see* 2555
Haworth Lignan *see* 5539
Haws *see* 2555
Hawthorn *see* 2555
Haxifal [Erempharma] *see* 1914
Hay-Crom [Ivax] *see* 2581
Haynon [Rosemont] *see* 2186
Hayo *see* 2438
Hazol *see* 7065
HB-419 *see* 4514
HBBL *see* 3652
HBCD *see* 4712
HBF-386 *see* 1611
HBK *see* 760
HBTU, 4654
HBW-023 *see* 4755
4-HC *see* 7276
HC-3 *see* 4678
HC-58 *see* 3585
HC-064 *see* 6626
HC-803 *see* 5291
HC-20511 *see* 5354
25-HCC *see* 1641
HCFU *see* 1844
HCG *see* 2221
γ-HCH *see* 5556
HCS, 4655
HCT-3012 *see* 6498
Hcy *see* 4772
HDNF *see* 6572
HDP *see* 7039
HE-69 *see* 6308
HE-781 *see* 9398
Headline [BASF] *see* 8064
Healon [Pfizer] *see* 4796
Healonid [Pfizer] *see* 4796
Healthied [Hokko] *see* 7173
Heartcin [Ohta] *see* 10025
Heartgard [Merial] *see* 5296
Heavy Hydrogen *see* 2941
Heavy Spar *see* 990
Heavy Water *see* 2942
HEC-5725 *see* 4216
Hecameg [Vegatec], 4656
Hecogenin, 4657
Hectorite, 4658
Hectorol [Bone Care] *see* 3484
Hederagenin, 4659
α-Hederin *see* 4660
Hedex [GSK] *see* 47
Hedione [Firmenich] *see* 6124
Hedonal MCPP [Bayer] *see* 5855
HEDTA *see* 10162
Hefasolon [Sanavita] *see* 7841
Heferol [Alkaloid] *see* 4073
Heidelberry *see* 1221
Heitrin [Abbott] *see* 9297
Helenalin, 4661
Helenien *see* 10261
Helenin *see* 200
Helenynolic Acid, 4662
Helfergin [Promonta] *see* 5851
Helianthine B *see* 6178
Helicin, 4663
Helicocerin *see* 2003
Heliosupine, 4664
Heliotropin *see* 7588
Helium, 4665
Helixin C *see* 6625
Helixin (the Saponin) *see* 4660
Helmatac [SK & F] *see* 7139
Helmet Flower *see* 8550
Helmetina *see* 7364
Helmex [Infectopharm] *see* 8066

Helmezine [Allen & Hanburys] *see* 7576
Helmifren *see* 7576
Helminal *see* 5324
Helmintox [Innothéra] *see* 8066
Helmitol [Bayer] *see* 6038
Helodermin *see* 3955
Helogaphen [Spitzner] *see* 2085
Helonias, 4666
Helospectin *see* 3955
Helvamox [Helvepharm] *see* 574
Helvecyclin [Helvepharm] *see* 9341
Helvetia Blue *see* 6102
Helveticoside, 4667
Helvolic Acid, 4668
Hem *see* 4674
Hemabate [Pfizer] *see* 1824
Hemanthine *see* 1497
Hematein, 4669
Hematin, 4670
β-Hematin *see* 4684
Hematin-protein *see* 2788
Hematite *see* 4055
Hematoporphyrin, 4671
Hematoporphyrin IX *see* 4671
Hematoxiline *see* 4672
Hematoxylin, 4672
Hematoxylon, 4673
Heme, 4674
Hemel *see* 316
Hementin, 4675
Hemerven [Interdelta] *see* 3325
Hemerythrin, 4676
Hemicelluloses, 4677
Hemicholinium, 4678
Hemicholinium-3 *see* 4678
Hemiglobin *see* 6037
Hemin, 4679
Hemineurin [AstraZeneca] *see* 2375
Heminevrin [AstraZeneca] *see* 2375
Hémipralon [Ipsen] *see* 7953
Hemipyocyanine, 4680
Hemisulfur Mustard, 4681
Hemlock *see* 2490
Hemocaprol [Sanofi-Aventis] *see* 427
Hémoclar [Sanofi Winthrop] *see* 7247
Hemocoagulase *see* 1008
Hemocuprein *see* 9131
Hemocuron [Takeda] *see* 9769
Hemocyanins, 4682
Hemodex [Alstoe] *see* 2950
Hemofil-M [Baxter] *see* 3962
Hemoglobin, 4683
Hemokine [Novartis] *see* 5039
Hemolidione [Sintetico] *see* 7348
Hemometina [Cusi] *see* 3616
Hemo-Pak [J & J] *see* 7038
Hemopoietine *see* 3744
Hemotrope [Andromaco] *see* 1513
Hemozoin, 4684
Hemp *see* 1752
Hempa *see* 4763
Henbane *see* 4898
2-Hendecanone *see* 6177
10-Hendecenoic Acid *see* 10033
Hendrickson's Reagent, 4685
Heneicosafluorotripropylamine *see* 4215
Henequem *see* 8687
Henna, 4686
HEOD *see* 3114
Hepadial [Biocodex] *see* 3226
Hepadist *see* 4716
Hepalande [Delalande] *see* 5902
Hepalidine [3M Pharma] *see* 9602
Hepa-Merz [Merz] *see* 6969
Heparamine, 4687

Hepar Calcis *see* 1709
Heparegen [Syntex] *see* 9602
Heparexine *see* 7474
Heparides *see* 4687
Heparin, 4688
β-Heparin *see* 2219
Heparinic Acid *see* 4688
Heparin Sodium *see* 4688
Hepation [Nippon Chemiphar] *see* 5776
Hepatocatalase *see* 1900
Hepatocyte Stimulating Factor *see* 5042
Hepatolite [DuPont] *see* 3400
Hepatoxane [Pharm. l'Esplanade] *see* 9649
HEPES, 4689
HEPP *see* 7241
Hepsal [Weddel] *see* 4688
Hepsera [Gilead] *see* 139
Heptabarb *see* 4690
Heptabarbital, 4690
Heptachlor, 4691
1,4,5,6,7,8,8-Heptachloro-3a,4,7,7a-tetra-
  hydro-4,7-methano-1*H*-indene *see*
  4691
Heptadecanoic Acid *see* 5816
(2*E*,8*E*,10*E*,14*R*)-2,8,10-Heptadecatriene-
  4,6-diyne-1,14-diol *see* 3624
(8*E*,10*E*,12*E*,14*R*)-8,10,12-Heptadecatri-
  ene-4,6-diyne-1,14-diol *see* 2270
2-(8*Z*)-(8-Heptadecen-1-yl)-4,5-dihydro-
  1*H*-imidazole-1-ethanol *see* 6925
(*Z*)-2-(8-Heptadecenyl)-2-imidazoline-1-
  ethanol *see* 6925
3-Heptadecyl-1,2-benzenediol *see* 10076
3-Heptadecylcatechol *see* 10076
cis-5-Heptadecylene-1-carboxylic Acid
  see 7304
rel-(2*R*,3*S*,5*R*,6*R*)-3-(1*E*,3*E*)-1,3-Hepta-
  dien-1-yl-5,6-dihydroxy-2-(hydroxymethyl)cyclohexanone *see* 7088
6-(1,3-Heptadien-1-yl)-3,4-dihydroxy-2-
  oxocyclohexanecarboxaldehyde *see*
  4299
Heptadon [Ebewe] *see* 6016
Heptadorm *see* 4690
1,1,2,2,3,3,3-Heptafluoro-*N*,*N*-bis(heptafluoropropyl)-1-propanamine *see* 4215
Heptakis-*O*-(4-sulfobutyl)-β-cyclodextrin
  see 2711
Heptaldehyde *see* 4693
2-(1-*N*,*N*-Heptamethylenimino)ethylgua-
  nidine *see* 4595
(*all-E*)-2-(3,7,11,15,19,23,27-Heptameth-
  yl-2,6,10,14,18,22,26-octacosaheptaen-
  yl)-3-methyl-1,4-naphthalenedione *see*
  5901
Heptaminol, 4692
Heptamul *see* 4691
Hept-a-myl [Delalande] *see* 4692
Heptanal, 4693
2-Heptanamine *see* 9982
Heptane *see* 4694
n-Heptane, 4694
1,7-Heptanedicarboxylic Acid *see* 898
Heptanedioic Acid *see* 7543
Heptanoic Acid, 4695
Heptanoic Acid Ethyl Ester *see* 3889
1-Heptanol, 4696
2-Heptanol, 4697
(±)-2-Heptanol *see* 4697
Heptanon [Pliva] *see* 6016
2-Heptanone, 4698
4-Heptanone *see* 3386
17β-Heptanoyloxy-1-methyl-5α-androst-
  1-en-3-one *see* 6039

Heptedrine [Bellon] *see* 9982

D-*arabino*-Hept-2-enonic Acid γ-Lactone *see* 4486

Heptenophos, 4699

n-Heptoic Acid *see* 4695

D-*glycero*-D-*gulo*-Heptonic Acid *see* 4491

Heptoxime, 4700

D-*manno*-Heptulose, 4701

D-*manno*-2-Heptulose *see* 4701

n-Heptyl Alcohol *see* 4696

Heptylaldehyde *see* 4693

N-Heptylcarbamic Acid (3a*S*,8a*R*)-1,2,3,-3a,8,8a-Hexahydro-1,3a,8-trimethyl-pyrrolo[2,3-*b*]indol-5-yl Ester *see* 3695

6-*O*-(*N*-Heptylcarbamoyl)-methyl-α-D-glucopyranoside *see* 4656

2-Heptyl-4-hydroxyquinoline *N*-Oxide *see* 4790

19-Heptyl-10-hydroxy-1,5,10,14-tetraaza-cyclononadecan-15-one *see* 7622

n-Heptylic Acid *see* 4695

2,2′-[3-[2-(3-Heptyl-4-methyl-2(3*H*)-thia-zolylidene)ethylidene]-1-propene-1,3-diyl]bis(3-heptyl-4-methylthiazolium) Iodide (1:2) *see* 7645

Heptylon [Delalande] *see* 4692

Heptylphysostigmine *see* 3695

2-Heptyl-4-quinolinol 1-Oxide *see* 4790

Heptylstigmine *see* 3695

6-Heptyltetrahydro-2*H*-pyran-2-one *see* 3448

Hepzide [Merck & Co.] *see* 6655

Heraclene [Natrapharm] *see* 2435

Heraclin *see* 1160

Herbadon [Mavena] *see* 2944

Herbadox [BASF] *see* 7193

Herbisan [Roberts Chemicals] *see* 3428

Herceptin [Genentech] *see* 9736

Hercules 528 *see* 3330

Hercules 3956 *see* 9716

Herculon [Hercules] *see* 7701

Hercynine, 4702

Heritage [Syngenta] *see* 917

Hermesetas *see* 8445

Hermophényl *see* 5954

Hero [Bayer CropSci.] *see* 3808

Heroin *see* 2971

Herperal *see* 3404

Herpes-Gel [Master] *see* 4931

Herplex [Allergan] *see* 4931

Herqueinone, 4703

Herrmann-Beller Catalyst, 4704

Herrmann's Palladacycle *see* 4704

Herzberg's Paper *see* 2482

HES *see* 4708

Hesofen *see* 5913

Hespan [BMS] *see* 4708

Hespander [Kyorin] *see* 4708

Hesperetin, 4705

Hesperetin 7-Rhamnoglucoside *see* 4706

Hesperetin-7-rutinoside *see* 4706

Hesperidin, 4706

Hestar [Pisa] *see* 4708

Hestrium Chloride *see* 4724

Hetacillin, 4707

Hetacin-K [Fort Dodge] *see* 4707

Hetaphenone *see* 3765

Hetastarch, 4708

5-HETE, 4709

Heteroauxin *see* 5004

Heterophylline *see* 774

Hetol *see* 1304

Hetoxol [Heterene Chem.] *see* 7697

Hetoxol CS-20 [Heterene Chem.] *see* 7697

Hetoxol L-9 [Heterene Chem.] *see* 7675

Hetrazan [Lederle] *see* 3128

Hetrazeen [Heterochemical] *see* 5899

Heulandite *see* 1651

Hevyteck [Penreco] *see* 7301

Heweneural [Hevert] *see* 5535

Hexaaminecobalt Trichloride, 4710

(*OC*-6-11)-Hexaammine-cobalt(3+) Chloride (1:3) *see* 4710

Hexaammonium Molybdate (Mo₇O₂₄⁶⁻) *see* 530

Hexaaquochromium Triacetate *see* 2224

Hexaaquochromium Tribromide *see* 2225

Hexaaquochromium Trichloride *see* 2226

1,4,7,10,13,16-Hexaazacyclotricosane Cyclic Peptide Deriv *see* 3585

Hexabendin *see* 4741

Hexaborane(10), 4711

Hexaboron Decahydride *see* 4711

Hexabrix [Guerbet] *see* 5110

2,2′,4,4′,5,5′-Hexabromobiphenyl *see* 7680

Hexabromocyclododecane, 4712

1,2,5,6,9,10-Hexabromocyclododecane *see* 4712

(*OC*-6-11)-Hexacarbonylvanadate(1−) *see* 10105

Hexacarboxybenzene *see* 5893

Hexachlorobenzene, 4713

1,2,3,4,5,6-Hexachlorobenzene *see* 4713

1,2,3,4,7,7-Hexachlorobicyclo[2.2.1]-2-heptene-5,6-bisoxymethylene Sulfite *see* 3629

1,4,5,6,7,7-Hexachloro-*endo*-bicyclo[2.2.-1]hept-5-ene-2,3-dicarboxylic Anhydride *see* 2087

Hexachlorobutadiene, 4714

1,1,2,3,4,4-Hexachloro-1,3-butadiene *see* 4714

(1α,2α,3β,4α,5α,6β)-1,2,3,4,5,6-Hexachlorocyclohexane *see* 5556

Hexachlorodimethyl Carbonate *see* 9921

Hexachloroendomethylenetetrahydro-phthalic Anhydride *see* 2087

1,2,3,4,10,10-Hexachloro-6,7-epoxy-1,4,-4a,5,6,7,8,8a-octahydro-*endo*,*endo*-1,-4:5,8-dimethanonaphthalene *see* 3633

1,2,3,4,10,10-Hexachloro-6,7-epoxy-1,4,-4a,5,6,7,8,8a-octahydro-*endo*,*exo*-1,-4:5,8-dimethanonaphthalene *see* 3114

Hexachloroethane, 4715

1,1,1,2,2,2-Hexachloroethane *see* 4715

rel-(1*R*,4*S*,4a*S*,5*S*,8*R*,8a*R*)-1,2,3,4,10,10-Hexachloro-1,4,4a,5,8,8a-hexahydro-1,4:5,8-dimethanonaphthalene *see* 219

6,7,8,9,10,10-Hexachloro-1,5,5a,6,9,9a-hexahydro-6,9-methano-2,4,3-benzo-dioxathiepin 3-Oxide *see* 3629

1,4,5,6,7,7-Hexachloro-*endo*-5-norbor-nene-2,3-dicarboxylic Anhydride *see* 2087

1,4,5,6,7,7-Hexachloro-5-norbornene-2,3-dimethanol Cyclic Sulfite *see* 3629

rel-(1a*R*,2*R*,2a*R*,3*R*,6*S*,6a*S*,7*S*,7a*S*)-3,4,5,-6,9,9-Hexachloro-1a,2,2a,3,6,6a,7,7a-octahydro-2,7:3,6-dimethanonaphth-[2,3-*b*]oxirene *see* 3633

rel-(1a*R*,2*R*,2a*S*,3*S*,6*R*,6a*R*,7*S*,7a*S*)-3,4,5,-6,9,9-Hexachloro-1a,2,2a,3,6,6a,7,7a-octahydro-2,7:3,6-dimethanonaphth-[2,3-*b*]oxirene *see* 3114

(*OC*-6-11)-Hexachloroosmate(2−) Ammonium (1:2) *see* 533

(*OC*-6-11)-Hexachloroosmate(2−) Potassium (1:2) *see* 7756

Hexachloropentadiene Dimer *see* 6290

Hexachlorophene, 4716

(*OC*-6-11)-Hexachloroplatinate(2−) Ammonium (1:2) *see* 545

(*OC*-6-11)-Hexachloroplatinate(2−) Potassium (1:2) *see* 7757

(*OC*-6-11)-Hexachloroplatinate(2−) Sodium (1:2) *see* 8758

Hexachloroplatinic(IV) Acid *see* 7641

4,5,6,7,8,8-Hexachloro-3a,4,7,7a-tetrahy-dro-4,7-methanoisobenzofuran-1,3-di-one *see* 2087

α,α,α,α′,α′,α′-Hexachloro-*p*-xylene *see* 1304

Hexaconazole, 4717

Hexacyanoferrate(4−) Dicopper(2+) *see* 2627

Hexacycline [BMS] *see* 9341

7,11-Hexadecadien-1-ol 1-Acetate *see* 4563

Hexadecadrol *see* 2945

Hexadecaheptaenal *see* 8036

2,4,6,8,10,12,14-Hexadecaheptaenal *see* 8036

Hexadecahydro-5,2,1,6,3,4-[2,3]butane-diyl[1,4]diylidenedipentaleno[2,1,6-*cde*:2′,1′,6′-*gha*]pentalene *see* 3447

(2*E*,4*E*)-5-[(3*S*,4*S*,4a*R*,6a*S*,12a*R*,13*S*,-15*S*,16b*S*,16c*S*)-2,3,4,4a,5,6,6a,7,10,-12,12a,13,14,15,16b,16c-Hexadecahy-dro-3,13-dihydroxy-4,10,10,12,12,16b,-16c-heptamethyl-15-(1-methylethenyl)-14-oxo-1*H*-benz[6,7]indeno[1,2-*b*]py-rano[3′,4′:4,5]cyclopenta[1,2-*f*]pyr-rolo[3,2,1-*hi*]indol-4-yl]-2-methyl-2,4-pentadienoic Acid *see* 6756

(3*S*,3a*S*,6*S*,6a*R*,6b*S*,7*S*,7a*R*,8*S*,10a*S*,11*S*,-13a*S*,13b*R*,13c*R*,14b*S*)-3,3a,4,5,6,6a,-6b,7,7a,8,9,10,10a,13a,13c,14b-Hexa-decahydro-6,8-dihydroxy-3,6,8,11,-14,15-hexamethyl-2*H*-7,13b-etheno-pentaleno[1″,2″:6,7;5″,4″:6′,7′]dicyclo-hepta[1,2-b:1′,2′-b′]difuran-2,12-(11*H*)-dione *see* 13

(3*S*,4*R*,5*S*,8*R*,9*E*,12*S*,14*S*,15*R*,16*S*,18*R*,-19*R*,26a*S*)-5,6,8,11,12,13,14,15,16,17,-18,19,24,25,26,26a-Hexadecahydro-5,19-dihydroxy-3-[(1*E*)-2-[(1*R*,3*R*,4*R*)-4-hydroxy-3-methoxycyclohexyl]-1-methylethenyl]-14,16-dimethoxy-4,10,-12,18-tetramethyl-8-(2-propen-1-yl)-15,19-epoxy-3*H*-pyrido[2,1-*c*][1,4]-oxaazacyclotricosine-1,7,20,21(4*H*,-23*H*)-tetrone *see* 9155

(3*S*,6*R*,7*E*,9*R*,10*R*,12*R*,14*S*,15*E*,17*E*,19*E*,-21*S*,23*S*,26*R*,27*R*,34a*S*)-9,10,12,13,14,-21,22,23,24,25,26,27,32,33,34,34a-Hexadecahydro-9,27-dihydroxy-3-[(1*R*)-2-[(1*S*,3*R*,4*R*)-4-hydroxy-3-meth-oxycyclohexyl]-1-methylethyl]-10,21-dimethoxy-6,8,12,14,20,26-hexameth-3*H*-pyrido[2,1-*c*][1,4]oxaazacyclohen-triacontine-1,5,11,28,29(4*H*,6*H*,31*H*)-pentone 4′-[2,2-Bis(hydroxymethyl)-propionate] *see* 9285

(2′*S*,6*R*,7*R*,8*R*,9*S*,10*R*,13*S*,14*S*,15*S*,16*S*)-1,3′,4′,6,7,8,9,10,11,12,13,14,15,16,-20,21-Hexadecahydro-10,13-dimethyl-spiro[17*H*-dicyclopropa[6,7:15,16]-cyclopenta[*a*]phenanthrene-17,-2′(5′*H*)-furan]-3,5′(2*H*)-dione *see* 3502

(1*R*,4a*S*,4b*S*,6a*S*,8*R*,10a*R*,10b*S*,12a*S*)-Hexadecahydro-1,4b,6a,8,10a,12a-hexamethyl-8-(4-methyl-3-penten-1-yl)-2(1*H*)-chrysenone *see* 8620

(3S,3aS,6S,6aR,6bS,7R,7aR,8S,10aS,11S,-
13aS,13bS,13cR,14bS,15S)-3,3a,4,5,6,-
6a,6b,7,7a,8,9,10,10a,13a,13c,14b-
Hexadecahydro-6-hydroxy-3,6,8,11,-
14,15-hexamethyl-2H-8,15-epoxy-7,-
13b-ethanopentaleno[1″,2″:6,7;
5‴,4‴:6′,7′]dicyclohepta[1,2-b:1′,2′-b′]-
difuran-2,12(11H)-dione *see* 613
(2′R,3S,3′R,3′AS,6′S,6aS,6bS,7′aR,11aS,-
11bR)-2,3,3′a,4,4′,5′,6,6′,6a,6b,7,7′,-
7′a,8,11a,11b-Hexadecahydro-3-hy-
droxy-3′,6′,10,11b-tetramethylspiro-
[9H-benzo[a]fluorene-9,2′(3′H)-furo-
[3,2-b]pyridin]-11(1H)-one *see* 5312
1,2,3,3a,3b,4,5,5a,6,8,10,10a,10b,11,12,-
12a-Hexadecahydro-1,10a,12a-tri-
methylcyclopenta[7,8]phenanthro[2,3-
c]pyrazol-1-ol *see* 8921
Hexadeca-μ-hydroxytetracosahydroxy-
[μ₈-[[1,3,4,6-tetra-O-sulfo-β-D-fructo-
furanosyl α-D-Glucopyranoside Tetra-
kis(sulfato-κO′)](8−)]]hexadecaalu-
minum *see* 9010
Hexadecanoic Acid *see* 7097
Hexadecanoic Acid Aluminum Salt (3:1)
see 351
Hexadecanoic Acid Ammonium Salt
(1:1) *see* 535
Hexadecanoic Acid Calcium Salt (2:1)
see 1689
Hexadecanoic Acid 3-[2-[4-(6-Fluoro-1,2-
benzisoxazol-3-yl)-1-piperidinyl]ethyl]-
6,7,8,9-tetrahydro-2-methyl-4-oxo-4H-
pyrido[1,2-a]pyrimidin-9-yl Ester *see*
7087
Hexadecanoic Acid Hexadecyl Ester *see*
2030
Hexadecanoic Acid 1,1′,1″-(1,2,3-Pro-
panetriyl) Ester *see* 9908
1-Hexadecanol *see* 2027
1-Hexadecanol Lactate *see* 2029
N²⁶-(Hexadecanoyl-γ-glutamyl)-(34-argi-
nine)GLP-1-(7-37)-peptide *see* 5569
Hexadecanoyl Sphingomyelin *see* 8873
(7Z)-7-Hexadecen-1-ol 1-Acetate *see*
4723
cis-7-Hexadecenyl Acetate *see* 4723
n-Hexadecylcitric Acid *see* 178
N-Hexadecyl-N,N-dimethylbenzenemeth-
anaminium Chloride (1:1) *see* 2016
Hexadecyldimethylbenzylammonium
Chloride *see* 2016
d-α-(n-Hexadecyl)glycerol *see* 2060
Hexadecyl Hexadecanoate *see* 2030
Hexadecyl(2-hydroxycyclohexyl)dimeth-
ylammonium Bromide *see* 2017
N-Hexadecyl-2-hydroxy-N,N-dimethyl-
cyclohexanaminium Bromide (1:1) *see*
2017
Hexadecylic Acid *see* 7097
2-[[(Hexadecyloxy)hydroxyphosphinyl]-
oxy]-N,N,N-trimethylethanaminium
Inner Salt *see* 6279
2-(Hexadecyloxy)-6-methyl-4H-3,1-ben-
zoxazin-4-one *see* 2020
(2S)-3-(Hexadecyloxy)-1,2-propanediol
see 2060
Hexadecyl Palmitate *see* 2030
Hexadecylphosphocholine *see* 6279
n-Hexadecylphosphorylcholine *see* 6279
1-Hexadecylpyridinium Chloride (1:1) *see*
2031
Hexadecyl 2-(N,N,N-Trimethylamino)-
ethyl Phosphate *see* 6279
Hexadecyltrimethylammonium Bromide
see 2024

(19α)-3,4,5,6,16,17-Hexadehydro-16-
(methoxycarbonyl)-19-methyloxa-
yohimbanium *see* 8603
(19α,20α)-3,4,5,6,16,17-Hexadehydro-16-
(methoxycarbonyl)-19-methyloxa-
yohimbanium Inner Salt *see* 312
Hexadenol *see* 8848
2,4-Hexadienedioic Acid *see* 6385
(2E,4E)-2,4-Hexadienoic Acid *see* 8847
(2E,4E)-2,4-Hexadien-1-al *see* 8848
3-(3,5-Hexadien-1-ynyl)-6-(1-propynyl)-
1,2-dithiin *see* 9456
Hexadilat [Durascan] *see* 6613
Hexadimethrine Bromide, 4718
Hexadrin *see* 3633
Hexadrol [Organon] *see* 2945
Hexaflumuron, 4719
Hexafluorenium Bromide, 4720
Hexafluoroacetone, 4721
(OC-6-11)-Hexafluoroaluminate(3−)
Ammonium (1:3) *see* 521
Hexafluorobenzene, 4722
1,2,3,4,5,6-Hexafluorobenzene *see* 4722
Hexafluorodiethyl Ether *see* 4231
1,1,1,3,3,3-Hexafluoro-2-(fluorometh-
oxy)propane *see* 8614
(OC-6-11)-Hexafluorogallate(3−) Ammo-
nium (1:3) *see* 522
Hexafluorophosphate(1−) Ammonium
(1:1) *see* 523
1,1,1,3,3,3-Hexafluoro-2-propanone *see*
4721
Hexafluorosilicate(2−) Aluminum (3:2)
see 336
Hexafluorosilicate(2−) Ammonium (1:2)
see 524
Hexafluorosilicate(2−) Barium (1:1) *see*
972
Hexafluorosilicate(2−) Copper(2+) (1:1)
see 2631
Hexafluorosilicate(2−) Hydrogen (1:2)
see 4214
Hexafluorosilicate(2−) Lead(2+) (1:1)
see 5465
Hexafluorosilicate(2−) Magnesium (1:1)
see 5732
Hexafluorosilicate(2−) Potassium (1:2)
see 7759
Hexafluorosilicate(2−) Sodium (1:2) *see*
8759
Hexafluorosilicate(2−) Zinc (1:1) *see*
10337
Hexafluorosulfur *see* 9103
Hexafluorouranium *see* 10041
(OC-6-11)-Hexafluorozirconate(2−)
Potassium (1:2) *see* 7760
Hexafluosilicic Acid *see* 4214
Hexagastron [Durascan] *see* 9010
Hexahydroadiphenine *see* 3494
Hexahydroaniline *see* 2722
Hexahydro-2H-azepin-2-one *see* 1763
N-[2-[4-[[[[(Hexahydro-1H-azepin-1-yl)-
amino]carbonyl]amino]sulfonyl]phen-
yl]ethyl]-5-methyl-3-isoxazolecarbox-
amide *see* 4479
N-[[(Hexahydro-1H-azepin-1-yl)amino]-
carbonyl]-4-methylbenzenesulfon-
amide *see* 9665
1-[[4-[2-(Hexahydro-1H-azepin-1-yl)eth-
oxy]phenyl]methyl]-2-(4-hydroxy-
phenyl)-3-methyl-1H-indol-5-ol *see*
1011
(2S,5R,6R)-6-[[(Hexahydro-1H-azepin-1-
yl)methylene]amino]-3,3-dimethyl-7-
oxo-4-thia-1-azabicyclo[3.2.0]heptane-
2-carboxylic Acid *see* 383

(2S,5R,6R)-6-[[(Hexahydro-1H-azepin-1-
yl)methylene]amino]-3,3-dimethyl-7-
oxo-4-thia-1-azabicyclo[3.2.0]heptane-
2-carboxylic Acid (2,2-Dimethyl-1-
oxopropoxy)methyl Ester *see* 384
6-[(Hexahydro-1H-azepin-1-yl)methyl-
eneamino]penicillanic Acid *see* 383
1-(Hexahydro-1H-azepin-1-yl)-3-[[p-[2-
(5-methyl-3-isoxazolecarboxamido)-
ethyl]phenyl]sulfonyl]urea *see* 4479
1-(Hexahydro-1H-azepin-1-yl)-3-(p-tolyl-
sulfonyl)urea *see* 9665
N-[2-(Hexahydro-1(2H)-azocinyl)ethyl]-
guanidine *see* 4595
Hexahydrobenzene *see* 2716
Hexahydrobenzoic Acid *see* 2717
Hexahydrobenzyl Alcohol *see* 2724
7,7′,8,8′,11,12-Hexahydro-ψ,ψ-carotene
see 7501
N-[[(Hexahydrocyclopenta[c]pyrrol-
2(1H)-yl)amino]carbonyl]-4-methyl-
benzenesulfonamide *see* 4474
1-(Hexahydrocyclopenta[c]pyrrol-2(1H)-
yl)-3-(p-tolylsulfonyl)urea *see* 4474
Hexahydrodesoxyephedrine *see* 7972
5-[(Hexahydro-1H-1,4-diazepin-1-yl)sul-
fonyl]isoquinoline *see* 3979
1,5a,6,9,9a,9b-Hexahydro-4a(4H)-diben-
zofurancarboxaldehyde *see* 8205
(4aR,9S,10aS)-1,3,4,9,10,10a-Hexahydro-
5,6-dihydroxy-1,1-dimethyl-7-(1-meth-
ylethyl)-2H-9,4a-(epoxymethano)phen-
anthren-12-one *see* 1852
(4aR,10aS)-1,3,4,9,10,10a-Hexahydro-
5,6-dihydroxy-1,1-dimethyl-7-(1-meth-
ylethyl)-4a(2H)-phenanthrenecarbox-
ylic Acid *see* 1850
1,2,6,7,8,8a-Hexahydro-β,δ-dihydroxy-
2,6-dimethyl-8-(2-methyl-1-oxobut-
oxy)-1-naphthaleneheptanoic Acid δ-
Lactone *see* 5644
(4S)-4,5,6,7,8,9-Hexahydro-11,13-dihy-
droxy-4-methyl-2H-3-benzoxacyclodo-
decin-2,10(1H)-dione *see* 2666
(3S,11E)-3,4,5,6,9,10-Hexahydro-14,16-
dihydroxy-3-methyl-1H-2-benzoxa-
cyclotetradecin-1,7(8H)-dione *see*
10314
(1aS,1bR,2S,2′R,5R,6S,6aR,7aR,8R)-
Hexahydro-1b,6-dihydroxy-6a-methyl-
8-(1-methylethenyl)spiro[2,5-methano-
7H-oxireno[3,4]cyclopent[1,2-d]oxe-
pin-7,2′-oxiran]-3(2H)-one *see* 10008
(4bR,5S,11bS)-4b,5,6,11b,12,13-Hexahy-
dro-1,2-dimethoxy-12-methyl[1,3]-
benzodioxolo[5,6-c]phenanthridin-5-ol
see 4770
rel-(3R,11bR)-1,3,4,6,7,11b-Hexahydro-
9,10-dimethoxy-3-(2-methylpropyl)-
2H-benzo[a]quinolizin-2-one *see* 9326
6,7,8,9,10,11-Hexahydro-N,N-dimethyl-
5H-cyclooct[b]indole-5-propanamine
see 5119
(4S,4aS,7S,7aR)-Hexahydro-4,7-di-
methylcyclopenta[c]pyran-3(1H)-one
see 5133
(3aα,4β,7β,7aα)-Hexahydro-3a,7a-di-
methyl-4,7-epoxyisobenzofuran-1,3-di-
one *see* 1755
rel-(3aR,4S,7R,7aS)-Hexahydro-3a,7a-di-
methyl-4,7-epoxyisobenzofuran-1,3-di-
one *see* 1755
(1S,6S)-2,3,4,5,6,7-Hexahydro-1,4-di-
methyl-1,6-methano-1H-4-benzazonin-
10-ol *see* 3696

buten-1-yl]-2*H*-cyclohepta[*b*]furan-2-one *see* 10256

(4a*R*,5*S*,10b*R*,12*R*)-2,3,4,4a,5,6-Hexahydro-12-methyl-1*H*-5,10b-propano-1,7-phenanthroline *see* 5680

1,2,3,4,10,14b-Hexahydro-2-methylpyrazino[2,1-*a*]pyrido[2,3-*c*][2]benzazepine *see* 6291

(1*S*)-3-[(1*E*)-2-[(1*R*,3a*R*,7a*R*)-2,3,3a,6,7,-7a-Hexahydro-7a-methyl-1-[(1*R*,2*E*,-4*R*)-1,4,5-trimethyl-2-hexen-1-yl]-1*H*-inden-4-yl]ethenyl]-4-methyl-3-cyclohexen-1-ol *see* 9153

Hexahydronicotinic Acid *see* 6647

(3a*S*,4*S*,6a*R*)-Hexahydro-2-oxo-1*H*-thieno[3,4-*d*]imidazole-4-pentanoic Acid *see* 1236

[3a*S*-(3aα,4β,5β,6aα)]-Hexahydro-2-oxo-1*H*-thieno[3,4-*d*]imidazole-4-pentanoic Acid 5-Oxide *see* 1237

(3a*S*,4*S*,5*R*,6a*R*)-Hexahydro-2-oxo-1*H*-thieno[3,4-*d*]imidazole-4-pentanoic Acid 5-Oxide *see* 1237

cis-Hexahydro-2-oxo-1*H*-thieno[3,4]imidazole-4-valeric Acid *see* 1236

N [6]-[5-[(3a*S*,4*S*,6a*R*)-Hexahydro-2-oxo-1*H*-thieno[3,4-*d*]imidazol-4-yl]-1-oxopentyl]-L-lysine *see* 1232

Hexahydrophenol *see* 2718

Hexahydropicolinic Acid *see* 7570

4,5,7,8,9,12-Hexahydro-5*H*-pyrano[3,4-*d*]pyrrolo[3,2,1-*jk*][1]benzazepin-5-one *see* 731

Hexahydropyrazine *see* 7576

Hexahydropyridine *see* 7580

rel-(3a*R*,4*S*,7*R*,7a*S*)-Hexahydro-2-[4-[4-(2-pyrimidinyl)-1-piperazinyl]butyl]-4,7-methano-1*H*-isoindole-1,3(2*H*)-dione *see* 9183

2,3,4,6,7,8-Hexahydropyrrolo[1,2-*a*]pyrimidine *see* 2995

(−)-Hexahydro-1,3,4,5-tetrahydroxybenzoic Acid *see* 8175

(3a*S*,5*S*,8*S*,9*R*,11*R*,13b*S*,15*R*)-3,3a,5,8,-11,13b-Hexahydro-7,8,12,15-tetrahydroxy-5,9-dimethyl-8,11-ethanofuro-[2,3-*e*]naphtho[2,3-*c*:6,7-*c′*]dipyran-2,-6,13(9*H*)-trione *see* 4570

9,11,12,13,13a,14-Hexahydro-2,3,5,6-tetramethoxydibenzo[*f*,*h*]pyrrolo[1,2-*b*]isoquinoline *see* 10011

(13a*S*)-9,11,12,13,13a,14-Hexahydro-2,3,-6,7-tetramethoxydibenzo[*f*,*h*]pyrrolo-[1,2-*b*]isoquinoline *see* 10012

1,10,19,22,23,24-Hexahydro-2,7,13,17-tetramethyl-1,19-dioxo-3,18-divinylbiline-8,12-dipropionic Acid *see* 1222

Hexahydrothymol *see* 5905

Hexahydrotoluene *see* 6118

Hexahydro-*s*-triazaborine *see* 1337

(5*R*,6*R*,7*S*,8*R*,8a*S*)-Hexahydro-6,7,8-trihydroxy-5-(hydroxymethyl)imidazo-[1,2-*a*]pyridine-2,3-dione *see* 5357

(β*R*,δ*R*,1*S*,2*S*,6*S*,8*S*,8a*R*)-1,2,6,7,8,8a-Hexahydro-β,δ,6-trihydroxy-2-methyl-8-[(2*S*)-methyl-1-oxobutoxy]-1-naphthaleneheptanoic Acid Sodium Salt (1:1) *see* 7835

(4*R*,7*S*,7a*S*)-1,2,3,4,7,7a-Hexahydro-2,-4,7-trimethyl-6*H*-cyclopenta[*c*]pyridin-6-one *see* 9243

Hexahydro-α,3a,5-trimethyl-6,8-dioxo-1,4-methanoindan-1-acetic Acid *see* 8497

1,2,3,4,5,6-Hexahydro-3,6,11-trimethyl-2,6-methano-3-benzazocin-8-ol *see* 6005

(4a*S*,9a*S*)-2,3,4,4a,9,9a-Hexahydro-2,-4a,9-trimethyl-1,2-oxazino[6,5-*b*]indol-6-ol 6-(*N*-Methylcarbamate) *see* 4425

(3a*R*,4a*S*,6a*S*,7*S*,9a*S*,9b*R*)-5,6,6a,7,9a,9b-Hexahydro-1,4a,7-trimethyl-3*H*-oxireno[8,8a]azuleno[4,5-*b*]furan-8(4a*H*)-one *see* 761

[4*R*-(4α,7β,7aβ)]-1,2,3,4,7,7a-Hexahydro-2,4,7-trimethyl-6*H*-2-pyridin-6-one *see* 9243

(3a*S*,8a*R*)-1,2,3,3a,8,8a-Hexahydro-1,-3a,8-trimethylpyrrolo[2,3-*b*]indol-5-ol 5-(*N*-Methylcarbamate) *see* 7496

(3a*S*,8a*R*)-1,2,3,3a,8,8a-Hexahydro-1,-3a,8-trimethylpyrrolo[2,3-*b*]indol-5-ol 5-(*N*-Phenylcarbamate) *see* 7372

Hexahydro-1,3,5-trinitro-1,3,5-triazine *see* 2728

5-[(Hexahydro-2,4,6-trioxo-5-pyrimidinyl)imino]-2,4,6(1*H*,3*H*,5*H*)-pyrimidinetrione Ammonium Salt (1:1) *see* 6391

1,2,3,5,6,7-Hexahydroxy-9,10-anthracenedione *see* 8425

1,2,3,5,6,7-Hexahydroxyanthraquinone *see* 8425

(1β,3β,5β,11α)-1,3,5,11,14,19-Hexahydroxycard-20(22)-enolide *see* 6999

(2β,3β,5β,22*R*,)-2,3,14,20,22,25-Hexahydroxycholest-7-en-6-one *see* 3538

Hexahydroxycyclohexane *see* 5021

1,1′,6,6′,7,7′-Hexahydroxy-3,3′-dimethyl-5,5′-bis(1-methylethyl)[2,2′-binaphthalene]-8,8′-dicarboxaldehyde *see* 4564

4,5,7,4′,5′,7′-Hexahydroxy-2,2′-dimethylnaphthodianthrone *see* 4902

1,3,4,6,8,13-Hexahydroxy-10,11-dimethylphenanthro[1,10,9,8-*opqra*]perylene-7,14-dione Stereoisomer *see* 4902

Hexahydroxydisiloxane *see* 8629

3,3′,4,4′,5,7-Hexahydroxyflavane *see* 5504

3,3′,4′,5,5′,7-Hexahydroxyflavanone *see* 576

3,3′,4′,5,5′,7-Hexahydroxyflavone *see* 6417

3,3′,4′,5,6,7-Hexahydroxyflavone *see* 8149

3,3′,4′,5,5′,7-Hexahydroxyflavylium Chloride *see* 2885

5,6,9,17,19,21-Hexahydroxy-23-methoxy-2,4,12,16,18,20,22-heptamethyl-2,7-(epoxypentadeca[1,11,13]trienimino)-naphtho[2,1-*b*]furan-1,11(2*H*)-dione 21-Acetate *see* 8343

5,6,9,17,19,21-Hexahydroxy-23-methoxy-2,4,12,16,18,20,22-heptamethyl-8-[*N*-(4-methyl-1-piperazinyl)formimidoyl]-2,7-(epoxypentadeca[1,11,13]trienimino)naphtho[2,1-*b*]furan-1,11(2*H*)-dione 21-Acetate *see* 8341

3,3′,4′,5,5′,7-Hexahydroxy-2-phenylbenzopyrylium Chloride *see* 2885

2,2,4,4,6,6-Hexakis(1-aziridinyl)-2,2,4,4,-6,6-hexahydro-1,3,5,2,4,6-triazatriphosphorine *see* 715

Hexakis(1-aziridinyl)phosphonitrile *see* 715

2,2,4,4,6,6-Hexakis(1-aziridinyl)-2λ5,-4λ5,6λ5-1,3,5,2,4,6-triazatriphosphorine *see* 715

(*OC*-6-11)-Hexakis(cyano-*κC*)cobaltate(3−) Potassium (1:3) *see* 7758

(*OC*-6-11)-Hexakis(cyano-*κC*)ferrate(3−) Ammonium (1:3) *see* 516

(*OC*-6-11)-Hexakis(cyano-*κC*)ferrate(4−) Ammonium (1:4) *see* 517

(*OC*-6-11)-Hexakis(cyano-*κC*)ferrate(4−) Copper(2+) (1:2) *see* 2627

(*OC*-6-11)-Hexakis(cyano-*κC*)ferrate(4−) Sodium (1:4) *see* 8752

Hexakis(β,β-dimethylphenethyl)distannoxane *see* 3997

Hexakis(hydroxymethyl)melamine *see* 4729

(*OC*-6-11)-Hexakis(1-isocyano-2-methoxy-2-methylpropane)technetium(1+)-[99mTc] *see* 9239

Hexakis(2-methoxy-2-methylpropyl Isocyanide)[99mTc] Technetium(1+) *see* 9239

1,1,1,3,3,3-Hexakis(2-methyl-2-phenylpropyl)distannoxane *see* 3997

(*OC*-6-11)-Hexakis(nitrito-*κN*)cobaltate-(3−) Potassium (1:3) *see* 2416

(*OC*-6-11)-Hexakis(nitrito-*κN*)cobaltate-(3−) Sodium (1:3) *see* 8738

Hexakis(nitrito-*N*)cobaltate(3−) Tripotassium *see* 2416

Hexaldehyde *see* 1762

Hexalen [U.S. Bioscience] *see* 316

Hexalene *see* 4723

Hexalin *see* 2718

Hexalure, 4723

Hexalyse [DOMS] *see* 2914

Hexamarium *see* 3405

Hexametapol *see* 4763

Hexametazine *see* 3953

Hexamethone *see* 4724

Hexamethonium, 4724

3,3,6,6,9,9-Hexamethyl-1,4,7-cyclononatriperoxane *see* 9752

N [1],*N* [1],*N* [1],*N* [10],*N* [10],*N* [10]-Hexamethyl-1,10-decanediaminium Bromide (1:2) *see* 2851

N,*N*,*N*,*N′*,*N′*,*N′*-Hexamethyl-1,10-decanediaminium Dibromide *see* 2851

Hexamethyldiaminoisopropanol Diiodide *see* 7897

Hexamethyldisilazane, 4725

Hexamethyldisiloxane, 4726

1,1,1,3,3,3-Hexamethyldisiloxane *see* 4726

Hexamethylenamine *see* 6038

Hexamethylene *see* 2716

1,1′-Hexamethylenebis[5-(*p*-chlorophenyl)biguanide] *see* 2092

Hexamethylenebis(dimethyl-9-fluorenylammonium Bromide) *see* 4720

Hexamethylenebis[dimethyl[1-methyl-3-(2,2,6-trimethylcyclohexyl)propyl]ammonium Chloride] *see* 9818

1,1′-Hexamethylenebis[5-(2-ethylhexyl)-biguanide] *see* 226

Hexamethylenebis[9-fluorenyldimethylammonium Bromide] *see* 4720

α,α′-[Hexamethylenebis(iminomethylene)]bis[3,4-dihydroxybenzyl Alcohol] *see* 4743

Hexamethylenebis[*N*-methylcarbamic Acid Ester Bromomethylate] *see* 3405

Hexamethylenebis[methylcarbamic Acid] Ester of 3-Hydroxy-1-methylpyridinium Bromide *see* 3405

Hexamethylenebis[*N*-methylcarbaminoyl-1-methyl-3-hydroxypyridinium Bromide] *see* 3405

Hexamethylenebis(trimethylammonium) *see* 4724

Hoelon [Bayer CropSci.] *see* 3092
Hoffmann's Anodyne *see* 8883
Hoffmann's Drops *see* 8882
HOG *see* 7333
Hoggar [Stada] *see* 3489
Hogival *see* 3763
Hogpax *see* 577
Hog's Bean *see* 4898
Hogweed *see* 8540
Hokunalin [Hokuriku] *see* 9991
Holland Balsam *see* 5319
Holligold *see* 1715
Holly-leaved Barberry *see* 1157
Holmia *see* 4765
Holmium, 4765
Holocaine Hydrochloride [Winthrop] *see* 7317
Holomycin, 4766
Holopon [Altana] *see* 6077
Holoxan [Baxter] *see* 4937
Holy Thistle *see* 6274
Homapin [Mission Pharmacal] *see* 4768
Homarine, 4767
Homatrisol [Cooper] *see* 4768
Homatropine, 4768
Hombre [LiphaTech] *see* 3158
Homidium, 4769
Homobatrachotoxin *see* 1007
Homocaine *see* 2439
Homochelidonine, 4770
α-Homochelidonine *see* 4770
Homochlorcyclizine, 4771
Homoclomin [Eisai] *see* 4771
Homocodeine *see* 7441
Homocysteine, 4772
L-Homocysteine *see* 4772
Homocysteine S-Ethyl Ether *see* 3792
L-Homocystine, 4773
Homoeriodictyol, 4774
Homogentisic Acid, 4775
Homoharringtonine, 4776
Homomenthyl Salicylate *see* 4777
Homomycin *see* 4891
Homonal [SS Pharm.] *see* 5712
D-Homo-17a-oxaandrosta-1,4-diene-3,17-dione *see* 9323
D-Homopantothenic Acid *see* 4783
18a-Homo-pregna-4,9,11-trien-17β-ol-3-one *see* 9359
Homoproline *see* 7570
Homosalate, 4777
m-Homosalicylic Acid *see* 2569
o-Homosalicylic Acid *see* 2570
p-Homosalicylic Acid *see* 2571
Homoserine, 4778
L-Homoserine *see* 4778
Homosulfamine *see* 5712
4-Homosulfanilamide *see* 5712
Homotaurine *see* 9728
Homotrilobine *see* 9866
Homovanillic Acid, 4779
HON, 4780
Honduras Balsam *see* 942
Honduras Bark *see* 1879
Honokiol, 4781
Honvan [Baxter] *see* 4280
Honvol [Horner] *see* 4280
Hoodia, 4782
Hopantenate Calcium *see* 4783
Hopantenic Acid, 4783
Hopate [Tanabe] *see* 4783
Hopeite *see* 10351
Hops, 4784
Hordenine, 4785
Horehound, 4786

Horizon [Novartis] *see* 2390
Horizon [Bayer CropSci.] *see* 9229
Hormezon [Tobishi] *see* 1182
Hormoestrol *see* 4737
Hormofemin [Medo] *see* 3115
Hormofollin *see* 3763
Hormomed [Merckle] *see* 3762
Hormonisene [Temis Lostalo] *see* 2178
Hormovarine [Clin-Comar-Byla] *see* 3763
Horn Silver *see* 8643
Horse Balm *see* 2467
Horse Chestnut, 4787
Horse Chestnut Seed Extract *see* 4787
Horseheal *see* 5047
Horse Nettle *see* 8835
Horse-radish, 4788
Horseweed *see* 3724
Hostacortin [Hoechst] *see* 7842
Hostacyclin [Hoechst] *see* 9341
Hostaginan [Hoechst] *see* 7855
Hostalival [Hoechst] *see* 6760
Hostapon SCI [Clariant] *see* 8739
Hostaquick [Bayer CropSci.] *see* 4699
Hostathion [Bayer CropSci.] *see* 9767
HOSu *see* 4884
Houpo *see* 5761
HP-549 *see* 5284
HP-873 *see* 4941
HPA-23, 4789
HPC *see* 6279, 7056
HPEK-1 *see* 9397
5-HPETE *see* 4709
hpGRF *see* 8840
hpGRF(1-29)NH₂ *see* 8600
HPL *see* 4655
HPMC *see* 4880
HPMF *see* 3652
HPMP *see* 5388
(S)-HPMPC *see* 2271
HPP *see* 271
4-HPR *see* 4025
hPTH 1-34 acetate *see* 9309
HPTS *see* 8065
HQ-495 *see* 175
HQNO, 4790
HR *see* 9969
HR-111V *see* 1944
HR-158 *see* 5631
HR-221 *see* 1929
HR-376 *see* 2356
HR-756 *see* 1934
HR-810 *see* 1941
h-R3 *see* 6638
HRF [Intrapharm] *see* 4561
HRP-1 *see* 4757
HRP-3 *see* 4757
HRP-5 *see* 4757
HRS-16 *see* 3116
HS-592 *see* 2343
HS-902 *see* 9616
HSA *see* 8607
H₂sal₂en *see* 8458
HSP-2986 *see* 7827
HSR-803 *see* 5291
HSR-902 *see* 9616
5-HT *see* 8601
HTIB *see* 5369
L-5HTP *see* 4886
5-HTP *see* 4886
HTX *see* 4761
(−)-HTX 1 *see* 4761
HTX D *see* 4436
HTZ *see* 4749
HU-211 *see* 2946
Hu1124 *see* 3567

Huanghuahaosu *see* 807
Huckleberry *see* 1221
Humagel [Parke-Davis] *see* 7146
Humalog [Lilly] *see* 5023, 5030
Human Amylin *see* 603
Human β-Amyloid Peptide (1-40) *see* 608
Human β-Amyloid Peptide (1-42) *see* 608
Human α-ANP *see* 863
Human α-ANP (4-28) *see* 863
Human Atriopeptin (1-28) *see* 863
Human Atriopeptin (4-28) *see* 863
Human Atriopeptin (102-126) *see* 863
Human Brain Natriuretic Peptide-32 *see* 1365
Human Chorionic Gonadotropin *see* 2221
Human Chorionic Gonadotropin, Glycoform α *see* 2221
Human Chorionic Somatomammotropin *see* 4655
Human EGF-URO *see* 3574
Human Growth Hormone *see* 8842
Human Growth Hormone-releasing Factor(1-29)amide *see* 8600
Human IgE Pentapeptide *see* 7241
Human Insulin-like Growth Factor I *see* 5029
Humanized Anti-Tac *see* 2796
Human LFA-3-IgG₁ Fusion Protein *see* 220
Human Menopausal Gonadotropins *see* 4250
Human Pancreatic Somatoliberin(1-29)-amide *see* 8600
Human Placental Lactogen *see* 4655
Human Serum Albumin *see* 8607
Human Thymosin α1 *see* 9560
Human Transmembrane Activator and CAML Interactor (TACI) Immunoglobulin G₁ Fc Domain Fusion Protein (Fc5) *see* 848
Human Tumor Necrosis Factor Receptor p75 Fc Fusion Protein *see* 3767
Human Urodilatin *see* 10071
Humate-P [Centeon] *see* 3962
Humatin [Parke-Davis] *see* 7146
Humatrope [Lilly] *see* 8842
HuMax-CD20 [Genmab] *see* 6862
Humegon [Organon] *see* 4250
Humic Acids, 4791
Huminsulin [Lilly] *see* 5023
Humira [Abbott] *see* 136
Humorsol [Merck & Co.] *see* 2889
Humoryl [Delalande] *see* 9680
Humulene, 4792
α-Humulene *see* 4792
β-Humulene *see* 4792
Humulin [Lilly] *see* 5023
Humulon, 4793
Humulone *see* 4793
Hungarian Chamomile *see* 2040
Hünig's Base *see* 3219
Hunter Corrective Factor *see* 4934
Huntexil [NeuroSearch] *see* 7860
HUP *see* 4794
Huperzine A, 4794
Hurler Corrective Factor *see* 4935
Husar [Bayer CropSci.] *see* 5089
Hussar [Bayer CropSci.] *see* 5089
Hustazol [Yoshitomi] *see* 2384
HWA-285 *see* 7927
HWA-486 *see* 5486
HWG-1608 *see* 9229

Hydroquinol *see* 4845
Hydroquinone, 4845
Hydroquinone Benzyl Ether *see* 6334
Hydroquinone Calcium Sulfonate *see* 3439
Hydroquinone Compd with *p*-(Methylamino)phenol (1:2) *see* 6229
Hydroquinone-β-D-glucopyranoside *see* 763
Hydroquinone Glucose *see* 763
Hydroquinone Monobenzyl Ether *see* 6334
Hydro-rapid [Sanorania] *see* 4338
Hydroresorcinol *see* 3196
Hydroretrocortine *see* 7841
Hydrorhombinine *see* 5667
Hydrosaluric [Merck & Co.] *see* 4819
Hydrosarpan [Eutherapie] *see* 8237
Hydrosilicofluoric Acid *see* 4214
"Hydrosulfuric Acid" *see* 4837
Hydrotalcite *see* 5724
Hydrotrichlorosilane *see* 9811
Hydrotrichlorothiazide *see* 9789
Hydrotricine [RPR] *see* 10023
Hydrourushiol *see* 7219, 10076
Hydrous Lanolin *see* 5409
Hydroxamethocaine *see* 4885
Hydroxocobalamin, 4846
Hydroxocobemine *see* 4846
Hydroxyacetaldehyde *see* 4532
N-Hydroxyacetamide *see* 62
p-Hydroxyacetanilide *see* 47
4'-Hydroxyacetanilide *see* 47
2-Hydroxyacetic Acid *see* 4534
2-Hydroxyacetic Acid (3a*S*,4*R*,5*S*,6*S*,8*R*,-9*R*,9a*R*,10*R*)-6-Ethenyldecahydro-5-hydroxy-4,6,9,10-tetramethyl-1-oxo-3a,9-propano-3a*H*-cyclopentacyclo-octen-8-yl Ester *see* 7650
Hydroxyacetone *see* 64
p-Hydroxyacetophenone-D-glucoside *see* 7507
4-Hydroxy-*N*-acetylproline *see* 7002
1-[[(3-Hydroxy-1-adamantyl)amino]acetyl]-2-cyano-(*S*)-pyrrolidine *see* 10177
(*S*)-3-Hydroxyadamantylglycine-L-*cis*-4,5-methanoprolinenitrile *see* 8525
(2α,3β)-3-Hydroxy-A(1),28-dinorlup-20-(29)-ene-2,17-dicarboxylic Acid *see* 1911
4-Hydroxyaflatoxin B₁ *see* 173
4-Hydroxyaflatoxin B₂ *see* 173
β-Hydroxyalanine *see* 8599
2-Hydroxy-6-alkylbenzoic Acids *see* 614
17-Hydroxy-17α-allyl-4-estrene *see* 282
l-3-Hydroxy-*N*-allylmorphinan *see* 5510
Hydroxyaluminum Acetate *see* 359
α-Hydroxy-δ-aminobutylidenediphosphonic Acid *see* 223
γ-Hydroxy-β-aminobutyric Acid *see* 440
m-Hydroxy-α-(1-aminoethyl)benzyl Alcohol *see* 6003
5-Hydroxy-3-(β-aminoethyl)indole *see* 8601
2-Hydroxy-4-amino-5-fluoropyrimidine *see* 4156
[4-(*N*-Hydroxyamino)-2*R*-isobutyl-3*S*-(2-thienylthiomethyl)succinyl]-L-phenylalanine-*N*-methylamide *see* 1006
3-Hydroxy-5-aminomethylisoxazole *see* 6397
6-(*p*-Hydroxy-α-aminophenylacetamido)penicillanic Acid *see* 574
Hydroxyamphetamine, 4847
p-Hydroxyampicillin *see* 574

(17β)-17-Hydroxyandrosta-1,4-dien-3-one *see* 1327
(3α,5α)-3-Hydroxyandrostan-17-one *see* 633
(3β,5α)-3-Hydroxyandrostan-17-one *see* 3664
3α-Hydroxy-17-androstanone *see* 633
3β-Hydroxy-17-androstanone *see* 3664
(5α,17β)-17-Hydroxyandrostan-3-one *see* 8920
17β-Hydroxy-3-androstanone *see* 8920
3α-Hydroxy-5α-androst-16-ene *see* 632
4-Hydroxyandrost-4-ene-3,17-dione *see* 4268
(3β)-3-Hydroxyandrost-5-en-17-one *see* 2875
(17α)-17-Hydroxyandrost-4-en-3-one *see* 3679
(17β)-17-Hydroxyandrost-4-en-3-one *see* 9324
N-Hydroxyaniline *see* 7405
p-Hydroxyaniline *see* 458
2-Hydroxyaniline *see* 457
3-Hydroxyaniline *see* 456
p-Hydroxyanilinoacetic Acid *see* 4875
o-Hydroxyanisole *see* 4589
9-Hydroxyanthracene *see* 678
1-Hydroxy-2-anthraquinonyl 6-*O*-β-D-Xylopyranosyl-β-D-glucopyranoside *see* 8413
Hydroxyapatite *see* 3514
(3*R*)-Hydroxy-8'-apo-β,ψ-carotenal *see* 2323
1-Hydroxy-7-azabenzotriazole, 4848
p-Hydroxybenzaldehyde, 4849
2-Hydroxybenzaldehyde *see* 8462
4-Hydroxybenzaldehyde *see* 4849
2-Hydroxybenzaldehyde Oxime *see* 8464
2-Hydroxybenzaldehyde 2-Phenylhydrazone *see* 8463
o-Hydroxybenzal Isonicotinylhydrazone *see* 8472
2-Hydroxybenzamide *see* 8465
N-Hydroxybenzenamine *see* 7405
Hydroxybenzene *see* 7354
α-Hydroxybenzeneacetic Acid *see* 5781
α-Hydroxybenzeneacetic Acid Ammonium Salt (1:1) *see* 529
α-Hydroxybenzeneacetic Acid (3-*endo*)-8-Methyl-8-azabicyclo[3.2.1]oct-3-yl Ester *see* 4768
α-Hydroxybenzeneacetic Acid 3-Methylbutyl Ester *see* 5782
α-Hydroxybenzeneacetic Acid 3,3,5-Trimethylcyclohexyl Ester *see* 2697
α-Hydroxybenzeneacetonitrile *see* 5783
4-*p*-Hydroxybenzeneazo-1-*p*-chloro-*o*-sulfophenyl-3-methyl-5-hydroxypyrazole Toluene-*p*-sulfonyl Ester Sodium Salt *see* 7670
2-Hydroxybenzenecarbodithioic Acid *see* 3418
4-Hydroxy-1,3-benzenedicarboxylic Acid *see* 4866
4-Hydroxy-*m*-benzenedisulfonic Acid *see* 7355
4-Hydroxy-1,3-benzenedisulfonic Acid *see* 7355
4-Hydroxybenzeneethanol *see* 10022
2-Hydroxybenzenemethanol *see* 8461
N-Hydroxybenzenesulfonamide *see* 7538
4-Hydroxybenzenesulfonic Acid *see* 7359
4-Hydroxybenzenesulfonic Acid Calcium Salt (2:1) *see* 1692
4-Hydroxybenzenesulfonic Acid Mercury(2+) Sodium Salt (2:1:2) *see* 5954

4-Hydroxybenzenesulfonic Acid Zinc Salt (2:1) *see* 10350
2-Hydroxybenzhydroxamic Acid *see* 8468
1-Hydroxy-1,2-benziodoxol-3(1*H*)-one 1-Oxide *see* 5091
[2-(Hydroxy-κ*O*)benzoato(2−)-κ*O*]mercury *see* 5953
(2-Hydroxybenzoato-*O*¹,*O*²)oxobismuth *see* 1285
2-Hydroxy-4*H*-1,3,2-benzodioxabismin-4-one *see* 1285
6-Hydroxy-5-benzofuranacrylic Acid δ-Lactone *see* 8037
p-Hydroxybenzoic Acid, 4850
2-Hydroxybenzoic Acid *see* 8469
4-Hydroxybenzoic Acid *see* 4850
2-Hydroxybenzoic Acid Ammonium Salt (1:1) *see* 547
4-Hydroxybenzoic Acid Butyl Ester *see* 1587
2-Hydroxybenzoic Acid 2-Carboxyphenyl Ester *see* 8477
2-Hydroxybenzoic Acid Compd with 1*H*-Imidazole (1:1) *see* 4954
2-Hydroxybenzoic Acid Copper Salt *see* 2639
2-Hydroxybenzoic Acid 4-(1,1-Dimethylethyl)phenyl Ester *see* 1589
2-Hydroxybenzoic Acid Ethyl Ester *see* 3902
4-Hydroxybenzoic Acid Ethyl Ester *see* 3890
p-Hydroxybenzoic Acid Ethyl Ester 6-Guanidinohexanoate *see* 4350
2-Hydroxybenzoic Acid 2-Ethylhexyl Ester *see* 6860
2-Hydroxybenzoic Acid 2-Hydroxyethyl Ester *see* 4535
2-Hydroxybenzoic Acid Magnesium Salt *see* 5749
2-Hydroxybenzoic Acid 3-Methylbutyl Ester *see* 5170
2-Hydroxybenzoic Acid Methyl Ester *see* 6193
4-Hydroxybenzoic Acid Methyl Ester *see* 6180
2-Hydroxybenzoic Acid 1-Naphthalenyl Ester *see* 6495
2-Hydroxybenzoic Acid 2-Naphthalenyl Ester *see* 6496
4-Hydroxybenzoic Acid 2-[(5-Nitro-2-furanyl)methylene]hydrazide *see* 6619
p-Hydroxybenzoic Acid (5-Nitrofurfurylidene)hydrazide *see* 6619
2-Hydroxybenzoic Acid Phenyl Ester *see* 7420
2-Hydroxybenzoic Acid Phenylmethyl Ester *see* 1146
2-Hydroxybenzoic Acid Potassium Salt (1:1) *see* 7785
4-Hydroxybenzoic Acid Propyl Ester *see* 7977
2-Hydroxybenzoic Acid 3,3,5-Trimethylcyclohexyl Ester *see* 4777
2-Hydroxybenzoic-5-sulfonic Acid *see* 9094
7-Hydroxy-2*H*-1-benzopyran-2-one *see* 10032
Hydroxybenzopyridine *see* 4881
2-(6-Hydroxybenzothiazol-2-yl)-2-thiazoline-4-carboxylic Acid *see* 4117
(±)-*N*-Hydroxy-*N*-(1-benzo[*b*]thien-2-yl-ethyl)urea *see* 10323
N-Hydroxybenzotriazole *see* 4851

Name Index

7α-[1(R)-Hydroxy-1-methylbutyl]-6,14-*endo*-ethenotetrahydrooripavine *see* 3931

2-Hydroxy-2-methyl-3-butyne *see* 6108

N-(S)-2-Hydroxy-3-methylbutyryl-1-1-L-alaninyl-(S)-1-amino-3-methyl-2,3,4,5-tetrahydro-1H-3-benazazepin-2-one *see* 8580

3-Hydroxymethyl-β-carboline *see* 3834

3-Hydroxymethylchrysazin *see* 300

7-Hydroxy-4-methylcoumarin *see* 4894

3-Hydroxy-N-methyl-*cis*-crotonamide Dimethyl Phosphate *see* 6337

Hydroxymethylcyclohexane *see* 2724

2R-Hydroxymethyl-4R-(cytosin-1′-yl)-1,3-oxathiolane *see* 742

4-Hydroxy-N-methyldichloroacetanilide *see* 3223

N-(Hydroxymethyl)-N-(1,3-dihydroxy-methyl-2,5-dioxo-4-imidazolidinyl)-N′-(hydroxymethyl)urea *see* 3003

2-(Hydroxymethyl)-α,3-dimethylcyclopentaneacetic Acid δ-Lactone *see* 5133

1-(Hydroxymethyl)-5,5-dimethylhydantoin, 4870

1-(Hydroxymethyl)-5,5-dimethyl-2,4-imidazolidinedione *see* 4870

8-(Hydroxymethyl)-6,11-dimethyl-2H,-4H-oxazolo[5,4,3-ij]pyrido[3,2-g]quinoline-4,10(11H)-dione *see* 6820

2-(Hydroxymethyl)-1,1-dimethylpiperidinium Methyl Sulfate Benzilate *see* 1197

2-Hydroxymethyl-1,1-dimethylpyrrolidinium Methyl Sulfate Benzilate *see* 7671

(4R,9aS)-5-Hydroxy-4-methyl-6,10-dioxo-3,4,6,9,9a,10-hexahydro-2H-1-oxa-4a,8a-diazaanthracene-7-carboxylic Acid 2,4-Difluorobenzylamide *see* 3460

4-(Hydroxymethyl)-1,3-dioxolane *see* 4521

(−)-(2S,4S)-1-[2-(Hydroxymethyl)-1,3-dioxolan-4-yl]cytosine *see* 9968

3-(Hydroxymethyl)-5,5-diphenylhydantoin Phosphate Ester *see* 4285

9-Hydroxy-2-methylellipticinium Acetate *see* 3604

P,P′-(Hydroxymethylene)bisphosphonic Acid *see* 7039

(εR,1R,3aS,4E,7aR)-4-[(2Z)-2-[(5S)-5-Hydroxy-2-methylenecyclohexylidene]-ethylidene]octahydro-α,α,ε,7a-tetramethyl-1H-indene-1-pentanol *see* 1641

7-Hydroxy-5′,6′-methylenedioxybenzo-furano(3′,2′:3,4)coumarin *see* 5859

7-Hydroxy-11,12-(methylenedioxy)coumestan *see* 5859

7-Hydroxy-3′,4′-(methylenedioxy)isoflavone *see* 8019

1-Hydroxy-6,7-methylenedioxy-2-methyl-1,2,3,4-tetrahydroisoquinoline *see* 4807

(Hydroxymethylene)diphosphonic Acid *see* 7039

2-Hydroxymethylene-17α-methylandrostan-17β-ol-3-one *see* 7066

2-Hydroxymethylene-17α-methyldihydrotestosterone *see* 7066

2-Hydroxymethylene-17α-methyl-17β-hydroxy-5α-androstan-3-one *see* 7066

(1S,6R)-6-Hydroxy-5-methylene-1-[(1S,-2S)-1,2,3-trihydroxy-2-methylpropyl]-

2-oxa-7,9-diazabicyclo[4.2.2]decane-8,10-dione *see* 1207

(17β)-17-Hydroxy-17-methylestra-4,9,11-trien-3-one *see* 6203

(17β)-17-Hydroxy-17-methylestr-4-en-3-one *see* 6799

1-[6-Hydroxy-2-(1-methylethenyl)-5-benzofuranyl]ethanone *see* 3942

2-(2-Hydroxy-1-methylethenyl)-5-methylcyclopentanecarboxylic Acid Delta Lactone *see* 6554

4-[1-Hydroxy-2-[(1-methylethyl)amino]-butyl]-1,2-benzenediol *see* 5215

4-[1-Hydroxy-2-[(1-methylethyl)amino]-ethyl]-1,2-benzenediol *see* 5263

5-[1-Hydroxy-2-[(1-methylethyl)amino]-ethyl]-1,3-benzenediol *see* 6002

N-[4-[1-Hydroxy-2-[(1-methylethyl)amino]ethyl]phenyl]methanesulfonamide *see* 8854

4-[2-Hydroxy-3-[(1-methylethyl)amino]-propoxy]benzeneacetamide *see* 850

4-[2-Hydroxy-3-[(1-methylethyl)amino]-propoxy]benzenepropanoic Acid Methyl Ester *see* 3755

1-[7-[2-Hydroxy-3-[(1-methylethyl)amino]propoxy]-2-benzofuranyl]ethanone *see* 1021

4-[(2S)-2-Hydroxy-3-[(1-methylethyl)-amino]propoxy]phenol *see* 7853

N-[4-[2-Hydroxy-3-[(1-methylethyl)amino]propoxy]phenyl]acetamide *see* 7818

4-[2-Hydroxy-3-[(1-methylethyl)amino]-propoxy]-2,3,6-trimethylphenol 1-Acetate *see* 6216

2-Hydroxy-4-(1-methylethyl)-2,4,6-cycloheptatrien-1-one *see* 9544

N-[α-(Hydroxymethyl)ethyl]-D-lysergamide *see* 3712

(5′α,8α)-12′-Hydroxy-2′-(1-methylethyl)-5′-(2-methylpropyl)ergotaman-3′,-6′,18-trione *see* 3708

(5′α,10α)-12′-Hydroxy-2′-(1-methylethyl)-5′-[(1S)-1-methylpropyl]ergotaman-3′,6′,18-trione *see* 3708

(5′α)-12′-Hydroxy-2′-(1-methylethyl)-5′-[(1S)-1-methylpropyl]ergotaman-3′,-6′,18-trione *see* 3707

(5′α)-12′-Hydroxy-2′-(1-methylethyl)-5′-(2-methylpropyl)ergotaman-3′,6′,18-trione *see* 3707

(5′α)-12′-Hydroxy-2′-(1-methylethyl)-5′-(phenylmethyl)ergotaman-3′,6′,18-trione *see* 3706

4-(1-Hydroxy-1-methylethyl)-2-propyl-1-[[2′-(1H-tetrazol-5-yl)[1,1′-biphenyl]-4-yl]methyl]-1H-imidazole-5-carboxylic Acid 5-Methyl-2-oxo-1,3-dioxol-4-yl)methyl Ester *see* 6933

(1-Hydroxy-1-methylethyl)succinic Acid γ-Lactone *see* 9304

5-Hydroxymethyl-2-formylfuran *see* 4871

5-(Hydroxymethyl)-2-furaldehyde, 4871

2-Hydroxymethylfuran *see* 4334

5-(Hydroxymethyl)-2-furancarbonal *see* 4871

5-(Hydroxymethyl)-2-furancarboxaldehyde *see* 4871

5-(Hydroxymethyl)-2-furfural *see* 4871

3-Hydroxy-3-methylglutaraldehydic Acid *see* 6241

3-Hydroxy-3-methylglutaric Acid *see* 5877

3-C-(Hydroxymethyl)-D-glyceroaldotetrose *see* 722

2-Hydroxymethyl-3-hydroxy-6-(1-hydroxy-2-*tert*-butylaminoethyl)pyridine *see* 7602

2-Hydroxymethyl-5-hydroxy-γ-pyrone *see* 5365

3-Hydroxy-1-methyl-5,6-indolinedione *see* 161

3-Hydroxy-1-methyl-5,6-indolinedione 5-Semicarbazone *see* 1789

(2E)-N-Hydroxy-3-[4-[[[2-(2-methyl-1H-indol-3-yl)ethyl]amino]methyl]phenyl]-2-propenamide *see* 7113

4-Hydroxymethyl-2-iodomethyl-1,3-dioxolane *see* 3463

6-Hydroxymethyl-2-isopropylamino-methyl-7-nitro-1,2,3,4-tetrahydroquinoline *see* 7018

3-Hydroxy-5-methylisoxazole *see* 4895

2-Hydroxy-4-(methylmercapto)butyric Acid *see* 6048

(6R,7S)-3-(Hydroxymethyl)-7-methoxy-8-oxo-7-[2-(2-thienyl)acetamido]-5-thia-1-azabicyclo[4.2.0]oct-2-ene-2-carboxylic Acid Carbamate (Ester) *see* 1938

6-[(2S,3R,4R,6E)-3-Hydroxy-4-methyl-2-(methylamino)-6,8-nonadienoic Acid]-cyclosporin A *see* 10224

(2S,3R,4R,6E)-3-Hydroxy-4-methyl-2-(methylamino)-6-octenoic Acid *see* 5840

2-Hydroxy-6-methyl-3-(1-methylethyl)-benzoic Acid *see* 9563

2-Hydroxy-3-methyl-6-(1-methylethyl)-2-cyclohexen-1-one *see* 3326

(5S,8S,10S,11S)-10-Hydroxy-2-methyl-5-(1-methylethyl)-1-[2-(1-methylethyl)-4-thiazolyl]-3,6-dioxo-8,11-bis(phenyl-methyl)-2,4,7,12-tetraazatridecan-13-oic Acid 5-Thiazolylmethyl Ester *see* 8366

(3S,4S,6S,9S)-4-Hydroxy-12-methyl-9-(1-methylethyl)-13-[2-(1-methylethyl)-4-thiazolyl]-8,11-dioxo-3,6-bis(phenyl-methyl)-2,7,10,12-tetraazatridecanoic Acid 5-Thiazolylmethyl Ester *see* 8366

3β-(Hydroxymethyl)-2α-methyl-4β-[(9-methyl-9H-pyrido[3,4-b]indol-1-yl)-methyl]-2H-pyran-5-carboxylic Acid Methyl Ester *see* 311

(2S)-2-Hydroxy-3-methyl-N-[(1S)-1-methyl-2-oxo-2-[[(1S)-2,3,4,5-tetrahydro-3-methyl-2-oxo-1H-3-benzazepin-1-yl]amino]ethyl]butanamide *see* 8580

5-(Hydroxymethyl)-3-(3-methylphenyl)-2-oxazolidinone *see* 9680

N-Hydroxy-α-methyl-4-(2-methylpropyl)benzeneacetamide *see* 4920

(5′α)-12′-Hydroxy-2′-methyl-5′-(2-methylpropyl)ergotaman-3′,6′,18-trione *see* 3713

5-(Hydroxymethyl)-6-methyl-2,4(1H,3H)-pyrimidinedione *see* 7250

4-Hydroxy-2-methyl-N-(5-methyl-2-thiazolyl)-2H-1,2-benzothiazine-3-carboxamide 1,1-Dioxide *see* 5894

(2E)-N-[(1S)-1-(Hydroxymethyl)-2-[(R)-[(methylthio)methyl]sulfinyl]ethyl]-3-(1,2,3,4-tetrahydro-6-methyl-2,4-dioxo-5-pyrimidinyl)-2-propenamide *see* 8863

N-(Hydroxymethyl)-N′-methylthiourea *see* 6813

1-(Hydroxymethyl)-3-methyl-2-thiourea *see* 6813

2-[4-(Hydroxymethylureidosulfonyl)-phenylcarbamoyl]benzoic Acid *see* 9044

β-Hydroxy-β-methyl-δ-valerolactone *see* 6242

17-Hydroxy-13-methyl-17α-vinyl-1,2,3,-6,7,8,9,10,11,12,13,14,16,17-tetrade-cahydro-15H-cyclopenta[a]phenan-thren-3-one *see* 6806

(−)-3-Hydroxymorphinan *see* 6796

4-[(2S)-2-Hydroxy-3-[[2-[(4-morpholinyl-carbonyl)amino]ethyl]amino]propoxy]-benzenepropanoic Acid [(4S)-2,2-Di-methyl-1,3-dioxolan-4-yl]methyl Ester *see* 5405

Hydroxymycin *see* 7146

α-Hydroxynaphthalene *see* 6468

β-Hydroxynaphthalene *see* 6469

1-Hydroxy-2-naphthalenecarboxylic Acid *see* 4873

3-Hydroxy-2-naphthalenecarboxylic Acid *see* 4874

2-Hydroxy-1,4-naphthalenedione *see* 5448

5-Hydroxy-1,4-naphthalenedione *see* 5317

2-Hydroxynaphthalene-6,8-disulfonic Acid *see* 6473

3-Hydroxy-2,7-naphthalenedisulfonic Acid *see* 6472

4-Hydroxy-1,5-naphthalenedisulfonic Acid *see* 6471

7-Hydroxy-1,3-naphthalenedisulfonic Acid *see* 6473

6-Hydroxy-2-naphthalenepropanoic Acid *see* 251

2-Hydroxy-6-naphthalenepropionic Acid *see* 251

1-Hydroxy-2-naphthalenesulfonic Acid *see* 6475

4-Hydroxy-1-naphthalenesulfonic Acid *see* 6476

6-Hydroxy-2-naphthalenesulfonic Acid *see* 6477

7-Hydroxy-1-naphthalenesulfonic Acid *see* 2578

7-Hydroxy-2-naphthalenesulfonic Acid *see* 1892

4-[2-(2-Hydroxy-1-naphthalenyl)diazen-yl]benzenesulfonic Acid Sodium Salt (1:1) *see* 6953

4-[2-(4-Hydroxy-1-naphthalenyl)diazen-yl]benzenesulfonic Acid Sodium Salt (1:1) *see* 6952

N-Hydroxynaphthalimide Diethyl Phos-phate *see* 6438

1-Hydroxy-2-naphthoic Acid, 4873

3-Hydroxy-2-naphthoic Acid, 4874

2-Hydroxy-1,4-naphthoquinone *see* 5448

5-Hydroxy-1,4-naphthoquinone *see* 5317

8-Hydroxy-1,4-naphthoquinone *see* 5317

4-Hydroxy-α-naphthylamine *see* 449

(2S,5R,6R)-6-[[(2R)-2-[[(4-Hydroxy-1,5-naphthyridin-3-yl)carbonyl]amino]-2-phenylacetyl]amino]-3,3-dimethyl-7-oxo-4-thia-1-azabicyclo[3.2.0]heptane-2-carboxylic Acid *see* 710

2-Hydroxy-2′,6′-nicotinoxylidide *see* 5238

4-Hydroxy-3-nitrobenzenearsonic Acid *see* 8406

4-Hydroxy-3-nitrobenzoic Acid *see* 6717

2-Hydroxy-5-nitrobenzoic Acid *see* 6718

As-(4-Hydroxy-3-nitrophenyl)arsonic Acid *see* 8406

2-Hydroxy-5-[2-(4-nitrophenyl)diazenyl]-benzoic Acid *see* 247

2-Hydroxy-5-[2-(3-nitrophenyl)diazenyl]-benzoic Acid Sodium Salt (1:1) *see* 5989

4-Hydroxy-3-[1-(4-nitrophenyl)-3-oxobu-tyl]-2H-1-benzopyran-2-one *see* 30

3-(Hydroxynitrosoamino)-L-alanine *see* 199

N-Hydroxy-N-nitrosobenzenamine Am-monium Salt (1:1) *see* 2612

3-Hydroxy-4-nitroso-2,7-naphthalenedi-sulfonic Acid Sodium Salt (1:2) *see* 6731

2-Hydroxy-1,2,3-nonadecanetricarbox-ylic Acid *see* 178

1-Hydroxynonane *see* 6766

(2E)-4-Hydroxy-2-nonenal *see* 4764

17β-Hydroxy-19-norandrosta-4,9,11-trien-3-one *see* 9745

17β-Hydroxy-19-nor-4-androsten-3-one *see* 6450

m-Hydroxynorephedrine *see* 6003

(17α)-17-Hydroxy-19-norpregna-4,20-dien-3-one *see* 6806

(17α)-17-Hydroxy-19-norpregna-5(10),20-dien-3-one *see* 6791

(17α)-17-Hydroxy-19-norpregna-4,9,11-trien-20-yn-3-one *see* 6794

17-Hydroxy-19-norpregn-4-ene-3,20-di-one Hexanoate *see* 4448

(17α)-17-Hydroxy-19-norpregn-4-en-3-one *see* 6785

(17α)-17-Hydroxy-19-norpregn-4-en-20-yn-3-one *see* 6786

(17α)-17-Hydroxy-19-norpregn-5(10)-en-20-yn-3-one *see* 6787

17α-Hydroxy-19-norprogesterone Ca-proate *see* 4448

1-Hydroxyoctadecane *see* 8932

(9Z,12R)-12-Hydroxy-9-octadecenoic Acid *see* 8335

(9S,10E)-9-Hydroxy-10-octadecen-12-ynoic Acid *see* 4662

4-[[Hydroxy(octadecyloxy)phosphinyl]-oxy]-1,1-dimethylpiperidinium Inner Salt *see* 7280

N-Hydroxyoctanamide *see* 6839

1-Hydroxyoctane *see* 6840

2-Hydroxyoctane *see* 6841

5-Hydroxy-4-octanone *see* 1599

8-Hydroxy-2,4,6-octatriynamide *see* 185

[1S-[1α,3α(1E,3R*)4β(Z),5α]]-7-[3-(3-Hydroxy-1-octenyl)-2,6-dioxabicyclo-[3.1.1]hept-4-yl]-5-heptenoic Acid *see* 9542

2-Hydroxy-4-(octyloxy)benzophenone *see* 6831

[2-Hydroxy-4-(octyloxy)phenyl]phenyl-methanone *see* 6831

(3β)-3-Hydroxyolean-12-en-28-oic Acid *see* 6920

d-12-Hydroxyoleic Acid *see* 8335

6β-Hydroxy-4-oxo-10αH-ambrosa-2,11-(13)-dien-12-oic Acid γ-Lactone *see* 379

6-Hydroxy-2-oxo-1,3-benzoxathiole *see* 9612

γ-Hydroxy-β-oxobutane *see* 63

1-Hydroxy-4-oxo-2,5-cyclohexadiene-1-sulfonic Acid Compd with Diethylam-ine *see* 3777

N-[(2R)-3-Hydroxy-1-oxodecyl]-L-leucyl-D-α-glutamyl-D-allothreonyl-D-valyl-L-leucyl-D-seryl-L-leucyl-D-seryl-L-isoleu-cine (9 → 3)-Lactone *see* 10205

3-Hydroxy-22-oxokopsan-1-carboxylic Acid Methyl Ester *see* 5368

(17α)-17-Hydroxy-3-oxo-19-norpregna-4,9-diene-21-nitrile *see* 3117

δ-Hydroxy-γ-oxo-L-norvaline *see* 4780

5-Hydroxy-4-oxo-L-norvaline *see* 4780

N²-[(9Z,12Z,15Z,17S)-17-Hydroxy-1-oxo-9,12,15-octadecatrien-1-yl]-L-gluta-mine *see* 10227

3β-Hydroxy-11-oxoolean-12-en-30-oic Acid *see* 3644

(3β,20β)-3-Hydroxy-11-oxoolean-12-en-29-oic Acid *see* 3644

3β-Hydroxy-11-oxoolean-12-en-30-oic Acid Acetate *see* 74

3β-Hydroxy-11-oxoolean-12-en-30-oic Acid Hydrogen Succinate *see* 1794

8-Hydroxy-5-oxoophiobola-3,6,19-trien-25-oic Acid *see* 1976

4-Hydroxy-3-(3-oxo-1-phenylbutyl)-2H-1-benzopyran-2-one *see* 10236

(1α,2β,4β,5α,7β)-7-[(2S)-3-Hydroxy-1-oxo-2-phenylpropoxy]-9,9-dimethyl-3-oxa-9-azoniatricyclo[3.3.1.0²,⁴]nonane Bromide (1:1) *see* 6077

(3-endo,8-syn)-3-(3-Hydroxy-1-oxo-2-phenylpropoxy)-8-methyl-8-(1-methyl-ethyl)-8-azoniabicyclo[3.2.1]octane Bromide (1:1) *see* 5117

(17α)-17-Hydroxy-3-oxopregna-4,6-di-ene-21-carboxylic Acid γ-Lactone *see* 1753

3β-Hydroxy-20-oxo-5α-pregnane *see* 268

20α-Hydroxy-3-oxo-5α-pregnane *see* 269

20β-Hydroxy-3-oxo-5α-pregnane *see* 270

3-Hydroxy-4-oxo-4H-pyran-2,6-dicar-boxylic Acid *see* 5853

3-Hydroxy-4-oxo-1(4H)-pyridinealanine *see* 6280

1-Hydroxy-5-oxo-5H-pyrido[3,2-a]phen-oxazine-3-carboxylic Acid *see* 7603

4-Hydroxy-2-oxo-1-pyrrolidineacetamide *see* 7041

13-Hydroxy-3-oxo-13,17-secoandrosta-1,4-dien-17-oic Acid δ-Lactone *see* 9323

[2α,3β,5α]-14-Hydroxy-19-oxo-2,3-[[(2S,3S,4S,6R)-tetrahydro-3,4-dihy-droxy-6-methyl-2H-pyran-3,2-diyl]bis-(oxy)]card-20(22)-enolide *see* 1722

(2α,3β,5α)-14-Hydroxy-19-oxo-2,3-[[(2S,3R,6R)-tetrahydro-3-hydroxy-6-methyl-4-oxo-2H-pyran-3,2-diyl]bis-(oxy)]card-20(22)-enolide *see* 10077

6-(10-Hydroxy-6-oxo-trans-1-undecenyl)-β-resorcylic Acid Lactone *see* 10314

2′-Hydroxypelargidenolon 1522 *see* 6355

6-Hydroxypelargidenon 1465 *see* 8549

6-Hydroxypelargidenon-6,4′-dimethyl Ether 1467 *see* 7170

1-Hydroxypentachlorobenzene *see* 7218

15-Hydroxypentadecanoic Acid ε-Lac-tone *see* 3952

4-Hydroxy-2,4-pentadienoic Acid γ-Lac-tone *see* 8001

2-Hydroxypentane *see* 7231

4-Hydroxy-2-pentenoic Acid γ-Lactone *see* 640

4-Hydroxy-3-pentenoic Acid γ-Lactone *see* 640

α-Hydroxyphenazine *see* 4680

β-Hydroxyphenethylamine *see* 7399

4-Hydroxyphenethylamine *see* 10016

2-(β-Hydroxyphenethylamino)pyridine *see* 7432

2-[6-(β-Hydroxyphenethyl)-1-methyl-2-piperidyl]acetophenone *see* 5609

6-(*p*-Hydroxyphenethyl)salicylic Acid *see* 5666

7-Hydroxy-3*H*-phenoxazin-3-one 10-Oxide *see* 8261

3-[4-(*p*-Hydroxyphenoxy)-3,5-diiodophenyl]alanine *see* 3209

1-Hydroxy-2-phenoxyethane *see* 7369

(−)-(*S*)-1-(*p*-Hydroxyphenoxy)-3-(isopropylamino)-2-propanol *see* 7853

1-(4-Hydroxyphenoxy)-3-[2-(4-morpholinocarboxamido)ethylamino]-2-propanol *see* 10253

N-(4-Hydroxyphenyl)acetamide *see* 47

α-Hydroxyphenylacetic Acid *see* 5781

α-Hydroxy-α-phenylacetophenone *see* 1095

2-Hydroxy-2-phenylacetophenone Oxime *see* 1096

(6*R*,7*R*)-7-[[(2*R*)-2-Hydroxy-2-phenylacetyl]amino]-3-[[(1-methyl-1*H*-tetrazol-5-yl)thio]methyl]-8-oxo-5-thia-1-azabicyclo[4.2.0]oct-2-ene-2-carboxylic Acid *see* 1916

(6*R*,7*R*)-7-[[(2*R*)-2-Hydroxy-2-phenylacetyl]amino]-8-oxo-3-[[[1-(1-hydroxyphenoxy)-1*H*-tetrazol-5-yl]thio]methyl]-5-thia-1-azabicyclo[4.2.0]oct-2-ene-2-carboxylic Acid *see* 1930

β-[4-Hydroxyphenyl]acrylic Acid *see* 2548

β-(*p*-Hydroxyphenyl)alanine *see* 10020

3-Hydroxy-L-phenylalanine *see* 10021

p-Hydroxyphenylaminoacetic Acid *see* 4875

α-(4-Hydroxyphenyl)-β-aminoethane *see* 10016

1-(*m*-Hydroxyphenyl)-2-aminoethanol *see* 6788

1-(*p*-Hydroxyphenyl)-2-aminoethanol *see* 6848

α-(*m*-Hydroxyphenyl)-β-aminopropanol *see* 6003

1-(*m*-Hydroxyphenyl)-2-amino-1-propanol *see* 6003

4-Hydroxy-5-[2-[4-(phenylamino)-5-sulfo-1-naphthalenyl]diazenyl]-2,7-naphthalenedisulfonic Acid Sodium Salt (1:3) *see* 622

N-Hydroxy-3-[3-[(phenylamino)sulfonyl]phenyl]-2-propenamide *see* 1025

2-Hydroxy-*N*-phenylbenzamide *see* 8467

α-Hydroxy-α-phenylbenzeneacetic Acid *see* 1082

α-Hydroxy-α-phenylbenzeneacetic Acid 2-(Diethylamino)ethyl Ester *see* 1031

α-Hydroxy-α-phenylbenzeneacetic Acid 2-(Dimethylamino)-1,1-dimethylethyl Ester *see* 3152

α-Hydroxy-α-phenylbenzeneacetic Acid (3-*endo*)-8-Methyl-8-azabicyclo[3.2.1]-oct-3-yl Ester *see* 9961

α-Hydroxy-α-phenylbenzeneacetic Acid 2-(1-Piperidinyl)ethyl Ester *see* 7583

2-[2-(4-Hydroxyphenyl)-6-benzimidazoyl]-6-(1-methyl-4-piperazyl)benzimidazole *see* 1250

3-(4-Hydroxyphenyl)-2*H*-1-benzopyran-7-ol *see* 2876

p-Hydroxyphenyl Benzyl Ether *see* 6334

N-(4-Hydroxyphenyl)butanamide *see* 4853

4-(*p*-Hydroxyphenyl)-2-butanone Acetate *see* 2606

p-Hydroxyphenylbutazone *see* 7071

4-Hydroxy-α¹-[[[6-(4-phenylbutoxy)hexyl]amino]methyl]-1,3-benzenedimethanol *see* 8475

(±)-4-Hydroxy-α¹-[[[6-(4-phenylbutoxy)hexyl]amino]methyl]-*m*-xylene-α,α¹-diol *see* 8475

1-(*p*-Hydroxyphenyl)-2-butylaminoethanol *see* 947

3-Hydroxy-4-phenylcarbostyril *see* 10201

3-Hydroxy-2-phenylcinchoninic Acid *see* 7056

1-Hydroxy-α-phenylcyclopentaneacetic Acid 2-Diethylaminoethyl Ester *see* 2712

1-Hydroxy-α-phenylcyclopentaneacetic Acid 2-(Dimethylamino)ethyl Ester *see* 2741

2-(*p*-Hydroxyphenyl)-5,7-dihydroxychromone *see* 717

O-(4-Hydroxyphenyl)-3,5-diiodotyrosine *see* 3209

4-[(1*Z*,3*Z*)-4-(4-Hydroxyphenyl)-2,3-diisocyano-1,3-butadien-1-yl]-1,2-benzenediol *see* 10259

(−)-(*R*)-1-(4-Hydroxyphenyl)-2-(3,4-dimethoxyphenethylamino)ethanol *see* 2896

(3-Hydroxyphenyl)dimethylethylammonium Chloride *see* 3564

[[2(*S*)-[[4(*R*)-(3-Hydroxyphenyl)-3(*R*),4-dimethyl-1-piperidinyl]methyl]-1-oxo-3-phenylpropyl]amino]acetic Acid *see* 366

N-[(2*S*)-2-[[(3*R*,4*R*)-4-(3-Hydroxyphenyl)-3,4-dimethyl-1-piperidinyl]methyl]-1-oxo-3-phenylpropyl]glycine *see* 366

m-Hydroxyphenylethanolamine *see* 6788

p-Hydroxyphenylethanolamine *see* 6848

5-[(1*E*)-2-(4-Hydroxyphenyl)ethenyl]-1,3-benzenediol *see* 8279

2-*p*-Hydroxyphenylethylamine *see* 10016

α-(*m*-Hydroxyphenyl)-β-(ethylamino)ethanol *see* 3914

N-(*p*-Hydroxyphenylethyl)-4-hydroxynorephedrine *see* 8365

N-[2-(*p*-Hydroxyphenyl)ethyl]-*N*-[2-(*p*-hydroxyphenyl)-2-hydroxy-1-methylethyl]amine *see* 8365

2-[(2*R*,6*S*)-6-[(2*S*)-2-Hydroxy-2-phenylethyl]-1-methyl-2-piperidinyl]-1-phenylethanone *see* 5609

4-Hydroxyphenyl-β-D-glucopyranoside *see* 763

N-(4-Hydroxyphenyl)glycine, 4875

2-Hydroxyphenylhydroxamic Acid *see* 8468

1-(4-Hydroxyphenyl)-1-hydroxy-2-butylaminoethane *see* 947

1-(*p*-Hydroxyphenyl)-2-[[β-hydroxy-β-(3′,5′-dihydroxyphenyl)]ethyl]aminopropane *see* 4012

N-[2-(4-Hydroxyphenyl)-2-hydroxyethyl]-1-methyl-3-(4-hydroxyphenyl)propylamine *see* 8208

1-(4-Hydroxyphenyl)-2-[2-(4-hydroxyphenyl)ethylamino]propanol *see* 8365

4-Hydroxyphenyl Iodide *see* 5079

p-Hydroxyphenylisopropylamine *see* 4847

β-(*p*-Hydroxyphenyl)isopropylmethylamine *see* 7442

Hydroxyphenylmercury Mixt with (Nitrato-*O*)phenylmercury *see* 7413

1-Hydroxy-1-phenyl-2-methylaminoethane *see* 4635

l-1-(*m*-Hydroxyphenyl)-2-methylaminoethanol *see* 7398

1-(4-Hydroxyphenyl)-2-methylaminoethanol *see* 9142

α-(*p*-Hydroxyphenyl)-β-methylaminopropane *see* 7442

1-(4-Hydroxyphenyl)-2-methylaminopropanol *see* 4863

2-[(2-Hydroxyphenyl)methylene]hydrazide 4-Pyridinecarboxylic Acid *see* 8472

1-(4-Hydroxyphenyl)-2-[1-methyl-3-(4-hydroxyphenyl)-propylamino]ethanol *see* 8208

1-(*p*-Hydroxyphenyl)-2-(1-methyl-2-phenoxyethylamino)-1-propanol *see* 5285

α-(4-Hydroxyphenyl)-β-methyl-4-(phenylmethyl)-1-piperidineethanol *see* 4936

1-(*p*-Hydroxyphenyl)-2-(1′-methyl-3′-phenylpropylamino)-1-propanol *see* 6821

P-(Hydroxyphenylmethyl)phosphinic Acid *see* 4852

2′-(4-Hydroxyphenyl)-5-(4-methyl-1-piperazinyl)-2-5′-bi(1*H*-benzimidazole) *see* 1250

1-[4-(3-Hydroxyphenyl)-1-methyl-4-piperidinyl]-1-propanone *see* 5348

4-(*m*-Hydroxyphenyl)-1-methyl-4-piperidyl Ethyl Ketone *see* 5348

4-[2-[[3-(4-Hydroxyphenyl)-1-methylpropyl]amino]ethyl]-1,2-benzenediol *see* 3440

(±)-4-[2-[[3-(*p*-Hydroxyphenyl)-1-methylpropyl]amino]ethyl]pyrocatechol *see* 3440

(1*R*)-3-(4-Hydroxyphenyl)-1-methylpropyl β-D-Glucopyranoside *see* 8314

(2-Hydroxyphenyl)-4-morpholinyl Methanone *see* 8470

1-(2′-Hydroxyphenyl)-4-(3″-nitrophenyl)-1,2,3,6-tetrahydropyrimidine-2-one *see* 4925

*N*¹-Hydroxy-*N*⁸-phenyloctanediamide *see* 10232

α-(*p*-Hydroxyphenyl)-α-(4-oxo-2,5-cyclohexadien-1-ylidine)-*o*-toluic Acid *see* 7356

3α-Hydroxy-8-(*p*-phenylphenacyl)-1α*H*,5α*H*-tropanium Bromide (−)-Tropate *see* 4033

1-(*p*-Hydroxyphenyl)-2-phenyl-4-butyl-pyrazolidine-3,5-dione *see* 7071

β-(*p*-Hydroxyphenyl)phloropropiophenone *see* 7437

m-Hydroxyphenylpropanolamine *see* 6003

1-(4-Hydroxyphenyl)-1-propanone *see* 7149

3-(4-Hydroxyphenyl)-2-propenoic Acid *see* 2548

α-Hydroxy-α-phenylpropionic Acid *see* 864

2-Hydroxy-2-phenylpropionic Acid *see* 864

dl-1-*p*-Hydroxyphenyl-2-propylamine *see* 4847

L-(1-Hydroxy-1-phenyl-2-propylamino)-1-(*m*-methoxyphenyl)-1-propanone *see* 7061

4-Hydroxy-3-(1-phenylpropyl)-2*H*-1-benzopyran-2-one *see* 7371 γ-(p-Hydroxyphenyl)-α-propylene *see* 2049

1-[γ-Hydroxy-γ-phenylpropyl]-4-phenyl-4-carbethoxypiperidine *see* 7362

1-(3-Hydroxy-3-phenylpropyl)-4-phenyl-isonipecotic Acid Ethyl Ester *see* 7362

1-(3-Hydroxy-3-phenylpropyl)-4-phenyl-4-piperidinecarboxylic Acid Ethyl Ester *see* 7362

3-Hydroxy-2-phenyl-4-quinolinecarboxylic Acid *see* 7056

3-Hydroxy-4-phenyl-2(1H)-quinolinone *see* 10201

N-(4-Hydroxyphenyl)retinamide *see* 4025

N-(p-Hydroxyphenyl)salicylamide *see* 6984

N-Hydroxy-3-(3-phenylsulfamoylphenyl)acrylamide *see* 1025

4-[3-Hydroxy-3-phenyl-3-(2-thienyl)propyl]-4-methylmorpholinium Iodide (1:1) *see* 9587

[3S-(3R*,4aR*,8aR*,2′S*,3′S*)]-2-[2′-Hydroxy-3′-phenylthiomethyl-4′-aza-5′-oxo-5′-(2″-methyl-3″-hydroxyphenyl)pentyl]decahydroisoquinoline-3-N-t-butylcarboxamide *see* 6528

2-[N-(m-Hydroxyphenyl)-p-toluidino-methyl]imidazoline *see* 7376

3-(4-Hydroxyphenyl)-1-(2,4,6-trihydroxyphenyl)-1-propanone *see* 7437

β-(p-Hydroxyphenyl)-2,4,6-trihydroxy-propiophenone *see* 7437

(m-Hydroxyphenyl)trimethylammonium Bromide, Decamethylenebis(methylcarbamate) *see* 2889

4-p-Hydroxyphenyl-2,2,4-trimethylchro-man *see* 2991

Hydroxyphosphine Oxide *see* 4907

2-Hydroxy-3-(phosphonooxy)propanal *see* 4518

(S)-1-[3-Hydroxy-2-(phosphonylmethoxy)propyl]cytosine *see* 2271

N-Hydroxyphthalimide, 4876

[6-Hydroxy-3-[4-[2-(1-piperidinyl)ethoxy]phenoxy]2-(4-methoxyphenyl)]-benzo[b]thiophene *see* 810

rel-4-[(R)-Hydroxy-(2S)-2-piperidinylmethyl]-1,2-benzenediol *see* 8353

3-[3-[(2R,3S)-3-Hydroxy-2-piperidinyl]-2-oxopropyl]-4(3H)-quinazolinone *see* 3982

N-[(2R)-2-Hydroxy-3-(1-piperidinyl)propoxy]-3-pyridinecarboximidoyl Chloride 1-Oxide *see* 775

10-[3-(4-Hydroxy-1-piperidinyl)propyl]-10H-phenothiazine-2-carbonitrile *see* 7279

3-[3-(3-Hydroxy-2-piperidyl)acetonyl]-4(3H)-quinazolinone *see* 3982

2-Hydroxypiperitone *see* 3326

Hydroxypiracetam *see* 7041

12-Hydroxypodocarpa-8,11,13-trien-16-oic Acid *see* 7660

Hydroxypolyethoxydodecane *see* 7675

16α-Hydroxyprednisolone-16α,17-acetonide *see* 2929

(3α,5α)-3-Hydroxypregnane-11,20-dione *see* 228

(3α,5α)-3-Hydroxypregnan-20-one *see* 267

(3α,5β)-3-Hydroxypregnan-20-one *see* 7847

(3β,5α)-3-Hydroxypregnan-20-one *see* 268

3α-Hydroxy-5β-pregnan-20-one *see* 7847

(5α,20R)-20-Hydroxypregnan-3-one *see* 270

(5α,20S)-20-Hydroxypregnan-3-one *see* 269

20α-Hydroxy-5α-pregnan-3-one 269

20β-Hydroxy-5α-pregnan-3-one *see* 270

17-Hydroxypregn-4-ene-3,20-dione *see* 4877

21-Hydroxypregn-4-ene-3,20-dione *see* 2902

21-Hydroxypregn-4-ene-3,20-dione 21-Acetate *see* 2903

21-Hydroxypregn-4-ene-3,11,20-trione *see* 2873

(3β)-3-Hydroxypregn-5-en-20-one *see* 7851

17α-Hydroxypregn-4-en-20-yn-3-one *see* 3795

17α-Hydroxyprogesterone, 4877

21-Hydroxyprogesterone *see* 2902

17α-Hydroxyprogesterone 3-Cyclopentyl Enol Ether *see* 7225

17α-Hydroxyprogesterone Hexanoate *see* 4877

Hydroxyproline, 4878

L₅-Hydroxyproline *see* 4878

(−)-(2S,4R)-4-Hydroxyproline *see* 4878

(2S,4S)-4-Hydroxyproline *see* 4878

trans-4-Hydroxyproline *see* 4878

(4R)-4-Hydroxy-L-proline *see* 4878

m-Hydroxypropadrine *see* 6003

1-Hydroxypropane *see* 7955

2-Hydroxypropanedioic Acid *see* 9207

5,5′-[(2-Hydroxy-1,3-propanediyl)bis-(acetylimino)]bis[N,N′-bis(2,3-dihydroxypropyl)-2,4,6-triiodo-1,3-benzenedicarboxamide] *see* 5066

5,5′-[(2-Hydroxy-1,3-propanediyl)bis-(oxy)]bis[4-oxo-4H-1-benzopyran-2-carboxylic Acid] *see* 2581

3-Hydroxypropanenitrile *see* 3848

2-Hydroxy-2-propanesulfonic Acid Sodium Salt (1:1) *see* 68

2-Hydroxy-1,2,3-propanetricarboxylic Acid *see* 2325

2-Hydroxy-1,2,3-propanetricarboxylic Acid Ammonium Iron(3+) Salt (1:?:?) *see* 4044

2-Hydroxy-1,2,3-propanetricarboxylic Acid Ammonium Salt (1:2) *see* 509

2-Hydroxy-1,2,3-propanetricarboxylic Acid Bismuth(3+) Potassium Salt (2:1:3) *see* 2469

2-Hydroxy-1,2,3-propanetricarboxylic Acid Calcium Salt (2:3) *see* 1663

2-Hydroxy-1,2,3-propanetricarboxylic Acid Copper(2+) Salt (1:2) *see* 2626

2-Hydroxy-1,2,3-propanetricarboxylic Acid Gallium Salt (1:1) *see* 4379

2-Hydroxy-1,2,3-propanetricarboxylic Acid Iron(2+) Salt (1:?) *see* 4071

2-Hydroxy-1,2,3-propanetricarboxylic Acid Iron(3+) Salt (1:?) *see* 4048

2-Hydroxy-1,2,3-propanetricarboxylic Acid Lithium Salt (1:3) *see* 5586

2-Hydroxy-1,2,3-propanetricarboxylic Acid Magnesium Salt (1:?) *see* 5727

2-Hydroxy-1,2,3-propanetricarboxylic Acid Mixt. with Cyanamide Calcium Salt (1:1) *see* 1665

2-Hydroxy-1,2,3-propanetricarboxylic Acid Potassium Salt (1:1) *see* 7745

2-Hydroxy-1,2,3-propanetricarboxylic Acid Potassium Salt (1:3) *see* 7744

2-Hydroxy-1,2,3-propanetricarboxylic Acid Silver(1+) Salt (1:3) *see* 8650

2-Hydroxy-1,2,3-propanetricarboxylic Acid Sodium Salt (1:3) *see* 8737

2-Hydroxy-1,2,3-propanetricarboxylic Acid 1,2,3-Tributyl Ester *see* 1565

2-Hydroxy-1,2,3-propanetricarboxylic Acid Zinc Salt (2:3) *see* 10333

[2-(Hydroxy-κO)propanoato-κO]silver *see* 8656

2-Hydroxypropanoic Acid *see* 5384

(2R)-2-Hydroxypropanoic Acid *see* 5383

(2S)-2-Hydroxypropanoic Acid *see* 5385

3-Hydroxypropanoic Acid *see* 4801

2-Hydroxy-propanoic Acid Ammonium Salt (1:1) *see* 527

2-Hydroxypropanoic Acid Calcium Salt (2:1) *see* 1680

2-Hydroxypropanoic Acid 1-Carboxyethyl Ester *see* 5387

2-Hydroxypropanoic Acid Ethyl Ester *see* 3872

2-Hydroxypropanoic Acid Hexadecyl Ester *see* 2029

2-Hydroxypropanoic Acid Iron(2+) Salt (2:1) *see* 4077

2-Hydroxypropanoic Acid Magnesium Salt *see* 5736

2-Hydroxypropanoic Acid Methyl Ester *see* 6166

2-Hydroxypropanoic Acid Silver Salt *see* 8656

2-Hydroxypropanoic Acid Sodium Salt (1:1) *see* 8767

1-Hydroxy-2-propanone *see* 64

2-(2-Hydroxypropanoyloxy)propanoic Acid *see* 5387

4-Hydroxy-1-propenylbenzene *see* 670

4-[(1E)-3-Hydroxy-1-propen-1-yl]-2,6-dimethoxyphenyl-β-D-glucopyranoside *see* 9146

4-(3-Hydroxypropenyl)-2,6-dimethoxyphenyl-D-glucoside *see* 9146

(17β)-17-Hydroxy-17-(2-propen-1-yl)estra-4,9,11-trien-3-one *see* 315

4-(3-Hydroxy-1-propen-1-yl)-2-methoxyphenol *see* 2488

4-(3-Hydroxy-1-propen-1-yl)-2-methoxyphenyl β-Glucopyranoside *see* 2487

α-Hydroxypropionic Acid *see* 5384

β-Hydroxypropionic Acid *see* 4801

β-Hydroxypropionitrile *see* 3848

5-(α-Hydroxypropionylamino)-2,4,6-triiodoisophthalic Acid Di(1,3-dihydroxyisopropylamide) *see* 5100

p-Hydroxypropiophenone *see* 7149

4′-Hydroxypropiophenone *see* 7149

2β-(3-Hydroxypropoxy)-1α,25-dihydroxyvitamin D₃ *see* 3587

(1R,2R,3R,5Z)-2-(3-Hydroxypropoxy)-4-methylene-5-[(2E)-2-[(1R,3aS,7aR)-octahydro-1-[(1R)-5-hydroxy-1,5-dimethylhexyl]-7a-methyl-4H-inden-4-ylidene]ethylidene]-1,3-cyclohexanediol *see* 3587

(5Z,7E)-2β-(3-Hydroxypropoxy)-9,10-secocholesta-5,7,10(19)-triene-1α,-3β,25-triol *see* 3587

1-[2-[2-Hydroxy-3-(propylamino)propoxy]phenyl]-3-phenyl-1-propanone *see* 7908

2′-[2-Hydroxy-3-(propylamino)propoxy]-3-phenylpropiophenone *see* 7908

α-Hydroxypropylbenzene *see* 3823

Hydroxypropyl Cellulose, 4879

3-Hydroxypropylene Oxide *see* 4525

N-(2-Hydroxypropyl)ethylenediamine *see* 435

3-Hydroxypropyl-*N*-methionylcolony-stimulating Factor (Human) 1-Ether with *α*-Methyl-*ω*-hydroxypoly(oxy-1,2-ethanediyl) *see* 4573

6-[(2-Hydroxypropyl)methylamino]-3(2*H*)-pyridazinone Hydrazone *see* 7535

Hydroxypropyl Methylcellulose, 4880

1-(2-Hydroxypropyl)-2-methyl-5-nitro-imidazole *see* 8556

2-(*α*-Hydroxypropyl)piperidine *see* 2484

5-Hydroxy-2-propylpiperidine *see* 8022

5-(2-Hydroxypropyl)-5-(2-propen-1-yl)-2,4,6(1*H*,3*H*,5*H*)-pyrimidinetrione *see* 8013

[10-[2-(Hydroxy-*κO*)propyl]-1,4,7,10-tetraazacyclododecane-1,4,7-triaceta-to(3−)-*κN*¹,*κN*⁴,*κN*⁷,*κN*¹⁰,*κO*¹,*κO*⁴,-*κO*⁷]gadolinium *see* 4358

10-(2-Hydroxypropyl)-1,4,7,10-tetraaza-cyclododecane-1,4,7-triacetic Acid Gadolinium Complex *see* 4358

7-(2-Hydroxypropyl)theophylline *see* 8014

(−)-5-Hydroxy-2[*N*-*n*-propyl-*N*-2-(2-thi-enyl)ethylamino]tetralin *see* 8404

(2-Hydroxypropyl)trimethylammonium Chloride Acetate *see* 6012

(2-Hydroxypropyl)trimethylammonium Chloride Carbamate *see* 1187

4-Hydroxypyrazolo[3,4-*d*]pyrimidine *see* 271

8-Hydroxy-1,3,6-pyrenetrisulfonic Acid Sodium Salt (1:3) *see* 8065

1-Hydroxy-2(1*H*)-pyridinethione *see* 8107

2-Hydroxy-5-[2-[4-[(2-pyridinylamino)-sulfonyl]phenyl]diazenyl]benzoic Acid *see* 9073

P,*P*′-[1-Hydroxy-2-(3-pyridinyl)ethyli-dene]bisphosphonic Acid *see* 8360

β-[*N*-(3-Hydroxy-4-pyridone)]-*α*-amino-propionic Acid *see* 6280

1-Hydroxy-5*H*-pyrido[3,2-*a*]phenoxazin-5-one-3-carboxylic Acid *see* 7603

5-(*α*-Hydroxy-*α*-2-pyridylbenzyl)-7-(*α*-2-pyridylbenzylidene)-5-norbornene-2,3-dicarboximide *see* 6775

3(*S*)-*N*-Hydroxy-4-((4-((pyrid-4-yl)oxy)-benzenesulfonyl)-2,2-dimethyl)tetrahy-dro-2*H*-1,4-thiazine-3-carboxamide *see* 7869

4-Hydroxy-2(1*H*)-pyrimidinethione *see* 9520

2-Hydroxy-4(1*H*)-pyrimidinone *see* 10035

2-Hydroxy-4(3*H*)-pyrimidinone *see* 10035

4-Hydroxy-2(1*H*)-pyrimidinone *see* 10035

α-Hydroxypyrotartaric Acid *see* 2322

4(*R*)-Hydroxy-2(*S*)-pyrrolidinecarboxylic Acid *see* 4878

1-Hydroxy-2,5-pyrrolidinedione *see* 4884

2-(4-Hydroxypyrrolidin-2-on-1-yl)acet-amide *see* 7041

(1*R*,2*R*)-2-[(3*R*)-Hydroxypyrrolidinyl]-1-(3,4-dimethoxyphenethoxy)cyclohex-ane *see* 10157

N-[(1*S*)-2-[(3*S*)-3-Hydroxy-1-pyrrolidin-yl]-1-phenylethyl]-*N*-methyl-*α*-phenyl-benzeneacetamide *see* 823

4-Hydroxyquinaldic Acid *see* 5375

Hydroxyquinol *see* 1075

2-Hydroxyquinoline *see* 1826

8-Hydroxyquinoline, 4881

4-Hydroxy-2-quinolinecarboxylic Acid *see* 5375

8-Hydroxy-5,7-quinolinedisulfonic Acid Copper Derivative Compound with Diethylamine *see* 2660

8-Hydroxy-5-quinolinesulfonic Acid, 4882

3-Hydroxyquinuclidine *see* 8199

all-trans-*N*-4′-Hydroxyretinanilide *see* 4025

14-Hydroxy-3*β*-(rhamnosyloxy)bufa-4,-20,22-trienolide *see* 7985

5-Hydroxy-1-*β*-D-ribofuranosyl-1*H*-imid-azole-4-carboxamide *see* 6308

9-Hydroxyrisperidone *see* 7087

9-Hydroxyrisperidone Palmitate *see* 7087

4′-Hydroxysalicylanilide *see* 6984

5-Hydroxysalicylic Acid *see* 4433

8-Hydroxysantonin *see* 806

3-Hydroxy-16,17-secoestra-1,3,5(10)-trien-17-oic Acid *see* 3457

12-Hydroxysenecionan-11,16-dione *see* 8590

[15(20)*E*]-12-Hydroxysenecionan-11,16,-dione *see* 8590

3*α*-Hydroxyspiro[1*αH*,5*αH*-nortropane-8,1′-pyrrolidinium] Chloride Benzilate *see* 9966

(3*β*,5*α*,25*R*)-3-Hydroxyspirostan-12-one *see* 4657

(2*α*,3*β*,5*α*,25*R*)-2-Hydroxyspirostan-3-yl *O*-*β*-D-Glucopyranosyl-(1 → 2)-*O*-[*β*-D-xylopyranosyl-(1 → 3)]-*O*-*β*-D-gluco-pyranosyl-(1 → 4)-*β*-D-galactopyrano-side *see* 4467

2-Hydroxystilbamide *see* 4883

Hydroxystilbamidine, 4883

2-Hydroxy-4,4′-stilbenedicarboxamidine *see* 4883

(*E*)-5-(*p*-Hydroxystyryl)resorcinol *see* 8279

Hydroxysuccinic Acid *see* 5771

N-Hydroxysuccinimide, 4884

2-Hydroxy-5-sulfobenzoic Acid *see* 9094

5-Hydroxy-4-[(4-sulfo-1-naphthalenyl)-azo]-1-(4-sulfophenyl)-1*H*-pyrazole-3-carboxylic Acid 3-Ethyl Ester Di-sodium Salt *see* 6954

3-Hydroxy-4-[2-(4-sulfo-1-naphthalenyl)-diazenyl]-2,7-naphthalenedisulfonic Acid Sodium Salt (1:3) *see* 369

2-[[*α*-(2-Hydroxy-5-sulfophenylazo)-ben-zylidene]-hydrazino]-benzoic Acid *see* 10345

6-Hydroxy-5-[2-(4-sulfophenyl)diazenyl]-2-naphthalenesulfonic Acid Sodium Salt (1:2) *see* 9130

2-[2-[[2-(2-Hydroxy-5-sulfophenyl)di-azenyl]phenylmethylene]hydrazinyl]-benzoic Acid *see* 10345

o-[1-(2-Hydroxy-5-sulfophenyl)-3-phen-yl-5-formazane]benzoic Acid *see* 10345

2-[5-(2-Hydroxy-5-sulfophenyl)-3-phen-yl-1-formazyl]benzoic Acid *see* 10345

5-Hydroxy-1-(*p*-sulfophenyl)-4-[(*p*-sulfo-phenyl)azo]pyrazole-3-carboxylic Acid Trisodium Salt *see* 9206

3-Hydroxytamoxifen *see* 3495

4-Hydroxy-TEMPO *see* 9284

Hydroxytetracaine, 4885

Hydroxytetracycline *see* 7075

1-Hydroxytetradecane *see* 6421

(+)-(3*S*,4*S*)-7-Hydroxy-Δ⁶-tetrahydrocan-nabinol-1,1-dimethylheptyl *see* 2946

5-Hydroxy-3-[4-(1,2,3,6-tetrahydro-4-phenyl-1-pyridyl)butyl]indole *see* 8408

4-Hydroxy-3-[1,2,3,4-tetrahydro-3-[4-(4-trifluoromethylbenzyloxy)phenyl]-1-naphthyl] *see* 4134

4-Hydroxy-3-[1,2,3,4-tetrahydro-3-[4-[[4-(trifluoromethyl)phenyl]methoxy]phen-yl]-1-naphthalenyl]-2*H*-1-benzopyran-2-one *see* 4134

(±)-5-[4-[(6-Hydroxy-2,5,7,8-tetrameth-ylchroman-2-yl)methoxy]benzyl]-2,4-thiazolidinedione *see* 9948

4-Hydroxy-2,2,6,6-tetramethyl Piperi-dine *N*-Oxide *see* 9284

4-Hydroxy-2,2,6,6-tetramethylpiperi-dinooxy *see* 9284

4-Hydroxy-2,2,6,6-tetramethyl-1-piperi-dinyloxy *see* 9284

α-Hydroxytoluene *see* 1127

o-Hydroxytoluene *see* 2564

α-Hydroxy-*α*-toluic Acid *see* 5781

2-Hydroxy-*m*-toluic Acid *see* 2570

2-Hydroxy-*p*-toluic Acid *see* 2569

6-Hydroxy-*m*-toluic Acid *see* 2571

2-(*m*-Hydroxy-*N*-*p*-tolylanilinomethyl)-2-imidazoline *see* 7376

ω-(4-Hydroxy-4-*p*-tolylpiperidino)-*p*-fluorobutyrophenone *see* 6351

2-Hydroxy-5-tosyloxybenzenesulfonic Acid *see* 9124

[Hydroxy(tosyloxy)iodo]benzene *see* 5369

1-Hydroxytriacontane *see* 9753

3-Hydroxy-3*H*-1,2,3-triazolo[4,5-*b*]pyri-dine *see* 4848

β-Hydroxytricarballylic Acid *see* 2325

(2*S*)-1-[2-[(3-Hydroxytricyclo[3.3.1.1³,⁷]-dec-1-yl)amino]acetyl]-2-pyrrolidine-carbonitrile *see* 10177

2-Hydroxytriethylamine *see* 3123

ω-[4-Hydroxy-4-(*m*-trifluoromethylphen-yl)piperidino]-*p*-fluorobutyrophenone *see* 9852

7-Hydroxy-3-(3,4,5-trihydroxyphenyl)-4*H*-1-benzopyran-4-one *see* 953

1-Hydroxy-2,10,11-trimethoxyaporphine *see* 2533

11-Hydroxy-1,2,10-trimethoxyaporphine *see* 5207

3-Hydroxy-2,9,10-trimethoxyberbine *see* 2536

5-Hydroxy-6,7,8-trimethoxy-2-(3,4,5-tri-methoxyphenyl)-4*H*-1-benzopyran-4-one *see* 4399

(−)-*β*-Hydroxy-*γ*-trimethylaminobutyric Acid *see* 1849

6-Hydroxy-*β*,2,7-trimethyl-5-benzofu-ranacrylic Acid *δ*-Lactone *see* 9907

2′-Hydroxy-2,5,9-trimethyl-6,7-benzo-morphan *see* 6005

3-Hydroxy-1,7,7-trimethylbicyclo[2.2.1]-heptan-2-one *see* 4856

N-[[[(1*S*,2*S*,3*R*,4*R*)-3-Hydroxy-4,7,7-tri-methylbicyclo[2.2.1]hept-2-yl]amino]-carbonyl]-4-methylbenzenesulfon-amide *see* 4473

4-Hydroxy-*α*,*α*,4-trimethylcyclohexane-methanol *see* 9311

5-Hydroxy-*α*,*α*,4-trimethyl-3-cyclohex-ene-1-methanol *see* 8705

(2E,4E,6E,8E,10E,12E,14E,16E)-17-
[(4R)-4-Hydroxy-2,6,6-trimethyl-1-
cyclohexen-1-yl]-2,6,11,15-tetrameth-
yl-2,4,6,8,10,12,14,16-heptadecaoc-
taenal see 2323
5,5′-[(2-Hydroxytrimethylene)bis(acetylimino)]bis[N,N′-bis(2,3-dihydroxypro-
pyl)-2,4,6-triiodoisophthalamide] see
5066
(2-Hydroxytrimethylene)bis[trimethyl-
ammonium] Iodide see 7897
5,5′-[(2-Hydroxytrimethylene)dioxy]-
bis(4-oxo-4H-1-benzopyran-2-carbox-
ylic Acid) see 2581
5,5′-(2-Hydroxytrimethylenedioxy)bis(4-
oxochromene-2-carboxylic Acid) see
2581
2-Hydroxy-N,N,N-trimethylethanami-
nium see 2211
2-Hydroxy-N,N,N-trimethylethanami-
nium 2-[4-(4-Chlorobenzoyl)phenoxy]-
2-methylpropanoate (1:1) see 4009
2-Hydroxy-N,N,N-trimethylethanami-
nium 2-Hydroxybenzoate (1:1) see
2213
2-Hydroxy-N,N,N-trimethylethanami-
nium Salt with 3,9-Dihydro-1,3-di-
methyl-1H-purine-2,6-dione (1:1) see
2215
2-Hydroxy-N,N,N-trimethylethanami-
nium Salt with 2-Hydroxybenzoic
Acid (1:1) see 2213
2-Hydroxy-N,N,N-trimethylethanami-
nium (OC-6-44)-Triaqua[2-(hydroxy-
κO)-1,2,3-propanetricarboxylato(4−)-
κO¹,κO²]ferrate(1−) (1:1) see 4065
(2Z,4E)-5-[(1S)-1-Hydroxy-2,6,6-trimeth-
yl-4-oxo-2-cyclohexen-1-yl]-3-methyl-
2,4-pentadienoic Acid see 12
(7R)-4-Hydroxy-N,N,N-trimethyl-10-
oxo-7-[(1-oxohexadecyl)oxy]-3,5,9-
trioxa-4-phosphapentacosan-1-ami-
nium Inner Salt 4-Oxide see 2461
(7R)-4-Hydroxy-N,N,N-trimethyl-10-
oxo-7-[(1-oxohexadecyl)oxy]-3,5,9-
trioxa-4-phosphapentacosan-1-ami-
nium, Inner Salt, 4-Oxide Mixture
with Formaldehyde Polymer with Oxirane and 4-(1,1,3,3-Tetramethylbu-
tyl)phenol and 1-Hexadecanol see
3957
(5Z,11α,13E,15R)-15-Hydroxy-11,16,16-
trimethyl-9-oxoprosta-5,13-dien-1-oic
Acid see 9895
1-(4-Hydroxy-2,3,5-trimethylphenoxy)-3-
(isopropylamino)-2-propanol 4-Ace-
tate see 6216
3-Hydroxy-4-[2-(2,4,5-trimethylphenyl)-
diazenyl]-2,7-naphthalenedisulfonic
Acid Sodium Salt (1:2) see 7712
11β-Hydroxy-16α,17α,21-trimethylpreg-
na-1,4-diene-3,20-dione see 8352
9-Hydroxy-2,5,11-trimethyl-6H-pyrido-
[4,3-b]carbazolium Acetate (1:1) see
3604
1-Hydroxy-2,4,6-trinitrobenzene see
7522
Hydroxytriphenylmethane see 9912
Hydroxytriphenylstannane see 9920
Hydroxytriphenyltin see 9920
3β-Hydroxy-1αH,5αH-tropane-2β-car-
boxylic Acid see 3540
3β-Hydroxy-1αH,5αH-tropane-2β-car-
boxylic Acid Benzoate see 1116

3β-Hydroxy-1αH,5αH-tropane-2α-car-
boxylic Acid Methyl Ester Benzoate
see 8020
3β-Hydroxy-1αH,5αH-tropane-2β-car-
boxylic Acid Methyl Ester Benzoate
see 2440
2-Hydroxytropone see 9963
5-Hydroxytryptamine see 8601
5-Hydroxytryptophan, 4886
3-Hydroxytyramine see 3470
3-Hydroxytyrosine see 3469
3-Hydroxy-L-tyrosine see 5516
Hydroxytyrosol, 4887
Hydroxyurea, 4888
(3β)-3-Hydroxyurs-12-ene-27,28-dioic
Acid see 8192
(3β)-3-Hydroxyurs-12-en-28-oic Acid see
10075
(3α,4β)-3-Hydroxyurs-12-en-23-oic Acid
see 1358
17-Hydroxy-17α-vinyl-4-estren-3-one see
6806
17β-Hydroxy-17α-vinylestr-5(10)-en-3-
one see 6791
1α-Hydroxyvitamin D₂ see 3484
1α-Hydroxyvitamin D₃ see 4858
1β-Hydroxyvitamin D₃ see 4858
25-Hydroxyvitamin D₃ see 1641
8-Hydroxyxanthine see 10060
1-Hydroxy-2-[(6-O-β-D-xylopyranosyl-β-
D-glucopyranosyl)oxy]-9,10-anthra-
cenedione see 8413
(3β,16α,17α)-17-Hydroxyyohimban-16-
carboxylic Acid Methyl Ester see 8031
(16α,17α)-17-Hydroxyyohimban-16-car-
boxylic Acid Methyl Ester see 10300
(16β,17α)-17-Hydroxyyohimban-16-car-
boxylic Acid Methyl Ester see 2535
(16α,17α,20α)-17-Hydroxyyohimban-16-
carboxylic Acid Methyl Ester see
10302
(16β,17α,20α)-17-Hydroxyyohimban-16-
carboxylic Acid Methyl Ester see
10301
Hydroxyzine, 4889
Hydrozincite see 10330
Hygric Acid Methylbetaine see 8899
Hygrine, 4890
Hygromix [Elanco] see 4892
Hygromycin, 4891
Hygromycin A see 4891
Hygromycin B, 4892
Hygrophylline, 4893
Hygroton [Novartis] see 2200
Hykinone [Abbott] see 5899
Hylorel [Aventis] see 4594
Hymecromone, 4894
Hymecromone O,O-Diethyl Phosphoro-
thioate see 7724
Hymenidin see 6971
Hymexazol, 4895
Hyndarin see 9360
Hyodeoxycholic Acid, 4896
α-Hyodeoxycholic Acid see 4896
Hyodesoxycholic Acid see 4896
Hyonate [Bayer] see 4796
Hyoscine see 8543
Hyoscine-N-butyl Bromide see 1591
Hyoscine Methyl Bromide see 6077
Hyoscyamine, 4897
dl-Hyoscyamine see 866
l-Hyoscyamine see 4897
Hyoscyamus, 4898
Hyp see 4878
Hypadil [Kowa] see 6648

Hypalon [DuPont], 4899
Hypaphorine, 4900
Hypaque Meglumine [GE Healthcare] see
2993
Hypaque Sodium see 2993
Hyperdrol [Britcair] see 339
Hyperforin, 4901
Hyperglycemic-Glycogenolytic Factor
see 4481
Hyperici Herba see 4903
Hypericin, 4902
Hypericum, 4903
Hypericum Red see 4902
Hyperin see 8150
Hyperium [Biopharma] see 8347
Hyperol [Gedeon Richter] see 1784
Hyperoside see 8150
Hypersin [Zeria] see 1188
Hyperstat [Schering-Plough] see 3007
Hypertensin see 644
Hypertensin [Novartis] see 644
Hypertensinogen see 644
Hyphylline see 3526
Hypnodil [Janssen] see 6225
Hypnodin [Takeda] see 7293
Hypnomidate [Janssen-Cilag] see 3926
Hypnone see 71
Hypnorex [Sanofi-Synthelabo] see 5583
Hypnovel [Roche] see 6261
"Hypo" see 8821
Hypoca [Yamanouchi] see 998
Hypochlorous Acid, 4904
Hypochlorous Acid Calcium Salt (2:1)
see 1676
Hypochlorous Acid 1,1-Dimethylethyl
Ester see 1574
Hypochlorous Acid Sodium Salt (1:1)
see 8762
Hypochlorous Anhydride see 2099
Hypochylin [Recip] see 4505
Hypocretin-1 see 6961
Hypocretin-2 see 6961
Hypocretins see 6961
Hypocrine [Mitsubishi Tanabe] see 4561
Hypoglycin see 4905
Hypoglycin A see 4905
Hypoglycine A, 4905
Hypoglycine B see 4905
Hyponitric Acid Disodium Salt see 641
Hyponitrous Acid Anhydride see 6740
β-Hypophamine see 10129
Hypophosphoric Acid, 4906
Hypophosphorous Acid, 4907
Hypophyseal Growth Hormone see 8842
Hypostamine [Promedica] see 9942
Hypostat [Herbrand] see 7149
Hyposulfite of Gold and Sodium see
4554
Hypovase [Pfizer] see 7838
Hypoxanthine, 4908
Hypoxanthine Riboside see 5018
Hypoxanthine Riboside-5-phosphoric
Acid see 5020
Hypoxanthosine see 5018
Hyprenan [Astra] see 7853
Hypromellose see 4880
Hypurin [CP Pharm.] see 5023
Hypusine, 4909
Hyrazin [Kowa] see 9453
Hyryl see 4123
Hyskon [Pharmacia] see 2950
Hysron [Kyowa] see 5864
Hytakerol [Winthrop] see 3198
Hytone [Dermik] see 4824
Hytracin [Dainabot-Dainippon] see 9297

IL-18 *see* 1884
IL-17803A *see* 20
^{123}I Labeled IMP *see* 5092
Ilaris [Novartis] *see* 1739
Ildamen [Viatris] *see* 7061
Ile *see* 5225
α-Ile *see* 5225
Ileogastrone *see* 3651
Iletin II [Lilly] *see* 5023
Iliadin [Meda] *see* 7065
Ilimaquinone, 4939
Iliren [Provet] *see* 9578
Ilixathin *see* 8438
Illcut [Nippon Zoki] *see* 9660
Illoxan [Bayer CropSci.] *see* 3092
Illudin M *see* 4940
Illudins, 4940
Illudin S *see* 4940
Ilmenite *see* 9628
ILO-522 *see* 4941
Ilodecakin *see* 5043
Ilomedin [Schering AG] *see* 4942
Ilopan [Warren-Teed] *see* 2949
Iloperidone, 4941
Iloprost, 4942
Ilosone [Lilly] *see* 3740
Ilotycin [Lilly] *see* 3739
Ilvin [Merck KGaA] *see* 1453
^{123}I-M123 *see* 5092
Imadyl [Roche] *see* 1862
Imagent [Alliance] *see* 7270
Imagopaque [GE Healthcare] *see* 5102
Imagotan [Sandoz] *see* 9093
Imakol [Rhône-Poulenc] *see* 7045
Imap [McNeil] *see* 4237
Imatinib, 4943
Imaverol [Janssen] *see* 3637
Imavist [Schering AG] *see* 7271
Imazalil *see* 3637
Imazamethabenz, 4944
Imazamethabenz Methyl *see* 4944
Imazamox, 4945
Imazapyr, 4946
Imazaquin, 4947
Imazethabenz *see* 4944
Imazethapyr, 4948
Imazosulfuron, 4949
Imbun [Merckle] *see* 4919, 7071
[(Im-Bzl)-D-His⁶,Pro⁹-NEt]-gonadotro-
   pin-releasing Hormone *see* 4760
IMC-11F8 *see* 6520
IMC-C225 *see* 2026
Imdur [Key] *see* 5271
Imecromone *see* 4894
Imeron [Bracco] *see* 5097
Imeson [Taurus] *see* 6661
IMET-3393 *see* 1036
Imex [Merz] *see* 9341
Imexim [Cimex] *see* 9050
Imferon [Fisons] *see* 5138
IMI-28 *see* 3678
IMI-30 *see* 4927
Imibenconazole, 4950
Imidacloprid, 4951
Imidamine *see* 672
Imidan [Gowan] *see* 7448
Imidapril, 4952
Imidaprilat *see* 4952
Imidazole, 4953
1*H*-Imidazole *see* 4953
1*H*-Imidazole-1-carbonitrile *see* 2686
2,4-(3*H*,5*H*)-Imidazoledione *see* 4799
1*H*-Imidazole-5-ethanamine *see* 4756
4-Imidazoleethylamine *see* 4756
5-Imidazoleethylamine *see* 4756

Imidazole Salicylate, 4954
Imidazoletrione *see* 7126
2,4-Imidazolidinedione *see* 4799
2-Imidazolidinethione *see* 3858
2,4,5-Imidazolidinetrione *see* 7126
2-Imidazolidinone, 4955
D-α-[(Imidazolidin-2-on-1-yl)carbonyl-
   amino]benzylpenicillin *see* 908
Imidazolidinyl Urea 4957
2-Imidazolidone *see* 4955
Imidazoline-2-thiol *see* 3858
1-(2-Δ²-Imidazolinyl)-2,2-diphenylcyclo-
   propane *see* 2272
N-(2-Imidazolin-2-yl)-*N*-(4-indanyl)-
   amine *see* 4972
β-(4-Imidazolyl)acrylcholine *see* 6392
3-(Imidazol-1-yl)-7-chlorobenzo-1,2,4-tri-
   azine-1-oxide *see* 9768
α-Imidazol-4-yl-2,3-cresotamide *see* 6306
2-(4-Imidazolyl)ethylamine *see* 4756
2-(Imidazol-1-yl)-1-hydroxyethane-1,1-
   diphosphonic Acid *see* 10388
(*E*)-4-(Imidazol-1-ylmethyl)cinnamic
   Acid *see* 7079
(2*E*)-3-[4-(1*H*-Imidazol-1-ylmethyl)phen-
   yl]-2-propenoic Acid *see* 7079
2-[[3-(1*H*-Imidazol-4-yl)-1-oxo-2-propen-
   yl]oxy]-*N*,*N*,*N*-trimethylethanami-
   nium *see* 6392
7*H*-Imidazo[4,5-*d*]pyrimidine *see* 8052
Imidocarb, 4956
Imidodicarbonic Diamide *see* 1310
Imidodicarbonimidic Diamide *see* 1218
Imidol [Mitsubishi] *see* 4962
Imidole *see* 8128
Imidurea, 4957
ImIFN *see* 5036
Imiglucerase *see* 4958, 4958
Imigran [GSK] *see* 9126
Imiject [GSK] *see* 9126
Iminazole *see* 4953
1,1′-Iminobis[4-amino-9,10-anthracene-
   dione] *see* 1247
1,1′-Iminobis-9,10-anthracenedione *see*
   681
2,2′-Iminobis[benzoic Acid] *see* 3348
2,2′-Iminobisethanol *see* 3118
rel-(α*R*,α′*R*,2*R*,2′*S*)-α,α′-[Iminobis-
   (methylene)]bis[6-fluoro-3,4-dihydro-
   2*H*-1-benzopyran-2-methanol] *see*
   6516
1,1′-Iminobis-2-propanol *see* 3214
1,1′-Iminodiacetic Acid, 4959
1,1′-Iminodianthraquinone *see* 681
2,2′-Iminodiethanol *see* 3118
α,α′-(Iminodimethylene)bis[6-fluoro-2-
   chromanmethanol] *see* 6516
Iminodisuccinic Acid, 4960
2,2′-Iminodisuccinic Acid *see* 4960
*N*⁵-[Imino(methylamino)methyl]-L-orni-
   thine *see* 6096
3-Imino-7-(methylamino)-3*H*-phenothia-
   zine Hydrochloride *see* 922
N-[Imino(phosphonoamino)methyl]-*N*-
   methylglycine *see* 7452
β-(5-Imino-2-pyrrolidinecarboxamido)-
   propamidine *see* 6757
2-(2-Imino-4,5,6,7-tetrahydrobenzothia-
   zol-3-yl)-1-*p*-tolylethanone Hydrobromide *see* 7533
(6*R*,7*R*)-7-[2-(2-Imino-4-thiazolyl)glyox-
   ylamido]-8-oxo-5-thia-1-azabicyclo-
   [4.2.0]oct-2-ene-2-carboxylic Acid 7²-
   (*Z*)-(*O*-Methyloxime) *see* 1952

Iminourea *see* 4597
Imipem [Neopharmed] *see* 4961
Imipemide *see* 4961
Imipenem, 4961
Imipramine, 4962
Imipramine *N*-Oxide, 4963
Imiprex [Dumex] *see* 4963
Imiquimod, 4964
Imisopasem Manganese, 4965
Imitrex [GSK] *see* 9126
Imizad Equine Injection [Burroughs
   Wellcome] *see* 4956
Imizin *see* 4962
Imizocarb *see* 4956
Imizol *see* 4956
Immenoctal [Houdé] *see* 8557
Immobilon [Reckitt & Colman] *see* 3931
IMMU-hLL2 *see* 3687
Immukin [Boehringer, Ing.] *see* 5036
Immune IFN *see* 5036
Immune Protein P5 *see* 867
Immuneron [Biogen] *see* 5036
Immunoglobulin G1 (Synthetic Human
   Fc Domain Fragment) Fusion Protein
   with Angiopoietin 1/Angiopoietin 2-
   Binding Peptide (Synthetic) *see* 9741
Immunox [Cilag] *see* 9558
Immunozima [Salus] *see* 5701
ImmuRAID-MN3-Tc-99m *see* 9027
IMMU-4 ⁹⁹ᵐTc Fab′ *see* 764
Imnudorm *see* 5913
Imodium [Janssen] *see* 5628
Imosec [Janssen] *see* 5628
Imossel [Janssen-Solucal] *see* 5628
Imotryl [Cassenne] *see* 1125
Imovane [Aventis] *see* 10393
IMP *see* 5020, 5092, 8773
IMPA, 4966
Impact [Cheminova] *see* 4242
Impavido [Zentaris] *see* 6279
Impedex [DuPont] *see* 8798
Imperacin [AstraZeneca] *see* 7075
Imperan [Millet] *see* 6219
Imperatorin, 4967
Imperial Green *see* 2615
Imperialine, 4968
Imperial Purple *see* 3028
Impetreve [Merck KGaA] *see* 2276
Implanon [Organon] *see* 3928
Importal [Novartis] *see* 5389
Impral [Tanabe Seiyaku] *see* 2264
Imprelis [DuPont] *see* 432
Impromen [Janssen] *see* 1452
Improntal [Rottapharm] *see* 7619
Impruvol *see* 1550
Impugan [Dumex] *see* 4338
Impulse [Bayer] *see* 8889
Imtack [AstraZeneca] *see* 5271
Imunace [Shionogi] *see* 5038
Imunoviral [Newport] *see* 5019
Imuran [GSK] *see* 895
Imurek [GSK] *see* 895
Imurel [GSK] *see* 895
Imuthiol *see* 3421
IN-60 *see* 5305
IN-511 *see* 7432
IN-A8947 *see* 905
Inacid [Merck & Co.] *see* 5009
Inactin [Byk Gulden] *see* 9476
Inactive Limonene *see* 5546
Inadine [J & J] *see* 7815
Inalone [Lampugnani] *see* 1017
Inapetyl [Pharmacia & Upjohn] *see* 1121
Inapsine [McNeil] *see* 3499
Inavir [Daiichi Sankyo] *see* 5406

INA-X14 see 5025
Incadronate Sodium see 4969
Incadronic Acid, 4969
Incafolic see 4248
Incazan [USSR] see 6231
INCB-018424 see 8440
Incel [Vertex] see 1244
Incidal see 5839
Incital [Fabre Santé] see 5870
Incivek [Vertex] see 9256
Incivo [Janssen-Cilag] see 9256
Incoran [ITA] see 7855
Incorporation Factor see 4520
Incostop [SLA Pharma] see 7398
Increlex [Tercica] see 5844
^{111}In-CYT-103 see 8521
Indacaterol, 4970
Indacin [Merck & Co.] see 5009
Indaflex [Lampugnani] see 4975
Indalapril see 2881
Indamol [Aventis] see 4975
Indan, 4971
Indanazoline, 4972
Indanofan, 4973
Indanthrene, 4974
Indanthrene Blue R see 4974
Indanthrene Brilliant Blue FF see 4974
1,2,3-Indantrione Monohydrate see 6640
1-(5-Indanyl) N-(2-Carboxy-3,3-dimeth-yl-7-oxo-4-thia-1-azabicyclo[3.2.0]-hept-6-yl)-2-phenylmalonamate see 1839
O-(5-Indanyl) m,N-Dimethylthiocarbani-late see 9675
α-(5-Indanyloxycarbonyl)benzylpenicil-lin see 1839
(S)-cis-4-[1-[2-(5-Indanyloxycarbonyl)-3-(2-methoxyethoxy)propyl]-1-cyclopen-tanecarboxamido]-1-cyclohexanecar-boxylic Acid see 1745
6-[2-(5-Indanyloxycarbonyl)phenylacet-amido]-3,3-dimethyl-7-oxo-4-thia-1-azabicyclo[3.2.0]heptane-2-carboxylic Acid see 1839
Indapamide, 4975
Indar [Dow AgroSci.] see 3995
1H-Indazole, 4976
Indecainide, 4977
Indema see 7348
Indene, 4978
1H-Indene see 4978
Indenolol, 4979
(±)-1-[Inden-4(or 7)-yloxy]-3-(isopropyl-amino)-2-propanol see 4979
1-[1H-Inden-4(or 7)-yloxy]-3-[(1-methyl-ethyl)amino]-2-propanol see 4979
Inderal [Wyeth] see 7953
Inderite see 5722
Inderm [Dermapharm] see 3739
Indermil [Loctite] see 1566
Indian Aconite see 112
Indian Apple see 7664
Indian Arrow Wood see 3941
Indian Balm see 9865
Indian Balsam see 942
Indian Barberry see 1158
Indian Berry see 2441
Indian Dogbane see 730
Indian Grass Oil see 6888
Indian Gum see 4450
Indian Hemp see 1752
Indian Laburnum see 1894
Indian Lilac see 6522
Indian Melissa Oil see 6891
Indian Oil of Verbena see 6891

Indian Physic see 730
Indian Pink see 8875
Indian Poke see 10151
Indian Tobacco see 5608
Indian Tragacanth see 5332
Indian Turmeric see 4808
"Indian Yellow" see 2416
India Rubber see 8411
Indican, 4980
Indican see 5013
Indic Fluoride see 4999
Indigo, 4981
Indigo Blue see 4981
Indigo Carmine, 4982
Indigo Copper see 2645
Indigopurpurin see 4985
Indigo Red see 4985
Indigotin see 4981
Indigotine see 4982
Indigo Weed see 954
Indinavir, 4983
Indiplon, 4984
Indirubin, 4985
cis-Indirubin see 4985
Indisulam, 4986
Indium, 4987
Indium Antimonide, 4988
Indium Arsenide, 4989
Indium Bromide (InBr₃) see 4997
Indium Chloride (InCl₃) see 4998
Indium Fluoride (InF₃) see 4999
Indium Gallium Aluminum Phosphide, 4990
Indium In 111 Pentetreotide see 7238
Indium In 111 Satumomab Pendetide see 8521
Indium Mononitride see 4991
Indium Monophosphide see 4993
Indium Nitride, 4991
Indium Oxide, 4992
Indium Oxide (In₂O₃) see 4992
Indium Phosphide, 4993
Indium Selenide, 4994
Indium Sesquioxide see 4992
Indium Sesquisulfate see 4995
Indium Sesquitelluride see 4996
Indium Sulfate, 4995
Indium Telluride, 4996
Indium Telluride (In₂Te₃) see 4996
Indium Tribromide, 4997
Indium Trichloride, 4998
Indium Trifluoride, 4999
Indium Trisulfate see 4995
Indo-1, 5000
Indobufen, 5001
Indocid [Merck & Co.] see 5009
Indocid PDA [Merck & Co.] see 5009
Indocin [Merck & Co.] see 5009
Indocin I.V. [Merck & Co.] see 5009
Indocyanine Green, 5002
Indocybin [Sandoz] see 8035
Indoklon [Ohio Med.] see 4231
Indole, 5003
1H-Indole see 5003
Indoleacetic Acid, 5004
1H-Indole-3-acetic Acid see 5004
1H-Indole-3-butanoic Acid see 5005
Indolebutyric Acid, 5005
Indole-3-butyric Acid see 5005
1H-Indole-3-carboxylic Acid (3-endo)-8-Methyl-8-azabicyclo[3.2.1]oct-3-yl Ester see 9962
1H-Indole-3-carboxylic Acid Octahydro-3-oxo-2,6-methano-2H-quinolizin-8-yl Ester Stereoisomer see 3458

1H-Indole-2,3-dione see 5148
1H-Indole-3-ethanamine see 9976
1H-3-Indole-3-ethanol see 9978
Indolicidin, 5006
2,3-Indolinedione see 5148
Indolmycin, 5007
1H-Indol-3-ol 3-(Hydrogen Sulfate) see 5013
1H-Indol-3-ol Hydrogen Sulfate Ester see 5013
3-Indolylacetone, 5008
l-β-3-Indolylalanine see 9977
4-(3-Indolyl)butyric Acid see 5005
2-Indolyl(3)-ethanol see 9978
β-Indolylethyl Alcohol see 9978
2-(3-Indolyl)ethyl Alcohol see 9978
2-(3-Indolyl)ethylamine see 9976
(1R,5S)-5-(1-Indol-3-ylethyl)-2-(methyl-amino)-2-oxazolin-4-one see 5007
(5S)-5-[(1R)-1-(1H-Indol-3-yl)ethyl]-2-(methylamino)-4(5H)-oxazolone see 5007
N-[1-[2-(1H-Indol-3-yl)ethyl]-4-piperi-dinyl]benzamide see 5010
1H-Indol-3-yl-β-D-glucopyranoside see 4980
(S)-3-(1H-Indol-3-yl)-2-methylaminopro-pionic Acid see 11
3-Indolylmethylcarboxylic Acid see 5004
2-C-(1H-Indol-3-ylmethyl)-α-L-lyxo-3-hexulofuranosonic Acid γ-Lactone see 820
2-C-(1H-Indol-3-ylmethyl)-α-L-xylo-3-hexulofuranosonic Acid γ-Lactone see 820
1-(1H-Indol-4-yloxy)-3-[(1-methylethyl)-amino]-2-propanol see 7554
Indol-3-yl Potassium Sulfate see 5013
1-(1H-Indol-3-yl)-2-propanone see 5008
Indol-3-yl Sulfate see 5013
Indomed [Merck & Co.] see 5009
Indomee [Merck & Co.] see 5009
Indometacin see 5009
Indomethacin, 5009
Indomethacin Glucosamide see 4483
Indomod [Pfizer] see 5009
Indon [Parke-Davis] see 7348
Indonaphthene see 4978
Indoptic [Merck & Co.] see 5009
Indoptol [Merck & Co.] see 5009
Indoramin, 5010
Indorm [Wyeth] see 7936
Indospicine, 5011
Indoxacarb, 5012
Indoxamic Acid see 7015
Indoxen [Sigma-Tau] see 5009
Indoxyl-β-D-glucoside see 4980
Indoxyl Sulfate, 5013
3-Indoxylsulfuric Acid see 5013
[^{111}In-DTPA-D-Phe^1]-octreotide see 7238
Induchlor [PPG] see 1676
InductOs [Pfizer] see 1333
Indusil [Recordati] see 2435
INF see 4337
INF-1837 see 4163
INF-3355 see 5869
INF-4668 see 5850
Infacol [Pharmax] see 3237
Infasurf [Forest] see 1716
INFeD [Watson] see 5138
Inferax [Astellas] see 5034
Infergen [Amgen] see 5034
Inflaced [Dexo] see 7619
Inflamase [Novartis] see 7841
Inflanefran [Allergan] see 7841

Inflatine *see* 5609
Inflazon [Taisho] *see* 5009
Infliximab, 5014
Informational RNA *see* 8327
Infusorial Earth, 5015
InGaAlP *see* 4990
INGN-201 *see* 2495
INH *see* 5232
INH-G *see* 4536
Inhibace [Roche] *see* 2275
Inhibins, 5016
Inhibitor of CAD *see* 1885
Inhibostamin [Zyma] *see* 9942
Inhiston [Biomed. Foscama] *see* 7349
Inicardio *see* 6627
Inimur [ICN] *see* 6616
Iniparib, 5017
Iniprol [Sanofi Winthrop] *see* 746
Initiss *see* 2275
INN-00835 *see* 6530
Innohep [Leo Pharm] *see* 9606
InnoPran [Reliant] *see* 7953
Innovace [Merck & Co.] *see* 3623
Innovar *see* 3499
Innoxalon [Sanko] *see* 6444
Inocor [Sanofi Winthrop] *see* 587
Inofal [Sandoz] *see* 9093
Inokiten [Nippon Yakuhin] *see* 10025
Inolin [Tanabe Seiyaku] *see* 9749
Inopamil [Simes] *see* 4915
Inorganic Benzene *see* 1337
Inosie [Morishita] *see* 5018
Inosine, 5018
Inosine Acedobene Dimepranol *see* 5019
Inosine Compd with 1-(Dimethylamino)-
2-propanol 4-(Acetylamino)benzoate
(1:3:3) *see* 5019
Inosine Pranobex, 5019
Inosinic Acid, 5020
5-Inosinic Acid *see* 5020
5'-Inosinic Acid *see* 5020
5'-Inosinic Acid Homopolymer Complex
with 5'-Cytidylic Acid Polymer with
5'-Uridylic Acid (1:1) *see* 8357
Inosiplex *see* 5019
Inosite *see* 5021
Inositol, 5021
i-Inositol *see* 5021
meso-Inositol *see* 5021
myo-Inositol *see* 5021
myo-Inositol 1,2,3,4,5,6-Hexakis(dihydro-
gen Phosphate) *see* 7499
Inositol Hexanicotinate *see* 5022
meso-Inositol Hexanicotinate *see* 5022
Inositolhexaphosphoric Acid *see* 7499
myo-Inositol Hexa-3-pyridinecarboxylate
see 5022
Inositol Niacinate, 5022
Inotropin [Bago] *see* 3470
Inovan [Kyowa] *see* 3470
Inovelon [Eisai] *see* 8426
Inoxyl [Debat] *see* 7044
INPC *see* 7930
INS-365 *see* 3391
INS-37217 *see* 2898
Insane Root *see* 4898
Insariotoxin *see* 9981
Insecticide 1179 *see* 6054
Insecticide No. 497 *see* 3114
Insecticide No. 4049 *see* 5764
Insect Repellent 3535 *see* 3828
Insegar [Syngenta] *see* 4017
Insibrin [Byk-Liprandi] *see* 3710
Insidon [Novartis] *see* 6948
Insiswa *see* 9138

Insogen [Altana] *see* 2192
Insoluble Anhydrite *see* 1708
Insoluble Cyanuric Acid *see* 2672
Insoluble Prussian Blue *see* 8018
Insoluble Sodium Metaphosphate *see*
8773
Insoluble Sodium Polyphosphate *see*
8773
Insoral [Spofa] *see* 7345
Inspra [Pfizer] *see* 3681
Instalac [Virbac] *see* 9878
Insubeta *see* 5050
Insulamin [Iwaki] *see* 1481
Insulatard [Novo] *see* 5023
Insulin, 5023
Insulinase, 5024
Insulin Aspart, 5025
Insulin-degrading Enzyme *see* 5024
Insulin Detemir, 5026
Insulin (emp) *see* 5023
Insulin Glargine, 5027
Insulin Glulisine, 5028
Insulin ¹³¹I *see* 5023
Insulin-like Growth Factor I *see* 5029
Insulin-like Growth Factor I (Human)
see 5844
Insulin-like Growth Factor I (Human)
Complex with Insulin-like Growth
Factor-binding Protein IGFBP-3 (Hu-
man) *see* 5845
Insulin-like Growth Factor II *see* 5029
Insulin-like Growth Factors, 5029
Insulin Lispro, 5030
Insulinoma Amyloid Peptide *see* 603
Insulin (prb) *see* 5023
Insulin Protease *see* 5024
Insulton *see* 5920
Insulysin *see* 5024
Insuman Rapid [Sanofi-Aventis] *see* 5023
Insumin [Kyorin] *see* 4228
Insuven [Berenguer] *see* 3325
In-Synch [ProLabs] *see* 7989
Intal [Sanofi-Aventis] *see* 2581
Inteban [Sumitomo] *see* 5009
Intedanib, 5031
Integerrimine *see* 8590
Integrilin [COR] *see* 3698
Integrin (formerly) [Sterling] *see* 7070
Integrins, 5032
Intelence [Tibotec] *see* 3935
Intenkordin [Polfa] *see* 2243
Intensain [HMR] *see* 2243
Intensopan [Sandoz] *see* 1462
Intercept [Ortho] *see* 6764
Interceptor [Novartis] *see* 6270
Intercron [Zambon] *see* 2581
Interferon, 5033
Interferon-α, 5034
Interferon-αA (Human Leukocyte Pro-
tein Moiety) *see* 5034
IPBC, 5112
Interferon α2 (Human Clone pAD19B-
IFN Protein Moiety Reduced) *see*
5034
Interferon α2 (Human Leukocyte Clone
Hif-SN206 Protein Moiety Reduced)
see 5034
Interferon α2 (Human Leukocyte Clone
pM21 Protein Moiety Reduced) *see*
5034
Interferon-β, 5035
Interferon-β₂ *see* 5042
Interferon-γ, 5036
Interferon αA (Human Leukocyte)
Mono(N^2,N^6-dicarboxy-L-lysyl) Deriv.

Diester with α-Methyl-ω-hydroxy-
poly(oxy-1,2-ethanediyl) *see* 7175
Interferon Alfa-2a *see* 5034
Interferon Alfa-2b *see* 5034
Interferon Alfa-2c *see* 5034
Interferon Alfacon-1 *see* 5034
Interferon Alfa-n1 *see* 5034
Interferon Alfa-n3 *see* 5034
Interferon Beta-1a *see* 5035
Interferon Beta-1b *see* 5035
Interferon Beta (Human) *see* 5035
Interferon Gamma-1a *see* 5036
Interferon Gamma-1b *see* 5036
Interferon Gamma-2a *see* 5036
Interferon-γ-inducing Factor *see* 1884
Intergravin-orales [Werfft-Chemie] *see*
1661
Interleukin-1, 5037
Interleukin-1β *see* 1884
Interleukin-2, 5038
Interleukin-3, 5039
Interleukin-4, 5040
Interleukin-5, 5041
Interleukin-6, 5042
Interleukin-10, 5043
Interleukin-10 (Human Clone pH15C)
see 5043
Interleukin-11, 5044
2-178-Interleukin 11 (Human Clone
pXM/IL-11) *see* 5044
Interleukin-18 *see* 1884
Interleukin-1 Inhibitor Protein *see* 5045
Interleukin-1β Converting Enzyme *see*
1884
Interleukin 1β Precursor Proteinase *see*
1884
Interleukin-1 Receptor Antagonist, 5045
Interleukin-4 Receptor, 5046
Interleukin 1 Receptor Accessory Pro-
tein (Human Extracellular Domain
Fragment) Fusion Protein with Type I
Interleukin 1 Receptor (Human Ex-
tracellular Domain Fragment) Fusion
Protein with Immunoglobulin G1
(Human Fc Fragment), Homodimer
see 8348
Interleukin-1 Trap *see* 8348
Intermedin(e) *see* 6381
Intermezzo [Transcept Pharma] *see* 10391
Internally Compensated Tartaric Acid
see 9202
Interstitial Cell Stimulating Hormone
see 5674
Intertocine-S [Veterinaria] *see* 7078
Intestibar [Spofa] *see* 990
Intestin-Euvernil [Heyden] *see* 9044
Intrabilix *see* 5065
Intrabutazone [Organon] *see* 7390
Intradex [Dextran Prod. Ltd.] *see* 2950
Intradine [Norbrook] *see* 9047
Intrapan [USV] *see* 2949
Intrasite [Fisch] *see* 2576
Intrastigmina [Lusofarmaco] *see* 6549
Intraval Sodium [RPR] *see* 9503
Intrazone [Arnolds] *see* 7390
Intrepid [Dow AgroSci.] *see* 6066
Intrinsic Factor *see* 1897
Intromene [Teikoku] *see* 9789
Intron A [Schering-Plough] *see* 5034
Intruder [Nippon Soda] *see* 48
Intuniv [Shire] *see* 4596
Intybin *see* 5394
Inula, 5047
Inula Camphor *see* 200
Inulin, 5048

IPD [Taiho] see 9132
IPD-1151T see 9132
Ipecac, 5113
Ipecacuanha see 5113
Ipertrofan [SPA] see 5914
Iphosphamid(e) see 4937
IPI-504 see 8282
IPI-1040 see 8692
Ipilimumab, 5114
Iplex [Insmed] see 5845
IPNO see 4963
Ipnovel [Roche] see 6261
Ipodate, 5115
Ipolab [Finmedical] see 5377
Ipomea, 5116
Ipotensivo see 5842
Ipradol [CL Pharma] see 4743
Ipratropium Bromide, 5117
Ipren [Pharmacia] see 4919
Ipriflavone, 5118
Iprindole, 5119
Iproclozide, 5120
Iprodione, 5121
Ipronal [Biosedra] see 8013
Iproniazid, 5122
Ipronidazole, 5123
Ipropran [Roche] see 5123
Iprosten [Takeda] see 5118
Iprovalicarb, 5124
Iproveratril see 10144
Ipsapirone, 5125
IPTG, 5126
I.P.U. see 5264
IPy$_2$BF$_4$ see 1303
IQ see 4939
IR 3535 [Merck KGaA] see 3828
IR-5878 see 6981
IRAP see 5045
Irbesartan, 5127
trans-IrCl(CO)(PPh$_3$)$_2$ see 10128
Ircon [Kenwood] see 4073
[Ir(Cp*)Cl$_2$]$_2$ see 7226
Irenat [Troponwerke] see 8784
Irescein [Americal] see 4192
Iressa [AstraZeneca] see 4412
Irgacare MP [Ciba] see 9822
Irgafen see 9063
Irgalax [Novartis] see 4488
Irgamide [Dispersa] see 9036
Irgarol 1051 [Ciba], 5128
Irgasan BS200 see 9336
Irgasan CF3 see 2372
Irgasan DP 300 [Ciba] see 9822
Iridin see 5134
Iridina Due [Montefarmaco] see 6453
Iridine see 5134
Iridium, 5129
Iridium Chloride (IrCl$_3$) see 5132
(OC-6-11)-Iridium Fluoride (IrF$_6$) see 5130
Iridium Hexafluoride, 5130
trans-Iridium(I)bis(triphenylphosphine)-carbonyl Chloride see 10128
Iridium Oxide (Ir$_2$O$_3$) see 5131
Iridium Pentamethylcyclopentadienyl Dichloride Dimer see 7226
Iridium Sesquioxide, 5131
Iridium Trichloride, 5132
Iridocin [Bayer] see 3791
Iridomyrmecin, 5133
Iridomyrmexin see 5133
Irifan see 2704
Irigenin, 5134
Irinotecan, 5135
Iris Blue B see 5380

Irish Broom see 8540
Irisone [Givaudan] see 5099
Irium see 8768
Irofulven, 5136
Iromon [Gador] see 4074
Iron, 5137
Iron Alum see 515
Iron(III) Ammonium Citrate see 4044
Iron Ammonium Disulfate see 515
Iron(II) Ammonium Sulfate see 518
Ironate [Wyeth] see 4084
Iron Blue see 8018
Iron Bromide (FeBr$_2$) see 4069
Iron Bromide (FeBr$_3$) see 4045
Iron Carbonyl see 5143
(TB-5-11)-Iron Carbonyl (Fe(CO)$_5$) see 5143
Iron(III) Chloride see 4046
Iron Chloride (FeCl$_2$) see 4070
Iron Chloride (FeCl$_3$) see 4046
Iron Choline Citrate Complex see 4065
Iron Dextran, 5138
Iron Dibromide see 4069
Iron Dichloride see 4070
Iron Difluoride see 4072
Iron Dihydroxide see 4075
Iron Diiodide see 4076
Iron Dithiocyanate see 4086
Irone, 5139
α-Irone, 5140
β-Irone, 5141
γ-Irone, 5142
Iron Fluoride (FeF$_2$) see 4072
Iron Fluoride (FeF$_3$) see 4049
Iron Hemiphosphide see 4081
Iron(III) Hexacyanoferrate(4−) see 8018
(OC-6-11)-Iron(3+) (3:4) Hexakis(cyano-κC) Ferrate(4−) see 8018
Iron Hydroxide (Fe(OH)$_2$) see 4075
Iron Hydroxide Oxide (Fe(OH)O) see 4052
Iron(III) Hydroxide Sucrose Complex see 5145
Iron Iodide (FeI$_2$) see 4076
Iron Monoselenide see 4082
Iron Monoxide see 4079
Iron(III) Nitrate see 4054
Iron Orthophosphate see 4056
Iron Oxalate see 4078
Iron Oxide (FeO) see 4079
Iron Oxide (Fe$_2$O$_3$) see 4055
Iron Oxide (Fe$_3$O$_4$) see 4067
Iron Pentacarbonyl, 5143
Iron(2+) Phosphate see 4080
Iron Phosphide (Fe$_2$P) see 4081
Iron(III) Potassium Hexacyanoferrate-(4−) see 8018
Iron Saccharate see 5145
Iron Selenide (FeSe) see 4082
Iron Sorbitex, 5144
Iron Sorbitol see 5144
Iron Sorbitol Citrate see 5144
Iron Sucrose, 5145
Iron(II) Sulfate see 4084
Iron(2+) Sulfate see 4084
Iron Sulfate-glycine Complex see 4066
Iron Sulfide (FeS) see 4085
Iron Sulfuret see 4085
Iron Tetracarbonyl see 5143
Iron Vitriol see 4084
Iron Yttrium Oxide (Fe$_5$Y$_3$O$_{12}$) see 10295
Irox see 4074
Irritren [Byk Gulden] see 5623
Irrodan [Biomed. Foscama] see 1480

Irrorin see 7855
Irsogladine, 5146
Irtan [Aventis] see 6521
Irtonin [Takeda] see 9750
Iruxol [Knoll] see 2465
Irvalec [PharmaMar] see 3601
IS-370 see 9005
IS-401 see 3216
IS-499 see 7671
ISA-247 see 10224
Isacen see 7073
Isactid [Ferring] see 125
Isalon [Grelan] see 216
Isanic Acid, 5147
Isano Oil see 1331
Isaphen see 7073
Isapirone see 5125
Isatidine see 8290
Isatin, 5148
γ-Isatropaic Acid see 9971
ε-Isatropaic Acid see 9971
ISAtx-247 see 10224
Isavuconazole, 5149
Iscador [Weleda] see 6294
Iseganan, 5150
Isentress [Merck & Co.] see 8216
Isepacin [Toyo Jozo] see 5151
Isépalline [Schering] see 5151
Isepamicin, 5151
Isethion [Mect] see 4511
Isethionic Acid, 5152
ISF-2001 see 9937
ISF-2123 see 7535
ISF-2522 see 7041
ISF-09334 see 8427
Ishikawa Reagent, 5153
ISI-641A see 744
Isicom [Desitin] see 1798
Isinglass, 5154
Isipen [Krka] see 7575
ISIS-2302 see 237
ISIS-2922 see 4256
ISIS-3521 see 744
ISIS-301012 see 6288
Islet Amyloid Polypeptide see 603
Ismelin [Novartis] see 4595
Ismo [Roche] see 5271
Ismotic [Alcon] see 5270
Isoacetophorone see 5242
Isoalloxazine-adenine Dinucleotide see 4122
Isoamidone see 5230
Isoamidone I see 6798
Isoaminile, 5155
Isoamyl Acetate, 5156
Isoamyl Alcohol see 5241
dl-sec-Isoamyl Alcohol see 6105
Isoamylamine, 5157
Isoamyl Benzoate, 5158
Isoamyl Bromide, 5159
Isoamyl Butyrate, 5160
Isoamylcarbamic Acid Thymyl Ester see 9564
Isoamyl Chloride, 5161
Isoamyl Cyanide, 5162
β-Isoamylene see 601
Isoamyleneguanidine see 4371
8-Isoamylenoxypsoralen see 4967
4-(β-Isoamylenyl)-1,2-diphenyl-3,5-py-razolidinedione see 4039
Isoamyl Ether, 5163
5-Isoamyl-5-ethylbarbituric Acid see 567
Isoamyl Formate, 5164
Isoamyl 2-Hydroxybenzoate see 5170
Isoamyl Iodide, 5165

Isoamyl Isovalerate, 5166
Isoamyl Mandelate *see* 5782
Isoamyl Nitrate, 5167
Isoamyl Nitrite, 5168
Isoamyl Oxide *see* 5163
Isoamyl Phthalate, 5169
Isoamyl Salicylate, 5170
Isoamyl Valerianate *see* 5166
Isoandrosterone *see* 3664
Isoarteril [Isola-Ibi] *see* 8237
Isoascorbic Acid, 5171
Isobamate *see* 1842
Isobarbaloin *see* 303
Isobatrachotoxin *see* 1007
Isobebeerine *see* 5203
Iso-Benzacyl [Novartis] *see* 1118
Isobenzan, 5172
1,3-Isobenzofurandione *see* 7484
Isobide [Nikken] *see* 5270
Isobixin *see* 1312
Isoborneol, 5173
(±)-Isoborneol *see* 5173
(+)-Isoborneol *see* 5173
(−)-Isoborneol *see* 5173
d-Isoborneol *see* 5173
dl-Isoborneol *see* 5173
l-Isoborneol *see* 5173
6-Isobornyl-3,4-dimethylphenol *see* 10273
Isobornyl Thiocyanoacetate, 5174
6-(2-Isobornyl)-3,4-xylen-1-ol *see* 10273
Isobromyl [Clin-Comar-Byla] *see* 1404
Isobutanoic Acid *see* 5200
Isobutanol *see* 5176
1-[[(α-Isobutanoyloxyethoxy)carbonyl]-aminomethyl]-1-cyclohexaneacetic Acid *see* 4349
4-[[(1*S*)-Isobutanoyloxyisobutoxy]carbonylamino]-(3*R*)-(4-chlorophenyl)butanoic Acid *see* 758
Isobutene *see* 5186
Isobutenenitrile *see* 6014
Isobutol *see* 3775
1-Isobutoxy-2-pyrrolidino-3-*N*-benzyl-anilinopropane *see* 1152
3-Isobutoxy-2-pyrrolidino-*N*-phenyl-*N*-benzylpropylamine *see* 1152
Isobutyl Acetate, 5175
Isobutyl Alcohol, 5176
Isobutylaldehyde *see* 5199
5-Isobutyl-5-allylbarbituric Acid *see* 1514
Isobutylamine, 5177
Isobutyl *p*-Aminobenzoate, 5178
(*S*)-(+)-3-Isobutyl-γ-aminobutyric Acid *see* 7843
2-(Isobutylamino)ethanol *p*-Aminobenzoate (Ester) *see* 1526
2-(Isobutylamino)ethyl *p*-Aminobenzoate *see* 1526
Isobutylbenzene, 5179
Isobutyl Bromide, 5180
Isobutyl Butanoate *see* 5181
3-Isobutyl-6-*sec*-butyl-2-hydroxypyrazine 1-Oxide *see* 831
Isobutyl *n*-Butyrate, 5181
Isobutyl Carbamate, 5182
Isobutylcarbamic Acid Ethyl Ester *see* 5198
Isobutyl Carbinol *see* 5241
Isobutylcarbylamine *see* 5157
Isobutyl Chloride, 5183
Isobutyl Chlorocarbonate, 5184
Isobutyl Chloroformate *see* 5184
Isobutyl Cyanoacrylate, 5185

(*E*,*E*)-*N*-Isobutyl-2,4-decadienamide *see* 7182
(*E*,*E*,*Z*)-*N*-Isobutyl-2,6,8-decatrienamide *see* 170
N-Isobutyldeca-*trans*-2-*cis*-6-*trans*-8-trienamide *see* 170
Isobutyl 1,4-Dihydro-5-methoxycarbonyl-2,6-dimethyl-4-(2-nitrophenyl)-3-pyridinecarboxylate *see* 6651
Isobutylene, 5186
Isobutyl Ether, 5187
Isobutyl Formate, 5188
p-Isobutylhydratropic Acid *see* 4919
Isobutylhydrochlorothiazide *see* 1527
Isobutyl Iodide, 5189
Isobutyl Isobutanoate *see* 5190
Isobutyl Isobutyrate, 5190
Isobutyl Isopentanoate *see* 5191
Isobutyl Isovalerate, 5191
Isobutyl Kelo-form *see* 5178
Isobutyl Mercaptan, 5192
Isobutyl Methyl 1,4-Dihydro-2,6-dimethyl-4-(*o*-nitrophenyl)-3,5-pyridinedicarboxylate *see* 6651
Isobutyl Nitrate, 5193
Isobutyl Nitrite, 5194
Isobutyl Octadecanoate *see* 5196
2-(4-Isobutylphenyl)butyric Acid *see* 1529
(±)-2-(4-Isobutylphenyl)propionic Acid *see* 4919
dl-2-(4-Isobutylphenyl)propionohydroxamic Acid *see* 4920
Isobutyl Propionate, 5195
4-*N*-Isobutylspiropiperidylrifamycin S *see* 8338
Isobutyl Stearate, 5196
Isobutyl Sulfide, 5197
Isobutyl Thiol *see* 5192
Isobutyltrimethylmethane *see* 5239
Isobutyl Urethane, 5198
Isobutyl Valerate *see* 5191
Isobutyraldehyde, 5199
Isobutyric Acid, 5200
Isobutyric Aldehyde *see* 5199
Isobutyronitrile, 5201
3-Isobutyryl-2-isopropylpyrazolo[1,5-*a*]-pyridine *see* 4918
O-Isobutyrylthiamine Disulfide *see* 9449
Isocaine [Columbus] *see* 5178
Isocalm [Kaken] *see* 9681
Isocapronitrile *see* 5162
Isocaramidine *see* 2846
Isocarboxazid, 5202
Isocard [Schwarz] *see* 5271
Isocaryophyllene *see* 1875
Isocef [Recordati] *see* 1950
Isochavicine *see* 7584
Isochavicinic Acid *see* 7579
Isochondrodendrine, 5203
Isochrysene *see* 9913
Isocillin [Aventis] *see* 7209
Isocinchomeronic Acid, 5204
Isococaine *see* 8020
Isocolin [Isola-Ibi] *see* 7474
Isocolumbin *see* 2476
Isoconazole, 5205
Iso-Cornox [FBC] *see* 5855
Isocorybulbine, 5206
Isocorydine, 5207
Isocorypalmine, 5208
Isocrin [Warner-Chilcott] *see* 7073
Isocrotonic Acid, 5209
Isocrotyl Chloride *see* 2149
Isocumene *see* 7957

Isocyanatobenzene *see* 7407
4-Isocyanato-1,1′-biphenyl *see* 10271
1-Isocyanatobutane *see* 1578
Isocyanatomethane *see* 6161
1-Isocyanatonaphthalene *see* 6492
Isocyanic Acid, 5210
Isocyanic Acid 4-Biphenylyl Ester *see* 10271
Isocyanic Acid Butyl Ester *see* 1578
Isocyanic Acid Methyl Ester *see* 6161
Isocyanic Acid 1-Naphthyl Ester *see* 6492
1-[(Isocyanomethyl)sulfonyl]-4-methyl-benzene *see* 9715
Isocyanuric Acid *see* 2690
Isodiamylamine *see* 3211
Isodihydroperparine *see* 3503
Isodine [Blair] *see* 7815
Isodiprene *see* 1835
Isodrin *see* 219
Isodulcit *see* 8295
Isodurene, 5211
Isoendoxan *see* 4937
d-Isoephedrine *see* 8024
Isoequol *see* 3701
Isoestradiol, 5212
8-Isoestradiol-17β *see* 5212
Isoestragole *see* 636
8-Isoestrone, 5213
Iso E Super [Int. Flavors & Fragrances], 5214
Isoetam [Ferrer] *see* 3775
Isoetarine *see* 5215
Isoetharine, 5215
Isoeugenol, 5216
Isofagomine, 5217
Isofedrol [Boehringer, Mann.] *see* 3663
Isofenphos, 5218
(*S*)-(−)-4′,7-Isoflavandiol *see* 3701
Isoflav-3-ene-4′,7-diol *see* 2876
Isoflavone, 5219
Isofluorphate *see* 5222
Isoflupredone, 5220
Isoflurane, 5221
Isoflurophate, 5222
Isoforon *see* 5242
Isogel [Charwell] *see* 7631
Isoglaucon [Alcon] *see* 2380
L-Isoglutamine, 5223
Isoindazole *see* 4976
Isoindirubin *see* 4985
1*H*-Isoindole-1,3(2*H*)-dione *see* 7485
Isoinokosterone *see* 3538
Iso-Iodeikon *see* 7375
Iso-JH 0 *see* 5321
Iso-K [San Carlo] *see* 5352
Isokahalalide F *see* 3601
Isoket [Schwarz] *see* 5271
Isolan, 5224
Isoleucine, 5225
L-Isoleucine *see* 5225
3-Isoleucine-8-arginine Vasopressin *see* 10129
L-Isoleucyl-L-leucyl-L-arginyl-L-tryptophyl-L-prolyl-L-tryptophyl-L-tryprophyl-L-prolyl-L-tryptophyl-L-arginyl-L-arginyl-L-lysinamide *see* 6940
L-Isoleucyl-L-leucyl-L-prolyl-L-tryptophyl-L-lysyl-L-tryptophyl-L-prolyl-L-tryptophyl-L-tryptophyl-L-prolyl-L-tryptophyl-L-arginyl-L-argininamide *see* 5006
Isolevin *see* 5263
α-Isolupanine *see* 5667
Isolysergic Acid, 5226

IsoMack [Mack, Illert.] *see* 5271
Isomaltol, 5227
Isomaltulose, 5228
Isomenyl [Kaken] *see* 5263
Isomeride [Ardix] *see* 4004
Isometamidium Chloride, 5229
Isomethadone, 5230
Isometheptene, 5231
Isomist *see* 5263
Isomonat [Roche] *see* 5271
Isomycin [Werfft] *see* 8001
Isomycomycin *see* 6411
Isomytal [Nippon Shinyaku] *see* 567
Isonaphthoic Acid *see* 6467
Isonaphthol *see* 6469
Isonex [Pfizer] *see* 5232
Isoniazid, 5232
Isonicotinic Acid, 5233
Isonicotinic Acid 2-[2-(Benzylcarbam-
   oyl)ethyl]hydrazide *see* 6576
Isonicotinic Acid Diethylamide, 5234
Isonicotinic Acid Hydrazide *see* 5232
Isonicotinic Acid Hydrazide Hydrazone
   with Glucuronic Acid Lactone *see*
   4536
Isonicotinic Acid 2-Isopropylhydrazide
   see 5122
Isonicotinic Acid α-Methylfurfurylidene-
   hydrazide *see* 4337
Isonicotinic Acid Salicylidenehydrazide
   see 8472
Isonicotinic Acid *m*-Sulfobenzylidene
   Hydrazide *see* 9089
N-Isonicotinoyl-*N*'-[β-(*N*-benzylcarbox-
   amido)ethyl]hydrazide *see* 6576
Isonicotinoylhydrazine *see* 5232
Isonicotinoylhydrazone of D-Glucuronic
   Acid Lactone *see* 4536
1-Isonicotinoyl-2-isopropylhydrazine *see*
   5122
1-Isonicotinoyl-2-salicylidenehydrazine
   see 8472
Isonicotinylhydrazine *see* 5232
Isonicotinyl Hydrazonotoluene-*m*-sul-
   fonic Acid *see* 9089
1-Isonicotinyl-2-isopropylhydrazine *see*
   5122
Isonipecaine *see* 5916
Isonipecotic Acid, 5235
Isonitropropane *see* 6715
Isonitrosoacetone, 5236
Isonitrosoacetophenone, 5237
5-Isonitrosobarbituric Acid *see* 10196
β-Isonitrosopropane *see* 73
Isonixin, 5238
Isonootkatone *see* 10167
Isonorin [SM & P] *see* 5263
Isonovobiocin *see* 6811
Isooctane, 5239
Isooctanol *see* 5240
Isooctyl Alcohol, 5240
p-Isooctylpolyoxyethylenephenol Form-
   aldehyde Polymer *see* 10014
Isoolean-14-en-3β-ol *see* 9195
Iso-oosporein *see* 6945
Isopaque [Winthrop] *see* 6235
Isopelletierine *see* 7181
Isopentanal *see* 5275
Isopentanamide *see* 5276
Isopentanoic Acid *see* 5277
Isopentanoic Acid Methyl Ester *see* 6165
Isopentanoyl Chloride *see* 5278
4-(2-Isopentenyl)-1,2-diphenyl-3,5-pyraz-
   olidinedione *see* 4039
Isopentyl Alcohol, 5241

sec-Isopentyl Alcohol *see* 6105
Isopentylamine *see* 5157
Isopentyl 5,6-Dihydro-7,8-dimethyl-4,5-
   dioxo-4*H*-pyrano[3,2-*c*]quinoline-2-
   carboxylate *see* 8258
Isopentyl Formate *see* 5164
Isopentyl Iodide *see* 5165
Isopentyl 3-Methylbutanoate *see* 5166
Isopentyl Nitrate *see* 5167
Isopentyl Nitrite *see* 5168
Isophane Insulin *see* 5023
Isophen [Knoll] *see* 6020
Isophenphos *see* 5218
Isophorone, 5242
α-Isophorone *see* 5242
Isophosphamide *see* 4937
Isophthalic Acid, 5243
Isophytol, 5244
Isopilocarpine *see* 7536
Isopilosine, 5245
Isopimaric Acid, 5246
Isopiperine *see* 7584
Isopiperinic Acid *see* 7579
IsoPPC *see* 7930
Isopral *see* 9810
10α-Isopregnenone *see* 3521
Isoprenaline *see* 5263
Isoprene, 5247
Isoprinosin [Newport] *see* 5019
Isoprinosina [Newport] *see* 5019
Isoprinosine [Newport] *see* 5019
Isoprochin P [Merckle] *see* 7982
Isopromedol *see* 7899
Isopropamide Iodide, 5248
Isopropanol *see* 5254
Isopropene Cyanide *see* 6014
Isopropenyl Acetate, 5249
4-Isopropenyl-1-cyclohexene-1-carboxal-
   dehyde *see* 7282
2-Isopropenyl-2,3-dihydro-5-acetylben-
   zofuran *see* 9743
2-Isopropenyl-6-methoxy-5-benzofuranyl
   Methyl Ketone *see* 3942
cis-(+)-2-Isopropenyl-1-methylcyclobu-
   taneethanol *see* 4571
4-(5-Isopropenyl-2-methyl-1-cyclopen-
   ten-1-yl)-2-butanone, 5250
Isopropenyl Methyl Ether *see* 6073
Isopropenylnitrile *see* 6014
*N*¹-(4-Isopropoxybenzoyl)-*p*-aminoben-
   zenesulfonamide *see* 9067
*N*¹-(*p*-Isopropoxybenzoyl)sulfanilamide
   see 9067
5-Isopropoxycarbonylaminothiabenda-
   zole *see* 1728
5-Isopropoxycarbonylamino-2-(4-thiazol-
   yl)benzimidazole *see* 1728
(*E*)-*O*-2-Isopropoxycarbonyl-1-methylvi-
   nyl *O*-Methyl Ethylphosphoramidothioate *see* 7929
(±)-1-[*p*-(2-Isopropoxyethoxymethyl)-
   phenoxy]-3-(isopropylamino)-2-propa-
   nol *see* 1295
(±)-1-[[α-(2-Isopropoxyethoxy)-*p*-tolyl]-
   oxy]-3-(isopropylamino)-2-propanol
   see 1295
7-Isopropoxyisoflavone *see* 5118
Isopropoxymethylphosphoryl Fluoride
   see 8511
7-Isopropoxy-3-phenyl-4*H*-1-benzopy-
   ran-4-one *see* 5118
7-Isopropoxy-3-phenylchromone *see*
   5118
o-Isopropoxyphenyl *N*-Methylcarbamate
   see 7949

2-Isopropoxypropane *see* 5258
Isopropydrin *see* 5263
Isopropyl Acetate, 5251
Isopropylacetic Acid *see* 5277
Isopropyl Acetoacetate, 5252
Isopropylacetone, 5253
Isopropyl Alcohol, 5254
Isopropylamine, 5255
4-Isopropylaminoantipyrine *see* 8220
(±)-1-(Isopropylamino)-3-[*p*-(cyclopropylmethoxyethyl)phenoxy]-2-propanol
   see 1184
4-Isopropylamino-2,3-dimethyl-1-phenyl-
   3-pyrazolin-5-one *see* 8220
7-[2-(*N*-Isopropylamino)ethyl]-(20*S*)-
   camptothecin *see* 1028
(20*S*)-7-(2-Isopropylamino)ethylcampto-
   thecin *see* 1028
1-(Isopropylamino)-2-hydroxy-3-[*o*-(allyloxy)phenoxy]propane *see* 7051
(±)-1-(Isopropylamino)-3-[*p*-(β-methoxyethyl)phenoxy]-2-propanol *see*
   6228
Isopropylaminomethyl-(3,4-dihydroxy-
   phenyl)carbinol *see* 5263
1-(Isopropylamino)-3-[(2-methylindol-4-
   yl)oxy]-2-propanol *see* 5923
α-(Isopropylaminomethyl)protocate-
   chuyl Alcohol *see* 5263
1-(Isopropylamino)-3-(1-naphthyloxy)-2-
   propanol *see* 7953
Isopropylaminophenazone *see* 8220
9-[3-(Isopropylamino)propyl]-9-(aminocarboyl)fluorene *see* 4977
α-(1-Isopropylaminopropyl)protocate-
   chuyl Alcohol *see* 5215
1-[3-(Isopropylamino)-2-pyridyl]-4-[(5-
   methanesulfonamidoindol-2-yl)car-
   bonyl]piperazine *see* 2882
p-Isopropylaniline *see* 2610
4-Isopropylantipyrine *see* 7982
Isopropylarterenol *see* 5263
p-Isopropylbenzaldehyde *see* 2611
Isopropylbenzene *see* 2607
p-Isopropylbenzoic Acid *see* 2608
3-Isopropyl-1*H*-2,1,3-benzothiadiazin-
   4(3*H*)-one 2,2-Dioxide *see* 1054
Isopropyl 4-(2,1,3-Benzoxadiazol-4-yl)-
   1,4-dihydro-5-methoxycarbonyl-2,6-
   dimethyl-3-pyridinecarboxylate *see*
   5287
p-Isopropylbenzyl Alcohol *see* 2609
Isopropyl Bromide, 5256
N-4-Isopropylcarbamoylbenzyl-*N*'-meth-
   ylhydrazine *see* 7876
3-Isopropylcarbamylsulfonamido-4-(3'-
   methylphenyl)aminopyridine *see* 9711
Isopropyl Carbanilate *see* 7930
Isopropylcarbinol *see* 5176
Isopropyl Chloride, 5257
N-Isopropyl-α-chloroacetanilide *see*
   7907
Isopropyl [4'-(*p*-Chlorobenzoyl)-2-phen-
   oxy-2-methyl]propionate *see* 4009
Isopropyl-*m*-chlorocarbanilate *see* 2193
Isopropyl *N*-(3-Chlorophenyl)carbamate
   see 2193
Isopropyl-*o*-cresol *see* 1872
Isopropyl-*m*-cresyl Ester of Isoamylcar-
   bamic Acid *see* 9564
Isopropyl Cyanide *see* 5201
Isopropyl 6-Cyano-5-methoxycarbonyl-
   2-methyl-4-(3-nitrophenyl)-1,4-dihy-
   dropyridine-3-carboxylate *see* 6630
(−)-*N*-(*trans*-4-Isopropylcyclohexyl-1-
   carbonyl)-D-phenylalanine *see* 6510

(2-Isothiocyanatoethyl)benzene *see* 7340
Isothiocyanatomethane *see* 6164
1-Isothiocyanato-4-[(*R*)-methylsulfinyl]-
butane *see* 9092
1-Isothiocyanatonaphthalene *see* 6493
4-Isothiocyanato-4'-nitrodiphenylamine
see 571
1-Isothiocyanato-4-(4-nitrophenoxy)ben-
zene *see* 6719
4-Isothiocyanato-*N*-(4-nitrophenyl)ben-
zenamine *see* 571
1-(4-Isothiocyanatophenoxy)-4-nitroben-
zene *see* 6719
3-Isothiocyanato-1-propene *see* 286
Isothiocyanic Acid *see* 9481
Isothiocyanic Acid Allyl Ester *see* 286
Isothiocyanic Acid 1-Naphthyl Ester *see*
6493
Isothiocyanic Acid *p*-(*p*-Nitroanilino)-
phenyl Ester *see* 571
Isothiocyanic Acid *p*-(*p*-Nitrophenoxy)-
phenyl Ester *see* 6719
Isothiocyanic Acid *p*-Phenylene Ester *see*
1309
Isothiocyanic Acid Phenyl Ester *see* 7408
Isothipendyl, 5273
d-Isothujone *see* 9546
Isothymol *see* 1872
Isotrate [Warner-Chilcott] *see* 5271
Isotretinoin, 5274
Isotrex [Stiefel] *see* 5274
Isotrifoliin *see* 5267
Isotron 2 *see* 3073
Isovaleral *see* 5275
Isovaleraldehyde, 5275
Isovaleramide, 5276
Isovalerianic Acid *see* 5277
Isovaleric Acid, 5277
Isovaleric Acid Amide *see* 5276
Isovaleric Acid Chloride *see* 5278
Isovaleric Acid Diester with 8,9-Dihy-
dro-9-hydroxy-8-(1-hydroxy-1-methyl-
ethyl)-2*H*-furo[2,3-*h*]-1-benzopyran-2-
one *see* 851
Isovaleric Acid *l*-Menthyl Ester *see* 5909
Isovaleric Aldehyde *see* 5275
Isovaleryl Chloride, 5278
2-Isovaleryl-1,3-indandione *see* 5279
2-Isovalerylindane-1,3-dione, 5279
N-Isovaleryl-L-valyl-L-valyl-3-hydroxy-6-
methyl-γ-aminoheptanoyl-L-alanyl-3-
hydroxy-6-methyl-γ-aminoheptanoic
Acid *see* 7260
Isovaline, 5280
Isovex *see* 3783
Isoviral [Lenza] *see* 5019
Isovist [Schering AG] *see* 5108
Isovitamin C *see* 5171
Isovon *see* 5263
Isovue [BMS] *see* 5100
Isovyl [Rhovyl] *see* 7707
Isoxaben, 5281
Isoxadifen-ethyl, 5282
Isoxaflutole, 5283
Isoxepac, 5284
Isoxsuprine, 5285
Isoyohimbine *see* 10301
Isozid [Fatol] *see* 5232
ISP-1 *see* 6418
Ispaghula Husk *see* 7631
Ispan [Scott Med. Prod.] *see* 7273
Ispenoral [Rosen] *see* 7209
Isphamycin *see* 2198
Ispronicline, 5286
Isradipine, 5287

Israpafant, 5288
Isrodipine *see* 5287
Issium [Lifepharma] *see* 4175
Isteropac E.R. [Bracco] *see* 5055
Istin [Pfizer] *see* 487
Istizin [Winthrop] *see* 2814
Istodax [Astellas] *see* 8382
Istradefylline, 5289
Isuprel [Winthrop] *see* 5263
Isupren *see* 5263
IT-290 *see* 6207
IT-931 *see* 3411
ITA-104 *see* 7630
Itaconic Acid, 5290
Italprid [Teofarma] *see* 9576
Itavastatin *see* 7621
Itax [Axcan] *see* 5291
ITF-182 *see* 4954
Itobarbital *see* 1514
Itopride, 5291
Itraconazole, 5292
Itramin Tosylate, 5293
Itridal *see* 2704
Itrin [Abbott] *see* 9297
Itrizole [J & J] *see* 5292
Itrol *see* 8650
Itrop [Boehringer, Ing.] *see* 5117
Ituran [Promonta] *see* 6686
IUdR *see* 4931
Iuvacor [Sanofi-Synthelabo] *see* 10025
Ivabradine, 5294
Ivacaftor, 5295
Ivadal [Sanofi-Aventis] *see* 10391
Ivermectin, 5296
Ivomec [Merial] *see* 5296
Ivosit [Hoechst] *see* 3315
Ivy Block [Enviroderm] *see* 1058
Ixabepilone, 5297
Ixbut, 5298
Ixel [Fabre] *see* 6275
Ixempra [BMS] *see* 5297
Ixense [Takeda] *see* 733
Ixertol [Rottendorf] *see* 2306
Ixodin, 5299
Ixoten [Baxter] *see* 9947
Izilox [Bayer] *see* 6377
J-38 *see* 4273
J-867 *see* 824
J-107088 *see* 3560
JA *see* 5308
Jaborandi *see* 7537
Jaclacin [Medac] *see* 104
Jacobsen's Catalyst, 5300
Jacodine *see* 8591
Jacutin [Hermal] *see* 5556
Jaikin [Basotherm] *see* 8726
Jakafi [Incyte] *see* 8440
Jalap, 5301
Jalovis *see* 4797
Jamaica Pepper *see* 7544
Jamboo *see* 5302
Jambul, 5302
JA-ME *see* 5308
Jamestown Weed *see* 8950
Jamylène [Théraplix] *see* 3446
Janus Green B, 5303
Januvia [Merck & Co.] *see* 8690
Japan Agar *see* 176
Japan Camphor *see* 1734
Japanese Oil of Camphor *see* 6872
Japan Isinglass *see* 176
Japan Tallow *see* 5304
Japan Wax, 5304
Japonilure, 5305
Jasmolin I *see* 5306

Jasmolin II *see* 5306
Jasmolins, 5306
Jasmone, 5307
Jasmonic Acid, 5308
Jateorrhizine *see* 5309
Jatroneural [Procter & Gamble] *see* 9846
Jatropur [Procter & Gamble] *see* 9760
Jatrorrhizine, 5309
Jatrox [Procter & Gamble] *see* 1285
JAU-6476 *see* 7997
Jaundice Berry *see* 956
Jaune Brilliant *see* 1633
Javanicin, 5310
Java Pepper *see* 2602
Java Plum *see* 5302
Javelin [Certis] *see* 925
Javelle Water *see* 4904
Javlor [Fabre] *see* 10186
JB-11 *see* 9893
JB-251 *see* 8004
JB-305 *see* 7582
JB-323 *see* 7573
JB-8181 *see* 2921
JDL-464 *see* 9711
JE-049 *see* 4922
Jectofer [AstraZeneca] *see* 5144
Jecto-Sal [Mallard] *see* 9513
Jee Cocktail *see* 6035
Jeff *see* 6035
Jellin [Grünenthal] *see* 4181
Jenacaine *see* 7875
Jequirity *see* 10
Jeremejevite *see* 326
Jerusalem Artichoke, 5311
Jerusalem Tea *see* 6877
Jerva Acid *see* 2051
Jervasic Acid *see* 2051
Jervine, 5312
Jesaconitine, 5313
Jestmin [Kyowa] *see* 6984
Jestryl [Ankerpharm] *see* 1781
Jesuit's Balsam *see* 2501
Jesuit's Bark *see* 2287
Jesuit's Tea *see* 5824
Jevtana [Sanofi-Aventis] *see* 1604
Jeweler's Rouge *see* 4055
Jexin [Evans] *see* 9987
JF-6064 *see* 9251
JH *see* 5321
JH 0 *see* 5321
JH III Bisepoxide *see* 5321
Jimpson Weed *see* 8950
Jimson Weed *see* 8950
Jinofloxacin *see* 6430
JM-8 *see* 1823
JM-216 *see* 8520
JM-473 *see* 7517
JM-3100 *see* 7649
JNJ-10234094 *see* 1841
JO-1016 *see* 9641
Jockey [AgrEvo] *see* 4226
Jodairol *see* 5077
Jodid [Merck KGaA] *see* 7764
Jodobil *see* 5069
Jodomiron [Bracco] *see* 5055
Joghurt *see* 10299
Johannisbrotmehl *see* 5612
JohnPhos Phosphine Ligand *see* 1469
Jojoba Oil, 5314
Jomybel [Sarva] *see* 5315
Jonctum [Hoechst] *see* 7002
Jonit [Hoechst] *see* 1309
Jopamiro [Bracco] *see* 5100
Jordan Almond *see* 296
Jordapon ACI 30 [BASF] *see* 8739

Jordapon CI [BASF] see 8739
Joristen see 7287
Josacine [Bellon] see 5315
Josamina [Novag] see 5315
Josamy [Yamanouchi] see 5315
Josamycin, 5315
Josaxin [Boots] see 5315
JP-992 see 1041
JTK-303 see 3610
JTT-705 see 2801
Judean Pitch see 835
Judolor [Woelm] see 4340
Juglans, 5316
Juglone, 5317
Juice Bromelain see 1395
Julodin [Takeda] see 3757
Jumbul see 5302
Jumex [Chiesi] see 8565
Junipene see 5624
Juniper, 5318
Juniperberry Oil see 5318
Juniper Tar, 5319
Junovan [IDM] see 6265
Jurnista [Janssen-Cilag] see 4840
Justamil [Anphar-Rolland] see 9056
Justar [Intersan] see 2267
Justice [Dow AgroSci.] see 8877
Justicidin A see 5320
Justicidin B see 5320
Justicidins, 5320
Justor [Chiesi] see 2275
Juvenile Hormones, 5321
Juvenimicin A₃ see 8392
K III see 3305
K IV see 3305
K-17 see 9403
K-33 see 9241
K-218 see 7273
K-351 see 6648
K-386 see 9671
K-1875 see 1252
K-1900 see 6637
K-3917 see 9275
K-3920 see 5001
K-4024 see 4477
K-9321 see 102
K-11941 see 227
K-21060E see 3495
K-22023 see 3436
Ka-2547 see 5990
Kaà-hê-é see 8937
Kabi 2165 see 2804
Kabi 2213 see 7847
Kabi 2234 see 9684
Kabikinase [Pharmacia] see 8955
KAD-1229 see 6296
Kadaya see 5332
Kadian [Alpharma] see 6361
Kaemferitrin see 5322
Kaempferol, 5322
Kaempferol 3-Robinoside 7-Rhamnoside
see 8372
Kahalalide F, 5323
Kainic Acid, 5324
α-Kainic Acid see 5324
L₅-xylo-Kainic Acid see 5324
Kalazaquine see 8691
Kalbitor [Dyax] see 3536
Kaleorid [Leo Pharm] see 7742
Kaletra [Abbott] see 5630
Kalgut [Tanabe Seiyaku] see 2896
Kalignost see 8818
Kalimozan [Nikken] see 4492
Kalinite see 354
Kalirechin [Toho] see 5327

Kalitabs [Leo Pharm] see 7742
Kalium see 7725
Kalkitoxin, 5325
Kallidin, 5326
Kallidin I see 1364
Kallidin II see 5326
Kallidin-9 see 1364
Kallidin-10 see 5326
Kallidinogenase see 5327
Kallikrein, 5327
Kallikrein RK2 see 9705
Kallinin see 5401
Kalma [Fresenius] see 9977
Kalmegh see 626
Kalmopyrin see 1650
Kalodil [Fidia] see 3216
Kalsetal see 1650
Kalydeco [Vertex] see 5295
Kalymin [AWD Pharma] see 8090
Kamala, 5328
Kamassin see 8146
Kameela see 5328
Kamila see 5328
Kammerer's Porphyrin see 8006
Kamoran [Lilly] see 123
Kampfstoff "Lost" see 6401
Kamycine [BMS] see 5329
Kamynex see 5329
Kanacedin see 5329
Kanamycin, 5329
Kanamycin A see 5329
Kanamycin B see 5329
Kanamycin C see 5329
Kanamytrex [Basotherm] see 5329
Kanaqua [Andromaco] see 5329
Kanasig see 5329
Kanatrol [Lusofarmaco] see 5329
Kanechlor [Kanegafuchi] see 7682
Kanemite [Agro-Kanesho] see 34
Kanendomycin [Meiji] see 5329
Kanendos [Crinos] see 5329
Kanescin [Torlan] see 5329
Kanicin see 5329
Kankerbos see 9138
Kankohso 101 see 7645
Kannasyn see 5329
Kannit [Mect] see 10283
Kanochol [Taiyo] see 6984
Kanrenol [SPA] see 1753
Kantec [Daiichi] see 5776
Kantrex [Apothecon] see 5329
Kaolin, 5330
Kaolinite see 356, 4312
Kaon [Altana] see 4492
Kaon-Cl [Altana] see 7742
Kapanol [GSK] see 6361
Kaparlem see 5319
Kapidex [Takeda] see 5413
Kapilin [Glaxo] see 5898
Kapilon Injectable see 5900
Kappadione [Lilly] see 5898
Kappaxin [Sterling Winthrop] see 5899
Kapron [Klin] see 6823
Karanjin, 5331
Karate [Syngenta] see 5398
Karathane [Dow AgroSci.] see 3314
Karaya Gum, 5332
Karbam Black see 4040
Karbinon [Badrial] see 6479
Kardiamed [Medice] see 3186
Karenitecin [BioNumerik] see 2538
Karidium see 8754
Karil [Novartis] see 1646
Karmex see 3425
Karmex Monuron Herbicide see 6349

Karminomycin see 1870
Karsil, 5333
Karsivan [Provet] see 7927
Karstenite see 1708
Karsulphan see 9072
"Kastle-Meyer Reagent" see 7357
KAT-256 see 2360
Katadolon [AWD] see 4221
Katana [Ishihara] see 4129
Katchung Oil see 7165
Kathon [Rohm & Haas] see 6844
Kathon CG [Rohm & Haas] see 6163
Katilo see 5332
Katine see 6803
Katlex [Iwaki] see 4338
Katonil [Kali-Chemie] see 2105
Katoseran [Hishiyama] see 2306
Katovit [Thomae] see 7896
Kaurie see 2502
Kautschin see 5546
Kava, 5334
Kavahin see 6214
Kavain see 5335
Kava-kava see 5334
Kavatin see 6214
Kawa see 5334
Kawain, 5335
Kay-Cee-L [Geistlich] see 7742
Kaydol [Sonneborn] see 7301
Kayexalate [Sanofi-Aventis] see 8797
Kayhydrin see 7492
Kayquinone [Abbott] see 5899
Kaytwo [Eisai] see 5901
KB-227 see 9727
KB-1585 see 5491
KB-2115 see 3692
KB-2413 see 3614
KB-2796 see 5620
KB-3022 see 7103
KBR-2738 see 4005
KBR 3023 see 7506
KBT-1585 see 5491
KBT-3022 see 7103
KC-404 see 4918
KC-2547 see 5990
KC-8857 see 9246
KC-9147 see 9669
K-CIT-V [V.E.T.] see 7744
K-Contin [Mundipharma] see 7742
KD-136 see 4634
K-Dur [Schering-Plough] see 7742
Keal [Sinbio] see 9010
Kebuzone, 5336
Kedacillina [Bracco] see 9022
Kefadol [Lilly] see 1916
Kefir Fungi, 5337
Kefir Grains see 5337
Kefir Seeds see 5337
Keflex [Lilly] see 1974
Keflin [Sandoz] see 1981
Keflor [Alpharma] see 1914
Keforal [Lab Sciencex] see 1974
Kefzol [Lilly] see 1918
Keimax [Essex] see 1950
Keiperazon [Kaken] see 1919
Kelcogel [CP Kelco] see 4417
Kelecin see 5483
Kelene see 3837
Kelfer [Cipla] see 2862
Kelfiprim [Farmitalia] see 9042
Kelfizina [Farmitalia] see 9042
Kelfizine W [Farmitalia] see 9042
Kelgum [CP Kelco] see 10255
Kelnac [Sankyo] see 7647
Kelocyanor [Serb] see 3565

Kelp [Omni] *see* 846
Kelpware *see* 4311
Kelthane [Dow AgroSci.] *see* 3096
Keltrol [CP Kelco] *see* 10255
Kelzan [CP Kelco] *see* 10255
Kemadrin [Glaxo Wellcome] *see* 7881
Kemicetine [Pharmacia] *see* 2077
Kemithal *see* 9444
Kemsol [Axxess Pharma] *see* 3285
Kenacort [BMS] *see* 9757
Kenacort-A [BMS] *see* 9758
Kenaf, 5338
Kenalog [Apothecon] *see* 9758
Kendall's Compound A *see* 2873
Kendall's Compound B *see* 2524
Kendall's Compound C *see* 265
Kendall's Compound E *see* 2525
Kendall's Compound F *see* 4824
Kendall's Desoxy Compound B *see* 2902
Kenicef *see* 1929
Kentera [UCB] *see* 7054
Kenzen [Takeda] *see* 1742
Keoxifene *see* 8215
Kephalins *see* 1975
Kepinol [Pfleger] *see* 9050
Kepivance [Amgen] *see* 7086
Kepone *see* 2084
Keppra [UCB] *see* 5513
Keraphen *see* 5081
Keracyanin *see* 2677
Keral [Menarini] *see* 5352
Keralyt [BMS] *see* 8469
Keratin, 5339
Keratinamin [Kowa] *see* 10052
Keratinase, 5340
24-163-Keratinocyte Growth Factor
  (Human) *see* 7086
Keratol [Streuli] *see* 10359
Kerb [Dow AgroSci.] *see* 7983
Kerecid *see* 4931
Kerlocal [Pierre Fabre] *see* 8288
Kerlone [Synthelabo] *see* 1184
Kermesic Acid, 5341
Kernechtrot *see* 6573
Kerosene, 5342
Kerosine *see* 5342
Kerosine (Petroleum) *see* 5342
Kertasin [Merck KGaA] *see* 3914
Keselan [Sumitomo] *see* 4634
Kessar [Pharmacia] *see* 9180
Kessazulen *see* 4590
Kestin [Almirall] *see* 3531
Kestine [Almirall] *see* 3531
Kestrone [Cooper] *see* 3763
Ket [Irbi] *see* 5344
Ketalar [Pfizer] *see* 5343
Ketalgin [Amino] *see* 6016
Ketamine, 5343
Ketanest [Pfizer] *see* 5343
Ketanserin, 5344
Ketas [Kyorin] *see* 4918
Ketaset [Fort Dodge] *see* 5343
Ketavet [Gellini] *see* 5343
Ketazolam, 5345
Ketazon [Leciva] *see* 5336
Ketek [Aventis] *see* 9261
Ketene, 5346
Ketensin [Janssen] *see* 5344
Kethoxal, 5347
Ketobemidone, 5348
[3′-Keto-Bmt1]-[Val2]-cyclosporin *see*
  10103
β-Ketobutyranilide *see* 57
2-Ketobutyric Acid *see* 58
3-Ketobutyric Acid *see* 58

4-Keto-β-carotene *see* 3542
Ketochromin *see* 3200
Ketoconazole, 5349
Ketocycloheptane *see* 2715
Ketocyclopentane *see* 2738
Ketoderm [Janssen-Cilag] *see* 5349
3-Ketodesogestrel *see* 3928
Ketodestrin *see* 3763
(dl)-9-Keto-11α,15α-dihydroxy-16-phen-
  oxy-17,18,19,20-tetranorprosta-4,5,13-
  trans-trienoic Acid Methyl Ester *see*
  3645
Ketodur [Pfizer] *see* 5348
17-(1-Ketoethyl)androstane-3,17-diol *see*
  263
17β-(1-Ketoethyl)-Δ⁵-androsten-3β-ol
  see 7851
Ketofen [Fort Dodge] *see* 5352
Ketogan
Ketogan Novum [Pfizer] *see* 5348
3-Keto-D-glucoheptonofuranolactone *see*
  4486
α-Ketoglutaric Acid, 5350
β-Ketoglutaric Acid *see* 67
2-Keto-L-gulonic Acid, 5351
Ketoheptamethylene *see* 2715
D-manno-Ketoheptose *see* 4701
Ketohexamethylene *see* 2719
2-Ketohexamethylenimine *see* 1763
D-erythro-3-Ketohexonic Acid Lactone
  see 5171
D-ribo-2-Ketohexose *see* 8033
L-3-Keto-threo-hexuronic Acid Lactone
  see 819
Ketohydroxyestrin *see* 3763
17-(1-Keto-2-hydroxyethyl)androstane-
  3,11,17-triol *see* 265
17-(1-Keto-2-hydroxyethyl)-Δ⁴-andro-
  sten-3,11-dione *see* 2873
17-(1-Keto-2-hydroxyethyl)-4-androsten-
  17α-ol-3-one *see* 2933
3-[β-Keto-γ-(3-hydroxy-2-piperidyl)pro-
  pyl]-4-quinazolone *see* 3982
Ketoisdin [Isdin] *see* 5349
Ketomalonic Acid *see* 5983
γ-Keto-β-methoxy-δ-methylene-Δ^α-hex-
  enoic Acid *see* 7201
1-Keto-6-methoxy-1,2,3,4-tetrahydro-
  naphthalene *see* 6074
5-Keto-2-methoxy-5,6,7,8-tetrahydro-
  naphthalene *see* 6074
Ketopentamethylene *see* 2738
D-erythro-2-Ketopentose *see* 8331
Ketophenylbutazone *see* 5336
Ketoprofen, 5352
Ketopron [Alcon] *see* 5352
β-Ketopropane *see* 65
2-Ketopropionaldehyde *see* 6153
α-Ketopropionic Acid *see* 8135
2-Ketopyrrolidine *see* 8130
2-Ketopyrrolidine-1-ylacetamide *see*
  7600
Ketorax [Pfizer] *see* 5348
Ketorolac, 5353
Ketoscilium [Zambon] *see* 4033
Ketosuccinic Acid *see* 7008
1-Keto-3-(3′-sulfamyl-4′-chlorophenyl)-
  3-hydroxyisoindoline *see* 2200
Ketotifen, 5354
17-Ketotrilostane *see* 9867
2-Keto-1,7,7-trimethylnorcamphane *see*
  1734
Kevopril [Rhone-Poulenc] *see* 8200
Kew Tree *see* 4460
K-Exit [Omega] *see* 8797

Keyhole-limpet Hemocyanin *see* 4682
Keypyrone [Key] *see* 3390
KF-868 *see* 5703
Khas Khas Oil *see* 6908
Khat, 5355
Khellin, 5356
Khinocyde *see* 8183
Khloratsizin *see* 2068
Khus Oil *see* 6908
Kiditard [Belga] *see* 8176
Kidonax [Unipharm] *see* 6653
Kidrolase [OPi] *see* 826
Kieselguhr *see* 5015
Kieserite *see* 5717, 5755
KIF-230 *see* 1055
KIF-3535 *see* 5912
Kifunensine, 5357
KIH-485 *see* 8123
KIH-2023 *see* 1302
KIH-2031 *see* 8106
KIH-8921 *see* 8106
KIH-9201 *see* 4239
Kiku Oil, 5358
Killax *see* 9349
Kilmicen [Farmitalia] *see* 9669
Kilmite 40 *see* 9349
KIM-112 *see* 7892
Kimopsin [Eisai] *see* 2264
Kimoral [Nippon Chemiphar] *see* 2264
Kinaden *see* 4797
Kinadon [United Phosphorus] *see* 7449
Kinavet CA1 [AB Science] *see* 5822
Kinecid [Schering AG] *see* 9046
Kinedak [Ono] *see* 3660
Kineorl [Showa Shinyaku] *see* 9681
Kineret [Amgen] *see* 5045
Kinetin [Schering] *see* 4797
Kinetin, 5359
Kinevac [BMS] *see* 8681
King's Gold *see* 797
King's Yellow *see* 797, 5460
Kinic Acid *see* 8175
Kinichron [Biochimica] *see* 8176
Kinidin Durules [Astra] *see* 8176
Kininogens, 5360
Kino, 5361
Kinotomin [Toa Eiyo] *see* 2343
Kino-yellow *see* 5706
Kinupril [Bellon] *see* 8200
Kinzalmono [Bayer] *see* 9271
Kionex [Paddock] *see* 8797
Kir Richter [Lepetit] *see* 746
Kirappu [Bayer CropSci.] *see* 3793
Kirocid [Schering AG] *see* 9046
Kiron [Schering AG] *see* 9046
Kirromycin *see* 6311
Kisselo-mleko *see* 10299
Kistrin, 5362
Kistrin (Agkistrodon rhodostoma
  Reduced) *see* 5362
Kitasamycin *see* 5505
Kit Ligand *see* 8934
Kitol, 5363
Kixor [BASF] *see* 8451
KL4 *see* 8680
KL-255 *see* 1500
KL-373 *see* 1238
Klacid [Abbott] *see* 2338
Klaricid [Abbott] *see* 2338
Klavocin [Pliva] *see* 2339
Klebcil [SKB] *see* 5329
Kleenodyne *see* 5080
KLH *see* 4682
Klimax E [Fink] *see* 3762
Klimicin [Lek] *see* 2354

Klinit [Eisai] *see* 10283
Klinium [Janssen] *see* 5537
Klinomycin [Valeant] *see* 6283
Klion [Gedeon Richter] *see* 6236
Klismacort [Novartis] *see* 7841
Kloben Neburon [DuPont] *see* 6519
Klockmannite *see* 2641
Klonopin [Roche] *see* 2379
Klot [Warren-Teed] *see* 9679
Klotogen [Abbott] *see* 5899
KL₄-surfactant *see* 8680
Klucel [Hercules] *see* 4879
K-Lyte [Apothecon] *see* 7730
KMD-3213 *see* 8642
Knallquecksilber *see* 5967
Knight's Spur *see* 5424
Knob Root *see* 2467
Knoll H₇₅ *see* 7442
Ko-1173 *see* 6248
Ko-1366 *see* 1493
Koate-HP [Bayer] *see* 3962
Kochite *see* 356
Kodel [Eastman-Kodak], 5364
Kodocytochalasin-1 *see* 2787
Koerbl's Catalyst *see* 8663
Kofler's Quinone *see* 7637
Kogenate *see* 3962
Koglucoid [Panray] *see* 8238
Koji *see* 9162
Kojic Acid, 5365
Kokam *see* 4398
Kokam Butter *see* 4398
Kokum *see* 4398
Kola, 5366
Kollateral [Ursapharm] *see* 6374
Kollerdormfix *see* 2704
Kollidon [BASF] *see* 7814
Kollidon CL [BASF] *see* 7814
Kolpon [Organon] *see* 3763
Kolton [Promonta] *see* 3370
Kombetin [Boehringer, Mann.] *see* 8982
Kombiquens [Novo] *see* 5876
Komet [Zeneca Ag Prod] *see* 9251
Kompensan [Roerig] *see* 3202
Konakion [Roche] *see* 7492
Kondurangin *see* 2479
Konesta [Akzo] *see* 9792
Koninckite *see* 4056
Konjac, 5367
Konjac Glucomannan *see* 5367
Konjac Mannan *see* 5367
Konlax [Nippon Shinyaku] *see* 7859
Konnyaku *see* 5367
Kon Oil *see* 5704
Kontexin [Pfizer] *see* 7417
Konyne [Miles] *see* 3963
Kopsine, 5368
Korazol *see* 7254
Korbutone [Nippon Glaxo] *see* 1017
Korec [Sanofi-Aventis] *see* 8167
Korglykon *see* 2497
Korlan *see* 8387
Koromite [Sankyo] *see* 6269
Koroseal [Goodrich] *see* 7707
KOS-862 *see* 3684
Koser's Reagent, 5369
Kosins, 5370
Koso *see* 1371
Kosso *see* 1371
Kotetsu [Nippon Soda] *see* 2089
K-Othrine [Bayer CropSci.] *see* 2888
Kotoite *see* 5722
Kouso *see* 1371
Kousso *see* 1371
KP-102 *see* 7824

KP-363 *see* 1521
KPABA *see* 7727
KPB *see* 5336
KPU-2 *see* 7654
Kraton [Shell] *see* 8527
Krazy Glue [Toagosei] *see* 3842
Krebiozen, 5371
Krebon *see* 1481
Kredex [GSK] *see* 1873
Kremol [Sonneborn] *see* 7301
Kren *see* 4788
Kresoxim-methyl, 5372
Krestin [Kureha] *see* 7702
Kristallose *see* 8445
KRM-1648 *see* 8339
KRN-1493 *see* 2285
KRN-8601 *see* 4573
Kronitex TCP [FMC] *see* 9940
KRX-101 *see* 9116
KRX-0401 *see* 7280
Kryolith *see* 2595
Kryptocur *see* 4561
Kryptofix 222 [Merck KGaA] *see* 2596
Krypton, 5373
Krypton Difluoride *see* 5373
Kryptosterol *see* 5411
Krysid *see* 708
Krystexxa [Savient] *see* 7177
K-Selectride *see* 8564
K-Strophanthin *see* 8982
K-Strophanthin-α *see* 2758
K-Strophanthoside *see* 8982
KT-30 *see* 4262
KT-611 *see* 6440
KT-3777 *see* 5633
K-Thrombin [Fawns & McAllan] *see* 5899
KTU 3616 *see* 1863
K₂tu-PA *see* 385
KU-54 *see* 9970
KU-0059436 *see* 6915
Kubacron [Kayaku] *see* 9789
KUE 13032c *see* 3053
KUF-6201 *see* 5912
KUH-833 *see* 7892
KUH-911 *see* 1302
Kujimycin B *see* 5407
Kukoline *see* 8684
Kullo *see* 5332
Kupferkies *see* 2037
Kurchatovium *see* 8437
Kurnakovite *see* 5722
Kuromanin *see* 2677
Kuromatsuene *see* 5624
Kusso *see* 1371
Kusum Oil *see* 5704
Kuteera *see* 5332
Kutkasin [Lääkefarmos] *see* 9080
Kuvan [BioMarin] *see* 8505
Ku-Zyme [Schwarz] *see* 7108
KVX-478 *see* 585
KW-110 *see* 26
KW-1062 *see* 6260
KW-2228 *see* 4573
KW-2307 *see* 10187
KW-3049 *see* 1045
KW-4679 *see* 6934
KW-6002 *see* 5289
KWD-2183 *see* 946
KWG-4168 *see* 8889
Kyamepromazine *see* 2673
Kyanmethin, 5374
Kyanol *see* 652
Kylar [Uniroyal] *see* 2809
Kylit [Taiho] *see* 10283

Kymo-trypure *see* 2264
Kynapid [Astellas] *see* 10157
Kynurenic Acid, 5375
Kynurenine, 5376
Kytril [Roche] *see* 4572
Kyurinett [Zensei] *see* 9874
L1 *see* 2862
L-67 *see* 7862
L-86-8275 *see* 4126
L-104 *see* 2381
L-105 *see* 8345
L-141 *see* 3985
L-300 *see* 9657
L-554 *see* 9942
L-627 *see* 1201
L-1102 *see* 8470
L-1573 *see* 2776
L-1913 *see* 7270
L-2214 *see* 1067
L-2329 *see* 1085
L-3428 *see* 478
L-5458 *see* 2865
L-6257 *see* 7033
L-12717 *see* 5643
L-34314 *see* 3337
L-35355 *see* 3381
L-36352 *see* 9854
LA-1 *see* 6661
LA-956 *see* 7188
LA-1221 *see* 1513
LA-6023 *see* 6010
LA-III *see* 2997
LAAM *see* 5519
LAAO *see* 409
Lab *see* 8254
Labelol [ELEA] *see* 5377
Labetalol, 5377
Lab Ferment (German) *see* 8254
Labican [Boniscontro] *see* 2085
Labile Bixin *see* 1312
Labile Factor *see* 3960
Labitan [Labinca] *see* 3222
Labosept [L.A.B.] *see* 2914
Labrocol [Lagap] *see* 5377
Labroda [Specia] *see* 4138
Labrodax [Specia] *see* 4138
Labyrin *see* 2306
Lac *see* 8617
LAC-43 *see* 1499
Lacca *see* 8617
Lacca Coerulea *see* 5603
Laccaic Acid, 5378
Laccaic Acid A *see* 5378
Laccaic Acid A₁ *see* 5378
Laccaic Acid B *see* 5378
Laccaic Acid C *see* 5378
Laccaic Acid D *see* 5378
Lacca Musica *see* 5603
Lac Dye *see* 5378
Lacea [Mitsui Toatsu] *see* 5386
Lacflavin [Towa Yakuhin] *see* 8325
LACI *see* 9623
Lacidipine, 5379
Lacimen [Menarini] *see* 5379
Lacipil [GSK] *see* 5379
Lacirex [Guidotti] *see* 5379
Lacmoid, 5380
Lacmus *see* 5603
Lacolin *see* 8767
Lacosamide, 5381
Lacrimin [Santen] *see* 1049
Lacrisert [Merck & Co.] *see* 4879
Lacrypos [Alcon] *see* 2219
Lactacidogen *see* 4305
β-Lactamase *see* 7202

Lactamin [Teisan] *see* 7855
Lactaminic Acid *see* 8623
Lactate Dehydrogenase, 5382
D-Lactic Acid, 5383
D(−)-Lactic Acid *see* 5383
DL-Lactic Acid, 5384
L-Lactic Acid, 5385
L(+)-Lactic Acid *see* 5385
d-Lactic Acid *see* 5385
l-Lactic Acid *see* 5383
(*R*)-(−)-Lactic Acid *see* 5383
(*S*)-(+)-Lactic Acid *see* 5385
DL-Lactic Acid Ammonium Salt *see* 527
Lactic Acid Cetyl Ester *see* 2029
Lactic Acid Hexadecyl Ester *see* 2029
Lactic Acid Homopolymer, 5386
Lactic Acid Lactate, 5387
Lactic Acid Methyl Ester *see* 6166
Lactic Dehydrogenase *see* 5382
D-Lactide Homopolymer *see* 5386
Lactinium [Roland] *see* 6972
Lactisole, 5388
Lactit *see* 5389
Lactite *see* 5389
Lactit M *see* 5389
Lactitol, 5389
Lactobacillamide *see* 5390
Lactobacillic Acid, 5390
Lactobacillus bifidus Factor *see* 1215
Lactobacillus bifidus Growth Factor *see* 1215
Lactobacillus lactis Dorner Factor *see* 10212
Lactobionic Acid, 5391
Lactobionic δ-Lactone *see* 5391
Lactobiosit *see* 5389
Lactochrome *see* 8325
Lactoferrin *see* 9732
Lactoferrin (Recombinant Human LF00) *see* 9163
Lactoflavine *see* 8325
Lactogen *see* 7894
Lactogen (Human Placental) *see* 4655
Lactogenic Hormone (Placental Human) *see* 4655
Lactose, 5392
Lactosit *see* 5389
Lactositol *see* 5389
Lactosylceramide *see* 2791
Lactotransferrin *see* 9732
Lactovagan [Schwartzhaupt] *see* 5384
2-(Lactoyloxy)propanoic Acid *see* 5387
Lactron [Kanebo] *see* 5386
Lactucarium, 5393
Lactucerin *see* 9193
α-Lactucerol *see* 9193
Lactucin, 5394
Lactucon *see* 9193
Lactucopicrin *see* 5394
Lactuflor [Chephasaar] *see* 5395
Lactulose, 5395
Lacty [Shimadzu] *see* 5386
Lacty [CCA Biochem] *see* 5389
Ladakamycin *see* 884
Ladogal [Sanofi-Synthelabo] *see* 2811
Ladropen [Berk] *see* 4143
Lady's Slipper *see* 2767
Laetrile *see* 592
Laevilac [Fresenius] *see* 5395
Laevoral [Laevosan] *see* 4302
Laevosan [Laevosan] *see* 4302
Laevulinic Acid *see* 5526
LAF *see* 5037
LAF-237 *see* 10177
Lafutidine, 5396

Lagistase [Labaz] *see* 3602
Lagosin *see* 4319
Laguncurin *see* 5706
Laidlomycin, 5397
Laki-Lorand Factor *see* 3967
LAM-1 *see* 8563
Lamar [Tokyo Tanabe] *see* 9252
Lambda-cyhalothrin, 5398
Lambdamycin *see* 2044
Lambeth [Lambeth Rope] *see* 7701
Lamb Mint *see* 7258
Lamictal [GSK] *see* 5403
Lamifiban, 5399
Laminaran, 5400
Laminaran Hydrogen Sulfate *see* 5400
Laminarin *see* 5400
Laminin, 5401
Laminin M *see* 5401
Lamisil [Novartis] *see* 9299
Lamivudine, 5402
Lamoryl [Leo Pharm] *see* 4584
Lamotrigine, 5403
Lamoxactam *see* 6372
Lamp Black *see* 1808
Lampit [Bayer] *see* 6622
Lamprecid [Clariant] *see* 9401
Lampren(e) [Novartis] *see* 2367
Lampterol *see* 4940
Lamra [Merckle] *see* 2997
Lamuran [Boehringer, Mann.] *see* 8237
Lanacort [Combe] *see* 4824
Lanadigenin *see* 3185
Lanarkite *see* 5475
Lanatilin [Wabosan] *see* 3186
Lanatosides, 5404
Lanatoxin [Beiersdorf] *see* 3182
Lancetina [Farma Uno] *see* 4281
Landamycine [Delalande] *see* 8330
Landel [Zeria] *see* 3571
Landiolol, 5405
Landomycin *see* 6918
Landruma [Landerln] *see* 6615
Laneth *see* 7697
Lanexat [Roche] *see* 4166
Lanfrax [Henkel] *see* 5409
Langite *see* 2644
Langoran [Aventis] *see* 5271
Lanicor [Roche] *see* 3186
Lanimerck [Merck KGaA] *see* 5404
Lanimostim *see* 5710
Laninamivir, 5406
Laninamivir Octanoate *see* 5406
Lanirapid [Roche] *see* 3186
Lanitop [Roche] *see* 3186
Lankamycin, 5407
Lankavose *see* 2038
Lannate [DuPont] *see* 6054
Lanoconazole, 5408
Lanogene [Amerchol] *see* 5409
Lanolin, 5409
Lanolin Alcohols, 5410
Lanolin Oil *see* 5409
Lanolin Wax *see* 5409
(3β)-Lanosta-8,24-dien-3-ol *see* 5411
Lanosterol, 5411
Lanoxicaps [GSK] *see* 3186
Lanoxin [GSK] *see* 3186
Lanreotide, 5412
Lansfordite *see* 5724
Lansoprazole, 5413
Lansox [Takeda] *see* 5413
Lantadin [Lepetit] *see* 2865
Lantanon [Ravasini] *see* 6251
Lanthana *see* 5414
Lanthanum, 5414

Lanthanum Carbonate, 5415
Lanthanum Chloride (LaCl₃) *see* 5414
Lanthanum Hydroxide (La(OH)₃) *see* 5414
Lanthanum Oxide (La₂O₃) *see* 5414
Lanthanum Sesquicarbonate *see* 5415
Lanthanum Sesquioxide *see* 5414
Lanthanum Trioxide *see* 5414
Lanthionine, 5416
L-Lanthionine *see* 5416
Lantrol 1673 [Henkel] *see* 5409
Lantus [Aventis] *see* 5027
Lanvis [GSK] *see* 9490
Lanzor [Sanofi-Aventis] *see* 5413
Lanzyme [Nissui] *see* 5701
Lapachic Acid *see* 5417
Lapachol, 5417
Lapatinib, 5418
Lapatinib Ditosylate *see* 5418
Lapis Lazuli *see* 5450
Lappa, 5419
Lappaconitine, 5420
Lapudrine [Zeneca] *see* 2190
Lapyrium Chloride, 5421
Laquinimod, 5422
Laracor [Lagap] *see* 7051
Laradopa [Roche] *see* 5516
Laraflex [Lagap] *see* 6499
Laragon [Roemmers] *see* 6274
Larapam [Sandoz] *see* 7619
Laratrim [Lagap] *see* 9050
Larch Agaric *see* 177
Larch Turpentine *see* 10139
Lard, 5423
Lard-factor *see* 10211
Lard Oil *see* 5423
Laredo [Dow AgroSci.] *see* 6406
Larex [Geistlich] *see* 7694
Largactil [Aventis] *see* 2191
Large Fennel *see* 4008
Large-flowered Cereus *see* 1612
Largon [Wyeth] *see* 7936
Lariam [Roche] *see* 5872
Laricic Acid *see* 178
Laricin *see* 2487
Laridal [Elfar-Drag] *see* 846
Laristine [Roche] *see* 4758
Larixin *see* 1974
Larixinic Acid *see* 5777
Lark's-claw *see* 5424
Lark's-heel *see* 5424
Larkspur, 5424
Larnite *see* 1704
Larocaine [Roche] *see* 3242
Larodopa [Roche] *see* 5516
Laromustine, 5425
Laronidase *see* 4935
Laropiprant, 5426
Laroscorbine [Roche] *see* 819
Larostidin [Roche] *see* 4758
Larotid [Roche] *see* 574
Laroxyl [Roche] *see* 483
Larvacide 100 *see* 2159
Larvadex [Novartis] *see* 2772
Larvin [Bayer CropSci.] *see* 9482
LAS *see* 5698
LAS-11871 *see* 7390
LAS-17177 *see* 2295
LAS-31416 *see* 298
LAS-34273 *see* 106
Lasalocid A, 5427
Laser [Tosi] *see* 6499
Laserphyrin [Meiji] *see* 9166
Laserpitin, 5428
Lasilix [Aventis] *see* 4338

Lasiocarpine, 5429
Lasix [Aventis] see 4338
Lasma [Pharmax] see 9436
Lasofoxifene, 5430
Lasso [Monsanto] see 194
Lastet [Nippon Kayaku] see 3929
Lasurite see 5450
LAS-W-090 see 3531
LAT-1717 see 7552
LAT-A see 5433
Latamoxef see 6372
Latanoprost, 5431
LAT-B see 5433
Latibon [Bayer] see 3119
Latisse [Allergan] see 1225
Latocillin [Chassot] see 2403
Latoconazole see 5408
α-Latrotoxin, 5432
Latrunculin A see 5433
Latrunculin B see 5433
Latrunculins, 5433
Latuda [Dainippon] see 5673
Latumcidin see 8
Laubanite see 1651
Laudamonium [Henkel] see 1061
Laudanidine, 5434
dl-Laudanidine see 5435
Laudanine, 5435
l-Laudanine see 5434
Laudanosine, 5436
Laudis [Bayer CropSci.] see 9276
Laughing Gas see 6740
Laumontite see 1651
Laurabolin [Intervet] see 6450
Laurel, 5437
Laurel Berry Oil see 5437
Laurel Camphor see 1734
Laureline, 5438
Laurel Leaf Oil see 5437
Laurent's Acid see 6486
Laureth see 7697
Laureth 9 see 7675
Lauric Acid, 5439
Laurocapram, 5440
Laurodin see 5442
Lauroguadine, 5441
Laurolinium Acetate, 5442
Lauromicina [Dukron] see 3740
Laurostearic Acid see 5439
Laurotetanine, 5443
N-(Lauroylcolaminoformylmethyl)pyri-
   dinium Chloride see 5421
N-Lauroylsarcosine Sodium Salt see
   4401
Lauryl Alcohol see 3451
Lauryl Betaine see 3453
Lauryl Bromide, 5444
Lauryl Sulfate Salt of the Propionic
   Acid Ester of Erythromycin see 3740
Lautarite see 1678
Lauth's Violet see 9499
Lavandin see 5445
Lavandin Oil see 5445
Lavasept [Fresenius] see 7677
Lavender, 5445
Lavender Oil see 5445
Laverin [Lemmon] see 3783
Lawesson's Reagent, 5446
Lawrencite see 4070
Lawrencium, 5447
Lawsone, 5448
Lawsonite see 1651
Laxoberal [Boehringer, Ing.] see 7518
Laxoberon [Ferrosan] see 7518
Laxo-Isatin see 7073

Laxonalin see 1296
Layor Carang see 176
Lazabemide, 5449
Lazaroids see 9619
Lazurite, 5450
LB-46 see 7554
LB-502 see 4338
LB-20304 see 4423
LBH-589 see 7113
LC-33 see 5493
LC-44 see 4219
LCB-29 see 4933
LCR see 10183
LDA see 3215
LDP-03 see 222
LEA-103 see 8389
LEA29Y see 1023
Lead, 5451
Lead Acetate, 5452
Lead(II) Acetate see 5452
Lead(IV) Acetate see 5478
Lead Antimonate(V), 5453
Lead Arsenate, 5454
Lead Arsenite, 5455
Lead Azide, 5456
Lead Azide (Pb(N₃)₂) see 5456
Lead Benzoate see 1093
Lead Borate, 5457
Lead Bromide, 5458
Lead(II) Bromide see 5458
Lead Bromide (PbBr₂) see 5458
Lead Carbonate Hydroxide (Pb₃(CO₃)₂-
   (OH)₂) see 1002
Lead Chloride, 5459
Lead(2+) Chloride see 5459
Lead(II) Chloride see 5459
Lead Chloride (PbCl₂) see 5459
Lead Chromate(VI), 5460
Lead Chromate Oxide (Pb₂(CrO₄)O) see
   5461
Lead Chromate(VI) Oxide, 5461
Lead Chromium Oxide (PbCrO₄) see
   5460
Lead Difluoride see 5463
Lead Diformate see 5464
Lead Dioxide, 5462
Lead Distearate see 5473
Lead Fluoride, 5463
Lead Fluoride (PbF₂) see 5463
Lead Fluosilicate see 5465
Lead Formate, 5464
Lead(II) Formate see 5464
Lead Hexafluorosilicate, 5465
Lead Hydroarsenate see 5454
Lead Hydrogen Arsenate see 5454
Lead Iodide, 5466
Lead(II) Iodide see 5466
Lead Iodide (PbI₂) see 5466
Lead Metavanadate see 5482
Lead Molybdate(VI), 5467
Lead Molybdenum Oxide (PbMoO₄) see
   5467
Lead Monosubacetate see 5474
Lead Monoxide, 5468
Lead Nitrate, 5469
Lead Orthoplumbate see 5480
Lead Oxalate, 5470
Lead(II) Oxalate see 5470
Lead(II) Oxide see 5468
Lead(IV) Oxide see 5462
Lead Oxide Brown see 5462
Lead Oxide (PbO) see 5468
Lead Oxide (PbO₂) see 5462
Lead Oxide (Pb₂O₃) see 5472
Lead Oxide (Pb₃O₄) see 5480

Lead Oxide Red see 5480
Lead Oxide Yellow see 5468
Lead Peroxide see 5462
Lead Phosphate, 5471
Lead(II) Phosphate see 5471
Lead Protoxide see 5468
Lead Sesquioxide, 5472
Lead Silicofluoride see 5465
Lead-sodium Alloy see 8769
Lead Stearate, 5473
Lead Subacetate, 5474
Lead Subcarbonate see 1002
Lead Sulfate, 5475
Lead(II) Sulfate see 5475
Lead Sulfide, 5476
Lead(II) Sulfide see 5476
Lead Sulfide (PbS) see 5476
Lead Sulfocyanate see 5481
Lead Superoxide see 5462
Lead Telluride, 5477
Lead(2+) Telluride see 5477
Lead Telluride (PbTe) see 5477
Lead Tetraacetate, 5478
Lead Tetraethyl see 9347
Lead Tetrafluoride, 5479
Lead Tetroxide, 5480
Lead Thiocyanate, 5481
Lead(II) Thiocyanate see 5481
Lead Trioxide see 5472
Lead Vanadate(V), 5482
Lead Vanadium Oxide (PbV₂O₆) see
   5482
Leaf Alcohol see 4736
Leaf Green see 2238
Lealgin [Janssen] see 7362
Leandin see 2670
Leanol [Yoshitomi] see 4743
Leanstar [Pitman-Moore] see 8842
Lebaycid [Bayer CropSci.] see 4029
Leben Raib see 10299
Leblon [De Angeli] see 7604
LECAM-1 see 8563
Lecasol [Kaken] see 2343
Lecedil [Zdravlje] see 3971
Lecibis [Columbia] see 1150
Lecibral [Nezel] see 6580
Lecithins, 5483
Lecithol see 5483
Lecs [Bayer CropSci.] see 4034
Lectins, 5484
Lectopam [Roche] see 1392
Ledakrin [Polfa] see 6657
Ledclair [Sinclair] see 3565
Ledercort Cream [Lederle] see 9758
Ledercort (Tabl.) [Lederle] see 9757
Lederfen [Am. Cyanamid] see 3996
Lederfolat [Lederle] see 4249
Lederfolin [Lederle] see 4249
Lederkyn [Wyeth] see 9051
Lederlon [Lederle] see 9759
Ledermycin [Wyeth] see 2890
Lederspan [Lederle] see 9759
Ledertrexate [Wyeth] see 6057
Ledol, 5485
Ledopa [Lepetit] see 5516
"Ledum Camphor" see 5485
Lefax [Asche] see 3237
Leflunomide, 5486
Leftose [Nippon Shinyaku] see 5701
Legalon [Madaus] see 6274
Legederm [Schering] see 213
Legential see 4433
Leghemoglobin, 5487
Legoglobin see 5487
Legumex D see 2838

Legumin *see* 1883
Lehydan [Abigo] *see* 7433
LeIF *see* 5034
Lei Gong Teng *see* 9549
Leipzig Yellow *see* 5460
Lektinol [Madaus] *see* 6294
Lemakalim *see* 5512
Lemoflur *see* 8754
Lemonol *see* 4437
Lemon Peel, 5488
Lemon Walnut *see* 5316
Lemon Yellow *see* 968
Lemoran *see* 5524
Lenacil, 5489
Lenalidomide, 5490
Lenampicillin, 5491
Lenamycin *see* 1964
Lendorm [Boehringer, Ing.] *see* 1459
Lendormin [Boehringer, Ing.] *see* 1459
Lenicet *see* 359
Lenidolor [Menarini] *see* 5850
Lenigallol [Knoll] *see* 8112
Lenirit [Bonomelli] *see* 4824
Lenisarin *see* 3428
Lenitral [Besins-Iscovesco] *see* 6694
Lenograstim *see* 4573
Lenopect [Draco] *see* 7569
Lenoxin [GSK] *see* 3186
Lensan A [CIBA Vision] *see* 4835
Lentagran [Syngenta] *see* 8079
Lentaron [Novartis] *see* 4268
Lenthionine, 5492
Lentinacin *see* 3728
Lentinan, 5493
Lentisk *see* 5823
Lentizol [Lundbeck] *see* 483
Lentogest [Amsa] *see* 4877
Lento-Kalium [Teofarma] *see* 7742
Lentopenil [Grossman] *see* 7205
Lentotran [Farmopatria] *see* 2085
Lentrat *see* 7222
Lentysine *see* 3728
Lenvatinib, 5494
Lenvatinib Mesylate *see* 5494
Leo 640 *see* 5614
Leocillin [Leo Pharm] *see* 7194
Leodrine [Leo Pharm] *see* 4826
Leonurine, 5495
Leopard's Bane *see* 780
Leparan [Italfarmaco] *see* 4688
Lepargylic Acid *see* 898
Lepetan [Otsuka] *see* 1501
Lepicidin A *see* 8877
Lepidine, 5496
Lepidocrocite *see* 4052
Lepidolite *see* 2007, 5576, 8415
Lepidopterans *see* 1912
Lepirudin *see* 4755
Leponex [Novartis] *see* 2406
Leptanal [Janssen-Cilag] *see* 4028
Leptandra, 5497
Leptilan [Dolorgiet] *see* 10099
Leptin, 5498
Leptomycin B, 5499
Leptospermone *see* 5982
Leptryl [Bellon] *see* 7283
Lercanidipine, 5500
Lerdip [Recordati] *see* 5500
Lergefin [Larma] *see* 1800
Lergigan [Recip] *see* 7901
Lergine *see* 7881
Lergoban [3M Pharma] *see* 3370
Lergobine [3M Pharma] *see* 3370
Leritine [Merck & Co.] *see* 651
Lescodil [Zdravlje] *see* 6580

Lescol [Novartis] *see* 4245
Lesopitron, 5501
Lespedin *see* 5322
Lesser Centaury *see* 1970
Lestid [Pfizer] *see* 2459
Letairis [Gilead] *see* 378
Lethelmin *see* 7364
Lethidrone [Burroughs Wellcome] *see* 6446
Letrozole, 5502
LETS *see* 4104
Lettuce Opium *see* 5393
Letusin *see* 5523
Leu *see* 5503
Leu-8 *see* 8563
Leucaenine *see* 6280
Leucaenol *see* 6280
Leu-Cam Proteins *see* 5032
Leucenine *see* 6280
Leucenol *see* 6280
Leuchtenbergite *see* 345
Leucine, 5503
L-Leucine *see* 5503
23-L-Leucinecolony-Stimulating Factor 2 (Human Clone pHG25 Protein Moiety) *see* 4574
Leucine⁵-enkephalin *see* 3627
6-D-Leucine-9-(N-ethyl-L-prolinamide)-10-deglycinamideluteinizing Hormone-releasing Factor (Pig) *see* 5509
6-D-Leucine-9-(N-ethyl-L-prolinamide)-1-9-luteinizing Hormone-releasing Factor (Swine) *see* 5509
6-L-Leucine Sporidesmolide I *see* 8893
1-L-Leucine-2-L-threonine-63-desulfohirudin (*Hirudo medicinalis* Isoform HV1) *see* 4755
Leuco-4 *see* 140
Leucocyanidin, 5504
Leucocyanidol *see* 5504
Leucoharmine *see* 4649
Leucokinin *see* 9989
Leucoline *see* 8185
Leucomalachite Green *see* 5763
Leucomax [Schering-Plough] *see* 4574
Leucomycin, 5505
Leucomycin A₁ *see* 5505
Leucomycin A₃ *see* 5315
Leucomycin V 3-Acetate 4ᴮ-(3-Methylbutanoate) *see* 5315
Leucomycin V 4B-Butanoate 3B-Propanoate *see* 8379
Leucomycin V 3ᴮ,9-Diacetate 3,4ᴮ-Dipropanoate *see* 6286
Leucomycin V 3,4ᴮ-Dipropanoate *see* 6262
Leucophor R *see* 1316
Leucopterin, 5506
Leucosar [Adria] *see* 4249
Leucotrofina [Ellem] *see* 9557
Leucovorin *see* 4249
Leucovorin [Lederle] *see* 4249
(D-Leu⁶)-des-Gly¹⁰-LH-RH-ethylamide *see* 5509
Leu⁵-E *see* 3627
Leu⁵-Enkephalin *see* 3627
Leukaemomycin C *see* 2834
Leukeran [GSK] *see* 2073
Leukerin *see* 5938
Leukine [Immunex] *see* 4574
Leukochthol *see* 4924
Leukocyte Adhesion Molecules *see* 5032
Leukocyte Elastase *see* 3583
Leukocyte-Endothelial Cell Adhesion Molecule-1 *see* 8563

Leukocyte Integrins *see* 5032
Leukocyte Interferon *see* 5034
Leukocytic Endogenous Mediator *see* 5037
Leukoprol [Morinaga] *see* 5710
LeukoScan [Immunomedics] *see* 9027
Leukotriene A *see* 5507
Leukotriene A₄ *see* 5507
Leukotriene B *see* 5507
Leukotriene B₄ *see* 5507
Leukotriene C *see* 5507
Leukotriene C₁ *see* 5507
Leukotriene C₄ *see* 5507
Leukotriene D *see* 5507
Leukotriene D₄ *see* 5507
Leukotriene E *see* 5507
Leukotriene E₄ *see* 5507
Leukotrienes, 5507
Leunase [Kyowa] *see* 826
Leupeptin Ac-LL *see* 5508
Leupeptin Pr-LL *see* 5508
Leupeptins, 5508
Leuplin [Takeda] *see* 5509
Leuprolide, 5509
Leuprolide Acetate *see* 5509
Leuprorelin *see* 5509
Leurocristine *see* 10183
Leustatin [Ortho Biotech] *see* 2336
LeuTech [Palatin] *see* 9237
Leutrol [Abbott] *see* 10323
Levacecarnine Hydrochloride *see* 79
Levacide [Norbrook] *see* 5511
Levact [Napp] *see* 1036
Levadex [MAP] *see* 3191
Levadin [Univet] *see* 5511
Levalbuterol *see* 210
Levallorphan, 5510
Levamfetamine *see* 579
Levamisole, 5511
Levamphetamine *see* 579
Levant Wormseed *see* 8496
Levanxene [Montedison] *see* 9275
Levanxol [Montedison] *see* 9275
Levaquin [Ortho-McNeil] *see* 6863
Levarterenol *see* 6784
Levarterenol Bitartrate *see* 6784
Levasole [Mallinckrodt] *see* 5511
Levatol [Schwarz] *see* 7191
Levaxin [Nycomed] *see* 9570
Levcromakalim, 5512
Levemir [Novo] *see* 5026
Levetiracetam, 5513
Levitra [Bayer] *see* 10124
Levlen [Berlex] *see* 6793
Levobren [Knoll] *see* 9119
Levobunolol, 5514
Levobupivacaine *see* 1499
Levocabastine, 5515
Levocarbinoxamine *see* 1800
Levocarnil [Sigma-Tau] *see* 1849
Levocarnitine *see* 1849
Levocarnitine Chloride *see* 1849
Levocetirizine *see* 2021
Levocetirizine Hydrochloride *see* 2021
Levocitrol [Firmenich] *see* 2329
Levocycline *see* 8381
Levodopa, 5516
Levo-Dromoran *see* 5524
Levodropropizine *see* 3500
Levofloxacin *see* 6863
Levoglucosenone, 5517
Levoglutamina *see* 4507
Levoleucovorin Calcium *see* 4249
Levomenol *see* 1245
Levomeprazine *see* 5518

Levomepromazine, 5518
Levomethadone Hydrochloride see 6016
Levomethadyl Acetate, 5519
Levomethorphan, 5520
Levonelle [Schering AG] see 6793
Levonordefrin see 6781
Levonorgestrel see 6793
Levonorgestrel 3-Oxime see 6783
Levonova [Schering AG] see 6793
Levopa [ICN] see 5516
Levophacetoperane, 5521
Levophed [Winthrop] see 6784
Levophta [Chauvin] see 5515
Levopimaric Acid, 5522
Levopraid [Ravizza] see 9119
Levopropoxyphene, 5523
Levopropoxyphene Napsylate see 5523
Levopropylcillin see 7931
Levorenin see 3674
Levorin see 1743
Levorin A₂ see 1743
Levormeloxifene see 1972
Levorotatory Lactic Acid see 5383
Levorphan see 5524
Levorphanol, 5524
Levosalbutamol see 210
Levosimendan, 5525
Levosulpiride see 9119
Levotartaric Acid see 9203
Levothroid [Forest] see 9570
Levothym [Lundbeck] see 4886
Levothyrox [Merck-Clévenot] see 9570
Levothyroxine see 9570
Levothyroxine Sodium see 9570
Lévotonine [Pan Medica] see 4886
Levotuss [Dompé] see 3500
Levovermax [Virbac] see 5511
Levovist [Schering] see 4365
Levoxadrol Hydrochloride see 3327
Levoxan [Cutter] see 3327
Levoxyl [Jones] see 9570
Levsin [Kremers-Urban] see 4897
Levugen see 4302
Levulan [Dusa] see 443
Levulinic Acid, 5526
Levulinic Acid Calcium Salt see 1681
Levulose see 4302
Levynite see 1651
Lewisite see 3071
Lexapro [Forest] see 2317
Lexinor [Kyorin] see 6789
Lexipafant, 5527
Lexiscan [Astellas] see 8247
Lexiva [GSK] see 4276
Lexobene [Merckle] see 3091
Lexomil [Roche] see 1392
Lexone [DuPont] see 6233
Lexotan [Roche] see 1392
Lexotanil [Roche] see 1392
Lexus [DuPont] see 4225
LF00 see 9163
LF-153 see 4009
LF-178 see 4009
LFA-2043 see 5121
1-92-LFA-3 (Antigen) (Human) Fusion
  Protein with Immunoglobulin G1
  (Human Hinge-C_H2-C_H3 γ₁-Chain),
  Dimer see 220
LFA3TIP see 220
LG-152 see 5861
LG-11457 see 3765
LG-30158 see 8374
LGD-1057 see 241
LGD-1069 see 1198
LH see 5674

LH-RH see 4561
Lialda [Shire] see 5974
Liarozole, 5528
Liatris, 5529
Liazal [Janssen] see 5528
Libanil [APS] see 4514
Libeprosta [Pierre Fabre] see 8524
Liberty [Bayer CropSci.] see 7451
Libethenite see 2638
Libexin see 7854
Libratar [UCB] see 2081
Libritabs [Roche] see 2085
Librium [Roche] see 2085
Licain [DeltaSelect] see 5535
(S)-Licarbazepine see 3754
Licareol see 5550
Licarpin [Allergan] see 7536
Lichenan see 5531
Licheniformins, 5530
Lichenin, 5531
Licochalcone A see 5532
Licochalcone B see 5532
Licochalcone C see 5532
Licochalcone D see 5532
Licochalcones, 5532
Licofelone, 5533
Licorice see 4540
Lidakol [Avanir] see 3444
Lidamidine, 5534
Lidanar [Sandoz] see 5980
Lidanil [Sandoz] see 5980
Lidaprim [Nycomed] see 9052
Lidarral [Rorer] see 5534
Lidepran see 5521
Lidesthesin [Ritsert] see 5535
Lidex [Syntex] see 4182
Lidocaine, 5535
Lidoderm [Hind] see 5535
Lidofast [Angelini] see 5535
Lidofenin, 5536
Lidoflazine, 5537
Lidoject [Hexal] see 5535
Lidone [Abbott] see 6319
LidoPosterine [Kade] see 5535
Lidrian [Baxter] see 5535
Life [SIT] see 8109
Life Root see 8589
Lifril [Kissei] see 9252
Lighter [Sanonda] see 5218
Light Green N see 5763
Light Green SF Yellowish, 5538
Lightning Fast see 3670
Light Oil of Camphor see 6872
Light Spar see 1708
Lignans, 5539
Lignin, 5540
Lignite Wax see 6345
Lignocaine see 5535
Lignocaine Benzyl Benzoate see 2893
Lignoceric Acid, 5541
Lignum Vitae see 4588
Ligroin, 5542
Ligroine see 5542
Ligustrin see 9146
Likinozym [Kaigai] see 5701
Likuden [HMR] see 4584
Lilacillin [Takeda] see 9022
Lilacin see 9146
Lilly 35483 see 2752
Lilly 36352 see 9854
Lilly 53838 see 4011
Lilly 69323 see 4011
Lilly 99170 see 743
Lily of the Valley see 2496
Liman [Kali-Chemie] see 9291

Limaprost, 5543
Limas [Taisho] see 5583
Lima Wood see 7295
Limber [Sumitomo] see 4324
Limbial [Chiesi] see 7025
Limclair [Sinclair] see 3565
Lime see 1688
Limestone, 5544
Limettin, 5545
Limifen [Janssen-Cilag] see 229
Limonene, 5546
Limonin, 5547
Limonite see 4052
Limonoic Acid Di-δ-lactone see 5547
Limonoic Acid 3,19:16,17-Dilactone see
  5547
Limovan [Aventis] see 10393
Limpidex [Sigma-Tau] see 5413
LIN-1418 see 9123
Linaclotide, 5548
Linagliptin, 5549
Linalol see 5550
Linalool, 5550
d-Linalool see 5550
l-Linalool see 5550
Linalool Acetate see 5551
Linalyl Acetate, 5551
l-Linalyl Acetate see 5551
Linalyl Alcohol see 5550
Linamarin, 5552
Linarigenin-glucoside see 5553
Linarin, 5553
Linaris [R.A.N.] see 9050
Linatine, 5554
Lincocin [Pharmacia & Upjohn] see 5555
Lincoln Bean see 8856
Lincolnensin see 5555
Lincomix [Pharmacia & Upjohn] see 5555
Lincomycin, 5555
Lindane, 5556
Linden Tree see 1004
Lindlar Catalyst, 5557
Lindol [Stauffer] see 9940
Lindoxyl [Lindopharm] see 380
Linear Gramicidins see 4568
Lineatin, 5558
4,6,6-Lineatin see 5558
Linezolid, 5559
Lingraine [Sanofi-Synthelabo] see 3718
Lingzhi see 4397
Linnaeite see 2411
Linodil [Sanofi Winthrop] see 5022
Linoleic Acid, 5560
9,12-Linoleic Acid see 5560
Linoleic Acid Methyl Ester see 6167
Linolenic Acid, 5561
α-Linolenic Acid see 5561
γ-Linolenic Acid, 5562
Linolexamide see 5560
Linolic Acid see 5560
Linomide [Pfizer] see 8390
Linseed, 5563
Linton [Yoshitomi] see 4634
Lintrin see 6490
Linum see 5563
Linurex [Makhteshim-Agan] see 5564
Linuron, 5564
Linyl [Roussel Diamant] see 7374
Liometacen [Chiesi] see 5009
Lion's Tooth see 9192
Lioresal [Novartis] see 930
Liothyronine, 5565
Liothyronine Sodium see 5565
Lioxone see 8082
Lipamone see 3115

Lipan *see* 3305
Lipanor [Winthrop] *see* 2312
Lipanthyl [Solvay] *see* 4009
Lipantil [Solvay] *see* 4009
Lipase, 5566
Lipaxan [Italfarmaco] *see* 4245
Lipex [Elea] *see* 7672
Liphadione [Lipha] *see* 2155
Lipiarmycin A3 *see* 4108
Lipidil [Solvay] *see* 4009
Lipid-mobilizing Hormone *see* 5568
Lipiodol [Guerbet] *see* 5067
Lipitor [Parke-Davis] *see* 855
Liple [Mitsubishi] *see* 7987
Lipobay [Bayer] *see* 1992
Lipoclar [Crinos] *see* 4009
Lipoclin [Sumitomo] *see* 2355
Lipocol [Lipo] *see* 7697
Lipocol C-20 [Lipo] *see* 7697
Lipocortin *see* 668
Lipodel [Delalande] *see* 7115
Lipofene [Teofarma] *see* 4009
Lipoglutaren [Ausonia] *see* 5877
Lipo-Hepin [3M Pharma] *see* 4688
α-Lipoic Acid *see* 9479
Lipolytic Hormone *see* 5568
Lipo-Merz [Merz] *see* 3923
Liponorm [Gentili] *see* 8677
Lipopeg [Lipo] *see* 7698
Lipopeg 4-S [Lipo] *see* 7698
Lipoprotein-associated Coagulation
    Inhibitor *see* 9623
Lipoprotein Lipase, 5567
Liposolvin [Tosi] *see* 8675
Liposorb O-20 [Lipo] *see* 7703
Lipostat [BMS] *see* 7835
Lipotrophin *see* 5568
Lipotropic Hormone, 5568
Lipotropin *see* 5568
β-Lipotropin (61-91) *see* 3627
β-Lipotropin C-fragment *see* 3627
Lipozid [Pfizer] *see* 4422
Liprevil [Schwarz] *see* 7835
Liprinal [BMS] *see* 2370
Liprodène [Farlamedia] *see* 7246
Liprostin [Endovasc] *see* 7987
Lipsin [Caber] *see* 4009
Lipur [Pfizer] *see* 4422
Liquaemin Sodium [Organon] *see* 4688
Liquamar [Organon] *see* 7371
Liquamycin [Pfizer] *see* 7075, 9341
Liquémin [Roche] *see* 4688
Liquiband [Medlogic] *see* 1566
Liquid Butter of Antimony *see* 696
Liquid Glass *see* 8804
Liquid Lanolin *see* 5409
Liquid Paraffin *see* 7301
Liquid Rosin *see* 9172
Liquid Silver *see* 5966
Liquifilm [Allergan] *see* 7706
LiquiVent [Alliance] *see* 7270
Liquoid *see* 8796
Liquorice *see* 4540
LIR-1660 *see* 10142
Liraglutide, 5569
Liranaftate, 5570
Liranol [Wyeth] *see* 7898
Lironox *see* 2793
Lisagal *see* 7073
Lisdexamfetamine, 5571
Lisdexamfetamine Dimesylate *see* 5571
Liserdol [Farmitalia] *see* 6009
Lisino [Essex] *see* 5635
Lisinopril, 5572
Lisitril [Ecosol] *see* 5572

Liskantin [Desitin] *see* 7865
Liskonum [GSK] *see* 5583
Lisofylline, 5573
Lisomucil [Sanofi-Synthelabo] *see* 1802
Lisozima [SPA] *see* 5701
Lispamol [Specia] *see* 463
Lispine [Sawai] *see* 3402
Lissamine Green SF [Zeneca] *see* 5538
Lissapol *see* 4400
Lissephen [Abbott] *see* 5917
Lissolamine V *see* 2024
Listenon *see* 9006
Listrocol [Farmitalia] *see* 2763
Lisuride, 5574
Lita [Opta], 5575
Litalir [BMS] *see* 4888
Litarex [Dumex] *see* 5586
Litec [Novartis] *see* 7629
Litesse [Pfizer] *see* 7685
Lithamide *see* 5578
Lithane [Bayer] *see* 5583
Litharge *see* 5468
Lithicarb [Aspen] *see* 5583
Lithiodiisopropylamine *see* 3215
Lithiofor [Vifor] *see* 5597
Lithium, 5576
Lithium Acetate, 5577
Lithium Alanate *see* 344
Lithium Aluminohydride *see* 344
Lithium Aluminum Hydride *see* 344
Lithium Amide, 5578
Lithium Benzoate, 5579
Lithium Biborate *see* 5580
Lithium Borate, 5580
Lithium Borohydride, 5581
Lithium Bromide, 5582
Lithium Carbonate, 5583
Lithium Chloride, 5584
Lithium Chloride (LiCl) *see* 5584
Lithium Chromate(VI), 5585
Lithium Citrate, 5586
Lithium Diisopropylamide *see* 3215
Lithium Fluoborate *see* 5599
Lithium Fluoride, 5587
Lithium Fluoroboride *see* 5599
Lithium Formate, 5588
Lithium Hydrate *see* 5590
Lithium Hydride, 5589
Lithium Hydrido(9-BBN-nopol Benzyl
    Ether Adduct) *see* 6513
Lithium (*T*-4)-Hydrotris(1-methylpro-
    pyl)borate(1−) *see* 8564
Lithium Hydroxide, 5590
Lithium Hydroxide (Li(OH)) *see* 5590
Lithium Iodide, 5591
Lithium Metasilicate *see* 5596
Lithium Monobromide *see* 5582
Lithium Monofluoride *see* 5587
Lithium Monohydride *see* 5589
Lithium Monoiodide *see* 5591
Lithium Nitrate, 5592
Lithium Oxalate, 5593
Lithium Oxide, 5594
Lithium Perchlorate, 5595
Lithium Platinocyanide *see* 5598
Lithium Silicate, 5596
Lithium Sulfate, 5597
Lithium Tetraborate *see* 5580
Lithium Tetracyanoplatinate(II), 5598
Lithium Tetrafluoroborate, 5599
Lithium Tetrafluoroborate(1−) (1:1) *see*
    5599
Lithium Tetrahydroaluminate *see* 344
Lithium Tetrahydroborate(1−) (1:1) *see*
    5581

Lithium Tri-*sec*-butylborohydride *see*
    8564
Lithium Triethylborohydride, 5600
(*T*-4)-Lithium Triethylhydroborate(1−)
    (1:1) *see* 5600
Lithobid [Solvay] *see* 5583
Lithocholic Acid, 5601
Lithographic Stone *see* 5544
Lithol *see* 4924
Lithonate S [Rowell] *see* 5586
Lithopone, 5602
Lithostat [Mission Pharmacal] *see* 62
Litlure A *see* 7885
Litmomycin *see* 4570
Litmus, 5603
Litosmil [Evans] *see* 3325
Litsoeine *see* 5443
Little Water *see* 10225
Litursol [Stada] *see* 10074
Livalfa [Mitsubishi] *see* 9942
Livalo [Kowa] *see* 7621
Livazone [Specia] *see* 9442
Liver Cell Growth Factor *see* 4452
Liver *Lactobacillus casei* Factor *see* 4248
Livermorium *see* 3592
Liver of Lime *see* 1709
Liver of Sulfur *see* 9107
Liver Starch *see* 4531
Livetins, 5604
Livial [Organon] *see* 9582
Livonal *see* 3823
Livostin [Novartis] *see* 5515
Lixacol [Schering-Plough] *see* 5974
Lixiana [Daiichi Sankyo] *see* 3562
Lixidol [Roche] *see* 5353
Lixisenatide, 5605
Lixivaptan, 5606
LJ-48 *see* 5856
LJ-206 *see* 1802
LJC-10141 *see* 3985
LJC-10627 *see* 1201
LJP-394 *see* 6
L-Kininogen *see* 5360
LL-1558 *see* 9571
LL-1656 *see* 1480
LL-AV290 *see* 882
LLD Factor *see* 10212
LL-E33288 *see* 1717
LLF *see* 3967
LL-F-28249α *see* 6529
LM-001 *see* 3985
LM-91 *see* 2155
LM-123 *see* 4165
LM-192 *see* 10199
LM-208 *see* 8200
LM-209 *see* 5931
LM-280 *see* 5053
LM-427 *see* 8338
LM-637 *see* 1389
LM-2219 *see* 3158
LM-2717 *see* 2356
LMD *see* 2950
L-MTP-PE *see* 6265
LMWD *see* 2950
Lobak *see* 2106
Lobamine [Lobica] *see* 6047
Lobaplatin, 5607
Lobelia, 5608
Lobelidine *see* 5609
Lobeline, 5609
α-Lobeline *see* 5609
Lobenzarit, 5610
Lobivon [Menarini] *see* 6516
Locabiotal [Servier] *see* 4342
Locacorten [Novartis] *see* 4169

Localyn [Syntex] *see* 4181
Locapred [Fabre] *see* 2929
Locatop [Fabre] *see* 2929
Loceryl [Roche] *see* 570
Lochnericine, 5611
Locilex [Magainin] *see* 7308
Lock-On [Dow AgroSci.] *see* 2195
Locoid [Yamanouchi] *see* 4824
Locorten [Novartis] *see* 4169
Locron [Hoechst] *see* 339
Locust Bean Gum, 5612
Lodalès [Sanofi-Synthelabo] *see* 8677
Loderix [Egis] *see* 8611
Lodine [Wyeth] *see* 3920
Lodopin [Fujisawa] *see* 10397
Lodosyn [Merck & Co.] *see* 1798
Lodoxamide, 5613
Lodoxamide Ethyl *see* 5613
Lodoxamide Trometamol *see* 5613
Lodoxamide Tromethamine *see* 5613
Loestrin [Warner-Lambert] *see* 6786
Lofepramine, 5614
Lofetensin [Nattermann] *see* 5615
Lofexidine, 5615
Loflucarban, 5616
Lofoxin [Locatelli] *see* 4281
Loftan [GSK] *see* 210
Lofton [Abbott] *see* 1480
Loftyl [Abbott] *see* 1480
Loganin, 5617
Logiparin [Novo] *see* 9606
Logna [Tanabe] *see* 2346
Logran [Syngenta] *see* 9761
Logwood *see* 4673
Logynon [Schering AG] *see* 6793
Loline, 5618
Lomadine [M & B] *see* 7322
Lomebact [Shionogi] *see* 5619
Lomefloxacin, 5619
Lomerizine, 5620
Lomeron [Lilly] *see* 10254
Lomexin [Recordati] *see* 4032
Lomir [Novartis] *see* 5287
Lomont [Rosemont] *see* 5614
Lomotil [Searle] *see* 3343
Lomper [Esteve] *see* 5837
Lomudal [Sanofi-Aventis] *see* 2581
Lomupren [Sanofi-Aventis] *see* 2581
Lomusol [Sanofi-Aventis] *see* 2581
Lomustine, 5621
LON-798 *see* 4596
Lonactene [Ferring] *see* 1796
Lonafarnib, 5622
Lonasen [Dainippon] *see* 1320
Lonavar [Searle] *see* 7020
Lonazolac, 5623
Londax [DuPont] *see* 1053
Londomin [Taiyo] *see* 9147
Londomycin *see* 6015
Longastatin [Italfarmaco] *see* 6851
Longatren [Bayer] *see* 902
Longdigox [Trommsdorff] *see* 3186
Longheparin *see* 4688
Longicid [Chassot] *see* 3128
Longifene [UCB] *see* 1474
Longifolene, 5624
α-Longilobine *see* 8591
β-Longilobine *see* 8290
Longum [Farmitalia] *see* 9042
Lonidamine, 5625
Loniten [Pfizer] *see* 6285
Lonmiel [Teikoku] *see* 1040
Lonolox [Pfizer] *see* 6285
Lonomycins, 5626
Lontermin [Lek] *see* 7877

Lontrel [Dow AgroSci.] *see* 2389
Looplure, 5627
Looser [Kaken] *see* 1500
Lo/Ovral [Wyeth] *see* 6793
Lopantrol [Janssen] *see* 5637
Lopatol [Novartis] *see* 6719
Lopemid [Gentili] *see* 5628
Lopemin [Dainippon] *see* 5628
Loperamide, 5628
Loperyl [Zambeletti] *see* 5628
Lophotoxin, 5629
Lopid [Pfizer] *see* 4422
Lopinavir, 5630
Lopirin [BMS] *see* 1776
Lopitrex [BMS] *see* 1983
Lopramine *see* 5614
Loprazolam, 5631
Loprazolam Mesylate *see* 5631
Lopremone (rescinded USAN) *see* 9750
Lopresor [Novartis] *see* 6228
Lopressor [Novartis] *see* 6228
Lopril [BMS] *see* 1776
Loprinone, 5632
Loprox [HMR] *see* 2269
Lorabid [Lilly] *see* 5633
Loracarbef, 5633
Lorajmine, 5634
Loramet [Wyeth] *see* 5639
Loramyc [BioAlliance Pharma] *see* 6257
Loratadine, 5635
Lorax [Lilly] *see* 5633
Lorax [Wyeth] *see* 5636
Lorazepam, 5636
Lorcainide, 5637
Lorcaserin, 5638
Lorelco [HMR] *see* 7873
Lorestat [Recordati] *see* 9683
Loretin *see* 4865
Lorexane [AstraZeneca] *see* 5556
Lorfan [Roche] *see* 5510
Loricin *see* 9021
Lorinden [Polfa] *see* 4169
Lorisal [Seatrace] *see* 5749
Lorivox [Srbolek] *see* 5637
Lormetazepam, 5639
Lormin [Lilly] *see* 2102
Lornoxicam, 5640
Loron [Roche] *see* 2365
Lorothidol *see* 1307
Lorox [DuPont] *see* 5564
Lorqess [Arena] *see* 5638
Lorsban [Dow AgroSci.] *see* 2195
Lorsilan [Belupo] *see* 5636
Lortaan [Merck & Co.] *see* 5641
Lortab [UCB] *see* 4823
Lorzaar [Merck & Co.] *see* 5641
Losantin *see* 1676
Losaprex [Sigma-Tau] *see* 5641
Losartan, 5641
Losec [AstraZeneca] *see* 6939
Lospel [Isagro] *see* 9338
Lotemax [Bausch & Lomb] *see* 5642
Lotensin [Novartis] *see* 1033
Loteprednol Etabonate, 5642
Lotrifen, 5643
Lotrimin [Schering-Plough] *see* 2401
Lotronex [GSK] *see* 304
Loturine *see* 4648
Lotus [BASF] *see* 2294
Lotusate [Winthrop] *see* 9167
Lovan [Lilly] *see* 4217
Lovastatin, 5644
Lovaza [Reliant] *see* 6938
"Love" *see* 5834
Love-lies-bleeding *see* 370

Lovenox [Sanofi-Aventis] *see* 3642
Loverine [Isei] *see* 2945
Loviscol [Wyeth] *see* 1802
Low Acyl Purified Gellan Gum *see* 4417
Lowgan [Yamanouchi] *see* 572
Low Molecular Weight Kininogen *see* 5360
Lowpres [Chugai] *see* 6371
Lowpston [Maruko] *see* 4338
Loxapac [Eisai] *see* 5645
Loxapine, 5645
Loxeen [Tobishi] *see* 7859
Loxen [Novartis] *see* 6580
Loxiglumide, 5646
Loxitane [Watson] *see* 5645
Loxonin [Sankyo] *see* 5647
Loxoprofen, 5647
Loxuran [EGYT] *see* 3128
Lozol [Aventis] *see* 4975
LPH *see* 5568
β-LPH *see* 5568
γ-LPH *see* 5568
LPH (61-55) *see* 3627
LPH (61-76) *see* 3627
LPH (61-77) *see* 3627
LPH (61-91) *see* 3627
LS-11 *see* 9166
LS-74-783 *see* 4279
LS-121 *see* 6437
LS-519 *see* 7604
LS-519-Cl2 *see* 7604
LS-2616 *see* 8390
LS-860263 *see* 1456
LSD *see* 5695
LSD-25 *see* 5695
L-Selectin *see* 8563
L-Selectride *see* 8564
LT-1 *see* 9005
LT-31-200 *see* 1335
LTA₄ *see* 5507
LTB₄ *see* 5507
LTC₄ *see* 5507
LTD₄ *see* 5507
LTE₄ *see* 5507
LTG *see* 5403
LTH *see* 7894
LTs *see* 5507
LTX *see* 5629
Lu-02-030 *see* 4351
Lu-5-110 *see* 4219
Lu-10-171 *see* 2317
Lu-23-174 *see* 8605
Lu-26-054-0 *see* 2317
LU-1631 *see* 388
LU-47311 *see* 8293
LU-127043 *see* 2830
LU-135252 *see* 2830
LU-208075 *see* 378
Lubalix [Lubapharm] *see* 2404
Lubeluzole, 5648
Lubiprostone, 5649
Lucamide *see* 3786
Lucanal [Lucava] *see* 6442
Lucanthone, 5650
Lucassin [Orphan Ther.] *see* 9310
Lucayan [Corvi] *see* 9594
Lucebanol [Hormona] *see* 4929
Lucensomycin, 5651
Lucentis [Genentech] *see* 8226
Luciculine *see* 6452
Lucidol [Elf Atochem] *see* 1119
Lucidril [Dainippon] *see* 5851
Luciferin, 5652
D-(−)-Luciferin *see* 4117
Lucifer Yellow Carbohydrazide *see* 5653

Lucifer Yellow CH *see* 5653
Lucifer Yellows, 5653
Lucifer Yellow Vinylsulfone *see* 5653
Lucifer Yellow VS *see* 5653
Lucinactant *see* 8680
Lucinite *see* 352
Lucite *see* 6013
Lucofen [Lundbeck] *see* 2188
Lucosil [Lundbeck] *see* 9048
Lucrin [Abbott] *see* 5509
Ludiomil [Novartis] *see* 5813
Lufenuron, 5654
Lufyllin [Mallinckrodt] *see* 3526
Lugacin [Sandoz] *see* 4427
LüH6 *see* 6827
Lukadin [San Carlo] *see* 400
Luliberin *see* 4561
Lulicon [Pola Pharma] *see* 5655
Luliconazole, 5655
Lullamin [Reed & Carnrick] *see* 6031
Lullan [Sumitomo] *see* 7296
Lumazine, 5656
Lumbinon [Lichtenstein] *see* 4535
Lumefantrine, 5657
Lumichrome, 5658
Lumiflavine, 5659
Lumigan [Allergan] *see* 1225
Luminal [Desitin] *see* 7352
Luminol, 5660
Luminous Calcium Sulfide *see* 1709
Lumiracoxib, 5661
Lumirelax *see* 6052
Lumiren [Guerbet] *see* 4091
Lumisol RV *see* 1316
9α-Lumista-5,7,22-trien-3β-ol *see* 8110
Lumisterol, 5662
9α-Lumisterol *see* 8110
Lumitens [Solvay] *see* 10276
Lumizyme [Genzyme] *see* 100
Lumopaque [Winthrop] *see* 10018
Lumota [Thomae] *see* 710
Lunacridine, 5663
Lunacrine, 5664
Lunamycin (obsolete) *see* 4940
Lunar Caustic *see* 8657
Lunasine, 5665
Lunesta [Sepracor] *see* 10393
Lunetoron [Sankyo] *see* 1489
Lunipax [SKB] *see* 4228
Lunis [Valeas] *see* 4176
Lunularic Acid, 5666
Luostyl [UPSA] *see* 3152
Lupanine, 5667
(3β)-Lup-20(29)-ene-3,28-diol *see* 1192
Lup-20(30)-ene-3β,28-diol *see* 1192
(3β)-Lup-20(29)-en-3-ol *see* 5668
Lupeol, 5668
Lupinidine *see* 8864
Lupinine, 5669
(−)-Lupinine *see* 5669
epi-Lupinine *see* 5669
l-Lupinine *see* 5669
l-epi-Lupinine *see* 5669
Lupolen [BASF] *see* 7687
Lupron [Abbott] *see* 5509
Luprostiol, 5670
α-Lupulic Acid *see* 4793
β-Lupulic Acid *see* 5672
Lupulin, 5671
Lupulon, 5672
Lurasidone, 5673
Luret [Sanofi Winthrop] *see* 915
Luride-SF [Hoyt] *see* 8754
Luridine *see* 2211
Luronase *see* 4797

Lurselle [HMR] *see* 7873
Lusedra [Eisai] *see* 4286
Lustral [Pfizer] *see* 8606
Lutalyse [Pharmacia & Upjohn] *see* 7989
Luteanine *see* 5207
Lutecium *see* 5676
Lutein *see* 10261
Luteinizing Hormone, 5674
Luteinizing Hormone (Human α-Subunit Reduced) Complex with Luteinizing Hormone (Human β-Subunit Reduced) Glycoform α *see* 5674
Luteinizing Hormone-Releasing Factor *see* 4561
Luteinizing Hormone-Releasing Factor (Swine) *see* 4561
Luteinizing Hormone-Releasing Hormone *see* 4561
Luteocobaltic Chloride *see* 4710
Luteohormone *see* 7889
Luteolas [Serono] *see* 3909
Luteolin, 5675
Luteolin-4′-methyl Ether *see* 3324
Luteonorm [Serono] *see* 3909
Luteotropic Hormone *see* 7894
Luteotropin *see* 7894
Luteran [Cassenne] *see* 2102
Lutestral [Cassenne] *see* 2102
Lutetium, 5676
Lutetium Texaphyrin *see* 6368
Lu-Tex *see* 6368
α,α′-Lutidin *see* 5677
β-Lutidine *see* 3900
2,6-Lutidine, 5677
Luto-Metrodiol *see* 3909
Lutrelef [Ferring] *see* 4561
Lutrepulse [Ferring] *see* 4561
Lutrin [Pharmacyclics] *see* 6368
Luvatren [Cilag-Chemie] *see* 6351
Luveris [Serono] *see* 5674
Luviskol [BASF] *see* 7814
Luvox [Solvay] *see* 4246
Luxazone [Allergan] *see* 2945
Luxoben [Chinoin] *see* 9576
LVD *see* 2950
LY-141B *see* 7277
LY-12735 *see* 6372
LY-031537 *see* 8208
LY-048740 *see* 880
LY-61017 *see* 3985
LY-061188 *see* 1974
LY-097964 *see* 1924
LY-099094 *see* 10184
LY-109514 *see* 6427
LY-110140 *see* 4217
LY-122772 *see* 3655
LY-127809 *see* 7277
LY-135837 *see* 4977
LY-139037 *see* 6747
LY-139481 *see* 8215
LY-139603 *see* 854
LY-146032 *see* 2823
LY-156758 *see* 8215
LY-163892 *see* 5633
LY-170053 *see* 6914
LY-177370 *see* 9596
LY-177837 *see* 8842
LY-188011 *see* 4420
LY-188695 *see* 3614
LY-210448 *see* 2821
LY-231514 *see* 7186
LY-232105 *see* 8877
LY-237216 *see* 3394
LY-246708 *see* 10254
LY-246736 *see* 366

LY-248686 *see* 3512
LY-253351 *see* 9181
LY-275585 *see* 5030
LY-300164 *see* 9164
LY-303366 *see* 649
LY-315920 *see* 10126
LY-317615 *see* 3656
LY-333013 *see* 10126
LY-333328 *see* 6963
LY-333531 *see* 8423
LY-335348 *see* 2895
LY-335979 *see* 10396
LY-353381 *see* 810
LY-355703 *see* 2597
LY-450139 *see* 8580
LY-570310 *see* 9256
LY-640315 *see* 7833
LY-900003 *see* 744
LY-2148568 *see* 3955
LY-2484595 *see* 3947
Lyapolate Sodium, 5678
Lycaconitine *see* 5679
Lycanol *see* 4543
Lycine *see* 1181
Lycoctonine, 5679
Lycodine, 5680
Lycopene, 5681
(*all-trans*)-Lycopene *see* 5681
(*all-trans*)-Lycopene-16,16′-diol *see* 5682
(*all-trans*)-Lycopen-16-ol *see* 5688
Lycopersicin *see* 9704
Lycophyll, 5682
Lycopodine, 5683
Lycopodium, 5684
Lycopodium Seed (Spores) *see* 5684
Lycopus, 5685
Lycoramine, 5686
Lycoremine *see* 4370
Lycorine, 5687
Lycoxanthin, 5688
Lycra [DuPont] *see* 8860
Lyman [Drossapharm] *see* 305
Lymecycline, 5689
Lymphazurin [Pharmascience] *see* 9117
Lymphoblastoid Interferon *see* 5034
LymphoCide [Immunomedics] *see* 3687
Lymphocyte Activating Factor *see* 5037
Lymphocyte Adhesion Molecule-1 *see* 8563
Lymphocyte Mitogenic Factor *see* 5038
LymphoStat-B [Human Genome Sciences] *see* 1024
Lymphotoxin *see* 9992
Lynamine [Memphis] *see* 5356
Lyndiol [Organon] *see* 5690
Lynestrenol, 5690
Lynoral [Organon] *see* 3788
Lynx [Bayer CropSci.] *see* 9229
Lyogen [Promonta Lundbeck] *see* 4220
Lyphocin [Fujisawa] *see* 10116
Lypressin, 5691
Lyrica [Pfizer] *see* 7843
Lys *see* 5697
Lysal [Gellini] *see* 5698
Lysalbinic Acid, 5692
Lysalgo [SIT] *see* 5869
Lysanxia [Lynapharm] *see* 7836
[Lys(B3),Glu(B29)]-insulin (Human) *see* 5028
[Lys(B28),Pro(B29)]-insulin (Human) *see* 5030
Lysenyl [Spofa] *see* 5574
Lyseen [Novartis] *see* 7859
Lysergamide, 5693
Lysergic Acid, 5694

Magnesium Diethyl *see* 3138
Magnesium Difluoride *see* 5729
Magnesium Dihydride *see* 5733
Magnesium Diiodide *see* 5735
Magnesium Dimecrotate *see* 3226
Magnesium Dioxide *see* 5744
Magnesium Diphenyl *see* 3361
Magnesium Fluoride, 5729
Magnesium Fluosilicate *see* 5732
Magnesium Formate, 5730
Magnesium Germanide, 5731
Magnesium Glucoheptonate *see* 4491
Magnesium Glucomonocarbonate *see* 4491
Magnesium Gluconate *see* 4492
Magnesium Glucosemonocarbonate *see* 4491
Magnesium Glutamate Hydrobromide *see* 4505
Magnesium Heparinate *see* 4688
Magnesium Hexafluorosilicate, 5732
Magnesium Hydrate *see* 5734
Magnesium Hydride, 5733
Magnesium Hydrogen Citrate *see* 5727
Magnesium Hydrogen Phosphate *see* 5745
Magnesium Hydroxide, 5734
Magnesium Hyposulfite *see* 5758
Magnesium Iodide, 5735
Magnesium Lactate, 5736
Magnesium Mesotrisilicate *see* 5751
Magnesium Metaborate *see* 5722
Magnesium Metasilicate *see* 5751
Magnesium Monoperoxyphthalate, 5737
Magnesium Monoselenide *see* 5750
Magnesium Nitrate, 5738
Magnesium Oleate, 5739
Magnesium Orthosilicate *see* 5751
Magnesium Oxalate, 5740
Magnesium Oxide, 5741
Magnesium Oxide (MgO) *see* 5741
Magnesium Pemoline *see* 7188
Magnesium Perchlorate, 5742
Magnesium Perhydrol *see* 5744
Magnesium Permanganate, 5743
Magnesium Peroxide, 5744
Magnesium Phosphate, Dibasic, 5745
Magnesium Phosphate, Monobasic, 5746
Magnesium Phosphate, Tribasic, 5747
Magnesium Pyrophosphate, 5748
Magnesium Salicylate, 5749
Magnesium Selenide, 5750
Magnesium Silicates, 5751
Magnesium Silicide, 5752
Magnesium Silicofluoride *see* 5732
Magnesium Stannide, 5753
Magnesium Stearate, 5754
Magnesium Sulfate, 5755
Magnesium Sulfite, 5756
Magnesium Sulfocyanate *see* 5757
Magnesium Superoxol *see* 5744
Magnesium Thiocyanate, 5757
Magnesium Thiosulfate, 5758
Magnesium Trisilicate *see* 5751
Magneson, 5759
Magnespirin *see* 5719
Magnetic Iron Oxide *see* 4067
Magnetite *see* 4067
Magnetkies *see* 4085
Magnevist [Schering AG] *see* 4357
Magnipen [Clin-Comar-Byla] *see* 5996
Magnofenyl *see* 7056
Magnoflorine, 5760
Magnogene *see* 5726
Magnolol, 5761

Magnophenyl *see* 7056
Magnopyrol [Abbott] *see* 3390
Magnosil [Polfa] *see* 5751
Magnosulf [Normark] *see* 5758
Magrilon [Sintyal] *see* 5831
Ma Huang *see* 3662
Maidenhair Tree *see* 4460
Maiorad [Rottapharm] *see* 9622
Maipedopa [Maipe] *see* 5516
Maitansine *see* 5829
Maitotoxin, 5762
Maize Oil *see* 2520
Maizena [CPC] *see* 8925
Majeptil [RPR] *see* 9510
Majudin *see* 1160
Maki [Lipha] *see* 1389
Makon *see* 6764
Malabar Nut *see* 147
Malabar Tamarind *see* 4398
Malachite *see* 2621
Malachite Green, 5763
Malachite Green G *see* 1380
Malaquin [Ahn Gook] *see* 2165
Malaria Pigment *see* 4684
Malaridine *see* 8117
Malathion, 5764
Malathon *see* 5764
Malayan Camphor *see* 1340
Malazide *see* 5769
Malcotran [Fisons] *see* 4768
Maldison *see* 5764
Maleamic Acid, 5765
Maleanilic Acid, 5766
Male Fern *see* 838
Maleic Acid, 5767
Maleic Acid Diethyl Ester *see* 3139
Maleic Acid Hydrazide *see* 5769
Maleic Acid Monoamide *see* 5765
Maleic Anhydride, 5768
Maleic Hydrazide, 5769
Male Shield-fern *see* 838
Maleuric Acid, 5770
Malex [Ecosol] *see* 47
Malexil [Ferrosan] *see* 3989
Malexin [BASF] *see* 6499
Maleylurea *see* 5770
Maliasin [Knoll] *see* 7972
Malic Acid, 5771
Maliner Kren *see* 4788
Malipuran [Heumann] *see* 1479
Malix [Bayer CropSci.] *see* 3629
Mallebrin [Krewel] *see* 332
Mallorol [Novartis] *see* 9512
Mallotoxin *see* 8405
Mallow, 5772
Malocide [Aventis] *see* 8098
Malol *see* 10075
Malonaldehyde *see* 5773
Malonamide Nitrile *see* 2678
Malondialdehyde, 5773
Malonic Acid, 5774
Malonic Acid Cyclic Isopropylidene Ester *see* 5885
Malonic Acid Di-*tert*-butyl Ester *see* 3041
Malonic Acid Dimethyl Ester *see* 6168
Malonic Acid Ethyl Ester Nitrile *see* 3841
Malonic Ester *see* 3878
Malonic Mononitrile *see* 2679
Malononitrile, 5775
Malononitrile Dimer *see* 476
Malononitrile Hydrazide *see* 2670
5,5′-[Malonylbis(methylimino)]bis[*N*,*N*′-bis[2,3-dihydroxy-1-(hydroxymethyl)-

propyl]-2,4,6-triiodoisophthalamide] *see* 5108
Malonylurea *see* 958
Malotilate, 5776
Maltobiose *see* 5778
Maltol, 5777
Maltonic Acid *see* 4492
Maltos [Otsuka] *see* 5778
Maltose, 5778
Malt Sugar *see* 5778
Malt Sugar-dextrin *see* 2954
Malvidin Chloride, 5779
Malvin *see* 5779
Malvoside *see* 5779
Mamallet-A [Showa] *see* 470
Mammacillin [Stricker] *see* 7206
Mammex *see* 6687
Mammotropin *see* 7894
Mamyzin [Boehringer, Ing.] *see* 7194
Manage [Monsanto] *see* 4636
Manage [Hokko] *see* 4950
Mancef [Lafare] *see* 1916
Manchurian Bean *see* 8856
Mancona Bark *see* 8519
Mancozeb, 5780
Mandarin G *see* 6953
Mandate [Dow AgroSci.] *see* 9463
7-D-Mandelamido-3-(1-methyl-1,2,3,4-tetrazole-5-thiomethyl)-Δ³-cephem-4-carboxylic Acid *see* 1916
7-D-Mandelamido-3-[[(1-methyl-1*H*-tetrazol-5-yl)thio]methyl]-3-cephem-4-carboxylic Acid *see* 1916
(6*R*,7*R*)-7-[(*R*)-Mandelamido]-8-oxo-3-[[[1-(sulfomethyl)-1*H*-tetrazol-5-yl]-thio]methyl]-5-thia-1-azabicyclo-[4.2.0]oct-2-ene-2-carboxylic Acid *see* 1930
7-[(*R*)-Mandelamido]-3-(1-sulfomethyl-1*H*-tetrazol-5-ylthiomethyl)-3-cephem-4-carboxylic Acid *see* 1930
Mandelamine [Warner-Chilcott] *see* 6038
Mandelic Acid, 5781
dl-Mandelic Acid *see* 5781
Mandelic Acid Ammonium Salt *see* 529
Mandelic Acid Isoamyl Ester, 5782
Mandelic Acid Isopentyl Ester *see* 5782
Mandelic Acid Nitrile *see* 5783
Mandelic Acid 3,3,5-Trimethylcyclohexyl Ester *see* 2697
Mandelonitrile, 5783
Mandelonitrile-β-gentiobioside *see* 592
Mandelonitrile Glucoside, 5784
D-Mandelonitrile-β-D-glucosido-6-β-D-glucoside *see* 592
Mandelonitrile β-Glucuronide *see* 592
Mandelyltropeine *see* 4768
Mandenol *see* 3875
Mandipropamid, 5785
Mandokef [Lilly] *see* 1916
Mandol [Lilly] *see* 1916
Maneb, 5786
Maneon [Poli] *see* 403
Manerix [Roche] *see* 6312
Manex [Griffin] *see* 5786
Manexin *see* 5811
Mangafodipir, 5787
Manganblende *see* 5805
Manganese, 5788
Manganese Acetate, 5789
Manganese Benzoate *see* 1093
Manganese Binoxide *see* 5795
Manganese Bromide, 5790
Manganese Carbonate, 5791
Manganese Carbonyl, 5792

Manganese Chloride, 5793
Manganese Chloride (MnCl₂) see 5793
Manganese Dibromide see 5790
Manganese Dichloride see 5793
Manganese(II)dichloro[(4R,9R,14R,19R)-
3,10,13,20,26-pentaazatetracyclo-
[20.3.1.0.⁴·⁹0¹⁴·¹⁹]hexacosa-1(26),22-
(23),24-triene] see 4965
Manganese Difluoride, 5794
Manganese Diiodide see 5797
Manganese Dinitrate see 5798
Manganese Dioxide, 5795
Manganese Dipyridoxal Diphosphate see
5787
Manganese(II)-N,N′-dipyridoxylethyl-
enediamine-N,N′-diacetate-5,5-bis-
(phosphonate) see 5787
Manganese Ethylenebis(dithiocarba-
mate) (Polymeric) Complex with Zinc
Salt see 5780
Manganese Fluoride see 5794
Manganese Fluoride (MnF₂) see 5794
Manganese Fluoride (MnF₃) see 5806
Manganese Green see 977
Manganese Hypophosphite, 5796
Manganese Iodide, 5797
Manganese Monosulfide see 5805
Manganese Nitrate, 5798
Manganese Oxalate, 5799
Manganese Oxide, 5800
Manganese Oxide (MnO₂) see 5795
Manganese Oxide (Mn₂O₃) see 5802
Manganese Pentacarbonyl Dimer see
5792
Manganese Peroxide see 5795
Manganese Selenide, 5801
Manganese Sesquioxide, 5802
Manganese Silicate, 5803
Manganese Sulfate, 5804
Manganese Sulfide, 5805
Manganese Superoxide see 5795
Manganese Tetroxide see 5800
Manganese Trifluoride, 5806
Manganic Acid (H₂MnO₄) Barium Salt
(1:1) see 977
Manganic Acid (H₂MnO₄) Potassium
Salt (1:2) see 7765
Manganic Fluoride see 5806
Manganite see 5802
Manganjustite see 5803
Manganomanganic Oxide see 5800
Manganosite see 5788
Manganous Chloride see 5793
Manganous Ethylenebis[dithiocarba-
mate] see 5786
Manganous Fluoride see 5794
Mangostin, 5807
Manicol see 5810
Manidipine, 5808
Maninil [Berlin-Chemie] see 4514
Maniol [Morishita] see 3341
Manna, 5809
Manna Sugar see 5810
Mannentake see 4397
Mannidex see 5810
Mannite see 5810
Mannitol, 5810
D-Mannitol see 5810
D-Mannitol Hexanitrate, 5811
Mannitol Nitrate, 5811
Mannitrin see 5811
L-Mannomethylose see 8295
D-Mannose, 5812
Mannose-terminated Human Placental
Glucocerebrosidase see 4487

Manoplax [Boots] see 4142
Mansil [Pfizer] see 7018
Mantadan [Boehringer, Ing.] see 368
Mantadine [Endo] see 368
Mantadix [BMS] see 368
Man-Tan see 3200
Manzate [Griffin] see 5780
Manzeb see 5780
Manzin [Crystal Chem.] see 5780
MAO see 6333
Maolate [Upjohn] see 2185
MAP see 5864
MAP-0004 see 3191
Mapeg [PPG] see 7698
Mapeg 400 MS [PPG] see 7698
Mapharsen [Parke-Davis] see 7047
Maphenide see 5712
MAPO see 6008
Maposol [Procida] see 6021
Mappine see 1483
Maprofix see 4400
Maprotiline, 5813
Maratan [Ravizza] see 1296
Marathon [Olympic] see 4951
Maraviroc, 5814
Marbocyl [Vétoquinol] see 5815
Marbofloxacin, 5815
Marboran see 6049
Marcaine [AstraZeneca] see 1499
Marcoeritrex see 3740
Marcoumar [Roche] see 7371
Marcumar [Roche] see 7371
Maretin [Bayer] see 6438
Marevan [Goldshield] see 10236
Marezine [Himmel] see 2703
Marfanil see 5712
Margaric Acid, 5816
Margarite see 1651
Margatoxin, 5817
Margosa Tree see 6522
Marigold see 1715
Marijuana see 1752
Marimastat, 5818
Marinco C see 5724
Marinco H see 5734
Marindinin see 5335
Mariner [DuPont] see 1053
Marinol [Unimed] see 9354
Maripen [Biogal] see 7190
Markweed see 7666
Marlate [DuPont] see 6061
Marmelosin see 4967
Marograstim see 4573
Maropitant, 5819
Marplan [Roche] see 5202
Marquis [Bayer] see 7711
Marrubiin, 5820
Marshal [FMC] see 1827
Marsh Gas see 6024
Marshite see 2652
Marshmallow see 313
Marsh Trefoil see 5910
Marsilid [Roche] see 5122
Marsin [Ikapharm] see 7351
Marsthine [Towa] see 2343
Martin Sulfurane, 5821
Martos-10 [Otsuka] see 5778
Marucotol [Maruko] see 5851
Marvelon 150/30 [Akzo] see 2928
Marvinol [Naugatuck] see 7707
Mary-bud see 1715
Maryland Pink see 8875
Mary Thistle see 6274
Marzine [GlaxoWellcome] see 2703
Masacin [Otsuka] see 7877

Mascagnite see 552
Masdil [Esteve] see 3224
Masdiol see 6022
Masitinib, 5822
Masitinib Mesylate see 5822
Masivet [AB Science] see 5822
Masletine [Shioe] see 2343
Masmoran [Pfizer] see 4889
Masnidipine see 5500
Masoprocol see 6782
Masor [Formenti] see 8936
Maspiron [Sawai] see 6983
Massicot see 5468
Massoi Lactone see 2849
Mast Cell Growth Factor see 8934
Masterfen [Dompé] see 9134
Masterid [Grünenthal] see 3497
Masteril [Syntex] see 3497
Masterone [Syntex] see 3497
Mastic, 5823
Mastiche see 5823
Mastimyxin [Chassot] see 7693
Mastisol see 5823
Mastix see 5823
Maté, 5824
Matacil [Bayer] see 428
Mataperro see 2479
Matico, 5825
Matric [Sankyou Agro] see 2223
Matricaria see 2040
Matricaria Oil see 2040
Matricarin, 5826
Matridine see 5827
Matridin-15-one see 5827
Matrifen [Nycomed] see 4028
Matrine, 5827
Matrix [DuPont] see 8356
Matrol see 2102
Matromycin [Pfizer] see 6918
MATS see 6087
Matulane [Sigma-Tau] see 7876
Maveral [Farmades] see 4246
Maverick [Monsanto] see 9095
Mavid [Abbott] see 2338
Mavik [Abbott] see 9729
Mavrik [Zoecon] see 4244
4-MAX see 6092
Maxacalcitol, 5828
Maxair [3M Pharma] see 7602
Maxalt [Merck & Co.] see 8370
Maxaquin [Pharmacia] see 5619
Maxeran [Aventis] see 6219
Maxforce see 4804
Maxibolin [Organon] see 3860
Maxidex [Alcon] see 2945
Maxiflor [Allergan] see 3159
Maxilase [Sanofi Winthrop] see 594
Maxim [Syngenta] see 4160
Maxipen [Pfizer] see 7337
Maxipime [BMS] see 1923
Maxivate [Westwood] see 1182
Maxolon [GSK] see 6219
Maxtrex [Pharmacia] see 6057
Max-Uric [Labinca] see 1067
Maxus [Elanco] see 880
Mayapple see 7664
May Blossom see 2496
Maycor [Parke-Davis] see 5271
Maydol see 2520
Maygace [BMS] see 5876
May Lily see 2496
Maynar [Ferrer] see 134
May Pops see 7156
Maytansine, 5829
Maytansinoid DM1, 5830

Mazildene [Farmochim. Ital.] see 5831
Mazindol, 5831
Mazipredone, 5832
Mazola [CPC] see 2520
Mazun see 10299
MB-10064 see 1451
MB-10731 see 1451
MB-46030 see 4116
MBA see 5846
MBBT see 1294
MBC see 1792
4-MBC see 6101
MBI-226 see 6940
MBK see 6107
MBLA see 5890
MBOCA see 6131
MBP-8298 see 3396
MBR-12325 see 5873
MBT see 5935
MBTS see 3413
MBU see 6011
3-MC see 6115
MC 540 see 5969
MC-838 see 6371
MC-903 see 1644
MC-1275 see 5828
MC-2188 see 2103
MC-4379 see 1212
MCA see 2112, 6117
MCAA see 5634
Mch5 see 1886
Mch6 see 1887
MCI see 6163
MCI-186 see 6802
MCI-196 see 2458
MCI-2016 see 1210
MCI-9038 see 769
MCMN see 8385
MCMT see 6309
McN-485 see 10399
McN-742 see 472
McN-822 see 6092
McN-1025 see 6775
McN-1210 see 8102
McN-2559 see 9677
McN-2559-21-98 see 9677
McN-4853 see 9706
McN-A-2833 see 7285
McN-A-2833-109 see 7285
McN-JR-4584 see 1050
McN-JR-7094 see 5537
McN-JR-8299 see 5511
McN-JR-15403-11 see 3155
McN-R-73-Z see 1800
McN-R-726-47 see 7671
McN-R-1967 see 4025
McN-X-181 see 10098
MCNU see 8227
MCP see 5833
1-MCP see 6120
MCPA, 5833
MCPBA see 2154
MCPP see 5855
M-CSF see 5710
MCT see 6336
MCW-275 see 6809
MD-141 see 3777
MD-805 see 769
MD-1100 see 5548
MD-69276 see 9680
MD-370503 see 1020
MDA, 5834
MDE, 5835
MDEA see 5835
M-Det see 2859

MDGF see 4101
MDL-473 see 8344
MDL-507 see 9255
MDL-9918 see 9306
MDL-14042A see 5615
MDL-16455 see 4097
MDL-17043 see 3643
MDL-62198 see 8222
MDL-63397 see 2800
MDL-71754 see 10175
MDL-73147 see 3458
MDL-73147EF see 3458
MDMA, 5836
MDMH see 4870
MDP see 5863, 6390
MDP-Lys (L18) see 8384
MDS see 2953
MDX-010 see 5114
ME-1206 see 1922
ME-1207 see 1922
ME-2906 see 9166
ME-3202 see 7242
ME-3625 see 6603
ME-3737 see 9598
MEA see 2776
Meadow Anemone see 8048
Meadow Clover see 9858
Meadow Crocus see 2456
Meadow Saffron see 2456
Me₃Al see 9879
Me₂AlCl see 3249
MEAP see 7453
Meat Sugar see 5021
Meaverin [Woelm] see 5926
Mebadin [Glaxo] see 2874
Meballymal Sodium see 8557
Mebaral [Winthrop] see 5921
Me₂BBr see 1426
Mebendazole, 5837
Mebenvet [Janssen] see 5837
Meberyt see 3743
Mebeverine, 5838
Mebhydroline, 5839
MeBmt, 5840
Mebonat [Boehringer, Mann.] see 2365
(R,R)-Me-BPE see 1363
(S,S)-Me-BPE see 1363
Mebrofenin, 5841
Mebron [Daiichi] see 3677
Mebubarbital see 7243
Mebutamate, 5842
Mebutina [Formenti] see 5842
3-MECA see 6115
Mecadox [Pfizer] see 1782
Mecalmin [Yoshitomi] see 3341
Mecamine see 5843
Mecamylamine, 5843
Mecasermin Rinfabate, 5845
MeCbl see 6116
Mecca-galls see 6817
Mechiaron [Choseido] see 5985
Mechlorethamine, 5846
Mechlorethamine Oxide, 5847
Mechlorprop see 5855
Meciclin [Citobios] see 2890
Mecillinam see 383
Meclan [Ortho] see 5849
Meclastine see 2343
Meclizine, 5848
Meclocycline, 5849
Mecloderm see 5849
Meclodol [Parke-Davis] see 5850
Meclofenamic Acid, 5850
Meclofenoxane see 5851
Meclofenoxate, 5851

Meclomen [Parke-Davis] see 5850
Meclophenamic Acid see 5850
Meclopran [Lagap] see 6219
Mecloqualone, 5852
Meclosorb [Basotherm] see 5849
Meclozine see 5848
Meclutin [ABC] see 5849
Mecobalamin see 6116
Mecocyanin see 2677
Mecomec [PBI/Gordon] see 5855
Meconic Acid, 5853
Meconin, 5854
Meconinic Acid Lactone see 5854
Mecopar see 5855
Mecoprop, 5855
Mecoprop-P see 5855
Mecrilate see 6117
Mecrylate see 6117
Mectizan [Merck & Co.] see 5296
Mecysteine, 5856
MED-15 see 591
Medaject [Neda] see 7875
Medallion [Syngenta] see 4160
Medan [Kowa] see 1180
Medaron see 4330
Medax [BASF] see 7892
Medazepam, 5857
Medazepol [Farmasa] see 5857
Medemanol see 5811
Medemycin [Meiji] see 6262
Mederantil [Boehringer, Ing.] see 1459
Medetomidine, 5858
d-Medetomidine see 2948
MEDI-493 see 7089
Medialan LL-33 see 4401
Medialan LL-99 see 4401
Mediator [Biopharma] see 1041
Mediaven [Syntex] see 6479
Mediaxal [Servier] see 1041
Medicagol, 5859
Medicarpin, 5860
Medicoal [Lundbeck] see 1808
Medifoxamine, 5861
Medifuran see 4323
Medigoxin see 3186
Medihaler-Epi [3M Pharma] see 3674
Medilave [Farillon] see 2031
Medinal see 957
Medio-Contrix "38" see 5107
Medomin [Novartis] see 4690
Medopren [Malesci] see 6127
Medphalan see 5896
Medrate [Pharmacia & Upjohn] see 6184
Medrogestone, 5862
Medroglutaric Acid see 5877
Medrol [Pharmacia & Upjohn] see 6184
Medrone [Pharmacia] see 6184
Medronic Acid, 5863
Medroxyprogesterone, 5864
Medrysone, 5865
Medullipin, 5866
Medullipin I see 5866
Medullipin II see 5866
(R,R)-Me-DuPHOS see 3513
(S,S)-Me-DuPHOS see 3513
Meerschaum see 5751
Meerwein's Reagent, 5867
Meerzwiebel see 8898
Mefamide [Bayer] see 5712
Mefedina [Farmitalia] see 5916
Mefenacet, 5868
Mefenacid [Streuli] see 5869
Mefenamic Acid, 5869
Mefenorex, 5870
Mefenoxam see 5993

Mefenpyr-diethyl, 5871
Mefloquine, 5872
Mefluidide, 5873
Mefoxin [Merck & Co.] *see* 1938
Mefoxitin [Merck & Co.] *see* 1938
Mefruside, 5874
Mega *see* 5842
Megabion (Japanese) *see* 6022
Megace [BMS] *see* 5876
Megacert [Wakamoto] *see* 4350
Megacillin [Grünenthal] *see* 7209
Megacillin (susp.) [Frosst] *see* 7205
Megacillin (Tabl.) [Frosst] *see* 7203
Megacins, 5875
Megakaryocyte Potentiating Factor *see* 9540
Megalocin [Roche] *see* 4131
Megalone [Roche] *see* 4131
Megasedan [DIF] *see* 5857
Megasul [Am. Cyanamid] *see* 6705
Megestat [BMS] *see* 5876
Megestil [Boehringer, Mann.] *see* 5876
Megestrol Acetate, 5876
Megimide [Inibsa] *see* 1029
Megion [Sandoz] *see* 1955
Meglum [Bago] *see* 5511
Meglumine *see* 6150
Meglumine Amidotrizoate *see* 2993
Meglumine Diatrizoate *see* 2993
Meglumine Iothalamate *see* 5107
Meglutol, 5877
(4*S*)-MeHex-D-Val-Thr-Val-D-Val-D-Pro-Orn-D-*allo*-Ile-cyclo[D-*allo*-Thr-D-*allo*-Ile-D-Val-Phe-(*Z*)-Dhb-Val] *see* 3601
Meiact [Meiji] *see* 1922
Meicelin [Meiji] *see* 1928
Meilax [Meiji Seika] *see* 3876
Meionite *see* 1651
Meitnerium, 5878
MeJA *see* 5308
Mejoralito [GSK] *see* 47
MEK *see* 6143
Meksamin *see* 6076
Mel B *see* 5883
Meladinine [DB Pharma] *see* 6059
Melagatran, 5879
Melaleuca Oil *see* 9226
Melamine, 5880
2-(4-Melamin-2-ylphenyl)-4-hydroxy-methyl-1,3-dithia-2-arsolane *see* 5883
Melampyrin *see* 4362
Melampyrite *see* 4362
Melampyrum *see* 4362
Melaniline *see* 3358
Melanins, 5881
Melanocyte-stimulating Hormone *see* 6381
Melanocyte-stimulating Hormone Release-inhibiting Factor *see* 5882
Melanophore-affecting Hormone *see* 6381
Melanophore Dilating Hormone *see* 6381
Melanophore Expanding Hormone *see* 6381
Melanophore Hormone *see* 6381
Melanophore-stimulating Hormone *see* 6381
Melanosome-dispersing Hormone *see* 6381
Melanostatin, 5882
Melanostatin I (Ox) *see* 5882
Melanotan [Clinuvel] *see* 168
Melanotropic Hormone *see* 6381
Melanotropin *see* 6381

α-Melanotropin *see* 6381
β-Melanotropin *see* 6381
γ-Melanotropin *see* 6381
Melanotropin Inhibiting Factor *see* 5882
Melanterite *see* 4084
Melanthigenin *see* 4659
Melarsoprol, 5883
Melatonin, 5884
Meldrum's Acid, 5885
Melengestrol Acetate, 5886
Meletin *see* 8150
Melex [Sankyo] *see* 6245
Melezitose, 5887
Melfalan *see* 5896
Meliane [Schering AG] *see* 4447
Melibiase *see* 4366
Melibiose, 5888
Melilot, 5889
Melin *see* 8438
Melinamide, 5890
Melinonine G *see* 4125
Melipramine [EGIS] *see* 4962
Melissyl Alcohol *see* 9753
Melitose *see* 8213
Melitoxin *see* 3100
Melitracen, 5891
Melitriose *see* 8213
Melittin, 5892
Melittin (Honeybee) *see* 5892
Melittin I *see* 5892
Melium [Han Wha] *see* 6245
Melixeran [Lusofarmaco] *see* 5891
Mellaril [Novartis] *see* 9512
Melleretten [Novartis] *see* 9512
Melleril [Novartis] *see* 9512
Mellic Acid *see* 5893
Mellitic Acid, 5893
Mellitin *see* 5892
Melody [Bayer CropSci.] *see* 5124
Melon Tree *see* 7123
Melopat [Medopharm] *see* 1180
Meloxicam, 5894
Melperone, 5895
Melphalan, 5896
Melprex [Chimac-Agriphar] *see* 3455
Meltatox [BASF] *see* 3454
Melysin [Takeda] *see* 384
Memantine, 5897
Memary [Merz] *see* 5897
Memcor [Lilly] *see* 10254
Memento 400 [Volpino] *see* 7572
MeMet *see* 6170
Memine [Macfarlan Smith] *see* 7441
Memoq [Gödecke] *see* 6581
Memphenesin *see* 5917
Memric [GSK] *see* 8444
MEN-8029 *see* 10386
MEN-9063 *see* 385
Menadiol, 5898
Menadione, 5899
Menadione Carboxymethoxime Ammonium Salt *see* 5900
Menadoxime, 5900
Menaelle [Théramex] *see* 7889
Menamin [RPR] *see* 5352
Menaphthone *see* 5899
Menaphthone Carboxymethoxime Ammonium Salt *see* 5900
Menaquinone 4 *see* 5901
Menaquinone 6 *see* 5901
Menaquinone 7 *see* 5901
Menaquinones, 5901
Menatetrenone *see* 5901
Menazone *see* 4337
Menbutone, 5902

Mendelevium, 5903
Mendon [Dainippon] *see* 2392
Mendrin *see* 3633
Menetryl *see* 3764
Menformon [Organon] *see* 3763
Menhaden Oil, 5904
Menichlopholan *see* 6603
Menitazine [Towa Yakuhin] *see* 1180
Menocil [Cilag-Chemie] *see* 472
Menoctyl [Dong Wha] *see* 6997
Menoflavin [Melbrosin] *see* 9858
Menogon [Ferring] *see* 4250
Menopatol [Chemiphar] *see* 9681
Menophase [Syntex] *see* 5987
Menorest *see* 3758
Menostar [Berlex] *see* 3758
Menotrophin *see* 4250
Menotropins *see* 4250
Menova [Merck KGaA] *see* 2102
Mensiso [Menarini] *see* 8688
Menstridyl *see* 2102
Mentax [Kaken] *see* 1521
p-Mentha-1,5-diene *see* 7315
p-Mentha-1,7(2)-diene *see* 7316
p-Mentha-1,8-diene *see* 5546
p-Mentha-6,8-dien-2-one *see* 1874
trans-(−)-2-*p*-Mentha-1,8-dien-3-yl-5-pentylresorcinol *see* 1750
p-Menthane-1,8-diol *see* 9311
3-*p*-Menthanol *see* 5905
(1*R*,4*S*)-(−)-*p*-Menthan-3-one *see* 5906
2-[(*p*-Mentha-1,3,5-trien-2-yloxy)methyl]-2-imidazoline *see* 10015
1-*p*-Menthene-6,8-diol *see* 8705
p-Menth-6-ene-2,8-diol *see* 8705
p-Menth-1-en-8-ol *see* 9313
1-*p*-Menthen-2-ol-3-one *see* 3326
p-Menth-1-en-3-one *see* 7585
R-(+)-*p*-Menth-4(8)-en-3-one *see* 8046
Menthol, 5905
l-Menthol *see* 5905
l-Menthol Acetate *see* 5907
l-Menthone, 5906
Menthyl Acetate, 5907
Menthyl *o*-Aminobenzoate *see* 5908
Menthyl Anthranilate, 5908
Menthyl Isovalerate, 5909
l-Menthyl Valerate *see* 5909
Mentium [Guidotti] *see* 8104
Menyanthes, 5910
Menzol [Schwarz] *see* 6786
Meobentine, 5911
5-MeO-DIPT *see* 6062
MeOTf *see* 6204
Meothrin [Shell] *see* 4019
MEP *see* 4007
Mepacrine *see* 8160
Mepact [Takeda] *see* 6265
Mepanipyrim, 5912
Meparfynol, 5913
Mepartricin, 5914
Mepartricin A *see* 5914
Mepartricin B *see* 5914
Mepavlon [Zeneca] *see* 5929
Mepedyl [Glaxo] *see* 3370
Mepenzolate Bromide, 5915
Meperidine, 5916
Mepha-Butazon [Mepha] *see* 7390
Mephenamin [Boehringer, Mann.] *see* 6975
Mephenesin, 5917
Mephenhydramine *see* 6373
Mephenon [Federa] *see* 6016
Mephenoxalone, 5918
Mephentermine, 5919

Mercurous Sulfate, 5965
Mercury, 5966
Mercury and Sodium Phenolsulfonate
  see 5954
Mercury Bichloride see 5943
Mercury Biniodide see 5946
Mercury Bisulfate see 5955
Mercury Bromide (HgBr$_2$) see 5942
Mercury Chloride (HgCl$_2$) see 5943
Mercury Chloride (Hg$_2$Cl$_2$) see 5962
Mercury Cyanide (Hg(CN)$_2$) see 5944
Mercury Cyanide Oxide (Hg$_2$(CN)$_2$O)
  see 5951
Mercury Dicyanide see 5944
Mercury Difluoride see 5945
Mercury Fluoride (HgF$_2$) see 5945
Mercury Fulminate, 5967
Mercury Iodide (HgI$_2$) see 5946
Mercury Monochloride see 5962
Mercury Oxide (HgO) see 5949, 5950
Mercury Perchloride see 5943
Mercury Pernitrate see 5947
Mercury Potassium Cyanide (HgK$_2$-
  (CN)$_4$) see 5952
Mercury Protochloride see 5962
Mercury Protoiodide see 5963
Mercury Protonitrate see 5964
Mercury Subchloride see 5962
Mercury Subsalicylate see 5953
Mercury Sulfide (HgS) see 5956, 5957
Mereprine [Cassella-med] see 3489
Meresa [Dolorgiet] see 9119
Merfen [Zyma] see 7414
Merfene [Novartis] see 2092
Mergamma [Zeneca] see 7411
Meridia [Abbott] see 8624
Merilid see 2105
Merimepodib, 5968
Merislon [Eisai] see 1180
Merit [Bayer CropSci.] see 4951
Merital [Hoechst] see 6760
Merlin [Bayer CropSci.] see 5283
Merocet [HMR] see 2031
Merocyanine 540, 5969
Merodicein [Becton Dickinson] see 5932
Meronem [AstraZeneca] see 5970
Meropen [Sumitomo] see 5970
Meropenem, 5970
Merosin see 5401
Merpan see 1774
Merphalan see 5896
Merphenyl Nitrate see 7413
Merrem [AstraZeneca] see 5970
Mersalyl, 5971
Mersalyl Acid see 5971
Mersarex [Arpida] see 4926
Mertansine see 5830
Mertax see 5935
Mertect [Syngenta] see 9440
Mertestate [Sterling Winthrop] see 9324
Merthiolate [Lilly] see 9470
Mervaldin [Lannett] see 5917
Merzonin [Takeda] see 9470
MES, 5972
Mesaconic Acid, 5973
Mesalamine, 5974
Mesalazine, 5974
Mesantoin [Sandoz] see 5920
Mesasal [Sanofi-Synthelabo] see 5974
Mescaline, 5975
Mescaline Oxidase see 6333
Mesembrine, 5976
Mesentol see 3801
Mesidicaine see 9869
Me$_3$SiOSiMe$_3$ see 4726

Mesitylene, 5977
Mesityl Oxide, 5978
Mesna, 5979
Mesnex [Baxter] see 5979
Mesocaine see 9869
Mesoetioporphyrin see 3917
Mesoinosite see 5021
Mesokain see 9869
Mesonex see 5022
Mesontoin see 5920
Mesopin [Endo] see 4768
Mesoridazine, 5980
Mesoridazine Besylate see 5980
Mesosulfuron-methyl, 5981
Mesotal [Galma] see 5022
Mesotartaric Acid see 9202
Mesothorium I see 8210
Mesothorium II see 127
Mesotrione, 5982
Mesoxalic Acid, 5983
Mesoxalonitrile (−)-[p-[(R)-1,4,5,6-
  Tetrahydro-4-methyl-6-oxo-3-pyrid-
  azinyl]phenyl]hydrazone see 5525
Mesoxalylcarbamide see 273
Mesoxalylurea see 273
Mesoyohimbine see 10301
Mespafin [Merckle] see 3488
Mesquite Gum, 5984
Messenger [Eden Bioscience] see 4650
Messenger RNA see 8327
Mestanolone, 5985
Mestenediol see 6022
Mesterolone, 5986
Mestilbol see 3148
Mestinon [Valeant] see 8090
Mestoranum [Schering AG] see 5986
Mestranol, 5987
Mesudin see 5712
Mesudrin see 5712
Mesulfa [Triosol] see 9045
Mesulfen, 5988
Mesulid [LPB] see 6633
Mesulphen see 5988
Mesurol [Bayer CropSci.] see 6044
Mesuximide see 6078
Mesyl Chloride see 6027
Met see 6047
Metab-Auxil see 7221
Metabenzthiazuron see 6011
Metabolic Indican see 5013
Metabolin [Takeda] see 9447
Metabutethamine Hydrochloride see
  1526
Metacam [Boehringer, Ing.] see 5894
Metacetaldehyde see 5994
Metachlor see 194
Metachloral see 9791
Metachrome Yellow, 5989
Metacide [Bayer CropSci.] see 6181
Metaclazepam, 5990
Metacortandracin see 7842
Metacortandralone see 7841
Metacresol Acetate see 2572
Metacycline see 6015
[2.2]Metacyclophane, 5991
Metadate [UCB] see 6183
Metadelphene [Hercules] see 2859
Metadiazine see 8100
Metadomus [Med. Domus] see 6015
Metafar [Lafare] see 1927
Metaflumizone, 5992
Metaformaldehyde see 9906
Metahydrin [HMR] see 9789
Metalaxyl, 5993
Metalaxyl-M see 5993

Metalcaptase [Heyl] see 7197
β$_1$-Metal-combining Protein see 9732
Metaldehyde, 5994
Metallibure see 6018
Metallic Arsenic see 784
Metallum Problematum see 9263
Metalphthalein see 2566
Metalutin [Parke-Davis] see 6799
Metalyse [Boehringer, Ing.] see 9286
Metamide [Protea] see 6219
Metamin [Takeda] see 4219
Metamitron, 5995
Metamizol see 3390
Metampicillin, 5996
Metam Sodium see 6021
Metamucil [Procter & Gamble] see 7631
Metandiol see 6022
Metandren [Novartis] see 6197
Metanephrine, 5997
Metanilic Acid, 5998
Metanil Yellow, 5999
Metanyl see 6030
Metaoxedrin see 7398
Metaphanine, 6000
Metaphen [Abbott] see 6697
Metaphos see 6181
Metaphosphoric Acid (HPO$_3$) see 7457
Metaphosphoric Acid (HPO$_3$) Potassium
  Salt (1:1) see 7767
Metaphosphoric Acid (HPO$_3$) Sodium
  Salt (1:1) see 8773
Metaphosphoric Acid (H$_3$P$_3$O$_9$) Sodium
  Salt (1:3) see 8823
Metaplexan [Rhône-Poulenc] see 5931
Metapramine, 6001
Metaprel [Dorsey] see 6002
Metaproterenol, 6002
Metaradrine see 6003
Metaraminol, 6003
Metarsenobillon see 9072
Metasilicic Acid see 8629
Metastron [Amersham] see 8970
Meta-Systox see 6123
Metathion see 4007
Metatyrosine see 10021
Metavariscite see 352
Metaxan see 6030
Metaxanin see 5993
Metazachlor, 6004
Metazepium Iodide see 1603
Metazocine, 6005
α-(−)-Metazocine see 6005
α-(+)-Metazocine see 6005
β-(−)-Metazocine see 6005
β-(+)-Metazocine see 6005
cis-(−)-Metazocine see 6005
cis-(+)-Metazocine see 6005
trans-(−)-Metazocine see 6005
trans-(+)-Metazocine see 6005
Metconazole, 6006
Met^5-E see 3627
Metelilachlor see 6223
Meteloidine, 6007
Metenarin [Teikoku Zoki] see 6142
Metendiol see 6022
Metenix [Sanofi-Aventis] see 6224
Met^5-Enkephalin see 3627
Méténolone see 6039
Metepa, 6008
Meterazine see 7879
Meterfer [Sinclair] see 4073
Metergoline, 6009
Metflorylthiazidine see 4826
Metformin, 6010
Metformin Pamoate see 6010

5-Methoxyacetylamino-2,4,6-triiodoiso-
phthalic Acid [(2,3-Dihydroxy-*N*-
methylpropyl)-(2,3-dihydroxypropyl)]-
diamide *see* 5105
Methoxyamine, 6060
3-Methoxy-4-(4′-aminobenzenesulfon-
amido)-1,2,5-thiadiazole *see* 9052
4-Methoxy-4′-aminodiphenylamine *see*
6072
6-Methoxy-8-(4-aminopentylamino)quin-
oline *see* 8183
Methoxyammonium Chloride *see* 6060
2,2′-[3-Methoxy-4′-amyl-5′-methyl-5-(2″-
pyrryl)]dipyrrylmethene *see* 7884
3-Methoxyaniline *see* 660
4-Methoxy-6-[β-(*p*-anisyl)vinyl]-α-py-
rone *see* 10291
10-Methoxy-6aβ-aporphin-11-ol *see* 727
4-Methoxybenzaldehyde *see* 656
2-Methoxybenzenamine *see* 660
3-Methoxybenzenamine *see* 660
4-Methoxybenzenamine *see* 660
Methoxybenzene *see* 662
4-Methoxy-1,3-benzenediamine *see* 2978
4-Methoxybenzenemethanol *see* 658
4-Methoxybenzoic Acid *see* 659
2-(4-Methoxybenzoyl)benzoic Acid *see*
664
4-Methoxybenzoyl Chloride *see* 665
1-(4-Methoxybenzoyl)-2-pyrrolidinone
see 655
p-Methoxybenzyl Alcohol *see* 658
1-(*p*-Methoxybenzyl)-2,3-dimethylguani-
dine *see* 5911
N-*p*-Methoxybenzyl-*N*′,*N*′-dimethyl-*N*-
α-pyridylethylenediamine *see* 8097
α-(α-Methoxybenzyl)-4-(β-methoxy-
phenethyl)-1-piperazineethanol *see*
10370
2-(4-Methoxy[1,1′-biphenyl]-3-yl)hydra-
zinecarboxylic Acid 1-Methylethyl
Ester *see* 1211
2-Methoxy-4,6-bis(isopropylamino)-*s*-tri-
azine *see* 7903
5-Methoxy-*N*,*N*-bis(1-methylethyl)-1*H*-
indole-3-ethanamine *see* 6062
6-Methoxy-*N*²,*N*⁴-bis(1-methylethyl)-1,-
3,5-triazine-2,4-diamine *see* 7903
4-Methoxy-*N*¹,*N*³-bis(3-pyridinylmeth-
yl)-1,3-benzenedicarboxamide *see*
7519
4-Methoxy-*N*,*N*′-bis(3-pyridylmethyl)-
isophthalamide *see* 7519
2-(Methoxycarbonylamino)benzimid-
azole *see* 1792
2-[[[(Methoxycarbonyl)amino][[2-nitro-
5-(propylthio)phenyl]amino]methyl-
ene]amino]ethanesulfonic Acid *see*
6565
2-[[[(Methoxycarbonyl)amino][[2-nitro-
5-(propylthio)phenyl]imino]methyl]-
amino]ethanesulfonic Acid *see* 6565
3-Methoxycarbonyl-6-*n*-butyl-7-benzyl-
oxy-4-oxoquinoline *see* 6556
N-Methoxycarbonyl-*N*′-[2-nitro-5-(propylthio)phenyl]-*N*″-2-(ethylsulfonic
Acid)guanidine *see* 6565
4-(Methoxycarbonyl)-4-[(1-oxopropyl)-
phenylamino]-1-piperidinepropanoic
Acid Methyl Ester *see* 8251
N-(2-Methoxycarbonylphenylsulfonyl)-
N′-[4,6-bis(difluoromethoxy)pyrimi-
din-2-yl]urea *see* 7866
1-Methoxycarbonyl-1-propen-2-yl Dimethyl Phosphate *see* 6244

(Methoxycarbonylsulfamoyl)triethylam-
monium Hydroxide Inner Salt *see*
1505
*N*¹-Methoxycarbonylsulfanilamide *see*
847
Methoxychlor, 6061
N-[4-(β-(2-Methoxy-5-chlorobenzamido)-
ethyl)benzosulfonyl]-*N*′-cyclohexyl-
urea *see* 4514
*N*¹-[4-[β-(2-Methoxy-5-chlorobenzoyl-
amino)ethyl]benzenesulfonyl]-*N*²-cy-
clohexylurea *see* 4514
(8α,9*R*)-6′-Methoxycinchonan-9-ol *see*
8177
(8α,9*S*)-6′-Methoxycinchonan-9-ol *see*
3676
(9*R*)-6′-Methoxycinchonan-9-ol *see* 3675
(9*S*)-6′-Methoxycinchonan-9-ol *see* 8176
(8α)-6′-Methoxycinchonan-9-one *see*
8179
6-Methoxycinchoninic Acid *see* 8178
p-Methoxycinnamic Acid 2-Ethoxyethyl
Ester *see* 2310
Methoxyconiferine *see* 9146
9-Methoxycorynantheidine *see* 6303
2-Methoxy-*p*-cresol *see* 2559
4-Methoxy-2,6-cresotic Acid Methyl Ester *see* 8861
(1α,3α)-7-Methoxycrinan-1,3-diol *see*
1497
Methoxycyclopentane *see* 2742
Methoxycyclopropane *see* 2746
Methoxy-DDT *see* 6061
10-Methoxydeserpidine *see* 2919
Methoxydichlorobenzoate *see* 3050
2-Methoxy-3,6-dichlorobenzoic Acid *see*
3050
6-(α-Methoxy-3,4-dichlorophenylacet-
amido)penicillanic Acid *see* 2376
8-Methoxydictamnine *see* 3969
6-Methoxy-8-(6-diethylaminohexylam-
ine)-4-methylquinoline *see* 8691
6-Methoxy-8-(3-diethylaminopropylam-
ino)quinoline *see* 7636
[3-Methoxy-4-[(*N*,*N*-diethylcarbamido)-
methoxy]phenyl]acetic Acid *n*-Propyl
Ester *see* 7917
6-Methoxy-3,4-dihydro-1(2*H*)-naphtha-
lenone *see* 6074
4-Methoxy-2,2′-dihydroxybenzophenone
see 3332
5-Methoxy-*N*,*N*-diisopropyltryptamine,
6062
2-Methoxy-*N*,α-dimethylbenzeneethan-
amine *see* 6071
5-Methoxy-8,8-dimethyl-2*H*,8*H*-benzo-
[1,2-*b*:5,4-*b*′]dipyran-2-one *see* 10264
7-Methoxy-2,2-dimethyl-2*H*-1-benzopy-
ran *see* 7839
7-Methoxy-2,2-dimethylchromene *see*
7839
(8β)-10-Methoxy-1,6-dimethylergoline-8-
methanol 5-Bromo-3-pyridinecarbox-
ylate (Ester) *see* 6581
o-Methoxy-*N*,*N*,α-dimethylphenethylamine
see 6071
(*trans*)-1-[2-[*p*-(7-Methoxy-2,2-dimethyl-
3-phenyl-4-chromanyl)phenoxy]ethyl]-
pyrrolidine *see* 1972
2-Methoxy-3,5-dimethyl-6-[(2*R*,4*Z*)-tetra-
hydro-4-[(2*E*)-2-methyl-3-(4-nitro-
phenyl)-2-propen-1-ylidene]-2-furan-
yl]-4*H*-pyran-4-one *see* 870
Methoxydiuron *see* 5564
Methoxydon(e) *see* 5918

Methoxyethane *see* 3883
2-Methoxyethanol *see* 6063
2-Methoxyethanol 1-Acetate *see* 6064
2-(2-Methoxyethoxy)ethanol *see* 3136
N-[5-(2-Methoxyethoxy)-2-pyrimidinyl]-
benzenesulfonamide *see* 4543
2-Methoxyethyl Acetate, 6064
Methoxyethylbenzeneboronic Acid, 6065
2-Methoxyethyl 1,4-Dihydro-5-(isopro-
poxycarbonyl)-2,6-dimethyl-4-(3-nitro-
phenyl)-3-pyridinecarboxylate *see*
6636
1-[4-(2-Methoxyethyl)phenoxy]-3-[(1-
methylethyl)amino]-2-propanol *see*
6228
B-[2-[(1*R*)-1-Methoxyethyl]phenyl]bo-
ronic Acid *see* 6065
2-(2-Methoxyethyl)pyridine *see* 6239
11β-Methoxy-17α-ethynylestradiol *see*
6375
11β-Methoxy-17α-ethynyl-Δ^1,3,5(10)-estra-
triene-3,17β-diol *see* 6375
Methoxyfenozide, 6066
Methoxyflurane, 6067
"Methoxy Foxy" *see* 6062
8-Methoxy[furano-3′,2′:6,7-coumarin]
see 6059
4-Methoxy-7*H*-furo[3,2-*g*][1]benzopyran-
7-one *see* 1160
9-Methoxy-7*H*-furo[3,2-*g*][1]benzopyran-
7-one *see* 6059
8-Methoxy-4′,5′:6,7-furocoumarin *see*
6059
4-Methoxyfuro[2,3-*b*]quinoline *see* 3099
10-Methoxyharmalan, 6068
Methoxyhydrastine *see* 6807
4-Methoxy-2-hydroxybenzophenone *see*
7053
3-Methoxy-4-hydroxycinnamic Acid *see*
4089
3-Methoxy-4-hydroxymandelic Acid *see*
10121
8-(2-Methoxy-3-hydroxymercuripropyl)-
coumarin-3-carboxylic Acid *see* 5939
L-3-Methoxy-4-hydroxyphenylalanine
see 6128
3-Methoxy-4-hydroxyphenylglycol, 6069
3-Methoxy-10-[3-(4-hydroxypiperidyl)-2-
methylpropyl]phenothiazine *see* 7283
3-Methoxy-4-hydroxytoluene *see* 2559
12-Methoxyibogamine *see* 4914
13-Methoxyibogamine *see* 9150
12-Methoxyibogamine-18-carboxylic
Acid Methyl Ester *see* 10223
(α*Z*,3*R*)-α-(Methoxyimino)-1-azabicyclo-
[2.2.2]octane-3-acetonitrile *see* 8444
(*E*)-2-Methoxyimino-*N*-methyl-2-[2-(2,5-
dimethylphenoxymethyl)phenyl]acet-
amide *see* 3292
(α*E*)-α-(Methoxyimino)-2-[(2-methyl-
phenoxy)methyl]benzeneacetic Acid
Methyl Ester *see* 5372
(α*E*)-α-(Methoxyimino)-2-[[[(*E*)-[1-[3-
(trifluoromethyl)phenyl]ethylidene]-
amino]oxy]methyl]benzeneacetic Acid
Methyl Ester *see* 9843
(*E*,*E*)-Methoxyimino-[2-[1-(3-trifluoro-
methylphenyl)ethylideneaminooxy-
methyl]phenyl]acetic Acid Methyl
Ester *see* 9843
5-Methoxy-1*H*-indole-3-ethanamine *see*
6076
N-[2-(5-Methoxy-1*H*-indol-3-yl)ethyl]-
acetamide *see* 5884

1-(4-Methoxy-6-trifluoromethyl-1,3,5-triazin-2-yl)-3-(2-trifluoromethylbenzenesulfonyl)urea *see* 9943

5-Methoxy-4'-(trifluoromethyl)valerophenone (*E*)-*O*-(2-Aminoethyl)oxime *see* 4246

2-Methoxytrimeprazine *see* 5518

(3β,16β,17α,18β,20α)-17-Methoxy-18-[(3,4,5-trimethoxybenzoyl)oxy]yohimban-16-carboxylic Acid Methyl Ester *see* 2919

3-Methoxy-6-[(1*Z*)-2-(3,4,5,-trimethoxyphenyl)ethenyl]-1,2-benzenediol *see* 2477

2-Methoxy-5-[(1*Z*)-2-(3,4,5,-trimethoxyphenyl)ethenyl]phenol *see* 2477

2-Methoxy-5-[(1*Z*)-2-(3,4,5-trimethoxyphenyl)ethenyl]phenol 1-(Dihydrogen Phosphate) *see* 4277

(2*E*,4*E*)-11-Methoxy-3,7,11-trimethyl-2,4-dodecadienoic Acid 1-Methylethyl Ester *see* 6055

(β*R*)-2-Methoxy-*N*,*N*,β-trimethyl-10*H*-phenothiazine-10-propanamine *see* 5518

(2*E*,4*E*,6*E*,8*E*)-9-(4-Methoxy-2,3,6-trimethylphenyl)-3,7-dimethyl-2,4,6,8-nonatetraenoic Acid *see* 103

(*all-E*)-9-(4-Methoxy-2,3,6-trimethylphenyl)-3,7-dimethylnona-2,4,6,8-tetraen-1-oic Acid Ethyl Amide *see* 6370

(2*E*,4*E*,6*E*,8*E*)-9-(4-Methoxy-2,3,6-trimethylphenyl)-3,7-dimethyl-2,4,6,8-nonatetraenoic Acid Ethyl Ester *see* 3936

1-Methoxy-3-(trimethylsilyloxy)-1,3-butadiene, 6075

5-Methoxytryptamine, 6076

3-Methoxy-L-tyrosine *see* 6128

6-Methoxyumbelliferone *see* 8545

Methoxyverapamil *see* 4386

5-Methoxy-2-[(6-*O*-β-D-xylopyranosyl-β-D-glucopyranosyl)oxy]-benzoic Acid Methyl Ester *see* 7867

Methrazone [Boehringer, Ing.] *see* 4039

Methscopolamine Bromide, 6077

Methscopolamine Nitrate *see* 8543

Methsuximide, 6078

Methural *see* 7067

Methyclothiazide, 6079

Methycobal [Eisai] *see* 6116

Methyl Abietate, 6080

Methylacetaldehyde *see* 7937

Methyl 4-Acetamido-2-ethoxybenzoate *see* 3798

m-Methylacetanilide *see* 72

N-Methylacetanilide, 6081

Methyl Acetate, 6082

Methylacetic Acid *see* 7939

Methylacetic Anhydride *see* 7940

Methyl Acetoacetate, 6083

(+)-*N*-Methyl-2-acetonylpyrrolidine *see* 4890

Methylacetopyronone *see* 2869

6-α-Methyl-17α-acetoxyprogesterone *see* 5864

Methyl Acetylsalicylate, 6084

β-Methylacrolein *see* 2587

Methyl Acrylate, 6085

α-Methylacrylic Acid *see* 6013

β-Methylacrylic Acid *see* 2588

α-Methylacrylonitrile *see* 6014

α-Methyl-1-adamantanemethylamine *see* 8351

N-Methyladrenaline *see* 6141

3-*O*-Methyladrenaline *see* 5997

Methylal, 6086

Methylal Acetamide *see* 4270

2-Methylalanine *see* 442

8-(2-Methylalanine)-35-(2-methylalanine)-36-L-argininamide-7-36-glucagon-like Peptide I (Human) *see* 9209

3-Methyl-*N*,α-(β-alanyl)-L-histidine *see* 671

Methyl Alcohol *see* 6029

Methyl Aldehyde *see* 4263

2-(*p*-Methylallylaminophenyl)propionic Acid *see* 293

α-Methylallyl Chloride *see* 2131

β-Methylallylchloride *see* 2150

γ-Methylallyl Chloride *see* 2130

13β-Methyl-17α-allyl-Δ⁴,⁹,¹¹-gonatriene-17β-ol-3-one *see* 315

Methyl Allyl Trisulfide, 6087

Methyl (2*R*,3*R*)-2-[3-Amidinobenzyl]-3-[[4-(1-oxido-4-pyridinyl)benzoyl]amino]butanoate *see* 6996

Methylamine, 6088

N-Methylaminoacetic Acid *see* 8510

2-(Methylamino)-*m*-anisic Acid Methyl Ester *see* 2807

Methyl *p*-Aminobenzenesulfonylcarbamate *see* 847

Methyl 2-Aminobenzoate *see* 6094

2-(4-Methylaminobutoxy)diphenylmethane *see* 1210

N-[[(Methylamino)carbonyl]oxy]ethanimidothioic Acid Methyl Ester *see* 6054

2-[4-[(Methylamino)carbonyl]-1*H*-pyrazol-1-yl]adenosine *see* 8247

3-Methylamino-6-carboxamido-1,2,3,4-tetrahydrocarbazole *see* 4301

4-Methylaminocetopyrocatechol *see* 160

Methyl 6-Amino-5-chloro-2-cyclopropyl-4-pyrimidinecarboxylate *see* 432

4''-*epi*-(Methylamino)-4''-deoxyavermectin B₁ₐ *see* 3612

4''-*epi*-(Methylamino)-4''-deoxyavermectin B₁ᵦ *see* 3612

3-Methylamino-7-dimethylaminophenazathonium Chloride *see* 921

4-Methylamino-1,5-dimethyl-2-phenyl-3-pyrazolone Sodium Methanesulfonate *see* 3390

α-*dl*-6-(Methylamino)-4,4-diphenyl-3-heptanol Acetate *see* 6773

N-Methylaminodithioformic Acid Sodium Salt *see* 6021

N-Methylaminoethane Sodium Sulfonate *see* 6196

β-Methylaminoethane-α-sulfonic Acid *see* 6196

2-(Methylamino)ethanesulfonic Acid *see* 6196

Methylaminoethanoic Acid *see* 8510

2-(Methylamino)ethanol, 6089

l-Methylaminoethanolcatechol *see* 3674

m-Methylaminoethanolphenol *see* 7398

p-Methylaminoethanolphenol *see* 9142

p-Methylaminoethanolphenol Tartrate *see* 9142

4-[2-(Methylamino)ethyl]-1,2-benzenediol *see* 2904

(α*R*)-α-[(1*S*)-1-(Methylamino)ethyl]benzenemethanol *see* 3663

(α*S*)-α-[(1*S*)-1-(Methylamino)ethyl]benzenemethanol *see* 8024

α-(1-Methylaminoethyl)-*p*-hydroxybenzyl Alcohol *see* 4863

4-[2-(Methylamino)ethyl]-*o*-phenylene Diisobutyrate *see* 4915

2-[2-(Methylamino)ethyl]pyridine *see* 1180

4-[2-(Methylamino)ethyl]pyrocatechol *see* 2904

2-Methyl-6-aminoheptane *see* 6846

2-Methyl-6-amino-2-heptanol *see* 4692

6-Methyl-2-amino-6-heptanol *see* 4692

Methyl 3-Amino-4-hydroxybenzoate *see* 6979

2-Methyl-4-amino-1-hydroxynaphthalene *see* 10220

β-Methylamino-α-(4-hydroxyphenyl)-ethyl Alcohol *see* 9142

2-Methylamino-5α-(β-indolyl)ethyl-2-oxazolin-4-one *see* 5007

2-Methylaminoisocamphane *see* 5843

3-Methylaminoisocamphane *see* 5843

Methyl Aminolevulinate, 6090

Methyl 2-Amino-3-mercaptopropionate *see* 5856

(α*R*)-α-[(Methylamino)methyl]benzenemethanol *see* 4635

(−)-α-(Methylaminomethyl)benzyl Alcohol *see* 4635

1-(Methylaminomethyl)dibenzo[*b*,*e*]bicyclo[2.2.2]octadiene *see* 1089

9-(Methylaminomethyl)-9,10-dihydro-9,10-ethanoanthracene *see* 1089

6-Methylamino-2-methylheptene *see* 5231

Methylaminomethyl 4-Hydroxyphenyl Carbinol *see* 9142

1-Methylaminomethyl-2-methyl-2-phenylethane *see* 7418

2-Methylaminomethyl-2-methyl-1-phenylpropane *see* 5919

α-(Methylaminomethyl)vanillyl Alcohol *see* 5997

N-[4-(6-Methylamino-7-nitro-2-thia-5-aza-6-heptene-1-yl)-2-thiazolylmethyl]-*N*,*N*-dimethylamine *see* 6747

Methyl [[3-(Aminooxoacetyl)-1-benzyl-2-ethyl-1*H*-indol-4-yl]oxy]acetate *see* 10126

p-Methylaminophenol Sulfate, 6091

4-(Methylamino)phenol Sulfate (2:1) *see* 6091

2-Methylamino-1-phenylethanol *see* 4635

1-Methylamino-2-phenylpropane *see* 7418

(1*R*,2*S*)-2-Methylamino-1-phenylpropan-1-ol *see* 3663

(1*S*,2*S*)-2-Methylamino-1-phenylpropan-1-ol *see* 8024

L-*erythro*-2-(Methylamino)-1-phenylpropan-1-ol *see* 3663

2-(Methylamino)-1-phenyl-1-propanone *see* 6035

1-Methyl-4-amino-*N*-phenyl-*N*-(2-thenyl)piperidine *see* 9429

5-Methylamino-2-phenyl-4-[3-(trifluoromethyl)phenyl]-3(2*H*)-furanone *see* 4235

5-Methylamino-2-phenyl-4-(α,α,α-trifluoro-*m*-tolyl)furan-3(2*H*)-one *see* 4235

α-Methylaminopropiophenone *see* 6035

β-(Methylamino)propylcyclopentane *see* 2735

1-(3-Methylaminopropyl)dibenzo[*b*,*e*]bicyclo[2.2.2]octadiene *see* 5813

7-(3-Methylaminopropyl)-1,2:5,6-dibenzocycloheptatriene *see* 8010

Methyl *p*-Chlorobenzoate *see* 2126

3-Methyl-7-chloro-1,2,4-benzothiadiazine 1,1-Dioxide *see* 3007

N-Methyl-*N*-(4-chlorobenzoylmethyl)-3-(10,11-dihydro-5*H*-dibenzo[*b*,*f*]azepin-5-yl)propylamine *see* 5614

2-Methyl-1-(*p*-chlorobenzyl)benzimidazole *see* 2107

Methyl Chlorocarbonate, 6114

Methyl 3-Chloro-5-(4,6-dimethoxypyrimidin-2-ylcarbamoylsulfamoyl)-1-methylpyrazole-4-carboxylate *see* 4636

Methyl 3-Chloro-2-(5-ethoxy-7-fluoro[1,2,4]triazolo[1,5-*c*]pyrimidin-2-ylsulfonamido)benzoate *see* 2391

Methyl 3-Chloro-*N*-(5-ethoxy-7-fluoro[1,2,4]triazolo[1,5-*c*]pyrimidin-2-ylsulfonyl)anthranilate *see* 2391

Methyl 6-[[[(2-Chloroethyl)nitrosoamino]carbonyl]amino]-6-deoxy-α-D-glucopyranoside *see* 8227

4-Methyl-5-(β-chloroethyl)thiazole *see* 2375

5-Methyl-2-(2′-chloro-6′-fluoroanilino)phenylacetic Acid *see* 5661

Methylchloroform *see* 9801

Methyl Chloroformate *see* 6114

Methylchloroisothiazolinone *see* 6163

Methyl Chloromethyl Ether *see* 2148

3-Methyl-4-chlorophenol *see* 2133

2-Methyl-4-chlorophenoxyacetic Acid *see* 5833

N-(2-Methyl-3-chlorophenyl)anthranilic Acid *see* 9673

trans-4-Methyl-5-(4-chlorophenyl)-3-cyclohexylcarbamoyl-2-thiazolidone *see* 4749

Methyl (+)-(*S*)-α-(*o*-Chlorophenyl)-6,7-dihydrothieno[3,2-*c*]pyridine-5(4*H*)-acetate *see* 2385

1-Methyl-6-*o*-chlorophenyl-8-ethyl-4*H*-s-triazolo[3,4-*c*]thieno[2,3-*e*]-1,4-diazepine *see* 3919

[5-Methyl-3-(*o*-chlorophenyl)-4-isoxazolyl]penicillin *see* 2403

Methyl *N*-[2-[[1-(4-Chlorophenyl)-1*H*-pyrazol-3-yl]oxymethyl]phenyl]-*N*-methoxycarbamate *see* 8064

2-Methyl-3-(2-chlorophenyl)-4-quinazolone *see* 5852

1-Methyl-2-chloropyridinium Iodide *see* 6387

Methyl 7-Chloro-6,7,8-trideoxy-6-[[[(2*S*,4*R*)-4-ethyl-2-piperidinyl]carbonyl]amino]-1-thio-L-*threo*-α-D-*galacto*-octopyranoside *see* 7613

Methyl 7-Chloro-6,7,8-trideoxy-6-[[[(2*S*,4*R*)-1-methyl-4-propyl-2-pyrrolidinyl]carbonyl]amino]-1-thio-L-*threo*-α-D-*galacto*-octopyranoside *see* 2354

3-Methylcholanthrene, 6115

20-Methylcholanthrene *see* 6115

β-Methylcholine Chloride Urethan *see* 1187

3-Methylchromone *see* 9823

3-Methylchrysazin *see* 2262

5-[(*E*,*E*)-β-Methylcinnamylidene]-4-oxo-2-thioxo-3-thiazolidineacetic Acid *see* 3660

Methylcobalamin, 6116

Methylcobaz [Labaz] *see* 6116

Co-Methylcobinamide Dihydrogen Phosphate (Ester) Inner Salt 3′-Ester with 5,6-Dimethyl-1-α-D-ribofuranosyl-1*H*-benzimidazole-κ*N*[3] *see* 6116

6α-Methylcompactin *see* 5644

3-Methylcrotonaldehyde *see* 8587

(*E*)-2-Methylcrotonic Acid *see* 9589

γ-Methyl-α,β-crotonolactone *see* 640

γ-Methyl-β,γ-crotonolactone *see* 640

Methyl 7-Crotonylidenecyclopenta[*c*]pyran-1-(7*H*)-one-4-carboxylate *see* 4314

7-*O*-Methylcurine *see* 2218

Methyl Cyanide *see* 69

Methyl Cyanoacrylate, 6117

Methyl α-Cyanoacrylate *see* 6117

Methyl 2-Cyanoacrylate *see* 6117

Methyl 2-Cyano-3,12-dioxooleana-1,9-(11)-dien-28-oate *see* 959

Methyl (*E*)-2-[2-[6-(2-Cyanophenoxy)pyrimidin-4-yloxy]phenyl]-3-methoxyacrylate *see* 917

Methylcyclohexane, 6118

1-Methyl-cyclohexanecarboxylic Acid (2,3-Dichloro-4-hydroxy-phenyl)-amide *see* 4005

N-Methyl-*N*-[(2-cyclohexyl-2-phenyl-1,3-dioxolan-4-yl)-methyl]piperidinium Iodide *see* 2729

N-(2-Methylcyclohexyl)-*N*′-phenylurea *see* 8626

(3*R*)-3-Methylcyclopentadecanone *see* 6398

Methylcyclopentadienylmanganese Tricarbonyl *see* 6309

3′-Methyl-1,2-cyclopentenophenanthrene, 6119

1-Methylcyclopropene, 6120

Methyl Cymate *see* 10372

Methyl Cysteinate *see* 5856

Methyl Cysteine *see* 5856

Methylcytisine, 6121

N-Methylcytisine *see* 6121

12-Methylcytisine *see* 6121

5-Methylcytosine, 6122

4′-Methyldaidzin *see* 4271

Methyldamascenine *see* 2807

6α-Methyl-6-dehydro-16-methylene-17-acetoxyprogesterone *see* 5886

Methyl Demeton, 6123

Methyl *O*-2-Deoxy-6-*O*-sulfo-2-(sulfoamino)-α-D-glucopyranosyl-(1 → 4)-*O*-β-D-glucopyranuronosyl-(1 → 4)-*O*-2-deoxy-3,6-di-*O*-sulfo-2-(sulfoamino)-α-D-glucopyranosyl-(1 → 4)-*O*-2-*O*-sulfo-α-L-idopyranuronosyl-(1 → 4)-2-deoxy-2-(sulfoamino)-α-D-glucopyranoside 6-(Hydrogen Sulfate) Sodium Salt (1:10) *see* 4259

Methyl-*N*-depropionyldecorticasine *see* 5618

N-Methyl-*N*-desacetylcolchicine *see* 2891

N-Methyl-3,5-diacetamido-2,4,6-triiodobenzoic Acid *see* 6235

Methyl [4,6-Diamino-2-[1-(2-fluorobenzyl)-1*H*-pyrazolo[3,4-*b*]pyridin-3-yl]-pyrimidin-5-yl]methylcarbamate *see* 8358

Methyl Diazepinone *see* 2997

1-Methyl-4-(5-dibenzo[*a*,*e*]cyclohepta-trienylidene)piperidine *see* 2770

N-Methyl-5*H*-dibenzo[*a*,*d*]cycloheptene-5-propanamine *see* 8010

1-Methyl-4-(5*H*-dibenzo[*a*,*d*]cyclohepten-ylidene)piperidine *see* 2770

3-Methyl-4,6-di-*tert*-butylphenol *see* 2840

4-Methyl-2,6-di-*tert*-butylpyridine *see* 3042

N-Methyl-2,2′-dichlorodiethylamine *see* 5846

N-Methyl-2,2′-dichlorodiethylamine *N*-Oxide *see* 5847

Methyldi(2-chloroethyl)amine *see* 5846

Methyldi(2-chloroethyl)amine *N*-Oxide *see* 5847

Methyl 5-(2,4-Dichlorophenoxy)-2-nitrobenzoate *see* 1212

Methyl 2-[4-(2,4-Dichlorophenoxy)phenoxy]propionate *see* 3092

Methyl 6,8-Dideoxy-6-[[[(2*S*,4*R*)-1-methyl-4-propyl-2-pyrrolidinyl]carbonyl]amino]-1-thio-D-*erythro*-α-D-*galacto*-octopyranoside *see* 5555

Methyl 2-Diethylaminoacetamido-*m*-toluate *see* 9700

3-Methyl-2-diethylaminoacetylaminobenzoic Acid Methyl Ester *see* 9700

6-Methyl-8α-(diethylcarbamoylamino)-ergoline *see* 9307

3-Methyldigitoxose *see* 2759

Methyl Digol *see* 3136

β-Methyldigoxin *see* 3186

4‴-*O*-Methyldigoxin *see* 3186

o-Methyldihydroartemisinin *see* 805

(+)-5-Methyl-10,11-dihydro-5*H*-dibenzo-[*a*,*d*]cyclohepten-5,10-imine *see* 3430

(+)-Methyl Dihydroepijasmonate *see* 6124

Methyl Dihydrojasmonate, 6124

(+)-*cis*-Methyl Dihydrojasmonate *see* 6124

(−)-*trans*-Methyl Dihydrojasmonate *see* 6124

(+)-*trans*-Methyl Dihydrojasmonate *see* 6124

Methyldihydromorphinone *see* 6227

(*S*)-*N*-(1-Methyl-4,5-dihydroorotyl)-L-histidyl-L-prolinamide *see* 9177

6-Methyl-3,4-dihydro-1,2,3-oxathiazin-4-one 2,2-Dioxide *see* 38

N-Methyl-2,5-dihydropyrrole *see* 6191

1α-Methyl-5α-dihydrotestosterone *see* 5986

2α-Methyldihydrotestosterone *see* 3497

4-Methyl-6,7-dihydroxycoumarindisulfate *see* 9114

(±)-Methyl-7-[3,5-dihydroxy-2-[(*E*)-3-hydroxy-3-methyl-1-octenyl]cyclopentyl]-4,5-heptadienoate *see* 7991

N-Methyl-5,6-dihydroxyindoxyl *see* 163

L-α-Methyl-3,4-dihydroxyphenylalanine *see* 6127

Methyl[β-(3,4-dihydroxyphenyl)ethyl]-amine *see* 2904

Methyl 3-(Dimethoxyphosphinyloxy)crotonate *see* 6244

Methyl 2-[(4,6-Dimethoxypyrimidin-2-ylcarbamoyl)sulfamoyl]-6-(trifluoromethyl)nicotinate *see* 4225

Methyl 2-[3-(4,6-Dimethoxypyrimidin-2-yl)ureidosulfonyl]-4-methanesulfonamidomethylbenzoate *see* 5981

Methyl 2-[(4,6-Dimethoxypyrimidin-2-yl)ureidosulfonylmethyl]benzoate *see* 1053

Methyl 2,2-Dimethylacetate *see* 6160

4′-*O*-Methyl-8-γ,γ-dimethylallylkaempferol-3-rhamnoside-7-glucoside *see* 3672

5-Methyl-10β-dimethylaminoethyl-10,11-dihydro-11-oxodibenzo[*b*,*e*][1,4]diazepine *see* 3012

5-(2-Methyl-3-dimethylaminopropyl)dibenzo[*a*,*d*][1,4]cycloheptadiene *see* 1535

6-Methylionone *see* 5139

N-Methylisatin 3-Thiosemicarbazone *see* 6049

Methyl Isobutyl Ketone *see* 5253

Methyl Isobutyrate, 6160

N-Methyl-2-isocamphanamine *see* 5843

2-Methylisocrotonic Acid *see* 639

Methyl Isocyanate, 6161

N-(D-6-Methyl-8-isoergolenyl)-N,N-diethylurea *see* 5574

N-(D-6-Methyl-8-isoergolin-1-yl)-N',N'-diethylurea *see* 9307

Methylisomyn *see* 6020

Methylisooctenylamine *see* 5231

1-Methyl-4-isopropenyl-Δ⁶-cyclohexen-2-one *see* 1874

4-(2-Methyl-5-isopropenyl-1-cyclopenten-1-yl)-2-butanone *see* 5250

Methyl Isopropenyl Ether *see* 6073

Methylisopropylbenzene *see* 2760

1-Methyl-4-isopropylcyclohexan-3-one *see* 5906

1-Methyl-4-isopropyl-1-cyclohexen-2-ol-3-one *see* 3326

1-Methyl-4-isopropylidene-3-cyclohexanone *see* 8046

Methyl Isopropyl Ketone, 6162

Methyl 2-(4-Isopropyl-4-methyl-5-oxo-2-imidazolin-2-yl)-p-toluate *see* 4944

Methyl 6-(4-Isopropyl-4-methyl-5-oxo-2-imidazolin-2-yl)-m-toluate *see* 4944

L-(+)-2-Methyl-5-isopropyl-1,3-nonadien-8-one *see* 8834

5-Methyl-2-isopropyl-1-phenol *see* 9554

6-Methyl-3-isopropylsalicylic Acid *see* 9563

(2S,3S,5S)-5-[N-[N-[[N-Methyl-N-[(2-isopropyl-4-thiazolyl)methyl]amino]carbonyl]valinyl]amino]-2-[N-[(5-thiazolyl)methoxycarbonyl]amino]-1,6-diphenyl-3-hydroxyhexane *see* 8366

Methylisothiazolinone, 6163

2-Methyl-3(2H)-isothiazolone *see* 6163

Methyl Isothiocyanate, 6164

Methyl Isovalerate, 6165

4-[4-[β-(5-Methylisoxazole-3-carboxamido)ethyl]phenylsulfonyl]-1,1-hexamethylenesemicarbazide *see* 4479

5-Methyl-4-isoxazolecarboxylic Acid 2-(Phenylmethyl)hydrazide *see* 5202

5-Methylisoxazole-4-carboxylic Acid Trifluoromethylanilide *see* 5486

5-Methyl-3(2H)-isoxazolone *see* 4895

N¹-(5-Methyl-3-isoxazolyl)sulfanilamide *see* 9050

Methyl Jasmonate *see* 5308

Methyl Lactate, 6166

2-Methyllactonitrile *see* 66

Methyl Linoleate, 6167

N-Methyllorazepam *see* 5639

1-Methyllumilysergol 8-(5-Bromonicotinate) 10-Methyl Ether *see* 6581

(15R)-15-Methyllycopodan-5-one *see* 5683

1-Methyl-d-lysergic Acid Butanolamide *see* 6213

1-Methyl-d-lysergic Acid (+)-1-Hydroxy-2-butylamide *see* 6213

Methylmaleic Acid *see* 2320

2-Methylmalic Acid *see* 2322

Methyl Malonate, 6168

α-Methylmandelic Acid *see* 864

Methylmelubrin *see* 3390

Methylmercadone *see* 6616

Methyl Mercaptan *see* 6028

Methyl β-Mercaptoalanine *see* 5856

1-Methyl-2-mercaptoimidazole *see* 6043

5-(Methylmercaptomethyl)-3-(5-nitro-2-furfurylideneamino)-2-oxazolidinone *see* 6616

2-Methylmercapto-10-[2-(N-methyl-2-piperidyl)ethyl]phenothiazine *see* 9512

3-Methylmercapto-N-[2'-(N'-methyl-2-piperidyl)ethyl]phenothiazine *see* 9512

Methyl-mercaptophos *see* 6123

D-2-Methyl-3-mercaptopropanoyl-L-proline *see* 1776

Methyl Mercury *see* 3274

Methyl Mesylate *see* 6169

Methyl Methacrylate *see* 6013

N-Methylmethanamine *see* 3250

Methylmethane *see* 3778

Methyl Methanesulfonate, 6169

Methyl Methanesulfonic Acid *see* 6169

S-Methylmethionine, 6170

S-Methyl-L-methionine *see* 6170

Methylmethioninesulfonium Bromide *see* 6170

Methylmethioninesulfonium Chloride *see* 6170

Methyl N-(2-Methoxyacetyl)-N-(2,6-xylyl)-DL-alaninate *see* 5993

1-Methyl-6-methoxy-3,4-dihydro-2-carboline *see* 6068

1-Methyl-7-methoxy-3,4-dihydro-β-carboline *see* 4646

Methyl-(E)-methoxyimino[α-(o-tolyloxy)-o-tolyl]acetate *see* 5372

l-α-2-Methyl-8-methoxy-6,7-methylenedioxy-1-(6,7-dimethoxy-3-phthalidyl)-1,2,3,4-tetrahydroisoquinoline *see* 6807

Methyl 3-(3-(4-Methoxy-6-methyl-1,3,5-triazin-2-yl)ureidosulfonyl)thiophene-2-carboxylate *see* 9468

(+)-trans-1-Methyl-3-[p-(methoxy)phenoxymethyl]-4-phenylpiperidine *see* 3989

2-(3-Methyl-5-methoxy-1-pyrazolyl)-4-methoxy-6-methylpyrimidine *see* 3677

2-Methyl-4-methoxy-6-sulfanilamido-1,3-diazine *see* 9049

2-Methyl-4-methoxy-6-sulfanilamidopyrimidine *see* 9049

1-Methyl-6-methoxy-1,2,3,4-tetrahydro-2-carboline *see* 162

3-Methyl-8-methoxy-3H-1,2,5,6-tetrahydropyrazino[1,2,3-ab]-β-carboline *see* 6231

1-Methyl-6-(1-methylallyl)-2,5-dithiobiurea *see* 6018

1-[[N-Methyl-N-3-[(methylamino)acetoxymethyl]pyridin-2-yl]carbamoyl-oxy]ethyl-1-[(2R,3R)-2-(2,5-difluorophenyl)-2-hydroxy-3-[4-(4-cyanophenyl)thiazol-2-yl]butyl]-1H-[1,2,4]triazol-4-ium Chloride *see* 5149

2-Methyl-6-methylamino-2-heptene *see* 5231

Methyl 2-(Methylamino)-3-methoxybenzoate *see* 2807

6-[(2S,4R,6E)-4-Methyl-2-(methylamino)-3-oxo-6-octenoic Acid]cyclosporin D *see* 10103

6-[(2S,4R,6E)-4-Methyl-2-(methylamino)-3-oxo-6-octenoic Acid]-7-L-valine-cyclosporin A *see* 10103

1-Methyl-N-[(3-endo)-9-methyl-9-azabicyclo[3.3.1]non-3-yl]-1H-indazole-3-carboxamide *see* 4572

4-[[4-Methyl-6-(1-methyl-2-benzimidazolyl)-2-propyl-1-benzimidazolyl]methyl]-2-biphenylcarboxylic Acid *see* 9271

1-Methyl-4-[(3-methyl-2(3H)-benzothiazolylidene)methyl] Quinolinium *see* 9460

1-Methyl-5-(4-methylbenzoyl)-1H-pyrrole-2-acetic Acid *see* 9677

N-[2-[1-Methyl-5-(4-methylbenzoyl)-1H-pyrrol-2-yl]acetyl]glycine 2-Methoxyphenyl Ester *see* 591

Methyl 1-(α-Methylbenzyl)imidazole-5-carboxylate *see* 6225

Methyl 3-Methylbutanoate *see* 6165

3-Methyl-N-(3-methylbutyl)-1-butanamine *see* 3211

S-Methyl N-[(Methylcarbamoyl)oxy]-thioacetimidate *see* 6054

(2S,3R)-5-Methyl-3-[[(αS)-α-(methylcarbamoyl)phenethyl]carbamoyl]-2-[(2-thienylthio)methyl]hexanohydroxamic Acid *see* 1006

Methyl O-(Methylcarbamoyl)thiolacetohydroxamate *see* 6054

1-Methyl-2-[2-(α-methyl-p-chlorobenzhydryloxy)ethyl]pyrrolidine *see* 2343

1-Methyl-2-[2-(methyl-p-chlorodiphenylmethyloxy)ethyl]pyrrolidine *see* 2343

6-Methyl-2-(4-methyl-3-cyclohexen-1-yl)-5-hepten-2-ol *see* 1245

α-[(α-Methyl-3,4-methylenedioxyphenethylamino)methyl]protocatechuyl Alcohol *see* 8004

1-Methyl-2-(3,4-methylenedioxyphenyl)-ethyl Octyl Sulfoxide *see* 9097

N-Methyl-3,4-methylenedioxyphenylisopropylamine *see* 5836

N-Methyl-N-methylenemethanaminium Iodide *see* 3275

2-Methyl-6-methylene-1,7-octadiene *see* 6416

2-Methyl-6-methylene-2,7-octadiene *see* 6416

7-Methyl-3-methylene-1,6-octadiene *see* 6416

1-Methylmethylergonovine *see* 6213

(1R,2S)-1-Methyl-2-(1-methylethenyl)-cyclobutaneethanol *see* 4571

1-Methyl-4-(1-methylethenyl)cyclohexene *see* 5546

2-Methyl-5-(1-methylethenyl)-2-cyclohexen-1-one *see* 1874

2-[(1R,6R)-3-Methyl-6-(1-methylethenyl)-2-cyclohexen-1-yl]-5-pentyl-1,3-benzenediol *see* 1750

4-[2-Methyl-5-(1-methylethenyl)-1-cyclopenten-1-yl]-2-butanone *see* 5250

(2S,4E)-N-Methyl-5-[5-(1-methylethoxy)-3-pyridinyl]-4-penten-2-amine *see* 5286

Methyl (1-Methylethyl)benzene *see* 2760

1-Methyl-2-(1-methylethyl)benzene *see* 2760

1-Methyl-3-(1-methylethyl)benzene *see* 2760

1-Methyl-4-(1-methylethyl)benzene *see* 2760

4-Methyl-1-(1-methylethyl)bicyclo[3.1.0]hexan-3-one *see* 9546

2-Methyl-5-(1-methylethyl)-1,3-cyclohexadiene *see* 7315

(1R,2S,5R)-5-Methyl-2-(1-methylethyl)-cyclohexanol *see* 5905

(1R,2S,5R)-5-Methyl-2-(1-methylethyl)-cyclohexanol 1-Acetate *see* 5907

5-Methyl-2-(1-methylethyl)cyclohexanol 1-(2-Aminobenzoate) *see* 5908

(2*S*,5*R*)-5-Methyl-2-(1-methylethyl)cyclohexanone *see* 5906

3-Methyl-6-(1-methylethyl)-2-cyclohexen-1-one *see* 7585

1-Methyl-4-(1-methylethyl)-2,3-dioxabicyclo[2.2.2]oct-5-ene *see* 816

(5*R*)-5-Methyl-2-(1-methylethylidene)cyclohexanone *see* 8046

rel-(1*R*,2*S*,4*S*)-1-Methyl-4-(1-methylethyl)-2-[(2-methylphenyl)methoxy]-7-oxabicyclo[2.2.1]heptane *see* 2296

1-Methyl-2-(1-methylethyl)-5-nitro-1*H*-imidazole *see* 5123

(5*S*,6*E*)-8-Methyl-5-(1-methylethyl)-6,8-nonadien-2-one *see* 8834

2-Methyl-4-(1-methylethyl)-7-oxo-8-oxa-3-thia-2,4-diazadecanoic Acid 2,3-Dihydro-2,2-dimethyl-7-benzofuranyl Ester *see* 1044

2-Methyl-4-(1-methylethyl)-7-oxo-8-oxa-3-thia-2,4-diazadecanoic Acid 2,3-Dihydro-2,2-dimethyl-7-benzofuranyl-ester *see* 1044

1-Methyl-7-(1-methylethyl)phenanthrene *see* 8283

2-Methyl-5-(1-methylethyl)phenol *see* 1872

5-Methyl-2-(1-methylethyl)phenol *see* 9554

7-Methyl-1-(1-methylethyl)-4-phenyl-2(1*H*)-quinazolinone *see* 7984

2-Methyl-1-[2-(1-methylethyl)pyrazolo[1,5-*a*]pyridin-3-yl]-1-propanone *see* 4918

(3*R*,6*R*,9*S*)-12-Methyl-13-[2-(1-methylethyl)-4-thiazolyl]-9-[2-(4-morpholinyl)ethyl]-8,11-dioxo-3,6-bis(phenylmethyl)-2,7,10,12-tetraazatridecanoic Acid 5-Thiazolymethyl Ester *see* 2436

N-Methyl-*N*-(1-methylethyl)-*N*-[2-[(9*H*-xanthen-9-ylcarbonyl)oxy]ethyl]-2-propanaminium Bromide (1:1) *see* 7919

(1*R*,3a*R*,7a*R*)-7a-Methyl-1-((*R*)-6-methylheptan-2-yl)-octahydroinden-4-one *see* 4587

N-Methyl-*N*-[[4-[(2-methyl-1*H*-imidazo[4,5-*c*]pyridin-1-yl)methyl]phenyl]sulfonyl]-L-leucine Ethyl Ester *see* 5527

N-Methyl-*N*-[[α-(2-methyl-1*H*-imidazo[4,5-*c*]pyridin-1-yl)-*p*-tolyl]sulfonyl]-L-leucine Ethyl Ester *see* 5527

(2*Z*)-2-Methyl-5-[(1*S*,2*R*,4*R*)-2-methyl-3-methylenebicyclo[2.2.1]hept-2-yl]-2-penten-1-ol *see* 8495

2-Methyl-5-(2-methyl-3-methylene-2-norbornyl)-2-penten-1-ol *see* 8495

β-[1-Methyl-4-(1-methyl-4-(1-methyl-4-formylaminopyrrole-2-carboxamido)-pyrrole-2-carboxamido)pyrrole-2-carboxamido]propionamidine *see* 3404

d-3-Methyl-*N*-methylmorphinan *see* 3229

O-Methyl [2-(2-Methyl-5-nitroimidazol-1-yl)ethyl]thiocarbamate *see* 1848

2β-Methyl-8α-(2-methyl-1-oxobutoxy)-mevinic Acid Lactone *see* 6243

Methyl 2-[[[(4-Methyl-5-oxo-3-propoxy-4,5-dihydro-1*H*-1,2,4-triazol-1-yl)carbonyl]amino]sulfonyl]benzoate *see* 7951

α-*dl*-1-Methyl-5-(1-methyl-2-pentynyl)-5-allylbarbituric Acid *see* 6053

1-Methyl-5-(1-methyl-2-pentyn-1-yl)-5-(2-propen-1-yl)-2,4,6(1*H*,3*H*,5*H*)-pyrimidinetrione *see* 6053

8-[[2-[Methyl(α-methylphenethyl)amino]ethyl]amino]caffeine *see* 3999

(γ*R*)-*N*-Methyl-γ-(2-methylphenoxy)benzenepropanamine *see* 854

2-Methyl-4-[(2-methylphenyl)azo]benzenamine *see* 415

N-Methyl-*N*-(3-methylphenyl)carbamothioic Acid *O*-(2,3-Dihydro-1*H*-inden-5-yl) Ester *see* 9675

N-Methyl-*N*-(3-methylphenyl)carbamothioic Acid *O*-2-Naphthalenyl Ester *see* 9678

N-Methyl-*N*-(3-methylphenyl)carbamothioic Acid *O*-(1,2,3,4-Tetrahydro-1,4-methanonaphthalen-6-yl) Ester *see* 9669

2-Methyl-4-[2-(2-methylphenyl)diazenyl]benzenamine *see* 415

1-[2-[2-Methyl-4-[2-(2-methylphenyl)diazenyl]phenyl]diazenyl]-2-naphthalenol *see* 8530

N-[(1*S*)-2-Methyl-1-[[[1-(4-methylphenyl)ethyl]amino]carbonyl]propyl]carbamic Acid 1-Methylethyl Ester *see* 5124

(6*S*)-2-Methyl-6-(4-methylphenyl)-2-hepten-4-one *see* 10006

1-Methyl-*N*-[(1*S*)-2-[methyl(phenylmethyl)amino]-2-oxo-1-phenylethyl]-5-[[[4'-(trifluoromethyl)[1,1'-biphenyl]-2-yl]carbonyl]amino]-1*H*-indole-2-carboxamide *see* 3395

2-Methyl-1-(4-methylphenyl)-3-(1-piperidinyl)-1-propanone *see* 9681

2-Methyl-3-(2-methylphenyl)-4(3*H*)-quinazolinone *see* 6032

2-[[2-Methyl-8-[[(4-methylphenyl)sulfonyl]amino]-6-quinolinyl]oxy]acetic Acid *see* 10369

2-Methyl-4-(4-methyl-1-piperazinyl)-10*H*-thieno[2,3-*b*][1,5]benzodiazepine *see* 6914

N-Methyl-3-(1-methyl-4-piperidinyl)-1*H*-indole-5-ethanesulfonamide *see* 6502

(3β,16β,17α,22*S*)-17-Methyl-20-[(2*S*,5*S*)-5-methyl-2-piperidinyl]-18-norpregna-5,12-diene-3,16-diol *see* 10143

17-Methyl-20α-((2*S*,5*S*)-5-methyl-2-piperidyl)-18-nor-17α-pregna-5,12-diene-3β,16β-diol *see* 10143

Methyl 2-Methylpropanoate *see* 6160

α-Methyl-4-[(2-methyl-2-propen-1-yl)amino]benzeneacetic Acid *see* 293

*N*¹-Methyl-*N*²-(1-methyl-2-propen-1-yl)-1,2-hydrazinedicarbothioamide *see* 6018

Methyl 4-Methyl-3-(2-propylaminopropionamido)thiophene-2-carboxylate *see* 1869

α-Methyl-4-(2-methylpropyl)benzeneacetic Acid *see* 4919

2-Methyl-2-(1-methylpropyl)-1,3-propanediol 1,3-Dicarbamate *see* 5842

(3*R*,4*R*)-4-Methyl-3-(methyl-7*H*-pyrrolo[2,3-*d*]pyrimidin-4-ylamino)-β-oxo-1-piperidinepropanenitrile *see* 9661

3-[(3*R*,4*R*)-4-Methyl-3-[methyl-(7*H*-pyrrolo[2,3-*d*]pyrimidin-4-yl)amino]piperidin-1-yl]-3-oxopropanenitrile *see* 9661

N-[4-[2-[Methyl[2-[4-[(methylsulfonyl)amino]phenoxy]ethyl]amino]ethyl]phenyl]methanesulfonamide *see* 3456

Methyl *N*-Methyltetrahydronicotinate *see* 768

Methyl 1-Methyl-Δ³,⁴-tetrahydro-3-pyridinecarboxylate *see* 768

(2*R*,4*R*)-4-Methyl-1-[*N*²-(3-methyl-1,2,-3,4-tetrahydro-8-quinolinesulfonyl)-L-arginyl]-2-piperidinecarboxylic Acid *see* 769

4-Methyl-5-[*p*-(methylthio)benzoyl]-4-imidazolin-2-one *see* 3643

2-Methyl-2-(methylthio)propanal *O*-[(Methylamino)carbonyl]oxime *see* 215

2-Methyl-2-(methylthio)propionaldehyde *O*-(Methylcarbamoyl)oxime *see* 215

1-Methyl-3-[3-methyl-4-[4-[(trifluoromethyl)sulfonyl]phenoxy]phenyl]-1,-3,5-triazine-2,4,6(1*H*,3*H*,5*H*)-trione *see* 7711

1-Methyl-3-[3-methyl-4-[4-[(trifluoromethyl)thio]phenoxy]phenyl]-1,3,5-triazine-2,4,6(1*H*,3*H*,5*H*)-trione *see* 9685

Methyl 1-Methylvinyl Ether *see* 6073

25α-Methylmilbemycin B *see* 6271

N-Methylmitomycin C *see* 7719

1-Methylmocimycin *see* 873

17-Methylmorphinan-3-ol *see* 5524

Methylmorphine *see* 2448

N-Methylmorpholine *see* 6362

N-Methylmorpholine *N*-Oxide, 6171

4-Methylmorpholine 4-Oxide *see* 6171

(+)-1-(3-Methyl-4-morpholino-2,2-diphenylbutyryl)pyrrolidine *see* 2958

(3*S*)-3-Methyl-4-(4-morpholinyl)-2,2-diphenyl-1-(1-pyrrolidinyl)-1-butanone *see* 2958

1-[(3*S*)-3-Methyl-4-(4-morpholinyl)-1-oxo-2,2-diphenylbutyl]pyrrolidine *see* 2958

Methyl Mustard Oil *see* 6164

Methylnaltrexone Bromide, 6172

2-Methyl-1,4-naphthalenediol *see* 5898

2-Methyl-1,4-naphthalenediol Diphosphoric Acid Ester Tetrasodium Salt *see* 5898

2-Methyl-1,4-naphthalenedione *see* 5899

(+)-*N*-Methyl-3-(1-naphthalenyloxy)-3-(2-thienyl)propanamine *see* 3512

(γ*S*)-*N*-Methyl-γ-(1-naphthalenyloxy)-2-thiophenepropanamine *see* 3512

2-Methyl-1,4-naphthohydroquinone *see* 5898

2-Methyl-1,4-naphthoquinol *see* 5898

2-Methyl-1,4-naphthoquinone *see* 5899

Methyl β-Naphthyl Ether *see* 6070

trans-*N*-Methyl-*N*-(1-naphthylmethyl)-6,6-dimethylhept-2-en-4-ynyl-1-amine *see* 9299

N-Methyl-*N*-(1-naphthylmethyl)-3-phenylpropen-1-amine *see* 6439

(+)-(*S*)-*N*-Methyl-γ-(1-naphthyloxy)-2-thiophenepropylamine *see* 3512

*N*¹-Methylnicotinamide Chloride *see* 9862

Methyl Nicotinate, 6173

Methyl Nitrate, 6174

1-Methylnitrazepam *see* 6634

Methylnitrobenzene *see* 6737

Methyl *p*-Nitrobenzenesulfonate, 6175

1-Methylnitroethane *see* 6715

1-Methyl-7-nitro-5-(2-fluorophenyl)-3*H*-1,4-benzodiazepin-2(1*H*)-one *see* 4177

3-Methyl-*N*-[(5-nitro-2-furanyl)methylene]-4-thiomorpholinamine 1,1-Dioxide *see* 6622

N-(α-Methylphenethyl)nicotinamide *see* 7330

N-(α-Methylphenethyl)-2-phenylglyci-nonitrile *see* 580

N-[1-(α-Methylphenethyl)-4-piperidyl]-propionanilide *see* 6145

Methylphenidan *see* 6183

Methylphenidate, 6183

Methyl Phenidylacetate *see* 6183

Methylphenobarbital *see* 5921

2-Methylphenol *see* 2564

3-Methylphenol *see* 2564

4-Methylphenol *see* 2564

10-Methyl-10H-phenothiazine-2-acetic Acid *see* 6215

N-Methyl-3-phenothiazinylacetic Acid *see* 6215

(10-Methyl-2-phenothiazinyl)acetic Acid *see* 6215

2-[2-[4-[2-Methyl-3-(10H-phenothiazin-10-yl)propyl]-1-piperazinyl]ethoxy]-ethanol *see* 3429

α-Methyl-3-phenoxybenzeneacetic Acid *see* 4011

2-[1-Methyl-2-(4-phenoxyphenoxy)eth-oxy]pyridine *see* 8103

5-Methyl-5-(4-phenoxyphenyl)-3-(phen-ylamino)-2,4-oxazolidinedione *see* 3972

3-(2-Methylphenoxy)-1,2-propanediol *see* 5917

N-Methyl-N-phenylacetamide *see* 6081

N-(3-Methylphenyl)acetamide *see* 72

Methyl N-Phenylacetyl-N-2,6-xylyl-DL-alaninate *see* 1032

N-Methyl-N-phenylbenzenamine *see* 6126

N-Methyl-ω-phenyl-*tert*-butylamine *see* 5919

N-(3-Methylphenyl)carbamic Acid 3-[(Methoxycarbonyl)amino]phenyl Es-ter *see* 7350

N-Methyl-4-phenyl-4-carbethoxypiperi-dine *see* 5916

1-[2-(2-Methylphenyl)diazenyl]-2-naphthalenamine *see* 10294

1-Methyl-α-phenyl-O-(2-dimethylamino-ethyl)-1H-pyrazole-5-methanol *see* 2335

4,4′-[(4-Methyl-1,3-phenylene)bis(azo)]-bis[6-methyl-1,3-benzenediamine] Di-hydrochloride *see* 1256

5,5′-[(4-Methyl-*m*-phenylene)bis(azo)]-bis[toluene-2,4-diamine] Dihydrochlo-ride *see* 1256

α-[(1-Methyl-2-phenylethyl)amino]ben-zeneacetonitrile *see* 580

7-[β-(α-Methyl-β-phenylethylamino)eth-yl]theophylline *see* 4003

3-[(1-Methyl-2-phenylethyl)amino]pro-panenitrile *see* 4022

N-(1-Methyl-1-phenylethyl)-2-bromo-3,3-dimethylbutanamide *see* 1419

3-Methyl-5,5-phenylethylhydantoin *see* 5920

N-(1-Methyl-2-phenylethyl)-γ-phenyl-benzenepropanamine *see* 7855

N-[1-(1-Methyl-2-phenylethyl)-4-piperi-dinyl]-N-phenylpropanamide *see* 6145

1-(1-Methyl-2-phenylethyl)-4-(N-propan-ilido)piperidine *see* 6145

N-(1-Methyl-2-phenylethyl)-3-pyridine-carboxamide *see* 7330

Methyl DL-Phenylglycinate *see* 7402

N-Methyl-N-[(1S)-1-phenyl-2-((3S)-3-hy-droxy-1-pyrrolidino)ethyl]-2-diphenyl-acetamide *see* 823

1-Methyl-6-phenyl-1H-imidazo[4,5-b]-pyridin-2-amine *see* 7435

2-(4-Methylphenyl)imidazo[2,1b]-5,6,7,8-tetrahydrobenzothiazole *see* 7533

1-Methyl-4-phenylisonipecotic Acid Ethyl Ester *see* 5916

Methyl-β-phenylisopropylamine *see* 6020

6-(5-Methyl-3-phenyl-2-isoxazoline-4-carboxamido)penicillanic Acid *see* 7003

4-(5-Methyl-3-phenyl-4-isoxazolyl)ben-zenesulfonamide *see* 10086

5-Methyl-3-phenyl-4-isoxazolylpenicillin *see* 7003

N-[[4-(5-Methyl-3-phenyl-4-isoxazolyl)-phenyl]sulfonyl]propanamide *see* 7140

2-Methyl-N-(phenylmethylene)-2-pro-panamine N-Oxide *see* 7162

N-Methyl-4-[2-(phenylmethyl)phenoxy]-1-butanamine *see* 1210

1-[1-Methyl-2-[2-(phenylmethyl)phen-oxy]ethyl]piperidine *see* 1051

1-Methyl-2-(phenylmethyl)-4(1H)-quina-zolinone *see* 4538

3-Methyl-2-phenylmorpholine *see* 7351

1-Methyl-5-phenyl-7-nitro-1,3-dihydro-2H-1,4-benzodiazepin-2-one *see* 6634

(2-Methylphenyl)phenylmethanone *see* 7426

(4-Methylphenyl)phenylmethanone *see* 7426

1-Methyl-N-phenyl-N-(phenylmethyl)-4-piperidinamine *see* 949

1-Methyl-4-phenyl-4-piperidinecarbox-ylic Acid Ethyl Ester *see* 5916

Methyl α-Phenyl-α-(2-piperidyl)acetate *see* 6183

2-Methyl-1-phenylpropane *see* 5179

2-Methyl-2-phenylpropane *see* 1553

(5Z)-5-[(2E)-2-Methyl-3-phenyl-2-pro-pen-1-ylidene]-4-oxo-2-thioxo-3-thi-azolidineacetic Acid *see* 3660

N-Methyl-N-[(2E)-3-phenyl-2-propen-1-yl]-1-naphthalenemethanamine *see* 6439

1-Methyl-4-phenyl-3-(2-propen-1-yl)-4-piperidinol 4-Propanoate *see* 287

dl-N-Methyl-2-phenylpropylamine *see* 7418

N-(2-Methylphenyl)-2-(propylamino)pro-panamide *see* 7862

4-Methyl-N-phenyl-6-(1-propyn-1-yl)-2-pyrimidinamine *see* 5912

3-Methyl-1-phenyl-2-pyrazolin-5-one *see* 6802

3-Methyl-1-phenyl-5-pyrazolyl Dimethyl-carbamate *see* 8114

N¹-(3-Methyl-1-phenylpyrazol-5-yl)sulfa-nilamide *see* 9076

4-[2-[(4-Methyl-6-phenyl-3-pyridazinyl)-amino]ethyl]morpholine *see* 6281

N-(4-Methyl-6-phenyl-3-pyridazinyl)-4-morpholineethanamine *see* 6281

5-Methyl-1-phenyl-2(1H)-pyridinone *see* 7606

trans-1-(4-Methylphenyl)-1-(2-pyridyl)-3-pyrrolidinoprop-1-ene *see* 9922

1-Methyl-3-phenyl-2,5-pyrrolidinedione *see* 7373

1-(4-Methylphenyl)-2-(1-pyrrolidinyl)-1-pentanone *see* 8122

2-[(1E)-1-(4-Methylphenyl)-3-(1-pyrroli-dinyl)-1-propen-1-yl]pyridine *see* 9922

(2E)-3-[6-[(1E)-1-(4-Methylphenyl)-3-(1-pyrrolidinyl)-1-propen-1-yl]-2-pyri-dinyl]-2-propenoic Acid *see* 118

N-Methyl-α-phenylsuccinimide *see* 7373

N-Methyl-2-phenylsuccinimide *see* 7373

3-Methyl-1-phenyl-5-(sulfanilamido)pyr-azole *see* 9076

Methyl-5-(phenylsulfinyl)-2-benzimid-azolecarbamate *see* 7034

(4-Methylphenylsulfonyl)methyl Isocya-nide *see* 9715

5-Methyl-1-phenyl-1,3,4,6-tetrahydro-5H-benz[*f*]-2,5-oxazocine *see* 6525

1-(4-Methylphenyl)-2-(4,5,6,7-tetrahy-dro-2-imino-3(2H)-benzothiazolyl)eth-anone Hydrobromide (1:1) *see* 7533

3-Methyl-2-phenyltetrahydro-2H-1,4-ox-azine *see* 7351

2-Methyl-9-phenyl-2,3,4,9-tetrahydro-1-pyridindene *see* 7347

1-Methyl-4-phenyl-1,2,3,6-tetrahydro-pyridine *see* 6380

1-Methyl-4-[phenyl-(2-thenyl)amino]-piperidine *see* 9429

1-Methyl-N-phenyl-N-(2-thienylmethyl)-4-piperidinamine *see* 9429

Methyl 5-(Phenylthio)-2-benzimidazole-carbamate *see* 3994

1-[1-Methyl-2-[(α-phenyl-*o*-tolyl)oxy]eth-yl]piperidine *see* 1051

4-[5-(4-Methylphenyl)-3-(trifluorometh-yl)-1H-pyrazol-1-yl]benzenesulfon-amide *see* 1958

(±)-N-Methyl-3-phenyl-3-[(α,α,α-trifluo-ro-*p*-tolyl)oxy]propylamine *see* 4217

3-Methyl-2-phenylvaleric Acid 2-Diethyl-aminoethyl Ester Methyl Bromide *see* 10092

3-Methyl-2-phenylvaleric Acid Diethyl(3-hydroxyethyl)methylammonium Bro-mide Ester *see* 10092

P-Methylphosphonic Acid Dimethyl Es-ter *see* 3435

P-Methylphosphonic Acid Monoethyl Ester *see* 3620

P-Methylphosphonic Acid Mono(1-meth-ylethyl) Ester *see* 4966

P-Methylphosphonofluoridic Acid 1-Methylethyl Ester *see* 8511

P-Methylphosphonofluoridic Acid 1,2,2-Trimethylpropyl Ester *see* 6525

P-Methylphosphonothioic Acid S-[2-[Bis(1-methylethyl)amino]ethyl] O-Ethyl Ester *see* 10234

P-Methylphosphonothioic Acid O-Ethyl Ester *see* 3621

Methyl Phthalate *see* 3279

2-Methyl-2-phytyl-6-chromanol *see* 9652

2-Methyl-2-phytyl-6-hydroxychroman *see* 9652

2-Methyl-3-phytyl-1,4-naphthoquinone *see* 7492

N-Methylpicolinic Acid *see* 4767

dl-N-Methylpipecolic Acid 2,6-Dimethyl-anilide *see* 5926

1-Methyl-2′,6′-pipecoloxylidide *see* 5926

4-Methyl-1-piperazinecarboxylic Acid 6-(5-Chloro-2-pyridinyl)-6,7-dihydro-7-oxo-5H-pyrrolo[3,4-b]pyrazin-5-yl Es-ter *see* 10393

N-[5-[4-(4-Methylpiperazinomethyl)ben-zoylamido]-2-methylphenyl]-4-(3-pyri-dyl)-2-pyrimidineamine *see* 4943

Methyltheobromine *see* 1639

N^1-(5-Methyl-1,3,4-thiadiazol-2-yl)sulfa-
nilamide *see* 9048

(6*R*,7*R*)-3-[[(5-Methyl-1,3,4-thiadiazol-2-
yl)thio]methyl]-8-oxo-7-[[2-(1*H*-tetra-
zol-1-yl)acetyl]amino]-5-thia-1-azabi-
cyclo[4.2.0]oct-2-ene-2-carboxylic Acid
see 1918

4-Methyl-2-thiazolamine *see* 447

4-Methyl-5-thiazoleethanol, 6199

N-(3-Methyl-4-thiazolin-2-ylidene)-2,4-
xylidene *see* 2761

4-Methyl-2-thiazolylamine *see* 447

α-Methyl-4-(2-thienylcarbonyl)benzene-
acetic Acid *see* 9134

N-Methyl-*N*-[3-[3-(2-thienylcarbonyl)-
pyrazolo[1,5-*a*]pyrimidin-7-yl]phenyl]-
acetamide *see* 4984

γ-Methylthio-α-aminobutyric Acid *see*
6047

2-Methylthio-4,6-bis(isopropylamino)-*s*-
triazine *see* 7904

2-Methylthio-4,6-bis(monoethylamino)-*s*-
triazine *see* 8674

N-Methylthiocarbamoyl-*N'*-[(1-methylal-
lyl)thiocarbamoyl]hydrazine *see* 6018

7-(Methylthiocarbonyl)-benzo-1,2,3-thia-
diazole *see* 98

Methyl Thiocyanate, 6200

4-Methylthio-3,5-dimethylphenyl *N*-
Methylcarbamate *see* 6044

2-Methylthio-4-ethylamino-6-isopropyl-
amino-*s*-triazine *see* 387

O-[[*N*-[*N'*-(1-Methylthioethylideneimino-
oxycarbonyl)-*N'*-methylaminosul-
fenyl]-*N*-methylcarbamoyl]-*S*-methyl-
acetohydroximate *see* 9482

5-(2-Methylthioethyl)-5-(2-pentyl)-2-
thiobarbituric Acid *see* 6050

N-[2-(Methylthio)ethyl]-2-[(3,3,3-tri-
fluoropropyl)thio]-5'-adenylic Acid
Monoanhydride with (Dichloromethylene)bis[phosphonic Acid] *see* 1749

N^6-(2-Methylthio)ethyl)-2-(3,3,3-trifluoro-
propylthio)-β,γ-dichloromethylene
ATP *see* 1749

5-[(Methylthio)methyl]-3-[[(5-nitro-2-fu-
ranyl)methylene]amino]-2-oxazolidi-
none *see* 6616

5-[(Methylthio)methyl]-3-[(5-nitrofurfur-
ylidene)amino]-2-oxazolidinone *see*
6616

8-Methylthio-10-(4-methylpiperazino)-
10,11-dihydrodibenzo[*b,f*]thiepine *see*
6217

(8β)-8-[(Methylthio)methyl]-6-propyler-
goline *see* 7277

Methylthioninium Chloride *see* 6132

1-[(1*E*)-2-(Methylthio)-1-[2-(pentyloxy)-
phenyl]ethenyl]-1*H*-imidazole *see* 6563

Methylthiophenol *see* 9478

1-(Methylthio)propyl 1-Propenyl Disulfide *see* 811

Methylthiouracil, 6201

4-Methyl-2-thiouracil *see* 6201

6-Methyl-2-thiouracil *see* 6201

1-Methyl-3-(9*H*-thioxanthen-9-ylmethyl)-
piperidine *see* 6051

4-(Methylthio)-3,5-xylyl Methylcarbamate *see* 6044

Methylthymol Blue, 6202

5-Methyltocol *see* 9659

8-Methyltocol *see* 9656

1-Methyl-5-*p*-toluoylpyrrole-2-acetami-
doacetic Acid Guaicil Ester *see* 591

1-Methyl-5-*p*-toluoylpyrrole-2-acetic
Acid *see* 9677

N-[(1-Methyl-5-*p*-toluoylpyrrol-2-yl)ace-
tyl]glycine *o*-Methoxyphenyl Ester *see*
591

Methyl 3-(*m*-Tolylcarbamoyloxy)phenyl-
carbamate *see* 7350

Methyl-*p*-tolylcarbinol *see* 3259

Methyl *p*-Tolylcarbinol Camphorate Diethanolamine Salt *see* 9649

2-Methyl-6-*p*-tolyl-2-hepten-4-one *see*
10006

(−)-*N*-Methyl-3-(*o*-tolyloxy)-3-phenyl-
propylamine *see* 854

2-Methyl-3-*o*-tolyl-4(3*H*)-quinazolinone
see 6032

2-Methyl-3-*o*-tolyl-6-sulfamyl-7-chloro-
1,2,3,4-tetrahydro-4-quinazolinone *see*
6224

6-Methyl-1,2,4-triazine-3,5(2*H*,4*H*)-dione
see 896

Methyltrichloromethane *see* 9801

α-Methyltricyclo[3.3.1.1^{3,7}]decane-1-
methanamine *see* 8351

Methyltrienolone, 6203

Methyl-*N*-(triethylammoniumsulfonyl)-
carbamate *see* 1505

Methyl Triflate *see* 6204

2-[[[3-Methyl-4-(2,2,2-trifluoroethoxy)-2-
pyridinyl]methyl]sulfinyl]-1*H*-benz-
imidazole *see* 5413

2-Methyl-3-(β,β,β-trifluoroethylthio-
methyl)-6-chloro-7-sulfamyl-3,4-dihy-
dro-1,2,4-benzothiadiazine 1,1-Dioxide
see 7705

Methyl Trifluoromethanesulfonate, 6204

2-(2-Methyl-3-trifluoromethylanilino)nic-
otinic Acid *see* 4178

Methyl(trifluoromethyl)dioxirane, 6205

2-[[α-Methyl-*m*-(trifluoromethyl)phen-
ethyl]amino]ethanol Benzoate (Ester)
see 1041

N-Methyl-γ-[4-(trifluoromethyl)phen-
oxy]benzenepropanamine *see* 4217

dl-*N*-Methyl-3-(*p*-trifluoromethylphen-
oxy)-3-phenylpropylamine *see* 4217

2-[[2-Methyl-3-(trifluoromethyl)phenyl]-
amino]-3-pyridinecarboxylic Acid *see*
4178

N-Methyl-*N*-(4-trifluoromethylphenyl)-
1,2-dihydro-4-hydroxy-5-methoxy-1-
methyl-2-oxoquinoline-3-carboxamide
see 9210

2-[[1-Methyl-2-[3-(trifluoromethyl)phen-
yl]ethyl]amino]ethanol 1-Benzoate *see*
1041

5-Methyl-*N*-[4-(trifluoromethyl)phenyl]-
4-isoxazolecarboxamide *see* 5486

(α*R*)-α-Methyl-*N*-[3-[3-(trifluoromethyl)-
phenyl]propyl]-1-napthalenemethan-
amine *see* 2285

Methyl (*E*)-2-[2-[6-(Trifluoromethyl)pyr-
idin-2-yloxymethyl]phenyl]-3-meth-
oxyacrylate *see* 7520

1-Methyl-3-[4-[*p*-[(trifluoromethyl)thio]-
phenoxy]-*m*-tolyl]-*s*-triazine-2,4,6(1*H*,-
3*H*,5*H*)-trione *see* 9685

16-Methyl-1,11α,16*RS*-trihydroxyprost-
13*E*-en-9-one *see* 8359

N-Methyl-*N*-(2,4,6-triiodo-3-aminophen-
yl)glutaramidic Acid *see* 5096

N-Methyltrilobine *see* 9866

Methyl 3,4,5-Trimethoxycinnamoyl
Reserpate *see* 8262

5-Methyl-6-[[(3,4,5-trimethoxyphenyl)-
amino]methyl]-2,4-quinazolinediamine
see 9893

Methyl 1,1,*N*-Trimethyl-*N*-(3,3-diphenyl-
propyl)-2-aminoethyl 1,4-Dihydro-2,6-
dimethyl-4-(3-nitrophenyl)pyridine-
3,5-dicarboxylate *see* 5500

Methyltrimethylene Glycol *see* 1570

Methyl *O*-2,3,4-Tri-*O*-methyl-6-*O*-sulfo-
α-D-glucopyranosyl-(1 → 4)-*O*-2,3-di-
O-methyl-β-D-glucopyranuronosyl-
(1 → 4)-*O*-2,3,6-tri-*O*-sulfo-α-D-gluco-
pyranosyl-(1 → 4)-*O*-2,3-di-*O*-methyl-
α-L-idopyranuronosyl-(1 → 4)-α-D-
glucopyranoside 2,3,6-Tris(hydrogen
Sulfate) *see* 4932

2-Methyl-2-(4,8,12-trimethyltridecyl)-6-
chromanol *see* 9652

1-Methyl-2,4,6-trinitrobenzene *see* 9902

2-Methyl-1,3,5-trinitrobenzene *see* 9902

6α-Methyl-11β,17α,21-triol-1,4-pregna-
diene-3,20-dione *see* 6184

Methyltrioxorhenium, 6206

(*T*-4)-Methyltrioxorhenium *see* 6206

α-Methyltryptamine, 6207

N-Methyl-L-tryptophan *see* 11

α-Methyltyramine *see* 4847

α-Methyltyrosine *see* 6240

α-Methyl-*m*-tyrosine, 6208

α-Methyl-L-tyrosine *see* 6240

α-Methyl-*p*-tyrosine *see* 6240

N-Methyl-L-tyrosine *see* 9136

(2-*O*-Methyltyrosine)deamino-1-carba-
oxytocin *see* 1796

β-Methylumbelliferone *see* 4894

4-Methylumbelliferone *see* 4894

4-Methylumbelliferone *O*,*O*-Diethyl
Phosphorothioate *see* 7724

4-Methyluracil *see* 6209

5-Methyluracil *see* 9553

6-Methyluracil, 6209

Methylurethane *see* 6110

Methyl Vaccenate *see* 10084

N-Methyl-L-valyl-*N*-[(1*S*,2*R*)-4-[(2*S*)-2-
[(1*R*,2*R*)-3-[[(1*R*,2*S*)-2-hydroxy-1-
methyl-2-phenylethyl]amino]-1-meth-
oxy-2-methyl-3-oxopropyl]-1-pyrroli-
dinyl]-2-methoxy-1-[(1*S*)-1-methylpro-
pyl]-4-oxobutyl]-*N*-methyl-L-valinam-
ide *see* 6340

trans-8-Methyl-*N*-vanillyl-6-nonenamide
see 1770

Methyl Vinyl Ketone, 6210

13β-Methyl-13-vinylpodocarp-7-ene-15-
oic Acid *see* 5246

13α-Methyl-13-vinylpodocarp-8(14)-ene-
15-oic Acid *see* 7540

Methyl Viologen (2+) *see* 7135

Methyl Vitamin B$_{12}$ *see* 6116

1-Methyl-2-(2,6-xylyloxy)ethylamine *see*
6248

N-Methyl-*N'*-2,4-xylyl-*N*-(*N*-2,4-xylyl-
formimidoyl)formamidine *see* 482

Methyl Yellow *see* 3251

Methylzineb *see* 7933

Methymycin, 6211

Methypranol *see* 6216

Methyprylon, 6212

Methyridine *see* 6239

Methysergide, 6213

Methysergide Maleate *see* 6213

Methysticin, 6214

Metiazic Acid, 6215

Metiazinic Acid, 6215

Meticlorpindol *see* 2386

MIH *see* 7876
Mikado [Bayer CropSci.] *see* 9026
Mikametan [Mikasa] *see* 5009
Mikamycin IA *see* 7871
Mikavir [Salus] *see* 400
Mikelan [Otsuka] *see* 1866
Mikrotsid *see* 4496
Milban [Mallinckrodt] *see* 3454
Milbeknock [Sankyo] *see* 6269
Milbemectin, 6269
Milbemycin α₁ *see* 6269
Milbemycin α₃ *see* 6269
Milbemycin β₁ *see* 6271
Milbemycin A₃ *see* 6269
Milbemycin A₃ 5-Oxime *see* 6270
Milbemycin A₄ *see* 6269
Milbemycin A₄ 5-Oxime *see* 6270
Milbemycin D *see* 6271
Milbemycin Oxime, 6270
Milbemycins, 6271
Milchsäure *see* 5384
D-Milchsäure *see* 5383
L-Milchsäure *see* 5385
Milcopen [Leiras] *see* 7209
Milcurb [Zeneca] *see* 3239
Milcurb Super [Zeneca] *see* 3794
Mild Aconitine *see* 111
Mildiomycin, 6272
Mildison [Astellas] *see* 4824
Mild Mercury Chloride *see* 5962
Mild Silver Protein *see* 8666
Milestone [DuPont] *see* 887
Milezin [Zentiva] *see* 5518
Milfoil *see* 97
Milgo [Zeneca] *see* 3794
Milid [Opfermann] *see* 7891
Milide [Beytout] *see* 7891
Milk, 6273
Milk Ipecac *see* 729
Milk of Sulfur *see* 9107
Milk Sugar *see* 5392
Milk Thistle, 6274
Milkweed *see* 818
Millafol [Parke-Davis] *see* 4248
Millenium [DuPont] *see* 4225
Millerite *see* 6583
Millevit [Abbott] *see* 10212
Millicaine [Corbiere] *see* 1189
Milligynon [Schering AG] *see* 6786
Millisrol [Nippon Kayaku] *see* 6694
Millon's Base *see* 5966
Millophyline [Arnolds] *see* 3766
Milnacipran, 6275
Milontin [Warner-Lambert] *see* 7373
Milorganite [Milwaukee Met. Sewer.
    Dist.], 6276
Milori Blue *see* 8018
Miloxacin, 6277
Milrila [Yamanouchi] *see* 6278
Milrinone, 6278
Milstem [Zeneca] *see* 3794
Miltaun [Wallace Labs.] *see* 5929
Miltefosine, 6279
Miltex [Baxter] *see* 6279
Miltown [Wallace Labs.] *see* 5929
Milvane [Schering AG] *see* 4447
Mimedran [Esteve] *see* 9124
Mimetite *see* 5451
Mimic [Dow AgroSci.] *see* 9230
Mimosine, 6280
Mimpara [Amgen] *see* 2285
Minacalm [Tobishi] *see* 9681
Minalfene [Bouchara] *see* 293
Minalgin [Streuli] *see* 3390

Minaprine, 6281
Mindererus Spirit *see* 492
Mindiab [Pfizer] *see* 4477
Minedil [Formenti] *see* 8082
Mineral Carbon *see* 4575
Mineral Green *see* 2615
Mineral Oil *see* 7301, 7302
Mineral Orange *see* 5480
Mineral Pitch *see* 835
Mineral Red *see* 5480
Mineral Spirits, 6282
Mineral Wax *see* 1986
Mineral White *see* 1708
Mingin *see* 10067
Minicid *see* 3202
Minidiab [Pfizer] *see* 4477
Minihep [Leo Pharm] *see* 4688
Miniluteolas [Serono] *see* 3909
Minilyn [Organon] *see* 5690
Mini-Pe [Searle] *see* 6786
"Mini-pill" *see* 6786
Minipress [Pfizer] *see* 7838
Minirin [Ferring] *see* 2926
Mini-Sintrom [Novartis] *see* 30
Minitran [3M Pharma] *see* 6694
Minium *see* 5480
Minocin [Wyeth] *see* 6283
Minocycline, 6283
Minodronate *see* 6284
Minodronic Acid, 6284
Minolip [Master] *see* 1041
Minor Centaury *see* 1970
Minovlar [Schering AG] *see* 6786
Minoxidil, 6285
Minoximen [Menarini] *see* 6285
Minprog [Pharmacia] *see* 7987
Minprostin E₂ [Pharmacia] *see* 7988
Mintacol *see* 7134
Mintec [Shire] *see* 7258
Mintezol [Merck & Co.] *see* 9440
Mintic [Ayerst] *see* 6239
Minulet [Wyeth] *see* 4447
Minuric [R & C] *see* 1067
Minusin [Doetsch, Grether] *see* 6803
Mioblock [Sankyo] *see* 7109
Miocamycin [Meiji] *see* 6286
Miochol [Cooper] *see* 82
Miokamycin, 6286
Miolaxene [Lepetit] *see* 6052
Miolene [Lusofarmaco] *see* 8365
Mional [Eisai] *see* 3661
Miorel [Lederle] *see* 9477
Mio-Sed *see* 2106
Miostat [Alcon] *see* 1781
Miotolon [Daiichi] *see* 4329
MIP *see* 749
Mipafox, 6287
Mipax *see* 3279
Mipomersen, 6288
Mirabilite *see* 8809
Miracil D *see* 5650
Miracle Fruit *see* 6289
Miraclid [Mochida] *see* 10067
Miracol *see* 5650
Miraculin, 6289
Miradol [Mitsui] *see* 9119
Miradon [Schering] *see* 661
Miralin [Miralin] *see* 6289
Miranax [Syntex] *see* 6499
Mirapex [Boehringer, Ing.] *see* 7825
Mirapexin [Boehringer, Ing.] *see* 7825
Mirapront [Mack, Illert.] *see* 7374
Mirasept [CIBA Vision] *see* 4835
Mirbanil [Boehringer, Ing.] *see* 9119
Mircol [Rhône-Poulenc] *see* 5931

Mirena [Schering AG] *see* 6793
Miretilan [Sandoz] *see* 3632
Mirex, 6290
Mirfat [Merckle] *see* 4338
Miridacin [Taiho] *see* 7890
Mirimostim *see* 5710
Mirlon *see* 6823
Miropinic Acid *see* 5246
Mirsol [Permamed] *see* 10370
Mirtazapine, 6291
Misch Metal, 6292
Miso *see* 8856
Misodex [Lusofarmaco] *see* 6293
Misoprostol, 6293
Mistabron [UCB] *see* 5979
Mistabronco [UCB] *see* 5979
Mistarel *see* 5263
Mistel *see* 6294
Mistletoe, 6294
Mistletoe Lectins, 6295
Mistral [M & B] *see* 4021
Misulvan [Bernabo] *see* 9119
MIT *see* 6163
Mitaban [Pharmacia & Upjohn] *see* 482
Mitac [Bayer CropSci.] *see* 482
Mithracin *see* 7652
Mithramycin *see* 7652
Mitigal *see* 5988
Mitigan [Makhteshim-Agan] *see* 3096
Mitiglinide, 6296
Mitis Green *see* 2615
Mitobronitol, 6297
Mitochondrial Ferritin *see* 4063
Mitocin-C [BMS] *see* 6300
Mitocor [Zambon] *see* 10025
Mitoguazone, 6298
Mitolactol, 6299
Mitomen *see* 5847
Mitomycin A *see* 6300
Mitomycin B *see* 6300
Mitomycin C *see* 6300
Mitomycins, 6300
Mitoquinones *see* 10025
Mitotane, 6301
Mitoxana [Baxter] *see* 4937
Mitoxantrone, 6302
Mitragynine, 6303
Mitramycin *see* 7652
Mitratapide, 6304
Mitrinermine *see* 8322
Mitronal [Searle] *see* 2306
Mittoval [Schering AG] *see* 231
Mivacron [GSK] *see* 6305
Mivacurium Chloride, 6305
Mivazerol, 6306
Mixobar [Byk Gulden] *see* 990
Miyaura's Reagent *see* 1301
Mizolastine, 6307
Mizollen [Lorex] *see* 6307
Mizoribine, 6308
MJ-505 *see* 7432
MJ-1992 *see* 8855
MJ-1996 *see* 393
MJ-1999 *see* 8854
MJ-4309-1 *see* 7054
MJ-5190 *see* 393
MJ-9067 *see* 3625
MJ-10061 *see* 1067
MJ-13105-1 *see* 1471
MJ-13754-1 *see* 6523
MJF-9325 *see* 4937
MJF-12264 *see* 9252
MJF-11567-3 *see* 1915
MK 4 *see* 5901
MK 6 *see* 5901

MK-130 *see* 2706
MK-135 *see* 2209
MK-188 *see* 10319
MK-191 *see* 7626
MK-208 *see* 3971
MK-217 *see* 223
MK-223 *see* 4919
MK-231 *see* 9112
MK-239 *see* 9231
MK-240 *see* 8010
MK-243 *see* 4973
MK-244 *see* 3612
MK-264 *see* 4246
MK-306 *see* 1938
MK-351 *see* 6127
MK-360 *see* 9440
MK-366 *see* 6789
MK-383 *see* 9620
MK-397 *see* 3689
MK-401 *see* 2395
MK-421 *see* 3623
MK-422 *see* 3623
MK-0431 *see* 8690
MK-0462 *see* 8370
MK-476 *see* 6346
MK-486 *see* 1798
MK-507 *see* 3475
MK-0517 *see* 741
MK-0518 *see* 8216
MK-521 *see* 5572
MK-0524 *see* 5426
MK-0524A *see* 5426
MK-595 *see* 3772
MK-621 *see* 3572
MK-639 *see* 4983
MK-647 *see* 3165
MK-650 *see* 2527
MK-663 *see* 3930
MK-0663 *see* 3930
MK-733 *see* 8677
MK-781 *see* 6240
MK-787 *see* 4961
MK-791 *see* 2274
MK-801 *see* 3430
MK-803 *see* 5644
MK-0822 *see* 6861
MK-826 *see* 3731
MK-0859 *see* 615
MK-0869 *see* 741
MK-870 *see* 401
MK-905 *see* 1728
MK-906 *see* 4113
MK-0928 *see* 4351
MK-933 *see* 5296
MK-936 *see* 2
MK-950 *see* 9601
MK-954 *see* 5641
MK-955 *see* 4281
MK-965 *see* 826
MK-0966 *see* 8377
MK-0974 *see* 9259
MK-990 *see* 8214
MK-0991 *see* 1889
MK-7009 *see* 10122
MK-8669 *see* 8336
MKH 6562 *see* 4151
ML-236 B *see* 6243
ML-1024 *see* 9435
ML-1129 *see* 1154
ML-1229 *see* 1154
ML-3000 *see* 5533
ML-1785713 *see* 4118
ML-I *see* 6295
ML-II *see* 6295
ML-III *see* 6295

MM 4550 *see* 6929
MM 13902 *see* 6929
MM 14151 *see* 2339
MM 17880 *see* 6929
MM 22380 *see* 6929
MM 22381 *see* 6929
MM 22382 *see* 6929
MM 22383 *see* 6929
MM-416775 *see* 5548
MMAE *see* 6340
MMC *see* 6300
MMF *see* 6147
MMH *see* 6158
MMNO *see* 6171
4-MMPD *see* 2978
4-MMPDS *see* 2978
MMPP *see* 5737
MMS *see* 6169
MMSC *see* 6170
MMT, 6309
α-MMT *see* 6208
MN-10T *see* 9309
MN-1695 *see* 5146
MnDPDP *see* 5787
MNE *see* 6581
Mnesis [Takeda] *see* 4929
MNNG *see* 6176
αMNP, 6310
MNPA *see* 6499
MO-911 *see* 7142
MOB *see* 7053
Mobactam [Takeda] *see* 1871
Moban [Endo] *see* 6319
Mobec [Boehringer, Ing.] *see* 5894
Mobenol [Church & Dwight] *see* 9667
Mobic [Boehringer, Ing.] *see* 5894
Mobicox [Boehringer, Ing.] *see* 5894
Mobidin [Ascher] *see* 5749
Mobiflex [Roche] *see* 9291
Mobisyl [Ascher] *see* 9830
Mobutazon *see* 6317
Mobuzon *see* 6317
MOCA *see* 6131
Mocap [Bayer CropSci.] *see* 3799
Mocimycin, 6311
Moclamine [Roche] *see* 6312
Moclobemide, 6312
Moctanin [Ethitek] *see* 6338
Modacin [GSK] *see* 1948
Modacor [Houdé] *see* 7061
Modacrylic Fibers, 6313
Modafinil, 6314
Modalim [Sanofi] *see* 2312
Modalina [Sanofi-Synthelabo] *see* 9846
Modamide [Merck & Co.] *see* 401
Modane [Adria] *see* 2814
Modatrop [Nordmark] *see* 7246
Modecate [Sanofi Winthrop] *see* 4220
Modenol *see* 1527
Moderatan [Theranol] *see* 3145
Modicare *see* 3464
Modicon [Ortho] *see* 6786
Modimmunal [Ravizza] *see* 5019
Modinal *see* 4400
Moditen [Sanofi Winthrop] *see* 4220
Modivid [Sanofi-Aventis] *see* 1929
Modown [Mobil; Rhône-Poulenc] *see* 1212
Modrasone [Schering] *see* 213
Modrenal [Bioenvision] *see* 9867
Modulan [Amerchol] *see* 5409
Modulon [Pfizer] *see* 9868
Modumate [Abbott] *see* 771
Moduretic [Merck & Co.] *see* 401
Moduretik [Merck & Co.] *see* 401

Modustatina [Midy] *see* 8841
Moenocinol *see* 945
Moenomycin *see* 945
Moenomycin A *see* 945
Moexipril, 6315
Moexiprilat Hydrochloride *see* 6315
Mofarotene, 6316
Mofebutazone, 6317
Mofenar [C.E.P.A.] *see* 1479
Mofesal [Medice] *see* 6317
Mofezolac, 6318
Mogadan [Valeant] *see* 6661
Mogadon [Valeant] *see* 6661
Mogeton [Agro-Kanesho] *see* 8184
Mohr's Salt *see* 518
Molatoc *see* 3446
Molcer [Wallace] *see* 3446
Molded Silver Nitrate *see* 8657
Molevac [Parke-Davis] *see* 8136
Molgramostim *see* 4574
Molindone, 6319
Molipaxin [Sanofi-Aventis] *see* 9740
Mol-Iron [Schering] *see* 4084
Mol-Iron (obsolete) *see* 6324
Moloid [Südmedica] *see* 5811
Molsidolat [Aventis] *see* 6320
Molsidomine, 6320
(*T*-4)-Molybdate (MoO$_4$$^{2-}$) Calcium
(1:1) *see* 1683
(*T*-4)-Molybdate (MoO$_4$2−) Potassium
(1:2) *see* 7768
Molybdenite *see* 6322
Molybdenum, 6321
Molybdenum Disulfide, 6322
(*OC*-6-11)-Molybdenum Fluoride (MoF$_6$)
see 6323
Molybdenum Hexafluoride, 6323
Molybdenum(VI) Oxide *see* 6325
Molybdenum Oxide (Mo$_2$O$_3$) *see* 6324
Molybdenum Oxide (MoO$_3$) *see* 6325
Molybdenumphosphorus Hydroxide Oxide *see* 7454
Molybdenum Sesquioxide, 6324
Molybdenum Sulfide (MoS$_2$) *see* 6322
Molybdenum Trioxide, 6325
Molybdic(VI) Acid, 6326
Molybdic Anhydride *see* 6325
Molybdophosphoric Acid *see* 7454
Molycor-R [Mepha] *see* 6788
Molyhibit 100 [Amax] *see* 8776
Molysite *see* 4046
MOM *see* 6286
Mometasone Furoate, 6327
Momicine [Morrith] *see* 6262
MON-097 *see* 60
MON-0573 *see* 4547
MON-2139 *see* 4547
MON-7200 *see* 3417
MON-12000 *see* 4636
MON-12037 *see* 4636
MON-13200 *see* 9463
MON-15100 *see* 3417
MON-21200 *see* 2368
MON-21233 *see* 2368
MON-24000 *see* 9469
MON-37500 *see* 9095
Monacetin, 6328
Monacolin K *see* 5644
Monacrin [Sterling Winthrop] *see* 402
Monactin *see* 6762
Monalium Hydrate *see* 5714
Monapen [Fujisawa] *see* 9584
Monarda, 6329
Monardin *see* 7180
Monasirup [RPR] *see* 636

Monaspor [Novartis] *see* 1946
Monastrol, 6330
Monazan *see* 6317
Monazol *see* 4875
Monazoline O [Croda] *see* 6925
Moncut [Nihon Nohyaku] *see* 4241
Mondamin [CPC] *see* 8925
Mondus [Labinca] *see* 4175
Monellins, 6331
Monensic Acid (obsolete) *see* 6332
Monensin, 6332
Monetite *see* 1694
Mongarit [Sankyo] *see* 8673
Monicor [Fabre] *see* 5271
Moniflagon [Dieetzout] *see* 9290
Monistat [Janssen] *see* 6257
Monit [Lorex] *see* 5271
Monitan [Ives] *see* 7703
Monitor [Bayer CropSci.] *see* 6019
Monitor [Monsanto] *see* 9095
Monkey's Bread Tree *see* 950
Monkshood *see* 108
Monk's Pepper *see* 182
Monoverin [Merck KGaA] *see* 7827
Monoacetin *see* 6328
Monoacetylneriifolin *see* 6559
N^1-Monoacetyl Sulfisoxazole *see* 9082
Monoamine Oxidase, 6333
Mono-L-aspartyl Chlorin E$_6$ *see* 9166
Monobasic Lead Acetate *see* 5474
Mono-Baycuten *see* 2401
Monobenzone, 6334
Monobenzyl Hydroquinone *see* 6334
Monobromoacetanilide *see* 1406
Monobromobenzene *see* 1413
Monobromoethane *see* 3826
2-Monobromoisovalerylurea *see* 1404
Monobromomethane *see* 6103
Monobromosuccinic Acid *see* 1447
Monobutyl *see* 6317
Monobutyltin Trichloride *see* 1594
Monobutyrin, 6335
α-Monobutyrin *see* 6335
Monocaine [Novocol] *see* 1526
Monocalcium Bis[(3R,5S,6E)-7-[2-cyclo-propyl-4-(4-fluorophenyl)-3-quinolin-yl]-3,5-dihydroxy-6-heptenoate] *see* 7621
Monocalcium Orthophosphate *see* 1695
Monocalcium Phosphate *see* 1695
α-Monocaprylin *see* 6338
Mono-Cedocard [Altana] *see* 5271
Monochloroacetone *see* 2114
Monochlorethane *see* 3837
Monochloroacetaldehyde *see* 2109
Monochloroacetic Acid *see* 2112
Monochloroacetic Acid Anhydride *see* 2113
Monochloroacetone *see* 2114
17-Monochloroacetylajmaline *see* 5634
Monochlorobenzene *see* 2122
Monochlorodibromomethane *see* 2135
8-Monochloro-3-(β-diethylaminoethyl)-4-methyl-7-ethoxycarbonylmethoxy-coumarin *see* 2393
α-Monochlorohydrin *see* 2146
Monochloromethyl Ether *see* 2148
Monocid [GSK] *see* 1930
Monocillin [Chassot] *see* 7206
Monoclair [Hennig] *see* 5271
Monoclate-P [Centeon] *see* 3962
Monoclonal Antibody B72.3-GYK-DTPA-In-111 *see* 8521
Monocor [Am. Cyanamid] *see* 1295
Monocortin [Grünenthal] *see* 7133

Monocrotaline, 6336
Monocrotophos, 6337
Monoctanoin, 6338
Monoethanolamine *see* 3782
Monoethylamine *see* 3817
Monoethyl Tartrate, 6339
Monoferrous Acid Citrate Monohydrate *see* 4071
Monoflocet [Aventis] *see* 6863
Monofluoroacetamide *see* 4198
Monofluoroacetic Acid *see* 4199
Monofluorobenzene *see* 4201
Mono-Gesic [Central] *see* 8477
Monoglycerol p-Aminobenzoate *see* 4523
Monoglycol Salicylate *see* 4535
Monoglyme *see* 3245
Monogynol B *see* 5668
Monohydrated Selenium Dioxide *see* 8567
Mono(2-hydroxybenzoate)-1H-imidazole *see* 4954
Monohydroxymercuridiiodoresorcinsul-fonphthalein Sodium Salt *see* 5932
Monoiodotyrosine *see* 5090
Mono-Kay [Abbott] *see* 7492
Monoket [Schwarz] *see* 5271
Monolene *see* 435
Monolupine *see* 617
Mono Mack [Mack, Illert.] *see* 5271
(Monomethoxypolyethylene Glycol Suc-cinimidyl)$_{74}$-L-asparaginase *see* 826
Monomethylamine *see* 6088
Monomethyl-p-aminophenol Sulfate *see* 6091
N-Monomethylaniline *see* 6093
N^G-Monomethyl-L-arginine *see* 6096
Monomethylarsinic Acid *see* 6025
Monomethylauristatin E, 6340
Monomethyldiaminodiphenazothionium Chloride *see* 922
o-Monomethyldiphenhydramine *see* 6975
N-Monomethylformamide *see* 6147
Monomethylhydrazine *see* 6158
Monomethyl Mercury *see* 3274
Monomethyloldimethylhydantoin *see* 4870
Monomethylpropion *see* 6035
Monomethyl Sulfate *see* 6195
Monomethylthionine Chloride *see* 922
Monomycin [Grünenthal] *see* 3739
Monomycin A *see* 7146
Mononine [ZLB Behring] *see* 3963
Mononitrogen Monoxide *see* 6664
Mononuclear Cell Factor *see* 5037
Monooctadecyl Ether of Glycerol *see* 1009
Monooctanoin *see* 6338
Monoparin [CP Pharm.] *see* 4688
Monophenolase *see* 10019
Monophenol Monooxygenase *see* 10019
Monophenoloxidase *see* 10019
Monophenylbutazone *see* 6317
Monopina [Bioindustria] *see* 487
Monopotassium Citrate *see* 7745
Monopotassium Oxalate *see* 7732
Monopotassium Phosphate *see* 7780
Monopril [BMS] *see* 4283
Monopropionylerythromycin *see* 3742
Monorden, 6341
Monores [Valeas] *see* 2345
Monorhein *see* 8299
Monorheumetten [Benzon] *see* 6317
Monos [SKB] *see* 8427
Monosilane *see* 8627

Monosilicic Acid *see* 8629
Monosodium Glutamate, 6342
Monosodium Methylarsonate *see* 6025
Monosodium Orthophosphate *see* 8790
Monosodium N-Phenylglycinamide-p-ar-sonate *see* 9974
Monosorb [Dexcel] *see* 5271
Monostearin, 4524
Monosulfiram *see* 9080
Monotard [Novo] *see* 5023
Monothioethyleneglycol *see* 5936
α-Monothioglycerol *see* 9488
Monotrim [Duphar] *see* 9878
Monotropein, 6343
Monovent [AstraZeneca] *see* 9302
Monoxychlorosene *see* 7055
Monsel's Solution, 6344
Montan Wax, 6345
Montanyl Alcohol *see* 6833
Monteban [Elanco] *see* 6501
Montelukast, 6346
Monteplase, 6347
Montmorillonite, 6348
Montricin [SPA] *see* 5914
Monuril [Zambon] *see* 4281
Monurol [Zambon] *see* 4281
Monuron, 6349
Monuron TCA *see* 6349
Monydrin [AstraZeneca] *see* 7417
Monzal [Boehringer, Ing.] *see* 10168
Monzaldon [Boehringer, Ing.] *see* 10168
Moogrol [Burroughs Wellcome] *see* 2047
MoOPH *see* 10134
Moore's Ketene, 6350
5-MOP *see* 1160
8-MOP [Valeant] *see* 6059
Moperone, 6351
Moplen [Montecatini] *see* 7701
Mopral [AstraZeneca] *see* 6939
MOPS, 6352
Moracizine *see* 6354
Moranoline *see* 2906
Morantel, 6353
Morbicid *see* 4263
Morcap [Mayne] *see* 6361
Mordant Violet 25 *see* 4374
Mordant Yellow 3R *see* 247
Morestan [Bayer] *see* 7077
Morfazinamide *see* 6359
Morial [Takeda] *see* 6320
Moricizine, 6354
Morin, 6355
Morinamide *see* 6359
Morindin, 6356
β-Morindin *see* 6356
Morindone *see* 6356
Moringine *see* 1128
Morintannic Acid *see* 5706
Moriperan [Morishita] *see* 6219
Morison's Paste *see* 5755
Moritannic Acid *see* 5706
Morman Tea *see* 3662
Mornidine [Searle] *see* 7567
Morniflumate *see* 6615
Morning-glory *see* 5116
Moroctic Acid *see* 8931
Moroctocog Alfa, 6357
Moronal [BMS] *see* 6825
Moroxydine, 6358
Morpan T [A.B.M. Chem.] *see* 6422
Morphazinamide, 6359
Morphia *see* 6361
Morphinan, 6360
Morphinan-3-ol *see* 6796
(9α,13α,14α)-Morphinan-3-ol *see* 6796

Morphine, 6361
Morphine Bis(nicotinate)
Morphine Bis(pyridine-3-carboxylate) see 6605
Morphine Dinicotinate see 6605
Morphine Hemisulfate see 6361
Morphine 3-Methyl Ether see 2448
Morphine Monomethyl Ether see 2448
Morphium see 6361
Morpholine, 6362
4-Morpholinecarboximidoylguanidine see 6358
Morpholine *N,N'*-Disulfide see 3415
4-Morpholineethanesulfonic Acid see 5972
4-Morpholinepropanesulfonic Acid see 6352
2-(4-Morpholinoanilino)-N^6-cyclohex-yladenine see 8292
2-(4-Morpholinoanilino)-6-cyclohexylam-inopurine see 8292
(−)-3-Morpholino-4-(3-*tert*-butylamino-2-hydroxypropoxy)-1,2,5-thiadiazole see 9601
2-(*N*-Morpholino)ethanesulfonic Acid see 5972
3-(2-Morpholinoethyl)morphine see 7441
N-2-Morpholinoethyl-5-nitroimidazole see 6637
2-(4-Morpholino)ethyl Sulfonate see 5972
3-(Morpholinomethyl)-1-[(5-nitrofurfur-ylidene)amino]hydantoin see 6617
5-Morpholinomethyl-3-(5-nitrofurfuryli-deneamino)-2-oxazolidinone see 4323
N-Morpholinomethylpyrazinamide see 6359
3-(*N*-Morpholino)propanesulfonic Acid see 6352
10-(3-Morpholinopropionyl)phenothia-zine-2-carbamic Acid Ethyl Ester see 6354
γ-Morpholinopropyl 4-*n*-Butoxyphenyl Ether see 7829
(αS)-α-[[2-(4-Morpholinyl)acetyl]ami-no]benzenebutanoyl-L-leucyl-*N*-[(1S)-3-methyl-1-[[(2R)-2-methyl-2-oxiran-yl]carbonyl]butyl]-L-phenylalanin-amide see 1837
β-Morpholinylethylmorphine see 7441
3-[2-(4-Morpholinyl)ethyl]morphine see 7441
1-(2-*N*-Morpholinylethyl)-5-nitroimid-azole see 6637
3-(4-Morpholinylmethyl)-1-[[(5-nitro-2-furanyl)methylene]amino]-2,4-imidaz-olidinedione see 6617
5-(4-Morpholinylmethyl)-3-[[(5-nitro-2-furanyl)methylene]amino]-2-oxazolidi-none see 4323
N-(4-Morpholinylmethyl)-2-pyrazinecar-boxamide see 6359
N-[10-[3-(4-Morpholinyl)-1-oxopropyl]-10*H*-phenothiazin-2-yl]carbamic Acid Ethyl Ester see 6354
4-Morpholinyl(3,4,5-trimethoxyphenyl)-methanethione see 9937
4-Morpholinyl(3,4,5-trimethoxyphenyl)-methanone see 9892
Morsydomine see 6320
MORT1-associated CED-3 Homolog see 1886
Mortopal see 9349
Mosaic Gold see 8907
Mosapramine, 6363

Mosapride, 6364
Mosatil see 3565
Moscontin [Mundipharma] see 6361
Mosegor [Novartis] see 7629
(±)-Mosher's Acid see 6365
Mosher's Reagent, 6365
Mospilan [Nippon Soda] see 48
Mossbunker Oil see 5904
Moss Starch see 5531
Motazomin [Takeda] see 6320
Motens [Boehringer, Ing.] see 5379
Motesanib, 6366
Motexafin Gadolinium, 6367
Motexafin Lutetium, 6368
Mother-of-Pearl Sulfur see 9099
Motiax [Neopharmed] see 3971
Motifene [Sankyo] see 3091
Motilex [Guidotti] see 2342
Motilin, 6369
Motilin (Swine) see 6369
Motilium [Janssen] see 3467
Motilyn [Abbott] see 2949
Motilyo [Janssen] see 3467
Motosol [Europharma] see 380
Motox see 9716
Motretinide, 6370
Motrin [McNeil] see 4919
Mountain Balm see 3727
Mountain Grape see 1157
Mountain Tobacco see 780
Mouse Antialopecia Factor see 5021
Mouse-bane see 108
Mouse EGF-URO see 3574
Movalis [Boehringer, Ing.] see 5894
Movatec [Boehringer, Ing.] see 5894
Movecil [Farmitalia] see 8088
Movectro [Merck Serono] see 2336
Moveltipril, 6371
Movens [Inverni] see 5850
Movento [Bayer CropSci.] see 8888
Mover [Mitsubishi Kasei] see 124
Movergan [Orion] see 8565
Movyl [Montecatini] see 7707
Mowiol [Hoechst] see 7706
Moxadil [Lederle] see 573
Moxalactam, 6372
Moxastine, 6373
Moxaverine, 6374
Moxestrol, 6375
Moxeza [Alcon] see 6377
Moxidectin, 6376
Moxifloxacin, 6377
Moxisylyte, 6378
Moxivig [Alcon] see 6377
Moxonidine, 6379
Mozobil [Genzyme] see 7649
MP see 6190
4-MP see 4253
6MP see 5938
8-MP see 6059
MP-11 see 7293
MP-302 see 5110
MP-328 see 5109
MP-620 see 5054
MP-1177 see 4359
MP-1727 see 7238
MPC-1304 see 756
MPC-7869 see 9196
MPK see 6187
MPP+ see 6380
α-MPT see 6240
MPTP, 6380
MPV-785 see 5858
MPV-1248 see 852
MPV-1440 see 2948

M-Pyrol [ISP] see 6190
MQPA see 769
MRA see 9651
MRD-108 see 7597
MRE-0470 see 1229
MRIH see 5882
MRL-41 see 2377
MRNA see 8327
MRX-115 see 7273
MS Contin [Purdue Frederick] see 6361
MS-222 see 9786
MS-325 see 4355
MS-551 see 6614
MS-752 see 5650
MS-932 see 124
MS-1144 see 2751
MSA III-2 see 5029
MSG see 6342
MSH, 6381
MSH Release-inhibiting Factor I see 5882
MSH Release-inhibiting Hormone see 5882
MSI-78 see 7308
MSIR [Purdue Frederick] see 6361
MSMA see 6025
MST Continus [Mundipharma] see 6361
MST-16 see 8706
L-α-MT see 6240
MT-141 see 1928
MTB-51 see 6030
MTBE see 6106
MTF-752 see 7239
MTI-446 see 3317
MTI-500 see 3922
MTO see 6206
L-3-MTO see 6128
MTPA see 6365
(R)-MTPA-chloride see 6365
MTP-cephalin see 6265
MTP-PE see 6265
MTU see 6201
MTX see 5762, 6057
4-MU see 4894
MU-495 see 6607
Mucara see 5332
Mucic Acid see 4361
Muciclar [Pfizer] see 1802
Mucidin see 8964
Mucinex [Adams] see 4591
Mucinol [Sanofi Winthrop] see 637
Mucins, 6382
Mucitux [Organon] see 3688
Muclox [Sigma-Tau] see 3971
Mucocedyl [3M] see 83
Mucochloric Acid, 6383
Mucochloric Anhydride, 6384
Mucocis [So.Se.] see 1802
Mucoclear [Mundipharma] see 380
Mucodil [Valeas] see 8936
Mucodyne [Sanofi-Aventis] see 1802
Mucofar [Farmakos] see 380
Mucofluid [UCB] see 5979
Mucoglucoronan see 2810
Mucolase [Lampugnani] see 1802
Mucolator [Abbott] see 83
Mucolex [Warner-Lambert] see 1802
Mucolitico [Maggioni] see 3463
Mucomyst [BMS] see 83
Muconic Acid, 6385
Muconomycin A see 10160
Muconomycin B see 10160
Mucopront [Mack, Illert.] see 1802
Mucosan [Fher] see 380
Muco Sanigen [Thiemann] see 83

Mucosil [Dey] see 83
Mucosolvan [Thomae] see 380
Mucosta [Otsuka] see 8242
Mucovent [Byk Gulden] see 380
Mucret [Pharma-Stern] see 83
Muira Puama, 6386
Mukaiyama Reagent, 6387
Mulka see 6035
Mulsopaque [Alcon] see 5103
Multaq [Sanofi-Aventis] see 3498
Multergan [Specia] see 9458
Multhiomycin see 6808
Multicrom [Menarini] see 2581
Multi-CSF see 5039
MultiHance [Bracco] see 4352
Multilase [Sigma-Tau] see 666
Multiplication-stimulating Activity III-2
    see 5029
Multipotent Colony Stimulating Factor
    see 5039
Multum [Chephasaar] see 2085
Mundisal [Purdue Frederick] see 2213
Munobal [Aventis] see 3987
Muphoran [Servier] see 4289
Mupirocin, 6388
Muplestim see 5039
Muracil [Organon] see 6201
Muramic Acid, 6389
Muramidase see 5701
Muramyl Dipeptide, 6390
Muramyl Tripeptide Phosphatidylethan-
    olamine see 6265
Murazyme [Prospa] see 5701
Murexide, 6391
Murexine, 6392
Muriacite see 1708
Muriatic Acid see 4818
Murillo Bark see 8156
Muroctasin see 8384
Muromonab CD3, 6393
Murrayin see 8546
Musaril [Sanofi-Synthelabo] see 9391
Muscalm [Nippon Kayaku] see 9681
Muscalure, 6394
Muscarine, 6395
Muscatox [Bayer] see 2546
Muscazone, 6396
Muscimol, 6397
Muscle Adenylic Acid see 146
Muscle Inosinic Acid see 5020
Muscone, 6398
Musco-Ril [Inverni] see 9477
Musculamine see 8870
Musculax [Nippon Organon] see 10132
Muse [Meda] see 7987
Mushroom Sugar see 9742
Muskelat [Azupharma] see 9391
Muskone see 6398
Musk Root see 9127
Musks, 6399
Mussel Poison see 8526
Mustard, 6400
Mustard Chlorohydrin see 4681
Mustard Gas, 6401
"Mustard Oil" see 286
Mustargen [Merck & Co.] see 5846
Mustron see 5847
Mutabase [Schering-Plough] see 3007
Mutamycin [BMS] see 6300
Mutesa [Wyeth] see 7032
Muthesa [Wyeth] see 7032
Muthmann's Liquid see 9330
Mutilin 14-(exo-8-Methyl-8-azabicyclo-
    [3.2.1]oct-3-ylsulfanyl)acetate see 8281

Mutton Suet see 9017
MV-119A see 3411
MX-70906 see 4973
MY-301 see 4591
MY-5116 see 8258
Myacine see 6539
Myalex [ICI] see 4001
Myambutol [Wyeth] see 3775
Myanesin [BDH] see 5917
Myanol see 5917
Myarsenol see 9072
MYC-8003 see 6311
Mycamine [Astellas] see 6254
Mycaminose, 6402
Mycardol [Winthrop] see 7222
Mycarose, 6403
Mycelex-G [Miles] see 2401
Mycelianamide, 6404
Mycetins, 6405
Mycifradin [Upjohn] see 6539
Mycil [BDH] see 2184
Myclobutanil, 6406
Mycobacidin, 6407
Mycobacillin, 6408
Mycobactic Acid see 6409
Mycobactin P see 6409
Mycobactins, 6409
Mycobactyl see 4536
Mycobutin [Farmitalia] see 8338
Mycobutol [Cadila] see 3775
Mycodécyl Solution [Doms-Adrian] see
    10033
Mycofug [Hermal] see 2401
Mycoin C$_3$ see 7158
Mycolic Acids, 6410
Mycolutein see 870
(6-O-Mycolyl-α-D-glucopyranosyl)-6-O-
    mycolyl-α-D-glucopyranoside see 2513
Mycomycin, 6411
Mycophenolate Mofetil see 6412
Mycophenolate Sodium see 6412
Mycophenolic Acid, 6412
Mycophyt [Mycofarm] see 6509
Myco-protein see 8204
Mycosamine, 6413
Mycose see 9742
Mycospor [Bayer] see 1216
Mycosporan [Bayer] see 1216
Mycosporin [Bayer] see 2401
Mycostatin [BMS] see 6825
Mycoster [Fabre] see 2269
Mycotoxin T-2 see 9981
Mydecamycin see 6262
Mydetone see 9681
Mydfrin [Alcon] see 7398
Mydocalm [Labatec] see 9681
Mydplegic [Cooper] see 2741
Mydriacyl [Alcon] see 9959
Mydrial [Winzer] see 10016
Mydriaticum [Merck & Co.] see 9959
Mydrilate [Boehringer, Ing.] see 2741
Myebrol [Kyorin] see 6297
Myelostat [Stragen] see 4776
Myelobromol [Enzypharm] see 6297
Myelotrast [Winthrop] see 5053
Myfortic [Novartis] see 6412
Myfungar [Wyeth] see 7036
Mygdalon [DDSA] see 6219
Myk [Cassenne] see 9025
Mykrox [Celltech] see 6224
Mylabris, 6414
Mylar [DuPont] see 7689
Mylaxen see 4720
Mylepsinum [AWD] see 7865
Myleran [GSK] see 1508

Mylicon [J & J-Merck] see 3237
Mylis [Organon] see 2875
Mylodorm [G.P.] see 567
Mylosar (formerly) see 884
Mylotarg [Wyeth] see 4424
Myocet [Elan] see 3487
Myocholine [Glenwood] see 1187
Myochrysine [Merck & Co.] see 4553
Myocord [Ivax] see 850
Myodetensine see 5917
Myodigin see 3182
Myodil see 5103
Myoflex [Novartis] see 9830
Myohematin see 2788
Myolastan [Sanofi-Synthelabo] see 9391
Myolysin see 5917
Myonal [Eisai] see 3661
Myopan see 5917
Myopone see 10239
Myorexon [AstraZeneca] see 5271
Myosalvarsan see 9072
Myoscain [Sanochemia] see 4591
Myoserol see 5917
Myosin, 6415
Myoson [Strathmann] see 7859
Myoten [Central Pharm.] see 5917
Myotonine [Glenwood] see 1187
Myoview [GE Healthcare] see 9395
Myoxam [Menarini] see 6262
Myoxanthin see 3542
Myozyme [Genzyme] see 100
Myprozine [Am. Cyanamid] see 6509
α-Myrcene see 6416
β-Myrcene, 6416
Myreth see 7697
Myrica see 1010
Myricetin, 6417
Myricitrin see 6417
Myricyl Alcohol see 9753
Myriocin, 6418
Myristica see 6818
Myristic Acid, 6419
Myristic Acid Isopropyl Ester see 5261
Myristicin, 6420
Myristin see 9896
29^B-(N^6-Myristoyl-L-lysine)-30^B-de-L-
    threonineinsulin (Human) see 5026
Myristyl Alcohol, 6421
Myristyltrimethylammonium Bromide,
    6422
Myrizone [Squibb] see 9442
Myrj [ICI] see 7698
Myrj 45 [ICI] see 7698
Myrj 52 [ICI] see 7698
Myrj 52S [ICI] see 7698
Myronate Potassium see 8683
Myrrh, 6423
Myrticolorin see 8438
Myrtillin-a see 2885
Myrtle Wax see 1010
Myrtol, 6424
Myrtrimonium Bromide see 6422
Mysalfon [Ivax] see 9307
Myser [Nikken] see 3171
Myslee [Sanofi-Aventis] see 10391
Mysoline [Acorus] see 7865
Mytab [Zeeland] see 6422
Mytelase [Sanofi-Aventis] see 373
Myxin, 6425
Myxoviromycin see 394
N-5′ see 9731
N-021 see 9657
N-22 see 6318
N-0252 see 5440
N-0437 see 8404

1,4-Naphthoquinone, 6480
1,2-Naphthoquinone-4-sulfonic Acid
  Sodium Salt see 8777
β-Naphthoquinone-4-sulfonic Acid
  Sodium Salt see 8777
Naphthoresorcinol, 6481
Naphthosalol see 6496
β-Naphthoxyacetic Acid see 6482
2-Naphthoxyacetic Acid, 6482
2-(α-Naphthoxy)-N,N-diethylpropionam-
  ide see 6497
β-Naphthoxyethanol see 6494
Naphthylacetic Acid see 6456
[6-[3-(2-Naphthyl)-D-alanine]]gonadore-
  lin see 6435
α-Naphthylamine see 6483
β-Naphthylamine see 6484
1-Naphthylamine, 6483
2-Naphthylamine, 6484
Naphthylamine Blue see 9972
2-Naphthylamine-3,6-disulfonic Acid see
  396
2-Naphthylamine-6,8-disulfonic Acid see
  395
β-Naphthylamine-δ-sulfonic Acid see
  1893
1-Naphthylamine-4-sulfonic Acid, 6485
1-Naphthylamine-5-sulfonic Acid, 6486
1-Naphthylamine-6-sulfonic Acid see
  2348
1-Naphthylamine-7-sulfonic Acid see
  2349
1-Naphthylamine-8-sulfonic Acid, 6487
2-Naphthylamine-1-sulfonic Acid, 6488
2-Naphthylamine-5-sulfonic Acid, 6489
2-Naphthylamine-6-sulfonic Acid see
  1893
2-Naphthylamine-7-sulfonic Acid see
  1893
2-Naphthylamine-8-sulfonic Acid see 933
2-Naphthyl Benzoate, 6490
β-Naphthylbis(β-chloroethyl)amine see
  2108
1-Naphthyl Chloride see 2151
β-Naphthyldi(2-chloroethyl)amine see
  2108
O-2-Naphthyl m,N-Dimethylthiocarb-
  anilate see 9678
N-(1-Naphthyl)ethylenediamine, 6491
β-Naphthyl Ethyl Ether see 3806
N-(1-Naphthyl)-N'-(3-ethylphenyl)-N'-
  methylguanidine see 747
O-(2-Naphthyl)glycolic Acid see 6482
1-Naphthylisocyanate, 6492
1-Naphthylisothiocyanate, 6493
1-Naphthyl Mercaptan see 6463
2-Naphthyl Mercaptan see 6464
1-Naphthyl N-Methylcarbamate see 1788
2-(1-Naphthylmethyl)imidazoline see
  6453
2-Naphthyl N-Methyl-N-(3-tolyl)thiono-
  carbamate see 9678
2-(2-Naphthyloxy)ethanol, 6494
α-Naphthylphthalamic Acid see 6500
N-1-Naphthylphthalamic Acid see 6500
α-Naphthyl Salicylate see 6495
1-Naphthyl Salicylate, 6495
2-Naphthyl Salicylate, 6496
3-(1-Naphthyl)-2-tetrahydrofurfurylpro-
  pionic Acid 2-(Diethylamino)ethyl
  Ester see 6437
α-Naphthylthiocarbamide see 708
α-Naphthylthiourea see 708
N-1-Naphthylthiourea see 708
1-(1-Naphthyl)-2-thiourea see 708

Naples Yellow see 5453
NaPMP see 5388
NAPQI see 56
Naprelan [Wyeth-Ayerst] see 6499
Napren [Nycomed] see 6499
Naprilene [Sigma-Tau] see 3623
Naprius [Magis] see 6499
Napropamide, 6497
Naprosyn [Syntex] see 6499
Naprosyne [Syntex] see 6499
Naproxcinod, 6498
Naproxen, 6499
Naprux [Andromaco] see 6499
Naptalam, 6500
NAPVSIPQ see 2836
Naqua [Schering] see 9789
Naramig [GSK] see 6502
Naramycin A see 2721
Narasin, 6501
Narasin A see 6501
Naratriptan, 6502
Narbivolol see 6516
Narbomycin, 6503
Narbonolide see 6503
Narcan [Endo] see 6447
Narcanti [BMS] see 6447
Narcaricin [Heumann] see 1067
Narceine, 6504
Narcissine see 5687
Narcoren [Veterinaria] see 7243
Narcosan Soluble see 4740
Narcosine see 6807
Narcotile see 3837
Narcotine see 6807
dl-Narcotine see 6807
l-α-Narcotine see 6807
Narcotoline, 6505
Narcoxyl [Veterinaria] see 10277
Narcozep [Roche] see 4177
Narcyl see 6504
Nardelzine [Warner-Lambert] see 7335
Nardil [Warner-Lambert] see 7335
Naridan see 7072
Narigix [Taiyo] see 6444
Narilet [Boehringer, Ing.] see 5117
Naringenin, 6506
Naringenin-7-rhamnoglucoside see 6507
Naringen 7-Methyl Ether see 8455
Naringetol see 6506
Naringin, 6507
Narone [Ulmer] see 3390
Naropin [AstraZeneca] see 8389
Narphen see 7332
Narsis [Sumitomo] see 5857
Nartate see 3390
Nartograstim see 4573
Nasacort [Aventis] see 9758
Nasalcrom [Pharmacia & Upjohn] see
  2581
Nasalide [Syntex] see 4176
Nasanyl [Searle] see 6435
Nasea [Yamanouchi] see 8223
NaSF see 8807
Nasivin [Merck KGaA] see 7065
Nasofelin [Neo Quimica] see 4016
Nasonex [Schering-Plough] see 6327
Nastenon see 7066
NAT-333 see 4026
Natacillin [Banyu] see 4707
Natacyn [Alcon] see 6509
Natalizumab, 6508
Natamycin, 6509
Nateglinide, 6510
Naticardina [Viatris] see 8176
Natil [3M] see 2697

Native Calcium Sulfate see 1708
Native Gellan Gum see 4417
Natrecor [Scios] see 1365
Natrilix [Servier] see 4975
Natriphene see 7415
Natrite see 8731
Natrium see 8708
Natriuretic Factor-32 (Human Brain
  Clone λhBNP57) see 1365
Natroba [ParaPro Pharma] see 8877
Natrol see 1280
Natron see 8731
Natto see 8856
Natulan [Sigma-Tau] see 7876
Natural Calcium Carbonate see 5544
Natural Gas, 6511
Natural Tartaric Acid see 9205
Naturalyte [Dow AgroSci.] see 8877
Naturetin [BMS] see 1039
Naturlose [Spherix] see 9161
Naturon see 6079
Natyl [Interdelta] see 3388
Naucaine see 7875
Naumannite see 8568
Nausicalm [Brothier] see 3230
Nausidol [SK & F] see 7567
Nausilen [Baldacci] see 242
Nautamine [Sanofi-Aventis] see 3339
Navadel see 3330
Navane [Roerig] see 9519
Navelbine [Fabre] see 10187
Navidrex [Goldshield] see 2740
Navigator [Idexx] see 6653
Navilox [Univet] see 5285
Navisin see 7065
Navitoclax, 6512
Navoban [Novartis] see 9962
Navolin see 4491
NAX 31 [Prospec] see 8766
Naxcel [Pfizer] see 1951
Naxen [Syntex] see 6499
Naxofem [Farmitalia] see 6637
Naxogin [Erba] see 6637
Naxy [Cephalon] see 2338
Nazasetron see 893
NBA-061 see 4034
NBCA see 1566
NB-Enantride, 6513
NBI-34060 see 4984
NBI-56418 see 3580
NBS see 1448
9-NC see 8421
NC-14 see 738
NC-45 see 10132
NC-123 see 5980
NC-129 see 8078
NC-302 see 8203
NC-319 see 4636
NC-531 see 9728
NC-758 see 9728
NC-1667 see 9829
NC-1968 see 7284
NC-6897 see 1038
NC-8438 see 3796
NC-21314 see 2369
NCI-129 see 8078
NCI-96683 see 8203
NCI-C55243 see 1425
NCI-C55254 see 2135
NCI-C56462 see 6336
NCPI see 2775
NCS see 10367
NCSA see 2166
NCSac see 2166
NDEA see 6723

Name Index

NDELA *see* 6722
2'NDG *see* 4393
NDGA *see* 6782
NDR-304 *see* 3843
NDR-5998A *see* 4026
NE-10064 *see* 904
NE-58095 *see* 8360
Neamine, 6514
Neantine [Givaudan] *see* 7483
Neatsfoot Oil, 6515
Neazina [Centralvet] *see* 9047
Nebcin [Lilly] *see* 9647
Neberk [Morishita] *see* 9252
Nebicina [Lilly] *see* 9647
Nebilet *see* 6516
Nebilox [GSK] *see* 6516
Nebivolol, 6516
dl-Nebivolol *see* 6516
Nebramycin *see* 9647
Nebramycin Factor 2 *see* 739
Nebramycin Factor 6 *see* 9647
Nebularine, 6517
Nebulicina [Boehringer, Ing.] *see* 4016
Nebulin, 6518
NebuPent [Astellas] *see* 7227
Neburon, 6519
Necatorina *see* 1816
Necic Acids *see* 8588
"Necine" Bases *see* 8289
Necitumumab, 6520
Nectocyd [Pfizer] *see* 3412
Nedocromil, 6521
Needle Antimony *see* 701
Neem, 6522
Nefazodone, 6523
Nefco *see* 6687
Nefiracetam, 6524
Nefopam, 6525
Nefrocarnit [Medice] *see* 1849
Neftin *see* 4330
Negamycin, 6526
Negatol [Altana] *see* 7673
NegGram [Sanofi Winthrop] *see* 6444
Negram [Sanofi Winthrop] *see* 6444
Nehydrin [TAD] *see* 3710
Nelarabine, 6527
Nelbon [Sankyo] *see* 6661
Nelfinavir, 6528
Nelfinavir Mesylate *see* 6528
Nema [Parke-Davis] *see* 9335
Nemacur [Bayer CropSci.] *see* 3991
Nemadectin, 6529
Nemafax [Rhône-Mérieux] *see* 9505
Nemafume *see* 3025
Nemagon *see* 3025
Nemasol Sodium [Valeant] *see* 473
Nemathorin [Ishihara Sangyo] *see* 4288
Nematolyt *see* 7119
Nemazine *see* 7364
Nembutal [Abbott] *see* 7243
Nemestran [HMR] *see* 4449
Nemex [Pfizer] *see* 8066
Nemicide [Coopers] *see* 5511
Nemifitide, 6530
Nemifitide Ditriflutate *see* 6530
Nemispor [Isagro] *see* 5780
Nemonapride, 6531
Nemurel [Takeda] *see* 3757
Nendrin *see* 3633
Neo Heliopan 303 [Symrise] *see* 6845
Neo Heliopan 357 [Symrise] *see* 881
Neo Heliopan AV [Symrise] *see* 6859
Neo Heliopan BB [Symrise] *see* 7053
Neo Heliopan HMS [Symrise] *see* 4777
Neo Heliopan Hydro [Symrise] *see* 3647

Neo Heliopan MA [Symrise] *see* 5908
Neo Heliopan OS [Symrise] *see* 6860
Neo-Absentol *see* 3773
Neoamyl Alcohol *see* 6542
Neo-Antergan [Poulenc] *see* 8097
Neoantimosan *see* 8942
Neo-Arsoluin *see* 6532
Neoarsphenamine, 6532
Neo-Atromid *see* 2370
Neobar [Merck KGaA] *see* 990
Neo-Benodine *see* 6125
Neobiosamine B *see* 6539
Neobiosamine C *see* 6539
Neo-Bradoral [Novartis] *see* 3464
Neobrettin [Norbrook] *see* 6539
Neocaine *see* 7875
Neocarcinostatin *see* 10367
Neo-Cardiamine *see* 9348
Neocarzinostatin *see* 10367
Neo-Cebicure *see* 5171
Neo-Cebitate *see* 5171
Neocembrene, 6533
Neocembrene A *see* 6533
Neochanin *see* 4271
Neocidol [Zagro] *see* 2998
Neoclym [Poli] *see* 2713
Neo-Cobefrin [Cook-Waite] *see* 6781
Neocon 1/35 [Ortho-Cilag] *see* 6786
Neo-Corovas *see* 7222
Neocuproine, 6534
Neo-Cytamen [Medeva] *see* 4846
Neodaian [Nichiiko] *see* 9449
Neodalit [Hässle] *see* 3012
Neo-Dioxanin [Boehringer, Ing.] *see* 3186
Neo Dohyfral D₃ *see* 10216
Neodorm [Minden] *see* 7243
Neo-Douxan *see* 6704
Neodrenal *see* 5263
Neo-Duplamox [ISF] *see* 2339
Neodymium, 6535
Neodyne *see* 3843
Neo-Epinine [Burroughs Wellcome] *see* 5263
Neoergosterol, 6536
Neo-Erycinum [Schering AG] *see* 3740
Neofamid *see* 5712
Neo-Farmadol *see* 7071
Neofemergen *see* 3712
Neo-Fluimucil [Zambon] *see* 83
Neofocin [Medici] *see* 4281
Neogama [Hormosan] *see* 9119
Neogel [Hexal] *see* 1794
Neogest [Schering AG] *see* 6793
Neo-Gilurytmal [Teofarma] *see* 7820
Neo H2 [Boehringer, Ing.] *see* 8407
Neohesperidin DHC *see* 6537
Neohesperidin Dihydrochalcone, 6537
Neo-Hombreol (Amp.) [Organon] *see* 9324
Neohombreol M [Organon] *see* 6197
Neohydrazid *see* 2670
Neoisocodeine *see* 8021
Neoisuprel *see* 5215
Neo Kodan [Schülke & Mayr] *see* 6843
Neolamin [Nippon Kayaku] *see* 9449
Neolate *see* 6539
Neolignans *see* 5539
Neolinarin *see* 7170
Neoloid [Lederle] *see* 1898
Neolone 950 [Rohm & Haas] *see* 6163
Neo-Lotan [Neopharmed] *see* 5641
Neomagnol [Chinoin] *see* 2074
Neomark [NeoPharm] *see* 1461
Neomas *see* 6539
Neo-mercazole [Roche] *see* 1799

Neomestine [Taiyo] *see* 4258
Neo-Methiodal *see* 5083
Neomethymycin, 6538
Neomethynolide *see* 6538
Neomin *see* 6539
Neomix [Tuco] *see* 6539
Neomycin, 6539
Neomycin A *see* 6514
Neomycin B *see* 6539
Neomycin C *see* 6539
Neomycin E *see* 7146
Neomyson [Eisai] *see* 9453
Neon, 6540
Neo-Naclex [Goldshield] *see* 1039
Neonal [Abbott] *see* 1524
Neonicotine *see* 612
Neo-Nilorex [A.V.P.] *see* 7334
Neopax [Janssen] *see* 7518
Neo-Penil *see* 7194
Neopentaldehyde *see* 7624
Neopentane, 6541
Neopentanol *see* 6542
Neopentyl Alcohol, 6542
Neopentyl Glycol, 6543
Neopenyl [Grünenthal] *see* 7203
Neoperidole [Kyowa Hakko] *see* 4634
Neophyl Chloride, 6544
Neopine, 6545
Neoplatin [BMS] *see* 2316
Neoprene, 6546
Neoprotoveratrine *see* 8008
Neopsicaine *see* 8020
Neopterin, 6547
Neo-Pynamin [Sumitomo] *see* 9370
Neopyrithiamine *see* 8105
Neoquassin, 6548
Neoral [Novartis] *see* 2750
NeoRecormon [Roche] *see* 3744
Neoron [Syngenta] *see* 1445
Neosalvarsan [Hoechst] *see* 6532
Neosanamid II *see* 9061
Neoscan [Medi-Physics] *see* 4379
Neo-Skiodan *see* 5083
Neosorexa [Sorex] *see* 3153
Neospiran *see* 9348
Neostene [SM & P] *see* 6022
Neostenovasan *see* 3526
Neosteron *see* 6022
Neostigmine, 6549
Neosulf *see* 6539
Neosulfine *see* 5988
Neo-Synephrine *see* 7398
Neo-T *see* 6551
Neotame, 6550
Neoteben [Bayer] *see* 5232
NeoTect [Diatide] *see* 2912
Neo-Tenebryl *see* 5083
Neo-Tenebryl *see* 5083
Neotetrazolium Blue *see* 6551
Neotetrazolium Chloride, 6551
Neothyl *see* 6186
Neothylline [Lemmon] *see* 3526
Neo-Thyreostat [Herbrand] *see* 1799
Neotigason [Roche] *see* 103
Neotilina *see* 3526
Neotocopherol *see* 9654
Neoton [Alfa] *see* 7452
Neotran *see* 1252
Neo-Vasophylline *see* 3526
Neovitamin A, 6552
Neovitamin A Acid *see* 5274
Neo-Zine [Neo] *see* 7351
Nepafenac, 6553
Nepetalactone, 6554
Nepetalic Acid *see* 6554
Nephril [Pfizer] *see* 7705

Nephroflow [Medi-Physics] *see* 5077
Nephrotest [Biolog. Arbeitsgem.] *see* 439
Neporex [Novartis] *see* 2772
Nepresol [Novartis] *see* 3187
Nepresol Inject [Novartis] *see* 3187
Népressol [Novartis] *see* 3187
Neprotin *see* 5309
Neptal [Procter & Gamble] *see* 20
Neptunium, 6555
Nequinate, 6556
Neral *see* 2321
Neratinib, 6557
Neraval [Schering] *see* 6050
Nerbowdine *see* 1497
Nerdipina [Ferrer] *see* 6580
Neribas [Schering AG] *see* 3162
Nericur [Schering] *see* 1119
Neridronate Sodium *see* 6558
Neridronic Acid, 6558
Neriforte [Schering AG] *see* 3162
Neriifolin, 6559
Neriine *see* 2480
Neriodin [Nagase] *see* 3091
Neriolin *see* 6919
Nerisona [Schering AG] *see* 3162
Nerisone [Schering AG] *see* 3162
Nerixia [Abiogen] *see* 6558
Nerobol *see* 6023
Nerol, 6560
Nerolidol, 6561
Nerolin Bromelia *see* 3806
Nerolin "Old" *see* 6070
Neroli Oil (Artificial) *see* 6094
Nervanaid B *see* 3565
Nerve Growth Factor, 6562
Nerve Growth Factor-2 *see* 6572
Nerve Root *see* 2767
Nervinex [Menarini] *see* 1385
Nervonus [Orion] *see* 5929
Nesacaine [Astra] *see* 2160
Nesdonal Sodium [RPR] *see* 9503
Nesina [Takata] *see* 302
Nesiritide *see* 1365
NESP *see* 2825
Nestorone [Pop. Council] *see* 3586
Netamifide *see* 6530
Nethamine [HMR] *see* 3764
Neticonazole, 6563
Netillin [Schering] *see* 6564
Netilmicin, 6564
Netilyn [Schering] *see* 6564
Netobimin, 6565
Netromicine [Schering] *see* 6564
Netromycin [Schering] *see* 6564
Netropsin, 6566
Netsusarin [Ohta] *see* 470
Nettacin [Schering] *see* 6564
Nettle, 6567
Neuberg Ester *see* 4305
Neu5Ac *see* 8623
Neuer [Daiichi] *see* 2023
Neulactil [M & B] *see* 7279
Neulasta [Amgen] *see* 4573
Neuleptil [Aventis] *see* 7279
Neumega [Genetics Inst.] *see* 5044
Neupentedrin *see* 9142
Neupogen [Amgen] *see* 4573
Neuprex [Xoma] *see* 931
Neupro [Schwarz] *see* 8404
Neuquinon [Eisai] *see* 10025
Neuracen [Byk Gulden] *see* 1016
Neuractiv [Novartis] *see* 7041
Neuraminic Acid, 6568
Neuraxin *see* 6052
Neuridine *see* 8870

Neurine, 6569
Neuriplege [Génévrier] *see* 2189
NeuroBloc [Elan] *see* 1362
Neurocil (Inj.) [Bayer] *see* 5518
Neurocil (Tabl.) [Bayer] *see* 5518
Neurodye GH1-43 *see* 4247
Neurolene [Magis] *see* 6760
Neurolite [BMS] *see* 9236
Neuromet [SKB] *see* 7041
Neuronotrophin-3 *see* 6572
Neurontin [Parke-Davis] *see* 4348
Neuropeptide Tyrosine *see* 6570
Neuropeptide Y, 6570
Neurosedyn *see* 9403
Neurosin *see* 1672
Neurotensin, 6571
Neurotensin (Ox) Triacetate (Salt) *see* 6571
Neuroton [Berlin-Chemie] *see* 2318
Neurotoxin *see* 2590
Neurotrast *see* 5103
Neurotrophin-3 *see* 6572
Neurotrophins, 6572
Neurturin *see* 4410
Neustab [Boots] *see* 9442
Neutase [Sawai] *see* 5701
Neuthion [Senju] *see* 4511
Neutrafil *see* 3526
Neutral Acriflavine *see* 115
Neutral Aluminum Acetate *see* 320
Neutral Ammonium Chromate *see* 507
Neutral Ammonium Fluoride *see* 519
Neutral Cupric Chromate *see* 2624
Neutral Cysteine Proteinase Inhibitor *see* 2775
Neutral Endopeptidase *see* 863
Neutralized Sarin *see* 4966
Neutral Lead Acetate *see* 5452
"Neutral" Magnesium Phosphate *see* 5747
Neutral Potassium Chromate *see* 7743
Neutral Proflavine Sulfate *see* 7887
Neutral Protein Hagedorn Insulin *see* 5023
Neutral Red, 6573
Neutral Red Chloride *see* 6573
Neutral Spirits, 6574
Neutral Verdigris *see* 2614
Neutrapen [3M Pharma] *see* 7202
Neutraphylline [Dorsey] *see* 3526
Neutrexin [MedImmune] *see* 9893
Neutroflavine *see* 115
Neutrogin [Chugai] *see* 4573
Neutronyx 600's [Onyx] *see* 6764
Neutronyx 605 [Onyx] *see* 6850
Neutronyx 611 [Onyx] *see* 6764
Neutrormone *see* 6022
NeutroSpec [Palatin] *see* 9237
Neutrosteron *see* 6022
Neuvitan [Fujisawa] *see* 6849
Neuzyme [Eisai] *see* 5701
Nevanac [Alcon] *see* 6553
Nevax *see* 3446
Nevergor *see* 5634
Nevigramon [Chinoin] *see* 6444
Nevile and Winther's Acid *see* 6476
Nevin [Grünenthal] *see* 9056
Nevirapine, 6575
Nevite *see* 8731
Nevralgina [Climax] *see* 3390
Nevriton [RPR] *see* 53
Newberyite *see* 5745
Newtol [Mohan] *see* 10283
Newtonite *see* 356
New Victoria Green Extra I *see* 5763

New Victoria Green Extra II *see* 5763
New Victoria Green Extra O *see* 5763
Nexaband [Closure Medical] *see* 6855
Nexagan *see* 1442
Nexavar [Bayer] *see* 8846
Nexide [Cheminova] *see* 5398
Nexion [BASF] *see* 1442
Nexium [AstraZeneca] *see* 6939
Nexplanon [Organon] *see* 3928
Nexter [BASF] *see* 8078
Nexus *see* 1910
Nezeril [Draco] *see* 7065
NF 6 *see* 9647
NF-64 *see* 6626
NF-149 *see* 2753
NF-180 *see* 4330
NF-260 *see* 4323
NFBS *see* 4202
NFSI *see* 4202
NGF *see* 6562
β-NGF *see* 6562
NGF-2 *see* 6572
NH-502 *see* 4973
N-HATU *see* 4653
N-HBTU *see* 4654
NHDC *see* 6537
NHPI *see* 4876
NHS *see* 4884
NHSI *see* 4884
NI-25 *see* 48
Nia [Novo] *see* 5876
NIA-10242 *see* 1803
NIA-18739 *see* 1235
NIA-33297 *see* 7294
Niacin *see* 6610
Niacinamide *see* 6608
Niacor [Upsher-Smith] *see* 6610
Niagara 1240 *see* 3790
Niagara 4512 *see* 8829
Niagara 4556 *see* 3098
Niagara 4562 *see* 5333
Niagara 5006 *see* 3052
Niagara Blue *see* 9972
Niagestin [Novo] *see* 5876
Nialamide, 6576
Nialamid [Pfizer] *see* 6576
Niamidal *see* 6576
Niamine *see* 6627
Niaprazine, 6577
Niaproof 4 [Niacet] *see* 8817
Niaquitil *see* 6576
Niaspan [Kos] *see* 6610
Niazol *see* 6453
Nibiol *see* 6742
Nicabate [GSK] *see* 6609
Nicalex *see* 346
Nicamide *see* 6627
Nicangin [AstraZeneca] *see* 6610
Nicant [Piam] *see* 6580
Nicapress [Boehringer, Ing.] *see* 6580
Nicaragua Wood *see* 7295
Nicaraven, 6578
Nicarb [Merck & Co.] *see* 6579
Nicarbazin, 6579
Nicardal [Italfarmaco] *see* 6580
Nicardipine, 6580
Nicarpin [San Carlo] *see* 6580
Niccolite *see* 6583
Nicein *see* 5401
Nicelate [Toyo Jozo] *see* 6444
Nicergolent [Ramon] *see* 6581
Nicergoline, 6581
Niceritrol, 6582
Nicetile [Sigma-Tau] *see* 79
Nicholin [Wyeth] *see* 2318

Nicizina [Pharmacia] see 5232
Nickel, 6583
Nickel Acetate, 6584
Nickel Acetylacetonate, 6585
Nickel Ammonium Sulfate, 6586
Nickel Benzoate see 1093
Nickel Bromide, 6587
Nickel Carbonate Hydroxide, 6588
Nickel Carbonate Hydroxide (Ni$_3$(CO$_3$)-
 (OH)$_4$) see 6588
Nickel Carbonyl, 6589
(T-4)-Nickel Carbonyl (Ni(CO)$_4$) see
 6589
Nickel Chloride, 6590
Nickel Chloride (NiCl$_2$) see 6590
Nickel Cyanide, 6591
Nickel Cyanide (Ni(CN)$_2$) see 6591
Nickel Diacetate see 6584
Nickel Dibromide see 6587
Nickel Dichloride see 6590
Nickel Dicyanide see 6591
Nickel Difluoride see 6593
Nickel Diformate see 6594
Nickel Dihydroxide see 6595
Nickel Diiodide see 6596
Nickel Dimethylglyoxime, 6592
Nickel Fluoride, 6593
Nickel Formate, 6594
Nickel Hydroxide, 6595
Nickel Hydroxide (Ni(OH)$_2$) see 6595
Nickelic Oxide see 6601
Nickel Iodide, 6596
Nickel Iodide (NiI$_2$) see 6596
Nickel Monoxide, 6597
Nickel Nitrate, 6598
Nickelous Acetate see 6584
Nickelous Bromide see 6587
Nickelous Fluoride see 6593
Nickelous Formate see 6594
Nickelous Hydroxide see 6595
Nickelous Iodide see 6596
Nickelous Oxide see 6597
Nickelous Phosphate see 6600
Nickel Oxalate, 6599
Nickel(II) Oxide see 6597
Nickel Oxide (NiO) see 6597
Nickel Oxide (Ni$_2$O$_3$) see 6601
Nickel Phosphate, 6600
Nickel Potassium Cyanide see 7802
Nickel Protoxide see 6597
Nickel Sesquioxide, 6601
Nickel Sulfate, 6602
Nickel Tetracarbonyl see 6589
Niclocide [Miles] see 6604
Niclofolan, 6603
Niclosamide, 6604
Nicopatch [Fabre] see 6609
Nicobid [Aventis] see 6610
Nicobifen see 1354
Nicobion [Merck KGaA] see 6608
Nicodel [Mitsui] see 6580
Nicoderm CQ [GSK] see 6609
Nicolan [Elan] see 6609
Nicolar [Aventis] see 6610
Nicolen see 4330
Nicomorphine, 6605
Niconacid [Novartis] see 6610
Nicor see 6627
Nicorandil, 6606
Nicorette [GSK] see 6609
Nicorine see 6627
Nicorol [Lundbeck] see 4338
Nicoscorbine see 6608
Nicosulfuron, 6607
Nicotafuryl see 9550

Nicotergoline see 6581
Nicotinamide, 6608
Nicotinamide-adenine Dinucleotide see
 6429
Nicotinamide-adenine Dinucleotide
 Phosphate see 6432
Nicotinamide Ascorbate see 6608
Nicotinamide Chloromethylate see 9862
Nicotinamide Methyl Chloride see 9862
Nicotinamide Mononucleotide see 6750
Nicotine, 6609
Nicotinell [Novartis] see 6609
Nicotine Neutral Sulfate see 6609
Nicotine Tartrate see 6609
Nicotinic Acid, 6610
Nicotinic Acid Amide see 6608
Nicotinic Acid Benzyl Ester, 6611
Nicotinic Acid Diethylamide see 6627
Nicotinic Acid 2-Hydroxyethyl Ester 2-
 (p-Chlorophenoxy)-2-methylpropion-
 ate (Ester) see 3923
Nicotinic Acid N-Methylbetaine see 9863
Nicotinic Acid Methyl Ester see 6173
Nicotinic Acid Neopentanetetrayl Ester
 see 6582
Nicotinic Acid 1-Oxide see 7040
Nicotinic Acid β-Phenylisopropylamide
 see 7330
Nicotinic Acid Tetrahydrofurfuryl Ester
 see 9550
Nicotinic Alcohol see 6612
3-(Nicotinoyloxy)propyl p-Chlorophen-
 oxyisobutyrate see 8386
Nicotinoyl-β-phenylisopropylamine see
 7330
Nicotinyl Alcohol, 6612
Nicoumalone see 30
Nicoxin see 6579
Nicozid [Piam] see 5232
Nicrazin [Merck & Co.] see 6579
Nidanthel [Cooper] see 6683
Nidantin [Warner-Lambert] see 7044
Nide [Ibirn] see 6633
Nidol [Tosi] see 6633
Nidrafur see 6626
Nidran [Sankyo] see 6639
Nidrane [Phoenix] see 1016
Nidrel [Specia] see 6662
Nielsbohrium see 1324, 3509
Nifedicor [Alfa] see 6613
Nifedipine, 6613
Nifekalant, 6614
Nifelan [Elan] see 6613
Nifelat [TAD] see 6613
Nifensar [Aventis] see 6613
Niflan [Mitsubishi] see 7831
Niflumic Acid, 6615
Nifluril see 6615
Nifluril (Suppositories) [UPSA] see 6615
Nifos T [Monsanto] see 9349
NIF-T see 4574
Niftolid see 4238
Nifulidone see 4330
Nifuratel, 6616
Nifurfoline, 6617
Nifuroquine, 6618
Nifuroxazide, 6619
Nifuroxime, 6620
Nifurpirinol, 6621
Nifurtimox, 6622
Nifurtoinol, 6623
Nifurzide, 6624
Nifuzon see 6687
Nigakilactone D see 8142
Nigelline see 2807

Nigericin, 6625
Night-blooming Cereus see 1612
NIH-4185 see 9445
NIH-7440 see 287
NIH-7519 see 7332
NIH-7539 see 6796
NIH-7667 see 6773
NIH-8805 see 1501
Nihydrazone, 6626
Nikardin see 6627
Nikethamide, 6627
Nikoform see 4872
Nikos [Dow AgroSci.] see 4139
Nilandron [Aventis] see 6629
Nilatil [USV] see 5293
Nile Blue A Oxazone see 6628
Nile Red, 6628
Nilergex see 5273
Nilevar [Searle] see 6785
Nilodin see 5650
Niltuvin [EGYT] see 6612
Nilutamide, 6629
Nilvadipine, 6630
Nilverm [Mallinckrodt] see 5511
Nim see 6522
Nimbecetin see 5322
Nimbex [GSK] see 859
Nimbin, 6631
Nimbiol, 6632
Nimbisan [Solvay] see 1459
Nimergoline see 6581
Nimesulide, 6633
Nimetazepam, 6634
Nimicor [Formenti] see 6580
Nimidane, 6635
Nimodipine, 6636
Nimorazole, 6637
Nimotop [Bayer] see 6636
Nimotuzumab, 6638
Nimrod [Syngenta] see 1498
Nimustine, 6639
Ninhydrin, 6640
Ninopterin, 6641
Niobium, 6642
Niobium(V) Chloride see 6643
Niobium Chloride (NbCl$_5$) see 6643
(TB-5-11)-Niobium Fluoride (NbF$_5$) see
 6644
Niobium(V) Oxide see 6645
Niobium Oxide (Nb$_2$O$_5$) see 6645
Niobium Pentachloride, 6643
Niobium Pentafluoride, 6644
Niobium Pentoxide, 6645
Nioform see 5075
Nionate [Nion] see 4074
Niopam [Bracco] see 5100
Niotal [Sanofi-Aventis] see 10391
Nioxime, 6646
Nip [Nichay] see 7707
Nipagin A see 3890
Nipagin M see 6180
Nipantiox 1-F [Nipa] see 1549
Nipasol M [Nipa] see 7977
Nipaxon [Pharmacia] see 6807
Nipecotic Acid, 6647
Nipent [SuperGen] see 7248
Niperyt see 7222
Nipolept [Aventis] see 10397
Nippas Calcium [Tanabe] see 473
Nipradilol, 6648
Nipradolol see 6648
Nipride [Roche] see 8780
Nipruss [Schwarz] see 8780
NiQuitin CQ [GSK] see 6609
Niran see 2083

Niridazole, 6649
Nirvanil [SKB] see 10098
NIS see 5088
NISac see 5086
Nisaplin [Aplin & Barrett] see 6650
Nische [Kodama] see 216
Nisentil [Roche] see 306
Nisin, 6650
Nisin A see 6650
Nisin Z see 6650
Nisoldipine, 6651
Nisotin see 3791
Nissorun [Nippon Soda] see 4749
Nisulid [Wyeth] see 6633
Nisvastatin see 7621
Nitalapram see 2317
Nitarsone, 6652
Nitazoxanide, 6653
Nitenpyram, 6654
Niter see 7769
Nithiamide see 404
Nithiazide, 6655
Nithiocyamine see 571
Nitisinone, 6656
Nitobanil [Kyowa Yakuhin] see 9252
Nitogenin see 3323
Nitoman [Roche] see 9326
Niton see 8212
Nitorol [Eisai] see 5271
Nitossil [Zyma] see 2384
Nitracrine, 6657
Nitradisc [Pfizer] see 6694
Nitrados [APS] see 6661
Nitraldone see 4323
Nitramide, 6658
Nitramine, 6659
m-Nitraniline see 6668
o-Nitraniline see 6669
p-Nitraniline see 6670
Nitraphen see 6685
Nitrapyrin, 6660
2-Nitratoethylaminotoluene-p-sulfonate
 see 5293
(Nitrato-κO)phenylmercury see 7413
Nitrazepam, 6661
Nitrendipine, 6662
Nitric Acid, 6663
Nitric Acid Aluminum Salt (3:1) see 347
Nitric Acid Ammonium Salt (1:1) see
 531
Nitric Acid Barium Salt (2:1) see 979
Nitric Acid Beryllium Salt (2:1) see 1172
Nitric Acid Bismuth(3+) Salt (3:1) see
 1271
Nitric Acid Cadmium Salt (2:1) see 1627
Nitric Acid Calcium Salt (2:1) see 1684
Nitric Acid Cerium(3+) Salt (3:1) see
 1999
Nitric Acid Cesium Salt (1:1) see 2014
Nitric Acid Chromium(3+) Salt (3:1) see
 2230
Nitric Acid Cobalt(2+) Salt (2:1) see
 2428
Nitric Acid Copper(2+) Salt (2:1) see
 2633
Nitric Acid, Fuming see 6663
Nitric Acid Gallium Salt (3:1) see 4380
Nitric Acid Iron(3+) Salt (3:1) see 4054
Nitric Acid Lanthanum(3+) Salt (3:1)
 see 5414
Nitric Acid Lead(2+) Salt (2:1) see 5469
Nitric Acid Lithium Salt (1:1) see 5592
Nitric Acid Magnesium Salt (2:1) see
 5738
Nitric Acid Manganese(2+) Salt (2:1) see
 5798

Nitric Acid Mercury(1+) Salt (1:1) see
 5964
Nitric Acid Mercury(2+) Salt (2:1) see
 5947
Nitric Acid Methyl Ester see 6174
Nitric Acid 2-Methylpropyl Ester see
 5193
Nitric Acid Nickel(2+) Salt (2:1) see 6598
Nitric Acid Palladium(2+) Salt (2:1) see
 7093
Nitric Acid Potassium Salt (1:1) see 7769
Nitric Acid Propyl Ester see 7975
Nitric Acid Silver(1+) Salt (1:1) see 8657
Nitric Acid Silver(1+) Salt Mixt with Sil-
 ver Chloride (AgCl) see 8657
Nitric Acid Sodium Salt (1:1) see 8778
Nitric Acid Strontium Salt (2:1) see 8975
Nitric Acid Thallium(1+) Salt (1:1) see
 9414
Nitric Acid Thallium(3+) Salt (3:1) see
 9420
Nitric Acid Thorium(4+) Salt (4:1) see
 9530
Nitric Acid Zinc Salt (2:1) see 10343
Nitric Acid Zirconium(4+) Salt (4:1) see
 10379
Nitric Anhydride see 6692
Nitric Oxide, 6664
Nitric Oxide Synthase see 6664
5,5′-Nitrilodibarbituric Acid Monoam-
 monium Salt see 6391
Nitrilotriacetic Acid, 6665
2,2′,2″-Nitrilotrisethanol see 9830
Nitrimidazine see 6637
Nitrin, 6666
Nitritocobalamin see 4846
Co-(Nitrito-κO)-cobinamide Dihydrogen
 Phosphate (Ester) Inner Salt 3′-Ester
 with (5,6-Dimethyl-1-α-D-ribofurano-
 syl-1H-benzimidazole-κN³) see 4846
Nitroacetanilide, 6667
m-Nitroacetanilide see 6667
o-Nitroacetanilide see 6667
p-Nitroacetanilide see 6667
Nitro Acid Sulfite see 6735
3-Nitroalizarin see 246
4-Nitro-2-aminophenetole see 6704
1-Nitro-3-amino-4-phenyl Ethyl Ether
 see 6704
m-Nitroaniline, 6668
o-Nitroaniline, 6669
p-Nitroaniline, 6670
Nitroanisole, 6671
m-Nitroanisole see 6671
o-Nitroanisole see 6671
p-Nitroanisole see 6671
5-Nitrobarbituric Acid, 6672
Nitrobenzaldehyde, 6673
m-Nitrobenzaldehyde see 6673
o-Nitrobenzaldehyde see 6673
p-Nitrobenzaldehyde see 6673
2-Nitrobenzenamine see 6669
3-Nitrobenzenamine see 6668
4-Nitrobenzenamine see 6670
Nitrobenzene, 6674
4-Nitrobenzeneacetic Acid see 6709
4-Nitrobenzeneacetonitrile see 6678
p-Nitrobenzenearsonic Acid see 6652
p-Nitrobenzeneazochromotropic Acid
 Sodium Salt see 2244
p-Nitrobenzeneazosalicylic Acid see 247
4-Nitro-1,2-benzenediamine see 6710
3-Nitrobenzenepropenoic Acid see 6682
4-Nitrobenzenesulfonamide, 6675
p-Nitrobenzenesulfonic Acid Methyl
 Ester see 6175

4-Nitrobenzenesulfonic Acid Methyl
 Ester see 6175
Nitrobenzoic Acid, 6676
m-Nitrobenzoic Acid see 6676
o-Nitrobenzoic Acid see 6676
p-Nitrobenzoic Acid see 6676
4-Nitrobenzoic Acid Chloride see 6677
Nitrobenzol see 6674
p-Nitrobenzoyl Chloride see 6677
4-Nitrobenzoyl Chloride, 6677
4-Nitrobenzyl Cyanide see 6678
Nitro-Bid [Altana] see 6694
o-Nitrobiphenyl, 6679
p-Nitrobiphenyl, 6680
2-Nitro-1,1′-biphenyl see 6679
4-Nitrobiphenyl see 6680
4-Nitro-1,1′-biphenyl see 6680
9-Nitrocamptothecin see 8421
9-Nitro-(20S)-camptothecin see 8421
N-Nitrocarbamide see 6738
Nitrocarbol see 6698
Nitrocefin, 6681
Nitrocellulose see 8125
Nitrochlorobenzene see 2153
m-Nitrochlorobenzene see 2153
Nitrochloroform see 2159
7-Nitro-5-(2-chlorophenyl)-3H-1,4-ben-
 zodiazepin-2(1H)-one see 2379
Nitrocine [Schwarz] see 6694
3-Nitrocinnamic Acid, 6682
Nitrocobalamin see 4846
Nitrocontin [Mundipharma] see 6694
Nitrodan, 6683
Nitroderm [Novartis] see 6694
4-Nitro-1,2-diaminobenzene see 6710
o-Nitrodiphenyl see 6679
p-Nitrodiphenyl see 6680
Nitrodulcitol see 4362
Nitro-Dur [Schering-Plough] see 6694
Nitroerythrite see 3735
Nitroethane, 6684
5-Nitro-2-ethoxyaniline see 6704
Nitrofen, 6685
Nitrofene see 6685
Nitroform see 9901
Nitrofural see 6687
5-Nitro-2-furaldehyde Acetylhydrazone
 see 6626
5-Nitro-2-furaldehyde p-Hydroxybenzo-
 ylhydrazone see 6619
5-Nitro-2-furaldehyde Semicarbazone
 see 6687
anti-5-Nitro-2-furaldoxime see 6620
[C(Z)]-5-Nitro-2-furancarboxaldehyde
 Oxime see 6620
Nitrofurantoin, 6686
6-[2-(5-Nitro-2-furanyl)ethenyl]-2-pyri-
 dinemethanol see 6621
1-[[(5-Nitro-2-furanyl)methylene]amino]-
 2,4-imidazolidinedione see 6686
3-[[(5-Nitro-2-furanyl)methylene]amino]-
 2-oxazolidinone see 4330
2-[(5-Nitro-2-furanyl)methylene]hydra-
 zinecarboxamide see 6687
2-[3-(5-Nitro-2-furanyl)-1-[2-(5-nitro-2-
 furanyl)ethenyl]-2-propen-1-ylidene]-
 hydrazinecarboximidamide see 6741
4-(5-Nitro-2-furanyl)-2-quinolinecarbox-
 ylic Acid 1-Oxide see 6618
Nitrofurazone, 6687
1-(5-Nitro-2-furfurylidene)-2-acetylhy-
 drazine see 6626
1-(5-Nitro-2-furfurylideneamino)hydan-
 toin see 6686
N-(5-Nitro-2-furfurylidene)-1-aminohy-
 dantoin see 6686

5-Nitro-2-thiophenecarboxylic Acid 2-[3-(5-Nitro-2-furanyl)-2-propen-1-ylidene]hydrazide *see* 6624

5-Nitro-2-thiophenecarboxylic Acid [3-(5-Nitro-2-furyl)allylidene]hydrazide *see* 6624

Nitrotoluene, 6737

m-Nitrotoluene *see* 6737

o-Nitrotoluene *see* 6737

p-Nitrotoluene *see* 6737

2-Nitrotoluene *see* 6737

3-Nitrotoluene *see* 6737

4-Nitrotoluene *see* 6737

p-Nitro-α-toluic Acid *see* 6709

p-Nitro-α-tolunitrile *see* 6678

2-Nitro-3,4,6-trichlorophenol *see* 9807

2-[2-Nitro-4-(trifluoromethyl)benzoyl]-1,3-cyclohexanedione *see* 6656

4′-Nitro-3′-trifluoromethylisobutyranilide *see* 4238

4-Nitro-3-(trifluoromethyl)phenol *see* 9401

2-[Nitro(2,4,6-trinitrophenyl)amino]ethanol 1-Nitrate *see* 7252

Nitrourea, 6738

N-Nitrourea *see* 6738

Nitrous Acid, 6739

Nitrous Acid Barium Salt (2:1) *see* 980

Nitrous Acid Butyl Ester *see* 1585

Nitrous Acid *tert*-Butyl Ester *see* 1586

Nitrous Acid Calcium Salt (2:1) *see* 1685

Nitrous Acid 1,1-Dimethyl Ethyl Ester *see* 1586

Nitrous Acid Ethyl Ester *see* 3885

Nitrous Acid Isopropyl Ester *see* 5262

Nitrous Acid 3-Methylbutyl Ester *see* 5168

Nitrous Acid 1-Methylethyl Ester *see* 5262

Nitrous Acid 2-Methylpropyl Ester *see* 5194

Nitrous Acid Potassium Salt (1:1) *see* 7770

Nitrous Acid Propyl Ester *see* 7976

Nitrous Acid Silver(1+) Salt (1:1) *see* 8658

Nitrous Acid Sodium Salt (1:1) *see* 8779

Nitrous Anhydride *see* 3307

Nitrous Ether *see* 3885

Nitrous Oxide, 6740

Nitrovin, 6741

Nitroxanthic Acid *see* 7522

Nitroxoline, 6742

Nitroxy Fluoride *see* 4197

Nitroxyl Chloride *see* 6744

Nitroxylsulfuric Acid *see* 6735

Nitroxynil, 6743

Nitroxynil Eglumine *see* 6743

Nitroxynil Meglumine *see* 6743

Nitrozone [Century] *see* 6687

Nitryl Chloride, 6744

Nitryl Chloride ((NO₂)Cl) *see* 6744

Nitryl Fluoride, 6745

Nitryl Fluoride ((NO₂)F) *see* 6745

Nitryl Hypofluorite *see* 4197

Nivadil [Fujisawa] *see* 6630

Nivadipine *see* 6630

Nivaldipine *see* 6630

Nivalenol, 6746

Nivalin [Sopharma] *see* 4370

Nivaquine [Aventis] *see* 2165

Nivemycin [Boots] *see* 6539

Nix [Pharmygiène-Scat] *see* 7294

Nixyn [Teofarma] *see* 5238

Nizatidine, 6747

Nizax [Lilly] *see* 6747

Nizaxid [Norgine] *see* 6747

Nizin [Broemmel] *see* 9058

Nizofenone, 6748

Nizoral [Janssen-Cilag] *see* 5349

NK-19 *see* 7645

NK-104 *see* 7621

NK-421 *see* 10024

NK-631 *see* 7257

NK-1006 *see* 5329

NKK-105 *see* 5776

NKT-01 *see* 4617

[Nle⁴,D-Phe⁷]-α-MSH *see* 168

NM-441 *see* 8016

L-NMA *see* 6096

NMDA, 6749

NMF *see* 6147

NMMO *see* 6171

NMN, 6750

NMO *see* 6171

NMOR *see* 6727

NMP *see* 6190

NN-304 *see* 5026

NN-2211 *see* 5569

NN-29285 *see* 1502

NNC-05-0328 *see* 9572

NNC-11-0232 *see* 10254

NND-318 *see* 5408

NND-502 *see* 5655

NNF-136 *see* 4241

NNF-9425 *see* 4014

NNI-0001 *see* 4150

NNI-750 *see* 1502

NNI-850 *see* 4024

NNK-758 *see* 1502

No. 220 Sol *see* 5934

No. 356 *see* 2376

NO-05-0328 *see* 9572

NO-328 *see* 9572

NOA-446510 *see* 5785

Noah's Ark *see* 2767

Noan [Teofarma] *see* 2997

NOBA *see* 6720

Nobacter [Chantereau] *see* 9819

Nobelium, 6751

N-Oblivon *see* 5913

Nobrium [Roche] *see* 5857

Nocardamin, 6752

Nocardic Acids *see* 6410

Nocardicin A *see* 6753

Nocardicin B *see* 6753

Nocardicins, 6753

Nocardomycolic Acids *see* 6410

Nocertone [Labaz] *see* 7033

Nociceptin, 6754

Noctal [UCB] *see* 7910

Noctamid [Schering AG] *see* 5639

Noctan *see* 6212

Noctazepam [Hexal] *see* 7025

Noctec [BMS] *see* 2071

Noctivane *see* 4740

Noctivane Sodium *see* 4740

Noctone [GEA] *see* 8228

Noctosom *see* 4228

Nodakenetin Glucoside *see* 6755

Nodakenin, 6755

Nodapton *see* 4537

No-Doz [BMS] *see* 1639

Nodulisporic Acid, 6756

Nodulisporic Acid A *see* 6756

Noflo [Merck & Co.] *see* 6789

Noformicin, 6757

Noformycin *see* 6757

Nogalamycin, 6758

L-Nogalose *see* 6758

Nogram [Sanofi Winthrop] *see* 6444

Nolatrexed, 6759

Noleptan [Thomae] *see* 4255

Nolicin [Krka] *see* 6789

Nolipax [Salus] *see* 4009

Nolotil [Boehringer, Ing.] *see* 3390

Noludar [Roche] *see* 6212

Nolvadex [AstraZeneca] *see* 9180

Nolvasan [Fort Dodge] *see* 2092

Nometan *see* 7577

Nomifensine, 6760

Nomilin, 6761

Nominee [Kumiai] *see* 1302

Nomolt [BASF] *see* 9249

Nonactic Acid *see* 6762

Nonactin, 6762

Nonaethyleneglycol Monomethyl Ether *p-n*-Butylaminobenzoate *see* 1098

Nonahydropentaborane *see* 7213

(3*R*,4*S*,6*S*,8*S*,10*R*,12*R*,14*R*,15*R*,16*R*,17*E*,-19*E*,21*E*,23*E*,25*E*,27*S*,28*R*)-4,6,8,10,-12,14,15,16,27-Nonahydroxy-3-[(1*R*)-1-hydroxyhexyl]-17,28-dimethyloxacyclooctacosa-17,19,21,23,25-pentaen-2-one *see* 4319

Nonaisoprenol *see* 8830

Nonalol *see* 6766

(2*E*,6*E*,10*E*,14*E*,18*E*,22*E*,26*E*,30*E*,)-3,7,-11,15,19,23,27,31,35-Nonamethyl-2,6,-10,14,18,22,26,30,34-hexatriacontanonaen-1-ol *see* 8830

Nonanedioic Acid *see* 898

Nonanoic Acid *see* 7179

Nonanoic Acid Ethyl Ester *see* 3891

1-Nonanol *see* 6766

Nonanol Acetate *see* 6765

NO-naproxen *see* 6498

2-Nonenal, 6763

(2*E*)-2-Nonenal *see* 6763

trans-2-Nonenaldehyde *see* 6763

Nonipol NO [Sankyo] *see* 6764

Nonoic Acid *see* 7179

Nonoxinol *see* 6764

Nonoxynol, 6764

Nonoxynol-9 *see* 6764

Nonoxynol-11 *see* 6764

Nonulosaminic Acids *see* 8623

n-Nonyl Acetate, 6765

n-Nonyl Alcohol, 6766

Nonylcarbinol *see* 2858

2-Nonyldioxolane, 6767

2-Nonyl-1,3-dioxolane *see* 6767

n-Nonyl Ethanoate *see* 6765

Nonylic Acid *see* 7179

Nonyl Methyl Ketone *see* 6177

Nonyl Phenol, 6768

(*p*-Nonylphenoxy)acetic Acid, 6769

2-(4-Nonylphenoxy)acetic Acid *see* 6769

Nonylphenoxypolyethoxyethanol *see* 6764

α-(4-Nonylphenyl)-ω-hydroxypoly(oxy-1,2-ethanediyl) *see* 6764

Nonylphenyl Polyethyleneglycol Ether *see* 6764

Nootkatone, 6770

(+)-Nootkatone *see* 6770

Nootrop [UCB] *see* 7600

Nootropil [UCB] *see* 7600

Nootropyl [UCB] *see* 7600

Nopalcol [Henkel] *see* 7698

Nopar [Lilly] *see* 7277

Nopia [Daiichi] *see* 8384

Nopil [Mepha] *see* 9050

Nopinene *see* 7557

Nopol, 6771

Novarsan *see* 6532
Novarsenobenzol *see* 6532
Novarsenobillon *see* 6532
Novaruca [Bioglan] *see* 4508
Novasin [Danisco] *see* 6650
Novasmasol [Zambeletti] *see* 6002
Novastan [Tokyo Tanabe] *see* 769
Novastat W [Salsbury] *see* 9064
Novatec [Merck & Co.] *see* 5572
Novatrin [Ayerst] *see* 4768
Novatropine *see* 4768
Novazam [Génévrier] *see* 2997
Novedrin *see* 3764
Novel Erythropoiesis Stimulating Protein *see* 2825
Novemina [Lazar] *see* 3390
Noveril [Morishita] *see* 3012
Noverme [J & J] *see* 5837
Novesine [CIBA Vision] *see* 1049
Novex *see* 4031
Noviben [Merck & Co.] *see* 1728
Novidium *see* 4769
Novidroxin [Fatol] *see* 4846
Noviflumuron, 6810
Noviform [Dispersa] *see* 1203
Novil *see* 3390
Novismuth *see* 1284
Novobiocin, 6811
Novocain [Sanofi Winthrop] *see* 7875
Novocamid [Hoechst] *see* 7874
Novocef [Pliva] *see* 1956
Novodigal [Beiersdorf-Lilly] *see* 3186
Novodil [Augot] *see* 2697
Novodolan [HMR] *see* 4135
Novodrin *see* 5263
Novofluen [Engelhard KG] *see* 3710
Novogent [Temmler] *see* 4919
Novoldiamine, 6812
Novolin [Novo] *see* 5023
Novol Ketone *see* 6812
Novon *see* 2799
Novonal [Hoechst] *see* 10087
Novo-Nastizol A [Bago] *see* 846
Novonorm [Novo Nordisk] *see* 8256
Novophone *see* 2822
Novopulmon [Viatris] *see* 1476
Novorapid [Novo Nordisk] *see* 5025
Novoridazine [Novopharm] *see* 9512
Novorin [Polfa] *see* 10284
Novosparol [Merck KGaA] *see* 1209
Novotrone *see* 9098
Novotusil [Inpharzam] *see* 2384
Novrad [Lilly] *see* 5523
Noxaben [Unifa] *see* 3094
Noxafil [Schering-Plough] *see* 7723
Noxfire [AgrEvo] *see* 8403
Noxibiol *see* 6742
Noxigram [Firma] *see* 2309
Noxiurotan *see* 7576
Noxyflex-S [Geistlich] *see* 6813
Noxythiolin, 6813
Noxytiolin *see* 6813
Nozinan (Inj.) [Sanofi-Aventis] *see* 5518
Nozinan (Tabl.) [Sanofi-Aventis] *see* 5518
NP *see* 6705
NP-13 *see* 1029
NP-55 *see* 8612
NP-113 *see* 5019
NPA-3 *see* 6500
NPAB *see* 7820
NPe6 *see* 9166
NPH-1320 *see* 1451
NPH Insulin *see* 5023
NPI-0047 *see* 8474
NPI-0052 *see* 8474

NPI-2358 *see* 7654
Nplate [Amgen] *see* 8383
NPPB, 6814
NPT-10381 *see* 5019
NPY *see* 6570
NPYR *see* 6730
NRC-910 *see* 5121
NRDC-107 *see* 1235
NRDC-143 *see* 7294
NRDC-149 *see* 2764
NRDC-161 *see* 2888
NRP-104 *see* 5571
NS-2330 *see* 9322
NsA A *see* 6756
NSC-185 *see* 2721
NSC-739 *see* 468
NSC-1026 *see* 2727
NSC-1771 *see* 9525
NSC-2101 *see* 8406
NSC-3051 *see* 6147
NSC-3053 *see* 2797
NSC-3364 *see* 4110
NSC-3590 *see* 4249
NSC-5159 *see* 2044
NSC-5366 *see* 6807
NSC-5547 *see* 3522
NSC-8806 *see* 5896
NSC-9706 *see* 9836
NSC-10023 *see* 7842
NSC-11905 *see* 5417
NSC-12198 *see* 3497
NSC-13875 *see* 316
NSC-15200 *see* 4380
NSC-15780 *see* 592
NSC-17777 *see* 7432
NSC-19893 *see* 4213
NSC-20264 *see* 146
NSC-21626 *see* 7934
NSC-23436 *see* 3471
NSC-23759 *see* 9323
NSC-25154 *see* 7593
NSC-25614 *see* 7486
NSC-25855 *see* 9602
NSC-26805 *see* 3882
NSC-27640 *see* 4144
NSC-28693 *see* 6336
NSC-30152 *see* 3235
NSC-31712 *see* 4208
NSC-32946 *see* 6298
NSC-34462 *see* 10036
NSC-35051 *see* 5896
NSC-37095 *see* 10057
NSC-38297 *see* 1461
NSC-39084 *see* 895
NSC-39415 *see* 3464
NSC-39470 *see* 1182
NSC-40725 *see* 5270
NSC-43193 *see* 8921
NSC-45383 *see* 8957
NSC-45388 *see* 2795
NSC-47439 *see* 4223
NSC-47774 *see* 7594
NSC-49171 *see* 8477
NSC-52644 *see* 8127
NSC-52947 *see* 7083
NSC-59729 *see* 8863
NSC-60584 *see* 9113
NSC-64375 *see* 1122
NSC-68982 *see* 4593
NSC-69856 *see* 10367
NSC-70731 *see* 5555
NSC-70845 *see* 6758
NSC-71047 *see* 9907
NSC-75520 *see* 9855
NSC-77518 *see* 2997

NSC-77625 *see* 9760
NSC-78502 *see* 5849
NSC-79037 *see* 5621
NSC-82116 *see* 5347
NSC-83653 *see* 368
NSC-84223 *see* 4854
NSC-85998 *see* 8962
NSC-91523 *see* 7953
NSC-102816 *see* 884
NSC-104800 *see* 6299
NSC-105014-F *see* 2336
NSC-107429 *see* 2698
NSC-107430 *see* 7233
NSC-107433 *see* 9242
NSC-108166 *see* 9949
NSC-109229 *see* 826
NSC-109723 *see* 9947
NSC-109724 *see* 4937
NSC-112931 *see* 9397
NSC-114901 *see* 2921
NSC-115944 *see* 3635
NSC-118218 *see* 4157
NSC-118218-H *see* 4157
NSC-119875 *see* 2316
NSC-122819 *see* 9288
NSC-123127 *see* 3487
NSC-125066 *see* 1319
NSC-125973 *see* 7081
NSC-127716 *see* 2856
NSC-129943 *see* 8241
NSC-130044 *see* 9467
NSC-130181 *see* 9617
NSC-138780 *see* 9981
NSC-140865 *see* 10025
NSC-141540 *see* 3929
NSC-141549 *see* 589
NSC-145668 *see* 623
NSC-148958 *see* 9252
NSC-153858 *see* 5829
NSC-156303 *see* 589
NSC-164011 *see* 10395
NSC-165563 *see* 1463
NSC-169780 *see* 8241
NSC-178248 *see* 2183
NSC-180024 *see* 1870
NSC-181815 *see* 7276
NSC-182986 *see* 2999
NSC-192965 *see* 8885
NSC-208734 *see* 104
NSC-218321 *see* 7248
NSC-226080 *see* 8232
NSC-234714 *see* 714
NSC-239336 *see* 3640
NSC-241240 *see* 1823
NSC-245382 *see* 6639
NSC-245467 *see* 10184
NSC-246131 *see* 10101
NSC-249008 *see* 9893
NSC-249992 *see* 589
NSC-256439 *see* 4927
NSC-264137 *see* 3604
NSC-266046 *see* 7011
NSC-0270516 *see* 8227
NSC-279836 *see* 6302
NSC-286193 *see* 9580
NSC-296961 *see* 399
NSC-301467 *see* 3768
NSC-301739 *see* 6302
NSC-305263 *see* 4296
NSC-312887 *see* 4157
NSC-325319 *see* 3111
NSC-328002 *see* 4157
NSC-332598 *see* 8304
NSC-337766 *see* 1248
NSC-339140 *see* 5034

Name Index

Octyl Acetate, 6852
n-Octyl Bromide, 6853
sec-Octyl Bromide, 6854
[2-[2-(*p*-Octylcresoxy)ethoxy]ethyl]di-
  methylbenzylammonium Chloride *see*
  6097
Octyl Cyanoacrylate, 6855
n-Octyl α-Cyanoacrylate *see* 6855
Octyl 2-Cyanoacrylate *see* 6855
2-*n*-Octyl-4,5-dichloro-1-isothiazolin-3-
  one *see* 3080
Octyldodecanol, 6856
2-Octyl-1-dodecanol *see* 6856
Octylene *see* 1766
Octylene Glycol *see* 3797
Octyl-β-D-glucopyranoside *see* 6857
n-Octylglucoside *see* 6857
n-Octyl-β-D-glucoside, 6857
sec-Octyl Iodide, 6858
2-Octyl-4-isothiazolin-3-one *see* 6844
2-Octyl-3(2*H*)-isothiazolone *see* 6844
Octyl Methoxycinnamate, 6859
Octylonium Bromide *see* 6997
β-*p*-*tert*-Octylphenoxyethyldiethylbenzyl-
  ammonium Chloride *see* 7353
Octylphenoxy Polyethoxyethanol *see*
  6850
Octyl Salicylate, 6860
5-[2-(Octylsulfinyl)propyl]-1,3-benzodi-
  oxole *see* 9097
n-Octylsulfoxide of Isosafrole *see* 9097
Octyl Triazone *see* 3864
Ocucoat [Bausch & Lomb] *see* 4880
Ocufen [Allergan] *see* 4229
Ocuflox [Allergan] *see* 6863
Oculinum *see* 1362
Ocupress [Novartis] *see* 1866
Ocusert Pilo [Cusi] *see* 7536
Ocu-Tracin [Ocumed] *see* 927
Ocytocin *see* 7078
ODA-914 *see* 2844
Odanacatib, 6861
Odanon [Towa Yakuhin] *see* 1790
L-OddC *see* 9968
Odemase [Azupharma] *see* 4338
Odontalg [Giovanardi] *see* 5535
Odorigenin *see* 10083
ODPH *see* 3365
Odrik [Aventis] *see* 9729
Odylen [Winthrop] *see* 5988
Oedemex [Mepha] *see* 4338
Oekolp [Kade] *see* 3762
Oenanthal *see* 4693
Oenanthaldehyde *see* 4693
Oenanthic Acid *see* 4695
Oenanthic Ether *see* 3889
Oenanthol *see* 4693
Oenanthotoxin *see* 3624
Oenanthylic Acid *see* 4695
Oestrasid *see* 3115
Oestrin *see* 3763
Oestriol *see* 3762
Oestrodiene *see* 3115
Oestrone *see* 3763
Oestroperos *see* 3763
Oestroral *see* 3115
Ofatumumab, 6862
OFF! [SC Johnson] *see* 2859
Oflocet [Aventis] *see* 6863
Oflocin [GSK] *see* 6863
Ofloxacin, 6863
Ofloxacine *see* 6863
OFQ *see* 6754
Oftalar [Alcon] *see* 7831
Oftanol [Bayer] *see* 5218

OG *see* 6857
Ogast [Takeda] *see* 5413
Ogastoro [Takeda] *see* 5413
Ogen [Abbott] *see* 3763
Ogostal [Lilly] *see* 1759
OGT-918 *see* 6268
Ogyline [Roussel Diamant] *see* 6794
γ-OH *see* 4854
4-OHA *see* 4268
OHB₁₂ *see* 4846
1α-OH-CC *see* 4858
l-OHP *see* 7011
(+)-11-OH-Δ⁸-THC-DMH *see* 2946
Oil Anise, Japanese, 6864
Oil Bergamot, 6865
"Oil Garlic" *see* 288
Oil Green *see* 2238
Oil Nut *see* 5316
Oil of American Wormseed *see* 6877
Oil of Anise *see* 657
Oil of Anthemis *see* 2040
Oil of Balm, 6866
Oil of Basil, 6867
Oil of Bay, 6868
Oil of Bitter Almond, 6869
Oil of Bitter Orange, 6870
Oil of Cade *see* 5319
Oil of Cajeput, 6871
Oil of Camphor, Rectified, 6872
Oil of Canada Fleabane *see* 6886
Oil of Caraway *see* 1779
Oil of Cardamom *see* 1833
Oil of Cascarilla *see* 1881
Oil of Cashew Nut Shell, 6873
Oil of Cassia *see* 2300
Oil of Cedar Wood, 6874
Oil of Celery, 6875
Oil of Champaca, 6876
Oil of Chenopodium, 6877
Oil of Cherry Laurel, 6878
Oil of Chinese Cinnamon *see* 2300
Oil of Cinnamon, Ceylon *see* 2300
Oil of Cinnamon Leaf *see* 2300
Oil of Citronella, 6879
Oil of Clove *see* 2402
Oil of Copaiba *see* 2501
Oil of Cumin, 6880
Oil of Cypress, 6881
Oil of Dill, 6882
Oil of Dwarf Pine Needles, 6883
Oil of Egg Yolk *see* 3575
Oil of Erigeron *see* 6886
Oil of Eucalyptus *see* 3939
Oil of Fennel *see* 4008
Oil of Fir, 6884
Oil of Fir—Siberian, 6885
Oil of Fleabane, 6886
Oil of Geranium, 6887
Oil of Geranium—East Indian, 6888
Oil of Ginger *see* 4458
Oil of Grapes *see* 3889
Oil of Hedeoma *see* 7211
Oil of Hops *see* 4784
Oil of Hyssop, 6889
Oil of Jojoba *see* 5314
Oil of Juniper *see* 5318
Oil of Juniper Tar *see* 5319
Oil of Lavender *see* 5445
Oil of Lemon, 6890
Oil of Lemon Balm *see* 6866
Oil of Lemon Grass, 6891
Oil of Levant Wormseed, 6892
Oil of Linaloe, 6893
Oil of Marjoram, 6894
Oil of Melissa Balm *see* 6866

Oil of Mirbane *see* 6674
Oil of Monarda *see* 6329
Oil of Mountain Pine *see* 6883
Oil of Myrcia *see* 6868
Oil of Myrtle, 6895
Oil of Neem *see* 6522
Oil of Neroli *see* 6898
Oil of Niaouli, 6896
Oil of Niobe *see* 6098
Oil of Nutmeg *see* 6818
Oil of Orange, 6897
Oil of Orange Flowers, 6898
Oil of Origanum, 6899
Oil of Palma Christi *see* 1898
Oil of Patchouli, 6900
Oil of Pelargonium Geranium *see* 6887
Oil of Pennyroyal *see* 7211
Oil of Pepper *see* 1314
Oil of Peppermint *see* 7258
Oil of Pettigrain, 6901
Oil of Pimenta *see* 7544
Oil of Pine *see* 6885
Oil of Pine Needles, 6902
Oil of Pulegium *see* 7211
Oil of Rice Bran *see* 8332
Oil of Rose, 6903
Oil of Rose Geranium *see* 6887
Oil of Rosemary *see* 8395
Oil of Rue, 6904
Oil of Sassafras *see* 8518
Oil of Savin *see* 8523
Oil of Scotch Fir *see* 6902
Oil of Silver Fir *see* 6884
Oil of Silver Pine *see* 6884
Oil of Spearmint *see* 8865
Oil of Spike *see* 5445
Oil of Sweet Almond, 6905
Oil of Sweet Bay *see* 5437
Oil of Sweet Flag *see* 1640
Oil of Sweetwood Bark *see* 1881
Oil of Tansy, 6906
Oil of Thuja *see* 9543
Oil of Thyme, 6907
Oil of Turpentine *see* 10007
Oil of Valerian *see* 10089
Oil of Vetiver, 6908
Oil of Vitriol *see* 9104
Oil of White Cedar *see* 9543
Oil of Wild Marjoram *see* 6899
Oil of Wine *see* 3889
Oil of Wine, "Heavy", 6909
Oil of Wormwood *see* 10250
Oil of Yarrow *see* 97
Oil Palm, 6910
Oil Red XO *see* 10287
Oil Sweet Orange *see* 6897
Oil White Birch *see* 1242
Oizine *see* 8080
OK-174 *see* 1044
OK-1166 *see* 4129
Okacin [Novartis] *see* 5619
Okacyn [Novartis] *see* 5619
Okadaic Acid, 6911
Okenite *see* 1704
Oksafenamide *see* 6984
Oksilidin *see* 8199
OKT3 *see* 6393
Oktadin *see* 4595
Oktatenzin *see* 4595
OKY-046 *see* 7079
Olaflur, 6912
Olah's Reagent, 6913
Olamin [Sigamed] *see* 2306
Olane [Avi-Sun] *see* 7701
Olanzapine, 6914

Olaparib, 6915
Olaquindox, 6916
Olbemox [Pharmacia] see 102
Olbetam [Pharmacia] see 102
Olcadil [Sandoz] see 2404
Olcenon [Lederle] see 9657
Old Tuberculin see 9985
Old Yellow Enzyme, 6917
Olean [Procter & Gamble] see 9014
Oleandomycin, 6918
Oleandomycin 2″,4′,11-Triacetate see 9949
Oleandrin, 6919
(3β)-Olean-12-en-3-ol see 611
Oleanol see 6920
Oleanolic Acid, 6920
Oleate of Mercury see 5948
Olefiant Gas see 3845
Oleic Acid, 6921
Oleic Acid Ammonium Salt see 532
Oleic Acid Calcium Salt see 1686
Oleic Acid Copper(2+) Salt see 2634
Oleic Acid Potassium Salt see 7771
Olein see 9904
Oleocanthal, 6922
Oleocreosote see 2561
Oleo Stock see 9173
Oleovitamin A see 10211
Oleovitamin D$_2$ see 10215
Oleovitamin D$_3$ see 10216
Oleptan [Bender] see 4255
Oleptro [Labopharm] see 9740
Olestra see 9014
Oleth see 7697
Oleum see 9104
Oleum Abietis see 6885
Oleum Andropogonis Muricati see 6908
Oleum Rusci see 1242
Oleum Vitis Viniferae see 3889
Oleuropein, 6923
Oleyl Alcohol, 6924
Oleyl Hydroxyethyl Imidazoline, 6925
Olibanum, 6926
Olicard [Solvay] see 5271
Oligomycin A see 6927
Oligomycin B see 6927
Oligomycin C see 6927
Oligomycin D see 6927
Oligomycins, 6927
Oliprevin [BMS] see 7835
Olivacine, 6928
Olivanic Acids, 6929
Olive Oil, 6930
Oliver [Sanzen] see 6983
Olive Spurge see 6249
Olivil, 6931
Olivin see 6932
Olivomycin A see 6932
Olivomycins, 6932
Olivoretin D see 9260
Olizumab see 6936
Olmelin see 1231
Olmesartan, 6933
Olmesartan Medoxomil see 6933
Olmetec [Sankyo] see 6933
Olmifon [Lafon] see 159
Ololiuqui see 5116
Olopatadine, 6934
Olprinone see 5632
Olsalazine, 6935
Olux [Connetics] see 2358
Olymp [DuPont] see 4236
Olympus [Bayer CropSci.] see 7951
Olynth [Pfizer] see 10284
OM-518 see 5918

OM-805 see 769
OM-853 see 10182
Omacetaxine Mepesuccinate see 4776
Omacor [Pronova] see 6938
Omadine [Olin] see 8107
Omadine Disulfide [Olin] see 8107
Omal see 9809
Omalizumab, 6936
Omapatrilat, 6937
Omapro [ChemGenex] see 4776
Omca [BMS] see 4220
Omcilon [BMS] see 9757
OMD see 6128
OM-dopa see 6128
OMDS see 8107
Omebeta [Betapharm] see 6939
Omega-3 Acid Ethyl Esters, 6938
Omega-3 Polyunsaturated Fatty Acids Ethyl Esters see 6938
OmeLich [Winthrop] see 6939
Omelind [Lindopharm] see 6939
Omep [Hexal] see 6939
Omepral [AstraZeneca] see 6939
Omeprazen [Malesci] see 6939
Omeprazole, 6939
Ome-Puren [Alpharma] see 6939
Omeril [Bayer] see 5839
Omiganan, 6940
Omigard [Cadence] see 6940
Omite [Uniroyal] see 7922
Omnaris [Nycomed] see 2266
Omnes [Fumouze] see 6616
Omnibex [Bago] see 7374
Omnic [Astellas] see 9181
Omnicain [Daiichi] see 7875
Omnicef [Abbott] see 1921
Omniflox [Abbott] see 9274
Omnipaque [GE Healthcare] see 5095
Omnipen [Wyeth] see 583
Omniscan [GE Healthcare] see 4354
Omnitarg see 7298
Omnopon [Roche] see 7121
Omoconazole, 6941
OMPA see 8532
Omperan [Taiho] see 9119
OMS-29 see 1788
OMS-45 see 4007
OMS-75 see 6442
OMS-115 see 3436
OMS-658 see 1442
OMS-659 see 1442
OMS-968 see 4273
OMS-1696 see 4842
OMS-1804 see 3161
OMS-2004 see 7886
OMS-3002 see 3922
OMS-3023 see 4037
OMS-3048 see 9725
Onaka [Max Pharma] see 7531
Onapristone, 6942
ONB see 6679
Onbrez Breezhaler [Novartis] see 4970
Oncaspar [Enzon] see 826
Onclast [Merck & Co.] see 223
Oncol [Otsuka] see 1044
Onconase [Alfacell] see 8230
OncoScint [Cytogen] see 8521
1-25-Oncostatin M (Human Precursor) Fusion Protein with CTLA-4 (Antigen)(human) Fusion Protein with Immunoglobulin G1 (Human Heavy Chain Fragment), Bimol. (146 → 146′)-Disulfide see 4
Oncovin [Genus] see 10183
Ondansetron, 6943

Ondena [Bayer] see 2834
Ondero [Boehringer, Ing.] see 5549
Ondogyne [Roussel Diamant] see 2713
One-Alpha [Leo Pharm] see 4858
Onglyza [BMS] see 8525
Onicit [Helsinn] see 7098
Onion Oil, 6944
Oniria [Lifepharma] see 8145
Onkotrone [Baxter] see 6302
ONO-802 see 4421
ONO-1078 see 7830
ONO-1101 see 5405
ONO-1206 see 5543
ONO-1308 see 6970
ONO-2235 see 3660
ONO-5046 see 8696
ONO-5920 see 6284
ONO-RS-411 see 7830
Onoact [Ono] see 5405
Onokrein P [Ono] see 5327
Onon [Ono] see 7830
Ononin see 4271
Onsenal [Pfizer] see 1958
Onsukil [Otsuka] see 7877
Ontak [Ligand] see 2895
Ontosein [Diagnostic Data] see 9131
Onychomal [Hermal] see 10052
Onyxide see 3844
Oöcyan see 1223
Oolong Tea see 9225
Ooporphyrin see 8006
Oosporein, 6945
OP-1 see 1333
OP-1206 see 5543
Opacin see 5081
Opacist E.R. [Bracco] see 5055
Opalène [Théraplix] see 9892
Opalgyne [Innothéra] see 1125
Opalmon [Ono] see 5543
Opalwax see 1899
Opana [Endo] see 7068
Oparenol see 5083
Opatanol [Alcon] see 6934
Opaxio [Cell Therapeutics] see 7081
OPC-31 see 776
OPC-1085 see 1866
OPC-2009 see 7877
OPC-7251 see 6430
OPC-8212 see 10166
OPC-12759 see 8242
OPC-13013 see 2280
OPC-14597 see 776
OPC-17116 see 4580
OPC-41061 see 9699
Opcon [Bausch & Lomb] see 6453
Opebacan see 931
Opegan [Santen] see 4796
Operidine [Janssen] see 7362
Ophena [QuatRx] see 6991
Ophiobolin D see 1976
Ophtacalm [Chauvin] see 2581
Ophtagram [Chauvin] see 4427
Ophthaine [BMS] see 7921
Ophthalamin (obsolete) see 10211
Ophthalgan [Wyeth-Ayerst] see 4520
Ophthalmadine [SAS] see 4931
Ophthetic [Allergan] see 7921
Ophtorenin [Winzer] see 1500
Ophtosol [Winzer] see 1398
Opian see 6807
Opianic Acid, 6946
Opianine see 6807
Opianyl see 5854
Opidol [Mundipharma] see 4840
Opilon [Hansam] see 6378

Opiniazide, 6947
Opino [Bayropharma] see 6821
Opipramol, 6948
Opiran [Cassenne] see 7546
Opium, 6949
Opium Active Principles see 7121
Opium Alkaloids Hydrochlorides see
  7121
Opium Tincture see 6949
Opobalsam see 943
Opocarbyl see 1808
Oporia [Pfizer] see 5430
Oprelvekin see 5044
Opridan [Locatelli] see 1443
Opromazine, 6950
Opsins, 6951
α_2-Opsonins see 4104
OPT-80 see 4108
Optaflexx [Elanco] see 8208
Optal see 7955
Optanox [Byk Gulden] see 10190
Optenyl [Stroschein] see 7122
Op-Thal-Zin [Alcon] see 10359
Optica DP [Marks] see 3088
Opticrom [Sanofi-Aventis] see 2581
Optidase [Sanofi Winthrop] see 1900
Optiflex [Lilly] see 8842
Optilast [Viatris] see 899
OptiMARK [Mallinckrodt] see 4359
Optimax [Merck KGaA] see 9977
Optimine [Schering] see 894
Optimmune [Schering-Plough] see 2750
Optimycin [Novartis] see 6015
Optinate [Procter & Gamble] see 8360
Option [Bayer CropSci.] see 4261
Optipect [Thiemann] see 2914
Optipranolol [Bausch & Lomb] see 6216
Optiray [Mallinckrodt] see 5109
Optison [GE Healthcare] see 7273
Optocain [Bayer] see 5926
Optovit [Hermes] see 9653
Optovit-A [Hermes] see 10211
Optrin [Pharmacyclics] see 6368
Opturem [Kade] see 4919
Opus [BASF] see 3685
Opyrin [Taisho] see 4163
OR-611 see 3648
OR-1259 see 5525
Orabet [Berlin-Chemie] see 9667
Orabilex [Fougera] see 1490
Orabilix see 1490
Orabolin [Organon] see 3860
Oracef [Lilly] see 1974
Oracéfal [BMS] see 1915
Oracilline [Schwarz] see 7209
Oracon (obsolete) [Mead Johnson] see
  3240
Oradexon [Organon] see 2945
Oradol [Novartis] see 3464
Oragest [Pharmagyn] see 5864
Orageston [Organon] see 282
Oragrafin Calcium [BMS] see 5115
Oragrafin Sodium [BMS] see 5115
Oraldene [Pfizer] see 4739
Oramorph [Boehringer, Ing.] see 6361
Oranabol [Farmitalia] see 7064
Orange I, 6952
Orange II, 6953
Orange III see 6178
Orange IV see 9955
Orange B, 6954
Orange Crush [Cyanamid] see 1248
Orange GS see 9955
Orange N see 9955
Orange Peel, 6955

Orange Root see 4808
Oranixon [Organon] see 5917
Orap [Janssen] see 7546
Orapem [Replidyne] see 3976
Orapred [BioMarin] see 7841
Oraprim [A.T.I.] see 9050
Orarsan [Boots] see 51
Oraseptic [Warner-Lambert] see 4739
Orasone [Solvay] see 7842
Oraspor [Alfresa Pharma] see 1945
Orasthin [Aventis] see 7078
Oratane [Douglas] see 5274
Orathecin [SuperGen] see 8421
Oratrast [RPR] see 990
Oratrol [Alcon] see 3087
Oravir [SKB] see 3970
Oravue [BMS] see 5106
Oraxim [Malesci] see 1956
Orazamide, 6956
Orbax [Schering-Plough] see 6957
Orbenin [GSK] see 2403
Orbicin [Pfizer/Mack, Illert.] see 3008
Orbifloxacin, 6957
Orbinamon [Pfizer] see 9519
Orbit [BASF] see 2294
Orbit [Syngenta] see 7932
Orcanette see 248
Orcein, 6958
Orchil see 2605
Orchisterone [Negroni] see 9324
Orcin see 6959
Orcinol, 6959
Orcinolcarboxylic Acid see 6977
Orciprenaline see 6002
Orcl [Nippon Shinyaku] see 124
Ordeal Bark see 8519
Ordeal Bean see 7495
Ordiflazine see 5537
Ordimel [Lilly] see 61
Ordinary Lactic Acid see 5384
Ordinary Tartaric Acid see 9205
Oregon Grape Root see 1157
Oregovomab, 6960
Orelox [Sanofi-Aventis] see 1942
Orencia [BMS] see 4
Oreoselone Methyl Ether see 7306
Orestralyn [McNeil] see 3788
Oretic [Abbott] see 4819
Oreton [Schering-Plough] see 9324
Oreton Methyl [Schering] see 6197
Orexins, 6961
ORF-10131 see 6792
ORF-11676 see 6445
ORF-15244 see 9558
ORF-15817 see 3563
ORF-15927 see 8359
ORF-17070 see 4760
ORF-20485 see 9294
ORF-22164 see 856
Orfadin [Swedish Orphan] see 6656
Orfiril [Desitin] see 10099
Org-2969 see 2928
Org-3236 see 3928
Org-3770 see 6291
Org-5222 see 821
Org-6216 see 8352
Org-9426 see 8375
Org-9487 see 8231
Org-10172 see 2810
Org-25969 see 9019
Org-31540 see 4259
Org-32489 see 4250
Org-34006 see 4932
Org-36286 see 2518
Org-37462 see 4395

Org-GB-94 see 6251
Org-NA-97 see 7109
Org-NC-45 see 10132
Org-OD-14 see 9582
Orgabolin (obsolete) [Organon] see 3860
Orgaboral see 3860
Orgadrone [Han Wha] see 2945
Orgafol [Organon] see 4250
Orgalutran [Schering-Plough] see 4395
Orgametil see 5690
Orgametril [Organon] see 5690
Orgaran [Organon] see 2810
Orgasteron [Organon] see 6799
Orgastyptin [Organon] see 3762
Orgatrax [Organon] see 4889
Orgotein 9131
Oricillin [Grünenthal] see 7931
Oriconazole see 5292
Oricur see 2105
Oriental Berry see 2441
Orientomycin see 2749
Orimeten [Novartis] see 437
Orimon see 7364
Orinase [Pharmacia & Upjohn] see 9667
Oripavine, 6962
Oritavancin, 6963
Orix [Biomedica] see 6613
Orizaba Jalap Root see 5116
γ-Orizanol see 6983
Orkanet see 248
ORLAAM [BioDevelopment] see 5519
Orlipastat see 6964
Orlistat, 6964
Orlon [DuPont], 6965
Ormeloxifene see 1972
Ormetein (Rescinded) see 9131
Ormosanine see 6966
21-Ormosanin-20-yl Panamine see 6966
Ormosinine, 6966
Ornicetil [Logeais; Nordmark] see 5350
Ornidal [Selvi] see 6967
Ornidazole, 6967
Ornidyl [Aventis] see 3570
Ornipressin, 6968
L-Ornithine see 6969
Ornithine, 6969
L-Ornithine-L-aspartate see 6969
L(+)-Ornithine α-Ketoglutarate see 5350
Ornithine-vasopressin see 6968
8-L-Ornithinevasopressin see 6968
Ornoprostil, 6970
Orn[8]-vasopressin see 6968
Orobronze [Applipharm] see 1756
Orofungin [SK & F] see 5914
Oroidin, 6971
Oroken [Sanofi-Aventis] see 1925
Oropur see 6972
Orospray [Vétoquinol] see 2198
Orotic Acid, 6972
Orotic Acid Compd with 5(or 4)-Amino-
  imidazole-4(or 5)-carboxamide (1:1)
  see 6956
Orotidine, 6973
Orotric [SAT] see 9001
Oroturic see 6972
Orotyl see 6972
Oroxine [GSK] see 9570
Oroxylin, 6974
Oroxylin A, 6974
ORP-178 see 5388
Orphanin FQ see 6754
Orphanin FQ (Swine) see 6754
Orphenadrine, 6975
Orphol [Sanofi] see 3710
Orpiment see 797

Orplatna [GPC] *see* 8520
Orris, 6976
Orris Root Oil *see* 6976
Orsanil [Orion] *see* 9512
Orseilles *see* 2605
o-Orsellinic Acid, 6977
Orsellinic Acid Methyl Ester 4-Methyl
   Ether *see* 8861
Orsin *see* 7397
Orstanorm [Novartis] *see* 3191
Ortazol *see* 4330
Orthanilic Acid, 6978
Orthene [Valent] *see* 32
Ortho 9006 *see* 6019
Ortho 12420 *see* 32
Ortho Cyclen [Ortho] *see* 6792
Ortho Evra [Ortho] *see* 6783
Ortho Spotless [Ortho] *see* 3294
Ortho Tri-Cyclen [Ortho] *see* 6792
Orthoarsenic Acid *see* 785
[Orthoborato(3−)-*κO*]phenylmercurate-
   (2−) Hydrogen (1:2) *see* 7414
Orthoboric Acid *see* 1339
Orthocaine, 6979
Ortho-Cept [Ortho] *see* 2928
Orthocide-406 *see* 1774
Orthoclase *see* 7725
Orthoclone OKT3 [J & J] *see* 6393
Ortho-Creme [Ortho] *see* 6764
Orthodiazine *see* 8080
Orthodichlorobenzene *see* 3065
Orthoform [Winthrop] *see* 6979
Orthoformic Acid, 6980
Orthoform New *see* 6979
Ortho-Gynest [Cilag] *see* 3762
Ortho-Klor *see* 2083
Ortho-Novin 1/50 [Ortho-Cilag] *see* 6786
Ortho-Novum [Ortho] *see* 6786
Ortho-Novum 1/50 [Ortho] *see* 6786
Orthophosphoric Acid *see* 7456
Orthophthalic Acid Didiethylamide *see*
   9348
[*μ*12-[Orthosilicato(4−)-*κO*:*κO*:*κO*:
   κO′:*κO′*:*κO′*:*κO″*:*κO″*:*κO″*:*κO‴*:
   κO‴]]tetracosa-*μ*-oxododecaoxodode-
   catungstate(4−) Hydrogen (1:4) *see*
   8641
Orthosulfamuron, 6981
Orthotelluric Acid *see* 9262
Orthotitanic Acid *see* 9625
Orthoxenol *see* 7415
Orthoxine [Roberts] *see* 6071
Ortrel [Ortho] *see* 6792
Orudis [RPR] *see* 5352
Oruvail [RPR] *see* 5352
Orvagil [Galenika] *see* 6236
Oryvita [Ijaku] *see* 6983
Oryzaal [Sankei] *see* 6983
Oryzalin, 6982
γ-Oryzanol, 6983
Oryzanol A *see* 6983
Oryzanol C *see* 6983
OS-1897 *see* 3025
OS-2046 *see* 6244
Osalmid, 6984
Osaterone, 6985
Os-Cal [GSK] *see* 1659
Oscine *see* 8547
Oseltamivir, 6986
OSI-774 *see* 3730
Osiren [Probiomed] *see* 6939
Osmarins *see* 6988
Osmaron B, 6987
Osmic Acid *see* 6990
Osmitrol [Baxter] *see* 5810

Osmium, 6988
Osmium Ammonium Chloride *see* 533
(*OC*-6-11)-Osmium Fluoride (OsF₆) *see*
   6989
Osmium Hexafluoride, 6989
Osmium Octafluoride *see* 6989
(*T*-4)-Osmium Oxide (OsO₄) *see* 6990
Osmium Potassium Chloride *see* 7756
Osmium Tetroxide, 6990
Osmosal *see* 5810
Osnervan [Glaxo Wellcome] *see* 7881
Osonide [Ranbaxy] *see* 2266
Ospamox [Novartis] *see* 574
Ospemifene, 6991
Ospen [Biochemie] *see* 7209
Ospolot [Bayer] *see* 9122
Osprey [Bayer CropSci.] *see* 5981
Ossalin [Chemipharm] *see* 8754
Ossein *see* 2464
Osseor [Servier] *see* 8224
Ossin [Sulzbach-Neuweiler] *see* 8754
Ossiten [Boehringer, Mann.] *see* 2365
Ossopan *see* 3514
Ostac [Boehringer, Mann.] *see* 2365
Ostelin [Roemmers] *see* 10215
Osten [Takeda] *see* 5118
Osteocalcin, 6992
Osteo-F [Colgate-Hoyt] *see* 8754
Osteofix [Chiesi] *see* 5118
Osteofluor [Merck-Clévenot] *see* 8754
Osteogenic Protein-1 *see* 1333
Osteol [Yamanouchi] *see* 4411
Osteolite [DuPont] *see* 5863
Osteoscan-HDP [Mallinckrodt] *see* 7039
Ostepam [Nordic] *see* 7104
Osthole, 6993
Osto-D2 [Triton] *see* 10215
Ostreogrycin A *see* 10200
Ostreogrycin B *see* 7871
Ostruthin, 6994
Ostruthol, 6995
Osvan [Takeda] *see* 1061
Osyritin *see* 8438
Osyritrin *see* 8438
Osyrol [Aventis] *see* 8887
Otamixaban, 6996
Otavite *see* 1621
OTC *see* 7049
OTCA *see* 7049
Oterben [Chinoin] *see* 9667
Otifuril *see* 4323
Otilonium Bromide, 6997
Otobain, 6998
Otobite *see* 6998
Otokalixin *see* 5329
Otrasel [Cephalon] *see* 8565
Otreon *see* 1942
Otrinol [Novartis] *see* 8024
Otrivine [Novartis] *see* 10284
Otto of Rose *see* 6903
OU-1308 *see* 6970
Ouabagenin, 6999
Ouabain, 7000
Ourari *see* 2661
Oust [DuPont] *see* 9087
Outlook [BASF] *see* 3236
O-V Statin [BMS] *see* 6825
Ovaban [Schering] *see* 5876
Ovahormon [Teikoku Zoki] *see* 3758
Ovahormon Depot [Teikoku] *see* 3758
Ovalbumin, 7001
Ovalyse [Pharmacia & Upjohn] *see* 4088
Ovanon [Nourypharma] *see* 5690
Ovaras [Serono] *see* 3909
OvaRex [AltaRex] *see* 6960

Ovastol *see* 5987
Ovazyme *see* 4496
Ovcon [Warner-Chilcott] *see* 6786
Overal [Lusofarmaco] *see* 8409
Ovesterin *see* 3762
Ovestin [Organon] *see* 3762
Ovide [Taro] *see* 5764
Ovidrel [Serono] *see* 2221
Ovifollin *see* 3763
Oviol [Nourypharma] *see* 2928
Ovisot *see* 82
Ovitelmin [Cilag-Chemie] *see* 5837
Ovitrelle [Serono] *see* 2221
Ovoester *see* 845
Ovoresta [Organon] *see* 5690
Ovotransferrin *see* 9732
Ovo-Vinces [Wolff] *see* 3762
Ovral [Wyeth] *see* 6793
Ovran [Wyeth] *see* 6793
Ovranette [Wyeth] *see* 6793
Ovrette [Wyeth] *see* 6793
Ovulen *see* 3909
Ovuplant [Wyeth Animal Health] *see*
   2924
Ovysmen [Janssen-Cilag] *see* 6786
7-Oxabicyclo[2.2.1]heptane-2,3-dicarbox-
   ylic Acid *see* 3630
22-Oxacalcitriol *see* 5828
Oxacarbazepine *see* 7028
Oxaceprol, 7002
2-Oxachlormadinone Acetate *see* 6985
Oxacillin, 7003
Oxacyclohexadecan-2-one *see* 3952
Oxadiargyl, 7004
1,3,4-Oxadiazole, 7005
Oxadiazon, 7006
22-Oxa-1*α*,25-dihydroxyvitamin D₃ *see*
   5828
Oxadixyl, 7007
6-Oxa-1,1,2,3,3,8-hexamethyl-2,3,5,6,7,8-
   hexahydro-1*H*-benz[*f*]-indene *see*
   4751
Oxaine M [Wyeth] *see* 7032
Oxalacetic Acid, 7008
Oxalaldehyde *see* 4544
Oxalamic Acid *see* 7016
Oxalamide *see* 7017
Oxalato (1*R*,2*R*-Cyclohexanediamine)-
   platinum(II) *see* 7011
Oxalato (*trans-l*-1,2-Diaminocyclohex-
   ane)platinum(II) *see* 7011
Oxalatoplatin *see* 7011
Oxalatoplatinum *see* 7011
Oxalenediuramidoxime, 7009
Oxalic Acid, 7010
Oxalic Acid Bismuth Salt *see* 1272
Oxalic Acid Cerium(3+) Salt (3:2) *see*
   2000
Oxalic Acid Chloride *see* 7013
Oxalic Acid Diamide *see* 7017
Oxalic Acid Diethyl Ester *see* 3142
Oxalic Acid Dimethyl Ester *see* 6179
Oxalic Acid Dinitrile *see* 2681
Oxalic Acid Dipotassium Salt *see* 7772
Oxalic Acid Disodium Salt *see* 8781
Oxalic Acid Hemipotassium Salt *see*
   7806
Oxalic Acid Monopotassium Salt *see*
   7732
Oxalic Acid Nickel(2+) Salt (1:1) *see*
   6599
Oxalic Acid Tin(2+) Salt (1:1) *see* 8913
Oxalid [USV] *see* 7071
Oxaliplatin, 7011
Oxaloacetic Acid *see* 7008

Oxaloacetic Ester *see* 3141
Oxalomolybdic Acid, 7012
Oxaloyl Chloride *see* 7013
[Oxalylbis(iminoethylene)]bis[(*o*-chloro-benzyl)diethylammonium Chloride] *see* 373
Oxalyl Chloride, 7013
Oxalylurea *see* 7126
Oxamarin, 7014
Oxametacine, 7015
Oxamic Acid, 7016
Oxamic Acid Hydrazide *see* 8584
Oxamide, 7017
Oxamidic Acid *see* 7016
Oxaminozoline *see* 8347
Oxammonium Hydrochloride *see* 4867
Oxammonium Sulfate *see* 4867
Oxamniquine, 7018
Oxamycin [Merck & Co.] *see* 2749
Oxamyl, 7019
Oxandrolone, 7020
Oxantel, 7021
Oxantel Embonate *see* 7021
22-Oxa-1,25(OH)$_2$D$_3$ *see* 5828
Oxaphenamide *see* 6984
Oxaphor *see* 4856
Oxapium Iodide *see* 2729
Oxapropanium Iodide, 7022
Oxaprozin, 7023
Oxarmin *see* 7043
Oxarol [Chugai] *see* 5828
Oxatomide, 7024
Oxatone [Baxter] *see* 3200
Oxazacort *see* 2865
Oxazepam, 7025
Oxazocilline *see* 7003
Oxazolam, 7026
Oxazolazepam *see* 7026
Ox Bile Extract, 7027
Oxcarbazepine, 7028
Oxedix [Labaz] *see* 7033
Oxedrine *see* 9142
Oxeladin, 7029
Oxendolone, 7030
Oxenin, 7031
Oxepinac *see* 5284
Oxetacaine *see* 7032
Oxetane *see* 9886
2-Oxetanone *see* 7934
Oxethazaine, 7032
Oxethazine *see* 7032
Oxetorone, 7033
Oxeze [AstraZeneca] *see* 4272
Oxfendazole, 7034
Oxiamine [Made] *see* 5018
Oxibendazole, 7035
Oxibuprokain *see* 1049
Oxibutinina *see* 7054
Oxichlorochine *see* 4857
Oxiconazole, 7036
Oxidimethiin, 7037
Oxidized Cellulose, 7038
Oxidized Glutathione *see* 4511
Oxidronate Sodium *see* 7039
Oxidronic Acid, 7039
Oxilan [Cook Imaging] *see* 5111
Oxilapine *see* 5645
Oxilin [Allergan] *see* 7065
2-[4-(3-Oximinocyclohexyl)phenyl]pro-pionic Acid *see* 10275
Oxine *see* 4881
Oxiniacic Acid, 7040
Oxinofen *see* 7056
Oxinorm [Zambeletti] *see* 9131
Oxipendyl *see* 7069

Oxiracetam, 7041
Oxirane *see* 3856
2-Oxiranemethanol *see* 4525
Oxistat [GSK] *see* 7036
Oxitol *see* 3803
Oxitriptan *see* 4886
Oxitropium Bromide, 7042
Oxivent [Boehringer, Ing.] *see* 7042
2-Oxoacetic Acid *see* 4546
3-(3-Oxo-7α-acetylthio-17β-hydroxy-4-androsten-17α-yl)propionic Acid γ-Lactone *see* 8887
2-Oxo-acid Carboxylase *see* 8134
1-(2-Oxo-4-amino-1,2-dihydropyrimidin-1-yl)-2-deoxy-2,2-difluororibose *see* 4420
α-Oxobenzeneacetaldehyde Aldoxime *see* 5237
β-Oxobenzenepropanoic Acid Ethyl Ester *see* 3822
2-Oxo-2*H*-1-benzopyran-3-carboxylic Acid *see* 2551
4-Oxo-4*H*-1-benzopyran-2-carboxylic Acid *see* 2241
γ-Oxo-[1,1'-biphenyl]-4-butanoic Acid *see* 3996
2-Oxo-3,3-bis[*p*-aminophenyl]butane *see* 578
d-2-Oxo-3-bornanecarboxylic Acid *see* 1733
(+)-2-Oxo-10-bornanesulfonic Acid *see* 1736
2-Oxobutane *see* 6143
2-Oxobutanedioic Acid *see* 7008
2-Oxobutanedioic Acid 1,4-Diethyl Ester *see* 3141
3-Oxobutanoic Acid *see* 58
3-Oxobutanoic Acid Ethyl Ester *see* 3812
3-Oxobutanoic Acid Methyl Ester *see* 6083
3-Oxobutanoic Acid 1-Methylethyl Ester *see* 5252
N-(1-Oxobutyl)adenosine Cyclic 3',5'-(Hydrogen Phosphate) 2'-Butanoate *see* 1473
4-(3-Oxobutyl)-1,2-diphenyl-3,5-pyrazoli-dinedione *see* 5336
δ-Oxo-α-butylene *see* 6210
d-2-Oxo-3-camphanecarboxylic Acid *see* 1733
4-Oxo-β-carotene *see* 3542
4-Oxo-4*H*-chromene-2-carboxylic Acid *see* 2241
N-(4-Oxo-2,5-cyclohexadien-1-ylidene)-acetamide *see* 56
p-(3-Oxocyclohexyl)hydratropic Acid Oxime *see* 10275
(±)-*p*-[(2-Oxocyclopentyl)methyl]hydra-tropic Acid *see* 5647
(2*E*)-9-Oxo-2-decenoic Acid *see* 8148
7-Oxo-7-deoxy-28,29-didehydronystatin *see* 1744
3-Oxodesogestrel *see* 3928
9-Oxo-11α,15α-dihydroxy-17*S*,20-di-methylprosta-*trans*-2,*trans*-13-dienoic Acid *see* 5543
6-Oxo-17*S*,20-dimethyl-PGE$_1$ Methyl Ester *see* 6970
1-Oxo-2-(2,6-dioxopiperidin-3-yl)-4-ami-noisoindoline *see* 5490
Oxodiperoxymolybdenum(pyridine)-(hexamethylphosphoric Triamide) *see* 10134
16-Oxoeburnane *see* 3534

Oxoethanoic Acid *see* 4546
1-Oxo-2-[*p*-[(α-ethyl)carboxymethyl]-phenyl]isoindoline *see* 5001
2-Oxoglutaric Acid *see* 5350
3-Oxoglutaric Acid *see* 67
3-Oxo-L-gulofuranolactone *see* 819
2-Oxo-L-gulonic Acid *see* 5351
Oxogulonic Acid Diacetonide *see* 3220
2-Oxohexamethylenimine *see* 1763
μ-Oxohexaphenyldiphosphorus(2+) 1,-1,1-Trifluoromethanesulfonic Acid (1:2) *see* 4685
N-[(3*E*)-1-Oxo-3-hexen-1-yl]-somatoli-berin (Human Pancreatic Islet) *see* 9320
D-*erythro*-3-Oxohexonic Acid Lactone *see* 5171
1-(5-Oxohexyl)-3,7-dimethylxanthine *see* 7249
1-(5'-Oxohexyl)-3-methyl-7-propylxan-thine *see* 7927
17-[(1-Oxohexyl)oxy]-19-norpregn-4-ene-3,20-dione *see* 4448
1-(5-Oxohexyl)theobromine *see* 7249
17α-Oxo-D-homo-1,4-androstadiene-3,17-dione *see* 9323
3-Oxo-17β-hydroxy-1,4-androstadiene *see* 1327
3-(3-Oxo-17β-hydroxy-4,6-androstadien-17α-yl)propionic Acid γ-Lactone *see* 1753
3-Oxo-17β-hydroxyandrostane *see* 8920
5-Oxoimino-25-cyclohexyl Avermectin B$_1$ Monosaccharide *see* 8561
N,N'-[(3-Oxo-1(3*H*)-isobenzofuranyli-dene)bis[(6-hydroxy-5-methyl-3,1-phenylene)methylene]]bis[*N*-(carboxy-methyl)glycine] *see* 2566
2-Oxo-3-isobutyl-9,10-dimethoxy-1,2,3,-4,6,7-hexahydro-11b*H*-benzo[*a*]quinol-izine *see* 9326
(±)-2-[*p*-(1-Oxo-2-isoindolinyl)phenyl]-butyric Acid *see* 5001
Oxolamine, 7043
Oxole *see* 4325
Oxolinic Acid, 7044
Oxomalonic Acid *see* 5983
Oxomemazine, 7045
Oxomethane *see* 4263
γ-Oxo-4-methoxy-1-naphthalenebutyric Acid *see* 5902
Oxonic Acid, 7046
(2*R*,3*S*)-3-[(4*E*,7*E*)-1-Oxo-4,7-nonadien-1-yl]-2-oxiranecarboxamide *see* 2003
28-Oxooligomycin A *see* 6927
1-[2-Oxo-2-[[2-[(1-oxododecyl)oxy]ethyl]-amino]ethyl]pyridinium Chloride (1:1) *see* 5421
2-Oxopentanedioic Acid *see* 5350
2-Oxo-1,5-pentanedioic Acid *see* 5350
3-Oxopentanedioic Acid *see* 67
4-Oxopentanoic Acid *see* 5526
4-Oxopentanoic Acid Calcium Salt (2:1) *see* 1681
4-Oxopentanoic Acid Ethyl Ester *see* 3874
(1*S*,2*S*)-3-Oxo-2-(2'-*cis*-pentenyl)cyclo-pentan-1-acetate *see* 5308
(1*R*,2*R*)-3-Oxo-2-(2*Z*)-2-penten-1-ylcyclo-pentaneacetic Acid *see* 5308
3-Oxo-2-pentylcyclopentaneacetic Acid Methyl Ester *see* 6124
N-(1-Oxopentyl)-*N*-[[2'-(2*H*-tetrazol-5-yl)[1,1'-biphenyl]-4-yl]methyl]-L-valine *see* 10102

Oxophenarsine, 7047
4-Oxo-4-(phenylamino)butanoic Acid *see* 8998
(2Z)-4-Oxo-4-(phenylamino)-2-butenoic Acid *see* 5766
Oxophenylarsine, 7048
3-Oxo-*N*-phenylbutanamide *see* 57
2-Oxo-2-[(1-phenylethyl)amino]acetic Acid Hydrazide *see* 7400
5-Oxo-L-proline *see* 8113
5-Oxo-L-prolyl-L-alanyl-L-α-aspartyl-L-prolyl-L-asparaginyl-L-lysyl-L-phenylalanyl-L-tyrosylglycyl-L-leucyl-L-methioninamide *see* 7493
5-Oxo-L-prolyl-L-histidyl-L-prolinamide *see* 9750
5-Oxo-L-prolyl-L-histidyl-L-tryptophyl-L-seryl-L-tyrosyl-3-(2-naphthyl)-D-alanyl-L-leucyl-L-arginyl-L-prolylglycinamide *see* 6435
(*R*)-3-[(*S*)-5-Oxoprolyl]-4-thiazolidinecarboxylic Acid *see* 7531
2-Oxopropanal *see* 6153
2-Oxopropanal 1-Oxime *see* 5236
2-Oxopropanedioic Acid *see* 5983
2-Oxopropanoic Acid *see* 8135
2-Oxopropanol *see* 64
3-(2-Oxopropyl)indole *see* 5008
(2*S*)-2-[(4*R*)-2-Oxo-4-propylpyrrolidinyl]butanamide *see* 1384
2-Oxo-2*H*-pyran-5-carboxylic Acid *see* 2545
4-Oxo-1,4-pyran-2,6-dicarboxylic Acid *see* 2051
4-Oxo-4*H*-pyran-2,6-dicarboxylic Acid *see* 2051
N-[1-Oxo-4-(1-pyrenyl)butyl]-L-phenylalanine *see* 8063
2-Oxopyrrolidine *see* 8130
2-Oxo-1-pyrrolidineacetamide *see* 7600
2-Oxo-1-pyrrolidineaceto-2′,6′-xylidide *see* 6524
5-Oxo-2-pyrrolidinecarboxylic Acid *see* 8113
2-Oxo-1-pyrrolidinylacetic Acid 2,6-Dimethylanilide *see* 6524
2(*S*)-(2-Oxopyrrolidin-1-yl)butyramide *see* 5513
(4*R*)-3-[[(2*S*)-5-Oxo-2-pyrrolidinyl]carbonyl]-4-thiazolidinecarboxylic Acid *see* 7531
2-(2-Oxo-1-pyrrolidinyl)-*N*-(2,6-dimethylphenyl)acetamide *see* 6524
1-(2-Oxo-1-pyrrolidinyl)-4-(1-pyrrolidinyl)-2-butyne *see* 7050
Oxo(salicylato)bismuth *see* 1285
Oxosilylene *see* 8635
2-Oxosparteine *see* 5667
Oxosuccinic Acid *see* 7008
1-Oxo-3-(3-sulfamyl-4-chlorophenyl)-3-hydroxyisoindoline *see* 2200
Oxo[sulfato(2−)-κ*O*,κ*O*′]titanium *see* 9631
Oxo[sulfato(2−)-κ*O*]vanadium *see* 10113
5-Oxo-4-[(4-sulfo-1-naphthyl)azo]-1-(*p*-sulfophenyl)-2-pyrazoline-3-carboxylic Acid 3-Ethyl Ester Disodium Salt *see* 6954
(1*R*,2*S*,5*R*)-7-Oxo-6-sulfoxy-1,6-diazabicyclo[3.2.1]octane-2-carboxamide *see* 878
29ᴮ-[*N*⁶-(1-Oxotetradecyl)-L-lysine](1ᴬ-21ᴬ),(1ᴮ-29ᴮ)-insulin (Human) *see* 5026

2-[[2-Oxo-2-[(tetrahydro-2-oxo-3-thienyl)amino]ethyl]thio]acetic Acid *see* 3704
2-Oxo-1,2,3,4-tetrahydroquinoline *see* 4817
DL-*S*-[2-[*N*-3-(2-Oxotetrahydrothienyl)-acetamido]]thioglycolic Acid *see* 3704
N-[4-Oxo-2-(2*H*-tetrazol-5-yl)-4*H*-1-benzopyran-8-yl]-4-(4-phenylbutoxy)benzamide *see* 7830
2-Oxo-2-[[3-(2*H*-tetrazol-5-yl)phenyl]amino]acetic Acid Butyl Ester *see* 9217
L-2-Oxo-4-thiazolidinecarboxylic Acid, 7049
(4*R*)-2-Oxo-4-thiazolidinecarboxylic Acid *see* 7049
4-Oxo-2-thiazolidinehexanoic Acid *see* 6407
P-(2-Oxo-3-thiazolidinyl)phosphonothioic Acid *O*-Ethyl *S*-(1-Methylpropyl) Ester *see* 4288
4-Oxo-4-[[4-[(2-thiazolylamino)sulfonyl]phenyl]amino]butanoic Acid *see* 9008
N-[1-Oxo-2-[(2-thienylcarbonyl)thio]propyl]glycine *see* 8936
4-Oxo-2-thionothiazolidine *see* 8308
9-Oxothioxanthene *see* 9523
Oxotremorine, 7050
2-[4-Oxo-3-[5-(trifluoromethyl)benzothiazol-2-ylmethyl]-3,4-dihydrophthalazin-1-yl]acetic Acid *see* 10394
(2*R*)-4-Oxo-4-[3-(trifluoromethyl)-5,6-dihydro[1,2,4]triazolo[4,3-*a*]pyrazin-7(8*H*)-yl]-1-(2,4,5-trifluorophenyl)butan-2-amine *see* 8690
(2*S*)-1-[2-Oxo-2-(3,4,5-trimethoxyphenyl)acetyl]-2-piperidinecarboxylic Acid 4-(3-Pyridinyl)-1-[3-(3-pyridinyl)propyl]butyl Ester *see* 1244
2-Oxo-4,4,7a-trimethyl-2,4,5,6,6,7a-hexahydrobenzofuran *see* 3188
4-Oxovaleric Acid *see* 5526
22-Oxovincaleukoblastine *see* 10183
9-Oxoxanthene *see* 10260
Oxpentifylline *see* 7249
Oxphylline [Noventis] *see* 3924
Oxprenolol, 7051
Oxsol 100 [Occidental Chemical] *see* 2127
Oxsoralen [Valeant] *see* 6059
Oxtriphylline *see* 2215
Oxucide *see* 7576
Oxurasin *see* 7577
Oxy-5 [SKB] *see* 1119
Oxyacanthine, 7052
OxyBan *see* 4496
Oxybenzene *see* 7354
Oxybenzone, 7053
Oxybenzopyridine *see* 4881
1,1′-Oxybisbenzene *see* 3357
4,4′-Oxybisbenzenecarboximidamide *see* 7322
1,1′-Oxybisbutane *see* 1572
4,4′-Oxybis-2-butanol *see* 7059
1,1′-Oxybis[2-chloroethane] *see* 3075
Oxybis[chloromethane] *see* 3079
1,1′-Oxybis[1-chloromethane] *see* 3079
5,5′-Oxybis[3,4-dichloro-2(5*H*)-furanone] *see* 6384
1,1′-Oxybisethane *see* 3861
2,2′-Oxybisethanol *see* 3131
1,1′-Oxybisethene *see* 3427
1,1′-Oxybis[2-ethoxyethane] *see* 3132
1,1′-Oxybismethane *see* 3266
1,1′-Oxybis[2-methoxyethane] *see* 3184

1,1′-Oxybis[3-methylbutane] *see* 5163
1,1′-[Oxybis(methylene)]bis[benzene] *see* 1135
1,1′-[Oxybis(methylene)]bis[4-(hydroxyimino)methyl]pyridinium Chloride (1:2) *see* 6827
3,3′-[Oxybis(methylenecarbonylimino)]-bis[2,4,6-triiodobenzoic Acid] *see* 5094
1,1′-Oxybis[2-methylpropane] *see* 5187
2,2′-Oxybis[2-methylpropane] *see* 3040
3,3′-[Oxybis[(1-oxo-2,1-ethanediyl)imino]]bis[2,4,6-triiodobenzoic Acid] *see* 5094
1,1′-Oxybis[2,3,4,5,6-pentabromobenzene] *see* 2848
1,1′-Oxybispentane *see* 602
10,10′-Oxybis-10*H*-phenoxarsine *see* 7060
1,1′-Oxybispropane *see* 3385
2,2′-Oxybispropane *see* 5258
3,3′-Oxybis-1-propene *see* 283
1,1′-Oxybis[2,2,2-trifluoroethane] *see* 4231
Oxybuprocaine *see* 1049
Oxybutynin, 7054
Oxybutynin Chloride *see* 7054
Oxycamphor *see* 4856
Oxycel [Parke-Davis] *see* 7038
Oxychelidonic Acid *see* 5853
Oxychinolin *see* 4881
Oxychloroquine *see* 4857
Oxychlorosene, 7055
5-Oxychlorpromazine *see* 6950
Oxycinchophen, 7056
Oxyclozanide, 7057
Oxycodone, 7058
OxyContin [Purdue] *see* 7058
Oxydapatit *see* 1696
N,*N*-(Oxydiacetyl)bis[3-amino-2,4,6-triiodobenzoic Acid] *see* 5094
Oxydiazepam *see* 9275
4,4′-Oxydibenzamidine *see* 7322
4,4′-Oxydi-2-butanol, 7059
2,2′-Oxydiethanol *see* 3131
Oxydifluoride *see* 4196
Oxydiformic Acid Diethyl Ester *see* 3146
1,1′-(Oxydimethylene)bis[4-formylpyridinium]dichloride Dioxime *see* 6827
Oxydimethylquinizine *see* 703
10,10′-Oxydiphenoxarsine, 7060
2,4:2′,4′-*O*-(Oxydistibylidyne)bis[D-gluconic Acid] *Sb*,*Sb*′-Dioxide Trisodium Salt Nonahydrate *see* 692
Oxydol *see* 4835
Oxyephedrin *see* 4863
Oxyethylated Tertiary Octylphenol Formaldehyde Polymer *see* 10014
Oxyethyltheophylline *see* 3924
Oxyfan [Coli] *see* 4886
Oxyfedrine, 7061
Oxyfluorfen, 7062
Oxyfly [Novartis] *see* 5398
Oxygen, 7063
Oxygen Dichloride *see* 2099
Oxygen Fluoride (OF₂) *see* 4196
Oxygen Fluoride ((O₂)F₂) *see* 4195
Oxygent [Alliance] *see* 7270
3-Oxygermylpropionic Acid Polymer *see* 7909
Oxygeron [Will-Pharma] *see* 10180
Oxygesic [Mundipharma] *see* 7058
Oxyhemerythrin *see* 4676
Oxyhemocyanin *see* 4682
Oxyhemoglobin *see* 4683
Oxyjavanicin *see* 4344

Oxylidine *see* 8199
Oxylupanine *see* 4868
Oxymesterone, 7064
Oxymestrone *see* 7064
Oxymetazoline, 7065
Oxymetholone, 7066
Oxymethurea, 7067
Oxymethylene *see* 4263
Oxymorphone, 7068
Oxy-Mycin [AgriPharm] *see* 7075
Oxyneurine *see* 1181
Oxynorm [Mundipharma] *see* 7058
Oxypaat [Katwijk] *see* 7577
Oxypangam [Sanorania] *see* 3216
Oxypendyl, 7069
Oxypertine, 7070
Oxyphedrine *see* 7061
Oxyphenbutazone, 7075
Oxyphencyclimine, 7072
(+)-Oxyphene *see* 2062
Oxyphenisatin Acetate, 7073
Oxyphenonium Bromide, 7074
Oxypropylated Cellulose *see* 4879
Oxyquinoline *see* 4881
Oxysept [Allergan] *see* 4835
Oxytetracycline, 7075
Oxytetrin [Schering-Plough] *see* 7075
Oxythiamine, 7076
Oxythioquinox, 7077
Oxytocin, 7078
Oxytrimethylline *see* 2215
Oxyzin (Tabl.) *see* 7577
OYE *see* 6917
OZ *see* 6983
γ-OZ *see* 6983
Ozagrel, 7079
Ozex [Toyama] *see* 9714
Ozidia [Pfizer] *see* 4477
Ozolinone *see* 3934
Ozone, 7080
P$_{450}$ *see* 2789
P-4 *see* 7945
P-7 *see* 5441
p28^{sis} *see* 7639
P-30 Protein *see* 8230
P-50 *see* 583
P53 *see* 9395
P57 *see* 4782
P57AS3 *see* 4782
P-071 *see* 2021
P-113 *see* 8509
P-165 *see* 892
P-170 *see* 7310
P-253 *see* 3370
P-280 *see* 9235
P-286 *see* 5110
P-350 *see* 2268
P-607 *see* 2192
P-638 *see* 8054
P-652 *see* 4257
P-725 *see* 7265
P829 *see* 2912
P-1011 *see* 3094
P-1134 *see* 7549
P-1202 *see* 6090
P-1206 *see* 4731
P-1496 *see* 10319
P-1779 *see* 314
P-2647 *see* 1122
P-3693A *see* 3483
P-4000 *see* 6716
P-4599 *see* 3483
P-5604 *see* 5642
P-7138 *see* 6621
P-720549 *see* 5284

PA-93 *see* 6811
PA-94 *see* 2749
PA-105 *see* 6918
PA-155A *see* 5007
PA-248 *see* 7931
PA-457 *see* 1196
PA-457N *see* 1196
PA-103001 *see* 1196
PA2024 *see* 8685
PABA *see* 418
Pabal [Ferring] *see* 1796
Pabanol [Valeant] *see* 418
PABG *see* 421
Pacerone [Upsher-Smith] *see* 478
Paceum [Orion] *see* 2997
Pacinol [Schering] *see* 4220
Pacitane [Wyeth] *see* 9864
Pacitron [Rorer] *see* 9977
Paclitaxel, 7081
Paclitaxel Poliglumex *see* 7081
Paclobutrazol, 7082
Pactamycin, 7083
Padan [Takeda] *see* 1865
PADGEM *see* 8563
Padimate A *see* 3254
Padimate O *see* 3254
Padisal [Bayer] *see* 9458
Padreatin *see* 5327
Padrin [Fujisawa] *see* 7861
Padukrein [Bayropharm] *see* 5327
Padutin [Winthrop] *see* 5327
Paederine *see* 7171
ψ-Paederine *see* 7171
Paediathrocin [Abbott] *see* 3739
PAF *see* 7638
PAF-acether *see* 7638
Pafnol [Yoshitomi] *see* 5288
Pagano-Cor [Helopharm] *see* 3765
Paginol [Jugoremedija] *see* 7191
Pagoclone, 7084
PAH *see* 439
Paidomal [Malesci] *see* 9436
Paindor [Novartis Agro] *see* 2768
Paint White *see* 1284
Paka Oil *see* 5704
Palafer [GSK] *see* 4073
Palaprin [Nicholas] *see* 305
Palatinol A [Advanced Solvents] *see* 7483
Palatinol M *see* 3279
Palatinose [Südzucker] *see* 5228
Palatone *see* 5777
Palau'amine, 7085
Palavale [Otsuka] *see* 3549
Pale Catechu *see* 4389
Pale Gentian *see* 4428
Palexia [Grünenthal] *see* 9189
Palfium *see* 2958
Palifermin, 7086
Palinum *see* 2704
Paliperidone, 7087
Palisade [Syngenta] *see* 9897
Palitantin, 7088
Paliuroside *see* 8438
Palivizumab, 7089
Palladia [Pfizer] *see* 9650
Palladium, 7090
Palladium(II) Acetate *see* 7092
Palladium Chloride, 7091
Palladium Chloride (PdCl$_2$) *see* 7091
Palladium Diacetate, 7092
Palladium Hydroxide On Carbon *see* 7166
Palladium Hydroxide (Pd(OH)$_2$) *see* 7166
Palladium Monoxide *see* 7094

Palladium Nitrate, 7093
Palladium Oxide, 7094
Palladium-tetrakis(triphenylphosphine) *see* 9367
Palladon [Mundipharma] *see* 4840
Palladone [Napp] *see* 4840
Palladous Chloride *see* 7091
Palladous Nitrate *see* 7093
Palladous Oxide *see* 7094
Pallidin [Merck KGaA] *see* 9065
Palmarosa Oil *see* 6888
Palmatine, 7095
Palmidrol, 7096
Palmitic Acid, 7097
Palmitic Acid Aluminum Salt *see* 351
Palmitic Acid Ammonium Salt *see* 535
Palmitic Acid Calcium Salt *see* 1689
Palmitic Acid Hexadecyl Ester *see* 2030
Palmitic Acid Palmityl Ester *see* 2030
Palmitic Triglyceride *see* 9908
Palmitin *see* 9908
N-Palmitoylsphingomyelin *see* 8873
N-Palmitoyl-D-*erythro*-sphingosylphos-phorylcholine *see* 8873
Palmityl Alcohol *see* 2027
Palmityl Palmitate *see* 2030
Palm Kernel Oil *see* 6910
Palmofen [Zambon] *see* 4281
Palm Oil *see* 6910
Palm Olein *see* 6910
Palm Stearin *see* 6910
Palohex *see* 5022
Palonosetron, 7098
Palosein [Diagnostic Data] *see* 9131
Paludrine [AstraZeneca] *see* 2091
Palustric Acid, 7099
Paluther [RPR] *see* 805
Palux [Taisho] *see* 7987
Palytoxin, 7100
Palytoxin (C51-55 Hemiacetal) *see* 7100
L-PAM *see* 5896
2-PAM *see* 7822
2-PAM Chloride *see* 7822
Pamabrom, 7101
Pamaquine, 7102
Pamaquine Embonate *see* 7102
Pamaquine Naphthoate *see* 7102
Pameion [Simes] *see* 7122
Pamelor [Mallinckrodt] *see* 6804
Pamicogrel, 7103
Pamidronate Disodium *see* 7104
Pamidronic Acid, 7104
Pamifos [Vipharm] *see* 7104
Pamine [Bradley] *see* 6077
Pamisol [Mayne Pharma] *see* 7104
Pamisyl Sodium [Parke-Davis] *see* 473
Pamiteplase, 7105
Pamoic Acid, 7106
PAN, 7107
Panacef [Valeas] *see* 1914
Panacid [Dainippon] *see* 7618
Panacur [Intervet] *see* 3994
Panadol [GSK] *see* 47
Panaldine [Daiichi] *see* 9585
Panama Bark *see* 8156
Panamax [Sanofi-Aventis] *see* 47
Panamicin [Gramon] *see* 3008
Panamine *see* 6966
Panaquilins *see* 4462
Panatus [Krka] *see* 1516
Panax *see* 4462
Panaxosides *see* 4462
Panazon [Toyama] *see* 6741
Pancho [Nippon Soda] *see* 2753
Pancoral [Eisai] *see* 4006

Pancrease *see* 7108
Pancreas Powder *see* 7108
Pancreatic Basic Trypsin Inhibitor *see* 746
Pancreatic Desoxyribonuclease *see* 2907
Pancreatic Dornase *see* 2907
Pancreatic Extract, 7108
Pancreatic Trypsin Inhibitor (Kunitz) *see* 746
Pancreatin *see* 7108
Pancreozymin *see* 2205
Pancreozymin *C*-Terminal Octapeptide *see* 8681
Pancrex-V [Paines & Byrne] *see* 7108
Pancuronium Bromide, 7109
Pandermite *see* 1655
Panectyl [RPR] *see* 9873
Panergon [Mack, Illert.] *see* 7122
Panflavin [Chinosolfabrik] *see* 115
Panformin [Shionogi] *see* 1481
Panfungol [Janssen] *see* 5349
Pangamic Acid, 7110
Pangram [Virbac] *see* 4427
Panheprin [Abbott] *see* 4688
Panimit [Nattermann] *see* 1500
Panimycin [Meiji] *see* 3008
Panipenem, 7111
Panitumumab, 7112
Panmycin [Pharmacia & Upjohn] *see* 9341
Panobinostat, 7113
Panodil [GSK] *see* 47
Panoral [Lilly] *see* 1914
Panoxolin *see* 7027
PanOxyl [Stiefel] *see* 1119
Panpurol [Nippon Shinyaku] *see* 7583
Panretin [Ligand] *see* 241
Panrone [Panray] *see* 9442
Pansporin [Takeda] *see* 1936
Pansporin T [Takeda] *see* 1936
Pantecta [Pfizer] *see* 7117
Pantelmin [J & J] *see* 5837
Pantenyl [Kay] *see* 2949
Pantestone [Organon] *see* 9324
Pantetheine, 7114
Pantethine, 7115
Pantetina [Maggioni] *see* 7115
Panthecin [Sawai] *see* 7115
Panthenol *see* 2949
Pantherine *see* 6397
Panthoderm [USV] *see* 2949
Pantholin [Lilly] *see* 7118
Pantocaine [Hoechst] *see* 9333
Pantocid [Abbott] *see* 4625
Pantogam *see* 4783
Pantoic Acid γ-Lactone *see* 7116
Pantoic Lactone *see* 7116
Pantolactone, 7116
Pantomicina [Abbott] *see* 3743
Pantomin [Daiichi] *see* 7115
Pantopan [Pharmacia] *see* 7117
Pantopaque [Lafayette] *see* 5103
Pantopon [Roche] *see* 7121
Pantoprazole, 7117
Pantorc [Altana] *see* 7117
Pantosin [Daiichi] *see* 7115
Pantostrep *see* 3197
Pantothenic Acid, 7118
Pantothenol *see* 2949
Pantothenyl Alcohol *see* 2949
N-(Pantothenyl)-β-aminoethanethiol *see* 7114
Pantothenylol *see* 2949
Pantoyl Lactone *see* 7116

N-Pantoyl-3-propanolamine *see* 2949
Pantozol [Altana] *see* 7117
Panzid [Valda] *see* 1948
Panzytrat [Axcan] *see* 7108
Papain, 7119
Papaveraldine, 7120
Papaveretum, 7121
Papaverine, 7122
Papaw *see* 7123
Papaya, 7123
Papayotin *see* 7119
Paper Red 4B [Crompton & Knowles] *see* 1104
Papital T.R. [Zemmer] *see* 7122
Papoose Root *see* 1321
PAPP *see* 466
Papra *see* 7664
Paprika *see* 1772
Papuamine, 7124
Papulex [Euroderma] *see* 6608
PAR, 7125
PAR-101 *see* 4108
Parabanic Acid, 7126
Parabis [Dow] *see* 3081
Parabolan *see* 9745
Parabromdylamine *see* 1453
(+)-Parabromdylamine *see* 1453
Paracarbinoxamine *see* 1800
Paracefan [Boehringer, Ing.] *see* 2380
Paracetaldehyde *see* 7131
Paracetamol *see* 47
Parachloramine *see* 5848
Parachlorometacresol *see* 2133
Parachlorometaxylenol *see* 2182
Paracide *see* 3066
Paracodin *see* 3189
Paradiazine *see* 8068
Paradichlorobenzene *see* 3066
Paradione [Abbott] *see* 7132
Paradise Plant *see* 6249
Paradisone [Firmenich] *see* 6124
Paradormalene *see* 6031
Paraffin, 7127
Paraffin Chlorinated, 7128
Paraffin Jelly *see* 7300
Paraffin Oil *see* 7301
Paraffin Wax *see* 7127
Paraflex [McNeil] *see* 2201
Paraflu [Dainippon] *see* 4163
Paraform *see* 7129
Paraformaldehyde, 7129
Paraguay Tea *see* 5824
Paraherquamide, 7130
(−)-Paraherquamide *see* 7130
Parahexyl *see* 9143
Paral [Forest] *see* 7131
Paralactic Acid *see* 5385
Paralaudin *see* 2970
Paraldehyde, 7131
Paralergin [Vita] *see* 846
Paralest [Parachemie] *see* 9864
Paralgin *see* 3390
Paralytic Shellfish Poison *see* 8526
Paramandelic Acid *see* 5781
Paramenyl *see* 5712
Paramethadione, 7132
Paramethasone, 7133
Paramezone [Recordati] *see* 7133
Paramicina [Raglionieri] *see* 7146
Paramidin [Takeda] *see* 1475
Paraminyl [Cooper] *see* 8097
Paramisan Sodium [Smith & Nephew] *see* 473
Paramix [Liomont] *see* 6653
Paramorfan *see* 3194

Paramorphan *see* 3194
Paramorphine *see* 9427
Para-moth [Mann Lake] *see* 3066
Paraoxon, 7134
Paraoxypropiophenone *see* 7149
Paraplatin [BMS] *see* 1823
Paraquat, 7135
Paraquat I *see* 7135
Parasan *see* 1031
Parasepiolite *see* 5751
Parasiticol *see* 171
Parasorbic Acid, 7136
Parasympatol *see* 9142
Paratartaric Acid *see* 9204
Paratect [Pfizer] *see* 6353
Paratensiol [Yoshitomi] *see* 8264
Parathar [Rorer] *see* 9309
Parathiazan *see* 9452
Parathion, 7137
Parathion-methyl *see* 6181
Parathorm [Hormone-Chemie] *see* 7138
Parathormone *see* 7138
Para-thor-mone [Lilly] *see* 7138
Parathyroid Hormone, 7138
Paravermin *see* 7576
Parawollastonite *see* 1704
Paraxenol *see* 7416
Paraxin [Roche] *see* 2077
Para-zene [Reliable Chem.] *see* 3066
Parazine *see* 7576
Parbendazole, 7139
Parda [Parke-Davis] *see* 5516
Pardisol *see* 3800
Pardroyd [Parke-Davis] *see* 7066
Parecoxib, 7140
Paredrine [SK & F] *see* 4847
Paredrinol [SK & F] *see* 7442
Pareira, 7141
Pareira Brava *see* 7141
Parenabol *see* 1327
Parentamin [Pharmacia] *see* 7996
Parenzymol [Horner] *see* 9975
Parest *see* 6032
Parfenac [Lederle] *see* 1479
Parfenal [Cyanamid] *see* 1479
Parfuran [Parke-Davis] *see* 6686
Pargin [Gibipharma] *see* 3549
Pargitan [Abigo] *see* 9864
Pargonyl *see* 7146
Pargyline, 7142
Paricalcitol, 7143
Paricina [Archifar] *see* 7146
Parid [Cooper] *see* 7576
Pariet [Eisai] *see* 8206
Parietic Acid *see* 8299
Parietin *see* 3618
Parietrope [Marcofina] *see* 8150
Parigenin *see* 8517
Parinaric Acid, 7144
α-Parinaric Acid *see* 7144
β-Parinaric Acid *see* 7144
Pariprazole *see* 8206
Paris Blue *see* 8018
Paris Green *see* 2615
Paris Red *see* 5480
Paris White *see* 1659
Paris Yellow *see* 5460
Paritane [Berk] *see* 7051
Parizac [Lacer] *see* 6939
Parkazin *see* 3119
Parkemed [Parke-Davis] *see* 5869
Parkin [Yoshitomi] *see* 3800
Parkinane [Wyeth] *see* 9864
Parkinsan [Byk Gulden] *see* 1477
Park Lily *see* 2496

Park Nucleotide, 7145
Parkopan [Hexal] *see* 9864
Parks 12 [Novartis] *see* 7859
Parlay [Zeneca Ag Prod] *see* 7082
Parlef [Parke-Davis] *see* 4163
Parlodel [Novartis] *see* 1424
Parlodion [Mallinckrodt] *see* 8125
Parmal [Central] *see* 8097
Parmid [Lagap] *see* 6219
Parnate [GSK] *see* 9734
Parocin [Almirall] *see* 5894
Paroex [CSP] *see* 2092
Paroidin [Parke-Davis] *see* 7138
Paroleine *see* 7301
ParoLich [Winthrop] *see* 7148
Paromomycin, 7146
Paromomycin I *see* 7146
Parotid Basic Proteins *see* 4757
Parotin, 7147
Paroven [Zyma] *see* 9969
Paroxat [Hexal] *see* 7148
Paroxetine, 7148
Paroxetine Mesylate *see* 7148
Paroxon *see* 7149
Paroxypropione, 7149
Parphezein *see* 3800
Parphezin *see* 3800
Parrot Green *see* 2615
Parsidol [Warner-Chilcott; Specia] *see* 3800
Parsitan *see* 3800
Parsley, 7150
Parsley Apiole *see* 721
Parsley Camphor *see* 721
Parsley Seed Oil *see* 7150
Parsol 1789 [DSM] *see* 881
Parsol HMS [DSM] *see* 4777
Parsol HS [DSM] *see* 3647
Parsol MCX [DSM] *see* 6859
Parsotil *see* 3800
Partel [Lilly] *see* 3412
Parterol *see* 3198
Parzone [Mallinckrodt] *see* 3189
Parthenicin *see* 7151
Parthenin, 7151
Parthenolide, 7152
Partocon [Ferring] *see* 7078
Partricin, 7153
Partricin A *see* 7153
Partricin B *see* 7153
Partricin Methyl Ester *see* 5914
Partusisten [Boehringer, Ing.] *see* 4012
Parvaquone, 7154
Parvolex [Evans] *see* 83
Parzate *see* 6426
Parzone [Mallinckrodt] *see* 3189
PAS *see* 473
Pasalon [Bayer] *see* 473
PASER [Jacobus] *see* 473
Pasetocin [Kyowa] *see* 574
Pasiniazide *see* 5232
Pasireotide, 7155
Paskalium [Glenwood] *see* 473
Pasminox [Beta] *see* 6997
Pasmus [Daiichi] *see* 4138
Pasolind N [Stada] *see* 47
Paspaline-P *see* 2787
Paspertin [Kali-Chemie] *see* 6219
Pasque Flower *see* 8048
Passiflora, 7156
Passiflorin *see* 4648
Passion Flower *see* 7156
Passion Vine *see* 7156
Pastaron [Sato] *see* 10052
Pataday [Alcon] *see* 6934
Patanase [Alcon] *see* 6934

Patanol [Alcon] *see* 6934
Patchouli Alcohol, 7157
Patchouli Camphor *see* 7157
Patchouli Oil *see* 6900
Patent Alum *see* 360
Paternoite *see* 5722
Pathfinder [Dow AgroSci.] *see* 9821
Pathilon *see* 9826
Pathocidin *see* 889
Pathocil [Wyeth] *see* 3094
Patoran [Syngenta] *see* 6218
Patricin [Kaken] *see* 4128
Patronite *see* 10104
Pattonex [Makhteshim-Agan] *see* 6218
Patulin, 7158
Patupilone *see* 3684
Paucimycin *see* 7146
Pauly's Reagent *see* 3002
Pavabid [HMR] *see* 7122
Pavacen [Central Pharm.] *see* 7122
Pavadel [Canright] *see* 7122
Pavagen [Darby] *see* 7122
Pavakey [Key] *see* 7122
Pavased [Mallard] *see* 7122
Paviin *see* 4295
Pavoninins, 7159
Pavulon [Organon] *see* 7109
Paxate [BMS] *see* 2997
Paxeladine [Beaufour] *see* 7029
Paxil [GSK] *see* 7148
Paxipam [Schering] *see* 4624
Paxistil *see* 4889
Paylean [Elanco] *see* 8208
Payoff *see* 4155
Payzone [Am. Cyanamid] *see* 6741
Pazopanib, 7160
Pazufloxacin, 7161
Pazufloxacin Mesylate *see* 7161
Pazzul [Bial-Portela] *see* 3754
PB-89 *see* 4255
PB-94 *see* 8613
PB-868Cl *see* 5703
Pbb *see* 4757
PBBs *see* 7680
PBDEs *see* 7681
Pbe *see* 4757
PBN, 7162
PB-NOX [Roussel-Bio] *see* 7589
PBU *see* 7344
PBZ [Novartis] *see* 9910
PC *see* 7452
PC 1 *see* 4006
PC-904 *see* 710
PC-1020 *see* 4452
PC-1421 *see* 7574
PCB 1242 *see* 7682
PCB 1254 *see* 7682
PCB 1260 *see* 7682
PCBs *see* 7682
PCBTF *see* 2127
PCC *see* 8086
PCDFs *see* 7684
PCDP *see* 7394
PCE [Abbott] *see* 3739
PCI-0120 *see* 6367
PCI-0123 *see* 6368
PCNB *see* 8197
PCP *see* 7218, 7333
PCPP-SA *see* 7676
PCV *see* 7192
PD-93 *see* 7618
PD-107779 *see* 3641
PD-109452-2 *see* 8167
PD-114720 *see* 5499
PD-127391 *see* 2353

PD-131501 *see* 8862
PD-144723 *see* 7843
PD-148515 *see* 876
PDB *see* 3066
PDC *see* 8087
3-PDC *see* 7219
PDDB *see* 3464
PDGF *see* 7639
P-DHP *see* 233
Pd(OAc)$_2$ *see* 7092
PDP *see* 6285
Pd(PPh$_3$)$_4$ *see* 9367
PDX *see* 7821
PDX Chloride *see* 7674
PEA *see* 7339
Peach Oil, Expressed, 7163
Peak [Syngenta] *see* 7992
Peanut, 7164
Peanut Oil, 7165
Pearl Ash *see* 7740
Pearlman's Catalyst, 7166
Pearl White *see* 1274
Pear Oil *see* 5156
Pearson's Creolin *see* 2558
Peau d'Espagne *see* 3821
PEBC *see* 7167
PeBDE *see* 7681
β-PEBG *see* 7345
Pebulate, 7167
PECA *see* 3842
2,3,4,7,8-PeCDF *see* 7684
Pecilocin, 7168
Pecitrol Veinogène *see* 1233
Pecram [Novartis] *see* 461
Pect [Rentschler] *see* 380
Pectamol [BDH] *see* 7029
Pectamon *see* 7029
Pectenine *see* 1847
Pectin, 7169
Pectin Sugar *see* 750
Pectobloc [Kytta-Siegfried] *see* 7554
Pectolin [Faulding] *see* 7441
Pectolinarigenin, 7170
Pectolinarin *see* 7170
Pectolitan [3M Pharma] *see* 2066
Pectox [Italfarmaco] *see* 1802
Pedameth [Forest] *see* 6047
Pederin, 7171
Pederine *see* 7171
ψ-Pederine *see* 7171
Pedestal [Uniroyal] *see* 6809
PEDG *see* 7345
Pediamycin [Abbott] *see* 3739
Pedisafe [Sagitta] *see* 2401
Peflacine [Bellon] *see* 7172
Peflox [Formenti] *see* 7172
Pefloxacin, 7172
Pefloxacine *see* 7172
Pefloxacin Mesylate *see* 7172
Pefurazoate, 7173
PEG *see* 7688
Pegasyl [SweVet] *see* 4835
Peganine *see* 10127
Peganone [Abbott] *see* 3802
Pegaptanib Sodium, 7174
PEG-asparaginase *see* 826
Pegaspargase *see* 826
Pegasus [Novartis Agro] *see* 2972
Pegasys [Roche] *see* 7175
PEG Fatty Acid Esters *see* 7698
PEG-IFN-α2a *see* 7175
Peginterferon α2a *see* 7175
Peginterferon Alfa-2a, 7175
Peginterferon Alfa-2b, 7176
PegIntron [Schering-Plough] *see* 7176

3,3',4',5,7-Pentahydroxyflavone *see* 8150
3,3',4',5,7-Pentahydroxyflavone-3-gluco-side *see* 5267
3,3',4',5,7-Pentahydroxyflavone-7-D-glu-coside *see* 8151
3,3',4',5,7-Pentahydroxyflavone-3-ruti-noside *see* 8438
3,3',4',5,7-Pentahydroxyflavylium Chloride *see* 2677
(2S,16Z,18E,20S,21S,22R,23R,24R,25S,-26R,27S,28E)-5,12,21,23,25-Pentahy-droxy-10-(4-isobutyl-1-piperazinyl)-27-methoxy-2,4,16,20,22,24,26-hepta-methyl-2,7-(epoxypentadeca[1,11,13]-trienimino)-6H-benzofuro[4,5-a]phen-oxazine-1(2H),6,15-trione 25-Acetate *see* 8339
3,3',4',5,7-Pentahydroxy-5'-methoxyflav-ylium Chloride *see* 7305
3,3',4',5,7-Pentahydroxy-2-phenylbenzo-pyrylium Chloride *see* 2677
Pentakis (N²-Acetyl-L-glutaminato)-tetrahydroxytrialuminum *see* 26
3,3',4',5,7-Pentakis(benzyloxy)flavone *see* 8150
Pentalin *see* 7217
Pentam [Astellas] *see* 7227
Pentamethylcyclopentadienyliridium(III) Dichloride Dimer, 7226
Pentamethylene *see* 2736
Pentamethylene-1,5-bis(1-methylpyrroli-dinium) Hydrogen Tartrate *see* 7245
Pentamethylenediamine *see* 1614
4,4'-(Pentamethylenedioxy)dibenzami-dine *see* 7227
Pentamethylene Glycol *see* 7229
Pentamethylene Oxide *see* 9362
1,5-Pentamethylenetetrazole *see* 7254
Pentamethylpararosaniline Chloride *see* 4430
1,2,2,6,6-Pentamethylpiperidine *see* 7189
Pentamidine, 7227
Pentamidine Mesylate *see* 7227
Pentamycin *see* 4319
Pentanal *see* 10088
tert-Pentanal *see* 7624
1-Pentanamine *see* 593
Pentane, 7228
n-Pentane *see* 7228
Pentanedial *see* 4508
1,5-Pentanediamine *see* 1614
Pentane-1,1-dicarboxylic Acid *see* 1579
1,5-Pentanedicarboxylic Acid *see* 7543
3,3-Pentanedicarboxylic Acid *see* 3140
Pentanedinitrile *see* 4510
Pentanedioic Acid *see* 4509
1,5-Pentanediol, 7229
2,4-Pentanedione *see* 76
2,4-Pentanedione Nickel Complex *see* 6585
1,1'-(1,5-Pentanediyl)bis[1-methylpyrrol-idinium] (2R,3R)-2,3-Dihydroxybu-tanedioate (1:2) *see* 7245
4,4'-[1,5-Pentanediylbis(oxy)]bisbenzene-carboximidamide *see* 7227
(1R,1'R,2R,2'R)-2,2'-[1,5-Pentanediylbis-[oxy(3-oxo-3,1-propanediyl)]]bis[1-[(3,4-dimethoxyphenyl)methyl]-1,2,-3,4-tetrahydro-6,7-dimethoxy-2-meth-ylisoquinolinium] Benzenesulfonate (1:2) *see* 859
2,2'-[1,5-Pentanediylbis[oxy(3-oxo-3,1-propanediyl)]]bis[1-[(3,4-dimethoxy-phenyl)methyl]-1,2,3,4-tetrahydro-6,7-dimethoxy-2-methylisoquinolinium] Benzenesulfonate (1:2) *see* 859

Pentanenitrile *see* 10091
1,2,3,4,5-Pentanepentol *see* 751
xylo-Pentane-1,2,3,4,5-pentol *see* 10283
1-Pentanethiol *see* 604
Pentanitrine *see* 7222
Pentanoic Acid *see* 10090
tert-Pentanoic Acid *see* 7625
Pentanoic Acid 2-[(2S,4S)-1,2,3,4,6,11-Hexahydro-2,5,12-trihydroxy-7-meth-oxy-6,11-dioxo-4-[[2,3,6-trideoxy-3-[(2,2,2-trifluoroacetyl)amino]-α-L-*lyxo*-hexopyranosyl]oxy]-2-naphthacenyl]-2-oxoethyl Ester *see* 10101
Pentanoic Acid Zinc Salt (2:1) *see* 10364
sec-Pentanol *see* 7232
tert-Pentanol *see* 7253
1-Pentanol, 7230
2-Pentanol, 7231
3-Pentanol, 7232
2-Pentanone *see* 6187
3-Pentanone *see* 3137
2,5,8,11,14-Pentaoxapentadecane *see* 9353
Pentaphen *see* 7255
Pentaphene *see* 3015
Pentapyrrolidinium Bitartrate *see* 7245
Pentasa [Ferring] *see* 5974
Pentasodium 4,4'-[(3-Sulfo-4,4'-biphen-ylene)bis(azo)]bis(3-amino-2,7-naph-thalenedisulfonate) *see* 9973
Pentasodium Triphosphate *see* 8824
1,2,3,5,6-Pentathiepane *see* 5492
Pentazocine, 7233
α-(-)-Pentazocine *see* 7233
α-(+)-Pentazocine *see* 7233
cis-(-)-Pentazocine *see* 7233
cis-(+)-Pentazocine *see* 7233
Pentcillin [Toyama-Sankyo] *see* 7575
Pentedrin *see* 9142
Pentek *see* 7221
1-Pentene, 7234
2-Pentene, 7235
Pent-4-enyl-N-furfuryl-N-imidazol-1-yl-carbonyl-DL-homoalaninate *see* 7173
Pentetate Calcium Trisodium, 7236
Pentetic Acid, 7237
Pentetrazol *see* 7254
Pentetreotide, 7238
Penthamil [Geigy] *see* 7236
Penthiobarbital Sodium *see* 9503
Penthiopyrad, 7239
Penthrane [Abbott] *see* 6067
Penthrit *see* 7222
Penticainide *see* 7242
Penticort [Lederle] *see* 382
Pentids [BMS] *see* 7203
Pentifylline, 7240
Pentigetide, 7241
Pentilium *see* 7245
Pentione *see* 5250
Pentisomide, 7242
Pentitrate *see* 7222
Pentlandite *see* 6583
Pentobarbital, 7243
Pentobarbital Calcium *see* 7243
Pentobarbital Sodium *see* 7243
Pentobarbitone *see* 7243
Pentofuryl [Karlspharma] *see* 6619
1-Pentol, 7244
1'-Pentol *see* 7244
1''-Pentol *see* 7244
Pentolinium Tartrate, 7245
Pentolonium Bitartrate *see* 7245
Pentone [Faulding] *see* 7243
Pentorex, 7246

Pentosalen *see* 4967
Pentosan Polysulfate, 7247
Pentostam [GSK] *see* 692
Pentostatin, 7248
Pentothal Sodium [Abbott] *see* 9503
Pentoxifylline, 7249
Pentoxiverin *see* 1795
Pentoxyl, 7250
Pentoxyverine *see* 1795
Pentoyl [Morishita] *see* 3619
Pentral 80 *see* 7222
Pentrane [Abbott] *see* 6067
Pentrex [BMS] *see* 583
Pentrexyl [BMS] *see* 583
Pentrinitrol, 7251
Pentrite *see* 7222
Pentryate *see* 7222
Pentryl, 7252
D-*erythro*-2-Pentulose *see* 8331
threo-2-Pentulose *see* 10286
Pentyde [Immunetech] *see* 7241
Pentyl Alcohol *see* 7230
tert-Pentyl Alcohol, 7253
Pentylamine *see* 593
Pentylbenzene *see* 595
Pentylcarbinol *see* 4734
Pentyl 1-(5-Deoxy-β-D-ribofuranosyl)-5-fluoro-1,2-dihyro-2-oxo-4-pyrimidine-carbamate *see* 1757
Pentylenetetrazole, 7254
Pentyl Ether *see* 602
Pentyl Hexanoate *see* 599
[2R-[2α(Z),3α]]-11-(3-Pentyloxiranyl)-9-undecenoic Acid *see* 10159
[2S-[2α(Z),3α]]-11-(3-Pentyloxiranyl)-9-undecenoic Acid *see* 10159
11-(3-Pentyloxiranyl)-9-undecenoic Acid *see* 10159
p-tert-Pentylphenol, 7255
Pentymal *see* 567
Pen-Vee K [Biochemie] *see* 7209
Peon [Zeria] *see* 10310
Peonidin, 7256
Peonin, 7256
PEP *see* 7686
Pepcid [Merck & Co.] *see* 3971
Pepcidac [McNeil] *see* 3971
Pepcidine [Merck & Co.] *see* 3971
Pepdine [Merck & Co.] *see* 3971
Pepdul [Merck & Co.] *see* 3971
Pepleo Injection [Nippon Kayaku] *see* 7257
Pepleomycin *see* 7257
Peplomycin, 7257
Pepo *see* 8051
Pepperidge Bush *see* 956
Peppermint, 7258
Peppermint Camphor *see* 5905
Pepsamar [Sanofi Winthrop] *see* 338
Pepsin, 7259
Pepsin A *see* 7259
Pepstatin, 7260
Pepstatin A *see* 7260
Pepstatin B *see* 7260
Pepstatin C *see* 7260
Peptan [Merck & Co.] *see* 3971
Peptard [3M Pharma] *see* 4897
Peptarom [Stada] *see* 10074
Peptavlon [Ayerst] *see* 7224
Peptazol [Recordati] *see* 7117
Pepticum [Grünenthal] *see* 6939
Peptide T, 7261
Peptide Tyrosine-tyrosine *see* 7262
Peptide YY, 7262
Peptidoglycan *see* 6389

Peptilate [Dojin] *see* 216
Peptinimid [Iromedica] *see* 3801
Pepto-Bismol [Procter & Gamble] *see* 1285
Per-Abrodil *see* 5083
Peracetic Acid, 7263
Peracon *see* 5155
Peragit [AFI] *see* 9864
Peramivir, 7264
Peraprin [Taiyo] *see* 6219
Perazil [Burroughs Wellcome] *see* 2082
Perazine, 7265
Perazolin [Zenyaku Kogyo] *see* 8706
Perazyl *see* 2082
Perbenzoic Acid, 7266
Perboric Acid (HBO(O₂)) Sodium Salt (1:1) *see* 8783
Percapyl *see* 2105
Perchloracap [Mallinckrodt] *see* 7774
Perchloric Acid, 7267
Perchloric Acid Ammonium Salt (1:1) *see* 538
Perchloric Acid Barium Salt (2:1) *see* 983
Perchloric Acid Copper(2+) Salt (2:1) *see* 2637
Perchloric Acid Lithium Salt (1:1) *see* 5595
Perchloric Acid Magnesium Salt (2:1) *see* 5742
Perchloric Acid Potassium Salt (1:1) *see* 7774
Perchloric Acid Silver(1+) Salt (1:1) *see* 8662
Perchloric Acid Sodium Salt (1:1) *see* 8784
Perchloric Acid Trioxidochloryl Ester *see* 2097
Perchloric Acid Zinc Salt (2:1) *see* 10348
Perchloric Anhydride *see* 2097
Perchlorobenzene *see* 4713
Perchlorobutadiene *see* 4714
Perchloroethane *see* 4715
Perchloroethylene *see* 9335
Perchloromethane *see* 1816
Perchloropentacyclo[5.2.1.0²,⁶.0³,⁹.0⁵,⁸]-decane *see* 6290
Perchloryl Fluoride, 7268
Perclene *see* 9335
Percoccide [A.C.F.] *see* 9045
Percocet [Endo] *see* 7058
Percodan [Endo] *see* 7058
Percorten [Novartis] *see* 2903
Percutol [Pliva] *see* 6694
6-Perdeoxy-6-per(2-carboxyethyl)thio-γ-cyclodextrin *see* 9019
Perderm [Schering] *see* 213
Perdiem Fiber [RPR] *see* 7631
Perdilatal [Smith & Nephew] *see* 6821
Perdipina [Novartis] *see* 6580
Perdipine [Syntex] *see* 6580
Perdix [Schwarz] *see* 6315
Perdorm [Lagap] *see* 9275
Perduren *see* 9492
Perebral [Beytout] *see* 2697
Perebron [Angelini] *see* 7043
Pereirine *see* 4413
Pérénan [Sanofi Winthrop] *see* 3710
Perenum [Delalande] *see* 9680
Perequil [Lepetit] *see* 5929
Perezone, 7269
Perfalqan [BMS] *see* 47
Perfan [HMR] *see* 3643
Perfane [HMR] *see* 3643
Perfekthion [BASF] *see* 3241

Perflan [Dow AgroSci.] *see* 9232
Perfluamine *see* 4215
Perflubron, 7270
Perflunafene *see* 4215
Perfluoroacetic Acid *see* 9847
Perfluorobenzene *see* 4722
Perfluorocaprylic Acid *see* 7272
Perfluoroctyl Bromide *see* 7270
Perfluorocyclobutane *see* 6834
Perfluorodecalin *see* 4215
Perfluoroethene *see* 9350
Perfluorohexane, 7271
Perfluoro-*n*-hexane *see* 7271
Perfluorooctanoic Acid, 7272
Perfluorophenol *see* 7223
Perfluoropropane, 7273
Perfluorotripropylamine *see* 4215
Perflutren *see* 7273
Perforin, 7274
Performic Acid, 7275
Perfosfamide, 7276
Perfusamine [Medi-Physics] *see* 5092
Pergamid [Scios-Nova] *see* 7276
Pergitral *see* 7222
Pergolide, 7277
Pergolide Mesylate *see* 7277
Pergonal [Serono] *see* 4250
Pergotime [Serono] *see* 2377
Perhexilene *see* 7278
Perhexiline, 7278
N-(2-Perhydroazocin-1-ylethyl)guanidine *see* 4595
Perhydronaphthalene *see* 2850
Perhydrosqualene *see* 8895
Peri Acid *see* 6487
Periactin [Merck & Co.] *see* 2770
Periactine [Merck & Co.] *see* 2770
Perichthol *see* 4924
Periciazine *see* 7279
Periciclina *see* 2890
Periclase *see* 5741
Peri-Colace *see* 1878
Peri-Colace [Roberts] *see* 3446
Pericyazine, 7279
Peridamol [Boehringer, Ing.] *see* 3388
Peridex [Zila] *see* 2092
Peridon [Italchimici] *see* 3467
Péridys [Fabre] *see* 3467
Perifosine, 7280
Perifunal [Aristegui] *see* 7630
Perigen [Bayer CropSci.] *see* 7294
Perilax [Nordic Drugs] *see* 1246
Perilla Ketone, 7281
Perillaldehyde, 7282
l-Perillaldehyde α-*anti*-Oxime *see* 7282
l-Perillaldehyde α-*syn*-Oxime *see* 7282
Perillartine *see* 7282
"Perilla Sugar" *see* 7282
Perimetazine *see* 7283
Perimethazine, 7283
Perimycin, 7284
Perimycin A *see* 7284
Perin [Endo] *see* 7576
Perindopril, 7285
Perindoprilat *see* 7285
Perindopril Erbumine *see* 7285
Periodic Acid, 7286
Periodic Acid (HIO₄) Potassium Salt (1:1) *see* 7775
Periodic Acid (HIO₄) Sodium Salt (1:1) *see* 8772
Periodyl, 7287
Periograf *see* 3514
Periplanone A *see* 7288
Periplanone B *see* 7288

(−)-Periplanone B *see* 7288
Periplanones, 7288
Periplocin, 7289
Periplocoside *see* 7289
Periplocymarin, 7290
Periplogenin, 7291
Periplum [Italfarmaco] *see* 6636
Peripress [Pfizer] *see* 7838
Perisalol [Chugai] *see* 6606
Peristim [Mead Johnson] *see* 1878
Periston [Bayer] *see* 7814
Peritol [Medphano] *see* 2770
Peritrate [Warner-Chilcott] *see* 7222
Perityl *see* 7222
Perivine, 7292
Perizin [Bayer] *see* 2546
Perketan [Inverni] *see* 5344
Perlacton [Chassot] *see* 7078
Perlapine, 7293
Perlatan *see* 3763
Perlinganit [Schwarz] *see* 6694
Perlon *see* 6823
Perlopal *see* 5913
Perlutal [Boehringer, Ing.] *see* 233
Perlutan [Boehringer, Ing.] *see* 233
Perlutex [Leo Pharm] *see* 5864
Permanent Yellow *see* 968
Permanganic Acid (HMnO₄) Barium Salt *see* 984
Permanganic Acid (HMnO₄) Calcium Salt *see* 1690
Permanganic Acid (HMnO₄) Potassium Salt (1:1) *see* 7776
Permanganic Acid (HMnO₄) Silver(1+) Salt *see* 8663
Permanganic Acid (HMnO₄) Sodium Salt *see* 8785
Permapen [Roerig] *see* 7205
Permasect [Mitchell Cotts] *see* 7294
Permastril [Cassenne] *see* 3497
Permax [Lilly] *see* 7277
Permeability Glycoprotein *see* 7310
Permease *see* 4797
Permethanoic Acid *see* 7275
Permethrin, 7294
Permiran [Laphal] *see* 10199
Permit [Nissan] *see* 4636
Permitil [Schering] *see* 4220
Permixon [Pierre Fabre] *see* 8524
Pernambuco, 7295
Pernazene [Synthelabo] *see* 10015
Pernovin [Chinoin] *see* 7347
Perolysen [M & B] *see* 7189
Peroidin *see* 7774
Perospirone, 7296
Perovskite *see* 9628
1,4-Peroxido-*p*-menthene-2 *see* 816
Peroxinorm [Grünenthal] *see* 9131
Peroxyacetic Acid *see* 7263
Peroxybenzoic Acid *see* 7266
Peroxyderm [Chassot] *see* 1119
Peroxydicarbonic Acid Potassium Salt (1:2) *see* 7773
Peroxydisulfuric Acid ([(HO)S(O)₂]₂O₂) Ammonium Salt (1:2) *see* 539
Peroxydisulfuric Acid ([(HO)S(O)₂]₂O₂) Potassium Salt (1:2) *see* 7777
Peroxydisulfuric Acid ([(HO)S(O)₂]₂O₂) Sodium Salt (1:2) *see* 8787
Peroxyformic Acid *see* 7275
Peroxymonosulfuric Acid *see* 1853
Perparin [Chemopuro] *see* 3783
Perperine *see* 3783
Perphenan *see* 7297
Perphenazine, 7297

Perphenazine 3,4,5-Trimethoxybenzoate *see* 6221
Perprazole *see* 6939
Per-Radiographol *see* 5083
Persa-gel [J & J] *see* 1119
Persantine [Boehringer, Ing.] *see* 3388
Persedon [Roche] *see* 8108
Persephin *see* 4410
Persian Bark *see* 1880
Persian Berries *see* 4291
Persian Insect Powder *see* 8077
Persian Red *see* 5461
Persic Oil *see* 7163
Persio *see* 2605
Persistol [Chevron] *see* 9836
Persolv [Lepetit] *see* 10072
Perspex *see* 6013
Persulfuric Acid *see* 1853
Pertestis [Orma] *see* 9324
Pertofran [Novartis] *see* 2921
Pertofrane [Novartis] *see* 2921
Pertranquil [HMR] *see* 5929
Pertrombon *see* 3867
Pertscan [Abbott] *see* 8788
Pertuzumab, 7298
Peruphasmal *see* 3459
Peruvian Balsam *see* 942
Peruvian Bark *see* 2287
Peruviol *see* 6561
Pervetral [Viatris] *see* 7069
Pervincamine [Sanofi-Aventis] *see* 10180
Pervitin [Temmler] *see* 6020
Perviton *see* 7330
Perycit [Bofors] *see* 6582
Perylene, 7299
PES *see* 5678
Peson [Hoechst] *see* 5678
Pesos [Valeas] *see* 4004
Pestox III [FBC] *see* 8532
Pestox XIV *see* 3228
PET *see* 7689
Petalite *see* 5576
Peteha [Fatol] *see* 8000
Pethidine *see* 5916
Petidiol *see* 3773
Petidion [Gerot] *see* 3773
Petidon *see* 9875
Petimin *see* 5751
Petinimid *see* 3801
Petinutin [Parke-Davis] *see* 6078
Petisan [Gerot] *see* 3773
Petit Muguet *see* 10247
PETN *see* 7222
Petnidan [Desitin] *see* 3801
Petrin [Parke-Davis] *see* 7251
Petrohol *see* 5254
Petrolatum, 7300
Petrolatum, Liquid, 7301
Petrol (British) *see* 4405
Petroleum, 7302
Petroleum Benzin, 7303
Petroleum Jelly *see* 7300
Petroleum Naphtha *see* 7303
Petroleum Spirits *see* 6282
Petroselic Acid *see* 7304
Petroselinamide *see* 7304
Petroselinic Acid, 7304
Petrosulpho *see* 4924
Petty-morrel *see* 755
Petunidin, 7305
Petunidol *see* 7305
Petunin *see* 7305
Petylyl [Temmler] *see* 2921
Petzite *see* 4549
Peucedanin, 7306

Pevaryl [Cilag] *see* 3549
Pexelizumab, 7307
Pexeva [JDS] *see* 7148
Pexid [Sigma] *see* 7278
Pexiganan, 7308
Peyocactine *see* 4785
Peyote *see* 5975
Peyrone's Chloride *see* 2316
Peyrone's Salt *see* 2316
Pezetamid [Hefa-Frenon] *see* 8067
PF-1022-221 *see* 3617
PF-1593 *see* 1489
PF-804950 *see* 3560
PF-2341066 *see* 2577
PF-3512676 *see* 180
P-FAD *see* 4496
PFC-218 *see* 7273
Pfeiffer's Substance, 7309
Pfizer-E [Pfizer] *see* 3743
Pfizerpen [Roerig] *see* 7203
Pfizerpen-AS [Roerig] *see* 7206
PFOA *see* 7272
PFOB *see* 7270
PFP *see* 7274
PFT Roche [Roche] *see* 1056
PFT-α *see* 7533
PG *see* 7971
PG490 *see* 9923
PG490-88 *see* 9923
PGA *see* 4248
PGE₁ *see* 7987
PGE₂ *see* 7988
PGF₂α *see* 7989
PGG₂ *see* 7986, 9542
PGH₂ *see* 7986
PGI₂ *see* 7986
PGL-4001 *see* 10030
P-Glycoprotein, 7310
[P(GMA-co-EA)] *see* 2499
P-gp *see* 7310
PGT/1A *see* 7531
PG-TXL *see* 7081
PGX *see* 7986
PH-60-40 *see* 3161
PH-70-23 *see* 4154
PH-73FF *see* 9775
PH-1882 *see* 1429
PH-5776 *see* 6653
PHA-291639 *see* 9650
PHA-291639E *see* 9650
PHA-782615 *see* 3560
Phacetoperane *see* 5521
Phacetur *see* 7318
Phaenthine *see* 9379
Phaeomelanins *see* 5881
Phaeva [Schering AG] *see* 4447
Phalloidin, 7311
Phalloidine *see* 7311
Phaltan *see* 4251
Phanchinone *see* 7312
Phanodorm [Winthrop] *see* 2704
Phanodorn [Winthrop] *see* 2704
Phanquinone, 7312
Phanquone *see* 7312
Phanurane [Specia] *see* 1753
Phardol [Kreussler] *see* 4535
Pharmatex [Innothéra] *see* 1061
Pharorid [Wellmark] *see* 6055
Phaseolin, 7313
Phaseollin *see* 7313
Phaseolunatin *see* 5552
Phaseomannite *see* 5021
Phasil [Reed & Carnrick] *see* 3237
Phasin, 7314
Phasins *see* 5484

Phazyme [Reed & Carnrick] *see* 3237
Ph BC *see* 4006
PHD *see* 9542
Phe *see* 7382
Pheasant's Eye *see* 156
α-Phellandrene, 7315
β-Phellandrene, 7316
Phe³-Lys⁸-oxytocin *see* 5691
Phe²-Lys⁸-vasopressin *see* 3988
Phemeride [Parke-Davis] *see* 1076
Phe-Mer-Nite *see* 7413
Phemerol Chloride [Parke-Davis] *see* 1076
Phemiton [Pliva] *see* 5921
Phenacaine Hydrochloride, 7317
Phenacemide, 7318
Phenacetin, 7319
Phenacetolin, 7320
Phenacetylurea *see* 7318
Phenacide *see* 9716
Phenacite *see* 1163
Phenacylamine, 7321
Phenacyl Bromide *see* 1409
Phenacyl Chloride *see* 2115
Phenalzine *see* 7335
Phenamidine, 7322
Phenamin [Nycomed] *see* 2186
Phenamiphos *see* 3991
Phenamizole *see* 479
Phenampromid *see* 7323
Phenampromide, 7323
9,10-Phenanthraquinone *see* 7327
Phenanthrene, 7326
9,10-Phenanthrenedione *see* 7327
Phenanthrenequinone, 7327
o-Phenanthroline *see* 7328
1,10-Phenanthroline *see* 7328
4,5-Phenanthroline *see* 7328
4,7-Phenanthroline-5,6-dione *see* 7312
4,7-Phenanthroline-5,6-quinone *see* 7312
Phenantoin *see* 5920
Phenarsazine Chloride, 7329
Phenatine, 7330
Phenatox *see* 9716
Phenazacillin *see* 4707
Phenazarsine Chloride *see* 7329
Phenazine, 7331
Phenazine [Jenkins] *see* 7334
2,3-Phenazinediamine *see* 2984
1,6-Phenazinediol 5,10-Dioxide *see* 5064
Phenazine Methosulfate *see* 6182
1-Phenazinol *see* 4680
Phenazocine, 7332
(±)-α-Phenazocine *see* 7332
α-(−)-Phenazocine *see* 7332
Phenazoline *see* 672
Phenazone *see* 703
Phenazopyridine, 7324
Phenazoxine *see* 7365
Phencamine *see* 3999
Phencyclidine, 7333
Phendimetrazine, 7334
Phénégic *see* 7364
Phenelzine, 7335
[ν-Phenenyltris(oxyethylene)]tris[triethylammonium Triiodide] *see* 4373
Phenesin *see* 1525
Phenetharbital, 7336
Phenethicillin Potassium, 7337
Phenethyl Alcohol, 7338
Phenethylamine, 7339
4-[2-[[6-(Phenethylamino)hexyl]amino]ethyl]pyrocatechol *see* 3472
Phenethylazocine *see* 7332
1-Phenethylbiguanide *see* 7345

2-(Phenylamino)benzoic Acid *see* 7384
β-Phenyl-γ-aminobutyric Acid *see* 460
N-[3-[[(Phenylamino)carbonyl]oxy]phenyl]carbamic Acid Ethyl Ester *see* 2925
dl-α-Phenyl-β-aminoethanol Sulfate *see* 7399
2-Phenylamino-4-methyl-6-cyclopropylpyrimidine *see* 2769
8-(Phenylamino)-1-naphthalenesulfonic Acid *see* 654
p-Phenylaminonitrosobenzene *see* 6726
3-[2-[4-(Phenylamino)phenyl]diazenyl]benzenesulfonic Acid Sodium Salt (1:1) *see* 5999
4-[2-[4-(Phenylamino)phenyl]diazenyl]benzenesulfonic Acid Sodium Salt (1:1) *see* 9955
1-Phenyl-2-aminopropane *see* 579
(*S*)-1-Phenyl-2-aminopropane *see* 2956
1-Phenyl-1-amino-2-propanol *see* 444
Phenyl Aminosalicylate, 7383
Phenyl *p*-Aminosalicylic Acid *see* 7383
N-Phenylanthranilic Acid, 7384
Phenyl Arsenoxide *see* 7048
Phenylarsine Oxide *see* 7048
Phenylarsonic Acid *see* 1069
As-Phenylarsonic Acid *see* 1069
Phenylarsonium Chloride (Ph₄AsCl) *see* 9383
Phenyl Azide, 7385
p-(Phenylazo)aniline *see* 414
4-(Phenylazo)benzenamine *see* 414
β-Phenylazo-α,α′-diaminopyridine *see* 7324
1-(*p*-Phenylazophenylazo)-2-naphthol *see* 9015
4-Phenylazo-*m*-phenylenediamine Hydrochloride *see* 2261
4-Phenylazo-*m*-phenylenediamine Hydrochloride Citrate *see* 2261
(Phenylazo)thioformic Acid 2-Phenylhydrazide *see* 3420
N-Phenylbarbital *see* 7336
N-Phenylbenzaldimine *see* 1141
N-Phenylbenzamide *see* 1063
Phenylbenzene *see* 3344
α-Phenylbenzeneacetic Acid *see* 3346
α-Phenylbenzeneacetic Acid 2-(Diethylamino)ethyl Ester *see* 149
α-Phenylbenzeneacetic Acid 1-Ethyl-3-piperidinyl Ester *see* 7582
α-Phenylbenzeneacetic Acid (3-*endo*)-8-Methyl-8-azabicyclo[3.2.1]oct-3-yl Ester *see* 9952
N-Phenylbenzeneamine *see* 3347
α-Phenylbenzeneethanethioic Acid *S*-[2-(Diethylamino)ethyl]ester *see* 9524
α-Phenylbenzenemethanamine *see* 1078
N-Phenylbenzenemethanamine *see* 1129
α-Phenylbenzenemethanimine *see* 1102
α-Phenylbenzenemethanol *see* 1092
Phenyl Benzenethiosulfonate, 7386
2-Phenyl-1*H*-benzimidazole, 7387
Phenylbenzimidazole Sulfonic Acid *see* 3647
2-Phenyl-1*H*-benzimidazole-6-sulfonic Acid *see* 3647
2-Phenyl-1,2-benzisoselenazol-3(2*H*)-one *see* 3533
Phenyl Benzoate, 7388
2-Phenyl-4*H*-1-benzopyran-4-one *see* 4124
3-Phenyl-4*H*-1-benzopyran-4-one *see* 5219
2-Phenyl-γ-benzopyrone *see* 4124

2-Phenyl-1,4-benzopyrone *see* 4124
β-*p*-Phenylbenzoylpropionic Acid *see* 3996
N-Phenylbenzylamine *see* 1129
2-(*N*-Phenyl-*N*-benzylaminomethyl)imidazoline *see* 672
N-Phenyl-*N*-benzyl-4-amino-1-methylpiperidine *see* 949
p-(α-Phenylbenzylidene)-1,1-dimethylpiperidinium Methylsulfate *see* 3336
Phenyl Biguanide, 7389
1-Phenylbiguanide *see* 7389
Phenylbis[2-chloroethylamine] *see* 653
Phenylbis(2,2,2-trifluoroacetato-*κO*)iodine *see* 7406
N-Phenylbis(trifluoromethanesulfonimide) *see* 7427
Phenylboric Acid *see* 1070
Phenylboron Dihydroxide *see* 1070
Phenylboronic Acid *see* 1070
Phenylboroxide *see* 1070
Phenyl Bromide *see* 1413
1-Phenylbutane *see* 1551
2-Phenylbutane *see* 1552
N-Phenylbutanedioic Acid Monoamide *see* 8998
N-Phenylbutanimide *see* 8997
Phenylbutazone, 7390
4-Phenyl-3-buten-2-one *see* 1140
8-[4-(4-Phenylbutoxy)benzamido]-2-(tetrazol-5-yl)-4*H*-1-benzopyran-4-one *see* 7830
8-[*p*-(4-Phenylbutoxy)benzoyl]amino-2-(5-tetrazolyl)-4-oxo-4*H*-1-benzopyran *see* 7830
Phenyl-*tert*-butylamine *see* 7374
Phenylbutylcarbinol *see* 4006
Phenyl-*sec*-butyl Norsuprifen *see* 6821
2-Phenylbutyric Acid 2-[2-(Diethylamino)ethyl]ethyl Ester *see* 1516
2-Phenylbutyric Acid 2-(Diethylamino)ethyl Ester *see* 1525
N-(α-Phenylbutyryl)urea *see* 7344
(2-Phenylbutyryl)urea *see* 7344
Phenylcarbamic Acid Dodecyl Ester *see* 3451
N-Phenylcarbamic Acid Ethyl Ester *see* 7430
N-Phenylcarbamic Acid 1-Methylethyl Ester *see* 7930
Phenylcarbamide *see* 7429
5-(*N*-Phenylcarbamoylamino)-1,2,3-thiadiazole *see* 9465
Phenylcarbimide *see* 7407
Phenylcarbinol *see* 1127
Phenyl Carbonate *see* 3354
α-Phenyl(carboxymethylpenicillin) *see* 1793
d,*l*-2-Phenyl-3-[1-(4-chlorobenzoyl)-5-methoxy-2-methylindole-3-acetoxy]propionic Acid *see* 9957
Phenylchloroform *see* 1112
2-Phenyl-6-chlorophenol, 7391
4-Phenyl-2-chlorophenol, 7392
1-Phenyl-1-(*o*-chlorophenyl)-3-dimethylaminopropanol *see* 2066
1-Phenyl-2-(3-chloropropylamino)propane *see* 5870
2-Phenylchromone *see* 4124
3-Phenylchromone *see* 5219
2-Phenylcinchoninic Acid *see* 2291
α-Phenylcinnamic Acid, 7393
α-Phenyl-*o*-cresol *see* 1144
α-Phenyl-*p*-cresol *see* 1145
Phenyl Cyanide *see* 1099

α-Phenylcyclohexaneacetic Acid 2-(Diethylamino)ethyl Ester *see* 3494
α-Phenylcyclohexaneglycolic Acid 4-(Diethylamino)-2-butynyl Ester *see* 7054
α-Phenylcyclohexaneglycolic Acid Ester Diethyl(2-hydroxyethyl)methylammonium Bromide *see* 7074
α-Phenylcyclohexaneglycolic Acid 1-Methyl-2-tetrahydroxypyrimidylmethyl Ester *see* 7072
1-(2-Phenyl-2-cyclohexyl-1,3-dioxolan-4-yl)methyl-1-methylpiperidinium Iodide *see* 2729
Phenylcyclohexyloxyacetic Acid Diethylaminoethyl Ester Bromomethylate *see* 7074
1-(1-Phenylcyclohexyl)piperidine *see* 7333
1-Phenyl-1-cyclohexyl-3-piperidyl-1-propanol Hydrochloride *see* 9864
1-Phenylcyclopentane-1-carboxylic Acid Diethylaminoethoxyethyl Ester *see* 1795
1-Phenylcyclopentanecarboxylic Acid 2-[2-(Diethylamino)ethoxy]ethyl Ester *see* 1795
1-Phenylcyclopentanecarboxylic Acid 2-(Diethylamino)ethyl Ester *see* 1778
3-(2-Phenyl-2-cyclopentylglycoloyloxy)-1,1-dimethylpyrrolidinium Bromide *see* 4537
rel-(1*R*,2*S*)-2-Phenylcyclopropanamine *see* 9734
Phenyl Cyclopropyl Sulfide *see* 2747
2-Phenyldiazenecarbothioic Acid 2-Phenylhydrazide *see* 3420
2-Phenyldiazenecarboxylic Acid 2-Phenylhydrazide *see* 3353
4-(2-Phenyldiazenyl)benzenamine *see* 414
4-(2-Phenyldiazenyl)-1,3-benzenediamine Hydrochloride (1:1) *see* 2261
1-(2-Phenyldiazenyl)-2-naphthalenamine *see* 10293
1-[2-[4-(2-Phenyldiazenyl)phenyl]diazenyl]-2-naphthalenol *see* 9015
3-(2-Phenyldiazenyl)-2,6-pyridinediamine *see* 7324
3-Phenyl-5-(dibutylaminoethylamino)-1,2,4-oxadiazole *see* 1513
Phenyl Dichlorophosphate, 7394
1-Phenyl-1-(diethylaminocarbonyl)-2-(aminomethyl)cyclopropane *see* 6275
β-Phenyl-*o*-(diethylaminoethoxy)propiophenone *see* 3765
3-Phenyl-5-(β-diethylaminoethyl)-1,2,4-oxadiazole *see* 7043
1-Phenyl-5,5-diethylbarbituric Acid *see* 7336
1-Phenyl-3-(*O*,*O*-diethylthionophosphoryl)-1,2,4-triazole *see* 9767
Phenyl Diguanide *see* 7389
2-Phenyl-3,5-dihydroxy-4-butylpyrazolidine *see* 6317
1-Phenyl-4-(2,3-dihydroxypropyl)diethylenediamine *see* 3500
1-Phenyl-4-(2,3-dihydroxypropyl)piperazine *see* 3500
2-Phenyl-1,3-diketohydrindene *see* 7348
Phenyldimazone *see* 6798
5-Phenyl-2-(dimethylamino)-2-oxazolin-4-one *see* 9533
1-Phenyl-2,3-dimethyl-4-amino-4-pyrazolone *see* 413
1-Phenyl-2-(α,α-dimethylethanolamino)ethanol *see* 4038

Phloracetophenone 4,6-Dimethyl Ether *see* 10265
Phloretin, 7437
Phloretin-2'-β-glucoside *see* 7438
Phlorhizin *see* 7438
Phloridzin, 7438
Phlorizin *see* 7438
Phloroglucin *see* 7439
Phloroglucinol, 7439
Phlorol *see* 3892
Phloropropiophenone *see* 4138
Phlorrhizen *see* 7438
Phloxine B, 7440
Phloxine O *see* 7440
PHMB *see* 7677
Phoenicin *see* 7346
Phoenicochroite *see* 5460
Pholcodine, 7441
Pholedrine, 7442
Phol-Tux [Interdelta] *see* 3884
Phomin *see* 2787
Phorate, 7443
Phoratoxins *see* 10206
Phorbol, 7444
Phorone, 7445
Phorwite RN *see* 1316
Phos *see* 4852
Phosalone, 7446
Phosdrin [AMVAC] *see* 6244
Phoselit [Kytta-Siegfried] *see* 4852
Phosethyl Al *see* 4279
Phos-ex [Vitaline] *see* 1649
Phosgene, 7447
Phosgeniminium Chloride *see* 10174
Phosilite *see* 4852
Phoskill [United Phosphorus] *see* 6337
PhosLo [Nabi] *see* 1649
Phosmet, 7448
Phosote *see* 2561
Phosphacol *see* 7134
Phosphagen *see* 7452
Phosphaljel [Am. Home] *see* 352
Phosphalugel [Yamanouchi] *see* 352
Phosphamidon, 7449
Phosphate of Soda *see* 8789
"Phosphatidal Choline" *see* 7633
"Phosphatidal Ethanolamine" *see* 7633
Phosphatidic Acids *see* 4522
Phosphatidylcholine *see* 5483
Phosphatidylethanolamine *see* 1975
Phosphatidylglycerol *see* 8049
Phosphine, 7450
Phosphinic Acid *see* 4907
Phosphinic Acid Ammonium Salt (1:1) *see* 525
Phosphinic Acid Barium Salt (2:1) *see* 974
Phosphinic Acid Calcium Salt (2:1) *see* 1677
Phosphinic Acid Iron(3+) Salt *see* 4053
Phosphinic Acid Manganese(2+) Salt (2:1) *see* 5796
Phosphinic Acid Potassium Salt (1:1) *see* 7762
Phosphinic Acid Sodium Salt (1:1) *see* 8763
3,3',3''-Phosphinidynetrispropanoic Acid *see* 9222
1,1',1''-Phosphinothioylidynetrisaziridine *see* 9518
Phosphinothricin, 7451
1,1',1''-Phosphinylidynetrisaziridine *see* 9837
1,1',1''-Phosphinylidynetris[2-methylaziridine] *see* 6008

Phosphocol P 32 [Mallinckrodt] *see* 2231
Phosphocozymase *see* 6432
Phosphocreatine, 7452
Phosphocysteamine, 7453
3-Phosphoglyceraldehyde *see* 4518
Phospholeum *see* 7700
Phospholine Iodide [Wyeth] *see* 3548
Phospholipase A *see* 5892
Phosphomolybdic Acid, 7454
Phosphonic Acid *see* 7458
Phosphonic Acid Bis(phenylmethyl) Ester *see* 3019
Phosphonic Acid Calcium Salt (1:1) *see* 1698
Phosphonic Acid Monoethyl Ester Aluminum Salt (3:1) *see* 4279
Phosphonic Acid Potassium Salt (1:2) *see* 7782
Phosphonic Acid Sodium Salt (1:2) *see* 8793
Phosphonium Iodide, 7455
N-(Phosphonoamidino)sarcosine *see* 7452
Phosphonodithioimidocarbonic Acid Cyclic Propylene P,P-Diethyl Ester *see* 5922
(R)-9-(2-Phosphonomethoxypropyl)adenine *see* 9289
Phosphono-O-methyl-2,6-diisopropylphenol *see* 4286
N-(Phosphonomethyl)glycine *see* 4547
Phosphonomycin *see* 4281
2-(Phosphonooxy)benzoic Acid *see* 4282
Phosphonorm [Medice] *see* 339
O-Phosphono-L-serine *see* 7475
9-(2-Phosphonylmethoxyethyl)adenine *see* 139
Phosphoprotein Phosphatase *see* 1643
Phosphoramidothioic Acid O,S-Dimethyl Ester *see* 6019
Phosphorazidic Acid Diphenyl Ester *see* 3366
Phosphoric Acid, 7456
Phosphoric Acid Aluminum Salt (1:1) *see* 352
Phosphoric Acid Ammonium Cobalt(2+) Salt (1:1:1) *see* 510
Phosphoric Acid Ammonium Salt (1:1) *see* 541
Phosphoric Acid Ammonium Salt (1:2) *see* 540
Phosphoric Acid Barium Salt (1:1) *see* 986
Phosphoric Acid Bismuth(3+) Salt (1:1) *see* 1276
Phosphoric Acid Calcium Salt (1:1) *see* 1694
Phosphoric Acid Calcium Salt (2:1) *see* 1695
Phosphoric Acid Calcium Salt (2:3) *see* 1696
Phosphoric Acid 7-Chlorobicyclo[3.2.0]-hepta-2,6-dien-6-yl Dimethyl Ester *see* 4699
Phosphoric Acid 2-Chloro-1-(2,4-dichlorophenyl)ethenyl Diethyl Ester *see* 2090
Phosphoric Acid 2-Chloro-3-(diethylamino)-1-methyl-3-oxo-1-propen-1-yl Dimethyl Ester *see* 7449
Phosphoric Acid (Z)-2-Chloro-1-(2,4,5-trichlorophenyl)ethenyl Dimethyl Ester *see* 9337
Phosphoric Acid Chromium(3+) Salt (1:1) *see* 2231

Phosphoric Acid Cobalt(2+) Salt (2:3) *see* 2431
Phosphoric Acid Copper(2+) Salt (2:3) *see* 2638
Phosphoric Acid 1,2-Dibromo-2,2-dichloroethyl Dimethyl Ester *see* 6442
Phosphoric Acid 2,2-Dichloroethenyl Dimethyl Ester *see* 3089
Phosphoric Acid 2,2-Dichlorovinyl Dimethyl Ester *see* 3089
Phosphoric Acid Diethyl 4-Nitrophenyl Ester *see* 7134
Phosphoric Acid (1E)-3-(Dimethylamino)-1-methyl-3-oxo-1-propen-1-yl Dimethyl Ester *see* 3097
Phosphoric Acid Dimethyl Ester, Ester with 2-Chloro-N,N-diethyl-3-hydroxycrotonamide *see* 7449
Phosphoric Acid Dimethyl Ester, Ester with cis-3-Hydroxy-N,N-dimethylcrotonamide *see* 3097
Phosphoric Acid Dimethyl Ester, Ester with (E)-3-Hydroxy-N-methylcrotonamide *see* 6337
Phosphoric Acid Dimethyl (1E)-1-Methyl-3-(methylamino)-3-oxo-1-propenyl Ester *see* 6337
Phosphoric Acid Glycerol Esters *see* 4522
Phosphoric Acid Iron(2+) Salt (2:3) *see* 4080
Phosphoric Acid Iron(3+) Salt (1:1) *see* 4056
Phosphoric Acid Lead(2+) Salt (2:3) *see* 5471
Phosphoric Acid Magnesium Salt (1:1) *see* 5745
Phosphoric Acid Magnesium Salt (2:1) *see* 5746
Phosphoric Acid Magnesium Salt (2:3) *see* 5747
Phosphoric Acid, Meta, 7457
Phosphoric Acid Monophenyl Ester Sodium Salt (1:2) *see* 3399
Phosphoric Acid Nickel(2+) Salt (2:3) *see* 6600
Phosphoric Acid Potassium Salt (1:1) *see* 7779, 7780
Phosphoric Acid Potassium Salt (1:3) *see* 7781
Phosphoric Acid Sodium Salt (1:1) *see* 8790
Phosphoric Acid Sodium Salt (1:2) *see* 8789
Phosphoric Acid Sodium Salt (1:3) *see* 8792
Phosphoric Acid Tributyl Ester *see* 9781
Phosphoric Acid Triethyleneimide *see* 9837
Phosphoric Acid Triethyl Ester *see* 9838
Phosphoric Acid Triphenyl Ester *see* 9915
Phosphoric Acid Tris(2,3-dibromopropyl) Ester *see* 9927
Phosphoric Acid Tris(1,3-dichloro-2-propyl)ester *see* 9224
Phosphoric Acid Tris(methylphenyl) Ester *see* 9940
Phosphoric Acid Tris(2-methylphenyl) Ester *see* 9941
Phosphoric Acid Zinc Salt (2:3) *see* 10351
Phosphoric Anhydride *see* 7467
Phosphoric Bromide *see* 7462
Phosphoric Chloride *see* 7463

2-(1-Phthalazinyl)hydrazinecarboxylic Acid Ethyl Ester see 9660
o-Phthaldialdehyde see 7480
Phthalein Complexone see 2566
Phthalein Purple see 2566
Phthalic Acid, 7483
m-Phthalic Acid see 5243
p-Phthalic Acid see 9305
o-Phthalic Acid Bis[diethylamide] see 9348
Phthalic Acid Diamide see 7481
Phthalic Acid Dibutyl Ester see 3044
Phthalic Acid Dimethyl Ester see 3279
Phthalic Acid 1-Ethyl-1-methyl-2-propynyl Ester see 7486
Phthalic Acid Potassium Salt see 7733
Phthalic Anhydride, 7484
Phthalidyl D-α-Aminobenzylpenicillanate see 9165
Phthalidyl 2-(3-Trifluoromethylanilino)-nicotinate see 9176
Phthalidyl 2-(α,α,α-Trifluoro-m-toluidino)nicotinate see 9176
Phthalimide, 7485
α-Phthalimidoglutarimide see 9403
3-Phthalimidoglutarimide see 9403
Phthalin see 7357
(SP-4-1)-[29H,31H-Phthalocyaninato-(2−)-κN²⁹,κN³⁰,κN³¹,κN³²]copper see 2505
[[N,N',N'',N'''-[(29H,31H-Phthalocyanine-C,C,C,C-tetrayl-κN²⁹,κN³⁰,κN³¹,-κN³²)tetrakis[methylenethio[(dimethylamino)methylidyne]]]tetrakis[N-methylmethanaminiumato]](2−)]Copper(4+) Chloride (1:4) see 212
Phthalofyne, 7486
Phthalophos see 6438
Phthalophos (USSR) see 7448
Phthaloyl Chloride, 7487
N-Phthaloylglutamimide see 9403
Phthaloylsulfacetamide see 7488
Phthalthrin see 9370
2-(N⁴-Phthalylaminobenzenesulfonamido)thiazole see 7489
N-Phthalylglutamic Acid Imide see 9403
Phthalylsulfacetamide, 7488
Phthalylsulfacetimide see 7488
2-(N⁴-Phthalylsulfanilamido)thiazole see 7489
Phthalylsulfathiazole, 7489
Phthalylsulfonazole see 7489
Phthiocol, 7490
Phtorazisin [Medexport] see 4145
PhXA-41 see 5431
Phycite see 3733
Phycobilin see 7491
Phycobiliproteins, 7491
Phycobilisomes see 7491
Phycocyanins see 7491
Phycoerythrins see 7491
Phygon see 3054
Phygon Paste see 3054
Phygon XL see 3054
Phylletten [Arznei Müller-Rorer] see 2914
Phyllocontin [Purdue Frederick] see 461
α-Phyllohydroquinone see 7492
Phylloquinone, 7492
Phyllotemp [Mundipharma] see 461
Phyllyrin see 7434
Phyone see 8842
Physalaemin, 7493
Physalemin see 7493
Physalien see 10316
Physalin see 10316

Physcione see 3618
Physeptone [Glaxo Wellcome] see 6016
Physex [Leo Pharm] see 2221
Physiomycine see 6015
Physiotens [Giulini] see 6379
Physodalin see 7494
Physodic Acid, 7494
Physostigma, 7495
Physostigmine, 7496
Physostigmine Aminoxide see 4425
Physovenine, 7497
Phytane-1,2,3-triol see 7498
Phytantriol, 7498
Phytar [Monterey] see 1609
Phytat D. B. [Daniel-Brunet] see 7499
Phytic Acid, 7499
Phytin see 7499
Phytoagglutinins see 5484
Phytoalexins see 8037
Phytochlorin, 7500
Phytochlorin e see 7500
Phytochrome see 7491
Phytocor [Pharma Dynamics] see 7672
Phytoecdysteroids see 3538
Phytofluene, 7501
all-trans-Phytofluene see 7501
Phytohemagglutinin see 5484
Phytol, 7502
Phytolacca, 7503
Phytomelin see 8438
Phytomenadione see 7492
Phytomonic Acid see 5390
Phytonadiol see 7492
Phytonadiol Sodium Diphosphate see 7492
Phytonadione see 7492
α-Phytosterol see 8694
3-Phytylmenadione see 7492
Piazofolina see 6359
Piazolin see 6359
PIBCA see 5185
Pibenzimol see 1250
Piberaline, 7504
Piboserod, 7505
Picaridin, 7506
Piceid see 8279
Picein, 7507
Picene, 7508
Piceol see 7507
Piceoside see 7507
Picfume [Dow] see 2159
Picloram, 7509
Picloxydine, 7510
Pico [BASF] see 7511
Pi-Coli [Ist. Vitamine] see 7572
Picolinafen, 7511
α-Picoline, 7512
β-Picoline, 7513
γ-Picoline, 7514
4-Picoline see 7514
Picolinic Acid, 7515
γ-Picolinic Acid see 5233
Picolinic Acid N-Methylbetaine see 4767
7-(β-3'-Picolylaminoethyl)theophylline see 7542
2-Picolylidenebis(p-phenyl Sodium Sulfate) see 7518
4,4'-(2-Picolylidene)bis(phenylsulfuric Acid) Disodium Salt see 7518
N-(2-Picolyl)-N-phenyl-N-(2-piperidinoethyl)amine see 7516
Picoperidamine see 7516
Picoperine, 7516
Picoplatin, 7517
Pico-Salax [Ferring] see 7518

Picosulfate Sodium, 7518
Picosulfol see 7518
Picotamide, 7519
Picoxystrobin, 7520
Picragol [Wyeth] see 8665
Picramic Acid, 7521
Picrasmin see 5266
Picric Acid, 7522
Picric Acid Ammonium Salt see 544
Picric Acid Silver Derivative see 8665
Picrocine see 2930
Picrocrocin, 7523
Picrolonic Acid, 7524
Picromycin, 7525
Picronitric Acid see 7522
Picropodophyllic Acid see 7662
Picropodophyllin see 7663
Picropodophyllinic Acid Lactone see 7663
Picrosalvin see 1852
Picrotin, 7526
Picrotol see 8665
Picrotoxin, 7527
Picrotoxinin, 7528
Picryl Chloride, 7529
Picrylmethylnitramine see 6659
Picrylnitromethylamine see 6659
Pictol [Mallinckrodt] see 6091
Pictor [BASF] see 3292
Picumast, 7530
PID see 7348
Pidilat [Solvay] see 6613
Pidorubicin see 3678
Pidotimod, 7531
Pierami [Fournier] see 400
PIFA see 7406
Pifarnine, 7532
Pifatidine see 8407
Pifazin [Pierrel] see 7532
Pifithrin-α, 7533
Pigitil [Fidia] see 7531
Pigmentation Hormone see 6381
Pigmex see 6334
Pigweed see 370
Piketoprofen, 7534
Pikromycin see 7525
Pilagan [Allergan] see 7536
Pildralazine, 7535
Piliophen see 5081
Pill-bearing Spurge see 3945
Pilo [Novopharma] see 7536
Pilocarpine, 7536
β-Pilocarpine see 7536
Pilocarpus, 7537
Pilofrin [Allergan] see 7536
Pilogel [Alcon] see 7536
Pilopine HS [Alcon] see 7536
Piloral [Nippon Kayaku] see 2343
Pilosine see 5245
Pilostat [Bausch & Lomb] see 7536
Pilot [Schering] see 8203
Piloty's Acid, 7538
Pilsicainide, 7539
Pilzcin [Shionogi] see 2580
Pimafucin [Basotherm] see 6509
Pimagedine see 438
Pimaric Acid, 7540
α-Pimaric Acid see 7540
β-Pimaric Acid see 5522
d-Pimaric Acid see 7540
l-Pimaric Acid see 5522
Pimaricin see 6509
Pimavecort [Mycofarm] see 6539
Pimecrolimus, 7541
Pimefylline, 7542

Pimelic Acid, 7543
Pimelic Ketone *see* 2719
Pimenta, 7544
Pimenta Berry Oil *see* 7544
Pimenta Leaf Oil *see* 7544
Pimento Oil *see* 7544
Pimephylline *see* 7542
Pimexone [Formenti] *see* 5927
Pimobendan, 7545
Pimozide, 7546
Pimpernel *see* 7547
Pimpinella, 7547
Pimpinellin, 7548
Pinacidil, 7549
Pinacol, 7550
Pinacol Diborane *see* 1301
Pinacolin *see* 7551
Pinacolone, 7551
Pinacoloxymethylphosphoryl Fluoride
  see 8839
Pinacolyl Methylphosphonofluoridate
  see 8839
Pinacone *see* 7550
Pinacyanol Chloride *see* 8164
Pinakon *see* 4746
Pinang *see* 766
Pinaverium Bromide, 7552
Pinazepam, 7553
Pinbetol [Dolorgiet] *see* 7554
Pindac [Leo Pharm] *see* 7549
Pindione [Lipha] *see* 7348
Pindolol, 7554
Pindone, 7555
Pindoptan [Kanoldt] *see* 7554
"Pineapple Oil" *see* 3830
Pinene, 7556
Pinene [Diachem] *see* 7509
α-Pinene, 7556
β-Pinene, 7557
2-Pinene, 7556
Pinene Hydrochloride *see* 1342
(1*R*,5*R*)-(+)-2-Pinen-4-one *see* 10154
2-Pinen-7-one *see* 2256
Pine Oil, 7558
Pine Resin *see* 10007
Pineroro [Maruko] *see* 3341
Pine Tar, 7559
Pine Tar Oil *see* 9200
Pine Tulip *see* 2057
Pinguinain, 7560
Pinite *see* 7561
Pinitol, 7561
D-Pinitol *see* 7561
Pinkroot *see* 8875
Pinnacle [DuPont] *see* 9468
Pinnoite *see* 5722
Pinolenic Acid, 7562
Pinol Hydrate *see* 8705
Pinorubin [Nippon Kayaku] *see* 7601
Pinosylvin, 7563
Pinoxaden, 7564
Pinozan *see* 7576
Pinrou [Century] *see* 7576
Pinus Montana Oil *see* 6883
Pinus Pumilio Oil *see* 6883
PIO *see* 7188
Pioglitazone, 7565
Pipacycline, 7566
Pipadox [Dagra] *see* 7577
Pipamazine, 7567
Pipamperone, 7568
Pipanol [Winthrop] *see* 9864
Pipazethate, 7569
Pipeacid [SIT] *see* 7572
Pipecolic Acid, 7570

Pipecolinic Acid *see* 7570
Pipecurium Bromide, 7571
Pipecuronium Bromide *see* 7571
Pipedac [Teofarma] *see* 7572
Pipemid [Gentili] *see* 7572
Pipemidic Acid, 7572
Pipenale [Fuso] *see* 7583
Pipenzolate Bromide, 7573
Pipenzolate Methylbromide *see* 7573
Pipenzolone Bromide *see* 7573
Piperacetazine, 7574
Piperacillin, 7575
Piperamic Acid *see* 7572
Piperate *see* 7576
Piperazate [Leeming] *see* 7576
Piperazidine *see* 7576
Piperazine, 7576
Piperazine Adipate, 7577
Piperazine-1,4-bis(*N*,*N*'-diethylenephos-
  phonediamide) *see* 3376
1,4-Piperazinebis(ethanesulfonic Acid)
  see 7592
Piperazine-*N*,*N*'-bis(2-ethanesulfonic
  Acid) *see* 7592
Piperazine Calcium Edathamil *see* 7576
1,4-Piperazinediethanesulfonic Acid *see*
  7592
2,5-Piperazinedione, 7578
1,4-Piperazinediylbis[bis(1-aziridinyl)-
  phosphine Oxide] *see* 3376
1,1'-[1,4-Piperazinediylbis(imidocarbonyl)]bis[3-(*p*-chlorophenyl)guanidine]
  see 7510
1,1'-(1,4-Piperazinediyl)bis[3-[(methylsulfonyl)oxy]-1-propanone] *see* 7594
N,*N*'-[1,4-Piperazinediylbis(2,2,2-trichlo-
  roethylidene)]bisformamide *see* 9859
Piperazine Estrone Sulfate *see* 3763
Piperazine Sultosylate *see* 9124
Piperic Acid, 7579
Piperidic Acid *see* 425
Piperidine, 7580
2-Piperidinecarboxylic Acid *see* 7570
3-Piperidinecarboxylic Acid *see* 6647
4-Piperidinecarboxylic Acid *see* 5235
1-Piperidineethanol Benzilate *see* 7583
3-Piperidino-4'-butoxypropiophenone
  see 3520
6-Piperidino-2,4-diaminopyrimidine 3-
  Oxide *see* 6285
2-(2-Piperidinoethoxy)ethyl 10*H*-Pyrido-
  [3,2-*b*][1,4]benzothiadiazine-10-car-
  boxylate *see* 7569
2-(2-Piperidinoethoxy)ethyl 10-Thia-1,9-
  diazaanthracene-10-carboxylate *see*
  7569
2-(1-Piperidino)ethyl Benzilate *see* 7583
β-Piperidinoethyl-4-butoxyphenyl Ketone *see* 3520
2-Piperidinoethyl 3-Methylflavone-8-car-
  boxylate *see* 4128
2-Piperidinoethyl 3-Methyl-4-oxo-2-
  phenyl-4*H*-1-benzopyran-8-carboxyl-
  ate *see* 4128
N-(2-Piperidinoethyl)-*N*-(2-pyridylmeth-
  yl)aniline *see* 7516
2-Piperidinomethyl-1,4-benzodioxan *see*
  7591
1-Piperidino-2-methyl-3-(*p*-tolyl)-3-pro-
  panone *see* 9681

3-Piperidino-1-phenyl-1-bicycloheptenyl-
  1-propanol *see* 1238
3-Piperidino-1-phenyl-(Δ⁵-bicyclo[2.2.1]-
  hepten-2-yl)-1-propanol *see* 1238
1-Piperidino-2-(*N*-propionylanilino)pro-
  pane *see* 7323
rel-(α*S*)-α-(2*R*)-2-Piperidinyl-2,8-bis(trifluoromethyl)-4-quinolinemethanol *see*
  5872
N-(Piperidin-1-yl)-5-(4-chlorophenyl)-1-
  (2,4-dichlorophenyl)-4-methyl-1*H*-pyr-
  azole-3-carboxamide *see* 8355
N-(2-Piperidinylmethyl)-2,5-bis(2,2,2-tri-
  fluoroethoxy)benzamide *see* 4130
1-(2-Piperidinyl)-2-propanone *see* 7181
3-(2*S*)-2-Piperidinylpyridine *see* 612
6-(1-Piperidinyl)-2,4-pyrimidinediamine
  3-Oxide *see* 6285
Piperidione, 7581
Piperidolate, 7582
Piperidylamidone *see* 3377
α-(2-Piperidyl)benzhydrol *see* 7597
DL-*erythro*-α-2-Piperidyl-2,8-bis(trifluoromethyl)-4-quinolinemethanol *see*
  5872
3-(1-Piperidyl)-1-cyclohexyl-1-phenyl-1-
  propanol Hydrochloride *see* 9864
3-(*N*-Piperidyl)-1,1-diphenyl-1-propanol
  see 7859
β-Piperidylethyl Benzilate *see* 7583
2-(1-Piperidyl)ethyl *p*-Butoxyphenyl
  Ketone *see* 3520
2-(β-*N*-Piperidylethyl)-4,4-diphenyl-1,3-
  dioxolan-5-one Hydrochloride *see*
  7596
2-(1-Piperidylmethyl)-1,4-benzodioxan
  see 7591
Piperilate, 7583
Piperine, 7584
Piperinic Acid *see* 7579
Piperinsäure (German) *see* 7579
Piperitone, 7585
Piperlongumine, 7586
Piperocaine, 7587
Piperonal, 7588
Piperonil [Lusofarmaco] *see* 7568
Piperonylaldehyde *see* 7588
Piperonyl Butoxide, 7589
Piperonylic Acid, 7590
10-[(4-Piperonyl-1-piperazinyl)acetyl]-
  phenothiazine *see* 4013
2-(4-Piperonyl-1-piperazinyl)pyrimidine
  see 7607
1-Piperonyl-4-(3,7,11-trimethyl-2,6,10-
  dodecatrienyl)piperazine *see* 7532
Piperoxan, 7591
(*E*,*E*)-1-Piperoylpiperidine *see* 7584
(*Z*,*Z*)-1-Piperoylpiperidine *see* 2048
Piperverm [Wild] *see* 7576
Piperylene *see* 7220
cis-Piperylene *see* 7220
trans-Piperylene *see* 7220
1,1'-(1,4-Piperzinediyl)bis-[3-bromo-1-
  propanone] *see* 7593
PIPES, 7592
Pipethanate *see* 7583
Pipitzahoic Acid *see* 7269
Pipizan Citrate [Merck & Co.] *see* 7576
Piplartine *see* 7586
Pipobroman, 7593
Piportil [Specia] *see* 7595
Piportil L4 [Specia] *see* 7595
Piportil M2 [Specia] *see* 7595
Piposulfan, 7594
Pipothiazine *see* 7595

Plesmet [Link] see 4066
Plessy's Green see 2231
Pletal [Otsuka] see 2280
Pleurisy Root see 817
Pleuromutilin, 7650
Pleurotin see 7651
Pleurotine, 7651
Plexamine [Biodica] see 4758
Plexiglas see 6013
Plexol 201 see 1255
Plicamycin, 7652
Plicatic Acid, 7653
Plicera [Amicus] see 5217
Plictran [Dow] see 2757
Plinabulin, 7654
Plitican [Synthelabo] see 242
PLLA see 5386
PLP see 8092
PLS see 7717
Pluchine see 1181
Plumbagin, 7655
Plumbago see 4575
Plumbic Acetate see 5478
Plumbic Fluoride see 5479
Plumbic Oxide see 5462
Plumbous Acetate see 5452
Plumbous Bromide see 5458
Plumbous Chloride see 5459
Plumbous Chromate see 5460
Plumbous Fluoride see 5463
Plumbous Iodide see 5466
Plumbous Molybdate see 5467
Plumbous Nitrate see 5469
Plumbous Oxide see 5468
Plumbous Plumbate see 5472
Plumbous Sulfide see 5476
Plumericin, 7656
Plumieride, 7657
Pluracol E [BASF] see 7688
Plurexid [UCB] see 2092
Plurimen [Viatris] see 8565
Pluripoietin see 4573
Pluronic F68 [BASF] see 7679
Pluronic L62 see 7679
Pluronic L101 [BASF] see 7679
Pluropon [Boehringer, Ing.] see 6274
Plutonium, 7658
PLV-2 see 3988
PLX-4032 see 10138
PM-150 see 10133
PM-02734 see 3601
PM-92102 see 5323
PM-185184 see 8556
PMA see 7411
PMAC see 7411
PMAS see 7411
PMEA see 139
PMP see 5279, 5388, 8093
iPMPA see 4966
(R)-PMPA see 9289
PMSF, 7659
PMSG see 2221
PN see 8132
PN-200-110 see 5287
PN-205-033 see 5287
PN-205-034 see 5287
PNB see 6680
Pneumorel [Stroder] see 4026
PNU-100766 see 5559
PNU-140690 see 9615
PNU-151774E see 8450
PNU-166196 see 1458
PNU-180638E see 298
PNU-200583E see 9684
P.O. 12 [Biothérax] see 3644

Poast [BASF] see 8612
Pocan see 7503
Podocarpic Acid, 7660
Pododacric Acid, 7661
Podofilox see 7663
Podomexef [Daiichi Sankyo] see 1942
Podophyllic Acids, 7662
Podophyllin see 7664
Podophyllinic Acid see 7662
Podophyllinic Acid 2-Ethylhydrazide see 7662
Podophyllinic Acid Hydrazide see 7662
Podophyllinic Acid Lactone see 7663
Podophyllotoxin, 7663
Podophyllum, 7664
Podophyllum Resin see 7664
POE Alcohol Ethers see 7697
POE Fatty Acid Esters see 7698
POE (40) Monostearate see 7698
POE Sorbitan Esters see 7703
POE (20) Sorbitan Monooleate see 7703
POE (8) Stearate see 7698
P(OEt)₃ see 9840
Pogy Oil see 5904
Poi, 7665
Poison Black Cherry see 1026
Poison Elder see 7668
Poison Hemlock see 2490
Poison Ivy, 7666
Poison Nut see 6819
Poison Oak, 7667
Poison Parsley see 2490
Poison Sumac, 7668
Poison Tobacco see 4898
Poison Vine see 7666
Poke Root see 7503
L-Polamidon [HMR] see 6016
Polamin [Schering-Plough] see 2186
Polaprezinc, 7669
Polaramine [Schering-Plough] see 2186
Polaronil [Schering-Plough] see 2186
Polar Yellow [Ciba], 7670
Poldine Methylsulfate, 7671
Poleon [Sumitomo] see 6444
Policydal [Eisai] see 9042
Polibutin [Juste] see 9868
Policosanol, 7672
Policresulen, 7673
Polidexide, 7674
Polidocanol, 7675
Polifeprosan, 7676
Polifeprosan 20 see 7676
Polihexanide, 7677
Polimod [Poli] see 7531
Polishing Powder see 8905
Polistin T-Caps [Trommsdorff] see 1800
Pollakisu [Kodama] see 7054
Pollucite see 2007
Polmiror [Poli] see 6616
Polo [Novartis Agro] see 2972
Polonium, 7678
Poloxamer 182 see 7679
Poloxamer 188 see 7679
Poloxamer 331 see 7679
Poloxamers, 7679
Polyacrylonitrile see 6965
Poly(allylamin-co-N,N'-diallyl-1,3-di-amino-2-hydroxypropane) see 8613
Polyamide see 6822
Polyaminopropylbiguanide see 7677
Polyanetholesulfonic Acid Sodium Salt see 8796
Polyanhydroglucuronic Acid see 7038
Poly[1,3-bis(p-carboxyphenoxy)propane-sebacic Acid] see 7676

Polybrene see 4718
Polybrominated Biphenyls, 7680
Polybrominated Diphenyl Ethers, 7681
Polybromobiphenyls see 7680
Poly(butyl 2-Cyanoacrylate) see 1566
Poly(carbon Monofluoride) see 4576
Polycarbophil Calcium Salt see 1699
Poly-trans-[(2-carboxyethyl)germases-quioxane] see 7909
Polycell see 1830
Polychlorinated Biphenyls, 7682
Polychlorinated Dibenzofurans, 7684
Poly(2-chloro-1,3-butadiene) see 6546
Polychlorocamphene see 9716
Polychloroprene see 6546
Polychol [Croda] see 7697
Polychrome see 3753
Polycidine see 2914
Polycillin [BMS] see 583
Polycillin-N [BMS] see 583
Polyclar [ISP] see 7814
Polydatin see 8279
Polydextrose, 7685
Poly(dicarbon Monofluoride) see 4576
Poly[2-(diethylamino)ethyl] Polyglycer-ylene Dextran Hydrochloride see 7674
Poly[(dimethyliminio)-1,3-propanediyldi-methyliminio)-1,6-hexanediyl Bromide (1:2)] see 4718
Polydimethylsiloxane see 3237
Polyestradiol Phosphate, 7686
Polyetherin A see 6625
Poly(ethyl 2-Cyanoacrylate) see 3842
Polyethylene, 7687
Polyethylene Glycol, 7688
Polyethylene Glycol 200 see 7688
Polyethylene Glycol 400 see 7688
Polyethylene Glycol 600 see 7688
Polyethylene Glycol 1500 see 7688
Polyethylene Glycol 4000 see 7688
Polyethylene Glycol 6000 see 7688
Polyethylene Glycol Esters of Fatty Acids see 7698
Polyethylene Glycol Fatty Alcohol Ethers see 7697
Polyethylene Glycol p-Isooctylphenyl Ether see 6850
Polyethylene Glycol (9) Monododecyl Ether see 7675
Polyethylene Glycol 400 Monostearate see 7698
Polyethylene Glycol-polypropylene Gly-col Block Copolymer see 7679
Polyethyleneglycols Mono(nonylphenyl) Ether see 6764
Polyethylene Sodium Sulfonate see 5678
Polyethylene Terephthalates, 7689
Polyferon [Biogen] see 5036
Poly-G [Olin] see 7688
L-Polyglutamic Paclitaxel see 7081
Poly(Glycidyl Methacrylate-co-Ethyl Acrylate) see 2499
Polyglycol E [Dow] see 7688
Polygodial, 7690
Polygris [Essex] see 4584
PolyHeme [Northfield] see 7692
Polyhexamethylene Biguanide Hydro-chloride see 7677
Polyhexanide see 7677
Poly[imino-1,4-butanediylimino(1,6-di-oxo-1,6-hexanediyl)] see 6824
Poly(iminocarbonimidoyliminocarbonim-idoylimino-1,6-hexanediyl) Hydrochlo-ride (1:?) see 7677
Poly(iminocarbonylpentamethylene) see 6823

Poly[imino[1-[4-[[2-(4-carboxy-5,5-di-
methyl-2-thiazolidinyl)-1-oxo-2-
[(phenylacetyl)amino]ethyl]amino]bu-
tyl]-2-oxo-1,2-ethanediyl]] *see* 7210
Poly[imino(1-oxo-1,6-hexanediyl)] *see*
6823
PolyI:polyC12U *see* 8357
PolyI.poly($C_{12}U$) *see* 8357
Poly(isobutyl 2-Cyanoacrylate) *see* 5185
Poly(lactic Acid) *see* 5386
Poly[(+)-lactic Acid] *see* 5386
Poly-D-lactic Acid *see* 5386
Poly-L-lactic Acid *see* 5386
Poly-L-lactide *see* 5386
Polylysine, 7691
Polymannuronic Acid *see* 235
Polymerized Pyridoxylated Hemoglobin,
7692
Polymer of *N,N,N',N'*-Tetramethyl-
hexamethylenediamine and Trimeth-
ylene Bromide *see* 4718
Polymetatelluric Acid *see* 9262
Poly(methyl 2-Cyanoacrylate) *see* 6117
Polymorphonuclear Elastase *see* 3583
Polymox [BMS] *see* 574
Polymyxin, 7693
Polymyxin B *see* 7693
Polymyxin B$_1$ *see* 7693
Polymyxin B$_2$ *see* 7693
Polymyxin D *see* 7693
Polymyxin D$_1$ *see* 7693
Polymyxin D$_2$ *see* 7693
Polymyxin E$_1$ *see* 2463
Polymyxin E$_2$ *see* 2463
Polymyxin F *see* 7693
Polymyxin K *see* 7693
Polymyxin M *see* 7693
Polymyxin P *see* 7693
Polymyxin S$_1$ *see* 7693
Polymyxin T$_1$ *see* 7693
Polynoxylin, 7694
Poly(octyl Cyanoacrylate) *see* 6855
Polyoxidonium, 7695
Polyoxin A *see* 7696
Polyoxin AL *see* 7696
Polyoxin B *see* 7696
Polyoxins, 7696
Poly[1-(2-oxo-1-pyrrolidinyl)ethylene]
see 7814
Polyoxyaluminum Acetylsalicylate *see*
305
Polyoxyethylene Alcohols, 7697
Polyoxyethylene (20) Cetyl Ether *see*
7697
Polyoxyethylene Fatty Acid Esters, 7698
Polyoxyethylene Lauryl Ether *see* 7675
Polyoxyethylene(*n*)nonylphenyl Ether
see 6764
Polyoxyethylene Sorbitan Esters *see*
7703
Polyoxyethylene (20) Sorbitan Mono-
oleate *see* 7703
Polyoxyethylene (40) Stearate *see* 7698
Polyoxyl 8 Stearate *see* 7698
Polyoxyl 40 Stearate *see* 7698
Polyoxymethylene *see* 7129
Poly[oxy(1-methyl-2-oxo-1,2-ethanediyl)]
see 5386
Polypeptide *β*$_1$ *see* 9560
Polyphase P-100 [Kop-Coat] *see* 5112
Polyphenolase *see* 10019
Polyphenoloxidase *see* 10019
Polyphosphazenes, 7699
Polyphosphoric Acid, 7700
Polyphosphoric Acids Sodium Salts *see*
8773

Polyplasdone XL [ISP] *see* 7814
Polypodine A *see* 3538
Polypropylene, 7701
Poly(rI).poly[r($C_{12}U)_n$] *see* 8357
Polysaccharide B-1459 *see* 10255
Polysaccharide-K, 7702
Polysaccharide S-60 *see* 4417
Poly SFH-P *see* 7692
Poly-Solv DPM [Olin] *see* 3384
Polysorbate 80 *see* 7703
Polysorbates, 7703
Polystichalbin *see* 202
Polystichin *see* 836
Polystichocitrin *see* 4121
Polystictine *see* 2297
Polystyrene *see* 8990
Poly(styrenesulfonic Acid) Sodium Salt
see 8797
Polysulfide Rubber *see* 9492
Polytef *see* 7704
Polytergent B [Olin] *see* 6764
Polytergent G [Union Carbide] *see* 6850
Polytetrafluoroethylene, 7704
Polytetrafluoroethylene Resin *see* 7704
Poly(tetramethyleneadipamide) *see* 6824
Poly(*N,N,N',N'*-tetramethyl-*N*-trimeth-
ylenehexamethylenediammonium Di-
bromide) *see* 4718
Polythene [DuPont] *see* 7687
Polythiazide, 7705
Polytrim [Wellcome] *see* 7693
Polyvidone *see* 7814
Polyvinyl Alcohol, 7706
Polyvinyl Chloride, 7707
Polyvinylpolypyrrolidone *see* 7814
Poly(4-vinylpyridine-*N*-oxide) *see* 8059
Polyvinylpyrrolidone *see* 7814
Polyvinylpyrrolidone-iodine Complex
see 7815
Polyviol [Wacker] *see* 7706
Pomarsol [Bayer CropSci] *see* 9525
Pomarsol Z [Bayer] *see* 10372
POMC *see* 7905
Pomegranate, 7709
Ponalar [Parke-Davis] *see* 5869
Ponalide [Novartis] *see* 3810
Ponasterone A, 7710
Ponasterones *see* 7710
Ponazuril, 7711
Ponceau 3R, 7712
Ponceau SX, 7713
Poncho [Bayer CropSci.] *see* 2398
Poncuronium Bromide (rescinded
USAN) *see* 7109
Poncyl-FP [Takeda] *see* 4584
Pondinol [Roche] *see* 5870
Ponderal [Servier] *see* 4004
Ponderax [Selpharm] *see* 4004
Ponderex [Robins] *see* 4004
Pondimin [Robins] *see* 4004
Pondinil [Roche] *see* 5870
Pondocil [Leo Pharm] *see* 7626
Pondocillin [Leo Pharm] *see* 7626
Pondocillina [Sigma-Tau] *see* 7626
Ponoxylan [Berk] *see* 7694
Ponsinomycin *see* 6286
Ponstan [Pfizer] *see* 5869
Ponstel [Warner-Lambert] *see* 5869
Ponstyl [Parke-Davis] *see* 5869
Pontal [Parke-Davis] *see* 5869
Pontamine White BR *see* 1316
Ponticin *see* 8297
Pontocaine Hydrochloride [Sanofi Win-
throp] *see* 9333
P.O.P. *see* 7149

POP *see* 4685
Popotillo *see* 3662
Poppy Capsules, 7714
Poppy Heads *see* 7714
Poppy Oil, 7715
Poppy-seed Oil *see* 7715
Populin, 7716
Populnetin *see* 5322
Populoside *see* 7716
POR 8 [Novartis] *see* 6968
Poractant Alfa, 7717
Porcelain Clay *see* 5330
Porcilene [Syntex] *see* 4023
Porcine Somatotropin *see* 8842
Pore-forming Protein *see* 7274
Porfimer Sodium, 7718
Porfiromycin, 7719
Porofor BSH [Bayer] *see* 1074
Porofor-57 *see* 911
Porphin *see* 7720
Porphine, 7720
21*H*,23*H*-Porphine *see* 7720
Porphobilinogen, 7721
Porphyrins *see* 7720
Porphyropsin, 7722
Portolac [Zyma] *see* 5389
Posaconazole, 7723
Posedrine [Merck KGaA] *see* 1016
Posicor [Roche] *see* 6252
Posicycline [Alcon] *see* 7075
Posiformin [Ursapharm] *see* 1203
Posorutin [Ursapharm] *see* 9969
Posse [FMC] *see* 1827
Possipione [Recordati] *see* 7149
Postacton *see* 5691
Postafen [UCB] *see* 1474, 5848
Post-*γ*-globulin *see* 2775
Post-parotid Basic Proteins *see* 4757
Potaba [Glenwood] *see* 7727
Potasan [7724, 7724
Potash *see* 7740
Potasoral [Bago] *see* 4492
Potassa *see* 7761
Potassiject [Vetus] *see* 7742
Potassio-cuprous Cyanide *see* 2655
Potassium, 7725
Potassium Acetate, 7726
Potassium Acid Arsenate *see* 7728
Potassium Acid Carbonate *see* 7730
Potassium Acid Fluoride *see* 7731
Potassium Acid Oxalate *see* 7732
Potassium Acid Phosphate *see* 7780
Potassium Acid Phthalate *see* 7733
Potassium Acid Sulfate *see* 7734
Potassium Acid Tartrate *see* 7736
Potassium Alum *see* 354
Potassium Aminobenzoate, 7727
Potassium *p*-Aminobenzoate *see* 7727
Potassium Antimonyltartrate *see* 691
Potassium Arsenate, 7728
Potassium Arsenite, 7729
Potassium Arsenite Solution *see* 7729
Potassium Atractylate *see* 858
Potassium Aurichloride *see* 7800
Potassium Aurocyanide *see* 7749
Potassium Benzoate *see* 1093
Potassium Biborate *see* 7799
Potassium Bicarbonate, 7730
Potassium Bichromate *see* 7748
Potassium Bifluoride, 7731
Potassium Binoxalate, 7732
Potassium Biphosphate *see* 7780
Potassium Biphthalate, 7733
Potassium Bis(cyano-*κC*)argentate(1−)
see 7789

Pregneninolone see 3795
4-Pregnen-17α-ol-3,20-dione see 4877
4-Pregnen-21-ol-3,20-dione see 2902
Pregnenolone, 7851
Δ⁵-Pregnen-3β-ol-20-one see 7851
Δ⁴-Pregnen-21-ol-3,11,20-trione see 2873
Pregnesin [Serono] see 2221
Pregnyl [Organon] see 2221
Prehemataminic Acid see 6568
Prelay [Sankyo] see 9948
Prelis [Mibe] see 6228
Prelog-Djerassi Lactone, 7852
(+)-Prelog-Djerassi Lactonic Acid see 7852
Preludin [Boehringer, Ing.] see 7351
Premarin [Wyeth] see 2492
Premarrubiin see 5820
Premerge [Dow] see 3315
Premier [Bayer CropSci.] see 4951
Prempar [Duphar] see 8365
Premsyn PMS [Chattem] see 7101
Prenalex [Itherapia] see 9316
Prenalterol, 7853
Prenazone see 4039
Prenormine [AstraZeneca] see 850
Prenoxdiazine, 7854
Prenoxdiazine Hibenzate see 7854
Prenoxid [Khandelwal] see 7854
Prent [Bayer] see 20
Prentox [Prentiss] see 7589
Prenylamine, 7855
4-Prenyl-1,2-diphenyl-3,5-dioxopyrazoli-dine see 4039
4-Prenyl-1,2-diphenyl-3,5-pyrazolidinedi-one see 4039
Prep [Bayer CropSci.] see 3787
Pre-Par [Duphar] see 8365
Preparation K see 3428
Prepared Calcium Carbonate see 1659
Prepared Chalk see 1659
Pre-Pen [Kremers-Urban] see 7210
Prephenic Acid, 7856
Prepidil [Pharmacia & Upjohn] see 7988
Prepulsid [Janssen-Cilag] see 2315
Pres [Boehringer, Ing.] see 3623
Pre-Sate [Warner-Lambert] see 2188
Prescal [Novartis] see 5287
Presdate [Alfa] see 5377
Presenilin 1 see 7857
Presenilin 2 see 7857
Presenilins, 7857
Preservex [BMS] see 23
Presfersul see 4084
Presidon see 8108
Presinex [Pliva] see 2926
Presinol [Bayer] see 6127
Pressalolo [Locatelli] see 5377
Pressunic [Unipharm] see 3187
Pressural [Polifarma] see 4975
Pressyn [Ferring] see 10129
Prestara [Genelabs] see 2875
Prestige [Valent] see 4262
Prethcamide see 2582
Pretilachlor, 7858
Pretyrosine see 781
Prevacid [Takeda] see 5413
Prevage [Allergan] see 4929
Prevangor see 7222
Preventol G-D see 3081
Preveon [Gilead] see 139
Prevex [Schering] see 3987
Previcox [Merial] see 4118
Previcur [Bayer CropSci.] see 7912
Previscan [Procter & Gamble] see 4165
Prexan [Lafare] see 6499

Prexidil [Bioindustria] see 6285
Prexidine [Expanscience] see 2092
Prexige [Novartis] see 5661
Prezatide Copper Acetate see 4452
Prezios [Chugai] see 5828
Prezista [Tibotec] see 2829
PrGCD see 9169
Priadel [Sanofi-Aventis] see 5583
Prialt [Elan] see 10321
Priamide [Janssen] see 5248
Priaxim [Ravizza] see 4179
Priceite see 1655
Pridinol, 7859
Pridinol Mesylate see 7859
Pridopidine, 7860
Prifinium Bromide, 7861
Priftin [Aventis] see 8344
Prilagin [Sofar] see 6257
Prilan [IPESA] see 7920
Priligy [Janssen-Cilag] see 2821
Prilocaine, 7862
Prilon [Cassenne] see 7043
Prilosec [AstraZeneca] see 6939
Primacor [Sanofi-Synthelabo] see 6278
Primalan [Inava] see 5931
Primamycin see 4642
Primaquine, 7863
Primary Ammonium Phosphate see 541
Primary Calcium Phosphate see 1695
Primary Isoamyl Alcohol see 5241
Primary Magnesium Phosphate see 5746
Primary Sodium Phosphate see 8790
Primatene Mist [Wyeth] see 3674
Primaverine see 7827
Primaxin [Merck & Co.] see 4961
Primbactam [Menarini] see 918
Primcillin [AstraZeneca] see 7209
Primeral [Master] see 6499
Primeverose, 7864
Primexin [Wander] see 4245
Primicid see 7610
Primidone, 7865
Primisulfuron-methyl, 7866
Primobolan Tablets [Schering AG] see 6039
Primobolan-Depot [Schering AG] see 6039
Primofax [Rhône-Poulenc] see 6808
Primofenac [Streuli] see 3091
Primogonyl [Schering] see 2221
Primogyn C [Schering AG] see 3788
Primolut N [Schering AG] see 6786
Primolut-Nor [Schering AG] see 6786
Primonabol see 6039
Primonabol Depot see 6039
Primosiston [Schering AG] see 6786
Primostat [Bayer Schering] see 4448
Primoteston [Schering AG] see 9324
Primperan [Delagrange] see 6219
Primulaverin, 7867
Primulidin see 5779
Primulin see 5779
Primus [Dow AgroSci.] see 4139
Primycin, 7868
Prinadol see 7332
Princep [Syngenta] see 8672
Prince's Feather see 370
Prince's Pine see 2057
Principen [BMS] see 583
Prinil [Merck & Co.] see 5572
Prinivil [Merck & Co.] see 5572
Prinodolol see 7554
Prinomastat, 7869
Priodax [Schering] see 5069
Prioderm [Seaton] see 5764

Priomicina [San Carlo] see 4281
Priscol [Novartis] see 9666
Priscoline [Novartis] see 9666
Prism [Valent] see 2347
Pristacin see 2031
Pristane, 7870
Pristimerin see 1957
Pristinamycin, 7871
Pristinamycin IA see 7871
Pristinamycin IIₐ see 10200
Pristiq [Wyeth] see 2937
Pritor [GSK] see 9271
Privaprol [Lepetit] see 5643
Privenal see 4740
Privine [Novartis] see 6453
PRN see 7425
iPr₂NEt see 3219
(n-Pr₄N)(RuO₄) see 9386
Pro see 7895
PRO-132365 see 9736
Proaccelerin see 3960
Pro-Actidil [Wellcome] see 9922
ProAmatine [Roberts] see 6263
Proamipide see 8242
Proaxis [Dow AgroSci.] see 5398
Proazamine see 7901
Proazaphosphatrane see 10155
Probamyl [Continental Pharma] see 5929
Proban see 2782
Pro-Banthine [Concord] see 7919
Probecid [Astra] see 7872
Proben [SCS] see 7872
Probenecid, 7872
Probilin [Gödecke; Warner-Lambert] see 7598
Probolin [Tennessee Pharm.] see 6022
Probucol, 7873
Probucol Monosuccinate see 9002
Pro-Bumin [Probitas] see 8607
Probutylin [Rorer] see 7875
Procadil [Recordati] see 7877
Procainamide Hydrochloride, 7874
Procaine, 7875
Procaine Benzylpenicillinate see 7206
Procaine Penicillin G see 7206
Procalmadiol see 5929
Procalmidol see 5929
Procamide [Sintesa] see 7874
Procanbid [Parke-Davis] see 7874
Procan-SR [Parke-Davis] see 7874
Procapan [Panray] see 7874
Procaptan [Stroder] see 7285
Procarbazine, 7876
Procardia [Pfizer] see 6613
Procaspase-1 see 1884
Procaspase-3 see 1885
Procaspase-8 see 1886
Procaspase-9 see 1887
Procaterol, 7877
Procef [BMS] see 1943
Processine [Sanko] see 2306
Procetofen see 4009
Procetofene see 4009
Procetofenic Acid see 4009
Prochloraz, 7878
Prochlorpemazine see 7879
Prochlorperazine, 7879
Prochlorperazine Mesylate see 7879
Procholon [Squibb] see 2872
Prociclide [Crinos] see 2864
Proclaim [Syngenta] see 3612
Proclorperazine see 7879
Procodazole, 7880
Procollagen see 2464
Proconazole see 7932

Proconvertin *see* 3961
Procoralan [Servier] *see* 5294
Procorum [Knoll] *see* 4386
Procrit [Ortho Biotech] *see* 3744
Proctocort [Salix] *see* 4824
Proctofoam-NS [Schwarz] *see* 7829
Procuta [Expanscience] *see* 5274
Procyanidins *see* 1902
Procyclidine, 7881
Procylin [Kaken] *see* 1154
Procymidone, 7882
Procysteine [Baxter] *see* 7049
Procytox [Baxter] *see* 2743
Pro-Dafalgan [UPSA] *see* 7906
Prodectin [Gedeon Richter] *see* 8088
Prodel *see* 3521
Pro-Diaban [Bayer; Schering AG] *see* 4479
Prodiamine, 7883
Prodif [Pfizer] *see* 4153
Prodigiosin, 7884
Prodigiosine *see* 7884
Prodigy [Dow AgroSci.] *see* 6066
dl-α-Prodine *see* 306
(±)-β-Prodine Hydrochloride *see* 306
Prodium [Breckenridge] *see* 7324
Prodlure, 7885
Pro Dorm [Synthelabo] *see* 5636
Proendotel [Fidia] *see* 2393
Pro-Entra [Wellcome-Sumitomo] *see* 9922
Pro-Epanutin [Pfizer] *see* 4285
Profasi [Serono] *see* 2221
Profemin [Yamanouchi] *see* 4338
Profenamine *see* 3800
Profender [Bayer] *see* 3617
Profenid [RPR] *see* 5352
Profenofos, 7886
Profenon [Morishita] *see* 4138
Profenone *see* 7149
Profibrinolysin *see* 7635
Profil [Nippon Soda] *see* 48
Profilate [Alpha Ther.] *see* 3962
Profilnine [Alpha Ther.] *see* 3963
Proflavine, 7887
Proflavine Hemisulfate *see* 7887
Proflex [Novartis] *see* 4919
Proflox [Esteve] *see* 6377
Proformiphen *see* 7370
Profume [Dow AgroSci.] *see* 9111
Progabide, 7888
Progallin P [Nipa] *see* 7971
Proge [Mochida] *see* 4877
Pro-Gen [Vétoquinol] *see* 783
Progeril [Midy] *see* 3710
Progesic [Lilly] *see* 4011
Progestasert [Alza] *see* 7889
Progesterone, 7889
Progestogel [Besins-Iscovesco] *see* 7889
Progeston [Fuji] *see* 7889
Proglicem *see* 3007
Proglumetacin, 7890
Proglumide, 7891
Proglycem [Ivax] *see* 3007
Prograf [Fujisawa] *see* 9155
Program [Novartis] *see* 5654
Proguanil, 2091
Progynon B [Schering AG] *see* 3758
Progynon C [Schering AG] *see* 3788
Progynon Depot [Schering AG] *see* 3758
Progynova [Schering AG] *see* 3758
ProHance [BMS] *see* 4358
Proheart 6 [Fort Dodge] *see* 6376
Proheptatriene *see* 2706
Prohexadione, 7892

Prohormo [Pharmar] *see* 5985
Proin [PRN] *see* 7417
Proinsulin, 7893
Prokarbol *see* 3305
Prokayvit Oral [BDH] *see* 5898
Prokine [Hoechst-Roussel] *see* 4574
Prokrein [Tobishi] *see* 5327
Prolacria [Allergan] *see* 3391
Prolactin, 7894
Proladone [Abbott] *see* 7058
Prolastin [Bayer] *see* 705
Prolax *see* 5917
Prolene [Industrial Rayon] *see* 7701
Proleukin [Chiron] *see* 5038
Prolia [Amgen] *see* 2897
Prolift [Pfizer] *see* 8244
Proline, 7895
Proline [Bayer CropSci.] *see* 7997
L-Proline *see* 7895
25-L-Proline-28-L-proline-29-L-proline-amylin (Human) *see* 7828
Prolintane, 7896
Prolixan [Siegfried] *see* 713
Prolixin [Apothecon] *see* 4220
Prolonium Iodide, 7897
Prolopa [Roche] *see* 1052
Proloprim [Wellcome] *see* 9878
Proluton [Schering AG] *see* 7889
Proluton Depot [Schering AG] *see* 4877
L-Prolyl-L-leucylglycinamide *see* 5882
Promac [Zeria] *see* 7669
PromAce [Fort Dodge] *see* 33
Promacid *see* 2191
Promacta [GSK] *see* 3609
Promactil [Combiphar] *see* 2191
Promanide *see* 4499
Promaquid [Rhône-Poulenc] *see* 4258
Promaxol [Otsuka] *see* 7877
Promazine, 7898
Promecon [Searle] *see* 1122
Promedol, 7899
α-Promedol *see* 7899
γ-Promedol *see* 7899
Promegestone, 7900
Promensil [Novogen] *see* 9858
ProMeris [Fort Dodge] *see* 5992
Promethazine, 7901
Promethazine Teoclate *see* 7901
Promethestrol *see* 6040
γ-Promethestrol *see* 6040
Promethium, 7902
Prometon, 7903
Prometone *see* 7903
Prometryn, 7904
Promid [Kaken] *see* 7891
Promidione *see* 5121
Promin [Parke-Davis] *see* 4499
Prominal *see* 5921
Promintic [Ayerst] *see* 6239
Promit [Pharmacia] *see* 2950
Promixin [Profile] *see* 2463
Promizole [Parke-Davis] *see* 9461
Promotil [Boehringer, Ing.] *see* 7896
Promotin [Nippon Chemiphar] *see* 5327
Promwill [Will-Pharma] *see* 7898
Pronamid *see* 7983
Prondol [Wyeth] *see* 5119
Pronestyl [BMS] *see* 7874
Pronon [Yamanouchi] *see* 7908
Pronox [Reiss] *see* 2704
Prontamid [SIT] *see* 9031
Prontobario [Bracco] *see* 990
Prontosil *see* 9055
Prontosil Album *see* 9057

Prontosil Flavum *see* 9055
Prontosil Rubrum *see* 9055
Prontylin *see* 9057
Pro-opiocortin *see* 7905
Pro-Opiomelanocortin, 7905
Propabloc [Azupharma] *see* 7953
Propacetamol, 7906
Propachlor, 7907
Propacil [Chugai] *see* 7980
Propaderm [GSK] *see* 1017
1,2-Propadiene-1,3-dione *see* 1815
Propafenone, 7908
Propagermanium, 7909
Propal *see* 5263
Propalin [Vétoquinol] *see* 7417
Propallylonal, 7910
Propamidine, 7911
Propaminodiphen *see* 7827
Propamocarb, 7912
Propanal *see* 7937
Propanamide, 7938
1-Propanamine *see* 7956
2-Propanamine *see* 5255
Propane, 7913
1-Propanearsonic Acid, 7914
2-Propanecarboxaldehyde *see* 5199
Propanedial *see* 5773
1,2-Propanediamine *see* 7965
1,3-Propanedicarboxylic Acid *see* 4509
Propane-diethyl Sulfone *see* 9090
Propanedinitrile *see* 5775
Propanedioic Acid *see* 5774
Propanedioic Acid Bis(1,1-dimethylethyl) Ester *see* 3041
Propanedioic Acid 1,3-Bis(1,1-dimethylethyl) Ester *see* 3041
Propanedioic Acid 1,3-Bis[2-methyl-2-(phenylthioxomethyl)hydrazide] *see* 3596
Propanedioic Acid 1,3-Diethyl Ester *see* 3878
Propanedioic Acid 1,3-Dimethyl Ester *see* 6168
1,2-Propanediol *see* 7968
(*R*)-1,2-Propanediol *see* 7968
(*S*)-1,2-Propanediol *see* 7968
1,3-Propanediol *see* 9885
1,3-Propanediol Bis[α-(*p*-chlorophenoxy)isobutyrate] *see* 8675
1,3-Propanediol Bis[2-(4-chlorophenoxy)-2-methylpropionate] *see* 8675
1,2-Propanediol-3-(*p*-chlorophenoxy)-1-carbamate *see* 2185
1,3-Propanedithiol, 7915
1,1'-[1,3-Propanediylbis[(dimethylimino)-3,1-propanediyl]]bis[4-(3-methyl-2(3*H*)-benzothiazolylidene)methyl]-quinolinium Tetraiodide *see* 9460
1,1'-(1,3-Propanediyl)bis[1,1-diphenylphosphine] *see* 1254
4,4'-[1,3-Propanediylbis(oxy)]bisbenzenecarboximidamide *see* 7911
4,4'-[1,3-Propanediylbis(oxy)]bis(3-bromobenzenecarboximidamide) *see* 3029
Propanenitrile *see* 7941
n-Propane Phosphonic Acid Anhydride *see* 7978
Propanethial *S*-Oxide, 7916
1,2,3-Propanetricarboxylic Acid *see* 9787
(2*R*,2'*R*)-1,2,3-Propanetricarboxylic Acid 1,1'-[(1*S*,2*R*)-1-[(2*S*,4*R*,9*R*,11*S*,12*S*)-12-Amino-4,9,11-trihydroxy-2-methyltridecyl]-2-[(1*R*)-1-methylpentyl]-1,2-ethanediyl] Ester *see* 4318
1,2,3-Propanetriol *see* 4520

1,2,3-Propanetriol 1-(4-Aminobenzoate) *see* 4523
1,2,3-Propanetriol Diacetate *see* 2964
1,2,3-Propanetriol 1-(Dihydrogen Phosphate) *see* 4522
1,2,3-Propanetriol 2-(Dihydrogen Phosphate) *see* 4522
(2*R*)-1,2,3-Propanetriol 1-(Dihydrogen Phosphate) *see* 4522
(2*S*)-1,2,3-Propanetriol 1-(Dihydrogen Phosphate) *see* 4522
1,2,3-Propanetriol Monoacetate *see* 6328
1,2,3-Propanetriol Mono(dihydrogen Phosphate) Calcium Salt (1:1) *see* 1672
1,2,3-Propanetriol Mono(dihydrogen Phosphate) Sodium Salt (1:2) *see* 8757
1,2,3-Propanetriol 1,2,3-Triacetate *see* 9751
1,2,3-Propanetriol 1,2,3-Trinitrate *see* 6694
Propanidid, 7917
Propanil, 7918
Propanoic Acid *see* 7939
Propanoic Acid 1,1′-Anhydride *see* 7940
Propanoic Acid Butyl Ester *see* 1590
Propanoic Acid Ethyl Ester *see* 3899
Propanoic Acid Methyl Ester *see* 6185
Propanoic Acid 2-Methylpropyl Ester *see* 5195
Propanoic Acid Propyl Ester *see* 7979
Propanoic Acid Sodium Salt (1:1) *see* 8798
Propanoic Acid Zinc Salt (2:1) *see* 10353
Propanoic Anhydride *see* 7940
1-Propanol *see* 7955
2-Propanol *see* 5254
2-Propanol Aluminum Salt (3:1) *see* 342
β-Propanolamine *see* 464
Propanolide *see* 7934
2-Propanol Nitrite *see* 5262
2-Propanol Titanium(4+) Salt (4:1) *see* 9636
2-Propanone *see* 65
Propanone 1-Oxime *see* 5236
2-Propanone Oxime *see* 73
Propanoyl Chloride *see* 7942
Propantheline Bromide, 7919
Propaphenin [Rodleben] *see* 2191
Propaquizafop, 7920
Proparacaine, 7921
Propargil *see* 7922
Propargite, 7922
Propargyl Alcohol, 7923
N-Propargyl-1-(*R*)-aminoindan *see* 8234
Propargyl Chloride, 7924
10-Propargyl-10-deazaaminopterin *see* 7821
Propargylic Acid *see* 7935
Propat [Hishiyama] *see* 9660
Propatyl Nitrate, 7925
Propavan [Kabi] *see* 7936
Propaxoline *see* 8012
Propazine, 7926
Propazol *see* 7880
Propecia [Merck & Co.] *see* 4113
2-Propenal *see* 119
2-Propenamide *see* 120
2-Propen-1-amine *see* 278
2-Propen-1-amine Polymer with 2-(Chloromethyl)oxirane *see* 8613
1-Propene *see* 7962
trans-1-Propene-1,2-dicarboxylic Acid *see* 5973
1-Propene Homopolymer *see* 7701

2-Propenenitrile *see* 122
Propene Oxide *see* 7969
2-Propene-1-sulfinothioic Acid *S*-2-Propen-1-yl Ester *see* 253
1-Propene-1,2,3-tricarboxylic Acid *see* 109
2-Propenoic Acid *see* 121
2-Propenoic Acid Butyl Ester *see* 1541
2-Propenoic Acid Ethyl Ester *see* 3813
2-Propenoic Acid Methyl Ester *see* 6085
1-Propenol-3 *see* 277
2-Propen-1-ol *see* 277
1-Propen-2-ol 2-Acetate *see* 5249
Propentofylline, 7927
1-Propen-2-yl Acetate *see* 5249
2-Propenylacrylic Acid *see* 8847
trans-p-Propenylanisole *see* 636
5-(1-Propen-1-yl)-1,3-benzodioxole *see* 5269
5-(2-Propen-1-yl)-1,3-benzodioxole *see* 8453
[*S*(*S*)]-*S*-2-Propen-1-yl-L-cysteine *S*-Oxide *see* 254
(17*β*)-17-(2-Propen-1-yl)estr-4-en-17-ol *see* 282
4-Propenylguaiacol *see* 5216
17-(2-Propen-1-yl)morphinan-3-ol *see* 5510
p-Propenylphenol *see* 670
4-(1-Propen-1-yl)phenol *see* 670
4-(2-Propen-1-yl)phenol *see* 2049
2-(1-Propen-1-yl)piperidine *see* 2485
N-2-Propen-1-yl-2-propen-1-amine *see* 2974
2-Propen-1-yl [3-(2-Propen-1-ylsulfinyl)-1-propen-1-yl] Disulfide *see* 188
3-[(*S*)-2-Propenylsulfinyl]-L-alanine *see* 254
N-2-Propen-1-ylthiourea *see* 9515
N-2-Propen-1-ylurea *see* 291
Properdin, 7928
Propericiazine *see* 7279
Propess [Ferring] *see* 7988
Propetamphos, 7929
Propethonum Iodide *see* 9826
Propham, 7930
Prophenatin [Nipro] *see* 3091
Propheniramine *see* 7349
Prophenpyridamine *see* 7349
Prophenytoin *see* 4285
Prophyllen [Streuli] *see* 3526
Prophylux [Hennig] *see* 7953
Propicillin, 7931
Propiconazole, 7932
Propildazine *see* 7535
PropiMax [Dow AgroSci.] *see* 7932
Propine [Allergan] *see* 3378
Propineb, 7933
Propiocine [Hoechst] *see* 3742
β-Propiolactone, 7934
Propiolic Acid, 7935
Propiomazine, 7936
Propionaldehyde, 7937
Propionamide, 7938
Propione *see* 3137
Propionic Acid, 7939
Propionic Acid Amide *see* 7938
Propionic Acid Butyl Ester *see* 1590
Propionic Acid Calcium Salt (2:1) *see* 1700
Propionic Acid Sodium Salt *see* 8798
Propionic Anhydride, 7940
Propionitrile, 7941
β-Propionolactone *see* 7934
Propionopyrrothine *see* 871

Propionylbenzene *see* 7944
Propionyl Chloride, 7942
Propionylcholinesterase *see* 2214
3-Propionyl-10-(*γ*-dimethylaminopropyl)phenothiazine *see* 7943
Propionylerythromycin *see* 3742
Propionylerythromycin Lauryl Sulfate *see* 3740
3″-Propionylleucomycin A5 *see* 8379
Propionyl Oxide *see* 7940
p-Propionylphenol *see* 7149
Propionylpromazine, 7943
Propionylpromethazine *see* 7936
Propiophenone, 7944
Propiopromazine *see* 7943
Propisomide *see* 7242
Propitan [Eisai] *see* 7568
Propitocaine *see* 7862
Propiverine, 7945
Propizepine, 7946
Proplant [Agriphar] *see* 7912
Proplex T [Baxter] *see* 3963
Propofol, 7947
Propolis, 7948
Propolis Balsam *see* 7948
Propolis Wax *see* 7948
Proponex-Plus [Shell] *see* 5855
Propoxur, 7949
1-Propoxy-2-amino-4-nitrobenzene *see* 6716
5-Propoxy-2-benzimidazolecarbamic Acid Methyl Ester *see* 7035
N-(6-Propoxy-1*H*-benzimidazol-2-yl)carbamic Acid Methyl Ester *see* 7035
N-(6-Propoxy-2-benzothiazolyl)carbamic Acid Methyl Ester *see* 9611
Propoxycaine Hydrochloride, 7950
Propoxycarbazone, 7951
5-Propoxy-2-(carbomethoxyamino)benzimidazole *see* 7035
N-Propoxyethyl-*N*-chloroacetyl-2,6-diethylaniline *see* 7858
Propoxyphene, 7952
α-d-Propoxyphene *see* 7952
l-Propoxyphene *see* 5523
Propranolol, 7953
Propranur [Stegropharm] *see* 7953
Propulm [DMS] *see* 7877
Propycil [Kali-Chemie] *see* 7980
Propyl Acetate, 7954
Propylacetic Acid *see* 10090
N-Propylajmaline *see* 7820
*N*⁴-Propylajmalinium *see* 7820
Propyl Alcohol, 7955
Propylaldehyde *see* 7937
Propylamine, 7956
(*S*)-7-[2-(*N*-Propylamino)ethyl]camptothecin *see* 1028
α-Propylamino-2-methylpropionanilide *see* 7862
2-(Propylamino)-*o*-propionotoluidide *see* 7862
3-Propylamino-*α*-propionylamino-2-carbomethoxy-4-methylthiophene *see* 1869
N-(*α*-Propylaminopropionyl)-*o*-toluidine *see* 7862
As-Propylarsonic Acid *see* 7914
Propylbenzene, 7957
Propyl Bromide, 7958
Propyl Butanoate *see* 7959
S-Propyl Butylethylthiocarbamate *see* 7167
Propyl Butyrate, 7959
Propyl Carbinol *see* 1542

n-Propylcarbinyl Chloride *see* 1561
Propyl Chloride, 7960
N-Propyl-*N'*-(*p*-chlorobenzenesulfonyl)-urea *see* 2192
Propyl Chlorocarbonate *see* 7961
Propyl Chloroformate, 7961
Propyl Cyanide *see* 1601
Propyldazine *see* 7535
Propyl [4-[(Diethylcarbamoyl)methoxy]-3-methoxyphenyl]acetate *see* 7917
(±)-1-*n*-Propyl-3α-diethylsulfamoyl-amino-6-hydroxy-1,2,3,4,4a,5,10,-10aβ-octahydrobenzo[g]quinoline *see* 8161
Propyl (3-Dimethylaminopropyl)carbamate *see* 7912
Propylene, 7962
[Propylenebis(dithiocarbamato)]zinc *see* 7933
Propylene Chlorohydrin, 7963
sec-Propylene Chlorohydrin, 7964
Propylenediamine, 7965
Propylene Dibromide, 7966
Propylenedicarboxylic Acid *see* 5290
Propylene Dichloride, 7967
(±)-*N,N'*-Propylenedinicotinamide *see* 6578
1,3-Propylenediol *see* 9885
(±)-4,4'-Propylenedi-2,6-piperazinedione *see* 8241
Propylene Glycol, 7968
1,3-Propylene Glycol *see* 9885
Propylene Oxide, 7969
Propylene Polymer *see* 7701
Propyl Ethanoate *see* 7954
Propyl Ether *see* 3385
Propyl Ethyl-*n*-butylthiolcarbamate *see* 7167
Propylethylene *see* 7234
Propyl Formate, 7970
Propyl Gallate, 7971
n-Propyl Gallate *see* 7971
3-(4-Propylheptyl)-4-morpholineethanol *see* 2884
Propylhexedrine, 7972
Propyl Hydride *see* 7913
Propyl *p*-Hydroxybenzoate *see* 7977
Propylic Alcohol *see* 7955
Propylidene Chloride, 7973
Propyl Iodide, 7974
Propyl Methanoate *see* 7970
Propyl Methyl Ether *see* 6186
Propyl Methyl Ketone *see* 6187
D-6-*n*-Propyl-8β-methylmercaptomethyl-ergoline *see* 7277
Propyl Nitrate, 7975
Propyl Nitrite, 7976
19-Propylorvinol *see* 3931
5-[2-Propyloxy-5-(1-methyl-2-pyrrolidi-nylethylamidosulfonyl)phenyl]-1-methyl-3-propyl-1,6-dihydro-7*H*-pyra-zolo[4,3-*d*]pyrimidin-7-one *see* 10027
Propylparaben, 7977
Propyl Parasept [Tenneco] *see* 7977
2-Propylpentamide *see* 10100
2-Propylpentanamide *see* 10100
2-Propylpentanoic Acid *see* 10099
1-(α-Propylphenethyl)pyrrolidine *see* 7896
Propylphosphonic Anhydride, 7978
n-Propylphosphonic Cyclic Anhydride *see* 7978
l-N-*n*-Propylpipecolic Acid-2,6-xylidide *see* 8389
(*S*)-(−)-1-Propyl-2',6'-pipecoloxylidide *see* 8389

2-*n*-Propyl-Δ¹-piperidine *see* 2486
(2*S*)-2-Propylpiperidine *see* 2489
(3*S*-*trans*)-6-Propyl-3-piperidinol *see* 8022
(3*S*,6*S*)-6-Propyl-3-piperidinol *see* 8022
6-Propylpiperonyl Butyl Diethylene Glycol Ether *see* 7589
N-Propyl-1-propanamine *see* 3383
Propyl Propanoate, 7979
Propyl Propionate *see* 7979
2-Propyl-4-pyridinecarbothioamide *see* 8000
Propylsulfamic Acid [5-(4-Bromophen-yl)-6-[2-(5-bromopyrimidin-2-yloxy)-ethoxy]pyrimidin-4-yl]amide *see* 5705
Propyl Sulfide *see* 3387
2-*n*-Propyl-3,4,5,6-tetrahydropyridine *see* 2486
(−)-2-(*N*-Propyl-*N*-2-thienylethylamino)-5-hydroxytetralin *see* 8404
N-[6-(Propylthio)-1*H*-benzimidazol-2-yl]-carbamic Acid Methyl Ester *see* 203
2-Propyl-4-thiocarbamoylpyridine *see* 8000
5-(Propylthio)-2-carbomethoxyamino-benzimidazole *see* 203
2-Propylthioisonicotinamide *see* 8000
Propylthiouracil, 7980
6-Propyl-2-thiouracil *see* 7980
Propyl-Thyracil [Frosst] *see* 7980
1-[*N*-Propyl-*N*-[2-(2,4,6-trichlorophen-oxy)ethyl]carbamoyl]imidazole *see* 7878
N-Propyl-*N*-[2-(2,4,6-trichlorophenoxy)-ethyl]-1*H*-imidazole-1-carboxamide *see* 7878
(*E*)-10-Propyl-5,9-tridecadien-1-ol Acetate *see* 7981
10-Propyl-*trans*-5,9-tridecadienyl Acetate *see* 7981
Propylure, 7981
2-Propylvaleramide *see* 10100
2-Propylvaleric Acid *see* 10099
2-Propynoic Acid *see* 7935
2-Propynoic Acid Ethyl Ester *see* 3898
2-Propyn-1-ol *see* 7923
2-Propynyl (*R*)-2-[4-(5-Chloro-3-fluoro-2-pyridyloxy)phenoxy]propionate *see* 2364
(*R*)-*N*-2-Propynyl-1-indanamine *see* 8234
Propyon [Makhteshim-Agan] *see* 7949
Propyphenazone, 7982
Propyzamide, 7983
Proquamezine *see* 463
Proquazone, 7984
Prorenal [Dainippon] *see* 5543
Prorennin *see* 8254
Prosaponin B *see* 3751
Proscar [Merck & Co.] *see* 4113
Proscillan [Streuli] *see* 7985
Proscillaridin, 7985
Proscillaridin A *see* 7985
Proscillaridin-4-methyl Ether *see* 7985
Prosedar [Schering-Plough] *see* 8145
Proserine *see* 6549
Proserout [Funai] *see* 5851
Proseryl [Funai] *see* 5851
Prosolvin [Intervet] *see* 5670
ProSom [Abbott] *see* 3757
Pro-Sonil *see* 2704
Prosopis Gum *see* 5984
Prostacyclin, 7986
Prostaglandin E₁, 7987
Prostaglandin E₂, 7988
Prostaglandin F₂α, 7989

Prostaglandin I₂ *see* 7986
Prostaglandins, 7990
Prostaglandin X *see* 7986
Prostaherb [Cesra] *see* 6567
Prostal [Teikoku Zoki] *see* 2102
Prostalene, 7991
Prostandin [Ono] *see* 7987
Prostanoic Acid *see* 7990
Prostap [Wyeth] *see* 5509
Prostaphlin [Apothecon] *see* 7003
Prostar [Bayer CropSci.] *see* 4241
Prostarmon F [Ono] *see* 7989
Prostasal [TAD] *see* 8694
ProstaScint *see* 1764
Prostavet [Virbac] *see* 3918
Prostenoglycine *see* 8936
Prostetin [Takeda] *see* 7030
Prostianol *see* 5670
Prostide [Sigma-Tau] *see* 4113
Prostigmin (Amp.) [Valeant] *see* 6549
Prostigmin (Tabl.) [Valeant] *see* 6549
Prostin/15M [Pfizer] *see* 1824
Prostin E₂ [Pharmacia & Upjohn] *see* 7988
Prostin VR [Pharmacia & Upjohn] *see* 7987
Prostivas [Pharmacia & Upjohn] *see* 7987
Prostrumyl *see* 6201
Prosulfuron, 7992
Prosultiamine, 7993
Prosynap [Janssen] *see* 5648
Protacine *see* 7890
Protactinium, 7994
Protactyl [Wyeth] *see* 7898
Protagent [Alcon-Thilo] *see* 7814
Protalba *see* 8008
Protamine Zinc Insulin *see* 5023
Protamone-D [Agri-Tech] *see* 9568
Protanabol *see* 7066
Protangix [Lefrancq] *see* 3388
Protara [Gensia] *see* 17
Protargin *see* 8666
Protargol [Winthrop] *see* 8666
Protaxon [Opfermann] *see* 7890
α₁-Protease Inhibitor *see* 705
ProTec [Cell Therapeutics] *see* 5573
Protecadin [Taiho] *see* 5396
Protectins *see* 5484
α₁-Proteinase Inhibitor *see* 705
Protein C, 7995
Protein Hydrolysates, 7996
Protein P5 *see* 867
Protein P9 *see* 1912
Protein Phosphoserine/phosphothreonine Phosphatase *see* 1643
Protein S *see* 7995
Protein (Synthetic Human Plasma Kalli-krein-inhibiting) *see* 3536
Protelos [Servier] *see* 8224
Protenolon [Showa] *see* 5985
Proterciclina [Proter] *see* 8381
Proternol *see* 5263
Prothiaden [Knoll] *see* 3478
Prothil [Kali-Chemie] *see* 5862
Prothioconazole, 7997
Prothionamide *see* 8000
Prothipendyl, 7998
Prothrombase *see* 7999
Prothrombin, 7999
Protionamide, 8000
Protirelin *see* 9750
Protium *see* 4829
Protium [Altana] *see* 7117
Protoactinium *see* 7994
Protoanemonin, 8001

Protocatechualdehyde, 8002
Protocatechualdehyde Dimethyl Ether
see 10145
Protocatechuic Acid, 8003
Protocatechuic Acid Methylene Ether
see 7590
Protocatechuic Aldehyde see 8002
Protoenstatite see 5751
Protoescigenin see 3751
Protogen A see 9479
Protoheme see 4674
Protoheme IX see 4674
Protokylol, 8004
Protomin see 4499
Protonix [Wyeth] see 7117
Proton Sponge see 3432
Protopam [Wyeth] see 7822
Protopic [Fujisawa] see 9155
Protopine, 8005
Protoporphyrin IX, 8006
Protopyrin see 3786
(+)-Protoquercitol see 8152
Protos [Servier] see 8224
Protosol see 3200
Protostephanine, 8007
Protostib see 6150
Protoveratrine A see 8008
Protoveratrine B see 8008
Protoveratrines, 8008
Protoverine, 8009
Protriptyline, 8010
Protropin [Genentech] see 8842
Protugan [Makhteshim-Agan] see 5264
Pro-UK see 8011
Pro U-PA see 8011
Pro-Urokinase, 8011
Prourokinase (Enzyme-activating) see
8011
Prourokinase (Enzyme-activating) (Hu-
man Clone pUK4/pUK18 Protein
Moiety Reduced) see 8011
Proustite see 8643
Provado [Bayer CropSci.] see 4951
Provas [Sanol Schwarz] see 1480
Provatene [Microbio] see 1855
Provaunt [DuPont] see 5012
Provell [Lilly] see 8008
Provenal [Pulitzer] see 9116
Provenge [Dendreon] see 8685
Proventil [Schering-Plough] see 210
Provera [Pharmacia & Upjohn] see 5864
Provigil [Cephalon] see 6314
Proviodine [Rougier] see 7815
Proviron [Schering AG] see 5986
Provisc [Alcon] see 4796
Provitamin D₃ see 2871
Provitar [Clin-Comar-Byla] see 7020
Provocholine [Roche] see 6012
Prower Factor see 3964
Prowl [BASF] see 7193
Proxine [Del Saz & Filippini] see 6499
Proxalyoc [Cephalon] see 7619
Proxazocain see 7829
Proxazole, 8012
Proxen [Syntex] see 6499
Proxibarbal, 8013
Proxibarbital see 8013
Proxigermanium see 7909
Proxil [Rottapharm] see 7890
Proxitane [Solvay Interox] see 7263
Proxymetacaine see 7921
Proxyphylline, 8014
Prozac [Lilly] see 4217
Prozei [Kissei] see 585
Prozin [Lusofarmaco] see 2191

Prozinex [Makhteshim-Agan] see 7926
PRT-60128 see 3600
Prucalopride, 8015
Prulaurasin see 5784
Prulet [Mission Pharmacal] see 7073
Prulifloxacin, 8016
Prunasin see 5784
Prunetin, 8017
Prunetol see 4426
Prunitrin see 8017
Prunol see 10075
Prunusetin see 8017
Pruri-ex [Lumavita] see 4319
Prussian Blue, 8018
Prussic Acid see 4832
Pruv [Rettenmaier] see 8807
Pruvel [Optimer] see 8016
Pryleugan [Temmler] see 4962
PS1 see 7857
PS2 see 7857
PS-60 see 4417
PS-207 see 9241
PS-341 see 1353
PSC-833 see 10103
P-Selectin see 8563
Pseudobaptigenin, 8019
Pseudobaptisin see 8019
Pseudobatrachotoxin see 1007
Pseudobutylbenzene see 1553
Pseudo-butylene see 1523
Pseudocef [Grünenthal] see 1946
Pseudochelerythrine see 8493
Pseudocholestane see 2508
Pseudocholinesterase see 2214
Pseudocinchonine see 4821
Pseudococaine, 8020
Pseudocodeine, 8021
Pseudoconhydrine, 8022
Pseudocubebin see 8609
Pseudocumene, 8023
Pseudocumol see 8023
Pseudodigitoxin see 4469
Pseudoephedrine, 8024
Pseudofructose see 8033
Pseudohecogenin see 4657
Pseudoionone, 8025
Pseudolibethenite see 2638
Pseudomalachite see 2638
trans-Pseudomonic Acid see 6388
Pseudomonic Acid A see 6388
Pseudomonic Acid B see 8026
Pseudomonic Acid C see 8026
Pseudomonic Acid D see 8026
Pseudomonic Acid I see 8026
Pseudomonic Acids, 8026
Pseudomorphine, 8027
Pseudomorphine (C34 Alkaloid) see
8027
Pseudopaederine see 7171
Pseudopederin see 7171
Pseudopelletierine, 8028
Pseudopterosins, 8029
Pseudopunicine see 8028
3-Pseudotropanol see 8030
Pseudotropine, 8030
Pseudotropine Benzoate see 9953
Pseudo-wollastonite see 1704
Pseudoyohimbine, 8031
Psicaine see 8020
Psicaine N see 8020
Psichial [Lang] see 2085
Psicofuranine, 8032
9β-D-Psicofuranosyladenine see 8032
9-β-D-Psicofuranosyl-9H-purin-6-amine
see 8032

Psicopax [Schering] see 5636
Psicoperidol [Lusofarmaco] see 9852
D-Psicose, 8033
Psicosoma [Ferrer] see 4505
Psicoterina [Francia] see 2085
Psicronizer [Albert-Pharma] see 6760
Psilocin, 8034
Psilocybin, 8035
Psilocyn see 8034
Psiquium [Sintofarma] see 5857
Psitticofulvins, 8036
PSK [Kureha] see 7702
Psoil [Basotherm] see 9612
Psoraderm-5 [Sunlife] see 1160
Psoradrate [Procter & Gamble] see 676
Psoralen, 8037
Psorcon [Dermik] see 3159
Psorcutan [Schering AG] see 1644
Psorex [GSK] see 2358
Psoriderm [Istoria] see 676
Psorigel [Alcon] see 2409
P.S.P. see 7360
PSTMS see 7421
Psychoperidol [Ethnor] see 9852
Psychotrine, 8038
Psychoverlan [Verla] see 4505
Psyllium Husk see 7631
Psyllium Hydrocolloid see 7631
Psyllium Hydrophylic Mucilloid see 7631
Psyllium Seed see 7631
Psyllium Seed Gum see 7631
Psymion [TAD] see 5813
Psyquil [Sanofi-Synthelabo] see 9853
PT see 8039
PTA see 3965
Ptaquiloside, 8039
PTC see 3963
Pteridine, 8040
2,4(1H,3H)-Pteridinedione see 5656
Pterin HB₂ see 1234
Pterocarpin, 8041
Pterofen see 9760
Pteroic Acid, 8042
Pteroyl-γ-glutamyl-γ-glutamylglutamic
Acid see 8043
Pterophene see 9760
Pteropterin, 8043
Pteroyldi-γ-glutamylglutamic Acid see
8043
Pteroylglutamic Acid see 4248
N-[N-(N-Pteroyl-γ-glutamyl)-γ-gluta-
myl]glutamic Acid see 8043
Pteroylheptaglutamic Acid see 8044
Pteroylhexaglutamylglutamic Acid, 8044
Pteroyltriglutamic Acid see 8043
PTFE see 7704
PTGA see 8043
PTH see 7138
Ptimal [Egis] see 9875
PTK-787 see 10130
PTO see 8107
Pt(oxalato)(trans-l-dach) see 7011
PTX see 7100
Ptychotis Oil see 189
Puccoon Root see 8492
Pudding Pipe see 1894
Pudding-stick see 1894
Puerzym [Firma] see 7259
PUK see 8011
Pukateine, 8045
Pularin see 4688
Pulegone, 8046
Pullulan, 8047
Pulmadil [3M Pharma] see 8353
Pulmaxan [AstraZeneca] see 1476

Pulmicort [AstraZeneca] *see* 1476
PulmiDur [Pharma-Stern] *see* 9436
Pulmoclase [UCB] *see* 1802
Pulmotil [Elanco] *see* 9596
Pulmo-Timelets [Temmler] *see* 9436
Pulmovet [Solvay] *see* 461
Pulmozyme [Genentech] *see* 2907
Pulsan [Yamanouchi] *see* 4979
Pulsar [Medosan] *see* 2938
Pulsatilla, 8048
Pulsatilla Camphor *see* 635
Pulsor [Dow AgroSci.] *see* 9469
Pulsotyl [Chinoin] *see* 7442
Pulzium [Solvay] *see* 9246
Puma *see* 4015
Pumactant, 8049
Pumice, 8050
Pumpellyite *see* 1651
Pumpkin Seed, 8051
Punch [DuPont] *see* 4236
Punicin *see* 3028, 7180
Punicine *see* 7181
Punktyl [Krewel] *see* 5636
Purantix [Wander] *see* 2363
Purapuridine *see* 8836
Purapurine *see* 8837
Purecal [Wyandotte] *see* 1659
Puregon [Organon] *see* 4250
Purgaceen *see* 7073
Purging Agaric *see* 177
Purging Cassia *see* 1894
Puricase [Mountain View] *see* 7177
Purified Araroba *see* 757
Purified Goa Powder *see* 757
Purified Oxgall *see* 7027
Purified Ozokerite *see* 1986
Purified Poloxamer 188 *see* 7679
Purified Protein Derivative *see* 9985
Purified Rayon *see* 8240
Purim [Laphal] *see* 7618
9*H*-Purin-6-amine *see* 140
1*H*-Purine, 8052
9*H*-Purine *see* 8052
9*H*-Purine-2,6-diamine *see* 2987
2,6(1*H*,3*H*)-Purinedione *see* 10257
Purine-6-thiol *see* 5938
Purinethol [GSK] *see* 5938
Puri-Nethol [GSK] *see* 5938
Purine-2,6,8-triol *see* 10060
Purine-2,6,8(1*H*,3*H*,9*H*)-trione *see* 10060
Purin-6(1*H*)-one *see* 4908
Purlytin [Pharmacia & Upjohn] *see* 8401
Purmorphamine, 8053
Purochin [Sclavo] *see* 10072
Purodigin [Wyeth] *see* 3182
Puromycin, 8054
Purosin-TC [Tatsumi] *see* 7985
Purostrophan [Kali-Chemie] *see* 7000
Purothionin, 8055
α_1-Purothionin *see* 8055
α_2-Purothionin *see* 8055
β-Purothionin *see* 8055
Purothionin A I *see* 8055
Purothionin A II *see* 8055
Purple Clover *see* 9858
Purple Coneflower *see* 3541
Purple Foxglove *see* 3178
Purple of Cassius *see* 4555
Purple of the Ancients *see* 3028
Purpurid [Promonta] *see* 3182
Purpurin, 8056
Purpurocatechol *see* 9963
Purpurogallin, 8057
Purpurosamine *see* 4427
Pursennid [Novartis] *see* 8594

Purshiana Bark *see* 1880
Pursuit [BASF] *see* 4948
Putrescine, 8058
Putty Powder *see* 8905
PVA *see* 7706
PVC *see* 7707
PVNO, 8059
PVP *see* 7814
PVP-I *see* 7815
PVPNO *see* 8059
PVPP *see* 7814
PX-917 *see* 9940
PX-105684 *see* 1025
PXD-101 *see* 1025
Pybox Ligands *see* 8083
Pycaril *see* 6611
Pydrin [Shell] *see* 4037
Pyelosil *see* 5083
Pygeum Africanum Extract, 8060
Pykaryl *see* 6611
Pyknolepsinum [Rhein-Pharma] *see* 3801
Pylen [Toyo] *see* 7687
Pylon [BASF] *see* 2089
Pylorid [GSK] *see* 8228
Pylumbrin *see* 5083
Pymetrozine, 8061
Pynamin [Sumitomo] *see* 252
Pyocyanine, 8062
Pyoktanin [Merck KGaA] *see* 4430
Pyopen [GSK] *see* 1793
Pyoredol [Aventis] *see* 7433
Pyostacine [RPR] *see* 7871
Pyoxanthose *see* 4680
Py-Phe, 8063
β-Pyracine *see* 8094
Pyraclostrobin, 8064
Pyrafat [Fatol] *see* 8067
Pyrahexyl *see* 9143
Pyralene [Prodelec] *see* 7682
Pyralgin [Savage] *see* 3390
Pyralin *see* 1965
Pyramal [Roxane] *see* 8097
Pyramidon *see* 470
Pyramin *see* 9720
Pyramine *see* 9720
Pyranica [Mitsubishi Kasei] *see* 9231
Pyranine, 8065
Pyranisamine *see* 8097
Pyrantel, 8066
Pyrantel Embonate *see* 8066
Pyranton *see* 2967
Pyrargyrite *see* 8643
Pyrasanone *see* 7390
Pyrathyn [Flint] *see* 6031
Pyrazinamide, 8067
Pyrazine, 8068
N-(2-Pyrazine)carbonyl-L-phenylalanine-
  L-leucine Boronic Acid *see* 1353
2-Pyrazinecarboxamide *see* 8067
Pyrazine Carboxylamide *see* 8067
2-Pyrazinecarboxylic Acid *see* 8070
2,3-Pyrazinedicarboxylic Acid, 8069
Pyrazinemonocarboxylic Acid *see* 8070
Pyrazinobutazone *see* 7390
Pyrazinoic Acid, 8070
Pyrazinoic Acid Amide *see* 8067
Pyrazino[2,3-*d*]pyrimidine *see* 8040
N^1-2-Pyrazinylsulfanilamide *see* 9068
Pyrazole, 8071
1*H*-Pyrazole *see* 8071
1*H*-Pyrazole-3-ethanamine *see* 1185
2-Pyrazoline, 8072
1*H*-Pyrazolo[3,4-*d*]pyrimidin-4-ol *see*
  271
Pyrazophos, 8073

Pyrcon [Meuselbach] *see* 8136
Pyrene, 8074
Pyrethrin I *see* 8075
Pyrethrin II *see* 8075
Pyrethrins, 8075
Pyrethrosin, 8076
Pyrethrum Flowers, 8077
Pyribenzamine [Novartis] *see* 9910
Pyricarbate *see* 8088
Pyricardyl *see* 6627
Pyrictal *see* 7336
Pyridaben, 8078
Pyridacil [Cilag] *see* 7324
Pyridate, 8079
Pyridazine, 8080
Pyridin [DuPont] *see* 4037
2-Pyridinamine *see* 469
3-Pyridinamine *see* 469
4-Pyridinamine *see* 2802
2*H*-Pyrid[4,3-*b*]indole *see* 1805
Pyridine, 8081
3-Pyridineacetic Acid, 8082
2-Pyridine Aldoxime Methiodide *see*
  7822
2-Pyridine Aldoxime Methyl Chloride
  see 7822
Pyridine Bis(oxazoline) Ligands, 8083
3-Pyridinecarboxamide *see* 6608
Pyridine-β-carboxylic Acid *see* 6610
γ-Pyridinecarboxylic Acid *see* 5233
o-Pyridinecarboxylic Acid *see* 7515
2-Pyridinecarboxylic Acid *see* 7515
3-Pyridinecarboxylic Acid *see* 6610
4-Pyridinecarboxylic Acid *see* 5233
Pyridine-β-carboxylic Acid Benzyl Ester
  see 6611
3-Pyridinecarboxylic Acid 2,2-Bis[[(3-
  pyridinylcarbonyl)oxy]methyl]-1,3-
  propanediyl Ester *see* 6582
4-Pyridinecarboxylic Acid [(2-Carboxy-
  3,4-dimethoxyphenyl)methylene]hy-
  drazide *see* 6947
3-Pyridinecarboxylic Acid 2-[2-(4-Chlo-
  rophenoxy)-2-methyl-1-oxopropoxy]-
  ethyl Ester *see* 3923
3-Pyridinecarboxylic Acid 3,3'-[2-[2-(4-
  Chlorophenoxy)-2-methyl-1-oxopro-
  poxy]-1,3-propanediyl] Ester *see* 1228
3-Pyridinecarboxylic Acid 3-[2-(4-Chlo-
  rophenoxy)-2-methyl-1-oxopropoxy]-
  propyl Ester *see* 8386
3-Pyridinecarboxylic Acid Compd with
  3,7-Dihydro-7-[2-hydroxy-3-[(2-hy-
  droxyethyl)methylamino]propyl]-1,3-
  dimethyl-1*H*-purine-2,6-dione (1:1)
  see 10258
Pyridine-3-carboxylic Acid Diethylamide
  see 6627
Pyridine-4-carboxylic Acid Diethylamide
  see 5234
4-Pyridinecarboxylic Acid 2-[1-(2-Furan-
  yl)ethylidene]hydrazide *see* 4337
4-Pyridinecarboxylic Acid Hydrazide *see*
  5232
3-Pyridinecarboxylic Acid Hydroxymethylamide *see* 4872
3-Pyridinecarboxylic Acid Methyl Ester
  see 6173
4-Pyridinecarboxylic Acid 2-(1-Methyl-
  ethyl)hydrazide *see* 5122
Pyridine-3-carboxylic Acid *N*-Methyl-
  olamide *see* 4872
3-Pyridinecarboxylic Acid 1-Oxide *see*
  7040

Quinestrol, 8171
Quinethazone, 8172
Quinfamide, 8173
Quinhydrone, 8174
Quinic Acid, 8175
Quinic Acid 1,5-Dicaffeic Ester *see* 2763
Quinicardine [Procter & Gamble] *see* 8176
Quinicine *see* 10199
Quinidex Extentabs [Robins] *see* 8176
Quinidine, 8176
α-Quinidine *see* 2289
Quinidine Bisulfate *see* 8176
Quiniduran [Teva] *see* 8176
Quinine, 8177
β-Quinine *see* 8176
Quinine Bimuriate *see* 8177
Quinine Dichloride *see* 8177
Quininic Acid, 8178
Quininone, 8179
Quiniofon *see* 4865
Quinizarin, 8180
Quinizarin Green SS, 8181
Quinmerac, 8182
Quinocide, 8183
Quinoclamine, 8184
Quinodis [Roche] *see* 4131
Quinofop-ethyl *see* 8203
Quinol *see* 4845
Quinoline, 8185
8-Quinolineboronic Acid, 8186
2-Quinolinecarboxylic Acid *see* 8162
8-Quinolinecarboxylic Acid, 8187
Quinoline Yellow, 8188
Quinoline Yellow A *see* 8189
Quinoline Yellow Base *see* 8189
Quinoline Yellow Spirit Soluble, 8189
Quinolinic Acid, 8190
2-Quinolinol *see* 1826
8-Quinolinol *see* 4881
2(1*H*)-Quinolinone *see* 1826
B-8-Quinolinylboronic Acid *see* 8186
2-(2-Quinolinyl)-1*H*-indene-1,3-(2*H*)-dione *see* 8189
2(1*H*)-Quinolone *see* 1826
Quinomethionate *see* 7077
Quinomycin A *see* 3544
Quinone, 8191
p-Quinone *see* 8191
Quinone Monoxime *see* 6729
Quinone Oxime *see* 6729
Quinophthalone *see* 8189
Quinora [Key] *see* 8176
Quinotoxine *see* 10199
Quinotoxol *see* 10199
Quinova-bitter *see* 8193
Quinovaic Acid *see* 8192
Quinovatine *see* 774
Quinovic Acid, 8192
Quinovic Acid β-D-Glucoside *see* 8193
Quinovic Acid β-D-Quinovoside *see* 8193
Quinovin, 8193
Quinovose, 8194
Quinoxaline, 8195
"Quinoxaline Antibiotic" *see* 3544
N-(2-Quinoxalinylcarbonyl)-*O*-[*N*-(2-quinoxalinylcarbonyl)-D-seryl-L-alanyl-3-mercapto-*N*,*S*-dimethylcysteinyl-*N*-methyl-L-valyl]-D-seryl-L-alanyl-*N*-methylcysteinyl-*N*-methyl-L-valine (8 → 1)-Lactone Cyclic (3 → 7)-Thioester *see* 3544
3-(2-Quinoxalinylmethylene)carbazic Acid Methyl Ester *N*,*N*′-Dioxide *see* 1782

(2-Quinoxalinylmethylene)hydrazinecarboxylic Acid Methyl Ester *N*,*N*′-Dioxide *see* 1782
*N*¹-(2-Quinoxalinyl)sulfanilamide *see* 9070
*N*¹-(2-Quinoxalyl)sulfanilamide *see* 9070
Quinoxyfen, 8196
Quinoxyl [Burroughs Wellcome] *see* 4865
Quintec [Dow AgroSci.] *see* 8196
Quintozene, 8197
Quintrate *see* 7222
Quinuclidine, 8198
3-Quinuclidinol, 8199
3-Quinuclidinyl Acetate *see* 8199
3-Quinuclidinyl Benzilate *see* 8199
3-Quinuclidinyl Benzoate *see* 8199
10-(3-Quinuclidinylmethyl)phenothiazine *see* 5931
(1*S*,3′*R*)-3′-Quinuclidinyl-1-phenyl-1,2,-3,4-tetrahydro-2-isoquinolinecarboxylate *see* 8838
5δ-[(3*S*)-3-Quinuclidinyl]thiomethylpristinamycin IA *see* 8201
Quinupramine, 8200
Quinupristin, 8201
Quinuronium Sulfate *see* 3393
Quinvet [Neopharmed] *see* 5698
Quisqualic Acid, 8202
L-Quisqualic Acid *see* 8202
Quitaxon [Napalm Sarl] *see* 3483
Quixin [Santen] *see* 6863
Quizalofop-ethyl, 8203
Quorn [Marlow], 8204
Qutenza [NeurogesX] *see* 1770
Qvar [3M] *see* 1017
QZ-2 *see* 6032
R-3-ZON *see* 7390
R-11, 8205
R-32 *see* 3170
R-48 *see* 2108
R-75 *see* 5500
R-79 *see* 5248
R-100 *see* 381
R-134a *see* 4750
R-148 *see* 6032
R-218 *see* 7273
R-242 *see* 9118
R-400 *see* 7146
R-406 *see* 4287
R-658 *see* 1603
R-738 *see* 6525
R-773 *see* 8344
R-788 *see* 4287
R-798 *see* 8353
R-802 *see* 4168
R-805 *see* 6633
R-818 *see* 4130
R-835 *see* 4911
R-837 *see* 4964
R-848 *see* 8271
R-875 *see* 2958
R-1132 *see* 3343
R-1273 *see* 7298
R-1406 *see* 7362
R-1439 *see* 221
R-1504 *see* 7448
R-1513 *see* 906
R-1569 *see* 9651
R-1582 *see* 906
R-1583 *see* 9209
R-1608 *see* 3697
R-1625 *see* 4634
R-1658 *see* 6351
R-1881 *see* 6203
R-1910 *see* 1548

R-1929 *see* 891
R-2028 *see* 4146
R-2113 *see* 2931
R-2323 *see* 4449
R-2498 *see* 9852
R-2858 *see* 6375
R-3248 *see* 31
R-3345 *see* 7568
R-3365 *see* 7611
R-3746 *see* 1942
R-3763 *see* 1942
R-4263 *see* 4028
R-4318 *see* 4135
R-4584 *see* 1050
R-4749 *see* 3499
R-4845 *see* 1200
R-4929 *see* 1077
R-5020 *see* 7900
R-6128 *see* 4237
R-6238 *see* 7546
R-6700 *see* 5172
R-7315 *see* 6225
R-7465 *see* 6497
R-7904 *see* 5537
R-8299 *see* 5511
R-11333 *see* 1452
R-12563 *see* 5511
R-12564 *see* 5511
R-14827 *see* 3549
R-14889 *see* 6257
R-15403 *see* 3155
R-15454 *see* 5205
R-15889 *see* 5637
R-16341 *see* 7196
R-16470 *see* 2947
R-16659 *see* 3926
R-17147 *see* 2705
R-17635 *see* 5837
R-17889 *see* 4149
R-18553 *see* 5628
R-23979 *see* 3637
R-25061 *see* 9134
R-25788 *see* 3058
R-25831 *see* 1848
R-28644 *see* 634
R-30730 *see* 9018
R-31520 *see* 2396
R-33800 *see* 9018
R-33812 *see* 3467
R-35443 *see* 7024
R-39209 *see* 229
R-41400 *see* 5349
R-41468 *see* 5344
R-42470 *see* 9303
R-43512 *see* 846
R-49945 *see* 5344
R-50547 *see* 5515
R-51211 *see* 5292
R-51619 *see* 2315
R-55667 *see* 8363
R-64433 *see* 3090
R-64766 *see* 8361
R-65824 *see* 6516
R-66905 *see* 8499
R-67555 *see* 6516
R-68070 *see* 8337
R-75251 *see* 5528
R-76477 *see* 7087
R-76713 *see* 10233
R-83842 *see* 10233
R-85246 *see* 5528
R-87926 *see* 5648
R-93877 *see* 8015
R-103757 *see* 6304
R-115777 *see* 9614

R-125489 see 5406
R-138727 see 7833
R-151993 see 9251
R-152450 see 4242
R-165335 see 3935
R-207910 see 1018
R-278474 see 8349
R-403323 see 1018
R-935788 see 4287
RA-8 see 3388
Rabalan see 7488
Rabeprazole, 8206
Rabon [Wyeth] see 9337
Rabond [Wyeth] see 9337
Race-acetylmethadol see 6017
Racecadotril, 8207
Racemethionine see 6047
Racemethorphan see 5520
Racemethorphan Hydrobromide see 5520
Racemic Acid see 9204
Racemic Ephedrine see 3663
Racemic Lactic Acid see 5384
Racemic Mandelic Acid see 5781
Racemic Propoxyphene see 7952
Racemic Tartaric Acid see 9204
Racemomycin A see 8959
Racemomycins see 8959
Racemorphan see 5524
Raceophenidol see 9453
Racephedrine see 3663
Racephedrine Hydrochloride see 3663
Racephenicol see 9453
Racepinephrine see 3674
R Acid see 6472
Racobalamin-60 [Abbott] see 10213
Ractopamine, 8208
RAD-001 see 3950
Radacef [Vianex] see 1932
Radacefe [BMS] see 1943
Radanil [Roche] see 1086
Radapon see 2799
Raddeamine see 4968
Rad-e-cate [Vineland] see 8730
Radecol see 6612
Radedorm [AWD] see 6661
Radibud [Takeda] see 1461
Radical Weed see 8835
Radicicol see 6341
Radicinin, 8209
Radinyl [Roberts] see 3768
Radioactive Colloidal Gold see 4552
Radioactive Cyanocobalamin see 10213
Radiocyanocobalamin see 10213
Radiogardase [Heyl] see 8018
Radio-gold (^{198}Au) Colloid see 4552
Radiographol see 6045
Radio-iodinated Insulin see 5023
Radiopaque [Schering AG] see 990
Radiotetrane see 5081
Radium, 8210
Radium-223 Chloride, 8211
Radium Chloride (^{223}RaCl$_2$) see 8211
Radium Emanation see 8212
Radium F see 7678
Radium ^{223}Ra Chloride see 8211
Radon, 8212
Rafamebin see 5085
Raffinose, 8213
Rafoxanide, 8214
Raft [Rhone-Poulenc] see 7004
Ragadan [Intervet] see 4699
Ralabol [Mallinckrodt Vet.] see 10319
Ralenova [Wyeth] see 6302
Ralgro [Mallinckrodt Vet.] see 10319

Rally [Dow AgroSci.] see 6406
Ralone [Mallinckrodt Vet.] see 10319
Raloxifene, 8215
Raltegravir, 8216
Raltitrexed, 8217
Ramace [AstraZeneca] see 8221
Ramatroban, 8218
Ramelteon, 8219
Rametin [Bayer] see 6438
Ramifenazone, 8220
Ramik [Hacco] see 3335
Ramipril, 8221
Ramiprilat see 8221
Ramoplanin, 8222
Ramosetron, 8223
R/AMP see 8341
Rampage [Motomco] see 1396
Ramrod [Monsanto] see 7907
Ramycin see 4346
Rancho [Bayer] see 5868
Rancinamycin IV see 8002
Randa [Nippon Kayaku] see 2316
Randolectil [Bayer] see 1519
Ranelic Acid, 8224
Ranestol [Parke-Davis] see 1195
Ranexa [CV Therapeutics] see 8229
Raney Nickel [W. R. Grace], 8225
Raney Nickel Catalyst see 8225
Raniben [Firma] see 8228
Ranibizumab, 8226
Raniclor [Ranbaxy] see 1914
Ranide [Merck & Co.] see 8214
Ranidil [Menarini] see 8228
Ranimustine, 8227
Raniplex [Solvay] see 8228
Ranitidine, 8228
Ranitidine Bismutrex see 8228
Ranman [Ishihara Sangyo] see 2693
Ranolazine, 8229
Ranomustine see 8227
Ranpirnase, 8230
Rantudil [Bayer] see 28
Ranunculin see 8001
Ranvil [Gentili] see 6580
RAPA see 8232
Rapacuronium Bromide, 8231
Rapaflo [Watson] see 8642
Rapamune [Wyeth] see 8232
Rapamycin, 8232
Rapamycin 42-(Dimethylphosphinate) see 8336
Rapamycin 42-Ester with 2,2-Bis-(hydroxymethyl)propionic Acid see 9285
Rapamycin 42-[3-Hydroxy-2-(hydroxymethyl)-2-methylpropanoate] see 9285
Rapeseed Oil, 8233
Raphanus Rusticanus see 4788
Rapiacta [Shionogi] see 7264
Rapicidin see 10017
Rapifen [Janssen-Cilag] see 229
Rapilysin [Roche] see 8284
Rapinovet [Schering-Plough Vet.] see 7947
Rapiscan [Gilead] see 8247
Rapitil [Aventis] see 6521
Rapitux [De Angeli] see 3500
Raplon [Organon] see 8231
Rapostan [Mepha] see 7071
Raptiva [Genentech] see 3567
Raptor [BASF] see 4945
Rapynogen [Maruko] see 7953
Rasagiline, 8234
Rasagiline Mesylate see 8234
Rasapen see 4788
Rasburicase, 8235

Rasilez [Novartis] see 239
Raspberry, 8236
Rastinon [Aventis] see 9667
Ratak [Sorex] see 3153
Ratanhine see 9136
Rat Antispectacled Eye Factor see 5021
Rathimed [Pfleger] see 6236
Raticate see 6775
Rattan Yellow see 4390
Rattlesnake Root see 8592
Rattle Weed see 1313
Rattoff [Anim. Control Technol.] see 10352
Rattrack see 708
Raubasine, 8237
Raudixin [BMS] see 8238
Rauhimbine see 2535
Raunormine [Penick] see 2919
Raunova [Zambeletti] see 9147
Raupine see 8515
Rauserpol [Rhewie Pharma] see 8238
Rautensin [Sandoz] see 8238
Rauverid [OJ & F] see 8238
Rauwiloid [3M Pharma] see 8238
Rauwoldin [OJ & F] see 8238
Rauwolfemms [Truxton] see 8238
Rauwolfia Serpentina, 8238
Rauwolfine see 187
Rauwolscine see 10301
Ra-Valeas [Valeas] see 8238
Ravenil [Caber] see 8088
Ravenol [Caber] see 9116
Raviac [Lipha] see 2155
Ravocaine Hydrochloride [Cook-Waite] see 7950
Ravuconazole, 8239
Ravyon [Bayer CropSci.] see 1788
Raxar [Glaxo Wellcome] see 4580
Raxil [Bayer CropSci.] see 9229
Rayon, 8240
Razadyne [J & J] see 4370
Razoxane, 8241
RB-1509 see 5621
RBPI$_{21}$ see 931
RC-61-91 see 4961
RC-160 see 10123
RC-172 see 216
RC-27109 see 6619
RCL I see 8333
RCL II see 8333
RCL III see 8333
RCL IV see 8333
RD-1572 see 4769
RD-4593 see 5855
RD-13621 see 4919
rDSPA alpha 1 see 3506
RDX see 2728
RE-4355 see 6442
RE-40885 see 4235
RE-45601 see 2347
Reacid see 2670
Reactine [Pfizer] see 2021
Reactivan [Cascan] see 3998
Realgar see 790
Reasec [Janssen] see 3343
Reason [Bayer CropSci.] see 3990
Reazide see 2670
Rebamipide, 8242
Rebaudioside A see 8243
Rebaudiosides, 8243
Rebetol [Schering-Plough] see 8323
Rebiana see 8243
Rebif [Serono] see 5035
Reboxetine, 8244
Reboxetine Mesylate see 8244

Rec-7-0040 *see* 4128
Rec-7-0267 *see* 3227
Rec-15-0691 *see* 9581
Rec-15-1476 *see* 4032
Rec-15-2375 *see* 5500
Recalbon [Ono] *see* 6284
Recanescine *see* 2919
Receptal [Intervet] *see* 1506
Reclaim [Dow AgroSci.] *see* 2389
Reclast [Novartis] *see* 10388
Recognan [Asahi] *see* 2318
Recolip [Benzon] *see* 2370
Recombinant Human Activated Protein
    C *see* 7995
Recombinant Human Soluble Interleu-
    kin-4 Receptor *see* 5046
Recombinant Methionyl Human G-CSF
    see 4573
Recombinate *see* 3962
Reconcile [Lilly] *see* 4217
Reconox *see* 7364
Recormon [Roche] *see* 3744
Recruit [Dow AgroSci.] *see* 4719
Recruit II [Dow AgroSci.] *see* 4719
Recruit III [Dow AgroSci.] *see* 6810
Recruit IV [Dow AgroSci.] *see* 6810
Rectadione *see* 7348
Rectalad [Carter-Wallace] *see* 3446
Rectodelt [Trommsdorff] *see* 7842
Rectogesic [Strakan] *see* 6694
Rectovalone [Jouveinal] *see* 9641
Redactiv [Alfa Wassermann] *see* 8345
Red-Al [Aldrich] *see* 8702
Red Arrow *see* 4407
Red Arsenic Glass *see* 790
Red Arsenic Sulfide *see* 790
Red Bole *see* 1330
Red Clover *see* 9858
Red Copper Oxide *see* 2654
Redeema [Merz] *see* 10359
Redeptin [SK & F] *see* 4237
Redergin [Lek] *see* 3710
Red Fluorescent Protein, 8245
Redigo [Bayer CropSci.] *see* 7997
Redisol [Merck & Co.] *see* 10212
Redisol H [Merck & Co.] *see* 4846
Red Lead *see* 5480
Red Lead Chromate *see* 5461
Red Mustard *see* 6400
Red No. 2 *see* 369
Red Oil *see* 10004
Redomex [Lundbeck] *see* 483
Red Orpiment *see* 790
Redoxon [Roche] *see* 819
Red Pepper *see* 1772
Red Phosphorus *see* 7459
Red Precipitate *see* 5949
Red Prussiate of Potash *see* 7751
Red Puccoon *see* 8492
Red Quebracho *see* 8147
Red Raspberry *see* 8236
Redroot *see* 370
Red Root *see* 8492
Red Sandalwood *see* 8522
Red Tetrazolium *see* 9918
Reduced Hematin *see* 4674
Reduced Nicotinamide Adenine Dinu-
    cleotide Phosphate Dehydrogenase *see*
    6917
Reduced Ninhydrin *see* 4813
Reduced Vitamin B$_{12}$ *see* 2410
Reductic Acid, 8246
Reductil [Abbott] *see* 8624
Reducymol [SePRO] *see* 625
Redul [Schering] *see* 4543

Red Uranium Oxide *see* 10045
Redux [Wyeth-Ayerst] *see* 4004
Red-water Tree Bark *see* 8519
Redwood *see* 7295
Reelon [Sanwa] *see* 7618
Reevon [Reeves] *see* 7687
Refacto [Wyeth] *see* 6357
Refined Solvent Naphtha *see* 5542
Refkas [Maruko] *see* 3739
Reflex [Syngenta] *see* 4254
Refludan [Aventis] *see* 4755
Refobacin [Merck KGaA] *see* 4427
Refrigerant 134a *see* 4750
Refugal [Bayer] *see* 2066
Refungine [A.C.F.] *see* 9024
Regacholyl *see* 4874
Regadenoson, 8247
Regaine [Pfizer] *see* 6285
Regelan [Zeneca] *see* 2370
Regenon [Temmler] *see* 3145
Regianin *see* 5317
Regitine [Novartis] *see* 7376
Reglan [Wyeth] *see* 6219
Regletin [Teikoku Zoki] *see* 309
Reglone [Syngenta] *see* 3392
Regolax [Corvi] *see* 7358
Regonol [Novartis] *see* 8090
Regramostim *see* 4574
Regranex [Chiron] *see* 7639
Regulan [Procter & Gamble] *see* 7631
Regulin [Young] *see* 5884
Regulipid [Heilit] *see* 2371
Regulox *see* 5769
Regulton [Knoll] *see* 388
Regulus of Antimony *see* 683
Regu-Mate [Intervet] *see* 315
Regurin [Galen] *see* 9966
Regutol [Schering-Plough] *see* 3446
Rehibin [Thames] *see* 2713
Reicaf [San Carlo] *see* 8381
Reichstein's Substance C *see* 265
Reichstein's Substance E *see* 7848
Reichstein's Substance Fa *see* 2525
Reichstein's Substance G *see* 165
Reichstein's Substance H *see* 2524
Reichstein's Substance L *see* 263
Reichstein's Substance M *see* 4824
Reichstein's Substance Q *see* 2902
Reichstein's Substance R *see* 266
Reichstein's Substance S *see* 2933
Reichstein's Substance U *see* 7849
Reinecke Salt, 8248
Reishi *see* 4397
Rela [Schering] *see* 1842
Relafen [GSK] *see* 6428
Relane [Cutter] *see* 3327
Relanium [Polfa] *see* 2997
Relaspium [Isei] *see* 9966
Relaxan [Hexal] *see* 4373
Relaxar *see* 5917
Relaxil *see* 5917
Relaxil G [Pannon] *see* 4591
Relaxin, 8249
Reldan [Dow AgroSci.] *see* 2195
Releasin (formerly) [Warner-Chilcott] *see*
    8249
Relefact LH-RH *see* 4561
Relenza [GSK] *see* 10311
Relestat [Allergan] *see* 3673
Relestrid *see* 6052
Reliberan [Geymonat] *see* 2085
Relicor *see* 3765
Relifen [GSK] *see* 6428
Relifex [GSK] *see* 6428
Relistor [Progenics] *see* 6172

Reliveran [Finadiet] *see* 6219
Relpax [Pfizer] *see* 3597
Relvene [Pharmascience] *see* 9969
Rely [Bayer CropSci.] *see* 7451
Remacemide, 8250
Remantadin(e) *see* 8351
Remark [Nippon Zoki] *see* 1180
Remedy [Dow AgroSci.] *see* 9821
Remeflin [Recordati] *see* 3227
Remeron [Organon] *see* 6291
Remestan [Wyeth] *see* 9275
Remestyp [Ferring] *see* 9310
Remicade [Centocor] *see* 5014
Remid [TAD] *see* 271
Remifemin [Schaper & Brümmer] *see*
    1313
Remifentanil, 8251
Remifentanyl *see* 8251
Reminyl [J & J] *see* 4370
Remitch [Toray] *see* 6443
Remivox [Janssen] *see* 5637
Remnos [DDSA] *see* 6661
Remodulin [United Ther.] *see* 9748
Removab [Fresenius] *see* 1909
Remoxipride, 8252
Renagel [Genzyme] *see* 8613
Renal Natriuretic Peptide *see* 10071
Renarcol *see* 5917
Rencal [BMS] *see* 7499
Rendix [Boehringer, Ing.] *see* 2794
Renese [Pfizer] *see* 7705
Renex 20 [ICI] *see* 7698
Renex 600'S [Atlas] *see* 6764
Renin, 8253
Renin Substrate *see* 644
Renitec [Merck & Co.] *see* 3623
Reniten [Merck & Co.] *see* 3623
Renivace [Merck & Co.] *see* 3623
Rennase *see* 8254
Rennet *see* 8254
Rennin, 8254
Renografin [Bracco] *see* 2993
Renoquid [Glenwood] *see* 9034
Renormax [Schering-Plough] *see* 8880
Ren-o-sal [Salsbury] *see* 8406
Renounce [Bayer CropSci.] *see* 2754
Renova [Ortho] *see* 8288
Renovist [BMS] *see* 2993
Renovue-65 [BMS] *see* 5055
Renovue-DIP [BMS] *see* 5055
Renpress [Sandoz] *see* 8880
Rentylin [Shire] *see* 7249
Renvela [Genzyme] *see* 8613
Renzapride, 8255
Renzepin [Bergamon] *see* 7604
Reocorin [Farmochimica] *see* 7855
Reomucil [DOC] *see* 1802
ReoPro [Lilly] *see* 5
Reoxyl [Byk Gulden] *see* 4741
Repagermanium *see* 7909
Repaglinide, 8256
Repanidal [Zentiva] *see* 9957
Reparil [Madaus] *see* 3751
Repel [WPC] *see* 2859
Repeltin [Fabre-Medikosma] *see* 9873
Repinotan, 8257
Repirinast, 8258
Replagal [Shire] *see* 4366
Repocal [Desitin] *see* 7243
Repodral [Winthrop] *see* 8942
Repoise [Robins] *see* 1519
Reprodin [Bayer] *see* 5670
Repronex [Ferring] *see* 4250
Reproterol, 8259
Reptilase [Knoll] *see* 1008

Repulson [Mochida] see 746
Requip [GSK] see 8388
Resacetophenone, 8260
Resazoin see 8261
Resazurin, 8261
Resbenzophenone see 1109
Resbuthrin [Wellcome] see 1235
Rescinnamine, 8262
Rescriptor [Pharmacia & Upjohn] see 2882
Rescufolin [Nordic] see 4249
Rescula [Novartis] see 10034
Rescuvolin [Medac] see 4249
Resectisol [McGaw] see 5810
Reserpic Acid, 8263
Reserpilic Acid Dimethylaminoethyl Ester Dihydrochloride see 8264
Reserpiline, 8264
Reserpine, 8265
Reserpinine see 8262
Reserpinolic Acid see 8263
Resibufogenin, 8266
Resilin, 8267
Resimatil [Sanofi-Synthelabo] see 7865
Resin Benjamin see 4615
Resin Benzoin see 4615
Resin Copal see 2502
Resin Damar see 2806
Resin Guaiac see 4588
Resin Ipomea, 8268
Resin Jalap, 8269
Resin Kino see 5361
Resin of Mexican Scammony see 8268
Resin Scammony, 8270
Resin Tolu see 943
Resiquimod, 8271
Resistomycin [Bayer] see 5329
Resistomycin, 8272
Resitan [Takeda] see 10092
Resitox [BASF] see 2546
Résivit [Oberlin] see 5504
Reslizumab, 8273
Resloom M 75 [Monsanto] see 4729
Resmethrin, 8274
(+)-trans-Resmethrin see 1235
Resmit [Shionogi] see 5857
Resochin [Bayer] see 2165
Resolor [Movetis] see 8015
Resolvable Tartaric Acid see 9204
Resonium A [Sanofi-Synthelabo] see 8797
Resorantel, 8275
Resorcin see 8276
Resorcin Blue see 5380
Resorcinol, 8276
Resorcinolphthalein see 4192
Resorcinol Phthalein Sodium see 4192
Resorcinolphthalin see 4193
Resorcinol Yellow see 9954
β-Resorcylaldehyde, 8277
Resorcylam see 8275
β-Resorcylic Acid, 8278
Resorcylic Acid Lactones see 10314
Resotren [Bayer] see 4865
Resource [Valent] see 4173
Respacal [Bios-Coutelier] see 9991
Respbid [Boehringer, Ing.] see 9436
Respicort [Mundipharma] see 9758
Respifral see 5263
Respigon [Taisho] see 8266
Respiram [Modern Vet. Ther.] see 3481
Respirase [Gibipharma] see 10370
Respiride [Schiapparelli] see 4026
Respirot [Novartis] see 2582
Respontin [GSK] see 5117
Resporisan [Kowa] see 8264

Restandol [Organon] see 9324
Restanza [Adv. Life Sci.] see 2018
Restasis [Allergan] see 2750
Restenacht [Kabi] see 3615
Restenil [Recip] see 5929
Rest-On see 6031
Restoril [Novartis] see 9275
Restrol [Central Pharm.] see 3115
Restryl see 6031
Resulax [AstraZeneca] see 8851
Resulfon [Nordmark] see 9040
Resveratrol, 8279
cis-Resveratrol see 8279
trans-Resveratrol see 8279
Resyl [Novartis] see 4591
Retabolil [Richter Gedeon] see 6450
Retacrit [Hospira] see 3744
ReTain [Valent] see 434
Retalon see 3115
Retalon-Lingual see 4737
Retalon Oleosum see 4737
Retalon-Oral see 3115
Retamine, 8280
Retapamulin, 8281
Retarcyl [Delagrange] see 6362
Retardon [Chassot] see 9065
Retaspimycin, 8282
Retavase [Centocor] see 8284
Retcin [DDSA] see 3739
Retene, 8283
Retens [Chiesi] see 3488
Reteplase, 8284
Reticulin, 8285
Reticuline, 8286
Retigabine see 3959
Retin-A [Ortho] see 8288
Retina-derived Endothelial Cell Growth Factors see 4101
Retinal, 8287
trans-Retinal see 8287
Retinal 2 see 2878
11-cis-Retinal see 8287
Retinaldehyde see 8287
Retinene₁ see 8287
Retinene 2 see 2878
Retinoic Acid, 8288
all-trans-Retinoic Acid see 8288
6-cis-Retinoic Acid see 241
9-cis-Retinoic Acid see 241
13-cis-Retinoic Acid see 5274
Retinoic Acid (±)-(2R)-3,4-Dihydro-2,5,-7,8-tetramethyl-2-[(4R,8R)-4,8,12-trimethyltridecyl]-2H-1-benzopyran-6-yl Ester see 9657
Retinol see 10211
all-trans-Retinol see 10211
Retinol₂ see 2879
13-cis-Retinol see 6552
Retinol Binding Protein see 10211
Retinova [Janssen-Cilag] see 8288
4-(all-trans-Retinoyl)aminophenol see 4025
Retolen [Byk Elmu] see 846
Retractyl [Rhovyl] see 7707
Retrocortine see 7842
Retroid [Roche] see 9746
Retrone see 3521
Retronecine, 8289
Retrorsine, 8290
Retrorsine N-Oxide see 8290
Retrovir [GSK] see 10322
Reumalon [Tosi] see 7056
Reumatox see 6317
Reutol [Errekappa] see 9677
REV-6000A see 2881

Revanex [Yuhan] see 8291
Revanil [Roche] see 5574
Revaprazan, 8291
Revasc [Aventis] see 4755
Revatio [Pfizer] see 8628
Reverin [HMR] see 8381
Reversine, 8292
Revex [Ohmeda] see 6445
Rev-Eyes [Bausch & Lomb] see 2820
ReVia [BMS] see 6448
Reviparin Sodium, 8293
Revivon [Reckitt & Colman] see 3380
Revlimid [Celgene] see 5490
Revolade [GSK] see 3609
Revolution [Pfizer] see 8561
Revus [Syngenta] see 5785
Rewodina [AWD] see 3091
Rexalgan [Dompé] see 9291
Rexan see 2106
Rexigen [Bago] see 2357
Rexitene [LPD] see 4593
Rexort [Hoechst] see 2318
Rexulfa [Medici] see 9065
Reyataz [BMS] see 849
Reychler's Acid see 1736
Rezipas [BMS] see 473
Rezulin see 9948
RF-2535 see 4730
RFP see 8245
RG-270 see 5096
RG-1068 see 8558
RG-7204 see 10138
RGCR see 4958
R-Gene [Pharmacia & Upjohn] see 770
RGH-1106 see 7571
RGH-3332 see 4167
RGH-4405 see 10188
RH-315 see 7983
RH-0345 see 4631
RH-787 see 8101
RH-893 see 6844
RH-2485 see 6066
RH-2915 see 7062
RH-3866 see 6406
RH-5287 see 3080
RH-5992 see 9230
RH-6201 see 101
RH-7281 see 10398
RH-7592 see 3995
RH-7988 see 9762
RH-112485 see 6066
RH-123652 see 9463
RH-130753 see 9469
Rh 123 see 8306
Rhabarberone see 300
Rhamnetin, 8294
β-Rhamnocitrin see 8294
Rhamnol see 8694
Rhamnolutein see 5322
Rhamnose, 8295
L-Rhamnose see 8295
3β-Rhamnosido-14β-hydroxy-Δ^{4,20,22}-bufatrienolide see 7985
6-O-α-L-Rhamnosyl-D-glucose see 8439
Rhamnoxanthin see 4292
Rhamnus Cathartica, 8296
rhAPC see 7995
Rhaponticin see 8297
Rhapontin, 8297
rhASB see 809
rhBMP-2 see 1333
rhBMP-7 see 1333
rhCG see 2221
trans-RhCl(CO)(PPh₃)₂ see 1819
rhDNase I see 2907

Rheadine, 8298
Rheic Acid see 8299
Rhein, 8299
Rhenic Anhydride see 8303
Rhenium, 8300
Rhenium Blue see 8303
(*OC*-6-11)-Rhenium Fluoride (ReF₆) see 8302
Rhenium Heptoxide, 8301
Rhenium Hexafluoride, 8302
Rhenium Oxide (Re₂O₇) see 8301
Rhenium Oxide (ReO₃) see 8303
Rhenium Trioxide, 8303
Rhenocain see 4885
Rheochrysidin see 3618
Rheocyclan [Grünenthal] see 9190
Rheomacrodex [Medisan] see 2950
Rheumacin LA [CP Pharm.] see 5009
Rheumatism Root see 3321
Rheumatism Weed see 729, 2057
Rheumatrex see 6057
Rheum Emodin see 3618
Rheumon Gel [Troponwerke; Bayer] see 3921
Rheumox [Goldshield] see 713
Rhex [Hobein] see 5917
RhIGF-1 see 5844
RhIGF-I/rhIGFB-3 Complex see 5845
rhIL-10 see 5043
Rhinalair [Fabre] see 8024
Rhinalar [Syntex] see 4176
Rhinall [Scherer] see 7398
Rhinanthin see 868
Rhinantin see 6453
Rhinaspray [Boehringer, Ing.] see 9727
Rhinathiol [Sanofi-Synthelabo] see 1802
Rhinathiol Tusso [Sanofi-Aventis] see 7854
Rhinocort [AstraZeneca] see 1476
Rhinofrenol see 7065
Rhinogutt [Thomae] see 9727
Rhinolast [Viatris] see 899
Rhinolitan [3M Pharma] see 7065
Rhinoperd see 6453
Rhinopront [Pfizer] see 9364
Rhinospray [Thomae] see 9727
Rhizoxin, 8304
RhLF see 9163
Rh₂(OAc)₄ see 8311
Rhocya see 7807
Rhod-2, 8305
Rhodalline see 9515
Rhodamine 123, 8306
Rhodamine B, 8307
Rhodamine 110 Methyl Ester see 8306
Rhodanic Acid see 8308
Rhodanine, 8308
Rhodanwasserstoffsäure (German) see 9481
Rhodasurf [Rhone-Poulenc] see 7697
Rhodeose see 4307
Rhodialothan [Nattermann] see 4637
Rhodiatox see 7137
Rhodinal see 2328
Rhodinol, 8309
β-Rhodinol see 2329
Rhodium, 8310
Rhodium(II) Acetate, 8311
Rhodium(II) Acetate Dimer see 8311
trans-Rhodium Carbonyl Bis(triphenyl-phosphine) Chloride see 1819
Rhodium Carbonyl Chloride, 8312
Rhodium Chloride, 8313
Rhodium Chloride (RhCl₃) see 8313
Rhodium Diacetate Dimer see 8311

Rhodium Trichloride see 8313
Rhodizite see 2007, 8415
Rhodochrosite see 5791
Rhodocide [Cheminova] see 3790
Rhododendrin, 8314
Rhodol see 6091
Rhodomycin A see 8315
Rhodomycin B see 8315
β-Rhodomycin II see 8315
Rhodomycins, 8315
Rhodonite see 5803
Rhodopin, 8316
Rhodopsin, 8317
Rhodoquine see 7636
Rhodoquinone, 8318
Rhodoquinone-10 see 8318
Rhodotoxin see 4578
Rhodoviolascin, 8319
Rhodoxanthin, 8320
Rhoeadine see 8298
Rhombinin see 617
Rhomex see 7576
Rhonal [Aventis] see 841
Rhonavil Extra Pure [Rhône-Poulenc] see 10120
Rhothane see 3070
Rhovyl [Rhovyl] see 7707
rhPDGF-BB see 7639
rhPM-1 see 9651
rhSCF see 8934
rhSOD see 9131
Rhubarb, 8321
Rhubarb Yellow see 8299
Rhucin [Pharming] see 2481
Rhudane see 5080
rHuEPO see 3744
rhuFabV2 see 8226
rHuKGF see 7086
rhuMAb 2C4 see 7298
rhuMAb-E25 see 6936
rhuMab HER2 see 9736
rhuMab-VEGF see 1194
Rhumalgan [Sandoz] see 3091
Rhynchophylline, 8322
Rhyncophylline see 8322
Rhythmy [Shionogi] see 8346
Riabal [Fujisawa] see 7861
Riacen [Chiesi] see 7619
Riamet [Novartis] see 5657
RiaSTAP [CSL Behring] see 4100
Ribamidine see 9197
Ribavirin, 8323
α-Ribazole, 8324
Ribena [GSK] see 819
Ribex [Formenti] see 3500
Ribitol see 157
Ribo [Tennessee Pharm.] see 4123
D-2-Ribodesose see 2909
Riboflavin, 8325
Riboflavin 5′-Adenosine Diphosphate see 4122
Riboflavin 5′-(Dihydrogen Phosphate) see 4123
Riboflavin Monophosphate see 4123
Riboflavin 5′-(Trihydrogen Diphosphate) *P′* → 5′-Ester with Adenosine see 4122
D-Ribofuranose-5-phosphoric Acid see 8329
9-β-D-Ribofuranosidoadenine see 141
1-β-D-Ribofuranosylcytosine see 2783
9-β-D-Ribofuranosylhypoxanthine see 5018
2-β-D-Ribofuranosylmaleimide see 8621
3-β-D-Ribofuranosylorotic Acid see 6973

9-β-D-Ribofuranosyl-9*H*-purin-6-amine see 141
9-β-D-Ribofuranosyl-9*H*-purine see 6517
9-β-D-Ribofuranosyl-9*H*-purine-2,6-diol see 10263
9-β-D-Ribofuranosyl-9*H*-purine-2,6-(1*H*,-3*H*)-dione see 10263
N-(9-β-D-Ribofuranosyl-9*H*-purin-6-yl)-butyramide Cyclic 3′,5′-(Hydrogen Phosphate) 2′-Butyrate see 1473
3-β-D-Ribofuranosyl-1*H*-pyrrole-2,5-dione see 8621
7-β-D-Ribofuranosyl-7*H*-pyrrolo[2,3-*d*]-pyrimidin-4-amine see 9984
2-β-D-Ribofuranosyl-4-thiazolecarbox-amide see 9580
2-β-D-Ribofuranosyl-1,2,4-triazine-3,-52*H*,4*H*)-dione see 897
1-β-D-Ribofuranosyl-1*H*-1,2,4-triazole-3-carboxamide see 8323
1-β-D-Ribofuranosyl-1,2,4-triazole-3-car-boxamidine see 9197
1-β-D-Ribofuranosyl-1*H*-1,2,4-triazole-3-carboximidamide see 9197
1-β-D-Ribofuranosyluracil see 10062
9-β-D-Ribofuranosyl Xanthine see 10263
D-Ribohexulose see 8033
Ribolact [Taiho] see 8325
Ribomustin [Mundipharma] see 1036
Ribomycine [Delalande] see 8330
Ribonosine [Toyo Jozo] see 5018
Ribonuclease, 8326
Ribonuclease (*Rana pipiens* Reduced) see 8230
Ribonucleic Acid, 8327
D-Ribose, 8328
D-Ribose-5-phosphate see 8329
D-Ribose-5-phosphoric Acid, 8329
Ribosomal RNA see 8327
Ribostamin [Delalande] see 8330
Ribostamycin, 8330
Ribotrex [Fabre] see 907
Riboxamide see 9580
Ribrain [Yamanouchi] see 1180
D-Ribulose, 8331
Ricainide see 4977
Ricamycin [Toyo Jozo] see 8379
Rice Bran Oil, 8332
Rice Oil see 8332
Ricestar [Bayer CropSci.] see 4015
Richweed see 2467
Ricidine see 8334
Ricin, 8333
Ricin D see 8333
Ricinine, 8334
Ricinoleic Acid, 8335
Ricinolsulfuric Acid see 8335
Ricinstearolic Acid Diiodide see 7287
Ricinus Oil see 1898
Rickamicin see 8688
Ricobendazole see 203
Ricridene [Lipha] see 6624
Ridaforolimus, 8336
Ridaura [Astellas] see 869
Ridauran [Fabre] see 869
Ridazin [Taro] see 9512
Ridect [Pfizer] see 7294
Ridene [Syntex] see 6580
Ridinol see 7859
Ridogrel, 8337
Ridomil [Syngenta] see 5993
Ridomil Gold [Syngenta] see 5993
Ridzol [Merck & Co.] see 8385
Riegel's Paper see 2482
Rifa [Grünenthal] see 8341

Rifabutin, 8338
Rifacol [Formenti] *see* 8345
Rifadin(e) [Sanofi-Aventis] *see* 8341
Rifalazil, 8339
Rifaldazine *see* 8341
Rifaldin [Aventis] *see* 8341
Rifamastene [Uce] *see* 8343
Rifamicine SV *see* 8343
Rifamide, 8340
Rifampicin *see* 8341
Rifampin, 8341
Rifamycin AG *see* 8342
Rifamycin AMP *see* 8341
Rifamycin B *see* 8342
Rifamycin B Diethylamide *see* 8340
Rifamycin L 105 *see* 8345
Rifamycin O *see* 8342
Rifamycins, 8342
Rifamycin S *see* 8342
Rifamycin X *see* 8342
Rifapentine, 8344
Rifapiam [Piam] *see* 8341
Rifaprodin [Almirall] *see* 8341
Rifaxidin *see* 8345
Rifaximin, 8345
Rifit [Syngenta] *see* 7858
Rifloc [Avensa] *see* 5271
Rifocin [Lepetit] *see* 8343
Rifocin M [Lepetit] *see* 8340
Rifoldin [Aventis] *see* 8341
Rifomycins *see* 8342
Rifomycin SV *see* 8343
Rigelon [Dojin] *see* 9453
Rikamycin *see* 8379
Rikelta [UCB] *see* 1384
Riker 52G *see* 746
Rilansyl *see* 2106
Rilaquil *see* 2106
Rilassol *see* 2106
Rilaten [Guidotti] *see* 8374
Rilmazafone, 8346
Rilmenidine, 8347
Rilonacept, 8348
Rilpivirine, 8349
Rilutek [Aventis] *see* 8350
Riluzole, 8350
Rimactan(e) [Novartis] *see* 8341
Rimadyl [Pfizer] *see* 1862
Rimantadine, 8351
Rimaon *see* 3771
Rimatil [Santen] *see* 1470
Rimazole [Cheil Sugar] *see* 2401
Rimexel [Organon] *see* 8352
Rimexolone, 8352
Rimifon [Pharmion] *see* 5232
Rimiterol, 8353
Rimitsid *see* 5232
Rimocidin, 8354
Rimon [Makhteshim-Agan] *see* 6809
Rimonabant, 8355
Rimostil [Novogen] *see* 9858
Rimso-50 [Bioniche] *see* 3285
Rimsulfuron, 8356
Rinatec [Boehringer, Ing.] *see* 5117
Rinderon-DP [Shionogi] *see* 1182
Rindex [Sidus] *see* 6015
Rineton [Sanwa] *see* 9758
Rinlaxer [Taisho] *see* 2185
Rino-Clenil [Chiesi] *see* 1017
Rinogutt [Boehringer, Ing.] *see* 9727
Rintal [Bayer] *see* 3981
Rintatolimod, 8357
Riociguat, 8358
Riol [Toa Eiyo] *see* 9252
Riomitsin *see* 7075

Riopan [Ayerst] *see* 5714
Rioprostil, 8359
Ripcord [BASF] *see* 2764
Ripercol [Am. Cyanamid] *see* 5511
Ripirin [Kabi] *see* 3615
Ripon *see* 5714
Riposon *see* 5913
Riquent [La Jolla Pharmaceuticals] *see* 6
Riripen [Daiichi] *see* 1125
Risamol [Jansen] *see* 2315
Risatarun [Ravensberg] *see* 2845
Rischiaril [Piam] *see* 2845
Rise [Yoshitomi] *see* 2400
Risedronate Sodium *see* 8360
Risedronic Acid, 8360
Risolid [Dumex] *see* 2085
Risordan [Aventis] *see* 5271
Risperdal [J & J] *see* 8361
Risperidone, 8361
Ristocetin, 8362
Ristogen [Kowa Yakuhin] *see* 4163
Ristomycin *see* 8362
Ristomycin A *see* 8362
Ristomycin B *see* 8362
Ristosamine *see* 123
Risumic [Dainippon] *see* 388
Ritacetyl [R.I.T.A.] *see* 5409
Ritalan [R.I.T.A.] *see* 5409
Ritalin [Novartis] *see* 6183
Ritalmex [Valeant] *see* 6248
Ritanserin, 8363
Ritawax [R.I.T.A.] *see* 5410
Ritipenem, 8364
Ritipenem Acoxil *see* 8364
Ritmodan [Sanofi-Aventis] *see* 3402
Ritmos [Sanofi-Synthelabo] *see* 187
Ritmos Elle [Inverni] *see* 5634
Ritmusin [Gebro] *see* 743
Ritodrine, 8365
Ritonavir, 8366
Rituxan [Genentech] *see* 8367
Rituximab, 8367
RIV-2093 *see* 307
Rivadescin [Schaper & Brümmer] *see*
  8238
Rivanol [Winthrop] *see* 3771
Rivaroxaban, 8368
Rivasin [Solvay] *see* 8265
Rivastatin *see* 1992
Rivastigmine, 8369
Riversideite *see* 1704
Rivistel [Delagrange] *see* 307
Rivizor [J & J] *see* 10233
Rivoltan [Lichtwer] *see* 2944
Rivotril [Roche] *see* 2379
Rixapen [RIT] *see* 2376
Rizaben [Kissei] *see* 9731
Rizatriptan, 8370
Rize [Yoshitomi] *see* 2400
Rizen [Formenti] *see* 2400
Rizinsan K2 A2 *see* 4541
Rizolex [Sumitomo] *see* 9670
RJR-1734 *see* 5286
RL-50 *see* 6342
r-metHuG-CSF *see* 4573
r-metHuLeptin *see* 6232
RMI-17043 *see* 3643
RMI-71754 *see* 10175
RMI-71782 *see* 3570
RML-I *see* 6295
RNA *see* 8327
RNase *see* 8326
Ro-1-5130 *see* 8090
Ro-1-5431 *see* 5524
Ro-1-5431/7 *see* 5524

Ro-1-5470 *see* 5520
Ro-1-5470/5 *see* 2957
Ro-1-5470/6 *see* 5520
Ro-1-6794 *see* 5524
Ro-1-7788 *see* 5520
Ro-1-7977 *see* 8995
Ro-1-9569 *see* 9326
Ro-2-2985 *see* 5427
Ro-2-3198 *see* 3564
Ro-2-3308 *see* 8199
Ro-2-3773 *see* 2352
Ro-2-7113 *see* 287
Ro-2-9578 *see* 9877
Ro-2-9757 *see* 4213
Ro-2-9915 *see* 4156
Ro-2-9945 *see* 4208
Ro-3-4787 *see* 1485
Ro-3-7008 *see* 195
Ro-4-1544/6 *see* 8941
Ro-4-1575 *see* 483
Ro-4-1577 *see* 2706
Ro-4-3780 *see* 5274
Ro-4-3816 *see* 214
Ro-4-4393 *see* 9038
Ro-4-4602 *see* 1052
Ro-4-5282 *see* 5870
Ro-4-5360 *see* 6661
Ro-4-6467 *see* 7876
Ro-4-8347 *see* 9746
Ro-5-0810/1 *see* 9818
Ro-5-0831 *see* 5202
Ro-5-2180 *see* 6780
Ro-5-2807 *see* 2997
Ro-5-3307/1 *see* 2846
Ro-5-3350 *see* 1392
Ro-5-3438 *see* 4159
Ro-5-4023 *see* 2379
Ro-5-4200 *see* 4177
Ro-5-5345 *see* 9275
Ro-5-6901 *see* 4228
Ro-6-4563 *see* 4473
Ro-7-0207 *see* 6967
Ro-7-1051 *see* 1086
Ro-7-1554 *see* 5123
Ro-7-4488/1 *see* 6425
Ro-7-6145 *see* 3220
Ro-8-6837 *see* 6217
Ro-9-1978 *see* 1757
Ro-10-1670 *see* 103
Ro-10-5970 *see* 1387
Ro-10-6338 *see* 1489
Ro-10-9070 *see* 383
Ro-10-9359 *see* 3936
Ro-11-1163 *see* 6312
Ro-11-1430 *see* 6370
Ro-12-0068 *see* 9291
Ro-13-1042 *see* 5637
Ro-13-5057 *see* 655
Ro-13-5223 *see* 4017
Ro-13-8996 *see* 7036
Ro-13-9297 *see* 5640
Ro-13-9904/001 *see* 1955
Ro-14-3169/000 *see* 4021
Ro-14-4767/000 *see* 570
Ro-14-4767/002 *see* 570
Ro-15-1297 *see* 8096
Ro-15-1788 *see* 4166
Ro-15-8074 *see* 1924
Ro-15-8075 *see* 1924
Ro-17-2301 *see* 1871
Ro-17-3664 *see* 7920
Ro-18-0647 *see* 6964
Ro-19-5247 *see* 1949
Ro-19-6327 *see* 5449
Ro-20-5720/000 *see* 1862

Ro-20-7234 see 4191
Ro-21-3981/001 see 6261
Ro-21-3981/003 see 6261
Ro-21-5535 see 1647
Ro-21-5998 see 5872
Ro-21-6937 see 9895
Ro-21-8837 see 3761
Ro-21-9738 see 3485
Ro-22-7796 see 2272
Ro-22-8181 see 5034
Ro-22-9000 see 227
Ro-23-6019 see 5038
Ro-23-6240 see 4131
Ro-24-7375 see 2796
Ro-25-8310/000 see 7175
Ro-31-2848 see 2275
Ro-31-3113 see 2275
Ro-31-8959 see 8506
Ro-31-8959/003 see 8506
Ro-40-5967 see 6252
Ro-40-7592 see 9668
Ro 40-8757 see 6316
Ro-42-1611 see 804
Ro-44-9883 see 5399
Ro-47-0203 see 1355
Ro-61-0612 see 9400
Ro-61-1790 see 2340
Ro-63-9141 see 1953
Ro-65-5788 see 1953
Ro-107-9070/194 see 10093
RO-76477 see 7087
RO-92670 see 7087
RO-0094815 see 5149
RO-0098557 see 5149
RO-4607381 see 2801
RO-5073031 see 9209
RO-5185426 see 10138
Roaccutane [Roche] see 5274
RoActemra [Roche] see 9651
Robac 22 [Robinson Bros.] see 3858
Robalate [Robins] see 3201
Robamol [Cenci] see 6052
Robamox [Wyeth] see 574
Robane [Robeco] see 8895
Robanul see 4537
Robaxin [Robins] see 6052
Robecote [Robeco] see 8616
Robenidine, 8371
Robenogatope I-131 [BMS] see 8393
Robenz [Am. Cyanamid] see 8371
Robenzidene see 8371
Robigenin see 5322
Robinin, 8372
Robinul [Robins] see 4537
Robison Ester see 4498
Robitet [Wyeth] see 9341
Robitussin [Wyeth] see 4591
Robizone-V [Robins] see 7390
Robnin [Nikken] see 10370
Robuoy [Robeco] see 7870
Rocain [Fuso] see 7875
Rocaltrol [Roche] see 1647
Roccal [Sanofi-Synthelabo] see 1061
Roccellic Acid, 8373
Rocefin [Roche] see 1955
Rocephin [Roche] see 1955
Rochelle Salt see 7790
Rociverine, 8374
Rock Oil see 7302
Rock Salt see 8734
Rocornal [Rodleben] see 9735
Rocuronium, 8375
Rodeo [Dow AgroSci.] see 4547
Rodewod see 634
Rodex [Hacco] see 10236

Rodinal see 458
Rodipal see 3800
Rodiuran see 4826
Rody [Sumitomo] see 4019
(+)-Roemerine see 735
Roentgenium, 8376
Roeridorm [Roerig] see 3784
Rofecoxib, 8377
Rofedex [Biofarm] see 2957
Roferon-A [Roche] see 5034
Roflual [Roche] see 8351
Roflumilast, 8378
Rogaine [Pfizer] see 6285
Rogitine [Novartis] see 7376
Rogojski's Salt see 2658
Rogor [Isagro] see 3241
Rohrbach's Soln see 978
Rohypnol [Roche] see 4177
Roinin [Mohan] see 7855
Roipnol [Roche] see 4177
Rökan [Intersan] see 4460
Rokital [Formenti] see 8379
Rokitamycin, 8379
Rolipram, 8380
Rolitetracycline, 8381
Roman Chamomile see 2040
Roman Chamomile Oil see 2040
Roman Laurel see 5437
Roman Vitriol see 2643
Romazicon [Roche] see 4166
Romensin [Elanco] see 6332
Romet [Mitsubishi] see 8258
Rometin see 5075
Romglizone see 9948
Romicil see 6918
Romidepsin, 8382
Romiplate [Kyowa] see 8383
Romiplostim, 8383
Rompun [Bayer] see 10277
Romurtide, 8384
RonaCare [Merck KGaA] see 6912
RonaCare Ectoin [Merck KGaA] see
  3552
Ronaxan [Merial] see 3488
Rondase see 4797
Rondomycin [Wallace] see 6015
Rongalite see 8755
Roniacol [Roche] see 6612
Roniacol Tartrate [Roche] see 6612
Ronicol see 6612
Ronidase see 4797
Ronidazole, 8385
Ronifibrate, 8386
Ronilan [BASF] see 10181
Ronnel, 8387
Ronok [Ono] see 6970
Ronoprost see 6970
Ronstar [Bayer CropSci.] see 7006
Rontyl [Leo Pharm] see 4826
ROP-500F see 5121
Ropinirole, 8388
Ropivacaine, 8389
Roptazol see 4330
Roquessine see 2480
Roquinimex, 8390
Roridan C see 9812
Roridins see 10160
Rosaced [Fabre] see 6236
Rosal [IBI] see 8391
Rosamicin see 8392
Rosampline [Tedis] see 583
Rosaniline Hydrochloride see 5715
Rosapin see 2918
Rosaprostol, 8391
Rosaramicin, 8392

Roscal [Reumert] see 3429
Roscoelite see 10104
Roscovitine see 8579
Rose Bengal, 8393
Rose Bengale B see 8393
Rose Bengal Extra see 8393
Rose Bengal Sodium I 131 see 8393
Rose Hips, 8394
Rosemary, 8395
Rosemide [Ono] see 4338
Rosenstiehl's Green see 977
Rose-pink see 1971
Rosiglitazone, 8396
Rosin, 8397
Rosin Oil, 8398
Rosinol see 8398
Rosmarinic Acid, 8399
p-Rosolic Acid see 872
Rosoxacin, 8400
Rossitrol [Corvi] see 8409
Rostaporfin, 8401
Rostil [Ortho] see 8359
Rosuvastatin, 8402
Rotacide [Roussel Uclaf] see 7589
Rotenone, 8403
Rotigotine, 8404
Rotondin [Casasco] see 4004
Rotoxamine see 1800
Rotramin [Glaxo-Allen] see 8409
Rottlerin, 8405
Rotundine see 9360
Rougoxin [Rougier] see 3186
Round-leafed Sundew see 3501
Roundup [Monsanto] see 4547
Rovalcyte [Roche] see 10093
Rovamicina [RPR] see 8879
Rovamycin [RPR] see 8879
Rovral [Bayer CropSci.] see 5121
Rowapraxin [Rowa-Wagner] see 7596
Rowasa [Solvay] see 5974
Roxanthin Red 10 [Roche] see 1756
Roxarsone, 8406
Roxatidine Acetate, 8407
Roxiam [Astra] see 8252
Roxiden [Pulitzer] see 7619
Roxindole, 8408
Roxindole Mesylate see 8408
Roxit [Aventis] see 8407
Roxithromycin, 8409
Roxomicina see 3740
Royal Demolition Explosive see 2728
Royal Jelly, 8410
Royal Jelly Acid see 8410
Royal Purple see 3028
Royline see 5679
Rozacrème [Biorga] see 6236
Rozagel [Biorga] see 6236
Rozerem [Takeda] see 8219
Rozex [Galderma] see 6236
Rozicel [BMS] see 1943
Rozol [Lipha] see 1389
Rozol see 2155
RP-40 see 6772
RP-103 see 2776
RP-217 see 9236
RP-866 see 8160
RP-2090 see 9074
RP-2168 see 6150
RP-2255 see 9075
RP-2259 see 4515
RP-2275 see 9040
RP-2512 see 7227
RP-2632 see 9045
RP-2786 see 8097
RP-2831 see 4692

RP-2987 *see* 3119
RP-3203 *see* 5083
RP-3276 *see* 7898
RP-3277 *see* 7901
RP-3356 *see* 3800
RP-3359 *see* 2091
RP-3377 *see* 2165
RP-3389 *see* 7901
RP-3554 *see* 9458
RP-3697 *see* 4373
RP-3735 *see* 5650
RP-3799 *see* 3128
RP-3828 *see* 463
RP-4482 *see* 9071
RP-4560 *see* 2191
RP-4753 *see* 8098
RP-4909 *see* 2189
RP-5015 *see* 5232
RP-5337 *see* 8879
RP-6140 *see* 7879
RP-6549 *see* 9873
RP-6847 *see* 7045
RP-7044 *see* 5518
RP-7162 *see* 9894
RP-7204 *see* 2673
RP-7293 *see* 7871
RP-7452 *see* 91
RP-7522 *see* 9051
RP-7746 *see* 9845
RP-7843 *see* 9510
RP-7891 *see* 4516
RP-8228 *see* 5521
RP-8595 *see* 3290
RP-8599 *see* 4258
RP-8823 *see* 6236
RP-8908 *see* 7279
RP-9159 *see* 7283
RP-9671 *see* 6808
RP-9715 *see* 2706
RP-9921 *see* 746
RP-9965 *see* 6226
RP-10192 *see* 2890
RP-11974 *see* 7446
RP-13057 *see* 2834
RP-13907 *see* 4138
RP-14539 *see* 8556
RP-16091 *see* 6215
RP-16272 *see* 1451
RP-17623 *see* 7006
RP-18429 *see* 9730
RP-19366 *see* 7595
RP-19551 *see* 7595
RP-19552 *see* 7595
RP-19560 *see* 6001
RP-19583 *see* 5352
RP-22050 *see* 10395
RP-22410 *see* 4479
RP-26019 *see* 5121
RP-27267 *see* 10393
RP-54274 *see* 8350
RP-54476 *see* 2803
RP-54563 *see* 3642
RP-54780 *see* 7011
RP-56976 *see* 3442
RP-57669 *see* 8201
RP-59500 *see* 2803
RP-62955 *see* 7084
r-PA *see* 8284
RPA-107382 *see* 3793
RPA-201772 *see* 5283
RPA-407213 *see* 3990
RPM *see* 8232
RPR-116258 *see* 1604
RPR-130673 *see* 6996
RR-32705 *see* 6927

RRNA *see* 8327
RS-533 *see* 7111
RS-1301 *see* 2883
RS-1320 *see* 4176
RS-3540 *see* 6499
RS-3650 *see* 6499
RS-3999 *see* 4176
RS-4691 *see* 2387
RS-5139 *see* 9279
RS-8852 *see* 203
RS-8858 *see* 7034
RS-9054 *see* 900
RS-9390 *see* 7991
RS-10085 *see* 6315
RS-10085-197 *see* 6315
RS-11988 *see* 5397
RS-21592 *see* 4393
RS-25259-197 *see* 7098
RS-26306 *see* 4395
RS-33295-198 *see* 10396
RS-35887 *see* 1531
RS-37619 *see* 5353
RS-43285 *see* 8229
RS-44872 *see* 9025
RS-61443 *see* 6412
RS-69216 *see* 6580
RS-79070-194 *see* 10093
RS-84043 *see* 4023
RS-84135 *see* 3645
RS-94991-298 *see* 6435
R Salt *see* 6472
RSD-1235 *see* 10157
r-VIII SQ *see* 6357
RSR-13 *see* 3568
RT *see* 9918
RTA-401 *see* 959
RTA-402 *see* 959
RTCA *see* 8323
RTI-3021-012 *see* 10030
rt-PA *see* 9624
RTY-319 *see* 4895
RU-43-715 *see* 7984
RU-0211 *see* 5649
RU-486 *see* 6266
RU-965 *see* 8409
RU-2267 *see* 315
RU-2323 *see* 4449
RU-5020 *see* 7900
RU-15060 *see* 9577
RU-15750 *see* 4135
RU-19110 *see* 4632
RU-22974 *see* 2888
RU-23908 *see* 6629
RU-24756 *see* 1934
RU-25474 *see* 9725
RU-28965 *see* 8409
RU-31158 *see* 5631
RU-38486 *see* 6266
RU-38702 *see* 117
RU-42173 *see* 10324
RU-44403 *see* 9729
RU-44570 *see* 9729
RU-66647 *see* 9261
Rubber, 8411
Rubeanic Acid, 8412
Rubene *see* 6454
Ruberythric Acid, 8413
Ruberythrinic Acid *see* 8413
Rubesol *see* 10212
Rubiadin, 8414
Rubian *see* 8413
Rubianic Acid *see* 8413
Rubiazol [Hoechst] *see* 9033
Rubiazol I *see* 9055
Rubichloric Acid *see* 834

Rubidazone [Bellon] *see* 10395
Rubidium, 8415
Rubidium Bromide, 8416
Rubidium Carnallite *see* 8415
Rubidium Chloride, 8417
Rubidium Hydroxide, 8418
Rubidium Iodide, 8419
Rubidium Monobromide *see* 8416
Rubidium Monochloride *see* 8417
Rubidium Monoiodide *see* 8419
Rubidomycin *see* 2834
Rubigan [Gowan] *see* 3992
Rubigervine *see* 8420
Rubijervine, 8420
Rubimycin [Nikken] *see* 6262
Rubinorm [IFI] *see* 8417
Rubitecan, 8421
Rubixanthin, 8422
Ruboxistaurin, 8423
Ruboxistaurin Mesylate Monohydrate
    see 8423
Rubozinc [Labcatal] *see* 4492
Rubramin PC [BMS] *see* 10212
Rubratope-57 [BMS] *see* 10213
Rubratope-60 [BMS] *see* 10213
Rubriment [Nordmark] *see* 6611
α-Rubromycin *see* 2466
Rubus, 8424
Ruby [Spencer] *see* 6604
Ruby Arsenic *see* 790
Ruby Wood *see* 8522
Ruconest [Pharming] *see* 2481
Rudilin [Darby] *see* 6821
Rudotel [OPW] *see* 5857
Rufast [AgrEvo] *see* 117
Rufigallic Acid *see* 8425
Rufigallol, 8425
Rufinamide, 8426
Rufloxacin, 8427
Rufol [Fournier] *see* 9048
Rugby [FMC] *see* 1636
Rugulovasine A *see* 8428
Rugulovasine B *see* 8428
Rugulovasines, 8428
Rulid [Hoechst] *see* 8409
Rumatel [Phibro] *see* 6353
Rumatral [Nicholas] *see* 305
Rumensin [Elanco] *see* 6332
Rumex, 8429
Runner [Bayer CropSci.] *see* 6066
Rupafin [Uriach] *see* 8430
Rupatadine, 8430
RuPhos Phosphine Ligand *see* 1469
Ruppert's Reagent *see* 9851
Rusalatide, 8431
Rusa Oil *see* 6888
Rusat *see* 1158
Ruswut *see* 1158
Rusyde [CP] *see* 4338
Rutaecarpine *see* 8432
Rutamycin *see* 6927
Rutecarpine, 8432
Ruthenium, 8433
Ruthenium Chloride (RuCl$_3$) *see* 8436
Ruthenium(VIII) Oxide *see* 8435
(*T*-4)-Ruthenium Oxide (RuO$_4$) *see* 8435
Ruthenium Oxychloride Ammoniated
    see 8434
Ruthenium Red, 8434
Ruthenium Tetroxide, 8435
Ruthenium Trichloride, 8436
Ruthenium Violet *see* 8434
Rutherfordium, 8437
Ruticina [Bernabo] *see* 5996
Rutile *see* 9628

Rutin, 8438
Rutinose, 8439
Rutoside *see* 8438
Ruven [Mepha] *see* 9969
Ruxolitinib, 8440
RV-144 *see* 3704
RV-12424 *see* 4179
RWJ-10553 *see* 6783
RWJ-17021-000 *see* 9706
RWJ-17070 *see* 4760
RWJ-20485 *see* 9294
RWJ-22164 *see* 856
RWJ-60235 *see* 7639
RWJ-270201 *see* 7264
RWJ-333369 *see* 1841
RX-5050M *see* 3380
RX-6029-M *see* 1501
RX-781094 *see* 4928
Ryanex *see* 8441
Ryania, 8441
Ryanicide [Penick] *see* 8441
Ryanodine, 8442
Ryanodol 3-(1*H*-Pyrrole-2-carboxylate)
see 8442
Rycarden [Syntex] *see* 6580
Rycoben [Novartis] *see* 203
Rycobendazole *see* 203
Rydar [Cutter] *see* 3327
Rydene [Syntex] *see* 6580
Rydrin [Kodama] *see* 6821
Ryegonovin [Aventis] *see* 6142
Rynacrom [Sanofi-Aventis] *see* 2581
Rynaxypyr [DuPont] *see* 2080
Rythmatine [Burroughs Wellcome] *see*
5911
Rythmodan [Sanofi-Aventis] *see* 3402
Rythmodul [Sanofi-Aventis] *see* 3402
Rythmol [Abbott] *see* 7908
Rytmobeta [Abbott] *see* 8854
Rytmonorm [Abbott] *see* 7908
S-2-676 *see* 9382
S-7 *see* 4031
S-23/46 *see* 664
S-041 *see* 4354
S-47 *see* 1419
S-73-4118 *see* 7605
S-77-0777 *see* 7840
S-94 *see* 4220
S-095 *see* 5787
S-145 *see* 3465
(+)-S-145 *see* 3465
S-165 *see* 3120
S-203 *see* 9869
S-222 *see* 3410
S-314 *see* 4342
S-596 *see* 782
S-640P *see* 1917
S-658 *see* 4324
S-768 *see* 4004
S-780 *see* 1041
S-805 *see* 5645
S-940 *see* 6438
S-992 *see* 1041
S-1006 *see* 1920
S-1108 *see* 1920
S-1320 *see* 1476
S-1452 *see* 3465
S-1520 *see* 4975
S-1530 *see* 6634
S-1574 *see* 9575
S-1605 *see* 3120
S-1694 *see* 403
S-1702 *see* 4474
S-1752 *see* 4029
S-1844 *see* 4037

S-1942 *see* 1442
S-2225 *see* 1442
S-2395 *see* 9316
S-2539 *see* 7325
S-2620 *see* 294
S-2703 *see* 2765
S-2703 Forte *see* 2765
S-3013 *see* 10126
S-3151 *see* 7294
S-3206 *see* 4019
S-3308-10 *see* 3294
S-3308L *see* 3294
S-3341 *see* 8347
S-3341-3 *see* 8347
S-3349 *see* 9670
S-3500 *see* 1527
S-4068 *see* 7823
S-4068SF *see* 7823
S-4084 *see* 2687
S-4347 *see* 1419
S-4522 *see* 8402
S-4661 *see* 3474
S-5602 *see* 4037
S-5602α *see* 4037
S-6059 *see* 6372
S-6810 *see* 5036
S-7131 *see* 7882
S-8527 *see* 2355
S-9318 *see* 8103
S-9490 *see* 7285
S-9490-3 *see* 7285
S-9780 *see* 7285
S-10036 *see* 4289
S-12911 *see* 8224
S-16257 *see* 5294
S-16257-2 *see* 5294
S-20098 *see* 183
S-21403 *see* 6296
S-23031 *see* 4173
S-25930 *see* 4911
S-26308 *see* 4964
S-28463 *see* 8271
S-31183 *see* 8103
S-32165 *see* 3120
S-50022 *see* 4463
S-53482 *see* 4174
SA *see* 10025
SA-79 *see* 7908
SA-96 *see* 1470
SA-97 *see* 4771
SA-504 *see* 9599
sab simplex [Pfizer] *see* 3237
Sabadilla, 8443
Sabcomeline, 8444
Saber [Intervet] *see* 5398
Sabidal *see* 2215
Sabiden [Szabo] *see* 7382
Sabril [HMR] *see* 10175
Sabromin [Sandoz] *see* 1460
Saccharase *see* 5049
Saccharated Ferric Oxide *see* 5145
Saccharic Acid *see* 4485
Saccharin, 8445
Saccharin Ammonium *see* 8445
Saccharin Insoluble *see* 8445
Saccharinol *see* 8445
Saccharinose *see* 8445
Saccharin Sodium *see* 8445
Saccharol *see* 8445
Saccharolactic Acid *see* 4361
Saccharolactone *see* 4485
L-Saccharopine, 8446
Saccharose *see* 9012
Saccharosonic Acid *see* 5171
Sacerno [EGYT] *see* 5920

Sacox [Intervet] *see* 8473
Sacred Bark *see* 1880
Sacrosidase, 8447
Sacysyl-cysteine *see* 2780
Sadopan *see* 4400
Safari [DuPont] *see* 9857
Safetray *see* 634
Saffan *see* 228
Safflor Carmine *see* 1867
Safflor Red *see* 1867
Safflower *see* 1868
Safflower Oil, 8448
Saffron, 8449
Saffron-bitter *see* 7523
Saffron Indian *see* 10005
Safinamide, 8450
Safinamide Mesylate *see* 8450
Saflufenacil, 8451
Safranal, 8452
Safranin B Extra *see* 7363
Safrole, 8453
Saga [Roussel-Uclaf] *see* 9725
Sagamicin [Kyowa] *see* 6260
Sagamicin (formerly) *see* 6260
Sage *see* 8480
Sagrotan [Schülke & Mayr] *see* 1061
SaH-42548 *see* 5831
SAHA *see* 10232
Saheli [Hindustan Latex] *see* 1972
SAICAR, 8454
Saigon Cinnamon *see* 2300
Saizen [Serono] *see* 8842
Sakarat X [Killgerm] *see* 10236
Sakura [Bayer CropSci.] *see* 8123
Sakuranetin, 8455
Sakuranin *see* 8455
Sal Acetosella *see* 7732
Saladjidi *see* 4924
Salagen [MGI] *see* 7536
Salamid [Hamilton] *see* 8465
Sal Ammoniac *see* 506
Salamol [IVAX] *see* 210
Salanil [Sato Yakuhin] *see* 4411
Salatrim, 8456
Salazopyrin [Pfizer] *see* 9073
Salazosulfadimidine, 8457
Salazosulfapyridine *see* 9073
Salbumol [GSK] *see* 210
Salbutamol *see* 210
Salcatonin *see* 1646
Salcyl *see* 8471
Salcylix *see* 8471
Saldox *see* 8464
Salen, 8458
Sal Enixum *see* 7734
Salep, 8459
Sal Ethyl *see* 3902
Salflex [Carnrick] *see* 8477
Salicain [Hoechst] *see* 4885
Salicin, 8460
Salicin Benzoate *see* 7716
Salicinerein *see* 7507
Salicoside *see* 8460
N-Salicoylaminophenol *see* 6984
Salicyclic Acid 2-Naphthyl Ester *see*
6496
Salicyl Alcohol, 8461
Salicyl Alcohol Glucoside *see* 8460
Salicylaldehyde, 8462
Salicylaldehyde β-D-Glucoside *see* 4663
Salicylaldehyde Phenylhydrazone, 8463
Salicylaldoxime, 8464
Salicylamide, 8465
Salicylamide *O*-Acetic Acid, 8466
Salicylamide *o*-Ethyl Ether *see* 3786

Salicylanilide, 8467
Salicylazosulfadimidine see 8457
Salicylazosulfamethazine see 8457
Salicylazosulfapyridine see 9073
Salicylhydroxamic Acid, 8468
Salicylic Acid, 8469
Salicylic Acid Acetate see 841
Salicylic Acid Acetate Calcium Salt see 1650
Salicylic Acid Acetate, Ester with 4-Hydroxyacetanilide see 1047
Salicylic Acid, Acid Sulfate see 8471
Salicylic Acid Benzyl Ester see 1146
Salicylic Acid Bimolecular Ester see 8477
Salicylic Acid p-tert-Butylphenyl Ester see 1589
Salicylic Acid Choline Salt see 2213
Salicylic Acid Dihydrogen Phosphate see 4282
Salicylic Acid Ethyl Ester see 3902
Salicylic Acid Isopropyl Ester O-Ester with O-Ethyl Isopropylphosphoramidothioate see 5218
Salicylic Acid Methyl Ester Benzoate see 6099
Salicylic Acid Monoammonium Salt see 547
Salicylic Acid 1-Naphthyl Ester see 6495
Salicylic Acid Sulfuric Acid Ester see 8471
Salicylic Acid 3,3,5-Trimethylcyclohexyl Ester see 4777
Salicylic Aldehyde see 8462
Salicylic Aldehyde Phenylhydrazone see 8463
Salicylic Ether see 3902
Salicyl Morpholide see 8470
Salicyloxysalicylic Acid see 8477
4-Salicyloylmorpholine, 8470
Salicyl Phosphate see 4282
Salicylsalicylic Acid see 8477
Salicylsulfonic Acid see 9094
Salicylsulfuric Acid, 8471
Salicyl-Vasogen see 547
Salicyl Yellow see 5989
Saligenin see 8461
Saligenin-β-D-glucopyranoside see 8460
Saligenol see 8461
Saliglutin [Streuli] see 8469
Salimid see 2740
Salinaphthol see 6496
Salinazid, 8472
Saliniazid see 8472
Salinidol [Doak] see 8467
Salinigrin see 7507
Salinomycin, 8473
Salinosporamide A see 8474
Salinosporamide B see 8474
Salinosporamide C see 8474
Salinosporamides, 8474
Salipran [Evans] see 1047
Salipurpol see 6506
Saliuzid see 6947
Salizell [Byk Gulden] see 8465
Salizell (Amp.) [Byk Gulden] see 8466
Salizolo see 4954
Salmetedur [Menarini] see 8475
Salmeterol, 8475
Salmeterol Xinafoate see 8475
Salmiac see 506
Salmine Sulfate, 8476
Salmine Sulfate (1:1) (Salt) see 8476
Salmotonin [Sanofi-Aventis] see 1646
Salocin [Intervet] see 8473

Salofalk [Falk] see 5974
Salol see 7420
Saloop see 8518
Salpetersäure (German) see 6663
Sal Polychrestum see 7793
Salsalate, 8477
Sal Soda see 8731
Salsoline, 8478
(+)-Salsoline see 8478
Salt see 8734
Salt Cake see 8809
Saltidin [Saltigo] see 7506
Salt of Lemon see 7732
Salt of Saturn see 7420
Salt of Sorrel see 7732, 7806
Salt of Tartar see 7740
Saltpeter see 7769
Saltucin [Boehringer, Mann.] see 1527
Salufer see 8759
Salures [Pfizer] see 1039
Saluric [Merck & Co.] see 2171
Salurin [Syntex] see 9789
Saluron [Apothecon] see 4826
Saluside see 6947
Salutaridine, 8479
Saluzid see 6947
Saluzide see 6947
Salvacard see 6627
Salvacyl [Ipsen] see 9924
Salvarsan see 801
Salvia, 8480
Salvin see 1850
Salvinin see 7180
Salvinorins, 8481
Sal Volatile see 505
Salymid [Biochemie] see 8465
Salyrgan [Sterling Winthrop] see 5971
Salyzoron [Hishiyama] see 1125
Salzburg Vitriol see 2643
Samaderin A see 8482
Samaderin B see 8482
Samaderin C see 8482
Samaderins, 8482
Samaderoside A see 8482
Samandarine, 8483
Samandarone see 8483
Samarium, 8484
Samarium Cobalt, 8485
Samarium Diiodide see 8486
Samarium 153 Ethylenediaminetetramethylenephosphonic Acid see 8487
Samarium Iodide, 8486
Samarium(II) Iodide see 8486
Samarium Pentacobalt see 8485
Samarium Sm 153 Lexidronam, 8487
Sambucin see 2677
Sambucus, 8488
Sambunigrin see 5784
SAMe see 144
S-Amet [Europharma] see 144
Samid [Knoll] see 8465
Samli see 4451
Samorin [M & B] see 5229
Sampatrilat, 8489
Samsca [Otsuka] see 9699
Samyr [Boehringer, Ing.] see 144
SAN-322I see 7929
SAN-371F see 7007
SAN-582H see 3236
SAN-619F see 2768
SAN-835H see 3164
SAN-1269H see 3164
SAN-9789 see 6790
Sanamycin [FBA Pharm] see 1611
Sanapert see 7073

Sanasthmax [GSK] see 1017
Sanazin see 8062
Sanceller 22 [Sanshin] see 3858
Sanctura [Indevus] see 9966
Sancuso [ProStrakan] see 4572
Sand see 8632
Sandalwood, 8490
Sandalwood Oil see 8490
Sandarac, 8491
Sand-brier see 8835
Sandimmun(e) [Novartis] see 2750
Sandofan [Syngenta] see 7007
Sandolanid [Sandoz] see 3186
Sandolase [Sandoz] see 8011
Sandomigran [Novartis] see 7629
Sandonorm [Novartis] see 1335
Sandopart [Novartis] see 2844
Sandopril [Sandoz] see 8880
Sandoptal [Sandoz] see 1514
Sandostatin [Novartis] see 6851
Sandostene [Sandoz] see 9429
Sandoz 43-715 see 7984
Sanecta [Holland Sweetener] see 829
Sanger's Reagent see 4204
Sangrel see 8602
Sanguicillin [Zdravlje] see 7626
Sanguinaria, 8492
Sanguinarine, 8493
Sanifer [Esseti] see 4063, 8751
Saniflor [Esseti] see 1125
Sanipirina [Bayer] see 47
Sanlit [Sankyo] see 8673
Sanmigran [Novartis] see 7629
Sanmite [Nissan] see 8078
Sanocrisin [Ferrosan] see 2713
Sanocrysin see 4554
Sanodin see 1794
Sanoma [Heilit] see 1842
Sanomigran [Novartis] see 7629
Sanorex [Novartis] see 5831
SANORG-34006 see 4932
Sanorin see 6453
Sanorin-Spofa see 6453
Sanotensin [Egis] see 4595
Sano-Tuss [Welti] see 3884
Sanoxit [Galderma] see 1119
Sansdolor see 5917
Sansert [Novartis] see 6213
α-Santalol, 8494
β-Santalol, 8495
Santavy's Substance F see 2891
Santemycin [Santen] see 6260
Santoflex see 3807
Santonica, 8496
Santonic Acid, 8497
α-Santonin, 8498
l-Santonin see 8498
Santophen 1 [Monsanto] see 2394
Santoquin [Novus] see 3807
Santovar A see 3334
Santyl [Knoll] see 2465
Sanvar [Debiopharm] see 10123
Sanzyme [Sankyo] see 9162
SAPEP see 399
Saperconazole, 8499
Saphris [Merck & Co.] see 821
Sapidan see 4400
l-Sapietic Acid see 5522
Sapilent [EGYT] see 9894
Sapoderm [Reckitt & Colman] see 9822
Saponaretin see 8501
Saponaria, 8500
Saponarin, 8501
Saponins, 8502
Saponite see 345

Saporin-6 *see* 8503
Saporins, 8503
Sapphire, 8504
Sapphire (Al$_2$O$_3$) *see* 8504
Sapphirine *see* 345
Sapresta [Taiho] *see* 756
Saprol [Shell Agrar] *see* 9859
Sapropterin, 8505
Saquinavir, 8506
Saquinavir Mesylate *see* 8506
Sar *see* 8510
Sarafem [Lilly] *see* 4217
Sarafloxacin, 8507
Sarafotoxins, 8508
Sarafotoxins S6 *see* 3631
1-Sar-8-ala-angiotensin II *see* 8509
Saralasin, 8509
Saralasin Acetate *see* 8509
Saran [Dow] *see* 10192
Sarasar [Schering-Plough] *see* 5622
Sarcine *see* 4908
Sarcolactic Acid *see* 5385
Sarcolysine *see* 5896
D-Sarcolysine *see* 5896
L-Sarcolysine *see* 5896
Sarcomycin *see* 8513
Sarcosine, 8510
Sarelon *see* 7164
Sarenin [Procter & Gamble] *see* 8509
Sargenor [Viatris] *see* 830
Sargramostim *see* 4574
Sarin, 8511
Sarizotan, 8512
Sarkin *see* 4908
Sarkomycin, 8513
Sarkomycin A *see* 8513
Sarmentose, 8514
Saroten [Bayer] *see* 483
Sarotex [Lundbeck] *see* 483
Sarpagan-17-al *see* 10137
Sarpagan-10,17-diol *see* 8515
Sarpagine, 8515
Sarpul [Roche] *see* 655
Sarsaparilla, 8516
Sarsasapogenin, 8517
Saruplase *see* 8011
Sarvinal *see* 704
Saryuurin [Nippon Shoji] *see* 6984
Sassafras, 8518
Sassolite *see* 1339
Sassy Bark, 8519
Sastridex [Lindopharm] *see* 4163
Sasulen [Faes] *see* 7619
Satanolon [Tatsumi] *see* 3341
Satinite *see* 1708
Satin Spar *see* 1708
Sativex [GW Pharma] *see* 1752
Satraplatin, 8520
Satumomab, 8521
Satumomab, N^6-[N-[2-[[2-[Bis(carboxy-
  methyl)amino]ethyl](carboxymethyl)-
  amino]ethyl]-N-(carboxymethyl)gly-
  cyl]-N^2-(N-glycyl-L-tyrosyl)-L-lysine
  Conjugate, Indium-^{111}In Chelate *see*
  8521
Saturn Red *see* 5480
Saucy Bark *see* 8519
Saunders, Red, 8522
Saurol *see* 4924
Savac [Am. Cyanamid] *see* 5083
Savella [Forest] *see* 6275
Savene [TopoTarget] *see* 8241
Saventrine [Pharmax] *see* 5263
Savey [Gowan] *see* 4749
Savin, 8523

Savinin *see* 8523
Saw Palmetto, 8524
Saxagliptin, 8525
Saxifrax *see* 8518
Saxin [Burroughs Wellcome] *see* 8445
Saxitoxin, 8526
Saxizon [Nikken] *see* 4824
Saxol *see* 7301
Saxoline *see* 7300
Saxon Bark *see* 8519
Saytex 102E [Albemarle] *see* 2848
Saytex CP-2000 [Albemarle] *see* 9328
Saytex HP-900 [Albemarle] *see* 4712
Sazio [Zyma] *see* 235
SB-58 *see* 8941
SB-75 *see* 2025
SB-5833 *see* 1727
SB-7505 *see* 4915
SB-202026 *see* 8444
SB-207266 *see* 7505
SB-207266A *see* 7505
SB-207499 *see* 2279
SB-209509 *see* 4301
SB-209509AX *see* 4301
SB-265805 *see* 4423
SB-275833 *see* 8281
SB-393229 *see* 9713
SB-480848 *see* 2824
SB-497115 *see* 3609
SB-659746A *see* 10176
SBA *see* 5484
SBB *see* 9016
SBB-I *see* 9016
SBB-II *see* 9016
SBE-β-CD *see* 2711
α_2-SB Glycoproteins *see* 4104
SBP-1382 *see* 8274
SBP-1390 *see* 1235
SBP-1513 *see* 7294
SBR Rubber, 8527
SC-046 *see* 2915
SC-0051 *see* 9026
SC-0224 *see* 4547
SC-2910 *see* 6030
SC-7031 *see* 3402
SC-7525 *see* 1325
SC-9387 *see* 7567
SC-9420 *see* 8887
SC-9880 *see* 4230
SC-11800 *see* 3909
SC-13957 *see* 3402
SC-18862 *see* 829
SC-29333 *see* 6293
SC-40230 *see* 1208
SC-47111 *see* 5619
SC-48334 *see* 6268
SC-54684A *see* 10269
SC-54701A *see* 10269
SC-58635 *see* 1958
SC-59046 *see* 2915
SC-59735 *see* 9623
SC-65872 *see* 10086
SC-66110 *see* 3681
SC-69124A *see* 7140
SC-72325 *see* 4965
SC-326421 *see* 6236
Scabecid [Stiefel] *see* 5556
Scabwort *see* 5047
Scala [Bayer CropSci.] *see* 8099
Scalibor [Intervet] *see* 2888
Scammony Root, 8528
Scandia *see* 8529
Scandicain [Bofors] *see* 5926
Scandine [Zambon] *see* 4915
Scandium, 8529

Scarlet Berry *see* 3510
Scarlet Red, 8530
scCO$_2$ *see* 1809
SCE-129 *see* 1946
SCE-963 *see* 1936
SCE-1365 *see* 1926
SCE-2174 *see* 1936
SCE-2787 *see* 1939
Scenesse [Clinuvel] *see* 168
Scepter [BASF] *see* 4947
SCF *see* 8934
Sch-1000 *see* 5117
Sch-3132 *see* 6050
Sch-3940 *see* 7297
Sch-4831 *see* 1182
Sch-10144 *see* 9678
Sch-10159 *see* 9820
Sch-10304 *see* 2381
Sch-10649 *see* 894
Sch-11460 *see* 1182
Sch-12041 *see* 4624
Sch-13475 *see* 8688
Sch-13521 *see* 4238
Sch-14714 *see* 4178
Sch-14947 *see* 8392
Sch-15719W *see* 5377
Sch-16134 *see* 8145
Sch-18020W *see* 1017
Sch-19927 *see* 5377
Sch-20569 *see* 6564
Sch-21420 *see* 5151
Sch-21480 *see* 9611
Sch-22219 *see* 213
Sch-25298 *see* 4141
Sch-28316Z *see* 4979
Sch-29851 *see* 5635
Sch-30500 *see* 5034
Sch-32088 *see* 6327
Sch-32481 *see* 6565
Sch-33844 *see* 8880
Sch-33861 *see* 8880
Sch-34117 *see* 2923
Sch-39300 *see* 4574
Sch-39720 *see* 1950
Sch-52000 *see* 5043
Sch-56592 *see* 7723
Sch-58235 *see* 3958
Sch-66336 *see* 5622
Sch-209579 *see* 5828
Sch-417690 *see* 10172
SCH-54031 *see* 7176
SCH-55700 *see* 8273
SCH-503034 *see* 1322
SCH-530348 *see* 10230
SCH-900962 *see* 2518
Schaeffer's α-Acid *see* 6475
Schaeffer's β-Acid *see* 6477
Schaeffer's Salt *see* 6477
Schardinger Dextrins *see* 2710
SCH-D *see* 10172
Scheele's Green *see* 2616
Scheelite *see* 1714
Schercozoline O [Lubrizol] *see* 6925
Schering 36268 *see* 2086
Schering 38584 *see* 7350
Scherisolon [Schering] *see* 7841
Schiff Bases, 8531
Schizophyllan *see* 8697
Schleimsäure (German) *see* 4361
Schmiedeberg's Digitalin *see* 3177
Scholar [Syngenta] *see* 4160
Schöllkopf's Acid *see* 6471
Schradan, 8532
Schultenite *see* 5454
Schwartz's Reagent, 8533

Schwefelstickstoff see 9389
Schweinfurt Green see 2615
Schweizer's Reagent, 8534
Schwesinger P$_4$ Base, 8535
SCI see 8739
Scifluorfen see 101
Scillaren, 8536
Scillaren A see 8536
Scillaren B see 8536
Scillarenin, 8537
Scillarenin 3β-Rhamnoside see 7985
Scilliroside, 8538
Scoburen [Renaudin] see 1591
Scoke see 7503
Scolecite see 1651
Scoline [Duncan Flockhart] see 9006
Scoparin, 8539
Scoparius, 8540
Scoparone, 8541
Scoparoside see 8539
Scopine Tropate see 8543
Scopoderm TTS [Novartis] see 8543
Scopola, 8542
Scopolamine, 8543
l-Scopolamine see 8543
Scopolamine Aminoxide see 8544
Scopolamine Bromobutylate see 1591
Scopolamine-N-butyl Bromide see 1591
Scopolamine Methobromide see 6077
Scopolamine Methyl Bromide see 6077
Scopolamine N-Oxide, 8544
Scopolammonium Bromide see 8543
Scopoletin, 8545
Scopolin, 8546
Scopoline, 8547
Score [Syngenta] see 3154
Scorprin [Fort Dodge] see 9035
Scotch Broom see 8540
Scotine [Unimed] see 2541
Scotophobin, 8548
Scotopsins see 6951
Scourge [Roussel Uclaf] see 7589
Scout [Roussel-Uclaf/AgrEvo] see 9725
SCP see 8732
S.C.T.Z. see 2375
Sculptra [Sanofi-Aventis] see 5386
Scu-PA see 8011
Scuroforme see 1515
Scutellarein, 8549
Scutellaria, 8550
Scutellarin see 8549
SD-271-12 see 2357
SD-1750 see 3089
SD-3211 see 8585
SD-3562 see 3097
SD-4402 see 5172
SD-7859 see 2090
SD-8447 see 9337
SD-9129 see 6337
SD-14114 see 3997
SD-15418 see 2675
SD-41706 see 4019
SD-43775 see 4037
SD-95481 see 2296
SD-208304 see 2088
S-Dimidine [Protea] see 9047
SDMH see 3272
SDS see 8768
Sdt-91 see 8942
SDX-105 see 1036
SDZ-205-502 see 8161
SDZ-212-713 see 8369
SDZ-215-811 see 7238
SDZ-ASM-981 see 7541
SDZ-CHI-621 see 1003

SDZ-DJN-608 see 6510
SDZ-ENA-713 see 8369
SDZ-HTF-919 see 9253
SDZ-ILE-964 see 5039
SDZ-PSC-833 see 10103
SDZ-RAD see 3950
SE-780 see 1041
SE-1520 see 4975
SE-2395 see 9316
Seaborgium, 8551
Sea-Nine [Rohm & Haas] see 3080
Sea-oak see 4311
Sea Onion see 8898
Searlequin see 5085
Sea Salt see 8734
Sea-wrack see 4311
Sebacic Acid, 8552
Sebacil [Bayer] see 7478
Sebacil, 8553
Sebacoin, 8554
Sebercim [GSK] see 6789
Sebivo [Novartis] see 9258
Sebizon [Schering] see 9031
Secaclavine see 2041
Secalan [Zyma] see 2092
Secale Cornutum see 3717
Secalip [Solvay] see 4009
Secalonic Acid A see 8555
Secalonic Acid B see 8555
Secalonic Acid C see 8555
Secalonic Acid D see 8555
Secalonic Acid E see 8555
Secalonic Acid F see 8555
Secalonic Acid G see 8555
Secalonic Acids, 8555
Secalysat [Ysatfabrik] see 2540
Secbutabarbital see 1509
Secbutobarbitone see 1509
Seccidin [Nippon Kayaku] see 7855
Secholex see 7674
Seclar [Andromaco] see 1016
Secletan [Ipsen] see 2267
Secnidazole, 8556
Secobarbital Sodium, 8557
(1α,3β,5Z,7E)-9,10-Secocholesta-5,7,10-
(19)-triene-1,3-diol see 4858
(3β,5Z,7E)-9,10-Secocholesta-5,7,10(19)-
triene-3,25-diol see 1641
(1α,3β,5Z,7E)-9,10-Secocholesta-5,7,10-
(19)-triene-1,3,25-triol see 1647
(1α,3β,5Z,7E,24R)-9,10-Secocholesta-5,-
7,10(19)-triene-1,3,24-triol see 9152
(3β,5Z,7E)-9,10-Secocholesta-5,7,10(19)-
trien-3-ol see 10216
(1S,1′E,3R,5Z,7E,20R)-9,10-Seco-20-(3′-
cyclopropyl-3′-hydroxyprop-1′-enyl)-
1,3-dihydroxypregna-5,7,10(19)-triene
see 1644
(1α,3β,5Z,7E,22E)-9,10-Secoergosta-5,7,-
10(19),22-tetraene-1,3-diol see 3484
(3β,5Z,7E,22E)-9,10-Secoergosta-5,7,10-
(19),22-tetraen-3-ol see 10215
(3β,6E,22E)-9,10-Secoergosta-5(10),6,-
8,22-tetraen-3-ol see 9153
(3β,5Z,7E)-9,10-Secoergosta-5,7,10(19)-
trien-3-ol see 10217
(3β,5E,7E,10α,22E)-9,10-Secoergosta-5,-
7,22-trien-3-ol see 3198
Secoiridoid see 6923
Seconal Sodium [Lilly] see 8557
Secondary Ammonium Phosphate see
540
Secondary Barium Phosphate see 986
Secondary Calcium Phosphate see 1694
Secondary Caprylic Alcohol see 6841

Secondary Magnesium Phosphate see
5745
Secondary Propyl Alcohol see 5254
Secondary Sodium Phosphate see 8789
Secotil see 6950
Secrebil [Isnardi] see 7598
Secreflo [Repligen] see 8558
γ-Secretase see 7857
Secretases see 608
Secretin, 8558
Secretin (Pig) see 8558
Secretin (Swine) see 8558
Secrosteron (obsolete) see 3240
Sectral [RPR] see 20
Securinan-11-one see 8559
Securinine, 8559
Securon [Abbott] see 10144
Securopen [Bayer] see 908
Securpres [Poli] see 4979
Sedaform see 2129
Sedagul [Wild] see 5535
Sedalande [J & J] see 4146
Sedalin [Chassot] see 33
Sedalito [Merck KGaA] see 47
Sedamyl [3M Pharma] see 22
Sedanoct [Woelm] see 9977
Sedantoinal [Sandoz] see 5920
Sedapain [Nichiiko] see 3696
Sedaplus [Rosen Pharma] see 3489
Sedapran [Kowa] see 7836
Sedatival [Raffo] see 5636
Sedator [Dechra] see 5858
Sedatromin [Takata] see 2306
Sedazin [Lagap] see 5636
Sedemesis see 2000
Sediel [Sumitomo] see 9183
Sednotic [Dista] see 567
Sedofarmolo [Centralvet] see 463
Sedolatan [Hoechst] see 7855
Sedoneural see 8729
Sedopretten [Schoening] see 3339
Sedotussin [UCB] see 1795
Sedoxil [Bial] see 6245
Sedrena [Daiichi] see 9864
Sedulon [Roche] see 7581
Sedural see 7324
Sedutain [Key] see 8557
Seduxen [Gedeon Richter] see 2997
Sefona see 4865
Seftem [Shionogi] see 1950
Séglor [UCB] see 3191
Segontin [Hoechst] see 7855
Segosin see 6045
Sehydrin see 4810
Seidlitz Mixture, 8560
Seignette Salt see 7790
Seki [Simes] see 2384
Sekisanine see 9219
Sekisanoline see 9219
Selacryn [SKB] see 9586
Selagine see 4794
Selagnine see 5684
Selamectin, 8561
Selbex [Eisai] see 9296
Seldane [Aventis] see 9306
Selecal [Toyama] see 9595
Selecron [Syngenta] see 7886
Select [Valent] see 2347
Selectfluor [Air Prods. Chem.], 8562
Selectin [BMS] see 7835
Selectins, 8563
Selectol [Pharmacia] see 1962
Selectomycin [Grünenthal] see 8879
Selectride [Aldrich], 8564
Seledat [Master] see 8565

Selegam [Hexal] *see* 8565
Selegiline, 8565
Selemicina [Italchemi] *see* 4281
Selenase [Biosyn] *see* 8802
Selenic Acid, 8566
Selenic Acid Ammonium Salt (1:2) *see* 548
Selenic Acid Copper(2+) Salt (1:1) *see* 2640
Selenic Acid Potassium Salt (1:2) *see* 7786
Selenic Acid Sodium Salt (1:2) *see* 8800
Seleninyl Chloride *see* 8572
Selenious Acid, 8567
Selenious Acid Diammonium Salt *see* 549
Selenious Acid Sodium Salt (1:2) *see* 8802
Selenious Anhydride *see* 8571
Selenite *see* 1708
Selenium, 8568
(T-4)-Selenium Bromide (SeBr$_4$) *see* 8574
Selenium Chloride, 8569
Selenium Chloride (Se$_2$Cl$_2$) *see* 8569
(T-4)-Selenium Chloride (SeCl$_4$) *see* 8575
Selenium Cysteine *see* 8577
Selenium Dioxide *see* 8571
(T-4)-Selenium Fluoride (SeF$_4$) *see* 8576
(OC-6-11)-Selenium Fluoride (SeF$_6$) *see* 8570
Selenium Hexafluoride, 8570
Selenium Hydride *see* 4836
Selenium Monochloride *see* 8569
Selenium Nitride *see* 6693
Selenium Oxide, 8571
Selenium Oxide (SeO$_2$) *see* 8571
Selenium Oxychloride, 8572
Selenium Sulfides, 8573
Selenium Tetrabromide, 8574
Selenium Tetrachloride, 8575
Selenium Tetrafluoride, 8576
Selenkupfer *see* 2656
Selenocysteine, 8577
Selenocystine *see* 8577
Selenomethionine, 8578
L-Selenomethionine *see* 8578
Selenomethionine Se 75 *see* 8578
Selenous Acid *see* 8567
3-Selenyl-L-alanine *see* 8577
Selepam [Schering-Plough] *see* 8145
Selepark [Betapharm] *see* 8565
Seles Beta [Schwarz] *see* 850
Selexid (inj.) [Leo Pharm] *see* 383
Selexid (susp.) [Leo Pharm] *see* 384
Selexid (tabl.) [Leo Pharm] *see* 384
Selezen [Italfarmaco] *see* 4954
Selgimed [Hennig] *see* 8565
Seliciclib, 8579
Selinon *see* 3305
Selipran [BMS] *see* 7835
Sellagen *see* 5751
Sellaite *see* 5729
Seloken [AstraZeneca] *see* 6228
Selopral [AstraZeneca] *see* 6228
Selozok [AstraZeneca] *see* 6228
Selsun [Abbott] *see* 8573
Selvigon [ICN-Galenika] *see* 7569
Selvjgon [Rorer] *see* 7569
Selzentry [Pfizer] *see* 5814
Semagacestat, 8580
Semap [Janssen] *see* 7196
Sematilide, 8581
Semduramicin, 8582

Semen Cydonia *see* 8170
Semesan [DuPont] *see* 4869
Semicarbazide Hydrochloride, 8583
Semicid [Am. Home] *see* 6764
Semi-mustard Gas *see* 4681
Seminose *see* 5812
Semioxamazide, 8584
Semisulfur Mustard *see* 4681
Semotiadil, 8585
Sempera [Janssen-Cilag] *see* 5292
Sempervirene *see* 8586
Sempervirine, 8586
Semprex [GSK] *see* 118
Semtex *see* 2728
Semtex A *see* 2728
Semtex H *see* 2728
Senarmontite *see* 699
Sencor [Bayer CropSci.] *see* 6233
Sencoral [Bayer CropSci.] *see* 6233
Sendoxan [Baxter] *see* 2743
Seneca Oil *see* 7302
Seneca Snakeroot *see* 8592
Senecialdehyde, 8587
Senecic Acid, 8588
Senecio, 8589
Senecioaldehyde *see* 8587
Senecionine, 8590
N^1-Senecioylsulfanilamide *see* 9036
Seneciphylline, 8591
Senega, 8592
Senega Snakeroot *see* 8592
Seniramin [Hokuriku] *see* 9147
Senna, 8593
Sennoside A *see* 8594
Sennoside B *see* 8594
Sennosides, 8594
Senociclin *see* 8381
Sensaval [Lundbeck] *see* 6804
Sensibion *see* 4671
Sensipar [Amgen] *see* 2285
Sensit [Thiemann] *see* 4002
Sensorcaine [AstraZeneca] *see* 1499
Sentiloc [Benzon] *see* 1195
Sentinel [Syngenta] *see* 2768
Seocalcitol, 8595
Seotalnatrium [Lilly] *see* 8557
SEP-0002093 *see* 3754
SEPA [MacroChem] *see* 6767
Sepamit [Organon] *see* 6613
Sepan [Janssen] *see* 2306
Sepatren [Dainippon Sumitomo] *see* 1940
Sepazon [Sankyo] *see* 2404
Sephadex 2-(Diethylamino)ethyl Ether *see* 7674
Sepia, 8596
Sepicide CI [Seppic] *see* 4957
Sepiolite *see* 5751
Sepiomelanin, 8597
Seponver [Ethnor] *see* 2396
Sepsinol *see* 4323
Septeal [Fabre] *see* 2092
Septicid *see* 5712
Septicide [Virbac] *see* 2313
Septiphene *see* 2394
Septopal [Merck KGaA] *see* 4427
Septosan [Mallard] *see* 9055
Septra [Monarch] *see* 9050
Septrin [GSK] *see* 9050
Septural [Grünenthal] *see* 7618
Sequel [Nihon] *see* 4024
Sequens *see* 2102
Sequestrene *see* 3565
Ser *see* 8599
Seradix [Rhone-Poulenc] *see* 5005
Seragon [Ferring] *see* 2221

Seral *see* 5913
Serastar [Yamanouchi] *see* 5009
Seratrodast, 8598
Serax [Wyeth] *see* 7025
D-Ser(Bu^t)^{6}Azgly^{10}-gonadorelin *see* 4562
[D-Ser(Bu^t)^6-des-Gly^{10}-NH$_2$]-LH-RH Ethylamide *see* 1506
Serc [Unimed] *see* 1180
Serdolect [Lundbeck] *see* 8605
Serecor [Houdé] *see* 4843
Serefrex [Janssen] *see* 5344
Seren Vita [Synthelabo] *see* 2085
Serenace [Searle] *see* 4634
Serenal [Sankyo] *see* 7026
Serenase [Lusofarmaco] *see* 4634
Serenesil [Abbott] *see* 3784
Serenid [Wyeth] *see* 7025
Serentil [Boehringer, Ing.] *see* 5980
Serepax [Wyeth] *see* 7025
Serepress [Formenti] *see* 5344
Sereprile [Synthelabo] *see* 9576
Sereprostat [Pierre Fabre] *see* 8524
Séresta [Biodim] *see* 7025
Serevent [GSK] *see* 8475
Serfolia [Hauck] *see* 8238
Sericin *see* 4102
Sericosol N *see* 4521
Seriel [Biogalenique] *see* 9662
Serine, 8599
L-Serine *see* 8599
L-Serine Diazoacetate (Ester) *see* 892
Serine Dihydrogen Phosphate (Ester) *see* 7475
125-L-Serine-2-133-interleukin 2 (Human Reduced) *see* 5038
Serine Phosphate *see* 7475
84-L-Serine-plasminogen Activator (Human Tissue-type Protein Moiety Reduced) *see* 6347
Serine 2-[(2,3,4-Trihydroxyphenyl)methyl] Hydrazide *see* 1052
173-L-Serine-174-L-tyrosine-175-L-glutamine-173-527-plasminogen Activator (Human Tissue-type) *see* 8284
[173-Serine,174-tyrosine,175-glutamine]-173-275-plasminogen Activator (Human Tissue-type Reduced) Fusion Protein with Urokinase (Human Urine β-Chain Reduced) *see* 385
Sermaka [Lilly] *see* 4227
Sermion [Farmitalia] *see* 6581
Sermorelin, 8600
Sernevin [Toho Yakuhin] *see* 9119
Sernyl *see* 7333
Sernylan *see* 7333
Serocion *see* 7909
Seroden [Allen & Hanburys] *see* 9442
Serofene [Serono] *see* 2377
Seromycin [Lilly] *see* 2749
Serophene [Serono] *see* 2377
Seropram [Lundbeck] *see* 2317
Seroquel [AstraZeneca] *see* 8155
Serotinex *see* 2370
Serotone [Yoshitomi; Japan Tobacco] *see* 893
Serotonin, 8601
Serotropin [Teikoku] *see* 2221
Seroxat [GSK] *see* 7148
Serozil [BMS] *see* 1943
Serozyme *see* 7999
Serpasil [Novartis] *see* 8265
Serpax [Wyeth] *see* 7025
Serpentaria, 8602
Serpentine *see* 815
Serpentine (Alkaloid), 8603

Serpentine (Mineral) *see* 5751
Serpina [Himalaya] *see* 8238
Sertaconazole, 8604
Sertan [Valeant] *see* 7865
Sertindole, 8605
Sertraline, 8606
Serum Albumin, 8607
Serum Gonadotropin *see* 2221
Serum Lactic Dehydrogenase *see* 5382
Serum Prothrombin Conversion Accelerator *see* 3961
Serum Spreading Factor *see* 10222
Serum Transferrin *see* 9732
Serum Tryptase *see* 7634
Serutan [Menley & James] *see* 7631
Servispor [Novartis] *see* 1974
Servizepam [Novartis] *see* 2997
L-Seryl-L-aspartyl-L-asparaginyl-L-asparaginyl-L-glutaminyl-L-glutaminylglycyl-L-lysyl-L-seryl-L-alanyl-L-glutaminyl-L-glutaminylglycylglycyl-L-tyrosinamide *see* 8548
N-(DL-Seryl)-*N'*-(2,3,4-trihydroxybenzyl) Hydrazine *see* 1052
L-Seryl-L-valyl-L-seryl-L-α-glutamyl-L-isoleucyl-L-glutaminyl-L-leucyl-L-methionyl-L-histidyl-L-asparaginyl-L-leucylglycyl-L-lysyl-L-histidyl-L-leucyl-L-asparaginyl-L-seryl-L-methionyl-L-α-glutamyl-L-arginyl-L-valyl-L-α-glutamyl-L-tryptophyl-L-leucyl-L-arginyl-L-lysyl-L-lysyl-L-leucyl-L-glutaminyl-L-α-aspartyl-L-valyl-L-histidyl-L-asparaginyl-L-phenylalanine Acetate Hydrate (1:?:?) *see* 9309
Serzone [BMS] *see* 6523
Sesame Oil, 8608
Sesamin, 8609
Sesamodil *see* 8585
Sesamolin, 8610
Sesden [Tanabe] *see* 9599
Setastine, 8611
Sethotope [BMS] *see* 8578
Sethoxydim, 8612
Sethyl *see* 4768
Setous [Shionogi] *see* 10397
Setrol [Baxter] *see* 7072
Sevelamer, 8613
Seven Barks *see* 4805
Sevin [Bayer CropSci.] *see* 1788
SevoFlo [Abbott] *see* 8614
Sevoflurane, 8614
Sevofrane [Maruishi] *see* 8614
Sevorane [Abbott] *see* 8614
Sevredol [Mundipharma] *see* 6361
Sexovid [Ferrosan] *see* 2713
Seyferth-Gilbert Reagent, 8615
SF-86-327 *see* 9299
SF-277 *see* 2963
SF-328 *see* 3396
SF-733 Antibiotic *see* 8330
SF-837 *see* 6262
SFAE *see* 8850
SG-75 *see* 6606
Sgd-12878 *see* 6941
Sgd-301-76 *see* 7036
S-GI *see* 2906
SGN-30 *see* 1375
SGN-35 *see* 1375
SH-100 *see* 2729
SH-261 *see* 3823
SH-406 *see* 9307
SH-437 *see* 5108
SH-582 *see* 4448
SH-617L *see* 5115

SH-714 *see* 2771
SH-717 *see* 4543
SH-770 *see* 4184
SH-818 *see* 2363
SH-863 *see* 2363
SH-881 *see* 2771
SH-926 *see* 5055
SH-80881 *see* 2771
SH B 331 *see* 4447
SH-E-222 *see* 5923
SH K 203 *see* 4183
SH U 454 *see* 4365
SH U 508A *see* 4365
SHA-486300 *see* 4842
Shadocol *see* 5081
Shark Liver Oil, 8616
Sharpen [BASF] *see* 8451
SHB-286 *see* 9120
SHCH-58 *see* 2611
Shekanin *see* 9245
Shellac, 8617
Shellfish Purple *see* 3028
Shellolic Acid, 8618
Sheridanite *see* 345
Shibagen [Ishihara] *see* 4129
Shigatox [Fawns & McAllan] *see* 9040
Shikalkin *see* 249
Shikimic Acid, 8619
Shikonin *see* 249
Shinbit [Nihon Schering] *see* 6614
Shiomarin [Shionogi] *see* 6372
Shionone, 8620
Shiosol [Shionogi] *see* 4553
Shirlan [Syngenta] *see* 4148
Shisonin A *see* 2677
SHL-451A *see* 4357
SHL-569B *see* 4360
Showdomycin, 8621
Showersan *see* 5080
Shoxin *see* 6775
Shvo Catalyst, 8622
Siagoside *see* 4394
Sialic Acids, 8623
Sialor [Solvay] *see* 637
Sibelium [Janssen] *see* 4175
Siberian Pine Needle Oil *see* 6885
SIB-S1 *see* 5513
Sibutol *see* 1306
Sibutramine, 8624
Siccanin, 8625
Sicoclor [Gramon] *see* 3999
Sicorten [Novartis] *see* 4633
Sideramines *see* 2863
Siderin Yellow *see* 4047
Siderite *see* 5137
Sideromycins *see* 2863
Siderophilin *see* 9732
Siderophores *see* 6409
Sideros [Sanofi-Synthelabo] *see* 4063
Siderotil *see* 4084
Siduron, 8626
Sieromicin [Sierochimica] *see* 7566
Siesta [BASF] *see* 5992
Sifrol [Boehringer, Ing.] *see* 7825
SIGA-246 *see* 9244
Sigamopen [Alpharma] *see* 574
Sigaperidol [Dumex] *see* 4634
Sigaprim [Dumex] *see* 9050
Sigmachrome [Svizera] *see* 1789
Sigmacort [Sigma] *see* 4824
Sigmafon [Fontana] *see* 5842
Sigmaform *see* 1291
Sigmart [Chugai] *see* 6606
Siklos [Addmedica] *see* 4888
SIL-4R *see* 5046

Silain [Wyeth] *see* 3237
Silane, 8627
3-Silapentane *see* 3147
Silapo [Stada] *see* 3744
Silarine [Vir] *see* 6274
Silbephylline [Berk] *see* 3526
Sildenafil, 8628
Silene *see* 1704
Silenor [Somaxon] *see* 3483
Silibin *see* 8671
Silibinin *see* 8671
Silibrin [Silesia] *see* 2085
Silica *see* 8632
Silicane *see* 8627
Siliceous Earth *see* 5015
Silicic Acid, 8629
Silicic Acid Aluminum Calcium Salt *see* 1651
Silicic Acid Calcium Salt *see* 1704
Silicic Acid (H_2SiO_3) *see* 8629
Silicic Acid (H_4SiO_4) *see* 8629
Silicic Acid ($H_6Si_2O_7$) *see* 8629
Silicic Acid (H_4SiO_4) Aluminum Magnesium Salt (2:2:1) *see* 345
Silicic Acid (H_2SiO_3) Lithium Salt (1:2) *see* 5596
Silicic Acid (H_2SiO_3) Sodium Salt (1:2) *see* 8804
Silicic Acid (H_4SiO_4) Tetraethyl Ester *see* 3903
Silicic Acid (H_4SiO_4) Zinc Salt (1:2) *see* 10357
Silicic Acid (H_4SiO_4) Zirconium(4+) Salt (1:1) *see* 10381
Silicic Acid Potassium Salt *see* 7788
Silicic Acid Sodium Salt *see* 8804
Silicic Anhydride *see* 8632
Silicobromoform *see* 9777
Silicobutane *see* 9387
Silicochloroform *see* 9811
Silicoethane *see* 3398
Silicofluoric Acid *see* 4214
Silicon, 8630
Silicon Bromide *see* 8638
Silicon Carbide, 8631
Silicon Carbide (SiC) *see* 8631
Silicon Chloride *see* 8639
Silicon Dioxide, 8632
Silicon Disulfide, 8633
Silicone Glycol Copolymers *see* 8634
Silicone Glycols *see* 8634
Silicone Polyethers *see* 8634
Silicones, 8634
Silicon Fluoride *see* 8640
Silicon Monoxide, 8635
Silicon Nitride, 8636
Silicon Nitride (Si_3N_4) *see* 8636
Silicon Sulfide (SiS_2) *see* 8633
Silicon Tetraacetate, 8637
Silicon Tetrabromide, 8638
Silicon Tetrachloride, 8639
Silicon Tetraethoxide *see* 3903
Silicon Tetrafluoride, 8640
Silicon Tetrahydride *see* 8627
Silicon Tetrahydroxide *see* 8629
Silicon Tetramethyl *see* 9377
Silicopropane *see* 9930
Silicotungstic Acid, 8641
Silirex [Lampugnani] *see* 6274
Silkis [Galderma] *see* 1647
Silkweed *see* 818
Sillimanite *see* 356
Silliver [Abbott] *see* 6274
Silmar [Hennig] *see* 6274
Silodosin, 8642

Silodyx [Recordati] *see* 8642
Silomat [Thomae] *see* 2360
Silosan [Zeneca] *see* 7610
Silubin *see* 1481
Silvadene [King] *see* 9035
Silver, 8643
Silver Acetate, 8644
Silverado [Bayer CropSci.] *see* 5981
Silver Balsam *see* 5319
Silver Benzoate *see* 1093
Silver Bromide, 8645
Silver Bromide (AgBr) *see* 8645
Silver Carbonate, 8646
Silver Chlorate, 8647
Silver Chloride, 8648
Silver Chloride (AgCl) *see* 8648
Silver Chromate(VI), 8649
Silver Citrate, 8650
Silver Cyanide, 8651
Silver Cyanide (Ag(CN)) *see* 8651
Silver Difluoride, 8652
Silver Drops *see* 5319
Silver Fluoride, 8653
Silver Fluoride (AgF) *see* 8653
Silver Fluoride (AgF$_2$) *see* 8652
Silver Iodate, 8654
Silver Iodide, 8655
Silver Iodide (AgI) *see* 8655
Silver Lactate, 8656
Silver Leaf *see* 8947
Silver Manganate(VII) *see* 8663
Silver Monofluoride *see* 8653
Silver Nitrate, 8657
Silver Nitrite, 8658
Silver Nucleate *see* 8666
Silver Nucleinate *see* 8666
Silver Orthophosphate *see* 8664
Silver Oxalate, 8659
Silver Oxide, 8660
Silver(II) Oxide, 8661
Silver Oxide (AgO) *see* 8661
Silver Oxide (Ag$_2$O) *see* 8660
Silver Perchlorate, 8662
Silver Permanganate, 8663
Silver Peroxide *see* 8661
Silver Phosphate, 8664
Silver Picrate, 8665
Silver Protein, 8666
Silver Proteinate *see* 8666
Silver Selenide, 8667
Silver Selenide (Ag$_2$Se) *see* 8667
Silver Suboxide *see* 8661
Silver Sulfate, 8668
Silver Sulfide, 8669
Silver Sulfide (Ag$_2$S) *see* 8669
Silver Tetraiodomercurate(II), 8670
Silver Trinitrophenolate *see* 8665
Silvisar *see* 8730
Silvol [Parke-Davis] *see* 8666
Silybin *see* 8671, 8671
Silybin A *see* 8671
Silybum Substance E$_6$ *see* 8671
Silycristin *see* 6274
Silydianin *see* 6274
Silymarin I *see* 8671
Simactil [Rorer] *see* 3710
Simanex [Makhteshim-Agan] *see* 8672
Simatin [Geistlich] *see* 3801
Simazine, 8672
Simdax [Orion] *see* 5525
Simeconazole, 8673
(*R*)-Simendan *see* 5525
Simethicone *see* 3237
Simetryn, 8674
Simfibrate, 8675

Simovil [Merck & Co.] *see* 8677
Simpadren *see* 9142
Simpalon *see* 9142
Simpatoblock *see* 4724
Simplesse [CP Kelco], 8676
Simplicef [Pharmacia & Upjohn] *see* 1942
Simplotan [Pfizer] *see* 9604
Simponi [Centocor] *see* 4560
Simulect [Novartis] *see* 1003
Simvastatin, 8677
SIN-10 *see* 6320
Sinalbin, 8678
Sinan *see* 5917
Sinanomycin *see* 6566
Sinapic Acid Choline Ester *see* 8679
Sinapine, 8679
Sinapine Glucosinalbate *see* 8678
Sinapsan [Rodleben] *see* 7600
Sinapultide, 8680
Sinartrol [SPA] *see* 7619
Sinbar [DuPont] *see* 9298
Sincalide, 8681
Sincaline *see* 2211
Sincodix [Beta] *see* 1516
Sincomen [Schering AG] *see* 8887
Sindiatil *see* 1481
Sinecod [Zyma] *see* 1516
Sinefungin, 8682
Sinemet [Merck & Co.] *see* 1798
Sineptina [Antibioticos] *see* 5505
Sinequan [Pfizer] *see* 3483
Sinerem [Guerbet] *see* 4092
Sinerol *see* 7065
Sinesalin [MIT Gesundheit] *see* 1039
Sinex [LaChartre] *see* 7065
Sinfibrate *see* 8675
Single-chain Pro-urokinase *see* 8011
Single-chain Urokinase-type Plasmino-
   gen Activator *see* 8011
Singulair [Merck & Co.] *see* 6346
Sinigrin, 8683
Sinigroside *see* 8683
Sinlestal [Daiichi Seiyaku] *see* 7873
Sinoaculine *see* 8479
Sinoflurol [Kaken] *see* 9252
Sinogan (Inj.) [Sanofi-Aventis] *see* 5518
Sinogan (Tabl.) [Sanofi-Aventis] *see* 5518
Sinomenine, 8684
Sinomin [Shionogi] *see* 9050
Sinorphan *see* 8207
Sinos *see* 2735
Sinosid [SIF] *see* 7146
Sinox *see* 3305
Sinox W [FMC] *see* 3315
Sinthrome [Novartis] *see* 30
Sintisone [Pharmacia] *see* 7841
Sintoclar [Pulitzer] *see* 2318
Sintomicetina [Lepetit] *see* 2077
Sintomodulina [Italfarmaco] *see* 9558
Sintonal [Europharma] *see* 1459
Sintosulfa [Afi] *see* 9065
Sintotrat [Bracco] *see* 4824
Sintrom [Novartis] *see* 30
Sinvacor [Merck & Co.] *see* 8677
Sipeimine *see* 4968
Sipiri Bark *see* 1015
Siplarol [Erba] *see* 4219
Siptazin [Isei] *see* 2306
Sipuleucel-T, 8685
SIPX *see* 8766
Siqualone [BMS] *see* 4220
SIR-8514 *see* 9844
Sirdalud [Novartis] *see* 9642
Siringina [Toyo Jozo] *see* 9147
Sirius Red, 8686

Sirius Red F3B *see* 8686
Sirolimus *see* 8232
Siros [Janssen-Cilag] *see* 5292
Siroxyl [Aventis] *see* 1802
Sirtal [Merck KGaA] *see* 1783
Sisal, 8687
Siscard [Boehringer, Ing.] *see* 2278
Siseptin [Schering] *see* 8688
Sisobiotic [Von Boch] *see* 8688
Sisolline [Schering-Plough] *see* 8688
Sisomicin, 8688
Sisomin [Max] *see* 8688
Sisosamine *see* 8688
Sissotrin *see* 1231
Sistalgin [Merck KGaA] *see* 7827
Sistan [DuPont] *see* 6021
Sisuril *see* 4826
Sitafloxacin, 8689
Sitagliptin, 8690
Sitagliptin Phosphate *see* 8690
Sitamaquine, 8691
Sitaxsentan, 8692
Sitofex [BASF] *see* 4262
Sito-Lande [Delalande] *see* 8694
β-Sitostanol *see* 8943
Sitosterin *see* 8694
α$_1$-Sitosterol, 8693
β-Sitosterol *see* 8694
γ-Sitosterol, 8695
Sivastin [Sigma-Tau] *see* 8677
Sivelestat, 8696
Sivlor [Sidus] *see* 5644
Sizofiran, 8697
SJO 0498 *see* 4151
SK-7 *see* 7240
SK-331-A *see* 10258
SK-818 *see* 7909
Skanitrol [Schering] *see* 6719
Skatole, 8698
Skeetal *see* 925
Skelid [Sanofi] *see* 9598
SKF-51 *see* 6846
SKF-385 *see* 9734
SKF-478 *see* 3341
SKF-478-A *see* 3341
SKF-1700-A *see* 6821
SKF-2601-A *see* 2191
SKF-4657 *see* 7879
SKF-5137 *see* 2958
SKF-5883 *see* 9510
SKF-7988 *see* 10200
SKF-8542 *see* 9760
SKF-18667 *see* 7679
SKF-29044 *see* 7139
SKF-30310 *see* 7035
SKF-38095 *see* 6912
SKF-39162 *see* 869
SKF-40383 *see* 1832
SKF-41558 *see* 1918
SKF-60771 *see* 1917
SKF-62698 *see* 9586
SKF-62979 *see* 203
SKF-75073 *see* 1930
SKF-82526 *see* 4010
SKF-82526-J *see* 4010
SKF-83088 *see* 1927
SKF-88373 *see* 1952
SKF-92334 *see* 2282
SKF-96022 *see* 7117
SKF-96148 *see* 2805
SKF-101468 *see* 8388
SKF-101468A *see* 8388
SKF-102362 *see* 6630
SKF-102886 *see* 4630
SKF-104864 *see* 9707

Sodium Nitrosylpentacyanoferrate(III) *see* 8780
Sodium Octadecanoate *see* 8806
Sodium Octadecyl (*E*)-Butendioate *see* 8807
Sodium Octadecyl Fumarate *see* 8807
Sodium Oxacillin *see* 7003
Sodium Oxalate, 8781
Sodium Oxide, 8782
Sodium Oxide (Na$_2$O) *see* 8782
Sodium Oxybate *see* 4854
Sodium γ-Oxybutyrate *see* 4854
Sodium α-Oxyhyponitrite *see* 641
Sodium Pentachlorophenate *see* 7218
Sodium Pentachlorophenoxide *see* 7218
Sodium (*OC*-6-22)-Pentakis(cyano-*κC*)-nitrosylferrate(2−) (2:1) *see* 8780
Sodium Pentosan Polysulfate *see* 7247
Sodium Perborate, 8783
Sodium Perchlorate, 8784
Sodium Periodate *see* 8772
Sodium Permanganate, 8785
Sodium Peroxide, 8786
Sodium Peroxide (Na$_2$(O$_2$)) *see* 8786
Sodium Peroxydisulfate *see* 8787
Sodium Persulfate, 8787
Sodium Pertechnetate Tc 99m, 8788
Sodium Phenolate *see* 7354
Sodium Phenoxide *see* 7354
Sodium *o*-Phenylphenate *see* 7415
Sodium 2-Phenylphenoxide *see* 7415
Sodium Phosphate, Dibasic, 8789
Sodium Phosphate, Monobasic, 8790
Sodium Phosphate (^{32}P) *see* 8791
Sodium Phosphate P 32, 8791
Sodium Phosphate, Tribasic, 8792
Sodium Phosphite, 8793
Sodium Phosphomolybdate, 8794
Sodium Phosphorothioate *see* 8820
Sodium Phosphotungstate, 8795
Sodium Phytate *see* 7499
Sodium Picosulfate *see* 7518
Sodium Pipesate *see* 7592
Sodium Platinichloride *see* 8758
Sodium Polyanethole Sulfonate, 8796
Sodium Polyanetholsulfonate *see* 8796
Sodium Polyethylene Sulfonate *see* 5678
Sodium Polymannuronate *see* 234
Sodium Polymetaphosphate *see* 8773
Sodium Polyphosphate Glass *see* 8773
Sodium Polystyrene Sulfonate, 8797
Sodium Propanoate *see* 8798
Sodium Propionate, 8798
Sodium Prussiate Yellow *see* 8752
Sodium Pteroylglutamate *see* 4248
Sodium Pyroborate *see* 8726
Sodium Pyrocatechol-2,4-disulfonate *see* 9621
Sodium Pyrophosphate *see* 9388
Sodium Pyrosulfate *see* 8722
Sodium Rhodanide *see* 9480
Sodium Rhodizonate, 8799
Sodium Ricinoleate *see* 8335
Sodium Rose Bengal I 131 *see* 8393
Sodium Salicylate *see* 8469
Sodium Salicylsulfate *see* 8471
Sodium Salt of Metanilylazodiphenylamine *see* 5999
Sodium Salt of 1-Nitroso-2-hydroxynaphthalene-3,6-disulfonic Acid *see* 6731
Sodium Selenate, 8800
Sodium Selenide, 8801
Sodium Selenide (Na$_2$Se) *see* 8801
Sodium Selenite, 8802

Sodium Sesquicarbonate, 8803
Sodium Silicate, 8804
Sodium Silicofluoride *see* 8759
Sodium Sorbate *see* 8847
Sodium Stannate(IV), 8805
Sodium Stearate, 8806
Sodium Stearyl Fumarate, 8807
Sodium Stibogluconate *see* 692
Sodium Succinate, 8808
Sodium Sulamyd [Schering] *see* 9031
Sodium Sulfate, 8809
Sodium Sulfate Sodium Salt (1:2) *see* 8809
Sodium Sulfhydrate *see* 8723
Sodium Sulfide, 8810
Sodium Sulfide (Na$_2$S) *see* 8810
Sodium Sulfide (Na(SH)) *see* 8723
Sodium Sulfite, 8811
Sodium Sulfocarbonate *see* 8825
Sodium Sulfocyanate *see* 9480
Sodium 2-Sulfonatoethyl Laurate *see* 8739
(5-[(3-Sodium-sulfopropyl-2(3*H*)-benzoxazolylidine)-2-butenylidene]-1,3-dibutyl-2-thiobarbituric Acid) *see* 5969
Sodium Sulfovinate *see* 8750
Sodium Sulfoxylate *see* 8747
Sodium Sulfuret *see* 8810
Sodium Superoxide *see* 8786
Sodium Tartrate, 8812
Sodium Taurocholate *see* 9212
Sodium Tellurate, 8813
Sodium Tellurate(IV) *see* 8814
Sodium Tellurate(VI) *see* 8813
Sodium Tellurite, 8814
Sodium L-Tenuazonate *see* 9292
Sodium Tetraborate *see* 8726
Sodium Tetrachloroaluminate, 8815
Sodium Tetrachloroaurate(III), 8816
Sodium Tetradecyl Sulfate, 8817
Sodium Tetrafluoroberyllate *see* 1176
Sodium Tetrafluoroborate *see* 8753
Sodium Tetraphenylborate, 8818
Sodium Tetraphenylborate(1−) (1:1) *see* 8818
Sodium 2-Thenoate *see* 9507
Sodium Theobromine Acetate *see* 9434
Sodium Thiacetarsamide *see* 9441
Sodium Thiocarbonate *see* 8825
Sodium Thiocyanate *see* 9480
Sodium Thioglycolate, 8819
Sodium 2-Thiophenecarboxylate *see* 9507
Sodium Thiophosphate, 8820
Sodium Thiosulfate, 8821
Sodium Tin Oxide *see* 8805
Sodium *p*-Toluenesulfonchloramide *see* 2075
Sodium Triacetoxyborohydride, 8822
Sodium 2,4,6-Tribromophenolate *see* 9775
Sodium Trichloroacetate *see* 9792
Sodium L-Triiodothyronine *see* 5565
Sodium Trimetaphosphate, 8823
Sodium Trioxodinitrate *see* 641
Sodium Trioxovanadate *see* 8774
Sodium Triphenyl-*p*-rosanilinetrisulfate *see* 6102
Sodium Triphosphate *see* 8824
Sodium Tripolyphosphate, 8824
Sodium Trithiocarbonate *see* 8825
Sodium Tungstate(VI), 8826
Sodium (*T*-4)-Tungstate (WO$_4{}^{2-}$) (2:1) *see* 8826
Sodium Tungstophosphate *see* 8795

Sodium Tyropanoate *see* 10018
Sodium Uranate(VI), 8827
Sodium Uranium Oxide (Na$_2$U$_2$O$_7$) *see* 8827
Sodium Valproate *see* 10099
Sodium Vanadate(V) *see* 8774
Sodium Vanadium Trioxide *see* 8774
Sodium Xylan Polysulfate *see* 7247
Sodothiol *see* 8821
Sofalcone, 8828
Sofarcid [Sofar] *see* 1930
Soframycin [HMR] *see* 6539
Softenon *see* 9403
Soja Bean *see* 8856
Solacen [Wallace Labs.] *see* 10010
Solan, 8829
Solancarpidine *see* 8836
Solanesol, 8830
(3β,12α)-Solanid-5-ene-3,12-diol *see* 8420
(3β)-Solanid-5-en-3-ol *see* 8831
(3β)-Solanid-5-en-3-yl *O*-6-Deoxy-α-L-mannopyranosyl-(1 → 2)-*O*-[6-deoxy-α-L-mannopyranosyl-(1 → 4)]-β-D-glucopyranoside *see* 8832
(3β)-Solanid-5-en-3-yl *O*-6-Deoxy-α-L-mannopyranosyl-(1 → 2)-*O*-[β-D-glucopyranosyl-(1 → 3)]-β-D-galactopyranoside *see* 8832
Solanidine, 8831
Solanidine-S *see* 8836
Solanine, 8832
α-Solanine *see* 8832
Solanine-S *see* 8837
Solanocapsin *see* 8833
Solanocapsine, 8833
Solanone, 8834
Solantal [Fujisawa] *see* 9579
Solanum, 8835
Solar [BASF] *see* 2294
Solar NP [Swift] *see* 6764
Solart [Bioindustria] *see* 28
Solaskil [RPR] *see* 5511
Solasod-5-en-3β-ol *see* 8836
Solasodine, 8836
Solasonine, 8837
Solaspin [Indian Health] *see* 1650
Solatene [Roche] *see* 1855
Solatubine *see* 8831
Solatunine *see* 8832
Solaxin *see* 2201
Solbrol A *see* 3890
Solbrol P [Bayer] *see* 7977
Solclot [Sumitomo] *see* 9624
Soldactone [Searle] *see* 1753
Soldesam [Farmacologico] *see* 2945
Solevar *see* 6785
Solfa [Takeda] *see* 486
Solganal [Schering] *see* 874
Solgol [BMS] *see* 6431
Solian [Synthelabo] *see* 481
Solicam [Syngenta] *see* 6790
Solid Crotonic Acid 2588
Solid Green *see* 1380
Solid Green O *see* 5763
Solifenacin, 8838
Solimidin [Selvi] *see* 10389
Soliris [Alexion] *see* 3553
Soliwax [Concept Pharm.] *see* 3446
Solketal 5259
Sol-Mycin [Roxane] *see* 3197
Solnhofen Stone 5544
Solocalm [Bernabó] *see* 7619
Solodelf [Cyanamid] *see* 9758

Solon [Taisho] *see* 8828
Solone [Abbott] *see* 7841
Solosa [Guidotti] *see* 4475
Solosin [Sanofi-Aventis] *see* 9436
Soloxine [Virbac] *see* 9570
Solozone *see* 8786
Sol Phenobarbital *see* 7352
Sol Phenobarbitone *see* 7352
Solprene [Phillips Petr.] *see* 8527
Solprin [Reckitt Benckiser] *see* 841
Solubacter [Pharmethic] *see* 9819
Solu-Biloptin [Schering AG] *see* 5115
Soluble Anhydrite *see* 1708
Soluble Aspirin *see* 1650
Soluble Barbital *see* 957
Soluble Fluorescein *see* 4192
Soluble Glass *see* 8804
Soluble Gun Cotton *see* 8125
Soluble Indigo Blue *see* 4982
Soluble Iodophthalein *see* 5081
Soluble Pentobarbital *see* 7243
Soluble Potash Glass *see* 7788
Soluble Potash Water Glass *see* 7788
Soluble Prussian Blue *see* 8018
Soluble RNA *see* 8327
Soluble Saccharin *see* 8445
Soluble Sulfacetamide *see* 9031
Soluble Sulfadiazine *see* 9035
Soluble Sulfamerazine *see* 9045
Soluble Sulfapyridine *see* 9069
Soluble Sulfathiazole *see* 9074
Soluble Tartar *see* 7796
Solucin *see* 6772
Solucort [Merck & Co.] *see* 7841
Solu-Cortef [Pharmacia & Upjohn] *see* 4824
SoluDecadron [Merck & Co.] *see* 2945
Solu-Decortin-H [Merck KGaA] *see* 7841
Soludillar [Syntex] *see* 7133
Solufen [SMB] *see* 4919
Solufilin *see* 3526
Solufilina [Merck KGaA] *see* 3766
Solufontamide *see* 9075
Solufyllin *see* 3526
Solumédine [Specia] *see* 9045
Solu-Medrol [Pharmacia & Upjohn] *see* 6184
Solu-Medrone [Pharmacia] *see* 6184
Solupemid [Recordati] *see* 7572
Solupen N [Winzer] *see* 2945
Solupred [Aventis] *see* 7841
Solupsan [UPSA] *see* 1650
Solurol [Delalande] *see* 9139
Soluseptasine *see* 6772
Soluseptazine *see* 6772
Solusetazine *see* 6772
Solustibosan [Bayer] *see* 692
Solutedarol [Specia] *see* 9758
Solu-Tracin [Alpharma] *see* 928
Solutrast [Bracco] *see* 5100
Solvay Soda *see* 8731
Solvazinc [Thames] *see* 10359
Solvazine [Solvay] *see* 10277
Solvent Naphtha *see* 5542
Solvezink [Tika] *see* 10359
Solvolan [Krka] *see* 380
Solvolip [Knoll] *see* 4022
SOM-230 *see* 7155
Soma [Wallace Labs.] *see* 1842
Somac [Pfizer] *see* 7117
Somacton [Ferring] *see* 8842
Somadril [Dumex] *see* 1842
Somagerol [Brenner-Efeka] *see* 5636
Somagrebove *see* 8842
Somalapor *see* 8842

Somalgen [Bago] *see* 9176
Somalgit [Inibsa] *see* 1842
Soman, 8839
Somatocrinin *see* 8840
Somatofalk [Falk] *see* 8841
SomatoKine [Insmed] *see* 5845
Somatoliberin, 8840
Somatoliberin (Human Pancreatic Islet) *see* 8840
Somatomedin *see* 5029
Somatomedin 1 *see* 5029
Somatomedin C *see* 5029
Somatonorm [Pfizer] *see* 8842
Somatostatin, 8841
Somatostatin (Sheep) *see* 8841
Somatotropic Hormone *see* 8842
Somatotropin, 8842
Somatotropin Release Inhibiting Factor *see* 8841
Somatrem *see* 8842
Somatropin *see* 8842
Somatuline LP [Ipsen] *see* 5412
Somatyl [Teofarma] *see* 1181
Somavert [Pharmacia & Upjohn] *see* 7178
Somavubove *see* 8842
Somazina [Ferrer] *see* 2318
Sombrevin [Gedeon Richter] *see* 7917
Sombucaps *see* 4740
Sombulex [3M Pharma] *see* 4740
Somelin [Sankyo] *see* 4638
Somenopor *see* 8842
Sometribove *see* 8842
Sometripor *see* 8842
Somfasepor *see* 8842
Somidobove *see* 8842
Sominex [GSK] *see* 3339
Somio *see* 2072
Somlan [Sintyal] *see* 4228
Somnafac *see* 6032
Somnalert [Warren-Teed] *see* 4740
Somnased [GSK] *see* 6661
Somnatrol [Takeda] *see* 3757
Somnesin [Central Pharm.] *see* 5913
Somnibel [UCB] *see* 6661
Somnite [Norgine] *see* 6661
Somnovit [Hosbon] *see* 5631
Somnurol [Synochem] *see* 1404
Somon [Atlas] *see* 2112
Somsanit [Köhler] *see* 4854
Sonacide [Wyeth-Ayerst] *see* 4508
Sonaform *see* 2704
Sonalan [Dow AgroSci.] *see* 3774
Sonalen [Dow AgroSci.] *see* 3774
Sonata [Wyeth] *see* 10309
Sone [Fawns & McAllan] *see* 7842
Soneryl [M & B] *see* 1524
Songar [Valeas] *see* 9765
Songorine, 8843
Sonicated Human Serum Albumin *see* 8607
Sonifilan [Kaken] *see* 8697
Sonilyn [Carter-Wallace] *see* 9032
Sonin [Lipha] *see* 5631
Sonora *see* 5984
Soothe-N-Seal [Colgate-Palmolive] *see* 6855
Sopental [Continental] *see* 7243
Sophidone [BMS] *see* 4840
Sophocarpidine *see* 5827
Sophorabioside, 8844
Sophoretin *see* 8150
Sophoricol *see* 4426
Sophoricoside *see* 4426
Sophorin *see* 8438
Sophorine *see* 2786

Sophorose, 8845
Sopivan [Aventis] *see* 10393
Soprol [Lederle] *see* 1295
Sorafenib, 8846
Sorbangil [Pharmacia] *see* 5271
Sorbic Acid, 8847
Sorbic Alcohol, 8848
Sorbichew [AstraZeneca] *see* 5271
Sorbic Oil *see* 7136
Sorbid SA [AstraZeneca] *see* 5271
Sorbidilat [AstraZeneca] *see* 5271
Sorbilande [Sanofi-Aventis] *see* 8851
Sorbilax [Vétoquinol] *see* 8851
Sorbin *see* 8852
Sorbinil, 8849
Sorbinol *see* 8848
Sorbinose *see* 8852
Sorbistat [Ono] *see* 3660
Sorbit *see* 8851
Sorbitan Esters, 8850
Sorbitan Fatty Acid Esters *see* 8850
Sorbitan Laurate *see* 8850
Sorbitan Monolaurate *see* 8850
Sorbitan Monooleate *see* 8850
Sorbitan Monostearate *see* 8850
Sorbitan Oleate *see* 8850
Sorbitan Stearate *see* 8850
Sorbitol, 8851
D-Sorbitol *see* 8851
Sorbitrate [AstraZeneca] *see* 5271
Sorbo [SPI] *see* 8851
Sorbose, 8852
L-Sorbose *see* 8852
Sorbostyl [Sanofi-Aventis] *see* 8851
Sorbsan [Maersk] *see* 235
Sordenac [Lundbeck] *see* 2383
Sordinol [Ayerst] *see* 2383
Sorensen's Phosphate *see* 8789
Sörensen's Potassium Phosphate *see* 7780
Sorensen's Sodium Phosphate *see* 8789
Soreton [Nippon Chemiphar] *see* 10310
Soriatane [Roche] *see* 103
Soricin *see* 8335
Soridermal [Rhône-Poulenc] *see* 6215
Soriflor [Nycomed] *see* 3159
Soripal [Specia] *see* 6215
Sorivudine, 8853
Sorlate [Abbott] *see* 7703
Sormetal *see* 3565
Sorot [Ravensberg] *see* 2914
Sortis [Pfizer] *see* 855
Sosegon [Sanofi-Aventis] *see* 7233
Sospitan [Kali-Chemie] *see* 8088
Sostril [Cascan] *see* 8228
Sotacor [BMS] *see* 8854
Sotal [Gramon] *see* 9613
Sotalex [BMS] *see* 8854
Sotalol, 8854
Soterenol, 8855
Sotradecol [Bioniche] *see* 8817
Soudan Coffee *see* 5366
Souframine *see* 7364
Sour-spine *see* 956
Soventol [Knoll] *see* 949
Sovran [BASF] *see* 5372
Sovrima [Santhera] *see* 4929
Sowberry *see* 956
Soxysympamine [Ferndale] *see* 6020
Soya Bean *see* 8856
Soybean, 8856
Soybean Agglutinin *see* 5484
Soybean Oil, 8857
Soy Sauce, 8858

SP *see* 8891, 8993
SP-54 [Bene-Chemie] *see* 7247
SP-63 *see* 6997
SP-281 *see* 10168
SP-732 *see* 7896
SP-1103 *see* 9370
Spabucol [Lagap] *see* 9868
Spaderizine [Kotobuki] *see* 2306
SPADNS, 8859
Spametrin-M [Yamanouchi Seiyaku] *see* 6142
Spamorin [Tanabe Seiyaku] *see* 4138
Span 20 [ICI] *see* 8850
Span 60 [ICI] *see* 8850
Span 80 [ICI] *see* 8850
Spandex, 8860
Spanidin [Nippon Kayaku] *see* 4617
Spanish Fly *see* 1754
"Spanish Fly" *see* 1755
Spanish Grass Wax *see* 3756
Spanish Saffron *see* 8449
Spanish Tea *see* 6877
Spanish White *see* 1284
Span-K [Aspen] *see* 7742
Spanon *see* 2086
Spanor [Bailleul] *see* 3488
Spantol [Nippon Chemiphar] *see* 7370
Spara [Dainippon] *see* 8862
Sparassol, 8861
Sparfloxacin, 8862
Sparine [Wyeth] *see* 7898
Sparsamycin A *see* 9984
Sparsomycin, 8863
(+)-Sparsomycin *see* 8863
Spartakon [Janssen] *see* 5511
Sparteine, 8864
l-Sparteine *see* 8864
Spartrix [Janssen] *see* 1848
SPA-S-132 *see* 7153
SPA-S-160 *see* 5914
SPA-S-222 *see* 5914
SPA-S-510 *see* 7619
Spasen [Menarini] *see* 6997
Spasfon-Lyoc [Cephalon] *see* 7439
Spasmaverine [Bellon] *see* 365
Spasmentral [Janssen] *see* 1077
Spasmex [Pfleger] *see* 9966
Spasmium *see* 1859
Spasmoctyl [Menarini] *see* 6997
Spasmocyclon [3M] *see* 2697
Spasmol *see* 5782
Spasmolyn *see* 5917
Spasmolysin [Kade] *see* 8014
Spasmolyt [Madaus] *see* 9966
Spasmolytin *see* 149
Spasmomen [Menarini] *see* 6997
Spasmonal [Norgine] *see* 365
Spasmo-Nit [Stroschein] *see* 7122
Spasmophen *see* 7074
Spasmopriv [Paillusseau] *see* 4013
Spasmoril *see* 4138
Spasmostenyl *see* 5782
Spassirex [Sanofi-Aventis] *see* 7439
Spasuret [Asche] *see* 4128
Spatomac *see* 1031
Spatonin *see* 3128
SP-B *see* 8680
SPC-97D *see* 902
SPCA *see* 3961
SPD-754 *see* 742
SPE *see* 9014
SPE-2792 *see* 661
Spearmint, 8865
Specifin [Bergamon] *see* 6444
Spectacillin [Sandoz] *see* 3669

Spectam [Abbott] *see* 8866
Spectamine [Medi-Physics] *see* 5092
Spectazole [Ortho] *see* 3549
Spectinomycin, 8866
Spectracef [Cornerstone] *see* 1922
Spectramast [Pfizer] *see* 1951
Spectra-Sorb UV 24 [Am. Cyanamid] *see* 3332
Spectra-Sorb UV 284 [Am. Cyanamid] *see* 9113
Spectra-Sorb UV 531 [Am. Cyanamid] *see* 6831
Spectrin, 8867
Spectro [Novartis Agro] *see* 3154
Spectrum [Sigma-Tau] *see* 1948
Speda [Byk Gulden] *see* 10190
"Speed" *see* 6020
Spergon *see* 2078
Spergualin *see* 4617
Spermaceti, 8868
Spermidine, 8869
Spermine, 8870
Spermine Oxidase *see* 6333
Spermine Phosphate *see* 8870
Sperm Oil, 8871
Spermwax [Robeco] *see* 8868
SPG *see* 8697
Sphaerocobaltite *see* 2420
Sphalerite *see* 10360
Sphene *see* 9626
Spheroidine *see* 9394
trans-4-Sphingenine *see* 8874
Sphingofungin C *see* 8872
Sphingofungins, 8872
Sphingomyelins, 8873
Sphingosine, 8874
(−)-D-*erythro*-Sphingosine *see* 8874
SPhos Phosphine Ligand *see* 1469
SP-I *see* 7662
SPI-0211 *see* 5649
Spice Berry *see* 755
Spigelia, 8875
Spignet *see* 755
Spike [Dow AgroSci.] *see* 9232
Spike Lavender *see* 5445
Spike Lavender Oil *see* 5445
Spikenard *see* 755
Spilanthol *see* 170
Spinacene *see* 8895
Spinacene *see* 8896
α-Spinasterin *see* 8876
α-Spinasterol, 8876
Spindle Tree *see* 3941
Spinnaker [Shell] *see* 9755
Spinosad *see* 8877
Spinosyn A *see* 8877
Spinosyn D *see* 8877
Spinosyns, 8877
SpinTor [Dow AgroSci.] *see* 8877
Spiperone, 8878
Spira 200 [RMB] *see* 8879
Spiramycin, 8879
Spiramycin Adipate *see* 8879
Spiramycin I *see* 8879
Spiramycin II *see* 8879
Spiramycin III *see* 8879
Spirapril, 8880
Spiraprilat *see* 8880
Spiraprilic Acid *see* 8880
Spirein *see* 4663
Spirilloxanthin *see* 8319
Spirit of Ammonia, Aromatic, 8881
Spirit of Ants *see* 4269
Spirit of Ether, 8882
Spirit of Ether Compound, 8883

Spirit of Glonoin *see* 6694
Spirit of Hartshorn *see* 491
Spirit of Hartshorn, Aromatic *see* 8881
Spirit of Nitroglycerin *see* 6694
Spirit of Nitrous Ether *see* 3885
Spirit of Turpentine *see* 10007
Spiriva *see* 9610
Spiro-32 [Unimed] *see* 8885
Spirobeta [Betapharm] *see* 8887
Spirocid [Hoechst] *see* 51
Spirocort [AstraZeneca] *see* 1476
Spiroctan [Ferlux] *see* 8887
Spirodiclofen, 8884
Spirodon *see* 9382
Spirofulvin *see* 4584
Spirogermanium, 8885
Spirolair [3M Pharma] *see* 7602
Spirolone [APS] *see* 8887
Spiromesifen, 8886
Spironolactone, 8887
Spironone [Dexo] *see* 8887
Spiropent [Boehringer, Ing.] *see* 2345
Spiropitan [Janssen] *see* 8878
(3β,5α,22β,25S)-Spirosolan-3-ol *see* 9703
(3β,5α,22β,25S)-Spirosolan-3-yl *O*-β-D-Glucopyranosyl-(1 → 2)-*O*-[β-D-xylo-pyranosyl-(1 → 3)]-*O*-β-D-glucopyran-osyl-(1 → 4)-β-D-galactopyranoside *see* 9704
(3β,22α,25R)-Spirosol-5-en-3-ol *see* 8836
(3β,22α,25R)-Spirosol-5-en-3-yl *O*-6-De-oxy-α-L-mannopyranosyl-(1 → 2)-*O*-[β-D-glucopyranosyl-(1 → 3)]-β-D-gal-actopyranoside *see* 8837
(2α,3β,5α,25R)-Spirostan-2,3-diol *see* 4466
(3β,5α,6α,25R)-Spirostan-3,6-diol *see* 2144
(3β,5α,15β,25R)-Spirostan-3,15-diol *see* 3173
(3β,5α,25R)-Spirostan-3-ol *see* 9591
(3β,5β,25R)-Spirostan-3-ol *see* 8703
(3β,5β,25S)-Spirostan-3-ol *see* 8517
(2α,3β,5α,15β,25R)-Spirostan-2,3,15-triol *see* 3179
(3β,25R)-Spirost-5-en-3-ol *see* 3323
(3β,25R)-Spirost-5-en-3-yl *O*-6-Deoxy-α-L-mannopyranosyl-(1 → 2)-*O*-[6-de-oxy-α-L-mannopyranosyl-(1 → 4)]-β-D-glucopyranoside *see* 3320
Spirotetramat, 8888
Spiroxamine, 8889
Spizef [Grünenthal] *see* 1936
Splenda [McNeil Nutritionals] *see* 9011
Splendil [AstraZeneca] *see* 3987
Splendipherin, 8890
Splenin, 8891
Splenopentin *see* 8891
Split Nut *see* 7495
Splotin [Taiyo] *see* 9119
SPM-907 *see* 4095
SPM-925 *see* 6315
SPM-927 *see* 5381
SPM-962 *see* 8404
SPM-8272 *see* 4095
Spodumene *see* 5576
Spofa 325 *see* 6373
Spondyvit [Wyeth] *see* 9653
Spongoadenosine *see* 10173
Sponsin [Farmasan] *see* 3710
Spoonwood *see* 5328
Sporanox [Janssen-Cilag] *see* 5292
Sporeine *see* 925
Sporidesmin *see* 8892
Sporidesmin A *see* 8892

Sporidesmins, 8892
Sporidesmolide I *see* 8893
Sporidesmolide II *see* 8893
Sporidesmolide III *see* 8893
Sporidesmolide IV *see* 8893
Sporidesmolides, 8893
Sporiline [Schering-Plough] *see* 9678
Sporostatin [Schering] *see* 4584
Sportak [Bayer CropSci.] *see* 7878
Sportak Delta [Aventis CropSci.] *see* 2768
Spotless [Sumitomo] *see* 3294
Spotlight [Dow AgroSci.] *see* 4233
Spotof [CCD] *see* 9730
Spotrete F [Cleary] *see* 9525
Spotted Alder *see* 4639
Spotted Cowbane *see* 2490
Spotted Hemlock *see* 2490
SPP-100 *see* 239
Spreading Dogbane *see* 729
Spreading Factor *see* 4797
Sprix [Roxro] *see* 5353
S-Protein *see* 10222
Sprout-Nip [PPG] *see* 2193
Sprycel [BMS] *see* 2831
SPS *see* 8796
SPS [Carolina] *see* 8797
Spurge Flax *see* 6249
Spurred Rye *see* 3717
S.Q. [Merck & Co.] *see* 9070
SQ-1489 *see* 9525
SQ-4918 *see* 4220
SQ-9453 *see* 3285
SQ-9538 *see* 9323
SQ-10496 *see* 9457
SQ-10733 *see* 4220
SQ-11302 *see* 3669
SQ-11436 *see* 1985
SQ-11725 *see* 6431
SQ-13050 *see* 3549
SQ-13396 *see* 5100
SQ-14055 *see* 9574
SQ-14225 *see* 1776
SQ-15101 *see* 233
SQ-15860 *see* 4542
SQ-16144 *see* 4220
SQ-16150 *see* 3758
SQ-18566 *see* 4626
SQ-19844 *see* 8681
SQ-21983 *see* 5106
SQ-22022 *see* 1985
SQ-22947 *see* 9574
SQ-26776 *see* 918
SQ-26991 *see* 10386
SQ-27519 *see* 4283
SQ-28555 *see* 4283
SQ-29852 *see* 1993
SQ-30217 *see* 9240
SQ-31000 *see* 7835
SQ-32692 *see* 4358
SQ-32756 *see* 8853
SQ-200475 *see* 3649
Squalamine, 8894
Squalane, 8895
Squalene, 8896
Squalestatin 1 *see* 10312
Squalestatins *see* 10312
Squalestatin S1 *see* 10312
Squalidine *see* 8590
Square-stem Rose-gentian *see* 1971
Squaric Acid, 8897
Squaw Root 1321
Squaw Weed *see* 8589
Squill, 8898
Squirrel Corn *see* 2532

SR-406 *see* 1774
SR-720-22 *see* 6224
SR-2508 *see* 3768
SR-4233 *see* 9617
SR-11247 *see* 1198
SR-25990 *see* 2385
SR-25990C *see* 2385
SR-29142 *see* 8235
SR-33589 *see* 3498
SR-33589B *see* 3498
SR-34006 *see* 4932
SR-41319B *see* 9598
SR-47436 *see* 5127
SR-57746A *see* 10252
SR-90107A *see* 4259
SR-141716 *see* 8355
SR-141716A *see* 8355
SRA-3886 *see* 3991
SRA-5172 *see* 6019
SRA-12869 *see* 5218
Srendam [Kayaku] *see* 9134
SRG-95213 *see* 3007
SRIF *see* 8841
SRIF-14 *see* 8841
SRIF-25 *see* 8841
SRIF-28 *see* 8841
SRIF-A *see* 8841
Srilane [Medicia] *see* 4933
sRNA *see* 8327
SRS *see* 5507
SRS-A *see* 5507
SRTx *see* 8508
SS-578 *see* 5085
SS-717 *see* 6563
SS-734 *see* 1178
SSB *see* 9016
SSB-I *see* 9016
SSB-II *see* 9016
SSF-129 *see* 3292
ST-21 *see* 7627
ST-37 *see* 4748
ST-52 [Baxter] *see* 4280
ST-155 *see* 2380
ST-200 *see* 79
ST-246 *see* 9244
ST-374 *see* 3999
ST-567 *see* 238
ST-679 *see* 591
ST-1085 *see* 6263
ST-1396 *see* 1962
ST-1435 *see* 3586
ST-1512 *see* 4743
ST-2121 *see* 3696
ST-4331 *see* 4891
ST-7090 *see* 4741
STA-4783 *see* 3596
Stabilene [Auclair] *see* 3825
Stable Bixin *see* 1312
Stablon [Ardix] *see* 9575
Stachydrine, 8899
Stadacain [Stada] *see* 1533
Stadadorm [Stada] *see* 567
Stadalax [Stada] *see* 1246
Stadol [BMS] *see* 1532
Stafac [SKB] *see* 10200
Stafoxil [Yamanouchi] *see* 4143
Stag Bush *see* 10170
Staggerweed *see* 5424
Stagid [Merck KGaA] *see* 6010
Stagural [Stada-Chemie] *see* 6788
Stalleril [Pharmacal] *see* 9512
Stallimycin *see* 3404
Stam [Dow AgroSci.] *see* 7918
Stamine [Tutag] *see* 8097
Stampede [Dow AgroSci.] *see* 7918

STAN *see* 8923
Stanazol *see* 8921
Stangen [Phys. Drug] *see* 8097
Stangyl (amp) [Rhone-Poulenc] *see* 9894
Stangyl (tabl) [Rhone-Poulenc] *see* 9894
Stanilo [Pharmacia] *see* 8866
Stannate (SnO_3^{2-}) Potassium (1:2) *see* 7791
Stannate (SnO_3^{2-}) Sodium (1:2) *see* 8805
Stannic Anhydride *see* 8905
Stannic Bromide, 8900
Stannic Chloride, 8901
Stannic Chromate(VI), 8902
Stannic Fluoride, 8903
Stannic Iodide, 8904
Stannic Oxide, 8905
Stannic Selenide, 8906
Stannic Sulfide, 8907
Stannochlor *see* 8910
Stannous Acetate, 8908
Stannous Bromide, 8909
Stannous Chloride, 8910
Stannous Fluoride, 8911
Stannous Iodide, 8912
Stannous Oxalate, 8913
Stannous Oxide, 8914
Stannous Pyrophosphate, 8915
Stannous Selenide, 8916
Stannous Sulfate, 8917
Stannous Sulfide, 8918
Stannous Tartrate, 8919
Stanolene [Amoco] *see* 7300
Stanolone, 8920
Stanozolol, 8921
Stanquinate *see* 5085
Stanyl [DSM] *see* 6824
Stapenor [Bayer] *see* 7003
Staphcillin [BMS] *see* 6041
Staphisagria, 8922
Staphlipen [Wyeth] *see* 4143
Staphylex [GSK] *see* 4143
Staphylokinase, 8923
Staphylokinase (Enzyme-activating) *see* 8923
Staphylomycin *see* 10200
Staphylomycine [SKB] *see* 10200
Staphylomycin M_1 *see* 10200
Staphylomycin S *see* 10200
Staple [DuPont] *see* 8106
STAR *see* 8923
Starane [Dow AgroSci.] *see* 4233
Star Anise, 8924
Starcef [Firma] *see* 1948
Starch, 8925
Starch Gum *see* 2955
Starch Polymer with 2-Propenenitrile *see* 2576
Starch, Soluble, 8926
Star Grass *see* 224
Staril [BMS] *see* 4283
Starkle [Mitsui] *see* 3317
Starlix [Novartis] *see* 6510
Starsis [Yamanouchi] *see* 6510
Starwort *see* 224, 4666
Starycide [Bayer] *see* 9844
Stas-Hustenlöser [Stada] *see* 380
Staticin [Westwood] *see* 3739
Statine, 8927
Statobex [Lemmon] *see* 7334
Statyl [Ayerst] *see* 6556
Stauffer 2061 *see* 7167
Stauroderm *see* 4228
Staurosporine, 8928
Stavesacre *see* 8922
Stavudine, 8929

Stayban [Boots] see 4229
Staycept [Syntex] see 6764
Stazepine [Polpharma] see 1783
St. Bartholomew's Tea see 5824
Steapyrium Chloride see 5421
Steareth see 7697
Stearic Acid, 8930
Stearic Acid Aluminum Salt see 358
Stearic Acid Ammonium Salt see 550
Stearic Acid Calcium Salt see 1705
Stearic Acid Ester with Lactate of Lactic
  Acid Calcium Salt see 1706
Stearic Acid Isobutyl Ester see 5196
Stearic Acid Lead Salt see 5473
Stearic Acid Potassium Salt see 7792
Stearic Acid Sodium Salt see 8806
Stearidonic Acid, 8931
Stearin see 9932
N-Stearoylsphingomyelin see 8873
N-Stearoyl-D-erythro-sphingosylphos-
  phorylcholine see 8873
Stearyl Alcohol, 8932
Steatite see 9168
Steclin [BMS] see 9341
Stedesa [Bial-Portela] see 3754
Stedicor [Procter & Gamble] see 904
Stédiril [Wyeth] see 6793
Steel Factor see 8934
Stefin A see 2775
Stefin B see 2775
Stefins see 2775
Steinbühl Yellow see 968
Stelabid [SK & F] see 5248
Steladone [Zagro] see 2090
Stelara [J & J] see 10079
Stelazine [GSK] see 9846
Stellamicina see 3740
Stellar [Staley], 8933
Stellarid [GSK] see 7985
Stellerite see 1651
Stem Bromelain see 1395
Stem Cell Factor, 8934
Stemetil [Aventis] see 7879
Stemex [Syntex] see 7133
Stemgen [Amgen] see 8934
Stem Peptide see 7145
Stemphylone see 8209
Stenbolone, 8935
Stenediol [Organon] see 6022
Stenobolone (rescinded USAN) see 8935
Stenol see 8932
Stenolon see 6023
Stenorol [Intervet] see 4632
Stephanine see 6000
Stepin [Basotherm] see 9612
Stepronin, 8936
Steranabol [Farmitalia] see 2397
Sterane [Pfizer] see 7841
Sterathal [Ward, Blenkinsop] see 7488
Sterax [Alcon] see 2929
Stercobilin see 10068
(−)-Stercobilin see 10068
Stercobilinogen see 10068
Stercorin see 2509
Sterculia see 5332
Stereomycine [Toraude; Ayerst] see 5505
Stereon [Firestone] see 8527
Sterinor [Heumann] see 9398
Sterisil [Warner-Lambert] see 4739
Steri/Sol [Warner-Lambert] see 4739
Steroderm [De Angeli] see 2929
Sterogyl [DB Pharma] see 10215
Sterosan [Novartis] see 2196
Sterox [Monsanto] see 6764
Steroxin [Novartis] see 2196

Sterretite see 352
Sterwin 904 see 226
Ster-Zac [Hough, Hoseason] see 9822
Stesolid [Alpharma] see 2997
Stevia, 8937
Stevioside, 8939
Stevioside A3 see 8243
Stevioside A4 see 8243
Steviosin see 8939
Steward [DuPont] see 5012
STH see 8842
STI-571 see 4943
Stibic Anhydride see 689
Stibine, 8940
Stibium see 683
Stibnal [Banyu] see 693
Stibnite see 701
Stibocaptate, 8941
Stibophen, 8942
Stickstofflost see 5846
Stickstoffwasserstoffsäure (German) see
  4812
Stiedex LP [Stiefel] see 2931
Stiemycin [Stiefel] see 3739
(3β)-Stigmasta-5,7-dien-3-ol see 2880
(3β,5α,22E)-Stigmasta-7,22-dien-3-ol see
  8876
(3β,5α,22E,24R)-Stigmasta-7,22-dien-3-
  ol see 2216
(3β,22E)-Stigmasta-5,22-dien-3-ol see
  8944
(3β,24E)-Stigmasta-5,24(28)-dien-3-ol
  see 4309
Stigmastanol, 8943
(3β,5α)-Stigmastan-3-ol see 8943
Δ5-Stigmasten-3β-ol see 8694
(3β)-Stigmast-5-en-3-ol see 8694
(3β,24S)-Stigmast-5-en-3-ol see 8695
Stigmasterol, 8944
St. Ignatius' Bean see 4938
Stilamin [Serono] see 8841
Stilalgin see 5917
Stilbamidine, 8945
Stilbene, 8946
[Stilbene-(4,4')]bis[ω-phenylurea]-2,2'-
  disulfonic Acid Disodium Salt see
  1316
Stilbene-α-carboxylic Acid see 7393
4,4'-Stilbenedicarboxamidine see 8945
E-3,5-Stilbenediol see 7563
3,4',5-Stilbenetriol see 8279
Stilbestrol see 3148
Stilbestrol Dimethyl Ether see 3234
Stilbestrol Diphosphate see 4280
Stilbetin [BMS] see 3148
Stilboestrol see 3148
Stilbostatin [Taro] see 4280
Stillacor [Wolff] see 3186
Stillingia see 8947
Stilnoct [Sanofi-Aventis] see 10391
Stilnox [Sanofi-Aventis] see 10391
Stilny [Will-Pharma] see 6780
Stilphostrol [Bayer] see 4280
Stilwet L [Union Carbide] see 8634
Stimate [ZLB Behring] see 2926
Stimol [Biocodex] see 2330
Stimolag [Lagap] see 7597
Stimsen [Lederle] see 9533
Stimulin see 6627
Stimulina see 4507
Stimu-TSH [HMR] see 9750
Stinger [Dow AgroSci.] see 2389
Stinging Nettle see 6567
Stingose see 360
Stinkweed see 8950

Stipolac [Burroughs Wellcome] see 5081
Stiptanon [Organon] see 3762
Stiripentol, 8948
Stirofos see 9337
Stivane [Beaufour] see 8104
St. John's Wort see 4903
StL-1106 see 4222
STMP see 8823
Stocrin [Merck & Co.] see 3569
Stoddard Solvent see 6282
Stogar [Fujirebio] see 5396
Stomamycin [Chassot] see 8879
Stomédine [GSK] see 2282
Stomp [BASF] see 7193
Stone Oil see 4924
Stone-root see 2467
Storax, 8949
Storax (Balsam) see 8949
Storinal see 1325
Storksbill see 4438
Storm [BASF] see 4134
Storocain [Eisai] see 7032
Stovaine see 607
Stovarsol [Abbott] see 51
Stoxil [SK & F] see 4931
Stozzon-Chlorophyll see 2158
STP see 3461
(+)-STP see 8948
(−)-STP see 8948
STPP see 8824
Strada [Isagro] see 6981
Straderm [ITA] see 4182
Stramonium, 8950
Stratene [Innothera; Sigma-Tau] see 2019
Strattera [Lilly] see 854
Strawberry Tree see 3941
Strengite see 4056
Strepogenin, 8951
Streptase [ZLB Behring] see 8955
Streptidine, 8952
Streptobrettin [Norbrook] see 8956
Streptocide [Frosst] see 9057
Streptococcal Deoxyribonuclease see
  8953
Streptococcal Fibrinolysin see 8955
Streptodornase, 8953
Streptogramin A see 10200
Streptogramin B see 7871
Streptogramins, 8954
Streptokinase, 8955
Streptokinase (Enzyme-activating) see
  8955
Streptolin see 8959
Streptomagma [Wyeth] see 3197
Streptomycin, 8956
Streptomycin A see 8956
Streptomycin Hydrochloride see 8956
Streptomycin Hydrochloride-calcium
  Chloride Complex see 8956
Streptomycin Sulfate see 8956
Streptonigrin, 8957
Streptonivicin see 6811
L-Streptose, 8958
Streptothenat [Grünenthal] see 8956
Streptothricin VI see 8959
Streptothricin F see 8959
Streptothricins, 8959
Streptovaricin, 8960
Streptovaricin A see 8960
Streptovaricin B see 8960
Streptovaricin C see 8960
Streptovaricin D see 8960
Streptovaricin E see 8960
Streptovaricin F see 8960
Streptovaricin G see 8960

Streptovaricin J *see* 8960
Streptovarycin *see* 8960
Streptovirudin, 8961
Streptozocin, 8962
Streptozon *see* 9055
Streptozotocin *see* 8962
Stresam [Beaufour] *see* 3913
Strese & Hofmann's Hectorite *see* 4658
Stresnil [Janssen] *see* 891
Stresson [Boehringer, Ing.] *see* 1493
Striadyne [Am. Home] *see* 143
Strictylon *see* 6453
Strigol, 8963
Striped Alder *see* 4639
Strobane-T *see* 9716
Strobilurin A *see* 8964
Strobilurins, 8964
Stroby [BASF] *see* 5372
Strodival [Herbert] *see* 7000
Stromba [Winthrop] *see* 8921
Strombaject [Winthrop] *see* 8921
Stromectol [Merck & Co.] *see* 5296
Strongarm [Dow AgroSci.] *see* 3093
Strongid [Pfizer] *see* 8066
Strong Silver Protein *see* 8666
Strontia *see* 8976
Strontianite *see* 8965, 8968
Strontium, 8965
Strontium Acetate, 8966
Strontium Bromide, 8967
Strontium Bromide (SrBr₂) *see* 8967
Strontium Carbonate, 8968
Strontium Chlorate, 8969
Strontium Chloride, 8970
Strontium Chloride Sr 89 *see* 8970
Strontium Chromate(VI), 8971
Strontium Dioxide *see* 8977
Strontium Fluoride, 8972
Strontium Fluoride (SrF₂) *see* 8972
Strontium Hydrate *see* 8973
Strontium Hydride *see* 8965
Strontium Hydroxide, 8973
Strontium Hydroxide (Sr(OH)₂) *see* 8973
Strontium Iodide, 8974
Strontium Iodide (SrI₂) *see* 8974
Strontium Monoxide *see* 8976
Strontium Nitrate, 8975
Strontium Nitride *see* 8965
Strontium Oxide, 8976
Strontium Oxide (SrO) *see* 8976
Strontium Peroxide, 8977
Strontium Peroxide (Sr(O₂)) *see* 8977
Strontium Ranelate *see* 8224
Strontium Sulfate, 8978
Strontium Sulfide, 8979
Strontium Sulfide (SrS) *see* 8979
Strontium Titanate, 8980
Strontium Titanium Oxide (SrTiO₃) *see* 8980
Strophadogenin *see* 8981
Strophanthidin, 8981
Strophanthidin α-L-Rhamnoside *see* 2497
Strophanthin, 8982
Strophanthus, 8983
Strophoperm [Sertürner] *see* 7000
Strotan [Strathmann] *see* 182
Structum [SKB] *see* 2219
Strumacil *see* 6201
Strumazol [Nycomed] *see* 6043
Struvite, 8984
Strychnidin-10-one *see* 8985
Strychnidin-10-one 19-Oxide *see* 8986
Strychnine, 8985

Strychnine N⁶-Oxide, 8986
C-Strychnotoxine *see* 9717
Stryker's Reagent, 8987
Stryphnasal [Sertürner] *see* 160
Stryphnon [Salus-Braumapharm] *see* 160
STS-557 *see* 3117
Stuart Factor *see* 3964
Stuart-Prower Factor *see* 3964
Stugeron [Janssen] *see* 2306
Stutgeron [Janssen] *see* 2306
Stutgin *see* 2306
STX *see* 8526
STX-209 *see* 930
Stylomycin *see* 8054
Stylophorin *see* 2052
Stylopine, 8988
Styphnic Acid, 8989
Styptic Collodion *see* 2468
Stypticin [Merck & Co.] *see* 2540
Styptochrome [Dr. Reddy's] *see* 1789
Styptocid [Stadmed] *see* 1789
Styptol *see* 2540
Styracin *see* 2305
Styrax *see* 8949
Styrene, 8990
Styrene-butadiene Rubber *see* 8527
Styrene Glycol, 8991
Styrol *see* 8990
Styrolene *see* 8990
Styryl Carbinol *see* 2303
5-Styrylresorcinol *see* 7563
Su-88 *see* 8828
Su-3118 *see* 9147
Su-4885 *see* 6238
Su-5864 *see* 4595
Su-6187 *see* 1527
Su-8341 *see* 2740
SU-11248 *see* 9129
SU-11654 *see* 9650
Suacron [Praemix] *see* 1780
Suadian [Schering AG] *see* 6439
Suanovil [Biokema] *see* 8879
Suavitil [Merck & Co.] *see* 1031
Subcutin [Ritsert] *see* 1088
Subdue [Syngenta] *see* 5993
Suberic Acid, 8992
Suberone *see* 2715
Suberoxime *see* 2715
Suberoylanilide Hydroxamic Acid *see* 10232
Subicard *see* 7222
Subitex [Hoechst] *see* 3315
Subitol *see* 4924
Sublimaze [Janssen-Cilag] *see* 4028
Sublimed Sulfur *see* 9107
Subose [BMS] *see* 4542
Substance P, 8993
Subtilin, 8994
Subutex [Schering-Plough] *see* 1501
Sucaryl [Abbott] *see* 8445
Sucaryl Calcium [Abbott] *see* 1667
Sucaryl Sodium *see* 2696
Success [Dow AgroSci.] *see* 8877
Succimal [Clin-Comar-Byla] *see* 3801
Succimer, 8995
Succinamide, 8996
Succinanil, 8997
Succinanilic Acid, 8998
Succinbromimide *see* 1448
Succinchlorimide *see* 2167
Succinic Acid, 8999
Succinic Acid Anhydride *see* 9000
Succinic Acid Bis[β-dimethylaminoethyl] Ester Dimethochloride *see* 9006
Succinic Acid Calcium Salt *see* 1707

Succinic Acid Diamide *see* 8996
Succinic Acid Di-*tert*-butyl Ester *see* 3046
Succinic Acid Dichloride *see* 9004
Succinic Acid 2,2-Dimethylhydrazide *see* 2809
Succinic Acid Dinitrile *see* 9003
Succinic Acid Monoanilide *see* 8998
Succinic Acid Monoester with 4-Butyl-4-(hydroxymethyl)-1,2-diphenyl-3,5-pyrazolidinedione *see* 9139
Succinic Acid Sodium Salt *see* 8808
Succinic Anhydride, 9000
Succinimide, 9001
Succiniodimide *see* 5088
Succinite *see* 374
Succino-AICAR *see* 8454
Succinobucol, 9002
Succinonitrile, 9003
4-Succinylamido-4'-aminodiphenylsulfone *see* 9009
N-Succinylanthranilic Acid Ester *see* 5679
Succinyl Chloride, 9004
Succinylcholine Bromide, 9005
Succinylcholine Chloride, 9006
Succinylcholine Iodide, 9007
Succinyl Dichloride *see* 9004
O,O-Succinyldicholine Iodide *see* 9007
Succinyl Oxide *see* 9000
2-(N⁴-Succinylsulfanilamido)thiazole *see* 9008
Succinylsulfathiazole, 9008
Succisulfone, 9009
Succosa [Astra Hässle] *see* 9010
Sucline [Soekami] *see* 8445
Suconox-4 *see* 4853
Sucostrin [Apothecon] *see* 9006
Sucraid [QOL Medical] *see* 8447
Sucralfate, 9010
Sucralfin [Inverni] *see* 9010
Sucralose, 9011
Sucrase *see* 5049
Sucrate [Lisapharma] *see* 9010
Sucrédulcor [Europ. Med.] *see* 8445
Sucrets [SKB] *see* 4748
Sucrol *see* 3511
Sucromat [Mayoly-Spindler] *see* 8445
Sucrosa *see* 2696
Sucrose, 9012
Sucrose Octaacetate, 9013
Sucrose Octakis(hydrogen Sulfate) Aluminum Complex *see* 9010
Sucrose Polyester, 9014
Sudafed [Pfizer] *see* 8024
Sudan III, 9015
Sudan IV *see* 8530
Sudan Black B, 9016
Sudan Red BK *see* 9015
Sudermo *see* 5988
Suet, Prepared, 9017
Sufenta [Janssen-Cilag] *see* 9018
Sufentanil, 9018
Sufentanyl *see* 9018
Sufrexal [Janssen] *see* 5344
Sugammadex, 9019
Sugammadex Sodium *see* 9019
Sugar *see* 9012
Sugar Cane Wax Alcohols *see* 7672
Sugar of Lead *see* 5452
Sugast [Alfa] *see* 9010
Suicalm [Janssen] *see* 891
Suiminth [Pfizer] *see* 6353
Suint, 9020
Sukor [Calif. Aromatics & Flavor] *see* 6537

Suxamethonium Chloride *see* 9006
Suxamethonium Iodide *see* 9007
Suxibuzone, 9139
Suxil [Eddé] *see* 9003
Suxilep *see* 3801
Suximal *see* 3801
Suxinutin [Parke-Davis] *see* 3801
Suzutolon [Tatsumi] *see* 1180
Swartziol *see* 5322
Sweet Bay *see* 5437
Sweet Birch Oil *see* 6193
Sweet Bugle *see* 5685
Sweet Cane *see* 1640
Sweet Clover *see* 5889
Sweet Elder *see* 8488
Sweet Fennel *see* 4008
Sweet Flag *see* 1640
Sweet Grass *see* 1640
Sweet Haw *see* 10170
Sweet Herb *see* 8937
Sweet Orange Peel *see* 6955
Sweet Oriental Gum *see* 8949
Sweet Root *see* 4540
Sweet Spirit of Niter *see* 3885
Sweet-wood Bark *see* 1881
Sweroside 2'-(3″,5″,3‴-Trihydroxydi-
   phenyl)-2″-carboxylic Acid Ester *see*
   371
Swertiamarin, 9140
Swing Gold [BASF] *see* 3292
Swiss Blue *see* 6132
Sword [Meiji] *see* 8016
SY-5555 *see* 3976
Sybron [Parke-Davis] *see* 4058
Sycrest [Merck & Co.] *see* 821
Sydnones, 9141
Sygen [Fidia] *see* 4394
Sykose *see* 8445
Sylatron [Schering-Plough] *see* 7176
Sylvanite *see* 4549
Sylvic Acid *see* 7
Sylvine *see* 7742
Sylvite *see* 7742
Symclosene *see* 9806
Symetra [Forest] *see* 7334
Symlin [Amylin] *see* 7828
Symmetrel [Endo] *see* 368
Sympatektoman *see* 9344
Sympathol *see* 9142
Sympectothion *see* 3720
Symptom 3 [Parke-Davis] *see* 1453
SYN-407855 *see* 7564
Synacid [Sterivet] *see* 4796
Synaclyn [Syntex] *see* 4176
Synacthen [Novartis] *see* 2539
Synadenylic Acid *see* 145
Synadrin [Hoechst] *see* 7855
Synagis [MedImmune] *see* 7089
Synalar [Syntex] *see* 4181
Synamol [Syntex] *see* 4181
Synandone [Syntex] *see* 4181
Synandrol [Pfizer] *see* 9324
Synanthic [Fort Dodge] *see* 7034
Synapause [Organon] *see* 3762
Synapse [Bayer CropSci.] *see* 4150
Synaptic Laminin *see* 5401
SynaptoGreen [Biotium] *see* 4247
Synapton [Forest] *see* 7496
Synarel [Searle] *see* 6435
Synasteron *see* 7066
Synatan [Ferring] *see* 2956
Synchrocept [Syntex] *see* 7991
Synchrocept B [Syntex] *see* 4023
Synchrodyn [Hoechst] *see* 310
Synchronate *see* 4230

Syncillin [Bayer] *see* 902, 7337
Syncortyl [Roussel Diamant] *see* 2903
Syncria [GSK] *see* 204
Syncro-Mate [Searle] *see* 4230
Syncuma [Philips Roxane] *see* 9649
Syncurine [Burroughs Wellcome] *see*
   2851
Syndeins *see* 8867
Syndrox [McNeil] *see* 6020
Synédil [Beytout] *see* 9119
Synemol [Syntex] *see* 4181
Synephrin [Neukonigsforder] *see* 9142
Synephrine, 9142
Syneptine [Toraude] *see* 5505
Synercid [Rhône-Poulenc] *see* 2803
Synestrol [Schering] *see* 3115
Synexin *see* 668
Synflex [Recordati] *see* 6499
Syngacillin [Wyeth] *see* 2695
Synhexyl, 9143
Synkamin [Parke-Davis] *see* 10220
Synkavit [Roche] *see* 5898
Synkayvite [Roche] *see* 5898
Synklor *see* 2083
Synnematin B *see* 7207
Synogil [Basotherm] *see* 6509
Synopen [Novartis] *see* 2164
Synotic [Syntex] *see* 4181
Synpen [Novartis] *see* 2164
Synpenin [Sankyo] *see* 583
Synperonic PE/F68 [ICI] *see* 7679
Synpitan [Werfft] *see* 7078
Synpol *see* 7695
Synsac [Syntex] *see* 4181
Synsorb 90 *see* 9144
Synsorb CD *see* 9144
Synsorb Pk *see* 9144
Synsorbs, 9144
Synstigmin *see* 6549
Syntaris [Syntex] *see* 4176
Syntarpen [Polfa] *see* 3094
Syntecort [Medicamenta] *see* 7133
Syntes 12a *see* 3565
Syntestan [Teofarma] *see* 2387
Syntetrin [BMS] *see* 8381
Synthenate *see* 9142
Synthovo *see* 4737
Synthroid [Knoll] *see* 9570
Syntocain [Sinteteca] *see* 7875
Syntocinon [Novartis] *see* 7078
Syntometrine *see* 3712
Syntons *see* 7990
Syntopressin [Sandoz] *see* 5691
Syntrogène *see* 4737
Synulox [Pfizer] *see* 2339
Synvinolin *see* 8677
Syprine [Merck & Co.] *see* 9828
Syprol [Rosemont] *see* 7953
SYR-322 *see* 302
Syraprim [Wellcome] *see* 9878
Syringaldehyde, 9145
Syringic Acid δ-Guanidinobutyl Ester
   see 5495
Syringic Aldehyde *see* 9145
Syringidin *see* 5779
Syringin, 9146
Syringopine *see* 9147
Syringoside *see* 9146
Syrosingopine, 9147
Syscor [Bayer] *see* 6651
Systamex [Schering-Plough] *see* 7034
Systemin, 9148
Systen [Janssen-Cilag] *see* 3758
Systhane [Dow AgroSci.] *see* 6406
Systox *see* 2892

Systral [Viatris] *see* 2187
Sytam [Murphy] *see* 8532
Sytasol [Murphy] *see* 3313
Szaibelyite *see* 5722
Szomolnikite *see* 4084
SZX-722 *see* 5124
T$_3$ *see* 5565
T$_4$ *see* 2728, 9570
2,4,5-T, 9149
T3P [Archimica] *see* 7978
T-20 *see* 3636
T 23P *see* 9927
T-34 *see* 1378
T-47 *see* 1378
T-113 *see* 9788
T-1220 *see* 7575
T-1258 *see* 9274
T-1384 *see* 6566
T-1551 *see* 1931
T-1824 *see* 3948
T-1982 *see* 1919
T-2525 *see* 1949
T-2588 *see* 1949
T-3262 *see* 9714
T-3761 *see* 7161
T-3762 *see* 7161
T-3811 *see* 4402
T-3811ME *see* 4402
T-91825 *see* 1947
TA-058 *see* 842
TA-064 *see* 2896
TA-870 *see* 3441
TA-903 *see* 1040
TA-0910 *see* 9177
TA -1790 *see* 875
TA-2711 *see* 3535
TA-3090 *see* 2346
TA-6366 *see* 4952
Tabalon [Teofarma] *see* 4919
Tabazur [Théraplix] *see* 6609
Tabe [Bernabo] *see* 7604
Tabernanthine, 9150
Table Salt *see* 8734
Table Spate *see* 1704
Tabloid [GSK] *see* 9490
Tabun, 9151
Tacalcitol, 9152
Tacaryl [Westwood] *see* 6036
Tace [HMR] *see* 2178
Tacef [Takeda] *see* 1926
Tachigaren [Sankyo] *see* 4895
Tachionin [Sana] *see* 9789
Tachmalin [AWD] *see* 187
Tachostyptan [Consolidated Chem.] *see*
   9539
Tachydaron [AWD] *see* 478
Tachyrol [Duphar] *see* 3198
Tachysterol, 9153
TACI-Fc5 *see* 848
TACI-Ig *see* 848
Tacitin [Novartis] *see* 1089
Tackle [Sankyo] *see* 8625
Tacosal [Stada] *see* 7433
Tacrine, 9154
Tacrolimus, 9155
Tacryl, 9156
Tadalafil, 9157
Tadenan [Debat] *see* 8060
t-Adenylic Acid *see* 146
Tadeonal *see* 7690
TAED, 9158
TAED 4303 *see* 9158
Tafenoquine, 9159
Tafethion [Rallis] *see* 3790
Tafil [Pharmacia] *see* 308

Taflotan [Santen] *see* 9160
Tafluprost, 9160
TAG-72 *see* 8521
Tagamet [GSK] *see* 2282
Tagathen [Lederle] *see* 2170
D-Tagatose, 9161
Tagilite *see* 2638
Tagonis [GSK] *see* 7148
Taidecanone [Taiyo] *see* 10025
Taiguic Acid *see* 5417
Tailed Pepper *see* 2602
TAK-375 *see* 8219
TAK-491 *see* 903
TAK-536 *see* 903
TAK-599 *see* 1947
Takacillin [Torii] *see* 5491
Taka-Diastase, 9162
Takanarumin [Takata] *see* 271
Takeoff [Takeda] *see* 4949
Takepron [Takeda] *see* 5413
Takesulin [Takeda] *see* 1946
Taketiam [Takeda] *see* 1936
Takizolite *see* 356
Taktic [Intervet] *see* 482
Takus [Montedison] *see* 2004
Talacen [Sanofi-Aventis] *see* 7233
Talactoferrin, 9163
Talactoferrin Alfa *see* 9163
Talampanel, 9164
Talampicillin, 9165
Talaporfin, 9166
Talaporfin Sodium *see* 9166
Talat [Polifarma] *see* 9165
Talbutal, 9167
Talc, 9168
Talcid [Bayer] *see* 5724
Talcum *see* 9168
Talecid [Schering] *see* 7488
Talerc [Boehringer, Ing.] *see* 3673
Taliglucerase Alfa, 9169
Talin [Tate & Lyle] *see* 9422
Talinolol, 9170
Talion [Tanabe] *see* 1151
Talipexole, 9171
Talis [Kali-Chemie] *see* 5990
Talisomycin *see* 9175
Talisomycin B *see* 9175
Talleol *see* 9172
Tall Oil, 9172
Tallol *see* 9172
Tallow, 9173
Tallow Alcohol, 9174
Tallow Shrub *see* 1010
Tallysomycin, 9175
Tallysomycin A *see* 9175
Tallysomycin B *see* 9175
Talniflumate, 9176
Talofen [Fournier] *see* 7898
Talon [Zeneca Ag Prod] *see* 1386
Taloxa [Schering-Plough] *see* 3984
Talpen [SKB] *see* 9165
Talpirid [Bell] *see* 1396
Talsis [Wyeth] *see* 1296
Talstar [FMC] *see* 1213
Taltirelin, 9177
Talusin [Abbott] *see* 7985
Talwin (Inj.) [Hospira] *see* 7233
Talwin (Tabl.) [Sanofi-Synthelabo] *see* 7233
Tamarind, 9178
Tamaron [Bayer CropSci.] *see* 6019
Tamas [Schering] *see* 1125
Tamatinib Fosdium *see* 4287
Tambocor [3M Pharma] *see* 4130
TAME *see* 605

Tamex [CFPI] *see* 1534
Tamibaro [Toko] *see* 9179
Tamibarotene, 9179
Tamiflu [Roche] *see* 6986
Tamik [IPRAD] *see* 3191
Tamofène [Aventis] *see* 9180
Tamoxifen, 9180
Tamsulosin, 9181
Tanacetin, 9182
Tanaclone [Amsa] *see* 3520
Tanadopa [Tanabe Seiyaku] *see* 3441
Tanafol *see* 2106
Tanakan [Ipsen] *see* 4460
Tanatril [Tanabe] *see* 4952
Tanbismuth *see* 1288
Tandearil [Novartis] *see* 7071
Tandix [Biofarma] *see* 4975
Tandospirone, 9183
Tanganil *see* 91
Tangantangan Oil *see* 1898
Tannic Acid, 9184
Tannic Acid Bismuth Derivative *see* 1288
Tannigen *see* 96
Tannins *see* 9184
Tannyl Acetate *see* 96
Tanorama *see* 3200
Tanphetamin *see* 2956
Tantalic Acid Anhydride *see* 9188
Tantalite *see* 6642, 9185
Tantalum, 9185
Tantalum Chloride (TaCl₅) *see* 9186
Tantalum Fluoride (TaF₅) *see* 9187
Tantalum Oxide (Ta₂O₅) *see* 9188
Tantalum Pentachloride, 9186
Tantalum Pentafluoride, 9187
Tantalum Pentoxide, 9188
Tantarone [Mohan] *see* 5985
Tan Tone *see* 3200
Tantum [Angelini] *see* 1125
TAO [Roerig] *see* 9949
TAP-031 *see* 4088
TAP-144 *see* 5509
Tapazol [Lilly] *see* 6043
Tapazole [King] *see* 6043
Tapentadol, 9189
Tapros [Santen] *see* 9160
Taprostene, 9190
TAPS, 9191
TAPSO *see* 9191
Taractan [Roche] *see* 2194
Tarasan [Roche] *see* 2194
Tarasyn [Syntex] *see* 5353
Taraxacum, 9192
Taraxast-20(30)-en-3β-ol *see* 9193
Taraxasterin *see* 9193
Taraxasterol, 9193
Taraxein, 9194
Taraxerol, 9195
Tar Camphor *see* 6455
Tarceva [OSI] *see* 3730
Tardak [Pfizer Animal Health] *see* 2883
Tardamide [Grünenthal] *see* 9056
Tardigal [Beiersdorf] *see* 3182
Tardisal [Sigma-Tau] *see* 8470
Tardocillin [Bayer] *see* 7205
Taredan [FMC] *see* 1636
Tareg [Novartis] *see* 10102
Tarenflurbil, 9196
Targa [Nissan] *see* 8203
Targinine *see* 6096
Targocid [Aventis] *see* 9255
Targosid [Aventis] *see* 9255
Targretin [Ligand] *see* 1198
Taribavirin, 9197

Tarichatoxin *see* 9394
Tariquidar, 9198
Tarivid [Aventis] *see* 6863
Tarka [Abbott] *see* 10144
Taroctyl [Taro] *see* 2191
Tarodyl *see* 4537
Tarodyn *see* 4537
Tar Oil, 9199
Tar Oil, Rectified, 9200
Tarragon, 9201
Tarragon Oil *see* 3760
Tartar Emetic *see* 691
(−)-Tartaric Acid *see* 9203
(+)-Tartaric Acid *see* 9205
d-Tartaric Acid *see* 9205
dl-Tartaric Acid *see* 9204
l-Tartaric Acid *see* 9203
meso-Tartaric Acid, 9202
D-Tartaric Acid, 9203
DL-Tartaric Acid, 9204
L-Tartaric Acid, 9205
Tartaric Acid Bismuth Complex Potassium Salt *see* 1278
Tartaric Acid Bismuth Complex Sodium Salt *see* 1280
Tartaric Acid Dipotassium Salt *see* 7796
L-Tartaric Acid Monoammonium Salt *see* 500
Tartaric Acid Tin(2+) Salt (1:1) *see* 8919
Tartarized Antimony *see* 691
Tartarus Vitriolatus *see* 7793
Tartrated Antimony *see* 691
Tartrazine, 9206
Tartrol *see* 1280
Tartronic Acid, 9207
Taselin [Eisai] *see* 6839
Taseron [Janssen] *see* 5344
Tasimelteon, 9208
Tasis [Wyeth] *see* 1296
Task [Boehringer, Ing.] *see* 3089
Tasmaderm [Roche] *see* 6370
Tasmar [Valeant] *see* 9668
Tasnon (elixir) [Tropon] *see* 7576
Tasnon (Tabl.) [Tropon] *see* 7576
Tasocitinib *see* 9661
Taspoglutide, 9209
Tasquinimod, 9210
Taste-modifying Protein *see* 6289
TAT-3 [Takeda] *see* 7516
TATBA *see* 9759
TATD *see* 6849
Tathiclon [Ohta] *see* 4511
Tathion [Yamanouchi] *see* 4511
Tationil [Boehringer Biochemia] *see* 4511
TATP *see* 9752
TAU-284 *see* 1151
Tauliz [HMR] *see* 7605
Taural [Roemmers] *see* 8228
Tauranol I-78 [Innospec] *see* 8739
Tauredon [Byk Gulden] *see* 4553
Taurine, 9211
Tauriscite *see* 4084
Taurocholic Acid, 9212
Tauroflex [Geistlich] *see* 9213
Taurolidine, 9213
Taurolin [HMR] *see* 9213
Tausonite *see* 8980
Tavanic [Aventis] *see* 6863
Tavegil [Novartis] *see* 2343
Tavegyl [Novartis] *see* 2343
Ta-Verm *see* 7576
Tavist [Novartis] *see* 2343
Tavocept [BioNumerik] *see* 3233
Tavor [Wyeth] *see* 5636
Taxicin-I *see* 9214

Teicoplanin A$_2$-1 *see* 9255
Teicoplanin A$_2$-2 *see* 9255
Teicoplanin A$_2$-3 *see* 9255
Teicoplanin A$_2$-4 *see* 9255
Teicoplanin A$_2$-5 *see* 9255
Teknar [Certis] *see* 925
Tektamer 38 [Lanxess] *see* 3026
Tektin A *see* 8867
Tekturna [Novartis] *see* 239
TEL *see* 9347
Telaprevir, 9256
Telar [DuPont] *see* 2197
Telavancin, 9257
Telazol [Fort Dodge] *see* 9592
Telbivudine, 9258
Telcagepant, 9259
Telcagepant Potassium *see* 9259
Telcyta [Telik] *see* 1748
Teldane [Aventis] *see* 9306
Teldanex [AstraZeneca] *see* 9306
Teldrin [SKB] *see* 2186
Telebar [Guerbet] *see* 990
Télémid [Procter & Gamble] *see* 9049
Teleocidin A *see* 9260
Teleocidin B *see* 9260
Teleocidin B$_4$ *see* 9260
Teleocidins, 9260
Telepaque [Winthrop] *see* 5101
Telepathine *see* 4649
Teleprim [Procter & Gamble] *see* 9050
Telesmin [Mitsubishi] *see* 1783
Telesol [Lasa] *see* 4886
Teletrast [Astra] *see* 5101
Telfast [Aventis] *see* 4097
Teli *see* 8519
Telithromycin, 9261
Tell [Syngenta] *see* 7866
Telluric(IV) Acid *see* 9270
Telluric(VI) Acid, 9262
Telluric Acid (H$_6$TeO$_6$) *see* 9262
Telluric Acid (H$_2$TeO$_3$) Potassium Salt
   (1:2) *see* 7798
Telluric Acid (H$_2$TeO$_4$) Potassium Salt
   (1:2) *see* 7797
Telluric Acid (H$_2$TeO$_3$) Sodium Salt
   (1:2) *see* 8814
Telluric Acid (H$_2$TeO$_4$) Sodium Salt
   (1:2) *see* 8813
Telluric Bromide *see* 9267
Telluric Chloride *see* 9268
Tellurite *see* 9265
Tellurium, 9263
(*T*-4)-Tellurium Bromide (TeBr$_4$) *see*
   9267
(*T*-4)-Tellurium Chloride (TeCl$_4$) *see*
   9268
Tellurium Dichloride, 9264
Tellurium Dioxide, 9265
(*OC*-6-11)-Tellurium Fluoride (TeF$_6$) *see*
   9266
Tellurium Hexafluoride, 9266
Tellurium Hexahydroxide *see* 9262
Tellurium Hydride *see* 4838
(*T*-4)-Tellurium Iodide (TeI$_4$) *see* 9269
Tellurium Oxide (TeO$_2$) *see* 9265
Tellurium Tetrabromide, 9267
Tellurium Tetrachloride, 9268
Tellurium Tetraiodide, 9269
Tellurobismuthite *see* 1289
Tellurous Acid, 9270
Tellurous Chloride *see* 9264
Telmicid [Lilly] *see* 3412
Telmid [Lilly] *see* 3412
Telmin [Janssen] *see* 5837
Telmisartan, 9271

Telodrin *see* 5172
Telomerase, 9272
Telomere Terminal Transferase *see* 9272
Telomycin, 9273
Telone II [Dow AgroSci.] *see* 3085
Telopar [Pfizer] *see* 7021
Telvar [DuPont] *see* 6349
Telzir [GSK] *see* 4276
TEM *see* 9836
Temac [Abbott] *see* 9274
Temafloxacin, 9274
Temaril [Allergan] *see* 9873
Temasept IV [Hexcel] *see* 9778
Temazepam, 9275
Tembotrione, 9276
TEMED, 9277
Tementil [Aventis] *see* 7879
Temephos, 9278
Temesta [Wyeth] *see* 5636
Temetex [Sauter] *see* 3162
Temgesic [Schering-Plough] *see* 1501
Temik [Bayer CropSci.] *see* 215
Temina *see* 10221
Temocapril, 9279
Temocaprilat *see* 9279
Temocillin, 9280
Temodal [Schering-Plough] *see* 9282
Temodar [Schering-Plough] *see* 9282
Temopen [SKB] *see* 9280
Temoporfin, 9281
Temovate [GSK] *see* 2358
Temozolomide, 9282
TEMPO, 9283
Tempo [Bayer] *see* 2754
TEMPOL, 9284
Temposil [Lederle] *see* 1665
Tempra [BMS] *see* 47
Temserin [Merck & Co.] *see* 9601
Temsirolimus, 9285
Temur *see* 9378
Tenacid [Sigma-Tau] *see* 4961
Tenathan [Robins] *see* 1188
Tendor [Chinoin] *see* 2846
Tenecteplase, 9286
Tenelid [Helsinn] *see* 4593
Tenesdol [Hishiyama] *see* 3341
Tenex [Reddy] *see* 4596
Teng Huang *see* 4390
Tenicid *see* 5069
Tenidap, 9287
Teniposide, 9288
Tennecetin *see* 6509
Tenoblock [Leiras] *see* 850
Tenofovir, 9289
Tenofovir DF *see* 9289
Tenofovir Disoproxil *see* 9289
Tenonitrozole, 9290
Tenorite *see* 2636
Tenormal [Zeneca] *see* 7189
Tenormin [AstraZeneca] *see* 850
Tenovil [Schering-Plough] *see* 5043
Tenox BHA [Eastman Kodak] *see* 1549
Tenox BHT [Eastman Kodak] *see* 1550
Tenox PG [Eastman] *see* 7971
Tenoxicam, 9291
Tensatrin [Gedeon Richter] *see* 8008
Tensilon [Valeant] *see* 3564
Tensinol [Farmaroma] *see* 7189
Tensipine [Genus] *see* 6613
Tensobon [Schwarz] *see* 1776
Tensofin [BMS] *see* 4220
Tensoral [Diana] *see* 7189
Tensozide [Mead Johnson] *see* 4283
Tenstaten [Ipsen] *see* 2267
Tenuate [Aventis] *see* 3145

Tenuazonic Acid, 9292
Tenyl [Nippon Kayaku] *see* 1180
Teonicon [Neopharmed] *see* 7542
Teoptic [Novartis] *see* 1866
Teorema [Panchemie] *see* 4483
Teoremac [Farmades] *see* 4483
Teosona [Phoenix] *see* 9436
TEPA *see* 9837
Tepadina [Adienne] *see* 9518
Tepanil [3M Pharma] *see* 3145
Tephroite *see* 5803
Tephrosin, 9293
Tephthol *see* 9305
Tepic [Nissan] *see* 9860
Tepilta [Wyeth] *see* 7032
Tepirubicin *see* 7601
Tepoxalin, 9294
TEPP *see* 9349
Tepraloxydim, 9295
Teprenone, 9296
Tequin [BMS] *see* 4409
TER-286 *see* 1748
Téralithe [Aventis] *see* 5583
Teramine [Darby] *see* 2188
Terazol [Cilag] *see* 9303
Terazosin, 9297
Terbacil, 9298
Terbasmin [AstraZeneca] *see* 9302
Terbia *see* 9300
Terbinafine, 9299
Terbium, 9300
Terbuconazole *see* 9229
Terbufos, 9301
Terbul [AstraZeneca] *see* 9302
Terbutaline, 9302
Terbutaline Bisdimethylcarbamate *see*
   946
Terbutrazole *see* 9229
Tercian [Théraplix] *see* 2673
Terconazole, 9303
Tercospor [Cilag] *see* 9303
Terebic Acid, 9304
Terebinic Acid *see* 9304
Terenol [Hoechst] *see* 8275
Terephthalic Acid, 9305
Terephthalylidene-3,3'-dicamphor-10,-
   10'-disulfonic Acid *see* 3537
Terfenadine, 9306
Terfenadine Carboxylate *see* 4097
Terfex [Leiras] *see* 9306
Terflurane *see* 9250
Terfluzine [Aventis] *see* 9846
Tergal [Rhône-Poulenc] *see* 7689
Tergavon *see* 4400
Tergitol [Union Carbide] *see* 7697
Tergitol 4 [Union Carbide] *see* 8817
Tergitol NP [Union Carbide] *see* 6764
Tergitol TP-9 [Union Carbide] *see* 6764
Terguride, 9307
Teriam [Roussel Diamant] *see* 9760
Teridax [Schering] *see* 5104
Teriflunomide, 9308
Terion [Lusofarmaco] *see* 4255
Teripatide Acetate, 9309
Terital [Montecatini] *see* 7689
Terlenka [Kunsztijde] *see* 7689
Terlipressin, 9310
Termidor [BASF] *see* 4116
Termierite *see* 356
Termitin *see* 10221
Tern [Syngenta] *see* 4020
Ternadin [Cantabria] *see* 9306
Ternelin [Novartis] *see* 9642
Teronac [Novartis] *see* 5831
Teropterin *see* 8043

Tetraphosphorus Trisulfide, 9385
Tetrapotassium Diphosphate see 7783
Tetrapotassium Hexakis(cyano-C)ferrate(4−) see 7752
Tetrapotassium Pyrophosphate see 7783
Tetrapropylammonium Perruthenate, 9386
Tetrapropylammonium Tetraoxoruthenate see 9386
TetraSan [Valent] see 3933
Tetrasilane, 9387
Tetrasilicane see 9387
Tetrasilicobutane see 9387
Tetrasilicon Decahydride see 9387
Tetrasodium 3,3′-[(3,3′-Dimethyl-4,4′-biphenylene)bis(azo)]bis(5-amino-4-hydroxy-2,7-naphthalenedisulfonate) see 9972
Tetrasodium Hexakis(cyano-C)ferrate-(4−) see 8752
Tetrasodium Pyrophosphate, 9388
Tetrasulfur Tetranitride, 9389
Tetra-Tablinen [Lichtenstein] see 7075
Tetrathiafulvalene, 9390
1,4,5,8-Tetrathiafulvalene see 9390
2,4,6,8-Tetrathia-1,3,5,7-tetraarsatricyclo[3.3.0.0³,⁷]octane see 790
1λ⁴,3,5λ⁴,7-Tetrathia-2,4,6,8-tetrazacycloocta-1,4,5,8-tetraene see 9389
1λ⁴δ², 5λ⁴δ²-1,3,5,7,2,4,6,8-Tetrathiatetrazocine see 9389
Tetrathiin see 7037
(T-4)-Tetrathioxomolybdate(2−) Ammonium (1:2) see 557
Tetraverin [Polfa] see 8381
7,8,9,10-Tetrazabicyclo[5.3.0]-8,10-decadiene see 7254
1,2,3,3a-Tetrazacyclohepta-8a,2-cyclopentadiene see 7254
Tetrazepam, 9391
Tetrazobenzene-β-naphthol see 9015
Tetrazolium Blue, 9392
N-[p-(o-1H-Tetrazol-5-ylphenyl)benzyl]-N-valeryl-L-valine see 10102
Tetrex [BMS] see 9341
Tetridin see 8108
Tetrin, 9393
Tetrin A see 9393
Tetrin B see 9393
Tetrodontoxin see 9394
Tetrodotoxin, 9394
Tetrofosmin, 9395
Tetrole see 4325
Tetron see 9349
Tetronasin, 9396
Tetropil see 9335
Tetroquinone, 9397
Tetroxoprim, 9398
L-glycero-Tetrulose see 3750
Tetryl see 6659
Tetryl Formate see 5188
Tetryzoline see 9364
Tetsol [Novartis] see 9341
Tetterwort see 8492
Tetucur [Teikoku Zoki] see 4084
Teturamin see 3407
Tevcodyne [Tevcon] see 7390
Teveten [Solvay] see 3691
Tevilon [Teikoku Rayon] see 7707
Texanol Isobutyrate, 9399
Texmeten [Roche] see 3162
Texodil [Grünenthal] see 1936
Texsolve S [Texaco] see 6282
Texsolve S-2 [Texaco] see 6282
Tezosentan, 9400

TF-1169 see 4147
TFD see 6205
TFDO see 6205
TFE see 9350, 9848
TFM, 9401
TFMSA see 9842
Tf₂NPh see 7427
Tf₂O see 9850
TFPI see 9623
TFT Thilo [Alcon-Thilo] see 9855
Tg see 9421
TGB see 9572
T/Gel [Farillon] see 2409
TGF-β see 9733
TGIC see 9860
TGS see 9011
TGV see 9549
TH-152 see 6002
TH-913 see 4949
TH-1165 see 4012
TH-1165a see 4012
TH-1395 see 4516
TH-2602 see 2673
TH-6040 see 3161
TH-9507 see 9320
THA [Woods] see 9154
Thacapzol [Recip] see 6043
Thalamonal see 3499
Thalamyd [Schering] see 7488
Thalazole [M & B] see 7489
Thalicarpine, 9402
Thalictrine see 5760
Thalidomide, 9403
Thalitone [Boehringer, Ing.] see 2200
Thallic Acetate see 9406
Thallic Oxide see 9417
Thallium, 9404
Thallium Acetate, 9405
Thallium(III) Acetate, 9406
Thallium Bromide, 9407
Thallium Carbonate, 9408
Thallium Chloride, 9409
Thallium Cyanide, 9410
Thallium Fluoride, 9411
Thallium Hydroxide, 9412
Thallium Iodide, 9413
Thallium Mononitrate, 9414
Thallium(I) Nitrate see 9414
Thallium(III) Nitrate see 9420
Thallium Oxide, 9415
Thallium Oxide (Tl₂O₃) see 9417
Thallium Peroxide see 9417
Thallium Selenide, 9416
Thallium Sesquioxide, 9417
Thallium Sulfate, 9418
Thallium Sulfide, 9419
Thallium Triacetate see 9406
Thallium Trinitrate, 9420
Thallous Acetate see 9405
Thallous Bromide see 9407
Thallous Carbonate see 9408
Thallous Chloride see 9409
Thallous Cyanide see 9410
Thallous Fluoride see 9411
Thallous Hydroxide see 9412
Thallous Iodide see 9413
Thallous Nitrate see 9414
Thallous Oxide see 9415
Thallous Selenide see 9416
Thallous Sulfate see 9418
Thallous Sulfide see 9419
Thalomid [Celgene] see 9403
THAM [Fisher Sci.] see 9951
Thanite [Hercules] see 5174
Thapsigargin, 9421

Thapsigargine see 9421
Thaumatin, 9422
Thaumatins Proteins see 9422
R-THBP see 8505
Δ¹-THC see 9354
Δ⁶-THC see 9354
Δ⁸-THC see 9354
Δ⁹-THC see 9354
(3R,4R)-Δ⁶-THC-DMH-7-oic Acid see 191
THE see 9355
Theaflavine, 9423
Theal Ampules [Boehringer, Mann.] see 3526
Theanine, 9424
Thearubigin see 9425
Thearubigins, 9425
Thebacon, 9426
Thebaine, 9427
Thebainone, 9428
Thebainone-A see 9428
Theelin [Parke-Davis] see 3763
Theelol [Parke-Davis] see 3762
Thefylan see 3526
Thein see 1639
Theiomycetin see 1162
Thelestrin [Carnrick] see 3763
Thelin [Encysive] see 8692
Thelmesan see 3522
Thelykinin see 3763
Themalon [Burroughs Wellcome] see 9445
Thenaldine, 9429
Thenalidine see 9429
Thenardite see 8809
Thenclor see 2170
Thenfadil [Winthrop] see 9431
Thenitrazole see 9290
2-Thenoic Acid see 9507
3-Thenoic Acid, 9430
Thenophenopiperidine see 9429
2-(α-Thenoylamino)-5-nitrothiazole see 9290
p-(2-Thenoyl)hydratropic Acid see 9134
2-(α-Thenoylthio)propionylglycine see 8936
Thenyldiamine, 9431
Thenylene [Abbott] see 6031
Thenylpyramine see 6031
Theochron [Forest] see 9436
Theobid [Cipla] see 9436
Theobroma Oil, 9432
Theobromine, 9433
1-Theobromineacetic Acid, 9434
Theocalcin [Knoll] see 9433
Theofibrate, 9435
Theograd [Abbott] see 9436
Theolair [3M] see 9436
Theon [Bock] see 8014
Theon [Klinge] see 9436
Theophyllamine see 461
Theophylline, 9436
7-Theophyllineacetic Acid see 25
Theophylline Cholinate see 2215
Theophylline Compd with Ethylenediamine see 461
Theophyllineethylamphetamine see 4003
Theophylline Ethylenediamine see 461
2-(7′-Theophyllinemethyl)-1,3-dioxolane see 3486
1-(Theophyllin-7-yl)ethyl 2-(p-Chlorophenoxy)isobutyrate see 9435
Theostat [Fabre] see 9436
Thephorin [Sinclair] see 7347
Theprubicin [Aventis] see 7601

Therabloat [Pfizer] *see* 7679
Theraderm [BMS] *see* 1119
Therafectin [BLS] *see* 480
Theralene [Rhône-Poulenc] *see* 9873
Theraloc [Oncoscience] *see* 6638
Theranabol *see* 7064
Therapas [Smith & Nephew] *see* 1118
Therapav [Berlex] *see* 7122
Therarubicin [Meiji Seika] *see* 7601
Theratuss [Squibb] *see* 7569
Thermolysin, 9437
Thermonatrite *see* 8731
Thermorubin, 9438
Thermorubin A *see* 9438
Thermovyl [Rhovyl] *see* 7707
Thermozymocidin *see* 6418
Thesodate *see* 9433
Thevetigenin *see* 3181
Thevetin A, 9439
THFA *see* 9358
THG *see* 9359
1-Thia-3-azaindene *see* 1110
Thiabendazole, 9440
Thiabendazole Hypophosphite *see* 9440
Thiabenzazonium Iodide *see* 9581
Thiabutazide *see* 1527
Thiacetarsamide, 9441
Thiacetazone, 9442
Thiacetic Acid *see* 9473
Thiacloprid, 9443
Thiactin *see* 9517
Thiacyclopentane *see* 9363
1-Thia-3-cyclopentene 1,1-Dioxide *see* 9086
Thiadiazuron *see* 9465
3'-Thia-2',3'-dideoxycytidine *see* 5402
Thialbarbital, 9444
Thialbarbitone *see* 9444
Thiamazole *see* 6043
Thiambutene, 9445
Thiamcol [Morishita] *see* 9453
Thiamenidine *see* 9573
Thiamethoxam, 9446
Thiameturon-methyl *see* 9468
Thiamine, 9447
Thiamine Chloride Hydrochloride *see* 9447
Thiamine Dichloride *see* 9447
Thiamine Diphosphate, 9448
Thiamine Diphosphate Ester Chloride *see* 9448
Thiamine Disulfide, 9449
Thiamine 8-(Methyl 6-Acetyldihydro-thioctate) Disulfide *see* 6849
Thiamine Monochloride *see* 9447
Thiamine Monophosphate *see* 9448
Thiamine Monophosphate Chloride *see* 9448
Thiamine Monophosphate Disulfide *see* 9449
Thiamine Propyl Disulfide *see* 7993
Thiamine Pyrophosphate *see* 9448
Thiamine Tetrahydrofurfuryl Disulfide *see* 4340
Thiamine Trihydrogen Pyrophosphate (Ester) *see* 9448
Thiamine Triphosphate, 9450
Thiamine Triphosphoric Acid Ester *see* 9450
Thiaminium Chloride *see* 9447
Thiaminogen [Ohta] *see* 6983
Thiamiprine, 9451
Thiamorpholine *see* 9452
Thiamphenicol, 9453
Thiamutilin *see* 9574

Thiamylal, 9454
Thianaphthene, 9455
Thianeuron [RPR] *see* 53
Thiantan *see* 3119
1-Thia-4,7,10,13,16-pentaazacycloeico-sane Cyclic Peptide Deriv *see* 1796
Thiapride *see* 9576
Thiarubrine A *see* 9456
Thiarubrine B *see* 9456
Thiarubrines, 9456
Thiasine *see* 3720
Thiaton [Hokuriku Seiyaku] *see* 9616
Thiaxanthene *see* 9522
Thiaxanthone *see* 9523
Thiazamide [May & Baker] *see* 9074
1,4-Thiazan *see* 9452
Thiazenone *see* 9457
Thiazesim, 9457
Thiazinamium Methylsulfate, 9458
2-Thiazolamine *see* 475
Thiazole, 9459
Thiazole Orange, 9460
Thiazolesulfone *see* 9461
4-Thiazolidinecarboxylic Acid *see* 9602
4-Thiazolidine-2-caproic Acid *see* 6407
ε-[2-(4-Thiazolidone)]hexanoic Acid *see* 6407
Thiazolinobutazone *see* 7390
Thiazolsulfone *see* 9461
Thiazol Yellow G, 9462
2-[[[4-[(2-Thiazolylamino)sulfonyl]phen-yl]amino]carbonyl]benzoic Acid *see* 7489
2-(4-Thiazolyl)-1*H*-benzimidazole *see* 9440
2-(4-Thiazolyl)-5-benzimidazolecarba-mate Isopropyl Ester *see* 1728
N-[2-(4-Thiazolyl)-1*H*-benzimidazol-6-yl]carbamic Acid 1-Methylethyl Ester *see* 1728
Thiazol-5-ylmethyl [(1*R*,4*R*)-1-Benzyl-4-[[(2*S*)-2-[[methyl-[[2-(1-methylethyl)-thiazol-4-yl]methyl]carbamoyl]amino]-4-(morpholin-4-yl)butanoyl]amino]-5-phenylpentyl]carbamate *see* 2436
4'-(2-Thiazolylsulfamoyl)succinanilic Acid *see* 9008
4'-(2-Thiazolylsulfamyl)phthalanilic Acid *see* 7489
p-2-Thiazolylsulfamylsuccinanilic Acid *see* 9008
*N*¹-2-Thiazolylsulfanilamide *see* 9074
Thiazopyr, 9463
Thiazosulfone *see* 9461
Thibenzazoline, 9464
Thibone *see* 9442
Thibutabarbital *see* 9476
Thidiazuron, 9465
Thienamycin, 9466
5-(2-Thienyl)-2,2'-bithiophene *see* 9317
Thienylic Acid *see* 9586
2-Thienylmethyl Isonicotinamide Cepha-losporin C_A *see* 1977
(15-*R*,*S*)-16-(3-Thienyloxy)-ω-tetranor-PGF_{2α} *see* 9578
Thiethylperazine, 9467
Thifensulfuron-methyl, 9468
Thifluzamide, 9469
Thiirane *see* 3857
Thilaven *see* 4924
Thimecil *see* 6201
Thimerosal, 9470
Thimet [AMVAC Chem] *see* 7443
Thiodelone [Shionogi] *see* 5925
Thioacetaldehyde, 9471

cis-Thioacetaldehyde *see* 9471
trans-Thioacetaldehyde *see* 9471
Thioacetamide, 9472
Thioacetazone *see* 9442
Thioacetic Acid, 9473
Thioacetic Acid *S*-Ester with *N*-[(4-Ami-no-2-methyl-5-pyrimidinyl)methyl]-*N*-4-hydroxy-2-mercapto-1-methyl-1-bu-tenyl)formamide Acetate *see* 53
Thioallyl Ether *see* 288
Thiobarbital, 9474
Thiobenzoic Acid *S*-Ester with *N*-[(4-Amino-2-methyl-5-pyrimidinyl)meth-yl]-*N*-(4-hydroxy-2-mercapto-1-meth-yl-1-butenyl)formamide *O*-Phosphate *see* 1043
Thiobenzyl Alcohol, 9475
2,2'-Thiobis[acetic Acid] *see* 9484
1,1'-Thiobisbenzene *see* 3371
1,1'-Thiobisbutane *see* 1593
1,1'-Thiobis[2-chloroethane] *see* 6401
2,2'-Thiobis[4-chlorophenol] *see* 4031
2,2'-Thiobis[4,6-dichlorophenol] *see* 1307
1,1'-Thiobisethane *see* 3905
2,2'-Thiobisethanol *see* 9483
1,1'-Thiobismethane *see* 3281
1,1'-[Thiobis(methylene)]bisbenzene *see* 1147
N,*N*'-[Thiobis[(methylimino)carbonyl-oxy]]bisethanimidothioic Acid Di-methyl Ester *see* 9482
1,1'-Thiobis[2-methylpropane] *see* 5197
1,1'-Thiobispropane *see* 3387
3,3'-Thiobis[propanoic Acid] *see* 9485
3,3'-Thiobis-1-propene *see* 288
Thiobutabarbital, 9476
Thiobutarit *see* 1470
Thiobutyl Alcohol *see* 1580
Thiocarbamide *see* 9521
Thiocarbanil *see* 7408
Thiocarbanilide *see* 3373
Thiocarbazil *see* 9442
Thiocarbonic Acid *O*-Ethyl Ester, *S*-Es-ter with *N*-[(4-Amino-2-methyl-5-py-rimidinyl)methyl]-*N*-(4-hydroxy-2-mercapto-1-methyl-1-butenyl)formam-ide Ethyl Carbonate (Ester) *see* 2022
Thiocolchicine, 9477
Thiocolchicoside *see* 9477
Thiocresol, 9478
Thioctacid [Viatris] *see* 9479
Thioctan [Katwijk] *see* 9479
Thioctic Acid, 9479
6,8-Thioctic Acid *see* 9479
Thiocuran [Sagitta] *see* 9050
Thiocyanate Sodium, 9480
2-Thiocyanatoacetic Acid *rel*-(1*R*,2*R*,4*R*)-1,7,7-Trimethylbicyclo[2.2.1]hept-2-yl Ester *see* 5174
Thiocyanic Acid, 9481
Thiocyanic Acid Aluminum Salt (3:1) *see* 363
Thiocyanic Acid Ammonium Salt (1:1) *see* 558
Thiocyanic Acid Barium Salt (2:1) *see* 994
Thiocyanic Acid (2-Benzothiazolylthio)-methyl Ester *see* 9223
Thiocyanic Acid Calcium Salt (2:1) *see* 1712
Thiocyanic Acid Cobalt(2+) Salt (2:1) *see* 2434
Thiocyanic Acid Copper(1+) Salt (1:1) *see* 2659

Thiocyanic Acid Iron(2+) Salt *see* 4086
Thiocyanic Acid Iron(3+) Salt *see* 4061
Thiocyanic Acid Lead(2+) Salt (2:1) *see* 5481
Thiocyanic Acid Mercury(2+) Salt (2:1) *see* 5958
Thiocyanic Acid Methyl Ester *see* 6200
Thiocyanic Acid Potassium Salt (1:1) *see* 7807
Thiocyanic Acid Sodium Salt (1:1) *see* 9480
Thiocyanic Acid Zinc Salt (2:1) *see* 10363
2-(Thiocyanomethylthio)benzothiazole *see* 9223
Thiocyl *see* 9513
Thiocymetin [Winthrop] *see* 9453
Thiodan [Bayer CropSci.] *see* 3629
Thiodemeton *see* 3408
Thioderon [Shionogi] *see* 5925
3,3′-Thiodi-L-alanine *see* 5416
Thiodicarb, 9482
2,2′-Thiodiethanol, 9483
Thiodiethylene Glycol *see* 9483
Thiodiglycol *see* 9483
Thiodiglycolic Acid, 9484
Thiodihydracrylic Acid *see* 9485
2-Thio-3,5-dimethyltetrahydro-1,3,5-thiadiazine *see* 2837
Thiodiphenylamine *see* 7364
O,*O*′-(Thiodi-4,1-phenylene)bis(*O*,*O*′-dimethylphosphorothioate) *see* 9278
O^P,*O*^P′-(Thiodi-4,1-phenylene)phosphorothioic Acid *O*^P,*O*^P,*O*^P′,*O*^P′-Tetramethyl Ester *see* 9278
Thiodiphosphoric Acid ([(HO)$_2$P(S)]$_2$O) *OP*,*OP*,*OP*′,*OP*′-Tetraethyl Ester *see* 9096
β,*β*-Thiodipropionic Acid *see* 9485
3,3′-Thiodipropionic Acid, 9485
Thiodrol [Shionogi] *see* 3680
Thioethanolamine *see* 2776
Thioethyl Alcohol *see* 3781
Thioethyl Ether *see* 3905
Thiofeed *see* 9504
Thiofide *see* 3413
Thioformamide, 9486
Thiofuran *see* 9506
Thiofurfuran *see* 9506
Thiogenal *see* 6050
1-Thio-D-glucitol *see* 9516
[1-(Thio-*κS*)-D-glucopyranosato-*κO*^{2}]-gold *see* 874
(1-Thio-*β*-D-glucopyranosato)(triethylphosphine)gold 2,3,4,6-Tetraacetate *see* 869
1-Thio-*β*-D-glucopyranose 1-[*N*-(Sulfooxy)-3-butenimidate] Potassium Salt (1:1) *see* 8683
[1-(Thio-*κS*)-*β*-D-glucopyranose-2,3,4,6-tetraacetato](triethylphosphine)gold *see* 869
5-Thio-D-glucose, 9487
Thioglycerin *see* 9488
Thioglycerol, 9488
Thioglycol *see* 5936
Thioglycolic Acid, 9489
Thioglycollic Acid *see* 9489
Thioglycollic-*β*-aminonaphthalide *see* 9497
Thioguanine, 9490
Thioguanosine, 9491
6-Thioguanosine *see* 9491
Thioimidodicarbonic Diamide ([(H$_2$N)C-(S)]$_2$NH) *see* 3414

2-Thio-4-ketothiazolidine *see* 8308
Thiokol [Thiokol], 9492
Thiokol A *see* 9492
Thiokol FA *see* 9492
Thiola [Mission] *see* 9609
Thiolacetic Acid *see* 9473
Thiolactic Acid, 9493
Thiole *see* 9506
2-Thiolhistidine, 9494
Thiolhistidine-betaine *see* 3720
Thiolin *see* 4924
2-Thiolpropionic Acid *see* 9493
Thiolutin, 9495
Thiomalic Acid, 9496
Thiomebumal Sodium *see* 9503
Thiomerin Sodium [Wyeth] *see* 5937
Thiomersalate *see* 9470
Thiomethyl Alcohol *see* 6028
Thiomorpholine *see* 9452
Thiomucase [Sanofi Winthrop] *see* 4797
Thionalide, 9497
Thionaphthene-2-carboxylic Acid, 9498
α-Thionaphthol *see* 6463
β-Thionaphthol *see* 6464
1-Thionaphthol *see* 6463
2-Thionaphthol *see* 6464
Thioneine *see* 3720
Thionembutal *see* 9503
Thionex [Makhteshim-Agan] *see* 3629
Thionic [UCB] *see* 10372
Thionicol [Mohan] *see* 9453
Thionine, 9499
5-Thio-2-nitrobenzoate Anion *see* 3606
Thiontan *see* 3119
Thionylan *see* 6031
Thionyl Bromide, 9500
Thionyl Chloride, 9501
Thionyl Difluoride *see* 9502
Thionyl Fluoride, 9502
2-Thio-4-oxo-6-propyl-1,3-pyrimidine *see* 7980
Thioparamizone [Smith & Nephew] *see* 9442
Thiopental Sodium, 9503
Thiopentone Sodium *see* 9503
Thiopeptin, 9504
Thioperazine *see* 9510
Thioperoxydicarbonic Acid ([(HO)C-(S)]$_2$S$_2$) *OC*,*OC*′-Diethyl Ester *see* 3428
Thiophanate, 9505
Thiophanate-methyl *see* 9505
Thiophan Sulfone *see* 9085
Thiophene, 9506
(2-Thiophenecarboxylato-*κO*^{2},*κS*^{1})copper *see* 2506
2-Thiophenecarboxylic Acid, 9507
3-Thiophenecarboxylic Acid *see* 9430
2,2′-(2,5-Thiophenediyl)bis[5-(1,1-dimethylethyl)benzoxazole] *see* 1012
β-Thiophenic Acid *see* 9430
Thiophenicol [Clin Comar] *see* 9453
Thiophenol, 9508
Thiophenylpyridylamino-10-carboxylic Acid Piperidinoethoxyethyl Ester *see* 7569
Thiophos *see* 7137
Thiophosphoric Anhydride *see* 7466
Thiophosphoryl Chloride *see* 7468
Thioproline *see* 9602
Thiopropanal *S*-Oxide *see* 7916
Thiopropazate, 9509
Thio-2-propene-1-sulfinic Acid *S*-Allyl Ester *see* 253
Thioproperazine, 9510

Thioproperazine Mesylate *see* 9510
Thiopropionaldehyde *S*-Oxide *see* 7916
2-Thio-6-propyl-1,3-pyrimidin-4-one *see* 7980
Thiopyrophosphoric Acid Tetraethyl Ester *see* 9096
Thioredoxin, 9511
Thioridazine, 9512
Thioridazine-2-sulfone *see* 9093
Thioridazine-2-sulfoxide *see* 5980
Thiorubber *see* 9492
Thiosalicylic Acid, 9513
Thiosarmine *see* 9072
Thioseconal *see* 9454
Thiosemicarbazide, 9514
Thioserine *see* 2778
Thiosinamine, 9515
Thiosol [Coop. Farm.] *see* 9609
1-Thiosorbitol, 9516
Thiostrepton, 9517
Thiosulfuric Acid (H$_2$S$_2$O$_3$) Ammonium Salt (1:2) *see* 559
Thiosulfuric Acid (H$_2$S$_2$O$_3$) Barium Salt (1:1) *see* 995
Thiosulfuric Acid (H$_2$S$_2$O$_3$) Magnesium Salt (1:1) *see* 5758
Thiosulfuric Acid (H$_2$S$_2$O$_3$) Potassium Salt (1:2) *see* 7808
Thiosulfuric Acid (H$_2$S$_2$O$_3$) Sodium Salt (1:2) *see* 8821
Thiotax *see* 5935
Thiotepa, 9518
Thiotepp *see* 9096
Thiotetrole *see* 9506
Thiothixene, 9519
2-Thiouracil *see* 9520
Thiourea, 9521
β-Thiovaline *see* 7197
Thioxamyl *see* 7019
Thioxanthene, 9522
9*H*-Thioxanthene *see* 9522
9*H*-Thioxanthen-9-one *see* 9523
Thioxanthone, 9523
2-Thioxo-3,5-dibenzyltetrahydro-1,3,5-thiadiazine *see* 9024
Thioxolone *see* 9612
2-Thioxo-4-thiazolidinone *see* 8308
Thiozin *see* 4924
THIP *see* 4351
Thiphenamil, 9524
Thiram, 9525
Thistle Saffron *see* 1868
Thomas Balsam *see* 943
Thomas Flour *see* 9526
Thomasin [Apogepha] *see* 3914
Thomas Phosphate, 9526
Thombolyse [Green Cross] *see* 8011
Thombran [Boehringer, Ing.] *see* 9740
Thonzylamine Hydrochloride, 9527
Thorazine [GSK] *see* 2191
Thoria *see* 9531
Thorium, 9528
Thorium Chloride, 9529
Thorium Dioxide *see* 9531
Thorium Nitrate, 9530
Thorium Oxide, 9531
Thorium Platinocyanide *see* 9532
Thorium Tetrachloride *see* 9529
Thorium Tetracyanoplatinate(II), 9532
Thorium(4+) Tetrakis(cyano-*C*)platinate(2−) (1:2) *see* 9532
Thorium X *see* 8210
Thorn Apple *see* 8950
Thoron *see* 8212
Thorotrast *see* 9531

Tilade [Aventis] *see* 6521
Tilarginine *see* 6096
Tilarin [Aventis] *see* 6521
Tilatil [Roche] *see* 9291
Tilavist [Aventis] *see* 6521
Tilcarex [Bayer] *see* 8197
Tilcotil [Roche] *see* 9291
Tildiem [Sanofi-Aventis] *see* 3224
Tiletamine, 9592
Tiliacorine, 9593
Tiliadin *see* 9195
Tilidine, 9594
Tilisolol, 9595
Tillam *see* 7167
Tillman's Reagent *see* 3078
Tilly Drops *see* 5319
Tilmapor [Novartis] *see* 1946
Tilmicosin, 9596
Tilorone, 9597
Tilt [Syngenta] *see* 7932
Tiludronate Disodium *see* 9598
Tiludronic Acid, 9598
Timabak [Théa Labs] *see* 9601
Timacar [Merck & Co.] *see* 9601
Timacor [Substipharm] *see* 9601
Timaxel [Rhône-Poulenc] *see* 6001
Timecef [Lepetit] *see* 1929
Timentin [SKB] *see* 2339
Timepidium Bromide, 9599
Timiperone, 9600
Timolol, 9601
Timonacic, 9602
Timonil [Desitin] *see* 1783
Timoped [Reckitt & Colman] *see* 9678
Timoptic [Merck & Co.] *see* 9601
Timoptol [Merck & Co.] *see* 9601
Timosina [Sclavo] *see* 9560
Timunox [Cilag] *see* 9558
Tin, 9603
Tin(II) Acetate *see* 8908
Tinactin [Schering-Plough] *see* 9678
Tinaderm [Schering-Plough] *see* 9678
Tin Ash *see* 8905
"Tin Bichloride" *see* 8901
Tin(II) Bromide *see* 8909
Tin(IV) Bromide *see* 8900
Tin Bromide (SnBr₂) *see* 8909
Tin Bronze *see* 8907
Tin(IV) Chloride *see* 8901
Tin Chloride (SnCl₂) *see* 8910
Tindamax [Presutti] *see* 9604
Tin Dibromide *see* 8909
Tin Dichloride *see* 8910
Tin Difluoride *see* 8911
Tin Diiodide *see* 8912
Tin Dioxide *see* 8905
Tin Diselenide *see* 8906
Tin Disulfide *see* 8907
Tin Ethyl Etiopurpurin Dichloride *see* 8401
Tin Etiopurpurin Dichloride *see* 8401
Tin(II) Fluoride *see* 8911
Tin(IV) Fluoride *see* 8903
Tin Fluoride (SnF₂) *see* 8911
Tinidazole, 9604
Tin(II) Iodide *see* 8912
Tin(IV) Iodide *see* 8904
Tin Iodide (SnI₂) *see* 8912
Tin Monoselenide *see* 8916
Tin Monosulfide *see* 8918
Tin Monoxide *see* 8914
Tinnitin [Sanova] *see* 1859
T-inosinic Acid *see* 5020
Tinosorb M [Ciba] *see* 1294
Tinosorb S [Ciba] *see* 1030

Tinostat *see* 3047
Tin(II) Oxalate *see* 8913
Tin(II) Oxide *see* 8914
Tin(IV) Oxide *see* 8905
Tin Oxide (SnO) *see* 8914
Tin Oxide (SnO₂) *see* 8905
Tin Phosphides, 9605
Tin Protochloride *see* 8910
Tin Protosulfide *see* 8918
Tin Protoxide *see* 8914
Tin(II) Selenide *see* 8916
Tin Selenide (SnSe) *see* 8916
Tin Selenide (SnSe₂) *see* 8906
Tinset [Janssen] *see* 7024
Tin(II) Sulfate *see* 8917
Tin(II) Sulfide *see* 8918
Tin(IV) Sulfide *see* 8907
Tin Sulfide (SnS) *see* 8918
Tin Sulfide (SnS₂) *see* 8907
Tin(2+) Tartrate *see* 8919
Tin Tetrabromide *see* 8900
Tin Tetrachloride *see* 8901
Tin Tetrafluoride *see* 8903
Tin Tetraiodide *see* 8904
Tintophen X *see* 1316
Tinuvin 360 [Ciba] *see* 1294
Tinuvin P [Ciba] *see* 3496
Tinzaparin, 9606
Tiobicina *see* 9442
Tiobutarit *see* 1470
Tioclomarol, 9607
Tioconazole, 9608
Tioctan [Fujisawa] *see* 9479
Tioctidasi [ISI] *see* 9479
Tio-Mid *see* 3791
Tiopronin, 9609
Tiorfan [Bioprojet] *see* 8207
Tioten [Mediolanum] *see* 8936
Tiotixene *see* 9519
Tiotropium Bromide, 9610
Tiovalon [Intersan] *see* 9641
Tiox [Schering-Plough] *see* 9611
Tioxidazole, 9611
Tioxolone, 9612
γ-TIP *see* 749
Tipedine *see* 9613
Tipepidine, 9613
Tipifarnib, 9614
T.I.P.P.S. *see* 5081
Tipranavir, 9615
Tiptor [Syngenta] *see* 2768
Tiquizium Bromide, 9616
Tirade [Sumitomo] *see* 4037
Tirapazamine, 9617
Tiratricol, 9618
Tirilazad, 9619
Tirilazad Mesylate *see* 9619
Tirodril [Estedi] *see* 6043
Tirofiban, 9620
Tiroidina *see* 9565
Tiron, 9621
Tiropramide, 9622
Tisercin *see* 5518
Tiserton [Janssen] *see* 8363
Tissue Factor *see* 9539
Tissue Factor Pathway Inhibitor, 9623
Tissue Plasminogen Activator, 9624
Tissue Thromboplastin *see* 9539
Tisuacryl [Centro de Biomateriales] *see* 1566
Titania *see* 9628
Titanic(IV) Acid, 9625
Titanic Hydroxide *see* 9625
Titanite *see* 9626
Titanium, 9626

Titanium Ammonium Oxalate *see* 560
Titanium Chloride (TiCl₂) *see* 9627
Titanium Chloride (TiCl₃) *see* 9637
Titanium Dichloride, 9627
Titanium Dihydride *see* 9629
Titanium Dioxide, 9628
Titanium Hydride, 9629
(T-4)-Titanium Hydroxide (Ti(OH)₄) *see* 9625
Titanium Isopropoxide *see* 9636
Titanium Isopropylate *see* 9636
Titanium Monochloride Triisopropoxide *see* 2174
Titanium Mononitride *see* 9630
Titanium Nitride, 9630
Titanium Oxide (TiO₂) *see* 9628
Titanium Oxysulfate, 9631
Titanium Potassium Oxalate *see* 7809
Titanium Sesquisulfate, 9632
Titanium Sulfate *see* 9631
Titanium Tetrabromide, 9633
Titanium Tetrachloride, 9634
Titanium Tetrafluoride, 9635
Titanium Tetrahydroxide *see* 9625
Titanium Tetraisopropoxide, 9636
Titanium Trichloride, 9637
Titanocene Dichloride, 9638
Titanous Chloride *see* 9637
Titanous Sulfate *see* 9632
Titan Yellow *see* 9462
Titanyl Ammonium Oxalate *see* 560
Titanyl Potassium Oxalate *see* 7809
Titanyl Sulfate *see* 9631
Titin, 9639
Titriplex III [Merck KGaA] *see* 3565
Titus [DuPont] *see* 8356
Tivantinib, 9640
Tixair [Madaus] *see* 83
Tixantone *see* 5650
Tixobar [Astra] *see* 990
Tixocortol, 9641
Tixogel VP [United Catalysts] *see* 1058
Tizanidine, 9642
Tizorol [Rudefsa] *see* 6043
TJN-318 *see* 5408
TLK-286 *see* 1748
Tl(OAc)₃ *see* 9406
TM-723 *see* 105
TM-906 *see* 9868
TMA *see* 9879
TMAH *see* 9371
TMAO *see* 9881
T-Maz 80 [Mazur] *see* 7703
TMC-114 *see* 2829
TMC-125 *see* 3935
TMC-278 *see* 8349
TMD, 9643
TMD-10 *see* 9344
TMEDA *see* 9277
TMG *see* 9375
TM-PGE₂ *see* 9895
TMQ *see* 9893
TMS *see* 9377
TMS-19Q *see* 8379
TMS-CF₃ *see* 105
TMSOTMS *see* 4726
TMTD *see* 9525
TMU *see* 9378
TMV *see* 9646
TMX-67 *see* 3983
TNAZ, 9644
TNB²⁻ *see* 3606
TNCA *see* 9498
TNF *see* 9900, 9992
TNF-β *see* 9992

Tomoxetine *see* 854
Tomtovok [Boehringer, Ing.] *see* 169
Tomudex [AstraZeneca] *see* 8217
Ton *see* 356
Tonamil [Ecobi] *see* 9527
Tonedron [Grimault] *see* 6020
Toness [Angelini] *see* 8012
Tonexol [Pharmaton] *see* 9333
Tonibral [Bouchara] *see* 2845
Tonin, 9705
Tonka Bean Camphor *see* 2550
Tonocalcin [Alfa] *see* 1646
Tonocard [AstraZeneca] *see* 9648
Tonocholin B *see* 81
Tonofolin [Zyma] *see* 4249
Tonofosfan [Hoechst] *see* 9672
Tonoftal [Essex] *see* 9678
Tonolift [Teisan] *see* 6788
γ-Tonoplast Intrinsic Protein *see* 749
Tonopres [Boehringer, Ing.] *see* 3191
Tonsillosan (Lösung) [Spitzner] *see* 5384
Tony Red *see* 9015
Tooler [BASF] *see* 9943
Topalgic [Nippon Shoji] *see* 9134
Topamax [Ortho] *see* 9706
Topaz *see* 336
Topazone *see* 4330
Topguard [Cheminova] *see* 4242
Topicain [Chugai] *see* 7032
Topical Thrombin *see* 9537
Topicort [HMR] *see* 2931
Topicycline [Roberts] *see* 9341
Topifug [Wolff] *see* 2929
Topik [Syngenta] *see* 2364
Topinambur *see* 5311
Topiramate, 9706
Topisolon [HMR] *see* 2931
Topline [Intervet] *see* 482
TopNotch [Dow AgroSci.] *see* 60, 3058
Topotecan, 9707
Toprec [RPR] *see* 5352
Toprek [RPR] *see* 5352
Toprol-XL [AstraZeneca] *see* 6228
Topsin [Nippon Soda] *see* 9505
Topsin-M [Nippon Soda] *see* 9505
Topstar [Rhone-Poulenc] *see* 7004
Topsym [Grünenthal] *see* 4182
Topsymin [Grünenthal] *see* 4182
Topsyn [Syntex] *see* 4182
Topsyne [Cassenne] *see* 4182
Toracizin *see* 4145
Toradiur [Roche] *see* 9711
Toradol [Roche] *see* 5353
Torasemide *see* 9711
Torbugesic [Fort Dodge] *see* 1532
Torbutrol [BMS] *see* 1532
Torcetrapib, 9708
Torch [Sanwa] *see* 10283
Tordon [Dow AgroSci.] *see* 7509
Torecan [Novartis] *see* 9467
Torem [Roche] *see* 9711
Toremifene, 9709
Torental [Sanofi-Aventis] *see* 7249
Toresten [Sankyo] *see* 9467
Torezolid *see* 9247
Torilin, 9710
Toripamide *see* 9909
Torisel [Wyeth] *see* 9285
Tornalate [Sterling Winthrop] *see* 1308
Torpex [Boehringer, Ing.] *see* 210
Torque [BASF] *see* 3997
Torrefaction Dextrin *see* 2955
Torsemide, 9711
Torularhodin, 9712
Torutilin *see* 10221

Torvast [Pfizer] *see* 855
Tosic Acid *see* 9692
Tositumomab, 9713
TosMIC *see* 9715
Tosmilen [Chemie Linz] *see* 2889
Tosnone [Nippon Shoji] *see* 1795
Tostram *see* 5293
Tosufloxacin, 9714
Tosufloxacin Tosylate *see* 9714
Tosuxacin [Dainabot] *see* 9714
Tosylchloramide Sodium *see* 2075
Tosyl Chloride *see* 9693
Tosylhydrazide *see* 9694
Tosylhydrazine *see* 9694
Tosylmethyl Isocyanide, 9715
D-3-*endo*-*p*-Tosylureidoborneol *see* 4473
Totabin *see* 6565
Totacef [BMS] *see* 1918
Totacillin N [GSK] *see* 583
Totalip [Guidotti] *see* 855
Totalon [Mallinckrodt] *see* 5511
Totapen [BMS] *see* 583
Totect [TopoTarget] *see* 8241
Totifen [Chiesi] *see* 5354
TOTO *see* 9460
Totocortin [Winzer] *see* 2945
Touchdown [Syngenta] *see* 4547
Touch Wood *see* 177
Tournesol *see* 5603
Tovalt [Biovail] *see* 10391
Tovasanib *see* 9741
Tovene [Kali-Chemie] *see* 3325
Toviaz [Pfizer] *see* 4095
Towk [Tanabe] *see* 9727
Toxakil *see* 9716
Toxalbumin *see* 10
Toxaphene, 9716
Toxicarol *see* 9293
Toxichlor *see* 2083
Toxiferine I, 9717
C-Toxiferine I *see* 9717
C-Toxiferine II *see* 9717
Toxiferine V *see* 9717
Toxiferine XI *see* 9717
Toxifren *see* 4121
Toxilic Acid *see* 5767
Toxilic Anhydride *see* 5768
Toxinal [Pharma-Selz] *see* 7075
Toxin HIL13-PE38QQR (Plasmid
    PhuIL13-Tx) *see* 2311
Toxin ML-I (Mistletoe Lectin I) (*Viscum
    album*) *see* 6295
Toxoflavin, 9718
Toxogonin [Merck KGaA] *see* 6827
Toxohormone, 9719
Toxopyrimidine, 9720
Toyocamycin, 9721
Toyolysom-DS [Maruko] *see* 5701
Toyomycin *see* 2242
Tozalinone *see* 9533
TP *see* 6551, 9559
TP-1 *see* 9562
TP-5 *see* 9558
TP-21 *see* 9512
TP-151 *see* 959
TP-155 *see* 959
TP508 *see* 8431
TPA *see* 7444, 9305, 9624
t-PA *see* 9624
TPAP *see* 9386
TPCM *see* 9944
TPD *see* 7993
TPMPA, 9722
TPN *see* 6432, 9318
TPN-12 *see* 9093

TPN$_Q$ *see* 9318
TPNH *see* 6432
TPO *see* 9540
TPP *see* 9448
TPP$^+$ *see* 9384
TPPDB *see* 9917
TPS-23 *see* 5980
TPTZ *see* 9918, 9926
TR-495 *see* 6032
TR-700 *see* 9247
TR-701 *see* 9247
TR-701 FA *see* 9247
TR-4698 *see* 8359
Trabectedin *see* 3550
Trabedersen, 9723
γ-Trace *see* 2775
Tracer [Dow AgroSci.] *see* 8877
Tracilon [Savage] *see* 9757
Tracker [DuPont] *see* 9725
Tracleer [Actelion] *see* 1355
Tracrium [GSK] *see* 859
Tractocile [Ferring] *see* 856
Tractur [Damor] *see* 7572
Tradaptive [Merck & Co.] *see* 5426
Tradjenta [Boehringer, Ing.] *see* 5549
Tradon [Lilly] *see* 7188
Tradonal [Viatris] *see* 9726
Trafuril [Novartis] *see* 9550
Tragacanth *see* 4616
Tragacanthin *see* 4616
Tral [Abbott] *see* 4742
Tralate [DuPont] *see* 9725
Tralin *see* 4742
Tralkoxydim, 9724
Tralomethrin, 9725
Tralox [Roussel-Uclaf AH] *see* 9725
Tramacin [J & J] *see* 9758
Tramadol, 9726
Tramal [Grünenthal] *see* 9726
Tramat [Bayer CropSci.] *see* 3796
Tramazoline, 9727
Tramfloc 904 [Tramfloc, Inc.] *see* 8825
Tramiprosate, 9728
Tramisol [Am. Cyanamid] *see* 5511
Trancalgyl *see* 3786
Trancolon [Fujisawa] *see* 5915
Trancopal [Winthrop] *see* 2106
Trancot [Truxton] *see* 5929
Trancote [Sawai] *see* 2106
Trandate [Allen & Hanburys] *see* 5377
Trandolapril, 9729
Trandolaprilat *see* 9729
Tranex [Malesci] *see* 9730
Tranexamic Acid, 9730
Tranexamic Acid *p*-(2-Carboxyethyl)-
    phenyl Ester *see* 2023
Trangorex [Sanofi-Aventis] *see* 478
Tranilast, 9731
Trankimazin [Pharmacia] *see* 308
Tran-Q [Pfizer] *see* 4889
Tranquilan [Bonin] *see* 5929
Tranquilax [Hokuriku] *see* 5857
Tranquinal [Bago] *see* 308
Tranquit [Promonta] *see* 7026
Tranquizine *see* 4889
Transacalm [Norgine] *see* 9868
Transamin [Daiichi] *see* 9730
Transanate [Teikoku Zoki] *see* 2106
Transbilix [Guerbet] *see* 5065
Transbronchin [Viatris] *see* 1802
Trans-citral *see* 2321
Transcobalamins *see* 10212
Transcop [Recordati] *see* 8543
Transcortin *see* 4824
Transcycline [HMR] *see* 8381

Transderm Scop [Novartis] *see* 8543
Transderm-Nitro [Novartis] *see* 6694
Transdihydrolisuride *see* 9307
Transducin *see* 4566
Transene [Clin-Comar-Byla] *see* 2392
Transferrins, 9732
Transfer RNA *see* 8327
Transforming Growth Factor-β, 9733
Transiderm-Nitro [Novartis] *see* 6694
Transline [Dow AgroSci.] *see* 2389
Translon [Daiichi] *see* 6524
Transthyretin *see* 9570
Transvaalin *see* 8536
Trantoin [McKesson] *see* 6686
Tranvet [Syntex] *see* 7943
Tranxene [Abbott] *see* 2392
Tranxilène [Clin-Comar-Byla] *see* 2392
Tranxilium [Mack, Illert.] *see* 2392
Tranylcypromine, 9734
TRAP-508 *see* 8431
Trapanal [Byk Gulden] *see* 9503
Trapex *see* 6164
Trapidil, 9735
Trapymin *see* 9735
Trasacor [Novartis] *see* 7051
Trasentine [Ciba-Geigy] *see* 149
Trasicor [Novartis] *see* 7051
Traslan [Gramon] *see* 2102
Trastuzumab, 9736
Trastuzumab Emtansine *see* 9736
Trastuzumab-MCC-DM1 *see* 9736
Trasulphane *see* 4924
Trasylol [Bayer] *see* 746
Traubensäure (German) *see* 9204
Traumacut [Brenner-Efeka] *see* 6052
Traumanase [Sanofi-Aventis] *see* 1395
Traumasept [Wolff] *see* 7815
Traumatic Acid, 9737
Traumatociclina [Biomed. Foscama] *see* 5849
Traumon Gel [Troponwerke] *see* 3921
Trausabun [Promonta] *see* 5891
Travamin [Baxter] *see* 7996
Travatan [Alcon] *see* 9738
Travel-Gum [Meda] *see* 3230
Travert [Baxter] *see* 5050
Travogen [Schering AG] *see* 5205
Travogyn [Keymer] *see* 5205
Travoprost, 9738
Traxam [Lederle] *see* 3985
Traxanox, 9739
Trazinin [Tobishi] *see* 746
Trazodone, 9740
Trazolan [Pfizer] *see* 9740
Trazyl [Angelini] *see* 4915
Treanda [Cephalon] *see* 1036
Trebananib, 9741
Trebiace [Mitsubishi] *see* 4973
Trecalmo [Bayer] *see* 2400
Trédémine [RPR] *see* 6604
Tree of Life *see* 9543
Treevix [BASF] *see* 8451
Treflan [Dow AgroSci.] *see* 9854
Trehalose, 9742
α,α-Trehalose *see* 9742
Trehalose 6,6′-Dimycolate *see* 2513
Tre-Hold [Amvac] *see* 6456
Trelibet [EGYT] *see* 7504
Trelstar [Watson] *see* 9924
Tremaril [Wander] *see* 6051
Tremarit [Wander] *see* 6051
Tremblex [Janssen] *see* 2947
Tremetone, 9743
Tremin [Schering-Plough] *see* 9864
Tremolite *see* 815

Tremonil [Wander] *see* 6051
Tremoquil *see* 6051
Tremorine, 9744
Trenbolone, 9745
Trenbolone Hexahydrobenzylcarbonate *see* 9745
Trengestone, 9746
Trenimon *see* 9764
Trentadil [Evans] *see* 948
Trental [Sanofi-Aventis] *see* 7249
Trepibutone, 9747
Trepidan [Max Farma] *see* 7836
Treprostinil, 9748
Trerief [Dainippon] *see* 10392
Tresanil [ISF] *see* 9937
Trescatyl [M & B] *see* 3791
Trescillin *see* 7931
Tresortil *see* 6052
Trest [Dorsey] *see* 6051
Tretamine *see* 9836
Trethylene *see* 9804
Tretinoin *see* 8288
Tretinoin Tocoferil *see* 9657
Tretoquinol, 9749
Trevintix [Théraplix] *see* 8000
Trevira [Hoechst] *see* 7689
TRF *see* 5037, 5041, 9750
TRH, 9750
Tri-4 [BASF] *see* 9854
Triac [Aché] *see* 9618
Triacana [Marcofina] *see* 9618
Triacetin, 9751
Triacetone Triperoxide, 9752
Triacetoxyperiodinane *see* 2934
1,2,3-Triacetoxypropane *see* 9751
1,3,4-Tri-*O*-acetyl-*N*-acetyl-6-desoxy-β-D-glucosamine *see* 2932
Triacetyl Glycerine *see* 9751
Triacetyloleandomycin *see* 9949
2-(2,3,5-Tri-*O*-acetyl-β-D-ribofuranosyl)-*as*-triazine-3,5(2*H*,4*H*)-dione *see* 897
Triacetylsphingosine *see* 8874
Triaconazole *see* 9303
1-Triacontanol *see* 9753
Triact [Misemer] *see* 5749
Triacylglycerol Lipase *see* 5566
Triadenyl [Henning] *see* 143
Triadimefon, 9754
Triadimenol, 9755
Trialkylaluminum *see* 322
Triallate, 9756
Triam [Lichtenstein] *see* 9758
Triamcinolone, 9757
Triamcinolone Acetonide, 9758
Triamcinolone 16α,17-Acetonide *see* 9758
Triamcinolone Acetonide *tert*-Butyl Acetate *see* 9759
Triamcinolone Hexacetonide, 9759
Triamelin *see* 9836
2,4,7-Triamino-6-phenylpteridine *see* 9760
2,4,6-Triamino-*s*-triazine *see* 5880
Triammonium Dodecatungstophophate-(3−) *see* 561
Triammonium Hexatriacontaoxo[phosphato(3−)]dodecatungstate(3−) *see* 561
Triammonium Pentafluorodioxouranate *see* 564
Triamterene, 9760
2,4,6-Trianilino-*p*-(carbo-2′-ethylhexyl-1′-oxy)-1,3,5-triazine *see* 3864
Tri-*p*-anisylchloroethylene *see* 2178
Triaquochromium Trifluoride *see* 2227

Triasporin [Italfarmaco] *see* 5292
Triasulfuron, 9761
Triatec [Aventis] *see* 8221
Triatomic Oxygen *see* 7080
Triatop [Janssen-Cilag] *see* 5349
s-Triazaborane *see* 1337
Triazamate, 9762
5,8,14-Triazatetracyclo[10.3.1.0²,¹¹.0⁴,⁹]-hexadeca-2(11)-3,5,7,9-pentaene *see* 10125
3,6,9-Triaza-3,6,9-tris(carboxymethyl)-4-(4-ethoxybenzyl)-undecanedioic Acid Gadolinium Complex *see* 4360
2,2′-(1-Triazene-1,3-diyldi-4,1-phenylene)bis[6-methyl-7-benzothiazolesulfonic Acid] Sodium Salt (1:2) *see* 9462
1,3,5-Triazido-2,4,6-trinitrobenzene *see* 9903
s-Triazine, 9763
1,3,5-Triazine *see* 9763
s-Triazine-2,4-dione-6-carboxylic Acid *see* 7046
1,3,5-Triazine-2,4,6-triamine *see* 5880
sym-Triazinetriol *see* 2690
1,3,5-Triazine-2,4,6(1*H*,3*H*,5*H*)-trione *see* 2690
4,4′,4″-(1,3,5-Triazine-2,4,6-triyltriimino)trisbenzoic Acid 1,1′,1″-Tris(2-ethylhexyl) Ester *see* 3864
1,1′,1″,1‴,1⁗,1⁗′-(1,3,5-Triazine-2,4,6-triyltrinitrilo)hexakismethanol *see* 4729
(*s*-Triazine-2,4,6-triyltrinitrilo)hexamethanol *see* 4729
Triaziquone, 9764
2,3,5-Tri-1-aziridinyl-2,5-cyclohexadiene-1,4-dione *see* 9764
Triazobenzene *see* 7385
Triazoic Acid *see* 4812
Triazolam, 9765
1*H*-1,2,4-Triazol-5-amine *see* 485
1*H*-1,2,4-Triazole, 9766
1*H*-1,2,4-Triazole-3,5(2*H*,4*H*)-dione *see* 10051
1,2,4-Triazolidine-3,5-dione *see* 10051
1,2,4-Triazol-1-yl-isothioacetic Acid 2′,4′-Dichloroanilide *S-p*-Chlorobenzyl Ether *see* 4950
4,4′-(1*H*-1,2,4-Triazol-1-ylmethylene)bis-benzonitrile *see* 5502
2β-[(1,2,3-Triazol-1-yl)methyl]-2α-methylpenam-3α-carboxylic Acid 1,1-Dioxide *see* 9220
2,2′-[5-(1*H*-1,2,4-Triazol-1-ylmethyl)-1,3-phenylene]di(2-methylpropionitrile) *see* 619
Triazophos, 9767
Triazoxide, 9768
Triazure [Calbiochem] *see* 897
Tri-Ban *see* 7555
Tribavirin *see* 8323
Tribenoside, 9769
Tribenzobicyclo[2.2.2]octatriene *see* 9925
Triborine Triamine *see* 1337
Triboron Nitride *see* 1337
Tribrissen [Schering] *see* 9035
2,2,2-Tribromoacetaldehyde *see* 1390
Tribromoacetaldehyde Hydrate *see* 1391
Tribromoacetic Acid, 9770
2,2,2-Tribromoacetic Acid *see* 9770
2,4,6-Tribromoaniline, 9771
2,4,6-Tribromoanisole, 9772
2,4,6-Tribromobenzenamine *see* 9771
Tribromobismuthine *see* 1262
Tribromoborane *see* 1349

Tribromo-*tert*-butyl Alcohol, 9773
2,4,6-Tribromo-*m*-cresol, 9774
2,2,2-Tribromo-1,1-ethanediol *see* 1391
Tribromohydrin *see* 9776
2,4,6-Tribromo-3-hydroxytoluene *see* 9774
Tribromoindium *see* 4997
Tribromomethane *see* 1430
1,3,5-Tribromo-2-methoxybenzene *see* 9772
2,4,6-Tribromo-3-methylphenol *see* 9774
1,1,1-Tribromo-2-methyl-2-propanol *see* 9773
Tribromomonosilane *see* 9777
2,4,6-Tribromophenol, 9775
Tribromophenolbismuth *see* 1291
2,4,6-Tribromophenol Bismuth(3+) Salt (3:1) *see* 1291
sym-Tribromopropane *see* 9776
1,2,3-Tribromopropane, 9776
3,4′,5-Tribromosalicylanilide *see* 9778
Tribromosilane, 9777
Tribromsalan, 9778
Tribufos, 9779
Tribunil [Bayer] *see* 6011
Triburon [Roche] *see* 9818
Tributoxyphosphine Oxide *see* 9781
Tri-*n*-butylallyltin *see* 289
Tributylamine, 9780
N,N,N-Tributyl-1-butanaminium Fluoride (1:1) *see* 9332
Tributylchlorostannane *see* 9782
Tributylchlorotin *see* 9782
Tributyl Citrate *see* 1565
Tributylmethoxystannane *see* 9784
Tributylmethoxytin *see* 9784
Tributyl Phosphate, 9781
Tributyl Phosphate *see* 9781
S,S,S-Tributyl Phosphorotrithioate *see* 9779
Tributyl-2-propen-1-ylstannane *see* 289
Tributylstannane *see* 9783
Tributylstannyl Chloride *see* 9782
Tributyltin Chloride, 9782
Tributyltin Hydride, 9783
Tri-*n*-butyltin Hydride *see* 9783
Tributyltin Methoxide, 9784
Tributyrin, 9785
Tributyrylglycerol *see* 9785
Tricaine, 9786
Tricaine Methanesulfonate *see* 9786
Tricalcium Arsenate *see* 1652
Tricalcium Orthophosphate *see* 1696
Tricalcium Phosphate *see* 1696
Tricandil [SPA] *see* 5914
Tricangine [SK & F] *see* 5914
Tricarballylic Acid, 9787
Tricarbon Dioxide *see* 1815
Tricarbonic Acid *C,C′*-Bis(1,1-dimethylethyl)ester *see* 3049
Tricarbonyl[(1,2,3,4,5-η)-1-methyl-2,4-cyclopentadien-1-yl]manganese *see* 6309
Tricarbonyl(η⁵-methylcyclopentadienyl)-manganese(I) *see* 6309
1,2,4-Tricarboxybenzene *see* 9871
Tri(carboxymethyl)amine *see* 6665
Trichloren *see* 9804
Trichlorfon, 9788
Trichlorine Nitride *see* 6689
Trichlormethiazide, 9789
Trichlormethine *see* 9804
Trichloroacetaldehyde, 9791
2,2,2-Trichloroacetaldehyde *see* 9791
Trichloroacetaldehyde Monoethylacetal *see* 2069

Trichloroacetaldehyde Monohydrate *see* 2071
Trichloroacetic Acid, 9792
2,2,2-Trichloroacetic Acid *see* 9792
Trichloroacetonitrile, 9793
2,2,2-Trichloroacetonitrile *see* 9793
S-2,3,3-Trichloroallyl Diisopropylthiocarbamate *see* 9756
2,4,6-Trichloroanisole, 9794
sym-Trichlorobenzene *see* 9797
unsym-Trichlorobenzene *see* 9796
vic-Trichlorobenzene *see* 9795
1,2,3-Trichlorobenzene, 9795
1,2,4-Trichlorobenzene, 9796
1,3,5-Trichlorobenzene, 9797
N,2,6-Trichlorobenzoquinone Imine *see* 4456
2,4,6-Trichlorobenzoyl Chloride, 9798
1,1,1-Trichloro-2,2-bis(*p*-chlorophenyl)-ethane *see* 2843
1,1,1-Trichloro-2,2-bis(*p*-fluorophenyl)-ethane *see* 2960
1,1,1-Trichloro-2,2-bis(*p*-methoxyphenyl)ethane *see* 6061
Trichlorobismuthine *see* 1264
Trichloroborane *see* 1350
2,2,3-Trichlorobutanal *see* 9799
β,β,β-Trichloro-*tert*-butyl Alcohol *see* 2129
Trichlorobutyltin *see* 1594
α,α,β-Trichloro-*n*-butyraldehyde, 9799
3,4,4′-Trichlorocarbanilide *see* 9819
1,2,4-Trichloro-5-[(4-chlorophenyl)sulfonyl]benzene *see* 9343
2,4,6-Trichloro-*m*-cresol, 9800
2,3,5-Trichloro-*N*-(3,5-dichloro-2-hydroxyphenyl)-6-hydroxybenzamide *see* 7057
2,2,2-Trichloro-1-(dimethoxyphosphinoyl)ethyl Butyrate *see* 9788
Trichloroethanal *see* 9791
1,1,1-Trichloroethane, 9801
1,1,2-Trichloroethane, 9802
2,2,2-Trichloro-1,1-ethanediol *see* 2071
2,2,2-Trichloroethanol, 9803
2,2,2-Trichloroethanol Dihydrogen Phosphate *see* 9820
1,1,2-Trichloroethene *see* 9804
2,2,2-Trichloro-1-ethoxyethanol *see* 2069
Trichloroethyl Alcohol *see* 9803
Trichloroethylene, 9804
2,2,2-Trichloroethyl *β*-D-Glucopyranosiduronic Acid *see* 10069
2,2,2-Trichloroethyl *β*-D-Glucosiduronic Acid *see* 10069
β,β,β-Trichloroethyl-D-glucuronide *see* 10069
1,1′-(2,2,2-Trichloroethylidene)bis[4-chlorobenzene] *see* 2843
1,1′-(2,2,2-Trichloroethylidene)bis[4-fluorobenzene] *see* 2960
1,1′-(2,2,2-Trichloroethylidene)bis[4-methoxybenzene] *see* 6061
1,2-*O*-[(1*R*)-2,2,2-Trichloroethylidene]-*α*-D-glucofuranose *see* 2072
Trichloroethyl Phosphate *see* 9820
Trichlorofluoromethane, 9805
1′,4,6′-Trichlorogalactosucrose *see* 9011
Trichlorogallium *see* 4378
1,3,5-Trichloro-2-hydroxybenzene *see* 9809
2,4,4′-Trichloro-2′-hydroxydiphenyl Ether *see* 9822
N-(2,2,2-Trichloro-1-hydroxyethyl)formamide *see* 2070

P-(2,2,2-Trichloro-1-hydroxyethyl)phosphonic Acid Dimethyl Ester *see* 9788
3,3,3-Trichloro-2-hydroxypropanenitrile *see* 2134
3,3,3-Trichloro-2-hydroxypropionitrile *see* 2134
2,4,6-Trichloro-3-hydroxytoluene *see* 9800
Trichloroiminocyanuric Acid *see* 9806
Trichloroindium *see* 4998
Trichloroiodine *see* 5063
Trichloroiridium *see* 5132
Trichloroisocyanuric Acid, 9806
Trichloroisopropanol *see* 9810
1,1,1-Trichloroisopropyl Alcohol *see* 9810
3,3,3-Trichlorolactonitrile *see* 2134
Trichloromethane *see* 2142
1,1,1-Trichloromethanol 1,1′-Carbonate *see* 9921
Trichloromethiazide *see* 9789
1,3,5-Trichloro-2-methoxybenzene *see* 9794
(Trichloromethyl)benzene *see* 1112
Trichloromethyl Carbonate *see* 9921
Trichloromethylcarbonochloridate *see* 3374
Trichloromethylchloroformate *see* 3374
Trichloromethyl Fluoride *see* 9805
3-Trichloromethylhydrochlorothiazide *see* 9241
N-Trichloromethylmercapto-4-cyclohexene-1,2-dicarboximide *see* 1774
N-(Trichloromethylmercapto)phthalimide *see* 4251
N-(Trichloromethylmercapto)-Δ⁴-tetrahydrophthalimide *see* 1774
Trichloromethylmethane *see* 9801
Trichloromethylnitrile *see* 9793
2,4,6-Trichloro-3-methylphenol *see* 9800
1,1,1-Trichloro-2-methyl-2-propanol *see* 2129
N-(Trichloromethylthio)-4-cyclohexene-1,2-dicarboximide *see* 1774
2-[(Trichloromethyl)thio]-1*H*-isoindole-1,3(2*H*)-dione *see* 4251
N-(Trichloromethylthio)phthalimide *see* 4251
N-Trichloromethylthio-3a,4,7,7a-tetrahydrophthalimide *see* 1774
Trichloromonofluoromethane *see* 9805
Trichloromonosilane *see* 9811
Trichloronitromethane *see* 2159
2,4,5-Trichloro-6-nitrophenol *see* 9807
3,4,6-Trichloro-2-nitrophenol, 9807
(*T*-4)-Trichlorooxovanadium *see* 10114
2,4,5-Trichlorophenol, 9808
2,4,6-Trichlorophenol, 9809
2-(2,4,5-Trichlorophenoxy)acetic Acid *see* 9149
1,1,1-Trichloro-2-propanol, 9810
2,3,3-Trichloro-2-propene-1-thiol Diisopropylcarbamate *see* 9756
S-(2,3,3-Trichloro-2-propenyl) Bis(1-methylethyl)carbamothioate *see* 9756
2-[(3,5,6-Trichloro-2-pyridinyl)oxy]acetic Acid *see* 9821
N,2,6-Trichloroquinoneimine *see* 4456
N,2,6-Trichloro-*p*-quinoneimine *see* 4456
Trichlorosilane, 9811
Trichlorostibine *see* 696
Trichlorotitanium *see* 9637
α,α,α-Trichlorotoluene *see* 1112
ω,ω,ω-Trichlorotoluene *see* 1112
Trichloro-*s*-triazine *see* 2691

1-Trifluoromethyl-1,3-dihydro-3,3-di-
methyl-1,2-benziodoxole *see* 9663
5-Trifluoromethyl-*N*-[3(*R*)-[1-[5,6-dihy-
dro-4-hydroxy-2-oxo-6(*R*)-(2-phenet-
yl)-6(*R*)-*n*-propyl-2*H*-pyran-3-yl]pro-
pyl]phenyl]-2-pyridinesulfonamide *see*
9615
6-Trifluoromethyl-3,4-dihydro-7-sulfam-
oyl-2*H*-1,2,4-benzothiadiazine 1,1-Di-
oxide *see* 4826
2-Trifluoromethyl-10-(γ-dimethylaminopropyl)phenothiazine *see* 9853
3′-Trifluoromethyldiphenylamine-2-car-
boxylic Acid *see* 4163
Trifluoromethylhydrothiazide *see* 4826
2-Trifluoromethyl-9-[3-[4-(β-hydroxyethyl)-1-piperazinyl]propylidene]thio-
xanthene *see* 4219
2-(Trifluoromethyl)-10-[3-[1-(β-hydroxyethyl)-4-piperazinyl]propyl]phenothia-
zine *see* 4220
Trifluoromethyl Iodide *see* 9849
o-Trifluoromethyl-*m*′-isopropoxybenzoic
Anilide *see* 4241
α,α,α-Trifluoro-5-methyl-4-isoxazolecar-
boxy-*p*-toluidide *see* 5486
2-Trifluoromethyl-10-[3′-(1-methyl-4-
piperazinyl)propyl]phenothiazine *see*
9846
3-Trifluoromethyl-4-nitrophenol *see*
9401
1-(3′-Trifluoromethyl-4′-nitrophenyl)-
4,4-dimethylimidazoline-2,5-dione *see*
6629
α,α,α-Trifluoro-2-methyl-4′-nitro-*m*-
propionotoluidide *see* 4238
4-Trifluoromethyl-*N*-(3,3a,4,4a,5,5a,6,6a-
octahydro-1,3-dioxo-4,6-ethenocyclo-
prop[*f*]isoindol-2(1*H*)-yl)benzamide
see 9244
4-[3-[2-(Trifluoromethyl)-10*H*-pheno-
thiazin-10-yl]propyl]-1-piperazineeth-
anol *see* 4220
(+)-16-[3-(Trifluoromethyl)phenoxy]-17,-
18,19,20-tetranorprostaglandin F$_{2α}$
Isopropyl Ester *see* 9738
2-[[3-(Trifluoromethyl)phenyl]amino]-
benzoic Acid *see* 4163
2-[[3-(Trifluoromethyl)phenyl]amino]-
benzoic Acid 2-(2-Hydroxyethoxy)-
ethyl Ester *see* 3921
2-[[3-(Trifluoromethyl)phenyl]amino]-3-
pyridinecarboxylic Acid *see* 6615
2-[[3-(Trifluoromethyl)phenyl]amino]-3-
pyridinecarboxylic Acid 1,3-Dihydro-
3-oxo-1-isobenzofuranyl Ester *see*
9176
1-(*m*-Trifluoromethylphenyl)-2-(β-benzoyloxyethyl)aminopropane *see* 1041
N-(3-Trifluoromethylphenyl)-*N*′,*N*′-di-
methylurea *see* 4185
1-(3-Trifluoromethylphenyl)-4-methyl-
amino-5-chloropyridazone *see* 6790
(*R*)-*N*-(3-(3-(Trifluoromethyl)phenyl)pro-
pyl)-1-(1-napthyl)ethylamine *see* 2285
2-[4-[[5-(Trifluoromethyl)-2-pyridinyl]-
oxy]phenoxy]propanoic Acid Butyl
Ester *see* 4147
2-[[8-(Trifluoromethyl)-4-quinolinyl]ami-
no]benzoic Acid 2,3-Dihydroxypropyl
Ester *see* 4135
N-[8-(Trifluoromethyl)-4-quinolyl]an-
thranilic Acid 2,3-Dihydroxypropyl
Ester *see* 4135
6-Trifluoromethyl-7-sulfamoyl-4*H*-1,2,4-
benzothiadiazine 1,1-Dioxide *see* 4170

6-Trifluoromethyl-7-sulfamyl-1,2,4-ben-
zothiadiazine 1,1-Dioxide *see* 4170
Trifluoromethylthiazide *see* 4170
4-[3-[2-(Trifluoromethyl)-9*H*-thioxan-
then-9-ylidene]propyl]-1-piperazine-
ethanol *see* 4219
Trifluoromethyltrimethylsilane, 9851
α,α,α-Trifluoro-4-nitro-*m*-cresol *see*
9401
(*T*-4)-Trifluoro[1,1′-oxybis[ethane]]bo-
ron *see* 1352
1,1,1-Trifluoro-*N*-phenyl-*N*-[(trifluoromethyl)sulfonyl]methanesulfonamide
see 7427
1-[2-(3,3,3-Trifluoropropyl)phenylsulfon-
yl]-3-(4-methoxy-6-methyl-1,3,5-tria-
zin-2-yl)urea *see* 7992
Trifluorostibine *see* 697
α,α,α-Trifluorothymidine *see* 9855
α,α,α-Trifluorotoluene *see* 1113
2-(α,α,α-Trifluoro-*m*-toluidino)nicotinic
Acid *see* 6615
N-(α,α,α-Trifluoro-*m*-tolyl)anthranilic
Acid *see* 4163
N-(α,α,α-Trifluoro-*m*-tolyl)anthranilic
Acid 2-(2-Hydroxyethoxy)ethyl Ester
see 3921
Trifluoro-1,3,5-triazine *see* 2692
2,4,6-Trifluoro-1,3,5-triazine *see* 2692
2-(α³,α³,α³-Trifluoro-2,3-xylidino)nico-
tinic Acid *see* 4178
Trifluperidol, 9852
Triflupromazine, 9853
Trifluralin, 9854
Triflurex [Makhteshim-Agan] *see* 9854
Trifluridine, 9855
Trifluron *see* 9844
Triflusal, 9856
Triflusulfuron-methyl, 9857
Trifoliin *see* 5267
Trifolitin *see* 5322
Trifolium, 9858
Triforine, 9859
Triformol *see* 7129
Triftazin *see* 9846
Trigard [Syngenta] *see* 2772
Trigonelline *see* 9863
Trigger [Polifarma] *see* 8228
Triglobe [AstraZeneca] *see* 9035
Triglycidyl Isocyanurate, 9860
Triglycine *see* 6665
Triglycine Sulfate *see* 4527
Triglycol Dimercaptan *see* 3329
Triglycollamic Acid *see* 6665
Triglycyl-lysine-vasopressin *see* 9310
Triglyme, 9861
Trigonellamide Chloride, 9862
Trigonelline, 9863
Trigonyl [Hoyer] *see* 9050
Trigot [BMS] *see* 3710
3,3′,3″-Triheptyl-4,4′,4″-trimethyl-7-(2″-
methylthiazolyl-2,2′-trimethine)thia-
zolcyanine 3,3″-Diiodide *see* 7645
Trihexyphenidyl Hydrochloride, 9864
Trihistan *see* 2082
(5*Z*,7*E*,22*E*,24*E*)-24a,26a,27a-Trihomo-
9,10-secocholesta-5,7,10(19),22,24-pen-
taene-1α,3β,25-triol *see* 8595
1,1,5-Trihydroperfluoropentanol *see*
6835
(*T*-4)-Trihydro(tetrahydrofuran)boron
see 1336
(*T*-4)-Trihydro[thiobis[methane]]boron
see 1336
2′,3′,4′-Trihydroxyacetophenone *see*
4372

11β,17α,21-Trihydroxyallopregnane-
3,20-dione *see* 4803
1,2,3-Trihydroxy-9,10-anthracenedione
see 675
1,2,4-Trihydroxy-9,10-anthracenedione
see 8056
1,2,3-Trihydroxyanthraquinone *see* 675
1,2,4-Trihydroxyanthraquinone *see* 8056
3′,4′,6-Trihydroxyaurone *see* 9102
3′,4′,6-Trihydroxybenzalcoumaranone
see 9102
1,2,3-Trihydroxybenzene *see* 8112
1,3,5-Trihydroxybenzene *see* 7439
3,4,5-Trihydroxy-1,2-benzenedicarboxal-
dehyde *see* 4252
3,4,5-Trihydroxybenzoic Acid *see* 4375
3,4,5-Trihydroxybenzoic Acid, (2*R*,3*R*)-
3,4-Dihydro-5,7-dihydroxy-2-(3,4,5-
trihydroxyphenyl)-2*H*-1-benzopyran-
3-yl Ester *see* 3573
3,4,5-Trihydroxybenzoic Acid Propyl
Ester *see* 7971
2-*C*-[[(3,4,5-Trihydroxybenzoyl)oxy]-
methyl]-D-ribose 5-(3,4,5-Trihydroxy-
benzoate) *see* 4640
[4a*S*-(4aα,5β,6α)]-3,3′,5-Trihydroxy-
[1,1′-biphenyl]-2-carboxylic Acid 2-
Ester with 5-Ethenyl-6-(β-D-glucopyranosyloxy)-4,4a,5,6-tetrahydro-
1*H*,3*H*-pyrano[3,4-*c*]pyran-1-one *see*
371
(6*R*)-3,5,6-Trihydroxy-4,6-bis(3-methyl-
2-buten-1-yl)-2-(3-methyl-1-oxobutyl)-
2,4-cyclohexadien-1-one *see* 4793
(3β,5β,11α)-3,11,14-Trihydroxybufa-
20,22-dienolide *see* 4388
3β,14,16β-Trihydroxy-5β-bufa-20,22-
dienolide 16-Acetate *see* 1482
(2*R*,3*R*)-2,3,4-Trihydroxybutanal *see*
3746
(2*R*,3*S*)-2,3,4-Trihydroxybutanal *see*
9536
(2*S*,3*R*)-2,3,4-Trihydroxybutanal *see*
9535
(2*S*,3*S*)-2,3,4-Trihydroxybutanal *see*
3747
(3*S*)-1,3,4-Trihydroxy-2-butanone *see*
3750
3β,22,23-Trihydroxy-24-(carboxymethylene)cholest-5-en-7-one 23,24-Lac-
tone *see* 673
(3β,5β)-3,5,14-Trihydroxycard-20(22)-
enolide *see* 7291
(3β,5β,12β)-3,12,14-Trihydroxycard-20-
(22)-enolide *see* 3185
(3β,5β,16β)-3,14,16-Trihydroxycard-20-
(22)-enolide *see* 4468
3β,12β,14-Trihydroxy-5β-card-20(22)-
enolide-3-(4‴-*O*-methyltridigitoxoside)
see 3186
9α,11α,15-Trihydroxy-16-*m*-chlorophen-
oxy-13-thia-17,18,19,20-tetranor-5-
prostenoic Acid *see* 5670
(3α,5β,7α,12α)-3,7,12-Trihydroxycho-
lan-24-oic Acid *see* 2210
Trihydroxycyanidine *see* 2690
(3*R*,4*S*,5*R*)-3,4,5-Trihydroxy-1-cyclohex-
ene-1-carboxylic Acid *see* 8619
3,4,5-Trihydroxy-2,2-dimethyl-6-chro-
manacrylic Acid δ-Lactone 4-Acetate
3-(2-Methylbutyrate) *see* 10208
2-(4,5-*cis*)-Trihydroxy-4,6-dimethylte-
trahydropyran *see* 6403
3,16α,17β-Trihydroxy-Δ^{1,3,5}-estratriene
see 3762

Triisopropoxytitanium Chloride *see* 2174

Triisopropylproazaphosphatrane *see* 10155

2,8,9-Triisopropyl-2,5,8,9-tetraaza-1-phosphabicyclo[3.3.3]undecane *see* 10155

3,7,12-Triketocholanic Acid *see* 2872

Triketohydrinden Hydrate *see* 6640

Trilafon [Schering] *see* 7297

Trilan [Esseti] *see* 9119

Trilene [Zeneca] *see* 9804

Trileptal [Novartis] *see* 7028

Tri-Levlen [Berlex] *see* 6793

Trilifan *see* 7297

Triline *see* 9804

Trilipix [Abbott] *see* 4009

Trilit *see* 9902

Trillekamin [Crookes] *see* 9790

Trillium, 9865

Trilobamine *see* 2819

Trilobine, 9866

Trilon A *see* 6665

Trilon B [BASF] *see* 3565

Trilophosphamide *see* 9947

Trilostane, 9867

Triludan [Aventis] *see* 9306

Trimagnesium Dicitrate *see* 5727

Trimagnesium Phosphate *see* 5747

Trimanganese Tetroxide *see* 5800

Trimangol [Cerexagri] *see* 5786

Trimanyl [Tosse] *see* 9878

Trimar *see* 9804

Trimaton *see* 6021

Trimax [Bayer CropSci.] *see* 4951

Trimebutine, 9868

Trimecaine, 9869

Trimedlure, 9870

Trimellitic Acid, 9871

Trimellitic Acid 1,2-Anhydride *see* 9872

Trimellitic Anhydride, 9872

Trimeperidine *see* 7899

Trimepranol *see* 6216

Trimeprazine, 9873

Trimeprimine *see* 9894

Trimeproprimine *see* 9894

Trimeric Acetone Peroxide *see* 9752

Trimesulf [LPB] *see* 9050

Trimet [Tessenderlo Kerley] *see* 8825

Trimetaphan Camphorsulfonate *see* 9876

Trimetaphosphoric Acid Trisodium Salt *see* 8823

Trimetazidine, 9874

Trimethadione, 9875

Trimethaphan Camphorsulfonate *see* 9876

Trimethaphan Camsylate, 9876

Trimethobenzamide, 9877

Trimethoprim, 9878

Trimethoquinol *see* 9749

1,2,10-Trimethoxy-6aα-aporphin-11-ol *see* 5207

2,10,11-Trimethoxy-6aα-aporphin-1-ol *see* 2533

3-(3,4,5-Trimethoxybenzamido)piperidine *see* 9970

3,4,5-Trimethoxybenzeneethanamine *see* 5975

3,4,5-Trimethoxybenzoic Acid 2-[4-[3-(2-Chloro-10*H*-phenothiazin-10-yl)propyl]-1-piperazinyl]ethyl Ester *see* 6221

3,4,5-Trimethoxybenzoic Acid Diester with 3,3′-[Ethylenebis(methylimino)]-di-1-propanol *see* 4741

3,4,5-Trimethoxybenzoic Acid Diester with Tetrahydro-1*H*-1,4-diazepine-1,4-(5*H*)-dipropanol *see* 3222

3,4,5-Trimethoxybenzoic Acid 2-(Dimethylamino)-2-phenylbutyl Ester *see* 9868

3,4,5-Trimethoxybenzoic Acid 1,1′-[1,2-Ethanediylbis[(methylimino)-3,1-propanediyl]] Ester *see* 4741

3,4,5-Trimethoxybenzoic Acid 1,1′-[2,6-Pyridinediylbis(methylene)]ester *see* 7620

3,4,5-Trimethoxybenzoic Acid 1,1′-[(Tetrahydro-1*H*-1,4-diazepine-1,4-(5*H*)-diyl)di-3,1-propanediyl] Ester *see* 3222

3,4,5-Trimethoxybenzoyl Methyl Reserpate *see* 8265

4-(3,4,5-Trimethoxybenzoyl)morpholine *see* 9892

N-[β-(3,4,5-Trimethoxybenzoyloxy)ethyl]-*N*′-[γ-(3-chloro-10-phenothiazinyl)propyl]piperazine *see* 6221

N-(3,4,5-Trimethoxybenzoyl)tetrahydro-1,4-oxazine *see* 9892

1-(2,3,4-Trimethoxybenzyl)-4-[bis(4-fluorophenyl)methyl]piperazine *see* 5620

1-(3′,4′,5′-Trimethoxybenzyl)-6,7-dihydroxy-1,2,3,4-tetrahydroisoquinoline *see* 9749

2,9,10-Trimethoxyberbin-3-ol *see* 2536

3,9,10-Trimethoxy-13aα-berbin-2-ol *see* 5208

Trimethoxyboron *see* 9882

3,4,5-Trimethoxycinnamic Acid Ester of Methyl Reserpate *see* 8262

6,7,8-Trimethoxycoumarin *see* 4294

6,6′,7-Trimethoxy-2,2′-dimethylberbaman-12-ol *see* 1155

6,6′,7-Trimethoxy-2,2′-dimethyloxyacanthan-12′-ol *see* 7052

(1β)-6,6′,7′-Trimethoxy-2,2′-dimethyltubocuraran-12′-ol *see* 2218

7′,10,11-Trimethoxyemetan-6′-ol *see* 1973

4,7,8-Trimethoxyfuro[2,3-*b*]quinoline *see* 8699

3,4,5-Trimethoxy-3′-hydroxy-4′-methoxy-(Z)-stilbene *see* 2477

Trimethoxymethane *see* 6980

(1α,6α,14α,16β)-1,6,16-Trimethoxy-4-(methoxymethyl)-20-methylaconitane-8,13,14-triol 8-Acetate 14-Benzoate *see* 2886

3,9,10-Trimethoxy-13α-methyl-13aβ-berbin-2-ol *see* 5206

6,6′,12′-Trimethoxy-2-methyloxyacanthan-7-ol *see* 2819

1,2,10-Trimethoxy-6aα-noraporphin-9-ol *see* 5443

1,1,1-Trimethoxypentane *see* 9888

3,4,5-Trimethoxyphenethylamine *see* 5975

1-[(2,3,4-Trimethoxyphenyl)methyl]-piperazine *see* 9874

5-[(3,4,5-Trimethoxyphenyl)methyl]-2,4-pyrimidinediamine *see* 9878

(2,4,6-Trimethoxyphenyl)-(3-pyrrolidinopropyl) Ketone *see* 1480

3,4,5-Trimethoxy-*N*-3-piperidinylbenzamide *see* 9970

1,2,4-Trimethoxy-5-(1-propenyl)benzene *see* 813

2,4,5-Trimethoxy-1-propenylbenzene *see* 813

2′,4′,6′-Trimethoxy-4-(1-pyrrolidinyl)butyrophenone *see* 1480

4-(3,4,5-Trimethoxythiobenzoyl)morpholine *see* 9937

Trimethylacetaldehyde *see* 7624

Trimethylacetic Acid *see* 7625

Trimethyl-β-acetoxypropylammonium Chloride *see* 6012

2-Trimethylacetyl-1,3-indandione *see* 7555

Trimethylallylsilane *see* 290

Trimethylaluminum, 9879

Trimethylamine, 9880

Trimethylamine N-Oxide, 9881

α,α′,α″-Trimethylaminetricarboxylic Acid *see* 6665

1-Trimethylammonio-3-(3-indolyl)propionate *see* 4900

Trimethylammonium Chloride *see* 9880

Trimethylanilinium Iodide *see* 7428

N,N,N-Trimethylbenzaminium Iodide (1:1) *see* 7428

sym-Trimethylbenzene *see* 5977

1,2,4-Trimethylbenzene *see* 8023

1,3,5-Trimethylbenzene *see* 5977

α,α,β-Trimethylbenzeneethanamine *see* 7246

N,α,α-Trimethylbenzeneethanamine *see* 5919

7,8,10-Trimethylbenzo[*g*]pteridine-2,4-(3*H*,10*H*)-dione *see* 5659

(±)-7-(3,5,6-Trimethyl-1,4-benzoquinon-2-yl)-7-phenylheptanoic Acid *see* 8598

(±)-4-(α,2,3-Trimethylbenzyl)imidazole *see* 5858

rel-(1*R*,2*R*,4*R*)-1,7,7-Trimethylbicyclo[2.2.1]heptan-2-ol *see* 5173

rel-(1*R*,2*S*,4*R*)-1,7,7-Trimethylbicyclo[2.2.1]heptan-2-ol *see* 1340

endo-1,7,7-Trimethylbicyclo[2.2.1]heptan-2-ol *see* 1340

exo-1,7,7-Trimethylbicyclo[2.2.1]heptan-2-ol *see* 5173

rel-(1*R*,2*S*,4*R*)-1,7,7-Trimethylbicyclo[2.2.1]heptan-2-ol 2-Acetate *see* 1341

(1*S*,4*R*)-1,3,3-Trimethylbicyclo[2.2.1]heptan-2-one *see* 4000

1,7,7-Trimethylbicyclo[2.2.1]heptan-2-one *see* 1734

2,6,6-Trimethylbicyclo[3.1.1]hept-2-ene *see* 7556

3,7,7-Trimethylbicyclo[4.1.0]hept-3-ene *see* 1835

(1*R*,5*R*)-4,6,6-Trimethylbicyclo[3.1.1]-hept-3-en-2-one *see* 10154

2,7,7-Trimethylbicyclo[3.1.1]hept-2-en-6-one *see* 2256

Trimethyl Borate, 9882

Trimethylbromo-IDA *see* 5841

Trimethylbromomethane *see* 1557

Trimethyl Carbinol *see* 1544

Trimethylchloromethane *see* 1563

Trimethylchlorosilane *see* 2180

p-(2,2,4-Trimethyl-4-chromanyl)phenol *see* 2991

trans-2,6,6-Trimethyl-1-crotonylcyclohexa-1,3-diene *see* 2808

2,2,6-Trimethyl-4,6-cyclohexadien-1-aldehyde *see* 8452

2,6,6-Trimethyl-1,3-cyclohexadiene-1-carboxaldehyde *see* 8452

(2*E*)-1-(2,6,6-Trimethyl-1,3-cyclohexadien-1-yl)-2-buten-1-one *see* 2808

3,3,5-Trimethylcyclohexanol α-Phenyl-α-hydroxyacetate *see* 2697

Triptide *see* 4511
Triptil [Merck & Co.] *see* 8010
Triptizol [SIT] *see* 483
Tript-Oh [Sigma-Tau] *see* 4886
Triptolide, 9923
Triptorelin, 9924
Triptycene, 9925
2,4,6-Tri-2-pyridinyl-*s*-triazine *see* 9926
2,4,6-Tri-2-pyridinyl-1,3,5-triazine *see* 9926
Tripyridyltriazine *see* 9926
2,4,6-Tripyridyl-*s*-triazine, 9926
TRIS *see* 9951
(*T*-4)-Tris(acetato-κ*O*)hydroborate(1−) Sodium (1:1) *see* 8822
Tris(acetato)thallium *see* 9406
1,1,1-Tris(acetyloxy)-1,1-dihydro-1,2-benziodoxol-3(1*H*)-one *see* 2934
Trisamine *see* 9951
2,3,5-Tris(aziridino)-1,4-benzoquinone *see* 9764
2,3,5-Tris(1-aziridinyl)-*p*-benzoquinone *see* 9764
Tris(1-aziridinyl)phosphine Oxide *see* 9837
Tris(1-aziridinyl)phosphine Sulfide *see* 9518
2,4,6-Tris(1-aziridinyl)-1,3,5-triazine *see* 9836
Tris-BP, 9927
Tris Buffer *see* 9951
Triscabol [Elf Atochem] *see* 10372
(*HB*-8-22-111′1′1″)-Tris[carbonato-(2−)-κ*O*,κ*O*′]dioxouranate(4−) Ammonium (1:4) *see* 563
(*T*-4)-Tris[carbonato(2−)-κ*O*]hydroxyzirconate(3−) Ammonium (1:3) *see* 566
Tris(2-carboxyethyl)phosphine *see* 9222
N-[2-[4,7,10-Tris(carboxymethyl)-1,4,7,10-tetraazacyclododec-1-yl]acetyl]-D-phenylalanyl-L-cysteinyl-L-tyrosyl-D-tryptophyl-L-lysyl-L-threonyl-*N*-[(1*R*,2*R*)-2-hydroxy-1-(hydroxymethyl)propyl]-L-cysteinamide Cyclic (2 → 7)-Disulfide *see* 3561
Tris[2-chloro-1-(chloromethyl)ethyl]phosphate *see* 9224
Tris(β-chloroethyl)amine *see* 9790
N,*N*,*N*′-Tris(2-chloroethyl)-*N*′,*O*-propylene Phosphoric Acid Ester Diamide *see* 9947
N,*N*,3-Tris(2-chloroethyl)tetrahydro-2*H*-1,3,2-oxazaphosphorin-2-amine 2-Oxide *see* 9947
Tris(2,3-dibromopropyl) Phosphate *see* 9927
Tris(1,3-dichloroisopropyl)phosphate *see* 9224
1,2,3-Tris(2-diethylaminoethoxy)benzene Tris(ethyl Iodide) *see* 4373
Tris[7-[4-(dimethylamino)phenyl]-*N*-hydroxy-4,6-dimethyl-7-oxo-2,4-heptadienamidato-*O*^*N*^*O*^1^]iron *see* 9814
2,4,6-Tris(dimethylamino)-*s*-triazine *see* 316
(*OC*-6-11)-Tris(*N*,*N*-dimethylcarbamodithioato-κ*S*,κ*S*′)iron *see* 4040
(*OC*-6-11)-Tris(dimethylcarbamodithioato-*S*,*S*′)iron *see* 4040
Tris(dimethyldithiocarbamato)iron *see* 4040
3,5,7-Triselena-1,2,4,6-tetraphosphatricyclo[2.2.1.0^2,6^]heptane *see* 7473
Trisenox [Cell Therapeutics] *see* 795

Tris(2,3-epoxypropyl) Isocyanurate *see* 9860
(*OC*-6-11)-Tris(1,2-ethanediamine-κ*N*^1^,κ*N*^2^)cadmium(2+) Hydroxide (1:2) *see* 9928
(*OC*-6-11)-Tris[ethanedioato(2−)-κ*O*^1^,κ*O*^2^]ferrate(3−) Ammonium (1:3) *see* 514
Tris(ethylenediamine)cadmium Dihydroxide, 9928
Tris(ethylenediamine)cadmium Hydroxide *see* 9928
2,3,5-Tris(ethyleneimino)benzoquinone *see* 9764
2,4,6-Tris(ethylenimino)-*s*-triazine *see* 9836
Tris(hydroxyethyl)amine *see* 9830
N,*N*,*N*′-Tris(2-hydroxyethyl)-*N*′-octadecyl-1,3-diaminopropane Dihydrofluoride *see* 6912
7,3′,4′-Tris[*O*-(2-hydroxyethyl)]rutin *see* 9969
Tris(hydroxymethyl)aminomethane *see* 9951
N-Tris(hydroxymethyl)methyl-2-aminoethanesulfonic Acid *see* 9319
N-Tris[(hydroxymethyl)methyl]-3-aminopropanesulfonic Acid *see* 9191
N-[Tris(hydroxymethyl)methyl]glycine *see* 9816
Tris(hydroxymethyl)nitromethane, 9929
Tris[2-(hydroxy-κ*O*)propanoato-κ*O*]aluminum *see* 343
Tris(8-hydroxyquinoline)aluminium *see* 364
Trisilane, 9930
Trisilicane *see* 9930
Trisilicon Octahydride *see* 9930
Trisilicon Tetranitride *see* 8636
Trisilicopropane *see* 9930
Tris[μ-[methanedisulfonato(2−)]]-dialuminum *see* 6046
Tris(*p*-methoxyphenyl)chloroethylene *see* 2178
Tris[2-methyl-1-aziridinyl]phosphine Oxide *see* 6008
Tris[1-methylethylene]phosphoric Triamide *see* 6008
N-[2-[2,4,6-Tris(1-methylethyl)phenyl]acetyl]sulfamic Acid 2,6-Bis(1-methylethyl)phenyl Ester *see* 876
2,8,9-Tris(1-methylethyl)-2,5,8,9-tetraaza-1-phosphabicyclo[3.3.3]undecane *see* 10155
Tris(2-methyl-2-propanolato)aluminum *see* 329
2,8,9-Tris(2-methylpropyl)-2,5,8,9-tetraaza-1-phosphabicyclo[3.3.3]undecane *see* 10155
1,1,1-Tris(nitratomethyl)propane *see* 7925
Trisodium 4′-Anilino-8-hydroxy-1,1′-azonaphthalene-3,6,5′-trisulfonate *see* 622
Trisodium Calcium Diethylenetriaminepentaacetate *see* 7236
Trisodium Carboxyphosphate *see* 4278
Trisodium Citrate *see* 8737
Trisodium (*OC*-6-11)-Hexakis(nitrito-κ*N*)cobaltate(3−) *see* 4528
Trisodium *N*-Hydroxyethylethylenediaminetriacetate *see* 10162
Trisodium Orthophosphate *see* 8792
Trisodium Phosphate *see* 8792
Trisodium Phosphonoformate *see* 4278

Trisodium 2-(*p*-Sulfophenylazo)-1,8-dihydroxynaphthalene-3,6-disulfonate *see* 8859
Trisodium Trimetaphosphate *see* 8823
Trisomin [Lilly] *see* 5751
Trisoralen [Valeant] *see* 9907
Trisorcin [Merckle] *see* 7197
Tris(oxalato)dialuminum *see* 349
Tris(oxalato)dicerium *see* 2000
1,3,5-Tris(2-oxiranylmethyl)-1,3,5-triazine-2,4,6(1*H*,3*H*,5*H*)-trione *see* 9860
Tris(2,3,4,5,6-pentafluorophenyl)borane *see* 9931
Tris(pentafluorophenyl)boron, 9931
Tris(2-pyridinecarboxylato-*N*^1^,*O*^2^)chromium *see* 2240
Tris(8-quinolinolato-κ*N*^1^,κ*O*^8^)aluminum *see* 364
Tri-Star [Nippon Soda] *see* 48
Tristat [Salsbury] *see* 6699
Tri-Stat IU [Tri-K] *see* 4957
Tristearin, 9932
Tris(2,4,6-tribromophenoxy)bismuthine *see* 1291
Tris(tribromophenoxy)triazine, 9933
2,4,6-Tris(2,4,6-tribromophenoxy)-1,3,5-triazine *see* 9933
Tris(tribromophenyl)cyanurate *see* 9933
1,2,3-Tris(2-triethylammonium Ethoxy)benzene Triiodide *see* 4373
Tris(trimethylsilyl)silane, 9934
Tristriphenylphosphine Rhodium Carbonyl Hydride, 9935
Trisulfurated Phosphorus *see* 9385
Tritace [Aventis] *see* 8221
Tritan *see* 9914
Tritanol *see* 9912
Tritec [GSK] *see* 8228
Triteren [Sumitomo] *see* 9760
Triterium *see* 9939
Tritheon [Ortho] *see* 404
4,5,9-Trithiadodeca-1,6,11-triene 9-Oxide *see* 188
3,5,7-Trithia-1,2,4,6-tetraphosphatricyclo[2.2.1.02,6]heptane *see* 9385
Trithioacetaldehyde *see* 9471
Trithioanethole *see* 637
Trithiocarbonic Acid, 9936
Trithiocarbonic Acid Disodium Salt *see* 8825
Trithio-*p*-methoxyphenylpropene *see* 637
Trithiozine, 9937
Triticum, 9938
Tritiozine *see* 9937
Tritium, 9939
Tritocaline *see* 9942
Tritoftorol [Cerexagri] *see* 10365
Tritolyl Phosphate, 9940
Tri-*o*-tolyl Phosphate, 9941
Triton A-20 [Rohm & Haas] *see* 10014
Triton N [Rohm & Haas] *see* 6764
Triton WR-1339 [Rohm & Haas] *see* 10014
Triton X [Rohm & Haas] *see* 6850
Triton X-100 [Rohm & Haas] *see* 6850
Tritopine *see* 5434
Tritoqualine, 9942
Tritosulfuron, 9943
Trittico [Angelini] *see* 9740
Trityl Chloride, 9944
Triuret, 9945
Trivastal [Itherapia] *see* 7607
Trizina [Francia Farm] *see* 1917
Trizinc Diphosphide *see* 10352
Trizma [Sigma-Aldrich] *see* 9951

d-Tubocurarine Iodide Dimethyl Ether *see* 6220

d-Tubocurine *see* 2217

Tubulin, 9988

α-Tubulin *see* 9988

β-Tubulin *see* 9988

Tuclase [Pfizer] *see* 1795

Tuduranine *see* 9980

Tuff-Lite [Indust. Plastic] *see* 7701

Tuftsin, 9989

Tulathromycin, 9990

Tulisan [Chiesi] *see* 6807

Tulobuterol, 9991

Tumenol *see* 4924

Tumeric *see* 10005

Tumil-K [Virbac] *see* 4492

Tumor Necrosis Factor, 9992

1-235-Tumor Necrosis Factor Receptor (Human) Fusion Protein with 236-467-Immunoglobulin G1 (Human γ₁-Chain Fc Fragment) *see* 3767

Tums Smooth Dissolve [GSK] *see* 1659

Tung Oil, 9993

Tungstate (W₁₂(OH)₂O₄₀¹⁰⁻) Ammonium (1:10) *see* 536

(*T*-4)-Tungstate (WO₄²⁻) Hydrogen (1:2) *see* 9998

(*T*-4)-Tungstate (WO₄²⁻) Potassium (1:2) *see* 7811

Tungsten, 9994

Tungsten Carbide, 9995

(*OC*-6-11)-Tungsten Fluoride (WF₆) *see* 9996

Tungsten Hexafluoride, 9996

Tungsten Hydroxide Oxide Phosphate *see* 7476

Tungsten Oxide (WO₃) *see* 9997

Tungsten Trioxide, 9997

Tungstic(VI) Acid, 9998

Tungstic Acid (H₂WO₄) Copper(2+) Salt (1:1) *see* 2647

Tungstic Anhydride *see* 9997

5-Tungsto-2-antimonate *see* 4789

21-Tungsto-9-antimonate *see* 4789

Tungstophosphoric Acid *see* 7476

Tungstophosphoric Acid (H₃PW₁₂O₄₀) Triammonium Salt *see* 561

Tungstosilicic Acid *see* 8641

Tunicamine *see* 9999

Tunicamycin, 9999

Tunichrome B-1, 10000

Tunichromes *see* 10000

Tunol *see* 2449

Tupersan [DuPont] *see* 8626

Turanose, 10001

Turbine [ISK] *see* 4137

Turbocalcin [SKB] *see* 3585

Turexan Crème [Turimed] *see* 10033

Turflon [Dow AgroSci.] *see* 9821

Turimycin P3 *see* 6262

Turinal [Gedeon Richter] *see* 282

Turixin [GSK] *see* 6388

Turkey Corn *see* 2532

Turkey-galls *see* 6817

Turkey-Red Oil, 10004

Turkish Geranium Oil *see* 6888

Turkish Laurel *see* 5437

Turlington's Balsam *see* 944

Turmeric, 10005

Turmeric Yellow *see* 2663

ar-Turmerone, 10006

Turnsole *see* 5603

Turoptin [Dispersa] *see* 6216

Turpentine, 10007

"Turpentine Camphor" *see* 1342

Turpentine Oil *see* 10007

Turpentine Substitutes *see* 6282

Tussafug [Robugen] *see* 1051

Tusseval [Marco Viti] *see* 7581

Tussirama [Serpero] *see* 4255

Tussistop *see* 2066

Tuss-Ornade [SKB] *see* 1778

Tutane *see* 1546

Tutin, 10008

Tuttomycin *see* 6539

TV-02 *see* 9152

TV-485 *see* 3921

TV-1322 *see* 28

TVP-1012 *see* 8234

TVX Q 7821 *see* 5125

TW 260/240 *see* 8867

Tween 80 [ICI] *see* 7703

TWHF *see* 9549

Twist [Bayer CropSci.] *see* 9843

Twistane, 10009

Twiston [McNeil] *see* 1800

TWSb *see* 8941

Tx 60 *see* 10234

TXA₂ *see* 9542

TXB₂ *see* 9542

TXD-258 *see* 1604

TXIB [Eastman Chemical] *see* 9399

Tybamate, 10010

Tybatran [Robins] *see* 10010

Tycerb [GSK] *see* 5418

Tydantil [Polichimica Sap] *see* 6616

Tygacil [Wyeth] *see* 9588

Tykerb [GSK] *see* 5418

Tylan [Elanco] *see* 10013

Tylcalsin *see* 1650

Tylciprine [Théraplix] *see* 9734

Tylenol [McNeil] *see* 47

Tylocrebrine, 10011

Tylonolide *see* 10013

Tylophorine, 10012

Tylose M [Clariant] *see* 6111

Tylose MGA *see* 1830

Tylosin, 10013

Tyloxapol, 10014

Tyloxypal *see* 10014

Tymazoline, 10015

Tymelyt [Lundbeck] *see* 5614

Tympanol *see* 6194

Tymtran [Adria] *see* 2004

Tyr *see* 10020

Tyraminase *see* 6333

Tyramine, 10016

Tyrene *see* 9794

Tyri 10 [Lindopharm] *see* 10023

Tyrimide *see* 5248

Tyrindoxyl Sulfate *see* 3028

Tyrisa [GSK] *see* 2824

Tyriverdin *see* 3028

Tyrocidine, 10017

Tyrocidine A *see* 10017

Tyrocidine B *see* 10017

Tyrocidine C *see* 10017

Tyropanoate Sodium, 10018

Tyropaque [Torii] *see* 10018

Tyrosamine *see* 10016

Tyrosinase, 10019

Tyrosine, 10020

L-Tyrosine *see* 10020

m-Tyrosine, 10021

[29-Tyrosine,104-glutamic Acid] CTLA-4 (Antigen) (Human Extracellular Domain-containing Fragment) Fusion Protein with Immunoglobulin G1 (Human Monoclonal Fc Domain-containing Fragment), Bimol. (120 → 120′)-Disulfide *see* 1023

[9-L-Tyrosine]heat-stable Enterotoxin (*Escherichia coli*)-(6-19)-peptide *see* 5548

Tyrosol, 10022

Tyrosur [Engelhard KG] *see* 10023

L-Tyrosylglycylglycyl-L-phenylalanyl-L-leucine *see* 3627

L-Tyrosylglycylglycyl-L-phenylalanyl-L-methionine *see* 3627

Tyrothricin, 10023

Tyrrlen *see* 1519

Tysabri [Elan] *see* 6508

Tyvaso [United Ther.] *see* 9748

Tyverb [GSK] *see* 5418

Tyzeka [Novartis] *see* 9258

Tyzine [Pfizer] *see* 9364

TZB *see* 7390

TZP-101 *see* 10029

TZP-4238 *see* 6985

TZU-0460 *see* 8407

U-27 *see* 7532

U-1363 *see* 3335

U-2032 *see* 5347

U-2043 *see* 9667

U-4527 *see* 2721

U-5956 *see* 4110

U-6013 *see* 5220

U-6591 *see* 6811

U-6987 *see* 1831

U-7800 *see* 4223

U-8344 *see* 10036

U-8471 *see* 5865

U-9586 *see* 8032

U-9889 *see* 8962

U-10071 *see* 9984

U-10136 *see* 7987

U-10149 *see* 5555

U-10858 *see* 6285

U-10997 *see* 6253

U-12062 *see* 7988

U-14583 *see* 7989

U-14743 *see* 7719

U-15167 *see* 6758

U-15800 *see* 7083

U-17835 *see* 9665

U-18496 *see* 884

U-19183 *see* 8863

U-19920 *see* 2781

U-21251 *see* 2354

U-22550 *see* 1724

U-24973A *see* 5891

U-26225A *see* 9726

U-26452 *see* 4514

U-26597A *see* 2459

U-27182 *see* 4229

U-28288D *see* 4594

U-28508 *see* 2354

U-28774 *see* 5345

U-31889 *see* 308

U-32070E *see* 1641

U-32921 *see* 1824

U-32921E *see* 1824

U-33030 *see* 9765

U-34865 *see* 3159

U-36059 *see* 482

U-36384 *see* 1824

U-41123 *see* 148

U-42585E *see* 5613

U-42718 *see* 5613

U-42842 *see* 759

U-53217 *see* 7986

U-53217A *see* 7986

U-57903E *see* 7613

U-62840 *see* 9748
U-64279E *see* 1951
U-67279A *see* 1951
U-69689E *see* 4088
U-70226E *see* 4921
U-72107A *see* 7565
U-74006F *see* 9619
U-76252 *see* 1942
U-76253 *see* 1942
U-76253A *see* 1942
U-90152 *see* 2882
U-90152S *see* 2882
U-100766 *see* 5559
Ub *see* 10026
Ubenimex, 10024
UBI-A1335 *see* 4154
Ubicardio [Tosi] *see* 10025
Ubichromenol 50 *see* 10025
Ubicor [Magis] *see* 10025
Ubidecarenone *see* 10025
Ubidenone [Esseti] *see* 10025
Ubifactor [San Carlo] *see* 10025
Ubimaior [Chiesi] *see* 10025
UBI-N252 *see* 7037
UBIP *see* 10026
Ubiquinone 10 *see* 10025
Ubiquinone 50 *see* 10025
Ubiquinones, 10025
Ubiquitin, 10026
Ubiquitous Immunopoietic Polypeptide
    see 10026
Ubisan [Leben's] *see* 10025
Ubiten-50 *see* 10025
Ubivis [AGIPS] *see* 10025
Ubretid [Nycomed] *see* 3405
UC-7744 *see* 1788
UC-21149 *see* 215
UC-51762 *see* 9482
Ucarcide [Union Carbide] *see* 4508
UCB-1474 *see* 2081
UCB-1967 *see* 3500
UCB-2543 *see* 1795
UCB-3412 *see* 3429
UCB-3983 *see* 5979
UCB-4445 *see* 1474
UCB-4492 *see* 4889
UCB-5062 *see* 5848
UCB-6215 *see* 7600
UCB-22073 *see* 6306
UCB-28556 *see* 2021
UCB-34714 *see* 1384
UCB-L059 *see* 5513
UCC 974 *see* 2837
UDCA *see* 10074
UD-CG 115 *see* 7545
UD-CG 115 BS *see* 7545
Udekinon [Tobishi] *see* 10025
Udenafil, 10027
Udicil [Pharmacia & Upjohn] *see* 2781
UDMH *see* 3271
Udolac [Mallinckrodt] *see* 2822
UDP *see* 10063
UDPG *see* 10064
UDP-Glucose *see* 10064
UDP-MurNAc-L-Ala-γ-D-Glu-X-D-Ala-D-
    Ala *see* 7145
Ue-5908 *see* 3146
U4Euh *see* 6092
UF-021 *see* 10034
Ugurol [Rottapharm] *see* 9730
UHF-8615 *see* 7173
UIC-94017 *see* 2829
Uintahite *see* 4457
Uintaite *see* 4457
UK-4271 *see* 7018

UK-14304 *see* 1381
UK-14304-18 *see* 1381
UK-20349 *see* 9608
UK-33274 *see* 3482
UK-33274-27 *see* 3482
UK-48340 *see* 487
UK-48340-11 *see* 487
UK-48340-26 *see* 487
UK-49858 *see* 4153
UK-61689 *see* 8582
UK-61689-2 *see* 8582
UK-67994 *see* 3473
UK-68798 *see* 3456
UK-73967 *see* 1745
UK-79300 *see* 1745
UK-81252 *see* 8489
UK-88525 *see* 2826
UK-92480 *see* 8628
UK-109496 *see* 10231
UK-116044 *see* 3597
UK-427857 *see* 5814
Ukidan [Serono] *see* 10072
Ularitide *see* 10071
Ulcar [Houdé] *see* 9010
Ulcedin [AGIPS] *see* 2282
UlcerGard [AstraZeneca] *see* 6939
Ulcerlmin [Chugai] *see* 9010
Ulcerone [Yamanouchi] *see* 2469
Ulcesium [Zambon] *see* 4033
Ulcex [Guidotti] *see* 8228
Ulcimet [Polipharm] *see* 2282
Ulcogant [Merck KGaA] *see* 9010
Ulcosan [Dompé] *see* 7604
Ulcuforton [Plantorgan] *see* 7604
Ulcus-Tablinen [Sanorania] *see* 1794
Uldumont *see* 6030
Ulexine *see* 2786
Ulfamid [Krka] *see* 3971
UL-FS-49 *see* 10313
Ulgut [Shionogi] *see* 1040
Ulifloxacin *see* 8016
Uliginosin A *see* 10028
Uliginosin B *see* 10028
Uliginosins, 10028
Ulimorelin, 10029
Ulinastatin *see* 10067
Uliprisnil Acetate *see* 10030
Ulipristal *see* 10030
Ulipristal Acetate, 10030
Uliron C *see* 9061
ULO [3M Pharma] *see* 2066
Ulobetasol Propionate *see* 4629
Ulone [3M Pharma] *see* 2066
Uloric [Teijin Pharma] *see* 3983
Ulsanic [Ferrer] *see* 3532
Ulstar [Roche] *see* 9895
Ulstron [ICI] *see* 7701
Ultane [Abbott] *see* 8614
Ultidine [GSK] *see* 8228
Ultiva [Glaxo Wellcome] *see* 8251
Ultracain [HMR] *see* 1869
Ultracarbon *see* 1808
Ultracef [BMS] *see* 1915
Ultracillin [Grünenthal] *see* 2695
Ultracorten [Novartis] *see* 7842
Ultracortenol [Novartis] *see* 7841
Ultradol [Procter & Gamble] *see* 3920
Ultrafur *see* 4323
Ultralan Oral [Schering AG] *see* 4184
Ultralanum [Schering AG] *see* 4184
Ultram [Ortho-McNeil] *see* 9726
Ultramarine, 10031
Ultramarine Green *see* 2238
Ultramarine Yellow *see* 968
Ultra-Mg [Melisana] *see* 4492

Ultramicina [Lisapharma] *see* 4281
Ultrapen *see* 7931
Ultraquinine *see* 2613
Ultrase [Axcan] *see* 7108
Ultrasulfon [Streuli] *see* 9037
Ultrasüss *see* 6716
Ultratard [Novo] *see* 5023
Ultra-Technekow [Mallinckrodt] *see* 8788
Ultravate [BMS] *see* 4629
Ultravist [Schering AG] *see* 5105
Ultrax [Chemie Linz] *see* 9046
UM-792 *see* 6448
UM-952 *see* 1501
Umatrope [Lilly] *see* 8842
Umbellatine *see* 1156
Umbelliferone, 10032
Umbelliferone Glucoside *see* 8700
Umbradil *see* 5083
UML-491 *see* 6213
UMP *see* 10066
Unacid PD Oral [Pfizer] *see* 9121
Unacil [Firma] *see* 3488
Unacim [Pfizer] *see* 9021, 9121
Unagen *see* 3390
Unakalm [Pharmacia & Upjohn] *see* 5345
Unal *see* 458
Unalmes [Andromaco] *see* 233
Unasyn [Pfizer] *see* 9121
Unasyn (Inj.) [Pfizer] *see* 9021
Unat [Roche] *see* 9711
Undecahydropentaborane *see* 7214
1-Undecanecarboxylic Acid *see* 5439
2-Undecanone *see* 6177
10-Undecenoic Acid *see* 10033
Undecylenic Acid, 10033
9-Undecylenic Acid *see* 10033
Unden [Bayer] *see* 7949
Unergol [Poli] *see* 3710
Ungernine *see* 9219
Unicide U-13 [Induchem] *see* 4957
Unicordium [Riom] *see* 1152
Unidasa *see* 4797
Unidone [Schering] *see* 661
Unidrox [Angelini] *see* 8016
Unifer [Tosi] *see* 4063
Uniflox [Bayer] *see* 2313
Unifur *see* 4323
Unifyl [Mundipharma] *see* 9436
Unihep [Leo Pharm] *see* 4688
Uniloc [Nycomed] *see* 850
Unipen [Wyeth-Ayerst] *see* 6436
Uniphyl [Purdue Frederick] *see* 9436
Uniphyllin [Mundipharma] *see* 9436
Unipril [Zambon] *see* 8221
Unipyranamide [Unichem] *see* 8067
Unisan [United Phosphorus] *see* 7411
Unisedil [Merck KGaA] *see* 2997
Unisept [Unichem] *see* 2092
Unisom [Pfizer] *see* 3489
Unisom Sleepgels [Chattem] *see* 3339
Unisomnia [Alliance] *see* 6661
Unispiran *see* 9348
Unistat *see* 6699
Unistat [Salsbury] *see* 9064
Unistat-3 [Salsbury] *see* 6699
Unitane *see* 9628
Unitiol *see* 3232
UNITOP [Roche] *see* 6425
Univasc [Schwarz] *see* 6315
Univer [Zeneus] *see* 10144
Unix [Syngenta] *see* 2769
Unixime [Firma] *see* 1925
Unizole *see* 3290
Unnatural Tartaric Acid *see* 9203
Unnilennium *see* 5878

Unnilhexium *see* 8551
Unniloctium *see* 4651
Unnilpentium *see* 3509
Unnilquadium *see* 8437
Unnilseptium *see* 1324
Unoprostone, 10034
Unresolvable Tartaric Acid *see* 9202
Ununbium *see* 2503
Ununhexium *see* 3592
Ununnilium *see* 2828
Ununoctium *see* 3594
Ununpentium *see* 3591
Ununquadium *see* 3590
Ununseptium *see* 3593
Ununtrium *see* 3589
Unununium *see* 8376
Unusual Tartaric Acid *see* 9203
Unwele *see* 9138
UP-33-901 *see* 2272
UP-83 *see* 6615
UP-106 *see* 7946
UP-34101 *see* 7906
Upbeet [DuPont] *see* 9857
UP$_4$dC *see* 2898
Upixon [Bayer] *see* 7576
Uplyso [Pfizer] *see* 9169
Uprima [TAP] *see* 733
Upside-down Tree *see* 950
UP$_4$U *see* 3391
UR-0003 *see* 7173
UR-112 *see* 2371
UR-1501 *see* 9856
UR-1521 *see* 4282
UR-4056 *see* 4243
UR-12592 *see* 8430
Urab [Hopkins] *see* 4036
Uracid [Hässle] *see* 338
Uracil, 10035
Uracil-6-carboxylic Acid *see* 6972
Uracil Deoxyriboside *see* 2911
Uracil Mustard, 10036
Uracil Riboside *see* 10062
Uractyl *see* 9062
Uragon [Makhteshim-Agan] *see* 1388
Uralenic Acid *see* 3644
Uralgin [Ceccarelli] *see* 6444
Uramid [Spofa] *see* 9062
l-Uramidoglutaric Acid *see* 1838
Uramil, 10037
Uramide [DuPont] *see* 10053
Uramustine *see* 10036
Uramycin B *see* 9721
Uranediol, 10038
Uranic Acid Diammonium Salt *see* 562
Uranic Oxide *see* 10045
Uranin(e) *see* 4192
Uranine Yellow *see* 4192
Uraninite *see* 10040
Uranium, 10039
Uranium Ammonium Carbonate *see* 563
Uranium Ammonium Fluoride *see* 564
Uranium Barium Oxide *see* 997
Uranium(IV) Chloride *see* 10042
Uranium Chloride (UCl$_4$) *see* 10042
Uranium Dioxide, 10040
Uranium Dioxydichloride *see* 10047
(*T*-4)-Uranium Fluoride (UF$_4$) *see* 10043
(*OC*-6-11)-Uranium Fluoride (UF$_6$) *see* 10041
Uranium Hexafluoride, 10041
Uranium Oxide Orange *see* 7812
Uranium Oxide (UO$_2$) *see* 10040
Uranium Oxide (UO$_3$) *see* 10045
Uranium Oxide Yellow *see* 8827
Uranium Tetrachloride, 10042

Uranium Tetrafluoride, 10043
Uranium Trichloride, 10044
Uranium Trioxide, 10045
Uranium X$_2$ *see* 7994
Uranium Yellow *see* 8827
Uranium Z *see* 7994
Uranous Oxide *see* 10040
Urantoin [DDSA] *see* 6686
Uranyl Acetate, 10046
Uranyl Ammonium Carbonate *see* 563
Uranyl Ammonium Fluoride *see* 564
Uranyl Benzoate *see* 1093
Uranyl Chloride, 10047
Uranyl Nitrate, 10048
Uranyl Sulfate, 10049
Urao *see* 8803
Urapidil, 10050
Uraprene [IBI] *see* 10050
Urari *see* 2661
Urate Oxidase *see* 10061
Urate Oxidase (*Aspergillus flavus* Clone 9C/9A Reduced) *see* 8235
Urate Oxidase (Synthetic *Sus scrofa* Variant Pig*KS*-Δ*N* Subunit) Homotetramer Amide with α-Carboxy-ω-methoxypoly(oxy-1,2-ethanediyl) *see* 7177
Urate:oxygen Oxidoreductase *see* 10061
Urazole, 10051
Urbac [Merck-Clévenot] *see* 6617
Urbadan [HMR] *see* 2356
Urbanyl [HMR] *see* 2356
Urbilat [Hor-Fer-Vit] *see* 5929
Urdes [Errekappa] *see* 10074
Urea, 10052
Urea Amidohydrolase *see* 10055
Urea Calcium Acetylsalicylate *see* 1650
Urea Compd with Hydrogen Peroxide (H$_2$O$_2$) (1:1) *see* 1784
Ureaform, 10053
Urea Hydrogen Peroxide *see* 1784
Urea Nitrate, 10054
Urea Nitrate (1:1) *see* 10054
Ureaphil [Abbott] *see* 10052
Urea Polymer with Formaldehyde *see* 10053
Urease, 10055
Urea Stibamine, 10056
Urecholine [Merck & Co.] *see* 1187
Uredepa, 10057
p-Ureidobenzenearsonic Acid *see* 1787
Ureidoglutaric Acid *see* 1838
5-Ureidohydantoin *see* 250
δ-Ureidonorvaline *see* 2330
4-Ureido-1-phenylarsonic Acid *see* 1787
Urem [Kade] *see* 4919
Urena, 10058
Urenil *see* 9062
Urepearl [Otsuka] *see* 10052
Urethan, 10059
Urethane *see* 10059
Urethimine *see* 10057
Urethylane *see* 6110
Uretrim [Bastian] *see* 9878
Urex [Mochida] *see* 4338
Urex [3M Pharma] *see* 6038
6,6'-Ureylenebis[1,1'-dimethylquinolinium] Sulfate *see* 3393
6,6'-Ureylenebis(1-methylquinolinium)-bis(methosulfate) *see* 3393
Urfadyn [Zambon] *see* 6623
Urfamycine [Inpharzam] *see* 9453
Uriben [Rosemont] *see* 6444
Uric Acid, 10060
Uric Acid Oxidase *see* 10061

Uricase, 10061
Uriclar [Crosara] *see* 6444
UriCon [ProLabs] *see* 7417
Uricovac [Sanofi Winthrop] *see* 1067
Uridine, 10062
P^1-(Uridine 5')-P^4-(2'-Deoxycytidine 5')-Tetraphosphate *see* 2898
Uridine 5'-Diphosphate, 10063
Uridine Diphosphate Glucose, 10064
Uridine-5'-diphosphoglucose *see* 10064
Uridine 5'-Monophosphate *see* 10066
Uridine 5'-(Pentahydrogen Tetraphosphate) P''' → 5'-Ester with 2'-Deoxycytidine *see* 2898
Uridine 5'-(Pentahydrogen Tetraphosphate) P''' → 5'-Ester with Uridine *see* 3391
Uridine 5'-Phosphoric Acid *see* 10066
Uridine 5'-Pyrophosphate *see* 10063
Uridine 5'-Pyrophosphate Glucose Ester *see* 10064
Uridine-5-pyrophosphoric Acid *see* 10063
Uridine 5'-(Tetrahydrogen Triphosphate) *see* 10065
Uridine 5'-(Trihydrogen Diphosphate) *see* 10063
Uridine 5'-(Trihydrogen Diphosphate) P'-α-D-Glucopyranosyl Ester *see* 10064
Uridine 5'-Triphosphate, 10065
5'-Uridylic Acid, 10066
Urief [Kissei] *see* 8642
UriKoxidase *see* 10061
Urimeth [North Am. Pharm.] *see* 6047
Urinary HCG *see* 2221
Urinary Indican *see* 5013
Urinary Trypsin Inhibitor *see* 10067
Urinastatin, 10067
Urinorm [Torii] *see* 1067
Uriodone *see* 5083
Urion [Zambon] *see* 231
Urisal [Sanofi-Synthelabo] *see* 8737
Urispas [SKB] *see* 4128
Uritrate [Warner-Lambert] *see* 7044
Uritrol *see* 6742
Urizept [RPR] *see* 6686
Uroalpha [Debat] *see* 6378
Uroanthelone *see* 3574
Urobilin *see* 10068
i-Urobilin *see* 10068
Urobilin IXα *see* 10068
Urobilinogen *see* 10068
Urobilins, 10068
Urocanylcholine *see* 6392
Urocaudal [Jorba] *see* 9760
Urochloralic Acid, 10069
Urocit-K [Mission Pharmacal] *see* 7744
Urocoli *see* 6742
Urocortin, 10070
Urodie [Abbott] *see* 9297
Urodilatin, 10071
Urodin [Streuli] *see* 6686
Urodixin [Italchimici] *see* 6444
Uroenterone *see* 3574
Urofollitrophin *see* 4250
Urofollitropin *see* 4250
Urogastrone *see* 3574
β-Urogastrone *see* 3574
γ-Urogastrone *see* 3574
Urografin [Schering AG] *see* 2993
Urokinase, 10072
Urokinase (Enzyme-activating) *see* 10072
Urolene Blue [Star] *see* 6132

Urolong [Thiemann] see 6686
Urolucosil [Lundbeck] see 9048
Uromaline see 5781
Uroman [Toho Kogyo] see 6444
Uromiro [Bracco] see 5055
Uromitexan [Baxter] see 5979
Uronamin [Sumitomo] see 6038
Uronase [Mochida] see 10072
Uroneg [Ibirn] see 6444
Uronorm [Alfa] see 2309
Uropan [Tiber] see 6444
Urophenil [Nippon Seiyaku] see 9453
Uropimid [CT] see 7572
Uropir [Sarm] see 7618
Uroplus [Shionogi] see 9050
Urorec [Recordati] see 8642
Uro-Ripirin [Kabi] see 3615
Urosin [Roche] see 271
Urosten [Bonomelli] see 7572
Urosulfan see 9062
Uro-Tablinen [Sanorania] see 6686
Urothion, 10073
Urotractan [Klinge] see 6038
Urotrate [Warner-Lambert] see 7044
Uroval [Firma] see 7572
Urox [Allied] see 6349
Urox B [Hopkins] see 1388
Urox D [Hopkins] see 3425
UroXatral [Sanofi-Synthelabo] see 231
Uroxin see 274
Ursacol [Zambon] see 10074
(3β)-Urs-12-en-3-ol see 610
(3β,18α,19α)-Urs-20(30)-en-3-ol see
   9193
Ursin see 763
Urso [Axcan] see 10074
Ursochol [Advanced Pharma] see 10074
Ursocholanic Acid see 2203
Ursodeoxycholic Acid see 10074
Ursodiol, 10074
Ursofalk [Falk] see 10074
Ursol D see 7397
Ursol P see 458
Ursolic Acid, 10075
Ursolvan [Sanofi-Aventis] see 10074
Urson see 10075
Urtosal [Lifepharma] see 8465
Urupan [Merckle] see 2949
Urushenol see 10076
Urushiol, 10076
(15:0)-Urushiol see 10076
Urushiol Monoene see 10076
USB-3153 see 7883
Uscharidin, 10077
Usevir [Yamasa Shoyu] see 8853
Uskan [Desitin] see 7025
Usnein see 10078
Usnic Acid, 10078
Usninic Acid see 10078
USR-604 see 3054
Ustekinumab, 10079
Ustilagic Acid, 10080
Ustimon see 4741
Ustizeain B see 10080
UT-15 see 9748
Utemerin [Kissei] see 8365
Uteplex [Am. Home] see 10065
Uteracon [National] see 7078
Uteroverdine see 1223
UTI see 10067
Uticort [Pfizer] see 1182
Utinor [Merck & Co.] see 6789
Utopar [Ferrosan] see 8365
Utovlan [Searle] see 6786
UTP see 10065

Utrogestan [Besins-Iscovesco] see 7889
Utrophin, 10081
Uval [Dorsey] see 9113
Uvaleral see 1404
Uvalysat [Ysatfabrik] see 10082
Uva Ursi, 10082
Uvesterol-D [Crinex] see 10215
Uvic Acid see 9204
Uvilon [Bayer] see 7576
Uvinul 400 [BASF] see 1109
Uvinul 3039 [BASF] see 6845
Uvinul D49 [BASF] see 1101
Uvinul M-40 [BASF] see 7053
Uvinul MC 80 [BASF] see 6859
Uvinul MS-40 [BASF] see 9113
Uvinul T 150 [BASF] see 3864
Uvistat [Boehringer, Ing.] see 6246
Uvitex OB [Ciba] see 1012
UX₂ see 7994
UZ see 7994
Uzarigenin see 10083
Uzarin, 10083
Uzone [Douglas] see 7390
V-7 see 9892
V-10029 see 1302
V-23031 see 4173
V-53482 see 4174
VA-2914 see 10030
VAA-I see 6295
Vabrocid see 6687
Vaccenic Acid, 10084
Vacor [Rohm & Haas] see 8101
Vadilex [Sanofi-Synthelabo] see 4936
Vadosilan [BMS] see 5285
Vagamin see 6030
Vagantin see 6030
Vagifem see 3758
Vagilen [Alfa] see 6236
Vagimid [Apogepha] see 6236
Vagisil [Combe] see 5535
Vagistat [BMS] see 9608
Vagran [UPSA] see 7946
Val see 10095
Valaciclovir see 10085
ValACV see 10085
Valacyclovir, 10085
Valamina [Schering] see 4220
Val⁵-angiotensin II-asp¹-β-amide see 644
Valbazen [Pfizer] see 203
Valbon [Kumai Chemical] see 1055
Valcote [Abbott] see 10099
Val²-cyclosporine see 2750
Valcyte [Roche] see 10093
Valdecoxib, 10086
Valdetamide, 10087
Valdoxan [Biopharma] see 183
Valecid [Medifarma] see 1930
Valemicina [Farmochim. Ital.] see 4281
Valentinite see 699
Valeral see 10088
n-Valeraldehyde, 10088
Valerian, 10089
Valerianic Acid see 10090
n-Valeric Acid, 10090
Valeric Aldehyde see 10088
Valeronitrile, 10091
Valethamate Bromide, 10092
Valette [Jenapharm] see 3117
Valganciclovir, 10093
Validacin [Takeda] see 10094
Validamycin A see 10094
Validamycins, 10094
Validol [Farmak] see 5909
Valifol [Valdecasas] see 5232
Valimon [Takeda] see 10094

Valine, 10095
L-Valine see 10095
L-Valine 2-[(2-Amino-1,6-dihydro-6-oxo-
   9H-purin-9-yl)methoxy]ethyl Ester see
   10085
L-Valine 2-[(2-Amino-1,6-dihydro-6-oxo-
   9H-purin-9-yl)methoxy]-3-hydroxy-
   propyl Ester see 10093
5-Valine-angiotensin II Amide see 644
L-Valine Ester with 9-[(2-Hydroxyeth-
   oxy)methyl]guanine see 10085
Valinomycin, 10096
Valiquid [Roche] see 2997
Valisone [Schering-Plough] see 1182
Valium [Roche] see 2997
Vallene 5842
Vallergan [M & B] see 9873
Vallesine see 840
Valmethamide see 10098
Valnemulin, 10097
Valnoctamide, 10098
Valodin [Ferrer] see 3394
Valoid [Glaxo Wellcome] see 2703
Valone see 5279
Valopride [Vita] see 1443
Valor [Valent] see 4174
Valoron [Gödecke] see 9594
Valproate Semisodium see 10099
Valproic Acid, 10099
Valpromide, 10100
Valrubicin, 10101
Valsartan, 10102
Valspodar, 10103
Valstar [Medeva] see 10101
Valsyn [Norwich] see 4323
Valtomicina [Midy] see 7566
Valtrex [GSK] see 10085
Valverde [Novartis] see 182
Valzin see 3511
Vampire Bat Salivary Plasminogen Acti-
   vator see 3506
Vanadate (VO₃¹⁻) Ammonium (1:1) see
   565
Vanadate (VO₃¹⁻) Sodium (1:1) see 8774
Vanadic Acid (HVO₃) Ammonium Salt
   see 565
Vanadic Anhydride see 10107
Vanadic Oxide see 10109
Vanadinite see 10104
Vanadium, 10104
Vanadium Carbonyl, 10105
Vanadium Fluoride (VF₅) see 10106
Vanadium Hexacarbonyl see 10105
Vanadium Oxide (V₂O₃) see 10109
Vanadium Oxide (V₂O₅) see 10107
Vanadium Oxydichloride see 10112
Vanadium Oxysulfate see 10113
Vanadium Oxytrichloride see 10114
Vanadium Pentafluoride, 10106
Vanadium Pentoxide, 10107
Vanadium Sesquioxide see 10109
Vanadium Sesquisulfide see 10110
Vanadium Sulfide (V₂S₃) see 10110
Vanadium(V) Trichloride Oxide see
   10114
Vanadium Trifluoride, 10108
Vanadium Trioxide, 10109
Vanadium Trisulfide, 10110
Vanadocene, 10111
Vanadyl Dichloride, 10112
Vanadyl Sulfate, 10113
Vanadyl Trichloride, 10114
Vanaspati, 10115
Vancenase [Schering-Plough] see 1017
Vanceril [Schering-Plough] see 1017

Ventolin [GSK] *see* 210
Ventox *see* 122
Ventramine *see* 6627
Ventrazol *see* 7254
Venturicidin A *see* 10141
Venturicidin B *see* 10141
Venturicidin B 3′-Carbamate *see* 10141
Venturicidins, 10141
Venzar [DuPont] *see* 5489
Venzoquimpe [Quimpe] *see* 5996
Vepesid [BMS] *see* 3929
Vepicombin [Nycomed] *see* 7209
VER-002 *see* 649
Veracillin [Ayerst] *see* 3094
Veracim [Cimex] *see* 10144
Veracur [Typharm] *see* 4263
Veradol [Schering AG] *see* 6499
Veraflox [Bayer] *see* 7819
Veralcamine *see* 10143
Veralipride, 10142
Veralkamine, 10143
Veramex [Sanofi-Synthelabo] *see* 10144
Veramix [Pharmacia] *see* 5864
Verapamil, 10144
Veraptin [Lagap] *see* 10144
Veratetrine *see* 8008
Veratraldehyde, 10145
Veratramine, 10146
Veratran [Latema] *see* 2400
Veratric Acid, 10147
Veratric Acid 4-[Ethyl(*p*-methoxy-α-
   methylphenethyl)amino]butyl Ester
   see 5838
Veratric Aldehyde *see* 10145
Veratridine, 10148
Veratrine *see* 2032
Veratrine (Mixture), 10149
Veratrole, 10150
3-Veratroylveracevine *see* 10148
Veratrum Viride, 10151
Verax [Tosi] *see* 1125
Verbascose, 10152
Verbenalin, 10153
Verbenalol *see* 10153
d-Verbenone, 10154
Vercyte [Abbott] *see* 7593
Verel [Eastman Kodak] *see* 6313
Verelan [Elan] *see* 10144
Vergentan [Synthelabo] *see* 242
Vergonil *see* 4826
Veritol [Knoll] *see* 7442
Verkade's Superbases, 10155
Verladyn [Verla] *see* 3191
Vermicidin [Rivopharm] *see* 5837
Vermicompren (Tabl.) [Merck KGaA]
   see 7577
Vermiculite, 10156
Vermilion *see* 5957
Verminum [Squibb] *see* 7139
Vermiparin *see* 5782
Vermirax [Biosintetica] *see* 5837
Vermitin *see* 7364
Vermizym *see* 7119
Vermouth *see* 10250
Vermox [Janssen] *see* 5837
Vernakalant, 10157
Vernamycin A *see* 10200
Vernamycin B_α *see* 7871
Vernine *see* 4601
Vernitest Reagent *see* 8164
Vernolepin, 10158
Vernolic Acid, 10159
Verol *see* 6091
Veronal *see* 957
Veronal Sodium *see* 957

Verospiron [Richter Gedeon] *see* 8887
Veroxil [Baldacci] *see* 4975
Verrucarin A *see* 10160
Verrucarin B *see* 10160
Verrucarin J *see* 10160
Verrucarin K *see* 10160
Verrucarins, 10160
Verrucarol *see* 10160
Verrugon [Pickles] *see* 8469
Versalide [Givaudan-Delawanna], 10161
Versapen [BMS] *see* 4707
Versatis [Grünenthal] *see* 5535
Versatrex [BMS] *see* 4707
Versed [Roche] *see* 6261
Versene-9 [Dow] *see* 3565
Versene 100 [Dow] *see* 3565
Versene Acid [Dow] *see* 3565
Versene Na₂ [Dow] *see* 3565
Versenol [Dow], 10162
Versus [Angelini] *see* 1037
Vertab [Chiesi] *see* 10144
Verteporfin, 10163
Verticillin A *see* 10164
Verticillin B *see* 10164
Verticillin C *see* 10164
Verticillins, 10164
Verticine, 10165
Vertigon [GSK] *see* 7879
Vertisan [DuPont] *see* 7239
Verucasep [Galen] *see* 4508
Verutal [Stiefel] *see* 4508
Verv-Ca *see* 1706
Verxite *see* 10156
Very Fast Death Factor *see* 621
Very Late Activation Antigens *see* 5032
Vesadin [Rhône-Mérieux] *see* 9047
Vésadol [Janssen] *see* 1603
Vesanoid [Roche] *see* 8288
Vesdil [AstraZeneca] *see* 8221
Vesicare [Yamanouchi] *see* 8838
Vesistol [Inverni] *see* 587
Vesnarinone, 10166
Vesperone [UCB] *see* 1366
Vesprin [BMS] *see* 9853
Vessel [Alfa] *see* 9116
Vestalin *see* 6791
Vestra [Pharmacia & Upjohn] *see* 8244
Vesulong [Novartis] *see* 9076
Vesuprim *see* 9035
Vesuvianite *see* 1651
Vesuvine *see* 1256
Vetacortyl [Vétoquinol] *see* 6184
Veta-K₁ [Sanofi] *see* 7492
Vetalar [Bioniche] *see* 5343
Vetalgina [ICC] *see* 5698
Vetalog [Solvay] *see* 9758
Vetalpha [A.T.I.] *see* 4858
Vetamox [Am. Cyanamid] *see* 52
Vetarsenobillon *see* 6532
Vetbond [3M] *see* 1566
Vetdectin [Fort Dodge] *see* 6376
Veteusan [Veterinaria] *see* 2585
Vetibenzamina [Novartis] *see* 9910
Vetiprost [Virbac] *see* 3918
Vetisulid [Fort Dodge] *see* 9032
Vetiver Oil *see* 6908
Vetiver Oil Haiti *see* 6908
Vetiver Oil Java *see* 6908
Vetiver Oil Reunion (Bourbon) *see* 6908
Vetivert Oil *see* 6908
α-Vetivone *see* 10167
β-Vetivone *see* 10167
Vetivones, 10167
Vetkelfizina [Centralvet] *see* 9042
Vetmedin [Boehringer, Ing.] *see* 7545

Vetoryl [Dechra] *see* 9867
Vetrabutine, 10168
Vetranquil [Sanofi] *see* 33
Vetrazin [Novartis] *see* 2772
Vetstrep [Merck & Co.] *see* 8956
Vetsulin [Intervet] *see* 5023
Vetyver Oil *see* 6908
Vexol [Alcon] *see* 8352
Vfend [Pfizer] *see* 10231
Viaben [Schrholz] *see* 1443
Viacutan [Boehringer, Ing.] *see* 5562
Viadent [Vipont] *see* 8493
Viadil [Pharma Investi] *see* 1443
Viadur [Alza] *see* 5509
Viagra [Pfizer] *see* 8628
Vianol *see* 1550
Viansin [Farmocosmici] *see* 2085
Viarespan [Servier] *see* 4026
Viarex [Schering-Plough] *see* 1017
Viarox [Byk Gulden] *see* 1017
Vibativ [Astellas] *see* 9257
Vibeline [Bellon] *see* 10208
Vibramycin [Pfizer] *see* 3488
Vibra-Tabs [Pfizer] *see* 3488
Vibraveineuse [Pfizer] *see* 3488
Vibravenös [Pfizer] *see* 3488
Vibriomycin *see* 3197
Viburnum Opulus, 10169
Viburnum Prunifolium, 10170
Vi-Cad [Vineland] *see* 1622
Vicard [Abbott] *see* 9297
Viccillin [Meiji] *see* 583
Viceton [Bimeda] *see* 2077
Vicilan [Zeneca] *see* 10178
Vicin *see* 10171
Vicine, 10171
Vicioside *see* 10171
Vicodin [Abbott] *see* 4823
Vicoprofen [Abbott] *see* 4823
Vicriviroc, 10172
Victan [Sanofi Winthrop] *see* 3876
Victoria Green B *see* 5763
Victoria Green W *see* 5763
Victoza [Novo] *see* 5569
Victrelis [Schering Corp.] *see* 1322
Vidarabine, 10173
Vidaza [Pharmion] *see* 884
Vi-De-3 [Novartis] *see* 10216
Videobil [Bracco] *see* 5106
Videophel *see* 5081
Viderpen [Sintex] *see* 5996
Videx [BMS] *see* 3109
Vidine *see* 2211
Vidopen [APS] *see* 583
Vidora [Wyeth] *see* 5010
Viehe's Salt, 10174
Vienna Green *see* 2615
Vigabatrin, 10175
Vigamox [Alcon] *see* 6377
Vigantol [Merck KGaA] *see* 10216
Vigazoo [Am. Cyanamid] *see* 9023
Vigilor [Bouchard] *see* 4115
Vigold [Bayer CropSci.] *see* 4216
Viibryd [Forest Labs.] *see* 10176
Vikane [Dow AgroSci.] *see* 9111
Vilan [Lannacher] *see* 6605
Vilazodone, 10176
Vildagliptin, 10177
Villiaumite *see* 8754
Viloxazine, 10178
Viminalol *see* 610
Vimpat [Schwarz] *see* 5381
Vinarol [Gen. Dyes] *see* 7706
Vinblastine, 10179
Vincafor [Pharmafarm] *see* 10180

Vincagil [Sanofi-Aventis] *see* 10180
Vincaleukoblastine *see* 10179
Vincamine, 10180
Vincaminine *see* 10180
Vincamone *see* 3534
Vincanorine *see* 3534
Vincare [Kumai Chemical] *see* 1055
Vincimax [Sanofi-Aventis] *see* 10180
Vincine *see* 10180
Vincinine *see* 10180
Vincit [Cheminova] *see* 4242
Vinclozolin, 10181
Vinconate, 10182
Vincristine, 10183
Vindesine, 10184
Vindolidine *see* 10185
Vindoline, 10185
Vindorosine *see* 10185
Vinegar Naphtha *see* 3811
Vinesthene *see* 3427
Vinethene *see* 3427
Vinetine *see* 7052
Vinflunine, 10186
Vingsal [Stroschein] *see* 6788
Vinol [Air Prods. Chem.] *see* 7706
Vinorelbine, 10187
Vinpocetine, 10188
Vinyl Acetate, 10189
Vinylacetonitrile *see* 281
γ-Vinyl-γ-aminobutyric Acid *see* 10175
Vinylbenzene *see* 8990
Vinylbital, 10190
Vinyl Carbinol *see* 277
Vinyl Chloride, 10191
Vinyl Cyanide *see* 122
α-Vinyldiacetonalkamine Benzoate *see* 3937
Vinylestrenolone *see* 6791
17α-Vinyl-5(10)-estren-17β-ol-3-one *see* 6791
Vinyl Ether *see* 3427
Vinylethylene *see* 1512
Vinylformic Acid *see* 121
γ-Vinyl GABA *see* 10175
Vinylidene Chloride, 10192
5-Vinyl-5-(1-methylbutyl)barbituric Acid *see* 10190
Vinyl Methyl Ketone *see* 6210
17α-Vinyl-19-nortestosterone *see* 6806
(−)-5-Vinyl-2-oxazolidinethione *see* 4548
β-Vinylpyridine *see* 3785
N-Vinyl-2-pyrrolidinone *see* 7814
1-Vinyl-2-pyrrolidinone Polymers *see* 7814
1-Vinyl-2-pyrrolidinone Polymers, Iodine Complex *see* 7815
5-Vinyl-2-thiooxazolidone *see* 4548
Vinyl Trichloride *see* 9802
Vinyzene *see* 7060
Viocin [Pfizer] *see* 10197
Vioform [Novartis] *see* 5075
Viokase [Axcan] *see* 7108
Violacein, 10193
Violanin *see* 2885
Violaquercitrin *see* 8438
Violaxanthin, 10194
Viologen, 10195
Violuric Acid, 10196
Viomycin, 10197
Viosterol *see* 10215
Vio-Thene [Rowell] *see* 7072
Vioxx [Merck & Co.] *see* 8377
Viozene *see* 8387
VIP, 10198
Viper [Dow AgroSci.] *see* 7212

Vipicil [Wyeth] *see* 2695
Viprinex [Knoll] *see* 624
Viprynium Chloride *see* 8136
Viprynium Embonate *see* 8136
Viquidil, 10199
Vira-A [Warner-Lambert] *see* 10173
Viracept [Agouron] *see* 6528
Viraferon [Schering-Plough] *see* 5034
Viramid [Alfa] *see* 8323
Viramidine [Valeant] *see* 9197
Viramune [Roxane] *see* 6575
Viras [Mect] *see* 8325
Virazide [Valeant] *see* 8323
Virazole [Valeant] *see* 8323
Viread [Gilead] *see* 9289
Virgan [Théa] *see* 4393
Virgimycin *see* 10200
Virginiamycin, 10200
Virginiamycin M₁ *see* 10200
Virginiamycin S₁ *see* 10200
Viridicatin, 10201
Viridin, 10202
Virilon [Star] *see* 9324
Virlix [Sanofi-Synthelabo] *see* 2021
Virmen [Menarini] *see* 134
Virofral [Novo] *see* 368
Virophta [Dulcis] *see* 9855
Viroptic [Burroughs Wellcome] *see* 9855
Virormone [Ferring] *see* 9324
Virosterone [Endo] *see* 9324
Virudermin [Robugen] *see* 10359
Virudox [Bioglan] *see* 4931
Virulizin [Lorus], 10203
Virulizin-2γ *see* 10203
Viru-Merz [Merz] *see* 9950
Viruseen [Hommel] *see* 134
Viruserol [Zyma] *see* 9950
Virusmin [Sumitomo] *see* 6358
Virustat [Delagrange] *see* 6358
Viruxan [Newport] *see* 5019
Viruzona *see* 6049
Visammin *see* 5356
Visano [Kade] *see* 5929
Visceralgina [SIT] *see* 9587
Visclair [Ranbaxy] *see* 5856
β-Viscol *see* 5668
Viscoleo [Lundbeck] *see* 4219
Viscose, 10204
Viscosin, 10205
Viscotoxin, 10206
Viscotoxin A₁ *see* 10206
Viscotoxin A₂ *see* 10206
Viscotoxin A₃ *see* 10206
Viscotoxin B *see* 10206
Viscum album Agglutinin I *see* 6295
Viscumin *see* 6295
Visha *see* 112
Visine [Pfizer] *see* 9364
Visipaque [GE Healthcare] *see* 5066
Viskaldix [Novartis] *see* 2382
Visken [Novartis] *see* 7554
Vismodegib, 10207
Visnadine, 10208
Visnagin, 10209
Visnamine [Chinoin] *see* 10208
Visor [Dow AgroSci.] *see* 9463
Vista [AgrEvo] *see* 4226
Vista [Dow AgroSci.] *see* 4233
Vistagan [Allergan] *see* 5514
Vista-Methasone [Richard Daniel] *see* 1182
Vistamycin [Meiji] *see* 8330
Vistar [3M] *see* 5873
Vistaril Pamoate [Pfizer] *see* 4889
Vistaril Parenteral [Pfizer] *see* 4889

Vistide [Gilead] *see* 2271
Visual Purple *see* 8317
Visubeta *see* 1182
Visubutina *see* 7071
Visudyne [Novartis] *see* 10163
Vita D [Austral] *see* 10215
Vitabact [Faure] *see* 7510
Vitaberin [Taisho] *see* 9449
Vitacimin [Takeda] *see* 819
Vitacin [Towa Yakuhin] *see* 819
Vital Red, 10210
Vital Red Evans *see* 10210
Vitamin A, 10211
2-*cis*-Vitamin A *see* 6552
Vitamin A Acid *see* 8288
2-*cis*-Vitamin A Acid *see* 5274
Vitamin A Alcohol *see* 10211
Vitamin A Aldehyde *see* 8287
Vitamin A₁ *see* 10211
Vitamin A₂ *see* 2879
Vitamin A₂ Aldehyde *see* 2878
Vitamin B₁ *see* 9447
Vitamin B₁ *O*,*S*-Diacetate *see* 53
Vitamin B₁ Disulfide *see* 9449
Vitamin B₁ Propyl Disulfide *see* 7993
Vitamin B₂ *see* 8325
Vitamin B₂ Phosphate *see* 4123
Vitamin B₃ *see* 6608
Vitamin B₄ *see* 140
Vitamin B₅ *see* 7118
Vitamin B₁₂, 10212
Vitamin B₁₂-*Co*(II) *see* 2410
Vitamin B₁₂ Coenzyme *see* 2435
Vitamin B₁₂, Radioactive, 10213
Vitamin B₁₂ₐ *see* 4846
Vitamin B₁₂ᵦ *see* 4846
Vitamin B₁₂ᵧ *see* 4846
Vitamin B₁₂ᵩ *see* 4846
Vitamin B₁₂ₚ *see* 3102
Vitamin B₁₂ᵣ *see* 2410
Vitamin B₁₂ₛ *see* 2410
Vitamin B₁₃ *see* 6972
Vitamin B₁₅ *see* 7110
Vitamin B₁₇ *see* 592
Vitamin Bₜ *see* 1849
Vitamin Bₓ *see* 418
Vitamin Bc *see* 4248
Vitamin Bc Conjugate *see* 8044
Vitamin C *see* 819
Vitamin D₁, 10214
Vitamin D₂, 10215
Vitamin D₃, 10216
Vitamin D₄, 10217
Vitamin E, 10218
Vitamin G *see* 8325
Vitamin H *see* 1236
Vitamin K, 10219
Vitamin K₁ *see* 7492
Vitamin K₁ Epoxide *see* 7492
Vitamin K₁ Oxide *see* 7492
Vitamin K₂ *see* 5901
Vitamin K₂₍₀₎ *see* 5899
Vitamin K₂₍₂₀₎ *see* 5901
Vitamin K₂₍₃₀₎ *see* 5901
Vitamin K₂₍₃₅₎ *see* 5901
Vitamin K₃ *see* 5899
Vitamin K₄ *see* 5898
Vitamin K₅, 10220
Vitamin L₁ *see* 417
Vitamin M *see* 4248
Vitamin P Complex *see* 1233
Vitamin PP *see* 6608
Vitamin T, 10221
Vitamin T Goetsch *see* 10221
Vitamin U *see* 6170

Vitamogen [Tanabe] *see* 9449
Vitaneurin [Mepha] *see* 9447
Vitanevril [Sanofi Winthrop] *see* 1043
Vitarubin [Streuli] *see* 10212
Vitascorbol [Aventis] *see* 819
Vitaseptol [Faure] *see* 9470
Vitasprint B$_{12}$ [Brenner-Efeka] *see* 7475
VitaStain [Arapahoe] *see* 9918
Vitas-U [Kaken] *see* 6170
Vitavax [Uniroyal] *see* 1828
Vitavel K [Vitamins Ltd] *see* 5898
Vitazechs *see* 8092
Vitellin *see* 5483
Vitrasert *see* 4393
Vitravene [Novartis] *see* 4256
Vitride [Zeeland] *see* 8702
Vitronectin, 10222
Vivactil [Merck & Co.] *see* 8010
Vival [Alpharma] *see* 2997
Vivalan [Zeneca] *see* 10178
Vivando [BASF] *see* 6230
Vivarint [Zeneca] *see* 10178
Vivatec [Merck & Co.] *see* 5572
Vivelle [Novartis] *see* 3758
Vivelle-Dot [Novartis] *see* 3758
Viviant [Wyeth] *see* 1011
Vividrin [Mann] *see* 2581
Vividyl [Lilly] *see* 6804
Viviful [Kumiai] *see* 7892
Vivitrol [Cephalon] *see* 6448
VK-55 *see* 9039
VLA Antigens *see* 5032
VLA Integrins *see* 5032
VLB *see* 10179
Vleminckx's Lotion *see* 9107
VM-26 *see* 9288
VMA *see* 10121
VML-251 *see* 4301
VML-588 *see* 2340
V.M.&P. Naphtha *see* 5542
VNP-40101M *see* 5425
Voacamine, 10223
Voacangine *see* 10223
Voacanginine *see* 10223
Voclosporin, 10224
Vodka, 10225
Voelicherite *see* 1696
Vogalene [Thraplix] *see* 6226
Vogan [Merck KGaA] *see* 10211
Vogesensäure (German) *see* 9204
Voglibose, 10226
Volatile Oil of Laurel *see* 5437
Volatile Oil of Mustard *see* 286
Voldal [Novartis] *see* 3091
Volibris [GSK] *see* 378
Volicitin, 10227
Volidan [BDH; Searle] *see* 5876
Volmax [GSK] *see* 210
Volon [BMS] *see* 9757
Volon A [BMS] *see* 9758
Volonimat [BMS] *see* 9758
Volpo [Croda] *see* 7697
Voltaren [Novartis] *see* 3091
Voltarol [Novartis] *see* 3091
Vomex A [Astellas] *see* 3230
Vomicine, 10228
Vomit Nut *see* 6819
Vomitoxin, 10229
Vomit Wort *see* 5608
Vonamycin Powder V *see* 6539
Vondozeb [Cerexagri] *see* 5780
Vonedrine *see* 7418
Vontil [SKB] *see* 9510
Vontrol [SK & F] *see* 3341
Vonum [Sanofi-Synthelabo] *see* 5009

Vorapaxar, 10230
Voriconazole, 10231
Vorinostat, 10232
Vorlan [Mallinckrodt] *see* 10181
Vorozole, 10233
Votrient [GSK] *see* 7160
Votubia [Novartis] *see* 3950
VP-16-213 *see* 3929
VP-63843 *see* 7648
v-PA *see* 3506
VPA-985 *see* 5606
V-Pen [Orion] *see* 7209
VPM *see* 6021
VPN-4090CE *see* 5425
Vpriv [Shire] *see* 10135
Vraap [Inverni] *see* 10180
V-Tablopen [Viatris] *see* 7209
Vucine *see* 3771
VUFB-6453 *see* 6216
VUFB-6638 *see* 9307
Vulgarobufotoxin *see* 1484
Vulkacit NPV/C [Mobay] *see* 3858
Vulkazit *see* 3358
Vulklor *see* 2078
Vumon [BMS] *see* 9288
VX, 10234
VX-175 *see* 4276
VX-478 *see* 585
VX-497 *see* 5968
VX-710 *see* 1244
VX-770 *see* 5295
VX-950 *see* 9256
Vydate [DuPont] *see* 7019
Vyrene [U.S. Rubber] *see* 8860
Vyvanse [Shire] *see* 5571
W-37 *see* 1481
W-45 *see* 462
W-483 *see* 3800
W-554 *see* 3984
W-583 *see* 5842
W-1544a *see* 7335
W-2197 *see* 7251
W-2900A *see* 3934
W-2946M *see* 8259
W-2979M *see* 899
W-3566 *see* 8171
W-3699 *see* 7598
W-4020 *see* 7836
W-4565 *see* 7044
W-5219 *see* 7891
W-5759A *see* 9594
W-5975 *see* 1182
W-6309 *see* 3171
W-6421A *see* 5514
W-7000A *see* 5514
W-19053 *see* 3911
W-36095 *see* 9648
WAC-104 *see* 1228
Wahoo *see* 3941
Wakaflavin L [Wakamoto] *see* 8325
Wakoboku *see* 5761
WAL-801 *see* 3673
WAL-801CL *see* 3673
Walconesin *see* 5917
Waldmeister (German) *see* 10247
Wandonorm [Novartis] *see* 1335
Wansar [Hoei] *see* 3341
Warburganal, 10235
Warfarin, 10236
WARF Compound 42 *see* 10236
Warfilone [Merck & Co.] *see* 10236
Warfotox [Aimco] *see* 10236
Warrior [Syngenta] *see* 5398
Wartec [Kabi] *see* 7663
Warticon [Perstorp] *see* 7663

Waruzol [Farmanic] *see* 846
Washed Sulfur *see* 9107
Washing Soda *see* 8731
Water, 10237
Water-d_2 *see* 2942
Water Bugle *see* 5685
Water Channel of Collecting Duct *see* 749
Water Glass *see* 8804
Water Hemp *see* 370
Watermelon, 10238
Water Shamrock *see* 5910
Wavellite *see* 352
Wax-O-Sol [Life Sci.] *see* 8895
Waxsol [Norgine] *see* 3446
WAY-140424 *see* 1011
WAY-173606 *see* 1357
WAY-179272 *see* 6557
WAY-GPA-748 *see* 7824
WAY-TSE-424 *see* 1011
WBA-8119 *see* 1386
WCH-CD *see* 749
WE-941-BS *see* 1459
WEB-2086 *see* 709
Weedazol [CFPI] *see* 485
Weed Hoe [Vineland] *see* 6025
Weedone [Nufarm] *see* 2793
Weifacodine [Weifa] *see* 7441
Weinsäure (German) *see* 9205
dl-Weinsäure (German) *see* 9204
Weinstein *see* 772
Weinsteinsäure (German) *see* 9205
Weiselfuttersaft (German) *see* 8410
Weisspiessglanz *see* 699
Welchol [Sankyo] *see* 2457
Welfurin [GAP] *see* 6686
Wellbutrin [GSK] *see* 1503
Wellcome 33-A-74 *see* 859
Wellcome 248U *see* 134
Wellcome Prepn 47-83 *see* 2703
Wellcoprim [Wellcome] *see* 9878
Wellcovorin [Burroughs Wellcome] *see* 4249
Welldorm [Smith & Nephew] *see* 3055
Wellferon [GSK] *see* 5034
Wellvone [GSK] *see* 857
Welvic [European Vinyls] *see* 7707
Wemid [Bernabo] *see* 3743
Weradys [Weimer] *see* 5876
Wescodyne [West Chemical] *see* 5080
Wespuril [Spitzner] *see* 3081
Westamine X *see* 5080
Westcort [Westwood] *see* 4824
Western Poison Oak *see* 7667
Westrosol *see* 9804
Weymouth Pine *see* 10241
WF-1360 *see* 8304
WG-253 *see* 8353
WG-696 *see* 9812
WGA *see* 5484
Wh-7286 *see* 10277
Wheat Germ Agglutinin *see* 5484
Wheat Germ Oil, 10239
Whey Factor *see* 6972
Whip *see* 4015
Whipcide *see* 7486
Whiskey *see* 10240
Whisky, 10240
White Agaric *see* 177
White Arsenic *see* 795
White Beeswax *see* 1019
White Bole *see* 5330
White Cedar *see* 9543
White Cinnamon *see* 1746
White Dextrin *see* 2955

White Flag *see* 6976
White Lead *see* 1002
White Mineral Oil *see* 7301
White Mustard *see* 6400
White Oak *see* 8154
White Oil of Camphor *see* 6872
White Phenolphthalein *see* 7356
White Phosphorus *see* 7459
White Pine, 10241
White Saunders *see* 8490
White Spirits *see* 6282
White Tin Oxide *see* 8905
White Vitriol *see* 10359
White Walnut *see* 5316
White Wax *see* 1019
Whiting *see* 1659
Whitlockite *see* 1696
Whorehouse Tea *see* 3662
Whortleberry *see* 1221
WHR-1142A *see* 5534
Widdrene *see* 9547
Widecillin [Meiji] *see* 574
WideMatch S [Dow AgroSci.] *see* 4233
Wieland-Gumlich Aldehyde, 10242
Wijs' Chloride *see* 5060
Wild Bergamot *see* 6329
Wild Black Cherry Bark *see* 10243
Wild Canilla *see* 1746
Wild Chamomile *see* 2040
Wild Cherry, 10243
Wild Cinnamon *see* 1746
Wild Cotton *see* 818
Wildfire Toxin, 10244
Wild Ginger *see* 814
Wild Indigo *see* 954
Wild Ipecac *see* 729
Wild Pepper *see* 6249
Wild Saffron *see* 2456
Wild Tobacco *see* 5608
Wild Woodbine *see* 4419
Wild Yam *see* 3321
Wilkinite *see* 1057
Wilkinson's Catalyst, 10245
Willemite *see* 10357
Wilpo [Dorsey] *see* 7374
Wilprafen [Mack, Illert.] *see* 5315
Wilzin [Orphan Europe] *see* 10327
Win [Bayer] *see* 1863
Win-1539 *see* 5348
Win-2848 *see* 9431
Win-3046 *see* 5215
Win-5063-2 *see* 9453
Win-8077 *see* 373
Win-8851-2 *see* 10018
Win-9317 *see* 7925
Win-11450 *see* 1047
Win-11831 *see* 5634
Win-13146 *see* 9242
Win-14833 *see* 8921
Win-17757 *see* 2811
Win-18320 *see* 6444
Win-18446 *see* 4087
Win-18501-2 *see* 7070
Win-20228 *see* 7233
Win-20740 *see* 2698
Win-21904 *see* 226
Win-22005 *see* 10072
Win-22118 *see* 9131
Win-24540 *see* 9867
Win-32729 *see* 3683
Win-32784 *see* 1308
Win-35213 *see* 8400
Win-35833 *see* 2312
Win-39103 *see* 6234
Win-39424 *see* 5095

Win-40014 *see* 8173
Win-40680 *see* 587
Win-41464 *see* 6843
Win-41464-2 *see* 6843
Win-41464-6 *see* 6843
Win-47203 *see* 6278
Win-59010 *see* 5787
Win-59075 *see* 9617
Win-63843 *see* 7648
Win-AM-13146 *see* 9242
Wincef [Astellas] *see* 1933
Wincoram [Sanofi Winthrop] *see* 587
Windaus-Grundmann Ketone *see* 4587
Wind Flower *see* 8048
Wine Ether *see* 3891
Wine Lees *see* 772
Win-Kinase [Winthrop] *see* 10072
Winobanin [Sanofi-Synthelabo] *see* 2811
Winstrol [Zambon] *see* 8921
Winter Bloom *see* 4639
Wintergreen Oil *see* 6193
Wintermin [Shionogi] *see* 2191
Wintersteiner's Compound D *see* 265
Wintersteiner's Compound F *see* 2525
Wintersteiner's Compound G *see* 263
Wintomylon [Daiichi] *see* 6444
Winuron [Sanofi Winthrop] *see* 8400
Wisdom [Amvac] *see* 1213
Witch Hazel *see* 4639
Withaferin A, 10246
Witherite *see* 965
WL-140 *see* 1699
WL-19805 *see* 2675
WL-41706 *see* 4019
WL-43775 *see* 4037
WL-108366 *see* 4134
WL-115110 *see* 4164
WL-136184 *see* 6006
WL-145158 *see* 9762
Wolfina [OJ & F] *see* 8238
Wolfram *see* 9994
Wolframite *see* 9994
Wolf's Bane *see* 108, 780
Wollastonite *see* 1704
Wood Alcohol *see* 6029
Wood Creosote *see* 2561
Wood Oil *see* 940
Woodruff, 10247
Woodsour *see* 956
Wood Spirit *see* 6029
Wood Sugar *see* 10285
Wood Vinegar *see* 8115
Woodward Herb *see* 10247
Woodward's Reagent K, 10248
Woody Nightshade *see* 3510
Wool Alcohols *see* 5410
Wool Fat *see* 5409
Woollins' Reagent, 10249
Woorali *see* 2661
Woorari *see* 2661
Worm Away [Robins] *see* 7576
Worm Guard [SK & F] *see* 7139
Worm-Agen [Merz] *see* 4748
Wormgrass *see* 8875
Wormwood, 10250
Wortmannin, 10251
Wourara *see* 2661
WP-833 *see* 9217
WR-638 *see* 7453
WR-2721 *see* 399
WR-6026 *see* 8691
WR-142490 *see* 5872
WR-152520 *see* 4208
WR-171669 *see* 4630
WR-238605 *see* 9159

WRC-0470 *see* 1229
Wrightine *see* 2480
WS3 *see* 3880
WS23 *see* 9887
WS-4545 Antibiotic *see* 1207
WSM-5008 *see* 2377
WSX-8365 *see* 3315
Wulfenite *see* 5467
Wurster's Blue *see* 9376
Wurster's Reagent *see* 9376
Wurtzite *see* 10360
WV-569 *see* 6848
WX-2412 *see* 7284
Wy-806 *see* 7032
Wy-1094 *see* 7898
Wy-1359 *see* 7936
Wy-3263 *see* 5119
Wy-3277 *see* 6436
Wy-3467 *see* 2997
Wy-3475 *see* 6774
Wy-3478 *see* 4854
Wy-3498 *see* 7025
Wy-3707 *see* 6793
Wy-3917 *see* 9275
Wy-4036 *see* 5636
Wy-4082 *see* 5639
Wy-4508 *see* 2695
Wy-8138 *see* 1296
Wy-8678 *see* 4593
Wy-16225 *see* 2959
Wy-20788 *see* 7190
Wy-21743 *see* 7023
Wy-21901 *see* 5010
Wy-22811 *see* 5930
Wy-42422 *see* 9924
Wy-42462 *see* 9924
Wy-44635 *see* 1940
Wy-45030 *see* 10140
Wy-45233 *see* 2937
Wy-45651 *see* 10140
Wy-45655 *see* 10140
Wy-47663 *see* 863
WY-49605 *see* 3976
WY-90493-RD *see* 765
Wyamine [Wyeth] *see* 5919
Wycillin [Wyeth] *see* 7206
Wydase [Wyeth] *see* 4797
Wydora [Wyeth] *see* 5010
Wymox [Wyeth] *see* 574
Wynestron [Wyeth] *see* 3763
Wyovin [Wyeth] *see* 3108
Wypax [Yamanouchi] *see* 5636
Wypresin [Wyeth] *see* 5010
Wytensin [Wyeth] *see* 4593
Wytrion [Wyeth] *see* 9949
Wyvital [Wyeth] *see* 2695
X-340 *see* 8272
X-537A *see* 5427
X-1497 *see* 6041
X-5108 *see* 873
X-550007 *see* 6810
Xagrid [Shire] *see* 616
Xalatan [Pharmacia & Upjohn] *see* 5431
Xaliproden, 10252
Xalkori [Pfizer] *see* 2577
Xamamina [Bracco] *see* 3230
Xametina [Zambeletti] *see* 9877
Xamoterol, 10253
Xanax [Pharmacia & Upjohn] *see* 308
Xanbon [Kissei] *see* 7079
Xanef [Merck & Co.] *see* 3623
Xanelim [Genentech] *see* 3567
Xanomeline, 10254
Xanomeline Tartrate *see* 10254
Xanteline *see* 6030

Name Index

Xanthaline *see* 7120
Xanthan Gum, 10255
Xanthatin, 10256
9*H*-Xanthen-9-one *see* 10260
Xanthine, 10257
Xanthine Riboside *see* 10263
Xanthinol Niacinate, 10258
Xanthinol Nicotinate *see* 10258
Xanthium [Galephar] *see* 9436
Xanthocillin, 10259
Xanthocillin X *see* 10259
Xanthocillin Y$_1$ *see* 10259
Xanthocillin Y$_2$ *see* 10259
Xanthokermesic Acid *see* 5378
Xanthone, 10260
Xanthophyll, 10261
Xanthopterin, 10262
Xanthopuccine *see* 1738
Xanthosine, 10263
Xanthothricin *see* 9718
Xanthotoxin *see* 6059
Xanthoxime *see* 10260
Xanthoxyletin, 10264
Xanthoxylin, 10265
Xanthoxylin N *see* 10264
Xanthoxylin S *see* 812
Xanthoxyloin *see* 10264
Xanthurenic Acid, 10266
Xanthyletin, 10267
Xantium [Cyanamid] *see* 9750
Xantocillin *see* 10259
Xantphos, 10268
Xantural [CP Kelco] *see* 10255
Xarelto [Bayer] *see* 8368
Xatral [Sanofi-Synthelabo] *see* 231
Xavin [Chinoin] *see* 10258
Xcytrin [Pharmacyclics] *see* 6367
XDE-007 *see* 6810
XDE-105 *see* 8877
XDE-537 *see* 2755
XDE-564 *see* 3093
XDE-565 *see* 2391
XDE-742 *see* 8124
XE-779L *see* 3294
Xefo [Nycomed] *see* 5640
Xeloda [Roche] *see* 1757
Xemilofiban, 10269
Xenalon [Mepha] *see* 8887
Xenar [Alfa] *see* 6499
Xenazine [Cambridge] *see* 9326
Xeneisol Xe 133 [Mallinckrodt] *see* 10270
Xenetix [Guerbet] *see* 5052
"Xenic Acid" *see* 10270
Xenical [Hoffmann-La Roche] *see* 6964
Xenid [RPG] *see* 3091
Xenon, 10270
Xenon Difluoride *see* 10270
Xenon Hexafluoride *see* 10270
Xenon Platinum Hexafluoride *see* 10270
Xenon Tetrafluoride *see* 10270
Xenon Trioxide *see* 10270
Xenon Xe 133 *see* 10270
Xenylamine *see* 1239
p-Xenylcarbimide, 10271
Xeplion [Janssen] *see* 7087
Xerac [Person & Covey] *see* 333
Xerac BP [Person & Covey] *see* 1119
Xerene [Martinet] *see* 5918
Xeroform [Warner-Chilcott] *see* 1291
X-gal, 10272
Xgeva [Amgen] *see* 2897
Xibornol, 10273
Xibrom [ISTA] *see* 1397
XIFAM *see* 10275
Xifaxan [Alfa Wassermann] *see* 8345

Xigris [Lilly] *see* 7995
Xilopar [Cephalon] *see* 8565
Ximelagatran, 10274
Ximoprofen, 10275
Ximovan [Aventis] *see* 10393
Xinlay [Abbott] *see* 861
Xipamide, 10276
Xitix [Woelm] *see* 819
XK-62-2 *see* 6260
XL-7 *see* 1307
XL-90 *see* 4591
Xobaline [Delano] *see* 6116
Xolair [Novartis] *see* 6936
Xonaltite *see* 1704
Xonotlite *see* 1704
Xopenex [Sepracor] *see* 210
X-Otag [Tutag] *see* 6975
XP-13512 *see* 4349
XP-19986 *see* 758
XPhos Phosphine Ligand *see* 1469
XR-9576 *see* 9198
XRD-473 *see* 4719
XRP-0673 *see* 6996
XRP-6258 *see* 1604
"XTC" *see* 5836
XU 62-320 *see* 4245
Xumbradil [Astra Pharm.] *see* 5083
Xuret [Galen] *see* 6224
Xusal [UCB] *see* 2021
Xymelin [Nycomed] *see* 10284
Xyduril [Dorsch] *see* 2370
Xylamide *see* 7891
Xylan Hydrogen Sulfate *see* 7247
Xylan Polysulfate *see* 7247
Xylans *see* 4677
Xylapan [Chassot] *see* 10277
Xylasol [Graeub] *see* 10277
Xylazine, 10277
Xylene, 10278
m-Xylene *see* 10278
o-Xylene *see* 10278
p-Xylene *see* 10278
Xylenol, 10279
as-m-Xylenol *see* 10279
as-o-Xylenol *see* 10279
p-Xylenol *see* 10279
sym-m-Xylenol *see* 10279
vic-m-Xylenol *see* 10279
vic-o-Xylenol *see* 10279
Xylenol Blue, 10280
Xylenol Orange, 10281
p-Xylenolsulfonephthalein *see* 10280
Xylidine, 10282
2,3-Xylidine *see* 10282
2,4-Xylidine *see* 10282
2,5-Xylidine *see* 10282
2,6-Xylidine *see* 10282
3,4-Xylidine *see* 10282
3,5-Xylidine *see* 10282
Xylite *see* 10283
Xylitol, 10283
Xyliton [Morishita] *see* 10283
Xylocaine [AstraZeneca] *see* 5535
Xylocard [AstraZeneca] *see* 5535
Xylocitin [Jenapharm] *see* 5535
Xyloidin *see* 8125
Xylol *see* 10278
Xylomed [Bio-Medical] *see* 10285
Xylometazoline, 10284
Xylo-Mucine *see* 1830
Xylonest [AstraZeneca] *see* 7862
Xyloneural [Strathmann] *see* 5535
Xylo-Pfan [Savage] *see* 10285
6-*O*-β-D-Xylopyranosyl-D-glucose *see* 7864

Xylose, 10285
D-Xylose *see* 10285
Xylosecarboxylic Acid *see* 4612
6-(β-D-Xylosido)-D-glucose *see* 7864
Xylotocan [AstraZeneca] *see* 9648
o-Xylotocopherol *see* 9655
p-Xylotocopherol *see* 9654
Xylulose, 10286
N-(2,3-Xylyl)anthranilic Acid *see* 5869
1-Xylylazo-2-naphthol, 10287
1-(2,5-Xylylazo)-2-naphthol *see* 10287
Xylyl Bromide, 10288
m-Xylyl Bromide *see* 10288
o-Xylyl Bromide *see* 10288
p-Xylyl Bromide *see* 10288
[[(2,6-Xylylcarbamoyl)methyl]imino]di-
acetic Acid *see* 5536
Xylyl Chloride, 10289
m-Xylyl Chloride *see* 10289
o-Xylyl Chloride *see* 10289
p-Xylyl Chloride *see* 10289
1-(2,6-Xylyloxy)-2-propylamine *see* 6248
Xymiazole *see* 2761
Xyotax [Cell Therapeutics] *see* 7081
Xyrem [Orphan Med.] *see* 4854
Xyzal [UCB] *see* 2021
XZ-450 *see* 907
Y-516 *see* 6363
Y-4153 *see* 2361
Y-6047 *see* 2400
Y-7131 *see* 3919
Y-8004 *see* 7831
Y-9179 *see* 6748
Y-12141 *see* 9739
Y-24180 *see* 5288
Y-25130 *see* 893
Yageine *see* 4649
Yal [Ritter] *see* 3446
Yamacillin [Yamanouchi] *see* 9165
Yamaful [Yamanouchi] *see* 1844
Yamaguchi Reagent *see* 9798
Yambolap [Chinoin] *see* 5118
Yam, Mexican, 10290
Yangonin, 10291
Yara Yara *see* 6070
Yarmor [Hercules] *see* 7558
Yarrow *see* 97
Yartsa Gunbu *see* 2515
Yarvitan [Janssen Animal Health] *see*
6304
Yasmin [Schering AG] *see* 3502
Yatren [Bayer] *see* 4865
Yaw Root *see* 8947
Yaxci *see* 8687
Yaz [Berlex] *see* 3502
Yazumycin A *see* 8959
Yazumycins *see* 8959
Y2B8-MX-DTPA *see* 4917
YB-2 *see* 4979
YC-93 *see* 6580
YCs *see* 1729
90Y-DOTA-hLL2 *see* 3687
Yeast, 10292
Yeast Adenylic Acid *see* 145
Yeast Nucleic Acid *see* 8327
Yectames [Valeant] *see* 233
Yellow AB, 10293
Yellow Arsenic Sulfide *see* 797
Yellow Beeswax *see* 1019
Yellow Cameleons *see* 1729
Yellow Copper *see* 2037
Yellow Cross Liquid *see* 6401
Yellow Cuprocide *see* 2654
Yellow Dextrin *see* 2955
Yellow Dock *see* 8429

Yellow Gentian *see* 4428
Yellow Indigo *see* 954
Yellow Jasmine *see* 4419
Yellow Jessamine *see* 4419
Yellow Melilot *see* 5889
Yellow Mercury Iodide *see* 5963
Yellow Moccasin Flower *see* 2767
Yellow Mustard *see* 6400
Yellow OB, 10294
Yellow Phenolphthalein *see* 7356
Yellow Precipitate *see* 5950
Yellow Prussiate of Potash *see* 7752
Yellow Prussiate of Soda *see* 8752
Yellow Puccoon *see* 4808
Yellow Resin *see* 8397
Yellow Root *see* 4808
Yellow Saunders *see* 8490
Yellow Sweet Clover *see* 5889
Yellow T *see* 9954
Yellow Ultramarine *see* 1662
Yellox [Bausch & Lomb] *see* 1397
Yentreve [Lilly] *see* 3512
Yerba Dulce *see* 8937
Yerba Maté *see* 5824
Yerba Santa *see* 3727
Yermonil [Novartis] *see* 5690
Yervoy [BMS] *see* 5114
YH-1 *see* 7188
YH-1885 *see* 8291
YI-5301 *see* 3933
Yig, 10295
Yingzhao *see* 10296
Yingzhaosu, 10296
Yingzhaosu A *see* 10296
Yingzhaosu C *see* 10296
(+)-Yingzhosu A *see* 10296
Yinyanghuo *see* 3672
YK-FH312 *see* 1196
YKP-509 *see* 1841
YL-704B1 *see* 6262
Ylangene, 10297
α-Ylangene *see* 10297
Ylang-Ylang Oil, 10298
YM-026 *see* 6510
YM-060 *see* 8223
YM-087 *see* 2491
YM-175 *see* 4969
YM-177 *see* 1958
YM-443 *see* 113
YM-529 *see* 6284
YM-617 *see* 9181
YM-866 *see* 7105
YM-905 *see* 8838
YM-08310 *see* 399
YM-09151-2 *see* 6531
YM-09330 *see* 1935
YM-09538 *see* 572
YM-09730-5 *see* 998
YM-11170 *see* 3971
YM-12617-1 *see* 9181
YM-14090 *see* 5034
YM-19020 *see* 3397
YM-35087 *see* 2491
YN-72 *see* 8853
Yobine [Lloyd] *see* 10300
Yobir [Maruko] *see* 309
Yochinol *see* 4865
Yocon [Palisades] *see* 10300
Yodoxin [Searle] *see* 5085
Yoghurt *see* 10299
Yogurt, 10299
Yohimbine, 10300
α-Yohimbine, 10301
δ-Yohimbine *see* 8237
allo-Yohimbine, 10302

Yohimex [Kramer] *see* 10300
Yohydrol [Riedel-Zabinka] *see* 10300
Yomesan [Bayer] *see* 6604
Yondelis [PharmaMar] *see* 3550
Yoristen [Kodama] *see* 4258
Yoshicol [Yoshitomi] *see* 6984
Yoshimilon [Kowa Yakuhin] *see* 9874
Youthwort *see* 3501
Yovital [Kenwood] *see* 10300
Yperite *see* 6401
YRC-2388 *see* 4034
⁹⁰Y-SMT-487 *see* 3561
YTR-830 *see* 9220
YTR-830H *see* 9220
Ytterbia *see* 10303
Ytterbium, 10303
Ytterbium Triflate, 10304
Ytterbium(III) Trifluoromethanesulfon-
ate *see* 10304
Yttria *see* 10305
Yttrium, 10305
Yttrium Iron Garnet *see* 10295
Yttrium Y 90 Edotreotide *see* 3561
Yttrium Y 90 Epratuzumab Tetraxetan
see 3687
Yttrium Y 90 Ibritumomab Tiuxetan *see*
4917
Yutopar [Astra] *see* 8365
Yxin [Pfizer] *see* 9364
Z-103 *see* 7669
Z-326 *see* 4033
Z-338 *see* 113
Z-876 *see* 3156
Z-905 *see* 7553
Z-4828 *see* 9947
Z-4942 *see* 4937
ZA-1296 *see* 5982
ZA-1963 *see* 7520
Zabromin [Sandoz] *see* 1460
Zacin [Cephalon] *see* 1770
Zackal [SS Pharm.] *see* 4411
Zactima [AstraZeneca] *see* 10117
Zacutex [Brit. Biotech] *see* 5527
Zadaxin [SciClone] *see* 9560
Zadine [Schering] *see* 894
Zadipina [Zambeletti] *see* 6651
Zaditen [Novartis] *see* 5354
Zaditor [Novartis] *see* 5354
Zadorin [Mepha] *see* 3488
Zadstat [Wyeth] *see* 6236
Zafirlukast, 10306
Zagam [Specia] *see* 8862
Zalain [Robert] *see* 8604
Zalcitabine, 10307
Zaldaride, 10308
Zaleplon, 10309
Zaltoprofen, 10310
Zamalin [Rafarm] *see* 1943
Zamanil *see* 7072
Zamocilline [Inpharzam] *see* 574
Zamudol [Viatris] *see* 9726
Zanaflex [Athena] *see* 9642
Zanamivir, 10311
Zanidip [Napp] *see* 5500
Zanil *see* 7057
Zanosar [Pharmacia & Upjohn] *see* 8962
Zantac [GSK] *see* 8228
Zantic [GSK] *see* 8228
Zantrène [Lederle] *see* 1248
Zappit [PPG] *see* 1676
Zaragozic Acids, 10312
Zaratite *see* 6588
Zarnestra [Janssen] *see* 9614
Zarontin [Parke-Davis] *see* 3801
Zaroxolyn [Teofarma] *see* 6224

Zarzio [Sandoz] *see* 4573
Zasten [Novartis] *see* 5354
Zatebradine, 10313
Zavedos [Pharmacia & Upjohn] *see* 4927
Zavesca [Actelion] *see* 6268
ZD-443 *see* 3731
ZD-473 *see* 7517
ZD-1033 *see* 619
ZD-1694 *see* 8217
ZD-1839 *see* 4412
ZD-4054 *see* 10320
ZD-4522 *see* 8402
ZD-6474 *see* 10117
Z-DNA *see* 2908
ZE-101 *see* 6747
Zeal [Valent] *see* 3933
α-Zearalanol *see* 10319
Zearalenone, 10314
Zeatin, 10315
trans-Zeatin *see* 10315
Zeaxanthin, 10316
Zeaxanthin Diepoxide *see* 10194
Zeaxanthol *see* 10316
Zebedassite *see* 345
Zebeta [Lederle] *see* 1295
Zebinix [Bial-Portela] *see* 3754
Zeclar [Abbott] *see* 2338
Zecovir [Guidotti] *see* 1385
Zeefra [Bouchara-Recordati] *see* 1985
Zeffix [GSK] *see* 5402
Zefnart [Tosoh] *see* 5570
Zeftera [J & J] *see* 1953
Zegerid [Santarus] *see* 6939
Zein, 10317
Zelapar [Valeant] *see* 8565
Zelboraf [Genentech] *see* 10138
Zeldox [Zeneca] *see* 4749
Zeldox [Pfizer] *see* 10371
Zelis [Pharma Italia] *see* 7619
Zelmac [Novartis] *see* 9253
Zelmid [Astra] *see* 10325
Zelnorm [Novartis] *see* 9253
Zemaira [ZLB Behring] *see* 705
Zemide [Alpharma] *see* 9180
Zemplar [Abbott] *see* 7143
Zemuron [Akzo] *see* 8375
Zen [Prophin] *see* 7619
Zenapax [Roche] *see* 2796
Zenas [EUSA] *see* 398
Zeniquin [Pfizer] *see* 5815
Zentel [GSK] *see* 203
Zentonil [Vétoquinol] *see* 144
Zentropil [Novartis] *see* 7433
Zenusin [Mepha] *see* 6613
Zeolites, 10318
Zepelin [De Angeli] *see* 4039
Zepharovicht *see* 352
Zephiran [Sanofi-Synthelabo] *see* 1061
Zephirol [Bayer] *see* 1061
Zephyr [Syngenta] *see* 2
Zepolas [Mikasa] *see* 4229
Zeranol, 10319
Zerit [BMS] *see* 8929
Zestril [AstraZeneca] *see* 5572
Zesulan [Toyo Jozo] *see* 5931
Zetamicin [Menarini] *see* 6564
Zeta Protein *see* 4104
Zetia [Merck/Schering-Plough] *see* 3958
Zettyn *see* 2016
Zevalin [IDEC] *see* 4917
Ziac [Am. Cyanamid] *see* 1295
Ziagen [GSK] *see* 1
Ziavetine *see* 1481
Zibeth *see* 2333
Zibotentan, 10320

Zibren [Sigma-Tau] *see* 79
Ziconotide, 10321
Zidoval [3M Pharma] *see* 6236
Zidovudine, 10322
Zienam [Merck & Co.] *see* 4961
Zildasac [Chugai] *see* 1037
Zilden [Pharmacia] *see* 3224
Zileuton, 10323
Zilmax [Intervet] *see* 10324
Zilpaterol, 10324
Zimeldine, 10325
Zimelidine *see* 10325
Zimovane [Aventis] *see* 10393
Zimox [Pfizer] *see* 574
Zinacef [GSK] *see* 1956
Zinamide [Merck & Co.] *see* 8067
Zinapar [Ziopharm] *see* 2827
Zinat [GSK] *see* 1956
Zinc, 10326
Zinc Acetate, 10327
Zinc Acexamate *see* 46
Zinc Ammonium Chloride *see* 537
Zincaps [RPR] *see* 10359
Zincate [Paddock] *see* 10359
"Zincates" *see* 10326
Zinc Bacitracin *see* 929
Zinc Bis(dimethylthiocarbamoyl) Disulfide *see* 10372
Zinc Blende *see* 10360
Zinc Bromide, 10328
Zinc Caprylate, 10329
Zinc Carbonate, 10330
Zinc Carbonate Hydroxide *see* 10330
Zinc L-Carnosine *see* 7669
Zinc Chloride, 10331
Zinc Chromate(VI) Hydroxide, 10332
Zinc Citrate, 10333
Zinc Cyanide, 10334
Zinc Diethyl *see* 3151
Zinc Dimethyl *see* 3289
Zinc Dimethyldithiocarbamate *see* 10372
Zinc Ethylenebis(dithiocarbamate) *see* 10365
Zinc Fluoride, 10335
Zinc Fluosilicate *see* 10337
Zinc Formate, 10336
Zinc Gluconate *see* 4492
Zinc Hexafluorosilicate, 10337
Zinc *p*-Hydroxybenzenesulfonate *see* 10350
Zinc Insulin *see* 5023
Zinc Iodate, 10338
Zinc Iodide, 10339
Zinc Iodide-Starch, 10340
Zincite *see* 10347
Zinc Lactate, 10341
Zinc Manganese Ethylenebisdithiocarbamate *see* 5780
Zinc Meta-arsenite, 10342
Zinc Nitrate, 10343
Zinc Oleate, 10344
Zincomed [Schwarz] *see* 10359
Zincon, 10345
Zinc Orthosilicate *see* 10357
Zinc Oxalate, 10346
Zinc Oxide, 10347
Zinc Perchlorate, 10348
Zinc Peroxide, 10349
Zinc *p*-Phenolsulfonate, 10350
Zinc Phosphate, 10351
Zinc Phosphide, 10352
Zinc Propionate, 10353
Zinc 1,2-Propylene Bisdithiocarbamate *see* 7933
Zinc Pyridinethione *see* 8107

Zinc Pyrithione *see* 8107
Zinc Pyrophosphate, 10354
Zinc Salicylate, 10355
Zinc Selenide, 10356
Zinc Silicate, 10357
Zinc Silicofluoride *see* 10337
Zincspar *see* 10330
Zinc Stearate, 10358
Zinc Subcarbonate *see* 10330
Zinc Sulfanilate Tetrahydrate *see* 9058
Zinc Sulfate, 10359
Zinc Sulfide, 10360
Zinc Sulfocarbolate *see* 10350
Zinc Sulfocyanate *see* 10363
Zinc Sulfophenate *see* 10350
Zinc Superoxide *see* 10349
Zinc Tartrate, 10361
Zinc Telluride, 10362
Zinc Thiocyanate, 10363
Zinc Undecylenate *see* 10033
Zinc Valerate, 10364
Zinc Vitriol *see* 10359
Zinc White *see* 10347
Zinc Yellow *see* 10332
Zindaclin [Epitan] *see* 2354
Zineb, 10365
Zinecard [Pharmacia & Upjohn] *see* 8241
Zingerone, 10366
Zingherone *see* 10366
Zingiberone *see* 10366
Zinnat [GSK] *see* 1956
Zinostatin, 10367
Zinpyr, 10368
Zinpyr-1 *see* 10368
Zinpyr-2 *see* 10368
Zinquin, 10369
Zinquin A *see* 10369
Zinquin E *see* 10369
Zinviroxime *see* 3655
ZIO-101 *see* 2827
Zipeprol, 10370
Ziprasidone, 10371
Ziram, 10372
Zircon *see* 10381
Zirconia *see* 10380
Zirconic Anhydride *see* 10380
Zirconium, 10373
Zirconium Chloride, 10374
(*T*-4)-Zirconium Chloride (ZrCl₄) *see* 10374
Zirconium Dioxide *see* 10380
Zirconium Fluoride, 10375
(*T*-4)-Zirconium Fluoride (ZrF₄) *see* 10375
Zirconium Hydride, 10376
Zirconium Hydroxide, 10377
(*T*-4)-Zirconium Hydroxide (Zr(OH)₄) *see* 10377
Zirconium Iodide, 10378
Zirconium Nitrate, 10379
Zirconium Orthosilicate *see* 10381
Zirconium Oxide, 10380
Zirconium Oxychloride *see* 10384
Zirconium Potassium Fluoride *see* 7760
Zirconium Silicate, 10381
Zirconium Sulfate, 10382
Zirconium Tetrachloride *see* 10374
Zirconium Tetrafluoride *see* 10375
Zirconium Tetraiodide *see* 10378
Zirconyl Acetate, 10383
Zirconyl Chloride, 10384
Zirgan [Théa] *see* 4393
Zirtek [UCB] *see* 2021
Zispin [Organon] *see* 6291
Zitazonium [Servier] *see* 9180

Zithromax [Pfizer] *see* 907
Zitoxil [Lifepharma] *see* 10370
Zitromax [Pfizer] *see* 907
Zixoryn [Gedeon Richter] *see* 4167
Zizanal *see* 6908
ZK-30595 *see* 3502
ZK-31224 *see* 9307
ZK-36374 *see* 4942
ZK-39482 *see* 5108
ZK-57671 *see* 9120
ZK-62711 *see* 8380
ZK-93035 *see* 4357
ZK-98299 *see* 6942
ZK-222584 *see* 10130
ZL-101 *see* 6747
ZMA *see* 10342
ZN-6 *see* 4346
Zoalene [Dow] *see* 3295
Zoamix [Dow] *see* 3295
Zoapatanol, 10385
Zoaquin *see* 5085
Zocor [Merck & Co.] *see* 8677
Zocord [Merck & Co.] *see* 8677
Zofac [AirPharma] *see* 8049
Zofenil [Menarini] *see* 10386
Zofenopril, 10386
Zofenoprilat *see* 10386
Zofran [GSK] *see* 6943
Zoisite *see* 1651
Zoladex [AstraZeneca] *see* 4562
Zolazepam, 10387
Zoledronate Disodium *see* 10388
Zoledronate Trisodium *see* 10388
Zoledronic Acid, 10388
Zoleptil [Abbott] *see* 10397
Zolicef [BMS] *see* 1918
Zolim [Schwarz] *see* 6307
Zolimidine, 10389
Zolinza [Merck & Co.] *see* 10232
Zoliparin [Mann] *see* 134
Zoliridine *see* 10389
Zolmitriptan, 10390
Zoloft [Pfizer] *see* 8606
Zolone *see* 7446
Zolpidem, 10391
Zolpidem Tartrate *see* 10391
Zoltum [AstraZeneca] *see* 6939
Zolvera [Rosemont] *see* 10144
Zolyse [Alcon] *see* 2264
Zomacton [Ferring] *see* 8842
Zometa [Novartis] *see* 10388
Zomig [AstraZeneca] *see* 10390
Zondel [Grelan] *see* 6788
Zonegran [Dainippon] *see* 10392
Zongorine *see* 8843
Zoniden [Irbi] *see* 9608
Zonisamide, 10392
Zooecdysteroids *see* 3538
"Zoomelanoidic" Acids *see* 4924
Zophren [GSK] *see* 6943
Zopiclone, 10393
Zopolrestat, 10394
Zorac [Allergan] *see* 9218
Zorial [Syngenta] *see* 6790
Zoroxin [Merck & Co.] *see* 6789
Zortress [Novartis] *see* 3950
Zorubicin, 10395
Zostex [Berlin-Chemie] *see* 1385
Zostrix [GenDerm] *see* 1770
Zosuquidar, 10396
Zosyn [Wyeth] *see* 9220
Zotepine, 10397
Zothelone *see* 3393
Zoton [Wyeth] *see* 5413
Zovant [Lilly] *see* 7995

Zovir [GSK] *see* 134
Zovirax [GSK] *see* 134
Zoxamide, 10398
Zoxan [Pfizer] *see* 3482
Zoxazolamine, 10399
Zoxium [Dow AgroSci.] *see* 10398
ZP-1 *see* 10368
ZP-2 *see* 10368
ZP-10A *see* 5605
ZPO *see* 10349
ZR-512 *see* 4842
ZR-515 *see* 6055
ZR-2458 *see* 6055
ZR-3210 *see* 4244
Z Span [Goldshield] *see* 10359
Zubrin [Schering-Plough] *see* 9294
Zuclomiphene *see* 2377

Zuclopenthixol *see* 2383
Zumenon [Solvay] *see* 3758
Zunden [Luitpold] *see* 7619
ZY-15021 *see* 1114
ZY-17617B *see* 10308
Zyban [GSK] *see* 1503
Zybrestat [Oxigene] *see* 4277
Zyclara [3M Pharma] *see* 4964
Zydena [Dong-A] *see* 10027
Zydol [Grünenthal] *see* 9726
Zydone [Endo] *see* 4823
Zyflo [Abbott] *see* 10323
Zygomycin A *see* 7146
Zyklolat [Mann] *see* 2741
Zylonite *see* 1965
Zyloprim [GSK] *see* 271
Zyloric [GSK] *see* 271

Zymafluor [Zyma] *see* 8754
Zymar [Allergan] *see* 4409
Zymofren [Specia] *see* 746
Zymoplastic Substance *see* 9539
Zymosan, 10400
Zymosans *see* 10400
ZypAdhera [Lilly] *see* 6914
Zyprexa [Lilly] *see* 6914
Zyrlex [UCB] *see* 2021
Zyrtec [Pfizer] *see* 2021
Zytiga [Centocor] *see* 9
Zytrim [Merckle] *see* 895
Zytron [Dow] *see* 3436
Zyvox [Pharmacia & Upjohn] *see* 5559
Zyvoxid [Pharmacia & Upjohn] *see* 5559
Zyxorin [Gedeon Richter] *see* 4167

NOTES

NOTES

NOTES

NOTES

NOTES

NOTES

NOTES

NOTES

NOTES

NOTES

NOTES

NOTES

NOTES

NOTES

NOTES

NOTES

NOTES

NOTES

ATOMIC WEIGHTS

(Alphabetical Order)

Element	Symbol	Atomic number	Atomic weight	Element	Symbol	Atomic number	Atomic weight
Actinium	Ac	89	227.0278**	Iridium	Ir	77	192.217
Aluminum	Al	13	26.9815386	Iron	Fe	26	55.845
Americium	Am	95	243.0614**	Krypton	Kr	36	83.798
Antimony	Sb	51	121.760	Lanthanum	La	57	138.90547
Argon	Ar	18	39.948	Lawrencium	Lr	103	262.1096**
Arsenic	As	33	74.92160	Lead	Pb	82	207.2
Astatine	At	85	209.9871**	Lithium	Li	3	[6.938; 6.997]
Barium	Ba	56	137.327				6.94*
Berkelium	Bk	97	247.0703**	Livermorium	Lv	116	292.200**
Beryllium	Be	4	9.012182	Lutetium	Lu	71	174.9668
Bismuth	Bi	83	208.98040	Magnesium	Mg	12	24.3050
Bohrium	Bh	107	267.1277**	Manganese	Mn	25	54.938045
Boron	B	5	[10.806; 10.821]	Meitnerium	Mt	109	276.151**
			10.81*	Mendelevium	Md	101	258.0984**
Bromine	Br	35	79.904	Mercury	Hg	80	200.59
Cadmium	Cd	48	112.411	Molybdenum	Mo	42	95.96
Calcium	Ca	20	40.078	Neodymium	Nd	60	144.242
Californium	Cf	98	251.0796**	Neon	Ne	10	20.1797
Carbon	C	6	[12.0096; 12.0116]	Neptunium	Np	93	237.0482**
			12.011*	Nickel	Ni	28	58.6934
Cerium	Ce	58	140.116	Niobium	Nb	41	92.90638
Cesium	Cs	55	132.9054519	Nitrogen	N	7	[14.00643; 14.00728]
Chlorine	Cl	17	[35.446; 35.457]				14.007*
			35.45*	Nobelium	No	102	259.1010**
Chromium	Cr	24	51.9961	Osmium	Os	76	190.23
Cobalt	Co	27	58.933195	Oxygen	O	8	[15.99903; 15.99977]
Copernicium	Cn	112	285.174**				15.999*
Copper	Cu	29	63.546	Palladium	Pd	46	106.42
Curium	Cm	96	247.0704**	Phosphorus	P	15	30.973762
Darmstadtium	Ds	110	281.162**	Platinum	Pt	78	195.084
Dubnium	Db	105	268.125**	Plutonium	Pu	94	244.0642**
Dysprosium	Dy	66	162.500	Polonium	Po	84	208.9824**
Einsteinium	Es	99	252.0830**	Potassium	K	19	39.0983
Erbium	Er	68	167.259	Praseodymium	Pr	59	140.90765
Europium	Eu	63	151.964	Promethium	Pm	61	144.9127**
Fermium	Fm	100	257.0951**	Protactinium	Pa	91	231.03588
Flerovium	Fl	114	289.187**	Radium	Ra	88	226.0254**
Fluorine	F	9	18.9984032	Radon	Rn	86	222.0176**
Francium	Fr	87	223.0197**	Rhenium	Re	75	186.207
Gadolinium	Gd	64	157.25	Rhodium	Rh	45	102.90550
Gallium	Ga	31	69.723	Roentgenium	Rg	111	280.164**
Germanium	Ge	32	72.63	Rubidium	Rb	37	85.4678
Gold	Au	79	196.966569	Ruthenium	Ru	44	101.07
Hafnium	Hf	72	178.49	Rutherfordium	Rf	104	265.1167**
Hassium	Hs	108	277.150**	Samarium	Sm	62	150.36
Helium	He	2	4.002602	Scandium	Sc	21	44.955912
Holmium	Ho	67	164.93032	Seaborgium	Sg	106	271.133**
Hydrogen	H	1	[1.00784; 1.00811]	Selenium	Se	34	78.96
			1.008*	Silicon	Si	14	[28.084; 28.086]
Indium	In	49	114.818				28.085*
Iodine	I	53	126.90447	Silver	Ag	47	107.8682